CHILTON®

GENERAL MOTORS
MECHANICAL SERVICE
2006 EDITION

THOMSON

★

DELMAR LEARNING™

Australia • Canada • Mexico • Singapore • Spain • United Kingdom • United States

THOMSON

DELMAR LEARNING

Chilton®
General Motors
Mechanical Service
2006 Edition

**Vice President,
Technology Professional
Business Unit:**

Gregory L. Clayton

**Publisher,
Professional Business Unit:**

David Koontz

Production Director:

Mary Ellen Black

Marketing Director:

Beth A. Lutz

Marketing Specialist:

Brian McGrath

Marketing Coordinator:

Marissa Maiella

Marketing Assistant:

Jennifer Stall

Sr. Production Editor:

Elizabeth Hough

Editorial Assistant:

Christine Wade

Publishing Coordinator:

Paula Baillie

Editors:

Dennis Bailey

Terry Blomquist

Thomas A. Mellon

Richard J. Rivele

Jon Wallace

Cover Design:

Melinda Possinger

© 2006 Thomson Delmar Learning,
a part of The Thomson Corporation.
Thomson, the Star logo, and
Delmar Learning are trademarks
used herein under license.

Printed in the United States of
America

1 2 3 4 5 XX 07 06 05

For more information contact
Thomson Delmar Learning
Executive Woods
5 Maxwell Drive, PO Box 8007,
Clifton Park, NY 12065-8007
Or find us on the World Wide Web
at
www.delmarlearning.com

ISBN: 1-4180-0602-5

ISSN: 1548-0887

NOTICE TO THE READER

Table of Contents

Sections

1   Alero, Grand Am

2   Astro, Safari

3   Aurora

4   Avalanche, Express, Savana, Sierra, Silverado

5   Aveo

6   Aztek, Rendezvous

7   Blazer, Jimmy, S10, Sonoma

8   Bonneville, Le Sabre, Park Avenue

9   Bravada, Envoy, Rainier, Trailblazer

10   Canyon, Colorado

11   Cavalier, Sunfire

12   Century, Grand Prix, Impala, Intrigue, Monte Carlo, Regal

13   Cobalt

14   Corvette

15   CTS, CTS-V

16   DeVille, Eldorado, Seville

17   Equinox

18   Escalade, Suburban, Tahoe, Yukon, Yukon Denali, Yukon XL

19   G6

20   GTO

21   ION, L Series, S Series

22   LaCrosse

23   Malibu, Malibu Maxx

24   Montana, Silhouette, Venture

25   Montana SV6, Relay, Terraza, Uplander

26   Solstice

27   SRX

28   STS

29   Tracker

30   Vibe

31   Vue

Model Index

Model	Section No.	Model	Section No.	Model	Section No.
A		**G**		**S**	
Alero	1-1	G6	19-1	S Series	21-1
Astro	2-1	Grand Am	1-1	S10	7-1
Aurora	3-1	Grand Prix	12-1	Safari	2-1
Avalanche	4-1	GTO	20-1	Savana	4-1
Aveo	5-1	**I**		Seville	16-1
Aztek	6-1	Impala	12-1	Sierra	4-1
B		Intrigue	12-1	Silhouette	24-1
Blazer	7-1	ION	21-1	Silverado	4-1
Bonneville	8-1	**J**		Solstice	26-1
Bravada	9-1	Jimmy	7-1	Sonoma	7-1
C		**L**		SRX	27-1
Canyon	10-1	L Series	21-1	STS	28-1
Cavalier	11-1	LaCrosse	22-1	Suburban	18-1
Century	12-1	Le Sabre	8-1	Sunfire	11-1
Cobalt	13-1	**M**		**T**	
Colorado	10-1	Malibu	23-1	Tahoe	18-1
Corvette	14-1	Malibu Maxx	23-1	Terraza	25-1
CTS	15-1	Montana	24-1	Tracker	29-1
CTS-V	15-1	Montana SV6	25-1	Trailblazer	9-1
D		Monte Carlo	12-1	**U**	
DeVille	16-1	**P**		Uplander	25-1
E		Park Avenue	8-1	**V**	
Eldorado	16-1	**R**		Venture	24-1
Envoy	9-1	Rainier	9-1	Vibe	30-1
Equinox	17-1	Regal	12-1	Vue	31-1
Escalade	18-1	Relay	25-1	**Y**	
Express	4-1	Rendezvous	6-1	Yukon	18-1
				Yukon Denali	18-1
				Yukon XL	18-1

USING THIS INFORMATION

Organization

To find where a particular model section or procedure is located, look in the Table of Contents. Main topics are listed with the page number on which they may be found. Following the main topics is an alphabetical listing of all of the procedures within the section and their page numbers.

Manufacturer and Model Coverage

This product covers 2002–2006 General Motors models that are produced in sufficient quantities to warrant coverage, and which have technical content available from the vehicle manufacturers before our publication date. Although this information is as complete as possible at the time of publication, some manufacturers may make changes which cannot be included here. While striving for total accuracy, the publisher cannot assume responsibility for any errors, changes, or omissions that may occur in the compilation of this data.

Part Numbers & Special Tools

Part numbers and special tools are recommended by the publisher and vehicle manufacturer to perform specific jobs. Before substituting any part or tool for the one recommended, you must be completely satisfied that neither your personal safety, nor the performance of the vehicle will be endangered.

ACKNOWLEDGEMENT

Portions of materials contained herein have been reprinted with permission from General Motors Corporation, Service and Parts Operations under License Agreement #0510757. The publisher would like to express appreciation to General Motors Corporation for its assistance in producing this publication. No further reproduction or distribution of the material in this manual is allowed without the expressed written permission of the publisher.

PRECAUTIONS

Before servicing any vehicle, please be sure to read all of the following precautions, which deal with personal safety, prevention of component damage, and important points to take into consideration when servicing a motor vehicle:

- Always wear safety glasses or goggles when drilling, cutting, grinding or prying.
- Steel-toed work shoes should be worn when working with heavy parts. Pockets should not be used for carrying tools. A slip or fall can drive a screwdriver into your body.
- Work surfaces, including tools and the floor should be kept clean of grease, oil or other slippery material.
- When working around moving parts, don't wear loose clothing. Long hair should be tied back under a hat or cap, or in a hair net.
- Always use tools only for the purpose for which they were designed. Never pry with a screwdriver.
- Keep a fire extinguisher and first aid kit handy.
- Always properly support the vehicle with approved stands or lift.
- Always have adequate ventilation when working with chemicals or hazardous material.
- Carbon monoxide is colorless, odorless and dangerous. If it is necessary to operate the engine with vehicle in a closed area such as a garage, always use an exhaust collector to vent the exhaust gases outside the closed area.
- When draining coolant, keep in mind that small children and some pets are attracted by ethylene glycol antifreeze, and are quite likely to drink any left in an open container, or in puddles on the ground. This will prove fatal in sufficient quantity. Always drain the coolant into a sealable container.
- To avoid personal injury, do not remove the coolant pressure relief cap while the engine is operating or hot. The cooling system is under pressure; steam and hot liquid can come out forcefully when the cap is loosened slightly. Failure to follow these instructions may result in personal injury. The coolant must be recovered in a suitable, clean container for reuse. If the coolant is contaminated it must be recycled or disposed of correctly.
- When carrying out maintenance on the starting system be aware that heavy gauge leads are connected directly to the battery. Make sure the protective caps are in place when maintenance is completed. Failure to follow these instructions may result in personal injury.
- Do not remove any part of the engine emission control system. Operating the engine without the engine emission control system will reduce fuel economy and engine ventilation. This will weaken engine performance and shorten engine life. It is also a violation of Federal law.
- Due to environmental concerns, when the air conditioning system is drained, the refrigerant must be collected using refrigerant recovery/recycling equipment. Federal law requires that refrigerant be recovered into appropriate recovery equipment and the process be conducted by qualified technicians who have been certified by an approved organization, such as MACS, ASI, etc. Use of a recovery machine dedicated to the appropriate refrigerant is necessary to reduce the possibility of oil and refrigerant incompatibility concerns. Refer to the instructions provided by the equipment manufacturer when removing refrigerant from or charging the air conditioning system.
- Always disconnect the battery ground when working on or around the electrical system.
- Batteries contain sulfuric acid. Avoid contact with skin, eyes, or clothing. Also, shield your eyes when working near batteries to protect against possible splashing of the acid solution. In case of acid contact with skin or eyes, flush immediately with water for a minimum of 15 minutes and get prompt medical attention. If acid is swallowed, call a physician immediately. Failure to follow these instructions may result in personal injury.
- Batteries normally produce explosive gases. Therefore, do not allow flames, sparks or lighted substances to come near the battery. When charging or working near a battery, always shield your face and protect your eyes. Always provide ventilation. Failure to follow these instructions may result in personal injury.
- When lifting a battery, excessive pressure on the end walls could cause acid to spew through the vent caps, resulting in personal injury, damage to the vehicle or battery. Lift with a battery carrier or with your hands on opposite corners. Failure to follow these instructions may result in personal injury.
- Observe all applicable safety precau-

tions when working around fuel. Whenever servicing the fuel system, always work in a well-ventilated area. Do not allow fuel spray or vapors to come in contact with a spark, open flame, or excessive heat (a hot drop light, for example). Keep a dry chemical fire extinguisher near the work area. Always keep fuel in a container specifically designed for fuel storage; also, always properly seal fuel containers to avoid the possibility of fire or explosion. Do not smoke or carry lighted tobacco or open flame of any type when working on or near any fuel-related components.

• Fuel injection systems often remain pressurized, even after the engine has been turned OFF. The fuel system pressure must be relieved before disconnecting any fuel lines. Failure to do so may result in fire and/or personal injury.

• The evaporative emissions system contains fuel vapor and condensed fuel vapor. Although not present in large quantities, it still presents the danger of explosion or fire. Disconnect the battery ground cable from the battery to minimize the possibility of an electrical spark occurring, possibly causing a fire or explosion if fuel vapor or liquid fuel is present in the area. Failure to follow these instructions can result in personal injury.

• The EPA warns that prolonged contact with used engine oil may cause a number of skin disorders, including cancer! You should make every effort to minimize your exposure to used engine oil. Protective gloves should be worn when changing oil. Wash your hands and any other exposed skin areas as soon as possible after exposure to used engine oil. Soap and water, or waterless hand cleaner should be used.

• Some vehicles are equipped with an air bag system, often referred to as a Supplemental Restraint System (SRS) or Supplemental Inflatable Restraint (SIR) system. The system must be disabled before performing service on or around system components, steering column, instrument panel components, wiring and sensors. Failure to follow safety and disabling procedures could result in accidental air bag deployment, possible personal injury and unnecessary system repairs.

• Always wear safety goggles when working with, or around, the air bag system. When carrying a non-deployed air bag, be sure the bag and trim cover are pointed away from your body. When placing a non-deployed air bag on a work surface, always face the bag and trim cover upward, away from the surface. This will reduce the motion of the module if it is accidentally deployed.

• Electronic modules are sensitive to electrical charges. The ABS module can be damaged if exposed to these charges.

• Brake pads and shoes may contain asbestos, which has been determined to be a cancer-causing agent. Never clean brake surfaces with compressed air. Avoid inhaling brake dust. Clean all brake surfaces with a commercially available brake cleaning fluid.

• When replacing brake pads, shoes, discs or drums, replace them as complete axle sets.

• When servicing drum brakes, disassemble and assemble one side at a time, leaving the remaining side intact for reference.

• Brake fluid often contains polyglycol ethers and polyglycols. Avoid contact with the eyes and wash your hands thoroughly after handling brake fluid. If you do get brake fluid in your eyes, flush your eyes with clean, running water for 15 minutes. If eye irritation persists, or if you have taken brake fluid internally, immediately seek medical assistance.

• Clean, high quality brake fluid from a sealed container is essential to the safe and proper operation of the brake system. You should always buy the correct type of brake fluid for your vehicle. If the brake fluid becomes contaminated, completely flush the system with new fluid. Never reuse any brake fluid. Any brake fluid that is removed from the system should be discarded. Also, do not allow any brake fluid to come in contact with a painted or plastic surface; it will damage the paint.

• Never operate the engine without the proper amount and type of engine oil; doing so will result in severe engine damage.

• Timing belt maintenance is extremely important! Many models utilize an interference-type, non-freewheeling engine. If the timing belt breaks, the valves in the cylinder head may strike the pistons, causing potentially serious (also time-consuming and expensive) engine damage.

• Disconnecting the negative battery cable on some vehicles may interfere with the functions of the on-board computer system(s) and may require the computer to undergo a relearning process once the negative battery cable is reconnected.

• Steering and suspension fasteners are critical parts because they affect performance of vital components and systems and their failure can result in major service expense. They must be replaced with the same grade or part number or an equivalent part if replacement is necessary. Do not use a replacement part of lesser quality or substitute design. Torque values must be used as specified during reassembly to ensure proper retention of these parts.

OLDSMOBILE AND PONTIAC

Alero • Grand Am

1

ENGINE REPAIR 1-12
BRAKES 1-56
DRIVE TRAIN 1-42
FUEL SYSTEM 1-39
SPECIFICATIONS AND
  MAINTENANCE CHARTS 1-2
Engine and Vehicle Identification 1-2
General Engine Identification 1-2
Engine Tune-Up Specifications 1-3
Firing Orders 1-3
Accessory Drive Belt Routing 1-4
Capacities 1-5
Valve Specifications 1-5
Crankshaft and Connecting Rod
  Specifications 1-6
Piston and Ring Specifications 1-7
Torque Specifications 1-8
Wheel Alignment 1-8
Tire, Wheel and Ball Joint
  Specifications 1-9
Brake Specifications 1-9
Scheduled Maintenance
  Intervals...................................1-10
STEERING AND
  SUSPENSION 1-48

A
Air Bag..1-48
  Disarming 1-48
  Precautions 1-48
  Rearming 1-48
Alternator 1-12
  Installation 1-12
  Removal....................................1-12

B
Brake Caliper 1-56
  Removal & Installation................1-56
Brake Drums..................................1-56
  Removal & Installation...............1-56
Brake Shoes..................................1-56
  Removal & Installation...............1-56

C
Camshaft and Valve Lifters 1-26
  Removal & Installation...............1-26

Clutch 1-43
  Removal & Installation..............1-43
Coil Spring.....................................1-51
  Removal & Installation..............1-51
CV-Joints 1-45
  Overhaul 1-45
Cylinder Head 1-18
  Removal & Installation..............1-18

D
Disc Brake Pads.............................1-56
  Removal & Installation..............1-56

E
Engine Assembly 1-12
  Removal & Installation..............1-12
Exhaust Manifold 1-24
  Removal & Installation..............1-24

F
Fuel Filter 1-39
  Removal & Installation..............1-39
Fuel Injector...................................1-41
  Removal & Installation..............1-41
Fuel Pump 1-39
  Removal & Installation..............1-39
Fuel System Pressure 1-39
  Relieving....................................1-39
Fuel System Service
  Precautions..............................1-39

H
Halfshaft..1-44
  Removal & Installation..............1-44
Heater Core....................................1-17
  Removal & Installation..............1-17
Hydraulic Clutch System 1-44
  Bleeding....................................1-44

I
Ignition Timing 1-12
  Adjustment................................1-12
Intake Manifold 1-21
  Removal & Installation..............1-21

L
Lower Ball Joint..............................1-52
  Removal & Installation..............1-52
Lower Control Arm 1-53

Control Arm Bushing
  Replacement............................1-53
  Removal & Installation..............1-53

O
Oil Pan..1-29
  Removal & Installation..............1-29
Oil Pump 1-31
  Removal & Installation..............1-31

P
Piston and Ring 1-38
  Positioning 1-38
Power Rack and Pinion
  Steering Gear 1-48
  Removal & Installation..............1-48

R
Rear Main Seal 1-33
  Removal & Installation..............1-33
Rocker Arms 1-21
  Removal & Installation..............1-21

S
Starter Motor 1-29
  Removal & Installation..............1-29
Strut..1-49
  Removal & Installation..............1-49

T
Timing Chain, Sprockets,
  Front Cover and Seal...................1-34
  Removal & Installation..............1-34
Torsion Bars 1-52
  Removal & Installation..............1-52
Transaxle.......................................1-42
  Removal & Installation..............1-42

V
Valve Lash 1-29
  Adjustment................................1-29

W
Water Pump 1-15
  Removal & Installation..............1-15
Wheel Bearings..............................1-54
  Adjustment................................1-54
  Removal & Installation..............1-54

SPECIFICATIONS AND MAINTENANCE CHARTS

ENGINE AND VEHICLE IDENTIFICATION

		Engine						Model Year	
Code ①	Liters (cc)	Cu. In. (cc)	Cyl.	Fuel Sys.	Engine Type	Eng. Mfg.		Code ②	Year
E	3.4 (3350)	207	6	MFI	OHV	BOC		2	2002
F	2.2 (2180)	134	4	MFI	OHV	BOC		3	2003
T	2.4 (2392)	146	4	MFI	DOHC	CUS		4	2004
								5	2005
								6	2006

BOC: Buick/Oldsmobile/Cadillac
CUS: Chevrolet/United States
MFI: Multi-point Fuel Injection

① 8th position of VIN
② 10th position of VIN

06025 -NBOD-C01

GENERAL ENGINE IDENTIFICATION

Year	Model	Engine Displacement Liters (VIN)	Net Horsepower @ rpm	Net Torque @ rpm (ft. lbs.)	Bore x Stroke (in.)	Com- pression Ratio	Oil Pressure @ rpm
2002	Alero	2.4 (T)	150@5600	155@4400	3.54x3.70	9.5:1	30@3000
	Alero	3.4 (E)	170@4800	200@4000	3.62x3.31	9.5:1	15@1100
	Alero	2.2 (F)	148@5600	150@4000	3.50x3.46	10.0:1	50-80@1000
	Grand Am	2.4 (T)	150@5600	155@4400	3.54x3.70	9.5:1	30@3000
	Grand Am	2.2 (F)	148@5600	150@4000	3.50x3.46	10.0:1	50-80@1000
	Grand Am	3.4 E)	170@4800	200@4000	3.62x3.31	9.5:1	15@1100
2003	Alero	2.4 (T)	150@5600	155@4400	3.54x3.70	9.5:1	30@3000
	Alero	3.4 (E)	170@4800	200@4000	3.62x3.31	9.5:1	15@1100
	Alero	2.2 (F)	148@5600	150@4000	3.50x3.46	10.0:1	50-80@1000
	Grand Am	2.4 (T)	150@5600	155@4400	3.54x3.70	9.5:1	30@3000
	Grand Am	2.2 (F)	148@5600	150@4000	3.50x3.46	10.0:1	50-80@1000
	Grand Am	3.4 E)	170@4800	200@4000	3.62x3.31	9.5:1	15@1100
2004	Alero	2.4 (T)	150@5600	155@4400	3.54x3.70	9.5:1	30@3000
	Alero	3.4 (E)	170@4800	200@4000	3.62x3.31	9.5:1	15@1100
	Alero	2.2 (F)	148@5600	150@4000	3.50x3.46	10.0:1	50-80@1000
	Grand Am	2.4 (T)	150@5600	155@4400	3.54x3.70	9.5:1	30@3000
	Grand Am	2.2 (F)	148@5600	150@4000	3.50x3.46	10.0:1	50-80@1000
	Grand Am	3.4 E)	170@4800	200@4000	3.62x3.31	9.5:1	15@1100
2005	Grand Am	2.2 (F)	148@5600	150@4000	3.38x3.72	10.0:1	50-80@1000
	Grand Am	3.4 E)	175@4800	205@4000	3.62x3.31	9.6:1	60@1800

MFI: Multi-point Fuel Injection

06025 -NBOD-C02

ENGINE TUNE-UP SPECIFICATIONS

Year	Engine Displacement Liters (VIN)	Spark Plugs Gap (in.)	Ignition Timing (deg.) MT	Ignition Timing (deg.) AT	Fuel Pump (psi)	Idle Speed (rpm) MT	Idle Speed (rpm) AT	Valve Clearance In.	Valve Clearance Ex.
2002	2.4 (T)	0.050	①	①	41-47	①	①	HYD	HYD
	2.2 (F)	0.060	①	①	50-60	①	①	HYD	HYD
	3.4 (E)	0.060	①	①	41-47	①	①	HYD	HYD
2003	2.4 (T)	0.050	①	①	41-47	①	①	HYD	HYD
	2.2 (F)	0.060	①	①	50-60	①	①	HYD	HYD
	3.4 (E)	0.060	①	①	41-47	①	①	HYD	HYD
2004	2.4 (T)	0.050	①	①	41-47	①	①	HYD	HYD
	2.2 (F)	0.060	①	①	50-60	①	①	HYD	HYD
	3.4 (E)	0.060	①	①	41-47	①	①	HYD	HYD
2005	2.2 (F)	0.040-0.046	①	①	50-60	①	①	HYD	HYD
	3.4 (E)	0.060	①	①	52-59	①	①	HYD	HYD

NOTE: The Vehicle Emission Control Information label often reflects specification changes made during production. The label figures must be used if they differ from those in this chart.

HYD: Hydraulic

① Refer to Vehicle Emission Control Information label

06025 -NBOD-C03

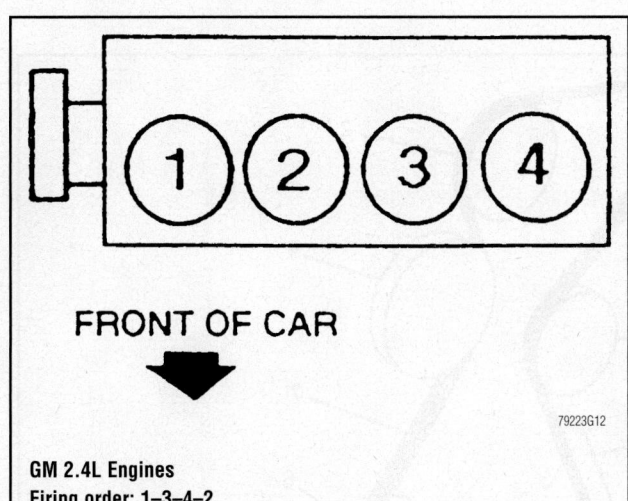

FRONT OF CAR

79223G12

GM 2.4L Engines
Firing order: 1–3–4–2
Distributorless ignition system

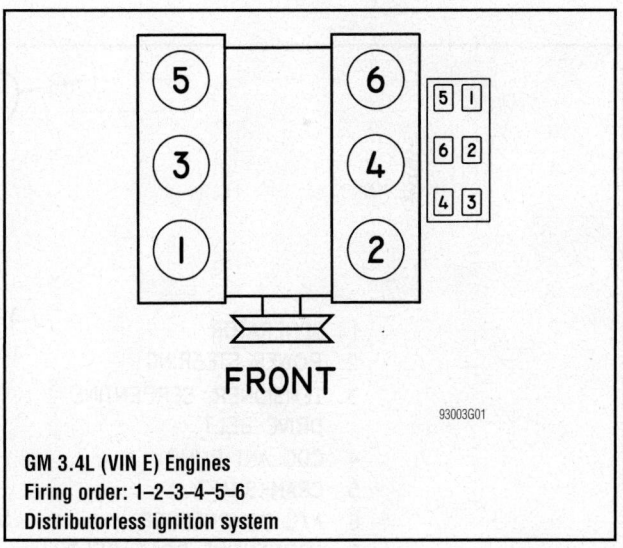

FRONT

93003G01

GM 3.4L (VIN E) Engines
Firing order: 1–2–3–4–5–6
Distributorless ignition system

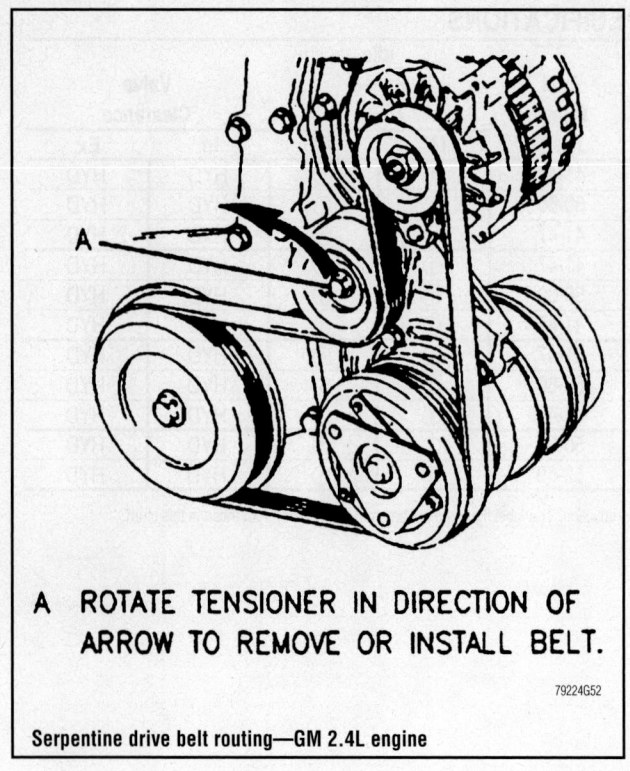

A ROTATE TENSIONER IN DIRECTION OF ARROW TO REMOVE OR INSTALL BELT.

79224G52

Serpentine drive belt routing—GM 2.4L engine

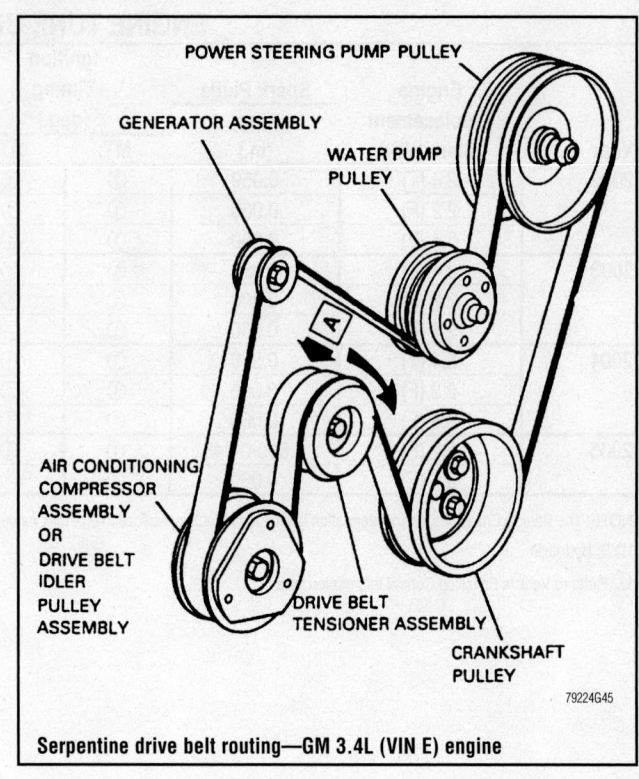

POWER STEERING PUMP PULLEY

GENERATOR ASSEMBLY

WATER PUMP PULLEY

AIR CONDITIONING COMPRESSOR ASSEMBLY OR DRIVE BELT IDLER PULLEY ASSEMBLY

DRIVE BELT TENSIONER ASSEMBLY

CRANKSHAFT PULLEY

79224G45

Serpentine drive belt routing—GM 3.4L (VIN E) engine

1 GENERATOR
2 POWER STEERING
3 TENSIONER, SERPENTINE DRIVE BELT
4 COOLANT PUMP
5 CRANKSHAFT
6 A/C COMPRESSOR
7 SERPENTINE DRIVE BELT
8 SERPENTINE DRIVE BELT ROTATION

79224G53

Serpentine drive belt routing—GM 3.1L engine

CAPACITIES

Year	Model	Engine Displacement Liters (VIN)	Engine ID/VIN	Engine Oil with Filter (qts.)	Transmission (pts.)		Fuel Tank (gal.)	Cooling System (qts.)
					Manual	Auto.		
2002	Alero	2.4 (T)	T	4.0	3.6	13.8	14.3	11.3
	Alero	3.4 (E)	E	4.5	3.6	13.8	14.3	13.6
	Alero	2.2 (F)	F	4.0	3.6	13.8	14.3	11.3
	Grand Am	2.2 (F)	F	4.0	3.6	13.8	14.3	11.3
	Grand Am	2.4 (T)	T	4.0	3.6	13.8	14.3	11.3
	Grand Am	3.4 (E)	E	4.5	3.6	13.8	14.3	13.6
2003	Alero	2.4 (T)	T	4.0	3.6	13.8	14.3	11.3
	Alero	3.4 (E)	E	4.5	3.6	13.8	14.3	13.6
	Alero	2.2 (F)	F	4.0	3.6	13.8	14.3	11.3
	Grand Am	2.2 (F)	F	4.0	3.6	13.8	14.3	11.3
	Grand Am	2.4 (T)	T	4.0	3.6	13.8	14.3	11.3
	Grand Am	3.4 (E)	E	4.5	3.6	13.8	14.3	13.6
2004	Alero	2.4 (T)	T	4.0	3.6	13.8	14.3	11.3
	Alero	3.4 (E)	E	4.5	3.6	13.8	14.3	13.6
	Alero	2.2 (F)	F	4.0	3.6	13.8	14.3	11.3
	Grand Am	2.2 (F)	F	4.0	3.6	13.8	14.3	11.3
	Grand Am	2.4 (T)	T	4.0	3.6	13.8	14.3	11.3
	Grand Am	3.4 (E)	E	4.5	3.6	13.8	14.3	13.6
2005	Grand Am	2.2 (F)	F	5.0	3.6	13.8	14.1	8.6
	Grand Am	3.4 (E)	E	4.5	3.6	13.8	14.1	13.6

NOTE: All capacities are approximate. Add fluid gradually and ensure a proper fluid level is obtained.

06025 -NBOD-C04

VALVE SPECIFICATIONS

Year	Engine Displacement Liters (VIN)	Seat Angle (deg.)	Face Angle (deg.)	Spring Test Pressure (lbs. @ in.)	Spring Installed Height (in.)	Stem-to-Guide Clearance (in.)		Stem Diameter (in.)	
						Intake	Exhaust	Intake	Exhaust
2002	2.4 (T)	45	46	50-55@ 1.437	1.437	0.0009- 0.0025	0.0016- 0.0032	0.2331- 0.2339	0.2326- 0.2334
	2.2 (F)	45	46	NA	NA	0.2362- 0.2367	0.2362- 0.2367	0.2344- 0.2355	0.2337- 0.2343
	3.4 (E)	45	45	75@ 1.701	1.701	0.0010- 0.0027	0.0010- 0.0027	NA	NA
2003	2.4 (T)	45	46	50-55@ 1.437	1.437	0.0009- 0.0025	0.0016- 0.0032	0.2331- 0.2339	0.2326- 0.2334
	2.2 (F)	45	46	NA	NA	0.2362- 0.2367	0.2362- 0.2367	0.2344- 0.2355	0.2337- 0.2343
	3.4 (E)	45	45	75@ 1.701	1.701	0.0010- 0.0027	0.0010- 0.0027	NA	NA
2004	2.4 (T)	45	46	50-55@ 1.437	1.437	0.0009- 0.0025	0.0016- 0.0032	0.2331- 0.2339	0.2326- 0.2334
	2.2 (F)	45	46	NA	NA	0.2362- 0.2367	0.2362- 0.2367	0.2344- 0.2355	0.2337- 0.2343
	3.4 (E)	45	45	75@ 1.701	1.701	0.0010- 0.0027	0.0010- 0.0027	NA	NA
2005	2.2 (F)	45	46	NA	NA	0.0012- 0.0022	0.0020- 0.0026	0.2344- 0.2355	0.2337- 0.2343
	3.4 (E)	45	46	75@ 1.701	1.701	0.0010- 0.0027	0.0010- 0.0027	NA	NA

NA: Not available

06025 -NBOD-C05

CRANKSHAFT AND CONNECTING ROD SPECIFICATIONS
All measurements are given in inches.

Year	Engine Displacement Liters (VIN)	Crankshaft				Connecting Rod		
		Main Brg. Journal Dia.	Main Brg. Oil Clearance	Shaft End-play	Thrust on No.	Journal Diameter	Oil Clearance	Side Clearance
2002	2.4 (T)	2.3622-2.3631	0.0004-0.0023	0.0034-0.0095	3	1.8887-1.8897	0.0004-0.0026	0.0059-0.0177
	2.2 (F)	2.2045-2.2050	0.0012-0.0026	0.0012-0.0150	3	2.0519-2.0525	0.0004-0.0026	0.0028-0.0146
	3.4 (E)	2.6473-2.6483	0.0008-① 0.0025	0.0024-0.0083	3	1.9987-1.9994	0.0007-0.0024	0.0070-0.0170
2003	2.4 (T)	2.3622-2.3631	0.0004-0.0023	0.0034-0.0095	3	1.8887-1.8897	0.0004-0.0026	0.0059-0.0177
	2.2 (F)	2.2045-2.2050	0.0012-0.0026	0.0012-0.0150	3	2.0519-2.0525	0.0004-0.0026	0.0028-0.0146
	3.4 (E)	2.6473-2.6483	0.0008-① 0.0025	0.0024-0.0083	3	1.9987-1.9994	0.0007-0.0024	0.0070-0.0170
2004	2.4 (T)	2.3622-2.3631	0.0004-0.0023	0.0034-0.0095	3	1.8887-1.8897	0.0004-0.0026	0.0059-0.0177
	2.2 (F)	2.2045-2.2050	0.0012-0.0026	0.0012-0.0150	3	2.0519-2.0525	0.0004-0.0026	0.0028-0.0146
	3.4 (E)	2.6473-2.6483	0.0008-① 0.0025	0.0024-0.0083	3	1.9987-1.9994	0.0007-0.0024	0.0070-0.0170
2005	2.2 (F)	2.2045-2.2050	0.0012-0.0026	0.0012-0.0150	3	1.9291-1.9297	0.0004-0.0026	0.0028-0.0146
	3.4 (E)	2.6473-2.6483	0.0008-① 0.0025	0.0024-0.0083	3	1.9987-1.9994	0.0007-0.0024	0.0100-0.0150

① Thrust bearing: 0.0012 - 0.0030

06025 -NBOD-C06

PISTON AND RING SPECIFICATIONS

All measurements are given in inches.

Year	Engine Displacement Liters (VIN)	Piston Clearance	Ring Gap			Ring Side Clearance		
			Top Compression	Bottom Compression	Oil Control	Top Compression	Bottom Compression	Oil Control
2002	2.4 (T)	0.0006-0.0015	0.006-0.012	0.010-0.016	0.010-0.030	0.0016-0.0031	0.0012-0.0028	0.0005-0.0089
	2.2 (F)	0.0004-0.0016	0.008-0.016	0.014-0.022	0.010-0.0300	0.0015-0.0031	0.0012-0.0028	0.0035-0.0052
	3.4 (E)	0.0013-0.0027	0.006-0.014	0.0197-0.0280	0.0098-0.0500	0.0020-0.0033	0.0020-0.0035	0.0080
2003	2.4 (T)	0.0006-0.0015	0.006-0.012	0.010-0.016	0.010-0.030	0.0016-0.0031	0.0012-0.0028	0.0005-0.0089
	2.2 (F)	0.0004-0.0016	0.008-0.016	0.014-0.022	0.010-0.0300	0.0015-0.0031	0.0012-0.0028	0.0035-0.0052
	3.4 (E)	0.0013-0.0027	0.006-0.014	0.0197-0.0280	0.0098-0.0500	0.0020-0.0033	0.0020-0.0035	0.0080
2004	2.4 (T)	0.0006-0.0015	0.006-0.012	0.010-0.016	0.010-0.030	0.0016-0.0031	0.0012-0.0028	0.0005-0.0089
	2.2 (F)	0.0004-0.0016	0.008-0.016	0.014-0.022	0.010-0.0300	0.0015-0.0031	0.0012-0.0028	0.0035-0.0052
	3.4 (E)	0.0013-0.0027	0.006-0.014	0.0197-0.0280	0.0098-0.0500	0.0020-0.0033	0.0020-0.0035	0.0080
2005	2.2 (F)	0.0004-0.0016	0.008-0.016	0.014-0.022	0.010-0.0300	0.0015-0.0031	0.0012-0.0027	0.0035-0.0042
	3.4 (E)	①	0.006-0.014	0.0188-0.0291	0.0098-0.0303	0.0020-0.0033	0.0020-0.0031	0.0028-0.0037

① Production 1-4: 0.006-0.0020
Production 5-6: -0.0003-0.0018

06025 -NBOD-C07

TORQUE SPECIFICATIONS
All readings in ft. lbs.

Year	Engine Displacement Liters (VIN)	Cylinder Head Bolts	Main Bearing Bolts	Rod Bearing Bolts	Crankshaft Damper Bolts	Flywheel Bolts	Manifold Intake	Manifold Exhaust	Spark Plug	Oil Pan Drain Plug
2002	2.4 (T)	①	②	③	④	⑤	⑥	⑦	13	18
	2.2 (F)	⑧	⑨	⑩	⑪	⑫	⑬	⑭	15	18
	3.4 (E)	⑮	⑯	⑰	⑱	52	⑲	12	15	18
2003	2.4 (T)	①	②	③	④	⑤	⑥	⑦	13	18
	2.2 (F)	⑧	⑨	⑩	⑪	⑫	⑬	⑭	15	18
	3.4 (E)	⑮	⑯	⑰	⑱	52	⑲	12	15	18
2004	2.4 (T)	①	②	③	④	⑤	⑥	⑦	13	18
	2.2 (F)	⑧	⑨	⑩	⑪	⑫	⑬	⑭	15	18
	3.4 (E)	⑮	⑯	⑰	76	52	⑰	12	20	18
2005	2.2 (F)	⑧	⑨	⑩	⑪	⑫	⑬	⑭	15	18
	3.4 (E)	⑮	⑯	⑰	⑱	52	⑲	12	15	18

① Nos. 1-8: 40 ft. lbs.
Nos. 9-10: 30 ft. lbs.
Tighten all bolts an additional 90 degrees
② 15 ft. lbs. plus 90 degrees
③ 18 ft. lbs. plus 80 degrees
④ 129 ft. lbs. plus 90 degrees
⑤ 22 ft. lbs. plus 45 degrees
⑥ Nuts: 19 ft. lbs.
Studs: 97 inch lbs.
⑦ Nuts: 11 ft. lbs.
Studs: 97 inch lbs.
⑧ First pass: 15 ft. lbs.
Second pass: 155 degrees
⑨ 15 ft. lbs. plus 70 degrees
⑩ 18 ft. lbs. plus 100 degrees

⑪ First pass: 74 ft. lbs.
Second pass: 125 degrees
⑫ 39 ft. lbs. plus 25 degrees
⑬ Manifold nut/bolt: 89 inch lbs.
Manifold stud: 53 inch lbs.
⑭ Manifold stud: 89 inch lbs.
Manifold nuts 124 inch lbs.
⑮ 44 ft. lbs. plus 95 degrees
⑯ 37 ft. lbs. plus 77 degrees
⑰ 15 ft. lbs. plus 75 degrees
⑱ 52 ft. lbs. plus 72 degrees
⑲ Lower: 115 inch lbs.
Upper: 18 ft. lbs.

06025 -NBOD-C08

WHEEL ALIGNMENT

Year	Model		Caster Range (+/-Deg.)	Caster Preferred Setting (Deg.)	Camber Range (+/-Deg.)	Camber Preferred Setting (Deg.)	Toe-in (in.)
2002	Alero	F	1.00	+4.10	1.00	-0.20	0.05 +/- 0.12
		R	—	—	0.50	-0.20	0.03 +/- 0.04
	Grand Am	F	1.00	+4.10	1.00	-0.20	0.05 +/- 0.13
		R	—	—	0.50	-0.20	0.03 +/- 0.10
2003	Alero	F	1.00	+4.10	1.00	-0.20	0.05 +/- 0.12
		R	—	—	0.50	-0.20	0.03 +/- 0.04
	Grand Am	F	1.00	+4.10	1.00	-0.20	0.05 +/- 0.13
		R	—	—	0.50	-0.20	0.03 +/- 0.10
2004	Alero	F	1.00	+4.10	1.00	-0.20	0.05 +/- 0.12
		R	—	—	0.50	-0.20	0.03 +/- 0.04
	Grand Am	F	1.00	+4.10	1.00	-0.20	0.05 +/- 0.13
		R	—	—	0.50	-0.20	0.03 +/- 0.10
2005	Grand Am	F	1.00	+4.10	0.75	-0.20	0.05 +/- 0.20
		R	—	—	0.50	-0.20	0.10 +/- 0.20

06025 -NBOD-C09

TIRE, WHEEL AND BALL JOINT SPECIFICATIONS

| Year | Model | OEM Tires | | Tire Pressures (psi) | | Wheel Size | Ball Joint Inspection | Lug Nut |
		Standard	Optional	Front	Rear			
2002	Alero GX, GL	P215/60R15	P225/50HR15	30	30	6-JJ	①	100
	Alero GL1, GL2, GL3	P225/50VR16	None	30	30	6-JJ	①	100
	Alero GLS	P225/50R16	None	30	30	6-JJ	①	100
	Grand AM SE/SE1	P215/60R15	None	30	30	6-JJ	0.125 in.	100
	Grand Am SE2	P225/50R16	None	30	30	6-JJ	0.125 in.	100
	Grand Am GT, GT1	P225/50R16	None	30	30	6-JJ	0.125 in.	100
2003	Alero GX, GL	P215/60R15	P225/50HR15	30	30	6-JJ	①	100
	Alero GL1, GL2, GL3	P225/50VR16	None	30	30	6-JJ	①	100
	Alero GLS	P225/50R16	None	30	30	6-JJ	①	100
	Grand AM SE/SE1	P215/60R15	None	30	30	6-JJ	0.125 in.	100
	Grand Am SE2	P225/50R16	None	30	30	6-JJ	0.125 in.	100
	Grand Am GT, GT1	P225/50R16	None	30	30	6-JJ	0.125 in.	100
2004	Alero GX, GL	P215/60R15	P225/50HR15	30	30	6-JJ	①	100
	Alero GL1, GL2, GL3	P225/50VR16	None	30	30	6-JJ	①	100
	Alero GLS	P225/50R16	None	30	30	6-JJ	①	100
	Grand AM SE/SE1	P215/60R15	None	30	30	6-JJ	0.125 in.	100
	Grand Am SE2	P225/50R16	None	30	30	6-JJ	0.125 in.	100
	Grand Am GT, GT1	P225/50R16	None	30	30	6-JJ	0.125 in.	100
2005	Grand AM SE/SE1	P215/60R15	None	30	30	6-JJ	0.125 in.	100
	Grand Am SE2	P225/50R16	None	30	30	6-JJ	0.125 in.	100
	Grand Am GT, GT1	P225/50R16	None	30	30	6-JJ	0.125 in.	100

OEM: Original Equipment Manufacturer

PSI: Pounds Per Square Inch

STD: Standard

OPT: Optional

① Replace if any measurable movement is found.

06025 -NBOD-C10

BRAKE SPECIFICATIONS
All measurements in inches unless noted

| Year | Model | | Brake Disc | | | Brake Drum Diameter | | | Minimum Lining Thickness | | Brake Caliper | |
			Original Thickness	Minimum Thickness	Maximum Runout	Original Inside Diameter	Max. Wear Limit	Maximum Machine Diameter	Front	Rear	Bracket Bolts (ft. lbs.)	Mounting Bolts (ft. lbs.)
2002	Alero	F	1.031	0.972	0.003	—	—	—	0.030	—	85	23
		R	0.430	0.410	—	8.863	8.920	8.909	—	①	85	81
	Grand Am	F	1.031	0.972	0.003	—	—	—	0.030	—	85	23
		R	0.430	0.410	—	8.863	8.920	8.909	—	①	85	81
2003	Alero	F	1.031	0.972	0.003	—	—	—	0.030	—	85	23
		R	0.430	0.410	—	8.863	8.920	8.909	—	①	85	81
	Grand Am	F	1.031	0.972	0.003	—	—	—	0.030	—	85	23
		R	0.430	0.410	—	8.863	8.920	8.909	—	①	85	81
2004	Alero	F	1.031	0.972	0.003	—	—	—	0.030	—	85	23
		R	0.430	0.410	—	8.863	8.920	8.909	—	①	85	81
	Grand Am	F	1.031	0.972	0.003	—	—	—	0.030	—	85	23
		R	0.430	0.410	—	8.863	8.920	8.909	—	①	85	81
2005	Grand Am	F	1.031	0.972	0.001	—	—	—	0.030	—	85	23
		R	0.433	0.354	0.001	8.868	8.909	8.889	—	①	85	81

NA: Not available

① 0.030 over rivet head; If bonded lining, use 0.030 from shoe

06025 -NBOD-C11

SCHEDULED MAINTENANCE INTERVALS
GM N BODY—2002-04 OLDSMOBILE ALERO & PONTIAC GRAND AM

TO BE SERVICED	TYPE OF SERVICE	VEHICLE MILEAGE INTERVAL (x1000)												
		7.5	15	22.5	30	37.5	45	52.5	60	67.5	75	82.5	90	97.5
Engine oil & filter	R	✓	✓	✓	✓	✓	✓	✓	✓	✓	✓	✓	✓	✓
Automatic transaxle fluid & filter ①	S/I	✓	✓	✓	✓	✓	✓	✓	✓	✓	✓	✓	✓	✓
Brake hoses	S/I	✓	✓	✓	✓	✓	✓	✓	✓	✓	✓	✓	✓	✓
Chassis lubrication	S/I	✓	✓	✓	✓	✓	✓	✓	✓	✓	✓	✓	✓	✓
Coolant level, hoses & clamps	S/I	✓	✓	✓	✓	✓	✓	✓	✓	✓	✓	✓	✓	✓
Driveshaft boots & front suspension components	S/I	✓	✓	✓	✓	✓	✓	✓	✓	✓	✓	✓	✓	✓
Exhaust system	S/I	✓	✓	✓	✓	✓	✓	✓	✓	✓	✓	✓	✓	✓
Lubricate chassis & suspension	S/I	✓	✓	✓	✓	✓	✓	✓	✓	✓	✓	✓	✓	✓
Lubricate steering linkage & transaxle shift linkage	S/I	✓	✓	✓	✓	✓	✓	✓	✓	✓	✓	✓	✓	✓
Lubricate parking brake cable guides, underbody contact points & linkage	S/I	✓	✓	✓	✓	✓	✓	✓	✓	✓	✓	✓	✓	✓
Manual transaxle oil	S/I	✓	✓	✓	✓	✓	✓	✓	✓	✓	✓	✓	✓	✓
Throttle linkage	S/I	✓	✓	✓	✓	✓	✓	✓	✓	✓	✓	✓	✓	✓
Brake linings	S/I	✓		✓		✓		✓		✓		✓		✓
Rotate tires	S/I	✓		✓		✓		✓		✓		✓		✓
Air filter element & PCV filter	R				✓				✓				✓	
Engine coolant ②	R													
Spark plugs ③	R				✓				✓				✓	
Accessory drive belt(s)	S/I				✓				✓				✓	
EGR & fuel systems	S/I				✓				✓				✓	
Ignition cables	S/I				✓				✓				✓	

R: Replace

S/I: Service or Inspect

① Automatic transaxle fluid & filter: replace at 100,000 miles (if not changed previously).

② Engine coolant: replace every 100,000 miles. Use O.E. specified (DEX-COOL™) coolant only. If any silicate coolant is used, the service interval is every 30,000 miles.

③ Platinum tip spark plugs: replace every 100,000 miles.

FREQUENT OPERATION MAINTENANCE (SEVERE SERVICE) ADDITIONS

If a vehicle is operated under any of the following conditions it is considered severe service:

- Towing a trailer or using a camper or car-top carrier.

- Extensive idling or low-speed driving for long distances as in heavy commercial use, such as delivery, taxi or police cars.

- Operating on rough, muddy or salt-covered roads.

- Operating on unpaved or dusty roads.

- 50% or more of the vehicle operation is in 32°C (90°F) or higher temperatures, or constant operation in temperatures below 0°C (32°F).

Engine oil and filter: change every 3000 miles or 3 months, whichever occurs first.

Wheels and tires: inspect and rotate every 6000 miles.

Air cleaner element: inspect every 15,000 miles and replace or clean as needed. Replace it at least every 30,000 miles.

Automatic transaxle fluid & filter: replace every 50,000 miles.

SCHEDULED MAINTENANCE INTERVALS
GM N BODY—2005 PONTIAC GRAND AM

TO BE SERVICED	TYPE OF SERVICE	VEHICLE MILEAGE INTERVAL (x1000)												
		7.5	15	22.5	30	37.5	45	52.5	60	67.5	75	82.5	90	97.5
Engine oil & filter ①	R	✓	✓	✓	✓	✓	✓	✓	✓	✓	✓	✓	✓	✓
Automatic transaxle fluid & filter ②	S/I	✓	✓	✓	✓	✓	✓	✓	✓	✓	✓	✓	✓	✓
Brake hoses	S/I	✓	✓	✓	✓	✓	✓	✓	✓	✓	✓	✓	✓	✓
Chassis lubrication	S/I	✓	✓	✓	✓	✓	✓	✓	✓	✓	✓	✓	✓	✓
Coolant level, hoses & clamps	S/I	✓	✓	✓	✓	✓	✓	✓	✓	✓	✓	✓	✓	✓
Driveshaft boots & front suspension components	S/I	✓	✓	✓	✓	✓	✓	✓	✓	✓	✓	✓	✓	✓
Exhaust system	S/I	✓	✓	✓	✓	✓	✓	✓	✓	✓	✓	✓	✓	✓
Lubricate chassis & suspension	S/I	✓	✓	✓	✓	✓	✓	✓	✓	✓	✓	✓	✓	✓
Lubricate steering linkage & transaxle shift linkage	S/I	✓	✓	✓	✓	✓	✓	✓	✓	✓	✓	✓	✓	✓
Lubricate parking brake cable guides, underbody contact points & linkage	S/I	✓	✓	✓	✓	✓	✓	✓	✓	✓	✓	✓	✓	✓
Manual transaxle oil	S/I	✓	✓	✓	✓	✓	✓	✓	✓	✓	✓	✓	✓	✓
Throttle linkage	S/I	✓	✓	✓	✓	✓	✓	✓	✓	✓	✓	✓	✓	✓
Brake linings	S/I	✓		✓		✓		✓		✓		✓		✓
Rotate tires	S/I	✓		✓		✓		✓		✓		✓		✓
Air filter element & PCV filter	R				✓				✓				✓	
Engine coolant ③	R													
Spark plugs ④	R				✓				✓				✓	
Accessory drive belt(s)	S/I				✓				✓				✓	
EGR & fuel systems	S/I				✓				✓				✓	
Ignition cables	S/I				✓				✓				✓	

R: Replace

S/I: Service or Inspect

① When the change engine oil message in the Driver Information Center (DIC) comes on, it means that service is required for your vehicle. Have your vehicle serviced as soon as possible within the next 600 miles (1 000 km). It is possible that, if you are driving under the best conditions, the engine oil life system may not indicate that vehicle service is necessary for over a year. However, your engine oil and filter must be changed at least once a year and at this time the system must be reset.

After the oil has been changed, the CHANGE ENGINE OIL message must be reset. To reset the message, do the following:

Turn the ignition to ON, with the engine off.

Fully press and release the accelerator pedal slowly three times within five seconds. The reset is complete when you hear the chimes and the CHANGE OIL light goes out. However, if the light stays on Turn the key to OFF.

If the CHANGE OIL light comes back on when you start your vehicle, the engine oil life system has not reset. Repeat the procedure.

② Automatic transaxle fluid & filter: replace at 100,000 miles (if not changed previously).

③ Engine coolant: replace every 100,000 miles. Use O.E. specified (DEX-COOL™) coolant only. If any silicate coolant is used, the service interval is every 30,000 miles.

④ Platinum tip spark plugs: replace every 100,000 miles.

FREQUENT OPERATION MAINTENANCE (SEVERE SERVICE) ADDITIONS

If a vehicle is operated under any of the following conditions it is considered severe service:

- Towing a trailer or using a camper or car-top carrier.
- Extensive idling or low-speed driving for long distances as in heavy commercial use, such as delivery, taxi or police cars.
- Operating on rough, muddy or salt-covered roads.
- Operating on unpaved or dusty roads.
- 50% or more of the vehicle operation is in 32°C (90°F) or higher temperatures, or constant operation in temperatures below 0°C (32°F).

Engine oil and filter: change every 3000 miles or 3 months, whichever occurs first.

Wheels and tires: inspect and rotate every 6000 miles.

Air cleaner element: inspect every 15,000 miles and replace or clean as needed. Replace it at least every 30,000 miles.

Automatic transaxle fluid & filter: replace every 50,000 miles.

ENGINE REPAIR

Alternator

REMOVAL

2.2L Engine

1. Before servicing the vehicle, refer to the precautions section.
2. Remove or disconnect the following:
   - Negative battery cable
   - Air cleaner assembly
   - Oil dipstick tube bolt and position the tube aside
   - Drive belt
   - Electrical connectors from the alternator
   - Alternator bolts
   - Alternator

2.4L Engine

1. Before servicing the vehicle, refer to the precautions section.
2. Remove or disconnect the following:
   - Negative battery cable
   - Accessory drive belt
   - Alternator mounting bolts
   - Alternator electrical connectors
   - Alternator

3.4L Engine

1. Before servicing the vehicle, refer to the precautions section.
2. Remove or disconnect the following:
   - Negative battery cable
   - Accessory drive belt
   - Alternator electrical connectors

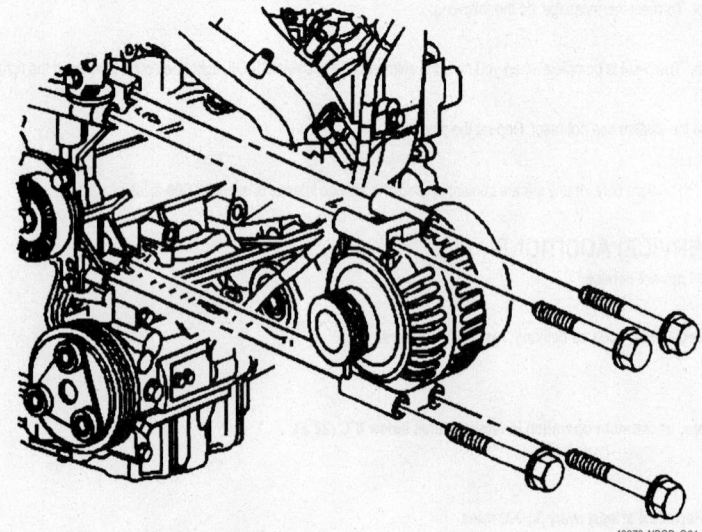

Alternator mounting—2.2L engine

42372-NB0D-G01

   - Power steering-to-alternator line clip
   - Alternator mounting nuts and bolts
   - Alternator

INSTALLATION

2.2L Engine

1. Install or connect the following:
   - Alternator
   - Alternator bolts and tighten to 15 ft. lbs. (20 Nm)
   - Electrical connectors to the alternator
   - Drive belt
   - Oil dipstick tube
   - Air cleaner assembly
   - Negative battery cable

2.4L Engine

1. Install or connect the following:
   - Alternator
   - Alternator electrical connectors
   - Alternator mounting bolts. Torque the bolts to 37 ft. lbs. (50 Nm).
   - Accessory drive belt
   - Negative battery cable

3.4L Engine

1. Install or connect the following:
   - Alternator. Torque the nuts to 22 ft. lbs. (30 Nm) and the bolts to 37 ft. lbs. (50 Nm).
   - Power steering-to-alternator line clip
   - Alternator electrical connectors

   - Accessory drive belt
   - Negative battery cable

Ignition Timing

ADJUSTMENT

The ignition timing is not adjustable, and is set electronically according to engine demand.

Engine Assembly

REMOVAL & INSTALLATION

2.2L Engine

MANUAL TRANSAXLE

1. Before servicing the vehicle, refer to the precautions section.
2. Remove or disconnect the following:
   - Hood
   - Negative battery cable
   - Air inlet duct and resonator
   - Accelerator and cruise control cable
   - Hose from the brake booster
   - Power steering pump bolts and set the pump aside
   - Fuel lines
   - Transmission shift control cables
   - Transmission shift control cables from the bracket
   - Clutch actuator cylinder from the transmission
3. Drain the cooling system.
   - Radiator inlet hose
   - Hose from the surge tank to the cylinder head
   - Outlet hose from the surge tank to the radiator
   - Bolt retaining the surge tank outlet hose to the intake manifold
   - Radiator outlet hose
   - Heater hoses
4. Disconnect the following electrical connectors:
   - Idle Air Control (IAC) motor
   - Throttle Position (TPS) sensor
   - Manifold Absolute Pressure (MAP) sensor
   - Crankshaft Position (CKP) sensor
   - Camshaft Position (CMP) sensor
   - Oil pressure sensor
   - Purge solenoid
   - Ignition coil and module assembly
   - Oxygen (O$_2$S) sensor
   - Vehicle speed sensor

- Engine Coolant Temperature (ECT) sensor
- Back-up lamp switch
- Electrical harness from the engine and set the harness aside

5. Raise the vehicle.
6. Remove or disconnect the following:
- Front suspension crossmember
- Drive axles
- Engine drive belt
- AC compressor bolts and set the compressor aside.
- Alternator and starter electrical connectors

7. Drain the engine oil.
- Front exhaust pipe from the exhaust manifold

8. Use a block of wood to support the front of the engine at the front of the oil pan.
9. Lower the vehicle onto an engine support table.
- Front engine mount
- Bolts which secure the transmission mounts to the frame

10. Raise the vehicle away from the engine and transmission assembly.
11. Install an engine hoist to the engine.
- Transmission bellhousing bolts
- Engine and the transmission

To install:
12. Installation is the reverse of removal. Please note the following torques:
- Engine to the transaxle. Torque the bolts to 66 ft. lbs. (90 Nm).
- Transaxle mount. Torque the bolt to 81 ft. lbs. (110 Nm).
- Engine mount nuts and bolts to 49 ft. lbs. (66 Nm) and the engine mount bolts to 81 ft. lbs. (110 Nm)
- A/C compressor. Torque the bolts to 16 ft. lbs. (22 Nm).
- Power steering pump. Torque the bolts to 25 ft. lbs. (34 Nm).

13. Refill the cooling system and engine oil.
14. Start the engine and check for proper operation.

AUTOMATIC TRANSAXLE

1. Before servicing the vehicle, refer to the precautions section.
2. Remove or disconnect the following:
- Hood
- Negative battery cable
- Air inlet duct and resonator
- Accelerator and cruise control cable
- Hose from the brake booster
- Power steering pump bolts and set the pump aside
- Fuel lines
- Transmission shift control cables

- Transmission shift control cables from the bracket
- Clutch actuator cylinder from the transmission

3. Drain the cooling system.
- Radiator inlet hose
- Hose from the surge tank to the cylinder head
- Outlet hose from the surge tank to the radiator
- Bolt retaining the surge tank outlet hose to the intake manifold
- Radiator outlet hose
- Heater hoses

4. Disconnect the following electrical connectors:
- Idle Air Control (IAC) motor
- Throttle Position (TPS) sensor
- Manifold Absolute Pressure (MAP) sensor
- Crankshaft Position (CKP) sensor
- Camshaft Position (CMP) sensor
- Oil pressure sensor
- Purge solenoid
- Ignition coil and module assembly
- Oxygen (O_2S) sensor
- Vehicle speed sensor
- Engine Coolant Temperature (ECT) sensor
- Back-up lamp switch
- Electrical harness from the engine and set the harness aside

5. Remove or disconnect the following:
- Upper transmission bellhousing bolts

6. Raise the vehicle.
- Engine drive belt
- AC compressor bolts and set the compressor aside
- Crankshaft balancer
- Alternator and starter electrical connectors

7. Drain the engine oil.
- Front exhaust pipe from the exhaust manifold
- Starter
- Flywheel-to-torque convertor bolts
- Lower transmission bellhousing bolts
- Transmission to engine brace

8. Lower the vehicle.
9. Install an engine hoist to the engine.
- Front engine mount
- Upper transmission bellhousing bolts
- Engine and the transmission
- Engine

To install:
10. Installation is the reverse of removal. Please note the following torques:
- Upper bell housing bolts and tighten to 66 ft. lbs. (90 Nm)

- Front engine mount and tighten the bolts to 81 ft. lbs. (110 Nm)
- Lower bell housing bolts and tighten to 66 ft. lbs. (90 Nm)
- Torque convertor bolts and tighten to 46 ft. lbs. (62 Nm)
- Brace from the transmission to the engine bolts to 53 ft. lbs. (72 Nm)
- A/C compressor. Torque the bolts to 16 ft. lbs. (22 Nm).
- Power steering pump bolts to 19 ft. lbs. (25 Nm)

11. Refill the cooling system and engine oil.
12. Start the engine and check for proper operation.

2.4L Engine

AUTOMATIC TRANSAXLE

1. Before servicing the vehicle, refer to the precautions section.
2. Properly drain the cooling system.
3. Relieve the fuel system pressure.
4. Drain the engine oil.
5. Remove or disconnect the following:
- Fuel rail assembly
- Air intake duct and bracket
- Ignition coil assembly
- Camshaft Position (CMP) sensor
- Power steering pump. DO NOT remove the lines.
- Oil pressure sending switch
- Cruise control assembly

6. Install an engine support fixture.
- Engine mount assembly
- Fuel pressure regulator vacuum line

7. Raise the engine with the engine support fixture J-hook.
- Engine mount bracket
- Accessory drive belt
- Front wheels
- Right front splash shield
- Crankshaft balancer
- Manifold Absolute Pressure (MAP) sensor electrical connector
- Intake Air Temperature (IAT) sensor electrical connector
- Exhaust Gas Recirculation (EGR) sensor electrical connector
- Evaporator canister electrical connector
- Alternator
- Accelerator cable and bracket
- Starter motor
- Exhaust manifold heat shield
- Exhaust pipe from the manifold
- Engine intake coolant pipe
- Torque converter cover
- Torque converter bolts
- Oxygen Sensor (O_2S)

- A/C compressor. DO NOT remove the hoses.
- Oil pan-to-bell housing bolts
- Transaxle mount
- Transaxle bolts

8. Install an engine lifting device and a transaxle support.

9. Raise the engine slightly and separate it from the transaxle to remove it from the vehicle.

To install:

✳✳ WARNING

Be sure the retaining bolts are in their correct locations. If not, engine damage may occur.

10. Install or connect the following:
- Engine to the transaxle. Torque the bolts to 75 ft. lbs. (100 Nm).
- Transaxle mount. Torque the through bolt to 55 ft. lbs. (75 Nm).
- Oil pan-to-bell housing bolts. Torque the bolts to 17 ft. lbs. (23 Nm).
- A/C compressor. Torque the bolts to 37 ft. lbs. (50 Nm).
- O_2S
- Torque converter bolts. Torque the bolts to 46 ft. lbs. (62 Nm).
- Torque converter cover. Torque the bolts to 115 inch lbs. (13 Nm).
- Intake coolant pipe
- Exhaust pipe to the manifold. Torque the bolts to 26 ft. lbs. (35 Nm).
- Exhaust manifold heat shield. Torque the bolts to 124 inch lbs. (14 Nm).
- Starter motor. Torque the bolts to 66 ft. lbs. (90 Nm).
- Accelerator cable and bracket
- Alternator
- Fuel pressure regulator vacuum line
- Evaporator canister electrical connector
- EGR sensor electrical connector
- IAT sensor electrical connector
- MAP sensor electrical connector
- Crankshaft balancer. Torque the bolt to 129 ft. lbs. (175 Nm).
- Right front splash shield
- Front wheels
- Accessory drive belt
- Engine mount bracket. Torque the bolts to 96 ft. lbs. (130 Nm).
- Engine mount assembly. Torque the bolts to 49 ft. lbs. (66 Nm).
- Cruise control assembly
- Oil pressure sending switch

- Power steering pump. Torque the bolts to 19 ft. lbs. (26 Nm).
- CMP sensor
- Ignition coil assembly
- Air intake assembly
- Fuel rail assembly

11. Refill the cooling system and engine oil.

12. Start the engine and check for proper operation.

MANUAL TRANSAXLE

1. Before servicing the vehicle, refer to the precautions section.

2. Properly drain the cooling system.

3. Relieve the fuel system pressure.

4. Drain the engine oil.

5. Discharge and recover the refrigerant.

6. Remove or disconnect the following:
- Air cleaner duct
- Air cleaner
- Underhood fuse block
- Air cleaner bracket
- Shift cables at the shift control
- Shift cables from the bracket and remove the bracket
- Back up lamp switch electrical connection
- Vehicle Speed Sensor (VSS) electrical connection
- Engine controls harness
- Vacuum lines
- Fuel line quick disconnect fittings
- Radiator hoses
- Heater hoses from the core
- Slave cylinder hydraulic line
- Evaporative (EVAP) solenoid
- Ground cables at the rear of the engine block
- Power steering pump with lines attached and position the pump aside
- Electrical connector from the A/C compressor and the Crankshaft Position (CKP) sensor and position the harness aside
- Starter with the wires attached and position it aside
- Surge tank bypass hose from the engine
- Cruise and accelerator cables from the throttle body
- Cruise control module

7. Tie the radiator to the hood latch panel with mechanics wire

8. Raise and support the vehicle. Safety strap the front of the vehicle to the hoist.
- Oil filter
- Front splash shields
- Front closeout panel
- Lower radiator support

- Right front brake hose
- Retaining nut from the Brake Pressure Modulator Valve (BPMV) at the mounting bracket
- Wheel Speed Sensors (WSS)
- WSS harnesses from the control arm retainers and the frame retainers and position them aside
- Driveshafts
- Ball joints from the control arms
- Outer tie rod ends from the control arms
- Catalytic converter from the exhaust manifold
- A/C compressor hose from the A/C compressor
- Bolt from the power steering pressure line retainer

9. Lower the vehicle until the front suspension crossmember rests on the support table. Position a three inch block of wood between the front of the oil pan and the crossmember.
- Front engine mount-to-bracket bolts
- Front suspension crossmember retaining bolts

10. Carefully raise the vehicle off of the engine/transaxle assembly. Install the engine hoist to the engine/transaxle assembly.
- Front and rear transaxle mount through-bolts
- 2 side transaxle mount lower nuts
- Engine/transaxle assembly off of the front suspension crossmember
- Transaxle from the engine
- Clutch drive plate and clutch driven plate
- Flywheel

11. Mount the engine on a suitable engine stand.

To install:

12. Remove the engine from the engine stand

13. Install or connect the following:
- Flywheel
- Clutch driven plate and clutch drive plate
- Transaxle to the engine

14. Lower the engine/transaxle assembly on to the front suspension crossmember.
- 2 side transaxle mount lower nuts and tighten the nuts to 60 ft. lbs. (44 Nm)
- Front and rear transaxle mount through-bolts. Tighten the through-bolts to 75 ft. lbs. (55 Nm).

15. Remove the engine hoist from the engine/transaxle assembly.

16. Carefully lower the vehicle on to the engine/transaxle assembly.

- Front suspension crossmember retaining bolts
- Front engine mount-to-bracket bolts
- Power steering pressure line retainer bolt.
- A/C compressor hose to the A/C compressor
- Catalytic converter to the exhaust manifold
- Outer tie rod ends to the control arms
- Ball joints to the control arms
- Driveshafts
- WSS harnesses to the frame retainers and the control arm retainers
- Wheel speed sensors
- Retaining nut to the BPMV at the mounting bracket
- Right front brake hose to the vehicle
- Lower radiator support
- Front closeout panel and the panel fasteners
- Front splash shields
- New oil filter

17. Remove the safety straps from the front of the vehicle and the hoist. Lower the vehicle. Untie the radiator from the hood latch panel.

- Cruise control module
- Cruise and accelerator cables to the throttle body
- Surge tank bypass hose to the engine
- Starter
- Connectors to the A/C compressor and the CKP sensor
- Power steering pump
- Ground cables at the rear of the engine block
- EVAP solenoid
- Slave cylinder hydraulic line
- Heater hoses to the heater core
- Radiator hoses
- Fuel line fittings
- Vacuum lines
- Engine controls harness
- VSS
- Back up lamp switch
- Shift cable bracket and the shift cables to the bracket
- Shift cables at the shift control
- Air cleaner bracket
- Underhood fuse block
- Air cleaner
- Air cleaner duct
- Negative battery cable

18. Refill the cooling system.
19. Refill the crankcase.
20. Recharge the A/C system.
21. Start the vehicle and verify no leaks.

22. Check and/or adjust the wheel alignment.

3.4L Engine

1. Before servicing the vehicle, refer to the precautions section.
2. Disconnect the negative battery cable.
3. Drain the engine coolant.
4. Remove or disconnect the following:

- Air cleaner assembly
- Hood
- Accessory drive belt
- Hoses from the surge tank
- Cruise control module
- Upper wiring harness from the engine
- Throttle and cruise control cables
- Starter motor
- A/C compressor and move it aside with the lines attached
- Lower wiring harness from the engine components
- Catalytic converter from the rear exhaust manifold
- Torque converter cover
- Torque converter-to-flywheel bolts
- Engine splash shields
- Transaxle-to-engine brace and the 2 outer transaxle mounting bolts
- Upper and lower radiator hoses
- Fuel lines from the engine
- Vacuum hose from the power brake booster
- Heater hoses

5. Install an engine hoist and raise the engine slightly.
6. Remove or disconnect the following:

- Engine mount and bracket
- Power steering pump
- Transaxle-to-engine bolts
- Engine from the vehicle

To install:

7. Install or connect the following:

- Engine. Torque the upper transaxle-to-engine bolts to 66 ft. lbs. (90 Nm).
- Engine support fixture and remove the engine hoist
- Power steering pump. Torque the bolts to 25 ft. lbs. (34 Nm).
- Engine mount bracket and mount. Torque the bolts to 43 ft. lbs. (58 Nm) and the nuts to 35 ft. lbs. (47 Nm).
- Heater hoses
- Fuel lines
- Vacuum hose to the power brake booster
- Upper and lower radiator hoses
- Both outer transaxle to engine

bolts. Torque the bolts to 66 ft. lbs. (90 Nm).
- Transaxle to the engine brace. Torque the bolts to 32 ft. lbs. (43 Nm).
- Engine splash shields
- Flexplate to the torque converter. Torque the bolts to 46 ft. lbs. (62 Nm).
- Torque converter cover. Torque the bolts to 89 inch lbs. (10 Nm).
- Catalytic converter to the exhaust manifold. Torque the bolts to 25 ft. lbs. (34 Nm).
- A/C compressor. Torque the bolts to 37 ft. lbs. (50 Nm).
- Starter motor. Torque the bolts to 32 ft. lbs. (43 Nm).
- Lower and upper wiring harnesses

✳✳ CAUTION

To avoid personal injury and/or damage to the vehicle, replace the throttle cable with a new one any time the engine has been removed from the vehicle.

- Cruise control and throttle cables
- Cruise control module
- Accessory drive belt
- Hoses to the surge tank
- Hood. Torque the bolts to 13 ft. lbs. (17 Nm).
- Air cleaner assembly

8. Refill the cooling system.
9. Refill the crankcase.
10. Start the engine and inspect for coolant and/or oil leaks.
11. Stop the engine and recheck the fluid levels after the engine has cooled.

Water Pump

REMOVAL & INSTALLATION

2.2L Engine

1. Before servicing the vehicle, refer to the precautions section.
2. Disconnect the negative battery cable.
3. Drain the engine coolant.
4. Remove or disconnect the following:

- Exhaust manifold, if equipped with an automatic transaxle
- Right front wheel
- Splash shield
- Water pump sprocket access plate from the timing cover

5. Attach Water Pump Sprocket Holding Tool J 43651 to the sprocket using the

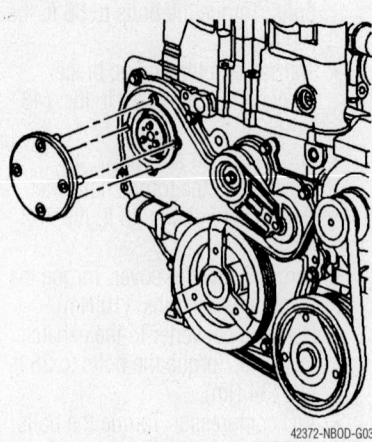

Remove the water pump sprocket access plate from the timing cover —2.2L engine

42372-NBOD-G03

access plate bolts to secure the tool to the engine front cover.

- Sprocket to water pump bolts
- Engine block to water pump bolt
- Engine front cover to water pump bolt
- Thermostat housing to water pump feed pipe
- 2 water pump to engine block bolts
- Water pump

To install:

6. Use a threaded stud in the water pump hub to align the hub to the water pump sprocket.

- Water pump and bolts
- Thermostat housing to water pump feed pipe
- Egine front cover bolt and engine block to water pump bolt. Tighten the bolts to 15 ft. lbs. (20 Nm)
- 2 of the water pump sprocket to water pump bolts

7. Remove the threaded stud.

- Last water pump sprocket to water pump bolt and tighten the sprocket bolts to 89 inch lbs. (10 Nm).

8. Remove the tool.

- Water pump sprocket access plate to the timing cover and tighten the bolts to 89 inch lbs. (10 Nm)

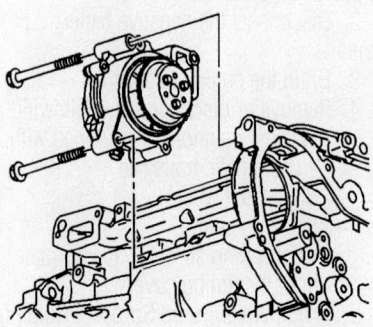

Water pump mounting—2.2L engine

42372-NBOD-G04

- Splash shield
- Right front wheel
- Exhaust manifold, if removed

9. Fill the cooling system.

10. Connect the negative battery cable.

2.4L Engine

1. Before servicing the vehicle, refer to the precautions section.

2. Disconnect the negative battery cable.

3. Drain the cooling system.

4. Remove or disconnect the following:

- Oxygen Sensor (O₂S) electrical connector
- Exhaust manifold heat shield
- Coolant inlet housing bolt through the exhaust manifold
- Exhaust manifold brace-to-manifold bolt
- Manifold-to-exhaust pipe studs
- Coolant inlet housing-to-water pump cover bolt
- Exhaust pipe from the exhaust manifold by pulling it downward

✳✳ WARNING

Do not rotate the flex coupling more than 4 degrees or damage may occur.

- Coolant inlet pipe from the oil pan
- Brake vacuum pipe from the camshaft housing
- Exhaust manifold from the cylinder head
- Heater hose from the heater outlet pipe
- Timing chain cover and tensioner
- Water pump cover-to-engine bolts
- 3 water pump-to-timing chain housing nuts
- Water pump and cover assembly

- Water pump cover from the pump

To install:

5. Install or connect the following:

- Water pump with a new gasket to the cover and tighten the bolts finger-tight
- Water pump cover-to-engine bolts and tighten finger-tight
- Water pump-to-timing chain housing nuts and tighten finger-tight
- Coolant inlet pipe into the water pump cover and tighten the bolts finger-tight

6. With all gaps closed, torque the bolts, in the following sequence, to the proper values:

   a. Pump assembly-to-chain housing nuts to 19 ft. lbs. (26 Nm).

   b. Pump cover-to-pump assembly to 124 inch lbs. (14 Nm).

   c. Water pump cover-to-engine bolts to 19 ft. lbs. (26 Nm).

   d. Coolant inlet pipe assembly-to-water pump cover bolts to 124 inch lbs. (14 Nm).

7. Install or connect the following:

- Exhaust manifold to the cylinder head. Torque the nuts to 11 ft. lbs. (15 Nm).
- Brake vacuum pipe to the camshaft housing
- Coolant inlet pipe to the oil pan. Torque the nuts to 19 ft. lbs. (25 Nm).
- Timing chain tensioner. Torque the bolts to 89 inch lbs. (10 Nm.).
- Front cover. Torque the bolts to 106 inch lbs. (12 Nm).
- Exhaust pipe to the manifold. Torque the bolts to 26 ft. lbs. (35 Nm).
- Exhaust manifold brace to the man-

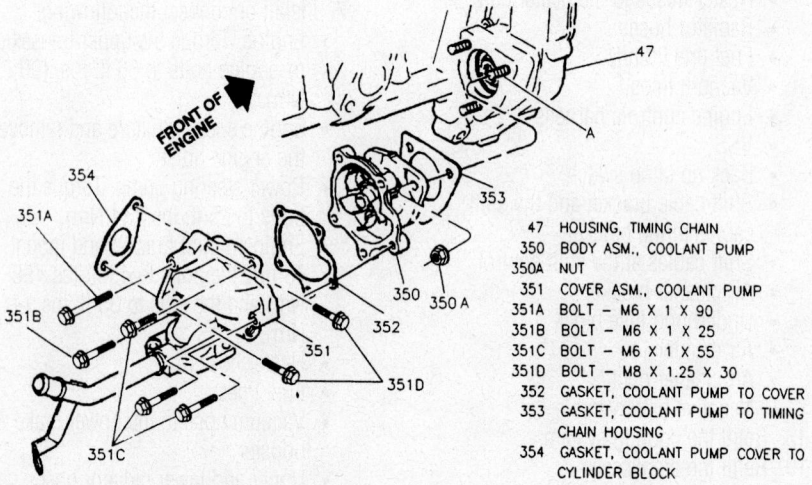

47 HOUSING, TIMING CHAIN
350 BODY ASM., COOLANT PUMP
350A NUT
351 COVER ASM., COOLANT PUMP
351A BOLT – M6 X 1 X 90
351B BOLT – M6 X 1 X 25
351C BOLT – M6 X 1 X 55
351D BOLT – M8 X 1.25 X 30
352 GASKET, COOLANT PUMP TO COVER
353 GASKET, COOLANT PUMP TO TIMING CHAIN HOUSING
354 GASKET, COOLANT PUMP COVER TO CYLINDER BLOCK

7922Z302

Exploded view of the water pump mounting—2.4L engine

ifold. Torque the bolt to 41 ft. lbs. (56 Nm) and the nuts to 19 ft. lbs. (26 Nm).
- Exhaust manifold to the exhaust pipe. Torque the nuts to 26 ft. lbs. (35 Nm).
- Heater hose to the heater pipe
- Exhaust manifold heat shield. Torque the bolts to 124 inch lbs. (14 Nm).
- O₂S electrical connector
- Negative battery cable

8. Refill the cooling system.
9. Start the engine and check for leaks.

3.4L Engines

1. Before servicing the vehicle, refer to the precautions section.
2. Disconnect the negative battery cable.
3. Drain the cooling system.
4. Remove or disconnect the following:
- Accessory drive belt
- Water pump pulley
- Water pump

To install:
5. Install or connect the following:
- Water pump with a new gasket. Torque the bolts to 89 inch lbs. (10 Nm).

- Water pump pulley. Torque the bolts to 18 ft. lbs. (25 Nm).
- Accessory drive belt
- Negative battery cable

6. Refill and bleed the cooling system.
7. Test run the engine and check for leaks.

Heater Core

REMOVAL & INSTALLATION

1. Before servicing the vehicle, refer to the precautions section.
2. Drain the engine cooling system.
3. Disable the Supplemental Inflatable Restraint (SIR) system.
4. Recover the A/C refrigerant.
5. Remove or disconnect the following:
- Evaporator hose from the evaporator
- Heater hoses
- Drain tube elbow
- Heater case plate for the heater pipes
- Heater case plate for the evaporator block
- End caps
- Sound insulators
- Instrument panel compartment

- Passenger inflator module
- Tilt steering lever
- Upper and the lower steering column covers
- Hazard warning switch
- Instrument panel cluster trim plate
- Instrument panel cluster
- Driver inflator module
- Steering wheel
- Multifunction lever
- Accessory trim plate
- Radio
- Cup holder
- Gearshift handle
- Ignition switch bolts
- Instrument panel upper bolt covers
- Electrical junction box connections
- Fog lamp switch
- Windshield side upper garnish molding
- Console from the instrument panel
- Instrument panel from the tie bar
- Air distribution duct
- Floor air duct
- Tie bar
- Daytime Running Lights (DRL) sensor wiring harness from the Heating, Ventilation, Air Conditioning (HVAC) module
- Blower motor electrical connectors
- Temperature actuator electrical connector
- HVAC module from the vehicle
- Heater core case cover
- Heater core bracket
- Heater core

To install:
6. Install or connect the following:
- Heater core
- Heater core bracket. Torque the screw to 9 inch lbs. (1 Nm).
- Heater core case cover. Torque the screws to 9 inch lbs. (1 Nm).
- HVAC module
- Temperature actuator electrical connector
- Blower motor electrical connectors
- DRL wiring harness
- Tie bar. Torque the bolts to 15 ft. lbs. (20 Nm).
- Floor air duct. Torque the bolts to 18 inch lbs. (2 Nm).
- Air distribution duct. Torque the bolts to 27 inch lbs. (3 Nm).
- Instrument panel
- Console to the instrument panel
- Windshield side upper garnish molding
- Fog lamp switch
- Electrical junction box connections
- Instrument panel upper bolt covers
- Ignition switch bolts

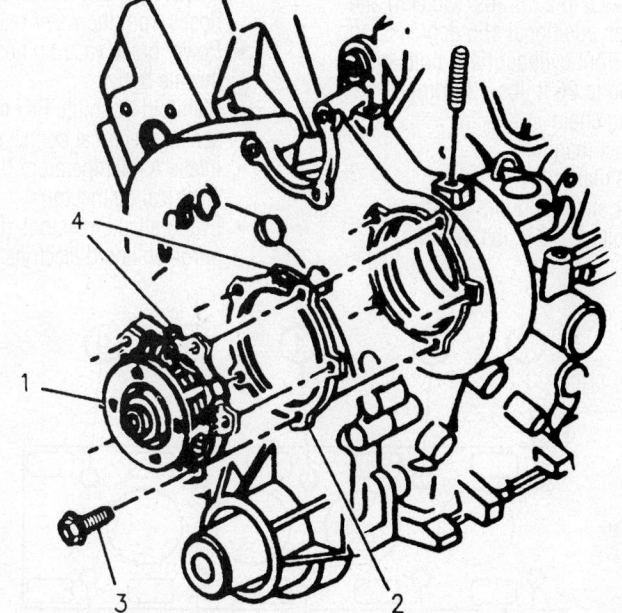

1   COOLANT PUMP
2   GASKET
3   BOLT – 10 N·m (89 LBS. IN.)
4   LOCATOR (MUST BE VERTICAL)

7922Z303

Exploded view of the water pump mounting—3.4L engines

- Gearshift handle
- Cup holder
- Radio. Torque the bolts to 27 inch lbs. (3 Nm).
- Accessory trim plate
- Multifunction lever. Torque the screw to 35 inch lbs. (4 Nm).
- Steering wheel. Torque the nut to 27 ft. lbs. (37 Nm).
- Driver inflator module. Torque the screw to 89 inch lbs. (10 Nm).
- Instrument panel cluster
- Instrument panel cluster trim plate
- Hazard warning switch
- Upper and the lower steering column covers. Torque the screw to 18 inch lbs. (2 Nm).
- Tilt steering lever
- Passenger inflator module. Torque the fasteners to 89 inch lbs. (10 Nm).
- Instrument panel compartment
- Sound insulators
- End caps
- Heater case plate for the evaporator block. Torque the nuts to 89 inch lbs. (10 Nm).
- Heater case plate for the heater pipes
- Drain tube elbow
- Heater hoses
- Evaporator hose. Torque the fitting to 18 ft. lbs. (25 Nm).

7. Fill the engine cooling system.
8. Evacuate and recharge the A/C system.

Cylinder Head

REMOVAL & INSTALLATION

2.2L Engine

1. Before servicing the vehicle, refer to the precautions section.
2. Remove or disconnect the following:

- Negative battery cable
- Intake manifold

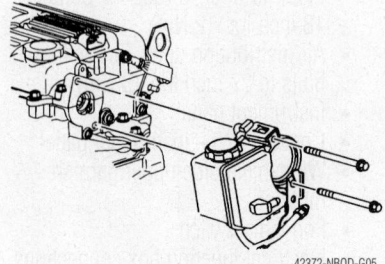

Remove the power steering pump bolts and set pump aside—2.2L engine

42372-NBOD-G05

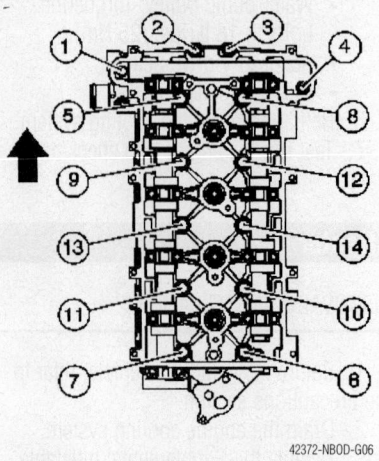

42372-NBOD-G06

Remove the cylinder head bolts in sequence and discard —2.2L engine

- Power steering pump bolts and set pump aside
- Exhaust manifold
- Timing chain
3. Drain the cooling system.
- Cylinder head bolts in sequence and discard
- Cylinder head and gasket

To install:
4. Install or connect the following:

- Cylinder head and gasket
- NEW cylinder head bolts (except the front bolts) and tighten in sequence to 22 ft. lbs. (30 Nm) and then an additional 155 degrees
- NEW front cylinder head bolts and tighten to 26 ft. lbs. (35 Nm)
- Timing chain
- Exhaust manifold
- Intake manifold
- Power steering pump and tighten the bolts to 19 ft. lbs. (25 Nm).

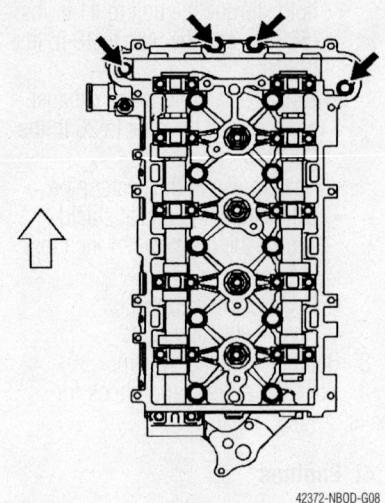

42372-NBOD-G08

Install the front cylinder head bolts—2.2L engine

5. Fill the cooling system.
6. Connect the negative battery cable.

2.4L Engine

1. Before servicing the vehicle, refer to the precautions section.
2. Relieve the fuel system pressure.
3. Drain the cooling system.
4. Remove or disconnect the following:

- Throttle body air intake duct
- Heater inlet and throttle body heater hoses from the water outlet
- Power brake vacuum hose from the throttle body
- Manifold Absolute Pressure (MAP) sensor electrical connector
- Intake Air Temperature (IAT) sensor electrical connector
- Evaporative Emissions (EVAP) purge solenoid electrical connector

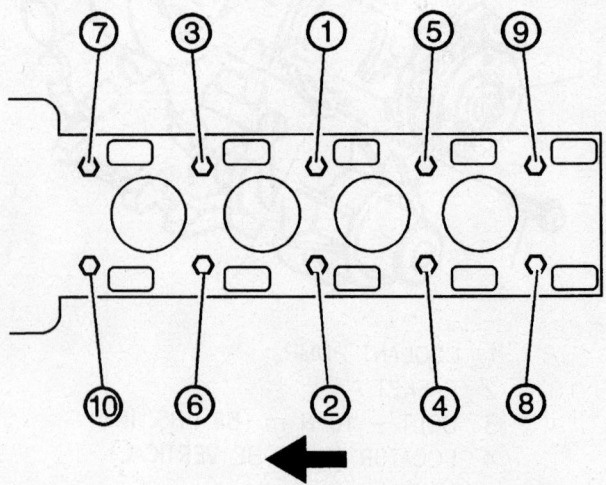

Install the cylinder head bolts in sequence (except the front bolts)—2.2L engine

42372-NBOD-G07

Cylinder head bolt tightening sequence—2.4L engine

- Camshaft Position (CMP) sensor electrical connector
- Stud-ended bolt from the alternator
- Intake manifold

5. Install the stud-ended alternator bolt back into the engine.

6. Install an engine support fixture.

7. Remove or disconnect the following:
- Exhaust manifold
- Ignition coil and module assembly
- Power steering pump
- Vacuum line from the fuel pressure regulator
- Fuel injector electrical connectors
- Fuel line clamp from the intake camshaft housing bracket
- Fuel rail without disconnecting the fuel lines
- Timing chain and sprocket
- Water pump
- Electrical connection from the oil switch
- Transaxle fluid level indicator tube assembly from the exhaust camshaft cover and move it aside, if equipped with an automatic transaxle

➡ **Any time the camshaft housing-to-cylinder head bolts are loosened or removed, the camshaft housing-to-cylinder head gasket must be replaced.**

- Intake camshaft housing

➡ **Turn the camshaft housing upside down as soon as it is removed from the cylinder head; otherwise, the lifters may fall out.**

- Exhaust camshaft housing
- Upper radiator hose
- Engine Coolant Temperature (ECT) sensor electrical connector
- Cylinder head bolts by reversing of the tightening sequence
- Cylinder head

To install:

8. Install or connect the following:

- Cylinder head with a new gasket
- New cylinder head bolts lubricated with clean engine oil

9. Torque the cylinder head bolts, in sequence, as follows:
  a. Step 1: Bolts 1 through 8 to 40 ft. lbs. (65 Nm).
  b. Step 2: Bolts 9 and 10 to 30 ft. lbs. (40 Nm).
  c. Step 3: All bolts an additional 90 degree (¼) turn.

10. Install the intake and exhaust camshaft housings and torque the bolts as follows:
  a. Long bolts: 11 ft. lbs. (15 Nm) plus an additional 90 degree turn.
  b. Short bolts: 11 ft. lbs. (15 Nm) plus an additional 30 degree turn.

11. Install or connect the following:
- Water pump. Torque the pump-to-cover bolts to 124 inch lbs. (14 Nm) and the water pump cover-to-block bolts to 19 ft. lbs. (26 Nm).
- Timing chain and sprocket. Torque the sprocket bolt to 52 ft. lbs. (70 Nm).
- Upper radiator hose to the water outlet
- ECT sensor connector

- Fuel rail
- Fuel injector electrical connectors
- Fuel pressure regulator vacuum line
- MAP sensor electrical connector
- IAT sensor electrical connector
- EVAP purge solenoid electrical connector
- CMP sensor electrical connector
- Oil switch electrical connector
- Exhaust manifold. Torque the nuts to 110 inch lbs. (13 Nm).
- Intake manifold. Torque the bolts to 18 ft. lbs. (24 Nm).
- Power steering pump. Torque the bolts to 19 ft. lbs. (26 Nm).
- Throttle body air intake duct
- Oil fill tube into the engine
- Heater inlet and throttle body heater hoses to the water outlet
- Power brake vacuum hose
- Ignition coil and module assembly

12. Remove the engine support fixture.

13. Fill all fluids to their proper levels.

➡ **An oil and filter change is recommended.**

14. Connect the negative battery cable. Start the vehicle and verify no leaks.

3.4L Engines

LEFT (FRONT) CYLINDER HEAD

1. Before servicing the vehicle, refer to the precautions section.

2. Relieve the fuel system pressure.

3. Drain the crankcase.

4. Drain the cooling system.

5. Remove or disconnect the following:
- Upper half of the air cleaner assembly
- Throttle body air inlet duct
- Exhaust crossover pipe heat shield

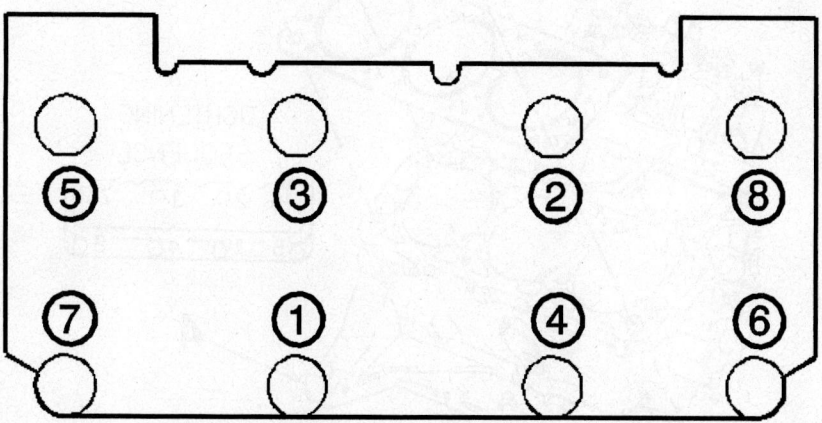

Cylinder head bolt loosening sequence—2005 3.4L engines

- Crossover pipe
- Spark plug wires from the spark plugs
- Rocker arm covers
- Upper intake plenum and lower intake manifold
- Left side exhaust manifold
- Oil level indicator tube

➡**When removing the valvetrain components, keep them in order for installation purposes.**

- Rocker arms and pushrods
- Cylinder head

To install:

6. Install the cylinder head with a new gasket.

7. Torque the cylinder head bolts to 33 ft. lbs. (45 Nm) plus an additional 90 degree turn.

8. Install or connect the following:
- New intake manifold gasket
- Pushrods and rocker arms. Torque the bolts to 89 inch lbs.

(10 Nm) plus an additional 30 degree turn.
- Lower intake manifold. Torque the bolts to 115 inch lbs. (13 Nm).
- Upper intake plenum. Torque the bolts to 18 ft. lbs. (25 Nm).
- Rocker arm covers. Torque the bolts to 89 inch lbs. (10 Nm).
- Oil level indicator tube
- Left side exhaust manifold. Torque the nuts to 12 ft. lbs. (16 Nm).
- Spark plug wires
- Exhaust crossover pipe. Torque the bolts to 18 ft. lbs. (25 Nm).
- Crossover pipe heat shield. Torque the bolts to 89 inch lbs. (10 Nm).
- Upper half of the air cleaner assembly
- Throttle body air inlet duct
- Negative battery cable

9. Refill the cooling system.
10. Refill the crankcase.

➡**A filter change is recommended.**

11. Start the engine and verify no leaks.

RIGHT (REAR) CYLINDER HEAD

1. Before servicing the vehicle, refer to the precautions section.
2. Relieve the fuel system pressure.
3. Drain the crankcase.
4. Drain the cooling system.
5. Remove or disconnect the following:
- Upper half of the air cleaner assembly
- Throttle body air inlet duct
- Exhaust crossover pipe heat shield
- Crossover pipe
- Oxygen Sensor (O_2S) electrical connector
- Exhaust pipe from the exhaust manifold
- Right side exhaust manifold
- Spark plug wires from the spark plugs
- Rocker arm covers
- Upper intake plenum
- Lower intake manifold

➡**When removing the valvetrain components keep them in order for installation purposes.**

- Rocker arms and pushrods
- Cylinder head

To install:

6. Install the cylinder head with a new gasket.

7. Torque the cylinder head bolts to 33 ft. lbs. (45 Nm) plus an additional 90 degree turn.

8. Install or connect the following:
- New intake manifold gasket
- Pushrods and rocker arms. Torque the bolts to 89 inch lbs. (10 Nm) plus an additional 30 degree turn.
- Lower intake manifold. Torque the bolts to 115 inch lbs. (13 Nm).
- Upper intake plenum. Torque the bolts to 18 ft. lbs. (25 Nm).
- Rocker arm covers. Torque the bolts to 89 inch lbs. (10 Nm).
- Spark plug wires
- Exhaust manifold. Torque the nuts to 12 ft. lbs. (16 Nm).
- Exhaust pipe to the exhaust manifold. Torque the nuts to 33 ft. lbs. (45 Nm).
- O_2S electrical connector
- Exhaust crossover pipe. Torque the bolts to 18 ft. lbs. (25 Nm).
- Crossover pipe heat shield. Torque the nuts to 89 inch lbs. (10 Nm).
- Upper half of the air cleaner assembly
- Throttle body air inlet duct
- Negative battery cable

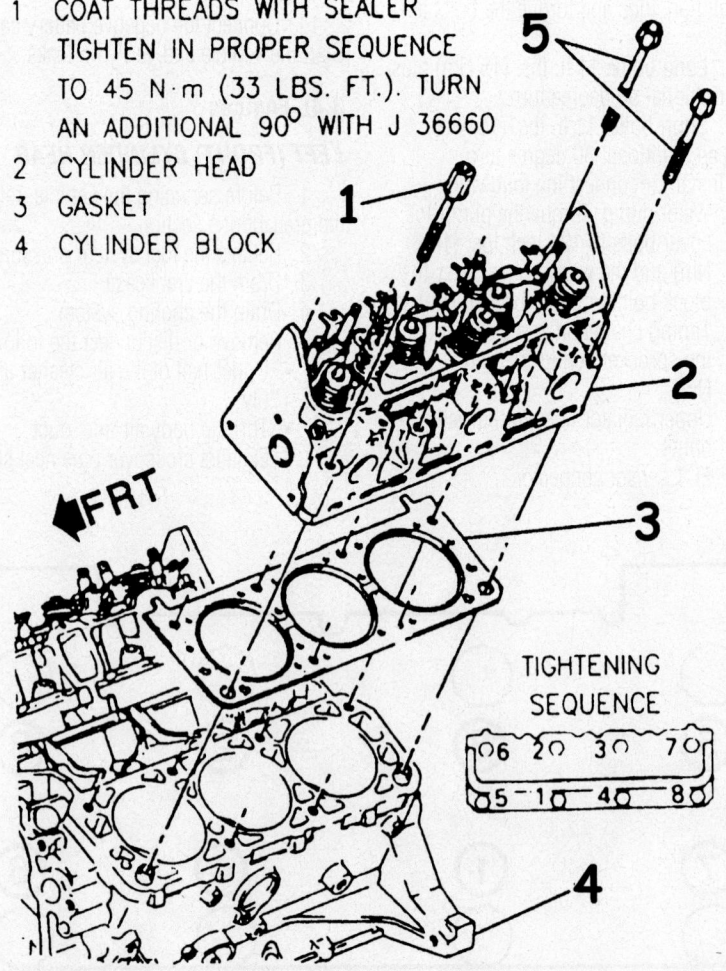

1 COAT THREADS WITH SEALER TIGHTEN IN PROPER SEQUENCE TO 45 N·m (33 LBS. FT.). TURN AN ADDITIONAL 90° WITH J 36660
2 CYLINDER HEAD
3 GASKET
4 CYLINDER BLOCK

FRT

TIGHTENING SEQUENCE

6 2 3 7
5 1 4 8

79222306

Cylinder head bolt tightening sequence—2002-04 3.4L engines

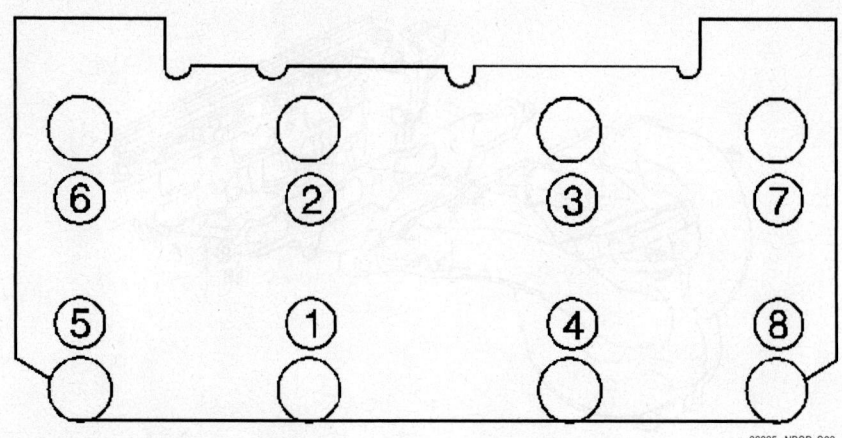

06025 -NBOD-G02

Cylinder head bolt tightening sequence—2005 3.4L engines

9. Refill the cooling system.
10. Refill the crankcase.

➡**An oil filter change is recommended.**

11. Start the engine and verify no leaks.

Rocker Arms

REMOVAL & INSTALLATION

2.2L Engine and 2.4L Engines

These engines are not equipped with rocker arms. The camshafts directly actuate the valves.

3.4L Engines

LEFT SIDE

1. Before servicing the vehicle, refer to the precautions section.
2. Drain the cooling system to a level below the coolant pipe on the front of the engine.
3. Remove or disconnect the following:
   - Negative battery cable
   - Spark plug wires
   - Heater bypass pipe
   - Positive Crankcase Ventilation (PCV) valve and hose
   - Rocker arm cover

➡**Keep the pushrods in order. Intake pushrods are 5¾ inches long and exhaust pushrods are 6 inches long.**

   - Rocker arms and pushrods

To install:

4. Lubricate all the valvetrain components with engine oil.
5. Install or connect the following:
   - Pushrods and the rocker arms. Torque the bolts to 24 ft. lbs. (32 Nm).
   - Rocker arm cover using a new gas-

ket. Torque the rocker cover bolts to 89 inch lbs. (10 Nm).
   - PCV valve and hose
   - Heater bypass pipe. Torque the screw at the water pump to 106 inch lbs. (12 Nm), the bolt at the cylinder head corner to 18 ft. lbs. (25 Nm) and the nut to 18 ft. lbs. (25 Nm).
   - Spark plug wires
   - Negative battery cable
6. Refill the cooling system.
7. Start the vehicle and verify no leaks.

RIGHT SIDE

1. Before servicing the vehicle, refer to the precautions section.
2. Remove or disconnect the following:
   - Negative battery cable
   - Alternator bracket on 2005 models
   - Spark plug wires from the spark plugs and the upper intake plenum wire retainer
   - Power brake booster vacuum pipe from the intake plenum
   - Accessory drive belt
   - Alternator, if necessary
   - Ignition coil assembly and Evapo-

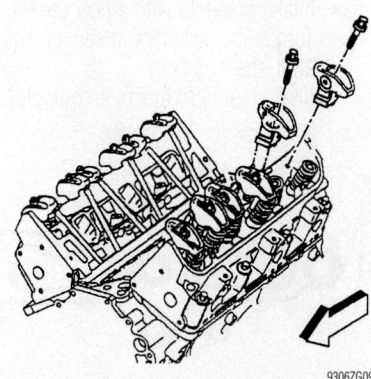

9306ZG09

Rocker arm components—3.4L engines

rative Emissions (EVAP) canister purge solenoid as an assembly
   - Rocker arm cover

➡**Keep the pushrods in order. Intake pushrods are 5¾ inches long and exhaust pushrods are 6 inches long.**

   - Rocker arms and pushrods

To install:

3. Lubricate all the valvetrain components with engine oil.
4. Install or connect the following:
   - Pushrods and the rocker arms. Torque the bolts to 24 ft. lbs. (32 Nm) plus an additional 30 degree turn.
   - Rocker arm cover using a new gasket. Torque the rocker cover bolts to 89 inch lbs. (10 Nm).
   - Ignition coil and EVAP solenoid assembly
   - Alternator, if removed. Torque the bolts to 37 ft. lbs. (50 Nm).
   - Accessory drive belt
   - Power brake booster vacuum pipe to the plenum
   - Spark plug wires
   - Negative battery cable
5. Start the vehicle and verify no leaks.

Intake Manifold

REMOVAL & INSTALLATION

These vehicles were filled at the factory with an antifreeze/coolant called GM Goodwrench DEX-COOL®. When adding coolant to vehicles, it is important that you use GM Goodwrench DEX-COOL (orange-colored, silicate-free) coolant. **Propylene glycol is not recommended for use in GM vehicles.** A 50/50 mixture of DEX-COOL and clean water will provide all the recommended protection. **DO NOT mix DEX-COOL with any other type of antifreeze.**

2.2L Engine

1. Before servicing the vehicle, refer to the precautions section.
2. Remove or disconnect the following:
   - Air inlet duct and resonator
   - Idle Air Control (IAC) electrical connector
   - Throttle Position (TP) sensor electrical connector
   - Manifold Absolute Pressure (MAP) sensor electrical connector
   - Evaporative Emissions (EVAP) hose
   - Positive Crankcase Ventilation (PCV) hose

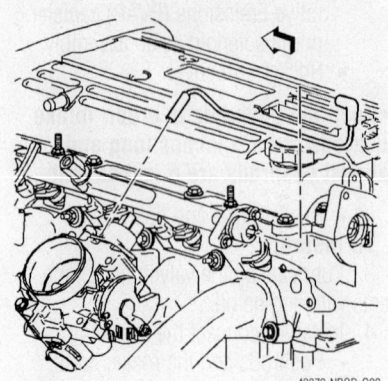

Remove the throttle body—2.2L engine

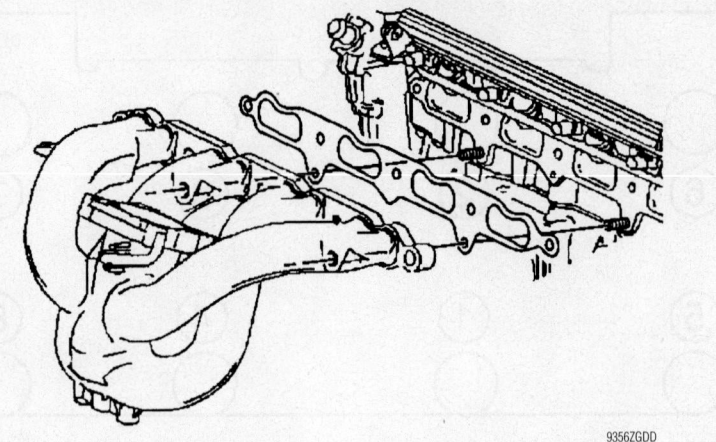

Exploded view of the intake manifold mounting—2.4L engine

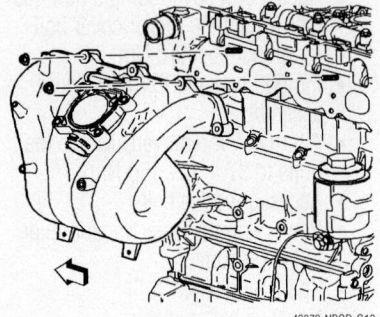

Remove the intake manifold—2.2L engine

- Purge solenoid tube
- Brake booster hose
- Oil level indicator tube bolt
- Accelerator and cruise control cables
- Throttle body
- Fuel rail
- Knock Sensor (KS) connector from the intake manifold
- Intake manifold nuts and bolts
- Intake manifold and gasket. The gasket is reusable if the gasket is not damaged.

To install:

3. Install or connect the following:
- Intake manifold gasket
- Intake manifold
- Intake manifold nuts and bolts and tighten to 89 inch lbs. (10 Nm)
- KS connector
- Fuel rail
- Throttle body
- IAC, TP, and MAP electrical connectors
- EVAP hose
- PCV hose
- Purge solenoid tube
- Brake booster hose
- Oil level indicator tube bolt
- Accelerator and the cruise control cables
- Air inlet duct and resonator

2.4L Engine

1. Before servicing the vehicle, refer to the precautions section.
2. Relieve the fuel system pressure.
3. Drain the cooling system to a level below the intake manifold.
4. Remove or disconnect the following:
- Air cleaner duct
- Manifold Absolute Pressure (MAP) sensor electrical connector
- Intake Air Temperature (IAT) sensor electrical connector
- Evaporative Emissions (EVAP) canister purge solenoid electrical connector
- Fuel injector electrical connectors
- Vacuum hoses from the intake manifold, fuel pressure regulator and EVAP canister purge solenoid
- Accelerator control cable bracket
- Coolant lines from the throttle body
- Stud-end alternator mount bolt
- Exhaust Gas Recirculation (EGR) pipe from the EGR adapter
- Intake manifold support brace, if equipped
- Intake manifold

To install:

5. Install or connect the following:
- Intake manifold with a new gasket. Torque the fasteners in sequence to 18 ft. lbs. (25 Nm).
- Intake manifold brace, if equipped.

Torque the bolts to 19 ft. lbs. (26 Nm).

➡**The brace-to-block bolts must be tightened first, then the brace-to-manifold bolt.**

- EGR pipe to the EGR adapter
- Stud-end alternator mount bolt
- Coolant lines to the throttle body
- Accelerator control cable bracket
- Vacuum hoses to the intake manifold, fuel pressure regulator and EVAP canister purge solenoid
- Electrical connectors to the fuel injectors
- MAP sensor electrical connector
- IAT sensor electrical connector
- EVAP canister purge solenoid electrical connector
- Air cleaner duct
6. Refill the cooling system.
7. Connect the negative battery cable.

3.4L Engines

1. Before servicing the vehicle, refer to the precautions section.
2. Relieve the fuel system pressure.
3. Drain the cooling system.
4. Remove or disconnect the following:
- Air cleaner assembly
- Accessory drive belt
- Exhaust Gas Recirculation (EGR) valve

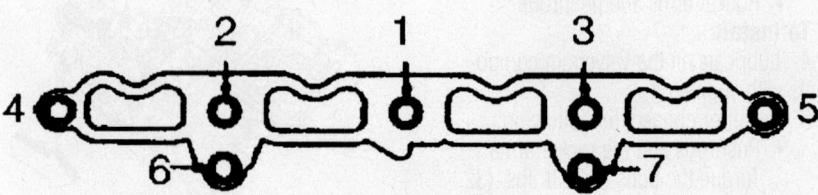

Intake manifold tightening sequence—2.4L engine

- Brake vacuum pipe at the intake plenum
- Fuel pressure regulator vacuum line
- Spark plug wires from the spark plugs and the intake plenum retainers
- Ignition coil assembly and the Evaporative Emissions (EVAP) canister purge solenoid as an assembly
- Throttle Position Sensor (TPS) electrical connector
- Idle Air Control (IAC) sensor electrical connector
- Fuel injector electrical connectors
- Engine Coolant Temperature (ECT) sensor electrical connector
- Camshaft Position (CMP) sensor electrical connector

- Vacuum modulator
- Manifold Absolute Pressure (MAP) sensor
- Coolant hoses from the throttle body
- Control cables from the throttle body and intake plenum bracket
- Upper intake plenum
- Both rocker arm covers
- Fuel lines from the fuel rail and fuel line bracket
- Fuel rail with the injectors
- Inlet cooling pipe from the outlet housing
- Heater bypass hose from the water pump and the cylinder head
- Upper radiator hose at the thermostat housing
- Thermostat housing
- Lower intake manifold

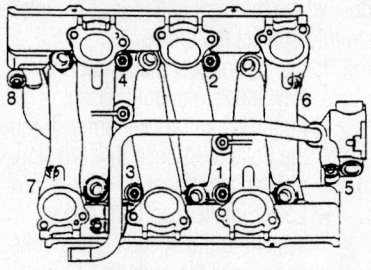

Lower intake manifold tightening sequence—3.4L engines

➡**When removing the valvetrain components, keep them in order for installation purposes.**

- Rocker arms and pushrods

To install:

5. Place a 3mm bead of RTV, on each

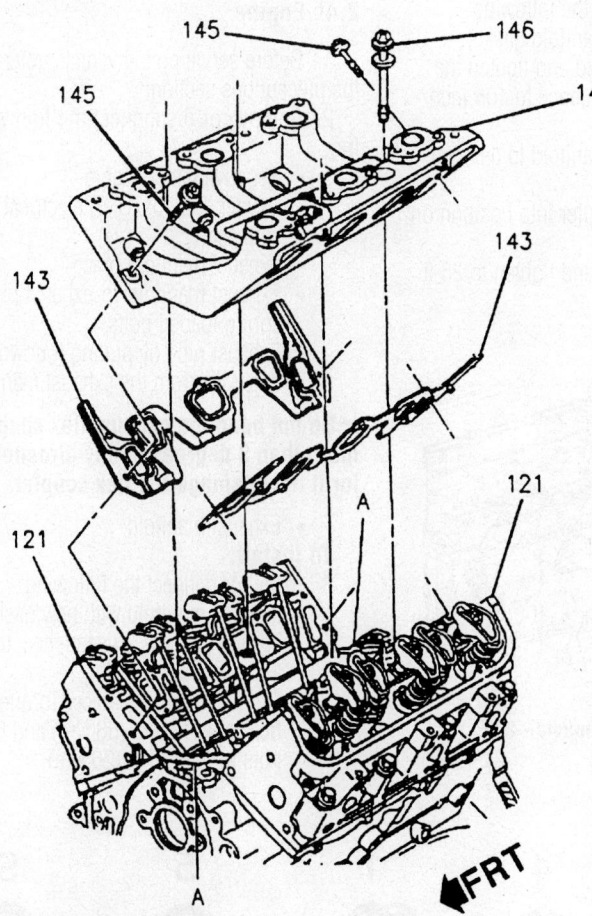

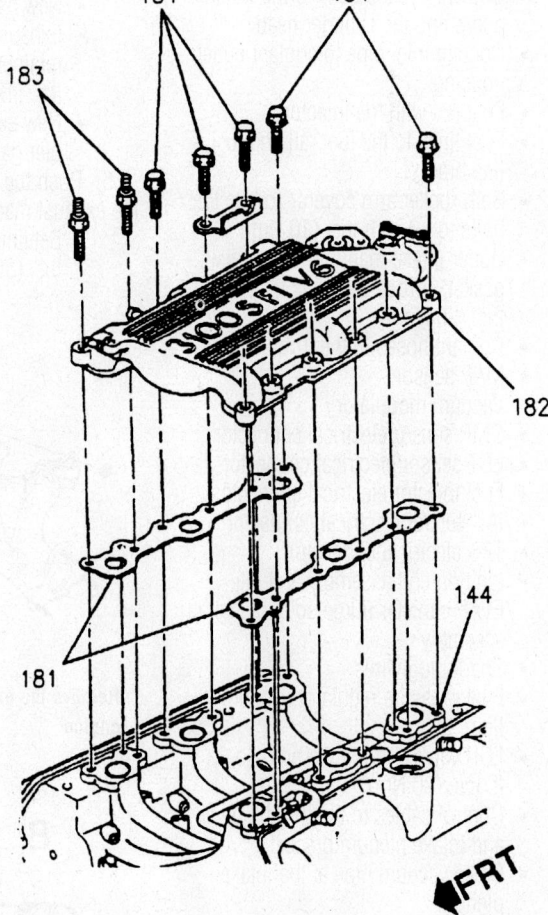

A   APPLY SEALANT
121   HEAD ASSEMBLY, CYLINDER
143   GASKET, LOWER INTAKE MANIFOLD
144   BOLT, LOWER INTAKE
145   BOLT, LOWER INTAKE MANIFOLD
146   BOLT, LOWER INTAKE MANIFOLD

144   MANIFOLD, LOWER INTAKE
181   GASKET, UPPER INTAKE MANIFOLD
182   MANIFOLD, UPPER INTAKE
183   STUD, UPPER INTAKE MANIFOLD
184   BOLT, UPPER INTAKE MANIFOLD

Exploded view of the upper and lower intake—3.4L engines

ridge, where the front and rear of the intake manifold contact the block.

6. Install or connect the following:
   - New intake manifold gasket
   - Pushrods and rocker arms. Torque the bolts to 89 inch lbs. (10 Nm) plus an additional 30 degree turn.
   - Lower intake manifold. Apply sealant to the threads of the bolts and torque the bolts to 115 inch lbs. (13 Nm).

✳✳ WARNING

In order to prevent oil leaks at the intake manifold, tighten the vertical bolts before the diagonal bolts.

- Thermostat housing with a new gasket. Torque the bolts to 19 ft. lbs. (26 Nm).
- Upper radiator hose at thermostat housing
- Coolant bypass hose to the water pump and the cylinder head
- Coolant inlet pipe to coolant outlet housing
- Fuel rail with the injectors
- Fuel lines to the fuel rail and fuel line bracket
- Both rocker arm covers. Torque the bolts to 89 inch lbs. (10 Nm).
- Upper intake manifold with a new gasket. Torque the bolts to 18 ft. lbs. (25 Nm).
- Coolant hoses to the throttle body
- MAP sensor
- Vacuum modulator
- CMP sensor electrical connector
- ECT sensor electrical connector
- Fuel injector electrical connectors
- IAC sensor electrical connector
- TPS electrical connector
- Ignition coil assembly and the EVAP canister purge solenoid as an assembly
- Spark plug wires
- Fuel pressure regulator vacuum line
- EGR valve. Torque the bolts to 22 ft. lbs. (30 Nm).
- Control cables to the throttle body and intake plenum bracket
- Brake vacuum pipe at the intake plenum
- Accessory drive belt
- Air cleaner assembly
- Negative battery cable

7. Refill the cooling system.

➡**An engine oil and filter change is recommended.**

8. Start the vehicle and verify no leaks.

Exhaust Manifold

REMOVAL & INSTALLATION

2.2L Engine

1. Before servicing the vehicle, refer to the precautions section.
2. Remove or disconnect the following:
   - Negative battery cable
   - Exhaust manifold heat shield
   - Oxygen (O2S) sensor
   - Manifold to the exhaust flex decoupler retainers
3. Pull down and back on the exhaust pipe in order to disconnect the pipe from the exhaust manifold.
   - Exhaust manifold to cylinder head retaining nuts
   - Exhaust manifold and gasket

To install:

4. Install or connect the following:
   - New exhaust manifold gasket
   - Exhaust manifold and tighten the retainers in sequence to 106 inch lbs. (12 Nm)
   - New exhaust manifold to flex coupler gasket
5. Push the flex coupler into position on the exhaust manifold.
   - Retaining nuts and tighten to 26 ft. lbs. (35 Nm)

Remove the exhaust manifold—2.2L engine

42372-NBOD-G11

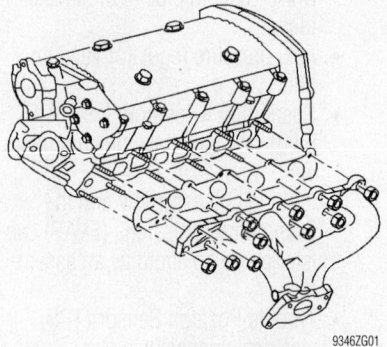

9346ZG01

Exploded view of the exhaust manifold assembly mounting—2.4L engine

- O2S sensor
- Exhaust manifold heat shield and tighten the bolts to 18 ft. lbs. (25 Nm)
- Negative battery cable

2.4L Engine

1. Before servicing the vehicle, refer to the precautions section.
2. Remove or disconnect the following:
   - Negative battery cable
   - Oxygen Sensor (O2S) electrical connector
   - Exhaust manifold brace
   - Exhaust manifold-to-exhaust pipe spring loaded bolts
   - Exhaust pipe by pulling it down and back from the exhaust manifold

➡**Do not bend the exhaust flex coupler more than 3 degrees in any direction, for it may damage the flex coupler.**

   - Exhaust manifold

To install:

3. Install or connect the following:
   - Exhaust manifold with new gaskets. Torque the nuts, in sequence, to 110 inch lbs. (13 Nm).
   - Exhaust manifold brace. Torque the bolt to 41 ft. lbs. (56 Nm) and the nuts to 19 ft. lbs. (26 Nm).

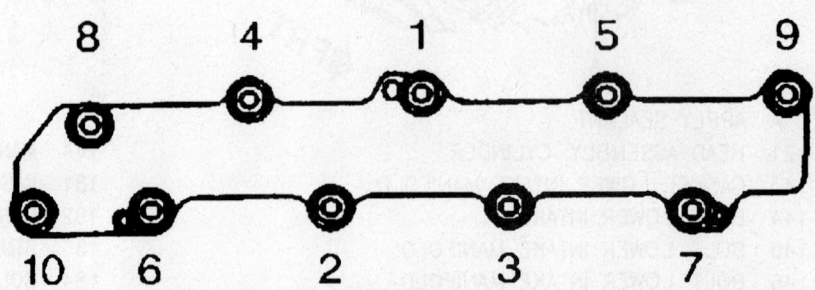

42372-NBOD-G12

Tighten the exhaust manifold retainers in this sequence—2.2L engine

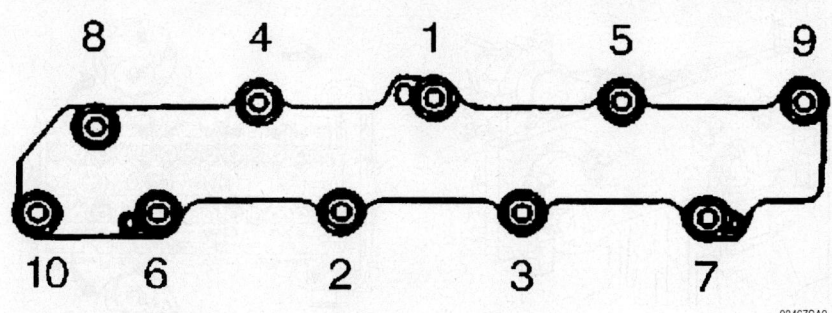

Exhaust manifold assembly torque sequence— 2.4L engine

- Exhaust pipe. Torque the bolts evenly to 26 ft. lbs. (35 Nm).
- O₂S electrical connector
- Negative battery cable

4. Start the vehicle and verify no exhaust leaks.

3.4L Engines

LEFT SIDE

1. Before servicing the vehicle, refer to the precautions section.

2. Partially drain the cooling system.
3. Remove or disconnect the following:

- Negative battery cable
- Air cleaner assembly
- Throttle body duct
- Exhaust crossover heat shield
- Exhaust crossover pipe from the manifold
- Radiator hose from the thermostat housing
- Spark plug wires

- Exhaust manifold heat shield
- Exhaust manifold

To install:

4. Install or connect the following:

- Exhaust manifold with a new gasket. Torque the nuts to 12 ft. lbs. (16 Nm).
- Exhaust manifold heat shield. Torque the nuts to 89 inch lbs. (10 Nm).
- Exhaust crossover pipe. Torque the bolts to 18 ft. lbs. (25 Nm).
- Crossover pipe heat shield. Torque the bolts to 89 inch lbs. (10 Nm).
- Spark plug wires
- Radiator hose to the coolant outlet housing
- Air cleaner assembly
- Throttle body duct
- Negative battery cable

RIGHT SIDE

1. Before servicing the vehicle, refer to the precautions section.
2. Remove or disconnect the following:

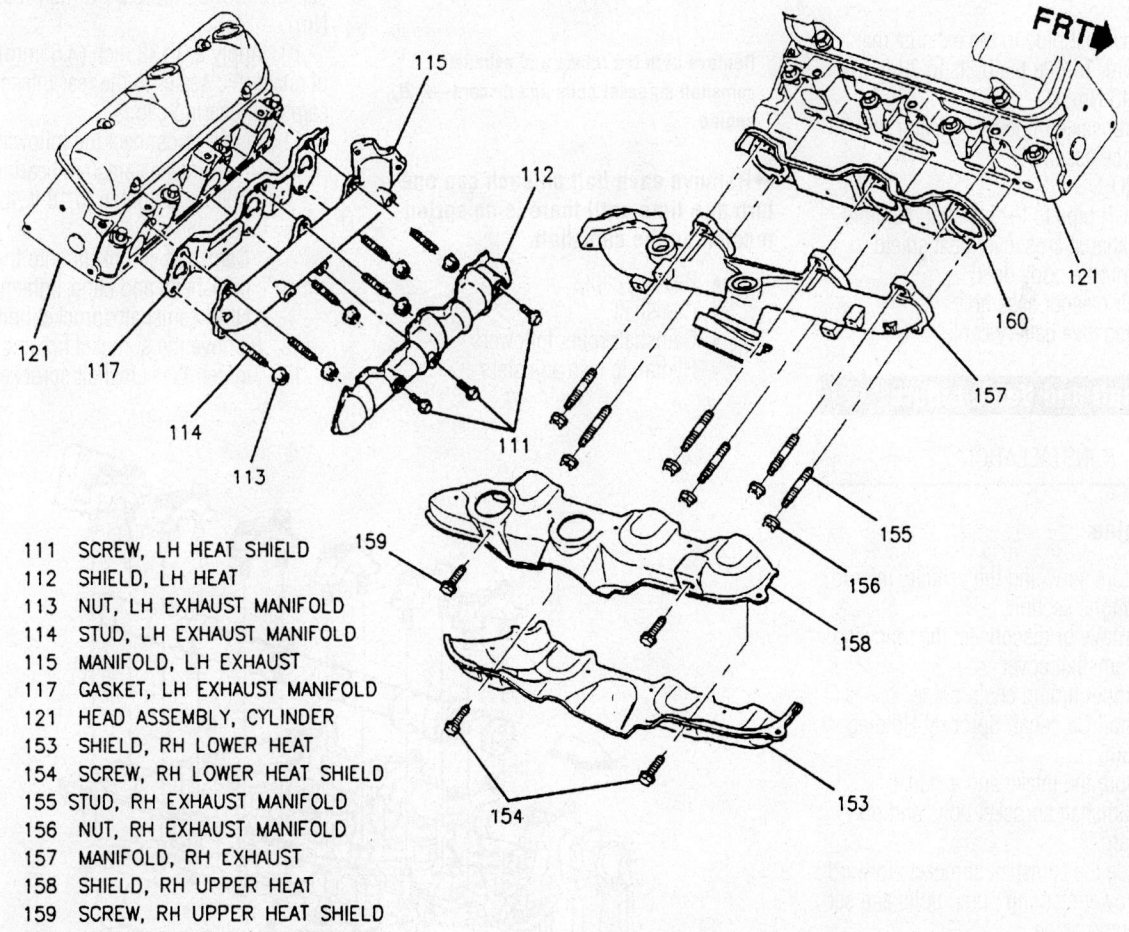

111	SCREW, LH HEAT SHIELD
112	SHIELD, LH HEAT
113	NUT, LH EXHAUST MANIFOLD
114	STUD, LH EXHAUST MANIFOLD
115	MANIFOLD, LH EXHAUST
117	GASKET, LH EXHAUST MANIFOLD
121	HEAD ASSEMBLY, CYLINDER
153	SHIELD, RH LOWER HEAT
154	SCREW, RH LOWER HEAT SHIELD
155	STUD, RH EXHAUST MANIFOLD
156	NUT, RH EXHAUST MANIFOLD
157	MANIFOLD, RH EXHAUST
158	SHIELD, RH UPPER HEAT
159	SCREW, RH UPPER HEAT SHIELD
160	GASKET, RH EXHAUST MANIFOLD

Exploded view of the exhaust manifold mounting—3.4L engines

- Negative battery cable
- Air cleaner assembly
- Throttle body duct
- Heated Oxygen Sensor (HO₂S)
- Exhaust crossover heat shield
- Exhaust Gas Recirculation (EGR) pipe from the exhaust manifold
- Exhaust crossover pipe
- Exhaust manifold heat shield
- Transaxle oil fill tube
- Front exhaust pipe from the exhaust manifold
- Wires from the spark plugs
- Exhaust manifold

To install:

3. Install or connect the following:
- Exhaust manifold with a new gasket. Torque the nuts to 12 ft. lbs. (16 Nm).
- Exhaust manifold heat shield. Torque the nuts to 89 inch lbs. (10 Nm).
- Exhaust crossover pipe to the manifold. Torque the bolts to 18 ft. lbs. (25 Nm).
- Exhaust crossover pipe heat shield. Torque the bolts to 89 inch lbs. (10 Nm).
- Exhaust pipe to the exhaust manifold. Torque the bolts to 33 ft. lbs. (45 Nm).
- Transaxle oil level indicator and fill tube assembly
- HO₂S
- EGR pipe to the exhaust manifold
- Exhaust crossover heat shield
- Throttle body duct
- Air cleaner assembly
- Negative battery cable

Camshaft and Valve Lifters

REMOVAL & INSTALLATION

2.2L Engine

1. Before servicing the vehicle, refer to the precautions section.
2. Remove or disconnect the following:
- Camshaft cover
- Upper timing chain guide
3. Install Camshaft Sprocket Holding Tool J 43665.
- Both the intake and exhaust camshaft sprocket bolts and discard
4. Slide the camshaft sprockets forward.
- Power steering pump bolts and set pump aside
5. Mark bearing caps to ensure they are installed in the original position.

42372-NBOD-G13

Install Camshaft Sprocket Holding Tool J 43665—2.2L engine

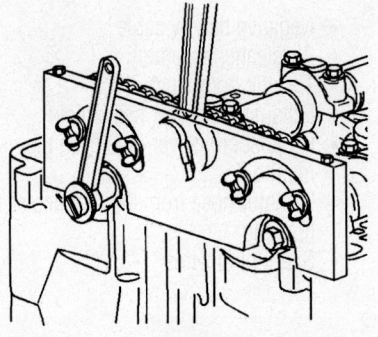

42372-NBOD-G14

Remove both the intake and exhaust camshaft sprocket bolts and discard—2.2L engine

➡**Remove each bolt on each cap one turn at a time until there is no spring tension on the camshaft.**

- Bearing caps
- Camshaft
- Camshaft roller followers
- Hydraulic lash adjusters

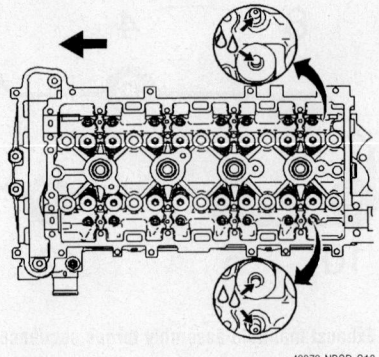

42372-NBOD-G16

Lubricate the valve tips—2.2L engine

To install:

6. Lubricate the valve tips.
7. Install or connect the following:
- Hydraulic lash adjusters
- Camshaft roller followers
8. Make sure that the alignment notches are aligned with the camshaft sprocket.
- Camshaft
- Camshaft bearing caps
9. Tighten the camshaft bearing cap bolts in 3 steps to 89 inch lbs. (10 Nm).
10. Apply a 0.138 inch (3.55mm) bead of anaerobic sealer to the rear intake camshaft bearing cap.
11. Install or connect the following:
- Rear intake camshaft bearing cap bolts and tighten to 18 ft. lbs. (25 Nm)
- Camshaft sprockets onto the camshafts and hand-tighten the NEW camshaft sprocket bolts
12. Remove the sprocket holding tool.
13. Tighten the camshaft sprocket bolts

42372-NBOD-G15

Remove the bearing caps and camshaft—2.2L engine

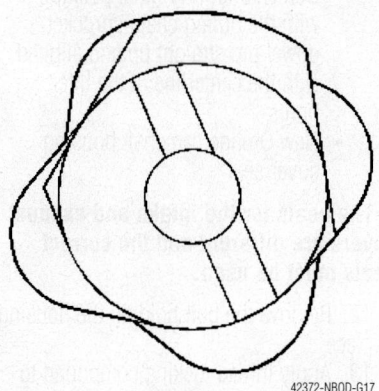

Make sure that the alignment notches are aligned with the camshaft sprocket—2.2L engine

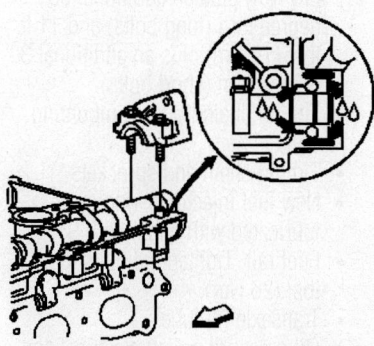

Apply a 0.197 inch (5mm) bead of anaerobic sealer97 in) bead to the rear intake camshaft bearing cap—2.2L engine

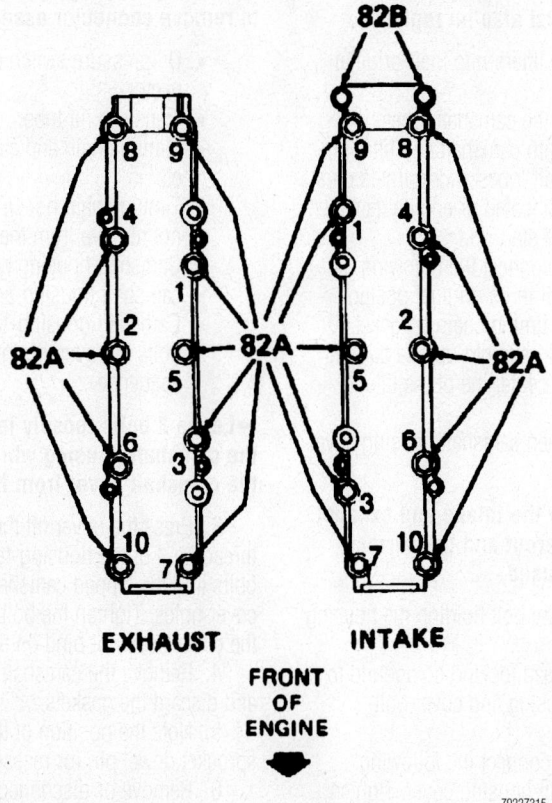

Camshaft housing bolt tightening sequence—2.4L engine

to 63 ft. lbs. (85 Nm), then an additional 30 degrees.

- Upper timing chain guide and tighten to 89 in. lbs. (10 Nm).
- Power steering pump and tighten the bolts to 19 ft. lbs. (25 Nm).
- Camshaft cover

2.4L Engine

INTAKE SIDE

➡**Anytime the camshaft housing-to-cylinder head bolts are loosened or removed, the camshaft housing-to-cylinder head gasket must be replaced.**

1. Before servicing the vehicle, refer to the precautions section.
2. Remove or disconnect the following:

- Negative battery cable
- Ignition assembly electrical connector
- 4 ignition assembly-to-camshaft housing bolts and the assembly by pulling it straight up

➡**Use a spark plug boot wire remover to remove connector assemblies.**

- Camshaft Position (CMP) sensor electrical connector
- Power steering pump and move it aside without disconnecting the lines
- Vacuum line from fuel pressure regulator
- Fuel injector wiring harness
- Both fuel line-to-intake camshaft housing clamps
- Fuel rail-to-camshaft housing bolts
- Fuel rail from the cylinder head and move it aside with the fuel lines attached
- Timing chain and camshaft sprockets
- Timing chain housing bolts but do not remove from the engine
- Camshaft housing cover-to-camshaft housing bolts
- Camshaft housing-to-cylinder head bolts by reversing of the tightening sequence

➡**Leave 2 bolts loosely in place to hold the camshaft housing while separating the camshaft cover from housing.**

3. Press the cover off the housing by threading 4 of the housing-to-cylinder head

bolts into the tapped camshaft housing cover holes. Tighten the bolts in evenly so the cover does not bind on the dowel pins.

4. Remove the camshaft housing cover and discard the gaskets.
5. Note the position of the chain sprocket dowel pin for reassembly.
6. Remove or disconnect the following:

- Camshaft
- Camshaft oil seal from camshaft and discard it

➡**The camshaft seal must be replaced any time the housing and cover are separated.**

➡**Store the valve lifters in order so that they may be installed in the their locations.**

- Valve lifters from the camshaft housing
- Camshaft carrier from the cylinder head and discard the gasket

To install:

7. Clean all the gasket surfaces completely.
8. Install or connect the following:

- New camshaft housing gasket
- Camshaft housing

➡**Install 1 bolt loosely to hold the housing in place.**

➡️**If the camshaft was replaced the valve lifters must also be replaced.**

9. Install the lifters into their original bores.

10. Lubricate the camshaft lobes, journals and lifters with camshaft and lifter pre-lube. The camshaft lobes and journals must be adequately lubricated or engine damage could occur upon start up.

11. Install or connect the following:
- Camshaft in its original position with the timing chain sprocket dowel pin straight up and aligned with the centerline of the lifter bores.
- New Green camshaft housing cover seal

➡️**The seals for the intake and exhaust covers are different and the correct seals must be used.**

12. Remove the bolt holding the housing in place.

13. Apply thread locking compound to the camshaft housing and cover bolt threads.

14. Install or connect the following:
- Camshaft housing cover. Tighten the bolts, in sequence, to 11 ft. lbs. (15 Nm) plus an additional 90 degree turn (long bolts) and 11 ft. lbs. (15 Nm) plus an additional 30 degree turn (short bolts).
- Timing chain housing bolts
- Timing chain and sprockets
- New fuel injector O-ring seals lubricated with engine oil
- Fuel rail. Tighten the bolts to 19 ft. lbs. (26 Nm).
- Ignition assembly on the camshaft housing. Tighten the bolts to 11 ft. lbs. (15 Nm), plus an additional 30 degree turn.
- Ignition assembly electrical connector
- Negative battery cable

15. Start the vehicle and verify proper operation and no leaks.

EXHAUST SIDE

➡️**Anytime the camshaft housing-to-cylinder head bolts are loosened or removed, the camshaft housing-to-cylinder head gasket must be replaced.**

1. Before servicing the vehicle, refer to the precautions section.

2. Remove or disconnect the following:
- Negative battery cable
- Ignition assembly electrical connector
- 4 ignition assembly-to-camshaft housing bolts and the assembly by pulling it straight up

➡️**Use a spark plug boot wire remover to remove connector assemblies.**

- Oil pressure switch electrical connector
- Transaxle fill tube
- Timing chain and camshaft sprockets
- Timing chain housing bolts but do not remove from the engine
- Camshaft housing cover-to-camshaft housing bolts
- Camshaft housing-to-cylinder head bolts by reversing of the tightening sequence

➡️**Leave 2 bolts loosely in place to hold the camshaft housing while separating the camshaft cover from housing.**

3. Press the cover off the housing by threading 4 of the housing-to-cylinder head bolts into the tapped camshaft housing cover holes. Tighten the bolts in evenly so the cover does not bind on the dowel pins.

4. Remove the camshaft housing cover and discard the gaskets.

5. Note the position of the chain sprocket dowel pin for reassembly.

6. Remove or disconnect the following:
- Camshaft
- Camshaft oil seal from camshaft and discard it

➡️**The camshaft seal must be replaced any time the housing and cover are separated.**

➡️**Store the valve lifters in order so that they may be installed in the their locations.**

- Valve lifters from the camshaft housing
- Camshaft carrier from the cylinder head and discard the gasket

To install:

7. Clean all the gasket surfaces completely.

8. Install or connect the following:
- New camshaft housing gasket
- Camshaft housing

➡️**Install 1 bolt loosely to hold the housing in place.**

➡️**If the camshaft was replaced the valve lifters must also be replaced.**

9. Install the lifters into their original bores.

10. Lubricate the camshaft lobes, journals and lifters with camshaft and lifter pre-lube. The camshaft lobes and journals must be adequately lubricated or engine damage could occur upon start up.

11. Install or connect the following:

- Camshaft in it original position with the timing chain sprocket dowel pin straight up and aligned with the centerline of the lifter bores.
- New Orange camshaft housing cover seal

➡️**The seals for the intake and exhaust covers are different and the correct seals must be used.**

12. Remove the bolt holding the housing in place.

13. Apply thread locking compound to the camshaft housing and cover bolt threads.

14. Install or connect the following:
- Camshaft housing cover. Tighten the bolts, in sequence, to 11 ft. lbs. (15 Nm) plus an additional 90 degree turn (long bolts) and 11 ft. lbs. (15 Nm) plus an additional 30 degree turn (short bolts).
- Timing chain housing mounting bolts
- Timing chain and sprockets
- New fuel injector O-ring seals lubricated with engine oil
- Fuel rail. Tighten the bolts to 19 ft. lbs. (26 Nm).
- Transaxle fill tube
- Oil pressure switch electrical connector
- Ignition assembly on the camshaft housing. Tighten the bolts to 11 ft. lbs. (15 Nm), plus an additional 30 degree turn.
- Ignition assembly electrical connector
- Negative battery cable

15. Start the vehicle and verify proper operation and no leaks.

3.4L Engines

1. Before servicing the vehicle, refer to the precautions section.

2. Relieve the fuel system pressure.

3. Remove or disconnect the following:
- Engine assembly

✳✳ WARNING

When removing valvetrain components they must be marked for installation in their original location.

- Rocker arm covers
- Intake manifold
- Rocker arms and pushrods
- Lifter guides
- Valve lifter(s) from the bores
- Crankshaft balancer and front cover

- Timing chain and sprockets
- Oil pump driven gear
- Camshaft thrust plate
- Camshaft

✳✳ WARNING

Avoid damaging the camshaft bearing surfaces.

To install:

4. Coat the camshaft with camshaft lubricant.

5. Install or connect the following:
- Camshaft
- Camshaft thrust plate. Torque the bolts to 89 inch lbs. (10 Nm).
- Oil pump driven gear. Torque the bolt to 27 ft. lbs. (36 Nm).
- Timing chain and sprocket. Torque the sprocket bolt to 103 ft. lbs. (140 Nm).
- Front cover. Torque the bolts to 15 ft. lbs. (21 Nm).
- Crankshaft balancer. Torque the bolt to 76 ft. lbs. (103 Nm).

6. Lubricate the bearing surfaces with Molykote®.

7. Install or connect the following:
- Lifters in their original locations
- Lifter guides. Torque the guide bolts to 89 inch lbs. (10 Nm).
- Pushrods and rocker arms. Torque the nuts to 89 inch lbs. (10 Nm) plus an additional 30 degree turn.
- Intake manifold
- Rocker arm covers. Torque the bolts to 89 inch lbs. (10 Nm).
- Engine assembly
- Negative battery cable

8. Adjust the valves, as required. Start the engine and verify no oil leaks.

Valve Lash

ADJUSTMENT

The engines are equipped with hydraulic valve lifters that do not require periodic valve lash adjustment. Adjustment to zero lash is maintained automatically by hydraulic pressure in the lifters.

Starter Motor

REMOVAL & INSTALLATION

2.2L Engine

1. Before servicing the vehicle, refer to the precautions section.
2. Remove or disconnect the following:

Starter mounting—2.2L engine

- Negative battery cable
- Starter electrical connectors
- Starter motor

To install:
3. Install or connect the following:
- Starter motor. Torque the bolts to 30 ft. lbs. (40 Nm).
- Starter electrical connectors
- Negative battery cable

2.4L Engine

1. Before servicing the vehicle, refer to the precautions section.
2. Remove or disconnect the following:
- Negative battery cable
- Air inlet duct from the throttle body
- Upper starter bolt
- Lower closeout panel
- Lower starter bolt
- Starter electrical connectors
- Starter motor

To install:
3. Install or connect the following:
- Starter electrical connectors
- Starter motor. Torque the bolts to 66 ft. lbs. (90 Nm).
- Lower closeout panel
- Air inlet duct to the throttle body
- Negative battery cable

3.4L Engine

1. Before servicing the vehicle, refer to the precautions section.
2. Remove or disconnect the following:
- Negative battery cable
- Flywheel inspection cover
- Starter electrical connectors
- Starter motor

To install:
3. Install or connect the following:
- Starter motor. Torque the bolts to 32 ft. lbs. (43 Nm).
- Starter electrical connectors

- Flywheel inspection cover. Torque the bolts to 89 inch lbs. (10 Nm).
- Negative battery cable

Oil Pan

REMOVAL & INSTALLATION

2.2L Engine

1. Before servicing the vehicle, refer to the precautions section.
2. Drain the engine oil.
3. Remove or disconnect the following:
- Engine mount strut bracket
- Drive belt
- Lower then upper AC compressor bolts
- Oil pan bolts
- Oil pan

To install:
4. Apply a 2mm bead of RTV sealant around the perimeter of the oil pan and the oil suction port. Do not over apply the RTV sealant.

5. Install or connect the following:
- Oil pan and tighten the bolts to 18 ft. lbs. (25 Nm)
- AC compressor bolts
- Engine mount bracket
- Drive belt

6. Refill the crankcase.
7. Start the vehicle and verify no leaks.

2.4L Engine

1. Before servicing the vehicle, refer to the precautions section.
2. Drain the engine oil.
3. Drain the cooling system.
4. Remove or disconnect the following:
- Negative battery cable
- Flywheel/converter cover
- Right wheel
- Right wheel well splash shield
- Accessory drive belt
- Air conditioning compressor lower bolts

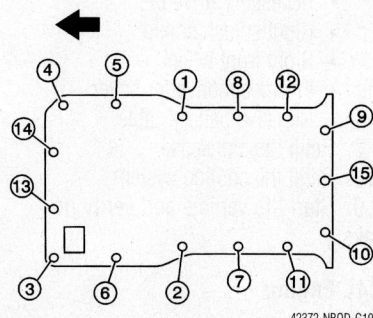

Remove the oil pan bolts in this sequence—2.2L engine

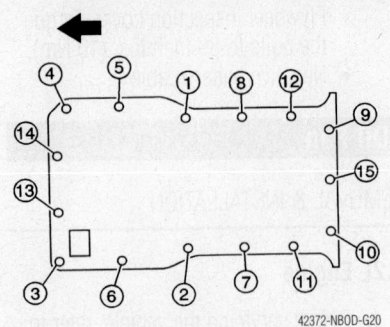

Install the oil pan bolts using this sequence—2.2L engine

- Transaxle-to-engine brace
- Engine mount strut bracket
- Radiator outlet pipe bolts
- Radiator outlet pipe from the oil pan
- Oil pan to the flywheel cover bolt and nut
- Flywheel cover stud for clearance
- Radiator outlet pipe from the lower radiator hose and oil pan
- Oil level sensor connector
- Oil pan

To install:

5. Inspect the oil pan gasket; it is reusable if not damaged.

6. Install or connect the following:
- Oil pan with the gasket. Torque the M8 bolts to 18 ft. lbs. (24 Nm) and the M6 bolts to 106 inch lbs. (12 Nm).
- Oil pan to the transaxle nut
- Oil level sensor connector
- Radiator outlet pipe to the lower radiator hose and oil pan
- Exhaust manifold brace
- Radiator outlet pipe. Torque the bolts to 124 inch lbs. (14 Nm).
- Engine mount strut bracket. Torque the bolts to 55 ft. lbs. (75 Nm).
- Transaxle to the engine brace
- Air conditioning compressor lower bolts. Torque the bolts to 37 ft. lbs. (50 Nm).
- Accessory drive belt
- Right splash shield
- Right front wheel
- Flywheel/converter cover
- Negative battery cable

7. Refill the crankcase.
8. Refill the cooling system.
9. Start the vehicle and verify no leaks.

3.4L Engine

1. Before servicing the vehicle, refer to the precautions section.

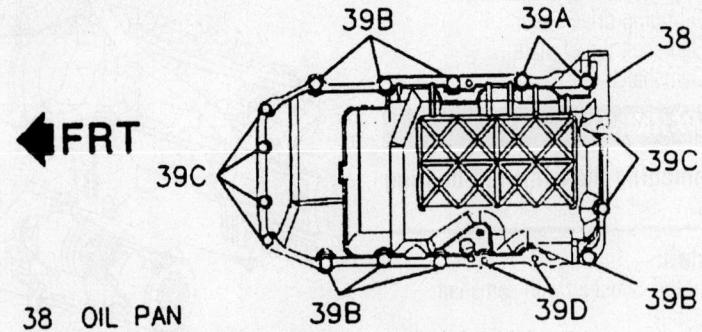

◀FRT

38   OIL PAN
39A BOLT, OIL PAN (M8 X 1.25 X 80)
     24 N•m (18 LB. FT.)
39B BOLT, OIL PAN (M8 X 1.25 X 22)
     24 N•m (18 LB. FT.)
39C BOLT, OIL PAN (M6 X 1.00 X 25)
     12 N•m (106 LB. IN.)
39D BOLT, STUD END OIL PAN
     26 N•m (19 LB. FT.)

Oil pan mounting bolt locations—2.4L engine

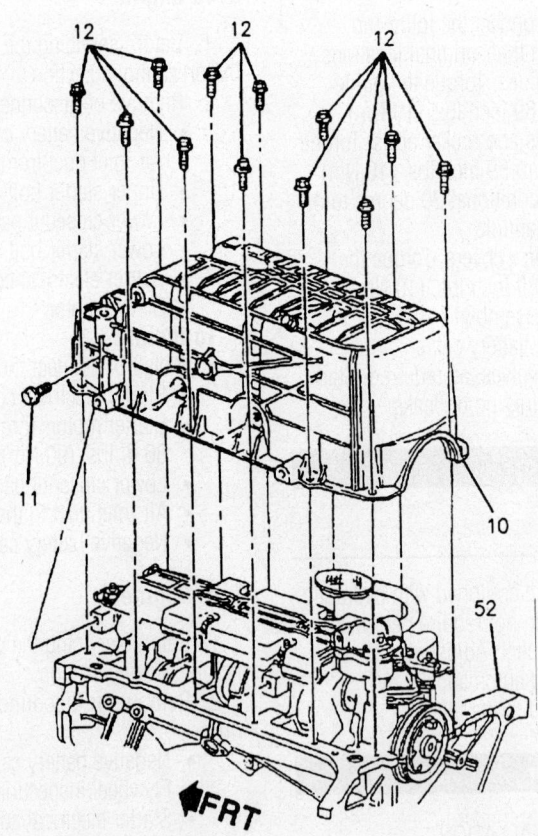

◀FRT

10 PAN, OIL
11 BOLT, OIL PAN SIDE
12 BOLT, OIL PAN RETAINING
52 BLOCK, ENGINE

Exploded view of the oil pan mounting—3.4L engines

2. Drain the engine oil.

3. Evacuate the A/C system.

4. Remove or disconnect the following:
   - Negative battery cable
   - Accessory drive belt
   - Right front wheel
   - Right splash shield
   - Anti-lock Brake System (ABS) Wheel Speed (WSS) sensor from the right subframe
   - Right lower ball joint from the steering knuckle
   - Right outer tie rod end
   - A/C compressor without disconnecting the lines
   - Evaporator-to-accumulator A/C line
   - Flywheel cover
   - Right side engine cradle bolts
   - Crankshaft balancer
   - Starter motor
   - Oil pan

To install:

5. Apply silicone sealer to the portion of the pan that contacts the rear of the block.

6. Install or connect the following:
   - Oil pan with a new gasket. Torque the flange bolts to 18 ft. lbs. (25 Nm) and the side bolts to 37 ft. lbs. (50 Nm).
   - Starter motor. Torque the bolts to 32 ft. lbs. (43 Nm).
   - Flywheel cover. Torque the bolts to 89 inch lbs. (10 Nm).
   - Crankshaft balancer. Torque the bolt to 76 ft. lbs. (103 Nm).
   - Engine cradle bolts. Torque the bolts to 84 ft. lbs. (115 Nm) plus an additional 120 degree turn.
   - A/C compressor. Torque the bolts to 37 ft. lbs. (50 Nm).
   - Evaporator-to-accumulator line
   - Ball joint. Torque the nut to 48 ft. lbs. (60 Nm).
   - Outer tie rod end. Torque the nut to 44 ft. lbs. (60 Nm).
   - WSS electrical connector
   - Wheel well splash shield
   - Right front wheel
   - Accessory drive belt
   - Negative battery cable

7. Fill the crankcase.

8. Evacuate and recharge the A/C system.

9. Start the engine and check for leaks.

➡**Whenever the vehicle subframe is removed or lowered, the wheel alignment should be checked.**

10. Check and/or adjust the front end alignment.

Oil Pump

REMOVAL & INSTALLATION

2.2L Engine

1. Before servicing the vehicle, refer to the precautions section.

2. Remove the engine front cover as follows:
   a. Remove the front wheel.
   b. Remove the engine splash shield.
   c. Remove the drive belt.
   d. Use crankshaft holding tool J 38122-A to prevent the crankshaft from rotating while loosening the crankshaft balancer bolt.
   e. Remove the crankshaft balancer bolt and discard. Remove the balancer.
   f. Remove the drive belt tensioner.
   g. Remove the water pump-to-front cover bolts.
   h. Remaining front cover-to-engine bolts, the cover and gasket.

3. Remove or disconnect the following:
   - Oil pressure relief valve
   - Oil pump cover
   - Oil pump gears

To install:

4. Install or connect the following:
   - Oil pump gears
   - Oil pump cover and tighten the bolts to 53 inch lbs. (6 Nm)
   - Oil pressure relief valve and tighten the plug to 30 ft. lbs. (40 Nm).

5. Install the engine front cover as follows:
   a. Install front cover gasket, cover and cover-to-engine bolts.
   b. Install the water pump-to-front cover bolts. Tighten all the cover bolts to 15 ft. lbs. (20 Nm).

c. Install the drive belt tensioner and tighten the bolt to 33 ft. lbs. (45 Nm).

d. Install the crankshaft balancer and bolt.

e. Use crankshaft holding tool J 38122-A to prevent the crankshaft from rotating while tightening the crankshaft balancer bolt. Tighten the bolt to 74 ft. lbs. (100 Nm) plus an additional 75 degrees.

f. Install the engine splash shield.

g. Install the drive belt.

h. Install the front wheel.

2.4L Engine

1. Before servicing the vehicle, refer to the precautions section.

2. Disconnect the negative battery cable.

3. Install an engine support fixture.

4. Properly drain the engine oil.

5. Remove or disconnect the following:
   - Oil pan
   - Balance shaft chain cover
   - Balance shaft chain tensioner
   - Oil pump cover
   - Oil pump assembly from the balance shaft assembly, by pulling the housing to disconnect the pump gear from the balance shaft

To install:

6. Lubricate the gears with clean engine oil.

7. Assemble the geroter gear into the housing.

➡**Fill the oil pump cavities with petroleum jelly prior to installation. This seals the pump and acts like a "prime" so the pump will draw oil as soon as the engine begins to turn. This will ensure that there is oil pressure immediately on start-up and will prevent engine damage.**

Oil pump assembly—2.2L engine

42372-NB0D-G21

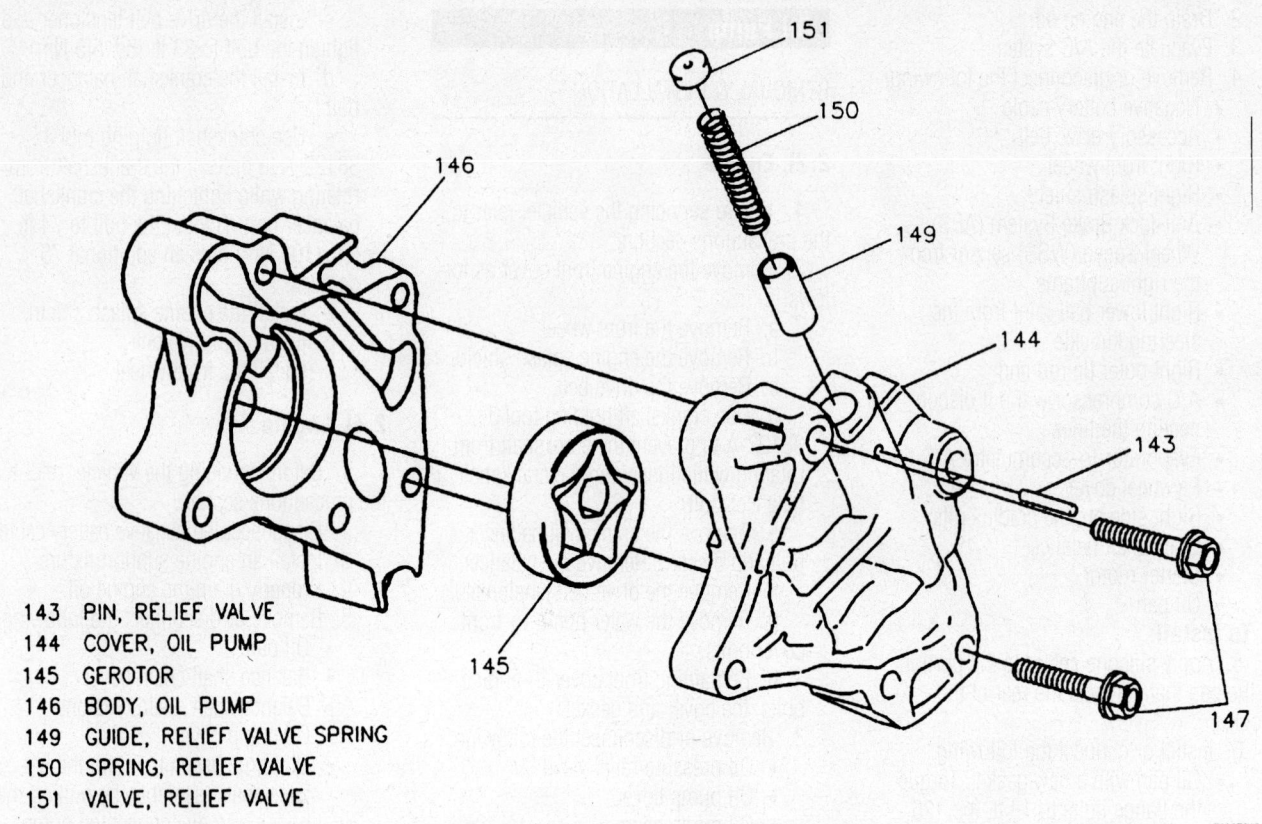

143 PIN, RELIEF VALVE
144 COVER, OIL PUMP
145 GEROTOR
146 BODY, OIL PUMP
149 GUIDE, RELIEF VALVE SPRING
150 SPRING, RELIEF VALVE
151 VALVE, RELIEF VALVE

7922Z318

Exploded view of the oil pump components—2.4L engine

8. Install or connect the following:
- Oil pump to the balance shaft assembly. Torque the bolts to 40 ft. lbs. (54 Nm).
- Oil pump cover. Torque the bolts to 40 ft. lbs. (54 Nm).
- Balance shaft chain tensioner with the bolts finger tight

9. Adjust the chain tension inserting a 0.40 in. (1mm) brass feeler, between the chain guide and the chain.

➡A brass feeler gauge must be used to ensure that correct measurements are obtained. If a steel gauge is used, it will not bend to conform to the guide and will allow for incorrect measurements.

10. Press the guide against the chain using about 10 lbs. of force. Torque the chain tensioner fastener to 89 inch lbs. (10 Nm).

11. Install or connect the following:
- Balance shaft chain cover. Torque the nut/bolt to 10 ft. lbs. (13 Nm).
- Oil pan. Torque the bolts to 18 ft. lbs. (24 Nm).
- Negative battery cable

12. Fill the crankcase.

➡An oil filter change is recommended.

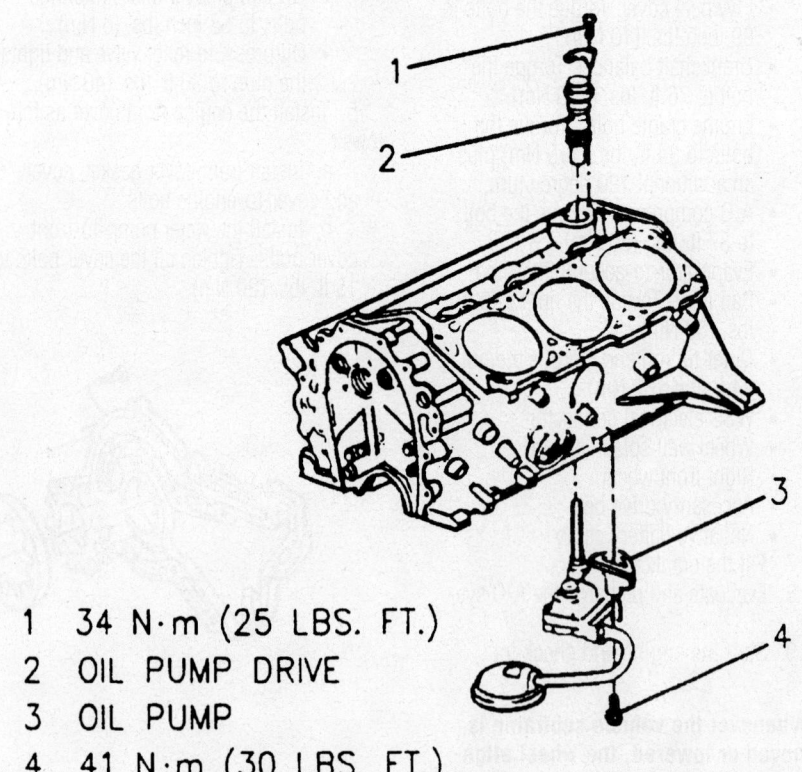

1   34 N·m (25 LBS. FT.)
2   OIL PUMP DRIVE
3   OIL PUMP
4   41 N·m (30 LBS. FT.)

7922Z319

Exploded view of the oil pump mounting—3.4L engines

13. Start the engine and verify oil pressure and no leaks.

3.4L Engines

1. Before servicing the vehicle, refer to the precautions section.
2. Disconnect the negative battery cable.
3. Drain the engine oil.
4. Remove or disconnect the following:
   • Oil pan
   • Oil pump and pump driveshaft

To install:
5. Install or connect the following:
   • Oil pump by engaging the oil pump driveshaft. Torque the oil pump bolts to 30 ft. lbs. (41 Nm).
   • Oil pan. Torque the bolts to 18 ft. lbs. (25 Nm).
   • Negative battery cable
6. Fill the crankcase.

➡**An oil filter change is recommended.**

7. Start the engine, check the oil pressure and check for leaks.

Rear Main Seal

REMOVAL & INSTALLATION

2.2L Engine

1. Before servicing the vehicle, refer to the precautions section.
2. Support the engine.
3. Remove or disconnect the following:
   • Transaxle

   • Flywheel
   • Rear main seal using a prytool

❋❋ WARNING

Be careful not to damage the crankshaft sealing surface.

To install:
4. Install or connect the following:
   • Seal using tool J 42067
   • Flywheel. Tighten the flywheel-to-crankshaft bolts to 39 ft. lbs. (53 Nm) plus an additional 25 degrees.
   • Transaxle
5. Start the engine and check for leaks.

2.4L Engine

1. Before servicing the vehicle, refer to the precautions section.
2. Remove or disconnect the following:
   • Negative battery cable
   • Transaxle
   • Pressure plate and clutch disc, if equipped
   • Flywheel
   • Oil pan-to-seal housing bolts
   • Seal housing
   • Rear main seal from the housing

❋❋ WARNING

Be careful not to damage the seal housing sealing surface; damage may result in an oil leak.

To install:
3. Install the new rear main seal into the housing.

4. Inspect the oil pan gasket inner silicone bead for damage and repair using a silicone sealant, if necessary.
5. Lubricate the lip of the seal with clean engine oil.
6. Install or connect the following:
   • Seal housing with a new gasket. Torque the bolts to 106 inch lbs. (12 Nm).
   • Flywheel. Torque the bolts to 22 ft. lbs. (30 Nm) plus an additional 45 degree turn.
   • Clutch, pressure plate and clutch cover assembly, if equipped with a manual transaxle
   • Transaxle
   • Negative battery cable
7. Start the engine and check for leaks.

3.4L Engines

1. Before servicing the vehicle, refer to the precautions section.
2. Support the engine.
3. Remove or disconnect the following:

   • Transaxle
   • Flywheel
   • Rear main seal using a prytool

❋❋ WARNING

Be careful not to damage the crankshaft sealing surface.

To install:
4. Lubricate the seal and bore with engine oil.
5. Install the new seal by sliding it over the mandrel of rear main seal installer tool J 34686 until the dust lip bottoms squarely against the tool collar.
6. Align the dowel pin of the tool with the dowel pin hole in the crankshaft and attach the tool to the crankshaft. Tighten the attaching screws to 24–60 inch lbs. (2.7–6.8 Nm).
7. Tighten the tool T-handle to press the seal into the bore until the tool collar is flush against the engine.
8. Remove the installation tool.

➡**Make sure that the seal is squarely seated in the bore.**

9. Install or connect the following:
   • Flywheel. Tighten the flywheel-to-crankshaft bolts to 61 ft. lbs. (83 Nm).
   • Transaxle
10. Start the engine and check for leaks.

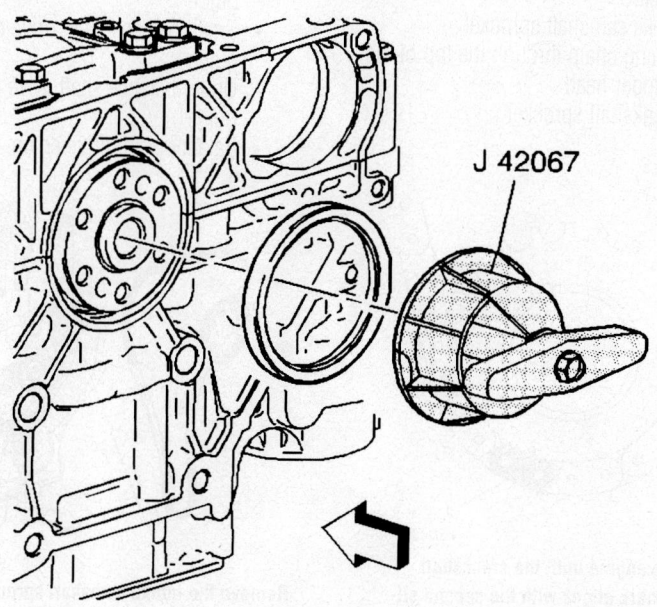

J 42067

42372-NB0D-G22

Install rear main seal using tool J 42067—2.2L engine

Timing Chain, Sprockets, Front Cover and Seal

REMOVAL & INSTALLATION

2.2L Engine

1. Before servicing the vehicle, refer to the precautions section.
2. Remove the valve cover.
3. Remove the engine front cover as follows:
   a. Remove the front wheel.
   b. Remove the engine splash shield.
   c. Remove the drive belt.
   d. Use crankshaft holding tool J 38122-A to prevent the crankshaft from rotating while loosening the crankshaft balancer bolt.
   e. Remove the crankshaft balancer bolt and discard. Remove the balancer.
   f. Remove the drive belt tensioner.
   g. Remove the water pump-to-front cover bolt.
   h. Remaining front cover-to-engine bolts, the cover and gasket.
4. Rotate the engine until the crankshaft

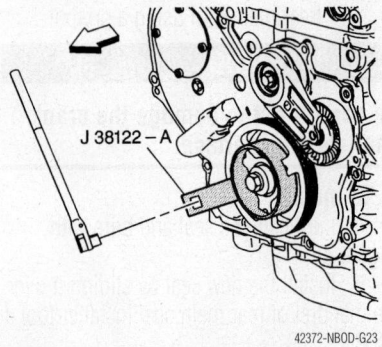

42372-NBOD-G23

Use crankshaft holding tool J 38122-A to prevent the crankshaft from rotating while loosening the crankshaft balancer bolt— 2.2L engine

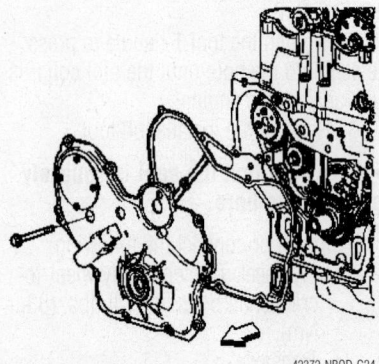

42372-NBOD-G24

Remove the water pump-to-front cover bolt—2.2L engine

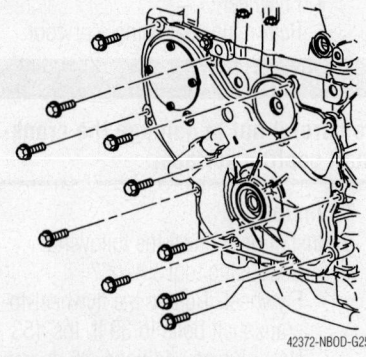

42372-NBOD-G25

Remove the front cover bolts and cover— 2.2L engine

sprocket mark aligns with the second silver link (2) at the 5 o'clock position. Refer to the illustration.

5. Make sure the INT diamond on the intake camshaft sprocket is aligned with the copper link at (1) at the 2 o'clock position. Refer to the illustration.

6. Make sure the EXH triangle on the exhaust camshaft sprocket is aligned with the silver link (3). Refer to the illustration.

7. Remove or disconnect the following:
   • Timing chain tensioner
   • Timing chain tensioner guide
   • Fixed timing chain guide access plug
   • Fxed timing chain guide
   • Upper timing chain guide

8. Use a 24 mm wrench to hold the camshafts from turning.
   • Exhaust camshaft sprocket bolt and discard
   • Exhaust camshaft sprocket
   • Intake camshaft sprocket bolt and discard
   • Intake camshaft sprocket
   • Timing chain through the top of the cylinder head
   • Crankshaft sprocket

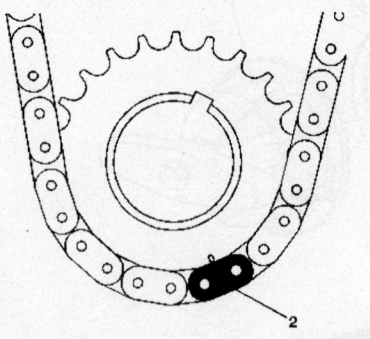

42372-NBOD-G26

Rotate the engine until the crankshaft sprocket mark aligns with the second silver link (2) at the 5 o'clock position—2.2L engine

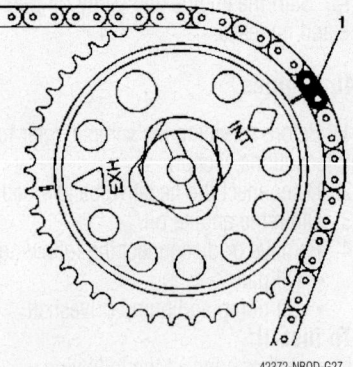

42372-NBOD-G27

Make sure the INT diamond on the intake camshaft sprocket is aligned with the copper link at (1) at the 2 o'clock position— 2.2L engine

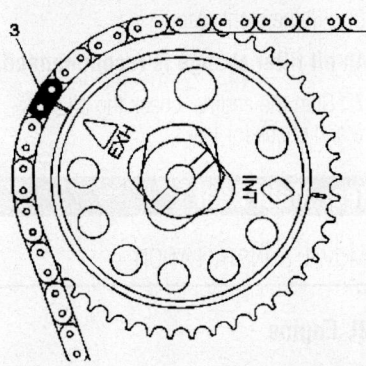

42372-NBOD-G28

Make sure the EXH triangle on the exhaust camshaft sprocket is aligned with the silver link (3)—2.2L engine

   • Balance shaft drive chain tensioner
   • Adjustable balance shaft chain guide
   • Small balance shaft drive chain guide
   • Upper balance shaft drive chain guide
   • Balance shaft drive chain.

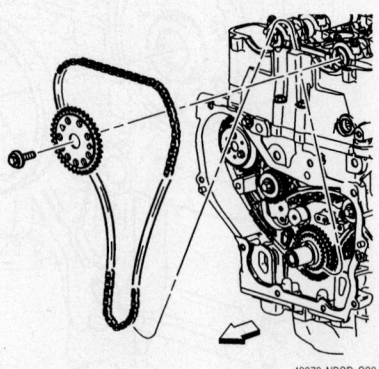

42372-NBOD-G29

Remove the intake camshaft sprocket, then the chain through the top of the head—2.2L engine

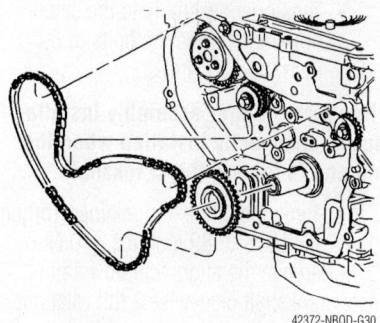

Remove the upper balance shaft chain guide, then the chain—2.2L engine

To install:

9. Install or connect the following:
- Upper balance shaft chain guide and tighten to 89 inch lbs. (10 Nm)
- Small balance shaft chain guide and bolts and tighten to 89 inch lbs. (10 Nm)
- Adjustable balance shaft drive chain guide and bolts and tighten to 89 inch lbs. (10 Nm)

10. Turn the tensioner plunger 90 degrees in its bore and compress the plunger until a paper clip can be inserted through the hole in the plunger body and into hole in the tensioner plunger.
- Timing chain tensioner and bolts and tighten to 89 inch lbs. (10 Nm)

11. Remove the paper clip from the balance shaft drive chain tensioner.

12. Install the crankshaft sprocket with timing mark at the 5 o'clock position.

13. Lower the timing chain through the opening in the top of the cylinder head. Carefully ensure that the chain goes around both sides of the cylinder block bosses.

14. Install or connect the following:
- Intake camshaft sprocket with the INT diamond at the 2 o'clock position

Install the crankshaft sprocket with timing mark at the 5 o'clock position—2.2L engine

15. Hand tighten a NEW intake camshaft sprocket bolt.

16. Route the timing chain around the crankshaft sprocket with the second silver link aligning with the timing mark.

17. Route the timing chain around the intake camshaft sprocket with the copper colored link aligning with the INT diamond.
- Timing chain tensioner guide through the opening in the top of the cylinder head and tighten the bolts to 89 inch lbs. (10 Nm).
- Exhaust camshaft sprocket with the timing chain silver link at EXH triangle aligned at the 10 o'clock position.

18. Use a 24 mm wrench to rotate the camshaft slightly, until exhaust sprocket aligns with the camshaft.

19. Hand tighten the NEW exhaust camshaft sprocket bolt.

20. Install the fixed timing chain guide and tighten the bolts to 89 inch lbs. (10 Nm).

21. Apply sealant, GM P/N 12345382 compound to thread and install the timing chain guide bolt access hole plug. Tighten the access hole plug to 30 ft. lbs. (40 Nm).
- Timing chain upper guide and bolts and tighten to 89 inch lbs. (10 Nm)

22. Measure the timing chain tensioner. In a fully compressed, non-active state, the tensioner will measure 2.83 in (72mm). A tensioner in the active state will measure 3.35 in (85mm). To put the tensioner in a non-active state, hold the flat end of the tensioner with a wrench and rotate the piston clockwise for slightly less than one full turn.
- Timing chain tensioner and tighten the tensioner to 55 ft. lbs. (75 Nm)

23. Use a suitable tool with a rubber tip on the end. Feed the tool down through the camshaft drive chant to rest on the timing chain. Then give a sharp jolt diagonally downwards to release the tensioner.

24. Use a 24 mm wrench to hold the camshaft and tighten the new bolts to 63 ft. lbs. (85 Nm) plus an additional 30 degrees.

25. Install the valve cover.

26. Install the engine front cover as follows:

a. Install front cover gasket, cover and cover-to-engine bolts.

b. Install the water pump-to-front cover bolts. Tighten all the cover bolts to 15 ft. lbs. (20 Nm).

c. Install the drive belt tensioner and tighten the bolt to 33 ft. lbs. (45 Nm).

d. Install the crankshaft balancer and bolt.

e. Use crankshaft holding tool J

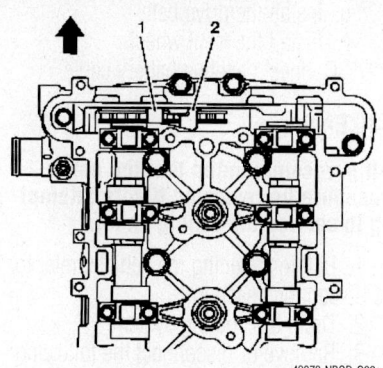

Lower the timing chain through the opening in the top of the cylinder head. Carefully ensure that the chain goes around both sides of the cylinder block bosses (1 & 2)—2.2L engine

38122-A to prevent the crankshaft from rotating while tightening the crankshaft balancer bolt. Tighten the bolt to 74 ft. lbs. (100 Nm) plus an additional 75 degrees.

f. Install the engine splash shield.

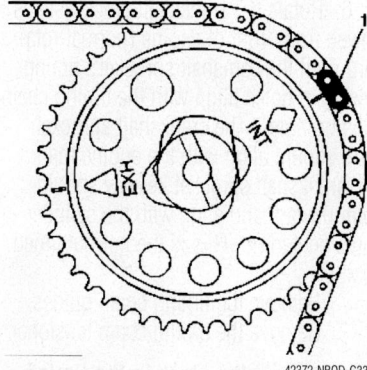

Install the intake camshaft sprocket with the INT diamond at the 2 o'clock position—2.2L engine

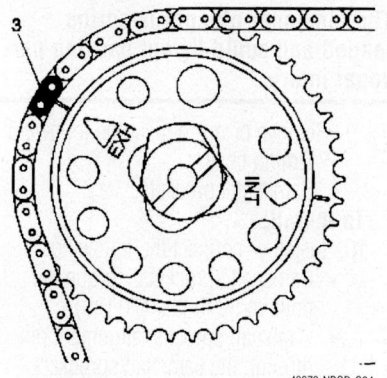

Install the exhaust camshaft sprocket with the timing chain silver link at EXH triangle aligned at the 10 o'clock position—2.2L engine

g. Install the drive belt.

h. Install the front wheel.

27. Connect negative battery cable.

2.4L Engine

➡️**It is recommended that the entire procedure be reviewed before attempting to service the timing chain.**

1. Before servicing the vehicle, refer to the precautions section.

2. Drain the cooling system.

3. Remove or disconnect the following:
   - Negative battery cable
   - Coolant surge tank
   - Accessory drive belt
   - Alternator

4. Install an engine support.

5. Remove or disconnect the following:
   - Upper cover fasteners
   - Front cover vent hose
   - Right engine mount and bracket
   - Right front wheel
   - Right lower splash shield
   - Crankshaft balancer
   - Lower cover fasteners
   - Front cover

6. Rotate the crankshaft clockwise, as viewed from front of engine (normal rotation) until the camshaft sprocket's timing dowel pin holes align with the timing chain housing holes. The crankshaft sprocket mark should align with the engine mark. The crankshaft sprocket keyway should point upward and align with the cylinder bores centerline. This is the normal timed position.

7. Remove the timing chain guides.

8. Remove the timing chain tensioner.

➡️**Be sure all the slack in the timing chain is above the tensioner assembly when removing it.**

❋❋ CAUTION

The tensioner plunger is spring loaded and could fly out causing personal injury.

9. Remove or disconnect the following:
   - Timing chain
   - Camshaft sprockets

To install:

10. Install or connect the following:
    - Camshaft sprockets. Torque the bolts to 52 ft. lbs. (70 Nm).
    - Camshaft sprocket alignment pin through the camshaft sprockets holes into the timing chain housing holes to position the camshafts for timing.

11. If the camshafts are out of position

and must be rotated more than ⅛ turn in order to install the alignment dowel pins, perform the following:

a. Rotated the crankshaft 90 degrees clockwise off Top Dead Center (TDC) in order to give the valves adequate clearance to open.

b. Once the camshafts are positioned and the dowels installed, rotate the crankshaft counterclockwise back to TDC.

❋❋ WARNING

Do not rotate the crankshaft clockwise to TDC or valve and piston damage may occur.

12. Install the timing chain over the exhaust camshaft sprocket, around the idler sprocket and around the crankshaft sprocket.

13. Remove the alignment dowel pin from the intake camshaft. Using a dowel pin remover tool, rotate the intake camshaft sprocket counterclockwise enough to slide the timing chain over the intake camshaft sprocket. Release the camshaft sprocket wrench. The length of chain between the 2 camshaft sprockets will tighten.

➡️**If properly timed, the intake camshaft alignment dowel pin should slide in easily. If the dowel pin does not fully index, the camshafts are not timed correctly and the procedure must be repeated.**

14. Leave the alignment dowel pins installed.

15. With slack removed from chain between intake camshaft sprocket and crankshaft sprocket, the timing marks on the crankshaft and the cylinder block should be aligned. If marks are not aligned, move the chain 1 tooth forward or rearward, remove slack and recheck the marks.

16. Tighten the chain housing to engine stud. The stud is installed under the timing chain. Torque it to 19 ft. lbs. (26 Nm).

17. Reload the timing chain tensioner as follows:

a. Form a keeper from heavy gauge wire.

b. Slightly, compress the shoe plunger and insert a small screwdriver into the access hole.

c. Release the ratchet pawl and compress the plunger completely into the hole.

d. Insert the keeper between the access hole and the blade.

18. Install or connect the following:

- Tensioner assembly to the chain housing. Torque the bolts to 89 inch lbs. (10 Nm).

➡️**Recheck plunger assembly installation. It is correctly installed when the long end is toward the crankshaft.**

- Tensioner shoe and retainer. Torque the bolts to 89 inch lbs. (10 Nm).

19. Remove the alignment dowel pins. Rotate crankshaft clockwise 2 full rotations. Align the crankshaft timing mark with mark on cylinder block and reinstall alignment dowel pins. Alignment dowel pins will slide in easily if engine is timed correctly.

❋❋ WARNING

If the engine is not correctly timed, severe engine damage could occur.

20. Install or connect the following:
    - Timing chain guides
    - New seal into the front cover by lubricating the seal lip and tapping it into place
    - Front cover and gaskets. Torque the nuts and bolts to 106 inch lbs. (12 Nm).
    - Crankshaft balancer. Torque the bolt to 129 ft. lbs. (175 Nm).
    - Right front lower splash shield
    - Front wheel. Torque the nuts to 100 ft. lbs. (140 Nm).
    - Right engine mount bracket. Torque the bolts to 81 ft. lbs. (110 Nm) plus an additional 90 degree turn.
    - Right engine mount. Torque the bolt to 49 ft. lbs. (60 Nm).
    - Upper cover vent hose

21. Remove the engine support.

22. Install or connect the following:
    - Alternator. Torque the bolts to 37 ft. lbs. (50 Nm).
    - Accessory drive belt
    - Coolant surge tank
    - Negative battery cable

23. Refill the cooling system.

24. Start the engine and check for leaks.

3.4L Engines

1. Before servicing the vehicle, refer to the precautions section.

2. Disconnect the negative battery cable.

3. Drain the cooling system.

4. Drain the engine oil.

5. Discharge and recover the A/C refrigerant.

6. Install an engine support fixture.

7. Remove or disconnect the following:
   - Front engine mount and bracket

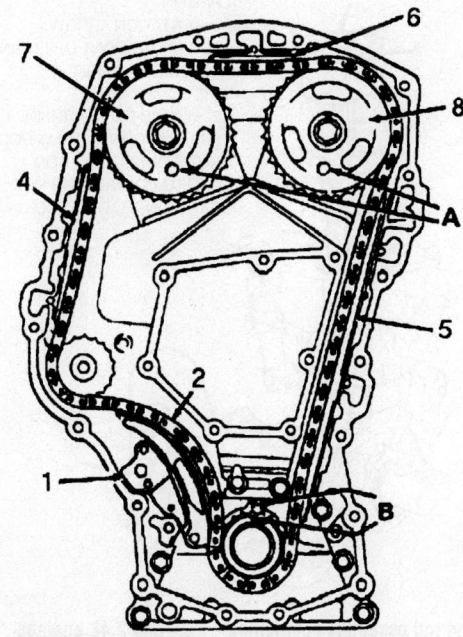

A. Camshaft timing alignment pin locations
B. Crankshaft gear timing marks
1. Shoe asm. timing chain tensioner
2. Timing chain
3. Timing chain tensioner
4. R.H. timing chain guide
5. L.H. timing chain guide
6. Upper timing chain guide
7. Exhaust camshaft sprocket
8. Intake camshaft sprocket

Timing chain and sprocket alignment positions—2.4L DOHC engine

- Accessory drive belt
- Air cleaner assembly
- Air intake duct
- 2 upper air conditioning compressor mounting bolts
- Power steering pump
- Alternator and bracket
- Right front wheel
- Right wheel well splash shield
- Crankshaft balancer
- Drive belt tensioner
- Right Wheel Speed Sensor (WSS) harness at the suspension support
- Lower ball joint
- Stabilizer bar from the control arm and suspension support
- Suspension support
- A/C compressor-to-oil pan bolts
- Oil filter and adapter
- Starter motor
- Oil pan
- Crankshaft Position (CKP) sensor
- Lower front cover bolts
- Coolant bypass hose
- Upper radiator hose

- Engine front cover

8. Place the number one piston at Top Dead Center (TDC).

9. Remove or disconnect the following:

- Camshaft sprocket
- Timing chain
- Crankshaft sprocket

To install:

10. Install the crankshaft sprocket.

➡**Be sure the timing mark on the crankshaft sprocket is pointing toward the mark on the chain damper.**

11. Place the timing chain over the camshaft sprocket and hold the sprocket so the timing mark is pointing down and the timing chain is hanging off the sprocket.

12. Loop the timing chain under the crankshaft sprocket and install the camshaft sprocket on the camshaft.

13. Verify that the marks are aligned; the camshaft sprocket will be at the 6 o'clock position and the crankshaft sprocket at the 12 o'clock position.

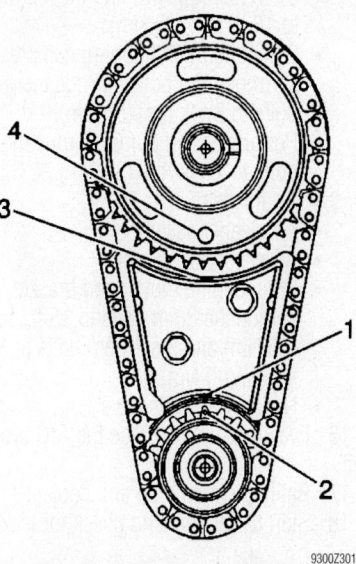

Be sure to align the damper mark (1) with the crankshaft mark (2) and the damper mark (3) with the camshaft sprocket mark (4)—3.4L engines

➡The No. 1 piston will be at TDC and the No. 4 piston will also be at TDC but on the compression stroke.

14. Torque the camshaft sprocket bolt to 103 ft. lbs. (140 Nm).

15. Install or connect the following:

- Front cover, using a new gasket and seal. Torque the small bolts to 15 ft. lbs. (21 Nm) and the long bolts to 35 ft. lbs. (47 Nm).
- Upper radiator hose
- Coolant bypass hose
- CKP sensor
- Oil pan. Torque the bolts to 18 ft. lbs. (25 Nm).
- Starter motor. Torque the bolts to 32 ft. lbs. (43 Nm).
- Oil filter adapter and filter
- A/C compressor-to-oil pan bolts. Torque the bolts to 37 ft. lbs. (50 Nm).
- Suspension support. Torque the bolts to 61 ft. lbs. (82 Nm).
- Stabilizer bar. Torque the bolts to 13 ft. lbs. (17 Nm).
- Ball joint. Torque the nuts to 41 ft. lbs. (55 Nm).
- WSS electrical connector
- Drive belt tensioner. Torque the bolts to 33 ft. lbs. (45 Nm).
- Crankshaft balancer. Torque the bolt to 76 ft. lbs. (103 Nm).
- Right wheel well splash shield
- Right front wheel
- Alternator. Torque the front bolt to

37 ft. lbs. (50 Nm) and the rear bolt to 18 ft. lbs. (25 Nm).
- 2 upper air conditioning compressor mounting bolts. Torque the bolts to 37 ft. lbs. (50 Nm).
- Power steering pump. Torque the bolts to 25 ft. lbs. (34 Nm).
- Air intake duct
- Air cleaner assembly
- Accessory drive belt
- Front engine mount and bracket. Torque the 8mm bolts to 15 ft. lbs. (20 Nm) and the 12mm bolts to 30 ft. lbs. (40 Nm).
- Negative battery cable

16. Evacuate and recharge the A/C system.
17. Refill the engine oil and coolant.
18. Start the engine and check for leaks.

Piston and Ring

POSITIONING

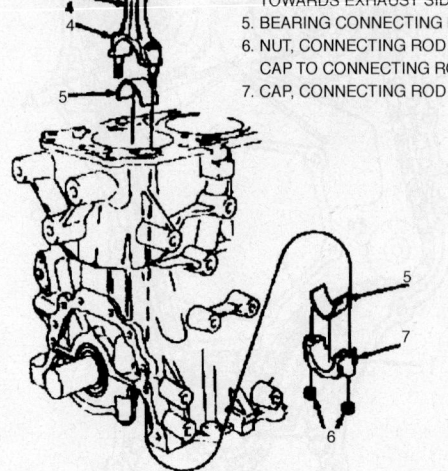

1. PISTON
2. ORIENTATION ARROW - TOWARDS FRONT OF ENGINE
3. CONNECTING ROD
4. OIL SQUIRT HOLE - TOWARDS EXHAUST SIDE
5. BEARING CONNECTING ROD
6. NUT, CONNECTING ROD CAP TO CONNECTING ROD
7. CAP, CONNECTING ROD

7922AG49

Piston and connecting rod assembly positioning—2.2L and 2.4L engines

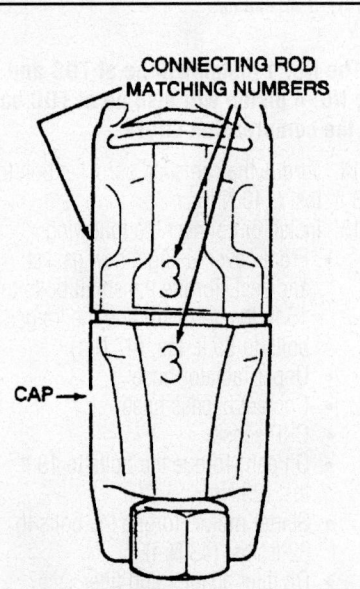

CONNECTING ROD MATCHING NUMBERS

CAP

7922AG51

Connecting rod and cap installation. Be sure to matchmark the cap and rod prior to disassembly, as shown—All engines

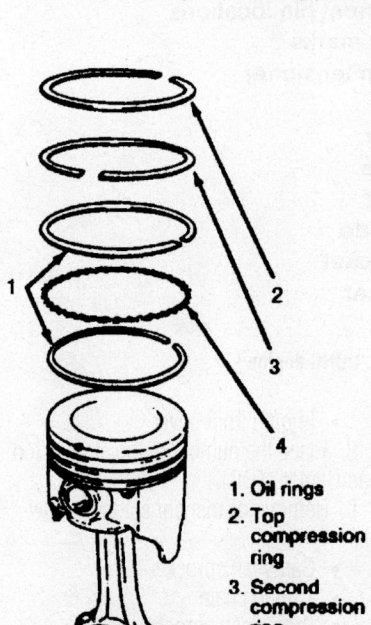

1. Oil rings
2. Top compression ring
3. Second compression ring
4. Expander

7922AG48

Piston ring positioning—3.4L engines

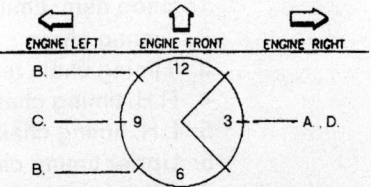

ENGINE LEFT | ENGINE FRONT | ENGINE RIGHT

A. OIL RING SPACER GAP (TANG IN HOLE OR SLOT WITH ARC)
B. OIL RING RAIL GAPS
C. 2ND COMPRESSION RING GAP
D. TOP COMPRESSION RING GAP

7922AG46

Piston ring end-gap spacing—3.4L engines

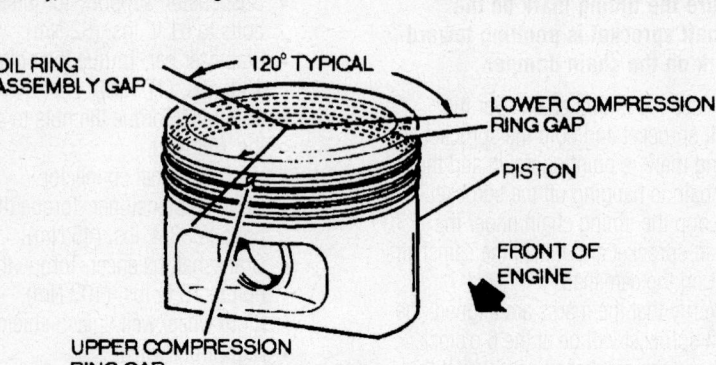

OIL RING ASSEMBLY GAP

120° TYPICAL

LOWER COMPRESSION RING GAP

PISTON

FRONT OF ENGINE

UPPER COMPRESSION RING GAP

7922AG50

Piston ring end-gap spacing—2.2L and 2.4L engines

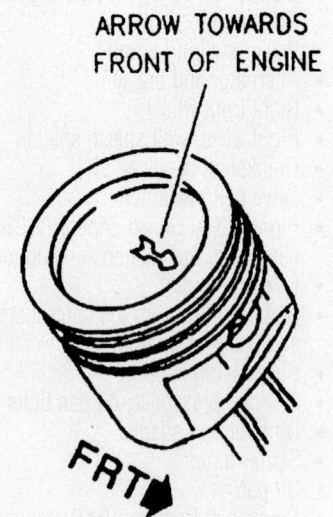

ARROW TOWARDS FRONT OF ENGINE

FRT

7922AG47

Piston positioning. Often the arrow is replaced by a notch, which also must face the front of the engine—3.4L engines

FUEL SYSTEM

Fuel System Service Precautions

Safety is the most important factor when performing not only fuel system maintenance but any type of maintenance. Failure to conduct maintenance and repairs in a safe manner may result in serious personal injury or death. Maintenance and testing of the vehicle's fuel system components can be accomplished safely and effectively by adhering to the following rules and guidelines.

• To avoid the possibility of fire and personal injury, always disconnect the negative battery cable unless the repair or test procedure requires that battery voltage be applied.

• Always relieve the fuel system pressure prior to disconnecting any fuel system component (injector, fuel rail, pressure regulator, etc.), fitting or fuel line connection. Exercise extreme caution whenever relieving fuel system pressure, to avoid exposing skin, face and eyes to fuel spray. Please be advised that fuel under pressure may penetrate the skin or any part of the body that it contacts.

• Always place a shop towel or cloth around the fitting or connection prior to loosening to absorb any excess fuel due to spillage. Ensure that all fuel spillage (should it occur) is quickly removed from engine surfaces. Ensure that all fuel soaked cloths or towels are deposited into a waste container.

• Always keep a dry chemical (Class B) fire extinguisher near the work area.

• Do not allow fuel spray or fuel vapors to come into contact with a spark or open flame.

• Always use a backup wrench when loosening and tightening fuel line connection fittings. This will prevent unnecessary stress and torsion to fuel line piping. Always follow the proper torque specifications.

• Always replace worn fuel fitting O-rings with new. Do not substitute fuel hose or equivalent, where fuel pipe is installed.

Fuel System Pressure

RELIEVING

2.4L Engine

1. Before servicing the vehicle, refer to the precautions section.

2. Loosen the fuel filler cap in order to relieve the pressure in the tank (do not tighten at this time).

3. Detach the fuel pump electrical connector.

4. Start and run the vehicle until it stalls, then engage the starter for an additional 3 seconds to ensure the relief of any remaining pressure.

5. Disconnect the negative battery cable.

6. Once the tests or repairs are completed, reattach the fuel pump electrical connector.

7. Connect the negative battery cable.

8. Tighten the fuel filler cap.

9. prime the fuel system by cycling the ignition switch **ON** for 2 seconds, **OFF** for 10 seconds, then **ON** again. Repeat, if necessary to build system pressure.

2.2L and 3.4L Engines

1. Before servicing the vehicle, refer to the precautions section.

2. Disconnect the negative battery cable in order to avoid possible fuel discharge if an accidental attempt is made to start the engine.

3. Loosen the fuel tank filler cap in order to relieve fuel tank pressure.

4. Connect a fuel pressure gauge (with bleed hose) to the fuel pressure test port connection. Wrap a towel around the fuel pressure connection when installing the fuel pressure gauge in order to avoid fuel spillage.

5. Install the bleed hose into an approved container and open the valve in order to bleed the fuel system pressure. The fuel pipe connections are now safe for servicing.

6. Drain any fuel remaining in the fuel pressure gauge into an approved container.

Fuel Filter

REMOVAL & INSTALLATION

1. Before servicing the vehicle, refer to the precautions section.

2. Relieve the fuel system pressure.

3. Remove or disconnect the following:
   • Fuel line from the filter, using a backup wrench
   • Quick-connect fitting from the fuel filter by compressing the tabs while pulling outward on the line
   • Fuel filter from the mounting bracket

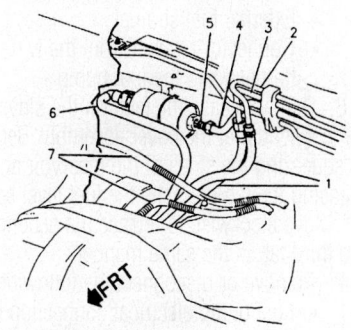

1 BODY
2 FUEL FILTER BRACKET
3 FUEL FILTER
4 SCREW – FULLY DRIVEN, SEATED AND NOT STRIPPED

79222Z323

Exploded view of the fuel filter mounting

1 HOSE, PART OF FUEL SENDER
2 FUEL VAPOR PIPE
3 FUEL RETURN PIPE
4 FUEL FEED PIPE
5 FUEL FEED PIPE NUT 27 N•m (20 LBS. FT.)
6 HOSE, PART OF FUEL SENDER
7 ABS AND FUEL SENDER HARNESS

79222Z324

Fuel filter mounting location and component identification

To install:

4. Install or connect the following:
   • Fuel filter to the mounting bracket
   • Fuel line using a backup wrench. Torque the fitting to 20 ft. lbs. (27 Nm).
   • Quick-connect fitting to the fuel filter
   • Negative battery cable

5. Pressurize the fuel system and verify no leaks.

Fuel Pump

REMOVAL & INSTALLATION

1. Before servicing the vehicle, refer to the precautions section.

2. Relieve the fuel system pressure.

3. Drain the fuel tank.

4. Remove or disconnect the following:

5. While holding the modular fuel sender assembly down, remove the snapring, if equipped or if equipped with a

cam lock ring use Fuel Sender Lock Nut Tool J-39765 to press down and rotate the cam lock ring to remove it.

❄ WARNING

If the modular fuel sender assembly is retained by a snapring, it may spring up from its position. When removing the modular fuel sender from the tank, be aware that the reservoir bucket is full of fuel. It must be tipped slightly during removal to avoid damage to the float.

- Fuel sender assembly
- External fuel strainer
- Connector retainer from the wiring harness and the fuel pump

6. Gently release the tabs on the sides of the fuel sender at the cover assembly. Begin by squeezing the sides of the reservoir and releasing the tab opposite the fuel level sensor. Move clockwise to release the second and third tab in the same manner.

7. Remove or disconnect the following:
- Fuel pump electrical connection by lifting the cover assembly
- Baffle and pump assembly from the retainer by rotating the fuel pump baffle counterclockwise
- Fuel pump outlet by sliding it out of slot
- Fuel pump outlet seal

To install:

8. Install or connect the following:
- Fuel pump outlet with a new seal by sliding it in the reservoir cover slots
- Fuel pump and baffle assembly onto the reservoir retainer by rotating it clockwise until seated
- Lower retainer assembly partially into the reservoir by aligning all 3 sleeve tabs and pressing the retainer onto the reservoir making sure all 3 tabs are firmly seated

➡ **Gently pull on the fuel pump reservoir to assure it is secure. If not secure, replace the entire fuel sender.**

- Connector retainer to the wiring harness and the fuel pump
- External fuel strainer
- Snapring, if equipped to the retainer slots while holding the modular fuel sender assembly down or use Fuel Sender Lock Nut Tool J-39765 in order to install the cam lock ring
- Fuel tank

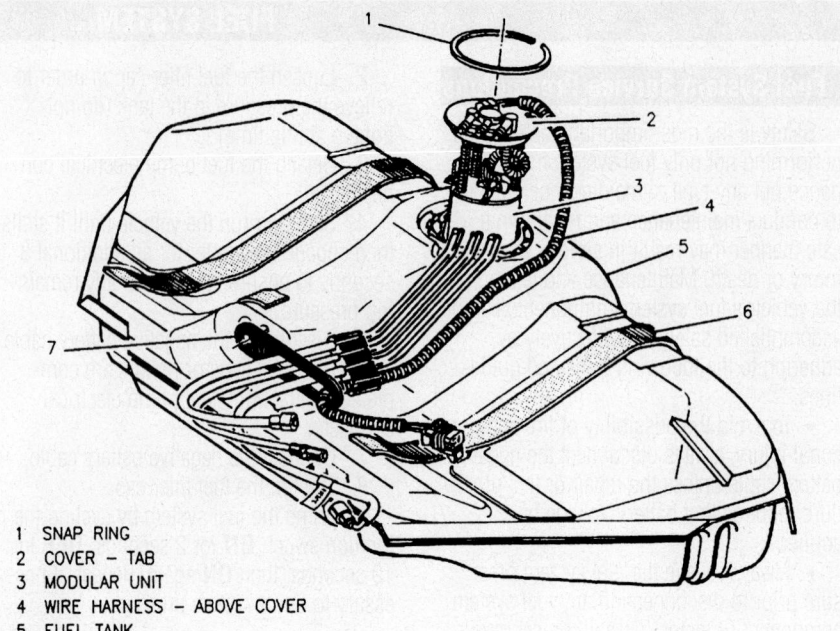

1 SNAP RING
2 COVER – TAB
3 MODULAR UNIT
4 WIRE HARNESS – ABOVE COVER
5 FUEL TANK
6 TANK ISOLATION STRIPS (3)
7 RUBBER ISOLATOR

7922Z325

Exploded view of the fuel sender assembly mounting to the tank

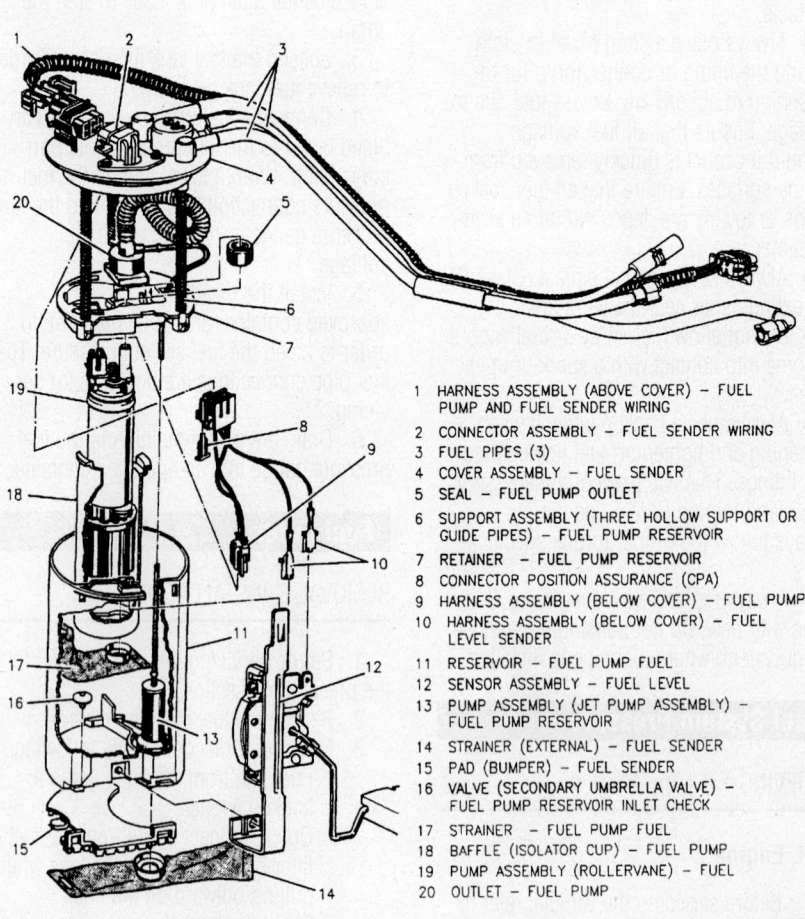

1 HARNESS ASSEMBLY (ABOVE COVER) – FUEL PUMP AND FUEL SENDER WIRING
2 CONNECTOR ASSEMBLY – FUEL SENDER WIRING
3 FUEL PIPES (3)
4 COVER ASSEMBLY – FUEL SENDER
5 SEAL – FUEL PUMP OUTLET
6 SUPPORT ASSEMBLY (THREE HOLLOW SUPPORT OR GUIDE PIPES) – FUEL PUMP RESERVOIR
7 RETAINER – FUEL PUMP RESERVOIR
8 CONNECTOR POSITION ASSURANCE (CPA)
9 HARNESS ASSEMBLY (BELOW COVER) – FUEL PUMP
10 HARNESS ASSEMBLY (BELOW COVER) – FUEL LEVEL SENDER
11 RESERVOIR – FUEL PUMP FUEL
12 SENSOR ASSEMBLY – FUEL LEVEL
13 PUMP ASSEMBLY (JET PUMP ASSEMBLY) – FUEL PUMP RESERVOIR
14 STRAINER (EXTERNAL) – FUEL SENDER
15 PAD (BUMPER) – FUEL SENDER
16 VALVE (SECONDARY UMBRELLA VALVE) – FUEL PUMP RESERVOIR INLET CHECK
17 STRAINER – FUEL PUMP FUEL
18 BAFFLE (ISOLATOR CUP) – FUEL PUMP
19 PUMP ASSEMBLY (ROLLERVANE) – FUEL
20 OUTLET – FUEL PUMP

7922Z326

Exploded view of the fuel pump assembly

- Negative battery cable
9. Pressurize the fuel system and verify no leaks.

Fuel Injector

REMOVAL & INSTALLATION

2.2L Engine

1. Before servicing the vehicle, refer to the precautions section.
2. Relieve the fuel system pressure.
3. Remove or disconnect the following:
   - Air cleaner outlet resonator
   - Vacuum pipe from the fuel pressure regulator
   - Engine fuel supply and return pipes
   - Fuel injector harness connectors
   - Fuel rail attaching studs.

➡**Be careful when removing the fuel rail so that you do not damage the injector tips or electrical connections.**

4. Remove the fuel rail as follows:
   a. Pull the fuel rail back and upward to remove the fuel injectors from the ports.
   b. Rotate the fuel rail to position the injectors downward.
   c. Remove the fuel rail.
5. Remove or disconnect the following:
   - Fuel injector retainer clip
   - Fuel injector from the fuel rail
6. Inspect the fuel injector in order to determine if the upper O-ring was also removed. If the upper O-ring in not removed, remove the O-ring from the fuel rail assembly.
   - Fuel injector O-rings and discard

To install:

7. Install or connect the following:
   - O-rings on the fuel injector
   - Fuel injector clip on the fuel injector

➡**The fuel injector will click when the injector is installed correctly.**

   - Fuel injector to the fuel rail with the connector facing upward
8. Install the fuel rail as follows:
   a. With the fuel injectors positioned downward, lower the fuel injectors into the ports.
   b. Align the injectors by rotating the fuel rail forward.
   c. Carefully push the fuel injectors into the cylinder head ports.
   d. Install and tighten the fuel rail retainers to 89 inch lbs. (10 Nm).
   - Fuel injector harness connectors.

Gently pull on the connectors to make sure they are fully engaged.
   - Fuel supply and return pipes
   - Vacuum pipe to the fuel pressure regulator
   - Air cleaner outlet resonator
   - Negative battery cable
9. Inspect for fuel leaks using the following procedure:
   a. Turn ON the ignition, with the engine OFF for 2 seconds.
   b. Turn OFF the ignition for 10 seconds.
   c. Turn ON the ignition.
   d. Inspect for fuel leaks.

2.4L Engine

1. Before servicing the vehicle, refer to the precautions section.
2. Relieve the fuel system pressure.
3. Remove or disconnect the following:
   - Air cleaner outlet resonator
   - Fuel pressure regulator vacuum hose
   - Camshaft Position (CMP) sensor electrical connector
   - Fuel injector electrical connectors
   - Fuel inlet pipe at the fuel rail
   - Fuel return pipe bracket screw and separate the fuel pipe from the pressure regulator
   - Fuel rail assembly
   - Fuel injector retaining clip
   - Fuel injector

To install:
4. Lubricate the O-rings with engine oil prior to installation.
5. Install or connect the following:

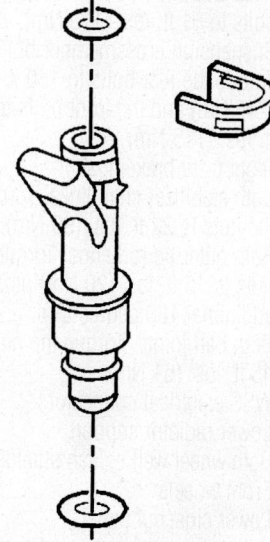

Exploded view of the fuel injector—2.4L engine

   - Fuel injector with new O-rings
   - Fuel injector retaining clip
   - Fuel rail assembly. Torque the bolts to 19 ft. lbs. (26 Nm).
   - New fuel pipe O-rings lubricated with engine oil
   - Fuel inlet pipe. Torque the fitting to 22 ft. lbs. (30 Nm).
   - Fuel return pipe to the pressure regulator. Torque the screw to 53 inch lbs. (6 Nm).
   - Fuel injector electrical connectors
   - CMP sensor electrical connector
   - Fuel pressure regulator vacuum hose
   - Air cleaner outlet resonator
   - Negative battery cable
6. Re-pressurize the fuel system and check for leaks.

3.4L Engines

1. Before servicing the vehicle, refer to the precautions section.

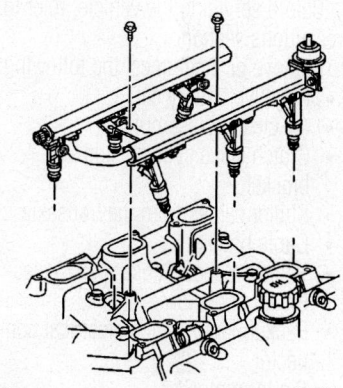

Exploded view of the fuel rail assembly—3.4L engine

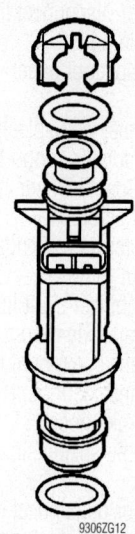

Exploded view of the fuel injector—3.4L engine shown

2. Relieve the fuel system pressure.

3. Remove or disconnect the following:

- Upper intake manifold
- Fuel feed pipe
- Fuel return pipe from the pressure regulator
- Main fuel injector wiring harness electrical connector
- Engine Coolant Temperature (ECT) sensor electrical connector
- Fuel rail assembly
- Fuel injector electrical connectors
- Fuel injector-to-fuel rail clip
- Fuel injector

To install:

✴✴ WARNING

If the fuel injector O-rings are color coded, install the black O-ring in the upper position and the brown O-ring in the lower position.

4. Lubricate the O-rings with engine oil prior to installation.

5. Install or connect the following:

- New fuel injector O-rings
- Fuel injector
- Fuel injector retaining clip
- Fuel injector electrical connectors

- Fuel rail assembly. Torque the bolts to 89 inch lbs. (10 Nm).
- ECT sensor electrical connector
- Main fuel injector wiring harness electrical connector
- New fuel pipe O-rings lubricated with engine oil
- Fuel feed pipe at the fuel rail. Torque the pipe nut to 13 ft. lbs. (17 Nm).
- Fuel return pipe to the pressure regulator. Torque the pipe nut to 13 ft. lbs. (17 Nm).
- Upper intake manifold

6. Re-pressurize the fuel system and check for leaks.

DRIVE TRAIN

Transaxle

REMOVAL & INSTALLATION

Manual

1. Before servicing the vehicle, refer to the precautions section.

2. Remove or disconnect the following:

- Negative battery cable
- Air cleaner assembly
- Clutch slave cylinder line and bracket
- Shifter cables from the transaxle
- Cable bracket
- Vehicle Speed Sensor (VSS) electrical connector
- Backup lamp switch electrical connector
- Starter motor

3. Tie the radiator to the upper hood latch panel.

4. Install an engine support fixture.

5. Remove or disconnect the following:

- Transaxle bolts
- Lower closeout panel
- Wheels
- Both wheel well splash shields
- Lower radiator support
- Wheel Speed sensor (WSS) electrical connectors
- Both outer tie rod ends
- Both ball joints
- Both stabilizer shaft links
- Front transaxle mount bolts
- Side transaxle mount nuts
- Rear transaxle mount stud
- Right brake hose

6. Support the suspension crossmember.

7. Remove or disconnect the following:

- Suspension crossmember bolts
- Rack and pinion bolts

8. Raise the vehicle off of the front suspension crossmember.

9. Remove or disconnect the following:

- Front transaxle mount
- Rear transaxle mount
- Halfshafts from the vehicle

10. Lower the engine and transaxle assembly enough to clear the left side inner body panel.

11. Support and remove the transaxle from the vehicle.

To install:

12. Install or connect the following:

- Transaxle. Torque the bolts to 66 ft. lbs. (90 Nm).
- Halfshafts
- Rear transaxle mount
- Front transaxle mount
- Rear transaxle mount bolt. Torque the bolt to 81 ft. lbs. (110 Nm).
- Front transaxle mount bolts. Torque the bolts to 84 ft. lbs. (115 Nm).
- Rack and pinion bolts. Torque the bolts to 81 ft. lbs. (110 Nm).
- Suspension crossmember bolts. Torque the rear bolts to 180 ft. lbs. (245 Nm) and the front bolts to 84 ft. lbs. (115 Nm).
- Right front brake hose
- Both stabilizer shaft links. Torque the nuts to 22 ft. lbs. (30 Nm).
- Both outer tie rod ends. Torque the nuts to 15 ft. lbs. (20 Nm) plus an additional 180 degree turn.
- Both ball joints. Torque the nuts to 45 ft. lbs. (61 Nm).
- WSS electrical connectors
- Lower radiator support
- Both wheel well splash shields
- Front wheels
- Lower closeout panel

13. Remove the engine support fixture.

14. Untie the radiator support.

15. Install or connect the following:

- Starter motor. Torque the bolts to 66 ft. lbs. (90 Nm) on 2.4L engines and 32 ft. lbs. (43 Nm) on 3.4L engines.
- Backup lamp switch electrical connector
- VSS electrical connector
- Shifter cables
- Clutch slave cylinder bracket and line
- Air cleaner assembly
- Negative battery cable

Automatic

1. Before servicing the vehicle, refer to the precautions section.

2. Disconnect the negative battery cable.

3. Drain the transaxle.

4. Remove or disconnect the following:

- Air inlet duct hose from the intake plenum
- Wiring harness from the transaxle and the Park Neutral Position (PNP) switch
- Upper transaxle-to-engine bolts and stud

5. Install an engine support fixture.

- Front wheels
- Both front fender liners
- Steering gear mounting bolts and secure the steering gear with mechanics wire
- Wheel Speed Sensor (WSS) wires from the front wheels and unclip them from the frame
- Ball joints from the steering knuckles
- Antilock Brake System (ABS) sensor from the WSS and frame
- Brake modulator assembly from the support

- Tie rod ends from the steering knuckles
- Front and rear transaxle mount bracket bolts
- Brake lines from the retainers on the crossmember

6. Lower the vehicle until the suspension crossmember rests on the jack stands.

7. Remove or disconnect the following:
- Front suspension crossmember support bolts
- Rear suspension crossmember support bolts
- Suspension crossmember-to-body bolts

8. Raise the vehicle off of the suspension
- Frame
- Bolts from the transaxle brace
- Shift cable from the shift linkage
- Cable from the bracket
- Flywheel inspection cover
- Starter
- Torque converter-to-flywheel bolts
- Transaxle cooler lines by removing the nut holding the bracket to the transaxle case
- Vehicle Speed Sensor (VSS) wiring harness from the sensor
- Drive shafts from the transaxle
- Body-to-transaxle mount bolts

9. Lower the transaxle with the engine support fixture enough to remove the transaxle.

10. Support the transaxle with a suitable jack.

11. Remove or disconnect the following:
- Transaxle-to-engine nut
- Engine from the transaxle
- Transaxle

To install:

12. Install or connect the following:
- Transaxle
- Lower transaxle-to-engine bolts and nuts and tighten to 66 ft. lbs. (90 Nm)
- Transaxle cooler pipes to the transaxle and tighten to 71 inch lbs. (8 Nm)
- Torque converter-to-flywheel bolts tighten to 46 ft. lbs. (62 Nm)
- Drive shafts
- VSS wiring
- Starter
- Flywheel inspection cover bolts tighten to 97 inch lbs. (11 Nm)

13. Use the engine support fixture to raise the engine and transaxle assembly.

14. Remove or disconnect the following:
- Transaxle mount-to-body bolts tighten to 66 ft. lbs. (90 Nm)
- Frame and hand-tighten the bolts

15. Once all the bolts have been installed tighten them as follows:
- Crossmember rear bolts to 180 ft. lbs. (245 Nm) plus an additional 180 degrees
- Front crossmember bolts to 84 ft. lbs. (115 Nm) plus an additional 120 degrees
- Front crossmember-to-body bolts to 81 ft. lbs. (110 Nm)

16. Raise the vehicle and support with jack stands
- Brake lines to the retainers on the crossmember
- Front and rear transaxle mount bracket bolts
- Power steering gear mounting bolts to 81 ft. lbs. (110 Nm)
- Tie rod ends to the steering knuckles
- Brake modulator assembly to the support bracket
- Lower ball joints to the steering knuckles
- ABS sensor to the WSS and frame
- Transaxle-to-engine brace bolts and tighten to 53 ft. lbs. (72 Nm)
- Both front fender liners
- Front wheels
- Upper transaxle-to-engine bolts and stud and tighten to 66 ft. lbs. (90 Nm)

17. Remove the engine support fixture.
- Shift linkage
- PNP switch and transaxle connections
- Air duct hose to the intake plenum
- Negative battery cable

18. Refill the transaxle with fluid.

19. Refill and bleed the power steering system.

Clutch

REMOVAL & INSTALLATION

1. Before servicing the vehicle, refer to the precautions section.

2. Remove or disconnect the following:
- Negative battery cable
- Hydraulic line from the clutch actuator (slave cylinder)
- Transaxle from the vehicle

3. If any clutch components are to be reused, use the following procedure:

a. Matchmark the clutch pressure plate to the flywheel. This is to retain the balance of the original parts. If all parts are to be replaced with new, this step is not necessary.

b. If the pressure plate is to be reused, loosen the pressure plate mounting bolts by turning each bolt 1 full turn until all the spring pressure is removed. This helps avoid warping the pressure plate.

c. Remove the clutch disc and pressure plate.

To install:

4. Apply a small amount of high-temperature grease to the pilot bearing as well as the tip of the transaxle input shaft and the clutch splines.

5. Install or connect the following:
- Clutch alignment tool into the flywheel
- Clutch disc onto the tool
- Pressure plate

➡**New pressure plate-to-flywheel replacement bolts are recommended.**

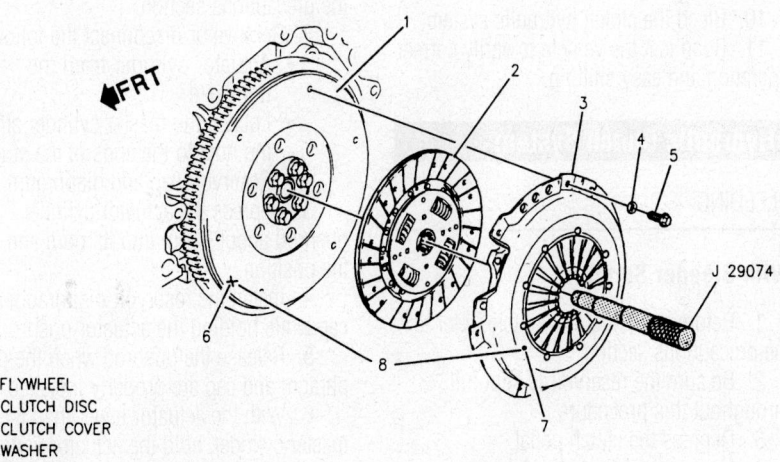

J 29074

1 FLYWHEEL
2 CLUTCH DISC
3 CLUTCH COVER
4 WASHER
5 BOLT
6 FLYWHEEL "HEAVY SIDE" IDENTIFICATION
7 CLUTCH COVER "LIGHT SIDE" IDENTIFICATION
8 ALIGN IDENTIFICATION MARKS ON ASSEMBLY

79222328

Exploded view of the clutch components—showing the clutch disk alignment tool

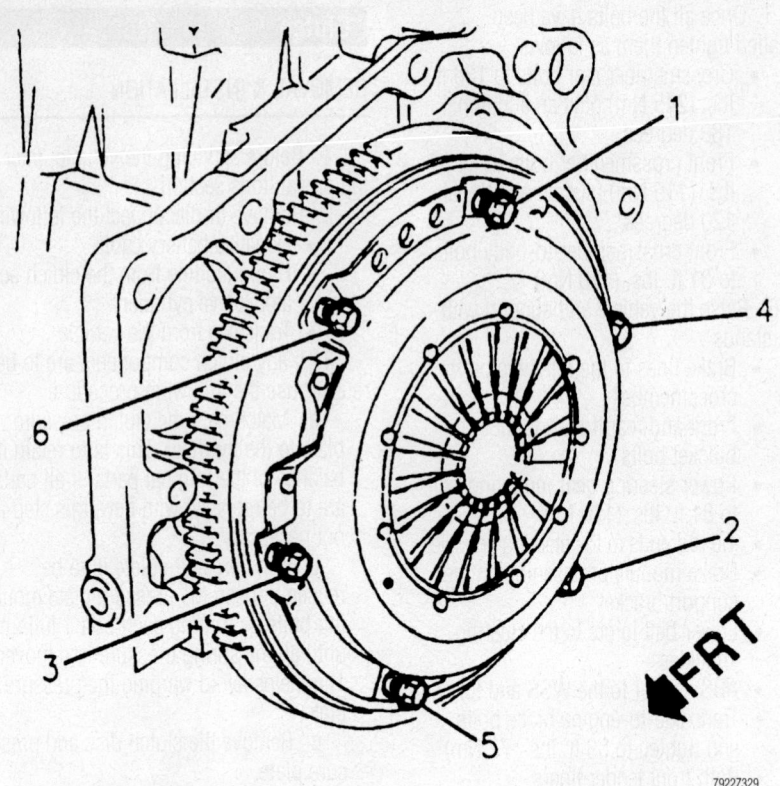

79222329

Clutch cover tightening sequence

6. Torque the pressure plate-to-flywheel bolts, in sequence to 12 ft. lbs. (16 Nm).

7. Remove the clutch alignment tool.

8. Lubricate the inside diameter of the actuator (slave cylinder/throw out bearing) with clutch bearing lubricant.

9. Install or connect the following:
- Transaxle
- Hydraulic line to the actuator (slave cylinder)
- Negative battery cable

10. Bleed the clutch hydraulic system.

11. Road test the vehicle to verify correct operation and easy shifting.

Hydraulic Clutch System

BLEEDING

With Bleeder Screw

1. Before servicing the vehicle, refer to the precautions section.

2. Be sure the reservoir is kept full throughout this procedure.

3. Depress the clutch pedal.

4. Loosen the bleed screw, located on the actuator cylinder body next to the inlet connection.

5. Torque the bleeder screw to 17 inch lbs. (2 Nm).

6. Repeat previous 3 steps until all air is removed from the system.

7. Refill the fluid reservoir.

8. To check the system, start the engine and wait 10 seconds.

9. Depress the clutch pedal and shift into **R**. If there is any gear clash, air may still be present.

Without Bleeder Screw

1. Before servicing the vehicle, refer to the precautions section.

2. Remove or disconnect the following:
- Actuator cylinder from the transaxle
- Loosen the master cylinder attaching nuts to the ends of the studs
- Reservoir cap and diaphragm

3. Depress the actuator cylinder pushrod about ¾ in. into its bore and hold the position.

4. Install the reservoir diaphragm and cap while holding the actuator pushrod.

5. Release the pushrod when the diaphragm and cap are properly installed.

6. With the actuator lower than the master cylinder, hold the actuator vertically with the pushrod end facing the ground.

7. Press the actuator pushrod into its bore with ½ in. strokes. Check the reservoir for bubbles. Continue until no bubbles enter the reservoir.

8. Install the master cylinder and actuator. Refill the fluid reservoir.

9. To check the system, start the engine and wait 10 seconds.

10. Depress the clutch pedal and shift into reverse. If there is any gear clash, air may still be present.

Halfshaft

REMOVAL & INSTALLATION

Left and Right Halfshafts

1. Before servicing the vehicle, refer to the precautions section.

2. Remove or disconnect the following:
- Negative battery cable
- Wheel
- Hub nut and washer
- Lower ball joint from the steering knuckle
- Anti-lock Brake System (ABS) electrical connector
- Sway bar link kit

3. Press the halfshaft from the hub/bearing assembly.

4. Pull the halfshaft from the transaxle using a slide hammer and axle removal tool.

To install:

5. Install or connect the following:
- Halfshaft into the transaxle or intermediate shaft using a brass drift positioned in the inboard joint groove and tap the joint in until it is seated

➡**Verify the joint is seated properly by grasping the inboard joint and pulling on it firmly.**

✳✳ WARNING

DO NOT pull on the halfshaft or damage to the inner joint may result.

- Halfshaft into the hub assembly
- Lower ball joint to the steering knuckle. Torque the nut to 44 ft. lbs. (60 Nm).

➡**If necessary, tighten the nut up to 60 degree (⅙) turn additional rotation to align the cotter pins. NEVER loosen the nut to make the holes align.**

- New cotter pin
- Washer and hub nut. Torque the hub nut to 284 ft. lbs. (385 Nm) on nuts that are colored black, or 183 ft. lbs. (235 Nm) on nuts that are colored gray.

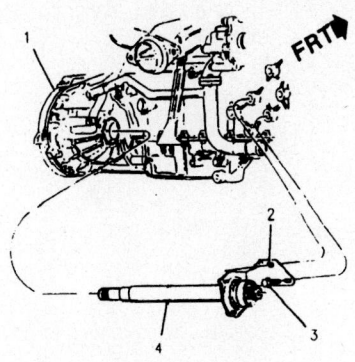

1 TRANSAXLE
2 INTERMEDIATE SHAFT SUPPORT BRACKET
3 BOLT
4 INTERMEDIATE SHAFT

79222330

Intermediate shaft components

- Sway bar link kit. Torque the nut to 13 ft. lbs. (17 Nm).
- Wheel. Torque the nuts to 100 ft. lbs. (140 Nm).
- Negative battery cable

6. Check the transaxle fluid level and top off as necessary.

Intermediate Shaft

1. Before servicing the vehicle, refer to the precautions section.
2. Install an engine support fixture.
3. Remove or disconnect the following:

- Right side wheel
- Sway bar link kit
- Lower ball joint from the steering knuckle
- Halfshaft from the intermediate shaft
- Intermediate shaft support bracket-to-engine bolts
- Intermediate shaft from the transaxle

To install:

4. Install or connect the following:

- Intermediate shaft into the transaxle. Torque the intermediate shaft support bracket-to-engine bolts to 49 ft. lbs. (66 Nm). Coat the splines of the intermediate shaft with chassis grease.
- Halfshaft to the intermediate shaft
- Lower ball joint to the steering knuckle. Torque the castle nut to 44 ft. lbs. (60 Nm).

➡**If necessary, tighten the nut up to 60 degrees (⅙) turn additional rotation to align the cotter pins. NEVER loosen the nut to make the holes align.**

- New cotter pin
- Sway bar link kit. Torque the nut to 13 ft. lbs. (17 Nm).
- Wheel. Torque the nuts to 100 ft. lbs. (140 Nm).

5. Remove the engine support fixture.
6. Connect the negative battery cable.
7. Check the transaxle fluid level and top off as necessary.

CV-Joints

OVERHAUL

Outer CV-Joint

1. Before servicing the vehicle, refer to the precautions section.
2. Remove or disconnect the following:

- Front wheel
- Halfshaft, position it in a vise
- Large CV-joint boot clamp
- Small CV-joint boot clamp
- CV-joint boot and slide it back on the shaft
- Outer race from the halfshaft, by spreading retaining ring
- Retaining ring from the halfshaft
- CV-joint boot from the halfshaft and discard it if damaged

3. Disassemble the chrome alloy balls from the CV-joint cage as follows:

a. Position a brass drift against the CV-joint cage and tap it with a hammer to tilt the cage.
b. Remove the 1st chrome alloy ball from the cage.
c. Tilt the cage in the opposite direction.
d. Remove the opposite chrome alloy ball.
e. Repeat the procedure until all 6 balls are removed.

4. Disassemble the CV-joint cage and inner race as follows:

a. Pivot the cage and race 90 degrees to the center line of the outer race.
b. Align the cage windows with outer race lands.
c. Remove the cage from the outer race.
d. Rotate the inner race upward and remove it from the cage.

To install:

5. Lubricate the parts with a light coat of grease.
6. Assemble the CV-joint cage and inner race, as follows:

a. Rotate the inner race 90 degrees to the cage centerline.

b. Align the cage windows with inner race lands.
c. Insert the inner race into the cage by rotating the inner race downward.
d. Insert the cage/inner race into the outer race.

7. Assemble the chrome alloy balls into the CV-joint cage, as follows:

a. Position a brass drift against the CV-joint cage and tap it with a hammer to tilt the cage.
b. Insert the 1st chrome alloy ball into the cage.
c. Tilt the cage in the opposite direction.
d. Insert the opposite chrome alloy ball.
e. Repeat the procedure until all 6 balls are inserted.

8. Install ½ of the grease provided, into the CV-joint.
9. Install or connect the following:

- Small ring clamp on the CV boot
- New retaining ring on the halfshaft
- Large ring clamp on the CV boot
- Outer race assembly onto the halfshaft until the ring engages the halfshaft groove

10. Slide the small end of the CV-joint boot/clamp into place, with the seal lip in the halfshaft groove.

➡**Make sure the boot lies flat against the halfshaft.**

11. Using a crimp tool, a torque wrench and a breaker bar, crimp the small CV-joint boot clamp to 100 ft. lbs. (136 Nm).
12. Check the clamp gap dimension; if it is not 0.085 in. (2.15mm), continue tightening the clamp until it is.
13. Install ½ kit grease into the CV-joint boot.
14. Measure approximately 0.687 in. (17.5mm) up from the bottom edge of the outer CV-joint assembly.
15. Slide the large end of the CV boot/clamp into place, with the seal lip in place over the outer race.

➡**Make sure the boot lies flat against the outer race.**

16. Using a crimp tool, a torque wrench and a breaker bar, crimp the large CV-joint boot clamp to 130 ft. lbs. (176 Nm).
17. Check the clamp gap dimension; if it is not 0.102 in. (2.60mm), continue tightening the clamp until it is.
18. Install the halfshaft and the front wheel.

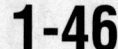

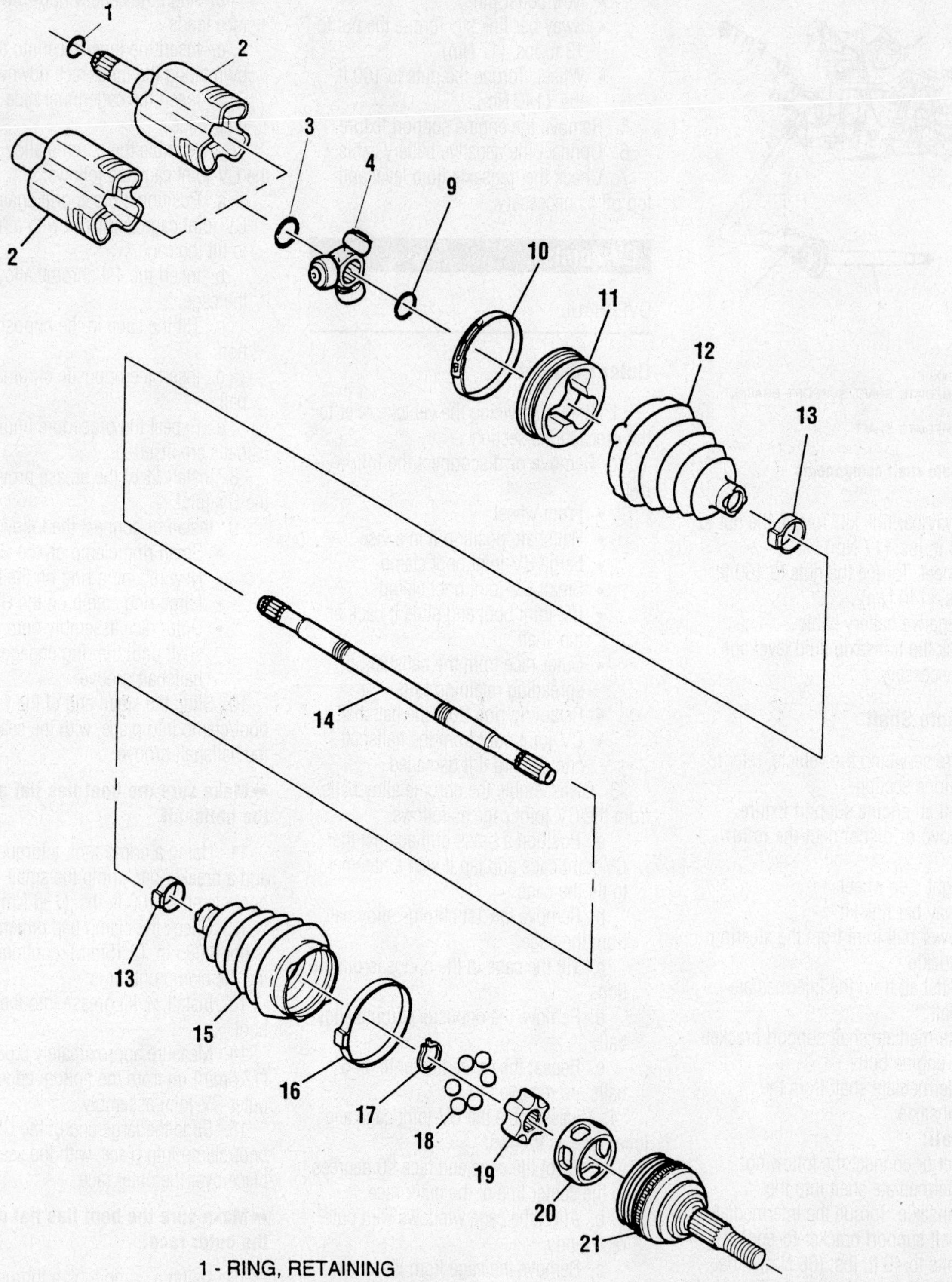

1 - RING, RETAINING
2 - HOUSING ASM, RETAINER &
3 - RING, SHAFT RETAINING
4 - SPIDER, TRIPOT JOINT
9 - RING, SPACER
10 - CLAMP, SEAL RETAINING
11 - BUSHING, TRILOBAL TRIPOT
12 - SEAL, DRIVE AXLE INBOARD
13 - CLAMP, SEAL RETAINING

14 - SHAFT, AXLE (RH SHOWN, LH SIMILAR)
15 - SEAL, DRIVE AXLE OUTBOARD
16 - CLAMP, SEAL RETAINING
17 - RING, RACE RETAINING
18 - BALL, CHROME ALLOY
19 - RACE, C/V JOINT INNER
20 - CAGE, C/V JOINT
21 - RACE, C/V JOINT OUTER

Exploded view of the halfshaft assembly

9306ZG13

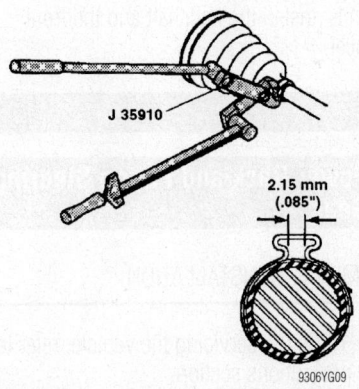

J 8059

9306XG19

Disconnecting the outer CV-joint from the axle shaft

J 35910

2.15 mm (.085")

9306YG09

Crimping the small boot clamp—Outer CV-joint

J 35910

2.60 mm (.102")

9306YG11

Crimping the large boot clamp—Outer CV-joint

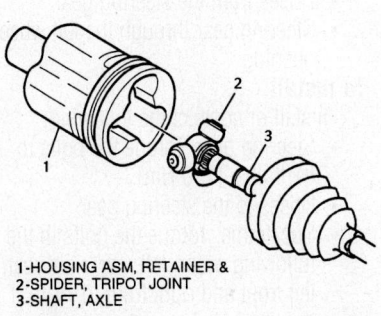

1-HOUSING ASM, RETAINER &
2-SPIDER, TRIPOT JOINT
3-SHAFT, AXLE

9306YG14

Exploded view of the inner (tri-pod) joint

Inner (Tri-Pod) Joint

1. Before servicing the vehicle, refer to the precautions section.
2. Remove or disconnect the following:
   - Front wheel
   - Halfshaft and place it in a vise
   - Small CV-joint boot clamp
   - Large CV-joint boot clamp
   - CV-joint boot by sliding it away from the tri-pod joint
   - Tri-pod housing from the tri-pod spider
   - Inboard spacer ring and slide it rearward on the shaft, using snapring pliers
   - Outboard retaining ring
   - Tri-pod joint spider assembly by tapping it from the halfshaft with a brass drift
   - Inboard tri-pod spider retaining ring
   - Trilobal tri-pod bushing from the housing
   - CV-joint boot
3. Thoroughly clean and inspect all parts.

To install:
4. Install or connect the following:
   - Small boot clamp

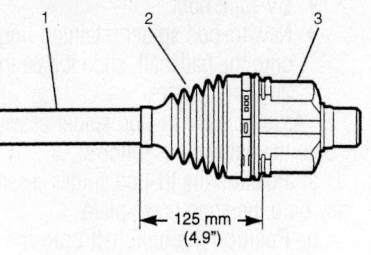

125 mm (4.9")

1-SHAFT, AXLE
2-SEAL, DRIVE AXLE INBOARD
3-HOUSING ASM, RETAINER

9306YG17

CV-joint boot measurement—Inner (tri-pod) joint

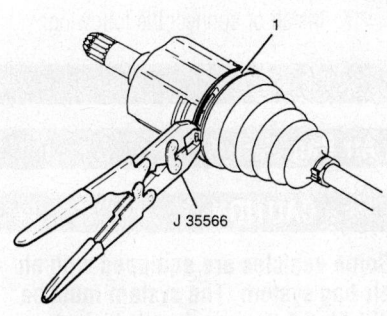

J 35566

1-CLAMP, SEAL RETAINING

9306YG18

Latching the large CV-joint boot ring—Inner (tri-pod) joint

INNER EAR GAP (A)

SEAL MOUNTING AREA

1   2   3

SEAL GROOVE   SITE GROOVE

J 35910   13   12,15

BREAKER BAR

TORQUE WRENCH

13-CLAMP, SEAL RETAINING
14-SHAFT, AXLE
12,15-SEAL, DRIVE AXLE OUTBOARD

9306YG16

Crimping the small CV-joint boot ring—Inner (tri-pod) joint

- CV-joint boot
- New tri-pod spider retaining ring onto the halfshaft, slide it past the 2nd ring groove

5. Assemble the tri-pod spider assembly onto the halfshaft as follows:

   a. Position the tri-pod spider assembly onto the shop press plate.

   b. Position the halfshaft onto the tri-pod spider assembly, in the shop press.

   c. Press the halfshaft into the tri-pod spider assembly until it passes the 2nd ring groove.

6. Remove the halfshaft from the shop press.

7. Install or connect the following:

- Outboard retaining ring into the axle shaft groove using snapring pliers
- Tri-pod joint spider assembly, slide it against the outboard retaining ring
- Inboard spacer ring, seat it in the groove
- ½ of the kit grease into the boot
- ½ of the kit grease into the tri-pod housing
- Trilobal tri-pod bushing flush with the tri-pod housing face
- New large seal clamp onto the CV-joint boot
- Tri-pod housing, slide it over the tri-pod joint spider assembly

- CV-joint boot/clamp, slide it into place, over the trilobal tri-pod bushing with the seal lip in the groove

➡**Make sure the boot lies flat against the trilobal bushing.**

8. Position the CV-joint boot so it measures 4.9 in. (125mm).

9. Using a crimp tool, a torque wrench and a breaker bar, crimp the small CV-joint boot clamp to 100 ft. lbs. (136 Nm).

10. Using a crimp tool, latch the large CV-joint boot clamp.

11. Install the halfshaft and the front wheel.

STEERING AND SUSPENSION

Air Bag

✻✻ CAUTION

Some vehicles are equipped with an air bag system. The system must be disabled before performing service on or around system components, steering column, instrument panel components, wiring and sensors. Failure to follow safety and disabling procedures could result in accidental air bag deployment, possible personal injury and unnecessary system repairs.

PRECAUTIONS

Several precautions must be observed when handling the inflator module to avoid accidental deployment and possible personal injury.

- Never carry the inflator module by the wires or connector on the underside of the module.
- When carrying a live inflator module, hold securely with both hands, and ensure that the bag and trim cover are pointed away.
- Place the inflator module on a bench or other surface with the bag and trim cover facing up.
- With the inflator module on the bench, never place anything on or close to the module which may be thrown in the event of an accidental deployment.

DISARMING

✻✻ CAUTION

The Supplemental Restraint System (SRS) must be disarmed before per-forming service procedures around the air bag or SRS wiring. Failure to do so may cause accidental deployment of the air bag, resulting in unnecessary SRS repairs and/or personal injury.

1. Disconnect the negative battery cable.

2. Turn the steering wheel so the vehicle's wheels are pointing straight ahead.

3. Turn the ignition switch to the **LOCK** position and remove the key.

4. Remove the **AIR BAG** fuse from the fuse block.

5. Remove the left sound insulator.

6. Remove the Connector Position Assurance (CPA) clip from the yellow 2-way connector at the base of the steering column, and detach the connector. If equipped with a passenger's side air bag, remove the CPA and detach the yellow 2-way connector from the passenger air bag lead.

REARMING

1. Turn the ignition switch to the **LOCK** position and remove the key.

2. Attach the yellow 2-way connector at the base of steering column and secure it with the Connector Position Assurance (CPA) clip. If equipped with a passenger's side air bag, attach the yellow 2-way connector at the passenger air bag lead and secure it with the CPA clip.

3. Install the left sound insulator.

4. Install the **AIR BAG** fuse in the fuse block.

5. Turn the ignition switch to the **RUN** position and verify that the **AIR BAG** warning lamp flashes 7 times, then turns **OFF**.

6. Connect the negative battery cable.

Power Rack and Pinion Steering Gear

REMOVAL & INSTALLATION

1. Before servicing the vehicle, refer to the precautions section.

2. Remove or disconnect the following:

- Negative battery cable
- Both front wheels
- Stabilizer shaft links from the control arms
- Tie rod ends from the steering knuckles
- Intermediate shaft lower pinch bolt
- Through-bolt from the rear transaxle mount
- Power steering line bracket from the crossmember
- Stabilizer bar

3. Support the rear of the sub-frame with jacks.

4. Remove the rear sub-frame mounting bolts and loosen the front ones.

5. Lower the sub-frame about 3 inches using the jacks.

6. Remove or disconnect the following:

- Hoses from the steering gear
- Steering gear through the left wheel opening

To install:

7. Install or connect the following:

- Steering gear. Torque the bolts to 81 ft. lbs. (110 Nm).
- Hoses to the steering gear
- Sub-frame. Torque the bolts in the following order: left rear, right rear, left front and right front to 71 ft. lbs. (110 Nm).

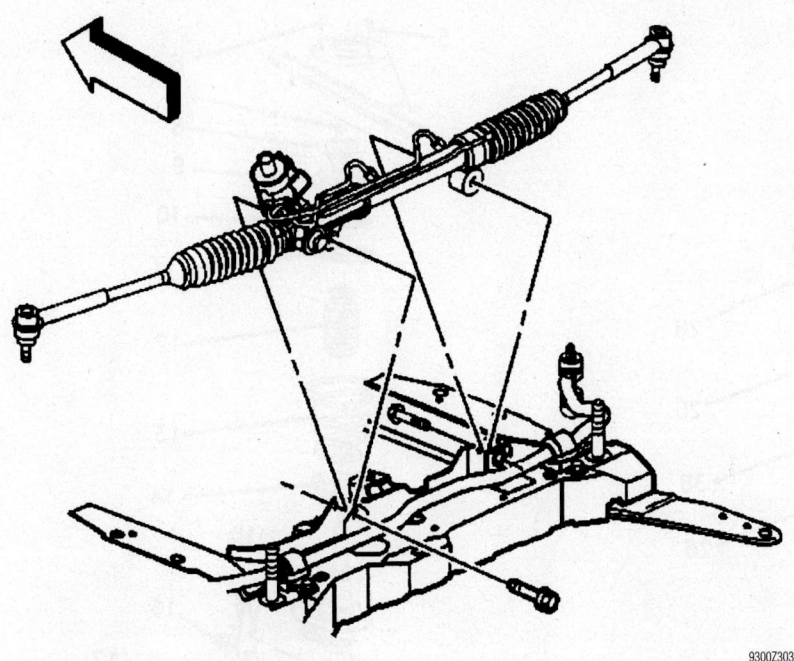

Power steering gear mounting

- Stabilizer bar bracket. Torque the bolts to 49 ft. lbs. (66 Nm).
- Power steering line bracket
- Rear transaxle mount. Torque the bolt to 89 ft. lbs. (120 Nm).
- Intermediate shaft. Torque the pinch bolt to 15 ft. lbs. (20 Nm).
- Tie rod ends. Torque the nuts to 15 ft. lbs. (20 Nm) plus an additional 180 degree turn.
- Stabilizer bar links. Torque the nuts to 13 ft. lbs. (17 Nm).
- Negative battery cable
8. Verify all steering hose fittings are tight.
9. Install the front wheels.
10. Fill the steering reservoir.

11. Install an adapter cap on the fluid reservoir with a vacuum pump attached to it.

12. Apply about 20 inches of vacuum to the system and wait 5 minutes. Typical vacuum drop is 2–3 inches of vacuum or)7–10 kPa). If the vacuum drop is greater, there may be a leak in the system allowing air to enter.

13. Remove the tools and install the reservoir cap.

14. Start the engine and allow it to idle.

15. Turn the engine **OFF** and check the fluid level. Do this until the fluid level stabilizes.

16. Start the engine and allow it to idle.

17. Turn the steering wheel in both directions 180–360 degrees 5 times.

18. Turn the engine **OFF** and check the fluid level.

19. Install an adapter cap on the fluid reservoir with a vacuum pump attached to it once again.

20. Apply about 20 inches of vacuum to the system and wait 5 minutes.

21. Remove the tools and check the fluid level. Install the cap.

Strut

REMOVAL & INSTALLATION

Front

1. Before servicing the vehicle, refer to the precautions section.
2. Support the front crossmember.
3. Remove or disconnect the following:
- Upper strut mounting fasteners
- Front wheel
- Brake line bracket from the strut
4. Scribe reference marks on the front strut and steering knuckle for installation purposes.
5. Remove the strut lower mounting bracket nut and through-bolts.
6. Remove the strut from the vehicle.
   To install:
7. Install or connect the following:
- Strut assembly
- Upper strut plate mounting nuts finger tight
- Lower strut bracket to the steering knuckle. Torque the through-bolt/nuts to 133 ft. lbs. (180 Nm) with the reference marks in alignment.
8. Torque the upper mounting fasteners to 18 ft. lbs. (25 Nm).
9. Install or connect the following:
- Brake line bracket to the strut.

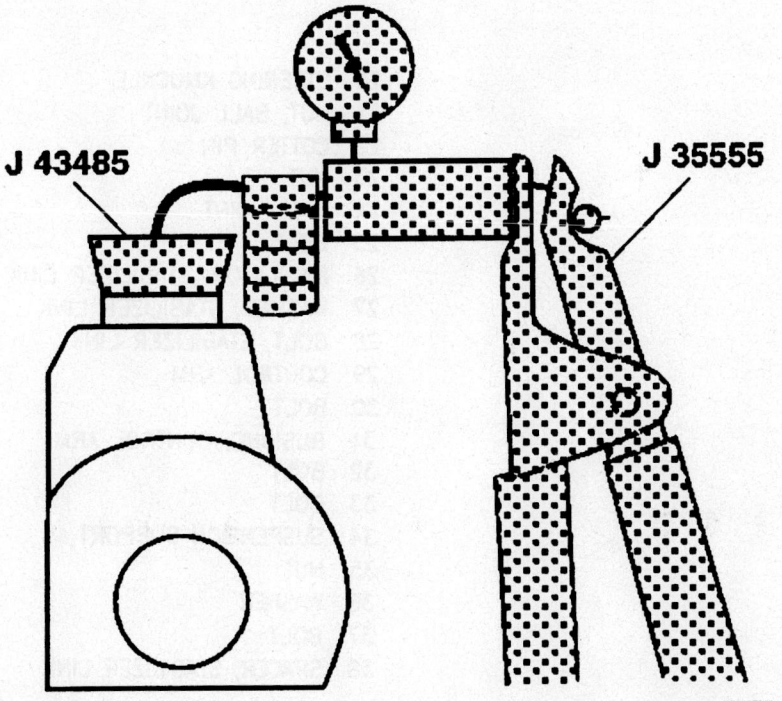

J 43485   **J 35555**

Install the special tools as shown on the steering fluid reservoir when bleeding the system

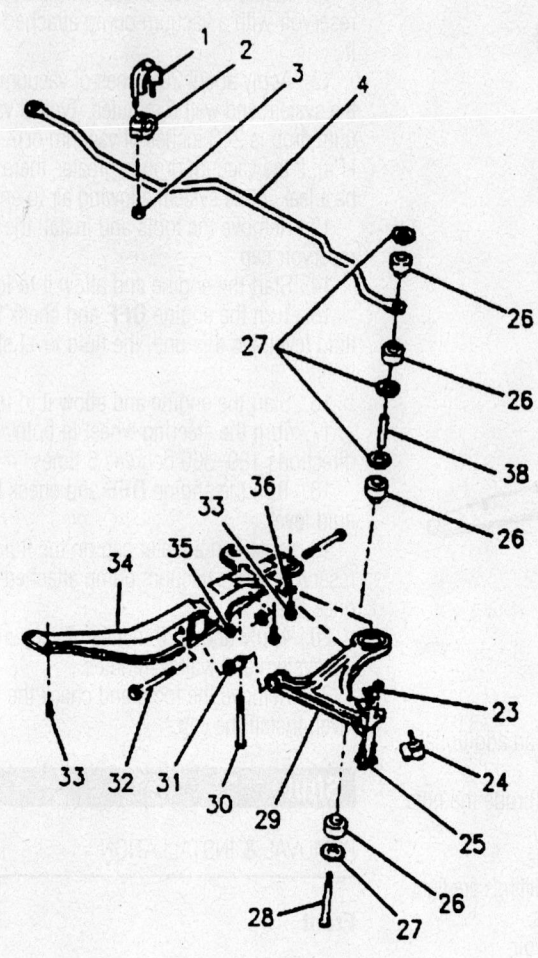
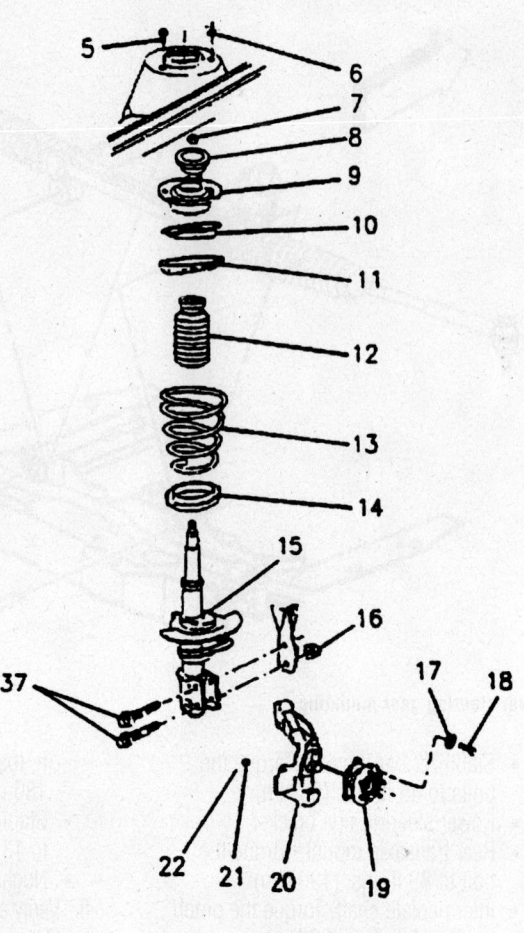

1 CLAMP, STABILIZER SHAFT	20 STEERING KNUCKLE
2 INSULATOR, STABILIZER SHAFT	21 NUT, BALL JOINT
3 NUT	22 COTTER PIN
4 STABILIZER SHAFT	23 NUT
5 BOLT	24 BALL JOINT
6 NUT	25 BOLT
7 NUT, STRUT DAMPENER SHAFT	26 INSULATOR, STABILIZER LINK
8 RATE WASHER	27 WASHER, STABILIZER LINK
9 STRUT MOUNT	28 BOLT, STABILIZER LINK
10 UPPER SPRING SEAT	29 CONTROL ARM
11 UPPER SPRING INSULATOR	30 BOLT
12 DUST TUBE ASSEMBLY	31 BUSHING, CONTROL ARM
13 SPRING	32 BOLT
14 LOWER SPRING INSULATOR	33 BOLT
15 STRUT	34 SUSPENSION SUPPORT
16 NUT	35 NUT
17 WASHER	36 WASHER
18 BOLT	37 BOLT
19 HUB AND BEARING ASSEMBLY	38 SPACER, STABILIZER LINK

Exploded view of the front suspension

79222332

Torque the bolt to 10 ft. lbs. (14 Nm).
- Wheel

10. Check and/or adjust the front end alignment.

Rear

1. Before servicing the vehicle, refer to the precautions section.
2. Remove the rear wheel.
3. Scribe a mark indicating the position of the strut on the knuckle.
4. Remove or disconnect the following:
- Strut nuts from inside the trunk
- Strut bolts from the wheel well area
- Strut-to-knuckle bolts
- Strut from the vehicle

To install:
5. Install or connect the following:
- Strut. Torque the upper bolts and nuts to 18 ft. lbs. (25 Nm).
- Strut on the knuckle by aligning the scribe marks made earlier. Torque the bolts to 89 ft. lbs. (120 Nm).
- Wheel

6. Check and/or adjust the wheel alignment.

Coil Spring

REMOVAL & INSTALLATION

Front & Rear

1. Before servicing the vehicle, refer to the precautions section.
2. Remove the strut assembly from the vehicle.
3. Mount the strut compressor in a holding fixture.
4. Mount the strut assembly into the compressor. Note that the strut compressor has strut mounting holes drilled for specific vehicle lines.
5. Compress the strut approximately ½ its height after initial contact with the top cap.

> **⁜ WARNING**
>
> **Never bottom the spring or damper rod.**

6. Remove the nut from the strut damper shaft and place alignment/guiding rod J-34013-27 on top of the damper shaft. Use the rod to guide the damper shaft straight down through the spring cap while compressing the spring. Remove the components.

To install:
7. Install the bearing cap into the strut compressor, if removed.

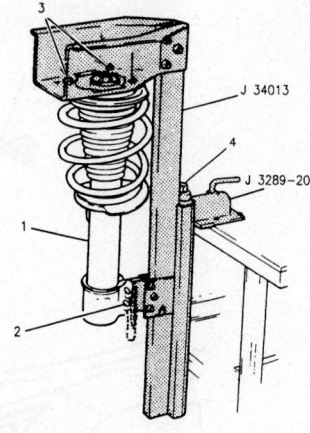

1. STRUT ASSEMBLY
2. INSTALL LOCKING PINS THROUGH STRUT ASSEMBLY
3. TIGHTEN NUTS UNTIL FLUSH WITH STRUT COMPRESSOR
4. COMPRESSOR FORCING SCREW

79222334

View of the strut assembly mounted in a compressor

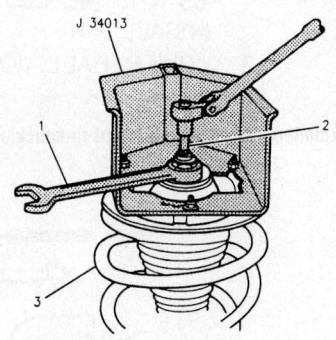

1. WRENCH
2. SOCKET
3. STRUT ASSEMBLY

79222335

Use a socket and a wrench to remove the damper shaft nut spring cap while compressing the spring

8. Mount the strut assembly in strut compressor, using bottom locking pin only. Extend the damper shaft and install clamp J-34013-20 on the damper shaft.
9. Install the spring over the damper and swing the assembly up so the upper locking pin can be installed.
10. Install all shields, bumpers and insulators on the spring seat. Install the spring seat on top of the spring. Be sure the flat on the upper spring seat is facing in the proper direction. The spring seat flat should be facing the same direction as the centerline of the strut assembly spindle.
11. Install the guiding rod and turn the forcing screw while the guiding rod centers

1. STRUT COMPRESSOR
2. STRUT ASSEMBLY

79222336

Install the rod to guide the damper shaft straight down through the spring cap while compressing the spring

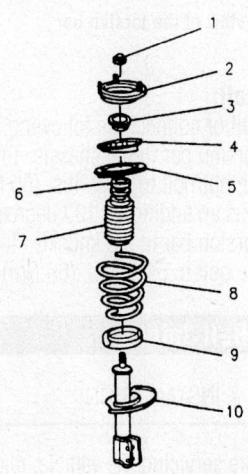

1. STRUT MOUNT NUT
2. STRUT MOUNT
3. RATE WASHER
4. SPRING SEAT
5. SPRING UPPER INSULATOR
6. JOUNCE BUMPER
7. STRUT DUST SHIELD
8. SPRING
9. SPRING LOWER INSULATOR
10. STRUT

79222337

Exploded view of the front strut assembly

the assembly. When the threads on the damper shaft are visible, remove the guiding rod and install the nut. On rear assemblies, tighten the nut to 52 ft. lbs. (70 Nm) on 2002–04 models or 34–47 Nm) on 2005 models. On front assemblies, tighten to 55 ft. lbs. (75 Nm). Use a crow's foot line wrench while holding the damper shaft with a socket.
12. Remove the clamp.

Torsion Bars

REMOVAL & INSTALLATION

1. Before servicing the vehicle, refer to the precautions section.
2. Remove or disconnect the following:
   - Torsion bar-to-knuckle bolt, washer and bushing
   - Torsion bar-to-chassis bolt
   - Torsion bar from the chassis

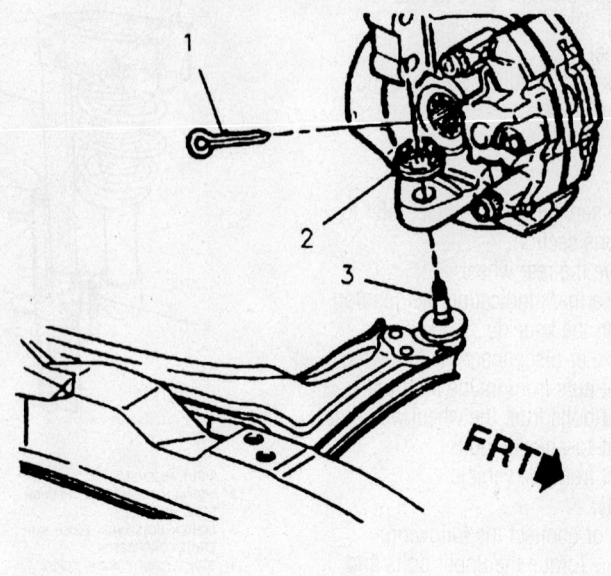

1 PIN
2 NUT – 55 N·m (41 LBS. FT.) MINIMUM TORQUE
  65 N·m (48 LBS. FT.) MAXIMUM TORQUE TO
  INSTALL PIN
3 LOWER BALL JOINT

79222339

Exploded view of the ball joint-to-knuckle mounting

Exploded view of the torsion bar

9306ZG14

To install:

3. Install or connect the following:
   - Torsion bar to the chassis. Torque the nut/bolt to 48 ft. lbs. (65 Nm) plus an additional 120 degree turn.
   - Torsion bar to the knuckle. Torque the bolt to 51 ft. lbs. (69 Nm).

Lower Ball Joint

REMOVAL & INSTALLATION

1. Before servicing the vehicle, refer to the precautions section.
2. Remove or disconnect the following:
   - Wheel
   - Wiring harness from the control arm
   - Stabilizer shaft link
   - Lower ball joint from the steering knuckle
3. Drill a ⅛ in. pilot hole in the center of each of the 3 ball joint mounting rivets.
4. Using a ½ in. drill bit, drill the heads off the rivets.
5. With a hammer and punch, knock the rivets out of the control arm. Remove the sway bar link kit.
6. Pull the control arm down so the ball stud clears the steering knuckle. Slide the ball joint out of the control arm.

To install:

7. Position the new ball joint into the lower control arm and install the bolts. The nuts must be on top of the control arm.

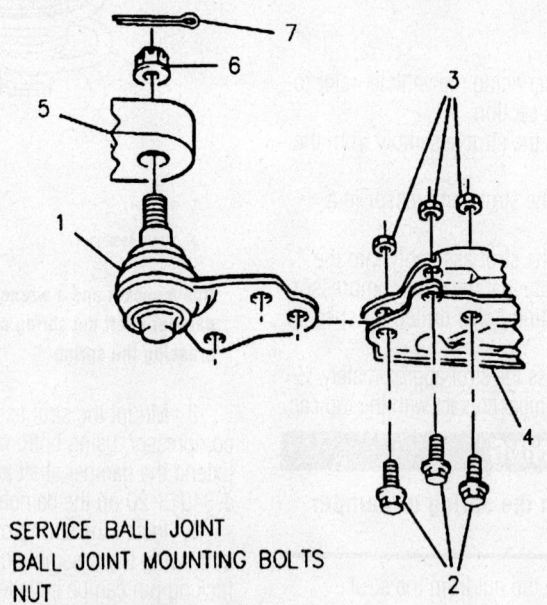

1 SERVICE BALL JOINT
2 BALL JOINT MOUNTING BOLTS
3 NUT
4 LOWER CONTROL ARM
5 STEERING KNUCKLE
6 NUT – 55 N·m (41 LBS. FT.) MINIMUM TORQUE
       65 N·m (48 LBS. FT.) MAXIMUM TORQUE
       TO INSTALL PIN
7 PIN

79222338

Exploded view of the replacement ball joint mounting

Tighten the mounting bolts to the specification provided with the ball joint service kit.

8. Connect the ball joint to the steering knuckle and torque the castle nut to 41 ft. lbs. (55 Nm) on 2002–04 models, or 48 ft. lbs. (60 Nm) on 2004–05 models.

➥**If necessary to align the cotter pin holes tighten the nut up to 60 degree (⅙) turn additional rotation. NEVER loosen the nut to make the holes align.**

9. Install or connect the following:
   • New cotter pin
   • Stabilizer shaft link. Torque the nut to 22 ft. lbs. (30 Nm) on 2002–03 models or 13 ft. lbs. (17 Nm) on 2004–05 models.
   • Wiring harness to the control arm
   • Wheel. Torque the nuts to 100 ft. lbs. (140 Nm).

Lower Control Arm

REMOVAL & INSTALLATION

Front

1. Before servicing the vehicle, refer to the precautions section.
2. Remove or disconnect the following:

• Front wheel
• Stabilizer link
• Anti-lock Brake System (ABS) harness from the lower control arm, if equipped
• Lower ball joint from the steering knuckle
• Lower control arm-to-suspension crossmember bolts
• Lower control arm

To install:

3. Install or connect the following:
   • Lower control arm-to-suspension crossmember and hand-tighten the bolts
4. Connect the ball joint to the steering knuckle and torque the castle nut to 41 ft. lbs. (55 Nm) on 2002–04 models, or 48 ft. lbs. (60 Nm) on 2004–05 models.
   • Stabilizer shaft link. Torque the nut to 22 ft. lbs. (30 Nm) on 2002–03 models or 13 ft. lbs. (17 Nm) on 2004–05 models.
   • ABS harness to the lower control arm, if equipped
   • Front wheel
5. Position the vehicle at curb height.
6. Tighten the bolts, with the vehicle at curb height as follows:
   a. Front lower control arm-to-sus-

pension crossmember bolt to 45 ft. lbs. (60 Nm), plus an additional 120 degree turn.
   b. Rear lower control arm-to-suspension crossmember bolt to 74 ft. lbs. (100 Nm), plus an additional 180 degree turn.
7. Check and/or adjust the front alignment.

CONTROL ARM BUSHING REPLACEMENT

Front Bushing

1. Before servicing the vehicle, refer to the precautions section.
2. Remove the lower control arm and place it in a vise.
3. Lubricate the threads of Screw J-21474-19 with high pressure lubricant.
4. Assemble Tools Screw J-21474-19, Remover/Installer J-41397-1A, Receiver J-41397-2A and J-21474-18 onto the front control arm bushing.
5. Tighten Tool J-21474-18 until the front bushing is pressed from the control arm.
6. Disassemble the tools.
To install:
7. Lubricate the new front bushing outer casing.

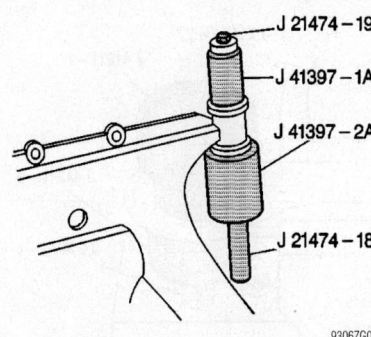

9306ZG05

Removing the front bushing from the lower control arm

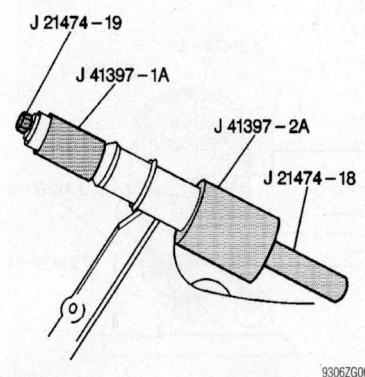

9306ZG06

Installing the front bushing to the lower control arm

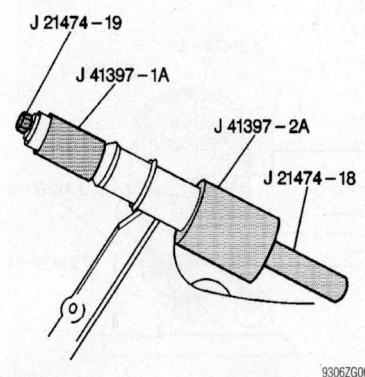

9306ZG04

Exploded view of the lower control arm and related components

8. Insert the new bushing into the control arm.

9. Assemble Tools Screw J-21474-19, Remover/Installer J-41397-1A, Receiver J-41397-2A and J-21474-18 onto the front control arm bushing.

10. Tighten Screw J-21474-19 until the front bushing is pressed into the control arm.

11. Disassemble the tools.

12. Install the lower control arm.

Rear Bushing

1. Before servicing the vehicle, refer to the precautions section.

2. Remove the lower control arm and place it in a vise.

3. Assemble Tools Screw J-21474-27, Remover/Installer J-41211-1, Receiver J-41211-3 and J-21474-4 onto the rear control arm bushing.

4. Tighten Tool J-21474-27 until the rear bushing is pressed from the control arm.

5. Disassemble the tools.

To install:

6. Insert the new bushing into the control arm.

7. Assemble Tools Screw J-21474-27, Remover/Installer J-41211-1, Receiver J-41211-3 and J-21474-4 onto the rear control arm bushing.

8. Tighten Screw J-21474-4 until the rear bushing is pressed into the control arm.

9. Disassemble the tools.

10. Install the lower control arm.

Wheel Bearings

ADJUSTMENT

These vehicles are equipped with sealed hub and bearing assemblies. The hub and bearing assemblies are non-serviceable. If the assembly is damaged, the complete unit must be replaced.

REMOVAL & INSTALLATION

Front

1. Before servicing the vehicle, refer to the precautions section.

2. Remove or disconnect the following:
   - Front wheel
   - Halfshaft nut and washer
   - Caliper from the steering knuckle and support it aside

✳✳ WARNING

DO NOT allow the caliper to hang unsupported from the brake hose.

- Brake rotor
- Anti-lock Brake System (ABS) connector, if equipped
- 3 hub/bearing assembly bolts
- Halfshaft from the hub/bearing assembly
- Hub/bearing assembly

To install:

3. Install or connect the following:
   - Hub/bearing assembly onto the halfshaft, making sure the splines engage smoothly
   - Hub/bearing assembly to the steering knuckle. Torque the bolts to 70 ft. lbs. (95 Nm).
   - Anti-lock Brake System (ABS) connector, if equipped
   - Brake rotor
   - Caliper onto the steering knuckle. Torque the bolts to 38 ft. lbs. (51 Nm).
   - Halfshaft nut. Torque the hub nut to 284 ft. lbs. (385 Nm) on nuts colored black, or 173 ft. lbs. (235 Nm) on gray colored nuts.

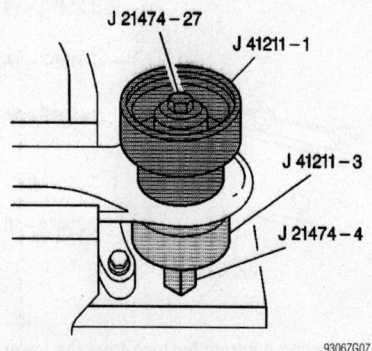

Removing the rear bushing from the lower control arm

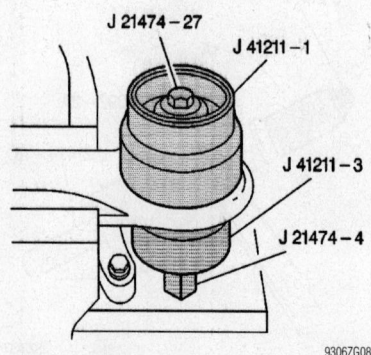

Installing the rear bushing to the lower control arm

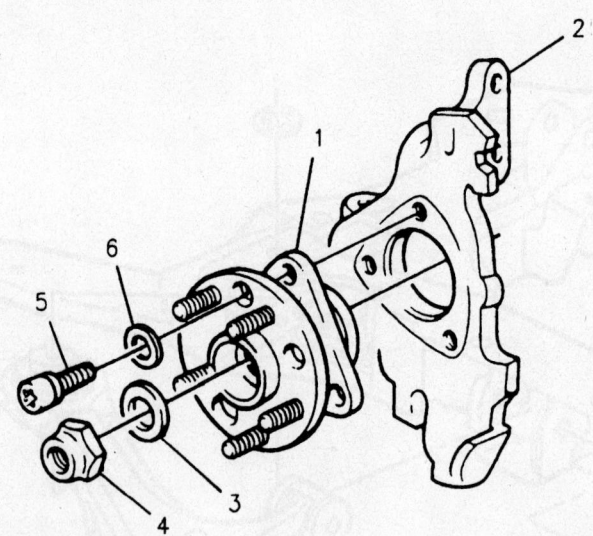

1 HUB AND BEARING ASSEMBLY
2 STEERING KNUCKLE
3 WASHER
4 DRIVE AXLE NUT – 260 N·m (192 LBS. FT.)
5 HUB AND BEARING RETAINING BOLT
6 WASHER

Exploded view of the front hub/bearing assembly

• Front wheel. Torque the nuts to 100 ft. lbs. (140 Nm).

Rear

DRUM BRAKE

1. Before servicing the vehicle, refer to the precautions section.
2. Remove or disconnect the following:
  • Wheel
  • Brake drum
  • 4 hub/bearing assembly to knuckle nuts

➡**The top rear bolt will not clear the brake shoes and must be removed with the bearing assembly.**

  • Anti-lock Brake System (ABS) speed sensor wire from the hub/bearing assembly
  • Hub/bearing assembly

To install:

3. Install or connect the following:
  • Hub/bearing assembly on the knuckle
  • ABS wheel speed sensor
  • Hub/bearing nuts. Torque the nuts to 44 ft. lbs. (60 Nm).
  • Brake drum
  • Wheel. Torque the nuts to 100 ft. lbs. (140 Nm).

DISC BRAKE

1. Before servicing the vehicle, refer to the precautions section.
2. Remove or disconnect the following:
  • Wheel
  • Brake rotor
  • Parking brake cable from the lever
  • Wheel Speed Sensor (WSS) electrical connector
  • Hub/bearing assembly bolts from the knuckle
  • Torx® bolts from the rear of the hub/bearing assembly
  • Hub/bearing assembly from the backing plate

To install:

3. Install or connect the following:
  • Hub/bearing assembly to the backing plate. Tighten the Torx® bolts to 89 inch lbs. (10 Nm).
  • Hub/bearing assembly on the knuckle. Torque the bolts to 62 ft. lbs. (85 Nm).
  • WSS electrical connector
  • Parking brake cable to the lever
  • Brake rotor
  • Wheel assembly

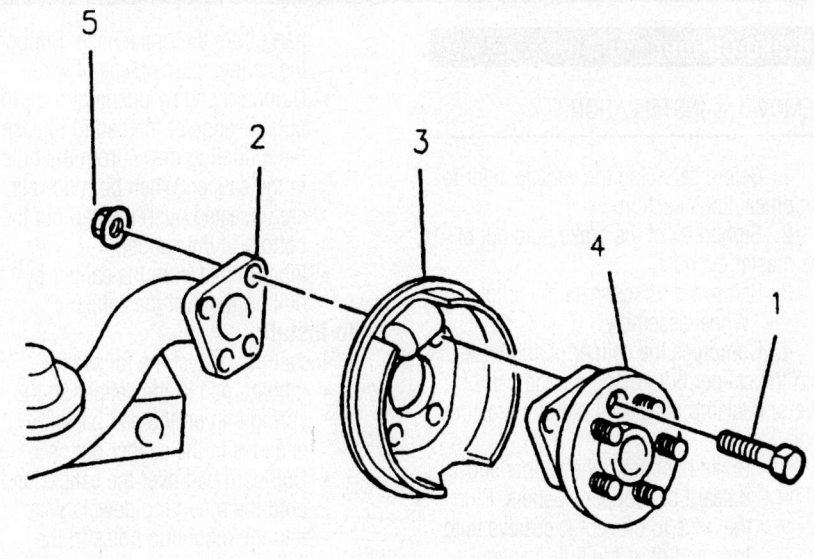

1 BOLT
2 REAR AXLE ASSEMBLY
3 BACKING PLATE
4 HUB AND BEARING ASSEMBLY
5 LOCKNUT

79222341

Rear hub and bearing assembly on models equipped with drum brakes

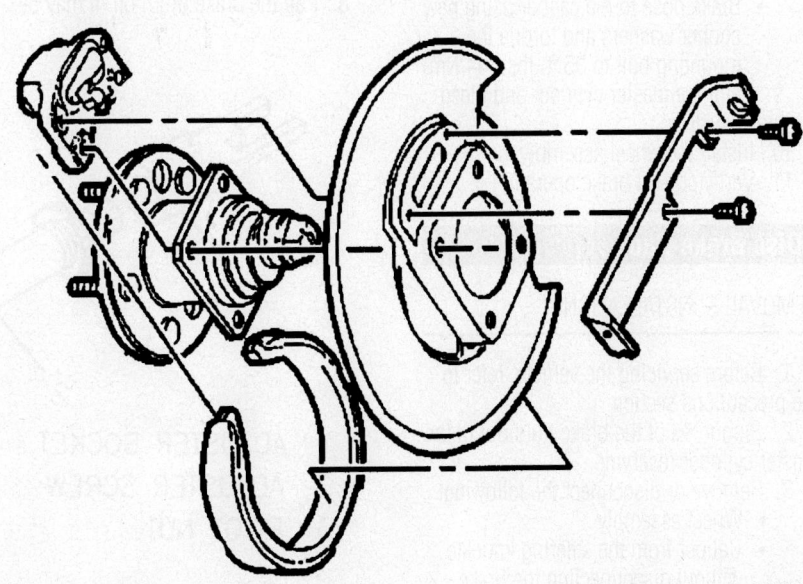

9300Z304

Rear wheel bearing assembly with disc brakes

BRAKES

Brake Caliper

REMOVAL & INSTALLATION

1. Before servicing the vehicle, refer to the precautions section.
2. Siphon ⅔ of the brake fluid out of the master cylinder.
3. Remove or disconnect the following:
   • Wheel assembly
4. Compress the caliper piston back into the caliper bore using a large pair of pliers, C-clamp or special piston retracting tool.

   • Brake hose from the caliper and discard the copper washers. Plug the hose to prevent excessive fluid loss and possible fluid contamination.
   • Caliper mounting bolts and the caliper from the knuckle
   • Brake pads from the caliper, if the caliper is being replaced

To install:
5. Inspect the condition of the caliper support for rust and corrosion that will hinder the travel of the caliper.
6. Inspect the caliper mounting hardware. New bolts are usually recommended.
7. Lubricate the mounting bushings and sleeves with silicone grease as required.
8. Install or connect the following:
   • Brake pads in the caliper
   • Caliper on the steering knuckle
   • Mounting bolts and torque to 40 ft. lbs. (51 Nm)
   • Brake hose to the caliper using new copper washers and torque the mounting bolt to 35 ft. lbs. (44 Nm)
9. Refill the master cylinder and bleed the brake system.
10. Install the wheel assembly.
11. Verify correct brake operation.

Disc Brake Pads

REMOVAL & INSTALLATION

1. Before servicing the vehicle, refer to the precautions section.
2. Siphon ⅔ of the brake fluid out of the master cylinder reservoir.
3. Remove or disconnect the following:
   • Wheel assembly
   • Caliper from the steering knuckle without disconnecting the brake hose. DO NOT allow the caliper to

hang from the brake hose. Support the caliper with a piece of wire.
   • Outboard pad by pushing in on the outside edge of the pad to release the mounting dowel from the hole in the caliper. When both dowels are unseated push the pad out the bottom of the caliper.
   • Inboard pad from the caliper by pulling it out of the caliper

To install:
4. Install or connect the following:
   • Inboard pad in the caliper so the spring clip on the pad back engages in the caliper piston
   • Outboard pad over the caliper end until the mounting dowels snap into the mounting holes in the caliper
   • Caliper over the rotor onto the steering knuckle
   • Caliper mounting bolts and sleeves and torque to 40 ft. lbs. (51 Nm)
   • Wheel assembly
5. Pump the brake pedal several times to seat the pads against the rotor before attempting to move the vehicle.
6. Check the master cylinder level and add fluid as necessary.

Brake Drums

REMOVAL & INSTALLATION

1. Before servicing the vehicle, refer to the precautions section.
2. Remove the wheel.
3. Pull the brake drum off. It may be

necessary to gently tap the rear edge of the drum to start it off the studs. If extreme resistance to removal is encountered, it will be necessary to retract the brake shoe self-adjuster screw, sometimes called the star-wheel. Knock out the access hole in the brake drum and turn the adjuster to retract the linings from the drum. Install a replacement hole cover before reinstalling the drum.

➡ **DO NOT hammer on the brake drum to remove it.**

To install:
4. Inspect the inside of the brake drum. If worn, heavily grooved or if the opening is distorted, the drum should be refinished or replaced. If refinishing, observe the maximum drum diameter specification.
5. Inspect the wheel cylinder for signs of brake fluid leakage. Inspect the brake shoe springs and self-adjuster mechanism. The adjuster should usually be disassembled, cleaned and lubricated when the drum is removed for brake service.
6. Install the drum over the brake shoes.
7. Install the wheel. Adjust the brakes.
8. Check brake operation.

Brake Shoes

REMOVAL & INSTALLATION

2002 Models

1. Before servicing the vehicle, refer to the precautions section.
2. Remove or disconnect the following:

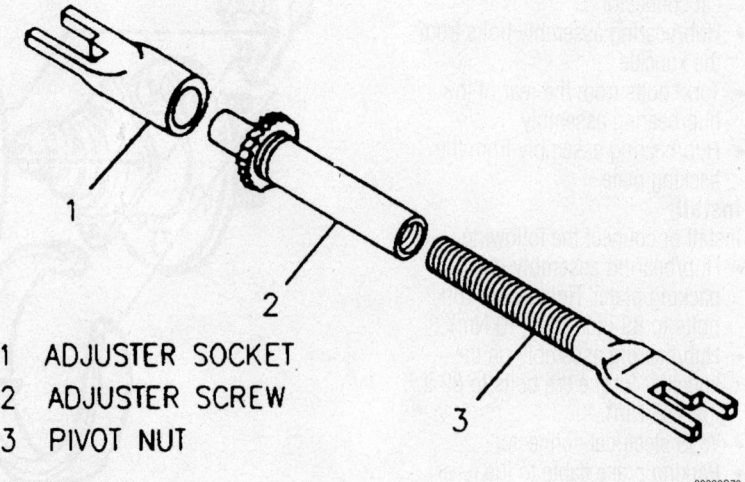

1 ADJUSTER SOCKET
2 ADJUSTER SCREW
3 PIVOT NUT

Brake adjuster—

93006G73

- Wheel assembly
- Brake drum
- Upper return springs from the shoes using tool J-8057 brake spring tool
- Hold-down springs using J-8049 brake spring tool
- Shoe hold-down pins from behind the brake backing plate

3. Lift up the actuator lever for the self-adjusting mechanism and remove the actuating link. Remove the actuator lever, pivot, and the pivot return spring.

4. Spread the shoes apart to clear the wheel cylinder pistons and remove the parking brake strut and spring.

- Parking brake cable from the lever
- Shoes, still connected by their adjusting screw spring

5. With the shoes removed, note the position of the adjusting spring and remove the spring and adjusting screw.

- C-clip from the parking brake lever and the lever from the secondary shoe

6. Use a damp cloth to remove all dirt and dust from the backing plate and brake parts.

To install:

7. Check the backing plate attaching bolts to make sure they are tight. Use fine emery cloth to clean all rust and dirt from the shoe contact surfaces on the plate and lubricate with brake grease. Check the wheel cylinder for signs of leakage.

8. Clean all parts completely in brake solvent and air dry. Clean the backing plate shoe contact points.

9. Inspect the inside of the brake drum. If worn, heavily grooved or if the opening is distorted, the drum should be refinished or replaced.

10. Inspect the brake shoe springs and self-adjuster mechanism. Disassemble the adjuster mechanism and clean the threads and coat with grease. Make sure the adjuster assembly turns freely before installing in the vehicle.

11. Install or connect the following:

- Parking brake lever on the secondary shoe and secure with C-clip
- Adjusting screw and spring on the shoes, connecting them together. The coils of the spring must not be over the starwheel on the adjuster. The left and right hand springs are not interchangeable.

12. Spread the shoe assemblies and connect the parking brake cable.

- Shoes on the backing plate, engag-

ing the shoes at the top temporarily with the wheel cylinder pistons. Make sure the starwheel on the adjuster is lined up with the adjusting hole in the backing plate, if equipped.

13. Spread the shoes slightly and install the parking brake strut and spring. Make sure the end of the strut without the spring engages the parking brake lever. The end with the spring engages the primary shoe (the one with the shorter lining).

- Actuator pivot, lever and return spring
- Actuating link in the shoe retainer.

Lift up the actuator lever and hook the link into the lever.

- Hold-down pins through the back of the plate using J-8057
- Lever pivots and hold-down springs. Install the shoe return springs using J-8049. Be careful not to stretch or distort the springs.

14. Make sure the linings are in the right place, the self-adjusting mechanism is correctly installed, and the parking brake parts are hooked up.

15. Measure the distance from the edge of the primary lining to the edge secondary lining, then measure the inside width of the drum. Adjust the linings by means of the

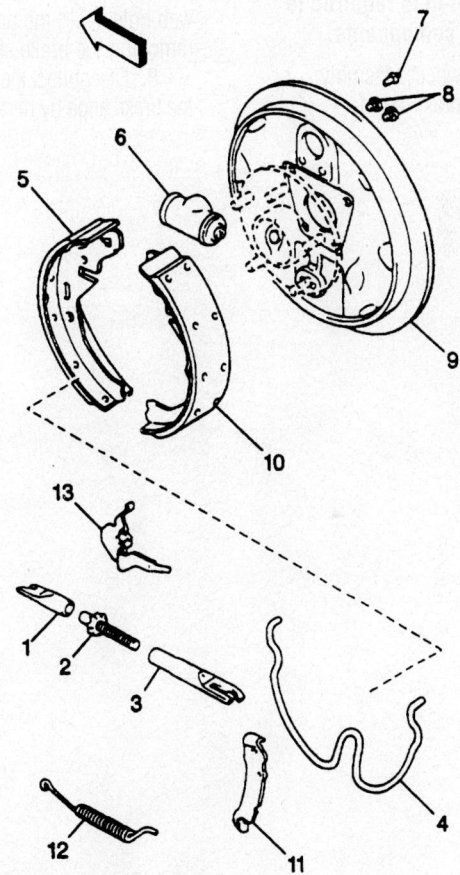

(1) Socket, Brake Adjuster
(2) Screw, Brake Adjuster
(3) Nut, Brake Pivot
(4) Spring, Retractor
(5) Brake Shoe and Lining
(6) Cylinder, Wheel Brake
(7) Valve, Bleeder
(8) Bolts, Wheel Cylinder
(9) Plate, Brake Backing
(10) Brake Shoe and Lining
(11) Lever, Park Brake
(12) Spring
(13) Adjuster Actuator

93006G78

Drum brake components, exploded view

adjuster so the drum will fit onto the linings.

16. Install the hub and bearing assembly onto the axle if removed. Torque the retaining bolts to 35 ft. lbs. (55 Nm).

17. Install the drum and the wheels.

18. Adjust the brakes. Install a rubber hole cover in the adjustment knock-out hole after the adjustment is complete. Adjust the parking brake.

19. Road test the vehicle and verify proper brake operation.

2003–05 Models

1. Before servicing the vehicle, refer to the precautions section.

➡ **Brake shoe spanner and spring removal tool J 38400 is required to remove the brake components.**

2. Remove the wheel assembly.
3. Remove the brake drum.

➡ **Be advised to repair one rear drum brake shoe assembly at a time. This enables you to use the other drum brake shoe assembly for visual reference.**

4. Remove the adjuster spring.

5. Use the brake tool to spread the top of the brake shoes to remove the adjuster assembly and the adjuster lever from the brake shoe.

6. Position the hook end of the brake tool under the universal spring and lightly pull the universal spring end out of the shoe web hole. Hold the universal spring while removing the brake shoe.

7. Position the hook end of the brake tool under the universal spring and lightly pull the universal spring end out of the shoe web hole. Hold the universal spring while removing the brake shoe.

8. Disconnect the park brake lever from the brake shoe by removing the retaining clip.

To install:

9. Connect the park brake lever to the brake shoe and install the retaining clip.

10. Position the hook end of the brake tool under the universal spring and lightly pull the universal spring end out while installing the brake shoe. Ensure that the universal spring properly engages the brake shoe web hole.

11. Position the hook end of the brake tool under the universal spring and lightly pull the universal spring end out while installing the brake shoe. Ensure that the universal spring engages the brake shoe web hole.

12. Use the brake tool to spread the top of the brake shoes to install the adjuster assembly and the adjuster lever to the brake shoe.

13. Install the adjuster spring.
14. Adjust the drum brake system.
15. Install the brake drum.
16. Install the wheel assembly.
17. Road test the vehicle.

CHEVROLET AND GMC

Astro • Safari

BRAKES 2-39
DRIVE TRAIN 2-22
ENGINE REPAIR.................2-9
FUEL SYSTEM 2-20
SPECIFICATIONS AND
  MAINTENANCE CHARTS 2-2
Engine and Vehicle Identification 2-2
General Engine Identification 2-2
Gasoline Engine Tune-Up
  Specifications 2-2
Firing Orders 2-3
Accessory Drive Belt Routing 2-3
Capacities 2-4
Valve Specifications.......................2-4
Crankshaft and Connecting Rod
  Specifications 2-5
Piston and Ring Specifications........2-5
Torque Specifications 2-6
Wheel Alignment 2-6
Tire, Wheel and Ball Joint
  Specifications 2-7
Brake Specifications 2-7
Scheduled Maintenance
  Intervals................................2-8
STEERING AND
  SUSPENSION 2-27

A

Air Bag......................................2-27
  Arming...................................2-27
  Disarming 2-27
  Precautions 2-27
Alternator 2-9
  Installation 2-9
  Removal 2-9
Automatic Transmission
  Assembly..................................2-22
  Removal & Installation...............2-22
Axle Shaft, Bearing and Seal........2-26
  Removal & Installation 2-26

B

Ball Joints...................................2-32
  Removal & Installation...............2-32
Brake Caliper 2-39
  Removal & Installation 2-39
Brake Drums 2-41
  Removal & Installation 2-41
Brake Shoes 2-41
  Removal & Installation..............2-41

C

Camshaft and Valve Lifters 2-16
  Removal & Installation..............2-16
Coil Springs 2-30
  Removal & Installation..............2-30
CV-Joints...................................2-24
  Overhaul 2-24
Cylinder Head 2-12
  Removal & Installation..............2-12

D

Disc Brake Pads...........................2-40
  Removal & Installation..............2-40
Distributor..................................2-9
  Installation 2-9
  Removal 2-9

E

Engine Assembly 2-9
  Removal & Installation...............2-9
Exhaust Manifold 2-16
  Removal & Installation..............2-16

F

Fuel Filter 2-20
  Removal & Installation..............2-20
Fuel Injector................................2-21
  Removal & Installation..............2-21
Fuel Pump 2-21
  Removal & Installation..............2-21
Fuel System Pressure 2-20
  Relieving 2-20
Fuel System Service
  Precautions.............................2-20

H

Halfshaft....................................2-23
  Removal & Installation..............2-23
Heater Core................................2-12
  Removal & Installation..............2-12

I

Ignition Timing 2-9
  Adjustment 2-9
Intake Manifold 2-14
  Removal & Installation..............2-14

L

Leaf Springs 2-30
  Removal & Installation..............2-30
Lower Control Arm 2-35
Control Arm Bushing
  Replacement 2-36
  Removal & Installation..............2-35

O

Oil Pan......................................2-17
  Removal & Installation..............2-17
Oil Pump 2-18
  Removal & Installation..............2-18

P

Pinion Seal 2-26
  Removal & Installation..............2-26
Piston and Ring 2-20
  Positioning 2-20
Power Steering Gear
  (Recirculating Ball).....................2-27
  Removal & Installation..............2-27

R

Rear Main Seal 2-18
  Removal & Installation..............2-18
Rocker Arms 2-13
  Removal & Installation..............2-13

S

Shock Absorbers 2-28
  Removal & Installation..............2-28
Starter Motor 2-17
  Removal & Installation..............2-17

T

Timing Chain, Sprockets, Front
  Cover and Seal 2-18
  Removal & Installation..............2-18
Torsion Bar 2-31
  Removal & Installation..............2-31
Transfer Case Assembly.................2-23
  Removal & Installation..............2-23

U

Upper Control Arm 2-34
Control Arm Bushing
  Replacement 2-35
  Removal & Installation..............2-34

V

Valve Lash 2-17
  Adjustment..............................2-17

W

Water Pump................................2-11
  Removal & Installation..............2-11
Wheel Bearings...........................2-36
  Adjustment 2-36
  Removal & Installation..............2-36

SPECIFICATIONS AND MAINTENANCE CHARTS

ENGINE AND VEHICLE IDENTIFICATION

Engine							Model Year	
Code ①	Liters (cc)	Cu. In.	Cyl.	Fuel Sys.	Engine Type	Eng. Mfg.	Code ②	Year
X	4.3 (4293)	263	6	MFI	OHV	CPC	2	2002
							3	2003
							4	2004
							5	2005

CPC: Chevrolet/Pontiac/Canada

MFI: Multi-port Fuel Injection

① 8th position of VIN

② 10th position of VIN

06025-ASTR-C01

GENERAL ENGINE SPECIFICATIONS

All measurements are given in inches.

Year	Model	Engine Displacement Liters	Engine Series VIN	Net Horsepower @ rpm	Net Torque @ rpm (ft. lbs.)	Bore x Stroke (in.)	Compression Ratio	Oil Pressure @ rpm
2002	Astro/Safari	4.3	X	190@4400	250@2800	4.00x3.48	9.2:1	18@2000
2003	Astro/Safari	4.3	X	190@4400	250@2800	4.00x3.48	9.2:1	18@2000
2004	Astro/Safari	4.3	X	190@4400	250@2800	4.00x3.48	9.2:1	18@2000
2005	Astro/Safari	4.3	X	190@4400	250@2800	4.00x3.48	9.2:1	18@2000

06025-ASTR-C02

GASOLINE ENGINE TUNE-UP SPECIFICATIONS

Year	Engine Displacement Liters	Engine VIN	Spark Plugs Gap (in.)	Ignition Timing (deg.)	Fuel Pump (psi)	Idle Speed (rpm) MT	Idle Speed (rpm) AT	Valve Clearance In.	Valve Clearance Ex.
2002	4.3	X	0.060	①	58-64 ②	600	625	HYD	HYD
2003	4.3	X	0.060	①	58-64 ②	600	625	HYD	HYD
2004	4.3	X	0.060	①	58-64 ②	600	625	HYD	HYD
2005	4.3	X	0.060	①	58-64 ②	600	625	HYD	HYD

NOTE: The Vehicle Emission Control Information label often reflects specification changes made during production.

The label figures must be used if they differ from those in this chart.

HYD: Hydraulic

① Ignition timing is preset and cannot be adjusted

② With key ON and engine OFF

06025-ASTR-C03

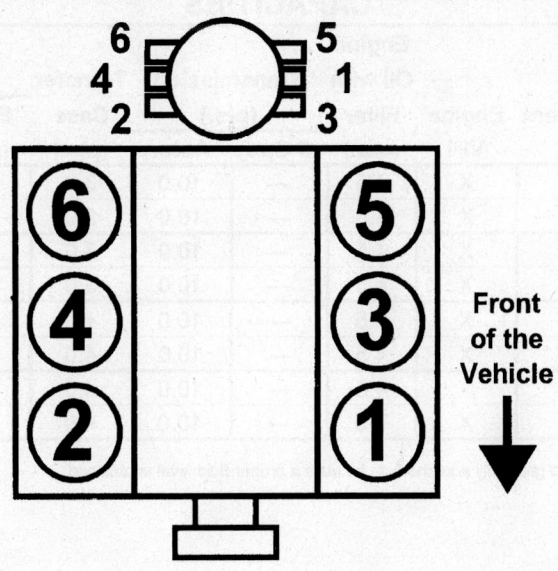

4.3L Engine
Firing order: 1–6–5–4–3–2
Distributorless ignition system

79243G61

Front
of the
Vehicle

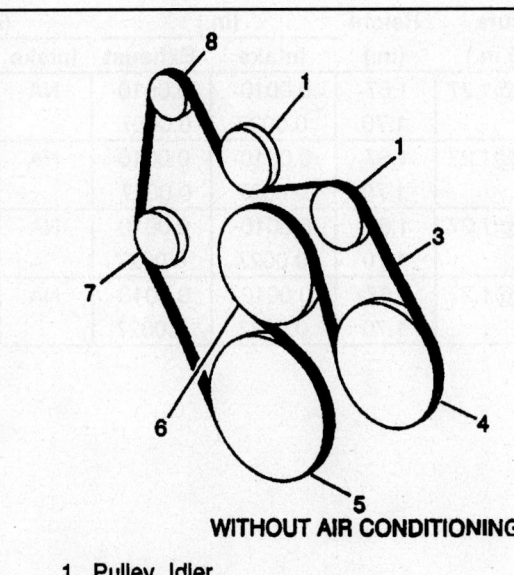

WITHOUT AIR CONDITIONING

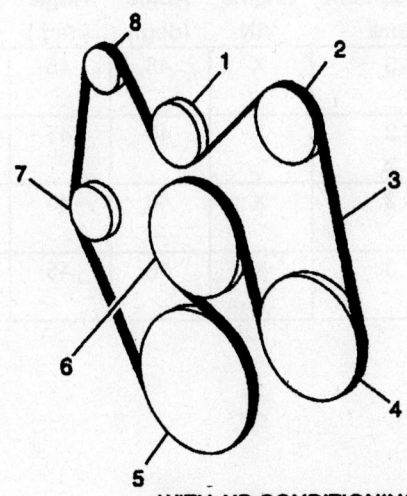

WITH AIR CONDITIONING

1. Pulley, Idler
2. Pulley, AC Compressor
3. Belt, Drive
4. Pulley, Power Steering Pump

5. Pulley, Crankshaft
6. Pulley, Water Pump
7. Pulley, Drive Belt Tensioner
8. Pulley, Generator

79244G24

Accessory serpentine belt routing—4.3L engine

CAPACITIES

Year	Model	Engine Displacement Liters	Engine VIN	Engine Oil with Filter (qts.)	Transmission (pts.) 5-Spd	Transmission (pts.) Auto.	Transfer Case (pts.)	Drive Axle Front (pts.)	Drive Axle Rear (pts.)	Fuel Tank (gal.)	Cooling System (qts.)
2002	Astro	4.3	X	4.5	—	10.0	3.0	2.6	3.5	27.0	①
	Safari	4.3	X	4.5	—	10.0	3.0	2.6	3.5	27.0	①
2003	Astro	4.3	X	4.5	—	10.0	3.0	2.6	3.5	27.0	①
	Safari	4.3	X	4.5	—	10.0	3.0	2.6	3.5	27.0	①
2004	Astro	4.3	X	4.5	—	10.0	4.0	2.6	3.5	27.0	①
	Safari	4.3	X	4.5	—	10.0	4.0	2.6	3.5	27.0	①
2005	Astro	4.3	X	4.5	—	10.0	4.0	2.6	3.5	27.0	①
	Safari	4.3	X	4.5	—	10.0	4.0	2.6	3.5	27.0	①

NOTE: All capacities are approximate. Add fluid gradually and check to be sure a proper fluid level is obtained.

① With rear heater: 16.5 qts.
Without rear heater: 13.5 qts.

06025-ASTR-C04

VALVE SPECIFICATIONS

Year	Engine Displacement Liters	Engine VIN	Seat Angle (deg.)	Face Angle (deg.)	Spring Test Pressure (lbs. @ in.)	Spring Installed Height (in.)	Stem-to-Guide Clearance (in.) Intake	Stem-to-Guide Clearance (in.) Exhaust	Stem Diameter (in.) Intake	Stem Diameter (in.) Exhaust
2002	4.3	X	46	45	187-203@1.27	1.67-1.70	0.0010-0.0027	0.0010-0.0027	NA	NA
2003	4.3	X	46	45	187-203@1.27	1.67-1.70	0.0010-0.0027	0.0010-0.0027	NA	NA
2004	4.3	X	46	45	187-203@1.27	1.67-1.70	0.0010-0.0027	0.0010-0.0027	NA	NA
2005	4.3	X	46	45	187-203@1.27	1.67-1.70	0.0010-0.0027	0.0010-0.0027	NA	NA

NA: Not Available

06025-ASTR-C05

CRANKSHAFT AND CONNECTING ROD SPECIFICATIONS

All measurements are given in inches.

Year	Engine Displacement Liters	Engine VIN	Crankshaft				Connecting Rod		
			Main Brg. Journal Dia.	Main Brg. Oil Clearance	Shaft End-play	Thrust on No.	Journal Diameter	Oil Clearance	Side Clearance
2002	4.3	X	①	②	0.0020-0.0080	4	2.2487-2.2497	0.0015 0.0031	0.0060-0.0170
2003	4.3	X	①	②	0.0020-0.0080	4	2.2487-2.2497	0.0015 0.0031	0.0060-0.0170
2004	4.3	X	①	②	0.0020-0.0080	4	2.2487-2.2497	0.0015 0.0031	0.0060-0.0170
2005	4.3	X	①	③	0.0020-0.0080	4	2.2487-2.2497	0.0015 0.0031	0.0060-0.0170

① No. 1: 2.4488-2.4495
  Nos. 2, 3: 2.4485-2.4494
  No. 4: 2.4480-2.4489

② No. 1: 0.0008-0.0020
  Nos. 2, 3: 0.0011-0.0023
  No. 4: 0.0017-0.0032

③ No. 1: 0.0008-0.0020
  Nos. 2, 3, 4: 0.0011-0.0023

06025-ASTR-C06

PISTON AND RING SPECIFICATIONS

All measurements are given in inches.

Year	Engine Displ. Liters	Engine VIN	Piston Clearance	Ring Gap			Ring Side Clearance		
				Top Compression	Bottom Compression	Oil Control	Top Compression	Bottom Compression	Oil Control
2002	4.3	X	0.0007-0.0024	0.010-0.016	0.015-0.023	0.010-0.029	0.0012-0.0027	0.0015-0.0031	0.0020-0.0070
2003	4.3	X	0.0007-0.0024	0.010-0.016	0.015-0.023	0.010-0.029	0.0012-0.0027	0.0015-0.0031	0.0020-0.0070
2004	4.3	X	0.0007-0.0024	0.010-0.016	0.015-0.023	0.010-0.029	0.0012-0.0027	0.0030-0.0110	0.0018-0.0077
2005	4.3	X	0.0007-0.0024	0.010-0.016	0.015-0.023	0.010-0.029	0.0012-0.0027	0.0030-0.0110	0.0018-0.0077

06025-ASTR-C07

TORQUE SPECIFICATIONS
All readings in ft. lbs.

Year	Engine Displacement Liters	Engine VIN	Cylinder Head Bolts	Main Bearing Bolts	Rod Bearing Bolts	Crankshaft Damper Bolts	Flywheel Bolts	Manifold Intake	Manifold Exhaust	Spark Plugs	Oil Pan Drain Plug
2002	4.3	X	①	77	②	70	74	③	④	11	18
2003	4.3	X	①	77	②	70	74	③	④	11	18
2004	4.3	X	①	77	②	70	74	③	④	11	18
2005	4.3	X	①	77	②	70	74	③	④	11	18

① 1st pass: 22 ft. lbs.
   2nd pass:
   Short bolts: Plus 55 degrees
   Medium bolts: Plus 65 degrees
   Long bolts: Plus 75 degrees

② 20 ft. lbs. plus 70 degrees

③ Lower intake manifold:
   1st pass: 27 inch lbs.
   2nd pass: 106 inch lbs.
   Final pass: 11 ft. lbs.
   Upper manifold bolts:
   1st pass: 44 inch lbs.
   2nd pass: 80 inch lbs.

④ Tighten bolts to 12 ft. lbs.

06025-ASTR-C08

WHEEL ALIGNMENT

Year	Model		Caster Range (+/-Deg.)	Caster Preferred Setting (Deg.)	Camber Range (+/-Deg.)	Camber Preferred Setting (Deg.)	Toe-in (in.)
2002	RWD	Left	1.00	+3.00	1.00	+0.60	0+/-0.20
		Right	1.00	+3.50	1.00	+0.60	0+/-0.20
	AWD	Left	1.00	+3.50	1.00	0	0+/-0.20
		Right	1.00	+4.50	1.00	0	0+/-0.20
2003	RWD	Left	1.00	+3.00	1.00	+0.60	0+/-0.20
		Right	1.00	+3.50	1.00	+0.60	0+/-0.20
	AWD	Left	1.00	+3.50	1.00	0	0+/-0.20
		Right	1.00	+4.50	1.00	0	0+/-0.20
2004	RWD	Left	1.00	+3.00	1.00	+0.60	0+/-0.20
		Right	1.00	+3.50	1.00	+0.60	0+/-0.20
	AWD	Left	1.00	+3.50	1.00	0	0+/-0.20
		Right	1.00	+4.50	1.00	0	0+/-0.20
2005	RWD	Left	1.00	+3.00	1.00	+0.60	0+/-0.20
		Right	1.00	+3.50	1.00	+0.60	0+/-0.20
	AWD	Left	1.00	+3.50	1.00	0	0+/-0.20
		Right	1.00	+4.50	1.00	0	0+/-0.20

06025-ASTR-C09

TIRE, WHEEL AND BALL JOINT SPECIFICATIONS

| Year | Model | OEM Tires | | Tire Pressures (psi) | | Wheel Size | Ball Joint Inspection | Lug Nut Torque (ft. lbs.) |
		Standard	Optional	Front	Rear			
2002	Astro/Safari	P215/75R15	None	36	36	6.5–JJ	U: 0.125 in. L ①	100
2003	Astro/Safari	P215/75R15	None	36	36	6.5–JJ	U: 0.125 in. L ①	100
2004	Astro/Safari	P215/75R15	None	36	36	6.5–JJ	U: 0.125 in. L ①	100
2005	Astro/Safari	P215/75R15	None	36	36	6.5–JJ	U: 0.125 in. L ①	100

OEM: Original Equipment Manufacturer

PSI: Pounds Per Square Inch

STD: Standard

OPT: Optional

L: Lower

U: Upper

① Do not lift truck. Inspect the boss into which the grease fitting is threaded. Replace if the boss is flush or receded below the surface of the ball joint

06025-ASTR-C10

BRAKE SPECIFICATIONS

All measurements in inches unless noted

| Year | Model | | Brake Disc | | | Brake Drum Diameter | | | Minimum Lining Thickness | Brake Caliper | |
			Original Thickness	Minimum Thickness	Maximum Runout	Original Inside Diameter	Max. Wear Limit	Maximum Machine Diameter		Bracket Bolts (ft. lbs.)	Mounting Bolts (ft. lbs.)
2002	Astro	F	①	②	0.004	—	—	—	0.030	NA	38
		R	—	—	—	9.50	9.59	9.56	0.030	NA	—
	Safari	F	①	②	0.004	—	—	—	0.030	NA	38
		R	—	—	—	9.50	9.59	9.56	0.030	NA	—
2003	Astro	F	①	②	0.004	—	—	—	0.030	NA	38
		R	—	—	—	9.50	9.59	9.56	0.030	NA	—
	Safari	F	①	②	0.004	—	—	—	0.030	NA	38
		R	—	—	—	9.50	9.59	9.56	0.030	NA	—
2004	Astro	F	1.142	1.102	0.005	—	—	—	0.030	129	80
		R	0.787	0.748	0.005	—	—	—	0.030	148	31
	Safari	F	1.142	1.102	0.005	—	—	—	0.030	129	80
		R	0.787	0.748	0.005	—	—	—	0.030	148	31
2005	Astro	F	1.142	1.102	0.005	—	—	—	0.030	129	80
		R	0.787	0.748	0.005	—	—	—	0.030	148	31
	Safari	F	1.142	1.102	0.005	—	—	—	0.030	129	80
		R	0.787	0.748	0.005	—	—	—	0.030	148	31

NA: Not Available

① Available with 1.040" and 1.250"rotors

② 1.040" rotors: 0.980
   1.250" rotors: 1.230"

06025-ASTR-C11

SCHEDULED MAINTENANCE INTERVALS
CHEVROLET—ASTRO & GMC—SAFARI

TO BE SERVICED	TYPE OF SERVICE	VEHICLE MILEAGE INTERVAL (x1000)															
		3	6	9	12	15	18	21	24	27	30	33	36	39	42	45	48
Accessory drive belt	S/I	Every 150,000 miles															
Air cleaner filter	R					✓					✓					✓	
Automatic transmission fluid	R					✓					✓					✓	
Brake system ①	S/I	✓	✓	✓	✓	✓	✓	✓	✓	✓	✓	✓	✓	✓	✓	✓	✓
Chassis & suspension grease points	L	✓	✓	✓	✓	✓	✓	✓	✓	✓	✓	✓	✓	✓	✓	✓	✓
CV-joint boots & axle seals	S/I	✓	✓	✓	✓	✓	✓	✓	✓	✓	✓	✓	✓	✓	✓	✓	✓
Engine coolant system ②	S/I	Every 150,000 miles															
Engine oil & filter	R	✓	✓	✓	✓	✓	✓	✓	✓	✓	✓	✓	✓	✓	✓	✓	✓
Front wheel bearings	S/I & L					✓					✓					✓	
Fuel filter	R	Every 30,000 miles															
Fuel tank, cap & lines	S/I								✓								✓
PCV valve	S/I	Every 100,000 miles															
Rear/front axle fluid level	S/I	✓	✓	✓	✓	✓	✓	✓	✓	✓	✓	✓	✓	✓	✓	✓	✓
Rotate tires	S/I		✓		✓		✓		✓		✓		✓		✓		✓
Spark plug wires	S/I	Every 100,000 miles															
Spark plugs	R	Every 100,000 miles															

R: Replace   S/I: Inspect and service, if necessary   L: Lubricate

① This should be performed when the tires are removed for rotation.

② Drain, flush and refill the cooling system, inspect the system hoses, and clean the radiator and condenser.

06025- ASTR-C12

ENGINE REPAIR

Distributor

REMOVAL

1. Before servicing the vehicle, refer to the Precautions Section.
2. Disconnect negative battery cable.
3. Disconnect spark plug wires and the coil leads from the distributor.
4. Disconnect electrical connector from the distributor.
5. Remove distributor cap fasteners and the cap.
6. Using a marker, matchmark the rotor-to-housing and housing-to-intake manifold positions so that they can be matched during installation.
7. Remove distributor hold-down bolt.
8. Remove distributor from the engine.
9. As the distributor is being removed from the engine the rotor will move in a counterclockwise direction about 42˚. This will appear as slightly more than one clock position.
10. Place a second mark on the distributor to mark the position of the rotor segment. This will help to ensure the correct rotor alignment when installing the distributor.

INSTALLATION

Engine Not Disturbed

1. If installing a new distributor, place two marks on the new distributor housing in the same position as the marks on the old distributor housing.
2. Align the rotor with the second mark made on the distributor.
3. Install the distributor in the engine making sure that the mounting hole in the distributor hold-down base is aligned over the mounting hole in the intake manifold.
4. As you are installing the distributor, watch the rotor move in a clockwise direction about 42˚.
5. Once the distributor is fully seated, the rotor should be aligned with the first mark made on the distributor housing. If the rotor is not aligned with the first mark made on the housing, the distributor and camshaft teeth have meshed one or more teeth out of alignment. If this is the case, remove the distributor and reinstall it so that all the marks are aligned.

6. Install hold-down bolt and tighten the bolt to 18 ft. lbs. (25 Nm).
7. Install distributor cap and engage the electrical connector to the distributor.
8. Install spark plug wires and coil leads.
9. Connect negative battery cable.

Engine Disturbed

1. Remove the No. 1 cylinder spark plug. Turn the engine using a socket wrench on the large bolt on the front of the crankshaft pulley. Place a finger near the No. 1 spark plug hole and turn the crankshaft until the piston reaches Top Dead Center (TDC). As the engine approaches TDC, you will feel air being expelled by the No. 1 cylinder. If the position is not being met, turn the engine another full turn (360 degree). Once the engine position is correct, install the spark plug.
2. Align the cast arrow in the distributor housing, the driven gear roll pin and the pre-drilled indent hole in the distributor driven gear. If the driven gear is installed correctly, the dimple will be approximately 180˚ opposite the rotor segment when it is installed in the distributor.

➡**Installing the distributor 180˚ out of alignment, or locating the rotor in the wrong holes, may cause a no start condition or can cause premature engine damage and wear.**

3. Make sure the rotor is pointing to the cap hold-down mount nearest the flat side of the housing.
4. Using a long screwdriver, align the oil pump drive shaft in the engine in the mating drive tab in the distributor.
5. Install the distributor in the engine. Make sure the spark plug towers are perpendicular to the centerline of the engine.
6. When the distributor is fully seated, the rotor segment should be aligned with the pointer cast in the distributor base. The pointer will have a "6" cast into it indicating a 6 cylinder engine. If the rotor segment is not within a few degrees of the pointer, the distributor gear may be off a tooth or more. If this is the case repeat the process until the rotor aligns with the pointer.
7. Install the cap and fasten the mounting screws.
8. Tighten the distributor mounting bolt to 18 ft. lbs. (25 Nm).
9. Engage the electrical connections and the spark plug wires.

Alternator

REMOVAL

1. Before servicing the vehicle, refer to the Precautions Section.
2. Disconnect negative battery cable.
3. Remove air intake assembly.
4. Remove accessory belt.
5. Remove heater hose brace, if necessary.
6. Disconnect wires.
7. Remove mounting bolts.
8. Remove alternator.

INSTALLATION

1. Install alternator and loosely install the mounting bolts.
2. Tighten the rear bolt to 37 ft. lbs. (50 Nm) and the front bolt to 18 ft. lbs. (25 Nm).
3. Tighten the brace-to-alternator and brace-to-intake retainers to 18 ft. lbs. (25 Nm). Tighten the brace-to-engine stud nut to 37 ft. lbs. (50 Nm).
4. Connect wires and the battery feed wire nut and tighten to 89 inch lbs.
5. Install heater hose bracket.
6. Install accessory belt.
7. Install air intake assembly.
8. Connect negative battery cable.

Ignition Timing

ADJUSTMENT

The ignition timing is preset and cannot be adjusted.

Engine Assembly

REMOVAL & INSTALLATION

➡**The engine assembly is removed from the bottom of the vehicle. A special engine lifting table is necessary to perform the following procedure.**

1. Before servicing the vehicle, refer to the Precautions Section.
2. Discharge the air conditioning refrigerant.
3. Drain the coolant.
4. Drain the crankcase.
5. Disconnect the battery cables.
6. Remove engine cover.
7. Remove battery.

8. Remove air intake assembly

9. Disconnect throttle cable and the cruise control cable (if equipped) from the throttle body.

10. Disconnect air conditioning lines at the condenser and accumulator.

11. Remove radiator.

12. Remove power steering reservoir and drain the fluid.

13. Disconnect lines from the Hydroboost unit.

14. Remove master cylinder from the Hydroboost unit and secure it to the oil fill tube.

15. Separate steering shaft from the steering gear.

16. Disconnect heater hoses and vacuum lines from the engine.

17. Remove fuse box and wiring harness from the bulkhead connector.

➡The engine/transmission assembly is removed from the bottom of the vehicle. Raise the vehicle so the rear of the vehicle is slightly higher than the front. When the frame bolts are removed, the body will be lifted away from the engine/transmission assembly.

18. Remove driveshaft. Matchmark it for reassembly prior to removal.

19. Remove starter and the starter opening cover.

20. Remove torque converter bolts through the starter opening.

21. Disconnect shift linkage from the transmission.

22. Remove exhaust pipe from the rear of the catalytic converter.

23. Remove parking brake bracket from the frame.

24. Disconnect rear brake line from the Brake Pressure Modulator Valve (BPMV).

25. Remove front bumper and the power steering cooler from the front air deflector.

26. Remove Supplemental Inflatable Restraint (SIR) connector.

27. Remove splash shields from the wheel openings.

28. Remove rear air conditioning lines at the rear crossmember, if equipped, leave the lines attached to the engine assembly.

29. Disconnect fuel lines at the filter and pull them through the crossmember.

30. Disconnect fuel tank electrical connector.

31. Remove transfer case vent tube, on all wheel drive models.

32. Make sure all lines and connections are free between the engine/transmission assembly and the body.

33. If using a twin post lift (side lift), perform the following:

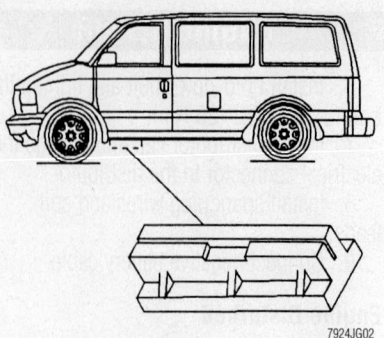

7924JG02

Attach the body protection pads to the pinch welds on both sides before raising the vehicle

a. Step 1: Lower the vehicle to the floor.

b. Step 2: Install the body protection lift adapter tool J 41602 pads to the pinch welds on both sides of the vehicle behind the front wheels.

c. Step 3: Position the front lifting arms of the lift under the body protection adapters.

d. Step 4: Be sure the rear of the vehicle will be slightly higher than the front when the lift is raised.

e. Step 5: Raise the lift about halfway up.

f. Step 6: Place jackstands under the

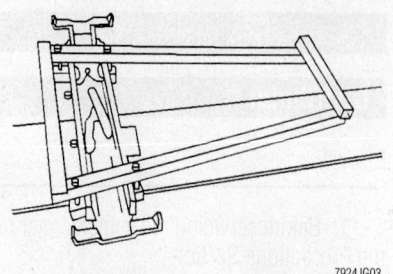

7924JG03

Install the engine lifting adapter to the front cylinder of the twin cylinder lift if applicable

frame attached to the engine/transmission assembly and remove the frame mounting bolts.

g. Step 7: Raise the vehicle to clear the engine/transmission assembly.

34. If using a dual cylinder (1 front and 1 rear) lift do the following:

a. Step 1: Install the body protection lift adapter tool J 41602 pads to the pinch welds on both sides of the vehicle behind the front wheels.

b. Step 2: Install stands under the body protection lift adapters and the rear of the vehicle.

c. Step 3: Lower the front cylinder of the lift and install the engine lifting adapter tool J 41617 to the lift.

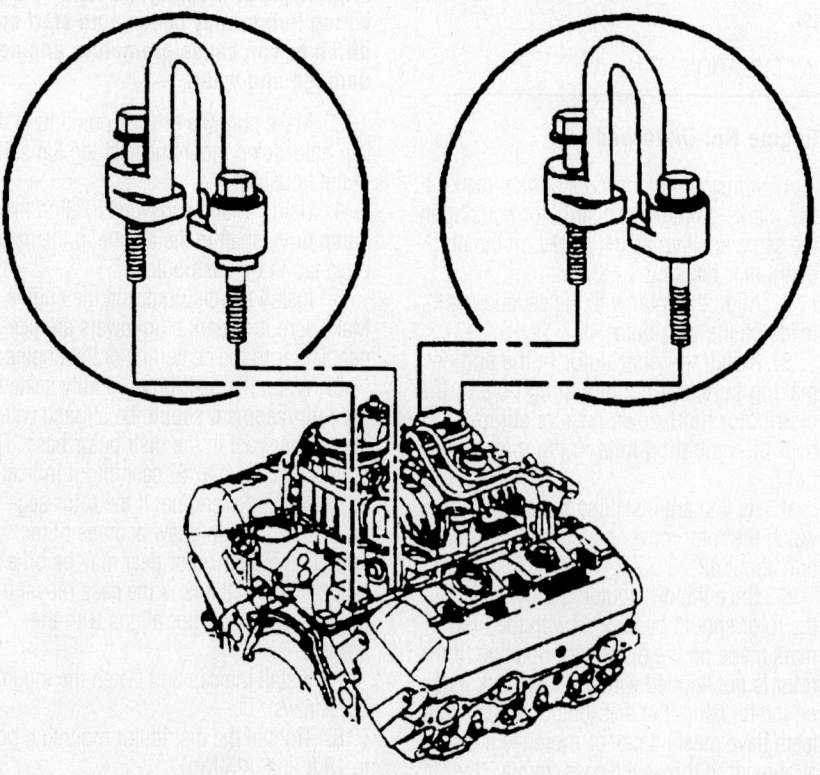

7924JG04

Install the engine lifting brackets at the right rear and left front of the intake manifold

d. Step 4: Raise the front cylinder with the adapter attached until it touches the engine/transmission assembly.

e. Step 5: Remove the frame bolts and lower the engine/transmission assembly from the vehicle.

35. Remove the 2 right rear and 2 left front intake manifold bolts.

36. Install the Engine Lifting Bracket tools J 41427 on the intake manifold to provide lifting points for an engine hoist. Install the hoist and raise the engine/transmission assembly.

37. Remove transmission from the engine.

38. Remove engine from the frame.

To install:

39. Install engine onto the frame and the transmission to the engine. Tighten the engine mount through-bolts to 74 ft. lbs. (100 Nm).

40. Install engine/transmission assembly on suitable stands or on the engine lifting adapter.

41. Remove the engine lifting brackets from the intake manifold and reinstall the bolts.

42. Position the engine/transmission assembly in the vehicle. Tighten the frame bolts in the following order:

a. Step 1: Right center bolt: 114 ft. lbs. (155 Nm).

b. Step 2: Left center bolt: 114 ft. lbs. (155 Nm).

c. Step 3: Right front bolt: 66 ft. lbs. (90 Nm).

d. Step 4: Left rear bolt: 66 ft. lbs. (90 Nm).

e. Step 5: Left front bolt: 66 ft. lbs. (90 Nm).

f. Step 6: Right rear bolt: 66 ft. lbs. (90 Nm).

43. Remove the stands or the engine lifting adapter. If the engine lifting adapter was used, raise the vehicle and remove the stands.

44. Remove the body protection adapter from the pinch welds.

45. Install splash shields in the wheel openings.

46. Connect steering shaft to the steering gear.

47. Install power steering cooler.

48. Connect lines to the Hydroboost unit.

49. Connect hose to the power steering reservoir.

50. Connect wiring harness to the bulkhead and fuse box.

51. Install heater hoses and the SIR connector.

52. Connect throttle cable and the cruise control cable (if equipped).

53. Install radiator and air cleaner assembly.

54. Install master cylinder to the booster.

55. Connect rear brake line to the BMPV.

56. Install parking brake bracket to the frame.

57. Connect fuel lines and the air conditioning lines (if equipped) at the rear crossmember.

58. Install transfer case vent tube, on all wheel drive models.

59. Install front bumper and transmission shift linkage.

60. Install torque converter bolts.

61. Install starter and driveshaft.

62. Install exhaust pipe.

63. Install engine cover and battery. Connect the positive battery cable first.

64. Refill the power steering, engine crankcase, brake system, cooling system and transmission.

65. Discharge the air conditioning system.

66. Bleed the brake system.

67. Start the engine and check for leaks.

Water Pump

REMOVAL & INSTALLATION

1. Before servicing the vehicle, refer to the Precautions Section.

2. Disconnect the negative battery cable.

3. Drain the engine cooling system.

4. Remove or disconnect the following:

- Air intake assembly
- Mass Air Flow (MAF) sensor
- Upper fan shroud
- Drive belt
- Fan and clutch assembly
- Water pump pulley using special tool J 41240

- Coolant hoses
- Water pump

➡On some engines, the pump retaining bolts will vary in size and thread. Be sure to note the positioning of all bolts during removal to assure proper installation.

5. Clean gasket mounting surface.

To install:

6. Install or connect the following:

- Water pump. Torque bolts to 33 ft. lbs. (45 Nm).
- Coolant hoses using new clamps
- Water pump pulley. Tighten bolts to 18 ft. lbs. (25 Nm).
- Fan and clutch assembly
- Drive belt
- Fan shroud
- MAF sensor
- Air intake assembly
- Negative battery cable

7. Refill the cooling system.

8. Start the engine and check for leaks.

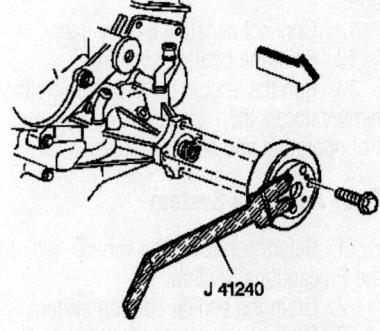

J 41240

06025-ASTR-G01

Removing the water pump pulley using special tool J-41240

Water pump mounting—2002 shown

9358LG01

Heater Core

REMOVAL & INSTALLATION

Front System

1. Before servicing the vehicle, refer to the Precautions Section.
2. Drain the engine cooling system.
3. Disconnect negative battery cable.
4. Remove heater hoses from the heater core.
5. Remove heater core cover-to-heater assembly screws and the cover.
6. Remove heater core-to-heater assembly strap screws and the straps.
7. Remove heater core.

To install:

8. Install heater core.
9. Install heater core straps and the straps-to-heater assembly screws, then tighten the screws to 18 inch lbs. (2 Nm).
10. Install heater core cover and the cover-to-heater assembly screws, then tighten the screws to 18 inch lbs. (2 Nm).
11. Install heater hoses to the heater core.
12. Connect negative battery cable.
13. Refill the cooling system.
14. Run the engine to normal operating temperatures; then, check the climate control operation and check for leaks.

Rear Auxiliary System

1. Before servicing the vehicle, refer to the Precautions Section.
2. Drain the engine cooling system.
3. Disconnect negative battery cable.
4. Remove body side front lower interior trim panel.
5. Loosen clamps from the rear auxiliary heater hoses.
6. Disconnect heater hoses from the rear auxiliary heater case.
7. Remove screws and band clamp from the right side of the heater core.
8. Remove screw and band clamp from the left side of the heater core.

➡**Place a cloth on the floor to catch any coolant that may spill from the heater core.**

9. Remove heater core from the rear auxiliary case assembly.
10. Remove seals from the heater core.

To install:

11. Install new seals to the heater core.
12. Install heater core to the rear auxiliary case assembly.
13. Install screws and band clamps to

the right and left sides of the heater core, then tighten to 18 inch lbs. (2 Nm).
14. Connect heater hoses to the rear auxiliary heater case.
15. Tighten clamps to the rear auxiliary heater hoses.
16. Install body side front lower interior trim panel.
17. Connect negative battery cable.
18. Refill the engine cooling system.
19. Run the engine to normal operating temperatures; then, check the climate control operation and check for leaks.

Cylinder Head

REMOVAL & INSTALLATION

Right Side

1. Before servicing the vehicle, refer to the Precautions Section.
2. Properly relieve the fuel system pressure.
3. Drain the cooling system.
4. Remove or disconnect the following:

- Negative battery cable
- Engine cover
- Engine cooling fan
- Air intake assembly
- PCV hose from valve cover
- Valve cover
- Intake manifold
- Spark plugs
- Spark plug wire support
- Exhaust manifold
- Alternator mounting bracket
- Alternator mounting bracket stud from cylinder head
- Engine wiring harness bolt/clip at the rear of the cylinder head
- Rocker arms
- Pushrods

➡**If valve train components, such as the rocker arms or pushrods, are to be reused, they must be tagged or arranged to insure installation in their original locations.**

5. Remove cylinder head bolts by loosening them in the reverse of the torque sequence.
6. Remove cylinder head.

To install:

7. Clean and inspect the gasket mounting surfaces.

➡**Do not apply sealer to composition steel/asbestos gaskets. If using a steel only gasket, apply a thin and even coat of sealer to both sides of the gaskets.**

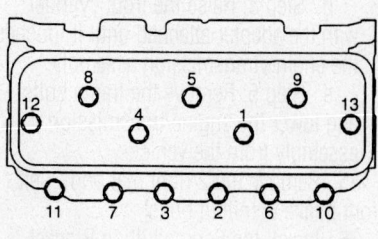

7924JG08

Cylinder head bolt torque sequence—4.3L engine

8. Install new gasket over the dowel pins with the bead or the words **This Side Up** facing upwards, as applicable.
9. Install cylinder head.
10. Using new bolts, coat the bolts with GM sealer 12346004 , then install the bolts and tighten in sequence to 22 ft. lbs. (30 Nm). The bolts must then be tightened again in sequence in the following order:
   a. Short length bolts: (11, 7, 3, 2, 6, 10) 55 degrees.
   b. Medium length bolts: (12, 13) 65 degrees.
   c. Long length bolts: (1, 4, 8, 5, 9) 75 degrees.
11. Install or connect the following:

- Pushrods
- Rocker arms
- Exhaust manifold
- Spark plug wire harness and support. Tighten the bolts to 106 inch lbs. (12 Nm).
- Intake manifold
- Alternator mounting bracket stud and tighten to 15 ft. lbs. (20 Nm)
- Alternator mounting bracket
- Engine wiring harness to the rear of the head. Tighten mounting bolt to 27 ft. lbs. (36 Nm).
- Valve cover
- PCV hose
- Air intake assembly
- Engine cooling fan
- Engine cover
- Negative battery cable

12. Refill the engine cooling system.
13. Start the engine to check for leaks
14. Check and/or adjust the ignition timing.

Left Side

1. Before servicing the vehicle, refer to the Precautions Section.
2. Properly relieve the fuel system pressure.
3. Drain the engine cooling system.
4. Properly discharge the air conditioning system, if equipped.

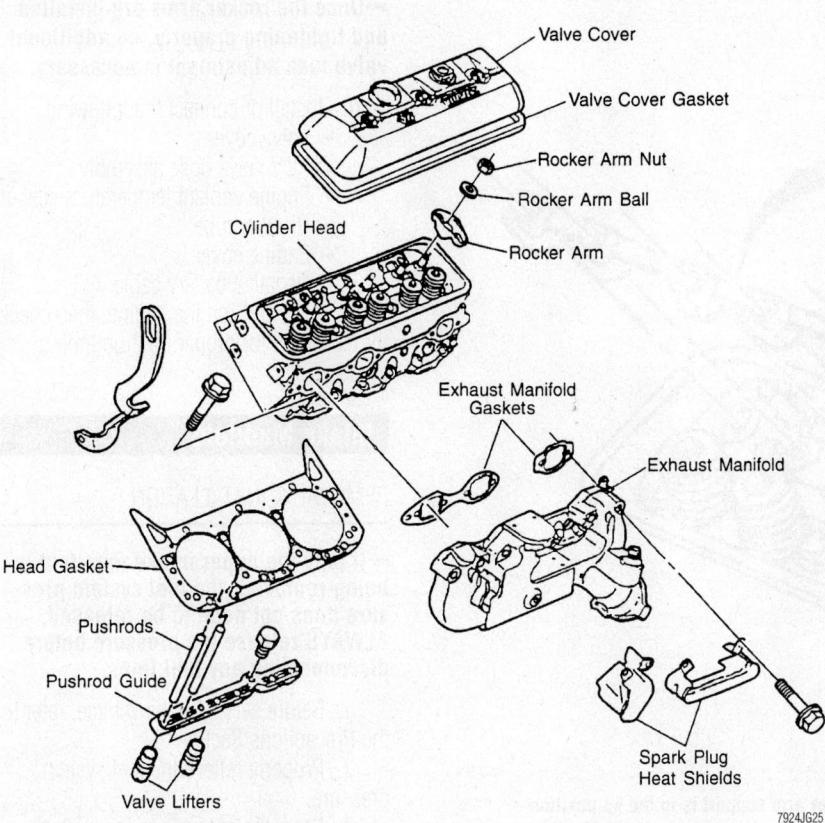

Cylinder head and related components—4.3L engine

5. Remove or disconnect the following:
- Negative battery cable
- Engine cover
- Engine cooling fan
- Accessory drive belt
- Air conditioning compressor, if equipped

➡**The power steering pump can stay mounted on the bracket and the hoses should not be disconnected from the pump.**

- Nut holding the power steering pump rear bracket to the engine
- Power steering pump mounting bracket stud from the cylinder head.
- Intake manifold
- Exhaust manifold
- Ground wires at the rear of the left cylinder head
- A/C pipe bracket nut, if equipped
- Fuel pipe bracket bolt from the rear of the cylinder head
- Engine coolant temperature sensor
- Spark plugs
- Spark plug wire support
- Valve cover
- Pushrods by loosening the rocker arms

➡**If valve train components, such as the rocker arms or pushrods, are to be reused, they must be tagged or arranged to insure installation in their original locations.**

6. Remove cylinder head bolts by loosening them in the reverse of the torque sequence.
7. Remove cylinder head.
To install:
8. Clean and inspect the gasket mounting surfaces.

➡**Do not apply sealer to composition steel/asbestos gaskets. If using a steel only gasket, apply a thin and even coat of sealer to both sides of the gaskets.**

9. Install new gasket over the dowel pins with the bead or the words **This Side Up** facing upwards, as applicable.
10. Install the cylinder head.
11. Coat the bolts with GM sealer 12346004 , then install the bolts and tighten in sequence to 22 ft. lbs. (30 Nm). The bolts must then be tightened again in sequence in the following order:
   a. Short length bolts: (11, 7, 3, 2, 6, 10) 55 degrees.
   b. Medium length bolts: (12, 13) 65 degrees.

   c. Long length bolts: (1, 4, 8, 5, 9) 75 degrees.
12. Install pushrods, secure the rocker arms and adjust the valves.
13. Install or connect the following:
- Valve cover
- Spark plug wire support
- Spark plugs
- Engine coolant temperature sensor
- Ground wires and bolt at the rear of the head. Tighten the bolt to 26 ft. lbs. (35 Nm).
- Fuel pipe bracket and stud. Tighten the stud to 24 ft. lbs. (35 Nm).
- A/C pipe bracket and nut, and tighten to 26 ft. lbs. (35 Nm).
- Exhaust manifold
- Intake manifold
- Power steering pump mounting bracket stud. Tighten to 15 ft. lbs. (20 Nm).
- Power steering pump mounting bracket. Tighten the nut to 30 ft. lbs. (41 Nm).
- Air conditioning compressor, if equipped.
- Accessory drive belt
- Engine cooling fan
- Engine cover
- Negative battery cable
14. Refill the engine cooling system.
15. Start the engine to check for leaks.
16. Check and/or adjust the ignition timing.

Rocker Arms

REMOVAL & INSTALLATION

➡**Make sure to keep all valvetrain components in order as you remove them. They must be reinstalled in their original locations.**

1. Before servicing the vehicle, refer to the Precautions Section.
2. Remove or disconnect the following:
- Negative battery cable
- Engine cover
- Oil filler tube
- Engine coolant temperature sensor
- PCV valve hose assembly
- Valve cover
- Rocker arm
- Rocker arm supports
- Pushrod(s)
To install:
3. Inspect and replace components if worn or damaged.
4. Install pushrods making sure they seat properly in the lifter.
5. Install rocker arm supports.

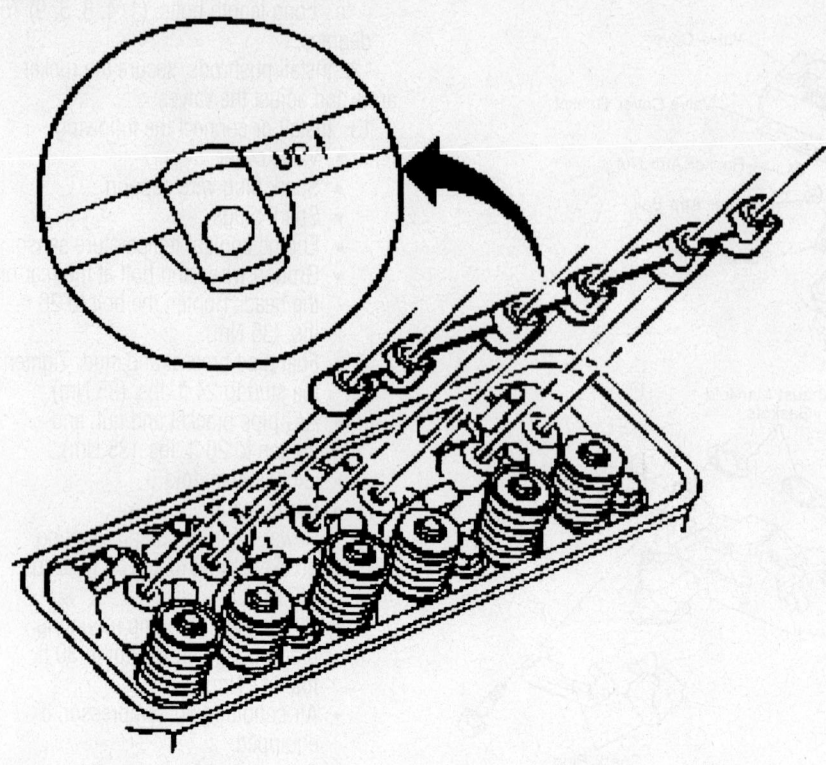

During installation, make sure the arrow on the rocker arm support is in the up position

9358LG03

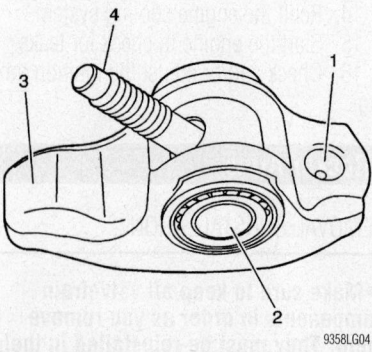

Prelube application locations on the rocker arm

9358LG04

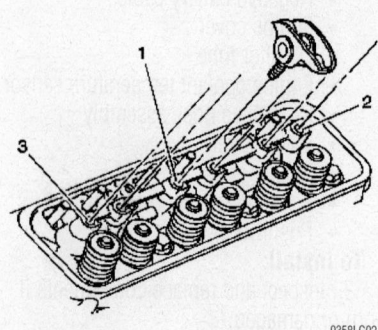

Proper rocker arm installation—2002–03 vehicles

9358LG02

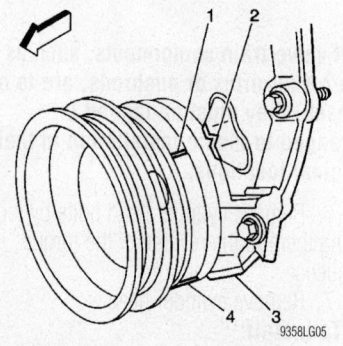

Proper crankshaft balancer alignment—2002–03 vehicles

9358LG05

6. Apply a suitable prelube to the following rocker arm contact surfaces:
   a. Valve pushrod socket (1).
   b. Roller pivot (2).
   c. Valve stem tip (3).
7. Install the rocker arms as follows:
   a. Finger-start bolt at locations 1, 2 and then 3.
   b. Finger-start the remaining rocker arm bolts.
8. Rotate the crankshaft balancer to position the alignment mark (1) 57–63 degrees clockwise or counterclockwise from the engine front cover alignment tab.
9. Tighten the rocker arm bolts to 22 ft. lbs. (30 Nm).

➡Once the rocker arms are installed and tightening properly, no additional valve lash adjustment is necessary.

10. Install or connect the following:
    • Valve cover
    • PCV valve hose assembly
    • Engine coolant temperature sensor
    • Oil filler tube
    • Engine cover
    • Negative battery cable
11. Start and run the engine, then check for leaks and for proper ignition timing adjustment.

Intake Manifold

REMOVAL & INSTALLATION

➡If only the upper intake manifold is being removed, the fuel system pressure does not need to be released. ALWAYS release the pressure before disconnecting any fuel lines.

1. Before servicing the vehicle, refer to the Precautions Section.
2. Properly relieve the fuel system pressure.
3. Drain the engine cooling system.
4. Remove or disconnect the following:
   • Negative battery cable
   • Engine cover
   • Air intake assembly
   • Throttle body
   • Throttle linkage from the upper intake manifold
   • Positive Crankcase Ventilation (PCV) hose at the rear of the upper intake manifold
   • Engine wiring harness bracket
   • A/C compressor clutch connector
   • A/C high pressure cutoff switch connector
   • Throttle position (TP) sensor
   • Idle air control (IAC) motor connector
   • Fuel meter body assembly connector
   • Manifold absolute pressure (MAP) sensor
   • EVAP canister purge solenoid valve
   • Ground strap
   • Vacuum hoses from both the front and rear of the upper intake
   • Automatic transmission dipstick tube
   • Brake booster vacuum hose at the upper intake manifold
   • Upper intake manifold
   • Lower manifold wiring harness
   • Distributor or High Voltage Switch (HVS) assembly

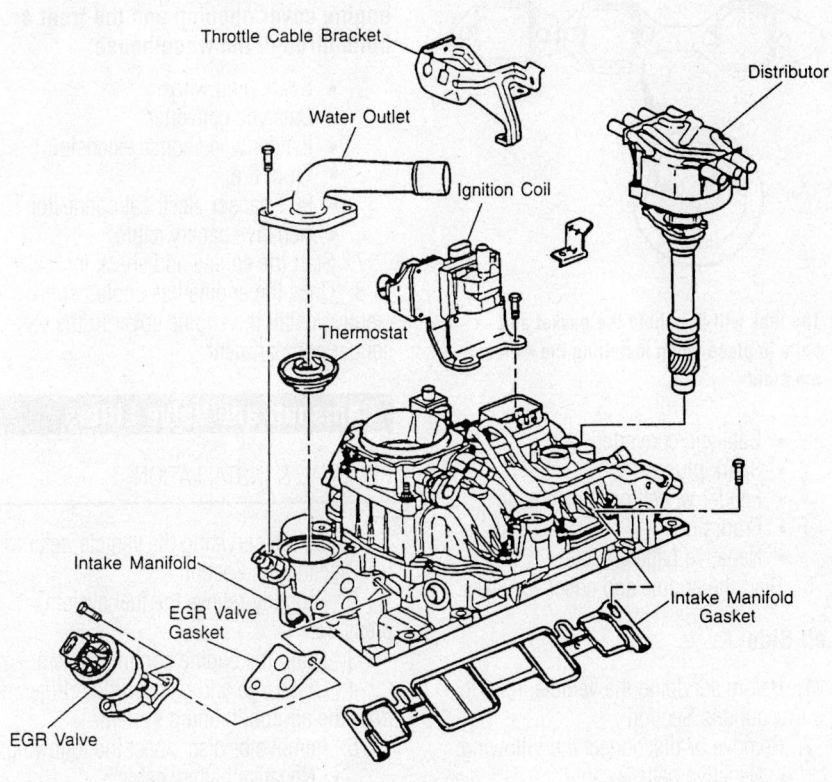

Intake manifold and related components—4.3L engine

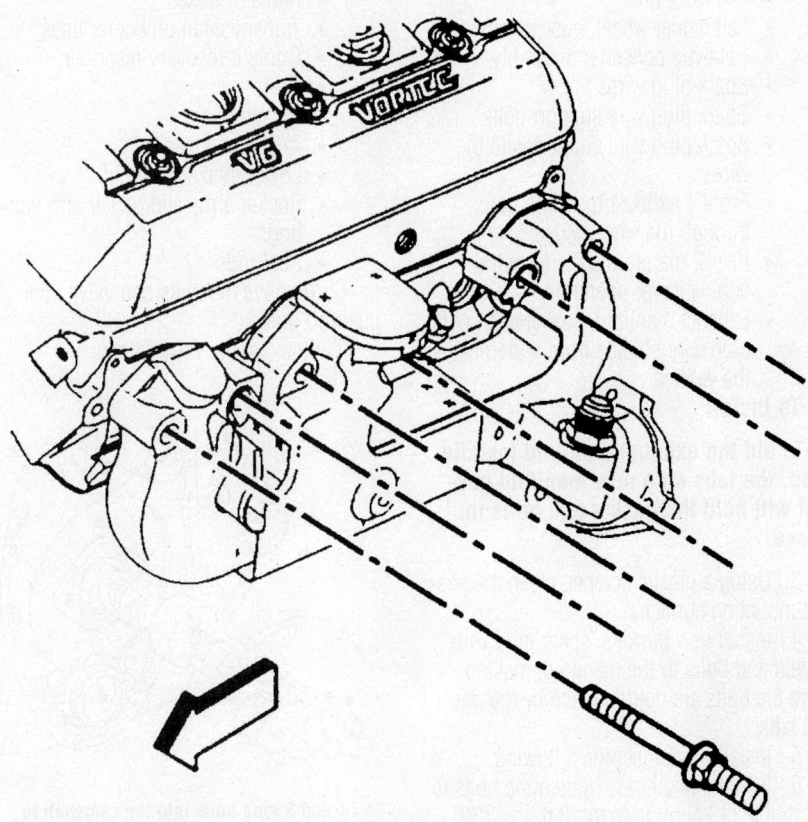

Lower intake manifold tightening sequence—4.3L engines

- Fuel lines and bracket from the rear of the lower intake manifold
- Ignition coil
- Upper radiator hose at the thermostat housing
- Heater hose at the lower intake manifold
- Air conditioning compressor bracket-to-lower intake manifold pencil brace
- Alternator bracket bolts near the thermostat housing

5. Loosen the power steering pump rear bracket and slide the power steering pump forward.

6. Remove the lower intake manifold.

To install:

7. Clean the gasket mounting surfaces. Be sure to inspect the manifold for warpage and/or cracks. If necessary, replace it.

8. Position the gaskets on the cylinder head with the port blocking plates to the rear and the **This Side Up** stamps facing upward. Then apply a ³⁄₁₆ in. (5mm) bead of RTV sealant on the front and rear of the engine block at the block-to-manifold mating surface. Extend the bead ½ in. (13mm) up each cylinder head to seal and retain the gaskets.

9. Install the lower intake manifold. Tighten the bolts in sequence and in 3 steps, as follows:
   a. Step 1: 26 inch lbs. (3 Nm).
   b. Step 2: 106 inch lbs. (12 Nm).
   c. Step 3: 11 ft. lbs. (15 Nm).

10. Install or connect the following:
- Power steering pump
- Alternator bracket bolt near the thermostat housing
- Wiring harness to the lower manifold components, including the injector, EGR valve and ECT sensor
- Air conditioning compressor bracket-to-the lower intake manifold pencil braces

11. Install transmission oil dipstick tube, if necessary.
- Fuel supply and return lines to the rear of the lower intake

12. Temporarily reattach the negative battery cable, then pressurize the fuel system (by cycling the ignition without starting the engine) and check for leaks.

13. Disconnect the negative battery cable.

14. Install or connect the following:
- Heater hose to the lower intake
- Upper radiator hose to the thermostat housing
- Distributor assembly and engage the wiring
- Ignition coil

- Connect vacuum hoses to the upper and lower intake manifold.
- New upper intake manifold gasket, making sure the green sealing lines are facing upward
- Upper intake manifold
- Manifold retainers. Tighten them to 88 inch lbs. (10 Nm) using two passes.
- Purge solenoid and bracket
- Brake booster vacuum hose at the upper intake manifold
- PCV hose to the rear of the upper intake manifold
- Vacuum hoses to both the front and rear of the manifold assembly
- Throttle body
- Throttle linkage to the upper intake
- Wiring to the upper intake components including the TP sensor, IAC motor, MAP sensor and the fuel meter
- Engine cover
- Air intake assembly

15. Connect negative battery cable.
16. Refill the engine cooling system.
17. Start the engine and check for leaks.

Exhaust Manifold

REMOVAL & INSTALLATION

Right Side

1. Before servicing the vehicle, refer to the Precautions Section.
2. Remove or disconnect the following:
   - Negative battery cable
   - Right front tire
   - Right fender wheelhouse extension
   - Spark plug wires
   - Catalytic converter
   - Exhaust manifold bolts through the wheelhouse
   - Exhaust manifold, with the gaskets and spark plug wire shields

To install:

➡**To aid the exhaust manifold installation, the tabs on a new manifold gasket will hold the gasket and bolts in place.**

3. Using a plastic scraper, clean the gasket mounting surfaces.
4. Install new gaskets, spark plug wire shield and bolts to the manifold, making sure the bolts are held in place by the gasket tabs.
5. Install or connect the following:
   - Exhaust manifold. Tighten the bolts to 11 ft. lbs. (15 Nm), then to 22 ft. lbs. (30 Nm).

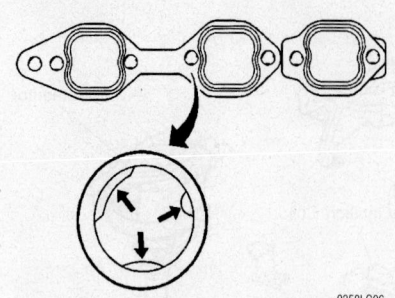

9358LG06

The tabs will help hold the gasket and bolts in place when installing the exhaust manifold

- Catalytic converter
- Spark plug wires
- Fender wheelhouse extension
- Front tire
- Negative battery cable

6. Start the engine and check for leaks.

Left Side

1. Before servicing the vehicle, refer to the Precautions Section.
2. Remove or disconnect the following:
   - Negative battery cable
   - Engine cover
   - Engine Coolant Temperature (ECT) sensor connector
   - Left front tire
   - Left fender wheelhouse extension
   - Catalytic converter assembly
   - Spark plug wires
   - Spark plug wire support bolts
   - Spark plug wire support with the wires
   - Front 4 exhaust manifold bolts through the wheelhouse
   - Rear 2 manifold bolts from the engine cover opening
   - Exhaust manifold, gasket and spark plug wire shields from underneath the vehicle

To install:

➡**To aid the exhaust manifold installation, the tabs on a new manifold gasket will hold the gasket and bolts in place.**

3. Using a plastic scraper, clean the gasket mounting surfaces.
4. Install new gaskets, spark plug wire shield and bolts to the manifold, making sure the bolts are held in place by the gasket tabs.
5. Install or connect the following:
6. Exhaust manifold. Tighten the bolts to 11 ft. lbs. (15 Nm), then to 22 ft. lbs. (30 Nm).

➡**Tighten the 2 rear bolts through the engine cover opening and the front 4 bolts through the wheelhouse.**

- Spark plug wires
- Catalytic converter
- Fender wheelhouse extension
- Front tire
- ECT sensor electrical connector
- Negative battery cable

7. Start the engine and check for leaks.
8. Once the engine has cooled sufficiently, install the engine cover to the passenger compartment.

Camshaft and Valve Lifters

REMOVAL & INSTALLATION

1. Before servicing the vehicle, refer to the Precautions Section.
2. Properly relieve the fuel system pressure.
3. Drain the engine cooling system.
4. Discharge and recover the refrigerant from the air conditioning system.
5. Remove or disconnect the following:
   - Negative battery cable
   - Engine cover
   - Air intake assembly
   - Engine cooling fan and shroud
   - Radiator hoses
   - Transmission oil cooler lines
   - Coolant recovery reservoir
   - Grille
   - Radiator
   - A/C condenser
   - Lower intake manifold
   - Rocker arms and rocker arm supports
   - Pushrods

6. Remove the bolts and valve lifter pushrod guide.
7. Remove the valve lifters.

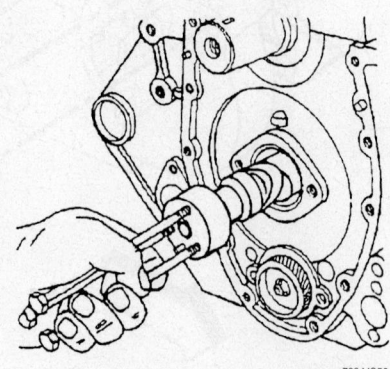

7924JG50

Thread 3 long bolts into the camshaft to use as a handle, then withdraw it from the engine

8. Remove or disconnect the following:
- Timing chain
- Camshaft sprocket
- Balance shift drive gear, if equipped
- Thrust plate

9. Install three 5/16 x 4 inch bolts into the engine camshaft front bolt holes.

10. Using the bolts as a handle, carefully rotate and pull the camshaft out of the camshaft bearings.

➡**Take care not to damage the camshaft bearings when removing the camshaft.**

To install:

11. Lubricate the camshaft journals with clean engine oil or a suitable pre-lube.

12. Install camshaft being extremely careful not to contact the bearings with the cam lobes.

13. Install or connect the following:
- Thrust plate. Torque the bolts to 106 inch lbs. (12 Nm).
- Balance shaft drive gear. Tighten the bolts to 15 ft. lbs. (20 Nm) plus 35 degrees.
- Camshaft sprocket
- Timing chain
- Valve lifters
- Valve lifter pushrod guide and tighten the bolts to 12 ft. lbs. (16 Nm)
- Pushrods
- Rocker arms
- Lower intake manifold assembly
- A/C condenser
- Radiator
- Grille
- Coolant recovery reservoir
- Transmission oil cooler lines
- Radiator hoses
- Engine cooling fan and shroud
- Air intake assembly
- Engine cover
- Negative battery cable

14. Refill the engine cooling system.

15. Start the engine and check for leaks.

Valve Lash

ADJUSTMENT

The 4.3L engines are equipped with screw-in rocker arm studs with positive stop shoulders. Because the shoulders that allow the rocker arms to be tightened into proper position, no adjustments are necessary or possible. If a valve train problem is suspected, check that the rocker arm nuts are tightened to 18 ft. lbs. (24 Nm). When valve lash falls out of specification (valve tap is heard), replace the rocker arm, pushrod and hydraulic lifter on the offending cylinder.

Starter Motor

REMOVAL & INSTALLATION

1. Before servicing the vehicle, refer to the Precautions Section.

2. Disconnect the negative battery cable.

3. Remove starter motor mounting bolts.

4. Remove shims, if necessary.

5. Disconnect starter electrical connections, after partially lowering the starter.

6. Remove starter motor.

To install:

7. Connect the starter leads to the starter as follows:

   a. Loosely install the starter enable relay lead (1) to the starter solenoid terminal. Align the terminal retaining tab to the starter solenoid.

   b. Loosely install alternator output (BAT) lead (3) and the positive battery cable (2) to the solenoid terminal. Align the positive battery cable terminal retaining tab to the starter solenoid.

   c. Tighten the starter enable relay lead nut to 18 inch lbs. (2 Nm) and the positive battery cable nut to 14 ft. lbs. (19 Nm).

8. Place starter motor into position.

9. Install starter motor mounting bolts but do not tighten it at this time.

10. Install starter motor shims, if equipped.

11. Install outboard starter motor bolt. Tighten the bolts to 32 ft. lbs. (43 Nm).

12. Connect the negative battery cable.

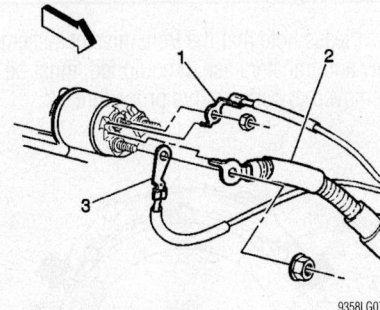

Starter motor wiring

Oil Pan

REMOVAL & INSTALLATION

1. Before servicing the vehicle, refer to the Precautions Section.

2. Drain the engine oil.

3. Remove or disconnect the following:
- Negative battery cable
- Oil level indicator
- Oil filter
- Bolt holding the oil cooler pipes bracket to the oil pan
- Oil filter adapter
- Bolt holding the bracket for the starter wiring harness and the transmission cooler pipes
- Crankshaft Position (CKP) sensor wiring harness from the retainer
- Starter
- Transmission cover
- Inner axle shaft housing support bracket-to-frame nuts and washers (AWD only)
- Front differential carrier upper and lower mounting nuts and bolts (AWD only)
- Lower the front differential carrier assembly only enough for removal of the oil pan
- Access plugs for the oil pan rear nuts
- Transmission-to-oil pan bolts
- Oil pan

To install:

➡**Any time the transmission and the engine oil pan are off of the engine at the same time, install the transmission before the oil pan. This is to allow for the proper oil pan alignment. Failure to achieve the correct oil pan alignment can result in transmission failure.**

4. Clean all sealing surfaces on the engine and the oil pan.

5. Install oil pan to the engine.

6. Install transmission to oil pan bolts and nuts. Tighten the bolts to 34 ft. lbs. (47 Nm).

7. Install access plugs for the oil pan rear nuts.

8. Install carrier upper and lower mounting bolts and nuts (AWD only).

9. Install inner axle shaft housing sup-

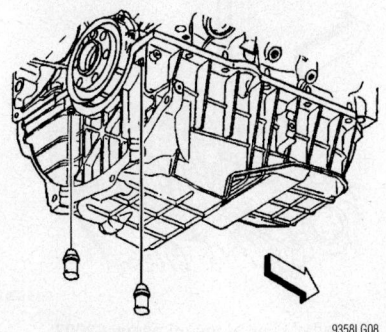

Oil pan mounting

port bracket to frame washers and nuts (AWD only).

10. Install transmission cover. Tighten the bolts to 106 inch lbs. (12 Nm).

11. Install starter motor.

12. Install bolt holding the bracket for the starter wiring harness and the transmission cooler pipes. Tighten the bolts to 89 inch lbs. (10 Nm).

13. Install CKP sensor wiring harness in the retainer.

14. Install oil filter adapter.

15. Install oil cooler pipes bracket bolt. Tighten the bolts to 89 inch lbs. (10 Nm).

16. Install oil filter.

17. Install oil pan drain plug and tighten 18 ft. lbs. (25 Nm).

18. Connect the negative battery cable.

19. Fill the crankcase with engine oil.

20. Start the engine and check for leaks.

Oil Pump

REMOVAL & INSTALLATION

1. Before servicing the vehicle, refer to the Precautions Section.

2. Remove oil pan.

3. Remove oil pump bolt, if necessary.

4. Remove oil pump and the pickup tube/shaft, if equipped.

➡ **Be careful not to crack the retainer.**

To install:

5. Ensure that the pump pickup tube is tight in the pump body. If the tube should come loose, oil pressure will be lost and oil starvation will occur. If the pickup tube is loose it should be replaced.

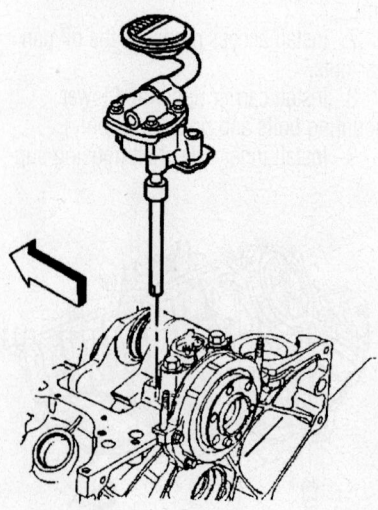

Exploded view of the oil pump—2002 model shown, others similar

9358LG09

6. If the pump has been disassembled and is being replaced or for any reason oil has been removed, it must be primed. It can either be filled with oil before installing the cover plate and oil kept within the pump during handling or the entire pump cavity can be filled with petroleum jelly.

❄ WARNING

If the pump is not primed, the engine could be damaged upon start up.

7. Install oil pump by aligning the pump shaft with the distributor drive gear as necessary. Tighten oil pump/pickup tube retainer(s) to 65 ft. lbs. (90 Nm).

➡**If the oil pump does not build up oil pressure almost immediately, remove the pan and check for a loose oil pump-to-pickup tube attachment. If necessary dismantle the pump and pack the pump cavity with petroleum jelly.**

8. Install oil pan.

9. Refill the crankcase.

10. Disable the ignition system; crank engine for approximately 10 seconds to aid in priming the oil pump and reducing the risk of engine damage.

❄ WARNING

Running the engine without measurable oil pressure will cause extensive damage.

Rear Main Seal

REMOVAL & INSTALLATION

Please note that the transmission assembly and transfer case, if equipped, must be removed to perform this procedure.

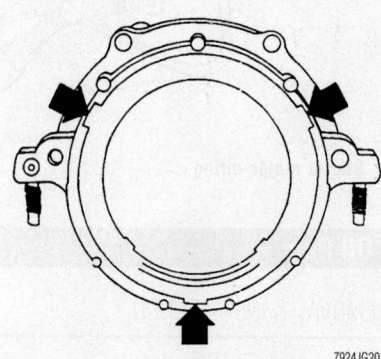

Carefully pry the rear main seal out of the retainer—4.3L engine

7924JG20

1. Before servicing the vehicle, refer to the Precautions Section.

2. Disconnect negative battery cable.

3. Remove transfer case, if equipped.

4. Remove transmission.

5. Remove clutch assembly/flywheel or flexplate.

6. Remove the crankshaft rear oil seal by inserting a suitable prying tool into the notches provided in the seal retainer and prying the seal out. Take care not to damage the crankshaft sealing surface.

To install:

7. Inspect the crankshaft for grit, rust or burrs and correct as necessary.

8. Clean the running surface of the crankshaft with a non-abrasive cleaner.

9. Install new rear seal lubricated with engine oil and a seal installer.

10. Install flywheel and clutch or flexplate.

11. Install transmission.

12. Install transfer case, if equipped.

13. Connect negative battery cable.

14. Start the engine and verify no oil leaks.

Timing Chain, Sprockets, Front Cover and Seal

REMOVAL & INSTALLATION

Front Cover and Seal

1. Before servicing the vehicle, refer to the Precautions Section.

2. Drain the cooling system.

3. Drain the engine oil.

4. Remove or disconnect the following:
- Negative battery cable
- Oil pan
- Crankshaft balancer
- Water pump assembly
- Crankshaft Position (CKP) sensor
- Front cover bolts and the reinforcements, if equipped
- Front cover

5. Pry the seal out of the front cover using a small prytool. Be very careful not to distort the front cover or to score the end of the crankshaft.

To install:

❄ WARNING

Anytime the front cover is removed, the cover must be replaced upon reassembly. If you reuse the old cover, oil leaks may develop.

6. Clean the gasket mating surfaces of the engine and cover of all remaining gasket

or sealer material. Be careful not to score or damage the surfaces.

→**The manufacturer suggests you wait until the front cover is mounted to the engine before you install the replacement crankshaft oil seal. This assures the cover is properly supported.**

7. Install new front cover gasket to the engine or cover using gasket cement to hold it in position. Lubricate the front of the oil pan seal with engine oil to aid in reassembly.

8. Install new front cover to the engine. Take care while engaging the front of the oil pan seal with the bottom of the cover. Apply sealer 12346141 to the oil pan rail where it contacts the timing cover-to-block joint (front) and the crankshaft rear seal retainer-to-block joint (rear). Continue the bead of sealant about 1 inch (25mm) in both directions from each of the four corners.

9. Install front cover retaining bolts and tighten to 106 inch lbs. (12 Nm).

10. Lightly coat the lips of the replacement crankshaft seal with clean engine oil, then position the seal with the open end facing inward the engine. Use a suitable seal installation driver to position the seal in the front cover.

11. Install or connect the following:
- Oil pan
- CKP sensor O-ring and the sensor
- Water pump
- Crankshaft balancer
- Negative battery cable

12. Properly refill the engine cooling system.

13. Run the engine until normal operating temperature has been reached, then check for leaks.

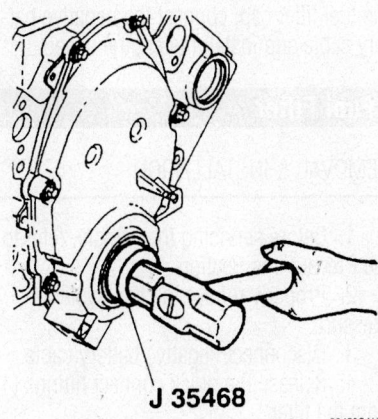

J 35468

88453GAU

Installing the crankshaft front oil seal—4.3L engines

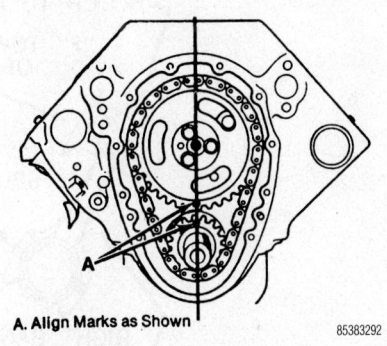

A. Align Marks as Shown

85383292

Timing mark alignment—4.3L engine

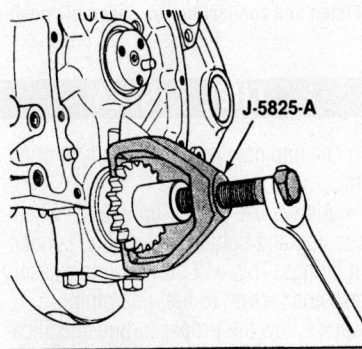

J-5825-A

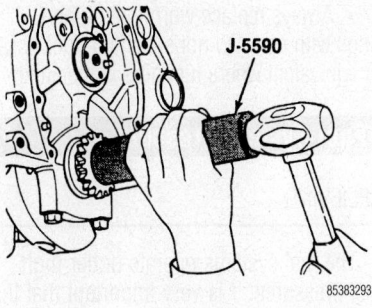

J-5590

85383293

Removal (top) and installation of the crankshaft timing gear

Timing Chain and Sprockets

→**The following procedure requires the use of the Crankshaft Sprocket Removal tool No. J-5825-A and the Crankshaft Sprocket Installation tool No. J-5590.**

1. Before servicing the vehicle, refer to the Precautions Section.

2. Remove the front cover from the engine.

3. Rotate the crankshaft until the No. 4 cylinder is on the Top Dead Center (TDC) of its compression stroke and the camshaft sprocket mark aligns with the mark on the

crankshaft sprocket (facing each other at a point closest together in their travel) and in line with the shaft centers.

4. Remove crankshaft Position (CKP) sensor reluctor ring, if equipped.

5. Remove camshaft sprocket-to-camshaft nut and/or bolts.

6. Remove camshaft sprocket (along with the timing chain). If the sprocket is difficult to remove, use a plastic mallet to bump the sprocket from the camshaft.

→**The camshaft sprocket (located by a dowel) is lightly pressed onto the camshaft and should come off easily. The chain comes off with the camshaft sprocket.**

7. If necessary use J-5825-A crankshaft sprocket removal tool to free the timing sprocket from the crankshaft.

8. If necessary, remove the crankshaft sprocket key.

To install:

9. Inspect the timing chain and the timing sprockets for wear or damage, replace the damaged parts as necessary.

10. Using a putty knife, clean the gasket mounting surfaces. Using solvent, clean the oil and grease from the gasket mounting surfaces.

11. Install crankshaft sprocket key, if removed.

12. Install crankshaft sprocket onto the crankshaft using J-5590 crankshaft sprocket installation tool and a hammer without disturbing the position of the engine.

→**During installation, coat the thrust surfaces lightly with Molykote, or an equivalent pre-lube.**

13. Install timing chain over the camshaft sprocket. Arrange the camshaft sprocket in such a way that the timing marks will align between the shaft centers and the camshaft locating dowel will enter the dowel hole in the cam sprocket.

14. Install timing chain under the crankshaft sprocket, then place the cam sprocket, with the chain still mounted over it, in position on the front of the camshaft.

15. Tighten camshaft sprocket-to-camshaft retainers to 18 ft. lbs. (25 Nm).

16. With the timing chain installed, turn the crankshaft two complete revolutions, then check to make certain that the timing marks are in correct alignment between the shaft centers.

17. Install CKP sensor reluctor ring, if equipped.

18. Install timing cover.

Piston and Ring

POSITIONING

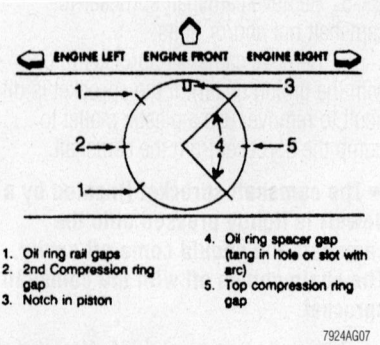

1. Oil ring rail gaps
2. 2nd Compression ring gap
3. Notch in piston
4. Oil ring spacer gap (tang in hole or slot with arc)
5. Top compression ring gap

7924AG07

Piston ring end-gap spacing—GM 4.3L engines

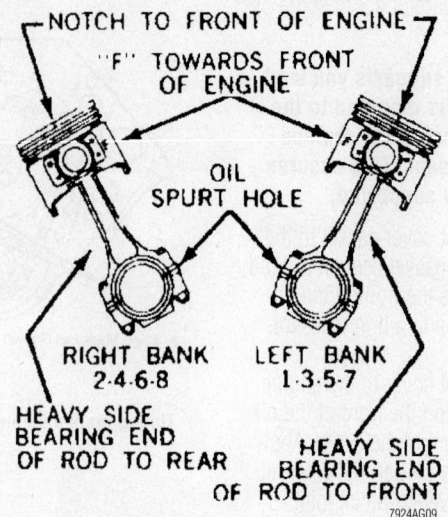

NOTCH TO FRONT OF ENGINE
"F" TOWARDS FRONT OF ENGINE
OIL SPURT HOLE
RIGHT BANK 2·4·6·8
HEAVY SIDE BEARING END OF ROD TO REAR
LEFT BANK 1·3·5·7
HEAVY SIDE BEARING END OF ROD TO FRONT

7924AG09

Piston and connecting rod assembly positioning—GM 4.3L engine

FUEL SYSTEM

Fuel System Service Precautions

Safety is the most important factor when performing not only fuel system maintenance but also any type of maintenance. Failure to conduct maintenance and repairs in a safe manner may result in serious personal injury or death. Maintenance and testing of the vehicle's fuel system components can be accomplished safely and effectively by adhering to the following rules and guidelines.

• To avoid the possibility of fire and personal injury, always disconnect the negative battery cable unless the repair or test procedure requires that battery voltage be applied.

• Always relieve the fuel system pressure prior to disconnecting any fuel system component (injector, fuel rail, pressure regulator, etc.), fitting or fuel line connection. Exercise extreme caution whenever relieving fuel system pressure, to avoid exposing skin, face and eyes to fuel spray. Please be advised that fuel under pressure may penetrate the skin or any part of the body that it contacts.

• Always place a shop towel or cloth around the fitting or connection prior to loosening to absorb any excess fuel due to spillage. Ensure that all fuel spillage (should it occur) is quickly removed from engine surfaces. Ensure that all fuel soaked cloths or towels are deposited into a suitable waste container.

• Always keep a dry chemical (Class B) fire extinguisher near the work area.

• Do not allow fuel spray or fuel vapors to come into contact with a spark or open flame.

• Always use a back-up wrench when loosening and tightening fuel line connection fittings. This will prevent unnecessary stress and torsion to fuel line piping. Always follow the proper torque specifications.

• Always replace worn fuel fitting O-rings with new. Do not substitute fuel hose or equivalent where fuel pipe is installed.

Fuel System Pressure

RELIEVING

The fuel systems operate under high fuel pressures. It is very important that the pressure be properly relieved prior to servicing the system or any of its components.

A Schrader valve is provided on these fuel systems to conveniently test or release the system pressure. A fuel pressure gauge and adapter will be necessary to connect the gauge to the fitting. Most of the MFI systems utilize a service valve on one end of the fuel rail assembly.

1. Before servicing the vehicle, refer to the Precautions Section.
2. Remove the engine cover.
3. Disconnect the negative battery cable to assure the prevention of fuel spillage if the ignition switch is accidentally turned **ON** while a fitting is still detached.
4. Loosen the fuel filler cap to release the fuel tank pressure.
5. Be sure the release valve on the fuel gauge is closed, then connect the fuel gauge to the pressure fitting located on the inlet fuel pipe fitting.

✻✻ CAUTION

When connecting the gauge to the fitting, be sure to wrap a rag around the fitting to avoid spillage. After repairs, place the rag in an approved container.

6. Install the bleed hose portion of the fuel gauge assembly into an approved container, then open the gauge release valve and bleed the fuel pressure from the system.
7. When the gauge is removed, be sure to open the bleed valve and drain all fuel from the gauge assembly.
8. When fuel service is finished, tighten the fuel filler cap, connect the negative battery cable and install the engine cover.

Fuel Filter

REMOVAL & INSTALLATION

1. Before servicing the vehicle, refer to the Precautions Section.
2. Properly relieve the fuel system pressure.
3. Disconnect negative battery cable.
4. Release the quick connect fittings (1) from the filter.
5. Loosen filter feed nut and the clamp bolt (2).
6. Remove filter (4) and the clamp (3) from the vehicle.

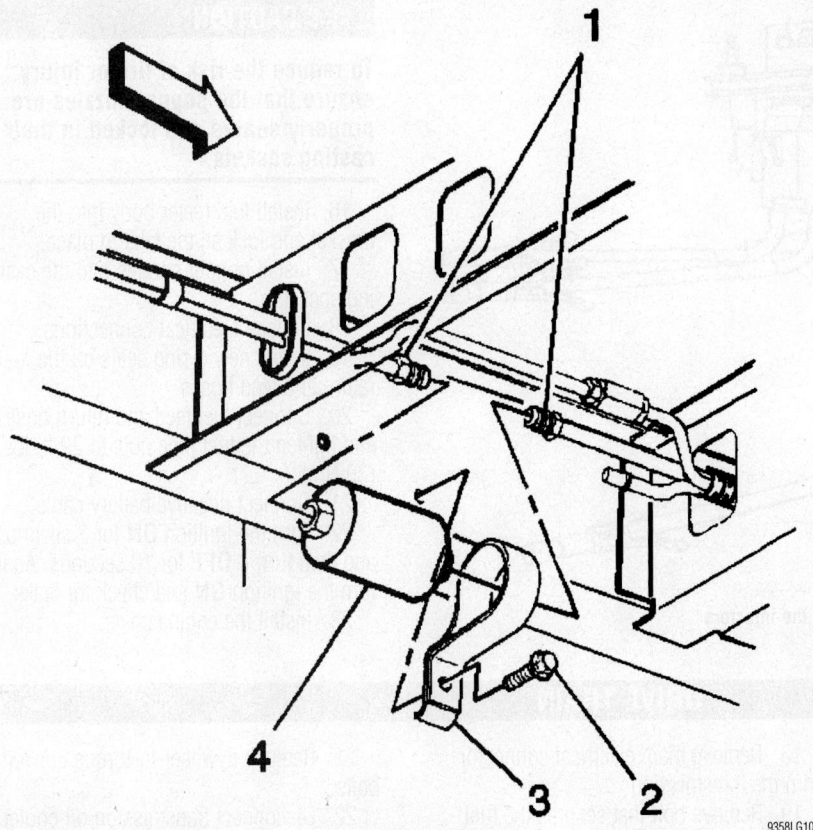

Exploded view of the fuel filter location along frame rail

To install:

7. Install filter and clamp with the directional arrow facing away from the fuel tank, towards the throttle body.

➡**The filter has an arrow (fuel flow direction) on the side of the case, be sure to install it correctly in the system, the with arrow facing away from the fuel tank.**

8. Tighten the fuel feed nut.
9. Tighten the filter clamp assembly bolt.
10. Install fuel quick disconnect fittings to the filter.
11. Connect negative battery cable.
12. Start the engine and check for leaks.

Fuel Pump

REMOVAL & INSTALLATION

1. Before servicing the vehicle, refer to the Precautions Section.
2. Properly relieve the fuel system pressure.
3. Drain the fuel tank.
4. Support the fuel tank.
5. Disconnect negative battery cable.
6. Disconnect filler neck from the tank.

7. Remove shield from tank and tank straps.
8. Disconnect fuel lines and vapor hose from pump.
9. Disconnect electrical connection from fuel pump.
10. Remove fuel tank.
11. Remove fuel pump/sending unit assembly by turning the locking ring

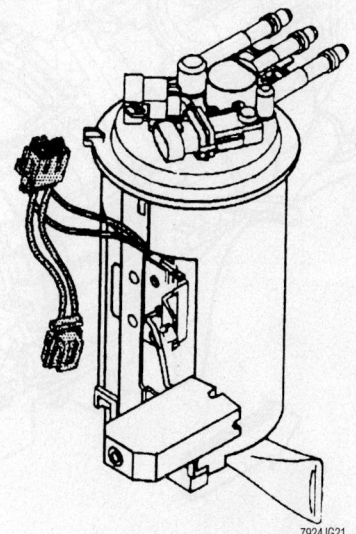

View of the in-tank fuel pump assembly

(located on top of the fuel tank) counter-clockwise using a spanner wrench.

12. Remove fuel pump from the fuel lever sending device.

To install:

13. Install fuel pump in tank with new seal around opening.
14. Raise tank and connect fuel lines and vapor hose.
15. Install tank to the frame. Torque the fasteners to 33 ft. lbs. (45 nm).
16. Install shield.
17. Install fuel filler neck and clamp.
18. Connect negative battery cable.
19. Refill the tank.
20. Run the engine and check for leaks.

Fuel Injector

REMOVAL & INSTALLATION

1. Before servicing the vehicle, refer to the Precautions Section.
2. Relieve the fuel system pressure. Refer to the fuel system relief procedure in this section.
3. Remove engine cover.
4. Disconnect negative battery cable.
5. Remove upper manifold assembly.
6. Disconnect fuel meter body electrical connection and the fuel feed and return hoses from the engine fuel pipes.
7. Remove poppet nozzle out of the casting socket by squeezing the nozzle locking tabs together.
8. Remove fuel meter body by releasing the locktabs.

➡**Each injector is calibrated. When replacing the fuel injectors, be sure to replace it with the correct injector.**

9. Remove lower hold-down plate and nuts.
10. While pulling the poppet nozzle tube downward, push with a small prytool down between the injector terminals and remove the injectors.

To install:

11. Lubricate the new injector O-ring seats with engine oil.
12. Install o-rings on the injector.
13. Install fuel injector into the fuel meter body injector socket.
14. Install lower hold-down plate and nuts. Torque the nuts to 27 inch lbs. (3 Nm).
15. Install fuel meter body assembly into the intake manifold. Torque the fuel meter bracket retainer bolts to 88 inch. lbs. (10 Nm).

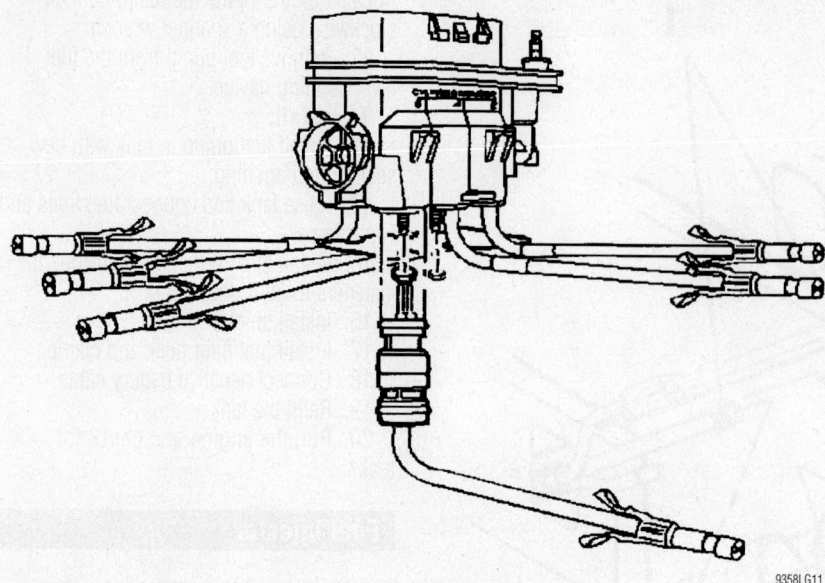

Exploded view of the fuel meter assembly, including the injectors

9358LG11

To reduce the risk of fire or injury ensure that the poppet nozzles are properly seated and locked in their casting sockets.

16. Install fuel meter body into the bracket and lock all the tabs in place.

17. Install poppet nozzles into the casting sockets.

18. Engage electrical connections.

19. Install new o-ring seals on the fuel return and feed hoses.

20. Connect fuel feed and return hoses and tighten the fuel pipe nuts to 22 ft. lbs. (30 Nm).

21. Connect negative battery cable.

22. Turn the ignition **ON** for 2 seconds and then turn it **OFF** for 10 seconds. Again turn the ignition **ON** and check for leaks.

23. Install the engine cover.

DRIVE TRAIN

Automatic Transmission Assembly

REMOVAL & INSTALLATION

1. Disconnect the negative battery cable.

2. Raise the vehicle.

3. Drain the transmission fluid.

4. Remove the rear propeller shaft.

5. Support the transmission with a transmission jack.

6. Remove the nut (2WD) or nuts (AWD) securing the transmission mount to the transmission support.

7. Raise the transmission slightly and remove the transmission support from the vehicle.

8. Remove front exhaust pipe assembly.

9. Remove starter motor.

10. Lower the transmission to gain access to the top and sides of the transmission.

11. Remove vent tube hose and the electrical connections from the transfer case, if equipped.

12. Remove transmission mount.

13. Remove transfer case, if equipped.

14. Disconnect range selector cable end from the transmission range selector lever ball stud and bracket.

15. Remove transmission heat shield.

16. Remove transmission vent hose.

17. Disconnect Park/Neutral Position (PNP) switch connector.

18. Remove main electrical connector from the transmission.

19. Remove bolt that secures the fuel line bracket to the left side of the transmission.

20. Remove torque converter access plug.

21. Remove flywheel-to-torque converter bolts.

22. Disconnect transmission oil cooler pipes from the transmission, then plug the openings in the transmission case.

23. Remove stud and the bolt securing the transmission to the engine.

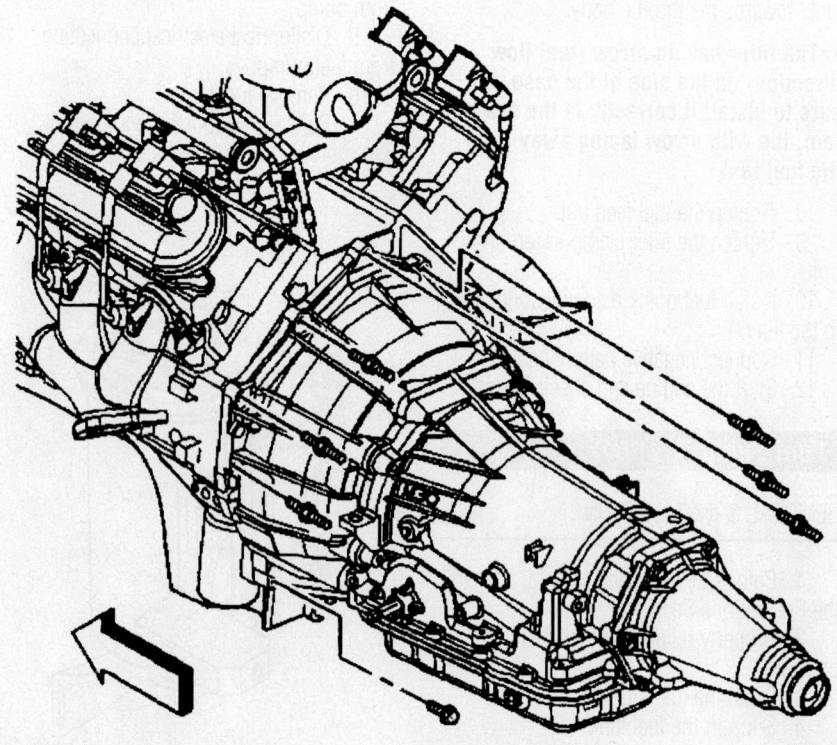

Exploded view of the automatic transmission—2002 vehicle shown

9358LG13

24. Remove six studs and one bolt securing the transmission to the engine.

25. Install tool J 21366 converter holding strap onto the transmission bell housing to retain the torque converter.

26. Pull the transmission straight back.

27. Remove the transmission from the vehicle while simultaneously removing the fluid level indicator tube.

To install:

28. Install the converter holding strap onto the transmission bell housing to retain the torque converter.

29. Support the transmission with a transmission jack.

30. Raise the transmission into place while simultaneously installing the fluid indicator tube.

31. Remove the converter holding strap from the transmission.

32. Slide the transmission straight onto the locating pins while lining up the marks on the flywheel and the torque converter. The torque converter must rotate freely by hand.

33. Install the fasteners securing the transmission to the engine. Tighten the studs and the bolt to 35 ft. lbs. (47 Nm).

34. Install flywheel-to-torque converter bolts and tighten to 46 ft. lbs. (63 Nm).

35. Install torque converter access plug.

36. Install transmission vent hose, fuel lines, and the wiring harness.

37. Install heat shield.

38. Connect r
ange selector cable.

39. Install transfer case, if equipped.

40. Install transmission mount to the transmission (2WD) or the transfer case adapter (AWD).

41. Connect vent hose and electrical connectors to the transfer case, if equipped.

42. Install starter motor.

43. Install front exhaust pipe assembly.

44. Install transmission support.

45. Install transmission mount to transmission support nut (2WD) or nuts (AWD). Tighten the nut(s) to 29 ft. lbs. (40 Nm).

46. Remove the transmission jack.

47. Install rear propeller shaft.

➡**Flush the transmission oil cooler and oil cooler pipes at this time, if necessary.**

48. Connect oil cooler pipes to the transmission.

49. Lower the vehicle.

50. Connect the negative battery cable.

51. Fill the transmission to the proper level.

52. Test the vehicle for proper operation and check for leaks.

Transfer Case Assembly

REMOVAL & INSTALLATION

1. Raise the vehicle.
2. Remove engine brace mounting bolts.
3. Remove engine brace.
4. Remove transfer case shield.
5. Remove rear propeller shaft.
6. Remove front propeller shaft.
7. Disconnect motor/encoder electrical connector.
8. Remove transfer case electrical wiring harness.
9. Disconnect transfer case vent hose.

✳✳ WARNING

To perform the following, the transmission mount must be removed for access to the bottom retaining nut of the transfer case. Do not support the transmission by the oil pan.

10. Support the transmission with a suitable transmission jack.
11. Remove transmission mount retaining nuts.

➡**Raise the transmission assembly enough to clear the rear crossmember.**

12. Remove transmission mount bolts and mount assembly.
13. Remove transfer case bottom retaining nut.

➡**The transmission mount must be replaced in its original position to support the transmission. When reinstalling the transmission mount, it is not necessary reinstall the mounting nuts and bolts. The weight of the transmission and engine will hold the transmission mount in place until this service procedure is completed.**

14. Install the transmission mount.
15. Remove the transmission jack stand from the transmission.
16. Remove the transfer case upper retaining nuts.
17. Install the transmission jack to the transfer case.
18. Remove the transfer case assembly.

✳✳ WARNING

DO NOT reuse the old gasket if it is damaged. DO NOT use any type of sealer in place of the gasket.

19. Remove the transfer case gasket.

To install:

➡**Make sure that the locating tab on the gasket is in the proper location. The locator tab should be facing up.**

20. Install new transfer case gasket.
21. Install transfer case assembly.
22. Install transfer case retaining nuts and tighten them to 40 ft. lbs. (55 Nm).
23. Remove the transmission jack from the transfer case.
24. Position transmission jack to the transmission.
25. Install transmission mount.
26. Install transmission mount bolts and tighten them to 35 ft. lbs. (47 Nm).
27. Lower the transmission into place.
28. Remove the transmission jack stand.
29. Install transmission mount retaining nuts and tighten to 29 ft. lbs. (40 Nm).
30. Connect motor/encoder electrical connector.
31. Install transfer case electrical wiring harness.
32. Connect transfer case vent hose.
33. Install rear propeller shaft.
34. Install front propeller shaft.
35. Install engine brace. Tighten the mounting bolts to 37 ft. lbs. (50 Nm).
36. Inspect the transfer case fluid level.
37. Install the transfer case shield.

Halfshaft

REMOVAL & INSTALLATION

1. Before servicing the vehicle, refer to the Precautions Section.
2. Unlock the steering column so the steering linkage is free to move.
3. Disconnect negative battery cable.
4. Remove front wheels.

➡**Place a drift through the caliper into the edge of the rotor to keep the rotor from turning when the nut is removed.**

5. Remove cotter pin, retainer, nut and washer.
6. Remove brake caliper and support it with a piece of wire to avoid damaging the brake hose.
7. Remove brake rotor.
8. Remove brake line support bracket and Anti-lock Brake System (ABS) wire bracket from the upper control arm.
9. Place a jackstand or jack under the lower control arm.
10. Separate axle shaft from the hub by placing a block of wood against the outer edge of the axle (to protect the threads), then strike the block of wood sharply with a

hammer. Do not remove the axle at this time.

11. Separate tie rods from the steering knuckles.

12. Remove lower shock absorber bolts.

13. Separate upper ball joint from the steering knuckle and suspend the steering knuckle on a wire.

14. Remove skid plate, if equipped.

15. Remove halfshaft-to-axle tube bolts.

16. Remove halfshaft by moving it forward and supporting it away from the frame.

17. Remove halfshaft from the hub and bearing assembly.

18. Separate halfshaft from the differential using a block of wood and a hammer.

To install:

➡️**It is essential that the differential carrier and axle seals are not lubricated or damaged during installation. Prior to shaft installation, cover the shock mounting bracket, lower control arm ball stud and ALL other sharp edges with a cloth or rag to help protect the boot.**

19. Install the axle into the carrier. With both hands on the tripod housing, align the splines on the shaft with the carrier. Then center the axle into the carrier seal and push the shaft straight into the carrier until the snapring is properly seated.

➡️**Be careful when supporting the lower control arm that any components are damaged with the supporting device.**

20. Raise the lower control arm using a jackstand or jack until the full weight of the arm is supported.

➡️**It is necessary to slightly start the knuckle onto the axle while at the same time guiding the lower ball joint into position on the knuckle.**

21. Install lower ball joint, the lower shock absorber and the upper ball joint.

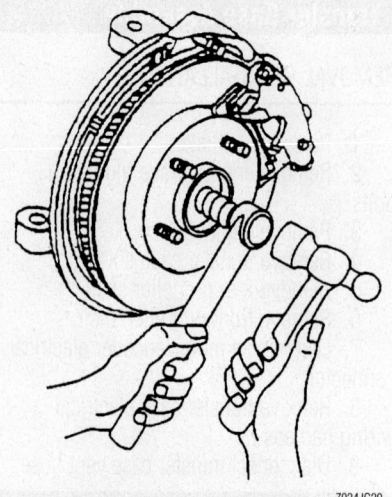

7924JG29

Tap the halfshaft out of the hub without damaging the threads

22. Install axle washer and nut. Tighten the nut to 103 ft. lbs. (140 Nm).

23. Install ABS and brake line brackets to the top of the upper control arm.

24. Install caliper and rotor.

25. Install tire and wheel assembly.

26. Install differential carrier shield.

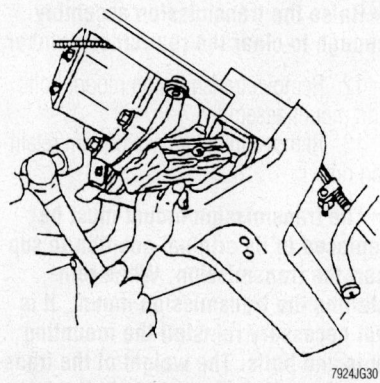

7924JG30

Using a block of wood and a mallet, disengage the halfshaft from the differential assembly

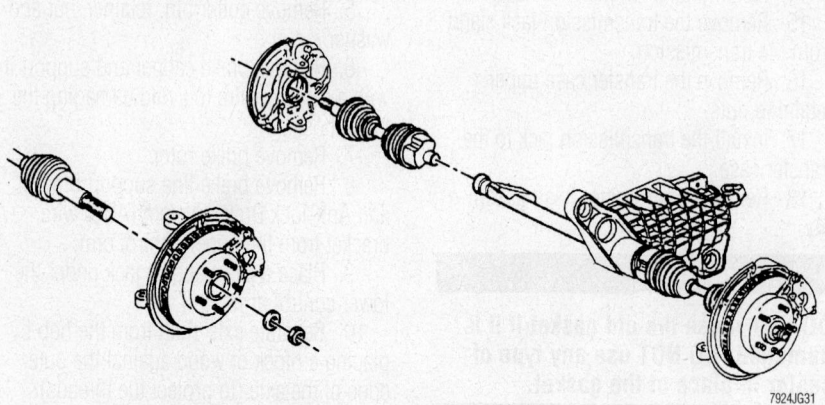

7924JG31

Halfshafts and related components

CV-Joints

OVERHAUL

Outer CV-Joint

1. Before servicing the vehicle, refer to the Precautions Section.

2. Remove front wheel.

3. Remove halfshaft and position it in a vise.

4. Remove large CV-joint boot clamp and discard it.

5. Remove small CV-joint boot clamp and discard it.

6. Remove CV-joint boot and slide it back on the shaft.

7. Remove outer race from the halfshaft, by spreading the outer race-to-halfshaft retaining ring, using Snapring Pliers J-8059.

8. Remove retaining ring from the halfshaft and discard it.

9. CV-joint boot from the halfshaft and discard it, if damaged.

10. Disassemble the chrome alloy balls from the CV-joint cage as follows:

a. Position a brass drift against the CV-joint cage and tap it with a hammer to tilt the cage.

b. Remove the 1st chrome alloy ball from the cage.

c. Tilt the cage in the opposite direction.

d. Remove the opposite chrome alloy ball.

e. Repeat the procedure until all 6 balls are removed.

11. Disassemble the CV-joint cage and inner race as follows:

a. Pivot the cage and race 90 degrees to the center line of the outer race.

b. Align the cage windows with outer race lands.

c. Remove the cage from the outer race.

d. Rotate the inner race upward and remove it from the cage.

12. Thoroughly clean and inspect all parts.

To install:

13. Lubricate the parts with a light coat of grease.

14. Assemble the CV-joint cage and inner race, as follows:

a. Rotate the inner race 90 degrees to the cage centerline.

b. Align the cage windows with inner race lands.

c. Insert the inner race into the cage by rotating the inner race downward.

d. Insert the cage/inner race into the outer race.

15. Assemble the chrome alloy balls into the CV-joint cage, as follows:

a. Position a brass drift against the CV-joint cage and tap it with a hammer to tilt the cage.

b. Insert the 1st chrome alloy ball into the cage.

c. Tilt the cage in the opposite direction.

d. Insert the opposite chrome alloy ball.

e. Repeat the procedure until all 6 balls are inserted.

16. Install ½ kit grease into the CV-joint.

17. Install small ring clamp on the CV boot.

18. Install new retaining ring on the half-shaft.

19. Install large ring clamp on the CV boot.

20. Install outer race assembly onto the halfshaft until the ring engages the halfshaft groove.

21. Slide the small end of the CV-joint

boot/clamp into place, with the seal lip in the halfshaft groove

➡**Make sure the boot lies flat against the halfshaft.**

22. Using the Crimp tool J-35910, a torque wrench and a breaker bar, crimp the small CV-joint boot clamp to 100 ft. lbs. (136 Nm).

23. Check the clamp gap dimension; if it is not 0.085 in. (2.15mm), continue tightening the clamp until it is.

24. Install ½ kit grease into the CV-joint boot.

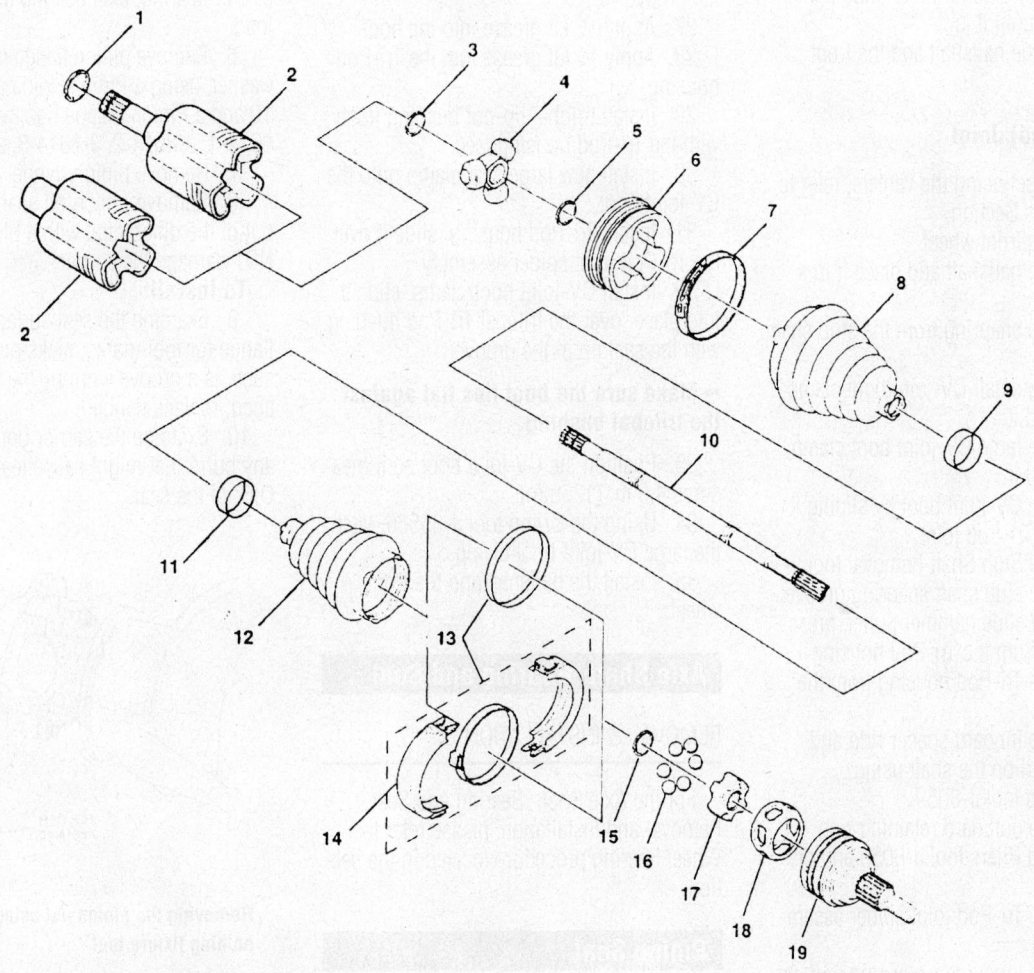

(1) Differential Shaft Ring	(11) Halfshaft Swage Ring
(2) Tripot Housing Assembly	(12) CV Joint Boot
(3) Spacer Ring	(13) Swage Ring
(4) Tripot Joint Spider Assembly	(14) Clamp Protector
(5) Spacer Ring	(15) Race Retaining Ring
(6) Tripot Bushing	(16) Ball
(7) Boot Retaining Clamp	(17) CV Joint Inner Race
(8) Tripot Joint Boot	(18) CV Joint Cage
(9) Halfshaft Swage Ring	(19) CV Joint Outer Race
(10) Halfshaft Bar	

9308JG09

Exploded view of the CV-Joint Assembly

25. Measure approximately 0.687 in. (17.5mm) up from the bottom edge of the outer CV-joint assembly.

26. Slide the large end of the CV boot/clamp into place, with the seal lip in place over the outer race.

➥**Make sure the boot lies flat against the outer race.**

27. Using the Crimp tool J-35910, a torque wrench and a breaker bar, crimp the large CV-joint boot clamp to 130 ft. lbs. (176 Nm).

28. Check the clamp gap dimension; if it is not 0.102 in. (2.60mm), continue tightening the clamp until it is.

29. Install the halfshaft and the front wheel.

Inner (Tri-Pod) Joint

1. Before servicing the vehicle, refer to the Precautions Section.
2. Remove front wheel.
3. Remove halfshaft and place it in a vise.
4. Remove snapring from the stub shaft and discard it.
5. Remove small CV-joint boot clamp, cut and discard it.
6. Remove large CV-joint boot clamp, cut and discard it.
7. Remove CV-joint boot by sliding it away from the Tri-Pod joint.
8. Install a Stub Shaft Removal tool J-38868-A to the stub shaft snapring groove.
9. Using a slide hammer puller, press the stub shaft from the Tri-Pod housing.
10. Remove Tri-Pod housing from the Tri-Pod spider.
11. Remove inboard spacer ring and slide it rearward on the shaft using Snapring Pliers tool J-8059.
12. Remove outboard retaining ring using Snapring Pliers tool J-8059 and discard it.
13. Remove Tri-Pod joint spider assembly.
14. Remove inboard spacer ring and discard it.
15. Remove CV-joint boot.
16. Remove trilobal Tri-Pod bushing from the housing.
17. Thoroughly clean and inspect all parts.

To install:
18. Install new snapring onto the stub shaft.
19. Install small boot clamp.
20. Install CV-joint boot.
21. Using the Crimp tool J-35910, a torque wrench and a breaker bar, crimp the small CV-joint boot clamp to 100 ft. lbs. (136 Nm).

22. Install inboard spacer ring slide it rearward on the shaft using Snapring Pliers tool J-8059, past the 2nd groove.

23. Install Tri-Pod joint spider assembly onto the shaft until it passes the 2nd groove.

24. Install outboard retaining ring into the axle shaft groove using Snapring Pliers tool J-8059.

25. Install Tri-Pod joint spider assembly, slide it against the outboard retaining ring.

26. Install inboard spacer ring, seat it in the groove.

27. Apply ½ kit grease into the boot.

28. Apply ½ kit grease into the Tri-Pod housing.

29. Install trilobal tip-pot bushing flush with the Tri-Pod housing face.

30. Install new large seal clamp onto the CV-joint boot.

31. Install Tri-Pod housing, slide it over the Tri-Pod joint spider assembly.

32. Install CV-joint boot/clamp, slide it into place, over the trilobal Tri-Pod bushing with the seal lip in the groove.

➥**Make sure the boot lies flat against the trilobal bushing.**

33. Position the CV-joint boot so it measures 4.9 in. (125mm).

34. Using the Crimp tool J-35566, latch the large CV-joint boot clamp.

35. Install the halfshaft and the front wheel.

Axle Shaft, Bearing and Seal

REMOVAL & INSTALLATION

For the Axle Shaft, Bearing and Seal, Removal and Installation, please refer to Wheel Bearing procedure located in the section.

Pinion Seal

REMOVAL & INSTALLATION

1. Before servicing the vehicle, refer to the Precautions Section.

➥**The following procedure requires the use of the Pinion Holding tool J-8614-10, the Pinion Flange Removal tool J-8614-1, J-8614-2, J-8614-3 and the Pinion Seal Installation tool J-23911.**

2. Remove driveshaft from the pinion flange. Matchmark the driveshaft prior to removal.

3. Remove driveshaft from the rear axle pinion flange and support the shaft up in body tunnel by wiring it to the exhaust pipe.

➥**If the U-joint bearings are not retained by a retainer strap, use a piece of tape to hold bearings on their journals.**

4. Mark the position of the pinion stem, flange and nut for reference.

5. Use an inch lbs. torque wrench to measure the amount of torque necessary to turn the pinion, then note this measurement as it is the combined pinion bearing, seal, carrier bearing, axle bearing and seal preload.

6. Remove pinion flange nut and washer, using a Pinion Holding tool J-8614-10 and a Pinion Flange Removal tool J-8614-1, J-8614-2, J-8614-3, as applicable.

7. Remove pinion flange.

8. Remove pinion oil seal by driving it out of the differential with a blunt chisel; DO NOT damage the carrier.

To install:
9. Examine the seal surface of pinion flange for tool marks, nicks or damage, such as a groove worn by the seal. If damaged, replace flange.

10. Examine the carrier bore and remove any burrs that might cause leaks around the O.D. of the seal.

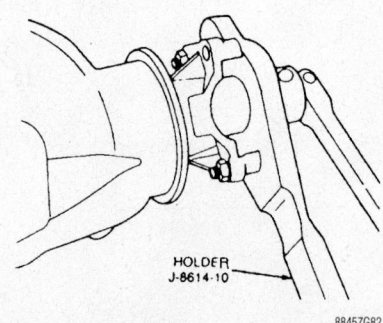

Removing the pinion nut using a pinion holding fixture tool

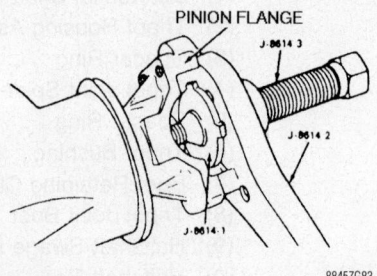

A puller and adapter should be used to withdraw the pinion from the housing

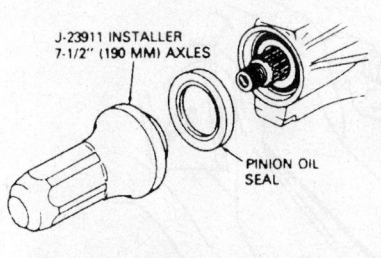

J-23911 INSTALLER
7-1/2" (190 MM) AXLES

PINION OIL
SEAL

88457G84

Use the appropriately sized installation
tool to drive the new seal into position.

11. Apply GM seal lubricant 1050169 to
the outside diameter of the pinion flange
and sealing lip of new seal.

12. Install new pinion oil seal using a
seal installer tool.

13. Install pinion flange and tighten nut
to the same position as marked earlier.
Tighten the nut a little at a time and turn the
pinion flange several times after each tightening in order to set the rollers.

14. Measure the torque necessary to turn
the pinion and compare this to the reading
taken during removal. Tighten the nut addi-

tionally, as necessary to achieve the same
preload as measured earlier.

➡**If fluid was lost from the differential
housing during this procedure, be sure
to check and add additional fluid, as
necessary.**

15. Remove the support then align and
secure the driveshaft assembly to the pinion
flange.

➡**The original matchmarks MUST be
aligned to assure proper shaft balance
and prevent vibration.**

STEERING AND SUSPENSION

Air Bag

✳✳ CAUTION

**Some vehicles are equipped with an
air bag system, also known as the
Supplemental Inflatable Restraint
(SIR) system. The system must be
disabled before performing service on
or around system components, steering column, instrument panel components, wiring and sensors. Failure to
follow safety and disabling procedures
could result in accidental air bag
deployment, possible personal injury
and unnecessary system repairs.**

PRECAUTIONS

Several precautions must be observed
when handling the inflator module to avoid
accidental deployment and possible personal injury.

• Never carry the inflator module by the
wires or connector on the underside of the
module.

• When carrying a live inflator module,
hold securely with both hands, and ensure
that the bag and trim cover are pointed
away.

• Place the inflator module on a bench
or other surface with the bag and trim cover
facing up.

• With the inflator module on the bench,
never place anything on or close to the
module, that may be thrown in the event of
an accidental deployment.

DISARMING

➡**With the AIR BAG fuse removed and
the ignition switch ON, the AIR BAG
warning lamp will be on. This is normal and does not indicate any system
malfunction.**

1. Turn the steering wheel so that the
vehicle's wheels are pointing straight ahead.

2. Turn the ignition switch to **LOCK**,
remove the key, then disconnect the negative battery cable.

3. Remove the AIR BAG fuse from the
fuse block.

4. Remove the steering column filler
panel.

5. Disengage the Connector Position
Assurance (CPA) and the yellow two way
connector located at the base of the steering
column.

6. Connect the negative battery cable.

ARMING

1. Disconnect the negative battery cable.

2. Turn the ignition switch to **LOCK**,
then remove the key.

3. Engage the yellow SIR connector and
CPA located at the base of the steering column.

4. Install the steering column filler
panel.

5. Install the AIR BAG fuse to the fuse
block.

6. Connect the negative battery cable.

7. Turn the ignition switch to **RUN** and
make sure that the AIR BAG warning lamp
flashes seven times and then shuts off. If
the warning lamp does not shut off, make
sure that the wiring is properly connected. If
the light remains on, take the vehicle to a
reputable repair facility for service.

Power Steering Gear
(Recirculating Ball)

REMOVAL & INSTALLATION

2WD Vehicles

➡**The wheels of the vehicle must be
straight ahead and the steering column**

in the LOCK position before disconnecting the steering column or intermediate
shaft from the steering gear. Failure to
do so will cause the coil assembly in
the steering column to become uncentered which will cause damage to the
coil assembly.

1. Before servicing the vehicle, refer to
the Precautions Section.

2. Remove lower fan shroud.

3. Remove intermediate shaft pinch bolt
from the intermediate shaft coupling.

4. Disconnect intermediate shaft from
the power steering gear.

5. Place a drain pan under the steering
gear.

➡**Cap or tape the ends of the hoses
and the gear fittings to prevent spillage
and contamination.**

6. Disconnect power brake booster outlet hose from the steering gear.

7. Disconnect power steering cooler
pipe from the steering gear.

8. Raise the vehicle.

9. Remove pitman arm from the steering gear.

10. Remove steering gear mounting
bolts and washers.

11. Remove steering gear from the vehicle.

To install:

12. Install steering gear to the vehicle.

13. Install steering gear washers and
bolts. Tighten to 55 ft. lbs. (75 Nm).

14. Install pitman arm to the steering
gear.

15. Lower the vehicle.

16. Connect power steering cooler pipe
to the steering gear and tighten to 20 ft. lbs.
(27 Nm).

17. Connect power brake booster outlet
hose to the steering gear and tighten the fitting to 20 ft. lbs. (27 Nm).

18. Remove the drain pan from under the
vehicle.

19. Connect intermediate shaft to the power steering gear.

20. Install intermediate shaft pinch bolt to the intermediate shaft coupling and tighten to 30 ft. lbs. (41 Nm).

21. Install the lower fan shroud.

22. Bleed the power steering system

4WD Vehicles

1. Before servicing the vehicle, refer to the Precautions Section.

2. Raise and support the vehicle.

3. Remove tire and the wheel.

4. Remove pitman arm-to-connecting rod retaining nut.

5. Remove pitman arm from the connecting rod.

6. Remove stabilizer links.

7. Rotate the stabilizer shaft downward.

8. Lower the vehicle.

9. Install steering column lock pin no. J 42640 into the lower steering column trim cover in order to lock the steering column.

10. Remove lower fan shroud.

11. Remove intermediate shaft-to-power steering gear pinch bolt.

12. Remove intermediate shaft from the power steering gear.

13. Place a drain pan under the vehicle.

14. Disconnect power steering cooler hose from the power steering gear.

15. Remove hydraulic brake booster from the power steering gear.

16. Raise the vehicle.

17. Remove power steering gear mounting bolts, then lower the vehicle.

18. Remove power steering gear.

19. Remove pitman arm from the power steering gear.

To install:

20. Install pitman arm to the power steering gear.

21. Install power steering gear to the vehicle.

22. Install pitman arm to the connecting rod.

23. Using the aid of an assistant, position the power steering gear to the frame and install the power steering mounting bolts. Tighten the power steering mounting bolts to 106 ft. lbs. (143 Nm).

24. Connect power brake booster outlet hose to the power steering gear and tighten to 20 ft. lbs. (27 Nm).

25. Connect power steering cooler hose to the power steering gear and tighten to 20 ft lbs. (27 Nm).

26. Install intermediate shaft to the power steering gear.

27. Install intermediate shaft to power

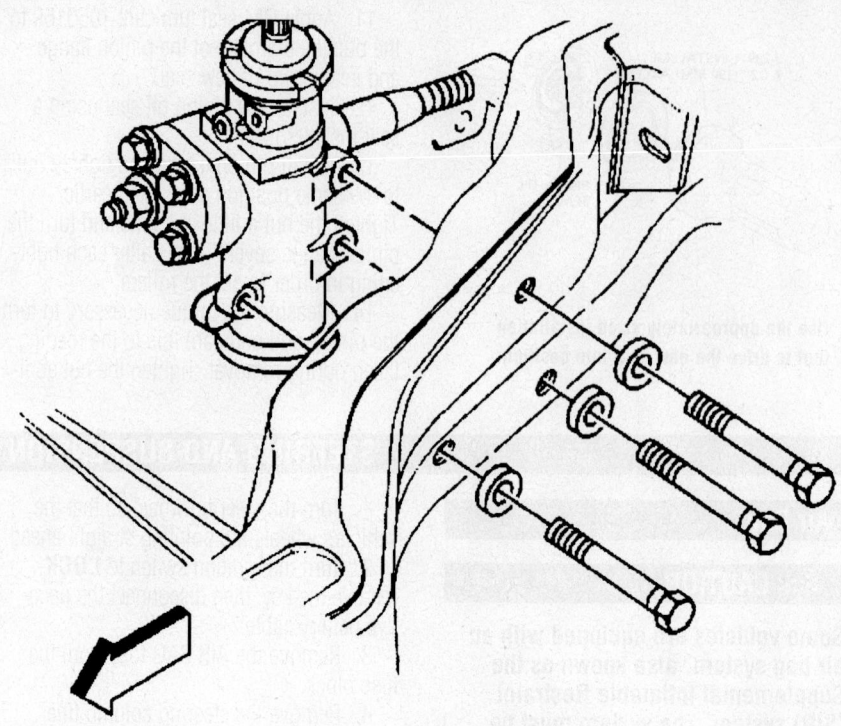

Exploded view of the power steering gear mounting—2002 vehicle shown, others similar

9358LG14

steering gear pinch bolt and tighten to 30 ft. lbs. (41 Nm).

28. Install lower fan shroud.

29. Remove the lock pin from the steering column.

30. Install pitman arm-to-connecting rod retaining nut and tighten to 48 ft. lbs. (62 Nm).

31. Rotate the stabilizer shaft to the correct position.

32. Install stabilizer links.

33. Install tire and the wheel.

34. Lower the vehicle.

35. Bleed the power steering system.

Shock Absorbers

REMOVAL & INSTALLATION

Front

2WD

1. Remove the tire and wheel.

2. Remove the shock absorber mounting nut, the retainer, and the insulator. Hold the shock absorber upper stem in order to keep the stem from turning. A hex is provided on the end of the stem for this purpose.

3. Remove the shock absorber mounting bolts. The nuts will remain attached to the lower control arm.

4. Pull the shock absorber out through the hole in the lower control arm.

To install:

5. Install the shock absorber, fully extended, up through the hole in the lower control arm.

6. Install the upper insulator, the retainer, and the shock absorber mounting nut over the shock absorber upper stem.

7. Install the shock absorber mounting bolts through the holes in the lower control arm. Tighten as follows:

  a. Tighten the shock absorber mounting nut to 15 ft. lbs. (20 Nm) while holding the upper stem of the shock absorber.

  b. Tighten the shock absorber mounting bolts to 18 ft. lbs. (25 Nm).

8. Install the tire and wheel.

4WD

1. Before servicing the vehicle, refer to the Precautions Section.

2. Support the lower control arm (front) or axle assembly (rear).

3. Remove wheel.

4. Remove inner wheel well splash shield, if removing the front shock absorber.

5. Remove lower nut, washer and bolt.

6. Remove upper nut, washer and bolt.

➡**Compress the front shock absorber to make removal easier.**

7. Remove shock absorber.

To install:

8. Compress the front shock absorber to make installation easier.

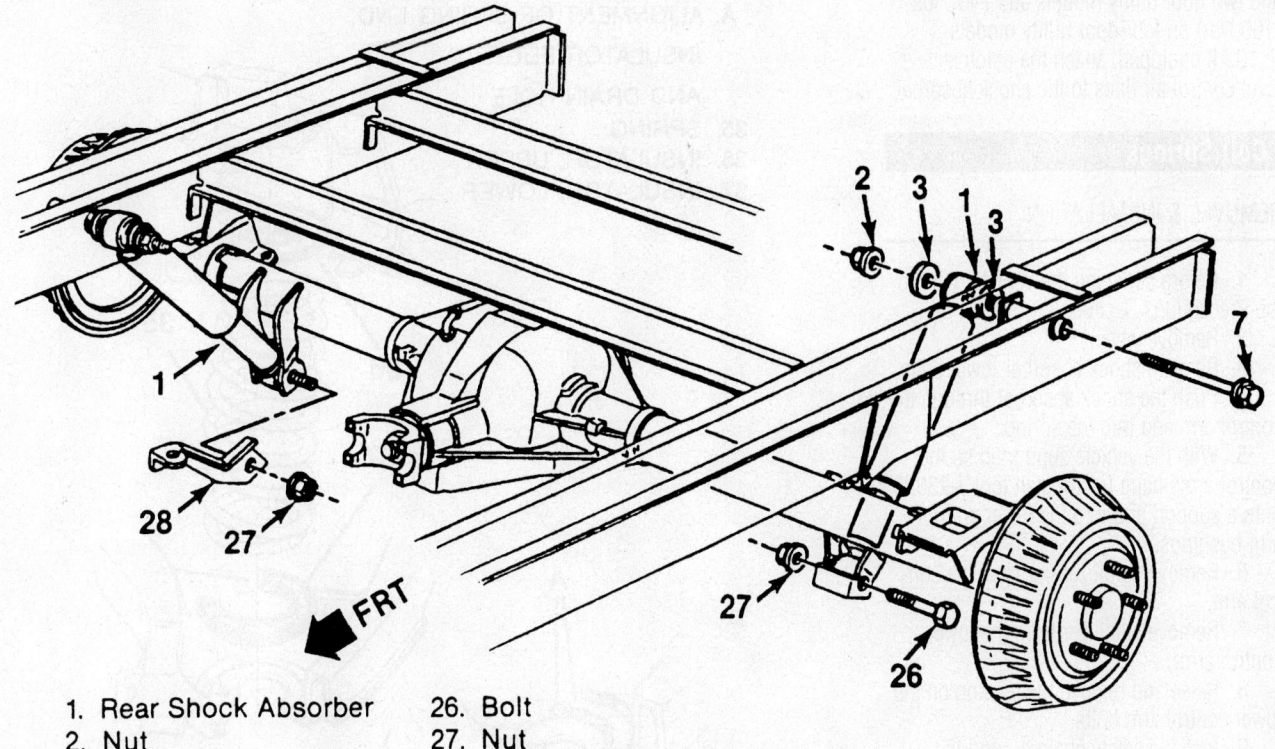

1. Rear Shock Absorber
2. Nut
3. Washer
7. Bolt

26. Bolt
27. Nut
28. Parking Brake Bracket

7924JG36

Rear shock absorber mounting

9. Install shock absorber.
10. Install lower nut. Torque the nut to 62 ft. lbs. (84 Nm).
11. Install upper bolt. Torque the bolts to 18 ft. lbs. (25 Nm).
12. Install wheel.

Rear

1. Before servicing the vehicle, refer to the Precautions Section.
2. Properly support the rear axle assembly.
3. Disconnect automatic level control air lines from the shock absorber, if equipped.
4. Remove shock absorber-to-frame retainers at the top of the shock.
5. Remove shock-to-axle retainers at the bottom of the shock.
6. Remove shock absorber.

To install:

7. Install the shock in the vehicle and loosely install the upper mounting fasteners to retain it.
8. Align the lower-end of the shock absorber with the axle mounting, then loosely install the retainers.
9. Tighten the upper shock retainers to 18 ft. lbs. (25 Nm). Tighten the lower shock retainers to 62 ft. lbs. (84 Nm) on pick-up

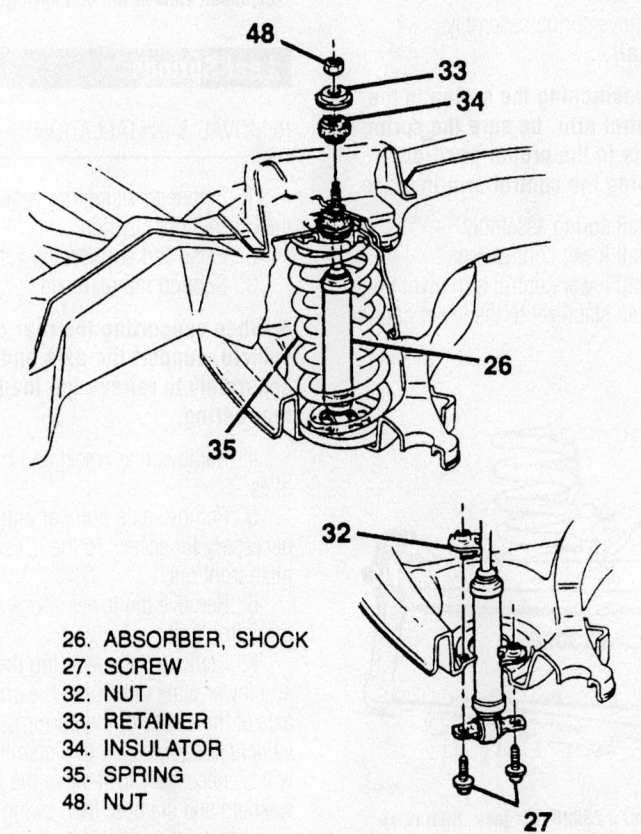

26. ABSORBER, SHOCK
27. SCREW
32. NUT
33. RETAINER
34. INSULATOR
35. SPRING
48. NUT

Front shock absorber mounting

7924JG37

and two door utility models and 74 ft. lbs. (100 Nm) on four door utility models.

10. If equipped, attach the automatic level control air lines to the shock absorber.

Coil Springs

REMOVAL & INSTALLATION

1. Before servicing the vehicle, refer to the Precautions Section.
2. Remove wheel.
3. Remove shock absorber lower bolts.
4. Push the shock absorber through the control arm and into the spring.
5. With the vehicle supported so the control arms hang free, install tool J-23028, onto a support and into the lower control arm bushings.
6. Remove stabilizer bar from the control arm.
7. Remove stabilizer from the lower control arm.
8. Raise and remove the tension on the lower control arm bolts.
9. Install a safety chain around the spring and through the lower control arm.
10. Remove lower control arm pivot bolts, the rear first.
11. Remove lower control arm and allow it to hang free.
12. Remove spring assembly.

To install:

➡**When positioning the spring in the lower control arm, be sure the spring insulator is in the proper position before lifting the control arm in place.**

13. Install spring assembly.
14. Install lower control arm.
15. Install lower control arm pivot bolts.
16. Install stabilizer to the lower control arm.

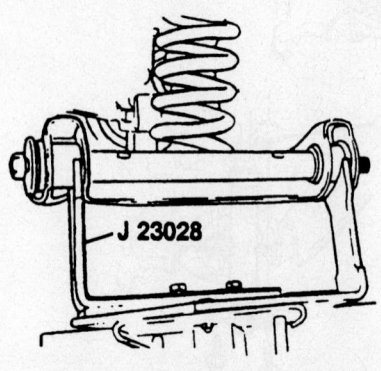

Secure tool J 23028 to a jack, then raise the jack to remove the tension on the lower control arm bolts

A. ALIGNMENT OF SPRING END, INSULATOR EDGE AND DRAIN HOLE
35. SPRING
36. INSULATOR, UPPER
37. INSULATOR, LOWER

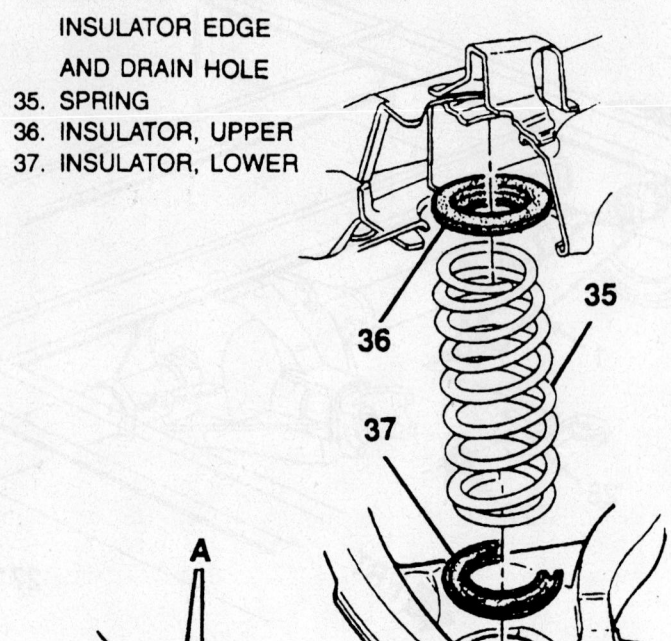

Exploded view of the coil spring mounting

Leaf Springs

REMOVAL & INSTALLATION

1. Before servicing the vehicle, refer to the Precautions Section.
2. Raise and support the vehicle.
3. Support the rear axle.

➡**When supporting the rear of the vehicle, support the axle and the body separately to relieve the load on the rear spring.**

4. Remove rear wheel and tire assemblies.
5. Remove axle bumper and retainer, if necessary for access to the lower spring plate front nut.
6. Remove the lower shock absorber mounting bolt.
7. Remove nuts securing the U-bolt and lower plate (attaching the spring to the axle at the center of the spring). If the vehicle is equipped with a stabilizer bar it will be necessary to remove the lower nuts, washers and clamps, then swing the stabilizer bar down to obtain clearance when lowering the axle assembly.

8. Remove U-bolt, lower plate and anchor plate, then CAREFULLY lower the axle away from the spring.

✳✳ WARNING

DO NOT let the axle hang by the brake hose at any point during the procedure or the hose may be severely damaged.

9. Remove shackle nut and bolt, then disengage the spring from the shackle, at the rear of the fiberglass spring.
10. Remove hanger nut and bolt, then the spring from the hanger, at the front of the fiberglass spring.

To install:

➡**To assure proper seating and attachment of the anchor plate over the spring end and the axle, the installation procedure must be followed closely.**

11. Install spring to the hanger, then loosely install the retaining nut and bolt.
12. Install spring to the shackle, then loosely install the retaining nut and bolt.
13. CAREFULLY raise the axle until it contacts the spring.

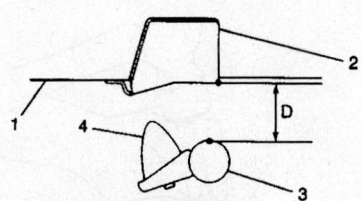

1. FRAME
2. AXLE STOP BRACKET
3. REAR AXLE (END VIEW)
4. BUMPER
5. TRIM HEIGHT 135–145MM
   (5.3–5.7 INCHES)

88268GB5

Measuring the rear suspension trim height

14. Apply rubber lubricant to the isolator on the spring in order to aid installation of the anchor plate, then install the anchor plate to the top of the spring.

15. Install lower plate and U-bolt around the axle and through the anchor plate. If your van is equipped with a stabilizer bar it will be necessary to install the clamps, washers and nuts.

16. Install nuts to the lower plate and U-bolts. Starting with the inner (lower plate side) nuts, gradually tighten the 4 nuts so the anchor plate moves uniformly, side-to-side, over the spring. Tighten the nuts to 41 ft. lbs. (56 Nm).

➡**After tightening the fasteners to specification, there should be no gap between the anchor plate, axle tube bracket and the lower plate. A metal-to-metal contact should exist.**

17. Install the lower shock absorber mounting bolt.

18. Raise the axle so the vehicle's weight is supported by the spring. The rear suspension height should be approximately 5.3–5.7 in. (135–145mm). With the suspension at normal ride height, tighten the shackle and hanger retainers to 74 ft. lbs. (100 Nm).

19. Install axle bumper and tighten the nut to 33 ft. lbs. (45 Nm), if removed.

20. Install tire and wheel assemblies.

Torsion Bar

Instead of the coil spring used on the front suspension of 2WD vehicles, the 4WD vehicles are equipped with a torsion bar.

REMOVAL & INSTALLATION

1. Before servicing the vehicle, refer to the Precautions Section.

2. Loosen adjustment assemblies on the torsion bar.

3. Mark the adjustment bolt setting.

4. Use tool J 36202 to increase the tension on the adjustment arm.

5. Remove adjustment bolt and the retaining nut, then move the tool aside.

6. Remove torsion bar adjustment arm by sliding the bar forward.

7. Slide the torsion bar partially back through the crossmember.

➡**The front end of the torsion bar are marked with a left and a right because there are different bars for both sides.**

8. Lower the front of the torsion bar down and slide it forward.

To install:

9. Lubricate the adjuster arm and the bolt with axle grease.

10. Install or torsion bar adjuster arm. The rear face of the torsion bar should be within 0.04-0.10 inch (1.0-2.0 mm) of the rear face of the adjuster arm when both are fully installed.

11. Install adjustment retaining nut and the adjustment bolt. Using tool J 36202, increase the tension on the torsion bar

12. Install retaining nut and adjustment bolt.

13. Place the adjustment bolt to the marked setting.

14. Use tool J 36202 to release the ten-

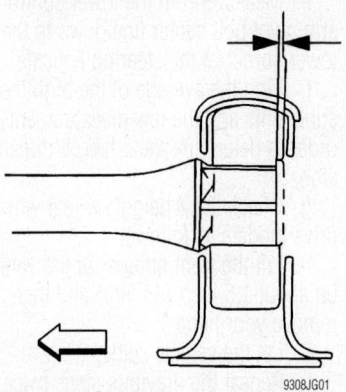

9308JG01

The rear face of the torsion bar should be within 0.04-0.10 inch (1.0-2.0 mm) of the rear face of the adjuster arm when fully installed

sion on the torsion bar until the load is taken up by the adjustment bolt, then remove the tool.

   a. Check the Z height on rear wheel drive models as follows:

   b. Lift the front bumper of the vehicle up about 1.5 inch (38 mm) and then remove your hands.

   c. Let the vehicle settle.

   d. Repeat the previous steps twice until you have lifted the bumper 3 times.

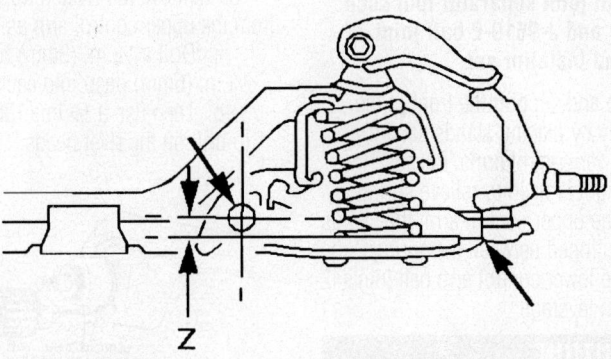

9308JG02

Z height measurement–2WD

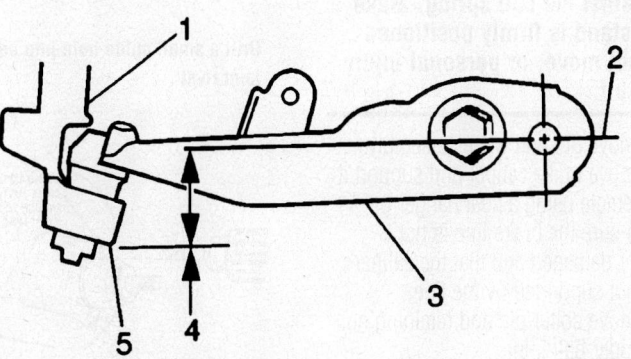

9308JG03

Z height measurement–4WD

e. Measure from the lower control arm pivot bolt center line down to the lower corner of the steering knuckle.

f. Find the average of the high measurements and the low measurements in order to determine the Z height dimension.

g. Check the Z height on rear wheel drive models as follows:

h. Lift the front bumper of the vehicle up about 1.5 inch (38 mm) and then remove your hands.

i. Let the vehicle settle.

j. Repeat the previous steps twice until you have lifted the bumper 3 times.

k. Measure from the lower control arm pivot bolt center line down to the lower corner of the steering knuckle for the Z height measurement.

Ball Joints

REMOVAL & INSTALLATION

2WD Vehicles

UPPER

1. Before servicing the vehicle, refer to the Precautions Section.

➡The following procedure requires the use of a ball joint separator tool such as J-23742 and J-9519-E ball joint remover and installer set.

2. Raise and support the front of the vehicle safely by placing stands securely under the lower control arms. Because the vehicle's weight is used to relieve spring tension on the upper control arm, the stands must be positioned between the spring seats and the lower control arm ball joints for maximum leverage.

✳✳ CAUTION

With components unbolted, the stand is holding the lower control arm in place against the coil spring. Make sure the stand is firmly positioned and cannot move, or personal injury could result.

3. Remove tire and wheel assembly.

4. Remove brake caliper and support it from the vehicle using a coat hanger or wire. Make sure the brake line is not stretched or damaged and that the caliper's weight is not supported by the line.

5. Remove cotter pin and retaining nut from the upper ball joint.

6. Remove anti-lock brake sensor wire bracket, if equipped.

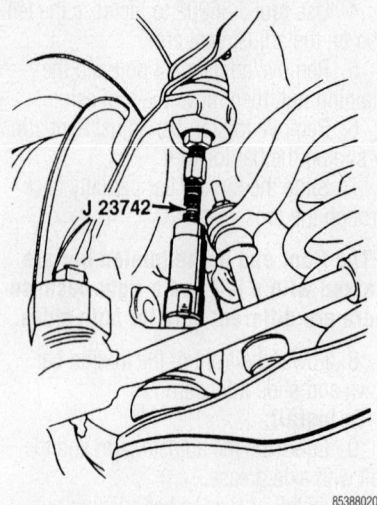

85388020

Use a ball joint separator tool to drive the upper ball joint from the steering knuckle

7. Remove upper ball joint from the steering knuckle using tool J-23742 and pull the steering knuckle free of the ball joint.

➡**After separating the steering knuckle from the upper ball joint, be sure to support the steering knuckle/hub assembly to prevent damaging the brake hose.**

8. Remove the riveted upper ball joint from the upper control arm as follows:

a. Drill a ⅛ in. (3mm) hole, about ¼ in. (6mm) deep into each rivet.

b. Then use a ½ in. (13mm) drill bit, to drill off the rivet heads.

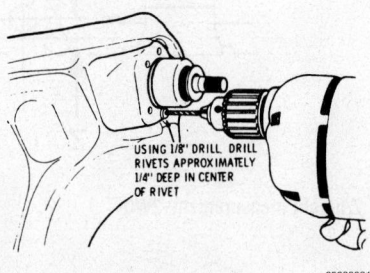

USING 1/8" DRILL DRILL RIVETS APPROXIMATELY 1/4" DEEP IN CENTER OF RIVET

85388021

Drill a small guide hole into each ball joint rivet

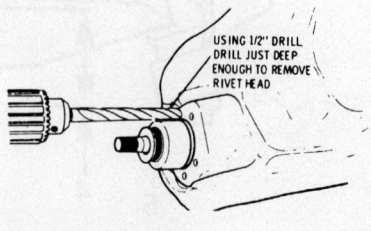

USING 1/2" DRILL DRILL JUST DEEP ENOUGH TO REMOVE RIVET HEAD

85388022

Then drill off the rivet heads

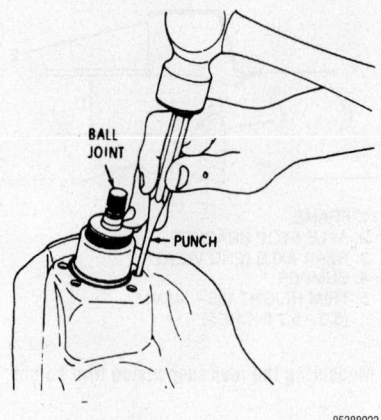

85388023

Punch the rivets out and remove the ball joint

c. Using a pin punch and the hammer, drive out the rivets in order to free the upper ball joint from the upper control arm assembly, then remove the upper ball joint.

9. Clean and inspect the steering knuckle hole. Replace the steering knuckle if the hole is out of round.

To install:

10. Install ball joint in the upper control arm.

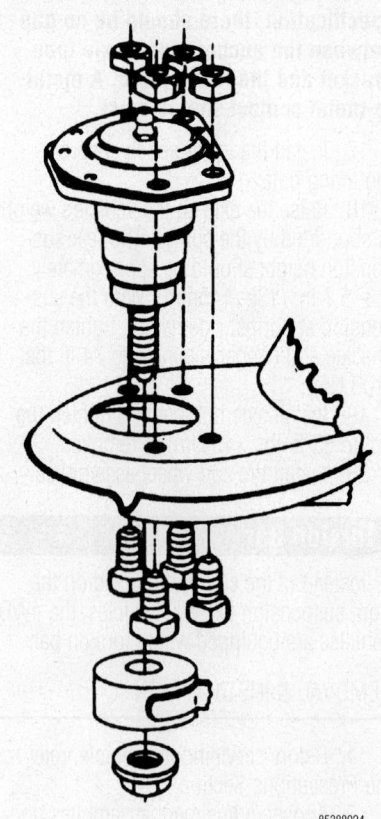

85388024

Service ball joints are bolted to the control arm

11. Install ball joint retaining nuts and bolts. Position the bolts threaded upward from under the control arm. Tighten the ball joint retainers to 22 ft. lbs. (30 Nm).

12. Install anti-lock brake sensor wire bracket, if removed.

13. Install ball joint to the knuckle. Make sure the joint is seated, then install the stud nut and tighten to 61 ft. lbs. (83 Nm). Insert a new cotter pin.

➡**When installing the cotter pin, never loosen the castle nut to expose the cotter pin hole.**

14. Thread the grease fitting into the ball joint. Use a grease gun to lubricate the upper ball joint until grease appears at the seal.

15. Install brake caliper.

16. Install tire and wheel assembly.

17. Check and adjust the front end alignment, as necessary.

LOWER

1. Before servicing the vehicle, refer to the Precautions Section.

➡**The following procedure requires the use of a ball joint remover/installer set (the particular set may vary upon application but must include a clamping-type tool with the appropriately sized adapters) and a ball joint separator tool, such as J-23742.**

2. Remove the wheel.

3. Position a jack under the spring seat of the lower control arm, then raise the jack to support the arm.

❊❊ CAUTION

The jack MUST remain under the lower control arm, during the removal and installation procedures, to retain the arm and spring positions. Make sure the jack is securely positioned and will not slip or release during the procedure or personal injury may result.

4. Remove brake caliper and support it aside using a hanger or wire. Make sure the brake line is not stressed or damaged.

5. Remove lower ball joint cotter pin and discard.

6. Remove ball joint stud nut.

7. Remove lower ball joint from the steering knuckle using tool J-23742.

8. Carefully guide the lower control arm out of the opening in the splash shield using a putty knife. Position a block of wood between the frame and upper control arm to keep the knuckle out of the way.

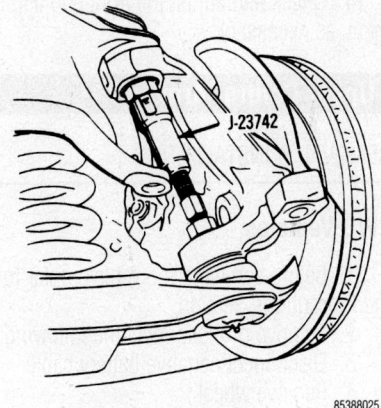

Use a ball joint separator to drive the lower joint from the knuckle

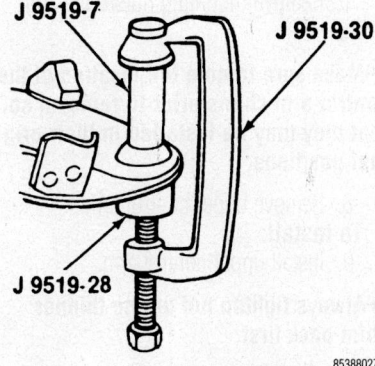

Driving the lower joint from the control arm

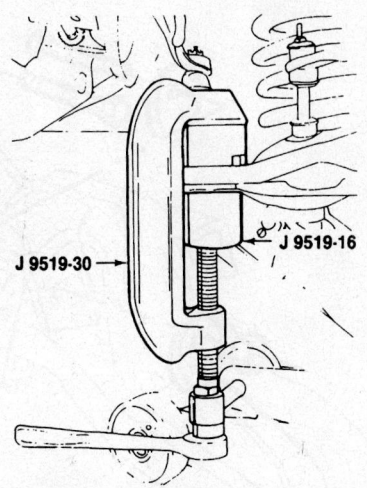

Installing a new ball joint

9. Remove grease fitting.

10. Remove ball joint from the control arm using the ball joint remover set along with the appropriate adapters.

To install:

11. Clean the tapered hole in the steering knuckle of any dirt or foreign matter,

then check the hole to see if it is out of round, deformed or otherwise damaged. If a problem is found, then knuckle must be replaced.

12. Press the new ball joint (with grease fitting pointing inward) until it bottoms in the control arm using a suitable installation set. Make sure the grease seal is facing inboard.

13. Install ball joint stud into the steering knuckle.

14. Install ball joint retaining nut and tighten to 90 ft. lbs. (125 Nm).

➡**When installing the cotter pin, never loosen the castle nut to expose the cotter pin hole.**

15. Install grease fitting into the ball joint, if not already installed.

16. Use a grease gun to lubricate the joint until grease appears at the seal.

17. Install brake caliper.

18. Install tire and wheel assembly.

19. Check and adjust the front end alignment, as necessary.

4WD Vehicles

On 4WD vehicles both the upper and lower ball joints are removed in the same manner. Once the joint is separated from the steering knuckle the rivets are drilled and punched to free the joint from the control arm. Service joints are bolted into position with the retaining bolts threaded upward from beneath the control arm. In this manner, the joint is replaced in an almost identical fashion to the upper joints on 2WD vehicles.

1. Remove tire and wheel assembly.

2. Remove wheel speed sensor wiring connector from the upper control arm, if removing the upper ball joint.

3. Remove the brake hose bracket from the upper control arm, if necessary.

4. Remove cotter pin from the ball joint, then loosen the retaining nut.

5. Position a suitable ball joint separator tool such as J-36607, then carefully loosen the joint in the steering knuckle. Remove the tool and the retaining nut, then separate the joint from the knuckle.

➡**After separating the steering knuckle from the upper ball joint, be sure to support the steering knuckle/hub assembly to prevent damaging the brake hose.**

6. Remove the riveted ball joint from the control arm:

  a. Drill a ⅛ in. (3mm) hole, about ¼ in. (6mm) deep into each rivet.

b. Then use a ½ in. (13mm) drill bit, to drill off the rivet heads.

c. Using a pin punch and the hammer, drive out the rivets in order to free the ball joint from the control arm assembly, then remove the ball joint.

To install:

7. Install ball joint in the control arm.

8. Install ball joint retaining nuts and bolts. Position the bolts threaded upward from under the control arm. Tighten the ball joint retainers to 17 ft. lbs. (23 Nm).

9. Install ball joint to the knuckle. Make sure the joint is seated, tighten the lower nut to 79 ft. lbs. (108 Nm) and the upper nut to 61 ft. lbs. (83 Nm). Install a new cotter pin.

➡**When installing the cotter pin, never loosen the castle nut to expose the cotter pin hole, but DO NOT tighten more than an additional ⅙ turn.**

10. Use a grease gun to lubricate the upper ball joint.

11. Install the brake hose brake on the upper control arm, if necessary.

12. Install wheel speed sensor wiring connector to the upper control arm, if the upper ball joint was removed.

13. Install tire and wheel assembly.

14. Check and adjust the front end alignment, as necessary.

Upper Control Arm

REMOVAL & INSTALLATION

2WD Vehicles

1. Before servicing the vehicle, refer to the Precautions Section.

2. Remove or disconnect the following:

3. Disconnect negative battery cable.

4. Remove wheel.

5. Remove wheel speed sensor harness bracket retaining bolt and nut, if equipped.

6. Remove steering knuckle from upper control arm ball joint.

7. Remove mounting nuts/bolts and shims.

➡**Make sure to note the location of the control arm shims prior to removal so that they may be installed in their original positions.**

8. Remove upper control arm.

To install:

9. Install upper control arm.

➡**Always tighten nut on the thinner shim pack first.**

10. Install mounting nuts/bolts and shims. Torque the nuts to 81 ft. lbs. (110 Nm).

11. Install steering knuckle to upper control arm ball joint.

12. Install new cotter pin.

➡**Tighten the nut to align the hole never loosen.**

13. Install wheel speed sensor harness bracket retaining bolt and nut, if equipped.

14. Install wheel.

4WD Vehicles

1. Before servicing the vehicle, refer to the Precautions Section.

2. Support the control arm with jackstands.

3. Remove tire and wheel assembly.

4. Remove brake hose bracket nut and bolt.

5. Remove speed sensor bracket nut and bolt.

6. Remove upper ball joint from the steering knuckle.

7. Remove upper control arm cam hardware nuts, cams, and bolts.

8. Remove upper control arm.

To install:

9. Install upper control arm.

10. Install upper control arm cam hardware bolts and the cams, making sure the bolt heads are opposed inside the bracket and that the cam lobes point down.

➡**Tighten the nuts with the control arm at Z height.**

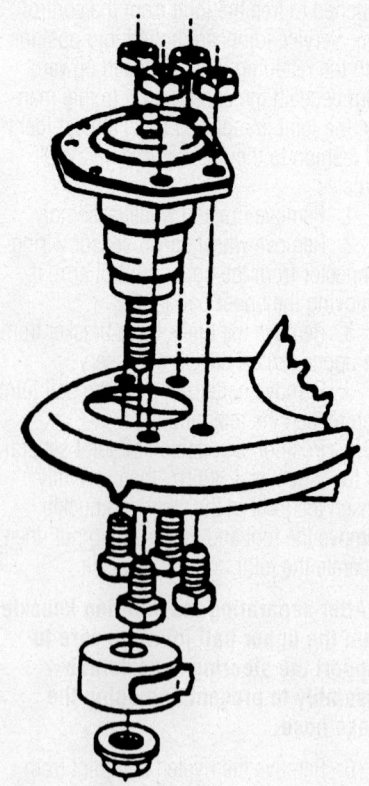

The replacement ball joint comes with nuts and bolts for installation

7924JG40

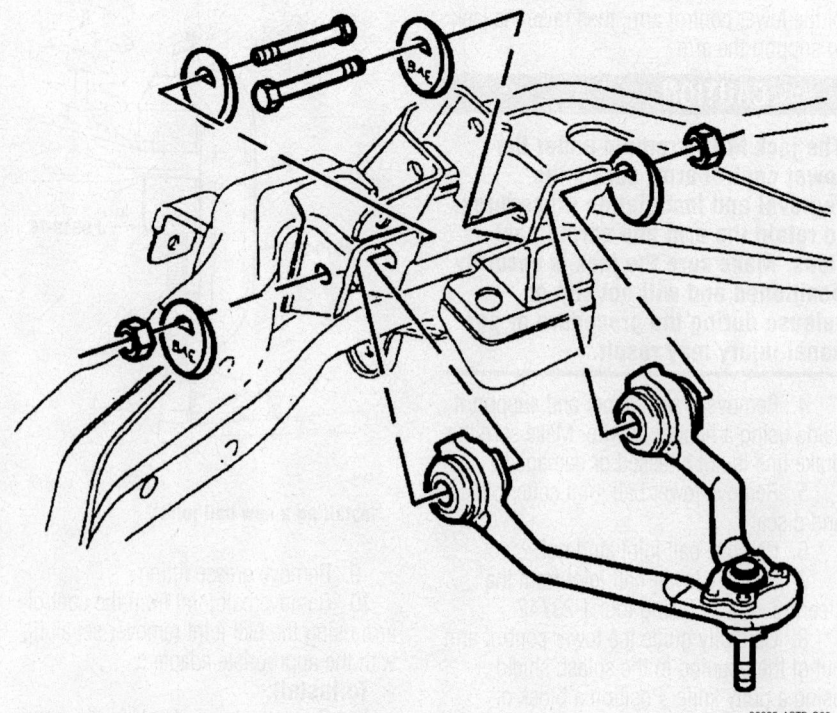

Exploded view of the upper control arm mounting bolts.

06025-ASTR-G02

11. Install new upper control arm cam hardware nuts. Tighten the front nut first, then the rear nut to 103 ft. lbs. (140 Nm).

12. Install the remaining components and check the wheel alignment.

CONTROL ARM BUSHING REPLACEMENT

2WD Vehicles

1. Before servicing the vehicle, refer to the Precautions Section.

2. Remove upper control arm and place it in a vice.

3. Remove upper control arm shaft nuts and retainers.

4. Remove upper control arm bushings using tool J 22269-1, a slotted washer and a short piece if pipe that is slightly larger than the bushing.

5. Remove upper control arm shaft.

To install:

6. Install upper control arm shaft.

7. Install upper control arm bushings using tool J 22269-1, a slotted washer and a short piece if pipe that is slightly larger than the bushing.

8. Tighten J 22269-1 until the bushing is positioned on the shaft and the control arm as shown in the accompanying illustration. The measurement should be 0.48–0.52 inch (12.8–13.8mm) at both sides when the properly installed.

9. Install upper control arm shaft nuts and retainers. Tighten to 85 ft. lbs. (115 Nm).

10. Install upper control arm.

4WD Vehicles

1. Before servicing the vehicle, refer to the Precautions Section.

2. Remove upper control arm.

3. Remove bushings from the arm.

4. Installation is the reverse of removal.

Lower Control Arm

REMOVAL & INSTALLATION

2WD Vehicles

1. Before servicing the vehicle, refer to the Precautions Section.

2. Remove coil spring.

3. Disconnect lower ball joint from the steering knuckle using special tool J–23742.

4. Remove lower control arm from the vehicle.

To install:

5. Install lower control arm.

6. Install lower ball joint stud into the steering knuckle.

7. Install ball joint-to-steering knuckle nut and tighten to 90 ft. lbs. (125 Nm).

8. Install new cotter pin to the lower ball joint stud.

9. Install coil spring.

10. Align the vehicle.

4WD Vehicles

1. Before servicing the vehicle, refer to the Precautions Section.

➡**Tools Needed: universal tie rod separator J–24319–B, torsion bar unloader J–36202, upper ball joint separator J–36607, lower control arm bushing service kit J–36618 (if the control arm bushing are being replaced) and ball joint C-clamp J–9519–23. Parts Needed: whether or not the control arm or bushing are being replaced, NEW control arm retaining nut should be used once the old ones have been loosened and removed.**

2. Remove front wheels.

3. Remove 2 bolts from the front splash shield and pivot it in order to gain access to the tie rod.

4. Remove the torsion bar.

5. Remove stabilizer bar from the control arm (keeping all of the link hardware sorted for proper installation). If necessary, completely remove the bar from the vehicle for access.

6. Remove shock absorber.

7. Remove the brake rotor.

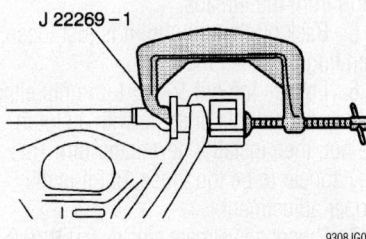

J 22269 – 1

9308JG04

Removing the upper control arm bushings using tool J22269-1—2WD

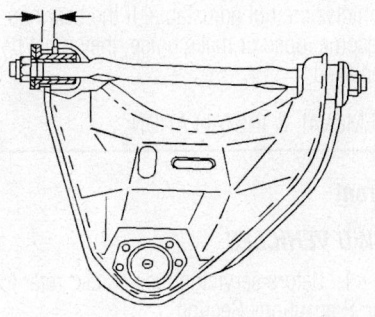

9308JG05

The bushing should be positioned as follows when correctly installed—2WD

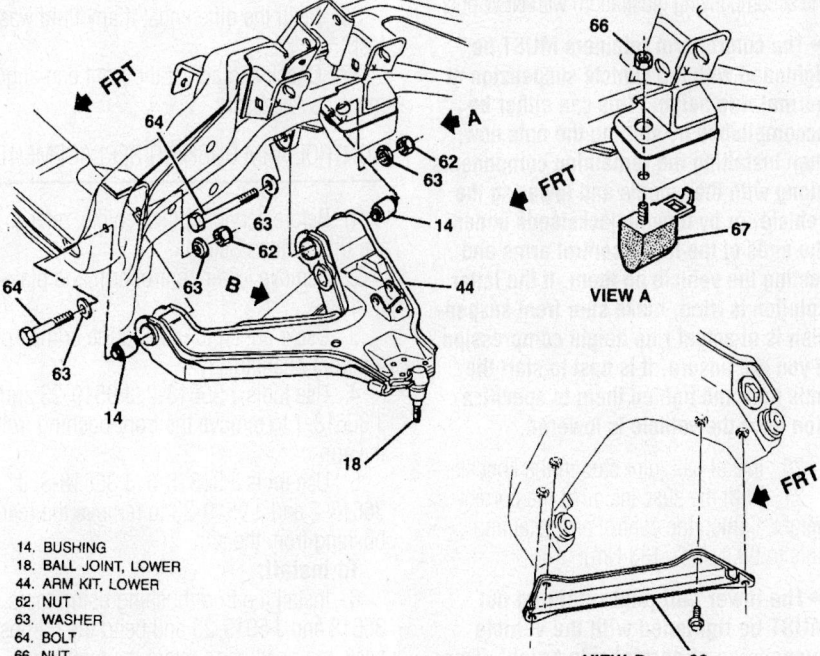

14. BUSHING
18. BALL JOINT, LOWER
44. ARM KIT, LOWER
62. NUT
63. WASHER
64. BOLT
66. NUT
67. BUMPER
68. BRACE

Exploded view of the lower control arm assembly mounting

88268GB1

8. Remove outer halfshaft nut and washer.

9. Disengage the halfshaft from the from the wheel hub and bearing.

10. Remove the outer tie rod cotter pin and retaining nut.

11. Disconnect the wheel speed sensor electrical connector.

12. Remove the wheel speed sensor bracket retaining bolts at the frame and upper control arm.

13. Disconnect the wheel speed sensor wiring harness clips form the upper control arm.

14. Remove the upper ball joint cotter pin and retaining nut.

15. Install special tool J-36607 to the upper control arm and steering knuckle and disconnect the upper control arm from the steering knuckle.

16. Remove lower ball joint cotter pin, nut and ball joint from the control arm using a ball joint separator.

17. Remove nuts and bolts and lower control arm. Note the direction which the control arm retaining bolts are facing for installation purposes.

To install:

18. Install lower control arm brace to the lower control arm and place the assembly into the vehicle. Position the front leg of the lower control arm into the crossmember before installing the rear leg into the frame bracket.

19. Install control arm bolts (facing in the direction as noted during removal or shown in the accompanying illustration) with NEW nuts.

➡**The control arm retainers MUST be tightened with the vehicle suspension at normal ride height. This can either be accomplished by starting the nuts now, then installing the remaining components along with the wheels and lowering the vehicle, or by moving jackstands under the ends of the lower control arms and resting the vehicle on them. If the latter solution is tried, make sure front suspension is at actual ride height compression. If you are unsure, it is best to start the nuts now and tighten them to specification once the vehicle is lowered.**

20. Install ball joint stud in the knuckle.

21. With the suspension at the correct height, tighten the control arm retaining nuts to 98 ft. lbs. (133 Nm).

➡**The lower ball joint retaining nut MUST be tightened with the vehicle suspension at normal ride height. This can either be accomplished by starting the nut now, then installing the remaining components along with the wheels**

and lowering the vehicle, or by moving jackstands under the ends of the lower control arms and resting the vehicle on them. If the latter solution is tried, make sure the FULL WEIGHT of the vehicle front end is on the suspension.

22. Install lower ball joint-to-control arm nut, then tighten the nut to 95 ft. lbs. (128 Nm) with the suspension at normal ride height and compression.

23. Install new cotter pin to the castellated nut. Tighten the nut (but no more than an additional 1/6 turn) in order to align the cotter pin. DO NOT loosen the nut from the specified torque.

24. Install adjuster arm by sliding the adapter forward, over the torsion bar to install the sides of the nut. Load the torsion bar and install the adjuster bolt aligning the installation mark.

25. Install drive axle through the hub and bearing assembly.

26. Tighten the hub and bearing assembly retaining bolts.

27. Install halfshaft nut and washer.

28. Install inner tie rod end to the relay rod.

29. Install shock absorber.

30. Install stabilizer bar, if removed.

31. Install stabilizer link(s) to the control arm(s).

32. Install torsion bar.

33. Install splash shield.

34. Install front wheels.

35. Recheck all fasteners for proper torque and installation before road testing.

36. Refill the differential if any fluid was lost.

37. Check and adjust the front end alignment, as necessary.

CONTROL ARM BUSHING REPLACEMENT

1. Before servicing the vehicle, refer to the Precautions Section.

2. Remove lower control arm and place it in vise.

3. Use a punch to unbend the crimps on the front bushing.

4. Use tools J 36618-2, J 9519-23 and J 36618-1 to remove the front bushing from the arm.

5. Use tools J 36618-5, J 36618-3, J 36618-2 and J 9519-23 to remove the rear bushing from the arm.

To install:

6. Install the front bushing using tools J 36618 and J 5919-23 and bend the crimps back into position to retain the bushing.

7. Install the rear bushing using tools J 36618 and J 5919-23 .

8. Install the lower control arm.

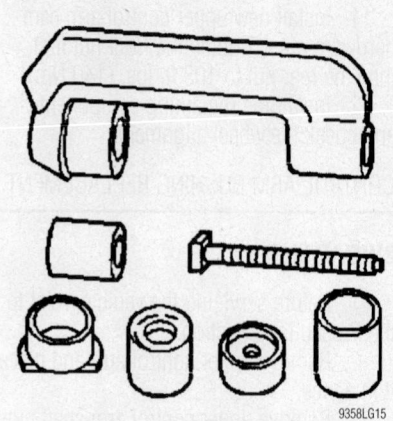

You need a variety of special tools to replace the control arm bushing

Wheel Bearings

ADJUSTMENT

2WD Vehicles

1. Before servicing the vehicle, refer to the Precautions Section.

2. If equipped, remove the wheel/hub cover for access, then remove the dust cap from the hub.

3. Remove the cotter pin and loosen the spindle nut.

4. Spin the wheel forward by hand and torque the nut to 12 ft. lbs. (16 Nm) in order to fully seat the bearings and remove any burrs from the threads.

5. Back off the nut until it is just loose, then finger-tighten the nut.

6. Loosen the nut 1/4–1/2 turn until either hole in the spindle lines up with a slot in the nut, then install a new cotter pin. This may appear to be too loose, but it is the proper adjustment.

7. Proper adjustment creates 0.001–0.005 in. (0.025–0.127mm) end-play.

4WD Vehicles

The front wheel bearings on the 4WD vehicles are not adjustable. If the bearings become loose or make noise, they must be replaced.

REMOVAL & INSTALLATION

Front

2WD VEHICLES

1. Before servicing the vehicle, refer to the Precautions Section.

2. Remove wheel.

3. Remove brake caliper with the pads without disconnecting the brake line.

4. Remove grease cap.

5. Remove cotter pin, spindle nut and washer.

6. Remove hub.

✳ WARNING

Be careful not to drop the outer wheel bearing. As the hub is pulled forward, the outer wheel bearings will often fall forward and they may easily be removed at this time.

7. Remove outer roller bearing assembly.

8. Remove inner seal by prying it out of the hub and discard it.

9. Remove inner bearing assembly.

To install:

10. Clean all parts in solvent and allow to air dry, then check for excessive wear or damage. Inspect all of the parts for scoring, pitting or cracking and replace if necessary.

➡**DO NOT remove the bearing races from the hub, unless they show signs of damage.**

11. If it is necessary to remove the wheel bearing races, use the GM front bearing race removal tool J-29117 to drive the races from the hub/disc assembly. A hammer and brass drift may also be used to drive the races from the hub, but the race removal tool is quicker.

12. If the bearing races were removed, position the replacement races in the freezer for a few minutes and then install them to the hub:

a. Lightly lubricate the inside of the hub/disc assembly using wheel bearing grease.

b. Using the GM seal installation tools J-8092 and J-8850, drive the inner bearing race into the hub/disc assembly until it seats. Be sure the race is properly seated against the hub shoulder and is not cocked.

➡**When installing the bearing races, be sure to support the hub/disc assembly with GM tool J-9746-02.**

c. Using the GM seal installation tools J-8092 and J-8457, drive the outer race into the hub/disc assembly until it seats.

13. Using a high melting point wheel bearing grease, lubricate the bearings, races and spindle; be sure to place a gob of grease (inside the hub/disc assembly) between the races to provide an ample supply of lubricant.

➡**To lubricate each bearing, place a gob of grease in the palm of the hand,** then scoop the bearing through the grease until it is well lubricated.

14. Place the inner bearing in the hub, then apply a thin coating of grease to the sealing lip and install a new inner seal, making sure the seal flange faces the bearing cup.

➡**Although a seal installation tool is preferable, a section of pipe with a smooth edge or a suitably sized socket may be used to drive the seal into position. Be sure the seal is flush with the outer surface of the hub assembly.**

15. Install wheel hub over the spindle.

16. Install outer bearing into the hub by hand.

17. Install spindle washer and nut.

18. Install brake caliper.

19. Install wheel.

20. Properly adjust the wheel bearings.

21. Install new cotter pin.

22. Install dust cap.

23. Install wheel cover.

4WD VEHICLES

1. Before servicing the vehicle, refer to the Precautions Section.

2. Install Torsion Bar Unloading tool J 36202 on the torsion bar adjusting bolt and remove the bolt. To aid during installation, count the number of turns required to remove the bolt.

3. Remove the wheel.

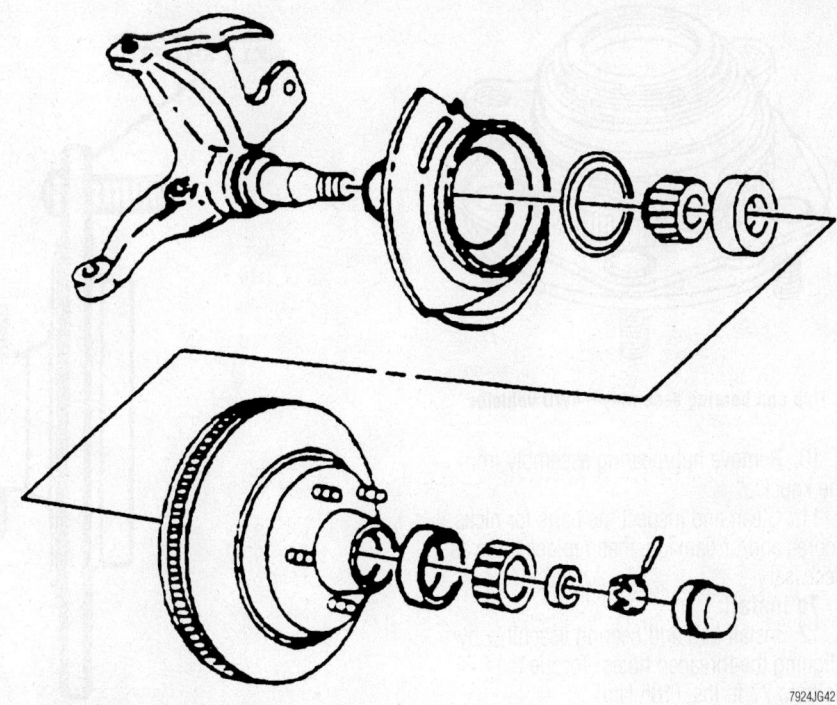

Wheel bearings, races and related components—2WD vehicles

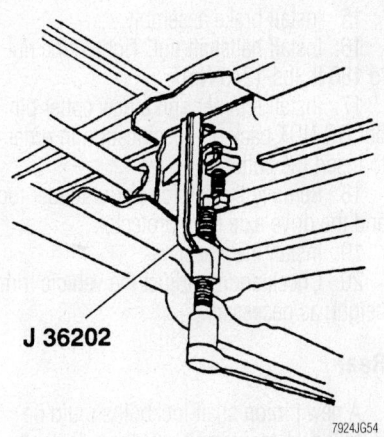

J 36202

Use Torsion Bar Unloading tool J 36202 to remove the adjusting bolt and unload the torsion bar

4. Install an axle shaft boot seal protector to the Tri-Pod axle joint.

5. Remove cotter pin and retainer.

6. Remove castle nut and the thrust washer.

7. Remove brake caliper and support it aside using wire or a coat hanger.

➡**Be sure the brake line is not stretched or damaged.**

8. Remove brake disc from the wheel hub.

9. Remove halfshaft from the hub/bearing assembly, using a Spindle Remover tool J-28733-A to prevent damage to the shaft or hub/bearing assembly.

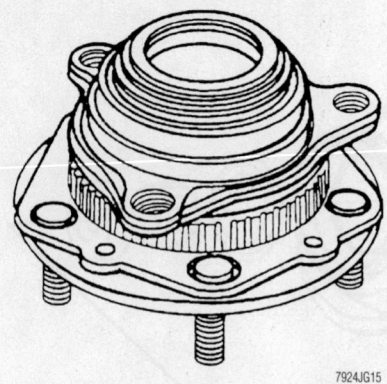

Hub and bearing assembly—4WD vehicles

10. Remove hub/bearing assembly from the knuckle.

11. Clean and inspect the parts for nicks, scores and/or damage, then replace them as necessary.

To install:

12. Install hub and bearing assembly by aligning the threaded holes. Torque the bolts to 77 ft. lbs. (105 Nm).

13. Install tie rod end to the steering knuckle using the retaining nut.

14. Install new cotter pin.

15. Install brake assembly.

16. Install halfshaft nut. Tighten the nut to 180 ft. lbs. (245 Nm).

17. Install retainer and a new cotter pin but DO NOT back off specification in order to insert the cotter pin.

18. Remove the torsion bar unloader tool and the drive axle boot protector.

19. Install the wheel.

20. Check and/or adjust the vehicle trim height, as necessary.

Rear

A new pinion shaft lockbolt should be installed whenever either of the axle shafts is removed.

The axle shaft and seal may be removed and replaced without disturbing the bearing or seal but it is highly recommended to replace the seals when removing the axle shaft.

1. Before servicing the vehicle, refer to the Precautions Section.

2. Remove rear wheels.

3. Remove brake drums.

4. Using a wire brush, clean the dirt/rust from around the rear axle cover.

5. Drain the fluid.

6. Remove rear pinion shaft lockbolt and the pinion shaft.

7. Remove C-lock from the button end of the axle shaft by pushing the axle shaft inward.

8. Remove axle shaft from the axle housing.

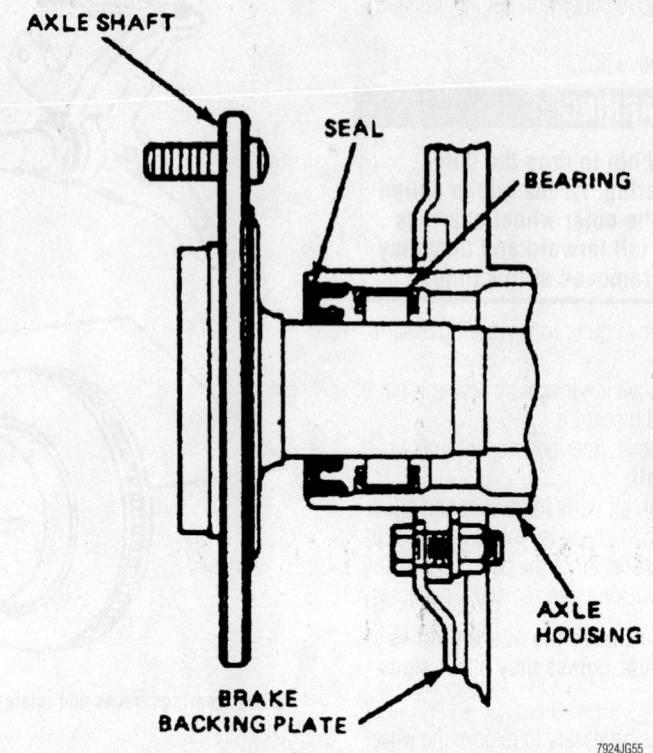

Cross-sectional view of the rear axle, bearing and seal assembly

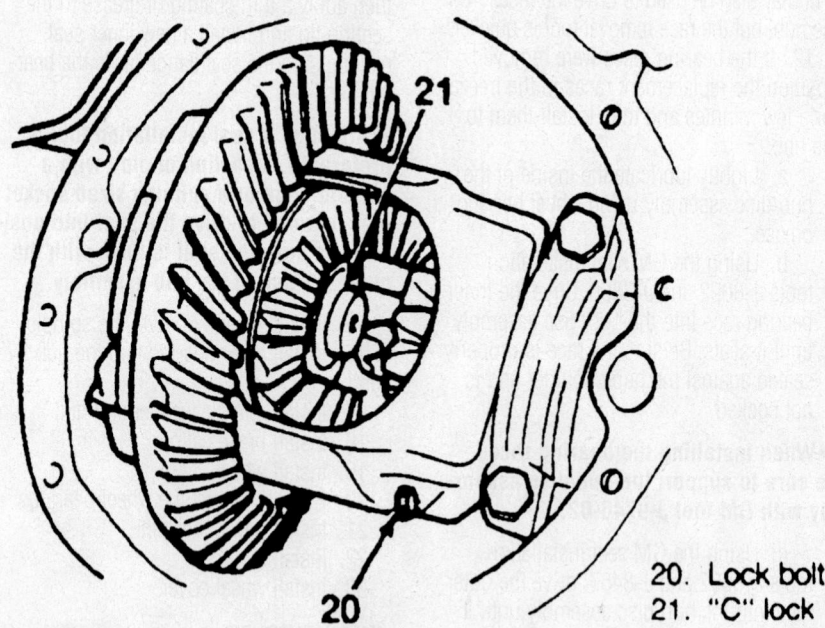

Pinion shaft lockbolt and axle C-lock locations, inside the differential

20. Lock bolt
21. "C" lock

➡ Be careful not to damage the oil seal.

✳✳ WARNING

If equipped with an Anti-Lock Brake System (ABS), be careful not to damage the reflector ring on the axle shaft or the speed sensor bolted to the backing plate, immediately adjacent to the shaft.

9. Remove oil seal by prying the it from the end of the rear axle housing.

DO NOT damage the housing oil seal surface.

10. Remove wheel bearing using the GM Slide Hammer tool J-2619, the GM Adapter tool J-2619-4 and the GM Axle Bearing Puller tool J-22813-01.

To install:

11. Clean and inspect the components for excessive wear or damage and replace them, if necessary.

12. Install new or reused bearing, coated with gear lubricant, using the Axle Shaft Bearing Installer tool J-34974 to drive the bearing in until it bottoms against the seat.

Be sure the bearing installer does not contact and damage the speed sensor on ABS equipped vehicles.

13. Install new seal lubricated with gear oil using the GM Axle Shaft Seal Installer tool J-33782 to seat it in the housing until it is flush with the axle tube.

➡**Be sure the seal installer does not contact and damage the speed sensor on ABS equipped vehicles.**

14. Install axle shaft into the housing by engaging the splines.

15. Install C-lock retainer on the axle shaft button end.

BE CAREFUL not to damage the wheel bearing seal.

16. Install axle shaft by pulling it outward to seat the C-lock retainer in the counterbore of the side gears.

17. Install pinion shaft through the case and the pinions. Tighten the new lockbolt to 27 ft. lbs. (36 Nm).

18. Install new rear axle cover gasket.

19. Install housing cover.

20. Install brake drums.

21. Install wheels.

22. Refill the housing.

BRAKES

Brake Caliper

REMOVAL & INSTALLATION

Front

1. Before servicing the vehicle, refer to the Precautions Section.

2. Remove ⅔ of the brake fluid from the master cylinder reservoir.

3. Remove tire and wheel assembly.

4. Disconnect brake fluid line and plug it.

5. Remove bolts retaining the caliper to the rotor.

6. Remove caliper from the rotor.

7. Remove disc brake pads from the caliper.

8. Remove disc brake pad retaining clips from inside the caliper.

To install:

9. Clean and lubricate the sleeves and bushings with silicon grease.

10. Install sleeves and bushings.

11. Install pads in the caliper.

12. Install caliper in position over the rotor.

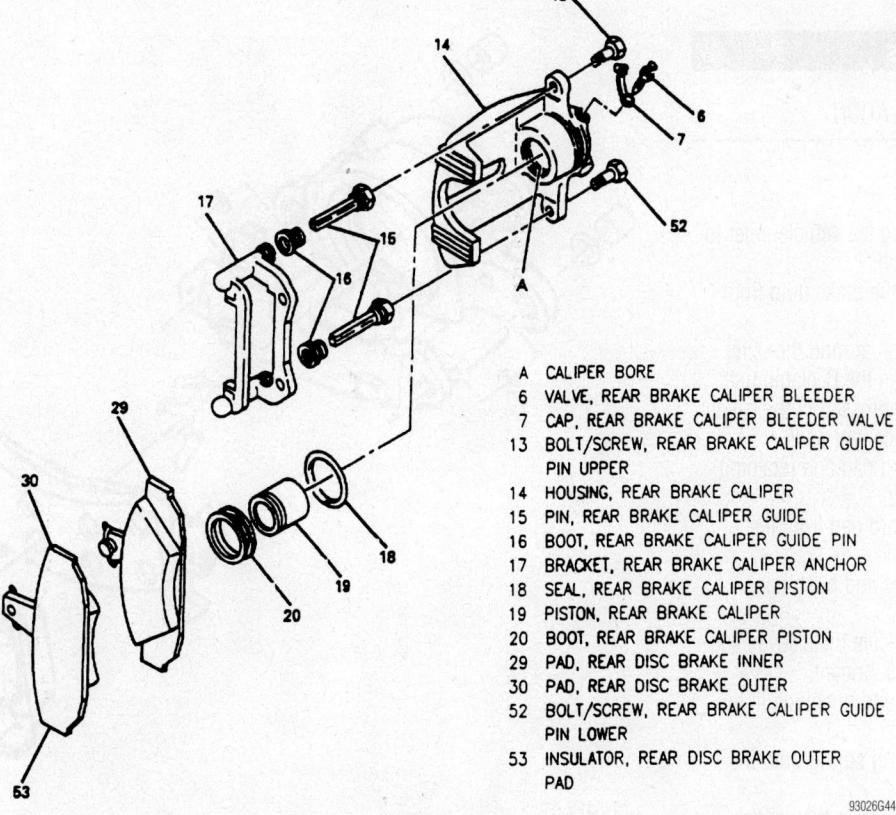

A CALIPER BORE
6 VALVE, REAR BRAKE CALIPER BLEEDER
7 CAP, REAR BRAKE CALIPER BLEEDER VALVE
13 BOLT/SCREW, REAR BRAKE CALIPER GUIDE PIN UPPER
14 HOUSING, REAR BRAKE CALIPER
15 PIN, REAR BRAKE CALIPER GUIDE
16 BOOT, REAR BRAKE CALIPER GUIDE PIN
17 BRACKET, REAR BRAKE CALIPER ANCHOR
18 SEAL, REAR BRAKE CALIPER PISTON
19 PISTON, REAR BRAKE CALIPER
20 BOOT, REAR BRAKE CALIPER PISTON
29 PAD, REAR DISC BRAKE INNER
30 PAD, REAR DISC BRAKE OUTER
52 BOLT/SCREW, REAR BRAKE CALIPER GUIDE PIN LOWER
53 INSULATOR, REAR DISC BRAKE OUTER PAD

93026G44

Rear brake caliper

13. Install mounting bolts and tighten the bolts to 38 ft. lbs. (51 Nm) for single piston calipers; 85 ft. lbs. (115 Nm) for dual piston calipers.

14. Connect fluid lines to the caliper, if disconnected, and tighten to 33 ft. lbs. (45 Nm).

15. Install wheel and tire assembly.

16. Refill the master cylinder to the correct level. Bleed the brake system if the fluid lines were disconnected from the caliper.

Rear

1. Before servicing the vehicle, refer to the Precautions Section.

2. Remove rear wheels.

3. Disconnect brake hose and cap line.

4. Remove retainers from caliper and remove caliper.

To install:

5. Install brake pads if removed.

6. Install caliper over rotor, and onto mounts.

7. Install retainers, and tighten to 23 ft. lbs. or (31 Nm).

8. Connect brake hose, and tighten to 20 ft. lbs. (27 Nm).

9. Bleed brake system.

10. Install tires and wheels.

11. Refill the master cylinder and pump pedal to attain full brake pedal before Road-testing the vehicle.

Disc Brake Pads

REMOVAL & INSTALLATION

Front

1. Before servicing the vehicle, refer to the Precautions Section.

2. Remove ⅔ of the brake fluid from the master cylinder.

3. Place a C-clamp around the outer pad and caliper; tighten the C-clamp until the piston is fully compressed in the caliper.

4. Remove brake caliper pads.

5. Remove inboard pad and retaining spring from the caliper.

6. Remove outboard pad from the caliper.

7. Remove sleeves and bushings.

To install:

8. Clean and lubricate the sleeves and bushing with silicone lubricant.

9. Install sleeves and bushings in the caliper.

10. Clip the retaining spring onto the inboard pad.

11. Install inboard pad in the caliper.

12. Install outboard pad into the caliper.

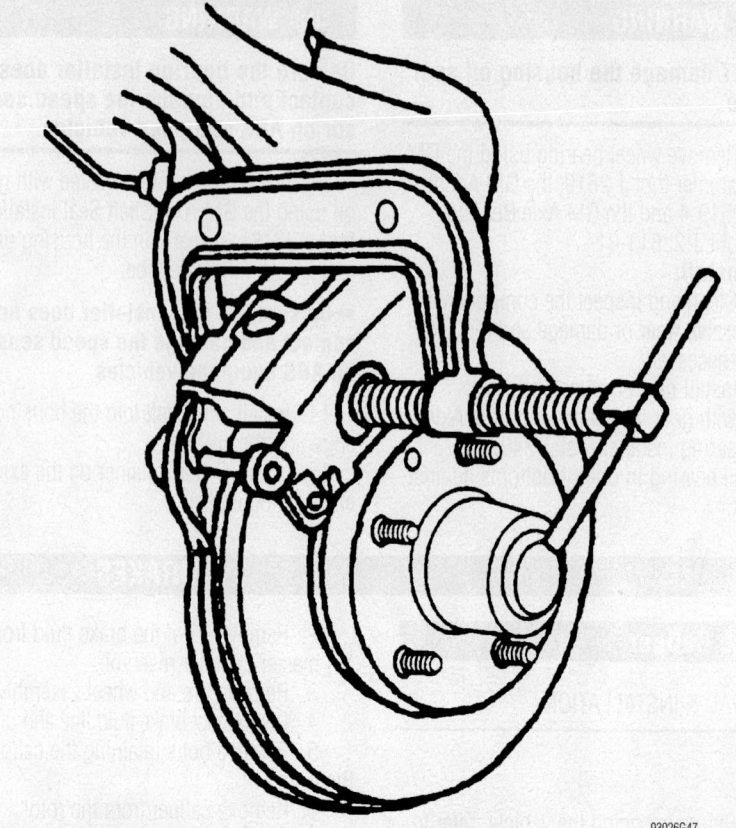

Compressing the caliper piston with a C-clamp—Astro and Safari

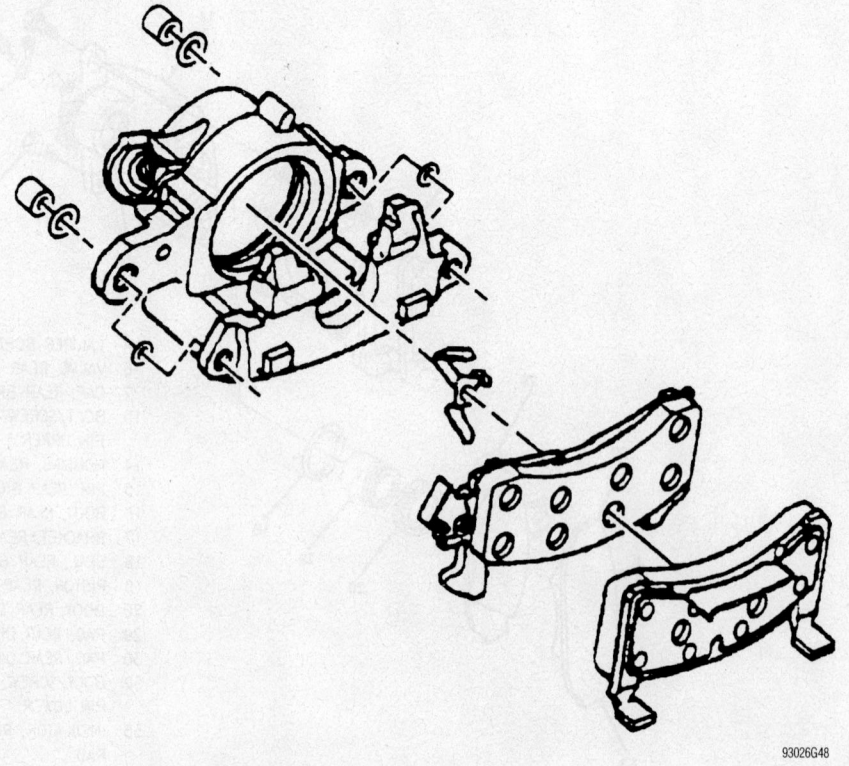

Exploded view of the disc brake assembly—Astro and Safari

13. Install caliper in position over the rotor.

14. Install mounting bolts. Bend the tabs, on the outboard brake pad, over the caliper.

15. Install wheel and tire assemblies.

16. Refill the master cylinder and pump pedal to attain full brake pedal before Road-testing the vehicle.

Rear

1. Before servicing the vehicle, refer to the Precautions Section.

2. Remove ⅔ of the brake fluid from the master cylinder.

3. Remove wheels and tires.

4. Place a C-clamp around the outer pad and caliper; tighten the C-clamp until the piston is fully compressed in the caliper. Remove top caliper retainer, and rotate caliper away from rotor.

5. Remove inboard pad and retaining spring from the caliper.

6. Remove outboard pad from the caliper.

To install:

7. Clean and lubricate the sleeves and bushing with silicone lubricant.

8. Install sleeves and bushings in the caliper.

9. Clip the retaining spring onto the inboard pad.

10. Install inboard pad in the caliper.

11. Install outboard pad into the caliper.

12. Install caliper in position over the rotor.

13. Install mounting bolts.

14. Install wheel and tire assemblies.

15. Refill the master cylinder and pump pedal to attain full brake pedal before Road-testing the vehicle.

Brake Drums

REMOVAL & INSTALLATION

1. Before servicing the vehicle, refer to the Precautions Section.

2. Remove wheel and tire assembly.

3. Remove brake drum. If the drum will not pull of the axle, use a rubber mallet and tap it around the edge.

To install:

4. Install drum on the axle.

5. Install wheel and tire assembly.

6. Refill the master cylinder and pump pedal to attain full brake pedal before road-testing the vehicle.

Brake Shoes

REMOVAL & INSTALLATION

1. Before servicing the vehicle, refer to the Precautions Section.

2. Remove wheel and tire assembly.

3. Remove brake drum.

4. Remove return springs from the brake shoes.

5. Remove shoe guide.

6. Remove hold-down springs and pins.

7. Remove actuator lever and pivot.

8. Remove lever return spring.

9. Remove actuator link.

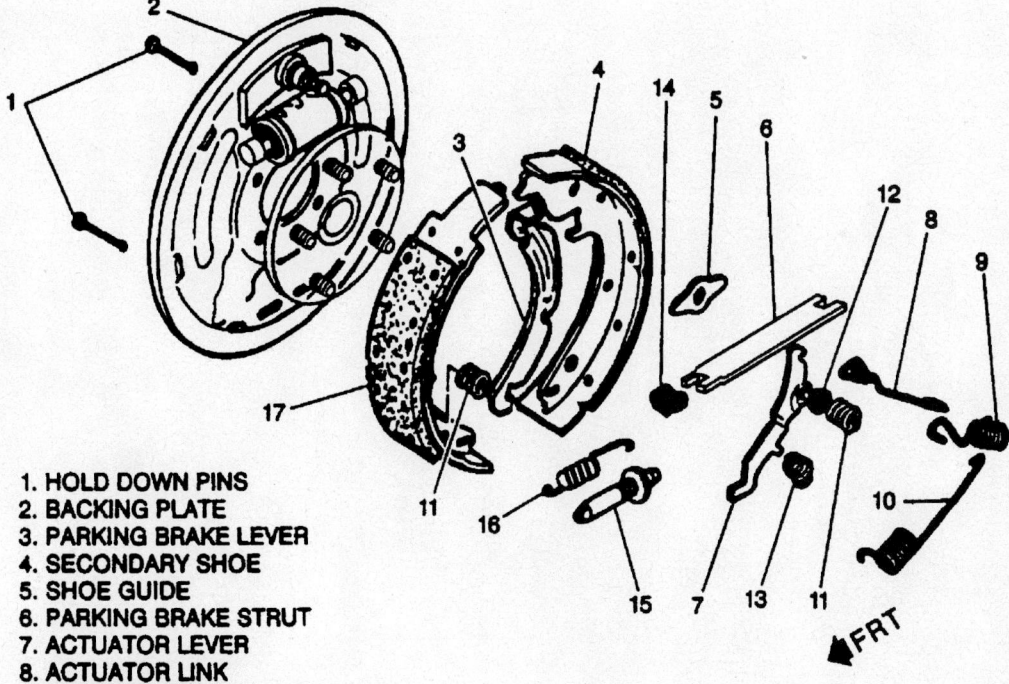

1. HOLD DOWN PINS
2. BACKING PLATE
3. PARKING BRAKE LEVER
4. SECONDARY SHOE
5. SHOE GUIDE
6. PARKING BRAKE STRUT
7. ACTUATOR LEVER
8. ACTUATOR LINK
9. RETURN SPRING
10. RETURN SPRING
11. HOLD DOWN SPRING
12. LEVER PIVOT
13. LEVER RETURN SPRING
14. STRUT SPRING
15. ADJUSTING SCREW ASSEMBLY
16. ADJUSTING SCREW SPRING
17. PRIMARY SHOE

93026G51

Exploded view of the drum brake components—Astro & Safari

10. Remove parking brake strut and spring.

11. Remove parking brake lever.

12. Remove brake shoes and the adjuster assembly.

To install:

13. Lubricate the contact points on the backing plate and the adjuster with lithium grease.

14. Install parking brake lever, adjusting screw and spring assembly.

15. Install shoe assembly onto the backing plate.

16. Install parking brake lever, strut and strut spring.

17. Install actuator lever and lever pivot.

18. Install actuator link.

19. Install lever spring, the hold-down pins and springs.

20. Install shoe guide.

21. Install return springs and install the brake drum in position.

22. Adjust the brakes as follows:

a. Remove the knockout area in the backing plate, behind the adjuster assembly.

b. Ensure the parking brake system is adjusted properly with no tension on the cables or parking brake lever. The tops of the shoes should be firmly seated against the upper spring retaining anchor, if not as specified, loosen the parking brake cables.

c. Install the drum and turn the brake adjuster until the wheels can just be turned by hand.

d. Then, back the adjuster off 24 notches. No brake drag should be felt after 12 notches.

e. Install an adjusting hole plug in the backing plate to prevent dirt and moisture from entering.

f. Readjust the parking brake cable as necessary.

23. Install the wheel and tire assemblies.

24. Refill the master cylinder and pump pedal to attain full brake pedal before Road-testing the vehicle.

OLDSMOBILE

Aurora

3

BRAKES 3-46
DRIVE TRAIN 3-37
ENGINE REPAIR................3-9
FUEL SYSTEM 3-34
SPECIFICATION AND
MAINTENANCE CHARTS 3-2
Engine and Vehicle Identification 3-2
General Engine Specifications 3-2
Engine Tune-Up Specifications 3-3
Firing Order 3-3
Accessory Drive Belt Routing 3-4
Capacities 3-4
Crankshaft and Connecting Rod
  Specifications 3-5
Valve Specifications 3-5
Piston and Ring Specifications 3-6
Torque Specifications 3-6
Wheel Alignment 3-7
Tire, Wheel and Ball Joint
  Specifications 3-7
Brake Specifications 3-7
Scheduled Maintenance
  Intervals.............................3-8
STEERING AND
SUSPENSION 3-40
A
Air Bag.................................3-40
  Arming.................................3-41
  Disarming.............................3-40
  Precautions...........................3-40
Alternator 3-9
  Installation 3-9
  Removal 3-9
B
Brake Caliper 3-46
  Removal & Installation..............3-46
C
Camshaft and Valve Lifters 3-20
  Removal & Installation..............3-20
Coil Spring 3-42
  Removal & Installation..............3-42

CV-Joint 3-38
  Removal & Installation..............3-38
Cylinder Head 3-14
  Removal & Installation..............3-14
D
Disc Brake Pads.........................3-49
  Removal & Installation..............3-49
E
Engine Assembly 3-9
  Removal & Installation..............3-9
Exhaust Manifold 3-18
  Removal & Installation..............3-18
F
Fuel Filter 3-34
  Removal & Installation..............3-34
Fuel Injector 3-35
  Removal & Installation..............3-35
Fuel Pump 3-35
  Removal & Installation..............3-35
Fuel System Pressure 3-34
  Relieving.............................3-34
Fuel System Service
  Precautions...........................3-34
H
Halfshaft...............................3-38
  Removal & Installation..............3-38
Heater Core 3-13
  Removal & Installation..............3-13
I
Ignition Timing 3-9
  Adjustment............................3-9
Intake Manifold 3-16
  Removal & Installation..............3-16
L
Lower Ball Joint........................3-43
  Removal & Installation..............3-43
Lower Control Arm 3-43
  Control Arm Bushing
    Replacement 3-44
  Removal & Installation..............3-43

O
Oil Pan.................................3-24
  Removal & Installation..............3-24
Oil Pump 3-25
  Removal & Installation..............3-25
P
Piston and Ring 3-32
  Positioning 3-32
Power Rack and Pinion Steering
  Gear.................................3-41
  Removal & Installation..............3-41
R
Rear Main Seal 3-27
  Removal & Installation..............3-27
Rocker Arms 3-16
  Removal & Installation..............3-16
S
Shock Absorber 3-42
  Removal & Installation..............3-42
Starter Motor 3-24
  Removal & Installation..............3-24
Strut 3-42
  Removal & Installation..............3-42
T
Timing Chain, Sprockets, Front
  Cover and Seal 3-28
  Removal & Installation..............3-28
Transmission Assembly 3-37
  Removal & Installation..............3-37
V
Valve Lash 3-24
  Adjustment............................3-24
W
Water Pump 3-12
  Removal & Installation..............3-12
Wheel Bearings.........................3-45
  Adjustment............................3-45
  Removal & Installation..............3-45

SPECIFICATION CHARTS

ENGINE AND VEHICLE IDENTIFICATION

Engine								Model Year	
Code ①	Liters (cc)	Cu. In.	Cyl.	Fuel Sys.	Engine Type	Eng. Mfg.		Code ②	Year
H	3.5 (3475)	212	6	SFI	DOHC	BOC		1	2001
1 ③	3.8 (3785)	231	6	MFI	OHV	BOC		2	2002
C	4.0 (3995)	244	8	MFI	DOHC	BOC		3	2003
								4	2004

NOTE: Oldsmobile did not market an Aurora for 2000.

DOHC: Double Overhead Camshafts

OHV: Overhead Valves

MFI: Multi-point Fuel Injection

SFI: Sequential Fuel Injection

BOC: Buick/Oldsmobile/Cadillac

① 8th position of VIN

② 10th position of VIN

③ Supercharged Engine

42372-AUR-C01

GENERAL ENGINE SPECIFICATIONS

Year	Model	Engine Displacement Liters (cc)	Engine Series (ID/VIN)	Fuel System	Net Horsepower @ rpm	Net Torque @ rpm (ft. lbs.)	Bore x Stroke (in.)	Com- pression Ratio	Oil Pressure @ rpm
2001	Aurora	3.5 (3475)	H	SFI	215@5600	230@4400	3.52x3.62	9.3:1	30@2000
		4.0 (3995)	C	SFI	250@5600	260@4000	3.43x3.31	10.2:1	35@2000
2002	Aurora	3.5 (3475)	H	SFI	215@5600	230@4400	3.52x3.62	9.3:1	30@2000
		4.0 (3995)	C	SFI	250@5600	260@4000	3.43x3.31	10.2:1	35@2000
2003	Aurora	4.0 (3995)	C	SFI	250@5600	260@4400	3.43x3.31	10.2:1	35@2000

MFI: Multi-point Fuel Injection

SFI: Sequential Fuel Injection

① Supercharged engine

42372-AUR-C02

ENGINE TUNE-UP SPECIFICATIONS

Year	Engine Displacement Liters (cc)	Engine ID/VIN	Spark Plug Gap (in.)	Ignition Timing (deg.)	Fuel Pump (psi)	Idle Speed (rpm)	Valve Clearance	
							Intake	Exhaust
2001	3.5 (3475)	H	0.050	①	41-47	②	HYD	HYD
	4.0 (3995)	C	0.050	①	41-47	②	HYD	HYD
2002	3.5 (3475)	H	0.050	①	41-47	②	HYD	HYD
	4.0 (3995)	C	0.050	①	41-47	②	HYD	HYD
2003	4.0 (3995)	C	0.050	①	41-47	②	HYD	HYD

NOTE: The Vehicle Emission Control Information label often reflects specification changes made during production.

The label figures must be used if they differ from those in the chart.

HYD: Hydraulic

① DIS Ignition System timing is not adjustable

② Idle speed is maintained by the ECM. There is no recommended adjustment procedure

42372-AUR-C03

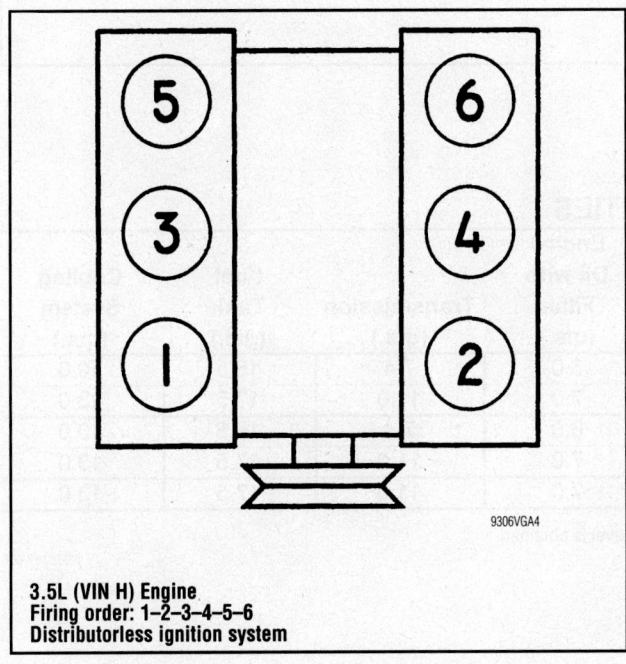

9306VGA4

3.5L (VIN H) Engine
Firing order: 1–2–3–4–5–6
Distributorless ignition system

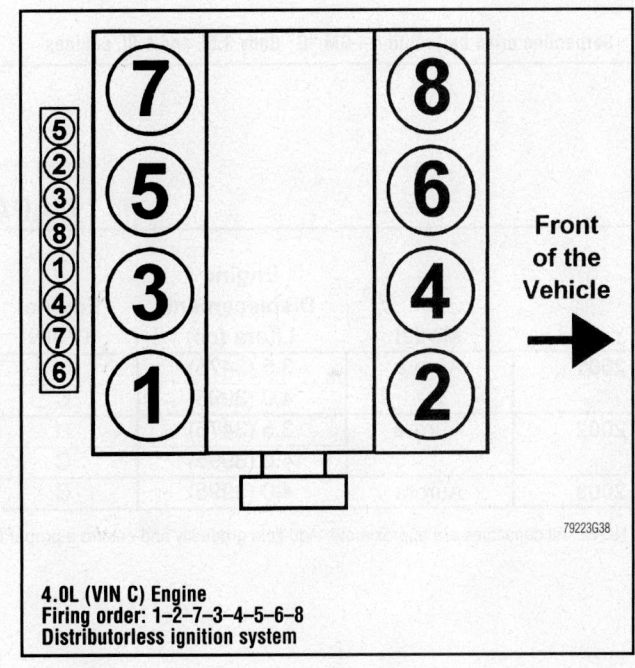

Front of the Vehicle
→

79223G38

4.0L (VIN C) Engine
Firing order: 1–2–7–3–4–5–6–8
Distributorless ignition system

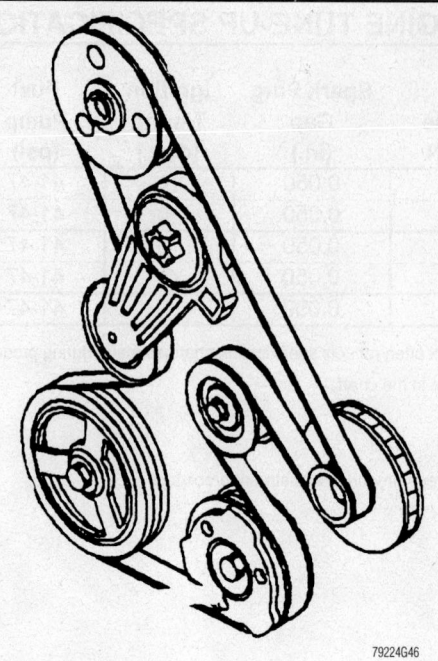

Serpentine drive belt routing—GM "G" Body 3.5L and 4.0L engines

CAPACITIES

Year	Model	Engine Displacement Liters (cc)	Engine ID/VIN	Engine Oil with Filter (qts.)	Transmission (qts.)	Fuel Tank (gal.)	Cooling System (qts.)
2001	Aurora	3.5 (3475)	H	6.0	7.4	18.5	10.0
		4.0 (3995)	C	7.0	11.0	17.5	13.0
2002	Aurora	3.5 (3475)	H	6.0	7.4	18.5	10.0
		4.0 (3995)	C	7.0	11.0	17.5	13.0
2003	Aurora	4.0 (3995)	C	7.0	11.0	17.5	13.0

NOTE: All capacities are approximate. Add fluid gradually and ensure a proper fluid level is obtained.

42372-AUR-C04

VALVE SPECIFICATIONS

Year	Engine Displacement Liters (cc)	Engine ID/VIN	Seat Angle (deg.)	Face Angle (deg.)	Spring Test Pressure (lbs. @ in.)	Spring Installed Height (in.)	Stem-to-Guide Clearance (in.)		Stem Diameter (in.)	
							Intake	Exhaust	Intake	Exhaust
2001	3.5 (3475)	H	45.75	45	136@0.964	1.377	0.0010-0.0030	0.0020-0.0040	0.233-0.234	0.233-0.234
	4.0 (3995)	C	45.75	45	136@0.964	1.190	0.0010-0.0030	0.0020-0.0040	0.233-0.234	0.233-0.234
2002	3.5 (3475)	H	45.75	45	136@0.964	1.377	0.0010-0.0030	0.0020-0.0040	0.233-0.234	0.233-0.234
	4.0 (3995)	C	45.75	45	136@0.964	1.190	0.0010-0.0030	0.0020-0.0040	0.233-0.234	0.233-0.234
2003	4.0 (3995)	C	45.75	45	136@0.964	1.138	0.0010-0.0030	0.0020-0.0040	0.233-0.234	0.233-0.234

NA: Not Available

42372-AUR-C05

CRANKSHAFT AND CONNECTING ROD SPECIFICATIONS
All measurements are given in inches.

Year	Engine Displacement Liters (cc)	Engine ID/VIN	Crankshaft				Connecting Rod		
			Main Brg. Journal Dia.	Main Brg. Oil Clearance	Shaft End-play	Thrust on No.	Journal Diameter ①	Oil Clearance	Side Clearance
2001	3.5 (3475)	H	2.7550-2.7560	0.0006-0.0021	0.0050-0.0200	3	2.1829-2.1835	0.0009-0.0025	0.0040-0.0130
	4.0 (3995)	C	2.5335-2.5337	0.0006-0.0025	0.0020-0.0200	2	2.2490	0.0010-0.0030	0.0080-0.0200
2002	3.5 (3475)	H	2.7550-2.7560	0.0006-0.0021	0.0050-0.0200	3	2.1829-2.1835	0.0009-0.0025	0.0040-0.0130
	4.0 (3995)	C	2.5335-2.5337	0.0006-0.0025	0.0020-0.0200	2	2.2490	0.0010-0.0030	0.0080-0.0200
2003	4.0 (3995)	C	2.5335-2.5341	0.0006-0.0025	0.0020-0.0197	2	2.1239-2.1245	0.0010-0.0030	0.0079-0.0197

① Large end of connecting rod ID

42372-AUR-C06

PISTON AND RING SPECIFICATIONS

All measurements are given in inches.

Year	Engine ID/VIN	Engine Displacement Liters (cc)	Piston Clearance	Ring Gap			Ring Side Clearance		
				Top Compression	Bottom Compression	Oil Control	Top Compression	Bottom Compression	Oil Control
2001	3.5 (3475)	H	0.0010-0.0025	0.008-0.018	0.014-0.020	0.010-0.030	0.0016-0.0037	0.0016-0.0037	Side-sealing
	4.0 (3995)	C	0.0008-0.0020	0.0098 0.0157	0.0138 0.0020	0.0098 0.0299	0.0016-0.0037	0.0016-0.0037	Side-sealing
2002	3.5 (3475)	H	0.0010-0.0025	0.008-0.018	0.014-0.020	0.010-0.030	0.0016-0.0037	0.0016-0.0037	Side-sealing
	4.0 (3995)	C	0.0008-0.0020	0.0098 0.0157	0.0138 0.0020	0.0098 0.0299	0.0016-0.0037	0.0016-0.0037	Side-sealing
2003	4.0 (3995)	C	0.0008-0.0020	0.0098 0.0157	0.0138 0.0020	0.0098 0.0299	0.0016-0.0037	0.0016-0.0037	Side-sealing

42372-AUR-C07

TORQUE SPECIFICATIONS

All readings in ft. lbs.

Year	Engine Displacement Liters (cc)	Engine ID/VIN	Cylinder Head Bolts	Main Bearing Bolts	Rod Bearing Bolts	Crankshaft Damper Bolts	Flywheel Bolts	Manifold		Spark Plugs	Lug Nuts
								Intake	Exhaust		
2001	3.5 (3475)	H	①	②	③	④	⑤	5	18	11	100
	4.0 (3995)	C	⑥	⑦	⑧	⑨	⑤	7.5	18	11	100
2002	3.5 (3475)	H	①	②	③	④	⑤	5	18	11	100
	4.0 (3995)	C	⑥	⑦	⑧	⑨	⑤	7.5	18	11	100
2003	4.0 (3995)	C	⑥	⑦	⑧	⑨	⑤	7.5	18	11	100

① M11 bolts:
   Step 1: 30 ft. lbs.
   Step 2: 100 degrees
   Step 3: 100 degrees
   M6 bolts: 22 ft. lbs.
② Cap Bolts
   Step 1: 15 ft. lbs.
   Step 2: 70 degrees
   Perimeter bolts: 22 ft. lbs.
③ Step 1: 22 ft. lbs.
   Step 2: Loosen completely
   Step 3: 18 ft. lbs.
   Step 4: 110 degrees
④ Step 1: 37 ft. lbs.
   Step 2: 120 degrees
⑤ Step 1: 11 ft. lbs.
   Step 2: 50 degrees

⑥ M11 bolts:
   Step 1: 30 ft. lbs.
   Step 2: 70 degrees
   Step 3: 60 degrees
   Step 4: 60 degrees
   M6 bolts: 106 inch lbs.
⑦ Step 1: 15 ft. lbs.
   Step 2: 65 degrees
⑧ Step 1: 22 ft. lbs.
   Step 2: Loosen completely
   Step 3: 18 ft. lbs.
   Step 4: 110 degrees
⑨ Step 1: 37 ft. lbs.
   Step 2: 120 degrees
⑪ New cylinder head 1st time installation: 20 ft. lbs.
   All others: 11 ft. lbs.

42372-AUR-C08

WHEEL ALIGNMENT

Year	Model		Caster Range (+/-Deg.)	Caster Preferred Setting (Deg.)	Camber Range (+/-Deg.)	Camber Preferred Setting (Deg.)	Toe-in (in.)	Steering Axis Inclination (Deg.)
2001	Aurora	F	+0.50	+6.00	+0.50	+0.20	0 +/- 0.09	12.50
		R	—	—	+0.50	-0.31	0.09 +/- 0.09	—
2002	Aurora	F	+0.50	+6.00	+0.50	+0.20	0 +/- 0.09	12.50
		R	—	—	+0.50	-0.31	0.09 +/- 0.09	—
2003	Aurora	F	+0.50	+5.00	+0.50	-0.20	0.20 +/- 0.20	NA
		R	—	—	+0.50	-0.30	0.20 +/- 0.20	—

F: Front

R: Rear

42372-AUR-C09

TIRE, WHEEL AND BALL JOINT SPECIFICATIONS

Year	Model	OEM Tires Standard	OEM Tires Optional	Tire Pressures (psi) Front	Tire Pressures (psi) Rear	Wheel Size	Ball Joint Inspection
2001	Aurora	P235/60R16	P235/60VR16	30	30	7-J	①
2002	Aurora	P235/60R16	P235/60VR16	30	30	7-J	①
2003	Aurora	P235/55HR17	NA	②	②	NA	①

OEM: Original Equipment Manufacturer

PSI: Pounds Per Square Inch

① Replace if any measurable movement is found.

② See vehicle placard

42372-AUR-C10

BRAKE SPECIFICATIONS
All measurements in inches unless noted

Year	Model		Brake Disc Original Thickness	Brake Disc Minimum Thickness	Brake Disc Maximum Run-out	Minimum Lining Thickness	Brake Caliper Mounting Bolt (ft. lbs.)
2001	Aurora	F	1.267	1.224	0.002	0.030	63
		R	0.433	0.423	0.002	0.030	20
2002	Aurora	F	1.267	1.224	0.002	0.030	63
		R	0.433	0.423	0.002	0.030	20
2003	Aurora	F	1.267	1.224	0.002	0.030	63
		R	0.433	0.423	0.002	0.030	20

F: Front

R: Rear

42372-AUR-C11

SCHEDULED MAINTENANCE INTERVALS
GM G BODY—OLDSMOBILE AURORA

TO BE SERVICED	TYPE OF SERVICE	VEHICLE MILEAGE INTERVAL (x1000)												
		7.5	15	22.5	30	37.5	45	52.5	60	67.5	75	82.5	90	97.5
Engine oil & filter	R	✓	✓	✓	✓	✓	✓	✓	✓	✓	✓	✓	✓	✓
Coolant level, hoses & clamps	S/I	✓	✓	✓	✓	✓	✓	✓	✓	✓	✓	✓	✓	✓
Driveshaft boots & front suspension components	S/I	✓	✓	✓	✓	✓	✓	✓	✓	✓	✓	✓	✓	✓
Exhaust system & brake hoses	S/I	✓	✓	✓	✓	✓	✓	✓	✓	✓	✓	✓	✓	✓
Lubricate chassis and suspension	S/I	✓	✓	✓	✓	✓	✓	✓	✓	✓	✓	✓	✓	✓
Lubricate steering linkage and transaxle linkage	S/I	✓	✓	✓	✓	✓	✓	✓	✓	✓	✓	✓	✓	✓
Lubricate parking brake cable guides, underbody contact points & linkage	S/I	✓	✓	✓	✓	✓	✓	✓	✓	✓	✓	✓	✓	✓
Throttle linkage	S/I	✓	✓	✓	✓	✓	✓	✓	✓	✓	✓	✓	✓	✓
Brake linings	S/I	✓		✓		✓		✓		✓		✓		✓
Rotate tires	S/I	✓		✓		✓		✓		✓		✓		✓
Air filter element	R				✓				✓				✓	
Engine coolant ①	R													
Spark plugs ②	R				✓				✓				✓	
Accessory drive belt(s)	S/I				✓				✓				✓	
Automatic transaxle fluid A8 & filter ③	S/I													
Fuel system	S/I				✓				✓				✓	
Ignition cables	R				✓				✓				✓	
Inspect throttle body bore & throttle plate for deposits	S/I				✓				✓				✓	
Supercharger oil	S/I				✓				✓				✓	

R: Replace

S/I: Service or Inspect

① Engine coolant: replace every 100,000 miles. Use O.E. specified (DEX-COOL™) coolant only. If any silicate coolant is used, the service interval is every 30,000

② Platinum tip spark plugs: replace every 100,000 miles.

③ Replace fuid every 50, 000 miles if driven in etxreme traffic or in places where the temperature exceeds 90 degrees F

FREQUENT OPERATION MAINTENANCE (SEVERE SERVICE)

If a vehicle is operated under any of the following conditions it is considered severe service:

- Extremely dusty areas.

- 50% or more of the vehicle operation is in 32°C (90°F) or higher temperatures, or constant operation in temperatures below 0°C (32°F).

- Prolonged idling (vehicle operation in stop and go traffic).

- Frequent short running periods (engine does not warm to normal operating temperatures).

- Police, taxi, delivery usage or trailer towing usage.

CV joints & front suspension components: service or inspect every 3000 miles.

Engine oil & filter change: change every 3000 miles.

Brake linings: check every 6000 miles.

Chassis lubrication: lubricate every 6000 miles.

Suspension, steering linkage, transaxle shift linkage, parking cable guides, underbody contact points: lubricate every 6000 miles.

Throttle body mount bolt torque: tighten at 6000 miles.

Air filter element: service or inspect every 15,000 miles.

Inspect throttle body bore & throttle plate for deposits: clean as required every 15,000 miles.

Rotate tires at 6000 miles, then every 15,000 miles.

ENGINE REPAIR

Alternator

REMOVAL

3.5L Engine

1. Before servicing the vehicle, refer to the precautions in the beginning of this section.
2. Remove or disconnect the following:
   - Negative battery cable
   - Drive belt
   - Engine cooling fan assembly
   - Alternator electrical connectors
   - Idler pulley
   - 2 upper alternator mounting bolts
   - A/C compressor
   - Alternator

4.0L Engine

1. Before servicing the vehicle, refer to the precautions in the beginning of this section.
2. Remove or disconnect the following:
   - Rear seat cushion
   - Negative battery cable
   - Radiator
   - Alternator electrical connectors
   - Alternator bolts
   - Alternator

INSTALLATION

3.5L Engine

1. Install or connect the following:
   - Alternator. Do not torque the bolts at this time.

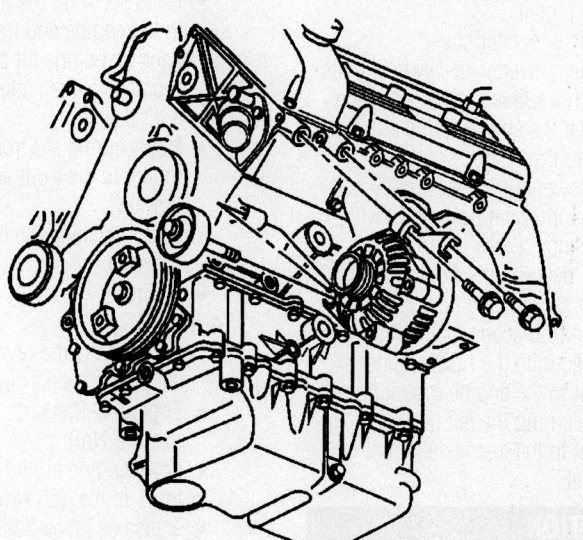

9306XG01

View of the alternator mounting—3.5L engine

- A/C compressor. Torque the bolts and the front nut to 37 ft. lbs. (50 Nm) and the rear nut to 18 ft. lbs. (25 Nm).
- Idler pulley. Torque the bolt to 37 ft. lbs. (50 Nm).

2. Torque the alternator bolts to 37 ft. lbs. (50 Nm).
3. Install or connect the following:
   - Alternator electrical connectors. Torque the positive battery terminal to 111 inch lbs. (12.5 Nm).
   - Engine cooling fan assembly. Torque the bolts to 53 inch lbs. (6 Nm).
   - Drive belt
   - Negative battery cable

4.0L Engine

1. Install or connect the following:
   - Alternator. Torque the bolts to 37 ft. lbs. (50 Nm).
   - Alternator electrical connectors
   - Radiator
   - Negative battery cable
   - Rear seat cushion

Ignition Timing

ADJUSTMENT

The engines are equipped with a Distributorless Ignition System (DIS). The system consists of 2 Crankshaft Position (CKP) sensors, crankshaft reluctor ring,

Camshaft Position (CMP) sensor, Ignition Control Module (ICM), 4 ignition coils, spark plug wires and spark plugs, Knock Sensor (KS) and the Powertrain Control Module (PCM).

The PCM controls spark advance under all driving conditions. The PCM incorporates a permanent spark control override, which electronically lowers the base timing if spark knock (detonation) is encountered during normal operation due to the use of low octane fuel.

Engine Assembly

REMOVAL & INSTALLATION

3.5L Engine

1. Before servicing the vehicle, refer to the precautions in the beginning of this section.
2. Drain the crankcase and cooling system.
3. Recover the air conditioning system refrigerant.
4. Remove or disconnect the following:
   - Negative battery cable
   - Vacuum brake booster hose from the vacuum brake booster and secure to the top of the engine
   - Fuel inlet and return quick-connect fittings at the fuel rail and secure to the air inlet grille
   - Hose from the evaporative emission canister purge valve and secure to the air inlet grille
   - Air cleaner assembly
   - 2 fuel injector sight shield nuts at the front of the fuel injector sight shield
   - Fuel injector sight shield (lift up) at the front and slide out of the engine bracket
   - Battery positive cable from the remote positive terminal and secure to the top of the engine
   - Secondary AIR relay from the relay bracket and secure to the top of the engine

5. Disconnect and secure the following wiring harness electrical connectors to the top of the engine:
   - Powertrain Control Module (PCM)
   - Connection C101
   - Engine electrical harness

6. Remove or disconnect the following:

- Engine ground cable from the right side body frame rail

✳✳ CAUTION

To avoid possible injury or vehicle damage, always replace the accelerator control cable with a NEW cable whenever you remove the engine from the vehicle. Also to avoid cruise control cable damage, position the cable out of the way while you remove or install the engine. Do not pry or lean against the cruise control cable and do not kink the cable. You must replace a damaged cable.

- Cruise control cable from the throttle body bracket and lever
- Accelerator control cable from the throttle body
- Transaxle range selector cable terminal from the transaxle manual shift lever pin
- Transaxle range selector cable from the bracket and position the cable aside
- Radiator inlet hose from the water housing crossover
- Radiator outlet hose from the thermostat housing
- Surge tank inlet hose from the surge tank
- Heater outlet hose from the thermostat housing
- Heater inlet hose from the water crossover
- Two master cylinder brake pipes from the Brake Pressure Modulator Valve (BPMV) and plug the lines
- Transaxle oil cooler pipe bracket bolt from the fan shroud
- Plastic cap off the transaxle oil cooler pipe quick connect fittings
- Upper and lower transaxle oil cooler pipes from the radiator

➥Make sure the wheels of the vehicle are in the straight position ahead and the steering column in the LOCK position before disconnecting the steering column or intermediate shaft from the steering gear. Failure to do so will cause the SIR coil assembly to become uncentered, which may cause damage to the coil assembly.

7. Lock the steering column by installing tool J42460 into the underside of the steering column.
- Right and left side strut tower bolts

- Transaxle oil cooler pipe bracket bolt from the lower tie bar
- Exhaust manifold pipe
- Front wheels
- Front wheel speed sensor electrical leads from the frame rail
- Front air deflector
- Front fascia extensions
- Secondary AIR inlet hose from the secondary AIR pump
- 2 nuts securing the front brake pipe frame brackets to the frame rails
- Front brake pipes from the retainers at the frame rails
- Front brake pipes away from the body frame rails by carefully pulling them
- 2 rear brake pipes at the rear of the engine frame
- A/C pressure sensor
- A/C discharge hose from the compressor and secure to the cooling fan assembly
- A/C suction hose from the compressor and secure to the cooling fan assembly

✳✳ CAUTION

Failure to disconnect the intermediate shaft from the rack and pinion stub shaft can result in damage to the steering gear and/or damage to the intermediate shaft. This damage may cause loss of steering control which could result in personal injury.

- Intermediate shaft pinch bolt
- Steering gear from the intermediate shaft
- Post Heated Oxygen Sensor (HO₂S) at the pigtail
- Torque converter cover
- Torque converter-to-flywheel bolts

8. Position a suitable powertrain support dolly under the engine frame and lower the vehicle onto the dolly,

9. If the powertrain support dolly is unavailable. Support the powertrain with 4 suitable jackstands. Place a 2 in x 4 in block of wood between the front of the engine oil pan and the engine frame.

10. Remove or disconnect the following:
- Nut securing the right engine mount to the engine mount bracket
- Nut securing the left transaxle mount to the transaxle mount bracket

✳✳ CAUTION

To avoid any vehicle damage, serious personal injury or death when major

components are removed from the vehicle and the vehicle is supported by a hoist, support the vehicle with jack stands at the opposite end from which the components are being removed.

11. Secure the front hoist pads to the vehicle.
- 6 frame to body mounting bolts (3).

12. Make clearance is maintained between the powertrain assembly and the following components:
- A/C accumulator hose
- A/C compressor hose
- Brake pipes
- Heater hoses
- Radiator hoses
- Wheel speed sensor leads
- Wiring harnesses

13. Carefully raise the vehicle in order to clear the supported engine/transaxle assembly.

14. Remove or disconnect the following:
- Secondary air injection crossover pipe/hose
- Engine wiring harness from the engine
- Power steering return hose from the power steering pump reservoir
- Power steering return hose retaining bolt from the cylinder head
- Power steering pressure hose from the power steering pump
- Bolt securing the power steering pressure hose to the secondary AIR valve bracket
- 3 bolts securing the right engine mount bracket to the engine
- Right engine mount bracket
- Bolts securing the transaxle brace to the engine and transaxle

15. Attach an engine lift chain to the engine lift brackets, then attach the chain to an engine lift device.
- Nut securing the front engine mount to the front engine mount bracket
- Bolts attaching the engine to the transaxle
- Engine

To install:

16. Installation is the reverse of removal, please note the following torques:
- Engine-to-transaxle bolts to 55 ft. lbs. (75 Nm)
- Front engine mount-to-bracket nut to 41 ft. lbs. (55 Nm)
- Transaxle brace bolts to 37 ft. lbs. (50 Nm)
- Right engine mount bracket bolts to 44 ft. lbs. (60 Nm)

- Power steering pressure hose to power steering pump to 22 ft. lbs. (30 Nm)
- Power steering pressure hose retaining bolt to 80 inch lbs. (9 Nm)
- Power steering return hose retaining bolt to 13 ft. lbs. (17 Nm)
- Frame mounting bolts to 141 ft. lbs. (191 Nm)
- Engine mount nuts to 59 ft. lbs. (80 Nm)
- Flywheel-to-torque converter bolts to 48 ft. lbs. (65 Nm)
- Intermediate shaft to steering gear pinch bolt to 33 ft. lbs. (45 Nm)
- A/C hoses to 15 ft. lbs. (20 Nm)

17. Connect the battery negative cable.
18. Fill the engine with oil.
19. Fill the cooling system.
20. Bleed the brake system.
21. Recharge the A/C system.
22. Bleed the power steering system.
23. Check the wheel alignment.
24. Complete the following procedure after the engine is installed in the vehicle:

    a. With the ignition **OFF** or disconnected, crank the engine several times. Listen for any unusual noises or evidence that any parts are binding.

    b. Start the engine and listen for abnormal conditions.

    c. Check the vehicle oil pressure gauge or light and confirm that the engine has acceptable oil pressure.

    d. Run the engine at approximately 1000 RPM until the engine reaches normal operating temperature.

    e. While the engine continues to idle raise and support the vehicle.

    f. Inspect for oil, coolant and exhaust leaks while the engine is idling.

    g. Lower the vehicle.

    h. Perform the Crankshaft Position (CKP) system variation learn procedure.

    i. Perform a final inspection for the proper engine oil and coolant levels.

    j. Road test the vehicle.

4.0L Engine

1. Before servicing the vehicle, refer to the precautions in the beginning of this section.
2. Drain the crankcase and cooling system.
3. Recover the air conditioning system refrigerant.
4. Remove or disconnect the following:
   - Negative battery cable
   - Vacuum brake booster hose from the vacuum brake booster

- Fuel inlet and return quick-connect fittings at the fuel rail and secure to the air inlet grille
- Hose from the evaporative emission canister purge valve and secure to the air inlet grille
- Air cleaner assembly
- 2 fuel injector sight shield nuts at the front of the fuel injector sight shield
- Fuel injector sight shield (lift up) at the front and slide out of the engine bracket
- Battery positive cable from the remote positive terminal and secure to the top of the engine
- Secondary AIR relay from the relay bracket and secure to the top of the engine

5. Disconnect and secure the following wiring harness electrical connectors to the top of the engine:
   - Powertrain control Module (PCM)
   - Connection C101
   - Engine electrical harness
6. Remove or disconnect the following:
   - Engine ground cable from the right side body frame rail

✳✳ CAUTION

To avoid possible injury or vehicle damage, always replace the accelerator control cable with a NEW cable whenever you remove the engine from the vehicle.

Also to avoid cruise control cable damage, position the cable out of the way while you remove or install the engine. Do not pry or lean against the cruise control cable and do not kink the cable. You must replace a damaged cable.

- Cruise control cable from the throttle body bracket and lever
- Accelerator control cable from the throttle body
- Transaxle range selector cable terminal from the transaxle manual shift lever pin
- Transaxle range selector cable from the bracket and position the cable aside
- Radiator inlet hose from the water housing crossover
- Radiator outlet hose from the thermostat housing
- Surge tank inlet hose from the surge tank
- Heater hoses

- Two master cylinder brake pipes from the Brake Pressure Modulator Valve (BPMV) and plug the lines
- Transaxle oil cooler pipe bracket bolt from the fan shroud
- Plastic cap off the transaxle oil cooler pipe quick connect fittings
- Upper and lower transaxle oil cooler pipes from the radiator

➡ **Make sure the wheels of the vehicle are in the straight position ahead and the steering column in the LOCK position before disconnecting the steering column or intermediate shaft from the steering gear. Failure to do so will cause the SIR coil assembly to become uncentered, which may cause damage to the coil assembly.**

7. Lock the steering column by installing tool J42460 into the underside of the steering column.
   - Right and left side strut tower bolts
   - Exhaust manifold pipe
   - Front wheels
   - Front wheel speed sensor electrical leads from the frame rail
   - Front air deflector
   - Front fascia extensions
   - Secondary AIR inlet hose from the secondary AIR pump
   - 2 nuts securing the front brake pipe frame brackets to the frame rails
   - Front brake pipes from the retainers at the frame rails
   - Front brake pipes away from the body frame rails by carefully pulling them
   - 2 rear brake pipes at the rear of the engine frame
   - A/C pressure sensor
   - A/C discharge hose from the compressor and secure to the cooling fan assembly
   - A/C suction hose from the compressor and secure to the cooling fan assembly

---

✳✳ CAUTION

Failure to disconnect the intermediate shaft from the rack and pinion stub shaft can result in damage to the steering gear and/or damage to the intermediate shaft. This damage may cause loss of steering control which could result in personal injury.

- Intermediate shaft pinch bolt
- Steering gear from the intermediate shaft

- Post Heated Oxygen Sensor (HO2S) at the pigtail
- Brace between the engine oil pan and the transaxle case
- Torque converter cover
- Torque converter-to-flywheel bolts

8. Position a suitable powertrain support dolly under the engine frame and lower the vehicle onto the dolly.

9. If the powertrain support dolly is unavailable. Support the powertrain with 4 suitable jackstands. Place a 2 in x 4 in block of wood between the front of the engine oil pan and the engine frame.

10. Remove or disconnect the following:
- Nut securing the right engine mount to the engine mount bracket
- Nut securing the left transaxle mount to the transaxle mount bracket

✳✳ CAUTION

To avoid any vehicle damage, serious personal injury or death when major components are removed from the vehicle and the vehicle is supported by a hoist, support the vehicle with jack stands at the opposite end from which the components are being removed.

11. Secure the front hoist pads to the vehicle.
- 6 frame to body mounting bolts (3).

12. Make clearance is maintained between the powertrain assembly and the following components:
- A/C accumulator hose
- A/C compressor hose
- Brake pipes
- Heater hoses
- Radiator hoses
- Wheel speed sensor leads
- Wiring harnesses

13. Carefully raise the vehicle in order to clear the supported engine/transaxle assembly.

14. Remove or disconnect the following:
- Heater pipes
- Intermediate hose from the secondary AIR valve at bank 2
- Nut securing the intermediate hose to the secondary AIR valve at bank 1
- Nut securing the coil cassette ground wire to the right cylinder head
- Engine wiring harness from the engine
- Power steering return hose from the power steering pump reservoir
- Power steering return hose retaining bolt from the cylinder head

- Power steering pressure hose from the power steering pump
- Bolt securing the power steering pressure hose to the to the right engine mount bracket
- Bolts securing the right engine mount bracket to the engine
- Right engine mount bracket
- Bolts/nuts securing the transaxle brace's to the engine and transaxle

15. Attach an engine lift chain to the engine lift brackets, then attach the chain to an engine lift devise.
- Nut securing the front engine mount to the front engine mount bracket
- Bolts attaching the engine to the transaxle
- Engine

To install:

16. Installation is the reverse of removal, please note the following torques:
- Engine-to-transaxle bolts to 55 ft. lbs. (75 Nm)
- Front engine mount-to-bracket nut to 52 ft. lbs. (70 Nm)
- Transaxle brace bolts to 37 ft. lbs. (50 Nm)
- Right engine mount bracket bolts to 37 ft. lbs. (50 Nm)
- Power steering pressure hose to power steering pump to 22 ft. lbs. (30 Nm)
- Power steering pressure hose retaining bolt to 80 inch lbs. (9 Nm)
- Power steering return hose retaining bolt to 37 ft. lbs. (50 Nm)
- Frame mounting bolts to 141 ft. lbs. (191 Nm)
- Engine mount nuts to 59 ft. lbs. (80 Nm)
- Flywheel-to-torque converter bolts to 44 ft. lbs. (60 Nm)
- Intermediate shaft to steering gear pinch bolt to 33 ft. lbs. (45 Nm)
- A/C hoses to 15 ft. lbs. (20 Nm)

17. Connect the battery negative cable.
18. Fill the engine with oil.
19. Fill the cooling system.
20. Bleed the brake system.
21. Recharge the A/C system.
22. Bleed the power steering system.
23. Check the wheel alignment.
24. Complete the following procedure after the engine is installed in the vehicle:

a. With the ignition **OFF** or disconnected, crank the engine several times. Listen for any unusual noises or evidence that any parts are binding.

b. Start the engine and listen for abnormal conditions.

c. Check the vehicle oil pressure gauge or light and confirm that the engine has acceptable oil pressure.

d. Run the engine at approximately 1000 RPM until the engine reaches normal operating temperature.

e. While the engine continues to idle raise and support the vehicle.

f. Inspect for oil, coolant and exhaust leaks while the engine is idling.

g. Lower the vehicle.

h. Perform the Crankshaft Position (CKP) system variation learn procedure.

i. Perform a final inspection for the proper engine oil and coolant levels.

j. Road test the vehicle.

Water Pump

REMOVAL & INSTALLATION

3.5L Engine

1. Before servicing the vehicle, refer to the precautions in the beginning of this section.
2. Drain the cooling system.
3. Remove or disconnect the following:
- Water pump pulley bolts, loosen but do not remove at this time
- Drive belt

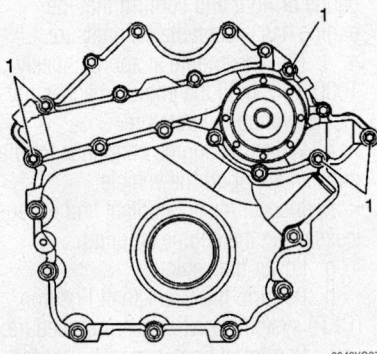

Be sure to install the 5 long water pump bolts (1) in the correct locations—3.5L engine

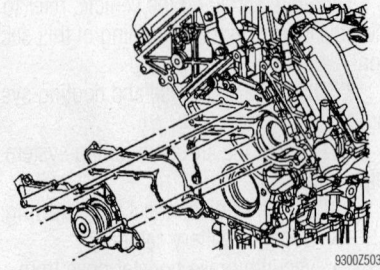

Water pump mounting—3.5L engine

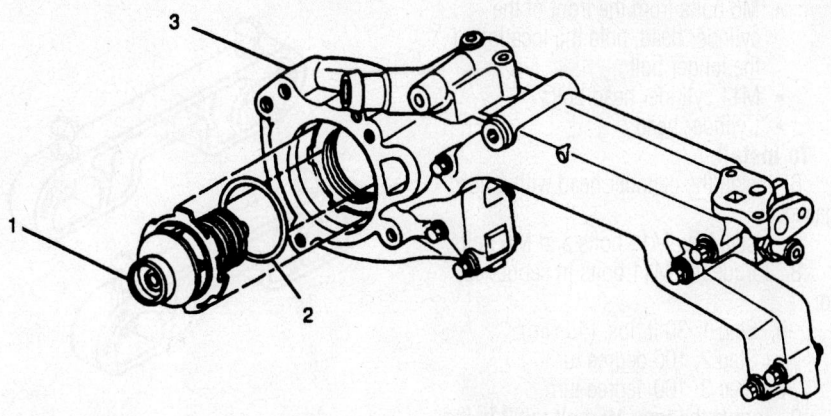

1 WATER PUMP ASM.   3 WATER PUMP HOUSING ASM.
2 O-RING SEAL

7922XG02

Exploded view of the water pump housing and water pump—4.0L engine

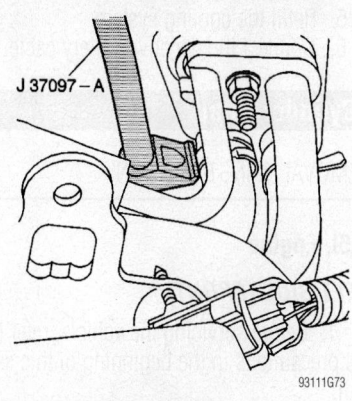

Remove the heater hoses clamp

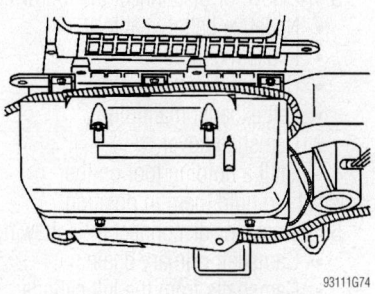

View of the heater core cover

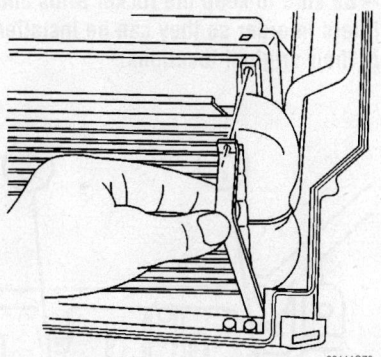

View of the heater core retaining straps

- Idler pulley
- Water pump pulley

➡ **The water pump is attached to the engine with both long and short bolts, be sure to note their locations.**

- Water pump

To install:

4. Install or connect the following:
  - Water pump using a new gasket. Torque the bolts to 124 inch lbs. (14 Nm).
  - Water pump pulley. Torque the bolts to 106 inch lbs. (12 Nm).
  - Idler pulley. Torque the bolt to 37 ft. lbs. (50 Nm).
  - Drive belt
5. Refill the cooling system.
6. Start the engine and check for leaks.

4.0L Engine

1. Before servicing the vehicle, refer to the precautions in the beginning of this section.
2. Drain the cooling system.
3. Remove or disconnect the following:
  - Negative battery cable
  - Air cleaner assembly
  - Secondary AIR injection control valve, if necessary
  - Oil level indicator tube nut, if necessary
  - Water pump drive belt cover
  - Water pump drive belt
  - Lower radiator hose
  - Bypass hose
  - Water pump cover
  - Water pump from the water pump housing, by turning the locking ring with a Water Pump Remover/Installer tool J-38816

To install:

4. Install or connect the following:
  - Water pump with new o-ring. Torque the pump to 73 ft. lbs. (100 Nm).
  - Water pump cover. Torque the bolts to 89 inch lbs. (10 Nm).
  - Lower radiator hose
  - Coolant bypass hose
  - Water pump drive belt
  - Water pump drive belt cover
  - Oil level indicator tube nut, if necessary
  - Secondary AIR injection control valve, if necessary
  - Air cleaner assembly
  - Negative battery cable
5. Refill and bleed the cooling system.

Heater Core

REMOVAL & INSTALLATION

1. Disconnect the negative battery cable.
2. Drain the cooling system into a clean container for reuse.
3. Remove or disconnect the following:
  - Heater hoses from the heater core
  - Center console assembly
  - Auxiliary air duct connector
  - Right sound insulator
  - Heater core heat shield-to-heating, HVAC module screws and the shield
  - Heater core cover-to-HVAC module screws and the cover
  - Heater core-to-HVAC module screws and straps
  - Heater core

To install:

4. Install or connect the following:
  - Heater core
  - Heater core-to-HVAC module screws and straps, then, tighten to 18 inch lbs. (1.5 Nm)
  - Heater core cover and the cover-to-HVAC module screws, then, tighten to 18 inch lbs. (1.5 Nm)
  - Heater core heat shield and the shield-to-heating, HVAC module screws, then, tighten to 18 inch lbs. (1.5 Nm)
  - Right sound insulator
  - Auxiliary air duct connector
  - Center console assembly
  - Heater hoses to the heater core

5. Refill the cooling system.
6. Connect the negative battery cable.

Cylinder Head

REMOVAL & INSTALLATION

3.5L Engine

LEFT SIDE (FRONT)

1. Before servicing the vehicle, refer to the precautions in the beginning of this section.
2. Drain the crankcase and the cooling system.
3. Remove or disconnect the following:
   - Negative battery cable
   - Intake manifold
   - Coolant crossover pipe
   - Left exhaust manifold
   - Camshaft cover
4. Install a holding tool on the camshafts to hold them in position.
5. Remove or disconnect the following:
   - Camshaft primary chain
   - Camshafts from the left cylinder head
   - Rocker arms and valve lifters

➡ **Be sure to keep the rocker arms and lifters in order so they can be installed in their original locations.**

- M6 bolts from the front of the cylinder head, note the location of the longer bolt
- M11 cylinder head bolts
- Cylinder head

To install:

6. Install the cylinder head with a new gasket.
7. Install new M11 bolts and M6 bolts.
8. Torque the M11 bolts in sequence to:
   a. Step 1: 30 ft. lbs. (40 Nm).
   b. Step 2: 100 degree turn.
   c. Step 3: 100 degree turn.
9. Torque the long M6 bolt to 22 ft. lbs. (30 Nm).
10. Torque both shorter M6 bolts to 106 inch lbs. (12 Nm).
11. Install or connect the following:
    - Lifters
    - Rocker arms
    - Camshafts
    - Primary chain
12. Remove the timing chain holding tool.
13. Install or connect the following:
    - Camshaft covers. Torque the bolts to 80 inch lbs. (9 Nm).
    - Exhaust manifold. Torque the bolts to 22 ft. lbs. (30 Nm).
    - Coolant crossover pipe. Torque the bolts to 18 ft. lbs. (25 Nm).

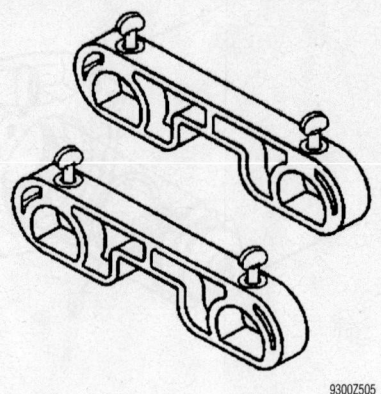

Camshaft holding fixture J-42038—3.5L engine

- Intake manifold. Torque the bolts to 62 inch lbs. (7 Nm).
- New oil filter
- Negative battery cable
14. Refill the crankcase.
15. Refill the cooling system.
16. Start the engine and check for leaks.

RIGHT SIDE (REAR)

1. Before servicing the vehicle, refer to the precautions in the beginning of this section.
2. Drain the crankcase and the cooling system.
3. Remove or disconnect the following:
   - Negative battery cable
   - Intake manifold
   - Coolant crossover pipe

➡ **Do not remove the rear exhaust manifold. Detach it from the cylinder head and the connection from the front manifold; then, move it aside.**

- Rear exhaust manifold from the cylinder head and front manifold
- Camshaft cover
4. Install a holding tool on the camshafts to hold them in position.
5. Remove or disconnect the following:
   - Primary camshaft chain
   - Right side camshafts
   - Rocker arms and lifters

➡ **Keep the rocker arms and lifters in order so they can be installed in their original positions.**

- Engine Coolant Temperature (ECT) sensor from the cylinder head
- M6 bolts from the front of the cylinder head, note the location of the longer bolt
- M11 cylinder head bolts
- Cylinder head

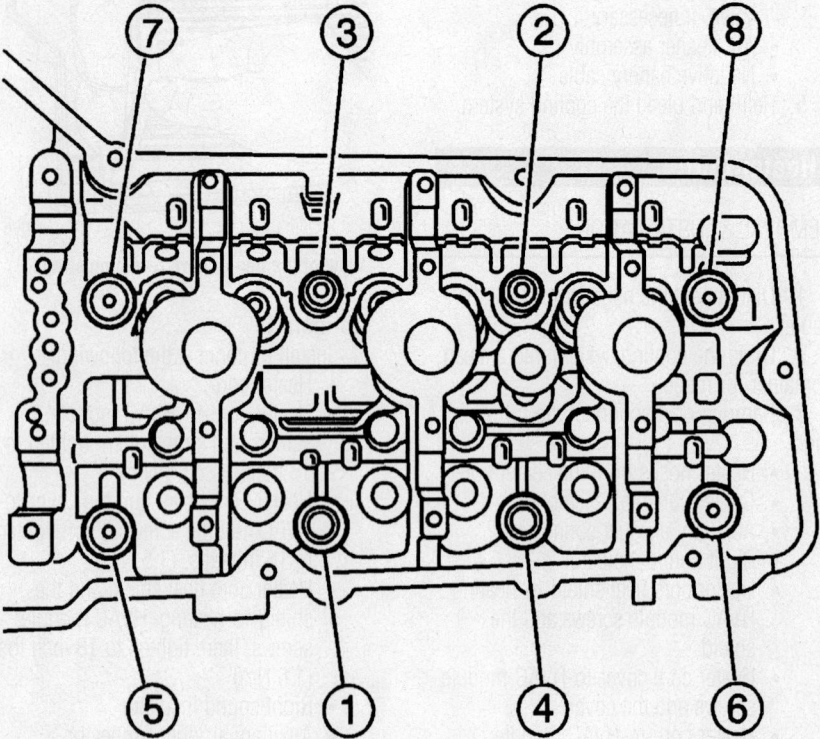

Front cylinder head bolt torque sequence—3.5L (VIN H) engine

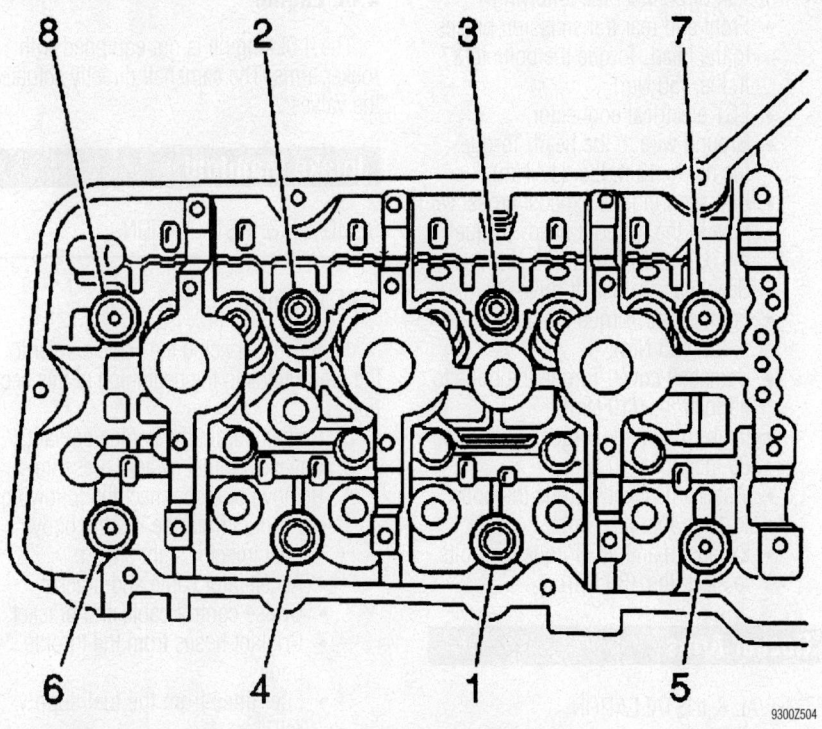

Rear cylinder head bolt torque sequence—3.5L (VIN H) engine

To install:

6. Install the cylinder head with a new gasket.

7. Install new M11 bolts.

8. Install M6 bolts in the front of the cylinder head.

9. Tighten the M11 bolts in sequence using the following steps:
   a. Step 1: 30 ft. lbs. (40 Nm).
   b. Step 2: 100 degree turn.
   c. Step 3: 100 degree turn.

10. Torque the long M6 bolt to 22 ft. lbs. (30 Nm).

11. Torque both short M6 bolts to 106 inch lbs. (12 Nm).

12. Install or connect the following:
   - ECT sensor. Torque the sensor to 15 ft. lbs. (20 Nm).
   - Lifters and rocker arms in their original positions
   - Camshafts
   - Primary chain
   - Camshaft covers. Torque the bolts to 80 inch lbs. (9 Nm).
   - Exhaust manifold on the cylinder head. Torque the bolts to 22 ft. lbs. (30 Nm).
   - Coolant crossover pipe. Torque the bolts to 18 ft. lbs. (25 Nm).
   - Intake manifold. Torque the bolts to 62 inch lbs. (7 Nm).
   - New oil filter
   - Negative battery cable

13. Refill the crankcase.

14. Refill the cooling system.

15. Start the engine and check for leaks.

4.0L Engine

LEFT SIDE

1. Before servicing the vehicle, refer to the precautions in the beginning of this section.

2. Remove or disconnect the following:
   - Exhaust manifold
   - Alternator
   - Water crossover
   - Intake manifold
   - Camshaft cover
   - Front cover
   - Secondary camshaft drive chain
   - Power steering hose retaining bolt from the head
   - Cylinder head

To install:

3. Install the cylinder head with a new gasket.

4. Install new M11 head bolts and M6 head bolts.

5. Torque the M11 bolts in sequence as follows:
   a. Step 1: 30 ft. lbs. (40 Nm).
   b. Step 2: 70 degree turn.
   c. Step 3: 60 degree turn.
   d. Step 4: 60 degree turn.

6. Torque the M6 bolts to 106 inch lbs. (12 Nm).

7. Install or connect the following:
   - Power steering hose retaining bolt. Torque the bolt to 37 ft. lbs. (50 Nm).
   - Secondary camshaft drive chain
   - Front cover. Torque the bolts to 89 inch lbs. (10 Nm).
   - Camshaft cover. Torque the bolts to 89 inch lbs. (10 Nm).
   - Intake manifold. Torque the bolts to 89 inch lbs. (10 Nm).
   - Water crossover. Torque the bolts to 18 ft. lbs. (25 Nm).

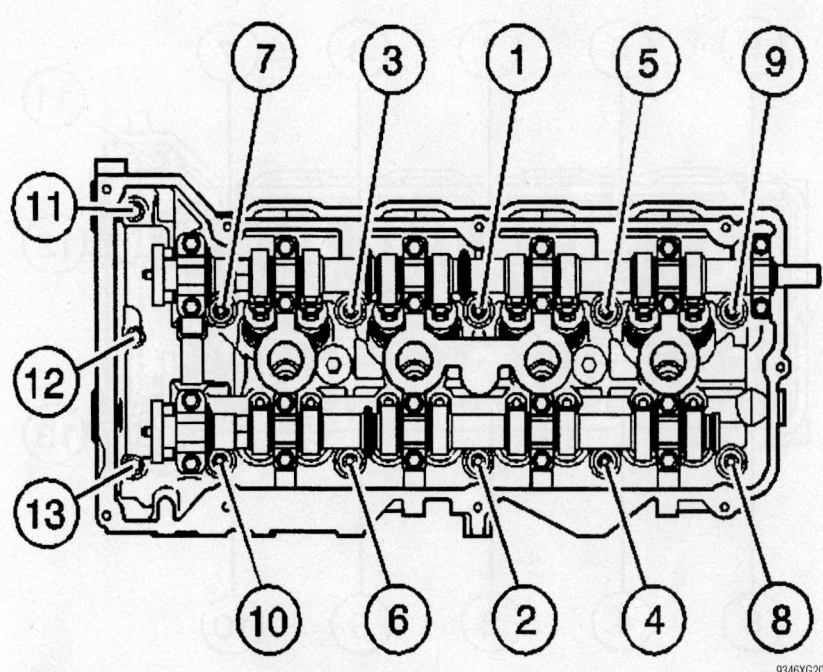

Left cylinder head bolt torque sequence—4.0L models

- Alternator. Torque the bolts to 37 ft. lbs. (50 Nm).
- Exhaust manifold. Torque the bolts to 18 ft. lbs. (25 Nm).

RIGHT SIDE

1. Before servicing the vehicle, refer to the precautions in the beginning of this section.
2. Remove or disconnect the following:
   - Exhaust manifold
   - Water crossover
   - Intake manifold
   - Camshaft cover
   - Front cover
   - Secondary camshaft drive chain
   - Engine Coolant Temperature (ECT) sensor electrical connector
   - Ground wire from the head
   - Bolt securing the exhaust crossover pipe to the cylinder head
   - Front and rear transmission braces from the head
   - Cylinder head

To install:

3. Install the cylinder head with a new gasket.
4. Install new M11 head bolts and M6 head bolts.
5. Torque the M11 head bolts in sequence as follows:
   a. Step 1: 30 ft. lbs. (40 Nm).
   b. Step 2: 70 degree turn.
   c. Step 3: 60 degree turn.
   d. Step 4: 60 degree turn.
6. Torque the M6 head bolts to 106 inch lbs. (12 Nm).

7. Install or connect the following:
   - Front and rear transmission braces to the head. Torque the bolts to 37 ft. lbs. (50 Nm).
   - ECT electrical connector
   - Ground wire to the head. Torque the nut to 13 ft. lbs. (17 Nm).
   - Bolt securing the exhaust crossover pipe to the cylinder head. Torque the bolt to 18 ft. lbs. (25 Nm).
   - Secondary camshaft drive chain
   - Front cover. Torque the bolts to 18 ft. lbs. (25 Nm).
   - Camshaft cover. Torque the bolts to 89 inch lbs. (10 Nm).
   - Intake manifold. Torque the bolts to 89 inch lbs. (10 Nm).
   - Water crossover. Torque the bolts to 18 ft. lbs. (25 Nm).
   - Exhaust manifold. Torque the bolts to 22 ft. lbs. (30 Nm).

Rocker Arms

REMOVAL & INSTALLATION

➡**All valve train components should be kept in the order that they were removed, so that they can be reinstalled in their original position.**

3.5L Engine

Refer to the camshaft removal and installation procedure for rocker arm service.

4.0L Engine

The 4.0L engine is not equipped with rocker arms. The camshaft directly actuates the valves.

Intake Manifold

REMOVAL & INSTALLATION

3.5L Engine

1. Before servicing the vehicle, refer to the precautions in the beginning of this section.
2. Partially drain the engine coolant.
3. Relieve the fuel system pressure.
4. Remove or disconnect the following:
   - Air duct from the throttle body
   - Fuel injector sight shield
   - Accelerator cable and bracket
   - Cruise control cable and bracket
   - Coolant hoses from the throttle body
   - Fuel lines from the fuel supply rail
   - Fuel vapor line from the Evaporative Emission (EVAP) canister purge solenoid
   - Brake booster vacuum hose
   - Air conditioning vacuum hose from the engine
   - Surge tank inlet pipe retainer from the fuel supply rail
   - Fuel injector electrical connectors
   - Throttle Position Sensor (TPS) electrical connector
   - Idle Air Control (IAC) valve electrical connector
   - Evaporative Emission (EVAP) canister purge solenoid connector
   - Manifold Absolute Pressure (MAP) sensor connector
   - Wiring harness from the camshaft covers
   - Vacuum hose from the fuel pressure regulator and throttle body
   - Positive Crankcase Ventilation (PCV) tubes from both camshaft covers and the intake manifold
   - Exhaust Gas Recirculation (EGR) valve outlet pipe
   - Fuel supply rail with injectors

➡**Disengage the snap-lock retainers by pushing toward the camshaft covers and lifting.**

   - Intake manifold

➡**The manifold-to-cylinder head seals are reusable unless cut or damaged.**

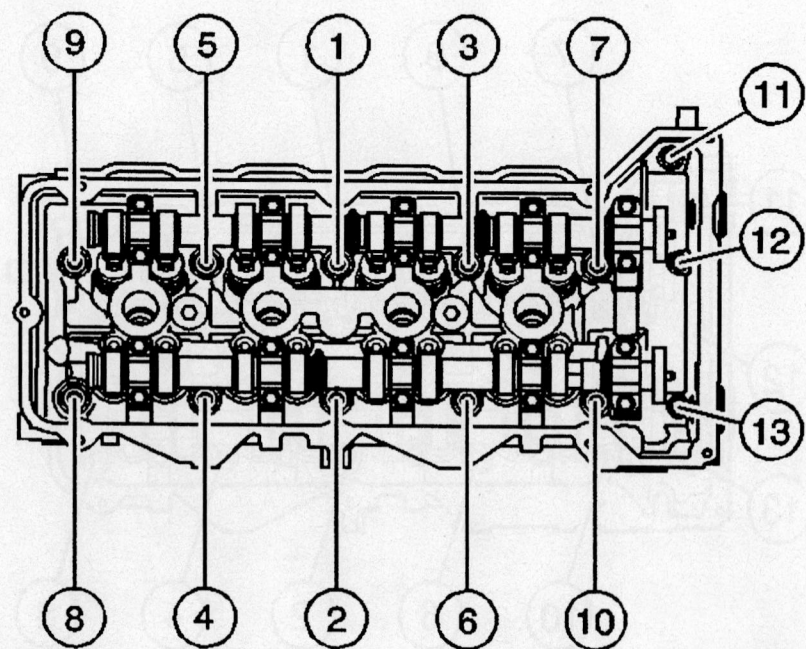

9346XG21

Right cylinder head bolt torque sequence—4.0L models

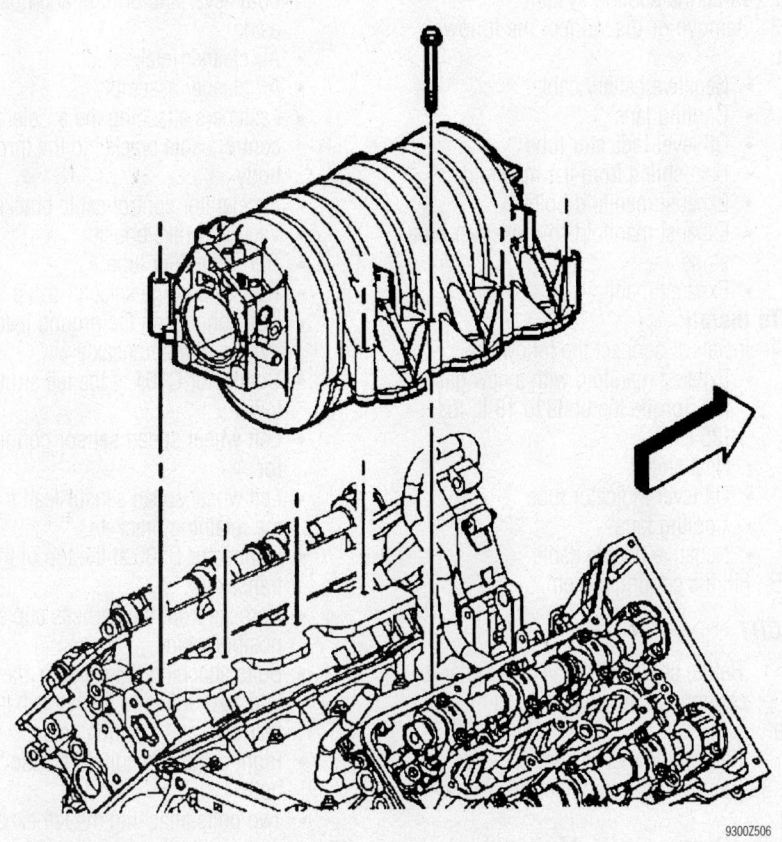

Intake manifold assembly—3.5L engine

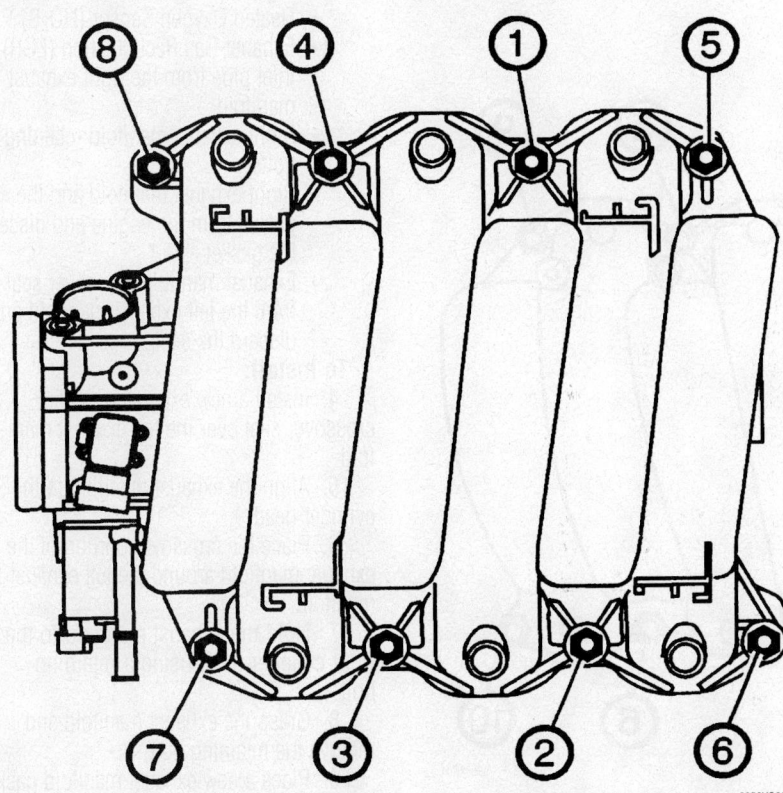

Intake manifold torque sequence—3.5L engine

To install:

5. Install or connect the following:
   - Intake manifold with new gaskets (if necessary). Torque the bolts in sequence to 89 inch lbs. (10 Nm).
   - New O-rings on the fuel injectors
   - Fuel supply rail with the injectors
   - EGR pipe. Torque the intake manifold bolts to 89 inch lbs. (10 Nm) and the coolant crossover bolt to 18 ft. lbs. (24 Nm).
   - PCV valve and feed tubes
   - Fuel pressure regulator vacuum hose
   - Engine wiring harness to the camshaft covers. Torque the bolts to 89 inch lbs. (10 Nm).
   - TPS electrical connector
   - IAC valve electrical connector
   - EVAP solenoid electrical connector
   - MAP sensor electrical connector
   - Fuel injector electrical connectors
   - Surge tank pipe retainer to the fuel supply rail
   - A/C vacuum hose
   - Brake booster vacuum hose
   - Vapor line to the EVAP canister purge solenoid
   - Fuel lines to the fuel supply rail
   - Coolant hoses to the throttle body
   - Cruise control cable and bracket
   - Accelerator cable and bracket
   - Fuel injector sight shield
   - Air duct to the throttle body
   - Negative battery cable
6. Refill the cooling system.
7. Start the engine and check for leaks.

4.0L Engine

1. Before servicing the vehicle, refer to the precautions in the beginning of this section.
2. Relieve the fuel system pressure.
3. Remove or disconnect the following:

   - Fuel injector sight shield
   - Ignition module connectors
   - Positive Crankcase Ventilation (PCV) valve
   - PCV fresh air tube
   - Fuel regulator vacuum tube
   - Secondary Air Injection (AIR) solenoid vacuum tubes
   - Fuel lines at the fuel rail
   - Engine coolant heater wire
   - Fuel injector electrical connectors
   - Fuel rail with injectors
   - Plenum duct clamp
   - Intake manifold

To install:

4. Install or connect the following:
- Intake manifold with a new gasket. Torque the bolts in sequence to 89 inch lbs. (10 Nm).
- Torque the plenum duct clamp to 20 inch lbs. (2.25 Nm)
- Fuel rail with the injectors. Torque the bolts to 89 inch lbs. (10 Nm).
- Fuel injector electrical connec-tors
- Engine coolant heater wire
- Fuel lines to the fuel rail
- AIR solenoid vacuum tubes
- Fuel regulator vacuum tube
- PCV fresh air tube
- PCV valve
- Ignition module connectors

5. Pressurize the fuel system and check for leaks.

6. Install the fuel injector sight shield. Torque the bolts to 27 inch lbs. (3 Nm).

Exhaust Manifold

REMOVAL & INSTALLATION

3.5L Engine

LEFT

1. Before servicing the vehicle, refer to the precautions in the beginning of this section.

2. Drain the cooling system.

3. Remove or disconnect the follow-ing:
- Negative battery cable
- Cooling fans
- Oil level indicator tube
- Heat shield from the manifold
- Exhaust manifold bolts
- Exhaust manifold-to-crossover pipe studs
- Exhaust manifold

To install:

4. Install or connect the following:
- Exhaust manifold with a new gas-ket. Torque the bolts to 18 ft. lbs. (25 Nm).
- Heat shield
- Oil level indicator tube
- Cooling fans
- Negative battery cable

5. Fill the cooling system.

RIGHT

1. Before servicing the vehicle, refer to the precautions in the beginning of this section.

2. Relieve the fuel system pressure.

3. Remove or disconnect the follow-ing:
- Cruise control cable from the throttle body lever and bracket and position aside
- Accelerator cable from the throttle body lever and bracket and position aside
- Air cleaner intake duct
- Air cleaner assembly
- Fasteners attaching the accelerator control cable bracket to the throttle body
- Accelerator control cable bracket
- Vacuum brake booster
- Transaxle filler tube
- Transaxle range selector cable
- Bolt connecting the ground lead to the top of the transaxle
- Connector C101 at the left strut tower
- Left wheel speed sensor connec-tor
- Left wheel speed sensor lead from the retaining brackets
- Connector C100 at the top of the transaxle
- Harness from the harness clip and position aside
- Bolts attaching the exhaust mani-fold heat shield to the left exhaust manifold and position aside
- Right exhaust manifold crossover bolts
- Two bolts attaching the left exhaust manifold to the right exhaust mani-fold crossover
- Exhaust manifold pipe
- Right exhaust manifold studs
- Heated Oxygen Sensor (HO$_2$S)
- Exhaust Gas Recirculation (EGR) inlet pipe from the right exhaust manifold
- Right exhaust manifold retaining bolts
- Right exhaust manifold and the gasket from the engine and discard the gasket
- Exhaust manifold crossover seal from the left exhaust manifold and discard the seal

To install:

4. Install a new exhaust manifold crossover seal over the left exhaust mani-fold

5. Align the exhaust manifold to the cylinder head.

6. Place the crossover portion of the exhaust manifold around the left exhaust manifold.

7. Hold the exhaust manifold to the right cylinder head using a retaining bolt.

8. Grasp the exhaust manifold and remove the retaining bolt.

9. Place a new exhaust manifold gasket between the exhaust manifold and the cylin-der head.

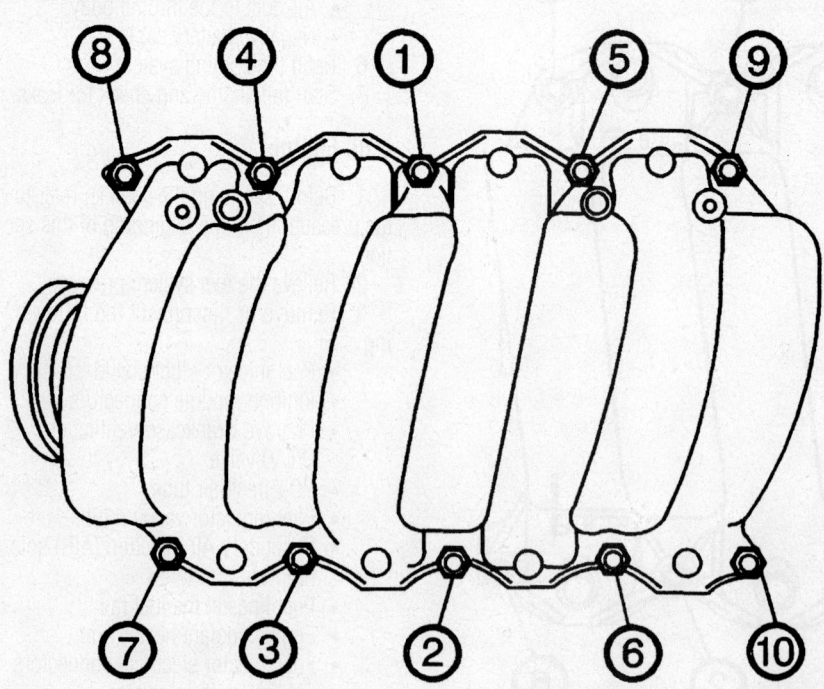

9356XG01

Intake manifold torque sequence—4.0L engine

10. Install or connect the following:
- Right exhaust manifold bolts and tighten to 18 ft. lbs. (25 Nm)
- EGR inlet pipe to the exhaust manifold nut to 44 ft. lbs. (60 Nm)
- Right exhaust manifold studs and tighten to 80 inch. lbs. (Nm)
- HO_2S
- Exhaust manifold pipe
- Two bolts attaching the left exhaust manifold to the right exhaust manifold crossover and tighten to 18 ft. lbs. (25 Nm)
- Heat shield to the left exhaust manifold and tighten the bolts to 80 inch. lbs. (9 Nm)
- Harness to the harness clip
- C100 to the transaxle
- Left wheel speed sensor connector
- Left wheel speed sensor lead to the retaining brackets
- C101 at the left strut tower
- Bolt connecting the ground lead to the top of the transaxle and tighten to 13 ft. lbs. (17 Nm)
- Transaxle range selector cable
- Transaxle filler tube
- Accelerator control cable bracket to the throttle body
- Fasteners attaching the accelerator control cable bracket to the throttle body. Tighten the accelerator control cable bracket bolt and to 115 inch. lbs. (13 Nm) and the accelerator control cable bracket nuts to 89 inch lbs. (10 Nm).
- Air cleaner assembly
- Air cleaner intake duct

➡**Make sure the throttle should operate freely without binding between full closed and wide open throttle.**

- Accelerator cable to the throttle body lever and bracket

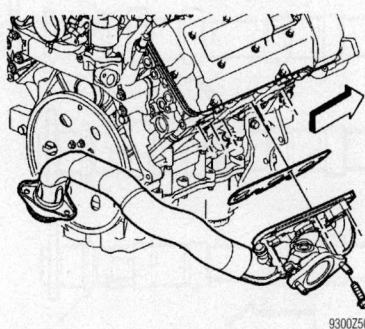

Exploded view of the right exhaust manifold—3.5L engine

9300Z507

- Cruise control cable to the throttle body lever and bracket

4.0L Engine

LEFT SIDE (FRONT)

1. Before servicing the vehicle, refer to the precautions in the beginning of this section.
2. Remove or disconnect the following:
- Negative battery cable
- Two nuts attaching the secondary AIR tube to the exhaust manifold
- Secondary AIR tube and the gasket from the exhaust manifold and discard the gasket
- Front engine mount bracket
- Oxygen Sensor (O_2S), if necessary
- Two bolts attaching the left exhaust manifold to the front exhaust manifold pipe
- Exhaust manifold retaining bolts
- Exhaust manifold and the gasket from the engine and discard the gasket
- Front exhaust manifold pipe seal from the left exhaust manifold and discard the seal
- Exhaust manifold flange seal retainer

To install:

3. Install or connect the following:
Retainer and a new front exhaust manifold pipe seal over the exhaust manifold
- New exhaust manifold gasket between the exhaust manifold and the cylinder head
- Exhaust manifold into the front exhaust manifold pipe and up to the cylinder head
- Left exhaust manifold bolts and tighten to 18 ft. lbs. (25 Nm)
- Two bolts attaching the left exhaust manifold to the front exhaust manifold pipe and tighten to 18 ft. lbs. (25 Nm)
- O_2S
- New gasket to the secondary AIR tube
- Two nuts attaching the secondary AIR tube to the exhaust manifold and tighten to 106 inch lbs. (12 Nm)

RIGHT SIDE (REAR)

1. Before servicing the vehicle, refer to the precautions in the beginning of this section.
2. Relieve the fuel system pressure.
3. Remove or disconnect the following:
- Cruise control cable from the throttle body lever and bracket and position aside

- Accelerator cable from the throttle body lever and bracket and position aside
- Air cleaner intake duct
- Air cleaner assembly
- Fasteners attaching the accelerator control cable bracket to the throttle body
- Accelerator control cable bracket
- Vacuum brake booster
- Transaxle filler tube
- Transaxle range selector cable
- Bolt connecting the ground lead to the top of the transaxle
- Connector C101 at the left strut tower
- Left wheel speed sensor connector
- Left wheel speed sensor lead from the retaining brackets
- Connector C100 at the top of the transaxle
- Harness from the harness clip and position aside
- Bolts attaching the exhaust manifold heat shield to the left exhaust manifold and position aside
- Right exhaust manifold crossover bolts
- Two bolts attaching the left exhaust manifold to the right exhaust manifold crossover
- Exhaust manifold pipe
- Right exhaust manifold studs
- Heated Oxygen Sensor (HO_2S)
- Exhaust Gas Recirculation (EGR) inlet pipe from the right exhaust manifold
- Right exhaust manifold retaining bolts
- Right exhaust manifold and the gasket from the engine and discard the gasket
- Exhaust manifold crossover seal from the left exhaust manifold and discard the seal

To install:

4. Install a new exhaust manifold crossover seal over the left exhaust manifold
5. Align the exhaust manifold to the cylinder head.
6. Place the crossover portion of the exhaust manifold around the left exhaust manifold.
7. Hold the exhaust manifold to the right cylinder head using a retaining bolt.
8. Grasp the exhaust manifold and remove the retaining bolt.
9. Place a new exhaust manifold gasket between the exhaust manifold and the cylinder head.
10. Install or connect the following:

- Right exhaust manifold bolts and tighten to 18 ft. lbs. (25 Nm)
- EGR inlet pipe to the exhaust manifold nut to 44 ft. lbs. (60 Nm)
- Right exhaust manifold studs and tighten to 80 inch. lbs. (Nm)
- HO$_2$S
- Exhaust manifold pipe
- Two bolts attaching the left exhaust manifold to the right exhaust manifold crossover and tighten to 18 ft. lbs. (25 Nm)
- Heat shield to the left exhaust manifold and tighten the bolts to 80 inch. lbs. (9 Nm)
- Harness to the harness clip
- C100 to the transaxle
- Left wheel speed sensor connector
- Left wheel speed sensor lead to the retaining brackets
- C101 at the left strut tower
- Bolt connecting the ground lead to the top of the transaxle and tighten to 13 ft. lbs. (17 Nm)
- Transaxle range selector cable
- Transaxle filler tube
- Accelerator control cable bracket to the throttle body
- Fasteners attaching the accelerator control cable bracket to the throttle body. Tighten the accelerator control cable bracket bolt and to 115 inch. lbs. (13 Nm) and the accelerator control cable bracket nuts to 89 inch lbs. (10 Nm).
- Air cleaner assembly
- Air cleaner intake duct

➡**Make sure the throttle should operate freely without binding between full closed and wide open throttle.**

- Accelerator cable to the throttle body lever and bracket
- Cruise control cable to the throttle body lever and bracket

Camshaft and Valve Lifters

REMOVAL & INSTALLATION

➡**All valve train components should be kept in the order that they were removed, so that they can be reinstalled in their original position.**

3.5L Engine

LEFT SIDE (FRONT)

1. Before servicing the vehicle, refer to the precautions in the beginning of this section.
2. Disconnect the negative battery cable.

3. Remove or disconnect the following:
- Fuel injector sight shield
- Oil level indicator tube
- Positive Crankcase Ventilation (PCV) valve
- Spark plug wires
- Ignition coil assembly
- Fuel injector wiring harness
- Engine wiring harness
- Camshaft cover

4. Install the camshaft holding fixture (J 42038) on the camshafts.

5. Remove or disconnect the following:
- Camshaft sprockets with the secondary drive chains
- Camshaft bearing caps
- Camshaft holding fixture
- Camshafts
- Rocker arms
- Lifters

To install:

6. Coat the lifters with clean engine oil and place them into position in the engine.

7. Coat the rocker arms with clean engine oil and set them in place in the head.

8. Coat the camshafts with clean engine oil and place them in their proper position in the head.

9. Install or connect the following:
- Camshaft holding fixture
- Camshaft bearing caps. Torque the bolts to 71 inch lbs. (8 Nm) plus an additional 22 degree turn.
- Camshaft sprockets with the secondary drive chains. Torque the bolts to 18 ft. lbs. (25 Nm) plus an additional 45 degree turn.

10. Remove the camshaft holding fixture from the head.

11. Install or connect the following:

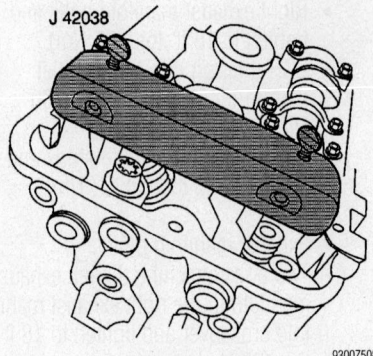

Camshaft holding fixture J-42038 installed on the camshafts—3.5L engine

- Camshaft cover (seals are reusable if they are not damaged). Torque the bolts to 80 inch lbs. (9 Nm).
- Engine wiring harness
- Fuel injector wiring harness
- Ignition coil assembly
- Spark plug wires
- PCV valve and feed tube
- Oil level indicator tube
- Fuel injector sight shield. Torque the nuts to 27 inch lbs. (3 Nm).
- Negative battery cable

RIGHT SIDE (REAR)

1. Before servicing the vehicle, refer to the precautions in the beginning of this section.

2. Remove or disconnect the following:
- Negative battery cable
- Transmission filler tube
- Fuel injector sight shield
- Positive Crankcase Ventilation (PCV) feed tube

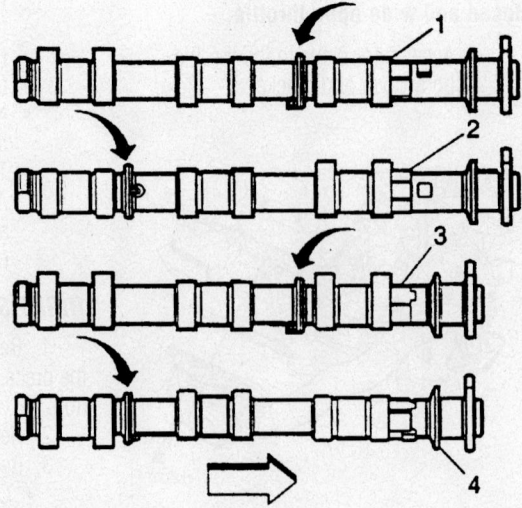

1. Left intake
2. Left exhaust
3. Right intake
4. Right exhaust

Camshaft identification—3.5L engine

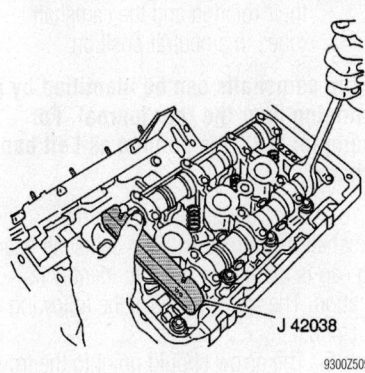

J 42038

9300Z509

Use the flats on the camshaft if rotation is necessary for installation of the holding tool—3.5L engine

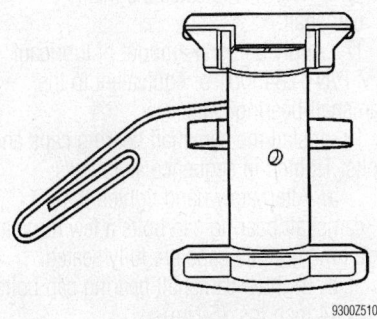

9300Z510

Before installation, compress the tensioner and lock it in place with a piece of wire—3.5L engine

- Engine wiring harness clips
- Oxygen Sensor (O2S) electrical connector
- Spark plug wires
- Ignition coil assembly
- Camshaft cover

3. Install the camshaft holding fixture (tool J-42038).

4. Remove or disconnect the following:
- Camshaft position sensor
- Camshaft sprockets with the secondary drive chains
- Camshaft bearing caps
- Camshaft holding fixture
- Camshafts
- Rocker arms
- Lifters

To install:

5. Coat the lifters with clean engine oil and place them into position in the engine.

6. Coat the rocker arms with clean engine oil and set them in place in the head.

7. Coat the camshafts with clean engine oil and place them in position in the head.

8. Install or connect the following:
- Camshaft holding fixture
- Camshaft bearing caps. Torque the

bolts to 71 inch lbs. (8 Nm). plus an additional 22 degree turn.
- Camshaft sprockets with secondary drive chains. Torque the bolts to 18 ft. lbs. (25 Nm) plus an additional 45 degree turn.
- Camshaft position sensor

9. Remove the camshaft holding fixture.

10. Install or connect the following:
- Camshaft cover (seals are reusable if undamaged). Torque the bolts to 80 inch lbs. (9 Nm).
- Ignition coil assembly
- Spark plug wires
- O2S electrical connector
- Engine wiring harness clips
- PCV feed tube
- Fuel injector sight shield. Torque the nuts to 27 inch lbs. (3 Nm).
- Transmission filler tube
- Negative battery cable

4.0L Engine

LEFT SIDE (FRONT)

1. Before servicing the vehicle, refer to the precautions in the beginning of this section.

2. Drain the cooling system.

3. Remove or disconnect the following:
- Negative battery cable

- Fuel injector sight shield
- Inlet radiator hose from the water housing crossover
- Positive Crankcase Ventilation (PCV) fresh air tube from the left side camshaft cover
- Ignition coil cassette and spark plug boots
- 2 pushnuts securing the engine coolant heater wire, if equipped and position aside
- Surge tank pipe from the fuel rail studs and carefully position aside
- Cable harness clips at the front of the camshaft cover and position the cable harness aside
- AIR valve bracket nut closest to the center of the engine. Pry outward slightly on the AIR valve bracket in order to gain clearance to remove the water pump drive belt shield nut.
- Water pump drive belt shield and belt
- Water pump belt tensioner
- 3 camshaft seal retainer bolts

➡**DO NOT reuse the camshaft seal.**

- Camshaft seal
- Camshaft cover

4. Rotate the crankshaft to Top Dead

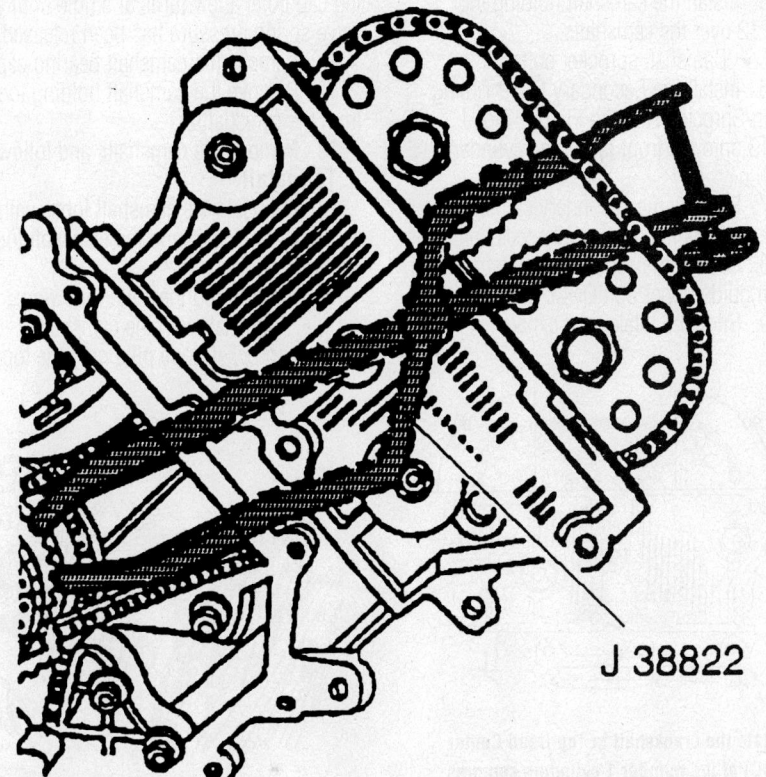

J 38822

9300XG05

Camshaft chain holding tool J-38822—4.0L engine

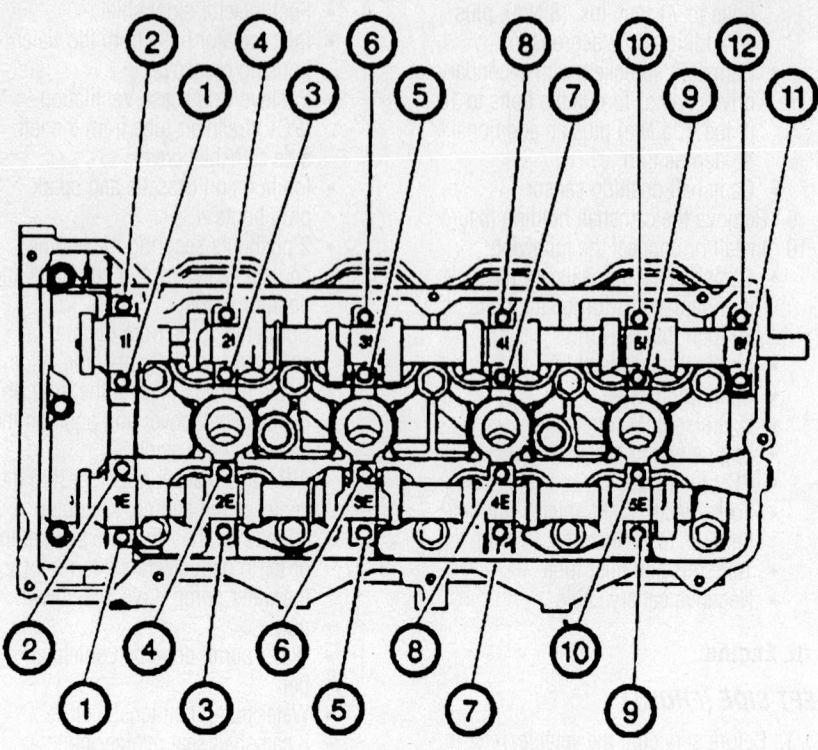

Left cylinder head camshaft bearing cap tightening sequence—4.0L (VIN C) engine

their rotation and the camshaft lobes in a neutral position.

➡ **The camshafts can be identified by a stamping near the rear journal. For example: L-EXH is defined as Left bank Exhaust.**

16. Observe the markings on the camshaft bearing caps. Each camshaft bearing cap is marked in order to identify its location. The markings have the following meanings:

　a. The arrow should point to the front of the engine.

　b. The number indicates the position from the front of the engine.

　c. The "E" indicates the exhaust camshaft.

　d. The "I" indicates the intake camshaft.

17. Apply a liberal amount of lubricant GM P/N 12345001 or equivalent to the camshaft bearing caps.

18. Install the camshaft bearing caps and bolts. Tighten in sequence as follows:

　a. Alternately hand tighten the camshaft bearing cap bolts a few turns at a time until all caps are fully seated.

　b. Tighten camshaft bearing cap bolts to 44 inch lbs. (5 Nm).

　c. Tighten camshaft bearing cap bolts an additional 30 degrees.

19. Align the camshafts.

20. Install the camshaft holding fixture.

➡ **Ensure the camshaft sprockets properly engage the camshaft sprocket drive pins and camshafts.**

21. Slide the intake and exhaust camshaft sprockets off the pins of the Secondary Drive Timing chain/Sprocket Holding Fixture Tool J 44213 and onto the pins of the camshafts.

22. Tighten the secondary camshaft drive

Center (TDC) of the number 1 cylinders compression stroke, both camshaft sprocket drive pins should be at the top of their rotation.

5. Install the camshaft holding tool J44212 over the camshafts.

　• Camshaft sprocket bolts

6. Install the Secondary Drive Timing Chain/Sprocket Holding Fixture Tool J 44213 onto the front rail of the cylinder head.

7. Remove the secondary camshaft drive chain guide upper bolt access plug.

8. Loosen the secondary camshaft drive chain guide upper bolt ONLY two turns.

9. Slide the intake and exhaust

camshaft sprockets onto the pins of the Secondary Drive Timing chain/Sprocket Holding Fixture Tool J 44213.

10. Alternately loosen the camshaft bearing cap bolts a few turns at a time until all valve spring pressure has been released.

11. Remove the camshaft bearing caps.

12. Remove the camshaft holding tool from the camshafts.

13. Remove the camshafts and followers.

To install:

14. Lubricate the camshaft lobes with assembly lubricant, and the camshaft journals with engine oil.

15. Install or connect the following:

　• Camshaft with the camshaft sprocket drive pins near the top of

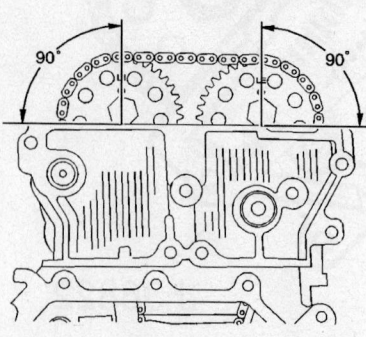

Rotate the crankshaft to Top Dead Center (TDC) of the number 1 cylinders compression stroke, both camshaft sprocket drive pins should be at the top of their rotation

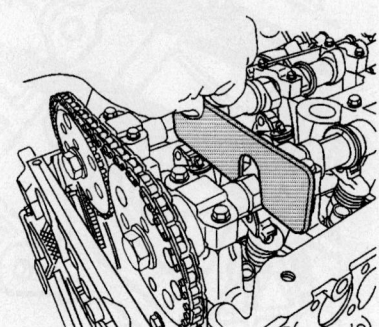

Install the camshaft holding tool J44212 over the camshafts

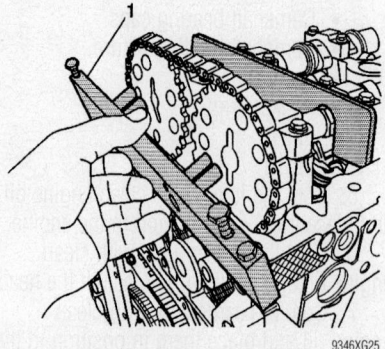

Install Secondary Drive Timing Chain/Sprocket Holding Fixture Tool J 44213 onto the front rail of the cylinder head

chain guide upper bolt to 18 ft. lbs. (25 Nm).

23. Install the secondary camshaft drive chain guide upper bolt access plug and tighten to 39 inch lbs. (4.5Nm).

24. Remove the Secondary Drive Timing chain/Sprocket Holding Fixture Tool J 44213 while carefully holding the camshaft sprockets against the camshafts.

25. Install the camshaft sprocket bolts and tighten to 89 ft. lbs. (120 Nm).

26. Verify the camshaft sprocket alignment.

27. Remove the camshaft holding tool.

28. Install or connect the following:
- Camshaft cover and tighten the bolts to 89 inch lbs. (10 Nm)
- Camshaft seal lips lubricated with clean oil
- Camshaft seal retainers coated with sealer GM P/N 1052080 and tighten to 27 inch lbs. (3 Nm)
- Water pump belt tensioner
- Water pump drive belt and shield
- AIR valve bracket nut to 80 inch lbs. (9Nm)
- Cable harness clips at the front of the camshaft cover
- 2 pushnuts securing the engine coolant heater wire, if equipped
- Inlet radiator hose to the water housing crossover
- Fuel injector sight shield
- Surge tank pipe to the fuel rail studs
- Ignition coil cassette and spark plug boots
- PCV fresh air tube to the left side camshaft cover
- Fuel injector sight shield
- Negative battery cable

29. Refill the cooling system.

RIGHT SIDE (REAR)

1. Before servicing the vehicle, refer to the precautions in the beginning of this section.

2. Drain the cooling system.

3. Remove or disconnect the following:
- Negative battery cable
- Fuel injector sight shield
- Positive Crankcase Ventilation (PCV) valve from the camshaft cover
- Vacuum tubes from the AIR vent solenoid
- AIR vent solenoid electrical connector
- AIR control valve bracket
- Nut securing the AIR tube
- Ignition coil cassette and spark plug boots

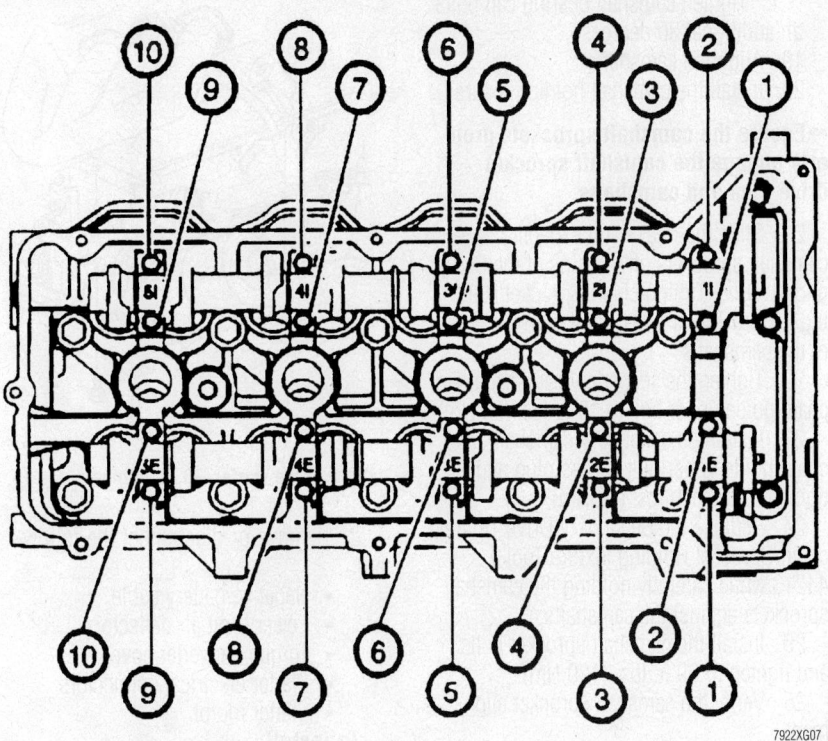

Right cylinder head camshaft bearing cap tightening sequence—4.0L engine

7922XG07

- Cable harness clips at the front of the camshaft cover and position the cable harness aside
- Camshaft cover

4. Rotate the crankshaft to Top Dead Center (TDC) of the number 1 cylinders compression stroke, both camshaft sprocket drive pins should be at the top of their rotation.

5. Install the camshaft holding tool J44212 over the camshafts.
- Camshaft sprocket bolts

6. Install the Secondary Drive Timing Chain/Sprocket Holding Fixture Tool J 44213 onto the front rail of the cylinder head.

7. Remove the secondary camshaft drive chain guide upper bolt access plug.

8. Loosen the secondary camshaft drive chain guide upper bolt ONLY two turns.

9. Slide the intake and exhaust camshaft sprockets onto the pins of the Secondary Drive Timing chain/Sprocket Holding Fixture Tool J 44213.

10. Alternately loosen the camshaft bearing cap bolts a few turns at a time until all valve spring pressure has been released.

11. Remove the camshaft bearing caps.

12. Remove the camshaft holding tool from the camshafts.

13. Remove the camshafts and followers.

To install:

14. Lubricate the camshaft lobes with assembly lubricant, and the camshaft journals with engine oil.

15. Install or connect the following:
- Camshaft with the camshaft sprocket drive pins near the top of their rotation and the camshaft lobes in a neutral position.

➥**The camshafts can be identified by a stamping near the rear journal. For example: L-EXH is defined as Left bank Exhaust.**

16. Observe the markings on the camshaft bearing caps. Each camshaft bearing cap is marked in order to identify its location. The markings have the following meanings:

a. The arrow should point to the front of the engine.

b. The number indicates the position from the front of the engine.

c. The "E" indicates the exhaust camshaft.

d. The "I" indicates the intake camshaft.

17. Apply a liberal amount of lubricant GM P/N 12345001 or equivalent to the camshaft bearing caps.

18. Install the camshaft bearing caps and bolts. Tighten in sequence as follows:

a. Alternately hand tighten the camshaft bearing cap bolts a few turns at a time until all caps are fully seated.

b. Tighten camshaft bearing cap bolts to 44 inch lbs. (5 Nm).

c. Tighten camshaft bearing cap bolts an additional 30 degrees.

19. Align the camshafts.

20. Install the camshaft holding fixture.

➥Ensure the camshaft sprockets properly engage the camshaft sprocket drive pins and camshafts.

21. Slide the intake and exhaust camshaft sprockets off the pins of the Secondary Drive Timing chain/Sprocket Holding Fixture Tool J 44213 and onto the pins of the camshafts.

22. Tighten the secondary camshaft drive chain guide upper bolt to 18 ft. lbs. (25 Nm).

23. Install the secondary camshaft drive chain guide upper bolt access plug and tighten to 39 inch lbs. (4.5Nm).

24. Remove the Secondary Drive Timing chain/Sprocket Holding Fixture Tool J 44213 while carefully holding the camshaft sprockets against the camshafts.

25. Install the camshaft sprocket bolts and tighten to 89 ft. lbs. (120 Nm).

26. Verify the camshaft sprocket alignment.

27. Remove the camshaft holding tool.

28. Install or connect the following:
- Camshaft cover and tighten the bolts to 89 inch lbs. (10 Nm)
- Cable harness clips at the front of the camshaft cover
- Ignition coil cassette and spark plug boots
- Nut securing the AIR tube
- AIR control valve bracket
- AIR vent solenoid electrical connector
- Vacuum tubes from the AIR vent solenoid
- Positive Crankcase Ventilation (PCV) valve to the camshaft cover
- Fuel injector sight shield
- Negative battery cable

29. Refill the cooling system.

Valve Lash

ADJUSTMENT

The valve lash in these models cannot be adjusted.

Starter Motor

REMOVAL & INSTALLATION

3.5L Engine

1. Before servicing the vehicle, refer to the precautions in the beginning of this section.

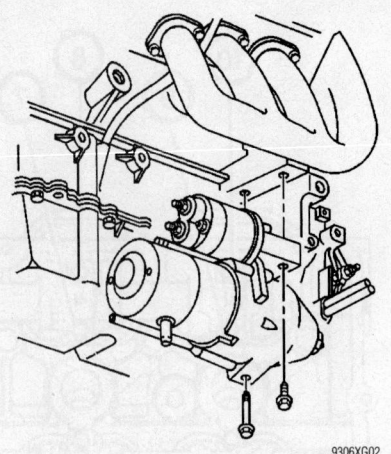

View of the starter—3.5L engine

2. Remove or disconnect the following:
- Negative battery cable
- Lower front air deflector
- Torque converter cover
- Starter electrical connectors
- Starter motor

To install:

3. Install or connect the following:
- Starter motor. Torque the bolts to 37 ft. lbs. (50 Nm).
- Starter electrical connectors. Torque the positive battery terminal nut to 89 inch lbs. (10 Nm) and the "S" terminal nut to 30 inch lbs. (3.5 Nm).
- Torque converter cover
- Lower front air deflector
- Negative battery cable

4.0L Engine

1. Before servicing the vehicle, refer to the precautions in the beginning of this section.

2. Remove or disconnect the following:
- Negative battery cable
- Intake manifold
- Starter electrical connectors
- Starter motor

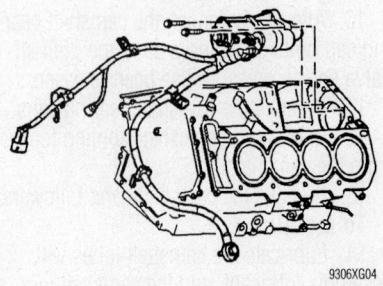

Exploded view of the starter—4.0L engine

To install:

✴✴ WARNING

Before installing the starter, torque the inner solenoid and battery terminal nuts to 70 inch lbs. (8 Nm). If not properly tightened, the starter may fail, due to terminal or cap damage.

3. Install or connect the following:
- Starter motor. Torque the bolts to 22 ft. lbs. (30 Nm).
- Starter electrical connectors. Torque the positive battery terminal nut to 89 inch lbs. (10 Nm) and the "S" terminal nut to 35 inch lbs. (4 Nm).
- Intake manifold
- Negative battery cable

Oil Pan

REMOVAL & INSTALLATION

3.5L Engine

1. Before servicing the vehicle, refer to the precautions in the beginning of this section.

2. Drain the crankcase.

3. Remove or disconnect the following:
- Oil filter
- Oil level sensor
- Transmission brace
- Oil pan

To install:

4. Install or connect the following:
- Oil pan with a new gasket. Do not tighten the bolts at this time.
- Transmission brace on the engine block only. Torque the bolts to 37 ft. lbs. (50 Nm).
- Oil level sensor. Torque the bolt to 80 inch lbs. (9 Nm).

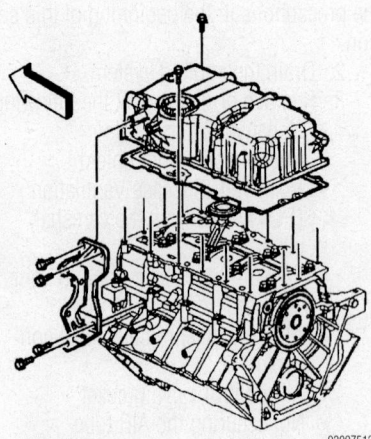

Exploded view of the oil pan—3.5L engine

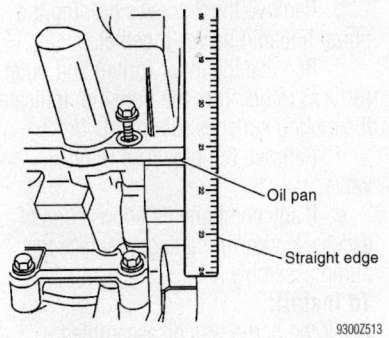

Use a straight edge to align the rear of the oil pan to the rear of the engine—3.5L engine

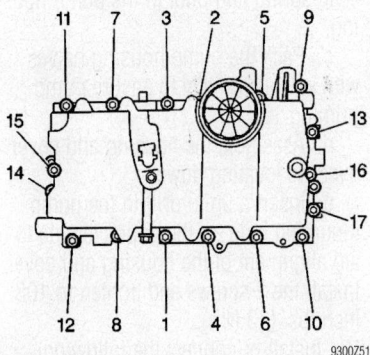

Oil pan mounting bolt tightening sequence—3.5L engine

- Brace-to-oil pan bolts, loosely install them

5. Align the rear of the oil pan flush with the rear of the engine block. Use a straight edge for reference.

6. Press the front of the oil pan against the transmission brace; then torque the brace-to-oil pan bolts to 18 ft. lbs. (25 Nm). Be sure to keep the rear of the pan flush with the rear of the engine.

7. Torque the oil pan bolts to 18 ft. lbs. (25 Nm).

8. Install or connect the following:
- Brace-to-transmission bolts. Torque them to 37 ft. lbs. (50 Nm).
- Oil level sensor electrical connector
- Drain plug. Torque it to 15 ft. lbs. (20 m).
- New oil filter. Torque the cap to 18 ft. lbs. (25 Nm).

9. Refill the crankcase.

10. Start the engine and inspect for leaks.

4.0L Engine

1. Before servicing the vehicle, refer to the precautions in the beginning of this section.

2. Raise the rear seat cushion to access the battery.

3. Disconnect the negative battery cable.

4. Drain the crankcase.

5. Remove or disconnect the following:

- Transmission
- Exhaust crossover pipe
- Oil pan

➡**The oil pan gasket is reusable unless it is damaged. Do not remove the gasket from the oil pan groove unless gasket replacement is required.**

To install:

6. Install or connect the following:
- Oil pan. Torque the bolts in sequence to 89 inch. lbs. (10 Nm).
- Exhaust crossover pipe. Torque the bolts to 18 ft. lbs. (25 Nm).
- Transmission
- Oil pan drain plug. Torque it to 15 ft. lbs. (20 Nm).
- Negative battery cable

7. Refill the crankcase.

Oil Pump

REMOVAL & INSTALLATION

3.5L Engine

1. Before servicing the vehicle, refer to the precautions in the beginning of this section.

2. Remove or disconnect the following:
- Front cover
- Rocker arm covers

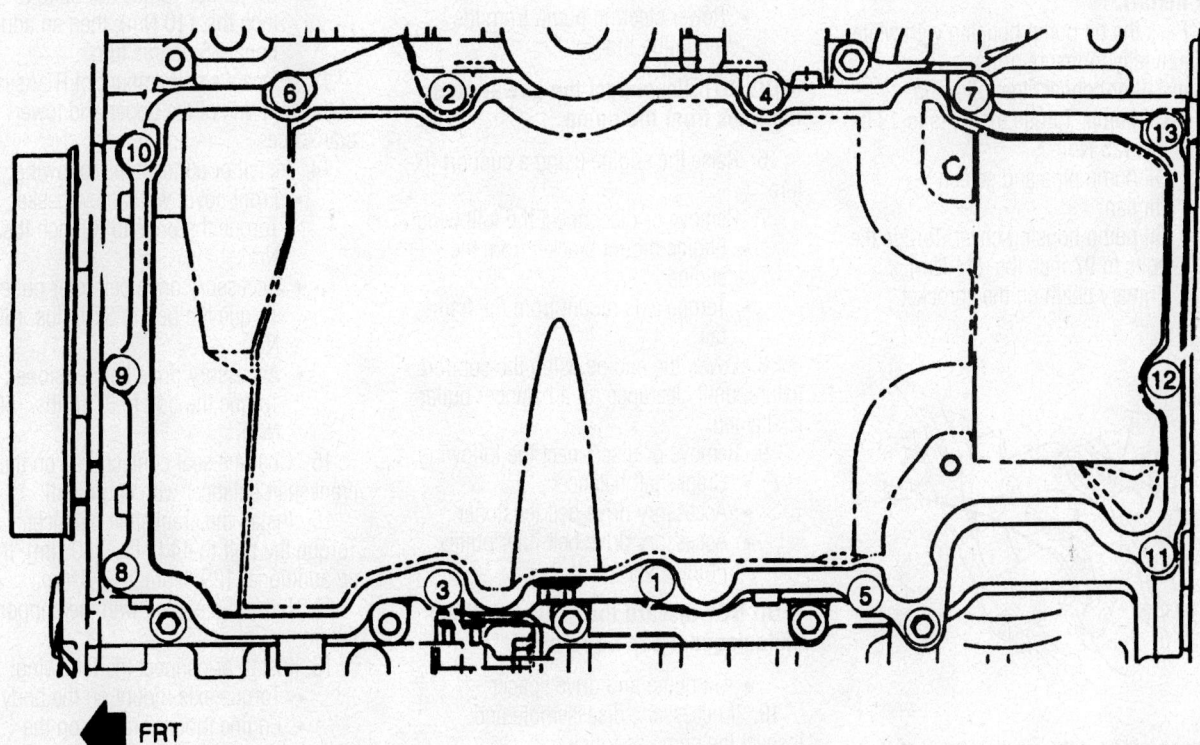

Oil pan bolt tightening sequence—4.0L engine

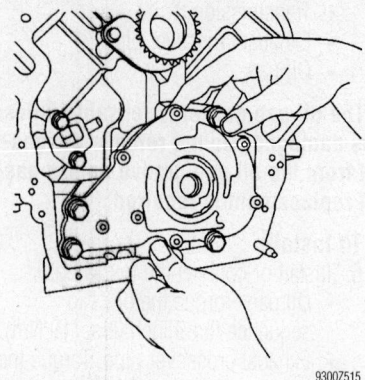

9300Z515

The oil pump is mounted on the front of the engine and driven by the crankshaft—3.5L engine

3. Install camshaft holding fixtures on both sets of camshafts.

4. Remove or disconnect the following:
- Primary chain tensioner
- Primary chain from the drive sprocket
- Oil pan
- Oil pump pipe and screen
- Oil pump by sliding it off the crankshaft

➡The internal parts of the oil pump are not serviced separately. The oil pump may be opened for inspection. If damage or wear is noted, replace the entire pump as an assembly.

To install:

5. Pack the oil pump housing with white petroleum jelly to insure priming.

6. Install or connect the following:
- Oil pump. Torque the bolts to 18 ft. lbs. (25 Nm).
- Oil pump pipe and screen
- Oil pan
- Oil pump housing cover. Torque the bolts to 97 inch lbs. (11 Nm).
- Primary chain on the sprocket

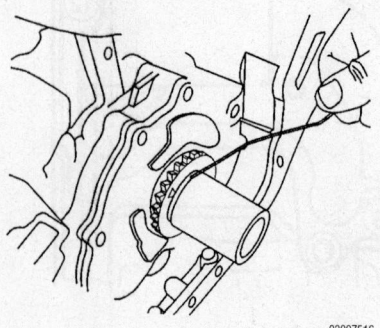

9300Z516

Correct position of the crankshaft sprocket when the oil pump is installed correctly—3.5L engine

➡Be sure to maintain correct timing.

- Chain tensioner

7. Remove the camshaft holding tools.

8. Install or connect the following:
- Rocker arm covers. Torque the bolts to 80 inch lbs. (9 Nm).
- Front cover. Torque the bolts to 124 inch lbs. (14 Nm).

4.0L Engine

1. Before servicing the vehicle, refer to the precautions in the beginning of this section.

2. Disconnect the negative battery cable.

3. Drain the engine oil.

4. Install an engine support fixture.

5. Remove or disconnect the following:
- Engine mount-to-frame through-bolt and nut
- Engine mount-to-engine through-bolt and nut
- Front wheels
- Right inner fender well splash shield
- Lower center air deflector
- Left transmission mount-to-frame through-bolt
- Power steering line retainer from the bracket
- Accessory drive belt from the power steering pump pulley
- Fuel injector sight shield from the intake manifold
- Power steering pump from the mounting bracket

➡DO NOT disconnect the power steering lines from the pump.

6. Raise the engine using a support fixture.

7. Remove or disconnect the following:
- Engine mount bracket from the engine
- Torque axis mount from the frame rail

8. Lower the engine using the support fixture, until clearance for a balancer puller is attained.

9. Remove or disconnect the following:
- Crankshaft balancer
- Accessory drive belt tensioner
- Accessory drive belt idler pulley
- Front cover

➡DO NOT discard the gasket, if it is undamaged it can be re-used.

- Oil pump and drive spacer

10. If necessary, disassemble and inspect the pump as follows:
   a. Remove the drive spacer from the pump housing.

   b. Remove the 2 screws holding the pump housing halves together.

   c. Remove the inner (drive) and outer (driven) rotors from the housing. Indicate the mating surfaces (dimples).

   d. Remove the pressure relief valve.

   e. If any components show signs of excessive wear or damage, replace the pump assembly.

To install:

11. If the pump was disassembled, reassemble as follows:
   a. Install the inner and outer rotors to the pump cover in the same orientation as removed.

   b. Install the pressure relief valve seat, spring and pilot in the pump housing.

   c. Pack the pump housing halves with petroleum jelly to ensure pump priming.

   d. Assemble the housing and cover over the locating dowel.

   e. Insert a 9mm drill in the pump mounting hole on the opposite side to aid alignment of the housing and cover. Install the 2 screws and tighten to 108 inch lbs. (12 Nm).

12. Install or connect the following:
- Oil pump drive spacer into the oil pump from the rear so the drive flat engages the pump rotor
- Oil pump. Torque the bolts to 89 inch lbs. (10 Nm); then an additional 35 degree turn.

13. Place a small amount of RTV sealant at the split line of the upper and lower crankcases.

14. Install or connect the following:
- Front cover with a new gasket. Torque the bolts to 89 inch lbs. (10 Nm).
- Accessory drive belt idler pulley. Torque the bolt to 37 ft. lbs. (50 Nm).
- Accessory drive belt tensioner. Torque the bolt to 37 ft. lbs. (50 Nm).

15. Coat the seal contact area on the crankshaft balancer with engine oil.

16. Install the crankshaft balancer. Torque the bolt to 44 ft. lbs. (60 Nm); then an additional 120 degrees (⅔) turn.

17. Raise the engine with the support fixture.

18. Install or connect the following:
- Torque axis mount on the body
- Engine mount bracket on the engine. Torque the nuts to 30 ft. lbs. (40 Nm) and the bolts to 41 ft. lbs. (55 Nm).

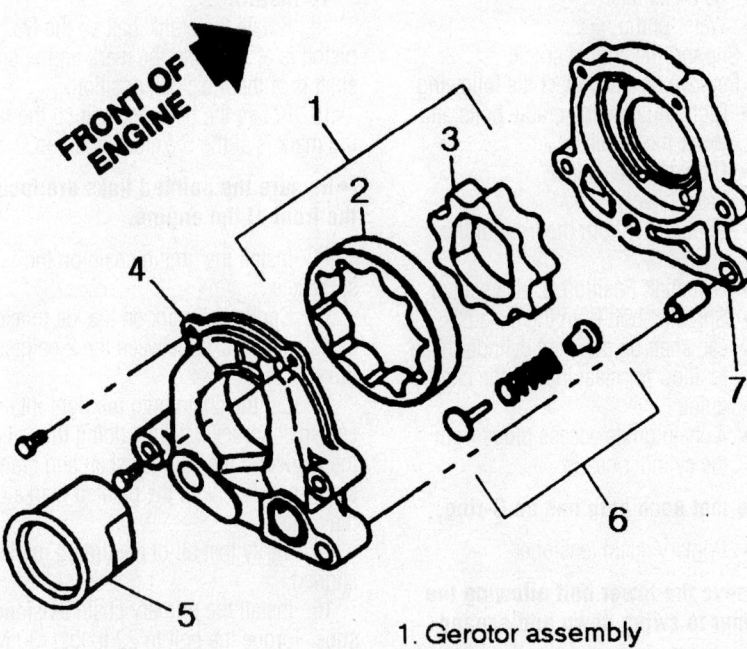

1. Gerotor assembly
2. Outer gear
3. Inner gear
4. Housing
5. Drive spacer
6. Relief valve
7. Cover

7922XG14

Exploded view of the oil pump—4.0L engine

19. Lower the engine support fixture until the engine is at its normal height.
20. Install or connect the following:
   • Power steering pump to the bracket
   • Fuel injector sight shield
   • Accessory drive belt
   • Power steering line retainer to the engine mount bracket
   • Left transmission mount-to-frame through-bolt. Torque the bolt to 63 ft. lbs. (85 Nm).
   • Lower center air deflector
   • Right inner fender well splash shield
   • Front wheels. Torque the nuts to 100 ft. lbs. (140 Nm).
   • Engine mount-to-engine bracket. Torque the through-bolt to 70 ft. lbs. (95 Nm).
   • Engine mount-to-frame. Torque the through-bolt to 37 ft. lbs. (50 Nm).
   • Negative battery cable
21. Remove the engine support fixture.
22. Replace the oil filter and refill the crankcase.
23. Run the engine and check for leaks and proper engine operation.

Rear Main Seal

REMOVAL & INSTALLATION

3.5L and 4.0L Engines With the Cartridge Type Seal

1. Before servicing the vehicle, refer to the precautions in the beginning of this section.
2. Remove or disconnect the following:
   • Engine and transmission assembly
   • Transmission from the engine
   • Flexplate
3. Remove the seal as follows:
   a. Place Seal Removal tool J-42841 onto the crankshaft.
   b. Install 8, 1-inch self starting screws through the guide holes of the tool and into the seal. A variable speed drill will be helpful for installing the screws.
   c. Tighten the center screw of the tool to remove the seal.
4. Clean out the drain at the bottom of the seal bore with a piece of wire or pipe cleaner.

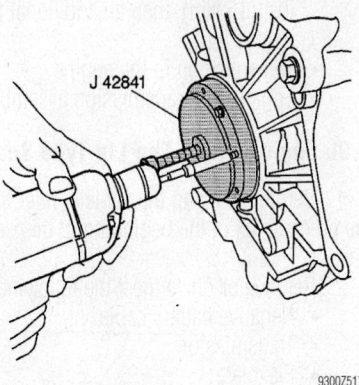

9300Z517

Use the guide holes in tool J-42841 to install the screws in the seal—3.5L and 4.0L engines

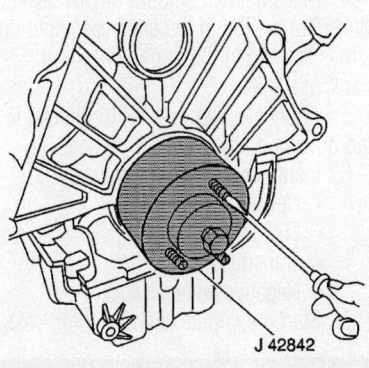

9300Z518

Use rear main seal installer J-42842 to install the seal—3.5L engine

To install:
5. Apply a small amount of RTV gasket maker at the crankcase split line across the end of the upper and lower crankcase seal.
6. Coat the outer diameter of the rear crankshaft seal area with engine oil.
7. Wipe the outer diameter of the crankshaft flexplate flange with a lint-free cloth.
8. Lubricate the outer rubber surface of the seal with clean engine oil. Do not apply any oil to the green coating, pre-applied to the inner diameter of the seal.
9. Loosen the center bolt of the seal installer until the center hub protrudes about ½ inch past the outer plate.
10. Thread the 3 mounting bolts into the crankshaft flange until the tool is firmly mounted on the crankshaft.
11. Install the seal by tightening the center bolt until the tool bottoms against the crankshaft.
12. Remove the tool and verify that the seal is installed evenly.
13. Install or connect the following:
   • Flexplate. Torque the bolts to 11 ft.

lbs. (15 Nm); then an additional 50 degree turn.
- Transmission to the engine
- Engine and transmission assembly

4.0L Engines With The Lip Type Seal

1. Before servicing the vehicle, refer to the precautions in the beginning of this section.
2. Remove or disconnect the following:
   - Negative battery cable
   - Transmission
   - Flexplate
3. Carefully pry the oil seal from the housing using a suitable prying tool, taking care not to damage the housing or the crankshaft sealing surface.

To install:

4. Place a small amount of GM Gasket Maker® at the top of the crankcase split line across the end of the upper and lower crankcase seal.
5. Apply clean engine oil to the inside and outside diameter of the oil seal.
6. Install or connect the following:
   - New seal
   - Flexplate
   - Transmission
   - Negative battery cable
7. Start the engine and check for leaks.

Timing Chain, Sprockets, Front Cover and Seal

REMOVAL & INSTALLATION

3.5L Engine

PRIMARY CHAIN

1. Before servicing the vehicle, refer to the precautions in the beginning of this section.
2. Remove or disconnect the following:
   - Negative battery cable
   - Rocker arm covers
3. Rotate the crankshaft so the No. 1 piston is at Top Dead Center (TDC) and the flats on the rear of the camshafts are parallel with the camshaft cover sealing surface.
4. Install camshaft holding fixtures on both sets of camshafts.
5. Drain the cooling system.
6. Remove or disconnect the following:
   - Front diagonal brace
   - Battery and tray
   - Washer and coolant reservoirs
   - Underhood accessory wiring junction block and move it aside
   - Drive belt
   - Power steering pump pulley
   - Idler pulley

- Belt tensioner
- Water pump
7. Support the engine cradle
8. Remove or disconnect the following:
   - Right side engine cradle bolts and lower the cradle
   - Crankshaft balancer
   - Front cover
   - Lift bracket from the front of the engine
   - Camshaft Position (CMP) sensor
   - Sprocket bolt from the exhaust camshaft on the right cylinder head to allow for clearance of the chain guide
   - 4 chain guide access plugs from the cylinder heads

➡ **Note that each plug has an O-ring.**

- Primary chain tensioner

➡ **Remove the lower bolt allowing the tensioner to swing down and expand.**

- Primary chain tensioner shoe by removing the bolt, pushing the guide downward slightly and pulling it up through the cylinder head
- Primary chain from the right camshaft, allowing it to fall into the oil pump area
- Primary chain

To install:

9. Rotate the crankshaft so the No. 1 piston is at TDC and the mark on the crankshaft is at the 4 o'clock position.
10. Rotate the balance shaft so the timing mark is at the 5 o'clock position.

➡ **Be sure the painted links are facing the front of the engine.**

11. Install the timing chain on the sprockets.
12. Center the mark on the left intake camshaft sprocket between the 2 painted links.
13. Lift the chain onto the right intake camshaft sprocket. While doing this, align the marks on the balance shaft and crankshaft sprockets with the painted marks on the chain.
14. Verify that all of the timing marks are aligned.
15. Install the primary chain tensioner shoe. Torque the bolt to 22 ft. lbs. (30 Nm).
16. Compress the primary chain tensioner using the following steps:
   a. Rotate the ratchet release lever counterclockwise and hold it.
   b. Press the tensioner shoe in and hold it.
   c. Release the ratchet lever and slowly release the pressure on the shoe.

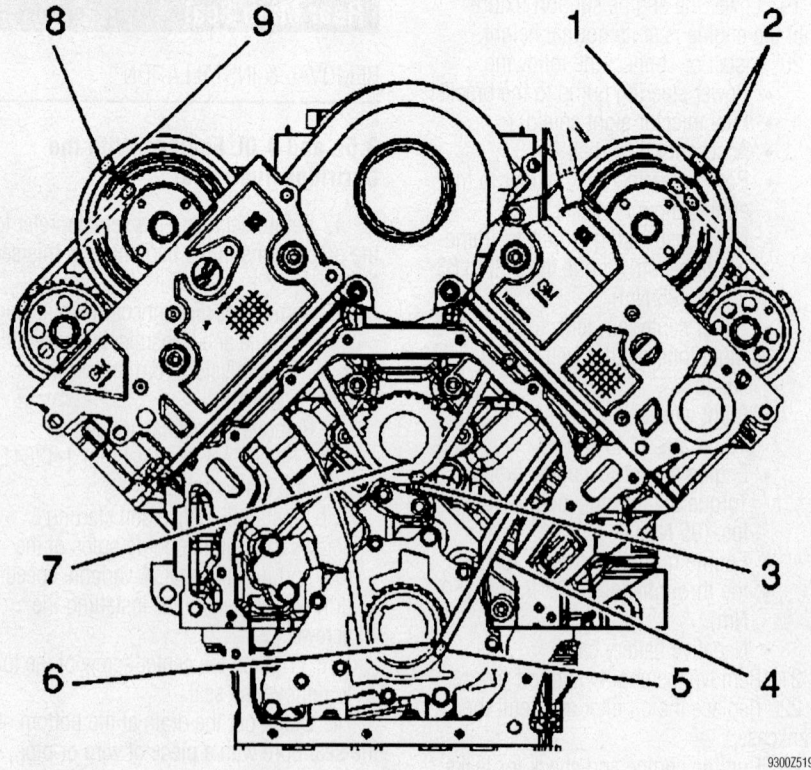

Primary timing chain alignment marks—3.5L engine

93002519

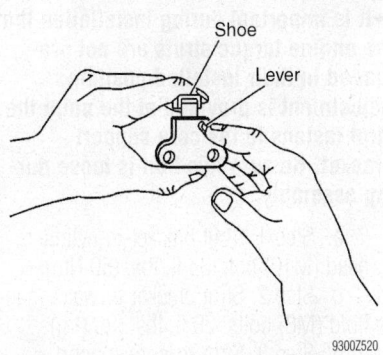

Compressing the primary chain tensioner—3.5L engine

d. Insert a pin through the hole in the lever as the lever moves to the first click. The ratchet should hold the shoe in the compressed position.

➡**Be sure the lever on the tensioner is facing you when installed.**

17. Install or connect the following:
- Primary chain tensioner. Torque the bolts to 18 ft. lbs. (25 Nm); then remove the chain tensioner pin.
- 4 chain guide access plugs. Torque the plugs to 44 inch lbs. (5 Nm).
- Front engine lift bracket. Torque the hex head bolt to 37 ft. lbs. (50 Nm)

and the internal drive bolt to 18 ft. lbs. (25 Nm).
- CMP sensor. Torque the bolts to 80 inch lbs. (9 Nm).
18. Remove the camshaft holding tools.
19. Place a small bead of RTV sealant on the 3 areas indicated in the diagram.
20. Install or connect the following:
- Front cover with a new gasket. Torque the bolts to 124 inch lbs. (14 Nm) and the coolant drain plug to 89 inch lbs. (10 Nm).
- Crankshaft balancer. Torque the bolt to 37 ft. lbs. (50 Nm); then an additional 120 degree turn.
21. Raise the engine cradle and install new bolts. Torque the bolts to 133 ft. lbs. (180 Nm).
22. Coat the sub-frame bushings with rubber lubricant.
23. Install or connect the following:
- Water pump with a new gasket. Torque the bolts to 124 inch lbs. (14 Nm).
- Belt tensioner. Torque the bolts to 37 ft. lbs. (50 Nm).
- Idler pulley. Torque the bolt to 37 ft. lbs. (50 Nm).
- Power steering pump pulley
- Drive belt

- Underhood accessory wiring junction block
- Washer and coolant reservoirs
- Battery and tray
- Front diagonal brace
- Rocker arm covers. Torque the bolts to 80 inch lbs. (9 Nm).
- Negative battery cable
24. Refill the cooling system.
25. Refill the crankcase.
26. Start the engine and verify no leaks.

SECONDARY TIMING CHAIN

1. Before servicing the vehicle, refer to the precautions in the beginning of this section.
2. Disconnect the negative battery cable.
3. Remove the rocker arm cover and install camshaft holding fixture J-42038.
4. Remove the camshaft sprocket bolts and install the timing chain holding fixture J-42042 on the cylinder head.
5. Remove or disconnect the following:
- Sprockets and chain
- Secondary timing sprocket and chain

To install:
6. Install or connect the following:
- Secondary timing chain on the sprockets, with the drive pins at the 12 o'clock positions

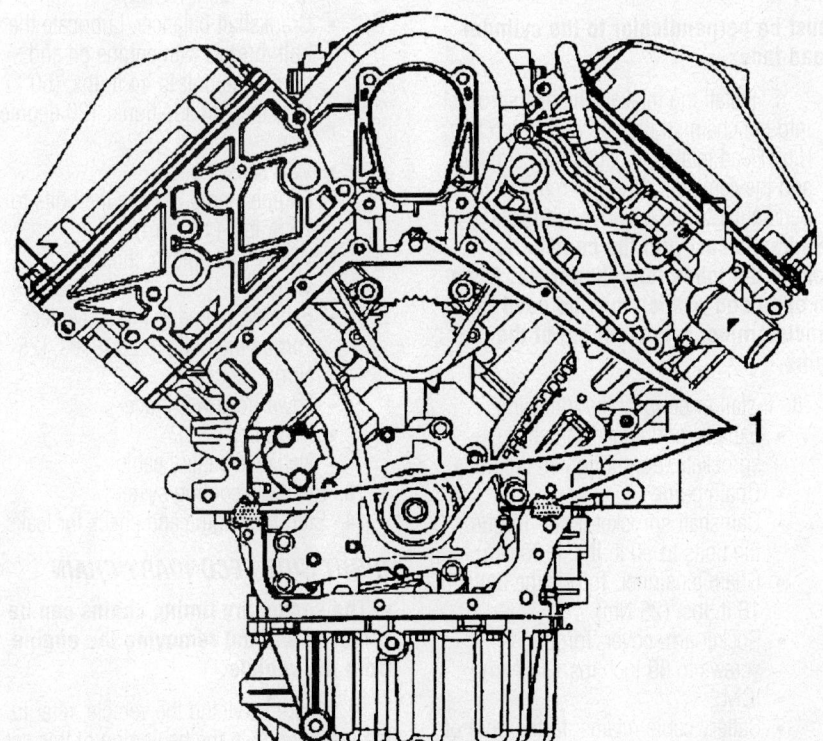

Apply RTV sealant to the 3 areas indicated before installing the front cover and gasket—3.5L engine

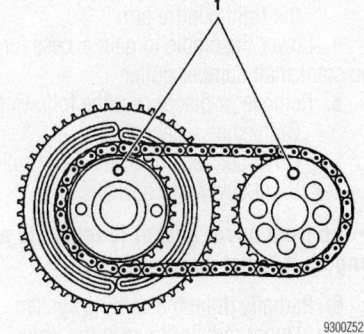

Correct sprocket alignment for the left secondary timing chain—3.5L engine

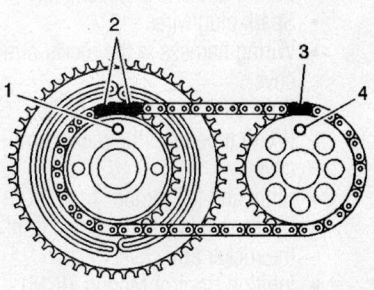

Correct sprocket alignment for the right secondary timing chain—3.5L engine

- Sprockets and chain assembly onto the camshafts, with the chain properly aligned on the tensioner
7. Remove the timing chain holding fixture
8. Install the sprocket bolts. Torque the bolts to 18 ft. lbs. (25 Nm); then an additional 45 degree turn.
9. Remove the camshaft holding fixture.
10. Install or connect the following:
   - Rocker arm cover
   - Negative battery cable

4.0L Engine

LEFT SIDE SECONDARY CHAIN

➡The secondary timing chains can be removed without removing the engine from the vehicle.

1. Before servicing the vehicle, refer to the precautions in the beginning of this section.
2. Remove or disconnect the following:
   - Negative battery cable
   - Drive belt
   - Tower-to-tower brace
   - Exhaust Y-pipe at the converter
   - Front wheels
   - Wheel well splash shields
   - Crankshaft balancer bolt
3. Support the engine cradle.
   - 3 right-side engine cradle bolts
   - Vehicle Speed Sensor (VSS) from the right control arm
4. Lower the cradle to gain access for the crankshaft damper puller.
5. Remove or disconnect the following:
   - Crankshaft balancer
   - Drive belt tensioner and idler pulley
   - Front cover and gasket

➡The front cover gasket is reusable as long as it is not damaged.

6. Partially drain the cooling system.
   - Upper radiator hose at the water crossover
   - Spark plug wires
   - Right side fan
   - Battery cable at the alternator
   - Spark plug wires
   - Wiring harness at the rocker arm cover
   - Positive Crankcase Ventilation (PCV) tube from the rocker arm cover
   - Right and left torque struts
   - Battery cable retainer at the front of the rocker arm cover
   - Ignition Control Module (ICM)
   - Rocker arm cover by pivoting it around the water pump driveshaft

➡Continue moving the cover upward and pivoting so that the edge of the cover closely follows the left edge of the intake manifold cover. The gasket is reusable as long as it is not damaged.

   - Left side secondary chain tensioner
   - Left side chain guide

➡Access the upper chain guide mounting bolt through the hole in the cylinder head capped with the plastic plug.

   - Secondary drive chain and sprockets

To install:

➡Correct timing exists when the crankshaft and intermediate shaft sprocket timing marks are in alignment and all 4 camshaft drive pins are perpendicular (90 degrees) to the cylinder head surface.

7. Assemble the left side secondary timing chain as follows:
   a. Route the timing chain over the intermediate sprocket teeth outer row.
   b. Route the timing chain over the chain guide and install the exhaust camshaft sprocket so the **LE** (Left Head Exhaust) pin engages the sprocket notch.

➡There should be no slack in the lower section of the timing chain and the camshaft drive pin

must be perpendicular to the cylinder head face.

   c. Install the intake camshaft sprocket into the chain so the sprocket notch **LI** (Left Head Intake) engages the camshaft and the camshaft drive pin remains perpendicular to the cylinder head face.

➡A hex is cast into the camshafts behind the lobes for cylinder No. 2, so an open end wrench may be used to provide minor repositioning of the cams.

8. Install or connect the following:
   - Exhaust and intake camshaft sprockets, do not tighten the bolts
   - Chain guide
   - Camshaft sprocket bolts. Torque the bolts to 90 ft. lbs. (120 Nm).
   - Chain tensioner. Torque the bolts to 18 ft. lbs. (25 Nm).
   - Rocker arm cover. Torque the screws to 89 inch lbs. (10 Nm).
   - ICM
   - Battery cable retainer to the front of the rocker arm cover
9. Install the right and left torque struts and torque the bolts as follows:

➡It is important during installation that the engine torque struts are not preloaded in their installed position. Adjustment is provided at the point the strut fastens to the core support bracket. Be sure this bolt is loose during assembly.

   a. Step 1: Strut bracket-to-cylinder head (M10) bolt: 35 ft. lbs. (50 Nm).
   b. Step 2: Strut bracket-to-water manifold (M8) bolts: 20 ft. lbs. (25 Nm).
   c. Step 3: Strut-to-core support bracket bolt: 45 ft. lbs. (60 Nm).
10. Install or connect the following:
   - PCV fresh air tube to the rocker arm cover
   - Wiring harness to the cover
   - Spark plug wires
   - Battery cable at the alternator
   - Right side fan
   - Upper radiator hose to the water crossover
   - Front cover with a new gasket. Torque the bolts to 89 inch lbs. (10 Nm).
11. Apply a dab of RTV to the split line between the upper and lower crankcase assemblies.
12. Install or connect the following:
   - Drive belt idler pulley. Torque the bolt to 35 ft. lbs. (47 Nm).
   - Drive belt tensioner. Torque the nut to 35 ft. lbs. (47 Nm).
   - Crankshaft balancer. Lubricate the bolt threads with engine oil and torque the bolt to 44 ft. lbs. (60 Nm) then an additional 120 degree turn.
   - VSS sensor
   - Engine cradle. Torque the bolts to 75 ft. lbs. (102 Nm).
   - Wheel well splash shields
   - Wheels
   - Exhaust Y-pipe to the converter. Torque the bolts to 20 ft. lbs. (25 Nm).
   - Tower-to-tower brace
   - Drive belt
   - Negative battery cable
13. Refill the cooling system.
14. Start the engine and check for leaks.

RIGHT SIDE SECONDARY CHAIN

➡The secondary timing chains can be removed without removing the engine from the vehicle.

1. Before servicing the vehicle, refer to the precautions in the beginning of this section.
2. Remove or disconnect the following:
   - Exhaust Y-pipe from the converter

90°
90°
90°
90°
90°
B
A
3
VIEW A
VIEW B
1
2
1 INTAKE POSITION
2 EXHAUST POSITION
3 TIMING MARKS
7922XG18

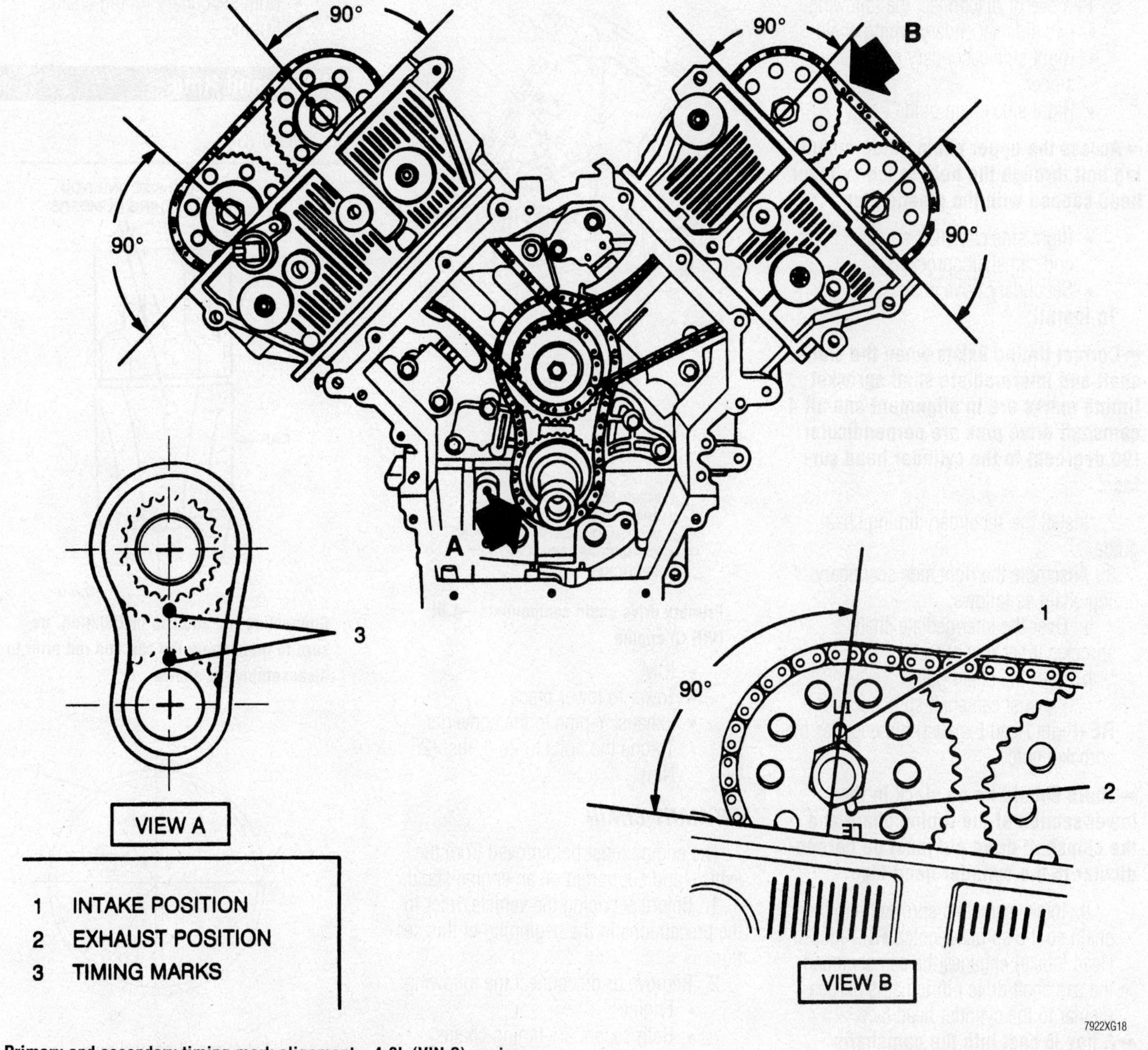

Primary and secondary timing mark alignment—4.0L (VIN C) engine

SECOND
PUSH DOWN
FIRST
PRESS IN
7922XG26

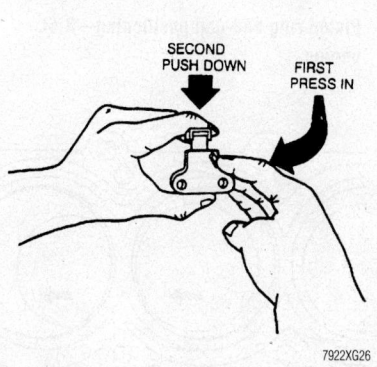

Rotating tensioner release lever—4.0L (VIN C) engine

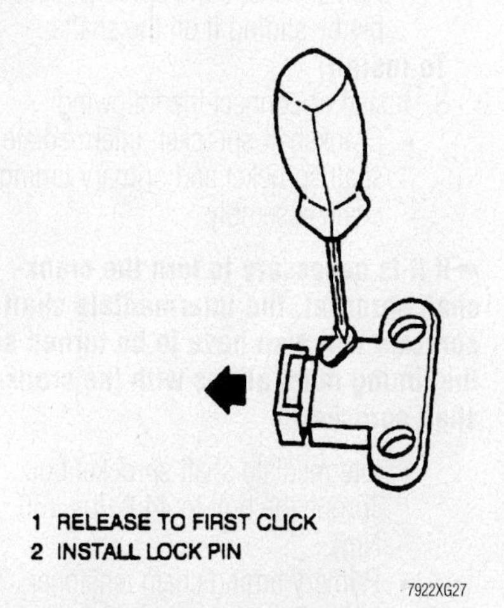

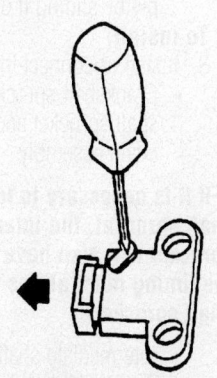

1 RELEASE TO FIRST CLICK
2 INSTALL LOCK PIN
7922XG27

Locking the tensioner into position—4.0L (VIN C) engine

- Tower-to-tower brace
- Ignition Control Module (ICM)
- Spark plug wires from the right bank
- Positive Crankcase Ventilation (PCV) valve
- Purge canister solenoid from the rear of the cover
- Rocker arm cover

3. Safely support the front of the engine cradle.
4. Remove or disconnect the following:
   - 2 mounting bolts at the front of the cradle
   - Right and left torque struts
5. Lower the engine cradle (or raise the vehicle) to provide clearance at the rear of the engine compartment.

6. Remove or disconnect the following:
- Left side secondary timing chain
- Right side secondary chain tensioner
- Right side chain guide

➡**Access the upper chain guide mounting bolt through the hole in the cylinder head capped with the plastic plug.**

- Right side camshaft sprocket bolts and camshaft sprockets
- Secondary drive chain

To install:

➡**Correct timing exists when the crankshaft and intermediate shaft sprocket timing marks are in alignment and all 4 camshaft drive pins are perpendicular (90 degrees) to the cylinder head surface.**

7. Install the secondary timing chain guide.

8. Assemble the right side secondary timing chain as follows:

a. Over the intermediate shaft sprocket inner row of teeth.

b. Over the chain guide.

c. Exhaust camshaft sprocket so the **RE** (Right Head Exhaust) pin engages the sprocket notch.

➡**There should be no slack in the lower section of the timing chain and the camshaft drive pin must be perpendicular to the cylinder head face.**

d. Intake camshaft sprocket into the chain so the sprocket notch **RI** (Right Head Intake) engages the camshaft and the camshaft drive pin remains perpendicular to the cylinder head face.

➡**A hex is cast into the camshafts behind the lobes for cylinder No. 1, so an open end wrench may be used to provide minor repositioning of the cams.**

9. Install or connect the following:
- Exhaust and intake camshaft sprockets, do not tighten
- Timing chain tensioner. Torque the bolts to 18 ft. lbs. (25 Nm).
- Camshaft sprocket bolts. Torque the bolts to 90 ft. lbs. (120 Nm).
- Left side secondary timing chain
- Torque struts
- Engine cradle. Torque the bolts to 75 ft. lbs. (102 Nm).
- Rocker arm cover. Torque the screws to 84 inch lbs. (10 Nm).
- Purge canister solenoid
- PCV valve
- Spark plug wires

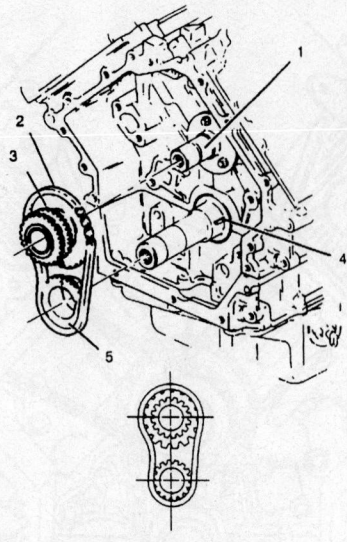

1 INTERMEDIATE SHAFT
2 PRIMARY CHAIN
3 INTERMEDIATE SHAFT SPROCKET
4 CRANKSHAFT SPROCKET KEY
5 SPROCKET

7922XG25

Primary drive chain components—4.0L (VIN C) engine

- ICM
- Tower to tower brace
- Exhaust Y-pipe to the converter. Torque the bolts to 20 ft. lbs. (25 Nm).

PRIMARY CHAIN

The engine must be removed from the vehicle and supported on an engine stand.

1. Before servicing the vehicle, refer to the precautions in the beginning of this section.

2. Remove or disconnect the following:
- Engine
- Both secondary timing chains
- Primary timing chain tensioner
- Intermediate shaft sprocket
- Primary chain and sprocket assembly by sliding it off the shafts

To install:

3. Install or connect the following:
- Crankshaft sprocket, intermediate shaft sprocket and primary timing chain assembly

➡**If it is necessary to turn the crankshaft sprocket, the intermediate shaft sprocket will also have to be turned so the timing mark aligns with the crankshaft sprocket.**

- Intermediate shaft sprocket bolt. Torque the bolt to 44 ft. lbs. (60 Nm).
- Primary timing chain tensioner bolts. Torque them to 18 ft. lbs. (25 Nm).

- Both secondary timing chains
- Engine

Piston and Ring

POSITIONING

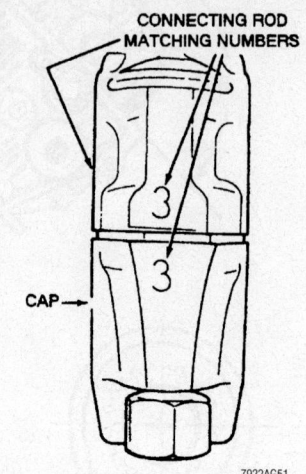

7922AG51

Connecting rod and cap installation. Be sure to matchmark the cap and rod prior to disassembly, as shown

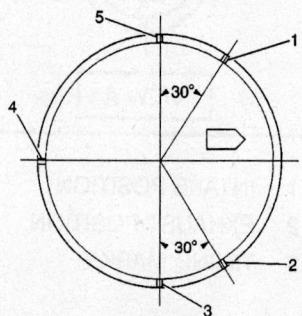

1. Lower oil control ring
2. Upper oil control ring
3. Top Ring
4. Oil control ring expander
5. Second ring

9306XG05

Piston ring end-gap positioning—3.5L engine

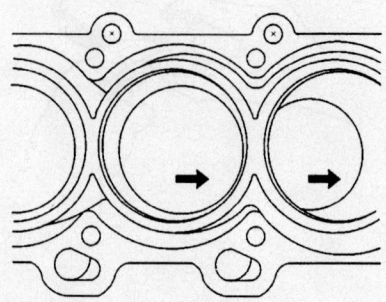

9306XG06

Piston positioning—3.5L engine

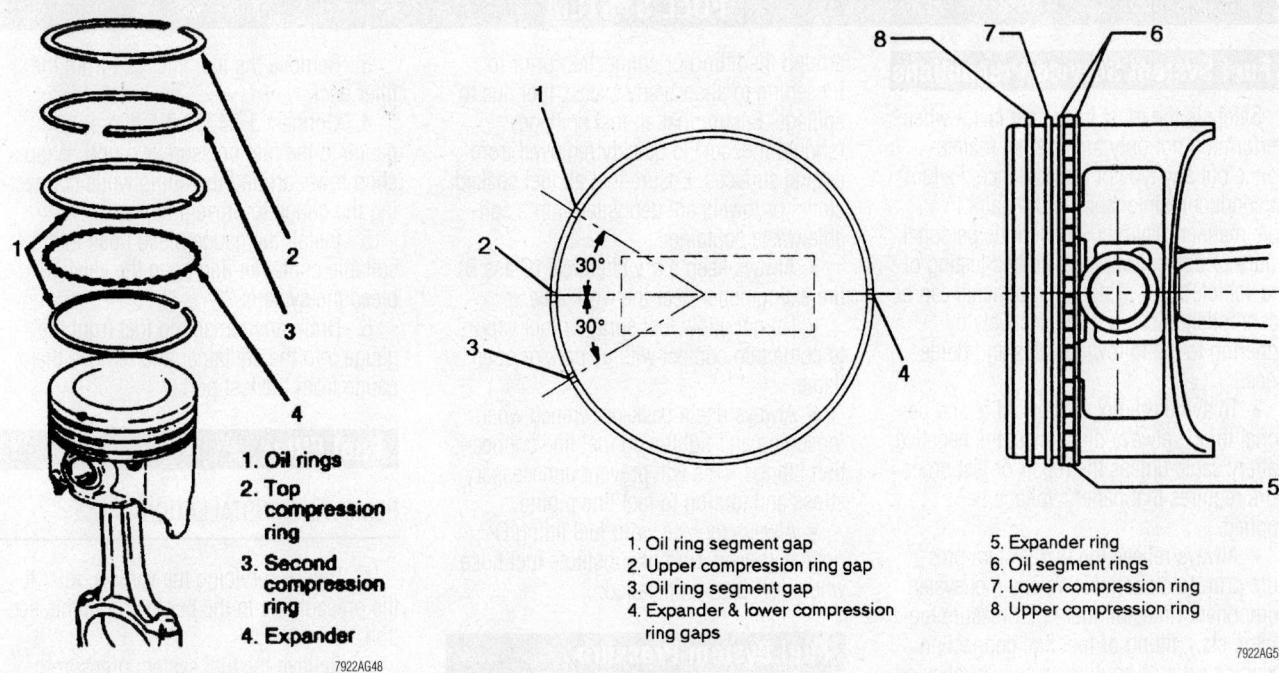

1. Oil rings
2. Top compression ring
3. Second compression ring
4. Expander

7922AG48

Piston ring positioning—3.5L engines

1. Oil ring segment gap
2. Upper compression ring gap
3. Oil ring segment gap
4. Expander & lower compression ring gaps

5. Expander ring
6. Oil segment rings
7. Lower compression ring
8. Upper compression ring

7922AG52

Piston ring and end-gap positioning—4.0L engine

LEFT BANK

FRT

B   A

RIGHT BANK

BOTTOM VIEW
(PAN - SIDE UP)

PISTON ARROW TOWARD
CHAINCASE ON BOTH SIDES

RIGHT BANK

FRT

1   3   5   7

LEFT BANK

6   8

4

2

TOP VIEW
(PAN - SIDE DOWN)

ROD CAPS

PISTON

FRT

LOCATOR LUGS
INDICATE PISTON
FRONT TOWARDS
ENGINE FRONT

ROD CAP

BEARING CAP NOTCHES
POINT TOWARD EACH
OTHER ON PAIRED RODS

FRT

PISTON

ROD CAP

BEARING CAP NOTCHES
POINT TOWARD EACH
OTHER ON PAIRED RODS

VIEW A

VIEW B

7922AG53

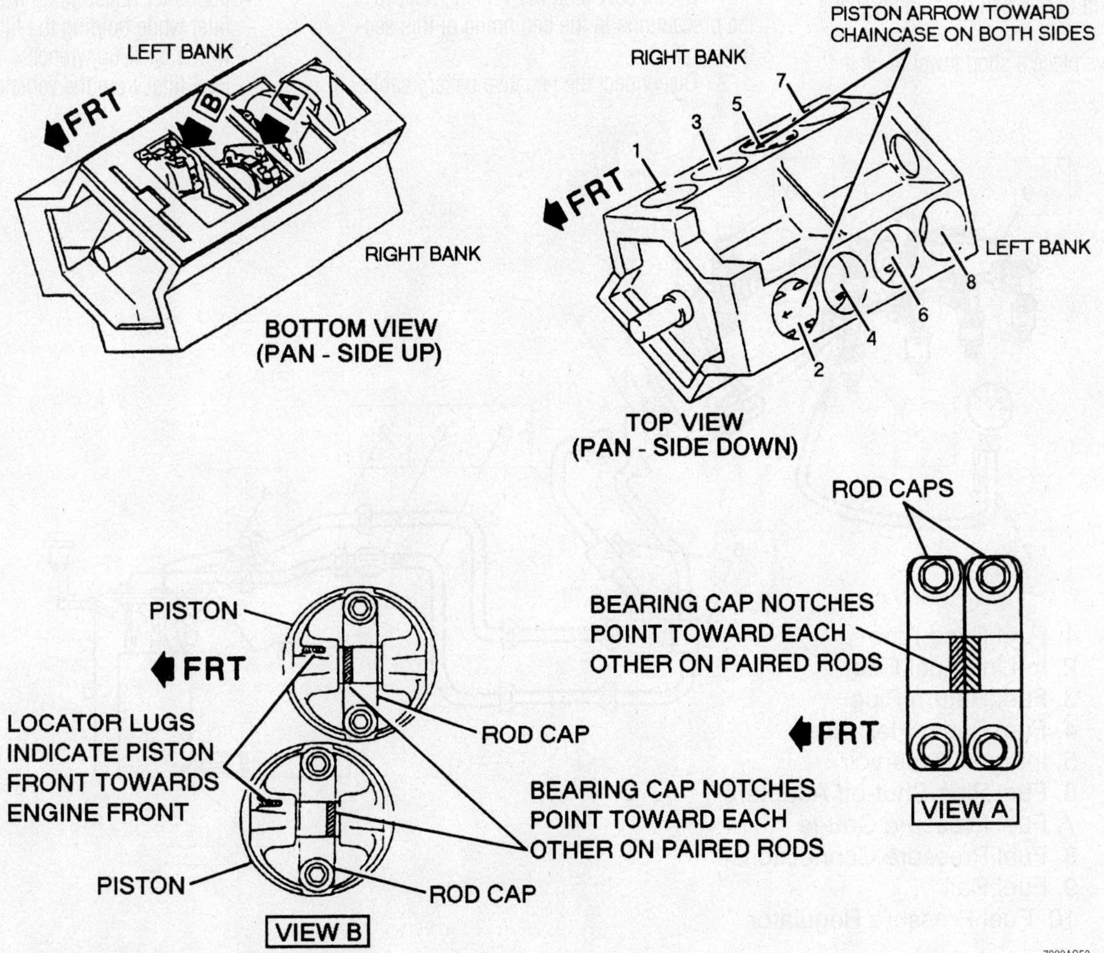

Piston and connecting rod assembly positioning—4.0L engine

FUEL SYSTEM

Fuel System Service Precautions

Safety is the most important factor when performing not only fuel system maintenance but any type of maintenance. Failure to conduct maintenance and repairs in a safe manner may result in serious personal injury or death. Maintenance and testing of the vehicle's fuel system components can be accomplished safely and effectively by adhering to the following rules and guidelines.

• To avoid the possibility of fire and personal injury, always disconnect the negative battery cable unless the repair or test procedure requires that battery voltage be applied.

• Always relieve the fuel system pressure prior to disconnecting any fuel system component (injector, fuel rail, pressure regulator, etc.), fitting or fuel line connection. Exercise extreme caution whenever relieving fuel system pressure, to avoid exposing skin, face and eyes to fuel spray. Please be advised that fuel under pressure may penetrate the skin or any part of the body that it contacts.

• Always place a shop towel or cloth around the fitting or connection prior to loosening to absorb any excess fuel due to spillage. Ensure that all fuel spillage (should it occur) is quickly removed from engine surfaces. Ensure that all fuel soaked cloths or towels are deposited into a suitable waste container.

• Always keep a dry chemical (Class B) fire extinguisher near the work area.

• Do not allow fuel spray or fuel vapors to come into contact with a spark or open flame.

• Always use a back-up wrench when loosening and tightening fuel line connection fittings. This will prevent unnecessary stress and torsion to fuel line piping.

• Always replace worn fuel fitting O-rings with new. Do not substitute fuel hose where fuel pipe is installed.

Fuel System Pressure

RELIEVING

1. Before servicing the vehicle, refer to the precautions in the beginning of this section.
2. Disconnect the negative battery cable.

3. Remove the fuel filler cap from the filler neck.
4. Connect J-34730-1 fuel pressure gauge to the fuel pressure test port. Wrap a shop towel around the fitting while connecting the gauge to prevent fuel spillage.
5. Install the gauge bleed hose into a suitable container and open the valve to bleed the system.
6. Drain any remaining fuel from the gauge into the container and remove the gauge from the test port.

Fuel Filter

REMOVAL & INSTALLATION

1. Before servicing the vehicle, refer to the precautions in the beginning of this section.
2. Relieve the fuel system pressure.
3. Remove or disconnect the following:
   • Quick connect fitting at the fuel filter inlet
   • Fuel filter outlet fitting from the fuel filter while holding the filter fitting with a back-up wrench
   • Fuel filter from the vehicle

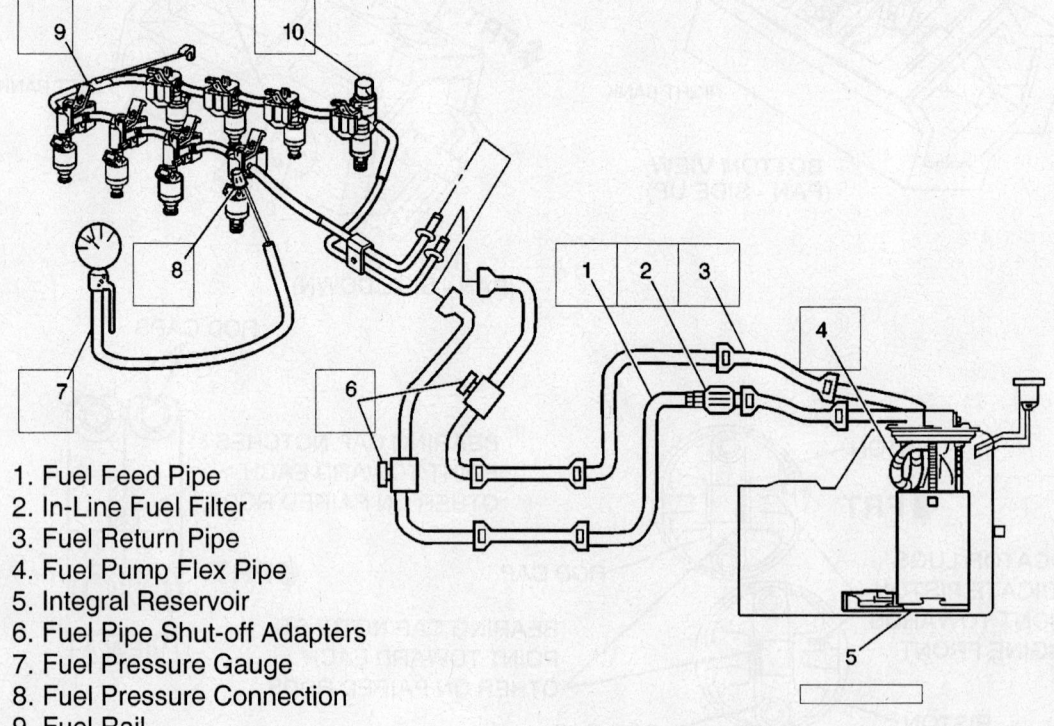

1. Fuel Feed Pipe
2. In-Line Fuel Filter
3. Fuel Return Pipe
4. Fuel Pump Flex Pipe
5. Integral Reservoir
6. Fuel Pipe Shut-off Adapters
7. Fuel Pressure Gauge
8. Fuel Pressure Connection
9. Fuel Rail
10. Fuel Pressure Regulator

9300XG07

Fuel system circuit showing the pressure test port on the supply rail—4.0L engine

To install:

4. Position the new fuel filter in the bracket.

5. Install a new plastic retainer on the fuel inlet line.

6. Apply a drop of oil on the fuel filter inlet fitting and snap the fitting onto the fuel filter.

7. Reconnect the fuel outlet line to the filter by holding the filter with a back-up wrench and tightening the line fitting to 22 ft. lbs. (30 Nm).

8. Pressurize the fuel system and verify no leaks.

Fuel Pump

REMOVAL & INSTALLATION

1. Before servicing the vehicle, refer to the precautions in the beginning of this section.

2. Relieve the fuel system pressure.

3. Drain the fuel tank.

4. Remove or disconnect the following:
   - Spare tire and jack
   - Floor trunk liner by pulling it back
   - Fuel sender access panel
   - Fuel sender assembly quick connect fittings
   - Fuel sender assembly electrical connector

✹✹ WARNING

When the lock-ring is removed from the fuel sender, the sender assembly will spring up. Downward pressure should be kept on the assembly and slowly released to ensure the sender assembly does not get damaged.

   - Lock-ring from the fuel sender

✹✹ CAUTION

The reservoir bucket on the fuel sender assembly will be full of fuel when it is removed from the tank. Be sure to have a catch pan nearby to drain the sender into.

   - Sender assembly from the tank

To install:

5. Install or connect the following:
   - New O-ring on top of the tank
   - Sender assembly into the tank
   - Retainer on top of the fuel tank, compress the sender until the retainer can be engaged; then lock it in place
   - Quick connect fittings to the fuel sender assembly

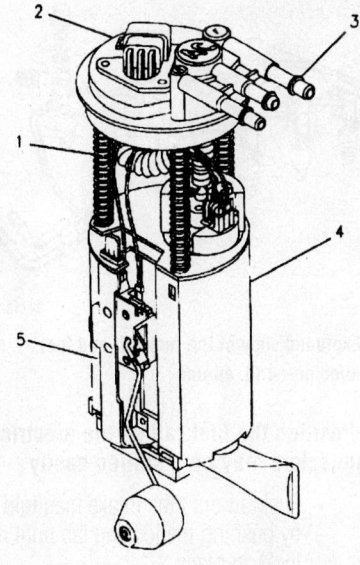

1	SUPPORT ASSEMBLY – FUEL SENDER
2	COVER ASSEMBLY – FUEL SENDER
3	FUEL PIPES (ABOVE COVER)
4	RESERVOIR – FUEL PUMP FUEL
5	SENSOR ASSEMBLY – FUEL LEVEL

7922XG19

Fuel pump and sending unit module assembly

   - Sender assembly electrical connector
   - Negative battery cable

6. Pressurize the fuel system and verify no leaks.

7. Install or connect the following:
   - Fuel sender access panel
   - Trunk liner
   - Spare tire and jack

Fuel Injector

REMOVAL & INSTALLATION

3.5L Engine

1. Before servicing the vehicle, refer to the precautions in the beginning of this section.

2. Relieve the fuel system pressure.

3. Remove or disconnect the following:
   - Fuel injector sight shield
   - Fuel line fittings
   - Fuel injector electrical connectors
   - All necessary vacuum hoses
   - Wiring harness from the fuel rail, if necessary
   - Fuel rail from the intake manifold by pushing the locking tab away

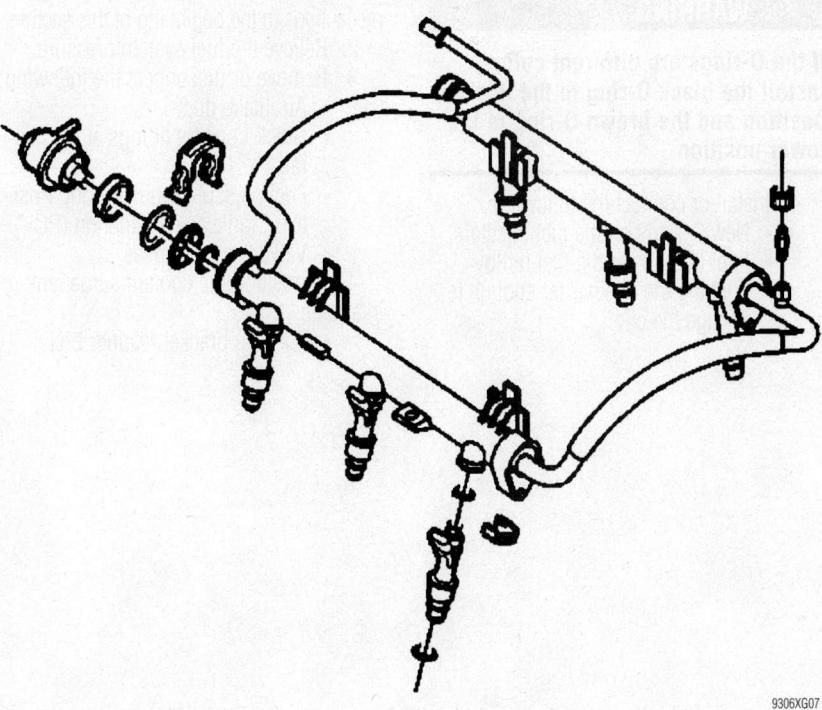

9306XG07

Exploded view of the fuel rail assembly–3.5L engine

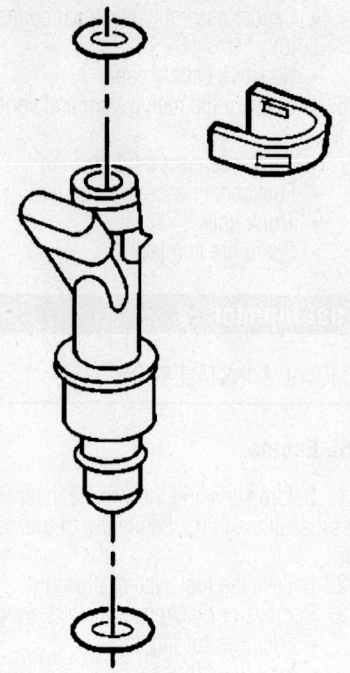

Exploded view of the fuel injector–3.5L engine

from the center of the intake manifold
- Fuel injector from the fuel rail and by spreading the retainer clip to release it

To install:

✳✳ WARNING

If the O-rings are different colors, install the black O-ring in the upper position and the brown O-ring in the lower position.

4. Install or connect the following:
- New O-rings on the fuel injectors
- Fuel injector on the fuel rail by pushing the retainer far enough to engage the clip

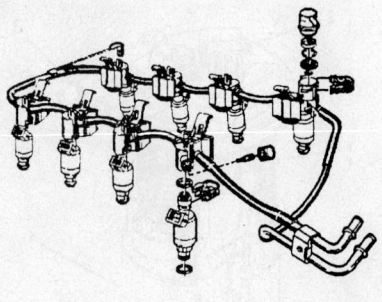

Exploded view of the fuel rail and fuel injector—4.0L engine

➡**Position the fuel rail so the electrical connectors may be installed easily**

- Fuel rail onto the intake manifold by pushing the locking tab until it locks in place
- Wiring harness to the fuel rail, if necessary
- All vacuum hoses
- Fuel injector electrical connectors
- Fuel line fittings
- Negative battery cable

5. Pressurize the fuel system and check for leaks.

6. Install the fuel injector sight shield. Torque the bolts to 27 inch lbs. (3 Nm).

4.0L Engine

1. Before servicing the vehicle, refer to the precautions in the beginning of this section.
2. Relieve the fuel system pressure.
3. Remove or disconnect the following:
- Air intake duct
- Quick-connect fittings at the fuel rail
- Fuel pressure regulator and Positive Crankcase Ventilation (PCV) Valve vacuum hoses

4. Reposition the coolant surge tank inlet pipe.
- Fuel rail bracket retainer bolt

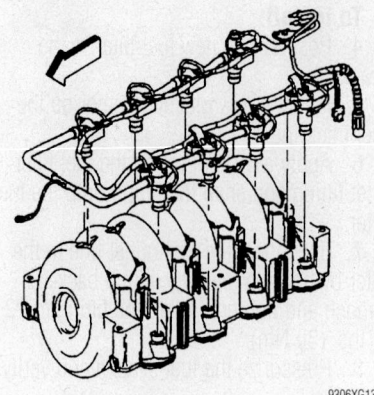

Exploded view of the fuel rail and intake manifold—4.0L engine

- Fuel injector electrical connectors
- Fuel rail bolts
- Rail and injectors as an assembly from the manifold
- Fuel injector from the fuel rail by spreading the retainer
- Fuel injector

To install:

5. Lubricate the new O-rings with engine oil.

6. Install or connect the following:
- Fuel injector using a new retainer clip and new O-rings
- Rail and injectors as an assembly into the manifold until fully seated
- Fuel rail bolts and tighten to 89 inch lbs. (10 Nm)
- Fuel injector electrical connectors
- Fuel rail bracket retainer bolt and tighten to 89 inch lbs. (10 Nm)
- Fuel pressure regulator and PCV vacuum hoses
- Coolant surge tank inlet pipe
- Quick-connect fittings at the fuel rail
- Air intake duct
- Negative battery cable

7. Pressurize the system and check for leaks.

DRIVE TRAIN

Transmission Assembly

REMOVAL & INSTALLATION

4T80-E Transmission

1. Before servicing the vehicle, refer to the precautions in the beginning of this section.

2. Remove or disconnect the following:
- Negative battery cable
- Air intake duct
- Range selector cable from the transmission
- Upper transmission oil cooler line from the radiator
- Lower transmission oil cooler line from the transmission
- Both heater tube retainers from the upper case side cover
- Coolant temperature sensor electrical connector
- Ground wire at the rear of the transmission
- Left front Wheel Speed Sensor (WSS) electrical connector
- Vehicle Speed Sensor (VSS) from the transmission
- Rack and pinion electrical connector
- Transmission vent hose
- Upper transmission bolts

3. Install an engine support fixture.

4. Remove or disconnect the following:
- Vacuum reservoir
- Ball joints from the steering knuckles
- Antilock Brake System (ABS) module
- Secondary Air Injection (AIR) pumps
- Left and right splash shields
- Air deflector
- Power steering line brackets at the frame
- Front and rear transmission mount nuts
- Rack and pinion from the frame

5. Support the engine frame.

6. Remove the 6 frame bolts and lower the frame from the body.

7. Remove or disconnect the following:
- Left and right halfshafts
- Exhaust heat shield
- Transmission-to-engine brace
- Front engine-to-transmission pencil brace
- Three ground connections at the front of the transmission
- Transmission main wiring harness
- Engine to transmission bracket
- Flexplate cover
- Torque converter bolts
- Right transmission bracket

8. Support the transmission.

9. Remove the lower transmission bolts and lower the transmission towards the left so it can clear the starter motor.

To install:

10. Install or connect the following:
- Transmission. Torque the transmission-to-engine bolts to 55 ft. lbs. (75 Nm).
- Right transmission bracket. Torque the bolt to 55 ft. lbs. (75 Nm).
- Torque converter bolts. Torque the bolts to 47 ft. lbs. (63 Nm).
- Flexplate cover. Torque the bolts to 20 ft. lbs. (27 Nm).
- Transmission brace to the engine. Torque the bolts to 44 ft. lbs. (60 Nm).
- Transmission main wiring harness
- Three ground connections to the transmission
- Front engine-to-transmission pencil brace
- Engine-to-transmission bracket
- Exhaust heat shield
- Left and right halfshafts
- Engine frame. Torque the bolts to 142 ft. lbs. (192 Nm).

11. Remove the frame support fixture.

12. Install or connect the following:
- Rack and pinion. Torque the bolts to 48 ft. lbs. (65 Nm).
- Front and rear transmission mounts. Torque the nuts to 48 ft. lbs. (65 Nm).
- Power steering line brackets. Torque the bolts to 53 inch lbs. (6 Nm).
- Air deflector
- Left and right splash shields
- AIR pumps
- ABS module
- Lower ball joints

13. Remove the engine support fixture.

14. Install or connect the following:
- Vacuum reservoir
- Upper transmission bolts. Torque the bolts to 55 ft. lbs. (75 Nm).
- Transmission vent hose
- Rack and pinion electrical connector
- VSS to the transmission
- WSS electrical connector
- Ground wire at the rear of the transmission
- Coolant temperature sensor electrical connector
- Both heater tube retainers from the upper case side cover
- Lower transmission oil cooler line
- Upper transmission oil cooler line
- Range selector cable and bracket
- Air intake duct
- Negative battery cable

15. Check and adjust the transmission fluid level.

16. Check the front end alignment.

4T65-E Transmission

1. Before servicing the vehicle, refer to the precautions in the beginning of this section.

➡ **Make sure the wheels of the vehicle must are in the straight ahead position and the steering column in the LOCK position before disconnecting the steering column or intermediate shaft from the steering gear. Failure to do so will cause the SIR coil assembly to become uncentered, which may cause damage to the coil assembly.**

2. Lock the steering column by installing tool J 42640 into the underside of the steering column.

3. Remove or disconnect the following:
- Negative battery cable
- Air cleaner assembly
- Range selector cable from the range selector lever
- Range selector cable with bracket from the transmission case and set aside
- Nut and bolt from the AIR pipe
- Bolt from the AIR pipe
- Ground cable bolt from the transaxle
- Transaxle electrical connector
- Wiring harness from the wiring harness retainer on the transaxle

4. Install the engine support fixture.
- Upper engine-to-transaxle case bolts
- Front wheels
- Both of the left facia extensions
- Front air deflector
- Intermediate shaft lower pinch bolt
- Intermediate shaft from the power steering gear
- Power steering gear heat shied
- Power steering gear mounting bolts
- Power steering line retainers from the frame

5. Secure the power steering gear to the exhaust manifold.

6. Loosen the two mounting nuts in order to allow removal of the Brake Pressure Modulator Valve (BPMV) from the bracket.

7. Remove or disconnect the following:
- Brake line retainers from the frame
- Left transaxle mount
- Frame
- Right and left drive axles from the transaxle
- Transmission oil cooler hoses from the transaxle
- Transaxle fluid filler tube
- Torque converter cover
- Flywheel-to-torque converter bolts

8. Support transaxle using an appropriate transaxle jack.
- Vehicle Speed Sensor (VSS) electrical connector
- Transaxle brace-to-transaxle bolts
- Transaxle brace-to-engine bolts
- Transaxle brace
- Engine-to-transaxle case bolt which can be accessed through the right wheel opening
- Remaining transaxle-to-engine bolt
- Transaxle from vehicle using an appropriate transaxle jack
- Rear transaxle bracket from the transaxle
- Left transaxle bracket from the transaxle

To install:

9. Installation is the reverse of removal, please note the following specifications:
- Left transaxle bracket bolts to 81 ft. lbs. (110 Nm)
- Rear transaxle bracket bolts to 46 ft. lbs. (63 Nm)
- Lower transaxle bolts to 55 ft. lbs. (75 Nm)
- Transaxle brace-to-engine bolts to 48 ft. lbs. (65 Nm)
- Transaxle brace-to-transaxle bolts to 26 ft. lbs. (36 Nm)
- Flywheel-to-torque converter bolts to 47 ft. lbs. (63 Nm)
- Power steering gear mounting bolts to 70 ft. lbs. (95 Nm)
- Intermediate shaft lower pinch bolt to 33 ft. lbs. (45 Nm)
- Upper transaxle case-to-engine bolts to 55 ft. lbs. (75 Nm)
- Bolt to AIR pipe to 17 ft. lbs. (23 Nm)
- Negative battery cable

10. Check and adjust the transmission fluid level.

11. Check the front end alignment.

Halfshaft

REMOVAL & INSTALLATION

1. Before servicing the vehicle, refer to the precautions in the beginning of this section.

2. Remove or disconnect the following:
- Wheel
- Outer tie rod end from the steering knuckle

3. Insert a drift or punch into the brake rotor and against the brake caliper in order to prevent the wheel hub and bearing from turning.
- Wheel driveshaft spindle nut and discard
- Stabilizer shaft link
- Electrical connector from the Wheel Speed Sensor (WSS) and position the wiring harness away from the ball joint
- Lower ball joint from the steering knuckle

4. Install wheel hub removal tool J 42129 onto the wheel hub and secure with wheel nuts.
- Wheel driveshaft from the wheel hub and bearing using the removal tool and support the wheel driveshaft.

5. Assemble tool J 2619-01 slide hammer with adapter, J 29794 extension, and J 33008-A wheel driveshaft removal

6. Using the assembled tools, separate the shaft from the transaxle.

7. Remove the wheel driveshaft from the vehicle.

To install:

8. Install or connect the following:
- Driveshaft to the transaxle

9. Verify that the driveshaft is properly engaged to the transaxle by grasping the inner tripod housing and pulling outward. Do not pull on the driveshaft bar. The driveshaft will remain firmly in place when properly engaged

10. Remove or disconnect the following:
- Driveshaft to the hub and bearing
- Ball joint to the steering knuckle
- Wheel speed sensor electrical connector
- Stabilizer shaft link

11. Insert a drift or punch into the rotor and against the caliper in order to prevent the hub and bearing from turning.

12. Remove or disconnect the following:
- A new driveshaft spindle nut and tighten to 118 ft. lbs. (160 Nm)
- Outer tie rod end to the steering knuckle

- Wheel

13. Check the alignment.

CV-Joint

REMOVAL & INSTALLATION

Inner (Tri-Pod) Joint

1. Before servicing the vehicle, refer to the precautions in the beginning of this section.

2. Remove or disconnect the following:
- Front wheel
- Halfshaft
- Swage ring
- Large CV-joint boot clamp
- CV-joint boot by sliding it away from the tri-pod joint
- Tri-pod housing from the tri-pod spider
- Trilobal tri-pod bushing from the housing
- Inboard spacer ring slide, it rearward on the shaft
- Outboard retaining ring
- Tri-pod joint spider assembly
- Inboard spacer ring and CV-joint boot

To install:

3. Install or connect the following:
- Swage ring clamp
- CV-joint boot

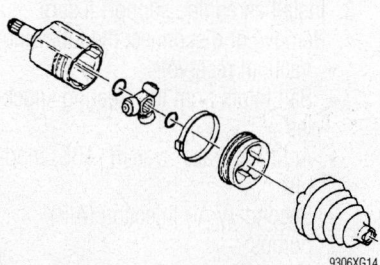

9306XG14

Exploded view of the inner (tri-pod) joint

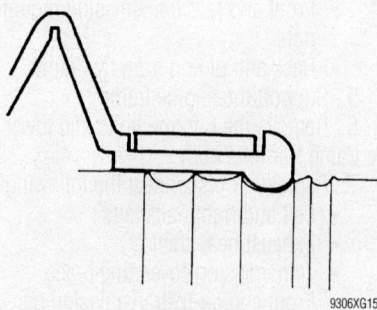

9306XG15

Positioning the inner CV-joint boot seal and swage ring—Inner (tri-pod) joint

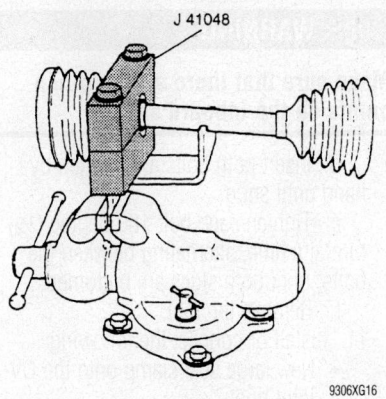

View of the swage ring crimping tool— Inner (tri-pod) joint

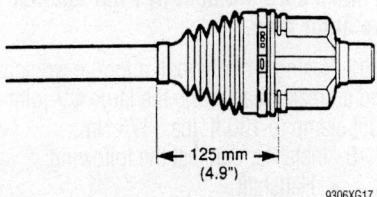

Boot measurement—Inner (tri-pod) joint

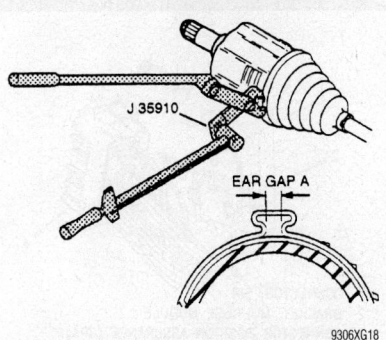

Crimping the large CV-joint boot ring— Inner (tri-pod) joint

4. Position the CV-joint boot seal into the axle shaft's joint seal groove and align the swage ring clamp on the boot.

5. Secure the swage ring clamp as follows:

   a. Mount the lower half of tool J-41048 in a vise.

   b. Position the outboard of the half-shaft in the tool.

   c. Position the upper end of tool J-41048 onto the lower half.

✳✳ WARNING

Make sure that there are no pinch points on the inboard seal.

   d. Insert both bolts and tighten by hand until snug.

   e. Tighten each bolt 180 degree (½) turn at a time, alternating between the bolts, until both sides are bottomed.

   f. Remove the tool.

6. Install or connect the following:

- Inboard spacer ring slide it rearward on the shaft
- Tri-pod joint spider assembly onto the shaft
- Outboard retaining ring into the axle shaft groove
- Tri-pod joint spider assembly, slide it against the outboard retaining ring
- Inboard spacer ring
- ½ kit grease into the boot
- ½ kit grease into the tri-pod housing
- Trilobal tri-pod bushing flush with the tri-pod housing face
- New large seal clamp onto the CV-joint boot
- Tri-pod housing, slide it over the tri-pod joint spider assembly
- CV-joint boot/clamp, slide it into place over the trilobal tri-pod bushing with the seal lip in the groove

➡**Make sure the boot lies flat against the trilobal bushing.**

7. Position the CV-joint boot so it measures 4.9 in. (125mm).

8. Using a crimp tool, a torque wrench and a breaker bar, crimp the large CV-joint boot clamp to 130 ft. lbs. (176 Nm).

9. Install or connect the following:

- Halfshaft
- Front wheel

Outer Joint

1. Before servicing the vehicle, refer to the precautions in the beginning of this section.

2. Remove or disconnect the following:

- Front wheel
- Halfshaft
- Swage ring
- Large boot clamp
- CV-joint boot, slide it away from the CV-joint
- CV-joint assembly by spreading the inner race-to-axle shaft retaining ring ears
- CV-joint boot from the axle shaft

3. Disassemble the chrome alloy balls from the CV-joint cage as follows:

   a. Position a brass drift against the CV-joint cage and tap it with a hammer to tilt the cage.

   b. Remove the 1st chrome alloy ball from the cage.

   c. Tilt the cage in the opposite direction.

   d. Remove the opposite chrome alloy ball.

   e. Repeat the procedure until all 6 balls are removed.

4. Disassemble the CV-joint cage and inner race as follows:

   a. Pivot the cage and race 90 degrees to the center line of the outer race.

   b. Align the cage windows with outer race lands.

   c. Remove the cage from the outer race.

   d. Rotate the inner race upward and remove it from the cage.

To install:

5. Lubricate the parts with a light coat of grease.

6. Assemble the CV-joint cage and inner race, as follows:

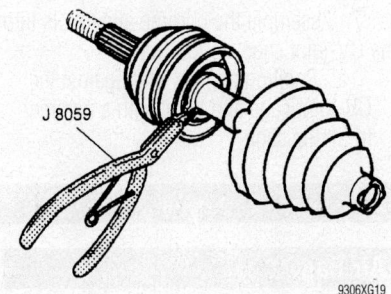

Disconnecting the outer CV-joint from the axle shaft

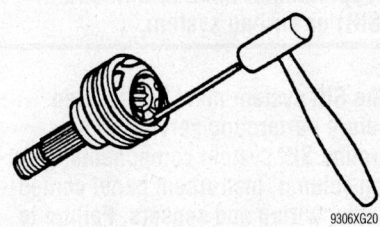

Tilting the cage—Outer CV-joint

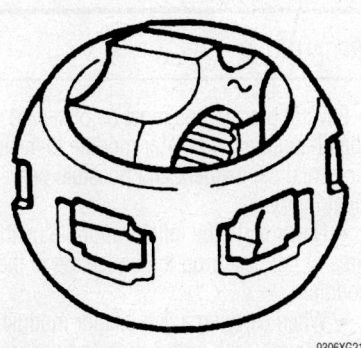

View the cage and inner race—Outer CV-joint

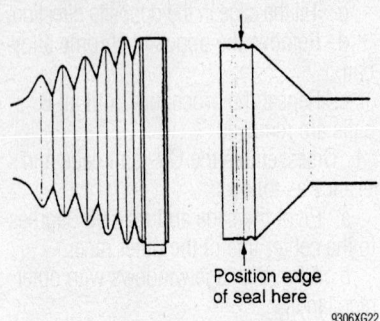

Positioning the boot—Outer CV-joint

a. Rotate the inner race 90 degrees to the cage centerline.

b. Align the cage windows with inner race lands.

c. Insert the inner race into the cage by rotating the inner race downward.

d. Insert the cage/inner race into the outer race.

7. Assemble the chrome alloy balls into the CV-joint cage, as follows:

a. Position a brass drift against the CV-joint cage and tap it with a hammer to tilt the cage.

b. Insert the 1st chrome alloy ball into the cage.

c. Tilt the cage in the opposite direction.

d. Insert the opposite chrome alloy ball.

e. Repeat the procedure until all 6 balls are inserted.

8. Install or connect the following:
- ½ kit grease into the CV-joint boot
- ½ kit grease into the CV-joint
- Swage ring clamp
- CV-joint boot
- CV-joint onto the axle shaft until the retaining ring seats into the groove

9. Position the CV-joint boot seal into the axle shaft's joint seal groove and align the swage ring clamp on the boot.

10. Secure the swage ring clamp as follows:

a. Mount the lower half of tool J-41048 in a vise.

b. Position the outboard of the half-shaft in the tool.

c. Position the upper end of tool J-41048 onto the lower half.

✷✷ WARNING

Make sure that there are no pinch points on the inboard seal.

d. Insert both bolts and tighten by hand until snug.

e. Tighten each bolt 180 degree (½) turn at a time, alternating between the bolts, until both sides are bottomed.

f. Remove the tool.

11. Install or connect the following:
- New large seal clamp onto the CV-joint boot
- CV-joint boot/clamp, slide it into place over the outer race with the seal lip in the groove

➡**Make sure the boot lies flat against the outer race.**

12. Using a crimp tool, a torque wrench and a breaker bar, crimp the large CV-joint boot clamp to 130 ft. lbs. (176 Nm).

13. Install or connect the following:
- Halfshaft
- Front wheel

STEERING AND SUSPENSION

Air Bag

✷✷ CAUTION

The vehicles are equipped with the Supplemental Inflatable Restraint (SIR) or air bag system.

The SIR system must be disabled before performing service on or around SIR system components, steering column, instrument panel components, wiring and sensors. Failure to follow safety and disabling procedures could result in accidental air bag deployment, possible personal injury and unnecessary SIR system repairs.

PRECAUTIONS

Several precautions must be observed when handling the inflator module to avoid accidental deployment and possible personal injury.

- Never carry the inflator module by the wires or connector on the underside of the module.
- When carrying a live inflator module, hold securely with both hands and ensure that the bag and trim cover are pointed away.

- Place the inflator module on a bench or other surface with the bag and trim cover facing up.
- With the inflator module on the bench, never place anything on or close to the module, which may be thrown in the event of an accidental deployment.

DISARMING

1. Turn the steering wheel so that the vehicle's wheels are pointing straight ahead.

2. Turn the ignition switch to the **OFF** position.

3. Remove the key from the ignition switch.

➡**With the SIR fuse removed and the ignition switch in the ON position, The AIR BAG warning lamp illuminates. This is normal operation, and does not indicate an SIR system malfunction.**

4. Remove the rear seat.

5. Remove the SIR fuse from the rear fuse block located under the rear seat.

6. Remove the driver sound insulator.

7. Remove the Connector Position Assurance (CPA) from the driver yellow connector located next to steering column.

8. Disconnect the driver frontal air bag yellow connector from the vehicle harness yellow connector.

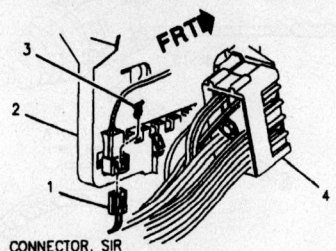

1 CONNECTOR, SIR
2 BRACKET, MULTIUSE MODULE
3 CONNECTOR POSITION ASSURANCE (CPA)
4 CONNECTOR, STEERING COLUMN WIRING HARNESS

SIR 2-way connector location—driver's side

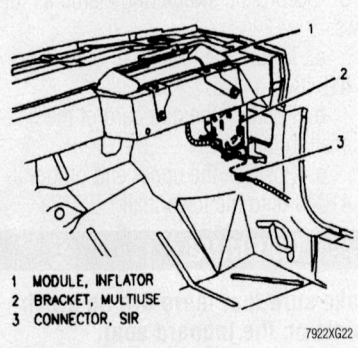

1 MODULE, INFLATOR
2 BRACKET, MULTIUSE
3 CONNECTOR, SIR

SIR 2-way connector location—passenger's side

9. Remove the passenger sound insulator.

10. Remove the CPA from the passenger yellow connector located above the passenger sound insulator.

11. Disconnect the passenger (IP) frontal air bag yellow connector from the vehicle harness yellow connector.

12. Remove both CPA locks from the driver side (seat) air bag and pretensioner yellow connector located under the driver seat.

13. Disconnect the driver side air bag and pretensioner yellow connector from the vehicle harness yellow connector.

14. Remove both CPA locks from the passenger side (seat) air bag and pretensioner yellow connector located under the passenger seat.

15. Disconnect the passenger side air bag and pretensioner yellow connector from the vehicle harness yellow connector.

16. The above procedures will disable the SIR system. If vehicle is equipped with optional Rear Air Bags the following steps must be done, to completely disable the SIR system:

  a. Remove the rear seat back.

  b. Remove the CPA from the passenger rear side air bag yellow connector.

17. Disconnect the passenger rear side air bag yellow connector from the vehicle harness yellow connector.

18. Remove the CPA from the driver rear side air bag yellow connector.

19. Disconnect the driver rear side air bag yellow connector from the vehicle harness yellow connector.

ARMING

1. Remove the key from the ignition switch.

2. Connect the passenger rear side air bag yellow connector to the vehicle harness yellow connector.

3. Install the (CPA) to the passenger rear side air bag yellow connector.

4. Connect the driver rear side air bag yellow connector to the vehicle harness yellow connector.

5. Install the CPA to the driver rear side air bag yellow connector.

6. Install the rear seat back.

7. Connect the driver side (seat) air bag and pretensioner yellow connector to the vehicle harness yellow connector.

8. Install both CPA locks to the driver side (seat) air bag and pretensioner yellow connector located under the driver seat.

9. Connect the passenger side (seat) air

bag and pretensioner yellow connector to the vehicle harness yellow connector.

10. Install both CPA locks to the passenger side (seat) air bag and pretensioner yellow connector located under the passenger seat.

11. Connect the passenger (IP) frontal air bag yellow connector to the vehicle harness yellow connector located above the passenger sound insulator.

12. Install the CPA to the passenger yellow connector.

13. Install the passenger sound insulator.

14. Connect the driver frontal air bag yellow connector to the vehicle harness yellow connector located next to steering column.

15. Install the CPA to the driver yellow connector.

16. Install the driver sound insulator.

17. Install the SIR fuse to the rear fuse block.

18. Install the rear seat.

19. Turn the ignition switch to the **RUN** position and verify that the **AIR BAG** warning lamp flashes 7 times, then turns **OFF**.

Power Rack and Pinion Steering Gear

REMOVAL & INSTALLATION

1. Before servicing the vehicle, refer to the precautions at the beginning of this section.

❋❋ CAUTION

Make sure the wheels of the vehicle are straight ahead and the steering column in the LOCK position before disconnecting the steering column or intermediate shaft from the steering gear. Failure to do so will cause the coil assembly in the steering column to become uncentered which will cause damage to the coil assembly.

2. Lock the steering column by installing the Steering Column Anti Rotation Pin tool J 42640 into the underside of the steering column.

3. Disconnect negative battery cable.

4. Remove or disconnect the following:
  • Intermediate shaft lower coupling
  • Outer tie rod end from the steering knuckle
  • Pressure and return pipes from the rack and pinion gear
  • Electrical connector, if equipped

  • Stabilizer shaft links at lower control arms

5. Rotate the stabilizer shaft to access steering gear bolts.
  • Rack and pinion attaching bolts
  • Rack and pinion assembly

To install:

6. Install or connect the following:
  • Rack and pinion assembly and tighten the bolts to 70 ft. lbs. (95 Nm)
  • Electrical connector, if equipped
  • Pressure and return lines to the steering gear and tighten the fittings to 20 ft. lbs. (27 Nm)
  • Heat shield and bolts. Tighten the bolts to 7 ft. lbs. (10 N)

7. Align stabilizer shaft.
  • Links to lower control arms
  • Tie rod ends to the steering knuckles. Torque the castle nuts to 52 ft. lbs. (70 Nm).
  • Intermediate shaft to the steering gear and tighten the pinch bolt to 35 ft. lbs. (47 Nm)

8. Remove the steering column anti rotation pin from the steering column.

9. Fill and bleed the power steering system.

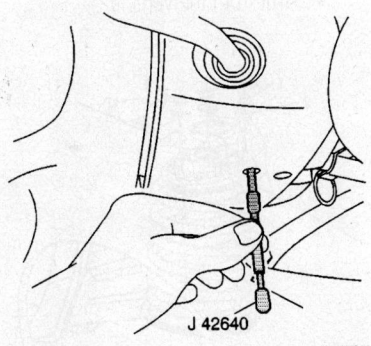

J 42640

9346UG08

Lock the steering column by installing the Steering Column Anti Rotation Pin tool J 42640 into the underside of the steering column

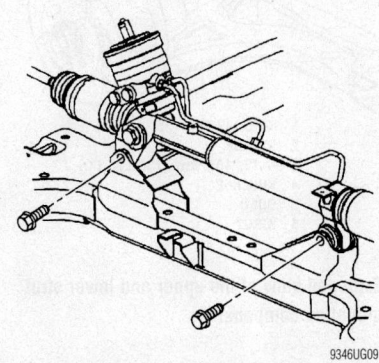

9346UG09

Steering gear mounting bolts

Strut

REMOVAL & INSTALLATION

Front

1. Before servicing the vehicle, refer to the precautions in the beginning of this section.
2. Remove or disconnect the following:

- Negative battery cable
- Front wheel
- Anti-lock Brake System (ABS) wheel speed sensor electrical connector
- ABS speed sensor bracket from the strut
- Brake line bracket, if removing the left strut

3. Matchmark the lower strut bracket to the steering knuckle.
4. Support the steering knuckle.
5. Remove or disconnect the following:

- Lower strut bracket nut and through-bolts
- 3 upper strut plate mounting nuts and washers
- Strut from the vehicle

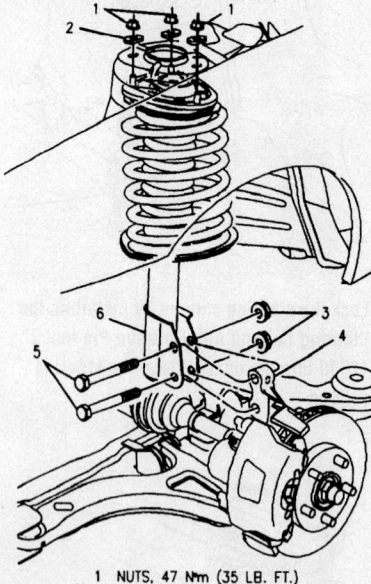

1 NUTS, 47 N·m (35 LB. FT.)
2 WASHER
3 NUTS, 185 N·m (136 LB. FT.)
4 KNUCKLE
5 BOLT
6 STRUT

7922XG23

Exploded view of the upper and lower strut mounting components

To install:

6. Install or connect the following:

- Strut, but do not tighten the upper strut plate-to-body nuts yet
- Lower strut through-bolts by aligning the matchmarks. Torque the nuts to 136 ft. lbs. (185 Nm).

7. Remove the steering knuckle support.
8. Install or connect the following:

- Brake line bracket to the strut, if removed
- ABS wheel speed sensor bracket on the strut
- ABS sensor electrical connector
- Front wheel. Torque the nuts to 100 ft. lbs. (140 Nm).
- Upper strut plate nuts. Torque the nuts to 35 ft. lbs. (47 Nm).
- Negative battery cable

9. Check the front end alignment and adjust as necessary.

Shock Absorber

REMOVAL & INSTALLATION

Rear

1. Before servicing the vehicle, refer to the precautions in the beginning of this section.
2. Support the lower control arm at such a height that the upper shock bolts will still be accessible.
3. Remove or disconnect the following:

- Rear wheel
- Electronic Level Control (ELC) air tube from the shock
- Both lower shock mount bolts
- Trunk trim panel to access the upper shock mount bolts
- Upper shock cap
- Both upper shock mount nuts and reinforcement
- Shock

To install:

4. Install or connect the following:

- Shock
- Reinforcement and upper shock mounting nuts. Torque the nuts to 15 ft. lbs. (20 Nm).
- Upper shock cap and inner trunk trim
- Lower shock bolts. Torque the bolts to 18 ft. lbs. (24 Nm).
- ELC air tube to the shock
- Rear wheel. Torque the nuts to 100 ft. lbs. (140 Nm).

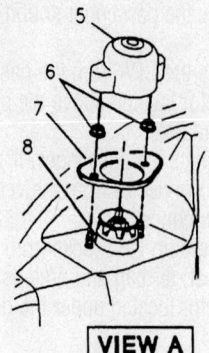

VIEW A

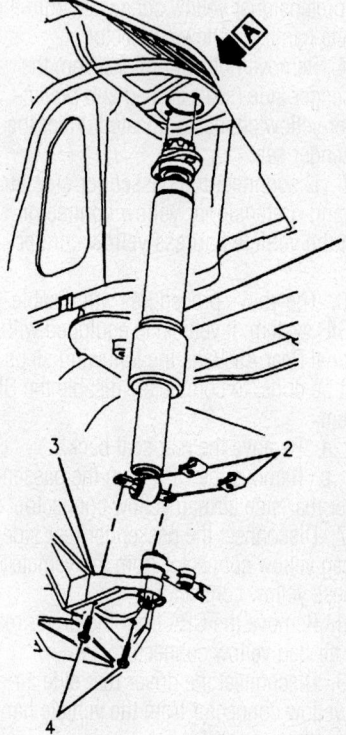

1 SHOCK
2 U-NUTS
3 CONTROL ARM
4 BOLTS 24 N·m (18 LB. FT.)
5 COVER
6 NUTS 20 N·m (15 LB. FT.)
7 REINFORCEMENT
8 MOUNT, UPPER

7922XG34

Exploded view of the rear shock mounting

Coil Spring

REMOVAL & INSTALLATION

Front

1. Remove the strut from the vehicle.
2. Disassemble the strut as follows:
   a. Step 1: Place the strut assembly

into compressor tool, to compress the coil spring.

b. Step 2: Compress the spring slightly.

c. Step 3: Hold the strut shaft from turning using a No. 50 Torx® socket and remove the 24mm nut on the top end of the strut.

d. Step 4: Install Rod tool J 34013-38 to help guide the strut shaft from the upper mount assembly.

e. Step 5: Loosen the spring compressor tool until the coil spring and mount can be removed as an assembly. Remove the lower spring insulator, if equipped.

To install:

3. Assemble the strut as follows:

a. Step 1: Place the strut in compressor tool.

b. Step 2: Install the coil spring over the strut.

c. Step 3: Compress the coil spring while guiding strut shaft through the top of the strut assembly.

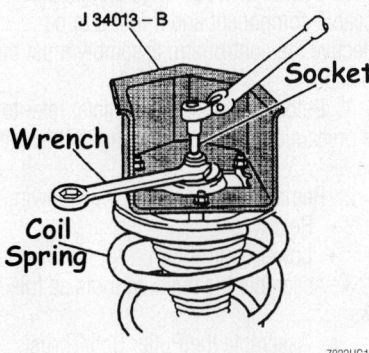

Use a Torx® socket to keep the piston rod from turning while removing the upper nut—front strut shown

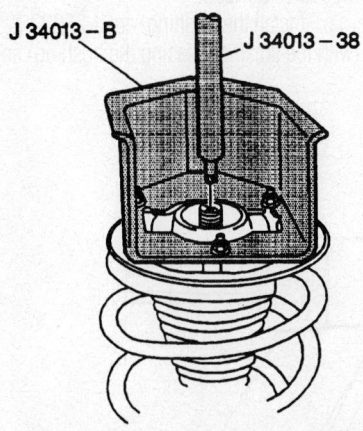

Install Rod J 34013-38 to help guide the strut shaft from the upper mount assembly—front strut shown

d. Step 4: Install the top strut nut. Torque the nut to 55 ft. lbs. (75 Nm).

e. Step 5: Remove the strut from the compressor.

Rear

1. Before servicing the vehicle, refer to the precautions in the beginning of this section.

2. Support the lower control arm.

3. Remove or disconnect the following:
- Rear wheel
- Electronic Level Control (ELC) air tube from the shock absorber
- Both lower shock absorber-to-control arm bolts
- Adjustment link outer ball stud nut
- Ball stud from the knuckle

4. Lower the control arm until the arm bottoms out on the rear suspension support.

5. Using a suitable prying tool, pry under the lower spring insulator to unseat it from the control arm.

6. Remove or disconnect the following:
- Insulator and coil spring
- Upper spring insulator, if necessary

To install:

7. Install or connect the following:
- Upper spring insulator, if removed

➡**Engage the retainer on the back of the insulator in the upper mount hole.**

- Coil spring and lower insulator

➡**Be sure to seat the lower insulator in the control arm hole.**

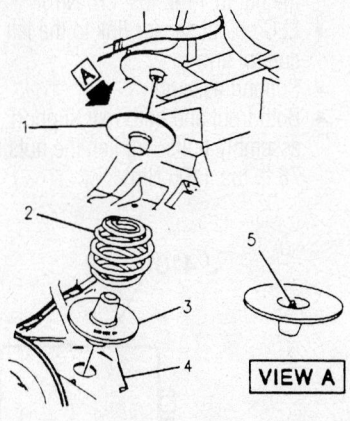

1	JOUNCE BUMPER
2	SPRING
3	LOWER SPRING INSULATOR
4	CONTROL ARM
5	RETAINER

Exploded view of the rear coil spring and related components

8. Raise the control arm into position.

9. Install or connect the following:
- Lower shock absorber-to-control arm bolts. Torque both bolts to 18 ft. lbs. (24 Nm).
- Adjustment link ball stud to the knuckle. Torque the castle nut to 55 ft. lbs. (70 Nm).

➡**If necessary to align the cotter pin holes, tighten the nut up to an additional 60 degree (⅙) turn. NEVER loosen the castle nut to align the cotter pin holes.**

- ELC air tube to the shock absorber
- Rear wheel. Torque the nuts to 100 ft. lbs. (140 Nm).

Lower Ball Joint

REMOVAL & INSTALLATION

The ball joint is an integral part of the control arm. If defective the control arm must be replaced.

Lower Control Arm

REMOVAL & INSTALLATION

Front

1. Before servicing the vehicle, refer to the precautions in the beginning of this section.

2. Remove or disconnect the following:
- Front wheel
- Sway bar link kit

➡**Take note of the positions of the washers and insulators for installation purposes.**

- Lower ball joint from the steering knuckle

✱✱ WARNING

Be careful not to overextend the half-shaft tri-pod joint.

- Lower control arm-to-engine frame nuts and bolts
- Lower control arm

To install:

3. Install or connect the following:
- Lower control arm, do not tighten the nuts and bolts
- Lower ball joint to the steering knuckle. Torque the castle nut to 41 ft. lbs. (55 Nm).

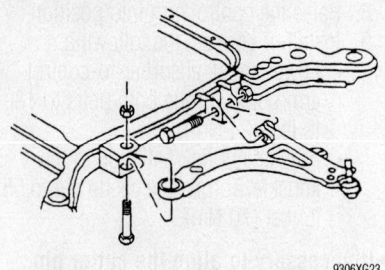

Exploded view of the lower control arm—front suspension

➡ **If necessary, tighten the castle nut up to an additional 60 degree (⅙) turn to align the cotter pin holes. NEVER loosen the nut to make the alignment.**

- Sway bar link kit. Torque the nut to 13 ft. lbs. (17 Nm).
- Front wheel. Torque the nuts to 100 ft. lbs. (140 Nm).

4. Lower the vehicle in order to achieve proper trim height.

5. On all except Aurora, torque the lower control arm-to-frame rear bolt to 117 ft. lbs. (158 Nm) and the lower control arm-to-frame front nut to 93 ft. lbs. (126 Nm).

6. On Aurora, torque the lower control arm-to-frame rear bolt to 117 ft. lbs. (158 Nm) and the lower control arm-to-frame front nut to 117 ft. lbs. (158 Nm).

Rear

1. Before servicing the vehicle, refer to the precautions in the beginning of this section.

2. Remove or disconnect the following:
- Rear wheels
- Exhaust system
- Coil springs
- Parking brake cable from the brake calipers
- Brake calipers from the control arms
- Parking brake cable from the rear suspension support assembly
- Support assembly electrical connectors
- Electronic Level Control (ELC) electrical connector and vent hose
- ELC air tube from the air compressor

3. Support the rear suspension support assembly.

4. Remove or disconnect the following:
- 3 support bracket-to-chassis bolts at each side
- Both front and both rear support assembly bolts
- Support assembly

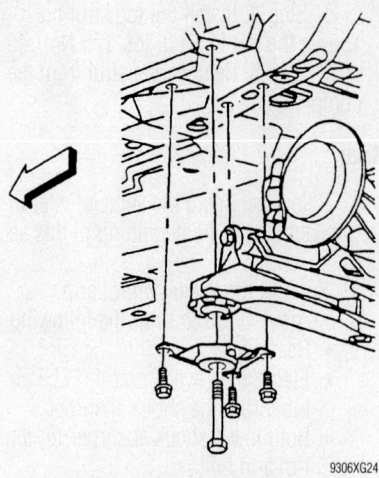

Exploded view of the support assembly—rear suspension

- ELC height sensor link from the left control arm
- Stabilizer link bolt and nut
- Anti-lock Brake System (ABS) electrical connector
- Wheel/hub assembly
- Lower control arm-to-rear suspension support assembly nuts and bolts
- Lower control arm

To install:

5. Install or connect the following:
- Lower control arm
- Lower control arm-to-rear suspension support assembly nuts and bolts, do not tighten
- Wheel/hub assembly
- ABS electrical connector
- Stabilizer link nut and bolt. Torque the nut to 11 ft. lbs. (15 Nm).
- ELC height sensor link to the left control arm
- Support assembly
- Both front and both rear support assembly bolts. Tighten the nuts to 78 ft. lbs. (106 Nm).

- 3 support bracket-to-chassis bolts at each side. Torque the bolts to 63 ft. lbs. (86 Nm).
- ELC air tube to the air compressor
- ELC electrical connector and vent hose
- Support assembly electrical connectors
- Parking brake cable to the rear suspension support assembly
- Brake calipers to the control arms
- Parking brake cable to the brake calipers
- Coil springs
- Exhaust system
- Rear wheels. Torque the nuts to 100 ft. lbs. (140 Nm).

6. Lower the vehicle in order to achieve proper trim height.

7. Torque the lower control arm nuts to 78 ft. lbs. (106 Nm).

CONTROL ARM BUSHING REPLACEMENT

Front

The control arm bushings are not a serviceable component and if found to be defective the control arm assembly must be replaced.

1. Before servicing the vehicle, refer to the precautions in the beginning of this section.

2. Remove or disconnect the following:
- Rear wheel
- Lower control arm

3. Assemble the bushing tools as follows:

a. Assemble the Puller Bolt/Thrust Bearing tool J-21474-19 through the Bushing Receiver tool J-14014-2 over the bushing against the control arm.

b. Lubricate the bolt threads with high pressure lubricant.

c. Install the Bushing tool J-22222-2 (with the small end facing the bushing) and

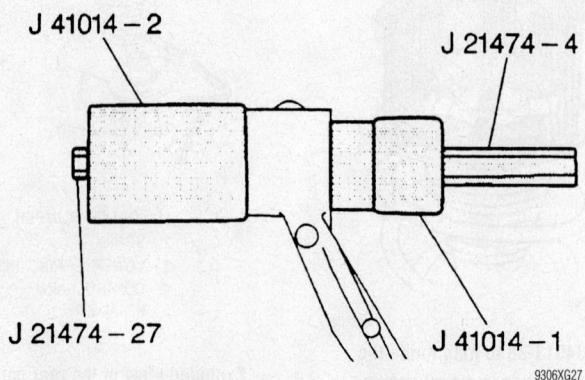

Removing the bushing from the lower control arm—rear suspension

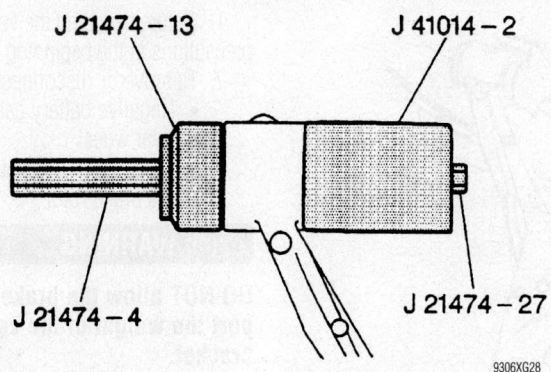

J 21474 – 13   J 41014 – 2

J 21474 – 4   J 21474 – 27

9306XG28

Installing the bushing to the lower control arm—rear suspension

the Long Nut J-21474-18 onto the Puller Bolt/Thrust Bearing tool J-21474-19.

4. Tighten the Long Nut J-21474-18 until the bushing is pressed from the control arm.

5. Remove the bushing tools.

To install:

6. Start the bushing onto the control arm.

➡**Position the bushing flat vertical and rearward.**

7. Assemble the bushing tools as folows:

a. Position the Puller Bolt/Thrust Bearing tool J-21474-19 through the Bushing Receiver tool J-14014-2 over the bushing against the control arm.

b. Lubricate the bolt threads with high pressure lubricant.

c. Install the Bushing Installer tool J-28685 (large end facing the bushing) and Long Nut J-21474-18 onto the Puller Bolt/Thrust Bearing tool J-21474-19.

8. Tighten the Puller Bolt/Thrust Bearing tool J-21474-19 until the bushing is fully seated in the control arm.

9. Remove the tools.

10. Install the lower control arm.

11. Install the front wheel.

Wheel Bearings

ADJUSTMENT

The wheel bearings are not adjustable. If the wheel bearings are defective, the hub and bearing assembly must be replaced.

REMOVAL & INSTALLATION

Front

The front wheel bearings are not serviced separately. If the front wheel bearings are defective, the hub and bearing assembly must be replaced.

1. Before servicing the vehicle, refer to the precautions in the beginning of this section.

2. Remove or disconnect the following:
- Negative battery cable
- Front wheel

3. Lubricate the threads on the halfshaft with clean engine oil.

4. Remove or disconnect the following:
- Halfshaft hub nut
- Caliper from the steering knuckle

✳✳ WARNING

DO NOT allow the brake hose to support the weight of the caliper.

- Brake rotor
- Anti-lock Brake System (ABS) sensor from the backing plate
- Hub and bearing assembly from the backing plate
- Hub and bearing assembly from the halfshaft
- Hub and bearing assembly

To install:

5. Install or connect the following:
- Hub and bearing assembly onto the halfshaft
- Hub and bearing assembly to the backing plate. Torque the 3 bolts alternately and evenly to 70 ft. lbs. (95 Nm).
- ABS sensor
- Brake rotor
- Caliper on the steering knuckle. Torque the bolts to 38 ft. lbs. (51 Nm).
- Halfshaft nut. Torque it to 118 ft. lbs. (160 Nm).
- Front wheel. Torque the nuts to 100 ft. lbs. (140 Nm).
- Negative battery cable

Rear

The rear wheel bearings are not serviced separately. If the rear wheel bearings are defective, the hub and bearing assembly must be replaced.

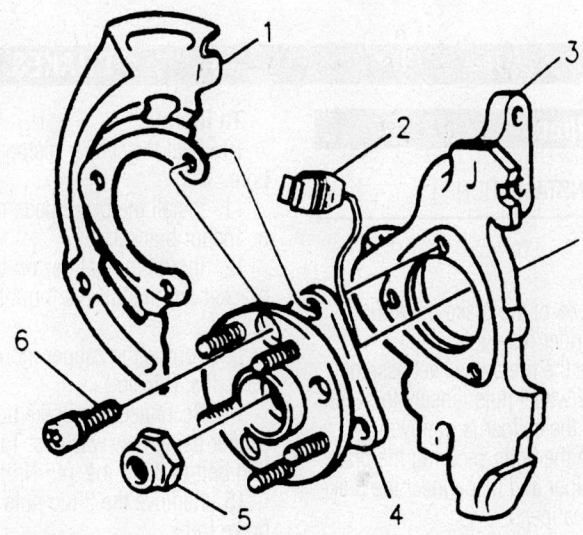

1 DUST SHIELD
2 WHEEL SPEED SENSOR CONNECTOR
3 STEERING KNUCKLE
4 HUB AND BEARING
5 NUT, DRIVE AXLE, 145 N·m (107 LB. FT.)
6 RETAINING BOLT, 95 N·m (75 LB. FT.)

7922XG36

Exploded view of the front hub mounting and related components

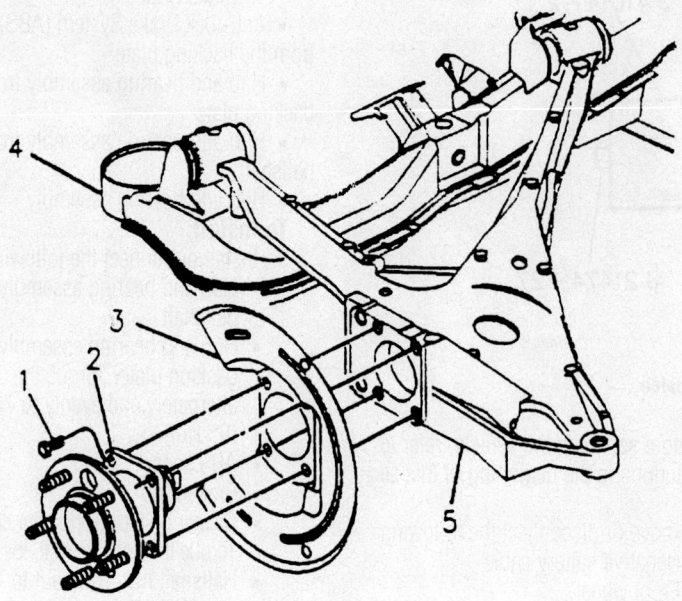

1 BOLT
2 HUB & BEARING
3 BRAKE SHIELD
4 REAR SUSPENSION SUPPORT
   ASSEMBLY
5 CONTROL ARM

7922XG37

Exploded view of the rear hub mounting and related components

1. Before servicing the vehicle, refer to the precautions in the beginning of this section.
2. Remove or disconnect the following:
   - Negative battery cable
   - Rear wheel
   - Caliper and bracket assembly from the brake rotor

✳✳ WARNING

DO NOT allow the brake hose to support the weight of the caliper and bracket.

   - Brake rotor
   - Anti-lock Brake System (ABS) sensor electrical connector
   - Hub and bearing assembly from the rear control arm
   To install:
3. Install or connect the following:
   - Hub and bearing assembly onto the control arm, loosely install the bolts
   - ABS sensor electrical connector
   - Hub and bearing bolts. Torque the bolts alternately and evenly to 52 ft. lbs. (70 Nm).
   - Brake rotor
   - Caliper and bracket assembly. Torque the bolts to 35 ft. lbs. (48 Nm).
   - Rear wheel. Torque the nuts to 100 ft. lbs. (140 Nm).
   - Negative battery cable

BRAKES

Brake Caliper

REMOVAL & INSTALLATION

Front

1. Siphon ⅔ of the brake fluid out of the master cylinder reservoir.
2. Remove the tire and wheel assembly.
3. Install 2 wheel nuts loosely to secure the rotor when the caliper is removed.
4. Remove the bolts securing the brake hose to the caliper and disconnect the brake hose from the caliper.
5. Plug the hose to prevent excessive fluid loss and possible fluid contamination.
6. Compress the piston into the caliper bore to provide clearance for removal.
7. Remove the caliper mounting bolts.
8. Remove the caliper from the anchor bracket.
9. If the caliper is being replaced remove the brake pads from the caliper or anchor bracket.

To install:
10. Seat the caliper piston fully in its bore.
11. Install the brake pads in the caliper or anchor bracket.
12. Install the caliper on the anchor bracket and install the mounting bolts.
13. Torque the caliper mounting bolts to 63 ft. lbs. (85 Nm).
14. Reconnect the brake hose to the caliper using new washers. Torque the fitting bolt to 33 ft. lbs. (45 Nm).
15. Remove the 2 lug nuts securing the brake rotor.
16. Install the wheel and tire assembly and torque the wheel nuts to 100 ft. lbs. (140 Nm).
17. Refill the master cylinder with fluid and bleed the brake system.
18. Pump the brake pedal several times to seat the brake pads against the rotor. Verify no hydraulic leaks and a good firm brake pedal.

Rear

1. Siphon ⅔ of the brake fluid out of the master cylinder reservoir.
2. Remove the tire and wheel assembly.
3. Install 2 wheel nuts loosely to secure the rotor when the caliper is removed.
4. Remove the bolt securing the brake hose to the caliper and disconnect the brake hose from the caliper.
5. Plug the hose to prevent excessive fluid loss and possible fluid contamination.
6. Remove the park brake cable
7. Remove upper and lower caliper bolts and lift caliper from mounting bracket.
To install:
8. Seat the caliper piston fully in its bore.
9. Make sure the notches in the caliper piston are at 6 and 12 o'clock.
10. Mount caliper onto anchor bracket. Install the 2 mounting bolts and torque bolts to 20 ft. lbs.
11. Reconnect the brake hose to the

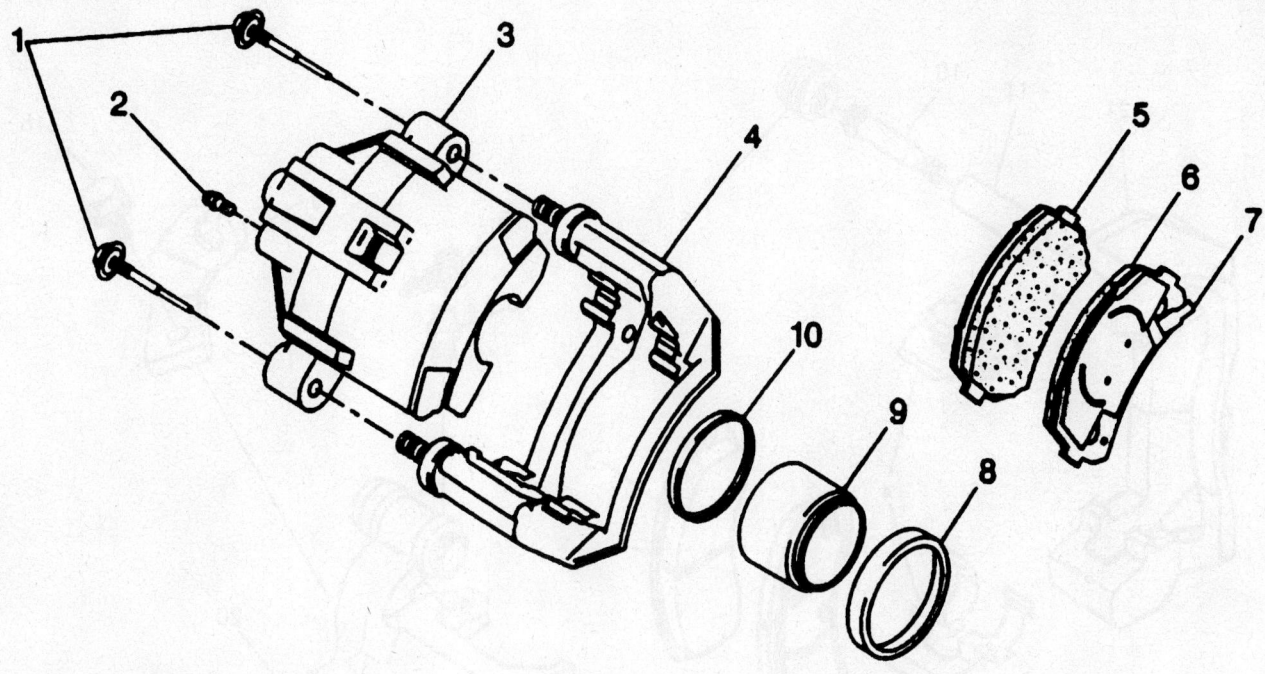

Front caliper and brake pad components

(1) Pin Bolts
(2) Outboard Pad
(3) Caliper Boot
(4) Inboard Pad
(5) Piston Seal

(6) Piston
(7) Anchor bracket
(8) Bleeder Valve
(9) Caliper Housing
(10) Caliper Anchor Bracket

93006G87

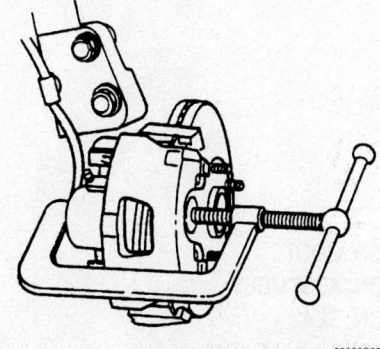

Compressing the front caliper piston

caliper using new copper washers. Torque the mounting bolt to 33 ft. lbs. (45 Nm).

12. Install the park brake cable.

13. Remove the wheel nuts securing the rotor.

14. Reinstall the tire and wheel assembly and torque the wheel nuts to 100 ft. lbs. (140 Nm).

15. Refill the master cylinder with fluid and bleed the brake system.

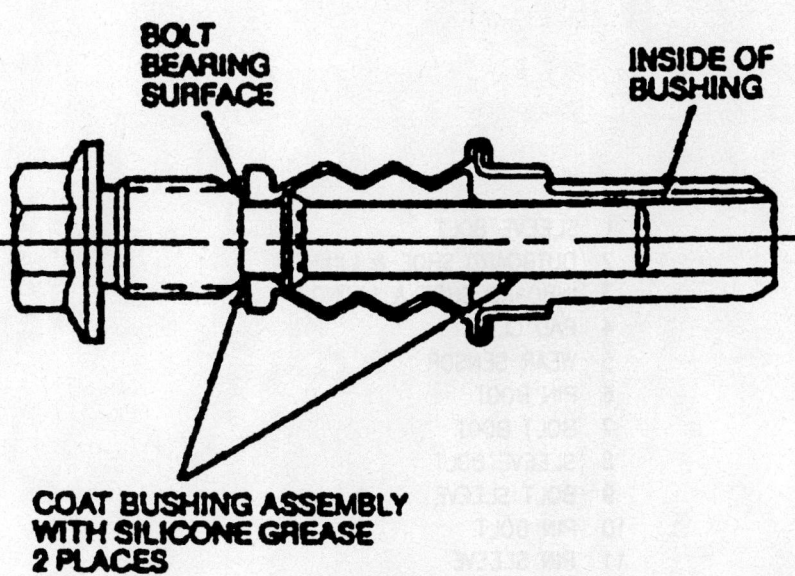

BOLT BEARING SURFACE

INSIDE OF BUSHING

COAT BUSHING ASSEMBLY WITH SILICONE GREASE 2 PLACES

93006G89

Caliper mounting bolt and sleeve lubrication points

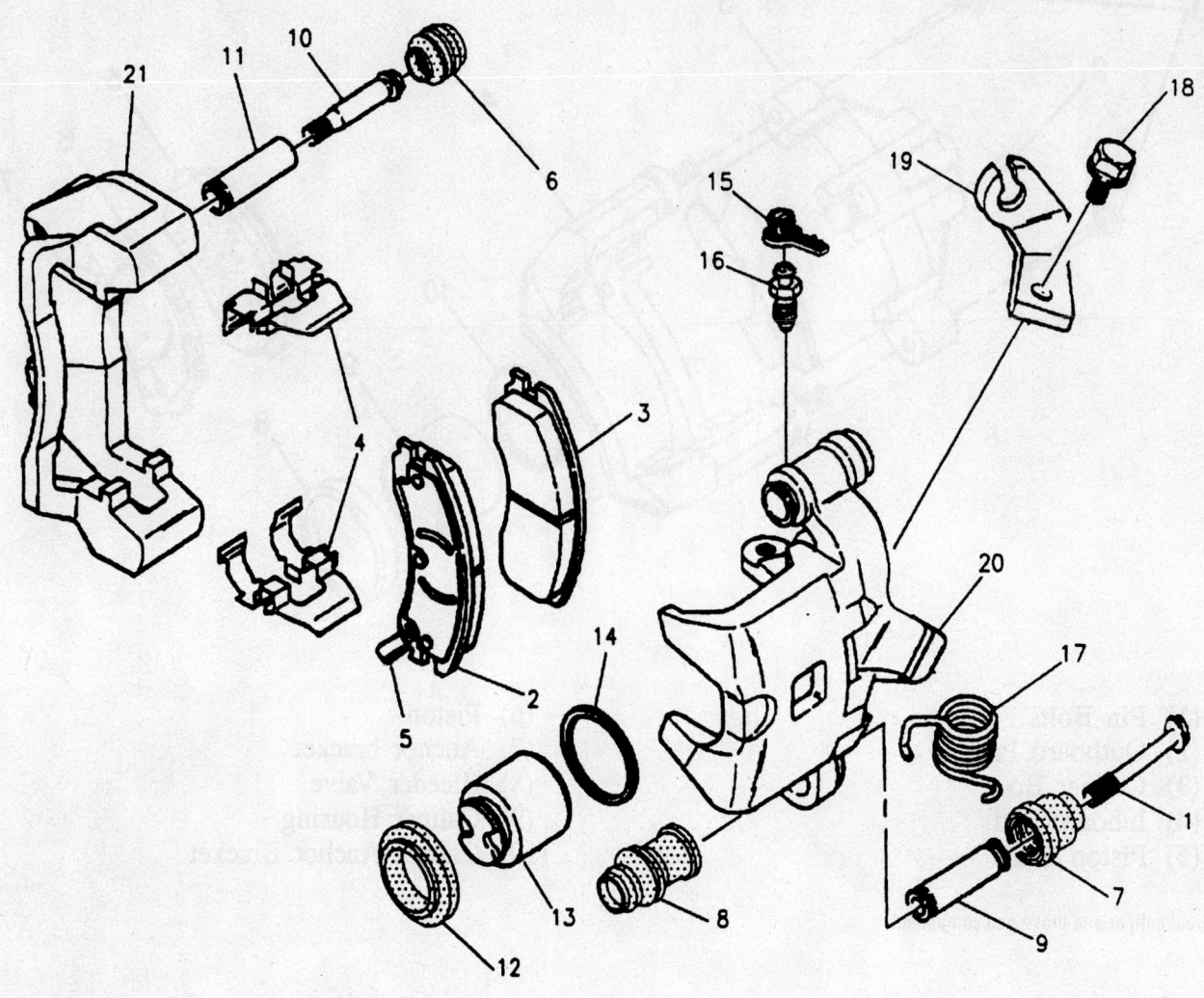

1	SLEEVE BOLT	12	PISTON BOOT
2	OUTBOARD SHOE & LINING	13	PISTON ASSEMBLY
3	INBOARD SHOE & LINING	14	PISTON SEAL
4	PAD CLIP	15	BLEEDER VALVE CAP
5	WEAR SENSOR	16	BLEEDER VALVE
6	PIN BOOT	17	LEVER RETURN SPRING
7	BOLT BOOT	18	BOLT AND WASHER
8	SLEEVE BOLT	19	CABLE SUPPORT BRACKET
9	BOLT SLEEVE	20	CALIPER BODY ASSEMBLY
10	PIN BOLT	21	CALIPER SUPPORT
11	PIN SLEEVE		

93006G90

Rear caliper and brake pad components

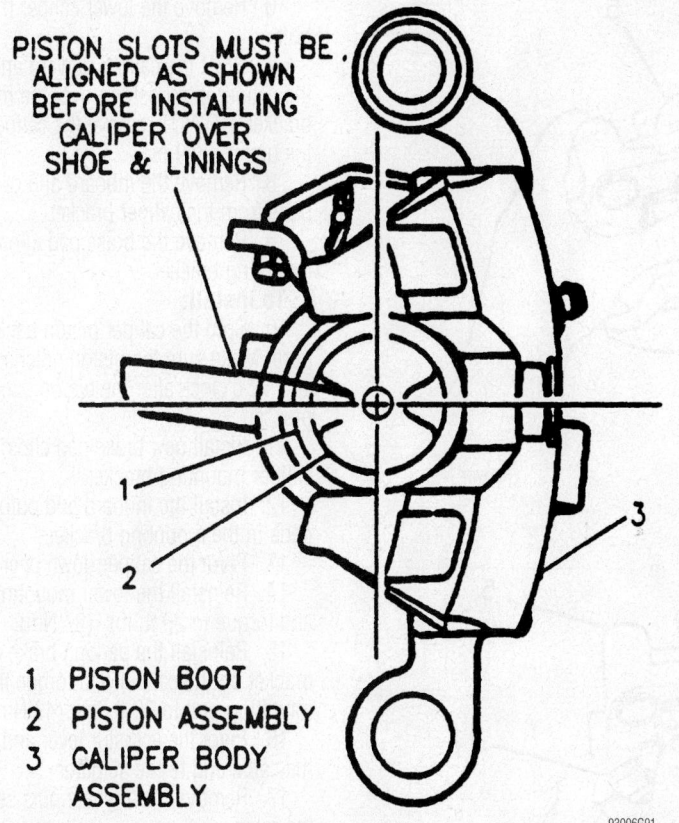

PISTON SLOTS MUST BE
ALIGNED AS SHOWN
BEFORE INSTALLING
CALIPER OVER
SHOE & LININGS

1 PISTON BOOT
2 PISTON ASSEMBLY
3 CALIPER BODY
   ASSEMBLY

93006G91

Aligning the notches in the piston

16. Pump the brake pedal several times to seat the brake pads against the rotor. Verify no hydraulic leaks and a good firm brake pedal.

Disc Brake Pads

REMOVAL & INSTALLATION

Front

1. Siphon ⅔ of the brake fluid from the master cylinder.
2. Remove the tire and wheel assembly.
3. Install 2 wheel nuts to secure the rotor on the hub.
4. Remove the caliper mounting bolts and sleeves.
5. Remove the caliper from the mounting bracket and support the caliper. DO NOT allow the caliper to hang unsupported from the brake hose.
6. Remove the outboard brake pad from the caliper by pushing the pad inward toward the piston to unseat the buttons on the back of pad from the holes in the caliper. Once the buttons are unseated push the pad out of the caliper.
7. Remove the inboard pad from the caliper by pulling the top of the pad away

from the piston and disengaging the spring clip from the caliper.

8. Compress the piston back into the bore using a C-Clamp.

To install:

9. Install the inboard pad into the caliper so the spring clip seats in the pis-

ton. Make sure lower edge of the spring clip is engaged in the piston and the pad is even against the base of the piston. Push the pad flat against caliper piston.

10. Install the outboard pad into the caliper so the wear indicator is at the trailing edge of the pad during forward wheel rotation. Push the pad the straight down into the caliper so the spring clips rode along the outside of the caliper. The buttons on the pad will snap into the mounting holes when the pad is correctly installed.

11. Reinstall the caliper onto the steering knuckle.

12. Remove the 2 wheel nuts securing the rotor.

13. Reinstall the tire and wheel assembly and torque the wheel nuts to 100 ft. lbs. (140 Nm).

14. Refill the master cylinder. Pump the brake pedal several times to seat the pads against the rotor.

15. Check the master cylinder level and add fluid as necessary.

REAR

1. Siphon ⅔ of the brake fluid from the master cylinder.
2. Remove the tire and wheel assembly.
3. Install 2 wheel nuts to secure the rotor on the hub.
4. Pivot the parking brake actuator level and disconnect the parking brake cable from the lever.
5. Remove the mounting bolt securing the parking brake cable bracket to the caliper and position the cable and bracket out of the way.

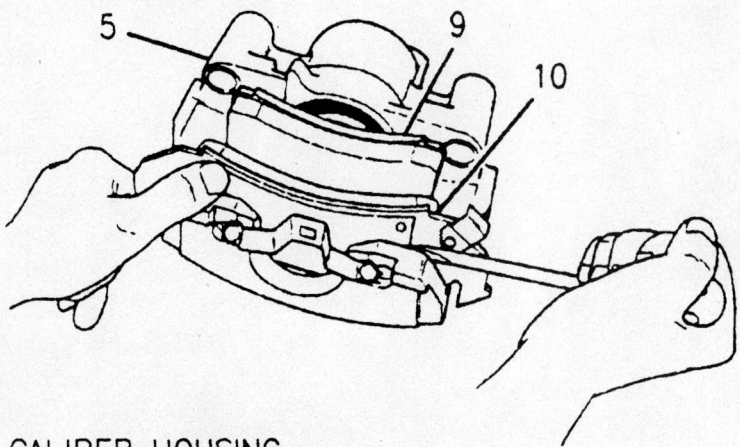

5 CALIPER HOUSING
9 INBOARD SHOE AND LINING
10 OUTBOARD SHOE AND LINING

93006G93

Removing the outboard front pad

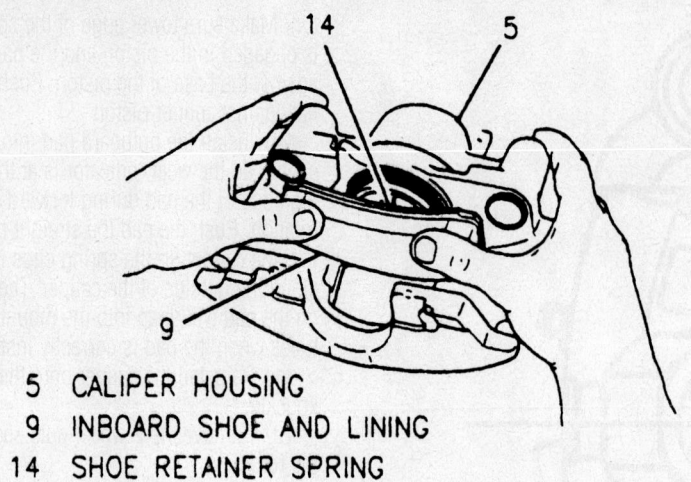

5 CALIPER HOUSING
9 INBOARD SHOE AND LINING
14 SHOE RETAINER SPRING

93006G94

Installing the inboard front pad

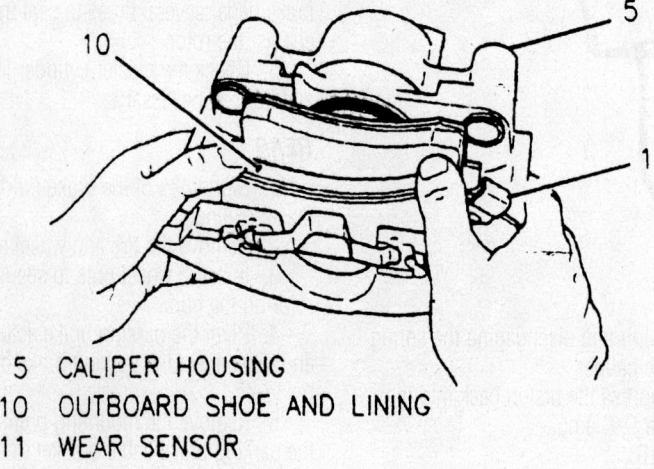

5 CALIPER HOUSING
10 OUTBOARD SHOE AND LINING
11 WEAR SENSOR

93006G95

Installing the outboard front pad

6. Remove the lower caliper mounting bolt.

7. Pivot the caliper upward and secure the caliper in a position over the mounting bracket. DO NOT remove the caliper from the upper pivot pin.

8. Remove the inboard and outboard pads from the caliper bracket.

9. Remove the brake pad clips from the mounting bracket.

To install:

10. Spin the caliper piston back into the bore. Make sure the piston notches are at 6 and 12 o'clock after the piston is compressed.

11. Install new brake pad clips in the caliper mounting bracket.

12. Install the inboard and outboard pads in the mounting bracket.

13. Pivot the caliper down over the pads.

14. Reinstall the lower mounting bolt and torque to 20 ft. lbs. (27 Nm).

15. Reinstall the parking brake cable bracket on the caliper and torque the mounting bolt to 32 ft. lbs. (43 Nm).

16. Pivot the actuator lever and connect the cable end to the actuator.

17. Remove the 2 wheel nuts securing the rotor.

18. Reinstall the tire and wheel assembly and torque the wheel nuts to 100 ft. lbs. (140 Nm).

19. Refill the master cylinder. Pump the brake pedal several times to seat the pads against the rotor.

20. Check the master cylinder level and add fluid as necessary.

CHEVROLET AND GMC

Avalanche • Express • Savana • Sierra • Silverado

BRAKES 4-150
DIESEL ENGINE REPAIR 4-76
DIESEL FUEL SYSTEM 4-105
DRIVE TRAIN 4-109
GASOLINE ENGINE REPAIR...4-33
GASOLINE FUEL SYSTEM 4-98
SPECIFICATIONS AND
  MAINTENANCE CHARTS 4-3
Engine and Vehicle Identification 4-3
General Engine Specifications 4-3
Engine Tune-Up Specifications........4-4
Firing Orders 4-5
Accessory Drive Belt Routing 4-6
Capacities 4-8
Valve Specifications....................4-12
Crankshaft and Connecting Rod
  Specifications 4-13
Piston and Ring Specifications 4-16
Torque Specifications 4-19
Wheel Alignment 4-21
Tire, Wheel and Ball Joint
  Specifications 4-24
Brake Specifications 4-25
Scheduled Maintenance
  Intervals...........................4-27
STEERING AND
  SUSPENSION 4-131

A
Air Bag...............................4-131
  Disarming & Arming..............4-131
  Precautions......................4-131
Alternator (Diesel Engine Repair)...4-76
  Removal & Installation..............4-76
Alternator (Gasoline Engine
  Repair) 4-36
  Removal 4-36
B
Brake Caliper 4-150
  Removal & Installation...........4-150
Brake Drums.........................4-154
  Removal & Installation...........4-154
Brake Shoes.........................4-154
  Removal & Installation...........4-154
C
Camshaft and Valve Lifters
(Diesel Engine Repair)................4-90
  Removal & Installation..............4-90

Camshaft and Valve Lifters
  (Gasoline Engine Repair)..............4-61
  Removal & Installation..............4-61
Clutch 4-116
  Adjustments 4-116
  Removal & Installation...........4-116
Coil Springs 4-136
  Removal & Installation...........4-136
CV-Joints............................4-118
  Overhaul 4-118
Cylinder Head (Diesel
  Engine Repair) 4-81
  Removal & Installation..............4-81
Cylinder Head (Gasoline
  Engine Repair) 4-48
  Removal & Installation..............4-48
D
Diesel Injection Pump.................4-106
  Removal & Installation...........4-106
Disc Brake Pads......................4-152
  Removal & Installation...........4-152
Distributor...........................4-33
  Installation 4-33
  Removal 4-33
E
Engine Assembly (Diesel
  Engine Repair) 4-76
  Removal & Installation..............4-76
Engine Assembly (Gasoline
  Engine Repair) 4-37
  Removal & Installation..............4-37
Exhaust Manifold (Diesel Engine
  Repair) 4-88
  Removal & Installation..............4-88
Exhaust Manifold (Gasoline
  Engine Repair) 4-59
  Removal & Installation..............4-59
F
Front Axle Shaft, Bearing
  and Seal 4-122
  Removal & Installation...........4-122
Front Drive Axle Differential
  Carrier 4-128
  Removal & Installation...........4-128
Front Drive Axle Pinion Seal.........4-126
  Removal & Installation...........4-126

Fuel Filter (Diesel Engine
  Repair).............................4-105
  Removal & Installation............4-105
Fuel Filter (Gasoline Engine
  Repair).............................4-99
  Removal & Installation............4-99
Fuel Injector (Diesel Engine
  Repair).............................4-107
  Removal & Installation............4-107
Fuel Injectors (Gasoline Engine
  Repair).............................4-101
  Removal & Installation............4-101
Fuel Pump 4-100
  Removal & Installation............4-100
Fuel System Air 4-105
  Bleeding............................4-105
Fuel System Pressure (Diesel
  Engine Repair) 4-105
  Relieving...........................4-105
Fuel System Pressure (Gasoline
  Engine Repair) 4-98
  Relieving...........................4-98
Fuel System Service
  Precautions.........................4-98
G
Glow Plugs 4-81
  Removal & Installation..............4-81
Glow Plugs 4-109
  Removal & Installation...........4-109
H
Halfshaft.............................4-117
  Removal & Installation...........4-117
Heater Core (Diesel Engine
  Repair).............................4-81
  Removal & Installation..............4-81
Heater Core (Gasoline Engine
  Repair).............................4-45
  Removal & Installation..............4-45
Hydraulic Clutch System 4-116
  Bleeding............................4-116
I
Idle Speed...........................4-105
  Adjustment........................4-105
Ignition Timing 4-37
  Adjustment........................4-37
Intake Manifold (Diesel
  Engine Repair) 4-87
  Removal & Installation..............4-87

Intake Manifold (Gasoline
  Engine Repair)..........................4-55
  Removal & Installation.............4-55

L

Leaf Springs 4-138
  Removal & Installation...........4-138
Lower Ball Joint...........................4-141
  Removal & Installation...........4-141
Lower Control Arm and
  Bushing...................................4-144
  Control Arm Bushing
    Replacement.......................4-145
  Removal & Installation...........4-144

O

Oil Pan (Diesel Engine Repair) 4-93
  Removal & Installation.............4-93
Oil Pan (Gasoline Engine
  Repair)......................................4-64
  Removal & Installation.............4-64
Oil Pump (Diesel Engine
  Repair)......................................4-95
  Removal & Installation.............4-95
Oil Pump (Gasoline Engine
  Repair)......................................4-68
  Removal & Installation.............4-68

P

Piston and Ring (Diesel Engine
  Repair)......................................4-98
  Positioning 4-98
Piston and Ring (Gasoline Engine
  Repair)......................................4-75
  Positioning 4-75
Power Steering Pump 4-132
  Removal & Installation...........4-132

R

Rack & Pinion Steering Gear 4-133
  Removal & Installation...........4-133
Rear Axle Shaft, Bearing
  and Seal 4-125
  Removal & Installation...........4-125

Rear Drive Axle Housing..............4-129
  Removal & Installation...........4-129
Rear Drive Axle Pinion Seal........4-127
  Removal & Installation...........4-127
Rear Main Seal (Diesel
  Engine Repair)...........................4-96
  Removal & Installation.............4-96
Rear Main Seal (Gasoline
  Engine Repair)...........................4-70
  Removal & Installation.............4-70
Recirculating Ball Power
  Steering Gear...........................4-132
  Removal & Installation...........4-132
Rocker Arms (Gasoline
  Engine Repair)...........................4-53
  Removal & Installation.............4-53
Rocker Arms/Shaft (Diesel
  Engine Repair)...........................4-83
  Removal & Installation.............4-83

S

Shock Absorber 4-134
  Removal & Installation...........4-134
Stabilizer Bar 4-148
  Removal & Installation...........4-148
Starter Motor (Diesel Engine
  Repair)......................................4-83
  Removal & Installation.............4-83
Starter Motor (Gasoline Engine
  Repair)......................................4-63
  Removal & Installation.............4-63
Steering Knuckle.........................4-149
  Removal & Installation...........4-149

T

Timing Chain, Sprockets,
  Front Cover and Seal
  (Diesel Engine Repair)................4-96
  Removal & Installation.............4-96
Timing Chain, Sprockets,
  Front Cover and Seal
  (Gasoline Engine Repair).............4-71
  Removal & Installation.............4-71

Timing Gears, Front Cover
  and Seal 4-97
  Removal & Installation.............4-97
Torsion Bars 4-139
  Removal & Installation...........4-139
Transfer Case Assembly..............4-116
  Removal & Installation...........4-116
Transfer Case Encoder Motor.......4-121
  Removal & Installation...........4-121
Transmission Assembly 4-109
  Removal & Installation...........4-109
Turbocharger...............................4-86
  Removal & Installation.............4-86

U

Upper Ball Joint...........................4-141
  Removal & Installation...........4-141
Upper Control Arm 4-144
  Control Arm Bushing
    Replacement.......................4-144
  Removal & Installation...........4-144

V

Valve Lash (Diesel Engine
  Repair)......................................4-92
  Adjustment...............................4-92
Valve Lash (Gasoline Engine
  Repair)......................................4-63
  Adjustment...............................4-63

W

Water Pump (Diesel Engine
  Repair)......................................4-79
  Removal & Installation.............4-79
Water Pump (Gasoline Engine
  Repair)......................................4-44
  Removal & Installation.............4-44
Wheel Bearings...........................4-146
  Adjustment.............................4-146
Wheel Hub, Bearings and
  Seal 4-146
  Removal & Installation...........4-146

SPECIFICATIONS AND MAINTENANCE CHARTS

ENGINE AND VEHICLE IDENTIFICATION

Code [1]	Liters (cc)	Cu. In.	Cyl.	Fuel Sys.	Engine Type	Eng. Mfg.
					Engine	
W	4.3 (4293)	263	6	MFI	OHV	CPC
X	4.3 (4293)	263	6	MFI	OHV	CPC
V	4.8 (4802)	293	8	MFI	OHV	CPC
M	5.0 (4999)	305	8	MFI	OHV	CPC
T	5.3 (5327)	325	8	MFI	OHV	CPC
T	5.3 (5327)	325	8	MFI	OHV	CPC
Z	5.3 (5327)	325	8	MFI	OHV	CPC
R	5.7 (5735)	350	8	MFI	OHV	CPC
N	6.0 (5966)	364	8	MFI	OHV	CPC
U	6.0 (5966)	364	8	MFI	OHV	CPC
F	6.5 (6473)	395	8	DSL	OHV	CPC
1	6.6 (6599)	364	8	DSL	OHV	CPC
2	6.6 (6599)	364	8	DSL	OHV	CPC
G	8.1 (8128)	496	8	MFI	OHV	CPC

Code [2]	Year
	Model Year
1	2001
2	2002
3	2003
4	2004
5	2005
6	2006

CPC: Chevrolet/Pontiac/Canada

DSL: Diesel

MFI: Multi-port Fuel Injection

[1] 8th position of VIN

[2] 10th position of VIN

06025-AVAL-C01

GENERAL ENGINE SPECIFICATIONS

Engine Displ. Liters	Engine VIN	Models	Years	Net Horsepower @ rpm	Net Torque @ rpm (ft. lbs.)	Bore x Stroke (in.)	Compression Ratio	Oil Pressure @ rpm
4.3	W	G-1500/2500 Sierra/Silverado	2002 2002	200@4600	260@2800	4.00x3.48	9.2:1	18@2000
4.3	X	Express/Savana Sierra/Silverado	2002-06 2002-06	200@4600	260@2800	4.00x3.48	9.2:1	18@2000
4.8	V	Avalanche Express/Savana Sierra/Silverado	2003-06 2003-06 2003-06	210@4400	290@3250	3.78x3.27	9.5:1	18@2000
5.0	M	G1500/2500	2002	220@4600	280@2800	3.74x3.48	9.4:1	18@2000
5.3	B	Sierra/Silverado	2005-06	285@4000	360@4000	3.78x3.62	9.9:1	18@2000
5.3	T	Avalanche Express/Savana Sierra/Silverado	2002-06 2003-06 2002-06	285@4000	360@4000	3.78x3.62	9.5:1	18@2000
5.3	Z	Avalanche Sierra/Silverado	2002-06 2002-06	285@4000	360@4000	3.78x3.62	9.5:1	18@2000
5.7	R	G1500/2500/3500	2002	255@4600	335@2800	4.00x3.48	9.4:1	18@2000
6.0	N	Express/Savana Sierra/Silverado	2003-06 2003-06	300@4400	360@4000	4.00x3.62	9.4:1	18@2000
6.0	U	Avalanche Express/Savana Sierra/Silverado	2003-06 2003-06 2003-06	300@4400	360@4000	4.00x3.62	9.4:1	18@2000
6.5	F	G2500/3500	2002	195@3400	430@1800	4.05x3.80	21.5:1	30-43@2000
6.6	1	Sierra/Silverado	2003-04	300@3000	520@1800	4.00x3.90	17.5:1	57@3250
6.6	2	Sierra/Silverado	2005-06	300@3000	520@1800	4.05x3.90	17.5:1	42@1800
8.1	G	Avalanche Sierra/Silverado	2003-06 2002-06	340@4200	455@3200	4.25x4.37	9.1:1	10@2000

06025-AVAL-C02

GASOLINE ENGINE TUNE-UP SPECIFICATIONS

Year	Engine Displacement Liters	Engine ID/VIN	Spark Plugs Gap (in.)	Ignition Timing (deg.) MT	Ignition Timing (deg.) AT	Fuel Pump (psi)	Idle Speed (rpm) MT	Idle Speed (rpm) AT	Valve Clearance In.	Valve Clearance Ex.
2002	4.3	W	0.060	①	①	58-64 ②	600	625	HYD	HYD
	4.8	V	0.060	①	①	55-62 ②	③	③	HYD	HYD
	5.0	M	0.060	①	①	60-66 ②	650 ④	550	HYD	HYD
	5.3	T	0.060	①	①	55-62 ②	③	③	HYD	HYD
	5.3	Z	0.060	①	①	55-62 ②	③	③	HYD	HYD
	5.7	R	0.060	①	①	60-66 ②	660 ⑤	525	HYD	HYD
	6.0	U	0.060	①	①	55-62 ②	③	③	HYD	HYD
	8.1	G	0.060	①	①	55-62 ②	③	③	HYD	HYD
2003	4.3	X	0.060	①	①	58-64 ②	600	625	HYD	HYD
	4.8	V	0.060	①	①	55-62 ②	③	③	HYD	HYD
	5.3	T	0.060	①	①	55-62 ②	③	③	HYD	HYD
	5.3	Z	0.060	①	①	55-62 ②	③	③	HYD	HYD
	6.0	N	0.060	①	①	55-62 ②	③	③	HYD	HYD
	6.0	U	0.060	①	①	55-62 ②	③	③	HYD	HYD
	8.1	G	0.060	①	①	55-62 ②	③	③	HYD	HYD
2004	4.3	X	0.060	①	①	58-64 ②	600	625	HYD	HYD
	4.8	V	0.060	①	①	55-62 ②	③	③	HYD	HYD
	5.3	T	0.060	①	①	55-62 ②	③	③	HYD	HYD
	5.3	Z	0.060	①	①	55-62 ②	③	③	HYD	HYD
	6.0	N	0.060	①	①	55-62 ②	③	③	HYD	HYD
	6.0	U	0.060	①	①	55-62 ②	③	③	HYD	HYD
	8.1	G	0.060	①	①	55-62 ②	③	③	HYD	HYD
2005	4.3	X	0.060	①	①	55-62 ②	600	625	HYD	HYD
	4.8	V	0.060	①	①	55-62 ②	③	③	HYD	HYD
	5.3	B	0.060	①	①	55-62 ②	③	③	HYD	HYD
	5.3	T	0.060	①	①	55-62 ②	③	③	HYD	HYD
	5.3	Z	0.060	①	①	55-62 ②	③	③	HYD	HYD
	6.0	N	0.060	①	①	55-62 ②	③	③	HYD	HYD
	6.0	U	0.060	①	①	55-62 ②	③	③	HYD	HYD
	8.1	G	0.060	①	①	55-62 ②	③	③	HYD	HYD
2006	4.3	X	0.060	①	①	55-62 ②	600	625	HYD	HYD
	4.8	V	0.060	①	①	55-62 ②	③	③	HYD	HYD
	5.3	B	0.060	①	①	55-62 ②	③	③	HYD	HYD
	5.3	T	0.060	①	①	55-62 ②	③	③	HYD	HYD
	5.3	Z	0.060	①	①	55-62 ②	③	③	HYD	HYD
	6.0	N	0.060	①	①	55-62 ②	③	③	HYD	HYD
	6.0	U	0.060	①	①	55-62 ②	③	③	HYD	HYD
	8.1	G	0.060	①	①	55-62 ②	③	③	HYD	HYD

NOTE: The Vehicle Emission Control Information label often reflects specification changes made during production.

The label figures must be used if they differ from those in this chart.

HYD: Hydraulic

① Ignition timing is preset and cannot be adjusted
② With key ON and engine OFF
③ Idle speed is maintained by the Powertrain Control Module (PCM)
④ Under 8500 GVW
⑤ Over 8500 GVW

06025-AVAL-C03

DIESEL ENGINE TUNE-UP SPECIFICATIONS

Year	Engine Displacement Liters	Engine ID/VIN	Valve Clearance		Intake Valve Opens (deg.)	Injection Pump Setting (deg.)	Injection Nozzle Pressure (psi)		Idle Speed (rpm)	Cranking Compression Pressure (psi)
			Intake (in.)	Exhaust (in.)			New	Used		
2002	6.5	F	HYD	HYD	①	①	1800	1700	①	380-400
	6.6	1	HYD	HYD	①	①	1800	1700	①	380-400
2003	6.6	1	HYD	HYD	①	①	1800	1700	①	380-400
2004	6.6	1	HYD	HYD	①	①	1800	1700	①	380-400
2005	6.6	2	HYD	HYD	①	①	NA	NA	①	380-400
2006	6.6	2	HYD	HYD	①	①	NA	NA	①	380-400

NOTE: The Vehicle Emission Control Information label often reflects specification changes made during production.

The label figures must be used if they differ from those in this chart.

HYD: Hydraulic

NA: Not Available

① Refer to Vehicle Emission Control Information label

06025-AVAL-C04

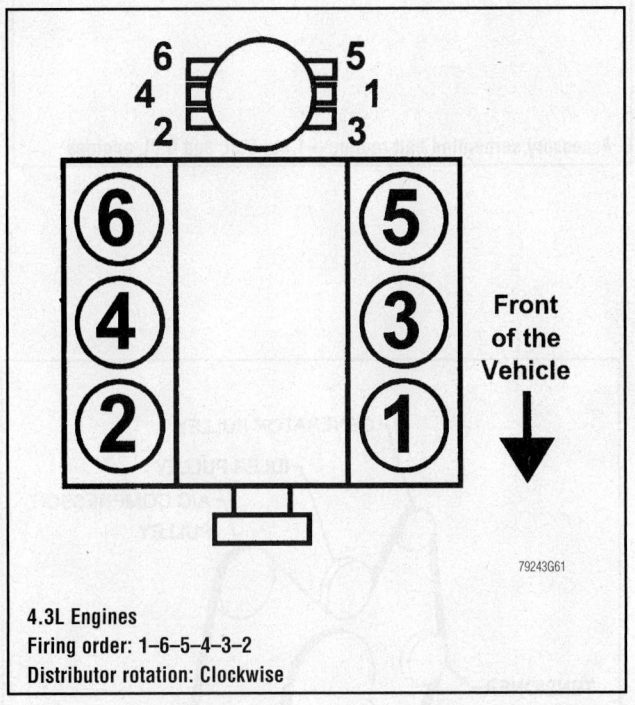

4.3L Engines
Firing order: 1–6–5–4–3–2
Distributor rotation: Clockwise

79243G61

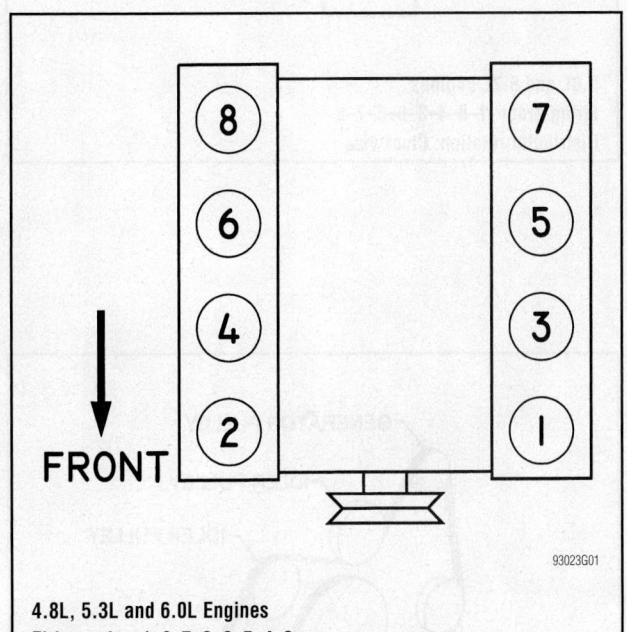

4.8L, 5.3L and 6.0L Engines
Firing order: 1–8–7–2–6–5–4–3
Distributorless ignition system (one coil on each cylinder)

93023G01

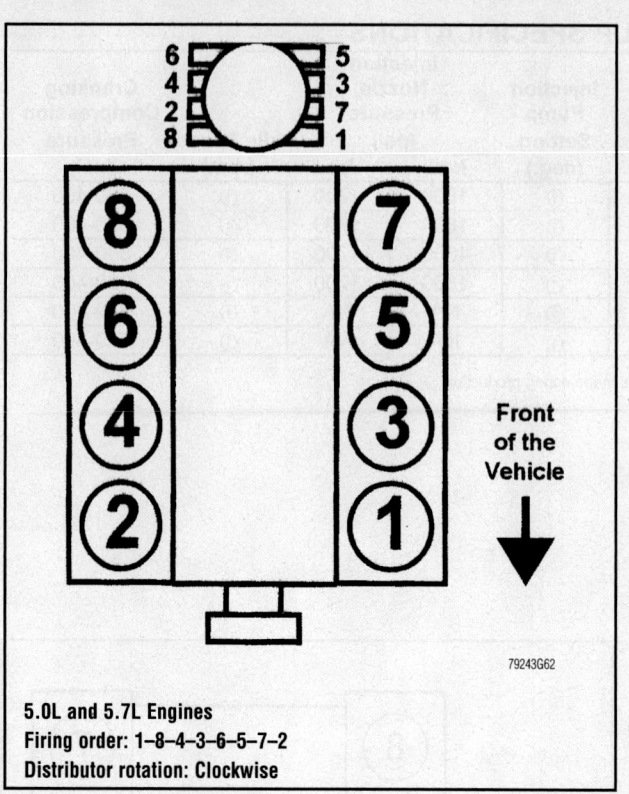

5.0L and 5.7L Engines
Firing order: 1–8–4–3–6–5–7–2
Distributor rotation: Clockwise

79243G62

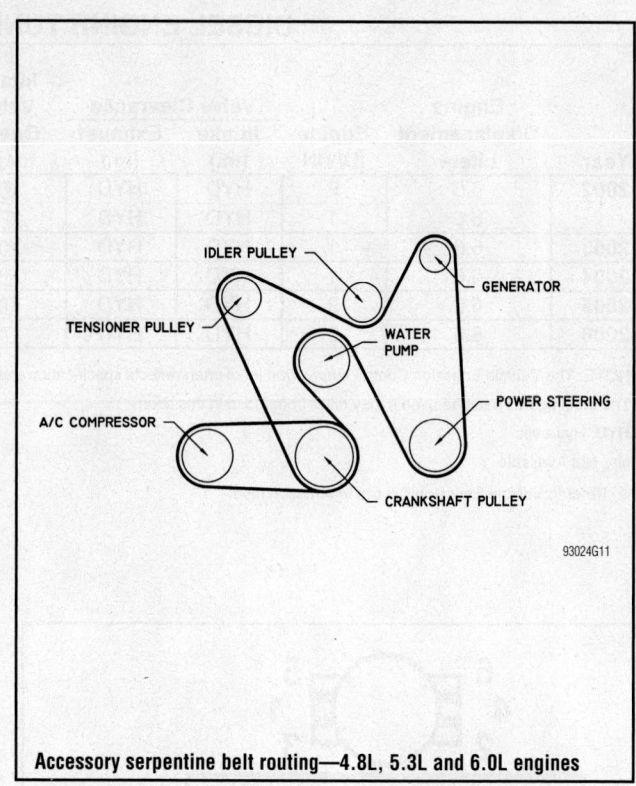

Accessory serpentine belt routing—4.8L, 5.3L and 6.0L engines

93024G11

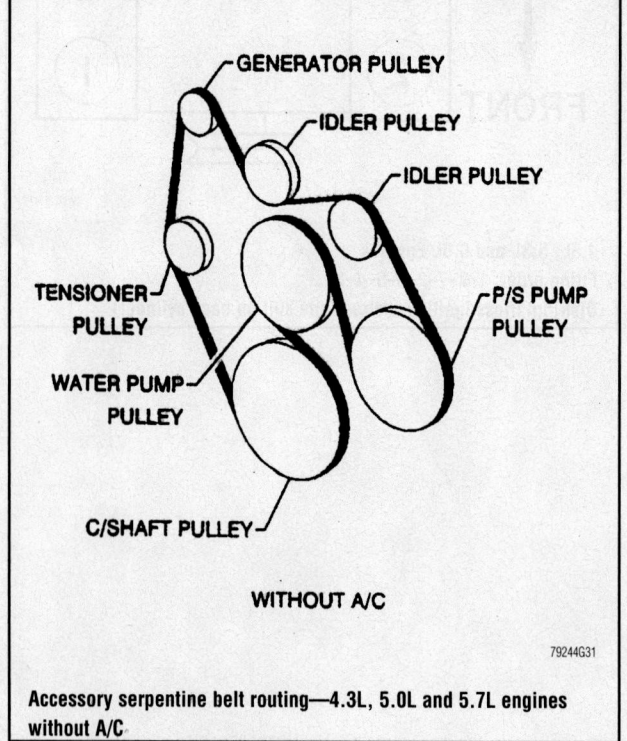

Accessory serpentine belt routing—4.3L, 5.0L and 5.7L engines without A/C

79244G31

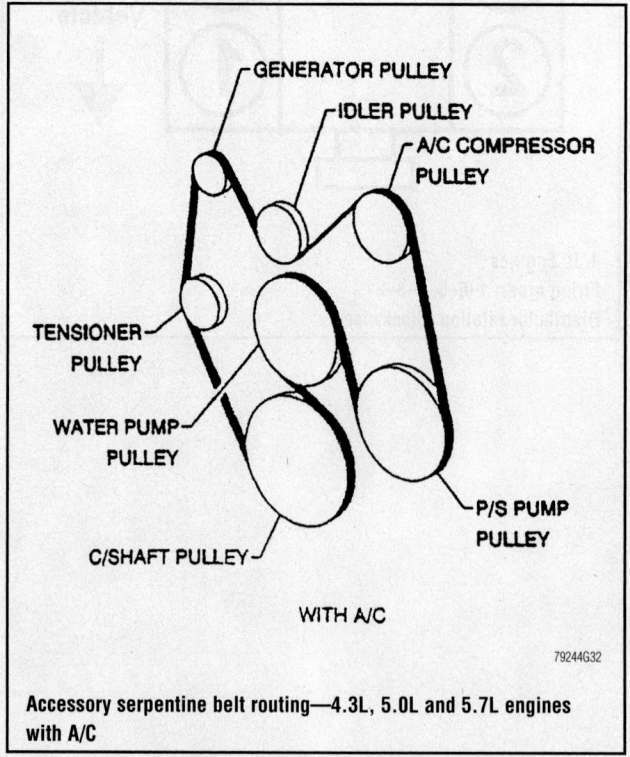

Accessory serpentine belt routing—4.3L, 5.0L and 5.7L engines with A/C

79244G32

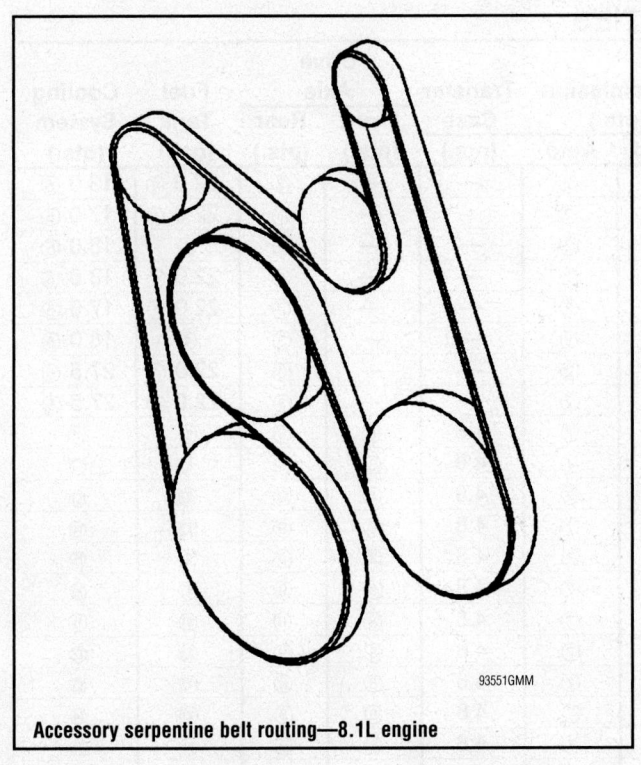

Accessory serpentine belt routing—8.1L engine

93551GMM

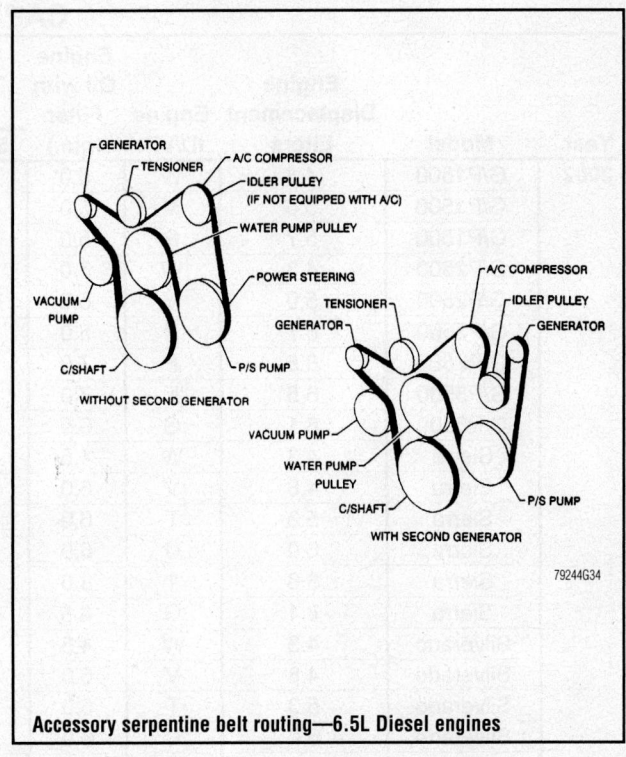

Accessory serpentine belt routing—6.5L Diesel engines

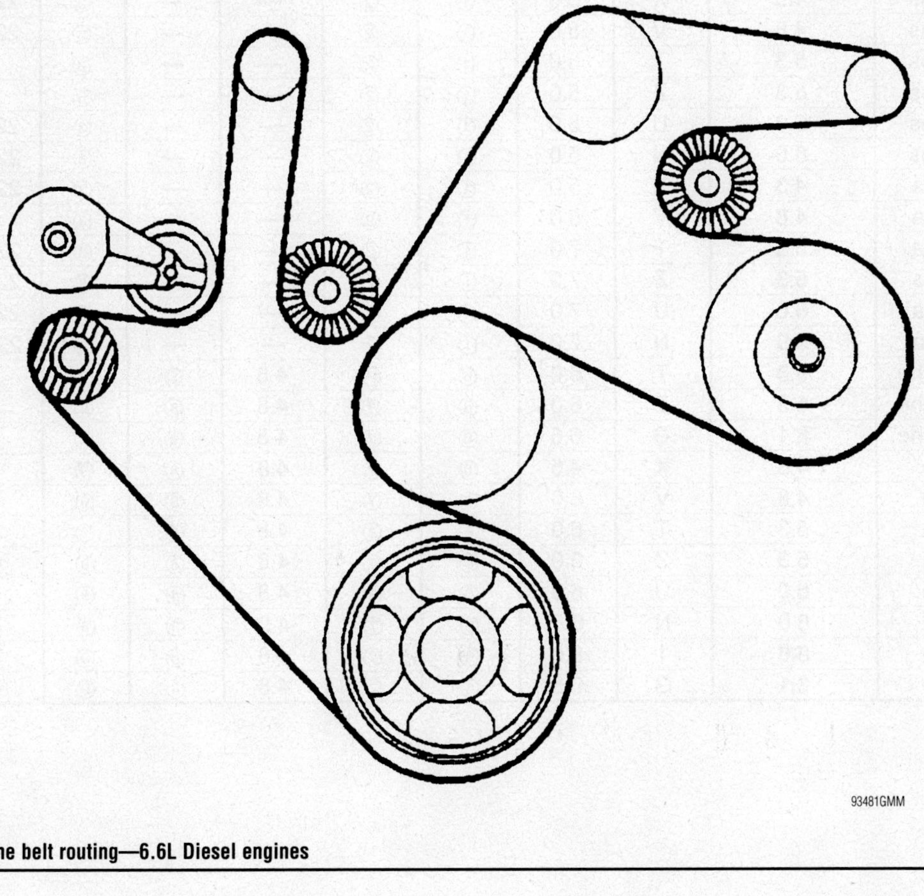

WITHOUT SECOND GENERATOR

- GENERATOR
- TENSIONER
- A/C COMPRESSOR
- IDLER PULLEY (IF NOT EQUIPPED WITH A/C)
- WATER PUMP PULLEY
- POWER STEERING
- VACUUM PUMP
- C/SHAFT
- P/S PUMP

WITH SECOND GENERATOR

- TENSIONER
- A/C COMPRESSOR
- GENERATOR
- IDLER PULLEY
- GENERATOR
- VACUUM PUMP
- WATER PUMP PULLEY
- C/SHAFT
- P/S PUMP

79244G34

Accessory serpentine belt routing—6.6L Diesel engines

93481GMM

CAPACITIES

Year	Model	Engine Displacement Liters	Engine ID/VIN	Engine Oil with Filter (qts.)	Transmission (pts.) 5-Spd	Transmission (pts.) Auto.	Transfer Case (pts.)	Drive Axle Front (pts.)	Drive Axle Rear (pts.)	Fuel Tank (gal.)	Cooling System (qts.)
2002	G/P1500	4.3	W	5.0	①	②	—	—	③	22.0 ④	13.0 ⑤
	G/P1500	5.0	M	5.0	①	②	—	—	③	22.0 ④	17.0 ⑤
	G/P1500	5.7	R	5.0	①	②	—	—	③	④	18.0 ⑤
	G/P2500	4.3	W	5.0	①	②	—	—	③	22.0 ④	13.0 ⑤
	G/P2500	5.0	M	5.0	①	②	—	—	③	22.0 ④	17.0 ⑤
	G/P2500	5.7	R	5.0	①	②	—	—	③	④	18.0 ⑤
	G/P2500	6.5	F	7.0	①	②	—	—	③	22.0 ④	27.5 ⑤
	G/P3500	6.5	F	7.0	①	②	—	—	③	22.0 ④	27.5 ⑤
	G/P3500	8.1	G	6.5	⑥	⑦	4.8	⑧	⑨	⑩	⑪
	Sierra	4.3	W	4.5	⑥	⑦	4.8	⑧	⑨	⑩	⑫
	Sierra	4.8	V	6.0	⑥	⑦	4.8	⑧	⑨	⑩	⑬
	Sierra	5.3	T	6.0	⑥	⑦	4.8	⑧	⑨	⑩	⑭
	Sierra	6.0	U	6.0	⑥	⑦	4.8	⑧	⑨	⑩	⑮
	Sierra	6.6	1	8.0	⑥	⑦	4.8	⑧	⑨	⑩	⑯
	Sierra	8.1	G	6.5	⑥	⑦	4.8	⑧	⑨	⑩	⑪
	Silverado	4.3	W	4.5	⑥	⑦	4.8	⑧	⑨	⑩	⑫
	Silverado	4.8	V	6.0	⑥	⑦	4.8	⑧	⑨	⑩	⑬
	Silverado	5.3	T	6.0	⑥	⑦	4.8	⑧	⑨	⑩	⑭
	Silverado	6.0	U	6.0	⑥	⑦	4.8	⑧	⑨	⑩	⑮
	Silverado	6.6	1	8.0	⑥	⑦	4.8	⑧	⑨	⑩	⑯
	Silverado	8.1	G	6.5	⑥	⑦	4.8	⑧	⑨	⑩	⑪
2003	Express	4.3	X	5.0	①	②	—	—	③	22.0 ④	13.0 ⑤
	Express	4.8	V	5.0	①	②	—	—	③	22.0 ④	17.0 ⑤
	Express	5.3	T	5.0	①	②	—	—	③	④	18.0 ⑤
	Express	5.3	Z	5.0	①	②	—	—	③	④	18.0 ⑤
	Express	6.0	U	5.0	①	②	—	—	③	22.0 ④	13.0 ⑤
	Express	6.0	N	5.0	①	②	—	—	③	22.0 ④	13.0 ⑤
	Savana	4.3	X	5.0	①	②	—	—	③	22.0 ④	17.0 ⑤
	Savana	4.8	V	5.0	①	②	—	—	③	④	18.0 ⑤
	Savana	5.3	T	7.0	①	②	—	—	③	22.0 ④	27.5 ⑤
	Savana	5.3	Z	7.0	①	②	—	—	③	22.0 ④	27.5 ⑤
	Savana	6.0	U	7.0	①	②	—	—	③	22.0 ④	27.5 ⑤
	Savana	6.0	N	7.0	①	②	—	—	③	22.0 ④	27.5 ⑤
	Avalanche	5.3	T	6.0	⑥	⑦	4.8	⑧	⑨	⑩	⑭
	Avalanche	5.3	Z	6.0	⑥	⑦	4.8	⑧	⑨	⑩	⑭
	Avalanche	8.1	G	6.5	⑥	⑦	4.8	⑧	⑨	⑩	⑪
	Sierra	4.3	X	4.5	⑥	⑦	4.8	⑧	⑨	⑩	⑫
	Sierra	4.8	V	6.0	⑥	⑦	4.8	⑧	⑨	⑩	⑬
	Sierra	5.3	T	6.0	⑥	⑦	4.8	⑧	⑨	⑩	⑭
	Sierra	5.3	Z	6.0	⑥	⑦	4.8	⑧	⑨	⑩	⑭
	Sierra	6.0	U	6.0	⑥	⑦	4.8	⑧	⑨	⑩	⑮
	Sierra	6.0	N	6.0	⑥	⑦	4.8	⑧	⑨	⑩	⑮
	Sierra	6.6	1	8.0	⑥	⑦	4.8	⑧	⑨	⑩	⑯
	Sierra	8.1	G	6.5	⑥	⑦	4.8	⑧	⑨	⑩	⑪

CAPACITIES

Year	Model	Engine Displacement Liters	Engine ID/VIN	Engine Oil with Filter (qts.)	Transmission (pts.) 5-Spd	Auto.	Transfer Case (pts.)	Drive Axle Front (pts.)	Rear (pts.)	Fuel Tank (gal.)	Cooling System (qts.)
2003 Cont.	Silverado	4.3	X	4.5	[6]	[7]	4.8	[8]	[9]	[10]	[12]
	Silverado	4.8	V	6.0	[6]	[7]	4.8	[8]	[9]	[10]	[13]
	Silverado	5.3	T	6.0	[6]	[7]	4.8	[8]	[9]	[10]	[14]
	Silverado	5.3	Z	6.0	[6]	[7]	4.8	[8]	[9]	[10]	[14]
	Silverado	6.0	U	6.0	[6]	[7]	4.8	[8]	[9]	[10]	[15]
	Silverado	6.0	N	6.0	[6]	[7]	4.8	[8]	[9]	[10]	[15]
	Silverado	6.6	1	8.0	[6]	[7]	4.8	[8]	[9]	[10]	[16]
	Silverado	8.1	G	6.5	[6]	[7]	4.8	[8]	[9]	[10]	[11]
2004	Express	4.3	X	5.0	[1]	[2]	—	—	[3]	22.0 [4]	13.0 [5]
	Express	4.8	V	5.0	[1]	[2]	—	—	[3]	22.0 [4]	17.0 [5]
	Express	5.3	T	5.0	[1]	[2]	—	—	[3]	[4]	18.0 [5]
	Express	5.3	Z	5.0	[1]	[2]	—	—	[3]	[4]	18.0 [5]
	Express	6.0	U	5.0	[1]	[2]	—	—	[3]	22.0 [4]	13.0 [5]
	Express	6.0	N	5.0	[1]	[2]	—	—	[3]	22.0 [4]	13.0 [5]
	Savana	4.3	X	5.0	[1]	[2]	—	—	[3]	22.0 [4]	17.0 [5]
	Savana	4.8	V	5.0	[1]	[2]	—	—	[3]	[4]	18.0 [5]
	Savana	5.3	T	7.0	[1]	[2]	—	—	[3]	22.0 [4]	27.5 [5]
	Savana	5.3	Z	7.0	[1]	[2]	—	—	[3]	22.0 [4]	27.5 [5]
	Savana	6.0	U	7.0	[1]	[2]	—	—	[3]	22.0 [4]	27.5 [5]
	Savana	6.0	N	7.0	[1]	[2]	—	—	[3]	22.0 [4]	27.5 [5]
	Sierra	4.3	X	4.5	[6]	[7]	4.8	[8]	[9]	[10]	[12]
	Sierra	4.8	V	6.0	[6]	[7]	4.8	[8]	[9]	[10]	[13]
	Sierra	5.3	T	6.0	[6]	[7]	4.8	[8]	[9]	[10]	[14]
	Sierra	5.3	Z	6.0	[6]	[7]	4.8	[8]	[9]	[10]	[14]
	Sierra	6.0	U	6.0	[6]	[7]	4.8	[8]	[9]	[10]	[15]
	Sierra	6.0	N	6.0	[6]	[7]	4.8	[8]	[9]	[10]	[15]
	Sierra	6.6	1	8.0	[6]	[7]	4.8	[8]	[9]	[10]	[16]
	Sierra	8.1	G	6.5	[6]	[7]	4.8	[8]	[9]	[10]	[11]
	Avalanche	5.3	T	6.0	[6]	[7]	4.8	[8]	[9]	[10]	[14]
	Avalanche	5.3	Z	6.0	[6]	[7]	4.8	[8]	[9]	[10]	[14]
	Avalanche	8.1	G	6.5	[6]	[7]	4.8	[8]	[9]	[10]	[11]
	Silverado	4.3	X	4.5	[6]	[7]	4.8	[8]	[9]	[10]	[12]
	Silverado	4.8	V	6.0	[6]	[7]	4.8	[8]	[9]	[10]	[13]
	Silverado	5.3	T	6.0	[6]	[7]	4.8	[8]	[9]	[10]	[14]
	Silverado	5.3	Z	6.0	[6]	[7]	4.8	[8]	[9]	[10]	[14]
	Silverado	6.0	U	6.0	[6]	[7]	4.8	[8]	[9]	[10]	[15]
	Silverado	6.0	N	6.0	[6]	[7]	4.8	[8]	[9]	[10]	[15]
	Silverado	6.6	1	8.0	[6]	[7]	4.8	[8]	[9]	[10]	[16]
	Silverado	8.1	G	6.5	[6]	[7]	4.8	[8]	[9]	[10]	[11]
2005	Express	4.3	X	4.5	—	[17]	—	—	NA	31.0 [4]	11.0 [5]
	Express	4.8	V	6.0	—	[17]	—	—	NA	31.0 [4]	13.4 [5]
	Express	5.3	T	6.0	—	[17]	—	—	[3]	31.0 [4]	13.4 [5]
	Express	6.0	N	6.0	—	[17]	—	—	[3]	31.0 [4]	14.8 [5]
	Savana	4.3	X	4.5	—	[17]	—	—	NA	31.0 [4]	11.0 [5]
	Savana	4.8	V	6.0	—	[17]	—	—	NA	31.0 [4]	13.4 [5]
	Savana	5.3	T	6.0	—	[17]	—	—	[3]	31.0 [4]	13.4 [5]
	Savana	6.0	N	6.0	—	[17]	—	—	[3]	31.0 [4]	14.8 [5]

06025-AVAL-C06

CAPACITIES

Year	Model	Engine Displacement Liters	Engine ID/VIN	Engine Oil with Filter (qts.)	Transmission (pts.) 5-Spd	Auto.	Transfer Case (pts.)	Drive Axle Front (pts.)	Rear (pts.)	Fuel Tank (gal.)	Cooling System (qts.)
2005 Cont.	Sierra	4.3	X	4.5	⑥	⑦	4.0	⑧	⑨	⑩	⑱
	Sierra	4.8	V	6.0	⑥	⑦	4.0	⑧	⑨	⑩	⑲
	Sierra	5.3	B	6.0	⑥	⑦	4.0	⑧	⑨	⑩	⑲
	Sierra	5.3	T	6.0	⑥	⑦	4.0	⑧	⑨	⑩	⑲
	Sierra	5.3	Z	6.0	⑥	⑦	4.0	⑧	⑨	⑩	⑲
	Sierra	6.0	U	6.0	⑥	⑦	4.8	⑧	⑨	⑩	⑳
	Sierra	6.0	N	6.0	⑥	⑦	4.8	⑧	⑨	⑩	⑳
	Sierra	6.6	2	10.0	⑥	⑦	4.8	⑧	⑨	⑩	⑯
	Sierra	8.1	G	6.5	⑥	⑦	4.8	⑧	⑨	⑩	㉑
	Silverado	4.3	X	4.5	⑥	⑦	4.0	⑧	⑨	⑩	⑱
	Silverado	4.8	V	6.0	⑥	⑦	4.8	⑧	⑨	⑩	⑲
	Silverado	5.3	B	6.0	⑥	⑦	4.0	⑧	⑨	⑩	⑲
	Silverado	5.3	T	6.0	⑥	⑦	4.0	⑧	⑨	⑩	⑲
	Silverado	5.3	Z	6.0	⑥	⑦	4.0	⑧	⑨	⑩	⑲
	Silverado	6.0	U	6.0	⑥	⑦	4.8	⑧	⑨	⑩	⑳
	Silverado	6.0	N	6.0	⑥	⑦	4.8	⑧	⑨	⑩	⑳
	Silverado	6.6	2	10.0	⑥	⑦	4.8	⑧	⑨	⑩	⑯
	Silverado	8.1	G	6.5	⑥	⑦	4.8	⑧	⑨	⑩	㉑
	Avalanche	4.8	V	6.0	⑥	⑦	4.8	⑧	⑨	⑩	⑲
	Avalanche	5.3	T	6.0	⑥	⑦	4.8	⑧	⑨	⑩	⑲
	Avalanche	5.3	Z	6.0	⑥	⑦	4.8	⑧	⑨	⑩	⑲
	Avalanche	6.0	U	6.0	⑥	⑦	4.8	⑧	⑨	⑩	⑳
	Avalanche	8.1	G	6.5	⑥	⑦	4.8	⑧	⑨	⑩	㉑
2006	Express	4.3	X	4.5	—	⑰	—	—	NA	31.0 ④	11.0 ⑤
	Express	4.8	V	6.0	—	⑰	—	—	NA	31.0 ④	13.4 ⑤
	Express	5.3	T	6.0	—	⑰	—	—	③	31.0 ④	13.4 ⑤
	Express	6.0	N	6.0	—	⑰	—	—	③	31.0 ④	14.8 ⑤
	Savana	4.3	X	4.5	—	⑰	—	—	NA	31.0 ④	11.0 ⑤
	Savana	4.8	V	6.0	—	⑰	—	—	NA	31.0 ④	13.4 ⑤
	Savana	5.3	T	6.0	—	⑰	—	—	③	31.0 ④	13.4 ⑤
	Savana	6.0	N	6.0	—	⑰	—	—	③	31.0 ④	14.8 ⑤
	Sierra	4.3	X	4.5	⑥	⑦	4.0	⑧	⑨	⑩	⑱
	Sierra	4.8	V	6.0	⑥	⑦	4.0	⑧	⑨	⑩	⑲
	Sierra	5.3	B	6.0	⑥	⑦	4.0	⑧	⑨	⑩	⑲
	Sierra	5.3	T	6.0	⑥	⑦	4.0	⑧	⑨	⑩	⑲
	Sierra	5.3	Z	6.0	⑥	⑦	4.0	⑧	⑨	⑩	⑲
	Sierra	6.0	U	6.0	⑥	⑦	4.8	⑧	⑨	⑩	⑳
	Sierra	6.0	N	6.0	⑥	⑦	4.8	⑧	⑨	⑩	⑳
	Sierra	6.6	2	10.0	⑥	⑦	4.8	⑧	⑨	⑩	⑯
	Sierra	8.1	G	6.5	⑥	⑦	4.8	⑧	⑨	⑩	㉑

06025-AVAL-C07

CAPACITIES

Year	Model	Engine Displacement Liters	Engine ID/VIN	Engine Oil with Filter (qts.)	Transmission (pts.) 5-Spd	Transmission (pts.) Auto.	Transfer Case (pts.)	Drive Axle Front (pts.)	Drive Axle Rear (pts.)	Fuel Tank (gal.)	Cooling System (qts.)
2006 Cont.	Silverado	4.3	X	4.5	⑥	⑦	4.0	⑧	⑨	⑩	⑱
	Silverado	4.8	V	6.0	⑥	⑦	4.8	⑧	⑨	⑩	⑲
	Silverado	5.3	B	6.0	⑥	⑦	4.0	⑧	⑨	⑩	⑲
	Silverado	5.3	T	6.0	⑥	⑦	4.0	⑧	⑨	⑩	⑲
	Silverado	5.3	Z	6.0	⑥	⑦	4.0	⑧	⑨	⑩	⑲
	Silverado	6.0	U	6.0	⑥	⑦	4.8	⑧	⑨	⑩	⑳
	Silverado	6.0	N	6.0	⑥	⑦	4.8	⑧	⑨	⑩	⑳
	Silverado	6.6	2	10.0	⑥	⑦	4.8	⑧	⑨	⑩	⑯
	Silverado	8.1	G	6.5	⑥	⑦	4.8	⑧	⑨	⑩	㉑
	Avalanche	4.8	V	6.0	⑥	⑦	4.8	⑧	⑨	⑩	⑲
	Avalanche	5.3	T	6.0	⑥	⑦	4.8	⑧	⑨	⑩	⑲
	Avalanche	5.3	Z	6.0	⑥	⑦	4.8	⑧	⑨	⑩	⑲
	Avalanche	6.0	U	6.0	⑥	⑦	4.8	⑧	⑨	⑩	⑳
	Avalanche	8.1	G	6.5	⑥	⑦	4.8	⑧	⑨	⑩	㉑

NOTE: All capacities are approximate. Add fluid gradually and check to be sure a proper fluid level is obtained.

① New Venture gear 4500: 8.0 pts.
New Venture gear 5LM60: 4.4 pts.

② 4L60E trans.: 10.0 pts.
4L80E trans.: 14.5 pts.

③ 8.5 in. ring gear: 4.2 pts.
9.5 in. ring gear: 6.5 pts.
9.75 in. ring gear: 6.0 pts.
10.5 in. ring gear: 6.5 pts.

④ Optional 31 and 40 gallon on 2002-04
Optional 33 and 57 gallon on 2005-06

⑤ Add three qts. with rear heater

⑥ New Venture Gear 3500: 4.8 pts.
New Venture Gear 4500: 8.0 pts.
6-speed M/T: 12.6 pts.

⑦ 4L60-E: 10 pts.
4L80-E: 15.4 pts.
Allison: 14.8 pts.

⑧ 8.25 in ring gear: 3.5 pts.
9.25 ring gear: 3.7 pts.

⑨ 8.6 in. ring gear: 4.3 pts.
9.5 & 10.5 in. ring gear: 5.5 pts.
9.75 in. ring gear: 6.0 pts.
11.5 in. ring gear: 6.3 pts.

⑩ Short bed: 26 gals.
Long bed: 34 gals.

⑪ With A/T: 20.7 qts.
With M/T: 21.1 qts.

⑫ With A/T: 12.6 qts.
With M/T: 12.9 qts.

⑬ With A/T: 13.4 qts.
With M/T: 13.7 qts.
With A/T and front A/C: 14.4 qts.
With A/T and front and rear A/C: 15.8 qts.

⑭ With A/T: 13.4 qts.
With A/T and optional A/C: 14.9 qts.
With A/T and front A/C: 14.4 qts.
With A/T and front and rear A/C: 15.8 qts.

⑮ With A/T: 14.8 qts.
With A/T and optional oil cooler: 14.4 qts.
With M/T: 15.2 qts.
With M/T and optional oil cooler: 15.8 qts.

⑯ With A/T: 20.3 qts.
With M/T: 20.7 qts.

⑰ 4L60-E: 17.6 pts. with 245 converter
4L60-E: 19.4 pts. with 258 converter
4L60-E: 22.4 pts. with 298 converter
4L80-E: 15.4 pts.

⑱ With A/T non electric fan: 14.8 qts.
With A/T electric fan: 16.5 qts.
With M/T non-electric fan: 15.1 qts.
With M/T electric fan: 16.6 qts.

⑲ With A/T non electric fan: 15.2 qts.
With A/T electric fan: 16.8 qts.
With M/T non-electric fan: 15.5 qts.
With M/T electric fan: 17.0 qts.

⑳ With A/T: 16.4 qts.
With A/T electric fan: 16.8 qts.

㉑ With A/T Light Duty: 26.9 qts.
With A/T Heavy Duty: 25.0 qts.

06025-AVAL-C08

VALVE SPECIFICATIONS

Engine Displ. Liters	Engine VIN	Years	Seat Angle (deg.)	Face Angle (deg.)	Spring Test Pressure (lbs. @ in.)	Spring Installed Height (in.)	Stem-to-Guide Clearance (in.)		Stem Diameter (in.)	
							Intake	Exhaust	Intake	Exhaust
4.3	W	2002	46	45	187-203@1.27	1.67-1.71	0.0010-0.0027	0.0010-0.0027	NA	NA
4.3	X	2003-06	46	45	187-203@1.27	1.67-1.71	0.0010-0.0027	0.0010-0.0027	NA	NA
4.8	V	2002-06	46	45	220@1.32	1.800	0.0010-0.0026	0.0010-0.0026	0.3132-0.3140	0.3132-0.3140
5.0	M	2002	46	45	187-203@1.27	1.69-1.71	0.0010-0.0027	0.0010-0.0027	NA	NA
5.3	B	2005-06	46	45	220@1.32	1.800	0.0010-0.0026	0.0010-0.0026	0.3130-0.3140	0.3130-0.3140
5.3	T	2002-06	46	45	220@1.32	1.800	0.0010-0.0026	0.0010-0.0026	0.3130-0.3140	0.3130-0.3140
5.3	Z	2003-06	46	45	220@1.32	1.800	0.0010-0.0026	0.0010-0.0026	0.3130-0.3140	0.3130-0.3140
5.7	R	2002	46	45	187-203@1.27	1.69-1.71	0.0010-0.0027	0.0010-0.0027	NA	NA
6.0	N	2002-06	46	45	230@1.40	1.800	0.0010-0.0026	0.0010-0.0026	0.3130-0.3140	0.3130-0.3140
6.0	U	2002-06	46	45	230@1.40	1.800	0.0010-0.0027	0.0010-0.0027	0.3130-0.3140	0.3130-0.3140
6.5	F	2002	46	45	220@1.32	1.800	0.0010-0.0026	0.0010-0.0026	NA	NA
6.6	1	2002-04	45	45	NA	1.610	0.0012-0.0025	0.0015-0.0028	0.2737-0.2744	0.2734-0.2741
6.6	2	2005-06	45	45	NA	1.610	0.0012-0.0025	0.0015-0.0028	0.2737-0.2744	0.2734-0.2741
8.1	G	2002-06	46	45	216-236@1.34	1.81-1.84	0.0010-0.0029	0.0012-0.0031	0.3715-0.3722	0.3713-0.3720

06025-AVAL-C09

CRANKSHAFT AND CONNECTING ROD SPECIFICATIONS

All measurements are given in inches.

Year	Engine Displacement Liters	Engine ID/VIN	Crankshaft				Connecting Rod		
			Main Brg. Journal Dia.	Main Brg. Oil Clearance	Shaft End-play	Thrust on No.	Journal Diameter	Oil Clearance	Side Clearance
2002	4.3	W	①	②	0.0020-0.0070	4	2.2487-2.2497	0.0013-0.0035	0.0060-0.0140
	4.8	V	2.5580-2.5593	0.0007-0.0021	0.0015-0.0078	5	2.0990-2.1000	0.0006-0.0030	0.0043-0.0200
	5.0	M	③	④	0.0020-0.0080	5	2.0978-2.0998	0.0013-0.0035	0.0060-0.0140
	5.3	T	2.5580-2.5593	0.0007-0.0021	0.0015-0.0078	5	2.0990-2.1000	0.0006-0.0030	0.0043-0.0200
	5.7	R	③	④	0.0020-0.0080	5	2.0978-2.0998	0.0013-0.0035	0.0060-0.0140
	6.0	U	2.5580-2.5593	0.0007-0.0021	0.0015-0.0078	5	2.0990-2.1000	0.0006-0.0030	0.0043-0.0200
	6.5	F	⑤	⑥	0.0039-0.0100	3	⑦	0.0018-0.0039	0.0067-0.0248
	6.6	1	3.1459-3.1466	0.0015-0.0028	0.0016-0.0081	NA	2.4789-2.4795	0.0014-0.0030	0.0122-0.0193
	8.1	G	2.7482-2.7489	⑧	0.0050-0.0110	NA	2.1990-2.1996	0.0008-0.0025	0.0151-0.0270
2003	4.3	X	①	②	0.0020-0.0070	4	2.2487-2.2497	0.0013-0.0035	0.0060-0.0140
	4.8	V	2.5580-2.5593	0.0007-0.0021	0.0015-0.0078	5	2.0990-2.1000	0.0006-0.0030	0.0043-0.0200
	5.3	T	2.5580-2.5593	0.0007-0.0021	0.0015-0.0078	5	2.0990-2.1000	0.0006-0.0030	0.0043-0.0200
	5.3	Z	2.5580-2.5590	0.0008-0.0021	0.0015-0.0078	5	2.0991-2.0999	0.0006-0.0030	0.0043-0.2000
	6.0	U	2.5580-2.5593	0.0007-0.0021	0.0015-0.0078	5	2.0990-2.1000	0.0006-0.0030	0.0043-0.0200
	6.0	N	2.5580-2.5590	0.0008-0.0021	0.0015-0.0078	5	2.0991-2.0999	0.0006-0.0030	0.0043-0.2000
	6.6	1	3.1459-3.1466	0.0015-0.0028	0.0016-0.0081	NA	2.4789-2.4795	0.0014-0.0030	0.0122-0.0193
	8.1	G	2.7482-2.7489	⑧	0.0050-0.0110	NA	2.1990-2.1996	0.0008-0.0025	0.0151-0.0270

06025-AVAL-C10

CRANKSHAFT AND CONNECTING ROD SPECIFICATIONS

All measurements are given in inches.

Year	Engine Displacement Liters	Engine ID/VIN	Crankshaft				Connecting Rod		
			Main Brg. Journal Dia.	Main Brg. Oil Clearance	Shaft End-play	Thrust on No.	Journal Diameter	Oil Clearance	Side Clearance
2004	4.3	X	①	②	0.0020-0.0070	4	2.2487-2.2497	0.0013-0.0035	0.0060-0.0140
	4.8	V	2.5580-2.5593	0.0007-0.0021	0.0015-0.0078	5	2.0990-2.1000	0.0006-0.0030	0.0043-0.0200
	5.3	T	2.5580-2.5593	0.0007-0.0021	0.0015-0.0078	5	2.0990-2.1000	0.0006-0.0030	0.0043-0.0200
	5.3	Z	2.5580-2.5590	0.0008-0.0021	0.0015-0.0078	5	2.0991-2.0999	0.0006-0.0030	0.0043-0.2000
	6.0	U	2.5580-2.5593	0.0007-0.0021	0.0015-0.0078	5	2.0990-2.1000	0.0006-0.0030	0.0043-0.0200
	6.0	N	2.5580-2.5590	0.0008-0.0021	0.0015-0.0078	5	2.0991-2.0999	0.0006-0.0030	0.0043-0.2000
	6.6	1	3.1459-3.1466	0.0015-0.0028	0.0016-0.0081	NA	2.4789-2.4795	0.0014-0.0030	0.0122-0.0193
	8.1	G	2.7482-2.7489	⑧	0.0050-0.0110	NA	2.1990-2.1996	0.0008-0.0025	0.0151-0.0270
2005	4.3	X	①	⑨	0.0020-0.0078	4	2.2487-2.2497	0.0015-0.0031	0.0060-0.0173
	4.8	V	2.5580-2.5593	0.0008-0.0021	0.0015-0.0078	5	2.0990-2.1000	0.0009-0.0025	0.0043-0.0200
	5.3	B	2.5580-2.5590	0.0008-0.0021	0.0015-0.0078	5	2.0990-2.1000	0.0006-0.0030	0.0043-0.0200
	5.3	T	2.5580-2.5593	0.0008-0.0021	0.0015-0.0078	5	2.0987-2.0999	0.0009-0.0025	0.0043-0.0200
	5.3	Z	2.5580-2.5590	0.0008-0.0021	0.0015-0.0078	5	2.0991-2.0999	0.0009-0.0030	0.0043-0.2000
	6.0	N	2.5580-2.5590	0.0008-0.0021	0.0015-0.0078	5	2.0991-2.0999	0.0009-0.0025	0.0043-0.2000
	6.0	U	2.5580-2.5593	0.0008-0.0021	0.0015-0.0078	5	2.0990-2.1000	0.0009-0.0025	0.0043-0.0200
	6.6	1	3.1459-3.1466	0.0015-0.0028	0.0016-0.0081	NA	2.4764-2.4772	0.0014-0.0030	0.0122-0.0193
	8.1	G	2.7482-2.7489	⑧	0.0050-0.0138	NA	2.1990-2.1996	0.0013-0.0027	0.0151-0.0270

06025-AVAL-C11

CRANKSHAFT AND CONNECTING ROD SPECIFICATIONS

All measurements are given in inches.

Year	Engine Displacement Liters	Engine ID/VIN	Crankshaft				Connecting Rod		
			Main Brg. Journal Dia.	Main Brg. Oil Clearance	Shaft End-play	Thrust on No.	Journal Diameter	Oil Clearance	Side Clearance
2006	4.3	X	①	⑨	0.0020-0.0078	4	2.2487-2.2497	0.0015-0.0031	0.0060-0.0173
	4.8	V	2.5580-2.5593	0.0008-0.0021	0.0015-0.0078	5	2.0990-2.1000	0.0009-0.0025	0.0043-0.0200
	5.3	B	2.5580-2.5590	0.0008-0.0021	0.0015-0.0078	5	2.0990-2.1000	0.0006-0.0030	0.0043-0.0200
	5.3	T	2.5580-2.5593	0.0008-0.0021	0.0015-0.0078	5	2.0987-2.0999	0.0009-0.0025	0.0043-0.0200
	5.3	Z	2.5580-2.5590	0.0008-0.0021	0.0015-0.0078	5	2.0991-2.0999	0.0009-0.0030	0.0043-0.2000
	6.0	N	2.5580-2.5590	0.0008-0.0021	0.0015-0.0078	5	2.0991-2.0999	0.0009-0.0025	0.0043-0.2000
	6.0	U	2.5580-2.5593	0.0008-0.0021	0.0015-0.0078	5	2.0990-2.1000	0.0009-0.0025	0.0043-0.0200
	6.6	1	3.1459-3.1466	0.0015-0.0028	0.0016-0.0081	NA	2.4764-2.4772	0.0014-0.0030	0.0122-0.0193
	8.1	G	2.7482-2.7489	⑧	0.0050-0.0138	NA	2.1990-2.1996	0.0013-0.0027	0.0151-0.0270

NA - Not Available

① No. 1: 2.4488 in.-2.4495 in.
   Nos. 2, 3: 2.4485 in.-2.4494 in.
   No. 4: 2.4480 in.-2.4489 in.

② No. 1: 0.0008 in.-0.0020 in.
   Nos. 2, 3: 0.0011 in.-0.0023 in.
   No. 4: 0.0017 in.-0.0032 in.

③ No. 1: 2.4484 in.-2.4493 in.
   Nos. 2, 3, 4: 2.4481 in.-2.4490 in.
   No. 5: 2.4479 in.-2.4488 in.

⑥ No. 1, 2, 3, 4: 0.0018 in.-0.0033 in.
   No. 5: 0.0022 in.-0.0037 in.

⑦ 2.399 in.-2.400 in. (Green)
   2.400 in.-2.401 in. (Yellow)

⑧ No. 1, 2, 3, 4: 0.0008-0.0020 in.
   No. 5: 0.0014-0.0026 in.

⑨ No. 1: 0.0008-0.0020 in.
   No. 2, 3, 4: 0.0011-0.00236 in.

06025-AVAL-C12

PISTON AND RING SPECIFICATIONS
All measurements are given in inches.

Year	Engine Displacement Liters	Engine ID/VIN	Piston Clearance	Ring Gap			Ring Side Clearance		
				Top Compression	Bottom Compression	Oil Control	Top Compression	Bottom Compression	Oil Control
2002	4.3	W	0.0007-0.0017	0.010-0.030	0.018-0.026	0.065 Max.	0.0042 Max.	0.0042 Max.	0.0020-0.0070
	4.8	V	0.0010-0.0024	0.009-0.015	0.017-0.025	0.007-0.027	0.0016-0.0033	0.0016-0.0031	0.0004-0.0087
	5.0	M	0.0007-0.0021	0.010-0.020	0.018-0.026	0.010-0.030	0.0012-0.0032	0.0012-0.0032	0.0020-0.0070
	5.3	T	0.0010-0.0024	0.009-0.015	0.017-0.025	0.007-0.027	0.0016-0.0033	0.0016-0.0031	0.0004-0.0087
	5.7	R	0.0007-0.0021	0.010-0.020	0.018-0.026	0.010-0.030	0.0012-0.0032	0.0012-0.0032	0.0020-0.0070
	6.0	U	0.0010-0.0024	0.009-0.015	0.017-0.025	0.007-0.027	0.0016-0.0033	0.0016-0.0031	0.0004-0.0087
	6.5	F	①	0.010-0.020	0.030-0.039	0.010-0.020	0.0015-0.0031	0.0015-0.0031	0.0016-0.0035
	6.6	1	0.0002-0.0007	0.012-0.018	0.020-0.026	0.006-0.014	0.0030-0.0067	0.0004-0.0012	0.0004-0.0012
	8.1	G	②	0.012-0.018	0.017-0.025	0.010-0.030	0.0012-0.0029	0.0012-0.0029	0.002-0.008
2003	4.3	X	0.0007-0.0017	0.010-0.030	0.018-0.026	0.065 Max.	0.0042 Max.	0.0042 Max.	0.0020-0.0070
	4.8	V	0.0010-0.0024	0.009-0.015	0.017-0.025	0.007-0.027	0.0016-0.0033	0.0016-0.0031	0.0004-0.0087
	5.3	T	0.0010-0.0024	0.009-0.015	0.017-0.025	0.007-0.027	0.0016-0.0033	0.0016-0.0031	0.0004-0.0087
	5.3	Z	-0.0014 0.0006	0.009-0.002	0.017-0.027	0.007-0.029	0.0016-0.0033	0.0016-0.0031	0.0005-0.0087
	6.0	U	0.0010-0.0024	0.009-0.015	0.017-0.025	0.007-0.027	0.0016-0.0033	0.0016-0.0031	0.0004-0.0087
	6.0	N	-0.0009 0.0012	0.012-0.020	0.020-0.030	0.012-0.034	0.0014-0.0031	0.0013-0.0030	0.0005-0.0008
	6.6	1	0.0002-0.0007	0.012-0.018	0.020-0.026	0.006-0.014	0.0030-0.0067	0.0004-0.0012	0.0004-0.0012
	8.1	G	②	0.012-0.018	0.017-0.025	0.010-0.030	0.0012-0.0029	0.0012-0.0029	0.002-0.008

06025-AVAL-C13

PISTON AND RING SPECIFICATIONS

All measurements are given in inches.

Year	Engine Displacement Liters	Engine ID/VIN	Piston Clearance	Ring Gap			Ring Side Clearance		
				Top Compression	Bottom Compression	Oil Control	Top Compression	Bottom Compression	Oil Control
2004	4.3	X	0.0007-0.0017	0.010-0.030	0.018-0.026	0.065 Max.	0.0042 Max.	0.0042 Max.	0.0020-0.0070
	4.8	V	0.0010-0.0024	0.009-0.015	0.017-0.025	0.007-0.027	0.0016-0.0033	0.0016-0.0031	0.0004-0.0087
	5.3	T	0.0010-0.0024	0.009-0.015	0.017-0.025	0.007-0.027	0.0016-0.0033	0.0016-0.0031	0.0004-0.0087
	5.3	Z	-0.0014 0.0006	0.009-0.002	0.017-0.027	0.007-0.029	0.0016-0.0033	0.0016-0.0031	0.0005-0.0087
	6.0	U	0.0010-0.0024	0.009-0.015	0.017-0.025	0.007-0.027	0.0016-0.0033	0.0016-0.0031	0.0004-0.0087
	6.0	N	-0.0009 0.0012	0.012-0.020	0.020-0.030	0.012-0.034	0.0014-0.0031	0.0013-0.0030	0.0005-0.0008
	6.6	1	0.0002-0.0007	0.012-0.018	0.020-0.026	0.006-0.014	0.0030-0.0067	0.0004-0.0012	0.0004-0.0012
	8.1	G	②	0.012-0.018	0.017-0.025	0.010-0.030	0.0012-0.0029	0.0012-0.0029	0.002-0.008
2005	4.3	X	0.0007-0.0024	0.010-0.020	0.015-0.031	0.0002 0.0035	0.0012 0.0033	0.0012 0.0033	0.0030-0.0079
	4.8	V	-0.0014 0.0006	0.0015-0.0033	0.015-0.0031	0.0005-0.0078	0.0090-0.0196	0.00173-0.0031	0.0070-0.0320
	5.3	B	-0.0014 0.0006	0.0090-0.0196	0.0173-0.030	0.007-0.032	0.0016-0.0033	0.0016-0.0031	0.0005-0.0078
	5.3	T	-0.0014 0.0006	0.0090-0.0196	0.0173-0.030	0.007-0.032	0.0016-0.0033	0.0016-0.0031	0.0005-0.0078
	5.3	Z	-0.0014 0.0006	0.0090-0.0196	0.0173-0.030	0.007-0.032	0.0016-0.0033	0.0016-0.0031	0.0005-0.0078
	6.0	N	-0.0009 0.0012	0.012-0.020	0.020-0.030	0.012-0.034	0.0014-0.0031	0.0013-0.0030	0.0005-0.0008
	6.0	U	-0.0009 0.0012	0.012-0.023	0.020-0.033	0.012-0.037	0.0015-0.0031	0.0015-0.0031	0.0006-0.0078
	6.6	2	NA	0.0118-0.018	0.0197-0.026	0.0059-0.014	0.0030-0.0067	0.0004-0.0012	0.0004-0.0012
	8.1	G	②	0.012-0.018	0.017-0.025	0.010-0.030	0.0012-0.0029	0.0012-0.0029	0.002-0.008

06025-AVAL-C14

PISTON AND RING SPECIFICATIONS
All measurements are given in inches.

Year	Engine Displacement Liters	Engine ID/VIN	Piston Clearance	Ring Gap			Ring Side Clearance		
				Top Compression	Bottom Compression	Oil Control	Top Compression	Bottom Compression	Oil Control
2006	4.3	X	0.0007-0.0024	0.010-0.020	0.015-0.031	0.0002 0.0035	0.0012 0.0033	0.0012 0.0033	0.0030-0.0079
	4.8	V	-0.0014 0.0006	0.0015-0.0033	0.015-0.0031	0.0005-0.0078	0.0090-0.0196	0.00173-0.0031	0.0070-0.0320
	5.3	B	-0.0014 0.0006	0.0090-0.0196	0.0173-0.030	0.007-0.032	0.0016-0.0033	0.0016-0.0031	0.0005-0.0078
	5.3	T	-0.0014 0.0006	0.0090-0.0196	0.0173-0.030	0.007-0.032	0.0016-0.0033	0.0016-0.0031	0.0005-0.0078
	5.3	Z	-0.0014 0.0006	0.0090-0.0196	0.0173-0.030	0.007-0.032	0.0016-0.0033	0.0016-0.0031	0.0005-0.0078
	6.0	N	-0.0009 0.0012	0.012-0.020	0.020-0.030	0.012-0.034	0.0014-0.0031	0.0013-0.0030	0.0005-0.0008
	6.0	U	-0.0009 0.0012	0.012-0.023	0.020-0.033	0.012-0.037	0.0015-0.0031	0.0015-0.0031	0.0006-0.0078
	6.6	2	NA	0.0118-0.018	0.0197-0.026	0.0059-0.014	0.0030-0.0067	0.0004-0.0012	0.0004-0.0012
	8.1	G	②	0.012-0.018	0.017-0.025	0.010-0.030	0.0012-0.0029	0.0012-0.0029	0.002-0.008

① 1-6: 0.0037-0.0047 in.
  7-8: 0.0042-0.0052 in.

② Interference fit (coated piston)

06025-AVAL-C15

TORQUE SPECIFICATIONS
All readings in ft. lbs.

Year	Engine Displacement Liters	Engine ID/VIN	Cylinder Head Bolts	Main Bearing Bolts	Rod Bearing Bolts	Crankshaft Damper Bolts	Flywheel Bolts	Manifold Intake *	Exhaust	Spark Plugs	Oil Pan Drain Plug
2002	4.3	W	①	77	②	70	74	③	④	11	18
	4.8	V	⑤	⑥	⑦	⑧	⑨	⑩	⑪	11	18
	5.0	M	①	⑫	⑬	70	74	⑭	④	11	18
	5.3	T	⑤	⑥	⑦	⑧	⑨	⑩	⑪	11	18
	5.7	R	①	⑫	⑬	70	74	⑭	④	11	18
	6.0	U	⑤	⑥	⑦	⑧	⑨	⑩	⑪	11	18
	6.5	F	⑮	⑯	48	200	65	31	26	—	18
	6.6	1	⑰	⑱	⑲	278	⑳	15	25	—	62
	8.1	G	㉑	㉒	㉓	189	㉔	㉕	㉖	15	21
2003	4.3	X	①	77	②	70	74	③	④	11	18
	4.8	V	⑤	⑥	⑦	⑧	⑨	⑩	⑪	11	18
	5.3	T	⑤	⑥	⑦	⑧	⑨	⑩	⑪	11	18
	5.3	Z	⑤	⑥	⑦	⑧	⑨	⑩	⑪	11	18
	6.0	U	⑤	⑥	⑦	⑧	⑨	⑩	⑪	11	18
	6.0	N	⑤	⑥	⑦	⑧	⑨	⑩	⑪	11	18
	6.6	1	⑰	⑱	⑲	278	⑳	15	25	—	62
	8.1	G	㉑	㉒	㉓	189	㉔	㉕	㉖	15	21
2004	4.3	X	①	77	②	70	74	③	④	11	18
	4.8	V	⑤	⑥	⑦	⑧	⑨	⑩	⑪	11	18
	5.3	T	⑤	⑥	⑦	⑧	⑨	⑩	⑪	11	18
	5.3	Z	⑤	⑥	⑦	⑧	⑨	⑩	⑪	11	18
	6.0	U	⑤	⑥	⑦	⑧	⑨	⑩	⑪	11	18
	6.0	N	⑤	⑥	⑦	⑧	⑨	⑩	⑪	11	18
	6.6	1	⑰	⑱	⑲	278	⑳	15	25	—	62
	8.1	G	㉑	㉒	㉓	189	㉔	㉕	㉖	15	21
2005	4.3	X	①	77	②	70	74	③	④	11	18
	4.8	V	⑤	⑥	⑦	⑧	⑨	⑩	⑪	11	18
	5.3	B	⑤	⑥	⑦	⑧	⑨	⑩	⑪	11	18
	5.3	T	⑤	⑥	⑦	⑧	⑨	⑩	⑪	11	18
	5.3	Z	⑤	⑥	⑦	⑧	⑨	⑩	⑪	11	18
	6.0	U	⑤	⑥	⑦	⑧	⑨	⑩	⑪	11	18
	6.0	N	⑤	⑥	⑦	⑧	⑨	⑩	⑪	11	18
	6.6	2	⑰	⑱	⑲	㉗⑦	⑳	15	25	—	62
	8.1	G	㉑	㉒	㉓	189	㉔	㉕	㉖	22	21

06025-AVAL-C16

TORQUE SPECIFICATIONS
All readings in ft. lbs.

Year	Engine Displacement Liters	Engine ID/VIN	Cylinder Head Bolts	Main Bearing Bolts	Rod Bearing Bolts	Crankshaft Damper Bolts	Flywheel Bolts	Manifold Intake *	Exhaust	Spark Plugs	Oil Pan Drain Plug
2006	4.3	X	①	77	②	70	74	③	④	11	18
	4.8	V	⑤	⑥	⑦	⑧	⑨	⑩	⑪	11	18
	5.3	B	⑤	⑥	⑦	⑧	⑨	⑩	⑪	11	18
	5.3	T	⑤	⑥	⑦	⑧	⑨	⑩	⑪	11	18
	5.3	Z	⑤	⑥	⑦	⑧	⑨	⑩	⑪	11	18
	6.0	U	⑤	⑥	⑦	⑧	⑨	⑩	⑪	11	18
	6.0	N	⑤	⑥	⑦	⑧	⑨	⑩	⑪	11	18
	6.6	2	⑰	⑱	⑲	㉗	⑳	15	25	—	62
	8.1	G	㉑	㉒	㉓	189	㉔	㉕	㉖	22	21

* NOTE: Applies to Lower Manifold only.

① Step 1: 22 ft. lbs.
　Step 2:
　Short bolt: Plus 55 degrees
　Medium bolt: Plus 65 degrees
　Long bolt: Plus 75 degrees

② 20 ft. lbs. plus 70 degrees

③ Lower intake manifold:
　Step 1: 27 in. lbs.
　Step 2: 106 in. lbs.
　Step 3: 11 ft. lbs.
　Upper manifold bolts:
　Step 1: 44 in. lbs.
　Step 2: 88 in. lbs.

④ Tighten bolts to 12 ft. lbs.
　Retorque to 22 ft. lbs.

⑤ 2002-04 Models: Step 1: 22 ft. lbs.
　Step 2: 90 degrees
　Step 3: 90 degrees,
　(except front and rear medium length bolts)
　Step 4: Tighten front and rear medium length bolts an additional 50 degrees
　2005-06 Models: M11 bolts Step 1: 22 ft. lbs.
　M11 bolts Step 2: 90 degrees
　M11 bolts Step 3: 70 degrees
　M8 bolts: 22 ft. lbs.

⑥ Inner bolts:
　Step 1: 15 ft. lbs.
　Step 2: 80 degrees
　Side Bolts: 15 ft. lbs.
　Outer bolts:
　Step 1: 15 ft. lbs.
　Step 2: 51 ft. lbs.

⑦ Step 1: 15 ft. lbs.
　Step 2 (early design): 60 degrees
　Step 2 (late design): 75 degrees

⑧ Installation pass: 240 ft. lbs.
　Step 1: Replace bolt with new bolt
　Step 2: 37 ft. lbs.
　Step 3: 140 degrees

⑨ Step 1: 15 ft. lbs.
　Step 2: 37 ft. lbs.
　Step 3: 74 ft. lbs.

⑩ Step 1: 44 in. lbs.
　Step 2: 89 in. lbs.

⑪ Step 1: 11 ft. lbs.
　Step 2: 18 ft. lbs.

⑫ Two bolt bearing cap:
　Step 1: 15 ft. lbs.
　Step 2: 73 degrees
　Four bolt bearing cap:
　Step 1: 15 ft. lbs.
　Step 2 (outboard bolt): 43 degrees
　Step 2 (inboard bolt): 73 degrees

⑬ Step 1: 20 ft. lbs.
　Step 2: 55 degrees

⑭ Step 1: 27 inch lbs.
　Step 2: 106 inch lbs.
　Step 3: 11 ft. lbs.

⑮ Step 1: 20 ft. lbs.
　Step 2: 55 ft. lbs
　Step 3: 55 ft. lbs.
　Final step: 90 degrees

⑯ Inner 12mm bolts:
　Step 1: 55 ft. lbs.
　Step 2: 90 degrees.
　Step 3: 90 degrees
　Outer 12mm bolts:
　Step 1: 15 ft. lbs.
　Step 2: 15 ft. lbs.
　Final step: 90 degrees
　Outer 10mm bolts: 30 ft. lbs.

⑰ 2002-04 M8 bolts: 18 ft. lbs.
　M12 bolts
　Step 1: 37 ft. lbs.
　Step 2: 59 ft. lbs.
　Step 3: Plus 90 degrees
　Step 4: Plus 75 degrees
　2005-06 M12 bolts: Step 1: 37 ft.
　Step 2: 59 ft. lbs.
　Step 3: Plus 60 degrees
　Step 4: Plus 90 degrees

⑱ 2002-04 Step 1: 72 ft. lbs.
　Step 2: 97 ft. lbs.
　Step 3: Plus 30 degrees
　2005-06 Step 1: 74 ft. lbs.
　Step 2: Plus 90 degrees

⑲ Step 1: 47 ft. lbs.
　Step 2: Plus 30 degrees
　Step 3: Plus 30 degrees

⑳ Step 1: 58 ft. lbs.
　Step 2: Plus 60 degrees
　Step 3: Plus 60 degrees

㉑ Step 1: 22 ft. lbs.
　Step 2: 22 ft. lbs.,
　Step 3: plus 120 degrees
　Step 4:
　Short bolt: Plus 60 degrees
　Med. bolt: Plus 45 degrees
　Long bolt: Plus 30 degrees

㉒ Inner bolts: 22 ft. lbs.,
　plus 90 degrees
　Outer studs: 22 ft. lbs.,
　plus 80 degrees

㉓ 22 ft. lbs., plus 90 degrees

㉔ Step 1: 59 ft. lbs.
　Step 2: 74 ft. lbs.

㉕ Steps 1 & 2: 44 inch lbs.
　Step 3: 89 inch lbs.
　Step 4: 106 inch lbs.

㉖ Center bolt: 26 ft. lbs.
　Nut: 12 ft. lbs.
　Stud: 15 ft. lbs.

㉗ 1st pass: 74 ft. lbs.
　2nd pass: Plus 90 degrees

06025-AVAL-C17

WHEEL ALIGNMENT

Year	Series	Model	Caster Range (+/-Deg.)	Caster Preferred Setting (Deg.)	Camber Range (+/-Deg.)	Camber Preferred Setting (Deg.)	Toe-in (Deg.)
2002	G-Series Vans	All models	1.00	L +3.75 R +3.75	0.60	+0.50	0.24+/-0.20
	Avalanche C15 w/ 16 inch tires	2wd/4wd	1.00	L +3.90 R +4.70	0.50	+0.25	0.10+/-0.20
	Avalanche C15/K15 w/ 17 inch tires	2wd/4wd	1.00	L +4.10 R +4.70	0.50	+0.25	0.10+/-0.20
	Avalanche K15 w/ 16 inch tires	2wd/4wd	1.00	L +3.80 R +4.50	0.50	+0.25	0.10+/-0.20
	Avalanche C25/K25	2wd/4wd	1.00	L +4.50 R +4.75	0.50	+0.25	0.10+/-0.20
	Sierra/Silverado C15 Series Reg. & Ext. Cab	2wd/4wd	1.00	L +3.75 R +4.00	1.00	+0.25	0.10+/-0.20
	Sierra/Silverado C15 Series Crew Cab	2wd/4wd	1.00	L +4.50 R +4.75	1.00	+0.25	0.10+/-0.20
	Sierra/Silverado K15 Series Reg. & Ext. Cab	2wd/4wd	1.00	L +3.40 R +4.25	1.00	+0.25	0.10+/-0.20
	Sierra/Silverado K15 Series Crew Cab	2wd/4wd	1.00	L +4.20 R +4.75	1.00	+0.25	0.10+/-0.20
	C25 LD	2wd/4wd	1.00	L +4.50 R +4.75	1.00	+0.25	0.10+/-0.20
	C25HD/C35HD/K25 LD	2wd/4wd	1.00	L +4.25 R +4.75	1.00	+0.25	0.10+/-0.20
	K25 HD/K35HD	2wd/4wd	1.00	L +4.00 R +4.75	1.00	+0.25	0.10+/-0.20
	K35 HD	2wd/4wd	1.00	L +4.00 R +4.75	1.00	+0.25	0.10+/-0.20
2003	Express/Savana w/ 6200, 7200 & 7300 GVW	2wd/4wd	1.00	L +4.20 R +4.50	0.50	+0.15	0.10+/-0.20
	Express/Savana 8500, 8600 & 9600 GVW	2wd/4wd	1.00	L +4.60 R +5.00	0.50	+0.25	0.10+/-0.20
	Express/Savana w/10000, 11,000, 11,500,12,000 & 12,300 GVW	2wd/4wd	1.00	L +4.60 R +4.90	0.50	+0.25	0.10+/-0.20
	H 1500/2500 (RV)	2wd/4wd	1.50	L +3.60 R +4.00	0.50	+0.10	0.30+/-0.20
	H 1500/2500 (except RV)	2wd/4wd	1.00	L +3.60 R +4.00	0.75	+0.10	0.20+/-0.20
	Avalanche C15 w/ 16 inch tires	2wd/4wd	1.00	L +3.90 R +4.70	0.50	+0.25	0.10+/-0.20
	Avalanche C15 w/ 17 inch tires	2wd/4wd	1.00	L +4.10 R +4.70	0.50	+0.25	0.10+/-0.20
	Avalanche K15 w/ 16 inch tires	2wd/4wd	1.00	L +3.50 R +4.50	0.50	+0.25	0.10+/-0.20
	Avalanche K15 w/ 17 inch tires	2wd/4wd	1.00	L +3.80 R +4.50	0.50	+0.25	0.10+/-0.20
	Avalanche C25/K25	2wd/4wd	1.00	L +4.50 R +4.75	0.50	+0.25	0.10+/-0.20

WHEEL ALIGNMENT

Year	Series	Model	Caster Range (+/-Deg.)	Caster Preferred Setting (Deg.)	Camber Range (+/-Deg.)	Camber Preferred Setting (Deg.)	Toe-in (Deg.)
2003 cont.	Silverado/Sierra C15 Series Reg. & Ext. Cab	2wd/4wd	1.00	L +3.50 R +4.00	0.50	+0.25	0.10+/-0.20
	Silverado/Sierra C15 Series Crew Cab	2wd/4wd	1.00	L +4.50 R +4.75	0.50	+0.25	0.10+/-0.20
	Silverado/Sierra K15 Series Reg. & Ext. Cab ①	2wd/4wd	1.00	②	0.50	+0.25	0.10+/-0.20
	Silverado/Sierra K15 Series Crew Cab	2wd/4wd	1.00	L +4.25 R +4.75	0.50	+0.25	0.10+/-0.20
	Silverado SS	2wd/4wd	1.00	L +4.10 R +5.10	0.50	+0.10	0.10+/-0.20
	C25 LD	2wd/4wd	1.00	L +4.50 R +4.75	1.00	+0.25	0.10+/-0.20
	C25HD/C35HD/K25 LD	2wd/4wd	1.00	L +4.25 R +4.75	1.00	+0.25	0.10+/-0.20
	K25 HD	2wd/4wd	1.00	L +4.00 R +4.75	1.00	+0.25	0.10+/-0.20
	K35 HD	2wd/4wd	1.00	L +4.00 R +4.75	1.00	+0.25	0.10+/-0.20
2004	Express/Savana1500 w/ 6200, 7200 & 2500 w/ 7300 GVW	2wd/4wd	1.00	L +4.20 R +4.50	0.50	+0.15	0.10+/-0.20
	Express/Savana2500 w/ 8500, 8600 &3500 w/ 8600, 9600 GVW	2wd/4wd	1.00	L +4.60 R +5.00	0.50	+0.25	0.10+/-0.20
	Express/Savana3500 w/10000,11,000 11, 500, 12,000 & 12,300 GVW	2wd/4wd	1.00	L +4.60 R +4.90	0.50	+0.25	0.10+/-0.20
	H 1500/2500	2wd/4wd	1.00	L +4.10 R +4.50	0.50	+0.15	0.10+/-0.20
	Avalanche C15 w/ 16 inch tires	2wd/4wd	1.00	L +3.90 R +4.70	0.50	+0.25	0.10+/-0.20
	Avalanche C15 w/ 17 inch tires	2wd/4wd	1.00	L +4.10 R +4.70	0.50	+0.25	0.10+/-0.20
	Avalanche K15 w/ 16 inch tires	2wd/4wd	1.00	L +3.60 R +4.40	0.50	+0.25	0.10+/-0.20
	Avalanche K15 w/ 17 inch tires	2wd/4wd	1.00	L +3.80 R +4.50	0.50	+0.25	0.10+/-0.20
	Avalanche C25/K25	2wd/4wd	1.00	L +4.50 R +4.75	0.50	+0.25	0.10+/-0.20
	Silverado/Sierra C15 Series w/ Reg.Ext./Crew Cab	2wd/4wd	1.00	L +3.75 R +4.00	0.50	+0.25	0.10+/-0.20
	Silverado/Sierra K15 Series Reg./Ext./Crew Cab	2wd/4wd	1.00	L +3.60 R +4.10	0.50	+0.25	0.10+/-0.20
	Silverado SS	2wd/4wd	1.00	L +4.10 R +5.10	0.50	+0.10	0.10+/-0.20
	C25 LD	2wd/4wd	1.00	L +4.50 R +4.75	1.00	+0.25	0.10+/-0.20
	C25HD/C35/K25 LD	2wd/4wd	1.00	L +4.25 R +4.75	1.00	+0.25	0.10+/-0.20
	K25HD/K35HD	2wd/4wd	1.00	L +4.00 R +4.75	1.00	+0.25	0.10+/-0.20

WHEEL ALIGNMENT

Year	Series	Model	Caster Range (+/-Deg.)	Caster Preferred Setting (Deg.)	Camber Range (+/-Deg.)	Camber Preferred Setting (Deg.)	Toe-in (Deg.)
2005-06	Express/Savana1500 w/ 6200, 7200 & 2500 w/ 7300 GVW	2wd/4wd	1.00	L +4.20 R +4.50	0.50	+0.15	0.10+/-0.20
	Express/Savana2500 w/ 8500, 8600 &3500 w/ 8600, 9600 GVW	2wd/4wd	1.00	L +4.60 R +5.00	0.50	+0.25	0.10+/-0.20
	Express/Savana3500 w/10000,11,000 11, 500, 12,000 & 12,300 GVW	2wd/4wd	1.00	L +4.60 R +4.90	0.50	+0.25	0.10+/-0.20
	H 1500/2500	2wd/4wd	1.00	L +4.10 R +4.50	0.50	+0.15	0.10+/-0.20
	Avalanche C15 w/ 16 inch tires	2wd/4wd	1.00	L +3.90 R +4.70	0.50	+0.25	0.10+/-0.20
	Avalanche C15 w/ 17 inch tires	2wd/4wd	1.00	L +4.10 R +4.70	0.50	+0.25	0.10+/-0.20
	Avalanche K15 w/ 16 inch tires	2wd/4wd	1.00	L +3.60 R +4.40	0.50	+0.25	0.10+/-0.20
	Avalanche K15 w/ 17 inch tires	2wd/4wd	1.00	L +3.80 R +4.50	0.50	+0.25	0.10+/-0.20
	Avalanche C25/K25	2wd/4wd	1.00	L +4.50 R +4.75	0.50	+0.25	0.10+/-0.20
	Silverado/Sierra C15 Series w/ Reg.Ext./Crew Cab	2wd/4wd	1.00	L +3.75 R +4.00	0.50	+0.25	0.10+/-0.20
	Silverado/Sierra K15 Series Reg./Ext./Crew Cab	2wd/4wd	1.00	L +3.60 R +4.10	0.50	+0.25	0.10+/-0.20
	Silverado SS	2wd/4wd	1.00	L +4.10 R +5.10	0.50	+0.10	0.10+/-0.20
	C25 LD	2wd/4wd	1.00	L +4.50 R +4.75	1.00	+0.25	0.10+/-0.20
	C25HD/C35/K25 LD	2wd/4wd	1.00	L +4.25 R +4.75	1.00	+0.25	0.10+/-0.20
	K25HD/K35HD	2wd/4wd	1.00	L +4.00 R +4.75	1.00	+0.25	0.10+/-0.20

① Vehicles built prior to the following VINs should be serviced with OLD specifications:

Vehicles built prior to the following VINs should be serviced with NEW specifications:

Fort Wayne: 1GECK19T53Z200018 (11/14/02)

Oshawa: 2GECK19TX31211735 (11/12/02)

Pontiac: 1GECK19T63E187691 (11/14/02)

② Old specification:

L +3.50

R +4.50

New specification:

L +3.80

R +4.10

06025-AVAL-C20

TIRE, WHEEL AND BALL JOINT SPECIFICATIONS

Year	Model	OEM Tires Standard	Optional	Tire Pressures (psi) Front	Rear	Wheel Size	Ball Joint Inspection	Lug Nut (ft. lbs.)
2002	1500 2wd	P235/75R15	None	36	36	6-JJ	L ①	③
	1500 4wd	P245/75R16	None	36	36	7-JJ	L ①	③
	2500	LT225/75R16D	LT245/75R16C	36	36	7-JJ	L ①	③
			LT245/75R16E	36	36			
	3500 SRW	LT245/75R16E	None	36	36	7-JJ	0.125 in.②	③
	3500 DRW	LT225/75R16D	LT215/85R16D	36	36	7-JJ	0.125 in.②	③
2003-04	1500 2wd	P235/75R15	None	36	36	6-JJ	L ①	③
	1500 4wd	P245/75R16	None	36	36	7-JJ	L ①	③
	2500	LT225/75R16D	LT245/75R16C	36	36	7-JJ	L ①	③
			LT245/75R16E	36	36			
	3500 SRW	LT245/75R16E	None	36	36	7-JJ	0.125 in.②	③
	3500 DRW	LT225/75R16D	LT215/85R16D	36	36	7-JJ	0.125 in.②	③
2005-06	1500 2wd	P235/75R15	None	36	36	6-JJ	L ①	③
	1500 4wd	P245/75R16	None	36	36	7-JJ	L ①	③
	2500	LT225/75R16D	LT245/75R16C	36	36	7-JJ	L ①	③
			LT245/75R16E	36	36			
	3500 SRW	LT245/75R16E	None	36	36	7-JJ	0.125 in.②	③
	3500 DRW	LT225/75R16D	LT215/85R16D	36	36	7-JJ	0.125 in.②	③

OEM: Original Equipment Manufacturer

PSI: Pounds Per Square Inch

STD: Standard

OPT: Optional

L: Lower

U: Upper

① Do not lift truck. Inspect the boss into which the grease fitting is threaded. Replace if the boss is flush or receded below the surface of the ball joint

② Applies to both upper and lower

③ Single wheels: 140 ft. lbs.
  Dual rear wheels: 175 ft. lbs.

06025-AVAL-C21

BRAKE SPECIFICATIONS
All measurements in inches unless noted

Year	Model		Brake Disc Original Thickness	Brake Disc Minimum Thickness	Brake Disc Maximum Runout	Brake Drum Diameter Original Inside Diameter	Brake Drum Diameter Max. Wear Limit	Brake Drum Diameter Max. Machine Diameter	Minimum Lining Thickness	Brake Caliper Bracket Bolts (ft. lbs.)	Brake Caliper Mounting Bolts (ft. lbs.)
2002	G/P1500	F	[1]	[2]	0.004	—	—	—	0.030	NA	38
		R	—	—	—	[3]	[4]	[5]	0.030	NA	—
	G/P2500	F	[1]	[2]	0.004	—	—	—	0.030	NA	38
		R	—	—	—	[3]	[4]	[5]	0.030	NA	—
	G/P3500	F	[1]	[2]	0.004	—	—	—	0.030	NA	38
		R	—	—	—	[3]	[4]	[5]	0.030	NA	—
	Avalanche	F	[6]	[7]	0.005	—	—	—	0.030	[8]	80
		R	[9]	[10]	0.005	—	—	—	0.030	[11]	[12]
	Sierra	F	[6]	[7]	0.005	—	—	—	0.030	[8]	80
		R	[9]	[10]	0.005	—	—	—	0.030	[11]	[12]
	Silverado	F	[6]	[7]	0.005	—	—	—	0.030	[8]	80
		R	[9]	[10]	0.005	—	—	—	0.030	[11]	[12]
2003	Express	F	[1]	[2]	0.004	—	—	—	0.030	NA	38
		R	—	—	—	[3]	[4]	[5]	0.030	NA	—
	Savana	F	[1]	[2]	0.004	—	—	—	0.030	NA	38
		R	—	—	—	[3]	[4]	[5]	0.030	NA	—
	Avalanche	F	[6]	[7]	0.005	—	—	—	0.030	[8]	80
		R	[9]	[10]	0.005	—	—	—	0.030	[11]	[12]
	Sierra	F	[6]	[7]	0.005	—	—	—	0.030	[8]	80
		R	[9]	[10]	0.005	—	—	—	0.030	[11]	[12]
	Silverado	F	[6]	[7]	0.005	—	—	—	0.030	[8]	80
		R	[9]	[10]	0.005	—	—	—	0.030	[11]	[12]
2004	Express	F	[1]	[2]	0.004	—	—	—	0.030	NA	38
		R	—	—	—	[3]	[4]	[5]	0.030	NA	—
	Savana	F	[1]	[2]	0.004	—	—	—	0.030	NA	38
		R	—	—	—	[3]	[4]	[5]	0.030	NA	—
	Avalanche	F	[6]	[7]	0.005	—	—	—	0.030	[8]	80
		R	[9]	[10]	0.005	—	—	—	0.030	[11]	[12]
	Sierra	F	[6]	[7]	0.005	—	—	—	0.030	[8]	80
		R	[9]	[10]	0.005	—	—	—	0.030	[11]	[12]
	Silverado	F	[6]	[7]	0.005	—	—	—	0.030	[8]	80
		R	[9]	[10]	0.005	—	—	—	0.030	[11]	[12]
2005	Express	F	[13]	[14]	0.005	—	—	—	—	[15]	[16]
		R	1.181	1.142	0.005	—	—	—	—	[15]	[16]
	Savana	F	[13]	[14]	0.005	—	—	—	—	[15]	[16]
		R	1.181	1.142	0.005	—	—	—	—	[15]	[16]
	Avalanche	F	[13]	[14]	0.005	—	—	—	—	[15]	[16]
		R	[17]	[18]	0.005	—	—	—	—	[15]	[16]
	Sierra	F	[19]	[20]	0.005	—	—	—	—	[21]	[22]
		R	[23]	[24]	0.005	—	—	—	—	[21]	[22]
	Silverado	F	[19]	[20]	0.005	—	—	—	—	[21]	[22]
		R	[23]	[24]	0.005	—	—	—	—	[21]	[22]

06025-AVAL-C22

BRAKE SPECIFICATIONS

All measurements in inches unless noted

Year	Model		Brake Disc			Brake Drum Diameter				Brake Caliper	
			Original Thickness	Minimum Thickness	Maximum Runout	Original Inside Diameter	Max. Wear Limit	Max. Machine Diameter	Minimum Lining Thickness	Bracket Bolts (ft. lbs.)	Mounting Bolts (ft. lbs.)
2006	Express	F	⑬	⑭	0.005	—	—	—	—	⑮	⑯
		R	1.181	1.142	0.005	—	—	—	—	⑮	⑯
	Savana	F	⑬	⑭	0.005	—	—	—	—	⑮	⑯
		R	1.181	1.142	0.005	—	—	—	—	⑮	⑯
	Avalanche	F	⑬	⑭	0.005	—	—	—	—	⑮	⑯
		R	⑰	⑱	0.005	—	—	—	—	⑮	⑯
	Sierra	F	⑲	⑳	0.005	—	—	—	—	㉑	㉒
		R	㉓	㉔	0.005	—	—	—	—	㉑	㉒
	Silverado	F	⑲	⑳	0.005	—	—	—	—	㉑	㉒
		R	㉓	㉔	0.005	—	—	—	—	㉑	㉒

NA: Not Available

① Available with 1.280 in. and 1.540 in. discs

② 1.2 in. disc: 1.230
  1.5 in. disc: 1.480

③ Available with 1 in., 11.15 in. and 13 in. drums

④ 10 in. drum: 10.05
  11.15 in. drum: 11.24
  1 in. drum: 13.09

⑤ 1 in. drum: 10.09
  11.15 in. drum: 11.21
  1 in. drum: 13.06

⑥ Vacuum: 1.14 in.
  Hydraulic: 1.50 in.

⑦ Vacuum: 1.10 in.
  Hydraulic: 1.46 in.

⑧ 15 series: 129 ft. lbs.
  25/35 series: 221 ft. lbs.

⑨ Vacuum: 0.787 in.
  Hydraulic: 1.14 in.

⑩ Vacuum: 0.748 in.
  Hydraulic: 1.10 in.

⑪ Vacuum: 148 ft. lbs.
  Hydraulic (9000 lbs.): 122 ft. lbs.
  Hydraulic (12,000 lbs.): 221 ft. lbs.

⑫ 15 series: 31 ft. lbs.
  25/35 series: 80 ft. lbs.

⑬ Available with 1.142 in. and 1.496 in. discs

⑭ 1.14 in. disc: 1.102
  1.49 in. disc: 1.457

⑮ Front 7200 GVW: 129 ft. lbs.
  Front & Rear all others: 221 ft. lbs.

⑯ Front: 80 ft. lbs.
  Rear: 7200 GVW: 31 ft. lbs., All others: 80 ft. lbs.

⑰ 7200 GVW: 1.181
  9900 GVW: 1.142

⑱ 7200 GVW: 1.142
  9900 GVW: 1.102

⑲ 6400/7000 GVW: 1.181
  7200 GVW: 1.142
  9900/12,300 GVW: 1.50

⑳ 6400/7000 GVW: 1.100
  7200 GVW: 1.100
  9900/12,300 GVW: 1.46

㉑ Light Duty: 133 ft. lbs. front, 148 ft. lbs. re
  Med/heavy duty: 221 ft. lbs. front and rea

㉒ Light Duty: 74 ft. lbs. front, 31 ft. lbs. rear
  Med/heavy duty: 80 ft. lbs. front and rear

㉓ 6400 GVW: 0.787
  7200/12,300 GVW: 1.181
  9900 GVW: 1.141

㉔ 6400 GVW: 0.784
  7200/12,300 GVW: 1.142
  9900 GVW: 1.102

06025-AVAL-C23

SCHEDULED MAINTENANCE INTERVALS - DIESEL ENGINES
2002-04 GENERAL MOTORS—AVALANCHE, SIERRA & SILVERADO—DIESEL

TO BE SERVICED	TYPE OF SERVICE	VEHICLE MILEAGE INTERVAL (x1000)																					
		5	8	10	15	20	23	25	30	35	38	40	45	50	53	55	60	65	68	70	75	80	83
Air intake system	S/I			✓		✓			✓			✓		✓			✓			✓		✓	
Automatic transmission fluid ①	R	Every 50,000 miles																					
Brake system	S/I	✓	✓		✓		✓		✓		✓		✓		✓		✓		✓		✓		✓
Chassis & suspension grease points	L	✓		✓	✓	✓		✓	✓	✓		✓	✓	✓		✓	✓	✓		✓	✓	✓	
Cooling fan operation	S/I			✓		✓			✓			✓		✓			✓			✓		✓	
Crankcase depression regular valve system hoses	S/I																✓						
CV-joint boots & axle seals	S/I	✓		✓	✓	✓		✓	✓	✓		✓	✓	✓		✓	✓	✓		✓	✓	✓	
EGR system ②	S/I																✓						
Engine coolant	R	Every 150,000 miles																					
Engine cooling system hoses & radiator	S/I & C	Initially at 100,000 miles, then every 50,000 miles																					
Engine oil & filter	R	✓		✓	✓	✓		✓	✓	✓		✓	✓	✓		✓	✓	✓		✓	✓	✓	
Front wheel bearings ③	S/I & L								✓					✓									
Fuel filter	R								✓					✓									
Rear/front axle fluid level	S/I	✓		✓	✓	✓		✓	✓	✓		✓	✓	✓		✓	✓	✓		✓	✓	✓	
Rotate tires	S/I	✓	✓		✓		✓		✓		✓		✓		✓		✓		✓		✓		✓
Shields & underhood insulation ①	S/I			✓		✓			✓			✓		✓			✓			✓		✓	

R: Replace   S/I: Inspect and service, if necessary   L: Lubricate   C: Clean

① Vehicles with a GVWR of 8500 lbs or more only.

② If equipped.

③ 2-wheel drive models only.

FREQUENT OPERATION MAINTENANCE (SEVERE SERVICE)

If a vehicle is operated under any of the following conditions it is considered severe service:

- Towing a trailer or using a camper or car-top carrier.
- Repeated short trips of less than 5 miles in temperatures below freezing, or trips of less than 10 miles in any temperature.
- Extensive idling or low-speed driving for long distances as in heavy commercial use, such as delivery, taxi or police cars.
- Operating on rough, muddy or salt-covered roads.
- Operating on unpaved or dusty roads.
- Driving in extremely hot (over 90°) conditions.

Engine oil & filter: replace every 2500 miles.

Chassis and suspension grease points: lubricate every 2500 miles

Rear/front axle fluid level: inspect initially at 5000 miles, then every 2500 miles thereafter.

Rotate tires: every 7500 miles.

Air cleaner filter: inspect every 15,000 miles.

Front wheel bearings (2-wheel drive only): clean, inspect and repack every 15,000 miles.

06025-AVAL-C24

SCHEDULED MAINTENANCE INTERVALS
2002-04 GENERAL MOTORS—AVALANCHE, SIERRA & SILVERADO—GASOLINE

TO BE SERVICED	TYPE OF SERVICE	VEHICLE MILEAGE INTERVAL (x1000)														
		7.5	15	22.5	30	37.5	45	52.5	60	67.5	75	82.5	90	97.5	100	150
Engine oil & filter	R	✓	✓	✓	✓	✓	✓	✓	✓	✓	✓	✓	✓	✓		
Chassis lubrication	S/I	✓	✓	✓	✓	✓	✓	✓	✓	✓	✓	✓	✓	✓		
Oil Life Monitor	S/I	✓	✓	✓	✓	✓	✓	✓	✓	✓	✓	✓	✓	✓		
Front/Rear Axle Fluid	S/I ①	✓	✓	✓	✓	✓	✓	✓	✓	✓	✓	✓	✓	✓		
CV joints & axle seals	S/I	✓	✓	✓	✓	✓	✓	✓	✓	✓	✓	✓	✓	✓		
Rotate tires	S/I	✓	✓	✓	✓	✓	✓	✓	✓	✓	✓	✓	✓	✓		
Passenger Compartment Air Filter	R		✓		✓		✓		✓		✓		✓				
Underhood Sound Shield	S/I	✓		✓		✓		✓		✓		✓		✓		
Fuel filter	R				✓				✓				✓			
Automatic transmission fluid & filter	R ②														✓	
Engine accessory drive belt	S/I								✓							
Fuel Tank, cap and lines	S/I								✓							
EGR System	S/I								✓							
EVAP System	S/I								✓							
Spark plugs	R														✓	
Spark Plug Wires	S/I														✓	
PCV Valve	S/I														✓	
Coolant	R															✓
Air cleaner filter	R				✓				✓				✓			

R: Replace   S/I: Service or Inspect

① If the vehicle is used for continuous trailer towing, change the fluid in the rear axle after the first 500 miles, then, every

② Vehicles over 8,600 lbs. GVWR: every 50,000 miles

FREQUENT OPERATION MAINTENANCE (SEVERE SERVICE)

If a vehicle is operated under any of the following conditions it is considered severe service:

- **Extremely dusty areas.**

- **50% or more of the vehicle operation is in 32°C (90°F) or higher temperatures, or constant operation**
  in temperatures below 0°C (32°F).

- **Prolonged idling (vehicle operation in stop and go traffic.**

- **Frequent short running periods (engine does not warm to normal operating temperatures).**

- **Police, taxi, delivery usage or trailer towing usage.**

Oil & oil filter change: change every 3000 miles

Lubricate chassis every 3000 miles

Drive axle: check every 3000 miles

Rotate tires every 6000 miles

Air cleaner filter: change every 24,000 miles

06025-AVAL-C25

SCHEDULED MAINTENANCE INTERVALS
2002-04 GENERAL MOTORS—G SERIES VAN, EXPRESS & SAVANA—GASOLINE

TO BE SERVICED	TYPE OF SERVICE	7.5	15	22.5	30	37.5	45	52.5	60	67.5	75	82.5	90	97.5	105	112.5	120
Accessory drive belt	S/I								✓								✓
Air cleaner filter	R				✓				✓				✓				✓
Automatic transmission fluid ①	R	Every 50,000 miles															
Chassis & suspension grease points	L	✓	✓	✓	✓	✓	✓	✓	✓	✓	✓	✓	✓	✓	✓	✓	✓
CV-joint boots & axle seals	S/I	✓	✓	✓	✓	✓	✓	✓	✓	✓	✓	✓	✓	✓	✓	✓	✓
EGR system	S/I								✓								✓
Engine coolant	R	Every 150,000 miles															
Engine oil & filter	R	✓	✓	✓	✓	✓	✓	✓	✓	✓	✓	✓	✓	✓	✓	✓	✓
EVAP system	S/I								✓								✓
Front wheel bearings	S/I & L				✓				✓				✓				✓
Fuel filter	R				✓				✓				✓				✓
Fuel system	S/I								✓								✓
PCV system	S/I	Every 100,000 miles															
Rear axle fluid level	S/I	✓	✓	✓	✓	✓	✓	✓	✓	✓	✓	✓	✓	✓	✓	✓	✓
Rotate tires	S/I	✓	✓	✓	✓	✓	✓	✓	✓	✓	✓	✓	✓	✓	✓	✓	✓
Shields & underhood insulation ①	S/I		✓		✓		✓		✓		✓		✓		✓		✓
Spark plugs	R	Every 100,000 miles															
Spark plug wires	S/I	Every 100,000 miles															

R: Replace   S/I: Inspect and service, if necessary   L: Lubricate

① Vehicles with a GVWR or 8500 lbs. or more only.

FREQUENT OPERATION MAINTENANCE (SEVERE SERVICE)

If a vehicle is operated under any of the following conditions it is considered severe service:

- Towing a trailer or using a camper or car-top carrier.

- Repeated short trips of less than 5 miles in temperatures below freezing, or trips of less than 10 miles in any temperature.

- Extensive idling or low-speed driving for long distances as in heavy commercial use, such as delivery, taxi or police cars.

- Operating on rough, muddy or salt-covered roads.

- Operating on unpaved or dusty roads.

- Driving in extremely hot (over 90°) conditions.

Engine oil & filter: replace every 3000 miles or 3 months, whichever occurs first.

Chassis and suspension grease points: lubricate every 3000 miles.

Rear/front axle fluid level: inspect every 3000 miles.

Rotate the tires ever 6000 miles.

Brake system components: inspect ever 6000 miles.

Front wheel bearings (2-wheel drive only): clean, inspect and repack every 15,000 miles.

Shields & underhood insulation (vehicles w/GVWR over 8500 lbs. Only): inspect every 15,000 miles

Cooling fan system hoses & connections: inspect every 15,000 miles.

Fuel filter: replace every 30,000 miles.

Air cleaner filter: inspect every 45,000 miles.

Automatic transmission fluid & filter: replace every 50,000 miles.

Accessory drive belt: inspect every 60,000 miles.

Fuel system tank, cap and lines: inspect every 60,000 miles.

EVAP system: inspect every 60,000 miles.

EGR system: inspect every 60,000 miles.

PCV system: inspect every 100,000 miles.

Engine cooling system components: inspect and clean every 150,000 miles.

06025-AVAL-C26

SCHEDULED MAINTENANCE INTERVALS
2002-04 GENERAL MOTORS—G SERIES VAN, EXPRESS & SAVANA—DIESEL

TO BE SERVICED	TYPE OF SERVICE	5	10	15	20	25	30	35	40	45	50	55	60	65	70	75	80	85	90	95	100	105	110	115	120
Air cleaner filter	R						✓						✓						✓						✓
Air intake system	S/I		✓		✓		✓		✓		✓		✓		✓		✓		✓		✓		✓		✓
Automatic transmission fluid ①	R										✓										✓				
Chassis & suspension grease points	L	✓	✓	✓	✓	✓	✓	✓	✓	✓	✓	✓	✓	✓	✓	✓	✓	✓	✓	✓	✓	✓	✓	✓	✓
Cooling fan, ducts & hoses	S/I												✓												✓
Crankcase depression regulator valve system hoses	S/I												✓												✓
CV-joint boots & axle seals	S/I	✓	✓	✓	✓	✓	✓	✓	✓	✓	✓	✓	✓	✓	✓	✓	✓	✓	✓	✓	✓	✓	✓	✓	✓
EGR system ②	S/I												✓												
Engine coolant	R	Every 100,000 miles																							
Engine cooling system hoses & radiator	S/I & C	Initially at 100,000 miles, then every 50,000 miles																							
Engine oil & filter ③	R	✓	✓	✓	✓	✓	✓	✓	✓	✓	✓	✓	✓	✓	✓	✓	✓	✓	✓	✓	✓	✓	✓	✓	✓
Front wheel bearings	S/I & L						✓						✓						✓						✓
Fuel filter	R												✓												✓
Rear axle fluid level	S/I	✓	✓	✓	✓	✓	✓	✓	✓	✓	✓	✓	✓	✓	✓	✓	✓	✓	✓	✓	✓	✓	✓	✓	✓
Rotate tires	S/I	✓	✓	✓	✓	✓	✓	✓	✓	✓	✓	✓	✓	✓	✓	✓	✓	✓	✓	✓	✓	✓	✓	✓	✓
Shields & underhood insulation	S/I						✓						✓						✓						✓

R: Replace   S/I: Inspect and service, if necessary   L: Lubricate   C: Clean

① For vehicles with a GVWR of 8500 lbs. or more.

② If equipped.

③ Perform at the mileage specified or every 3 months, whichever occurs first.

FREQUENT OPERATION MAINTENANCE (SEVERE SERVICE)

If a vehicle is operated under any of the following conditions it is considered severe service:

- Towing a trailer or using a camper or car-top carrier.

- Repeated short trips of less than 5 miles in temperatures below freezing, or trips of less than 10 miles in any temperature.

- Extensive idling or low-speed driving for long distances as in heavy commercial use, such as delivery, taxi or police cars.

- Operating on rough, muddy or salt-covered roads.

- Operating on unpaved or dusty roads.

- Driving in extremely hot (over 90°) conditions.

Engine oil & filter: replace every 2500 miles.

Chassis and suspension grease points: lubricate every 2500 miles.

Rear axle fluid level: inspect every 2500 miles.

CV-joint boots and axle seals: inspect for leakage every 2500 miles.

Rotate tires: every 7500 miles.

Air cleaner filter: inspect every 15,000 miles.

Front wheel bearings (2-wheel drive only): clean, inspect and repack every 15,000 miles.

Automatic transmission fluid & filter: replace every 50,000 miles.

06025-AVAL-C27

MAINTENANCE I AND II SERVICE SCHEDULES
2005-06 GENERAL MOTORS—AVALANCHE, EXPRESS, SAVANA, SIERRA, SILVERADO

When the CHANGE ENGINE OIL light appears, certain services and inspections are required.
Required services are described as Maintenance I and Maintenance II.
The first service on a vehicle should be Maintenance I, and the second service should be Maintenance II.
Alternate between the 2 thereafter. However, in some cases, Maintenance II may be required more often.
Maintenance I: Use Maintenance I if the CHANGE ENGINE OIL light comes on within 10 months
since vehicle was purchased or, if Maintenance II was performed.
Maintenance II: Use Maintenance II if the previous service performed was Maintenance I.
Always use Maintenance II whenever the CHANGE ENGINE OIL light comes on 10 months or more since the last
service, or, if the CHANGE ENGINE OIL light has not come on at all for one year.

Service	Maintenance I	Maintenance II
Change the engine oil and filter. Reset the oil life system.	✓	✓
Visually inspect the vehicle for leaks or damage. A fluid loss in the vehicle system could indicate a problem. Inspected, repair and add fluid to the system if necessary.	✓	✓
Inspect the engine air cleaner filter. If necessary, replace the filter.	✓	✓
Rotate the tires. Inspect the tire inflation pressures and the tire wear.	✓	✓
Visually inspect the brake lines and hoses for proper hook-up, binding, leaks, cracks, chafing, etc. Inspect the disc brake pads for wear and the rotors for surface condition. Inspect the drum brake linings for wear or cracks. Inspect other brake parts, including drums, wheel cylinders, calipers, parking brake, etc. Inspect the parking brake adjustment.	✓	✓
Inspect the engine coolant and the windshield washer fluid levels. Add fluid as needed.	✓	✓
Inspect the suspension and steering components. Inspect the front and rear suspension and the steering system for damaged, loose or missing parts, or signs of wear. Inspect the power steering lines and the hoses for proper hook-up, binding, leaks, cracks, chafing, etc.	--	✓
Visually inspect the coolant hoses and replace the hoses if they are cracked, swollen or deteriorated. Inspect all pipes, fittings and clamps; replace with GM parts as needed. To help ensure proper operation, a pressure test of the cooling system and pressure cap and cleaning the outside of the radiator and air conditioning condenser is recommended at least once a year.		✓
Inspect the wiper blades for wear or cracking.	--	✓
Inspect the restraint system components. Ensure the safety belt reminder light and all the belts, buckles, latch plates, retractors and anchorages are working properly. Look for any other loose or damaged safety belt system parts. If you see anything that might keep a safety belt system from working correctly, repair or replaced the damaged part. Replace torn or frayed safety belts, refer to Operational and Functional Checks in Seat Belts. Inspect for any opened or broken air bag coverings, and repair or replace as needed. The air bag system does require regular maintenance.	--	✓

06025-AVAL-C28

MAINTENANCE I AND II SERVICE SCHEDULES
2005-06 GENERAL MOTORS—AVALANCHE, EXPRESS, SAVANA, SIERRA, SILVERADO

Lubricate the body components.Lubricate all key lock cylinders, hood latch assemblies, secondary latches, pivots, spring anchor and release pawl, hood and door hinges, rear folding seats and liftgate hinges. Frequent lubrication may be required when exposed to a corrosive environment, refer to Fluid and Lubricant Recommendations . Applying dielectric silicone grease GM P/N 12345579 (Canadian P/N 1974984) or equivalent on the weatherstrips with a clean cloth.	--	✓
Inspect the transaxle fluid level and add fluid as needed.	--	✓
Inspect the suspension and steering components.Inspect the front and rear suspension and the steering system for damaged, loose or missing parts, or signs of wear. Inspect power steering lines and hoses for proper hook-up, binding, leaks, cracks, chafing, etc.	✓ --	
Inspect the throttle system for interference or binding and for damaged or missing parts. Replace the parts as needed. Replace any components that have high effort or excessive wear. Do not lubricate the accelerator or the cruise control cables.	✓ --	
Replace the passenger compartment air filter.	--	✓

06025-AVAL-C29

GASOLINE ENGINE REPAIR

Distributor

8.1L and 2005-06 4.8L, 5.3L and 6.0L engines use a distributorless ignition system.

REMOVAL

4.3L, 5.0L and 5.7L Engines

1. Remove or disconnect the following:
   - Negative battery cable
   - Spark plug wires and the coil leads from the distributor
   - Electrical connector at the base of the distributor
   - Distributor cap
2. Matchmark the rotor-to-housing and housing-to-engine block positions so that they can be matched during installation.
   - Distributor hold-down bolt
   - Distributor from the engine

4.8L, 5.3L and 6.0L Engines

➡ **If the Malfunction Indicator Lamp turns on, and a DTC code P1345 sets after installing the distributor, this indicates an incorrectly installed distributor. Engine damage or distributor damage may occur.**

1. Turn **OFF** the ignition switch.
2. Remove or disconnect the following:

- Spark plug wires from the distributor cap
- Electrical connector from the base of the distributor
- Two screws that hold the distributor cap to the housing. Discard the screws.
- Distributor cap from the housing

3. Use a grease pencil in order to note the position of the rotor in relation to the distributor housing.

4. Mark the distributor housing and the intake manifold with a grease pencil.

5. Remove or disconnect the following:

- Mounting clamp hold-down bolt
- Distributor

6. As the distributor is being removed from the engine, watch the rotor move in a counterclockwise direction about 42 degrees. This will appear as slightly more than the 1 o'clock position. Note the position of the rotor segment. Place a second mark on the base of the distributor. This will aid in achieving proper rotor alignment during the distributor installation.

INSTALLATION

4.3L, 5.0L and 5.7L Engines

TIMING NOT DISTURBED

1. Install or connect the following:
   - Distributor, aligning the match-marks are properly alignment
   - Distributor hold-down bolt
   - Distributor cap
   - Electrical connector at the base of the distributor
   - Spark plug wires and coil leads
   - Negative battery cable

TIMING DISTURBED

1. Remove the No. 1 cylinder spark plug. Turn the engine using a socket wrench on the large bolt on the front of the crankshaft pulley. Place a finger near the No. 1 spark plug hole and turn the crankshaft until the piston reaches Top Dead Center (TDC). As the engine approaches TDC, you will feel air being expelled through the No. 1 cylinder spark plug hole. The timing mark on the crankshaft pulley should now be aligned

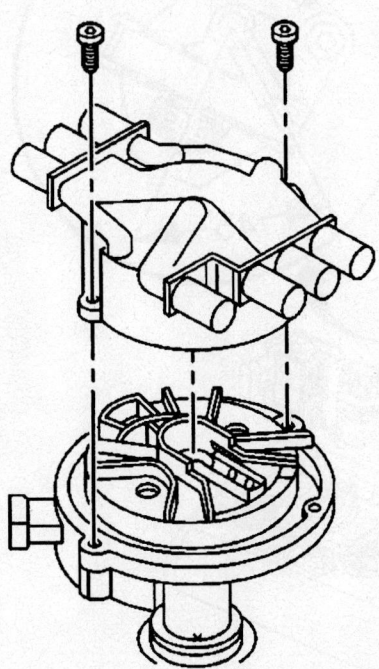

9308KG98

Distributor cap—4.3L, 4.8L, 5.3L and 6.0L engines

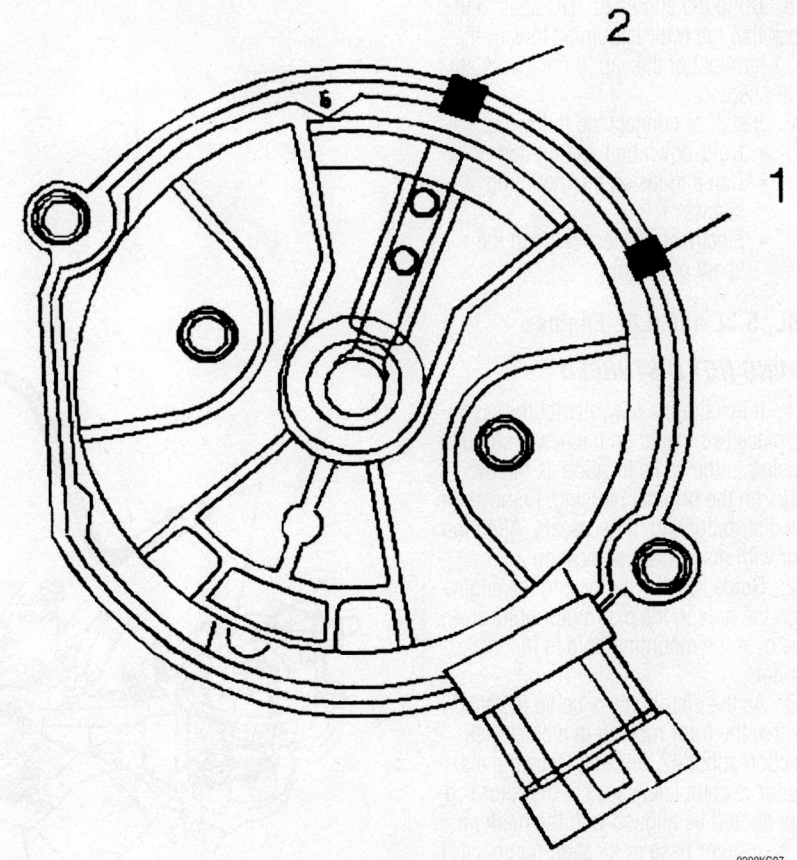

9308KG97

Distributor rotor starting point (1) and 42 degrees counterclockwise (2)—4.8L, 5.3L and 6.0L engines

with the **0** mark on the timing scale. If the position is not being met, turn the engine another full turn (360 degrees). Once the engines position is correct, install the spark plug.

➡ **Before installation, position the rotor so it points to the No. 2 terminal on the cap. As the distributor is lowered into the engine, the rotor will rotate clockwise and stop at the No. 1 terminal. This is the desired position.**

2. Turn the rotor so that it will point to the No. 1 terminal of the distributor cap when it is fully seated in the engine.

3. Install or connect the following:
  • Distributor. It may be necessary to turn the rotor a little in either direction, in order to engage the gears.

➡ **If the distributor will not seat completely in the engine, remove the distributor and align the groove on the top of the oil pump drive shaft with a long screwdriver to match the tab on the bottom of the distributor shaft. Reinstall the distributor.**

4. Tap the starter a few times to ensure that the oil pump shaft is mated to the distributor shaft.

5. Bring the engine to TDC again and check that the rotor is pointed toward the No. 1 terminal of the cap. If the marks are all aligned.

6. Install or connect the following:
  • Hold-down bolt and tighten
  • Cap and fasten the mounting screws
  • Electrical connections and the spark plug wires

4.8L, 5.3L and 6.0L Engines

TIMING NOT DISTURBED

1. If installing a new distributor assembly, place two marks on the new distributor housing in the same location as the two marks on the original housing. Remove the new distributor cap, if necessary. Align the rotor with mark made at location 2.

2. Guide the distributor into the engine. Align the hole in the distributor hold-down base over the mounting hole in the intake manifold.

3. As the distributor is being installed, observe the rotor moving in a clockwise direction about 42 degrees. Once the distributor is completely seated, the rotor segment should be aligned with the mark on the distributor base in location number 1. If the rotor segment is not aligned with the number 1 mark, the driven gear teeth and

the camshaft have meshed one or more teeth out of alignment.

4. Install or connect:
  • Distributor mounting clamp bolt and tighten to 18 ft. lbs. (25 Nm)

5. Distributor cap with NEW distributor cap screws. Tighten to 21 inch lbs. (2.4 Nm).

  • Electrical connector
  • Spark plug wires
  • Ignition coil wire

➡ **If the Malfunction Indicator lamp is turned on after installing the distributor, and a DTC P1345 is found, the distributor has been installed incorrectly.**

TIMING DISTURBED

1. Rotate the number 1 cylinder to TDC of the compression stroke. The engine front cover has 2 alignment tabs and the crankshaft balancer has 2 alignment marks (spaced 90 degrees apart) which are used for positioning number 1 piston at top dead center (TDC). With the piston on the compression stroke and at top dead center, the crankshaft balancer alignment mark must align with the engine front cover tab and the crankshaft balancer alignment mark must align with the engine front cover tab.

2. Align the white paint mark on the bottom stem of the distributor with the pre-drilled indent hole in the bottom of the gear. If the driven gear is installed incorrectly, the dimple will be approximately 180 degrees opposite of the rotor segment when it is installed in the distributor.

The OBD II ignition system distributor driven gear and rotor may be installed in multiple positions. In order to avoid mistakes, mark the distributor on the following components in order to ensure the same mounting position upon reassembly:
  • The distributor driven gear
  • The distributor shaft
  • The rotor holes

Installing the driven gear 180 degrees out of alignment, or locating the rotor in the wrong holes, will cause a no-start condition. Premature engine wear or damage may result.

3. Using a long screwdriver, align the oil pump drive shaft to the drive tab of the distributor. Guide the distributor into the engine. Ensure that the spark plug towers are perpendicular to the centerline of the engine.

Once the distributor is fully seated, the rotor segment should be aligned with the pointer cast into the distributor base.

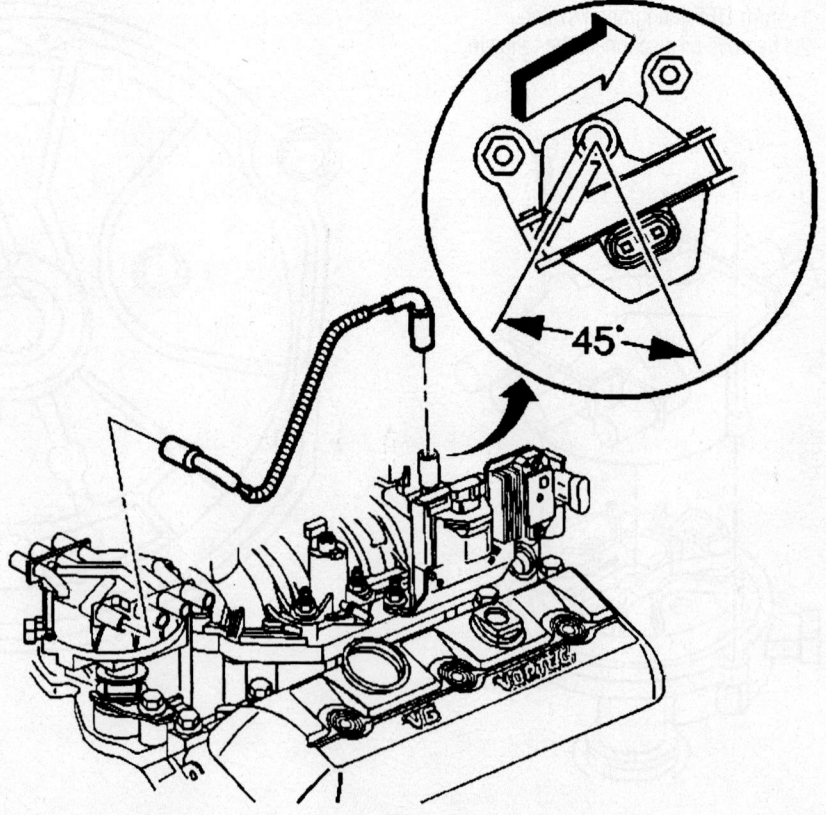

Distributor electrical connection—4.8L, 5.3L and 6.0L engines

9308KG96

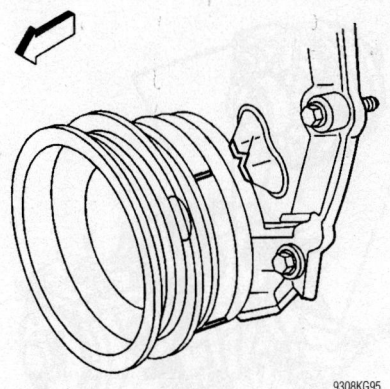

9308KG95

Engine at TDC compression—4.8L, 5.3L and 6.0L engines

This pointer may have a 6 cast into it, indicating that the distributor is to be used on a 6 cylinder engine or a 8 cast into it, indicating that the distributor is to be used on a 8 cylinder engine.

If the rotor segment does not come within a few degrees of the pointer, the gear mesh between the distributor and the camshaft may be off a tooth or more.

If this is the case, repeat the procedure again in order to achieve proper alignment.

➡ **Use the correct fastener in the correct location. Replacement fasteners must be the correct part number for that application. Fasteners requiring replacement or fasteners requiring the use of thread locking compound or sealant are identified in the service procedure. Do not use paints, lubricants, or corrosion inhibitors on fasteners or fastener joint surfaces unless specified. These coatings affect fastener torque and joint clamping force and may damage the fastener. Use the correct tightening sequence and specifications when installing fasteners in order to avoid damage to parts and systems.**

4. Install or connect the following:
   - Distributor mounting clamp bolt and tighten to 18 ft. lbs. (25 Nm)
   - Distributor cap with NEW distributor cap screws. Tighten to 21 inch lbs. (2.4 Nm).
   - Electrical connector
   - Spark plug wires
   - Ignition coil wire.

➡ **If the Malfunction Indicator lamp is turned on after installing the distributor, and a DTC P1345 is found, the distributor has been installed incorrectly.**

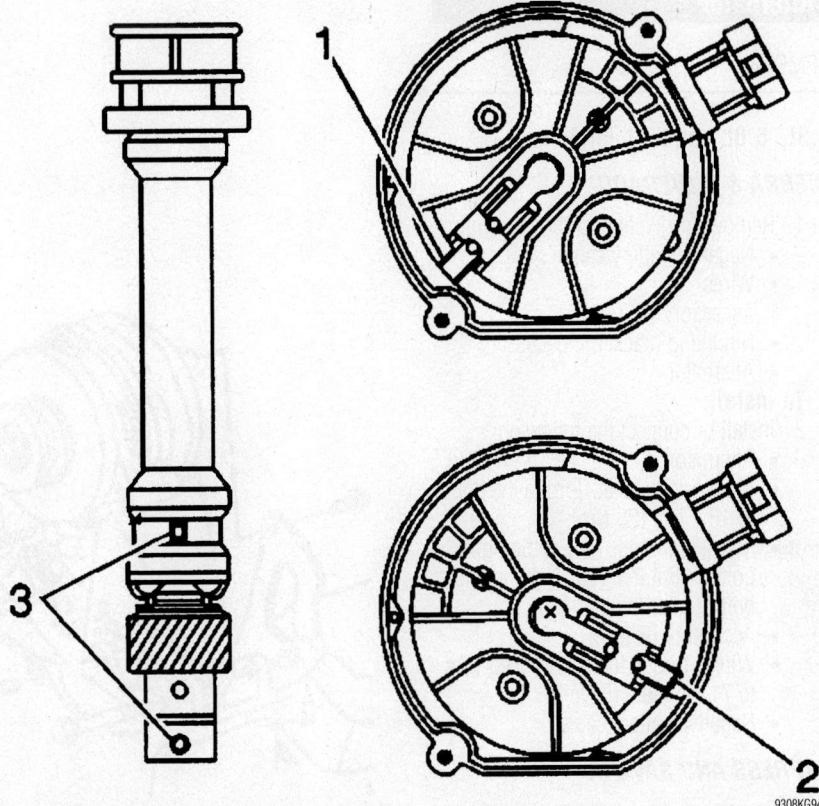

9308KG94

Distributor alignment. 1 is the starting point; 2 is installed; 3 are the shaft alignment marks—4.8L, 5.3L and 6.0L engines

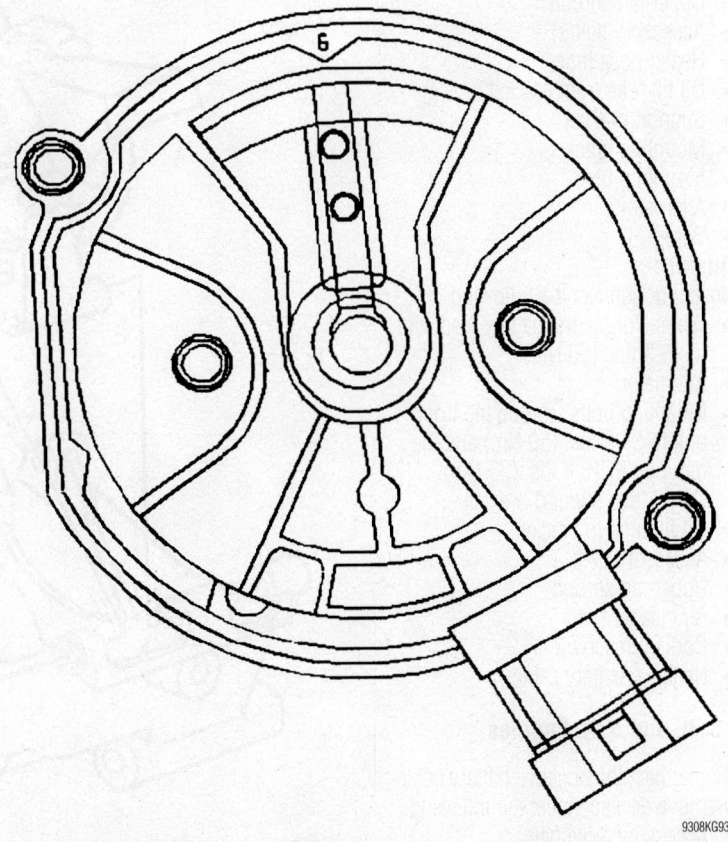

9308KG93

Distributor fully seated—4.8L, 5.3L and 6.0L

Alternator

REMOVAL

4.3L, 5.0L and 5.7L Engines

SIERRA & SILVERADO

1. Remove or disconnect the following:
   - Negative battery cable
   - Wires
   - Accessory belt(s)
   - Mounting bracket, if necessary
   - Alternator

To install:

2. Install or connect the following:
   - Alternator
   - Mounting bracket. Torque the bolts to 18 ft lbs. (25 Nm).
   - Mounting bolts. Torque the right bolt to 18 ft lbs. (25 Nm) and left bolt to 37 ft lbs. (50 Nm).
   - Accessory belt(s)
   - Wires. Torque the battery feed wire to 71 inch lbs. (8 Nm).
   - Negative battery cable

EXPRESS AND SAVANA

1. Remove or disconnect the following:
   - Negative battery cable
   - Coolant reservoir
   - Air cleaner
   - Upper fan shroud
   - Accessory belt(s)
   - Heater hose pipe
   - Oil fill tube from bracket
   - Support bracket
   - Mounting bracket
   - Mounting bolts
   - Alternator
   - Wires

To install:

2. Install or connect the following:
   - Wires. Torque the battery feed wire to 15 ft lbs. (20 Nm).
   - Alternator
   - Mounting bolts. Torque the front bolt to 37 ft lbs. (50 Nm) and the rear bolt to 18 ft lbs. (25 Nm).
   - Oil fill tube support bracket
   - Oil fill tube to bracket
   - Accessory belt(s)
   - Upper fan shroud
   - Air cleaner
   - Coolant reservoir
   - Negative battery cable

4.8L, 5.3L and 6.0L Engines

1. Disconnect the negative battery cable.
2. Remove or disconnect the following:
   - Accessory drive belt

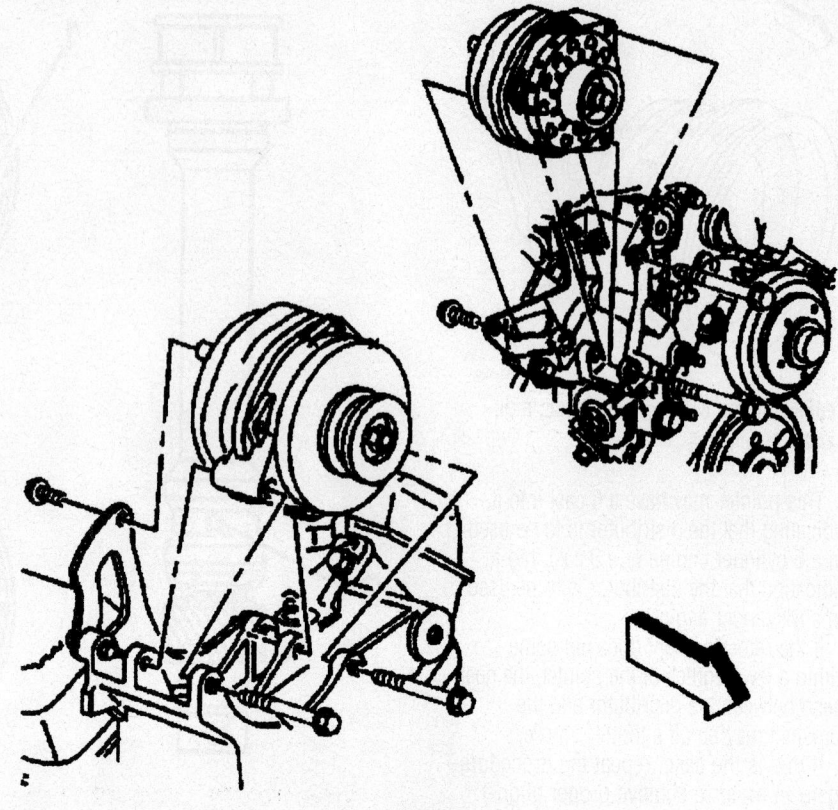

Exploded view of the alternator mounting—Truck shown

9308KG01

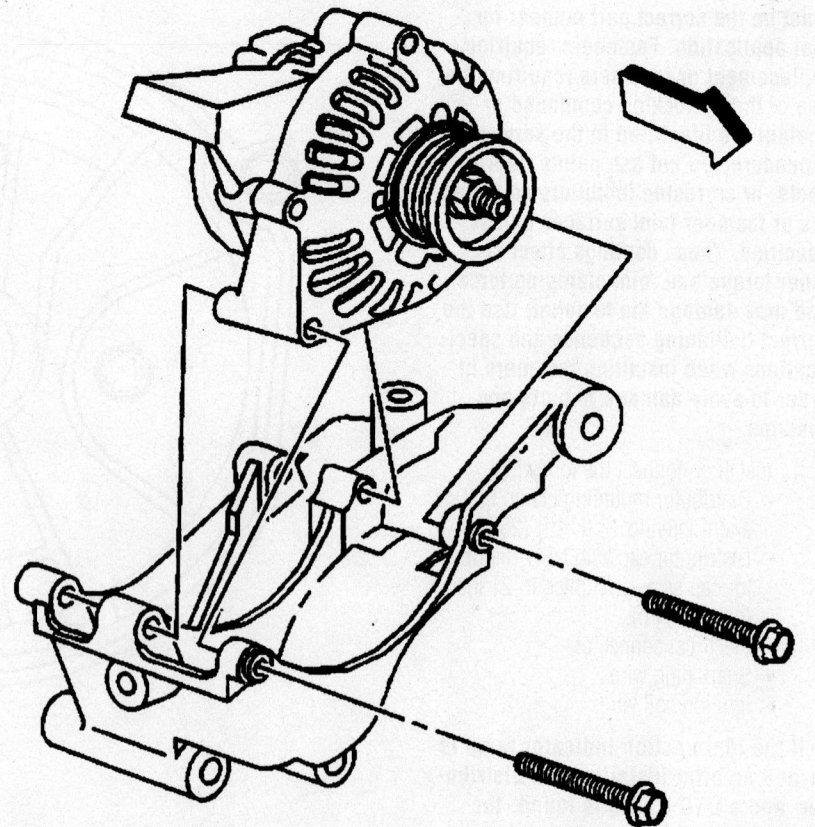

Exploded view of the alternator mounting—Van shown

9358KG02

- Engine sight shield, if necessary
- Electrical connections from the generator
- Mounting bolts
- Generator

To install:
3. Install the generator.
4. Install or connect the following:
   - Generator mounting bolts. Tighten the bolts to 37 ft. lbs. (50 Nm).
   - Electrical connections to the generator. Tighten the B+ nut to 13 ft. lbs. (18 Nm).
   - Engine sight shield, if removed
   - Accessory drive belt
5. Connect the negative battery cable. Tighten to bolt to 13 ft. lbs. (17 Nm).

8.1L Engine

1. Disconnect the negative battery cable.
2. Remove or disconnect the following:
   - Electrical connections from the generator
3. Remove the cable from the generator as follows:
   a. Slide the boot down, to reveal the terminal stud.
   b. Unfasten the cable nut from the stud, then remove the generator cable.
   - Accessory drive belt
   - Mounting bolts
   - Generator
   - Mounting bolts securing the generator to the brace and bracket

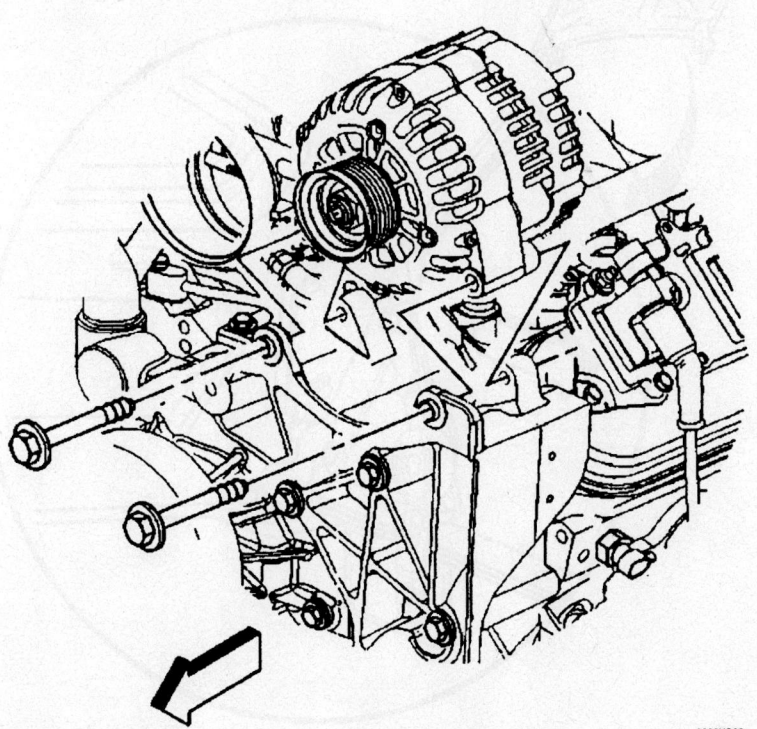

Alternator mounting—4.8L, 5.3L & 6.0L engines; 8.1L similar

- Generator

To install:
4. Install or connect the following:
   - Generator
   - Generator mounting bolts. Tighten the bolts to 37 ft. lbs. (50 Nm).
   - Accessory drive belt
5. Connect the generator cable, secure with the nut and tighten to 80 inch lbs. (9 Nm). Slide the boot back over the terminal stud.
   - Electrical connections to the generator
6. Connect the negative battery cable.

Ignition Timing

ADJUSTMENT

Always refer to the Vehicle Emissions Control Information label in the engine compartment for base ignition timing specification and adjustment procedures.

Engine Assembly

REMOVAL & INSTALLATION

4.3L (Exc. Silverado and Sierra), 5.0L and 5.7L Engines

1. Drain the cooling system.
2. Drain the engine oil.

3. Remove or disconnect the following:
   - Negative battery cable
   - Hood
   - Air cleaner
   - Accessory drive belt
   - Fan
   - Water pump pulley
   - Radiator and shroud
   - Heater hoses at the engine
   - Accelerator, cruise control and detent linkage if used
   - Air conditioning compressor, if used, and lay aside
   - Power steering pump, if used, and lay aside
   - Wiring from the engine
   - Fuel line
   - Vacuum lines from the intake manifold
   - Exhaust pipes from the manifold
   - Strut rods at the engine mountings, if used
   - Flywheel or torque converter cover
   - Wiring along the oil pan rail
   - Starter
   - Wire for the fuel gauge
   - Converter-to-flex plate bolts, if equipped with automatic transmission
4. Support the transmission
   - Bell housing to engine bolts
   - Rear engine mounting to frame bolts and the front through bolts and the engine

To install:
5. Lower the engine.
6. Install or connect the following:
   - Engine mounting bolts. Torque the rear engine mounting to frame bolts or nuts to 45 ft. lbs. (54 Nm), the front through-bolts to 70 ft. lbs. (97 Nm) and the front nuts to 50 ft. lbs. (67 Nm).
   - Bell housing to engine bolts and torque to 35 ft. lbs. (47 Nm)
7. Remove the transmission support.
   - Converter-to-flex plate bolts and tighten to 35 ft. lbs. (47 Nm)
   - Fuel gauge wiring
   - Starter
   - Flywheel or torque converter cover
   - Strut rods at the engine mountings, if used
   - Exhaust pipes at the manifold
   - Vacuum lines to the intake manifold
   - Fuel line
   - Engine wiring harness
   - Power steering pump, if used
   - Air conditioning compressor, if used
   - Accelerator, cruise control and detent linkage

9308KG99

- Heater hoses
- Radiator and shroud
- Accessory drive belts
- Hood
- Negative battery cable

8. Refill coolant and engine oil.

4.3L Silverado and Sierra

1. Remove or disconnect the following:

- Battery negative cable
- Coolant
- A/C refrigerant, if equipped
- Oil pan skid plate
- Engine shield
- Starter.
- Transmission cover
- Bolt holding the bracket for the starter cables and transmission cooler lines, if equipped
- Nuts at the catalytic converter pipe
- Exhaust pipes from the exhaust manifolds
- Bolts holding the brackets to the oil pan for both battery cables
- Crankshaft Position (CKP) sensor electrical connector and remove the harness from the retainer
- Low oil level sensor electrical connector and remove the wire harness from the retainer
- Bolt holding the battery negative cable and ground cable to the engine
- Torque converter-to-flywheel bolts, if equipped, through the starter opening
- Engine to transmission bolts

2. Move the hood hinge bolts to hold the hood in the service position.

3. Remove or disconnect the following:

- Positive Crankcase Ventilation (PCV) hose from the air cleaner outlet duct
- Air cleaner outlet duct from the throttle body and the air cleaner assembly
- Fan shroud
- Drive belt
- Engine cooling fan
- Radiator inlet and outlet hoses from the engine

❋❋ CAUTION

In order to avoid possible injury or vehicle damage, always replace the accelerator control cable with a NEW cable whenever you remove the engine from the vehicle.

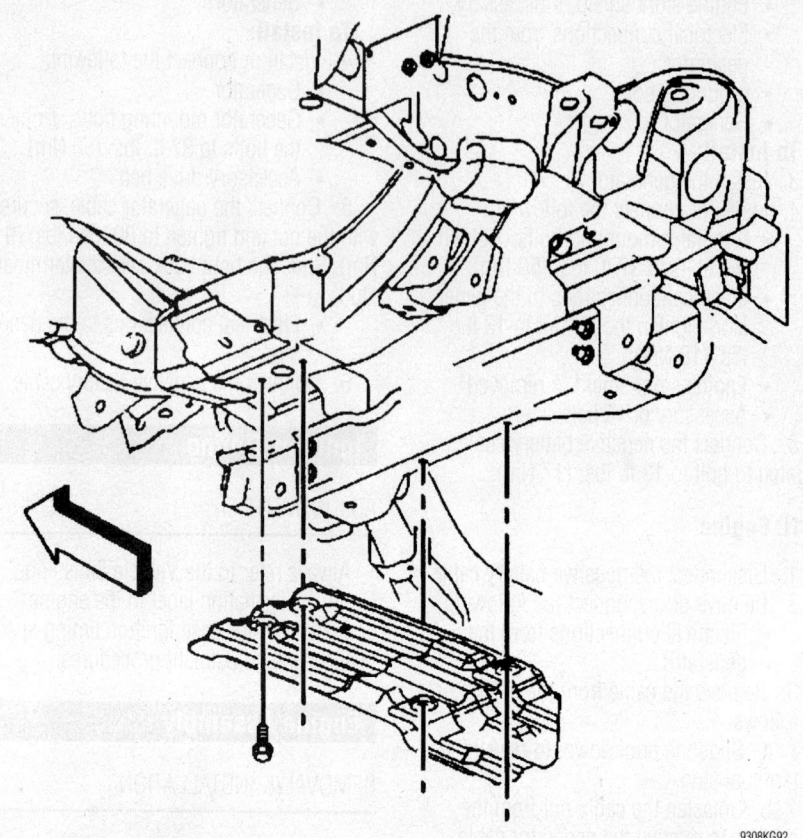

Engine shield removal

9308KG92

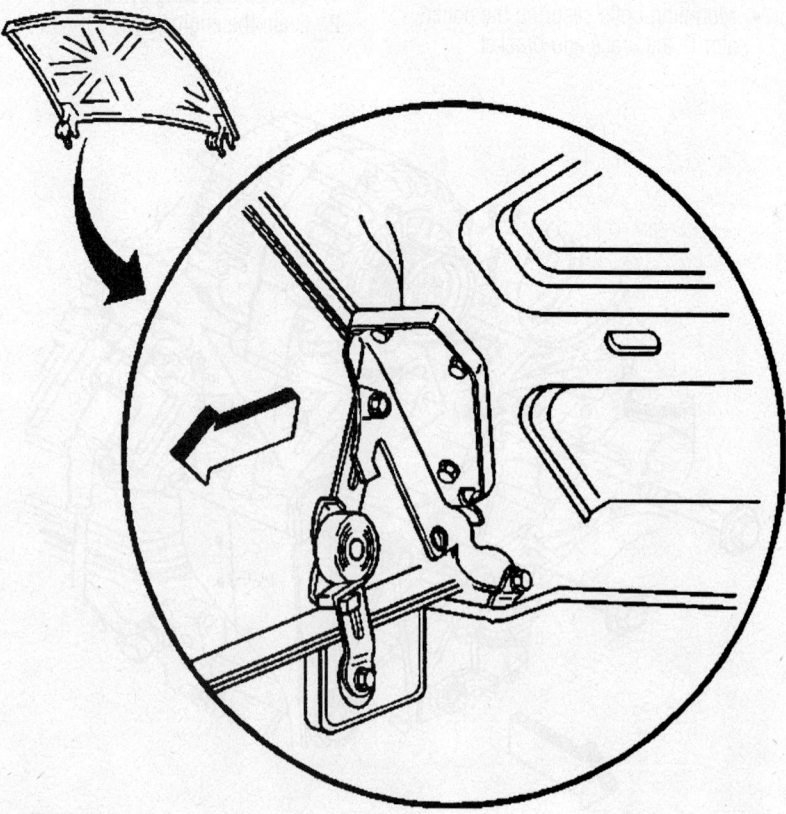

Hood in the service position

9308KG91

✳✳ WARNING

In order to avoid cruise control cable damage, position the cable out of the way while you remove or install the engine.

- Accelerator control cable
- Cruise control cable from the throttle body and the bracket on the throttle body and intake manifold, if equipped
- Engine wiring harness and clip from the accelerator control cable bracket
- Accelerator control cable bracket from the throttle body
- A/C hoses from the compressor and the accumulator, if equipped
- Secondary Air Injection (AIR) crossover pipe from the AIR pipe assemblies

➡**Remove the AIR pipes before engine removal. The AIR pipes can break or damage easily causing erratic engine operation.**

- AIR pipe assemblies from the left exhaust manifold, if equipped
- AIR pipe assembly from the AIR pump
- AIR pipe assembly from the right exhaust manifold, if equipped.
- A/C pressure switch, if equipped
- A/C compressor clutch, if equipped
- Exhaust Gas Recirculation (EGR) valve
- Battery positive cable
- Fuel meter body
- Idle Air Control (IAC) motor
- Throttle Position (TP) sensor
- Engine Coolant Temperature (ECT) sensor
- EVAP canister purge solenoid valve
- Manifold Absolute Pressure (MAP) sensor
- Ignition Control Module (ICM)
- Ignition coil
- Engine oil pressure gauge sensor
- Distributor
- Knock Sensor (KS)
- Nuts holding the bracket for the engine wiring harness to the intake manifold studs
- Bolt holding the engine wiring harness clip to the battery positive cable junction block bracket
- Bracket for the battery positive cable junction block from the power steering pump mounting bracket

- Battery positive and negative cables, if not done already
- Nut holding the ground wire to the stud at the rear of the right cylinder head
- Stud holding the engine wiring harness bracket to the rear of the right cylinder head
- Nut holding the engine wiring harness bracket to the stud for the EVAP canister purge solenoid valve
- Bolt holding the ground strap and the ground wire to the rear of the left cylinder head. Move the engine wiring harness aside.
- Both heater hoses from the engine and the cowl
- Distributor cap
- Fuel pipes at the rear of the engine
- Hose from the EVAP purge canister solenoid valve
- Power brake booster vacuum hose
- Nuts holding the power steering pump rear bracket to the side and front of the engine
- Three bolts and the nut holding the power steering pump mounting bracket to the engine

4. With the power steering pump and the A/C compressor still attached, slide the power steering pump mount bracket off the stud and set aside.

- Water outlet
- EGR valve inlet pipe from the intake and exhaust manifold

5. Attach the engine crane to the left front and right rear intake manifold mounting bolts.

6. Remove the engine motor mount to frame bracket bolts.

7. Support the transmission with a suitable jack.

8. Remove the engine.

To install:

9. Install or connect the following:

- Engine in the vehicle
- Engine mount to frame bracket bolts. Tighten the bolts to 50 ft. lbs. (65 Nm).

10. Remove the lifting device.

11. Apply thread lock GM P/N 12345382 or equivalent to the threads of the lower intake manifold bolts. Install the bolts and tighten as follows:

   a. 1st pass: 27 inch lbs. (3 Nm)
   b. 2nd pass: 106 inch lbs. (12 Nm)
   c. Final pass: 11 ft. lbs. (15 Nm)

12. Loosely install one transmission to engine bolt. Remove the support jack from under the transmission.

13. Install the EGR valve inlet pipe to the intake and the exhaust manifold and tighten as follows:

   a. Tighten the EGR valve inlet pipe intake nut to 18 ft. lbs. (25 Nm).
   b. Tighten the EGR valve inlet pipe exhaust nut to 22 ft. lbs. (30 Nm).
   c. Tighten the EGR valve inlet pipe clamp bolt to 18 ft. lbs. (25 Nm).

14. Install the water outlet.

15. Slide the power steering pump mounting bracket with the power steering pump and the A/C compressor on the stud.

16. Position the power steering pump rear bracket on the studs.

17. Install or connect the following:

- Power steering pump mounting bracket bolts and the nut
- Nut for the power steering pump rear bracket to the front of the engine. Tighten the power steering pump mounting bracket and the power steering pump rear bracket bolts and the nuts to 30 ft. lbs. (41 Nm).
- Fuel pipes
- Hose to the EVAP purge canister solenoid valve
- Vacuum brake booster hose to engine and the vacuum brake booster
- Distributor cap
- Both heater hoses to the engine and the cowl
- AIR pipe assembly with new gaskets to the right exhaust manifold, if equipped
- AIR pipe nuts and bracket bolt. Tighten the nuts to 18 ft. lbs. (25 Nm) and the bolt to 89 inch lbs. (10 Nm).
- AIR pipe assembly to the AIR pump
- AIR pipe assembly with new gaskets to the left exhaust manifold, if equipped. Install the AIR pipe nuts and bracket bolt. Tighten the to 18 ft. lbs. (25 Nm) and the bolt to 89 inch lbs. (10 Nm).
- AIR crossover pipe to the AIR pipe assemblies
- Engine wiring harness
- A/C pressure switch
- A/C compressor clutch
- EGR valve
- Generator battery positive cable
- Fuel meter body assembly
- IAC motor
- TP sensor
- ECT sensor
- EVAP canister purge solenoid valve
- MAP sensor
- ICM
- Ignition coil
- Engine oil pressure gauge sensor

- Distributor
- KS

18. Install the bolt holding the ground strap and the ground wire to the rear of the left cylinder head. Tighten the ground strap and ground wire bolt to 12 ft. lbs. (16 Nm).

19. Position the engine wiring harness bracket on the EVAP purge canister solenoid valve stud and install the nut.

20. Install the stud holding the wire harness bracket to the rear of the right cylinder head. Tighten the nut on the EVAP solenoid to 80 inch lbs. (9 Nm). Tighten the stud at rear of the cylinder head to 18 ft. lbs. (25 Nm).

21. Install the nut holding the ground wire on the stud at the rear of the right cylinder head. Tighten the ground wire nut to 12 ft. lbs. (16 Nm).

22. Position the battery positive and negative cables. Do not connect the negative battery cable to the battery.

23. Install or connect the following:
- Battery positive cable junction block bracket and bolt to the power steering pump mounting bracket. Tighten the bolt to 18 ft. lbs. (25 Nm).
- Bolt holding the engine wiring harness bracket to battery positive cable junction block bracket
- Engine wiring harness bracket on the intake manifold studs. Tighten the nuts to 106 inch lbs. (12 Nm) and the bolt to 80 inch lbs. (9 Nm).
- A/C hoses
- Accelerator control cable bracket. Tighten the nuts to 80 inch lbs. (9 Nm).
- Engine wire harness and clip to the accelerator control cable bracket

✳✳ CAUTION

In order to avoid possible injury or vehicle damage, always replace the accelerator control cable with a NEW cable whenever you remove the engine from the vehicle.

- NEW accelerator control cable
- Cruise control cable
- Radiator inlet and outlet hoses
- Engine cooling fan
- Drive belt
- Upper and lower radiator shroud
- Air cleaner outlet duct to the throttle body and the air cleaner assembly
- PCV hose to the air inlet duct

24. Move the hood hinge bolts from the service position to the normal operating position.

25. Raise the vehicle.
26. Install or connect the following:
- Remaining transmission to engine bolts except for the one where the transmission cover mounts
- Torque converter to flywheel
- Transmission cover and bolts. Tighten the transmission cover to oil pan bolt to 106 inch lbs. (12 Nm). Tighten the transmission cover to transmission bolt to 34 ft. lbs. (47 Nm).
- Bolt holding the bracket for the starter cables and the transmission cooler pipe and tighten to 80 inch lbs. (9 Nm)
- Positive and negative battery cable brackets-to-oil pan bolts and tighten to 106 inch lbs. (12 Nm).
- Bolt for the battery negative cable and ground wire to the front of the engine. Tighten the battery negative cable and ground wire bolt to 18 ft. lbs. (25 Nm).
- CKP sensor and install the harness in the retainer
- Low oil level sensor and install the wire harness in the retainer
- Exhaust pipe to the exhaust manifolds and tighten the nuts at the catalytic converter flange
- Starter motor
- Oil pan skid plate and tighten bolt to 15 ft. lbs. (20 Nm)
- Engine shield

27. Lower the vehicle.
- Battery negative cable
- Engine oil
- Coolant

28. Recharge the A/C system.

4.8L, 5.3L and 6.0L Engines

✳✳ CAUTION

Before servicing any electrical component, the ignition key must be in the OFF or LOCK position and all electrical loads must be OFF, unless instructed otherwise in these procedures.

1. Remove or disconnect the following:

- Negative battery cable
- Coolant
- A/C refrigerant

2. Raise the hood to the servicing position. Move the hood hinge bolt to hold the hood in the servicing position.

- Upper and the lower radiator hoses from the engine

- Air cleaner duct from the engine
- A/C condenser mounting bolts
- Radiator support from the vehicle
- A/C compressor
- Coolant hose from the throttle body
- Heater hoses from the engine and the cowl
- Engine sight shield from the intake manifold
- Accelerator control cable mounting bracket from the intake manifold

✳✳ CAUTION

In order to avoid possible injury or vehicle damage, always replace the accelerator control cable with a NEW cable whenever you remove the engine from the vehicle. In order to avoid cruise control cable damage, position the cable out of the way while you remove or install the engine.

- Accelerator control cable and the cruise control cable, if equipped, from the throttle shaft

3. Open the large electrical harness retainer. Remove one 10 mm nut in order to release the engine harness from the intake manifold.

4. Disconnect the electrical connectors from the following:

- Eight injectors
- Idle Air Control (IAC) motor
- Throttle Position (TP) sensor
- Evaporative Emissions (EVAP) canister purge solenoid
- Manifold Absolute Pressure (MAP) sensor
- Camshaft Position (CMP) sensor
- Ground splice at the rear of the right side of the block
- Ground splice and the ground strap at the rear of the left side of the block
- Coolant Temperature (CTS) sensor
- Oil pressure sensor/switch
- Electrical connector from intake and disconnect from harness
- Junction block bracket from alternator bracket

5. Set the electrical harness aside.
6. Remove or disconnect the following:

- EVAP canister purge solenoid vent tube from the solenoid by squeezing the retainer, then release the tube from the solenoid
- Battery negative cable from the engine block

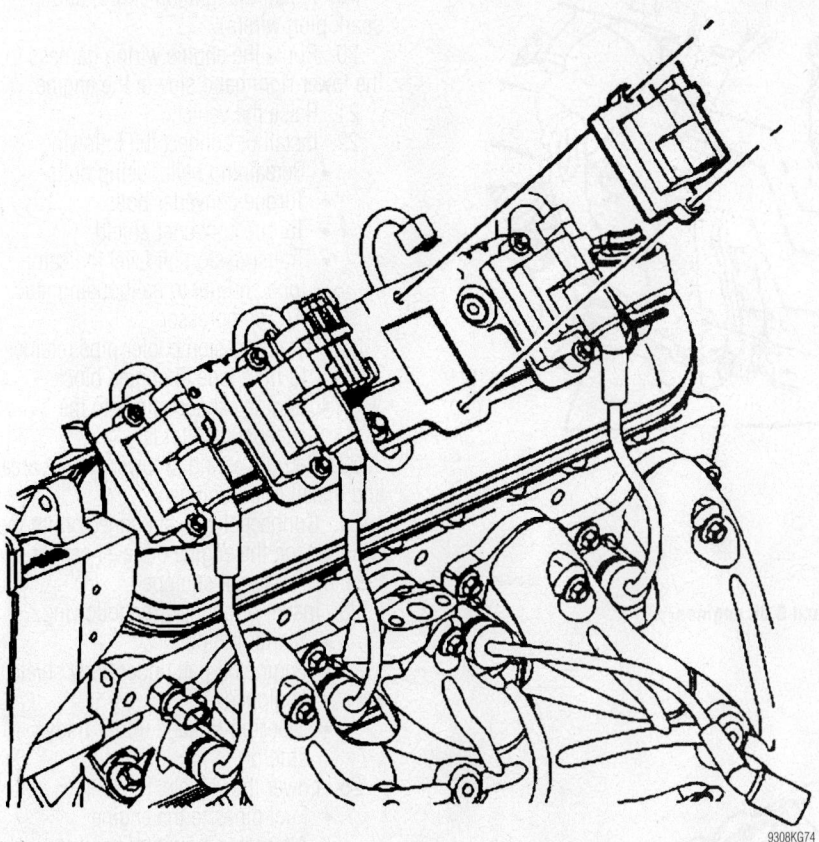

Ignition coil removal—4.8L, 5.3L and 6.0L engines

9308KG74

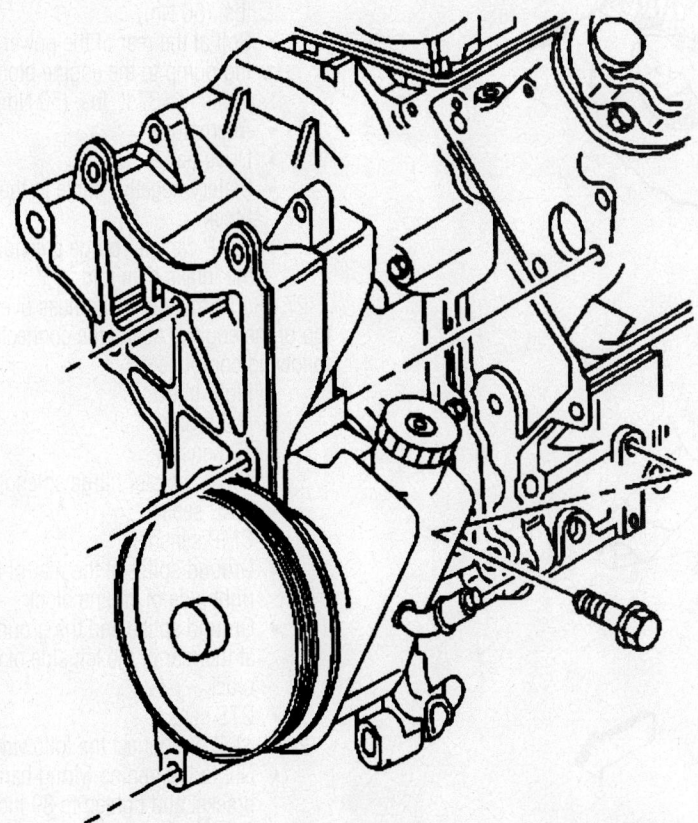

Power steering pump removal—4.8L, 5.3L and 6.0L engines

9308KG73

- Drive belt
- Bolts holding the alternator mounting bracket to the cylinder head and block
- Bolt behind the power steering pump to engine block
- Alternator mounting bracket. Position the bracket aside.
- Fuel pipes from the engine

7. Raise the vehicle.
- Steering linkage under body shield, if equipped
- Engine oil pan under body shield, if equipped
- Engine oil
- Starter motor

8. Disconnect the engine wiring harness from the following components:
- Crankshaft Position (CKP) sensor
- Engine oil level sensor
- Block heater, if equipped
- Oil pan wiring harness

9. Reposition wiring from the lower engine area.
- Exhaust pipes from the exhaust manifolds
- Transmission cooler pipe retainer from the right side of the engine block, if equipped
- Torque converter shield from the engine
- Torque converter bolts
- Nut and the transmission oil level indicator tube from the bellhousing stud
- Lower bellhousing studs from the engine

10. Lower the vehicle.
- Remaining bellhousing bolts
- Engine electrical harness aside
- Ignition coil(s)

11. Install an engine crane.

12. Install a floor jack or stands to transmission for support.

13. Remove the engine mount bolts.

➡**Use care while moving the engine assembly in order to avoid breaking the MAP sensor locating tabs. Broken MAP sensor tabs may result in decreased engine performance.**

14. Remove the engine from the vehicle.
To install:
15. Install or connect the following:
- Engine to the vehicle
- Engine mount bolts
- Upper bellhousing bolts

16. Remove transmission support apparatus.

17. Remove the lifting device.

18. Remove the lift brackets from both cylinder heads.

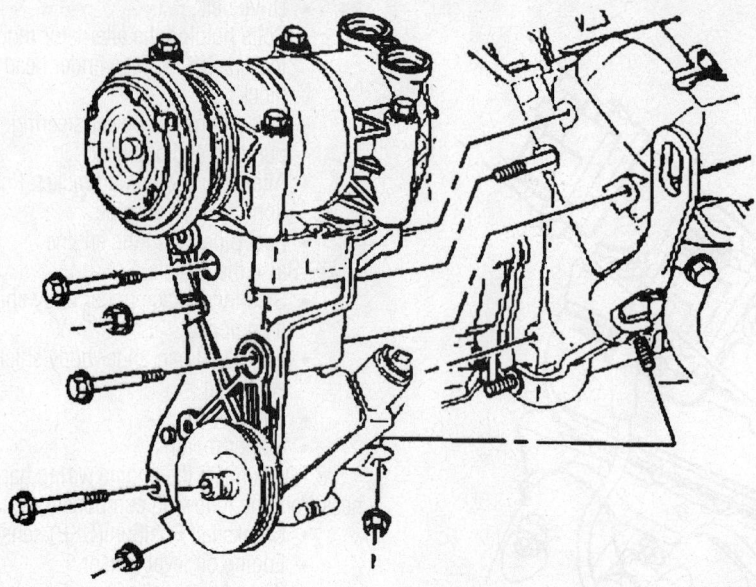

Power steering mount bracket removal—4.8L, 5.3L and 6.0L engines

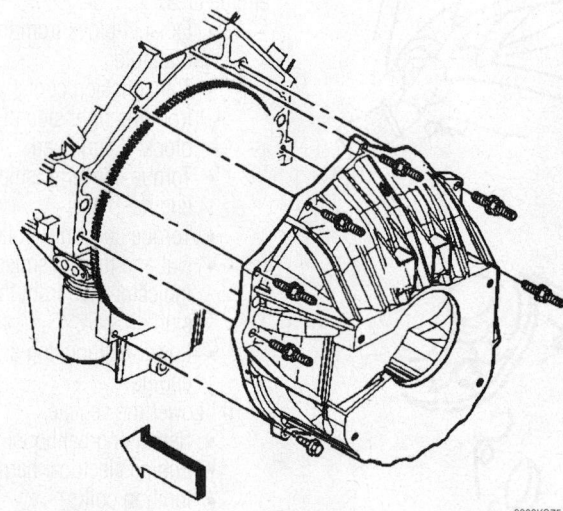

Bellhousing bolt removal—4.8L, 5.3L and 6.0L engines

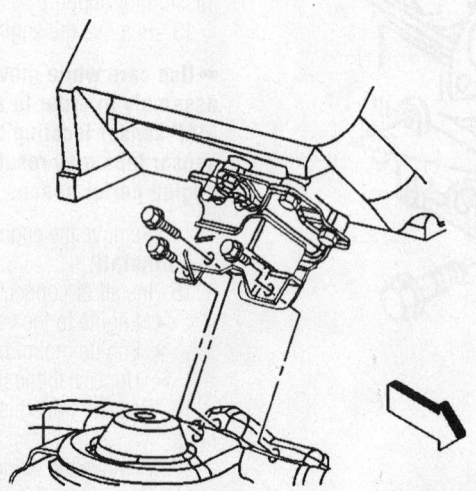

Engine mount disconnect—4.8L, 5.3L and 6.0L engines

19. Install the ignition coil(s) and the spark plug wire(s).

20. Route the engine wiring harness to the lower right hand side of the engine.

21. Raise the vehicle.

22. Install or connect the following:
- Remaining bellhousing bolts
- Torque converter bolts
- Torque converter shield
- Transmission oil level indicator tube and nut to bellhousing stud
- A/C compressor
- Transmission cooler pipe retainer to right side of engine block
- Engine exhaust pipes to the exhaust manifolds

23. Reroute wiring to lower engine area and install bolt to oil pan.

24. Connect electrical connectors to the CKP sensor, the engine oil level sensor and the block heater, if equipped.

25. Install or connect the following:
- Starter motor
- Engine oil pan under body shield, if equipped
- Steering linkage under body shield

26. Lower the vehicle.
- Fuel pipes to the engine
- Alternator mounting bracket to the cylinder head using the nuts and the bolts. Tighten the bolts to 37 ft. lbs. (50 Nm).
- Bolt at the rear of the power steering pump to the engine block and tighten to 37 ft. lbs. (50 Nm)
- Alternator
- Drive belt
- Battery negative cable to the engine block
- EVAP canister purge solenoid to the intake manifold

27. Route the engine harness over the top of the engine. Attach the connectors for following components:
- Eight injectors
- IAC motor
- TP sensor
- EVAP canister purge solenoid.
- MAP sensor
- CMP sensor
- Ground splice at the rear of the right side of engine block
- Ground splice and the ground strap at the rear of the left side of engine block
- CTS sensor

28. Install or connect the following:
- Nut to the engine wiring harness bracket and tighten to 89 inch lbs. (10 Nm)

> **⁂ CAUTION**
>
> In order to avoid possible injury or vehicle damage, always replace the accelerator control cable with a NEW cable whenever you remove the engine from the vehicle. In order to avoid cruise control cable damage, position the cable out of the way while you remove or install the engine.

- NEW accelerator control cable
- Cruise control cable, if equipped, to the throttle shaft
- Bolts for the accelerator control cable mounting bracket and tighten to 89 inch lbs. (10 Nm)
- Engine sight shield to the intake manifold
- Heater hoses to the cowl and the engine
- Coolant hose to the throttle body
- Radiator support in the vehicle
- A/C condenser mounting bolts
- Air cleaner duct
- Lower radiator hoses to the engine

29. Lower the hood.
30. Fill the engine with oil.
31. Fill the engine with coolant.
32. Connect the negative battery cable.

8.1L Engine

1. Raise the hood to the servicing position. Move the hood hinge bolt to hold the hood in the servicing position.
2. Release the fuel system pressure.
3. Remove or disconnect the following:
- Negative, then positive battery cables
- Coolant
- A/C refrigerant
- Engine oil cooler lines from the engine block
- Transmission-to-engine bolts
- Clutch pressure plate bolts, if equipped
- Torque converter bolts, if equipped
- Catalytic converter
- Exhaust manifold pipe
- Hoses from power steering pump, then plug the lines and ports
- Starter motor

4. Raise the vehicle.
- Engine electrical harness and tie aside
- Alternator
- Ground cable bolt from engine block
- Exhaust Gas Recirculation (EGR) valve adapter

- Vacuum lines (tag before removal)
- Throttle Actuator Control (TAC) module electrical connector

5. Install Engine Lift Brackets part No. J 36857, or equivalent, to the rear of the right cylinder head and the front of the left cylinder head.
6. Install the attaching bolt and washer. Use part No. 9428217 with 1560963. Tighten the bolts to 30 ft. lbs. (40 Nm).
7. Remove or disconnect the following:
- Engine mount heat shield bolt and shields
- Engine mount-to-engine mount bracket bolts
- Engine from the vehicle, using a suitable lifting device. Place on a suitable stand.
- A/C compressor/power steering pump bracket from the cylinder head
- Lift brackets from the cylinder head

To install:

8. Install Engine Lift Brackets part No. J 36857, or equivalent, to the rear of the right cylinder head and the front of the left cylinder head.
9. Install the attaching bolt and washer. Use part No. 9428217 with 1560963. Tighten the bolts to 30 ft. lbs. (40 Nm).
10. Install or connect the following:
- A/C compressor/power steering mounting bracket. Tighten the bolts and nut to 37 ft. lbs. (50 Nm).
- Alternator bracket
- Engine into the vehicle
- Engine mount-to-engine mount bracket bolts
- Engine mount heat shield and bolts

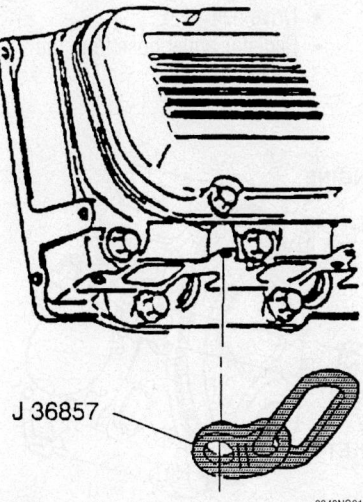

J 36857

9348NG01

Install suitable lift brackets to the rear of the right head and the front of the left head

11. Remove the lift hooks from the cylinder heads, then raise the vehicle.
- Engine oil cooler lines
- Transmission-to-engine bolts
- Clutch pressure plate bolts, if equipped
- Torque converter bolts, if equipped
- Catalytic converter
- Exhaust manifold pipe
- Hoses to the power steering pump
- Starter motor

12. Lower the vehicle.
- Engine electrical harness. Make sure the harness is properly routed.
- Alternator
- Ground cable bolt to engine block and tighten to 12 ft. lbs. (16 Nm)
- EGR valve adapter
- Vacuum lines, as tagged during removal
- TAC module electrical connector
- Radiator
- A/C compressor
- Fuel feed and return lines
- Ignition coils
- Positive, then negative battery cables
- Air cleaner outlet duct and secure with the clamp

13. Lower the hood from the service position.
14. Properly recharge the A/C system.
15. Fill the engine with oil.
16. Fill the engine with coolant.
17. Perform the Crankshaft Position (CKP) sensor variation learn procedure:

   a. Install a suitable scan tool and check for Diagnostic Trouble Codes (DTCs). If any DTCs, other than P1336 are set, resolve those codes first, before proceeding with this procedure.

   b. With the scan tool, select the crankshaft position variation learn procedure.

   c. Observe the fuel cut-off for the 8.1L engine.

   d. The scan tool will instruct you to perform certain steps, make sure you follow all directions given by the scan tool exactly.

   e. Enable the crankshaft position system variation learn procedure.

➡ **While the learn procedure is in progress, release the throttle immediately when the engine started to decelerate. The engine control is returned to the operator and the engine responds to throttle position after the learn procedure is complete.**

   f. Slowly increase the engine speed to the RPM that you observed.

g. Immediately release the throttle when fuel cut-out is reached.

h. The scan tool displays: Learn Status: Learned this ignition. If the scan tool does NOT display this message and not other DTCs set, you must perform further troubleshooting.

i. Turn the ignition **OFF** for 30 seconds after the learn procedure has been completed successfully.

18. Start and run the engine, then check for leaks.

Water Pump

REMOVAL & INSTALLATION

4.3L, 5.0L and 5.7L Engines

1. Drain the radiator.
2. Remove or disconnect the following:
   - Fan shroud
   - Negative battery cable
   - Drive belt(s)
   - Alternator and other accessories, if necessary
   - Fan, fan clutch and pulley
   - Accessory brackets that might interfere with water pump removal
   - Lower radiator hose from the water pump inlet
   - Heater hose from the nipple on the pump
   - Water pump assembly away from the timing cover

To install:

3. Clean all old gasket material from the timing chain cover.
4. Install or connect the following:
   - Pump assembly with a new gasket. Torque the bolts to 30 ft. lbs. (41 Nm).
   - Hose between the water pump inlet and the pump

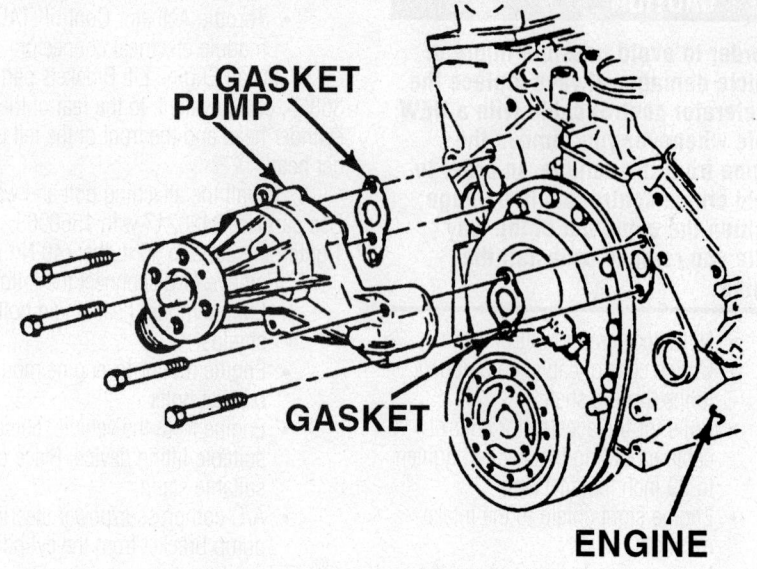

Exploded view of the water pump mounting—5.0L and 5.7L engines

   - Fan, fan clutch and pulley
   - Alternator and other accessories, if necessary
   - Drive belt(s)
   - Upper radiator shroud
5. Refill the cooling system.
6. Connect the battery.

4.8L, 5.3L and 6.0L Engines

1. Remove or disconnect the following:
   - Air outlet duct
   - Coolant
   - Inlet radiator hose from the water pump
   - Upper fan shroud
   - Cooling fan and clutch assembly
   - Drive belt
   - Radiator outlet hose from the coolant pump

   - Surge tank hose
   - Heater hose
   - Water pump

To install:

➡ **DO NOT use cooling system seal tabs (or similar compounds) unless otherwise instructed. The use of cooling system seal tabs (or similar compounds) may restrict coolant flow through the passages of the cooling system or the engine components. Restricted coolant flow may cause engine overheating and/or damage to the cooling system or the engine components/assembly.**

2. Install or connect the following:
   - Water pump. Install the water pump bolts. Tighten the water pump bolts first pass to 11 ft. lbs. (15 Nm); tighten the bolts final pass to 22 ft. lbs. (30 Nm).
   - Water pump drive belt pulley and bolts (if applicable). Tighten the pulley bolts first pass to 89 inch lbs. (10 Nm); tighten the bolts final pass to 18 ft. lbs. (25 Nm).
   - Surge tank hose
   - Heater hose
   - Outlet radiator hose to the coolant pump
   - Drive belt
   - Cooling fan and clutch assembly
   - Upper fan shroud
   - Inlet radiator hose to the water pump
   - Air inlet duct
   - Coolant

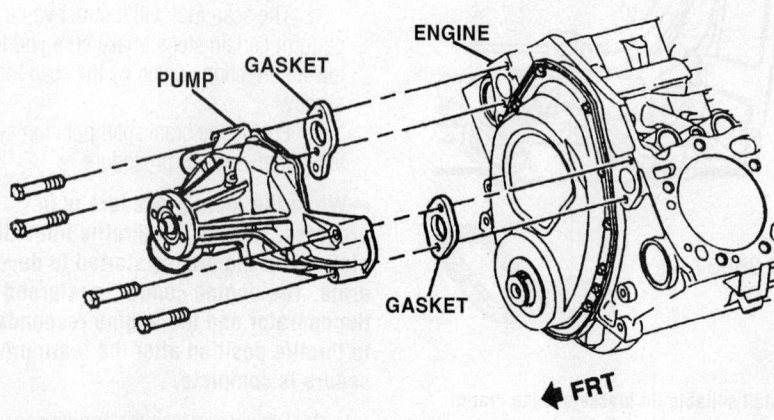

Exploded view of the water pump mounting—4.3L engine

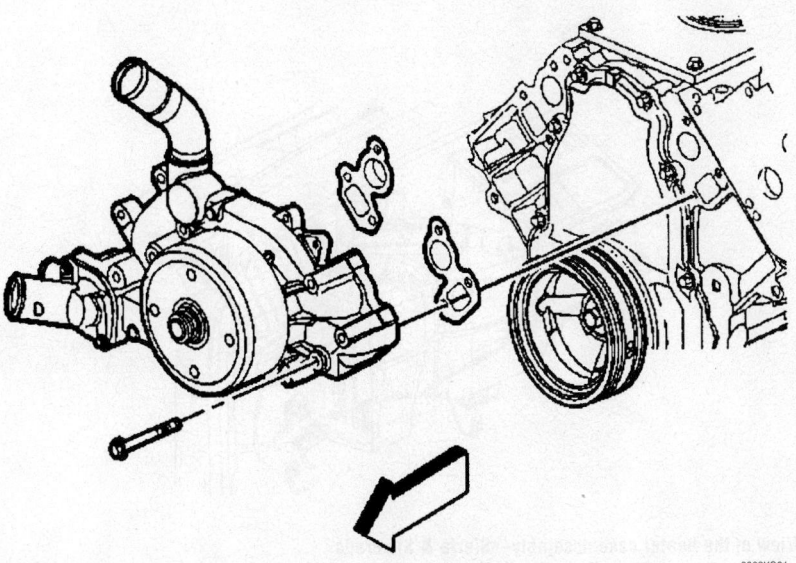

Exploded view of the water pump assembly—4.8L, 5.3L and 6.0L engines

8.1L Engines

1. Remove or disconnect the following:
   - Coolant
   - Drive belt
   - Fan clutch
   - Outlet hose clamp and hose
2. Reposition the bypass hose clamps at the water pump and water crossover
   - Bypass hose
   - Water pump bolt and pump. Discard the water pump gaskets.

To install:

3. Install or connect the following:
   - New water pump gaskets.
   - Water pump and bolts. Tighten the water pump bolts 37 ft. lbs. (50 Nm).
   - Bypass hose and clamps
   - Outlet hose and clamp
   - Fan clutch
   - Drive belt

- Surge tank hose
- Heater hose
- Outlet radiator hose to the coolant pump
- Drive belt
- Cooling fan and clutch assembly
- Upper fan shroud

- Inlet radiator hose to the water pump
- Air inlet duct
- Coolant

Heater Core

REMOVAL & INSTALLATION

Avalanche

➡This procedure requires the use of J 43181 Quick Connect Connector Removal Tool or equivalent.

1. Drain the engine cooling system into a clean container for reuse.
2. Remove or disconnect the following:
   - Negative battery cable
   - Air conditioning refrigerant, using the proper equipment
3. Using the J 43181 or equivalent quick connect tool, disconnect the inlet heater hose from the heater core.
   a. Install the tool to the heater core pipe.
   b. Close the tool around the heater core pipe.
   c. Firmly pull the tool into the quick connect end of the heater hose.

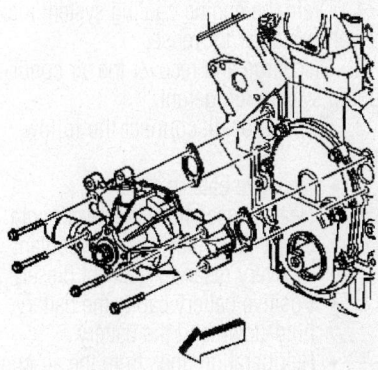

Exploded view of the water pump assembly—8.1L engine

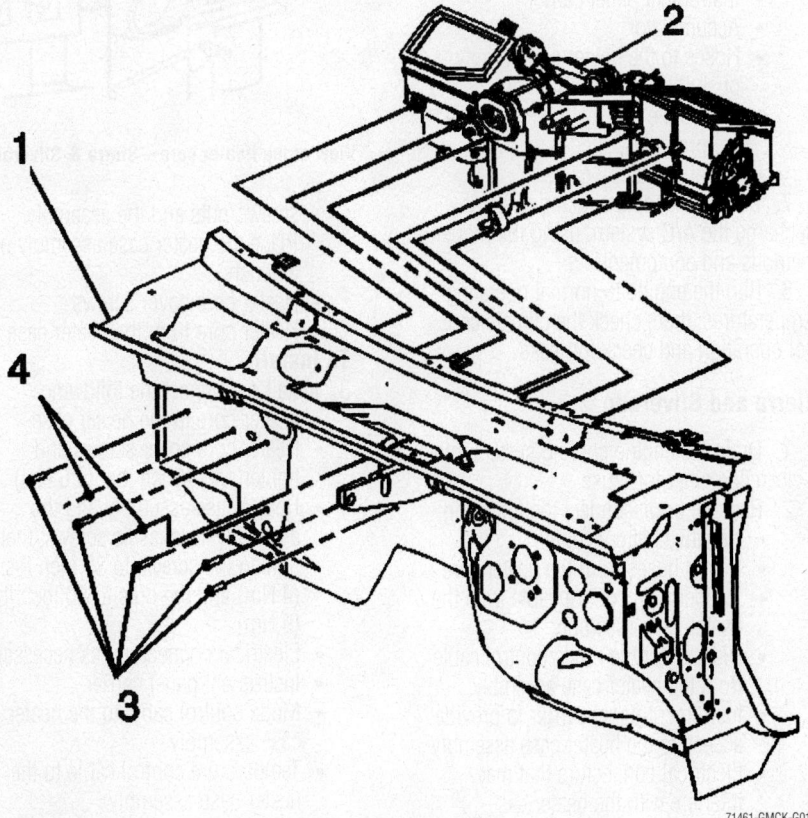

Remove the nuts (4), bolts (3), then remove the HVAC module (2) from the vehicle (1)—Avalanche

d. Firmly grasp the heater hose. Pull the heater hose forward in order to disengage the inlet hose from the heater core.

4. Disconnect the surge tank outlet hose from the heater core using tool J 43181.

5. Remove or disconnect the following:
- Accumulator
- Instrument panel carrier
- HVAC module drain hose
- Electrical harnesses and the ground connections from the HVAC module
- Retaining nuts and bolts and HVAC module
- Screws from the heater core cover
- Cover from the HVAC module
- Heater core cowl gasket from the heater core
- Heater core from the HVAC module assembly

To install:

6. Install or connect the following
- Heater core to the HVAC module
- Heater core cowl gasket
- HVAC module cover and secure with the retaining screws
- HVAC module into position. Torque the nuts to 90 inch lbs. (8 Nm) and the bolts to 35 inch lbs. (4 Nm).
- Electrical connectors to the module
- Instrument panel carrier
- Accumulator
- Hoses to the heater core. Firmly push the hoses on until you hear an audible "click" to be sure they are properly seated.
- Negative battery cable.

7. Refill the engine cooking system. Recharge the A/C system, using the proper methods and equipment.

8. Run the engine to normal operating temperatures; then, check the climate control operation and check for leaks.

Sierra and Silverado

1. Drain the engine cooling system into a clean container for reuse.

2. Remove or disconnect the following:
- Negative battery cable
- Heater hoses from the heater core
- Temperature control cable from the heater case assembly
- Disconnect the mode control cable from the heater case assembly
- Instrument panel carrier to provide access to the heater case assembly
- Electrical connectors that may interfere with the heater case assembly removal
- Heater case assembly-to-chassis

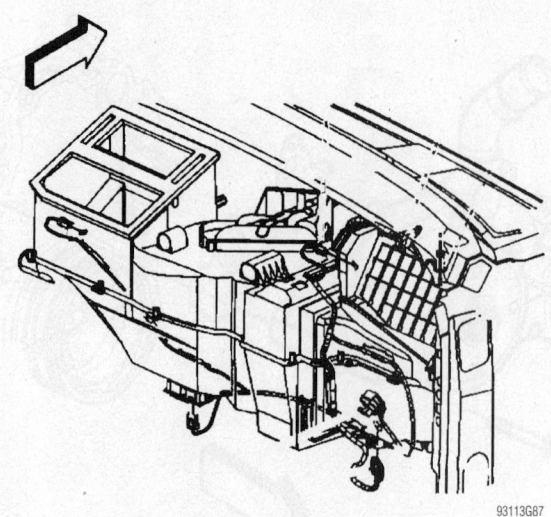

View of the heater case assembly—Sierra & Silverado

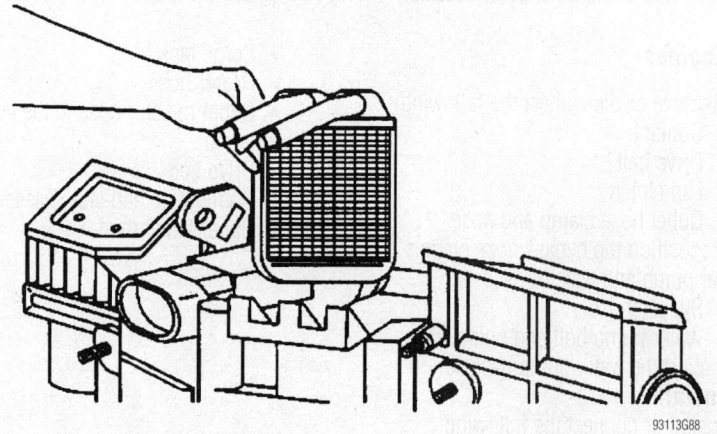

View of the heater core—Sierra & Silverado

screws/nuts and the assembly. Place the heater case assembly on a bench.
- Heater core cover screws
- Heater core from the heater case

To install:
3. Install or connect the following:
- Heater core to the heater case
- Heater core cover screws and tighten to 14 inch lbs. (1.6 Nm)
- Heater case assembly and the assembly-to-chassis screws, then, tighten the screws to 35 inch lbs. (4 Nm) and the nuts to 80 inch lbs. (9 Nm)
- Electrical connectors, as necessary
- Instrument panel carrier
- Mode control cable to the heater case assembly
- Temperature control cable to the heater case assembly
- Heater hoses to the heater core
- Negative battery cable.

4. Refill the engine cooking system.

5. Run the engine to normal operating temperatures; then, check the climate control operation and check for leaks.

Express and Savana

FRONT HEATER

1. Drain the engine cooling system into a clean container for reuse.

2. Discharge and recover the air conditioning system refrigerant.

3. Remove or disconnect the following:
- Negative battery cable
- Heater hoses from the heater core
- Surge tank (diesel) or the coolant recovery reservoir (except diesel).
- Positive battery cable, the battery hold-down and the battery
- Refrigeration lines from the air conditioning accumulator and discard the gaskets

- Air conditioning accumulator
- Lower right kick panel and the knee bolster, from the right side
- Lower outer floor air outlet duct
- Heater case screws

4. Carefully open the heater core access door.

- Heater core-to-heater case retainers and the heater core

To install:

5. Install or connect the following:

- Heater core and the heater core-to-heater case retainers. Carefully, close the heater core access door.
- Heater case screws
- Lower outer floor air outlet duct
- Knee bolster and the lower right kick panel
- Air conditioning accumulator.
- Refrigeration lines to the air conditioning accumulator, using new gaskets
- Battery, the battery hold-down and the positive battery cable
- Surge tank (diesel) or the coolant recovery reservoir (except diesel)
- Heater hoses to the heater core

6. Evacuate and charge the air conditioning system.

7. Refill the engine cooling system.

8. Connect the negative battery cable.

9. Run the engine to normal operating temperatures; then, check the climate control operation and check for leaks.

REAR AUXILIARY HEATER

1. Drain the engine cooling system into a clean container for reuse.

2. Remove or disconnect the following:

- Negative battery cable
- Heater hoses from the rear auxiliary heater core
- Rear interior quarter trim panel from the left rear side
- Blower motor from the rear auxiliary heater case
- Rear auxiliary heater case retainers and the case
- Lower auxiliary heater case retainers and the lower case
- Rear heater core from the case

To install:

3. Install or connect the following:

- Rear auxiliary heater core to the case
- Lower case and the lower auxiliary heater case retainers
- Rear auxiliary heater case retainers and the case
- Blower motor to the rear auxiliary heater case

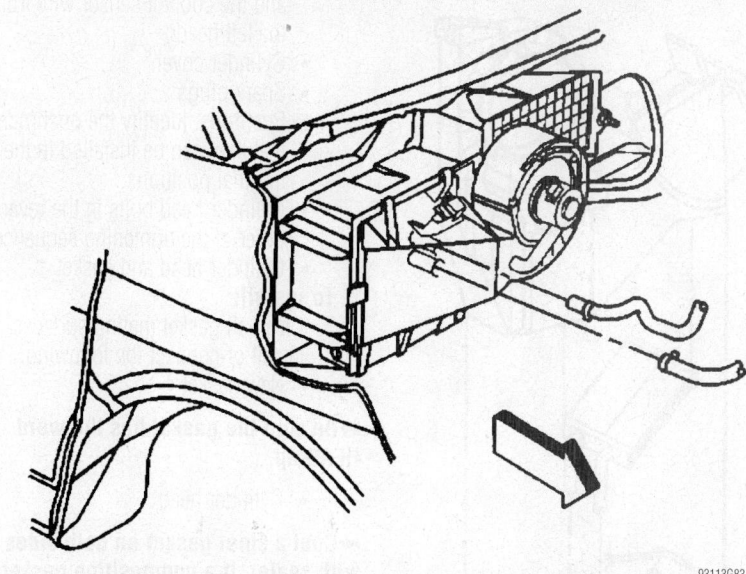

Underhood view of the heater assembly—Express and Savana

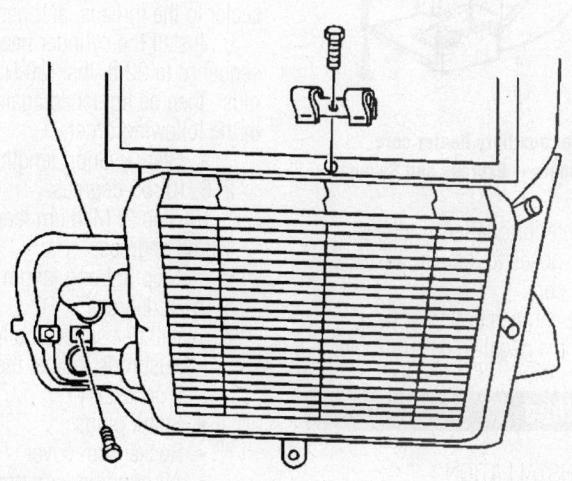

View of the heater case screws—Express and Savana

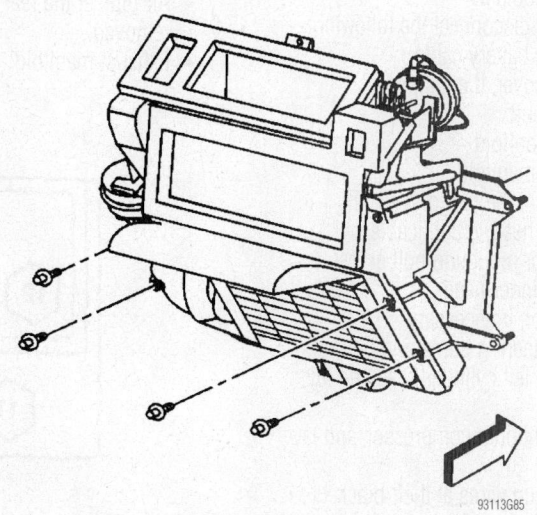

View of the heater core and retainers—Express and Savana

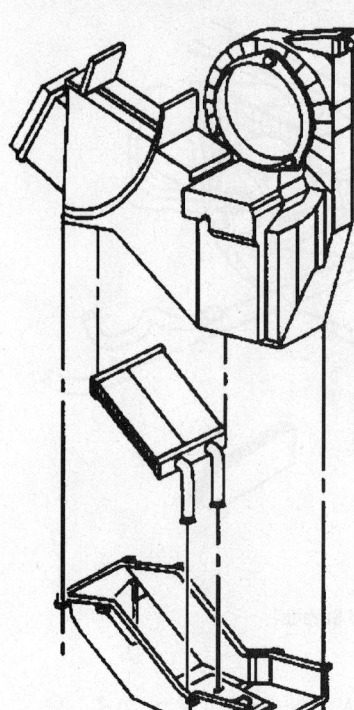

View of the rear auxiliary heater core, case and retainers— Express and Savana

- Rear interior quarter trim panel
- Heater hoses to the rear auxiliary heater core
4. Refill the engine cooling system.
5. Connect the negative battery cable.

Cylinder Head

REMOVAL & INSTALLATION

4.3L Engine—Express and Savana

1. Drain the coolant.
2. Remove or disconnect the following:
   - Negative battery cable
   - Engine cover, if equipped
   - Fan shroud
   - Intake manifold
   - Exhaust manifold
   - Air pipe at the rear of the right cylinder head, if applicable
   - Alternator mounting bolt at the right cylinder head
   - Alternator, if necessary
   - Power steering pump and brackets from the left cylinder head and lay aside
   - Air conditioner compressor, and lay aside
   - Spark plug wires at their brackets
   - Ground strap from the right side

and the coolant sensor wire from the left head
   - Cylinder cover
   - Spark plugs
   - Pushrods. Identify the pushrods so that they can be installed in their original positions.
   - Cylinder head bolts in the reverse order of the tightening sequence
   - Cylinder head and gasket

To install:
3. Clean all gasket mating surfaces.
4. Install or connect the following:
   - New gasket

➡**Be sure the gasket has the word HEAD up.**

   - Cylinder head

➡**Coat a steel gasket on both sides with sealer. If a composition gasket is used, do not use sealer.**

5. Clean the cylinder head bolts, apply sealer to the threads, and hand-tighten.
6. Install the cylinder head bolts in sequence to 22 ft. lbs. (30 Nm) The bolts must, then be tightened again in sequence in the following order:
   a. Step 1: Short length bolt: (11, 7, 3, 2, 6, 10) 55 degrees.
   b. Step 2: Medium length bolt: (12, 13) 65 degrees.
   c. Step 3: Long length bolts: (1, 4, 8, 5, 9) 75 degrees.
7. Install or connect the following
   - Pushrods (adjust the rocker arms, if necessary)
   - Spark plugs
   - Rocker arm cover
   - Air conditioner compressor
   - Power steering pump and brackets
   - Alternator or the alternator mounting bolt at the cylinder head
   - Air pipe at the rear of the head if removed
   - Exhaust manifold

   - Intake manifold
   - Engine cover, if removed
8. Refill the engine with coolant.
9. Connect the negative battery cable.

4.3L Engine—Silverado and Sierra

LEFT SIDE

1. Remove or disconnect the following:
   - Battery negative cable
   - Coolant
   - Accessory drive belt
   - Cooling fan assembly
   - Power steering pump mounting bracket
   - Power steering pump mounting bracket stud from the cylinder head
   - Lower intake manifold
   - Exhaust manifold
   - Spark plug wire harness and the spark plug wire support
   - Valve pushrods
   - Ground strap and ground wire bolt from the rear of the cylinder head
   - Engine Coolant Temperature (ECT) sensor (if applicable)
   - ECT gauge sensor (if applicable)
   - Spark plugs
   - Spark plug wire support
   - Cylinder head bolts
   - Cylinder head and the gasket

➡**Clean all dirt, debris, and coolant from the engine block cylinder head bolt holes. Failure to remove all foreign material may result in damaged threads, improperly tightened fasteners or damage to components.**

2. Clean the cylinder head bolts and the engine block bolt holes.
   To install:
3. Inspect the dowel pins (cylinder head locator) for proper installation.

➡**Do not use any type sealer on the cylinder head gasket (unless specified).**

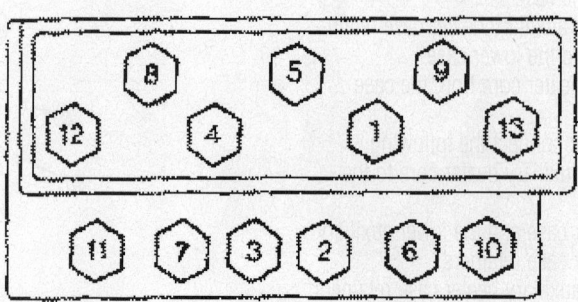

Cylinder head bolt tightening sequence—4.3L engine

4. Install or connect the following:
- NEW cylinder head gasket in position over the dowel pins (cylinder head locator)
- Cylinder head onto the engine block. Guide the cylinder head carefully into place over the dowel pins and the cylinder head gasket.
- Sealant GM P/N 12346004, or equivalent, to the threads of the cylinder head bolts
- Cylinder head bolts finger-tight

5. Tighten the cylinder head bolts in sequence:

   a. First pass: 22 ft. lbs. (30 Nm).

   b. Second pass: Long bolts (1, 4, 5, 8, and 9)—+ 75 degrees.

   c. Second pass: Medium bolts (12 and 13)—+ 65 degrees.

   d. Second pass: Short bolts (2, 3, 6, 7, 10, and 11)—+ 55 degrees.

6. Install or connect the following:
- Spark plug wire support and bolts. Tighten to 106 inch lbs. (12 Nm).
- Spark plugs. Tighten to 11 ft. lbs. (15 Nm), if USED; 22 ft. lbs. (30 Nm), if NEW.

7. If reusing the ECT gauge sensor (if applicable), apply sealant GM P/N 12346004 or equivalent to the threads of the ECT gauge sensor. Install the ECT gauge sensor (if applicable). Tighten the sensor to 15 ft. lbs. (20 Nm).

8. Install or connect the following:
- Ground strap and the ground wire bolt. Tighten the bolt to 12 ft. lbs. (16 Nm).
- Valve pushrods
- Lower intake manifold
- Exhaust manifold
- Stud for the power steering pump mounting bracket to the cylinder head. Tighten the power steering pump mounting bracket stud to 15 ft. lbs. (20 Nm).
- Power steering pump mounting bracket
- Engine cooling fan assembly
- Coolant
- Battery negative cable

RIGHT SIDE

1. Remove or disconnect the following:
- Battery negative cable
- Coolant
- Engine cooling fan assembly
- Alternator mounting bracket
- Alternator mounting bracket stud from the cylinder head
- Lower intake manifold
- Exhaust manifold

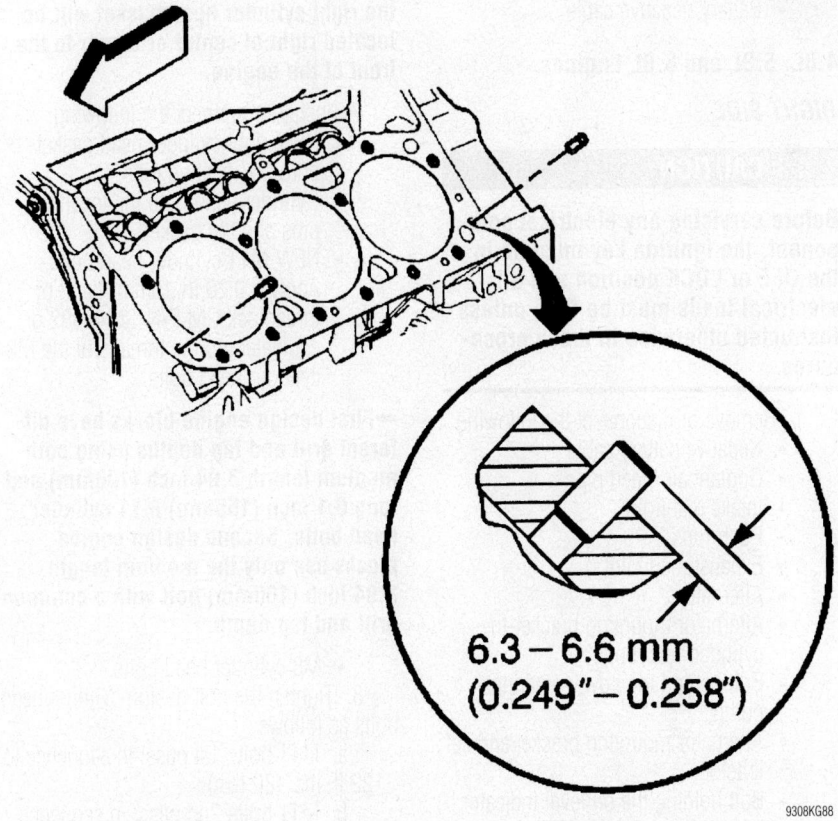

6.3 – 6.6 mm
(0.249" – 0.258")

9308KG88

Dowel pin installation—4.3L

- Spark plug wire harness and spark plug wire support
- Valve pushrods
- Cylinder head and the gasket

2. Clean the engine block and the cylinder head sealing surfaces.

To install:

3. Inspect the dowel pins (cylinder head locator) for proper installation.

➡**Do not use any type sealer on the cylinder head gasket (unless specified).**

4. Install or connect the following:
- NEW cylinder head gasket in position over the dowel pins (cylinder head locator)
- Cylinder head onto the engine block. Guide the cylinder head carefully into place over the dowel pins and the cylinder head gasket.
- Sealant GM P/N 12346004 or equivalent to the threads of the cylinder head bolts
- Cylinder head bolts finger-tight

5. Tighten the cylinder head bolts in sequence:

   a. First pass: 22 ft. lbs. (30 Nm).

   b. Second pass: Long bolts (1, 4, 5, 8, and 9)—+ 75 degrees.

   c. Second pass: Medium bolts (12 and 13)—+ 65 degrees.

   d. Second pass: Short bolts (2, 3, 6, 7, 10, and 11)—+ 55 degrees.

6. Install or connect the following:
- Spark plug wire support and bolts. Tighten only the rear support bolt to 106 inch lbs. (12 Nm).

➡**The front spark plug wire support bolt is used to fasten the oil level indicator tube, and will be installed within the oil level indicator tube installation procedure.**

- Front spark plug wire support bolt
- Spark plugs. Tighten to 11 ft. lbs. (15 Nm), if USED; 22 ft. lbs. (30 Nm), if NEW.
- Valve pushrods
- Lower intake manifold
- Spark plug wire harness and wire support. Tighten to 106 inch lbs. (12 Nm).
- Exhaust manifold
- Stud for the alternator mounting bracket. Tighten the alternator mounting bracket stud to 15 ft. lbs. (20 Nm).
- Alternator mounting bracket
- Engine cooling fan assembly

- Coolant
- Battery negative cable

4.8L, 5.3L and 6.0L Engines

RIGHT SIDE

✳✳ CAUTION

Before servicing any electrical component, the ignition key must be in the OFF or LOCK position and all electrical loads must be OFF, unless instructed otherwise in these procedures.

1. Remove or disconnect the following:
   - Negative battery cable
   - Coolant air bleed pipe
   - Intake manifold
   - Push rods
   - Exhaust manifold(s)
   - Alternator
   - Alternator mounting bracket-to-cylinder head bolts
   - Bolt behind the power steering pump
   - Alternator mounting bracket and set it aside
   - Bolt holding the oil level indicator tube to the right side cylinder head
   - Oil level indicator tube
   - Cylinder head(s) from the engine
   - Spark plugs

➡ **The M11 cylinder head bolts are NOT reusable. Install NEW M11 cylinder head bolts during reassembly.**

- Cylinder head bolts

➡ **After removal, place the cylinder head on two wood blocks to prevent damage.**

2. Remove the gasket. Discard the gasket. Discard the M11 cylinder head bolts.

To install:

➡ **Do not use any type sealant on the cylinder head gasket (unless specified). The cylinder head gaskets must be installed in the proper direction and position.**

3. Clean the engine block cylinder head bolt holes (if required). Thread repair tool J 42385-107 may be used to clean the threads of old threadlocking material.

4. Spray cleaner GM P/N 12346139, P/N 12377981, or equivalent into the hole.

5. Clean the cylinder head bolt holes with compressed air.

6. Check the cylinder head locating pins for proper installation.

➡ **When properly installed, the tab on the right cylinder head gasket will be located right of center or closer to the front of the engine.**

7. Install or connect the following:
   - NEW right cylinder head gasket onto the locating pins
   - Cylinder head onto the locating pins and the gasket
   - NEW M11 cylinder head bolts. Apply a 0.20 in. (5mm) band of threadlock GM P/N 12345382 or equivalent to the threads of the M8 cylinder head bolts.

➡ **First design engine blocks have different drill and tap depths using both medium length 3.94 inch (100mm) and long 6.1 inch (155mm) M11 cylinder head bolts. Second design engine blocks use only the medium length 3.94 inch (100mm) bolt with a common drill and tap depth.**

- M8 cylinder head bolts.

8. Tighten the first design cylinder head bolts as follows:
   a. M11 bolts 1st pass: in sequence to 22 ft. lbs. (30 Nm).
   b. M11 bolts 2nd pass: in sequence + 90 degrees.

   c. M11 bolts (1,2,3,4,5,6,7,8): + 90 degrees.
   d. M11 bolts (9 and 10): + 50 degrees.
   e. M8 cylinder head bolts (11,12,13,14,15) to 22 ft. lbs. (30 Nm). Begin with the center bolt (11) and alternating side-to-side, work outward tightening all of the bolts.

9. Tighten the second design cylinder head bolts as follows:
   a. M11 bolts (1-10) 1st pass: in sequence to 22 ft. lbs. (30 Nm).
   b. M11 bolts (1-10) 2nd pass: in sequence + 90 degrees.
   c. M11 bolts (1-10): + 70 degrees.
   d. M8 cylinder head bolts (11,12,13,14,15) to 22 ft. lbs. (30 Nm). Begin with the center bolt (11) and alternating side-to-side, work outward tightening all of the bolts.

➡ **The cylinder head gasket displacement can be verified by markings visible on the underside of the right gasket locating tab. Some 4.8/5.3L head gaskets may have 53 stamped onto the locating tab. Some 6.0L head gaskets may have 60 stamped onto the locating tab.**

10. Install or connect the following:

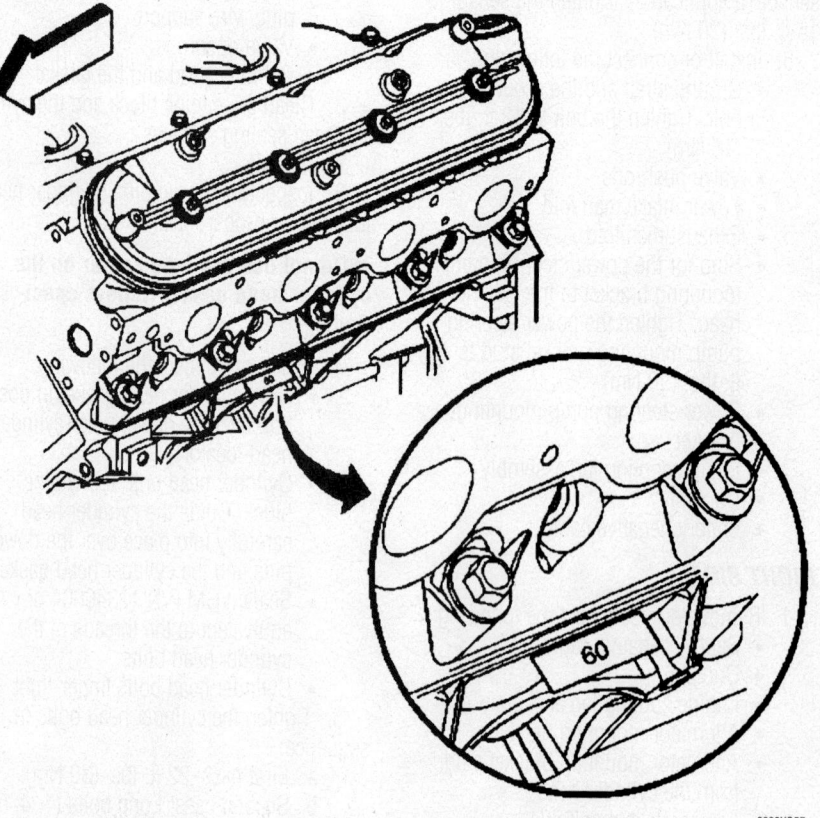

Locating tab—4.8L, 5.3L and 6.0L engines

9308KG57

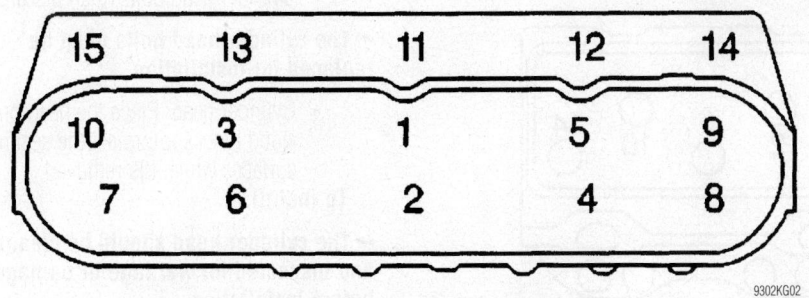

Cylinder head bolt tightening sequence—4.8L, 5.3L and 6.0L engines

- Alternator
- Exhaust manifold(s)
- Pushrods
- Intake manifold
- Negative battery cable

LEFT SIDE

⚡ CAUTION

Before servicing any electrical component, the ignition key must be in the OFF or LOCK position and all electrical loads must be OFF, unless instructed otherwise in these procedures.

1. Remove or disconnect the following:
   - Negative battery cable
   - Intake manifold
   - Push rods
   - Exhaust manifold(s)
   - Alternator
   - Alternator mounting bracket-to-cylinder head bolts
   - Bolt behind the power steering pump
   - Alternator mounting bracket and set it aside
   - Oil level indicator tube-to-cylinder head bolt
   - Oil level indicator tube
   - Cylinder head from the engine
   - Spark plugs

➡️**The M11 cylinder head bolts are NOT reusable. Install NEW M11 cylinder head bolts during assembly.**

2. Remove the cylinder head bolts.

➡️**After removal, place the cylinder head on two wood blocks to prevent damage.**

3. Remove the gasket. Discard the gasket. Discard the M11 cylinder head bolts.

To install:

➡️**Do not use any type sealant on the cylinder head gasket (unless specified). The cylinder head gaskets must** be installed in the proper direction and position.

4. Clean the engine block cylinder head bolt holes (if required). Thread repair tool J 42385-107 may be used to clean the threads of old threadlocking material.

5. Spray cleaner GM P/N 12346139, P/N 12377981, or equivalent into the hole.

6. Clean the cylinder head bolt holes with compressed air.

7. Check the cylinder head locating pins for proper installation.

➡️**When properly installed, the tab on the left cylinder head gasket will be located left of center or closer to the front of the engine.**

8. Install or connect the following:
   - NEW left cylinder head gasket onto the locating pins
   - Cylinder head onto the locating pins and the gasket
   - NEW M11 cylinder head bolts.

9. Apply a 0.20 in. (5mm) band of threadlock GM P/N 12345382 or equivalent to the threads of the M8 cylinder head bolts.
   - M8 cylinder head bolts

➡️**First design engine blocks have different drill and tap depths using both medium length 3.94 inch (100mm) and long 6.1 inch (155mm) M11 cylinder head bolts. Second design engine blocks use only the medium length 3.94 inch (100mm) bolt with a common drill and tap depth.**

   - M8 cylinder head bolts.

10. Tighten the first design cylinder head bolts as follows:
    a. M11 bolts 1st pass: in sequence to 22 ft. lbs. (30 Nm).
    b. M11 bolts 2nd pass: in sequence + 90 degrees.
    c. M11 bolts (1,2,3,4,5,6,7,8): + 90 degrees.
    d. M11 bolts (9 and 10): + 50 degrees.

e. M8 cylinder head bolts (11,12,13,14,15) to 22 ft. lbs. (30 Nm). Begin with the center bolt (11) and alternating side-to-side, work outward tightening all of the bolts.

11. Tighten the second design cylinder head bolts as follows:
    a. M11 bolts (1-10) 1st pass: in sequence to 22 ft. lbs. (30 Nm).
    b. M11 bolts (1-10) 2nd pass: in sequence + 90 degrees.
    c. M11 bolts (1-10): + 70 degrees.
    d. M8 cylinder head bolts (11,12,13,14,15) to 22 ft. lbs. (30 Nm). Begin with the center bolt (11) and alternating side-to-side, work outward tightening all of the bolts.

➡️**The cylinder head gasket displacement can be verified by markings visible on the top side of the left gasket locating tab. Some 4.8/5.3L head gaskets may have 53 stamped onto the locating tab. Some 6.0L head gaskets may have 60 stamped onto the locating tab.**

12. Install or connect the following:
    - Alternator mounting bracket. Tighten the four bolts to 37 ft. lbs. (50 Nm).
    - Bolt at the rear of the power steering pump and tighten to 37 ft. lbs. (50 Nm).
    - Exhaust manifold(s)
    - Pushrods
    - Intake manifold
    - Negative battery cable

5.0L and 5.7L Engines

1. Drain the coolant.
2. Remove or disconnect the following:
   - Negative battery cable
   - Engine cover
   - Coolant recovery reservoir, if applicable
   - Intake manifold
   - Exhaust manifolds and position them out of the way
   - Ground strap at the rear of the right AIR pipe, if equipped
3. For vans with A/C, remove the A/C compressor and the forward mounting bracket and lay the compressor aside. Do not disconnect any of the refrigerant lines.
   - Exhaust Gas Recirculation (EGR) inlet tube
4. On the right side cylinder head, disconnect the fuel pipe, spark plug wires and wiring harness bracket.
   - Nut and stud attaching the main accessory bracket to the cylinder head

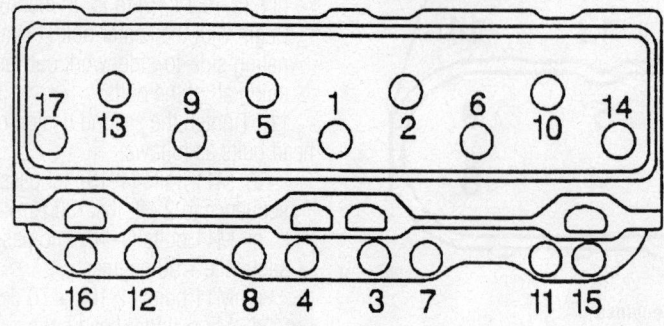

Cylinder head bolt tightening sequence—5.0L and 5.7L engines

➡ **You may have to loosen the remaining bolts and studs in order to remove the head.**

- Coolant sensor wire
- Spark plug wire bracket
- Cylinder head covers
- Spark plugs
- Pushrods. Identify the pushrods so that they can be installed in their original positions.
- Cylinder head bolts in the reverse order of the tightening sequence
- Heads

To install:

5. Inspect the cylinder head and block mating surfaces. Clean all old gasket material.

6. Install or connect the following:
- Cylinder heads using new gaskets. Install the gaskets with the word **HEAD** up.

➡ **Coat a steel gasket on both sides with sealer. If a composition gasket is used, do not use sealer.**

7. Clean the bolts, apply sealer to the threads, and hand-tighten.

8. Install the cylinder head bolts in sequence to 22 ft. lbs. (30 Nm). The bolts must be tightened once, then be tightened again in sequence in the following order:

a. Step 1: Short length bolt: (3, 4, 7, 8, 11, 12, 15, 16), plus 55 degrees.

b. Step 2: Medium length bolt: (14, 17), plus 65 degrees.

c. Step 3: Long length bolts: (1, 2, 5, 6, 9, 10, 13), plus 75 degrees.

9. Install or connect the following:
- Pushrods in their original positions
- Cylinder head covers
- Spark plugs
- Coolant sensor wire
- Spark plug wire bracket
- Main accessory bracket to the cylinder head
- EGR vent tube

- Fuel pipe
- Spark plug wires
- Wiring harness bracket
- A/C compressor and forward mounting bracket
- Ground strap to the rear of the right AIR pipe
- Exhaust manifolds
- Intake manifold
- Coolant recovery reservoir, if removed
- Engine cover
- Negative battery cable

10. Refill the engine with coolant.

8.1L Engine

LEFT SIDE

1. Drain the cooling system.
2. Remove or disconnect the following:
- Negative battery cable
- Water crossover
- Intake manifold
- Valve cover
- Rocker arms and pushrods, keeping them in order for installation
- Engine harness ground bolts

3. Reposition the engine harness grounds and ground straps from the cylinder head.
- Exhaust manifold

- Cylinder head bolts, then discard

➡ **The cylinder head bolts must be replaced for installation.**

- Cylinder head. Place the head on 2 wood blocks to protect the sealing surfaces while it is removed.

To install:

➡ **The cylinder head should be cleaned and inspected for warpage or damage before installation.**

4. Thoroughly clean the mating surfaces of the head and block. Clean the bolt holes thoroughly.

➡ **If a composition gasket is used, do not use sealer.**

5. Align the cylinder head gasket locating marks to face up. Make sure that the gasket tabs are located of the No. 1 and 2 cylinder for proper installation.

6. Install or connect the following:
- New cylinder head gasket
- Cylinder head
- Sealer to the threads of new cylinder head bolts, if not pre-applied

➡ **The long bolts are used in locations 1, 2, 3, 6, 7, 8, 9, 10, 11, 14, 16, and 17. The medium length bolts are used in locations 15 and 18. The short bolts are used in locations 4, 5, 12, and 13.**

7. Tighten the head bolts, in sequence, in 4 stages, as follows:

a. Step 1: 22 ft. lbs. (30 Nm).
b. Step 2: 22 ft. lbs. (30 Nm)
c. Step 3: Additional 120 degrees using a torque angle meter.
d. Step 4: Torque bolt numbers. 1, 2, 3, 6, 7, 8, 9, 10, 11, 14, 16 and 17 an additional 60 degrees.
e. Tighten bolts 15 and 18 an additional 45 degrees, and bolt numbers 4, 5, 12 and 13 an additional 30 degrees.

8. Install or connect the following:
- Exhaust manifold

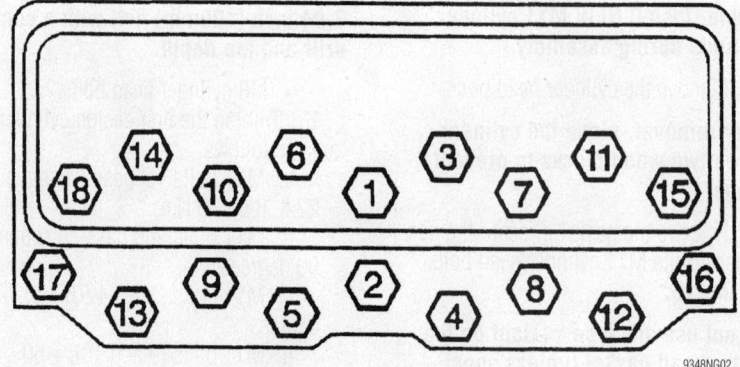

Cylinder head bolt tightening sequence—8.1L engine

- Water crossover
- Engine harness grounds and ground strap
- Rocker arms and pushrods
- Valve cover
- Intake manifold

9. Connect the battery cable and refill the cooling system.

RIGHT SIDE

1. Drain the cooling system.
2. Remove or disconnect the following:
   - Negative battery cable
   - Intake manifold
   - Valve cover
   - Rocker arms and pushrods, keeping them in order for installation
   - Engine Coolant Temperature (ECT) sensor clip from the bracket
   - ECT sensor
   - ECT sensor bracket bolt and bracket
   - Heater inlet and outlet hoses from the hose bracket
   - Water crossover
   - Exhaust manifold
   - Cylinder head bolts, then discard

➡**The cylinder head bolts must be replaced for installation.**

   - Cylinder head. Place the head on 2 wood blocks to protect the sealing surfaces while it is removed.

To install:

➡**The cylinder head should be cleaned and inspected for warpage or damage before installation.**

3. Thoroughly clean the mating surfaces of the head and block. Clean the bolt holes thoroughly.

➡**If a composition gasket is used, do not use sealer.**

4. Align the cylinder head gasket locating marks to face up. Make sure that the gasket tabs are located of the no. 1 and 2 cylinder for proper installation.
5. Install or connect the following:
   - New cylinder head gasket
   - Cylinder head
   - Sealer to the threads of new cylinder head bolts, if not pre-applied

➡**The long bolts are used in locations 1, 2, 3, 6, 7, 8, 9, 10, 11, 14, 16, and 17. The medium length bolts are used in locations 15 and 18. The short bolts are used in locations 4, 5, 12, and 13.**

6. Tighten the head bolts, in sequence, in 4 stages, as follows:
   a. Step 1: 22 ft. lbs. (30 Nm).

   b. Step 2: 22 ft. lbs. (30 Nm)
   c. Step 3: Additional 120 degrees using a torque angle meter.
   d. Step 4: Torque bolt numbers. 1, 2, 3, 6, 7, 8, 9, 10, 11, 14, 16 and 17 an additional 60 degrees.
   e. Tighten bolts 15 and 18 an additional 45 degrees, and bolt numbers 4, 5, 12 and 13 an additional 30 degrees.
7. Install or connect the following:
   - Exhaust manifold
   - Water crossover
   - Heater hose bracket and bolts. Tighten the bolts to 37 ft. lbs. (50 Nm).
   - ECT sensor bracket and bolt. Tighten to 37 ft. lbs. (50 Nm).
   - ECT sensor
   - ECT sensor clip
   - Rocker arms and pushrods
   - Valve cover
   - Intake manifold
8. Connect the battery cable and refill the cooling system.

Rocker Arms

REMOVAL & INSTALLATION

4.3L, 5.0L and 5.7L Engines

1. Remove or disconnect the following:
   - Engine cover

   - Cylinder head cover
   - Rocker arm nut. If you are only replacing the pushrod, back the nut off until you can swing the rocker out of the way.
   - Rocker arms and balls as a unit

➡**Always remove each set of rocker arms (1 set per cylinder) as a unit.**

   - Pushrods and pushrod guides

To install:

2. Install or connect the following:
   - Pushrods and their guides. Be sure that they seat properly in each lifter.
3. Position a set of rocker arms (for 1 cylinder) in the proper location.

➡**Install the rocker arms for each cylinder only when the lifters are off the cam lobe and both valves are closed.**

4. Coat the replacement rocker arm with Molykote® or its equivalent, and the rocker arm and pivot with SAE 90 gear oil, and install the pivots.
   - Nuts and finger tighten
5. On 4.3L engines, rotate the crankshaft balancer to position the crankshaft balancer alignment mark (1) 57-63 degrees clockwise or counterclockwise from the engine front cover alignment tab (2).
6. Tighten the rocker arm nuts to 22 ft. lbs. (30 Nm).

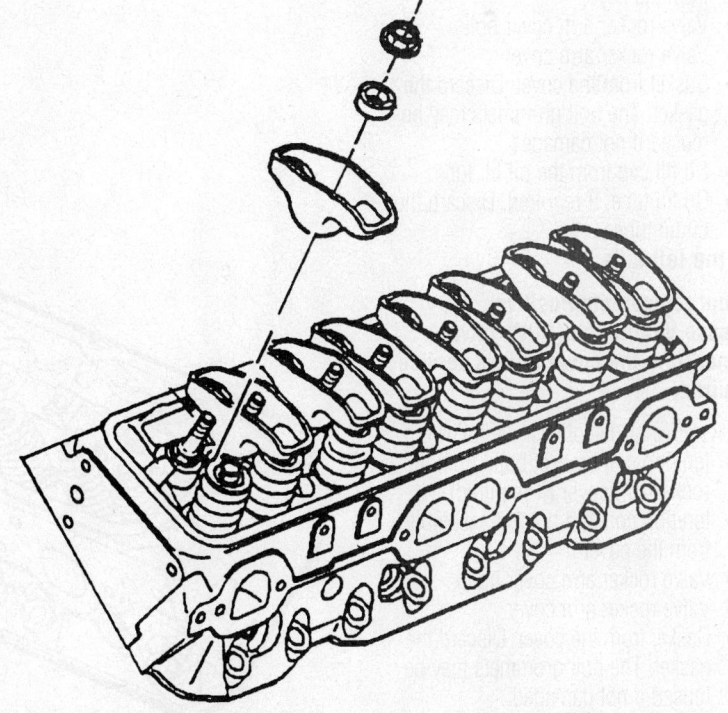

7924KG10

Exploded view to the rocker arm and related components—4.3L, 5.0L and 5.7L engines

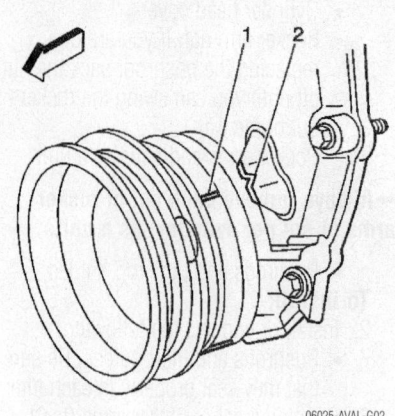

Positioning the crankshaft balancer alignment marks—4.3L engine

06025-AVAL-G02

7. Install the rocker arm cover.
8. Install the engine cover.

4.8L, 5.3L and 6.0L Engines

➡Do not remove the ignition coils from the valve rocker arm cover unless required. Do not remove the oil fill tube from the cover unless serice is required. If the oil fill tube has been removed from the cover, install a NEW tube during assembly.

On the right side:
1. Remove or disconnect the following:
   • Ignition coil bracket bolts from the rocker arm cover, if required
   • Ignition coil and bracket assembly from the cover
   • Valve rocker arm cover bolts
   • Valve rocker arm cover
   • Gasket from the cover. Discard the gasket. The bolt grommets may be reused if not damaged.
   • Oil fill cap from the oil fill tube
   • Oil fill tube, if required. Discard the oil fill tube.

On the left side:

➡Do not remove the Positive Crankcase Ventilation (PCV) valve grommet from the cover unless service is required.

2. Remove or disconnect the following:
   • Ignition coil bracket bolts from the rocker arm cover (if required)
   • Ignition coil and bracket assembly from the cover
   • Valve rocker arm cover bolts
   • Valve rocker arm cover
   • Gasket from the cover. Discard the gasket. The bolt grommets may be reused if not damaged.
   • Valve rocker arm bolts
   • Valve rocker arms

   • Valve rocker arm pivot support
   • Pushrods

To install:

➡Valve lash is built in. No valve adjustment is required.

3. Lubricate the valve rocker arms and pushrods with clean engine oil.
4. Lubricate the flange of the valve rocker arm bolts with clean engine oil.
5. Lubricate the flange or washer surface of the bolt that will contact the valve rocker arm.
6. Install or connect the following:
   • Valve rocker arm pivot support

➡Make sure that the pushrods seat properly to the valve lifter sockets.

   • Pushrods

➡Make sure that the pushrods seat properly to the ends of the rocker arms.

   • Rocker arms and bolts. DO NOT tighten the rocker arm bolts at this time

7. Rotate the crankshaft until number one piston is at top dead center of compression stroke. In this position, cylinder number one rocker arms will be off lobe lift, and the crankshaft sprocket key will be at the 1:30 position. If viewing from the rear of the engine, the additional crankshaft pilot hole (non-threaded) will be in the 10:30 position.

The engine firing order is 1, 8, 7, 2, 6, 5, 4, 3. Cylinders 1, 3, 5 and 7 are left bank. Cylinders 2, 4, 6, and 8 are right bank.

8. With the engine in the number one firing position, tighten the following valve rocker arm bolts:
   a. Tighten exhaust valve rocker arm bolts 1, 2, 7, and 8 to 22 ft. lbs. (30 Nm).
   b. Tighten intake valve rocker arm bolts 1, 3, 4, and 5 to 22 ft. lbs. (30 Nm).
9. Rotate the crankshaft 360 degrees. Tighten the following valve rocker arm bolts:
   a. Tighten exhaust valve rocker arm bolts 3, 4, 5, and 6 to 22 ft. lbs. (30 Nm).
   b. Tighten intake valve rocker arm bolts 2, 6, 7, and 8 to 22 ft. lbs. (30 Nm).

On the right side:

➡The valve rocker arm cover bolt grommets may be reused. If the oil fill tube has been removed from the valve rocker arm cover, install a NEW oil fill tube during assembly.

10. Lubricate the O-ring seal of the NEW oil fill tube with clean engine oil.
11. Install or connect the following:
   • NEW oil fill tube into the rocker arm cover and rotate the tube clockwise until locked in the proper position
   • Oil fill cap into the tube and rotate clockwise until locked in the proper position

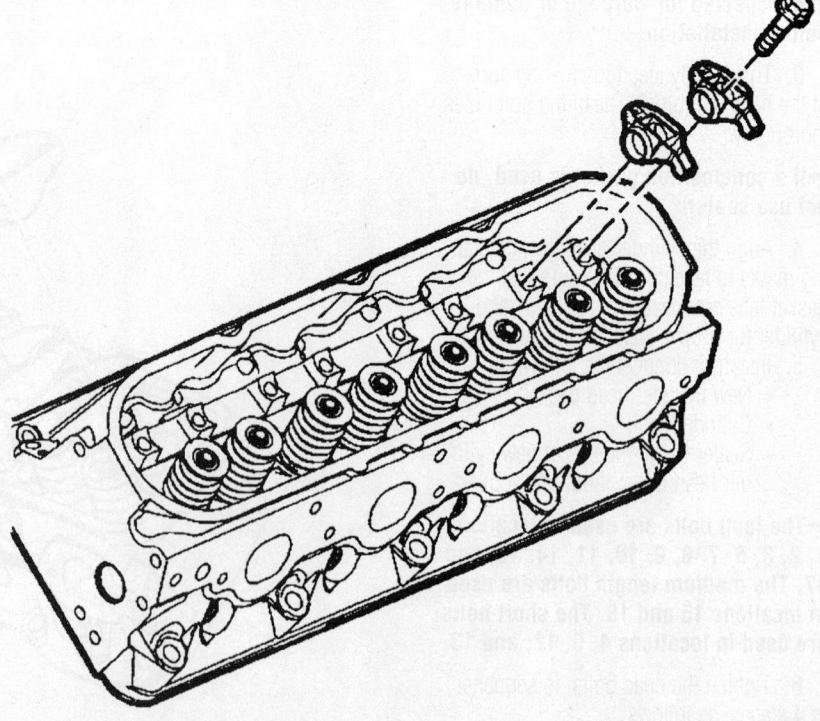

Rocker arm removal—4.8L, 5.3L and 6.0L engines

9308KG68

- NEW cover gasket into the valve rocker arm cover
- Valve rocker arm cover onto the cylinder head

12. Install the cover bolts with grommets. Tighten the valve rocker arm cover bolts to 106 inch lbs. (12 Nm).

13. Apply threadlock GM P/N 12345382 or equivalent to the threads of the bracket bolts. Install the ignition coil and bracket assembly and bolts. Tighten the ignition coil and bracket assembly studs to 106 inch lbs. (12 Nm).

On the left side:

➡**DO NOT reuse the valve rocker arm cover gasket. The valve rocker arm cover bolt grommets may be reused. If the vapor vent grommet has been removed from the valve rocker arm cover, install a NEW vapor vent gourmet during assembly.**

14. Install or connect the following:
- NEW cover gasket (1) into the valve rocker arm cover
- Valve rocker arm cover onto the cylinder head

15. Install the cover bolts with grommets. Tighten the valve rocker arm cover bolts to 106 inch lbs. (12 Nm).

16. Apply threadlock GM P/N 12345382 or equivalent to the threads of the bracket bolts. Install the ignition coils and bracket assembly and bolts. Tighten the ignition coil and bracket assembly bolts to 106 inch lbs. (12 Nm).

8.1L Engine

➡**Always make sure to keep all removed valve train components in order for reassembly. They must be installed in the same position from which they were removed.**

1. Remove or disconnect the following:
- Valve (rocker arm) cover
- Rocker arm nuts, balls and rocker arms

➡**The intake pushrods are shorter than the exhaust pushrods.**

- Pushrods
- Rocker arm guides and pushrod guides

2. Clean and inspect all components for damage.

To install:

3. Apply a suitable sealer to the rocker arm stud-to-cylinder head threads.

4. Install or connect the following:
- Pushrod guides and rocker arm studs. Tighten to 37 ft. lbs. (50 Nm).

- Pushrods

5. Coat the rocker arm and ball bearing surfaces with a suitable prelube.
- Rocker arms, balls and nuts. Tighten the nuts slowly to 18 ft. lbs. (25 Nm) on 2002–04 models, or 26 ft. lbs. (35 Nm) on 2005–06 engines, while guiding the tips of the rocker arms over the tips of the valves.
- Valve (rocker arm) cover

Intake Manifold

REMOVAL & INSTALLATION

4.3L Engine

1. Relieve the fuel system pressure.
2. Remove the fuel pipes/hoses.
3. Disconnect the cruise control cable from the throttle lever, if equipped.
4. Remove the cruise control cable from the accelerator control cable bracket, if equipped.
5. Remove the accelerator cable from the throttle body lever.
6. Remove the accelerator cable from the accelerator control cable bracket.
7. Disconnect the A/C compressor clutch and pressure switch connectors.
8. Remove the engine wiring harness clip from the accelerator control cable bracket.
9. Remove the accelerator control cable bracket with the cables attached, from the throttle body.
10. Reposition and secure the bracket and cables out of the way.
11. Disconnect the throttle position sensor, idle air control motor and control port injector connectors.
12. Remove the engine wiring harness clip bolt.
13. Disconnect the EVAP purge solenoid and Manifold absolute Pressure ((MAP) sensor connectors.

14. Disconnect the EVAP canister tube (1) from the purge solenoid valve.
15. Remove the engine wiring harness bracket nuts.
16. Remove the engine wiring harness bracket.
17. Remove the engine wiring harness ground nut and ground wire from the rear of the right cylinder head.
18. Remove the engine wiring harness rear bracket nut at the EVAP canister purge solenoid valve.
19. Remove the stud holding the engine wiring harness bracket.
20. Reposition the engine wiring harness with the bracket aside.
21. Remove the PCV valve hose from the valve and rocker cover.
22. Disconnect the power brake booster vacuum hose from the vacuum fitting.
23. Remove the accelerator cable bracket.
24. Remove the intake manifold upper studs.
25. Remove the front two throttle body studs.
26. Remove the upper intake manifold.
27. Remove the distributor housing and rotor, mark for reassembly.
28. Remove the radiator inlet hose from the thermostat housing.
29. Remove the drive belt.
30. Loosen the power steering (P/S) pump rear bracket nut.
31. Remove the P/S pump rear bracket front nut.
32. Remove the bolts and the nut for the P/S pump bracket.
33. Leave the A/C compressor, if equipped, and the P/S pump on the bracket.
34. Slide the P/S pump bracket forward to access the front intake manifold bolt.
35. Remove the lower intake manifold.
36. Remove and discard the intake manifold gaskets.
37. Clean and inspect the intake manifold.

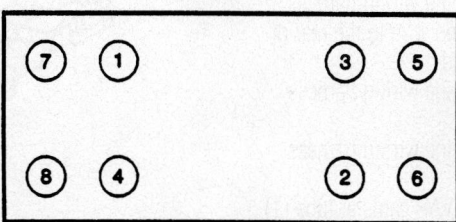

INTAKE SEQUENCE

7924KG14

Lower intake manifold bolt tightening sequence—4.3L engines

To install:

38. Clean all gasket mating surfaces thoroughly.

39. Position the new gaskets on the cylinder heads with the port blocking plates at the rear and the words **THIS SIDE UP** facing up.

40. Apply a ³⁄₁₆ inch (5mm) bead of RTV to the front and rear sealing surfaces on the engine block. Extend the bead ½ inch (13mm) up each cylinder head to retain the gasket.

41. Carefully position the lower intake manifold onto the engine.

42. Apply GM 1052080 or equivalent sealer to the lower intake manifold bolts

43. Torque the bolts using 3 steps in the sequence shown:
   a. Step 1: 24 inch lbs. (3 Nm).
   b. Step 2: 108 inch lbs. (12 Nm).
   c. Step 3: 11 ft. lbs. (15 Nm).

44. Slide the P/S pump bracket rearward.

45. Install the bolts and the nut for the P/S pump bracket.

46. Install the P/S pump rear bracket front nut.

47. Tighten the P/S pump rear bracket nut.

48. Install the drive belt.

49. Install the upper radiator and bypass hose to the thermostat housing.

50. Install the distributor housing and rotor.

51. Position the upper intake manifold gasket on the lower manifold.

52. Install the upper intake manifold. Torque the bolts and studs to 88 inch lbs. (10 Nm).

53. Install the accelerator cable bracket.

54. Disconnect the power brake booster vacuum hose from the vacuum fitting.

55. Install the PCV valve hose from the valve and rocker cover.

56. Reposition the engine wiring harness with the bracket aside.

57. Install the stud holding the engine wiring harness bracket.

58. Install the engine wiring harness rear bracket nut at the EVAP canister purge solenoid valve.

59. Install the engine wiring harness ground nut and ground wire to the rear of the right cylinder head.

60. Install the engine wiring harness bracket.

61. Install the engine wiring harness bracket nuts.

62. Connect the EVAP canister tube (1) from the purge solenoid valve.

63. Connect the EVAP purge solenoid and Manifold absolute Pressure ((MAP) sensor connectors.

64. Install the engine wiring harness clip bolt.

65. Connect the throttle position sensor, idle air control motor and control port injector connectors.

66. Install the accelerator control cable bracket with the cables attached, to the throttle body.

67. Install the engine wiring harness clip to the accelerator control cable bracket.

68. Connect the A/C compressor clutch and pressure switch connectors.

69. Install the accelerator cable to the accelerator control cable bracket.

70. Install the accelerator cable to the throttle body lever.

71. Install the cruise control cable to the accelerator control cable bracket, if equipped.

72. Connect the cruise control cable to the throttle lever, if equipped.

73. Connect the negative battery cable.

74. Refill and bleed the cooling system.

75. Pressurize the fuel system and check for leaks.

4.8L, 5.3L and 6.0L Engines

➡The intake manifold, throttle body, fuel injection rail, and fuel injectors may be removed as an assembly. If not servicing the individual components, remove the manifold as a complete assembly.

1. Remove or disconnect the following:
   - Positive Crankcase Ventilation (PCV) hose and valve
   - Manifold Absolute Pressure (MAP) sensor, if required
   - Engine coolant air bleed clamp and hose from the throttle body
   - Knock sensor connector.
   - Accelerator control cable bracket and bolts, if required
   - Fuel rail with injectors
   - EVAP solenoid, bolt, and isolator
   - Intake manifold bolts
   - Intake manifold with gaskets
   - Intake manifold-to-cylinder head gaskets from the manifold. Discard the intake manifold gaskets.
   - Throttle body and gasket

2. Clean the intake manifold in solvent.

3. Dry the intake manifold with compressed air.

4. Inspect the throttle body and wire harness studs and stud inserts for looseness or damaged threads.

5. Inspect the fuel rail bolt inserts for looseness or damaged threads.

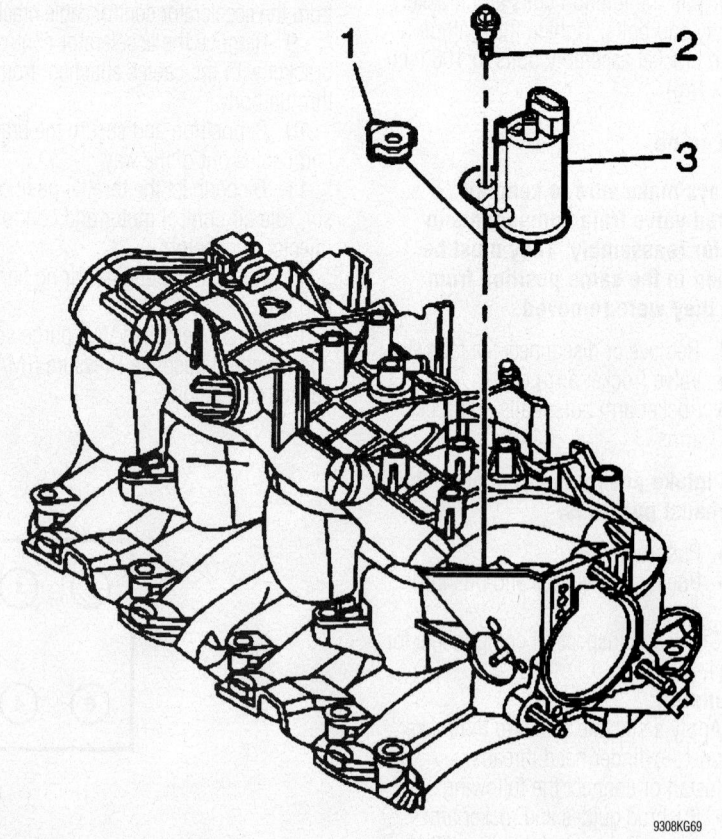

EVAP solenoid removal—4.8L, 5.3L and 6.0L engines

9308KG69

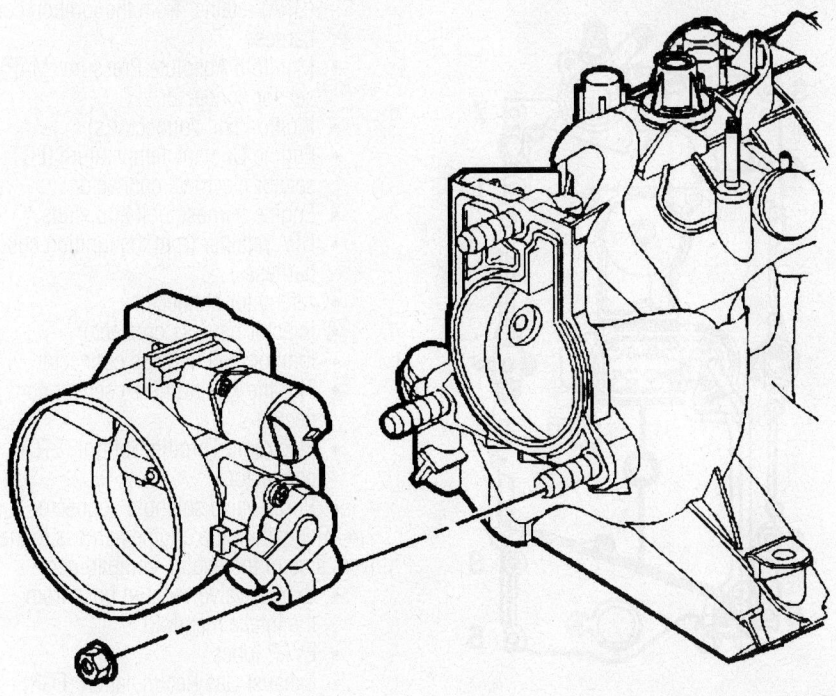

Throttle body removal—4.8L, 5.3L and 6.0L engines

- Air cleaner intake duct
- Coolant reservoir
- Wiring harness connectors and brackets
- Throttle linkage and bracket from the upper intake manifold
- Cruise control cable, if equipped
- Fuel lines and the bracket from the rear of the intake manifold
- Positive Crankcase Ventilation (PCV) valve and hoses
- Ignition coil and bracket
- Purge solenoid and bracket

➡ **Note the location of the manifold bolts and studs before removal for reassembly in their original positions.**

- Intake manifold bolts and studs
- Upper intake manifold

2. Clean the old gasket residue from both mating surfaces.

- Distributor
- Upper radiator hose from the thermostat housing
- Heater hose from the lower intake manifold
- Coolant bypass hose
- Exhaust Gas Recirculation (EGR) valve
- Fuel pressure and return lines from the lower intake manifold
- Wiring harnesses and brackets from the lower manifold
- Left side valve cover
- Transmission oil level indicator and tube, if equipped
- EGR tube, clamp and bolt
- Air conditioning compressor and bracket, but do not disconnect the lines

3. Loosen the compressor mounting bracket and slide it forward, but do not remove it.

- Power brake vacuum tube
- Lower intake manifold bolts and lower intake manifold

6. Inspect the intake manifold vacuum passages for debris or restrictions.

7. Inspect for damaged or broken vacuum fittings, damaged MAP sensor mounting bore, or broken MAP sensor retaining tabs.

8. Inspect the composite intake manifold assembly for cracks or other damage.

9. Inspect the areas between the intake runners. Inspect all the gasket sealing surfaces for damage.

10. Inspect the fuel injector bores for excessive scoring or damage. Inspect the intake manifold cylinder head deck for warpage.

11. Locate a straight edge across the intake manifold cylinder head deck surface. Position the straight edge across a minimum of two runner port openings.

12. Insert a feeler gauge between the intake manifold and the straight edge. A intake manifold with warpage in excess of 0.118 in. (3mm) over a 7.87 in. (200mm) area is warped and should be replaced.

To install:

13. Install or connect the following:

- MAP sensor
- EVAP solenoid, bolt, and isolator. Tighten the bolt to 89 inch lbs. (10 Nm).
- NEW intake manifold-to-cylinder head gaskets
- Intake manifold

14. Apply a 0.20 in. (5mm) band of

threadlock GM P/N 12345382 or equivalent to the threads of the intake manifold bolts.

- Intake manifold bolts. Tighten intake manifold bolts first pass in sequence to 44 inch lbs. (5 Nm). Tighten intake manifold bolts final pass in sequence to 89 inch lbs. (10 Nm).
- PCV valve and hose
- Coolant air bleed hose and clamp onto the throttle body
- Accelerator control cable bracket and bolts. Tighten the bolts to 89 inch lbs. (10 Nm).

5.0L and 5.7L Engines

1. Remove or disconnect the following:

- Negative battery cable
- Engine cover

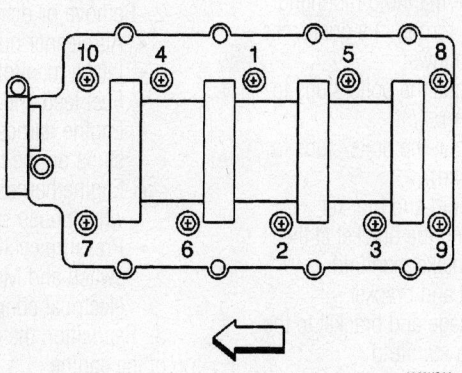

Lower intake manifold bolt tightening sequence—4.8L, 5.3L and 6.0L engines

To install:

4. Clean all gasket surfaces completely.

5. Install the intake manifold gaskets with the port blocking plates facing the rear. Factory gaskets should have the words **This Side Up** visible.

6. Apply gasket sealer to the front and rear sealing surfaces of the engine block. Extend the sealer approximately ½ inch (13mm) onto the heads.

7. Install the lower intake manifold.

8. Apply sealer to the lower intake manifold bolts prior to installation.

9. On the 5.0L and 5.7L engines, install the bolts and torque in sequence as follows:

a. Step 1: 71 inch lbs. (8 Nm).

b. Step 2: 106 inch lbs. (12 Nm).

c. Step 3: 11 ft. lbs. (15 Nm).

10. Install or connect the following:

- Power brake vacuum tube
- PCV valve and hose
- EGR tube, clamp and bolt
- Transmission oil level indicator and tube, if equipped
- Left side valve cover
- Wiring harnesses and brackets to the lower manifold
- Fuel pressure and return lines to the lower intake manifold
- EGR valve
- Coolant bypass hose
- Heater hose to the lower intake manifold
- Upper radiator hose to the thermostat housing
- Air conditioning compressor and bracket
- Distributor
- Upper intake manifold gasket
- Upper intake manifold

✳✳ WARNING

When installing the upper intake manifold be careful not to pinch the injector wires between the upper and lower intake manifolds.

- Upper intake manifold mounting bolts/studs, torque in a crisscross pattern as follows:

a. Step 1: Torque the bolts/studs to 44 inch lbs. (5 Nm).

b. Step 2: Torque the bolts/studs to 83 inch lbs. (10 Nm).

- Purge solenoid and bracket
- Fuel lines and the bracket at the rear of the intake manifold
- Ignition coil and bracket
- Throttle linkage and bracket to the upper intake manifold
- Throttle linkage cable

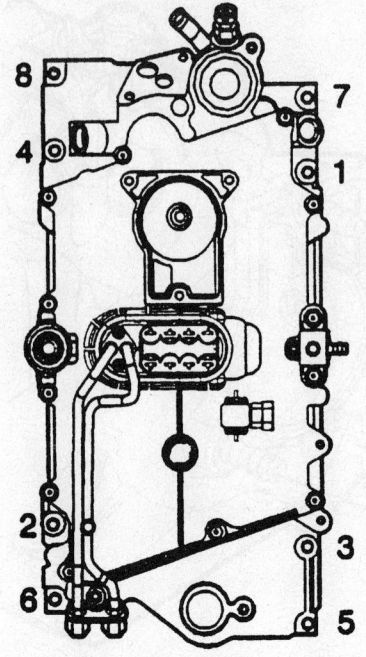

7924KG17

Lower intake manifold bolt tightening sequence—5.0L and 5.7L engines

- Cruise control cable, if equipped
- Wiring harness connectors and brackets
- Air cleaner intake duct
- Coolant recovery reservoir and the engine cover
- Negative battery cable

11. Start the vehicle and verify that there are no leaks.

8.1L Engine

➡The intake manifold, throttle body, fuel rail and injectors can be removed as an assembly. If you do not need to service these components individually, remove the manifold as a complete assembly.

1. Relieve the fuel system pressure and drain the cooling system.

2. Remove or disconnect the following:

- Air cleaner outlet duct
- Intake manifold sight shield
- Fuel feed and return pipes
- Engine harness clips from the studs on the front of the dash
- Engine harness clip from the wheelhouse splash shield
- Pressure cycling switch, surge tank switch and Mass Air Flow (MAF) electrical connectors

3. Reposition the engine harness to the top of the engine

- Connector Position Assurance

(CPA) retainer from the ignition coil harness

- Manifold Absolute Pressure (MAP) sensor connector
- Ignition coil connector(s)
- Engine Coolant Temperature (ECT) sensor electrical connector
- Engine harness bolt and studs
- CPA retainer from the ignition coil harness
- Alternator connector
- Injector harness connector
- Ignition coil harness connector
- Throttle Position (TP) sensor connector
- Electronic Throttle Control (ETC) connector
- Purge valve solenoid connector

4. Reposition the engine harness to the drivers side of the engine compartment.

- Bypass valve vacuum hose from the intake manifold
- EVAP tubes
- Exhaust Gas Recirculation (EGR) valve electrical connector
- EGR pipe bolts from the EGR adapter. Reposition the EGR pipe
- EGR valve pipe gasket and discard
- Secondary Air Injection (AIR) pipe nut from the fuel rail stud, if equipped
- Fuel pressure regulator vacuum hose
- Fuel rail studs and fuel rail, ONLY if replacing the manifold
- Intake manifold bolts

✳✳ WARNING

Do NOT try to remove the intake manifold by prying under the sealing surfaces.

- Intake manifold
- Intake manifold side gaskets and end seals and discard

➡The splash shield is reusable and secured using a snap-in fit. Do not distort the shield during removal.

- Splash shield

To install:

5. Clean all gasket surfaces completely.

6. Install or connect the following:

- Splash shield. Make sure the shield fits properly between the cylinder head.

➡Make sure the manifold gasket tabs align with the hole in the head gasket.

- New intake manifold end seals
- New intake manifold side gaskets onto the heads. Make sure the

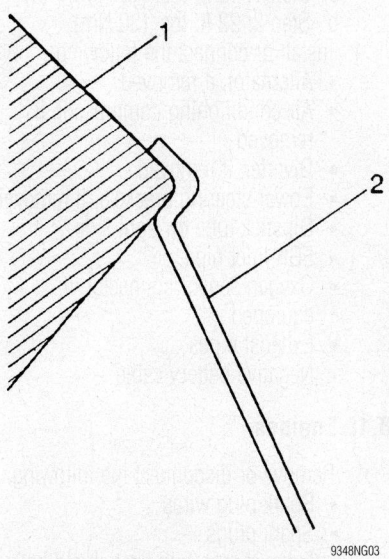

Make sure that the splash shield snap fits between the cylinder heads

stamped **This Side Up** is showing.
- Intake manifold to the block
- Apply a suitable thread locking material to at least 8 threads of the intake manifold bolts

7. On 2002–04 models, install the intake manifold bolts and tighten, in the sequence shown, in 4 passes:

    a. 1st pass: 44 inch lbs. (5 Nm).

    b. 2nd pass: 44 inch lbs. (5 Nm). Check the manifold joints for shifting and fix as necessary.

    c. 3rd pass: 89 inch lbs. (10 Nm).

    d. 4th pass: 106 inch lbs. (12 Nm).

8. On 2005–06 models, install the intake manifold bolts and tighten, in the sequence shown, in 4 passes:

    a. 1st pass: 44 inch lbs. (5 Nm).

    b. 2nd pass: 71 inch lbs. (8 Nm). Check the manifold joints for shifting and fix as necessary.

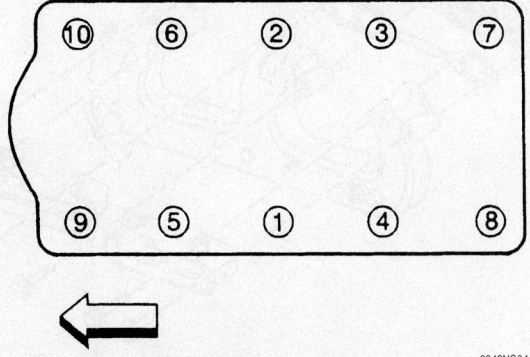

Intake manifold bolt tightening sequence—8.1L engine

    c. 3rd pass: 106 inch lbs. (12 Nm).

    d. 4th pass: 11 ft. lbs. (15 Nm).

9. Install the remaining components in the reverse order of the removal procedure.

10. Fill the cooling system, then connect the negative battery cable

11. Start the vehicle and verify that there are no leaks.

Exhaust Manifold

REMOVAL & INSTALLATION

4.3L Engine

1. Remove or disconnect the following:
- Negative battery cable
- Engine cover, if equipped
- Exhaust pipe from the exhaust manifold
- Spark plug wires from the plugs and the retaining clips
- Heat shields

    a. Remove the spark plugs, dipstick tube and wiring, if necessary.

2. Unbend the exhaust manifold bolt lock tangs.

3. Remove or disconnect the following:
- Exhaust manifold retaining bolts, washers and tab washers
- Exhaust manifold
- Old gaskets and discard

To install:

4. Clean the gasket mounting surfaces.

5. Inspect the exhaust manifold for distortion, cracks or damage; replace if necessary.

6. Install or connect the following:
- Exhaust manifold to the cylinder using a new gasket.

7. Tighten the exhaust manifold bolts and stud on the first pass to 11 ft. lbs. (5 Nm).

8. Tighten the exhaust manifold bolts and stud on the final pass to 22 ft. lbs. (30 Nm).

➡Once the bolts are tightened, bend the tabs on the washers back over the heads of all bolts in order to lock them in position.

- Spark plugs
- Dipstick tube
- Spark plug wires to the retainer clips and plugs
- Exhaust pipe to the manifold
- Engine cover, on van models
- Negative battery cable

4.8L, 5.3L and 6.0L Engines

1. Remove or disconnect the following:
- Spark plug wires from the spark plugs

➡Do not remove the spark plug wires from the ignition coils unless required.

- Exhaust manifold, bolts, and gasket. Discard the gasket.
- Heat shield and bolts from the manifold, if required

To install:

➡Do not reuse the exhaust manifold-to-cylinder head gaskets. Upon installation of the exhaust manifold, install a NEW gasket. A improperly installed gasket or leaking exhaust system may effect On-Board Diagnostics (OBD) II system performance.

2. Clean the exhaust manifold and heat shield in solvent. Dry the exhaust manifold with compressed air.

3. Use a straight edge and a feeler gauge and measure the exhaust manifold cylinder head deck for warpage. An exhaust manifold deck with warpage in excess of 0.01 in. (0.25mm) within the two front or two rear runners or 0.02 in. (0.5mm) overall, may cause an exhaust leak and may effect OBD II system performance. Exhaust manifolds not within specifications must be replaced.

4. Apply a 0.2 in. (5mm) wide band of threadlock GM P/N 12345493 or equivalent to the threads of the exhaust manifold bolts.

5. Install the exhaust manifold gasket and exhaust manifold

6. Install the exhaust manifold bolts and tighten, beginning with the center two bolts. Alternate from side-to-side, and work toward the outside bolts.

    a. Tighten the exhaust manifold bolts first pass to 11 ft. lbs. (15 Nm).

    b. Tighten the exhaust manifold bolts final pass to 18 ft. lbs. (25 Nm). Using a flat punch, bend over the exposed edge of the exhaust manifold gasket at the front of the right cylinder head.

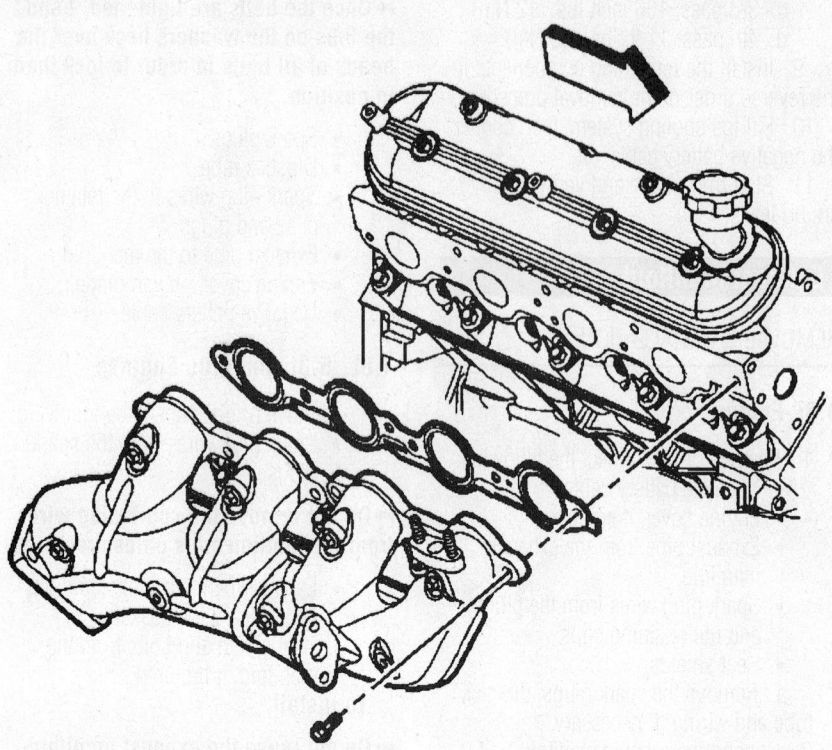

Right exhaust manifold removal—4.8L, 5.3L and 6.0L; left side similar

7. Install or connect the following:
- Heat shield and bolts and tighten to 80 inch lbs. (9 Nm)
- Spark plug wires

5.0L and 5.7L Engines

1. Remove or disconnect the following:
- Negative battery cable
- Engine cover
- Air cleaner, if needed
- Exhaust pipe at the manifold
- Oxygen Sensor (O_2S) wiring, if equipped
- AIR hose at the check valve
- Exhaust Gas Recirculation (EGR) valve, inlet pipe
- Heat stove pipe and the dipstick tube bracket, if working on the right side of the engine
- Power steering pump rear bracket at the manifold, if removing the left side manifold
- Loosen the alternator and remove the lower bracket, if necessary
- Air conditioner compressor rear bracket and the diverter valve and bracket. if needed

➥On models with air conditioning, it may be necessary to remove the compressor, do not disconnect the compressor lines.

- Manifold bolts and the manifold(s) Some models have lock tabs on the front and rear manifold bolts which must be removed before removing the bolts.

To install:

2. Clean gasket surfaces, and inspect manifold for cracks replace as necessary.

3. Install the manifold and torque it in the following steps:

a. Step 1: 15 ft. lbs. (20 Nm).
b. Step 2: 22 ft. lbs. (30 Nm).

4. Install or connect the following:
- Alternator, if removed
- Air conditioning compressor, if removed
- Diverter, if removed
- Power steering brackets, if removed
- Dipstick tube on right side
- EGR inlet pipe
- Oxygen sensor connector, if equipped
- Exhaust pipes
- Negative battery cable

8.1L Engines

1. Remove or disconnect the following:
- Spark plug wires
- Spark plugs
- Exhaust manifold heat shield bolts and shield
- Exhaust manifold bolt and nuts
- Exhaust manifold
- Exhaust manifold gasket and discard

To install:

2. Clean the mating surfaces and the retainer threads.

3. Install or connect the following:
- New exhaust manifold gasket
- Exhaust manifold
- Exhaust manifold bolt and nuts. Tighten the bolt to 26 ft. lbs. (35 Nm) and the nuts to 12 ft. lbs. (16 Nm).
- If removed, tighten the studs to 15 ft. lbs. (20 Nm).
- Heat shield. Tighten the retaining

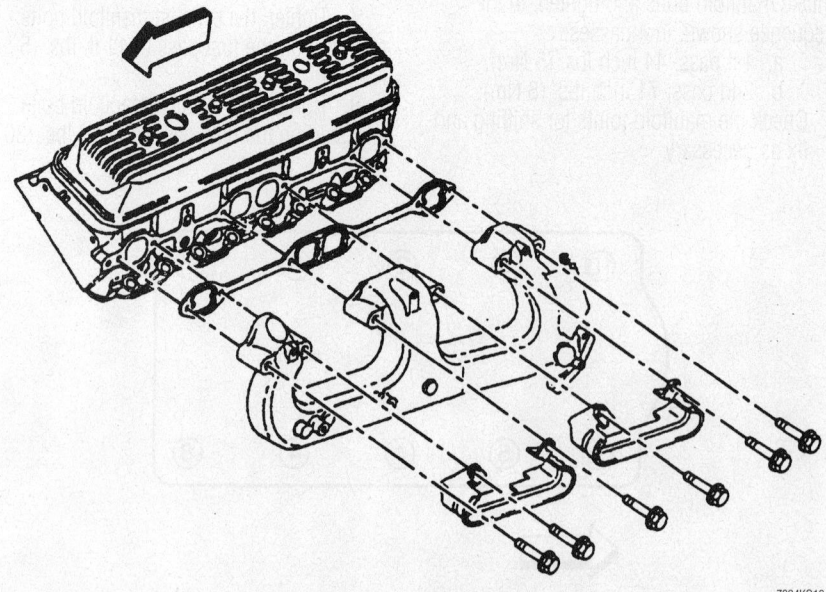

Exploded view of the left exhaust manifold, the right side is similar—5.0L and 5.7L engines

bolts and nuts to 18 ft. lbs. (25 Nm).
- Spark plugs and plug wires

Camshaft and Valve Lifters

REMOVAL & INSTALLATION

4.3L Engines

1. Properly relieve the fuel system pressure.
2. Drain the engine cooling system.
3. Remove or disconnect the following:
   - Negative battery cable
   - Radiator
   - Cooling fan
   - Water pump
   - Rocker arm covers from the engine
   - Intake manifold assembly
   - Rocker arms, pushrods and lifters
   - Crankshaft pulley and hub
   - Engine front cover
4. Align the timing marks on the crankshaft and camshaft sprockets.
   - Camshaft sprocket and timing chain
   - Balance shaft drive gear, if equipped
   - Camshaft thrust plate

➡Install the sprocket bolts or longer bolts of the same thread into the end of the camshaft as a handle.

   - Camshaft
   To install:
5. Lubricate the camshaft journals with clean engine oil or a suitable pre-lube.
6. Install or connect the following:
   - Camshaft
   - Camshaft thrust plate
   - Balance shaft drive gear, if equipped
   - Timing chain and camshaft sprocket
   - Engine front cover
   - Crankshaft pulley and hub
   - Valve lifters
   - Pushrods and rocker arms, properly adjust the valve clearance
   - Intake manifold assembly
   - Rocker arm covers to the engine
   - Radiator to the vehicle
   - Negative battery cable
7. Refill the engine cooling system.

4.8L, 5.3L and 6.0L Engines

1. Raise the hood to the servicing position and secure it. Move the hood hinge bolt to hold the hood in the servicing position.

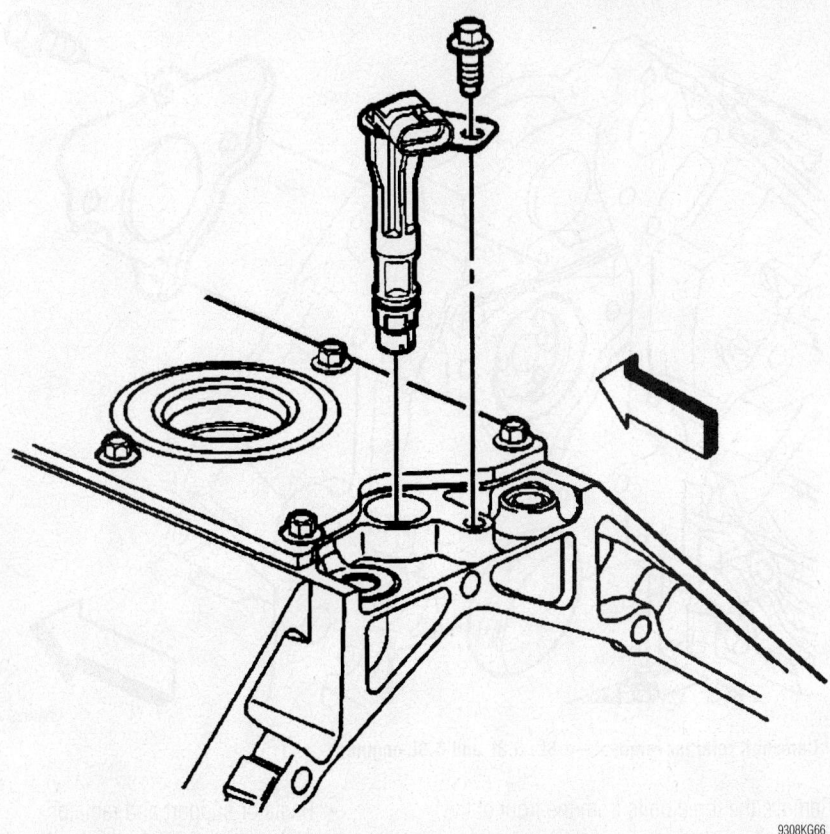

Camshaft sensor removal—4.8L, 5.3L and 6.0L engines

9308KG66

2. Remove or disconnect the following:
   - Battery negative cable
   - Coolant
   - Upper and lower radiator hoses from the engine
   - Air cleaner duct from the engine
   - A/C condenser mounting bolts, if equipped
   - Radiator support and radiator
   - Engine cooling fan
   - Drive belt
   - A/C drive belt, if equipped
   - Engine sight shield
   - Electrical wiring harness from the thermostat housing
   - Water pump
3. Raise the vehicle.
   - Starter motor
   - Right side closeout cover and bolt
   - Crankshaft balancer
   - Engine oil pan
   - Engine front cover
   - Cylinder heads from the engine
   - Valve lifters from the engine
4. Align the timing marks on the camshaft and crankshaft sprockets. Make sure that the number 1 piston is in the firing position.
   - Camshaft sprocket
   - Camshaft sensor bolt and sensor

   - Camshaft retainer bolts and retainer

➡All camshaft journals are the same diameter, so care must be used in removing or installing the camshaft to avoid damage to the camshaft bearings.

5. Install the three M8-1.25 x 100 mm bolts in the camshaft front bolt holes. Using the bolts as a handle, carefully rotate and pull the camshaft out of the engine block. Remove the bolts from the front of the camshaft.
6. Clean and inspect all sealing surfaces.
   To install:

➡If camshaft replacement is required, the valve lifters must also be replaced.

7. Lubricate the camshaft journals and the bearings with clean engine oil. Install three M8-1.25 x 100 mm (M8-1.25 x 4.0 in) bolts into the camshaft front bolt holes.

➡All camshaft journals are the same diameter, so care must be used in removing or installing the camshaft to avoid damage to the camshaft bearings.

8. Using the bolts as a handle, carefully install the camshaft into the engine block.

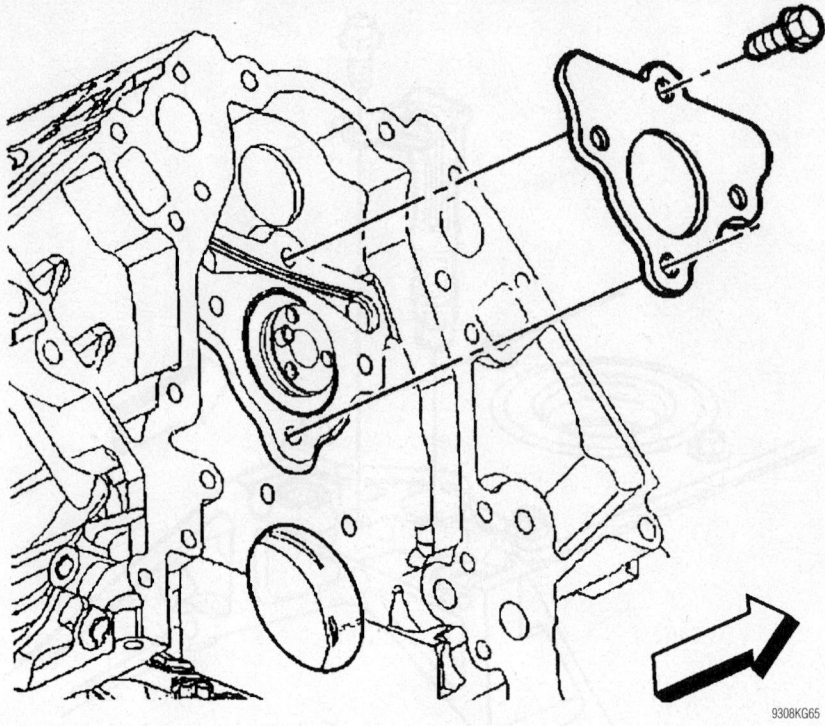

Camshaft retainer removal—4.8L, 5.3L and 6.0L engines

Remove the three bolts from the front of the camshaft.

➡ **Install the retainer plate with the sealing gasket facing the engine block. The gasket surface on the engine block should be clean and free of dirt or debris.**

9. Install or connect the following:
- Camshaft retainer and the bolts. Tighten the camshaft retainer bolts to 18 ft. lbs. (25 Nm).
10. Inspect the camshaft sensor O-ring seal. If the O-ring seal is not cut or damaged, it may be reused. Lubricate the O-ring seal with clean engine oil.
- Camshaft sensor and bolt. Tighten the bolt to 18 ft. lbs. (25 Nm).
- Camshaft sprocket and timing chain
- Valve lifters
- Cylinder heads
- Engine front cover to the engine
- Oil pan
- Right side closeout cover
- Starter motor
- Crankshaft balancer to the crankshaft
- Water pump
- Electrical wiring harness to the thermostat housing
- A/C drive belt, if equipped
- Drive belt
- Engine sight shield

- Radiator support and radiator
- A/C condenser mounting bolts
- Engine cooling fan
- Air cleaner duct
- Negative battery cable

5.0L and 5.7L Engines

1. Drain the cooling system.
2. Properly relieve the fuel system pressure.
3. Remove or disconnect the following:
- Engine cover, on van models
- Air cleaner
- Grille and center support, on van models
- Air conditioning condenser and swing the condenser forward from its mounting, if equipped
- Fan, the shroud and the radiator
- Valve covers
- Water pump assembly
4. Align the timing marks and remove the torsional damper.
- Timing chain cover
- Electrical and vacuum connections at the intake manifold
- Distributor assembly, mark the distributor rotor-to-housing location
- Intake manifold, pushrods and hydraulic lifters
- Camshaft sprocket bolts
- Camshaft sprocket and timing chain
- Crankshaft sprocket, as required

- Front engine mount through-bolts and raise the engine to gain sufficient clearance for camshaft removal, as required
5. Install 2 or 3 ⁵⁄₁₆–18 bolts 4–5 in. (102–127mm) long into the camshaft threaded holes.
- Camshaft

➡ **Inspect the shaft for signs of excessive wear or damage.**

To install:

➡ **Liberally coat camshaft and bearing with heavy engine oil or engine assembly lubricant.**

6. Install or connect the following:
- Camshaft, align the timing marks on the camshaft and crankshaft gears
- Engine mount through-bolts
- Camshaft sprocket and chain. Torque the bolts to 18 ft. lbs. (25 Nm).
- Hydraulic lifters and pushrods
- Distributor assembly
- Timing chain cover
- Torsional damper
- Water pump
- Valve covers
- Fan, the shroud and radiator
- Air conditioning condenser, if equipped
- Grille and center support, on van models
- Air cleaner
- Engine cover, on van models
- Negative battery cable
7. Refill the cooling system.

8.1L Engines

1. Properly discharge the air conditioning system.
2. Remove or disconnect the following:

- Grille
- A/C condenser
- Intake manifold
- Rocker arms and pushrods
- Valve lifter guide retainer bolts and retainer
- Valve lifter guides, keeping them in proper order for reassembly
- Valve lifters

➡ **If any lifters are stuck in their bores, use a suitable valve lifter to remove them.**

- Timing chain and sprocket
- Camshaft retaining bolts
- Camshaft retainer

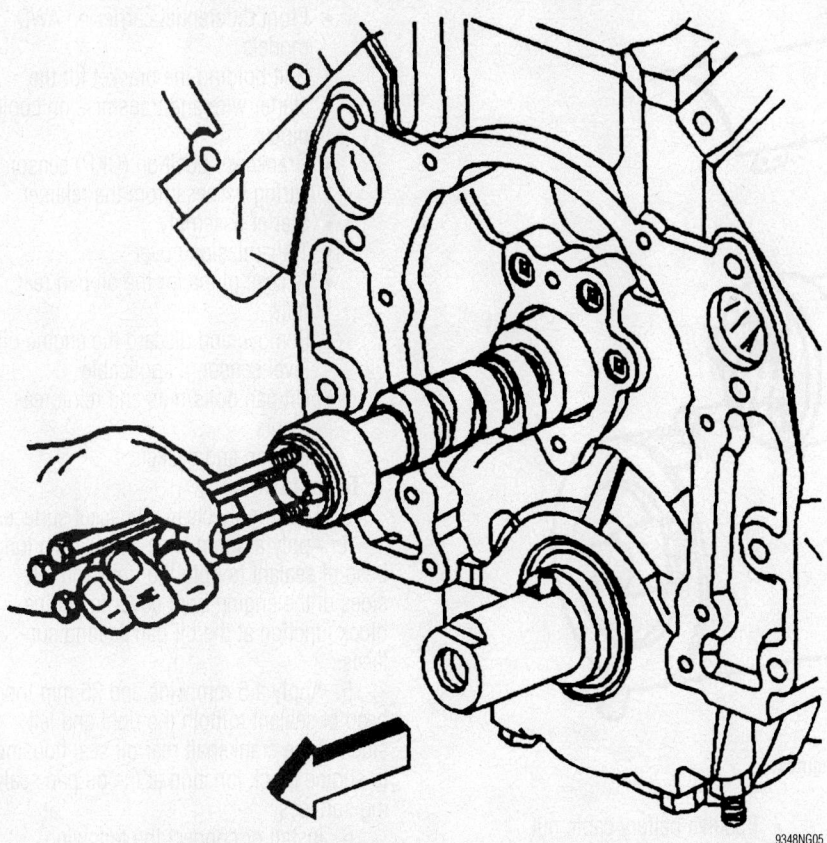

Use the 3 bolts as a handle to carefully remove and install the camshaft—8.1L engine shown

9348NG05

※ WARNING

All of the cam journals are the same size so be very careful when removing and installing the camshaft that you do not damage the bearings.

3. Install three 8-1.25 x 100mm bolts in the holes in the front of the camshaft and carefully pull the camshaft from the block.

4. Remove the bolts from the front of the camshaft.

5. Clean and inspect the camshaft for damage.

To install:

6. Liberally coat camshaft and bearings with heavy engine oil or engine assembly lubricant.

7. Install the camshaft, using the 3 bolts threaded into the camshaft bolt holes as a handle, then remove the bolts.

8. Install or connect the following:
  • Camshaft retainer and bolts. Tighten to 106 inch lbs. (12 Nm).

➡**If a new camshaft is installed, you MUST install new valve lifters.**

  • Timing chain and sprocket
  • Valve lifters
  • Valve lifter guides over the flats on

the lifters. Make sure the rollers of the lifters are properly aligned with the cam lobes.
  • Valve lifter guide retainer. Tighten the bolts to 18 ft. lbs. (25 Nm).
  • Rocker arms and pushrods
  • Intake manifold
  • A/C condenser
  • Grille

9. Recharge the A/C system.

Valve Lash

ADJUSTMENT

All engines use hydraulic lifters, which require no periodic adjustment.

Starter Motor

REMOVAL & INSTALLATION

4.3L and 5.0L Engines

1. Remove or disconnect the following:
  • Negative battery cable
  • Bracket and shield
  • Wires
  • Mounting bolts and shims
  • Starter

To install:

2. Install or connect the following:
  • Starter
  • Mounting bolts and shim. Torque the bolts to 33 ft lbs. (45 Nm).
  • Wires. Torque battery wire nut to 89 inch lbs. (10 Nm) and ignition nut to 18 inch lbs. (2 Nm).
  • Bracket and shield. Torque the nuts to 53 inch lbs. (6 Nm).
  • Negative battery cable

5.7L Engines

1. Remove or disconnect the following:
  • Negative battery cable
  • Mounting bolts and shims
  • Wires
  • Heat shield
  • Starter

To install:

2. Install or connect the following:
  • Starter
  • Wires. Torque battery wire nut to 89 inch lbs. (10 Nm), and ignition nut to 18 inch lbs. (2 Nm).
  • Heat shield. Torque the bolts to 53 inch lbs. (6 Nm) and the nuts to 35 inch lbs. (3 Nm).
  • Mounting bolts and shim. Torque the bolts to 33 ft lbs. (45 Nm).
  • Negative battery cable

4.8L, 5.3L and 6.0L Engines

1. Disconnect the negative battery cable.
2. Raise and support the vehicle.
3. Remove or disconnect the following:
  • Protective shields (as necessary)
  • Starter solenoid shield
  • Starter-to-transmission close out cover bolt
  • Engine oil level sensor connection
4. Slide the starter forward until the starter clears the transmission.
  • Starter transmission close out cover
  • Positive battery cable and wiring harness from the starter
  • Starter

To install:

5. Install or connect the following:
  • Starter
  • Positive battery cable.
  • Starter transmission close out cover
  • Mounting bolts to the engine block and tighten to 37 ft. lbs. (50 Nm)
  • Oil level sensor connection
  • Starter-to-transmission close out cover bolt
  • Starter solenoid shield
  • Protective shields (as necessary)

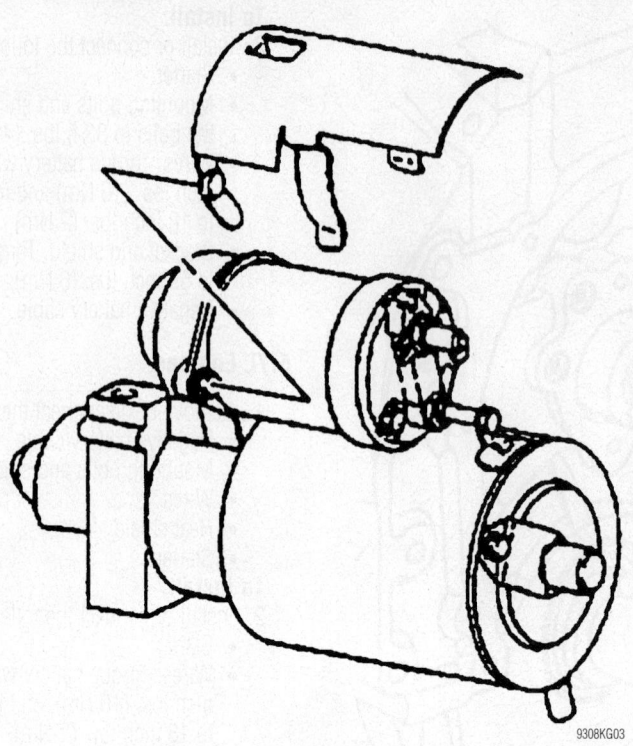

Exploded view of the starter motor—4.3L and 5.0L engines.

9308KG03

6. Remove the safety stands.
7. Lower the vehicle.
8. Connect the negative battery cable.

8.1L Engine

1. Remove or disconnect the following:
   • Negative battery cable

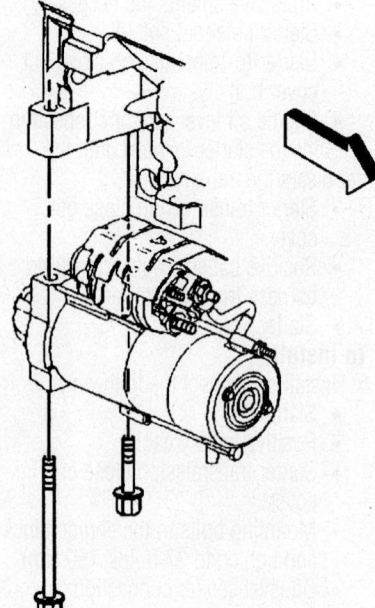

Starter removal—4.8L, 5.3L and 6.0L engines

9308KG00

   • Positive battery cable nut
   • Positive cable from the solenoid
   • Engine harness ground nut and ground from the solenoid
   • Mounting bolts and starter
   • Heat shield bolts, nut and shield, if necessary

To install:

2. Install or connect the following:
   • Heat shield, bolts and nut if removed. Tighten the bolts to 35 inch lbs. (3 Nm) and the nut to 44 inch lbs. (5 Nm).
   • Starter and bolts. Tighten to 37 ft. lbs. (50 Nm).
   • Ground wire and nut. Tighten to 30 inch lbs. (3.4 Nm).
   • Positive cable and nut. Tighten to 80 inch lbs. (9 Nm).
   • Negative battery cable

Oil Pan

REMOVAL & INSTALLATION

4.3L Engines

EXPRESS AND SAVANA

1. Drain the engine oil.
2. Remove or disconnect the following:
   • Negative battery cable
   • Oil filter
   • Oil filter adapter

   • Front differential carrier on AWD models
   • Bolt holding the bracket for the starter wire and transmission cooler pipes
   • Crankcase position (CKP) sensor wiring harness from the retainer.
   • Starter assembly
   • Transmission cover
   • Access plugs for the oil pan rear nuts
   • Remove and discard the engine oil level sensor, if applicable
   • Oil pan bolts/nuts and reinforcements
   • Oil pan and gaskets

To install:

3. Thoroughly clean all gasket surfaces,
4. Apply a 5 mm wide and 25 mm long bead of sealant to both the right and left sides of the engine front cover to engine block junction at the oil pan sealing surfaces.
5. Apply a 5 mm wide and 25 mm long bead of sealant to both the right and left sides of the crankshaft rear oil seal housing to engine block junction at the oil pan sealing surfaces.
6. Install or connect the following:
   • New gasket
   • Oil pan and new gasket
   • Install the oil pan bolts and nuts, but do not tighten
7. Measure the pan-to-transmission housing clearance using a feeler gage and a straight edge. Use a feeler gage to check the clearance between the oil pan-to-transmission housing measurement points. If the clearance exceeds 0.011 in. (0.3 mm) at any of the 3 oil pan-to-transmission housing measurement points (1), then repeat the step until the oil pan-to-transmission housing clearance is within the specification. The oil pan must always be forward of the rear face of the engine block.
8. Install the oil pan bolts, nuts and reinforcements. Torque bolts in sequence to 18 ft. lbs. (25 Nm).
9. Install or connect the following:
   • Install the engine oil level sensor, if applicable
   • Access plugs for the oil pan rear nuts
   • Transmission cover
   • Starter assembly
   • Crankcase position (CKP) sensor wiring harness to the retainer.
   • Bolt holding the bracket for the starter wire and transmission cooler pipes
   • Front differential carrier on AWD models

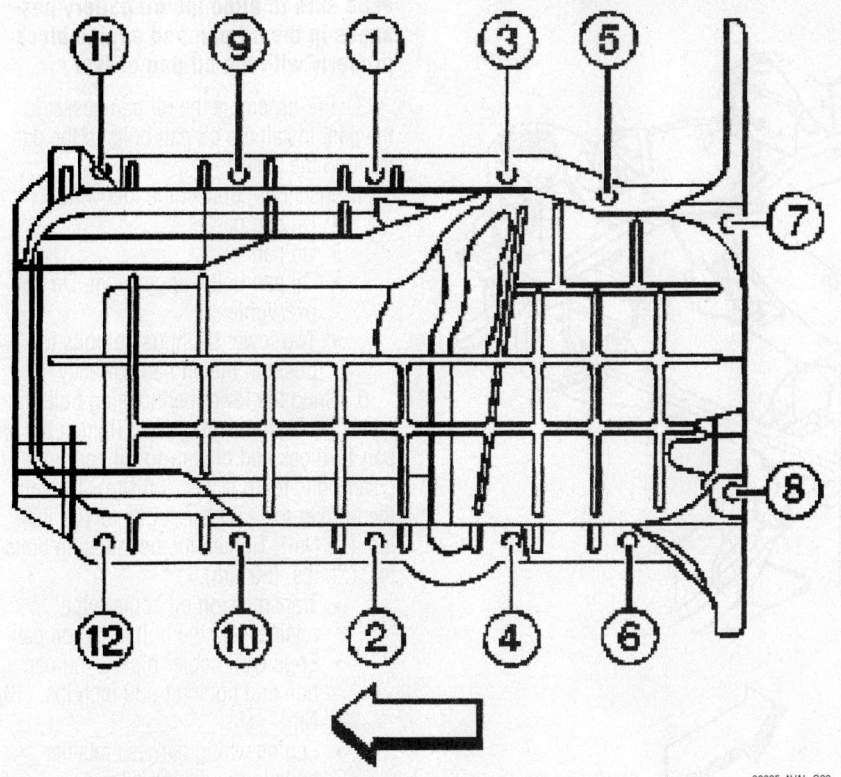

Oil pan bolt tightening sequence—4.3L engine

06025-AVAL-G03

- Oil filter adapter
- Oil filter
- Negative battery cable
10. Refill the engine with oil.

AVALANCHE, SIERRA AND SILVERADO

1. Drain the engine oil.
2. Disconnect the negative battery cable.
3. Raise and support the vehicle.
4. Remove or disconnect the following:
- Oil pan skid plate bolts and plate, if equipped
- Engine oil and filter
- Crossmember bolts and bar
- On 4WD, the front differential carrier
- Battery cable bracket bolts.
- Starter
- Transmission cover
- Positive battery cable clip bolt
- Oil level sensor electrical connector
- Transmission
- Oil level sensor and discard
- Oil pan bolts and oil pan
- Oil pan gasket
To install:
5. Thoroughly clean all gasket surfaces,
6. Apply a 5 mm wide and 25 mm long bead of sealant to both the right and left

sides of the engine front cover to engine block junction at the oil pan sealing surfaces.
7. Apply a 5 mm wide and 25 mm long bead of sealant to both the right and left sides of the crankshaft rear oil seal housing to engine block junction at the oil pan sealing surfaces.
8. Install or connect the following:
- Transmission
- New gasket
- Oil pan and new gasket
- Install the oil pan bolts and nuts, but do not tighten
9. Measure the pan-to-transmission housing clearance using a feeler gage and a straight edge. Use a feeler gage to check the clearance between the oil pan-to-transmission housing measurement points. If the clearance exceeds 0.011 in. (0.3 mm) at any of the 3 oil pan-to-transmission housing measurement points (1), then repeat the step until the oil pan-to-transmission housing clearance is within the specification. The oil pan must always be forward of the rear face of the engine block.
10. Install the oil pan bolts, nuts and reinforcements. Torque bolts in sequence to 18 ft. lbs. (25 Nm).
- Oil level sensor electrical connector
- Positive battery cable clip bolt
- Transmission cover

- Starter
- Battery cable bracket bolts.
- On 4WD, the front differential carrier
- Crossmember bolts and bar
- Engine oil and filter
- Oil pan skid plate bolts and plate, if equipped
- Negative battery cable
11. Refill the engine with oil.

4.8L, 5.3L and 6.0L Engines

➡ **The original oil pan gasket is retained and aligned to the oil pan by rivets. When installing a new gasket, it is not necessary to install new rivets. DO NOT reuse the oil pan gasket. When installing the oil pan, install a NEW oil pan gasket.**

1. Remove or disconnect the following:
- Negative battery cable
- Front differential if equipped with four wheel drive
- Under body shield from the vehicle
- Oil pan shield
- Cross brace if equipped
- Engine oil and filter
- Transmission-to-oil pan bolts
- Oil level sensor electrical connector
- Two front wiring harness retainer bolts
- Engine wiring harness retainer bolts from the engine oil pan
- Engine oil cooler pipe-to-oil pan bolt
- Transmission oil cooler pipe retainer and the bolt from the oil pan
- Closeout covers and bolts (one each side of engine)
- Engine mount bolts each side
- Oil pan
To install:

➡ **The alignment of the structural oil pan is critical. The rear bolt hole locations of the oil pan provide mounting points for the transmission bellhousing. To ensure the rigidity of the power-train and correct transmission alignment, it is important that the rear of the block and the rear of the oil pan must NEVER protrude beyond the engine block and transmission bellhousing plane.**

2. Apply a 0.20 in. (5mm) bead of sealant GM P/N 12378190 or equivalent 0.8 in. (20mm) long to the engine block. Apply the sealant directly onto the tabs of the front cover gasket that protrudes into the oil pan surface.

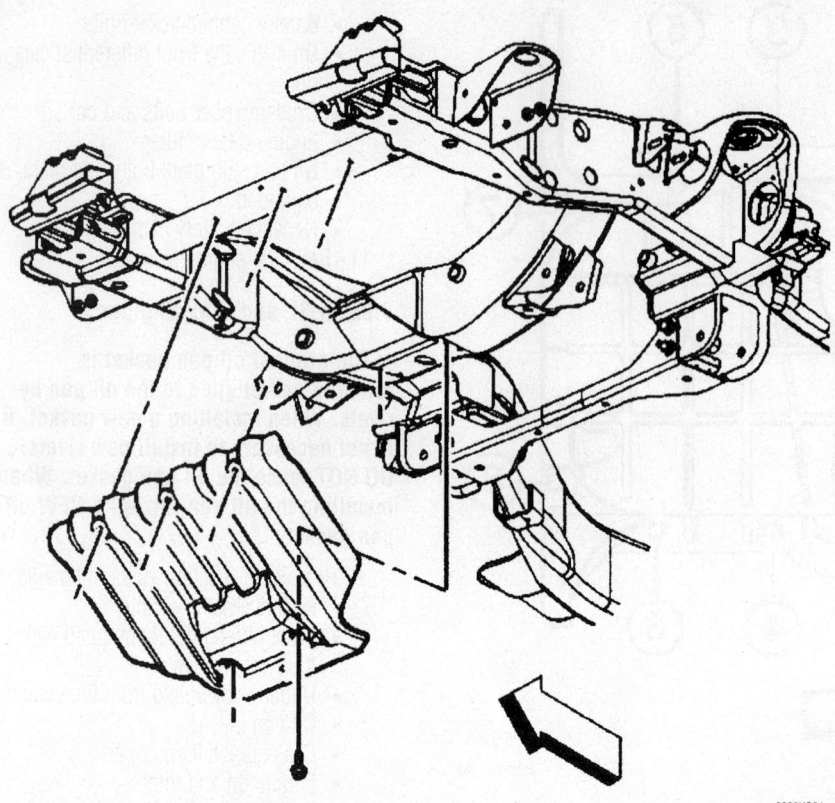

Oil pan shield—4.8L, 5.3L and 6.0L engines

9308KG81

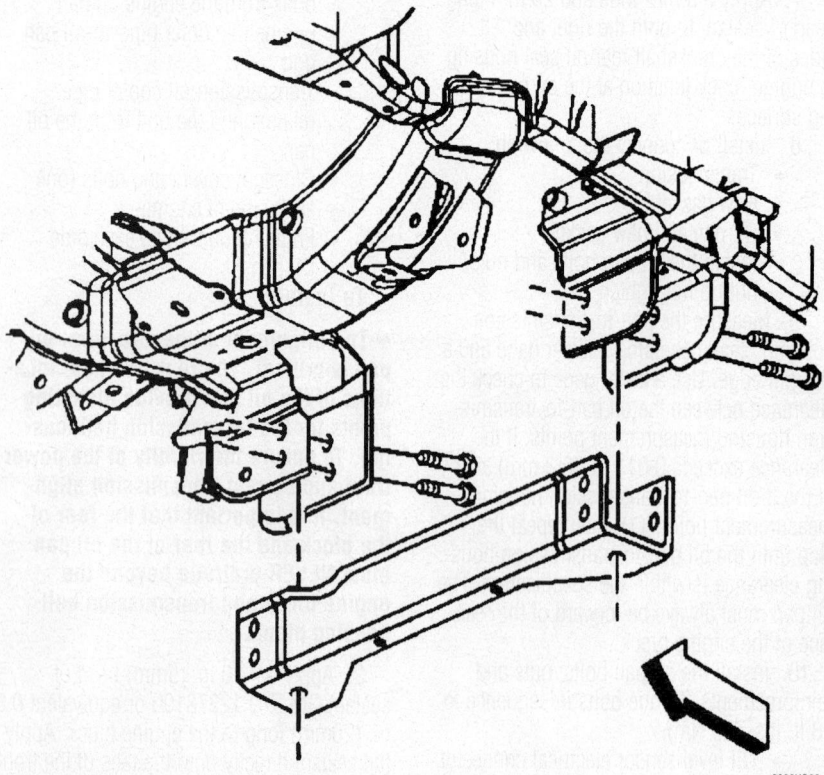

Cross brace—4.8L, 5.3L and 6.0L engines

9308KG80

➡ **Be sure to align the oil gallery passages in the oil pan and engine block properly with the oil pan gasket.**

3. Pre-assemble the oil pan gasket to the pan. Install the oil pan bolts to the pan through the gasket.

4. Install or connect the following:
- Oil pan gasket
- Oil pan
- Oil pan bolts, finger-tight. Do not overtighten.
- Two lower bellhousing bolts to position the oil pan correctly

5. Snug the lower bellhousing bolt finger-tight. Do not overtighten. Tighten the oil pan-to-block and oil pan-to-oil pan front cover bolts to 18 ft. lbs. (25 Nm). Tighten the oil pan-to-rear cover bolts to 106 inch lbs. (12 Nm). Tighten the bellhousing bolts to 37 ft. lbs. (50 Nm).
- Transmission oil cooler pipe retainer and the bolt to the oil pan
- Engine oil cooler pipe-to-oil pan bolt and tighten to 89 inch lbs. (10 Nm)
- Engine wiring harness retainer bolts to the engine oil pan
- Oil level sensor electrical connector
- Transmission-to-oil pan bolts and tighten to 41 ft. lbs. (55 Nm)
- Front differential, if equipped with four wheel drive
- Underbody shield

6. Lower the vehicle. Fill the engine with oil and install the engine oil filter.

7. Connect the negative battery cable.

5.0L and 5.7L Engines

1. Drain the engine oil.
2. Remove or disconnect the following:
- Negative battery cable
- Underbody protector shields
- Transmission and engine oil lines from guides
- Front driveshaft, if needed
- Front drive axles, if needed
- Exhaust crossover pipe
- Flywheel/flexplate or torque converter cover
- Oil filter and adapter
- Strut rods at the front engine mounting, if equipped
- Oil pan bolts, nuts and reinforcements
- Oil pan and gaskets

To install:

3. Thoroughly clean all gasket surfaces.
4. Install or connect the following:
- New gasket
- Oil pan
- Oil pan bolts, nuts and reinforce-

ments. Torque the bolts to 18 ft. lbs. (25 Nm).
- Strut rods at the front engine mounting
- Oil filter and adapter
- Torque converter or flywheel/flex-plate cover
- Exhaust crossover pipe
- Front drive axles and driveshaft, if removed
- Transmission and engine oil lines to guides
- Underbody protectors
- Negative battery cable

5. Fill the crankcase with oil.

8.1L Engine

➡**Removal of the transmission may be necessary on vans.**

1. Disconnect the negative battery cable and drain the engine oil.
2. Remove or disconnect the following:
   - Front differential, if equipped with 4WD
   - Starter motor
   - Oil pan skid plate bolts and plate
   - Crossbar bolt(s) and crossbar
   - Oil level dipstick
   - Oil level sensor electrical connector
   - Engine harness clip from the oil pan
   - Battery cable channel bolt
   - Battery cable channel and reposition
   - Oil pan bolts, oil pan and gasket

➡**You can reuse the oil pan gasket, if it is not damaged**

To install:

➡**You must install the oil pan within 5 minutes of applying the sealer.**

3. Before servicing the vehicle, refer to the precautions in the beginning of this section.
4. Apply sealant to the sides of the front and rear crankshaft bearing caps on the left and right sides.
5. Install or connect the following:
   - Oil pan gasket into the oil pan groove
   - Oil pan and bolts
6. Tighten the oil pan bolts, in sequence, as follows:
   a. 1st pass: 89 inch lbs. (10 Nm).
   b. 2nd pass: 18 ft. lbs. (25 Nm).
7. Install or connect the following:
   - Battery cable channel and bolt. Tighten to 80 inch lbs. (9 Nm).
   - Oil level sensor and tighten to 15 ft. lbs. (20 Nm)

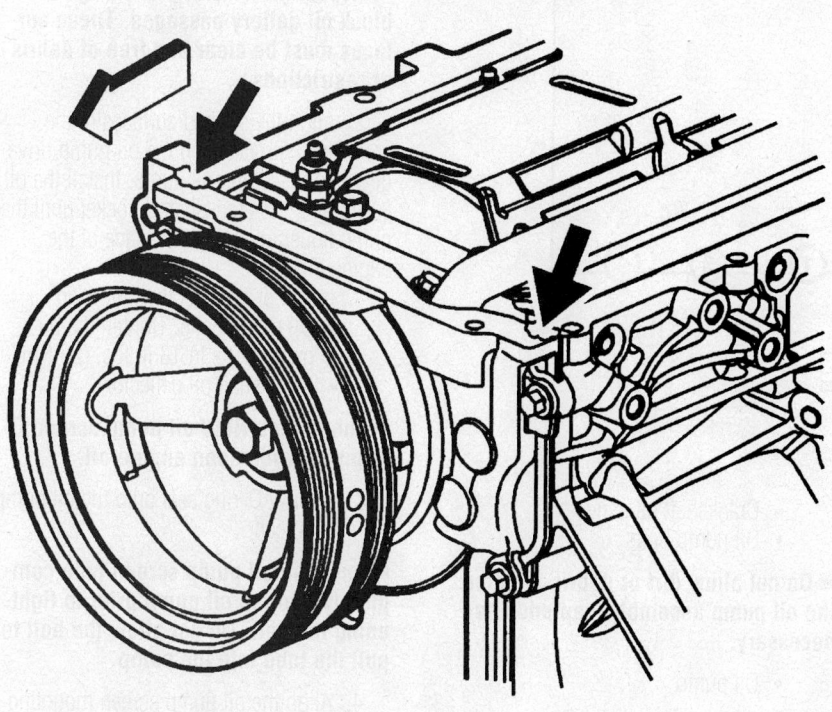

Apply sealant at these points at the front of the block—4.8L, 5.3L and 6.0L engines

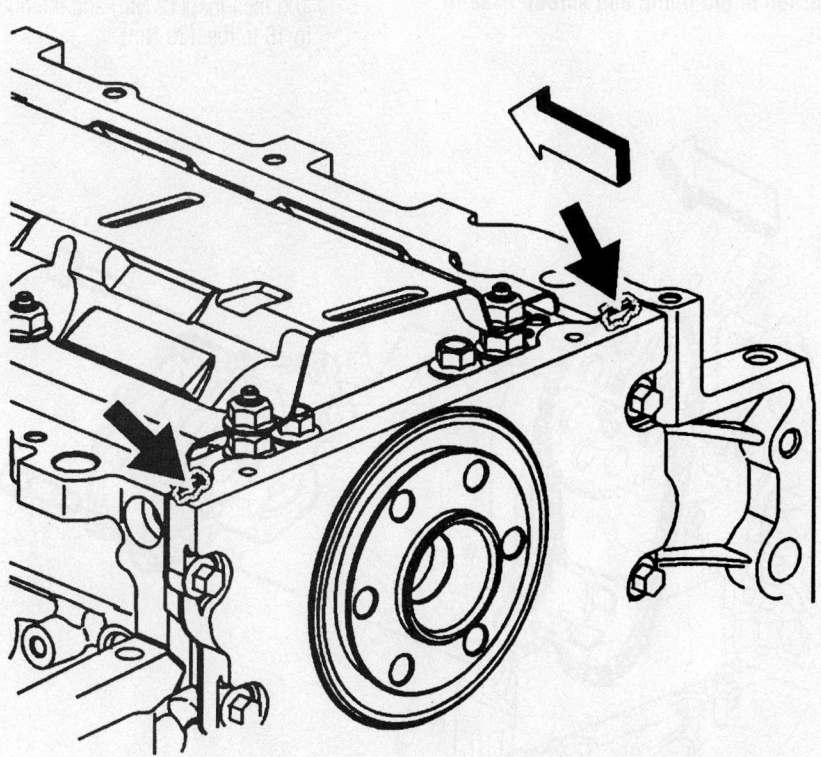

Apply sealant at these points at the rear of the block—4.8L, 5.3L and 6.0L engines

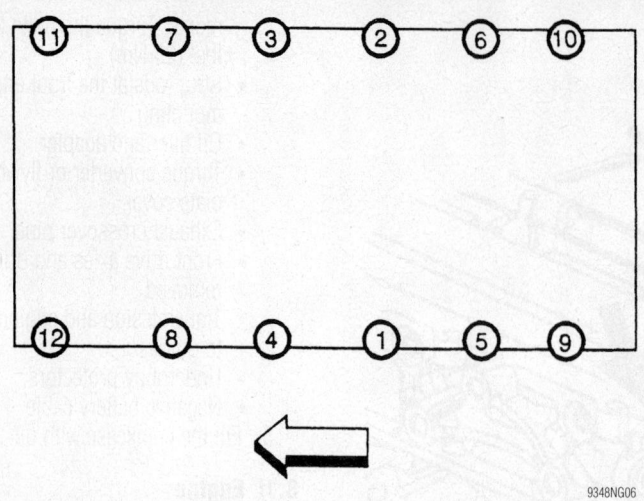

Oil pan bolt tightening sequence—8.1L engine

- Engine harness clip
- Oil level sensor connector
- Oil level dipstick
- Crossbar and bolt(s). Tighten to 74 ft. lbs. (100 Nm).
- Skid plate. Tighten the bolts to 15 ft. lbs. (20 Nm).
- Starter motor
- Front differential
- Negative battery cable
8. Fill the crankcase with oil.

Oil Pump

REMOVAL & INSTALLATION

4.3L Engines

1. Remove or disconnect the following:
   - Oil pan
   - Oil pump mounting bolt
   - Oil pump

To install:

2. Inspect the oil pump locator pins for damage, and replace if required.
3. Clean and inspect the oil pump.
4. Position the oil pump onto the locator pins.
5. Install the oil pump bolt and tighten the bolt to 66 ft. lbs. (90 Nm).
6. Install the oil pan.

4.8L, 5.3L and 6.0L Engines

1. Remove or disconnect the following:
   - Engine front cover
   - Oil pan
   - Oil pump screen bolt and nuts
   - Oil pump screen with O-ring seal.
   - O-ring seal from the pump screen. Discard the O-ring seal.
   - Remaining crankshaft oil deflector nuts.

- Crankshaft oil deflector
- Oil pump bolts

➡**Do not allow dirt or debris to enter the oil pump assembly, cap ends as necessary.**

   - Oil pump

➡**The internal parts of the oil pump assembly are not serviced separately (excluding the spring). If the oil pump components are worn or damaged, replace the oil pump as an assembly. Do not attempt to repair the wire mesh portion of the pump and screen assembly.**

To install:

➡**Inspect the oil pump and engine block oil gallery passages. These surfaces must be clear and free of debris or restrictions.**

2. Align the splined surfaces of the crankshaft sprocket and the oil pump drive gear and install the oil pump. Install the oil pump onto the crankshaft sprocket until the pump housing contacts the face of the engine block.
3. Install or connect the following:
   - Oil pump bolts. Tighten the oil pump bolts to 18 ft. lbs. (25 Nm).
   - Crankshaft oil deflector

➡**Lubricate a NEW oil pump screen O-ring seal with clean engine oil.**

   - NEW O-ring seal onto the oil pump screen

➡**Push the oil pump screen tube completely into the oil pump prior to tightening the bolt. Do not allow the bolt to pull the tube into the pump.**

4. Align the oil pump screen mounting brackets with the correct crankshaft bearing cap studs.
5. Install or connect the following:
   - Oil pump screen
   - Oil pump screen bolt and the deflector nuts. Tighten the bolt to 106 inch lbs. (12 Nm) and the nuts to 18 ft. lbs. (25 Nm).

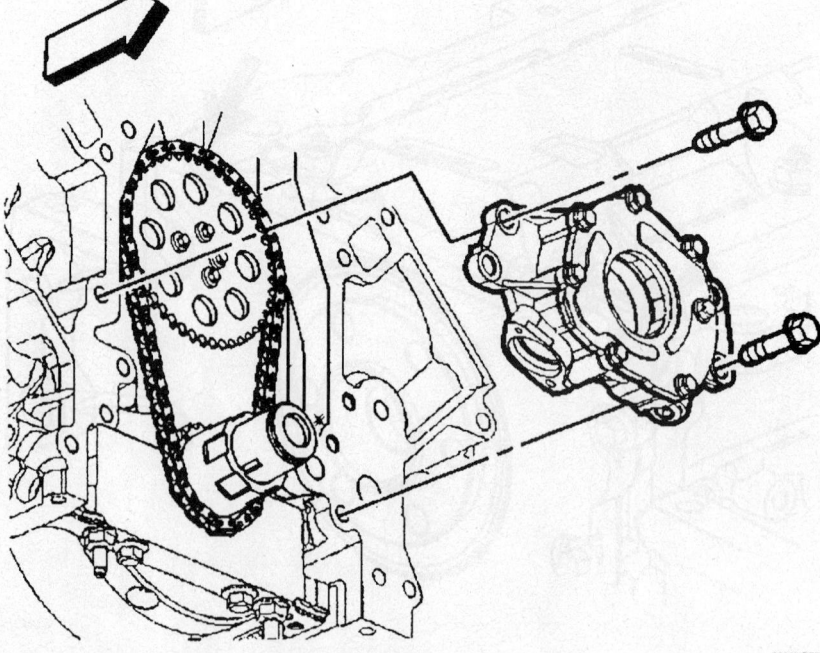

Oil pump removal—4.8L, 5.3L and 6.0L engines

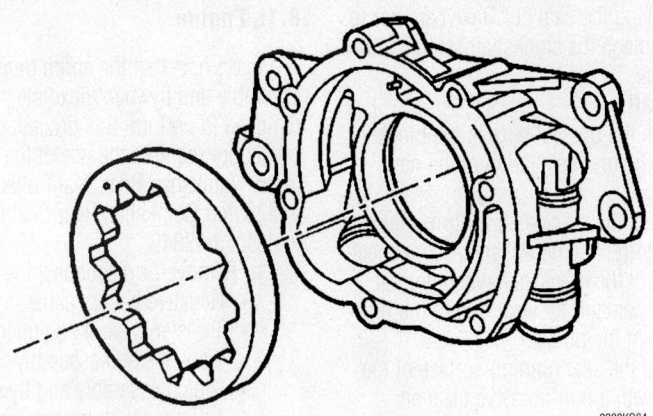

Oil pump disassembly—4.8L, 5.3L and 6.0L engines

- Oil pan
- Engine front cover

5.0L and 5.7L Engines

1. Remove or disconnect the following:

- Oil pan
- Oil pump attaching bolt, if equipped
- Pick-up tube nut/bolt
- Pump along with the pick-up tube and shaft, as necessary

2. Clean all sealing surfaces

To install:

3. Ensure that the pump pick-up tube is tight in the pump body. If the tube should come loose, oil pressure will be lost and oil starvation will occur. If the pick-up tube is loose it should be replaced.

4. If the pump has been disassembled and is being replaced or for any reason oil has been removed, it must be primed. It can either be filled with oil before installing the cover plate and oil kept within the pump during handling or the entire pump cavity can be filled with petroleum jelly.

➡If the pump is not primed, the engine could be damaged before it receives adequate lubrication when the engine is started.

5. Install or connect the following:
- Pump, aligning the pump shaft with the oil pump drive gear as necessary. Torque oil pump/pick-up tube retainer(s) to 65 ft. lbs. (90 Nm).
- Oil pan
6. Refill the engine crankcase
7. Disable the ignition system; crank engine for approximately 10 seconds to aid in priming the oil pump and reducing the risk of engine damage.

➡If the oil pump does not build up oil pressure almost immediately, remove the pan and check for a loose oil pump-to-pick-up tube attachment. If necessary dismantle the pump and pack the pump cavity with petroleum jelly. Running the engine without measurable oil pressure will cause extensive damage.

8.1L Engine

1. Remove or disconnect the following:

- Oil pan
- Oil pump screen bolt

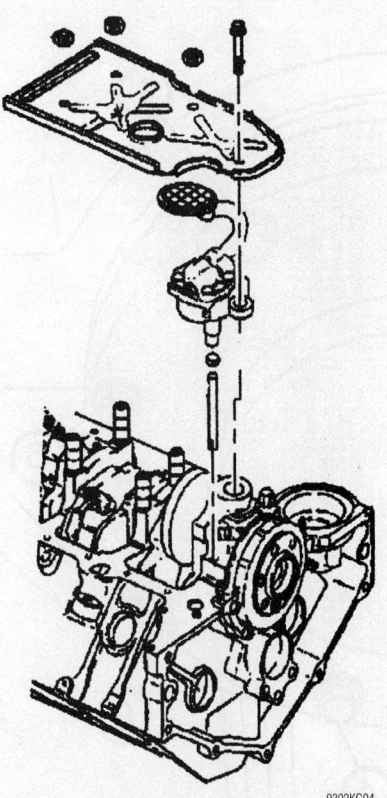

Exploded view of the oil pump mounting—
4.8L, 5.3L and 6.0L engines

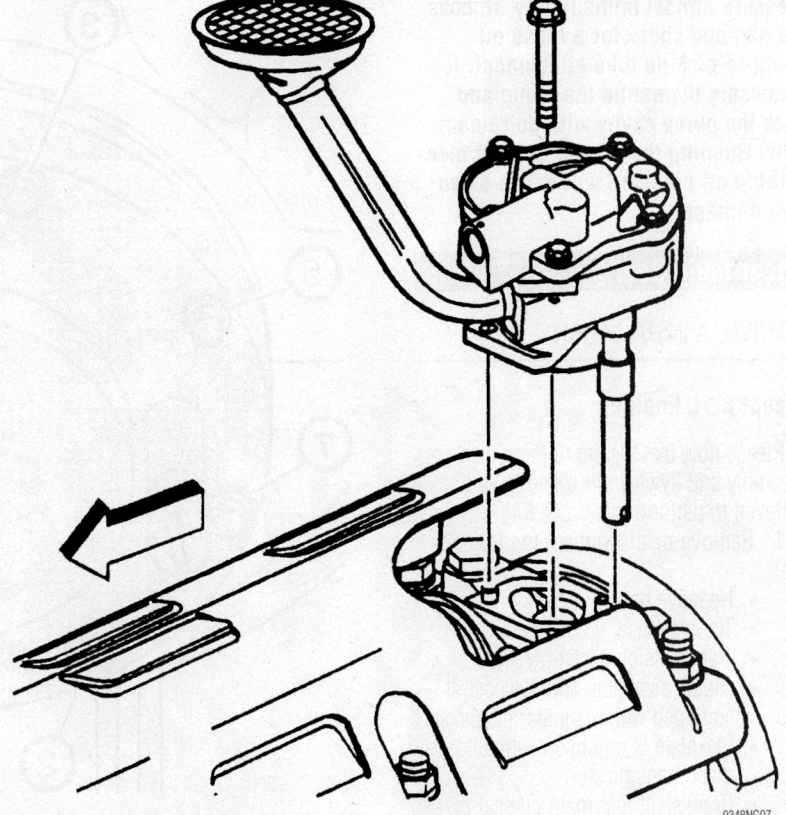

Oil pump removal—8.1L engine

- Oil pump, retainer and driveshaft. Discard the driveshaft retainer
- Crankshaft oil deflector nuts
- Crankshaft oil deflector
- Oil pump bolts
- Oil pump

2. Clean and inspect the oil pump

To install:

3. Install the crankshaft oil deflector. Tighten the nuts to 37 ft. lbs. (50 Nm).

➡**Always replace the retainer between the oil pump and the shaft, when installing the oil pump. During assembly, install a new oil pump driveshaft retainer. To ease installation, slightly heat the retainer to above room temperature.**

4. Assemble the oil pump, driveshaft and a new retainer.

5. Install or connect the following:
- Oil pump, positioning it on the locating pins
- Oil pump bolt and tighten to 56 ft. lbs. (75 Nm)
- Oil pan

6. Refill the engine crankcase

7. Disable the ignition system; crank engine for approximately 10 seconds to aid in priming the oil pump and reducing the risk of engine damage.

➡**If the oil pump does not build up oil pressure almost immediately, remove the pan and check for a loose oil pump-to-pick-up tube attachment. If necessary dismantle the pump and pack the pump cavity with petroleum jelly. Running the engine without measurable oil pressure will cause extensive damage.**

Rear Main Seal

REMOVAL & INSTALLATION

Except 8.1L Engine

Please note that the entire transmission assembly and flywheel/flexplate must be removed to perform this procedure.

1. Remove or disconnect the following:
- Negative battery cable
- Transfer case, if equipped
- Transmission assembly
- Clutch assembly and flywheel, if equipped with manual transmission
- Flexplate, if equipped with automatic transmission
- Crankshaft rear main oil seal by inserting a suitable prying tool and

prying the seal out. Take care not to damage the crankshaft sealing surface.

To install:

2. Clean the oil seal bore in the block thoroughly before installation of the new seal.

3. Inspect the crankshaft for grit, rust or burrs and correct as necessary. Also inspect the portion of the crankshaft where the oil seal makes contact, for wear due to the rubbing action of the oil seal.

4. Clean the seal running surface of the crankshaft with a non-abrasive cleaner.

5. Lubricate the inner diameter of the new seal and the outer diameter of the crankshaft with engine oil.

6. Install or connect the following:
- Rear main oil seal, using installation tool J 38841, J-35621-B or J-41479, until the tool bottoms against the block and crankshaft rear main bearing cap.
- Flywheel and clutch
- Flexplate, as required
- Transmission assembly
- Transfer case, if equipped
- Negative battery cable

7. Start the engine and verify no oil leaks.

8.1L Engine

Please note that the entire transmission assembly and flywheel/flexplate must be removed to perform this procedure. This procedure requires the use of the following tools: Crankshaft Rear Seal Puller tool No. J 43320 and Crankshaft Rear Seal Installer tool No. J 42849.

1. Remove or disconnect the following:
- Negative battery cable
- Transfer case, if equipped
- Transmission assembly
- Clutch assembly and flywheel, if equipped with manual transmission
- Flexplate, if equipped with automatic transmission

2. Install the guide pins from the Crankshaft Rear Sear Puller into the crankshaft.

3. Install the Rear Seal Puller over the guide pins.

4. Using a drill, insert 8 of the self-drilling sheet metal screws into the rear crankshaft seal, using a crisscross pattern as shown. The self tapping screws are included with the Crankshaft Rear Seal Puller.

5. Thread the center bolt of the Crankshaft Rear Seal Puller into the crankshaft to remove the seal.

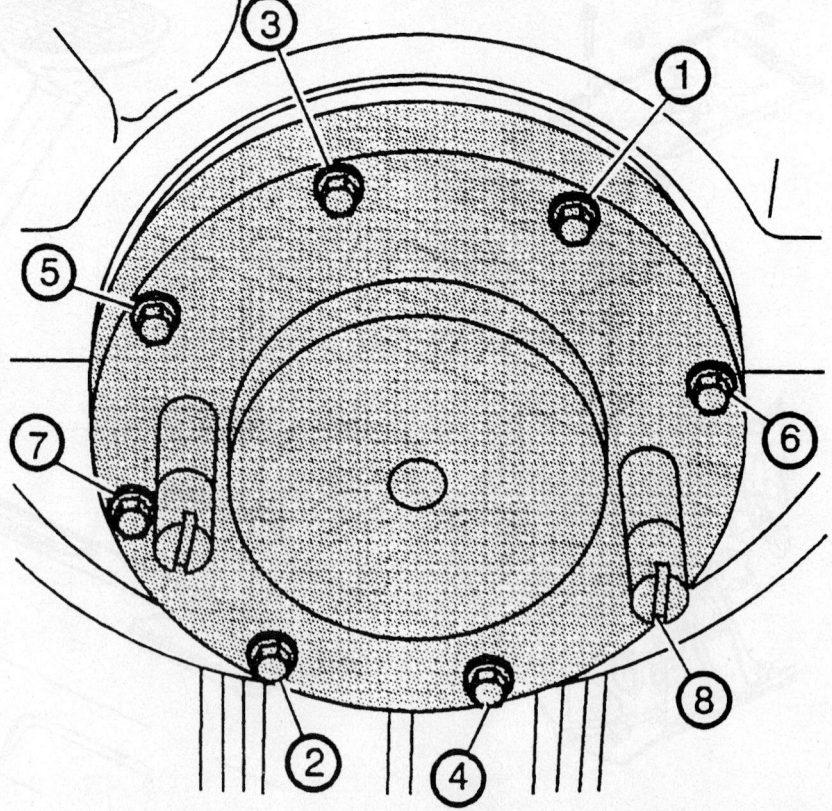

Drill the screws into the rear main seal using a crisscross pattern—8.1L engine

9348NG08

6. Remove the guide pins from the crankshaft.

To install:

7. Make sure there is no dirt, rust or loose burrs on the crankshaft.

8. Apply a light coating of engine oil to the crankshaft sealing surface. Do NOT get oil on the sealing surface of the engine block.

9. Install the new rear main seal onto the Crankshaft Rear Seal Installation Tool.

10. Position the Rear Seal Installation Tool against the crankshaft. Thread the attaching screws into the tapped holes in the crankshaft.

11. Use a screwdriver to tighten the screws securely to make sure the seal is squarely installed against the crankshaft.

12. Rotate the center nut until the installation tool bottoms, then remove the seal installation tool.

13. Install or connect the following:
- Flexplate, if equipped with automatic transmission
- Clutch assembly and flywheel, if equipped with manual transmission
- Transmission assembly
- Transfer case, if equipped
- Negative battery cable

Timing Chain, Sprockets, Front Cover and Seal

The manufacturer recommends that the front cover oil seal be replaced whenever the cover is removed.

REMOVAL & INSTALLATION

4.3L, 5.0L and 5.7L Engines

1. Drain the cooling system.
2. Remove or disconnect the following:
- Negative battery cable
- Fan shroud assembly
- Belts, pulleys and water pump assembly
- Crankshaft pulley and damper
- Oil pan-to-front cover bolts

➥**If equipped with a composite front cover, it must be replaced with a new one. Reusing the front cover may result in oil leaks.**

3. Remove the engine shield bolts and shield.

4. Disconnect the Crankshaft Position (CKP) sensor electrical connector, if equipped.

5. Remove the CKP sensor and discard the o-ring.

- Screws holding the timing chain cover to the block
- Cover and gaskets.

6. Remove the crankshaft position (CKP) sensor reluctor ring, if equipped.

7. Use a suitable tool to pry the old seal out of the front face of the cover.

8. Rotate the crankshaft until the timing marks on the camshaft and crankshaft sprockets are in proper alignment. This will put no. 4 cylinder at TDC.

9. Unsnap the timing chain tensioner shoe from the pin.

10. Remove or disconnect the following:
- Camshaft sprocket-to-camshaft nut and/or bolts
- Camshaft sprocket (along with the timing chain), if the sprocket is difficult to remove, use a plastic mallet to bump the sprocket from the camshaft.

➥**The camshaft sprocket (located by a dowel) is lightly pressed onto the camshaft and should come off easily. The chain comes off with the camshaft sprocket.**

11. If necessary use J-5825-A, or equivalent, crankshaft sprocket removal tool to free the timing sprocket from the crankshaft.

12. Remove the crankshaft balancer key.

13. If necessary, remove the timing chain tensioner bracket bolt and bracket.

To install:

14. Inspect the timing chain and the timing sprockets for wear or damage, replace the damaged parts as necessary.

15. Clean the gasket mounting surfaces of all remaining traces of old gasket.

➥**During installation, coat the thrust surfaces lightly with Molykote® or equivalent pre-lube.**

16. If necessary, install the timing chain tensioner bracket and bolt and tighten to 106 inch lbs. (12 Nm).

17. Install the key into the crankshaft keyway. The crankshaft balancer key should be parallel to the crankshaft or with a slight incline.

18. Install or connect the following:
- Crankshaft sprocket onto the crankshaft, use tool J-5590, crankshaft sprocket installation tool, and a hammer, without disturbing the position of the engine.
- Timing chain, arrange the camshaft sprocket in such a way that the timing marks will align between the shaft centers and the camshaft locating dowel will enter the dowel hole in the cam sprocket.

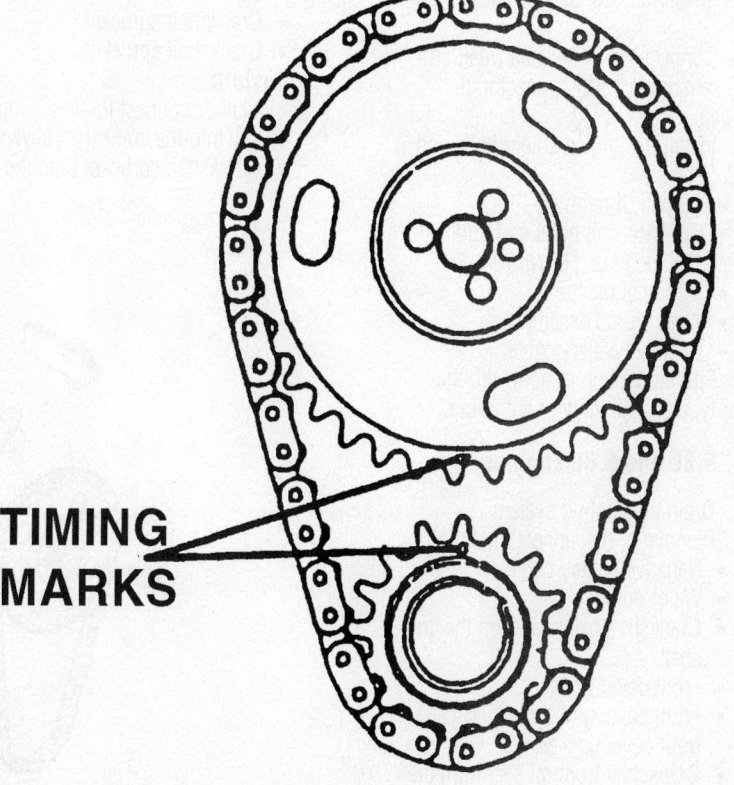

TIMING MARKS

7924KG21

Timing mark alignment for timing chain removal and installation—gasoline engines

• Cam sprocket, with the chain mounted under it in position on the front of the camshaft. Torque the camshaft sprocket-to-camshaft retainer bolts to 106 inch lbs. (12 Nm) on 5.0L and 5.7L, or 18 ft. lbs. (25 Nm) on 4.3L.

19. Install the timing chain tensioner shoe onto the bracket and position the top of the shoe under the tab at the top of the bracket.

20. With the timing chain installed, turn the crankshaft 2 complete revolutions, then check to make certain that the timing marks are in correct alignment between the shaft centers.

➡ **Coat the lip of the new seal with oil prior to installation.**

• New seal so that the open end is toward the inside of the cover, using seal driver J-22102
• New front pan seal, cutting the tabs off.

21. Coat a new cover gasket with adhesive sealer and position it on the block.

22. Apply a ⅛ in. (3mm) bead of RTV gasket material to the front cover.

23. Install the crankshaft position (CKP) sensor reluctor ring, if equipped.

24. Install the front cover carefully onto the locating dowels and tighten the attaching screws

25. Install the CKP sensor and new o-ring.

26. Connect the Crankshaft Position (CKP) sensor electrical connector, if equipped.

27. Install the engine shield bolts and shield.

• Oil pan, if removed
• Cover-to-pan bolts and tighten to 106 inch lbs. (12 Nm)
• Torsional damper
• Water pump assembly
• Negative battery cable

28. Fill the cooling system with the proper type and quantity of antifreeze.

4.8L, 5.3L and 6.0L Engines

1. Drain the cooling system.
2. Remove or disconnect the following:
• Negative battery cable
• Water pump
• Crankshaft balancer from the crankshaft
• Front cover bolts
• Front cover and gasket. Discard the front cover gasket.
• Crankshaft front oil seal from the cover

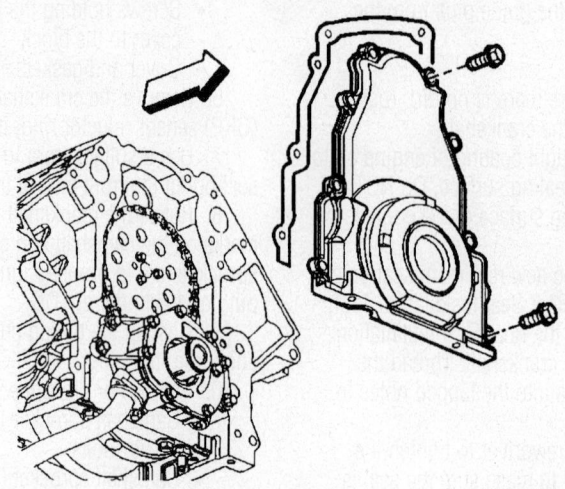

Front cover and gasket—4.8L, 5.3L and 6.0L engines

• Oil pump

3. Rotate the crankshaft until the timing marks on the crankshaft and the camshaft sprockets are aligned.

➡ **Do not turn the crankshaft assembly after the timing chain has been removed in order to prevent damage to the piston assemblies or the valves.**

4. Remove or disconnect the following:

• Camshaft sprocket bolts
• Camshaft sprocket and timing chain
• Crankshaft sprocket
• Crankshaft sprocket key

To install:

5. Install or connect the following:
• Key into the crankshaft keyway
• Crankshaft sprocket onto the front of the crankshaft. Align the crankshaft key with the crankshaft sprocket keyway. Rotate the crankshaft sprocket until the alignment mark is in the 12 o'clock position.
• Camshaft sprocket and timing chain. Locate the camshaft sprocket alignment mark in the 6 o'clock position.
• Camshaft sprocket bolts and tighten to 26 ft. lbs. (35 Nm)

➡ **Do not lubricate the oil seal sealing surface.**

6. Lubricate the outer edge of the oil seal with clean engine oil. Lubricate the front cover oil seal bore with clean engine oil.

7. Install the crankshaft front oil seal with an installer.

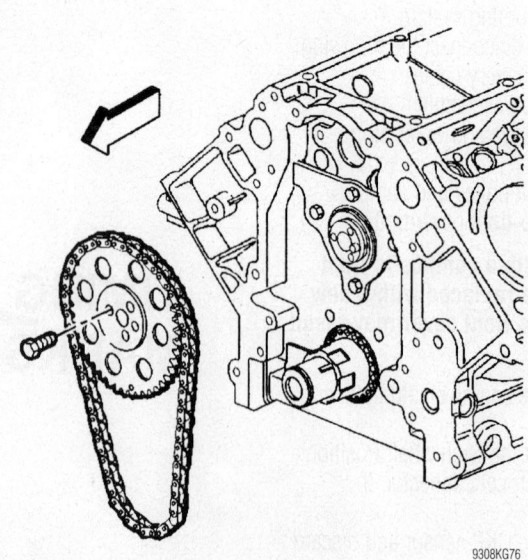

Sprocket and chain removal—4.8L, 5.3L and 6.0L engines

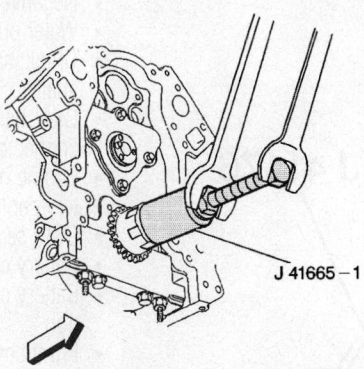

Crankshaft sprocket installation—4.8L, 5.3L and 6.0L engines

➡**Do not apply any type of sealant to the front cover gasket (unless specified). Special tools are used to properly align the engine front cover at the oil pan surface and to center the crankshaft front oil seal.**

8. Install the front cover gasket, cover, and bolts onto the engine. Tighten the cover bolts finger-tight. Do not overtighten.

9. Start the J41480 tool-to-front cover bolts. Don't tighten the bolts yet.

➡**Align the tapered legs of the tool with the machined alignment surfaces on the front cover.**

10. Install tool J41476 . Install the crankshaft balancer bolt. Tighten the crankshaft balancer bolt by hand until snug. Do

not overtighten. Tighten the J41480 bolts and front cover bolts to 18 ft. lbs. (25 Nm).

11. Remove the tools.

12. Install the used crankshaft balancer bolt and tighten to 240 ft. lbs. (330 Nm).

13. Remove the used bolt.

➡**The nose of the crankshaft should be recessed 2.4-4.48 mm (0.094-0.176 in) into the balancer bore.**

14. Install a NEW crankshaft balancer bolt and tighten to 37 ft. lbs. (50 Nm),, then tighten an additional 140 degrees.

15. Place a straight edge across the engine block and front cover oil pan sealing

surfaces. Avoid contact with the portion of the gasket that protrudes into the oil pan surface. Insert a feeler gauge between the front cover and the straight edge tool. The cover must be flush with the oil pan surface or no more than 0.02 in. (0.5mm) below flush. If the front cover-to-engine block oil pan surface alignment is not within specifications, repeat the cover alignment procedure. If the correct front cover-to-engine block alignment cannot be obtained, replace the front cover.

16. Install the crankshaft balancer bolt. Tighten the crankshaft balancer bolt by hand until snug. Do not overtighten the bolt.

17. Snug the oil pan-to-cover bolts in order to position the cover at the pan rail.

18. Tighten the oil pan-to-front cover bolts to 18 ft. lbs. (25 Nm).

19. Tighten the front cover bolts to 18 ft. lbs. (25 Nm).

20. Install the water pump.

8.1L Engine

➡**This procedure requires the use of Crankshaft Sprocket Installer tool No. J 22102 and Crankshaft Protector Button tool No. J 42846.**

1. Drain the cooling system.

2. Remove or disconnect the following:

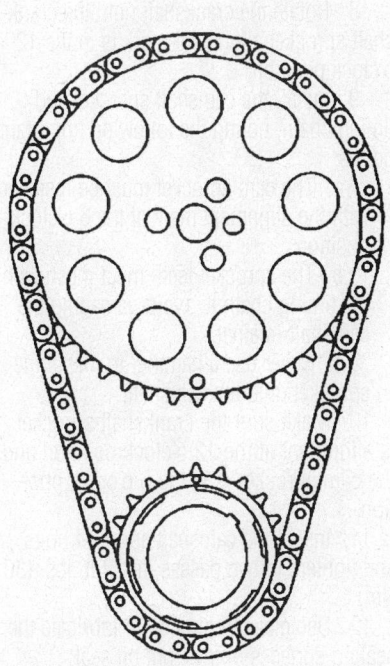

Timing mark alignment—4.8L, 5.3L and 6.0L engines

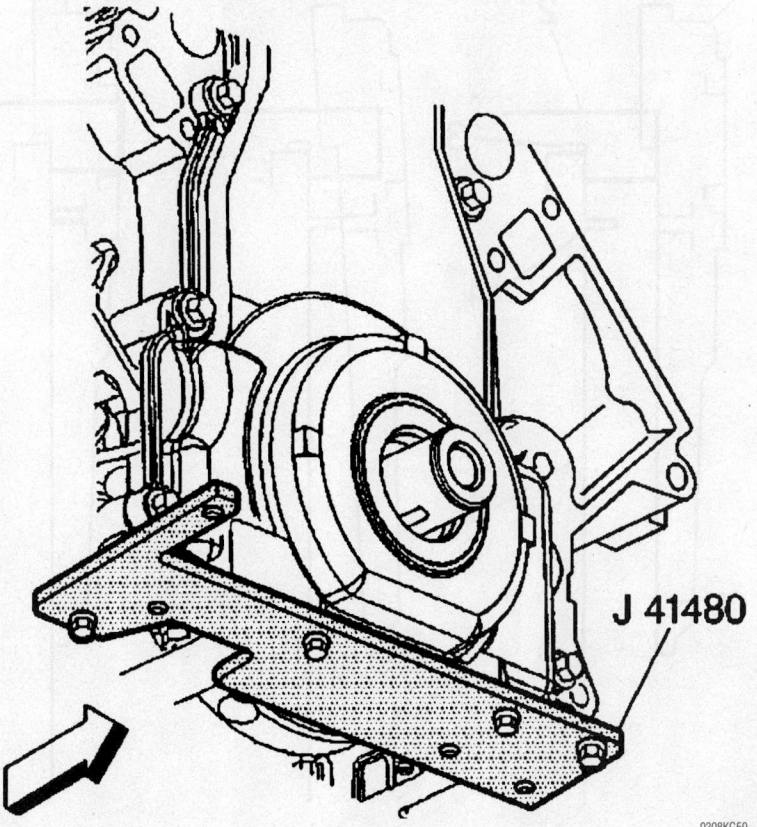

J 41480

J41480 installation—4.8L, 5.3L and 6.0L engines

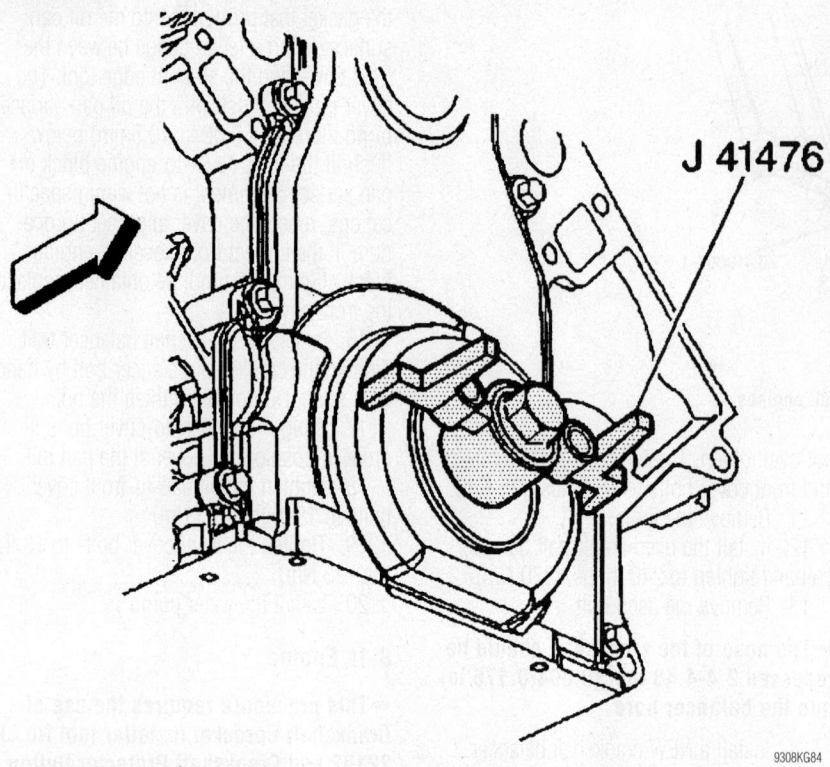

Seal alignment tool installation—4.8L, 5.3L and 6.0L engines

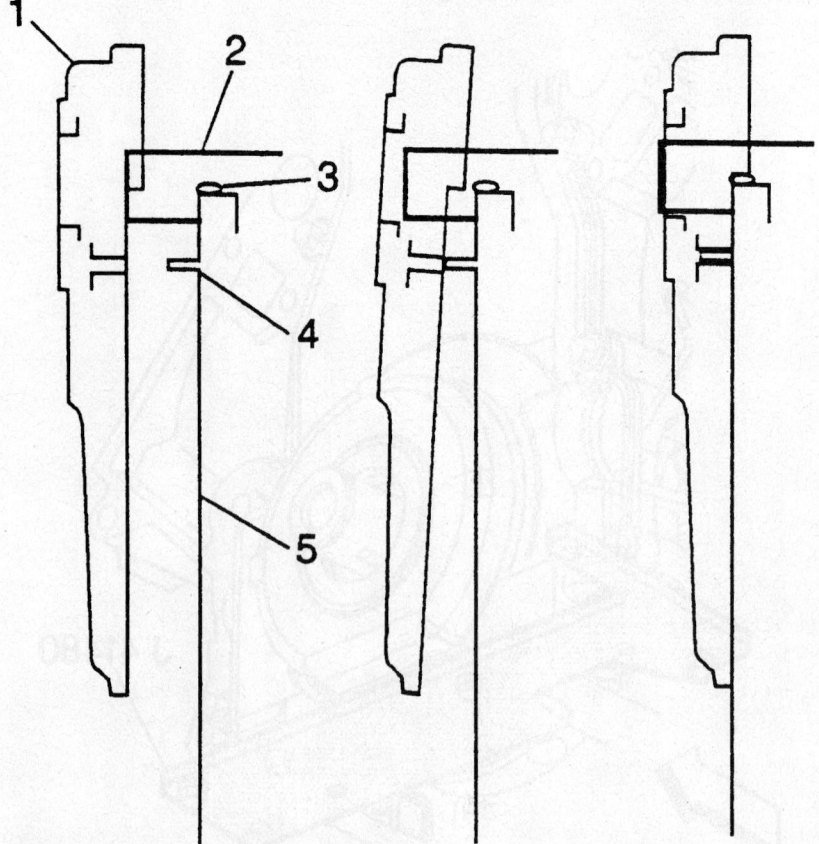

Proper front cover installation sequence—8.1L engine

- Negative battery cable
- Water pump
- Crankshaft balancer from the crankshaft
- Camshaft Position (CMP) sensor connector
- Engine harness clips from the battery cable channel
- CMP sensor bolt and sensor
- Battery cable channel bolt
- Battery cable channel and reposition
- Front cover bolts, front cover and gasket

➡ **The front cover gasket can be reused if it is not damaged.**

- Crankshaft front oil seal from the front cover

3. Align the timing marks on the camshaft and crankshaft sprockets.
- Camshaft sprocket bolts
- Camshaft sprocket and timing chain

4. Install Crankshaft Protector Button tool No. J 42846 into the end of the crankshaft and remove the crankshaft sprocket using a 3-jawed puller.

5. Clean and inspect the timing chain and sprockets.

To install:

6. Use the Crankshaft Sprocket Installer tool No. J 22102 to install the crankshaft sprocket. Align the keyway of the sprocket with the crankshaft pin.

7. Remove the installation tool.

8. Rotate the crankshaft until the crankshaft sprocket alignment mark is in the 12 o'clock position.

9. Install the camshaft sprocket and timing chain, noting the following important points:

   a. The cam sprocket must be installed with the alignment mark at the 6 o'clock position.

   b. The sprocket teeth must mesh with the timing chain to avoid damaging the camshaft retainer.

   c. Never use a hammer to install the sprocket onto the camshaft.

10. Make sure the crankshaft sprocket is alignment at the 12 o'clock position and the cam sprocket is at the 6 o'clock position.

11. Install the camshaft sprocket bolts and tighten, in two passes, to 22 ft. lbs. (30 Nm)

12. Use clean engine oil to lubricate the sealing surfaces of the front oil seal.

13. Install or connect the following:
- New seal into the front cover, using a suitable seal installation tool

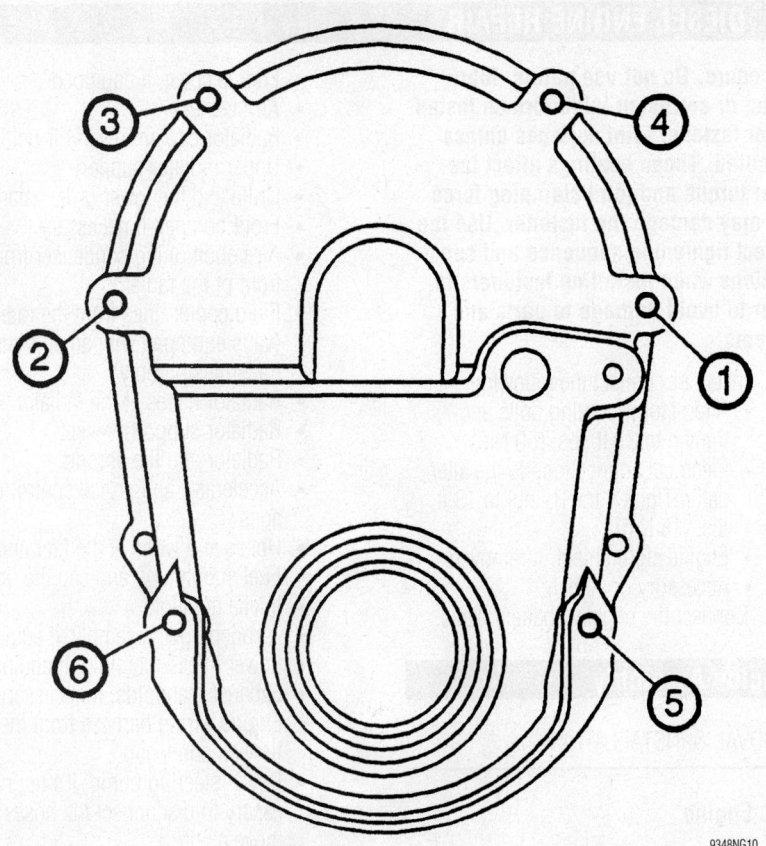

Engine front cover bolt tightening sequence—8.1L engine

➥**The front cover must be installed while the sealant is still wet to the touch.**

- Sealant to the 2 places on the engine block where the front cover meets the oil pan
- Front cover gasket into the cover

14. Install the front cover, referring to the accompanying figure and using the following steps only:

a. Hold the front cover (1) up to the crankshaft (2).

b. Lift the cover (1) while sliding the cover over the crankshaft (2).

c. Slide the front cover toward the engine block (5) while keeping the cover raised.

d. Lower the cover down over the dowel pin (4), allowing the front cover to rest on the sealant (3).

15. Install the front cover bolts and tighten, in sequence, as follows:

a. 1st pass: 53 inch lbs. (6 Nm)

b. 2nd pass: 106 inch lbs. (12 Nm)

16. Install or connect the following:

- Battery cable channel and bolt. Tighten to 80 inch lbs. (9 Nm).
- CMP sensor. Inspect the O-ring first, replace if necessary and coat with oil before installation
- CMP sensor bolt to 106 inch lbs. (12 Nm)
- Engine harness clips to the battery cable channel
- CMP sensor electrical connector
- Crankshaft balancer
- Water pump
- Negative battery cable.

17. Fill the cooling system with the proper type and quantity of antifreeze.

Piston and Ring

POSITIONING

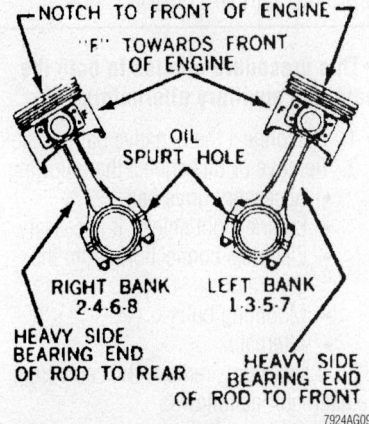

Piston and connecting rod assembly positioning—4.3L, 4.8L, 5.0L, 5.3L, 5.7L and 6.0L engines

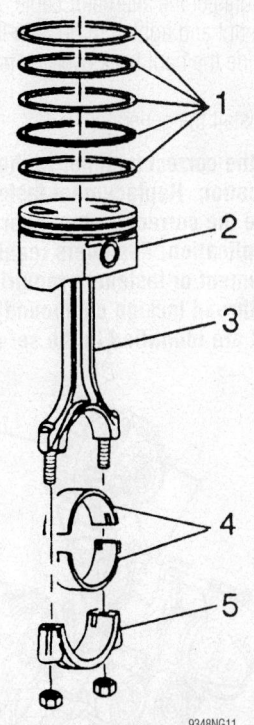

Piston rings (1), piston (2), connecting rod (3) and related components—8.1L engine

DIESEL ENGINE REPAIR

Alternator

REMOVAL & INSTALLATION

➡ **This procedure applies to both the main and auxiliary alternators.**

1. Disconnect the negative battery cable.
2. Remove or disconnect the following:
   - Accessory drive belt
   - Engine sight shield, if necessary
   - Electrical connections from the generator
   - Mounting bolts
   - Alternator
3. If necessary, remove the cable from the alternator as follows:

   a. Slide the boot down, to reveal the terminal stud.

   b. Unfasten the cable nut from the stud, then remove the alternator cable.

To install:

4. Connect the alternator cable, secure with the nut and tighten to 80 inch lbs. (9 Nm). Slide the boot back over the terminal stud.

5. Install the generator.

➡ **Use the correct fastener in the correct location. Replacement fasteners must be the correct part number for that application. Fasteners requiring replacement or fasteners requiring the use of thread locking compound or sealant are identified in the service procedure. Do not use paints, lubricants, or corrosion inhibitors on fasteners or fastener joint surfaces unless specified. These coatings affect fastener torque and joint clamping force and may damage the fastener. Use the correct tightening sequence and specifications when installing fasteners in order to avoid damage to parts and systems.**

6. Install or connect the following:
   - Alternator mounting bolts and tighten to 37 ft. lbs. (50 Nm)
   - Electrical connections to the alternator. Tighten the B+ nut to 13 ft. lbs. (18 Nm).
   - Engine sight shield, if removed
   - Accessory drive belt
7. Connect the negative battery cable.

Engine Assembly

REMOVAL & INSTALLATION

6.5L Engine

1. Drain the cooling system.
2. Discharge the air conditioning system and remove the air conditioning vacuum reservoir.
3. Drain the engine oil.
4. Remove or disconnect the following:
   - Battery cables
   - Engine cover, if equipped
   - Air cleaner
   - Radiator coolant reservoir bottle
   - Upper radiator support
   - Grille and the lower grille valance
   - Front bumper, if necessary
   - Air conditioning condenser from in front of the radiator
   - Fluid cooler lines from the radiator (vans equipped with an automatic transmission only)
   - Radiator hoses at the radiator
   - Radiator support bracket
   - Radiator and the shroud
   - Accelerator and cruise control linkages
   - Hoses and wires at the fuel unit
   - Fuel supply unit and cap the lines
   - Intake manifold
   - Turbocharger assembly, if equipped
   - Lower intake manifold, if equipped
   - Exhaust manifolds, if necessary
   - Engine wiring harness from the firewall connection
   - Power steering pump, it's not necessary to disconnect the hoses; just move aside
   - Heater hoses at the engine
   - Thermostat housing, if necessary
   - Oil filler and automatic transmission tubes
   - Cruise control servo, servo bracket and transducer, if equipped
   - Exhaust pipes at the manifolds
   - Driveshaft and plug the end of the transmission
   - Transmission shift linkage and the speedometer cable
   - Fuel line from the fuel tank and pump
   - Transmission mounting bolts
5. Support the transmission and engine.
6. Install lifting hooks J-41427 as follows:

   a. Remove the 2 right rear lower intake manifold retainers and install lifting hook J-41427 (the one marked "right"). Tighten the bolts to 11 ft. lbs. (15 Nm).

   b. Remove the air conditioning compressor and the accessory drive bracket.

   c. If equipped, disconnect the EGR tube and the 2 left lower bolts from the intake manifold.

   d. Install the lifting hook J-41427 (the one marked "left") and tighten the bolts to 11 ft. lbs. (15 Nm).

7. Remove or disconnect the following:

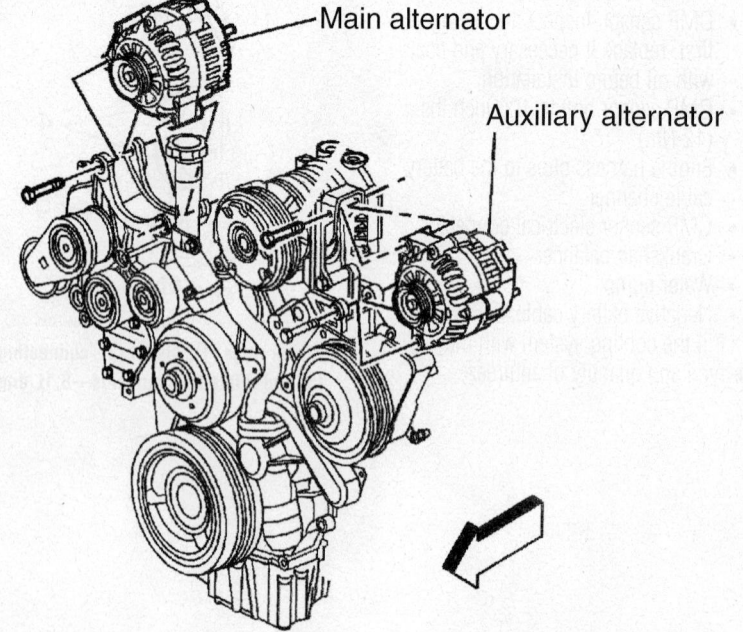

Main alternator

Auxiliary alternator

06025-AVAL-G04

Main and auxiliary alternator mounting

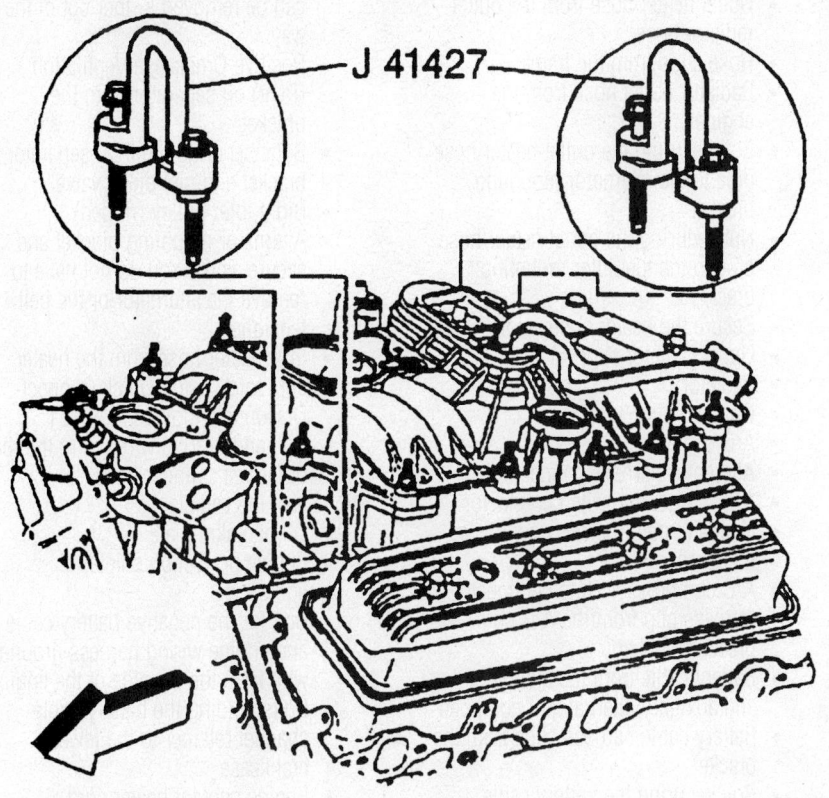

J 41427

7924KG01

For engine removal and installation, universal lift brackets should be installed in place of the proper intake manifold bolts

- Engine mount bracket-to-frame bolts
- Engine mount through-bolts

8. Raise the engine slightly and remove the engine mounts, support the engine with wood between the oil pan and the cross-member.

9. Remove the manual transmission and clutch as follows:

a. Remove the clutch housing rear bolts.

b. Remove the bolts attaching the clutch housing to the engine and remove the transmission and clutch as a unit.

➡**Support the transmission as the last bolt is being removed to prevent damaging the clutch.**

c. Remove the starter and clutch housing rear cover.

d. Loosen the clutch mounting bolts a little at a time to prevent distorting the disc until spring pressure is released. Remove all of the bolts, the clutch disc and the pressure plate.

10. Remove the automatic transmission as follows:

a. Lower the engine and support it on blocks.

b. Remove the starter and converter housing cover.

c. Remove the flexplate-to-converter attaching bolts.

d. Support the transmission on blocks.

e. Disconnect the detent cable on the Turbo Hydra-Matic.

f. Remove the transmission-to-engine mounting bolts.

11. Attach an engine crane to the engine.

12. Remove the blocks from the engine only and glide the engine away from the transmission.

To install:

13. Install the manual transmission and clutch as follows:

a. Install the clutch disc and the pressure plate. Tighten the clutch mounting bolts a little at a time to prevent distorting the disc.

b. Install the starter and clutch housing rear cover.

c. Install the bolts attaching the clutch housing to the engine and install the transmission and clutch as a unit. Tighten the bolts to specification.

d. Install the clutch housing rear bolts.

14. Install the automatic transmission as follows:

a. Position the transmission.

b. Install the transmission-to-engine mounting bolts.

c. Connect the throttle linkage and detent cable.

d. Install the flexplate-to-converter attaching bolts. Torque the bolts to 65 ft lbs. (90 Nm).

e. Install the starter and converter housing cover.

15. Install or connect the following:

- Engine mount through-bolts and tighten to 50 ft lbs. (68 Nm)
- Engine mount bracket-to-frame bolts. Torque the bolts to 44 ft lbs. (59 Nm).
- Clutch cross-shaft
- Transmission mounting bolts and tighten to 75 ft lbs. (100 Nm)
- Transmission shift linkage and the speedometer cable
- Driveshaft

16. Remove the lifting hooks.

- Compressor
- EGR valve tube
- Intake manifold retaining bolts
- Condenser
- Hood latch support
- Lower fan shroud and filler panel
- Transmission dipstick tube and the accelerator cable at the tube
- Coolant hose at the intake manifold and the PCV valve
- Cruise control servo, bracket and transducer
- Oil filler pipe and automatic transmission filler pipe
- Engine dipstick tube
- Thermostat housing, if removed
- Heater hoses at the engine
- Engine wiring harness to the firewall connection
- Radiator and the shroud
- Radiator support bracket
- Exhaust manifolds if removed
- Lower intake manifold if removed
- Intake manifold
- Fuel supply unit
- Lines to the fuel supply unit
- Accelerator and cruise control linkages
- Windshield wiper jar and bracket
- Air conditioning condenser
- Air conditioning vacuum reservoir, if equipped
- Fluid cooler lines at the radiator, if equipped with an automatic transmission
- Radiator coolant reservoir bottle
- Hoses at the radiator
- Upper radiator support
- Grille and the lower grille valance

- Air cleaner
- Air stove pipe
- Engine cover
- Battery cables

17. Refill the cooling system.
18. Recharge the air conditioning system.

6.6L Engine

➡**In order to remove the engine, the vehicle must be on a lift. the front tires also need to be removed. You will have to support the vehicle by its frame for tire removal.**

1. Drain the cooling system.
2. Discharge and recover the air conditioning system.
3. Drain the engine oil.
4. Raise the hood to the servicing position. Move the hood hinge bolt to hold the hood in the servicing position.
5. Disconnect the battery cables.
6. Remove the upper intake manifold sight shield as follows:

   a. Remove the retaining bolt in the front of the shield.

   b. Lift up on the front of the shield, then left the shield off the rear bracket.

7. Remove or disconnect the following:

➡**After you remove the duct, cover the turbocharger openings and ducts with tape to prevent foreign objects from entering.**

- Air cleaner outlet duct from the air cleaner and turbocharger.
- Mass Air Flow (MAF) switch connector
- A/C pressure cycling switch connector
- Surge tank switch
- Engine wire harness clip from the accumulator
- Engine wire harness clips from the wheelhouse inner panel and engine bracket
- Air cleaner assembly and bracket
- Surge tank

8. Raise the vehicle.
- Front tires and wheels
- Both front fender wheelhouse inner panels

9. Lower the vehicle.
- Charged air cooler pipes and hoses from the engine and charged air cooler
- Radiator inlet hose form the radiator and engine
- Upper and lower fan shrouds
- Radiator outlet hose from the radiator

- Outlet heater hose from the outlet radiator hose
- Hose clips from the frame
- Radiator outlet hose from the engine
- Bolt securing the outlet heater hose pipe to the alternator mounting bracket
- Nut securing the outlet heater hose pipe to the fuel filter mounting bracket
- Secure the heater hose aside
- Upper radiator support
- Radiator
- Charged air cooler
- A/C condenser
- Alternator harness connector
- A/C refrigerant switch connector
- Dual alternator harness connector, if equipped
- A/C compressor clutch connector
- Harness clip from the A/C compressor bracket
- Battery cable from the alternator and auxiliary alternator, if equipped
- Battery cable harness clips from the bracket
- Bolt securing the battery cable junction block from the power steering pump
- Move and secure the battery cables aside
- Both fuel injection control module harness connectors, by flipping the latch up
- Engine wire harness from the retainer
- Fuel lines at the engine
- Remove the nut and the fuel line bracket from the upper valve rocker arm cover stud
- Fuel lines aside
- Power supply cable from the glow plug relay
- Drive belt
- Suction hose from the accumulator. You can leave the compressor end on the compressor.
- A/C compressor bolts, then move the compressor, with the hoses attached, to the right side of the engine compartment
- Wiring harness to the left side of the engine and tie aside
- Bolts holding the power steering pump front bracket to the pump and A/C compressor mounting bracket
- A/C compressor and power steering pump bracket. Once the battery cables are removed from the engine, the power steering pump

can be removed further out of the way
- Positive Crankcase Ventilation (PCV) oil separator from the bracket
- Bolts securing the PCV separator bracket and fuel bleed valve
- Right idler pulley (ribbed)
- Alternator mounting bracket and secure aside. You do not have to remove the alternator or the belt tensioner
- Inlet heater hose from the heater core inlet, using Quick Connect-Disconnect tool No. J 43181
- Bolt and ground wires from the rear of the left cylinder head

10. Raise the vehicle
- Oil pan skid plate
- Engine protection shield, if equipped
- Bolt for the negative battery cable and engine wiring harness ground wire from the left side of the engine
- Bolts holding the battery cable channel retainer to the lower crankcase
- Engine coolant heater cord
- Starter motor
- Nut securing the battery cable bracket to the right side of the lower crankcase
- Bolt holding the auxiliary negative battery cable and the engine wiring harness ground wires to the right side of the engine
- Position the battery cables aside
- Exhaust pipe-to-exhaust outlet clamp
- Lower oil pan, if 4WD

11. If equipped with an automatic transmission, matchmark the installed position of the flywheel and torque converter.
- Torque converter bolts through the starter opening
- Transmission oil line clip nut if equipped with A/T
- Nuts securing the transmission fluid fill tube bracket, if equipped with A/T
- Transmission-to-engine stud and bolts. Note the location of the studs and any brackets attached to the studs

12. Lower the vehicle to work through the wheel opening
- Engine mount-to-frame bracket bolts

13. Lower the vehicle.
14. Install Engine Lifting Bracket tool No. J 36857 to the rear of the left cylinder head with a suitable bolt.

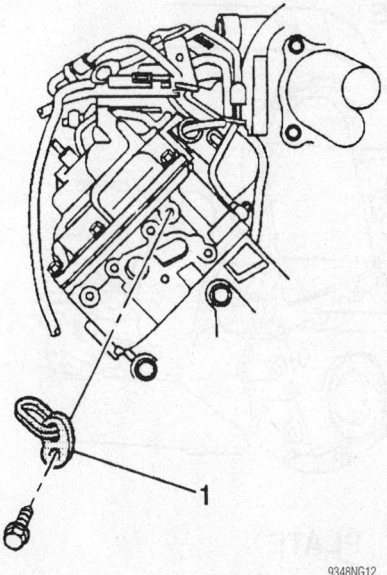

Install the engine lifting bracket to the rear of the left cylinder head

15. Install Engine Lifting Bracket tool No. J 36857 to the front of the right cylinder head with a suitable bolt.

16. Install a suitable lifting device. The engine will have to be angled to remove it. Use a load positioning sling to help in angling the engine.

17. Raise the vehicle off the engine mounts.

18. Remove the left and right engine mount frame brackets.

19. Remove the engine assembly from the vehicle.

20. Secure the engine on an engine stand by removing the following components:
  • Flywheel/flexplate
  • Rear main seal
  • Exhaust outlet
  • Oil pan
  • Flywheel housing

To install:

21. Install Engine Lifting Bracket tool No. J 36857 to the rear of the left cylinder head with a suitable bolt.

22. Install Engine Lifting Bracket tool No. J 36857 to the front of the right cylinder head with a suitable bolt.

23. Install a suitable lifting device. The engine will have to be angled to install it. Use a load positioning sling to help in angling the engine.

24. Install or connect the following:
  • Engine in the vehicle
  • 2 transmission-to-engine bolts, loosely
  • Left and right side engine mount frame brackets and tighten to 55 ft. lbs. (75 Nm)
  • Engine mount-to-frame bracket bolts and tighten to 50 ft. lbs. (65 Nm)

25. Remove the lifting brackets from the cylinder heads.
  • Transmission-to-engine bolts/studs and tighten to 37 ft. lbs. (50 Nm)
  • Torque converter bolts and tighten to 44 ft. lbs. (60 Nm)

26. Install the remaining components in the reverse of the removal procedure, noting the following important specifications and steps:
  • Transmission fluid tube bracket nuts: 13 ft. lbs. (18 Nm)
  • Transmission oil cooler line clip bolt: 80 inch lbs. (9 Nm)
  • Exhaust pipe clamp: 30 ft. lbs. (40 Nm)
  • Ground wire bolt: 25 ft. lbs. (34 Nm)
  • Battery cable bracket bolts: 106 inch lbs. (12 Nm)
  • Cable and ground wire bolts: 25 ft. lbs. (34 Nm)
  • Oil pan skid plate bolts: 15 ft. lbs. (20 Nm)
  • Engine protection shield bolts: 15 ft. lbs. (20 Nm)
  • Alternator bracket bolt: 37 ft. lbs. (50 Nm)
  • Idler pulley bolt: 32 ft. lbs. (43 Nm)
  • A/C compressor bolts: 37 ft. lbs. (50 Nm)

27. Refill the crankcase and the cooling system.

28. Recharge the air conditioning system.

Water Pump

REMOVAL & INSTALLATION

6.5L Engine

1. Drain the engine coolant.
2. Remove or disconnect the following:
  • Negative battery cables
  • Fan and fan shroud
  • Air conditioning hose bracket and/or the oil filler tube, as required
  • Accessory drive belt(s)
  • Vacuum pump mounting bracket nuts/bolt
  • Vacuum pump and bracket
  • Power steering pump and bracket
  • Coolant hoses from the pump
  • Water pump plate retaining bolts
  • Pump and plate assembly from the engine

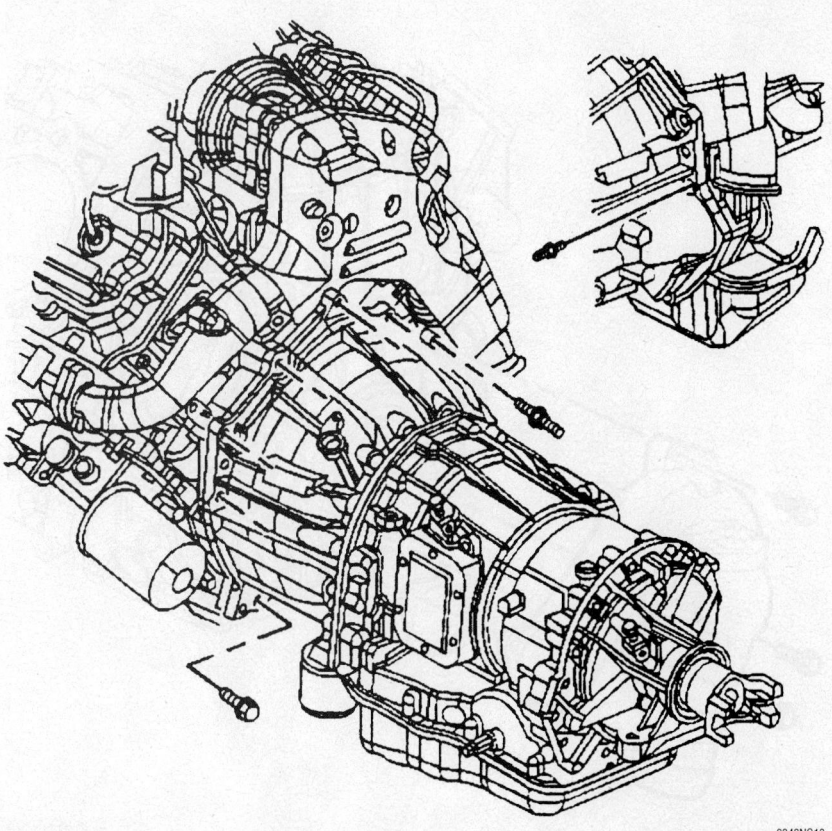

Transmission-to-engine mounting—6.6L engine with A/T

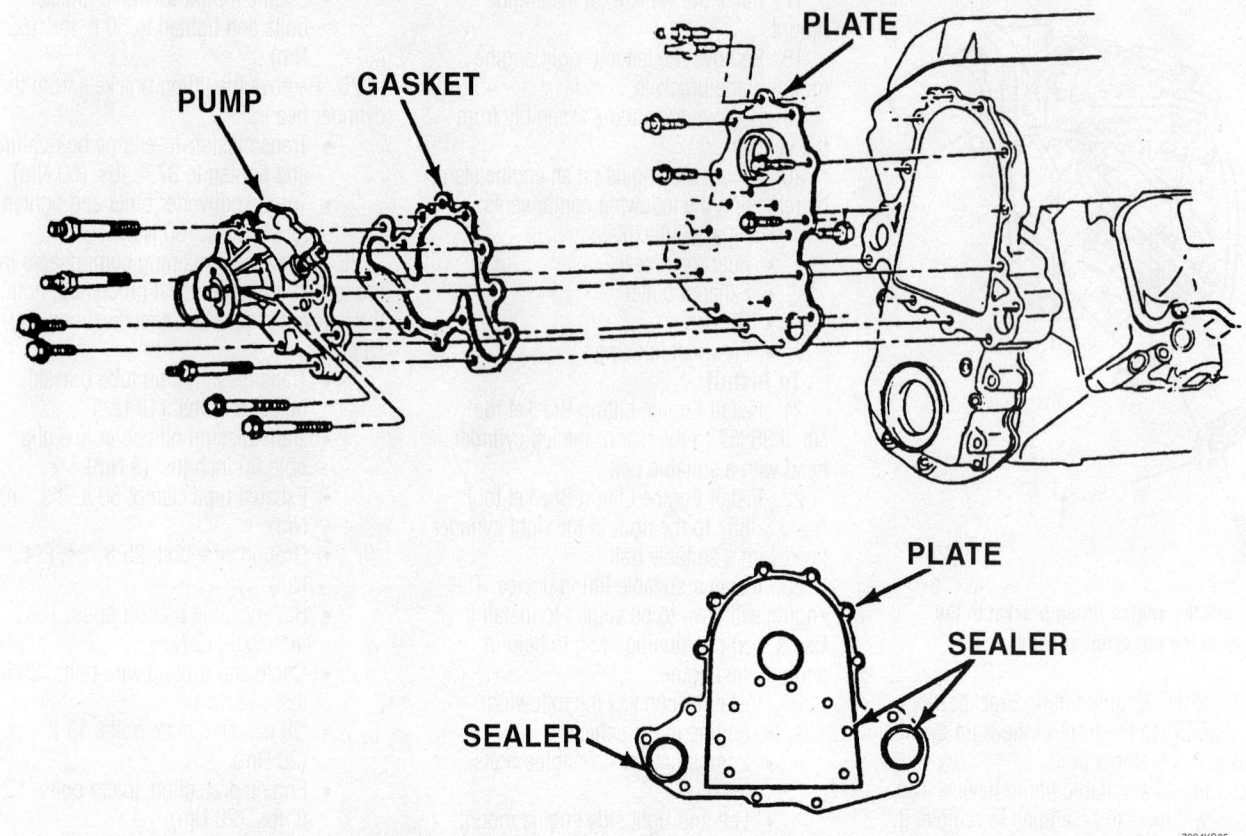

Exploded view of the water pump assembly and related components—6.5L diesel engines

➡ **Remove the bolt on the rear of the water pump plate.**

- Separate the pump and gasket from the plate

To install:

3. Install or connect the following:
   - Water pump and a new gasket to the plate. Torque the retaining bolt (at the rear of the plate) to 20 ft. lbs. (28 Nm).

4. Be sure the block mating surface and the plate flanges are free of oil. Apply an anaerobic sealer GM part 1052357 or equivalent.

➡ **The sealer must be wet to the touch when the bolts are tightened.**

- Water pump and plate assembly. Torque the bolts to 20 ft. lbs. (28 Nm).
- Coolant hoses to the pump assembly
- Power steering pump and bracket
- Vacuum pump and bracket, along with the bolt holding the pump and alternator
- Fan and pulley
- Accessory drive belt(s)
- Oil filler tube and/or air condi-

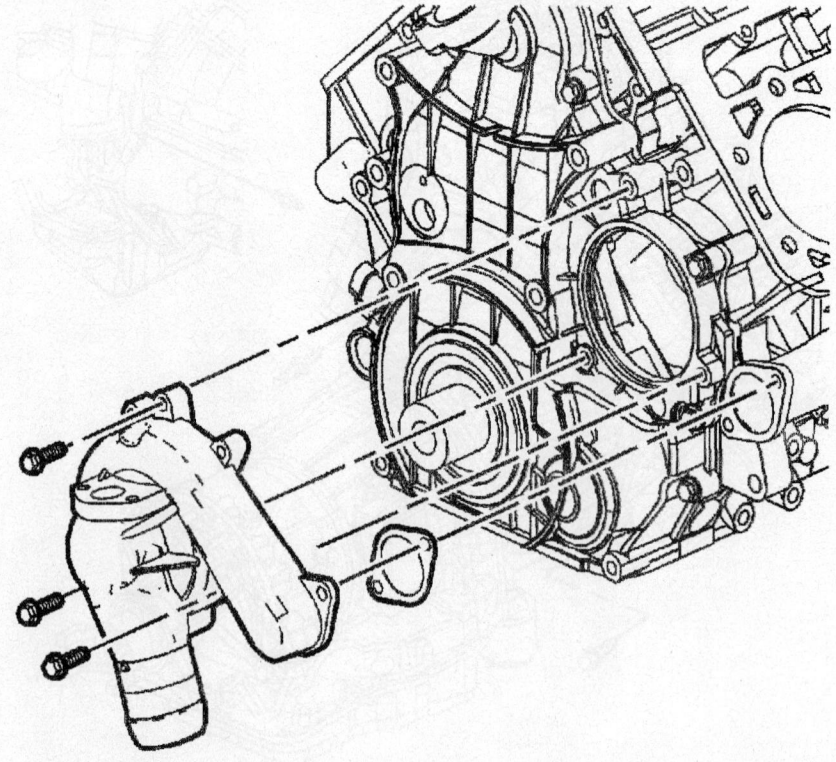

Exploded view of the water pump assembly and related components—6.6L diesel engines

tioning hose bracket nuts, if removed
- Fan shroud
- Batteries
5. Refill the radiator.

6.6L Engine

1. Remove the left front fender wheelhouse inner panel.
2. Drain the coolant.
3. Remove or disconnect the following:
- Thermostat housing crossover
- Fan clutch
- Crankshaft balancer
- Water pump outlet pipe-to-water pump nuts
- Engine wiring harness retainer front the inner stud
- Water pump bolts, noting their locations as they are different lengths
- Water pump and gasket

To install:
4. Lubricate the water pump O-ring with engine oil.
5. Install or connect the following:
- Water pump
- Water pump bolts and tighten to 15 ft. lbs. (20 Nm)
- Water pump-to-water pump outlet gasket
- Engine wiring harness retainer on the water pump outlet pipe inner stud
- Water pump-to-water pump outlet pipe nuts and tighten to 15 ft. lbs. (20 Nm)
- Thermostat housing crossover
- Crankshaft balancer
- Fan clutch

6. Fill the cooling system and install the left front fender wheelhouse inner panel.

Heater Core

REMOVAL & INSTALLATION

Please refer to Heater Core under Gasoline Engine Repair for this procedure.

Glow Plugs

REMOVAL & INSTALLATION

6.5L Engine

1. Remove or disconnect the following:
- Negative battery cables
- Glow plug lead wires
- Plugs

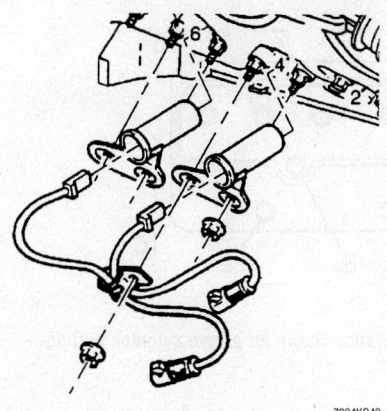

7924KG42

Exploded view of the heat shrouds and glow plug wiring—6.5L diesel engine

- Right front tire
- Inner splash shield from the fender well
- Lead wire from the plug at the No. 2 cylinder and the lead wires from plugs in the Nos. 4 and 6 cylinders at the harness connectors
- Heat shroud for the plug in the No. 4 and 6 cylinder. Slide the shrouds back just far enough to allow access so you can unplug the wires.
- Glow plugs in cylinders No. 2, 4 and 6.

2. Reach up under the vehicle and disconnect the lead wire at No. 8. Remove the glow plug.

➡**You may find that removing the exhaust pipe down might make this a bit easier when working on Nos. 6 and 8.**

To install:
3. Install or connect the following:
- Glow plugs and tighten to 16 ft. lbs. (22 Nm)
- Heat shrouds and electrical connection
- Exhaust pipe, if removed
- Splash shields, if removed
- Negative battery cable

6.6L Engine

BANK 1 OR 2

1. Remove or disconnect the following:
- Negative battery cables
- Front tire
- Inner splash shield from the fender well
- Air cleaner outlet duct
- Electrical nuts from the glow plug(s)

- Harness from the glow plug(s)

➡**On vehicles with Federal emissions systems, there is a buss bar connecting the glow plugs on each bank of the engine.**

- Glow plug(s)

To install:
2. Install or connect the following:
- Glow plug and tighten to 13 ft. lbs. (18 Nm)
- Buss bar and wiring
- Glow plug electrical nut and tighten to 13 inch lbs. (1.5 Nm)
- Air cleaner outlet duct
- Splash shield to the fender well
- Negative battery cables

Cylinder Head

REMOVAL & INSTALLATION

6.5L Engine

1. Relieve the fuel system pressure.
2. Drain the coolant system.
3. Discharge the air conditioning system.
4. Remove or disconnect the following:
- Negative battery cables
- Intake manifold
- Fan upper shroud
- Compressor assembly, if equipped
- Turbocharger, if equipped
- Exhaust manifold
- Valve cover
- Rocker arm assemblies and pushrods

➡**Mark all components so they may be returned to their original location.**

- Air cleaner resonator and bracket
- Transmission and oil dipstick tube; remove the oil fill tube from the coolant crossover pipe
- Heater, radiator and bypass hoses
- Alternator upper bracket
- Alternator
- Power steering pump
- Vacuum pump
- Fuel bleeder valve at the coolant crossover pipe
- Fuel return crossover line clamp bolts from both cylinder heads
- Wire connector from the sensor in the coolant crossover pipe
- Electrical connection and brackets from cylinder head
- Coolant crossover pipe/thermostat assembly
- Head bolts and the cylinder heads

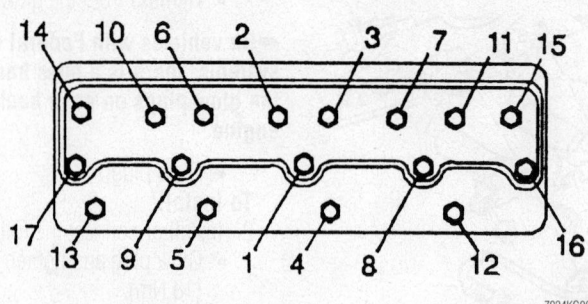

Tighten the cylinder head bolts according to the sequence shown for proper cylinder sealing—6.5L diesel engines

To install:

5. Clean the mating surfaces of the heads and block thoroughly.

6. Clean the head bolts thoroughly. Coat the threads of the head bolts with sealing compound GM part 1052080 or equivalent, before installation.

7. Install a new gaskets, and the cylinder head and bolts. Torque the bolts, in sequence, as follows:

   a. Step 1: 20 ft. lbs. (25 Nm)

   b. Step 2: 50 ft. lbs. (65 Nm).

   c. Step 3: An additional 90 degrees (¼ turn).

8. Install or connect the following:
- Coolant crossover pipe and thermostat
- Fuel valve
- Bypass hose
- Upper radiator hose
- Heater hoses at the head
- Transmission and oil dipstick tube
- Air cleaner resonator and bracket
- Pushrods, hardened ends facing up
- Rocker arm assemblies

9. Adjust the valves.
- Valve cover
- Alternator and upper bracket
- Exhaust manifolds. Torque bolts to 22 ft. lbs. (30 Nm).
- Upper fan shroud
- Intake manifold
- Turbocharger, if equipped
- Vacuum pump, if equipped
- Air conditioning compressor, if equipped
- Engine electrical connection
- Negative battery cables

10. Refill the cooling system with the proper type and quantity of antifreeze.

11. Evacuate and recharge the air conditioning system.

6.6L Engine

1. Relieve the fuel system pressure.
2. Drain the coolant system.

3. Remove or disconnect the following:
- Negative battery cables
- Left or right front splash shield from the fender well, as applicable
- Turbocharger
- Turbocharger charged air cooler inlet duct
- Thermostat housing crossover
- Left or right intake manifold, as necessary
- Upper left or right valve cover
- Fuel rail assembly
- Left or right exhaust manifold

- Bolt and ground straps from the rear of the cylinder head
- Lower left or right valve cover
- Rocker arm shaft assembly
- Glow plugs
- Fuel injector return pipe eye bolts and washers
- Fuel injector return pipe assembly
- Fuel injector bracket bolts
- Fuel injectors with the brackets, using a suitable removal tool
- Injector bracket pins
- Cylinder head bolts, in the proper sequence
- Cylinder head and gasket. Discard the gasket

To install:

4. Clean the mating surfaces of the heads and block thoroughly.

5. Position a new left or right side head gasket on the block. Note that the left and right side gaskets are NOT interchangeable.

➡The cylinder head bolts on these vehicles are pre-coated with an application of a molybdenum disulfide for thread lubrication. Do not remove the coating or add any additional lubrication.

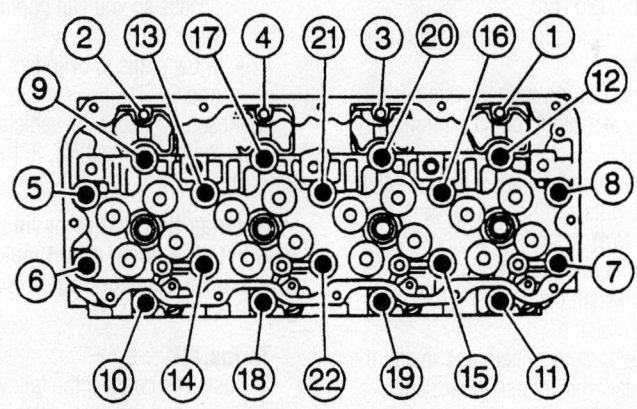

Cylinder head bolt loosening sequence—6.6L diesel engine

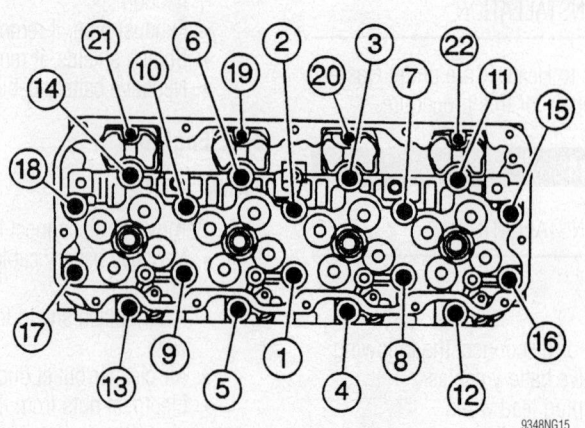

Cylinder head bolt tightening sequence—6.6L diesel engines

6. Install the cylinder head and bolts.

7. Tighten the cylinder head bolts, in sequence, as follows:

   a. Step 1: M12 bolts to 37 ft. lbs. (50 Nm).

   b. Step 2: M12 bolts to 59 ft. lbs. (80 Nm).

   c. Step 3: Tighten the M12 bolts an additional 90 degrees using a torque angle meter.

   d. Step 4: Tighten the M12 bolts an additional 75 degrees using a torque angle meter.

   e. Step 5: M8 bolts to 18 ft. lbs. (25 Nm).

8. Install or connect the following:
- New O-ring onto the fuel injectors after coating with clean engine oil
- New copper washer into the fuel injector bore in the cylinder head
- Fuel injector bracket pin

➡ **If you are reusing the old injectors, clean the carbon from the tips, but do not use a wire brush.**

- Fuel injector bracket bolt and tighten to 37 ft. lbs. (50 Nm)
- Fuel injector return pipe assembly
- Fuel injector return pipe-to-injector eye bolts and washers. Tighten to 11 ft. lbs. (15 Nm).
- Fuel return pipe-to-cylinder head eye bolts and washers. Tighten to 11 ft. lbs. (15 Nm).

- Bolt and ground straps to the rear of the cylinder head. Tighten to 18 ft. lbs. (25 Nm).
- Valve rocker shaft assembly
- Lower and upper valve covers
- Glow plugs
- Exhaust manifold
- Fuel rail assembly
- Intake manifold
- Thermostat housing crossover
- Turbocharger charged air cooler duct
- Clamp and hose to the charged air cooler. Tighten to 53 inch lbs. (6 Nm).
- Turbocharger
- Fender splash shield
- Negative battery cables

9. Refill the cooling system with the proper type and quantity of antifreeze.

10. Evacuate and recharge the air conditioning system.

Starter Motor

REMOVAL & INSTALLATION

1. Remove or disconnect the following:
- Negative battery cables
- Right front wheel and fender splash shield, on 6.6L engines
- Turbocharger exhaust pipe, on 6.6L engines

- Mounting bolts/nuts and shim, if used
- Starter
- Wires
- Heat shield and bracket

To install:

2. Install or connect the following:
- Heat shield and bracket. Torque the bolts to 13 ft lbs. (17 Nm).
- Wires. Torque battery wire nut to 89 inch lbs. (10 Nm), and ignition nut to 18 inch lbs. (2 Nm) on 6.5L engines. On 6.6L engines, tighten the solenoid nut to 30 inch lbs. (3.4 Nm) and the positive battery cable nut to 80 inch lbs. (9 Nm).
- Starter
- Mounting bolts/nuts and shim, if used. Torque the bolts to 33 ft lbs. (45 Nm) and the nut to 75 inch lbs. (8.5 Nm) for 6.5L engines. For 6.6L engines tighten the starter bolts to 58 ft. lbs. (78 Nm).
- Turbocharger exhaust pipe, on 6.6L engines
- Right front fender splash shield and wheel, on 6.6L engines
- Negative battery cables

Rocker Arms/Shaft

REMOVAL & INSTALLATION

6.5L Engine

1. Remove or disconnect the following:
- Engine cover

➡ **Rotate the engine until the mark on the crankshaft balancer is at the 2 o'clock position. Rotate the crankshaft counterclockwise 3½ in. (88mm) aligning the crankshaft balancer mark with the first lower water pump bolt, at about the 12:30 position. This will ensure that no valves are close to a piston crown.**

- Cylinder head cover
- Rocker shaft assembly

➡ **The rocker assemblies are mounted on 2 short rocker shafts per cylinder head, with each shaft operating 4 rockers. 2 bolts secure each rocker shaft assembly. Mark the shafts so they can be installed in their original locations.**

- Pushrods. The pushrods MUST be installed in the original direction! A paint stripe usually identifies the upper end of each rod, but if you can't see it, be sure to mark each rod yourself.

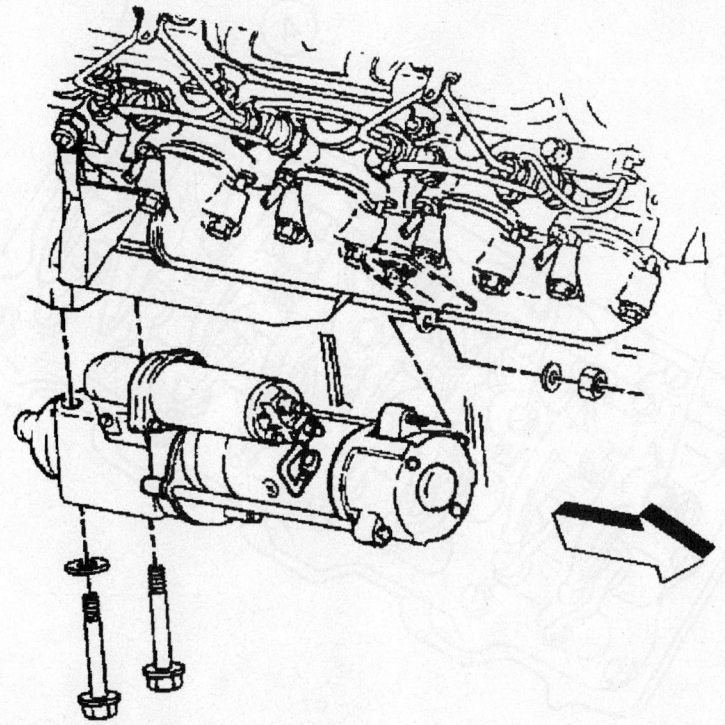

Exploded view of the starter motor—6.5L engine shown

06025-EXP-G01

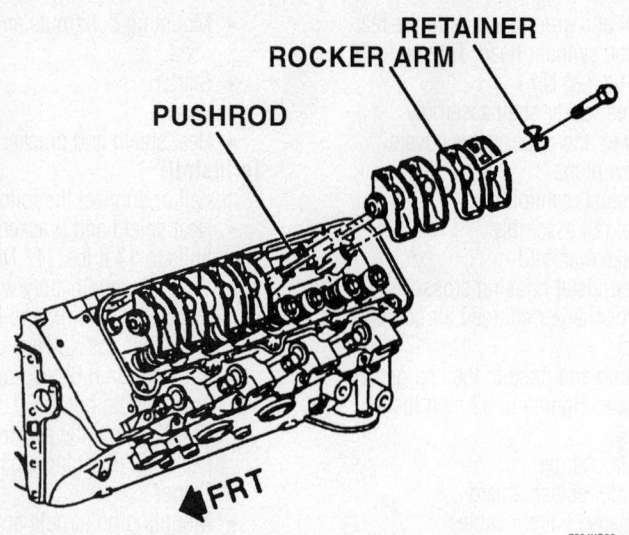

RETAINER
ROCKER ARM
PUSHROD

FRT

7924KG22

Rocker shaft assembly and related components—diesel engines

2. Insert a small prybar into the end of the rocker shaft bore and break off the end of the nylon retainers. Pull off the retainers with pliers, then slide off the rockers.

To install:

3. Be sure first that the rocker arms and springs go back on the shafts in the exact order in which they were removed. It's a good idea to coat them with engine oil.

4. Center the rockers on the corresponding holes in the shaft

5. Install or connect the following:
- New plastic retainers using a ½ in. (13mm) drift
- Pushrods with there marked ends up
- Rocker shaft assemblies and be sure that the ball ends of the pushrods seat themselves in the rockers

6. Rotate the engine clockwise until the mark on the torsional damper aligns with the **0** on the timing tab. Rotate the engine counterclockwise 3 ½ in. (88mm) measured at the damper. You can estimate this by checking that the mark on the damper is now aligned with the FIRST lower water pump bolt. BE CAREFUL! This ensures that the piston is away from the valves.
- Rocker shaft bolts and tighten to 40 ft. lbs. (55 Nm)
- Cylinder head cover
- Engine cover

6.6L Engine

1. Remove the lower valve (rocker arm) covers

2. Loosen the valve clearance lock nuts on each rocker arm

3. Loosen the valve clearance adjusting

screw on each rocker arm to relieve tension on the valve train

➡**The rocker arm bolts retain the rocker arms on the shaft. Do not remove the bolts from the rocker arm shaft brackets.**

4. Loosen the rocker arm shaft bolts in the proper sequence, leaving the bolts in the rocker arm shaft brackets.

5. Remove or disconnect the following:
- Rocker arm shaft assemblies from the cylinder head
- Valve bridge pins
- Valve bridges
- Valve push rods

6. Clean all parts in a suitable solvent. Disassemble the rocker arm shaft as necessary.

To install:

7. Lubricate the rocker arm shaft and the inside of the rocker arms with engine oil.

8. If disassembled, install or connect the following:
- Rocker arm bracket on one end of the rocker arm shaft with the bolt
- Rocker arm intake, spring, exhaust and the bracket with bolt. Continue in the same sequence to the last bracket.
- Push the bracket to compress the springs and then install the bolt

9. Lubricate the top of the valves, the valve bridge stem, the valve bridge and the valve bridge pins.

10. Install or connect the following:
- Valve bridge pins
- Valve bridges

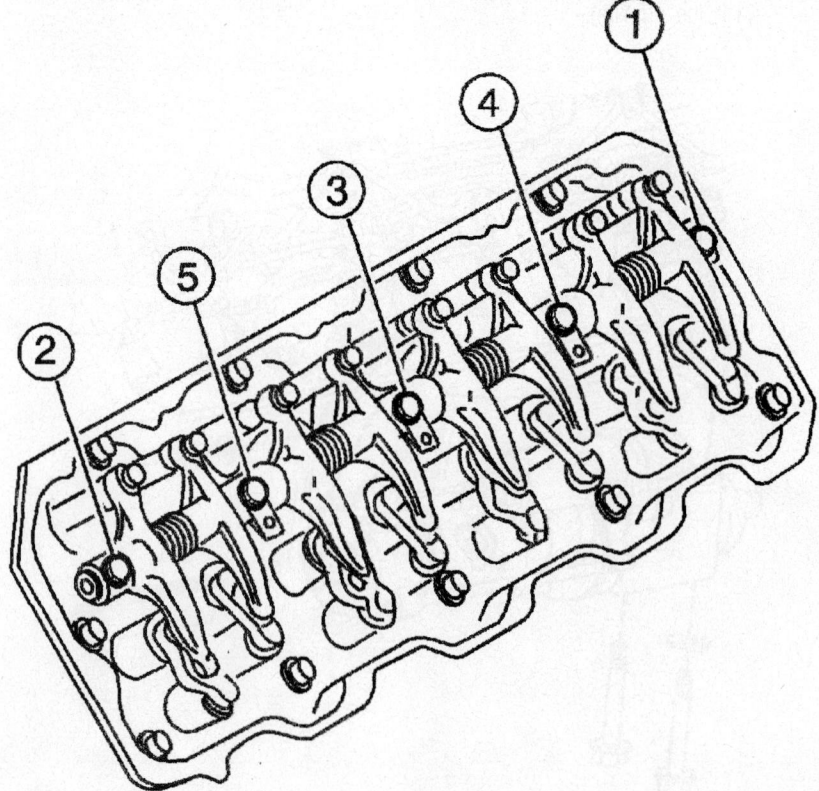

06025-AVAL-G06

Rocker arm shaft bolt loosening sequence—6.6L engine

- Pushrods. Make sure it is fully installed by gently pulling up on it. You should feel resistance from the pushrod trying to lift the valve lifter

11. Use clean engine oil to lubricate the rocker arm shaft bolt threads, tops of the push rods, rocker arms and rocker arm shaft.

- Rocker arm shaft assembly to the cylinder head
- Rocker arm shaft assembly bolts and tighten, in the proper sequence to 30 ft. lbs. (40 Nm)

12. Adjust the valve clearance, as follows:

a. Remove the fan clutch.

b. Remove both upper valve covers.

c. Rotate the engine in the normal direction and place the No. 1 piston at Top Dead Center (TDC) of the compression stroke. The No. 1 cylinder is at the right side front. While turning the engine, watch the intake valve to open and close. Align the mark on the crankshaft balancer with the pointer on the engine.

d. Loosen the valve clearance adjusting screws for the valve being adjusted.

e. Insert the feeler gauge between the tip of the rocker arm and the valve bridge.

f. Adjust the intake and the exhaust valve clearance to 0.012 in. (0.3mm) with the engine cold. Refer to the figure for the valves that can be adjusted TDC of the compression stroke.

g. Tighten the valve adjusting screw lock nut to 16 ft. lbs. (22 Nm).

h. Turn the engine one rotation in the normal direction and put the No. 1 piston at TDC of the exhaust stroke to adjust the remaining valve clearance. While turning the engine, watch the exhaust valve to open and close. Align the mark on the crankshaft balancer with the pointer on the engine.

i. Loosen the valve clearance adjusting screws for the valves being adjusted.

j. Insert the feeler gauge between the tip of the rocker arm and the valve bridge.

k. Adjust the intake and the exhaust valve clearance to 0.012 in. (0.3mm) with the engine cold. Refer to the figure for the valves that can be adjusted TDC of the exhaust stroke.

l. Tighten the valve adjusting screw lock nut to 16 ft. lbs. (22 Nm).

13. Install the upper and lower valve cover and fan clutch, as necessary.

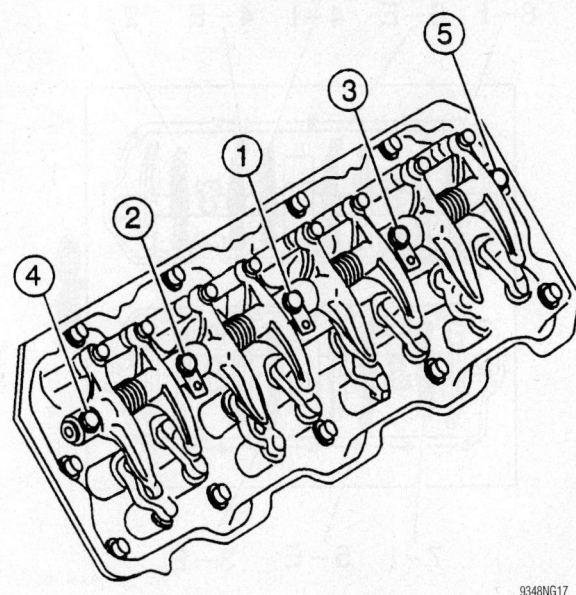

Rocker arm shaft tightening sequence—6.6L VIN 1 engines

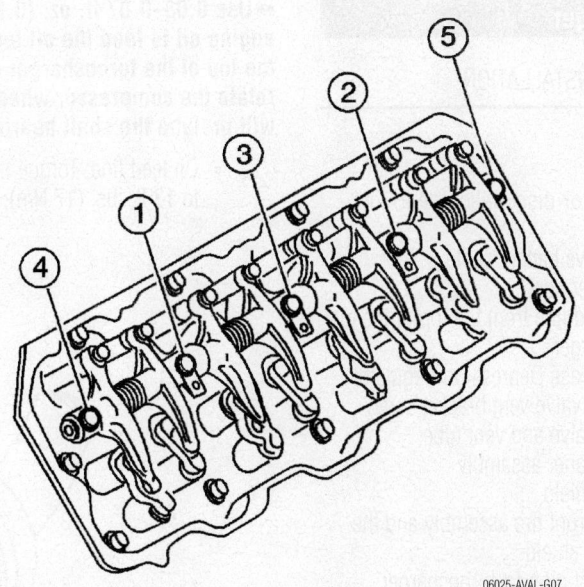

Rocker arm shaft tightening sequence—6.6L VIN 2 engines

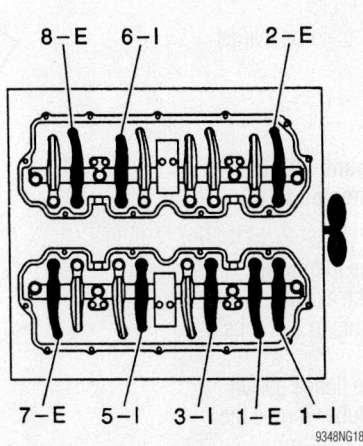

Location of the valves that are adjusted at TDC of the compression stroke—6.6L engine

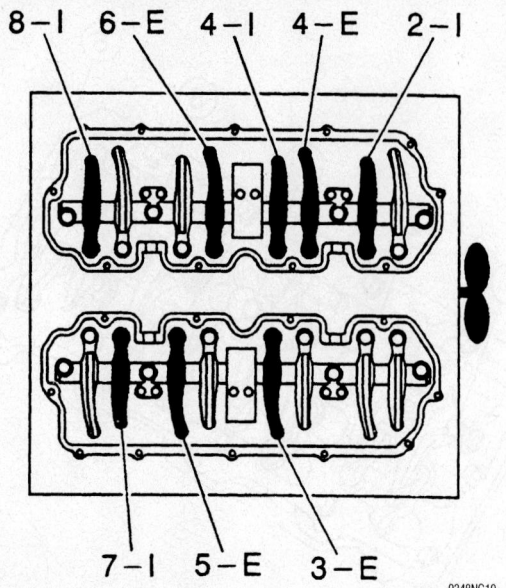

8–I 6–E 4–I 4–E 2–I

7–I 5–E 3–E

9348NG19

Location of the valves that are adjusted at TDC of the exhaust stroke—6.6L engine

Turbocharger

REMOVAL & INSTALLATION

6.5L Engine

1. Remove or disconnect the following:

- Negative battery cable
- Air inlet duct
- Oil feed line from the top of the turbocharger
- Crankcase Depression Regulator (CDR) valve vent bracket screw
- CDR valve and vent tube
- Air cleaner assembly
- Heat shield
- Right front tire assembly and the splash shield
- Exhaust pipe-to-turbocharger exhaust outlet elbow V-band clamp
- Oil drain tube-to-turbocharger center bearing bolts
- Exhaust manifold-to-turbocharger nuts
- Turbocharger

To install:

➡**Use anti-seize compound on all threaded fasteners connected to the turbocharger**

2. Install or connect the following:
- Turbocharger to the exhaust manifold. Torque the nuts to 37 ft. lbs. (50 Nm).
- New oil drain tube flange gasket and the oil drain tube. Torque the bolts to 19 ft. lbs. (26 Nm).

➡**Use 0.03–0.07 fl. oz. (0.88–2ml) of engine oil to feed the oil feed hole at the top of the turbocharger and hand rotate the compressor wheel/shaft. This will prelube the shaft bearings**

- Oil feed line. Torque the connection to 13 ft. lbs. (17 Nm).
- Exhaust pipe to the turbocharger exhaust elbow V-band clamp. Torque the clamp to 71 inch. lbs. (8 Nm).

3. Disengage the injection pump fuel shutdown solenoid connector and crank the engine for no more 15 seconds to prime the oil system. Do not let the engine start.
- Right front wheel and splash shield
- Heat shield. Apply Loctite® to the bolts and torque to 56 inch lbs. (6 Nm)
- Air cleaner assembly
- Turbocharger compressor outlet
- CDR valve, tube and bracket
- Air intake duct

➡**Operate the engine at idle for at least 3 minutes after installing the turbocharger**

6.6L Engine

1. Disconnect the negative battery cables.
2. Open the hood and move the hinge bolts to the service position.
3. Raise the vehicle.
4. Drain the coolant.
5. Remove or disconnect the following:
- Left and right wheelhouse liners
- Exhaust pipe-to-exhaust outlet

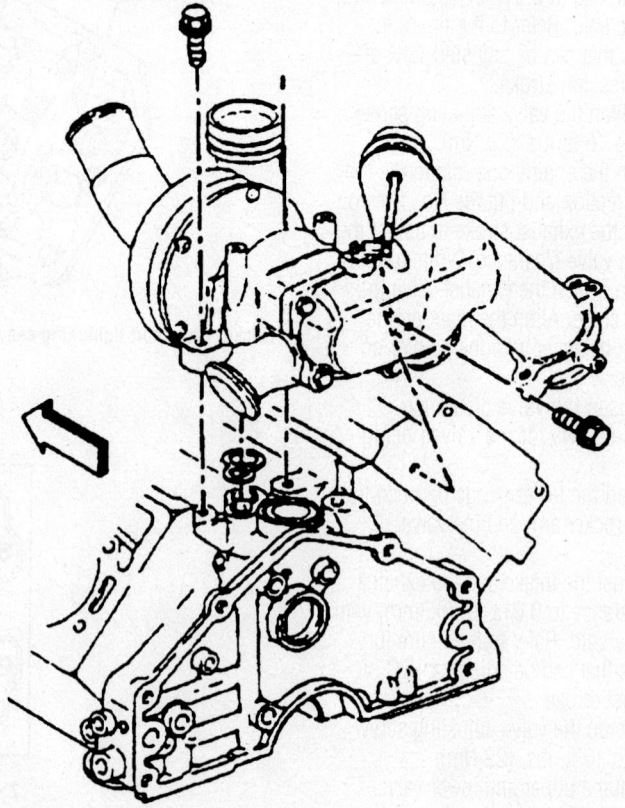

Turbocharger mounting—6.5L diesel engine

7924KG43

clamp. Move the clamp onto the exhaust pipe
- Transmission fluid fill tube-to-bell housing nuts if equipped with an A/T. Position the tube to the right side of the vehicle; it does not need to be removed from the transmission
- 3 nuts and left exhaust heat shield from the front of the lower dash panel
- Left exhaust pipe heat shield bolts

6. Position the left exhaust pipe heat shield to access the left exhaust pipe-to-manifold bolts. Do not remove the heat shield from the vehicle at this time.

➡ Do not bend the exhaust pipe at the expansion area.

- Left, then the right exhaust pipe-to-exhaust manifold bolts
- Gaskets and discard
- Lower bolt for the exhaust outlet shield

7. Lower the vehicle.
- Upper intake manifold sight shield front retaining bolt
- Sight shield
- Air cleaner outlet duct from the air cleaner and turbocharger. Cover the openings to prevent debris from entering
- Charged air cooler outlet duct-to-intake hose clamps (loosen only)
- Hose from the charged air cooler duct-to-intake manifold tube
- A/C compressor clutch electrical connector
- A/C cut-out switch connector
- Drive belt
- A/C compressor mounting bolts; position the compressor aside with the lines attached
- Turbocharger inlet coolant hose from the bypass valve
- Turbocharger outlet coolant hose from the turbocharger
- Crankcase hose from the left valve cover and position aside
- Wire connector from the intake heater
- Intake air heater relay, if equipped
- Heat shield-to-turbocharger bolts and heat shield
- Remaining 2 bolts from the exhaust outlet heat shield
- Exhaust outlet heat shield
- 4 bolts and 2 nuts from the exhaust outlet. You do not have to remove the outlet for turbocharger removal

8. Move the exhaust outlet to one side in order to access the right exhaust pipe-to-turbocharger bolts.

- Exhaust outlet gasket and discard
- Right exhaust pipe-to-turbocharger bolts
- Right exhaust pipe and gasket

9. Move the exhaust outlet to one side for access to the left pipe.
- Left exhaust pipe heat shield
- Left exhaust pipe-to-turbocharger bolts
- Left exhaust pipe and gasket
- Turbocharger oil supply hose eye bolt and washers. Move the hose aside
- Turbocharger oil drain pipe nuts from the flywheel housing
- Turbocharger mounting bolts
- Turbocharger with the oil drain pipe

10. If replacing the turbocharger, remove the oil drain pipe and coolant hose.

To install:

11. Thoroughly clean the gasket surfaces.

12. Install or connect the following:
- Turbocharger oil drain pipe and new gasket. Tighten the bolts to 16 ft. lbs. (21 Nm).
- Turbocharger inlet coolant hose
- Turbocharger oil supply hose to the engine block
- Turbocharger oil supply hose eye bolt and washers and tighten to 31 ft. lbs. (42 Nm)
- Turbocharger lower heat shield
- Turbocharger. Tighten the 3 mounting bolts to 80 ft. lbs. (108 Nm).
- New gasket for oil drain pipe
- Oil drain pipe nuts and tighten to 15 ft. lbs. (20 Nm)

13. If installing a new turbocharger, pour 4–5 oz. of clean engine oil into the turbocharger supply hose opening, while rotating the impeller.
- Oil supply hose, using new washers. Tighten the eye bolt to 31 ft. lbs. (42 Nm).

14. Install the remaining components in the reverse order of removal, noting the following important points:
- When installing the exhaust pipe, use new gaskets and align the tabs and make sure the proper pipe flange is towards the turbocharger, as they are different. Tighten the exhaust pipe-to-turbocharger bolts to 39 ft. lbs. (53 Nm).
- Tighten the turbocharger heat shield bolts to 80 inch lbs. (9 Nm)
- Tighten the A/C compressor bolts to 37 ft. lbs. (50 Nm)
- Tighten the exhaust pipe clamp to 30 ft. lbs. (40 Nm)

15. Fill the cooling system and connect the negative battery cables.

➡ Operate the engine at idle for at least 3 minutes after installing the turbocharger

Intake Manifold

REMOVAL & INSTALLATION

6.5L Engine

1. Recover the air conditioning refrigerant and reposition air conditioning lines.

2. Remove or disconnect the following:
- Negative battery cable
- Air cleaner assembly
- Fuel lines, and electrical connections
- Engine and transmission oil level tubes

✳✳ WARNING

Do not remove the center intake and side intakes as an assembly. Damage to the center intake and turbocharger may occur.

- Center intake assembly, and glow plug relay
- Side intake bolts and fuel retaining clips
- Side intakes

3. Clean all gasket surfaces.
To install:
4. Install or connect the following:
- Side intakes and new gaskets.

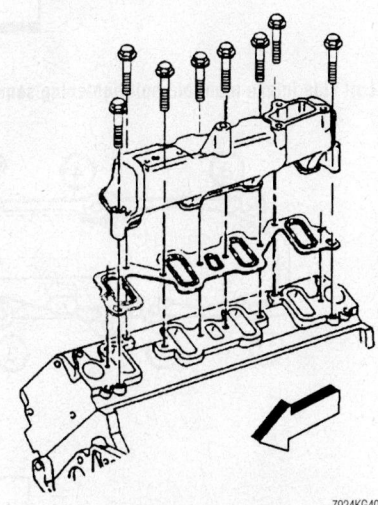

Exploded view of the side intake manifold mounting—6.5L engines

7924KG40

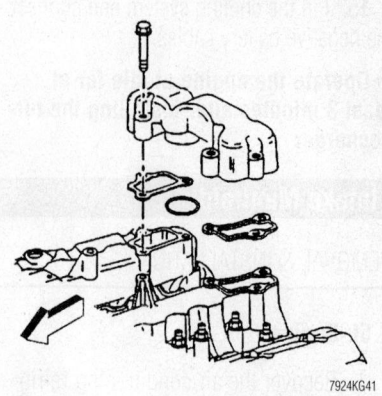

Center intake manifold mounting—6.5L engines

Torque the bolts to 31 ft. lbs. (42 Nm).
- Fuel lines retaining clips and electrical connection
- Center intake with new gaskets. Torque the bolts to 17 ft. lbs. (23 Nm).
- Engine oil and transmission oil level tubes
- Glow plug relay
- Air conditioning lines
- Air cleaner assembly

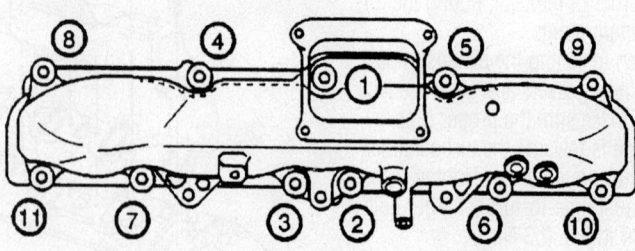

Left side intake manifold bolt tightening sequence—6.6L engine

Right side intake manifold bolt tightening sequence—6.6L engine

- Negative battery cable
5. Recharge air conditioning system.

6.6L Engines

CENTER MANIFOLD

1. Remove the Exhaust Gas Recirculation (EGR) valve cooler tube.
2. Remove the intake manifold tube.
3. Remove and discard the 2 intake manifold tube gaskets.
4. Remove the turbocharger.
5. Remove the center intake manifold bolts/nuts.
6. Pull-up the center intake manifold in order to remove.
7. Remove and discard the gaskets.
8. Clean the center intake manifold in cleaning solvent and air dry.

To install:

9. Install new center intake manifold gaskets.
10. Install the center intake manifold.
11. Install the center intake manifold bolts/nuts and tighten to 89 inch lbs. (10 Nm).
12. Install the turbocharger.
13. Install 2 new O-rings onto the intake manifold tube.

14. Lubricate the O-rings with clean engine oil to aid in the installation.
15. Install the intake manifold tube.
16. Install the EGR valve cooler tube.

LEFT AND RIGHT MANIFOLDS

1. Drain the cooling system.
2. Remove or disconnect the following:
- Batteries cables
- Center intake manifold
- Fuel junction block
- Left or right fuel rail
- Intake manifold tube
- 9 bolts and 2 nuts from the intake manifold. A bolt is located in the manifold opening.

➡The intake manifold uses sealer. If necessary, pry at the area by the common rail bolt holes and be careful to avoid damaging the sealing surfaces.

- Intake manifold from the head. Cover the head openings to prevent debris from entering.
3. Clean all gaskets surface.

To install:

4. Install or connect the following:
- A ⅛ in. (2–3mm) wide to ¹⁄₁₆ in (0.5–1.5mm) high bead of sealant to the sealing surface of the intake manifold

➡The left and right side manifolds are NOT interchangeable.

- Intake manifold
- Bolts and nuts. Tighten to 15 ft. lbs. (20 Nm), in sequence.
- Intake manifold tube
- Fuel rail
- Fuel junction block
- Turbocharger
- Negative battery cables
5. Fill cooling system.

Exhaust Manifold

REMOVAL & INSTALLATION

6.5L Engine

1. Remove or disconnect the following:
- Batteries cables
- Exhaust pipe from the manifold flange
- Engine oil and transmission oil fill tubes
- Engine cover and disconnect the glow plug wires
- Glow plugs
- Turbocharger assembly, as required
- Air conditioner compressor rear bracket, as required

- Manifold bolts and the manifold

To install:

2. Install or connect the following:
- Exhaust manifold. Torque the bolts to 26 ft. lbs. (35 Nm).
- Exhaust pipe
- Glow plugs and electrical connection
- Engine and transmission oil fill tubes
- Air conditioning compressor bracket
- Negative battery cable

6.6L Engine

LEFT SIDE

1. Raise the vehicle.
2. Remove or disconnect the following:
- Bolts securing the left exhaust pipe heat shield and move the heat shield aside
- Left exhaust pipe-to-manifold bolts
- Left front wheel
- Left front fender splash shield
- Charge air cooler duct
- Exhaust manifold heat shield bolts and shield
- 2 nuts and 6 bolts with the plain washer and bell view washer from the left manifold
- Exhaust manifold by removing it from the rear, then the front studs and sliding it out the bottom, past the oil filter
- Exhaust manifold gasket and discard

To install:

3. Installation is the reverse of the removal procedure. Tighten the retainers as follows:
   a. Exhaust manifold nuts and bolts, in sequence, in 2 passes to 25 ft. lbs. (34 Nm).
   b. Heat shield bolts: 71 inch lbs. (8 Nm).
   c. Exhaust pipe-to-manifold bolts: 39 ft. lbs. (59 Nm).

RIGHT SIDE

1. Raise the vehicle.
2. Remove or disconnect the following:
- Right front wheel
- Right front fender splash shield
- Exhaust manifold heat shield bolts and shield
- Right exhaust pipe-to-manifold bolts
- 2 nuts and 6 bolts with the plain washer and bell view washer from the left manifold
- Exhaust manifold by removing it

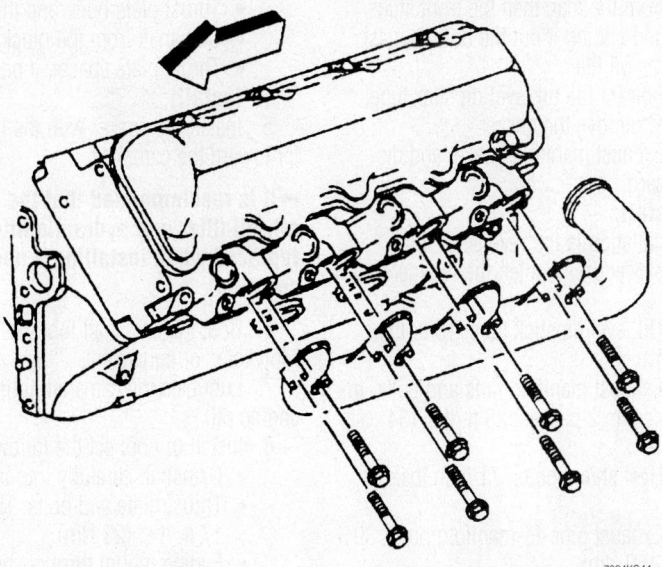

7924KG44

Exploded view of the left exhaust manifold mounting—6.5L diesel engines

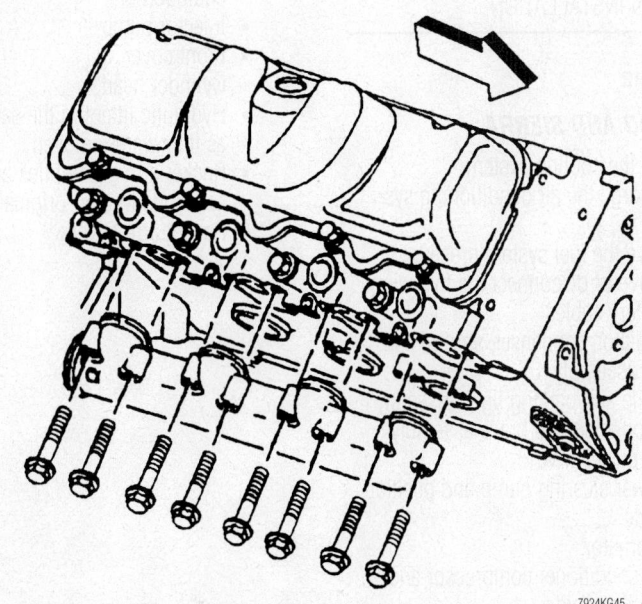

7924KG45

Exploded view of the right exhaust manifold mounting—6.5L diesel engines

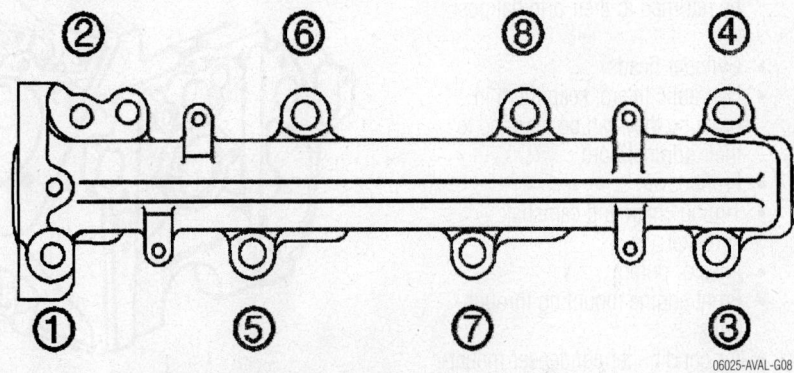

06025-AVAL-G08

Left and right side exhaust manifold bolt removal and installation sequence—6.6L engine

from the rear, then the front studs and sliding it out the bottom, past the oil filter
- Bolt for the oil level dipstick tube, to remove the gasket
- Exhaust manifold gasket and discard

To install:

3. Installation is the reverse of the removal procedure. Tighten the retainers as follows:

    a. Oil level dipstick tube: 15 ft. lbs. (20 Nm).

    b. Exhaust manifold nuts and bolts, in sequence, in 2 passes: 25 ft. lbs. (34 Nm).

    c. Heat shield bolts: 71 inch lbs. (8 Nm).

    d. Exhaust pipe-to-manifold bolts: 39 ft. lbs. (59 Nm).

Camshaft and Valve Lifters

REMOVAL & INSTALLATION

6.5L Engine

SILVERADO AND SIERRA

1. Drain the cooling system.
2. Discharge the air conditioning system.
3. Relieve the fuel system pressure.
4. Remove or disconnect the following:
- Battery cables
- Radiator, condenser, shroud and fan assembly
- Grille and parking light assembly
- Hood latch and brace assembly
- Oil pump drive
- Power steering pump and position aside
- Alternator
- Air conditioner compressor and position aside
- Rocker arm covers
- Rocker arm assemblies and pushrods. Mark them so they can be returned to their original position.
- Cylinder heads
- Hydraulic lifters; keep them in order so they can be returned to their original bore
- Front cover
- Timing chain and camshaft sprocket
- Injector pump
- Front engine mounting through-bolts
- Air conditioner condenser mounting bolts and lift the condenser out

- Thrust plate bolts and thrust plate
- Camshaft from the block
- Thrust plate spacer, if necessary

To install:

5. Install the spacer with the I.D. chamfer toward the camshaft.

➡It is recommended that the engine oil, oil filter and hydraulic lifters be replaced when installing a new camshaft.

6. Coat the camshaft lobes with Molykote® or equivalent.
7. Lubricate the camshaft journals with engine oil.
8. Install or connect the following:
- Camshaft carefully into the block
- Thrust plate and bolts. Tighten to 17 ft. lbs. (23 Nm).
- Engine mount through-bolts
- Timing chain and sprockets, align the timing marks
- Air conditioner condenser, if equipped
- Injector pump
- Front cover
- Cylinder head
- Hydraulic lifters in the same bore as they were removed
- Rocker arm assemblies and pushrods in their original locations
- Rocker arm covers

- Power steering pump, alternator and air conditioner compressor
- Oil pump drive
- Hood latch and brace
- Grille and parking light assembly
- Radiator, the shroud and fan assembly
- Negative battery cables

9. Refill the cooling system with the proper type and quantity of antifreeze.

EXPRESS AND SAVANA

1. Drain the cooling system.
2. Relieve the fuel system pressure.
3. Remove or disconnect the following:
- Battery cables
- Headlight bezels
- Grille, bumper and lower valance panel
- Hood latch
- Coolant recovery bottle
- Upper tie bar
- Air conditioner compressor
- Radiator and fan
- Oil pump drive
- Cylinder heads to gain clearance for lifter removal
- Alternator lower bracket
- Water pump
- Torsional damper
- Front cover

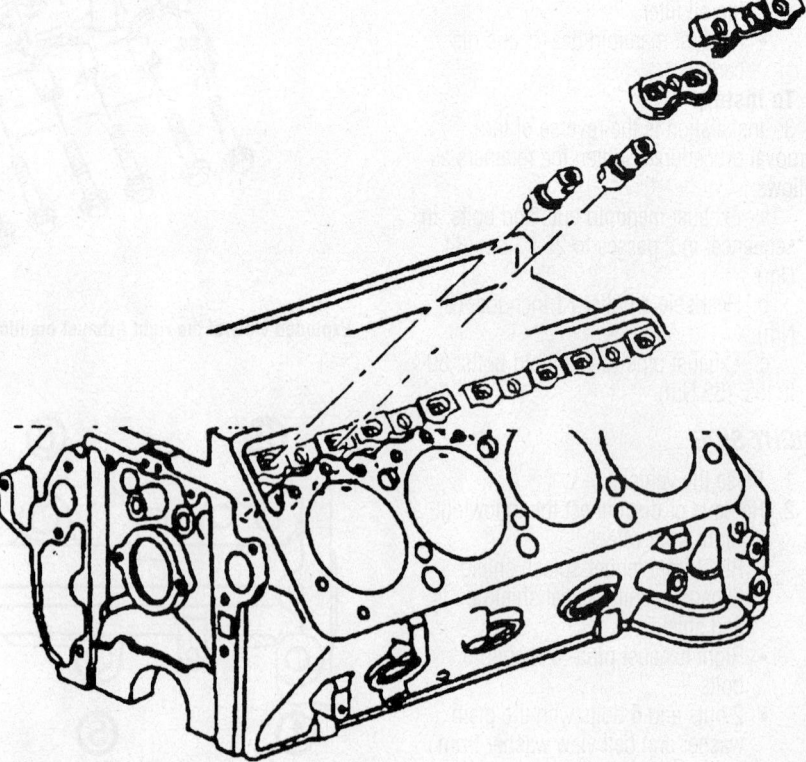

Exploded view of the lifter, guide plate and clamp—6.5L diesel engines

7924KG46

- Injection pump
- Rocker arm covers
- Rocker arm assemblies and pushrods. Mark them so they can be returned to their original position.
- Hydraulic lifters and keep them in order so they can be returned to their original bore.
- Timing chain and camshaft sprocket
- Thrust plate bolts and thrust plate
- Camshaft from the block
- Thrust plate spacer, if necessary

To install:

4. Install the spacer with the I.D. chamfer toward the camshaft.

➡**It is recommended that the engine oil, oil filter and hydraulic lifters be replaced when installing a new camshaft.**

5. Coat the camshaft lobes with Molykote®, or equivalent.
6. Lubricate the camshaft journals with engine oil.
7. Install or connect the following:
- Camshaft carefully into the block
- Thrust plate and bolts. Torque the bolts to 17 ft. lbs. (23 Nm).
- Timing chain and sprockets, align the timing marks
- Hydraulic lifters in the same bore as they were removed
- Rocker arm assemblies and pushrods in their original locations
- Rocker arm covers
- Fuel pump
- Front cover
- Torsional damper and water pump
- Alternator lower bracket
- Cylinder heads
- Oil pump drive
- Radiator and fan
- Air conditioner compressor
- Upper tie bar
- Coolant recovery bottle
- Hood latch
- Grille, bumper and lower valence panel
- Headlight bezels
- Battery cables
8. Refill the cooling system.
9. Evacuate and charge the air conditioner system.

6.6L Engines

➡**This procedure requires the use of the following special tools: Flywheel Holding Tool No. J 44643, Magnetic Base J 26900-13 and Dial Indicator J 26900-12.**

1. Properly discharge the A/C system.
2. Remove or disconnect the following:
- Both cylinder heads
- Valve lifter guide hold-down bracket bolts
- Valve lifter guide hold-down brackets
- Valve lifter guides
- Valve lifters
- Charged air cooler
- A/C condenser
- Starter

3. Install the Flywheel Holding Tool No. J 44643 in the starter opening. Make sure the tool is flush to the flywheel opening. The holding tool will be used to remove the crankshaft balancer bolt and camshaft drive gear bolt.
- Engine front cover
- Oil pump driven gear nut and gear

➡**The crankshaft reluctor and oil pump drive gear are timed together at the factory. Do NOT remove the reluctor from the oil pump drive gear.**

- Oil pump drive gear and crankshaft reluctor assembly. Do not remove the reluctor bolts or damage the reluctor teeth

4. Using the Magnetic Base J 26900-13 and Dial Indicator J 26900-12, measure the camshaft end-play. The production value is 0.002–0.0045 in. (0.050-0.114mm) and the service limit is 0.008 in. (0.20mm). Replace the cam gear or thrust plate if the measured value exceeds the service limit.
- Camshaft reluctor screws and reluctor

➡**Use the flywheel holding tool to hold the engine from turning while loosening the camshaft gear bolt.**

- Loosen the camshaft gear bolt and leave the bolt finger-tight
- Camshaft thrust plate bolts through the holes in the camshaft gear
- Camshaft with the gear attached
- Cam gear bolt and gear
- Thrust plate

5. Clean and inspect the camshaft and bearings.

To install:

6. Install or connect the following:
- Camshaft thrust plate
- Camshaft driven gear
- New driven gear bolt (finger-tight)
- Camshaft and gear assembly into the cylinder block. Align the gear to the crankshaft gear
- Threadlock to the thrust plate bolts
- Thrust plate bolts and tighten to 19 ft. lbs. (26 Nm)

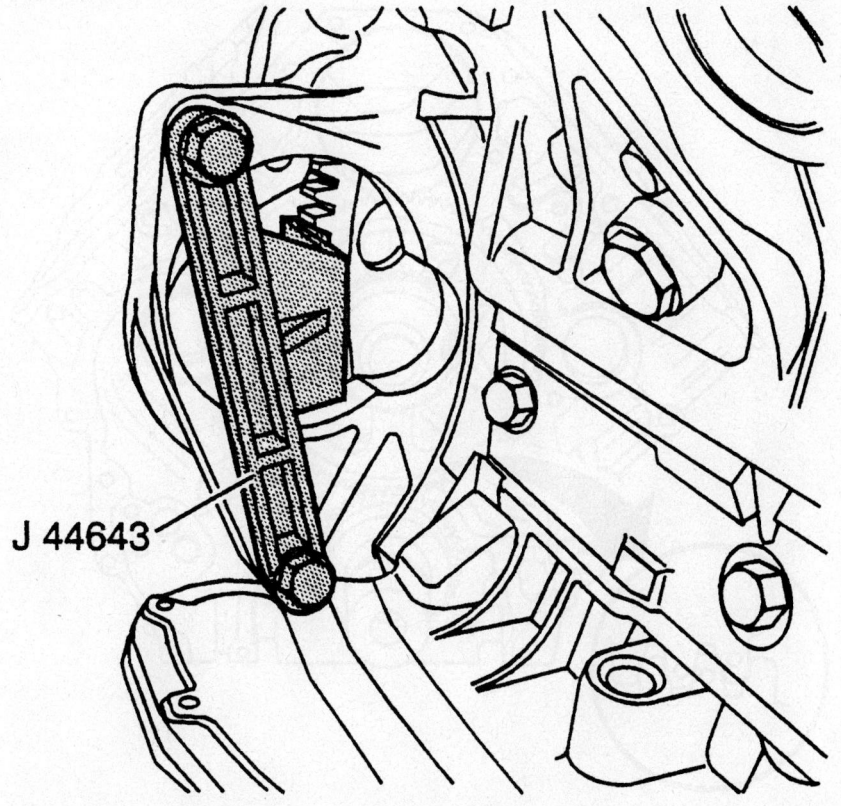

J 44643

Proper installation of the flywheel holding tool in the starter opening

9348NG25

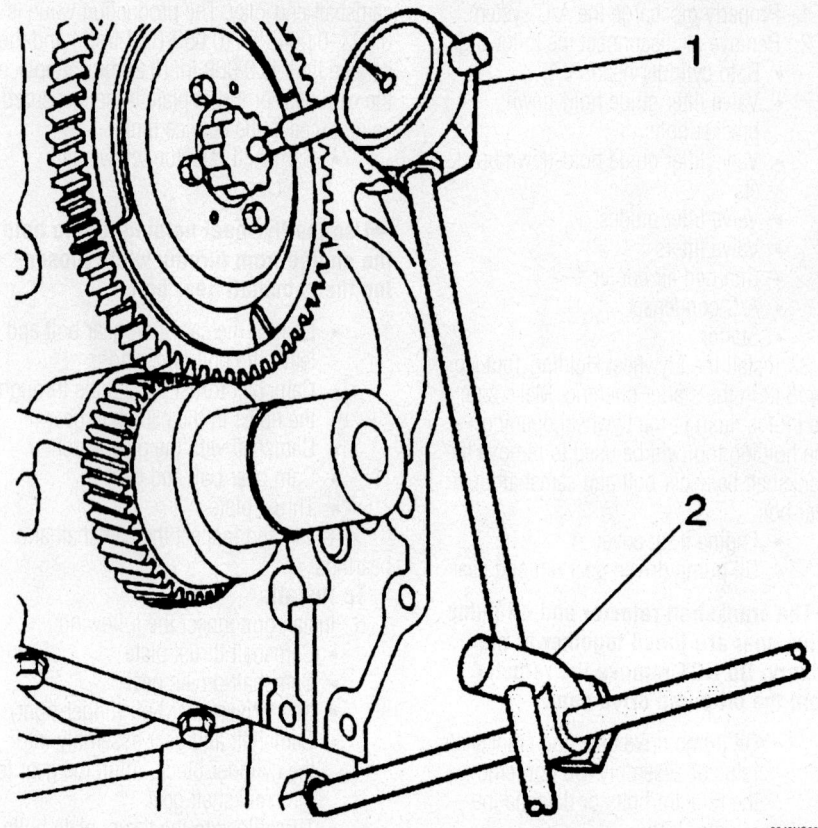

Use the dial indicator (1) and magnetic base (2) to measure the camshaft end-play

9348NG23

- Camshaft reluctor to the cam gear
- Reluctor bolts. Tighten to 80 inch lbs. (9 Nm) in a crisscross pattern.
- If removed, reinstall the flywheel holding tool in the starter opening
- Camshaft gear bolt and tighten to 173 ft. lbs. (234 Nm)

7. Using the Magnetic Base J 26900-13 and Dial Indicator J 26900-12, measure the camshaft end-play. The production value is 0.002–0.0045 in. (0.050-0.114mm) and the service limit is 0.008 in. (0.20mm). Replace the cam gear or thrust plate if the measured value exceeds the service limit.

- Oil pump drive gear and reluctor to the crankshaft. Do not damage the teeth of the reluctor.
- Oil pump driven gear and nut. Tighten to 74 ft. lbs. (100 Nm).
- Engine front cover
- A/C condenser
- Charged air cooler

8. Apply clean engine oil to the roller and outside of the lifters.

- Valve lifters
- Valve lifter guides
- Valve lifter guide hold-down brackets. Make sure that both tabs of the bracket are in the holes of the valve lifter guides.
- Valve lifter guide hold-down bracket bolts. Tighten to 97 inch lbs. (11 Nm).

Valve Lash

ADJUSTMENT

The 6.5L engine uses hydraulic lifters, which require no periodic adjustment.

6.6L Engine

1. Remove the fan clutch.
2. Remove both upper valve covers.
3. Rotate the engine in the normal direction and place the No. 1 piston at Top Dead Center (TDC) of the compression stroke. The No. 1 cylinder is at the right side front. While turning the engine, watch the intake valve to open and close. Align the mark on the crankshaft balancer with the pointer on the engine.
4. Loosen the valve clearance adjusting screws for the valve being adjusted.
5. Insert the feeler gauge between the tip of the rocker arm and the valve bridge.
6. Adjust the intake and the exhaust valve clearance to 0.012 in. (0.3mm) with the engine cold. Refer to the figure for the valves that can be adjusted TDC of the compression stroke.

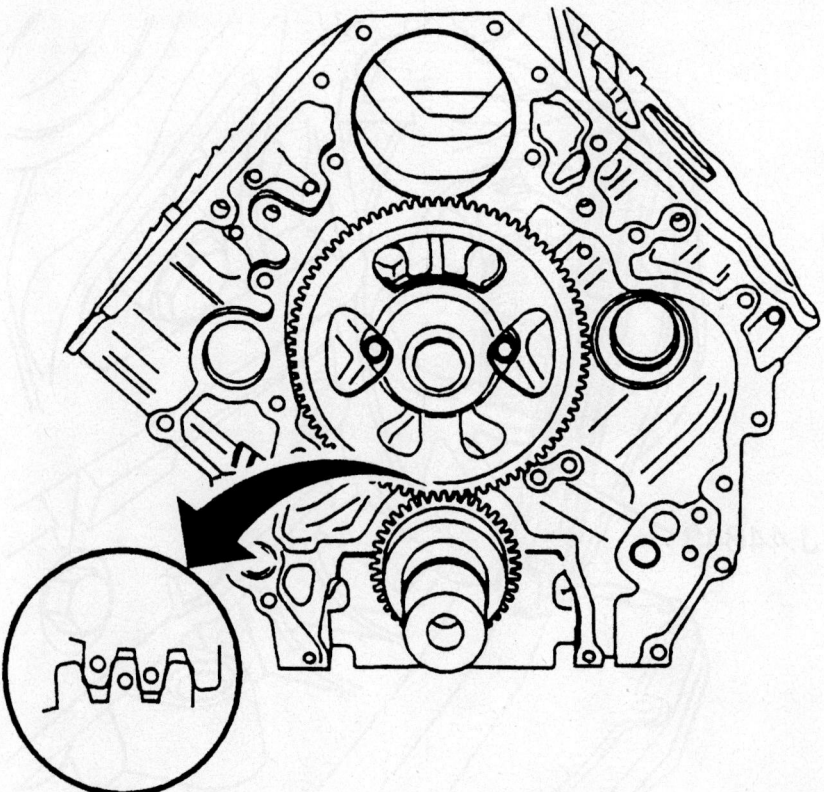

9348NG24

Camshaft and crankshaft gear alignment—6.5L engine

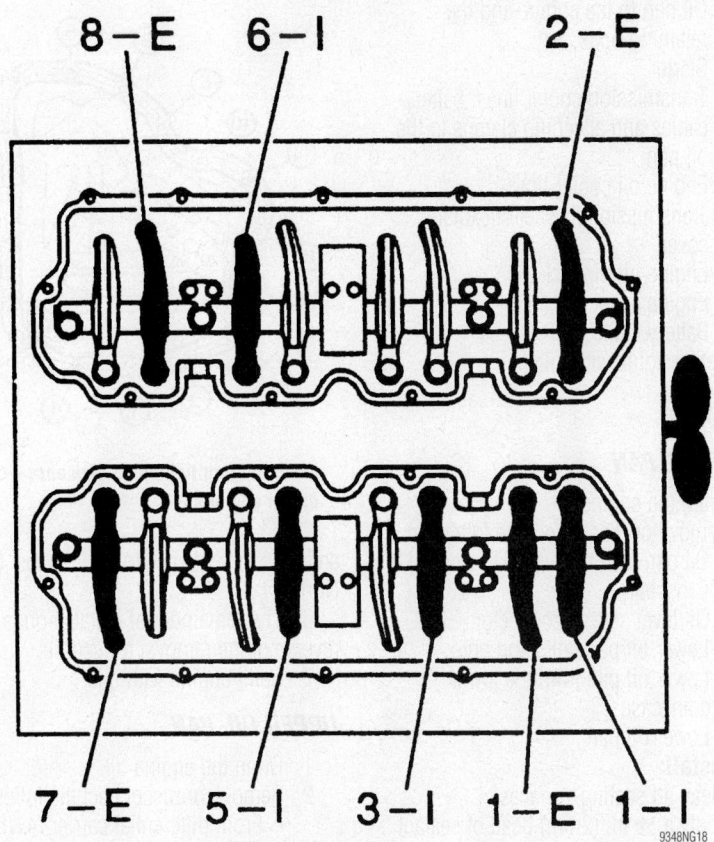

Location of the valves that are adjusted at TDC of the compression stroke—6.6L engine

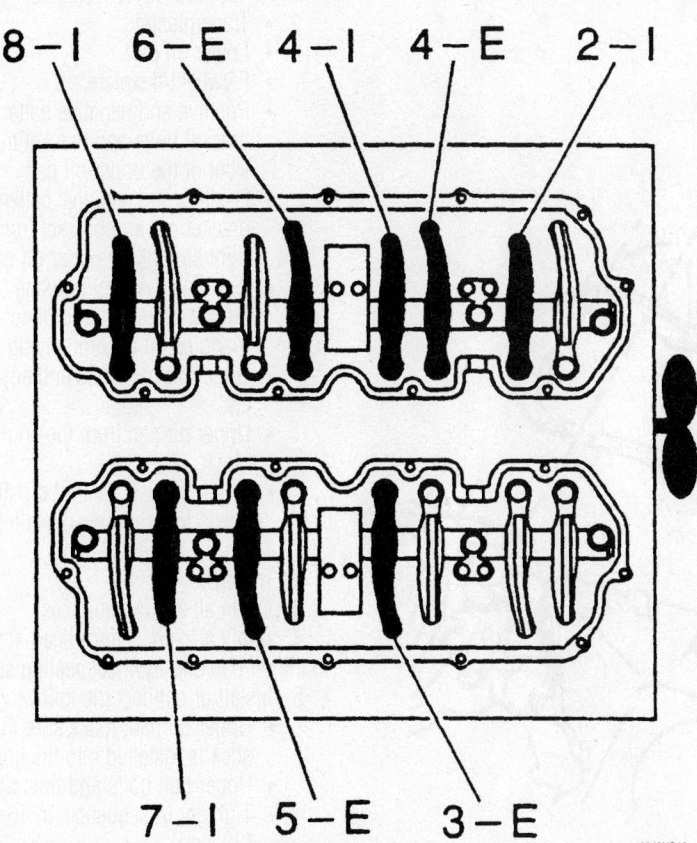

Location of the valves that are adjusted at TDC of the exhaust stroke—6.6L engine

7. Tighten the valve adjusting screw lock nut to 16 ft. lbs. (22 Nm).

8. Turn the engine one rotation in the normal direction and put the No. 1 piston at TDC of the exhaust stroke to adjust the remaining valve clearance. While turning the engine, watch the exhaust valve to open and close. Align the mark on the crankshaft balancer with the pointer on the engine.

9. Loosen the valve clearance adjusting screws for the valves being adjusted.

10. Insert the feeler gauge between the tip of the rocker arm and the valve bridge.

11. Adjust the intake and the exhaust valve clearance to 0.012 in. (0.3mm) with the engine cold. Refer to the figure for the valves that can be adjusted TDC of the exhaust stroke.

12. Tighten the valve adjusting screw lock nut to 16 ft. lbs. (22 Nm).

13. Install the upper and lower valve cover and fan clutch, as necessary.

Oil Pan

REMOVAL & INSTALLATION

6.5L Engine

SILVERADO AND SIERRA

1. Drain the engine oil.
2. Remove or disconnect the following:
   - Battery cables
   - Oil dipstick
   - Flywheel/flexplate cover
   - Oil cooler line guides
   - Front driveshaft
   - Front axle, if needed
   - Exhaust pipes from the manifolds
   - Front engine mount through-bolts
   - Oil pan bolts and the oil pan
   - Oil pan rear seal

To install:

3. Clean all sealing surfaces.
4. Apply a 3/16 in. (5mm) bead of RTV sealant to the oil pan sealing surface, inboard of the bolt holes. The sealant must be wet to the touch when the oil pan is to be installed.

5. Install or connect the following:
   - Oil pan rear seal
   - Oil pan to the engine. Torque all bolts except the rear 2 bolts to 84 inch lbs. (9.4 Nm). Tighten the rear bolts to 17 ft. lbs. (23 Nm).
   - Engine mounting through-bolt and nut
   - Front axles and front driveshaft, if removed
   - Oil cooler lines in guides
   - Oil dipstick

- Exhaust pipes to the manifolds
- Flywheel/flexplate cover
- Battery cables

6. Refill with the proper grade and quantity of oil.

EXPRESS AND SAVANA

1. Drain the engine oil.
2. Remove or disconnect the following:
   - Battery cables
   - Engine cover
   - Engine oil dipstick
   - Transmission flywheel/flexplate cover
   - Oil cooler lines at the block
   - Starter
   - Transmission cooler lines, battery cables and attaching clamps from the oil pan
   - Oil pan bolts
   - Oil pan and oil pan rear seal

To install:

3. Clean all sealing surfaces
4. Apply a ³⁄₁₆ in. (5mm) bead of RTV sealant to the oil pan sealing surface, inboard of the bolt holes. The sealant must be wet to the touch when the oil pan is to be installed.
5. Install or connect the following:
   - Oil pan rear seal

- Oil pan to the engine and the retaining bolts.
- Starter
- Transmission cooler lines, battery cables and attaching clamps to the oil pan
- Engine oil cooler lines
- Transmission flywheel/flexplate cover
- Engine oil dipstick tube
- Engine cover
- Battery cables

6. Refill engine with oil.

6.6L Engine

LOWER OIL PAN

1. Drain the engine oil.
2. Remove or disconnect the following:
   - Oil pan skid plate (2WD vehicles)
   - Crossbar
   - Oil level sensor connector
   - Lower oil pan bolts and nuts
   - Lower oil pan from the lower crankcase
   - Lower oil pan

To install:

3. Clean all sealing surfaces
4. Apply a ⅛ in. (2mm) bead of sealant to the oil pan sealing surface.
5. Install the oil pan. Tighten the bolts

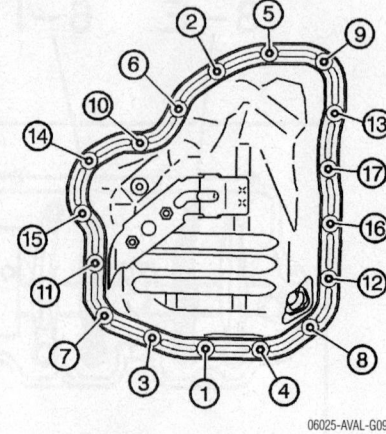

06025-AVAL-G09

Lower oil pan tightening sequence—6.6L diesel engines

and nuts in sequence to 89 inch lbs. (10 Nm)

6. The remainder of installation is the reverse of the removal procedure.
7. Refill engine with oil.

UPPER OIL PAN

1. Drain the engine oil.
2. Remove or disconnect the following:
   - Front differential carrier (4WD vehicles)
   - Relay rod from the pitman arm and idler arm (2WD vehicles)
   - Transmission
   - Lower oil pan
   - Flywheel/flexplate
   - Positive and negative battery cable bracket bolts and bracket from the front of the upper oil pan
   - Positive and negative battery cable bracket nut and bracket from the right side of the upper oil pan
   - 2 engine flywheel housing to upper oil pan bolts (refer to denoted black triangles on accompanying figure)
   - Upper oil pan bolts and any brackets
   - Upper oil pan from the engine block
   - Upper oil pan. The oil dipstick tube needs to be removed while lowering the upper oil pan.

To install:

3. Clean all sealing surfaces
4. Apply a ⅛ in. (2mm) bead of sealant to the oil pan and flywheel sealing surfaces.
5. Install or connect the following:
   - Upper oil pan; make sure the dipstick is installed into the upper pan
   - Upper pan bolts and brackets. Tighten, in sequence, to 15 ft. lbs. (20 Nm).
   - 2 engine flywheel housing to upper

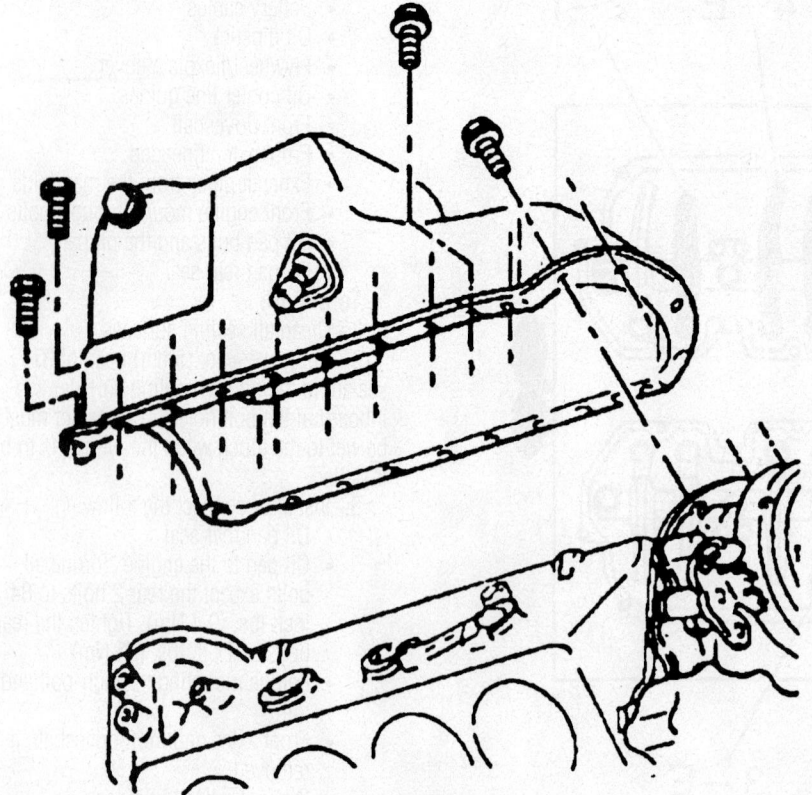

7924KG47

Exploded view of the oil pan mounting—6.5L diesel engines

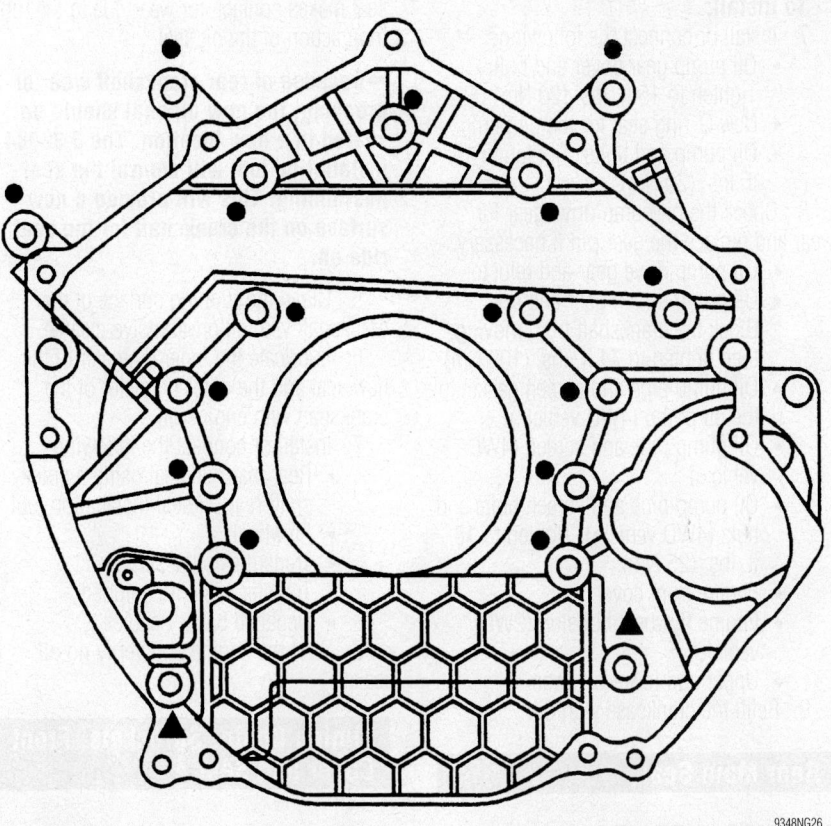

Remove only the flywheel housing-to-upper oil pan bolts designated with a black triangle—6.6L engine

9348NG26

oil pan bolts (refer to denoted black triangles on accompanying figure). Torque to 37 ft. lbs. (50 Nm).

6. The remainder of installation is the reverse of the removal procedure.

7. Refill engine with oil.

Oil Pump

REMOVAL & INSTALLATION

6.5L Engine

1. Drain the engine oil
2. Remove or disconnect the following:
   - Oil pan
   - Oil pump to crankshaft rear main bearing attaching bolt
   - Oil pump and hex drive

To install:

3. Inspect the oil pan pick up tube and screen for damage and the hex drive for cracks.

4. Install or connect the following:
   - Oil pump and extension shaft to the engine. Align the extension shaft hex with the drive hex, the oil pump should push easily into place.
   - Oil pump bolt and tighten to 65 ft. lbs. (90 Nm)

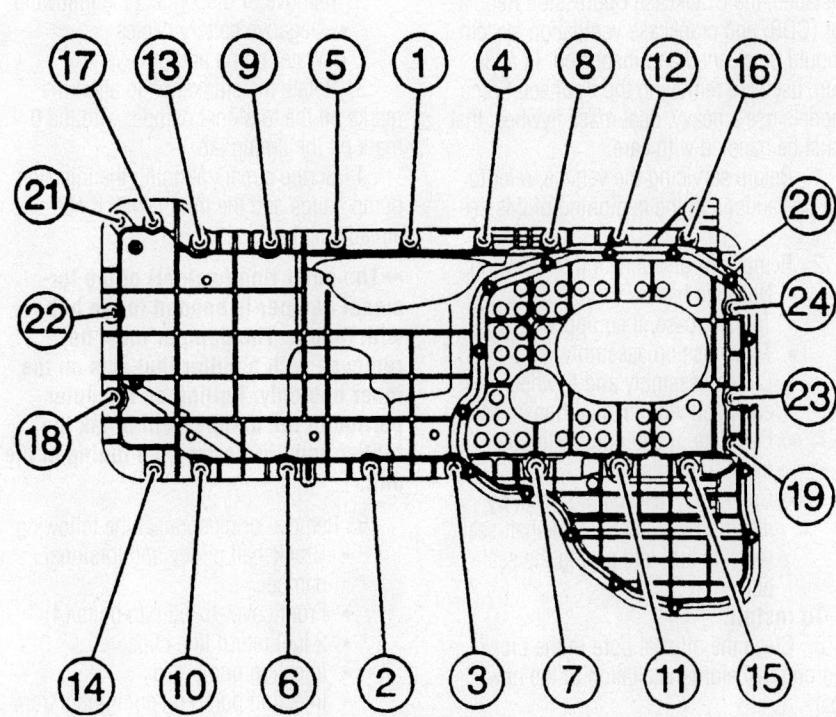

Upper oil pan bolt tightening sequence—6.6L engine

9348NG27

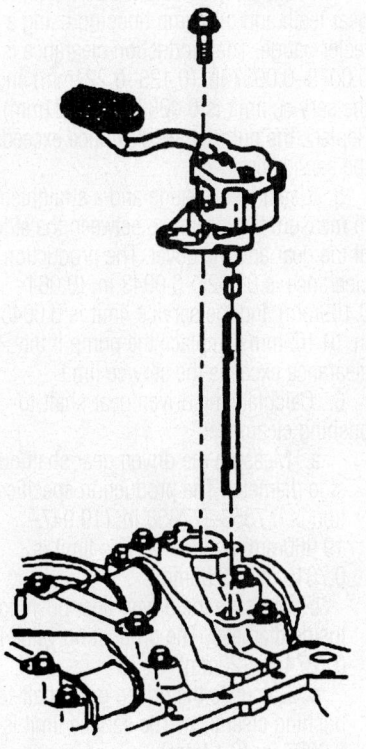

7924KG48

Exploded view of the oil pump mounting—6.5L diesel engines

- Oil pan
5. Refill the crankcase with oil.

6.6L Engine

1. Drain the engine oil
2. Remove or disconnect the following:
   - Engine flywheel housing (2WD vehicles)
   - Engine front cover
   - Lower and upper oil pans
   - Oil pump pipe and screen and gasket
3. Block the crankshaft from turning with a wooden dowel.
   - Oil pump driven gear nut
   - Oil pump driven gear

➡ **The crankshaft reluctor and oil pump drive gear are timed together at the factory. Do NOT remove the reluctor from the oil pump drive gear or damage the reluctor teeth.**

   - Oil pump drive gear and crankshaft reluctor assembly using a brass drift and tapping as close to the center of the reluctor assembly
   - 3 hex head and 1 Allen head bolt
   - Oil pump
   - Oil pump O-ring seal
   - Oil pump gear cover bolts and cover
4. Measure the clearance between the gear teeth and oil pump housing using a feeler gauge. The production clearance is 0.0049–0.0087 in. (0.125–0.221mm) and the service limit is 0.0087 in. (0.221mm). Replace the pump if the clearance exceeds the service limit.
5. Use a feeler gauge and a straightedge to measure the clearance between the side of the gear and the cover. The production clearance is 0.0025–0.0043 in. (0.064–0.109mm) and the service limit is 0.0043 in. (0.109mm). Replace the pump if the clearance exceeds the service limit.
6. Calculate the driven gear shaft-to-bushing clearance:
   a. Measure the driven gear shaft outside diameter. The production specification is 0.7853–0.7858 in. (19.947–19.960mm) and the service limit is 0.7819 in. (19.86mm).
   b. Measure the driven gear bushing inside diameter. The production value is 0.7874 in. (20mm).
   c. Calculate the driven gear shaft-to-bushing clearance. The service limit is 0.0055 in. (0.14mm).
   d. Replace the pump if the clearance exceeds the service limit.

To install:

7. Install or connect the following:
   - Oil pump gear cover and bolts. Tighten to 15 ft. lbs. (20 Nm).
   - New O-ring seal for the oil pump
   - Oil pump and bolts. Tighten to 15 ft. lbs. (20 Nm).
8. Check the oil pump drive gear for wear and replace the gear pin if necessary.
   - Oil pump drive gear and reluctor
   - Oil pump driven gear and nut. Block the crankshaft from moving, then tighten to 74 ft. lbs. (100 Nm)
   - Oil pump pipe and screen gasket to the oil pump (4WD vehicle)
   - Oil pump pipe and screen (4WD vehicle)
   - Oil pump pipe and screen bolts and nuts (4WD vehicle). Tighten to 18 ft. lbs. (25 Nm).
   - Engine front cover
   - Engine flywheel housing (2WD vehicle)
   - Upper and lower oil pans
9. Refill the crankcase with oil.

Rear Main Seal

REMOVAL & INSTALLATION

Please note that the entire transmission assembly must be removed before performing this procedure. Before a new seal is installed, the Crankcase Depression Regulator (CDR) and crankcase ventilation system should be cleaned and inspected. In addition, use care removing the flywheel. Some models use a heavy, dual mass flywheel that must be handled with care.

1. Before servicing the vehicle, refer to the precautions in the beginning of this section.
2. Remove or disconnect the following:
   - Negative battery cables
   - Transfer case, if equipped
   - Transmission assembly
   - Clutch assembly and flywheel, if equipped with manual transmission
   - Flexplate, if equipped with automatic transmission
   - Crankshaft rear main oil seal by inserting a suitable crankshaft seal removal tool and prying the seal out

To install:

3. Clean the oil seal bore in the block thoroughly before installation of the new seal.
4. Inspect the crankshaft for grit, rust or burrs and correct as necessary. Also inspect the portion of the crankshaft where the oil seal makes contact, for wear due to the rubbing action of the oil seal.

➡ **Because of rear crankshaft wear or grooving, the new oil seal should be seated in a new location. The J 39084 installation tool will control the seal positioning. This will provide a new surface on the crankshaft for the seal to ride on.**

5. Clean the running surface of the crankshaft with a non-abrasive cleaner.
6. Lubricate the inner diameter of the new seal and the outer diameter of the crankshaft with engine oil.
7. Install or connect the following:
   - Rear main oil seal using a crankshaft rear oil seal installation tool
   - Flywheel.
   - Transmission assembly
   - Transfer case, if equipped
   - Negative battery cables
8. Start the engine and verify no oil leaks.

Timing Chain, Sprockets, Front Cover and Seal

REMOVAL & INSTALLATION

6.5L Engine

1. Drain the cooling system.
2. Remove or disconnect the following:
   - Negative battery cables
   - Water pump and pulleys
3. Rotate the crankshaft to align the marks on the torsional damper with the **0** mark on the timing tab.
4. Scribe a mark aligning the injection pump flange and the front cover, if not already marked.

➡ **The outer ring (weight) of the torsional damper is bonded to the hub with rubber. The damper must be removed with a puller that acts on the inner hub only. Pulling on the outer portion of the damper will break the rubber bond or destroy the tuning of the unit.**

5. Remove or disconnect the following:
   - Crankshaft pulley and torsional damper
   - Front cover-to-oil pan bolts (4)
   - 2 fuel return line clips
   - Injection pump gear
   - Injection pump retaining nuts from the front cover
   - Crankshaft sensor
   - Baffle

- Cover bolts remaining and the front cover
- Injection pump gear

6. Align the camshaft timing gear marks

- Bolt and washer attaching the camshaft gear
- Camshaft sprocket with the timing chain
- Crankshaft sprocket.

To install:

7. Install or connect the following:

- Cam sprocket, timing chain and crankshaft sprocket as a unit, aligning the timing marks on the sprockets.

8. Rotate the crankshaft to align the injection pump and camshaft gears.

- Injection pump gear

9. If the front cover oil seal is to be replaced, it can now be pried out of the cover with a suitable prying tool. Press the new seal into the cover evenly.

10. Clean both sealing surfaces until all traces of old sealer are gone. Apply a ³⁄₃₂ in. (2mm) bead of GM sealant 1052357 or equivalent to the sealing surface. Apply a ³⁄₁₆ in. (5mm) bead of RTV type sealer to the bottom portion of the front cover which attaches to the oil pan.

- Front cover
- Baffle
- Injection pump. Torque the nuts to 31 ft. lbs. (42 Nm), making sure the scribe marks on the pump and front cover are aligned.
- Injection pump driven gear. Torque the injection pump gear bolts to 17 ft. lbs. (23 Nm), making sure the marks on the cam gear and pump are aligned.

➡**Verify that there is a minimum clearance of 0.040 in. (1.0mm) between the injection pump gear and baffle or noise may be result.**

- Fuel line clips
- Front cover-to-oil bolts
- Torsional damper
- Crankshaft pulley. Torque the bolts to 80 inch lbs. (9 Nm).
- Oil pan bolts. Torque the bolts to 106 inch lbs. (12 Nm).
- Water pump
- Pulley assembly
- Negative battery cables

11. Refill the cooling system with the proper type and quantity of antifreeze.

12. Inspect the engine for leaks.

Timing Gears, Front Cover and Seal

REMOVAL & INSTALLATION

6.6L Engine

➡**The 6.6L engine uses gears in place of a timing chain. For removal and installation, please see the Camshaft and Lifters procedure. This procedure covers the removal of the front cover and seal.**

1. Remove the upper intake manifold sight shield as follows:

   a. Remove the retaining bolt in the front of the shield.

   b. Lift up on the front of the shield, then lift the shield off the rear bracket.

2. Drain the cooling system.

3. Remove or disconnect the following:

- Negative battery cables

- Right front wheel
- Right front fender splash shield
- Upper fan shroud
- Fan clutch
- Drive belt
- Oil dipstick tube
- Thermostat housing crossover
- Crankshaft balancer
- Crankshaft front oil seal
- Water pump
- Camshaft sensor electrical connector
- Camshaft sensor bolt and sensor
- Crankshaft Position (CKP) sensor connector, bolt and sensor
- CKP sensor spacer bolts and spacer
- 5 bolts securing the upper oil pan to the front cover
- Bracket bolts and the bracket for the turbocharger outlet coolant pipe
- Engine front cover bolts
- Use a suitable seal cutter to separate the front cover from the cylinder block and upper oil pan

➡**Do not bend the turbocharger outlet pipe.**

- O-ring from the front cover
- Oil pressure relief valve from the front cover

To install:

4. Clean and inspect all sealing surfaces.

5. Install or connect the following:

- Oil pressure relief valve with a new O-ring. Tighten to 30 ft. lbs. (41 Nm).
- Apply a ⅛ in. (2–3mm) wide to ¹⁄₁₆ in. (0.5–1.5mm) high bead of sealant to the front cover sealing surfaces to the engine block and oil pan.
- New front cover O-ring after lubricating it with engine oil
- Front cover and bolts. Tighten to 18 ft. lbs. (25 Nm).
- Upper oil pan-to-front cover bolts. Tighten to 15 ft. lbs. (20 Nm).
- Turbocharger coolant outlet pipe bracket and bolts. Tighten to 15 ft. lbs. (20 Nm).
- Camshaft sensor and bolt. Tighten to 80 inch lbs. (9 Nm).
- Camshaft sensor connector

➡**The CKP sensor spacers are machined with different timing positions. If you have to replace a spacer, make sure it has the same part number.**

- CKP sensor spacer and spacer bolts. Tighten to 89 inch lbs. (10 Nm).

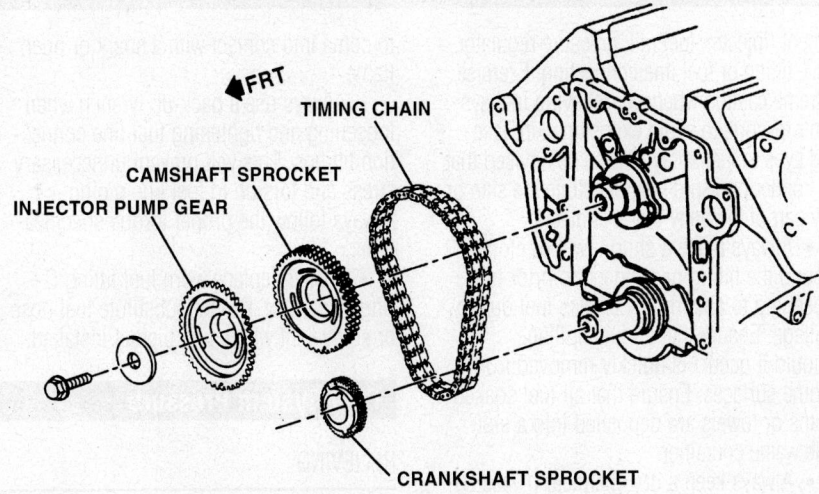

Timing chain and related components—6.5L diesel engines

7924KG23

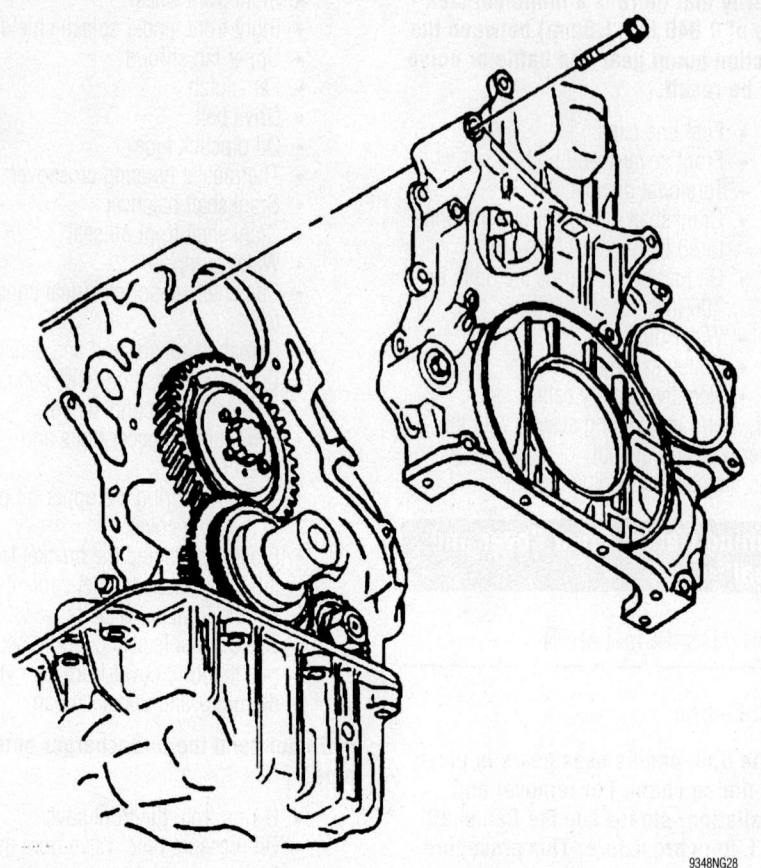

Engine front cover—6.6L engine

- CKP sensor and bolt. Tighten to 89 inch lbs. (10 Nm).
- Water pump
- Crankshaft front oil seal
- Crankshaft balancer
- Thermostat housing crossover
- Oil fill tube
- Drive belt

- Upper fan shroud
- Right front fender splash shield and wheel
- Negative battery cables
6. Refill the cooling system with the proper type and quantity of antifreeze.
7. Inspect the engine for leaks.

Piston and Ring

POSITIONING

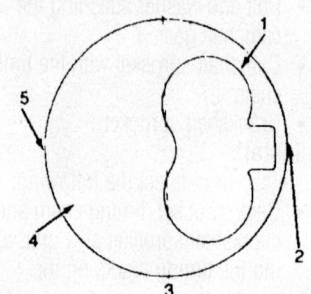

1	Oil control ring expander gap	3	Centerline of piston pin
2	Second compression ring gap	4	Oil control ring gap
		5	Top compression ring gap

7924AG11

Piston ring end-gap spacing —6.5L diesel engines

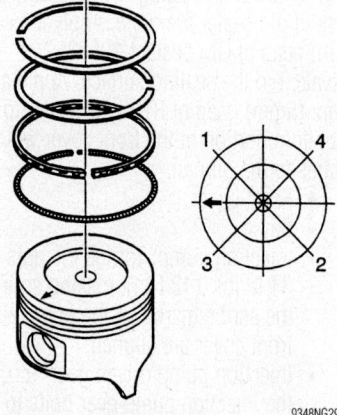

9348NG29

Piston ring positioning—6.6L diesel engines

GASOLINE FUEL SYSTEM

Fuel System Service Precautions

Safety is the most important factor when performing not only fuel system maintenance but any type of maintenance. Failure to conduct maintenance and repairs in a safe manner may result in serious personal injury or death. Maintenance and testing of the vehicle's fuel system components can be accomplished safely and effectively by adhering to the following rules and guidelines.

- To avoid the possibility of fire and personal injury, always disconnect the negative battery cable unless the repair or test procedure requires that battery voltage be applied.
- Always relieve the fuel system pressure prior to disconnecting any fuel system component (injector, fuel rail, pressure regulator, etc.), fitting or fuel line connection. Exercise extreme caution whenever relieving fuel system pressure, to avoid exposing skin, face and eyes to fuel spray. Please be advised that fuel under pressure may penetrate the skin or any part of the body that it contacts.

- Always place a shop towel or cloth around the fitting or connection prior to loosening to absorb any excess fuel due to spillage. Ensure that all fuel spillage (should it occur) is quickly removed from engine surfaces. Ensure that all fuel soaked cloths or towels are deposited into a suitable waste container.
- Always keep a dry chemical (Class B) fire extinguisher near the work area.
- Do not allow fuel spray or fuel vapors to come into contact with a spark or open flame.

- Always use a back-up wrench when loosening and tightening fuel line connection fittings. This will prevent unnecessary stress and torsion to fuel line piping. Always follow the proper torque specifications.
- Always replace worn fuel fitting O-rings with new. Do not substitute fuel hose or equivalent where fuel pipe is installed.

Fuel System Pressure

RELIEVING

A Schrader valve is provided on these fuel systems, in order to conveniently test or release the system pressure. A fuel pressure gauge and adapter will be necessary to connect the gauge to the fitting. Most of the MFI systems utilize a service valve on one end of the fuel rail assembly. The CMFI system covered here uses a valve located on the inlet pipe fitting, immediately before it enters the CMFI assembly (towards the rear of the engine)

1. Before servicing the vehicle, refer to the precautions in the beginning of this section.

2. Turn the ignition **OFF**.

3. Disconnect the negative battery cable.

4. Loosen the fuel filler cap in order to relieve the fuel tank vapor pressure.

5. Connect a fuel pressure gauge to the fuel pressure valve/fitting.

6. Wrap a shop towel around the fitting while connecting the gauge in order to avoid spillage.

7. Install the bleed hose of the gauge into an approved container.

8. Open the valve on the gauge to bleed the system pressure.

The fuel connections are now safe for servicing. Drain any fuel remaining in the gauge into an approved container.

Fuel Filter

REMOVAL & INSTALLATION

➡**On 2005-06 gasoline engines the fuel filter is integral with the fuel pump/sender assembly in the fuel tank.**

2002–04 Sierra, Silverado & Avalanche

EXCEPT 2004 5.3L ENGINE

1. Before servicing the vehicle, refer to the precautions in the beginning of this section.

2. Disconnect the negative battery cable.

3. Relieve the fuel system pressure.

4. Raise the vehicle.

5. Clean all the fuel filter connections and the surrounding areas before disconnecting the fuel pipes in order to avoid possible contamination of the fuel system.

6. Disconnect the threaded fittings from the fuel filter.

7. Cap the fuel pipes in order to prevent possible fuel system contamination.

8. Slide the fuel filter from the bracket located on the frame rail.

9. Inspect the fuel pipe O-rings for cuts, nicks, swelling, or distortion. Replace the O-rings if necessary.

To install:

10. Slide the fuel filter into the bracket. Remove the caps from the fuel pipes.

11. Connect the threaded fittings to the fuel filter. Tighten the fittings to 18 ft. lbs. (25 Nm).

12. Lower the vehicle.

13. Tighten the fuel filler cap.

14. Connect the negative battery cable.

15. Turn the ignition **ON** for 2 seconds.

16. Turn the ignition **OFF** for 10 seconds.

17. Turn the ignition **ON**.

18. Inspect for fuel leaks.

2004 5.3L ENGINE

➡**This procedure requires the use of tool J-46363, Fuel Line Release Tool, or equivalent.**

1. Before servicing the vehicle, refer to the precautions in the beginning of this section.

2. Disconnect the negative battery cable.

3. Relieve the fuel system pressure.

4. Raise the vehicle.

5. Remove the fuel composition sensor bracket nuts and reposition it aside.

6. Clean all the fuel filter connections and the surrounding areas before disconnecting the fuel pipes in order to avoid possible contamination of the fuel system.

7. Using a flare nut wrench and backup wrench disconnect the fuel line fitting from the fuel filter.

8. Use the Fuel Line Release Tool, J-46363, or equivalent to disconnect the quick connect fittings as follows:

   a. Pull the quick connect fitting back until the internal retainers hit the filter stop.

   b. Twist one end of J-46363 and insert it between the fitting and the filter

   c. Twist the opposite end of the tool and insert it between the fitting and filter

   d. Once tool J-46363 is inserted, make sure it is parallel to the filter and fitting.

   e. While holding the filter, push the fitting toward the filter in order to disconnect the quick connect fitting.

   f. Remove the fitting from the filter. Remove the tool from the filter.

9. Slide the fuel filter from the bracket.

10. Inspect the fuel pipe O-rings for cuts, nicks, swelling, or distortion. Replace the O-rings if necessary.

To install:

11. Slide a NEW fuel filter into the bracket.

12. While holding the filter, push the fitting towards the filter in order to connect the quick connect fitting.

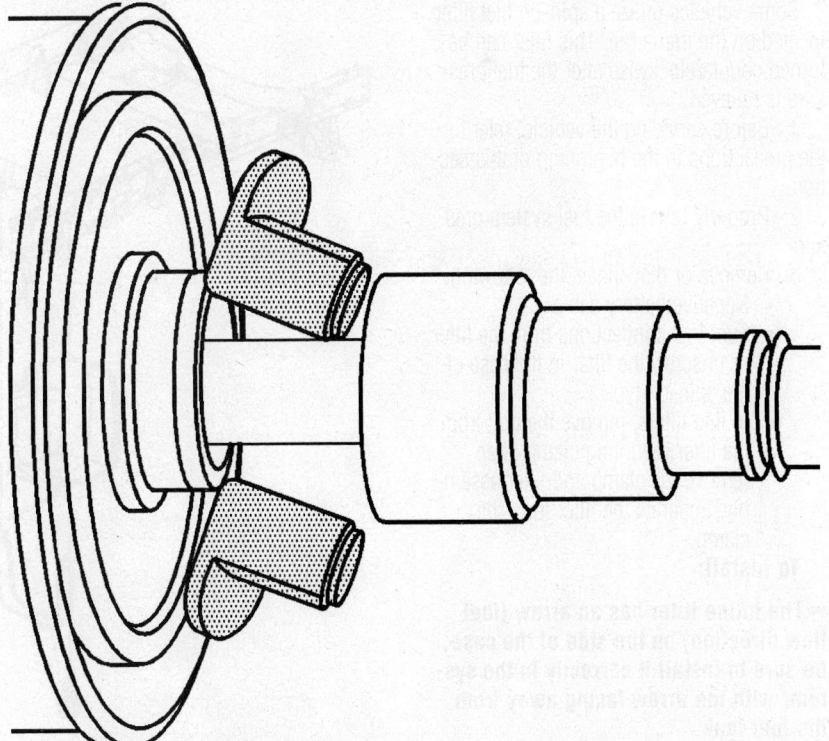

Twist one end of the fuel line release tool and insert it between the fitting and the fuel filter— 2004 5.3L engine

71461-GMCK-G02

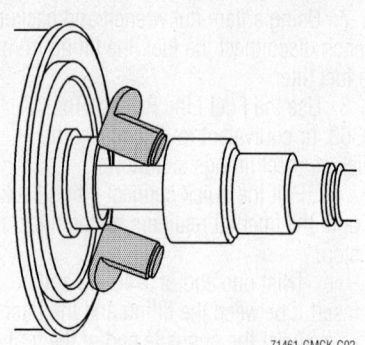

71461-GMCK-G02

The spin-on fuel filter is serviced in the same manner as a spin-on oil filter

13. Using a flare nut wrench and backup wrench connect the fuel line fitting to the fuel filter.

14. Position the fuel composition sensor/bracket into place and secure with the bracket nuts. Torque the nuts to 13 ft. lbs. (17 Nm).

15. Lower the vehicle

2002–04 Express and Savana

The fuel filter is normally located along the frame rail of the vehicle. On some vehicles however, it may have been relocated to the engine compartment. When in doubt, trace a fuel line from the engine backward or from the tank forward in order to locate the filter.

Some vehicles utilize a spin-on fuel filter located on the frame rail. This filter can be turned counterclockwise after the fuel pressure is relieved.

1. Before servicing the vehicle, refer to the precautions in the beginning of this section.

2. Properly relieve the fuel system pressure.

3. Remove or disconnect the following:
   • Negative battery cable
   • Fuel line connections from the filter or unscrew the filter in the case of the spin-on type
   • In line filters, remove the bolt from the filter mounting clamp, then remove the clamp and filter assembly. Separate the filter from the clamp.

To install:

➡ **The inline filter has an arrow (fuel flow direction) on the side of the case, be sure to install it correctly in the system, with the arrow facing away from the fuel tank.**

4. Install or connect the following:
   • In line filters place filter in clamp
   • Install filter and clamp

• Spin-on filters, lubricate the gasket before installation. Then tighten the filter an additional ¾ of a turn from the point when the gasket touches the filter adapter. Always check for leaks after a new filter is installed.

Fuel Pump

REMOVAL & INSTALLATION

1. Before servicing the vehicle, refer to the precautions in the beginning of this section.

2. Remove or disconnect the following:
   • Negative battery cable
3. Relieve the fuel system pressure.
4. Drain the fuel tank.
5. Remove or disconnect the following:
   • Fuel tank

✳✳ WARNING

Do not handle the fuel sender assembly by the fuel pipes. The amount of leverage generated by handling the fuel pipes could damage the joints.

• Fuel sender assembly retaining ring using a fuel tank sending unit

wrench. Remove the fuel sender assembly and the seal. Discard the seal.

6. Note the position of the fuel strainer on the fuel sender. Support the fuel sender assembly with one hand and grasp the strainer with the other hand. Pull the strainer off the fuel sender. Discard the strainer after inspection. Inspect the strainer. Replace a contaminated strainer and clean the fuel tank.
   • Fuel pump electrical connector
   • Electrical connector retaining clip from the fuel level sensor
   • Sensor electrical connector from under the fuel sender cover
   • Fuel level sensor retaining clip

7. Squeeze the locking tangs and remove the fuel level sensor.

8. Remove the fuel pressure sensor.

To install:

9. Install or connect the following:
   • Fuel pressure sensor
   • Fuel level sensor
   • Sensor retaining clip
   • Electrical connector to the fuel level sensor
   • Electrical connector retaining clip to the fuel level sensor
   • Fuel pump electrical connector

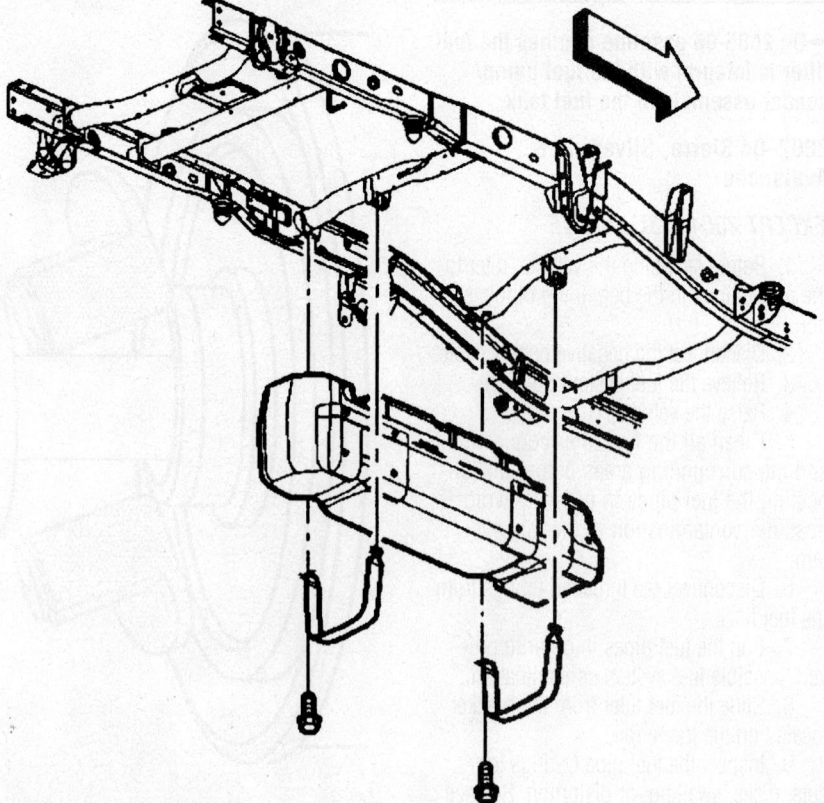

Fuel tank—Silverado shown

9308KG54

➥**Always install a new fuel strainer when replacing the fuel tank fuel pump module.**

- New fuel strainer in the same position as noted during disassembly. Push the strainer on the bottom of the fuel sender until the strainer is fully seated.
- New seal on the fuel tank

➥**The fuel pump strainer must be in a horizontal position when the fuel sender is installed in the tank. When installing the fuel sender assembly, assure that the fuel pump strainer does not block full travel of the float arm.**

- Fuel sender assembly into the fuel tank
- Fuel sender assembly retaining ring
- Fuel tank. Install the fuel tank strap attaching bolts. Tighten the bolts to 30 ft. lbs. (40 Nm).

10. Refill the fuel tank. Install the fuel filler cap. Connect the negative battery cable.

11. Turn the ignition **ON** for 2 seconds.

12. Turn the ignition **OFF** for 10 seconds.

13. Turn the ignition **ON**.

14. Inspect for fuel leaks.

Fuel Injector

REMOVAL & INSTALLATION

4.3L Engines

1. Before servicing the vehicle, refer to the precautions in the beginning of this section.

➥**Use care when removing the injectors to prevent damage to the electrical connector pins on the injector and the nozzle. The fuel injector is serviced as a complete assembly only. Since the injector is an electrical component, it should not be immersed in any type of cleaner.**

2. Relieve the fuel system pressure.

3. Remove the upper intake manifold.

4. Remove and discard the fuel meter body seal.

5. Before removal, clean the fuel meter body with a spray type engine cleaner, if necessary. Follow the package instructions. Do not soak the fuel meter body in liquid cleaning solvent.

➥**When disconnecting the injectors, remember the sequence in order to ensure correct injector placement to each cylinder.**

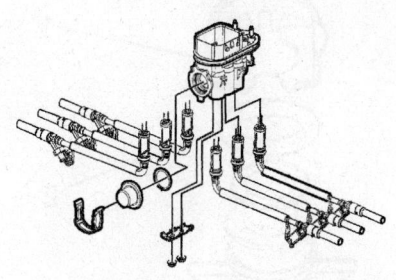

06025-AVAL-G10

Exploded view of the fuel injector assembly—4.3L engine

6. Remove the injectors from the lower intake manifold bores. Lightly pull on the injector tube and using a small pocket screwdriver, remove the injector from the manifold.

7. Remove the MFI (multi-port flexible injection) fuel meter body from the MFI bracket.

8. Remove the injector retainer lock nuts and retainer.

9. While pulling the fuel injector downward, push with a small tip punch down between the injector terminals until the injector is removed.

To install:

10. Lubricate the NEW injector O-ring seals with clean engine oil.

11. Install the fuel injector into the fuel meter body injector socket.

12. Install the injector retainer and the injector retainer lock nuts and tighten to 27 inch lbs. (3 Nm).

13. Position the fuel tubes of the MFI fuel meter body in the forward position.

14. Install the MFI fuel meter body into the fuel meter body bracket on the lower intake manifold. Push down firmly to lock the fuel meter body into the fuel meter body bracket.

15. Insert the injectors into the correct injector bore in the lower manifold.

16. Insure the electrical connectors on the injectors are positioned so that they do not interfere with each other and are pointing towards the center of the intake manifold. Rotate the electrical connectors inboard if necessary

17. Lubricate the upper intake manifold O-ring on the fuel meter body with clean engine oil.

18. Install the upper intake manifold.

19. Tighten the fuel fill cap.

20. Connect the negative battery cable.

21. Turn the ignition ON, with the engine OFF, for 2 seconds.

22. Turn the ignition OFF for 10 seconds.

23. Turn the ignition ON, with the engine OFF.

24. Inspect for fuel leaks.

5.0L and 5.7L Engines

1. Remove or disconnect the following:
- Negative battery cable
- Intake manifold plenum
- Fuel rail assembly
- Wiring harness
- Injector clip and discard it
- Injector O-ring seals from both

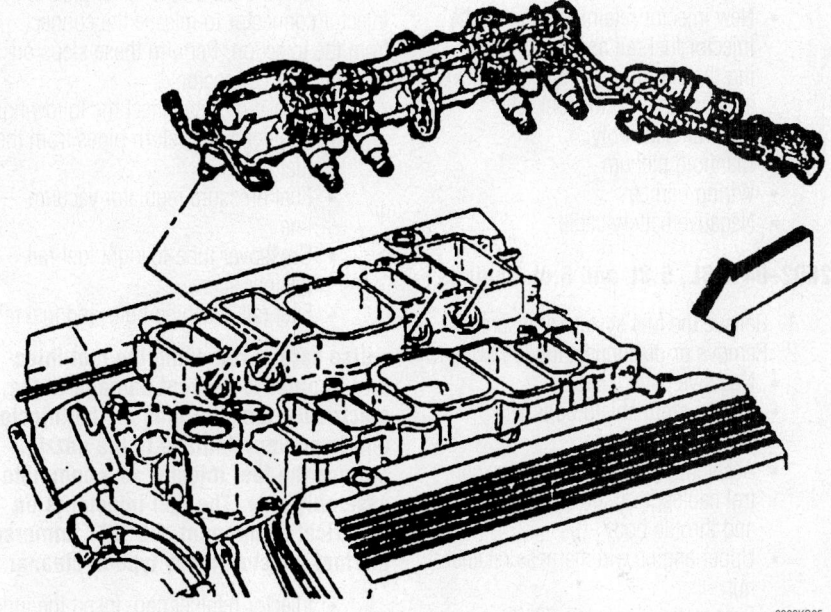

Exploded view of the fuel rail assembly—MFI systems

9308KG05

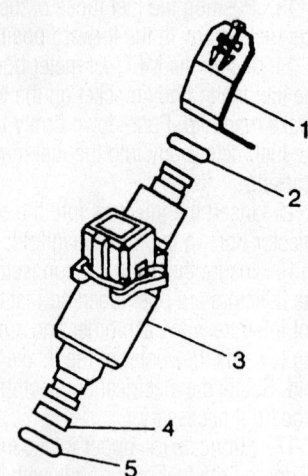

1. Clip - SFI Fuel Injector Retainer
2. O-ring - SFI Fuel Injector Upper
3. Injector Asm - SFI Fuel
4. O-ring - Backup
5. O-ring - SFI Fuel Injector Lower

9308KG06

Exploded view of the fuel injector assembly—MFI systems

ends of the injector. Save the O-ring backups for use on reassembly

To install:

2. Install or connect the following:
  • O-ring backups before installing the O-rings
3. Lubricate the new injector O-rings with clean engine oil
  • O-ring to injector assembly
  • Fuel injector into the fuel rail injector socket with the electrical connectors facing outward
  • New injector retaining clips on the injector fuel rail assembly by sliding the clip into the injector groove as it snaps onto the fuel rail
  • Fuel rail assembly
  • Manifold plenum
  • Wiring harness
  • Negative battery cable

2002–04 4.8L, 5.3L and 6.0L Engines

1. Relieve the fuel system pressure.
2. Remove or disconnect the following:
  • Negative battery cable
  • Engine sight shield bolts and bracket
  • Accelerator control and cruise control cables from the cable bracket and throttle body
  • Upper engine wire harness retainer nut
  • Evaporative Emission (EVAP) purge valve harness connector

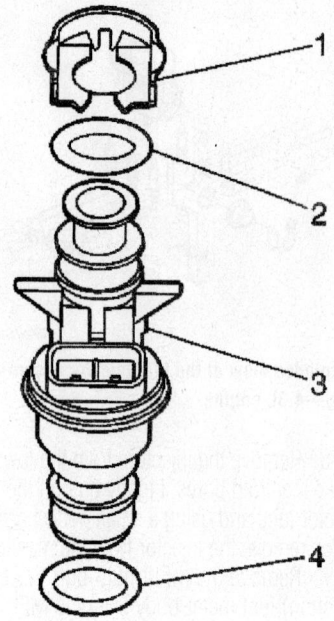

1. Retainer clip
2. O-ring
3. Injector
4. O-ring

06025-SAVA-G01

Fuel injector—4.8L, 5.3L and 6.0L engines

3. Position the upper engine wire harness aside
4. Tag the injector connectors for identification, then pull the top part of the injector connector up. Do not pull the top part of the connector past the top of the white portion.
5. Push the tab on the lower side of the injector connector to release the connect from the injection. Perform these steps on each injector connector.
6. Remove or disconnect the following:
  • Fuel feed and return pipes from the fuel rail
  • Fuel pressure regulator vacuum line
  • Crossover tube-to-right fuel rail retainer screw
  • Fuel rail attaching bolts and fuel rail

➡ **Use care in removing the fuel injectors in order to prevent damage to the electrical connector pins on the injector and to prevent damage to the nozzle. Service the fuel injector as a complete assembly only. The fuel injector is an electrical component. DO NOT immerse the fuel injector in any type of cleaner.**

  • Injector retainer clip. Insert the fork of a fuel injector assembly removal

tool behind the injector connector between the fuel rail pod and the 3 protruding retaining clip ledges. Use a prying motion while inserting the tool in order to force the injector out of the fuel rail pod.
  • Injector retainer clip
  • Injector O-ring seals from both ends of the injector. Discard the O-ring seals.

To install:

➡ **When ordering new fuel injectors, be sure to order the correct injector for the application being serviced. The fuel injector assembly is stamped with a part number identification.**

7. Lubricate the new injector O-ring seals with clean engine oil.
8. Install or connect the following:
  • New injector O-ring seals on the injector
  • New retainer clip on the injector
9. Push the fuel injector into the fuel rail injector socket with the electrical connector facing outward. The retainer clip locks on to a flange on the fuel rail injector socket.
10. Remove the crossover tube-to-right fuel rail retainer, then remove the crossover tube.
11. Replace the crossover tube O-ring with a new, lubricated one.
12. Install or connect the following:
  • Crossover tube and loosely install the retainer
  • Fuel rail to the intake manifold
  • Apply a 0.020 in. (5mm) band of threadlock to the fuel rail retaining bolts
  • Fuel rail bolts and tighten to 89 inch lbs. (10 Nm)
  • Crossover pipe retainer and tighten to 34 inch lbs. (3.8 Nm)
  • Fuel pressure regulator vacuum line
  • Fuel feed and return pipes
  • Fuel injector electrical connectors, as tagged. Rotate the injectors as necessary to avoid stretching the wire harness.
  • Upper engine wire harness
  • EVAP purge solenoid electrical connector
  • Upper engine wire harness retainer nut and tighten to 49 inch lbs. (5.5 Nm)
  • Accelerator control and cruise control cables
  • Engine sight shield mounting bracket and bolts
13. Tighten the fuel cap.

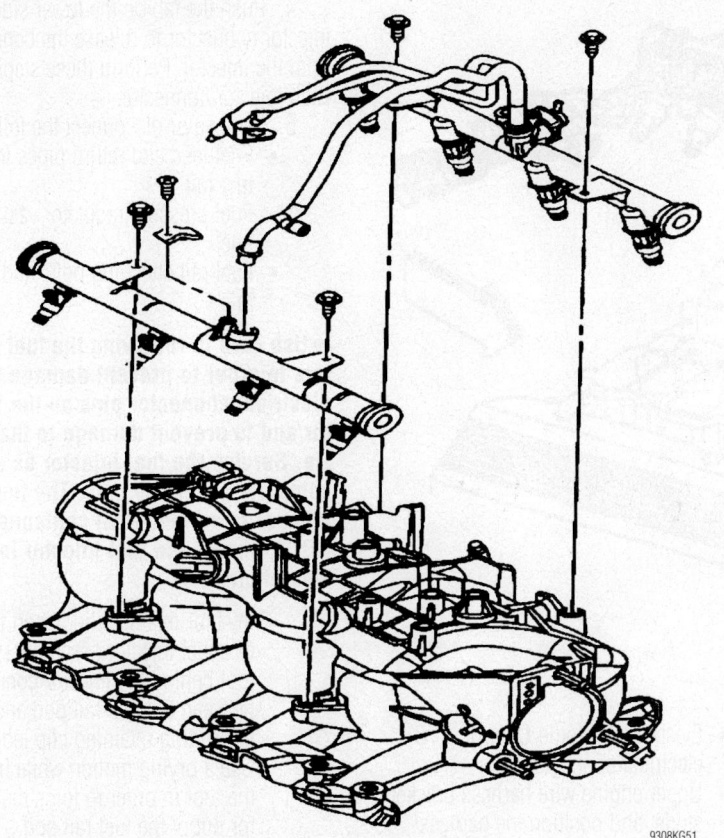

Fuel rail assembly—4.8L, 5.3L and 6.0L engines

14. Connect the negative battery cable.
15. Turn the ignition **ON** for 2 seconds.
16. Turn the ignition **OFF** for 10 seconds.
17. Turn the ignition **ON**.
18. Inspect for fuel leaks.
19. Install the engine sight shield. Tighten the engine sight shield bolts to 89 inch lbs. (10 Nm)

2005–06 4.8L, 5.3L and 6.0L Engines

1. Relieve the fuel system pressure.
2. Disconnect the negative battery cable.
3. Remove the wire harness bracket nut.
4. Disconnect the evaporative emission (EVAP) purge solenoid electrical connector.
5. Disconnect the generator electrical connector.
6. Disconnect the Manifold Absolute Pressure (MAP) sensor and the knock sensor connectors.
7. Remove the knock sensor harness connector from the intake manifold.
8. Disconnect the electronic throttle control (ETC) electrical connector by disengaging the gray retainer, push down the black clip and disconnect the connector.
9. Remove the connector position assurance (CPA) retainer.
10. Disconnect the main coil and fuel

injector connectors from the right side of the engine.
11. Remove the harness clips from the fuel rail
12. Remove the CPA retainer from the left side of the engine.
13. Disconnect the main coil and fuel injector connectors from the left side of the engine.
14. Remove the harness clips from the fuel rail.
15. Reposition the engine wire harness aside.
16. Mark the injector connectors to their corresponding injectors to ensure correct reassembly.
17. Pull the connector position assurance (CPA) retainer on the connector up 1 click.
18. Push the tab on the connector in.
19. Disconnect the fuel injector electrical connector. Repeat the steps for each injector electrical connector.
20. Remove the positive crankcase ventilation (PCV) hose.
21. Disconnect the fuel feed pipe from the fuel rail.
22. Remove the fuel rail.
23. Remove and discard the fuel injector lower O-ring seal from each injector.

To install:
24. Lubricate the new injector O-ring seals with clean engine oil.
25. Install the new injector O-ring seals onto the fuel injector.
26. Install a new retainer clip onto the fuel injector.
27. Push the fuel injector into the fuel rail injector socket with the electrical connector facing outward. The retainer clip locks onto a flange on the fuel rail injector socket.
28. Install the fuel rail.
29. Apply a 5 mm bead of threadlock to the threads of the fuel rail bolts.
30. Install the fuel rail bolts and tighten to 89 inch lbs. (10 Nm).
31. Connect the fuel feed pipe to the fuel rail.
32. Install the PCV hose.
33. Install the fuel injector electrical connectors to their corresponding injectors to ensure correct reassembly.
34. Connect the fuel injector electrical connector.
35. Push the CPA retainer on the connector in 1 click.
36. Repeat the steps for each injector electrical connector.
37. Position the engine wire harness.
38. Install the harness clips to the fuel rail.
39. Connect the main coil and fuel injector connectors to the left side of the engine.
40. Install the CPA retainer to the left side of the engine.
41. Install the harness clips from the fuel rail
42. Connect the main coil and fuel injector connectors to the right side of the engine.
43. Install the connector position assurance (CPA) retainer.
44. Connect the electronic throttle control (ETC) electrical connector.
45. Install the knock sensor harness connector to the intake manifold.
46. Connect the Manifold Absolute Pressure (MAP) sensor and the knock sensor connectors.
47. Connect the generator electrical connector.
48. Connect the evaporative emission (EVAP) purge solenoid electrical connector.
49. Install the wire harness bracket nut.
50. Connect the negative battery cable.

5.0L and 5.7L Engines

1. Remove or disconnect the following:
   • Negative battery cable
   • Intake manifold plenum

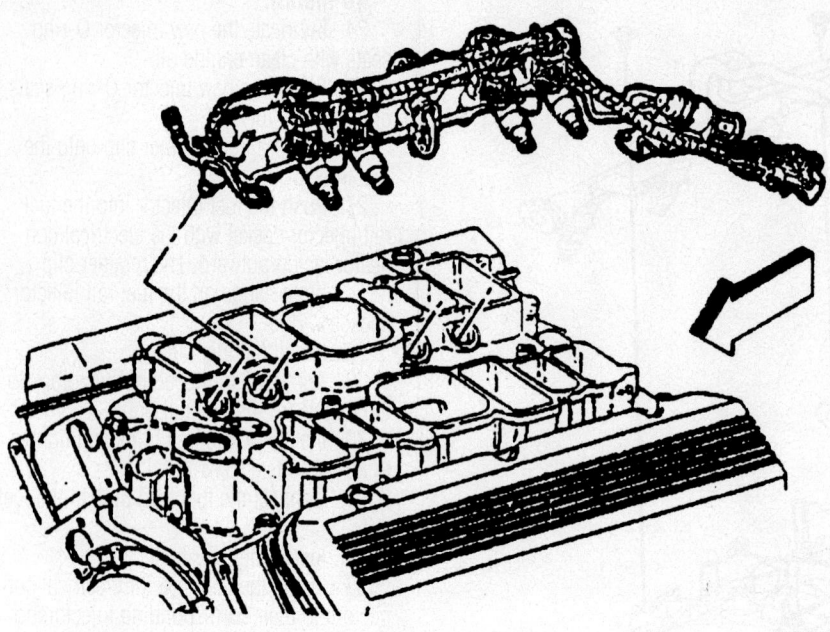

Exploded view of the fuel rail assembly—MFI systems

- Fuel rail assembly
- Wiring harness
- Injector clip and discard it
- Injector O-ring seals from both ends of the injector. Save the O-ring backups for use on reassembly

To install:

2. Install or connect the following:
  - O-ring backups before installing the O-rings
3. Lubricate the new injector O-rings with clean engine oil
  - O-ring to injector assembly
  - Fuel injector into the fuel rail injector socket with the electrical connectors facing outward
  - New injector retaining clips on the injector fuel rail assembly by sliding the clip into the injector groove as it snaps onto the fuel rail
  - Fuel rail assembly
  - Manifold plenum
  - Wiring harness
  - Negative battery cable

8.1L Engine

1. Relieve the fuel system pressure.
2. Remove or disconnect the following:
  - Negative battery cable
  - Engine sight shield nuts and bracket
  - Alternator harness connector
  - Evaporative Emission (EVAP) purge valve harness connector
  - Throttle Position (TP) sensor electrical connector
  - Electronic Throttle Control (ETC) electrical connector
  - Upper engine wire harness bracket studs, and position the harness aside
3. Tag the injector connectors for identification, then pull the top part of the injector connector up. Do not pull the top part of the connector past the top of the white portion.

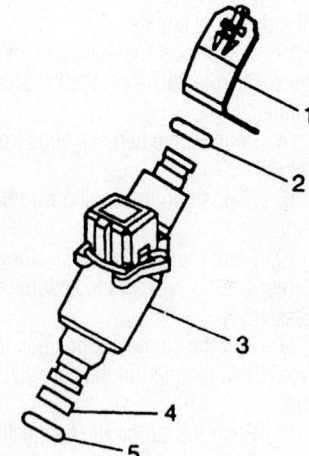

1. Clip - SFI Fuel Injector Retainer
2. O-ring - SFI Fuel Injector Upper
3. Injector Asm - SFI Fuel
4. O-ring - Backup
5. O-ring - SFI Fuel Injector Lower

Exploded view of the fuel injector assembly—MFI systems

4. Push the tab on the lower side of the injector connector to release the connect from the injector. Perform these steps on each injector connector.
5. Remove or disconnect the following:
  - Fuel feed and return pipes from the fuel rail
  - Fuel pressure regulator vacuum line
  - Fuel rail attaching bolts and fuel rail

➡**Use care in removing the fuel injectors in order to prevent damage to the electrical connector pins on the injector and to prevent damage to the nozzle. Service the fuel injector as a complete assembly only. The fuel injector is an electrical component. DO NOT immerse the fuel injector in any type of cleaner.**

- Injector retainer clip. Insert the fork of a fuel injector assembly removal tool behind the injector connector between the fuel rail pod and the 3 protruding retaining clip ledges. Use a prying motion while inserting the tool in order to force the injector out of the fuel rail pod.
- Injector retainer clip
- Injector from the fuel rail pod
- Injector O-ring seals from both ends of the injector. Discard the O-ring seals.

To install:

➡**When ordering new fuel injectors, be sure to order the correct injector for the application being serviced. The fuel injector assembly is stamped with a part number identification.**

6. Lubricate the new injector O-ring seals with clean engine oil.
7. Install or connect the following:
  - New injector O-ring seals on the injector
  - New retainer clip on the injector
8. Push the fuel injector into the fuel rail injector socket with the electrical connector facing outward. The retainer clip locks on to a flange on the fuel rail injector socket.
9. Install or connect the following:
  - Fuel rail to the intake manifold
  - Apply a 0.020 (5mm) band of threadlock to the fuel rail retaining bolts
  - Fuel rail bolts and tighten to 106 inch lbs. (12 Nm)
  - Fuel pressure regulator vacuum line
  - Fuel feed and return pipes

- Fuel injector electrical connectors, as tagged. Rotate the injectors as necessary to avoid stretching the wire harness
- Upper engine wire harness bracket
- Retainer studs to the upper engine wire harness and tighten the nut to 89 inch lbs. (10 Nm)

- Alternator electrical connector
- EVAP purge solenoid electrical connector
- TP and ETC sensor connectors
- Engine sight shield mounting bracket and bolts

10. Tighten the fuel cap.
11. Connect the negative battery cable.

12. Turn the ignition **ON** for 2 seconds.
13. Turn the ignition **OFF** for 10 seconds.
14. Turn the ignition **ON**.
15. Inspect for fuel leaks.
16. Install the engine sight shield. Tighten the engine sight shield bolts to 89 inch lbs. (10 Nm).

DIESEL FUEL SYSTEM

Fuel System Pressure

RELIEVING

Fuel system pressure can be released by wrapping a fuel fitting in a heavy shop towel and slightly loosening the fitting. NEVER perform this with any source of ignition nearby!

Fuel System Air

BLEEDING

1. Open the air bleed valve on the fuel manager/filter.
2. Connect a hose to the air bleed valve and place the other of the hose in a suitable container.

✳✳ CAUTION

The diesel/water mixture is flammable and may be hot. To avoid personal injury or property damage, do not allow the diesel/water mixture to come in contact with skin, open flame or a hot engine. Do not overfill the container holding the fuel mixture as heat from a warm engine or any another heat source may cause the fuel to expand and leak from the container that may lead to a fire.

3. Remove the F/SOL fuse from the fuse panel.
4. Crank the engine in short intervals of 10-to-15 seconds until clear fuel is observed at the air bleed hose (wait for 1 minute between cranking intervals)
5. Remove the hose and close the air bleed valve.
6. Install the F/SOL fuse and start the vehicle. Allow the vehicle to run at idle for 5 minutes.
7. Check for fuel leaks, and clear any Diagnostic Trouble Code's (DTC's)

Idle Speed

ADJUSTMENT

Idle speed and injection timing is controlled by the PCM. There is no provision for adjustment.

Fuel Filter

REMOVAL & INSTALLATION

6.5L Engine

1. Turn the ignition **OFF**. Remove the fuel tank cap to release any pressure or vacuum in the tank.

➡**It is not necessary to drain all the fuel from the header in order to change the element since the fuel will remain in the header's cavity.**

2. Remove or disconnect the following:
- Open the air bleed valve to relieve residual pressure

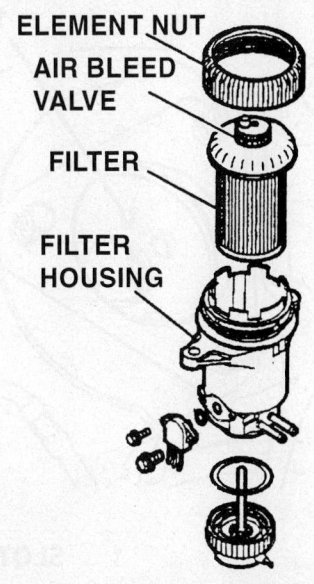

ELEMENT NUT
AIR BLEED VALVE
FILTER
FILTER HOUSING

7924KG26

Exploded view of the fuel filter assembly—6.5L diesel engines

- Element nut, turning it by hand to the left. If necessary, a strap wrench may be used to loosen the nut.
- Element by lifting straight up and out of the header assembly

To install:
3. Be sure the mating surface between the element assembly and the header assembly is clean.
4. Install or connect the following:
- New element by aligning the widest key slot located under the element assembly cap with the widest key in the header assembly.
5. Carefully push the element downward until the mating surfaces make contact.
- Element nut and tighten securely by hand
6. If not already done, open the air bleed valve on top of the fuel manager/filter assembly, then connect a length of hose placing the other end in a suitable container.

✳✳ CAUTION

Be extremely cautious when handling diesel fuel. Do not expose the fuel to sparks or open flames. Also, be cautious as the fuel coming out of the drain hose could be hot.

7. Disconnect the fuel injection pump shutdown solenoid wire.
8. Crank the engine for 10–15 seconds, then wait 1 minute for the starter motor to cool. Repeat until clear fuel is observed coming from the air bleed.
9. Close the air bleed valve, reconnect the injection pump solenoid wire and replace the fuel tank cap.
10. Start the engine, allow it to idle for 5 minutes and check the fuel manager/filter assembly for leaks.

6.6L Engine

1. Disconnect the negative battery cables.
2. Drain the fuel from the fuel filter as follows:

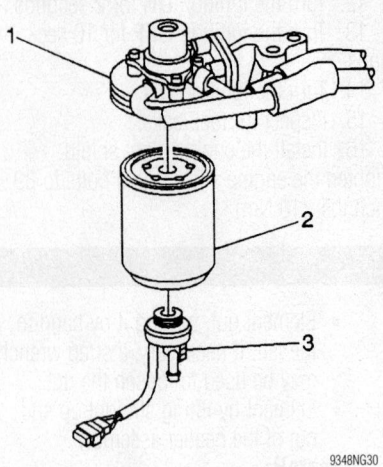

9348NG30

Exploded view of the fuel filter (2)—6.6L engine

   a. Install a hose on the water drain on the water-in-fuel sensor.

   b. Place the other end of the hose into an approved container.

   c. Drain as much fuel as possible from the fuel filter housing.

   d. Tighten the water drain on the water-in-fuel sensor.

3. Remove or disconnect the following:

- Water-in-fuel sensor harness connector
- Fuel filter from the fuel filter/heater element housing
- Water-in-fuel sensor from the fuel filter

To install:

4. Install or connect the following:

- Water-in-fuel sensor in the fuel filter

➡**Check the fuel filter/heater element housing and the filter for a dislocated filter seal or foreign debris. Contamination on the filter/heater housing may cause leakage at the fuel filter. Coat the seal with clean engine oil.**

- Fuel filter on the fuel filter/heater element housing
- Water-in-fuel harness connector
- Negative battery cables

5. Prime the fuel system:

   a. Pump the primer located on top of the fuel filter 30 times or until stiff.

   b. Try to start and run the engine. If the engine does not start, repeat the previous step.

   c. Allow the engine to run for 5 minutes at idle.

   d. Check for fuel leaks and clear all Diagnostic Trouble Codes (DTCs).

Diesel Injection Pump

All vehicles are equipped with an electronically controlled pump. The electronic pump is driven by gears and rotates at the same speed as the camshaft. An electronic stepper motor used to control injection timing and a fuel solenoid driver used to control the fuel injection solenoid on the electronic model.

REMOVAL & INSTALLATION

6.5L Engine

1. Relieve the fuel system pressure.
2. Remove or disconnect the following:

- Negative battery cables
- Intake manifold
- Fuel injection and inlet lines
- Cables wires, and hoses at the injection pump
- Fuel return line at the top of the injection pump
- Fuel feed line at the injection pump, if necessary
- Oil filler tube grommet

➡**Do not engage the starter in order to rotate the engine with the injection pump removed. The pump driven gear could jam in the front housing resulting in a sheared crankshaft or camshaft gear key and possible valvetrain damage.**

3. Scribe or paint a matchmark on the front cover and the injection pump flange.

4. Rotate the crankshaft by hand and remove the injection pump driven gear bolts, accessing the bolts through the oil filler neck hole.

5. Remove the injection pump-to-front cover attaching nuts. Remove the pump. Be sure to cap all open lines and nozzles in order to prevent system contamination and damage.

To install:

6. Align the locating pin on the pump hub with the slot in the injection pump driven gear (the SLOT not the hole in the gear) At the same time, align the timing marks.

7. Attach the injection pump to the front cover. Torque the nuts to 30 ft. lbs. (40 Nm) checking the timing marks before tightening.

8. Install or connect the following:

- Driven gear-to-injection pump bolts and tighten to 20 ft. lbs. (27 Nm)
- Grommet and oil fill tube
- Air conditioning bracket, if applicable
- Fuel feed line. Torque to 20 ft. lbs. (27 Nm).
- Fuel return line to the pump, if removed
- Cables, wires and hoses previously removed
- Injector lines
- Intake manifold

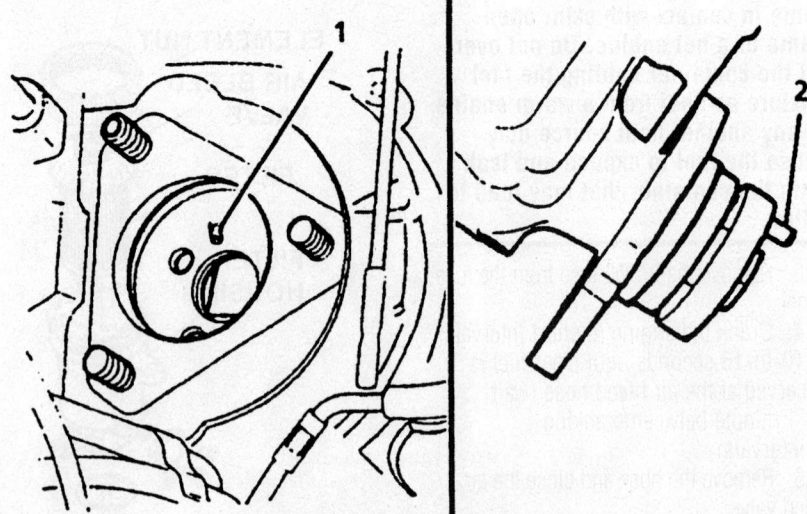

| 1 | SLOT IN DRIVEN GEAR |
| 2 | PUMP HUB |

7924KG27

Align the pin on the pump hub with the slot in the driven gear, NOT into the hole in the gear—diesel engines

- Negative battery cables
9. Start the engine and check for leaks.

6.6L Engine

1. Drain the cooling system
2. Remove or disconnect the following:
   - Negative battery cables
   - Air intake duct. Cover the end to prevent dirt from entering
   - Intake manifold cover
   - Fuel fill cap in order to relieve the fuel pressure
   - Fuel injection control module (FICM) electrical connectors
   - Fuel injection control module
   - Upper fan shroud
   - Fan blade assembly
   - Drive belt
   - Bolt holding the positive battery cable junction box and bracket and position aside
   - A/C compressor and power steering pump and position aside with the lines attached
   - Oil dipstick tube
   - A/C and power steering pump bracket
   - Drive belt tensioner and bolt.
   - Alternator
   - Thermostat housing bracket, wiring and fuel test port and 2 nuts
   - Positive Crankcase Ventilation (PCV) catch tank from the PCV bracket and the bolt below holding the lower line, then position aside
   - Alternator bracket
   - Turbo cooling hose return line clamp and hose
   - Upper radiator hose at the outlet pipe. Remove the bracket and support the bracket at the valve cover and swing out of the way
   - Bolt holding the wiring support bracket at the thermostat housing
3. Move the main wiring harness by disconnecting the following:
   a. The fuel pressure regulator connector on the fuel injection pump
   b. Fuel injection control module connectors
4. Flip the wire harness and harness tray towards the back and position aside.
5. Remove or disconnect the following:
   - Heater pipe bolt and temperature sensor wire from the thermostat housing
   - Air intake pipe
   - Water crossover assembly
   - Hose from the turbo water feed line

➡**Cap all open fuel connections to prevent contaminants from entering.**

- High pressure fuel lines and support pipe and hose at the fuel injection pump and junction block
- Fuel return hose from the fuel injection pump
- Y-junction banjo fitting from the junction block
- Bolts securing the fuel injection pump at the front cover and block

➡**When removing the pump, be careful not to damage any of the mating surfaces.**

- Fuel injection pump from the block using 2 prytools to work the pump from the block toward the rear of the engine, keeping the pump straight.
6. Prepare the fuel pump as follows:
   a. Hold the fuel pump by the drive gear in a vise with copper jaw liners.
   b. Loosen the gear nut until the nut is even with the end of the gear shaft.
   c. Separate the pump and adapter by removing the 3 bolts and spacers.
   d. Inspect the O-ring for damage on the pump adapter and replace if necessary. Lubricate the O-ring with clean engine oil.
   e. Clean all mating surfaces.
   f. Install the adapter on the pump
   g. Using the bolts and spacers,

reassemble the pump. Tighten the bolts to 15 ft. lbs. (20 Nm).
   h. Install the gear and nut and tighten to 52 ft. lbs. (70 Nm).

To install:

7. Installation is the reverse of removal, noting the following tightening specifications:
   - Fuel injection pump mounting bolts: 15 ft. lbs. (20 Nm)
   - Y-junction banjo fitting: 11 ft. lbs. (15 Nm)
   - High pressure fuel lines: 40 ft. lbs. (54 Nm)
   - Heater pipe bracket and water crossover bolts: 15 ft. lbs. (20 Nm)
   - Upper radiator hose mounting bolts: 89 inch lbs. (10 Nm)
   - A/C and power steering pump bracket bolts: 34 ft. lbs. (46 Nm)
8. Refill the cooling system, then start the engine and check for leaks.

Fuel Injectors

REMOVAL & INSTALLATION

6.5L Engine

➡**Special tool J–29873, or its equivalent, an injection nozzle socket, will be necessary for this procedure.**

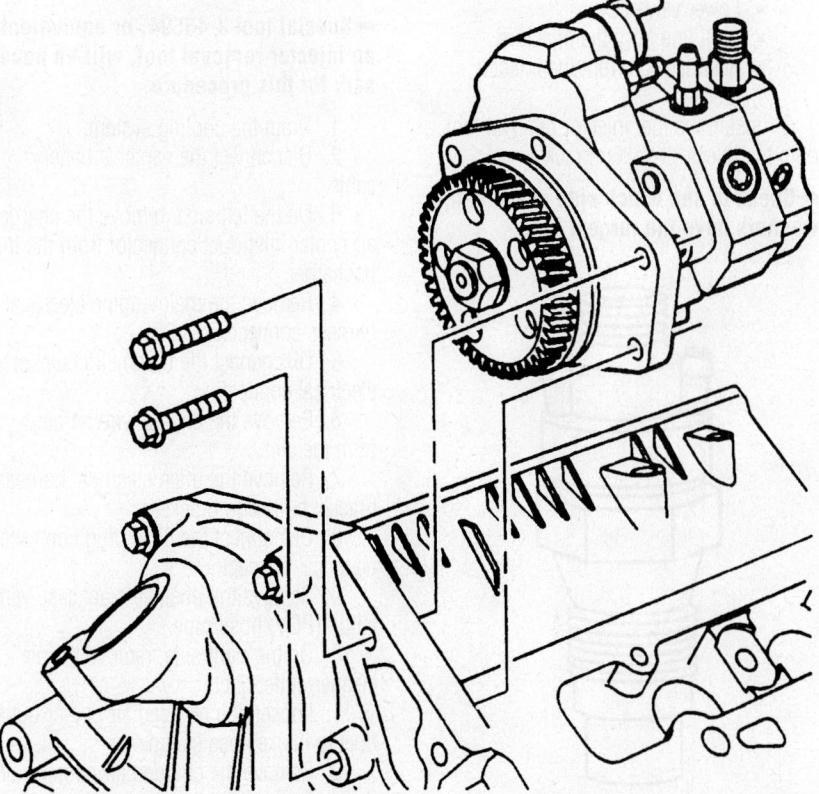

9348NG31

Diesel fuel injection pump—6.6L engine

1. Remove or disconnect the following:
   - Batteries.
   - Fuel line clip(s)
   - Fuel return hose
   - Fuel injection lines
2. Using GM special tool J–29873, remove the injector. Always remove the injector by turning the 30mm hex portion of the injector; turning the round portion will damage the injector. Always cap the injector and fuel lines when disconnected, to prevent contamination.

To install:

3. Always install the injector by turning the 30mm hex portion of the injector; turning the round portion will damage the injector.
4. Install or connect the following:
   - Injector with a new gasket. Torque to 50 ft. lbs. (70 Nm).
   - Injection line. Torque the nut to 20 ft. lbs. (25 Nm).
   - Fuel return hose
   - Fuel line clip(s)
   - Batteries

2002–04 6.6L Engine

➡**Special tool J 44639, or equivalent, an injector removal tool, will be necessary for this procedure.**

1. Remove or disconnect the following:
   - Negative battery cables
   - Lower valve cover
   - Spill line from the injectors
   - Retainer bolt from the injector bracket
2. Install the fuel injector removal tool onto the injector retainer bracket.

➡**Check to see which side of the banjo washers have the largest hole.**

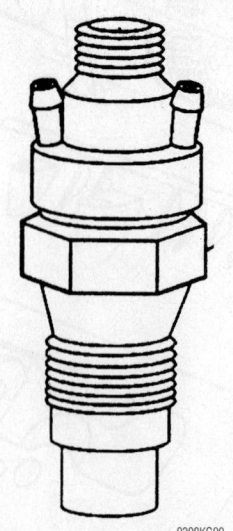

View of the fuel injector nozzle.

3. Install a wrench on the fuel injector removal tool and pry away from the injector.
4. Remove the faulty injectors.
5. Remove the copper compression washer from the injection hole if the washer did not come off with the injector.

To install:

6. If the injector sleeve is pulled from the cylinder head when removing the injector, the injector sleeve must be installed as follows:
   a. Set the new injector sleeve gaskets to the injector sleeve.
   b. Apply threadlock to the lower sealing area of the injector sleeve.
   c. Use a large brass drift to drive the injector sleeve into the cylinder head until fully seated.
7. Replace the copper compression washer. Assembly grease may be needed to hold the washer in place.
8. Install or connect the following:
   - New injectors
   - Retainer bolt on the injector bracket. Tighten to 36 ft. lbs. (46 Nm).
   - Spill line and tighten the bolts to 108 inch lbs. (12 Nm)
   - Lower valve cover
   - Negative battery cables

2005–06 6.6L Engine

➡**Special tool J-46594, or equivalent, an injector removal tool, will be necessary for this procedure.**

1. Drain the cooling system.
2. Disconnect the negative battery cable.
3. On the left side, remove the charged air cooler inlet duct connector from the turbocharger.
4. Remove the main engine electrical harness connectors.
5. Disconnect the barometric sensor electrical connector.
6. Remove the engine wire harness from the clip.
7. Remove the main electrical harness bracket bolts and bracket.
8. Disconnect the glow plug controller electrical connector.
9. Remove the positive crankcase ventilation (PCV) hose/pipe.
10. On the right side, remove the air cleaner outlet duct.
11. Loosen the charged air cooler outlet duct to intake hose clamp.
12. Remove the charged air cooler outlet duct from the intake.
13. Remove the fuel filter and bracket.

14. Remove the fuel injection control module.
15. Prior to removing the fuel injector pipes, use compressed air to blow any debris from between the injector line and fittings. Wipe the fittings clean of debris.
16. Remove the fuel injector pipes.
17. Remove the fuel return hose from the injectors.
18. Disconnect the fuel injector electrical connectors.
19. Remove the fuel injector bracket bolts.
20. Install the injector removal tool J-46594 into the bolt hole in the fuel injector bracket.
21. Install a flare nut wrench onto the tool and pull back away from the fuel injector, until the injector releases from its seat.
22. Remove the tool.
23. Remove the fuel injectors with brackets.
24. If necessary, remove the fuel injector bracket pins.
25. If necessary, remove and discard the copper washer from the fuel injector bore.
26. If necessary, remove and discard the O-ring from the fuel injector.

To install:

27. If necessary, install a new O-ring onto the fuel injector.
28. If necessary, install a new copper washer to the fuel injector bore.
29. If necessary, install the fuel injector bracket pins.
30. Install the fuel injectors with brackets.
31. Install the fuel injector bracket bolts and tighten to 22 ft. lbs. (30 Nm).
32. Connect the fuel injector electrical connectors.
33. Install the fuel return hose to the injectors.
34. Install the fuel return hose clips.
35. Install the fuel injector pipes and tighten to 30 ft. lbs. (40 Nm).
36. On the left side, install the positive crankcase ventilation (PCV) hose/pipe.
37. Connect the glow plug controller electrical connector.
38. Install the main electrical harness bracket bolts and bracket.
39. Install the engine wire harness to the clip.
40. Connect the barometric sensor electrical connector.
41. Install the main engine electrical harness connectors.
42. Install the charged air cooler inlet duct connector to the turbocharger.
43. On the right side, install the fuel injection control module.

9308KG09

44. Install the fuel filter and bracket.
45. Install the charged air cooler outlet duct to the intake.
46. Tighten the charged air cooler outlet duct to intake hose clamp.
47. Install the air cleaner outlet duct.
48. Connect the negative battery cable.
49. Refill the cooling system.

Glow Plugs

REMOVAL & INSTALLATION

6.5L Engine

1. Disconnect both battery cables.
2. Raise and support the vehicle.
3. On the left side, remove the left front tire.
4. Remove the splash shield from the left front wheel well.
5. Remove the lead wires from the glow plugs in cylinders 1 and 3.
6. Remove the glow plugs from cylinders 1 and 3.
7. Remove the engine cover.
8. From the inside of the vehicle, remove the lead wires from the glow plugs in cylinders 5 and 7.
9. Remove the glow plugs from cylinders 5 and 7.
10. On the right side, remove the right front tire.
11. Remove the splash shield from the right front wheel well.
12. Remove the lead wires from the glow plug in cylinder 2.
13. Remove the lead wires for the glow plugs in cylinders 4 and 6 at the harness connectors.
14. Remove the glow plugs from cylinders 2, 4, and 6.
15. Disconnect the lead wire at the glow plug from cylinder 8 from inside of the vehicle.
16. Remove the glow plug from cylinder 8.

To install:
17. On the right side, install the glow plug into cylinder 8 from inside of the vehicle and tighten to 13 ft. lbs. (17 Nm).
18. Connect the lead wire to the cylinder 8 glow plug.
19. Install the glow plugs into cylinders 2, 4, and 6 by reaching through the right front wheel well and tighten to 13 ft. lbs. (17 Nm).
20. Install the lead wire to glow plug 6.
21. Connect the wires for the glow plugs in cylinders 4 and 6 to the connectors at the wire harness.
22. Install the lead wire for the cylinder 2 glow plug.
23. Install the splash shield in the right front wheel well.
24. Inspect the wire routing, ensuring that the lead wires are not rubbing against the exhaust manifold or any part that may harm the wire insulation.
25. Install the right front tire.
26. On the left side, install the glow plugs into cylinders 1 and 3 by reaching through the left front wheel well and tighten to 13 ft. lbs. (17 Nm).
27. Install the lead wires for cylinder 1 and 3.
28. Install the splash shield in the left front wheel well.
29. Inspect the wire routing, ensuring that the lead wires are not rubbing against the exhaust manifold or any part that may harm the wire insulation.
30. Install the left front tire.
31. Install the glow plugs into cylinders 5 and 7 from inside of the vehicle and tighten to 13 ft. lbs. (17 Nm).
32. Connect the lead wires to the glow plugs.
33. Connect the negative battery cables.
34. Ensure that the PCM electrical harness is properly routed to avoid contact with the engine cover during installation.
35. Install the engine cover.

REMOVAL & INSTALLATION

6.6L Engine

1. Disconnect both battery cables.
2. Remove the left and right wheelhouse panels.
3. Remove the glow plug harness nut(s).
4. Remove the harness from the glow plug(s).
5. Remove the glow plug(s).

To install:
6. Install the glow plugs and tighten to 13 ft. lbs. (17 Nm).
7. Install the harness to the glow plugs.
8. Install the glow plug harness nut(s).
9. Install the left and right wheelhouse panels.
10. Connect both battery cables.

DRIVE TRAIN

Transmission Assembly

REMOVAL & INSTALLATION

Manual Transmission

SIERRA AND SILVERADO—NV3500

1. Shift the transmission into 3rd or 4th speed gear.
2. Remove or disconnect the following:
   - Shift lever
   - Shift tower
   - Transmission oil
   - If equipped with a transfer case, remove the front propeller shaft
   - Rear propeller shaft.
3. If equipped with a transfer case, remove or disconnect the following:
   - Two transfer case shields
   - Manual transfer case shift linkage
   - Bolt securing the left side support brace to the transmission
   - Bolt and stud securing the left side support brace to the transfer case
   - Bolt securing the right side support brace to the transmission
   - Bolt securing the right side support brace to the transfer case
4. Using tool J42371, push back on the white plastic sleeve on the quick connect in order to separate the hydraulic clutch line from the concentric slave cylinder quick connect.
5. Disconnect the wiring harness and connectors from the vehicle speed sensor, backup lamp switch, and transmission harness retainers.
6. If equipped with a 4.3L engine, remove the two bolts securing the clutch housing cover. Remove the transmission rear mount. Support the transmission with a transmission jack.
7. Remove or disconnect the following:
   - Bolts securing the bottom right side of the transmission to the engine
   - Stud securing the right side of the transmission to the engine
   - Bolt and six studs securing the transmission to the engine
8. Pull the transmission straight back on the clutch hub splines. Do not let the transmission hang from the clutch plate and the clutch cover.
   - Transmission from the vehicle
   - Clutch plate and the clutch cover from the engine flywheel, if required

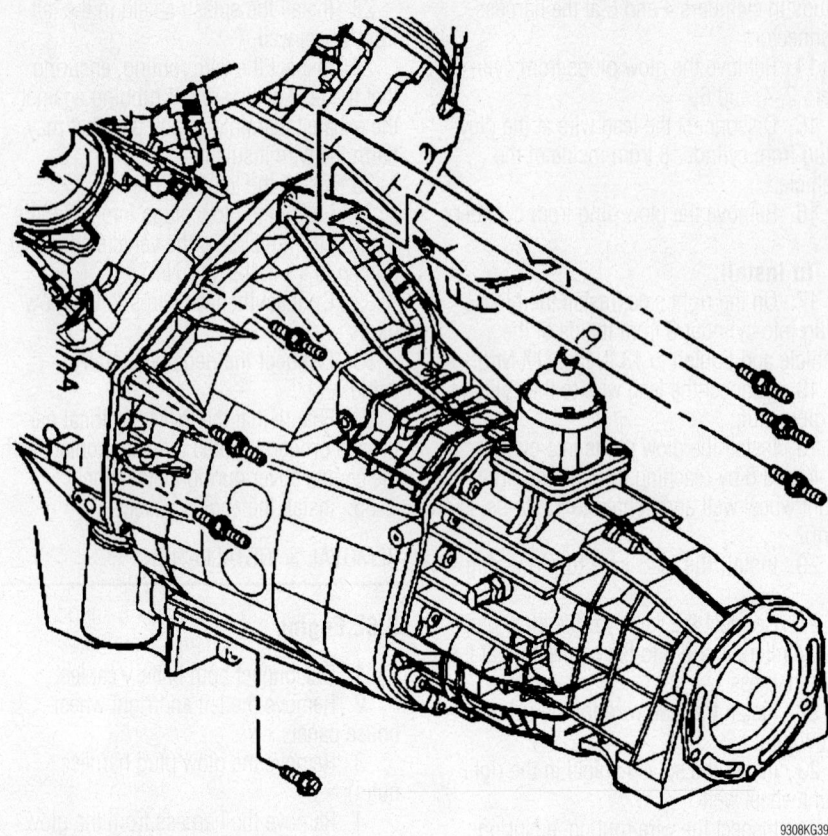

NV3500 removal—Sierra & Silverado

9308KG39

To install:

9. Install the clutch plate and the clutch cover to the engine flywheel if removed.

10. Ensure the transmission is positioned in the 3rd or 4th speed gear. Rotate the transmission clockwise onto the clutch hub splines. Install the bolt and the studs securing the transmission to the engine. Tighten the bolts to 37 ft. lbs. (50 Nm).

11. Install or connect the following:
   • Stud securing the right side of the transmission to the engine and tighten to 37 ft. lbs. (50 Nm)
   • Bolts securing the bottom right side of the transmission to the engine and tighten to 37 ft. lbs. (50 Nm).
   • Clutch housing cover using the two bolts (4.3L engine). Tighten the bolts to 10 ft. lbs. (14 Nm).
   • Transmission rear mount
   • Clutch line to the concentric slave cylinder

12. If equipped with a transfer case, install or connect the following:
   • Right side support brace-to-transmission bolt and tighten to 37 ft. lbs. (50 Nm)
   • Right side support brace-to-transfer case bolts and tighten to 37 ft. lbs. (50 Nm)

   • Left side support brace-to-transfer case bolt(s) and stud and tighten to 37 ft. lbs. (50 Nm)
   • Left side support brace-to-transmission bolts and tighten to 37 ft. lbs. (50 Nm)
   • Manual transfer case shift linkage.
   • Two transfer case shields
   • Front propeller shaft

13. Install or connect the following:
   • The rear propeller shaft.
   • The shift tower.
   • Transmission with transmission fluid
   • Shift lever

SIERRA AND SILVERADO—NV4500

1. Shift the transmission into 3rd or 4th speed gear.

2. Remove or disconnect the following:
   • The shift lever
   • The shift tower
   • The transmission oil
   • Front propeller shaft, 4WD only
   • Rear propeller shaft
   • Two transfer case shields
   • Manual transfer case shift linkage, if equipped
   • Two bolts securing the right side support bracket to the transmission

3. Using tool J42371, push back on the white plastic sleeve on the quick connect in order to separate the hydraulic clutch line from the concentric slave cylinder quick connect.
   • Vehicle Speed Sensor (VSS) connector and harness
   • Backup lamp switch connector and harness
   • Transmission harness retainers
   • Clutch housing cover-to-transmission bolts (4)
   • Left and right side transmission-to-engine cover bolts
   • Transmission rear mount. Support the transmission with a transmission jack.
   • Bolts and studs securing the transmission to the engine.

4. Pull the transmission straight back on the clutch hub splines. Do not let the transmission hang from the clutch plate and the clutch cover. Remove the transmission from the vehicle.

5. Remove the clutch plate and the clutch cover from the engine flywheel if required.

To install:

6. Install or connect the following:
   • Clutch plate and the clutch cover to the engine flywheel, if removed

7. Ensure the transmission is positioned in the 3rd or 4th speed gear. Rotate the transmission clockwise onto the clutch hub splines. Install the bolt and the studs securing the transmission to the engine. Tighten the bolts to 37 ft. lbs. (50 Nm).

8. Install or connect the following:
   • Right and left side transmission to engine cover bolts and tighten to 10 ft. lbs. (14 Nm)
   • Clutch cover-to-transmission bolts and tighten to 10 ft. lbs. (14 Nm)
   • Transmission rear mount
   • Clutch line to the slave cylinder
   • Right side support bracket-to-transmission bolts (2). Tighten to 37 ft. lbs. (50 Nm).
   • Manual transfer case shift linkage, if removed
   • Two transfer case shields, if equipped
   • Front propeller shaft, if equipped
   • Rear propeller shaft
   • Shift tower
   • Transmission with transmission fluid
   • Shift lever

SIERRA AND SILVERADO—ZF-S6-650

1. Shift the transmission into 3rd or 4th speed gear.

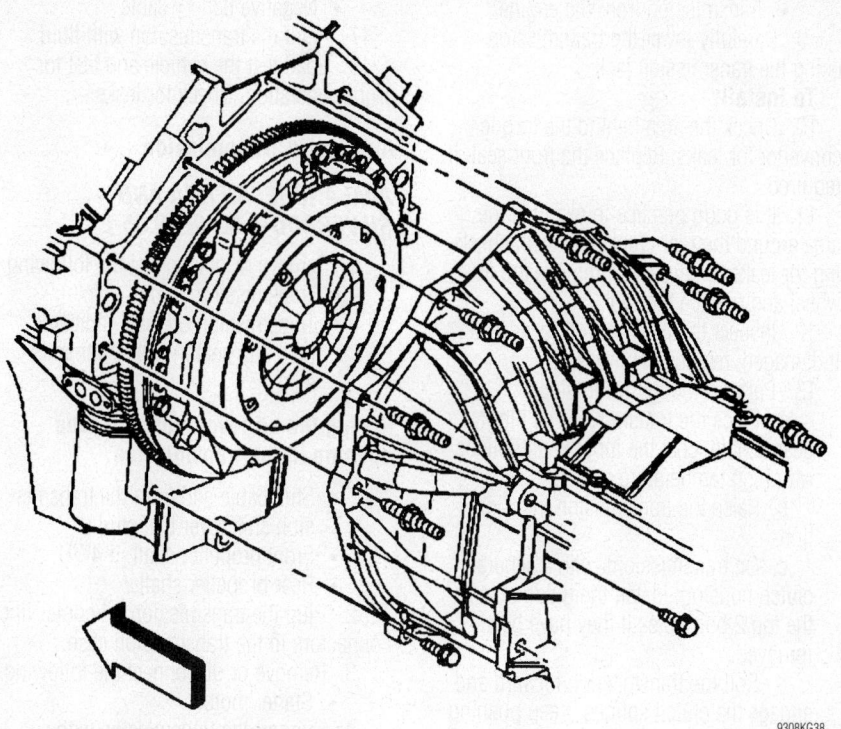

9308KG38

NV4500 removal—Sierra and Silverado

2. Remove the shift lever.

3. Raise and suitably support the vehicle.

4. If vehicle is a 2 wheel drive (2WD), remove the rear propeller shaft.

5. If equipped with 4 wheel drive (4WD), remove the transfer case.

6. Disconnect the clutch actuator cylinder hydraulic hose quick connect at the master cylinder.

7. Disconnect the clutch actuator cylinder line retaining clips.

8. If equipped with a 6.6L engine, disconnect the power take off, backup lamp and vehicle speed sensor connectors.

9. Remove the engine harness clips from the fuel feed/return brackets.

10. Remove the PTO connector from the fuel feed/return bracket.

11. Reposition the harness.

12. Remove the fuel lines from the fuel feed/return brackets.

13. If equipped with a 8.1L engine, disconnect the power take off, backup lamp, vehicle speed sensor and oxygen sensor connectors.

14. Remove the engine harness clips from the fuel feed/return brackets.

15. Remove the PTO connector from the fuel feed/return bracket.

16. Remove the oxygen sensors connectors from the fuel feed/return hose clip and transmission bracket.

17. Reposition the harness.

18. Remove the fuel lines from the fuel feed/return brackets.

19. If equipped with 4WD, remove the vent hose clip from the bracket.

20. Remove the vent hose clip nut and clip.

21. If equipped with 4WD, remove the vent hose clip from the fuel feed/return bracket.

22. Remove the starter.

23. Remove the exhaust pipe hanger bracket bolts and bracket.

24. Support the transmission with a suitable transmission jack.

25. Remove the transmission support

26. Remove the transmission bolts/studs.

27. Pull the transmission straight back on the clutch hub splines. Do not let the transmission hang from the clutch assembly.

28. Remove the transmission from the vehicle with the clutch hydraulic hose attached to the clutch actuator.

29. Remove the insulator from the top of the transmission.

To install:

30. Install the insulator to the top of the transmission.

➡**Connect the clutch actuator cylinder hose quick connect to the master cylinder before the transmission is fully installed. Failure to connect the hose could cause damage to the clutch pressure plate.**

31. Slowly feed the clutch hydraulic hose towards the clutch master cylinder and connect the clutch actuator cylinder hydraulic hose quick connect to the master cylinder.

32. Rotate the transmission clockwise onto the clutch hub splines.

33. Position the transmission to the engine. Do not use the transmission bolts to draw up the transmission.

34. Install the transmission studs and tighten to 37 ft. lbs. (50 Nm).

35. Slowly feed the clutch hydraulic hose towards the clutch master cylinder and connect the clutch actuator cylinder hydraulic hose quick connect to the master cylinder.

36. Install the transmission support.

37. Remove the transmission jack.

38. Install the exhaust pipe hanger bracket and bolts.

39. Install the starter.

40. If equipped with 4WD, install the vent hose clip to the fuel feed/return bracket.

41. Install the vent hose clip and nut.

42. Install the vent hose clip to the bracket.

43. Install the fuel lines to the fuel feed/return brackets.

44. Position the harness.

45. Install the oxygen sensors connectors to the fuel feed/return hose clip and transmission bracket.

46. Install the PTO connector to the fuel feed/return bracket.

47. Install the engine harness clips to the fuel feed/return brackets.

48. If equipped with a 8.1L engine, connect the power take off, backup lamp, vehicle speed sensor and oxygen sensor connectors.

49. Connect the clutch actuator cylinder line retaining clips.

50. If equipped with 4WD, install the transfer case.

51. If equipped with a 6.6L engine, connect the power take off, backup lamp and vehicle speed sensor connectors.

52. If vehicle is a 2WD, install the rear propeller shaft.

53. Lower the vehicle.

54. Install the shift lever.

55. Fill the transmission with fluid.

56. Bleed the clutch hydraulic system, if necessary.

57. Connect the battery cables.

EXPRESS AND SAVANA

1. Drain the transmission.
2. Remove or disconnect the following:
   - Negative battery cable
   - Shifter boot and lever
   - Exhaust pipes
   - Parking brake cables
   - Driveshaft, matchmark for reassembly
   - Transfer case, if equipped
   - Transmission to engine braces (vehicles equipped with diesel engines have only 1 brace)
   - Wiring harness at the transmission
3. Support the transmission with a transmission jack.
   - Nut securing the transmission mount to the crossmember
4. Position a transmission jack or equivalent, under the transmission for support.
   - Crossmember. Visually inspect to see if other equipment, brackets or lines, must be removed to permit removal of transmission.

➡**Mark position of crossmember when removing to prevent incorrect installation. The tapered surface should face the rear.**

5. Except the NV 3500, remove the top 2 transmission to housing bolts and insert 2 guide pins.

➡**The clutch housing on the NV 3500 is integral with the transmission as is the converter housing on the automatic transmissions.**

6. On the NV 3500, remove the bolts securing the clutch housing to the engine.

➡**The use of guide pins will not only support the transmission but will prevent damage to the clutch disc. Guide pins can be made by using 2 bolts, the same as those just removed only longer, and cutting off the heads. Make an adjustment slot. Be sure to support the clutch release bearing and support assembly during removal of the transmission.**

7. Remove the remaining bolts and slide transmission straight back from engine. Use care to keep the transmission drive gear straight in line with clutch disc hub.
8. Remove or disconnect the following:
   - Wiring, clips, tubes and brackets etc., which would interfere with the removal of the transmission

➡**Ensure that the engine is supported with a jack stand before detaching the transmission from the engine.**

   - Transmission from the engine
9. Carefully lower the transmission using the transmission jack.

To install:

10. Check the area behind the torque converter for leaks. Replace the front seal, if required.
11. It is good practice to examine the area around the rear crankshaft seal, checking for leaks. If necessary, remove the flywheel and replace the seal.
12. Inspect the flywheel ring gear teeth. If damaged, replace the flywheel.
13. Perform the following steps:
   a. Place the transmission in high gear. Lightly coat the input shaft splines with high temperature grease.
   b. Raise the transmission into position.
   c. On transmissions with a separate clutch housing, install the guide pins in the top 2 bolt holes if they have been removed.
   d. Roll the transmission forward and engage the clutch splines. Keep pushing the transmission forward until it mates with the engine.
   e. On transmissions with a separate clutch housing, remove the guide pins and install the bolts, tighten the bolts to 23 ft. lbs. (31 Nm).
   f. On the NV 3500, install the transmission-to-engine bolts. Tighten the bolts to 35 ft. lbs. (47 Nm)
14. When satisfied that the transmission is properly seated, install and tighten the transmission-to-engine bolts and/or studs to 34 ft. lbs. (47 Nm).
15. Install or connect the following:
   - Wiring, clips, tubes and brackets etc
   - Shifter cable
   - Starter
   - Exhaust pipes
   - Transmission crossmember. Torque the bolts to 56 ft. lbs. (77 Nm).
   - Transmission mount on the transmission. Torque the bolts to 35 ft. lbs. (47 Nm).
   - Nut and washer that secure the transmission mount to the crossmember. Torque the nut to 38 ft. lbs. (52 Nm).
   - Transmission to engine brace(s) Torque the bolts to 41 ft. lbs. (55 Nm) for gasoline engines and to 51 ft. lbs. (70 Nm) for diesel engines.
   - Transfer case, if equipped
16. Remove the transmission jack and engine support stands.
   - Driveshaft
   - Shifter lever and boot

   - Negative battery cable
17. Refill the transmission with fluid.
18. Road test the vehicle and test for proper operation. Check for leaks.

Automatic Transmission

AVALANCHE, SIERRA AND SILVERADO—4L60E/4L65-E

1. Remove or disconnect the following:
   - Transmission fluid
   - Transmission oil level indicator tube and seal from the transmission

➡**Plug the oil level indicator tube opening in the transmission.**

   - Shift cable end from the transmission shift lever ball stud
   - Front propeller shaft, if 4WD
   - Rear propeller shaft.
2. Plug the transmission oil cooler line connectors in the transmission case.
3. Remove or disconnect the following:
   - Starter motor
4. Support the transmission with a transmission jack.
5. Remove or disconnect the following:
   - Torque converter access plug
   - Flywheel-to-torque converter bolts
   - Transmission rear mount-to-transmission bolts and nut
   - Heat shield-to-transmission bolts
   - Transmission vent hose from the transmission
   - Fuel lines from the transmission
   - Wiring harness from the transmission
   - Transmission-to-engine stud and bolt
   - Six studs and bolt securing the transmission to the engine.
6. Install tool J21366 onto the transmission bell housing to retain the torque converter. Pull the transmission straight back.
7. The transmission from the vehicle
8. Flush the transmission oil cooler and cooling lines.

To install:

9. Install Tool J21366 onto the transmission bell housing to retain the torque converter.
10. Support the transmission with a transmission jack.
11. Raise the transmission into place and remove the tool from the transmission.
12. Slide the transmission straight onto the locating pins while lining up the marks on the flywheel and the torque converter. The torque converter must be flush onto the flywheel and rotate freely by hand.

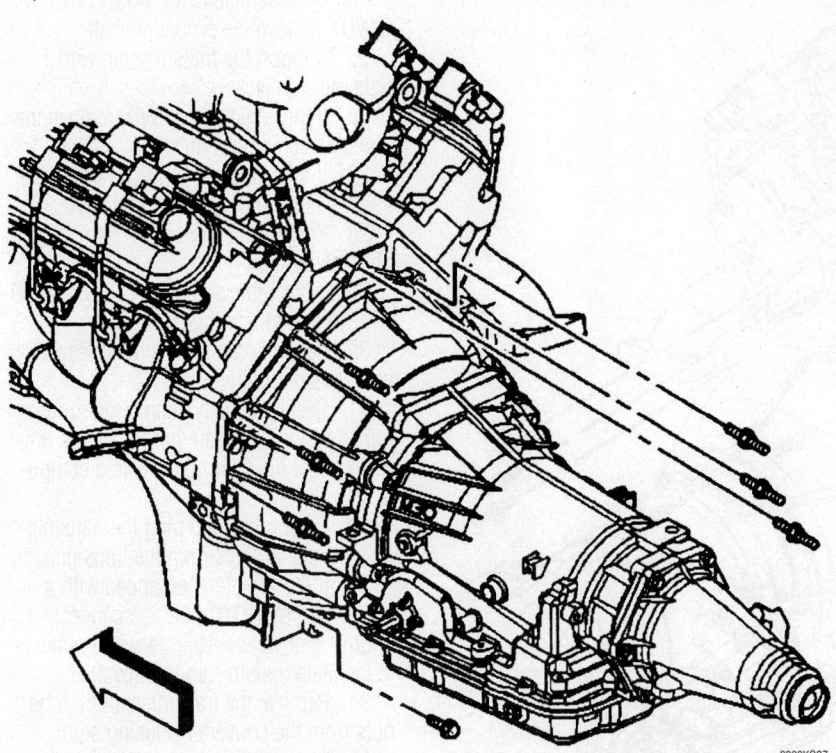

4L60E removal—Silverado shown

13. Install or connect the following:
- Studs and bolt securing the transmission to the engine. Tighten to 37 ft. lbs. (50 Nm).
- Flywheel to torque converter bolts
- Torque converter access plug
- Transmission vent hose to the transmission
- Fuel lines to the transmission
- Wiring harness to the transmission.
- Heat shield-to-transmission bolts and tighten to 13 ft. lbs. (17 Nm)
- Transmission rear mount-to-transmission bolt and nut and tighten to 18 ft. lbs. (25 Nm)

14. Remove the transmission jack from the transmission.

15. Unplug the transmission oil cooler line connectors in the transmission case.

16. Install or connect the following:
- Transmission oil cooler lines
- Front propeller shaft, if equipped
- Rear propeller shaft
- Shift cable end to the transmission shift lever ball stud

17. Unplug the oil level indicator tube opening in the transmission.

18. Install the transmission oil level indicator tube and seal to the transmission.

19. Tighten the oil pan bolts and fill the transmission with transmission fluid.

20. Lower the vehicle.

AVALANCHE, SIERRA AND SILVERADO—4L80E/4L85-E

1. Remove or disconnect the following:
- Transmission fluid
- Transmission oil level indicator tube and seal from the transmission

2. Plug the oil level indicator tube opening in the transmission.
- Shift cable from the transmission shift lever ball stud
- Front propeller shaft, if 4WD
- Rear propeller shaft.
- Transmission oil cooler lines, then plug thee openings in the transmission case
- Starter motor

3. Support the transmission with a transmission jack.
- Heat shield
- Transmission vent hose
- Fuel lines from the transmission
- Wiring harness from the transmission
- Transmission brace-to-engine bracket and transmission nut and bolt
- Torque converter cover
- Flywheel to torque converter bolts
- Transmission rear mount
- Stud and bolt on the right side

securing the transmission to the engine
- Remaining six studs and the bolt securing the transmission to the engine

4. Install Tool J21366 onto the transmission bell housing to retain the torque converter.

5. Pull the transmission straight back. Remove the transmission from the vehicle.

6. Flush the transmission oil cooler and cooling lines when you remove the transmission.

To install:

7. Install Tool J21366 onto the transmission bell housing to retain the torque converter.

8. Support the transmission with a transmission jack.

9. Raise the transmission into place and remove the tool from the transmission.

10. Slide the transmission straight onto the locating pins while lining up the marks on the flywheel and the torque converter. The torque converter must be flush onto the flywheel and rotate freely by hand.

11. Install or connect the following:
- Six studs and bolt securing the transmission to the engine. Tighten to 37 ft. lbs. (50 Nm).
- Stud and bolt on the right side securing the transmission to the engine. Tighten to 37 ft. lbs. (50 Nm).
- Flywheel-to-torque converter bolts and tighten to 44 ft. lbs. (60 Nm).
- Torque converter cover-to-engine bolts and tighten to 37 ft. lbs. (50 Nm)
- Torque converter cover-to-transmission stud and bolt and tighten to 24 ft. lbs. (33 Nm).
- Transmission vent hose
- Fuel lines
- Wiring harness
- Heat shield. Tighten the bolts to 13 ft. lbs. (17 Nm).
- Transmission rear mount-to-transmission nuts and bolt. Tighten to 18 ft. lbs. (25 Nm).
- Transmission brace. Tighten the bolts and nut to 37 ft. lbs. (50 Nm).

12. Remove the transmission jack from the transmission.
- Starter motor

13. Unplug the transmission oil cooler line connectors in the transmission case.

14. Connect the transmission oil cooler lines to the transmission.

15. Install or connect the following:
- Rear propeller shaft
- Front propeller shaft, if 4WD

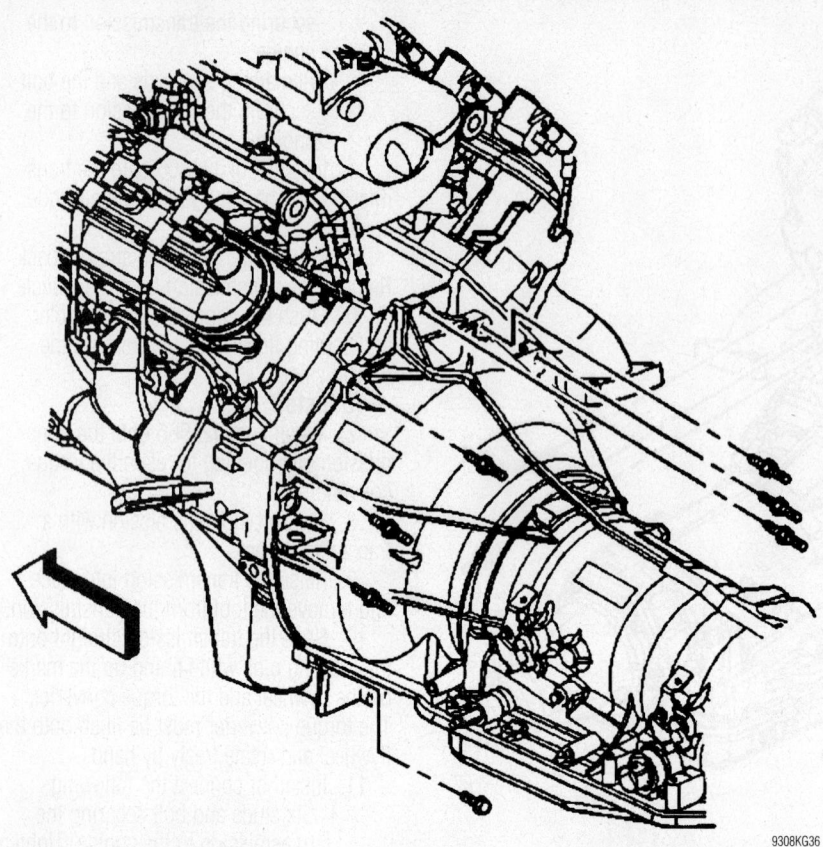

4L80E removal—Silverado shown

9308KG36

- Shift cable end to the transmission shift lever ball stud
16. Unplug the oil level indicator tube opening in the transmission.
17. Install the transmission oil level indicator tube and seal to the transmission.
18. Tighten the oil pan bolts and fill the transmission with transmission fluid.
19. Lower the vehicle.

SIERRA AND SILVERADO—ALLISON

1. Drain the transmission.
2. Disconnect both negative battery cables.
3. Remove the transmission fluid level indicator.
4. Remove the right front wheel and tire.
5. Remove the right front wheel house inner panel retainers.
6. Disconnect any harness retainers attached to the inner panel.
7. Remove the inner panel.
8. Remove the starter.
9. Remove the engine protection shield bolts and shield.
10. Rotate the engine clockwise, using the crankshaft bolt in order to access the torque converter bolts thru the starter open-

ing. Have an assistant rotate the engine while aligning the bolts.
11. Remove the torque converter bolts.
12. Completely raise the vehicle.
13. Disconnect the shift cable from the selector lever ball stud and remove the cable from the bracket.
14. Remove the shift cable bracket bolts and bracket from the transmission.
15. Reposition the bracket with the cable attached off to the side.
16. Remove the fuel line retainer bolts on the left side of the transmission.
17. Remove the fuel line bracket nut from the converter housing stud.
18. Disconnect the turbine speed sensor and input speed sensor electrical connectors.
19. Disconnect the output speed sensor electrical connector.
20. If the vehicle is equipped with 4 wheel drive (4WD), the output speed sensor is located on the transfer case and will be disconnected later.
21. Disconnect the transmission main electrical connector.
22. Disconnect the park/neutral position (PNP) switch electrical connector.
23. Remove the exhaust hanger bolts and reposition the hanger.

24. If the vehicle is a 2 wheel drive (2WD), remove the propeller shaft.
25. Support the transmission with a transmission jack.
26. If the vehicle is a 2WD, remove the transmission mount nuts.
27. If the vehicle is a 2WD, remove the transmission support bolts and nuts.
28. Remove the transmission mount bolts and mount.
29. If the vehicle is equipped with 4WD, remove the transfer case.
30. Reposition any wiring harness branches out of the way.
31. Secure a safety chain around the transmission. Use care not to overlap any wiring, fuel lines, or other related components.
32. Disconnect and plug the transmission oil cooler lines from the transmission.
33. If the vehicle is equipped with a power take off (PTO) unit , disconnect and/or remove any necessary components to facilitate transmission removal.
34. Remove the transmission fill tube nuts from the converter housing studs.
35. Remove the wire harness/vent tube bracket nut from the converter housing stud and reposition the bracket.
36. Remove the remaining converter housing bolts and studs.
37. Separate the transmission from the engine.
38. Install torque converter holding tool J-21366 to the converter housing in order to keep the torque converter from sliding off of the turbine shaft.
39. Carefully lower the transmission from the vehicle while simultaneously removing the fill tube.
40. Remove the holding tool.

To install:
41. Install torque converter holding tool J-21366 to the converter housing in order to keep the torque converter from sliding off of the turbine shaft.
42. Raise the transmission into place while simultaneously installing the transmission fill tube.
43. Remove the holding tool.
44. Align the transmission with the engine using the alignment dowels located at the rear of the engine.

➡**Ensure that the torque converter can be rotated before tightening the bolts and studs.**

45. Install the converter housing bolts and studs and tighten to 37 ft. lbs. (50 Nm).
46. Install the wire harness/vent tube bracket and nut to the converter housing stud.

47. Install the transmission fill tube and nuts to the converter housing studs.

48. If the vehicle is equipped with a PTO unit, connect and/or install the components at this time.

49. Remove the safety chain from around the transmission.

50. Install the transfer case, if the vehicle is equipped with 4WD.

51. If the vehicle is a 2WD, install the transmission mount and tighten to 37 ft. lbs. (50 Nm).

52. Install the transmission support and tighten to 70 ft. lbs. (95 Nm).

53. If the vehicle is a 2WD, install the transmission mount nuts and tighten to 30 ft. lbs. (40 Nm).

54. Remove the transmission jack.

55. If the vehicle is a 2WD, install the propeller shaft.

56. Position the exhaust hanger and install the bolts.

57. Position the wiring harness branches.

58. Connect the PNP switch electrical connectors.

59. Connect the transmission main electrical connector

60. Connect the output speed sensor electrical connector. If the vehicle is equipped with 4WD, the output speed sensor is located on the transfer case and has been connected during the transfer case installation.

61. Connect the turbine speed sensor and the input speed sensor electrical connectors.

62. Install the fuel line bracket and nut to the transmission converter housing stud.

63. Install the fuel line retainer and bolts to the left side of the transmission.

64. Install the shift cable bracket and bolts to the transmission.

65. Install the shift cable to the bracket and the selector lever ball stud.

66. Remove the access hole cover on the converter housing in order to rotate the converter and align the first torque converter bolt. If reusing the torque converter bolts, clean the bolt threads and apply Loctite® or equivalent to the threads prior to installation.

67. Install the torque converter bolts and tighten to 44 ft. lbs. (760 Nm).

68. Install the converter housing access hole cover.

69. Install the engine protection shield and bolts.

70. Position and install the starter.

71. Install the inner panel.

72. Connect any harness retainers to the inner panel.

73. Install the right front wheel house inner panel retainers.

74. Install the right front wheel and tire.

75. Remove the plugs from the transmission oil cooler line fittings in the transmission case, if necessary.

76. Flush the transmission oil cooler and lines, if necessary.

77. Connect the transmission oil cooler lines to the transmission.

78. Lower the vehicle.

79. Connect both negative battery cables.

80. Fill the transmission with new transmission fluid.

81. Install the transmission fluid level indicator.

82. If a replacement transmission was installed, perform the Fast Learn procedure using a scan tool.

EXPRESS AND SAVANA

1. Drain the transmission.
2. Remove or disconnect the following:
   - Negative battery cable
   - Shift cable, control lever and bracket
   - Exhaust pipes
   - Parking brake cables
   - Driveshaft, matchmark for reassembly
   - Transfer case, if equipped
   - Transmission to engine braces (vehicles equipped with diesel engines have only 1 brace)
   - Wiring harness at the transmission
3. Support the transmission with a transmission jack.
   - Nut securing the transmission mount to the crossmember
4. Position a transmission jack or equivalent, under the transmission for support.
   - Crossmember. Visually inspect to see if other equipment, brackets or lines, must be removed to permit removal of transmission.

➡ **Mark position of crossmember when removing to prevent incorrect installation. The tapered surface should face the rear.**

5. Perform the following:
   a. Remove the torque converter inspection cover.
   b. Mark the alignment of the torque converter to the flexplate.
   c. Remove the torque converter to flexplate bolts. Remove the dipstick tube and seal from the transmission. Plug the opening to avoid contamination.
   d. Disconnect both transmission lines at the transmission and plug them to avoid contamination and leakage.
   e. Position a J 21366 converter holding strap onto the transmission/torque converter to keep the torque converter from sliding off of the transmission turbine shaft.

➡ **The use of guide pins will not only support the transmission but will prevent damage to the clutch disc. Guide pins can be made by using 2 bolts, the same as those just removed only longer, and cutting off the heads. Make an adjustment slot.**

6. Remove the remaining bolts and slide transmission straight back from engine. Use care to keep the transmission drive gear straight in line with clutch disc hub.

7. Remove or disconnect the following:
   - Wiring, clips, tubes and brackets etc., which would interfere with the removal of the transmission

➡ **Ensure that the engine is supported with a jack stand before detaching the transmission from the engine.**

   - Transmission from the engine
8. Carefully lower the transmission using the transmission jack.

To install:

9. Check the area behind the torque converter for leaks. Replace the front seal, if required.

10. It is good practice to examine the area around the rear crankshaft seal, checking for leaks. If necessary, remove the flexplate and replace the seal.

11. Inspect the flexplate ring gear teeth. If damaged, replace the flexplate.

12. Perform the following steps:
   a. With tool J 21366 or equivalent, torque converter holding strap in place, raise the transmission into position with a transmission jack.
   b. Remove the torque converter holding strap and slide the transmission into place. Slide the transmission straight onto the locating pins while lining up the marks on the flexplate and the torque converter. Be sure the transmission is fully seated against the rear of the engine block and the locating pins are completely engaged.

❋❋ **WARNING**

DO NOT attempt to draw the transmission to the block with the mounting bolts. If the transmission is not properly seated, the bolts will break the transmission case.

➡**The torque converter must be flush with the flexplate and rotate freely by hand.**

13. When satisfied that the transmission is properly seated, install and tighten the transmission-to-engine bolts and/or studs to 34 ft. lbs. (47 Nm).

14. Perform the following steps:

a. Install the dipstick tube and seal.

b. Check the alignment marks on the torque converter and flexplate to be sure that they are properly aligned. Install the torque converter bolts. Finger-tighten the bolts to ensure proper converter seating. When the converter is properly seated, tighten the bolts to 46 ft. lbs. (63 Nm).

c. Install the torque converter cover. Tighten the retaining bolts to 24 ft. lbs. (33 Nm) on the 4.3L engines or 89 inch lbs. (10 Nm) on the V8 engines.

15. Install or connect the following:
- Wiring, clips, tubes and brackets etc
- Transmission cooling lines
- Shifter cable
- Starter
- Exhaust pipes
- Transmission crossmember. Torque the bolts to 56 ft. lbs. (77 Nm).
- Transmission mount on the transmission. Torque the bolts to 35 ft. lbs. (47 Nm).
- Nut and washer that secure the transmission mount to the crossmember. Torque the nut to 38 ft. lbs. (52 Nm).
- Transmission to engine brace(s) Torque the bolts to 41 ft. lbs. (55 Nm) for gasoline engines and to 51 ft. lbs. (70 Nm) for diesel engines.
- Transfer case, if equipped

16. Remove the transmission jack and engine support stands.
- Driveshaft
- Shifter lever and boot, on manual transmissions
- Negative battery cable

17. Refill the transmission with fluid.

18. Road test the vehicle and test for proper operation. Check for leaks.

Clutch

ADJUSTMENTS

The hydraulic clutch system requires no periodic adjustment.

REMOVAL & INSTALLATION

1. Remove or disconnect the following:
- Transmission
- Quick disconnect from the actuator cylinder

2. Install a clutch alignment tool.

3. Mark the flywheel and a clutch pressure plate lug for the installation alignment.

4. Remove the pressure plate bolts and the washers.

5. Secure the clutch pressure plate and the clutch driven plate to the flywheel.

6. Remove the clutch alignment tool.

To install:

7. Install the bolts and the washers securing the clutch pressure plate and the clutch driven plate to the flywheel.

8. Install the clutch alignment tool.

9. Align the marks made during removal or, if new align the lightest part of the clutch pressure plate identified by a yellow dot, to the heaviest part of the flywheel, identified by an "X". Tighten the clutch pressure plate to the flywheel bolts to 52 ft. lbs. (70 Nm) in a crisscross pattern.

10. Remove the clutch alignment tool.

11. Install the transmission.

12. Install the quick disconnect to the concentric slave cylinder.

Hydraulic Clutch System

BLEEDING

Bleeding air from the hydraulic clutch system is necessary whenever any part of the system has been disconnected or the fluid level (in the reservoir) has been allowed to fall so low, that air has been drawn into the master cylinder.

1. Fill master cylinder reservoir with new brake fluid conforming to DOT 3 specifications.

2. Have an assistant fully depress and hold the clutch pedal, then open the bleeder screw.

3. Close the bleeder screw and have your assistant release the clutch pedal.

4. Repeat the procedure until all of the air is evacuated from the system. Check and refill master cylinder reservoir as required to prevent air from being drawn through the master cylinder.

➡**Never release a depressed clutch pedal with the bleeder screw open or air will be drawn into the system.**

5. Test the clutch for proper operation.

Transfer Case Assembly

REMOVAL & INSTALLATION

Avalanche, Sierra and Silverado

1. Remove or disconnect the following:
- Transfer case shields
- Front propeller shaft
- Rear propeller shaft
- Shift rod from the transfer case
- Vent hose from the transfer case
- Vehicle Speed Sensor (VSS) electrical connectors
- All necessary wiring harnesses from the transfer case

2. Support the transfer case with a transmission jack.

3. If equipped with a NV3500 manual transmission, remove or disconnect the following:
- Bolt securing the left side support brace to the transmission
- Bolt and stud securing the left side support brace to the transfer case
- Two bolts securing the right side support brace to the transmission and transfer case

4. Remove or disconnect the following:
- Six nuts securing the transfer case and bracket to the transmission or transmission adapter, as applicable
- Transfer case
- Gasket, then discard

To install:

5. Install a new gasket to the transmission. Use Teflon pipe sealant GM P/N 12346004 in order to hold the gasket in place.

6. Raise and position the transfer case to the vehicle.

7. Install or connect the following:
- Six nuts securing the transfer case and bracket to the transmission adapter or transmission. Tighten to 37 ft. lbs. (50 Nm).

8. If equipped with a manual transmission, install or connect the following:
- Bolt securing the left side support brace to the transmission and tighten to 37 ft. lbs. (50 Nm)
- Bolt and stud securing the left side support brace to the transfer case and tighten to 37 ft. lbs. (50 Nm)
- Two bolts securing the right side support brace to the transmission and transfer case and tighten to 37 ft. lbs. (50 Nm)

9. Install or connect the following:
- Vent hose to the transfer case

10. Check the transfer case oil level.

11. VSS electrical connectors
- Wiring harness to the transfer case
- Shift rod to the transfer case
- Front and rear propeller shafts
- Transfer case shields

12. Lower the vehicle.

Express and Savana

1. Drain transfer case of lubricant.
2. Remove or disconnect the following:
- Negative battery cable
- Skid plate, if equipped
- Vent hose clamp at the transfer case
- Front driveshaft at the transfer case and support it aside
- Rear driveshaft and support it aside
- Electrical connections from the transfer case
- Transfer case shift linkage

3. Support the transfer case with a transmission jack.
- Transmission-to-transfer case bolts and spring washers
- Transfer case assembly and gasket

4. Carefully lower the transfer case.

To install:

5. Carefully raise the transfer case into position.

6. Install or connect the following:
- New gasket to the transmission, using gasket sealer to hold it in place

- Transfer case onto the transmission or transmission adapter. Torque the bolts to 33 ft. lbs. (45 Nm).
- Electrical harness connectors to the transfer case connections
- Transfer case shift linkage and make the proper adjustments
- Front and rear driveshafts

7. Refill the transfer case with DEXRON IIE automatic transmission fluid.
- Skid plate, if equipped
- Negative battery cable

8. Test drive for proper operation.

Halfshaft

REMOVAL & INSTALLATION

Avalanche, Sierra and Silverado

1. Remove or disconnect the following:
- Wheels

2. Insert a drift or a large screwdriver through the brake caliper into one of the brake rotor vanes in order to prevent the drive axle wheel drive shaft from turning.

3. Remove or disconnect the following:
- Nut and the washer from the hub

➡ **Do not reuse the hub nut. A new nut must be used when installing the wheel drive shaft.**

- Bolts (6) securing the wheel drive shaft inboard flange to the output shaft flange
- Drift from the rotor
- Stabilizer shaft link from the lower control arm

4. Wrap shop towels around both the inner and the outer wheel drive shaft boots in order to avoid damage to the boots during removal and installation.

5. Pull the wheel drive shaft through the lower control arm opening.

To install:

6. Wrap shop towels around both the inner and the outer wheel drive shaft boots in order to avoid damage to the boots during removal and installation.

➡ **Clean the steering knuckle and the wheel drive shaft splines and threads. These areas must be dry and free of grease, dirt, and contamination.**

7. Insert the wheel drive shaft splined shank into the knuckle hub.

➡ **Use only a genuine GM front wheel drive shaft nut. Installation of anything but an OEM front wheel drive shaft nut could cause damage to the vehicle.**

8. Install or connect the following:
- Washer and the new hub nut to the wheel driveshaft. Do not tighten.

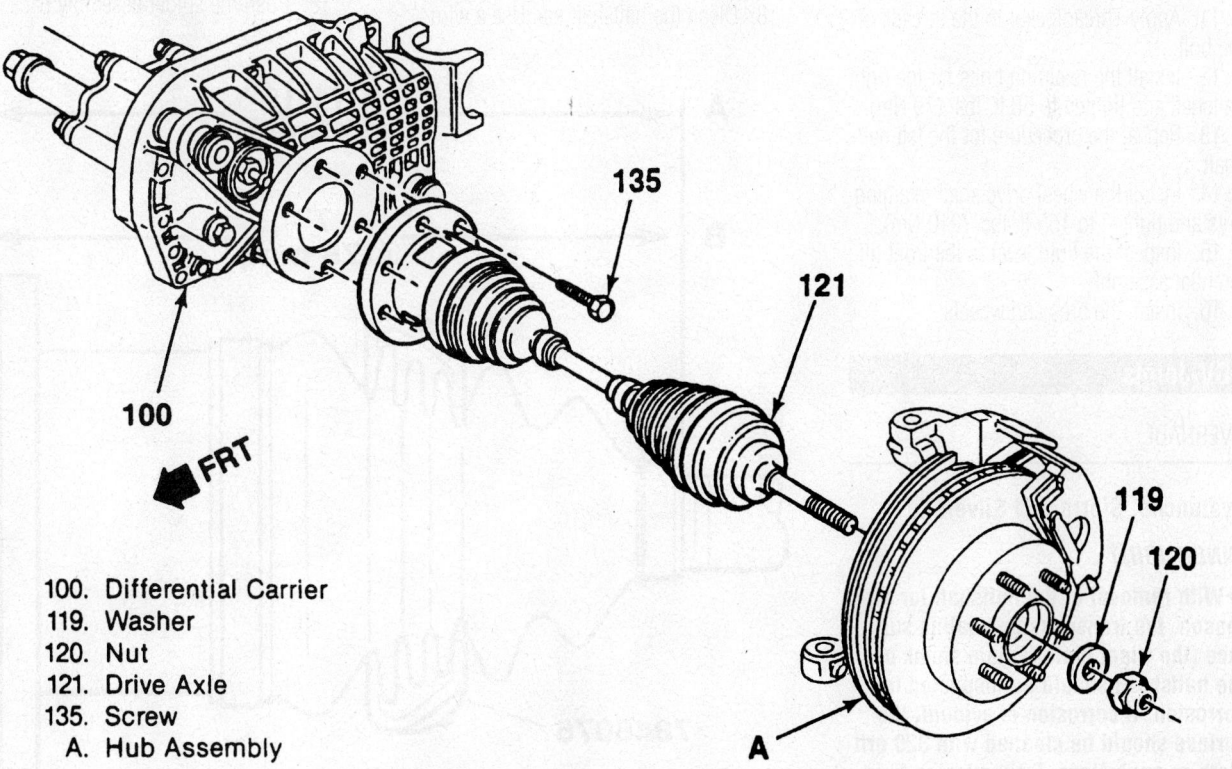

100. Differential Carrier
119. Washer
120. Nut
121. Drive Axle
135. Screw
A. Hub Assembly

The halfshaft is mounted to the flange on the differential and through the hub assembly—4-wheel drive models

7924KG29

- The wheel drive shaft inboard flange to the output shaft flange using the inboard flange bolts

9. Insert a drift or a large screwdriver through the brake caliper into 1 of the brake rotor vanes in order to prevent the wheel drive shaft from turning. Tighten the inboard flange bolts to 58 ft. lbs. (78 Nm). Tighten the hub nut to 177 ft. lbs. (240 Nm).

10. Remove the drift from the rotor.

11. Install the stabilizer shaft link.

12. Install the wheel and tire assembly.

Express and Savana

1. Remove the front wheel and tire assembly.

2. Remove the halfshaft nut.

3. Remove the halfshaft retaining bolts.

4. Remove the left and/or right ha;fshaft from the front differential flange.

5. Install axle remover tool J-45859 to the wheel hub.

6. Use the tool to remove the halfshaft from the steering knuckle.

7. Remove the left and/or right halfshaft

To install:

8. Install the right halfshaft in the steering knuckle.

9. Install the right halfshaft to the differential flange.

10. Clean the threads of the bolts with denatured alcohol or equivalent and allow to dry.

11. Apply Threadlocker to the threads of the bolt.

12. Install the retaining bolts for the right halfshaft and tighten to 58 ft. lbs. (79 Nm).

13. Repeat the procedure for the left halfshaft.

14. Install the wheel drive shaft retaining nuts and tighten to 155 ft. lbs. (210 Nm).

15. Inspect the fluid level of the front differential assembly.

16. Install the tires and wheels.

CV-Joints

OVERHAUL

Avalanche, Sierra and Silverado

INNER JOINT

➡ With removal of the halfshaft for any reason, the transmission sealing surface (the tripot male/female shank of the halfshaft) should be inspected for corrosion. If corrosion is evident, the surface should be cleaned with 320 grit cloth or equivalent. Transmission fluid may be used to clean off any remaining debris. The surface should be wiped dry and the halfshaft reinstalled free of any buildup.

1. Before servicing the vehicle, refer to the precautions in the beginning of this section.

2. Use a hand grinder in order to cut through the swage ring.

3. Remove the tripot housing from the halfshaft. Wipe the grease off of the tripot assembly roller bearings and the tripot housing. Thoroughly degrease the tripot housing. Allow the tripot housing to dry prior to assembly.

➡ Handle the tripot spider assembly with care. Tripot balls and needle rollers may separate from the spider trunnion if the tripot balls and needle rollers are not handled carefully.

4. Use side cutters to cut away the small boot clamp.

5. Compress the tripot boot up the halfshaft away from the tripot spider assembly toward the outboard (CV joint assembly) end of the halfshaft.

6. Spread the spider spacer ring with tool J8059, or equivalent.

7. Remove the following items from the halfshaft bar:
   a. The spacer ring.
   b. The spider assembly.
   c. The tripot boot.

8. Clean the halfshaft bar. Use a wire brush in order to remove any rust in the boot mounting area (grooves).

9. Inspect the needle rollers, needle bearings, and trunnion. Check the tripot housing for unusual wear, cracks, or other damage. Replace any damaged parts.

To assemble:

10. Place the new small boot clamp onto the small end of the joint boot.

11. Compress the joint boot and small boot clamp onto the halfshaft bar.

12. Position the small end of the joint boot into the joint boot groove on the halfshaft bar.

13. Secure the small boot clamp with tool J35910, or equivalent, a breaker bar, and a torque wrench. Tighten the small clamp (1) to 100 ft. lbs. (136 Nm).

14. Check the gap dimension on the clamp ear. Continue tightening until the gap dimension is reached.

➡ Assemble the CV joint with the convolute retainer in the correct position, as illustrated.

15. Install the convolute retainer over the inboard joint boot, being sure to capture three convolutions.

16. Install the tripot spider assembly onto the halfshaft bar with the counterbore towards the end of the halfshaft bar.

17. Install the spacer ring in the groove at the end of the halfshaft bar.

18. Push the spider assembly back

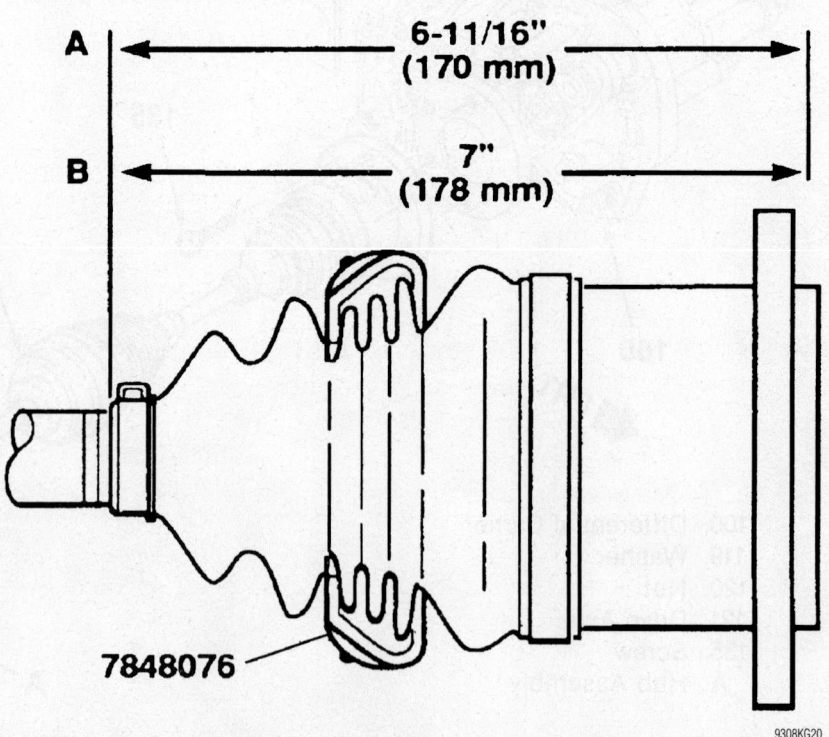

7848076

Assembled joint measurement—15 Series Pickups

9308KG20

toward the end of the halfshaft bar until the spacer ring is covered by the spider assembly counterbore.

19. Pack the tripot boot and the tripot housing with the grease supplied in the kit. The amount of grease supplied in this kit has been pre-measured for this application.

20. Reassemble the tripot housing and the tripot boot using the following procedure:

   a. Pinch the swage ring slightly by hand in order to distort it into an oval shape.

   b. Slide the distorted swage ring over the large diameter of the boot.

   c. Place the tripot housing over the spider assembly.

   d. Install the boot onto the tripot housing.

   e. Align the tripot boot with the swage ring in place, over the flat area on the tripot housing.

21. Mount tool J36652 in a vise. Install the bottom half of the split-plate swage clamp. For K15 models, use tool J36652-98. For K25 models, use tool J36652-1.

22. Check the inboard stroke position. Use measurement A for the K15 models. Use measurement B for the K25 models.

23. Position the inboard end (tripot end) of the halfshaft assembly in tool J36652. Install the top half of the proper size tool on the lower half of the tool. For K15 models, use tool J36652-98. For K25models, use tool J36652-1.

24. Align the swage ring and the swage ring clamp. Insert the bolts. Hand tighten the bolts in tool J36652 until the bolts are snug.

25. Align the following during this procedure:

   a. The tripot boot.

   b. The housing.

   c. The swage ring. Tighten each bolt 180 degrees at a time. Alternate between the bolts until both sides of the top half of J36652 touch the bottom half of the tool.

26. Loosen the bolts and remove the halfshaft assembly from J36652.

27. Remove the convolute retainer from the boot.

OUTER JOINT

1. Place protective covers over the vise jaws. Place the halfshaft in the vise.

2. Use a hand grinder to cut through the swage ring. Use side cutters to cut off the small boot clamp.

3. Slide the boot down the halfshaft bar and away from the CV-joint outer race. Wipe all grease away from the face of the CV joint.

4. Find the halfshaft bar retaining snap ring, which is located in the inner race.

5. Spread the snapring ears apart.

6. Pull the CV joint and the CV joint boot from the halfshaft bar. Discard the old CV joint boot.

7. Place a brass drift against the CV joint cage. Tap gently on the brass drift with a hammer in order to tilt the cage.

8. Remove the first chrome alloy ball when the CV joint cage tilts. Tilt the CV joint cage (1) in the opposite direction to remove the opposing chrome alloy ball. Repeat this process to remove all six of the balls.

9. Pivot the CV joint cage and the inner race 90 degrees to the center line of the outer race. At the same time, align the cage windows with the lands of the outer race. Lift out the cage and the inner race.

10. Remove the inner race from the cage by rotating the inner race upward. Clean the following items thoroughly with cleaning solvent. Remove all traces of old grease and any contaminates.

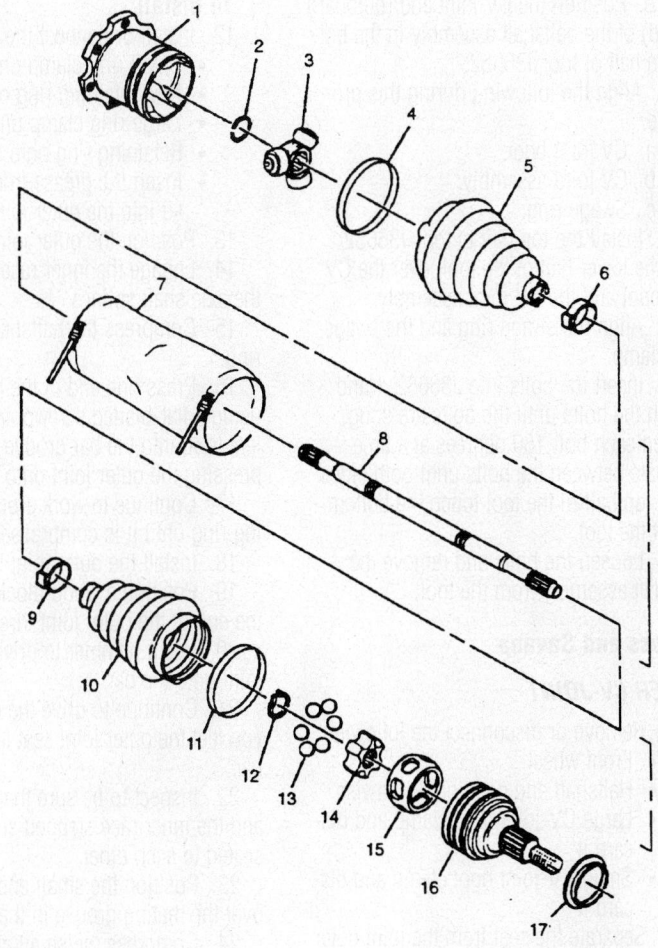

Legend

(1) Tripot Housing Assembly
(2) Spacer Ring
(3) Tripot Joint Spider Assembly
(4) Swage Ring
(5) Tripot Joint Seal
(6) Small Seal Retaining Clamp
(7) Drive Axle Seal Cover (Optional)
(8) Drive Axle Shaft

(9) CV Joint Seal
(10) Race Retaining Ring
(11) Ball
(12) CV Joint Inner Race
(13) CV Joint Cage
(14) CV Joint Outer Race
(15) Deflector Ring

9308KG10

Exploded view of the CV-Joint assembly

a. Inner and outer race assemblies.

b. CV joint cage.

c. Chrome alloy balls.

11. Dry all the parts. Check the CV joint assembly for unusual wear, cracks, or other damage. Replace any damaged parts. Clean the halfshaft bar. Use a wire brush to remove any rust in the boot mounting area (grooves).

To assemble:

12. Inspect all of the parts for unusual wear, cracks, or other damage. Replace the CV joint assembly if necessary. Put a light coat of the recommended grease on the inner and the outer race grooves.

13. Hold the inner race at 90 degrees to the centerline of the cage. Align the lands of the inner race with the windows of the cage. Insert the inner race into the cage by rotating the inner race downward.

14. Insert the cage and inner race into the outer race.

15. Place a brass drift against the CV joint cage. Tap gently on the brass drift with a hammer in order to tilt the cage. Install the first chrome alloy ball when the CV joint cage tilts. Tilt the CV joint cage in the opposite direction to install the opposing chrome alloy ball. Repeat this process in order to install all six of the balls.

16. Pack the CV joint boot and the CV joint assembly with the grease supplied in the kit. The amount of grease supplied in this kit has been pre-measured for this application.

17. Place the new small boot clamp onto the CV joint boot.

18. Slide the CV joint boot onto the halfshaft bar.

19. Position the small end of the CV joint boot into the joint boot groove on the halfshaft bar.

20. Secure the small boot clamp, a breaker bar, and a torque wrench. Tighten the small clamp (1) to 100 ft. lbs. (136 Nm).

21. Check the gap dimension on the clamp ear. Continue tightening until the gap dimension is reached.

22. Pinch the new swage ring slightly by hand to distort it into an oval shape. Slide the distorted swage ring over the large diameter of the boot.

➡ **Be sure that the retaining ring side of the CV joint inner race faces the half-shaft bar (3) before installation.**

23. Slide the CV joint onto the halfshaft bar. The retaining snap ring inside of the inner race engages in the halfshaft bar groove with a click when the CV joint is in the proper position.

24. Pull on the CV joint to verify engagement.

25. Slide the large diameter of the CV joint boot with the large swage ring in place, over the outside edge of the CV joint outer race.

26. Clamp the CV joint boot tightly to the CV joint outer race with the large swage ring, using the following procedure:

a. Mount tool J36652 in a vise.

b. Install the bottom half of the split-plate swage clamp. For K15 models, use tool J36652-98.

c. For K25 models, use tool J36652-1.

d. Position the CV joint end (outboard end) of the halfshaft assembly in the bottom half of tool J36652.

27. Align the following during this procedure:

a. CV joint boot.

b. CV joint assembly.

c. Swage ring.

28. Install the top half of tool J36652 onto the lower half of the tool, over the CV joint boot and the CV joint assembly.

29. Align the swage ring and the swage ring clamp.

30. Insert the bolts into J36652. Hand tighten the bolts until the bolts are snug. Tighten each bolt 180 degrees at a time. Alternate between the bolts until both sides of the top half of the tool touch the bottom half of the tool.

31. Loosen the bolts and remove the halfshaft assembly from the tool.

Express and Savana

OUTER CV-JOINT

1. Remove or disconnect the following:
- Front wheel
- Halfshaft and position it in a vise
- Large CV-joint boot clamp and discard it
- Small CV-joint boot clamp and discard it

2. Separate the seal from the joint outer race at the large end.

3. Position the seal behind the joint face.

4. Wipe the grease from the face of the joint inner race, cage, balls, etc.

5. Remove the outer joint from the half-shaft bar.

6. Have an assistant hold the joint housing.

7. Position a wood block between the seal and the joint (along the joint face).

8. Strike the wood block with a hammer to compress the axle shaft retaining ring.

9. Continue to strike the wood block to remove the outer joint from the bar.

10. Remove the retaining ring from the bar.

11. Remove the seal from the bar.

To install:

12. Install or connect the following:
- Small ring clamp on the CV boot
- New retaining ring on the halfshaft
- Large ring clamp on the CV boot
- Retaining ring onto the bar
- Insert the grease from the service kit into the outer joint

13. Position the outer joint horizontally.

14. Engage the inner race splines onto the axle shaft splines.

15. Compress the halfshaft retaining ring.

16. Press one end of the retaining ring, using a flat-bladed screwdriver or equivalent tool, into the bar groove while firmly pressing the outer joint onto the axle shaft.

17. Continue to work around the retaining ring until it is compressed.

18. Install the outer joint to the bar.

19. Position a wood block squarely over the end of the outer joint threaded shaft.

20. Use a hammer to drive the outer joint onto the bar.

21. Continue to drive the outer joint until you feel the outer joint seat fully onto the bar.

22. Inspect to be sure that the axle shaft and the inner race stepped surfaces are fully seated to each other.

23. Position the small end of the seal over the mating groove in the bar.

24. Compress the small clamp using crimping pliers until the crimp dimension is 0.02-0.06 in. (0.5-1.6 mm).

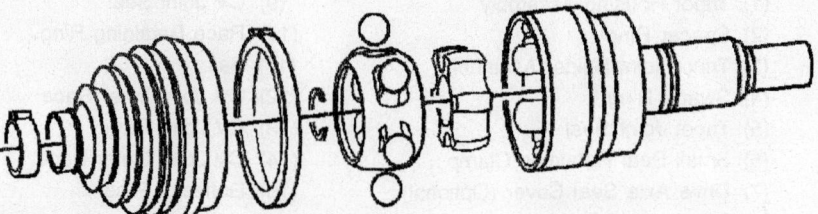

06025-AVAL-G11

Exploded view of the outer CV-Joint assembly—Express and Savana

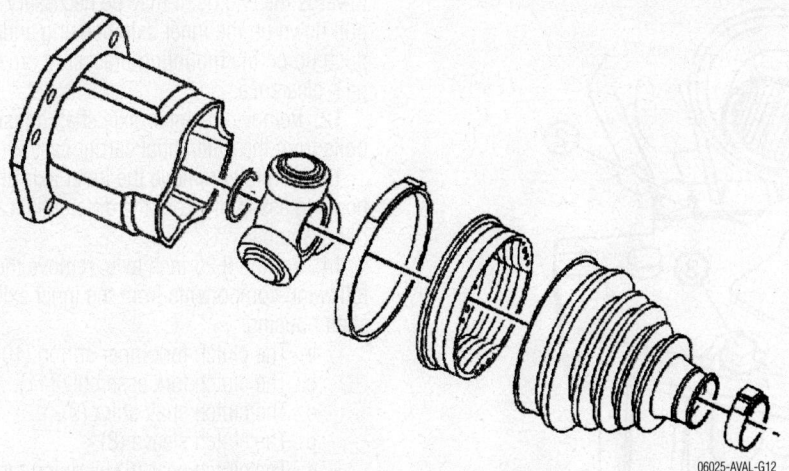

06025-AVAL-G12

Exploded view of the inner CV-Joint assembly—Express and Savana

25. Vent the joint to relieve excess trapped air.

26. Compress the small clamp using crimping pliers until the crimp dimension is 0.02-0.06 in. (0.5-1.6 mm).

27. Install the halfshaft and the front wheel.

INNER (TRI-POT) JOINT

1. Remove the front wheel.

2. Remove the halfshaft and place it in a vise.

3. Remove the large clamp and the small clamp. Use a side cutter or other suitable tool and discard the clamps.

4. Remove the tripot housing.

5. Compress the tripot seal up the bar, away from the tripot spider assembly toward the outboard end of the halfshaft assembly.

6. Spread the spider snap ring with snap ring pliers.

7. Remove and discard the snap ring.

8. Remove the spider assembly.

9. Remove the tripot bushing.

10. Remove the tripot seal.

11. Clean the bar with a wire brush in order to remove any rust in the seal mounting areas.

To install:

12. Slide the new clamps (4, 7) on the tripot seal (6).

13. Install the tripot seal (6) over the wheel drive shaft.

14. Install the tripot bushing (5) into the seal.

15. Slide the spider assembly onto the wheel drive shaft.

16. Install the new spider snap ring (2) with snap ring pliers.

17. Install the tripot housing (1) over the spider assembly and seat it in the bushing.

18. Seat the small end of the tripot seal (6) into the groove on the wheel drive shaft.

19. Compress the small clamp using the, until the gap dimension is 0.02-0.06 in. (0.5-1.6mm).

20. Fill the joint and seal cavity with the grease supplied in the service kit.

21. Seat the large end of the tripot seal (6) into the groove on the tripot bushing.

22. Check the stroke position of the joint, and set it so the dimension (A) is 6.92 in. (176mm).

23. Vent the joint to relive excess trapped air.

24. Secure the large clamp using a crimper.

25. Install the halfshaft and the front wheel.

Transfer Case Encoder Motor

REMOVAL & INSTALLATION

1. Remove the transfer case shield.

2. Remove the front propeller shaft.

3. Disconnect the transfer case switch electrical connector.

4. Disconnect the encoder motor electrical connector.

5. Remove the encoder motor bolts.

6. Remove the encoder motor.

7. Remove the actuator insulator gasket.

8. If replacing the encoder motor, remove the locating pins from the old motor.

To install:

➡**If the encoder motor is being replaced because it is defective, ensure that the transfer case is in the neutral position. Manually shift the transfer case at the shift shaft, using a crescent wrench if necessary. When installing the encoder motor, ensure that the encoder motor is indexed correctly and the motor is flat against the transfer case before tightening the bolts.**

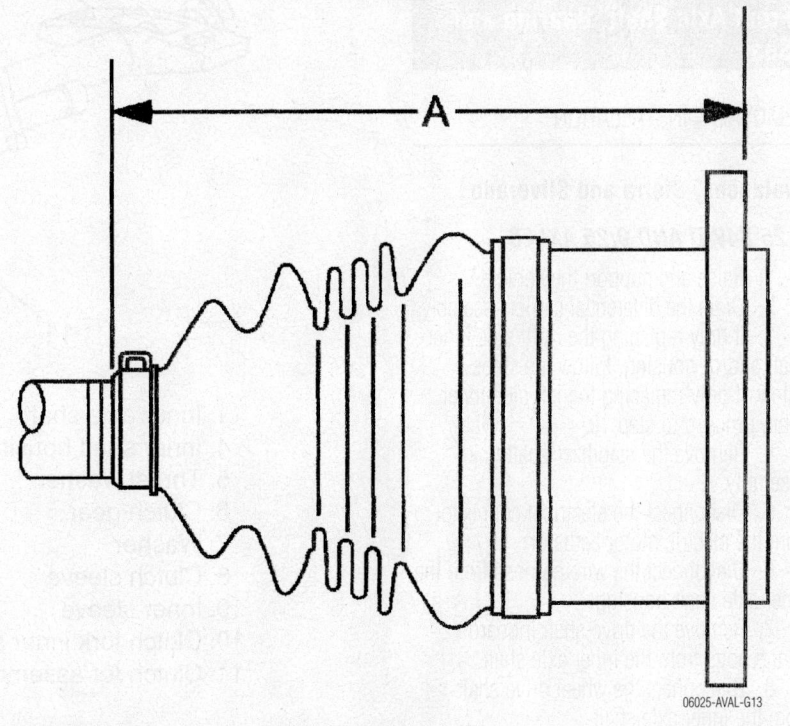

06025-AVAL-G13

Boot positioning dimension on inner joint—Express and Savana

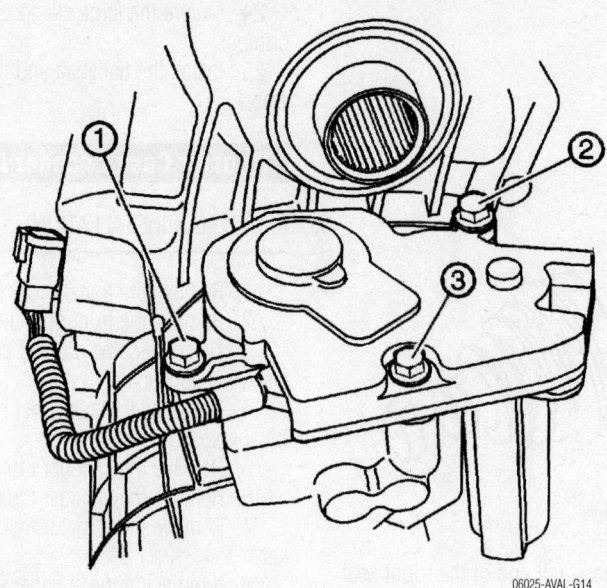

06025-AVAL-G14

Encoder motor tightening sequence

9. Install the locating pins to the new encoder motor.

10. Position a new actuator insulator gasket to the transfer case.

11. Install the encoder motor.

12. Install encoder motor bolts and tighten in sequence to 15 ft. lbs. (20 Nm).

13. Connect the encoder motor electrical connector.

14. Connect the transfer case switch electrical connector.

15. Install the front propeller shaft.

16. Install the transfer case shield.

Front Axle Shaft, Bearing and Seal

REMOVAL & INSTALLATION

Avalanche, Sierra and Silverado

8.25 S4WD AND 9.25 AXLES

1. Raise and support the vehicle.

2. Drain the differential carrier assembly.

3. If only replacing the right side inner shaft and/or housing, follow the steps below. If only replacing the left side inner shaft, proceed to step 19.

4. Remove the stabilizer shaft link assembly.

5. Disconnect the electrical connector from the electric motor actuator.

6. Disconnect the wire harness from the inner axle shaft housing.

7. Remove the drive shaft inboard flange bolts from the inner axle shaft.

8. Disconnect the wheel drive shaft from the inner axle shaft.

9. Remove the inner axle shaft housing nuts from the bracket.

10. For 25/35 series vehicles, remove the front axle mounting bracket to frame nuts.

11. Slide the front axle mounting bracket

towards the engine. It may be necessary to pull down on the inner axle housing and/or push up on the mounting bracket in order to gain clearance.

12. Remove the inner axle shaft housing bolts from the differential carrier case.

13. Carefully remove the inner axle shaft housing assembly from the differential carrier assembly.

14. For the 8.25 inch axle, remove the following components from the inner axle shaft housing:

a. The clutch fork inner spring (10).

b. The clutch fork assembly (11).

c. The clutch shaft shim (9).

d. The clutch sleeve (8).

e. The clutch gear (6) by doing the following:

f. Clamp the inner axle shaft housing (4) in a vise. Clamp only on the mounting flange.

g. Strike the inside surface of the shaft (1) flange with a hammer and a brass drift in order to dislodge the front drive axle clutch gear (6) from the inner axle shaft (1).

h. The thrust washer (5).

15. For the 9.25 inch axle, remove the

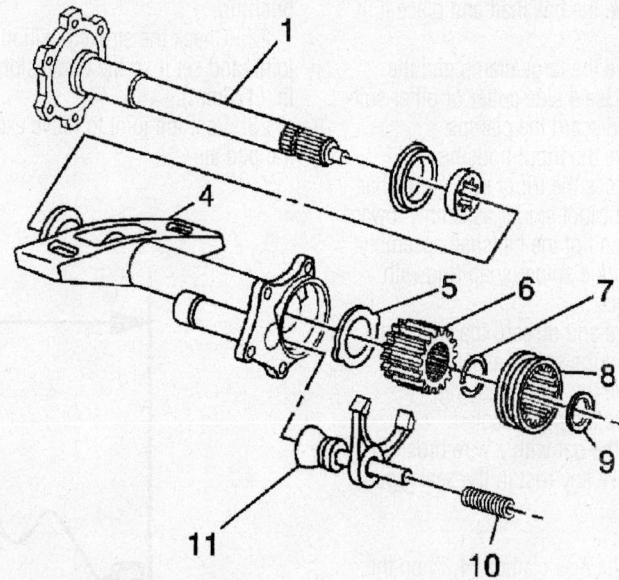

1. Inner axle shaft
4. Inner shaft housing
5. Thrust washer
6. Clutch gear
7. Washer
8. Clutch sleeve
9. Inner sleeve
10. Clutch fork inner spring
11. Clutch for assembly

06025-AVAL-G15

Exploded view of the front axle assembly—8.25 S4WD and 9.25 axles

following components from the inner axle shaft housing:

    a. The clutch fork inner spring (10).

    b. The clutch fork assembly (11).

    c. The clutch shaft shim (9).

    d. The clutch sleeve (8).

    e. The retainer ring (7).

    f. The thrust washers (5, 6).

16. Remove the inner axle shaft (2). Tap out the inner axle shaft with a soft-faced mallet, if necessary.

17. Remove the inner axle seal and the bearing from the axle housing.

18. If only replacing the left side inner axle shaft, remove the wheel drive shaft inboard flange bolts from the inner axle shaft. Disconnect the wheel drive shaft from the inner axle shaft.

19. Remove the inner axle shaft using a hammer and a brass drift.

20. Install the inner axle shaft housing into a vise. Clamp only on the mounting flange of the inner axle shaft housing.

21. Install the bushing and bearing removal tool J-29369-1, 8.25 inch axle, or J-29369-2, 9.25 inch axle, behind the inner axle shaft seal or the inner axle shaft bearing as necessary.

22. Install a slide hammer to the removal tool.

23. Remove the inner axle shaft seal and/or the inner axle shaft bearing using the slide hammer.

24. If only replacing the left side seal, place an alignment mark between the inner axle shaft and the wheel drive shaft.

25. Disconnect the wheel drive shaft from the inner axle shaft.

26. Remove the inner axle shaft using a hammer and a brass drift.

27. Remove the inner axle shaft seal using a suitable seal remover tool.

To install:

28. Install the right side bearing with the square shoulder in using and axle bearing tube installer and a universal driver handle.

29. Install the new axle shaft seal using the sane tools.

30. Install the inner axle shaft into the inner axle shaft housing. Carefully tap the inner axle shaft into place with a soft-faced mallet.

31. Install the inner axle shaft and clutch fork assembly components into the inner shaft housing.

32. If only the left side inner axle shaft was removed, install the shaft by performing the following steps:

33. Install the inner axle shaft into the differential case side gear using a soft-faced mallet until the retaining ring on the inner

axle shaft is fully seated within the groove in the differential case side gear.

34. Pull back on the inner axle shaft to ensure that the inner axle shaft is properly retained in the differential case side gear.

35. Connect the halfshaft to the inner axle shaft.

36. Install the halfshaft inboard flange to inner axle shaft bolts and tighten to 58 ft. lbs. (79 Nm).

37. If the right side inner axle shaft and/or housing was removed, install the shaft and/or housing using the following steps:

38. Install the new inner axle shaft bearing and the seal to the axle housing.

39. Install the inner axle shaft (2) into the inner axle shaft housing (1). Carefully tap the inner axle shaft into place with a soft-faced mallet.

40. Place the inner axle shaft housing on end so that the splines of the inner axle shaft is facing up.

41. For the 8.25 inch axle, install the following components into the inner axle shaft housing:

➡**Use chassis grease in order to hold the thrust washer in place.**

42. The thrust washer (5) Ensure the tabs on the thrust washer are aligned with the slots in the inner axle shaft housing (4).

43. The retainer ring (7) into the clutch gear (6).

44. The clutch gear (6) onto the inner axle shaft (1). Drive the clutch gear into place with a plastic hammer.

45. Install the original shim to the shaft. Use the chassis grease in order to hold the shim in place.

46. Install the inner axle housing assembly to the differential carrier case. Do not use sealer at this time.

47. Install the bolts and tighten to 30 ft. lbs. (40 Nm).

48. Install a dial indicator on the axle tube end. The plunger of the indicator must be at a right angle to the axle flange.

49. Move the shaft back and forth and read the end play. The correct end play is 0.001-0.020 in (0.03-0.51mm).

50. If the end play is incorrect, install a thicker or thinner shim as needed in order to bring the end play into the specified range.

51. Install the clutch gear shim (9). clutch sleeve (8), clutch fork assembly (11) and clutch fork inner spring (10).

52. For the 9.25 inch axle, install the following components into the inner axle shaft housing:

53. The thrust washer (5) Ensure the

tabs on the thrust washer are aligned with the slots in the inner axle shaft housing (4).

54. The second thrust washer (6).

55. The retainer ring (7) onto the inner axle shaft (1).

56. Determine the clutch gear shim thickness.

57. Install the clutch gear shim (9). clutch sleeve (8), clutch fork assembly (11) and clutch fork inner spring (10).

58. Apply sealant to the inner axle housing to differential carrier sealing surface.

59. Install the inner axle shaft housing assembly to the differential carrier assembly.

60. Install the inner axle shaft housing bolts and tighten to 30 ft. lbs. (40 Nm).

61. For 25/35 series vehicles, perform the following steps in order to install the front axle mounting bracket to the inner axle shaft housing:

62. Slide the front axle mounting bracket towards the frame. Install the front axle mounting bracket studs into the inner shaft housing mounting flange. It may be necessary to push up on the front axle mounting bracket and/or pull down on the inner axle housing in order to gain enough clearance to install the mounting bracket studs into the inner shaft housing.

63. Install the front axle mounting bracket to frame nuts.

64. Install the inner axle shaft housing washers and nuts to the bracket and tighten to 75 ft. lbs. (100 Nm).

65. Connect the wheel drive shaft inboard flange to the inner axle shaft and tighten to 30 ft. lbs. (40 Nm).

66. Install the wheel drive shaft inboard flange to the inner axle shaft bolts and tighten to 58 ft. lbs. (79 Nm).

67. Connect the wire harness to the inner axle shaft housing.

68. Connect the electrical connector to the front axle actuator.

69. Install the stabilizer shaft link assembly.

70. With either replacement procedure, fill the differential carrier assembly with axle lubricant.

71. Lower the vehicle.

8.25 F4WD AXLE

1. Raise and support the vehicle.

2. Drain the differential carrier assembly.

3. If only replacing the right side inner shaft and/or housing, follow the steps below. If only replacing the left side inner shaft, proceed to step 16.

4. Remove the stabilizer shaft link assembly.

5. Remove the wheel drive shaft inboard flange bolts from the inner axle shaft.

6. Disconnect the wheel drive shaft from the inner axle shaft.

7. Disconnect the inner axle shaft from the differential case side gear using a hammer and brass drift. Remove the inner axle shaft housing nuts from the bracket.

8. Remove the inner axle shaft housing bolts from the differential carrier assembly.

9. Remove the inner axle shaft and inner axle shaft housing from the vehicle.

10. Remove the inner axle shaft from the inner axle shaft housing.

11. Remove the inner axle shaft seal and the bearing from the inner axle shaft housing.

12. Install the inner axle shaft housing into a vise. Clamp only on the mounting flange of the inner axle shaft housing.

13. Install the bushing and bearing removal tool J-29369-1, 8.25 inch axle, or J-29369-2, 9.25 inch axle, behind the inner axle shaft seal or the inner axle shaft bearing as necessary.

14. Install a slide hammer to the removal tool.

15. Remove the inner axle shaft seal and/or the inner axle shaft bearing using the slide hammer.

16. If only replacing the left side seal, place an alignment mark between the inner axle shaft and the wheel drive shaft.

17. Disconnect the wheel drive shaft from the inner axle shaft.

18. Remove the inner axle shaft using a hammer and a brass drift.

19. Remove the inner axle shaft seal using a suitable seal remover tool.

To install:

20. Install the right side bearing with the square shoulder in using and axle bearing tube installer and a universal driver handle.

21. Install the new axle shaft seal using the sane tools.

22. Install the inner axle shaft into the inner axle shaft housing. Carefully tap the inner axle shaft into place with a soft-faced mallet.

23. Install the inner axle shaft and clutch fork assembly components into the inner shaft housing.

24. If only the left side inner axle shaft was removed, install the shaft by performing the following steps:

25. Install the inner axle shaft into the differential case side gear using a soft-faced mallet until the retaining ring on the inner axle shaft is fully seated within the groove in the differential case side gear.

26. Pull back on the inner axle shaft to ensure that the inner axle shaft is properly retained in the differential case side gear.

27. Connect the halfshaft to the inner axle shaft.

28. Install the halfshaft inboard flange to inner axle shaft bolts and tighten to 58 ft. lbs. (79 Nm).

29. If the right side inner axle shaft and/or housing was removed, install the shaft and/or housing using the following steps.

30. Install the new inner axle shaft bearing and the new seal to the inner axle shaft housing.

31. Install the inner axle shaft into the inner axle shaft housing. Do not install the inner axle shaft completely into the inner axle shaft housing at this time.

32. Apply sealant to the inner axle housing to differential carrier sealing surface.

33. Install the inner axle shaft and the inner axle shaft housing to the differential carrier assembly.

34. Install the inner axle shaft housing bolts and tighten to 30 ft. lbs. (40 Nm).

35. Install the inner axle shaft housing nuts to the bracket and tighten to 75 ft. lbs. (100 Nm).

36. Install the inner axle shaft into the differential case side gear by doing the following:

37. Turn the inner axle shaft and align the splines of the inner axle shaft with the splines on the differential side gear.

38. Install the inner axle shaft into the differential case side gear using a soft-faced mallet until the retaining ring on the inner axle shaft is fully seated within the groove in the differential case side gear.

39. Pull back on the inner axle shaft to ensure that the inner axle shaft is properly retained in the differential case side gear.

40. Install the wheel drive shaft inboard flange to the inner axle shaft.

41. Install the wheel drive shaft inboard flange to inner axle shaft bolts and tighten to 58 ft. lbs. (79 Nm).

42. Install the stabilizer shaft link assembly.

43. Fill the differential carrier assembly with axle lubricant

44. Lower the vehicle.

Express and Savana

1. Raise the vehicle.

2. Drain the differential carrier assembly.

3. Remove the engine protection shield.

4. Disconnect the wheel drive shaft from the inner axle shaft.

5. Place a pry bar between the inner axle shaft flange and the inner axle shaft housing.

6. Disconnect the inner axle shaft from the differential case side gear using the pry bar. Do not remove the inner axle shaft at this time.

7. Install a support jack underneath the differential carrier assembly.

8. Remove the upper inner shaft housing bushing bolt and nut.

9. Remove the lower inner shaft housing bushing bolt and nut.

10. Remove the inner axle shaft housing bolts from the differential carrier assembly.

11. Remove the inner axle shaft and inner axle shaft housing from the vehicle.

12. Remove the inner axle shaft from the inner axle shaft housing.

13. Remove the inner axle shaft seal and the bearing from the inner axle shaft housing using a bearing remover and slide hammer.

To install:

14. Install the new inner axle shaft bearing and the new seal to the inner axle shaft housing using a bearing installer and universal driver handle.

15. Install the inner axle shaft into the inner axle housing. Do not install the inner axle shaft completely into the inner axle shaft housing at this time.

16. Apply sealant to the inner axle housing to differential carrier sealing surface.

17. Install the inner axle shaft housing with the inner axle shaft to the differential carrier assembly.

18. Install the inner axle shaft housing bolts and tighten to 35 ft. lbs. (48 Nm).

19. Install the lower inner axle shaft housing bushing bolt and nut.

20. Install the upper inner axle shaft housing bushing bolt and nut and tighten to 63 ft. lbs. (85 Nm).

21. Remove the support jack.

22. Carefully guide the inner axle shaft through the inner axle housing until the retaining ring on the inner axle shaft contacts the differential case side gear.

23. Install the inner axle shaft into the differential case side gear by tapping the retaining ring into the retaining groove using a soft-faced mallet and until the retaining ring on the inner axle shaft is fully seated within the groove in the differential case side gear.

24. Pull back on the inner axle shaft to ensure that the inner axle shaft is properly retained in the differential case side gear.

25. Install the wheel drive shaft to the inner axle shaft.

26. Fill the differential carrier assembly with axle lubricant.

27. Lower the vehicle.

Rear Axle Shaft, Bearing and Seal

REMOVAL & INSTALLATION

8.5 Inch and 9.5 Inch Rear Axles

1. Raise and support the vehicle on a hoist.

2. Remove or disconnect the following:
   - Tire and wheel assembly
   - Brake caliper
   - Rear cover and the gasket
   - Pinion shaft locking screw.
   - Pinion shaft, on axles without locking differential

3. On axles with a locking differential, remove the shaft part way. Rotate the case until the pinion shaft touches the housing.

4. On axles with a locking differential, use a screwdriver, or a similar tool, in order to enter the differential case and rotate the lock until the lock aligns with the thrust block.

5. Push the flange of the axle shaft toward the differential. Remove the lock from the button end of the axle shaft.

➡**When removing the axle shaft, do not rotate the shaft. Rotating the shaft will misalign the gears. Misaligning the gears will make the assembly difficult.**

6. Remove the axle shaft from the housing.

7. If replacing only the axle shaft seal, remove the seal using a suitable seal removal tool.

8. Remove the bearing using a bearing remover.

9. Inspect all the parts for damage. Replace the parts as necessary.

To install:

10. Install a new bearing using a bearing installer.

11. Install new seal using a seal installer. Ensure the seal is fully seated in the axle tube.

➡**Carefully insert the axle shaft in order to not damage the seal.**

12. Install the axle shaft into the housing. Slide the axle shaft into place allowing the splines to engage the differential side gear.

13. On axles without a locking differential, place the lock on the button end of the axle shaft.

14. On axles with a locking differential, keep the pinion shaft partially withdrawn.

15. On axles with a locking differential, place the lock on the axle shaft so that the ends are flush with the thrust block. Pull the shaft flange outward in order to seat the lock in the differential gear.

➡**Anytime you remove a differential pinion shaft locking screw, coat the screw threads with Loctite® 242 before reinstalling the screws. The screw has an adhesive coating in order to prevent the screw from loosening in the case. Removing the screw removes the adhesive on the screw.**

16. Align the hole in the pinion shaft with the screw hole in the differential case.

17. Install or connect the following:
   - Pinion flange locking bolt and tighten to 25 ft. lbs. (34 Nm).
   - Rear cover and the gasket
   - Brake caliper
   - Tire and wheel assembly

18. Fill the rear axle.

19. Remove the supports and lower the vehicle.

8.6 Inch Rear Axles

1. Raise and support the vehicle on a hoist.

2. Remove the tire and wheel assembly.

3. Remove the brake caliper on disc brake models.

4. Remove the rear cover and gasket.

5. Remove the pinion shaft locking bolt.

6. On axles without a locking differential, remove the pinion shaft.

7. On axles with a locking differential, remove the shaft part way. Rotate the case until the pinion shaft touches the housing.

8. On axles with a locking differential, use a screwdriver, or a similar tool, in order to enter the differential case and rotate the lock until the lock aligns with the thrust block.

9. Remove the brake drum on drum brake models.

10. Push the flange of the axle shaft in toward the differential.

11. Remove the C-lock from the button end of the axle shaft.

12. When removing the axle shaft, do not rotate the shaft. Rotating the shaft will misalign the gears. Misaligning the gears will make assembly difficult.

13. Remove the axle shaft from the housing.

To install:

14. Install the axle shaft into the rear axle housing.

15. Slide the axle shaft into place allow-

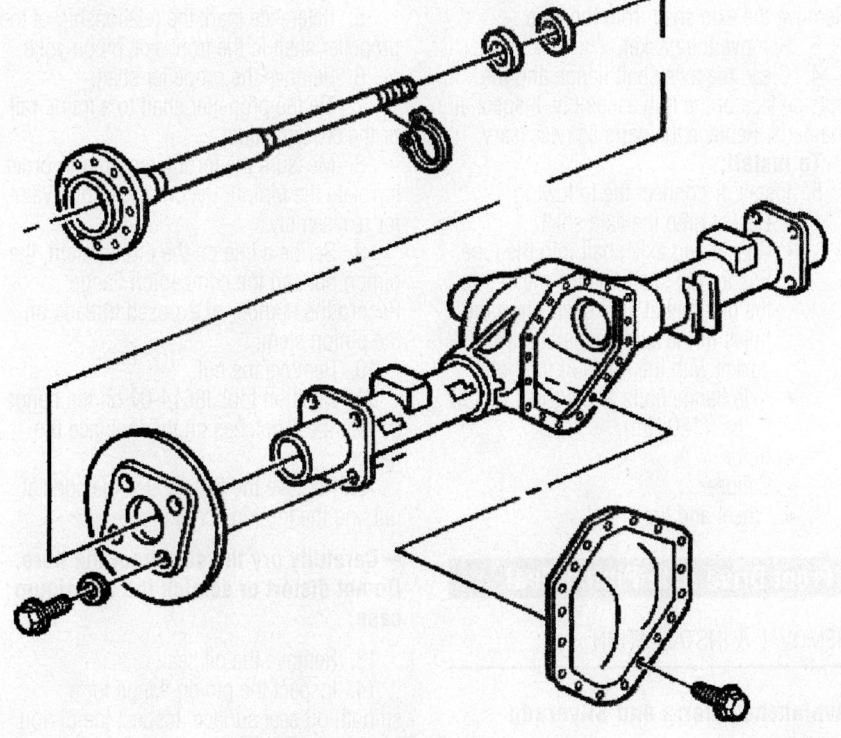

Exploded view of the rear axle—8.5/8.6/9.5 inch

06025-SAVA-G02

ing the splines to engage the differential side gear.

16. On axles without a locking differential, place the C-lock on the button end of the axle shaft.

17. On axles with a locking differential, keep the pinion shaft partially withdrawn.

18. Install the brake drum on drum brake models.

19. On axles with a locking differential, place the C-lock on the axle shaft so that the ends are flush with the thrust block.

20. Pull the shaft flange outward in order to seat the lock in the differential gear.

21. Align the hole in the pinion shaft with the bolt hole in the differential case.

22. Install the new pinion shaft locking bolt and tighten to 25 ft. lbs. (35 Nm).

23. Install the rear cover and the gasket.

24. Install the caliper on disc brake models.

25. Install the tire and wheel assembly.

26. Fill the rear axle, using the proper fluid.

27. Lower the vehicle.

9.75 Inch Rear Axles

1. Release the parking brake.
2. Raise and support the vehicle.
3. Remove the tire and wheel assembly.
4. Remove the rear steering gear assembly.
5. Remove the steering knuckle assembly.
6. Remove the lock clip from the axle shaft end. The lock clip is spring loaded and fits securely in the axle shaft slot and may need to be push off the shaft end with a screw driver or related tool. Pushing the axle shaft inwards towards the gears my help in removal of the lock clip.
7. When removing the axle shaft do not rotate the shaft. Rotating the shaft will cause the gears to move. Misalignment of the gears will make the assembly difficult.
8. Remove the axle shaft.

To install:

9. Install the axle shaft.
10. Install the spring loaded lock clip to the axle shaft end.
11. Install the steering knuckle assembly.
12. Install the rear steering gear assembly.
13. Install the tire and wheel assembly.
14. Lower the vehicle.

10.5 Inch Rear Axles

1. Remove or disconnect the following:
   - Tire and wheel
   - Brake caliper
   - Brake rotor
   - Flange bolts
2. Lightly rap the axle shaft with a soft-faced hammer in order to loosen the shaft.

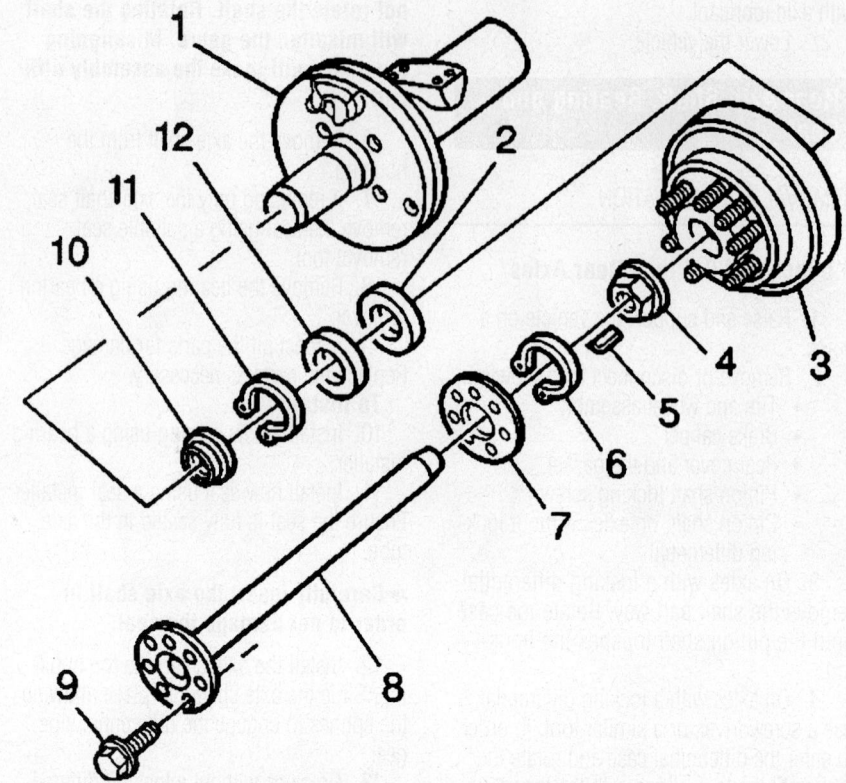

Rear axle shaft removal—10.5 inch 25 Series Silverado

9308KG16

Grip the rib on the axle shaft flange with a locking pliers. Twist the axle shaft flange in order to start the axle shaft removal. Remove the axle shaft from the tube.

3. Remove the gasket.
4. Clean the axle shaft flange and the outside face of the hub assembly. Inspect all the parts. Replace the parts as necessary.

To install:

5. Install or connect the following:
   - Gasket onto the axle shaft
   - Gasket and axle shaft into the tube. Ensure the shaft splines mesh into the differential side gear. Align the holes in the axle flange and the gasket with the holes in the hub.
   - Axle flange bolts and tighten to 110 ft. lbs. (150 Nm).
   - Rotor
   - Caliper
   - Wheel and tire

Front Drive Axle Pinion Seal

REMOVAL & INSTALLATION

Avalanche, Sierra and Silverado

1. Raise the vehicle on a hoist.
2. Remove the tire and wheel.

3. Remove the brake calipers.
4. Remove the differential carrier assembly shield, if equipped.
5. Reference mark the relationship of the propeller shaft to the front axle pinon yoke.
6. Remove the propeller shaft.
7. Tie the propeller shaft to a frame rail or the crossmember.
8. Measure the torque required in order to rotate the pinion. Record the torque value for reassembly.
9. Scribe a line on the pinion stem, the pinion nut and the companion flange. Record the number of exposed threads on the pinion stem.
10. Remove the nut.
11. Position tool J8614-01 on the flange so that the 4 notches on the tool face the flange.
12. Remove the flange. Use the special nut and the forcing screw.

➡Carefully pry the seal from the bore. Do not distort or scratch the aluminum case.

13. Remove the oil seal.
14. Inspect the pinion flange for a smooth oil seal surface. Inspect the pinion flange for worn drive splines. Replace the pinion flange if necessary.

15. Remove the dust deflector.

To install:

➡**Stake the new deflector at 3 new equally spaced positions. You must stake the new deflector in such a way that you do not damage the seal operating surface.**

16. Install and stake the dust deflector on the flange.

17. Position the oil seal in the bore. Then place a driver over the oil seal. Strike the driver with a hammer until the seal flange seats on the axle housing surface. Drive the seal in straight, not at an angle, as this will damage the aluminum housing.

➡**Do not hammer the pinion flange/ yoke onto the pinion shaft. Pinion components may be damaged if the pinion flange/yoke is hammered onto the pinion shaft.**

18. Install the flange onto the pinion using tool J8614-01. Place the washer and a new nut on the pinion threads. Tighten the nut to the original scribed position using the scribe marks and the exposed threads as reference.

19. Measure the rotating torque of the pinion. Compare the measurement with the rotating torque recorded earlier. Tighten the pinion nut by small increments until the torque required in order to rotate the pinion is 3 inch lbs. (0.35 Nm) greater than the original torque.

20. Install the propeller shaft.

21. Install the differential carrier assembly shield, if equipped.

22. Install the brake calipers

23. Install the tire and wheel.

24. Lower the vehicle.

Express and Savana

1. Raise and support the front end on jackstands.

2. Matchmark and disconnect the front driveshaft at the carrier.

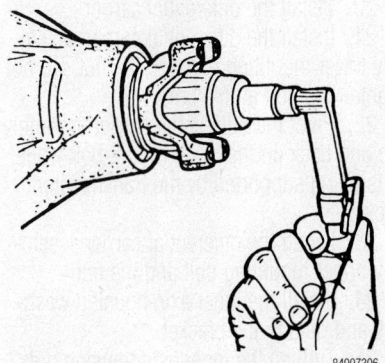

Measuring the pinion rotating torque

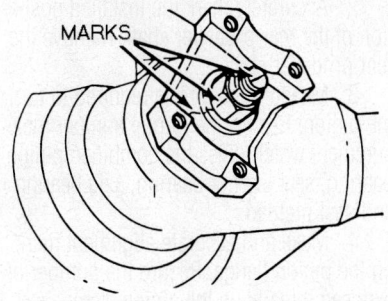

MARKS

84907307

View of the scribed marks

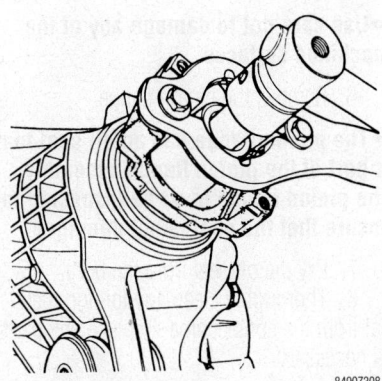

84907308

Removing the pinion nut

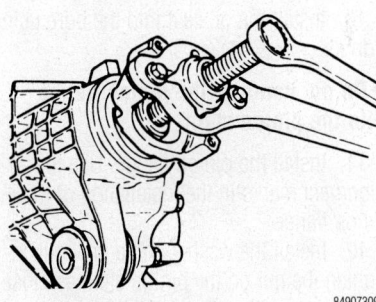

84907309

Removing the pinion flange

3. Remove or disconnect the following:
- Wheels
- Calipers and wire them up, out of the way

4. Position an inch pound torque wrench on the pinion nut. Measure the torque needed to rotate the pinion one full revolution. Record the figure.

5. Matchmark the pinion flange, shaft and nut. Count and record the number of exposed threads on the pinion shaft.
- Nut and washer, while holding the flange
- Flange, using a puller
- Seal from its bore by carefully prying it. Be careful to avoid scratching the seal bore.
- Deflector from the flange

To install:

6. Clean the seal bore thoroughly.

7. Remove any burrs from the deflector staking on the flange.

8. Tap the deflector onto the flange and stake it in three places.

9. Position the new seal in the carrier bore and drive it into place until flush. Coat the seal lips with wheel bearing grease.

10. Coat the outer edge of the flange neck with wheel bearing grease and slide it onto the pinion shaft.

11. Place a new nut and washer onto the pinion shaft and tighten it to the position originally recorded. That is, the alignment marks are aligned, and the recorded number of threads are exposed on the pinion shaft.

✳✳ WARNING

Never hammer the flange onto the pinion!

12. Measure the rotating torque of the pinion. Compare this to the original torque. Tighten the pinion nut, in small increments, until the rotating torque is 3 inch lbs. (0.35 Nm) GREATER than the original torque.

13. Install the driveshaft.

14. Install the calipers and install the wheels.

Rear Drive Axle Pinion Seal

REMOVAL & INSTALLATION

Avalanche, Sierra and Silverado

1. Raise the vehicle.

2. Remove the tire and wheel assemblies (8.6 inch, 9.5 inch axles).

3. Remove the rear brake calipers and rotors (8.6 inch, 9.5 inch axles).

4. Remove the axle shafts (10.5 inch, 11.5 inch axle).

5. Reference mark the rear propeller shaft to the rear axle pinion yoke.

6. Disconnect the propeller shaft from the axle.

7. Measure the torque required to turn the pinion. Record the torque number measurement which gives the combined pinion bearing, seal, carrier bearing, axle bearing and seal preload.

8. Make and accurate alignment mark on the pinion flange. Record the number of exposed threads on the pinion stem.

9. Remove the pinion flange nut and the washer. Use a container in order to catch any lubricant.

➡**Use care not to damage any of the machined surfaces.**

10. Remove the pinion flange.

➡ **The pinion flange has an oil seal that is part of the pinion flange assembly. The pinion flange must be inspected to ensure that the seal is not damaged.**

11. Pry the oil seal from the bore.

12. Thoroughly clean any foreign material from the contact area. Replace any parts as necessary.

To install:

13. Lubricate the cavity between the lips of the oil seal with wheel bearing lubricant.

14. Install the oil seal into the bore using a driver.

➡ **Do not hammer the pinion flange onto the pinion stem.**

15. Install the pinion flange. Use the alignment marks in the installation of the pinion flange.

16. Install the washer and a new nut. Tighten the nut on the pinion stem as close as possible to the alignment marks without going past the marks. Use the alignment marks and the thread count as a reference. Tighten the nut a little at a time. Turn the pinion flange several times after each tightening in order to seat the rollers.

17. Measure the torque required to rotate the pinion flange. Compare this to the original torque. Tighten the pinion nut, in small increments, until the rotating torque is 3 inch lbs. (0.35 Nm) GREATER than the original torque.

18. Align the propeller shaft with the alignment marks. Connect the propeller shaft.

19. Install the axle shafts (10.5 inch, 11.5 inch axle).

20. Install the rear brake calipers and rotors (8.6 inch, 9.5 inch axles).

21. Install the tire and wheel assemblies (8.6 inch, 9.5 inch axles).

Express and Savana

1. Raise the vehicle on a hoist.

➡ **Observe and accurately mark the positions of all driveline components relative to the propeller shaft and axles prior to disassembly. These components include the propeller shaft, drive axles, pinion flanges, output shafts, etc. Reassemble all components in the exact relationship the components had to each other during removal. Follow the specifications and the torque values. Follow any measurements made prior to removal.**

2. Accurately mark the installed position of the rear propeller shaft. Remove the rear propeller shaft.

3. Measure the torque required to turn the pinion. Record the torque number measurement which gives the combined pinion bearing, seal, carrier bearing, axle bearing and seal preload.

4. Make and accurate alignment mark on the pinion flange. Record the number of exposed threads on the pinion stem.

5. Remove the pinion flange nut and the washer. Use a container in order to catch any lubricant.

➡ **Use care not to damage any of the machined surfaces.**

6. Remove the pinion flange.

➡ **The pinion flange has an oil seal that is part of the pinion flange assembly. The pinion flange must be inspected to ensure that the seal is not damaged.**

7. Pry the oil seal from the bore.

8. Thoroughly clean any foreign material from the contact area. Replace any parts as necessary.

To install:

9. Lubricate the cavity between the lips of the oil seal with wheel bearing lubricant.

10. Install the oil seal into the bore using a driver.

➡ **Do not hammer the pinion flange onto the pinion stem.**

11. Install the pinion flange. Use the alignment marks in the installation of the pinion flange.

12. Install the washer and a new nut. Tighten the nut on the pinion stem as close as possible to the alignment marks without going past the marks. Use the alignment marks and the thread count as a reference. Tighten the nut a little at a time. Turn the pinion flange several times after each tightening in order to seat the rollers.

13. Measure the torque required to rotate the pinion flange. Compare this to the original torque. Tighten the pinion nut, in small increments, until the rotating torque is 3 inch lbs. (0.35 Nm) GREATER than the original torque.

14. Align the propeller shaft with the alignment marks. Connect the propeller shaft.

15. Install the retainers and the bolts. Tighten the bolts to 15 ft. lbs. (20 Nm).

16. Fill the rear axle.

17. Lower the vehicle.

Front Drive Axle Differential Carrier

REMOVAL & INSTALLATION

Avalanche, Sierra and Silverado

1. Turn the steering wheel all the way to the left.

2. Raise and support the vehicle.

3. Place jack or utility stands at the rear end of the vehicle.

4. Remove the engine protection shield.

5. Remove the front differential carrier assembly shield, if equipped.

6. Drain the differential carrier assembly, if necessary.

7. Disconnect the front propeller shaft from the differential carrier assembly.

8. Remove the relay rod.

9. Support the differential carrier assembly with a transmission jack.

10. Remove the wheel drive shaft inboard flange bolts from the inner axle shaft, both sides.

11. Disconnect the electrical connector from the front axle actuator, S4WD axle only.

12. Disconnect the wire harness from the inner axle shaft housing and differential, S4WD axle only.

13. Disconnect the vent hose from the differential carrier assembly.

14. Remove the inner axle housing nuts and washers from the bracket.

15. Remove the differential carrier assembly upper mounting bolt and the nut.

16. Pivot the differential carrier assembly forward and down on the lower mount bolt while it is being supported by the transmission jack.

17. Secure the differential carrier assembly to the jack.

18. Remove the differential carrier assembly lower mounting bolt and the nut.

19. Remove the differential carrier.

To install:

20. Install the differential carrier.

21. Install the differential carrier assembly lower mounting bolt and the nut. Do not tighten the bolt at this time.

22. Pivot the differential carrier assembly up and back on the lower mount bolt while it is being supported by the transmission jack.

23. Install the differential carrier assembly upper mounting bolt and the nut.

24. Install the inner axle housing washers and nuts to the bracket.

25. Tighten the inner axle housing nuts

and the upper and the lower differential carrier assembly bolts to 75 ft. lbs. (100 Nm).

26. Connect the vent hose to the differential carrier assembly.

27. Remove the transmission jack.

28. Connect the electrical connector from the front axle actuator, S4WD axle only.

29. Connect the wire harness from the inner axle shaft housing and differential, S4WD axle only.

30. Install the wheel drive shaft inboard flange to inner axle shaft bolts, both sides, and tighten to 58 ft. lbs. (79 Nm).

31. Install the relay rod.

32. Install the front propeller shaft to the differential carrier assembly.

33. Fill the differential carrier assembly.

34. Install the front differential carrier assembly shield, if equipped.

35. Install the engine protection shield.

36. Lower the vehicle.

Express and Savana

1. Raise and support the vehicle.

2. Place a jack or utility stands at the rear end of the vehicle.

3. Drain the fluid from the differential carrier assembly .

4. Disconnect the propeller shaft from the differential carrier assembly.

5. Support the propeller shaft as necessary.

6. Remove the engine protection shield.

7. Remove the front engine crossmember.

8. Place a transmission jack under the differential carrier assembly. Secure and support the differential carrier assembly to the transmission jack.

9. Place an alignment mark between the wheel drive shafts and the inner shaft flanges.

10. Remove the wheel drive shaft to inner shaft bolts.

11. Disconnect the left side wheel drive shaft from the inner shaft.

12. Disconnect the right side wheel drive shaft from the inner shaft.

13. Disconnect the differential carrier assembly vent hose from the differential carrier assembly.

14. Remove the upper inner shaft housing mounting bolt.

15. Remove the upper differential carrier assembly mounting bolt.

16. Remove the differential carrier assembly.

To install:

17. Install the differential carrier assembly to the vehicle.

18. Connect the vent hose to the differential carrier assembly.

19. Install the differential carrier assembly upper mounting bolt and the nut. Do not tighten the bolt to specification at this time.

20. Install the inner shaft upper mounting bolt and the nut. Do not tighten the bolt to specification at this time.

21. Install the front engine crossmember and tighten the bolts to 48 ft. lbs. (65 Nm).

22. Tighten the upper differential carrier assembly and inner shaft housing mounting bolts to 63 ft. lbs. (85 Nm).

23. Connect the left side wheel drive shaft to the inner shaft flange. Align the marks made during removal.

24. Clean the threads of the left and right side drive shaft bolt with denatured alcohol or equivalent and allow to dry.

25. Apply threadlocker to the threads of the bolts.

26. Install the right and left side wheel drive shaft to inner shaft bolts and tighten to 37 ft. lbs. (50 Nm).

27. Install the engine protection shield.

28. Connect the propeller shaft to the differential carrier assembly.

29. Fill the differential carrier assembly with the proper lubricant.

30. Remove the jack or utility stands.

31. Lower the vehicle.

Rear Drive Axle Housing

REMOVAL & INSTALLATION

Avalanche, Sierra and Silverado

2002–04 MODELS

1. Remove or disconnect the following:
   - Axle lubricant
   - Propeller shaft
   - Wheel assemblies
   - Parking brake cable
   - Brake calipers
   - Shock absorbers from the axle brackets
   - Vent hose from the rear axle vent fitting
   - Nuts and the washers from the U-bolts
   - U-bolts, the spring plates and the spacers from the axle assembly

2. Lower the axle assembly.

To install:

3. Place the rear axle assembly under the vehicle. Align the rear axle assembly with the springs. Connect the spacers, the spring plates and the U-bolts to the rear axle. Raise the rear axle assembly into position.

4. Install or connect the following:
   - Washers and nuts to the U-bolts. Tighten the nuts to 59 ft. lbs. (80 Nm) first, then to 89 ft. lbs. (120 Nm).
   - Vent hose to the rear axle vent fitting
   - Shock absorbers to the rear axle
   - Brake calipers
   - Parking brake cable
   - Wheel assemblies
   - Propeller shaft

5. Fill the rear axle.

6. Bleed the brake system.

7. Remove the supports and lower the vehicle.

2005–06 MODELS

1. Raise and support the vehicle.

2. Place jack stands at the front end of the vehicle.

3. Support the axle with jack stands.

4. Remove the tire and wheel assemblies.

5. Disconnect the upper stabilizer shaft link from the frame.

6. Reference mark the rear propeller shaft to the rear axle pinion yoke.

7. Disconnect the propeller shaft from the axle. Support the propeller shaft as necessary.

8. Disconnect the lower mount of the shock absorbers.

9. Disconnect the vent hose.

10. Disconnect the park brake cables.

11. Disconnect the junction block and brake pipe.

12. Remove and wire the calipers out of the way.

13. Remove the nuts and the washers from the spring assembly U-bolts.

14. Remove the U-bolts, the anchor plates and the spacers from the axle.

15. Remove the axle with the aid of a hydraulic assist.

16. Remove the stabilizer shaft U-bolt nuts and the U-bolts from the axle if necessary.

17. Remove the stabilizer shaft from the axle if necessary.

To install:

18. Install the stabilizer shaft to the axle if necessary.

19. Install the stabilizer shaft clamps, the U-bolts, and the nuts if necessary. Do not torque the stabilizer shaft U-bolt nuts at this time.

20. Place the axle under the vehicle.

21. Raise the axle to the springs with the aid of a hydraulic assist. Align the axle with the springs.

22. Install the spacers, the anchor plates and the U-bolts.

23. Install the washers if equipped and the nuts to the U-bolts and tighten in a crisscross pattern to 110 ft. lbs. (150 Nm).

24. Install the stabilizer shaft link to the frame if necessary.

25. Install the stabilizer shaft link bolt and the nut and tighten to 70 ft. lbs. (95 Nm).

26. Tighten the stabilizer shaft U-bolt nuts to 24 ft. lbs. (32 Nm).

27. Install the brake calipers.

28. Install the brake pipe fitting brackets.

29. Install the brake pipe.

30. Install the brake pipe junction block.

31. Connect the park brake cables.

32. Connect the vent hose to the axle vent fitting.

33. Install the shock absorbers to the lower mount bracket.

34. Install the shock absorber bolts and the nuts and tighten to 70 ft. lbs. (95 Nm).

35. Install the propeller shaft to the pinion yoke. Align the reference marks made during removal.

36. Install the propeller shaft yoke retaining clamps and the bolts and tighten to 18 ft. lbs. (25 Nm).

37. Install the tire and wheel assemblies.

38. Fill the axle with lubricant.

39. Remove the jack stands.

40. Lower the vehicle.

Express and Savana

2002–04 MODELS

1. Drain the lubricant from the axle housing

2. Remove or disconnect the following:
   - Driveshaft
   - Wheel, the brake drum or hub and the drum assembly
   - Parking brake cable from the lever and at the brake flange plate
   - Hydraulic brake lines from the connectors
   - Shock absorbers from the axle brackets
   - Vent hose from the axle vent fitting, if equipped
   - Height sensing and brake proportional valve linkage, if equipped
   - Stabilizer shaft, if equipped

3. Support the axle assembly with a jack.
   - U-bolts
   - Spring plates and spacers
   - Axle assembly

To install:

4. Raise the axle assembly into position.

5. Install or connect the following:
   - U-bolts
   - Spring plates and spacers
   - Nuts and washers on the U-bolts. Torque to 81 ft. lbs. (110 Nm).
   - Stabilizer shaft, if equipped
   - Height sensing and brake proportional valve linkage, if equipped
   - Vent hose at the axle vent fitting, if equipped
   - Shock absorbers at the axle brackets
   - Hydraulic brake lines
   - Parking brake cable
   - Wheels
   - Driveshaft

6. Fill the axle housing.

2005–06 MODELS

1. Drain the lubricant from the axle housing

2. Raise and support the vehicle.

3. Place jack stands at the front end of the vehicle.

4. Support the rear axle with jack or utility stands.

5. Remove the wheel and tire assemblies.

6. Disconnect the electrical connector from the rear wheel speed sensor, if equipped.

7. Remove the brake calipers from the rear axle. It is not necessary to disconnect the brake hose from the caliper. Support the brake calipers as necessary and set aside.

8. Disconnect the propeller shaft from the rear axle. Support the propeller shaft as necessary and set aside.

9. Disconnect the park brake cables.

10. Remove the rear axle vent hose from the axle.

11. Disconnect the lower portion of the shock absorber from the rear axle.

12. Remove the rear spring U-bolts, the rear spring anchor plates and the rear spring spacers from the vehicle.

13. Remove the rear axle from the vehicle.

To install:

14. Install the rear axle to the vehicle. Support the rear axle with jack or utility stands.

15. Install the rear spring U-bolts, the rear spring anchor plates and the rear spring spacers to the vehicle.

16. Install the leaf spring U-bolt nuts. Tighten the leaf spring U-bolt mounting nuts to 63 ft. lbs. (85 Nm) on 1500 models, 103–108 ft. lbs. (140–147 Nm) on 2500 Series and 3500 Series Cargo and Passenger Vans and Cutaways.

17. Tighten the front hanger bracket nut to 80 ft. lbs. (110 Nm), plus an additional 200 degrees.

18. Tighten the spring shackle mounting nuts to 66 ft. lbs. (90 Nm).

19. Connect the lower portion of the shock absorber to the rear axle.

20. Install the rear axle vent hose to the rear axle.

21. Connect the park brake cable to the rear axle.

22. Connect the propeller shaft to the rear axle.

23. Install the brake calipers.

24. Connect the electrical connector to the rear wheel speed sensor, if equipped.

25. Install the wheel and tire assemblies.

26. Fill the rear axle with the proper axle lubricant.

27. Remove the jack stands.

28. Lower the vehicle.

STEERING AND SUSPENSION

Air Bag

☀☀ CAUTION

Some vehicles are equipped with an air bag system. The system must be disabled before performing service on or around system components, steering column, instrument panel components, wiring and sensors. Failure to follow safety and disabling procedures could result in accidental air bag deployment, possible personal injury and unnecessary system repairs.

PRECAUTIONS

Several precautions must be observed when handling the inflator module to avoid accidental deployment and possible personal injury.

• Never carry the inflator module by the wires or connector on the underside of the module

• When carrying a live inflator module, hold securely with both hands, and ensure that the bag and trim cover are pointed away

• Place the inflator module on a bench or other surface with the bag and trim cover facing up

• With the inflator module on the bench, never place anything on or close to the module that may be thrown in the event of an accidental deployment

DISARMING & ARMING

2002 Models

1. Turn the front wheels to the straight-ahead position.

2. Turn the ignition switch to the **LOCK** position and remove the key.

➡ **If the key is in the RUN position when the Air Bag fuse is removed or open (blown), the Air Bag warning lamp in the dash will light up. This is normal operation, not a sign of a malfunction.**

3. Remove the Air Bag fuse from the fuse panel.

4. Remove the drivers side knee bolster and unplug the yellow 2-pin connector at the base of the steering column to disarm the driver's side Air Bag. Remove the passenger side knee bolster and unplug the

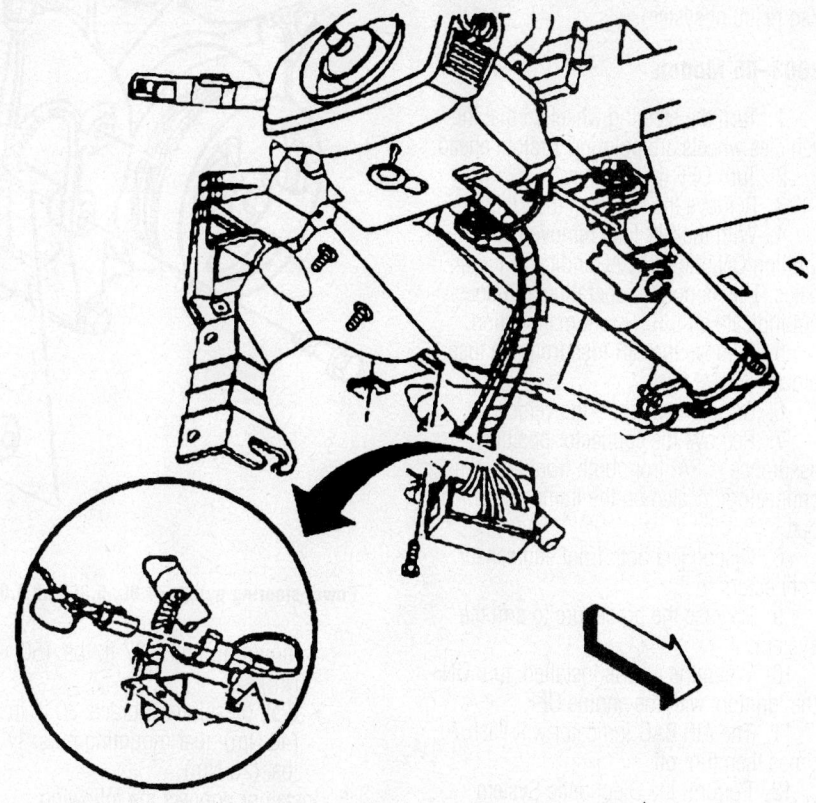

Typical air bag connector location—driver's side

7924KG30

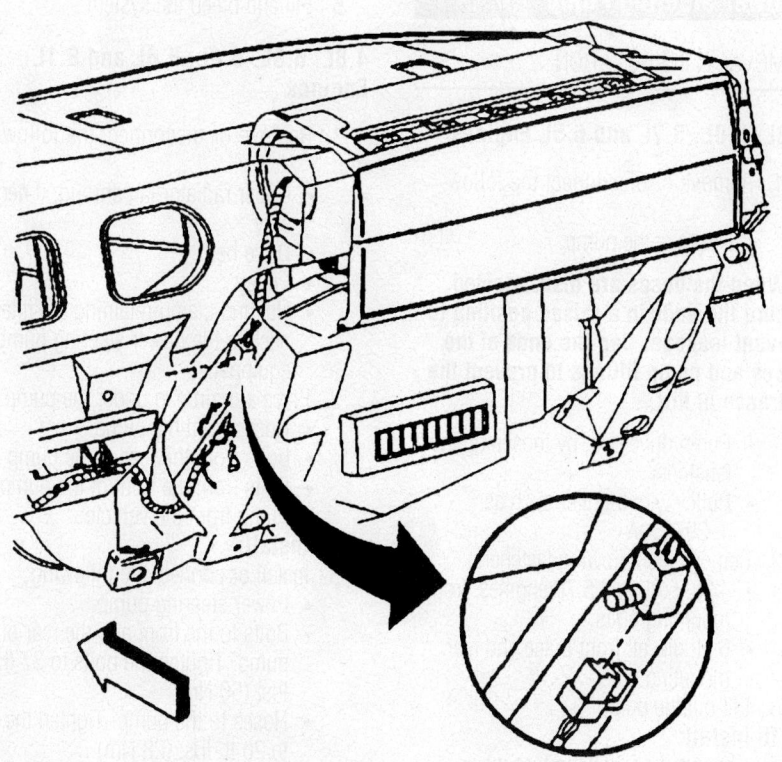

Typical air bag connector location—passenger's side

7924KG31

yellow 2-pin connector to disable the passenger's side Air Bag.

5. Reverse the procedure to arm the Air Bag restraint system.

2003–06 Models

1. Turn the steering wheel so that the vehicles wheels are pointing straight ahead.
2. Turn OFF the ignition.
3. Remove the key from the ignition.
4. With the SIR fuse removed and the ignition ON, the AIR BAG indicator illuminates. This is normal operation and does not indicate an SIR system malfunction.
5. Remove the SIR fuse from the fuse block.
6. Raise and support the vehicle.
7. Remove the connector position assurance (CPA) from both front end sensor connectors located on the frame crossmember.
8. Disconnect both front end sensor connectors
9. Reverse the procedure to arm the system.
10. When the fuse is installed, turn ON the ignition, with the engine OFF.
11. The AIR BAG indicator will flash 7 times then turn off.
12. Perform the Diagnostic System Check if the AIR BAG indicator does not operate as described.

Power Steering Pump

REMOVAL & INSTALLATION

4.3L, 5.0L, 5.7L and 6.5L Engines

1. Remove or disconnect the following:
   • Hoses at the pump.

➡**When the hoses are disconnected, secure the ends in a raised position to prevent leakage. Cap the ends of the hoses and pump fittings to prevent the entrance of dirt.**

   • Pump drive belt, by loosening the tensioner
   • Pulley with a puller such as J–29785–A
2. Remove the following fasteners:
   • 4.3L, 5.0L and 5.7L engines: front mounting bolts
   • 6.5L diesel: front brace and rear mounting nuts
3. Lift out the pump.

To install:
4. Observe the following torques:
   • 4.3L, 5.0L, and 5.7L engine front

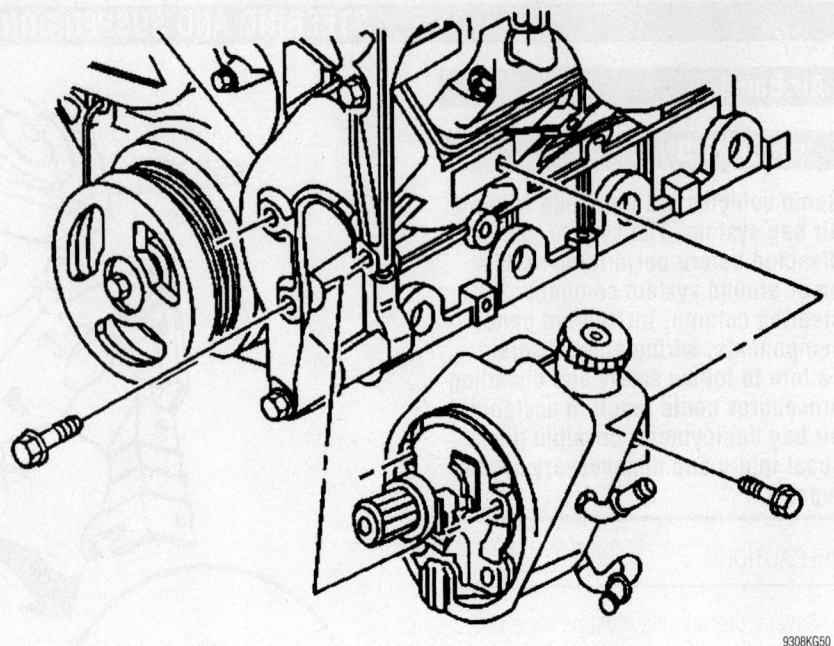

Power steering pump—4.8L, 5.3L and 6.0L engines

9308KG50

mounting bolts: 37 ft. lbs. (50 Nm)
   • 6.5L diesel, front brace: 30 ft. lbs. (40 Nm); rear mounting nuts: 17 ft. lbs. (23 Nm)
5. Install or connect the following:
   • Pulley with J–25033–B
   • Drive belt
   • Hoses
6. Fill and bleed the system.

4.8L, 5.3L, 6.0L, 6.6L and 8.1L Engines

1. Remove or disconnect the following:

   • Upper radiator fan shroud, if necessary
   • Drive belt
   • Pulley
   • Nut and clamp retaining the filler neck to the power steering pump, if equipped
2. Place a drain pan under the pump
   • Hoses from the pump
   • Bolts from the rear of the pump
   • Bolts from the front of the pump
   • Pump from the vehicle
To install:
3. Install or connect the following:
   • Power steering pump
   • Bolts to the front and the rear of the pump. Tighten the bolts to 37 ft. lbs. (50 Nm)
   • Hoses to the pump. Tighten the nut to 20 ft. lbs. (28 Nm)
   • Nut and clamp retaining the filler

neck to the power steering pump, if equipped
   • Pulley; install with 0.020 in. (0.5mm) play
   • Drive belt
   • Upper radiator shroud
4. Fill and bleed the power steering system.

Recirculating Ball Power Steering Gear

REMOVAL & INSTALLATION

1. Raise the vehicle.
2. Remove the shield.
3. Place a drain pan below the steering gear.
4. Remove or disconnect the following:
   • Hoses from the steering gear
   • Intermediate shaft from the steering gear
   • Pitman arm from the relay rod and steering gear
   • Steering gear frame bolts and the steering gear
To install:
5. Place the steering gear in position.
6. Install or connect the following:
   • Steering gear to the frame bolts and tighten to 110 ft. lbs. (150 Nm)
   • Pitman arm to steering gear and tighten to 184 ft. lbs. (250 Nm)
   • Intermediate shaft
7. Remove the plugs and the caps from the steering gear and the hoses.

3. Lower the vehicle.
4. Fill and bleed the power steering system.

Express and Savana

1. Raise and support the vehicle.
2. Place a drain pan under the vehicle.
3. Remove the front tires and wheels.
4. Remove the engine protection shield.
5. Remove the outer tie rods retaining nuts.
6. Disconnect the outer tie rods from the steering knuckles.
7. Disconnect the lower intermediate shaft from the power steering gear.
8. Disconnect the power brake booster outlet hose and power steering gear outlet hose or the power steering cooler hose from the power steering gear.
9. Cap the ends of the hoses and the power steering gear fittings in order to prevent the entrance of dirt.
10. Remove the power steering gear from the vehicle.

To install:

11. Install the power steering gear to the vehicle.
12. Install the power steering gear to the frame mounting bolts and tighten to 110 ft. lbs. (150 Nm).
13. Remove the caps from the steering gear and hoses.

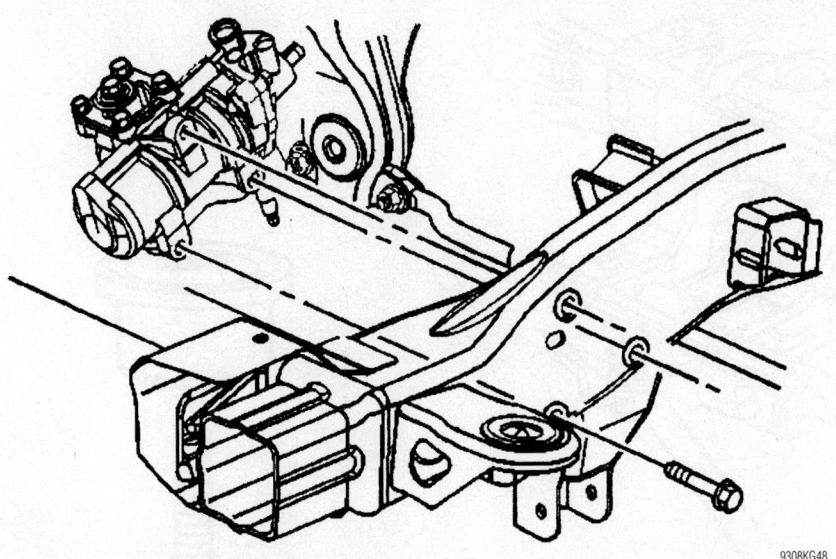

Recirculating ball steering gear—Silverado shown

- Hoses to the steering gear. Tighten the hose connection to 20 ft. lbs. (28 Nm).
- Shield
8. Fill and bleed the system.
9. Lower the vehicle.

Rack & Pinion Steering Gear

REMOVAL & INSTALLATION

Avalanche, Sierra and Silverado

1. Remove or disconnect the following:
   - Wheel assemblies
   - Engine shield, if equipped
   - Stabilizer shaft
   - Power steering high and low pressure lines
   - Coupler clamp bolt from the intermediate shaft
   - Outer tie rod ends from steering knuckle
   - Intermediate shaft from the rack and pinion assembly
   - Rack and pinion assembly mounting nuts, washers and bolts
   - Rack and pinion assembly from the vehicle

To install:
2. Install or connect the following:
   - Rack and pinion assembly into the vehicle
   - Rack and pinion assembly mounting bolts, washers and nuts. Tighten the nuts to 136 ft. lbs. (185 Nm).
   - Intermediate shaft to the rack and pinion assembly

- Coupler clamp bolt to the intermediate shaft. Tighten the bolt to 33 ft. lbs. (45 Nm).
- Low pressure hose
- High pressure hose. Tighten the hoses to 20 ft. lbs. (27 Nm).
- Outer tie rod ends
- Engine protection shield, if equipped
- Stabilizer shaft
- Wheels

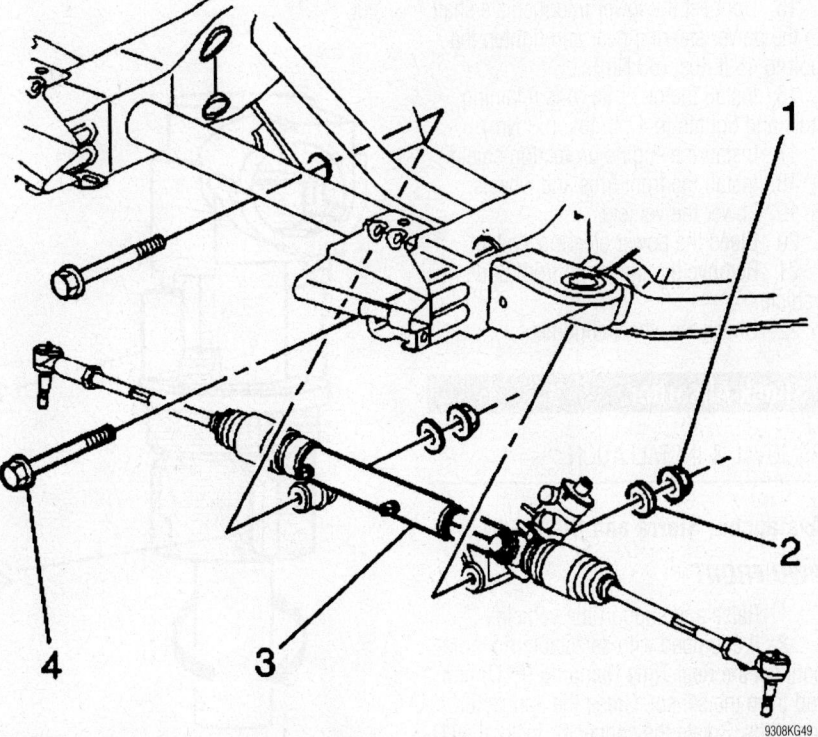

Rack and pinion steering gear—Silverado

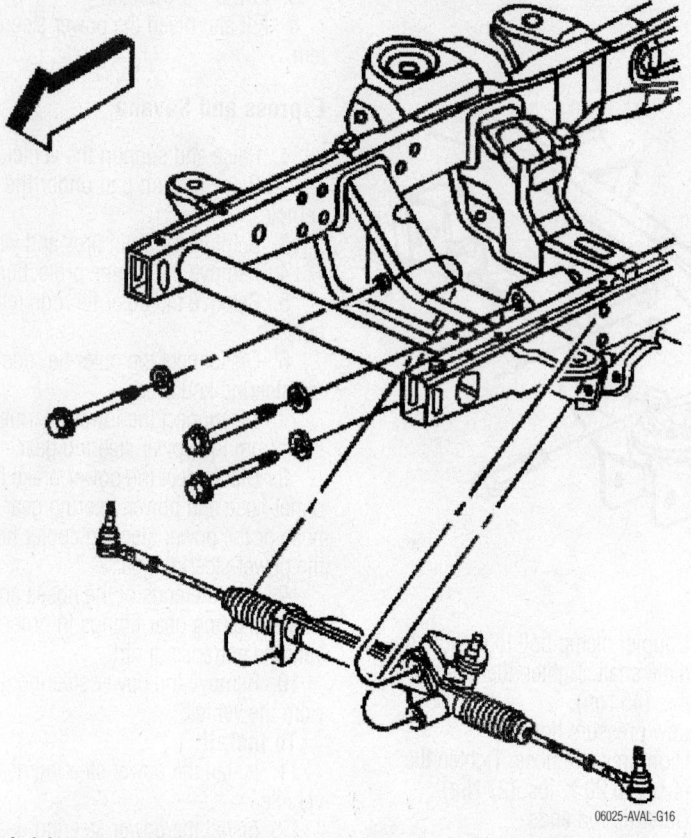

06025-AVAL-G16

Rack and pinion steering gear mounting—Express/Savana

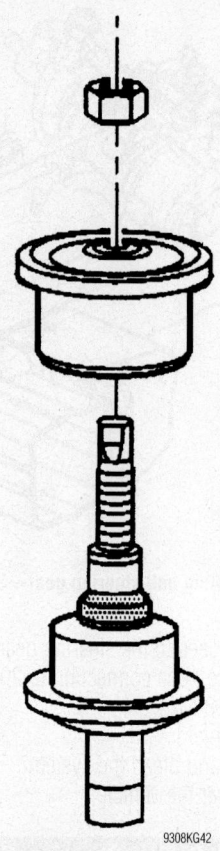

9308KG42

Upper shock insulator—Silverado

14. Connect the power brake booster outlet hose and power steering gear outlet hose or the power steering cooler hose to the power steering gear.

15. Connect the lower intermediate shaft to the power steering gear and tighten the bolt to 45 ft. lbs. (60 Nm).

16. Install the outer tie rods retaining nuts and tighten to 47 ft. lbs. (63 Nm).

17. Install the engine protection shield.

18. Install the front tires and wheels.

19. Lower the vehicle.

20. Bleed the power steering system.

21. Remove the drain pan from under the vehicle.

22. Check the wheel alignment.

Shock Absorber

REMOVAL & INSTALLATION

Avalanche, Sierra and Silverado

2WD FRONT

1. Raise and support the vehicle.

2. If equipped with selectable ride, disconnect the Real Time Damping (RTD) link rod from the sensor. Grasp the connector lock tabs. Rotate the connector lock tabs (1) and (2) counter-clockwise until the connector is unlocked. Disengage the connector from the tennon by firmly pulling the connector up. Hold the tennon end with a wrench while removing the nut. Remove the nut.

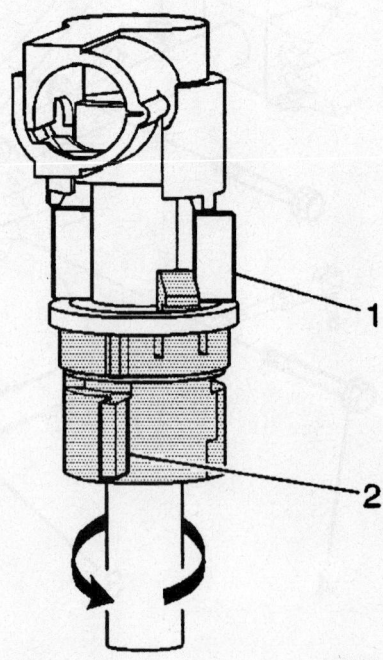

9308KG43

RTD connector—Silverado

3. Remove the upper insulator. Do not discard the plastic pilot ring.

4. Remove the shock absorber mounting bolts at the lower control arm. Remove the shock absorber through the lower control arm from below.

To install:

5. Support the lower control arm with a suitable jack in order to align the tennon with the mounting hole if equipped with selectable ride.

6. Install or connect the following:
- Shock absorber through the lower control arm from below
- Tennon through the mounting hole in the upper spring pocket

7. Align the shock absorber with the mounting holes in the lower control arm.
- Shock absorber mounting bolts to the lower control arm. Tighten to 18 ft. lbs. (25 Nm).

➡The upper insulators are substantially larger that the lower insulators. The upper insulator must be installed above the shock mounting bracket on the frame. The plastic pilot ring will assist the alignment of the isolators.

- Upper insulator to the shock absorber

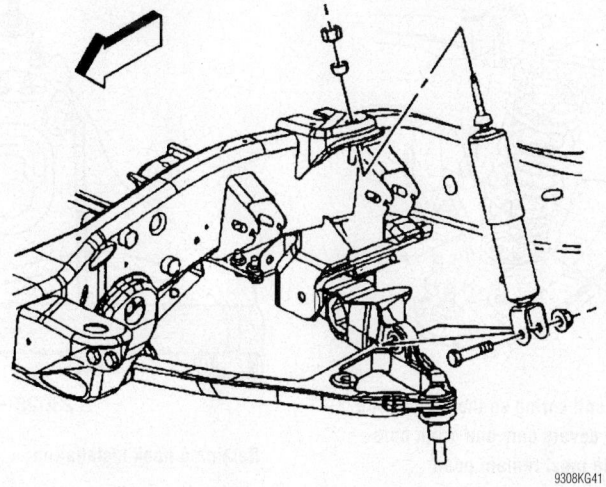

Shock absorber removal—4WD Silverado

9308KG41

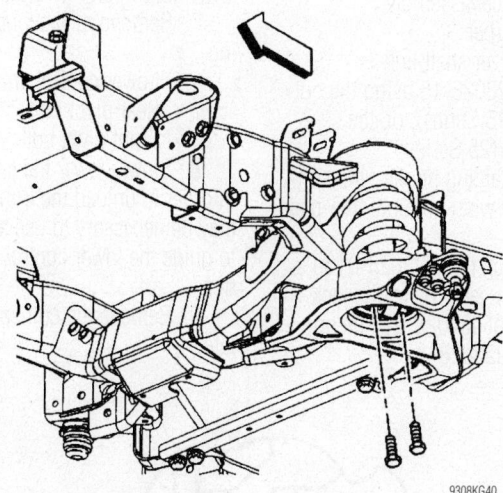

Shock absorber removal—2WD Silverado

9308KG40

- Nut to the tennon end. Do not tighten the nut.
- RTD link rod to the sensor (if equipped).

8. Remove the safety stands.

9. Lower the vehicle. Hold the tennon end with a wrench while torquing the nut. Tighten the nut to 15 ft. lbs. (20 Nm).

10. Connect the electrical connector using the following procedure:

   a. Verify that the connector is unlocked.

   b. Align the connector so that the tabs are perpendicular to the wrench flats on the tennon end.

   c. Engage the connector to the tennon by firmly pushing the connector down.

   d. Grasp the connector lock tabs. Rotate the connector counter clockwise.

11. The connector is locked into place when you hear an audible snap and the tabs are aligned.

4WD FRONT

1. Raise and support the vehicle.
2. Remove or disconnect the following:
   - Real Time Damping (RTD) link rod from the sensor, if equipped
   - Electrical connector, if equipped with selectable ride. Grasp the connector lock tabs. Rotate the connector tabs counter clockwise until the connector is unlocked. Disengage the connector from the tennon by firmly pulling the connector up. Hold the tennon end with a wrench while removing the nut. Remove the nut.
   - Upper insulator. Do not discard the plastic pilot ring.
   - Shock absorber mounting bolt at the lower control arm

➡**The lower shock mounting bushing is serviceable by driving the bushing out with the appropriate tool.**

- Shock absorber

To install:

3. Install the shock absorber. Insert the stem through the hole in the shock bracket on the frame. Align the shock absorber with the mounting holes in the lower control arm.

4. Install or connect the following:
   - Shock absorber through bolt to the lower control arm
   - Shock absorber through bolt nut and tighten to 59 ft. lbs. (80 Nm)

➡**The upper insulators are substantially larger that the lower insulators. The upper insulator must be installed above the shock mounting bracket on the frame. The plastic pilot ring will assist the alignment of the isolators.**

- Upper insulator to the shock absorber
- Nut to the tennon end. Do not tighten the nut
- RTD link rod to the sensor, if equipped

5. Remove the safety stands. Lower the vehicle. Hold the tennon end with a wrench while torquing the nut. Tighten the nut to 15 ft. lbs. (20 Nm).

6. Connect the electrical connector using the following procedure if equipped with selectable ride.

   a. Verify that the connector is unlocked.

   b. Align the connector so that the tabs (1) are perpendicular to the wrench flats on the tennon end.

   c. Engage the connector to the tennon by firmly pushing the connector down.

   d. Grasp the connector lock tabs (1, 2). Rotate the connector counter clockwise. The connector is locked into place when you hear an audible snap and the tabs are aligned.

REAR

1. Raise and support the vehicle.
2. Remove or disconnect the following:
   - Electrical connector, if equipped with selectable ride
   - Upper shock absorber nut and bolt
   - Lower shock absorber nut and bolt
   - Shock absorber

To install:

3. Installation is the reverse of removal. Tighten the nuts to 70 ft. lbs. (95 Nm).

4. Connect the electrical connector if equipped with Selectable Ride. Remove the safety stands. Lower the vehicle.

Express and Savana

FRONT

1. Support the front of the vehicle safely under the lower control arms.
2. Remove or disconnect the following:
   - Tire and wheel assembly
   - Upper and lower shock absorber retaining fastener(s)

➡ **Vehicles equipped with quad shocks have a spacer between them.**

   - Shock absorber

To install:

3. Install or connect the following:
   - Shock absorber and fastener(s)
4. On 2-wheel drive vehicles torque the upper bolt to 15 ft. lbs. (20 Nm) and the lower bolts to 18 ft. lbs. (25 Nm). On 4-wheel drive vehicles, torque the lower bolt to 59 ft. lbs. (80 Nm). and the upper bolt to 15 ft. lbs. (26 Nm). Be sure the bolts are inserted in the proper direction. The upper bolt head should be forward; the bottom bolt head should be rearward.
   - Tire and wheel

REAR

1. The vehicle's weight should rest on correctly placed safety stands located under the frame. Chock the front wheels to prevent vehicle movement.
2. Support the rear axle with a floor jack.
3. If the vehicle is equipped with air lift type shocks, bleed the air from the lines and disconnect the line from the shock absorber.
4. Remove or disconnect the following:
   - Shock absorber at the top by removing the 2 mounting bolts/nuts from the frame bracket
   - Nut, washers and bolt from the bottom mount
   - Shock

To install:

5. Install or connect the following:
   - Shock
   - Upper mounting nuts/bolts and tighten to 18 ft lbs. (25 Nm).
   - Lower mounting bolt/nuts and tighten to 60 ft lbs. (80 Nm).
6. Check that no parts such as exhaust components bind on the shock absorbers.

Coil Springs

REMOVAL & INSTALLATION

Avalanche, Sierra and Silverado

FRONT

1. Raise and support the vehicle.
2. Remove or disconnect the following:

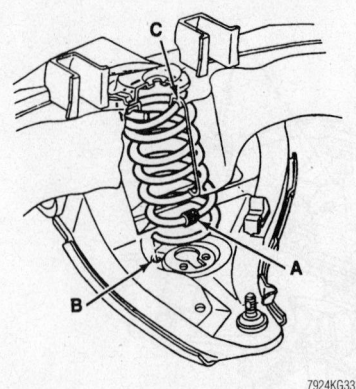

Position the coil spring so the bottom end of the spring covers only one drain hole—the other hole must remain open

   - Engine protection shield
   - Frame cross bar (25 series only)
   - Tire and wheel assembly
   - Shock absorber
   - Front stabilizer shaft link
3. Install tool J23028-15 using the outboard locating tab (15 Series), or, the inboard locating tab (25 Series).
4. Attach the retaining hook to the control arm. Tighten the wing nut until free-play is eliminated.
5. Securely attach tool J23028-01 to a suitable transmission jack. Raise the jack until the yokes of tool J23028-01 line up with the notches in J23028-15.

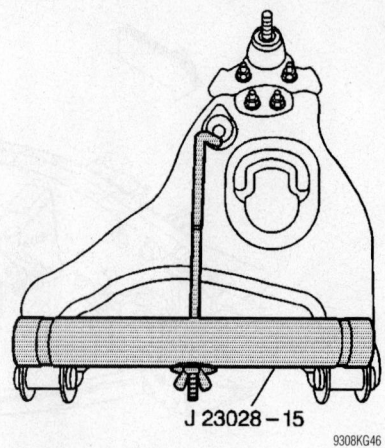

Retaining hook installation

6. Using the tools and the transmission jack, relieve the spring tension from the lower control arm pivot bolts.
7. Remove or disconnect the following:
   - Lower control arm pivot bolt nuts
   - Rear pivot bolt
   - Front pivot bolt
8. Slowly lower the transmission jack in order to unload the front coil spring. It may be necessary to use a pry bar in order to guide the lower control arm out of position.
9. Remove the coil spring and the insulator.

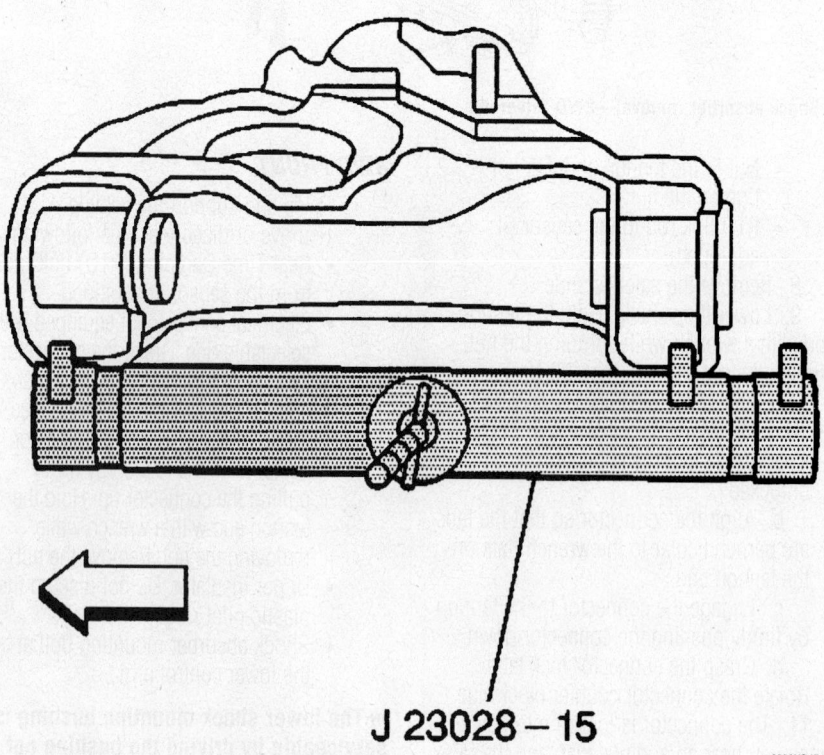

Installing J23028-15 on the 25 Series

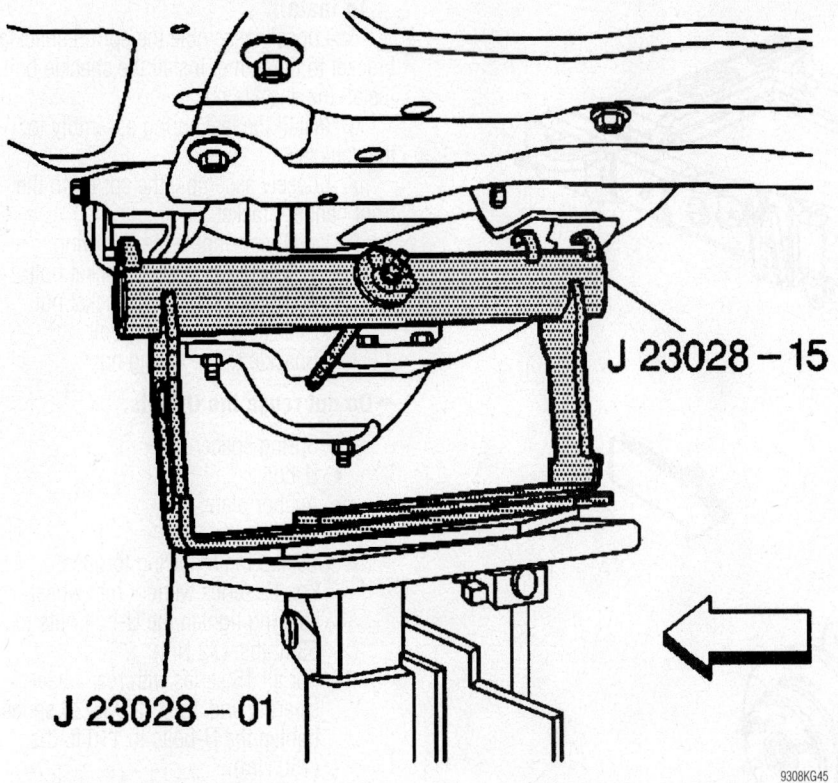

Tool attached to a jack

To install:

10. Install the coil spring and the insulator to the lower control arm.

11. Raise the transmission jack in order to compress the front coil spring. It may be necessary to use a pry bar in order to guide the lower control arm into position.

12. Install or connect the following:

- Front pivot bolt
- Rear pivot bolt
- Lower control arm pivot nuts. Tighten the pivot bolt nuts to 107 ft. lbs. (145 Nm).

13. Lower the jack. Remove the tool from the control arm.

- Front stabilizer shaft link

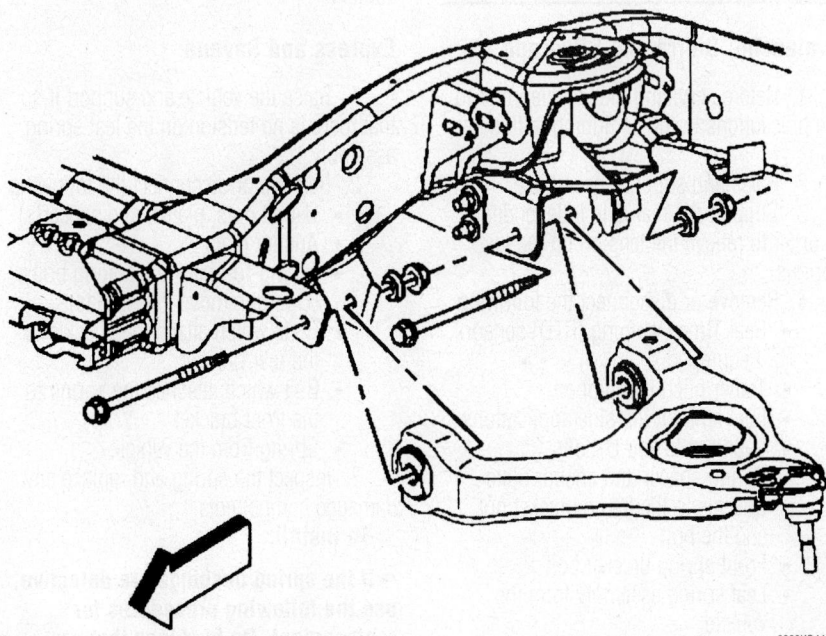

Lower control arm removal—Silverado

- Shock absorber
- Tire and wheel assembly
- Frame cross bar (25 series only). Tighten the nuts to 74 ft. lbs. (100 Nm).

14. Install the engine protection shield.

15. Remove the safety stands. Lower the vehicle.

REAR

1. Raise and support the vehicle.

2. Disconnect the Real Time Damping (RTD) sensor, if equipped.

3. Remove the lower shock absorber nuts and bolt from the rear axle.

4. Lower the rear axle until the springs are fully unloaded.

5. Remove the spring and the upper and lower insulators.

To install:

6. Position the spring and the upper and lower insulators.

7. Install the rear spring to the rear axle.

8. Raise the rear axle. Install the lower shock absorber nuts to the rear axle.

9. Connect the RTD sensor, if equipped.

10. Remove the rear axle support. Lower the vehicle.

Express and Savana

1. Support the vehicle safely under the frame rails. The control arms should hang freely.

2. Remove or disconnect the following:
- Wheel
- Shock absorber lower end mounting nut/bolt
- Stabilizer bar from the lower control arm

3. Support the lower control arm and install a spring compressor on the spring or chain the spring to the control arm as a safety precaution.

➡**If equipped with an air cylinder inside the spring, remove the valve core from the cylinder and expel the air by compressing the cylinder with a pry-bar. With the cylinder compressed, replace the valve core so the cylinder will stay in the compressed position. Push the cylinder as far as possible towards the top of the spring.**

4. Raise the front end to remove the tension from the lower control arm the bolts securing the control arm.

➡**The cross-shaft and lower control arm keeps the coil spring compressed. Use care when lowering the assembly.**

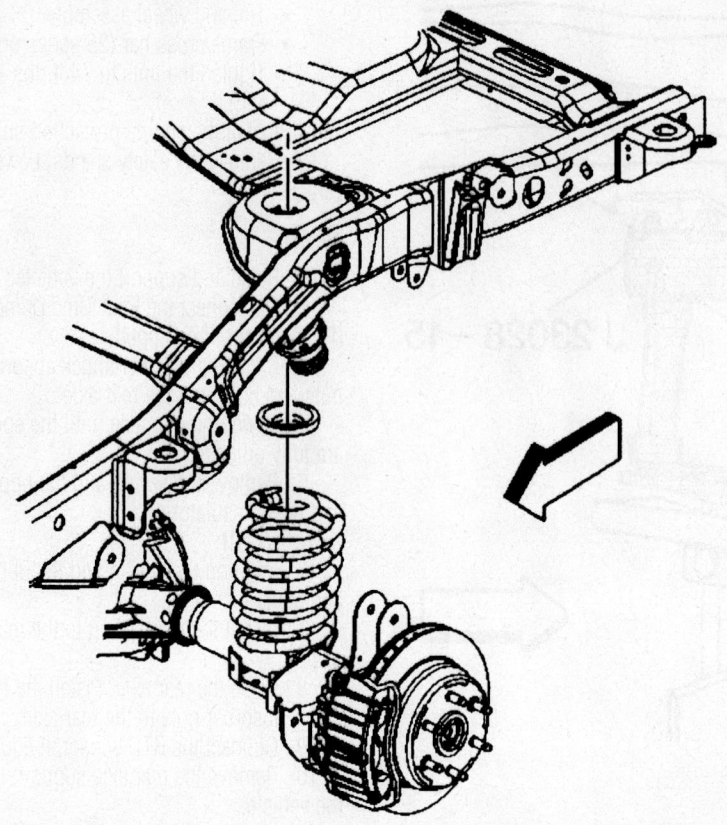

Rear coil spring removal—15 Series Silverado

9308KG24

5. Slowly lower the control arm until the spring can be removed. Be sure all compression is relieved from the spring.

6. If the coil spring was chained, remove the chain and spring. If a compressor was used, remove the spring and slowly release the compressor.

7. Remove the air cylinder, if equipped.

To install:

8. Install or connect the following:
- Air cylinder so the protector plate is toward the upper control arm. The Schrader valve should protrude through the hole in the lower control arm.
- Chain and spring or compress the spring and install the assembly

9. Slowly raise the control arm.
- Bolts securing the control arm. Tighten to 107 ft. lbs. (145 Nm).
- Stabilizer bar to the lower control arm. Torque the nuts to 24 ft. lbs. (34 Nm).
- Shock absorber at the lower end. Torque the nuts to 37 ft. lbs. (50 Nm).
- Air cylinders, inflate the cylinder to 60 psi (414 kpa) if equipped
- Wheel

10. Once the weight of the vehicle is on

the wheels, reduce the air cylinder pressure to 50 psi.

Leaf Springs

REMOVAL & INSTALLATION

Avalanche, Sierra and Silverado

1. Before servicing the vehicle, refer to the precautions in the beginning of this section.

2. Raise and support the vehicle.

3. Support the rear axle independently in order to relieve the tension on the leaf springs.

4. Remove or disconnect the following:
- Real Time Damping (RTD) sensors, if equipped
- Trailer hitch if equipped
- Fuel tank for left side applications
- U-bolt nuts and U-bolts
- Spring spacer and anchor plate
- Shackle to the frame bracket nut and the bolt
- Front spring bracket bolt
- Leaf spring assembly from the vehicle
- Shackle from the spring

To install:

5. Loosely assemble the spring shackle bracket to the frame. Install the shackle bolt. Install the shackle nut.

6. Install the leaf spring assembly to the vehicle.

7. Loosely assemble the spring to the front hanger bracket.

8. Install or connect the following:
- Front spring hanger bracket bolt
- Front spring hanger bracket nut
- Shackle to the spring bolt
- Shackle to the spring nut

➡**Do not reuse the U-bolts.**

- Spring spacer
- U-bolts
- Anchor plate
- U-bolt nuts

9. Observe the following torques:
- For 15 series without rear wheel steering tighten the U-bolt nuts to 53 ft. lbs. (72 Nm)
- For all 15 series with rear wheel steering and 15HD/25/35/36 series tighten the U-bolts to 110 ft. lbs. (150 Nm)
- Tighten the front hanger bracket nut to 110 ft. lbs. (150 Nm)
- Tighten the rear hanger bracket nut to 70 ft. lbs. (95 Nm)

10. Install the fuel tank for left side applications.

11. Install the trailer hitch if equipped.

12. Connect the RTD sensors, if equipped

13. Remove the rear axle support.

14. Remove the safety stands. Lower the vehicle.

Express and Savana

1. Raise the vehicle and support it so that there is no tension on the leaf spring assembly.

2. Remove or disconnect the following:
- U-bolt nuts, plates, and spacer(s)
- Anchor plate
- Spring-to-shackle retaining bolts. (Do not remove these bolts)
- Bolts which attach the shackle to the rear bracket
- Bolt which attaches the spring to the front bracket
- Spring from the vehicle

3. Inspect the spring and replace any damaged components.

To install:

➡**If the spring bushings are defective, use the following procedures for replacement. On bushings that are staked in place, the stakes must first**

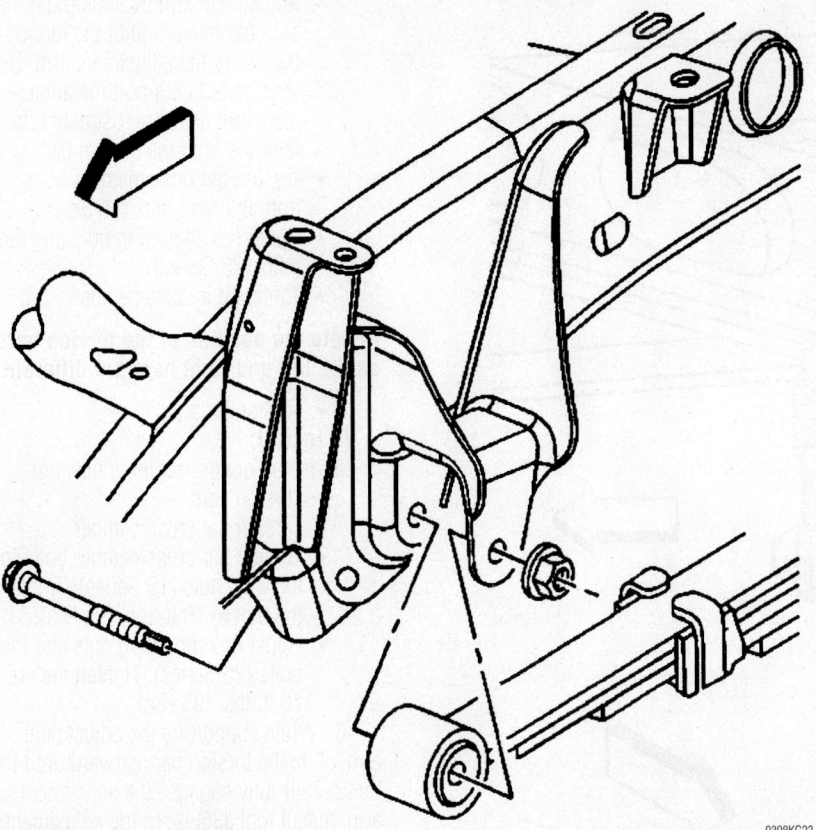

Rear leaf spring front shackle—Silverado

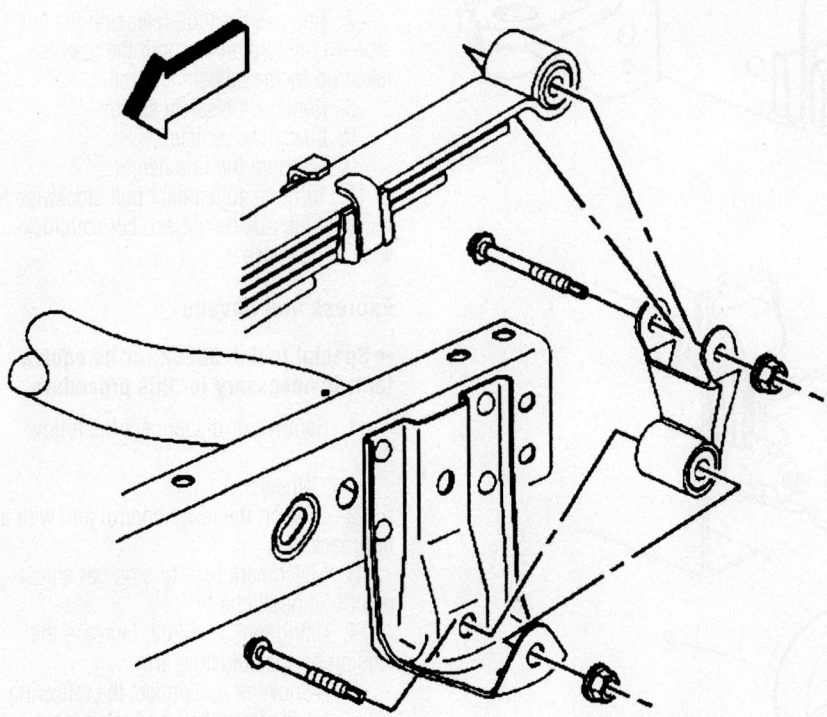

Rear leaf spring rear shackle—Silverado

be straightened. Using a press or vise, remove the bushing and install the new one. When a new, previously staked bushing is installed, stake it in 3 equally spaced locations.

4. Place the spring assembly onto the axle housing. Position the front and rear of the spring at the brackets. Raise the axle with a floor jack as necessary to make the alignments.

5. Install or connect the following:
- Front and rear brackets bolts loosely
- Spacers and spring plate
- New u-bolts, washers and nuts
- Anchor plate

6. Install the leaf spring U-bolt nuts. Tighten the leaf spring U-bolt mounting nuts to 63 ft. lbs. (85 Nm) on 1500 models, 103–108 ft. lbs. (140–147 Nm) on 2500 Series and 3500 Series Cargo and Passenger Vans and Cutaways.

7. Check that the hanger and shackle bolts are properly installed. All bolt heads should be inboard. Don't tighten them yet.

8. Using the floor jack, raise the axle until the distance between the bottom of the rebound bumper and its contact point on the axle is 182mm plus or minus 6mm.

9. When the spring is properly positioned, tighten all the hanger and shackle nuts to 70 ft. lbs. (95 Nm).

10. Tighten the following:
- Leaf spring-to-shackle nuts 15/25/35 series: 70 ft. lbs. (95 Nm)
- Leaf spring-to-shackle nuts C3HD series: 157 ft. lbs. (213 Nm)
- Shackle-to-bracket nuts C3HD series: 157 ft. lbs. (213 Nm)

Torsion Bars

REMOVAL & INSTALLATION

Avalanche, Sierra and Silverado

➥This procedure requires the removal of both torsion bars.

1. Raise and support the vehicle.
2. Mark the adjustment bolt setting. Install tool J36202 to the adjustment arm and the crossmember.
3. Increase the tension on the adjustment arm until the load is removed from the adjustment bolt and the adjuster nut.
4. Remove or disconnect the following:
- Adjustment bolt and the adjuster nut
- Tool, allowing the torsion bar to unload.

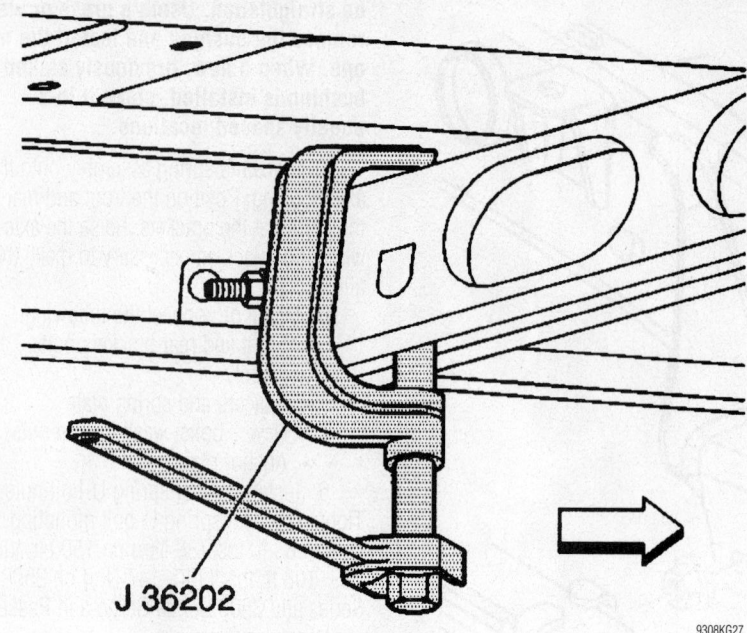

J 36202

Retainer installation—Silverado torsion bar

9308KG27

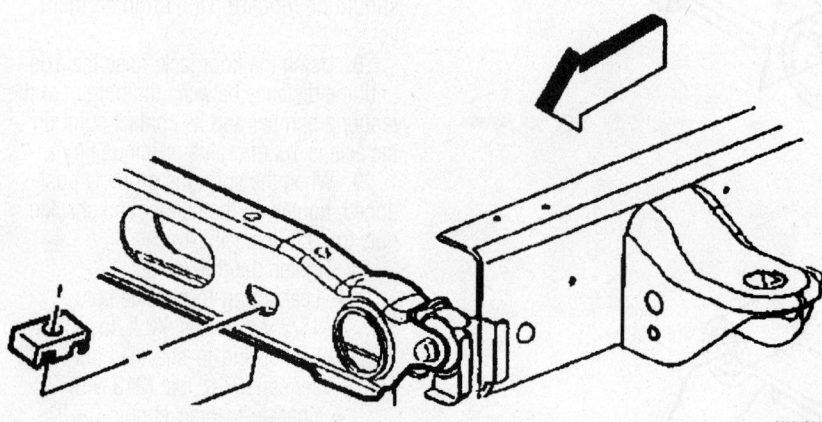

Adjuster nut removal—15 Series Silverado

9308KG26

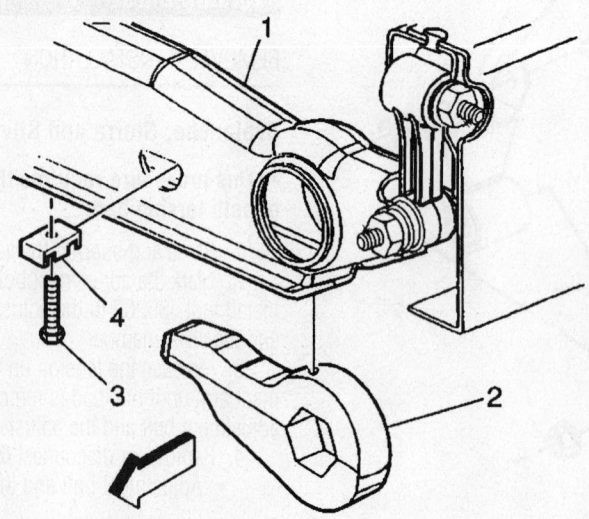

9308KG25

Adjuster bolt removal—15 Series Silverado

• Adjustment arm by sliding the torsion bar forward until the torsion bar clears the adjustment arm. Use your hand to support the adjustment arm as the adjustment arm releases from the torsion bar.
• Torsion bar crossmember bolts from the weld nuts (15 Series)
• Upper link mounting nuts and the bolts (25 Series)
• Torsion bar crossmember

➡ **Note the position of the torsion bars as the left and right bars are different.**

• Torsion bars

To install:

5. Install or connect the following:
   • Torsion bars
   • Torsion bar crossmember
   • Torsion bar crossmember bolts to the weld nuts (15 Series). Tighten the bolt to 70 ft. lbs. (95 Nm)
   • Upper link mounting nuts and the bolts (25 Series). Tighten the nut to 70 ft. lbs. (95 Nm)

6. While supporting the adjustment arm, slide the torsion bar rearward until the torsion bar fully engages the adjustment arm. Install tool J36202 to the adjustment arm and the crossmember. Increase the tension on the adjustment arm in order to load the torsion bar.
   • Adjustment bolt and the adjuster nut

7. Remove the tool, releasing the tension on the torsion bar until the load is taken up by the adjustment bolt.

8. Remove the safety stands.

9. Lower the vehicle.

10. Measure the ride height.

11. Turn the adjustment bolt clockwise to increase the ride height and counterclockwise to decrease it.

Express and Savana

➡ **Special tool J–36202, or its equivalent, is necessary for this procedure.**

1. Remove or disconnect the following:
   • Wheels

2. Support the lower control arm with a floor jack.

3. Matchmark both torsion bar adjustment bolt positions.

4. Using tool J–36202, increase the tension on the adjusting arm.

5. Remove or disconnect the following:
   • Adjustment bolt and retaining plate

6. Move the tool aside, and slide the torsion bars forward.
   • Adjusting arms

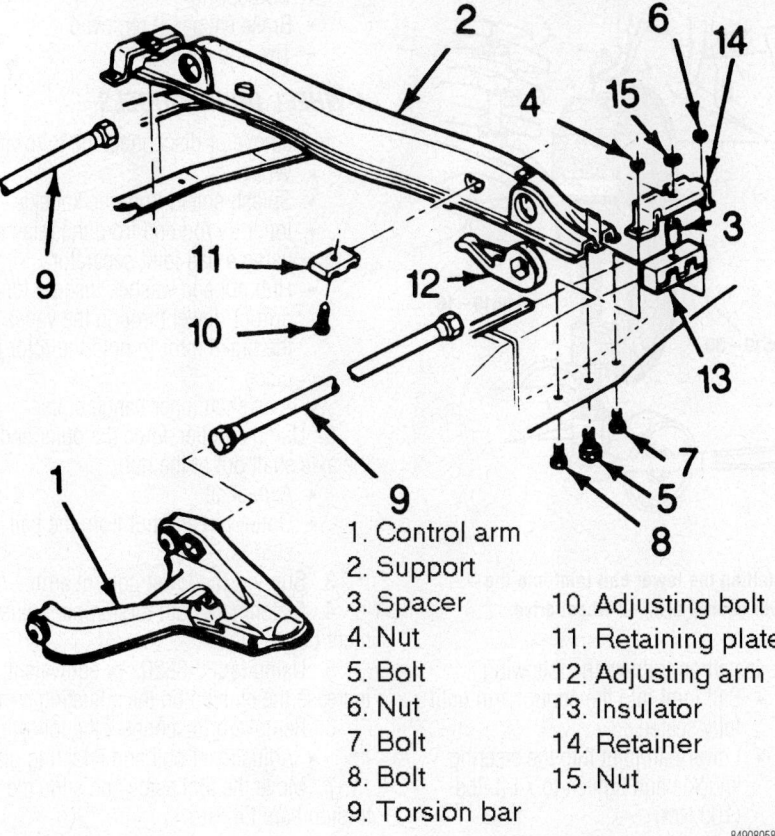

1. Control arm
2. Support
3. Spacer
4. Nut
5. Bolt
6. Nut
7. Bolt
8. Bolt
9. Torsion bar
10. Adjusting bolt
11. Retaining plate
12. Adjusting arm
13. Insulator
14. Retainer
15. Nut

84908058

Installing the torsion bar—K-Series

- Torsion bar support crossmember and slide the support crossmember rearwards
- Torsion bars, matchmark the position. They are not interchangeable.
- Support crossmember
- Retainer, spacer and bushing from the support crossmember

To install:
7. Install or connect the following:
  - Retainer, spacer and bushing to the support crossmember
  - Support assembly on the frame, out of the way
  - Torsion bars, sliding them forward until they are supported. Align the marks made when removed.
  - Support crossmember into position. Torque the bolts to 70 ft. lbs. (95 Nm).
  - Adjuster retaining plate and bolt on each torsion bar
8. Using tool J–36202, increase tension on both torsion bars.
9. Set the adjustment bolt to the marked position.
10. Release the tension on the torsion bar until the load is taken up by the adjustment bolt.
11. Install both wheels

Upper Ball Joint

REMOVAL & INSTALLATION

Avalanche, Sierra and Silverado

1. Raise and support the vehicle.
2. Remove or disconnect the following:
  - Tire and wheel assembly
  - Upper control arm
  - Upper ball joint, using a press

To install:

➡ **The ball joint must be installed with the flat edges or notches in the same position as the replaced ball joint. The ball joint is directional and damage will occur if this procedure is not followed.**

3. Install or connect the following:
  - Upper ball joint, using a press
  - Upper control arm
  - Tire and wheel assembly
4. Remove the safety stands.
5. Lower the vehicle.
6. Verify the wheel alignment.

Express and Savana

1. Remove or disconnect the following:
  - Wheel

- Brake hose bracket from the control arm
2. Using a ⅛ in. drill bit, drill a pilot hole through each ball joint rivet.
3. Drill out the rivets with a ½ in. drill bit. Punch out any remaining rivet material.
  - Cotter pin and nut from the ball stud
4. Support the lower control arm.
  - Stud from the knuckle, using a ball joint separator

To install:
5. Install or connect the following:
  - New ball joint on the control arm

➡ **Service replacement ball joints come with nuts and bolts to replace the rivets.**

  - Bolts and nuts. Torque the nuts to 17 ft. lbs. (23 Nm) for 15- and 25-Series, 52 ft. lbs. (70 Nm) for 35-Series.

➡ **The bolts are inserted from the bottom.**

6. Start the ball stud into the knuckle. Ensure it is squarely seated. Install the ball stud nut and pull the ball stud into the knuckle with the nut. Tighten the nut after the vehicle wheels are on the ground and the suspension is loaded.
7. Install the wheel.
8. Once the weight of the vehicle is on the wheels tighten the nut to 84 ft. lbs. (115 Nm).

Lower Ball Joint

REMOVAL & INSTALLATION

Avalanche, Sierra and Silverado

2WD MODELS

1. Raise and support the vehicle.
2. Remove or disconnect the following:
  - Tire and wheel assembly
  - Front coil spring
  - Lower control arm
3. Secure the lower control arm in a bench vise or equivalent.
4. Center punch the rivet heads.
5. Drill out the rivets.

To install:
6. Install or connect the following:
  - Ball joint to the lower control arm
  - Replacement bolts to the lower control arm
  - Nuts to the bolts. Tighten the nuts to 52 ft. lbs. (70 Nm).
7. Remove the lower control arm from the bench vise.

- Lower control arm
- Coil spring
- Tire and wheel tire assembly
8. Remove the safety stands.
9. Lower the vehicle.
10. Verify the wheel alignment.

4WD MODELS

1. Raise and support the vehicle.
2. Remove or disconnect the following:
   - Tire and wheel assembly
   - Lower control arm
3. Place the lower control arm in a bench vise.
4. Using a chisel, remove the 4 securing crimps from the ball joint body (15 series only).
5. Using a press, remove the ball joint from the lower control arm.

To install:

➡**Use the outer flange of the ball joint in order to press the ball joint into place.**

6. Install the new ball joint using a press.
7. Place the lower control arm in a bench vise.
8. Using a punch, install 4 crimps to the ball joint. Use the replaced ball joint as a reference (15 series only).
9. Install or connect the following:
   - Lower control arm
   - Tire and wheel assembly
10. Remove the safety stands.
11. Lower the vehicle.
12. Verify the wheel alignment.

Express and Savana

2-WHEEL DRIVE MODELS

1. Place jack under lower control arm, then raise the jack slightly.
2. Remove or disconnect the following:
   - Tire and wheel assembly
   - Brake caliper and position it to the side
   - Coil spring
   - Cotter pin and the lower ball joint retaining nut. Using the proper tool separate the ball joint from its mounting. Support the knuckle assembly so its weight will not damage the brake hose.
   - Ball joint out of the lower control arm, using tool J-9519-30-D or equivalent.

To install:

3. Start the new ball joint into the control arm. Position the bleed vent in the rubber boot facing inward.

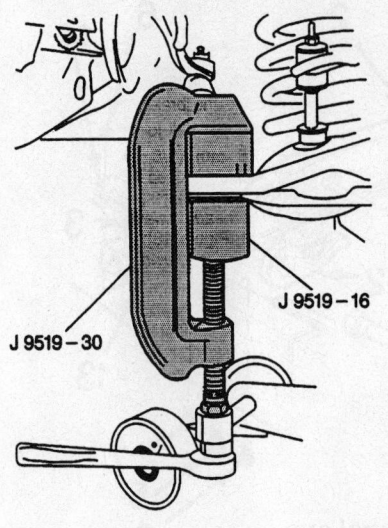

J 9519 – 16

J 9519 – 30

7924KG34

Installing the lower ball joint into the lower control arm—2-wheel drive

4. Install or connect the following:
   - Ball joint into the control arm until fully seated
   - Lower ball joint into the steering knuckle and tighten to 74 ft. lbs. (100 Nm)

- Coil spring
- Brake caliper, if removed
- Tire and wheel

4-WHEEL DRIVE MODELS

1. Remove or disconnect the following:
   - Wheel
   - Splash shield from the knuckle
   - Inner tie rod end from the relay rod using a ball joint separator
   - Hub nut and washer. Insert a long drift or dowel through the vanes in the brake rotor to hold the rotor in place.
   - Axle shaft inner flange bolts
2. Using a puller, force the outer end of the axle shaft out of the hub.
   - Axle shaft
   - Cotter pin and nut from the ball stud
3. Support the lower control arm.
4. Matchmark both torsion bar adjustment bolt positions.
5. Using tool J-36202 or equivalent, increase the tension on the adjusting arm.
6. Remove or disconnect the following:
   - Adjustment bolt and retaining plate
7. Move the tool aside and slide the torsion bars forward.

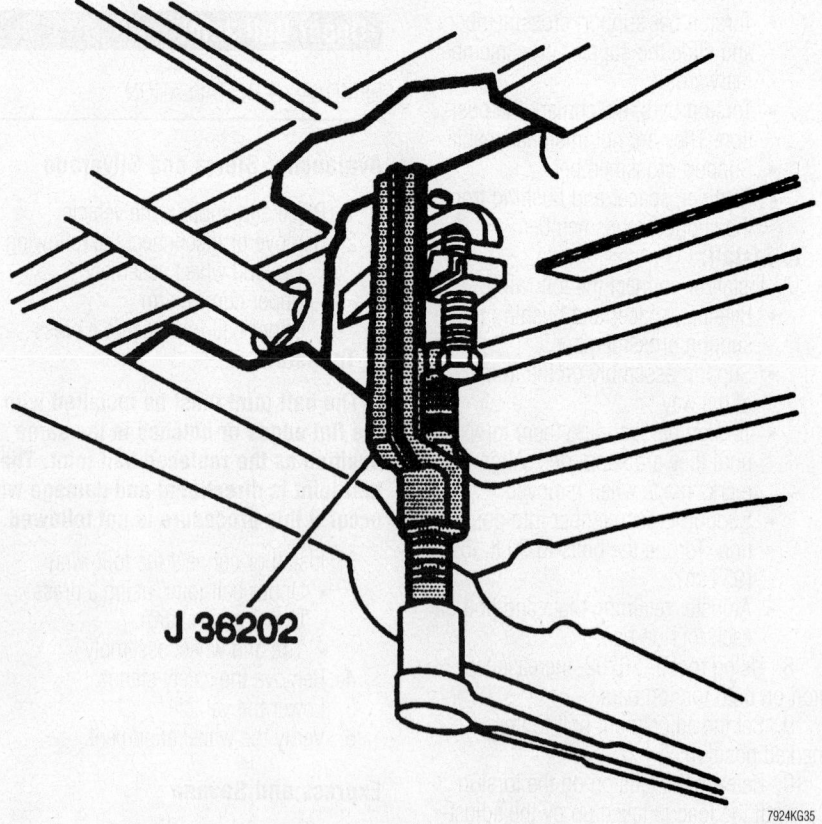

J 36202

7924KG35

A special tool is available for removing or installing the torsion bar adjusting bolt—4-wheel drive

- Ball joint from the knuckle, using a screw-type forcing tool
- Lower control arm
- Lower ball joint out of control arm with tool J-9519-E or equivalent ball joint press

To install:

8. Install or connect the following:
- New ball joint into the control arm with tool J-9519-E or equivalent
- Lower control arm

9. Using tool J-36202 or equivalent, increase tension on both torsion bars.
- Adjustment retainer plate and bolt on both torsion bars

10. Set the adjustment bolt to the marked position.

11. Release the tension on the torsion bar until the load is take up by the adjustment bolt and remove the tool.
- Shaft in the hub and the washer and hub nut. Leave the drift in the rotor vanes and torque the hub nut to 175 ft. lbs. (238 Nm).
- Flange bolts. Tighten them to 59 ft. lbs. (80 Nm), remove the drift.
- Inner tie rod end at the steering relay rod. Torque the nut to 35 ft. lbs. (48 Nm).
- Splash shield

- Wheel

12. Once the weight of the vehicle is on the wheels follow these steps:

a. Lift the front bumper about 1 ½ in. (38mm) and let it drop.

b. Repeat this procedure 2–3 more times.

c. Draw a line on the side of the lower control arm from the centerline of the control arm pivot shaft, dead level to the outer end of the control arm.

d. Measure the distance between the lowest corner of the steering knuckle and the line on the control arm, record the figure.

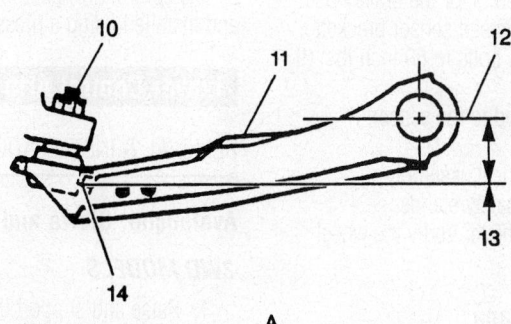

A

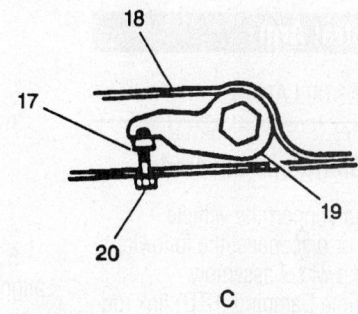

C

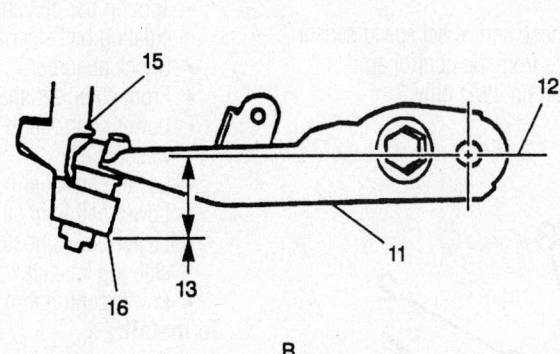

B

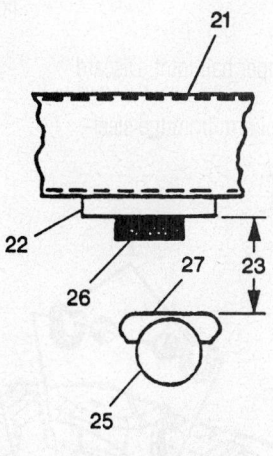

D

A. "C" MODEL	15. STEERING KNUCKLE
B. "K" MODEL	16. STEERING KNUCKLE LOWER CORNER
C. "K" MODEL TORSION BAR ADJUSTER	17. NUT
D. "CK" MODEL REAR SUSPENSION	18. TORSION BAR SUPPORT ASM.
10. LOWER BALL JOINT	19. TORSION BAR ADJUSTMENT ARM
11. LOWER CONTROL ARM	20. BOLT – ONE TURN EQUALS 6mm HEIGHT CHANGE
12. PIVOT BOLT CENTER LINE	21. FRAME
13. "Z" HEIGHT	22. BOTTOM SURFACE OF JOUNCE BRACKET
C 1,2,3 95.0 ± 6.0mm	23. "D" HEIGHT
K 1,2 157.0 ± 6.0mm	25. REAR AXLE
K 3 145.0 ± 6.0mm	26. JOUNCE BUMPER
14. LOWER BALL JOINT EXTRUSION	27. AXLE JOUNCE PAD

7924KG36

Use these specifications and diagrams to determine if the vehicle ride height is correct

e. Push down about 1½ in. (38mm) on the front bumper and let it return. Repeat the procedure 2–3 more times.

f. Re-measure the distance at the control arm.

g. Determine the average of the 2 measurements. This is the "Z" height measurement. The "Z" height should be as specified in the chart.

h. If the figure is correct, tighten the control arm pivot nuts to 94 ft. lbs. (128 Nm).

i. If the figure is not correct, tighten the pivot bolts to 94 ft. lbs. (128 Nm) and have the front end alignment corrected.

Upper Control Arm

REMOVAL & INSTALLATION

Avalanche, Sierra and Silverado

1. Raise and support the vehicle.
2. Remove or disconnect the following:
   - Tire and wheel assembly
   - Real Time Damping (RTD) link rod from the sensor, if equipped
   - Retaining bolt for the brake hose and the wheel speed sensor brackets
   - Halfshaft
   - Nut at the upper ball joint. Discard the nut
   - Upper control arm from the steering knuckle

- Upper control arm nuts and the adjustment cams
- Upper control arm bolts
- Upper control arm

To install:

3. Install or connect the following:
   - Upper control arm
   - Upper control arm bolts
   - Upper control arm nuts and the adjustment cams. Tighten the nuts to 140 ft. lbs. (190 Nm)
   - Upper control arm to the steering knuckle
   - Halfshaft
   - New nut to the upper ball joint stud. Tighten the nut to 37 ft. lbs. (50 Nm).
   - Retaining bolts for the brake hose and wheel speed sensor brackets. Tighten the bolts to 80 inch lbs. (9 Nm).
   - RTD link rod to the sensor, if equipped
   - Tire and wheel assembly
4. Remove the safety stands.
5. Lower the vehicle. Verify the wheel alignment.

Express and Savana

1. Support the lower control arm with a floor jack.
2. Remove or disconnect the following:
   - Wheel
   - Brake hose and wheel speed sensor brackets from the control arm
   - Halfshaft on 4WD only

- Ball joint from the knuckle
- Nuts and adjustment cams
- Control arm to the frame brackets

To install:

3. Install or connect the following:
   - Control arm in position
   - Nuts and adjustment cams and tighten to 129 ft. lbs. (174 Nm)
   - Ball joint to the knuckle. Torque the nut to 37 ft. lbs. (50 Nm).
   - Halfshaft on 4WD only
   - Brake hose and wheel speed sensor brackets from the control arm
   - Wheel

CONTROL ARM BUSHING REPLACEMENT

The control arm bushings are removed and installed using a press.

Lower Control Arm and Bushing

REMOVAL & INSTALLATION

Avalanche, Sierra and Silverado

2WD MODELS

1. Raise and support the vehicle.
2. Remove or disconnect the following:
   - Tire and wheel assembly
   - Coil spring on vehicles with rack and pinion steering
   - Torsion bar on vehicles with recirculating ball steering
   - Shock absorber
   - Front stabilizer shaft link
   - Lower control arm nuts and the washers
   - Lower control arm bolts
   - Lower ball joint stud nut
   - Lower ball joint stud from the steering knuckle
   - Lower control arm

To install:

3. Install or connect the following:
   - Lower control arm
   - Ball joint stud to the steering knuckle
   - Lower ball joint stud nut. Tighten the lower ball joint stud nut to 74 ft. lbs. (100 Nm)
   - Front coil spring or torsion bar
   - Lower control arm bolt
   - Lower control arm nuts and the washers. Tighten the nuts to 129 ft. lbs. (175 Nm)
   - Front stabilizer shaft link.
   - Shock absorber
   - Tire and wheel assembly
4. Remove the safety stands. Lower the vehicle. Verify the wheel alignment.

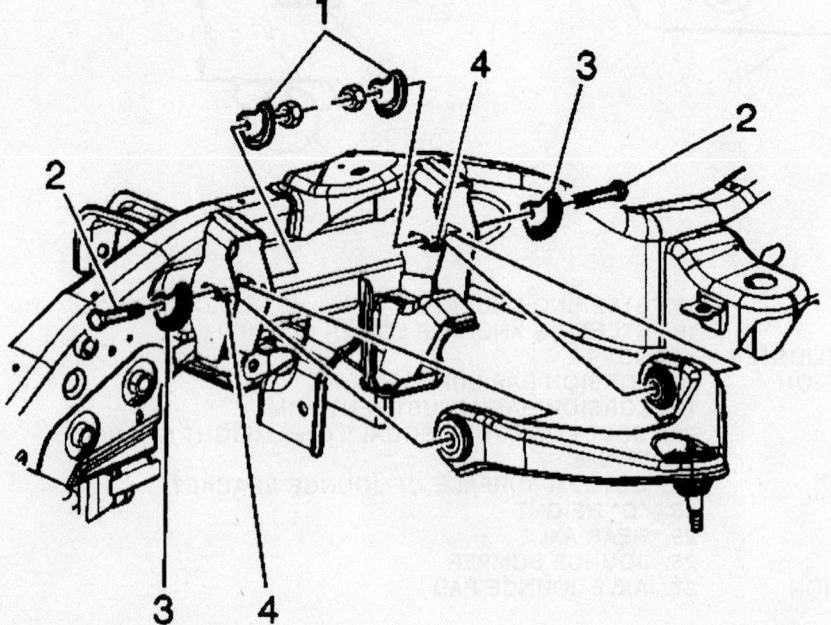

Upper control arm—Silverado

9308KG31

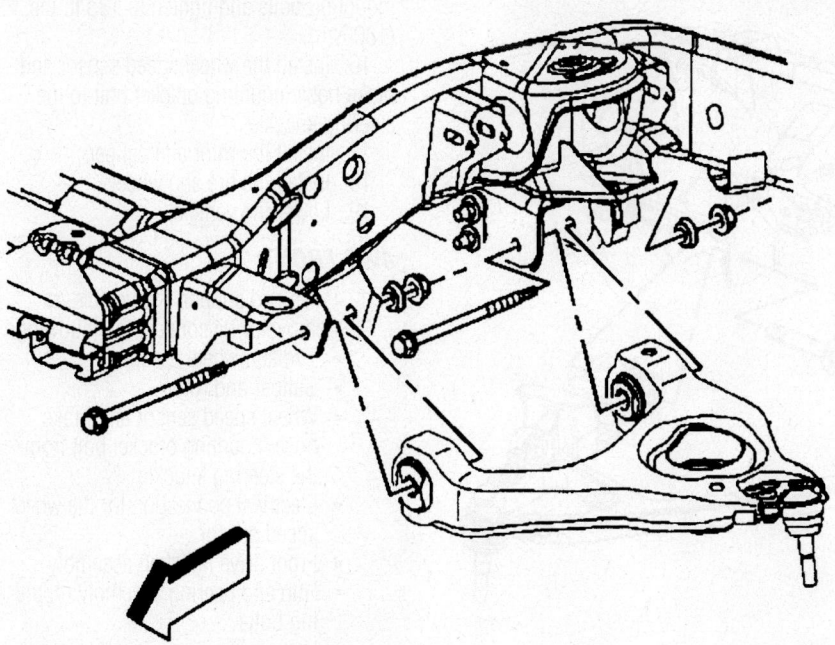

2WD lower control arm—15 Series Silverado

4WD

1. Raise and support the vehicle.
2. Remove or disconnect the following:

   • Tire and wheel assembly

   • Stabilizer shaft links from the lower control arm
   • Shock absorber nut and the bolt
   • Torsion bars
   • Halfshaft
   • Lower ball joint stud nut

   • Lower ball joint stud from the steering knuckle
   • Lower control arm nuts and the washers
   • Lower control arm bolts
   • Lower control arm

   To install:

   • Lower control arm
   • Lower control arm bolts
   • Washers with the shoulder facing the arm
   • Nuts and tighten to 129 ft. lbs. (175 Nm)
   • Halfshaft
   • Lower ball joint stud to the steering knuckle. Install the nut to the ball joint stud. Tighten the nut to 74 ft. lbs. (100 Nm).
   • Torsion bars
   • Shock absorber through nut and bolt
   • Stabilizer shaft links to the lower control arm
   • RTD link rod to the sensor (if equipped)
   • Tire and wheel assembly

3. Remove the safety stands. Lower the vehicle. Verify the wheel alignment.

Express and Savana

1. Raise and support the vehicle.
2. Remove the tire and wheel.
3. Remove the front coil spring on 2WD.
4. On 4WD, remove the torsion bar, stabilizer link, lower shock nut and halfshaft.
5. Remove the lower ball joint retaining nut.
6. Disconnect the lower ball joint from the steering knuckle using ball joint remover and separator.
7. Remove the lower control arm.

To install:

8. Install the lower control arm.
9. On 4WD, tighten the lower control arm mounting nuts and the washers to 107 ft. lbs. (145 Nm).
10. Connect the lower ball joint to the steering knuckle.
11. Install the lower ball joint retaining nut and tighten to 74 ft. lbs. (100 Nm).
12. On 4WD, install the torsion bar, stabilizer link, lower shock nut and halfshaft.
13. Install the front coil spring.
14. Install the tire and wheel.
15. Lower the vehicle.

CONTROL ARM BUSHING REPLACEMENT

➡ **Control arm bushings are not replaceable. If they are damaged, the control arm will have to be replaced.**

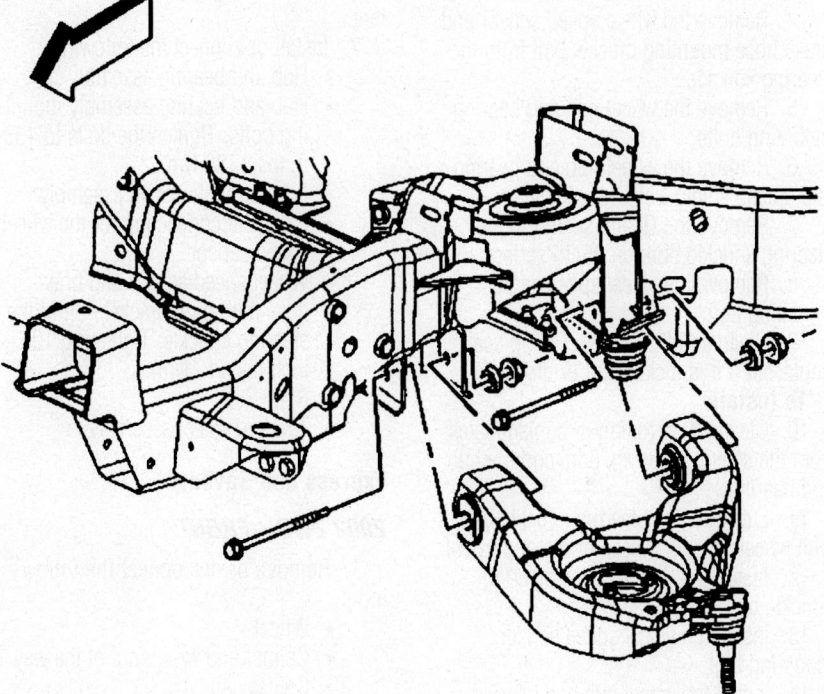

2WD lower control arm—25 Series Silverado

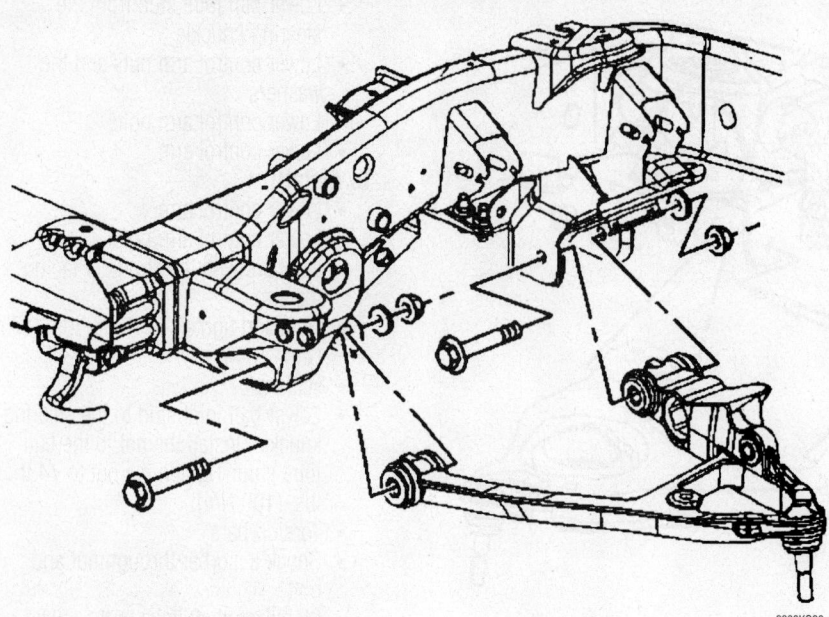

9308KG30

4WD lower control arm—15 Series Silverado

Wheel Bearings

ADJUSTMENT

2002 Express and Savana

1. Raise and support the vehicle.
2. Remove the wheel cover, if used.
3. Remove the dust cap from the hub.
4. Remove the cotter pin.
5. Tighten the hub/rotor bearing nut to 12 ft. lbs. (16 Nm), while turning the wheel forward by hand. This will seat the bearings.
6. Adjust the nut to so that it is just loose. Then, back the nut off until the hole in the spindle aligns with a slot in the nut.
7. Install a new cotter pin.
8. Make sure the cotter pin ends do not interfere with the cap.
9. Bend the ends of the cotter pin against the nut.
10. Attach a dial indicator and measure the endplay in the hub/rotor assembly. Proper endplay is 0.001-0.008 in. (0.03-0.20 mm).
11. Install the cap to the hub/rotor.
12. Install the wheel cover, if used.
13. Remove the safety stands.
14. Lower the vehicle.

➡On 2003–06 Express and Savanna and 2002–06 Avalanche, Sierra and Silverado, the wheel bearings on these vehicles are not adjustable.

Wheel Hub, Bearings and Seal

REMOVAL & INSTALLATION

Avalanche, Sierra and Silverado

2WD—FRONT

1. Raise and support the vehicle.
2. Remove the tire and wheel.
3. Remove the caliper rotor.
4. Remove the wheel speed sensor and brake hose mounting bracket bolt from the steering knuckle.
5. Remove the wheel hub and bearing mounting bolts.
6. Remove the wheel hub and bearing and splash shield from the vehicle.
7. Remove the O-ring seal from the steering knuckle bore on 25/35 series.
8. Remove the wheel speed sensor mounting bolt.
9. Clean and inspect the O-ring seal. Replace it if it is nicked, cut or dry.

To install:

10. Clean all corrosion or contaminates from the steering knuckle bore and the hub and bearing.
11. Lubricate the steering knuckle bore with wheel bearing grease or the equivalent.
12. Install the O-ring to the steering knuckle on 25/35 series.
13. Install the wheel speed sensor mounting bolt.
14. Install the wheel hub and bearing and splash shield.
15. Install the wheel hub and the bearing

mounting bolts and tighten to 133 ft. lbs. (180 Nm).
16. Install the wheel speed sensor and brake hose mounting bracket bolt to the steering knuckle.
17. Install the rotor and caliper.
18. Install the tire and wheel.
19. Lower the vehicle.

4WD FRONT

1. Raise and support the vehicle.
2. Remove or disconnect the following:
   - Tire and wheel assembly
   - Caliper and rotor
   - Wheel speed sensor and brake hose mounting bracket bolt from the steering knuckle
   - Electrical connection for the wheel speed sensor
   - Front drive halfshaft assembly
   - Hub and bearing assembly mounting bolts
   - Hub and bearing assembly
   - O-ring seal from the steering knuckle bore (25 Series)
3. Clean and inspect the O-ring seal (25 Series).

To install:

4. Clean all corrosion or contaminates from the steering knuckle bore and the hub and bearing assembly.
5. Install the O-ring to the steering knuckle (25 Series).
6. Lubricate the steering knuckle bore with wheel bearing grease or the equivalent.
7. Install or connect the following:
   - Hub and bearing assembly
   - Hub and bearing assembly mounting bolts. Tighten the bolts to 133 ft. lbs. (180 Nm).
   - Front drive halfshaft assembly
   - Electrical connection for the wheel speed sensor
   - Wheel speed sensor and brake hose mounting bracket bolt to the steering knuckle. Tighten to 106 inch lbs. (12 Nm).
   - Rotor
   - Tire and wheel assembly.

Express and Savana

2002 2WD—FRONT

1. Remove or disconnect the following:
   - Wheel
   - Caliper and wire it out of the way
   - Grease cap
   - Cotter pin, spindle nut, and washer
   - Hub.

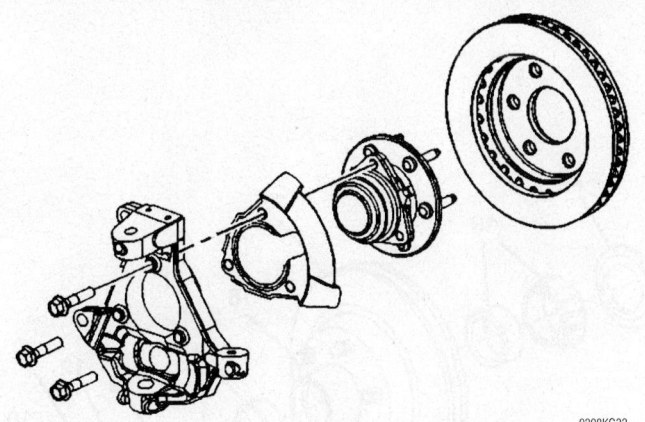

2WD/4WD front hub—15 Series Avalanche, Sierra and Silverado

9308KG33

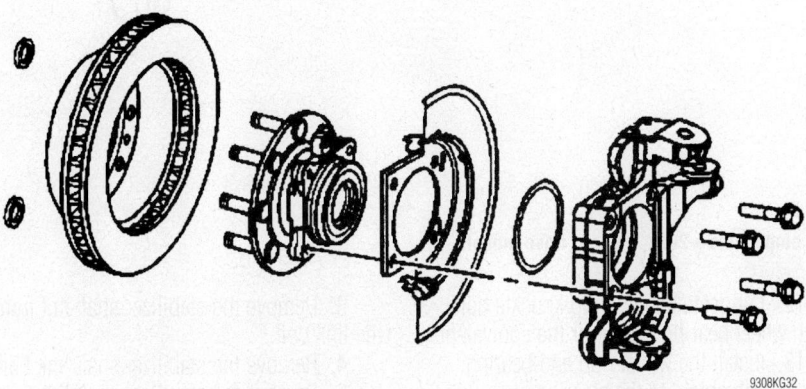

2WD/4WD front hub —25 Series Avalanche, Sierra and Silverado

9308KG32

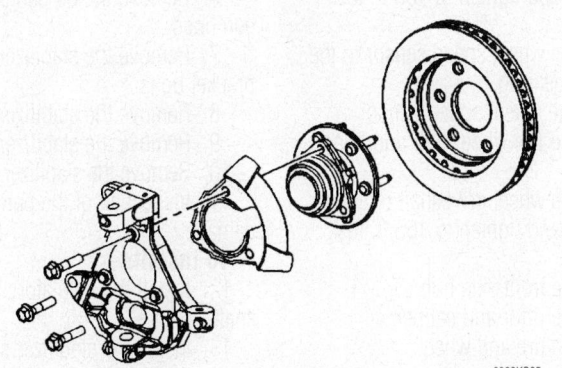

4WD front hub—15 Series Silverado

9308KG35

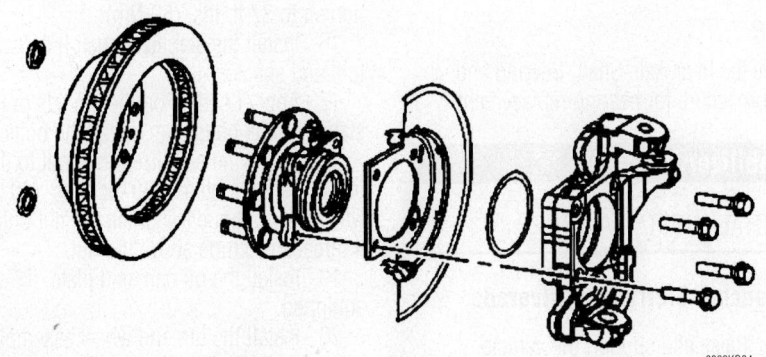

4WD front hub—25 Series Silverado

9308KG34

✳✳ WARNING

Do not drop the wheel bearings.

- Outer roller bearing assembly from the hub

2. The inner bearing assembly will remain in the hub and may be removed after prying out the inner seal. Discard the seal.

3. Clean all parts in a non-flammable solvent and let them air dry. Never spin-dry a bearing with compressed air! Check for excessive wear and damage.

4. If necessary for replacement, remove the bearing races from the hub using a hammer and drift. They are driven out from the inside out.

To install:

5. Install or connect the following:
- New bearing races, if required. When installing new races, ensure that they are not cocked and that they are fully seated against the hub shoulder.

6. Pack both wheel bearings using high melting point wheel bearing grease for disc brakes.

- Inner bearing in the hub and a new inner seal, making sure that the seal flange faces the bearing race.
- Wheel hub over the spindle
- Outer bearing into the hub
- Spindle washer and nut

7. Spin the wheel hub by hand and tighten the nut until it is just snug—12 ft. lbs. (16 Nm). Back off the nut until it is loose, then tighten it finger-tight. Loosen the nut until either hole in the spindle lines up with a slot in the nut and insert a new cotter pin. There should be 0.001–0.008 in. (0.025–0.200mm) end-play. This can be measured with a dial indicator.

8. Replace the dust cap, wheel and tire.

2003–06 2WD—FRONT

1. Raise and support the vehicle.

2. Remove the tire and wheel.

3. Remove the caliper and rotor.

4. Remove the wheel speed sensor mounting bolt.

5. Remove the wheel speed sensor from the wheel hub and bearing.

6. Remove the wheel hub and bearing mounting bolts.

7. Remove the wheel hub and bearing and splash shield from the steering knuckle.

To install:

8. Clean all corrosion or contaminates from the steering knuckle bore and the hub and bearing.

9. Lubricate the steering knuckle bore with wheel bearing grease or the equivalent.

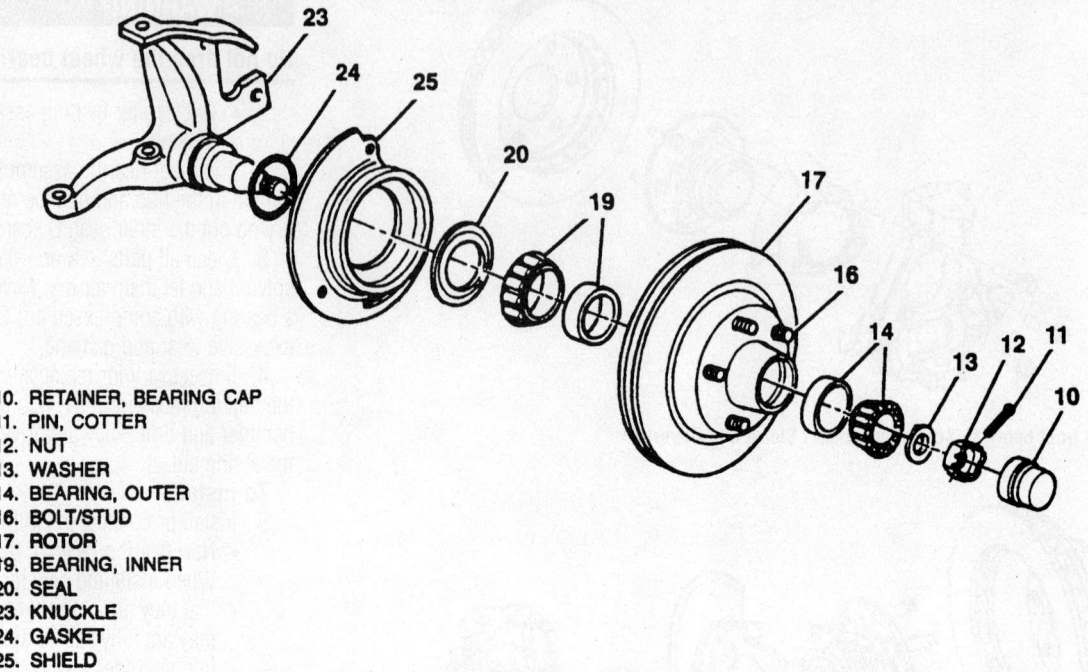

10. RETAINER, BEARING CAP
11. PIN, COTTER
12. NUT
13. WASHER
14. BEARING, OUTER
16. BOLT/STUD
17. ROTOR
19. BEARING, INNER
20. SEAL
23. KNUCKLE
24. GASKET
25. SHIELD

7924KG53

Exploded view of the front wheel bearing and related components—2002 2-wheel drive models

10. Install the wheel hub and bearing and splash shield to the steering knuckle.

11. Install the wheel hub and bearing mounting bolts and tighten to 133 ft. lbs. (180 Nm).

12. Install the wheel speed sensor to the wheel hub and bearing.

13. Install the wheel speed sensor retaining bolt.

14. Install the rotor and caliper.

15. Install the tire and wheel.

16. Lower the vehicle.

4WD—FRONT

1. Raise and support the vehicle.

2. Remove the tire and wheel.

3. Remove the caliper and rotor.

4. Remove the front axle hub cap.

5. Remove the halfshaft retaining nut and washer.

6. Using an axle remover, disengage the halfshaft from the hub and bearing.

7. Remove the wheel speed sensor mounting bolt from the wheel hub and bearing.

8. Remove the wheel speed sensor from the wheel hub and bearing.

9. Remove the wheel hub and bearing mounting bolts.

10. Remove the wheel hub and bearing and splash shield from the steering knuckle.

To install:

11. Clean all corrosion or contaminates from the steering knuckle bore and the hub and bearing.

12. Lubricate the steering knuckle bore with wheel bearing grease or the equivalent.

13. Install the wheel hub and bearing and splash shield to the steering.

14. Install the wheel hub and bearing mounting bolts and tighten to 133 ft. lbs. (180 Nm).

15. Install the wheel speed sensor to the wheel hub and bearing.

16. Install the wheel speed sensor mounting bolt to the wheel hub and bearing.

17. Install the wheel driveshaft retaining nut and washer and tighten to 155 ft. lbs. (210 Nm).

18. Install the front axle hub cap.

19. Install the rotor and caliper.

20. Install the tire and wheel.

21. Lower the vehicle .

22. Check and adjust the front end alignment and road test the vehicle.

REAR

See the Rear Axle Shaft, Bearing and Seal procedure for bearing replacement

Stabilizer Bar

REMOVAL & INSTALLATION

Avalanche, Sierra and Silverado

1. Raise and support the vehicle.

2. Remove the tire and wheel.

3. Remove the stabilizer shaft nut from the link bolt.

4. Remove the stabilizer shaft link bolt.

5. Remove the stabilizer shaft link insulators and spacers.

6. Remove the oil pan skid plate, if equipped.

7. Remove the stabilizer shaft insulator bracket bolts.

8. Remove the stabilizer shaft bracket.

9. Remove the stabilizer shaft.

10. Remove the stabilizer shaft insulators.

11. Inspect all of the parts for wear and damage.

To install:

12. Install the insulators to the stabilizer shaft.

13. Install the stabilizer shaft.

14. Install the brackets over the insulators and the stabilizer shaft.

15. Install insulator bracket bolts and tighten to 37 ft. lbs. (50 Nm).

16. Install the stabilizer shaft link insulators and spacers.

17. Apply Loctite® on the threads of the stabilizer link bolts then install the bolts.

18. Install the stabilizer shaft nut to the link bolt and tighten to 89 inch lbs. (10 Nm), and continue to tighten the nut until 2-4 threads protrude above the nut.

19. Install the oil pan skid plate, if equipped.

20. Install the tire and wheel assembly.

21. Remove the safety stands

22. Lower the vehicle.

Express and Savana

1. Raise and support the vehicle.
2. Remove the tire and wheel.
3. Remove the stabilizer shaft nut from the link bolt.
4. Remove the stabilizer shaft link bolt.
5. Remove the stabilizer shaft link insulators and spacers.
6. Remove the stabilizer shaft insulator clamps mounting bolts (15 and 25 series).
7. Remove the stabilizer shaft clamps (1500 and 2500 series w/7300 GVW).
8. Remove the stabilizer shaft (1500 and 2500 series w/7300 GVW).
9. Remove the stabilizer shaft insulators (1500 and 2500 series w/7300 GVW).
10. Remove the stabilizer shaft insulator clamps mounting nuts and/or studs (2500 series w/8500 and 8600 GVW and 3500 series).
11. Remove the stabilizer shaft clamps (2500 series w/8500 and 8600 GVW and 3500 series).
12. Remove the stabilizer shaft (2500 series w/8500 and 8600 GVW and 3500 series).
13. Remove the stabilizer shaft insulators (2500 series w/8500 and 8600 GVW and 3500 series).
14. Inspect all of the parts for wear and damage.

To install:

15. Install the stabilizer shaft insulator clamp studs, if removed (2500 series w/8500 and 8600 GVW and 3500 series) and tighten to 18 ft. lbs. (25 Nm).
16. Place the stabilizer shaft insulators on the stabilizer shaft with the slits facing toward the front of the vehicle.
17. Install the stabilizer shaft insulators to the stabilizer shaft (2500 series w/8500 and 8600 GVW and 3500 series).
18. Install the stabilizer shaft (2500 series w/8500 and 8600 GVW and 3500 series).
19. Install the stabilizer shaft clamps over the insulators and the stabilizer shaft (2) (2500 series w/8500 and 8600 GVW and 3500 series).
20. Install stabilizer shaft insulator clamps mounting nuts (2500 series w/8500 and 8600 GVW and 3500 series) and tighten the nuts to 34 ft. lbs. (46 Nm).

21. Install the stabilizer shaft insulators (1500 and 2500 series w/7300 GVW).
22. Install the stabilizer shaft (1500 and 2500 series w/7300 GVW).
23. Install the stabilizer shaft clamps (1500 and 2500 series w/7300 GVW).
24. Install the stabilizer shaft insulator clamps mounting bolts (1500 and 2500 series w/7300 GVW) and tighten to 18 ft. lbs. (25 Nm).
25.
26. Apply Loctite® on the threads of the stabilizer link bolts then install the bolts.
27. Install the stabilizer shaft nut to the link bolt and tighten to 97 inch lbs. (11 Nm).
28. Install the stabilizer shaft link insulators and spacers.
29. Install the tire and wheel.
30. Lower the vehicle.

Steering Knuckle

REMOVAL & INSTALLATION

Avalanche, Sierra and Silverado

1. Raise and support the vehicle.
2. Remove the tire and wheel.
3. Remove the wheel hub and bear-ing.
4. Support the lower control arm with a suitable jack.
5. Disconnect the outer tie rod from the knuckle.
6. Remove the brake hose bracket retaining bolt from the knuckle.
7. Remove the retaining nut and separate the upper and lower ball joints from the steering knuckle using a ball joint remover and adapters.
8. Remove the steering knuckle.

To install:

9. Clean all grease and contaminants from the tapered section and the threads of the upper ball joint, the lower ball joint, and the tie rod end.
10. Clean and inspect the taper holes and the mounting surfaces of the steering knuckle. If any of the tapered holes are elongated, out of round, or damaged, the replace the steering knuckle.
11. Install the steering knuckle.
12. Connect the lower ball joint to the steering knuckle and install the retaining nut and tighten to 74 ft. lbs. (100 Nm).
13. Connect the upper ball joint to the

steering knuckle and install the retaining nut and tighten to 37 ft. lbs. (50 Nm).
14. Install the brake hose bracket retaining bolt to the knuckle.
15. Connect the outer tie rod to the steering knuckle.
16. Install the wheel hub and bearing.
17. Install the tire and wheel.
18. Remove the safety stands.
19. Lower the vehicle.

Express and Savana

1. Raise and support the vehicle.
2. Remove the tire and wheel.
3. Remove the wheel hub and bearing.
4. Remove the outer tie rod-to-steering knuckle retaining nut.
5. Disconnect the outer tie rod from the steering knuckle.
6. Remove the brake hose bracket retaining bolt from the steering knuckle.
7. Support the lower control arm with a suitable jack.
8. Remove the upper and lower ball joint retaining nuts.
9. Separate the upper and lower ball joints from the steering knuckle using a ball joint remover and adapters.
10. Remove the steering knuckle.

To install:

11. Clean all grease and contaminants from the tapered section and the threads of the upper ball joint, the lower ball joint, and the tie rod end.
12. Clean and inspect the taper holes and the mounting surfaces of the steering knuckle. If any of the tapered holes are elongated, out of round, or damaged, the replace the steering knuckle.
13. Install the steering knuckle.
14. Connect the lower ball joint to the steering knuckle and install the retaining nut and tighten to 74 ft. lbs. (100 Nm).
15. Connect the upper ball joint to the steering knuckle and install the retaining nut and tighten to 37 ft. lbs. (50 Nm).
16. Install the brake hose bracket retaining bolt to the knuckle.
17. Connect the outer tie rod to the steering knuckle.
18. Install the outer tie rod to the steering knuckle retaining nut and tighten to 47 ft. lbs. (63 Nm).
19. Remove the safety stand.
20. Install the wheel hub and bearing.
21. Install the tire and wheel.

BRAKES

22. Lower the vehicle.

Brake Caliper

REMOVAL & INSTALLATION

Avalanche, Sierra and Silverado

FRONT

1. Remove or disconnect the following:
   - ⅔ of the brake fluid from the master cylinder
   - Tire and wheel assembly
2. Using a C-clamp or the equivalent, compress the caliper piston until the caliper piston bottoms in the bore.
   - Brake hose at caliper by removing the inlet fitting bolt. Plug the line.
   - Caliper mounting bolts
   - Caliper
3. Inspect the caliper assembly.

To install:

4. Install or connect the following:
   - Caliper
   - Caliper mounting bolts. Tighten the caliper guide pin bolts to 74 ft. lbs. (100 Nm) on 15 series or 80 ft. lbs. (108 Nm) on 25/35 series..
   - Brake hose at caliper by installing the inlet fitting bolt. Tighten the inlet fitting bolt to 33 ft. lbs. (45 Nm).
5. Bleed the brakes.
   - Tire and wheel assembly

REAR

1. Remove or disconnect the following:
   - ⅔ of the brake fluid from the master cylinder
   - Tire and wheel assembly
2. Using a C-clamp or the equivalent, compress the caliper piston until the caliper piston bottoms in the bore.
   - Brake hose at caliper by removing the inlet fitting bolt. Plug the line.
   - Caliper mounting bolts
   - Caliper
3. Inspect the caliper assembly.

To install:

4. Install or connect the following:
   - Caliper
5. Perform the following procedure before installing the caliper guide pin bolts (15 Series only).
   a. Remove all traces of the original adhesive patch.

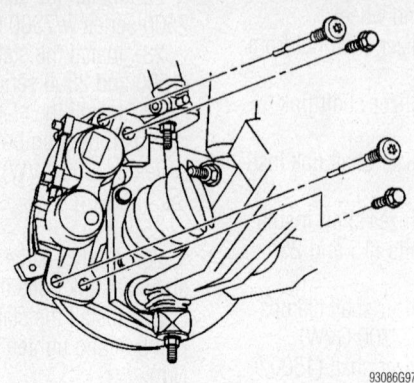

Front caliper removal—Avalanche, Sierra & Silverado

93086G97

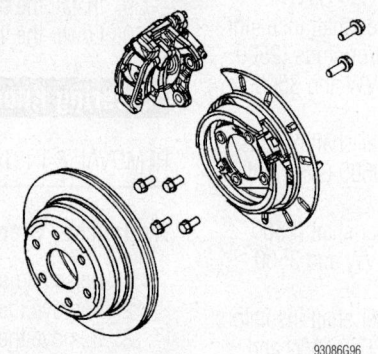

Rear caliper removal—15 Series Silverado/Sierra

93086G96

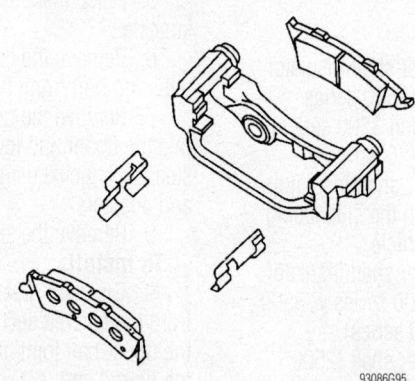

Rear pad removal—15 Series Silverado/Sierra

93086G95

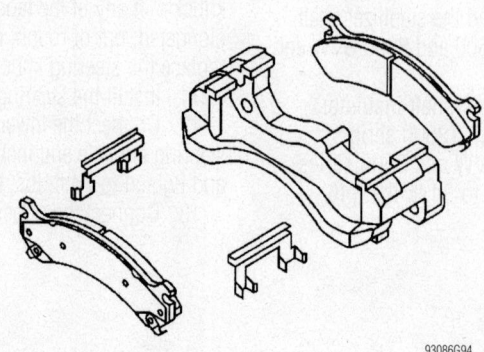

Rear pad removal—25 Series Silverado

93086G94

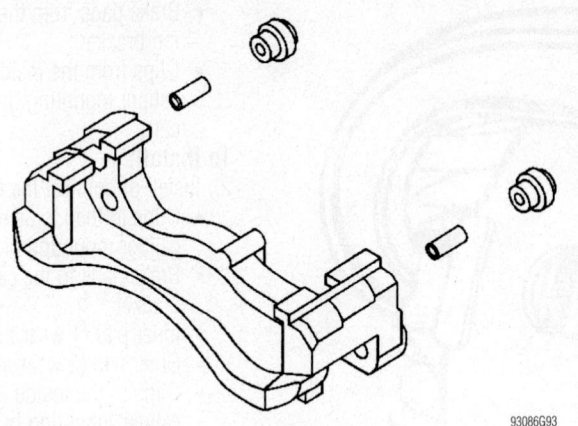

Sleeve and bushing removal—15 Series Silverado/Sierra

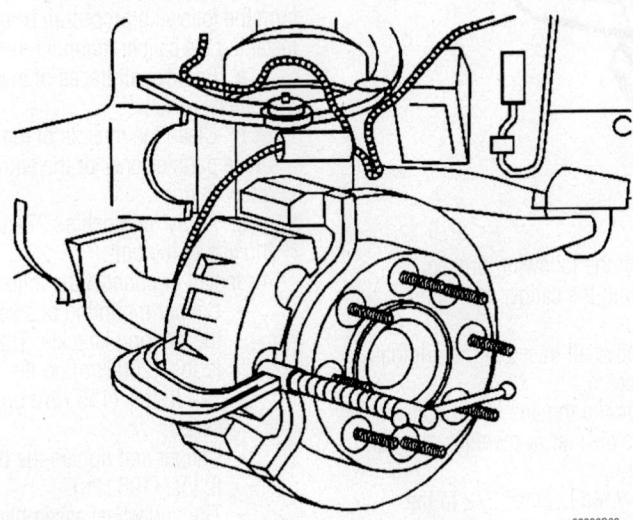

Compressing the rear caliper piston—15 Series Silverado/Sierra

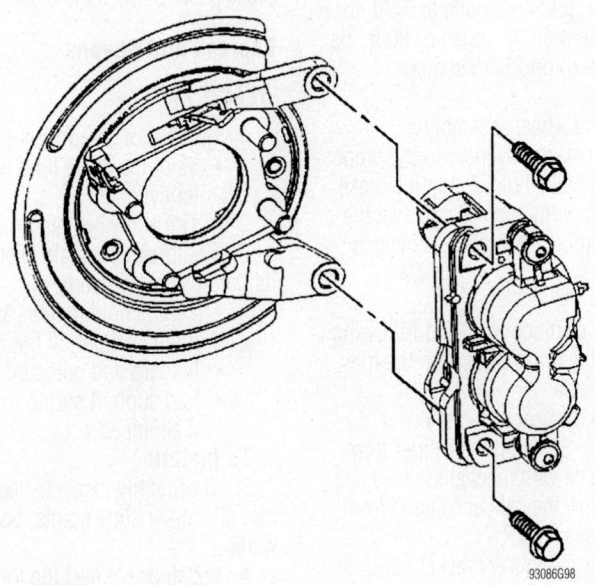

Rear caliper removal—25 Series Silverado

b. Clean the threads of the bolt with brake parts cleaner or the equivalent and allow to dry.

c. Apply Red Loctite® no. 272 to the threads of the bolt.

6. Install or connect the following:
- Caliper mounting bolts. Tighten the caliper guide pin bolts to 31 ft. lbs. (42 Nm) on the 15 series; 80 ft. lbs. (108 Nm) on the 25 series.
- Brake hose at the caliper by installing the inlet fitting bolt. Tighten the bolt to 33 ft. lbs. (45 Nm).

7. Bleed the brakes.
- Tire and wheel assembly

8. Refill the brake master cylinder to the proper level with fresh brake fluid.

Express and Savana

FRONT AND REAR

➡There are 2 caliper designs and they can be identified by the method used to secure the assembly to the spindle bracket. The Delco caliper is secured by a bolt and sleeve combination. The Bendix caliper assembly is secured by a slider, spring and bolt.

1. Remove or disconnect the following:
- ⅔ of the brake fluid from the master cylinder
- Tire and wheel assembly

2. Using a C-clamp or the equivalent, compress the caliper piston until the caliper piston bottoms in the bore.
- Brake hose at caliper by removing the inlet fitting bolt. Plug the line.
- Bolt and sleeve or bolt and slider assemblies that hold the caliper and then lift the caliper off the rotor

3. Inspect the caliper assembly.

To install:

4. Install or connect the following:
- Caliper onto the knuckle/rotor assembly and secure the assembly with the mounting bolts or sliders.

Tighten the front bolts to 80 ft. lbs. (108 Nm). Tighten the rear bolts to 31 ft. lbs. (30 Nm) on 15 series and 80 ft. lbs. (108 Nm) on 25/35 series.
- Brake line to the caliper

5. Bleed the brakes.

6. Pump the brake pedal and verify there is minimal brake pedal travel.

7. Check the brake fluid level. Install the tire and wheel assembly.

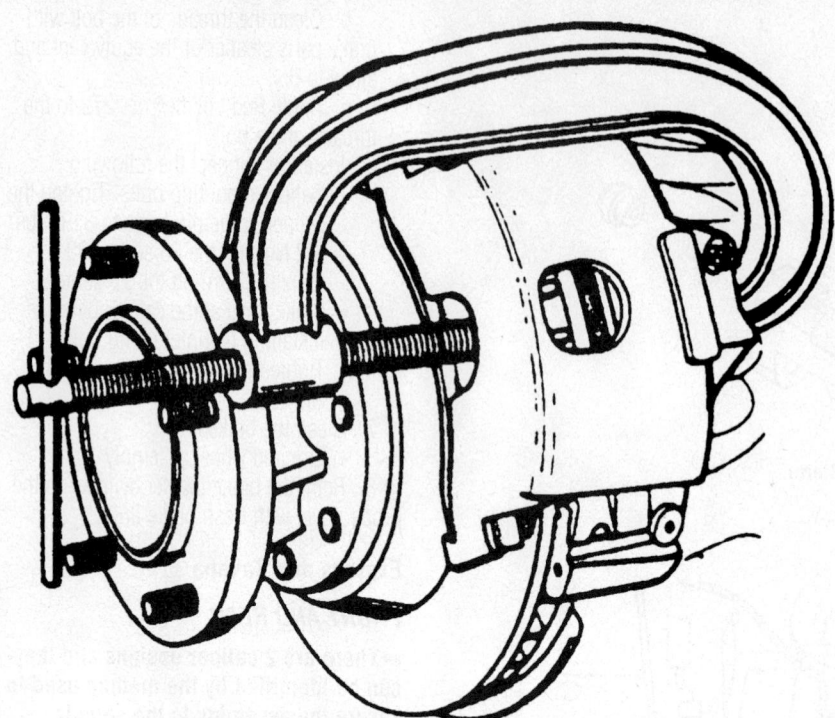

Compressing the caliper piston—Express & Savana

93026G45

Disc Brake Pads

REMOVAL & INSTALLATION

Avalanche, Sierra and Silverado

FRONT

1. Remove ⅔ of the brake fluid from the master cylinder.
2. Remove or disconnect the following:
   - Wheel
   - Caliper. Suspend the caliper from the frame with mechanic's wire. Do not allow the caliper to hang from the brake hose.
   - Caliper mounting bracket bolts
   - Caliper mounting bracket from the steering knuckle assembly
   - Brake pads from the caliper mounting bracket
   - Clips from the inside ends of the caliper mounting bracket and discard

To install:
3. Install or connect the following:
   - Clips to the inside ends of the caliper mounting bracket
   - Brake pads to the caliper mounting bracket
   - Inner pad (1 wear indicator)
   - Outer pad (2 wear indicators)
   - Caliper mounting bracket to the steering knuckle assembly

4. Perform the following procedure before installing the caliper mounting bracket bolts:
   a. Remove all traces of the original adhesive patch.
   b. Clean the threads of the bolt with brake parts cleaner or the equivalent and allow to dry.
   c. Apply red Loctite® 272 to the threads of the bolt.
5. Install or connect the following:
   - Caliper mounting bracket bolts to the steering knuckle. Tighten the caliper guide pin bolts to 74 ft. lbs. (100 Nm) on 15 series or 80 ft. lbs. (108 Nm) on 25/35 series.
   - Caliper
   - Tire and wheel assembly
6. Refill the master cylinder to the proper level with fresh brake fluid. Pump the brake pedal slowly and firmly in order to seat the brake pads. Burnish the brakes as needed.

REAR

1. Remove or disconnect the following:
   - ⅔ of the brake fluid from the master cylinder
   - Tire and wheel assembly
   - Caliper. Suspend the caliper from the frame with mechanic's wire. Do not allow the caliper to hang from the brake hose.
   - Caliper mounting bracket bolts from the backing plate

   - Brake pads from the caliper mounting bracket
   - Clips from the inside ends of the caliper mounting bracket and discard

To install:
2. Install or connect the following:
   - Clips to the inside ends of the caliper mounting bracket
   - Brake pads to the caliper mounting bracket
   - Inner pad (1 wear indicator)
   - Outer pad (2 wear indicators)
   - Clips to the inside ends of the caliper mounting bracket
   - Caliper mounting bracket to the backing plate assembly (15 series).
3. Install the caliper mounting bracket to the backing plate assembly (25 series). Perform the following procedure before installing the caliper mounting bracket bolts.
   a. Remove all traces of the original adhesive patch.
   b. Clean the threads of the bolt with brake parts cleaner or the equivalent and allow to dry.
   c. Apply red Loctite® 272 to the threads of the bolt.
4. Install or connect the following:
   - Caliper mounting bracket bolts to the steering knuckle. Tighten to 148 ft. lbs. (200 Nm) on the 15 series; 122 ft. lbs. (165 Nm) on the 25 series.
   - Caliper and tighten the bolts to 80 ft. lbs. (108 Nm).
   - Tire and wheel assembly
5. Refill the master cylinder to the proper level with fresh brake fluid. Pump the brake pedal slowly and firmly in order to seat the brake pads. Burnish the brakes as needed.

Express and Savana

DELCO TYPE

1. Remove or disconnect the following:
   - ⅔ of the brake fluid from the master cylinder
   - Tire and wheel assembly
2. Compress the brake piston back into its bore using a C-clamp.
   - 2 bolts holding the caliper and then lift the caliper off the disc
   - Inboard and outboard pads
   - Pad support spring from the piston, if equipped

To install:
3. Thoroughly inspect, clean and lubricate all caliper slide points, bolts and hardware.

4. Install or connect the following:
   - Retainer spring on the inner pad

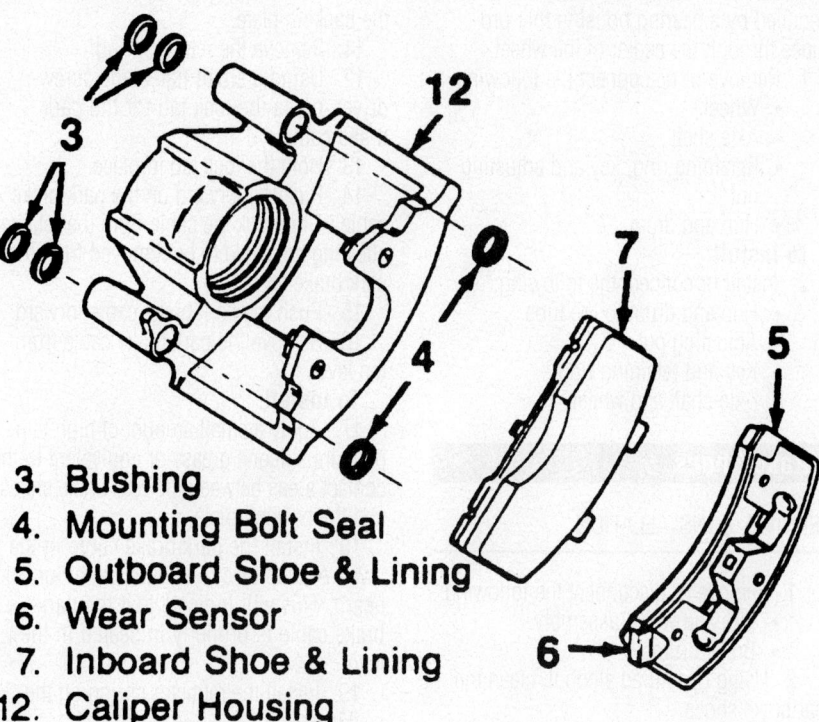

3. **Bushing**
4. **Mounting Bolt Seal**
5. **Outboard Shoe & Lining**
6. **Wear Sensor**
7. **Inboard Shoe & Lining**
12. **Caliper Housing**

93026G49

Replacing the disc brake pads—Delco type

and insert the assembly into the center cavity of the piston

5. Push down on the inner pad until it lays flat against the caliper. It is important to push the piston all the way into the caliper if new linings are installed or the caliper will not fit over the rotor.

- Outboard pad with the ears of the pad over the caliper ears and the tab at the bottom engaged in the caliper cutout
- Caliper over the brake disc and align the holes in the caliper with those of the mounting bracket
- Mounting bracket bolts through the sleeves in the inboard caliper ears and through the mounting bracket, making sure the ends of the bolts pass under the retaining ears on the inboard pad
- Mounting bolts to 80 ft. lbs. (108 Nm). After both calipers are mounted pump the brake pedal to seat the pad against the rotor. Use a pair of locking pliers to bend over the upper ears of the outer pad so it isn't loose.
- Wheels

6. Add fluid to the master cylinder reservoirs so they are ¼ in. (6.35mm) from the top.

7. Test the brake pedal by pumping it to obtain a hard pedal. Check the fluid level again and add fluid as necessary. Do not move the vehicle until a pedal is obtained.

BENDIX TYPE

1. Remove or disconnect the following:
- ⅔ of the brake fluid from the master cylinder
- Tire and wheel assembly

2. Compress the brake piston back into its bore using a C-clamp.
- Bolt at the caliper slider. Use a brass drift pin to remove the slider and spring.
- Caliper up and forward from the bottom and lift it off the caliper support. Tie the caliper out of the way with a piece of wire. Be careful not to damage the brake line.
- Inner shoe from the caliper support. Discard the inner shoe clip.
- Outer shoe from the caliper

To install:

3. Thoroughly clean, inspect and lubricate the caliper, slider and spring with silicone.

4. Install or connect the following:
- New inboard shoe clip on the shoe
- Lower end of the inboard shoe into the groove provided in the support. Slide the upper end of the shoe into position. Be sure the clip remains in position.
- Outboard shoe in the caliper, with the ears at the top of the shoe over the caliper ears and the tab at the bottom of the shoe engaged in the caliper cutout. If assembly is difficult, a C-clamp may be used. Be careful not to damage the lining.
- Caliper over the brake disc, top edge first. Rotate the caliper downward onto the support.
- Spring over the caliper support key, install the assembly between the support and lower caliper groove.

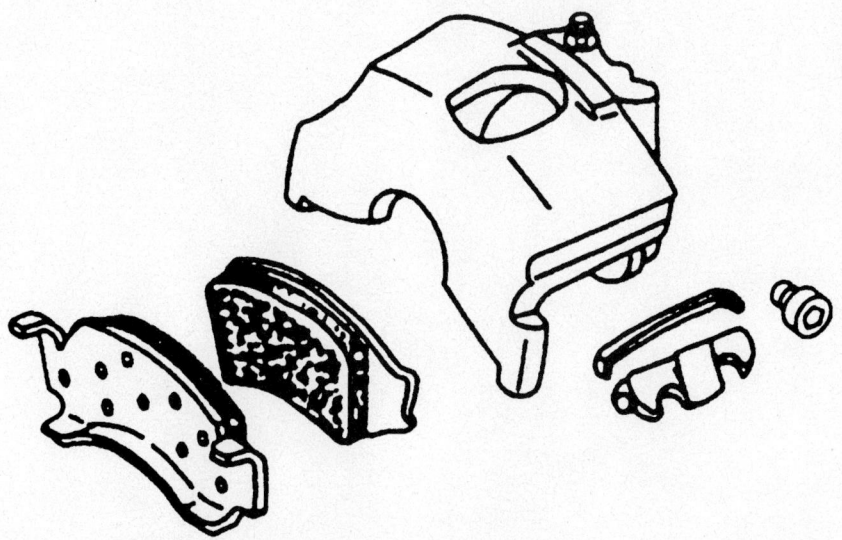

93026G50

Replacing the disc brake pads—Bendix type

Tap into place until the key retaining screw can be installed.

- Screw and torque to 15 ft. lbs. (20 Nm). The boss must fit fully into the circular cutout in the key.

5. Install the wheel and add brake fluid as necessary.

Brake Drums

REMOVAL & INSTALLATION

With Semi-Floating Axles

1. Raise and support the vehicle safely.
2. Mark the relationship of the wheel to the hub and remove the wheel.
3. Mark the relationship of the drum to the hub and pull the drum from the brake assembly. If the brake drums have been scored from worn linings, the brake adjuster must be backed off so the brake shoes will retract from the drum. The adjuster can be backed off by inserting a brake adjusting tool through the access hole provided. In some cases the access hole is provided in the brake drum. A metal cover plate is over the hole. This may be removed by using a hammer and chisel.

To install:

4. Align the mark on the drum to mark on hub and install drum
5. Align the mark on the wheel to mark on drum and install wheel
6. Adjust brake lining as needed. Pump brakes

With Full Floating Axles

To remove the drums from full floating rear axles, the axle shaft will have to be removed. Full-floating rear axles can be identified by a bearing housing that protrudes through the center of the wheel.

1. Remove or disconnect the following:
   - Wheel
   - Axle shaft
   - Retaining ring, key and adjusting nut
   - Hub and drum

To install:

2. Install or connect the following:
   - Hub and drum to the tube
   - Adjusting nut
   - Key and retaining ring
   - Axle shaft and wheel

Brake Shoes

REMOVAL & INSTALLATION

1. Remove or disconnect the following:
   - Tire and wheel assembly
   - Brake drums
2. Using denatured alcohol, clean the rear brake shoes.
3. Adjust the brake shoes to the lowest position. This will reduce the tension on the retractor spring.
4. Remove the adjuster spring.
5. Remove the brake adjuster lever.
6. Remove the adjuster assembly.
7. Using a pair of channel locks, remove the retractor spring from the secondary brake shoe.
8. Remove the secondary brake shoe from the backing plate.
9. Using a pair of channel locks, remove the retractor spring from the primary brake shoe.

10. Remove the primary brake shoe from the backing plate.
11. Remove the return spring.
12. Using a small flat-blade screwdriver, press the lock tab for the park brake cable.
13. Hold the lock tab in place.
14. Pushing forward on the park brake cable will unlock the cable from the retainer allowing the cable to be removed from the park brake lever.
15. Push the park brake cable forward.
16. Remove the park brake cable from the lever.

To install:

17. Apply a small amount of high temperature silicone grease or equivalent to the contact areas between the rear brake shoes and the backing plate.
18. Install the park brake cable in the lever. A snap or clip should be felt or heard. This will indicate that the park brake cable is properly in seated in the lever.
19. Install the retractor spring on the backing plate.
20. Using a pair of channel locks, install the retractor spring in the primary brake shoe.
21. Install the secondary brake shoe on the backing plate.
22. Using channel locks, install the retractor spring in the secondary brake shoe.
23. Install the adjuster spring.
24. Install the brake adjuster lever.
25. Install the adjuster assembly.
26. Adjust the rear brake shoes.
27. Install the rear brake drum.

CHEVROLET

Aveo

BRAKES 5-21
DRIVE TRAIN 5-15
ENGINE REPAIR................5-7
FUEL SYSTEM 5-14
SPECIFICATIONS AND
MAINTENANCE CHARTS 5-2
Vehicle and Engine Identification
  Chart....................................5-2
General Engine Specifications 5-2
Gasoline Engine Tune-Up
  Specifications 5-2
Accessory Drive Belt Routing 5-3
Capacities 5-3
Valve Specifications.....................5-3
Crankshaft and Connecting Rod
  Specifications 5-4
Piston and Ring Specifications........5-4
Torque Specifications 5-4
Wheel Alignment 5-5
Tire and Wheel Specifications..........5-5
Brake Specifications 5-5
Scheduled Maintenance Intervals 5-6
STEERING AND
SUSPENSION 5-17

A
Air Bag (Supplemental Restraint)
  System...................................5-17
  Arming The System 5-17
  Disarming The System...............5-17
  Service Precautions 5-17
Alternator 5-7
  Removal & Installation...............5-7

B
Brake Caliper 5-21
  Removal & Installation...............5-21
Brake Drums.............................5-21
  Removal & Installation...............5-21
Brake Pads...............................5-21
  Removal & Installation...............5-21

Brake Shoes.............................5-21
  Removal & Installation...............5-21

C
Camshafts................................5-11
  Removal & Installation...............5-11
Coil Springs 5-19
  Removal & Installation...............5-19
CV-Joint 5-16
  Overhaul 5-16
Cylinder Head............................5-8
  Removal & Installation...............5-8

E
Engine Assembly 5-7
  Removal & Installation...............5-7
Exhaust Manifold........................5-10
  Removal & Installation...............5-10

F
Front Suspension Crossmember ...5-17
  Removal & Installation...............5-17
Fuel Filter 5-14
  Removal & Installation...............5-14
Fuel Pump 5-14
  Removal & Installation...............5-14
Fuel Rail and Injectors 5-15
  Removal & Installation...............5-15
Fuel System Pressure 5-14
  Relieving 5-14
Fuel System Service Precautions....5-14

H
Halfshafts................................5-16
  Removal & Installation...............5-16
Heater Core..............................5-8
  Removal And Installation 5-8

I
Intake Manifold..........................5-9
  Removal & Installation...............5-9

L
Lower Ball Joint..........................5-19
  Removal & Installation...............5-19

Lower Control Arm 5-19
  Removal & Installation...............5-19
Lower Control Arm Bushing 5-20
  Removal & Installation...............5-20

O
Oil Pan 5-12
  Removal & Installation...............5-12
Oil Pump 5-12
  Removal & Installation...............5-12

P
Piston and Ring 5-13
  Positioning 5-13

R
Rack and Pinion Steering Gear 5-18
  Removal & Installation...............5-18
Rear Main Seal 5-12
  Removal & Installation...............5-12

S
Shock Absorber 5-18
  Removal & Installation...............5-18
Stabilizer Bar 5-19
  Removal & Installation...............5-19
Starter Motor 5-11
  Removal & Installation...............5-11
Steering Knuckle, Hub and Wheel
  Bearing 5-20
  Removal & Installation...............5-20
Strut......................................5-18
  Removal & Installation...............5-18

T
Timing Belt, Sprockets and Front
  Covers 5-12
  Removal & Installation...............5-12
Transaxle Assembly 5-15
  Removal & Installation...............5-15

W
Water Pump..............................5-7
  Removal & Installation...............5-7

SPECIFICATIONS AND MAINTENANCE CHARTS

VEHICLE AND ENGINE IDENTIFICATION CHART

		Engine						Model Year	
Code	Liters	Cu. In.	Cyl.	Fuel Sys.	Engine Type	Eng. Mfg.		Code	Year
6	1.6	97.5	4	MPI	DOHC	Daewoo		4	2004
								5	2005
								6	2006

MPI: Multi-port Fuel Injection

06025-AVEO-C01

GENERAL ENGINE SPECIFICATIONS

Year	Engine Displacement Liters	Engine VIN	Net Horsepower @ rpm	Net Torque @ rpm (ft. lbs.)	Bore x Stroke (in.)	Compression Ratio	Oil Pressure @ rpm
2004	1.6	6	103@6000	107@3600	3.10x3.21	9.5:1	①
2005	1.6	6	103@6200	107@3600	3.10x3.21	9.5:1	①

① Not available

06025-AVEO-C02

GASOLINE ENGINE TUNE-UP SPECIFICATIONS

Year	Engine Displacement Liters	Engine VIN	Spark Plugs Gap (in.)	Ignition Timing (deg.) MT	AT	Fuel Pump (psi)	Idle Speed (rpm) MT	AT	Valve Clearance In.	Ex.
2004	1.6	6	0.039-0.043	5	5	①	②	②	HYD	HYD
2005	1.6	6	0.039-0.043	5	5	①	②	②	HYD	HYD

NOTE: The Vehicle Emission Control Information label often reflects specification changes changes made during production.

The label figures must be used if they differ from those in this chart.

HYD: Hydraulic

① Not available.

② Controlled by the Powertrain Control Module (PCM) and cannot be manually adjusted.

06025-AVEO-C03

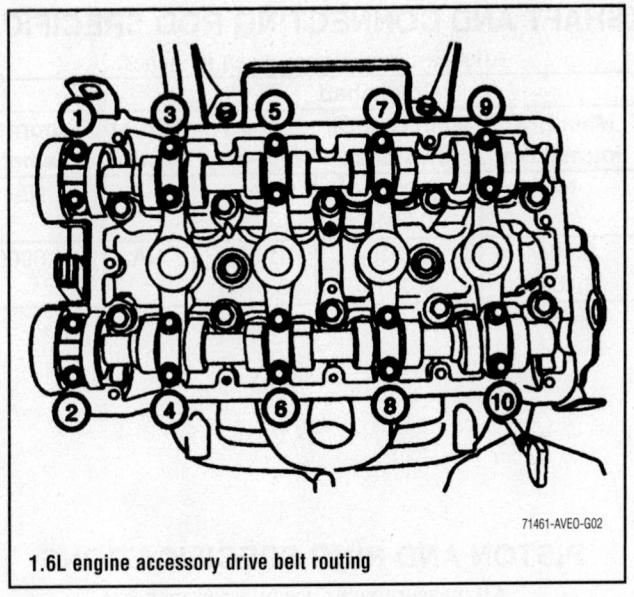

71461-AVEO-G02

1.6L engine accessory drive belt routing

CAPACITIES

Year	Model	Engine Displacement Liters	Engine ID/VIN	Engine Oil with Filter (qts.)	Transmission (pts.) 4-Spd	Transmission (pts.) 5-Spd	Transmission (pts.) Auto.	Drive Axle Front (pts.)	Drive Axle Rear (pts.)	Fuel Tank (gal.)	Cooling System (qts.)
2004	Aveo	1.6	6	3.9	—	3.8	12.40	①	—	11.9	7.0
2005	Aveo	1.6	6	3.9	—	3.8	12.40	①	—	11.9	7.0

NOTE: All capacities are approximate. Add fluid gradually and check to be sure a proper fluid level is obtained.

① Included in transaxle capacity

06025-AVEO-C04

VALVE SPECIFICATIONS

Year	Engine VIN	Engine Displacement Liters	Seat Angle (deg.)	Face Angle (deg.)	Spring Test Pressure (lbs. @ in.)	Spring Installed Height (in.)	Stem-to-Guide Clearance (in.) Intake	Stem-to-Guide Clearance (in.) Exhaust	Stem Diameter (in.) Intake	Stem Diameter (in.) Exhaust
2004	6	1.6	NA	NA	NA	NA	NA	NA	NA	NA
2005	6	1.6	44.5-45	45-45.25	NA	NA	NA	NA	0.234-0.235	0.2336-0.2342

06025-AVEO-C07

CRANKSHAFT AND CONNECTING ROD SPECIFICATIONS

All measurements are given in inches.

Year	Engine Displ. Liters	Engine VIN	Crankshaft				Connecting Rod		
			Main Brg. Journal Dia.	Main Brg. Oil Clearance	Shaft End-play	Thrust on No.	Journal Diameter	Oil Clearance	Side Clearance
2004	1.6	6	NA	0.0010-0.0018	NA	NA	NA	0.0007-0.0027	NA
2005	1.6	6	2.164-2.1650	0.0010-0.0016	0.0020-0.0110	NA	1.6900	0.0007-0.0027	0.0027-0.0090

NA: Not available

06025-AVEO-C06

PISTON AND RING SPECIFICATIONS

All measurements are given in inches.

Year	Engine Displ. Liters	Engine VIN	Piston Clearance	Ring Gap			Ring Side Clearance		
				Top Comp.	Bottom Comp.	Oil Control	Top Comp.	Bottom Comp.	Oil Control
2004	1.6	6	NA	NA	NA	NA	NA	NA	NA
2005	1.6	6	0.0008-0.0016	0.0060-0.0120	0.0120-0.0190	NA	0.0019-0.0031	0.0020-0.0030	NA

NA: Not available

06025-AVEO-C05

TORQUE SPECIFICATIONS

All readings in ft. lbs.

Year	Engine VIN	Engine Displacement Liters	Cylinder Head Bolts	Main Bearing Bolts	Rod Bearing Bolts	Crankshaft Damper Bolts	Flywheel Bolts	Manifold		Spark Plugs	Oil Pan Drain Plug
								Intake	Exhaust		
2004	6	1.6	①	②	③	④	⑤	18	18	18	26
2005	6	1.6	⑥	②	③	④	⑤	18	18	18	26

① Step 1: 18 ft. lbs.
Step 2: plus 70 degrees
Step 3: plus 70 degrees
Step 4: plus 50 degrees

② Step 1: 37 ft. lbs.
Step 2: plus 45 degrees
Step 3: plus 15 degrees

③ Step 1: 18 ft. lbs.
Step 2: plus 30 degrees
Step 3: plus 15 degrees

④ Step 1: 70 ft. lbs.
Step 2: plus 30 degrees
Step 3: plus 15 degrees

⑤ Step 1: 25 ft. lbs.
Step 2: plus 30 degrees
Step 3: plus 15 degrees

⑥ Step 1: 18 ft. lbs.
Step 2: plus 60 degrees
Step 3: plus 60 degrees
Step 4: plus 60 degrees
Step 4: plus 100 degrees

06025-AVEO-C08

WHEEL ALIGNMENT

Year	Model		Caster Range (+/-Deg.)	Caster Preferred Setting (Deg.)	Camber Range (+/-Deg.)	Camber Preferred Setting (Deg.)	Toe-in (Deg.)
2004	Aveo	Front	0.75	1.75	0.75	-0.40	4+/-0.167
		Rear	—	—	0.50	-0.50	0.25+/-0.33
2005	Aveo	Front	0.75	-0.40	0.75	-0.40	4+/-0.167
		Rear	—	—	0.50	-1.50	0.07+/-0.17

06025-AVEO-C09

TIRE AND WHEEL SPECIFICATIONS

Year	Model	OEM Tires Standard	OEM Tires Optional	Tire Pressures (psi) Front	Tire Pressures (psi) Rear	Wheel Size	Wheel Lug Nut Torque (Ft. Lbs.)
2004	Aveo	P185/60R14	—	30	30	①	88
2005	Aveo	P185/60R14	—	30	30	①	88

OEM: Original Equipment Manufacturer

PSI: Pounds Per Square Inch

① Not available

06025-AVEO-C10

BRAKE SPECIFICATIONS

All measurements in inches unless noted

Year	Model		Brake Disc Original Thickness	Brake Disc Minimum Thickness	Brake Disc Maximum Runout	Minimum Lining Thickness Front	Minimum Lining Thickness Rear	Brake Caliper Bracket Bolts (ft. lbs.)	Brake Caliper Mounting Bolts (ft. lbs.)
2004	Aveo	F	①	1.110	0.004	0.280	—	70	20
		R	②	②	②	—	0.020	②	②
2005	Aveo	F	0.945	0.866	0.002	0.280	—	70	20
		R	②	②	②	—	0.020	②	②

① Not available

② Drum brakes used on rear, specifications not available

06025-AVEO-C11

SCHEDULED MAINTENANCE INTERVALS
2004-05 CHEVROLET AVEO

TO BE SERVICED	TYPE OF SERVICE	VEHICLE MILEAGE INTERVAL (x1000)												
		7.5	15	22.5	30	37.5	45	52.5	60	67.5	75	82.5	90	97.5
Engine oil & filter	R	✓	✓	✓	✓	✓	✓	✓	✓	✓	✓	✓	✓	✓
Rotate tires	S/I	✓	✓	✓	✓	✓	✓	✓	✓	✓	✓	✓	✓	✓
Engine coolant strength hoses & clamps	S/I													
Air cleaner filter	R				✓				✓				✓	
Automatic transmission fluid & filter	R												✓	
Engine coolant	R				✓				✓				✓	
PCV valve	S/I				✓				✓				✓	
Spark plugs	R				✓				✓				✓	
Drive belts	S/I		✓		✓		✓		✓		✓		✓	
Front & rear brakes ①	S/I													
Fuel filter	R						✓						✓	
Passenger compartment air filter	R		✓		✓		✓		✓		✓		✓	
Timing belt	S/I				✓		✓		✓				✓	
Evaporative canister	S/I				✓		✓		✓				✓	

R: Replace   S/I: Service or Inspect

① Change clutch/brake fluid every 24 months.

FREQUENT OPERATION MAINTENANCE (SEVERE SERVICE)

If a vehicle is operated under any of the following conditions it is considered severe service:

- Extremely dusty areas.

- 50% or more of the vehicle operation is in 32°C (90°F) or higher temperatures, or constant operation in temperatures below 0°C (32°F).

- Prolonged idling (vehicle operation in stop and go traffic.

- Frequent short running periods (engine does not warm to normal operating temperatures).

- Police, taxi, delivery usage or trailer towing usage.

Engine oil & filter: replace every 3000 miles.

Rotate tires initially at 6000 miles and every 9000 miles thereafter.

Air cleaner filter: change every 15,000 miles.

Engine coolant strength, hoses & clamps: check every 15,000 miles.

Exhaust system: check every 15,000 miles.

Automatic transmission fluid & filter: change every 21,000 miles.

06025-AVEO-C12

ENGINE REPAIR

Alternator

REMOVAL & INSTALLATION

1. Disconnect the negative battery cable.

2. Disconnect the Intake Air Temperature (IAT) sensor connector.

3. Disconnect the air intake tube

4. Remove the accessory drive belt.

5. Remove the alternator wiring connector.

6. Remove the alternator.

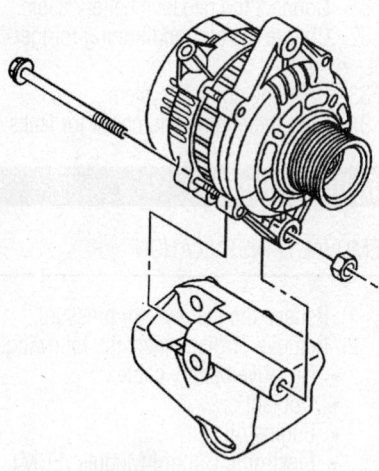

71461-AVEO-G03

Alternator mounting—1.6L engine

To install:

7. Position the alternator on the engine.

8. Install the alternator mounting bolts. Tighten the bolts to 18 ft. lbs. (25 Nm).

9. Tighten the positive cable nut to 11 ft. lbs. (15 Nm).

10. Connect the wiring connector.

11. Install and tension the accessory drive belt.

12. Install the air intake tube.

13. Connect the IAT sensor connector.

14. Connect the negative battery cable.

Engine Assembly

REMOVAL & INSTALLATION

1. Relieve the fuel system pressure.

2. Drain the engine coolant and engine oil.

3. Recover the air conditioning refrigerant, into a refrigerant recovery station

4. Remove the hood.

5. Remove or disconnect the following:
- Negative battery cable
- Air cleaner hose
- Valve cover breather tubes
- Accessory drive belt
- Right front wheel
- Right front wheel well splash shield
- Cooling fans
- Radiator hoses
- Radiator
- Power steering pressure and return lines
- Ignition coil connector
- Electronic Control Module (ECM) connector at intake manifold and starter
- Oxygen sensor connector
- Fuel injector connectors
- Engine sensor connectors
- Alternator
- All necessary vacuum lines
- Fuel feed line at fuel rail
- Throttle cable
- Throttle body cooling line
- Heater hoses from engine
- Surge tank coolant line
- Starter
- A/C compressor
- Exhaust pipe
- Crankshaft pulley
- EGR, knock sensor, oil pressure switch and charcoal canister connectors
- Torque converter bolts on automatic transmissions
- Bellhousing bolts
- Support the transaxle with a jack
- Install an engine lifting device
- Right engine mount bracket and engine mount

6. Carefully separate the engine from the transaxle and remove the engine.

To install:

7. Position the engine onto the transaxle and engage the alignment pins.

8. Install the bellhousing bolts. Torque the bolts to 55 ft. lbs. (75 Nm).

9. Install the right engine mount and mount bracket. Torque the bolts to 41 ft. lbs. (55 Nm).

10. Remove the engine lifting device and transaxle jack.

11. Install the torque converter bolts. Torque the bolts to 33 ft. lbs. (45 Nm).

12. Install or connect the following:
- EGR, knock sensor, oil pressure switch and charcoal canister connectors
- Crankshaft pulley. Torque the bolt

to 70 ft. lbs. (95 Nm), plus an additional 30°, then an additional 15°.
- Exhaust pipe. Torque the bolts to 26 ft. lbs. (35 Nm).
- Power steering pressure and return lines
- A/C compressor
- Alternator
- Starter
- Surge tank coolant line
- Heater hoses
- Throttle body cooling line
- Throttle cable
- Fuel feed line at fuel rail
- Vacuum lines
- Engine sensor connectors
- Fuel injector connectors
- Oxygen sensor connector
- ECM connector at intake manifold and starter
- Ignition coil connector
- Power steering pressure and return lines
- Radiator
- Radiator hoses
- Cooling fans
- Right front wheel well splash shield
- Right front wheel
- Accessory drive belt
- Valve cover breather tubes
- Air cleaner hose
- Negative battery cable
- Hood

13. Charge the air conditioning refrigerant using approved recycling equipment.

14. Fill the engine with coolant and engine oil.

15. Bleed the power steering system.

16. Start the engine and check for leaks.

Water Pump

REMOVAL & INSTALLATION

1. Remove or disconnect the following:
- Negative battery cable
- Coolant
- Radiator hoses
- Rear timing belt cover
- Water pump

To install:

2. Clean the gasket surfaces on the water pump and block.

3. Position a new water pump housing gasket on the water pump sealing surface using gasket sealant to hold the gasket in place.

4. Install the water pump to the block

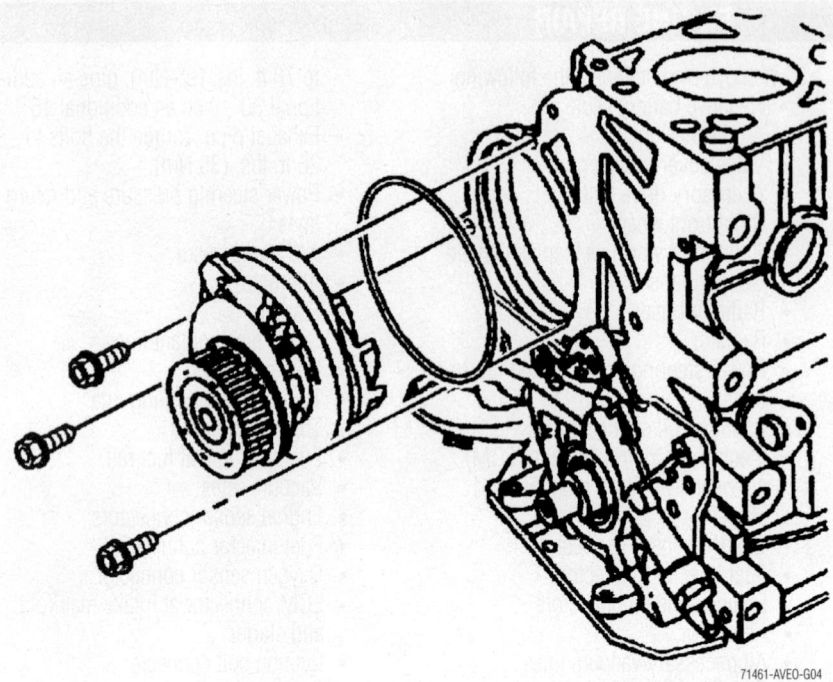

Water pump mounting—1.6L engine

71461-AVEO-G04

with the flange aligned with the recess of the rear timing belt cover.

5. Install the water pump and tighten the bolts to 89 inch lbs. (10 Nm).

6. Install the rear timing belt cover.

7. Install the radiator hoses.

8. Fill the engine with coolant.

9. Start the engine and check for leaks.

Heater Core

REMOVAL AND INSTALLATION

1. Recover the air conditioning refrigerant, into a refrigerant recovery station

2. Disconnect the negative battery cable.

3. Drain the cooling system into a clean container for reuse.

4. Disconnect the heater hoses at the firewall.

5. Remove 2 driver air bag module mounting bolts and discard them.

6. Disconnect the connector and remove the driver air bag.

7. With the front wheels in the straight ahead position, remove the steering wheel.

8. Remove the lower instrument trim panel.

9. Remove the steering column upper and lower trim panels.

10. Remove the turn signal and wiper switch.

11. Remove the A pillar garnish moldings.

12. Remove the instrument cluster.

13. Remove the instrument panel side covers, lower cover and upper and lower center covers.

14. Remove the ashtray, cigar lighter and cup holder.

15. Remove the A/C-heater control panel screws, pull out the control panel and disconnect the connectors and control cables, noting the cable locations.

16. Remove the A/C-heater control panel.

17. Remove the radio and clock.

18. Remove the glove box.

19. Disconnect the passenger side air bag connector, then remove the air bag.

20. Remove the instrument panel bolts behind the A/C-heater control panel opening.

21. Remove the instrument panel bolts below the steering column and at the ends of the instrument panel.

22. Remove the instrument panel end screws.

23. Remove the tie bar retaining bolts.

24. Remove the instrument panel.

25. Remove the A/C suction hose and liquid evaporator pipe connector block at the cowl.

26. From the firewall remove the screws that secure the heater/air distribution case.

assembly to the cowl.

27. Remove the heater/air distribution case.

28. Disconnect the control cables from the case.

29. Remove the heater core covers and clamps and remove the heater core.

To install:

30. Installation is the reverse of removal. Please note the following torque specifications:

- Tighten the instrument panel bolts to 15 ft. lbs. (20 Nm).
- Tighten the passenger air bag module mounting bolts to 97 inch lbs. (11 Nm).
- Tighten the driver air bag module mounting bolts to 71 inch lbs. (8 Nm).
- Tighten the steering wheel bolt to 28 ft. lbs. (38 Nm).

31. Connect the negative battery cable.

32. Charge the air conditioning refrigerant.

33. Refill the cooling system.

34. Start the engine and check for leaks.

Cylinder Head

REMOVAL & INSTALLATION

1. Relieve the fuel system pressure.

2. Remove or disconnect the following:

- Negative battery cable
- Coolant
- Engine oil
- Electronic Control Module (ECM) ground terminal from intake manifold
- Air cleaner housing
- Throttle cable
- Radiator hoses
- Intake manifold pressure sensor
- Intake Air Temperature (IAT) sensor
- Coolant temperature sensor
- Fuel feed line from fuel rail
- Intake manifold support bracket
- Intake manifold vacuum hoses
- Spark plug wires
- Direct Ignition Coil (DIS) and bracket
- Accessory drive belt
- Right front tire
- Right wheelwell splash shield
- Upper front timing cover
- Align the camshaft gear timing marks
- Crankshaft pulley
- Lower front timing belt cover
- Slightly loosen the water pump bolts
- Rotate the water pump counterclockwise to relieve the timing belt tension

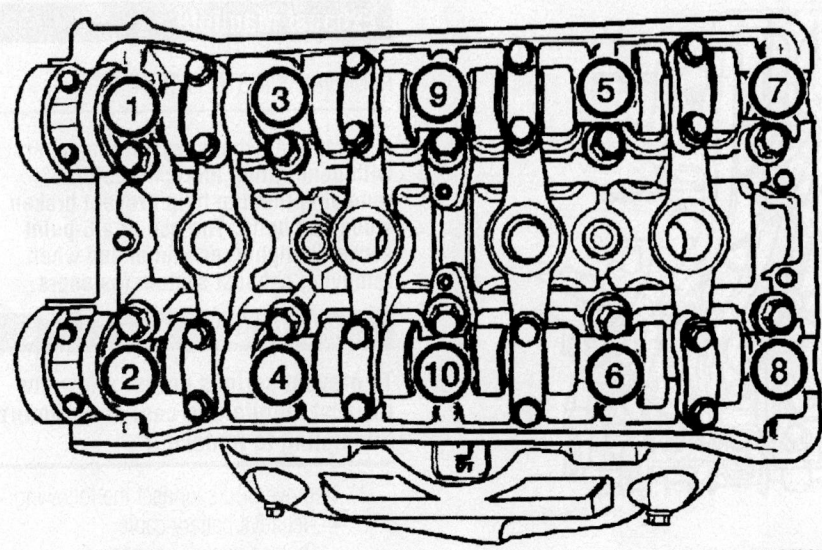

71461-AVEO-G05

Cylinder head bolt removal sequence—1.6L engine

- Remove the timing belt
3. Support the engine with a lifting device, then remove the engine mounting bracket
4. Remove the valve cover.
5. While holding the intake camshaft firmly in place, remove the intake camshaft gear bolt and gear.
6. While holding the exhaust camshaft firmly in place, remove the exhaust camshaft gear bolt and gear.
7. Remove or disconnect the following:
- Timing belt tensioner
- Idler pulley
- Right engine mount
- Rear timing belt cover
- Intake manifold
- Exhaust pipe
- Exhaust manifold
8. Remove the cylinder head bolts gradually in the sequence shown and remove the cylinder head.

To install:
9. The cylinder head should be cleaned and inspected prior to installation.
10. Lightly oil all bolt and stud bolt threads before installation.
11. Clean all gasket mating surfaces thoroughly.
12. Install a new head gasket on the cylinder block.

> ❋❋ **WARNING**
>
> **Always use new cylinder head bolts when installing the cylinder head or damage to the engine may occur.**

13. Install the cylinder head on the cylinder block.

14. Tighten the cylinder head bolts in reverse of the removal sequence. Tighten the 2004 cylinder head bolts to 18 ft. lbs. (25 Nm). Tighten an additional 70°, plus 70°, plus another 50°. Tighten 2005 bolts to 18 ft. lbs. (25 Nm). Tighten an additional 60°, plus 60°, plus another 10°.
15. Install the right engine mount. Tighten the engine mount attaching bolts to 22 ft. lbs. (30 Nm).
16. Install or connect the following:
- Intake manifold
- Intake manifold support bracket. Tighten the bolts to 18 ft. lbs. (25 Nm).
- Throttle cable
- Surge tank coolant hose
- Heater hose to cylinder head
- Fuel feed line at the fuel rail
- All vacuum hoses
- Exhaust manifold
- Rear timing belt cover
- Timing belt tensioner. Tighten the bolts to 18 ft. lbs. (25 Nm).
- Idler pulley. Tighten the bolts to 18 ft. lbs. (25 Nm).
- Intake and exhaust camshaft gears. Tighten the bolts to 49 ft. lbs. (68 Nm).
17. Apply a small amount of gasket sealant to the corners of the front camshaft caps and the top of the rear camshaft cover-to-cylinder head seal.
18. Install the valve cover using a new gasket
19. Align the timing marks on the camshaft gears.
20. Align the mark on the crankshaft gear with the notch at the bottom of the rear timing belt cover.

21. Install the timing belt and rotate the water pump clockwise to apply tension to the timing belt.
22. Tighten the water pump bolts to 89 inch lbs. (10 Nm).
23. Adjust the timing belt tension.
24. Install or connect the following:
- Spark plug wires
- Upper and lower timing belt covers
- Crankshaft pulley. Tighten the bolt to 70 ft. lbs. (95 Nm), plus an additional 30°, then another 15°.
- Accessory drive belt
- Air cleaner housing
- Radiator hoses
- Front splash shield
- Right front wheel
- Coolant temperature sensor
- Fuel feed line from fuel rail
- IAT sensor
- Intake manifold pressure sensor
- DIS coil connector
- Fuel injector connectors
- ECM ground terminal
- Negative battery cable
25. Fill the engine with oil and coolant.
26. Start the engine and check for leaks.

Intake Manifold

REMOVAL & INSTALLATION

1. Relieve the fuel system pressure.
2. Remove or disconnect the following:
- Negative battery cable
- Coolant
- Electronic Control Module (ECM) ground terminal from intake manifold
- Air cleaner housing
- Throttle body
- Throttle cable
- Radiator hoses
- Intake manifold pressure sensor
- Intake Air Temperature (IAT) sensor
- Coolant temperature sensor
- Fuel feed line from fuel rail
- EGR pipe
- Alternator
- Intake manifold support bracket
- Intake manifold vacuum hoses
- Intake manifold retaining bolts and in the sequence shown.

To install:
3. Install or connect the following:
- Intake manifold with new gasket. Tighten the retainers in sequence to 18 ft. lbs. (25 Nm).
- Fuel rail. Tighten the bolts to 18 ft. lbs. (25 Nm).

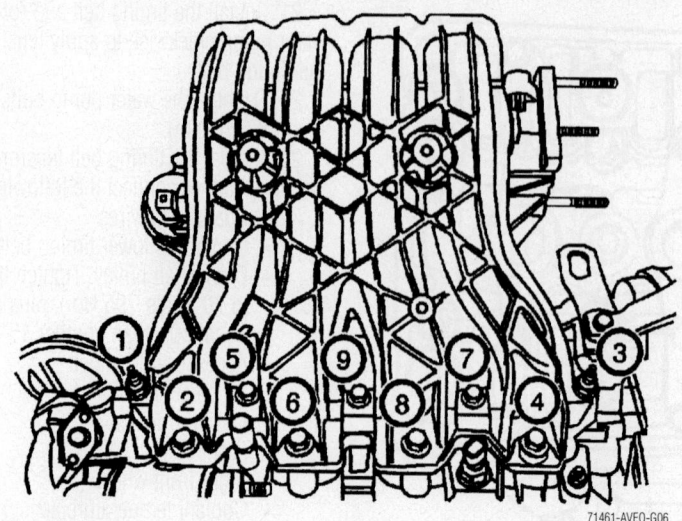

Intake manifold bolt removal sequence—1.6L engine

71461-AVEO-G06

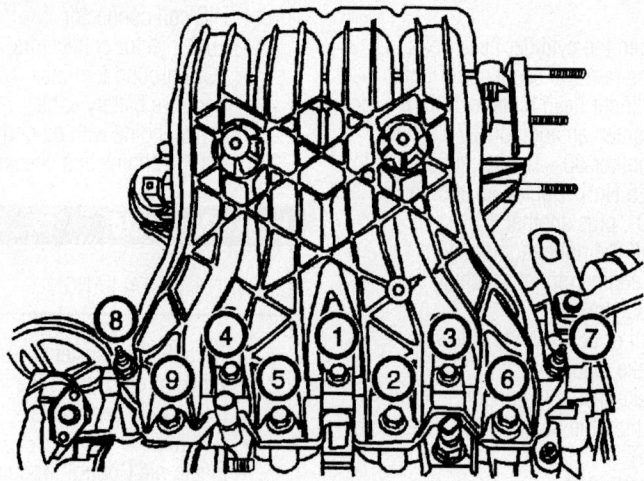

Intake manifold bolt tightening sequence—1.6L engine

71461-AVEO-G07

- EGR pipe
- Intake manifold support bracket. Tighten the upper bolts to 18 ft. lbs. (25 Nm) and the lower bolts to 33 ft. lb. (45 Nm).
- Alternator
- Intake manifold vacuum hoses
- IAT sensor
- Intake manifold pressure sensor
- Radiator hoses
- Throttle cable
- Throttle body
- Air cleaner housing
- ECM ground terminal from intake manifold
- Negative battery cable

4. Fill the engine with coolant.

5. Start the engine and check for leaks.

Exhaust Manifold

REMOVAL & INSTALLATION

➡Spray the exhaust system fasteners with penetrating lubricant before removing them to help prevent broken studs and bolts. The use of a 6-point socket is highly recommended when removing exhaust system fasteners.

✳✳ CAUTION

To prevent serious burns, allow the exhaust manifold to cool down before attempting to remove it.

1. Remove or disconnect the following:
   - Negative battery cable
   - Oxygen sensor connector
   - Manifold heat shield
   - Exhaust pipes
   - Exhaust manifold nuts in the sequence shown.
   - Exhaust manifold

To install:

2. Clean all gasket mating surfaces thoroughly.

3. Install a new exhaust manifold gasket and the exhaust manifold on the cylinder head. Start 2 nuts to hold the manifold in position. Tighten the nuts in sequence to 18 ft. lbs. (25 Nm).

4. Remove or disconnect the following:
   - Exhaust pipes. Tighten the nuts to 30 ft. lbs. (40 Nm).
   - Heat shield
   - Oxygen sensor
   - Negative battery cable

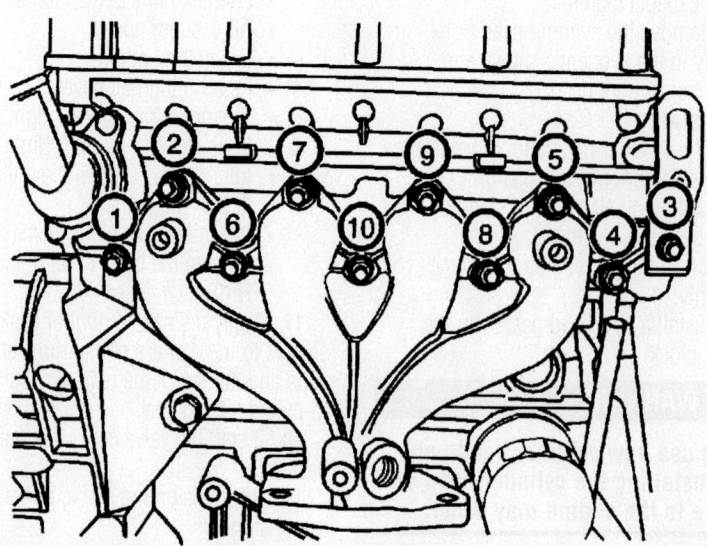

71461-AVEO-G08

Exhaust manifold bolt removal sequence—1.6L engine

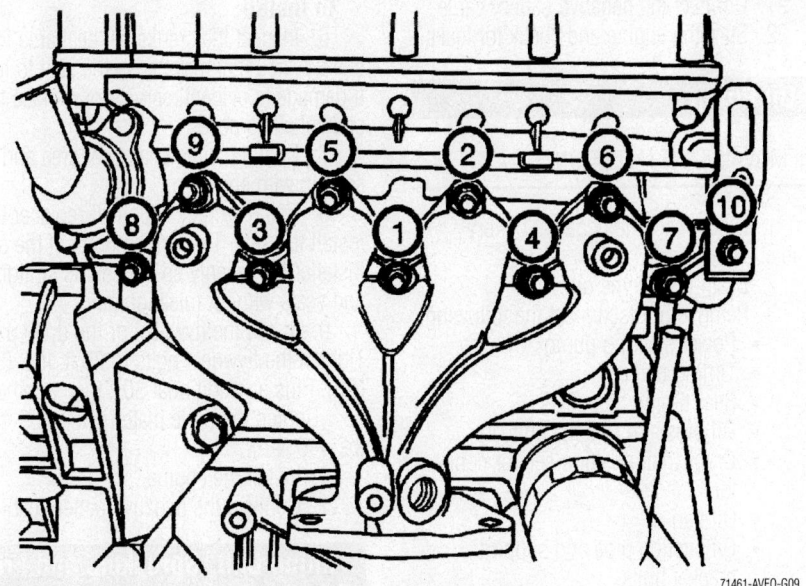

Exhaust manifold bolt installation sequence—2004 1.6L engine

71461-AVEO-G09

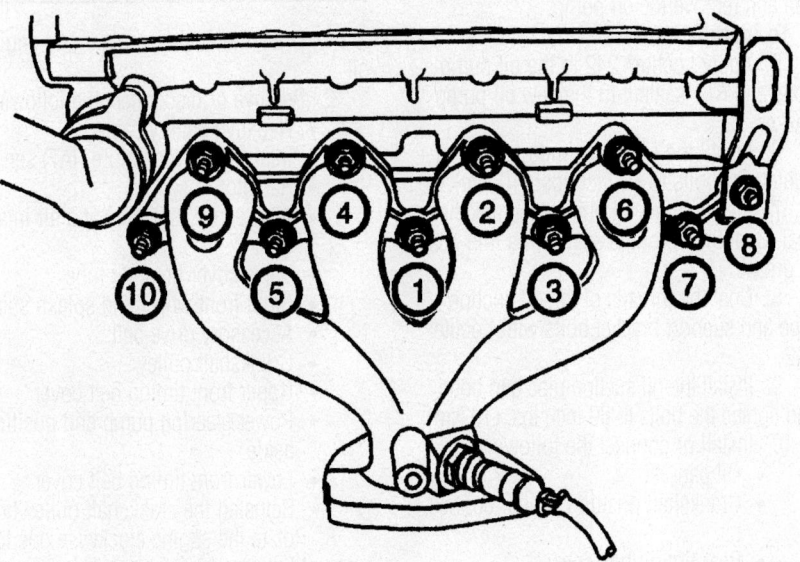

Exhaust manifold bolt installation sequence—2005 1.6L engine

06025-AVEO-G01

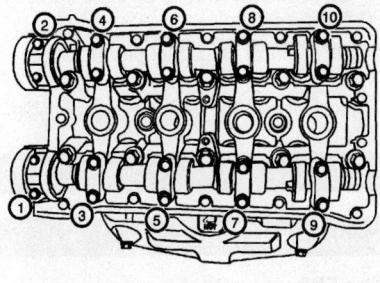

Camshaft bearing cap bolt removal and
installation sequence—2004 1.6L engine

71461-AVEO-G10

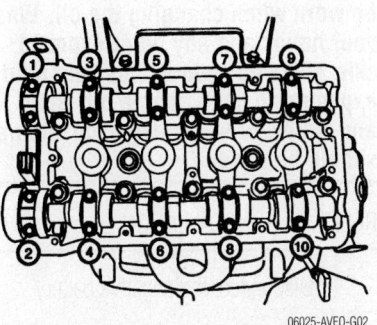

Camshaft bearing cap bolt removal and
installation sequence—2005 1.6L engine

06025-AVEO-G02

Camshafts

REMOVAL & INSTALLATION

1. Relieve the fuel system pressure.
2. Remove or disconnect the following:
   - Negative battery cable
   - Valve cover
   - Timing belt
3. While holding the intake camshaft firmly in place, remove the intake camshaft gear bolt and gear.
4. While holding the exhaust camshaft firmly in place, remove the exhaust camshaft gear bolt and gear.

➡Mark the camshaft cap positions for installation reference.

5. Remove the camshaft cap bolts gradually in the sequence shown.
6. Remove the exhaust and intake camshafts

To install:

7. Lubricate the camshaft journals and the camshaft caps with clean engine oil.

8. Install the intake/exhaust camshaft caps in their original positions, and tighten the cap bolts in reverse of the removal sequence to 12 ft. lbs. (16 Nm).

9. Install the camshaft gears.

10. While firmly holding the camshafts in place, tighten the camshaft gear bolts to 49 ft. lbs. (68 Nm).
11. Install the timing belt.
12. Install the valve cover.
13. Connect the negative battery cable.

Starter Motor

REMOVAL & INSTALLATION

1. Disconnect the negative battery cable.
2. Raise and support the vehicle safely.
3. Disconnect the starter electrical harness.
4. Remove the starter ground bolt.
5. Support the starter and remove the bolts.
6. Remove the starter from the vehicle.

To install:

7. Before servicing the vehicle, refer to the precautions in the beginning of this section.

8. Install the starter. Tighten the mounting bolts to 17 ft. lbs. (23 Nm) on 2004 models, or 32 ft. lbs. (43 Nm) on 2005 models. Tighten the solenoid nut to 11 ft. lbs. (15 Nm).

9. Install the ground bolt and tighten to 30 ft. lbs. (41 Nm).

10. Connect the starter electrical harness.

11. Connect the negative battery cable.

Oil Pan

✳✳ CAUTION

The EPA warns that prolonged contact with used engine oil may cause a number of skin disorders, including cancer! You should make every effort to minimize your exposure to used engine oil. Protective gloves should be worn when changing the oil. Wash your hands and any other exposed skin areas as soon as possible after exposure to used engine oil. Soap and water, or waterless hand cleaner, should be used.

REMOVAL & INSTALLATION

1. Disconnect the negative battery cable.

2. Raise and support the vehicle safely on jack stands.

3. Drain the engine oil.

4. Remove the right wheel.

5. Remove the right side splash shield.

6. Disconnect the heated oxygen sensor connector.

7. Remove the exhaust pipe and catalytic converter as a unit.

8. Remove the lower crossmember bracket, if necessary.

9. Remove the oil pan-to-transaxle case bolts.

10. Remove the oil pan-to-engine block bolts.

11. Remove the oil pan.

To install:

12. Clean the gasket mating surfaces thoroughly.

13. Coat the oil pan gasket with sealant.

14. Install the oil pan and tighten the bolts to 89 inch lbs. (10 Nm).

15. Tighten the oil-pan-to-transaxle bolts to 23 ft. lbs. (31 Nm).

16. Connect the exhaust pipe and catalytic converter. Tighten the nuts to 37 ft. lbs. (50 Nm). Tighten the front muffler nuts to 22 ft. lbs. (30 Nm).

17. Install the lower crossmember bracket, if removed.

18. Connect the oxygen sensor connector.

19. Install the splash shield and wheel.

20. Fill the engine with oil.

21. Connect the negative battery cable.

22. Start the engine and check for leaks.

Oil Pump

REMOVAL & INSTALLATION

1. Disconnect the negative battery cable.

2. Drain the engine oil.

3. Remove or disconnect the following:
   - Power steering pump
   - Timing belt
   - Rear timing belt cover
   - Oil pressure switch connector
   - Crankshaft position sensor connector
   - Oil pan
   - Oil suction pipe and support bracket bolts
   - Oil pump bolts

4. Carefully separate the oil pump and gasket from the engine block and the oil pan and remove the oil pump.

To install:

5. Apply Loctite® 242 to the oil pump bolts and RTV sealant to the new oil pump gasket.

6. Install the oil pump and gasket and tighten the bolts to 89 inch lbs. (10 Nm).

7. Install a new oil pump-to-crankshaft seal. Coat the lip of the seal with a thin coat of grease.

8. Coat the threads of the oil suction pipe and support bracket bolts with Loctite® 242.

9. Install the oil suction pipe and bolts and tighten the bolts to 89 inch lbs. (10 Nm).

10. Install or connect the following:
    - Oil pan
    - Crankshaft position sensor connector
    - Rear timing belt cover
    - Timing belt
    - Power steering pump
    - Negative battery cable

11. Fill the engine with oil.

12. Start the engine and check for leaks.

Rear Main Seal

REMOVAL & INSTALLATION

1. Disconnect the negative battery cable.

2. Remove the engine.

3. Remove the flywheel or drive plate bolts.

4. Remove the flywheel or drive plate

5. Remove the crankshaft rear oil seal.

To install:

6. Inspect the crankshaft seal area for any damage that may cause the seal to leak. If damage is evident, service or replace the crankshaft as necessary.

7. Coat the crankshaft seal area and the seal lip with engine oil.

8. Using a crankshaft seal replacer tool, install the seal. Tighten the bolts of the seal installer tool evenly so the seal is straight and seats without misalignment.

9. Install the flywheel or the drive plate. Tighten the flywheel bolts to 25 ft. lbs. (35 Nm). Plus an additional 30°, then another 15°. Tighten the drive plate bolts to 33 ft. lbs. (45 Nm).

10. Install the engine.

11. Connect the negative battery cable.

Timing Belt, Sprockets and Front Covers

REMOVAL & INSTALLATION

1. Drain the engine coolant and engine oil.

2. Remove or disconnect the following:
   - Negative battery cable
   - Intake Air Temperature (IAT) sensor
   - Radiator hoses
   - Air cleaner assembly and air intake duct
   - Valve cover breather tube
   - Right front wheel and splash shield
   - Accessory drive belt
   - Crankshaft pulley
   - Upper front timing belt cover
   - Power steering pump and position aside
   - Lower front timing belt cover
   - Reinstall the crankshaft pulley bolt

3. Rotate the engine clockwise one full turn until the mark on the crankshaft gear aligns with the notch at the bottom of the rear timing cover.

4. Align the camshaft gear timing marks.

5. Slightly loosen the water pump bolts.

6. Using tool J-42492-A, turn the water pump clockwise until the adjuster arm pointer of the timing belt tensioner is aligned with the notch in the tensioner bracket.

7. Tighten the water pump bolts.

8. Remove the timing belt.

9. Remove the crankshaft and camshaft gears.

10. Remove the timing belt tensioner.

11. Remove the idler pulley.

12. Remove the rear timing belt cover.

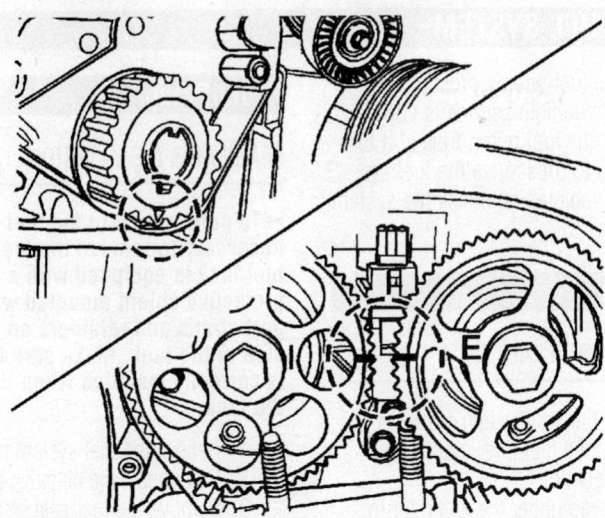

Camshaft gear timing marks
71461-AVEO-G11

Aligning the crankshaft and camshaft gear timing marks—1.6L engine

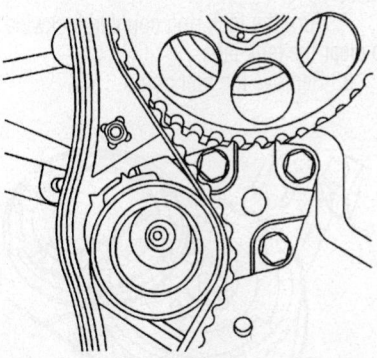

71461-AVEO-G12

**Aligning the timing belt tensioner pointer
with the notch in the bracket—1.6L engine**

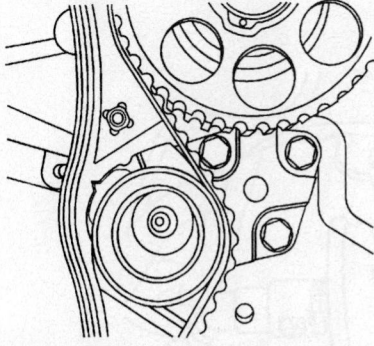

71461-AVEO-G13

**Aligning the timing belt tensioner pointer
with the pointer on the bracket—1.6L
engine**

- Air cleaner assembly and air intake duct
- Intake Air Temperature (IAT) sensor
- Radiator hoses
- Engine oil
- Coolant
- Negative battery cable

Piston and Ring

POSITIONING

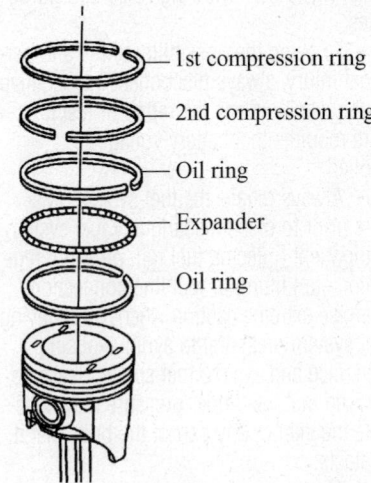

- 1st compression ring
- 2nd compression ring
- Oil ring
- Expander
- Oil ring

71461-AVEO-G14

Piston ring positioning—1.6L engine

To install:

13. Install the rear timing belt cover. Tighten the bolts to 89 inch lbs. (10 Nm).

14. Install the idler pulley. Tighten the bolt to 30 ft. lbs. (40 Nm).

15. Install the timing belt tensioner. Tighten the bolts to 18 ft. lbs. (25 Nm).

16. Install the crankshaft gear.

17. Install the camshaft gears.

18. While firmly holding the camshafts in place, tighten the camshaft gear bolts to 49 ft. lbs. (68 Nm).

19. Install the timing belt.

20. Rotate the crankshaft 2 full turns clockwise using the pulley bolt.

21. Slightly loosen the water pump bolts.

22. Using tool J-42492-A, turn the water pump until the adjuster arm pointer of the timing belt tensioner is aligned with the pointer on the tensioner bracket.

23. Tighten the water pump bolts. Tighten the bolts to 89 inch lbs. (10 Nm).

24. Remove the crankshaft pulley bolt.

25. Install the upper and lower timing belt covers. Tighten the bolts to 89 inch lbs. (10 Nm).

26. Install the crankshaft pulley bolt. Tighten the bolt to 70 ft. lbs. (95 Nm) plus an additional 15°.

27. Install or connect the following:
- Accessory drive belt
- Right front wheel and splash shield
- Valve cover breather tube

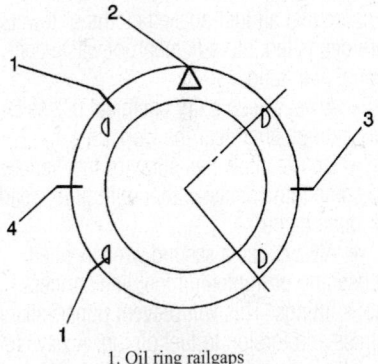

1. Oil ring railgaps
2. Piston notch
3. Compression ring gaps
4. Top compression ring

71461-AVEO-G15

Piston ring gap positioning—1.6L engine

FUEL SYSTEM

Fuel System Service Precautions

Safety is the most important factor when performing not only fuel system maintenance, but also any type of maintenance. Failure to conduct maintenance and repairs in a safe manner may result in serious personal injury or death. Work on a vehicle's fuel system components can be accomplished safely and effectively by adhering to the following rules and guidelines.

• To avoid the possibility of fire and personal injury, always disconnect the negative battery cable unless the repair or test procedure requires that battery voltage by applied.

• Always relieve the fuel system pressure prior to disconnecting any fuel system component (injector, fuel rail, pressure regulator, etc.) fitting or fuel line connection. Exercise extreme caution whenever relieving fuel system pressure, to avoid exposing skin, face and eyes to fuel spray. Please be advised that fuel under pressure may penetrate the skin or any part of the body that it contacts.

• Always place a shop towel or rag around the fitting or connection prior to loosening to absorb any excess fuel due to spillage. Ensure that all fuel spillage is quickly removed from engine surfaces. Ensure that all fuel-soaked cloths or towels are deposited into a flameproof waste container with a lid.

• Always keep a dry chemical (Class B) fire extinguisher near the work area.

• Do not allow fuel spray or fuel vapors to come into contact with a light bulb, spark or open flame.

• Always use a second wrench when loosening or tightening fuel line connections fittings. This will prevent unnecessary stress and torsion to fuel piping. Always follow the proper torque specifications.

• Always replace worn fuel fitting O-rings with new ones. Do not substitute fuel hose where rigid pipe is installed.

Fuel System Pressure

RELIEVING

Remove the fuel filler cap. Locate the fuel pump fuse EF10 in the engine compartment fuse box. Remove the fuel pump fuse. Start the engine and allow it to idle until it stalls. Crank the engine for 10 sec-onds to ensure fuel supply pressure is released. When vehicle service is complete, reinstall the fuel pump fuse and turn the ignition on to pressurize the fuel system. Start the vehicle and check the system for leaks.

Fuel Filter

REMOVAL & INSTALLATION

1. Relieve the fuel system pressure.
2. Remove the fuel filter mounting bracket assembly bolt.
3. Place a rag under the fuel filter to catch any residual fuel that may leak out when the filter is removed.
4. Remove the fuel filter cover.
5. Disconnect the inlet/outlet fuel lines by moving the line connector lock forward and pulling the hose off of the fuel filter tube.
6. Remove the fuel filter by pulling it from the bracket.

To install:

7. Install the fuel filter into the retaining clamp, ensuring proper direction of flow as noted earlier.
8. Connect the inlet and outlet lines and secure the connector lock.
9. Install the fuel filter cover.
10. Tighten the mounting bracket bolts to 35 inch lbs. (4 Nm).
11. Start the engine and check the filter connections for leaks by running the tip of your finger around each connection.

Fuel Pump

REMOVAL & INSTALLATION

➡To gain access to the fuel pump, it is necessary to remove the fuel tank. The fuel tank is equipped with a composite protective shield mounted with the support straps and retainers on the right side of the tank. Make sure the shield is correctly installed when installing the tank.

1. Relieve the fuel system pressure.
2. Disconnect the negative battery cable.
3. Remove the rear seat.
4. Remove the fuel pump access cover.
5. Disconnect the fuel pump electrical connector.
6. Disconnect the fuel inlet and return lines.
7. Turn the lock ring counterclockwise to clear the tank tabs.
8. Remove the fuel pump.

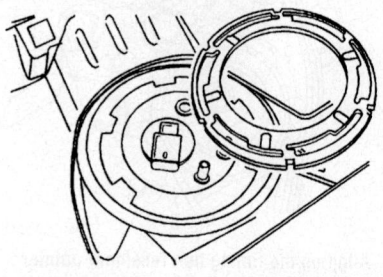

71461-AVEO-G17
Fuel pump locking ring—1.6L engine

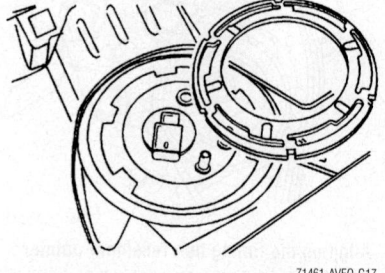

71461-AVEO-G16
Fuel filter mounting—1.6L engine

To install:

9. Clean the gasket mating surface, then install a new gasket.

10. Install the fuel pump in the same location as when removed.

11. Position the locking ring in place, then turn it clockwise to contact the stop.

12. Connect the fuel pump electrical connector.

13. Connect the fuel inlet and return lines.

14. Install the fuel pump access cover.

15. Install the rear seat.

16. Connect the negative battery cable.

17. Start the engine and check for leaks.

Fuel Rail and Injectors

REMOVAL & INSTALLATION

1. Disconnect the negative battery cable.

2. Relieve the fuel system pressure.

3. Remove the intake manifold support bracket.

4. Disconnect the fuel injector harness connectors.

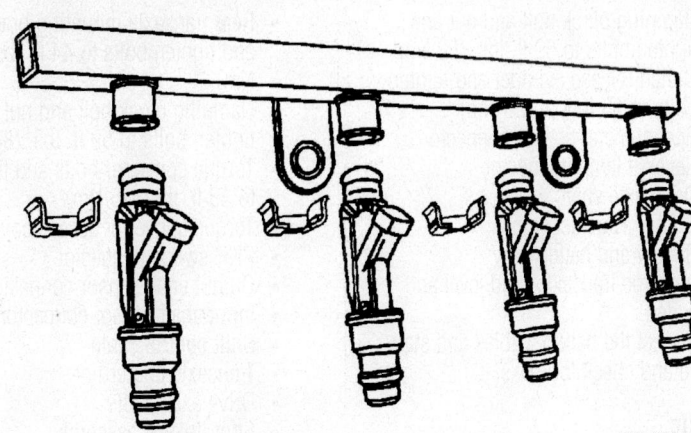

71461-AVEO-G18

Fuel injectors and fuel rail—1.6L engine

5. Disconnect the fuel feed line.

6. Remove the fuel rail, with injectors attached.

7. Remove the retaining clips and remove the injectors by pulling downward and out.

To install:

8. Lubricate new O-rings with clean engine oil and position onto the injectors.

9. Install the fuel injectors on the fuel rail with the injector terminals facing outward.

10. Install the retaining clips.

11. Install the fuel rail. Tighten the bolts to 18 ft. lbs. (25 Nm).

12. Connect fuel feed line.

13. Connect the fuel injector connectors.

14. Install the intake manifold bracket.

15. Start the engine and check for fuel leaks.

DRIVE TRAIN

Transaxle Assembly

REMOVAL & INSTALLATION

Manual

1. Remove or disconnect the following:
   - Battery and battery tray
   - Shift linkage assembly
   - Drive axle shaft
   - Reverse lamp connector
   - Speedometer sensor connector
   - Clutch release cylinder
   - Damping block connection bolt and nut
   - Damping block from front crossmember
   - Rear transaxle mounting bracket
   - Cage retaining bolts
   - Transaxle upper mounting bracket and cage
   - 3 transaxle-to-engine upper retaining bolts.

2. Support the transaxle with a transmission jack.

3. Remove the 7 lower transaxle-to-engine mounting bolts.

4. Remove the transaxle sideways away from the engine and remove the transaxle.

To install:

5. Position the transaxle sideways toward the engine until the input shaft is in the clutch disc and then position it to the engine block.

6. Install the 7 transaxle-to-engine lower bolts and tighten the bolts as follows:
   - Bolts no. 1: 54 ft. lbs. (73 Nm)
   - Bolts no. 2: 15 ft. lbs. (21 Nm)
   - Bolts no. 3: 23 ft. lbs. (31 Nm)

7. Install the 3 transaxle-to-engine upper bolts and tighten the bolts to 54 ft. lbs. (73 Nm).

8. Install or connect the following:
   - Cage and retaining bolts
   - Upper transaxle mounting bracket and tighten bolts to 44 ft. lbs. (60 Nm)
   - Rear damper block to crossmember and tighten bolts to 41 ft. lbs. (55 Nm)
   - Rear transaxle mounting bracket and tighten bolts to 59 ft. lbs. (80 Nm)

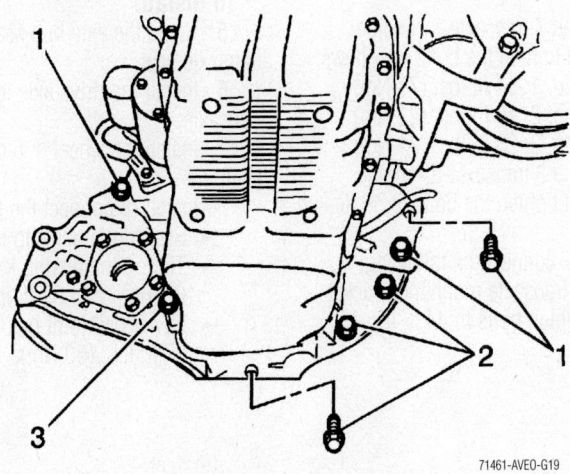

71461-AVEO-G19

Lower transaxle mounting bolt locations—Aveo

- Damping block bolt and nut and tighten bolts to 59 ft. lbs. (80 Nm)
- Clutch release cylinder and tighten bolts to 15 ft. lbs. (20 Nm)
- Speedometer sensor connector
- Reverse lamp connector
- Drive axle shaft
- Shift linkage assembly
- Battery and battery tray

9. Check the transaxle fluid level and add as needed.

10. Connect the battery cables and start the vehicle and check for leaks.

Automatic

1. Remove or disconnect the following:
- Battery and battery tray
- Shift linkage assembly
- Drive axle shafts
- Transaxle cooler lines
- Shift control cable
- Input shaft sensor connector
- Output shaft sensor connector
- Park/Neutral Position (PNP) switch connector
- Torque converter access cover
- Torque converter bolts
- Damping block connection bolt and nut
- Rear transaxle mounting bracket
- Transaxle upper mounting bracket and cage
- 3 transaxle-to-engine upper retaining bolts.

2. Support the transaxle with a transmission jack.

3. Remove the 7 lower transaxle-to-engine mounting bolts.

4. Remove the transaxle sideways away from the engine and remove the transaxle.

To install:

5. Position the transaxle sideways toward the engine and then position it to the engine block.

6. Install the 7 transaxle-to-engine lower bolts and tighten the bolts as follows:
- Bolts no. 1: 54 ft. lbs. (73 Nm)
- Bolts no. 2: 15 ft. lbs. (21 Nm)
- Bolts no. 3: 23 ft. lbs. (31 Nm)

7. Install the 3 transaxle-to-engine upper bolts and tighten the bolts to 54 ft. lbs. (73 Nm).

8. Install or connect the following:
- Upper transaxle mounting bracket and tighten bolts to 44 ft. lbs. (60 Nm)

- Rear transaxle mounting bracket and tighten bolts to 44 ft. lbs. (60 Nm)
- Damping block bolt and nut and tighten bolts to 59 ft. lbs. (80 Nm)
- Torque converter bolts and tighten to 33 ft. lbs. (45 Nm)
- Torque converter access cover
- PNP switch connector
- Output shaft sensor connector
- Input shaft sensor connector
- Shift control cable
- Transaxle cooler lines
- Drive axle shafts
- Shift linkage assembly
- Battery and battery tray

9. Check the transaxle fluid level and add as needed.

10. Connect the battery cables and start the vehicle and check for leaks.

Halfshafts

REMOVAL & INSTALLATION

1. Raise and support the vehicle.

2. Remove or disconnect the following:
- Front wheels
- Underbody splash shields
- Brake caliper and wire aside
- Brake rotor
- Axle shaft nut
- Lower ball joint nut
- Separate steering knuckle from ball joint
- Tie rod nut
- Separate the tie rod from the steering knuckle

3. Push the drive axle away from the wheel hub

4. Separate the axle from the transaxle and remove the axle.

To install:

5. Clean the axle hub seal and the transaxle seal.

6. Install the drive axle into the transaxle.

7. Install the wheel hub onto the axle shaft.

8. Install or connect the following:
- Steering knuckle to lower ball joint
- Tie rod to steering knuckle. Tighten tie rod nut to 33 ft. lbs. (45 Nm).
- Lower ball joint nut. Tighten nut to 37 ft. lbs. (50 Nm).

9. Loosely install a new axle shaft nut.

10. Install the brake rotor and caliper.

11. Install the road wheels and tighten the lug nuts snug.

12. Lower the vehicle to the ground and tighten the lug nuts to 88 ft. lbs. (120 Nm).

13. Tighten the axle shaft nut to 221 ft. lbs. (300 Nm).

14. Install the underbody splash shields.

15. Check the transaxle fluid level and add as needed.

CV-Joint

OVERHAUL

Outer Joint

1. Remove the large and small boot clamps and discard the clamps.

2. Remove the boot.

3. Remove the grease from the joint.

4. Compress the snap ring and remove the joint from the shaft.

5. Remove the seal from the axle shaft.

6. Install a new seal on the axle shaft

7. Install a new joint to the shaft.

8. Fill the joint seal with 3.9–4.6 ounces of grease.

9. Install the new boot.

10. Install a new large and small clamp on the boot and crimp the clamps.

Inner Joint

1. Remove the large and small boot clamps and discard the clamps.

2. Separate the joint housing from the boot.

3. Remove the grease from the tripod.

4. Remove the axle shaft retaining ring.

5. Remove the tripod joint retaining ring.

6. Remove the tripod joint seal.

7. Install a new small seal clamp on the seal.

8. Install the seal onto the axle shaft.

9. Install the axle shaft retaining ring.

10. Fill the tripod joint with 6.9–7.6 ounces of grease on auto. trans. models, or 4.2–4.96 ounces of grease on manual transaxle models.

11. Install the new boot.

12. Install a new large and small clamp on the boot and crimp the clamps.

STEERING AND SUSPENSION

Air Bag (Supplemental Restraint) System

The Supplemental Restraint System (SRS) is designed to work in conjunction with the standard 3-point safety belts to reduce injury in a head-on collision.

✳✳ CAUTION

The SRS can actually cause physical injury or death if the safety belts are not used, or if the manufacturer's warnings are not followed. The manufacturer's warnings can be found in your owner's manual, or, in some cases, on your sun visor.

The SRS is comprised of the following components:
• Driver's side air bag module
• Passenger's side air bag module
• Right-hand and left-hand primary crash front air bag sensors
• Air bag diagnostic monitor computer
• Electrical wiring

The SRS primary crash front air bag sensors are hard-wired to the air bag modules and determine when the air bags are deployed. During a frontal collision, the sensors quickly inflate the 2 air bags to reduce injury by cushioning the driver and front passenger from striking the dashboard, windshield, steering wheel and any other hard surfaces. The air bag inflates so quickly (in a fraction of a second) that in most cases it is fully inflated before you actually start to move during a collision.

Since the SRS is a complicated and essentially important system, its components are constantly being tested by a diagnostic computer. The computer illuminates the air bag indicator light on the instrument cluster for approximately 6 seconds when the ignition switch is turned to the **RUN** position when the SRS is functioning properly. After being illuminated for the 6 seconds, the indicator light should then turn off.

If the air bag light does not illuminate at all, stays on continuously, or flashes at any time, a problem has been detected by the diagnostic computer.

✳✳ CAUTION

If at any time the air bag light indicates that the computer has noted a problem, immediately diagnose the problem. A faulty SRS can cause severe physical injury or death.

SERVICE PRECAUTIONS

Whenever working around, or on, the air bag supplemental restraint system, ALWAYS adhere to the following warnings and cautions.

• Always wear safety glasses when servicing an air bag vehicle and when handling an air bag module.
• Carry a live air bag module with the bag and trim cover facing away from your body, so that an accidental deployment of the air bag will have a small chance of personal injury.
• Place an air bag module on a table or other flat surface with the bag and trim cover pointing up.
• Wear gloves, a dust mask and safety glasses whenever handling a deployed air bag module. The air bag surface may contain traces of sodium hydroxide, a byproduct of the gas that inflates the air bag and which can cause skin irritation.
• Ensure to wash your hands with mild soap and water after handling a deployed air bag.
• All air bag modules with discolored or damaged cover trim must be replaced, not repainted.
• All component replacement and wiring service must be made with the negative and positive battery cables disconnected from the battery for a minimum of 1 minute prior to attempting service or replacement.
• NEVER probe the air bag electrical terminals. Doing so could result in air bag deployment, which can cause serious physical injury.
• If the vehicle is involved in a fender-bender that results in a damaged front bumper or grille, the air bag sensors should be inspected to ensure that they were not damaged.
• If at any time, the air bag light indicates that the computer has noted a problem, immediately diagnose the problem. A faulty SRS can cause severe physical injury or death.

DISARMING THE SYSTEM

1. Turn the steering wheel to the straight-ahead position.
2. Turn the ignition switch to LOCK and remove the key.
3. Remove the air bag fuse F8 in the instrument panel fuse block and wait more than 1 minute for the SIR capacitor to discharge.

ARMING THE SYSTEM

1. Install air bag fuse F8 in the fuse block.
2. Turn the ignition switch on and verify that the air bag indicator flashes 7 times and turns off.

Front Suspension Crossmember

REMOVAL & INSTALLATION

1. Remove or disconnect the following:
• Power steering fluid
• Front wheels
• Lower control arm ball joint and stabilizer bar link nut
• Tie rod end ball joint
• Engine mounting reaction rod bolts.
• Steering gear feed and return lines
• Steering column intermediate shaft lower joint
• Crossmember mounting bolts and crossmember.

2. If the crossmember is being replaced, remove the stabilizer bar, steering gear and control arm from the crossmember.

To install:

3. If removed, install the stabilizer bar, steering gear and control arm to the crossmember.

4. Install the crossmember and tighten the bolts to 111 ft. lbs. (150 Nm)

5. Install or connect the following:
• Intermediate shaft lower joint
• Power steering lines
• Engine reaction rod bolts and

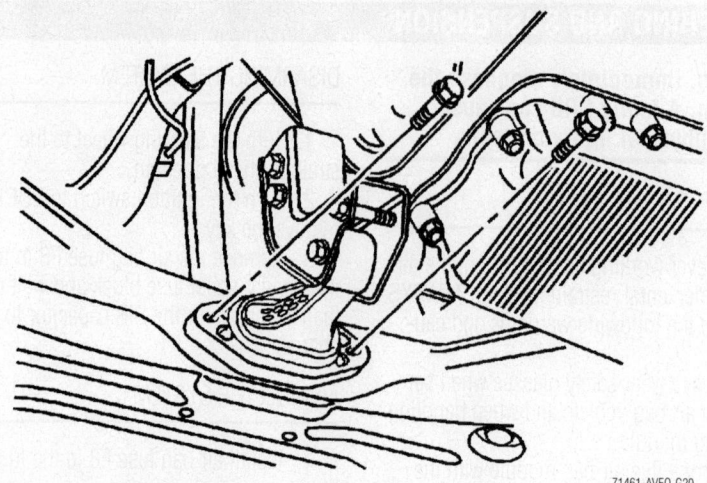

Removing engine reaction rod bolts—Aveo

71461-AVEO-G20

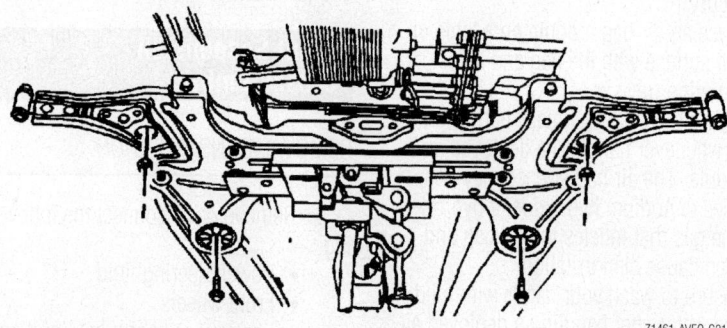

Removing front suspension crossmember—Aveo

71461-AVEO-G21

tighten the bolts to 44 ft. lbs. (60 Nm)
- Tie rod end ball joint
- Lower control arm ball joint and stabilizer bar link nut
- Front wheels
- Power steering fluid

Rack and Pinion Steering Gear

REMOVAL & INSTALLATION

1. Place the front wheels in the straight ahead position.
2. Drain the power steering fluid from the steering gear.
3. Remove or disconnect the following:
- Front wheels
- Steering gear feed and return lines
- Intermediate shaft from steering column
- Outer tie rod hex nuts
- Ball joint hex nuts
- Stabilizer bar shaft link assembly
- Stabilizer bar from steering knuckle
- Front crossmember assembly
- Steering gear retaining bracket
- Steering gear from crossmember

To install:

4. Install the steering gear and the retaining bracket. Tighten the bracket nuts to 37 ft. lbs. (50 Nm).
5. Install the front crossmember and tighten the nuts and bolts to 111 ft. lbs. (150 Nm).
6. Install the ball joint hex nuts and connect the stabilizer bar to the knuckle by tightening the bolt with the stabilizer shaft link assembly. Tighten the ball joint hex nuts to 33 ft. lbs. (45 Nm). Tighten the stabilizer bar shaft to 33 ft. lbs. (45 Nm).
7. Install the outer tie rod hex nuts and tighten to 33 ft. lbs. (45 Nm).
8. Connect the steering gear feed and return lines and tighten the fittings to 16 ft. lbs. (22 Nm).
9. Install the front wheels.
10. Connect the intermediate shaft to the steering column
11. Connect the negative battery cable.
12. Fill the power steering system and then bleed the system.

Strut

REMOVAL & INSTALLATION

1. Loosen the top strut mounting-to-body bolt in the engine compartment.
2. Raise and support the vehicle so the weight of the vehicle rests on the stands, not the control arms.
3. Remove or disconnect the following:
- Front wheels
- Axle hub nut
- Brake caliper and wire aside
- Anti-lock Brake System (ABS) speed sensor connector
- Ball joint-to-steering knuckle strut nut
- Separate the ball joint from the knuckle
- Outer tie rod from the steering knuckle
- Drive axle from wheel hub and wire the axle to the body
- Lower strut from the hub
- Strut-to-body nuts
- Strut

To install:

4. Install the strut to the body and tighten the nuts to 44 ft. lbs. (60 Nm).
5. Connect the drive axle to the front hub.
6. Install or connect the following:
- Outer tie rod to steering knuckle
- Ball joint to steering knuckle
- Ball joint to knuckle/strut nut and tighten to 74 ft. lbs. (100 Nm)
- ABS speed sensor connector
- Brake caliper
- Front wheels

7. Lower the vehicle to the ground and tighten the lug nuts to 88 ft. lbs. (120 Nm).
8. Tighten the axle shaft nut to 221 ft. lbs. (300 Nm).

Shock Absorber

REMOVAL & INSTALLATION

➡**Remove only one shock at a time. Do not suspend the rear axle by the brake hoses only.**

1. Remove the upper shock absorber-to-body bolts.
2. Raise and support the vehicle under the rear axle.
3. Remove the lower shock-to-axle bolts.
4. Remove the shock absorber.

To install:

5. Attach the shock absorber to the axle and install the bolt.

6. Lower the vehicle and guide the shock into the upper mounting location and install the bolts.

7. Tighten the lower shock bolt to 52 ft. lbs. (70 Nm), and the upper bolt to 37 ft. lbs. (50 Nm).

Coil Springs

REMOVAL & INSTALLATION

✳✳ CAUTION

When removing the rear springs, do not use a twin-post hoist. The swing arch tendency of the rear axle when some fasteners are removed may cause it to slip from the hoist causing personal injury.

1. Raise and safely support the vehicle so the rear control arms are supported.

2. Remove the rear wheels.

3. Remove the shock absorbers.

4. Slowly lower the axle and remove the springs and the insulators.

To install:

5. Position the upper spring insulators to the body and hold them in place with an adhesive.

6. Seat the lower spring bumper into position.

7. Install the springs and raise the axle.

8. Install the shock absorbers by bringing the axle assembly to trim height and tighten the shock absorber bolts.

9. Install the rear wheels.

10. Lower the vehicle.

Stabilizer Bar

REMOVAL & INSTALLATION

1. Raise and safely support the vehicle allowing the suspension to hang free.

2. Remove the wheels.

3. Remove the stabilizer bar-to-knuckle nut and the bar-to-link nut.

4. Remove the stabilizer bar links.

5. Remove the front crossmember assembly.

6. Remove the stabilizer bar from the crossmember by removing the U-clamp bolts.

To install:

7. Install or connect the following:
- Stabilizer bar and U-clamps. Tighten the clamp bolts to 18 ft. lbs. (25 Nm).

- Install the front crossmember assembly.

8. Install the stabilizer bar links.

9. Install the stabilizer bar-to-knuckle nut and the bar-to-link nut and tighten the nuts to 37 ft. lbs. (50 Nm).

10. Install the wheels.

11. Lower the vehicle.

Lower Ball Joint

REMOVAL & INSTALLATION

1. Raise and safely support the vehicle so the weight of the vehicle rests on the stands, not the control arms.

2. Remove the wheels.

3. Remove the lower control arm.

4. Remove the ball joint nuts and remove the ball joint.

To install:

5. Install or connect the following:
- Lower control arm
- Ball joint and tighten the bolts to 47 ft. lbs. (64 Nm)
- Wheels
- Lower the vehicle

Lower Control Arm

REMOVAL & INSTALLATION

1. Raise and support the vehicle so the weight rests on the stands, not on the control arms.

2. Remove the front wheels.

3. Remove the control arm link bolt and disconnect the stabilizer bar from the control arm.

4. Remove the ball joint-to-steering knuckle nut.

5. Separate the ball joint from the steering knuckle.

6. Remove the control arm mounting bolts and the bracket.

7. Remove the control arm.

To install:

8. Install the control arm.

9. Connect the front of the control arm to the body with the front mounting bolt and washer, but do not tighten.

10. Apply thread sealer to the control arm rear bolts.

11. Install the control arm rear bolts using a new self-locking nuts, but do not tighten the nuts.

12. Install the stabilizer bar link bolt.

13. Install the ball joint to the steering knuckle and tighten the nut to 74 ft. lbs. (100 Nm).

14. Connect the retaining clip to the ball joint stud.

15. Install the wheels.

16. Raise the vehicle and place jack stands under the control arms to bear the weight of the vehicle.

17. Tighten the control arm mounting bolts to 81 ft. lbs. (110 Nm).

18. Remove the jack stands and lower the vehicle.

71461-AVEO-G22

Lower control arm mounting bolts locations—Aveo

Lower Control Arm Bushing

REMOVAL & INSTALLATION

1. Remove the control arm.
2. The front and rear control arm bushings can be pressed out using suitable press tools.

To install:

3. Before installing new bushings, coat the control arm bushing openings with a multi-purpose lubricant.
4. Press the rear bushing into the arm so the flat of the bushing is on the top side the same as the ball joint.
5. Press the front bushing in from the back to the front so it is centered in the arm.
6. Install the control arm.

Steering Knuckle, Hub and Wheel Bearing

REMOVAL & INSTALLATION

1. Raise and support the vehicle.
2. Separate the front axle shaft from the front wheel hub.
3. Remove the backing plate.
4. Remove the front strut bolts and remove the steering knuckle.
5. Remove the inner snap ring from the knuckle/hub.
6. Using tools J-37105-1, -2, -3 and 500-2, press out the wheel hub as shown.
7. Remove the outer snap ring.
8. Using tools J-37105-1 and -2, 500-2 and J-36661-2, press out the wheel bearing as shown.
9. Clean the steering knuckle bore.

To install:

10. Install the outer snap ring.
11. Using the same tools as removal, press the new wheel bearing into position.
12. Install the inner snap ring.
13. Using the same tools as removal, press the new wheel hub into position.
14. Install the steering knuckle and install front strut bolts and tighten to 74 ft. lbs. (100 Nm).
15. Install the front axle shaft to the front wheel hub.
16. Install the backing plate.
17. Install the wheels.

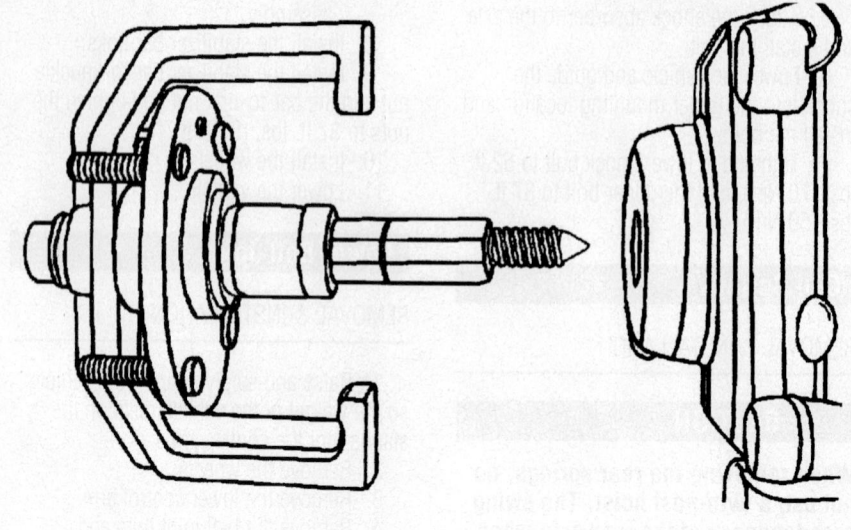

Removing the front wheel hub—Aveo

71461-AVEO-G23

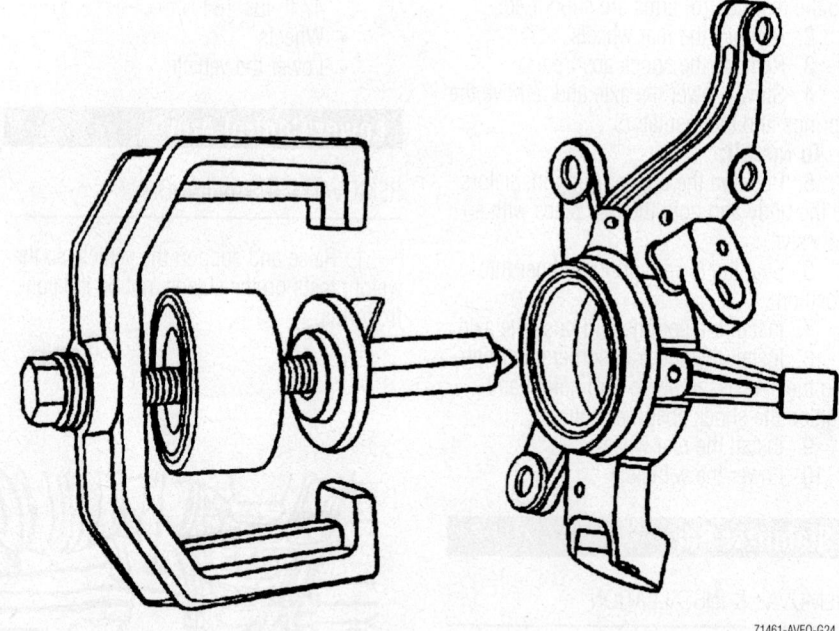

Removing the front wheel bearing—Aveo

71461-AVEO-G24

BRAKES

Brake Caliper

REMOVAL & INSTALLATION

Front

1. Raise and safely support the vehicle.
2. Remove the wheel and tire assembly.
3. Remove the 2 caliper mounting bolts.
4. Disconnect the brake hose and plug the openings.
5. Remove the caliper.

To install:

6. Install the caliper and tighten the bolts to 70 ft. lbs. (95 Nm).
7. Connect the brake hose and tighten the fittings to 30 ft. lbs. (40 Nm).
8. Install the wheel and tire.

9. Lower the vehicle.
10. Bleed the brake system.

Brake Pads

REMOVAL & INSTALLATION

Front

1. Remove the front wheels.
2. Remove the lower guide pin bolt and rotate the caliper upward.
3. Remove the brake pads.

To install:

4. Compress the caliper piston into the bore.
5. Fit the pads into the caliper.
6. Install the caliper and tighten the mounting bolt to 20 ft. lbs. (27 Nm).
7. Install the wheel and tire.
8. Lower the vehicle.

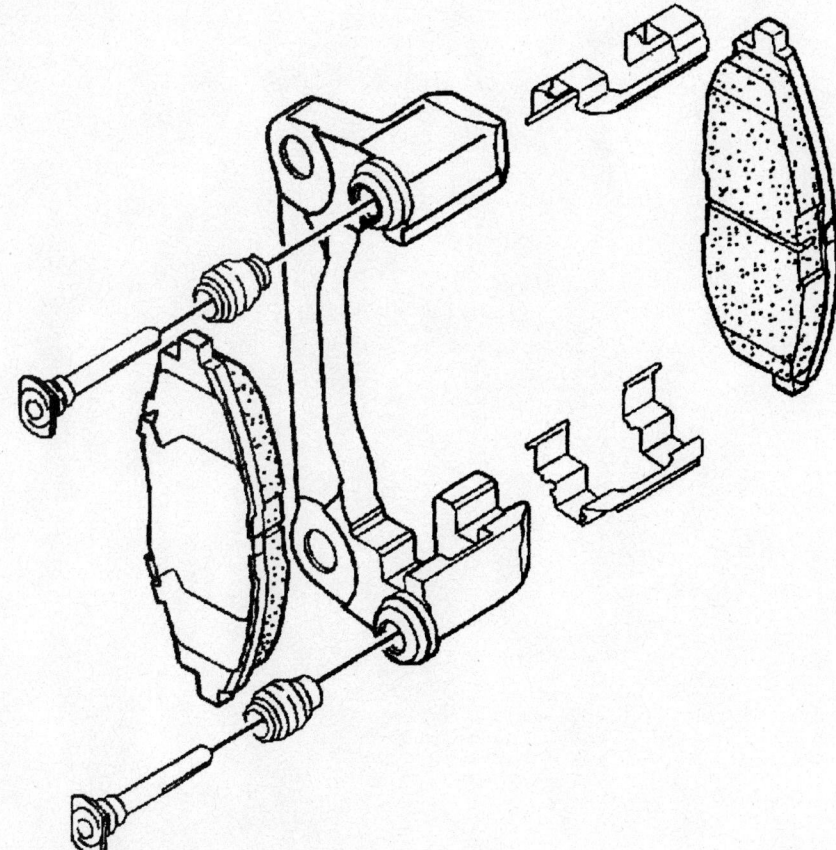

Exploded view of front brake caliper—Aveo

71461-AVEO-G25

Brake Drums

REMOVAL & INSTALLATION

Rear

1. Raise and support the vehicle.
2. Remove the rear wheels.
3. Loosen the parking brake cable.
4. Remove the axle nut.
5. Remove the brake drum.

To install:

6. Ensure the brake adjuster nut is drawn all the way against the stop.
7. Install the brake drum.
8. Install the axle nut and tighten to 147 ft. lbs. (200 Nm).
9. Tighten the parking brake cable.
10. Install the wheels.
11. Adjust the rear brakes and the parking brake.

Brake Shoes

REMOVAL & INSTALLATION

Rear

1. Remove the rear wheels.
2. Remove the brake drum.
3. Loosen the leading shoe hold-down return spring.
4. Disconnect the upper link of the connecting link spring of the leading shoe to relieve tension on the upper return spring.
5. Remove the upper return spring and the adjuster.
6. Disconnect the trailing shoe return spring.
7. Remove the trailing shoe and lining.
8. Disconnect the lower return spring.

To install:

9. Clean the adjuster assembly and apply Molykote® 111 grease to the brake shoe contact points.
10. Inspect the threads of the adjuster for smooth rotation.
11. Install the trailing shoe and lining assembly with the hold-down spring, the washer and the pin.
12. Properly route the parking brake cable and attach it to the shoe lever.
13. Install the lower return spring on the shoe.

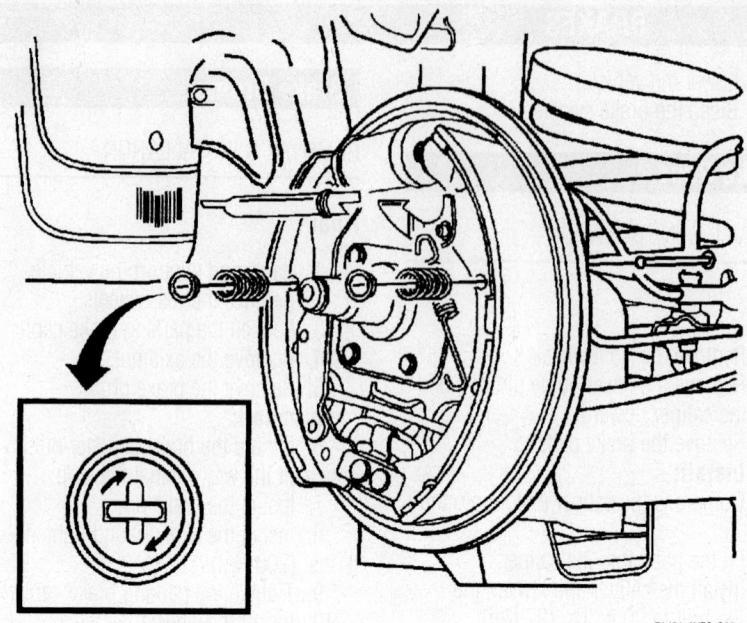

71461-AVEO-G26

Removing the brake shoe hold-down springs—Aveo

14. Install the leading shoe and adjuster assembly against the backing plate.

15. Install the lower return spring to the leading shoe.

16. Install the adjuster assembly and turn it in as far as possible.

17. Position the spring clip toward the backing plate.

18. Install the leading shoe with the hold-down spring.

19. Attach the leading shoe upper link spring connection to apply tension to the return spring.

20. Install the upper return spring.

21. Ensure the adjuster assembly nut is drawn all the way to the stop.

22. Install the brake drum.

23. Install the axle nut and tighten the nut to 147 ft. lbs. (200 Nm).

24. Install the wheels.

25. Adjust the rear brakes and the parking brake.

26. Check the brake fluid level.

BUICK AND PONTIAC

Aztek • Rendezvous

6

DISC BRAKES...................6-66
DRIVE TRAIN 6-52
DRUM BRAKES................6-69
ENGINE REPAIR................6-9
FUEL SYSTEM 6-48
SPECIFICATIONS CHARTS......6-2
Engine and Vehicle Identification 6-2
General Engine Specifications 6-2
Engine Tune-Up Specifications.......6-2
Firing Orders 6-3
Accessory Drive Belt Routing 6-3
Capacities 6-4
Valve Specifications.....................6-4
Crankshaft and Connecting Rod
  Specifications 6-5
Piston and Ring Specifications........6-5
Torque Specifications 6-6
Wheel Alignment 6-6
Tire, Wheel and Ball Joint
  Specifications 6-7
Brake Specifications 6-7
Scheduled Maintenance Intervals 6-8
STEERING AND
  SUSPENSION 6-59

A
Air Bag..................................6-59
  Disarming............................6-59
  Precautions..........................6-59
Alternator..................................6-9
  Removal & Installation...............6-9
Axle Shaft, Bearing and Seal.........6-57
  Removal & Installation.............6-57

B
Brake Drum..............................6-69
  Removal & Installation.............6-69
Brake Pads..............................6-66
  Removal & Installation.............6-66
Brake Shoes.............................6-69
  Removal & Installation.............6-69

C
Caliper...................................6-66
  Removal & Installation.............6-66
Camshaft and Valve Lifters 6-23

Removal & Installation..............6-23
Coil Spring................................6-62
  Removal & Installation.............6-62
CV-Joints.................................6-57
  Overhaul 6-57
Cylinder Head 6-13
  Removal & Installation.............6-13

D
Differential Carrier Assembly 6-58
  Removal & Installation.............6-58

E
Engine Assembly 6-9
  Removal & Installation...............6-9
Exhaust Manifold 6-20
  Removal & Installation.............6-20

F
Front Cover and Seal, Timing
  Chain and Sprockets 6-36
  Removal & Installation.............6-36
Fuel Filter 6-49
  Removal & Installation.............6-49
Fuel Injector.............................6-51
  Removal & Installation.............6-51
Fuel Pump...............................6-49
  Removal & Installation.............6-49
Fuel System Pressure 6-49
  Relieving 6-49
Fuel System Service Precaution 6-48

H
Halfshaft..................................6-55
  Removal & Installation.............6-55
Heater Core 6-13
  Removal & Installation.............6-13

I
Intake Manifold 6-16
  Removal & Installation.............6-16

L
Lower Ball Joint.........................6-63
  Removal & Installation.............6-63
Lower Control Arm 6-63
  Control Arm Bushing
    Replacement 6-64
  Removal & Installation............6-63

O
Oil Pan...................................6-32
  Removal & Installation.............6-32
Oil Pump 6-34
  Removal & Installation.............6-34

P
Parking Brake Shoes....................6-67
  Adjustment...........................6-68
  Removal & Installation.............6-67
Pinion Seal...............................6-58
  Removal & Installation.............6-58
Pistons and Ring 6-48
  Postioning 6-48
Power Steering Rack and Pinion....6-59
  Removal & Installation.............6-59

R
Rear Main Seal 6-35
  Removal & Installation.............6-35
Rocker Arms 6-15
  Removal & Installation.............6-15

S
Shock Absorber 6-61
  Removal & Installation.............6-61
Starter Motor 6-31
  Removal & Installation.............6-31
Strut......................................6-60
  Removal & Installation.............6-60

T
Transaxle Assembly 6-52
  Removal & Installation.............6-52
Transfer Case Assembly................6-53
  Removal & Installation.............6-53

V
Valve Lash 6-31
  Adjustment...........................6-31

W
Water Pump..............................6-12
  Removal & Installation.............6-12
Wheel Bearings..........................6-64
  Adjustment...........................6-64
  Removal & Installation.............6-64

SPECIFICATION CHARTS

ENGINE AND VEHICLE IDENTIFICATION

		Engine							Model Year	
Code ①	Liters (cc)	Cu. In.	Cyl.	Fuel Sys.	Engine Type	Eng. Mfg.		Code ②		Year
E	3.4 (3350)	207	6	SFI	OHV	CPC		2		2002
7	3.6 (3556)	217	6	SFI	DOHC	CPC		3		2003

SFI: Sequential Fuel Injection

OHV: Overhead Valves

DOHC: Dual overhead camshafts

CPC: Chevrolet/Pontiac/Canada

① 8th position of VIN

② 10th position of VIN

Code ②	Year
4	2004
5	2005
6	2006

06025-AZTEK-C01

GENERAL ENGINE SPECIFICATIONS
All measurements are given in inches.

Year	Model	Engine Displacement Liters	Engine Series VIN	Net Horsepower @ rpm	Net Torque @ rpm (ft. lbs.)	Bore x Stroke (in.)	Com-pression Ratio	Oil Pressure @ rpm
2002	Aztek	3.4	E	185@5200	210@4000	3.62x3.31	9.6:1	15@1100
	Rendezvous	3.4	E	185@5200	210@4000	3.62x3.31	9.6:1	15@1100
2003	Aztek	3.4	E	185@5200	210@4000	3.62x3.31	9.6:1	15@1100
	Rendezvous	3.4	E	185@5200	210@4000	3.62x3.31	9.6:1	15@1100
2004	Aztek	3.4	E	185@5200	210@4000	3.62x3.31	9.6:1	15@1100
	Rendezvous	3.4	E	185@5200	210@4000	3.62x3.31	9.6:1	15@1100
2005	Aztek	3.4	E	185@5200	210@4000	3.62x3.31	9.6:1	15@1100
	Rendezvous	3.4	E	185@5200	210@4000	3.62x3.31	9.6:1	15@1100
		3.6	7	245@6000	235@3200	3.70x3.37	10.2:1	20@2000

06025-AZTEK-C02

ENGINE TUNE-UP SPECIFICATIONS

Year	Engine Displacement Liters	Engine VIN	Spark Plug Gap (in.)	Ignition Timing (deg.)	Fuel Pump (psi)	Idle Speed (rpm)	Valve Clearance Intake	Valve Clearance Exhaust
2002	3.4	E	0.060	①	41-47	②	HYD	HYD
2003	3.4	E	0.060	①	41-47	②	HYD	HYD
2004	3.4	E	0.060	①	41-47	②	HYD	HYD
2005	3.4	E	0.060	①	41-47	②	HYD	HYD
	3.6	7	0.044	①	NA	②	HYD	HYD

NA: Information not available

NOTE: The Vehicle Emissions Control Information label often reflects specification changes made during production.

The label figures must be used if they differ from those in the chart.

HYD: Hydraulic

① Refer to underhood label for exact setting.

② Idle speed is maintained by the PCM.

06025-AZTEK-C03

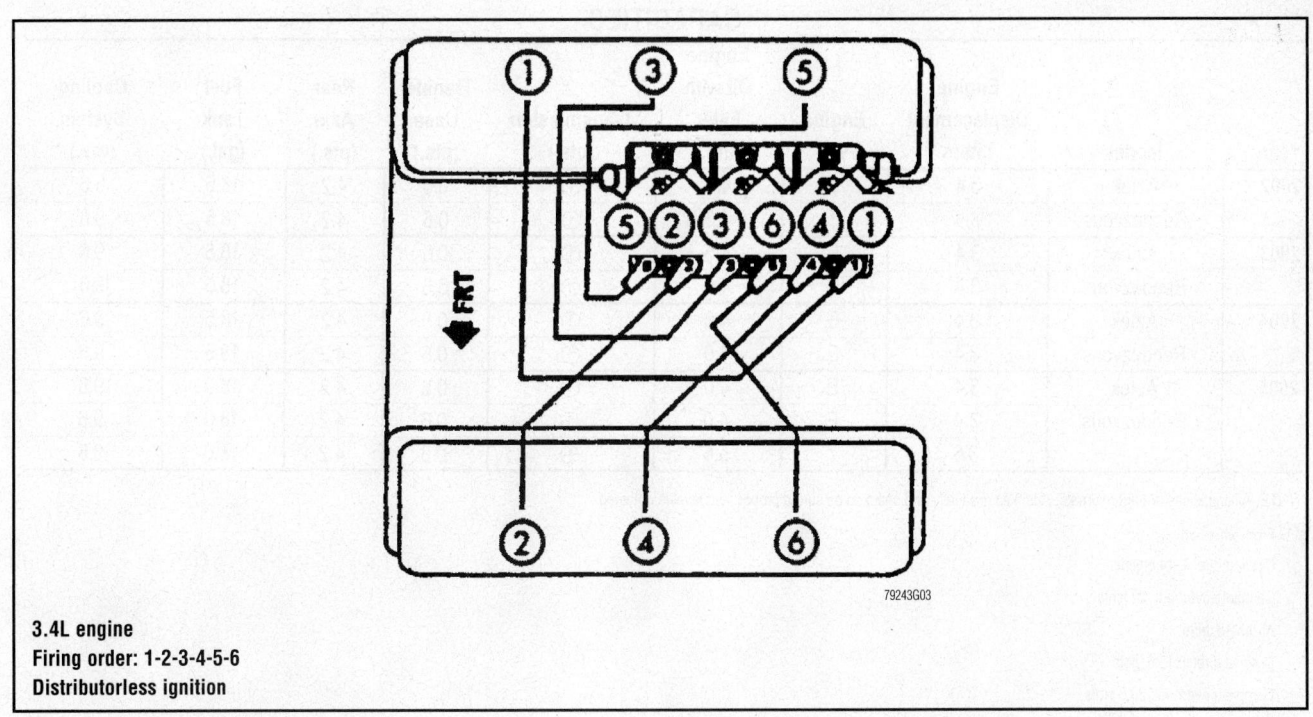

3.4L engine
Firing order: 1-2-3-4-5-6
Distributorless ignition

79243G03

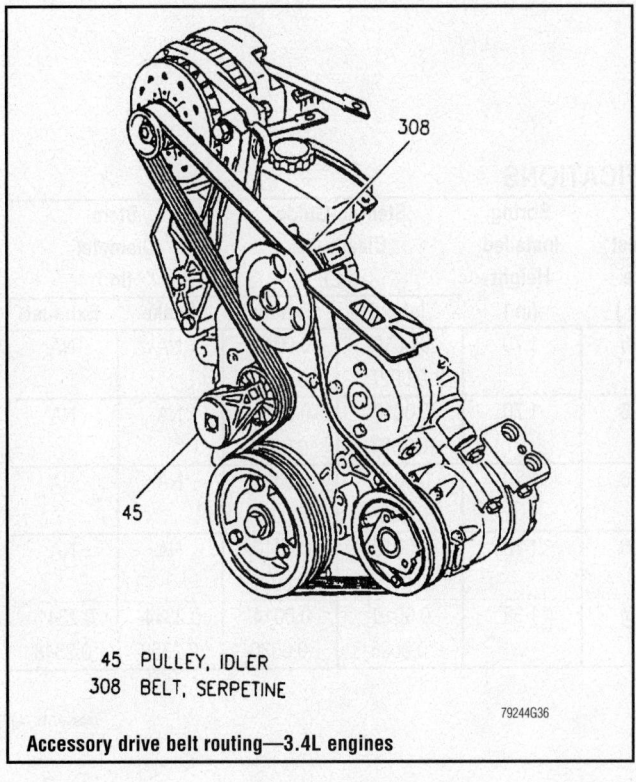

45   PULLEY, IDLER
308   BELT, SERPETINE

Accessory drive belt routing—3.4L engines

79244G36

06025-AZTEK-G80

Accessory drive belt routing—3.6L engines

CAPACITIES

Year	Model	Engine Displacement Liters	Engine VIN	Engine Oil with Filter (qts.)	Transmission (pts.)	Transfer Case (pts.)	Rear Axle (pts.)	Fuel Tank (gal.)	Cooling System (qts.)
2002	Aztek	3.4	E	4.5	①	0.6	4.2	18.5	9.6
	Rendezvous	3.4	E	4.5	①	0.6	4.2	18.5	9.6
2003	Aztek	3.4	E	4.5	①	0.6	4.2	18.5	9.6
	Rendezvous	3.4	E	4.5	①	0.6	4.2	18.5	9.6
2004	Aztek	3.4	E	4.0	①	0.6	4.2	18.5	9.6
	Rendezvous	3.4	E	4.0	①	0.6	4.2	18.5	9.6
2005	Aztek	3.4	E	4.0	①	0.6	4.2	18.0	9.6
	Rendezvous	3.4	E	4.0	①	0.6	4.2	18.0	9.6
		3.6	7	5.5	①	0.6	4.2	18.0	9.6

NOTE: All capacities are approximate. Add fluid gradually and check to be sure a proper fluid level is obtained.

① Front wheel drive:

    Drain and refill: 14.8 pints

    Complete overhaul: 20 pints

    All wheel drive:

    Drain and refill: 15.6 pints

    Complete overhaul: 20.8 pints

06025-AZTEK-C04

VALVE SPECIFICATIONS

Year	Engine Displacement Liters	Engine VIN	Seat Angle (deg.)	Face Angle (deg.)	Spring Test Pressure (lbs. @ in.)	Spring Installed Height (in.)	Stem-to-Guide Clearance (in.) Intake	Stem-to-Guide Clearance (in.) Exhaust	Stem Diameter (in.) Intake	Stem Diameter (in.) Exhaust
2002	3.4	E	46	45	230@1.26	1.70	0.0010-0.0027	0.0010-0.0027	NA	NA
2003	3.4	E	46	45	230@1.26	1.70	0.0010-0.0027	0.0010-0.0027	NA	NA
2004	3.4	E	46	45	230@1.26	1.70	0.0010-0.0027	0.0010-0.0027	NA	NA
2005	3.4	E	46	45	230@1.26	1.70	0.0010-0.0027	0.0010-0.0027	NA	NA
	3.6	7	44.25	45	134-149@0.9449	1.38	0.0010-0.0026	0.0014-0.0030	0.2344-0.2352	0.2341-0.2348

NA: Information not Available

06025-AZTEK-C05

CRANKSHAFT AND CONNECTING ROD SPECIFICATIONS

All measurements are given in inches.

Year	Engine Displacement Liters	Engine VIN	Crankshaft				Connecting Rod		
			Main Brg. Journal Dia.	Main Brg. Oil Clearance	Shaft End-play	Thrust on No.	Journal Diameter	Oil Clearance	Side Clearance
2002	3.4	E	2.6473-2.6483	0.0008-0.0023	0.0024-0.0083	3	1.9987-1.9994	0.0007-0.0024	0.007-0.017
2003	3.4	E	2.6473-2.6483	0.0008-0.0023	0.0024-0.0083	3	1.9987-1.9994	0.0007-0.0024	0.007-0.017
2004	3.4	E	2.6473-2.6483	0.0008-0.0023	0.0024-0.0083	3	1.9987-1.9994	0.0007-0.0024	0.007-0.017
2005	3.4	E	2.6473-2.6483	0.0008-0.0023	0.0024-0.0083	3	1.9987-1.9994	0.0007-0.0024	0.007-0.017
	3.6	7	2.6768-2.6775	0.0004-0.0024	0.0039-0.0130	3	2.2044-2.2050	0.0004-0.0028	0.0037-0.0140

06025-AZTEK-C06

PISTON AND RING SPECIFICATIONS

All measurements are given in inches.

Year	Engine Displ. Liters	Engine VIN	Piston Clearance	Ring Gap			Ring Side Clearance		
				Top Compression	Bottom Compression	Oil Control	Top Compression	Bottom Compression	Oil Control
2002	3.4	E	0.0013-0.0027	0.008-0.019	0.021-0.034	NA	0.0020-0.0034	0.0020-0.0035	NA
2003	3.4	E	0.0013-0.0027	0.008-0.019	0.021-0.034	NA	0.0020-0.0034	0.0020-0.0035	NA
2004	3.4	E	0.0013-0.0027	0.008-0.019	0.021-0.034	NA	0.0020-0.0034	0.0020-0.0035	NA
2005	3.4	E	0.0013-0.0027	0.008-0.019	0.021-0.034	NA	0.0020-0.0034	0.0020-0.0035	NA
	3.6	7	0.0010-0.0021	0.0059-0.0118	0.0110-0.0189	0.0059-0.0236	0.0012-0.0026	0.0006-0.0024	0.0012-0.0067

NA: Information not available

06025-AZTEK-C07

TORQUE SPECIFICATIONS
All readings in ft. lbs.

Year	Engine Displacement Liters	Engine VIN	Cylinder Head Bolts	Main Bearing Bolts	Rod Bearing Bolts	Crankshaft Damper Bolts	Flywheel Bolts	Manifold Intake*	Manifold Exhaust	Spark Plugs	Oil Pan Drain Plug
2002	3.4	E	①	②	③	④	52	⑤	12	15	18
2003	3.4	E	①	②	③	④	52	⑤	12	15	18
2004	3.4	E	⑥	②	③	⑦	52	⑧	12	15	18
2005	3.4	E	⑥	②	③	⑦	52	⑧	12	15	18
	3.6	7	⑨	⑩	⑪	⑫	⑬	17	15	13	18

* Lower

① 37 ft. lbs. plus 90 degrees

② 37 ft. lbs. plus 77 degrees

③ 15 ft. lbs. plus 75 degrees

④ 52 ft. lbs. plus 85 degrees

⑤ 1st step: 62 inch lbs.
  2nd step: 115 inch lbs.

⑥ 44 ft. lbs. Plus 95 degrees

⑦ 52 ft. lbs. Plus 72 degrees

⑧ Center bolts:
  Step 1: 62 inch lbs.
  Step 2: 115 inch lbs.
  Corner bolts:
  Step 1: 115 inch lbs.
  Step 2: 18 ft. lbs.

⑨ M8 bolts:
  Step 1: 10 ft. lbs.
  Step 2: plus 60 degrees
  M11 bolts:
  Step 1: 33 ft. lbs.
  Step2: plus 120 degress

⑩ Inner
  Step 1: 15 ft. lbs.
  Step 2: plus 80 degrees
  Outer:
  Step 1: 10 ft. lbs.
  Step 2: plus 110 degrees
  Side
  Step 1: 22 ft. lbs.
  Step 2: plus 60 degrees

⑪ Step 1: 22 ft. lbs.
  Step 2: back off to 0
  Step 3: 18 ft. lbs.
  Step 4: plus 110 degrees

⑫ Step 1: 74 ft. lbs.
  Step 2: plus 150 degrees

⑬ Step 1: 22 ft. lbs.
  Step 2: plus 45 degrees

06025-AZTEK-C08

WHEEL ALIGNMENT

Year	Model		Caster Range (+/-Deg.)	Caster Preferred Setting (Deg.)	Camber Range (+/-Deg.)	Camber Preferred Setting (Deg.)	Toe-in (in.)
2002	All	F	0.75	+3.10	0.50	-0.70	0 +/-0.20
	FWD	R	—	—	0.30	-1.00	0 +/-0.30
	AWD	R	—	—	0.25	-0.25	0 +/-0.15
2003	All	F	0.45	+3.10	0.50	-0.70	0 +/-0.20
	FWD	R	—	—	0.40	0	0 +/-0.20
	AWD	R	—	—	0.25	-0.25	0 +/-0.20
2004	FWD w/twist axle	F	0.50	+2.39	0.50	-0.60	0+/-0.75
		R	—	—	0.40	0	0 +/-0.20
	AWD and I.R.S.	F	0.50	+2.39	0.50	-0.60	0+/-0.75
		R	—	—	0.60	-0.25	0 +/-0.20
2005	FWD	F	0.75	+2.40	0.50	-0.65	0+/-0.20
		R	—	—	0.50	0	0 +/-0.30
	AWD and I.R.S.	F	0.75	+2.40	0.75	-0.65	0+/-0.20
		R	—	—	0.60	-0.30	0 +/-0.20

All alignment figures based on nominal ride height and standard tires

06025-AZTEK-C09

TIRE, WHEEL AND BALL JOINT SPECIFICATIONS

Year	Model	OEM Tires		Tire Pressures (psi)		Wheel Size	Ball Joint Inspection	Lugnut Torque (ft. lbs.)
		Standard	Optional	Front	Rear			
2002	All	P215/70R16	P215/70R16/ P235/55R17	std: 35 opt: 32	std.: 35 opt.: 32	6-JJ	U ① L: 0.090 in.	100
2003	All	P215/70R16	P215/70R16/ P235/55R17	std: 35 opt: 32	std.: 35 opt.: 32	6-JJ	U ① L: 0.090 in.	100
2004	All	P215/70R16	P215/70R16	②	②	6-JJ	U ① L: 0.090 in.	100
2005	All	P215/70R16	P215/70R16	②	②	6-JJ	U ① L: 0.090 in.	100

OEM: Original Equipment Manufacturer

PSI: Pounds Per Square Inch

STD: Standard

OPT: Optional

L: Lower

U: Upper

① Replace if any movement is noted or if stud can be moved by hand

② See placard on vehicle

06025-AZTEK-C10

BRAKE SPECIFICATIONS
All measurements in inches unless noted

Year	Model		Brake Disc			Brake Drum Diameter			Minimum Lining Thickness		Brake Caliper	
			Original Thickness	Minimum Thickness	Maximum Runout	Original Inside Diameter	Max. Wear Limit	Maximum Machine Diameter	Front	Rear	Bracket Bolts (ft. lbs.)	Mounting Bolts (ft. lbs.)
2002	Aztek	F	1.181	1.063	0.002	—	—	—	NA	—	137	26
		R	0.430	0.350	0.002	8.86	8.92	8.90	—	0.030	92	33
	Rendezvous	F	1.181	1.063	0.002	—	—	—	NA	—	137	26
		R	0.043	0.350	0.002	8.86	8.92	8.90	—	0.030	92	33
2003	Aztek	F	1.181	1.063	0.002	—	—	—	NA	—	137	26
		R	0.430	0.350	0.002	—	9.90	9.882	—	0.030	92	33
	Rendezvous	F	1.181	1.063	0.002	—	—	—	NA	—	137	26
		R	0.043	0.350	0.002	—	9.90	9.882	—	NA	92	33
2004	Aztek	F	1.181	1.063	0.002	—	—	—	NA	—	137	26
		R	0.430	0.350	0.002	—	9.90	9.882	—	NA	96	33
	Rendezvous	F	1.181	1.063	0.002	—	—	—	NA	—	137	26
		R	0.043	0.350	0.002	—	9.90	9.882	—	NA	96	33
2005	Aztek	F	1.181	1.063	0.002	—	—	—	NA	—	137	26
		R	0.430	0.350	0.002	—	9.90	9.882	—	NA	96	33
	Rendezvous	F	1.181	1.063	0.002	—	—	—	NA	—	137	26
		R	0.043	0.350	0.002	—	9.90	9.882	—	NA	96	33

NA: Information not available

06025-AZTEK-C11

SCHEDULED MAINTENANCE INTERVALS
2002-04 BUICK—RENDEZVOUS, PONTIAC—AZTEK

TO BE SERVICED	TYPE OF SERVICE	VEHICLE MILEAGE INTERVAL (x1000)															
		7.5	15	22.5	30	37.5	45	52.5	60	67.5	75	82.5	90	97.5	105	112	120
Accessory drive belt	I								✓								✓
Air cleaner filter	R								✓								✓
Air distributor air filter	R		✓		✓		✓		✓		✓		✓		✓		✓
Brake system	I	✓	✓	✓	✓	✓	✓	✓	✓	✓	✓	✓	✓	✓	✓	✓	✓
Engine coolant	R	Every 150,000 miles															
Engine oil & filter	S/I	✓	✓	✓	✓	✓	✓	✓	✓	✓	✓	✓	✓	✓	✓	✓	✓
Fuel tank, cap & lines	I								✓								✓
Transmission fluid ①	R																
Rotate tires	S/I	✓	✓	✓	✓	✓	✓	✓	✓	✓	✓	✓	✓	✓	✓	✓	✓
Spark plug wires	S/I	Every 100,000 miles															
Spark plugs	R	Every 100,000 miles															

R: Replace   I: Inspect   S: Service

① Change the transmission fluid every 50, 000 miles if the vehicle meets any of the criteria outlined in the severe service list below.

FREQUENT OPERATION MAINTENANCE (SEVERE SERVICE)

If a vehicle is operated under any of the following conditions it is considered severe service:

- Towing a trailer or using a camper or car-top carrier.

- Repeated short trips of less than 5 miles in temperatures below freezing, or trips of less than 10 miles in any temperature.

- Extensive idling or low-speed driving for long distances as in heavy commercial use, such as delivery, taxi or police cars.

- Operating on rough, muddy or salt-covered roads.

- Operating on unpaved or dusty roads.

- Driving in extremely hot (over 90°) conditions.

Automatic transaxle fluid and filter: replace every 50,000 miles.

Tires: rotate every 6000 miles.

Brake system: inspect every 6000 miles.

Air distributor air filter: replace every 12,000 miles.

06025-AZTEK-C12

SCHEDULED MAINTENANCE INTERVALS
2005 BUICK—RENDEZVOUS, PONTIAC—AZTEK

TO BE SERVICED	TYPE OF SERVICE	VEHICLE MILEAGE INTERVAL (x1000)															
		5	10	15	20	25	30	35	40	45	50	55	60	65	70	75	80
Accessory drive belt	I/R	Every 150,000 miles															
Air cleaner filter	R					✓					✓					✓	
Air distributor air filter	R		✓		✓		✓		✓		✓		✓		✓		✓
Brake system	I	✓	✓	✓	✓	✓	✓	✓	✓	✓	✓	✓	✓	✓	✓	✓	✓
Engine coolant	R	Every 150,000 miles															
Engine oil & filter	S/I	✓	✓	✓	✓	✓	✓	✓	✓	✓	✓	✓	✓	✓	✓	✓	✓
Exhaust system	I					✓					✓					✓	
Fuel tank, cap & lines	I					✓					✓					✓	
Transmission fluid	R										✓						
Rotate tires	S/I	✓	✓	✓	✓	✓	✓	✓	✓	✓	✓	✓	✓	✓	✓	✓	✓
Spark plug wires	S/I	Every 100,000 miles															
Spark plugs	R	Every 100,000 miles															

R: Replace   I: Inspect   S: Service

06025-AZTEK-C13

ENGINE REPAIR

Alternator

REMOVAL & INSTALLATION

3.4L

1. Before servicing the vehicle, refer to the Precautions Section.
2. Remove or disconnect the following:
   - Negative battery cable
3. Rotate the engine forward.
   - Alternator terminal nut, lead and electrical connector
   - Serpentine belt
   - Front bolts and two rear bolts
   - Alternator from the bracket; position it above the drive axle
   - Serpentine belt tensioner
   - Bracket
   - Power steering pipes from the retainer
   - Fuel pressure test port cap from the injector rail

➡ **Do not disconnect the power steering pipes from the pump**

   - Power steering pump and reposition it to gain access to the alternator
   - Alternator

To install:

4. Install or connect the following:
   - Alternator
   - Electrical harness to the right fender well retainer
   - Power steering pump. Torque the bolts to 25 ft. lbs. (34 Nm).
   - Fuel pressure test port cap to the fuel rail
   - Power steering pipes to the retainer. Torque the fastener to 54 inch lbs. (6 Nm).
   - Alternator bracket. Torque the bolt to 37 ft. lbs. (50 Nm).
   - Serpentine belt tensioner
   - Alternator to the bracket. Torque the bolts to 37 ft. lbs. (50 Nm).
   - Serpentine belt
   - Alternator electrical connector, lead and nut. Torque the nut to 115 inch lbs. (13 Nm).
5. Rotate the engine to its original position.
   - Negative battery cable
6. Perform a charging system test and verify the proper operation of the system.

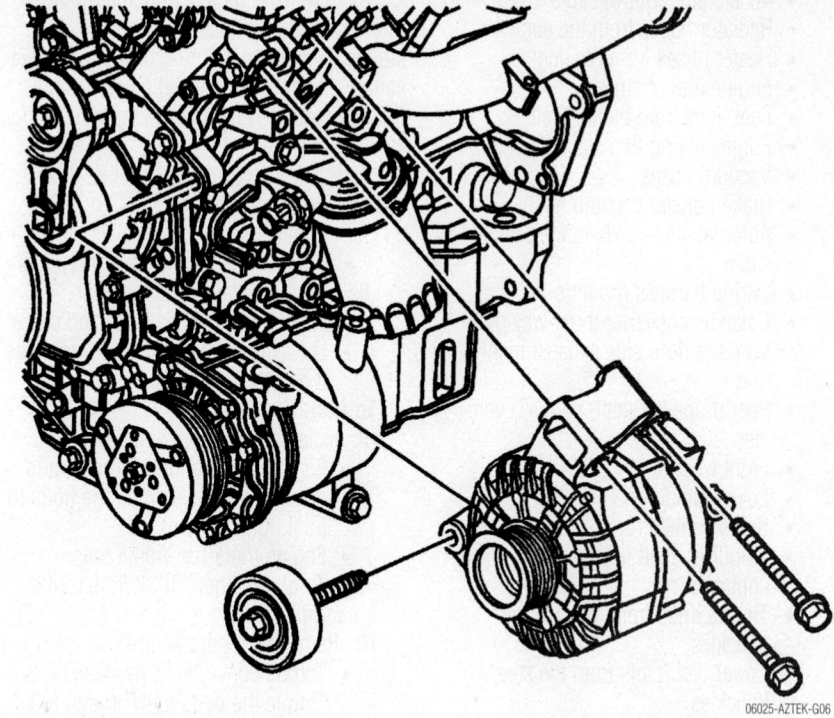

Alternator mounting—3.6L engine

06025-AZTEK-G06

3.6L

1. Before servicing the vehicle, refer to the Precautions Section.
2. Disconnect the battery ground (negative) cable from the battery
3. Remove the torque struts and rotate the engine in order to provide access to remove the alternator.
4. Remove the bolts and the nuts from the engine mount struts brackets on the engine.
5. Remove the bolt and the nut from the engine mount strut at the engine mount strut bracket on the upper radiator support.
6. Remove the engine mount strut.
7. Remove the alternator B+ terminal nut and the battery cable from the alternator.
8. Disconnect the alternator electrical connector.
9. Remove the drive belt from the alternator.
10. Remove the idler pulley.
11. Remove the alternator bolts.
12. Remove the alternator from the vehicle.

To install:

13. Install the alternator to the vehicle.
14. Install the alternator bolts. Tighten the alternator bolts to 50 Nm (37 ft. lbs.).
15. Install the idler pulley.
16. Install the drive belt.

17. Connect the alternator electrical connector.
18. Install the battery cable and the alternator B+ terminal nut to the alternator. Tighten the alternator B+ terminal nut 20 Nm (15 ft. lbs.).
19. Install the torque struts.
20. Install the bolt and the nut to the engine mount strut at the engine mount strut bracket on the upper radiator support. HAND TIGHTEN ONLY.
21. Install the bolt and the nut to the engine mount strut at the engine mount strut bracket on the engine.
22. Tighten engine mount strut nuts to 48 Nm (35 ft. lbs.).
23. Connect the battery ground (negative) cable to the battery.

Engine Assembly

REMOVAL & INSTALLATION

3.4L Engine

1. Before servicing the vehicle, refer to the Precautions Section.
2. Drain the cooling system.
3. Drain the engine oil.
4. Relieve the fuel system pressure.
5. Remove or disconnect the following:
   - Negative battery cable

- Throttle body air inlet duct
- Cruise control cable
- Accelerator control cable
- Radiator hoses from the engine
- Heater hoses from the engine
- Engine mount struts
- Fuel lines from the fuel rail
- Engine wiring harness connectors
- Vacuum hoses
- Brake booster vacuum hose
- Automatic transaxle range selector cable
- Wiring harness grounds
- Catalytic converter three-way pipe from the right side exhaust manifold
- Rear propeller shaft, on AWD vehicles
- Front wheels
- Lower radiator baffle
- Splash shields
- Stabilizer shaft links from the lower control arms
- Tie rod ends from the steering knuckles
- Lower ball joints from the steering knuckles
- Cooler lines and bracket from the transaxle
- A/C compressor bolts and position compressor aside
- Axles from the transaxle and secure them to the steering knuckle/struts

✳✳ CAUTION

Failure to remove the intermediate shaft from the steering gear may result in damage to the gear or intermediate shaft and may cause a loss of steering control.

- Intermediate shaft from the steering gear
6. Lower the vehicle until the frame is in contact with a suitable transaxle table/engine stand (such as J 39580). Make certain that an engine stand (such as J 39580) is aligned below the engine.
   - Frame bolts
7. Raise the vehicle to separate the assembly from the vehicle.
   - Starter
   - Flywheel-to-torque converter bolts
8. Install a suitable engine hoist
   - Engine to transaxle bolts and studs
   - Engine from the transaxle and place it on the engine stand

To install:

9. Install or connect the following:
   - Engine to the transaxle/frame and install the bolts. Torque the bolts to 55 ft. lbs. (75 Nm).
   - Engine mount-to-frame nuts. Torque the nuts to 32 ft. lbs. (43 Nm).
10. Remove the engine hoist.
   - Torque converter to flywheel bolts. Torque the bolts to 47 ft. lbs. (63 Nm).
   - Starter
11. Position the transaxle table with powertrain/frame under the vehicle. Lower the vehicle until the frame contacts the transaxle table.
   - New frame to body bolts. Torque the front bolts to 111 ft. lbs. (150 Nm) and the rear bolts to 122 ft. lbs. (165 Nm). Remove the transaxle table.
   - Intermediate shaft to the steering gear

✳✳ CAUTION

When installing the intermediate shaft be certain that the shaft is seated properly before installing the pinch bolt. If the pinch bolt is inserted into the coupling before the shaft, the mating surfaces disengage. Disengagement of the two shafts may lead to a loss of steering control.

- Pinch bolt at the intermediate shaft. Torque the bolt 35 ft. lbs. (48 Nm).
- Drive axles to the transaxle
- Cooler lines and bracket to the transaxle. Torque the fasteners to 17 ft. lbs. (23 Nm).
- Lower ball joints to the steering knuckles. Torque to 40 ft. lbs. (55 Nm).
- Tie rod ends to the steering knuckles
- Stabilizer shaft links to the lower control arms. Torque the bolts 17 ft. lbs. (23 Nm).
- Inner fender splash shield. Torque the fasteners to 18 inch lbs. (2 Nm).
- Front wheels
- Catalytic converter pipe to the right side exhaust manifold. Torque the nuts to 25 ft. lbs. (34 Nm).
- Wiring harness grounds
- Rear propeller shaft, if removed
- Brake booster vacuum hose
- Vacuum hoses to the engine
- Range selector cable. Torque the screw to 14 ft. lbs. (20 Nm).
- Engine wiring harness connectors
- Fuel lines to the fuel rail. Torque the fasteners to 13 ft. lbs. (17 Nm).
- Throttle body brackets and cables. Torque the fasteners to 18 ft. lbs. (25 Nm).
- Engine mount strut. Torque the bolt to 35 ft. lbs. (48 Nm).
- Heater hoses
- Radiator hoses

✳✳ CAUTION

Whenever the engine has been removed from the vehicle it is necessary to install a new accelerator control cable to avoid damage or personal injury.

- New accelerator control cable
- Cruise control cable
- Throttle body air inlet duct
- Negative battery cable

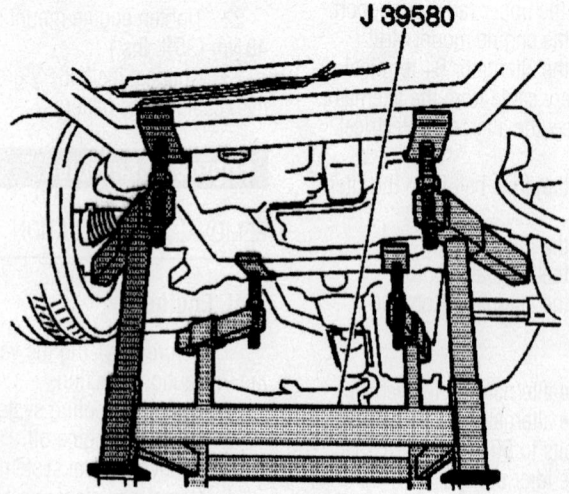

J 39580

9358KG02

Use a engine stand to support the assembly during frame removal—3.4L Engine

12. Fill the engine with oil.

13. Fill the engine with coolant.

14. Inspect the transmission fluid level and top off, if necessary.

15. Turn the ignition to the **ON** position several times to pressurize the fuel system.

16. Start the engine and inspect for leaks, repair if necessary.

17. Check and top off the fluid levels if required.

3.6L Engine

1. Before servicing the vehicle, refer to the Precautions Section.

2. Disconnect the battery negative cable.

3. Remove the throttle body air inlet duct.

➡**Do not disconnect the battery negative cable from the vehicle.**

4. Disconnect the battery negative cable from the engine block.

➡**Do not disconnect the battery positive cable from the vehicle, underhood electrical center or the battery.**

5. Disconnect the battery positive cable from the alternator and the starter.

6. Drain the cooling system.

7. Disconnect the radiator hoses from the engine.

8. Disconnect the heater hoses from the engine.

9. Remove the engine mount struts.

✳✳ WARNING

Relieve the fuel pressure.

10. Disconnect the fuel pressure and evaporative emission (EVAP) pipes from the engine.

11. Remove the ECM chassis (outboard) side electrical connector from the ECM.

12. Remove the wiring harness ground from the transmission.

13. Remove the vacuum brake booster hose from the intake manifold.

14. Properly evacuate the air conditioning system.

15. Disconnect the discharge hose from the condenser.

16. Disconnect the suction hose from the evaporator outlet tube.

17. Secure the discharge and suction hoses to the powertrain.

18. Remove the transmission electrical connector.

19. Raise and support the vehicle.

20. If you will be separating the engine from the transmission, remove the torque converter bolts.

21. Drain the engine oil.

22. If the vehicle is equipped with all wheel drive (AWD), remove rear propeller shaft.

23. Remove the catalytic converter.

24. Remove the front tires and wheels.

25. Remove lower radiator air baffle.

26. Remove the engine splash shields.

27. Disconnect the vehicle speed sensor (VSS) electrical connector and secure the wiring harness to the vehicle.

28. Remove the front wheel speed sensor wiring harnesses from the lower control arms and the frame.

29. Remove the tie rod ends from the steering knuckles.

30. Remove the lower ball joints from the knuckles.

31. Disconnect the drive axles from the transaxle.

32. Rotate the struts and reposition the drive axles toward the rear of the vehicle in order to provide clearance for the powertrain to be removed.

✳✳ CAUTION

Failure to disconnect the intermediate shaft from the rack and pinion steering gear stub shaft can result in damage to the steering gear and/or intermediate shaft. This damage may cause loss of steering control which could result in an accident and possible personal injury

33. Separate the intermediate steering shaft from the steering gear.

34. Remove the engine mount lower nuts.

35. Remove the transmission mount lower nuts.

36. Position a powertrain lift table below the powertrain.

37. Lower the vehicle until the powertrain is supported by the powertrain lift table.

38. Remove the frame bolts.

39. Carefully raise the vehicle or lower the powertrain table in order to remove the powertrain from the vehicle.

40. Remove the exhaust crossover pipe.

41. Remove the coolant inlet pipe.

42. Perform the following steps if it is necessary to separate the engine from the transmission.

➡**Do not disconnect the power steering pipes or drain the power steering fluid.**

43. Remove the power steering pressure pipe/hose from the water outlet.

44. Remove the accessory drive belt.

45. Remove the power steering reservoir and reposition to provide access.

46. Remove and reposition the power steering pump with the power steering pipe/hose.

47. Remove the transfer case brace.

48. Remove the transmission lower brace.

49. Remove the transmission upper brace nut located behind the power steering pump.

50. Remove the engine to transmission (bell housing) bolts.

51. Use 4 M10x1.5x40 GM P/N 11519182, or equivalent bolts to install the EN 46114 engine lift brackets to the left rear and right front cylinder heads.

52. Tighten the lift bracket bolts to 65 Nm (48 ft. lbs.).

53. Use an engine hoist in order to separate the engine from the transmission and the frame.

54. Install the engine to a suitable engine stand.

To install:

➡**Use an engine hoist in order to remove the engine from the engine stand.**

55. Install the engine to the transmission and the frame at the powertrain lift table.

56. Remove the EN 46114 engine lift brackets.

57. Install the engine to transmission (bell housing) bolts.

58. Install the transmission upper brace nut located behind the power steering pump.

59. Tighten the transmission upper brace nut to 50 Nm (37 ft. lbs.).

60. Install the transmission lower brace.

61. Install the transfer case brace.

62. Install the power steering pump with the power steering pipe/hose.

63. Install the power steering reservoir and reposition to provide access.

64. Install the accessory drive belt.

65. Install the power steering pressure pipe/hose to the water outlet.

66. Install the coolant inlet pipe.

67. Install the exhaust crossover pipe.

68. Carefully lower the vehicle or raise the powertrain table in order to install the powertrain to the vehicle.

69. Install the frame bolts.

70. Raise the vehicle and remove the powertrain lift table.

71. Install the transmission mount lower nuts. Torque the bolt to 35 ft. lbs. (48 Nm).

72. Install the engine mount lower nuts.

❊❊ CAUTION

Failure to disconnect the intermediate shaft from the rack and pinion steering gear stub shaft can result in damage to the steering gear and/or intermediate shaft. This damage may cause loss of steering control which could result in an accident and possible personal injury

73. Install the intermediate steering shaft to the steering gear.

74. Rotate the struts and install the drive axles to the transaxle.

75. Install the lower ball joints to the knuckles.

76. Install the tie rod ends to the steering knuckles.

77. Install the front wheel speed sensor wiring harnesses to the lower control arms and the frame.

78. Connect the VSS electrical connector and secure the wiring harness to the vehicle.

79. Install the engine splash shields.

80. Install lower radiator air baffle.

81. Install the front tires and wheels.

82. Install the catalytic converter.

83. If the vehicle is equipped with AWD, install rear propeller shaft.

84. Install the torque converter bolts as necessary.

85. Lower the vehicle.

86. Fill the engine oil as necessary.

87. Install the transmission electrical connector.

88. Connect the suction hose to the evaporator outlet pipe.

89. Connect the discharge hose to the condenser.

90. Recharge the air conditioning system.

91. Install the brake booster vacuum hose to the intake manifold.

92. Install the transmission ground wire and the bolt.

93. Tighten the transmission ground bolt 75 Nm (55 ft. lbs.).

94. Install the ECM chassis (outboard) side electrical connector to the ECM.

95. Connect the fuel pressure and EVAP pipes to the engine.

96. Install the engine mount struts.

97. Connect the heater hoses to the engine.

98. Connect the radiator hoses to the engine.

99. Fill the cooling system.

100. Connect the battery positive cable to the alternator and the starter.

101. Connect the battery negative cable to the engine block.

102. Install the throttle body air inlet duct.

103. Connect the battery negative cable to the battery.

Water Pump

REMOVAL & INSTALLATION

3.4L Engine

1. Before servicing the vehicle, refer to the Precautions Section.

2. Drain the coolant from the engine.

3. Remove or disconnect the following:
- Negative battery cable
- Serpentine drive belt guard
- Loosen the water pump pulley bolts
- Serpentine drive belt
- Water pump pulley
- Water pump
- Water pump gasket

To install:

4. Clean the gasket mounting surfaces.

5. Install or connect the following:
- Gasket
- Water pump. Torque the bolts to 89 inch lbs. (10 Nm).
- Water pump pulley and hand-tighten the bolts at this time
- Serpentine drive belt
- Water pump pulley bolts to 18 ft. lbs. (25 Nm)
- Serpentine drive belt guard

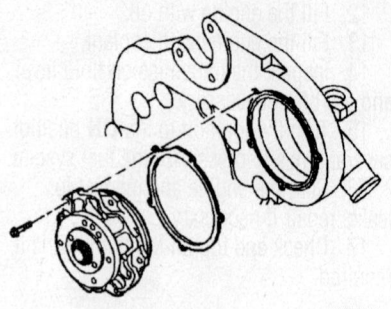

Exploded view of the water pump—3.4L engine

9358KG01

6. Fill the cooling system.

7. Start the engine and check for leaks, repair if necessary.

8. Road test the vehicle and verify there is no air in the cooling system.

3.6L Engine

1. Before servicing the vehicle, refer to the Precautions Section.

2. Install the EN 46104 onto the water pump pulley.

3. Remove the water pump pulley bolts.

4. Remove the water pump pulley.

5. Remove the water pump bolts.

6. Remove the water pump from the front cover.

7. Remove the water pump gasket.

To install:

8. Ensure that the engine front cover and water pump are clear of old gasket material.

EN 46104 installed—3.6L engine

06025-AZTEK-G07

9. Place a new water pump gasket on the water pump.

10. Place the water pump in position on the front cover.

11. Install the water pump bolts. Tighten the water pump bolts to 10 Nm (89 inch lbs.).

12. Install the water pump pulley.

13. Loosely install the water pump pulley bolts.

14. Install the EN 46104 onto the water pump pulley. Tighten the water pump pulley bolts to 10 Nm (89 inch lbs.).

Heater Core

REMOVAL & INSTALLATION

1. Before servicing the vehicle, refer to the Precautions Section.

2. Drain the cooling system.

3. Remove or disconnect the following:
- Wiper module
- Brake booster vacuum hose
- Air cleaner cover and intake tube
- Throttle and cruise control cables, from the TBI bracket, also
- Transmission filler tube
- Hoses from the heater core
- Right and left instrument panel insulators
- Center floor air outlet
- Heater outlet duct
- Tie strap holding the instrument panel harness to the heater core cover
- Heater core cover
- Heater core retaining screw
- Heater core

To install:

4. Install or connect the following:
- Heater core
- Heater core retaining screw. Torque to 8 inch lbs.
- Heater core cover. Torque to 8 inch lbs.
- Heater outlet duct. Torque to 8 inch lbs.
- Tie strap holding the instrument panel harness to the heater core cover
- Center floor air outlet
- Right and left instrument panel insulators
- Hoses to the heater core
- Transmission filler tube
- Throttle and cruise control cables
- Air cleaner cover and intake tube
- Brake booster vacuum hose
- Wiper module

Cylinder Head

REMOVAL & INSTALLATION

3.4L Engine

This engine uses aluminum cylinder heads. Use care when working with light alloy parts. Valve guides are pressed in. Roller rocker arms are located on a pedestal in a slot in the cylinder head and are retained on individual threaded bolts.

The cylinder heads are retained by torque-to-yield bolts. A torque angle meter is required for proper torque during assembly. New replacement head bolts are recommended.

Before removing the cylinder head(s) from the engine and before disassembling the valve mechanism, perform a compression test and note the results. During disassembly, be sure that the valvetrain components are kept together and identified so that they can be installed in their original locations.

LEFT (FRONT) SIDE

1. Before servicing the vehicle, refer to the Precautions Section.

2. Relieve the fuel system pressure using the recommended procedure.

3. Drain the cooling system.

4. Drain the oil.

5. Remove or disconnect the following:
- Negative battery cable
- Upper intake manifold
- Lower intake manifold
- Valve rocker arms and pushrods
- Exhaust crossover pipe
- Thermostat bypass pipe
- Right side engine mount strut bracket
- Oil level indicator tube
- Left side spark plug wires and spark plugs
- Left side exhaust manifold
- Left side cylinder head and gasket

To install:

6. Clean all parts well. Clean all gasket surfaces. Carefully remove all varnish soot and carbon to the bare metal. DO NOT use a motorized wire brush on any gasket surface since the soft aluminum will be damaged. If necessary, the head can be disassembled for thorough inspection and reconditioning.

7. Inspect the cylinder head for cracks. Do not attempt to weld the cylinder head. If cracked, replace it. Check the cylinder head deck, intake and exhaust manifold mating surfaces for flatness. These surfaces may be reconditioned by milling. If the surfaces are warped more than 0.005 in. (0.127mm), the surface should be milled. If more than 0.010 in. (0.251mm) of metal must be removed from the head, the head should be replaced.

8. Clean the cylinder head bolts and the bolt holes. Check the head bolts for damaged threads or stretching. New replacement head bolts are recommended.

9. Install or connect the following:
- New cylinder head gasket which is marked which side is **UP**
- Cylinder head by aligning it with the dowel pins
- New cylinder head bolts coated with a sealant (such as GM 1052080). Torque the bolts in the proper sequence (1–8) to 37 ft. lbs. (50 Nm) for 2001–03; 44 ft. lbs. (60 Nm) for 2004–05. Using a torque angle meter turn the bolts 90 degrees for 2001–03; 95 degrees for 2004–05, in the proper sequence.
- Left side exhaust manifold. Torque the bolts to 12 ft. lbs. (16 Nm).
- Spark plugs. Torque the plugs 11 ft. lbs. (15 Nm).
- Spark plug wires
- Oil level indicator tube. Torque the fastener to 18 ft. lbs. (25 Nm).
- Right side mount strut bracket. Torque the fastener to 37 ft. lbs. (50 Nm).

Cylinder head bolt torque sequence—3.4L engine

7924LG03

- Thermostat bypass pipe. Torque it to 18 ft. lbs. (25 Nm).
- Exhaust crossover pipe. Torque the fastener to 18 ft. lbs. (25 Nm).
- Valve rocker arms and pushrods. Torque the fastener to 89 inch lbs. (10 Nm). Using a torque angle meter torque the fastener an additional 30 degrees.
- Lower intake manifold. Torque the bolts to 115 inch lbs. (13 Nm).
- Upper intake manifold. Torque the bolts to 18 ft. lbs. (25 Nm).
- Negative battery cable

10. Refill the coolant system.

11. Change the oil filter and fill the engine with clean oil.

12. Turn the ignition to the **ON** position several times to pressurize the fuel system. Start the engine and inspect for any leaks, repair if necessary. Check and top off the fluid levels if required.

RIGHT (REAR) SIDE

1. Before servicing the vehicle, refer to the Precautions Section.

2. Relieve the fuel system pressure using the recommended procedure.

3. Drain the coolant system.

4. Drain the oil from the engine.

5. Remove or disconnect the following:
- Negative battery cable
- Upper intake manifold
- Lower intake manifold
- Valve rocker arms and pushrods
- Exhaust crossover pipe
- Right side spark plug wires
- Right side exhaust manifold
- Right side cylinder head and gasket
- Right side spark plugs from the cylinder head

To install:

6. Clean all parts well. Clean all gasket surfaces. Carefully remove all varnish soot and carbon to the bare metal. DO NOT use a motorized wire brush on any gasket surface since the soft aluminum will be damaged. If necessary, the head can be disassembled for thorough inspection and reconditioning.

7. Inspect the cylinder head for cracks. Do not attempt to weld the cylinder head. If cracked, replace it. Check the cylinder head deck, intake and exhaust manifold mating surfaces for flatness. These surfaces may be reconditioned by milling. If the surfaces are "out of flat" by more than 0.005 inch, the surface should be milled. If more than 0.010 inch of metal must be removed from the head, the head should be replaced.

8. Clean the cylinder head bolts and the bolt holes. Check the head bolts for damaged threads or stretching. New replacement head bolts are recommended.

9. Install or connect the following:
- New cylinder head gasket
- Cylinder head on top of the gasket and make certain it is lined up properly with the dowel pins
- New cylinder head bolts coated with a sealant (such as GM 1052080). Torque the bolts in the proper sequence (1-8) to 37 ft. lbs. (50 Nm). Using a torque angle meter turn the bolts 90 degrees in the proper sequence.
- Right side exhaust manifold. Torque the bolts to 12 ft. lbs. (16 Nm).
- Spark plugs. Torque the plugs 11 ft. lbs. (15 Nm).
- Spark plug wires
- Exhaust crossover pipe. Torque the fastener to 18 ft. lbs. (25 Nm).
- Valve rocker arms and pushrods. Torque the fastener to 89 inch lbs. (10 Nm). Using a torque angle meter torque the fastener an additional 30 degrees.
- Lower intake manifold. Torque the bolts to 115 inch lbs. (13 Nm).
- Upper intake manifold. Torque the bolts to 18 ft. lbs. (25 Nm).
- Negative battery cable

10. Refill the coolant system.

11. Change the oil filter and fill the engine with clean oil.

12. Start the vehicle and verify no leaks, abnormal noises and correct engine operation.

13. Check the fluid levels and top off if necessary.

3.6L Engine

LEFT SIDE

1. Before servicing the vehicle, refer to the Precautions Section.

2. Remove the upper and lower intake manifolds.

3. Remove the camshafts.

4. Remove the exhaust manifold.

5. Remove the two front M8 left cylinder head bolts.

6. Remove the left cylinder head bolts.

7. Remove the left cylinder head.

8. Remove and discard the left cylinder head gasket.

To install:

✳✳ WARNING

Ensure that the crankshaft is in the stage one timing drive assembly position using the EN 46111.

9. Ensure the cylinder head locating pins are securely mounted in the cylinder block deck face.

10. Install a NEW left cylinder head gasket using the deck face locating pins for retention.

11. Align the left cylinder head with the deck face locating pins. Place the left cylinder head in position on the deck face.

➡**DO NOT allow oil on the cylinder head bolt bosses. DO NOT reuse the old M11 cylinder head bolts.**

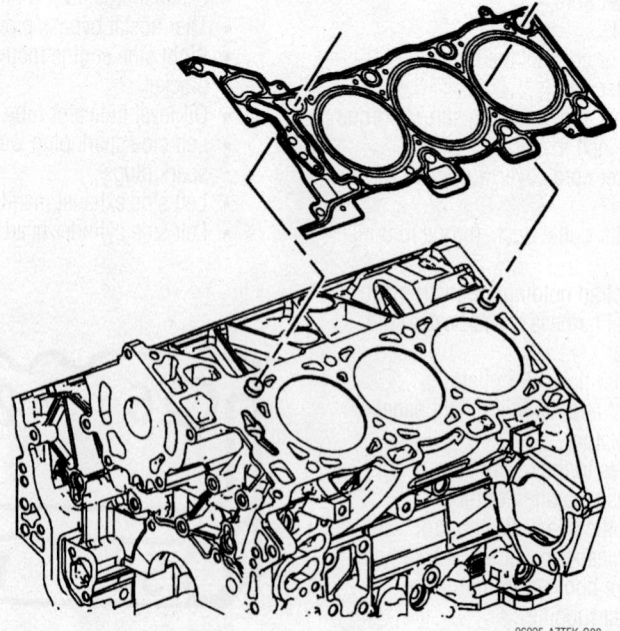

Cylinder head locating pins—3.6L engine

06025-AZTEK-G08

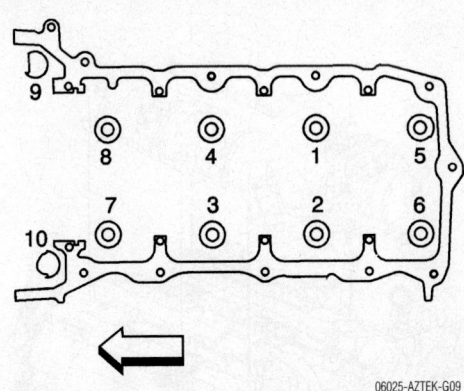

Left cylinder head bolt torque sequence—3.6L engine

06025-AZTEK-G09

12. Install new M11 cylinder head bolts.

a. Tighten the M11 cylinder head bolts a first pass in sequence to 45 Nm (33 ft. lbs.).

b. Tighten the M11 cylinder head bolts a second pass in sequence an additional 120 degrees.

13. Install the 2 front M8 left cylinder head bolts.

a. Tighten the M8 cylinder head bolts a first pass to 15 Nm (11 ft. lbs.).

b. Tighten the M8 cylinder head bolts a second pass in sequence an additional 60 degrees.

14. The remainder of installation is the reverse of removal.

RIGHT SIDE

1. Before servicing the vehicle, refer to the Precautions Section.

2. Remove the upper and lower intake manifolds.

3. Remove the camshafts.

4. Remove the exhaust manifold.

5. Remove the right cylinder head bolts.

6. Remove the right cylinder head.

7. Remove and discard the right cylinder head gasket.

To install:

➡**Ensure that the crankshaft is in the timing drive assembly position using the EN 46111.**

8. Ensure the cylinder head locating pins are securely mounted in the cylinder block deck face.

9. Install a NEW right cylinder head gasket using the deck face locating pins for retention.

10. Align the right cylinder head with the deck face locating pins.

11. Place the right cylinder head in position on the deck face.

➡**DO NOT allow oil on the cylinder head bolt bosses. DO NOT reuse the old M11 cylinder head bolts.**

12. Install new M11 cylinder head bolts.

a. Tighten the M11 cylinder head bolts a first pass in sequence to 45 Nm (33 ft. lbs.).

b. Tighten the M11 cylinder head

bolts a second pass in sequence an additional 120 degrees using the J 45059.

13. The remainder of installation is the reverse of removal.

Rocker Arms

REMOVAL & INSTALLATION

3.4L Engine

➡**Valve train components which are to be reused must be installed in their original positions. If removed, be sure to tag or arrange all rocker arms and pushrods to assure proper installation.**

1. Before servicing the vehicle, refer to the Precautions Section.

2. Remove or disconnect the following:
• Negative battery cable
• Rocker arm cover
• Rocker arm bolts
• Rocker arms

➡**Place the valve train parts in order to ensure they are installed in the proper location. Intake pushrods are yellow and measure 5.68 inches (144.18mm). Exhaust pushrods are green and measure 6.0 inches (152.51mm). When removing the pushrods, make certain they do not fall into the lifter valley.**

• Pushrods

To install:

3. Inspect and replace components if worn or damaged. Clean all old thread locking material from the pedestal bolts.

4. Coat the bearing surface of the rocker arms, pushrods and rocker arm bolts with a prelube (such as GM 1052365). Make certain to install the components in their original position.

5. Install or connect the following:
• Intake valve pushrods which are 5.68 inches (144.18mm) long
• Exhaust valve pushrods which are 6.0 inches (152.51mm) long
• Rocker arms. Torque the bolt to 14 ft. lbs. (19 Nm) plus 30 degrees.
• Rocker arm cover. Torque the bolt to 89 inch lbs. (10 Nm).
• Negative battery cable

6. Start the engine and verify the vehicle is running properly.

3.6L Engine

See the procedure under Camshaft Removal and Installation.

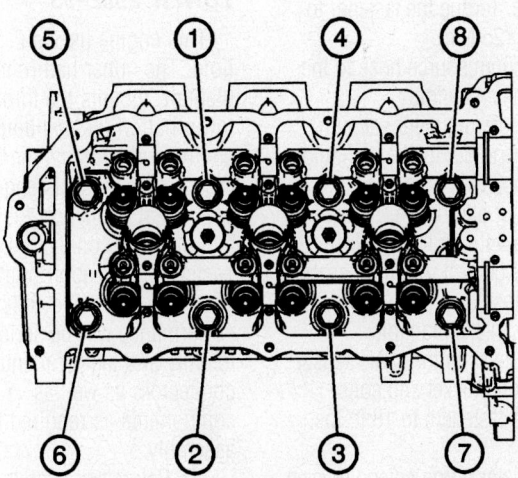

06025-AZTEK-G10

Right side cylinder head bolt torque sequence—3.6L engine

Intake Manifold

REMOVAL & INSTALLATION

3.4L Engine

UPPER

This engine uses a 2-piece intake manifold. The upper half (often called a plenum) mounts the throttle body. The lower half of the manifold bolts to the engine and contains the fuel injectors. Please note that this engine uses a sequential multi-port fuel injection system. Injector connectors must be connected to their appropriate fuel injector assembly or engine emissions and engine performance will be seriously affected. Identify and tag for identification all wiring connectors as well as vacuum and other components as required to assure correct assembly.

1. Before servicing the vehicle, refer to the Precautions Section.
2. Drain the engine coolant. Remove the coolant recovery bottle.
3. Relieve the fuel system pressure using the recommended procedure.
4. Remove or disconnect the following:
   - Negative battery cable
   - Throttle body air inlet duct
   - Accelerator and cruise control cables and bracket from the throttle body
   - Throttle Position (TP) sensor connector from the throttle body
   - Idle Air Control (IAC) valve connector from the throttle body
   - Left side spark plug wires
   - Left side spark plug wire harness clip and harness
   - Throttle body heater hoses
   - Evaporative emissions (EVAP) canister purge solenoid valve vacuum hoses
   - EVAP canister purge solenoid valve
   - Ignition coil bracket and coils
   - Wire harness for the Manifold Air Pressure (MAP) sensor
   - Vacuum harness from the MAP sensor and upper intake manifold
   - Emissions control vacuum harness
   - Brake booster vacuum hose from the upper intake manifold
   - Vacuum hose connection for the Heater Vent Air Conditioning (HVAC) source hose
   - Vacuum hose connection for the fuel pressure regulator
   - Exhaust Gas Recirculation (EGR) valve
   - MAP sensor and bracket

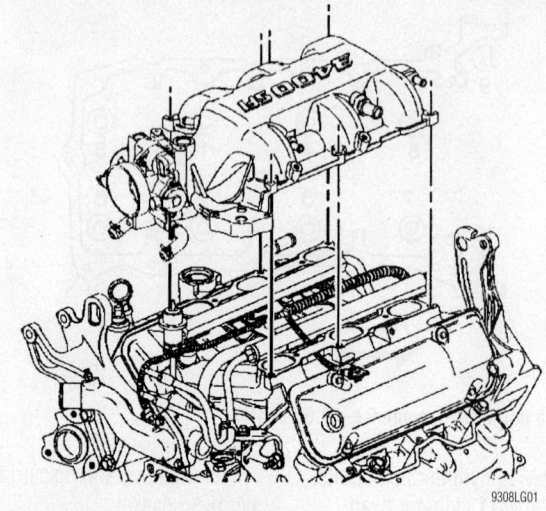

Removing the upper intake manifold—3.4L engine

   - Alternator through-bolt bracket
   - Upper intake manifold
   - Upper intake manifold gasket
   - Throttle body, if replacing the manifold

To install:

5. Clean all parts well. Use care in cleaning old gasket material from the machined aluminum surfaces on the plenum and manifold as sharp tools may damage sealing surfaces.
6. Clean the mating surfaces to the upper intake manifold and engine block. Remove any loose pieces of RTV sealer.
7. Install or connect the following:
   - Throttle body to the upper intake manifold (if removed). Torque the bolts to 18 ft. lbs. (25 Nm).
   - Upper intake manifold gasket
   - Upper intake manifold
   - MAP sensor and bracket. Torque the bolt to 44 inch lbs. (5 Nm).
   - Upper intake manifold bolts. Torque the bolts to 18 ft. lbs. (25 Nm).
   - EGR valve. Torque the fastener to 18 ft. lbs. (25 Nm).
   - HVAC vacuum source hose to the upper intake manifold
   - Fuel pressure regulator vacuum hose to the upper intake manifold
   - Brake booster vacuum hose
   - MAP sensor and bracket. Torque the bolt to 44 inch lbs. (5 Nm).
   - Emissions control vacuum harness
   - Vacuum hose for the MAP sensor and upper intake manifold
   - Wiring harness to the MAP sensor
   - Ignition coil bracket and coils. Torque the fasteners to 18 ft. lbs. (25 Nm).
   - EVAP canister purge solenoid valve

   - Vacuum hoses to the EVAP canister purge solenoid valve
   - Throttle body heater hoses
   - Left side spark plug wire harness clip
   - Spark plugs wires
   - TP sensor wire harness connector to the throttle body
   - IAV valve wire harness connector to the throttle body
   - Accelerator and cruise control cables and bracket to the throttle body. Torque the fasteners to 106 inch lbs. (12 Nm).
   - Throttle body air inlet duct
   - Negative battery cable

8. Fill the coolant system.
9. Fill the engine with new oil.
10. Turn the ignition to the **ON** position several times to pressurize the fuel system.
11. Start the engine and check for any leakage and repair if necessary.
12. Check and top off all fluid levels if needed.

LOWER 2002–03

This engine uses a 2-piece intake manifold. The upper half (often called a plenum) mounts the throttle body. The lower half of the manifold bolts to the engine and contains the fuel injectors. Please note that this engine uses a sequential multi-port fuel injection system. Injector connectors must be connected to their appropriate fuel injector assembly or engine emissions and engine performance will be seriously affected. Identify and tag for identification all wiring connectors as well as vacuum and other components as required to assure correct assembly.

1. Before servicing the vehicle, refer to the Precautions Section.

2. Drain the engine coolant.

3. Relieve the fuel system pressure using the recommended procedure.

4. Remove or disconnect the following:

- Negative battery cable
- Upper intake manifold
- Left and right side valve covers
- Wire harness from the Engine Coolant Temperature (ECT) sensor
- Fuel injector, Manifold Absolute Pressure (MAP) and ECT wire harness
- Fuel feed and return pipe from the injector rail
- Fuel injector rail
- Power steering pump from the front cover

➡**Do not disconnect the power steering pipes or hoses from the steering pump.**

- Heater inlet pipe with the heater hose from the lower intake manifold
- Radiator inlet hose
- Thermostat bypass hose from the manifold
- Lower intake manifold
- Pushrods after loosening the rocker arms
- Lower intake manifold gasket and seals
- ECT sensor, if replacing the manifold
- Thermostat and housing, if replacing the manifold

To install:

5. Clean the gasket mounting surfaces.

6. Inspect the intake manifold for cracks or damage, replace if necessary.

7. Install or connect the following:

- ECT sensor, if removed. Torque the sensor to 17 ft. lbs. (23 Nm).
- Thermostat and housing, if removed. Torque the fastener to 18 ft. Lbs. (25 Nm).
- Thin bead of RTV sealer (such as GM 12345739) on the ridge of the engine block where the lower intake manifold makes contact
- Lower intake manifold gaskets
- Pushrods and tighten the rocker arms. Torque the arms to 14 ft. lbs. (19 Nm) plus 30 degrees.
- Lower intake manifold. Torque the bolts to first to 62 inch lbs. (7 Nm) then to 115 inch lbs. (13 Nm) after applying a sealant (such as GM 12345382) to the threads of the bolts.
- Thermostat bypass hose to the lower intake manifold pipe. Torque the fastener to 18 ft. lbs. (25 Nm).

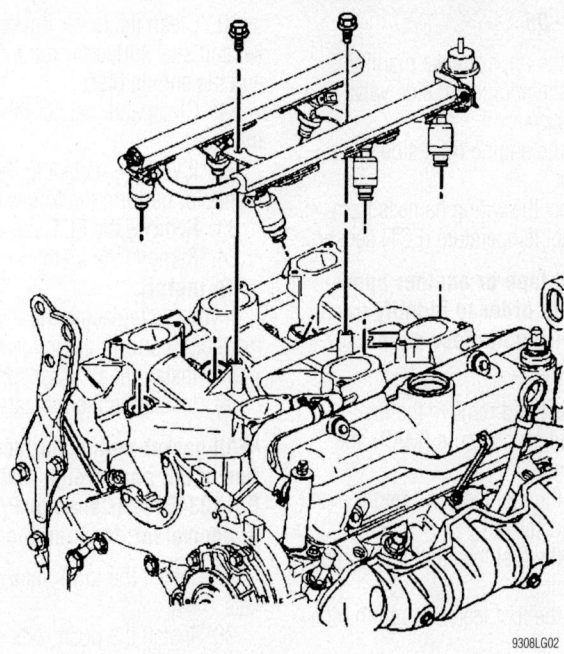

Remove the fuel injector rail—2002–03 3.4L engine

- Radiator inlet hose from the engine
- Power steering pump to the front cover
- Fuel injector rail. Torque the fastener to 89 inch lbs. (10 Nm).
- Fuel feed and return pipes to the injector rail. Torque the fasteners to 13 ft. lbs. (17 Nm).
- Wire harness for the fuel injector, MAP sensor and the ECT sensor
- Right side rocker arm cover. Torque the bolts to 89 inch lbs. (10 Nm).

- Left side rocker arm cover. Torque the bolts to 89 inch lbs. (10 Nm).
- Upper intake manifold. Torque the bolts to 18 ft. lbs. (25 Nm).
- Negative battery cable

8. Fill the coolant system.

9. Turn the ignition to the **ON** position several times to pressurize the fuel system.

10. Start the engine and check for any leakage and repair if necessary.

11. Check and top off all fluid levels if needed.

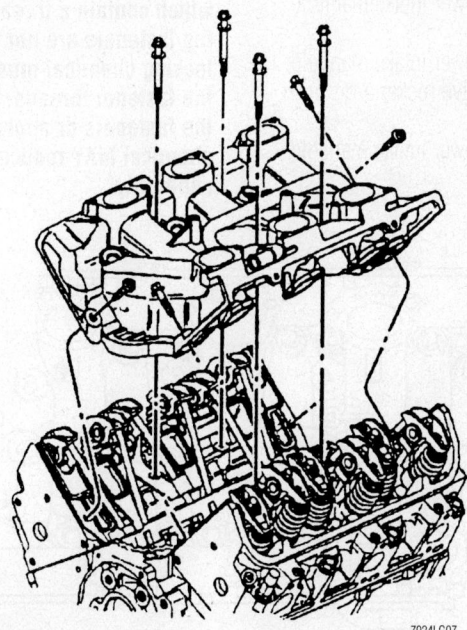

Lower intake manifold assembly—2002–03 3.4L engine

LOWER 2004–05

12. Remove the upper intake manifold.
13. Remove the engine left side valve rocker arm cover.
14. Remove the engine right side valve rocker arm cover.
15. Disconnect the wiring harness from the engine coolant temperature (ECT) sensor.

➡**Use masking tape or another appropriate method in order to identify each fuel injector wiring harness connector cylinder number.**

16. Disconnect and remove the fuel injector, manifold air pressure (MAP) and ECT wiring harness.
17. Remove the fuel feed and return pipe from the fuel injector rail.
18. Remove the fuel feed and return pipe retaining clip bolt.
19. Remove the fuel feed and return pipe retaining clip.
20. Remove the fuel injector rail bolts.
21. Remove the fuel injector rail with the fuel injectors.

➡**Do NOT disconnect the power steering pipes or hoses from the power steering pump.**

22. Remove the power steering pump from the front engine cover and reposition.
23. Disconnect the heater inlet pipe with heater hose from the lower intake manifold and reposition.
24. Disconnect the inlet radiator hose from the engine.
25. Disconnect the thermostat bypass hose from the lower intake manifold pipe.
26. Remove the lower intake manifold bolts.
27. Remove the lower intake manifold.
28. Loosen the valve rocker arms and remove the push rods.
29. Remove the lower intake manifold gaskets and seals.

30. Clean the lower intake manifold gasket and seal surfaces on the cylinder heads and the engine block.
31. Clean and inspect the lower intake manifold.
32. If you are replacing the lower intake manifold, perform the following steps:
33. Remove the ECT sensor.
34. Remove the thermostat.
To install:
If you are replacing the lower intake manifold, perform the following steps.
35. Install the ECT sensor.
36. Install the thermostat.

➡**All gasket-mating surfaces need to be free of oil, and foreign material. Use GM P/N 12346139 (Canadian P/N 10953463), or equivalent, to clean the surfaces.**

37. Install the lower intake manifold gaskets.
38. Install the push rods and tighten the valve rocker arms.
39. With the gaskets in place apply a small drop 8–10 mm (0.31–0.39 in.) of RTV sealer GM P/N 12346286 (Canadian P/N 10953472) or equivalent to the 4 corners of the intake manifold to block joint.
40. Connect the 2 small drops with a bead of RTV sealer that is between 8–10 mm (0.31–0.39 in.) wide and 3.0–5.0 mm (0.12–0.20 in.) thick.
41. Install the lower intake manifold.
42. Apply sealer GM P/N 12345382, (Canadian P/N 10953489) or the equivalent to the lower intake manifold bolt threads.

➡**Maximum gasket performance is achieved when using new fasteners, which contain a thread-locking patch. If the fasteners are not replaced, a thread locking chemical must be applied to the fastener threads. Failure to replace the fasteners or apply a thread-locking chemical MAY reduce gasket sealing capability.**

➡**All lower intake manifold bolts need to be cleaned, free of any foreign material, and reused only if new bolts are unavailable. Use GM P/N 12345382 (Canadian P/N 10953489) or equivalent and apply to the old intake manifold bolt threads.**

❋❋ WARNING

The manufacturer recommends the center bolts be fully torqued before the diagonal bolts to assure proper seal ability.

➡**Lower intake manifold bolts in location 6 and 7 should be torqued to specification using a crow's foot type tool.**

43. Install the lower intake manifold bolts.
44. Tighten the lower intake manifold bolts in sequence to 7 Nm (62 inch lbs.) on the first pass.
45. Tighten the lower intake manifold bolts (1, 2, 3, 4) in sequence to 13 Nm (115 inch lbs.) on the final pass.
46. Tighten the lower intake manifold bolts (5, 6, 7, 8) in sequence to 25 Nm (18 ft. lbs.) on the final pass.
47. Connect the thermostat bypass hose to the lower intake manifold pipe.
48. Connect the inlet radiator hose to the engine.
49. Connect the heater inlet pipe to the lower intake manifold.
50. Install the power steering pump to the front engine cover.
51. Install the fuel injector rail.
52. Connect the fuel feed and return pipe to the fuel injector rail.
53. Install the fuel feed and return pipe retaining clip.
54. Install the fuel feed and return pipe retaining clip bolt. Tighten the fuel feed and return pipe retaining clip bolt to 8 Nm (71 inch lbs.).

➡**The fuel injector wiring harness connectors must be connected to their respective fuel injectors. Failure to connect the fuel injector connectors to their respective fuel injectors may result in excessive exhaust emissions and poor engine performance.**

55. Install and connect the fuel injector, MAP and ECT wiring harness.
56. Install the wiring harness to the ECT sensor.
57. Install the engine right side valve rocker arm cover.
58. Install the engine left side valve rocker arm cover.
59. Install the upper intake manifold.

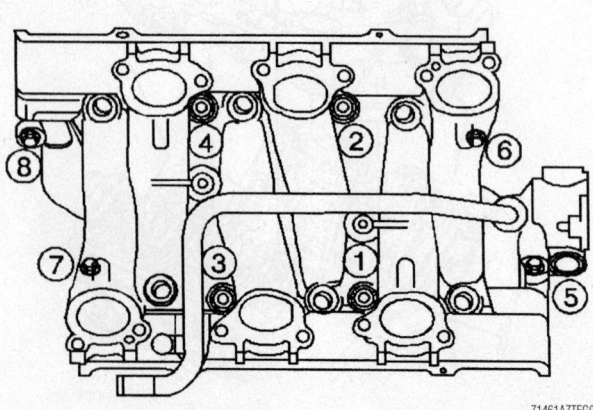

71461AZTEG01

Lower intake manifold torque sequence—3.4L engine

3.6L Engine

UPPER

1. Before servicing the vehicle, refer to the Precautions Section.
2. Turn the ignition OFF.
3. Remove the air inlet duct.
4. Relieve the fuel system pressure.
5. Disconnect the fuel pressure and evaporative emission (EVAP) hoses from the engine.
6. Disconnect the BARO sensor electrical connector.
7. Remove the purge line from the purge line retainer.
8. Remove the fuel feed hose bracket bolt and reposition the fuel feed hose.

➡**Do NOT disconnect the engine control module (ECM) electrical connectors. Do NOT remove the ECM from the ECM bracket.**

9. Remove the ECM bracket with the ECM and reposition aside.
10. Raise and support the vehicle.
11. Disconnect the intermediate steering shaft from the steering gear.
12. Remove and reposition the front portion of the front fender liners in order to gain access to the frame front bolts.
13. Lower the vehicle.
14. Position a floor jack at the front center section of the frame in order to support the powertrain.
15. Remove the frame front bolts.
16. Carefully lower the powertrain or raise the vehicle enough to provide access.
17. Disconnect the purge solenoid electrical connector.
18. Remove the wiring harness from the right side of the intake manifold.
19. Disconnect the fuel injector electrical connector.
20. Remove the fuel injector electrical connector from the fuel injector electrical connector bracket.
21. Disconnect the intake manifold runner control solenoid electrical connector.
22. Disconnect the throttle body electrical connector.
23. Remove the brake booster vacuum hose and check valve from the intake manifold.
24. Remove the positive crankcase ventilation (PCV) hose from the cylinder head and the intake manifold.
25. Remove the intake manifold bolts (1-6).
26. Remove the upper intake manifold.
27. Remove and discard the upper intake manifold gasket.

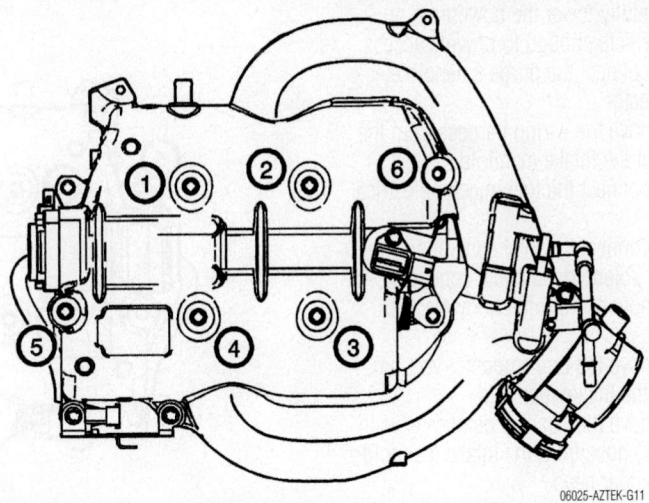

Upper intake manifold bolt loosening/tightening sequence—3.6L engine

06025-AZTEK-G11

28. Disassemble the intake manifold if necessary.
29. Clean and inspect the intake manifold and the sealing surfaces.

To install:

30. Assemble the intake manifold if necessary.
31. Install the intake manifold.
32. Install the intake manifold bolts (1-6). Tighten the bolts in the order shown to 23 Nm (17 ft. lbs.).
33. Install the PCV hose to the intake manifold and the cylinder head.
34. Install the brake booster vacuum hose and check valve to the brake booster.
35. Install the throttle body electrical connector.
36. Connect the intake manifold runner control solenoid electrical connector.
37. Install the fuel injector electrical connector to the fuel injector electrical connector bracket.
38. Install the fuel injector bracket.
39. Connect the fuel injector electrical connector.
40. Install the wiring harness and bracket to the right side of the intake manifold. Tighten the bracket bolts to 10 Nm (89 inch lbs.).
41. Connect the purge solenoid electrical connector.
42. Carefully raise the powertrain or lower the vehicle in order to install the frame bolts.
43. Install the frame front bolts. Torque to 114 ft. lbs. (155 Nm).
44. Remove the floor jack.
45. Raise and support the vehicle.
46. Install the front fender liners.
47. Connect the intermediate steering shaft to the steering gear.

48. Lower the vehicle.
49. Install the ECM bracket with the ECM.
50. Install the fuel feed hose bracket and the bracket bolt. Tighten the bracket bolt to 10 Nm (89 inch lbs.).
51. Connect the BARO sensor electrical connector.
52. Install the purge line to the purge line retainer.
53. Connect the fuel pressure and EVAP hoses to the engine.
54. Install the air inlet duct.

LOWER

1. Before servicing the vehicle, refer to the Precautions Section.
2. Turn the ignition OFF.
3. Remove the air inlet duct.
4. Relieve the fuel pressure.
5. Disconnect the fuel pressure and evaporative emission (EVAP) hoses from the engine.
6. Disconnect the barometric pressure (BARO) sensor electrical connector.

➡**Do NOT disconnect the engine control module (ECM) electrical connectors.**

7. Remove the ECM bracket with the ECM and reposition aside.
8. Raise and support the vehicle.
9. Disconnect the intermediate steering shaft from the steering gear.
10. Remove and reposition the front portion of the front fender liners in order to gain access to the frame front bolts.
11. Lower the vehicle.
12. Position a floor jack at the front center section of the frame in order to support the powertrain.
13. Remove the frame front bolts.

14. Carefully lower the powertrain or raise the vehicle enough to provide access.

15. Disconnect the purge solenoid electrical connector.

16. Remove the wiring harness from the right side of the intake manifold.

17. Disconnect the fuel injector electrical connector.

18. Disconnect the intake manifold runner control solenoid electrical connector.

19. Remove the throttle body electrical connector.

20. Remove the brake booster vacuum hose from the intake manifold.

21. Remove the positive crankcase ventilation (PCV) hose from the intake manifold and the cylinder head.

22. Remove the intake manifold bolts (1-6).

23. Remove the intake manifold.

24. Disassemble the intake manifold as necessary.

25. Clean and inspect the intake manifold and the sealing surfaces.

To install

26. Assemble the intake manifold as necessary.

27. Install the intake manifold. Tighten the bolts in the order shown to 23 Nm (17 ft. lbs.).

28. Install the PCV hose to the intake manifold and the cylinder head.

29. Install the brake booster vacuum hose to the intake manifold.

30. Connect the throttle body electrical connector.

31. Connect the intake manifold runner control solenoid electrical connector.

32. Connect the fuel injector electrical connector.

33. Install the wiring harness to the right side of the intake manifold. Tighten the bracket bolts to 10 Nm (89 inch lbs.).

34. Connect the purge solenoid electrical connector.

35. Carefully raise the powertrain or lower the vehicle in order to install the frame bolts.

36. Install the frame front bolts. Torque to 114 ft. lbs. (155 Nm).

37. Remove the floor jack.

38. Raise and support the vehicle.

39. Install the front fender liners.

40. Connect the intermediate steering shaft to the steering gear.

41. Lower the vehicle.

42. Install the ECM bracket with the ECM.

43. Connect the BARO sensor electrical connector.

44. Connect the fuel pressure and EVAP hoses to the engine.

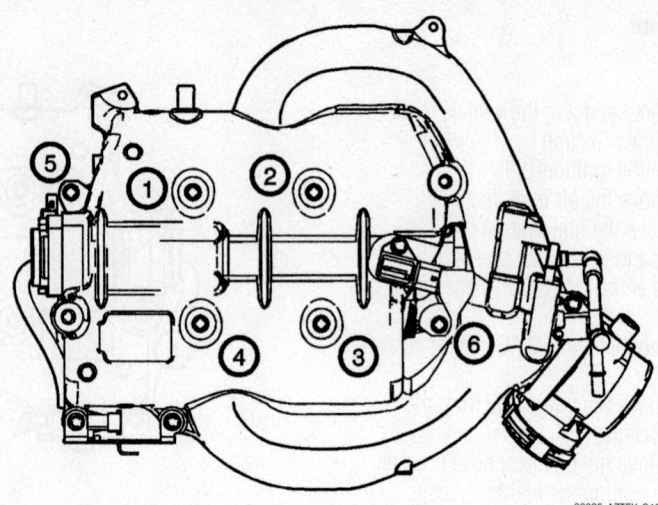

06025-AZTEK-G12

Lower intake manifold bolt torque sequence—3.6L engine

45. Install the air inlet duct.

Exhaust Manifold

REMOVAL & INSTALLATION

3.4L Engine

The exhaust manifolds are conventional iron castings. The left and right manifolds are connected by a crossover pipe. Use care with the exhaust manifold-to-cylinder head fasteners. The cylinder heads are aluminum.

LEFT (FRONT) SIDE

1. Before servicing the vehicle, refer to the Precautions Section.

2. Drain the cooling system.

3. Remove or disconnect the following:
   • Negative battery cable
   • Throttle body air inlet duct
   • Right side engine mount strut bracket

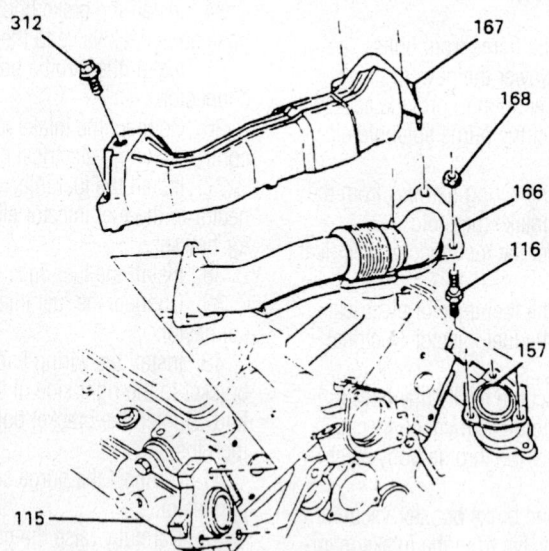

115	MANIFOLD, LEFT HAND EXHAUST
116	STUD, EXHAUST CROSSOVER
157	MANIFOLD, RIGHT HAND EXHAUST
166	CROSSOVER PIPE, EXHAUST
167	SHIELD, EXHAUST CROSSOVER UPPER HEAT
168	NUT, EXHAUST CROSSOVER
312	BOLT/SCREW, EXHAUST CROSSOVER UPPER HEAT SHIELD

7924LG28

Exploded view of the exhaust crossover and heat shield mounting

- Radiator inlet hose
- Thermostat bypass pipe
- Exhaust crossover pipe heat shield
- Exhaust crossover pipe
- Left side exhaust manifold heat shield
- Left side exhaust manifold and discard the gasket

To install:

4. Clean the gasket mounting surfaces.
5. Install or connect the following:
   - Left side exhaust manifold gasket
   - Left side exhaust manifold. Torque the nuts to 12 ft. lbs. (16 Nm).
   - Left side exhaust manifold heat shield. Torque the fasteners to 89 inch lbs. (10 Nm).
   - Exhaust crossover pipe. Torque the bolts to 18 ft. lbs. (25 Nm).
   - Exhaust crossover pipe heat shield. Torque the bolts to 89 inch lbs. (10 Nm).
   - Thermostat bypass pipe. Torque the fastener to 18 ft. lbs. (25 Nm).
   - Radiator inlet hose to the engine
   - Right side engine mount strut bracket. Torque the bolts to 35 ft. lbs. (48 Nm).
   - Throttle body air inlet duct
   - Negative battery cable
6. Fill the cooling system
7. Start the vehicle and check for leaks, repair if necessary.
8. Check and top off all fluid levels if necessary.

RIGHT (REAR) SIDE

1. Before servicing the vehicle, refer to the Precautions Section.
2. Remove or disconnect the following:
   - Negative battery cable
   - Throttle body air inlet duct
   - Accelerator cable bracket from the throttle body
   - Manifold Absolute Pressure (MAP) sensor
   - Exhaust Gas Recirculation (EGR) valve
3. Rotate the engine for access.
   - Ignition module
   - Ignition coils and bracket
   - Spark plug wires from the right bank spark plugs
   - Heated Oxygen (HO2) sensor electrical connector
   - EVAP solenoid bracket
   - Fasteners for the crossover pipe from the right bank exhaust manifold
   - Catalytic converter
   - Right side exhaust manifold heat shields
   - Right side exhaust manifold

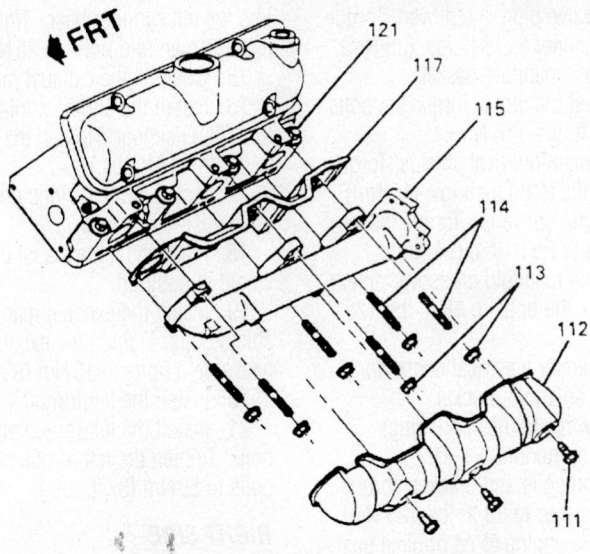

111 SCREW, LH EXHAUST MANIFOLD HEAT SHIELD
112 SHIELD LH EXHAUST MANIFOLD
113 NUT, LH EXHAUST MANIFOLD
114 STUD, LH EXHAUST MANIFOLD
115 MANIFOLD, LH EXHAUST
117 GASKET, LH EXHAUST MANIFOLD
121 HEAD, LH CYLINDER

7924LG29

Left exhaust manifold mounting—3.4L engine

- Right side exhaust manifold gasket
- EGR valve pipe, if replacing the exhaust manifold
- HO2 sensor, if replacing the exhaust manifold

To install:

4. Clean the gasket mounting surfaces.
5. Install or connect the following:
   - HO2 sensor, if removed. Torque the sensor to 31 ft. lbs. (42 Nm).

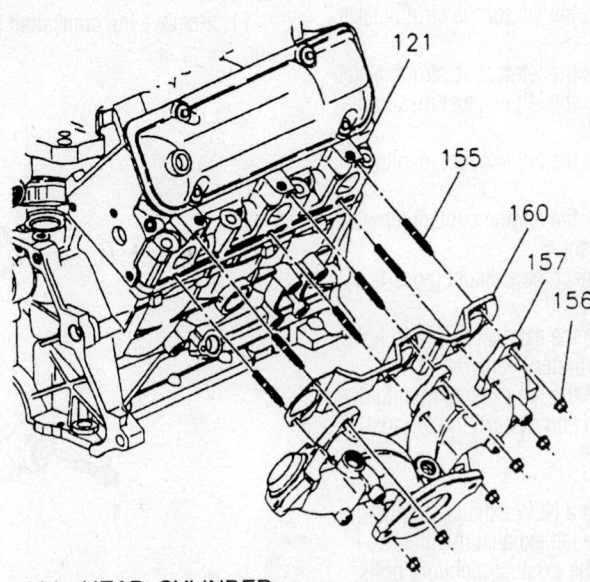

121 HEAD, CYLINDER
155 STUD, EXHAUST MANIFOLD
156 NUT, EXHAUST MANIFOLD
157 MANIFOLD RIGHT EXHAUST
160 GASKET, RIGHT EXHAUST MANIFOLD

7924LG30

Exploded view of the right exhaust manifold mounting—3.4L engine

- EGR valve pipe, if removed. Torque the fastener to 18 ft. lbs. (25 Nm).
- Exhaust manifold gasket
- Exhaust manifold. Torque the bolts to 12 ft. lbs. (16 Nm).
- Both manifold heat shields. Torque the bolts to 89 inch lbs. (10 Nm).
- Catalytic converter. Torque the fasteners to 25 ft. lbs. (34 Nm).
- Exhaust manifold crossover pipe. Torque the bolts to 18 ft. lbs. (25 Nm).
- HO_2 sensor electrical connector
- EVAP solenoid bracket
- Plug wires to the spark plugs
- Ignition module
- Ignition coils and bracket. Torque the fastener to 18 ft. lbs. (25 Nm).

6. Rotate the engine to its original position.

7. Install or connect the following:
- EGR valve to the intake manifold
- MAP sensor
- Accelerator cable bracket to the throttle body. Torque the bolt to 89 inch lbs. (10 Nm).
- Throttle body air inlet duct
- Negative battery cable

8. Start the engine and inspect for leaks, repair if necessary.

3.6L Engine

LEFT SIDE

1. Before servicing the vehicle, refer to the Precautions Section.

2. Remove the left torque strut bracket bolts.

3. Remove the left torque strut bracket.

4. Remove the left exhaust manifold heat shield bolts.

5. Remove the left exhaust manifold heat shield.

6. Remove the engine coolant temperature (ECT) sensor.

7. Disconnect the exhaust pipes from the manifold.

8. Remove the exhaust manifold bolts from the left cylinder head.

9. Remove the left exhaust manifold.

10. Remove and discard the exhaust manifold gasket.

To install:

11. Position a NEW exhaust manifold gasket onto the left exhaust manifold.

12. Install the exhaust manifold bolts into the left exhaust manifold.

13. Place the left exhaust manifold, exhaust manifold gasket and bolts as an assembly in position on the left cylinder head.

14. Install the exhaust manifold bolts into the left cylinder head. Tighten the exhaust manifold bolts to 25 Nm (18 ft. lbs.).

15. Connect the exhaust pipes.

16. Install the engine coolant temperature (ECT) sensor. Tighten the ECT sensor to 22 Nm (16 ft. lbs.).

17. Install NEW O-rings on the crankshaft position sensor.

18. Place the left exhaust manifold heat shield in position.

19. Install the exhaust manifold heat shield bolts. Tighten the exhaust manifold heat shield bolts to 10 Nm (89 inch lbs.).

20. Install the left torque strut bracket.

21. Install the left torque strut bracket bolts. Tighten the left torque strut bracket bolts to 50 Nm (37 ft. lbs.).

RIGHT SIDE

1. Before servicing the vehicle, refer to the Precautions Section.

2. Remove the right exhaust manifold heat shield bolts.

3. Remove the right exhaust manifold heat shield.

4. Remove the exhaust manifold bolts from the right cylinder head.

5. Remove the right exhaust manifold.

6. Remove and discard the exhaust manifold gasket.

7. Remove the block heater cartridge, if equipped.

8. Remove the right knock sensor bolt.

9. Remove the right knock sensor.

10. Remove the crankshaft position sensor bolt.

11. Remove the crankshaft position sensor.

12. Remove and discard the crankshaft position sensor O-ring, if damaged.

To install

13. Install a NEW O-ring on the crankshaft position sensor, if damaged.

14. Position the crankshaft position sensor into the cylinder block.

15. Install the crankshaft position sensor bolt. Tighten the crankshaft position sensor bolt to 10 Nm (89 inch lbs.).

16. Position the right knock sensor to the cylinder block as shown.

17. Install the knock sensor bolt. Tighten the knock sensor bolt to 23 Nm (17 ft. lbs.). Ensure proper knock sensor orientation.

18. Install the block heater cartridge, if equipped.

19. Position a NEW exhaust manifold gasket onto the right exhaust manifold.

20. Install the exhaust manifold bolts into the right exhaust manifold.

21. Place the right exhaust manifold, exhaust manifold gasket and bolts as an assembly in position on the right cylinder head.

22. Connect the exhaust pipes.

23. Install the exhaust manifold bolts into the right cylinder head. Tighten the exhaust manifold bolts to 25 Nm (18 ft. lbs.).

24. Place the right exhaust manifold heat shield in position.

25. Install the exhaust manifold heat shield bolts. Tighten the exhaust manifold heat shield bolts to 10 Nm (89 inch lbs.).

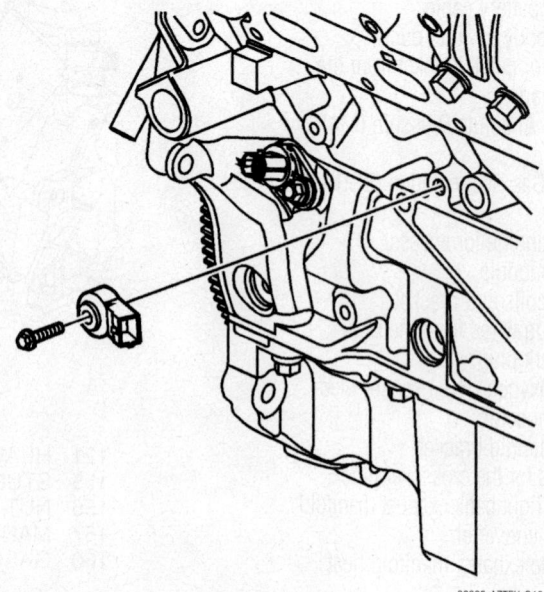

06025-AZTEK-G13

Proper knock sensor installation—3.6L engine

Camshaft and Valve Lifters

REMOVAL & INSTALLATION

3.4L Engine

1. Before servicing the vehicle, refer to the Precautions Section.
2. Relieve the fuel system pressure.
3. Remove or disconnect the following:
   • Engine assembly

✳✳ WARNING

When removing valvetrain components they must be marked for installation in their original location. When the camshaft is being replaced, the valve lifters must also be replaced.

   • Right and left side valve covers
   • Upper and lower intake manifold
   • Rocker arm bolts, balls, rocker arms and pushrods
   • Oil splash shield
   • Lifter guide bolts and the guide
   • Valve lifter(s) from the bores
   • Crankshaft balancer and front cover
   • Timing chain and sprockets
   • Oil pump driven gear bolt and gear
   • Camshaft thrust plate
   • Camshaft, using a large screwdriver inserted in the bolt hole to carefully rotate and pull the camshaft out of the bearings

✳✳ WARNING

Avoid damaging the camshaft bearing surfaces.

To install:
4. Coat the camshaft with Prelube.
5. Install or connect the following:
   • Camshaft
   • Camshaft thrust plate. Tighten the bolts to 89 inch lbs. (10 Nm).
   • Oil pump driven gear. Tighten the bolt to 27 ft. lbs. (36 Nm).
   • Timing chain and sprocket
   • Camshaft thrust button and front cover
   • Crankshaft balancer
6. Lubricate the bearing surfaces with Molykote.

➡**Installation of a new camshaft or a wear pattern on the old valve lifter will require the replacement of the camshaft and lifters together. If camshaft replacement is not necessary, be sure to install the used valve lifters in their original position.**

7. Install or connect the following:
   • Lifters in their original locations
   • Lifter guide. Tighten the guide bolts to 89 inch lbs. (10 Nm).
   • Oil splash shield
   • Pushrods, rocker arms, balls and bolts. Tighten the nuts to 89 inch lbs. (10 Nm) plus an additional 30 degree turn.
   • Lower and upper intake manifold
   • Right and left side valve covers
   • Engine assembly
   • Negative battery cable
8. Adjust the valves, as required. Start the engine and verify no oil leaks.

3.6L Engine

LEFT SIDE

1. Before servicing the vehicle, refer to the Precautions Section.
2. Remove the upper intake manifold with the lower intake manifold.
3. Disconnect the ignition coil electrical connectors.
4. Remove the wiring harness from the side of the camshaft cover by sliding the conduit down and outboard.
5. Remove the wiring conduit retainers from the camshaft cover by rotating the wiring harness conduit retainers counterclockwise.

➡**It is not necessary to disconnect the engine front cover electrical connectors.**

6. Remove the wiring harness from the front of the camshaft cover.
7. Reposition and secure the wiring harnesses away from the camshaft cover in order to provide clearance.
8. Remove the ignition coils.
9. Loosen the left engine strut bracket.
10. Loosen the left engine strut bracket-to-cylinder head bolts.
11. Remove the camshaft cover bolts and camshaft cover.
12. Remove and discard the camshaft cover seal and grommets. DO NOT reuse.
13. Remove the camshaft sensors.
14. Remove the camshaft position actuator solenoid.
15. Remove the crankshaft balancer.
16. Rotate the crankshaft with the EN 46111 until the camshafts are in a neutral (low tension) position. The camshaft flats will be parallel with the camshaft cover rail.

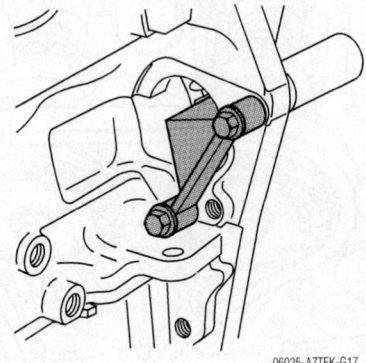

06025-AZTEK-G17

Flywheel locking tool in place—3.6L engine

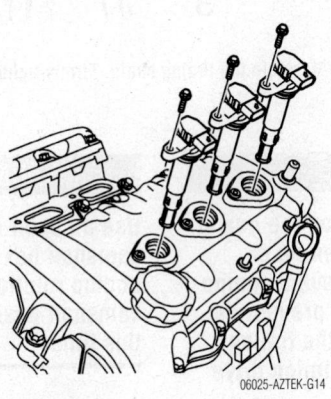

06025-AZTEK-G14

Ignition coil removal—3.6L engine

06025-AZTEK-G18

Rotate the crankshaft with the EN 46111 until the camshafts are in a neutral (low tension) position. The camshaft flats will be parallel with the camshaft cover rail—3.6L engine

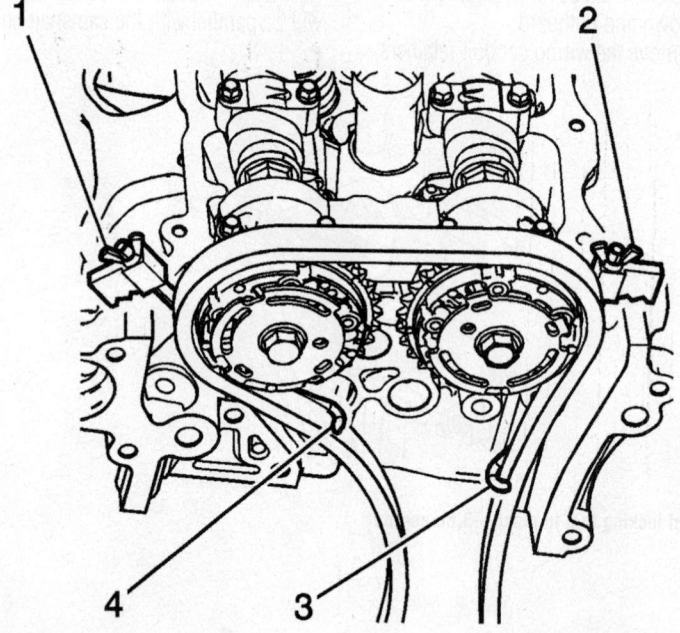

06025-AZTEK-G19

Install the EN 46108 in order to retain the timing chain. Firmly tighten the EN 46108 nuts—3.6L engine

✳✳ WARNING

A wrench must be used on the hex of the camshaft when loosening or tightening in order to prevent component damage. Failure to prevent the torque reaction against the timing drive chain can lead to timing drive chain failure.

✳✳ WARNING

Use an open-end wrench at the camshaft hex to prevent camshaft/engine rotation. DO NOT remove the camshaft position actuator bolt at this time.

17. Loosen the camshaft position actuator bolt.

➡ **Ensure that the tips of the EN 46108 are fully engaged into the timing chain.**

18. Install the EN 46108 in order to retain the timing chain. Firmly tighten the EN 46108 nuts.

➡ **Ensure that the camshaft timing chain and the camshaft position actuators are marked for proper assembly.**

19. Mark the timing chain and the respective locations on the camshaft position actuators.

20. Remove the camshaft position actuator bolt.

21. Remove the EN 46105-2 from the left camshafts.

22. Position the camshaft lobes in a neutral position.

23. Observe the markings on the bearing caps. Each bearing cap is marked in order to identify its location. The markings have the following meanings:

- The raised feature must always be oriented toward the center of the cylinder head.
- The I indicates the intake camshaft.
- The E indicates the exhaust camshaft.
- The number indicates the journal position from the front of the engine.

24. Remove the camshaft bearing cap bolts.

25. Remove the camshaft bearing caps.

26. Remove the camshafts.

27. Replace the camshaft bearing caps and bolts.

28. Remove the valve rocker arms, camshaft followers, from the left cylinder head.

✳✳ WARNING

Do not stroke/cycle the stationary hydraulic lash adjuster plunger without oil in the lower pressure chamber. Do not allow the stationary hydraulic lash adjuster to tip over, plunger down, after the oil fill.

29. Remove the valve lifters, stationary hydraulic lash adjuster, (SHLAs) from the left cylinder head.

To install:

30. Fill the stationary hydraulic lash adjuster (SHLA) with clean engine oil GM P/N 12378006 or equivalent. Take precautions to prevent scratching the pivot sphere area (1) of the SHLA.

31. Lubricate the SHLA bores in the

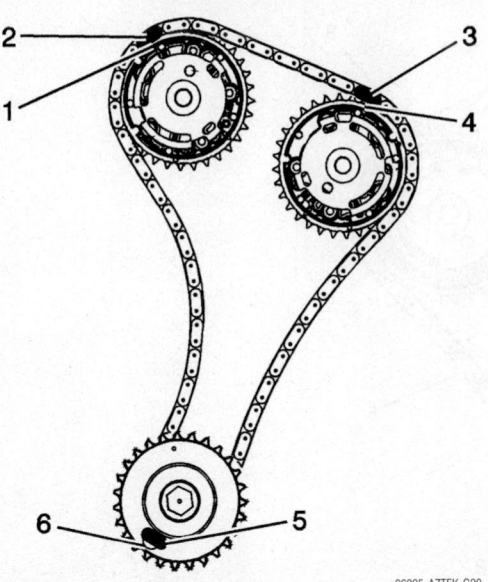

06025-AZTEK-G20

Mark the timing chain and the respective locations on the camshaft position actuators—3.6L engine

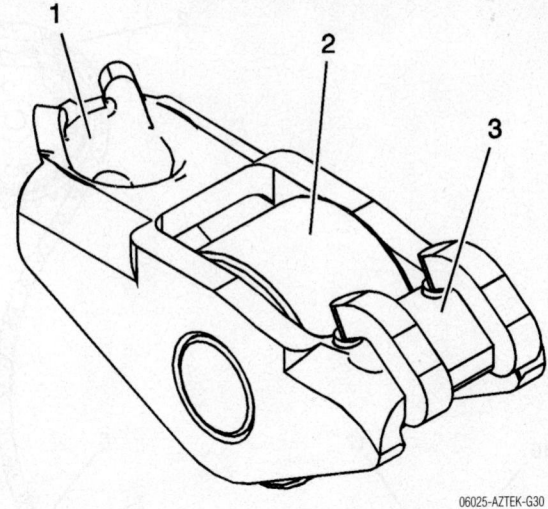

06025-AZTEK-G30

Rocker arm—3.6L engine

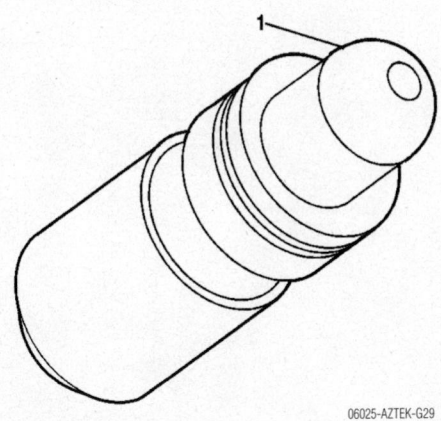

06025-AZTEK-G29

Stationary hydraulic lash adjuster—3.6L engine

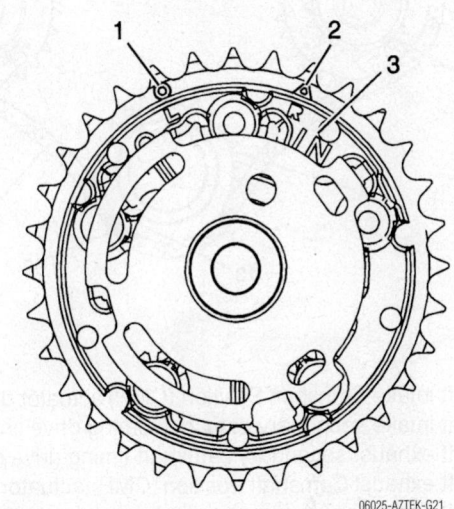

06025-AZTEK-G21

CMP actuator—3.6L engine

cylinder head with clean engine oil GM P/N 12378006 or equivalent.

32. Install the SHLAs in the cylinder head.

33. Apply a liberal amount of lubricant GM P/N 12345501 (Canadian P/N 992704) or equivalent to the SHLA pivot spheres.

34. Apply a liberal amount of lubricant GM P/N 12345501 (Canadian P/N 992704) or equivalent to the pivot pocket (1), roller (2) and valve slot (3) areas of the camshaft followers.

❈❈ WARNING

The follower must be positioned squarely on the valve tip so that the full width of the roller will completely contact the camshaft lobe. If the followers are being reused you must put them back in their original location.

35. Place the camshaft followers in position on the valve tip and stationary hydraulic lash adjuster (SHLA). The rounded head end of the follower goes on the SHLA while the flat end goes on the valve tip.

36. Clean the camshaft journals and carriers with a clean, lint-free cloth.

➡**Ensure that the marks on the camshaft position actuator and the timing chain are aligned. DO NOT tighten the camshaft position actuator bolt at this time.**

37. Locate the camshafts to the cylinder head and assemble the camshaft actuators to the camshafts.

38. Ensure that the crankshaft is in the stage one timing drive assembly position using the EN 46111.

39. Ensure that the camshaft sealing rings are in place in the camshaft grooves.

40. Select the proper camshaft for the particular installation location. The ring placement is defined as follows: The number 4 identification ring for the left intake camshaft is machined off. The number 5 identification ring for the left exhaust camshaft is machined off.

41. Apply a liberal amount of lubricant GM P/N 12345501 (Canadian P/N 992704) or equivalent to the camshaft journals and the left cylinder head camshaft carriers.

42. Place the left intake and left exhaust camshafts in position in the left cylinder head.

43. Position the camshaft lobes in a neutral position with the flats on the back of the camshafts up and parallel with the left cylinder head camshaft cover rail.

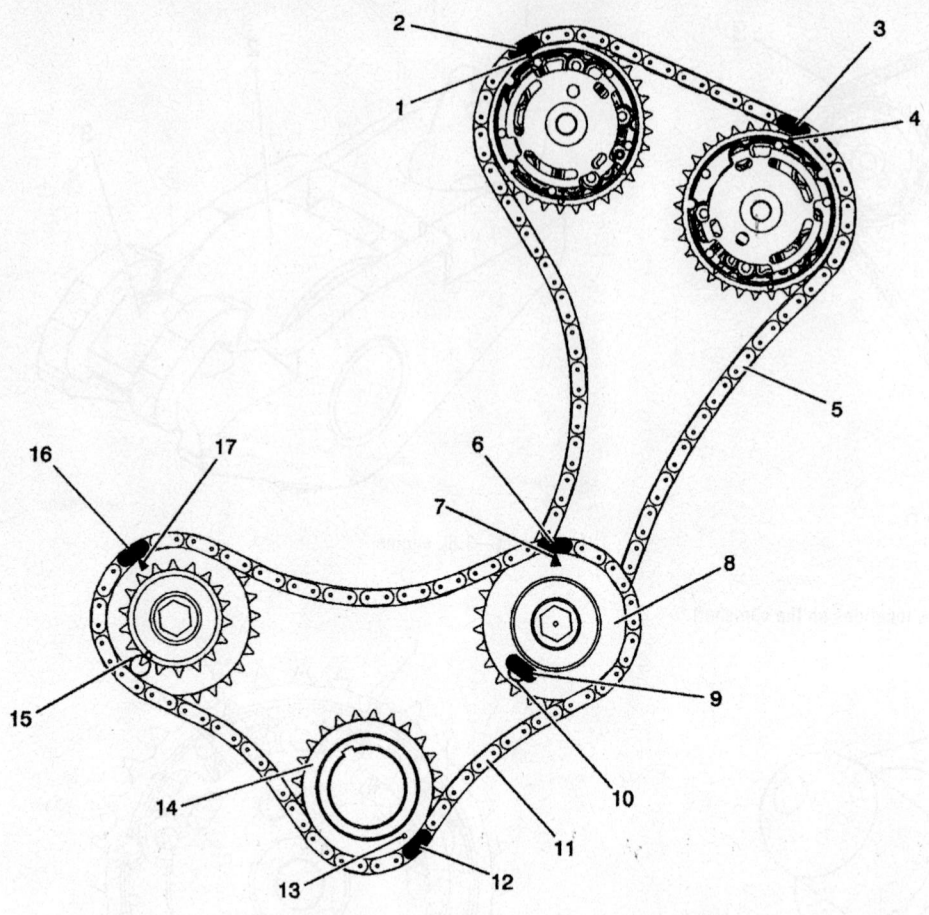

1. Left intake Camshaft Position (CMP) actuator timing mark
2. Left intake secondary camshaft timing drive chain timing link
3. Left exhaust secondary camshaft timing drive chain timing link
4. Left exhaust Camshaft Position (CMP) actuator timing mark
5. Left secondary camshaft timing drive chain
6. Primary camshaft drive chain timing link for the left primary camshaft intermediate drive chain sprocket
7. Left primary camshaft intermediate drive chain sprocket timing mark for the primary camshaft drive chain
8. Left primary camshaft intermediate drive chain sprocket
9. Left secondary camshaft timing drive chain timing link for the left primary camshaft intermediate drive chain sprocket
10. Left primary camshaft intermediate drive chain sprocket timing window for the left secondary camshaft timing drive chain timing link
11. Primary camshaft drive chain
12. Primary camshaft drive chain timing link for the crankshaft sprocket
13. Crankshaft sprocket timing mark
14. Crankshaft sprocket
15. Right primary camshaft intermediate drive chain sprocket
16. Primary camshaft drive chain timing link for the right primary camshaft intermediate drive chain sprocket
17. Right primary camshaft intermediate drive chain sprocket timing mark

06025-AZTEK-G22

Ensure that the crankshaft is in the stage one timing drive assembly position using the EN 46111—3.6L engine

44. Observe the markings on the left cylinder head camshaft bearing caps. Each bearing cap is marked in order to identify its location. The markings have the following meanings:
   • The raised feature must always be oriented toward the center of the cylinder head.
   • The I indicates the intake camshaft.
   • The E indicates the exhaust camshaft.
   • The number 2, 4, 6 indicates the cylinder position from the front of the engine.

45. Apply a liberal amount of lubricant GM P/N 12345501 (Canadian P/N 992704) or equivalent to the camshaft bearing caps.

46. Install the camshaft bearing thrust

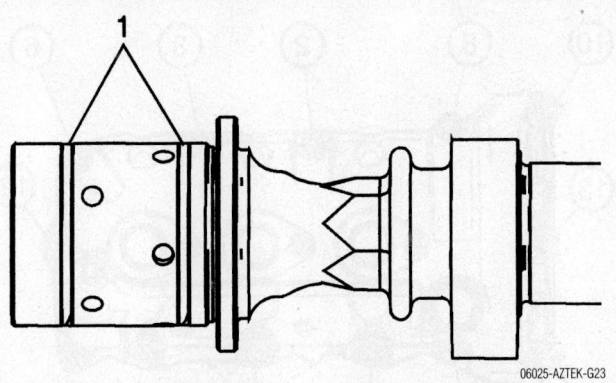

1

Ensure that the camshaft sealing rings are in place in the camshaft grooves—3.6L engine

06025-AZTEK-G23

cap in the first journal of the left cylinder head.

47. Install the remaining bearing caps with their orientation mark toward the center of the cylinder head.

48. Hand start all the camshaft bearing cap bolts.

49. Tighten the camshaft bearing cap bolts in the sequence shown to 10 Nm (89 inch lbs.).

50. Loosen the center intake camshaft bearing cap bolts 1, 2 and the center exhaust camshaft bearing cap bolts 3, 4.

51. Retighten the center camshaft bearing cap bolts 1, 2, 3, 4 to 10 Nm (89 inch lbs.).

52. Remove the EN 46108.

✳✳ WARNING

Use an open-end wrench at the camshaft hex to prevent camshaft/engine rotation.

53. Observe the body of the camshaft position actuator for the "IN" marking. The marking is for an intake camshaft position actuator.

54. Ensure the proper timing mark is used. Observe the outer ring of the camshaft position actuator for the "L" and circle marking. The marking is for alignment to the highlighted timing chain link on the left side of the engine.

55. Use an open wrench on the hex cast into the camshaft in order to prevent camshaft rotation when tightening the camshaft position actuator bolt.

56. Install the left intake camshaft position actuator.

57. Install the camshaft position actuator bolt. Tighten the camshaft position actuator bolt to 58 Nm (43 ft. lbs.).

58. Observe the body of the camshaft position actuator for the "EX" marking. The marking is for an exhaust camshaft position actuator.

59. Ensure the proper timing mark is used. Observe the outer ring of the camshaft position actuator for the "L" and circle marking. The marking is for alignment to the highlighted timing chain link on the left side of the engine.

60. Use an open wrench on the hex cast into the camshaft in order to prevent camshaft rotation when tightening the camshaft position actuator bolt.

61. Install the left exhaust camshaft position actuator.

62. Install the camshaft position actuator bolt. Tighten the camshaft position actuator bolt to 58 Nm (43 ft. lbs.).

63. Install the CMP actuator valve.

64. Install the CMP actuator valve bolt. Tighten the CMP actuator valve bolt to 10 Nm (89 inch lbs.).

65. Install the camshaft sensors. Torque to 89 inch lbs. (10 Nm).

➡**The EN 46106 must be installed onto the flywheel.**

66. Use the J 41998-B, nut, bearing and washer to install the crankshaft balancer.

➡**Do not lubricate the crankshaft front oil seal or crankshaft balancer sealing surfaces. The crankshaft balancer is installed into a dry seal.**

67. Apply lubricant to the inside of the crankshaft balancer hub bore.

68. Place the crankshaft balancer in position on the crankshaft.

69. Thread the J 41998-B in the crankshaft. Ensure you engage at least 10 threads of the J 41998-B before pressing the crankshaft balancer in place.

70. Push the crankshaft balancer into position by tightening the nut on the J 41998-B until the large washer bottoms out on the crankshaft end.

71. Remove the J 41998-B.

72. Install the crankshaft balancer bolt. Tighten the crankshaft balancer bolt to 100 Nm (74 ft. lbs.). Tighten the crankshaft balancer bolt an additional 150 degrees using the J 45059.

73. Remove the EN 46106.

74. Install a NEW camshaft cover seal and NEW grommets.

75. Remove the EN 46105-2 from the rear of the left camshafts.

76. Install the EN 46101 onto the spark plug tubes of the left cylinder head.

77. Install the camshaft cover bolt grommets prior to installing the camshaft cover bolts.

78. Wipe the camshaft cover sealing surface on the left cylinder head with a clean, lint-free cloth.

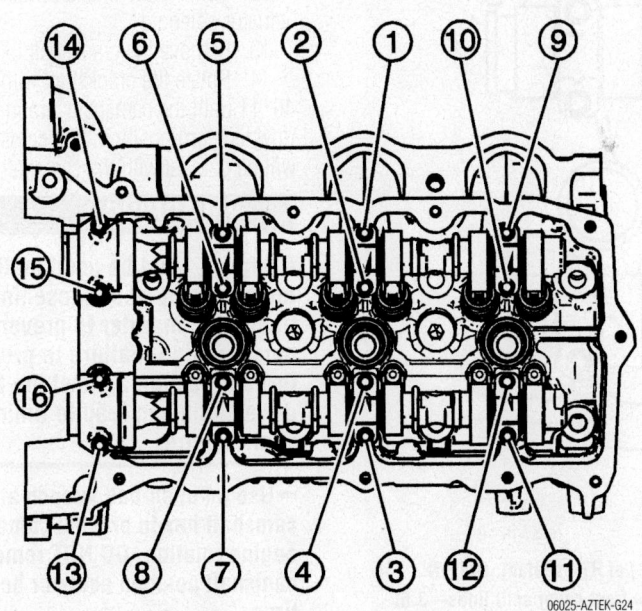

Left side camshaft bearing cap torque sequence—3.6L engine

06025-AZTEK-G24

79. Place a bead 8 mm (0.3150 in.) in diameter by 4 mm (0.1575 in.) in height of RTV sealant, GM P/N 12378521 (Canadian P/N 88901148) or equivalent, on the engine front cover split lines.

80. Place the left camshaft cover into position onto the left cylinder head.

81. Loosely install the left camshaft cover bolts.

82. Tighten the left camshaft cover bolts in the sequence shown to 10 Nm (89 inch lbs.).

83. Remove the EN 46101 from the spark plug tubes of the left cylinder head.

84. Install the NEW spark plugs into the left cylinder head. Tighten the spark plugs to 20 Nm (15 ft. lbs.).

85. Install each ignition coil through the left camshaft cover into the spark plug tube taking care not to damage the spark plug and/or the seal in the left camshaft cover.

86. Install each ignition coil bolt. Tighten the ignition coil bolt to 10 Nm (89 inch lbs.).

87. Tighten the left engine strut bracket-to-cylinder head bolts to 50 Nm (37 ft. lbs.).

88. Install the ignition coils.

89. Install the wiring harness to the front of the camshaft cover.

90. Install the wiring harness conduit retainers to the wiring harness conduit.

91. Install the wiring harness to the side of the camshaft cover.

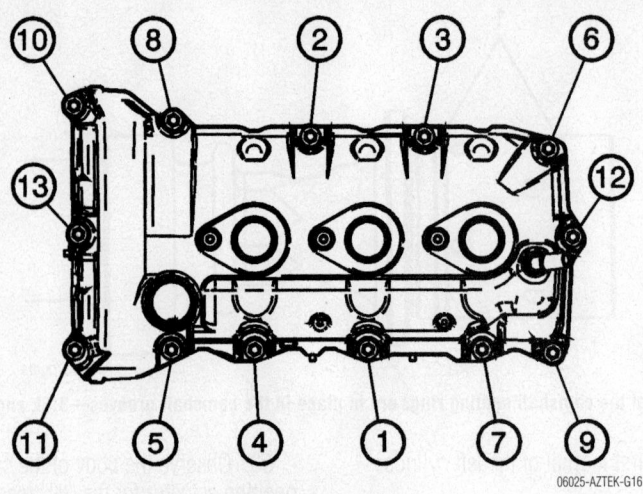

06025-AZTEK-G16

Left camshaft cover torque sequence—3.6L engine

92. Connect the ignition coil electrical connectors.

93. Install the upper intake manifold with the lower intake manifold.

RIGHT SIDE

1. Before servicing the vehicle, refer to the Precautions Section.

2. Remove the upper intake manifold with the lower intake manifold.

3. Disconnect the ignition coil electrical connectors.

4. Remove the wiring harness from the side of the camshaft cover by sliding the conduit down and outboard.

5. Remove the wiring conduit retainers from the camshaft cover by rotating the wiring harness conduit retainers counter-clockwise.

➡ **It is not necessary to disconnect the engine front cover electrical connectors.**

6. Remove the wiring harness from the front of the camshaft cover.

7. Reposition and secure the wiring harnesses away from the camshaft cover in order to provide clearance.

8. Remove the ignition coils.

9. Remove the camshaft cover.

10. Remove and discard the camshaft cover seal and grommets.

11. Remove the camshaft sensors.

12. Remove the intake camshaft position actuator solenoid.

13. Remove the crankshaft balancer.

14. Rotate the crankshaft with the EN 46111 until the camshafts are in a neutral (low tension) position. The camshaft flats will be parallel with the camshaft cover rail.

❋❋ WARNING

A wrench must be used on the hex of the camshaft when loosening or tightening in order to prevent component damage. Failure to prevent the torque reaction against the timing drive chain can lead to timing drive chain failure.

➡ Use an open-end wrench at the camshaft hex to prevent camshaft/engine rotation. DO NOT remove the camshaft position actuator bolt at this time.

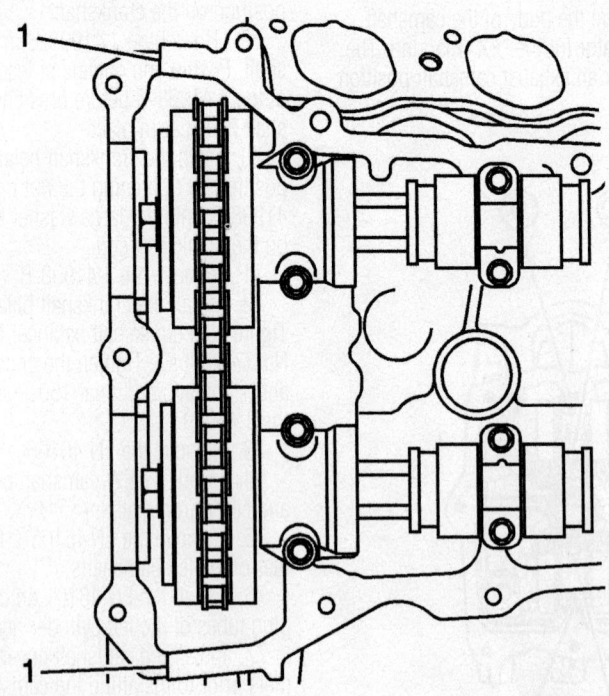

06025-AZTEK-G15

Place a bead 8 mm (0.3150 in.) in diameter by 4 mm (0.1575 in.) of RTV sealant, GM P/N 12378521 (Canadian P/N 88901148) or equivalent, on the engine front cover split lines—3.6L engine

15. Loosen the camshaft position actuator bolt.

➡**Ensure that the tips of the EN 46108 are fully engaged into the timing chain.**

16. Install the EN 46108 in order to retain the timing chain. Firmly tighten the EN 46108 nuts.

➡**Ensure that the camshaft timing chain and the camshaft position actuators are marked for proper assembly.**

17. Mark the timing chain and the respective locations on camshaft position actuators.

18. Remove the camshaft position actuator bolt.

19. Remove the EN 46105-1 from the right camshafts.

20. Position the camshaft lobes in a neutral position.

21. Observe the markings on the bearing caps. Each bearing cap is marked in order to identify its location.

22. The markings have the following meanings:
- The raised feature must always be oriented toward the center of the cylinder head.
- The I indicates the intake camshaft.
- The E indicates the exhaust camshaft.
- The number indicates the journal position from the front of the engine.

23. Remove the camshaft bearing cap bolts.

24. Remove the camshaft bearing caps.

25. Remove the camshafts.

26. Replace the camshaft bearing caps and bolts.

✳✳ WARNING

Do not stroke/cycle the stationary hydraulic lash adjuster plunger without oil in the lower pressure chamber. Do not allow the stationary hydraulic lash adjuster to tip over, plunger down, after the oil fill.

27. Remove the valve lifters, stationary hydraulic lash adjuster, (SHLAs) from the left cylinder head.

To install:

28. Fill the stationary hydraulic lash adjuster (SHLA) with clean engine oil GM P/N 12378006 or equivalent. Take precautions to prevent scratching the pivot sphere area (1) of the SHLA.

29. Lubricate the SHLA bores in the cylinder head with clean engine oil GM P/N 12378006 or equivalent.

30. Install the SHLAs in the cylinder head.

31. Apply a liberal amount of lubricant GM P/N 12345501 (Canadian P/N 992704) or equivalent to the SHLA pivot spheres.

32. Apply a liberal amount of lubricant GM P/N 12345501 (Canadian P/N 992704) or equivalent to the pivot pocket (1), roller (2) and valve slot (3) areas of the camshaft followers.

✳✳ WARNING

The follower must be positioned squarely on the valve tip so that the full width of the roller will completely contact the camshaft lobe. If the followers are being reused you must put them back in their original location.

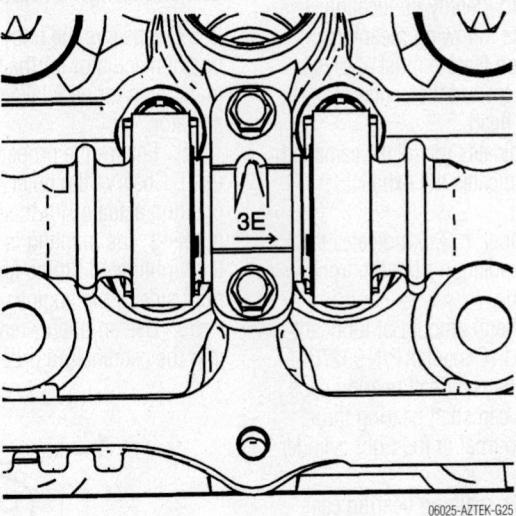

Camshaft cap markings—3.6L engine

33. Place the camshaft followers in position on the valve tip and stationary hydraulic lash adjuster (SHLA). The rounded head end of the follower goes on the SHLA while the flat end goes on the valve tip.

34. Clean the camshaft journals and carriers with a clean, lint-free cloth.

➡**Ensure that the marks on the camshaft position actuators and the timing chain (15-18) are aligned. DO NOT tighten the camshaft position actuator bolt at this time.**

35. Locate the camshafts to the cylinder head and assemble the camshaft actuators to the camshafts.

36. Ensure that the crankshaft is in the stage one timing drive assembly position using the EN 46111.

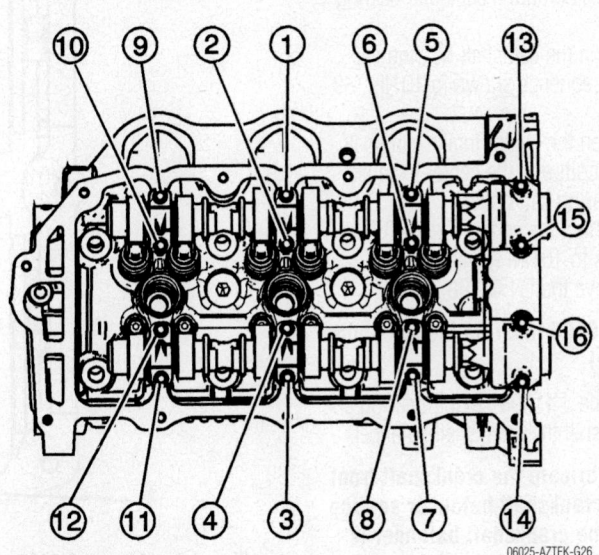

Right side camshaft cap torque sequence—3.6L engine

37. Ensure that the camshaft sealing rings are in place in the camshaft grooves.

38. Select the proper camshaft for the particular installation location. The ring placement is defined as follows: The number 2 identification ring for the right exhaust camshaft is machined off. The number 3 identification ring for the right intake camshaft is machined off.

39. Apply a liberal amount of lubricant GM P/N 12345501 (Canadian P/N 992704) or equivalent to the camshaft journals and the right cylinder head camshaft carriers. Place the right intake and right exhaust camshafts in position in the right cylinder head.

40. Position the camshaft lobes in a neutral position with the flats on the back of the camshafts up and parallel with the right cylinder head camshaft cover rail. Observe the markings on the right cylinder head camshaft bearing caps. Each bearing cap is marked in order to identify its location. The markings have the following meanings:

- The raised feature must always be oriented toward the center of the cylinder head.
- The I indicates the intake camshaft.
- The E indicates the exhaust camshaft.
- The number 1, 3, 5 indicates the cylinder position from the front of the engine.

41. Apply a liberal amount of lubricant GM P/N 12345501 (Canadian P/N 992704) or equivalent to the camshaft bearing caps.

42. Install the camshaft bearing thrust caps in the first journal of the right cylinder head.

43. Install the remaining bearing caps with their orientation mark toward the center of the cylinder head.

44. Hand start all the camshaft bearing cap bolts.

45. Tighten the camshaft bearing cap bolts in the sequence shown to 10 Nm (89 inch lbs.).

46. Loosen the center intake camshaft bearing cap bolts and the center exhaust camshaft bearing cap bolts.

47. Retighten the center camshaft bearing cap bolts to 10 Nm (89 inch lbs.).

48. Remove the EN 46108.

➡**The EN 46106 must be installed onto the flywheel.**

49. Use the J 41998-B, nut, bearing and washer to install the crankshaft balancer.

➡**Do not lubricate the crankshaft front oil seal or crankshaft balancer sealing surfaces. The crankshaft balancer is installed into a dry seal.**

50. Apply lubricant to the inside of the crankshaft balancer hub bore.

51. Place the crankshaft balancer in position on the crankshaft.

52. Thread the J 41998-B in the crankshaft. Ensure you engage at least 10 threads of the J 41998-B before pressing the crankshaft balancer in place.

53. Push the crankshaft balancer into position by tightening the nut on the J 41998-B until the large washer bottoms out on the crankshaft end.

54. Remove the J 41998-B.

55. Tighten the crankshaft balancer bolt to 100 Nm (74 ft. lbs.). Tighten the crankshaft balancer bolt an additional 150 degrees using the J 45059.

56. Remove the EN 46106.

➡**Use an open-end wrench at the camshaft hex to prevent camshaft/engine rotation.**

57. Observe the body of the camshaft position actuator for the "IN" marking. The marking is for an intake camshaft position actuator.

58. Ensure the proper timing mark is used. Observe the outer ring of the camshaft position actuator for the "R" and triangle marking. The marking is for alignment to the highlighted timing chain link on the right side of the engine.

59. Use an open wrench on the hex cast into the camshaft in order to prevent camshaft rotation when tightening the camshaft position actuator bolt.

60. Install the right intake camshaft position actuator.

61. Install the camshaft position actuator bolt. Tighten the camshaft position actuator bolt to 58 Nm (43 ft. lbs.).

62. Observe the body of the camshaft position actuator for the "EX" marking. The marking is for an exhaust camshaft position actuator.

63. Ensure the proper timing mark is used. Observe the outer ring of the camshaft position actuator for the "R" and triangle marking. The marking is for alignment to the highlighted timing chain link on the right side of the engine.

64. Use an open wrench on the hex cast into the camshaft in order to prevent camshaft rotation when tightening the camshaft position actuator bolt.

65. Install the right exhaust camshaft position actuator.

66. Install the camshaft position actuator bolt. Tighten the camshaft position actuator bolt to 58 Nm (43 ft. lbs.).

67. Install the CMP actuator valve.

68. Install the CMP actuator valve bolt. Tighten the CMP actuator valve bolt to 10 Nm (89 inch lbs.). Connect the CMP actuator valve electrical connector.

69. Install the power steering pressure hose bracket and nut as necessary.

70. Install the CMP sensor.

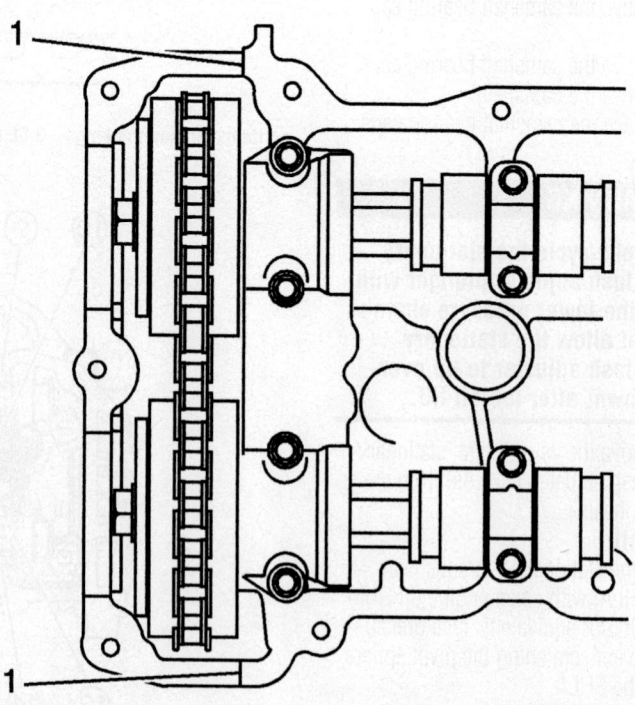

06025-AZTEK-G27

Place a bead 8 mm (0.3150 in) in diameter by 4 mm (0.1575 in) in height of RTV sealant, GM P/N 12378521 (Canadian P/N 88901148) or equivalent, on the engine front cover split lines

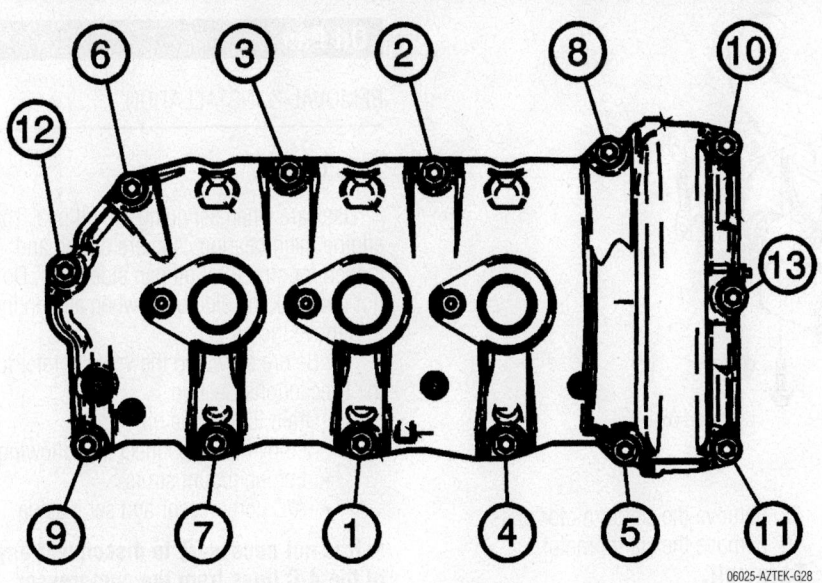

Right side camshaft cover torque sequence—3.6l engine

06025-AZTEK-G28

71. Install the CMP sensor bolt. Tighten the CMP sensor bolt to 10 Nm (89 inch lbs.).

72. Connect the CMP sensor electrical connector.

73. Install the power steering pressure hose as necessary.

74. Install a NEW camshaft cover seal and NEW grommets.

75. Remove the EN 46105-1 from the rear of the right camshafts.

76. Install the EN 46101 onto the spark plug tubes of the right cylinder head.

77. Install the camshaft cover bolt grommets prior to installing the camshaft cover bolts.

78. Wipe the camshaft cover sealing surface on the right cylinder head with a clean, lint-free cloth.

79. Place a bead 8 mm (0.3150 in) in diameter by 4 mm (0.1575 in) in height of RTV sealant, GM P/N 12378521 (Canadian P/N 88901148) or equivalent, on the engine front cover split lines (1).

80. Place the right camshaft cover into position onto the right cylinder head.

81. Loosely install the right camshaft cover bolts.

82. Tighten the right camshaft cover bolts in the sequence shown to 10 Nm (89 inch lbs.).

83. Remove the EN 46101 from the spark plug tubes of the right cylinder head.

84. Install the NEW spark plugs into the right cylinder head. Tighten the spark plugs to 20 Nm (15 ft. lbs.).

85. Install each ignition coil through the right camshaft cover into the spark plug tube taking care not to damage the spark plug and/or the seal in the right camshaft cover.

86. Install each ignition coil bolt. Tighten the ignition coil bolt to 10 Nm (89 inch lbs.).

87. Install the upper intake manifold with the lower intake manifold.

Valve Lash

ADJUSTMENT

3.4L Engine

Because the rocker arm fasteners are secured and tightened, valve lash is not adjustable. If a valve train problem is suspected, check that the rocker arm pedestals

bolts are tightened to specification. During initial installation the bolts are coated with thread locking compound. If they are sufficiently loosened to cause valvetrain noise, they should be removed and thoroughly cleaned. Apply thread locking compound to the rocker arm pedestal bolts. Tighten the bolts to 14 ft. lbs. (19 Nm) plus 30 degrees.

When valve lash falls out of specification (valve tap is heard) and tightening the bolts does not solve the problem, replace the rocker arm, pushrod and hydraulic lifter on the offending cylinder.

3.6L Engine

See the procedure under Camshaft and Lifters.

Starter Motor

REMOVAL & INSTALLATION

3.4L Engine

1. Before servicing the vehicle, refer to the Precautions Section.

2. Remove or disconnect the following:
   - Negative battery cable
   - Radiator air baffle assembly
   - Electrical connections
   - Torque converter cover
   - Starter

To install:

3. Install or connect the following:
   - Starter. Torque the bolts to 35 ft. lbs. (47 Nm).

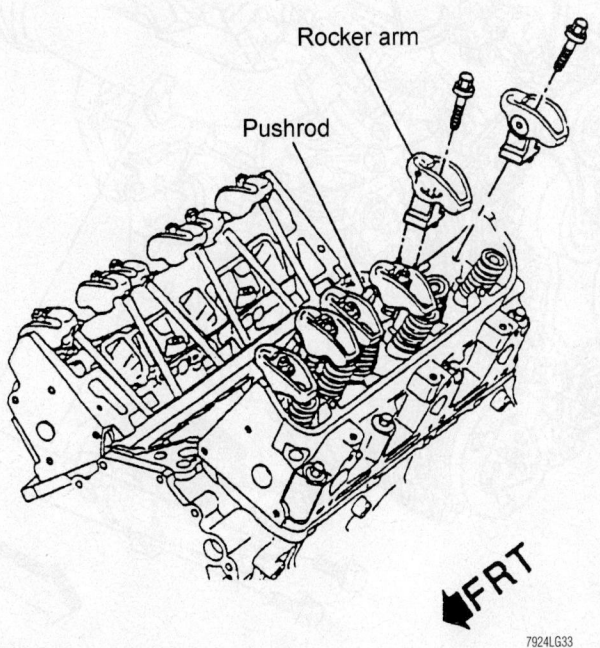

Rocker arm

Pushrod

FRT

7924LG33

Valve rocker arm and related components—3.4L engine

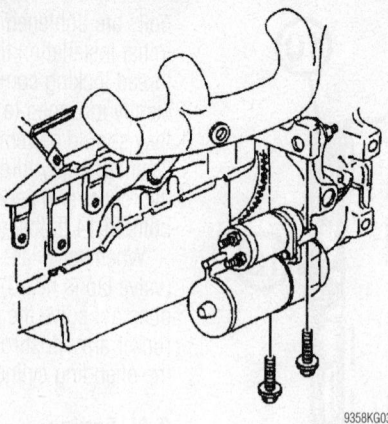

Starter motor mounting—3.4L engine

- Torque converter cover
- Solenoid "BAT" terminal. Torque the nut to 89 inch lbs. (10 Nm).
- Solenoid "S" terminal. Torque the nut to 27 inch lbs. (3 Nm).
- Radiator air baffle assembly
- Negative battery cable

4. Perform a charging system test and verify the starter is operating properly.

3.6L Engine

1. Before servicing the vehicle, refer to the Precautions Section.
2. Disconnect the battery ground (negative) cable from the battery.
3. Raise and support the vehicle.
4. Remove the radiator air baffle.
5. Remove the starter motor BAT terminal nut and electrical leads.

6. Remove the starter motor bolts.
7. Remove the starter motor.

To install:

8. Install the starter motor.

➥**Use the correct fastener in the correct location.**

9. Install the starter motor bolts. Tighten the starter motor bolts to 50 Nm (37 ft. lbs.).
10. Install the starter motor S terminal electrical connector.
11. Install the battery positive cable and the BAT terminal nut to the starter motor BAT terminal. Tighten the starter motor BAT terminal nut to 13 Nm (115 inch lbs.).
12. Install radiator air baffle assembly.
13. Lower the vehicle.
14. Install the battery ground (negative) cable to the battery.

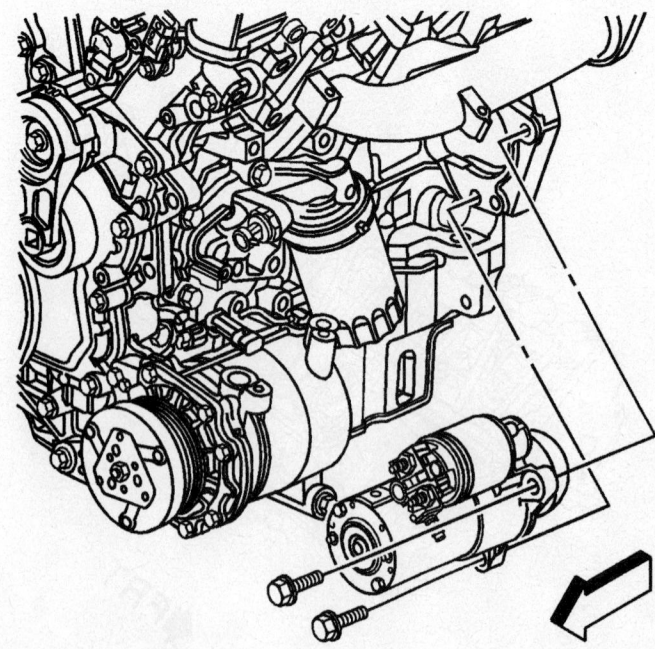

Starter motor mounting—3.6L engine

Oil Pan

REMOVAL & INSTALLATION

3.4L Engine

Use care when servicing the oil pan. The engine main bearing caps are drilled and tapped for structural oil pan side bolts. Do not overlook the side bolts when attempting to remove the oil pan.

1. Before servicing the vehicle, refer to the Precautions Section.
2. Drain the engine oil.
3. Remove or disconnect the following:
   - Engine mount struts
   - A/C compressor and set it aside

➥**It is not necessary to disconnect any of the A/C lines from the compressor.**

4. Install an engine support fixture.
5. Remove or disconnect the following:
   - Catalytic converter pipe from the right side exhaust manifold
   - Frame bolts and make certain that an engine stand (such as J 39580) is aligned below the engine, lower the frame
   - Oil level sensor wiring harness connector
   - Starter
   - Transaxle brace from the oil pan
   - Transaxle mount lower nuts
   - Engine mount lower nuts and raise the engine with the support fixture
   - Engine mount and bracket from the oil pan

➥**You will need to use a torque wrench adapter to remove and install the right side oil pan bolts.**

   - Oil pan and gasket

To install:

6. Clean the gasket mounting surfaces.
7. Apply a small amount of sealer GM 1234579 on both sides of the bearing cap.
8. Install or connect the following:
   - Oil pan gasket
   - Oil pan. Tighten the bottom bolts to 18 ft. lbs. (25 Nm) and the side bolts, using a torque wrench adapter (tool J 39505), to 37 ft. lbs. (50 Nm).
   - Engine mount and bracket to the oil pan. Torque the bolt to 43 ft. lbs. (58 Nm).
   - Lower the engine into position
   - Transaxle lower nuts. Torque the nuts to 90 inch lbs. (10 Nm).
   - Transaxle brace to the oil pan. Torque the bolts to 32 ft. lbs. (43 Nm).

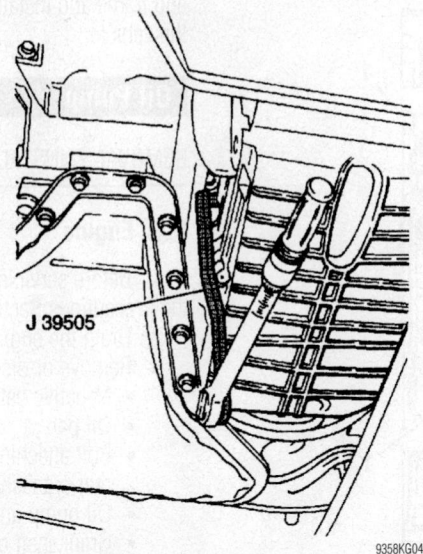

Use a torque wrench adapter to tighten the oil pan side bolts—3.4L engine

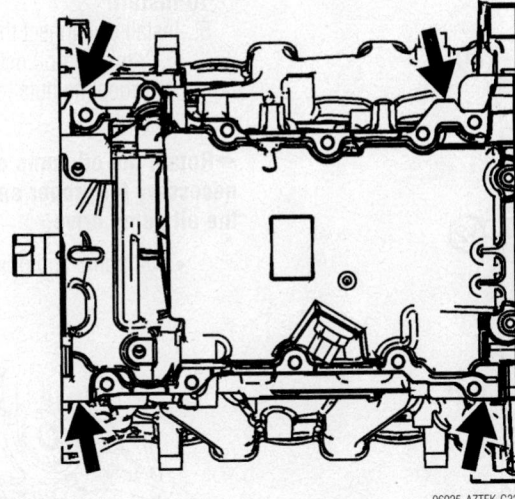

Oil pan pry points—3.6L engine

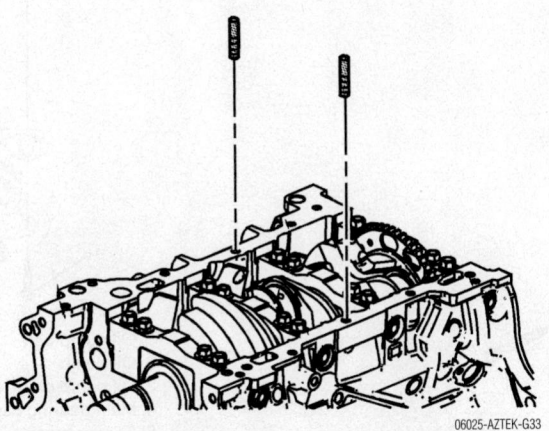

Install 8 mm (0.315 in) guides pins into the center oil pan rail bolt hole on each side of the engine block—3.6L engine

- Starter. Torque the bolts to 35 ft. lbs. (47 Nm).
- Catalytic converter pipe to the right side exhaust manifold. Torque the bolts to 26 ft. lbs. (35 Nm).
- Frame and use new frame to body bolts. Torque the front bolts to 111 ft. lbs. (150 Nm) and the rear bolts to 122 ft. lbs. (165 Nm).

9. Remove the engine support fixture.
10. Install or connect the following:
- A/C compressor. Torque the bolts to 37 ft. lbs. (50 Nm).
- Engine mount struts. Torque the fasteners to 52 ft. lbs. (70 Nm).
- Negative battery cable

11. Fill the engine with oil.
12. Start the vehicle and inspect for leaks, repair if necessary.

3.6L Engine

1. Before servicing the vehicle, refer to the Precautions Section.
2. Remove the engine, transaxle and sub-frame from the vehicle and install the engine on an engine stand.
3. Remove the A/C compressor with the hoses from the engine.
4. Remove the engine front cover.
5. Remove the oil pan bolts and, using the pry points, remove the pan.

To install:

6. Install 8 mm (0.315 in) guides pins into the center oil pan rail bolt hole on each side of the engine block.
7. Place a 3 mm (0.118 in) bead of RTV sealant, GM P/N 12378521 (Canadian P/N 88901148) or equivalent, on the block pan rail and the crankshaft rear oil seal housing.
8. Position the oil pan onto the block.
9. Remove the 8 mm (0.315 in) guides from the engine block.
10. Loosely install the oil pan bolts.
11. Tighten the oil pan bolts in sequence shown. Tighten the 8 mm bolts (1-11) to 23 Nm (17 ft. lbs.). Tighten the 6 mm bolts (12, 13) to 10 Nm (89 inch lbs.).
12. Install the engine front cover.
13. Place the A/C compressor in position to the cylinder block.
14. Loosely install the A/C compressor bolts.
15. Tighten the A/C compressor front upper bolt to 50 Nm (37 ft. lbs.).
16. Tighten the A/C compressor rear upper bolt to 50 Nm (37 ft. lbs.).
17. Tighten the A/C compressor rear lower bolt to 50 Nm (37 ft. lbs.).
18. Tighten the A/C compressor front lower bolt to 22 Nm (16 ft. lbs.).
19. Install the engine to the transmission

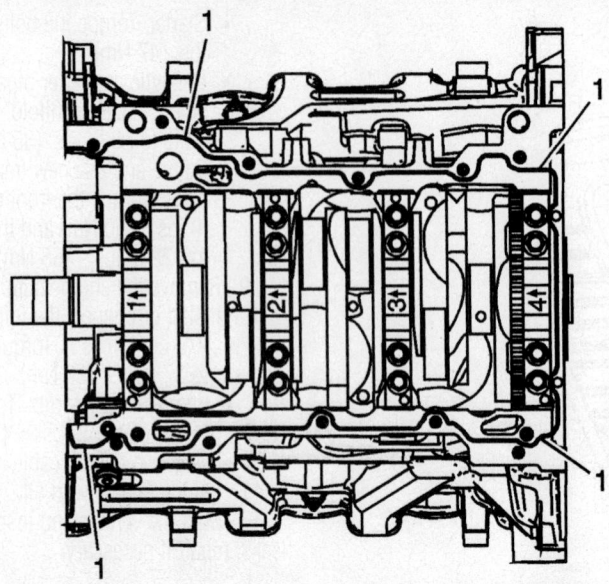

06025-AZTEK-G34

Place a 3 mm (0.118 in) bead of RTV sealant on the block pan rail and the crankshaft rear oil seal housing—3.6L engine

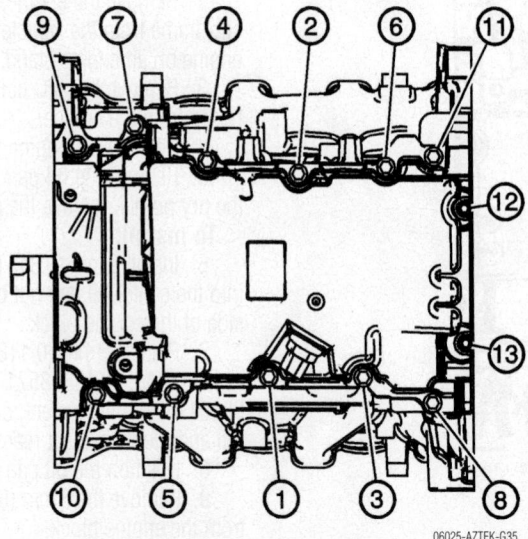

06025-AZTEK-G35

Oil pan bolt torque sequence—3.6L engine

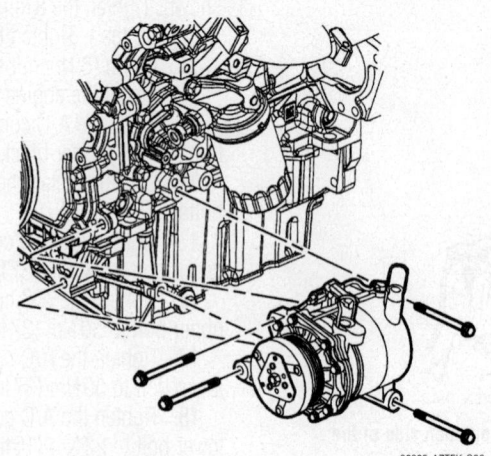

06025-AZTEK-G36

Compressor mounting—3.6L engine

and frame and install the assembled unit in the vehicle.

Oil Pump

REMOVAL & INSTALLATION

3.4L Engine

1. Before servicing the vehicle, refer to the Precautions Section.
2. Drain the engine oil
3. Remove or disconnect the following:
   • Negative battery cable
   • Oil pan
   • Bolt attaching the oil pump to the rear crankshaft bearing cap
   • Oil pump and driveshaft
   • Crankshaft oil deflector nuts and deflector, if necessary
4. Inspect the oil pump and oil pump driveshaft.
   To install:
5. Install or connect the following:
   • Crankshaft oil deflector, if removed. Torque the nuts to 18 ft. lbs. (25 Nm).

➡ **Rotate the oil pump driveshaft as necessary for proper engagement with the oil pump drive.**

   • Oil pump and driveshaft

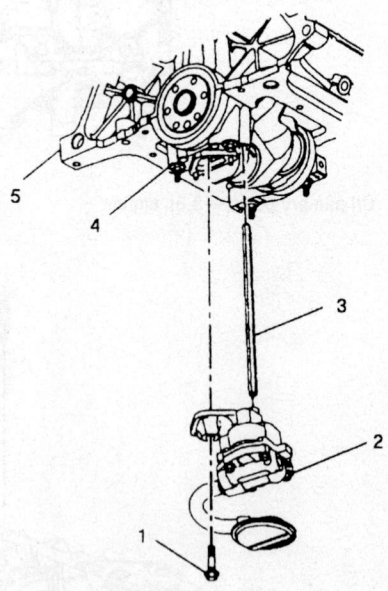

1. Oil pump bolt
2. Oil pump
3. Oil pump drive rod
4. Main bearing cap
5. Engine block

7924LG37

Exploded view of the oil pump mounting—3.4L engine

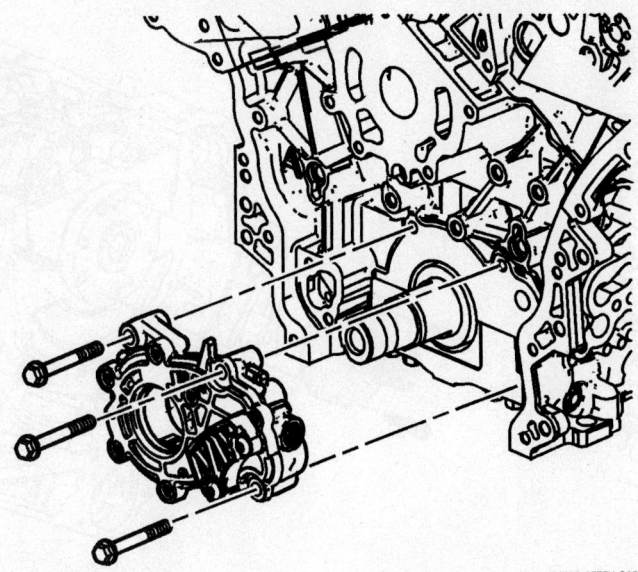

Oil pump installation—3.6L engine

06025-AZTEK-G37

- Oil pump to the rear crankshaft bearing cap. Torque the bolt to 30 ft. lbs. (41 Nm).
- Oil pan
- Negative battery cable
6. Fill the engine with oil.
7. Start the vehicle and inspect for leaks, repair if necessary.

3.6L Engine

1. Before servicing the vehicle, refer to the Precautions Section.

✳✳ WARNING

Do not remove the left bank idler sprocket.

2. Remove the primary timing chain.
3. Remove the crankshaft sprocket.
4. Remove the oil pump bolts and the oil pump.

To install:

➡There are no serviceable components within the oil pump. Disassemble the pump only to diagnose an oiling concern. A disassembled oil pump must not be reused. A disassembled oil pump must be replaced.

5. Align the oil pump gerotor with the crankshaft flats and install the oil pump to the engine block.
6. Align the pump body with the mounting holes in the cylinder block.
7. Install the oil pump bolts. Tighten the oil pump bolts to 23 Nm (17 ft. lbs.).
8. Install the crankshaft sprocket.
9. Install the primary timing chain.

Rear Main Seal

REMOVAL & INSTALLATION

3.4L Engine

The transaxle assembly must be removed to perform this service. This requires special tooling to support the engine assembly while the transaxle and sub-frame are lowered from under the vehicle.

1. Before servicing the vehicle, refer to the Precautions Section.
2. Remove or disconnect the following:
- Negative battery cable
- Transmission assembly
- Engine flywheel
- Oil seal

✳✳ WARNING

When removing the seal, use care so that no damage occurs to the crankshaft. Once the seal is removed, inspect the crankshaft surface for any nicks or burrs. Repair or replace crankshaft as necessary.

To install:
3. Install or connect the following:
- New oil seal lubricated with engine oil, using an Oil Seal Installer tool J 34686 until it is seated properly over the crankshaft
- Flywheel
- Transmission assembly
- Negative battery cable
4. Start the vehicle and check for leaks, repair if necessary.

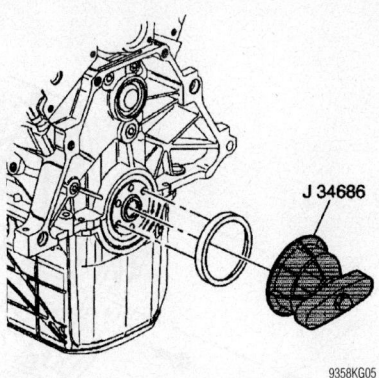

9358KG05

Use the correct installation tool when replacing the rear main seal—3.4L engine

3.6L Engine

1. Before servicing the vehicle, refer to the Precautions Section.
2. Remove the engine/transaxle assembly.
3. Separate the transaxle from the engine.
4. Remove the flywheel.
5. Remove the crankshaft rear oil seal housing bolts. Using the pry points located at the edge of the crankshaft rear oil seal housing, shear the RTV sealant.
6. Remove and discard the crankshaft rear oil seal housing.

To install:
7. Install the 6 mm (0.236 in) guide pins into the 2 crankshaft rear oil seal housing corner bolt holes of the engine block.
8. Install the EN-47839 with the J-42183 (1, 2) onto the rear of the crankshaft flange.
9. Place a 3 mm (0.118 in) bead of RTV sealant, GM P/N 12378521 (Canadian P/N 88901148) or equivalent, to the NEW crankshaft rear oil seal housing as shown (1).

➡DO NOT allow any engine oil on the area where the crankshaft rear oil seal housing is to be installed.

10. Install the crankshaft rear oil seal housing to the engine block.
11. Remove the 6 mm (0.236 in) guides from the engine block.
12. Install the crankshaft rear oil seal housing bolts. Tighten the crankshaft rear oil seal housing bolts to 10 Nm (89 inch lbs.).
13. Remove the EN-47839 and J-42183 (1, 2) from the crankshaft flange.
14. Place the engine flywheel in position on the crankshaft.
15. Install 2 NEW bolts in location at the top and bottom of the engine flywheel bolt pattern allowing the engine flywheel to hang in position.

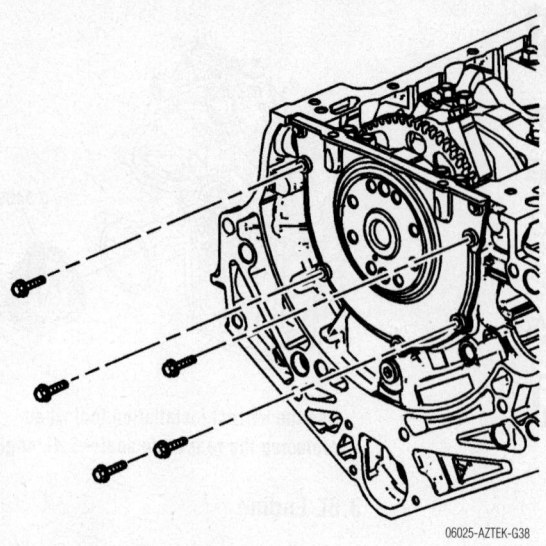

Rear main seal housing bolts—3.6L engine

06025-AZTEK-G38

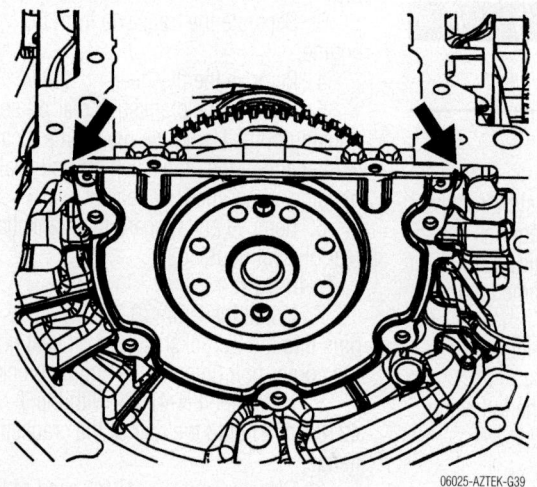

Rear main seal housing pry points—3.6L engine

06025-AZTEK-G39

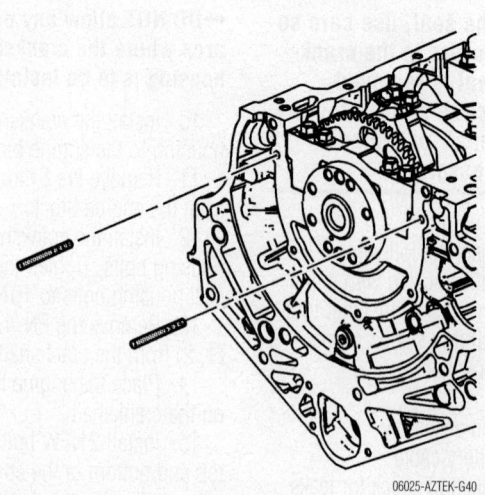

Install the 6 mm (0.236 in) guide pins into the 2 crankshaft rear oil seal housing corner bolt holes of the engine block—3.6L engine

06025-AZTEK-G40

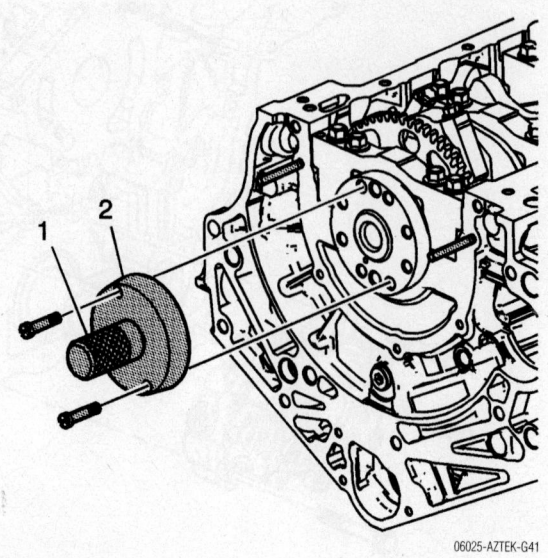

Install the EN-47839 with the J-42183 (1, 2) onto the rear of the crankshaft flange—3.6L engine

06025-AZTEK-G41

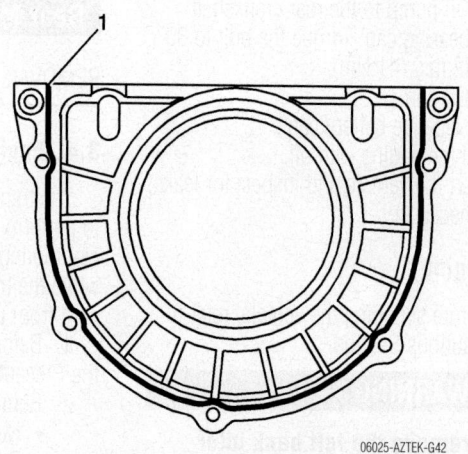

Place a 3 mm (0.118 in) bead of RTV sealant, GM P/N 12378521 (Canadian P/N 88901148) or equivalent, to the NEW crankshaft rear oil seal housing as shown (1)—3.6L engine

06025-AZTEK-G42

16. Install an engine holding tool.

17. Install the remaining NEW engine flywheel bolts. Tighten the NEW engine flywheel bolts to 30 Nm (22 ft. lbs.). Tighten the NEW engine flywheel bolts an additional 45 degrees.

18. Install the engine/transaxle.

Front Cover and Seal, Timing Chain and Sprockets

REMOVAL & INSTALLATION

3.4L Engine

1. Before servicing the vehicle, refer to the Precautions Section.

2. Drain the engine oil.

3. Drain the coolant.

4. Remove or disconnect the following:
- Negative battery cable
- Crankshaft balancer
- Drive belt tensioner
- Power steering pump and lines. Do not disconnect the lines from the pump.
- Thermostat bypass pipe from the front cover
- Radiator outlet hose from the water pump
- Water pump pulley
- Upper and lower Crankshaft Position (CKP) sensor wire harness bracket from the front cover
- CKP sensor from the front cover
- Front cover and gasket

5. Rotate the crankshaft until the timing marks are aligned in the following locations:
- Camshaft alignment pin (1)
- Timing chain damper (2) to the crankshaft sprocket (3)
- Crankshaft key (4)
- Timing chain damper (5) to the camshaft sprocket locator hole (6)

6. Remove or disconnect the following:
- Camshaft sprocket bolt
- Timing chain, timing chain sprockets and damper
- Front oil seal

To install:

7. Install or connect the following:
- New front oil seal by making certain the seal is fully seated

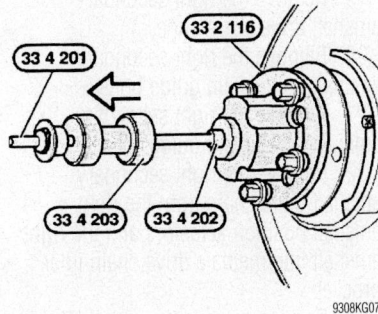

Crankshaft timing mark locations—3.4L engine

- Timing chain damper. Torque the bolts to 15 ft. lbs. (21 Nm).
- Timing chain to the camshaft sprocket
- Crankshaft sprocket
- Timing chain to the crankshaft sprocket by making certain the chain is fully seated

8. Align the crankshaft timing mark to the bottom mark on the damper.

9. Align the timing mark on the camshaft gear center line of the locator hole with the timing mark on the top of the damper.

10. Align the dowel in the camshaft with the dowel hole in the camshaft sprocket.

11. Install or connect the following:
- Camshaft sprocket bolt. Torque the bolt to 103 ft. lbs. (140 Nm).

12. Apply a 0.20 inch (5mm) bead of sealer to both sides of the lower tabs of the engine front cover gasket.
- Front cover. Refer to the accompanying figure and torque bolts (2) to 15 ft. lbs. (21 Nm), bolts (3) to 41 ft. lbs. (55 Nm) and bolts (1) to 35 ft. lbs. (47 Nm).
- Water pump to the front cover. Torque the bolts to 89 inch lbs. (10 Nm).
- Water pump pulley. Torque the bolt to 18 ft. lbs. (25 Nm).
- CKP sensor to the front cover
- Upper/lower CKP wire harness brackets to the front cover
- Radiator outlet hose to the water pump
- Thermostat bypass pipe to the front cover
- Power steering pump and lines
- Drive belt tensioner
- Crankshaft balancer
- Negative battery cable

13. Fill the engine with oil.
14. Fill the coolant system.
15. Start the vehicle and verify that the engine is running properly.

3.6L Engine

1. Before servicing the vehicle, refer to the Precautions Section.

2. Remove the engine/transaxle assembly.

3. Remove the alternator, power steering pump and A/C compressor.

4. Remove the crankshaft balancer.

5. Remove the camshaft position actuator valve bolts.

6. Remove the camshaft position actuator valves from the front cover.

7. Remove the engine front cover bolts.

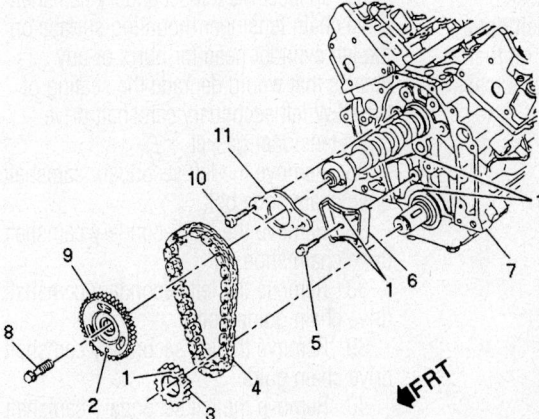

1. Timing alignment marks
2. Locator hole
3. Crankshaft sprocket
4. Timing chain
5. Timing chain dampener bolt
6. Timing chain dampener
7. Engine block
8. Camshaft sprocket bolt
9. Camshaft sprocket
10. Thrust plate bolt
11. Thrust plate

Exploded view of the timing chain assembly—3.4L engine

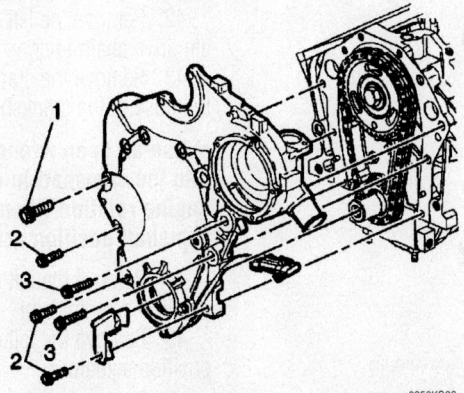

Timing chain front cover bolt tightening specifications—3.4L engine

➡Do not pry between the engine front cover and the camshaft position sensors or the camshaft position actuators in order to shear the RTV. Use the pry points and a bolt in the jackscrew hole in order to remove the engine front cover. Damage to the camshaft position sensors or the camshaft position actuators may occur if the camshaft position sensors or the camshaft position actuators are used to pry against in order to remove the engine front cover.

8. Loosely install a 10 x 1.5 mm bolt in the "jackscrew" hole (1).

9. Using the pry points (2) located at the edge of the front cover and the "jackscrew", shear the RTV sealant.

10. Remove the engine front cover.

11. Using tool EN 46111, rotate the crankshaft until the left cylinder head camshafts align with tool EN 46105-2 and the right cylinder head camshafts align with tool EN 46105-1.

12. Install tool EN 46105-1 to the right camshafts.

13. Install tool EN 46105-2 to the left camshafts.

14. Remove the right secondary camshaft drive chain tensioner bolts.

15. Remove the right secondary camshaft drive chain tensioner.

16. Remove and discard the right secondary camshaft drive chain tensioner gasket.

17. Inspect the right secondary camshaft drive chain tensioner mounting surface on the right cylinder head for burrs or any defects that would degrade the sealing of the NEW right secondary camshaft drive chain tensioner gasket.

18. Remove the right secondary camshaft drive chain shoe bolt.

19. Remove the right secondary camshaft drive chain shoe.

20. Remove the right secondary camshaft drive chain guide bolts.

21. Remove the right secondary camshaft drive chain guide.

22. Remove the right secondary camshaft drive chain from the right camshaft position actuators and the right camshaft intermediate drive chain idler sprocket.

23. Remove the primary camshaft drive chain tensioner bolts.

24. Remove the primary camshaft drive chain tensioner.

25. Remove and discard the primary camshaft drive chain tensioner gasket.

26. Inspect the primary camshaft drive chain tensioner mounting surface on the engine block for burrs or any defects that would degrade the sealing of the NEW primary camshaft drive chain tensioner gasket.

27. Remove the primary camshaft drive chain upper guide bolts.

28. Remove the primary camshaft drive chain upper guide.

➡Do not remove the primary camshaft drive chain lower guide. The primary camshaft drive chain lower guide is not serviceable separately. If the primary camshaft drive chain lower guide must be replaced, the oil pump must be replaced.

29. Remove the primary camshaft drive chain.

30. Remove the right camshaft intermediate drive chain idler bolt.

31. Remove the right camshaft intermediate drive chain idler.

32. Remove the left secondary camshaft drive chain tensioner bolts.

33. Remove the left secondary camshaft drive chain tensioner.

34. Remove and discard the left secondary camshaft drive chain tensioner gasket.

35. Inspect the left secondary camshaft drive chain tensioner mounting surface on the left cylinder head for burrs or any defects that would degrade the sealing of the NEW left secondary camshaft drive chain tensioner gasket.

36. Remove the left secondary camshaft drive chain shoe bolt.

37. Remove the left secondary camshaft drive chain shoe.

38. Remove the left secondary camshaft drive chain guide bolts.

39. Remove the left secondary camshaft drive chain guide.

40. Remove the left secondary camshaft drive chain from the left camshaft position actuators and the left camshaft intermediate drive chain idler sprocket.

41. Remove the left camshaft intermediate drive chain idler bolt.

42. Remove the left camshaft intermediate drive chain idler.

43. Remove the crankshaft sprocket from the nose of the crankshaft.

➡Use an open wrench on the hex cast into the camshaft in order to prevent engine rotation when loosening the camshaft position actuator bolt.

44. Remove the left exhaust camshaft position actuator bolt.

45. Remove the left exhaust camshaft position actuator.

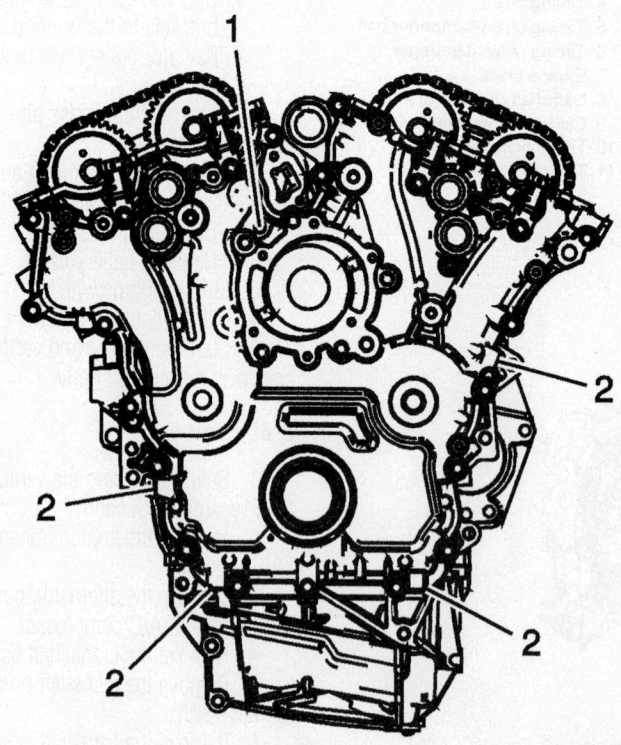

06025-AZTEK-G70

Loosely install a 10 x 1.5 mm bolt in the "jackscrew" hole (1). Using the pry points (2) located at the edge of the front cover and the "jackscrew", shear the RTV sealant—3.6L engine

➡A wrench must be used on the hex of the camshaft when loosening or tightening in order to prevent component damage. Failure to prevent the torque reaction against the timing drive chain can lead to timing drive chain failure.

46. Use an open wrench on the hex cast into the camshaft in order to prevent engine rotation when loosening the camshaft position actuator bolt.

47. Remove the left intake camshaft position actuator bolt.

48. Remove the left intake camshaft position actuator.

➡Use an open wrench on the hex cast into the camshaft in order to prevent engine rotation when loosening the camshaft position actuator bolt.

49. Remove the right exhaust camshaft position actuator bolt.

50. Remove the right exhaust camshaft position actuator.

➡Use an open wrench on the hex cast into the camshaft in order to prevent engine rotation when loosening the camshaft position actuator bolt.

51. Remove the right intake camshaft position actuator bolt.

52. Remove the right intake camshaft position actuator.

To install:

53. Observe the body of the camshaft position actuator for the "IN" marking (3). The marking is for an intake camshaft position actuator.

54. Ensure the proper timing mark is used. Observe the outer ring of the camshaft

position actuator for the "R" and triangle marking (2). The marking is for alignment to the highlighted timing chain link on the right side of the engine.

55. Use an open wrench on the hex cast into the camshaft in order to prevent camshaft rotation when tightening the camshaft position actuator bolt.

56. Install the right intake camshaft position actuator.

57. Install the camshaft position actuator bolt. Tighten the camshaft position actuator bolt to 58 Nm (43 ft. lbs.).

58. Observe the body of the camshaft position actuator for the "EX" marking (1). The marking is for an exhaust camshaft position actuator.

59. Ensure the proper timing mark is used. Observe the outer ring of the camshaft position actuator for the "R" and triangle marking (2). The marking is for alignment to the highlighted timing chain link on the right side of the engine.

➡Use an open wrench on the hex cast into the camshaft in order to prevent camshaft rotation when tightening the camshaft position actuator bolt.

60. Install the right exhaust camshaft position actuator.

61. Install the camshaft position actuator bolt. Tighten the camshaft position actuator bolt to 58 Nm (43 ft. lbs.).

62. Observe the body of the camshaft position actuator for the "IN" marking (3). The marking is for an intake camshaft position actuator.

63. Ensure the proper timing mark is used. Observe the outer ring of the camshaft position actuator for the "L" and circle

marking (1). The marking is for alignment to the highlighted timing chain link on the left side of the engine.

➡Use an open wrench on the hex cast into the camshaft in order to prevent camshaft rotation when tightening the camshaft position actuator bolt.

64. Install the left intake camshaft position actuator.

65. Install the camshaft position actuator bolt. Tighten the camshaft position actuator bolt to 58 Nm (43 ft. lbs.).

66. Observe the body of the camshaft position actuator for the "EX" marking (1). The marking is for an exhaust camshaft position actuator.

67. Ensure the proper timing mark is used. Observe the outer ring of the camshaft position actuator for the "L" and circle marking (3). The marking is for alignment to the highlighted timing chain link on the left side of the engine.

➡Use an open wrench on the hex cast into the camshaft in order to prevent camshaft rotation when tightening the camshaft position actuator bolt.

68. Install the left exhaust camshaft position actuator.

69. Install the camshaft position actuator bolt. Tighten the camshaft position actuator bolt to 58 Nm (43 ft. lbs.).

70. Ensure the crankshaft sprocket is installed with the timing mark (1) visible. Install the crankshaft sprocket on to the nose of the crankshaft.

71. Align the notch in the crankshaft sprocket with the pin in the crankshaft.

72. Slide the crankshaft sprocket on the crankshaft nose until the crankshaft sprocket contacts the step in the crankshaft.

73. If necessary, align the crankshaft sprocket to the stage one timing drive assembly position using tool EN 46111.

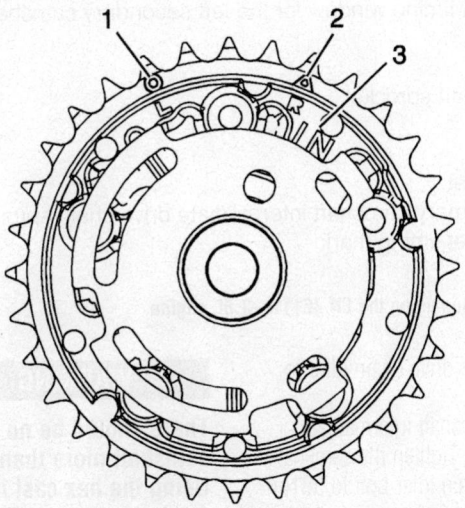

Right intake CMP actuator—3.6L engine

06025-AZTEK-G43

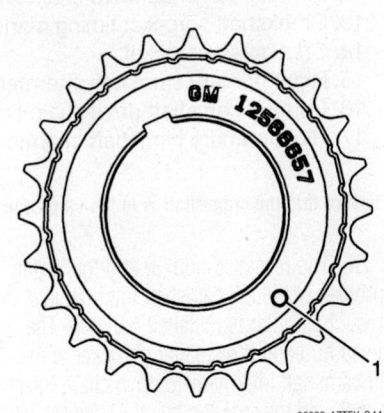

Crankshaft sprocket—3.6L engine

06025-AZTEK-G44

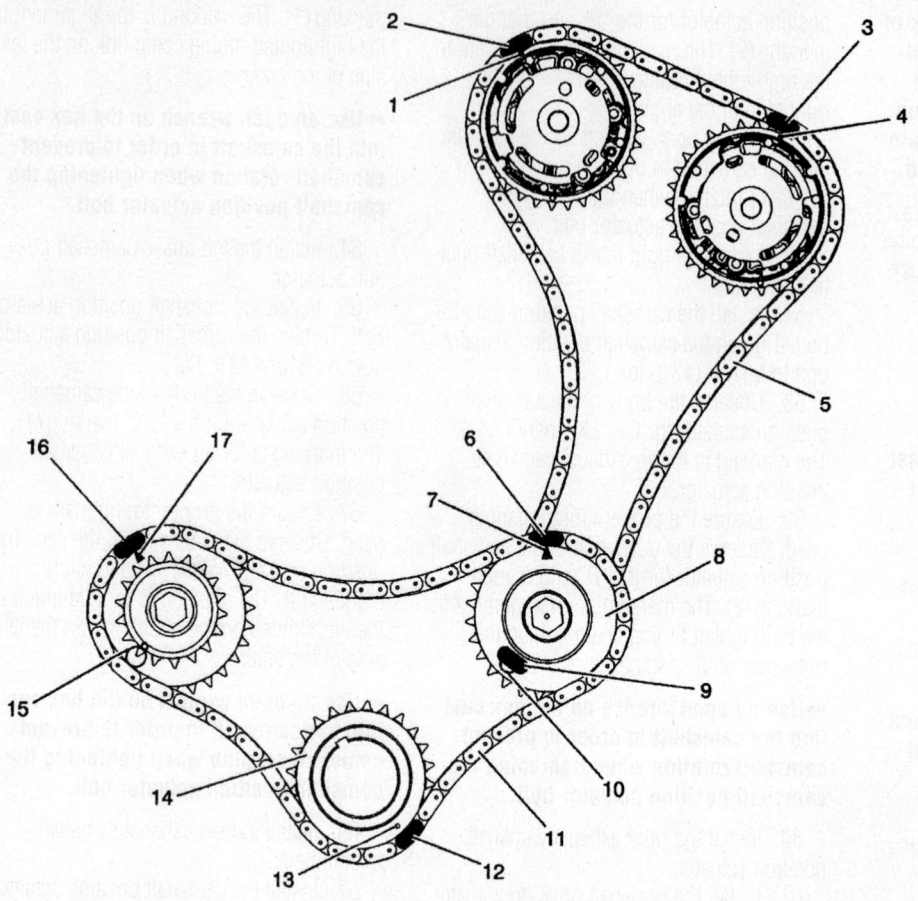

1. Left intake Camshaft Position (CMP) actuator timing mark
2. Left intake secondary camshaft timing drive chain timing link
3. Left exhaust secondary camshaft timing drive chain timing link
4. Left exhaust Camshaft Position (CMP) actuator timing mark
5. Left secondary camshaft timing drive chain
6. Primary camshaft drive chain timing link for the left primary camshaft intermediate drive chain sprocket
7. Left primary camshaft intermediate drive chain sprocket timing mark for the primary camshaft drive chain
8. Left primary camshaft intermediate drive chain sprocket
9. Left secondary camshaft timing drive chain timing link for the left primary camshaft intermediate drive chain sprocket
10. Left primary camshaft intermediate drive chain sprocket timing window for the left secondary camshaft timing drive chain timing link
11. Primary camshaft drive chain
12. Primary camshaft drive chain timing link for the crankshaft sprocket
13. Crankshaft sprocket timing mark
14. Crankshaft sprocket
15. Right primary camshaft intermediate drive chain sprocket
16. Primary camshaft drive chain timing link for the right primary camshaft intermediate drive chain sprocket
17. Right primary camshaft intermediate drive chain sprocket timing mark

06025-AZTEK-G22

Ensure that the crankshaft is in the stage one timing drive assembly position using the EN 46111—3.6L engine

74. The recessed hub (3) and the larger sprocket of the left camshaft intermediate drive chain idler is installed outward. The raised hub and the smaller sprocket of the left camshaft intermediate drive chain idler is installed towards the block. Place the left camshaft intermediate drive chain idler to the cylinder block.

75. Install the camshaft intermediate drive chain idler bolt. Tighten the camshaft intermediate drive chain idler bolt to 58 Nm (43 ft. lbs.).

⁂ WARNING

There should be no need to rotate the camshaft more than 10 degrees. Using the hex cast into the camshaft rotate the camshaft in order to install tool EN 46105.

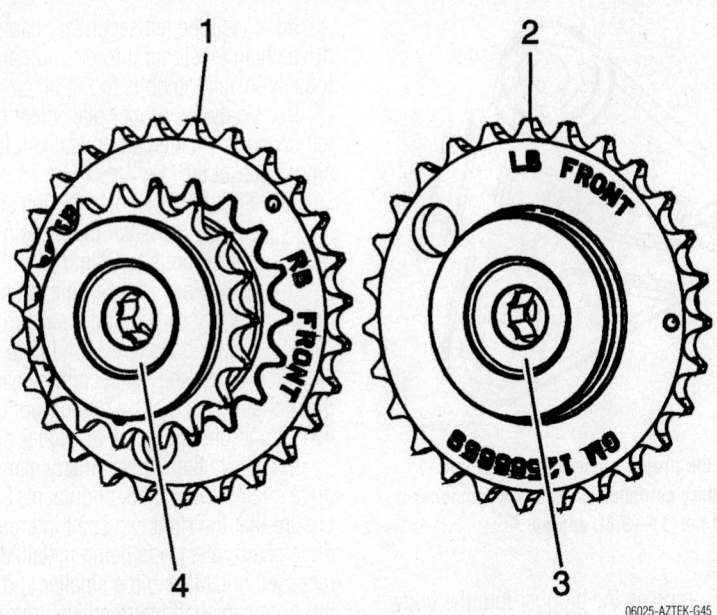

06025-AZTEK-G45

Left camshaft intermediate drive chain idler—3.6L engine

76. Install tool EN 46105-1 onto the rear of the left camshafts.

✳✳ WARNING

All camshafts must be locked in place before installation of any camshaft drive chains.

77. Ensure that tool EN 46105-1 is fully seated onto the camshafts.

78. Ensure that the crankshaft is in the stage one timing drive assembly position using tool EN 46111.

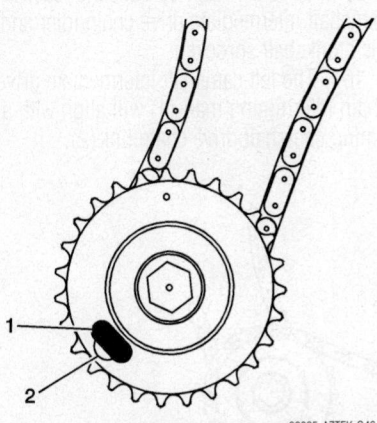

06025-AZTEK-G46

Place the left secondary camshaft drive chain around the inner sprocket of the left camshaft intermediate drive chain idler with the timing camshaft drive chain link (1) aligned to the alignment access hole (2) made in the left camshaft intermediate drive chain idler outer sprocket—3.6L engine

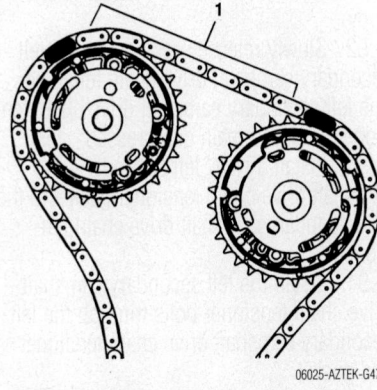

06025-AZTEK-G47

Ensure there are 7 links (1) between the timing camshaft drive chain links for the camshaft position actuator sprockets— 3.6L engine

79. Install the left secondary camshaft drive chain.

80. Place the left secondary camshaft drive chain around the inner sprocket of the left camshaft intermediate drive chain idler with the timing camshaft drive chain link (1) aligned to the alignment access hole (2) made in the left camshaft intermediate drive chain idler outer sprocket. Wrap the secondary camshaft drive chain around both left actuator drive sprockets.

81. Ensure there are 7 links (1) between the timing camshaft drive chain links for the camshaft position actuator sprockets.

82. Align the left exhaust camshaft position actuator sprocket alignment circle mark (2) with the timing camshaft drive chain link (1).

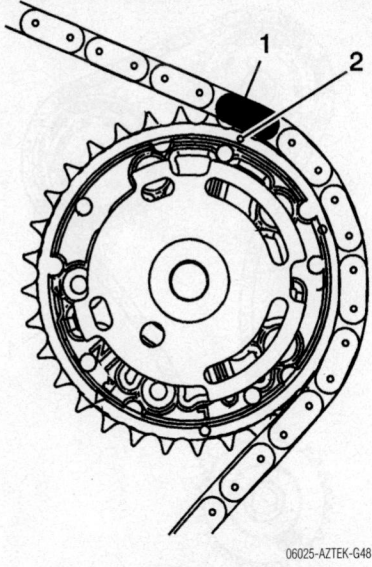

06025-AZTEK-G48

Align the left exhaust camshaft position actuator sprocket alignment circle mark (2) with the timing camshaft drive chain link (1)—3.6L engine

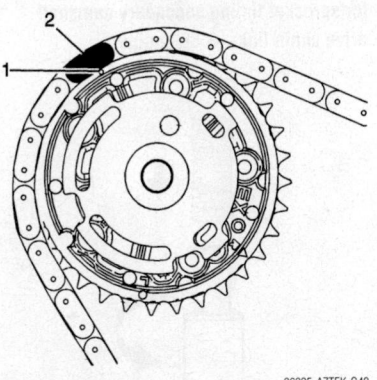

06025-AZTEK-G49

Align the left intake camshaft position actuator sprocket alignment circle mark (1) with the timing camshaft drive chain link (2)—3.6L engine

83. Align the left intake camshaft position actuator sprocket alignment circle mark (1) with the timing camshaft drive chain link (2).

84. There will be 18 links (1) between the left camshaft intermediate drive chain idler timing secondary camshaft drive chain link and each left camshaft position actuator sprocket timing secondary camshaft drive chain link.

85. Position the left secondary camshaft drive chain guide.

86. Install the secondary camshaft drive chain guide bolts. Tighten the secondary camshaft drive chain guide bolts to 23 Nm (17 ft. lbs.).

87. Position the left secondary camshaft drive chain shoe.

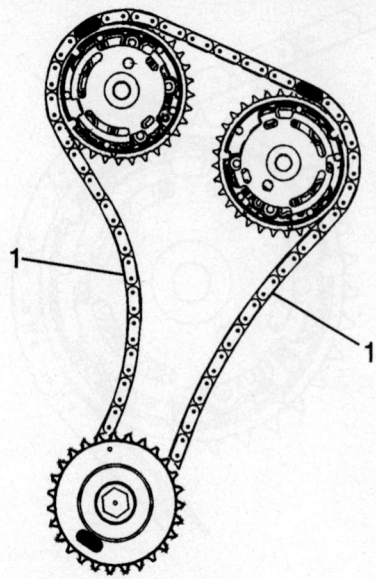

06025-AZTEK-G50

There will be 18 links (1) between the left camshaft intermediate drive chain idler timing secondary camshaft drive chain link and each left camshaft position actuator sprocket timing secondary camshaft drive chain link—3.6L engine

06025-AZTEK-G51

Using tool J 45027 reset the left secondary camshaft drive chain tensioner plunger—3.6L engine

88. Install the secondary camshaft drive chain shoe bolt. Tighten the secondary camshaft drive chain shoe bolt to 23 Nm (17 ft. lbs.).

89. Using tool J 45027 reset the left secondary camshaft drive chain tensioner plunger.

90. Install the plunger into the left secondary camshaft drive chain tensioner body.

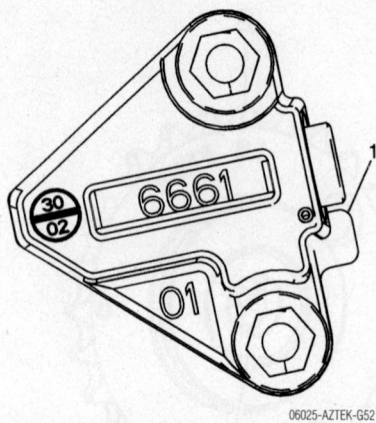

06025-AZTEK-G52

Verify the proper placement of the left secondary camshaft drive chain tensioner gasket tab (1)—3.6L engine

91. Compress the plunger into the body and lock the left secondary camshaft drive chain tensioner by inserting tool EN 46112 into the access hole in the side of the left secondary camshaft drive chain tensioner body.

92. Slowly release pressure on the left secondary camshaft drive chain tensioner. The left secondary camshaft drive chain tensioner should remain compressed.

93. Install a NEW left secondary camshaft drive chain tensioner gasket to the left secondary camshaft drive chain tensioner.

94. Install the left secondary camshaft drive chain tensioner bolts through the left secondary camshaft drive chain tensioner and gasket.

95. Ensure the left secondary camshaft drive chain tensioner mounting surface on the left cylinder head does not have any burrs or defects that would degrade the sealing of the NEW left secondary camshaft drive chain tensioner gasket.

96. Place the left secondary camshaft drive chain tensioner into position and loosely install the bolts to the block.

97. Verify the proper placement of the left secondary camshaft drive chain tensioner gasket tab (1).
- First Pass: Tighten the left secondary camshaft drive chain tensioner bolts to 5 Nm (44 inch lbs.).
- Final Pass: Tighten the left secondary camshaft drive chain tensioner bolts to 23 Nm (17 ft. lbs.).

98. Release the left secondary camshaft drive chain tensioner by pulling out tool EN 46112 and unlocking the tensioner plunger.

99. Verify the left secondary camshaft drive chain timing mark alignments (1-6). Ensure that the right camshaft intermediate drive chain idler (1) is being installed. The recessed hub (4) and the smaller sprocket of the right camshaft intermediate drive chain idler is installed outward. The raised hub and the larger sprocket of the right camshaft intermediate drive chain idler is installed towards the block.

100. Install the right camshaft intermediate drive chain idler.

101. Install the camshaft intermediate drive chain idler bolt. Tighten the camshaft intermediate drive chain idler bolt to 58 Nm (43 ft. lbs.).

➡ **Ensure that the crankshaft is in the stage one timing drive assembly position.**

102. Install the primary camshaft drive chain.

103. Wrap the primary camshaft drive chain around the large sprockets of each camshaft intermediate drive chain idler and the crankshaft sprocket.

104. The left camshaft intermediate drive chain idler timing mark (1) will align with a timing camshaft drive chain link (2).

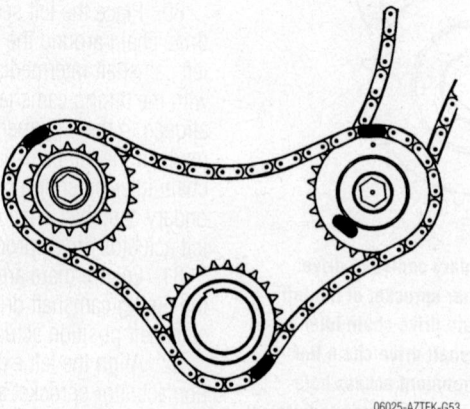

06025-AZTEK-G53

Wrap the primary camshaft drive chain around the large sprockets of each camshaft intermediate drive chain idler and the crankshaft sprocket—3.6L engine

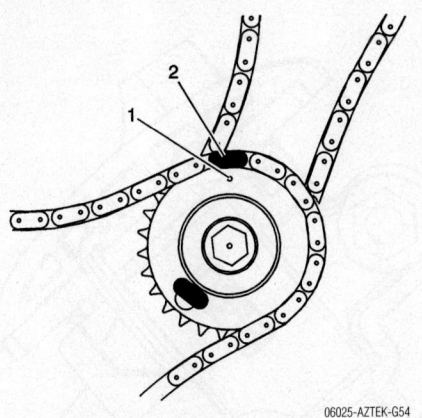

06025-AZTEK-G54

The left camshaft intermediate drive chain idler timing mark (1) will align with a timing camshaft drive chain link (2)—3.6L engine

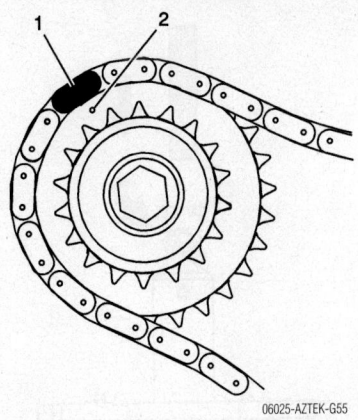

06025-AZTEK-G55

The right camshaft intermediate drive chain idler timing mark (2) will align with a timing camshaft drive chain link (1)—3.6L engine

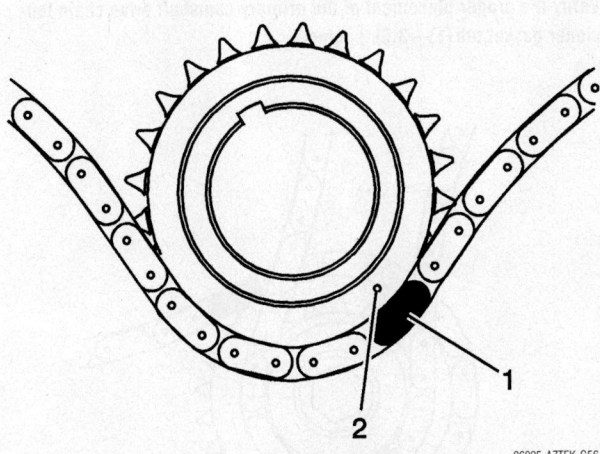

06025-AZTEK-G56

The crankshaft sprocket timing mark (2) will align with a timing camshaft drive chain link (1)—3.6L engine

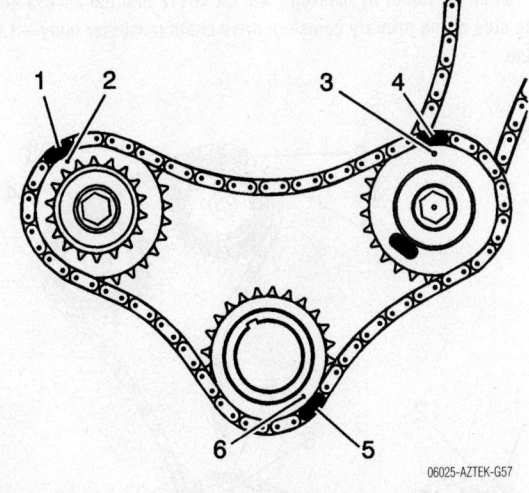

06025-AZTEK-G57

Ensure all the timing marks (2, 3, 6) are properly aligned with the timing camshaft drive chain links (1, 4, 5)—3.6L engine

105. The right camshaft intermediate drive chain idler timing mark (2) will align with a timing camshaft drive chain link (1).

106. The crankshaft sprocket timing mark (2) will align with a timing camshaft drive chain link (1).

107. Ensure all the timing marks (2, 3, 6) are properly aligned with the timing camshaft drive chain links (1, 4, 5).

108. Install the primary camshaft drive chain upper guide.

109. Install the primary camshaft drive chain upper guide bolts. Tighten the primary camshaft drive chain upper guide bolts to 23 Nm (17 ft. lbs.).

110. Using tool J 45027 reset the primary camshaft drive chain tensioner plunger.

111. Install the plunger into the primary camshaft drive chain tensioner body.

112. Compress the plunger into the body and lock the primary camshaft drive chain tensioner by inserting tool EN 46112 into

the access hole in the side of the primary camshaft drive chain tensioner body.

113. Slowly release pressure on the primary camshaft drive chain tensioner. The primary camshaft drive chain tensioner should remain compressed.

114. Install a NEW primary camshaft drive chain tensioner gasket to the primary camshaft drive chain tensioner.

115. Install the primary camshaft drive chain tensioner bolts through the primary camshaft drive chain tensioner and gasket.

116. Ensure the primary camshaft drive chain tensioner mounting surface on the engine block does not have any burrs or defects that would degrade the sealing of the NEW primary camshaft drive chain tensioner gasket.

117. Place the primary camshaft drive chain tensioner into position and loosely install the bolts to the block.

118. Verify the proper placement of the

primary camshaft drive chain tensioner gasket tab (1).

- First Pass: Tighten the primary camshaft drive chain tensioner bolts to 5 Nm (44 inch lbs.).
- Final Pass: Tighten the primary camshaft drive chain tensioner bolts to 23 Nm (17 ft. lbs.).

119. Release the primary camshaft drive chain tensioner by pulling out tool EN 46112 and unlocking the tensioner plunger.

120. Verify the primary and left secondary camshaft drive chain timing mark alignments (1-12).

121. Remove tool EN 46105-1 from the rear of the left camshafts.

122. Using tool EN 46111 rotate the crankshaft and crankshaft sprocket from the stage one alignment position (1) to the stage two alignment position (2), 115 crankshaft degrees, in order to install the right secondary camshaft drive chain components.

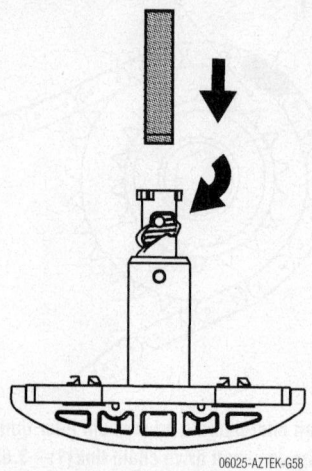

Compress the plunger into the body and lock the primary camshaft drive chain tensioner by inserting tool EN 46112 into the access hole in the side of the primary camshaft drive chain tensioner body—3.6L engine

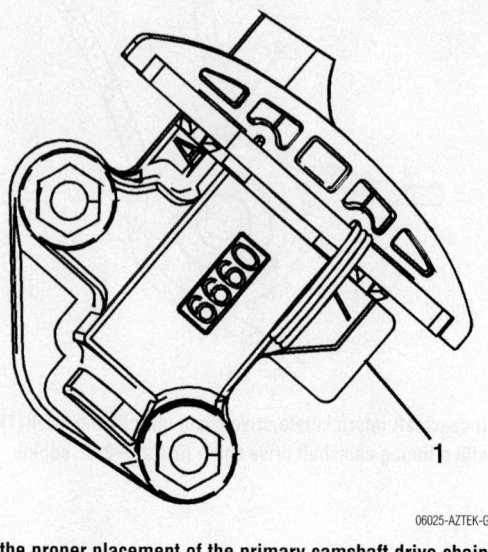

Verify the proper placement of the primary camshaft drive chain tensioner gasket tab (1)—3.6L engine

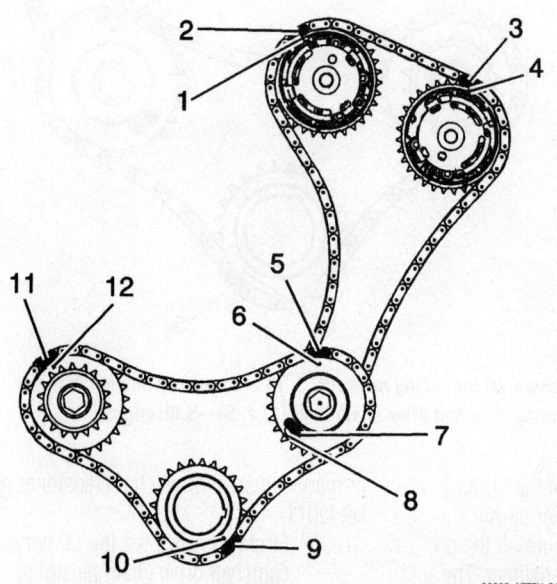

Verify the primary and left secondary camshaft drive chain timing mark alignments (1-12)—3.6L engine

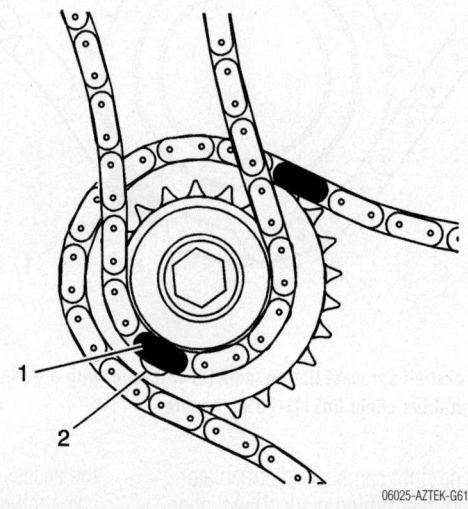

Place the secondary camshaft drive chain around the right camshaft intermediate drive chain idler outer sprocket, aligning the timing camshaft drive chain link (1) with the alignment access hole (2) made in the right camshaft intermediate drive chain idler inner sprocket—3.6L engine

123. Install tool EN 46105-2 onto the rear of the left camshafts.

124. Install tool EN 46105-1 onto the rear of the right camshafts.

125. Ensure that the crankshaft is in the stage two timing drive assembly position.

126. Install the right secondary camshaft drive chain.

127. Place the secondary camshaft drive chain around the right camshaft intermediate drive chain idler outer sprocket, aligning the timing camshaft drive chain link (1) with the alignment access hole (2) made in the right camshaft intermediate drive chain idler inner sprocket.

128. Wrap the secondary camshaft drive chain around both right actuator drive sprockets.

129. Ensure there are 7 links (1) between the timing camshaft drive chain links for the camshaft position actuator sprockets.

130. Align the right exhaust camshaft position actuator sprocket alignment triangle mark (1) with the timing camshaft drive chain link (2).

131. Align the right intake camshaft position actuator sprocket alignment triangle mark (2) with the timing camshaft drive chain link (1).

132. There will be 18 links (1) between the right camshaft intermediate drive chain idler timing camshaft drive chain link and each right camshaft position actuator sprocket timing camshaft drive chain link.

133. Position the right secondary camshaft drive chain guide.

134. Install the secondary camshaft drive chain guide bolts. Tighten the secondary camshaft drive chain guide bolts to 23 Nm (17 ft. lbs.).

135. Position the right secondary camshaft drive chain shoe.

136. Install the secondary camshaft drive chain shoe bolt. Tighten the secondary

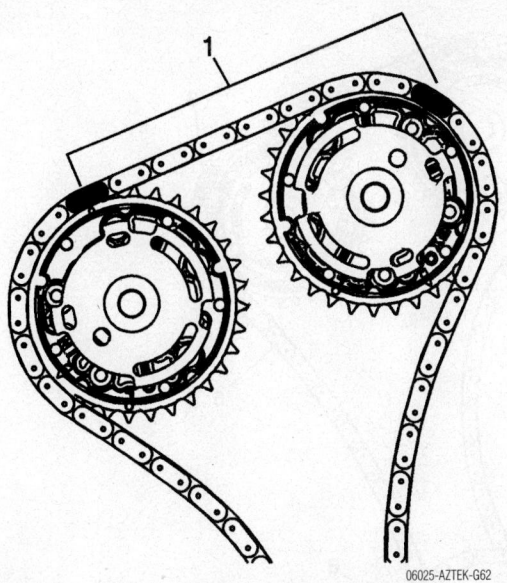

06025-AZTEK-G62

Ensure there are 7 links (1) between the timing camshaft drive chain links for the camshaft position actuator sprockets—3.6L engine

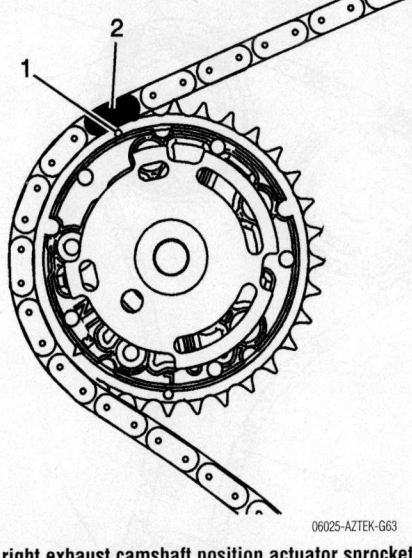

06025-AZTEK-G63

Align the right exhaust camshaft position actuator sprocket alignment triangle mark (1) with the timing camshaft drive chain link (2)—3.6L engine

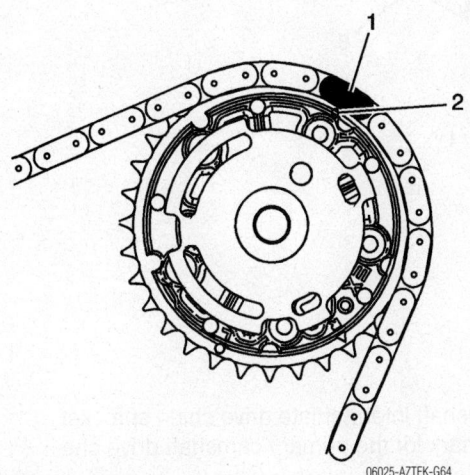

06025-AZTEK-G64

Align the right intake camshaft position actuator sprocket alignment triangle mark (2) with the timing camshaft drive chain link (1)—3.6L engine

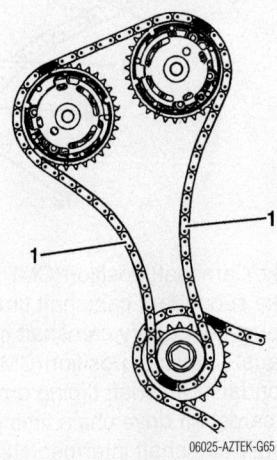

06025-AZTEK-G65

There will be 18 links (1) between the right camshaft intermediate drive chain idler timing camshaft drive chain link and each right camshaft position actuator sprocket timing camshaft drive chain link—3.6L engine

camshaft drive chain shoe bolt to 23 Nm (17 ft. lbs.).

137. Using tool J 45027 reset the right secondary camshaft drive chain tensioner plunger.

138. Install the plunger into the right secondary camshaft drive chain tensioner body.

139. Compress the plunger into the body and lock the right secondary camshaft drive chain tensioner by inserting tool EN 46112 into the access hole in the side of the right secondary camshaft drive chain tensioner body.

140. Slowly release pressure on the right secondary camshaft drive chain tensioner.

The right secondary camshaft drive chain tensioner should remain compressed.

141. Install a NEW right secondary camshaft drive chain tensioner gasket to the right secondary camshaft drive chain tensioner.

142. Install the right secondary camshaft drive chain tensioner bolts through the right secondary camshaft drive chain tensioner and gasket.

143. Ensure the right secondary camshaft drive chain tensioner mounting surface on the right cylinder head does not have any burrs or defects that would degrade the sealing of the NEW right secondary camshaft drive chain tensioner gasket.

144. Place the right secondary camshaft drive chain tensioner into position and loosely install the bolts to the block.

145. Verify the proper placement of the right secondary camshaft drive chain tensioner gasket tab (1).

- First Pass: Tighten the right secondary camshaft drive chain tensioner bolts to 5 Nm (44 inch lbs.).
- Final Pass: Tighten the right secondary camshaft drive chain tensioner bolts to 23 Nm (17 ft. lbs.).

146. Release the right camshaft drive chain tensioner by pulling out tool EN 46112 and unlocking the tensioner plunger.

147. Verify all primary and secondary

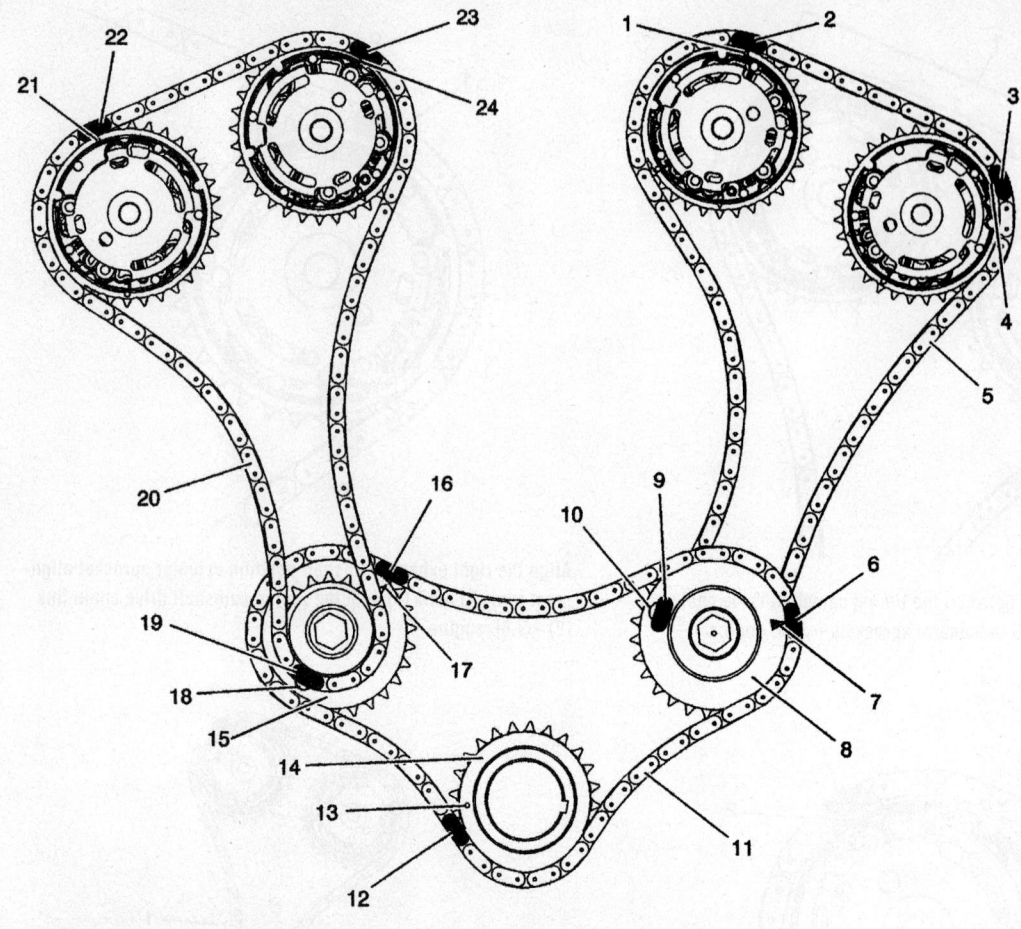

1. Left intake Camshaft Position (CMP) actuator timing mark
2. Left intake secondary camshaft timing drive chain timing link
3. Left exhaust secondary camshaft timing drive chain timing link
4. Left exhaust Camshaft Position (CMP) actuator timing mark
5. Left secondary camshaft timing drive chain
6. Primary camshaft drive chain timing link for the left primary camshaft intermediate drive chain sprocket
7. Left primary camshaft intermediate drive chain sprocket timing mark for the primary camshaft drive chain
8. Left primary camshaft intermediate drive chain sprocket
9. Left secondary camshaft timing drive chain timing link for the left primary camshaft intermediate drive chain sprocket
10. Left primary camshaft intermediate drive chain sprocket timing window
11. Primary camshaft drive chain
12. Primary camshaft drive chain timing link for the crankshaft sprocket
13. Crankshaft sprocket timing mark
14. Crankshaft sprocket
15. Right primary camshaft intermediate drive chain sprocket
16. Primary camshaft drive chain timing link for the right primary camshaft intermediate drive chain sprocket
17. Right primary camshaft intermediate drive chain sprocket timing mark for the primary camshaft drive chain
18. Right primary camshaft intermediate drive chain sprocket timing mark/window for the right secondary camshaft timing drive chain
19. Right secondary camshaft timing drive chain timing link for the right primary camshaft intermediate drive chain sprocket
20. Right secondary camshaft timing drive chain
21. Right exhaust Camshaft Position (CMP) actuator timing mark
22. Right exhaust secondary camshaft timing drive chain timing link
23. Right intake Camshaft Position (CMP) actuator timing mark
24. Right intake Camshaft Position (CMP) actuator timing mark

06025-AZTEK-G71

Verify all primary and secondary camshaft drive chain timing mark alignments (1-18) Stage Two—3.6L engine

camshaft drive chain timing mark alignments (1-18) Stage Two.

148. Install the 8 mm (0.315 in) guide pins from tool EN 46109 into the cylinder block positions as shown.

149. Install the NEW engine front cover to cylinder block seal.

150. Place a 3 mm (0.118 in) bead of RTV sealant, GM P/N 12378521 (Canadian P/N 88901148) or equivalent, on the engine front cover as shown (1).

151. Place the engine front cover onto tool EN 46109 and slide into position.

152. Remove tool EN 46109 from the cylinder block.

153. Hand start all the front cover bolts.

154. Tighten the engine front cover bolts in the sequence shown. Tighten the engine front cover bolts in sequence to 23 Nm (17 ft. lbs.).

155. Install NEW O-rings on the camshaft position sensor.

156. Place the camshaft position sensors in position on the front cover.

157. Install the camshaft position sensor bolts. Tighten the camshaft position sensor bolts to 10 Nm (89 inch lbs.).

158. Place the camshaft position actuator valves in position on the front cover.

159. Install the camshaft position actuator valve bolts. Tighten the camshaft position actuator valve bolts to 10 Nm (89 inch lbs.).

160. The EN 46106 must be installed onto the flywheel.

161. Use tool J 41998-B, nut, bearing and washer to install the crankshaft balancer.

➡ **Do not lubricate the crankshaft front oil seal or crankshaft balancer sealing surfaces. The crankshaft balancer is installed into a dry seal.**

162. Apply lubricant to the inside of the crankshaft balancer hub bore.

163. Place the crankshaft balancer in position on the crankshaft.

164. Thread tool J 41998-B in the crankshaft. Ensure you engage at least 10 threads of tool J 41998-B before pressing the crankshaft balancer in place.

165. Push the crankshaft balancer into position by tightening the nut on tool J 41998-B until the large washer bottoms out on the crankshaft end.

166. Remove tool J 41998-B.

167. Install the crankshaft balancer bolt.

168. Tighten the crankshaft balancer bolt. Tighten the crankshaft balancer bolt to 100 Nm (74 ft. lbs.).

169. Tighten the crankshaft balancer bolt

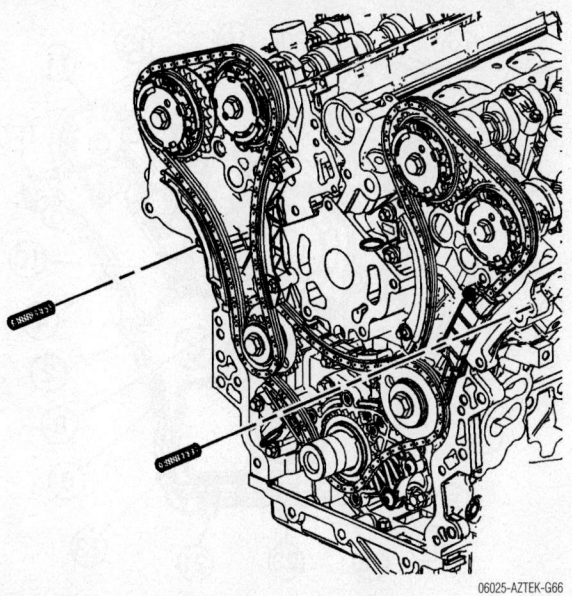

06025-AZTEK-G66

Install the 8 mm (0.315 in) guide pins from tool EN 46109 into the cylinder block positions as shown—3.6L engine

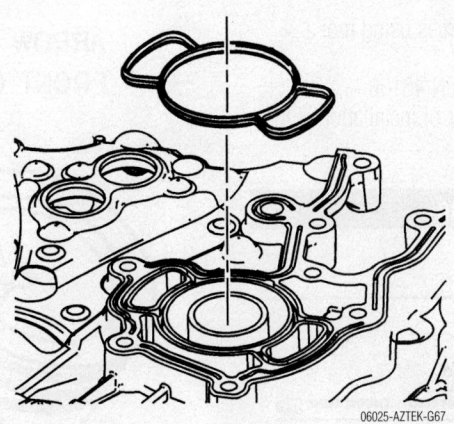

06025-AZTEK-G67

Install the NEW engine front cover to cylinder block seal—3.6L engine

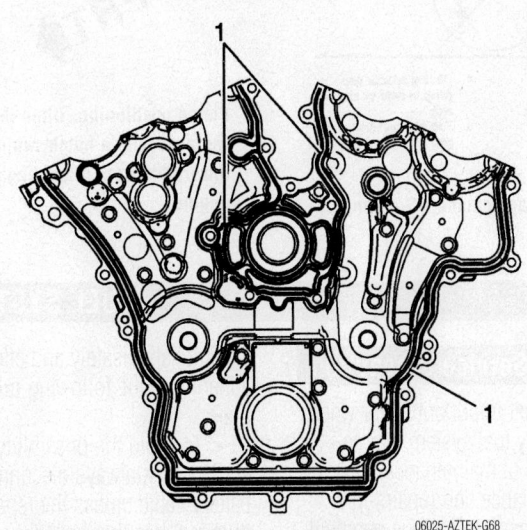

06025-AZTEK-G68

Place a 3 mm (0.118 in) bead of RTV sealant, GM P/N 12378521 (Canadian P/N 88901148) or equivalent, on the engine front cover as shown (1)—3.6L engine

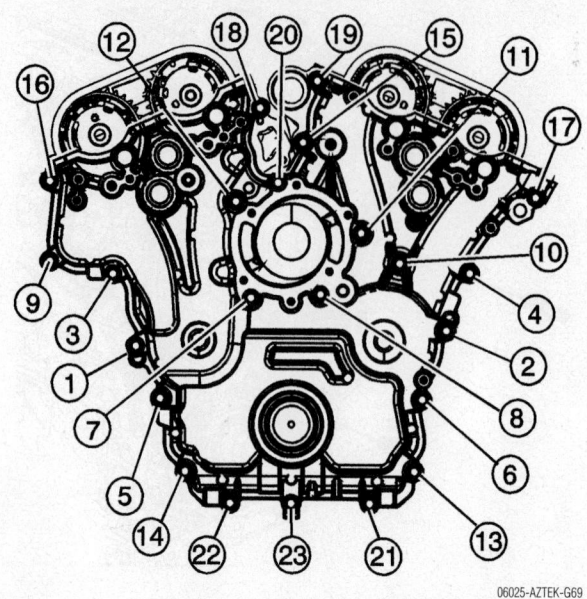

Tighten the engine front cover bolts in the sequence shown—3.6L engine

06025-AZTEK-G69

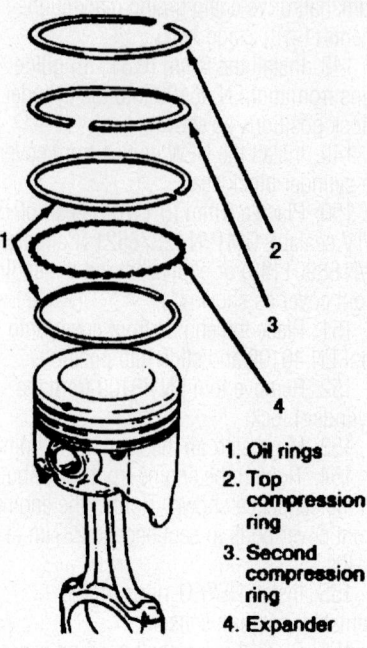

1. Oil rings
2. Top compression ring
3. Second compression ring
4. Expander

7922AG48

Piston ring positioning—3.4L engine

an additional 150 degrees using tool J 45059.

170. Remove tool EN 46106.

171. The remainder of installation is the reverse of removal.

Pistons and Ring

POSTIONING

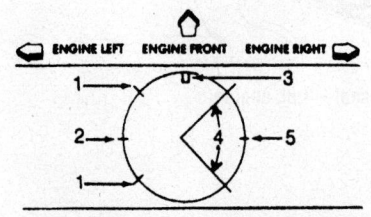

1. Oil ring rail gaps
2. 2nd Compression ring gap
3. Notch in piston
4. Oil ring spacer gap (tang in hole or slot with arc)
5. Top compression ring gap

7924AG07

Piston ring end-gap spacing—3.4L engine

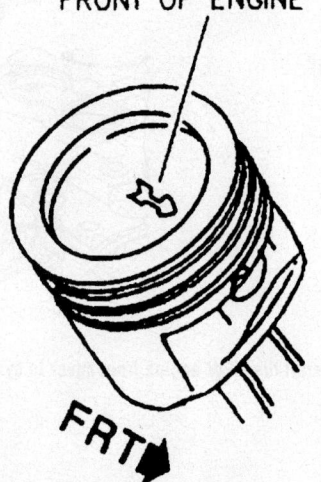

ARROW TOWARDS FRONT OF ENGINE

FRT

7922AG47

Piston positioning. Often the arrow is replaced with a notch, which must face toward the front of the engine—3.4L engine

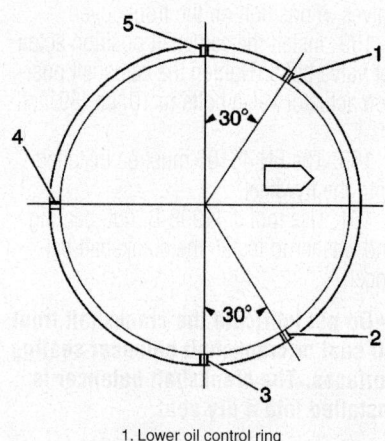

1. Lower oil control ring
2. Upper oil control ring
3. Top ring
4. Oil control ring expander
5. Second ring

06025-AZTEK-G72

Piston ring gap positioning—3.6L engine

FUEL SYSTEM

Fuel System Service Precaution

Safety is the most important factor when performing not only fuel system maintenance but any type of maintenance. Failure to conduct maintenance and repairs in a safe manner may result in serious personal injury or death. Maintenance and testing of the vehicle's fuel system components can be accomplished safely and effectively by adhering to the following rules and guidelines.

• To avoid the possibility of fire and personal injury, always disconnect the negative battery cable unless the repair or test procedure requires that battery voltage be applied.

• Always relieve the fuel system pressure prior to disconnecting any fuel system component (injector, fuel rail, pressure regulator, etc.), fitting or fuel line connection. Exercise extreme caution whenever relieving fuel system pressure, to avoid exposing skin, face and eyes to fuel spray. Please be advised that fuel under pressure may penetrate the skin or any part of the body that it contacts.

• Always place a shop towel or cloth around the fitting or connection prior to loosening to absorb any excess fuel due to spillage. Ensure that all fuel spillage (should it occur) is quickly removed from engine surfaces. Ensure that all fuel soaked cloths or towels are deposited into a suitable waste container.

• Always keep a dry chemical (Class B) fire extinguisher near the work area.

• Do not allow fuel spray or fuel vapors to come into contact with a spark or open flame.

• Always use a back-up wrench when loosening and tightening fuel line connection fittings. This will prevent unnecessary stress and torsion to fuel line piping. Always follow the proper torque specifications.

• Always replace worn fuel fitting O-rings with new ones. Do not substitute fuel hose, or equivalent, where fuel pipe is installed.

Fuel System Pressure

RELIEVING

3.4L Engine

A Schrader valve is provided on these fuel systems to conveniently test or release the system pressure. A fuel pressure gauge and adapter will be necessary to connect the gauge to the fitting. Most of the SFI systems utilize a service valve on one end of the fuel rail assembly.

1. Before servicing the vehicle, refer to the Precautions Section.
2. Disconnect the negative battery cable
3. Loosen the fuel filler cap to relieve tank vapor pressure.
4. Connect a fuel pressure gauge to the connector. Wrap a shop towel around the fittings to prevent spillage.
5. Install the bleed hose into an approved container and open the valve.
6. Drain any remaining fuel from the pressure gauge.
7. When fuel service is finished, tighten the fuel filler cap and connect the negative battery cable.

3.6L Engine

1. Before servicing the vehicle, refer to the Precautions Section.
2. Turn the ignition OFF.
3. Remove the fuel pump fuse and the fuel pump relay.
4. Loosen the fuel filler cap to relieve the fuel tank vapor pressure.

5. Attempt to start the engine and allow the engine to run until it stops.
6. Remove the fuel pressure test port cap.

> **❋❋ CAUTION**
>
> **Wrap a shop towel around the fuel pressure connection in order to reduce the risk of fire and personal injury. The towel will absorb any fuel leakage that occurs during the connection of the fuel pressure gage. Place the towel in an approved container when the connection of the fuel pressure gage is complete.**

7. Wrap a shop towel around the fuel pressure test port and use a small flat-bladed tool in order to depress (open) the fuel pressure test port valve.
8. Place the shop towel in an approved container.
9. Install the fuel pressure test port cap.
10. Tighten the fuel filler cap.

Fuel Filter

REMOVAL & INSTALLATION

3.4L Engine

1. Before servicing the vehicle, refer to the Precautions Section.
2. Relieve fuel system pressure.
3. Remove or disconnect the following:
   • Negative battery cable
   • Quick connect fittings at the inlet/outlet sides of the in-pipe fuel filter
   • Fuel filter mounting bracket nut
   • Fuel filter and drain any remaining fuel

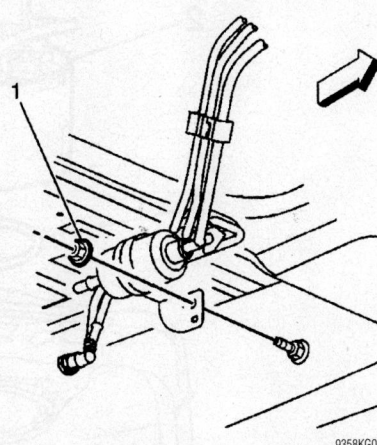

The fuel filter is located on the frame rail near the tank—with 3.4L engine

To install:
4. Install or connect the following:
   • Fuel filter to the bracket
   • Fuel filter assembly to the side rail near the fuel tank. Torque the nut to 89 inch lbs. (10 Nm).
   • Inlet/outlet quick connectors to the fuel filter
   • Negative battery cable
5. Start the vehicle and checks for leaks, repair if necessary.

3.6L Engine

See the Fuel Pump procedure.

Fuel Pump

REMOVAL & INSTALLATION

3.4L Engine

1. Before servicing the vehicle, refer to the Precautions Section.
2. Properly relieve the fuel system pressure.
3. Drain and remove the fuel tank from the vehicle
4. Remove or disconnect the following:
   • Negative battery cable
   • Fuel tank
   • Fuel sender shield
   • Quick connect fittings at the fuel pump
   • Fuel Tank Pressure (FTP) electrical connector
   • Fuel pump assembly lockring

> **❋❋ WARNING**
>
> **Do NOT pick up the fuel pump by the fuel pipes. Doing this could cause the joints to be damaged.**

   • Fuel pump assembly from the fuel tank and discard the O-ring
To install:
5. Install or connect the following:
   • New O-ring on the fuel tank
   • Fuel pump into the fuel tank making certain not to fold or twist the strainer and that it does not interfere with the full travel of the float arm
   • Fuel pump locking nut
   • Quick connect fittings at the fuel pump
   • FTP sensor electrical connector
   • Fuel sender shield
   • Fuel tank and fill the tank
   • Negative battery cable
6. Prime the fuel system as follows:

a. Turn the ignition switch **ON** for two seconds.

b. Turn the ignition switch **OFF** for 10 seconds.

c. Turn the ignition switch **ON** and checks for leaks. Repair if necessary.

3.6L Engine

1. Before servicing the vehicle, refer to the Precautions Section.

2. Relieve the fuel system fuel pressure.

3. Drain the fuel tank.

4. Raise the vehicle.

5. Disconnect the fuel tank wiring harness connector.

6. Disconnect the fuel evaporative emission (EVAP) pipes.

7. Remove the EVAP canister bracket nut.

8. Loosen the fuel tank fill pipe hose clamp.

9. Disconnect the fuel tank fill pipe hose from the fuel tank.

10. Disconnect the EVAP vent pipe near the fill pipe hose.

➡ **Do not bend the fuel tank straps. Bending the fuel tank straps may damage the straps.**

11. With the aid of an assistant or a transmission jack, support the fuel tank and remove the fuel tank strap attaching bolts.

12. Remove the fuel tank from the vehicle and place the fuel tank in a suitable work area.

13. Disconnect the fuel sender module electrical connectors.

➡ **Do Not handle the fuel sender assembly by the fuel pipes. The amount of leverage generated by handling the fuel pipes could damage the joints.**

14. Disconnect the fuel pipes from the fuel sender.

➡ **Avoid damaging the lock ring. Use only J-45722 to prevent damage to the lock ring.**

✱✱ WARNING

Do NOT use impact tools. Significant force will be required to release the lock ring. The use of a hammer and screwdriver is not recommended. Secure the fuel tank in order to prevent fuel tank rotation.

15. Use tool J 45722 and a long breaker-bar in order to unlock the fuel sender lock ring. Turn the lock ring in a counterclockwise direction.

16. Remove the fuel sender lock ring and the fuel sender from the fuel tank.

17. Remove and discard the fuel sender seal.

18. Remove the fuel level sensor from the fuel sender module.

➡ **Some lock rings were manufactured with "DO NOT REUSE" stamped into them. These lock rings may be reused if they are not damaged or warped.**

➡ **Inspect the lock ring for damage due to improper removal or installation procedures. If damage is found, install a NEW lock ring.**

➡ **Check the lock ring for flatness.**

19. Place the lock ring on a flat surface. Measure the clearance between the lock ring and the flat surface using a feeler gage at 7 points.

20. If warpage is less than 0.41 mm (0.016 in.), the lock ring does not require replacement. If warpage is greater than 0.41 mm (0.016 in.), the lock ring must be replaced.

To install:

21. Install the fuel level sensor to the fuel sender module.

22. Clean the fuel sender sealing flange.

➡ **Always replace the fuel sender seal when installing the fuel sender assem-**

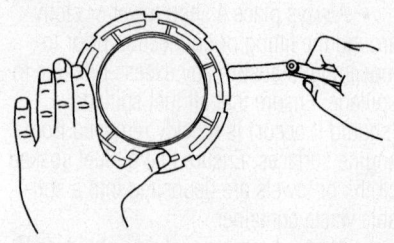

Measuring lock ring for flatness—with 3.6L engine

bly. Replace the lock ring if necessary. DO NOT apply any type of lubrication to the seal groove.

23. Install the NEW fuel sender seal to the fuel tank seal groove.

24. Install the fuel sender and the fuel sender lock ring.

25. Use tool J 45722 in order to install the fuel sender lock ring. Turn the lock ring in a clockwise direction.

26. Install the fuel pipes to the fuel sender.

27. Install the fuel sender sensor electrical connectors.

28. Install the EVAP canister.

29. Install the EVAP vent solenoid valve.

30. Install the fuel pipes to the fuel tank.

31. Install the fuel tank wiring harness to the fuel tank.

32. Install the fuel sender.

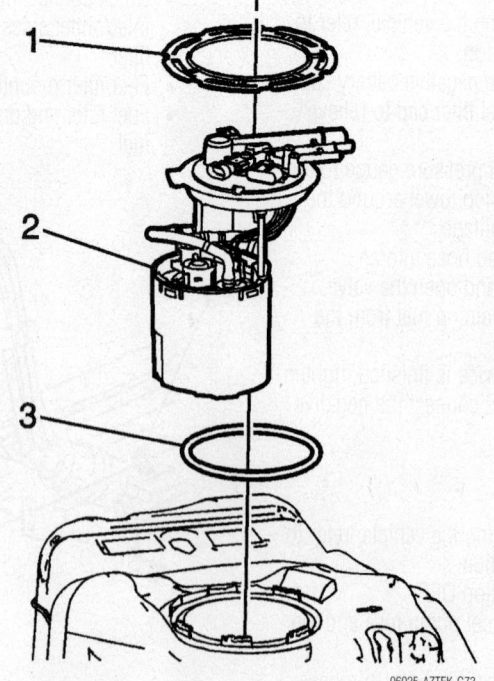

Fuel pump module—with 3.6L engine

33. With the aid of an assistant or a transmission jack, position and support the fuel tank.

➥**Do not bend the fuel tank straps. Bending the fuel tank straps may damage the straps.**

34. Install the fuel tank strap bolts. Tighten the fuel tank strap bolts to 47 Nm (35 ft. lbs.).

35. Install the EVAP canister bracket nut. Tighten the EVAP canister bracket nut to 6 Nm (53 inch lbs.).

36. Install the EVAP vent pipe near the fill pipe hose.

37. Install the fuel tank filler pipe hose to the fuel tank. Tighten the fuel tank filler pipe hose clamp to 2.5 Nm (22 inch lbs.).

38. Connect the fuel pipes.

39. Install the fuel tank wiring harness connector.

40. Lower the vehicle.

41. Add fuel as necessary and install the fuel tank filler pipe cap.

42. Inspect for fuel leaks. Perform the following steps:
- Turn ON the ignition for 2 seconds.
- Turn OFF the ignition for 10 seconds.
- Turn ON the ignition.
- Inspect for fuel leaks.

Fuel Injector

REMOVAL & INSTALLATION

3.4L Engine

1. Before servicing the vehicle, refer to the Precautions Section.

2. Relieve the fuel system pressure.

3. Remove or disconnect the following:
- Negative battery cable
- Upper intake manifold
- Engine fuel feed pipe at the fuel rail
- Fuel return line from the regulator
- Main injector harness and individual injector electrical connectors
- Electrical harness from the fuel rail
- Coolant Temperature Sensor (CTS) electrical connector
- Fuel rail retaining bolts
- Fuel rail
- Fuel injector retaining clips and injectors
- O-rings and discard them

To install:

➥**When replacing the fuel injector O-rings install the brown O-ring in the lower position. The lower O-ring uses a nylon collar to properly position it on**

the injector. **Be sure to install the O-ring backup or the sealing O-ring may move when the injector is installed to the fuel rail. If the sealing ring is not seated properly, a vacuum leak is possible thus causing drivability complaints.**

4. Install or connect the following:
- Upper O-ring to the fuel injector
- Lower O-ring backup to the injector
- Lower O-ring to the injector

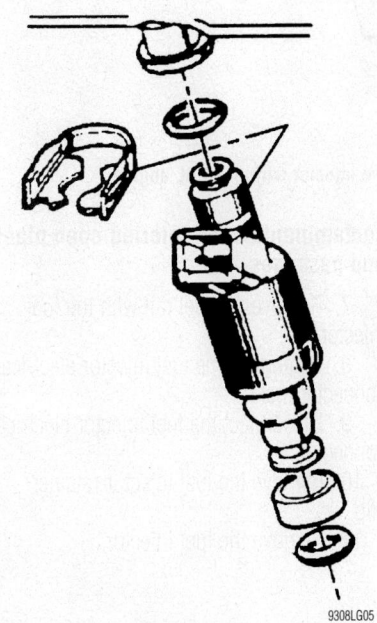

Fuel injector—3.4L engine

9308LG05

- Fuel injector to the fuel rail
- Fuel rail into the intake manifold. Tilt the rail to install the injectors.
- Fuel rail retaining bolts. Torque the bolts to 89 inch lbs. (10 Nm).
- CTS electrical connector
- Injector electrical harness
- Injector electrical connectors. Make sure to push the slide locks into position.
- Main injector harness connector
- Fuel feed and return pipe, using new O-rings
- Fuel feed pipe. Torque the nut to 13 ft. lbs. (17 Nm).
- Fuel return pipe to the pressure regulator. Torque the nut to 13 ft. lbs. (17 Nm).
- Upper intake manifold
- Negative battery cable

5. Prime the fuel system as follows:

a. Turn the ignition switch **ON** for two seconds.

b. Turn the ignition switch **OFF** for 10 seconds.

c. Turn the ignition switch **ON** and checks for leaks. Repair if necessary.

3.6L Engine

1. Before servicing the vehicle, refer to the Precautions Section.

2. Relieve the fuel system pressure.

3. Remove the upper intake manifold.

4. Remove the fuel pipe retaining clip.

5. Use compressed air in order to

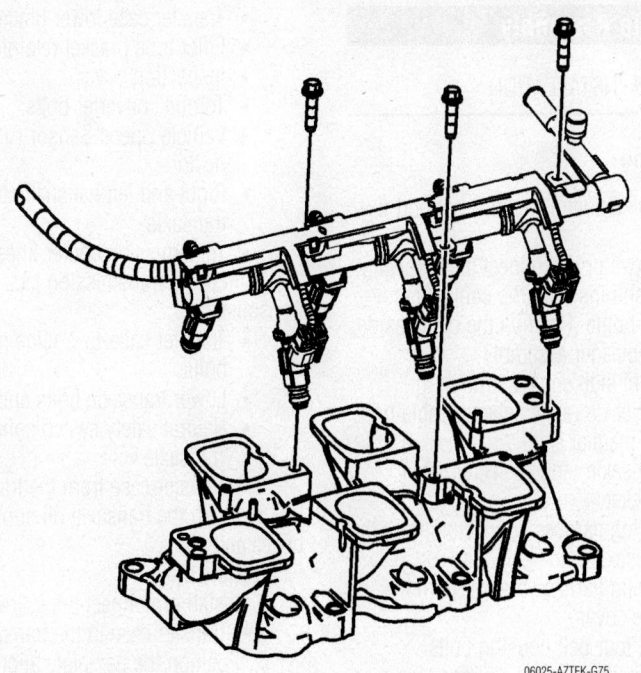

Fuel rail removal—3.6L engine

06025-AZTEK-G75

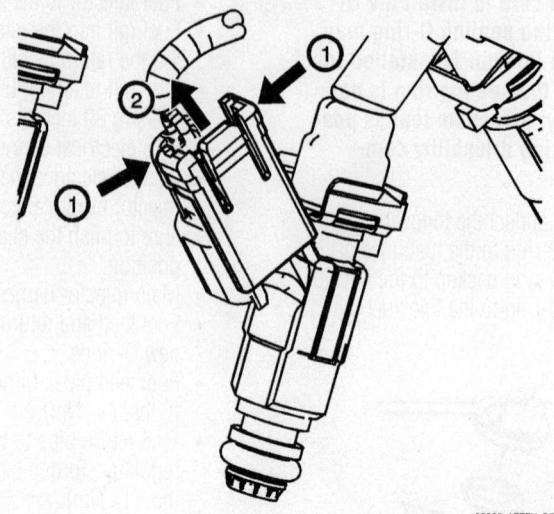

06025-AZTEK-G76

Squeeze the tabs (1) and pull up (2) to disconnect the injector wiring—3.6L engine

remove debris from the area where the fuel injectors enter the intake manifold.

6. Remove the fuel rail bolts.

➡️**Remove the fuel rail assembly carefully in order to prevent damage to the injector electrical connector terminals and the injector spray tips. Support the fuel rail after the fuel rail is removed in order to avoid damaging the fuel rail components. Cap the fittings and plug the holes when servicing the fuel system in order to prevent dirt and other contaminants from entering open pipes and passages.**

7. Remove the fuel rail with the fuel injectors.

8. Disengage the fuel injector electrical connector lock.

9. Disconnect the fuel injector electrical connector.

10. Remove the fuel injector retainer clip.

11. Remove the fuel injector.

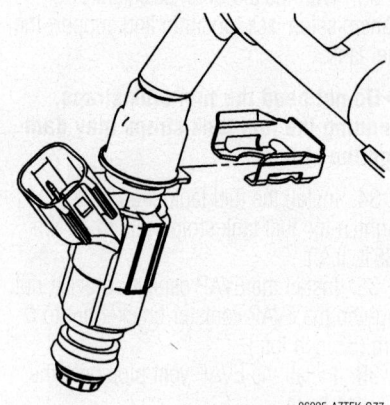

06025-AZTEK-G77

Fuel injector retaining clip—3.6L engine

12. Remove and discard the fuel injector seals.

To install:

13. Install NEW fuel injector seals.

14. Install the fuel injector.

15. Install the fuel injector retainer clip.

16. Install the fuel injector electrical connector.

17. Engage the fuel injector electrical connector lock.

18. Install the fuel rail with the fuel injectors.

19. Install the fuel rail bolts. Tighten the fuel rail bolts to 10 Nm (89 inch lbs.).

20. Install the upper intake manifold.

DRIVE TRAIN

Transaxle Assembly

REMOVAL & INSTALLATION

3.4L Engine

1. Install a suitable engine support fixture.

2. Remove or disconnect the following:
   - Push pins from the coolant recovery bottle. Position the bottle aside.
   - Air cleaner assembly
   - Right side engine strut
   - Transaxle range selector cable from the manual shaft
   - Transaxle range selector cable bracket
   - Wiring harness connectors from the transaxle
   - Wiring harness bracket from the side cover
   - Top four bell housing bolts
   - Propeller shaft, if equipped
   - Frame
   - Transfer case lower brace
   - Filler tube bracket retaining bolt
   - Inspection cover
   - Torque converter bolts
   - Vehicle Speed Sensor (VSS) connector
   - Right and left halfshafts from the transaxle
   - Transmission cooler lines

3. Install a transmission jack under the transmission.
   - Transfer case-to-engine mount bolts
   - Lower transaxle bolts and stud
   - Neutral safety switch connector
   - Transaxle
   - Transfer case from the transaxle

4. Flush the transaxle oil coolers, hoses and pipes.

To install:

5. Install or connect the following:
   - Transfer case to the transaxle

6. Position the flex plate alignment hole to the seven o'clock position.

7. Align the transaxle filler tube to the transmission and install the transaxle into the vehicle.

8. Install or connect the following:
   - Lower transaxle bolt and stud. Torque to 55 ft. lbs. (75 Nm).
   - Wiring harness bracket to the side cover
   - Neutral safety switch connector
   - Transaxle brace
   - Transmission cooler lines
   - Right and left halfshafts
   - VSS connector
   - Transfer case lower brace. Torque bolts to 35 ft. lbs. (47 Nm).
   - Torque converter bolts. Tighten the torque converter bolts to 46 ft. lbs. (63 Nm).
   - Torque converter cover
   - Filler tube bracket retaining bolt

➡️**Thoroughly clean and apply LOCTITE® DRI-LOC 201® (GM P/N 12345493, or equivalent) to the bolt threads prior to assembly.**

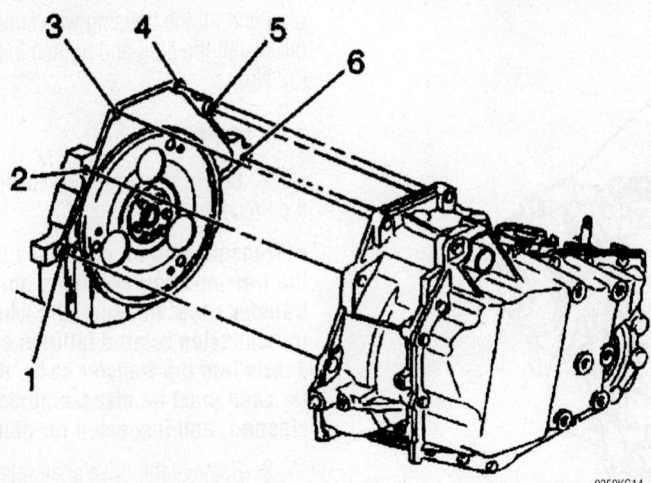

Transaxle mounting and bolt locations—3.4L engine

- Propeller shaft, if equipped. Ensure the special washer is in place on each pair of bolts. Tighten the bolts to 24 ft. lbs. (33 Nm).
- Front wheels

9. Lower vehicle and remove the engine support fixture

- Right side engine strut
- Upper transaxle bolts and stud. Torque to 55 ft. lbs. (75 Nm).
- Wiring harness the transaxle
- Range selector cable bracket
- Range selector cable on the manual shaft
- Air cleaner assembly
- Coolant recovery bottle.

10. Check and adjust the fluid level. Inspect for fluid leaks

3.6L Engine

1. Before servicing the vehicle, refer to the Precautions Section.

➡ **Transmission oil circulates between the transmission assembly and the transfer case. In situations where transmission related failures circulate debris into the transfer case, the transfer case must be disassembled, cleaned, and inspected for damage.**

2. Raise and support the vehicle.

3. Remove the transmission lower bell housing bolt located between the engine oil pan and the transmission oil pan.

4. Remove the powertrain (engine, transmission and frame) from the vehicle.

5. Remove the filler tube.

6. Remove the transmission bell housing bolts.

7. If the vehicle is equipped with all wheel drive (AWD), remove the transfer case brace.

8. Remove the coolant inlet pipe.

9. Remove the transmission mount with the transmission mount bracket.

10. Remove the starter motor.

11. Remove the torque converter bolts.

12. Remove the transmission lower brace.

13. Remove the transmission upper brace nut located behind the power steering pump and above the transfer case.

14. Remove the transmission bolt near the crank sensor and the engine coolant (block) heater.

15. Separate the transmission from the engine.

16. If the vehicle is equipped with AWD, remove the transfer case.

17. Flush the transaxle oil cooler and the transaxle oil cooler hoses.

To install:

18. If the vehicle is equipped with AWD, install the transfer case.

19. Use the dowel locator pins in order to align and install the transmission to the engine.

20. Install the transmission bell housing bolts. Tighten the transmission bell housing bolts to 50 Nm (37 ft. lbs.).

21. Install the transmission bell housing bolt near the crank sensor and the engine coolant (block) heater.

22. Tighten the transmission bell housing bolt to 50 Nm (37 ft. lbs.).

23. Install the transmission upper brace nut located behind the power steering pump and above the transfer case. Tighten the transmission upper brace nut to 50 Nm (37 ft. lbs.).

24. Install the transmission lower brace. Tighten the transmission upper brace nut to 50 Nm (37 ft. lbs.).

25. Install the torque converter bolts. Tighten the bolts to 63 Nm (47 ft. lbs.).

26. Install the starter motor.

27. Install the transmission mount with the transmission mount bracket. Start with the forward lower bolt and work clockwise to tighten the transaxle mount bracket bolts to 95 Nm (70 ft. lbs.).

28. Install the coolant inlet pipe.

29. If the vehicle is equipped with AWD, install the transfer case brace. Tighten the transfer case brace bolts to 50 Nm (37 ft. lbs.). Tighten the transfer case brace nuts to 50 Nm (37 ft. lbs.).

30. Install the filler tube.

31. Install the powertrain (engine, transmission and frame) to the vehicle.

32. Install the transmission lower bell housing bolt located between the engine oil pan and the transmission oil pan. Tighten the transmission bell housing bolt to 50 Nm (37 ft. lbs.).

33. Inspect and adjust the transaxle fluid level as needed.

Transfer Case Assembly

REMOVAL & INSTALLATION

3.4L Engine

1. Before servicing the vehicle, refer to the Precautions Section.

2. Remove or disconnect the following:

- Propeller shaft
- Gear oil from the extension housing
- Transmission fluid from the case
- Speed sensor electrical connector
- Transfer case lower brace bolts and brace
- Clamp from the extension housing vent hose coupling
- Vent hose bracket-to-transfer case bolt
- Vent hose coupling, vent hose and bracket
- Transaxle
- Output shaft retaining ring

3. Rotate the transaxle 90 degrees.

- Transfer case lower brace bolt

4. Rotate the transfer case 90 degrees back to the installed position

- Transfer case side brace bolts and side brace
- Transfer case-to-transaxle bolts

✳✳ **CAUTION**

The transfer case weighs about 60 pounds. Be sure to lift the case properly, to avoid injury.

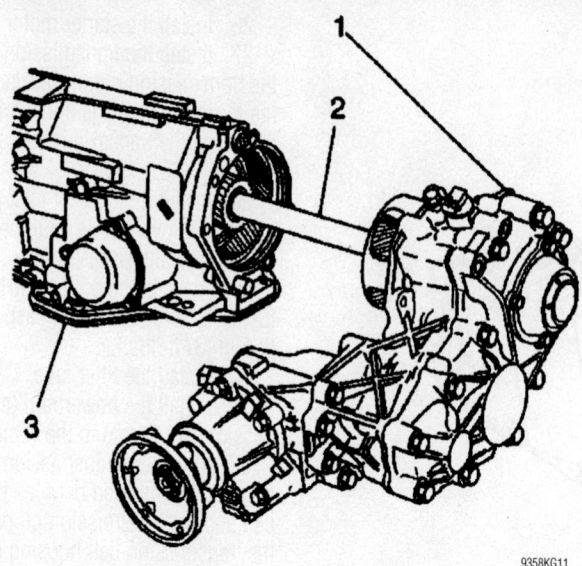

Transfer case (1), output shaft (2) and transaxle (3)—3.4L engine

9358KG11

- Transfer case from the transaxle. Note that the output shaft withdraws from the transaxle with the transfer case.
- Output shaft
- Transfer case O-ring from the transfer case

To install:

5. Install or connect the following:
   - O-ring seal to the transfer case
6. Rotate the transaxle so the bottom pan is facing the floor.
   - Transfer case to the transaxle
7. Torque the transfer case bolts, in sequence, as follows:
   a. Bolts 1 and 2: 26 ft. lbs. (35 Nm), plus an additional 160 degrees.
   b. Bolt 3: 26 ft. lbs. (35 Nm), plus an additional 70 degrees.
   c. Bolts 4 and 5: 30 ft. lbs. (40 Nm).
   - Transfer case side brace. Tighten the bolts, in sequence, to 35 ft. lbs. (47 Nm).

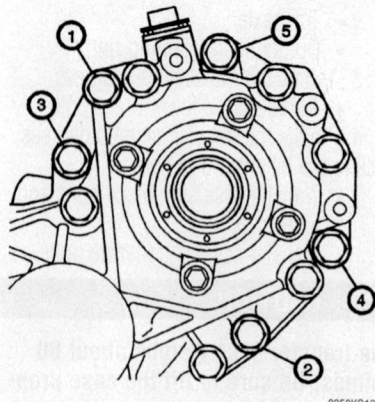

Transfer case bolt tightening sequence—3.4L engine

9358KG12

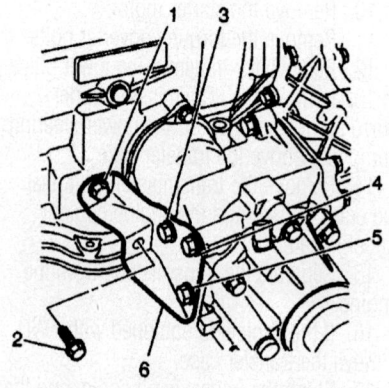

Transfer case side brace bolt tightening sequence—3.4L engine

9358KG13

8. Rotate the transaxle 90 degrees.
   - Transfer case lower brace-to-transaxle bolt. Torque the bolt to 35 ft. lbs. (47 Nm).
   - Output shaft
   - New output shaft retaining ring
   - Transaxle
   - Vent hose and coupling to the extension housing. Secure with the clamp.
   - Vent hose bracket and bolt/stud. Tighten to 106 inch lbs. (10 Nm).
   - Speed sensor electrical connector
   - Propeller shaft
   - Transfer case lower brace. Tighten the bolts to 35 ft. lbs. (47 Nm).
   - Drain plugs and gaskets to the transfer case and extension housing. Torque the plugs to 24 ft. lbs. (32 Nm).
9. Check the transaxle fluid level.
10. Remove the extension housing fill

plug and fill the housing with suitable gear oil. Install the plug and tighten to 24 ft. lbs. (32 Nm).

3.6L Engine

1. Before servicing the vehicle, refer to the Precautions Section.

➡**Transmission oil circulates between the transmission assembly and the transfer case. In situations where transmission related failures circulate debris into the transfer case, the transfer case must be disassembled, cleaned, and inspected for damage.**

2. Remove the drive shaft retaining ring.
3. Rotate the transaxle 90 degrees.
4. Remove the transfer case lower brace bolt.
5. Rotate the transaxle 90 degrees.
6. Remove the transfer case side brace bolts. Remove the side brace.
7. Remove the transfer case to case bolts.

✷✷ CAUTION

This component weighs approximately 60 lbs. Personal injury may result if you lift the component improperly.

➡**During removal of the transfer case/output shaft, do not use excessive force or damage to the bushings may occur.**

8. Remove the transfer case assembly from the transmission case.
9. Remove the transfer case lower brace bolts.
10. Remove the transfer case lower brace.
11. Remove the case extension seal from the transfer case.
12. Position the transaxle so that the case side cover is facing down.
13. Remove the oil dam from the transaxle.
14. Install the transfer case assembly onto a work stand.

To install:

15. Install the case extension seal.

➡**The oil dam must be installed with the oil passage notch aligned to the passage in the case. Incorrect alignment will cause oil flow stoppage and damage to the transmission.**

16. Line up the notch to the hole in the case and install the oil dam.
17. Install the transfer case lower brace.

18. Install the transfer case lower brace bolts. Hand tighten.

➡ The park gear thrust bearing must be retained in the park gear when installing the transfer case to the transmission, or damage may occur.

➡ When the transfer case is installed onto the transmission, there should be no gap between these parts. If a gap exists, check the park gear thrust bearing for proper retention to the park gear.

19. Install the transfer case assembly onto the transmission case.

20. Install the 5 transfer case to case bolts.

➡ Do not use air powered tools in order to assemble or disassemble transmissions. Use hand tools in order to properly determine bolt tightness. Improper bolt torque can contribute to transmission repair conditions, and this information, which is vital to diagnosis, can only be detected when using hand tools.

➡ Use the correct fastener in the correct location. Replacement fasteners must be the correct part number for that application. Fasteners requiring replacement or fasteners requiring the use of thread locking compound or sealant are identified in the service procedure. Do not use paints, lubricants, or corrosion inhibitors on fasteners or fastener joint surfaces unless specified. These coatings affect fastener torque and joint clamping force and may damage the fastener. Use the correct tightening sequence and specifications when installing fasteners in order to avoid damage to parts and systems.

21. Torque the bolts in the following sequence:
- Tighten 2 transfer case bolts (1, 2): First pass 35 Nm (26 ft. lbs.). Final Pass an additional 160 degrees
- Tighten 1 transfer case bolt (3): First pass 35 Nm (26 ft. lbs.). Final Pass an additional 70 degrees
- Tighten 2 transfer case bolts (4, 5): Tighten transfer case bolts (4, 5) to 40 Nm (30 ft. lbs.).

22. Install the transfer case side brace to case.

23. Install the transfer case side brace to case bolts. Hand tighten the bolts.

24. Torque the bolts in the following

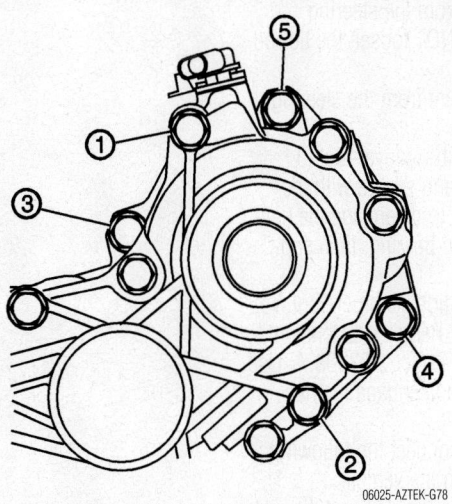

Transfer case bolt torque sequence—with 3.6L engine

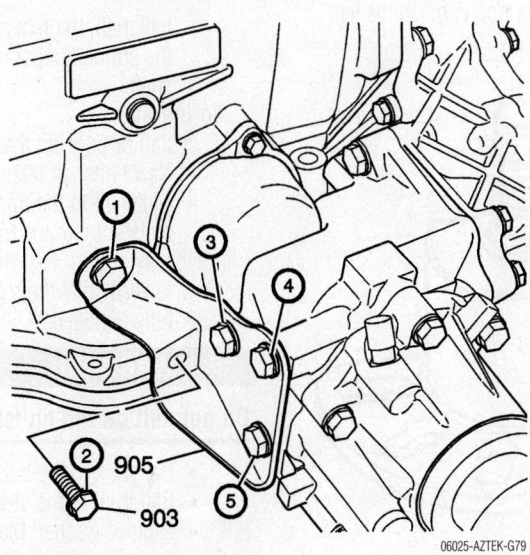

Transfer case side bolt torque sequence—with 3.6L engine

sequence (1, 2, 3, 4, 5). Tighten the transfer case side brace bolts to 31 Nm (23 ft. lbs.).

25. Install the transfer case lower brace to case bolt. Tighten the transfer case lower brace bolt to 56 Nm (42 ft. lbs.).

26. Tighten the transfer case lower brace to case bolts. Tighten the transfer case lower brace bolts to 31 Nm (23 ft. lbs.).

Halfshaft

REMOVAL & INSTALLATION

These procedures requires the use of the following special tools. Slide Hammer Tool No. J 2619-01, Axle Shaft Remover Extension J 29794, Axle Shaft Puller J 33008-A and Wheel Hub Remover J 42129.

Front

1. Before servicing the vehicle, refer to the Precautions Section.

2. Remove or disconnect the following:
- Front wheel
- Splash shield
- Stabilizer shaft link
- Wheel Speed Sensor (WSS) electrical connector

3. Insert a drift into the caliper and into the rotor to prevent the rotor from turning.

➡ Note the installed position of the 2-piece washer before removing it. The ramped sides must face each other when installing the washer.

- Halfshaft nut and the ramped 2-piece washer. Discard the nut.

- Tie rod end from the steering knuckle; Do NOT loosen the tie rod jam nut
- Lower ball joint from the steering knuckle

4. Install wheel hub removal tool (J 42129) onto the hub and secure with the lug nuts. Use the tool to disengage the half-shaft from the hub and bearing, then support the shaft.

5. Assemble the Slide Hammer Tool No. J 2619-01, Axle Shaft Remover Extension J 29794 and Axle Shaft Puller J 33008-A. Use the assembled tool to disengage the half-shaft from the transaxle.

6. Remove or disconnect the following:
- Halfshaft from the vehicle
- Halfshaft retaining ring and discard. On the right halfshaft, the retaining ring is on the splined shaft of the inner tripot housing. On the left

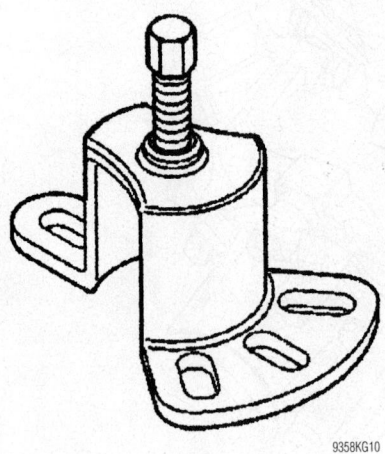

Wheel hub removal tool

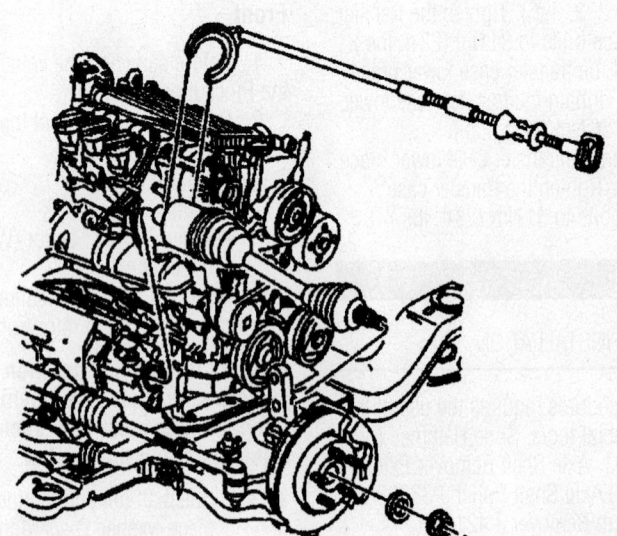

Assemble the special tools to separate the halfshaft from the transaxle

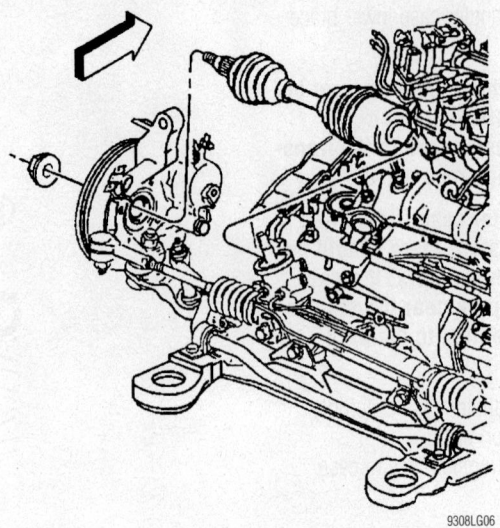

Install the left side halfshaft

halfshaft, the retaining ring is on the splined transmission output shaft.

To install:
7. Install or connect the following:
- New halfshaft retaining ring
- Halfshaft to the transaxle. Push the halfshaft into the transaxle until it is fully seated. Pull on the tri-pot joint to verify that the retaining ring is fully engaged.

✳✳ WARNING

Do not pull on the halfshaft bar.

- Halfshaft to the hub and bearing.
- Ball joint to the steering knuckle
- 2-piece washer. Make sure the ramped sides of the washer are facing each other.

- Drift through the caliper and into the rotor to prevent it from turning
- New halfshaft nut. Torque it to 192 ft. lbs. (260 Nm).
- Stabilizer shaft link. Torque the nut to 17 ft. lbs. (23 Nm).
- WSS electrical connector
- Splash shield
- Front wheel

8. Check the transaxle fluid level. Check and adjust the alignment as needed.

9. Road test the vehicle and check for any abnormal noise.

Rear

1. Before servicing the vehicle, refer to the Precautions Section.
2. Apply the parking brake.
3. Remove or disconnect the following:
- Rear wheel
- Halfshaft nut and discard
4. Release the parking brake.
- Brake caliper bracket and support it with a piece of wire
- Parking brake cable routing bracket nut, if necessary
- Rear tie rod end-to-knuckle bolt; DO NOT loosen the tie rod end jam nut
- Parking brake cable bracket-to-knuckle bolts; loosen only.
- Wheel Speed Sensor (WSS) electrical connector

5. Install the Wheel Hub Removal Tool No. K 42129 onto the wheel hub and secure with wheel nuts. Begin to disengage the halfshaft from the hub and bearing.

- Bolt and nut securing the upper control arm to the suspension knuckle

- Halfshaft completely from the hub and bearing
6. Reposition the suspension knuckle toward the rear of the vehicle.
- Tool from the wheel hub

➡**Support the halfshaft until it is completely removed from the vehicle.**

7. Assemble the Slide Hammer Tool No. J 2619-01, Axle Shaft Remover Extension J 29794 and Axle Shaft Puller J 33008-A. Install the Axle Shaft Puller evenly onto the rear beveled surface of the halfshaft inner joint housing.
- Halfshaft from the rear axle differential using the assembled tools
- Halfshaft from the vehicle

➡**The differential output shaft oil seal must be replaced when removing the rear wheel drive shaft.**

- Halfshaft oil seal
To install:

❄❄ WARNING

Support the wheel drive shaft until it is completely installed.

8. Install or connect the following:
- Halfshaft to the differential output shaft. Carefully align and guide the halfshaft onto the differential output shaft.

❄❄ WARNING

Be careful not to damage the differential output shaft oil seal.

- Halfshaft fully onto the differential output shaft using light force. Make sure it is fully seated on the differential output shaft retaining ring by grasping the inner tripot housing and pulling outward. Do not pull on the wheel drive shaft bar. The halfshaft will remain firmly in place when properly engaged.
9. Begin to position the suspension knuckle to the halfshaft. Align and carefully guide the halfshaft into the hub and bearing but do not seat fully.
- Suspension knuckle to the upper control arm
- Bolt and nut to the upper control arm/suspension knuckle assembly. Torque the bolt to 63 ft. lbs. (85 Nm).
- Parking brake cable bracket.
- Tie rod to the knuckle. Tighten the bolt to 63 ft. lbs. (85 Nm).
- Brake caliper bracket

- WSS electrical connector
- Parking brake cable routing bracket, if removed. Tighten the nuts to 89 inch lbs. (10 Nm).
10. Set the park brake.
- New halfshaft nut. Tighten to 192 ft. lbs. (260 Nm).
- Wheel
11. Lower the vehicle and release the parking brake.

CV-Joints

OVERHAUL

1. Before servicing the vehicle, refer to the Precautions Section.
2. Remove or disconnect the following:
- Halfshaft
- Large seal retaining clamp from the CV-joint and discard the clamp
- Swage ring by using a hand grinder to cut through it. Do not damage the axle shaft with the grinder
- Halfshaft outboard seal from the CV-joint outer race and slide the seal away from the joint
- CV-joint and boot from the halfshaft and discard the boot
3. Place a brass drift against the CV-joint cage and gently tap on it until it tilts. Remove the chrome alloy ball. Tilt the cage in the opposite direction and remove the ball. Continue to rotate the cage until all six alloy balls have been removed.
4. Pivot the cage and inner race 90 degrees to the center line of the of the outer race. Align the cage windows with the outer race lands. Lift the cage and the inner race out of the CV-joint.
5. Remove the inner race from the cage by rotating the race upward.
6. Clean the grease and contaminates with cleaning solvent from the inner/outer races; CV-joint cage and the alloy balls.
7. Remove any rust from the boot mounting area and clean the halfshaft bar.
To assemble:
8. Coat the inner and outer race grooves with grease and align the inner race with the windows of the cage.
9. Insert the inner race to the cage by rotating the race downward.
10. Insert the cage and inner race into the outer race.
11. Install the six alloy balls into the cage by tilting the cage. Repeat this process until all the balls are in place.
12. Pack the CV boot and joint with the grease supplied in the kit.

13. Install or connect the following:
- New boot clamp onto the boot
- CV boot on to the halfshaft bar and position the small end of the boot into the groove on the halfshaft bar. Secure the clamp to the boot with a Seal Clamp tool J 35910. Torque the clamp to 100 ft. lbs. (136 Nm).
- Swage ring over the large diameter of the boot by pinching the ring into an oval shape

➡**Make certain that the retaining ring side of the inner race faces the halfshaft bar before installation.**

14. Slide the joint onto the halfshaft with the retaining snapring inside of the inner race. The race is properly seated when it snaps into place. Pull on the CV-joint to verify full engagement.
15. Install the large diameter of the boot with the large swage ring in place over the outside edge of the joint outer race.
16. Clamp the boot tightly to the outer race with the large swage ring by mounting Split Plate Swage Clamp tool J 36652 in a vise.
17. Position the outboard end of the halfshaft in the bottom of the tool.
18. Align the CV boot, joint and swage ring.
19. Install the top half of the tool and align the swage ring and clamp. Install the bolts to the top of the tool and tighten snugly. Tighten each bolt an additional 180 degrees. Alternate between the bolts until both sides of the top portion of the tool touch the bottom half.
20. Loosen the bolts and remove the split plate swage clamp tool.
21. Install the halfshaft.
22. Road test the vehicle and make certain there are no abnormal noises in the front end.
23. Check and adjust the alignment if necessary.

Axle Shaft, Bearing and Seal

REMOVAL & INSTALLATION

1. Before servicing the vehicle, refer to the Precautions Section.
2. Remove or disconnect the following:
- Rear tire and wheel assembly
- Rear halfshaft
- Differential output shaft oil seal. Do not damage the differential sealing surfaces.

To install:

3. Install or connect the following:
- Left axle shaft oil seal (1), using the J 44809

➡**Inspect the sealing surface of the wheel drive shaft inner tri-pot housing to ensure it is free of corrosion. Use a crocus cloth in order to remove any light corrosion and clean the sealing surface with denatured alcohol, or equivalent.**

4. Lubricate the wheel drive shaft sealing surface of the oil seal with synthetic gear oil.
- Left wheel halfshaft
- Tire and wheel assembly

5. Inspect the differential lubricant level.

Pinion Seal

REMOVAL & INSTALLATION

1. Before servicing the vehicle, refer to the Precautions Section.
2. Drain the differential lubricant.
3. Remove or disconnect the following:
- Torque tube assembly.
- Drive pinion oil seal from the drive pinion housing and discard
- Pinion housing-to-differential bolts
- Drive pinion and housing assembly
- Drive pinion housing shim from the differential
- O-ring seal from the drive pinion housing and discard

To install:

4. Lubricate the O-ring seal with synthetic gear oil.

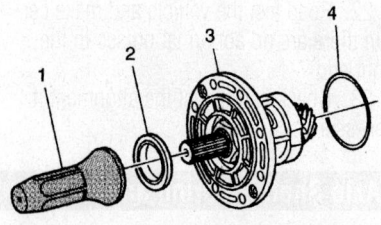

9358KG20

View of the seal installer (1), drive pinion housing shim (2), differential (3) and O-ring (4)

5. Install or connect the following:
- Oil seal to the drive pinion housing, using a seal installation tool
- O-ring seal to the drive pinion housing
- Drive pinion housing assembly and shim to the differential carrier
- Pinion housing-to-differential mounting bolts. Torque the bolts to 21 ft. lbs. (28 Nm).

6. Lubricate the sealing surfaces of the drive pinion oil seal with synthetic gear oil.
- Torque tube assembly.

7. Fill the differential with synthetic gear oil.

Differential Carrier Assembly

REMOVAL & INSTALLATION

1. Before servicing the vehicle, refer to the Precautions Section.
2. Set the parking brake.
3. Drain the rear differential gear oil.
4. Remove or disconnect the following:
- Right rear tire and wheel
- Electrical connector from the clutch pump check valve
- Right rear halfshaft
- Front propeller shaft

5. Place an adjustable support beneath the torque tube.
- Torque tube-to-bracket through bolt and nut; loosen only
- Bolts from the torque tube bracket
- Differential carrier-to-cradle mounting bolts, nuts, washers, and mounts from the differential

➡**During the removal of the wheel drive shaft, the differential output shaft may become disengaged from the differential. If this occurs, firmly grasp and separate the output shaft from the wheel drive shaft. Align the splines on the output shaft to the differential and reposition the output shaft to the differential.**

6. While simultaneously moving the differential assembly to the right side of the vehicle, disengage the left halfshaft from the differential.

- Rear differential and torque tube as an assembly
- Torque tube from the differential

To install:

7. Install or connect the following:
- Torque tube to the differential.
- Rear differential and torque tube assembly to the suspension cradle. Simultaneously guide the left wheel drive shaft onto the differential output shaft while positioning the differential assembly to the suspension cradle.

8. Place an adjustable support under the torque tube. Ensure that the left wheel drive shaft is fully engaged to the differential output shaft.
- Differential carrier mounts, washers, bolts, and nuts to the differential. Torque the bolts to 37 ft. lbs. (50 Nm).
- Torque tube bracket-to-body bolts. Torque the torque tube bracket-to-body bolts to 41 ft. lbs. (55 Nm) and the torque tube-to-bracket through bolt and nut to 47 ft. lbs. (64 Nm).
- Front propeller shaft
- Right rear wheel halfshaft
- Tire and wheel assembly
- Differential drain plug and gasket. Tighten the drain plug to 22 ft. lbs. (30 Nm).

9. Fill the axle with synthetic gear oil. Check the differential oil level to ensure it is even with, to no lower than, 0.25 in. (6mm) below the opening of the fill hole.
- Differential fill plug and gasket. Tighten the fill plug to 22 ft. lbs. (30 Nm).
- Clutch pump check valve connector

10. Remove the adjustable support from the torque tube.

11. Operate the vehicle making tight left, then right turns in order to engage the All-Wheel-Drive (AWD) system and distribute the gear oil throughout the differential.

12. Fill the axle with synthetic gear oil. Check the differential oil level to ensure it is even with, to no lower than, 0.25 in. (6mm) below the opening of the fill hole.

STEERING AND SUSPENSION

Air Bag

❊❊ CAUTION

All models are equipped with a Supplemental Inflatable Restraint (SIR) system. Before attempting any work on or near the steering column, ALWAYS disarm the air bag to prevent a costly and possibly dangerous accidental deployment.

PRECAUTIONS

Several precautions must be observed when handling the inflator module to avoid accidental deployment and possible personal injury.

• Never carry the inflator module by the wires or connector on the underside of the module

• When carrying a live inflator module, hold securely with both hands, and ensure that the bag and trim cover are pointed away from your body

• Place the inflator module on a bench or other surface with the bag and trim cover facing up

• With the inflator module on the bench, never place anything on or close to the module that may be thrown in the event of an accidental deployment

DISARMING

1. Turn the wheels to the straight-ahead position, then turn the ignition switch to **LOCK**.
2. Remove the console accessory wiring junction block access hole cover.
3. Remove the "AIR BAG" or "SIR" fuse from the block, as applicable.
4. Remove the left-hand sound insulator, for access to the SIR wiring harness.
5. Remove the Connector Position Assurance (CPA) device, then disengage the yellow 2-way connector at the base of the steering column.
6. Remove the right hand insulator panel.
7. Remove the CPA from the inflatable restraint instrument panel module connector located behind the RH insulator panel.
8. Disconnect the instrument panel module connector
9. If equipped with side air bags, remove the CPA from the inflatable restraint side impact module—left front connector located under the driver seat.

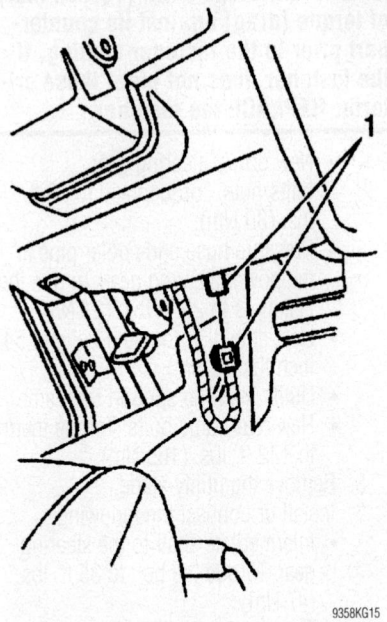

9358KG15

Driver's side air bag connector location

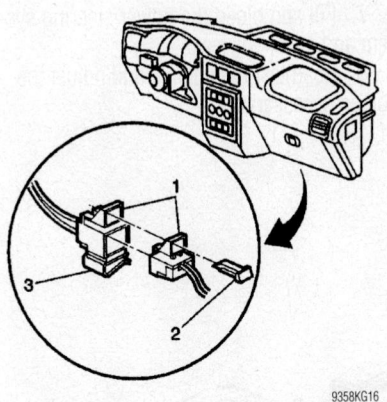

9358KG16

Passenger's side air bag connector location

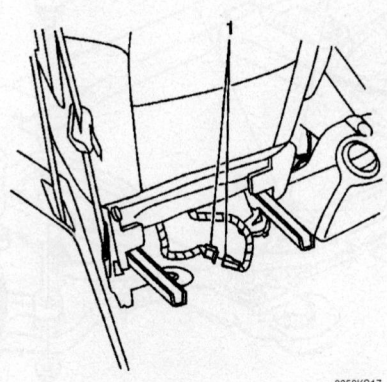

9358KG17

Side air bag connector location

10. Disconnect the side impact module—left front connector.
11. Remove the CPA from the inflatable restraint side impact module—right front connector located under the front passenger seat.
12. Disconnect the side impact module—right front connector.

➡ With the fuse removed, the AIR BAG or SIR light will illuminate if the ignition switch is turned ON at any time. This is normal and does not indicate a problem when the system is disarmed.

To enable:
13. Be sure the ignition is in the **LOCK** position.
14. Connect the inflatable restraint side impact module—right front connector located under the front passenger seat.
15. Install the CPA to the side impact module—right front connector.
16. Connect the inflatable restraint side impact module—left front connector located under the driver seat.
17. Install the CPA to the side impact module—left front connector.
18. Engage the yellow SIR connector, then secure using the CPA device for both the drivers and passenger sides.
19. Install the sound insulator panel.
20. Install the SIR system fuse to the fuse block.
21. Turn the ignition switch to the **ON** position and verify that the AIR BAG indicator light flashes 7 times, then extinguishes. If it does not go out, troubleshoot the SIR system fault.
22. Install the instrument panel lower extension.

Power Steering Rack and Pinion

REMOVAL & INSTALLATION

1. Before servicing the vehicle, refer to the Precautions Section.
2. Remove or disconnect the following:
   • Negative battery cable
   • Left front wheel
   • Stabilizer shaft
   • Tie rod ends from the steering knuckle
   • Intermediate shaft from the steering gear
3. Support the frame using a suitable utility stand.
   • Frame rear bolts and discard them
   • Power steering gear heat shield

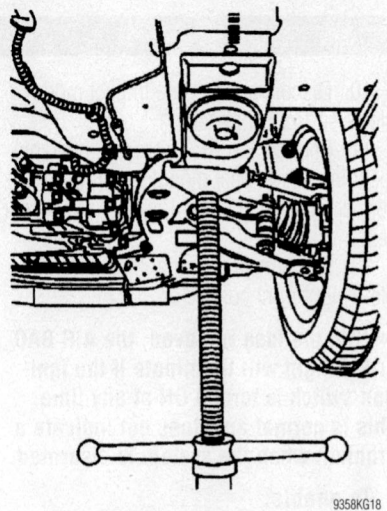

Use a utility stand to support the frame

- Cooler pipe from the power steering gear
- Pressure hose from the power steering gear
- Power steering gear through the left wheel opening

To install:

4. Install or connect the following:
- Power steering gear through the left wheel opening

✷✷ WARNING

This is a prevailing torque type fastener. This fastener may be reused

---

ONLY if: The fastener and its counterpart are clean and free from rust. The fastener develops 2 Nm (18 inch lbs.) of torque (drag) against its counterpart prior to the fastener seating. If the fastener does not meet these criteria, REPLACE the fastener.

- New power steering gear bolts/nuts. Torque them to 59 ft. lbs. (80 Nm).
- Pressure hose and cooler pipe to the power steering gear. Torque the fasteners to 20 ft. lbs. (27 Nm).
- Heat shield. Torque the bolts to 54 inch lbs. (6 Nm).
- Utility stand to support the frame
- New rear frame bolts. Torque them to 122 ft. lbs. (165 Nm).

5. Remove the utility stand.
6. Install or connect the following:
- Intermediate shaft to the steering gear. Torque the bolt to 35 ft. lbs. (47 Nm).
- Tie rod ends to the steering knuckle
- Stabilizer shaft. Torque the bolt to 17 ft. lbs. (23 Nm).
- Left front wheel

7. Fill and bleed the power steering system and check for leaks.
8. Road test the vehicle and adjust the toe as necessary.

Strut

REMOVAL & INSTALLATION

✷✷ CAUTION

Do not remove the top center nut from the strut assembly. This nut should only be removed when the strut assembly is out of the vehicle, mounted in a holding fixture and the coil spring is in a compressed position using the proper coil spring compressor.

1. Before servicing the vehicle, refer to the Precautions Section.
2. Remove or disconnect the following:
- Wiper module
- Three upper strut nuts

➡**Use a frame contact lift to raise the vehicle. DO NOT use a suspension contact lift.**

- Tire and wheel
- Lower strut bolts and nuts after marking the position of the strut to the knuckle

➡**The strut to steering knuckle position must be marked so that the camber angle will not change. If the angle is change, the wheel alignment will also be affected.**

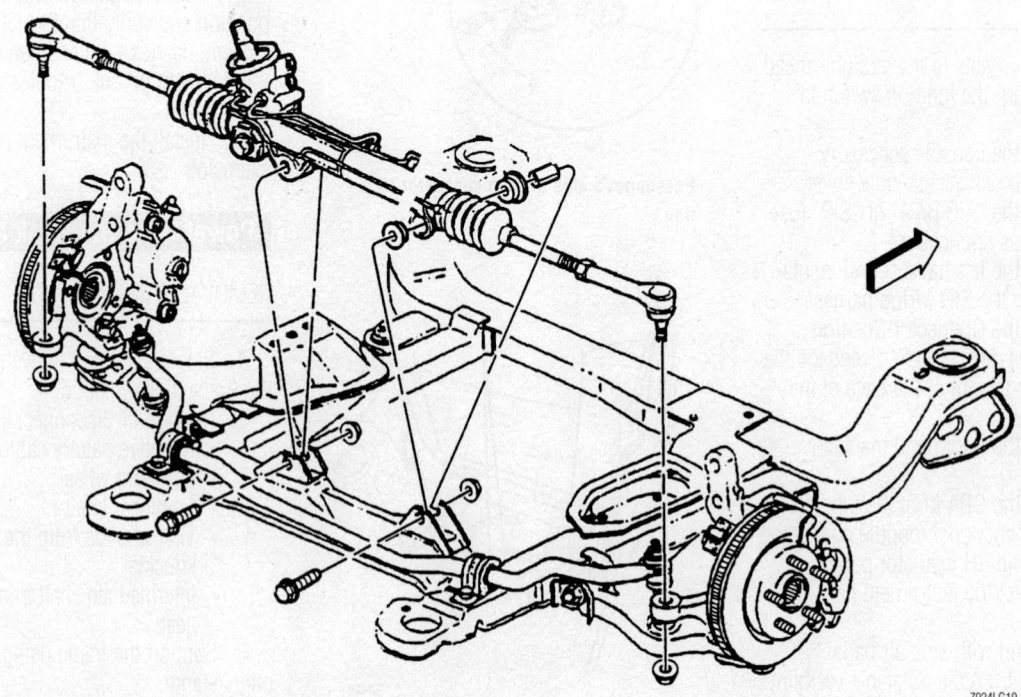

The rack and pinion steering gear is bolted to the rear of the subframe, as shown

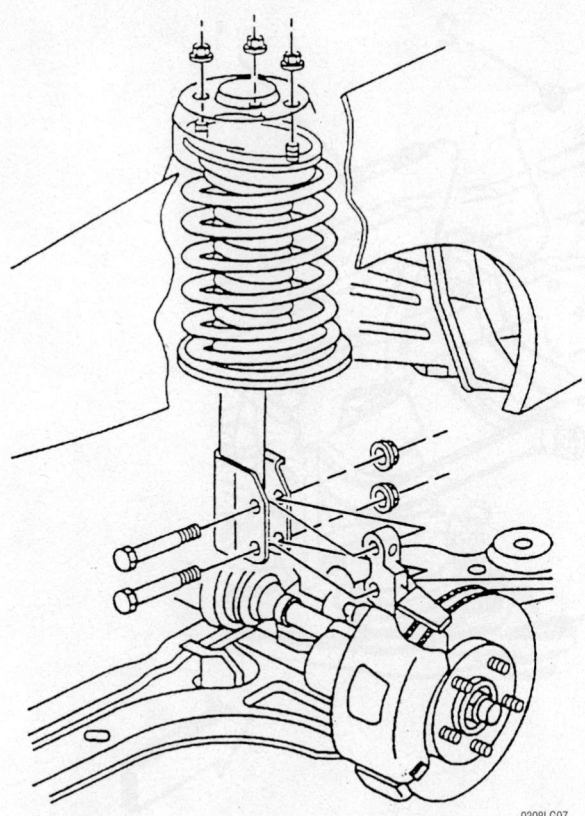

Strut assembly mounting

9308LG07

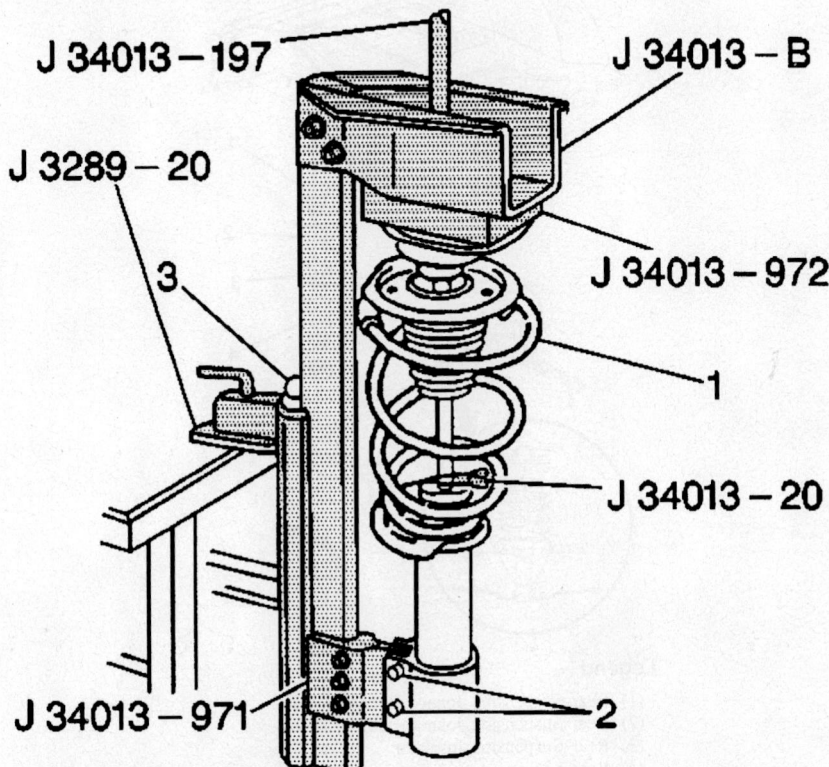

9358LG07

Disassembling the strut assembly

- Strut
- Nut from the top of the strut by placing the assembly in a Strut Compressor tool J 34013-B and Damper Rod Clamp J 34013-20. Turn the compressor forcing screw until the spring compresses slightly
- Strut mount
- Spring from the strut assembly

To install:

3. Install or connect the following:
- Spring over the strut in the proper position
- Strut mount
- Compressor screw and start turning the screw clockwise until the strut shaft threads are visible through the top of the strut. Torque the nut to 63 ft. lbs. (85 Nm).
- Strut and the upper nuts. Torque the nuts to 30 ft. lbs. (41 Nm) for 2002–04; 22 ft. lbs. (30 Nm) for 2005.
- Wiper module
- Lower strut bolts and nuts by aligning the strut to the steering knuckle. Torque the nuts to 90 ft. lbs. (123 Nm).
- Front wheel

4. Road test the vehicle and check the front end alignment and adjust as needed.

Shock Absorber

REMOVAL & INSTALLATION

1. Before servicing the vehicle, refer to the Precautions Section.

2. Raise and support the vehicle.

3. Support the rear axle using a utility stand to slightly compress the coil spring and relieve the shock absorber tension.

4. If the vehicle is equipped with automatic level control, disconnect the air tube connector from the shock absorber.

5. Remove or disconnect the following:
- Shock absorber upper bolt and nut
- Shock absorber lower bolt and/or nut, as applicable
- Shock absorber from the brackets by compressing it slightly

To install:

6. Inspect the shock absorber, upper and lower mounting brackets and the frame mounting hole for cracks excessive wear and burrs.

7. Install or connect the following:
- Shock absorber
- Shock absorber bolts and nuts. Torque the nuts to 63 ft. lbs. (85 Nm) for 2002–04; 66 ft. lbs. (90 Nm) for 2005.

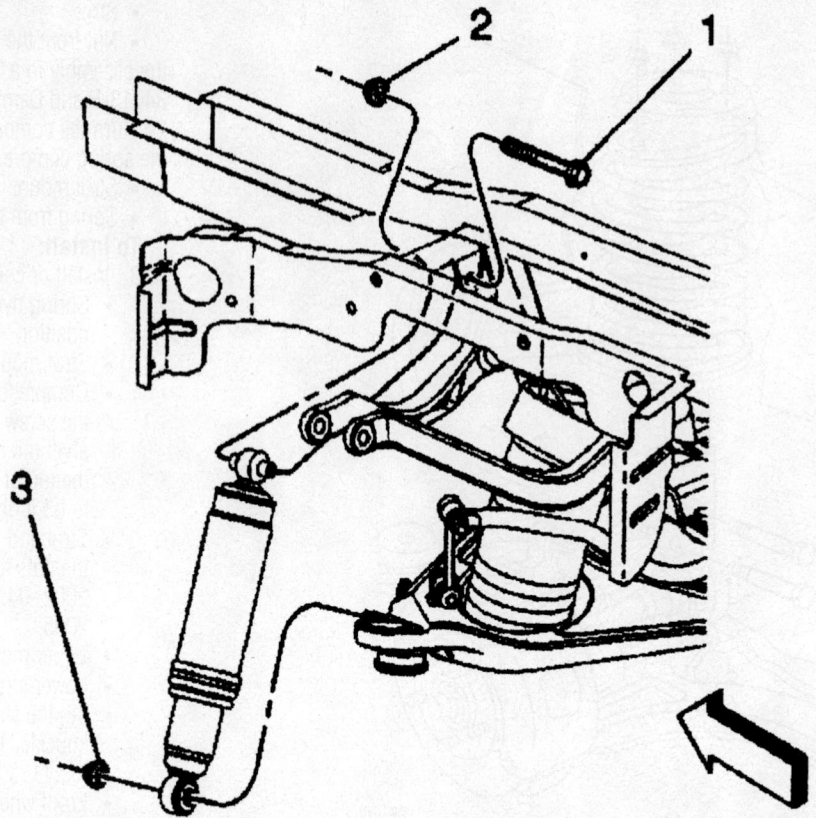

Rear shock absorber mounting—AWD vehicle shown

- Air tube connector to the shock absorber, if equipped with automatic level control
8. Remove the support from the rear axle.
9. Road test the vehicle.

Coil Spring

REMOVAL & INSTALLATION

Front

The service procedure for the front coil springs is covered under MacPherson Strut removal and installation.

Rear

1. Before servicing the vehicle, refer to the Precautions Section.
2. Remove or disconnect the following:
   - Brake hose bracket screw from the control arm
   - Shock absorber lower bolt while using a utility stand to support the rear axle
   - Tie rod from the rear axle
   - Spring and insulators after lowering the rear axle

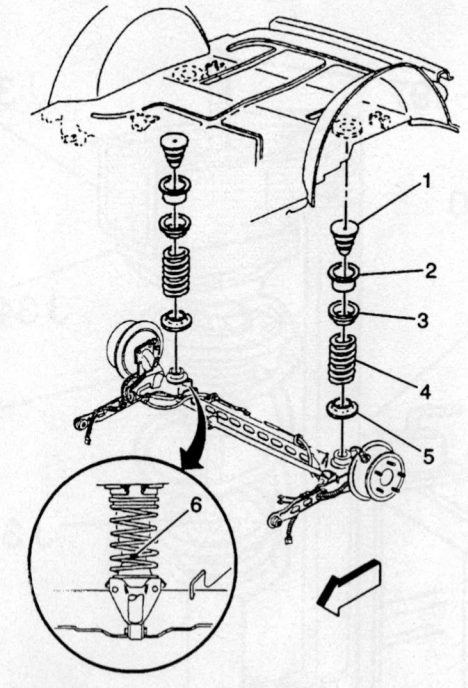

Legend

(1) Rear Suspension Jounce Bumper
(2) Rear Suspension Jounce Bumper Retainer
(3) Rear Suspension Insulator
(4) Rear Spring
(5) Rear Spring Insulator
(6) Paint Stripe

Exploded view of the coil spring assembly

To install:

3. Install or connect the following:
- Insulators and springs on the rear axle with the paint stripe is facing rearward
- Rear axle tie rod after raising the rear axle into its proper position. Torque the nut to 92 ft. lbs. (125 Nm).
- Shock absorbers to the rear axle. Torque the upper and lower nuts to 63 ft. lbs. (85 Nm) for 2002–04; 66 ft. lbs. (90 Nm) for 2005.
- Brake hose bracket to the control arm. Torque the bolt to 33 ft. lbs. (44 Nm).

4. Remove the axle supports.

Lower Ball Joint

REMOVAL & INSTALLATION

1. Before servicing the vehicle, refer to the Precautions Section.
2. Remove the lower control arm.
3. Place the control arm in a vise.
4. Drill or grind off the ball stud rivet heads and use a punch to remove the rivets.
5. Remove the ball joint from the lower control arm

To install:

6. Install the ball joint using new fasteners facing down away from the joint and tighten to 50 ft. lbs. (63 Nm)
7. Install the lower control arm.
8. Road test the vehicle and check the front wheel alignment and adjust if necessary.

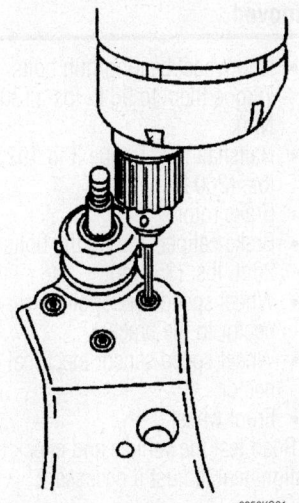

9358KG21

Drill off the rivet heads

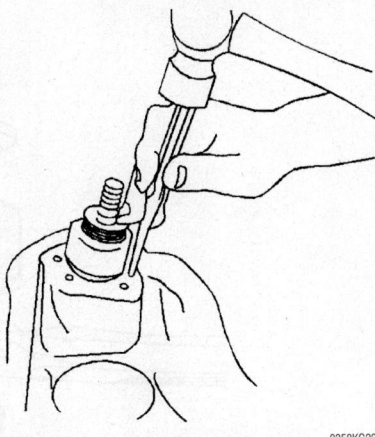

9358KG22

Use a punch and hammer to remove the rivets

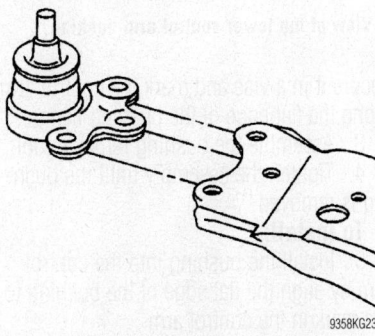

9358KG23

Remove the ball joint from the control arm

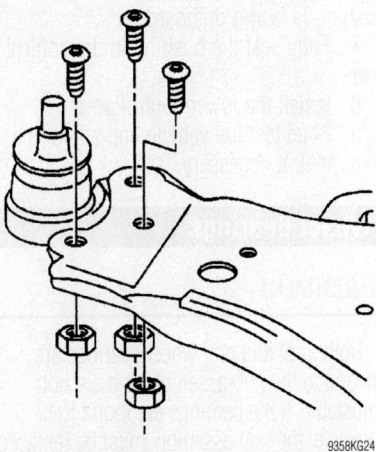

9358KG24

Install the ball joint fasteners facing down away from the joint

Lower Control Arm

REMOVAL & INSTALLATION

2002–04

1. Before servicing the vehicle, refer to the Precautions Section.

2. Remove or disconnect the following:
- Front wheel
- Anti-lock Brake System (ABS) wheel speed sensor connector and jumper harness
- Stabilizer shaft link
- Cotter pin from the ball joint stud and loosen the nut
- Ball Joint from the steering knuckle
- Lower control arm

To install:

3. Install or connect the following:
- Lower control arm
- Ball joint stud to the knuckle

➡Align the ball stud cotter pin hole parallel to the knuckle to ease the pin installation.

- Ball joint stud castle nut. Torque it to 40 ft. lbs. (55 Nm).
- New cotter pin
- Stabilizer shaft link. Torque the nut to 17 ft. lbs. (23 Nm).
- ABS jumper harness to the retainer clips
- ABS sensor connector
- Lower control arm nuts. Torque them to 83 ft. lbs. (113 Nm).
- Front wheel

4. Road test the vehicle and check the front end alignment, adjust if necessary.

2005

✳✳ WARNING

Use only the recommended tools for separating the ball joint from the knuckle. Do NOT hammer or pry the ball joint from the knuckle. Failure to use the recommended tools may cause damage to the ball joint and seal.

➡Use the ignition key in order to unlock the steering column.

1. Turn the steering wheel in order to move the front of the applicable wheel to the outboard most position in order to allow for tool access to the lower control arm ball stud nut.

➡Use only a frame-contact type vehicle lift or a floor jack at the recommended lift points. Do NOT use a suspension-contact type vehicle lift. Do NOT lift the front of the vehicle by the lower control arms.

2. Raise and support the vehicle.
3. Remove the tire and wheel.
4. Disconnect the wheel speed sensor wiring harness from the lower control arm.

5. Remove the stabilizer shaft link.
6. Remove the cotter pin from the ball stud.
7. Loosen the ball stud nut. Do not remove the ball stud nut.
8. Install tool J 41820 over the ball stud nut and the steering knuckle.
9. Rotate the ball stud nut counterclockwise in order to separate the ball stud from the steering knuckle.
10. Remove the lower control arm bolts and nuts.
11. Remove the lower control arm.

To install:
12. Install the lower control arm.
13. Install the control arm bolts and nuts. Hand tighten only.

➡**Align the ball stud cotter pin hole parallel to the knuckle in order to ease the cotter pin installation.**

14. Install the ball stud to the knuckle.
15. Install the ball stud castle nut.
16. Tighten the ball stud castle nut to 30 Nm (22 ft. lbs.) plus 135 degrees.

➡**Do not loosen the ball stud nut in order to align the ball stud nut slots to the ball stud cotter pin hole. If necessary, tighten the ball stud castle nut in order to align the ball stud castle nut slot to the ball stud cotter pin hole.**

➡**Ensure that the cotter pin ends do not contact the ABS sensor connector or the drive axle.**

17. Install a new cotter pin and bend the ends as shown in either example.
18. Install the stabilizer shaft link.
19. Install the wheel speed sensor harness to the lower control arm.

➡**This is a prevailing torque type fastener. This fastener may be reused only if: The fastener and its counterpart are clean and free from rust. The fastener develops 3 Nm (27 inch lbs.) of torque (drag) against its counterpart prior to the fastener seating. If the fastener does not meet these criteria, replace the fastener.**

20. Install the lower control arm nuts.
21. Tighten the lower control arm nuts to 98 Nm (72 ft. lbs.).
22. Install the tire and wheel.
23. Lower the vehicle.

CONTROL ARM BUSHING REPLACEMENT

1. Before servicing the vehicle, refer to the Precautions Section.
2. Remove the lower control arm and

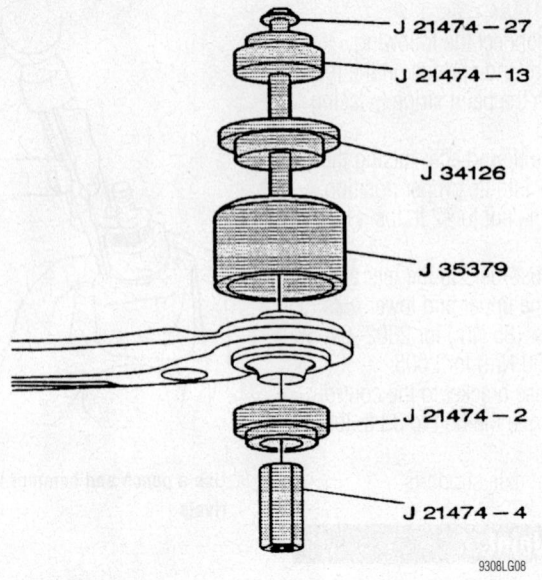

View of the lower control arm bushing

secure it in a vise and mark the control arm along the flat edge of the bushing flange.
3. Assemble the bushing removal tool.
4. Tighten the assembly until the bushing is removed.

To install:
5. Install the bushing into the control arm by align the flat edge of the bushing to the mark in the control arm.
6. Make certain that the flat edge of the bushing is 30 degrees from the centerline of the control arm and the thin slot in the bushing is facing outboard.
7. Fully seat the bushing in the control arm.
8. Install the lower control arm.
9. Road test the vehicle and adjust the alignment, if necessary.

Wheel Bearings

ADJUSTMENT

Both front and rear wheel bearings are integral to the hub assembly and are not adjustable. If the bearings are found to be defective, the hub assembly must be replaced.

REMOVAL & INSTALLATION

Front

1. Before servicing the vehicle, refer to the Precautions Section.
2. Remove or disconnect the following:
   • Front wheel
   • Wheel speed sensor electrical connector and the connector from the bracket

   • Brake caliper and bracket
   • Brake rotor
   • Halfshaft nut
3. Attach a front hub spindle removal tool to the wheel bearing/hub.
4. Push the halfshaft out of the wheel bearing hub assembly.
5. Remove or disconnect the following:
   • Wheel bearing/hub bolts and discard them
   • Wheel bearing/hub assembly

To install:
6. Install or connect the following:
   • Wheel bearing/hub assembly

✳✳ CAUTION

The wheel bearing/hub bolts must be replaced whenever they are loosened or removed.

   • New wheel bearing/hub bolts. Torque them to 96 ft. lbs. (130 Nm).
   • Halfshaft nut. Torque it to 192 ft. lbs. (260 Nm).
   • Brake rotor
   • Brake caliper. Torque the bolts to 26 ft. lbs. (35 Nm).
   • Wheel speed sensor electrical connector to the bracket
   • Wheel speed sensor electrical connector
   • Front wheel
7. Road test the vehicle and check the front alignment, adjust if necessary.

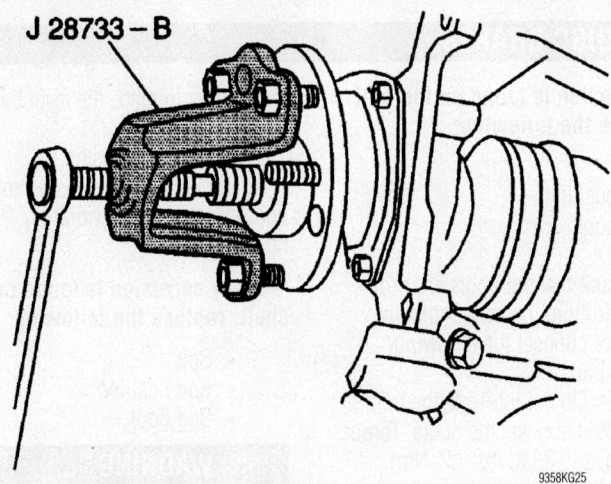

J 28733 – B

9358KG25

Use a pullet to separate the hub from the halfshaft

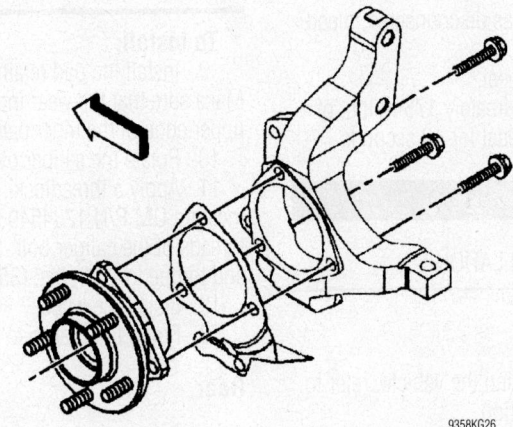

9358KG26

Front hub and bearing assembly mounting

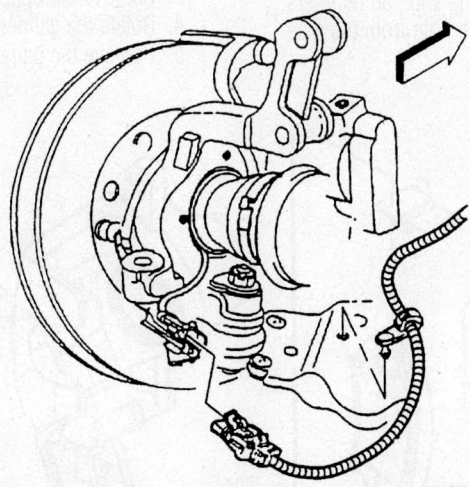

9308LG09

View of the front steering assembly

Rear

FRONT WHEEL DRIVE

1. Before servicing the vehicle, refer to the Precautions Section.

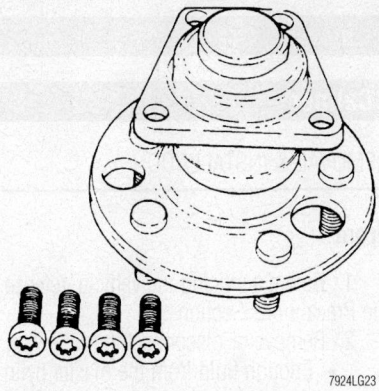

7924LG23

The rear wheel hub is mounted with 4 Torx head bolts

To install:

3. Install or connect the following:
- Bearing/hub to the axle beam. Torque the bolts to 63 ft. lbs. (85 Nm) for drum brakes; 96 ft. lbs. (130 Nm) with disc brakes.
- Wheel speed sensor electrical connector to the bearing/hub assembly
- Brake drum
- Rear wheel

ALL WHEEL DRIVE

1. Before servicing the vehicle, refer to the Precautions Section.

2. Remove or disconnect the following:
- Rear wheel
- Brake caliper and support it with a piece of wire
- Brake caliper bracket
- Rotor
- Wheel speed sensor connector from the sensor and bracket
- Halfshaft from the hub and bearing
- Bearing/hub bolts and bearing/hub

To install:

3. Install or connect the following:
- Bearing/hub. Tighten the bolts to 63 ft. lbs. (85 Nm) for drum brakes; 96 ft. lbs. (130 Nm) with disc brakes.
- Halfshaft
- Wheel speed sensor connector to the bracket and sensor
- Rotor
- Caliper bracket
- Caliper
- Tire and wheel

2. Remove or disconnect the following:
- Rear wheel
- Brake drum
- Wheel speed sensor
- Bearing/hub assembly

DISC BRAKES

Caliper

REMOVAL & INSTALLATION

Front

1. Before servicing the vehicle, refer to the Precautions Section.
2. Remove or disconnect the following:
   - Enough fluid from the master cylinder to place the level between Full and Minimum
   - Wheel
3. If the caliper is being repaired or replaced, remove the brake hose, cap the end, and discard the washers.
4. Remove the caliper bolts and lift the caliper off the bracket.
5. Brake pads

➡**If any corrosion is found on the bolt shaft, replace the following:**

   - Bolt
   - Bolt bushing
   - Bolt boot

To install:

6. Make sure that the boots are properly installed. Bottom the caliper piston.
7. Install or connect the following:
   - Brake pads
   - Caliper. Do not lubricate the bolt threads. Lubricate the boots. Torque the bolts to 26 ft. lbs. (35 Nm).
   - Brake hose, using new washers. Torque the bolt to 40 ft. lbs. (54 Nm).
8. If the hose was disconnected, bleed the system.
9. Install the wheel.
10. Apply approximately 175 ft. lbs. of force to the brake pedal for 10 seconds.

Rear

1. Before servicing the vehicle, refer to the Precautions Section.
2. Remove or disconnect the following:
   - Enough fluid from the master cylinder to place the level between Full and Minimum
   - Wheel
3. If the caliper is being repaired or replaced, remove the brake hose, cap the end, and discard the washers.
4. Remove the caliper bolts and lift the caliper off the bracket.
5. Brake pads

➡**If any corrosion is found on the bolt shaft, replace the following:**

   - Bolt
   - Bolt bushing
   - Bolt boot

To install:

6. Make sure that the boots are properly installed. Bottom the caliper piston.
7. Install or connect the following:
   - Brake pads
   - Caliper. Do not lubricate the bolt threads. Lubricate the boots. Torque the bolts to 33 ft. lbs. (45 Nm).
   - Brake hose, using new washers. Torque the bolt to 40 ft. lbs. (54 Nm).
8. If the hose was disconnected, bleed the system.
9. Install the wheel.
10. Apply approximately 175 ft. lbs. of force to the brake pedal for 10 seconds.

Brake Pads

REMOVAL & INSTALLATION

Front

1. Before servicing the vehicle, refer to the Precautions Section.
2. Remove the wheel.
3. Remove the lower caliper bolt.
4. Rotate the caliper up.
5. Remove the pads and pad retainers.
6. Remove enough fluid from the master cylinder to place the level between Full and Minimum
7. Bottom the piston.
8. Inspect the bolt boots and piston boot for tears or deterioration. Replace as necessary.

➡**If any corrosion is found on the bolt shaft, replace the following:**

   - Bolt
   - Bolt bushing
   - Bolt boot

❉❉ WARNING

Never attempt to polish away the corrosion.

To install:

9. Install the pad retainers and pads. Make sure that the wear indicator is at the upper edge of the inner pad.
10. Rotate the caliper over the pads.
11. Apply a threadlocking compound meeting GM P/N 12345493 specs to the threads of the caliper bolt. Install the bolt and torque to 26 ft. lbs. (35 Nm).
12. Install the wheel.
13. Pump the brakes to seat the pads.

Rear

1. Before servicing the vehicle, refer to the Precautions Section.
2. Remove the wheel.
3. Remove the upper caliper bolt.
4. Rotate the caliper down.
5. Remove the pads and pad retainers.

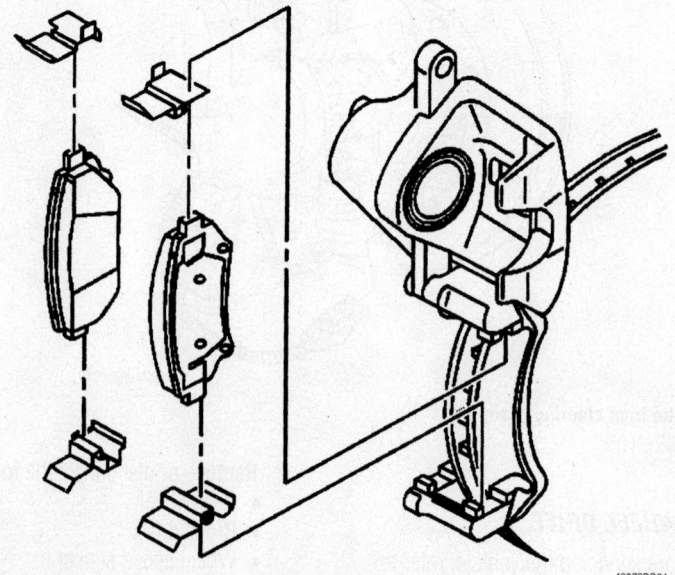

Front pads and retainers

42372BG01

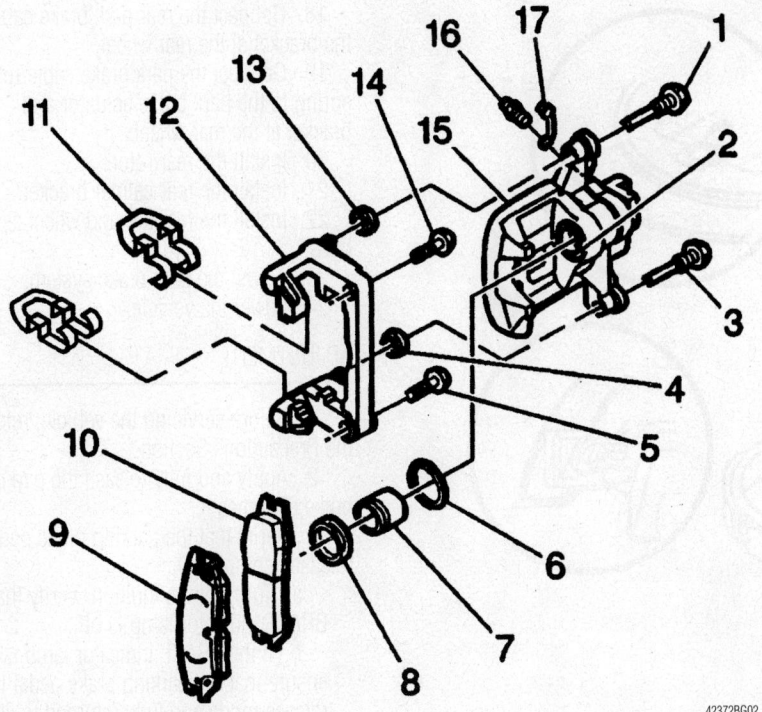

Rear disc brake parts

42372BG02

5. Remove the rear caliper bracket.
6. Remove the rear rotor.
7. Disconnect and remove the rear park brake cable from the bracket at the rear wheel.
8. Disconnect the park brake cable return spring from the park brake actuator and bracket at the rear wheel.
9. Remove the rear hub. The rear hub, backing plate, park brake cable bracket, and park brake actuator will be removed as an assembly.
10. Remove the 2 retainers and the park brake cable bracket from the park brake actuator. Position the rear hub aside.
11. Remove the park brake shoe and actuator from the backing plate.
12. Separate the park brake shoe from the actuator.

To install:
13. Assemble the park brake shoe to the actuator.
14. Install the park brake shoe and actuator onto the backing plate.
15. Position the park brake shoe, actuator, and backing plate over the rear hub.
16. Install the park brake cable bracket and the two retainers.
17. Install the rear hub. The rear hub, backing plate, park brake cable bracket, and park brake actuator will be installed as an assembly.

6. Remove enough fluid from the master cylinder to place the level between Full and Minimum
7. Bottom the piston.
8. Inspect the bolt boots and piston boot for tears or deterioration. Replace as necessary.

➡️If any corrosion is found on the bolt shaft, replace the following:

- Bolt
- Bolt bushing
- Bolt boot

❄❄ **WARNING**

Never attempt to polish away the corrosion.

To install:
9. Install the pad retainers and pads. Make sure that the wear indicator is at the downward edge of the outer pad.
10. Rotate the caliper over the pads.
11. Install the bolt and torque to 33 ft. lbs. (45 Nm).
12. Install the wheel.
13. Pump the brakes to seat the pads.

Parking Brake Shoes

REMOVAL & INSTALLATION

1. Before servicing the vehicle, refer to the Precautions Section.

2. Raise and support the vehicle.
3. Remove the rear tire and wheel assemblies.
4. Relieve the park brake system tension at the equalizer assembly.

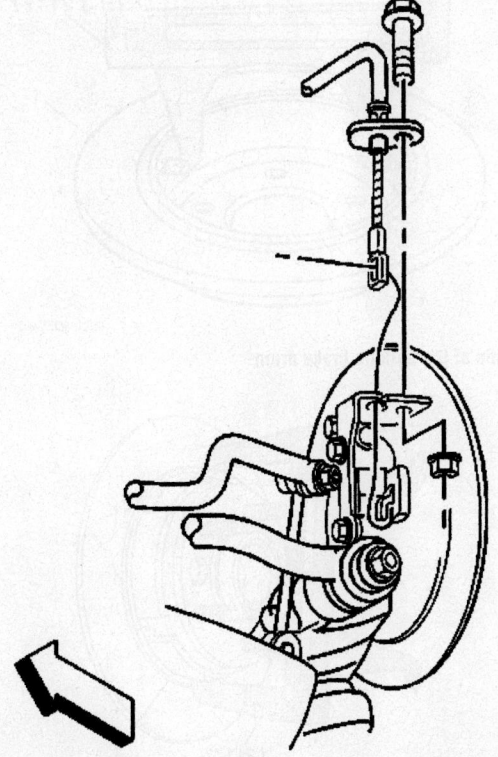

06025-AZTEK-G03

Parking brake cable attachment—rear disc brakes

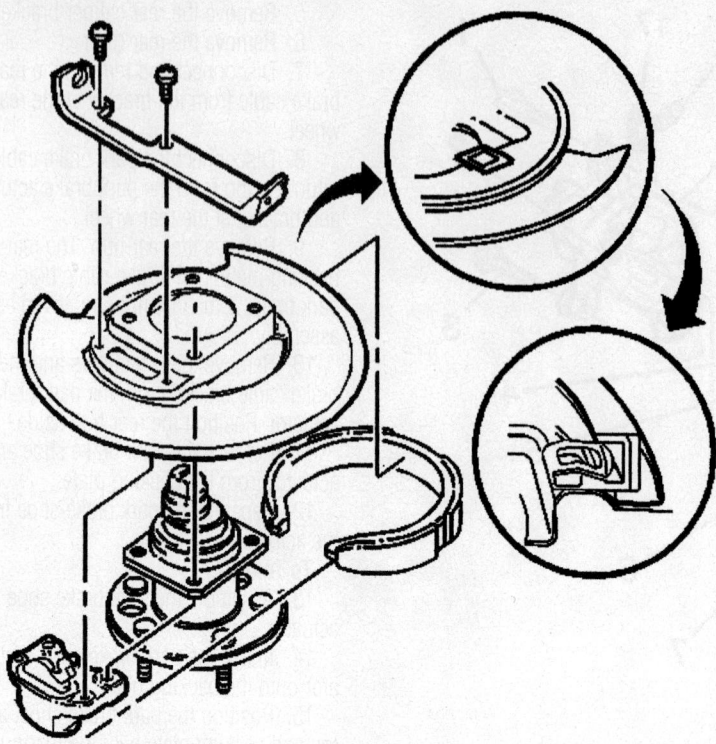

06025-AZTEK-G02

Parking brake shoes and related parts—rear disc brakes

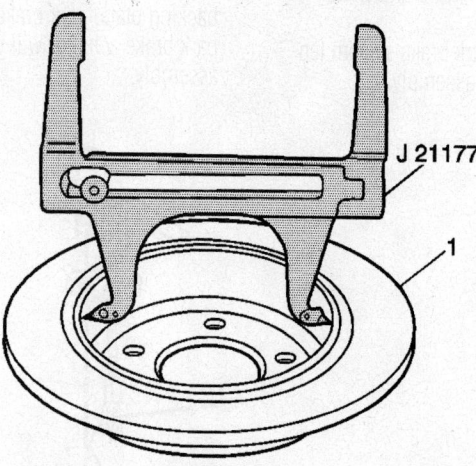

06025-AZTEK-G05

Measuring the inside of the parking brake drum

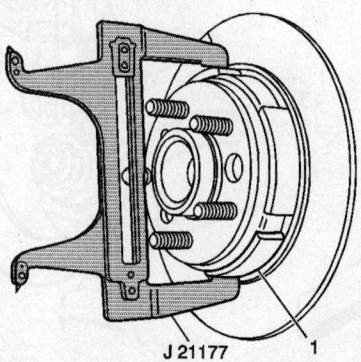

06025-AZTEK-G04

Measuring the parking brake shoes

18. Connect the rear park brake cable to the bracket at the rear wheel.

19. Connect the park brake cable return spring to the park brake actuator and bracket at the rear wheel.

20. Install the rear rotor.

21. Install the rear caliper bracket.

22. Install the rear tire and wheel assemblies.

23. Adjust the park brake system.

24. Lower the vehicle.

ADJUSTMENT

1. Before servicing the vehicle, refer to the Precautions Section.

2. Apply and fully release the parking brake six times.

3. Verify that the parking brake pedal releases completely.

   a. Turn ON the ignition. Verify that the BRAKE indicator lamp is off.

   b. If the BRAKE indicator lamp is on, ensure that the parking brake pedal is in release mode and fully returned to stop.

   c. Remove the slack in the front parking brake cable by pulling downward on the cable.

4. Raise and suitably support the vehicle.

5. Remove the rear wheels and tires.

6. Relieve tension on the park brake system at the park brake equalizer.

7. Remove both rear caliper bracket bolts.

✳✳ WARNING

Support the brake caliper with heavy mechanic's wire, or equivalent, whenever it is separated from its mount and the hydraulic flexible brake hose is still connected. Failure to support the caliper in this manner will cause the flexible brake hose to bear the weight of the caliper, which may cause damage to the brake hose and in turn may cause a brake fluid leak.

8. Remove brake caliper and bracket as one assembly. Do not disconnect the brake hose from the caliper.

9. Support the assembly with mechanic's wire, or equivalent.

10. Remove rear brake rotor.

11. Set the J 21177-A inside of the park brake drum at the widest point. Tighten the set screw on the tool to ensure the proper measurement when removing the tool from the drum.

12. Position the J 21177-A over the park brake shoe at the widest point.

13. Turn the adjuster on the actuator until the park brake shoe just contacts the J 21177-A.

14. Repeat these steps for the opposite side.

15. Install both rear brake rotors.

16. Install both rear brake calipers and brackets.

➡ **To ensure that the proper clamp load will be present when installed. It is imperative that the threads on the caliper bracket bolts, as well as the mounting holes in the knuckle, be cleaned of all debris and inspected before proceeding with installation.**

17. Clean and visually inspect threads of the caliper bracket bolts and mounting holes in the knuckle.

18. Apply THREADLOCKER, GM P/N 12345493 (Canadian P/N 10953488), or equivalent to the threads of the brake caliper bracket bolts.

✳ WARNING

Use the correct fastener in the correct location. Replacement fasteners must be the correct part number for that application. Fasteners requiring replacement or fasteners requiring the use of thread locking compound or sealant are identified in the service procedure. Do not use paints, lubricants, or corrosion inhibitors on fasteners or fastener joint surfaces unless specified. These coatings affect fastener torque and joint clamping force and may damage the fastener. Use the correct tightening sequence and specifications when installing fasteners in order to avoid damage to parts and systems.

19. Install the caliper bracket bolts. Tighten the brake caliper bracket bolts to 130 Nm (96 ft. lbs.).

20. Install the rear wheels and tires.

21. Adjust the parking brake cable by turning the nut at the equalizer while spinning both rear wheels. When either rear wheel starts to drag, back off the nut one full turn.

22. Lower the vehicle to curb height.

23. Apply the parking brake, then inspect for rotation of the rear wheels. If the rear wheels rotate during this inspection, readjust the parking brake cable.

24. Release the parking brake. Verify that the wheels rotate freely.

25. Lower the vehicle.

DRUM BRAKES

Brake Drum

REMOVAL & INSTALLATION

1. Before servicing the vehicle, refer to the Precautions Section.

2. Release the parking brake.

3. Raise and support the vehicle.

4. Remove the tire and wheel.

5. Mark the relationship of the drum to the hub.

6. Remove and discard the retaining clip (if applicable).

7. Remove the brake drum.

8. If the brake drum does not come off easily, perform the following steps:

  a. Loosen the parking brake cable.

  b. Remove the access hole plug from the backing plate.

  c. Insert a flat-bladed tool through the backing plate access hole in order to disengage the self adjuster.

  d. Insert another flat-bladed tool through the same backing plate access hole in order to loosen the adjuster screw.

  e. Install the access hole plug in order to prevent dirt or contamination from entering the drum brake.

  f. Apply a small amount of penetrating oil around the brake drum center hole.

  g. Remove the brake drum.

To install:

➡ **Align the marks on the brake drum and the hub made during the removal procedure.**

9. Install the brake drum.

10. Inspect the brake to shoe adjustment.

11. Install the tire and wheel.

12. Lower the vehicle.

Brake Shoes

REMOVAL & INSTALLATION

1. Before servicing the vehicle, refer to the Precautions Section.

2. Raise and support the vehicle.

3. Remove the brake drum.

➡ **Do not over stretch the adjuster spring. Damage can occur if the spring is over stretched.**

4. Disengage the adjuster spring hook end from the tab on the adjuster actuator.

5. Remove the straight end of the adjuster spring from the brake shoe.

6. Remove the adjuster actuator from the brake shoe.

7. Remove the return spring from the brake shoes.

8. Remove the park brake cable from the park brake actuator lever.

9. Remove the brake shoe hold-down springs and retainers from the brake shoes.

10. Remove the adjuster from the brake shoes and the park brake actuator lever.

11. Remove the horseshoe clip retaining the park brake actuator lever to the brake shoe.

12. Remove the park brake actuator lever and wave washer from the brake shoe.

13. Clean all of the drum brake system components with denatured alcohol.

14. Inspect all of the drum brake system components.

15. Replace drum brake system components as necessary.

16. Inspect the wheel cylinder for the following conditions:

- Brake fluid leakage
- Worn or damaged dust boots

17. Replace damaged or leaking wheel cylinders as necessary.

To install:

18. Apply GM P/N 1052196 (Canadian P/N 5264008) brake lubricant, or equivalent, to the following areas:

- The brake shoe contact points on the backing plate
- The adjuster screw threads
- The inside diameter of the adjuster socket

19. Install the park brake actuator lever and wave washer to the brake shoe.

20. Install the horseshoe clip to the park brake actuator lever pivot pin.

21. Install the brake shoes to the brake backing plate.

22. Install the brake shoe hold-down pins, springs and retainers to the brake shoes.

23. Install the park brake cable to the park brake actuator lever.

➡**Ensure that the adjuster engages the brake shoe and the park brake actuator properly.**

24. Install the adjuster screw to the brake shoe and the park brake actuator.

25. Apply GM P/N 1052196 (Canadian P/N 5264008) brake lubricant or equivalent to the adjuster actuator/brake shoe interface.

26. Install the adjuster actuator to the brake shoe.

➡**Do not over stretch the adjuster spring. Damage can occur if the spring is over stretched.**

27. Install the straight end of the adjuster spring to the brake shoe.

28. Install the adjuster spring hook end to the tab on the adjuster actuator.

29. Install the return spring to the brake shoes.

➡**Ensure that the adjuster operates properly.**

30. Move the park brake actuator lever in order to spread the brake shoes apart. The adjuster actuator lever should move downward, then upward as the park brake actuator lever is released, forcing the adjuster wheel to rotate. If the adjuster does not operate properly, remove then reinstall the adjuster.

31. Adjust the brake shoes.

32. Adjust the park brake cable.

33. Install the brake drum.

34. Lower the vehicle.

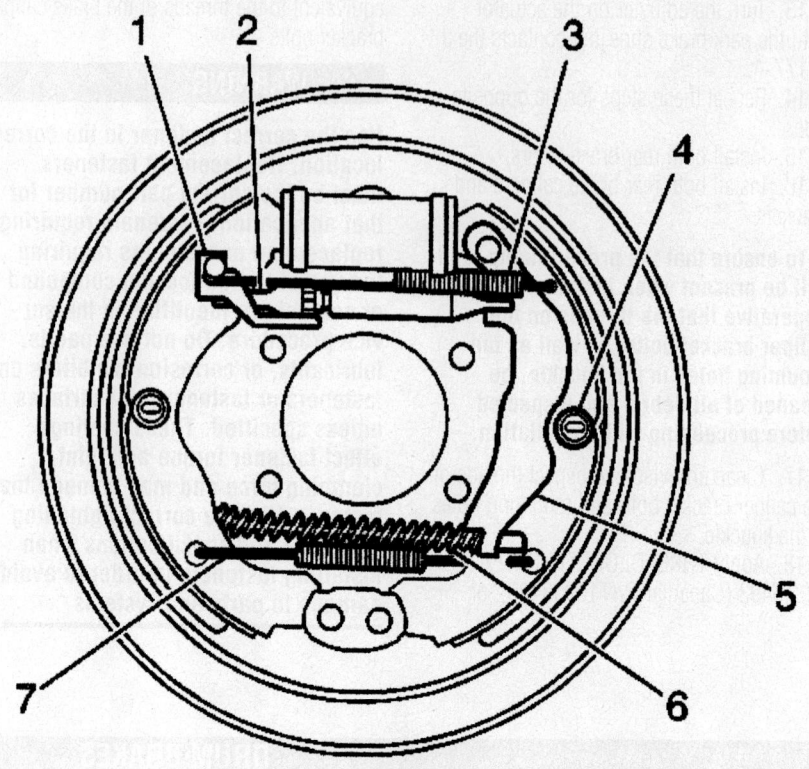

1. Adjuster actuator
2. Adjuster
3. Upper spring
4. Hold-down springs
5. Parking brake lever
6. Parking brake cable
7. Lower spring

06025-AZTEK-G01

Drum brakes

CHEVROLET AND GMC

7

Blazer • Jimmy • S10 • Sonoma

BRAKES **7-56**
DRIVE TRAIN **7-37**
ENGINE REPAIR **7-13**
FUEL SYSTEM **7-33**
**SPECIFICATIONS AND
   MAINTENANCE CHARTS** **7-3**
Engine and Vehicle
   Identification 7-3
General Engine Specifications 7-3
Engine Tune-Up Specifications 7-4
Firing Order 7-4
Accessory Drive Belt Routing 7-5
Capacities 7-6
Valve Specifications.....................7-7
Crankshaft and Connecting
   Rod Specifications 7-7
Piston and Ring Specifications 7-8
Torque Specifications 7-8
Wheel Alignment 7-9
Tire, Wheel and Ball Joint
   Specifications 7-10
Brake Specifications 7-11
Scheduled Maintenance
   Intervals............................7-12
**STEERING AND
   SUSPENSION** **7-43**
A
Air Bag...................................7-43
   Arming (2002) 7-43
   Arming (2003–05) 7-44
   Disarming (2002) 7-43
   Disarming (2003–05) 7-44
   Precautions 7-43
Alternator 7-13
   Installation 7-13
   Removal 7-13
Automatic Transmission 7-37
   Removal & Installation.............7-37
Axle Shaft, Bearing and Seal........7-42
   Removal & Installation.............7-42
B
Ball Joints...............................7-48
   Removal & Installation.............7-48
Brake Caliper 7-56
   Removal & Installation.............7-56

Brake Drums...........................7-57
   Removal & Installation.............7-57
Brake Shoes............................7-57
   Removal & Installation.............7-57
C
Camshaft and Valve Lifters 7-25
   Removal & Installation.............7-25
Clutch 7-38
   Removal & Installation.............7-38
Coil Springs 7-46
   Removal & Installation.............7-46
CV-Joints................................7-40
   Overhaul 7-40
Cylinder Head 7-17
   Removal & Installation.............7-17
D
Disc Brake Pads........................7-57
   Removal & Installation.............7-57
Distributor...............................7-13
   Removal 7-13
E
Engine Assembly 7-13
   Removal & Installation.............7-13
Exhaust Manifold 7-23
   Removal & Installation.............7-23
F
Fuel Filter 7-33
   Removal & Installation.............7-33
Fuel Injector 7-34
   Removal & Installation.............7-34
Fuel Pump 7-34
   Removal & Installation.............7-34
Fuel System Pressure 7-33
   Relieving 7-33
Fuel System Service
   Precautions........................7-33
H
Halfshaft.................................7-39
   Removal & Installation.............7-39
Heater Core.............................7-15
   Removal & Installation.............7-15
Hydraulic Clutch System 7-38
   Bleeding............................7-38
I
Ignition Timing 7-13
   Adjustment........................7-13

Intake Manifold.........................7-21
   Removal & Installation.............7-21
L
Leaf Springs 7-47
   Removal & Installation.............7-47
Lower Control Arm 7-51
   Control Arm Bushing
      Replacement 7-52
   Removal & Installation.............7-51
M
Manual Transmission 7-37
   Removal & Installation.............7-37
O
Oil Pan..................................7-27
   Removal & Installation.............7-27
Oil Pump 7-29
   Removal & Installation.............7-29
P
Pinion Seal 7-42
   Removal & Installation.............7-42
Piston and Ring 7-32
   Positioning 7-32
Power Steering Gear 7-44
   Removal & Installation.............7-44
R
Rear Main Seal 7-29
   Removal & Installation.............7-29
Rocker Arms 7-20
   Removal & Installation.............7-20
S
Shock Absorbers 7-45
   Removal & Installation.............7-45
Stabilizer Bar 7-48
   Removal & Installation.............7-48
Starter Motor 7-26
   Removal & Installation.............7-26
Steering Knuckle........................7-47
   Removal & Installation.............7-47
T
Timing Chain, Sprockets,
   Front Cover and Seal...................7-30
   Removal & Installation.............7-30
Torsion Bar 7-47
   Removal & Installation.............7-47
Transfer Case Assembly................7-39
   Removal & Installation.............7-39

U

Upper Control Arm 7-51
   Control Arm Bushing
     Replacement 7-51
   Removal & Installation 7-51

V

Valve Lash 7-26
   Adjustment 7-26

W

Water Pump 7-15
   Removal & Installation 7-15

Wheel Bearings 7-53
   Adjustment 7-53
   Removal & Installation 7-53

SPECIFICATIONS AND MAINTENANCE CHARTS

ENGINE AND VEHICLE IDENTIFICATION

	Engine						Model Year	
Code ①	Liters (cc)	Cu. In.	Cyl.	Fuel Sys.	Engine Type	Eng. Mfg.	Code ②	Year
4	2.2 (2189)	134	4	MFI	OHV	CPC	2	2002
W	4.3 (4293)	263	6	MFI	OHV	CPC	3	2003
							4	2004
							5	2005
							6	2006

CPC: Chevrolet/Pontiac/Canada

MFI: Multi-port Fuel Injection

① 8th position of VIN

② 10th position of VIN

06025-S10P-C01

GENERAL ENGINE SPECIFICATIONS

All measurements are given in inches.

Year	Model	Engine Displacement Liters	Engine Series (ID/VIN)	Net Horsepower @ rpm	Net Torque @ rpm (ft. lbs.)	Bore x Stroke (in.)	Compression Ratio	Oil Pressure @ rpm
2002	Blazer	2.2	4	120@5000	140@3600	3.50x3.46	9.0:1	56@3000
		4.3	W	230@4600	285@2800	4.00x3.48	9.4:1	18@2000
	Jimmy	2.2	4	120@5000	140@3600	3.50x3.46	9.0:1	56@3000
		4.3	W	230@4600	285@2800	4.00x3.48	9.4:1	18@2000
	S10	2.2	4	120@5000	140@3600	3.50x3.46	9.0:1	56@3000
		4.3	W	230@4600	285@2800	4.00x3.48	9.4:1	18@2000
	Sonoma	2.2	4	118@5200	130@2800	3.50x3.46	9.0:1	56@3000
		4.3	W	230@4600	285@2800	4.00x3.48	9.4:1	18@2000
	Xtreme	2.2	4	120@5000	140@3600	3.50x3.46	9.0:1	56@3000
		4.3	W	230@4600	285@2800	4.00x3.48	9.4:1	18@2000
2003	Blazer	2.2	4	120@5000	140@3600	3.50x3.46	9.0:1	56@3000
		4.3	W	230@4600	285@2800	4.00x3.48	9.4:1	18@2000
	Jimmy	2.2	4	120@5000	140@3600	3.50x3.46	9.0:1	56@3000
		4.3	W	230@4600	285@2800	4.00x3.48	9.4:1	18@2000
	S10	2.2	4	120@5000	140@3600	3.50x3.46	9.0:1	56@3000
		4.3	W	230@4600	285@2800	4.00x3.48	9.4:1	18@2000
	Sonoma	2.2	4	118@5200	130@2800	3.50x3.46	9.0:1	56@3000
		4.3	W	230@4600	285@2800	4.00x3.48	9.4:1	18@2000
	Xtreme	2.2	4	120@5000	140@3600	3.50x3.46	9.0:1	56@3000
		4.3	W	230@4600	285@2800	4.00x3.48	9.4:1	18@2000
2004	Blazer	4.3	W	190@4400	250@2800	4.00x3.48	9.4:1	18@2000
	Jimmy	4.3	W	230@4600	285@2800	4.00x3.48	9.4:1	18@2000
	S10	4.3	W	230@4600	285@2800	4.00x3.48	9.4:1	18@2000
	Sonoma	4.3	W	230@4600	285@2800	4.00x3.48	9.4:1	18@2000
	Xtreme	4.3	W	190@4400	250@2800	4.00x3.48	9.2:1	18@2000
2005	Blazer	4.3	W	190@4400	250@2800	4.00x3.48	9.4:1	18@2000

06025-S10P-C02

GASOLINE ENGINE TUNE-UP SPECIFICATIONS

Year	Engine Displacement Liters	Engine ID/VIN	Spark Plugs Gap (in.)	Ignition Timing (deg.) MT	Ignition Timing (deg.) AT	Fuel Pump (psi)	Idle Speed (rpm) MT	Idle Speed (rpm) AT	Valve Clearance In.	Valve Clearance Ex.
2002	2.2	4	0.040	①	①	41-47	②	②	HYD	HYD
	4.3	W	0.060	①	①	58-64 ③	600	625	HYD	HYD
2003	2.2	4	0.040	①	①	41-47	②	②	HYD	HYD
	4.3	W	0.060	①	①	58-64 ③	600	625	HYD	HYD
2004	4.3	W	0.060	①	①	58-64 ③	600	625	HYD	HYD
2005	4.3	W	0.060	①	①	58-64 ③	600	625	HYD	HYD

NOTE: The Vehicle Emission Control Information label often reflects specification changes made during production.

The label figures must be used if they differ from those in this chart.

HYD: Hydraulic

① Ignition timing is preset and cannot be adjusted

② Idle speed is maintained by the PCM

③ With key ON and engine OFF

06025-S10P-C03

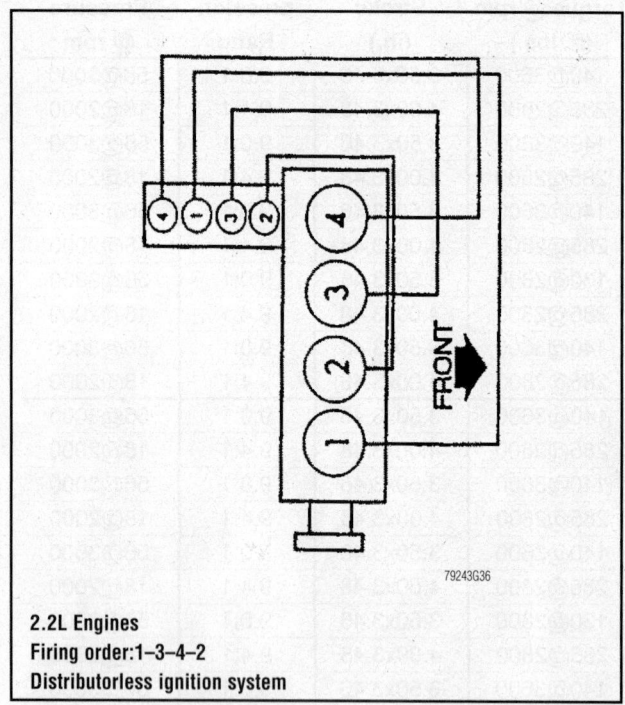

2.2L Engines
Firing order: 1–3–4–2
Distributorless ignition system

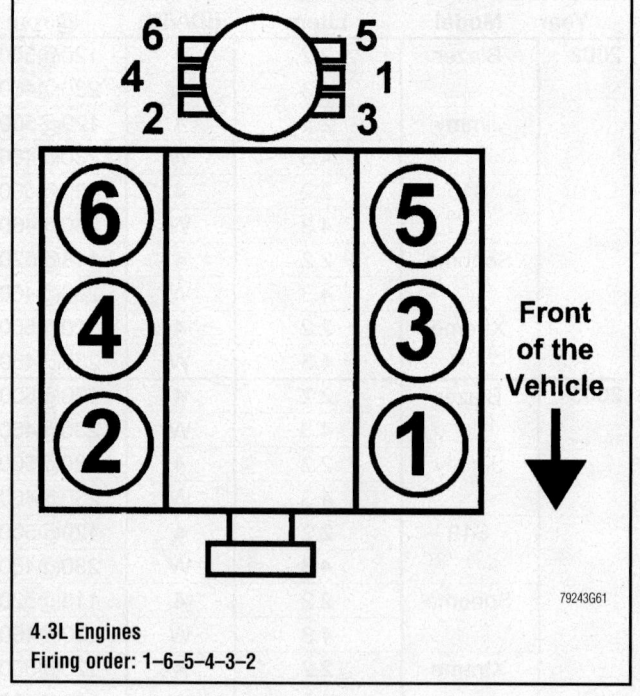

Front of the Vehicle

4.3L Engines
Firing order: 1–6–5–4–3–2

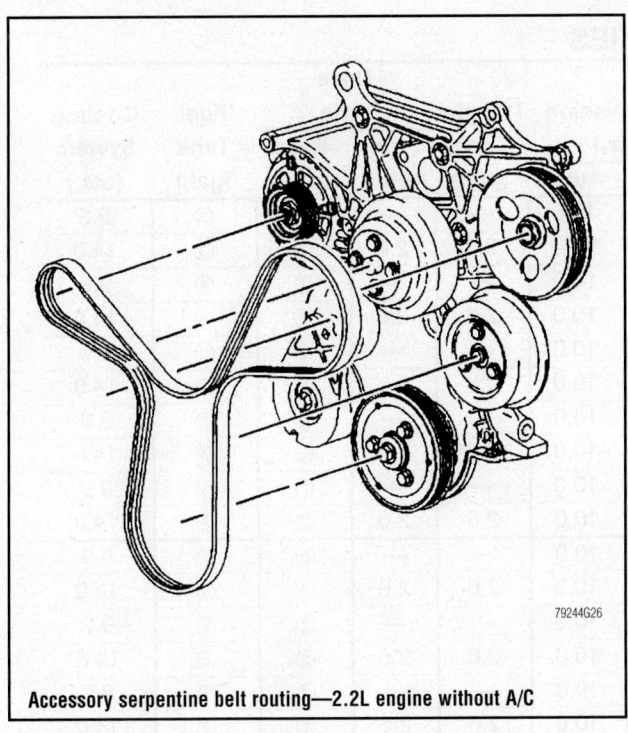

Accessory serpentine belt routing—2.2L engine without A/C

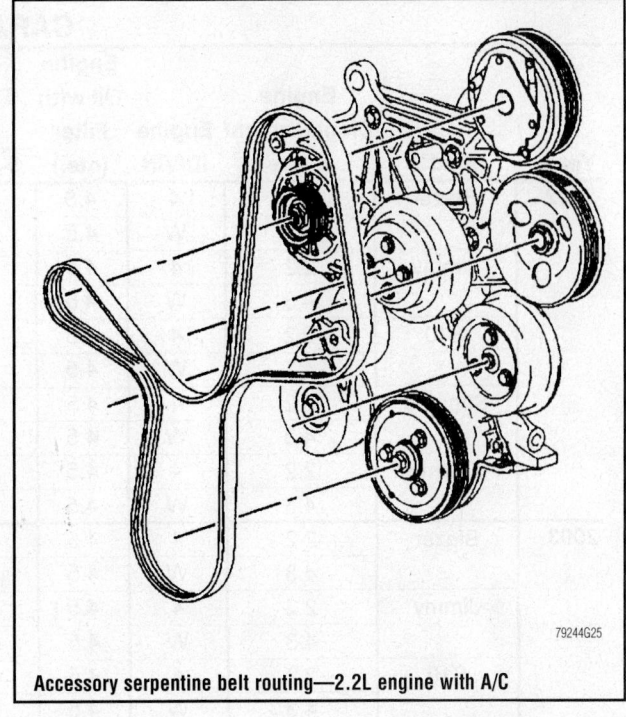

Accessory serpentine belt routing—2.2L engine with A/C

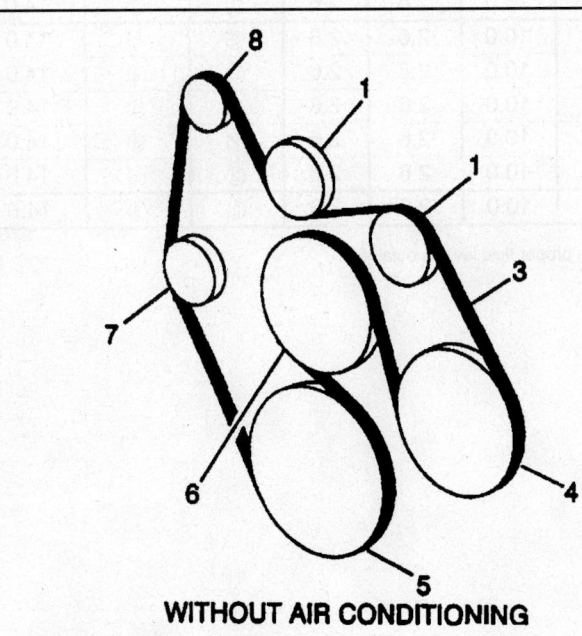

WITHOUT AIR CONDITIONING

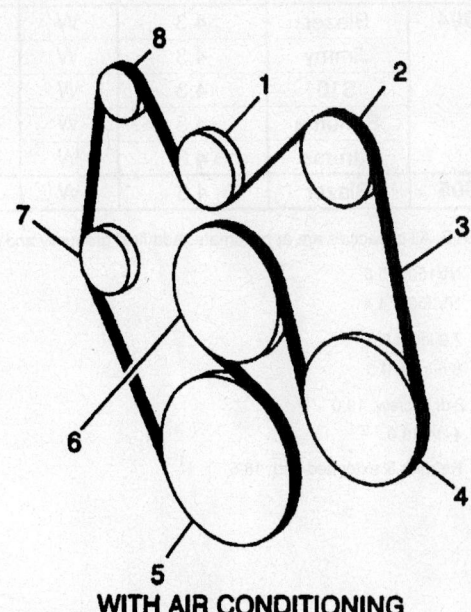

WITH AIR CONDITIONING

1. Pulley, Idler
2. Pulley, AC Compressor
3. Belt, Drive
4. Pulley. Power Steering Pump

5. Pulley, Crankshaft
6. Pulley, Water Pump
7. Pulley, Drive Belt Tensioner
8. Pulley, Generator

Accessory serpentine belt routing—4.3L engine

CAPACITIES

Year	Model	Engine Displacement Liters	Engine ID/VIN	Engine Oil with Filter (qts.)	Transmission (pts.) 5-Spd	Transmission (pts.) Auto.	Transfer Case (pts.)	Drive Axle Front (pts.)	Drive Axle Rear (pts.)	Fuel Tank (gal.)	Cooling System (qts.)
2002	Blazer	2.2	4	4.5	①	10.0	—	—	②	③	9.9
		4.3	W	4.5	①	10.0	2.6	2.6	②	③	14.0
	Jimmy	2.2	4	4.5	①	10.0	—	—	②	③	9.9
		4.3	W	4.5	①	10.0	2.6	2.6	②	③	14.0
	S10	2.2	4	4.5	①	10.0	—	—	②	③	9.9
		4.3	W	4.5	①	10.0	2.6	2.6	②	③	14.0
	Sonoma	2.2	4	4.5	①	10.0	—	—	②	③	9.9
		4.3	W	4.5	①	10.0	2.6	2.6	②	③	14.0
	Xtreme	2.2	4	4.5	①	10.0	—	—	②	③	9.9
		4.3	W	4.5	①	10.0	2.6	2.6	②	③	14.0
2003	Blazer	2.2	4	4.5	①	10.0	—	—	②	③	9.9
		4.3	W	4.5	①	10.0	2.6	2.6	②	③	14.0
	Jimmy	2.2	4	4.5	①	10.0	—	—	②	③	9.9
		4.3	W	4.5	①	10.0	2.6	2.6	②	③	14.0
	S10	2.2	4	4.5	①	10.0	—	—	②	③	9.9
		4.3	W	4.5	①	10.0	2.6	2.6	②	③	14.0
	Sonoma	2.2	4	4.5	①	10.0	—	—	②	③	9.9
		4.3	W	4.5	①	10.0	2.6	2.6	②	③	14.0
	Xtreme	2.2	4	4.5	①	10.0	—	—	②	③	9.9
		4.3	W	4.5	①	10.0	2.6	2.6	②	③	14.0
2004	Blazer	4.3	W	4.5	①	10.0	2.6	2.6	②	③	14.0
	Jimmy	4.3	W	4.5	①	10.0	2.6	2.6	②	③	14.0
	S10	4.3	W	4.5	①	10.0	2.6	2.6	②④	③	14.0
	Sonoma	4.3	W	4.5	①	10.0	2.6	2.6	②	③	14.0
	Xtreme	4.3	W	4.5	①	10.0	2.6	2.6	②	③	14.0
2005	Blazer	4.3	W	4.5	①	10.0	2.6	2.6	②	③	14.0

NOTE: All capacities are approximate. Add fluid gradually and check to be sure a proper fluid level is obtained.

① NV1500: 5.8
   NV3500: 4.4
② 7.6 inch: 3.6
   8.6 inch: 4.0
③ 2-dr & crew: 19.0
   4-dr: 18.0
   Regular & extended cab: 18.5

06025-S10P-C04

VALVE SPECIFICATIONS

Year	Engine Displacement Liters	Engine ID/VIN	Seat Angle (deg.)	Face Angle (deg.)	Spring Test Pressure (lbs. @ in.)	Spring Installed Height (in.)	Stem-to-Guide Clearance (in.)		Stem Diameter (in.)	
							Intake	Exhaust	Intake	Exhaust
2002	2.2	4	46	45	201-215@1.18	1.71	0.0007-0.0020	0.0014-0.0030	NA	NA
	4.3	W	46	45	187-203@1.27	1.69-1.71	0.0010-0.0027	0.0010-0.0027	NA	NA
2003	2.2	4	46	45	201-215@1.18	1.71	0.0007-0.0020	0.0014-0.0030	NA	NA
	4.3	W	46	45	187-203@1.27	1.69-1.71	0.0010-0.0027	0.0010-0.0027	NA	NA
2004	4.3	W	46	45	187-203@1.27	1.69-1.71	0.0010-0.0027	0.0010-0.0027	NA	NA
2005	4.3	W	46	45	187-203@1.27	1.69-1.71	0.0010-0.0027	0.0010-0.0027	NA	NA

NA: Not Available

06025-S10P-C05

CRANKSHAFT AND CONNECTING ROD SPECIFICATIONS

All measurements are given in inches.

Year	Engine Displacement Liters	Engine ID/VIN	Crankshaft				Connecting Rod		
			Main Brg. Journal Dia.	Main Brg. Oil Clearance	Shaft End-play	Thrust on No.	Journal Diameter	Oil Clearance	Side Clearance
2002	2.2	4	2.4945-2.4954	0.0006-0.0019	0.0020-0.0070	4	1.9983-1.9994	0.0010-0.0031	0.0039-0.0149
	4.3	W	①	②	0.0020-0.0070	4	2.2487-2.2497	0.0013-0.0035	0.0060-0.0140
2003	2.2	4	2.4945-2.4954	0.0006-0.0019	0.0020-0.0070	4	1.9983-1.9994	0.0010-0.0031	0.0039-0.0149
	4.3	W	①	②	0.0020-0.0070	4	2.2487-2.2497	0.0013-0.0035	0.0060-0.0140
2004	4.3	W	①	②	0.0020-0.0070	4	2.2487-2.2497	0.0013-0.0035	0.0060-0.0140
2005	4.3	W	①	②	0.0020-0.0070	4	2.2487-2.2497	0.0013-0.0035	0.0060-0.0140

① No. 1: 2.4488-2.4495
  Nos. 2, 3: 2.4485-2.4494
  No. 4: 2.4480-2.4489

② No. 1: 0.0008-0.0020
  Nos. 2, 3: 0.0011-0.0023
  No. 4: 0.0017-0.0032

06025-S10P-C06

PISTON AND RING SPECIFICATIONS

All measurements are given in inches.

Year	Engine Displacement Liters	Engine ID/VIN	Piston Clearance	Ring Gap			Ring Side Clearance		
				Top Compression	Bottom Compression	Oil Control	Top Compression	Bottom Compression	Oil Control
2002	2.2	4	0.0007-0.0017	0.010-0.020	0.012-0.018	0.010-0.03	0.0020-0.0035	0.0008-0.0031	0.0005-0.0087
	4.3	W	0.0007-0.0024	0.010-0.016	0.015-0.023	0.010-0.029	0.0012-0.0027	0.0015-0.0031	0.0020-0.0070
2003	2.2	4	0.0007-0.0017	0.010-0.020	0.012-0.018	0.010-0.03	0.0020-0.0035	0.0008-0.0031	0.0005-0.0087
	4.3	W	0.0007-0.0024	0.010-0.016	0.015-0.023	0.010-0.029	0.0012-0.0027	0.0015-0.0031	0.0020-0.0070
2004	4.3	W	0.0007-0.0024	0.010-0.016	0.015-0.023	0.010-0.029	0.0012-0.0027	0.0015-0.0031	0.0020-0.0070
2005	4.3	W	0.0007-0.0024	0.010-0.016	0.015-0.023	0.010-0.029	0.0012-0.0027	0.0015-0.0031	0.0020-0.0070

06025-S10P-C07

TORQUE SPECIFICATIONS

All readings in ft. lbs.

Year	Engine Displacement Liters	Engine ID/VIN	Cylinder Head Bolts	Main Bearing Bolts	Rod Bearing Bolts	Crankshaft Damper Bolts	Flywheel Bolts	Manifold		Spark Plugs	Oil Pan Drain Plug
								Intake *	Exhaust		
2002	2.2	4	①	70	38	77	55	17	10	11	33
	4.3	W	②	77	③	74	74	④	⑤	11	18
2003	2.2	4	①	70	38	77	55	17	10	11	33
	4.3	W	②	77	③	74	74	④	⑤	11	18
2004	4.3	W	②	77	③	74	74	④	⑤	11	18
2005	4.3	W	②	77	③	74	74	④	⑤	11	18

* NOTE: Applies to Lower Manifold only.

① Short bolts: 43 ft. lbs. plus 90 degrees
   Long bolts: 46 ft. lbs. plus 90 degrees

② 1st pass: 22 ft. lbs.
   2nd pass:
   Short bolt: Plus 55 degrees
   Medium bolt: Plus 65 degrees
   Long bolt: Plus 75 degrees

③ 20 ft. lbs. plus 70 degrees

④ Lower intake manifold:
   1st pass: 27 inch lbs.
   2nd pass: 106 inch lbs.
   Final pass: 11 ft. lbs.
   Upper manifold bolts:
   1st pass: 44 inch lbs.
   2nd pass: 88 inch lbs.

⑤ Tighten bolts to 12 ft. lbs.
   Retorque to 22 ft. lbs.

06025-S10P-C08

WHEEL ALIGNMENT

Year	Model			Caster		Camber		Toe-in (in.)
				Range (+/-Deg.)	Preferred Setting (Deg.)	Range (+/-Deg.)	Preferred Setting (Deg.)	
2002	Exc. ZQ8/Z87	Left		1.0	+2.8	1.0	0	0.10+/-0.10
		Right		1.0	+3.3	—	—	—
	ZQ8/Z87	Left		1.0	+4.7	1.0	0	0.10+/-0.10
		Right		1.0	+5.2	—	—	—
2003	Exc. ZQ8/Z87	Left		1.0	+2.8	1.0	0	0.10+/-0.10
		Right		1.0	+3.3	—	—	—
	ZQ8/Z87	Left		1.0	+4.7	1.0	0	0.10+/-0.10
		Right		1.0	+5.2	—	—	—
2004	Exc. ZQ8/Z87	Left		1.0	+2.8	1.0	0	0.10+/-0.20
		Right		1.0	+3.3	—	—	—
	ZQ8/Z87	Left		1.0	+4.7	1.0	0	0.10+/-0.20
		Right		1.0	+5.2	—	—	—
2005	Exc. ZQ8/Z87	Left		1.0	+2.8	1.0	0	0.10+/-0.20
		Right		1.0	+3.3	—	—	—
	ZQ8/Z87	Left		1.0	+4.7	1.0	0	0.10+/-0.20
		Right		1.0	+5.2	—	—	—

ZQ8: Sport chassis package

Z87: Low rider chassis package

06025-S10P-C09

TIRE, WHEEL AND BALL JOINT SPECIFICATIONS

Year	Model	OEM Tires		Tire Pressures (psi)		Wheel Size	Ball Joint Inspection	Lug Nut (ft. lbs.)
		Standard	Optional	Front	Rear			
2002	2wd, base	P205/70R15	P235/70R15	36	36	6-JJ	U: 0.125 in. L ①	100
	2wd, Sport	P215/65R15	None	36	36	6-JJ	U: 0.125 in. L ①	100
	4wd, Reg. Cab, w/117.9 WB	P235/70R15	P235/75R15	36	36	6-JJ	U: 0.125 in. L ①	100
	4wd, all others	P235/75R15	None	36	36	6-JJ	U: 0.125 in. L ①	100
2003	2wd, base	P205/70R15	P235/70R15	36	36	6-JJ	U: 0.125 in. L ①	100
	2wd, Sport	P215/65R15	None	36	36	6-JJ	U: 0.125 in. L ①	100
	4wd, Reg. Cab, w/117.9 WB	P235/70R15	P235/75R15	36	36	6-JJ	U: 0.125 in. L ①	100
	4wd, all others	P235/75R15	None	36	36	6-JJ	U: 0.125 in. L ①	100
2004	2wd, base	P205/70R15	P235/70R15	36	36	6-JJ	U: 0.125 in. L ①	100
	2wd, Sport	P215/65R15	None	36	36	6-JJ	U: 0.125 in. L ①	100
	4wd, Reg. Cab, w/117.9 WB	P235/70R15	P235/75R15	36	36	6-JJ	U: 0.125 in. L ①	100
	4wd, all others	P235/75R15	None	36	36	6-JJ	U: 0.125 in. L ①	100
2005	2wd, base	P205/70R15	P235/70R15	36	36	6-JJ	U: 0.125 in. L ①	100
	2wd, Sport	P215/65R15	None	36	36	6-JJ	U: 0.125 in. L ①	100
	4wd, Reg. Cab, w/117.9 WB	P235/70R15	P235/75R15	36	36	6-JJ	U: 0.125 in. L ①	100
	4wd, all others	P235/75R15	None	36	36	6-JJ	U: 0.125 in. L ①	100

OEM: Original Equipment Manufacturer

PSI: Pounds Per Square Inch

STD: Standard

OPT: Optional

L: Lower

U: Upper

① Do not lift truck. Inspect the boss into which the grease fitting is threaded. Replace if the boss is flush or receded below the surface of the ball joint

06025-S10P-C10

BRAKE SPECIFICATIONS
All measurements in inches unless noted

Year	Model		Brake Disc			Brake Drum Diameter			Minimum Lining Thickness	Brake Caliper	
			Original Thickness	Minimum Thickness	Maximum Runout	Original Inside Diameter	Max. Wear Limit	Maximum Machine Diameter		Bracket Bolts (ft. lbs.)	Mounting Bolts (ft. lbs.)
2002	Blazer	F	1.140	1.130	0.002	—	—	—	0.030	①	②
		R	0.787	0.735	0.002	9.50	9.59	9.56	0.030	①	23
	Jimmy	F	1.140	1.130	0.002	—	—	—	0.030	①	②
		R	0.787	0.735	0.002	9.50	9.59	9.56	0.030	①	23
	S10	F	1.140	1.130	0.002	—	—	—	0.030	①	②
		R	0.787	0.735	0.002	9.50	9.59	9.56	0.030	①	23
	Sonoma	F	1.140	1.130	0.002	—	—	—	0.030	①	②
		R	0.787	0.735	0.002	9.50	9.59	9.56	0.030	①	23
	Xtreme	F	1.140	1.130	0.002	—	—	—	0.030	①	②
		R	0.787	0.735	0.002	9.50	9.59	9.56	0.030	①	23
2003	Blazer	F	1.140	1.130	0.002	—	—	—	0.030	①	②
		R	0.787	0.735	0.002	9.50	9.59	9.56	0.030	①	23
	Jimmy	F	1.140	1.130	0.002	—	—	—	0.030	①	②
		R	0.787	0.735	0.002	9.50	9.59	9.56	0.030	①	23
	S10	F	1.140	1.130	0.002	—	—	—	0.030	①	②
		R	0.787	0.735	0.002	9.50	9.59	9.56	0.030	①	23
	Sonoma	F	1.140	1.130	0.002	—	—	—	0.030	①	②
		R	0.787	0.735	0.002	9.50	9.59	9.56	0.030	①	23
	Xtreme	F	1.140	1.130	0.002	—	—	—	0.030	①	②
		R	0.787	0.735	0.002	9.50	9.59	9.56	0.030	①	23
2004	Blazer	F	1.140	1.130	0.002	—	—	—	0.030	①	②
		R	0.787	0.735	0.002	9.50	9.59	9.56	0.030	①	23
	Jimmy	F	1.140	1.130	0.002	—	—	—	0.030	①	②
		R	0.787	0.735	0.002	9.50	9.59	9.56	0.030	①	23
	S10	F	1.140	1.130	0.002	—	—	—	0.030	①	②
		R	0.787	0.735	0.002	9.50	9.59	9.56	0.030	①	23
	Sonoma	F	1.140	1.130	0.002	—	—	—	0.030	①	②
		R	0.787	0.735	0.002	9.50	9.59	9.56	0.030	①	23
	Xtreme	F	1.140	1.130	0.002	—	—	—	0.030	①	②
		R	0.787	0.735	0.002	9.50	9.59	9.56	0.030	①	23
2005	Blazer	F	1.140	1.130	0.002	—	—	—	0.030	①	②
		R	0.787	0.735	0.002	9.50	9.59	9.56	0.030	①	23

NA: Not Available

① Dual piston caliper-to-knuckle: 133 ft. lbs

② Single piston: 38 ft. lbs.
　Dual piston 85 ft. lbs.

06025-S10P-C11

SCHEDULED MAINTENANCE INTERVALS
GENERAL MOTORS—S-SERIES BLAZER, JIMMY, S10 PICKUP, SONOMA & XTREME

TO BE SERVICED	TYPE OF SERVICE	VEHICLE MILEAGE INTERVAL (x1000)															
		7.5	15	22.5	30	37.5	45	52.5	60	67.5	75	82.5	90	97.5	105	112.5	120
Accessory drive belt	S/I								✓								✓
Air cleaner filter	R				✓				✓				✓				✓
Automatic transmission fluid	R	Every 50,000 miles															
Brake system ①	S/I	✓	✓	✓	✓	✓	✓	✓	✓	✓	✓	✓	✓	✓	✓	✓	✓
Chassis & suspension grease points	L	✓	✓	✓	✓	✓	✓	✓	✓	✓	✓	✓	✓	✓	✓	✓	✓
CV-joint boots & axle seals	S/I	✓	✓	✓	✓	✓	✓	✓	✓	✓	✓	✓	✓	✓	✓	✓	✓
Engine coolant system ②	S/I	Every 150,000 miles															
Engine oil & filter	R	✓	✓	✓	✓	✓	✓	✓	✓	✓	✓	✓	✓	✓	✓	✓	✓
Front wheel bearings	S/I & L				✓				✓				✓				✓
Fuel filter	R				✓				✓				✓				✓
Fuel tank, cap & lines	S/I								✓								✓
PCV valve	S/I	Every 100,000 miles															
Rear/front axle fluid level	S/I	✓	✓	✓	✓	✓	✓	✓	✓	✓	✓	✓	✓	✓	✓	✓	✓
Rotate tires	S/I	✓	✓	✓	✓	✓	✓	✓	✓	✓	✓	✓	✓	✓	✓	✓	✓
Spark plug wires	S/I	Every 100,000 miles															
Spark plugs	R	Every 100,000 miles															

R: Replace   S/I: Inspect and service, if necessary   L: Lubricate

① This should be performed when the tires are removed for rotation.

② Drain, flush and refill the cooling system, inspect the system hoses, and clean the radiator and condenser.

③ 2-wheel drive models only.

FREQUENT OPERATION MAINTENANCE (SEVERE SERVICE)

If a vehicle is operated under any of the following conditions it is considered severe service:

- Towing a trailer or using a camper or car-top carrier.

- Repeated short trips of less than 5 miles in temperatures below freezing, or trips of less than 10 miles in any temperature.

- Extensive idling or low-speed driving for long distances as in heavy commercial use, such as delivery, taxi or police cars.

- Operating on rough, muddy or salt-covered roads.

- Operating on unpaved or dusty roads.

- Driving in extremely hot (over 90°) conditions.

Engine oil & filter: replace every 3000 miles or 3 months, whichever occurs first.

Chassis and suspension grease points: lubricate every 3000 miles.

Rear/front axle fluid level: inspect every 3000 miles.

Rotate the tires ever 6000 miles.

Brake system components: inspect ever 6000 miles.

Front wheel bearings (2-wheel drive only): clean, inspect and repack every 15,000 miles.

Air cleaner filter: inspect every 15,000 miles.

Automatic transmission fluid & filter: replace every 15,000 miles.

06025-S10P-C12

ENGINE REPAIR

Distributor

REMOVAL

The 2.2L engine is equipped with a distributorless ignition.

4.3L Engine

1. Remove or disconnect the following:
   - Negative battery cable
   - Air cleaner assembly
   - Spark plug and coil wires
   - Electrical connector
   - Distributor cap
2. Use a grease pencil in order to note the position of the rotor in relation to the distributor housing.
3. Mark the distributor housing and the intake manifold location with a grease pencil.
4. Remove the mounting clamp hold down bolt and the distributor.

➡**As the distributor is being removed from the engine, watch the rotor move in a counter-clockwise direction about 42 degrees. This will appear as slightly more than the 1 o'clock position. Note the position of the rotor segment. Place a second mark on the base of the distributor. This will aid in achieving proper rotor alignment during the distributor installation.**

To install:

5. If installing a new distributor assembly, place 2 marks on the new distributor housing in the same location as the 2 marks on the original housing.
6. Align the rotor with the second mark.
7. Guide the distributor into the engine.
8. Align the hole in the distributor hold-down base over the mounting hole in the intake manifold.
9. As the distributor is being installed, observe the rotor moving in a clockwise direction about 42 degrees.
10. Once the distributor is completely seated, the rotor segment should be aligned with the mark on the distributor base.
11. If the rotor segment is not aligned with the number 1 mark, the driven gear teeth and the camshaft have meshed one or more teeth out of alignment. In order to correct this condition, remove and reinstall the distributor.
12. Install the distributor mounting clamp bolt and tighten to 18 ft. lbs. (25 Nm).

13. Install or connect the following:
    - Distributor cap
    - Electrical connector
    - Spark plug and coil wires
    - Air cleaner assembly
    - Negative battery cable

Alternator

REMOVAL

2.2L Engine

1. Remove or disconnect the following:
   - Passenger side wheel assembly
   - Alternator brace-to-block bolt, the brace-to-intake nut and the brace-to-engine stud
   - Alternator wiring
   - Accessory belt
   - Mounting bolts
   - Alternator

4.3L Engine

1. Remove or disconnect the following:
   - Negative battery cable
   - Air inlet duct, if necessary
   - Accessory belt
   - Heater hose brace
   - Wires
   - Mounting bolts
   - Alternator

INSTALLATION

2.2L Engine

Install or connect the following:
- Alternator
- Mounting bolts. Torque the left bolt to 22 ft. lbs. (30 Nm) and the right bolt to 32 ft. lbs. (43 Nm).
- Wires. Torque the battery feed wire nut to 71 inch lbs. (8 Nm).
- Alternator brace. Torque the nuts and bolts to 22 ft. lbs. (30 Nm).
- Accessory belt
- Negative battery cable

4.3L Engine

Install or connect the following:
- Alternator and loosely install the mounting bolts
- Tighten the rear bolt to 37 ft. lbs. (50 Nm) and the front bolt to 18 ft. lbs. (25 Nm)
- Tighten the brace-to-alternator and brace-to-intake retainers to 18 ft. lbs.

(25 Nm). Tighten the brace-to-engine stud nut to 37 ft. lbs. (50 Nm).
- Wires and the battery feed wire nut
- Heater hose bracket
- Accessory belt
- Negative battery cable

Ignition Timing

ADJUSTMENT

The ignition timing is preset and cannot be adjusted.

Engine Assembly

REMOVAL & INSTALLATION

2.2L Engine

➡**In certain cases on some models the A/C system will have to be evacuated because the compressor may need to be removed from the vehicle to allow clearance for engine removal. On other models you maybe able to set the compressor and lines to one side and still have enough clearance to remove the engine. In this case the system does not have to be evacuated because the lines do not have to be disconnected from the compressor. To check if your system has to be evacuated, unplug the electrical connectors from the compressor, then unbolt the compressor assembly. Unfasten any brackets holding the refrigerant lines and try to set the components aside so that you will have enough clearance for engine removal. If there is not enough clearance for engine removal you must recover the refrigerant from the A/C system with an approved recovery station before attempting to remove the engine from your vehicle. DO NOT attempt this without the proper equipment. R-134a should NOT be mixed with R-12 refrigerant and, depending on your local laws, attempting to service this system could be illegal.**

1. Disconnect the negative battery cable and properly relieve the fuel system pressure.
2. Drain the engine cooling system and the engine oil into separate drain pans.
3. Remove or disconnect the following:
   - Hood
   - Oxygen Sensor (O2S) electrical connection
   - Exhaust pipe from the manifold

➡**On some models it may also be necessary to disconnect the catalytic converter from the exhaust pipe.**

- Braces from the engine and the transmission, if equipped
- Starter motor
- Transmission and separate it from the engine or, if necessary, remove it from the vehicle
- Alternator rear brace by unfastening the bolt and nuts
- Ground straps from the engine block
- Drive belt
- A/C compressor and bracket. If possible, set the compressor and bracket to one side without disconnecting the lines.
- Hoses and transmission coolant lines engaged to the radiator
- Radiator
- Power steering pump and cap the power steering lines to avoid contamination
- Heater hoses from the heater core
- 12 volt supply from the mega fuse, if necessary
- All electrical connections and wiring harnesses
- All vacuum lines
- Throttle cable, and if equipped the cruise control cable
- Exhaust Gas Recirculation (EGR) pipe and the EGR valve
- Fuel lines

4. Install a suitable lifting device to the engine.

5. Remove the engine mount bolts and carefully lift the engine from the vehicle. Pause several times while lifting the engine to make sure no wires or hoses have become snagged.

To install:

6. Carefully lower the engine into the vehicle and install the engine mount bolts. Remove the engine lifting device.

7. Install or connect the following:
- Fuel lines
- 12 volt supply to the mega fuse, if removed
- All vacuum lines, electrical connections and wiring harnesses
- EGR valve and pipe, if removed
- Throttle and if equipped, the cruise control cable
- Heater hoses to the heater core
- Power steering pump and attach the lines
- A/C compressor
- Radiator, all hoses and fluid cooler lines

- Water pump, if removed
- Drive belt
- Ground strap to the engine
- Alternator rear brace and tighten the bolt and nuts, if removed
- Transmission to the engine
- Starter motor, if removed
- Braces to the engine and the transmission, if equipped
- Exhaust pipe to the manifold
- Catalytic converter to the exhaust pipe, if removed
- O_2S electrical connection
- Battery
- Hood

8. Check all powertrain fluid levels and add, as necessary. Be sure to properly fill the engine crankcase with clean engine oil.

9. Connect the battery cables and properly fill the engine cooling system.

10. Start and run the engine, then check for leaks.

4.3L Engine

1. Drain the engine cooling system
2. Drain the engine oil.
3. Remove or disconnect the following:
- Negative battery cable
- Fuel system pressure
- Vacuum reservoir and/or the underhood light from the hood, as equipped
- Outer cowl vent grilles
- Hood
- Oxygen Sensor (O_2S) and/or wiring
- Exhaust pipes at the manifolds and loosen the hanger at the catalytic converter. This is necessary to remove the rear catalytic converter cushion mounts for removal of the exhaust assembly.
- Skid plate, if equipped
- Engine-to-transmission pencil braces
- Slave cylinder and position aside, if equipped
- Line clamp at the bell housing
- Wiring from the starter
- Starter
- Transfer case
- Oil filter
- Engine mount through bolts
- Rear engine mount crossbar, nut and washer
- Bell housing bolts, except the upper left.
- Battery ground (negative) cable from the engine
- Front drive axle bolts and roll the axle downward, on 4WD vehicles
- Air cleaner assembly

- Upper radiator shroud
- Fan assembly
- Drive belt assembly
- Water pump pulley
- Upper radiator hose
- Air conditioning compressor, if equipped, and position aside with the lines intact
- Lower radiator hose
- Oil cooler and overflow lines from the radiator, plug the openings to prevent system contamination or excessive fluid loss
- Radiator and lower radiator shroud
- Power steering hoses from the steering gear, then cap the openings to prevent system contamination or excessive fluid loss
- Heater hoses from the intake manifold and the water pump
- Wiring harness and vacuum lines from the engine
- Throttle cables
- Remaining bell housing bolt
- Fuel lines and the bracket
- Ground strap(s) from the rear of the cylinder head
- Front body mount bolts, on 4WD vehicles

4. Support the transmission.
5. Install a lifting device and lift the engine.

To install:

6. Install or connect the following:
- Engine into the vehicle
- Front body mount bolts, on 4WD vehicles
- Ground strap(s) to the rear of the cylinder head
- Fuel lines and the bracket
- Upper left bell-housing bolt
- Throttle cables
- Vacuum lines and wiring harness connectors
- Heater hoses
- Power steering hoses
- Lower shroud and radiator
- Oil cooler lines to the radiator and overflow hose
- Lower radiator hose
- Air conditioning compressor to the engine, if equipped
- Upper radiator hose
- Water pump pulley
- Drive belt assembly
- Fan assembly
- Upper radiator shroud
- Air cleaner assembly
- Front drive axle, for 4WD vehicles
- Battery ground strap to the engine block
- Remaining bell housing bolts

- Engine mount through-bolts. Torque them to 49 ft. lbs. (66 Nm).
- Rear engine mount crossbar nut and washer. Tighten the nut to 33 ft. lbs. (45 Nm).
- Oil filter
- Starter motor
- Flywheel cover
- Clutch slave cylinder, if equipped
- Pencil brace and the skid plate, as equipped
- Catalytic converter Y-pipe assembly and hangers
- Hood
- Outer cowl vent grilles
- Vacuum reservoir and/or the under-hood light to the hood, as equipped
- Negative battery cable

7. Check all powertrain fluid levels and add, as necessary.

8. Refill the engine crankcase.

9. Refill the engine cooling system.

10. Start and run the engine, then check for leaks.

Water Pump

REMOVAL & INSTALLATION

1. Disconnect the negative battery cable.
2. Drain the engine cooling system.
3. Relieve the belt tension and remove the accessory drive belts or the serpentine drive belt, as applicable.
4. Remove or disconnect the following:

- Upper fan shroud
- Fan or fan and clutch assembly, as applicable
- Water pump pulley
- Coolant hose(s) from the water pump

➡ **For the hoses on some engines, removal may be easier if the hose is left attached until the pump is free from the block. Once the pump is removed from the engine, the pump may be pulled (giving a better grip and greater leverage) from the tight hose connection.**

- Water pump retainers
- Water pump from the engine

❋❋ WARNING

Note the positions of all retainers as some engines will utilize different length fasteners in different locations and/or bolts and studs in different locations.

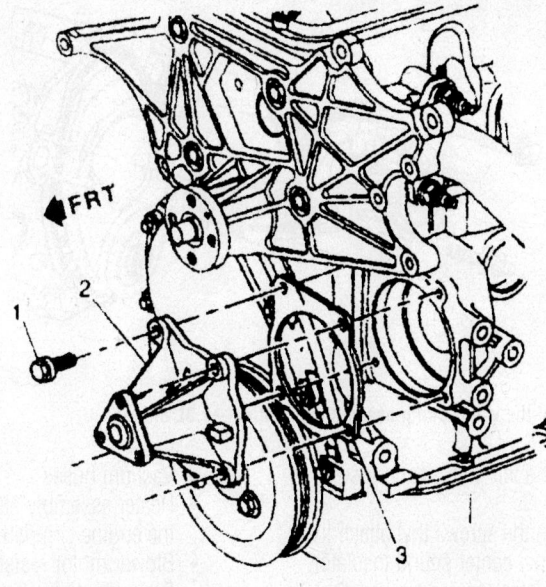

1. BOLT
2. PUMP, COOLANT
3. GASKET

7924JG05

Exploded view of the water pump mounting—2.2L engine

To install:

5. Clean the gasket mounting surfaces.

➡ **The water pumps on some of the earlier engines covered may have been installed using sealer only, no gasket, at the factory. If a gasket is supplied with the replacement part, it should be used. Otherwise, a ⅛ in. (3mm) bead of RTV sealer should be used around the sealing surface of the pump.**

6. Apply sealant to the water pump retainer threads.

7. Install or connect the following:

- Water pump using a new gasket. Tighten the water pump retainers to 18 ft. lbs. (25 Nm) for 2.2L engine or to 30 ft. lbs. (41 Nm) for 4.3L engine.
- Coolant hose(s)
- Water pump pulley
- Fan or fan and clutch assembly
- Serpentine drive belt (if equipped) by positioning the belt over the pulleys and carefully allow the tensioner back into contact with the belt
- V-belts (if equipped) and adjust the tension
- Upper fan shroud
- Negative battery cable

8. Refill the engine cooling system.

9. Run the engine and check for leaks.

Heater Core

REMOVAL & INSTALLATION

1. Disconnect the negative battery cable.
2. Drain the engine cooling system.
3. Remove or disconnect the following:

- Heater hoses from the heater core

4. Remove the instrument panel as follows:

a. Disable the air bag system.

b. Set the parking brake and block the wheels.

c. Disconnect the parking brake release cable from the parking brake lever.

d. Unfasten the screws that retain the DLC instrument panel left side sound insulator. Feed the DLC through the hole in the sound insulator.

e. Unfasten the right side sound insulator panel screws and remove the panel.

f. Unfasten the screws that attach the instrument panel left side sound insulator to the knee bolster and cowl panel.

g. Unfasten the nut that attaches the left side sound insulator to the accelerator pedal bracket.

h. Unplug the remote control door lock receiver module electrical connector.

i. Remove the door lock receiver module from the left side sound insula-

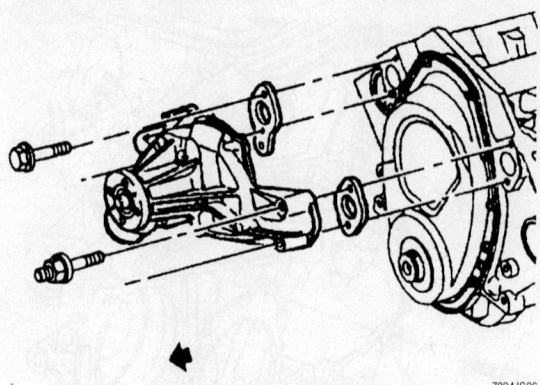

7924JG06

Exploded view of the water pump assembly mounting—4.3L engine

tor. Remove the left side sound insulator.

j. Unfasten the screws that attach the instrument panel center sound insulator to the knee bolster, instrument panel, heater assembly and floor duct.

k. Remove the center sound insulator.

l. Unfasten the screws that attach the courtesy lamp to the knee bolster.

m. Unfasten the screws that attach the knee bolster to the instrument panel.

n. Disconnect the lap cooler duct from the knee bolster.

o. Unplug the lighter electrical connection and remove the knee bolster.

p. Unfasten the steering column-to-instrument panel nuts and lower the column.

q. Unfasten the screws that attach the instrument panel accessory trim plate to the instrument panel.

r. Remove the trim plate and unplug all necessary electrical connection.

s. Remove the heater and/or air conditioning control assembly.

t. Remove the radio and the storage compartment assembly (if equipped).

u. If necessary, remove the instrument cluster.

v. Unfasten the left and right instrument panel pivot bolts and the panel lower support bolt.

w. Unfasten the speaker grilles retaining screws and remove the speaker grilles.

x. Remove the windshield defroster grille using a flat-bladed prytool. Start at one end of the grille and work your way down the grille.

y. Unfasten the 4 instrument panel upper support screws.

z. Tag and unplug all necessary electrical connections.

aa. Remove the instrument panel from the vehicle.

5. Remove or disconnect the following:
- Air inlet assembly, if equipped

- Vacuum hoses
- Heater assembly studs, from inside the engine compartment
- Blower motor resistor
- From inside the heater case assem-

bly, the stud; the stud is located behind the blower motor resistor
- Heater assembly-to-chassis screws
- Heater assembly from the vehicle
- Access cover screws and cover from the heater assembly
- Heater core from the heater case assembly

To install:
6. Install or connect the following:
- Heater core to the heater case assembly
- Access cover to the heater assembly and the cover screws
- Heater assembly to the vehicle
- Heater assembly-to-chassis screws and torque them to 40 inch lbs. (4.5 Nm)
- The stud, working inside the heater case assembly; the stud is located behind the blower motor resistor

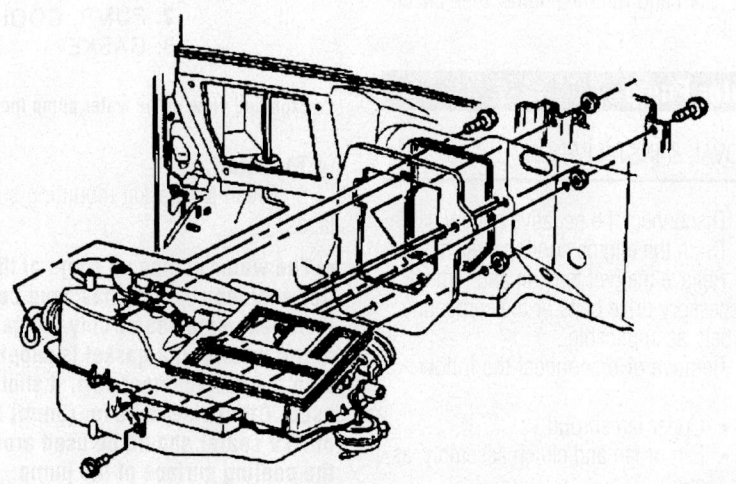

93113G77

View of the heater case assembly

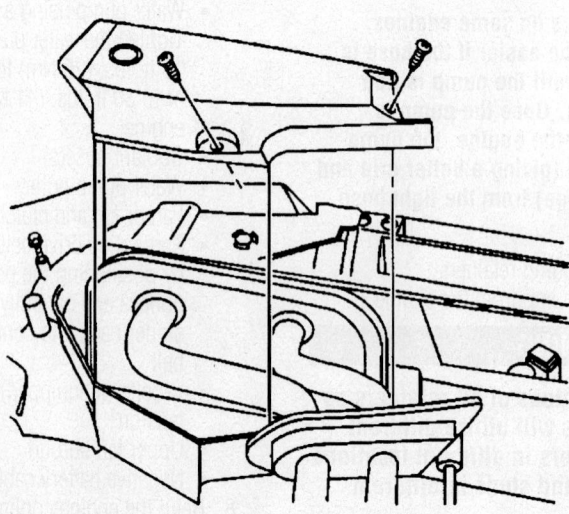

93113G78

View of the heater case cover

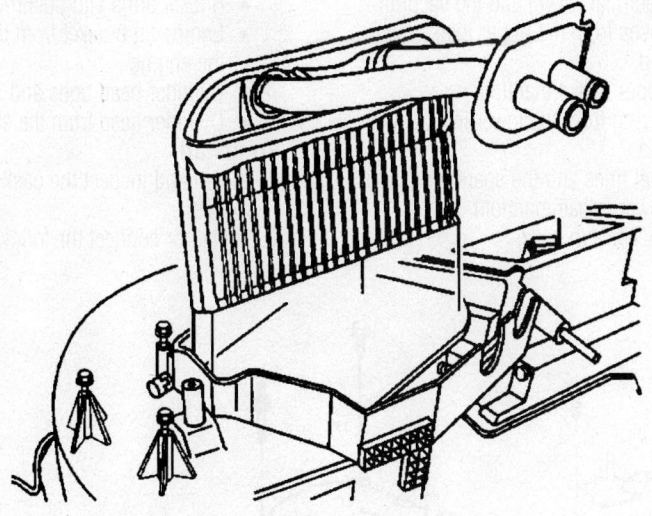

93113G79

View of the heater core

- Blower motor resistor
- Heater assembly studs, working inside the engine compartment, and torque them to 17 inch lbs. (1.9 Nm)
- Vacuum hoses
- Air inlet assembly, if equipped

7. Install the instrument panel as follows:

a. Rest the instrument panel on the lower pivot studs.

b. Attach the electrical connections.

c. Install but do not tighten the 4 upper instrument panel support screws.

d. Install the left and right panel pivot bolts. Tighten the bolts to 102 inch lbs. (11.5 Nm).

e. Install the panel lower support bolt. Tighten the bolt to 102 inch lbs. (11.5 Nm).

f. Tighten the upper support screws to 17 inch lbs. (1.9 Nm).

g. Install the windshield defroster grille and the speaker grilles.

h. Install the radio and storage compartment assembly (if equipped).

i. If removed, install the instrument cluster.

j. Install the heater and/or air conditioning control assembly.

k. Attach the electrical connections to the instrument panel accessory trim plate.

l. Place the trim plate in position and install its retaining screws. Tighten the screws to 17 inch lbs. (1.9 Nm).

m. Place the steering column into position and install its retaining nuts. Tighten the nuts to 22 ft. lbs. (30 Nm).

n. Attach the lighter electrical connection and the lap cooler duct to the knee bolster.

o. Place the knee bolster into position and install its retaining screws. Tighten the Torx® head screws to 80 inch lbs. (9

Nm) and the hex head screws to 17 inch lbs. (1.9 Nm).

p. Place the courtesy lamp in position and install its screws. Tighten the screws to 17 inch lbs. (1.9 Nm).

q. Place the instrument panel center sound insulator in position. Install the screws that attach the center sound insulator to the knee bolster, instrument panel and the floor duct. Tighten the screws to 17 inch lbs. (1.9 Nm).

r. Install the screw that attaches the center sound insulator to the heater assembly. Tighten the screw to 13 inch lbs. (1.5 Nm).

s. Install the remote control door lock receiver module to the instrument panel left side sound insulator.

t. Attach the door lock receiver electrical connection.

u. Install the nut that attaches the left side sound insulator to the accelerator pedal bracket. Tighten the nut to 35 inch lbs. (4 Nm).

v. Install the screw that attaches the left side sound insulator to cowl panel. Tighten the screw to 13 inch lbs. (1.5 Nm).

w. Install the screws that attach the left side sound insulator to knee bolster. Tighten the screw to 17 inch lbs. (1.9 Nm).

x. Feed the DLC through the hole in the sound insulator, place the DLC in position and install its retaining screws. Tighten the screws to 21 inch lbs. (2.4 Nm).

y. Install the right side sound insulator and tighten the screws

z. Connect the parking brake release cable to the lever.

aa. Enable the air bag system.

8. Install the heater hoses to the heater core.

9. Refill the cooling system.

10. Connect the negative battery cable.

11. Run the engine to normal operating temperatures; then, check the climate control operation and check for leaks.

Cylinder Head

REMOVAL & INSTALLATION

2.2L Engine

1. Relieve the fuel system pressure.
2. Disconnect the negative battery cable.
3. Drain the engine cooling system.
4. Remove or disconnect the following:

- Air duct from the air inlet
- Upper radiator hose and upper fan shroud
- Radiator assembly
- Lower fan shroud
- Fan assembly
- Drive belt assembly
- Water pump pulley
- Heater hose from the intake manifold and the thermostat housing
- Thermostat housing
- Alternator support brace and the alternator wiring
- Air conditioning compressor with brackets, if equipped, move it aside without disconnecting the lines
- Accessory bracket along with the alternator and power steering pump

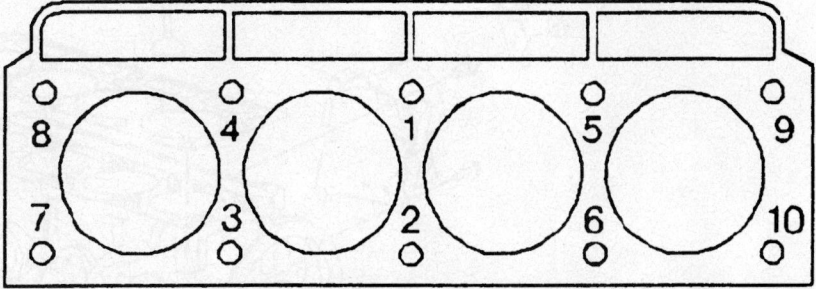

7924JG07

Cylinder head bolt torque sequence—2.2L engine

still attached. Be careful not to damage the steering pump lines.
- Throttle cable and cable support linkage
- Heater hose from the water pump
- Oil fill tube
- Exhaust pipe
- Oxygen Sensor (O2S)
- Exhaust manifold

- Electrical wiring and the vacuum hoses from the upper intake manifold
- Upper intake manifold
- Wiring from the lower intake manifold
- Fuel lines and the spark plug wires
- Lower intake manifold
- Rocker arm cover

- Rocker arms and pushrods
- Engine lift bracket from the rear of the engine
- Cylinder head bolts and studs
- Cylinder head from the engine

To install:
5. Clean and inspect the gasket mounting surfaces.
6. Install or connect the following:

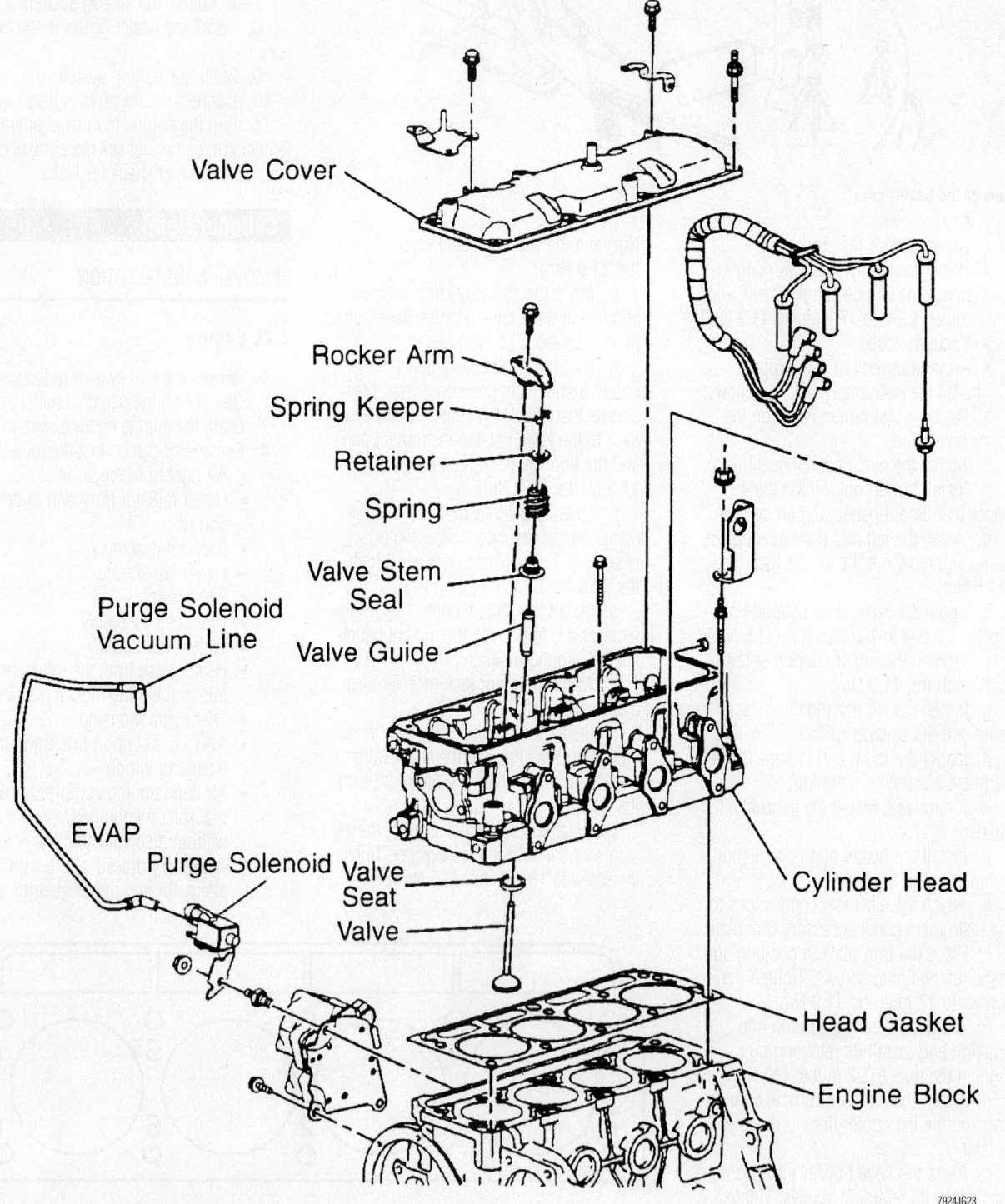

Valve Cover

Rocker Arm

Spring Keeper

Retainer

Spring

Valve Stem Seal

Valve Guide

Purge Solenoid Vacuum Line

EVAP

Purge Solenoid

Valve Seat

Valve

Cylinder Head

Head Gasket

Engine Block

Cylinder head and related components—2.2L engine

7924JG23

- Cylinder head using a new gasket
- Cylinder head bolt threads coated with sealer 1052080. Tighten the bolts within 15 minutes of sealer application, in sequence, to 46 ft. lbs. (63 Nm) for long bolts and to 43 ft. lbs. (58 Nm) for short bolts; then, tighten all bolts an additional 90 degree turn using a torque angle meter.
- Engine lift bracket
- Rocker arms and pushrods
- Rocker arm cover
- Lower intake manifold
- Spark plug wires and the fuel lines
- Lower intake manifold and wiring
- Upper intake manifold
- Vacuum hoses and electrical wiring to the upper intake
- Oil fill tube assembly
- Exhaust manifold
- Exhaust pipe and O$_2$S
- Heater hose to the water pump
- Throttle cable support and throttle cable

- Accessory support bracket and components
- Air conditioning compressor, if equipped
- Power steering support brace
- Alternator support brace and wiring
- Thermostat housing and the heater hose
- Water pump pulley and drive belt assembly
- Fan assembly
- Radiator and the lower fan shroud
- Upper fan shroud and upper radiator hose
- Air inlet ductwork
- Negative battery cable

7. Refill the engine cooling system and check for leaks.

4.3L Engine

1. Properly relieve the fuel system pressure, then disconnect the negative battery cable.

2. Drain the engine cooling system.
3. Remove or disconnect the following:

- Intake manifold
- Exhaust manifold
- Alternator and bracket, if removing the right cylinder head
- Cooling fan assembly
- Air conditioning compressor (position it aside with the refrigerant lines attached)
- Air pipe bracket and nut from the rear of the power steering pump if removing the left cylinder head
- Engine accessory bracket with power steering pump (position the pump aside with the lines attached) and brackets, if removing the left cylinder head
- Wiring harness and clip from the rear of the cylinder head
- Coolant sensor wire
- Wiring from the spark plugs
- Spark plugs, if necessary
- Ground wires and if necessary, the

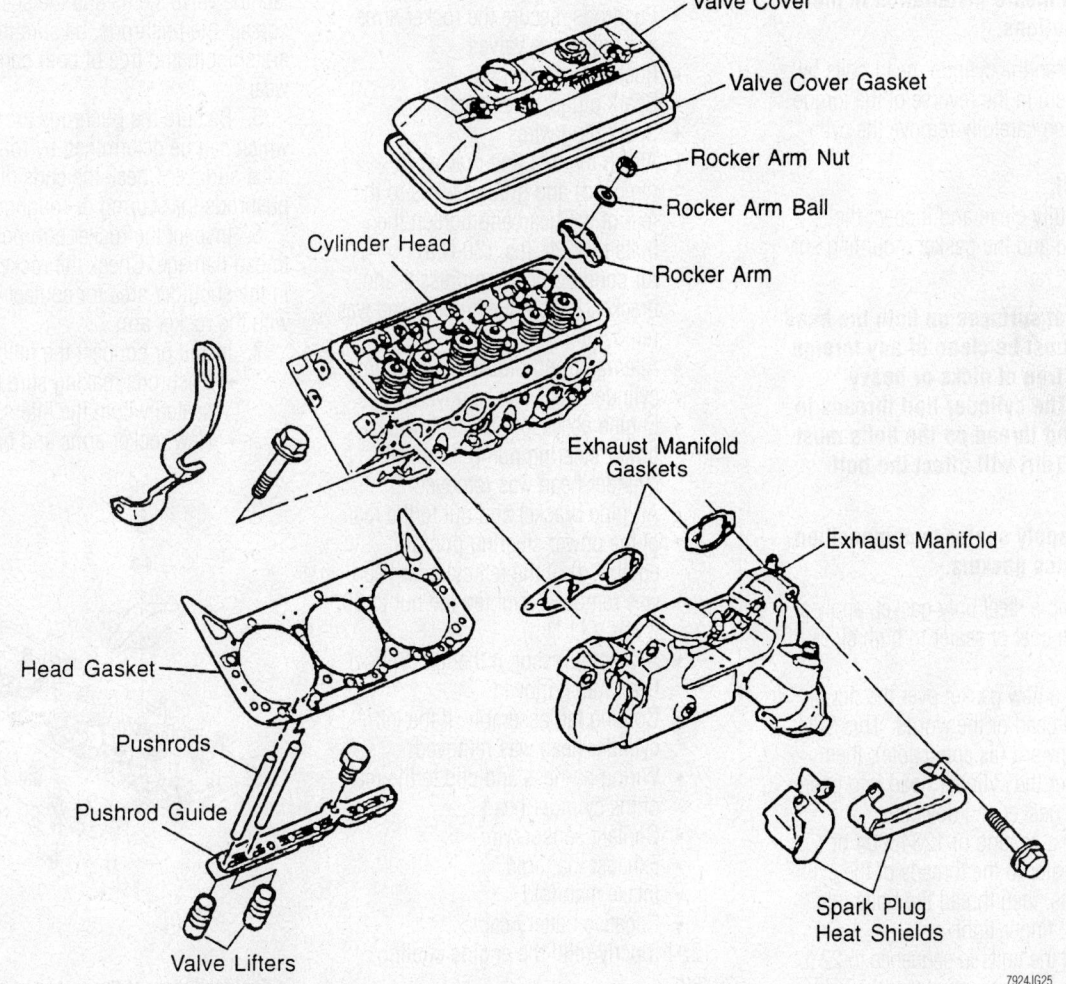

Cylinder head and related components—4.3L engine

7924JG25

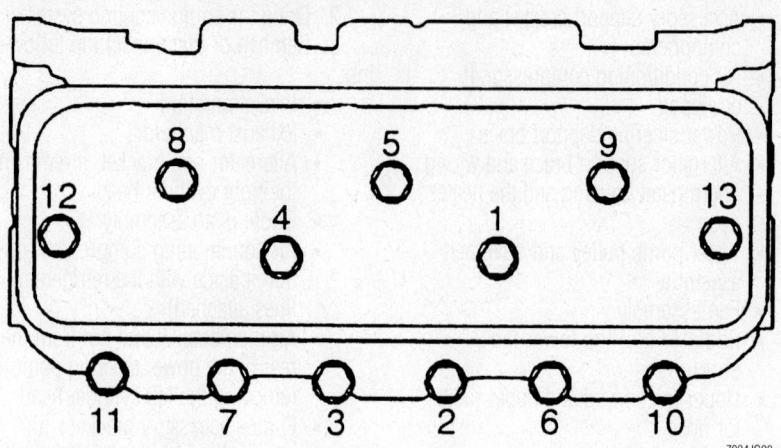

Cylinder head bolt torque sequence—4.3L engine

fuel line bracket from the rear of the cylinder head
- Rocker arm cover

4. Loosen the rocker arms and remove the pushrods.

➡**If valve train components, such as the rocker arms or pushrods, are to be reused, they must be tagged or arranged to insure installation in their original locations.**

5. Unfasten the cylinder head bolts by loosening them in the reverse of the torque sequence, then carefully remove the cylinder head.

To install:

6. Carefully clean and inspect the cylinder head and the gasket mounting surfaces.

➡**The gasket surfaces on both the head and block must be clean of any foreign matter and free of nicks or heavy scratches. The cylinder bolt threads in the block and thread on the bolts must be cleaned (dirt will affect the bolt torque).**

➡**DO NOT apply sealer to composition steel-asbestos gaskets.**

7. If using a steel only gasket, apply a thin and even coat of sealer to both sides of the gaskets.

8. Place a new gasket over the dowel pins with the bead or the words "This Side Up" facing upward (as applicable), then carefully lower the cylinder head into position over the gasket and dowels.

9. Apply a coating of 12346004 or equivalent sealer to the threads of the cylinder head bolts, then thread the bolts into position until finger-tight.

10. Install the bolts in sequence to 22 ft. lbs. (30 Nm). The bolts must then be tight-

ened again in sequence in the following order:
  a. Short length bolts: (11, 7, 3, 2, 6, 10) 55 degrees.
  b. Medium length bolts: (12, 13) 65 degrees.
  c. Long length bolts: (1, 4, 8, 5, 9) 75 degrees.

11. Install or connect the following:
- Pushrods, secure the rocker arms and adjust the valves
- Rocker arm cover
- Spark plugs, if removed
- Spark plug wires
- Attach the fuel line bracket (if removed) and ground wires to the rear of the head and tighten the bolts to 22 ft. lbs. (30 Nm)
- Air conditioning compressor and bracket, if the left cylinder head was removed
- Alternator and bracket, if the right cylinder head was removed
- Engine accessory bracket with power steering pump if the left cylinder head was removed
- Air pipe bracket and nut to the rear of the power steering pump (if equipped), if the left cylinder head was removed. Tighten the nut to 30 ft. lbs. (41 Nm).
- A/C compressor, if the left cylinder head was removed
- Cooling fan assembly, if the left cylinder head was removed
- Wiring harness and clip to the rear of the cylinder head
- Coolant sensor wire
- Exhaust manifold
- Intake manifold
- Negative battery cable

12. Properly refill the engine cooling system.

13. Run the engine to check for leaks.

Rocker Arms

REMOVAL & INSTALLATION

2.2L Engine

1. Remove or disconnect the following:
- Rocker arm cover
- Rocker arm retaining nut, arm and ball
- Pushrod, if necessary

➡**Valvetrain components, being reused, must be installed in their original positions. If removed, be sure to tag or arrange all rocker arms and pushrods to assure proper installation.**

To install:

2. Inspect the rocker arms, balls and pushrods for damage or wear and replace as necessary.

3. Check the rocker arms, balls and their mating surfaces. Be sure the surfaces are smooth and free from scoring or other damage.

4. Check the rocker arm areas that contact the valve stems and the sockets that contact the pushrods, be sure these areas are smooth and free of both damage and wear.

5. Be sure the pushrods are not bent which can be determined by rolling them on a flat surface. Check the ends of the pushrods for scoring or roughness

6. Inspect the rocker arm bolts for thread damage. Check the rocker arm bolts in the shoulder area for contact damage with the rocker arm.

7. Install or connect the following:
- Pushrods making sure they are seated within the lifters, if removed
- New rocker arms and balls by coat-

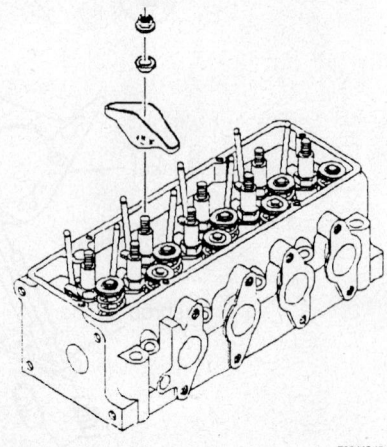

Exploded view of the rocker arm assembly—2.2L engine

ing the friction surfaces using Dri-Slide Molykote® or equivalent pre-lube

✳ WARNING

When tightening a rocker arm retainer, be sure the lifter for that valve is resting on the base circle of the camshaft and not on the lobe, otherwise the valve train can be damaged. Do not over-tighten the retainers.

- Rocker arms and ball. Tighten the nuts to 18 ft. lbs. (25 Nm).
- Rocker arm cover

8. Start and run the engine to check for leaks.

4.3L Engine

1. Remove or disconnect the following:
- Rocker arm cover(s)
- Rocker arm nut, rocker arm and ball washer

➡ **If only the pushrod is to be removed, loosen the rocker arm nut, swing the rocker arm to the side and remove the pushrod.**

- Pushrod(s)

To install:

2. Inspect and replace components if worn or damaged.

3. Coat the bearing surfaces of the rocker arms and the rocker arm ball washers with Molykote® or equivalent pre-lube.

4. Install or connect the following:
- Pushrods making sure they seat properly in the lifter
- Rocker arms, ball washers and the nuts

➡ **The 4.3L engines are equipped with screw-in rocker arm studs with positive stop shoulders.**

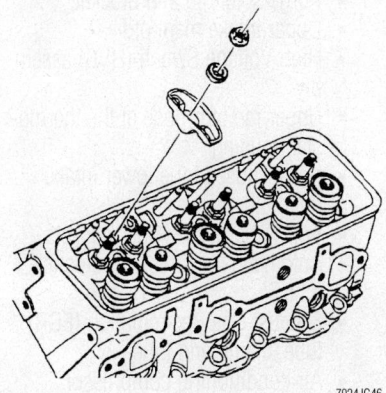

Exploded view of the rocker arm assembly—4.3L engine

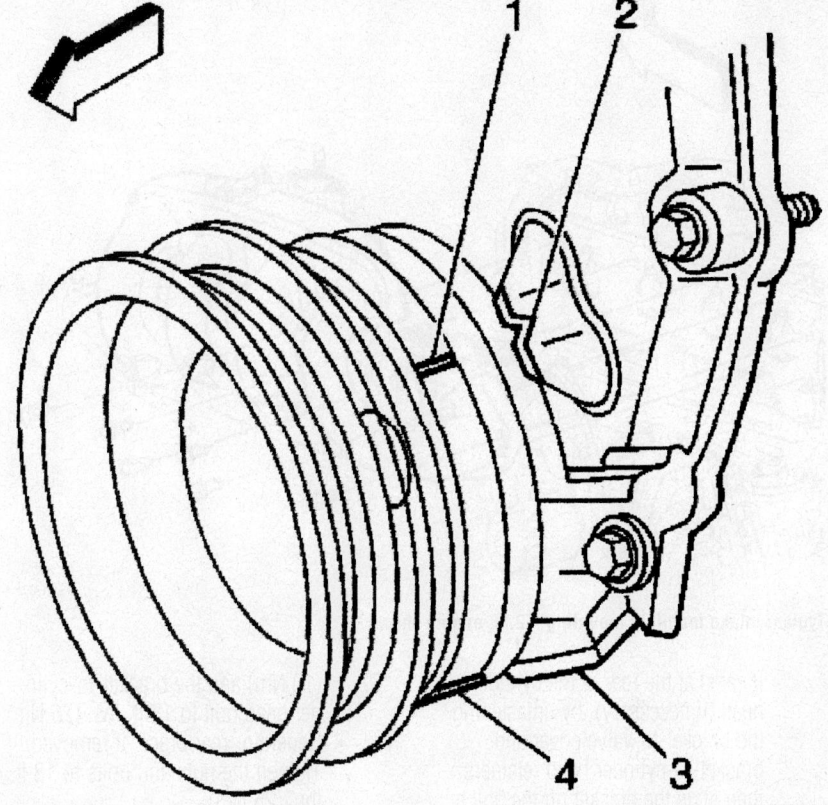

Positioning the crankshaft balancer—4.3L engine

06025-S10P-G01

- Rocker arm adjusting nuts and hand-tighten

5. Rotate the crankshaft balancer to position the crankshaft balancer alignment mark (1) 57-63 degrees clockwise or counterclockwise from the engine front cover alignment tab (2).

6. Tighten the rocker arm bolts to 22 ft. lbs. (30 Nm). No further adjustment is necessary or possible.

7. Install the rocker arm cover(s).

8. Start and run the engine, then check for leaks and for proper ignition timing adjustment.

Intake Manifold

REMOVAL & INSTALLATION

2.2L Engine

1. Remove or disconnect the following:

- Negative battery cable and remove the air cleaner resonator.
- Three vacuum hoses from the throttle body
- Throttle cable support bracket and the throttle body assembly
- Upper fan shroud and disconnect

the vacuum brake booster hose, if necessary

- Exhaust Gas Recirculation (EGR) pipe-to-manifold bolts and the EGR pipe-to-EGR adapter bolt, and the EGR pipe.
- EGR adapter
- Idle Air Control (IAC) motor connector
- Idle Air Control (IAC) motor
- Manifold Absolute Pressure (MAP) sensor connector
- Throttle Position (TP) sensor connector
- Fuel injector harness connector
- Right fender wheelhouse extension
- Retainers from the engine harness bracket, the transmission filler tube (if equipped) and the fuel system evaporator pipe
- Fuel pipes from the fuel rail
- Accelerator cable and if equipped, the cruise control cable
- Spark plug wires from the plugs
- Spark plug wire harness retainer from the heater hose pipe and set aside the harness
- Alternator rear brace by accessing the retaining nuts and bolts through the wheelhouse, if necessary
- Engine wiring harness bracket

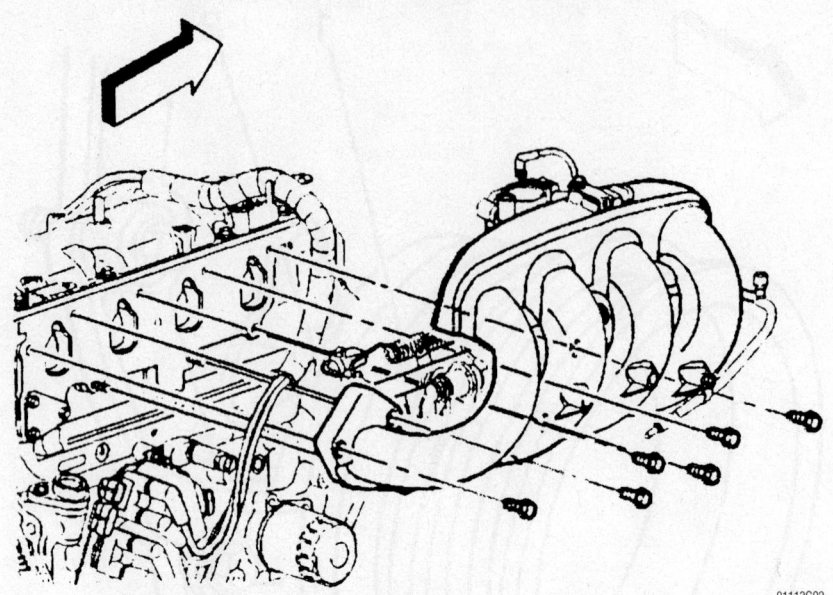

Typical intake manifold mounting—2.2L engine shown

located at the rear of the cylinder head (if necessary), by unfastening the bracket-to-valve cover and bracket-to-cylinder head retainers, then slide the bracket off the bolt at the rear of the cylinder head
- Intake manifold bolts
- Fuel rail bracket
- Intake manifold and gasket

To install:

2. Carefully remove all traces of gasket material from the mating surfaces. Check the EGR passage to be sure it is free of excessive carbon deposits and clean, as necessary.

3. Install or connect the following:
- Lower intake manifold using a new gasket, then tighten the retaining bolts to 17 ft. lbs. (24 Nm) using the sequence illustrated
- Engine wiring harness bracket, if removed. Tighten the bracket-to-valve cover bolts to 88 inch lbs.

(10 Nm) and the bracket-to-cylinder head bolt to 18 ft. lbs. (25 Nm).
- Generator rear brace, if removed. Tighten the nuts and bolts to 18 ft. lbs. (25 Nm).
- Spark plug wire harness and retainer and attach the spark plug wires to the plugs
- Throttle body assembly, if removed and the throttle cable support bracket
- Three vacuum lines to the throttle body
- Accelerator cable and if equipped, the cruise control cable
- Fuel lines
- Retainers to the engine harness bracket, the transmission filler tube (if equipped) and the fuel system evaporator pipe
- IAC motor connector
- MAP sensor connector
- TP sensor connector

- Fuel injector harness connector
- Wheelhouse extension
- EGR adapter and tighten the retainers to 97 inch lbs. (11 Nm)
- EGR pipe to the EGR adapter and tighten the bolt to 18 ft. lbs. (25 Nm)
- EGR pipe-to-intake manifold bolts and tighten the bolts to 89 inch lbs. (10 Nm)
- Upper fan shroud and the brake booster hose
- Air cleaner resonator and connect the negative battery cable.

4. Start the engine and check for leaks.

4.3L Engine

➡️If only the upper intake manifold is being removed, the fuel system pressure does not need to be released. ALWAYS release the pressure before disconnecting any fuel lines.

1. Remove the engine cover, if equipped

2. Properly relieve the fuel system pressure.

3. Drain the engine cooling system.

4. Remove or disconnect the following:

- Negative battery cable
- Air cleaner and air inlet duct
- Wiring harness connectors and brackets
- Throttle linkage from the upper intake manifold
- Ignition coil
- Fuel lines and bracket from the rear of the lower intake manifold
- Brake booster vacuum hose at the upper intake manifold
- Positive Crankcase Ventilation (PCV) hose at the rear of the upper intake manifold
- Vacuum hoses from both the front and rear of the upper intake
- Purge solenoid and bracket
- Upper intake manifold
- High Voltage Switch (HVS) assembly
- Upper radiator hose at the thermostat housing
- Heater hose at the lower intake manifold
- Wiring harnesses and brackets.
- Automatic transmission dipstick tube
- Exhaust Gas Recirculation (EGR) tube, clamp and tube
- Air conditioning compressor bracket-to-lower intake manifold pencil brace

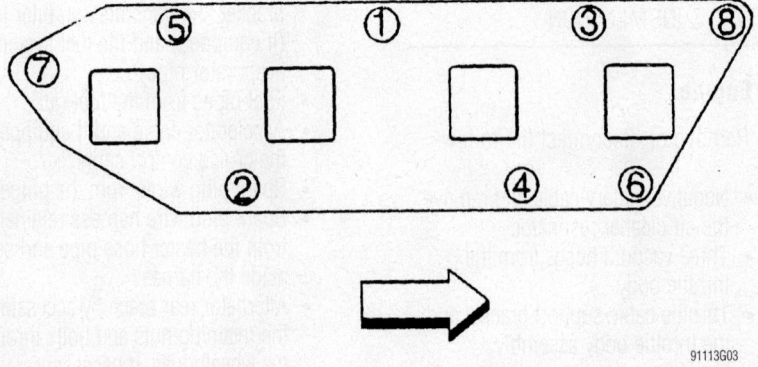

Intake manifold bolt tightening sequence—2.2L engine

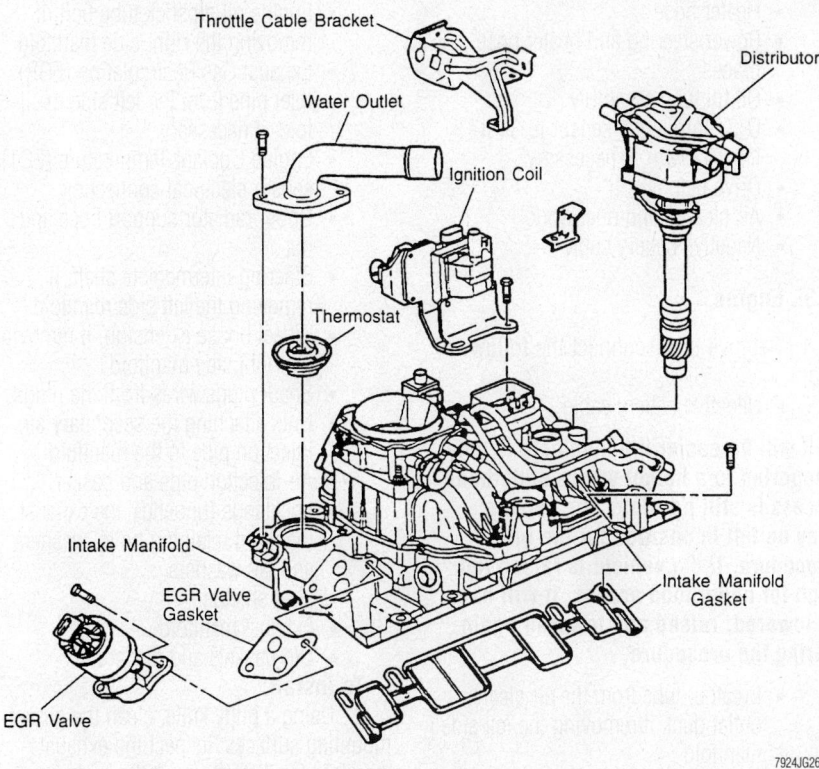

Intake manifold and related components—4.3L engine

- Alternator bracket bolts near the thermostat housing
- Lower intake manifold

5. Insert clean rags into the openings in the cylinder head to prevent dirt and debris from entering the engine.

6. Clean the gasket mounting surfaces. Be sure to inspect the manifold for warpage and/or cracks. If necessary, replace it.

To install:

7. Remove the rags from the cylinder heads.

8. Position the gaskets on the cylinder head with the port blocking plates to the rear and the **This Side Up** stamps facing upward. Then apply a ³⁄₁₆ in. (5mm) bead of RTV sealant on the front and rear of the engine block at the block-to-manifold mating surface. Extend the bead ½ in. (13mm) up each cylinder head to seal and retain the gaskets.

9. Install the lower intake manifold. Tighten the bolts in sequence and in 3 steps, as follows:
   a. Step 1: 26 inch lbs. (3 Nm).
   b. Step 2: 106 inch lbs. (12 Nm).
   c. Step 3: 11 ft. lbs. (15 Nm).

10. Install or connect the following:
- Alternator bracket bolt near the thermostat housing
- EGR tube, clamp and bolt
- Wiring harness to the lower manifold components, including the injector, EGR valve and ECT sensor
- Air conditioning compressor bracket-to-the lower intake manifold pencil braces
- Transmission oil dipstick tube, if necessary
- Fuel supply and return lines to the rear of the lower intake

11. Temporarily reattach the negative battery cable, then pressurize the fuel system (by cycling the ignition without starting the engine) and check for leaks.

12. Disconnect the negative battery cable.

13. Install or connect the following:
- Heater hose to the lower intake
- Upper radiator hose to the thermostat housing
- Vacuum hoses to the upper and lower intake manifold
- New upper intake manifold gasket, making sure the green sealing lines are facing upward
- Upper intake manifold being careful not to pinch the fuel injector wires between the manifolds
- Manifold retainers. Tighten them to 88 inch lbs. (10 Nm) using two passes.
- Purge solenoid and bracket
- Brake booster vacuum hose at the upper intake manifold.
- PCV hose to the rear of the upper intake manifold
- Vacuum hoses to both the front and rear of the manifold assembly
- Throttle linkage to the upper intake
- Ignition coil
- Wiring to the upper intake components including the TP sensor, IAC motor, MAP sensor and the IMTV.
- Plastic cover
- Air cleaner and air inlet duct
- Negative battery cable

14. Refill the engine cooling system.

Exhaust Manifold

REMOVAL & INSTALLATION

2.2L Engine

1. Remove or disconnect the following:
- Negative battery cable

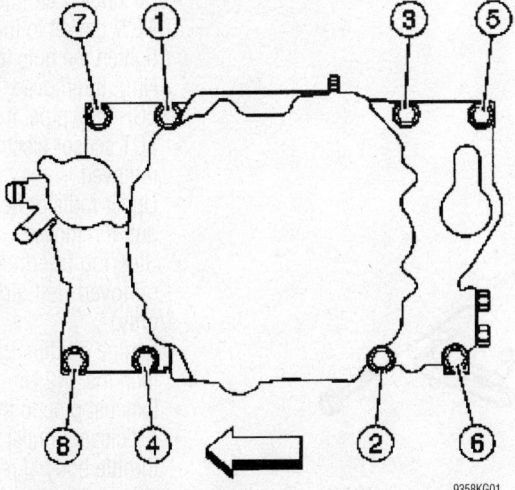

Lower intake manifold tightening sequence—4.3L engines

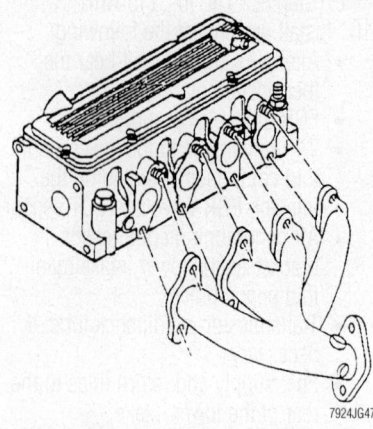

Exhaust manifold mounting—2.2L engine

- Air cleaner and duct work
- Oxygen Sensor (O_2S) from the manifold, if replacing it
- Drive belt
- Oil fill tube assembly
- Heater hose brace
- Power steering brace and set the pump aside
- Air conditioning pencil and rear braces. Set the compressor aside without disconnecting the lines
- Exhaust manifold nuts
- Exhaust manifold

To install:

2. Clean the exhaust manifold retainer threads and the gasket the mating surfaces.

3. Install or connect the following:
- Exhaust manifold using a new gasket. Torque the nuts to 115 inch lbs. (13 Nm).
- Exhaust pipe to the manifold
- Air conditioning pencil and rear braces

- Heater hose
- Power steering and heater hose braces
- Oil fill tube assembly
- O_2S. Torque the sensor to 31 ft. lbs. (42 Nm), if necessary.
- Drive belt
- Air cleaner and duct work
- Negative battery cable

4.3L Engine

1. Remove or disconnect the following:

- Negative battery cable

➡**It will be easier if the vehicle is only supported to a height where underhood access is still possible, the vehicle may be left in position for the entire procedure. If the vehicle is raised too high for underhood access, it will have to lowered, raised and lowered again during the procedure.**

- Breather tube from the air cleaner outlet duct, if removing the left side manifold
- Air cleaner outlet duct retaining wingnut, if removing the left side manifold
- Intake Air Temperature (IAT) sensor harness connector, if removing the left side manifold
- Air cleaner outlet duct from the throttle body, if removing the left side manifold
- Exhaust pipe from the exhaust manifold. It may be necessary to remove the tires to gain access to the rear manifold bolts.

- Engine oil dipstick tube bolt, if removing the right side manifold
- Exhaust Gas Recirculation (EGR) inlet pipe from the left side manifold, if necessary
- Engine Coolant Temperature (ECT) sensor electrical connection
- Upper radiator support hose and nut
- Steering intermediate shaft, if removing the left side manifold
- Wheel house extension, if removing the right side manifold
- Spark plugs wires from the plugs
- Nuts attaching the secondary air injection pipe to the manifold
- Air injection pipe and gasket
- Locktangs (unbend), the exhaust manifold retaining bolts, washers and tab washers
- Heat shields
- Exhaust manifold
- Old gaskets and discard

To install:

2. Using a putty knife, clean the gasket mounting surfaces. Inspect the exhaust manifold for distortion, cracks or damage; replace if necessary.

3. Apply a threadlock such as GM 12345493 to the threads of the manifold retainers prior to installation.

4. Install or connect the following:
- Exhaust manifold to the cylinder using a new gasket, then tighten the to 11 ft. lbs. (15 Nm) and then to 22 ft. lbs. (30 Nm). Once the bolts are tightened, bend the tabs on the washers back over the heads of all bolts in order to lock them in position.
- Spark plug wires to the plugs
- Fender wheelhouse extension and the tire assembly, if removed
- Secondary air injection pipe with a NEW gasket to the manifold and tighten the nuts to 18 ft. lbs. (25 Nm), if removed
- EGR inlet pipe, if removed
- ECT sensor electrical connection, if removed
- Upper radiator hose support and nut, if removed
- Steering intermediate shaft, if removed (left side manifold only)
- Engine oil dipstick tube bolt to 106 inch lbs. (12 Nm), if removed
- Exhaust pipe to the manifold
- Air cleaner outlet duct to the throttle body, if removed
- IAT sensor harness connector, if removed

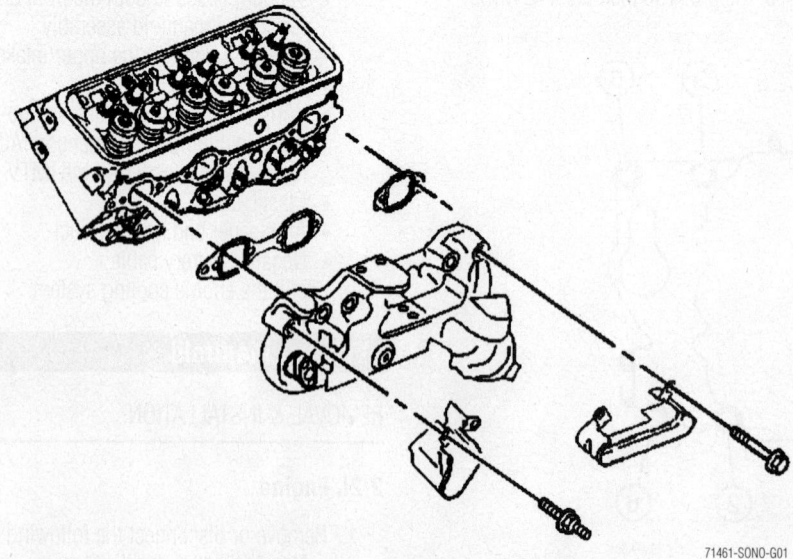

Exploded view of the exhaust manifold mounting—4.3L engine, left side shown

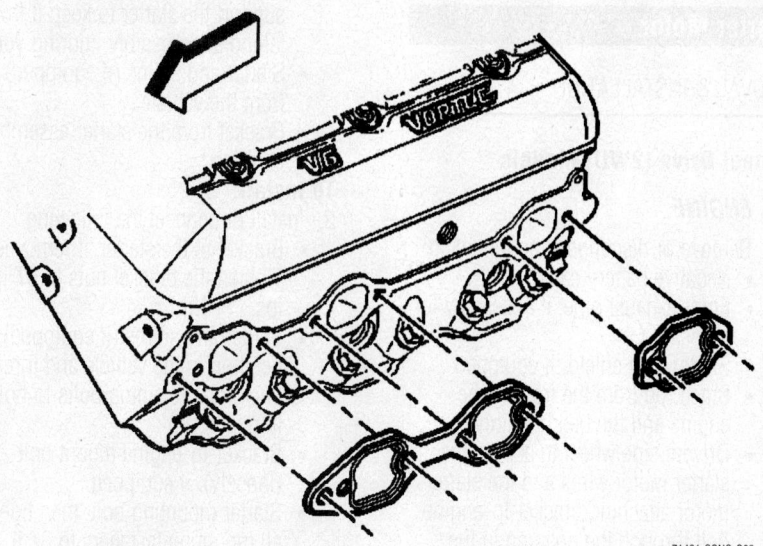

71461-SONO-G02

The new exhaust manifold gasket may have tab, which will help hold the gasket and bolts in place

- Air cleaner outlet duct retaining wingnut, if removed
- Breather tube to the air cleaner outlet duct, if removed
- Negative battery cable

Camshaft and Valve Lifters

REMOVAL & INSTALLATION

2.2L Engine

1. Properly relieve the fuel system pressure.
2. Disconnect the negative battery cable.
3. Drain the engine cooling system and the engine oil.
4. Remove or disconnect the following:
- Radiator
- Rocker arm cover
- Cylinder head
- Anti-rotation bracket bolts and brackets
- Valve lifters
- Oil pump drive retaining bolt and the drive by lifting and twisting
- Camshaft Position (CMP) sensor, if equipped
- Crankshaft pulley and hub
- Drive belt idler pulley
- Timing cover from the engine
- Timing chain and camshaft sprocket
- Camshaft thrust plate
- Camshaft by pulling it straight out of the engine, while turning it slightly as it is withdrawn and taking care not to damage the bearings.

To install:

5. Inspect the camshaft, journals and lobes for wear and replace, if necessary.
6. If removed, use the camshaft bearing tool to install a new set of bearings.
7. Coat the camshaft lobes and journals with a high viscosity oil with zinc such as GM 12345501.
8. Install or connect the following:
- Camshaft by turning it slightly from side-to-side as it is inserted
- Thrust plate. Torque the bolts to 106 inch lbs. (12 Nm).
- Timing chain and camshaft sprocket
- Timing cover
- Serpentine drive belt idler pulley
- Crankshaft pulley and hub

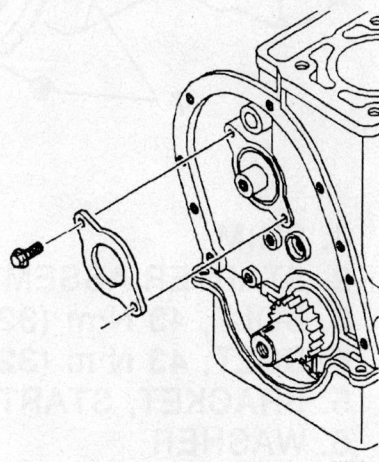

7924JG49

Remove the camshaft thrust plate and withdraw the camshaft from the engine—2.2L engine

- Oil pump drive by inserting while twisting. Torque the fasteners to 18 ft. lbs. (25 Nm).
- Valve lifters and the anti-rotation brackets
- Cylinder head. Torque the bolts to 46 ft. lbs. (62 Nm) plus an additional 90 degrees turn.
- Rocker arm cover
- Radiator
- Negative battery cable
9. Refill the engine cooling system.

4.3L Engine

1. Properly relieve the fuel system pressure.
2. Disconnect the negative battery cable.
3. Drain the engine cooling system.
4. Discharge and recover the refrigerant from the air conditioning system.
5. Remove or disconnect the following:

- Radiator
- Air conditioning condenser
- Rocker arm covers
- Intake manifold assembly
- Rocker arms, pushrods and lifters
- Crankshaft pulley and hub
- Engine front (timing) cover

6. Align the timing marks on the crankshaft and camshaft sprockets.
7. Remove or disconnect the following:
- Camshaft sprocket and timing chain
- Balance shaft drive gear, if equipped
- Camshaft thrust plate
- Camshaft by installing the sprocket bolts or longer bolts the camshaft end to act as a handle; then, remove the camshaft while turning slightly from side to side, as necessary.

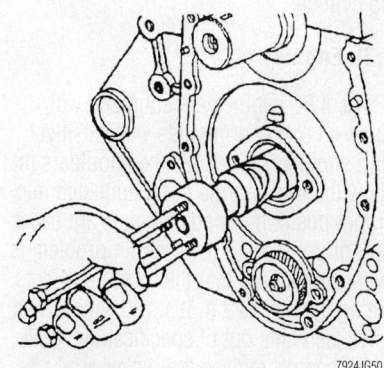

7924JG50

Thread 3 long bolts into the camshaft to use as a handle, then withdraw it from the engine

➡Take care not to damage the camshaft bearings when removing the camshaft.

To install:

8. Lubricate the camshaft journals with clean engine oil or a suitable pre-lube.

9. Install or connect the following:
   - Camshaft being extremely careful not to contact the bearings with the cam lobes
   - Thrust plate. Torque the bolts to 106 inch lbs. (12 Nm).
   - Balance shaft drive gear, if equipped
   - Timing chain and camshaft sprocket
   - Engine front (timing) cover
   - Crankshaft pulley and hub
   - Valve lifters, pushrods and rocker arms. Adjust the valve clearance.
   - Intake manifold assembly
   - Rocker arm covers
   - Radiator
   - Negative battery cable

10. Refill the engine cooling system.

Valve Lash

ADJUSTMENT

2.2L Engine

Because the rocker arm fasteners are secured and tightened, valve lash is not adjustable on the 2.2L engine. If a valve-train problem is suspected, check that the rocker arm nuts are tightened to 18 ft. lbs. (25 Nm). Be very careful not to over-tighten the rocker arm nuts. ONLY tighten the nuts when the hydraulic lifter is resting on the base circle of the camshaft and not when it is held upward on the lobe. When valve lash falls out of specification (valve tap is heard), replace the rocker arm, pushrod and hydraulic lifter on the offending cylinder.

4.3L Engine

The 4.3L engines are equipped with screw-in rocker arm studs with positive stop shoulders. Because the shoulders that allow the rocker arms to be tightened into proper position, no adjustments are necessary or possible. If a valvetrain problem is suspected, check that the rocker arm nuts are tightened to 22 ft. lbs. (30 Nm). When valve lash falls out of specification (valve tap is heard), replace the rocker arm, pushrod and hydraulic lifter on the offending cylinder.

Starter Motor

REMOVAL & INSTALLATION

2 Wheel Drive (2WD) Models

2.2L ENGINE

1. Remove or disconnect the following:
   - Negative battery cable
   - Front exhaust pipe, if necessary for access
   - Starter heat shield, if equipped
   - Brace rod from the front of the engine and the bell housing
   - Drivers side wheel to access the starter motor wires and the starter motor attaching bracket-to-engine bolt through the opening in the wheel well
   - Wires from the starter solenoid
   - Attaching bracket-to-engine mount bolt
   - Starter-to-engine block bolts. When removing the last bolt, be sure to support the starter to keep it from falling and possibly injuring you.
   - Starter and shims (if equipped) from the vehicle
   - Bracket from the starter assembly, if equipped

To install:

2. Install or connect the following:
   - Bracket to the starter, if removed. Tighten the bracket nuts to 97 inch lbs. (11 Nm).
   - Starter and shims (if equipped) into position in the vehicle and thread one of the retaining bolts to hold it in position.
   - Bracket-to-engine mount bolt (loosely), if equipped
   - Starter mounting bolt, then tighten all mounting fasteners to 32 ft. lbs. (43 Nm)
   - Wiring to the solenoid
   - Brace rod and tighten the retainers
   - Front exhaust pipe and tighten the fasteners, if removed
   - Starter heat shield, if equipped

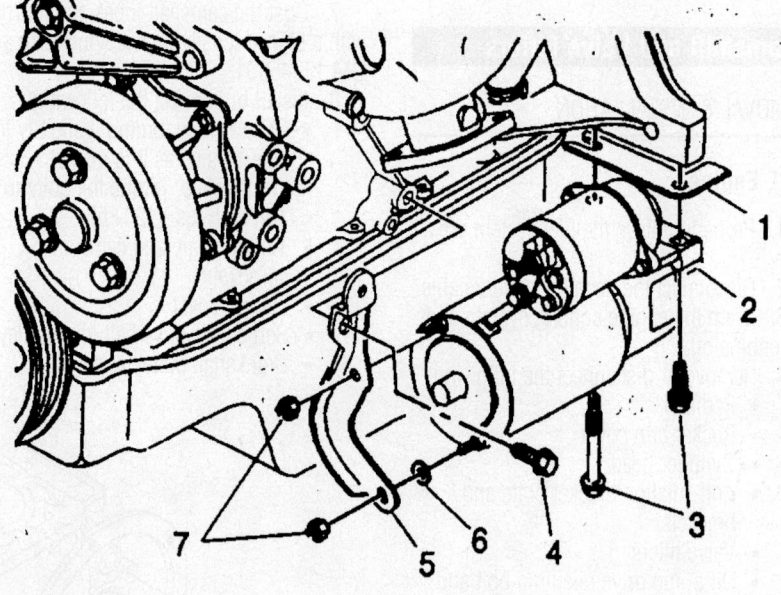

1. SHIM
2. STARTER ASSEMBLY
3. BOLT, 43 N·m (32 LBS. FT.)
4. BOLT, 43 N·m (32 LBS. FT.)
5. BRACKET, STARTER MOTOR
6. WASHER
7. NUT, 11 N·m (97 LBS. IN.)

88452G08

Starter motor and related components—2.2L engine

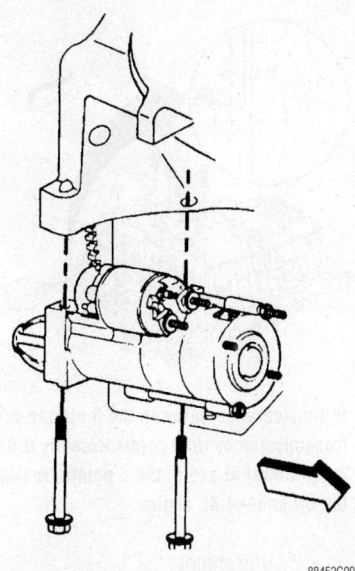

The starter motor on 4.3L engines is retained by two long bolts

- Driver's side wheel, if removed
- Negative battery cable

4.3L MODELS

1. Remove or disconnect the following:
   - Negative battery cable
   - Wires from the starter solenoid
   - Starter motor mounting bolts
   - Starter motor and if equipped, the shims

To install:

2. Install or connect the following:
   - Starter motor into position
   - Starter motor inboard bolt but do not tighten it at this time
   - Starter motor shims, if equipped
   - Outboard starter motor bolt. Tighten the bolts to 32 ft. lbs. (43 Nm).
   - Wires to the solenoid
   - Negative battery cable

4 Wheel Drive (4WD) Models

1. Remove or disconnect the following:
   - Negative battery cable

➡ **In some cases it may be easier to access the starter motor bolts if you raise the vehicle and remove the wheel assembly.**

- Wheel assembly, if necessary
- Engine mounts
- Transmission mount and support the transmission assembly
- Starter-to-engine bolts and support the starter

2. Rotate the starter as necessary for access, then tag and disconnect the solenoid wiring.

3. Carefully lower the starter and shims (if equipped) from the vehicle. Note the location of any shims for installation purposes.

4. If necessary, remove the shield from the starter assembly.

To install:

5. Install or connect the following:
   - Starter into position in the vehicle along with any shims (making sure they are in their original positions), then tighten the mounting bolts to 37 ft. lbs. (50 Nm).
   - Shield to the starter assembly and tighten the retaining nuts to 106 inch lbs. (12 Nm), if removed
   - Wiring to the solenoid
   - Transmission mount and remove the supports
   - Secure the engine mounts, then remove the lifting device
   - Wheel assembly, if removed

6. Connect the negative battery cable.

Oil Pan

REMOVAL & INSTALLATION

2.2L Engine

1. Drain the engine oil.
2. Remove or disconnect the following:
   - Engine
   - Clutch pressure plate and disc, if equipped
   - Flywheel
   - Oil pan retainers and the pan

To install:

3. Clean the gasket mating surfaces
4. Install or connect the following:
   - New gasket and seal onto the oil pan using a thin bead of sealant at either side of the seal
   - Oil pan. Torque the bolts to 89 inch. lbs. (10 Nm).
   - Flywheel

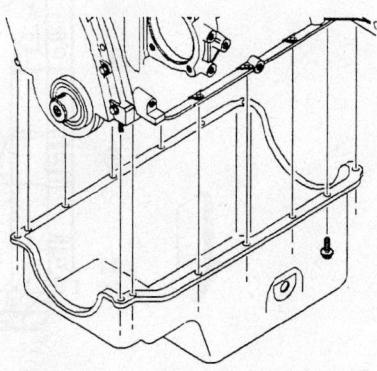

Oil pan mounting—2.2L engine

- Pressure plate and disc, if equipped
- Engine

4.3L Engine

2WD MODELS

1. Drain the engine oil.
2. Remove or disconnect the following:
   - Engine
   - Oil level sensor, if equipped, and discard
   - Oil pan retainers (nuts, studs and/or bolts) and rail reinforcements, if equipped
   - Oil pan
   - Rubber bell housing plugs and gasket

To install:

3. Clean the gasket mounting surfaces.

➡ **The alignment between the rear of the oil pan and the rear of the block is critical. The oil pan must be flush or slightly forward of the rear of the block to allow for proper alignment with the transmission housing. Use a feeler gauge to measure the clearance between the 3 oil pan-to-transmission contact points. If the clearance exceeds 0.011 in. (0.3mm) at any of the 3 points, realign the oil pan.**

4. Apply sealant to the oil pan rail where it contacts the timing cover-to-block joint (front) and the crankshaft rear seal retainer-to-block joint (rear). Continue the bead of sealant about 1 in. (25mm) in both directions from each of the 4 corners.

5. Install or connect the following:
   - Rubber bell housing plugs, if equipped
   - Oil pan using a new gasket

➡ **The alignment between the rear of the pan and rear of the block is critical. The two surfaces must be flush to allow for proper alignment with the transmission housing.**

6. Use a feeler gauge to check the clearance between the oil pan-to-transmission contacts. If clearance exceeds 0.011 inch (0.3mm) at any of the three contact points, readjust the pan until the clearance is within specification.

7. Once the pan is in its correct position tighten the retainers to 18 ft. lbs. (25 Nm) using the proper sequence.

8. Install a new oil level sensor, if used and tighten to 115 inch lbs. (13 Nm).

9. Install the engine into the vehicle. Refill the crankcase with fresh oil. Start the

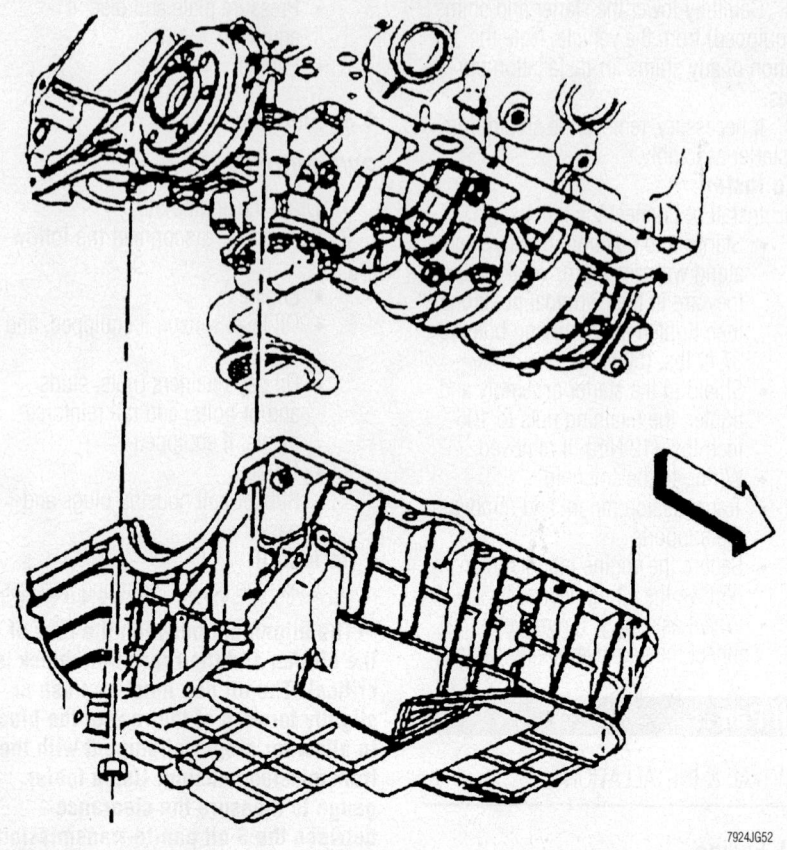

Oil pan mounting—4.3L engine

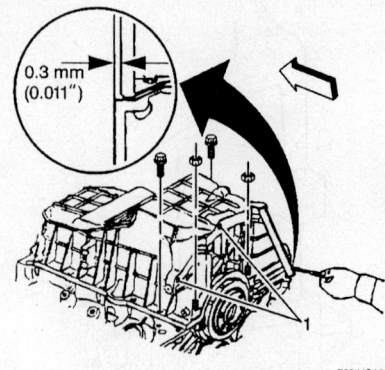

If the clearance between the 3 oil pan-to-transmission contact points exceeds 0.011 in. (0.3mm) at any of the 3 points, realign the oil pan—4.3L engine

- Starter motor
- Oil pan bolts, nuts and reinforcements
- Oil pan and discard the gasket

To install:

6. Clean the gasket mounting surfaces.

➡The alignment between the rear of the oil pan and the rear of the block is critical. The oil pan must be flush or slightly forward of the rear of the block to allow for proper alignment with the transmission housing. Use a feeler gauge to measure the clearance between the 3 oil pan-to-transmission contact points. If the clearance exceeds 0.011 in. (0.3mm) at any of the 3 points, realign the oil pan.

engine, establish normal operating temperatures and check for leaks.

4WD MODELS

1. Disconnect the negative battery cable.
2. Drain the engine crankcase oil.
3. Remove or disconnect the following:
   - Dipstick
   - Drivebelt splash shield, the front axle shield and the transfer case shield
   - Front skid plate and the flywheel cover
   - Left and right engine mount through-bolts
4. Raise the engine using a lifting device and block in position. This may be accomplished using large wooden blocks between the motor mounts and brackets.

➡Use extreme caution when blocking the engine in position. Get out from under the vehicle and rock the engine slightly once the blocks are in place to be sure the engine is properly supported.

5. Remove or disconnect the following:
   - Oil cooler line
   - Pitman arm bolt and pitman arm

- Idler arm bolts and idler arm
- Front differential through-bolts
- Front driveshaft, if necessary
- Differential assembly by rolling it forward for clearance

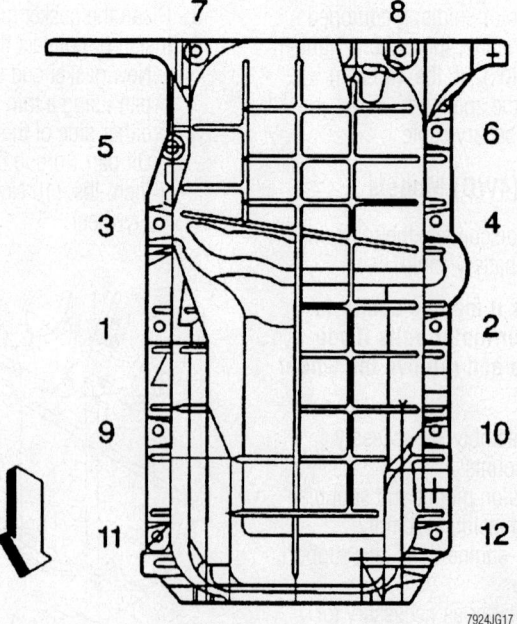

Tighten the bolts in sequence to prevent warping the sealing surface of the oil pan—4.3L engine

7. Apply sealant to the oil pan rail where it contacts the timing cover-to-block joint (front) and the crankshaft rear seal retainer-to-block joint (rear). Continue the bead of sealant about 1 in. (25mm) in both directions from each of the 4 corners.

8. Install or connect the following:
- Oil pan, using a new gasket. Tighten the retainers, in sequence, to 18 ft. lbs. (25 Nm).
- Starter motor
- Differential by rolling it back into position
- Front driveshaft
- Front differential through-bolts
- Idler arm and secure using the retaining bolts
- Pitman arm and secure using the bolts
- Transfer case shield
- Flywheel cover
- Front skid plate
- Front axle shield
- Drive belt splash shield
- Dipstick
- Negative battery cable

9. Refill the engine crankcase.

10. Start the engine and check for leaks.

Oil Pump

REMOVAL & INSTALLATION

1. Remove or disconnect the following:
- Oil pan
- Oil pump and the pickup tube/shaft, if equipped
- Extension shaft and retainer from

the pump, if necessary for the 2.2L engine

➡**Be careful not to crack the retainer.**

To install:

2. Inspect the pins (oil pump locator) for damage, and replace the pins if required.

To install:

3. For the 2.2L engine, if the extension shaft was removed, heat the extension shaft retainer in hot water, then install the shaft and retainer to the oil pump. Be sure the retainer does not crack during installation.

4. Ensure that the pump pickup tube is tight in the pump body. If the tube should come loose, oil pressure will be lost and oil starvation will occur. If the pickup tube is loose it should be replaced.

5. If the pump has been disassembled and is being replaced or for any reason oil has been removed, it must be primed. It can either be filled with oil before installing the cover plate and oil kept within the pump during handling or the entire pump cavity can be filled with petroleum jelly.

> ❈❈ **WARNING**
>
> **If the pump is not primed, the engine could be damaged upon start up.**

➡**Do not reuse the oil pump driveshaft retainer. During assembly, install a NEW oil pump driveshaft retainer.**

6. Install or connect the following:
- Oil pump. Tighten oil pump/pickup tube retainer(s) to 32 ft. lbs. (44 Nm) for the 2.2L engine or to 65 ft. lbs. (90 Nm), for the 4.3L engine.

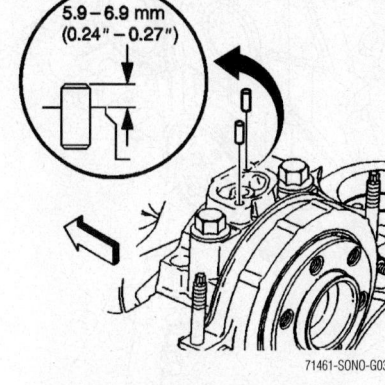

71461-SONO-G03

On 4.3L engines, inspect the pins (oil pump locator) for damage, and replace the pins if required

➡**If the oil pump does not build up oil pressure almost immediately, remove the pan and check for a loose oil pump-to-pickup tube attachment. If necessary dismantle the pump and pack the pump cavity with petroleum jelly.**

- Oil pan
7. Refill the crankcase.
8. Disable the ignition system; crank engine for approximately 10 seconds to aid in priming the oil pump and reducing the risk of engine damage.

> ❈❈ **WARNING**
>
> **Running the engine without measurable oil pressure will cause extensive damage.**

Rear Main Seal

REMOVAL & INSTALLATION

2.2L Engine

Please note that the transmission assembly and transfer case, if equipped, must be removed to perform this procedure.

1. Remove or disconnect the following:
- Negative battery cable
- Transmission assembly and transfer case, if equipped
- Flexplate, if equipped
- Clutch assembly and flywheel, if equipped
- Crankshaft seal by prying it from out

➡**Be careful not to damage the crankshaft seal surface with the prying tool.**

To install:

2. Install the new rear seal by lubricating it with engine oil and using a seal tool J-34686.

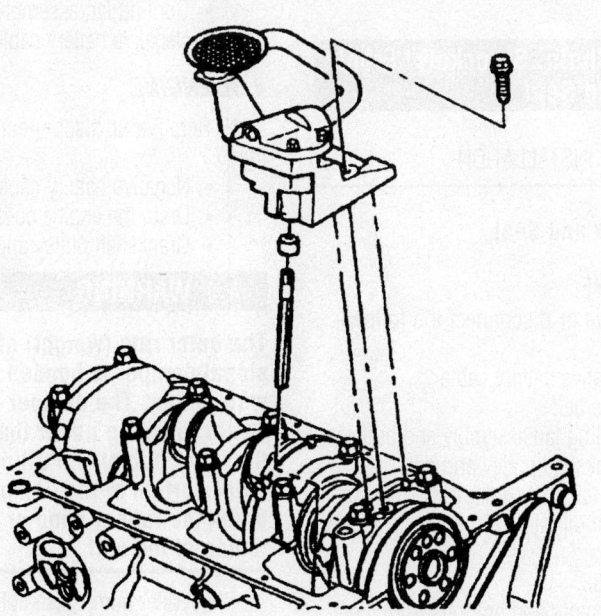

7924JG18

Exploded view of the oil pump mounting—2.2L engine

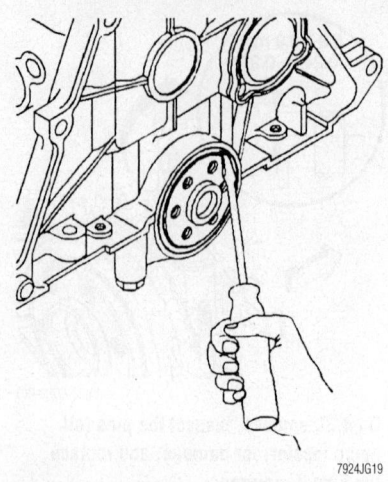

Carefully pry the rear main oil seal out of
its bore—2.2L engine

3. Slide the seal over the mandrel until
the dust lip bottoms squarely against the
tool collar.

4. Align the dowel pin of the tool with
the dowel pinhole in the crankshaft and
attach the tool to crankshaft.

5. Tighten the T-handle of the tool to
push the seal into the bore. Continue until
the tool collar is flush against the block.

6. Loosen the T-handle completely.
Remove the attaching screws and tool.
Check to be sure the seal is seated squarely
in the bore.

7. Install or connect the following:
  • Flywheel/clutch assembly or flex-
    plate
  • Transmission assembly and trans-
    fer case, if equipped
  • Negative battery cable
8. Start the engine and check for leaks.

4.3L Engine

Please note that the transmission assembly and transfer case, if equipped, must be
removed to perform this procedure.

1. Remove or disconnect the following:

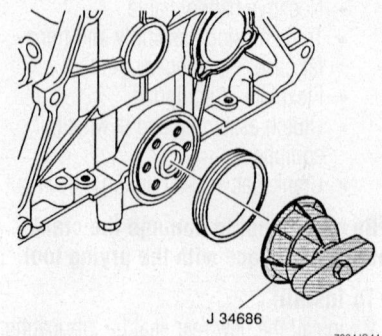

Rear main oil seal installation using tool
J-34686—2.2L engine

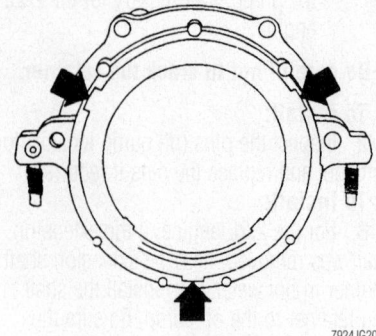

Carefully pry the rear main seal out of the
retainer—4.3L engine

  • Negative battery cable
  • Transfer case, if equipped
  • Transmission
  • Clutch assembly/flywheel or flex-
    plate
2. Remove the crankshaft rear oil seal by
inserting a suitable prying tool into the
notches provided in the seal retainer and
prying the seal out. Take care not to damage
the crankshaft sealing surface.

To install:
3. Inspect the crankshaft for grit, rust or
burrs and correct as necessary.
4. Clean the running surface of the
crankshaft with a non-abrasive cleaner.
5. Install or connect the following:
  • New rear seal lubricated with
    engine oil and a seal installer
  • Flywheel and clutch or flexplate
  • Transmission
  • Transfer case, if equipped
  • Negative battery cable
6. Start the engine and verify no oil
leaks.

Timing Chain, Sprockets, Front Cover and Seal

REMOVAL & INSTALLATION

Front Cover and Seal

2.2L ENGINE

1. Remove or disconnect the following:

  • Negative battery cable
  • Drive belt
  • Cooling fan assembly and pulley
  • Crankshaft pulley and hub
  • Belt tensioner/idler pulley assembly
  • Front oil pan-to-front cover nuts or
    studs
  • Starter
  • Alternator and brackets from the
    engine, then position them aside

  • Oil pan bolts, loosen but do not
    remove
  • Crankcase (timing) front cover
    bolts and the cover. Make sure all
    bolts are removed and be careful
    not to force and damage the cover.
  • Old crankshaft seal from the cover
    using a suitable prytool. Be very
    careful not to distort the front cover
    or to score the end of the crank-
    shaft.

To install:
2. Carefully remove all traces of gasket
or sealant from the mating surfaces.
3. Lubricate the lips of a new seal with
clean engine oil, then use a seal centering
tool (such as J-35468) to install the seal to
the front cover. Leave the tool in position in
the seal until the cover is installed.
4. Apply a ⅜ in. (10mm) wide by ⁵⁄₁₆
(5mm) thick bead of RTV sealer to the oil
pan at the front crankcase cover sealing surface. Then apply a ¼ in. (6mm) by ⅛ in.
(3mm) thick bead of RTV to the crankcase
front cover at the block sealing surface.
5. Install or connect the following:
  • Crankcase front cover to the engine
    using the seal tool to assure it is
    properly centered and prevent dam-
    age to the hub. Tighten the cover
    retaining bolts to 97 inch lbs. (11
    Nm), then remove the seal center-
    ing tool.
  • Oil pan bolts
  • Starter
  • Alternator with brackets
  • Belt tensioner/idler pulley assembly
  • Belt assembly
  • Crankshaft pulley and hub
  • Cooling fan assembly and pulley
  • Negative battery cable

4.3L ENGINE

1. Remove or disconnect the following:

  • Negative battery cable
  • Drain the engine cooling system.
  • Crankshaft pulley and damper

❊❊ WARNING

**The outer ring (weight) of the tor-
sional damper is bonded to the hub
with rubber. The damper must be
removed with a puller that acts on
the inner hub only. Pulling on the
outer portion of the damper will
break the rubber bond or destroy the
tuning of the unit.**

  • Water pump assembly
  • Crankshaft Position (CKP) sensor

➡**Depending upon the year of your truck, you may just need to loosen the oil pan bolts, or you may need to remove it completely.**

- Oil pan or loosen bolts, as applicable
- Crankshaft Position (CKP) sensor, if equipped
- Front cover bolts and the reinforcements, if equipped
- Front cover from the engine

2. Pry the seal out of the front cover using a small prytool. Be very careful not to distort the front cover or to score the end of the crankshaft.

To install:

➡**Anytime the front cover is removed, the cover must be replaced upon reassembly. If you reuse the old cover, oil leaks may develop.**

3. Clean the gasket mating surfaces of the engine and cover of all remaining gasket or sealer material. Be careful not to score or damage the surfaces.

➡**The manufacturer suggests you wait until the front cover is mounted to the engine before you install the replacement crankshaft oil seal. This assures the cover is properly supported.**

4. Install or connect the following:
- New front cover gasket to the engine or cover using gasket cement to hold it in position. Lubricate the front of the oil pan seal with engine oil to aid in reassembly.
- Front cover to the engine. Take care while engaging the front of the oil pan seal with the bottom of the cover. Apply sealer 12346141 to the oil pan rail where it contacts the timing cover-to-block joint (front) and the crankshaft rear seal retainer-to-block joint (rear). Continue the bead of sealant about 1 in. (25mm) in both directions from each of the four corners.
- Front cover retaining bolts and tighten to 106 inch. lbs. (12 Nm)

5. Lightly coat the lips of the replacement crankshaft seal with clean engine oil, then position the seal with the open end facing inward the engine. Use a suitable seal installation driver to position the seal in the front cover.
- CKP sensor O-ring and the sensor, if equipped
- Oil pan, if removed
- Tighten the oil pan bolts

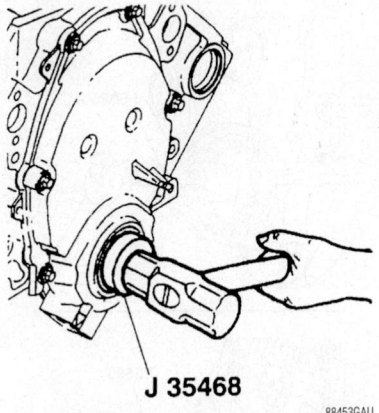

J 35468

88453GAU

Installing the crankshaft front oil seal— 4.3L engines

- Water pump
- Crankshaft damper and pulley
- Negative battery cable

6. Properly refill the engine cooling system.

7. Run the engine until normal operating temperature has been reached, then check for leaks.

Timing Chain and Sprockets

2.2L ENGINE

1. Remove or disconnect the following:
- Negative battery cable

- Crankcase (timing) front cover from the engine

2. Turn the crankshaft until the timing marks on the sprockets are in alignment. The marks should also be in alignment with the tabs on the tensioner.
- Tensioner retaining bolts
- Camshaft sprocket retaining bolts
- Camshaft sprocket and timing chain at the same time
- Tensioner assembly
- Crankshaft Position (CKP) sensor, if equipped
- Crankshaft sprocket using J-22888-20, if equipped

To install:

3. Install or connect the following:
- Crankshaft sprocket using a suitable installer such as J-5590, if removed. Make sure the sprocket is fully seated against the crankshaft.

4. Compress the tensioner spring and insert a cotter pin or nail in the hole provided to hold the tensioner in position.
- Tensioner retaining bolts
- Camshaft sprocket in the timing chain, position the chain under the crankshaft sprocket and the camshaft sprocket to the camshaft

5. Verify that the timing marks are all properly aligned, then loosely install the camshaft sprocket bolt.

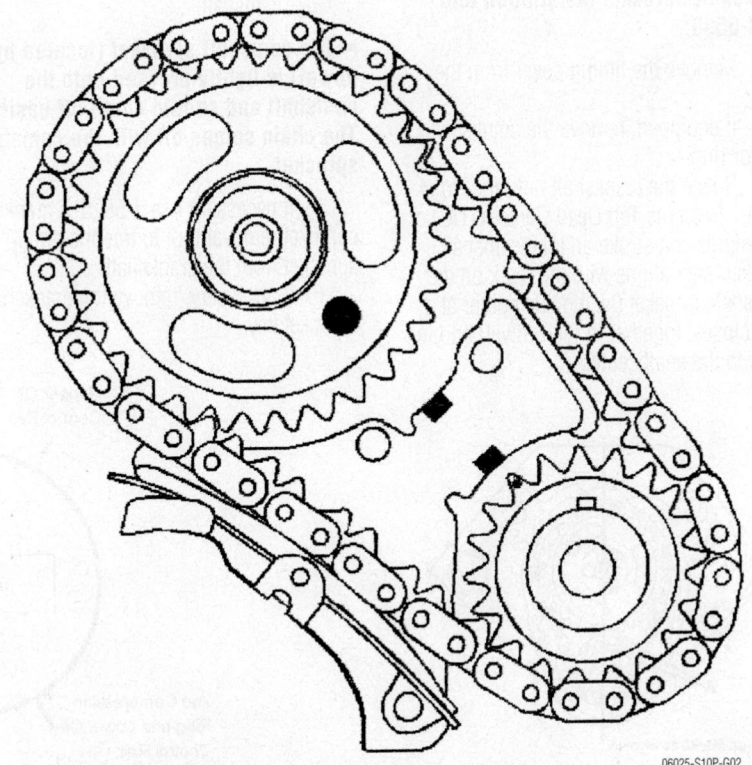

06025-S10P-G02

Timing chain, sprocket and camshaft alignment marks—2.2L engine

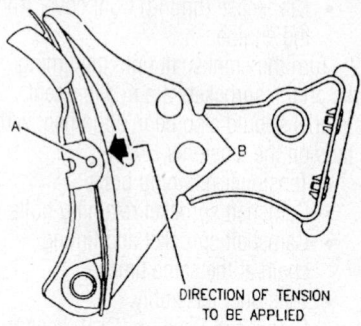

A INSERT PIN AFTER TENSION HAS BEEN APPLIED
B TABS, USED FOR CAMSHAFT AND CRANKSHAFT
   ALIGNMENT

85383290

Locking the timing chain tensioner into position for chain installation—2.2L engine

- CKP sensor, if equipped
- Tighten the tensioner bolts to 18 ft. lbs. (24 Nm), then tighten the camshaft sprocket bolt to 96 ft. lbs. (130 Nm)

6. Remove the cotter pin or nail holding the tensioner in position off the chain.

- Timing cover to the engine

4.3L ENGINE

➡The following procedure requires the use of the Crankshaft Sprocket Removal tool No. J-5825-A and the Crankshaft Sprocket Installation tool No. J-5590.

1. Remove the timing cover from the engine.

2. If equipped, remove the crankshaft reluctor ring.

3. Rotate the crankshaft until the No. 4 cylinder is on the Top Dead Center (TDC) of its compression stroke and the camshaft sprocket mark aligns with the mark on the crankshaft sprocket (facing each other at a point closest together in their travel) and in line with the shaft centers.

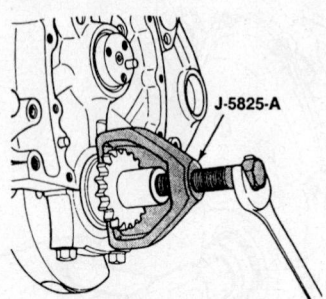

J-5825-A

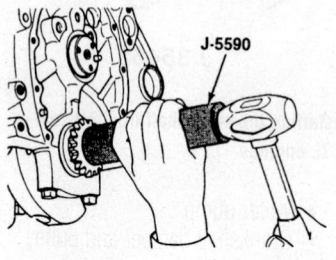

J-5590

85383293

Removal (top) and installation of the crankshaft timing gear

4. Remove or disconnect the following:
- Crankshaft Position (CKP) sensor reluctor ring, if equipped
- Camshaft sprocket-to-camshaft nut and/or bolts
- Camshaft sprocket (along with the timing chain). If the sprocket is difficult to remove, use a plastic mallet to bump the sprocket from the camshaft.

➡The camshaft sprocket (located by a dowel) is lightly pressed onto the camshaft and should come off easily. The chain comes off with the camshaft sprocket.

5. If necessary use J-5825-A crankshaft sprocket removal tool to free the timing sprocket from the crankshaft.

6. If necessary, remove the crankshaft sprocket key.

To install:

7. Inspect the timing chain and the timing sprockets for wear or damage, replace the damaged parts as necessary.

8. Using a putty knife, clean the gasket mounting surfaces. Using solvent, clean the oil and grease from the gasket mounting surfaces.

9. Install or connect the following:
- Crankshaft sprocket key, if removed
- Crankshaft sprocket onto the crankshaft using J-5590 crankshaft sprocket installation tool and a hammer without disturbing the position of the engine

➡During installation, coat the thrust surfaces lightly with Molykote® or an equivalent pre-lube.

- Timing chain over the camshaft sprocket. Arrange the camshaft sprocket in such a way that the timing marks will align between the shaft centers and the camshaft locating dowel will enter the dowel hole in the cam sprocket.
- Timing chain under the crankshaft sprocket, then place the cam sprocket, with the chain still mounted over it, in position on the front of the camshaft
- Camshaft sprocket-to-camshaft retainers to 18 ft. lbs. (25 Nm)

10. With the timing chain installed, turn the crankshaft two complete revolutions, then check to make certain that the timing marks are in correct alignment between the shaft centers.

- CKP sensor reluctor ring, if equipped
- Timing cover

Piston and Ring

POSITIONING

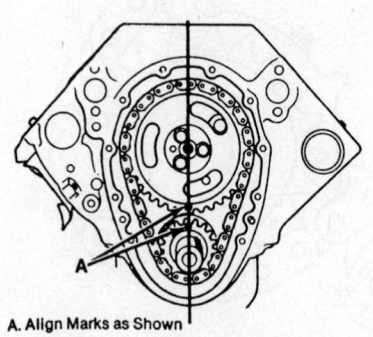

A. Align Marks as Shown

85383292

Timing mark alignment—4.3L engine

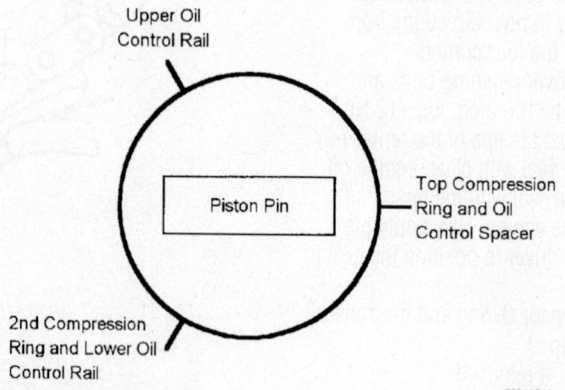

Upper Oil
Control Rail

Piston Pin

Top Compression
Ring and Oil
Control Spacer

2nd Compression
Ring and Lower Oil
Control Rail

7924AG12

Piston ring end-gap spacing—2.2L engine

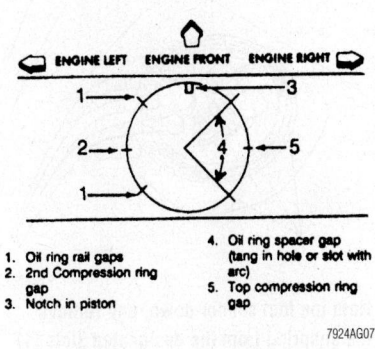

Piston ring end-gap spacing—4.3L engines

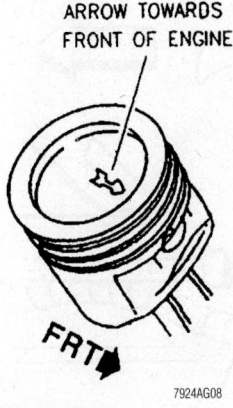

Piston/connecting rod-to-engine positioning—2.2L engines

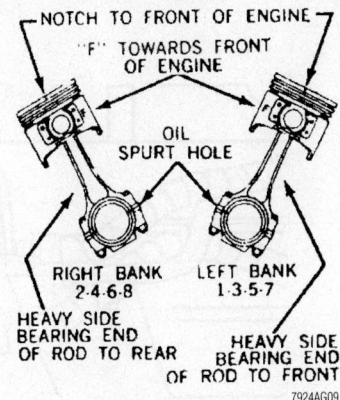

Piston and connecting rod assembly positioning—4.3L engine

FUEL SYSTEM

Fuel System Service Precautions

Safety is the most important factor when performing not only fuel system maintenance but also any type of maintenance. Failure to conduct maintenance and repairs in a safe manner may result in serious personal injury or death. Maintenance and testing of the vehicle's fuel system components can be accomplished safely and effectively by adhering to the following rules and guidelines.

• To avoid the possibility of fire and personal injury, always disconnect the negative battery cable unless the repair or test procedure requires that battery voltage be applied.

• Always relieve the fuel system pressure prior to disconnecting any fuel system component (injector, fuel rail, pressure regulator, etc.), fitting or fuel line connection. Exercise extreme caution whenever relieving fuel system pressure, to avoid exposing skin, face and eyes to fuel spray. Please be advised that fuel under pressure may penetrate the skin or any part of the body that it contacts.

• Always place a shop towel or cloth around the fitting or connection prior to loosening to absorb any excess fuel due to spillage. Ensure that all fuel spillage (should it occur) is quickly removed from engine surfaces. Ensure that all fuel soaked cloths or towels are deposited into a suitable waste container.

• Always keep a dry chemical (Class B) fire extinguisher near the work area.

• Do not allow fuel spray or fuel vapors to come into contact with a spark or open flame.

• Always use a back-up wrench when loosening and tightening fuel line connection fittings. This will prevent unnecessary stress and torsion to fuel line piping.

Always follow the proper torque specifications.

• Always replace worn fuel fitting O-rings with new. Do not substitute fuel hose or equivalent where fuel pipe is installed.

Fuel System Pressure

RELIEVING

Multi-Port Fuel Injection and Central Port Injection Systems

The fuel systems operate under high fuel pressures. It is very important that the pressure be properly relieved prior to servicing the system or any of its components.

A Schrader valve is provided on these fuel systems to conveniently test or release the system pressure. A fuel pressure gauge and adapter will be necessary to connect the gauge to the fitting. Most of the MFI systems utilize a service valve on one end of the fuel rail assembly.

1. Before servicing the vehicle, refer to the precautions in the beginning of this section.

2. Disconnect the negative battery cable to assure the prevention of fuel spillage if the ignition switch is accidentally turned **ON** while a fitting is still detached.

3. Loosen the fuel filler cap to release the fuel tank pressure.

4. Be sure the release valve on the fuel gauge is closed, then connect the fuel gauge to the pressure fitting located on the inlet fuel pipe fitting.

❈ CAUTION

When connecting the gauge to the fitting, be sure to wrap a rag around the fitting to avoid spillage. After

repairs, place the rag in an approved container.

5. Install the bleed hose portion of the fuel gauge assembly into an approved container, then open the gauge release valve and bleed the fuel pressure from the system.

6. When the gauge is removed, be sure to open the bleed valve and drain all fuel from the gauge assembly.

7. When fuel service is finished, tighten the fuel filler cap and connect the negative battery cable.

Fuel Filter

REMOVAL & INSTALLATION

1. Before servicing the vehicle, refer to the precautions in the beginning of this section.

2. Properly relieve the fuel system pressure.

3. Remove or disconnect the following:
   • Negative battery cable
   • Fuel filler cap
   • Quick connect fittings from the filter
   • Filter feed nut and the clamp bolt
   • Filter and the clamp from the vehicle

To install:

4. Install or connect the following:
   • Filter and clamp with the directional arrow facing away from the fuel tank, toward the throttle body

→The filter has an arrow (fuel flow direction) on the side of the case, be sure to install it correctly in the system, the with arrow facing away from the fuel tank.

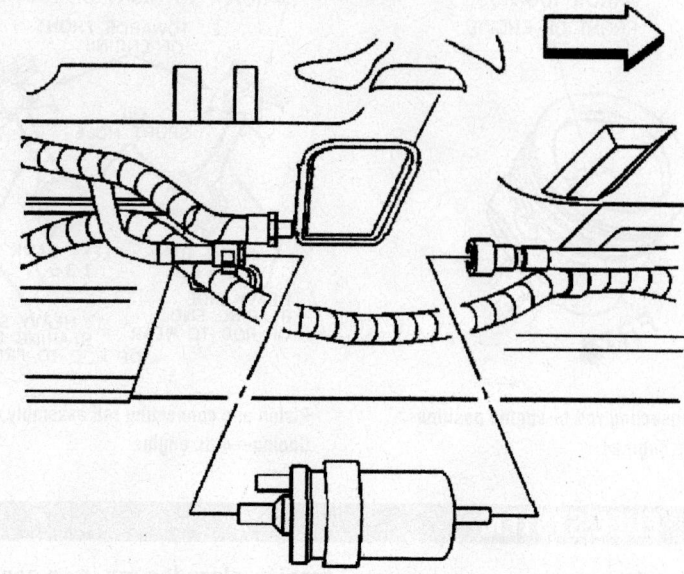

Fuel filter location along frame rail—2.2L engine

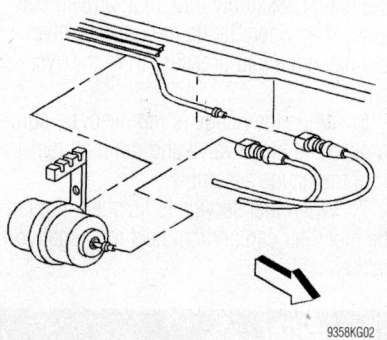

Fuel filter location along frame rail—4.3L engine

- Fuel feed nut
- Filter clamp assembly bolt
- Fuel quick disconnect fittings to the filter
- Fuel filler cap
- Negative battery cable

5. Check for fuel leaks as follows:
   a. Turn the ignition **ON** for 2 seconds.
   b. Turn the ignition **OFF** for 10 seconds.
   c. Turn the ignition **ON**.
   d. Check for leaks.

Fuel Pump

REMOVAL & INSTALLATION

1. Properly relieve the fuel system pressure.
2. Disconnect the negative battery cable.

3. Lower the spare tire.
4. Remove or disconnect the following:
   - Rear tail lamp assemblies
   - Frame-to-pickup box bolts
   - Wiring harness ground wire from the frame
   - License plate lamp
   - Fuel filler neck-to-pickup box ground wire and screws
   - Fuel tank on crew cab models
   - Pickup box
   - Fuel sender electrical connectors
   - Fuel sender and Evaporative Emission (EVAP) pipes

✳✳ WARNING

The fuel sender assembly may spring up from the fuel tank. When removing the sender from the fuel tank, keep in mind that the reservoir bucket is full of fuel, so you must tip the sender slightly during removal to avoid damaging the float. Discard the fuel sender O-ring and replace with a new on during installation.

5. While holding the module fuel sender down, remove the snapring from the designated slots (1) found on the retainer.

To install:

6. Install a new O-ring on the fuel sender to the tank.
7. Align the tab on the front of the sender with the slot on the front of the retainer snapring.
8. Slowly apply pressure to the top of the spring-loaded sender until it aligns flush with the retainer on the tank.

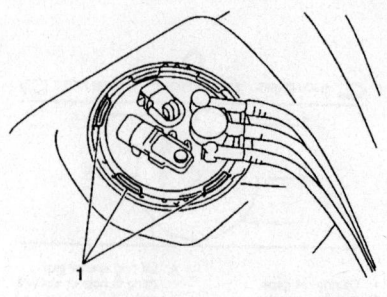

Hold the fuel sender down, and remove the snapring from the designated slots (1) found on the retainer

➡**Make sure that the snapring is properly and fully seated in the tab slots.**

9. Install or connect the following:
   - Snapring into the proper slots
   - Fuel and EVAP pipes
   - Electrical connectors
   - Negative battery cable
10. Check for fuel leaks as follows:
    a. Turn the ignition **ON** for 2 seconds.
    b. Turn the ignition **OFF** for 10 seconds.
    c. Turn the ignition **ON**.
    d. Check for leaks.
11. Install or connect the following:
    - Pickup box to the truck
    - Filler neck-to-pickup box screws and ground wire
    - License plate lamp
    - Wiring harness ground wire to the frame
    - Frame-to-pickup box bolts and tighten to 52 ft. lbs. (70 Nm)
    - Rear tail lamp assemblies

Fuel Injector

REMOVAL & INSTALLATION

2.2L Engine

1. Relieve the fuel system pressure.
2. Remove or disconnect the following:
   - Negative battery cable
   - Intake manifold, if necessary
   - Fuel injector electrical connections by pushing in the wire connector clip and gently pulling on the connector
   - Fuel feed inlet pipe from the rail

➡**Use a back-up wrench on the fuel rail return fitting to prevent it from turning.**

   - Fuel return pipe from the fuel pressure regulator

- Fuel pressure regulator
- Fuel rail attaching bolts and lift the fuel rail assembly from the cylinder head
- Fuel rail by moving the rail toward the front of the engine
- Fuel injector retaining clip
- Fuel injector

➡**Because each injector is calibrated for a specific flow rate, make sure you only replace fuel injectors using an IDENTICAL part number to the old injectors.**

To install:

➡**When installing the injector care should be taken not to tear or misalign O-rings.**

3. Lubricate the injector O-ring seals with clean engine oil and install them injector.

4. Install or connect the following:
- Upper O-ring, lower back-up O-ring and lower O-ring
- Fuel injector to the fuel rail
- Fuel injector retaining clip
- Fuel rail and insert it into the cylinder head
- Fuel rail retaining bolts and tighten to 18 ft. lbs. (25 Nm)
- Fuel pressure regulator

➡**Use a back-up wrench on the fuel rail return fitting to prevent it from turning.**

- Return pipe to the fuel pressure regulator. Tighten the fuel pipe nut to 22 ft. lbs. (30 Nm).
- Fuel feed inlet pipe to rail

➡**Rotate the fuel injectors as necessary to avoid stretching the wire harness.**

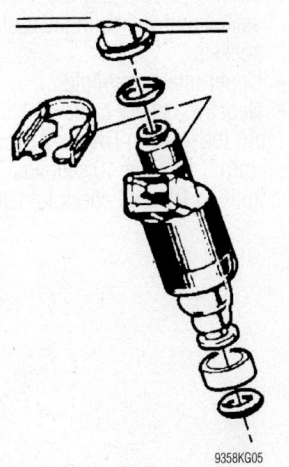

9358KG05

Exploded view of a typical fuel injector—2.2L engine

- Injector electrical connections
- Intake manifold, if removed
- Negative battery cable

5. Inspect for leaks as follows:
   a. Turn the switch to the **ON** position for 2 seconds.
   b. Turn the ignition switch **OFF** for 10 seconds.
   c. Turn the ignition switch to the **ON** position and check for leaks.

4.3L Engine

2002 VEHICLES

1. Relieve the fuel system pressure. Refer to the fuel system relief procedure in this section.
2. Relieve the fuel system pressure.
3. Remove or disconnect the following:
- Negative battery cable
- Fuel meter body electrical connection and the fuel feed and return hoses from the engine fuel pipes
- Upper manifold assembly

➡**When disconnecting the poppet nozzles, note the sequence for correct poppet nozzle placement to each cylinder.**

- Poppet nozzle out of the casting socket by squeezing the poppet nozzle locking tabs together while lifting the nozzle out of the casting socket
- Fuel meter body by releasing the locktabs

➡**Each injector is calibrated. When replacing the fuel injectors, be sure to replace it with the correct injector.**

- Lower hold-down plate and nuts

4. While pulling the poppet nozzle tube downward, push with a small prytool down between the injector terminals and remove the injectors.

To install:

5. Lubricate the new injector O-ring seats with engine oil.

6. Install or connect the following:
- O-rings on the injector
- Fuel injector into the fuel meter body injector socket.
- Lower hold-down plate and nuts. Torque the nuts to 27 inch lbs. (3 Nm).
- Fuel meter body assembly into the intake manifold. Torque the fuel meter bracket retainer bolts to 88 inch. lbs. (10 Nm).

To reduce the risk of fire or injury ensure that the poppet nozzles are

properly seated and locked in their casting sockets

- Fuel meter body into the bracket and lock all the tabs in place

➡**The fuel meter body assemblies are numbered to indicate poppet nozzle order.**

- Poppet nozzles into the casting sockets
- Electrical connections
- New o-ring seals on the fuel return and feed hoses
- Fuel feed and return hoses and tighten the fuel pipe nuts to 22 ft. lbs. (30 Nm)
- Negative battery cable

7. Turn the ignition **ON** for 2 seconds and then turn it **OFF** for 10 seconds. Again turn the ignition **ON** and check for leaks.

8. Install the manifold plenum.

2003–05 VEHICLES

1. Relieve the fuel system pressure. Refer to the fuel system relief procedure in this section.
2. Relieve the fuel system pressure.
3. Remove or disconnect the following:
- Negative battery cable
- Upper manifold assembly

Do not use any solvent that contains Methyl Ethyl Ketone (MEK). This solvent may damage fuel system components.

➡**Before removal, clean the fuel meter body assembly with a spray type engine cleaner, GM X-30A or the equivalent, if necessary. Follow the package instructions. DO NOT soak fuel meter body assemblies in liquid cleaning solvent.**

4. Cover the fuel injector sockets to prevent dirt and other debris from entering the open fuel passages.

➡**When disconnecting the fuel injectors, note the sequence to ensure correct injector placement to each cylinder.**

5. Lightly pull on the injector tube and use a small screwdriver to carefully release the injector from the manifold.

6. Remove the fuel meter body from the bracket by releasing the lock tabs on the bracketRemove the injector retainer lock nuts and retainer.

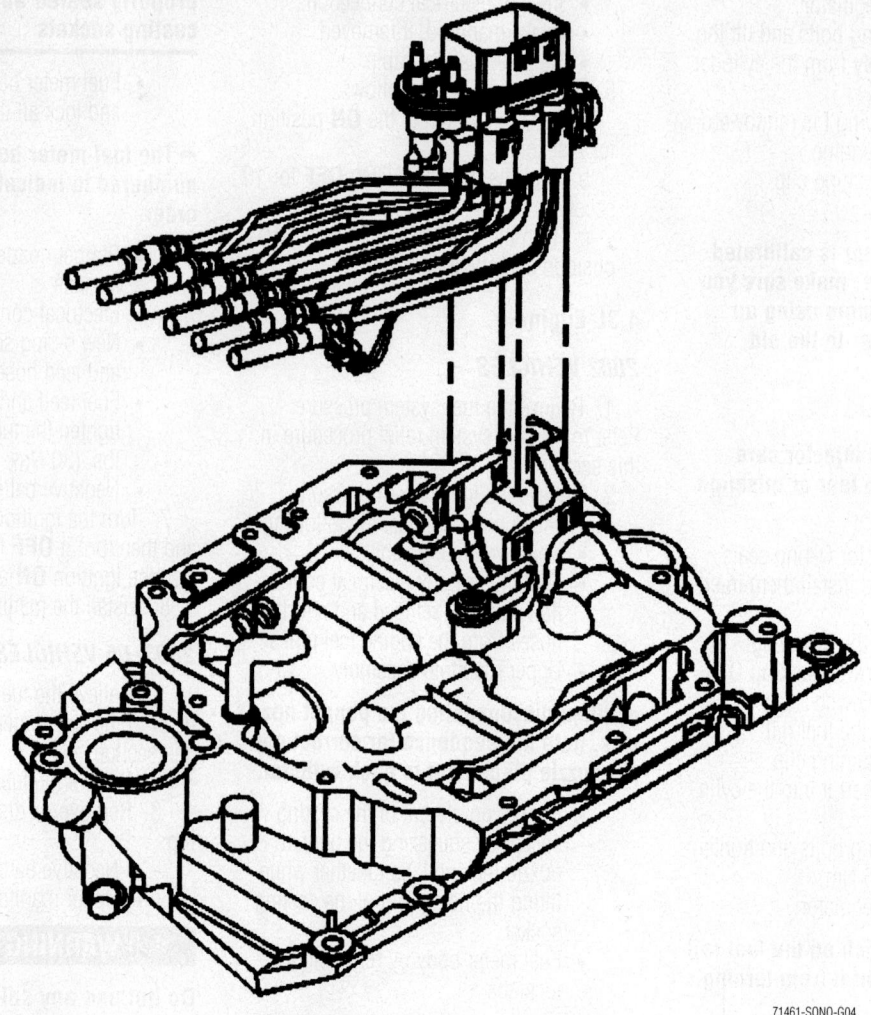

71461-SONO-G04

Exploded view of the fuel meter assembly, including the injectors

➡Be careful when removing the fuel injectors to avoid damaging the electrical connector terminals. The fuel injector is serviced as a complete assembly only. Also, since the injectors are electrical components, these injectors should not be immersed in any type of liquid solvent or cleaner as damage may occur. Fuel injector cleaning is not recommended.

7. While pulling the fuel injector downward, push with a small tip punch down between the injector terminals until the injector is removed.

To install:

8. Lubricate the new injector O-ring seats with engine oil.

9. Install or connect the following:
- O-rings on the injector
- Fuel injector into the fuel meter body injector socket.
- Retainer and the injector retainer lock nuts. Tighten the nuts to 27 inch lbs. (3 Nm).
- Fuel meter body in the intake manifold. Push the fuel meter body into the bracket. Make sure all of the tabs are locked into place

➡The fuel meter body assemblies are numbered to indicate poppet nozzle order.

- Poppet nozzles into the casting sockets. Inspect the injectors in order to ensure they are firmly seated and locked in the casting sockets.
- Upper intake manifold
- Negative battery cable

10. Turn the ignition **ON** for 2 seconds and then turn it **OFF** for 10 seconds. Again turn the ignition **ON** and check for leaks.

DRIVE TRAIN

Manual Transmission

REMOVAL & INSTALLATION

1. Shift the transmission into 3rd or 4th gear position.
2. Remove or disconnect the following:
   - Negative battery cable
   - Shift lever and the if necessary, the shift housing
   - Parking brake cable for clearance
   - Propeller shaft
   - Side plate, if equipped
   - Transfer case and shift lever, on 4WD models
   - All wiring harness that would interfere with transmission removal
   - Fuel line retainers from the rear crossmember
   - Muffler from the catalytic converter
   - Exhaust pipes from the exhaust manifold
   - Catalytic converter hangar, if necessary
   - Exhaust section
   - Bolts and nuts attaching any transmission braces to the engine and transmission
3. Disconnect the hydraulic clutch quick-connect from the concentric slave cylinder following 1 of the 2 steps:
   a. Use 2 small prytools at 180 degrees from each other to depress the white plastic sleeve on the quick connect to separate the clutch line from the concentric slave cylinder quick connect.
   b. Use special tool J–36221 to depress the white plastic sleeve on the quick connect to separate the clutch line end from the concentric slave cylinder quick connect.
4. Remove or disconnect the following:
   - Bolts securing the clutch housing cover to the transmission, if equipped
   - Clutch plate and clutch cover, if necessary
5. Support the transmission with a suitable jack.
   - Rear crossmember from the frame rail
   - Wiring harness from the front crossmember, if equipped. Move the wiring harness away from the transmission oil pan. Lower the transmission enough to gain access to the top of the transmission.
   - Fuel line retainers or wiring harnesses from the top of the transmission
   - Bolt, washer, and nut securing the wiring harness ground wires to the engine block
   - Bolts retaining the transmission to the engine. Pull the transmission straight back on the clutch hub splines.
6. Lower the transmission using the transmission jack.

To install:

Installation is the reverse of removal, but please note the following important steps.

7. Place a THIN coat of high-temperature grease on the main drive gear (input shaft) splines.
8. Secure the transmission to the floor jack and raise the transmission into position.

➡ **On some models, it may be necessary to rotate the transmission clockwise while inserting it into the clutch hub.**

9. Slowly insert the input shaft through the clutch. Rotate the output shaft slowly to engage the splines of the input shaft into the clutch while pushing the transmission forward into place. Do not force the transmission into position, the transmission should easily fall into place once everything is properly aligned.
10. Tighten the transmission mounting bolts to 35 ft. lbs. (47 Nm).
11. Do not remove the transmission jack until the crossmembers have been installed.
12. Check the transmission fluid level and replenish as necessary.

Automatic Transmission

REMOVAL & INSTALLATION

1. Remove or disconnect the following:
   - Negative battery cable
2. Drain the transmission fluid.
   - Driveshaft from the transmission (2WD) and transfer case, if equipped (4WD)
3. Support the transmission with a suitable transmission jack.
   - Shift cable from the transmission control lever and bracket
   - Nut and washer securing the transmission mount to the crossmember
   - Bolts and washers securing the mount to the transmission
   - Exhaust pipe from the exhaust manifold(s)
   - Bolts securing the converter pan cover to the transmission, if equipped
   - 3 bolts securing the torque converter to the flywheel
   - Bolt, clip, and strap securing the three fuel lines and transmission vent hose to the transmission case
   - Bolts and nut securing the transmission to the engine
   - Oil filler tube and seal from the transmission
   - Transmission cooler lines from the transmission. Plug the lines and the ports in the transmission.
   - Wiring harness connectors from the transmission.
4. Inspect for any other wiring, brackets etc. which may interfere with the removal of the transmission.
5. Since the transmission acts as a rear engine mount, properly support the rear of the engine with an underbody support or other suitable support before attempting to remove the transmission. Otherwise the rear of the engine may pitch downward and components on the rear of the engine and on the firewall may be damaged.
6. Remove the transmission from the engine by pulling the transmission rearward to disengage it from the locator dowel pins on the back of the block. Carefully lower the transmission from the vehicle. Use care that the torque converter does not fall out of the front of the transmission.

➡ **Use converter holding strap tool No. J-21366, to secure the torque converter to the transmission during removal and installation procedures.**

To install:

Installation is the reverse of removal, but please note the following important steps.

7. Make sure the torque converter is fully seated in the pump drive. If not, the transmission will not fit tightly to the rear of the engine block.
8. Raise the transmission into position and remove the torque converter holding strap and carefully. Slide the transmission forward until the dowel pins are engaged.
9. The torque converter should be flush with the flywheel and turn freely by hand.
10. Install the transmission–to–engine bolts. Tighten the bolts to 34 ft. lbs. (47 Nm).
11. Tighten the torque converter-to-flywheel bolts to 37 ft. lbs. (50 Nm).

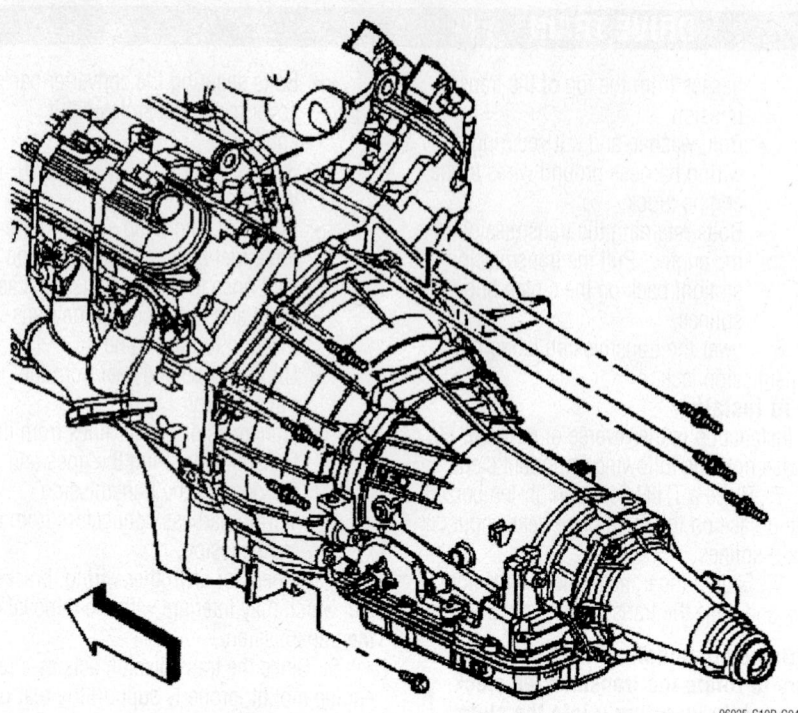

Automatic transmission mounting—4.3L engine shown: 2.2L similar

12. If equipped, tighten the converter pan cover to the transmission bolts to 37 ft. lbs. (50 Nm)

13. Tighten the bolts and washers securing the transmission mount to 35 ft. lbs. (47 Nm).

14. Tighten the nut and washer securing the transmission mount to the crossmember to 38 ft. lbs. (52 Nm).

15. Refill the transmission with the proper amount and type of fluid.

16. Connect the negative battery cable. Start the vehicle and allow to warm while checking for leaks. Road test the vehicle to check for shift quality.

Clutch

REMOVAL & INSTALLATION

1. Remove or disconnect the following:
   • Negative battery cable
   • Transmission

2. Install a clutch alignment tool or a used transmission input shaft to support the clutch.

3. If the clutch assembly is going to be reused, mark the flywheel, clutch cover and a pressure plate lug for alignment when installing.

4. Remove or disconnect the following:
   • Clutch cover bolts and washers
   • Clutch cover assembly and the clutch plate
   • Clutch alignment tool

5. Clean all parts and inspect for damage.
To install:
6. Install or connect the following:
   • Clutch alignment tool, to support the clutch

• Clutch cover by aligning the matchmarks or, if new, align the lightest part of the cover, identified by a yellow dot, with the heaviest part identified by an **X**.
• Clutch plate/clutch cover assembly to the flywheel. Tighten the bolts to 33 ft. lbs. (45 Nm) for 2.2L engines or to 29 ft. lbs. (40 Nm) for 4.3L engines.

➡**Tighten each screw 1 turn at a time to avoid warping the clutch cover.**

7. Remove the clutch alignment tool.
8. Install or connect the following:
   • Transmission
   • Negative battery cable

Hydraulic Clutch System

Bleeding air from the hydraulic clutch system is necessary whenever any part of the system has been disconnected or the fluid level (in the reservoir) has been allowed to fall so low, that air has been drawn into the master cylinder.

BLEEDING

1. Fill master cylinder reservoir with new brake fluid conforming to DOT 3 specifications.

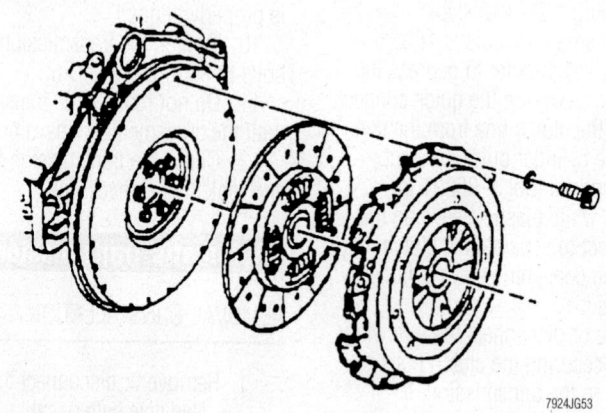

Exploded view of the clutch disc and related components

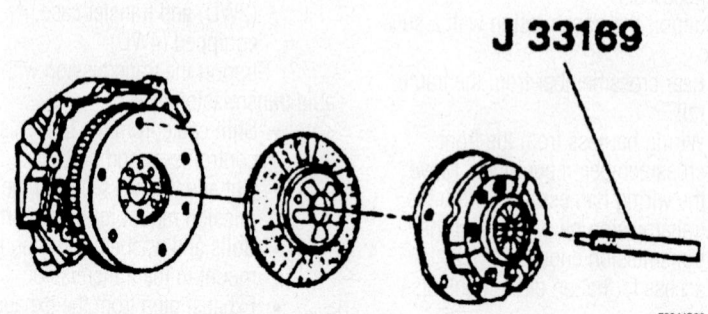

J 33169

Use the clutch alignment tool to center and support the clutch disc during installation

✳✳ CAUTION

Always use new fluid from a sealed container. Never, under any circumstances, use fluid that has been bled from a system to fill the reservoir as it may be aerated, have too much moisture content and possibly be contaminated.

2. Have an assistant fully depress and hold the clutch pedal, then open the bleeder screw.

3. Close the bleeder screw and have your assistant release the clutch pedal.

4. Repeat the procedure until all of the air is evacuated from the system. Check and refill master cylinder reservoir as required to prevent air from being drawn through the master cylinder.

➡**Never release a depressed clutch pedal with the bleeder screw open or air will be drawn into the system.**

5. If the previous steps do not result in satisfactory pedal feel, remove the reservoir cap and pump the clutch pedal very fast for 30 seconds. Stop to let the air escape, then repeat the procedure as necessary to purge all remaining air.

6. Test the clutch for proper operation.

Transfer Case Assembly

REMOVAL & INSTALLATION

1. Disconnect the negative battery cable.
2. Shift the transfer case into the **4HI** range.
3. Drain the transfer case fluid.
4. Support the transfer case.
5. Remove or disconnect the following:
   - Skid plate
   - Front and rear driveshafts from the transfer case. Matchmark the shafts prior to removal.
   - Vacuum lines and/or the electrical connectors, as equipped
   - Transfer case shift rod/cable from the case, if applicable
   - Support brace-to-transfer case bolts, if applicable
   - Transfer case

6. Remove all traces of old gasket material from the mating surfaces.

To install:

7. Install or connect the following:
   - New gasket using sealer to hold it in position
   - Transfer case. Torque the bolts to 33–35 ft. lbs. (45–47 Nm).
   - Support brace bolts. Torque the

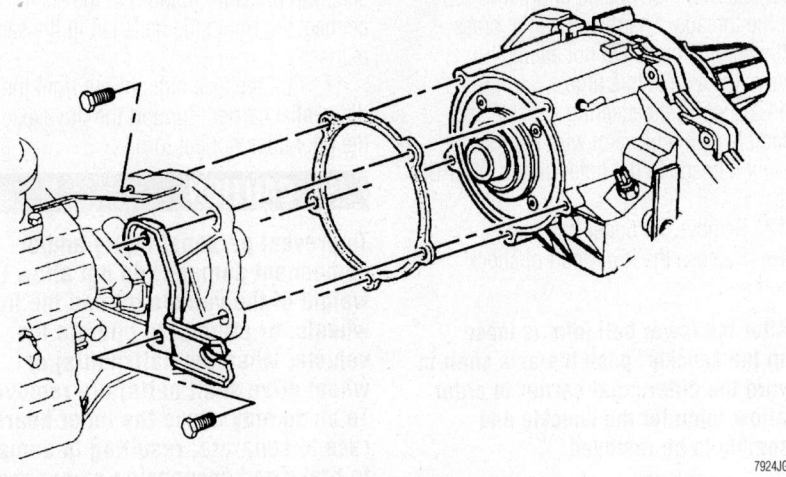

Typical transfer case-to-manual transmission mounting

7924JG27

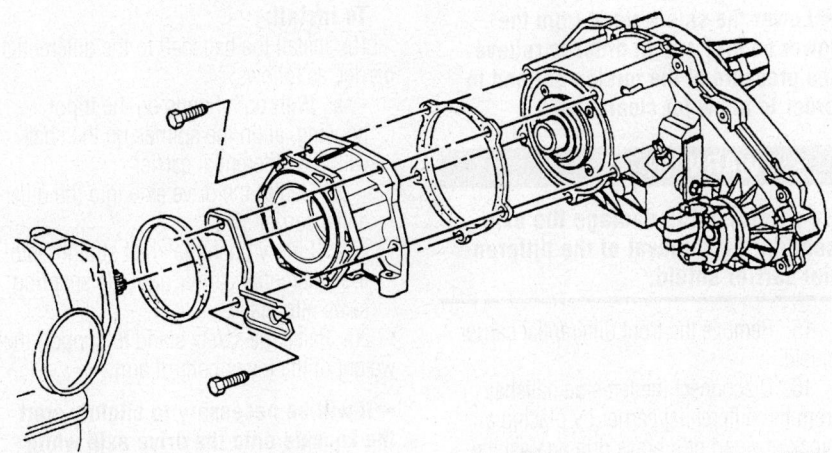

Typical transfer case-to-automatic transmission mounting

7924JG28

bolts to 35–37 ft. lbs. (47–50 Nm), if equipped.
   - Shift rod to the case, if equipped
   - Vacuum lines and/or electrical connections, as necessary
   - Front and rear driveshafts by aligning the matchmarks

8. Refill the transfer case.
   - Skid plate, if equipped
   - Negative battery cable

Halfshaft

REMOVAL & INSTALLATION

1. Unlock the steering column so the steering linkage is free to move.

2. Remove the front wheel and tire assemblies.

3. Insert a drift through the brake caliper and into one of the rotor vanes to prevent the drive axle from turning.

4. Remove the axle nut and washer. Remove the drift from the brake rotor.

5. Remove the front brake rotors and support the caliper with a piece of wire in order to prevent damage to the brake hose.

6. Remove the brackets from the upper control arm holding both the Anti-lock Brake System (ABS) wire and the brake hose.

7. Remove the ABS bracket located on the top of the upper control arm ball joint.

8. Strap the frame securely to the hoist in order to prevent movement.

➡**Be careful that the jackstand does not damage or bend any components it may contact.**

9. Position a jackstand under the lower control arm. Support the weight of the steering knuckle assembly and lower control arm with a jackstand.

10. Disengage the wheel drive shaft from the hub by placing a brass drift against the

outer end of the drive axle in order to protect the threads. Sharply strike the brass drift with a hammer. Do not attempt to remove the axle at this time.

11. Support the steering knuckle and assembly with a piece of wire in order to prevent damage to the outer tie rod and ABS wire.

12. Remove the upper ball joint.

13. Remove the lower part of shock absorber.

➡ **After the lower ball joint is loose from the knuckle, push the axle shaft in toward the differential carrier in order to allow room for the knuckle and assembly to be removed.**

14. Remove the lower ball joint.
   • Remove the axle from the steering knuckle assembly.

➡ **Lower the safety stand from the lower control arm in order to relieve the pressure of the torsion bar and in order to allow for clearance.**

✸✸ WARNING

Be careful not to damage the axle seal during removal of the differential carrier shield.

15. Remove the front differential carrier shield.

16. Disconnect the left side halfshaft from the differential carrier by placing a block of wood or a brass drift against the tripot housing. Firmly strike the block of wood outward from the case with a hammer. Strike hard enough to overcome the snapring pressure holding in the shaft. Disconnect the right side halfshaft in the same manner.

17. Pull the axle straight out from the differential carrier. Support the drive axle so the boot does not get torn.

✸✸ CAUTION

To prevent personal injury and/or component damage, do not allow the weight of the vehicle to load the front wheels, or attempt to operate the vehicle, when the halfshaft(s) or wheel drive shaft nut(s) are removed. To do so may cause the inner bearing race to separate, resulting in damage to brake and suspension components and loss of vehicle control.

18. Remove the halfshaft (drive axle).

To install:

19. Install the halfshaft to the differential carrier, as follows:

   a. With both hands on the tripot housing, align the splines on the shaft with the differential carrier.

   b. Center the drive axle into the differential carrier seal.

   c. Firmly push the shaft straight into the differential carrier until the snapring seats into place.

20. Raise the safety stand to support the weight of the lower control arm.

➡ **It will be necessary to slightly start the knuckle onto the drive axle while simultaneously guiding the lower ball stud to its proper location on the steering knuckle.**

21. Install the lower ball joint.

22. Install the lower part of the shock absorber.

23. Install the upper ball joint.

24. Install the halfshaft/axle shaft washer and nut. Tighten to 103 ft. lbs. (140 Nm).

25. Install the ABS bracket located on the top of the upper control arm ball joint.

26. Install the brackets from the upper control arm holding both the ABS wire and the brake hose.

27. Install the front brake rotors.

28. Install the front differential carrier shield.

29. Remove the strap from the frame.

30. Install the front tire and wheel assemblies.

CV-Joints

OVERHAUL

Outer CV-Joint

➡ **This procedure requires the use of the following: J 35910 Seal Clamp Tool, J 41048 Swage Clamp Tool and J 8059 Snap Ring Pliers, or equivalent tools.**

1. Remove or disconnect the following:
   • Front wheel
   • Halfshaft and place it in a vise. Use protective covers on the vise to avoid damaging the halfshaft.

2. Use a hand grinder to cut through the swage rings. Do not damage the outer race.

3. Compress the seal on the halfshaft and away from the CV-joint outer race. Wipe all grease away from the face of the CV-joint.

4. Find the halfshaft retaining snapring, which is in the inner race. Spread the snapring ears apart using Snapring Pliers J-8059, or equivalent.

5. Pull the outer CV-joint from the halfshaft. Discard the old seal.

6. Disassemble the chrome alloy balls from the CV-joint cage as follows:

   a. Position a brass drift against the CV-joint cage and tap it with a hammer to tilt the cage.

   b. Remove the 1st chrome alloy ball from the cage.

   c. Tilt the cage in the opposite direction.

   d. Remove the opposite chrome alloy ball.

   e. Repeat the procedure until all 6 balls are removed.

7. Disassemble the CV-joint cage and inner race as follows:

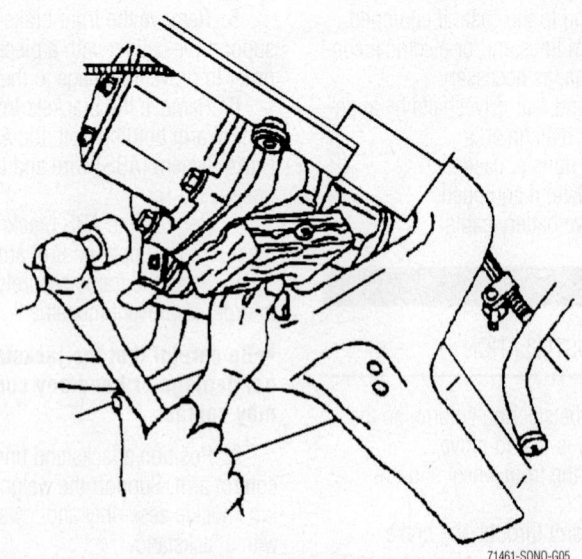

71461-SONO-G05

To separate the halfshaft, place a block of wood against the tripot housing. Firmly strike the block of wood outward from the case with a hammer

a. Pivot the cage and race 90 degrees to the center line of the outer race.

b. Align the cage windows with outer race lands.

c. Remove the cage from the outer race.

d. Rotate the inner race upward and remove it from the cage.

8. Thoroughly clean and inspect all parts.

To install:

9. Lubricate the parts with a light coat of grease.

10. Assemble the CV-joint cage and inner race, as follows:

a. Rotate the inner race 90 degrees to the cage centerline.

b. Align the cage windows with inner race lands.

c. Insert the inner race into the cage by rotating the inner race downward.

d. Insert the cage/inner race into the outer race.

11. Assemble the chrome alloy balls into the CV-joint cage, as follows:

a. Position a brass drift against the CV-joint cage and tap it with a hammer to tilt the cage.

b. Insert the 1st chrome alloy ball into the cage.

c. Tilt the cage in the opposite direction.

d. Insert the opposite chrome alloy ball.

e. Repeat the procedure until all 6 balls are inserted.

12. Pack the CV-joint seal and the CV-joint assembly with the grease supplied in the kit. The amount of grease supplied in this kit has been pre-measured for this application.

13. Place the new small swage clamp onto the CV-joint seal.

14. Place the large retaining clamp on the seal.

15. Position the small end of the CV-joint seal into the joint seal groove on the halfshaft bar.

16. Position the outboard end of the halfshaft assembly in Swage Clamp Tool J 41048, or equivalent. Align the swage clamp within the clamp tool.

17. Place the top half of the Swage Clamp Tool J 41048 on the bottom half. Check to make sure there are no pinch points on the seal before proceeding with procedures. Insert the bolts and tighten by hand until snug.

18. Align the seal, the halfshaft bar and swage clamp. Tighten each bolt 180 degrees at a time, using a ratchet wrench. Alternate between each bolt until both sides are bottomed.

19. Loosen the bolts. Separate the dies. Check the swage clamp for any "lip" deformities. If the deformities exist, place the swage clamp back into the Swage Clamp Tool J 41048.

20. Make sure the retaining ring side of the CV-joint inner race faces the halfshaft bar before installation.

21. Place the retaining snapring into the CV-Joint inner race.

➡**The retaining snapring inside of the inner race will engage in the halfshaft bar groove with a "click" when the CV-Joint is in the proper position.**

22. Slide the CV-Joint onto the halfshaft bar. Pull on the CV-Joint to be sure it is properly engaged.

23. Slide the large diameter of the CV-Joint seal, with the large retaining ring in place, over the outside edge of the CV-Joint outer race.

24. Position the lip of the CV-Joint seal into the groove on the CV-Joint outer race. Make sure to remove any excess air from the CV-Joint seal.

➡**Make sure the boot lies flat against the outer race.**

25. Using the Crimp tool J-35910, a torque wrench and a breaker bar, crimp the large CV-joint boot clamp to 130 ft. lbs. (176 Nm).

26. Check the clamp gap dimension; if it is not 0.085 in. (2.15mm), continue tightening the clamp until it is.

27. Install the halfshaft and the front wheel.

Inner (Tri-Pot) Joint

➡**This procedure requires the use of the following: J 35566 Drive Axle Seal Clamp Pliers, J 41048 Swage Clamp Tool and J 8059 Snap Ring Pliers, or equivalent tools.**

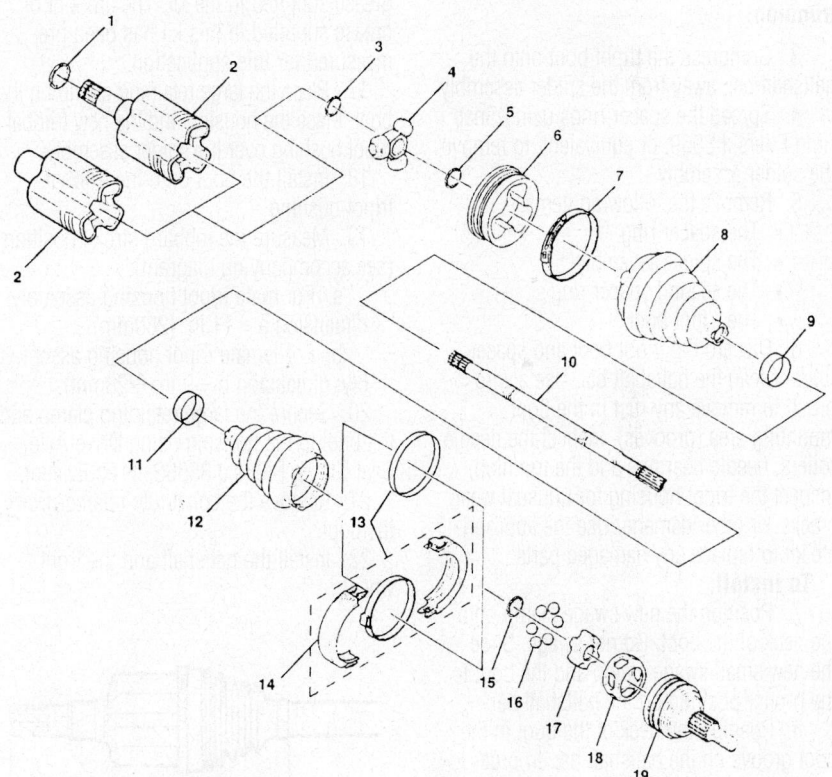

(1) Differential Shaft Ring
(2) Tripot Housing Assembly
(3) Spacer Ring
(4) Tripot Joint Spider Assembly
(5) Spacer Ring
(6) Tripot Bushing
(7) Boot Retaining Clamp
(8) Tripot Joint Boot
(9) Halfshaft Swage Ring
(10) Halfshaft Bar
(11) Halfshaft Swage Ring
(12) CV Joint Boot
(13) Swage Ring
(14) Clamp Protector
(15) Race Retaining Ring
(16) Ball
(17) CV Joint Inner Race
(18) CV Joint Cage
(19) CV Joint Outer Race

Exploded view of the CV-Joint Assembly

9308JG09

1. Remove or disconnect the following:
   • Front wheel
   • Halfshaft and place it in a vise
   • Clamp from the boot with a pair of side cutters.

➡**Be careful not to damage the tripot housing.**

2. Use a hand grinder to cut through the swage ring.
   • Tripot housing and the trilobal tripot bushing from the halfshaft bar. Thoroughly degrease the housing and the spider assembly. Discard the tripot bushing. Use 320 grit 3M cloth (or equivalent) to remove any evident corrosion in the transmission sealing surface. Allow the housing and the spider assembly to dry .

➡**Handle the tripot spider assembly with care or the tripot balls and needle rollers may separate from the spider trunnion.**

3. Compress the tripot boot onto the halfshaft bar, away from the spider assembly

4. Spread the spacer rings using Snap Ring Pliers J 8059, or equivalent, to remove the spider assembly.

5. Remove the following items:
   • The spacer ring
   • The spider assembly
   • The second spacer ring
   • The tripot boot

6. Discard the tripot boot and spacer rings. Clean the halfshaft bar. Use a wire brush to remove any rust in the boot mounting area (grooves). Inspect the needle rollers, needle bearings and the trunnion. Inspect the tripot housing for unusual wear, cracks, or other damage. Use the appropriate kit to replace any damaged parts.

To install:

7. Position the new swage clamp onto the neck of the boot. Do not swage. Slide the new small swage clamp and the boot to the proper position on the halfshaft bar.

8. Position the neck of the boot in the boot groove on the halfshaft bar. In order to swage the swage clamp, position the inboard end of the halfshaft assembly in Swage Clamp Tool J 41048, or equivalent. align the swage clamp within J 41048.

9. Place the top half of the Swage Clamp Tool J 41048 on the bottom half. Check to make sure there are no pinch points on the boot before continuing. Insert the bolts and tighten by hand until snug.

10. Align the boot, the halfshaft bar and the swage clamp. Tighten each bolt 180 degrees at a time, using a ratchet wrench.

Alternate between each bolt until both sides are bottomed.

11. Loosen the bolts and separate the dies.

➡**If deformities exist in the swage clamp, place the swage clamp back into the Swage Clamp Tool. Make sure the swage clamp covers the whole swaging area. Re-swage the swage clamp.**

12. Inspect the swage clamp for any "lip" deformities.

13. Assemble the joint with the convolute retainer in the correct position. Assemble the joint to meet the specified dimension to avoid boot damage.

14. Install the convolute retainer over the boot capturing four convolutions

15. Install the spacer ring and spider assembly onto the halfshaft bar. Install the other spacer ring in the groove at the end of the halfshaft bar. Ensure that the rings are fully seated.

16. Pack the boot and housing with the grease supplied in the kit. The amount of grease supplied in this kit has been pre-measured for this application.

17. Place the large retaining clamp on the boot. Place the housing and the new trilobal tripot bushing over the spider assembly.

18. Install the boot onto the trilobal tripot bushing.

19. Measure the inboard stroke position (see accompanying diagram):
   a. For male tripot housing assembly: dimension a = 11 in. (280mm).
   b. For female tripot housing assembly: dimension b = 9 in. (228mm).

20. Secure the large retaining clamp and the boot to the housing using Drive Axle Seal Clamp Pliers J 35566, or equivalent.

21. Remove the convolute retainer from the boot.

22. Install the halfshaft and the front wheel.

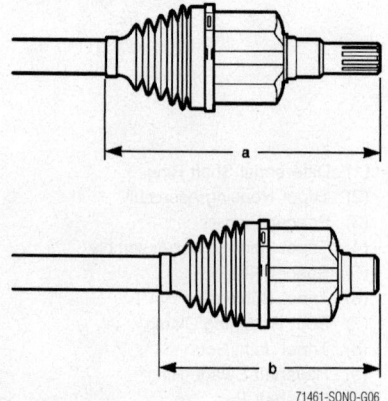

Measure the inboard stroke position as shown

71461-SONO-G06

Axle Shaft, Bearing and Seal

REMOVAL & INSTALLATION

For the Axle Shaft, Bearing and Seal, Removal and Installation, please refer to Wheel Bearing procedure located in the section.

Pinion Seal

REMOVAL & INSTALLATION

➡**The following procedure requires the use of the Pinion Holding tool J-8614-10, the Pinion Flange Removal tool J-8614-1, J-8614-2, J-8614-3 and the Pinion Seal Installation tool J-23911.**

1. Remove or disconnect the following:
   • Driveshaft from the pinion flange. Matchmark the driveshaft prior to removal.
   • Driveshaft from the rear axle pinion flange and support the shaft up in body tunnel by wiring it to the exhaust pipe.

➡**If the U-joint bearings are not retained by a retainer strap, use a piece of tape to hold bearings on their journals.**

2. Mark the position of the pinion stem, flange and nut for reference.

3. Use an inch lbs. torque wrench to measure the amount of torque necessary to turn the pinion, then note this measurement as it is the combined pinion bearing, seal, carrier bearing, axle bearing and seal preload.

4. Remove or disconnect the following:
   • Pinion flange nut and washer, using a Pinion Holding tool J-8614-10 and a Pinion Flange Removal tool J-8614-1, J-8614-2, J-8614-3, as applicable
   • Pinion flange
   • Pinion oil seal by driving it out of the differential with a blunt chisel; DO NOT damage the carrier

To install:

5. Examine the seal surface of pinion flange for tool marks, nicks or damage, such as a groove worn by the seal. If damaged, replace flange.

6. Examine the carrier bore and remove any burrs that might cause leaks around the O.D. of the seal.

7. Apply GM seal lubricant 1050169 to the outside diameter of the pinion flange and sealing lip of new seal.

8. Install or connect the following:
   • New pinion oil seal using a seal installer tool

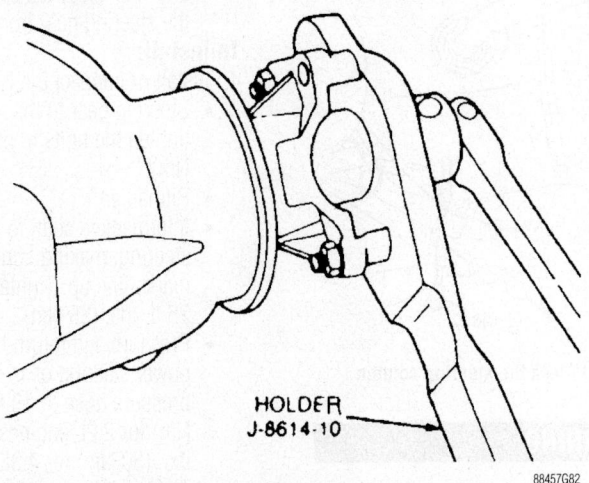

Removing the pinion nut using a pinion holding fixture tool

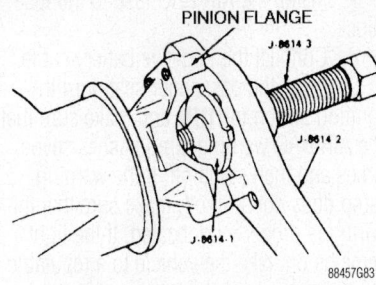

A puller and adapter should be used to withdraw the pinion from the housing

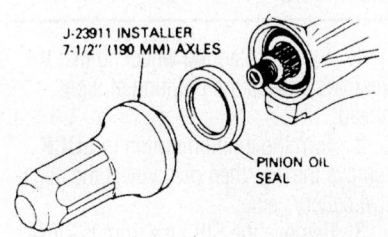

Use the appropriately sized installation tool to drive the new seal into position.

➡**The original matchmarks MUST be aligned to assure proper shaft balance and prevent vibration.**

- Pinion flange and tighten nut to the same position as marked earlier. Tighten the nut a little at a time and turn the pinion flange several times after each tightening in order to set the rollers.

9. Measure the torque necessary to turn the pinion and compare this to the reading taken during removal. Tighten the nut additionally, as necessary to achieve the same preload as measured earlier,

then tighten an additional 3–5 inch lbs. (0.34–0.56 Nm).

➡**If fluid was lost from the differential housing during this procedure, be sure to check and add additional fluid, as necessary.**

10. Remove the support then align and secure the driveshaft assembly to the pinion flange.

STEERING AND SUSPENSION

Air Bag

❊❊ CAUTION

Some vehicles are equipped with an air bag system, also known as the Supplemental Inflatable Restraint (SIR) system. The system must be disabled before performing service on or around system components, steering column, instrument panel components, wiring and sensors. Failure to follow safety and disabling procedures could result in accidental air bag deployment, possible personal injury and unnecessary system repairs.

PRECAUTIONS

Several precautions must be observed when handling the inflator module to avoid accidental deployment and possible personal injury.

- Never carry the inflator module by the wires or connector on the underside of the module.
- When carrying a live inflator module,

hold securely with both hands, and ensure that the bag and trim cover are pointed away.
- Place the inflator module on a bench or other surface with the bag and trim cover facing up.
- With the inflator module on the bench, never place anything on or close to the module, that may be thrown in the event of an accidental deployment.

2002 Vehicles

DISARMING

1. Turn the steering wheel so that the vehicle's wheels are pointing straight ahead.
2. Turn the ignition switch to **LOCK**, remove the key, then disconnect the negative battery cable.
3. Remove the AIR BAG fuse from the fuse block.
4. Remove the steering column filler panel or knee bolster.
5. Unplug the Connector Position Assurance (CPA) and yellow two way connector at the base of the steering column.
6. Remove the Connector Position Assurance (CPA) from the passenger yellow

two way connector located behind the glove box.
7. Unplug the yellow two way connector located behind the glove box.
8. Connect the negative battery cable.

➡**With the AIR BAG fuse removed, the battery cable connected and the ignition in the ON position, the AIR BAG warning lamp will be ON. This is normal and does not indicate a system malfunction.**

ARMING

1. Disconnect the negative battery cable.
2. Attach the yellow two way connector located behind the glove box.
3. Install the Connector Position Assurance (CPA) to the passenger yellow two way connector located behind the glove box.
4. Turn the ignition switch to **LOCK**, then remove the key.
5. Attach the two way connector at the base of the steering column and the Connector Position Assurance (CPA).
6. Install the steering column filler panel or knee bolster.

7. Install the AIR BAG fuse to the fuse block.

8. Connect the negative battery cable.

9. From the passenger seat, turn the ignition switch to **RUN** and make sure that the AIR BAG warning lamp flashes seven times and then shuts off. If the warning lamp does not shut off, make sure that the wiring is properly connected. If the light remains on, take the vehicle to a reputable repair facility for service.

2003–05 Vehicles

DISARMING

1. Turn the steering wheel so that the vehicle's wheels are pointing straight ahead.

2. Turn the ignition switch to **LOCK**, remove the key, then disconnect the negative battery cable.

3. Remove the SIR fuse from the fuse block.

4. Raise and support the vehicle.

5. Remove the Connector Position Assurance (CPA) from both inflatable restraints front end discriminating sensor connectors located on the frame crossmember.

6. Disconnect the inflatable restraints front end discriminating sensor connectors.

ARMING

1. Disconnect the negative battery cable.

2. Connect the inflatable restraint front end discriminating sensor connectors to the inflatable restraints front end discriminating sensor.

3. Install the CPA to the inflatable restraints front end discriminating sensor connectors.

4. Install the SIR fuse into the fuse block.

5. Staying well away from all air bags, turn ON the ignition, with the engine OFF.

6. The AIR BAG indicator will flash 7 times.

7. The AIR BAG indicator will then turn OFF.

8. Perform the SIR-Diagnostic System Check if the AIR BAG indicator does not operate as described.

Power Steering Gear

REMOVAL & INSTALLATION

2 Wheel Drive (2WD) Models

1. Position a fluid catch pan under the power steering gear.

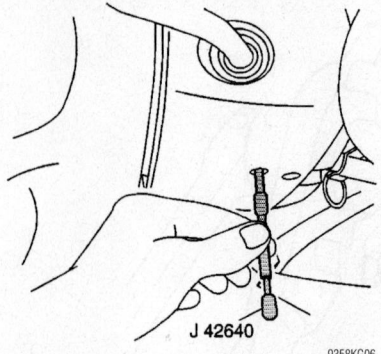

Use J 42640 to lock the steering column

✳✳ WARNING

Do NOT rotate the steering shaft after the steering column has been removed.

2. Lock the steering column through the access hole in the steering column lower trim cover using steering column anti-rotation pin J 42640.

3. Remove or disconnect the following:
- Air cleaner assembly
- Intermediate shaft from the steering gear
- Feed and return fluid hoses from the steering gear. Immediately cap or plug all openings to prevent system contamination or excessive fluid loss.
- Intermediate shaft lower coupling shield, if equipped
- Lower intermediate shaft coupling bolt
- Matchmark the lower intermediate shaft coupling and the steering shaft
- Lower intermediate shaft coupling from the steering shaft
- Pitman arm from the gear pitman shaft
- Power steering gear-to-frame bolts

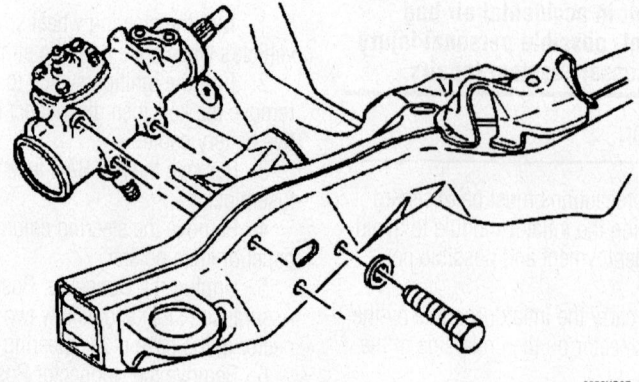

Power steering gear mounting

and washers, then carefully remove the steering gear from the vehicle

To install:

4. Install or connect the following:
- Steering gear to the vehicle and tighten the bolts to 55 ft. lbs. (75 Nm)
- Pitman arm
- Intermediate shaft to the power steering, making sure the match-marks line up. Tighten the bolt to 26 ft. lbs. (35 Nm).
- Pressure and return hoses to the power steering gear. Tighten the pressure hose to 18 ft. lbs. (25 Nm) for 2.2L engines and 22 ft. lbs. (30 Nm) for 4.3L engines. Tighten the return hose to 18 ft. lbs. (25 Nm).
- Shield over the intermediate shaft lower coupling, if equipped
- Air cleaner assembly

5. Remove the steering column lock pin

6. Bleed the power steering system

4 Wheel Drive (4WD) Models

This procedure requires the use of the following special tools: J 24319-B Steering Linkage and Tie Rod Puller, J 42640 Steering Column Anti-Rotation Pin, J 29193 Steering Linkage Installer (12mm), J 29194 Steering Linkage Installer (14mm).

1. Position a fluid catch pan under the power steering gear.

✳✳ WARNING

DO NOT rotate the steering shaft after the steering column has been removed.

2. Lock the steering column through the access hole in the steering column lower trim cover using steering column anti-rotation pin J 42640.

3. Remove or disconnect the following:

- Air cleaner assembly
- Intermediate shaft lower coupling shield, if equipped
- Wiring harness clip from the power steering return hose at the power steering gear
- Feed and return fluid hoses from the steering gear. Immediately cap or plug all openings to prevent system contamination or excessive fluid loss.
- Lower intermediate shaft coupling bolt
- Matchmark the lower intermediate shaft coupling and the steering shaft
- Lower intermediate shaft coupling from the steering shaft
4. Raise the vehicle.
- Steering linkage shield
- Differential carrier shield mounting bolts
- Differential carrier shield
- Pitman arm ball stud cotter pin and nut at the relay rod
- Pitman arm from relay rod using a suitable puller
- Steering gear mounting bolts and the washers from the frame
- Steering gear
- Pitman arm

To install:
5. Install or connect the following:
- Pitman arm
- Steering gear
- Power steering gear to the frame washers and mounting bolts. Tighten the bolts to 55 ft. lbs. (75 Nm).
- Relay rod to the pitman arm ball stud. Ensure the seal is on the stud
6. On 2002–03 vehicles, seat the taper using tool J 29193 or J 29194 and tighten the tool to 48 ft. lbs. (62 Nm).
7. Remove the special tool from the pitman arm ball stud, if necessary.
8. Install or connect the following:
- New nut and cotter pin to the pitman arm ball stud at the relay rod and tighten the pitman arm ball stud nut at the relay rod to 61 ft. lbs. (83 Nm)
- Differential carrier shield
- Differential carrier shield mounting bolts
- Steering linkage shield
9. Lower the vehicle.
- Intermediate shaft to the power steering, making sure the matchmarks line up. Tighten the bolt to 26 ft. lbs. (35 Nm).
- Pressure and return hoses to the power steering gear. Tighten the

pressure hose to 22 ft. lbs. (30 Nm) and the return hose to 18 ft. lbs. (25 Nm).
- Wiring harness clip to the power steering return hose at the power steering gear
- Shield over the intermediate shaft lower coupling, if equipped
- Air cleaner assembly
10. Remove the steering column lock pin
11. Bleed the power steering system

Shock Absorbers

REMOVAL & INSTALLATION

Front

2WD MODELS

1. Remove or disconnect the following:
- Wheel
- Mounting nut

➡**Hold the shock absorber stem with a wrench while backing the nut off.**

- Retaining nut and grommet
- Shock absorber-to-lower control arm bolts
- Shock absorber
- Replace the parts, as necessary.

To install:
2. Fully extend the shock absorber stem, then push it up through the lower control arm and spring so that the upper stem passes through the mounting hole in the upper control arm frame bracket.
3. Install or connect the following:
- Retaining nut and grommet on the stem. Tighten the nut to 106 inch lbs. (12 Nm).
- Shock absorber-to-lower control

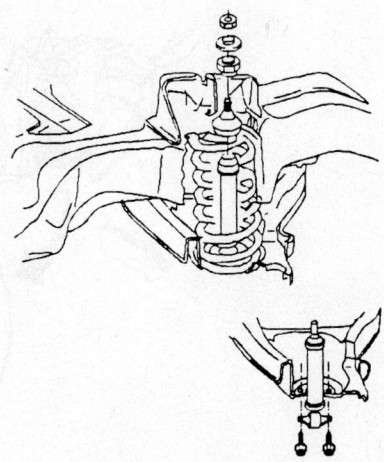

7924JG33

Front shock absorber mounting—2WD vehicles

arm bolts and tighten to 22 ft. lbs. (30 Nm)
- Wheel

4WD MODELS

1. Remove or disconnect the following:
- Wheel
- Lower nut/bolt and collapse the shock absorber
- Shock absorber upper nut and bolt
- Shock absorber

To install:
2. Install or connect the following:
- Shock absorber to the bracket. Tighten the nuts/bolts to 54 ft. lbs. (73 Nm).
- Wheel

Rear

1. Properly support the rear axle assembly.
2. Remove or disconnect the following:
- Automatic level control air lines from the shock absorber, if equipped
- Shock absorber-to-frame retainers at the top of the shock
- Shock-to-axle retainers at the bottom of the shock
- Shock absorber

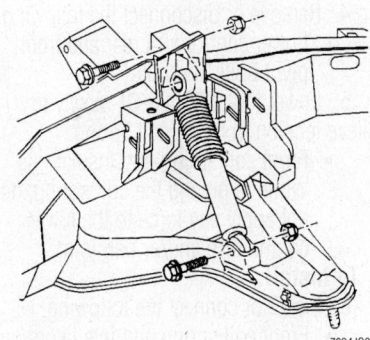

7924JG34

Front shock absorber mounting—4WD vehicles

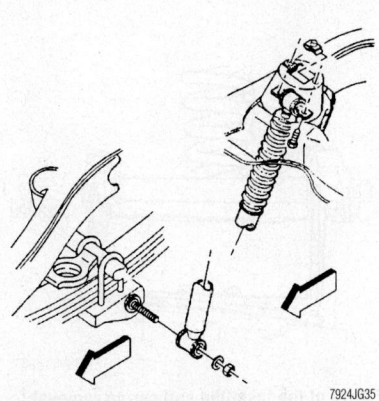

7924JG35

Rear shock absorber mounting

To install:

3. Install the shock in the vehicle and loosely install the upper mounting fasteners to retain it.

4. Align the lower-end of the shock absorber with the axle mounting, then loosely install the retainers.

5. Tighten the upper shock retainers to 18 ft. lbs. (25 Nm). Tighten the lower shock retainers to 62 ft. lbs. (84 Nm).

6. If equipped, attach the automatic level control air lines to the shock absorber.

Coil Springs

REMOVAL & INSTALLATION

1. Remove or disconnect the following:
   - Wheel
   - Stabilizer shaft link from the lower control arm
   - Shock absorber

2. Secure Coil Spring Remover & Installer tool J 23028-01, or equivalent to the end of a suitable jack. Cradle the lower control arms using the tool. Raise the jack to relieve tension on the lower control arm pivot bolts.

3. Turn the steering wheel to one side, to allow the steering linkage to clear the lower control arm front pivot bolt.

4. Remove or disconnect the following:
   - Lower control arm rear and front pivot bolts and nuts

5. Lower tool J 23028-01 slowly to relieve tension from the coil spring.
   - Front coil spring and insulators. While removing the coil spring, do not apply any force to the lower control arm and/or ball joint.

To install:

6. Install or connect the following:
   - Front coil spring and insulators on the lower control arm

➡**Make sure that the coil spring covers all or part of one inspection drain hole.**

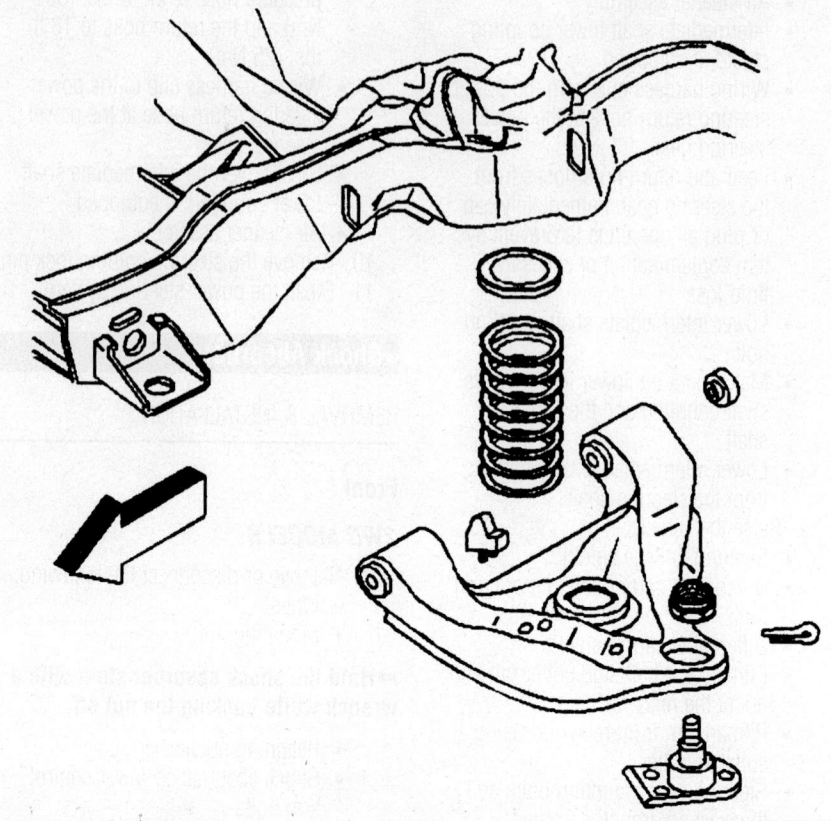

Exploded view of the coil spring removal

The other hole must be partly or completely uncovered. Rotate the coil spring as necessary.

7. Support the control arm using tool J 23028-01. Position the coil spring and insulator in the upper spring seat on the frame.

8. Raise the lower control arm using tool J 23028-01.

9. Install or connect the following:

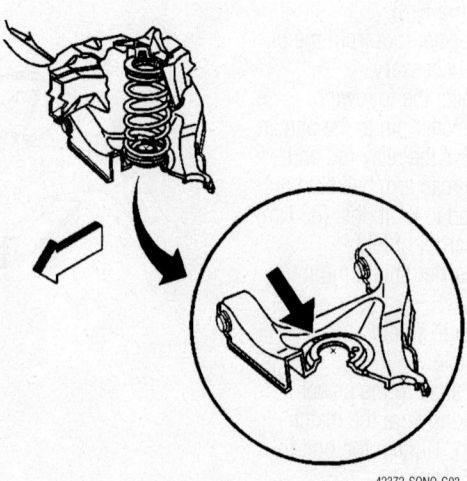

- Lower control arm to the frame

➡**You must install the bolts in the direction shown to keep proper steering linkage clearance.**

- Lower control arm front and rear pivot bolts with NEW nuts. With the suspension loaded, torque the front bolt to 85 ft. lbs. (115 Nm) and the rear bolt to 72 ft. lbs. (98 Nm). Remove tool J 23028-01.

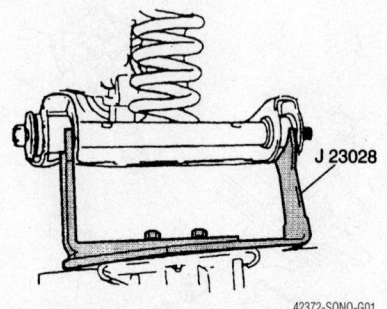

View of the installed coil spring removal tool

The coil spring must cover all or part of one inspection drain hole. The other hole must be partly or completely uncovered

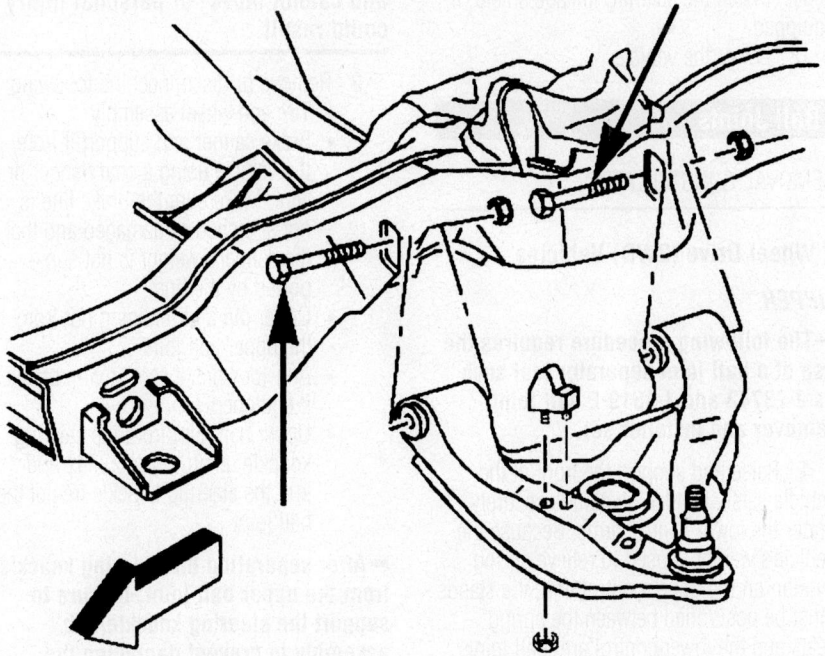

42372-SONO-G04

For steering linkage clearance, the control arm pivot bolts must be installed as shown

- Shock absorber
- Stabilizer shaft link to the lower control arm

10. Check and adjust the front wheel alignment.

Leaf Springs

REMOVAL & INSTALLATION

➡ **The following procedure requires the use of two sets of jackstands.**

1. Support the rear axle with jackstands, support the axle and the body separately in order to relieve the load on the rear spring.
2. Remove or disconnect the following:
   - Wheel
   - Shock absorber
   - U-bolt nuts, washers, anchor plate and bolts
   - Spare tire, if equipped
   - Rear exhaust hangers and lower the rear exhaust, if necessary
   - Shackle-to-frame bolt, washers and nut
   - Fuel tank, if necessary
   - Front bracket nut, washers and bolt
   - Spring
   - Shackle from the spring, if necessary

To install:

3. Install or connect the following:
   - Shackle to the rearward spring eye using the bolt, washers and nut, but do not fully tighten at this time.

- Spring assembly
- Spring to the front bracket using the bolt, washers and nut, but do not fully tighten at this time.
- Fuel tank, if removed
- Shackle-to-frame bolt, washers and nut, but do not fully tighten at this time. If used, remove the spring support.
- U-bolts, anchor plate, washers and U-bolt nuts. Torque the nuts using 2 passes of a diagonal sequence:
  a. Step 1: Torque to 18 ft. lbs. (25 Nm).
  b. Step 2: Torque to 73 ft. lbs. (100 Nm) in the sequence.

4. Position the axle to achieve an approximate gap of 6.46–6.94 in. (164–176mm) between the axle housing tube and the metal surface of the rubber frame bumper bracket. Measure from the housing between the U-bolts to the metal part of the rubber bump stop on the frame.

5. While supporting the axle in this position, tighten the front and rear spring mounting fasteners to 89 ft. lbs. (122 Nm).

6. Install or connect the following:
   - Rear exhaust in position and tighten the hangers
   - Spare tire
   - Shock absorber

Torsion Bar

Instead of the coil spring used on the front suspension of 2WD vehicles, the 4WD vehicles are equipped with a torsion bar.

REMOVAL & INSTALLATION

➡ **The following procedure requires the use of the Torsion Bar Unloader tool J-36202.**

1. Remove or disconnect the following:
   - Transmission shield, if equipped
   - Torsion bar unloader tool to relax the tension on the torsion bar adjusting arm screw; record the number of turns necessary to properly install the tool. Remove the adjusting screw and the unloader tool.
   - Lower link mount nut from one side
   - Torsion bars by disengaging them

➡ **Note the direction of the forward end and side of the torsion bar being removed**

   - Lower link nut from the opposite side
   - Lower link mount, upper link mount nut
   - Upper link mount
   - Torsion bar from the frame

To install:

2. Install or connect the following:
   - Torsion bar and support
   - Upper link mount. Torque the nut to 48 ft. lbs. (68 Nm).

3. Place a jack under the torsion bar to release tension.

4. Install or connect the following:
   - Lower link mount bushing and nut. Torque the nut to 37 ft. lbs. (50 Nm).
   - Torsion bar unloader tool. Tighten the tool against the adjusting arm the same number turns recorded earlier and remove the tool. This loads the torsion bars.
   - Transmission shield, if removed

Steering Knuckle

REMOVAL & INSTALLATION

1. Raise and support the vehicle.
2. Support the lower control arm with jack stands.
3. Remove the wheel hub and bearing.
4. Remove the bolts that attach the splash shield to the steering knuckle.
5. Remove the cotter pin.
6. Remove the tie rod end stud nut.
7. Disconnect the tie rod end from the steering knuckle
8. Remove the lower ball joint stud from the steering knuckle.

9. Remove the upper ball joint stud from the steering knuckle.

10. Raise the upper control arm. Disengage the ball joint stud from the steering knuckle.

11. Remove the steering knuckle from the lower ball joint stud.

To install:

12. Install the upper ball joint to the steering knuckle.

13. Install the lower ball joint to the steering knuckle.

14. Install the splash shield to the steering knuckle and tighten to 19 ft. lbs. (26 Nm).

15. Install the tie rod end to the steering knuckle. Ensure that the seal is on the stud.

16. Seat the taper using Steering Linkage Installer J-29193 and tighten the tool to 40 ft. lbs. (54 Nm).

17. Remove the tool.

18. Install the tie rod end retaining nut and tighten to 39 ft. lbs. (53 Nm).

19. Install the wheel hub and bearing.

20. Lower the vehicle.

21. Check the front wheel alignment.

Stabilizer Bar

REMOVAL & INSTALLATION

1. Raise and support the vehicle.

2. Remove the tire and wheel.

3. Remove the steering linkage shield, if equipped.

4. Remove the to the stabilizer shaft link retaining nut.

5. Remove the grommets and retainers from the stabilizer shaft link bolt.

6. Remove the stabilizer shaft bracket mounting bolts.

7. Remove the stabilizer shaft brackets.

8. Remove the stabilizer shaft.

9. Remove stabilizer shaft insulators from stabilizer shaft.

To install:

10. Install the stabilizer shaft insulators to the stabilizer shaft with the slits facing the front of the vehicle.

11. Install the stabilizer shaft.

12. Install the stabilizer shaft brackets over the stabilizer shaft insulators.

13. Install the stabilizer shaft bracket mounting bolts and tighten to 26 ft. lbs. (36 Nm) on 2WD vehicles, or 48 ft. lbs. (65 Nm) on 4WD vehicles..

14. Install the grommets and retainers from the stabilizer shaft link bolt.

15. Install the stabilizer shaft links and tighten to 13 ft. lbs. (18 Nm).

16. Install the wheel and tire.

17. Install the steering linkage shield, if equipped.

18. Lower the vehicle.

Ball Joints

REMOVAL & INSTALLATION

2 Wheel Drive (2WD) Vehicles

UPPER

➡**The following procedure requires the use of a ball joint separator tool such as J-23742 and J-9519-E ball joint remover and installer set.**

1. Raise and support the front of the vehicle safely by placing stands securely under the lower control arms. Because the vehicle's weight is used to relieve spring tension on the upper control arm, the stands must be positioned between the spring seats and the lower control arm ball joints for maximum leverage.

❊❊ CAUTION

With components unbolted, the stand is holding the lower control arm in place against the coil spring. Make sure the stand is firmly positioned and cannot move, or personal injury could result.

2. Remove or disconnect the following:
   - Tire and wheel assembly
   - Brake caliper and support it from the vehicle using a coat hanger or wire. Make sure the brake line is not stretched or damaged and that the caliper's weight is not supported by the line.
   - Cotter pin and retaining nut from the upper ball joint
   - Anti-lock brake sensor wire bracket, if equipped
   - Upper ball joint from the steering knuckle using tool J-23742 and pull the steering knuckle free of the ball joint

➡**After separating the steering knuckle from the upper ball joint, be sure to support the steering knuckle/hub assembly to prevent damaging the brake hose.**

3. Remove the riveted upper ball joint from the upper control arm as follows:
   a. Drill a ⅛ in. (3mm) hole, about ¼ in. (6mm) deep into each rivet.
   b. Then use a ½ in. (13mm) drill bit, to drill off the rivet heads.

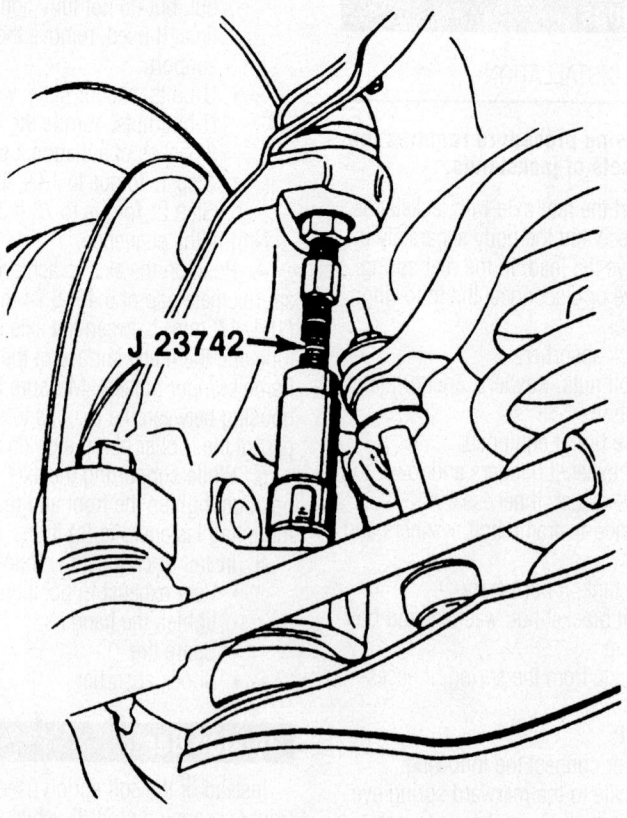

Use a ball joint separator tool to drive the upper ball joint from the steering knuckle

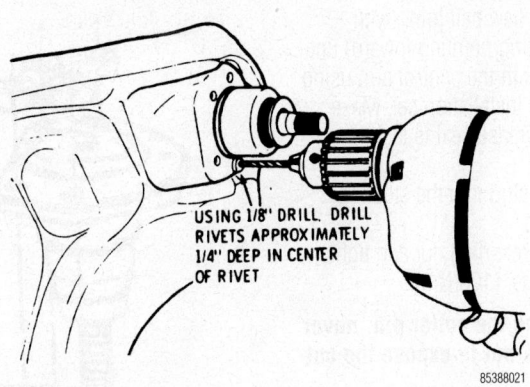

Drill a small guide hole into each ball joint rivet

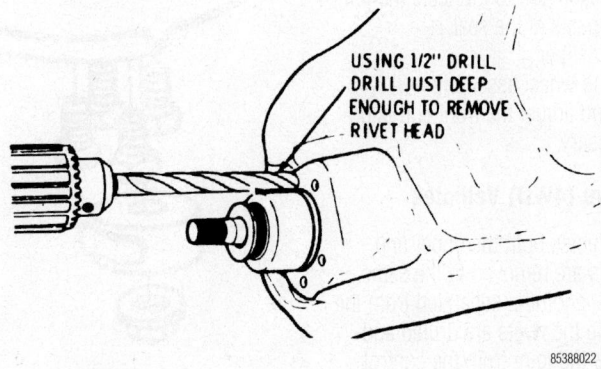

Then drill off the rivet heads

c. Using a pin punch and the hammer, drive out the rivets in order to free the upper ball joint from the upper control arm assembly, then remove the upper ball joint.

4. Clean and inspect the steering knuckle hole. Replace the steering knuckle if the hole is out of round.

To install:

5. Install or connect the following:
- Ball joint in the upper control arm
- Ball joint retaining nuts and bolts.

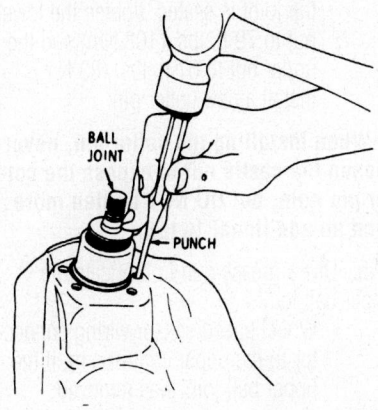

Punch the rivets out and remove the ball joint

Position the bolts threaded upward from under the control arm. Tighten the ball joint retainers to 17 ft. lbs. (23 Nm).
- Anti-lock brake sensor wire bracket, if removed

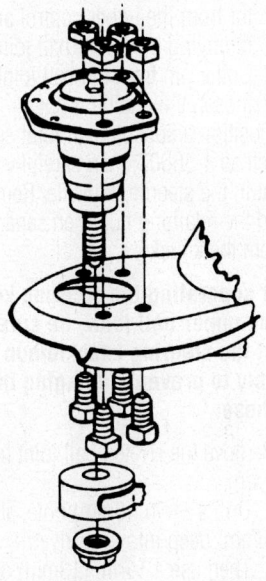

Service ball joints are bolted to the control arm

- Ball joint to the knuckle. Make sure the joint is seated, then install the stud nut and tighten to 61 ft. lbs. (83 Nm). Insert a new cotter pin.

➡ **When installing the cotter pin, never loosen the castle nut to expose the cotter pin hole.**

- Thread the grease fitting into the ball joint. Use a grease gun to lubricate the upper ball joint until grease appears at the seal.
- Brake caliper
- Tire and wheel assembly

6. Check and adjust the front end alignment, as necessary.

LOWER

➡ **The following procedure requires the use of a ball joint remover/installer set (the particular set may vary upon application but must include a clamping-type tool with the appropriately sized adapters) and a ball joint separator tool, such as J-23742.**

- Tire and wheel assembly

1. Position a jack under the spring seat of the lower control arm, then raise the jack to support the arm.

❋❋ CAUTION

The jack MUST remain under the lower control arm, during the removal and installation procedures, to retain the arm and spring positions. Make sure the jack is securely positioned and will not slip or release during the procedure or personal injury may result.

2. Remove or disconnect the following:
- Brake caliper and support it aside using a hanger or wire. Make sure

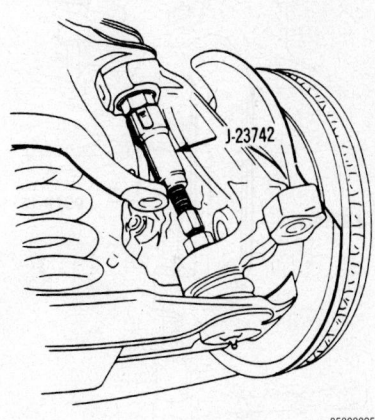

Use a ball joint separator to drive the lower joint from the knuckle

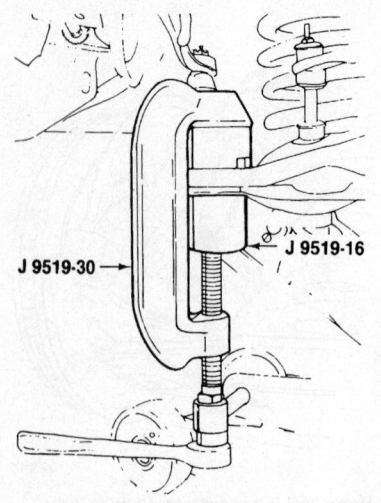

Driving the lower joint from the control arm

the brake line is not stressed or damaged.
- Lower ball joint cotter pin and discard
- Ball joint stud nut
- Lower ball joint from the steering knuckle using tool J-23742

3. Carefully guide the lower control arm out of the opening in the splash shield using a putty knife. Position a block of wood between the frame and upper control arm to keep the knuckle out of the way.
- Grease fitting
- Ball joint from the control arm using the ball joint remover set along with the appropriate adapters

To install:

4. Clean the tapered hole in the steering knuckle of any dirt or foreign matter, then check the hole to see if it is out of round, deformed or otherwise damaged. If a problem is found, then knuckle must be replaced.

5. Install or connect the following:

Installing a new ball joint

- Press the new ball joint (with grease fitting pointing inward) until it bottoms in the control arm using a suitable installation set. Make sure the grease seal is facing inboard.
- Ball joint stud into the steering knuckle
- Ball joint retaining nut and tighten to 79 ft. lbs. (108 Nm)

➡ **When installing the cotter pin, never loosen the castle nut to expose the cotter pin hole.**

- Grease fitting into the ball joint, if not already installed

6. Use a grease gun to lubricate the joint until grease appears at the seal.
- Brake caliper
- Tire and wheel assembly

7. Check and adjust the front end alignment, as necessary.

4 Wheel Drive (4WD) Vehicles

On 4WD vehicles both the upper and lower ball joints are removed in the same manner. Once the joint is separated from the steering knuckle the rivets are drilled and punched to free the joint from the control arm. Service joints are bolted into position with the retaining bolts threaded upward from beneath the control arm. In this manner, the joint is replaced in an almost identical fashion to the upper joints on 2WD vehicles.

1. Remove or disconnect the following:

- Tire and wheel assembly
- Wheel speed sensor wiring connector from the upper control arm, if removing the upper ball joint
- Cotter pin from the ball joint, then loosen the retaining nut

2. Position a suitable ball joint separator tool such as J-36607, then carefully loosen the joint in the steering knuckle. Remove the tool and the retaining nut, then separate the joint from the knuckle.

➡ **After separating the steering knuckle from the upper ball joint, be sure to support the steering knuckle/hub assembly to prevent damaging the brake hose.**

3. Remove the riveted ball joint from the control arm:

a. Drill a ⅛ in. (3mm) hole, about ¼ in. (6mm) deep into each rivet.

b. Then use a ½ in. (13mm) drill bit, to drill off the rivet heads.

c. Using a pin punch and the ham-

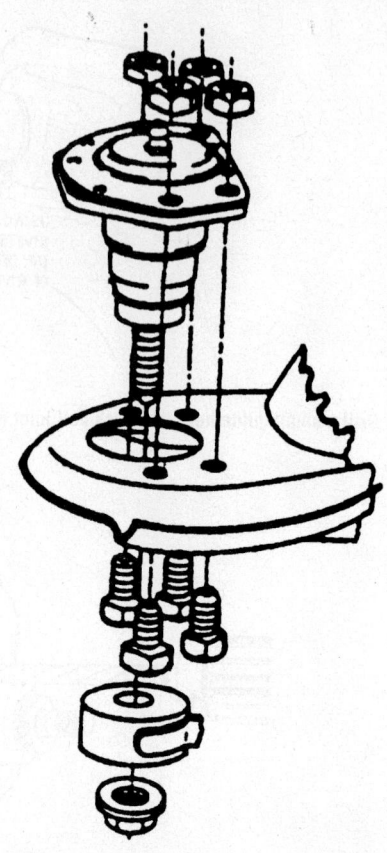

The replacement ball joint comes with nuts and bolts for installation

mer, drive out the rivets in order to free the ball joint from the control arm assembly, then remove the ball joint.

To install:

4. Install or connect the following:
- Ball joint in the control arm
- Ball joint retaining nuts and bolts. Position the bolts threaded upward from under the control arm. Tighten the ball joint retainers to 17 ft. lbs. (23 Nm).
- Ball joint to the knuckle. Make sure the joint is seated, tighten the lower nut to 79 ft. lbs. (108 Nm) and the upper nut to 61 ft. lbs. (83 Nm). Install a new cotter pin.

➡ **When installing the cotter pin, never loosen the castle nut to expose the cotter pin hole, but DO NOT tighten more than an additional ⅙ turn.**

5. Use a grease gun to lubricate the upper ball joint.
- Wheel speed sensor wiring connector to the upper control arm, if the upper ball joint was removed
- Tire and wheel assembly

6. Check and adjust the front end alignment, as necessary.

Upper Control Arm

REMOVAL & INSTALLATION

2 Wheel Drive (2WD) Vehicles

1. Remove or disconnect the following:
   - Negative battery cable
   - Wheel
   - Wheel speed sensor harness bracket retaining bolt and nut, if equipped
   - Support the lower control arm with a jack stand
   - Steering knuckle from upper control arm ball joint
   - Mounting nuts/bolts and shims

➡ **Make sure to note the location of the control arm shims prior to removal so that they may be installed in their original positions.**

   - Upper control arm

To install:

2. Install or connect the following:
   - Upper control arm

➡ **Always tighten nut on the thinner shim pack first.**

   - Mounting nuts/bolts and shims. Torque the nuts to 81–85 ft. lbs. (110–115 Nm).
   - Steering knuckle to upper control arm ball joint
   - New cotter pin

➡ **Tighten the nut to align the hole never loosen.**

   - Wheel speed sensor harness bracket retaining bolt and nut, if equipped
   - Wheel

4 Wheel Drive (4WD) Vehicles

1. Remove or disconnect the following:
   - Tire and wheel assembly
   - Support the lower control arm with a jack stand
   - Unload the torsion bar
   - Cotter pin from the ball joint, then loosen the retaining nut
   - Steering knuckle from the upper ball joint. Be sure to support the steering knuckle/hub assembly to prevent damaging the brake hose.

➡ **The 4WD vehicles do not use shims to adjust the front wheel alignment. Instead, the upper control arm bolts are equipped with cams, which are rotated to achieve caster and camber**

adjustments. In order to preserve adjustment and ease installation, matchmark the cams to the control arm before removal. If the control arm is being replaced, transfer the alignment marks to the new component before installation.

   - Front and rear nuts retaining the control arm retaining bolts to the frame
   - Outer cams from the bolts
   - Bolts and inner cams
   - Control arm from the vehicle
   - Retaining nut and the bumper from the control arm, if necessary

2. If the bushings are being replaced, use a suitable bushing service set to remove the bushings from the arm.

To install:

3. Install or connect the following:
   - Bushing service set to drive the new bushings into the control arm, if removed
   - Bumper and retaining nut to the control arm, if removed. Tighten the bumper retaining nut to 20 ft. lbs. (27 Nm).
   - Control arm, retaining bolts (from the inside of the frame brackets facing outward) and the inner cams. The inner cams must be positioned on the bolts before they are inserted through the control arm and frame brackets.
   - Outer cams over the retaining bolts, then the nuts to the ends of the bolts at the front and rear of the control arm

4. Align the cams to the reference marks made earlier, then tighten the end nuts to 85 ft. lbs. (115 Nm).
   - Ball joint to the knuckle
   - Load the torsion bar
   - Tire and wheel assembly

5. Check and adjust the front end alignment, as necessary.

CONTROL ARM BUSHING REPLACEMENT

2 Wheel Drive (2WD) Vehicles

1. Remove or disconnect the following:
   - Upper control arm and place it in a vise
   - Upper control arm shaft nuts and retainers
   - Upper control arm bushings using tool J 22269-1, a slotted washer and a short piece if pipe that is slightly larger than the bushing
   - Upper control arm shaft

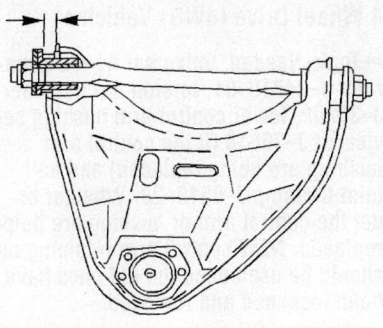

06025-S10P-G05

Control arm bushing installation measurement

To install:
   - Upper control arm shaft
   - Upper control arm bushings using tool J 22269-1, a slotted washer and a short piece if pipe that is slightly larger than the bushing

2. Tighten J 22269-1 until the bushing is positioned on the shaft and the control arm as shown in the accompanying illustration. The measurement should be 0.48–0.52 inch (12.8–13.8mm) at both sides when the properly installed.
   - Upper control arm shaft nuts and retainers. Tighten to 85 ft. lbs. (115 Nm).
   - Upper control arm

4 Wheel Drive (4WD) Vehicles

If the bushings require replacement, refer to the control arm removal and installation procedure for bushing replacement.

Lower Control Arm

REMOVAL & INSTALLATION

2 Wheel Drive (2WD) Vehicles

1. Remove or disconnect the following:
   - Coil spring
   - Lower ball joint from the steering knuckle
   - Lower control arm from the vehicle

To install:

2. Install or connect the following:
   - Lower control arm
   - Lower ball joint stud into the steering knuckle
   - Ball joint-to-steering knuckle nut and tighten to specification
   - New cotter pin to the lower ball joint stud
   - Coil spring

3. Align the vehicle.

4 Wheel Drive (4WD) Vehicles

➡**Tools Needed: universal tie rod separator J–24319–01, torsion bar unloader J–36202, lower control arm bushing service kit J–36618 (if the control arm bushing are being replaced) and ball joint C-clamp J–9519–23. Whether or not the control arm or bushing are being replaced, NEW control arm retaining nut should be used once the old ones have been loosened and removed.**

1. Raise and support the vehicle.
2. Remove the tire and wheel.
3. Remove the torsion bar from the lower control arm.
4. Remove the steering linkage shield.
5. Remove the stabilizer shaft link from the lower control arm.
6. Remove the stabilizer shaft bracket bolts.
7. Lower the stabilizer shaft.
8. Remove the shock absorber lower mounting bolt.
9. Compress the shock absorber.
10. Remove the steering knuckle.
11. Remove the lower control arm to the crossmember and the frame bracket mounting nuts and bolts.
12. Remove the lower control arm from the frame.

To install:

13. Install the front leg of the lower control arm into the crossmember before installing the rear leg into the frame bracket.
14. Install the lower control arm mounting bolts in the direction shown.

15. Tighten the nuts with the lower control arm at the proper trim height. Tighten the nuts and the bolts with the front suspension loaded.
16. Install the lower control arm mounting nuts and tighten to 81 ft. lbs. (110 Nm).
17. Install the steering knuckle.
18. Install the shock absorber lower mounting bolt and tighten to 54 ft. lbs. (73 Nm).
19. Raise the stabilizer shaft, install the stabilizer shaft bracket bolts and tighten to 48 ft. lbs. (65 Nm).
20. Install the stabilizer shaft link to the lower control arm.
21. Install the steering linkage shield.
22. Install the torsion bar.
23. Install the tire and wheel.
24. Lower the vehicle.
25. Check the front wheel alignment.

CONTROL ARM BUSHING REPLACEMENT

2 Wheel Drive (2WD) Vehicles

1. Remove lower control arm and place it in vise.
2. Install tools J 22269-01, 21474-8, 12 and 13 on the rear bushing and tighten until the bushing is removed.
3. Using a blunt chisel, drive the front bushing flare flush with the rubber part of the bushing.
4. Place a wedge or a spacer between the bushing housing to keep the housing from bending while removing or installing the bushing.

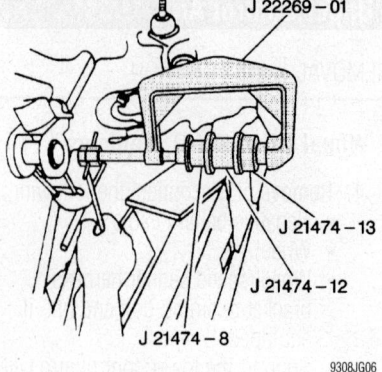

Removing the lower control arm rear bushing—all 2 wheel drive

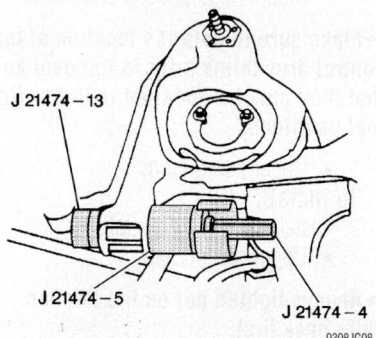

Installing the lower control arm front bushing—all 2 wheel drive

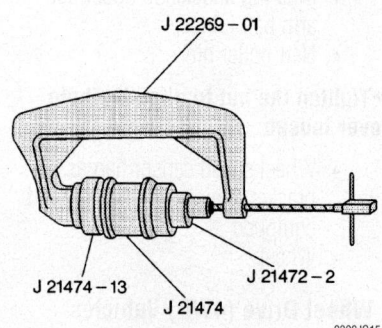

Installing the lower control arm rear bushing—all 2 wheel drive

5. Install tools J 21474-3, 4, 5 and 6 on the front bushing and tighten until the bushing is removed.

To install:

6. Install the front bushing into the control arm.
7. Install tools J 21474-4, 5 and 13. Tighten until the bushing is fully seated.
8. Install the rear bushing into the control arm
9. Install tools J 22269-01, J 21474-2 and 13. Tighten until the bushing is fully seated.
10. Install the lower control arm.

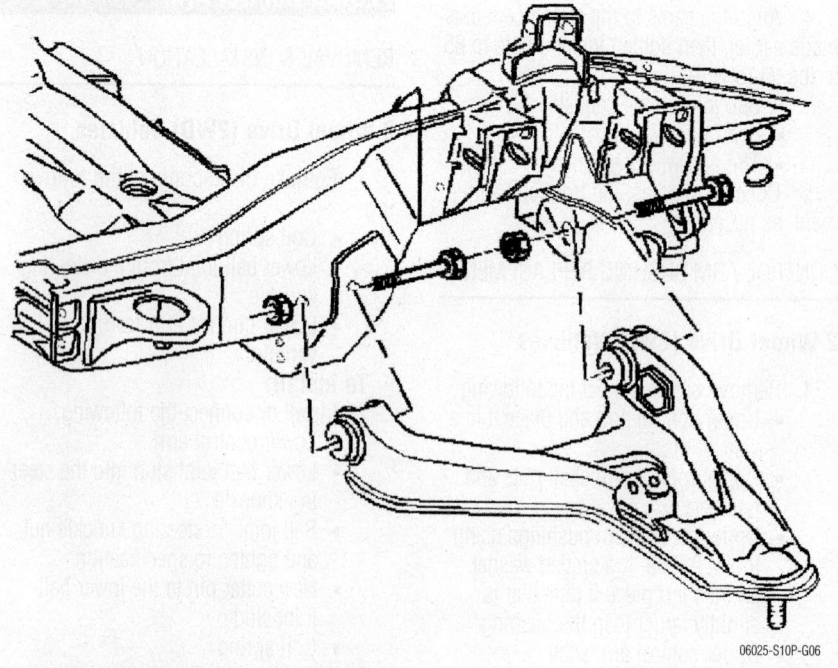

Lower control arm mounting—4WD models

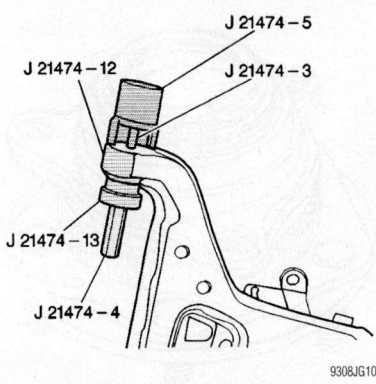

Removing the lower control arm front bushing—4 wheel drive

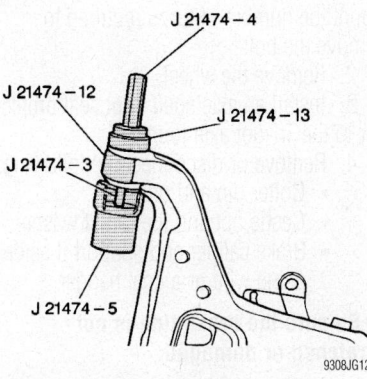

Installing the lower control arm front bushing—4 wheel drive

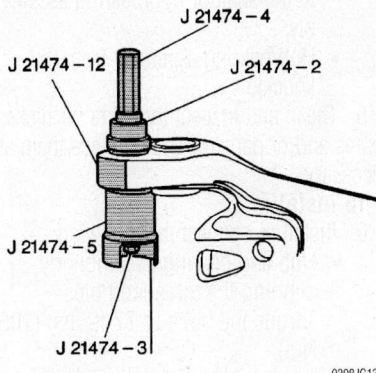

Installing the lower control arm rear bushing—4 wheel drive

4 Wheel Drive (4WD) Vehicles

1. Remove lower control arm and place it in vise.
2. Using bushing service set J 21474, remove the front and rear bushings.
To install:
3. Using bushing service set J 21474, install the front and rear bushings.
4. Install the lower control arm.

Wheel Bearings

ADJUSTMENT

2 Wheel Drive (2WD) Vehicles

1. If equipped, remove the wheel/hub cover for access, then remove the dust cap from the hub.
2. Remove the cotter pin and loosen the spindle nut.
3. Spin the wheel forward by hand and torque the nut to 12 ft. lbs. (16 Nm) in order to fully seat the bearings and remove any burrs from the threads.
4. Back off the nut until it is just loose, then finger-tighten the nut.
5. Loosen the nut ¼–½ turn until either hole in the spindle lines up with a slot in the nut, then install a new cotter pin. This may appear to be too loose, but it is the proper adjustment.
6. Proper adjustment creates 0.001–0.005 in. (0.025–0.127mm) endplay.

4 Wheel Drive (4WD) Vehicles

The front wheel bearings on the 4-wheel drive vehicles and 2004–05 2WD vehicles are not adjustable. If the bearings become loose or make noise, they must be replaced.

REMOVAL & INSTALLATION

Front

2WD MODELS 2002–2003

1. Remove or disconnect the following:
   - Wheel
   - Brake caliper with the pads without disconnecting the brake line
   - Grease cap
   - Cotter pin, spindle nut and washer
   - Hub

❊❊ WARNING

Be careful not to drop the outer wheel bearing. As the hub is pulled forward, the outer wheel bearings will often fall forward and they may easily be removed at this time.

   - Outer roller bearing assembly
   - Inner seal by prying it out of the hub and discard it
   - Inner bearing assembly

To install:
2. Clean all parts in solvent and allow to air dry, then check for excessive wear or damage. Inspect all of the parts for scoring, pitting or cracking and replace if necessary.

➡**DO NOT remove the bearing races from the hub, unless they show signs of damage.**

3. If it is necessary to remove the wheel bearing races, use the GM Front Bearing Race Removal tool J-29117 to drive the races from the hub/disc assembly. A hammer and brass drift may also be used to drive the races from the hub, but the race removal tool is quicker.
4. If the bearing races were removed, position the replacement races in the freezer for a few minutes and then install them to the hub:
   a. Lightly lubricate the inside of the hub/disc assembly using wheel bearing grease.
   b. Using the GM Seal Installation tools J-8092 and J-8850, drive the inner bearing race into the hub/disc assembly until it seats. Be sure the race is properly seated against the hub shoulder and is not cocked.

➡**When installing the bearing races, be sure to support the hub/disc assembly with GM tool J-9746-02.**

   c. Using the GM Seal Installation tools J-8092 and J-8457, drive the outer race into the hub/disc assembly until it seats.
5. Using a high melting point wheel bearing grease, lubricate the bearings, races and spindle; be sure to place a gob of grease (inside the hub/disc assembly) between the races to provide an ample supply of lubricant.

➡**To lubricate each bearing, place a gob of grease in the palm of the hand, then scoop the bearing through the grease until it is well lubricated.**

6. Place the inner bearing in the hub, then apply a thin coating of grease to the sealing lip and install a new inner seal, making sure the seal flange faces the bearing cup.

➡**Although a seal installation tool is preferable, a section of pipe with a smooth edge or a suitably sized socket may be used to drive the seal into position. Be sure the seal is flush with the outer surface of the hub assembly.**

7. Install or connect the following:
   - Wheel hub over the spindle
   - Outer bearing into the hub by hand
   - Spindle washer and nut
   - Brake caliper
   - Wheel
8. Properly adjust the wheel bearings

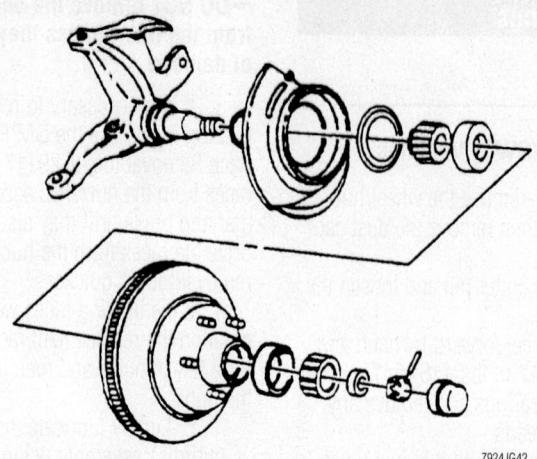

7924JG42

Wheel bearings, races and related components—2002-03 2WD vehicles

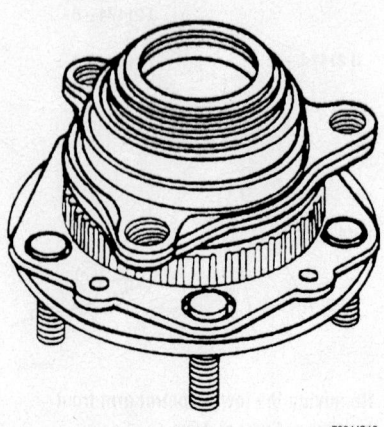

7924JG15

Hub and bearing assembly—4WD vehicles

9. Install or connect the following:
  • New cotter pin
  • Dust cap
  • Wheel cover

2WD MODELS 2004–05

1. Raise and support the vehicle.
2. Remove the tire and wheel.
3. Remove the brake rotor.
4. Remove the wheel speed sensor mounting bolt from the wheel hub and bearing.
5. Remove the wheel speed sensor from the wheel hub and bearing.
6. Remove the wheel hub and bearing-to-steering knuckle mounting bolts.
7. Remove the wheel hub and bearing from the steering knuckle.
8. Remove the splash shield from the steering knuckle.

9. Remove the wheel hub seal.

To install:

10. Install the wheel hub seal.
11. Install and align the splash shield to the steering knuckle.
12. Install the wheel hub and bearing assembly to the steering knuckle. Align the threaded holes.
13. Install the mounting bolts and tighten to 77 ft. lbs. (105 Nm).
14. Install the wheel speed sensor and tighten to 13 ft. lbs. (18 Nm).
15. Install the rotor.
16. Install the tire and wheel.
17. Lower the vehicle.

4WD MODELS

1. Install Torsion Bar Unloading tool J 36202 on the torsion bar adjusting bolt and remove the bolt. To aid during installation,

count the number of turns required to remove the bolt.
2. Remove the wheel.
3. Install an axle shaft boot seal protector to the Tri-pot axle joint.
4. Remove or disconnect the following:
  • Cotter pin and retainer
  • Castle nut and the thrust washer
  • Brake caliper and support it aside using wire or a coat hanger

➡ **Be sure the brake line is not stretched or damaged.**

  • Brake disc from the wheel hub
  • Halfshaft from the hub/bearing assembly, using a Spindle Remover tool J-28733-A to prevent damage to the shaft or hub/bearing assembly
  • Hub/bearing assembly from the knuckle

5. Clean and inspect the parts for nicks, scores and/or damage, then replace them as necessary.

To install:

6. Install or connect the following:
  • Hub and bearing assembly by aligning the threaded holes. Torque the bolts to 77 ft. lbs. (105 Nm).
  • Tie rod end to the steering knuckle using the retaining nut
  • New cotter pin
  • Brake assembly
  • Halfshaft nut. Tighten the nut to 103 ft. lbs. (140 Nm).
  • Retainer and a new cotter pin but DO NOT back off specification in order to insert the cotter pin.

7. Remove the torsion bar unloader tool and the drive axle boot protector.
8. Install the wheel.
9. Check and/or adjust the vehicle trim height, as necessary.

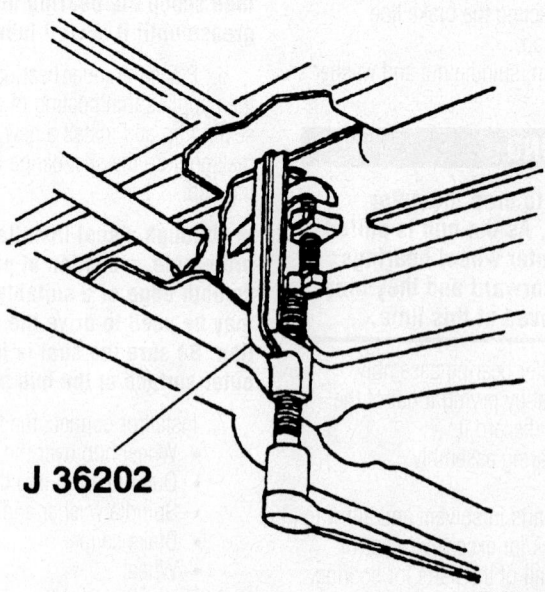

J 36202

7924JG54

Use Torsion Bar Unloading tool J 36202 to remove the adjusting bolt and unload the torsion bar

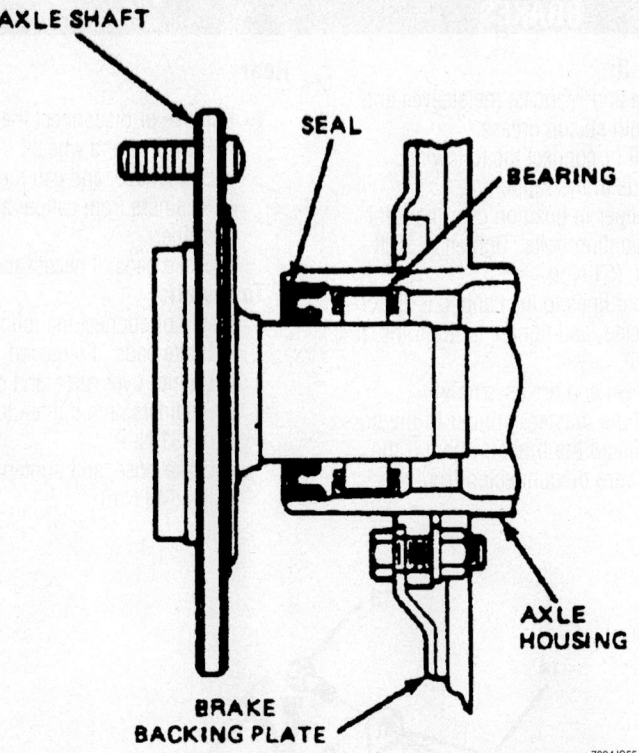

Cross-sectional view of the rear axle, bearing and seal assembly

Rear

A new pinion shaft lockbolt should be installed whenever either of the axle shafts is removed.

The axle shaft and seal may be removed and replaced without disturbing the bearing or seal but it is highly recommended to replace the seals when removing the axle shaft.

1. Remove or disconnect the following:
   - Rear wheels
   - Brake drums
2. Using a wire brush, clean the dirt/rust from around the rear axle cover.
3. Drain the fluid.

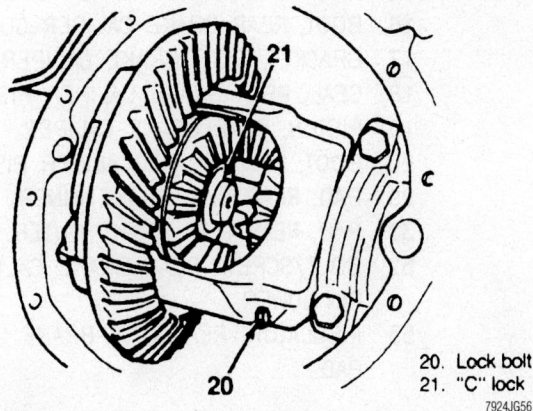

20. Lock bolt
21. "C" lock

Pinion shaft lockbolt and axle C-lock locations, inside the differential

4. Remove or disconnect the following:
   - Rear pinion shaft lockbolt and the pinion shaft
   - C-lock from the button end of the axle shaft by pushing the axle shaft inward
   - Axle shaft from the axle housing

➡ **Be careful not to damage the oil seal.**

✳✳ WARNING

If equipped with an Anti-Lock Brake System (ABS), be careful not to damage the reflector ring on the axle shaft or the speed sensor bolted to the backing plate, immediately adjacent to the shaft.

5. Remove or disconnect the following:
   - Oil seal by prying it from the end of the rear axle housing

✳✳ WARNING

DO NOT damage the housing oil seal surface.

   - Wheel bearing using the GM Slide Hammer tool J-2619, the GM Adapter tool J-2619-4 and the GM Axle Bearing Puller tool J-22813-01

To install:
6. Clean and inspect the components for excessive wear or damage and replace them, if necessary.
7. Install or connect the following:
   - New or reused bearing, coated with gear lubricant, using the Axle Shaft Bearing Installer tool J-34974 to drive the bearing in until it bottoms against the seat

✳✳ WARNING

Be sure the bearing installer does not contact and damage the speed sensor on ABS equipped vehicles.

   - New seal lubricated with gear oil using the GM Axle Shaft Seal Installer tool J-33782 to seat it in the housing until it is flush with the axle tube

➡ **Be sure the seal installer does not contact and damage the speed sensor on ABS equipped vehicles.**

   - Axle shaft into the housing by engaging the splines
   - C-lock retainer on the axle shaft button end

✳✳ WARNING

BE CAREFUL not to damage the wheel bearing seal.

   - Axle shaft by pulling it outward to seat the C-lock retainer in the counterbore of the side gears
   - Pinion shaft through the case and the pinions. Tighten the new lockbolt to 27 ft. lbs. (36 Nm).
   - New rear axle cover gasket
   - Housing cover
   - Brake drums
   - Wheels
8. Refill the housing.

BRAKES

Brake Caliper

REMOVAL & INSTALLATION

Front

1. Remove or disconnect the following:
   - ⅔ of the brake fluid from the master cylinder reservoir
   - Tire and wheel assembly
   - Brake caliper fluid line, then plug it
   - Bolts retaining the caliper to the rotor
   - Caliper from the rotor
   - Disc brake pads from the caliper
   - Brake pad retaining clips from inside the caliper

To install:

2. Clean and lubricate the sleeves and bushings with silicon grease.
3. Install or connect the following:
   - Pads in the caliper
   - Caliper in position over the rotor
   - Mounting bolts. Tighten to 38 ft. lbs. (51 Nm).
   - Fluid lines to the caliper, if disconnected, and tighten to 40 ft. lbs. (54 Nm)
   - Wheel and tire assembly
4. Refill the master cylinder to the correct level. Bleed the brake system if the fluid lines were disconnected from the caliper.

Rear

1. Remove or disconnect the following:
   - Rear tires and wheels
   - Brake hose; and cap the line
   - Retainers from caliper and remove caliper
   - Brake pads, if necessary

To install:

2. Install or connect the following:
   - Brake pads, if removed
   - Caliper over rotor, and onto mounts
   - Retainers, and tighten to 23 ft. lbs. or (31 Nm)
   - Brake hose, and tighten to 33 ft. lbs. (46 Nm)

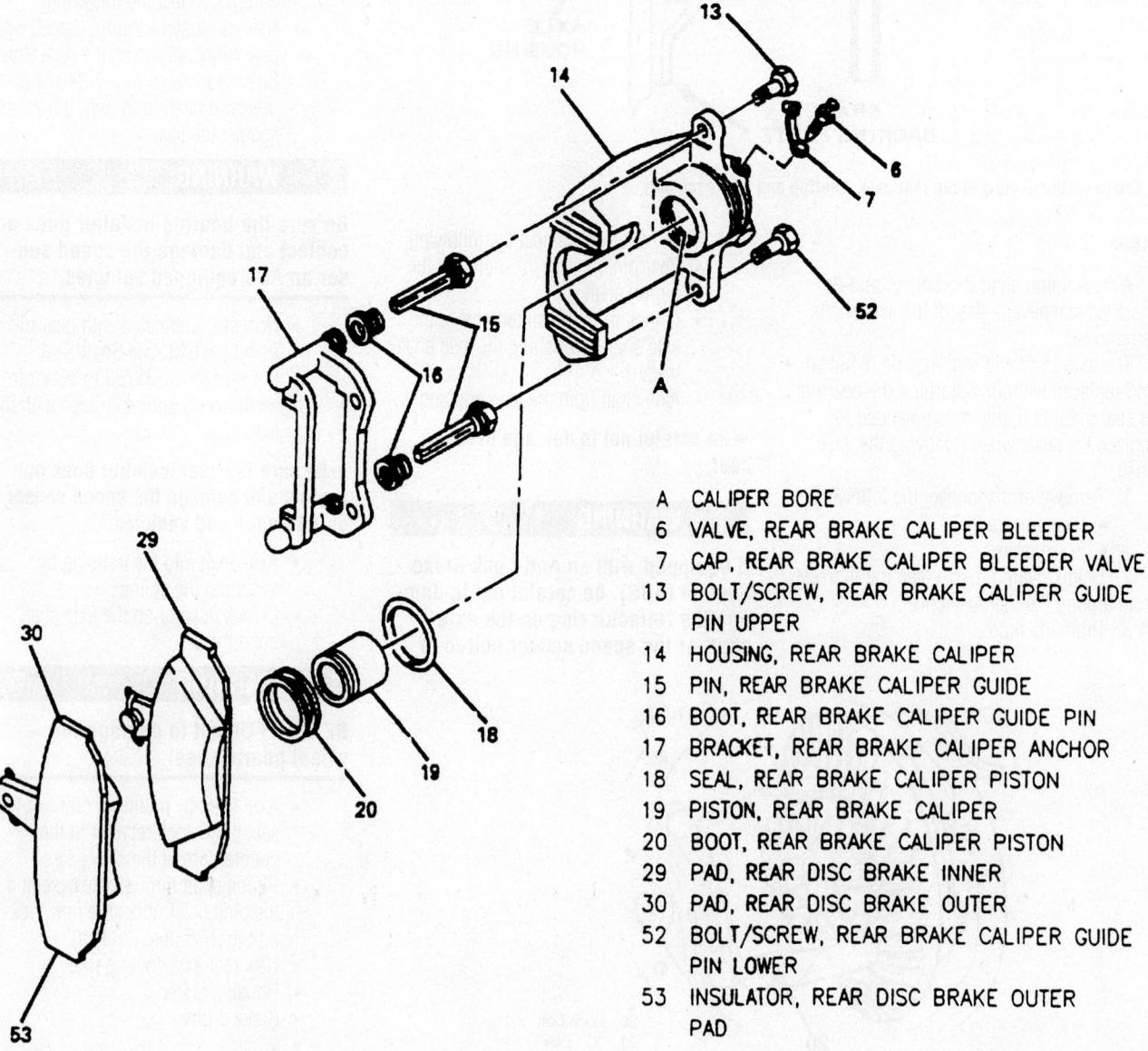

A CALIPER BORE
6 VALVE, REAR BRAKE CALIPER BLEEDER
7 CAP, REAR BRAKE CALIPER BLEEDER VALVE
13 BOLT/SCREW, REAR BRAKE CALIPER GUIDE PIN UPPER
14 HOUSING, REAR BRAKE CALIPER
15 PIN, REAR BRAKE CALIPER GUIDE
16 BOOT, REAR BRAKE CALIPER GUIDE PIN
17 BRACKET, REAR BRAKE CALIPER ANCHOR
18 SEAL, REAR BRAKE CALIPER PISTON
19 PISTON, REAR BRAKE CALIPER
20 BOOT, REAR BRAKE CALIPER PISTON
29 PAD, REAR DISC BRAKE INNER
30 PAD, REAR DISC BRAKE OUTER
52 BOLT/SCREW, REAR BRAKE CALIPER GUIDE PIN LOWER
53 INSULATOR, REAR DISC BRAKE OUTER PAD

93026G44

Rear brake caliper

• Rear tires and wheels

3. Refill the master cylinder to the correct level. Bleed the brake system if the fluid lines were disconnected from the caliper.

Disc Brake Pads

REMOVAL & INSTALLATION

Front

1. Remove or disconnect the following:
   • ⅔ of the brake fluid from the master cylinder
2. Place a C-clamp around the outer pad and caliper; tighten the C-clamp until the piston is fully compressed in the caliper.
   • Remove top caliper retainer, and rotate caliper away from rotor
   • Inboard pad and retaining spring from the caliper
   • Outboard pad from the caliper
   • Sleeves and bushings

To install:

3. Clean and lubricate the sleeves and bushing with silicone lubricant and install them in the caliper.
4. Install or connect the following:
   • Retaining spring onto the inboard pad
   • Inboard pad in the caliper
   • Outboard pad into the caliper
   • Caliper in position over the rotor
   • Caliper mounting bolts. Bend the tabs, on the outboard brake pad, over the caliper.
   • Wheel and tire assemblies
5. Refill the master cylinder and pump pedal to attain full brake pedal before Road-testing the vehicle.

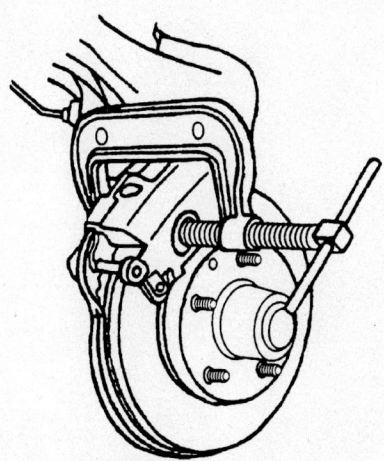

93026G47

Compressing the caliper piston with a C-clamp

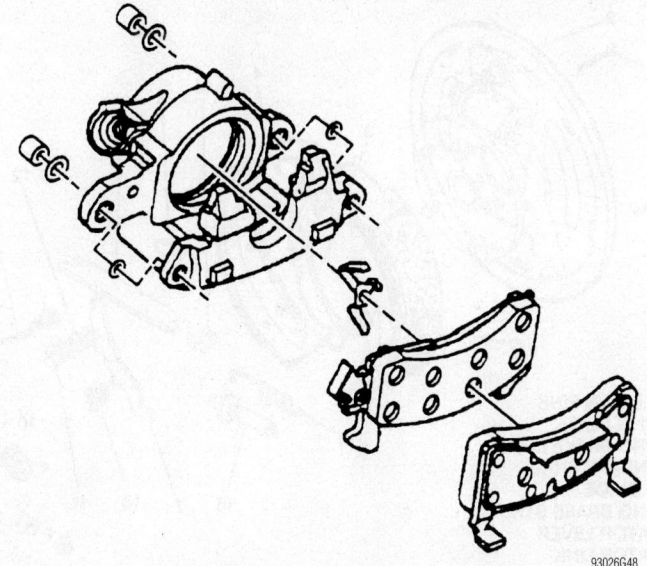

93026G48

Exploded view of the single piston front disc brake caliper

Rear

1. Remove or disconnect the following:
   • ⅔ of the brake fluid from the master cylinder
   • Wheels
2. Place a C-clamp around the outer pad and caliper; tighten the C-clamp until the piston is fully compressed in the caliper.
   • Top caliper retainer, and rotate caliper away from rotor
   • Inboard pad and retaining spring from the caliper
   • Outboard pad from the caliper

To install:

3. Clean and lubricate the sleeves and bushing with silicone lubricant and install them in the caliper.
4. Install or connect the following:
   • Retaining spring onto the inboard pad
   • Inboard pad in the caliper
   • Outboard pad into the caliper
   • Caliper in position over the rotor
   • Caliper mounting bolts
   • Wheel and tire assemblies
5. Refill the master cylinder and pump pedal to attain full brake pedal before Road-testing the vehicle.

Brake Drums

REMOVAL AND INSTALLATION

1. Remove or disconnect the following:
   • Wheel and tire assembly
   • Brake drum. If the drum will not pull of the axle, use a rubber mallet and tap it around the edge.

To install:

1. Install or connect the following:
   • Drum on the axle
   • Wheel and tire assembly
   • Refill the master cylinder and pump pedal to attain full brake pedal before road-testing the vehicle.

Brake Shoes

REMOVAL AND INSTALLATION

1. Remove or disconnect the following:
   • Wheel and tire assembly
   • Brake drum
   • Return springs from the brake shoes
   • Shoe guide
   • Hold-down springs and pins
   • Actuator lever and pivot
   • Lever return spring
   • Actuator link
   • Parking brake strut and spring
   • Parking brake lever
   • Brake shoes and the adjuster assembly

To install:

2. Lubricate the contact points on the backing plate and the adjuster with lithium grease.
3. Install or connect the following:
   • Parking brake lever, adjusting screw and spring assembly
   • Shoe assembly onto the backing plate
   • Parking brake lever, strut and strut spring
   • Actuator lever and lever pivot
   • Actuator link

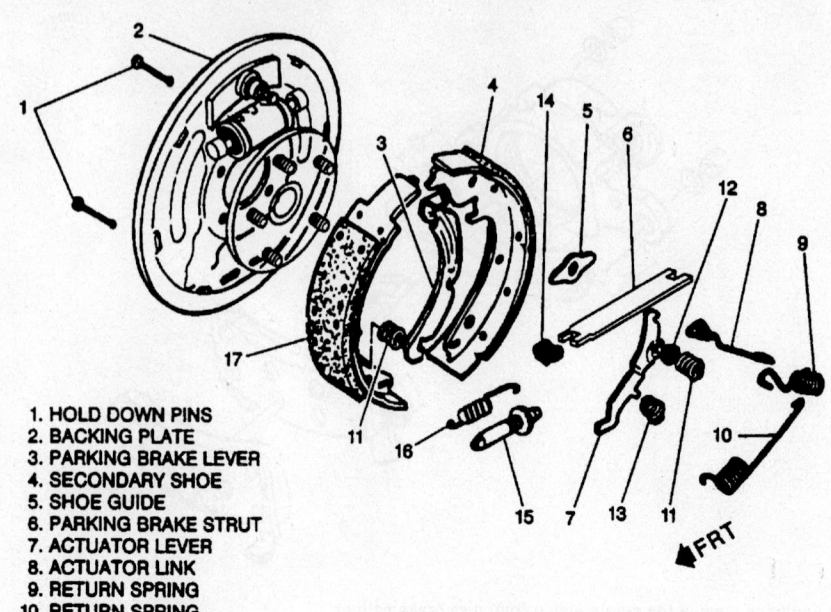

1. HOLD DOWN PINS
2. BACKING PLATE
3. PARKING BRAKE LEVER
4. SECONDARY SHOE
5. SHOE GUIDE
6. PARKING BRAKE STRUT
7. ACTUATOR LEVER
8. ACTUATOR LINK
9. RETURN SPRING
10. RETURN SPRING
11. HOLD DOWN SPRING
12. LEVER PIVOT
13. LEVER RETURN SPRING
14. STRUT SPRING
15. ADJUSTING SCREW ASSEMBLY
16. ADJUSTING SCREW SPRING
17. PRIMARY SHOE

93026G51

Exploded view of the drum brake components

- Lever spring, the hold-down pins and springs
- Shoe guide
- Return springs
- Brake drum in position

1. Adjust the brakes as follows:

a. Remove the knockout area in the backing plate, behind the adjuster assembly.

b. Ensure the parking brake system is adjusted properly with no tension on the cables or parking brake lever. The tops of the shoes should be firmly seated against the upper spring retaining anchor, if not as specified, loosen the parking brake cables.

c. Install the drum and turn the brake adjuster until the wheels can just be turned by hand.

d. Then, back the adjuster off 24 notches. No brake drag should be felt after 12 notches.

e. Install an adjusting hole plug in the backing plate to prevent dirt and moisture from entering.

f. Readjust the parking brake cable as necessary.

2. Install the wheel and tire assemblies.

3. Refill the master cylinder and pump pedal to attain full brake pedal before Road-testing the vehicle.

BUICK AND PONTIAC

Bonneville • Le Sabre • Park Avenue

8

BRAKES 8-56
DRIVE TRAIN 8-47
ENGINE REPAIR 8-12
FUEL SYSTEM 8-45
SPECIFICATIONS AND
  MAINTENANCE CHARTS 8-2
Engine and Vehicle Identification 8-2
General Engine Specifications 8-2
Engine Tune-up Specifications 8-3
Firing Order 8-3
Accessory Drive Belt Routing 8-4
Capacities 8-5
Valve Specifications......................8-5
Crankshaft and Connecting Rod
  Specifications 8-6
Piston and Ring Specifications........8-6
Torque Specifications 8-7
Wheel Alignment 8-7
Tire, Wheel and Ball Joint
  Specifications 8-8
Brake Specifications 8-9
Short Trip/City Scheduled
  Maintenance Intervals..................8-10
Long Trip/City Scheduled
  Maintenance Intervals..................8-11
STEERING AND
  SUSPENSION 8-50

A
Air Bag.............................8-50
  Arming............................8-50
  Disarming 8-50
  Precautions......................8-50
Alternator 8-12
  Removal & Installation.............8-12

B
Brake Caliper 8-56
  Removal & Installation............8-56

C
Camshaft and Valve Lifters 8-29
  Removal & Installation.............8-29
Coil Spring 8-53
  Removal & Installation............8-53

CV-Joints.........................8-48
  Overhaul 8-48
Cylinder Head 8-20
  Removal & Installation.............8-20

D
Disc Brake Pads...................8-58
  Removal & Installation.............8-58

E
Engine Assembly 8-12
  Removal & Installation.............8-12
Exhaust Manifold 8-26
  Removal & Installation.............8-26

F
Fuel Filter 8-45
  Removal & Installation.............8-45
Fuel Injectors 8-46
  Removal & Installation.............8-46
Fuel Pump 8-45
  Removal & Installation.............8-45
Fuel System Pressure 8-45
  Relieving........................8-45
Fuel System Service Precautions...8-45

H
Halfshaft.........................8-47
  Removal & Installation.............8-47
Heater Core......................8-18
  Removal & Installation.............8-18

I
Ignition Timing 8-12
Intake Manifold 8-24
  Removal & Installation.............8-24

L
Lower Ball Joint...................8-53
  Removal & Installation.............8-53
Lower Control Arm 8-54
  Control Arm Bushing
    Replacement 8-55
  Removal & Installation.............8-54

O
Oil Pan...........................8-36
  Removal & Installation.............8-36
Oil Pump 8-37
  Removal & Installation.............8-37

P
Piston and Ring 8-44
  Positioning 8-44
Power Rack and Pinion Steering
  Gear...........................8-51
  Removal & Installation.............8-51

R
Rear Main Seal 8-38
  Removal & Installation.............8-38
Rocker Arms 8-22
  Removal & Installation.............8-22

S
Shock Absorber 8-52
  Removal & Installation.............8-52
Starter Motor 8-36
  Removal & Installation.............8-36
Strut.............................8-51
  Removal & Installation.............8-51
Supercharger 8-22
  Removal & Installation.............8-22

T
Timing Chain, Sprockets, Front
  Cover and Seal 8-40
  Removal & Installation.............8-40
Transaxle Assembly 8-47
  Removal & Installation.............8-47

V
Valve Lash 8-36
  Adjustment......................8-36

W
Water Pump......................8-16
  Removal & Installation.............8-16
Wheel Bearings...................8-55
  Adjustment......................8-55
  Removal & Installation.............8-55

SPECIFICATIONS AND MAINTENANCE CHARTS

ENGINE AND VEHICLE IDENTIFICATION

Code ①	Liters (cc)	Cu. In.	Cyl.	Fuel Sys.	Engine Type	Eng. Mfg.
1 ③	3.8 (3785)	231	6	MFI	OHV	BOC
K	3.8 (3785)	231	6	MFI	OHV	BOC
Y	4.6 (4572)	279	8	MFI	DOHC	BOC

Code ②	Year
2	2002
3	2003
4	2004
5	2005

MFI: Multi-point Fuel Injection

BOC: Buick/Oldsmobile/Cadillac

OHV: Overhead Valves

① 8th position of VIN

② 10th position of VIN

③ Supercharged engine

06025-BONN-C01

GENERAL ENGINE SPECIFICATIONS

Year	Model	Engine Displacement Liters	Engine Series VIN	Net Horsepower @ rpm	Net Torque @ rpm (ft. lbs.)	Bore x Stroke (in.)	Compression Ratio	Oil Pressure @ rpm
2002	Bonneville	3.8	1	240@5000	275@3200	3.80x3.40	8.5:1	60@1850
	Bonneville	3.8	K	205@5200	230@4000	3.80x3.40	9.4:1	60@1850
	LeSabre	3.8	K	200@5200	230@4000	3.80x3.40	9.4:1	60@1850
	Park Avenue	3.8	1	240@5200	280@3600	3.80x3.40	8.5:1	60@1850
	Park Avenue	3.8	K	200@5200	230@4000	3.80x3.40	9.4:1	60@1850
2003	Bonneville	3.8	1	240@5000	275@3200	3.80x3.40	8.5:1	60@1850
	Bonneville	3.8	K	205@5200	230@4000	3.80x3.40	9.4:1	60@1850
	LeSabre	3.8	K	200@5200	230@4000	3.80x3.40	9.4:1	60@1850
	Park Avenue	3.8	1	240@5200	280@3600	3.80x3.40	8.5:1	60@1850
	Park Avenue	3.8	K	200@5200	230@4000	3.80x3.40	9.4:1	60@1850
2004	Bonneville	3.8	K	205@5200	230@4000	3.80x3.40	9.4:1	60@1850
	Bonneville	4.6	Y	275@5600	300@4000	3.66x3.31	10:01	35@2000
	LeSabre	3.8	K	200@5200	230@4000	3.80x3.40	9.4:1	60@1850
	Park Avenue	3.8	1	240@5200	280@3600	3.80x3.40	8.5:1	60@1850
	Park Avenue	3.8	K	200@5200	230@4000	3.80x3.40	9.4:1	60@1850
2005	Bonneville	3.8	K	205@5200	230@4000	3.80x3.40	9.4:1	60@1850
	Bonneville	4.6	Y	275@5600	300@4000	3.66x3.31	10:01	35@2000
	LeSabre	3.8	K	200@5200	230@4000	3.80x3.40	9.4:1	60@1850
	Park Avenue	3.8	1	240@5200	280@3600	3.80x3.40	8.5:1	60@1850
	Park Avenue	3.8	K	200@5200	230@4000	3.80x3.40	9.4:1	60@1850

06025-BONN-C02

ENGINE TUNE-UP SPECIFICATIONS

Year	Engine Displacement Liters	Engine VIN	Spark Plug Gap (in.)	Ignition Timing (deg.)	Fuel Pump (psi)	Idle Speed (rpm)	Valve Clearance	
							Intake	Exhaust
2002	3.8	1	0.060	①	41-47②	③	HYD	HYD
	3.8	K	0.060	①	41-47②	③	HYD	HYD
2003	3.8	1	0.060	①	41-47②	③	HYD	HYD
	3.8	K	0.060	①	41-47②	③	HYD	HYD
2004	3.8	1	0.060	①	41-47②	③	HYD	HYD
	3.8	K	0.060	①	41-47②	③	HYD	HYD
	4.6	Y	0.050	①	NA	③	HYD	HYD
2005	3.8	1	0.060	①	41-47②	③	HYD	HYD
	3.8	K	0.060	①	41-47②	③	HYD	HYD
	4.6	Y	0.050	①	NA	③	HYD	HYD

NOTE: The Vehicle Emission Control Information label often reflects specification changes made during production.

The label figures must be used if they differ from those in this chart.

HYD: Hydraulic

NA: Information not available

① DIS Ignition System timing not adjustable

② Pressure at fuel pump

③ Idle speed maintained by ECM. There is no recommended adjustment procedure

06025-BONN-C03

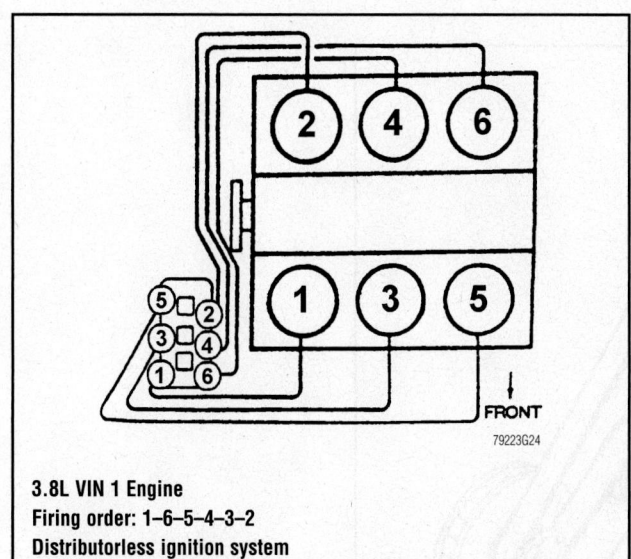

3.8L VIN 1 Engine
Firing order: 1–6–5–4–3–2
Distributorless ignition system

79223G24

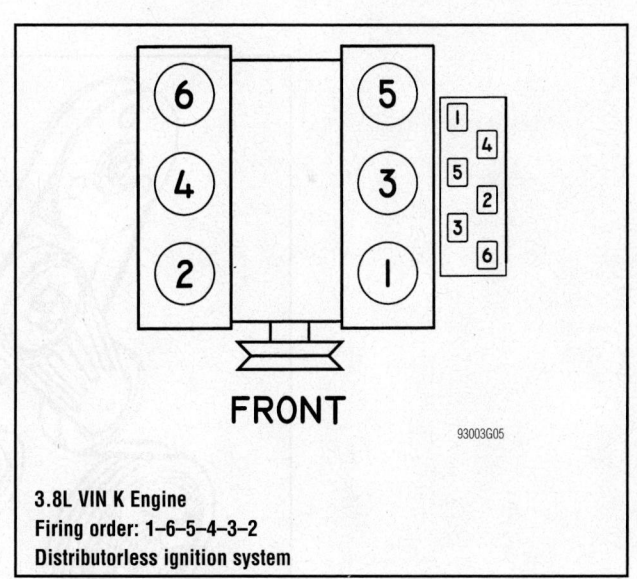

3.8L VIN K Engine
Firing order: 1–6–5–4–3–2
Distributorless ignition system

93003G05

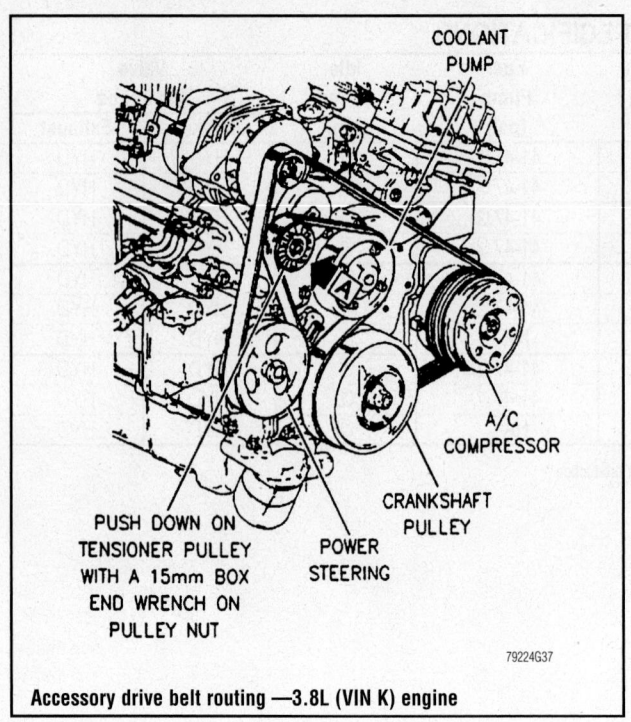

Accessory drive belt routing —3.8L (VIN K) engine

COOLANT PUMP

A/C COMPRESSOR

CRANKSHAFT PULLEY

POWER STEERING

PUSH DOWN ON TENSIONER PULLEY WITH A 15mm BOX END WRENCH ON PULLEY NUT

79224G37

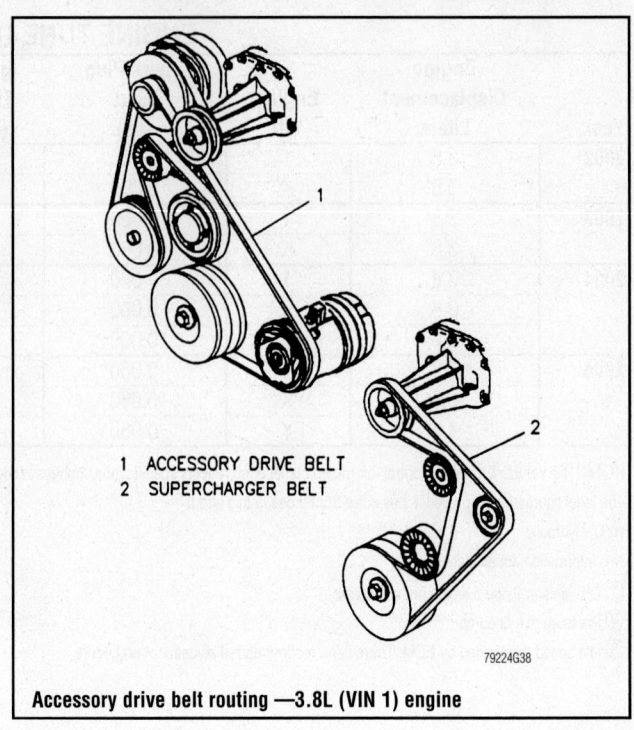

Accessory drive belt routing —3.8L (VIN 1) engine

1 ACCESSORY DRIVE BELT
2 SUPERCHARGER BELT

79224G38

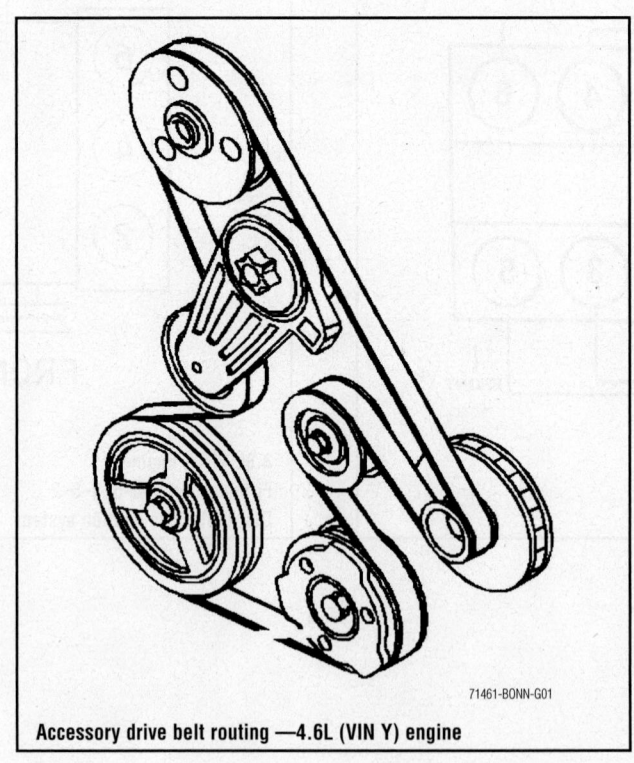

Accessory drive belt routing —4.6L (VIN Y) engine

71461-BONN-G01

CAPACITIES

Year	Model	Engine Displacement Liters	Engine VIN	Engine Oil with Filter (qts.)	Transmission (pts.)	Fuel Tank (gal.)	Cooling System (qts.)
2002	Bonneville	3.8	1& K	4.5	14.8	18.0	10.0
	LeSabre	3.8	1& K	4.5	14.8	18.0	10.0
	Park Avenue	3.8	1& K	4.5	12.0	18.0	13.0
2003	Bonneville	3.8	1& K	4.5	14.8	18.0	10.0
	LeSabre	3.8	K	4.5	14.8	18.0	10.0
	Park Avenue	3.8	1& K	4.5	12.0	18.0	13.0
2004	Bonneville	3.8	K	4.5	14.8	18.5	10.0
	Bonneville	4.6	Y	7.5	30.0	18.5	13.0
	LeSabre	3.8	K	4.5	14.8	18.5	10.0
	Park Avenue	3.8	1& K	4.5	12.0	18.5	13.0
2005	Bonneville	3.8	K	4.5	14.8	18.5	10.0
	Bonneville	4.6	Y	7.5	30.0	18.5	13.0
	LeSabre	3.8	K	4.5	14.8	18.5	10.0
	Park Avenue	3.8	1& K	4.5	12.0	18.5	13.0

NOTE: All capacities are approximate. Add fluid gradually and ensure a proper fluid level is obtained.

06025-BONN-C04

VALVE SPECIFICATIONS

Year	Engine Displacement Liters	Engine VIN	Seat Angle (deg.)	Face Angle (deg.)	Spring Test Pressure (lbs. @ in.)	Spring Installed Height (in.)	Stem-to-Guide Clearance (in.) Intake	Stem-to-Guide Clearance (in.) Exhaust	Stem Diameter (in.) Intake	Stem Diameter (in.) Exhaust
2002	3.8	1	45	45	80@1.750	1.690-1.720	0.0015-0.0032	0.0015-0.0032	0.3129-0.3136	0.3129-0.3136
	3.8	K	45	45	80@1.750	1.690-1.720	0.0015-0.0032	0.0015-0.0032	0.3129-0.3136	0.3129-0.3136
2003	3.8	1	45	45	80@1.750	1.690-1.720	0.0015-0.0032	0.0015-0.0032	0.3129-0.3136	0.3129-0.3136
	3.8	K	45	45	80@1.750	1.690-1.720	0.0015-0.0032	0.0015-0.0032	0.3129-0.3136	0.3129-0.3136
2004	3.8	1	45	45	80@1.750	1.690-1.720	0.0015-0.0032	0.0015-0.0032	0.3129-0.3136	0.3129-0.3136
	3.8	K	45	45	80@1.750	1.690-1.720	0.0015-0.0032	0.0015-0.0032	0.3129-0.3136	0.3129-0.3136
	4.6	Y	45.75	45	130-142@0.9646	1.378	0.0011-0.0027	0.0020-0.0039	0.2331-0.2339	0.2331-0.2339
2005	3.8	1	45	45	80@1.750	1.690-1.720	0.0015-0.0032	0.0015-0.0032	0.3129-0.3136	0.3129-0.3136
	3.8	K	45	45	80@1.750	1.690-1.720	0.0015-0.0032	0.0015-0.0032	0.3129-0.3136	0.3129-0.3136
	4.6	Y	45.75	45	130-142@0.9646	1.378	0.0011-0.0027	0.0020-0.0039	0.2331-0.2339	0.2331-0.2339

06025-BONN-C05

CRANKSHAFT AND CONNECTING ROD SPECIFICATIONS

All measurements are given in inches.

Year	Engine Displacement Liters	Engine VIN	Crankshaft				Connecting Rod		
			Main Brg. Journal Dia.	Main Brg. Oil Clearance	Shaft End-play	Thrust on No.	Journal Diameter	Oil Clearance	Side Clearance
2002	3.8	1	2.4988-2.4998	①	0.0030-0.0110	2	2.2487-2.2499	0.0005-0.0026	0.0040-0.0200
	3.8	K	2.4988-2.4998	①	0.0030-0.0110	2	2.2487-2.2499	0.0005-0.0026	0.0040-0.0200
2003	3.8	1	2.4988-2.4998	①	0.0030-0.0110	2	2.2487-2.2499	0.0005-0.0026	0.0040-0.0200
	3.8	K	2.4988-2.4998	①	0.0030-0.0110	2	2.2487-2.2499	0.0005-0.0026	0.0040-0.0200
2004	3.8	1	2.4988-2.4998	①	0.0030-0.0110	2	2.2487-2.2499	0.0005-0.0026	0.0040-0.0200
	3.8	K	2.4988-2.4998	①	0.0030-0.0110	2	2.2487-2.2499	0.0005-0.0026	0.0040-0.0200
	4.6	Y	2.5335-2.5341	0.0006-0.0022	0.0020-0.0197	NA	2.1239-2.1245	0.0010-0.0030	0.0079-0.0197
2005	3.8	1	2.4988-2.4998	①	0.0030-0.0110	2	2.2487-2.2499	0.0005-0.0026	0.0040-0.0200
	3.8	K	2.4988-2.4998	①	0.0030-0.0110	2	2.2487-2.2499	0.0005-0.0026	0.0040-0.0200
	4.6	Y	2.5335-2.5341	0.0006-0.0022	0.0020-0.0197	NA	2.1239-2.1245	0.0010-0.0030	0.0079-0.0197

NA: Information not available

① Journal 1: 0.0007 - 0.0016
Journals 2 and 3: 0.0010 - 0.0020
Journal 4: 0.0009 - 0.0018

06025-BONN-C06

PISTON AND RING SPECIFICATIONS

All measurements are given in inches.

Year	Engine Displacement Liters	Engine VIN	Piston Clearance	Ring Gap			Ring Side Clearance		
				Top Compression	Bottom Compression	Oil Control	Top Compression	Bottom Compression	Oil Control
2002	3.8	1	0.0004-0.0020	0.012-0.022	0.030-0.040	0.010-0.030	0.0013-0.0031	0.0013-0.0031	0.0009-0.0079
	3.8	K	0.0004-0.0020	0.012-0.022	0.030-0.040	0.010-0.030	0.0013-0.0031	0.0013-0.0031	0.0009-0.0079
2003	3.8	1	0.0004-0.0020	0.012-0.022	0.030-0.040	0.010-0.030	0.0013-0.0031	0.0013-0.0031	0.0009-0.0079
	3.8	K	0.0004-0.0020	0.012-0.022	0.030-0.040	0.010-0.030	0.0013-0.0031	0.0013-0.0031	0.0009-0.0079
2004	3.8	1	0.0004-0.0020	0.012-0.022	0.030-0.040	0.010-0.030	0.0013-0.0031	0.0013-0.0031	0.0009-0.0079
	3.8	K	0.0004-0.0020	0.012-0.022	0.030-0.040	0.010-0.030	0.0013-0.0031	0.0013-0.0031	0.0009-0.0079
	4.6	Y	NA	0.0098-0.0157	0.0020-0.0138	0.0098-0.0299	0.0016-0.0037	0.0016-0.0037	side sealing
2005	3.8	1	0.0004-0.0020	0.012-0.022	0.030-0.040	0.010-0.030	0.0013-0.0031	0.0013-0.0031	0.0009-0.0079
	3.8	K	0.0004-0.0020	0.012-0.022	0.030-0.040	0.010-0.030	0.0013-0.0031	0.0013-0.0031	0.0009-0.0079
	4.6	Y	NA	0.0098-0.0157	0.0020-0.0138	0.0098-0.0299	0.0016-0.0037	0.0016-0.0037	side sealing

NA: Information not available

06025-BONN-C07

TORQUE SPECIFICATIONS
All readings in ft. lbs.

Year	Engine Displacement Liters	Engine VIN	Cylinder Head Bolts	Main Bearing Bolts	Rod Bearing Bolts	Crankshaft Damper Bolts	Flywheel Bolts	Manifold		Spark Plugs	Oil Pan Drain Plug
								Intake	Exhaust		
2002	3.8	1	①	②	③	④	⑤	⑥	22	11	22
	3.8	K	①	⑦	③	④	⑤	11	38	11	22
2003	3.8	1	①	②	③	④	⑤	⑥	22	11	22
	3.8	K	①	⑦	③	④	⑤	11	38	11	22
2004	3.8	1	①	②	③	④	⑤	⑥	22	11	22
	3.8	K	①	⑦	③	④	⑤	11	38	11	22
	4.6	Y	⑧	⑨	⑩	⑪	⑫	7.5	18	15	15
2005	3.8	1	①	⑬	③	④	⑤	⑥	22	⑭	22
	3.8	K	①	⑬	③	④	⑤	11	22	⑭	22
	4.6	Y	⑧	⑨	⑩	⑪	⑫	7.5	18	11	15

① Step 1: Tighten all bolts to 37 ft. lbs.
Step 2: Turn all bolts 120 degrees

② Step 1: Tighten caps in equal increments to 52 ft. lbs.
Step 2: Loosen 360 degrees
Step 3: 15 ft. lbs.
Step 4: 54 ft. lbs.
Step 5: Plus three turns of 35 degrees for a total of 105 degrees

③ 20 ft. lbs. plus 50 degrees

④ 111 ft. lbs. plus 76 degrees

⑤ 11 ft. lbs. plus 50 degrees

⑥ Upper manifold: 8 ft. lbs.
Lower manifold: 11 ft. lbs.

⑦ 26 ft. lbs. plus 50 degrees

⑧ M11 bolts:
Step 1: 30 ft. lbs.
Step 2: +70 degrees
Step 3: +60 degrees
Step 4: +45 degrees
Do not exceed 175 degrees total
M6 bolts: 106 inch lbs.

⑨ Lower crankcase
M10x1.5
Step 1: 15 ft. lbs.
Step 2: +65 degrees
M8x1.25: 22 ft. lbs.

⑩ Step 1: 22 ft. lbs.
Step 2: back off to zero
Step 3: 18 ft. lbs.
Step 4: +110 degrees

⑪ Step 1: 37 ft. lbs.
Step 2: +120 degrees

⑫ Step 1: 11 ft. lbs.
Step 2: +50 degrees

⑬ Cap bolts: 30 ft. lbs. plus 110 degrees
Side bolts: 11 ft. lbs. plus 45 degrees

⑭ Initial installation: 20 ft. lbs.
Re-installation: 11 ft. lbs.

06025-BONN-C08

WHEEL ALIGNMENT

Year	Model		Caster		Camber		Toe-in (in.)
			Range (Deg.)	Preferred Setting (Deg.)	Range (Deg.)	Preferred Setting (Deg.)	
2002	All	F	0.50	+6.00	0.50	-0.20	0.20 +/- 0.20
		R	—	—	0.50	-0.30	0.20 +/- 0.20
2003	All	F	0.50	+6.00	0.50	-0.20	0.20 +/- 0.20
		R	—	—	0.50	-0.30	0.20 +/- 0.20
2004	All	F	0.50	+5.00	0.50	-0.20	0.20 +/- 0.20
		R	—	—	0.50	-0.30	0.20 +/- 0.20
2005	All	F	0.50	+5.00	0.50	-0.20	0.20 +/- 0.20
		R	—	—	0.50	-0.30	0.20 +/- 0.20

06025-BONN-C09

TIRE, WHEEL AND BALL JOINT SPECIFICATIONS
Pontiac

Year	Model	OEM Tires Standard	OEM Tires Optional	Tire Pressures (psi) Front	Tire Pressures (psi) Rear	Wheel Size	Ball Joint Inspection	Wheel lug Torque (ft. lbs.)
2002	Bonneville SE	P215/65R15	P225/60R16	30	30	6-JJ	①	100
	Bonneville SSE	P225/60R16	None	30	30	7-JJ	①	100
2003	Bonneville SE	P215/65R15	P225/60R16	30	30	6-JJ	①	100
	Bonneville SSE	P225/60R16	None	30	30	7-JJ	①	100
2004	Bonneville SE	P225/60R16	None	②	②	NA	①	100
	Bonneville SLE	P235/55R17	None	②	②	NA	①	100
	Bonneville GXP	P235/50WR18	None	②	②	NA	①	100
2005	Bonneville SE	P225/60R16	None	②	②	NA	①	100
	Bonneville SLE	P235/55R17	None	②	②	NA	①	100
	Bonneville GXP	P235/50WR18	None	②	②	NA	①	100

NA: Information not available

OEM: Original Equipment Manufacturer

PSI: Pounds Per Square Inch

① Do not lift car. Inspect the boss into which the grease fitting is threaded. Replace if the boss is flush or receded below the surface of the ball joint.

② See placard on vehicle

06025-BONN-C10

TIRE, WHEEL AND BALL JOINT SPECIFICATIONS
Buick

Year	Model	OEM Tires Standard	OEM Tires Optional	Tire Pressures (psi) Front	Tire Pressures (psi) Rear	Wheel Size	Ball Joint Inspection	Wheel lug Torque (ft. lbs.)
2002	LeSabre/Park Avenue	P205/70R15	P215/60R16	30	30	6-JJ	①	100
2003	LeSabre/Park Avenue	P205/70R15	P215/60R16	30	30	6-JJ	①	100
2004	LeSabre	P215/70R15	P225/60R16	②	②	NA	①	100
	Park Avenue	P225/60R16	P235/55R17	②	②	NA	①	100
2005	LeSabre	P215/70R15	P225/60R16	②	②	NA	①	100
	Park Avenue	P225/60R16	P235/55R17	②	②	NA	①	100

NA: Information not available

OEM: Original Equipment Manufacturer

PSI: Pounds Per Square Inch

① Do not lift car. Inspect the boss into which the grease fitting is threaded. Replace if the boss is flush or receded below the surface of the ball joint.

06025-BONN-C11

BRAKE SPECIFICATIONS

All measurements in inches unless noted

Year	Model		Brake Disc			Minimum Lining Thickness		Caliper Bracket Bolts	Caliper Mounting Bolts
			Original Thickness	Minimum Thickness	Maximum Runout	Front	Rear	(ft. lbs.)	(ft. lbs.)
2002	Bonneville	F	1.267	1.200	0.002	0.030	—	137	63
		R	0.433	0.374	0.002	—	0.030	94	20
	LeSabre	F	1.267	1.200	0.002	0.030	—	137	63
		R	0.433	0.374	0.002	—	0.030	94	20
	Park Avenue	F	1.267	1.200	0.002	0.030	—	137	63
		R	0.433	0.374	0.002	—	0.030	94	20
2003	Bonneville	F	1.267	1.200	0.002	0.030	—	137	63
		R	0.433	0.374	0.002	—	0.030	94	20
	LeSabre	F	1.267	1.200	0.002	0.030	—	137	63
		R	0.433	0.374	0.002	—	0.030	94	20
	Park Avenue	F	1.267	1.200	0.002	0.030	—	137	63
		R	0.433	0.374	0.002	—	0.030	94	20
2004	Bonneville	F	NA	NA	NA	NA	—	137 ①	NA
		R	NA	NA	NA	—	NA	94	NA
	LeSabre	F	1.267	1.200	0.002	0.030	—	137	63
		R	0.433	0.374	0.002	—	0.030	94	20
	Park Avenue	F	1.267	1.209	0.002	NA	—	137	63
		R	0.433	0.354	0.002	—	NA	94	20
2005	Bonneville	F	NA	②	NA	NA	—	③	④
		R	0.433	0.354	0.002	—	NA	94	20
	LeSabre	F	1.267	1.200	0.002	0.030	—	137	63
		R	0.433	0.374	0.002	—	0.030	94	20
	Park Avenue	F	1.267	1.209	0.002	NA	—	137	63
		R	0.433	0.354	0.002	—	NA	94	20

NA: Inforation not available

① Bonneville V8: 111 ft. lbs.

② Minimum thickness is cast into the rotor

③ L36: 137 ft. lbs.

   LD8: 111 ft. lbs.

④ L36: 63 ft. lbs.

   LD8: 25 ft. lbs.

06025-BONN-C12

SHORT TRIP/CITY SCHEDULED MAINTENANCE INTERVALS
BUICK LESABRE, PARK AVENUE
PONTIAC BONNEVILLE

TO BE SERVICED	TYPE OF SERVICE	VEHICLE MILEAGE INTERVAL (x1000)												
		3	6	9	12	15	18	21	24	27	30	33	36	39
Engine oil & filter	R	Change oil and filter every 3,000 miles or 3 months whichever occurs first												
Rotate tires	S/I		✓		✓		✓		✓		✓		✓	
Exhaust system & brake hoses	S/I	✓	✓	✓	✓	✓	✓	✓	✓	✓	✓	✓	✓	✓
Driveshaft boots & front suspension components	S/I	✓	✓	✓	✓	✓	✓	✓	✓	✓	✓	✓	✓	✓
Lubricate chassis, suspension, steering linkage, transaxle shift linkage, parking brake cable guides, underbody contact points & linkage	S/I	✓	✓	✓	✓	✓	✓	✓	✓	✓	✓	✓	✓	✓
Coolant level, hoses & clamps	S/I	✓	✓	✓	✓	✓	✓	✓	✓	✓	✓	✓	✓	✓
Throttle linkage	S/I	✓	✓	✓	✓	✓	✓	✓	✓	✓	✓	✓	✓	✓
Brake linings	S/I	✓	✓	✓	✓	✓	✓	✓	✓	✓	✓	✓	✓	✓
Supercharger oil	S/I								✓					
Accessory drive belts	S/I	Every 150,000 miles												
Engine coolant	R	every 5 years or 150,000 miles												
Spark plug wires (V6)	I	every 100,000 miles												
Spark plugs ②	R				✓				✓				✓	
Passenger compartment air filter	R		✓		✓		✓		✓		✓		✓	
Air filter element	S/I					✓					✓			
PCV filter	R				✓				✓				✓	
Ignition cables	S/I				✓				✓				✓	
EGR & fuel systems	S/I				✓				✓				✓	
Throttle body bore and valve plates (V8)	I										✓			
Automatic transaxle fluid & filter ③	R	every 100,000 miles												
Throttle body mount bolt torque	S/I													

R: Replace   S/I: Service or Inspect

① Engine coolant: replace every 100,000 miles. Use O.E. specified (DEX-COOL™) coolant only. If any silicate coolant is used, every 30,000 miles.

② Platinum tip spark plugs: replace every 100,000 miles.

③ Change at 50,000 miles if driven under one or more of the following conditions:

    a. heavy city traffic at temps. Above 90 degrees F

    b. hilly or mountainous terrain

    c. frequent trailer towing

    d. taxi, police or delivery service

06025-BONN-C13

LONG TRIP/HIGHWAY SCHEDULED MAINTENANCE INTERVALS
BUICK LESABRE, PARK AVENUE
PONTIAC BONNEVILLE

TO BE SERVICED	TYPE OF SERVICE	VEHICLE MILEAGE INTERVAL (x1000)												
		7.5	15	23	30	38	45	50	53	60	68	75	83	90
Engine oil & filter	R	✓	✓	✓	✓	✓	✓	✓	✓	✓	✓	✓	✓	✓
Rotate tires	S/I	✓	✓	✓	✓	✓	✓	✓	✓	✓	✓	✓	✓	✓
Tire inflation	I	once a month												
Restraint system	I	every 6 months												
Transmission fluid level	S/I	every 6 months												
Exhaust system & brake hoses	S/I	✓	✓	✓	✓	✓	✓	✓	✓	✓	✓	✓	✓	✓
Driveshaft boots & front suspension components	S/I	✓	✓	✓	✓	✓	✓	✓	✓	✓	✓	✓	✓	✓
Lubricate chassis, suspension, steering linkage, transaxle shift linkage, parking brake cable guides, underbody contact points & linkage	S/I	✓	✓	✓	✓	✓	✓	✓	✓	✓	✓	✓	✓	✓
Coolant level, hoses & clamps	S/I	✓	✓	✓	✓	✓	✓	✓	✓	✓	✓	✓	✓	✓
Throttle linkage	S/I	✓	✓	✓	✓	✓	✓	✓	✓	✓	✓	✓	✓	✓
Brake linings	S/I	✓	✓	✓	✓	✓	✓	✓	✓	✓	✓	✓	✓	✓
Supercharger oil	S/I								✓					
Accessory drive belts	S/I	Every 150,000 miles												
Engine coolant	R	every 5 years or 150,000 miles												
Spark plug wires (V6)	I	every 100,000 miles												
Spark plugs ②	R				✓				✓				✓	
Passenger compartment air filter	R		✓		✓		✓		✓		✓		✓	
Air filter element ④	I		✓		✓		✓		✓		✓		✓	
PCV filter	R				✓				✓				✓	
Ignition cables	S/I				✓				✓				✓	
EGR & fuel systems	S/I				✓				✓				✓	
Throttle body bore and valve plates (V8)	I				✓				✓				✓	
Automatic transaxle fluid & filter ③	R	every 100,000 miles												
Throttle body mount bolt torque	S/I													

R: Replace   S/I: Service or Inspect

① Engine coolant: replace every 100,000 miles. Use O.E. specified (DEX-COOL™) coolant only. If any silicate coolant is used, every 30,000 miles.

② Platinum tip spark plugs: replace every 100,000 miles.

③ Change at 50,000 miles if driven under one or more of the following conditions:
   a. heavy city traffic at temps. Above 90 degrees F
   b. hilly or mountainous terrain
   c. frequent trailer towing
   d. taxi, police or delivery service

④ Replace every 30,000 miles

06025-BONN-C14

ENGINE REPAIR

➡Disconnecting the negative battery cable on some vehicles may interfere with the operation of the onboard computer system. The computer may undergo a relearning process once the negative battery cable is reconnected.

Alternator

REMOVAL & INSTALLATION

3.8L Engine

1. Before servicing the vehicle, refer to the Precautions Section.
2. Remove or disconnect the following:
   - Negative battery cable
   - Supercharger belt, if equipped
   - Accessory drive belt
   - Fuel injector sight shield
   - Alternator brace
   - Electrical connections
   - Alternator bolts and the alternator

To install:

3. Install or connect the following:
   - Alternator and torque the bolts to 37 ft. lbs. (50 Nm)
   - Electrical connections and torque the nut to 111 inch lbs. (12.5 Nm)
   - Alternator brace. Torque the nut to 37 ft. lbs. (50 Nm) and the bolt to 22 ft. lbs. (30 Nm).
   - Fuel injector sight shield
   - Accessory drive belt
   - Supercharger belt, if equipped
   - Negative battery cable

4.6L Engine

1. Disconnect the battery negative cable.

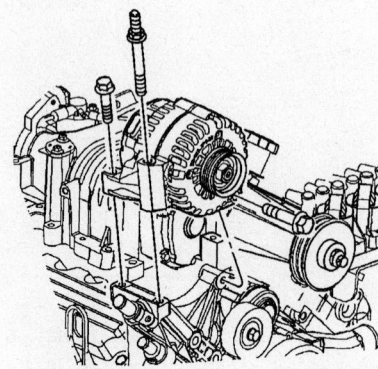

9346UGDB

Exploded view of the alternator mounting—supercharged models

2. Remove the drive belt.
3. Remove the engine cooling fans.
4. Disconnect the wiring harness connector from the generator.
5. Reposition the protective boot from the generator output BAT terminal for access.
6. Remove the generator output BAT terminal nut and disconnect the positive lead from the generator.
7. Loosen the lower generator bolt.
8. Remove the generator bolts from the generator.
9. Remove the generator from the vehicle.

To install:

10. Position the generator to the A/C compressor and engine.

➡Use the correct fastener in the correct location. Replacement fasteners must be the correct part number for that application. Fasteners requiring replacement or fasteners requiring the use of thread locking compound or sealant are identified in the service procedure. Do not use paints, lubricants, or corrosion inhibitors on fasteners or fastener joint surfaces unless specified. These coatings affect fastener torque and joint clamping force and may damage the fastener. Use the correct tightening sequence and specifications when installing fasteners in order to avoid damage to parts and systems.

11. Install the generator mounting bolts. Tighten the generator mounting bolts to 50 Nm (37 ft. lbs.).
12. Connect the wiring harness connector to the generator.
13. Connect the battery positive lead and install the generator output BAT terminal nut. Tighten the generator output BAT terminal nut to 12.5 Nm (111 inch lbs.).
14. Press the protective boot on to the generator output BAT terminal.
15. Install the engine cooling fans.
16. Install the drive belt.
17. Connect the battery negative cable.

Ignition Timing

The ignition timing is not adjustable, and is set according to engine demand electronically. The Powertrain Control Module (PCM) controls the ignition timing for all driving conditions.

Engine Assembly

REMOVAL & INSTALLATION

3.8L Engine

1. Before servicing the vehicle, refer to the Precautions Section.
2. Disconnect the negative battery cable.
3. Remove the hood.
4. Relieve the fuel system pressure.
5. Drain the coolant system and crankcase.
6. Remove or disconnect the following:
   - Negative cable
   - Fuel injector sight shield
   - Vacuum brake booster hose from the vacuum connections
   - Fuel feed and return lines from the fuel rail
   - Evaporative emission canister purge valve
   - Cruise control cable from the throttle body bracket and lever
   - Electrical connector from the cruise control module
   - Cruise control module from the mounting studs

➡Always replace the accelerator control cable with a NEW cable whenever you remove the engine from the vehicle.

   - Accelerator control cable
   - Drive belt
   - Bolt securing both the battery negative cable and the engine harness ground lead to the engine block
7. Disconnect the wiring harness connectors from the following components:
   - A/C compressor clutch
   - A/C pressure sensor
   - Knock (KS) sensor
   - Engine coolant block heater
   - Oil level sensor
8. Remove or disconnect the following:
   - Wiring harness from the harness clip at the rear of the A/C compressor
   - Torque converter cover
   - Starter motor
   - Bolts securing the flywheel to the torque converter
9. Disconnect then secure the following wiring harness electrical connectors to the cowl panel:

- Knock (KS) sensor number 2 which can be found behind the right exhaust manifold
- Oil pressure sensor
- Vehicle Speed Sensor (VSS)

10. Remove or disconnect the following:
- Bolts securing the transaxle brace to the transaxle
- Nuts attaching the exhaust manifold pipe to the right exhaust manifold
- Exhaust manifold pipe from the right exhaust manifold studs, allowing it to rest on top of the power steering gear heat shield
- Exhaust manifold pipe gasket and discard the gasket
- Right wheelhouse extension
- Font A/C compressor mounting nuts
- Rear A/C compressor mounting bolt
- Compressor off of the mounting studs and rest on top of the engine frame
- Bolt securing the Powertrain Control Module (PCM) ground located at the left front cylinder head

11. Disconnect the wiring harness electrical connectors from the following components on the left side of the engine:
- Fuel injectors
- Ignition harness
- Boost control solenoid (VIN I only)
- Engine Coolant Temperature (ECT) sensor
- Throttle Position (TP) sensor
- Idle Air Control (IAC) valve
- Mass Air Flow (MAF) sensor

12. Disconnect the wiring harness connectors from the following components on the right side of the engine:
- Fuel injectors
- Exhaust Gas Recirculation (EGR) valve
- Manifold Absolute Pressure (MAP) sensor
- Heated Oxygen (O₂S) sensor
- AIR solenoid, if equipped
- Alternator

13. Remove or disconnect the following:
- Alternator
- Air cleaner intake duct

14. Attach an engine support fixture.
- Front power steering pump mounting bolts
- Side power steering pump mounting bolt and piston the power steering pump against the cowl, allowing it to rest on top of the transaxle housing
- Right engine mount bracket

- Right lower engine-to-transaxle mounting bolt
- Coolant and heater hoses

15. Use a block of wood between a floor jack and the transaxle, support the transaxle at the pan.

16. Remove the engine support fixture.

17. Attach an engine lift chain to the engine lift brackets and attach to an engine lift device.

18. Remove all remaining engine-to-transaxle bolts

19. Remove the engine from the vehicle.

To install:

20. Installation is the reverse of removal, please note the following torques:
- 5 upper engine-to-transaxle mounting bolts to 55 ft. lbs. (75 Nm)
- Right lower engine-to-transaxle mounting bolt to 55 ft. lbs. (75 Nm)
- Power steering pump bolts to 20 ft. lbs. (27 Nm)
- Bolt attaching the PCM ground to the left front cylinder head and tighten to 37 ft. lbs. (50 Nm)
- A/C compressor bolts 37 ft. lbs. (50 Nm)
- Transaxle brace bolts to 48 ft. lbs.(65 Nm)
- Flywheel-to-torque converter bolts to 46 ft. lbs. (63 Nm)

21. Refill the crankcase.

22. Refill and bleed the engine cooling system.

23. Start the engine and check for leaks.

24. Road test the vehicle and check operation.

4.6L Engine

1. Disconnect the battery negative cable.

2. Recover the A/C refrigerant system.

3. Disconnect the vacuum brake booster hose from the vacuum connection and position aside.

4. Disconnect the fuel inlet and return quick-connect fittings at the fuel rail and secure to the air inlet grille.

5. Disconnect the hose from the evaporative emission canister purge valve and secure to the air inlet grille.

6. Remove the upper filler panel.

7. Remove the air cleaner assembly.

8. Remove the 2 nuts from the fuel injector sight shield .

9. Remove the fuel injector sight shield from the engine.

10. Remove the nut securing the battery positive cable to the remote positive terminal and secure to the top of the engine.

11. Disconnect the secondary AIR relay from the relay bracket and secure to the top of the engine.

12. Disconnect and secure the following wiring harness electrical connectors to the top of the engine:
- The PCM.
- The C101
- The engine electrical harness.

13. Remove the bolt securing the engine ground cable from the right side body frame rail.

⁜ WARNING

In order to avoid possible injury or vehicle damage, always replace the accelerator control cable with a NEW cable whenever you remove the engine from the vehicle. In order to avoid cruise control cable damage, position the cable out of the way while you remove or install the engine. Do not pry or lean against the cruise control cable and do not kink the cable. You must replace a damaged cable.

14. Push the lock release and remove the cruise control cable from the throttle body bracket and lever.

15. Remove the accelerator control cable from the throttle body. Push the lock release and remove the cruise control cable from the throttle body bracket and lever.

16. Remove the shift cable from the bracket and manual shift lever and position aside.

17. Drain the cooling system.

18. Position tool J 37097-A, or equivalent, to the clamp in order to remove the radiator inlet hose from the water housing crossover and position aside.

19. Position tool J 38185, or equivalent, to the clamp in order to remove the radiator outlet hose from the thermostat housing and position aside.

20. Disconnect the surge tank inlet hose from the water housing crossover.

21. Disconnect the surge tank outlet hose from the heater pipe.

22. Disconnect the heater hoses from the heater pipes.

⁜ CAUTION

Brake fluid may irritate eyes and skin.

➡Record the location of the brake pipes to the brake pressure modulator valve(BPMV) for use as an aid during installation.

23. Remove the 2 master cylinder brake pipes from the BPMV.

24. Plug the open outlet ports to prevent fluid loss and contamination.

25. Remove the upper transaxle oil cooler pipe retaining bolt from the fan shroud.

26. Slide the plastic cap off the upper transaxle oil cooler pipe quick connect fitting.

27. Disconnect the upper transaxle oil cooler pipe from the radiator using tool J 41623-B, or equivalent.

28. Disconnect the lower transaxle oil cooler pipe fitting from the radiator.

➡The wheels of the vehicle must be straight ahead and the steering column in the LOCK position before disconnecting the steering column or intermediate shaft from the steering gear. Failure to do so will cause the SIR coil assembly to become un-centered, which may cause damage to the coil assembly.

29. Lock the steering column by installing tool J 42640, or equivalent, into underside of the steering column.

30. Remove the right and left side strut tower bolts.

31. Raise and support the vehicle.

32. Remove the rear exhaust manifold pipe.

33. Remove the front wheels.

34. Disconnect the front electronic brake pad wear sensor electrical connectors.

35. Disconnect the front electronic brake pad wear sensor electrical leads from the strut brackets and secure to the body frame rails.

36. Disconnect the front wheel speed sensor electrical leads from the body frame rail.

37. Disconnect the road sensing suspension electrical connector at the body frame rail, if equipped.

38. Disconnect the road sensing suspension electrical leads from the body frame rail, if equipped.

39. Disconnect the electronic suspension position sensor links from the lower control arms, if equipped.

40. Remove the front air deflector.

41. Remove the front fascia extensions.

42. Disconnect the secondary AIR inlet hose from the secondary AIR pump.

43. Loosen the nuts securing the front brake pipe frame brackets to the body frame rails.

44. Disconnect the front brake pipes from the retainers at the body frame rails.

45. Carefully pull the front brake pipes away from the body frame rails.

46. Disconnect the 2 rear brake pipes at the rear of the engine frame. Plug the open outlet ports to prevent fluid loss and contamination.

47. Disconnect the A/C pressure sensor.

48. Disconnect the A/C discharge hose from the compressor and secure to the cooling fan assembly.

49. Disconnect the A/C suction hose from the compressor and secure to the cooling fan assembly.

✳✳ WARNING

Failure to disconnect the intermediate shaft from the rack and pinion stub shaft can result in damage to the steering gear and/or damage to the intermediate shaft. This damage may cause loss of steering control which could result in personal injury.

50. Remove the intermediate shaft pinch bolt.

51. Remove the steering gear from the intermediate shaft.

52. Disconnect the post HO$_2$S at the sensor pigtail.

53. Remove the engine oil cooler quick connect fittings from the engine oil filter adapter, with the oil pipes still attached, and position aside-if equipped.

54. Remove the brace between the engine oil pan and the transaxle case.

55. Remove the torque converter cover.

➡Mark the flywheel to torque converter relationship prior to removal of the bolts.

56. Remove the torque converter to the flywheel bolts.

57. Position tool J 39580, or equivalent, powertrain support dolly under the engine frame.

58. Lower the vehicle on to tool J 39580, or equivalent.

59. If the powertrain support dolly is unavailable, support the powertrain with 4 suitable jackstands.

60. Place a 2 x 4 in block of wood between the front of the engine oil pan and the engine frame.

61. Remove the nut securing the right engine mount to the right engine mount bracket.

62. Remove the nut securing the left transaxle mount to the left transaxle mount bracket.

✳✳ CAUTION

To avoid any vehicle damage, serious personal injury or death when major components are removed from the vehicle and the vehicle is supported by a hoist, support the vehicle with jack stands at the opposite end from which the components are being removed.

63. Secure the front hoist pads to the vehicle. Remove the 6 frame to body mounting bolts.

➡Ensure clearance is maintained between the engine/transaxle assembly and the following:

- The A/C accumulator hose
- The A/C compressor hose
- The brake pipes
- The electronic brake pad wear sensor leads
- The heater hoses
- The radiator hoses
- The wheel speed sensor leads
- The wiring harnesses

64. Carefully raise the vehicle in order to clear the supported engine/transaxle assembly.

65. Drain the engine oil.

66. Remove the heater pipes.

67. Disconnect the intermediate hose from the secondary AIR valve at bank 2.

68. Remove the nut securing the intermediate hose to the secondary AIR valve at bank 1.

69. Remove the nut securing the coil cassette ground wire to the right cylinder head.

70. Disconnect the engine wiring harness from the engine.

71. Disconnect the power steering hose from the power steering pump reservoir.

72. Remove the power steering return hose retaining bolt from the cylinder head.

73. Remove the power steering pressure hose from the power steering pump.

74. Remove the nut securing the power steering pressure hose to the right engine mount bracket.

75. Remove the 4 bolts securing the right engine mount bracket to the engine.

76. Remove the right engine mount bracket.

77. Remove the bolt securing the rear transaxle brace to the transaxle.

78. Remove the nuts securing the rear transaxle brace to the stud located on the right cylinder head.

79. Remove the rear transaxle brace.

80. Remove the bolts securing the front transaxle brace to the transaxle and right cylinder head.

81. Remove the nuts securing the vehicle speed sensor heat shield to the transaxle.

82. Remove the bolts securing the center transaxle brace to the engine and transaxle.

83. Install tool J 42504, or equivalent, to the cylinder head.

84. Install an engine lift chain to the engine lift brackets and attach to an engine lift device.

85. Remove the nut securing the front engine mount to the engine frame.

86. Remove the bolts attaching the engine to the transaxle.

87. Raise the engine from the supported frame and transaxle assembly.

88. Remove the front engine mount bracket.

To install:

89. Install the front engine mount bracket.

90. Carefully position the engine to the supported frame and transaxle assembly, aligning the engine dowels to the transaxle cover.

91. Install the bolts attaching the engine to the transaxle. Tighten the engine to transaxle mounting bolts to 75 Nm (55 ft. lbs.).

92. Install the nut securing the front engine mount to the engine frame. Tighten the front engine mount nut to 70 Nm (52 ft. lbs.).

93. Place a block of wood between the front of the engine oil pan and the engine frame.

94. Remove the engine lift chain from the engine lift brackets.

95. Remove tool J 42504, or equivalent, from the cylinder head.

96. Install the 4 bolts securing the center transaxle brace to the engine and transaxle. Tighten the center transaxle brace bolts to 50 Nm (37 ft. lbs.).

97. Install the retaining nuts securing the vehicle speed sensor heat shield to the transaxle. Tighten the vehicle speed sensor heat shield nuts to 50 Nm (37 ft. lbs.).

98. Install the bolts securing the front transaxle brace to the transaxle and right cylinder head. Tighten the front transaxle brace bolts to 50 Nm (37 ft. lbs.).

99. Position the rear transaxle brace over the studs located at the rear of the right cylinder head.

100. Loosely install the nuts securing the rear transaxle brace to the cylinder head.

101. Install the bolt securing the rear transaxle brace to the transaxle. Tighten the rear transaxle brace bolt to 50 Nm (37 ft. lbs.). Tighten the rear transaxle brace nuts to 50 Nm (37 ft. lbs.).

102. Position the right engine mount bracket to the engine.

103. Install the 4 bolts securing the right engine mount bracket to the engine. Tighten the right engine mount bracket bolts to 50 Nm (37 ft. lbs.).

104. Install the nut securing the power steering pressure hose to the right engine mount bracket. Tighten the power steering pressure hose retaining nut to 9 Nm (80 inch lbs.).

105. Install the power steering pressure hose to the power steering pump. Tighten the power steering hose to power steering pump to 30 Nm (22 ft. lbs.).

106. Connect the power steering hose to the power steering pump reservoir.

107. Install the power steering return hose retaining bolt to the cylinder head. Tighten the power steering return hose retaining bolt to 50 Nm (37 ft. lbs.).

108. Connect the engine wiring harness to the engine.

109. Install the nut securing the coil cassette ground wire to the cylinder head. Tighten the coil cassette ground wire nut to 17 Nm (13 ft. lbs.).

110. Install the nut securing the intermediate hose to the secondary AIR valve at bank 1. Tighten the intermediate hose to secondary AIR valve nut to 9 Nm (80 inch lbs.).

111. Connect the intermediate hose to the secondary AIR valve at bank 2.

112. Install the heater pipes.

113. Position the engine/transaxle assembly under the vehicle.

➡ **Ensure clearance is maintained between the engine/transaxle assembly and the following:**

- The A/C accumulator hose
- The A/C compressor hose
- The brake pipes
- The electronic brake pad wear sensor leads
- The heater hoses
- The radiator hoses
- The wheel speed sensor leads
- The wiring harnesses

114. Carefully lower the vehicle over the engine/transaxle assembly, aligning the struts to the strut towers.

115. Install the 6 frame mounting bolts retaining the frame to the vehicle. Using dowel pins in the alignment holes, align the engine frame with the vehicle. Tighten the frame mounting bolts to 191 Nm (141 ft. lbs.).

116. Install the nut securing the right engine mount to the right engine mount bracket. Tighten the right engine mount nut to 80 Nm (59 ft. lbs.).

117. Remove the block of wood between the front of the engine oil pan and the engine frame.

118. Install the nut securing the left transaxle mount to the left transaxle mount bracket. Tighten the left transaxle mount nut to 80 Nm (59 ft. lbs.).

119. Raise and support the vehicle.

120. Remove tool J 39580, or equivalent, powertrain support dolly from under the engine frame.

➡ **Line up the flywheel and converter, using the alignment marks made during disassembly.**

121. Install the bolts securing the flywheel to the torque converter. Tighten the flywheel to torque converter bolts to 60 Nm (44 ft. lbs.).

122. Install the torque converter cover.

123. Install the oil pan to transaxle brace. Tighten the oil pan to transaxle brace bolts to 50 Nm (37 ft. lbs.).

124. Install the engine oil cooler quick connect fittings to the engine oil filter adapter, if equipped.

125. Connect the post HO2S at the sensor pigtail.

❋❋ CAUTION

When installing the intermediate shaft make sure that the shaft is seated prior to pinch bolt installation. If the pinch bolt is inserted into the coupling before shaft installation, the two mating shafts may disengage. Disengagement of the two mating shafts will cause loss of steering control which could result in personal injury.

126. Connect the intermediate shaft to the steering gear.

127. Install the intermediate shaft to the steering gear pinch bolt. Tighten the intermediate shaft to steering gear pinch bolt to 45 Nm (33 ft. lbs.).

128. Connect the A/C suction hose to the compressor. Tighten the A/C suction hose nut to 20 Nm (15 ft. lbs.).

129. Connect the A/C discharge hose to the compressor. Tighten the A/C discharge hose nut to 20 Nm (15 ft. lbs.).

130. Connect the A/C pressure sensor.

131. Connect the 2 rear brake pipes at the rear of the engine frame. Tighten the brake pipes to 15 Nm (11 ft. lbs.).

132. Connect the front brake pipes to the retainers at the body frame rails.

133. Install the nuts securing the front brake pipe frame brackets to the body frame rails. Tighten the brake pipe frame bracket nuts to 15 Nm (11 ft. lbs.).

134. Connect the secondary AIR inlet hose to the secondary AIR pump.

135. Install the front fascia extensions.

136. Install the front air deflector.

137. Attach the front wheel speed sensor electrical leads to the body frame rail.

138. Attach the front electronic brake pad wear sensor electrical leads to the strut brackets.

139. Connect the front electronic brake pad wear sensor electrical connectors.

140. Connect the road sensing suspension electrical connector at the body frame rail, if equipped.

141. Connect the road sensing suspension electrical leads to the body frame rail, if equipped.

142. Connect the electronic suspension position sensor links to the lower control arms, if equipped.

143. Install the rear exhaust manifold pipe.

144. Install the front wheels.

145. Lower the vehicle ONLY enough to allow the threaded holes in the strut to align with the holes in the strut towers.

146. Install the right and left side strut tower bolts. Tighten the strut tower bolts to 60 Nm (44 ft. lbs.).

147. Lower the vehicle.

148. Remove tool J 42640, or equivalent, from the steering column.

149. Connect the lower transaxle oil cooler pipe fitting to the radiator.

➡**Ensure the lower transaxle oil cooler pipe is positioned upwards while tightening.**

150. Tighten the transaxle oil cooler pipe fitting to 35 Nm (26 ft. lbs.).

151. Install the upper transaxle oil cooler pipe retaining bolt to the fan shroud. Tighten the transaxle oil cooler pipe retaining bolt to 6 Nm (53 inch lbs.).

152. Push the upper transaxle oil cooler pipe into the radiator quick connect fitting, until a click is heard.

153. Tug gently on the cooler pipe to ensure proper retention.

154. Slide the plastic cap over the quick connect fitting.

155. Install the 2 master cylinder brake pipes to the BPMV using the location recorded during the removal procedure. Tighten the brake pipes to 15 Nm (11 ft. lbs.).

156. Connect the heater hoses to the heater pipes.

157. Connect the surge tank outlet hose to the heater pipe.

158. Connect the surge tank inlet hose to the water housing crossover.

159. Position tool J 38185, or equivalent, to the clamp in order to connect the radiator outlet hose to the thermostat housing.

160. Position tool J 37097-A, or equivalent, to the clamp in order to connect the radiator inlet hose to the engine.

161. Install the shift cable to the manual shift lever and the bracket.

❊❊ CAUTION

In order to avoid possible injury or vehicle damage, always replace the accelerator control cable with a NEW cable whenever you remove the engine from the vehicle. In order to avoid cruise control cable damage, position the cable out of the way while you remove or install the engine. Do not pry or lean against the cruise control cable and do not kink the cable. You must replace a damaged cable.

162. Install a NEW accelerator control cable to the throttle body.

163. Install the cruise control cable to the throttle body lever.

164. Slide the cruise control fully into the throttle body bracket until it snaps into place.

165. Install the bolt securing the engine ground cable to the right side body frame rail. Tighten the engine ground cable bolt to 50 Nm (37 ft. lbs.).

166. Connect the wiring harness electrical connectors to the following components:

- The PCM.
- The C101
- The engine electrical harness.

167. Install the nut securing the battery positive cable to the remote positive terminal. Tighten the battery positive cable to remote positive terminal nut to 12 Nm (106 inch lbs.).

168. Attach the secondary AIR relay to the relay bracket.

169. Install the air cleaner assembly.

170. Install the upper filler panel.

171. Connect the hose to the evaporative emission canister purge valve.

172. Connect the fuel inlet and return quick-connect fittings at the fuel rail.

173. Connect the vacuum brake booster hose to the vacuum connection.

174. Position the fuel injector sight shield to the engine.

175. Install the 2 fuel injector sight shield nuts. Tighten the fuel injector sight shield nuts to 3 Nm (27 inch lbs.).

176. Connect the battery negative cable.

177. Fill the engine with oil.

178. Fill the cooling system.

179. Bleed the hydraulic brake system.

180. Recharge the A/C refrigerant system.

181. Bleed the power steering system.

182. Measure the wheel alignment.

Complete the following procedure after the engine is installed in the vehicle:

183. With the ignition OFF or disconnected, crank the engine several times. Listen for any unusual noises or evidence that any parts are binding.

184. Start the engine and listen for abnormal conditions.

185. Check the vehicle oil pressure gage or light and confirm that the engine has acceptable oil pressure.

186. Run the engine at approximately 1,000 RPM until the engine reaches normal operating temperature.

187. While the engine continues to idle raise and support the vehicle.

188. Inspect for oil, coolant and exhaust leaks while the engine is idling.

189. Lower the vehicle.

190. Perform the CKP system variation learn procedure.

191. Perform a final inspection for the proper engine oil and coolant levels.

192. Road test the vehicle.

Water Pump

REMOVAL & INSTALLATION

3.8L (VIN K) Engine

1. Before servicing the vehicle, refer to the Precautions Section.

2. Drain the cooling system.

3. Remove or disconnect the following:

- Negative battery cable
- Accessory drive belt
- Coolant hoses from the water pump
- Water pump pulley bolts

➡**The long bolt can be removed by aligning the bolt head up with the hole in the frame rail.**

- Pulley
- Water pump bolts
- Water pump

To install:

4. Apply a thin bead of sealer around the outside edge of the water pump.

5. Install or connect the following:

- Water pump with new gasket. Torque the water pump short bolts

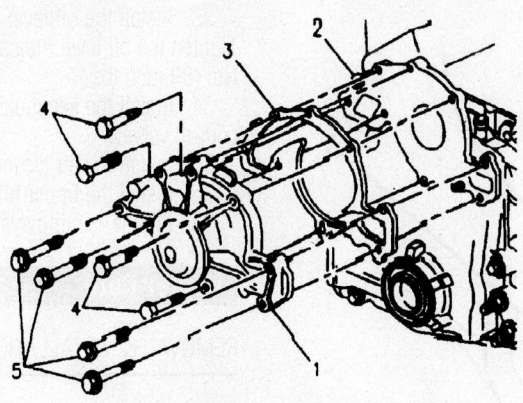

1. Coolant pump
2. Engine front cover
3. Gasket
4. 11 ft. lb.(15 Nm)
5. 22 ft. lb.(30 Nm)

7922UG01

Exploded view of the water pump—3.8L (VIN K and 1) engines

to 11 ft. lbs. (15 Nm) and the long bolts to 22 ft. lbs. (30 Nm).
- Water pump pulley. Torque the bolts to 115 inch lbs. (13 Nm).
- Coolant hoses to the water pump
- Drive belt
- Supercharger belt, if equipped

6. Refill and bleed the cooling system.
7. Run the engine and check for leaks.
8. Recheck the coolant level when the engine has cooled.

3.8L (VIN 1) Engine

1. Before servicing the vehicle, refer to the Precautions Section.
2. Drain the cooling system.
3. Support the engine using an engine support fixture.
4. Remove or disconnect the following:

- Negative battery cable
- Air conditioning compressor splash shield
- Supercharger and accessory drive belts
- Coil pack, if equipped
- Supercharger belt tensioner
- Front engine mount
- Power steering pump
- Engine mount bracket
- Idler pulley
- Water pump pulley
- Water pump

To install:

5. Apply a thin bead of sealer around the outside edge of the water pump.
6. Install or connect the following:

- Water pump with a new gasket. Torque the short bolts to 11 ft. lbs.

(15 Nm) and the long bolts to 22 ft. lbs. (30 Nm).
- Water pump pulley and torque the bolts to 115 inch lbs. (13 Nm)
- Engine mount bracket. Torque the nut closest to the a/c compressor to 33 ft. lbs. (45 Nm), and the others to 58 ft. lbs. (78 Nm).
- Idler pulley and torque the bolts to 37 ft. lbs. (50 Nm)
- Power steering pump and torque the bolts to 20 ft. lbs. (27 Nm)
- Front engine mount and torque the nuts to 58 ft. lbs. (78 Nm)
- Supercharger belt tensioner and torque the bolts to 37 ft. lbs. (50 Nm)
- Ignition coil pack assembly, if

equipped and torque the nuts to 70 in. lbs. (8 Nm)
- Supercharger drive belt
- Accessory drive belt
- Air conditioning compressor splash shield
- Negative battery cable

7. Refill and bleed the cooling system.
8. Run the engine and check for leaks.
9. Recheck the coolant level when the engine has cooled.

4.6L Engine

1. Drain the cooling system.
2. Remove the upper filler panel.
3. Remove the air cleaner assembly
4. Remove the secondary AIR injection control valve.
5. Remove the oil level indicator tube nut.
6. Remove the water pump belt shield fasteners. Remove the water pump belt shield.
7. Remove the water pump drive belt.
8. Remove the radiator outlet hose from the thermostat housing.
9. Remove the water pump cover bolts.
10. Remove the water pump cover.
11. Disconnect the heater return hose.
12. Position tool J 38816-A, or equivalent, to the water pump locking ears.
13. Fasten the support plate to the water housing crossover, to ensure proper engagement of the tool to the water pump locking ears.
14. Using tool J 38816-A, or equivalent, turn the pump clockwise in order to remove the pump from the housing.
15. Remove the support plate from the water housing crossover.

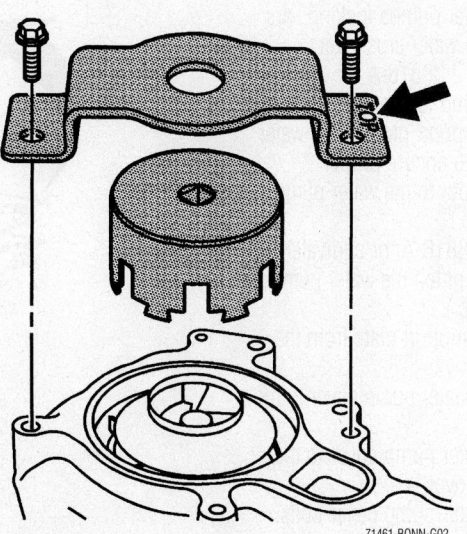

71461-BONN-G02

Tool J 38816-A used on the water pump—4.6L engine

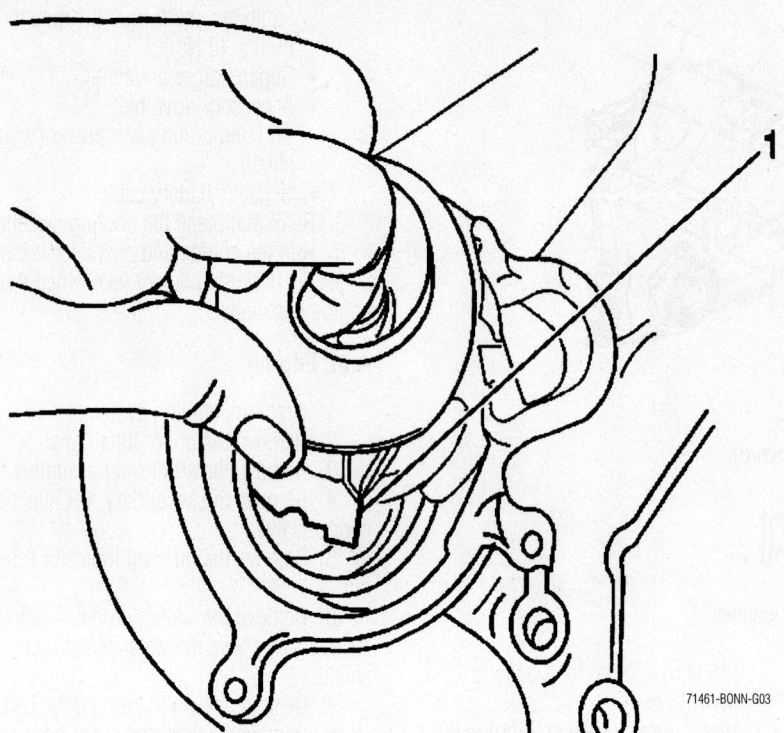

The notched locking ear (1) must be in the 7 o'clock position

16. Remove the water pump from the vehicle.

17. Remove the seal from the water crossover.

18. Clean the sealing surfaces of the water housing crossover and the pump cover.

19. Clean and inspect the water pump.

To install:

20. Insert the seal into the recessed portion of the water crossover.

➡**The notched locking ear (1) must be in the 7 o'clock position.**

21. Index the water pumps locking ears with the tangs in the water crossover.

22. Position tool J 38816-A, or equivalent, to the water pump locking ears.

23. Fasten the support plate to the water housing crossover, to ensure proper engagement of the tool to the water pump locking ears.

24. Turn tool J 38816-A, or equivalent, counterclockwise. Tighten the water pump to 100 Nm (74 ft. lbs.).

25. Remove the support plate from the water housing crossover.

26. Connect the water pump cover to the heater return hose.

27. Install the water pump cover to the water housing crossover.

28. Install the water pump cover bolts. Tighten the water pump cover bolts to 10 Nm (89 inch lbs.).

29. Connect the radiator outlet hose to the thermostat housing.

30. Install the water pump drive belt.

31. Place the water pump belt shield in position.

32. Install the water pump belt shield fasteners. Tighten the water pump belt shield fasteners to 10 Nm (89 inch lbs.).

33. Install the oil level indicator tube nut. Tighten the oil level indicator tube nut to 10 Nm (89 inch lbs.).

34. Install the secondary AIR injection control valve.

35. Install the air cleaner assembly.

36. Install the upper filler panel.

37. Fill the cooling system.

Heater Core

REMOVAL & INSTALLATION

Bonneville and LeSabre

1. Disconnect the negative battery cable.

2. Drain the cooling system into a clean container for reuse.

3. Remove or disconnect the following:
   - Heater hoses from the heater core
   - Right sound insulator
   - All necessary electrical connectors and remove the instrument panel compartment
   - Temperature valve actuator
   - Electrical connector and remove the HVAC programmer
   - Heater core cover-to-heater assembly screws and the cover
   - Heater core-to-heater assembly screws and the heater core

To install:

4. Install or connect the following:
   - Heater core and the heater core-to-heater assembly screws, then,

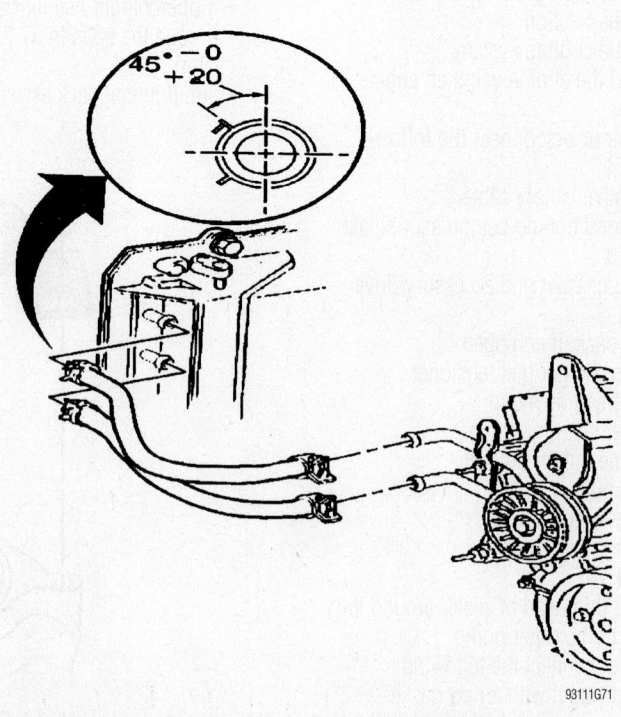

View of the heater hoses and clamp positions—Bonneville and LeSabre

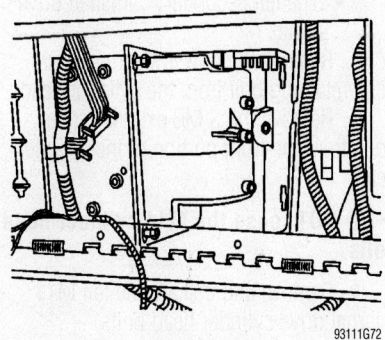

View of the heater core cover—Bonneville and LeSabre

tighten the screws to 12 inch lbs. (1.4 Nm)
- Heater core cover and the cover-to-heater assembly screws, then, tighten the screws to 12 inch lbs. (1.4 Nm)
- HVAC programmer and connect the electrical connector
- Temperature valve actuator
- Instrument panel compartment and connect any necessary electrical connectors
- Right sound insulator
- Heater hoses to the heater core

5. Refill the cooling system.
6. Connect the negative battery cable.
7. Operate the engine to normal operating temperatures; then, check the climate control operation and check for leaks.

Park Avenue

1. Disconnect the negative battery cable.
2. Drain the cooling system into a clean container for reuse.
3. Remove or disconnect the following:
- Heater hoses from the heater core
- Center console assembly
- Auxiliary air distribution duct
- Heater core heat shield cover from the HVAC module
- Air filter access cover and the heater core cover screws
- Heater core cover to obtain access to remove the heater core
- Heater core retaining bolts and straps
- Heater core

To install:

4. Install or connect the following:
- Heater core
- Heater core straps and retaining bolts, then, tighten to 18 inch lbs. (1.5 Nm)
- Heater core cover
- Air filter access cover and the

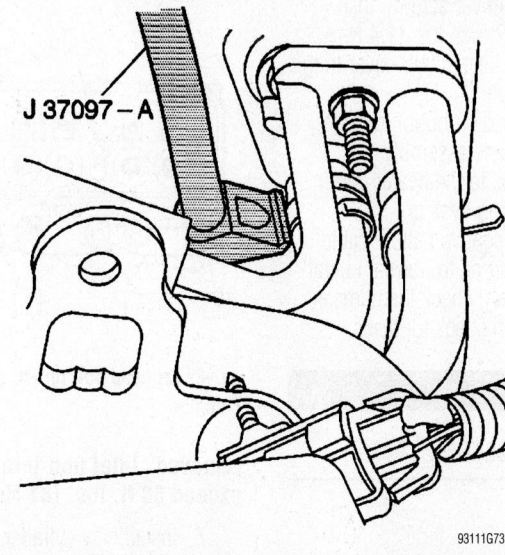

Using Hose Clamp Pliers tool J-37097-A remove the heater hoses clamp—Park Avenue

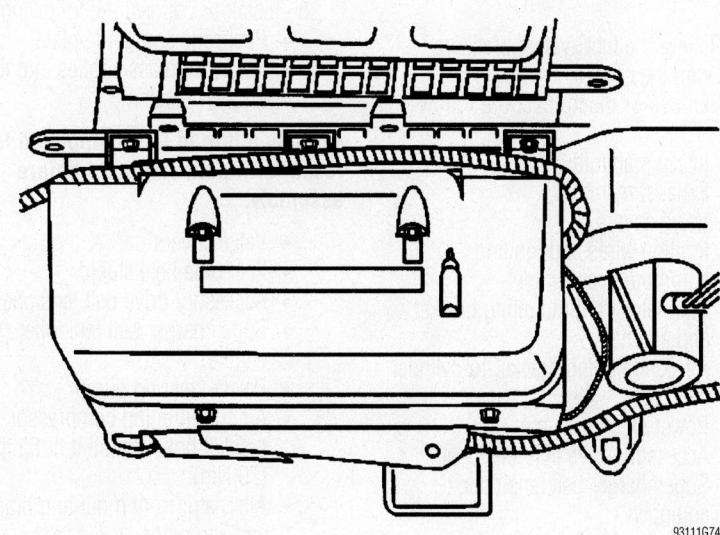

View of the heater core cover—Park Avenue

View of the heater core—Park Avenue

heater core cover screws, then tighten to 12 inch lbs. (1.4 Nm)
- Heater core heat shield cover to the HVAC module
- Auxiliary air distribution duct
- Center console assembly
- Heater hoses to the heater core

5. Refill the cooling system.
6. Connect the negative battery cable.
7. Operate the engine to normal operating temperatures; then, check the climate control operation and check for leaks.

Cylinder Head

REMOVAL & INSTALLATION

3.8L Engine

1. Before servicing the vehicle, refer to the Precautions Section.
2. Disconnect the negative battery cable.
3. Relieve the fuel system pressure.
4. Drain the cooling system.
5. Remove or disconnect the following:

- Intake manifold
- Exhaust manifold
- Valve covers
- Ignition wires and ignition coil/module assembly
- Alternator front mounting bracket and alternator
- Air conditioning bracket-to-cylinder head bolt
- Power steering pump
- Accessory drive belt tensioner
- Supercharger belt tensioner, if equipped
- Fuel pipe heat shield
- Rocker arm assemblies, note their original position
- Pushrods and guide plate
- Cylinder head bolts
- Cylinder head

To install:

6. Place the new cylinder head gasket on the engine block dowels with the note **THIS SIDE UP** facing the cylinder head and the arrow facing the front of the engine. Position the cylinder head on the engine block.

➡**The head gasket is identified by either a L or a R stamped on it next to the arrow.**

➡**This engine uses special torque-to-yield head bolts. The procedure must be followed carefully and new bolts must be used whenever the head is**

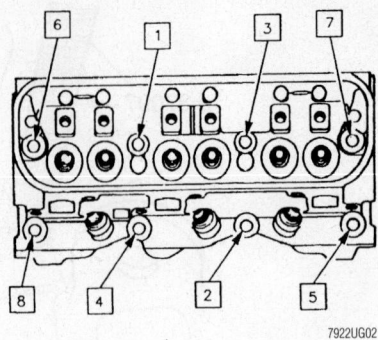

Cylinder head bolt torque sequence—3.8L engine

removed. **Total bolt torque should not exceed 60 ft. lbs. (81 Nm).**

7. Install new cylinder head bolts and torque them in sequence as follows:
   a. Step 1: 37 ft. lbs. (50 Nm).
   b. Step 2: Plus 120 degrees.
8. Install or connect the following:
   - Pushrods and guide plate
   - Rocker arm assemblies into their original location

➡**Apply a thread lock compound to the rocker arm pedestal bolts before assembly.**

- Valve covers
- Fuel pipe heat shield
- Accessory drive belt tensioner
- Supercharger belt tensioner (VIN 1 engine)
- Power steering pump
- Air conditioning compressor bracket bolt. Torque it to 52 ft. lbs. (70 Nm).
- Alternator front mounting bracket, and alternator
- Ignition coil/module assembly and spark plug wires
- Exhaust manifold. Torque the bolts to 22 ft. lbs. (30 Nm).
- Intake manifold
- Negative battery cable
9. Refill and bleed the cooling system.
10. Start the engine and check for leaks and proper operation.

4.6L Engine

LEFT SIDE

1. Remove the following subassemblies:

- The left exhaust manifold.
- The generator.
- The water crossover.
- The intake manifold.
- The camshaft cover.
- The engine front cover.

- The left secondary camshaft drive chain.
2. Remove the power steering return hose retaining bolt from the cylinder head.
3. Remove the 3 M6 external drive bolts from the front portion of the cylinder head.

➡**DO NOT reuse the M11 cylinder head bolts.**

4. Remove and discard the ten M11 internal drive cylinder head bolts.
5. Remove the left cylinder head. Make sure that no dowel guide pins are stuck in the cylinder head.

➡**You must clean the thread sealant material from the cylinder head bolt holes in the cylinder block. Failure to do so could cause false torque readings during reassembly. After removing the cylinder head, remove any remaining bolt thread sealant material from the threaded cylinder block holes.**

➡**DO NOT reuse the cylinder head gasket.**

6. Remove the left cylinder head gasket.
7. Remove all remaining gasket material from the cylinder head and cylinder block using tool J 28410.
8. Place the cylinder head on a flat, clean surface with the combustion chambers face-up in order to prevent damage to the deck face.
9. Clean and inspect the cylinder head.

To install:

10. Make sure all the cylinder head locating pins are securely mounted in the cylinder block deck face.

➡**Failure to remove all the old thread sealant material from the cylinder block could cause false torque readings.**

11. Make sure any old thread sealant material is removed from the cylinder head bolt holes in the cylinder block.
12. Install a new left cylinder head gasket using the deck face locating pins for retention.
13. Align the cylinder head with the deck face locating pins.
14. Place the cylinder head in position on the deck face.

➡**DO NOT reuse the old M11 cylinder head bolts.**

15. Install new M11 cylinder head bolts in the cylinder head.
16. Install the M6 cylinder head bolts at the front of the cylinder head.

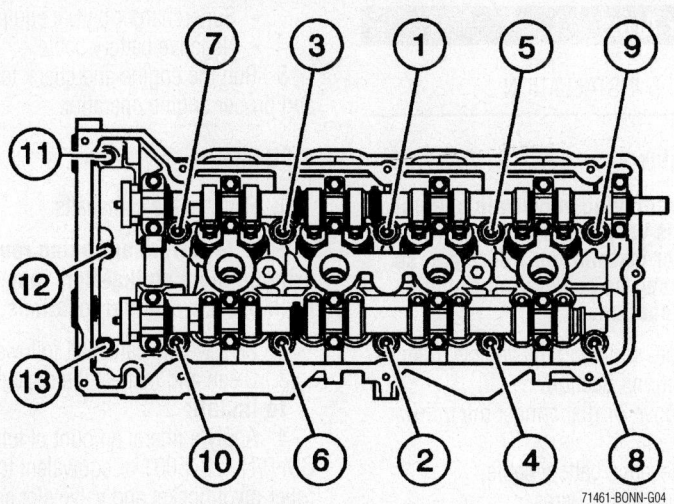

Left cylinder head bolt torque sequence—4.6L engine

71461-BONN-G04

17. Tighten the left cylinder head bolts in the sequence shown.

    a. First Pass: Tighten the left cylinder head M11 cylinder head bolts to 40 Nm (30 ft. lbs.) in the sequence shown.

    b. Second Pass: Tighten the left cylinder head M11 cylinder head bolts an additional 70 degrees in the sequence shown.

    c. Third Pass: Repeat the sequence turning each bolt another 60 degrees.

    d. Final Pass: Repeat the sequence again turning each bolt a final 45 degrees (total 175 degrees).

18. Tighten the M6 bolts at the front of the cylinder head to 12 Nm (106 inch lbs.).

19. Install the power steering return hose retaining bolt to the cylinder head. Tighten the power steering return hose retaining bolt to 50 Nm (37 ft. lbs.).

20. Install the following subassemblies:

- The left secondary camshaft drive chain.
- The engine front cover.
- The camshaft cover.
- The intake manifold.
- The water crossover.
- The generator.
- The left exhaust manifold.

RIGHT SIDE

1. Remove the following subassemblies:

- The right exhaust manifold.
- The water crossover.
- The intake manifold.
- The camshaft cover.
- The engine front cover.
- The right secondary camshaft drive chain.

2. Disconnect the electrical connector from the ECT sensor.

3. Remove the nut securing the coil cassette ground wire to the cylinder head.

4. Remove the bolt securing the exhaust crossover pipe to the cylinder head.

5. Raise and support the vehicle.

6. Remove the bolt securing the front transaxle brace to the cylinder head.

7. Loosen the bolts attaching the transaxle brace to the transaxle.

8. Remove the bolt securing the rear transaxle brace to the transaxle.

9. Lower the vehicle.

10. Remove the nuts securing the rear transaxle brace to the cylinder head.

11. Remove the rear transaxle brace.

12. Remove the 3 M6 external drive bolts from the front portion of the cylinder head.

➡**DO NOT reuse the M11 cylinder head bolts.**

13. Remove and discard the ten M11 internal drive cylinder head bolts.

14. Remove the right cylinder head.

Make sure that no dowel guide pins are stuck in the cylinder head.

➡**You must clean the thread sealant material from the cylinder head bolt holes in the cylinder block. Failure to do so could cause false torque readings during reassembly.**

15. After removing the cylinder head, remove any remaining bolt thread sealant material from the threaded cylinder block holes.

➡**DO NOT reuse the cylinder head gasket.**

16. Remove the right cylinder head gasket.

17. Remove all remaining gasket material from the cylinder head and cylinder block.

18. Place the cylinder head on a flat, clean surface with the combustion chambers face-up in order to prevent damage to the deck face.

19. Clean and inspect the cylinder head.

To install:

20. Make sure all the cylinder head locating pins are securely mounted in the cylinder block deck face.

➡**Failure to remove all the old thread sealant material from the cylinder block could cause false torque readings.**

21. Make sure any old thread sealant material is removed from the cylinder head bolt holes in the cylinder block.

22. Install a new right cylinder head gasket using the deck face locating pins for retention.

23. Align the cylinder head with the deck face locating pins.

24. Place the cylinder head in position on the deck face.

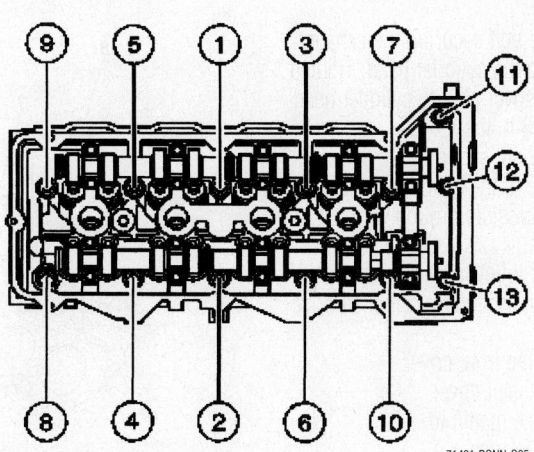

71461-BONN-G05

Right cylinder head torque sequence—4.6L engine

➡**DO NOT reuse the old M11 cylinder head bolts.**

25. Install new M11 cylinder head bolts in the cylinder head.

26. Install the M6 cylinder head bolts at the front of the cylinder head.

27. Tighten the right cylinder head bolts in the sequence shown.

  a. First Pass: Tighten the right cylinder head M11 cylinder head bolts to 40 Nm (30 ft. lbs.) in the sequence shown.

  b. Second Pass: Tighten the right cylinder head M11 cylinder head bolts an additional 70 degrees in the sequence shown.

  c. Third Pass: Repeat the sequence turning each bolt another 60 degrees.

  d. Final Pass: Repeat the sequence again turning each bolt a final 45 degrees (total 175 degrees).

28. Tighten the M6 bolts at the front of the cylinder head to 12 Nm (106 inch lbs.).

29. Position the rear transaxle brace over the studs located at the rear of the right cylinder head.

30. Loosely install the nuts securing the rear transaxle brace to the cylinder head.

31. Raise and support the vehicle.

32. Install the bolt securing the rear transaxle brace to the transaxle. Tighten the rear transaxle brace bolt to 50 Nm (37 ft. lbs.).

33. Install the bolt securing the front transaxle brace to the cylinder head. Tighten the front transaxle brace bolts to 50 Nm (37 ft. lbs.).

34. Lower the vehicle.

35. Tighten the rear transaxle brace to the cylinder head nuts to 50 Nm (37 ft. lbs.).

36. Connect the electrical connector to the ECT sensor.

37. Install the bolt securing the exhaust crossover pipe to the cylinder head. Tighten the exhaust crossover pipe to cylinder head bolt to 25 Nm (18 ft. lbs.).

38. Install the nut securing the coil cassette ground wire to the cylinder head. Tighten the coil cassette ground wire nut to 17 Nm (13 ft. lbs.).

39. Install the following subassemblies:
- The right secondary camshaft drive chain.
- The engine front cover.
- The camshaft cover.
- The intake manifold.
- The water crossover.
- The right exhaust manifold.

Rocker Arms

REMOVAL & INSTALLATION

3.8L Engine

➡**When removing valvetrain components, it is very important that they are marked for installation reference, so that they can be reinstalled in their original location.**

1. Before servicing the vehicle, refer to the Precautions Section.

2. Remove or disconnect the following:

- Negative battery cable
- Spark plug wires
- Rocker arm cover(s)
- Rocker arm pedestal bolts and assemblies
- Pushrods

To install:

3. Lubricate the pushrod tips and put them in their proper locations.

4. Lubricate the rocker arms and pedestals and install them. Be sure the pushrod tips are properly seated in the rocker arms.

5. Apply thread locking compound to the rocker arm pedestal bolt threads. Torque the bolts to 11 ft. lbs. (15 Nm) plus an additional 90 degree turn.

6. Apply suitable thread locking compound to the rocker arm cover bolts.

7. Install or connect the following:

- Rocker arm cover using a new gasket. Torque the bolts to 89 inch lbs. (10 Nm).
- Spark plug wires
- Accessory drive belt

- Supercharger belt, if equipped
- Negative battery cable

8. Run the engine and check for leaks and proper engine operation.

4.6L Engine

1. Remove the camshafts.

➡**If the followers are being reused, arrange them so that they may be put back in their original locations.**

2. Remove the camshaft followers.

3. Clean and inspect the rocker arms.

To install:

4. Apply a liberal amount of lubricant GM P/N 12345001 or equivalent to the roller pivot pocket and valve slot areas of the camshaft followers.

➡**The follower must be positioned squarely on the valve tip so that the full width of the roller will completely contact the camshaft lobe. If the followers are being reused you must put them back in their original location.**

5. Place the camshaft followers in position on the valve tip and the SHLA. The rounded head of the follower goes on the SHLA, while the flat end goes on the valve tip.

6. Install the camshafts.

Supercharger

REMOVAL & INSTALLATION

➡**A small amount of oil seepage through the front seal is normal. This seepage is caused by minute traces of oil escaping around the seal due to**

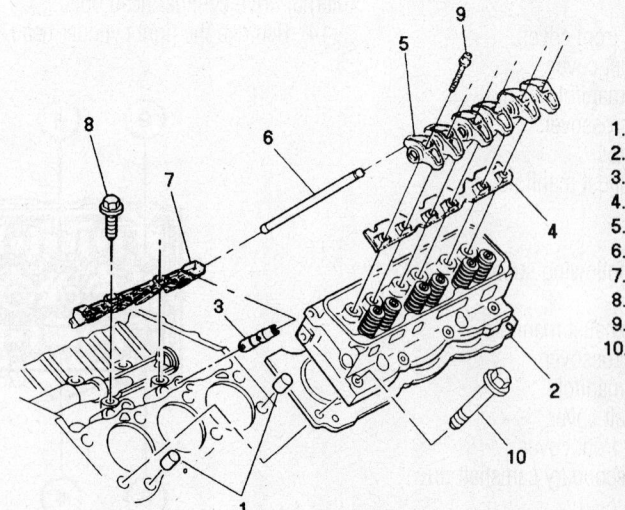

1. Dowel pin
2. Head gasket
3. Valve lifter
4. Pivot retainer
5. Rocker arm
6. Pushrod
7. Lifter guide
8. Bolt
9. Bolt
10. Head bolt

7922UG03

Exploded view of the rocker arms and related components mounting—3.8L engines

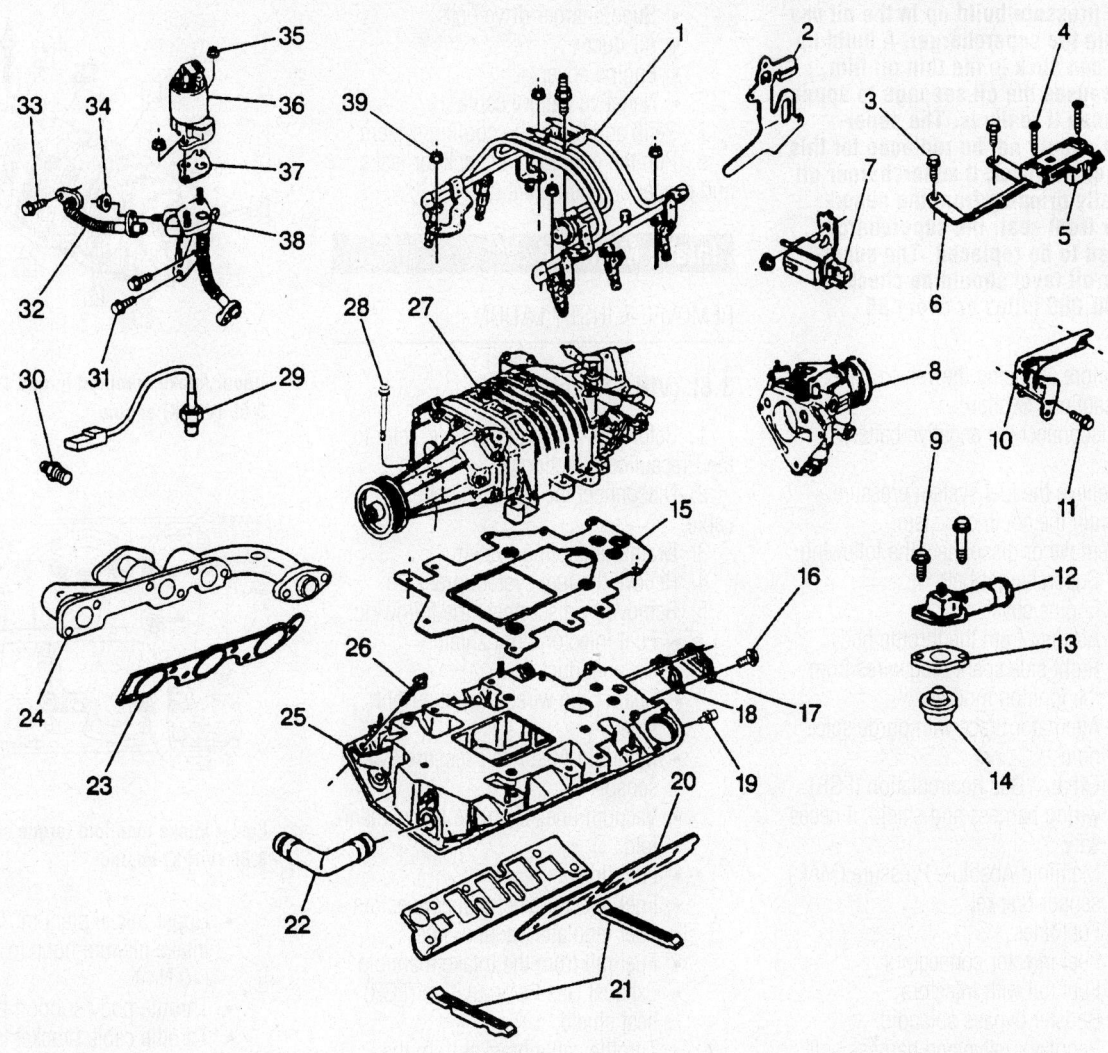

1	Fuel Injection Rail	21	Lower Intake Manifold Seal
2	Fuel Injector Sight Shield Bracket	22	Heater Inlet Pipe With Seal
3	MAP Sensor Bracket Bolt	23	Exhaust Manifold Gasket
4	MAP Sensor Bolt	24	Exhaust Manifold (Right)
5	MAP Sensor	25	Lower Intake Manifold
6	MAP Sensor Bracket	26	Lower Intake Manifold Bolt
7	Bypass Valve	27	Supercharger
8	Throttle Body	28	Supercharger Bolt
9	Water Outlet Bolt	29	Heated Oxygen Sensor
10	Accelerator Cable Control Bracket	30	Exhaust Manifold Bolt/Stud (Right)
11	Accelerator Cable Control Bracket Bolt	31	EGR Valve Adapter Bolt
12	Water Outlet	32	EGR Valve Outlet Pipe
13	Water Outlet Gasket	33	EGR Valve Outlet Pipe Bolt
14	Thermostat	34	EGR Valve Outlet Pipe Nut
15	Supercharger Gasket	35	EGR Valve Nut
16	Engine Coolant Manifold Bolt	36	EGR Valve
17	Engine Coolant Manifold	37	EGR Valve Gasket
18	Engine Coolant Manifold Gasket	38	EGR Valve Adapter
19	Coolant Temperature Sensor	39	Fuel Injection Rail Nut
20	Lower Intake Manifold Gasket		

9300UG01

Exploded view of the supercharger and related components—3.8L (VIN 1) engine

normal pressure build up in the oil cavity within the supercharger. A build up of dust can stick to the thin oil film, which causes the oil seepage to appear worse than it really is. The supercharger should not be replaced for this seepage. However, if supercharger oil is visually dripping from the supercharger front seal, the supercharger will need to be replaced. The supercharger oil level should be checked every 30,000 miles or every 36 months.

1. Before servicing the vehicle, refer to the Precautions Section.
2. Disconnect the negative battery cable.
3. Relieve the fuel system pressure.
4. Drain the cooling system.
5. Remove or disconnect the following:
   - Supercharger belt
   - Engine cover
   - Air duct from the throttle body
   - Right side spark plug wires from the ignition module
   - Alternator brace with purge solenoid
   - Exhaust Gas Recirculation (EGR) wiring harness and shield, if necessary
   - Manifold Absolute Pressure (MAP) sensor bracket
   - Fuel lines
   - Fuel injector connectors
   - Fuel rail with injectors
   - Booster bypass solenoid
   - Regulator valve and harness bolt
   - Boost control solenoid
   - Throttle body bolts and the assembly
   - Supercharger bolts and the assembly

To install:
6. Install or connect the following:
   - Supercharger with new gasket and torque the bolts, gradually and evenly, to 17 ft. lbs. (23 Nm).
   - MAP sensor bracket
   - Throttle body and torque the nuts to 89 inch lbs. (10 Nm)
   - Boost control solenoid and torque the nut to 72 inch lbs. (8 Nm)
   - Regulator valve and harness bolt
   - Booster bypass solenoid
   - Fuel rail with injectors
   - Fuel lines
   - Electrical connectors to the fuel injectors
   - Alternator brace with purge solenoid
   - EGR wiring harness and shield
   - Right side spark plug wires

- Supercharger drive belt
- Air duct
- Engine cover
- Negative battery cable
7. Refill and bleed the cooling system.
8. Run the engine and check for leaks and proper engine operation.

Intake Manifold

REMOVAL & INSTALLATION

3.8L (VIN K) Engine

1. Before servicing the vehicle, refer to the Precautions Section.
2. Disconnect the negative battery cable.
3. Drain the cooling system.
4. Relieve the fuel system pressure.
5. Remove or disconnect the following:
   - Fuel injector sight shield
   - Air inlet duct
   - Spark plug wires from the right side
   - Manifold Absolute Pressure (MAP) sensor
   - Vacuum lines from the intake manifold
   - Fuel lines
   - Fuel injector electrical connectors
   - Fuel regulator vacuum line
   - Fuel rail from the intake manifold
   - Exhaust Gas Recirculation (EGR) heat shield
   - Throttle cable bracket from the cylinder head mounting bracket and the throttle body cables
   - Throttle body support bracket
   - Upper intake plenum and gasket
   - Thermostat housing
   - Electrical connector from the Engine Coolant Temperature (ECT) sensor
   - Drive belt tensioner assembly
   - EGR valve outlet pipe
   - Lower intake manifold

To install:
6. Install or connect the following:
   - Intake manifold using new manifold gaskets. Torque the bolts in sequence to 11 ft. lbs. (15 Nm); then, re-torque to 11 ft. lbs. (15 Nm).
   - EGR valve outlet pipe
   - Drive belt tensioner assembly. Torque the tensioner bolts to 37 ft. lbs. (50 Nm).
   - Electrical connector to the ECT sensor
   - Thermostat housing

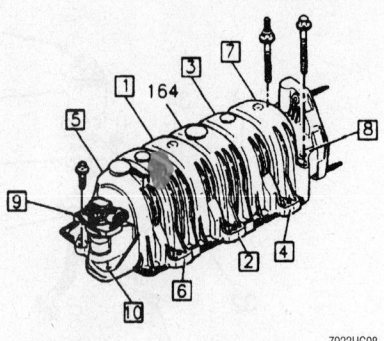

Upper intake manifold torque sequence—3.8L (VIN K) engine

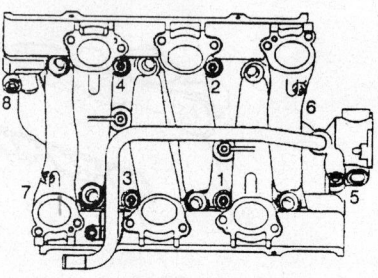

Lower intake manifold torque sequence—3.8L (VIN K) engine

   - Upper intake plenum. Torque the intake plenum bolts to 88 inch. lbs. (10 Nm).
   - Throttle body support bracket
   - Throttle cable bracket to the cylinder head mounting bracket and the cables to the throttle body lever
   - EGR heat shield
   - Fuel rail. Torque the fuel rail bolts to 88 inch. lbs. (10 Nm).
   - Fuel lines
   - Fuel regulator vacuum line
   - Fuel injector electrical connectors
   - Vacuum lines to the intake manifold
   - MAP sensor
   - Spark plug wires
   - Fuel injector sight shield and air inlet duct
   - Negative battery cable
7. Refill and bleed the cooling system.
8. Run the engine and check for leaks and proper engine operation.

3.8L (VIN 1) Engine

1. Before servicing the vehicle, refer to the Precautions Section.
2. Relieve the fuel system pressure.
3. Drain the cooling system.
4. Remove or disconnect the following:

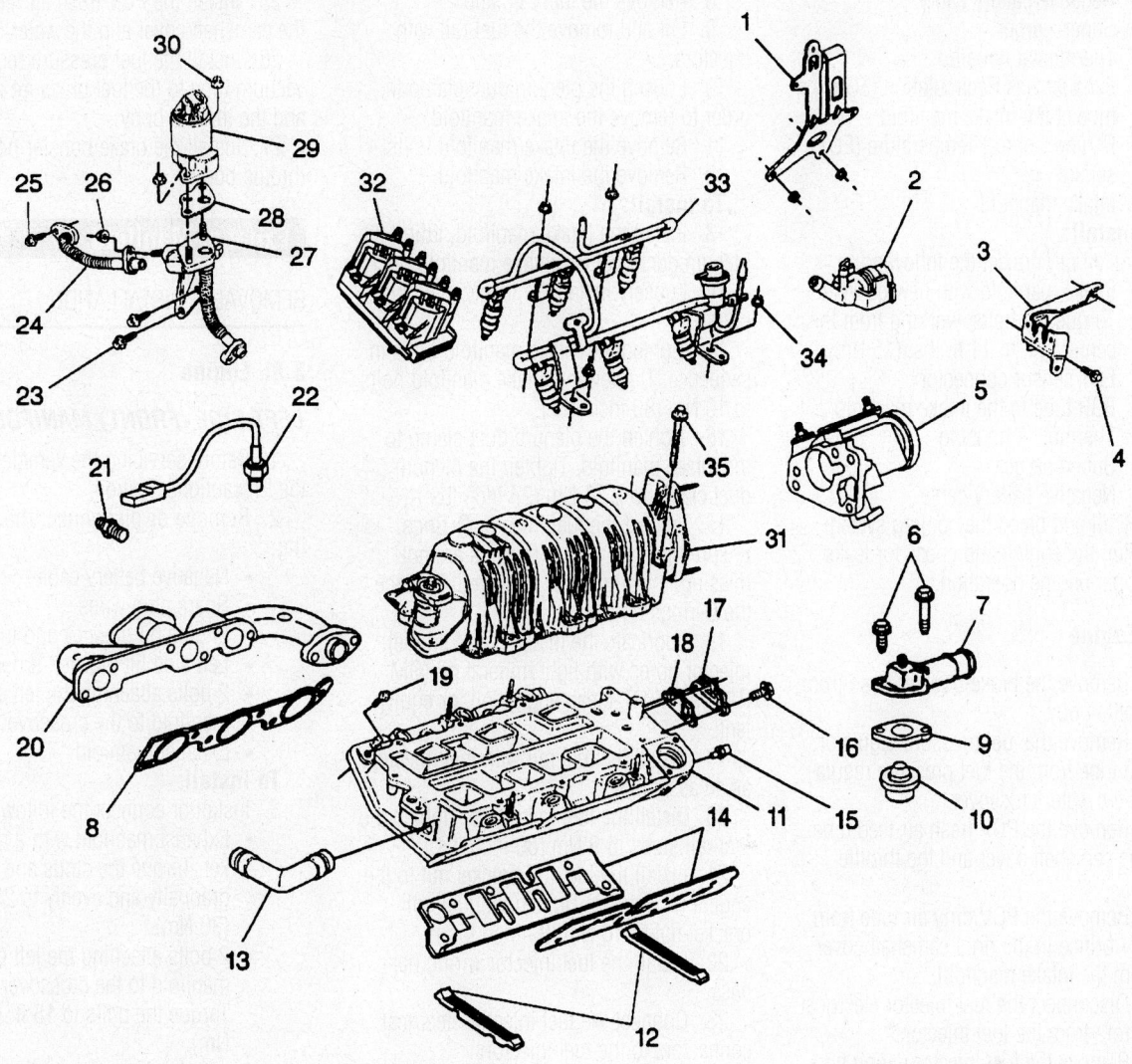

1	Fuel Injector Sight Shield Bracket	19	Lower Intake Manifold Bolt
2	Vacuum Source Manifold	20	Exhaust Manifold (Right)
3	Accelerator Cable Control Bracket	21	Exhaust Manifold Bolt/Stud
4	Throttle Body Support Bolt	22	Exhaust Oxygen Sensor
5	Throttle Body	23	EGR Valve Adapter Bolt
6	Water Outlet Bolt	24	EGR Valve Outlet Pipe
7	Water Outlet	25	EGR Valve Outlet Pipe Bolt
8	Exhaust Manifold Gasket	26	EGR Valve Outlet Pipe Nut
9	Water Outlet Gasket	27	EGR Valve Adapter
10	Thermostat	28	EGR Valve Gasket
11	Lower Intake Manifold	29	EGR Valve
12	Intake Manifold Seal	30	EGR Valve Nut
13	Heater Water Inlet Pipe	31	Upper Intake Manifold
14	Lower Intake Manifold Gasket	32	ICM
15	Coolant Temperature Sensor	33	Fuel Injection Rail
16	Engine Coolant Manifold Bolt	34	Fuel Injector Rail Nut
17	Engine Coolant Manifold	35	Upper Intake Manifold Bolt
18	Engine Coolant Manifold Gasket		

Exploded view of the intake manifold and related components—3.8L (VIN K) engine

- Negative battery cable
- Supercharger
- Thermostat housing
- Exhaust Gas Recirculation (EGR) tube at the intake manifold
- Engine Control Temperature (ECT) sensor
- Intake manifold

To install:

5. Install or connect the following:
   - Intake manifold with new gaskets. Torque the bolts, working from the center out, to 11 ft. lbs. (15 Nm).
   - ECT sensor connector
   - EGR tube to the intake manifold
   - Thermostat housing
   - Supercharger
   - Negative battery cable

6. Refill and bleed the cooling system.

7. Run the engine and check for leaks and proper engine operation.

4.6L Engine

1. Remove the brake booster hose from the throttle body.

2. Remove the fuel pressure regulator vacuum tube from the fuel pressure regulator and the water crossover.

3. Remove the PCV fresh air feed tube from the camshaft cover and the throttle body.

4. Remove the PCV dirty air tube from the PCV orifice in the right camshaft cover and from the intake manifold.

5. Disconnect the fuel injector electrical connectors from the fuel injectors.

6. Remove the fuel injector wiring harness.

7. Remove the fuel rail bracket nut from the engine lift bracket.

8. Remove the fuel rail studs.

9. Lift and remove the fuel rail with injectors.

10. Loosen the plenum duct clamp in order to remove the intake manifold.

11. Remove the intake manifold bolts.

12. Remove the intake manifold.

To install:

13. Install the intake manifold, fitting the plenum duct over the intake manifold duct.

14. Loosely install the intake manifold bolts.

15. Tighten the intake manifold bolts in sequence. Tighten the intake manifold bolts to 10 Nm (89 inch lbs.).

16. Tighten the plenum duct clamp to the intake manifold. Tighten the plenum duct clamp to 2.75 Nm (24 inch lbs.).

17. Inspect the fuel injector O-rings. Ensure the fuel injector O-rings are not missing, misaligned or damaged. Replace the O-rings if necessary.

18. Lubricate the intake manifold fuel injector bores with light mineral oil (GM P/N 9981704), clean engine oil, or equivalent.

19. Install the fuel rail with fuel injectors as an assembly.

20. Install the fuel rail studs. Tighten the fuel rail studs to 9 Nm (80 inch lbs.).

21. Install the fuel rail bracket nut to the engine lift bracket. Tighten the fuel rail bracket nut to 10 Nm (89 inch lbs.).

22. Install the fuel injector wiring harness.

23. Connect the fuel injector electrical connectors to the fuel injectors.

24. Install the PCV dirty air tube to the PCV orifice in the right camshaft cover and to the intake manifold.

25. Install the PCV fresh air feed tube to the camshaft cover and the water crossover.

26. Install the fuel pressure regulator vacuum tube to the fuel pressure regulator and the throttle body.

27. Install the brake booster hose to the throttle body.

Exhaust Manifold

REMOVAL & INSTALLATION

3.8L Engine

LEFT SIDE (FRONT) MANIFOLD

1. Before servicing the vehicle, refer to the Precautions Section.

2. Remove or disconnect the following:

   - Negative battery cable
   - Spark plug wires
   - Engine oil dipstick and tube
   - Left side lift bracket, if necessary
   - 2 bolts attaching the left exhaust manifold to the crossover pipe
   - Exhaust manifold

To install:

3. Install or connect the following:

   - Exhaust manifold with a new gasket. Torque the studs and bolts gradually and evenly to 22 ft. lbs. (30 Nm).
   - 2 bolts attaching the left exhaust manifold to the crossover pipe. Torque the bolts to 15 ft. lbs. (20 Nm).
   - Left side lift bracket, if removed
   - Engine oil dipstick and tube. Torque the bolts to 15 ft. lbs. (20 Nm).
   - Spark plug wires
   - Negative battery cable

4. Run the engine and check for exhaust leaks.

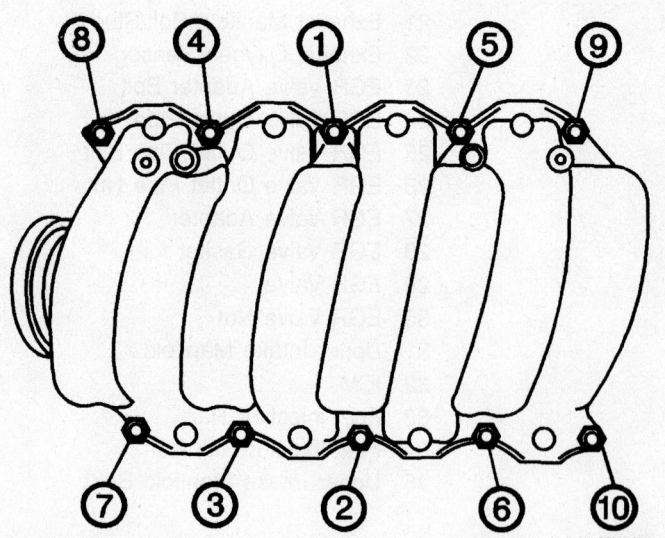

71461-BONN-G06

Intake manifold torque sequence—4.6L engine

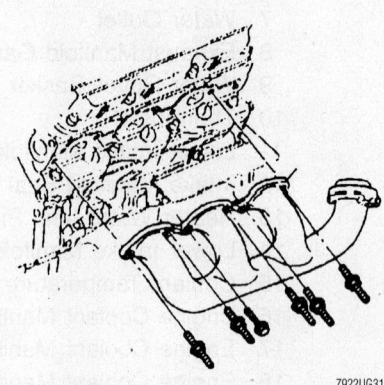

7922UG31

Exploded view of the left exhaust manifold mounting

RIGHT SIDE (REAR) MANIFOLD—MODELS WITHOUT NC8

1. Before servicing the vehicle, refer to the Precautions Section.

2. Remove or disconnect the following:

- Negative battery cable
- Fuel injector sight shield
- Air cleaner assembly
- Spark plug wires
- Brake booster heat shield
- Crossover pipe
- Engine harness from the right hand engine lift hook bracket
- Transaxle fluid dipstick and tube
- Oxygen (O_2S) sensor
- Exhaust Gas Recirculation (EGR) feed pipe bolt from the manifold
- Transmission oil level tube and seal
- Exhaust manifold flange nuts
- Front exhaust pipe
- Engine lift bracket
- Exhaust manifold

To install:

3. Install or connect the following:

- Manifold to the cylinder head and crossover pipe using new gaskets
- Manifold mounting studs. Torque the studs and bolts to 22 ft. lbs. (30 Nm), beginning at the center and working outwards.
- Engine lift bracket
- Front exhaust pipe
- Front exhaust pipe-to-manifold nuts. Torque the nuts to 22 ft. lbs. (30 Nm).
- Transmission dipstick tube seal and the tube
- EGR feed pipe to the manifold
- O_2S sensor
- Spark plug wires to the spark plugs
- Engine harness to the right hand engine lift hook bracket
- Crossover pipe
- Brake booster heat shield
- Air cleaner assembly
- Fuel injector sight shield
- Negative battery cable

4. Run the engine and check for exhaust leaks.

RIGHT SIDE (REAR) MANIFOLD—MODELS WITH NC8

1. Before servicing the vehicle, refer to the Precautions Section.

2. Remove or disconnect the following:

- Negative battery cable
- Air cleaner assembly
- Fuel injector sight shield

- Heated Oxygen (HO_2S) sensor electrical connection
- Spark plug wires
- Connector from the secondary AIR valve
- Hose from the secondary AIR valve
- Bolts attaching the secondary AIR tube to the exhaust manifold
- Secondary AIR tube and the gasket from the exhaust manifold and discard the gasket
- Bolt attaching the secondary AIR valve to the fuel injector sight shield bracket
- Secondary AIR valve from the fuel injector sight shield bracket
- Brake booster heat shield
- Crossover pipe
- Transaxle fluid dipstick and tube
- Exhaust manifold pipe
- Right exhaust manifold studs
- Fuel injector sight shield bracket retaining nuts and the bracket
- Right engine lift bracket retaining fasteners and the bracket
- Bolt attaching the Exhaust Gas Recirculation (EGR) inlet pipe to the right exhaust manifold
- Right exhaust manifold retaining fasteners
- Right exhaust manifold and the gasket from the engine and discard the gasket
- Exhaust crossover seal from the exhaust crossover and discard the seal

To install:

3. Install or connect the following:

- New exhaust crossover seal
- New manifold gasket
- Manifold to the cylinder head and crossover pipe
- Manifold mounting studs. Torque the studs and bolts to 22 ft. lbs. (30 Nm), beginning at the center and working outwards.
- Bolt attaching the Exhaust Gas Recirculation (EGR) inlet pipe to the right exhaust manifold and tighten to 22 ft. lbs. (30 Nm)
- Engine lift bracket
- Fuel injector sight shield bracket and tighten the retainers to 22 ft. lbs. (30 Nm)
- Right exhaust manifold studs and tighten to 80 inch lbs. (9 Nm)
- HO_2S sensor, if removed
- Exhaust manifold pipe
- Transmission dipstick tube
- Exhaust crossover to the right exhaust manifold bolts and tighten to 15 ft. lbs. (20 Nm)

- Brake booster heat shield
- Air cleaner assembly
- Secondary AIR valve to the fuel injector sight shield bracket and tighten the bolt to 18 ft. lbs. (25 Nm)
- New gasket on the secondary AIR tube
- Bolts attaching the secondary AIR tube to the exhaust manifold and tighten to 89 inch lbs. (10 Nm)
- Hose to the secondary AIR valve
- Connector to the secondary AIR valve
- Spark plug wires to the spark plugs
- HO_2S sensor connection
- Fuel injector sight shield
- Negative battery cable

4. Run the engine and check for exhaust leaks.

4.6L Engine

LEFT WITH RPO NC1 OR NF7

1. Remove the A.I.R. valve pipe nuts from the exhaust manifold.

2. Remove the A.I.R. valve bolt from the mounting bracket.

3. Remove the A.I.R. valve nuts from the mounting bracket.

4. Remove the A.I.R. valve.

➡ **DO NOT reuse the old A.I.R. valve pipe gasket.**

5. Remove and discard the A.I.R. valve pipe gasket.

6. Remove the oxygen sensor and inspect the sensor for excessive deposits or damage.

7. Replace the oxygen sensor if necessary.

8. Remove the exhaust manifold to crossover pipe bolts.

9. Remove the exhaust manifold bolts.

10. Remove the exhaust manifold.

➡ **DO NOT reuse the exhaust manifold to crossover pipe gasket.**

11. Remove and discard the exhaust manifold to crossover pipe gasket.

➡ **DO NOT reuse the exhaust manifold gasket.**

12. Remove and discard the exhaust manifold gasket.

13. Disconnect the EGR inlet tube nut from the exhaust crossover.

➡ **The EGR valve inlet pipe incorporates a crush seal connection at the water crossover. The EGR valve inlet pipe must be replaced if disconnected from the water crossover.**

14. Remove the EGR inlet pipe bolt and flange from the water crossover.

15. Remove the EGR inlet tube and discard.

16. Remove the exhaust crossover stud and bolt.

17. Remove the exhaust crossover.

18. If equipped, remove the left coolant heater as follows:

19. Remove the coolant heater bolt.

20. Remove the coolant heater from the cylinder block.

To install:

21. If the engine is equipped with coolant heaters, install the left side heater as follows:

    a. Place the coolant heater in position on the cylinder block.

    b. Install the coolant heater bolt. Tighten the coolant heater bolt to 10 Nm (89 inch lbs.).

➡**DO NOT reuse the exhaust manifold gasket.**

22. With the manifold still on the bench, position a new manifold gasket in place on the manifold sealing surface.

23. Install 2 outer manifold bolts in the manifold to retain the gasket.

24. Install a new manifold to intermediate pipe flange gasket.

25. Using 2 hands, place the manifold in the intermediate pipe and onto the cylinder head.

26. Install new manifold flange bolts for maximum joint integrity. Hand tighten. Do not torque the flange bolts to specification until the engine is mounted in the vehicle.

27. Hand tighten the 2 outer manifold bolts.

28. Install the remaining manifold bolts. Tighten the exhaust manifold bolts to 25 Nm (18 ft. lbs.).

29. Coat the oxygen sensor threads with high temperature anti-seize, GM P/N 12377953 or equivalent.

30. Install the oxygen sensor. Tighten the oxygen sensor to 40 Nm (30 ft. lbs.).

➡**DO NOT reuse the old A.I.R. valve pipe gasket.**

31. Install the NEW A.I.R. valve pipe gasket.

32. Install the A.I.R. valve.

33. Install the A.I.R. valve nuts to the mounting bracket. Tighten the A.I.R. valve nuts to 9 Nm (80 inch lbs.).

34. Install the A.I.R. valve bolt through the mounting bracket. Tighten the A.I.R. valve bolt to 9 Nm (80 inch lbs.).

35. Install the A.I.R. valve pipe nuts to the exhaust manifold. Tighten the A.I.R. valve pipe nuts to 9 Nm (80 inch lbs.).

LEFT W/O RPO, NC1 OR NF7

1. Remove the oxygen sensor and inspect the sensor for excessive deposits or damage.

2. Replace the oxygen sensor if necessary.

3. Remove the exhaust manifold to crossover pipe bolts.

4. Remove the exhaust manifold bolts.

5. Remove the exhaust manifold.

➡**DO NOT reuse the exhaust manifold to crossover pipe gasket.**

6. Remove and discard the exhaust manifold to crossover pipe gasket.

➡**DO NOT reuse the exhaust manifold gasket.**

7. Remove and discard the exhaust manifold gasket.

8. Disconnect the EGR inlet tube nut from the exhaust crossover.

➡**The EGR valve inlet pipe incorporates a crush seal connection at the water crossover. The EGR valve inlet pipe must be replaced if disconnected from the water crossover.**

9. Remove the EGR inlet pipe bolt and flange from the water crossover.

10. Remove the EGR inlet tube and discard.

11. Remove the exhaust crossover stud and bolt.

12. Remove the exhaust crossover.

13. If equipped, remove the left coolant heater as follows:

    a. Remove the coolant heater bolt.

    b. Remove the coolant heater from the cylinder block.

To install:

14. If the engine is equipped with coolant heaters, install the left side heater as follows:

    a. Place the coolant heater in position on the cylinder block.

    b. Install the coolant heater bolt. Tighten the coolant heater bolt to 10 Nm (89 inch lbs.).

15. With the manifold still on the bench, position a new manifold gasket in place on the manifold sealing surface.

16. Install 2 outer manifold bolts in the manifold to retain the gasket.

17. Install a new manifold to intermediate pipe flange gasket.

18. Using 2 hands, place the manifold in the intermediate pipe and onto the cylinder head.

19. Install new manifold flange bolts for maximum joint integrity. Hand tighten. Do not torque the flange bolts to specification until the engine is mounted in the vehicle.

20. Hand tighten the 2 outer manifold bolts.

21. Install the remaining manifold bolts. Tighten the exhaust manifold bolts to 25 Nm (18 ft. lbs.).

22. Coat the oxygen sensor threads with high temperature anti-seize, (GM P/N 12377953) or equivalent.

23. Install the oxygen sensor. Tighten the oxygen sensor to 40 Nm (30 ft. lbs.).

RIGHT WITH RPO NC1 OR NF7

1. Remove the A.I.R. valve pipe nuts from the exhaust manifold.

2. Remove the A.I.R. valve bolts from the mounting bracket.

3. Remove the A.I.R. valve.

➡**DO NOT reuse the A.I.R. valve pipe gasket.**

4. Remove and discard the A.I.R. valve pipe gasket.

5. Remove the oxygen sensor and inspect the sensor for excessive deposits or damage.

6. Replace the oxygen sensor if necessary.

➡**The stud must be reinstalled in the original location. The stud may remain attached to the nut when initially removed during exhaust manifold removal. In order to prevent exhaust leakage between the exhaust manifold and cylinder head the nut and stud combination must be reinstalled during exhaust manifold installation in the original location.**

7. Remove the right exhaust manifold retaining nuts.

➡**DO NOT reuse the exhaust manifold gasket.**

8. Remove the exhaust manifold and the gasket.

9. If equipped, remove the right coolant heater as follows:

10. Remove the coolant heater retaining bolt.

11. Remove coolant heater from the cylinder block.

To install:

12. If the engine is equipped with coolant heaters, install the right side heater as follows:

    a. Place the coolant heater in position on the cylinder block.

    b. Install the coolant heater bolt.

Tighten the coolant heater bolt to 10 Nm (89 inch lbs.).

13. Position a new manifold gasket in place on the cylinder head studs.

14. Using two hands, position the manifold onto the cylinder head.

15. Install two outer manifold nuts to hold the manifold in place.

16. Install the remaining manifold nuts. Tighten the exhaust manifold nuts to 25 Nm (18 ft. lbs.).

17. Coat the oxygen sensor threads with high temperature anti-seize, GM P/N 12377953 or equivalent.

18. Install the oxygen sensor. Tighten the oxygen sensor to 40 Nm (30 ft. lbs.).

➥**DO NOT reuse the old A.I.R. valve pipe gasket.**

19. Install the NEW A.I.R. valve pipe gasket.

20. Install the A.I.R. valve.

21. Install the A.I.R. valve bolts through the mounting bracket. Tighten the A.I.R. valve bolts to 9 Nm (80 inch lbs.).

22. Install the A.I.R. valve pipe nuts to the exhaust manifold. Tighten the A.I.R. valve pipe nuts to 9 Nm (80 inch lbs.).

23. Place the exhaust intermediate pipe in position.

24. Install the intermediate pipe stud at the cylinder head and the bolt at the lower crankcase. Tighten the intermediate pipe-to-lower crankcase bolt to 35 Nm (18 ft. lbs.). Tighten the intermediate pipe-to-cylinder head stud to 25 Nm (18 ft. lbs.).

➥**The EGR valve inlet pipe incorporates a crush seal connection at the water crossover. The EGR valve inlet pipe must be replaced if disconnected from the water crossover.**

25. Hand start the NEW EGR inlet tube to intermediate pipe nut to prevent cross-threading. Tighten the EGR inlet tube-to-intermediate pipe nut to 60 Nm (44 ft. lbs.).

26. Connect the EGR inlet pipe flange to the water crossover.

27. Install the EGR inlet pipe to water crossover flange bolt. Tighten the EGR inlet tube to water crossover bolt to 25 Nm (18 ft. lbs.).

RIGHT W/O RPO, NC1 OR NF7

1. Remove the oxygen sensor and inspect the sensor for excessive deposits or damage.

2. Replace the oxygen sensor if necessary.

➥**The stud must be reinstalled in the original location. The stud may remain**

attached to the nut when initially removed during exhaust manifold removal. In order to prevent exhaust leakage between the exhaust manifold and cylinder head the nut and stud combination must be reinstalled during exhaust manifold installation in the original location.

3. Remove the right exhaust manifold retaining nuts.

➥**DO NOT reuse the exhaust manifold gasket.**

4. Remove the exhaust manifold and the gasket.

5. If equipped, remove the right coolant heater as follows:

   a. Remove the coolant heater retaining bolt.

   b. Remove coolant heater from the cylinder block.

To install:

6. If the engine is equipped with coolant heaters, install the right side heater as follows:

   a. Place the coolant heater in position on the cylinder block.

   b. Install the coolant heater bolt. Tighten the coolant heater bolt to 10 Nm (89 inch lbs.).

7. Position a new manifold gasket in place on the cylinder head studs.

8. Using two hands, position the manifold onto the cylinder head.

9. Install two outer manifold nuts to hold the manifold in place.

10. Install the remaining manifold nuts. Tighten the exhaust manifold nuts to 25 Nm (18 ft. lbs.).

11. Coat the oxygen sensor threads with high temperature anti-seize, GM P/N 12377953 or equivalent.

12. Install the oxygen sensor. Tighten the oxygen sensor to 40 Nm (30 ft. lbs.).

13. Place the exhaust intermediate pipe in position.

14. Install the intermediate pipe stud at the cylinder head and the bolt at the lower crankcase. Tighten the intermediate pipe-to-lower crankcase bolt to 35 Nm (18 ft. lbs.). Tighten the intermediate pipe-to-cylinder head stud to 25 Nm (18 ft. lbs.).

➥**The EGR valve inlet pipe incorporates a crush seal connection at the water crossover. The EGR valve inlet pipe must be replaced if disconnected from the water crossover.**

15. Hand start the NEW EGR inlet tube to intermediate pipe nut to prevent cross-threading. Tighten the EGR inlet tube-to-intermediate pipe nut to 60 Nm (44 ft. lbs.).

16. Connect the EGR inlet pipe flange to the water crossover.

17. Install the EGR inlet pipe to water crossover flange bolt. Tighten the EGR inlet tube to water crossover bolt to 25 Nm (18 ft. lbs.).

Camshaft and Valve Lifters

REMOVAL & INSTALLATION

3.8L Engine

1. Before servicing the vehicle, refer to the Precautions Section.

2. Relieve the fuel system pressure.

3. Remove the engine and mount it on an engine stand.

4. Remove or disconnect the following:

- Negative battery cable
- Supercharger, if equipped
- Intake manifold
- Rocker arm covers
- Rocker arm assemblies
- Pushrods
- Lifters and guides

➥**A magnet may be helpful when pulling the lifters out of their bores. Identify all parts as they are removed, so they can be reinstalled in their original locations.**

- Crankshaft balancer
- Timing chain front cover

5. Set the engine to Top Dead Center (TDC) No. 1 cylinder (firing position) to align the timing marks, before disassembling the timing chain and sprockets.

❉❉ **WARNING**

Align the timing marks of the camshaft and crankshaft sprockets to avoid burring the camshaft journals by the crankshaft.

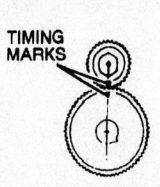

TIMING MARKS

BALANCE SHAFT GEAR TO BALANCE SHAFT DRIVE GEAR

TIMING MARKS

CAMSHAFT SPROCKET TO CRANKSHAFT SPROCKET

7922UG09

The timing marks should face each other when the chain and gears are installed properly

6. Remove or disconnect the following:
- Camshaft sprocket and timing chain
- Camshaft thrust plate
- Camshaft

✳✳ WARNING

If the camshaft was replaced the lifters must also be replaced. The old lifters have developed a wear pattern and will cause the new camshaft to wear prematurely.

To install:

7. Coat the camshaft lobes and bearings with camshaft break-in prelube prior to installation.

8. Install or connect the following:
- Camshaft
- Camshaft thrust plate. Torque the bolts to 10 ft. lbs. (14 Nm).
- Camshaft sprocket and timing chain with timing marks aligned. Torque the camshaft sprocket bolt to 74 ft. lbs. (100 Nm) plus an additional 90 degree (¼) turn.
- Timing chain front cover
- Crankshaft balancer. Torque the mounting bolt to 111 ft. lbs. (150 Nm). plus an additional 76 degree turn.

9. Coat the valve lifters with camshaft break-in prelube.

10. Install or connect the following:
- Valve lifters
- Lifter guides and lifter guide retainer. Torque the retainer mounting bolts to 22 ft. lbs. (30 Nm).
- Pushrods and rocker arms. Torque the rocker arm bolts to 11 ft. lbs. (15 Nm) plus an additional 90 degree turn.
- Rocker arm covers
- Intake manifold
- Supercharger, if equipped
- Engine
- Negative battery cable

11. Verify that all fluid levels are full and correct.

12. Start the engine and check for leaks. Check engine operation.

4.6L Engine

LEFT SIDE

Tools Required
- J 45059 Torque Angle/Meter
- J 44212 Camshaft Holding Tool
- EN 46327 Timing Chain Retention Tool
- J 38185 Hose Clamp Pliers
- J 38823 Water Pump Drive Pulley Installer

- J 38825 Water Pump Drive Pulley Remover

1. Remove the fuel injector sight shield.
2. Partially drain the cooling system.
3. Position tool J 38185 to the clamp in order to remove the radiator inlet hose from the water housing crossover.
4. Disconnect the PCV fresh air tube from the camshaft cover.
5. Remove the ignition coil cassette.
6. Remove the 4 spark plug boots.
7. Disconnect the cable harness clips at the front of the camshaft cover and position the cable harness aside.
8. Remove the secondary AIR valve bracket nut closest to the center of the engine.

9. Pry outward slightly on the secondary AIR valve bracket in order to gain clearance to remove the water pump drive belt shield nut.
10. Remove the water pump drive belt shield fasteners.
11. Remove the water pump drive belt shield.
12. Disconnect the water pump drive belt.
13. Loosen the 2 bolts attaching the water pump belt tensioner to the water crossover.
14. Remove the water pump belt tensioner.
15. Remove the plastic dust cap from the end of the intake camshaft.

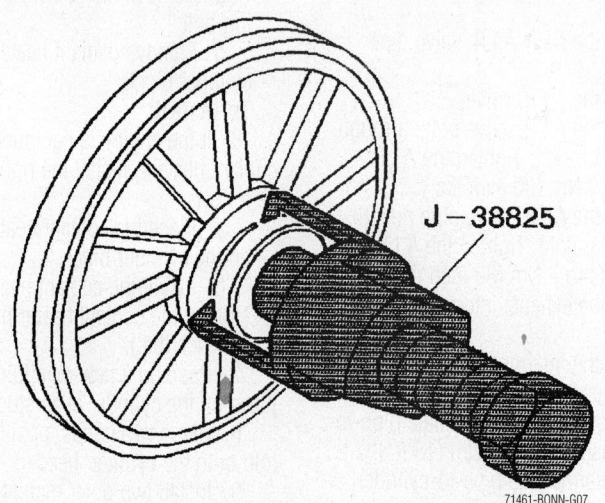

71461-BONN-G07

Remove the water pump drive pulley from the intake camshaft using tool J 38825—4.6L engine

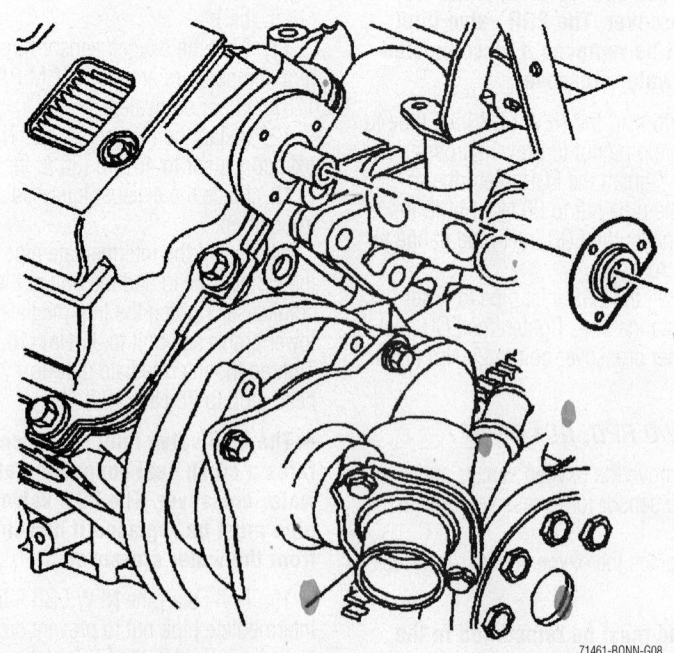

71461-BONN-G08

Remove the camshaft seal—4.6L engine

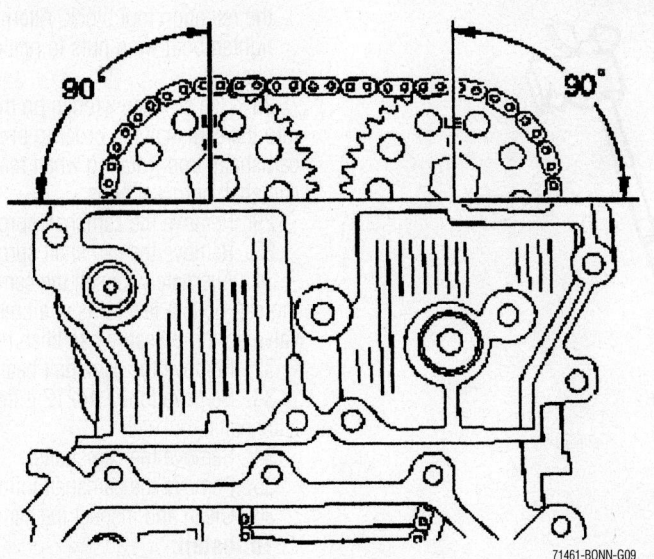

71461-BONN-G09

Rotate the crankshaft to TDC of the #1 cylinders compression stroke, both camshaft sprocket drive pins should be at the top of their rotation—4.6L engine

16. Remove the water pump drive pulley from the intake camshaft using tool J 38825.

17. Remove the 3 camshaft seal retainer bolts.

➡ **DO NOT reuse the camshaft seal.**

18. Remove the camshaft seal.

19. Remove the camshaft cover bolts.

20. Lift the camshaft drive end of the camshaft cover up.

21. Remove the camshaft cover reward to clear the water pump drive shaft.

22. Discard the camshaft cover perimeter seals and spark plug seals if there is any evidence of damage or if the seal comes out of the groove in the cover during removal.

23. Clean and inspect the camshaft cover.

24. Rotate the crankshaft to TDC of the #1 cylinders compression stroke, both camshaft sprocket drive pins should be at the top of their rotation.

✳✳ CAUTION

The camshaft holding tools must be installed on the camshafts to prevent camshaft rotation. When performing service to the valve train and/or timing components, valve spring pressure can cause the camshafts to rotate unexpectedly and can cause personal injury.

25. Install tool J 44212 over the camshafts.

26. Use a paint stick to create a mark on the timing chain link adjacent to each camshaft sprocket timing mark.

27. Install the timing chain retention tools using the procedure below:

a. Rotate the wing nut of the to the top of its travel.

b. Position the bottom retention tool on the cylinder head with the V-notch of the block adjacent to the left exhaust camshaft sprocket and chain.

c. Insert the hook end into a secondary timing chain link as shown.

d. Rotate the wing nut until it contacts the retention tool block. DO NOT tighten the wing nut at this time.

e. Rotate the wing nut to the top of its travel.

f. Position the top retention tool on the cylinder head with the V-notch of the block adjacent to the left intake camshaft sprocket and chain.

g. Insert the hook end into a secondary timing chain link as shown

h. Rotate the wing nut until it contacts

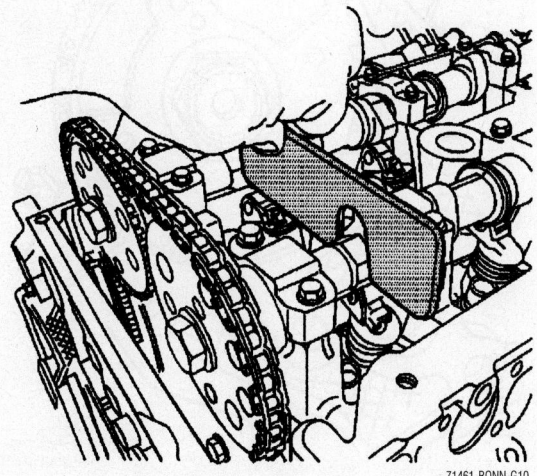

71461-BONN-G10

Install tool J 44212 over the camshafts—4.6L engine

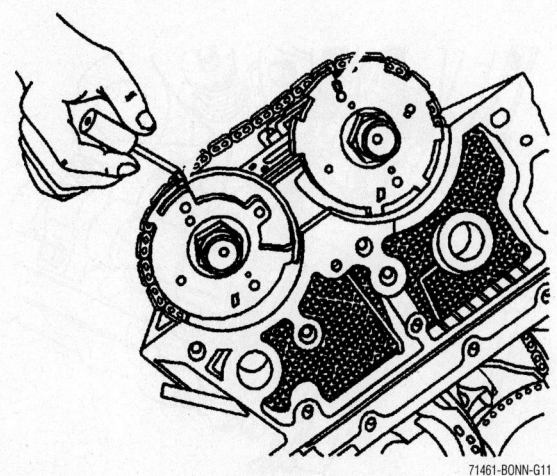

71461-BONN-G11

Use a paint stick to create a mark on the timing chain link adjacent to each camshaft sprocket timing mark—4.6L engine

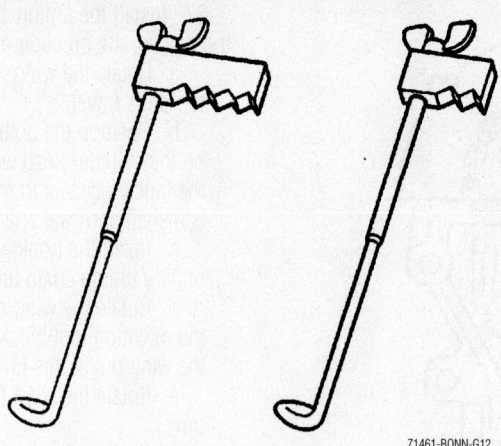

71461-BONN-G12

Timing chain retention tools—4.6L engine

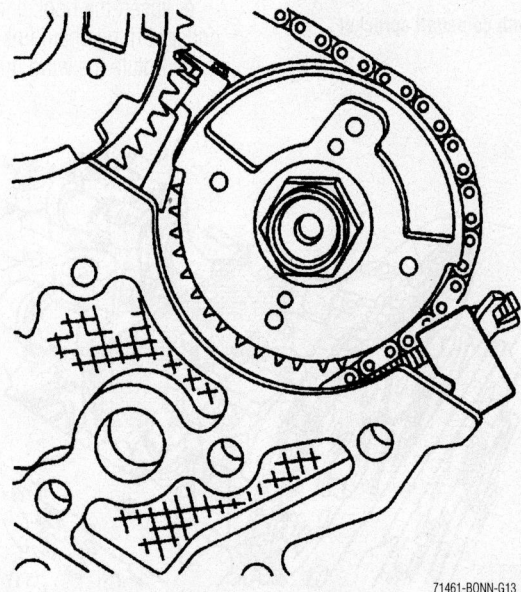

71461-BONN-G13

Position the bottom retention tool on the cylinder head with the V-notch of the block adjacent to the left exhaust camshaft sprocket and chain—4.6L engine

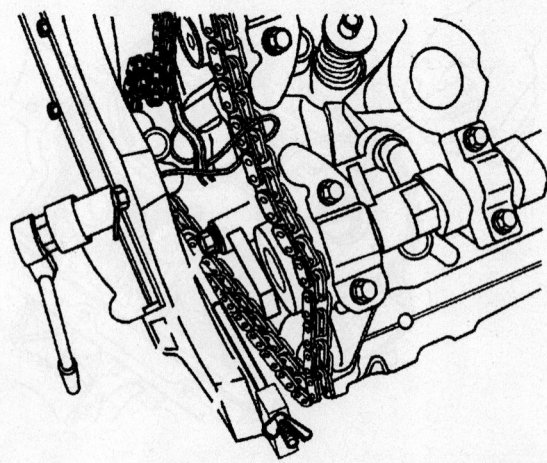

71461-BONN-G14

Insert the hook end into a secondary timing chain link as shown—4.6L engine

the retention tool block. Alternately tighten both wing nuts to retain the chain.

28. Use an open wrench on the hex cast into the camshafts in order to prevent the camshafts from rotating when removing the camshaft sprocket bolts.

29. Remove the camshaft sprocket bolts.

30. Remove the camshaft sprockets.

31. Alternately loosen the camshaft bearing cap bolts a few turns at a time until all valve spring pressure has been released.

32. Remove the camshaft bearing caps.

33. Remove tool J 44212 from the camshafts.

34. Remove the camshafts.

35. Remove the camshaft followers.

36. Clean and inspect the camshafts.

To install:

37. Apply a liberal amount of lubricant GM P/N 12345001 or equivalent to the roller pivot pocket and valve slot areas of the camshaft followers.

➡**The follower must be positioned squarely on the valve tip so that the full width of the roller will completely contact the camshaft lobe. If the followers are being reused you must put them back in their original location.**

38. Place the camshaft followers in position on the valve tip and the stationary hydraulic lash adjusters (SHLA). The rounded head of the follower goes on the SHLA, while the flat end goes on the valve tip.

39. Clean the camshaft carriers with a clean, lint-free cloth.

40. Apply a liberal amount of lubricant GM P/N 12345001 or equivalent to the camshaft carriers, camshaft lobes and the camshaft journals.

41. Place the camshaft in the camshaft carriers with the camshaft sprocket drive pins near the top of their rotation and the camshaft lobes in a neutral position. The camshafts can be identified by a stamping near the rear journal. For example: L-EXH is defined as Left bank Exhaust.

42. Observe the markings on the camshaft bearing caps. Each camshaft bearing cap is marked in order to identify its location. The markings have the following meanings:

- The arrow should point to the front of the engine.
- The number indicates the position from the front of the engine.
- The "E" indicates the exhaust camshaft.
- The "I" indicates the Intake camshaft.

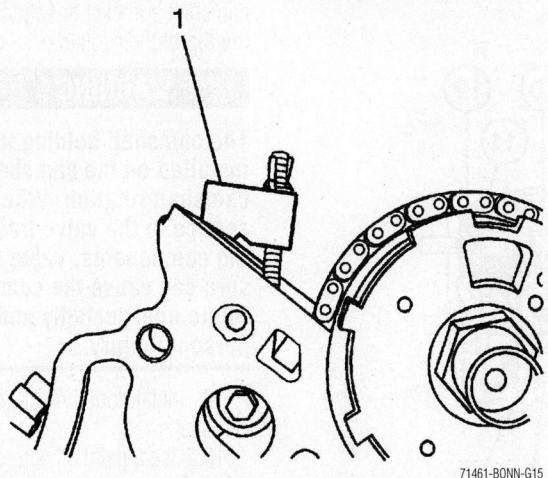

Position the top retention tool on the cylinder head with the V-notch of the block adjacent to the left intake camshaft sprocket and chain—4.6L engine

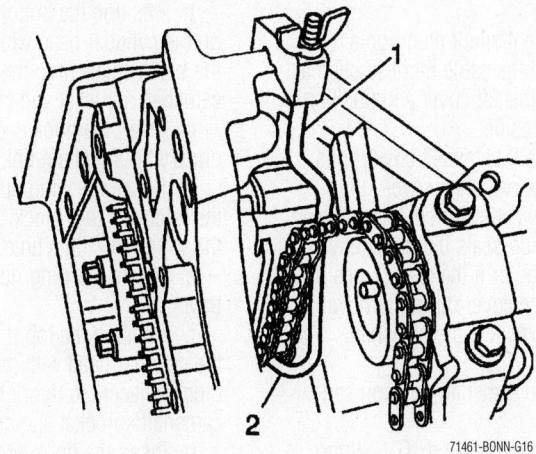

Insert the hook end into a secondary timing chain link as shown—4.6L engine

43. Apply a liberal amount of lubricant GM P/N 12345001 or equivalent to the camshaft bearing caps.

44. Install the camshaft bearing caps according to the identification marks.

➡ **Use the correct fastener in the correct location. Replacement fasteners must be the correct part number for that application. Fasteners requiring replacement or fasteners requiring the use of thread locking compound or sealant are identified in the service procedure. Do not use paints, lubricants, or corrosion inhibitors on fasteners or fastener joint surfaces unless specified. These coatings affect fastener torque and joint clamping force and may damage the fastener. Use the correct tightening sequence and specifications when installing fasteners in order to avoid damage to parts and systems.**

45. Install the camshaft bearing cap bolts in sequence.

   a. Alternately hand tighten the camshaft bearing cap bolts a few turns at a time until all caps are fully seated.

   b. Tighten the camshaft bearing cap bolts to 5 Nm (44 inch lbs.).

   c. Tighten the camshaft bearing cap bolts an additional 30 degrees using tool J 36660-A.

46. Align the camshafts.

✳✳ CAUTION

The camshaft holding tools must be installed on the camshafts to prevent camshaft rotation. When performing service to the valve train and/or timing components, valve spring pressure can cause the camshafts to rotate unexpectedly and can cause personal injury.

47. Install tool J 44212 over the camshafts.

➡ **Ensure the camshaft sprockets properly engage the camshaft sprocket drive pins and camshafts.**

48. Slide the intake and exhaust camshaft sprockets off the pins of tool J 44213 and onto the pins of the camshafts.

49. Use an open wrench on the hex cast into the camshafts in order to prevent the camshafts from rotating when tightening the camshaft sprocket bolts.

50. Install the camshaft sprocket bolts. Tighten the camshaft sprocket bolts to 120 Nm (89 ft. lbs.).

51. Verify the camshaft sprocket alignment.

52. Remove tool J 44212 from the camshafts.

53. Install the camshaft cover seal as required.

➡ **Be careful to prevent the exposed section of the camshaft cover seal from being damaged by the edge of the cylinder head casting.**

54. Insert the intake camshaft end through the hole in the camshaft cover.

55. Work the camshaft cover into position by pivoting the cover down and to the left allowing the cover to clear the camshaft drive chain and then aligning the bolt holes.

56. Install the 9 camshaft cover bolts. Tighten the camshaft cover bolts to 10 Nm (89 inch lbs.).

57. Install the NEW seal as follows:

58. Lubricate the camshaft seal lips with engine oil.

59. Push the camshaft seal into position around the intake camshaft using the protective sleeve supplied with the seal.

60. Coat the bolt threads with sealant GM P/N 1052080 (Canadian P/N 10953480) or equivalent.

61. Install the camshaft seal bolts. Tighten the camshaft seal bolts to 3 Nm (27 inch lbs.).

62. Place the water pump drive pulley in position on the intake camshaft.

63. Install the water pump pulley using tool J 38823. During installation, the tool will bottom out on the camshaft at the proper depth.

64. Install the plastic dust cap into the end of the camshaft.

65. Position the water pump belt tensioner to the water crossover. Tighten the water pump belt tensioner bolts to 10 Nm (89 inch lbs.).

66. Connect the water pump drive belt.

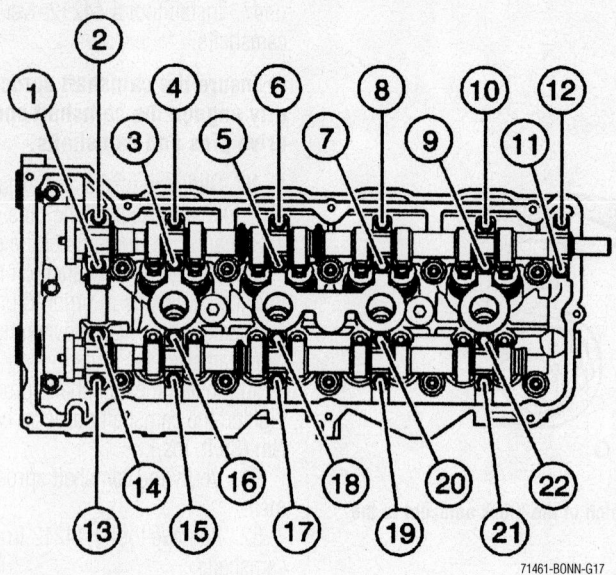

Left side camshaft bearing cap torque sequence—4.6L engine

71461-BONN-G17

67. Install the water pump drive belt shield.

68. Install the water pump drive belt shield fasteners. Tighten the water pump drive belt shield fasteners to 10 Nm (89 inch lbs.). Tighten the secondary AIR valve bracket nut to 9 Nm (80 inch lbs.).

69. Connect the cable harness clips to the cable harness at the front of the camshaft cover.

70. Install the spark plug boots onto the coil cassette. Ensure that the boots are fully seated against the cassette.

71. Install the ignition coil cassette.

72. Connect the PCV fresh air tube to the left camshaft cover.

73. Position tool J 38185 to the clamp in order to connect the radiator inlet hose to the water housing crossover.

74. Install the fuel injector sight shield.

75. Fill the cooling system.

RIGHT SIDE

1. Remove the 2 nuts from the intake manifold sight shield.

2. Remove the sight shield from the engine.

3. Disconnect the PCV dirty air tube from the camshaft cover.

4. Disconnect the oxygen sensor wire.

5. Disconnect the vacuum tubes from the secondary AIR vent solenoid.

6. Disconnect the secondary AIR vent solenoid electrical connector.

7. Remove the secondary AIR control valve bracket.

8. Remove the nut securing the secondary AIR tube.

9. Remove the ignition coil cassette.

10. Remove the 4 spark plug boots.

11. Disconnect the cable harness clips at the front of the camshaft cover and position the cable harness aside.

12. Remove the 9 camshaft cover bolts.

13. Remove the camshaft cover.

14. Discard the camshaft cover perimeter seals and spark plug seals if there is any evidence of damage or if the seal comes out of the groove in the cover during removal.

15. Clean and inspect the camshaft cover.

16. Remove the camshaft position sensor.

17. Rotate the crankshaft to TDC of the #1 cylinders compression stroke, both

camshaft sprocket drive pins should be at the top of their rotation.

✸✸ WARNING

The camshaft holding tools must be installed on the camshafts to prevent camshaft rotation. When performing service to the valve train and/or timing components, valve spring pressure can cause the camshafts to rotate unexpectedly and can cause personal injury.

18. Install tool J 44212 over the camshafts.

19. Use a paint stick to create a mark on the timing chain link adjacent to each camshaft sprocket timing mark.

20. Install the using the procedure below:

a. Rotate the wing nut to the top of its travel.

b. Position the bottom retention tool on the cylinder head with the V-notch of the block adjacent to the right exhaust camshaft sprocket and chain.

c. Insert the hook end into a secondary timing chain link as shown.

d. Rotate the wing nut until it contacts the retention tool block. DO NOT tighten the wing nut at this time.

e. Rotate the wing nut to the top of its travel.

f. Position the top retention tool on the cylinder head with the V-notch of the block adjacent to the right intake camshaft sprocket and chain.

g. Insert the hook end into a secondary timing chain link as shown.

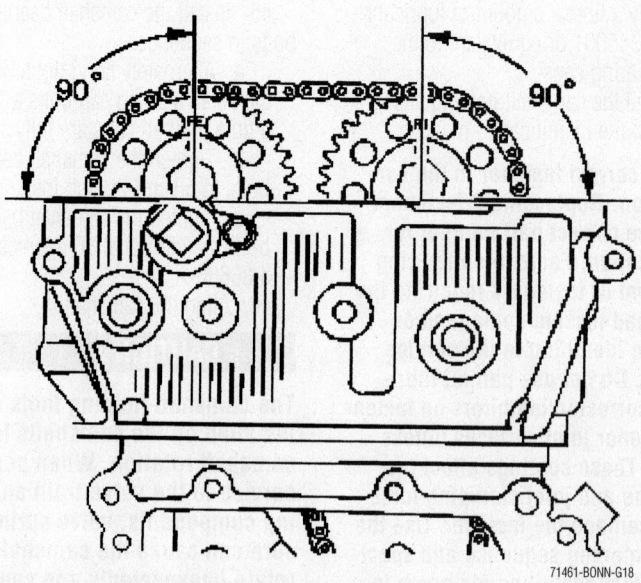

71461-BONN-G18

Rotate the crankshaft to TDC of the #1 cylinders compression stroke, both right side camshaft sprocket drive pins should be at the top of their rotation—4.6L engine

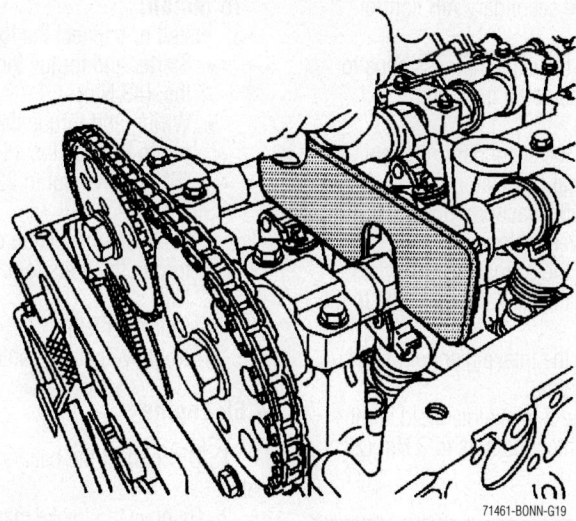

Install tool J 44212 over the camshafts

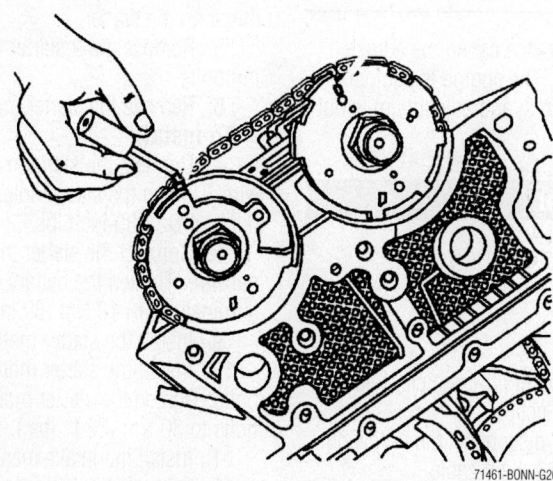

Use a paint stick to create a mark on the timing chain link adjacent to each camshaft sprocket timing mark

h. Rotate the wing nut until it contacts the retention tool block. Alternately tighten both wing nuts to retain the chain.

21. Use an open wrench on the hex cast into the camshafts in order to prevent the camshafts from rotating when removing the camshaft sprocket bolts.

22. Remove the camshaft sprocket bolts.

23. Alternately loosen the camshaft bearing cap bolts a few turns at a time until all valve spring pressure has been released.

24. Remove the camshaft bearing caps.

25. Remove tool J 44212 from the camshafts.

26. Remove the camshafts.

27. Remove the camshaft followers.

28. Clean and inspect the camshafts.

To install:

29. Apply a liberal amount of lubricant GM P/N 12345001 or equivalent to the

roller pivot pocket and valve slot areas of the camshaft followers.

➡**The follower must be positioned squarely on the valve tip so that the full width of the roller will completely contact the camshaft lobe. If the followers are being reused you must put them back in their original location.**

30. Place the camshaft followers in position on the valve tip and the stationary hydraulic lash adjusters (SHLA). The rounded head of the follower goes on the SHLA, while the flat end goes on the valve tip.

31. Clean the camshaft carriers with a clean, lint-free cloth.

32. Apply a liberal amount of lubricant GM P/N 12345001 or equivalent to the camshaft carriers, camshaft lobes and the camshaft journals.

33. Place the camshaft in the camshaft carriers with the camshaft sprocket drive pins near the top of their rotation and the camshaft lobes in a neutral position. The camshafts can be identified by a stamping near the rear journal. For example: R-EXH is defined as Right bank Exhaust.

34. Observe the markings on the camshaft bearing caps. Each camshaft bearing cap is marked in order to identify its location. The markings have the following meanings:

- The arrow should point to the front of the engine.
- The number indicates the position from the front of the engine.
- The "E" indicates the exhaust camshaft.
- The "I" indicates the Intake camshaft.

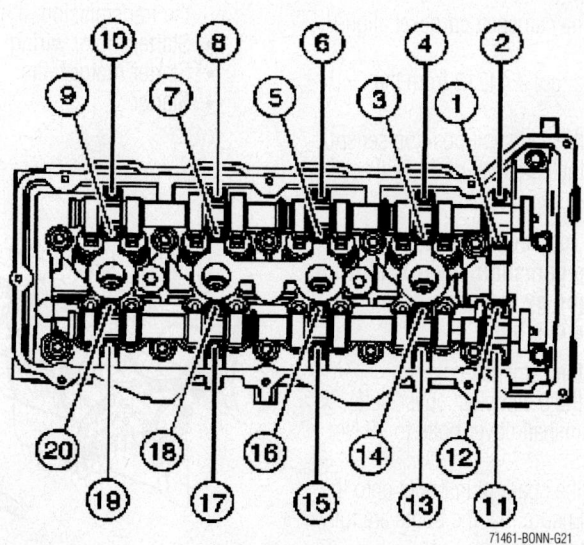

Right side camshaft bearing cap torque sequence—4.6L engine

35. Apply a liberal amount of lubricant GM P/N 12345001 or equivalent to the camshaft bearing caps.

36. Install the camshaft bearing caps according to the identification marks.

37. Install the camshaft bearing cap bolts in sequence. Alternately hand tighten the camshaft bearing cap bolts a few turns at a time until all caps are fully seated.

   a. Tighten the camshaft bearing cap bolts to 5 Nm (44 inch lbs.).

   b. Tighten the camshaft bearing cap bolts an additional 30 degrees using tool J 36660-A.

38. Align the camshafts.

✳ CAUTION

The camshaft holding tools must be installed on the camshafts to prevent camshaft rotation. When performing service to the valve train and/or timing components, valve spring pressure can cause the camshafts to rotate unexpectedly and can cause personal injury.

39. Install tool J 44212 over the camshafts.

➡**Ensure the camshaft sprockets properly engage the camshaft sprocket drive pins and camshafts.**

40. Slide the intake and exhaust camshaft sprockets off the pins of tool J 44213 and onto the pins of the camshafts.

41. Use an open wrench on the hex cast into the camshafts in order to prevent the camshafts from rotating when tightening the camshaft sprocket bolts.

42. Install the camshaft sprocket bolts. Tighten the camshaft sprocket bolts to 120 Nm (89 ft. lbs.).

43. Verify the camshaft sprocket alignment.

44. Remove tool J 44212 from the camshafts.

45. Install the camshaft position sensor.

46. Install the camshaft cover seal as required.

➡**Be careful to prevent the exposed section of the camshaft cover seal from being damaged by the edge of the cylinder head casting.**

47. Install the camshaft cover.

48. Install the 9 camshaft cover bolts. Tighten the camshaft cover bolts to 10 Nm (89 inch lbs.).

49. Install the spark plug boots onto the coil cassette. Ensure that the boots are fully seated against the cassette.

50. Install the ignition coil cassette.

51. Install the secondary AIR control valve bracket.

52. Connect the cable harness clips to the cable harness at the front of the camshaft cover.

53. Connect the secondary AIR vent solenoid electrical connector.

54. Connect the vacuum tubes to the secondary AIR vent solenoid.

55. Connect the oxygen sensor wire.

56. Connect the PCV dirty air tube to the camshaft cover.

57. Position the intake manifold sight shield to the engine.

58. Install the 2 intake manifold sight shield nuts. Tighten the nuts to 3 Nm (27 inch lbs.).

Valve Lash

ADJUSTMENT

The valve clearance cannot be adjusted on these engines. The engine is equipped with hydraulic lifters, and adjustment is not necessary.

Starter Motor

REMOVAL & INSTALLATION

3.8L Engine

1. Before servicing the vehicle, refer to the Precautions Section.

2. Remove or disconnect the following:
   - Negative battery cable
   - Flexplate inspection cover
   - Splash shield, if equipped
   - Electrical connectors
   - Transmission cooler line clip from the transmission, if necessary
   - Starter motor wiring
   - Starter motor bolts
   - Starter

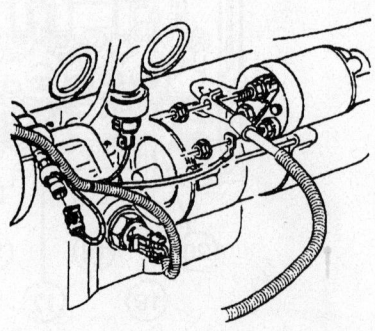

9306UG02

Starter in place with wiring

To install:

3. Install or connect the following:
   - Starter and torque the bolts to 32 ft. lbs. (43 Nm)
   - Wiring and torque the "B" terminal nut to 89 inch lbs. (10 Nm) and the "S" terminal nut to 22 inch lbs. (3 Nm).
   - Flexplate inspection cover and torque the bolts to 62 inch lbs. (7 Nm)
   - Splash shield
   - Negative battery cable

4.6L Engine

1. Disconnect the battery negative cable.

2. Remove the intake manifold.

3. Disconnect the BAT cable from the starter.

4. Disconnect the wire from the S terminal on the starter.

5. Remove the 2 starter motor mounting bolts.

6. Remove the starter motor.

To install:

7. Connect the starter motor S terminal wire. Tighten the starter solenoid S terminal nut to 4 Nm (35 inch lbs.).

8. Connect the starter motor BAT terminal wire. Tighten the battery cable to starter terminal nut to 10 Nm (89 inch lbs.).

9. Install the starter motor.

10. Install the starter motor mounting bolts. Tighten the starter motor mounting bolts to 30 Nm (22 ft. lbs.).

11. Install the intake manifold.

12. Connect the battery negative cable.

Oil Pan

REMOVAL & INSTALLATION

3.8L Engine

✳ WARNING

The oil level sensor, located in the oil pan, must be removed prior to removal of the oil pan.

If the oil pan is removed first, damage to the oil level sensor may occur.

1. Before servicing the vehicle, refer to the Precautions Section.

2. Drain oil into an approved container.

3. Remove or disconnect the following:
   - Negative battery cable

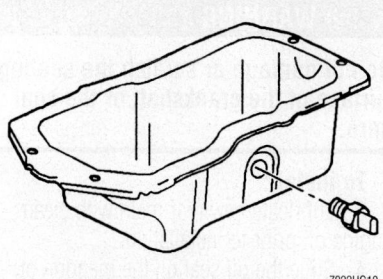

7922UG10

If equipped, be sure to remove the oil level sensor before removing the pan

- Right engine mount bracket, if necessary
- Flexplate cover
- Oil level sensor
- Oil filter
- Torque axis mount bracket bolts, if necessary
- Oil pan bolts
- Oil pan
- Oil pan gasket

4. Clean the oil pan and cylinder block mating surfaces.

To install:

5. Install or connect the following:
- Oil pan with a new gasket and torque the bolts to 125 inch lbs. (14 Nm)
- Torque axis mount bracket bolts, if removed
- Oil filter
- Flexplate cover
- Oil level sensor
- Oil drain plug and torque the plug to 30 ft. lbs. (40 Nm)
- Right engine mount bracket, if necessary
- Negative battery cable

6. Refill the crankcase.
7. Run the engine and check for leaks.

4.6L Engine

1. Drain the engine oil.
2. Remove the front exhaust manifold pipe.
3. Disconnect the electrical connector from the engine oil level sensor.
4. Remove the engine oil level sensor from the oil pan.
5. Remove the oil pan bolts.

➡**The oil pan gasket is reusable unless damaged. Do not remove the gasket from the oil pan groove unless replacement is required.**

6. Remove the oil pan.
7. Clean and inspect the oil pan.

To install:

8. If required, install a new oil pan seal using the following procedure:
   a. Clean any residual oil from the seal groove.
   b. Work the seal into the pan groove in both directions around the pan.

9. Position the oil pan to the crankcase. Install the oil pan retaining bolts. Tighten the oil pan bolts to 10 Nm (89 inch lbs) in the sequence shown.

10. Install the engine oil level sensor into the oil pan. Tighten the engine oil level sensor to 20 Nm (15 ft. lbs.).

11. Connect the electrical connector to the engine oil level sensor.

12. Install the front exhaust manifold pipe.

13. Fill the engine oil.

14. Inspect for oil leaks after engine start up.

Oil Pump

REMOVAL & INSTALLATION

3.8L Engine

1. Before servicing the vehicle, refer to the Precautions Section.

2. Support the engine using an engine support fixture.

3. Remove or disconnect the following:
- Negative battery cable
- Engine drive belts and tensioner assembly
- Drive belt idler pulley and bracket

4. Remove or disconnect the following:
- Torque axis mount bracket, if necessary
- Engine front cover assembly
- Oil filter adapter with pressure regulator valve and spring
- Oil pump cover
- Inner and outer pump gears

To install:

5. Lubricate the oil pump gears with petroleum jelly.

6. Install the gears into the oil pump housing.

7. Pack the gear cavity with petroleum jelly after the gears have been installed in the housing.

8. Install or connect the following:
- Oil pump cover. Torque the screws to 97 inch lbs. (11 Nm).
- Oil filter adapter with new gasket, pressure regulator valve and spring. Torque the bolts to 11 ft. lbs. (15 Nm).
- Front cover assembly
- Tensioner assembly
- Drive belt idler pulley and bracket, if removed
- Drive belts
- Torque axis mount bracket
- Negative battery cable

Oil pan bolt torque sequence—4.6L engine

71461-BONN-G22

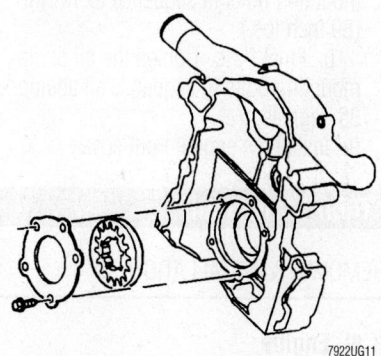

7922UG11

The oil pump is located inside the front engine cover—3.8L (VIN K and 1) engines

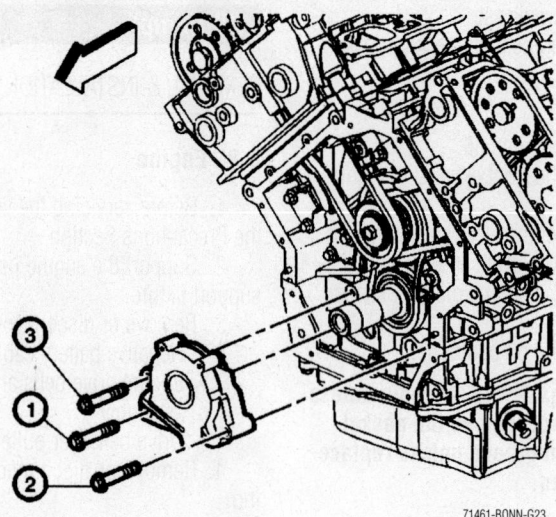

Oil pump installation—4.6L engine

71461-BONN-G23

9. Remove the engine support fixture.
10. Verify the correct engine oil level.
11. Start the vehicle and verify no leaks and proper oil pressure.

4.6L Engine

1. Remove the engine front cover.
2. Remove the 3 oil pump assembly retaining bolts identified by the larger head size.
3. Slide the oil pump assembly off the nose of the crankshaft with the drive collar in place.
4. Clean and inspect the oil pump.

To install

5. Install the oil pump drive spacer into the oil pump so that the drive flat engages the pump rotor.
6. Position the oil pump on the crankshaft.
7. Install the retaining bolts.
8. Apply upward pressure on the pump while tightening the 3 retaining bolts. Tighten the bolts in the sequence shown.

   a. First Pass: Tighten the oil pump mounting bolts in sequence to 10 Nm (89 inch lbs.).
   b. Final Pass: Tighten the oil pump mounting bolts in sequence an additional 35 degrees.
9. Install the engine front cover.

Rear Main Seal

REMOVAL & INSTALLATION

3.8L Engine

1. Before servicing the vehicle, refer to the Precautions Section.

2. Remove or disconnect the following:
- Transaxle assembly
- Flexplate from the crankshaft
- Rear main seal from engine block by inserting a small flat-bladed prytool through the dust lip at an angle, then pry out the crankshaft rear oil seal. Repeat as necessary around the seal until it is removed.

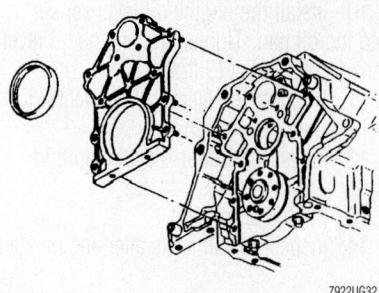

7922UG32

Rear main oil seal and rear cover

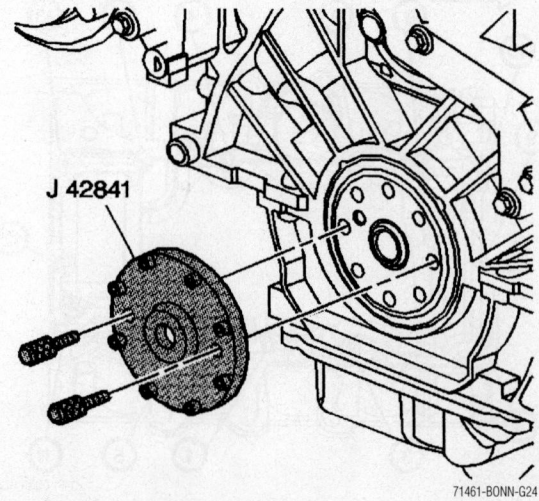

J 42841

71461-BONN-G24

Place the tool J 42841 onto the crankshaft

✳✳ WARNING

Do not damage or scratch the sealing surface of the crankshaft or the seal bore.

To install:

3. Lubricate new rear main with clean engine oil prior to installation.
4. Slide the oil seal on the mandrel of seal installer tool J-38196 until the back of the seal is seated squarely against the collar of the tool.
5. Attach the seal installer to the rear of the crankshaft with the 2 mounting bolts, then turn the T-handle until the oil seal is fully seated into the rear of the engine.
6. Loosen the T-handle of the tool completely.
7. Remove both bolts and the tool.
8. Install or connect the following:
- Flexplate. Torque the bolts to 11 ft. lbs. (15 Nm), plus an additional 50 degrees.
- Transaxle

4.6L Engine

1. Remove the transaxle assembly.

➡**Do not reuse the flywheel bolts.**

2. Remove the 8 mounting bolts.
3. Remove the flywheel and the reinforcement.
4. Place the tool J 42841 onto the crankshaft.
5. Install the tool J 42841 retaining bolts.
6. Using a drill motor, variable speed preferred, with a socket adapter, install eight 25 mm (1.0 in) self-drilling screws into the seal using the guide holes in the removal tool.

➡**When drilling, make sure you reduce the drill speed when the screw begins threading into the seal.**

7. With all 8 removal screws installed, remove the tool J 42841 retaining bolts.

8. Install the center forcing screw.

9. Tighten the center screw on the tool J 42841 to pull the seal assembly off the end of the crankshaft.

To install:

➡**Make sure the drain is clear before installing the new crankshaft rear oil seal. Failure to clear the drain could cause the crankshaft rear oil seal to leak.**

10. Clean any debris from the crankshaft rear oil seal drain using wire or an unbound plastic tie-wrap.

11. Place a small amount of Gasket Maker, GM P/N 1052942 (Canadian P/N 10953466), or equivalent, at the crankcase split line across the end of the upper/lower crankcase seal.

12. Coat the outer diameter of the cylinder block crankshaft rear oil seal area with clean engine oil GM P/N 12345501 (Canadian P/N 992704), or equivalent.

➡**DO NOT allow any engine oil on the area where the crankshaft rear oil seal is to be pressed onto the crankshaft. The green coating pre-applied to the inner diameter of the crankshaft rear oil seal must not be contaminated.**

13. Wipe the outer diameter of the flywheel flange clean with a lint-free cloth.

➡**DO NOT put any engine oil on the green coating pre-applied to the inner diameter of the crankshaft rear oil seal. This coating is a sealant that must not be contaminated.**

14. Lubricate the outer rubber surface of the crankshaft rear oil seal with clean engine oil GM P/N 12345501 (Canadian P/N 992704), or equivalent.

15. Loosen the center bolt of the tool J 45930 until the center hub protrudes approximately 13 mm (0.5 in.) beyond the outer plate. It is not necessary to completely unthread the center bolt and separate the 2 pieces of the tool J 45930.

16. Install the tool J 45930 to the rear of the crankshaft.

17. Thread the 2 mounting bolts into the crankshaft flange.

18. Tighten the bolts until the tool J 45930 is firmly mounted on the crankshaft.

19. Install the crankshaft rear oil seal by

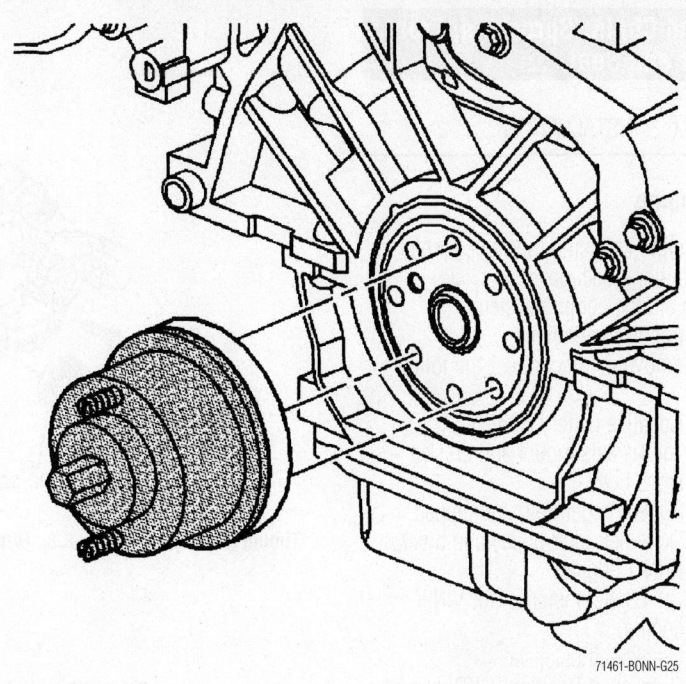

71461-BONN-G25

Tool J 45930

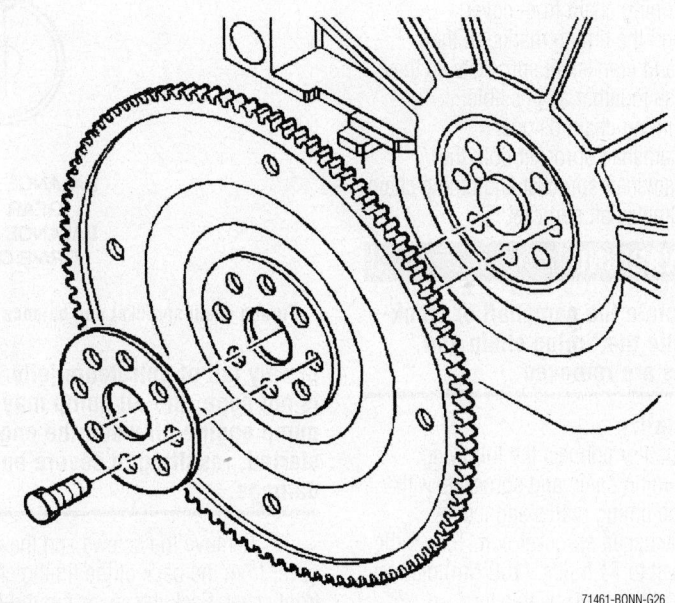

71461-BONN-G26

Position the flywheel and the reinforcement to the crankshaft

tightening the center bolt until the tool J 45930 bottoms against the crankcase.

20. Loosen the center bolt to release pressure on the crankcase.

21. Loosen the 2 mounting bolts.

22. Remove the tool J 45930 from the crankshaft flange.

23. Inspect to ensure the installation depth is equal around the crankshaft rear oil seal's circumference. If the depth is not equal reinstall the tool J 45930 and repeat the installation procedures.

24. Position the flywheel and the reinforcement to the crankshaft.

25. Apply sealant, GM P/N 12346004 (Canadian P/N 10953480) or equivalent, to the flywheel mounting bolts.

26. Install the 8 NEW mounting bolts.

a. First Pass: Tighten the flywheel mounting bolts to 15 Nm (11 ft. lbs.).

b. Final Pass: Tighten the flywheel mounting bolts an additional 50 degrees using the tool J 36660-A .

27. Install the transaxle assembly.

Timing Chain, Sprockets, Front Cover and Seal

REMOVAL & INSTALLATION

3.8L Engine

1. Before servicing the vehicle, refer to the Precautions Section.
2. Drain the cooling system.
3. Support the engine.
4. Remove or disconnect the following:

- Negative battery cable
- Torque axis mount and bracket
- Drive belt
- Supercharger belt, if equipped
- Drive belt idler pulley and bracket (VIN 1 engine)
- Drive belt tensioner (for VIN K engine)
- Crankshaft balancer
- Crankshaft Position (CKP) sensor shield and the CKP sensor
- Oil pan-to-front cover bolts
- Timing chain front cover

5. Align the timing marks on the camshaft and crankshaft sprockets so they are as close together as possible.

- Timing chain damper
- Camshaft sprocket bolt, the camshaft sprocket and timing chain
- Crankshaft sprocket

> ※※ **WARNING**
>
> **Do not rotate the camshaft or crankshaft while the timing chain and sprockets are removed.**

To install:

6. Install or connect the following:
- Timing chain and sprockets with the timing marks aligned
- Camshaft sprocket bolt. Torque the bolt to 74 ft. lbs. (100 Nm) plus an additional 90 degree turn.
- Timing chain damper. Torque the bolts to 16 ft. lbs. (22 Nm).

> ※※ **WARNING**
>
> **The oil pump is built into the front cover. When the cover is removed, oil drains from the pump. Since the pump "loses its prime" it may not establish oil pressure as soon as the engine starts. Therefore, it is important to remove the oil pump cover from the back of the timing chain front cover and pack the space around the oil pump gears com-**

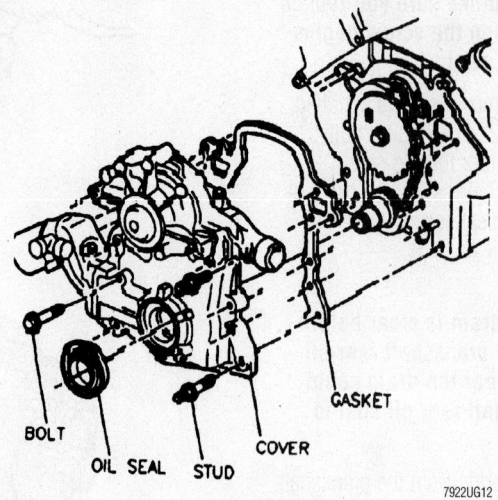

Timing chain front cover—3.8L (VIN K and 1) engines

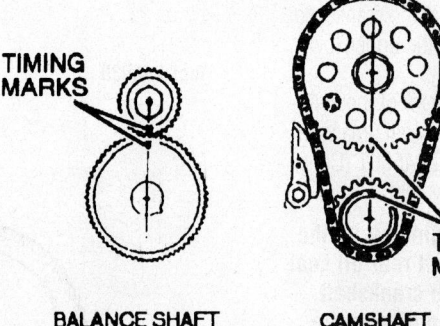

Timing chain sprocket and balance shaft gear alignment—3.8L (VIN K and 1) engines

pletely full of petroleum jelly. If this is not done, the oil pump may not pump engine oil when the engine is started, resulting in severe engine damage.

7. Remove the screws and the oil pump cover from the back of the timing chain front cover. Pack the space around the oil pump gears completely full of petroleum jelly. There must be no air space left inside the pump.

8. Install or connect the following:
- Pump cover with new gaskets. Torque the screws to 97 inch lbs. (11 Nm).
- Timing chain front cover. Torque the front cover-to-engine bolts to 11 ft. lbs. (15 Nm) plus an additional 40 degrees.
- Oil pan-to-front cover bolts. Torque the bolts to 125 inch lbs. (14 Nm).
- CKP sensor. Torque the bolts to 14–28 ft. lbs. (20–40 Nm).

- CKP sensor shield
- Crankshaft balancer. Torque the bolt to 111 ft. lbs. (150 Nm) plus an additional 76 degree turn.
- Drive belt tensioner assembly (VIN K engine)
- Drive belt idler pulley (VIN 1 engine)
- Right inner fender access panel and the right front wheel
- Drive belt(s)
- Engine mount
- Coolant hoses
- Negative battery cable

9. Remove the engine support fixture.
10. Refill and bleed the cooling system.
11. Start the vehicle and check for leaks and proper engine operation.

4.6L Engine

1. Remove the front cover perimeter bolts.
2. Remove the front cover and the gas-

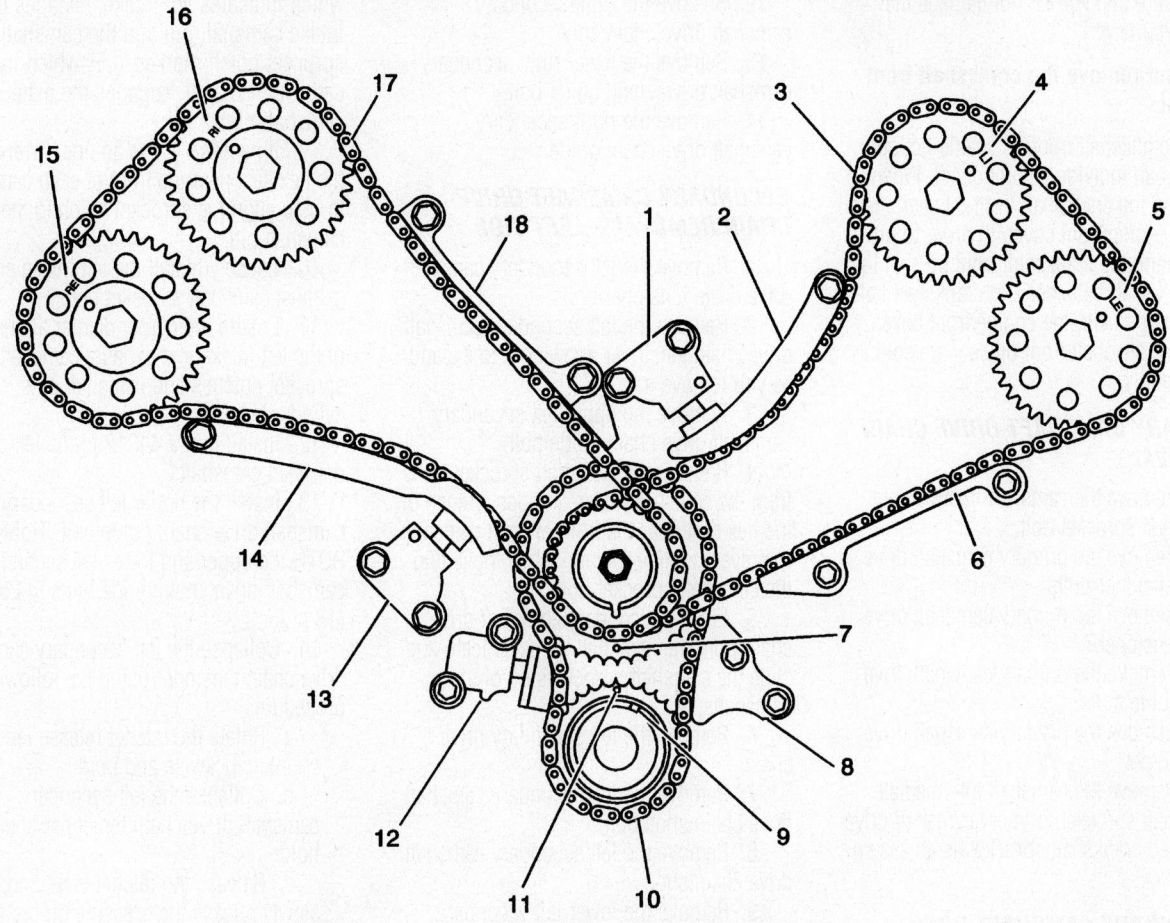

(1) Left Secondary Timing Chain Tensioner
(2) Left Secondary Timing Chain Shoe
(3) Left Secondary Timing Chain
(4) Left Intake Camshaft Sprocket Timing Mark
(5) Left Exhaust Camshaft Sprocket Timing Mark
(6) Left Secondary Timing Chain Guide
(7) Intermediate Sprocket Timing Mark
(8) Primary Timing Chain Guide
(9) Crankshaft Sprocket Pin Alignment Slot
(10) Primary Timing Chain
(11) Crankshaft Sprocket Timing Mark
(12) Primary Timing Chain Tensioner
(13) Right Secondary Timing Chain Tensioner
(14) Right Secondary Timing Chain Shoe
(15) Right Exhaust Camshaft Sprocket Timing Mark
(16) Right Intake Camshaft Sprocket Timing Mark
(17) Right Secondary Timing Chain
(18) Right Secondary Timing Chain Guide

71461-BONN-G27

Correct alignment of the primary and secondary timing chains—4.6L engine

ket. Ensure that you do not damage the sealing surface.

➡**Do not remove the crankshaft front oil seal.**

The crankshaft front oil seal is not serviced as an individual component. When replacing the crankshaft front oil seal, install a NEW engine front cover. In order to precisely align the crankshaft front oil seal to the crankshaft balancer and crankshaft balancer dust shield, the engine front cover and the crankshaft front oil seal are sold as an assembly.

PRIMARY CAMSHAFT DRIVE CHAIN REMOVAL

1. Remove the camshaft intermediate drive shaft sprocket bolt.
2. Remove the primary camshaft drive chain tensioner bolts.
3. Remove the primary camshaft drive chain tensioner.
4. Remove the primary camshaft drive chain guide bolts.
5. Remove the primary camshaft drive chain guide.
6. Remove the camshaft intermediate drive shaft sprocket, primary camshaft drive chain and crankshaft sprocket as an assembly.

SECONDARY CAMSHAFT DRIVE CHAIN REMOVAL—RIGHT SIDE

1. Remove the camshaft position sensor bolt.
2. Remove the camshaft position sensor.
3. Remove and discard the camshaft position sensor O-ring.
4. Remove the chain guide access plugs located in the cylinder heads. Ensure the O-ring seal is on each access plug.
5. Remove the right secondary drive chain tensioner bolts.
6. Remove the right secondary camshaft drive chain tensioner allowing it to expand as you remove it.
7. Remove the upper right secondary camshaft drive chain guide bolt.
8. Remove the camshaft sprocket bolts from the camshafts. Use an open wrench on the hex cast near the front of each camshaft to prevent engine rotation when loosening the camshaft sprocket bolts.
9. Lift the secondary camshaft drive chain from the camshaft sprocket teeth and slide the camshaft sprockets off of the camshafts.
10. Remove the right secondary camshaft drive chain.
11. Remove the right secondary camshaft drive chain shoe bolt.

12. Remove the right secondary camshaft drive chain shoe.
13. Remove the lower right secondary camshaft drive chain guide bolt.
14. Remove the right secondary camshaft drive chain guide.

SECONDARY CAMSHAFT DRIVE CHAIN REMOVAL—LEFT SIDE

1. Remove the left secondary camshaft drive chain tensioner bolts.
2. Remove the left secondary camshaft drive chain tensioner allowing it to expand as you remove it.
3. Remove the upper left secondary camshaft drive chain guide bolt.
4. Remove the camshaft sprocket bolts from the camshafts. Use an open wrench on the hex cast near the front of each camshaft to prevent engine rotation when loosening the camshaft sprocket bolts.
5. Lift the secondary camshaft drive chain from the camshaft sprocket teeth and slide the camshaft sprockets off of the camshafts.
6. Remove the left secondary drive chain.
7. Remove the left secondary camshaft drive chain shoe bolt.
8. Remove the left secondary camshaft drive chain shoe.
9. Remove the lower left secondary camshaft drive chain guide bolt.
10. Remove the left secondary camshaft drive chain guide.

To install:

SECONDARY CAMSHAFT DRIVE CHAIN INSTALLATION—LEFT SIDE

1. Install the left secondary camshaft drive chain guide.
2. Loosely install the lower left secondary camshaft drive chain guide bolt.
3. Install the left secondary camshaft drive chain shoe.
4. Install the left secondary camshaft drive chain shoe bolt. Tighten the left secondary camshaft drive chain shoe bolt to 25 Nm (18 ft. lbs.).
5. Install the left secondary camshaft drive chain by sliding the chain down through the left cylinder head and placing the chain on the end of the camshafts.
6. Route the left secondary camshaft drive chain around the inner row of the intermediate drive chain sprocket teeth.
7. Install the left intake and exhaust camshaft sprockets into the left secondary camshaft drive chain.
8. Install the left intake and exhaust camshaft sprockets onto the camshafts. The camshaft sprocket notch marked "LI"

which indicates left intake, engages the intake camshaft pin and the camshaft sprocket notch marked "LE" which indicates left exhaust, engages the exhaust camshaft pin.

9. If necessary, use an open wrench on the hex cast near the front of each camshaft to help align the sprocket notch to the camshaft pin.
10. Loosely install the left intake and exhaust camshaft sprocket bolts.
11. Ensure the perpendicular alignment of the left intake and exhaust camshaft sprocket notches and camshaft pins to the cylinder head.
12. Install tool J 44212 to the left cylinder head camshafts.
13. Install the upper left secondary camshaft drive chain guide bolt. Tighten BOTH the upper and lower left secondary camshaft drive chain guide bolts to 25 Nm (18 ft. lbs.).
14. Collapse the left secondary camshaft drive chain tensioner using the following procedure:

a. Rotate the ratchet release lever counterclockwise and hold.
b. Collapse the left secondary camshaft drive chain tensioner shoe and hold.
c. Release the ratchet lever and slowly release the pressure on the shoe.
d. When the ratchet lever moves to the first detent a click should be heard and felt.
e. Insert a pin through the hole in the release lever in order to lock the left secondary camshaft drive chain tensioner shoe in the collapsed position.

➡**Ensure the left secondary camshaft drive chain tensioner release lever is facing out.**

15. Install the left secondary camshaft drive chain tensioner.
16. Install the left secondary camshaft drive chain tensioner bolts. Tighten the left secondary camshaft drive chain tensioner bolts to 25 Nm (18 ft. lbs.).
17. Remove pin from left secondary camshaft drive chain tensioner lever.

SECONDARY CAMSHAFT DRIVE CHAIN INSTALLATION—RIGHT SIDE

1. Install the right secondary drive chain guide.
2. Loosely install the lower right secondary camshaft drive chain guide bolt.
3. Install the right secondary camshaft drive chain shoe.
4. Install the right secondary camshaft drive chain shoe bolt. Tighten the right sec-

ondary camshaft drive chain shoe bolt to 25 Nm (18 ft. lbs.).

5. Install the right secondary camshaft drive chain by sliding the chain down through the right cylinder head and placing the chain on the end of the camshafts.

6. Route the right secondary camshaft drive chain around the outer row of the intermediate drive chain sprocket teeth.

7. Install the right intake and exhaust camshaft sprockets into the right secondary camshaft drive chain.

8. Install the right intake and exhaust camshafts onto the camshafts. The camshaft sprocket notch marked "RI" which indicates right intake, engages the intake camshaft pin and the camshaft sprocket notch marked "RE" which indicates right exhaust, engages the exhaust camshaft pin.

9. If necessary, use an open wrench on the hex cast near the front of each camshaft to help align the sprocket notch to the camshaft pin.

10. Loosely install the right intake and exhaust camshaft sprocket bolts.

11. Ensure the perpendicular alignment of the right intake and exhaust camshaft sprocket notches and camshaft pins to the cylinder head.

12. Install tool J 44212 to the right cylinder head camshafts.

13. Install the upper right secondary camshaft drive chain guide bolt. Tighten BOTH the upper and lower right secondary camshaft drive chain guide bolts to 25 Nm (18 ft. lbs.).

14. Collapse the right secondary camshaft drive chain tensioner using the following procedure:

   a. Rotate the ratchet release lever counter-clockwise and hold.

   b. Collapse the right secondary camshaft drive chain tensioner shoe and hold.

   c. Release the ratchet lever and slowly release the pressure on the shoe.

   d. When the ratchet lever moves to the first detent a click should be heard and felt.

   e. Insert a pin through the hole in the release lever in order to lock the right secondary camshaft drive chain tensioner shoe in the collapsed position.

➡**Ensure the right secondary camshaft drive chain tensioner release lever is facing out.**

15. Install the right secondary camshaft drive chain tensioner.

16. Install the right secondary camshaft drive chain tensioner bolts. Tighten the right

secondary camshaft drive chain tensioner bolts to 25 Nm (18 ft. lbs.).

17. Remove pin from right secondary camshaft drive chain tensioner lever.

18. Ensure the correct alignment of all secondary timing components.

19. Ensure the correct alignment of all primary timing components.

20. Tighten ALL camshaft sprocket bolts. Use the hex cast into each camshaft to prevent engine rotation and provide leverage. Tighten ALL camshaft sprocket bolts to 120 Nm (90 ft. lbs.).

21. Install the chain guide access plugs located in the cylinder heads. Ensure the O-ring seal is on each access plug. Tighten the chain guide access plugs to 4.5 Nm (39 inch lbs.).

22. Install a NEW O-ring on the camshaft position sensor.

23. Lubricate the O-ring with clean engine oil.

24. Install the camshaft position sensor.

25. Install the camshaft position sensor bolt. Tighten the camshaft position sensor bolt to 10 Nm (89 inch lbs.).

PRIMARY CAMSHAFT DRIVE CHAIN INSTALLATION

1. Install the primary camshaft drive chain on the camshaft intermediate drive shaft sprocket and crankshaft sprocket.

2. Align the timing marks of the camshaft intermediate drive shaft sprocket and crankshaft sprocket. Ensure the marks are aligned vertically.

3. Ensure the number one piston is at Top Dead Center (TDC) and the crankshaft

pin is approximately at the one o'clock position using tool J 39946.

4. Install the primary camshaft drive chain, camshaft intermediate drive shaft sprocket and crankshaft sprocket as an assembly onto the camshaft intermediate drive shaft and the crankshaft.

5. Install the camshaft intermediate drive shaft sprocket bolt. Tighten the camshaft intermediate drive shaft sprocket bolt to 60 Nm (44 ft. lbs.).

6. Install the primary camshaft drive chain guide.

7. Install the primary camshaft drive chain guide bolts. Tighten the primary camshaft drive chain guide bolts to 25 Nm (18 ft. lbs.).

8. Collapse the primary camshaft drive chain tensioner using the following procedure:

   a. Rotate the ratchet release lever counterclockwise and hold.

   b. Collapse the primary camshaft drive chain tensioner shoe and hold.

   c. Release the ratchet lever and slowly release the pressure on the shoe.

   d. When the ratchet lever moves to the first detent a click should be heard and felt.

   e. Insert a pin through the hole in the release lever in order to lock the primary camshaft drive chain tensioner shoe in the collapsed position.

➡**Ensure the primary camshaft drive chain tensioner release lever is facing out.**

9. Install the primary camshaft drive chain tensioner.

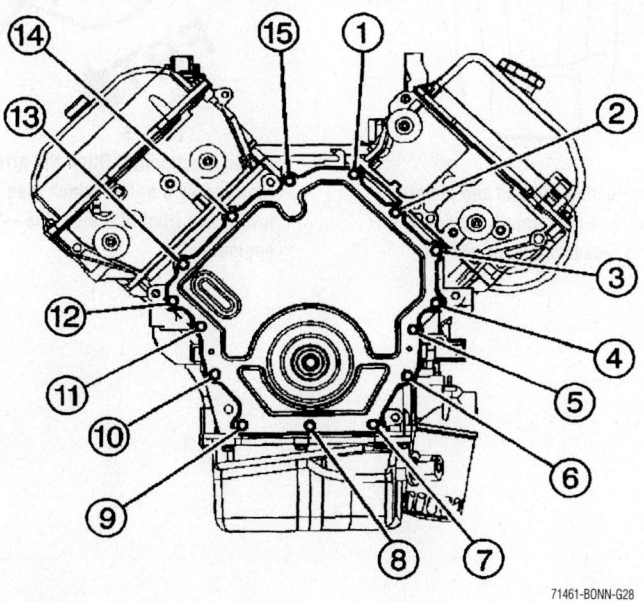

Front cover torque sequence—4.6L engine

71461-BONN-G28

10. Install the primary camshaft drive chain tensioner bolts. Tighten the primary camshaft drive chain tensioner bolts to 25 Nm (18 ft. lbs.).

11. Remove the pin in the release lever locking the primary camshaft drive chain tensioner.

12. Ensure the timing marks are aligned vertically.

13. Place a small amount of sealant GM P/N 12345739, (Canadian P/N 10953541), or equivalent at the split line of the upper and lower crankcases.

14. Place the front cover gasket over the crankcase dowel pins.

15. Place the front cover in position on the crankcase.

16. Install the front cover retaining bolts.

17. Tighten the front cover retaining bolts in the sequence shown. Tighten the front cover retaining bolts in proper sequence to 10 Nm (89 inch lbs.).

Piston and Ring

POSITIONING

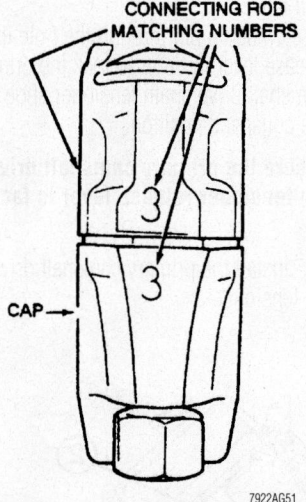

Engine connecting rod and cap installation. Be sure to matchmark the cap and rod prior to disassembly, as shown.

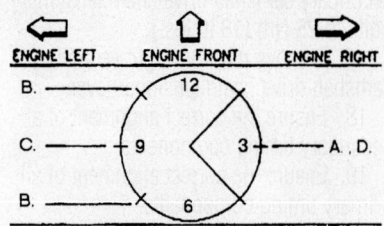

A. OIL RING SPACER GAP (TANG IN HOLE OR SLOT WITH ARC)
B. OIL RING RAIL GAPS
C. 2ND COMPRESSION RING GAP
D. TOP COMPRESSION RING GAP

Piston ring end–gap spacing—3.8L engines

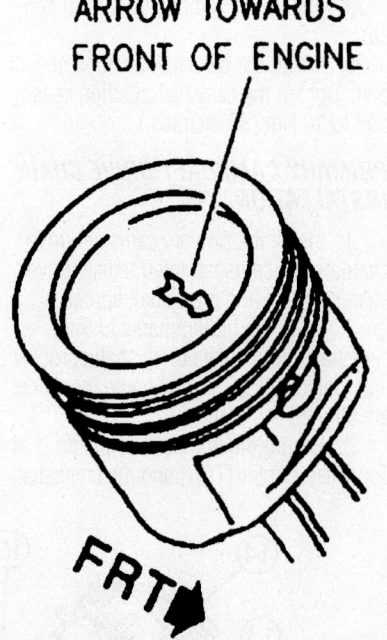

Piston positioning. Often the arrow is replaced by a notch, which also must face toward the front of the engine—3.8L engines

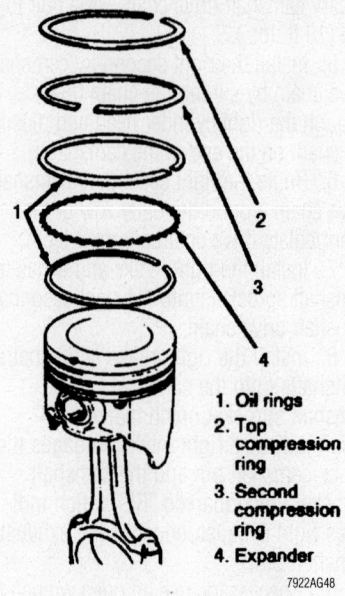

1. Oil rings
2. Top compression ring
3. Second compression ring
4. Expander

Piston ring positioning—3.8L engines

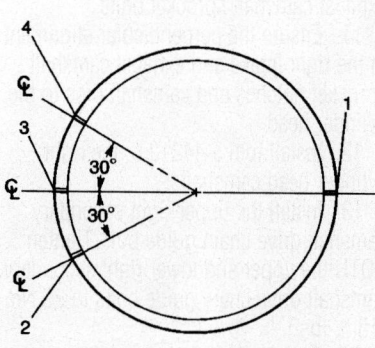

Correct piston ring installation—4.6L engine
1. Oil control ring and expander, and 2nd compression ring gaps
2. Upper oil control ring gap
3. Top compression ring gap
4. Lower oil control ring gap

FUEL SYSTEM

Fuel System Service Precautions

Safety is the most important factor when performing not only fuel system maintenance but any type of maintenance. Failure to conduct maintenance and repairs in a safe manner may result in serious personal injury or death. Maintenance and testing of the vehicle's fuel system components can be accomplished safely and effectively by adhering to the following rules and guidelines.

• To avoid the possibility of fire and personal injury, always disconnect the negative battery cable unless the repair or test procedure requires that battery voltage be applied.

• Always relieve the fuel system pressure prior to disconnecting any fuel system component (injector, fuel rail, pressure regulator, etc.), fitting or fuel line connection. Exercise extreme caution whenever relieving fuel system pressure, to avoid exposing skin, face and eyes to fuel spray. Please be advised that fuel under pressure may penetrate the skin or any part of the body that it contacts.

• Always place a shop towel or cloth around the fitting or connection prior to loosening to absorb any excess fuel due to spillage. Ensure that all fuel spillage (should it occur) is quickly removed from engine surfaces. Ensure that all fuel soaked cloths or towels are deposited into a suitable waste container.

• Always keep a dry chemical (Class B) fire extinguisher near the work area.

• Do not allow fuel spray or fuel vapors to come into contact with a spark or open flame.

• Always use a back-up wrench when loosening and tightening the fuel line connection fittings. This will prevent unnecessary stress and torsion to fuel line piping.

• Always replace worn fuel fitting O-rings with new. Do not substitute fuel hose or equivalent, where fuel pipe is installed.

Fuel System Pressure

RELIEVING

1. Disconnect the negative battery cable to avoid possible fuel discharge if an accidental attempt is made to start the engine.
2. Remove the fuel tank cap to relieve tank pressure. Do not tighten until the service procedure has been completed.

3. Connect a fuel pressure gauge with bleed valve to the fuel pressure test port. Wrap a shop towel around the fitting while connecting the gauge to catch any spilled fuel.
4. Install the bleed hose into an approved container and open the valve to bleed off the fuel system pressure.
5. Drain any fuel remaining in the gauge into an approved container.

✳✳ CAUTION

There may still be residual fuel in the system, and a small amount of fuel may be released when servicing fuel lines or connections. In order to reduce the chance of personal injury, cover the fuel line fittings with a shop towel before disconnecting to catch any fuel that may leak out.

Fuel Filter

REMOVAL & INSTALLATION

1. Before servicing the vehicle, refer to the Precautions Section.
2. Disconnect the negative battery cable.
3. Relieve the fuel system pressure.
4. Twist the quick connector ¼ turn in each direction to loosen any dirt that may have accumulated in the connector.
5. Use compressed air to remove any dirt in the connector.
6. Squeeze the plastic tabs of the male connector and pull apart.
7. Remove threaded connection from the filter inlet.
8. Remove the filter.

To install:
9. Position the fuel filter, making sure it is facing in the proper direction.
10. Attach the outlet quick connect line to the fuel filter as follows:
  a. Step 1: Be sure the connector is clean and that a new plastic retainer is used on the filter.
  b. Step 2: Apply a couple of drops of engine oil to the male pipe end of filter.
  c. Step 3: Push the fuel line onto the fuel filter until the plastic retainer snaps into place.
  d. Step 4: Check that the connector is locked into place by trying to pull the connector from the filter.
11. Install the threaded connection to the

inlet side of filter and tighten to 22 ft. lbs. (30 Nm).
12. Connect the negative battery cable.
13. Pressurize the fuel system by turning the ignition switch to the **ON** position for 2 seconds. Turn OFF the ignition switch for 10 seconds. Turn ON the ignition switch. Check for fuel leaks.

Fuel Pump

REMOVAL & INSTALLATION

1. Before servicing the vehicle, refer to the Precautions Section.
2. Relieve the fuel system pressure.
3. Drain the fuel tank.
4. Remove or disconnect the following:
  • Negative battery cable
  • Spare tire and jack
  • Trunk lining
  • Fuel sender access panel
  • Sender and quick connect fittings from the sender
  • Electrical connector from the sender and position harness and hoses aside

✳✳ CAUTION

When removing the fuel sender from the tank, the reservoir bucket is full of fuel. Use caution in containing the fuel.

  • Sender retaining ring
  • Sender and take note of its position
  • Fuel sender O-ring and discard it

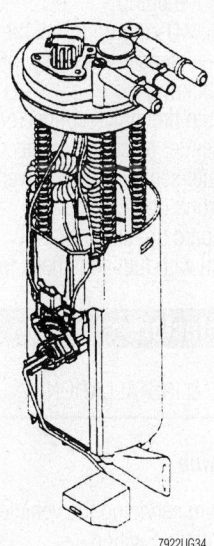

7922UG34

Fuel pump and level sender assembly

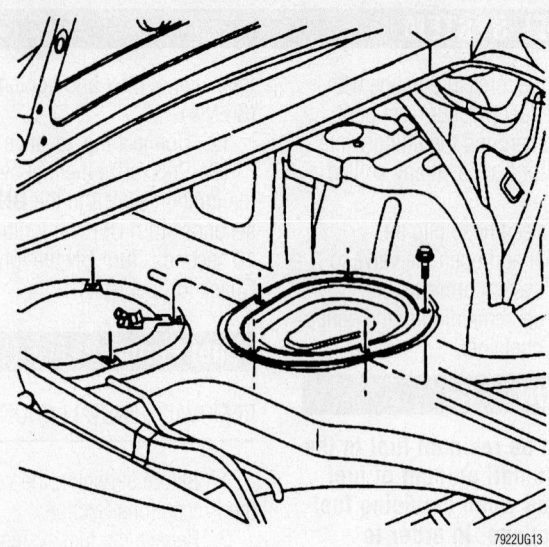

7922UG13

The fuel pump service cover is located in the luggage compartment under the spare tire—Park Avenue shown others similar

→**Note the direction the strainer is pointing.**

- Strainer from the pump by pulling it down and twisting
- Pump electrical wires and hoses
- Pump assembly out of the rubber connectors

To install:

5. Transfer any insulators and grommets from the old pump to the new one.

6. Connect the pump to the fuel hose and tilt the bottom of the pump into the mounting bracket.

7. Install or connect the following:
- New strainer on the pump so it points in the same direction as noted during removal
- Electrical connectors and fuel lines to the pump
- New O-ring on top of the fuel tank
- Fuel sender assembly into the tank
- Lockring
- Fuel line quick connectors
- Sender electrical connector
- Fuel sender access cover
- Trunk liner
- Spare tire and jack

8. Refill with fuel and check for leaks.

Fuel Injectors

REMOVAL & INSTALLATION

3.8L Engine

1. Before servicing the vehicle, refer to the Precautions Section.
2. Relieve the fuel system pressure.

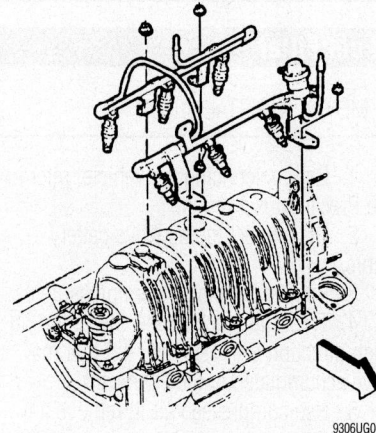

9306UG04

Exploded view of the fuel rail assembly—Vin 1 models

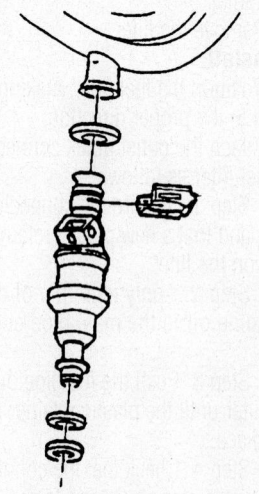

9306UG05

Exploded view of the fuel injector assembly

3. Remove or disconnect the following:
- Negative battery cable
- Fuel lines
- Fuel rail
- Injector retaining clips
- Fuel injector

To install:

4. Install or connect the following:
- Fuel injector with new O-rings, coat the o-rings with clean engine oil prior to installation
- Injector retaining clips
- Fuel rail and torque the bolts to 89 inch lbs. (10 Nm). On VIN 1 models, tighten the fuel rail hold-down stud to 18 ft. lbs. (25 Nm), if equipped.
- Fuel lines
- Negative battery cable

4.6L Engine

→**If the fuel injectors are leaking, the engine oil may be contaminated with fuel.**

1. Remove the fuel rail.
2. Remove the fuel injector by pushing the locking tab on the fuel rail toward the center of the fuel rail.
3. Remove the fuel injector upper O-ring.
4. Remove the fuel injector lower O-ring.
5. Discard the injector O-ring seals.

To install:

※ CAUTION

In order to reduce the risk of fire and personal injury that may result from a fuel leak, always install the fuel injector O-rings in the proper position. If the upper and lower O-rings are different colors (black and blue), be sure to install the blue O-ring in the upper position and the black O-ring in the lower position on the fuel injector. The O-rings are the same size but are made of different materials.

→**The fuel injector O-rings should always be replaced whenever the fuel injectors are serviced. Be sure to install the correct O-ring for the fuel injector. If the O-ring is not seated properly, a vacuum leak is possible and driveability complaints may occur.**

6. Lubricate the new upper and lower O-rings with clean engine oil.
7. Install the new upper and lower injector O-rings on the fuel injector.
8. Install the fuel injector on to the fuel rail.
9. Install the fuel rail.

DRIVE TRAIN

Transaxle Assembly

REMOVAL & INSTALLATION

1. Before servicing the vehicle, refer to the Precautions Section.

➡ **Make sure the wheels of the vehicle are in the straight ahead position and the steering column in the LOCK position before disconnecting the steering column or intermediate shaft from the steering gear. Failure to do so will cause the SIR coil assembly to become uncentered, which may cause damage to the coil assembly.**

2. Lock the steering column by installing tool J-42640 into the underside of the steering column.
3. Remove or disconnect the following:

- Negative battery cable
- Air cleaner
- Range selector cable from the range selector lever
- Range selector cable with the bracket from the transmission case
- Transaxle electrical connector
- Wiring harness from the wiring harness retainer
- Ground cable bolt from the transaxle

4. Install an engine support fixture.

- Upper engine-to-transaxle case bolts
- Front tire and wheel assembly
- Front fascia extensions from each side
- Front air deflector

✳✳ CAUTION

Failure to disconnect the intermediate shaft from the rack and pinion steering gear stub shaft can result in damage to the steering gear or to the intermediate shaft. This damage may cause loss of steering control, which could result in an accident and possible personal injury.

- Intermediate shaft lower pinch bolt
- Intermediate shaft from the power steering gear
- Power steering gear heat shield
- Power steering gear mounting bolts
- Power steering line retainers from the frame and attach the gear to the exhaust manifold

5. Loosen the two mounting nuts to allow removal of the brake pressure modulator valve from the bracket.

- Brake line retainers from the frame
- Left transaxle mount
- Frame
- Front transaxle mount bracket with mount attached
- Right and left drive axles from the transaxle
- Transmission oil cooler hoses from the transaxle
- Transaxle fluid filler tube
- Torque converter cover
- Flywheel-to-torque converter bolts

6. Support transaxle using an appropriate transaxle jack.

- Vehicle Speed Sensor (VSS) electrical connector
- Torque strut bracket-to-transaxle bolts
- Torque strut bracket-to-engine bolts
- Torque strut bracket
- Engine-to-transaxle case bolt
- Remaining transaxle-to-engine bolt
- Transaxle

To install:

7. Installation is the reverse of removal, please note the following specifications:

- Left transaxle bracket bolts to 81 ft. lbs. (110 Nm)
- Rear transaxle bracket bolts to 46 ft. lbs. (63 Nm)
- Lower transaxle bolts to 55 ft. lbs. (75 Nm)
- Torque strut bracket-to-engine bolts to 48 ft. lbs. (65 Nm)
- Torque strut bracket-to-transaxle bolts to 26 ft. lbs. (36 Nm)
- Flywheel-to-torque converter bolts to 46 ft. lbs. (63 Nm)
- Brake pressure modulator valve nuts 89 inch lbs. (10 Nm)
- Power steering gear mounting bolts to 95 ft. lbs. (70 Nm)
- Intermediate shaft lower pinch bolt to 33 ft. lbs. (45 Nm)
- Upper transaxle case-to-engine bolts to 55 ft. lbs. (75 Nm)
- Range selector cable bracket nuts to 18 ft. lbs. (25 Nm)

8. Check and adjust transaxle fluid level.
9. Road test vehicle and check for transaxle leaks.

Halfshaft

REMOVAL & INSTALLATION

✳✳ WARNING

Use care when removing the halfshaft to prevent the inner CV-joint from becoming over-extended. Over-extension of the joint could result in separation of internal components and possible joint failure.

1. Before servicing the vehicle, refer to the Precautions Section.
2. Install a boot protector on the outer CV-joint boot.
3. Remove or disconnect the following:

- Front wheel
- Speed sensor connector

4. Loosen the stabilizer shaft link assembly bolt (to accommodate ball joint separation).
5. Remove the ball joint cotter pin and nut. Loosen the joint.

➡ **The grease fitting may have to be removed from the ball joint for tool access.**

6. Separate the lower control arm from the joint.
7. Remove or disconnect the following:

- Hub nut
- Halfshaft from the hub using tool J 28733

8. Move the strut and knuckle rearward.
9. Remove the halfshaft from the transaxle using tool J 42129 Axle Shaft Remover and tool J 2619-01 Slide Hammer.

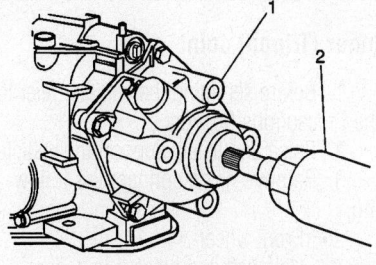

9346UG07

Remove the halfshaft from the transaxle using tool J 42129 Axle Shaft Remover and tool J 2619-01 Slide Hammer

❋❋ WARNING

If equipped with anti-lock brakes, care must be used to prevent damage to the toothed sensor ring on the halfshaft and the wheel speed sensor on the steering knuckle.

To install:

➡**If installing the right halfshaft, install a seal protector, so that it can be pulled out after the halfshaft is installed.**

10. Install the halfshaft into the transaxle by placing a drift pin or punch into the groove on the joint housing and tapping lightly until seated. Verify that the halfshaft is seated by grasping the inner joint housing and pulling. DO NOT pull on the halfshaft.

11. Install or connect the following:
- Halfshaft into the hub/bearing assembly with new hub nut and tighten to 118 ft. lbs. (160 Nm)
- Ball joint into the steering knuckle. Torque the nut to 88 inch lbs. (10 Nm), plus an additional 120 degree turn during which a torque of 41 ft. lbs. (55 Nm) must be obtained.

➡**Tighten the nut up to one more flat in order to align the slot with the hole in the stud.**

12. Install or connect the following:
- Stabilizer shaft link assembly. Torque the nut to 14 ft. lbs. (17 Nm).
- Speed sensor connector
- Front wheel

13. If a seal protector was installed, remove it by pulling in line with the handle.

14. Road test for proper operation.

CV-Joints

OVERHAUL

Inner (Tripod) Joint

1. Before servicing the vehicle, refer to the Precautions Section.
2. Raise and safely support the vehicle.
3. Remove or disconnect the following:

- Front wheel
- Halfshaft and place it in a vise
- Small CV-joint boot clamp, cut and discard it
- Large CV-joint boot clamp, cut and discard it

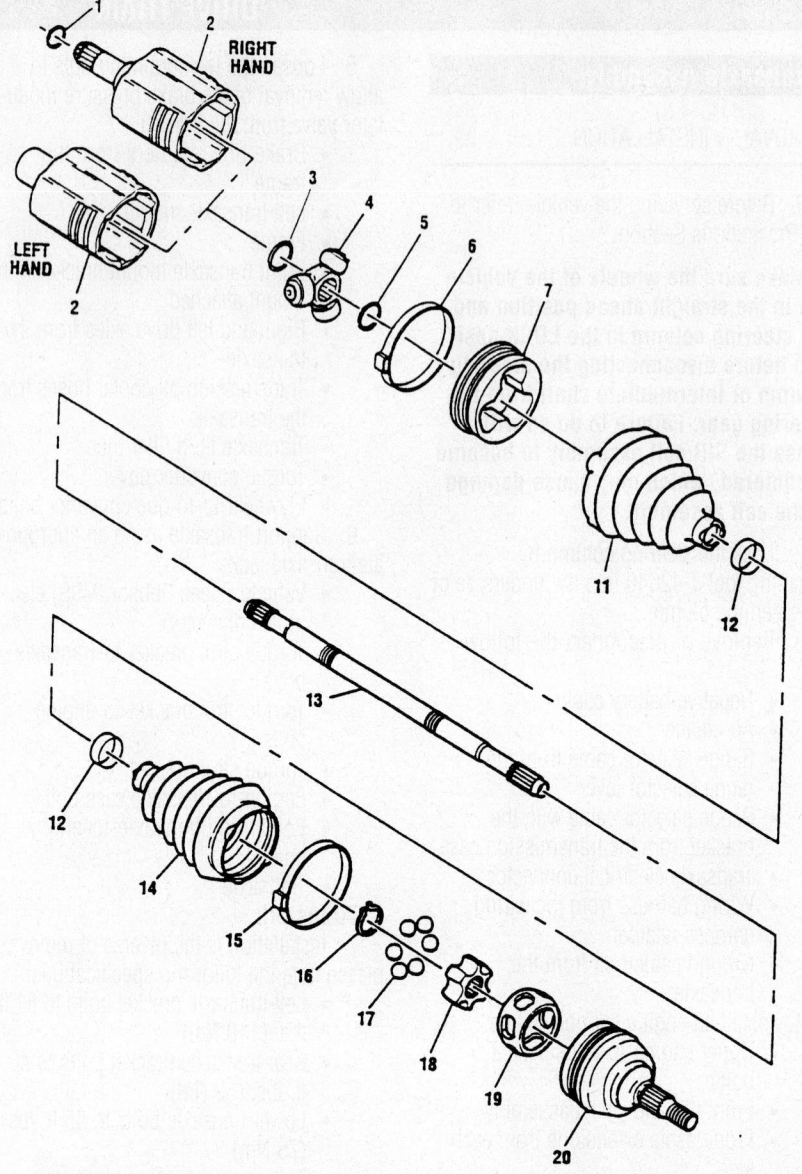

1 - RING, RETAINING
2 - HOUSING ASM, RETAINER &
3 - RING, SHAFT RETAINING
4 - SPIDER, TRIPOT JOINT
5 - RING, SPACER
6 - CLAMP, SEAL RETAINING
7 - BUSHING, TRILOBAL TRIPOT
11 - SEAL, DRIVE AXLE INBOARD
12 - RING, SWAGE

13 - SHAFT, AXLE (RH SHOWN, LH SIMILAR)
14 - SEAL, DRIVE AXLE OUTBOARD
15 - CLAMP, SEAL RETAINING
16 - RING, RACE RETAINING
17 - BALL, CHROME ALLOY
18 - RACE, C/V JOINT INNER
19 - CAGE, C/V JOINT
20 - RACE, C/V JOINT OUTER

9306UG06

Exploded view of the halfshaft assembly

- CV-joint boot by sliding it away from the tripod joint
- Tripod housing from the tripod spider
- Inboard spacer ring and slide it rearward on the shaft
- Outboard retaining ring
- Tripod joint spider assembly
- Inboard spacer ring and discard it

- Tripod joint spider assembly by tapping it from the halfshaft with a brass drift
- Tripod spider retaining ring and discard it
- Trilobal tripod bushing from the housing
- CV-joint boot

4. Thoroughly clean and inspect all parts.

To install:

5. Install or connect the following:
- Small boot clamp
- CV-joint boot
- New inboard spacer ring. Slide it rearward on the shaft past the 2nd groove
- Tripod joint spider assembly onto the shaft until it passes the 2nd groove

6. Assemble the tripod spider assembly onto the halfshaft as follows:

a. Position the tripod spider assembly onto the shop press plate.

b. Position the halfshaft onto the tripod spider assembly, in the shop press.

c. Press the halfshaft into the tripod spider assembly until the spider assembly passes the 2nd groove.

※※ WARNING

When assembling the tripod assembly onto the halfshaft, do not exceed 4,000 lbs. pressure.

7. Remove the halfshaft from the shop press and place it in vise.

8. Install or connect the following:
- New outboard retaining ring into the axle shaft groove
- Tripod joint spider assembly, slide it against the outboard retaining ring using a brass drift
- Inboard spacer ring, seat it in the groove

9. Use ½ of the grease supplied in the kit into the boot and the other ½ into the tripod housing.
- Trilobal tripod bushing flush with the tripod housing face
- New large seal clamp onto the CV-joint boot
- Tripod housing, slide it over the tripod joint spider assembly
- CV-joint boot/clamp, slide it into place, over the trilobal tripod bushing with the seal lip in the groove

➡ **Make sure the boot lies flat against the trilobal bushing.**

10. Using the crimp tool, a torque wrench and a breaker bar, crimp the small CV-joint boot clamp to 100 ft. lbs. (136 Nm).

11. Using the crimp tool, latch the large CV-joint boot clamp.

12. Install the halfshaft and the front wheel.

Outer CV-Joint

1. Before servicing the vehicle, refer to the Precautions Section.

2. Remove or disconnect the following:
- Front wheel
- Halfshaft
- Swage ring using a hand grinder
- Large boot clamp
- CV-joint boot, slide it away from the CV-joint
- CV-joint assembly by spreading the inner race-to-axle shaft retaining ring ears using Snapring Pliers
- CV-joint boot from the axle shaft

3. Disassemble the chrome alloy balls from the CV-joint cage as follows:

a. Position a brass drift against the CV-joint cage and tap it with a hammer to tilt the cage.

b. Chrome alloy ball from the cage.

c. Tilt the cage in the opposite direction.

d. Remove the opposite chrome alloy ball.

e. Repeat the procedure until all 6 balls are removed.

4. Disassemble the CV-joint cage and inner race as follows:

a. Pivot the cage and race 90 degrees to the center line of the outer race.

b. Align the cage windows with outer race lands.

c. Remove the cage from the outer race.

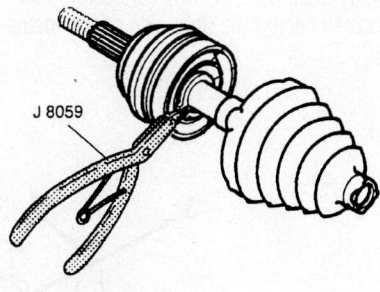

Disconnecting the outer CV-joint from the axle shaft

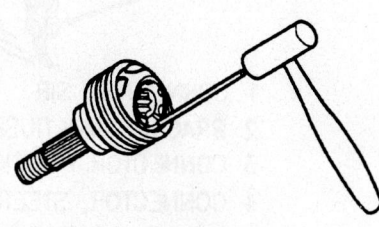

Tilting the cage—Outer CV-joint

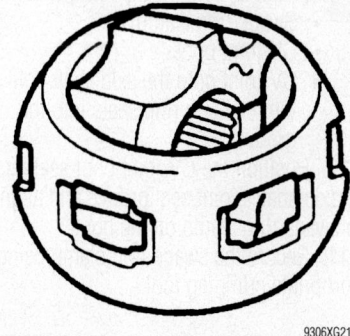

View the cage and inner race—Outer CV-joint

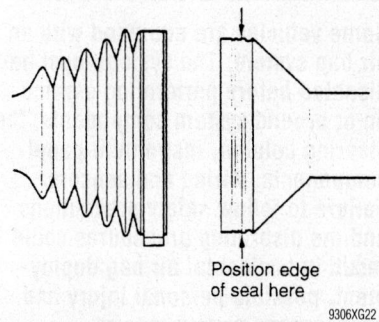

Positioning the boot—Outer CV-joint

d. Rotate the inner race upward and remove it from the cage.

5. Thoroughly clean and inspect all parts.

To install:

6. Lubricate the parts with a light coat of grease.

7. Assemble the CV-joint cage and inner race, as follows:

a. Rotate the inner race 90 degrees to the cage centerline.

b. Align the cage windows with inner race lands.

c. Insert the inner race into the cage by rotating the inner race downward.

d. Insert the cage/inner race into the outer race.

8. Assemble the chrome alloy balls into the CV-joint cage, as follows:

a. Position a brass drift against the CV-joint cage and tap it with a hammer to tilt the cage.

b. Insert the 1st chrome alloy ball into the cage.

c. Tilt the cage in the opposite direction.

d. Insert the opposite chrome alloy ball.

e. Repeat the procedure until all 6 balls are inserted.

9. Install or connect the following:
- Swage ring clamp
- CV-joint boot
- CV-joint onto the axle shaft until the retaining ring seats into the groove

10. Position the CV-joint boot seal into the axle shaft's joint seal groove and align the swage ring clamp on the boot.

11. Secure the swage ring clamp using appropriate crimping tool.

✳✳ WARNING

Make sure that there are no pinch points on the inboard seal.

12. Install or connect the following:
- ½ kit grease into the CV-joint boot
- ½ kit grease into the CV-joint
- New large seal clamp onto the CV-joint boot
- CV-joint boot/clamp, slide it into

place, over the outer race with the seal lip in the groove

➡ **Make sure the boot lies flat against the outer race.**

13. Using a Crimp tool, a torque wrench and a breaker bar, crimp the large CV-joint boot clamp to 130 ft. lbs. (176 Nm).

14. Install the halfshaft and the front wheel.

STEERING AND SUSPENSION

Air Bag

✳✳ CAUTION

Some vehicles are equipped with an air bag system. The system must be disabled before performing service on or around system components, the steering column, instrument panel components, wiring and sensors. Failure to follow safety precautions and the disarming procedures could result in accidental air bag deployment, possible personal injury and unnecessary system repairs.

PRECAUTIONS

Several precautions must be observed when handling the inflator module to avoid accidental deployment and possible personal injury.

- Never carry the inflator module by the wires or connector on the underside of the module.
- When carrying a live inflator module, hold it securely with both hands, and ensure that the bag and trim cover are pointed away.
- Place the inflator module on a bench or other surface with the bag and trim cover facing up.
- With the inflator module on the bench, never place anything on or close to the module which may be thrown in the event of an accidental deployment.

DISARMING

✳✳ CAUTION

The Supplemental Inflatable Restraint system (SIR) must be disarmed before performing service around the air bag or SIR wiring. Failure to do so may cause accidental deployment of the air bag, result-

ing in unnecessary SIR repairs and/or personal injury.

1. Turn the steering wheel so the front wheels are in the straight-ahead position.
2. Turn the ignition switch to the **LOCK** position.
3. Remove or disconnect the following:
- Negative battery cable
- Air bag fuse from the fuse panel

➡ **The position of the fuse on the panel varies according to model and year. Consult the vehicle owner's manual for fuse location.**

- Left-hand sound insulator trim panel under the instrument panel
- Connector retainer clip and the yellow 2-way connector at the base of the steering column
- Connector retainer and detach the passenger side yellow 2-way connector. Located behind the right side sound insulator.

ARMING

After the necessary repairs have been made, re-enable the air bag system as follows:

1. Turn the steering wheel so the front wheels are in the straight-ahead position.
2. Turn the ignition switch to the **LOCK** position.
3. Disconnect the negative battery cable.
4. Install or connect the following:
- Yellow 2-way connector at the base of the steering column and the connector retainer
- Left-hand sound insulator
- Yellow 2-way connector on the right side and the connector retainer
- Sound insulator and/or glove box
- Air bag fuse
- Negative battery cable

5. Turn the ignition switch to the **RUN** position. Verify that the INFLATABLE RESTRAINT indicator lamp flashes 7–9

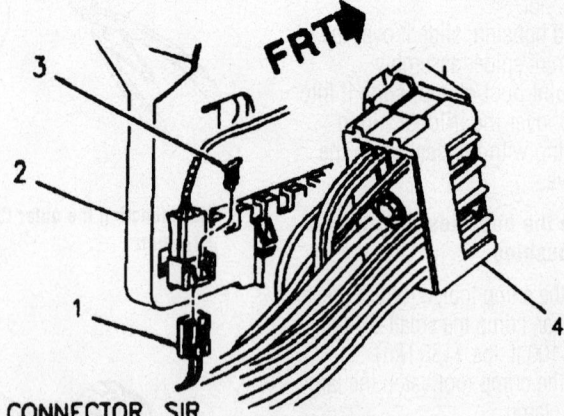

1 CONNECTOR, SIR
2 BRACKET, MULTIUSE MODULE
3 CONNECTOR POSITION ASSURANCE (CPA)
4 CONNECTOR, STEERING COLUMN WIRING HARNESS

7922UG35

Drivers side air bag connector

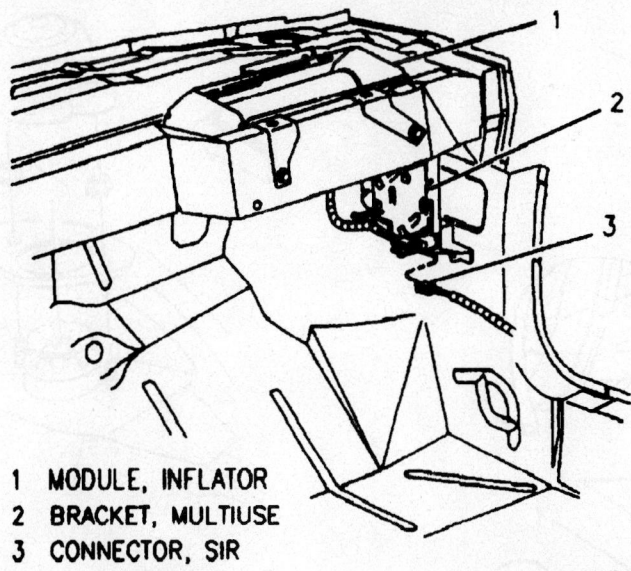

1 MODULE, INFLATOR
2 BRACKET, MULTIUSE
3 CONNECTOR, SIR

7922UG36

Passenger's side air bag connector

times, then remains OFF. If the lamp does not function as specified, there is a malfunction in the air bag system.

Power Rack and Pinion Steering Gear

REMOVAL & INSTALLATION

1. Before servicing the vehicle, refer to the Precautions Section.

✳✳ CAUTION

Make sure the wheels of the vehicle are straight ahead and the steering column in the LOCK position before disconnecting the steering column or intermediate shaft from the steering gear. Failure to do so will cause the coil assembly in the steering column to become uncentered which will cause damage to the coil assembly.

2. Lock the steering column by installing the Steering Column Anti Rotation Pin tool J 42640 into the underside of the steering column.
3. Disconnect negative battery cable.
4. Remove or disconnect the following:
- Front wheels
- Power steering gear heat shield

✳✳ CAUTION

Failure to disconnect the intermediate shaft from the rack and pinion stub shaft can result in damage to the steering gear and/or intermediate

shaft. This damage can cause loss of steering control which could result in personal injury.

- Intermediate shaft lower pinch bolt
- Intermediate shaft from the power steering gear
- Outer tie rod retaining nuts
- Outer tie rod from the steering knuckles
- Variable effort steering electrical connector from the power steering gear, if equipped
- Power steering gear pressure and return hoses from the gear
- Left stabilizer shaft insulator
- Power steering gear mounting bolts (lift from the mounting holes)
- Power steering gear through the left wheel opening

5. Transfer the outer tie rods if replacing the power steering gear.

To install:

6. Install or connect the following:
- Power steering gear through the left wheel opening
- Power steering gear mounting bolts and tighten to 48 ft. lbs. (65 Nm) for 2002–03; 55 ft. lbs. (75 Nm) for 2004–05.
- Left stabilizer shaft insulator
- Power steering gear pressure and return hoses to the gear and tighten to 20–22 ft. lbs. (27–30 Nm)
- Variable effort steering electrical connector, if equipped
- Outer tie rod end to the steering knuckles and tighten the nuts to 52–55 ft. lbs. (70–75 Nm) for

2002–04; 22 ft. lbs. plus 180 degrees for 2005.
- Intermediate shaft to the power steering gear
- Intermediate shaft lower pinch bolt and tighten to 33–35 ft. lbs. (45–47 Nm)
- Power steering gear heat shield
- Wheels

7. Remove tool J-42640 from the steering column.
8. Bleed the power steering system.
9. Inspect the power steering system for leaks.
10. Adjust the front toe.

Strut

REMOVAL & INSTALLATION

Front

✳✳ WARNING

The steering knuckle must be retained after the strut-to-steering knuckle bolts have been removed. Failure to observe this may cause ball joint and/or halfshaft damage.

1. Before servicing the vehicle, refer to the Precautions Section.
2. Matchmark the strut-to-steering knuckle location.
3. If equipped with electronic ride control, detach the electrical connection.
4. Remove or disconnect the following:
- 3 upper strut mount nuts and allow the control arms to hang free
- Anti-lock Brakes System (ABS) front wheel speed sensor
- Wheel speed sensor bracket from the strut
- Brake line bracket from the strut
- Strut-to-steering knuckle bolts and the strut

To install:
5. Install or connect the following:
- Strut. Torque the 3 nuts to 30–35 ft. lbs. (40–47 Nm).
- Electronic ride control electrical connector, if removed
- Strut-to-knuckle bolts. Torque the bolts to 135 ft. lbs. (185 Nm) for 2002–03; 108 ft. lbs. (147 Nm) for 2004–05.
- Brake line bracket to the strut
- Wheel speed sensor bracket to the strut
- ABS front wheel speed sensor connector

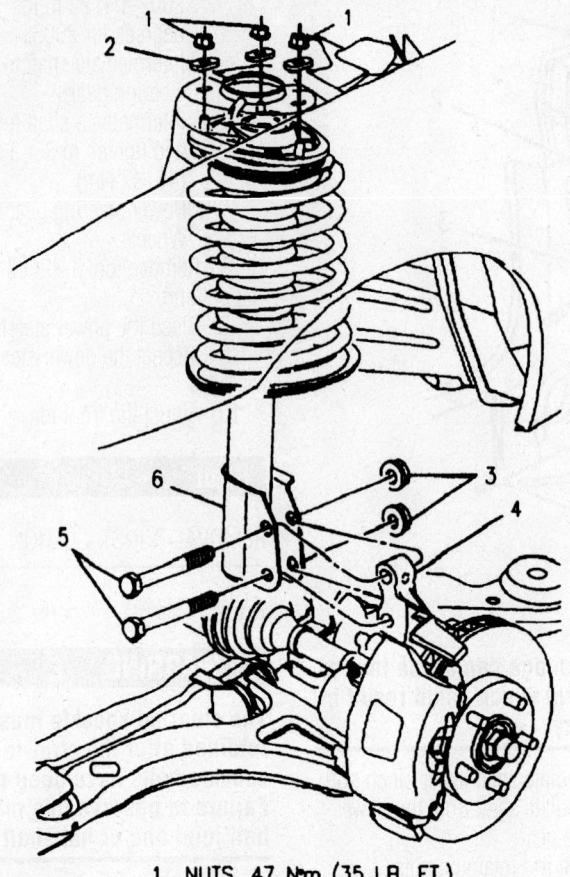

1 NUTS, 47 N·m (35 LB. FT.)
2 WASHER
3 NUTS, 185 N·m (136 LB. FT.)
4 KNUCKLE
5 BOLT
6 STRUT

7922UG17

The strut assembly is mounted between the steering knuckle and the body—front strut shown

• Front wheel. Torque the lug nuts to 100 ft. lbs. (140 Nm).
5. Check and adjust the wheel alignment.

Rear

1. Remove rear seat cushion and seatback to gain access to the strut tower mounting nuts.
2. Remove tire.
3. Support the control arm.
4. Remove or disconnect the following:
• Strut to knuckle bolts
• Upper strut nuts
• Strut from vehicle
To install:
5. Install or connect the following:
• Strut
• Upper strut nuts. Torque the nuts to 35 ft. lbs. (47 Nm).
• Strut to knuckle bolts. Torque the bolts to 140 ft. lbs. (190 Nm).

• Rear seatback and cushion
• Tire. Torque the lug nuts to 100 ft. lbs. (140 Nm).

Shock Absorber

REMOVAL & INSTALLATION

Rear

1. Before servicing the vehicle, refer to the Precautions Section.
2. Remove the tire and wheel assembly.
3. Support the control arm with a jack stand.
4. Remove or disconnect the following:

• Electronic Ride Control (ELC) air tube from the shock
• Two bolts securing the shock to the control arm

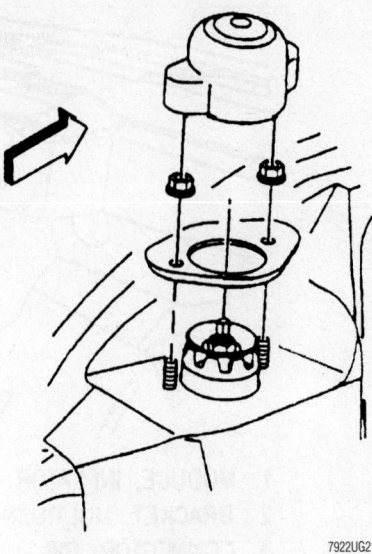

7922UG21

Remove the cover to gain access to the upper shock absorber mounting components—rear suspension

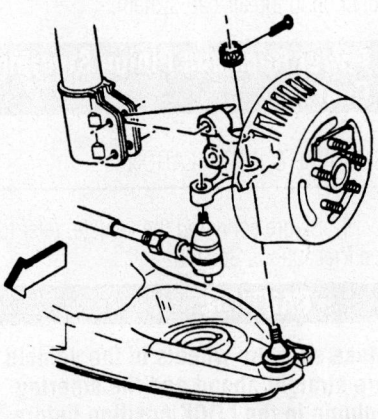

7922UG22

Exploded view of the lower shock absorber mounting to the knuckle assembly—rear suspension

• Trunk trim to gain access to the shock upper mounting nuts.
• Cover, the two nuts, and the reinforcement from the top of the shock
• Shock from the vehicle
To install:
5. Install or connect the following:
• Shock, reinforcement, and the two nuts. Tighten the mounting nuts to 15 ft. lbs. (20 Nm).
• Shock cover
• Trunk trim
• Shock-to-control arm bolts and tighten the bolts to 18 ft. lbs. (24 Nm)
• ELC air tube to the shock.
• Tire and wheel assembly

Coil Spring

REMOVAL & INSTALLATION

※※ CAUTION

The coil springs are under a considerable amount of tension. Be very careful when removing or installing them; they can exert enough force to cause very serious injury.

Front

1. Remove the strut from the vehicle.
2. Disassemble the strut as follows:

   a. Step 1: Place the strut assembly into compressor tool, to compress the coil spring.

   b. Step 2: Compress the spring slightly.

   c. Step 3: Hold the strut shaft from turning using a No. 50 Torx® socket and remove the 24mm nut on the top end of the strut.

   d. Step 4: Install Rod tool J 34013-38 to help guide the strut shaft from the upper mount assembly.

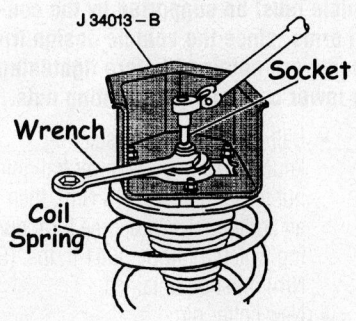

Use a Torx® socket to keep the piston rod from turning while removing the upper nut—front strut shown

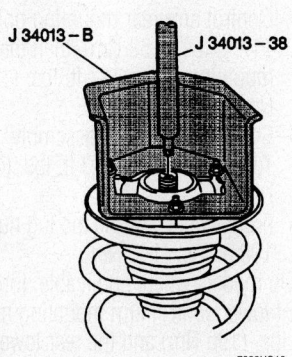

Install Rod J 34013-38 to help guide the strut shaft from the upper mount assembly—front strut shown

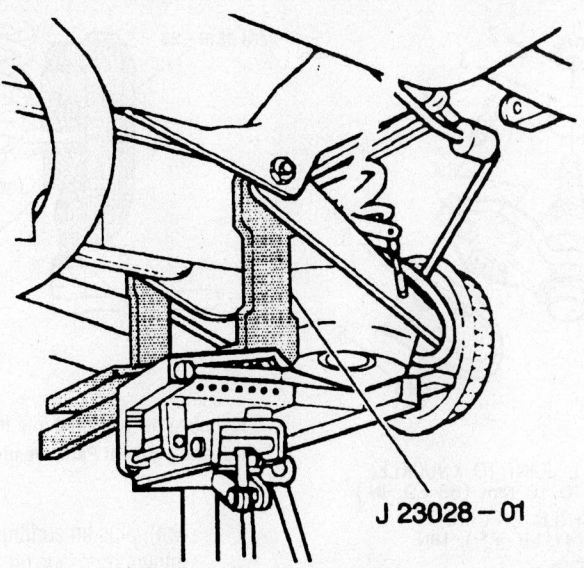

Use support bracket tool J-23028-01 mounted on a jack to support the rear lower control arm

   e. Step 5: Loosen the spring compressor tool until the coil spring and mount can be removed as an assembly. Remove the lower spring insulator, if equipped.

To install:

3. Assemble the strut as follows:

   a. Step 1: Place the strut in compressor tool.

   b. Step 2: Install the coil spring over the strut.

   c. Step 3: Compress the coil spring while guiding strut shaft through the top of the strut assembly.

   d. Step 4: Install the top strut nut. Torque the nut to 55 ft. lbs. (75 Nm).

   e. Step 5: Remove the strut from the compressor.

Rear

1. Before servicing the vehicle, refer to the Precautions Section.
2. Remove or disconnect the following:
   - Tire
   - Electronic Leveling Control (ELC) air tube
3. Support the control arm.
   - Two lower shock bolts
   - Trunk trim
   - Cover, 2 nuts and the reinforcement from the top of the shock
   - Cotter pin and hex nut on control arm
4. Separate the adjustment link from the knuckle.
5. Slowly lower the control arm.
6. Remove the spring with the lower insulator.

To install

7. Install or connect the following:
   - Coil spring with insulator
   - Lower shock bolts. Torque the bolts to 18 ft. lbs. (24 Nm).
   - Adjustment link. Torque the nut to 88 inch lbs. (10 Nm) 36 ft. lbs. (50 Nm). Tighten an additional turn to align the cotter pin.
   - ELC air tube
   - Tire. Torque the lug nuts to 100 ft. lbs. (140 Nm).

Lower Ball Joint

REMOVAL & INSTALLATION

Front

The ball joint is an integral part of the lower control arm and if found to be defective the control arm should be replaced.

Rear

1. Before servicing the vehicle, refer to the Precautions Section.
2. Remove or disconnect the following:
   - Rear wheel
   - Height sensor link from the right control arm
   - Parking brake cable retaining clip from the left control arm
   - Adjustment link from the knuckle
3. Support the control arm with a jack.
4. Remove the ball joint stud cotter pin and nut.
5. Reinstall the nut on the stud with the

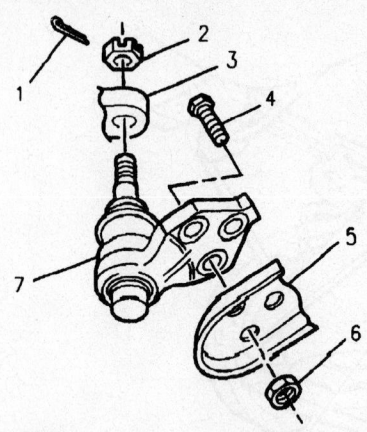

1 PIN
2 NUT, BALL JOINT TO KNUCKLE;
   TIGHTEN TO 10 N•m (88 LB. IN.)
   THEN TIGHTEN 2 FLATS TO
   55 N•m (41 LB. FT.), MIN.
3 KNUCKLE
4 BALL JOINT MOUNTING BOLTS MUST
   FACE DOWN
5 CONTROL ARM
6 BALL JOINT MOUNTING NUTS
   68 N•m (50 LB. FT.)
7 SERVICE BALL JOINT

7922UG25

The replacement ball joint should be
attached to the control arm using 3 bolts
and nuts—except Park Avenue

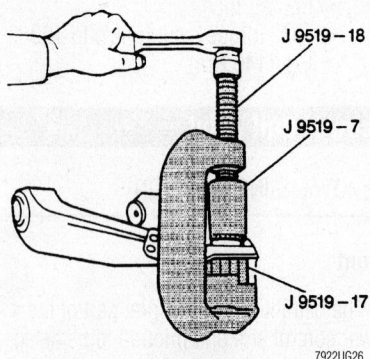

Removing the ball joint from the lower
control arm—except Park Avenue

flat side of the nut facing up; do not torque
the nut.
6. Remove or disconnect the following:
   • Stud from the knuckle
   • Slotted hex nut from the ball joint
     stud
7. Press the ball joint out of the control
arm
To install:
8. Install or connect the following:
   • New ball joint, press it into the
     control arm
   • Ball joint stud into the knuckle.
     Torque the nut to 88 inch lbs. (10

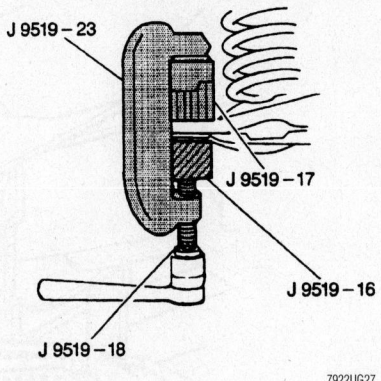

Installing the ball joint into the lower control arm—except Park Avenue

Nm), plus an additional 4 flats. The
minimum torque on the nut should
be 40 ft. lbs. (55 Nm).
   • New cotter pin

➡Tighten the nut up to one more flat in
order to align the slot with the hole in
the stud.

   • Adjustment link to the knuckle.
     Torque the slotted nut to 33 ft. lbs.
     (45 Nm).
   • New cotter pin

➡Tighten the nut up to one more flat in
order to align the slot with the hole in
the stud.

   • Height sensor link to the right con-
     trol arm
   • Parking brake cable retainer to the
     left control arm
   • Rear wheel

Lower Control Arm

REMOVAL & INSTALLATION

Front

1. Before servicing the vehicle, refer to
the Precautions Section.
2. Allow the control arms to hang free.
3. Remove or disconnect the following:
   • Front wheel
   • Stabilizer shaft link assembly from
     the control arm
   • Cotter pin from the lower ball joint
     and remove the nut
   • Ball joint from the steering
     knuckle
   • Lower control arm from the frame
To install:
4. Install or connect the following:
   • Lower control arm to the frame

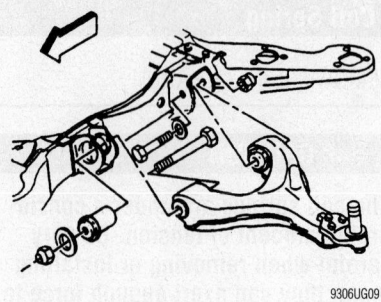

Lower control arm assembly—except Park
Avenue

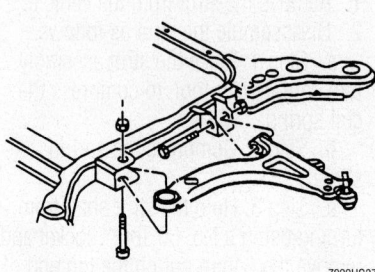

Lower control arm mounting—Park
Avenue

➡Do not tighten the lower control arm
nuts at this time. The weight of the
vehicle must be supported by the control arms, since the vehicle design trim
heights are obtained before tightening
the lower control arm mounting nuts.

   • Ball joint stud to the steering
     knuckle. Torque the lower ball joint
     nut to 88 inch lbs. (10 Nm), then
     an additional 120 degree turn, dur-
     ing which a torque of 41 ft. lbs. (55
     Nm) must be obtained.
   • New cotter pin

➡Align the slot in the nut with the hole
in the lower ball joint stud. Torque the
ball joint nut up to one more flat. Never
loosen the nut to align the slot.

   • Control arm rear mounting bolts to
     117 ft. lbs. (158 Nm), then the front
     mounting nut to 117 ft. lbs. (158
     Nm)
   • Stabilizer shaft link assembly.
     Torque the nuts to 13 ft. lbs. (17
     Nm).
   • Front wheel. Torque the lug nuts to
     100 ft. lbs. (140 Nm).
5. On models where applicable, torque
the front lower control arm mounting nut to
140 ft. lbs. (190 Nm) and the rear lower
control arm nut to 93 ft. lbs. (126 Nm).
6. Check and adjust the wheel alignment.

Rear

1. Before servicing the vehicle, refer to the Precautions Section.
2. Remove or disconnect the following:
   - Rear suspension support assembly
   - Electronic Ride Control height sensor
   - Tie rod
   - Stabilizer link assembly from the control arm
   - Antilock Brake System (ABS) electrical connector
   - Hub and bearing
   - Lower control arm
   - Adjustment link from the control arm, if necessary
   - Nuts and bolts and the control arm

To install:
3. Install or connect the following:
   - Lower control arm to the frame.

➡**Tighten the control arm nuts with the vehicle unsupported and resting on the wheels at the normal trim height.**

   - Bolts and the nuts
   - Hub and bearing
   - Adjustment link to the control arm, if necessary. tighten the nut to 36 ft. lbs. (50 Nm).
   - ABS electrical connector
   - Tie rod
   - Stabilizer link assembly to the control arm
   - Height sensor
   - Rear wheel

4. Lower the vehicle and tighten the control arm nuts to 78 ft. lbs. (106 Nm).

CONTROL ARM BUSHING REPLACEMENT

Except Park Avenue

FRONT

1. Before servicing the vehicle, refer to the Precautions Section.

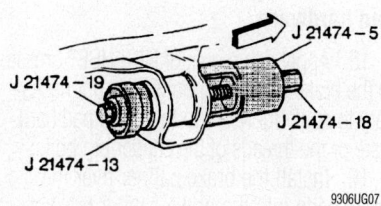

Removing the bushing from the control arm

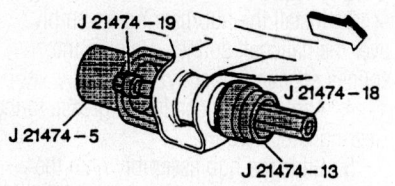

Installing the bushing into the control arm

2. Remove or disconnect the following:
   - Front wheel
   - Lower control arm
3. Press the control arm bushing out of the control arm.

To install:
4. Insert the new bushing into control arm.
5. Press the new bushing into the control arm.
6. Install the lower control arm.
7. Install the front wheel. Torque the lug nuts to 100 ft. lbs. (140 Nm).

REAR

1. Before servicing the vehicle, refer to the Precautions Section.
2. Remove the control arm.
3. Place tool J 21474-27 with the washer through tool J 41014-2 over the bushing against the control arm.
4. Lubricate the bolt threads with high pressure lubricant.
5. Install tool J 41014-1 (with small end facing the bushing), the thrust bearing and tool J 21474-4 onto tool J 21474-27.
6. Tighten the nut until the bushing is driven out of the control arm.
7. Remove the bushing tools.

To install:
8. Start the bushing into the control arm.
9. Make sure that the flat on the bushing is vertical and facing rearward. Place tool J 21474-27 with the washer through tool J41014-1 over the bushing against the control arm.

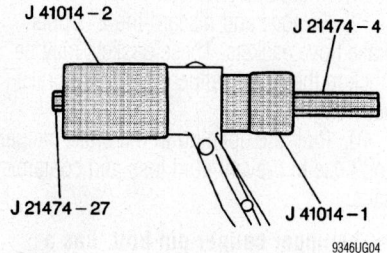

Removing the bushing from the rear control arm

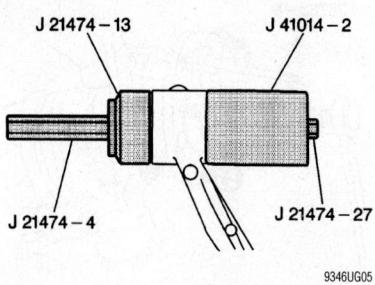

Installing the bushing into the rear control arm

10. Lubricate the bolt threads with high pressure lubricant.
11. Install tool J 21474-13 (making sure the large end is facing the bushing), the thrust bearing and tool J 21474-4 onto tool J 21474-27.
12. Tighten the bolt until the bushing is fully seated into the control arm.
13. Remove the bushing tools.
14. Install the control arm.

Park Avenue

The lower control arm is replaced as a unit. The bushing can not be removed.

Wheel Bearings

ADJUSTMENT

The wheel bearings are not adjustable. If a wheel bearing is out of specifications, it must be replaced. Using a dial indicator, check for looseness. If play exceeds 0.005 inch (0.127mm), the bearing wear is excessive and the hub/bearing should be replaced.

REMOVAL & INSTALLATION

Front

1. Before servicing the vehicle, refer to the Precautions Section.
2. Remove or disconnect the following:
   - Front wheel

➡**Insert a drift punch through the caliper and into the rotor cooling fins to prevent the rotor from turning.**

   - Halfshaft nut and washer
   - Brake caliper
   - Brake rotor
   - Anti-Lock Brake System (ABS) speed sensor
   - 3 hub/bearing assembly bolts

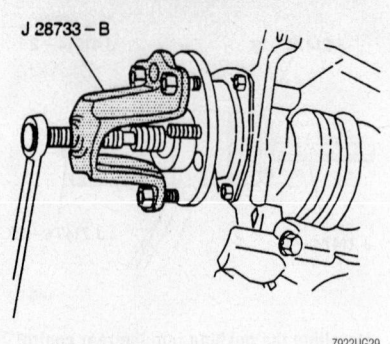

J 28733 – B

Use a puller such as J 28733-B to press the halfshaft from the hub assembly

7922UG29

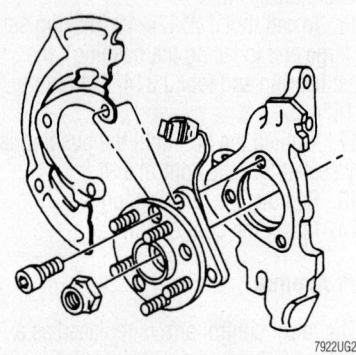

Exploded view of the front hub and wheel bearing assembly

7922UG28

- Dust shield
- Hub/bearing assembly from the halfshaft, using a puller
- Hub/bearing assembly from the steering knuckle

To install:

3. Install the hub/bearing assembly over the halfshaft splines. Be sure the splines engage smoothly.

4. Apply a light coating of grease to the steering knuckle bore.

5. Slide the hub assembly onto the halfshaft as far as possible. If the hub will not bottom out on the halfshaft, install the hub mounting bolts and use the halfshaft nut to draw the hub onto the halfshaft.

6. Once the hub is flush with the steering knuckle, remove the mounting bolts and install the dust shield.

7. Install or connect the following:

8. Install the mounting bolts. Torque the bolts to 70 ft. lbs. (95 Nm).

9. Place the transaxle in **N**.

10. Install or connect the following:
- ABS front wheel speed sensor connector, and clip to the dust shield
- Brake rotor
- Caliper.

11. Torque the halfshaft nut to 118 ft. lbs. (160 Nm) on models except Park Avenue.

12. Install the front wheels. Torque the lug nuts to 100 ft. lbs. (140 Nm).

13. Road test the vehicle.

Rear

1. Before servicing the vehicle, refer to the Precautions Section.

2. Remove or disconnect the following:

- Wheel and the tire
- Brake caliper

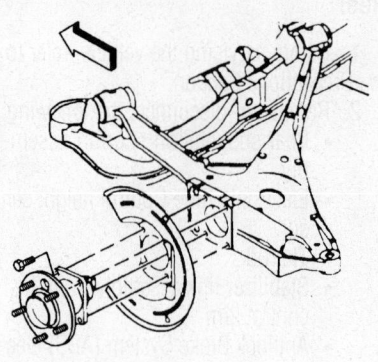

9346UG06

Exploded view of the rear hub/wheel bearing assembly

- Brake rotor
- ABS sensor wire connector.
- Four bolts from the control arm
- Hub and bearing from the control arm
- Brake shield from the control arm

To install:

3. Clean the control arm face and the bore before installing the hub and the bearing.

4. Install or connect the following:
- Brake shield and the hub and the bearing to the control arm with the four bolts. Reconnect the ABS sensor. Tighten the hub and bearing bolts to 52 ft. lbs. (70 Nm).
- Brake rotor
- Brake caliper
- Wheel and the tire

BRAKES

Brake Caliper

REMOVAL & INSTALLATION

Front

ALL EXC. BONNEVILLE V8

1. Inspect the fluid level in the brake master cylinder reservoir. If the brake fluid level is midway between the maximum-full point and the minimum allowable level, no brake fluid needs to be removed from the reservoir before proceeding. If the brake fluid level is higher than midway between the maximum-full point and the minimum allowable level, remove brake fluid to the midway point before proceeding.

2. Raise and support the vehicle.

3. Remove the tire and wheel assembly.

4. Install and hand tighten 2 wheel lug nuts to retain the rotor to the hub.

5. Install a large C-clamp over the body of the brake caliper with the C-clamp ends against the rear of the caliper body and against the outer brake pad.

6. Tighten the C-clamp until the caliper piston is compressed into the caliper bore enough to allow the caliper to slide past the brake rotor.

7. Remove the C-clamp from the caliper.

8. Remove the brake hose to caliper bolt from the brake caliper.

9. Remove and discard the 2 copper brake hose gaskets. These gaskets may be stuck to the brake caliper and/or the brake hose end.

10. Plug the opening in the brake caliper and hose to prevent fluid loss and contamination.

➡**The upper caliper pin bolt, has a bushing as part of the bolt. The lower caliper pin bolt, is of a solid design.**

11. Remove the caliper pin bolts. Note the location of the caliper pin bolts.

12. Remove the caliper from the caliper bracket.

13. Inspect the caliper pin bolt boots in the caliper bracket.

To install:

14. If reusing the brake caliper pin bolts, wipe the old grease from the brake caliper pin bolts with a clean shop towel.

➡**DO NOT apply lubricant to the brake pad hardware.**

15. Apply a thin coat of NIGLUBE® grease to the brake caliper pin bolts. It is not necessary to apply lubricant to the brake pad hardware or the threads of the caliper pin bolts.

16. Install the brake caliper over the brake pads into the brake caliper bracket.

➡**The upper caliper pin bolt, has a bushing as part of the bolt. The lower caliper pin bolt, is of a solid design.**

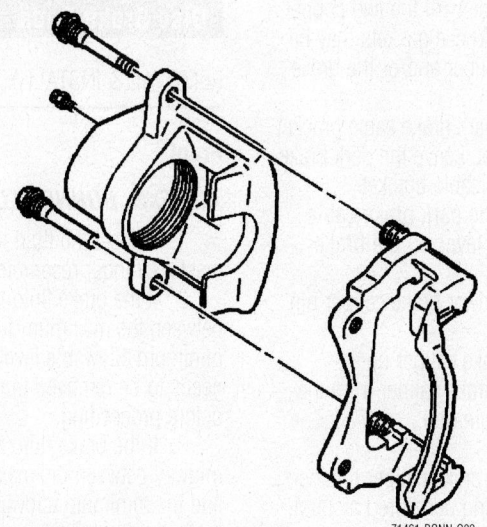

Front caliper—exc. Bonneville V8

71461-BONN-G32

17. Install the caliper pin bolts. Tighten the brake caliper pin bolts to 85 Nm (63 ft. lbs.).

➡**Install NEW copper brake hose gaskets.**

18. Assemble the brake hose bolt and the NEW copper brake hose gaskets to the brake hose.

19. Install the brake hose to caliper bolt to the brake caliper. Tighten the brake hose to caliper bolt to 44 Nm (32 ft. lbs.).

20. Bleed the brake system.

21. With the engine OFF, gradually apply the brake pedal to approximately ⅔ of its travel distance.

22. Slowly release the brake pedal.

23. Wait 15 seconds, then repeat until a firm brake pedal is obtained. This will properly seat the brake caliper pistons and brake pads.

24. Remove the 2 wheel lug nuts retaining the brake rotor to the hub.

25. Install the tire and wheel assembly.

26. Lower the vehicle.

BONNEVILLE V8

1. Inspect the fluid level in the brake master cylinder reservoir.

2. If the brake fluid level is midway between the maximum-full point and the minimum allowable level, no brake fluid needs to be removed from the reservoir before proceeding.

3. If the brake fluid level is higher than midway between the maximum-full point and the minimum allowable level, remove brake fluid to the midway point before proceeding.

4. Raise and suitably support the vehicle.

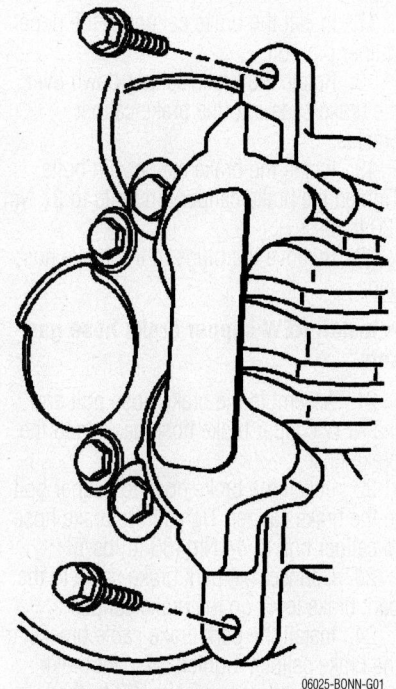

06025-BONN-G01

Bonneville V8 front caliper removal

5. Remove the tire and wheel assembly.

6. Install a large C-clamp over the body of the brake caliper with the C-clamp ends against the rear of the caliper body and against the outer brake pad.

7. Tighten the C-clamp until the caliper piston is compressed into the caliper bore enough to allow the caliper to slide past the brake rotor.

8. Remove the C-clamp from the caliper.

9. Remove the brake hose-to-caliper bolt attaching the brake hose to the brake caliper.

10. Remove the brake hose from the brake caliper.

11. Remove and discard the 2 copper brake hose gaskets. These gaskets may be stuck to the brake caliper and/or the brake hose end.

12. Plug the opening in the brake caliper and the brake hose to prevent fluid loss and contamination.

13. Remove the brake caliper pin bolts.

14. Remove the brake caliper from the brake caliper bracket.

To install:

➡**Ensure that the caliper guide pin boots are fully seated to the caliper guide pin retaining seat of the caliper guide pin. Ensure that the caliper guide pin boots are fully seated to the caliper boot seal retaining seat of the brake caliper mounting bracket.**

15. Inspect the caliper guide pin boots for cuts, tears, or deterioration. If damaged, replace the slides and boots.

➡**If reusing the brake caliper pin bolts, the threads of the caliper pin bolts and the threads of the caliper bracket mounting holes must be free of residue and debris prior to application of threadlocker in order to ensure proper adhesion and fastener retention.**

16. Prepare the bolts and the threaded holes for assembly: Thoroughly clean the residue from the bolt threads by using denatured alcohol or equivalent and allow to dry. Thoroughly clean the residue from the threaded holes by using denatured alcohol or equivalent and allow to dry.

17. Apply threadlocker GM P/N 12345493 (Canadian P/N 10953488) or equivalent, to ⅔ of the threaded length of the lower caliper bracket bolts. Ensure that there are no gaps in the threadlocker along the length of the filled area of the bolts.

18. Allow the threadlocker to cure approximately 10 minutes before installation.

19. Apply a thin coat of high temperature silicone brake lubricant to the brake caliper pin bolts.

20. Install the brake caliper pin bolts. Tighten the brake caliper pin bolts to 34 Nm (25 ft. lbs.).

21. Remove the plug from the brake caliper opening and the brake hose.

➡**Install NEW copper brake hose gaskets.**

22. Assemble the NEW copper brake hose gaskets, and the brake caliper bolt to the brake hose.

23. Install the brake hose and the brake caliper bolt to the brake caliper. Tighten the brake hose to caliper bolt to 44 Nm (32 ft. lbs.).

24. Bleed the hydraulic brake system.

25. With the engine OFF, gradually apply the brake pedal to approximately ⅔ of its travel distance.

26. Slowly release the brake pedal.

27. Wait 15 seconds, then repeat steps 11 and 12 until a firm brake pedal apply is obtained. This will properly seat the brake caliper pistons and brake pads.

28. Install the tire and wheel assembly.

29. Lower the vehicle.

Rear

Inspect the fluid level in the brake master cylinder reservoir. If the brake fluid level is midway between the maximum-full point and the minimum allowable level, no brake fluid needs to be removed from the reservoir before proceeding. If the brake fluid level is higher than midway between the maximum-full point and the minimum allowable level, remove brake fluid to the midway point before proceeding.

1. Raise the vehicle and suitably support.

2. Remove the tire and wheel assembly.

3. Pull down on the front park brake cable.

4. Remove the front park brake cable from the park brake cable connector.

➡When using a large C-clamp to compress a caliper piston into a caliper bore of a caliper equipped with an integral park brake mechanism, do not exceed more than 1 mm (0.039 in.) of piston travel. Exceeding this amount of piston travel will cause damage to the internal adjusting mechanism and/or the integral park brake mechanism.

5. Install a large C-clamp over the top of the brake caliper housing and against the back of the outboard brake pad. Compress the brake caliper piston into brake caliper bore to allow the piston enough clearance to slide the brake caliper off the brake rotor.

6. Remove the C-clamp.

➡Be sure to plug the opening in the brake caliper and brake hose to prevent fluid loss and/or contamination.

7. Remove brake hose to caliper bolt attaching the brake hose to the brake caliper.

8. Remove the brake hose from the brake caliper.

9. Remove and discard the two copper brake hose gaskets. These gaskets may be stuck to the brake caliper and/or the brake hose end.

10. Remove the park brake cable bracket from the brake caliper. Leave the park brake cable attached to the cable bracket.

11. Disconnect the park brake cable from the park brake lever on the brake caliper.

12. Remove the lower brake caliper pin bolt.

13. Rotate the brake caliper up.

14. Remove the brake caliper from the upper brake caliper pin bolt.

To install:

15. If reusing the brake caliper retainers, clean the sleeves using denatured alcohol, or equivalent.

16. Lubricate the brake caliper pin bolts with a thin coat of high temperature silicone lube.

17. Install the brake caliper to the upper caliper pin bolt.

18. Rotate the brake caliper down over the brake pads into the brake caliper bracket.

19. Install the brake caliper pin bolts. Tighten the brake caliper pin bolts to 27 Nm (20 ft. lbs.).

20. Remove the plugs in the brake hose end.

➡Install NEW copper brake hose gaskets.

21. Assemble the brake hose bolt and the NEW copper brake hose gaskets to the brake hose.

22. Install the brake hose to caliper bolt to the brake caliper. Tighten the brake hose to caliper bolt to 44 Nm (33 ft. lbs.).

23. Connect the park brake cable to the park brake lever on the brake caliper.

24. Install the park brake cable bracket to the brake caliper. Tighten the park brake cable bracket bolt to 43 Nm (32 ft. lbs.).

25. Install the front park brake cable to the park brake cable connector.

26. Bleed the brake system.

27. With the engine OFF, gradually apply the brake pedal to approximately ⅔ of its travel distance.

28. Slowly release the brake pedal.

29. Wait 15 seconds, then repeat until a firm brake pedal is obtained. This will properly seat the brake caliper pistons and brake pads.

30. Install the tire and wheel assembly.

31. Lower the vehicle.

Disc Brake Pads

REMOVAL & INSTALLATION

Front

ALL EXC. BONNEVILLE V8

1. Inspect the fluid level in the brake master cylinder reservoir.

2. If the brake fluid level is midway between the maximum-full point and the minimum allowable level, no brake fluid needs to be removed from the reservoir before proceeding.

3. If the brake fluid level is higher than midway between the maximum-full point and the minimum allowable level, remove brake fluid to the midway point before proceeding.

4. Raise the vehicle and suitably support.

5. Remove the tire and wheel assembly.

6. Install a large C-clamp over the body of the brake caliper with the C-clamp ends against the rear of the caliper body and against the outer brake pad.

7. Tighten the C-clamp until the caliper piston is compressed into the caliper bore enough to allow the caliper to slide past the brake rotor.

8. Remove the C-clamp from the caliper.

9. Remove the bottom brake caliper pin bolt.

10. Pivot the brake caliper body upward and secure out of the way with heavy mechanic's wire or equivalent. Do NOT disconnect the hydraulic brake flexible hose from the caliper.

11. Install a large C-clamp over the body of the brake caliper with the C-clamp ends against the rear of the caliper body and against an old inner brake pad or a wood block installed against the caliper piston.

12. Tighten the C-clamp until the caliper piston is compressed completely into the caliper bore.

13. Remove the C-clamp and the old brake pad or wood block from the caliper.

14. Remove the inboard and outboard brake pads from the brake caliper bracket.

15. Remove and inspect the brake pad retainers.

To install:

16. Inspect the caliper bolt suspension boots for cuts, tears, or deterioration. If damaged, replace the pin boots.

17. Inspect the caliper pin bolts for damage or corrosion. Replace if damaged or

corroded. Do not attempt to clean away corrosion. Corrosion is typically caused by damaged pin boots.

18. Inspect the caliper piston boot for deterioration, replace if damaged.

19. Lubricate the front brake caliper pin bolts with a thin coat of NIGLUBE® grease.

20. Install the brake pad retainers into the brake caliper bracket.

21. Install both brake pads into the brake caliper bracket.

22. Pivot the brake caliper down over the brake pads and into the brake caliper bracket.

23. Insert the lower brake caliper pin bolt. Tighten the brake caliper pin bolt to 85 Nm (63 ft. lbs.).

24. Install the tire and wheel assembly.

25. Lower the vehicle.

26. With the engine OFF, gradually apply the brake pedal to approximately ⅔ of its travel distance.

27. Slowly release the brake pedal.

28. Wait 15 seconds, then repeat until a firm brake pedal is obtained. This will properly seat the brake caliper pistons and brake pads.

29. Fill the brake master cylinder reservoir to the proper level.

30. Burnish the pads and rotors.

BONNEVILLE V8

✲✲ CAUTION

Two different designs of the front brake rotors and front brake pads are used on this vehicle. Do NOT interchange first design and second design parts, or a loss of braking and personal injury could occur.

1. Inspect the fluid level in the brake master cylinder reservoir.

2. If the brake fluid level is midway between the maximum-full point and the minimum allowable level, no brake fluid needs to be removed from the reservoir before proceeding.

3. If the brake fluid level is higher than midway between the maximum-full point and the minimum allowable level, remove brake fluid to the midway point before proceeding.

4. Raise and suitably support the vehicle.

5. Remove the tire and wheel assembly.

6. Install large C-clamp over the body of the brake caliper with the C-clamp ends against the rear of the caliper body and against the outboard brake pad.

7. Tighten the C-clamp evenly until the

caliper pistons are compressed into the caliper bores enough to allow the caliper to slide past the brake rotor.

8. Remove the C-clamp from the caliper.

9. To loosen the brake caliper lower pin bolt, hold the brake caliper guide pin with a wrench.

10. Remove the brake caliper pin bolt.

✲✲ WARNING

Support the brake caliper with heavy mechanic's wire, or equivalent, whenever it is separated from its mount and the hydraulic flexible brake hose is still connected. Failure to support the caliper in this manner will cause the flexible brake hose to bear the weight of the caliper, which may cause damage to the brake hose and in turn may cause a brake fluid leak.

11. Pivot the brake caliper body upward and secure the caliper out of the way with heavy mechanic's wire or equivalent. Ensure that there is no tension on the hydraulic brake flexible hose. Do NOT disconnect the hydraulic brake flexible hose from the caliper.

12. Remove the brake pads from the caliper bracket.

13. Remove and inspect the brake pad retainers from the caliper bracket.

To install:

➡**Ensure that the caliper guide pin boots are fully seated to the caliper guide pin retaining seat of the caliper pin. Ensure that the caliper guide pin boots are fully seated to the caliper boot seal retaining seat of the brake caliper mounting bracket.**

14. Inspect the brake caliper guide pin bolts. If damaged or corroded replace the brake caliper guide bolts.

15. Inspect the brake caliper guide pins. If damaged, or corroded replace the brake caliper guide pin. Do not attempt to clean away any corrosion.

16. Inspect the brake caliper guide pin boots for cuts, tears, or deterioration. If damaged, replace the brake caliper guide pin boots.

17. Carefully pull outward on the caliper guide pin to ensure that the caliper guide pin retaining seat is fully seated to the caliper guide pin boot.

18. Inspect the brake caliper piston boot for deterioration, replace if damaged.

19. Install a large C-clamp over the body

of the brake caliper, with the C-clamp ends against the rear of the caliper body and against an old inboard brake pad or a wood block installed against the caliper pistons.

20. Tighten the C-clamp evenly until the caliper pistons are compressed completely into the caliper bores.

21. Remove the C-clamp and the old brake pad or wood block from the caliper.

22. Install the brake pad retainers to the caliper bracket.

23. Install the brake pads to the caliper bracket.

24. Pivot the brake caliper downward, over the brake pads and into the caliper bracket.

➡**If reusing the lower caliper pin bolt, the threads of the lower caliper pin bolt and the threads of the caliper bracket mounting holes must be free of residue and debris prior to application of threadlocker in order to ensure proper adhesion and fastener retention.**

25. If reusing the caliper pin bolts prepare the bolt and the threaded hole for assembly: Thoroughly clean the residue from the bolt threads by using denatured alcohol or equivalent and allow to dry. Thoroughly clean the residue from the threaded holes by using denatured alcohol or equivalent and allow to dry.

26. Apply threadlocker GM P/N 12345493 (Canadian P/N 10953488), or equivalent to two-thirds of the threaded length of the lower caliper pin bolt. Ensure that there are no gaps in the threadlocker along the length of the filled area of the bolt.

27. Allow the threadlocker to cure approximately ten minutes before installation.

28. Apply a thin coat of Niglube® GM P/N 18046532 grease or equivalent, to the front brake caliper guide pin.

29. Install the lower brake caliper pin bolt. Hold the lower brake caliper guide pin with a wrench and tighten the lower brake caliper pin bolt to 34 Nm (25 lb ft).

30. Install the tire and wheel assembly.

31. Lower the vehicle.

32. With the engine OFF, gradually apply the brake pedal to approximately ⅔ of its travel distance.

33. Slowly release the brake pedal.

34. Wait 15 seconds, then repeat steps 15 and 16 until a firm brake pedal apply is obtained; this will properly seat the brake caliper pistons and brake pads.

35. Fill the brake master cylinder reservoir to the proper level.

36. Burnish the pads and rotors.

Rear

Inspect the fluid level in the brake master cylinder reservoir. If the brake fluid level is midway between the maximum-full point and the minimum allowable level, no brake fluid needs to be removed from the reservoir before proceeding. If the brake fluid level is higher than midway between the maximum-full point and the minimum allowable level, remove brake fluid to the midway point before proceeding.

Raise the vehicle and suitably support.

1. Remove the tire and wheel assembly.

➡When using a large C-clamp to compress a caliper piston into a caliper bore of a caliper equipped with an integral park brake mechanism, do not exceed more than 1 mm (0.039 in.) of piston travel. Exceeding this amount of piston travel will cause damage to the internal adjusting mechanism and/or the integral park brake mechanism.

2. Using a large C clamp, compress the brake caliper piston into the brake caliper bore to gain enough clearance to allow the brake caliper to pivot off the brake caliper bracket.

3. Compress the piston until resistance is felt, but no more than 1 mm (0.039 in) of piston travel.

4. Remove the park brake cable guide bolt from the lower control arm.

5. Remove the bottom brake caliper pin bolt.

➡Support the brake caliper with heavy mechanic's wire, or equivalent, whenever it is separated from its mount and the hydraulic flexible brake hose is still connected. Failure to support the caliper in this manner will cause the flexible brake hose to bear the weight of the caliper, which may cause damage to the brake hose and in turn may cause a brake fluid leak.

6. Pivot the brake caliper body upward and secure out of the way with heavy mechanic's wire. Do NOT disconnect the hydraulic brake flexible hose from the caliper.

7. Remove the inboard and outboard brake pads from the brake caliper bracket.

8. Remove and inspect the brake pad retainers.

To install:

9. Inspect the brake caliper bolt suspension boots for cuts, tears, or deterioration. If damaged, replace the brake caliper pin boots.

10. Inspect the brake caliper pin bolts for damage or corrosion. Replace if damaged or corroded. Do not attempt to clean away corrosion. Corrosion is typically caused by damaged pin boots.

11. Inspect the brake caliper piston boot for deterioration, repair or replace the brake caliper if damaged.

12. Retract the brake caliper piston into the brake caliper bore. Use a spanner type wrench to turn the piston clockwise until it bottoms in the brake caliper bore and align the piston.

13. Align the cutouts in the brake caliper piston to the alignment pins on the brake pads.

14. Apply a thin coat of high temperature silicone lube to the rear brake caliper bolts.

15. Install the brake pad retainers into the brake caliper bracket.

16. Install the inboard and outboard brake pads into the brake caliper bracket.

17. Pivot the brake caliper down over the brake pads and into the brake caliper bracket.

18. Insert the lower brake caliper pin bolt. Tighten the brake caliper pin bolt to 27 Nm (20 ft. lbs.).

19. Install the park brake cable guide bolt to the lower control arm. Tighten the park brake cable guide bolt to 24 Nm (18 ft. lbs.).

20. Install the tire and wheel assembly.

21. Lower the vehicle.

22. With the engine OFF, gradually apply the brake pedal to approximately ⅔ of its travel distance.

23. Slowly release the brake pedal.

24. Wait 15 seconds, then repeat until a firm brake pedal is obtained. This will properly seat the brake caliper pistons and brake pads.

25. Fill the brake master cylinder reservoir to the proper level.

26. Burnish the pads and rotors.

BRAKES **9-47**
DRIVE TRAIN **9-35**
ENGINE REPAIR **9-12**
FUEL SYSTEM **9-32**
SPECIFICATIONS AND
MAINTENANCE CHARTS **9-2**
Engine and Vehicle
  Identification 9-2
General Engine Specifications 9-2
Engine Tune-Up Specifications.......9-3
Accessory Drive Belt Routing 9-3
Capacities 9-4
Valve Specifications.................9-5
Crankshaft and Connecting
  Rod Specifications................9-5
Piston and Ring Specifications........9-6
Torque Specifications 9-6
Wheel Alignment 9-7
Tire, Wheel and Ball Joint
  Specifications 9-8
Brake Specifications 9-8
Scheduled Maintenance
  Intervals.........................9-9
STEERING AND
SUSPENSION **9-40**

A

Air Bag............................9-40
  Arming 9-41
  Disarming 9-40
  Precautions.......................9-40
Air Suspension 9-46
  Depressurizing....................9-46
  Removal & Installation............9-46
Alternator 9-12
  Installation 9-12
  Removal 9-12
Automatic Transmission
  Assembly.........................9-35
  Removal & Installation............9-35
Axle Housing 9-40
  Removal & Installation............9-40
Axle Shaft, Bearing and Seal........9-39
  Removal & Installation............9-39

B

Brake Caliper 9-47
  Removal & Installation............9-47

C

Camshaft and Valve Lifters 9-23
  Removal & Installation............9-23
Coil Springs 9-43
  Removal & Installation............9-43
CV-Joints..........................9-37
  Overhaul 9-37
Cylinder Head 9-17
  Removal & Installation............9-17

D

Disc Brake Pads....................9-47
  Removal & Installation............9-47
Distributor.........................9-12

E

Engine Assembly 9-13
  Removal & Installation............9-13
Exhaust Manifold 9-22
  Removal & Installation............9-22

F

Fuel Filter 9-33
  Removal & Installation............9-33
Fuel Injector.......................9-33
  Removal & Installation............9-33
Fuel Pump 9-33
  Removal & Installation............9-33
Fuel System Pressure 9-33
  Relieving 9-33
Fuel System Service Precautions...9-32

H

Halfshaft..........................9-36
  Removal & Installation............9-36
Heater Core.......................9-16
  Removal & Installation............9-16

I

Ignition Timing 9-12
  Adjustment.......................9-12
Intake Manifold 9-20
  Removal & Installation............9-20

L

Lower Ball Joints 9-44
  Removal & Installation............9-44
Lower Control Arm 9-44
  Removal & Installation............9-44

O

Oil Pan............................9-25
  Removal & Installation............9-25

Oil Pump

Oil Pump 9-27
  Removal & Installation............9-27

P

Pinion Seal 9-39
  Removal & Installation............9-39
Piston and Ring 9-32
  Positioning 9-32
Power Steering Gear 9-41
  Removal & Installation............9-41

R

Rear Main Seal 9-28
  Removal & Installation............9-28
Rocker Arms/Shafts 9-19
  Removal & Installation............9-19

S

Shock Absorbers 9-42
  Removal & Installation............9-42
Stabilizer Bar 9-43
  Removal & Installation............9-43
Starter Motor 9-24
  Removal & Installation............9-24
Steering Knuckle...................9-45
  Removal & Installation............9-45
Strut/Shock Module................9-42
  Removal & Installation............9-42

T

Timing Chain, Sprockets, Front
  Cover and Seal 9-29
  Removal & Installation............9-29
Transfer Case Assembly.............9-36
  Removal & Installation............9-36

U

Upper Ball Joints 9-44
  Removal & Installation............9-44
Upper Control Arm 9-44
  Removal & Installation............9-44

V

Valve Lash 9-24
  Adjustment.......................9-24

W

Water Pump 9-15
  Removal & Installation............9-15
Wheel Bearings and Hub 9-45
  Adjustment.......................9-45
  Removal & Installation............9-45

SPECIFICATIONS AND MAINTENANCE CHARTS

ENGINE AND VEHICLE IDENTIFICATION

Engine								Model Year	
Code ①	Liters (cc)	Cu. In.	Cyl.	Fuel Sys.	Engine Type	Eng. Mfg.		Code ②	Year
S	4.2 (4200)	256	6	MFI	DOHC	CPC		2	2002
P	5.3 (5326)	325	8	SFI	OHV	CPC		3	2003

CPC: Chevrolet/Pontiac/Canada

MFI: Multi-port Fuel Injection

① 8th position of VIN

② 10th position of VIN

Code ②	Year
2	2002
3	2003
4	2004
5	2005
6	2006

06025-BRAV-C01

GENERAL ENGINE SPECIFICATIONS

All measurements are given in inches.

Year	Model	Engine Displacemen Liters	Engine Series (ID/VIN)	Net Horsepower @ rpm	Net Torque @ rpm (ft. lbs.)	Bore x Stroke (in.)	Com- pression Ratio	Oil Pressure @ rpm
2002	Bradava	4.2	S	270@6000	275@3600	3.66x4.02	10.0:1	12@1200
	Envoy	4.2	S	270@6000	275@3600	3.66x4.02	10.0:1	12@1200
	TrailBlazer	4.2	S	270@6000	275@3600	3.66x4.02	10.0:1	12@1200
2003	Bravada	4.2	S	275@6000	275@3600	3.66x4.02	10.0:1	12@1200
		5.3	P	290@5300	325@4000	3.78x3.62	9.45:1	18@2000
	Envoy	4.2	S	275@6000	275@3600	3.66x4.02	10.0:1	12@1200
		5.3	P	290@5300	325@4000	3.78x3.62	9.45:1	18@2000
	TrailBlazer	4.2	S	270@6000	275@3600	3.66x4.02	10.0:1	12@1200
		5.3	P	290@5300	325@4000	3.78x3.62	9.45:1	18@2000
2004	Bravada	4.2	S	275@6000	275@3600	3.66x4.02	10.0:1	12@1200
		5.3	P	290@5300	325@4000	3.78x3.62	9.45:1	18@2000
	Envoy	4.2	S	275@6000	275@3600	3.66x4.02	10.0:1	12@1200
		5.3	P	290@5300	325@4000	3.78x3.62	9.45:1	18@2000
	Rainier	4.2	S	275@6000	275@3600	3.66x4.02	10.0:1	12@1200
		5.3	P	290@5300	325@4000	3.78x3.62	9.45:1	18@2000
	TrailBlazer	4.2	S	270@6000	275@3600	3.66x4.02	10.0:1	12@1200
		5.3	P	290@5300	325@4000	3.78x3.62	9.45:1	18@2000
2005	Envoy	4.2	S	275@6000	275@3600	3.66x4.02	10.0:1	12@1200
		5.3	P	300@5000	325@4000	3.78x3.62	9.95:1	18@2000
	Rainier	4.2	S	275@6000	275@3600	3.66x4.02	10.0:1	12@1200
		5.3	P	300@5300	325@4000	3.78x3.62	9.95:1	18@2000
	TrailBlazer	4.2	S	275@6000	275@3600	3.66x4.02	10.0:1	12@1200
		5.3	P	300@5300	325@4000	3.78x3.62	9.95:1	18@2000

06025-BRAV-C02

GASOLINE ENGINE TUNE-UP SPECIFICATIONS

Year	Engine Displacement Liters	Engine ID/VIN	Spark Plugs Gap (in.)	Ignition Timing (deg.) MT	Ignition Timing (deg.) AT	Fuel Pump (psi)	Idle Speed (rpm) MT	Idle Speed (rpm) AT	Valve Clearance In.	Valve Clearance Ex.
2002	4.2	S	0.050	—	①	50-57 ②	—	③	HYD	HYD
2003	4.2	S	0.050	—	①	50-57 ②	—	③	HYD	HYD
	5.3	P	0.060	①	①	55-62 ②	—	625	HYD	HYD
2004	4.2	S	0.050	—	①	50-57 ②	—	③	HYD	HYD
	5.3	P	0.060	①	①	55-62 ②	—	625	HYD	HYD
2005	4.2	S	0.042	—	①	50-57 ②	—	③	HYD	HYD
	5.3	P	0.040	①	①	55-62 ②	—	③	HYD	HYD

NOTE: The Vehicle Emission Control Information label often reflects specification changes made during production.

The label figures must be used if they differ from those in this chart.

HYD: Hydraulic

① Distributorless ignition, cannot be adjusted

② With key ON, engine OFF

③ Idle speed is maintained by the PCM

06025-BRAV-C03

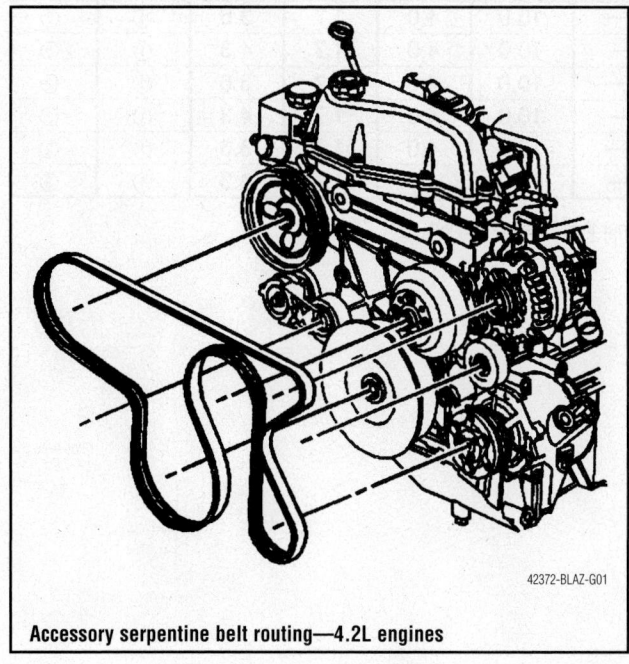

Accessory serpentine belt routing—4.2L engines

42372-BLAZ-G01

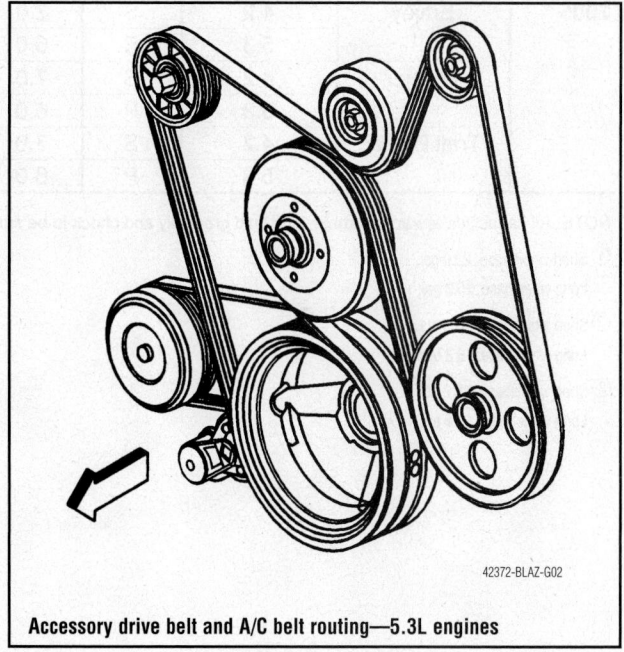

Accessory drive belt and A/C belt routing—5.3L engines

42372-BLAZ-G02

CAPACITIES

Year	Model	Engine Displacement Liters	Engine ID/VIN	Engine Oil with Filter (qts.)	Transmission (pts.) 5-Spd	Transmission (pts.) Auto.	Transfer Case (pts.)	Drive Axle Front (pts.)	Drive Axle Rear (pts.)	Fuel Tank (gal.)	Cooling System (qts.)
2002	Bravada	4.2	S	7.0	—	10.0	2.6	1.7	4.0	18.6	13.9
	Envoy	4.2	S	7.0	—	10.0	2.6	1.7	4.0	18.6	13.9
	TrailBlazer	4.2	S	7.0	—	10.0	2.6	1.7	4.0	18.6	13.9
2003	Bravada	4.2	S	7.0	—	10.0	4.0	1.7	3.6	①	②
		5.3	P	6.0	—	10.0	4.0	1.7	4.3	①	③
	Envoy	4.2	S	7.0	—	10.0	4.0	1.7	3.6	①	②
		5.3	P	6.0	—	10.0	4.0	1.7	4.3	①	③
	TrailBlazer	4.2	S	7.0	—	10.0	4.0	1.7	3.6	①	②
		5.3	P	6.0	—	10.0	4.0	1.7	4.3	①	③
2004	Bravada	4.2	S	7.0	—	10.0	4.0	1.7	3.6	①	②
		5.3	P	6.0	—	10.0	4.0	1.7	4.3	①	③
	Envoy	4.2	S	7.0	—	10.0	4.0	1.7	3.6	①	②
		5.3	P	6.0	—	10.0	4.0	1.7	4.3	①	③
	Rainier	4.2	S	7.0	—	10.0	4.0	1.7	3.6	①	②
		5.3	P	6.0	—	10.0	4.0	1.7	4.3	①	③
	Trail Blazer	4.2	S	7.0	—	10.0	4.0	1.7	3.6	①	②
		5.3	P	6.0	—	10.0	4.0	1.7	4.3	①	③
2005	Envoy	4.2	S	7.0	—	10.0	4.0	1.7	3.6	①	②
		5.3	P	6.0	—	10.0	4.0	1.7	4.3	①	③
	Rainier	4.2	S	7.0	—	10.0	4.0	1.7	3.6	①	②
		5.3	P	6.0	—	10.0	4.0	1.7	4.3	①	③
	Trail Blazer	4.2	S	7.0	—	10.0	4.0	1.7	3.6	①	②
		5.3	P	6.0	—	10.0	4.0	1.7	4.3	①	③

NOTE: All capacities are approximate. Add fluid gradually and check to be sure a proper fluid level is obtained.

① Short wheelbase: 22.0 gal.
   Long wheelbase: 25.3 gal.

② Short wheelbase: 13.9 qts.
   Long wheelbase: 15.3 qts.

③ Short wheelbase: 15.2 qts.
   Long wheelbase: 17.9 qts.

06025-BRAV-C04

VALVE SPECIFICATIONS

Year	Engine Displacement Liters	Engine ID/VIN	Seat Angle (deg.)	Face Angle (deg.)	Spring Test Pressure (lbs. @ in.)	Spring Installed Height (in.)	Stem-to-Guide Clearance (in.)		Stem Diameter (in.)	
							Intake	Exhaust	Intake	Exhaust
2002	4.2	S	NA	NA	130-142@1.26	NA	0.0011-0.0025	0.0015-0.0030	NA	NA
2003	4.2	S	NA	NA	130-142@1.26	NA	0.0011-0.0025	0.0015-0.0030	NA	NA
	5.3	P	46	45	220@1.32	1.80	0.0010-0.0026	0.0010-0.0026	0.313-0.314	0.313-0.314
2004	4.2	S	NA	NA	130-142@1.26	NA	0.0011-0.0025	0.0015-0.0030	NA	NA
	5.3	P	46	45	220@1.32	1.80	0.0010-0.0026	0.0010-0.0026	0.313-0.314	0.313-0.314
2005	4.2	S	NA	NA	130-142@1.26	NA	0.0011-0.0025	0.0015-0.0030	NA	NA
	5.3	P	46	45	220@1.32	1.80	0.0010-0.0026	0.0010-0.0026	0.313-0.314	0.313-0.314

NA: Not Available

06025-BRAV-C05

CRANKSHAFT AND CONNECTING ROD SPECIFICATIONS

All measurements are given in inches.

Year	Engine Displacement Liters	Engine ID/VIN	Crankshaft				Connecting Rod		
			Main Brg. Journal Dia.	Main Brg. Oil Clearance	Shaft End-play	Thrust on No.	Journal Diameter	Oil Clearance	Side Clearance
2002	4.2	S	2.7567-2.7574	0.0004-0.0025	0.0044-0.0153	4	2.2337-2.2342	0.0008-0.0025	0.0019-0.0137
2003	4.2	S	2.7567-2.7574	0.0004-0.0025	0.0044-0.0153	4	2.2337-2.2342	0.0008-0.0025	0.0019-0.0137
	5.3	P	2.5587-2.5593	0.0008-0.0021	0.0015-0.0078	4	2.0991-2.0999	0.0009-0.0025	0.0043-0.0200
2004	4.2	S	2.7567-2.7574	0.0004-0.0025	0.0044-0.0153	4	2.2337-2.2342	0.0008-0.0025	0.0019-0.0137
	5.3	P	2.5587-2.5593	0.0008-0.0021	0.0015-0.0078	4	2.0991-2.0999	0.0009-0.0025	0.0043-0.0200
2005	4.2	S	2.7567-2.7574	0.0004-0.0025	0.0044-0.0153	4	2.2337-2.2342	0.0008-0.0025	0.0019-0.0137
	5.3	P	2.5587-2.5593	0.0008-0.0021	0.0015-0.0078	4	2.0991-2.0999	0.0009-0.0025	0.0043-0.0200

06025-BRAV-C06

PISTON AND RING SPECIFICATIONS

All measurements are given in inches.

Year	Engine Displacement Liters	Engine ID/VIN	Piston Clearance	Ring Gap			Ring Side Clearance		
				Top Compression	Bottom Compression	Oil Control	Top Compression	Bottom Compression	Oil Control
2002	4.2	S	-0.0006 0.0014	0.0059- 0.0118	0.0142- 0.0201	0.0098- 0.0299	0.0017- 0.0037	0.0017- 0.0037	0.0023- 0.0085
2003	4.2	S	-0.0006 0.0014	0.0059- 0.0118	0.0142- 0.0201	0.0098- 0.0299	0.0017- 0.0037	0.0017- 0.0037	0.0023- 0.0085
	5.3	P	-0.0014 0.0006	0.0090- 0.0173	0.0173- 0.0275	0.007- 0.0295	0.0015- 0.0033	0.0015- 0.0031	0.0005- 0.0078
2004	4.2	S	-0.0006 0.0014	0.0059- 0.0118	0.0142- 0.0201	0.0098- 0.0299	0.0017- 0.0037	0.0017- 0.0037	0.0023- 0.0085
	5.3	P	-0.0014 0.0006	0.0090- 0.0173	0.0173- 0.0275	0.007- 0.0295	0.0015- 0.0033	0.0015- 0.0031	0.0005- 0.0078
2005	4.2	S	-0.0006 0.0014	0.0059- 0.0118	0.0142- 0.0201	0.0098- 0.0299	0.0017- 0.0037	0.0017- 0.0037	0.0023- 0.0085
	5.3	P	-0.0014 0.0006	0.0090- 0.0173	0.0173- 0.0275	0.0070- 0.0295	0.0015- 0.0033	0.0015- 0.0031	0.0005- 0.0078

06025-BRAV-C07

TORQUE SPECIFICATIONS

All readings in ft. lbs.

Year	Engine Displacement Liters	Engine ID/VIN	Cylinder Head Bolts	Main Bearing Bolts	Rod Bearing Bolts	Crankshaft Damper Bolts	Flywheel Bolts	Manifold Intake *	Exhaust	Spark Plugs	Oil Pan Drain Plug
2002	4.2	S	①	②	③	④	⑤	⑥	⑦	13	19
2003	4.2	S	①	②	③	④	⑤	⑥	⑦	13	19
	5.3	P	⑧	⑨	⑩	⑪	⑫	⑬	⑭	11	18
2004	4.2	S	①	②	③	④	⑤	⑥	⑦	13	19
	5.3	P	⑧	⑨	⑩	⑪	⑫	⑬	⑭	11	18
2005	4.2	S	①	②	③	④	⑤	⑥	⑦	13	19
	5.3	P	⑧	⑨	⑩	⑪	⑫	⑬	⑭	11	18

* NOTE: Applies to Lower Manifold only.

① Cylinder head bolts (14)
　1st pass: 22 ft. lbs.
　2nd pass: Plus 155 degrees
　2 short end bolts: 62 INCH lbs.
　2nd pass: plus 60 degrees
　1 long end bolt: 62 INCH lbs.
　2nd pass: plus 120 degrees

② 18 ft. lbs., plus 180 depress
　2nd pass: 106 inch lbs.

③ 18 ft. lbs., plus 180 degrees

④ 110 ft. lbs., plus 180 degrees

⑤ 18 ft. lbs., plus 50 degrees

⑥ 89 INCH lbs.

⑦ 1st pass: 15 ft. lbs.
　2nd pass: 15 ft. lbs.
　3rd pass: 15 ft. lbs.

⑧ M11 bolts: 22 ft. lbs.
　2nd pass: Plus 90 degrees22 ft. lbs.
　3rd pass: Plus 70 degrees
　M8 bolts: 22 ft. lbs.

⑧ 111 ft. lbs., plus 180 degrees

⑨ Inner bolts:
　1st pass: 15 ft. lbs.
　Final pass: Plus 80 degrees
　Outer bolts:
　1st pass: 15 ft. lbs.
　Final pass: Plus 51 degrees

⑩ 15 ft. lbs. plus 75 degrees

⑪ Installation pass: 240 ft. lbs. (discard bolt)
　First pass: 37 ft. lbs. (new bolt)
　Final pass: Plus 140 degrees

⑫ 1st pass: 15 ft. lbs.
　2nd pass: 37 ft. lbs.
　3rd pass: 74 ft. lbs.

⑬ 1st pass: 44 inch lbs.
　2nd pass: 89 inch lbs.

⑭ 1st pass: 11 ft. lbs.
　2nd pass: 15 ft. lbs.

06025-BRAV-C08

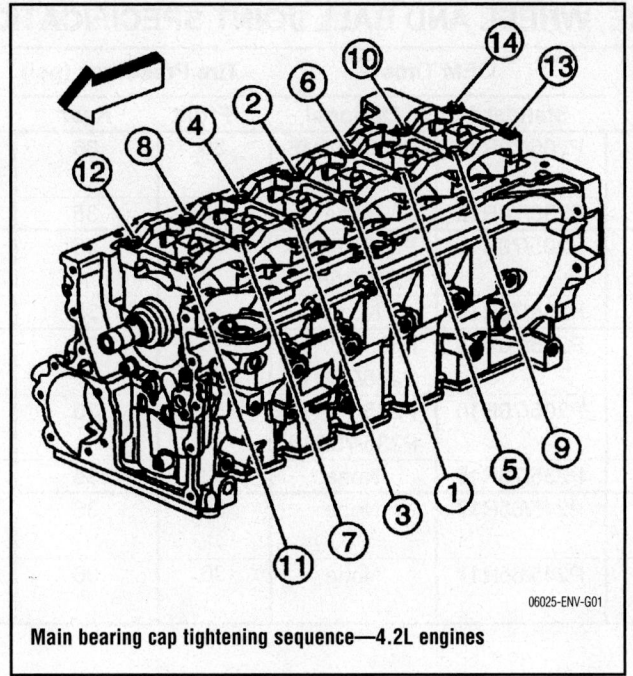

Main bearing cap tightening sequence—4.2L engines

06025-ENV-G01

WHEEL ALIGNMENT

Year	Model		Caster Range (+/-Deg.)	Caster Preferred Setting (Deg.)	Camber Range (+/-Deg.)	Camber Preferred Setting (Deg.)	Toe-in (in.)
2002	Bravada/Envoy/	2WD	1.00	+3.50	1.00	-0.5	-0.10+0.30
	TrailBlazer	4WD	1.00	+3.50	1.00	-0.5	-0.10+0.30
2003	Bravada/Envoy/	2WD	1.00	+3.50	1.00	-0.5	-0.10+0.30
	TrailBlazer	4WD	1.00	+3.50	1.00	-0.5	-0.10+0.30
2004	Bravada/Envoy/	2WD	0.50	+3.50	0.30	0.00	-0.10+/-0.20
	TrailBlazer	4WD	0.50	+3.50	0.30	0.00	-0.10+/-0.20
	Rainier	2WD	0.50	+3.50	0.30	0.00	-0.10+/-0.20
		4WD	0.50	+3.50	0.30	0.00	-0.10+/-0.20
2005	Bravada/Envoy/	①	0.60	+4.00	0.60	0.00	-0.10+/-0.20
	TrailBlazer	②	0.60	+4.25	0.60	0.00	-0.10+/-0.20
	Rainier	①	0.60	+4.00	0.60	0.00	-0.10+/-0.20
		②	0.60	+4.25	0.60	0.00	-0.10+/-0.20

① With rear coil spring suspension.

② With rear air spring suspension.

06025-BRAV-C09

TIRE, WHEEL AND BALL JOINT SPECIFICATIONS

Year	Model	OEM Tires Standard	OEM Tires Optional	Tire Pressures (psi) Front	Tire Pressures (psi) Rear	Wheel Size	Ball Joint Inspection	Lug Nut (ft. lbs.)
2002	Envoy TrailBlazer	P205/70R15	P235/70R15 P235/75R15	36	36	6-JJ	L ①	103
	Bravada	P235/70R15	None	33	35	7-JJ	②	103
2003	Envoy TrailBlazer	P205/70R15	P235/70R15 P235/75R15	36	36	6-JJ	L ①	103
	Bravada	P235/70R15	None	33	35	7-JJ	②	103
2004	Envoy TrailBlazer	P205/70R15	P235/70R15 P235/75R15	36	36	6-JJ	L ①	103
	Rainier	P205/70R15	P235/70R15 P235/75R15	36	36	6-JJ	L ①	103
	Bravada	P235/70R15	None	33	35	7-JJ	②	103
2005	Envoy TrailBlazer	P245/65R17	None	36	36	7-JJ	L ①	103
	Rainier	P245/65R17	None	36	36	7-JJ	L ①	103

OEM: Original Equipment Manufacturer

PSI: Pounds Per Square Inch

STD: Standard

OPT: Optional

L: Lower

U: Upper

① Do not lift truck. Inspect the boss into which the grease fitting is threaded. Replace if the boss is flush or receded below the surface of the ball joint

② Replace if any measurable movement is found.

06025-BRAV-C10

BRAKE SPECIFICATIONS
All measurements in inches unless noted

Year	Model	Front Brake Disc Original Thickness	Front Brake Disc Minimum Thickness	Front Brake Disc Max Run-out	Rear Brake Disc Original Thickness	Rear Brake Disc Minimum Thickness	Rear Brake Disc Max Run-out	Minimum Lining Thickness Front	Minimum Lining Thickness Rear	Brake Caliper Bracket Bolts (ft. lbs.)	Brake Caliper Mounting Bolts (ft. lbs.)
2002	Bravada	1.140	1.080	0.002	0.787	0.728	0.002	NA	NA	①	②
	Envoy	1.140	1.080	0.002	0.787	0.728	0.002	NA	NA	①	②
	TrailBlazer	1.140	1.080	0.002	0.787	0.728	0.002	NA	NA	①	②
2003	Bravada	1.140	1.080	0.002	0.787	0.728	0.002	NA	NA	①	②
	Envoy	1.140	1.080	0.002	0.787	0.728	0.002	NA	NA	①	②
	TrailBlazer	1.140	1.080	0.002	0.787	0.728	0.002	NA	NA	①	②
2004	Bravada	1.140	1.080	0.002	0.787	0.728	0.002	NA	NA	①	②
	Envoy	1.140	1.080	0.002	0.787	0.728	0.002	NA	NA	①	②
	Rainier	1.140	1.080	0.002	0.787	0.728	0.002	NA	NA	①	②
	TrailBlazer	1.140	1.080	0.002	0.787	0.728	0.002	NA	NA	①	②
2005	Envoy	1.140	1.080	0.002	0.787	0.728	0.002	NA	NA	①	②
	Rainier	1.140	1.080	0.002	0.787	0.728	0.002	NA	NA	①	②
	TrailBlazer	1.140	1.080	0.002	0.787	0.728	0.002	NA	NA	①	②

NA: Not Available

① Front: 110 ft. lbs.   ② Front: 31 ft. lbs.
   Rear: 100 ft. lbs.   Rear: 23 ft. lbs.

06025-BRAV-C11

SCHEDULED MAINTENANCE INTERVALS
2002-04 BRAVADA, ENVOY, RAINIER & TRAILBLAZER

TO BE SERVICED	TYPE OF SERVICE	VEHICLE MILEAGE INTERVAL (x1000)															
		7.5	15	22.5	30	37.5	45	52.5	60	67.5	75	82.5	90	97.5	105	112.5	120
Accessory drive belt	S/I									✓							✓
Air cleaner filter	R				✓					✓			✓				✓
Automatic transmission fluid	R	Every 50,000 miles															
Brake system ①	S/I	✓	✓	✓	✓	✓	✓	✓	✓	✓	✓	✓	✓	✓	✓	✓	✓
Chassis & suspension grease points	L	✓	✓	✓	✓	✓	✓	✓	✓	✓	✓	✓	✓	✓	✓	✓	✓
CV-joint boots & axle seals	S/I	✓	✓	✓	✓	✓	✓	✓	✓	✓	✓	✓	✓	✓	✓	✓	✓
Engine coolant ②	S/I	Every 150,000 miles															
Engine oil & filter	R	✓	✓	✓	✓	✓	✓	✓	✓	✓	✓	✓	✓	✓	✓	✓	✓
Front wheel bearings	S/I & L				✓					✓			✓				✓
Fuel filter	R				✓					✓			✓				✓
Fuel tank, cap & lines	S/I									✓							✓
PCV valve	S/I	Every 100,000 miles															
Rear/front axle fluid	S/I	✓	✓	✓	✓	✓	✓	✓	✓	✓	✓	✓	✓	✓	✓	✓	✓
Rotate tires	S/I	✓	✓	✓	✓	✓	✓	✓	✓	✓	✓	✓	✓	✓	✓	✓	✓
Spark plug wires	S/I	Every 100,000 miles															
Spark plugs	R	Every 100,000 miles															

R: Replace   S/I: Inspect and service, if necessary   L: Lubricate

① This should be performed when the tires are removed for rotation.

② Drain, flush and refill the cooling system, inspect the system hoses, and clean the radiator and condenser.

③ 2-wheel drive models only.

FREQUENT OPERATION MAINTENANCE (SEVERE SERVICE)

If a vehicle is operated under any of the following conditions it is considered severe service:

- Towing a trailer or using a camper or car-top carrier.

- Repeated short trips of less than 5 miles in temperatures below freezing, or trips of less than 10 miles in any temperature.

- Extensive idling or low-speed driving for long distances as in heavy commercial use, such as delivery, taxi or police cars.

- Operating on rough, muddy or salt-covered roads.

- Operating on unpaved or dusty roads.

- Driving in extremely hot (over 90°) conditions.

Engine oil & filter: replace every 3000 miles or 3 months, whichever occurs first.

Chassis and suspension grease points: lubricate every 3000 miles.

Rear/front axle fluid level: inspect every 3000 miles.

Rotate the tires ever 6000 miles.

Brake system components: inspect ever 6000 miles.

Front wheel bearings (2-wheel drive only): clean, inspect and repack every 15,000 miles.

Air cleaner filter: inspect every 15,000 miles.

Automatic transmission fluid & filter: replace every 15,000 miles.

06025-BRAV-C12

MAINTENANCE I AND II SERVICE SCHEDULES
2005 Envoy, Rainer & Trail Blazer

When the CHANGE ENGINE OIL light appears, certain services and inspections are required.
Required services are described as Maintenance I and Maintenance II.
The first service on a vehicle should be Maintenance I, and the second service should be Maintenance II.
Alternate between the 2 thereafter. However, in some cases, Maintenance II may be required more often.
Maintenance I: Use Maintenance I if the CHANGE ENGINE OIL light comes on within 10 months since vehicle was purchased or, if Maintenance II was performed.
Maintenance II: Use Maintenance II if the previous service performed was Maintenance I.
Always use Maintenance II whenever the CHANGE ENGINE OIL light comes on 10 months or more since the last service, or, if the CHANGE ENGINE OIL light has not come on at all for one year.

Service	Maintenance I	Maintenance II
Change the engine oil and filter. Reset the oil life system.	✓	✓
Visually inspect the vehicle for leaks or damage. A fluid loss in the vehicle system could indicate a problem. Inspected, repair and add fluid to the system if necessary.	✓	✓
Inspect the engine air cleaner filter. If necessary, replace the filter.	✓	✓
Rotate the tires. Inspect the tire inflation pressures and the tire wear.	✓	✓
Visually inspect the brake lines and hoses for proper hook-up, binding, leaks, cracks, chafing, etc. Inspect the disc brake pads for wear and the rotors for surface condition. Inspect the drum brake linings for wear or cracks. Inspect other brake parts, including drums, wheel cylinders, calipers, parking brake, etc. Inspect the parking brake adjustment.	✓	✓
Inspect the engine coolant and the windshield washer fluid levels. Add fluid as needed.	✓	✓
Inspect the suspension and steering components. Inspect the front and rear suspension and the steering system for damaged, loose or missing parts, or signs of wear. Inspect the power steering lines and the hoses for proper hook-up, binding, leaks, cracks, chafing, etc.	--	✓
Visually inspect the coolant hoses and replace the hoses if they are cracked, swollen or deteriorated. Inspect all pipes, fittings and clamps; replace with GM parts as needed. To help ensure proper operation, a pressure test of the cooling system and pressure cap and cleaning the outside of the radiator and air conditioning condenser is recommended at least once a year.		✓
Inspect the wiper blades.	--	✓
Inspect the restraint system components. Ensure the safety belt reminder light and all the belts, buckles, latch plates, retractors and anchorages are working properly. Look for any other loose or damaged safety belt system parts. If you see anything that might keep a safety belt system from working correctly, repair or replaced the damaged part. Replace torn or frayed safety belts, refer to Operational and Functional Checks in Seat Belts. Inspect for any opened or broken air bag coverings, and repair or replace as needed. The air bag system does require regular maintenance.	--	✓

MAINTENANCE I AND II SERVICE SCHEDULES
2005 Envoy, Rainer & Trail Blazer

Lubricate the body components.Lubricate all key lock cylinders, hood latch assemblies, secondary latches, pivots, spring anchor and release pawl, hood and door hinges, rear folding seats and liftgate hinges. Frequent lubrication may be required when exposed to a corrosive environment, refer to Fluid and Lubricant Recommendations . Applying dielectric silicone grease GM P/N 12345579 (Canadian P/N 1974984) or equivalent on the weatherstrips with a clean cloth.	--	✓
Inspect the transaxle fluid level and add fluid as needed.	--	✓
Inspect the suspension and steering components.Inspect the front and rear suspension and the steering system for damaged, loose or missing parts, or signs of wear. Inspect power steering lines and hoses for proper hook-up, binding, leaks, cracks, chafing, etc.	--	✓
Inspect the throttle system for interference or binding and for damaged or missing parts. Replace the parts as needed. Replace any components that have high effort or excessive wear. Do not lubricate the accelerator or the cruise control cables.	--	✓
Replace the passenger compartment air filter.	--	✓

06025-BRAV-C14

ENGINE REPAIR

Distributor

All models are equipped with distributor-less ignition systems.

Alternator

REMOVAL

4.2L Engine

1. Remove or disconnect the following:
   - Negative battery cable
   - Accessory belt
   - Positive battery cable nut from the generator
   - A/C line mounting bracket bolt at the engine lift hook
   - Right engine lift hook bolts
   - Engine lift hook
   - Mounting bolts
   - Alternator

5.3L Engine

1. Remove or disconnect the following:
   - Negative battery cable
   - Accessory belt
   - Electrical connector
   - Terminal stud nut, after sliding boot down
   - Alternator cable
   - Mounting bolts
   - Alternator

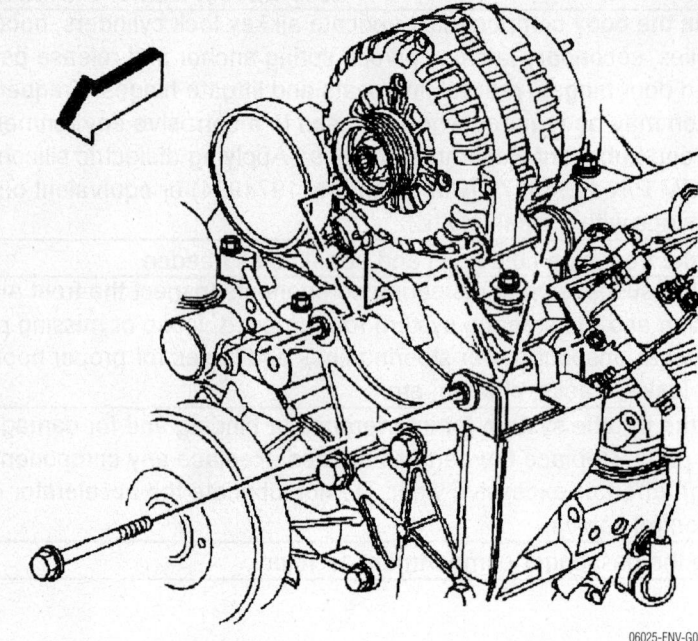

06025-ENV-G03

Alternator mounting—5.3L engines

INSTALLATION

4.2L Engine

1. Install or connect the following:
   - Alternator and loosely install the mounting blots
   - Tighten the alternator mounting bolts to 37 ft. lbs. (50 Nm)

- Positive battery cable and secure with the nut; tighten the nut to 80 inch lbs. (9 Nm)
- Engine lift hook and bolts; tighten the bolts to 37 ft. lbs. (50 Nm)
- A/C line bracket to the lift hook, then tighten the retaining bolt to 89 inch lbs. (10 Nm)
- Accessory belt
- Negative battery cable

5.3L Engine

1. Install or connect the following:
   - Alternator and loosely install the mounting bolts
   - Tighten the bolts to 37 ft. lbs. (50 Nm)
   - Alternator cable
   - Terminal stud nut and tighten to 80 inch lbs. (9 Nm)
   - Boot back over terminal stud
   - Electrical connector
   - Accessory belt
   - Negative battery cable

Ignition Timing

ADJUSTMENT

The ignition timing is preset and cannot be adjusted.

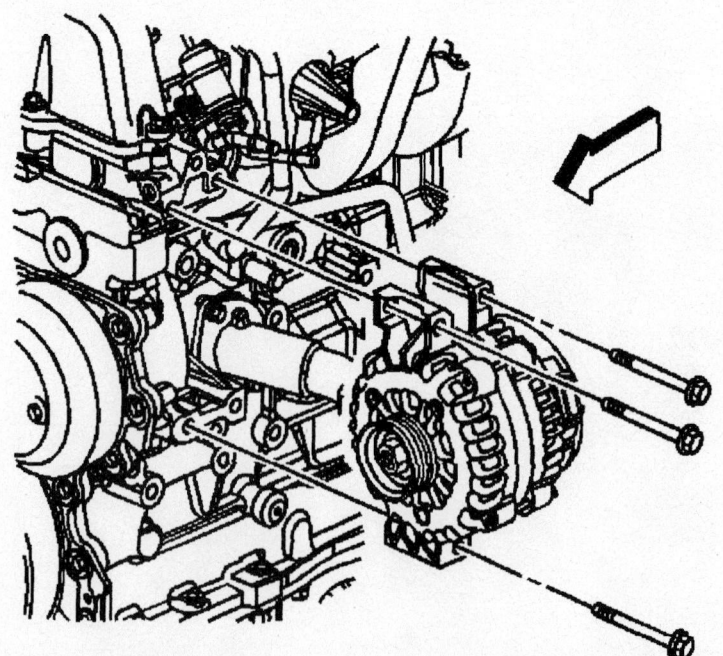

06025-ENV-G02

Alternator mounting—4.2L engines

Engine Assembly

REMOVAL & INSTALLATION

4.2L Engine

1. Drain the engine cooling system

➡ **Keep the oil drain plug removed during the engine removal and installation.**

2. Drain the engine oil. Install a suitable plug into the oil pan to prevent oil leakage during the remainder of the procedure.

3. Using the proper equipment, discharge and recover the refrigerant from the A/C system, if equipped.

4. Remove or disconnect the following:
- Hood
- Negative battery cable
- Fuel system pressure
- Air cleaner assembly
- Throttle body
- Manifold Absolute Pressure (MAP) sensor
- Windshield washer solvent container
- Air intake baffle
- Grille
- Headlight housing
- Radiator support brace
- Hood latch
- A/C lines from the condenser
- Transmission cooler lines from the engine, not the radiator

5. Remove the cooling fan and shroud, tilting the radiator forward, and the cooling fan and shroud rearward for clearance.
- Accessory belt
- Power steering pump bolts; position the pump aside
- Heater hoses from the heater core
- Transmission filler tube bracket nut from the Air Injector Reactor (AIR) adapter
- AIR adapter

6. Install a suitable lift hook to the AIR adapter

7. Remove or disconnect the following:
- Oxygen (O$_2$) sensor connector
- A/C line from the accumulator
- Front axle actuator electrical connector
- Camshaft phaser actuator valve electrical connector
- Transmission cooler lines from the clips on the right side of the engine block
- Ignition coil harness connectors
- Harness retainer from the clips

- Power brake hose from the booster
- Powertrain Control Module (PCM)
- Fuel lines from the fuel pressure regulator. Cap the lines to avoid excessive fuel leakage.
- All harnesses from the engine harness bracket
- Engine harness bracket bolt and bracket
- Starter electrical connections
- A/C pressure sensor and clutch electrical connector
- Alternator electrical connector and battery lead
- Knock Sensor (KS), Crankshaft Position (CKP) and Camshaft Position (CMP) sensor electrical connectors
- 4 grounds on the left side of the block

8. Raise and safely support the vehicle.
- Left and right side driveshafts
- Propeller shaft from the front axle pinion yoke
- Engine protection shield
- Exhaust pipe from the exhaust manifold. Slide the exhaust pipe backward slightly.
- Fuel tank shield, if equipped
- AIR lines from the AIR pump
- Torque converter access cover and bolts

9. Place a jack on the transmission fluid pan for support.

10. Remove the transmission support.

11. Lower the transmission enough to reach the top bell housing bolts.

12. Remove the top 4 bell housing bolts, there may be 2 harness clips that will need to be removed in order to have access to 2 of the top bolts.

13. Raise the transmission.

14. Reinstall the transmission support using only 2 through bolts.

15. Remove or disconnect the following:
- Remaining bell housing bolts (11 total)
- Left and right engine lower mount nuts
- Oil level sensor electrical connector
- Oil pressure switch electrical connector

16. Carefully lower the vehicle.

17. Remove the left, then the right upper engine mount nut.

18. Install a suitable engine hoist.

19. Raise the engine out of the compartment slowly, keeping the transmission supported.

20. Remove both engine mounts for clearance.

21. Continue raising the engine out of the vehicle.

22. Place the engine on a suitable engine stand.

To install:

23. Remove the engine from the engine stand.

24. Slowly install the engine into the engine compartment, aligning the engine mounts with the brackets.

25. When the engine mounts are aligned, install the engine mounts, putting the mount up through the engine mount brackets before inserting into the chassis mount brackets.

26. Lower the engine onto the mounts and install the upper engine mounting nuts. Tighten the nuts to 51 ft. lbs. (71 Nm).

27. Remove the engine hoist.

28. Lay the radiator into the radiator support, but do not install the radiator completely.

29. Raise and safely support the vehicle.

30. Install the lower bell housing bolts, except the top four.

31. Remove the 2 through bolts secure the transmission support, then lower the transmission.

32. Install the top 4 bell housing bolts and tighten all 11 bolts to 37 ft. lbs. (50 Nm).

33. Raise the transmission.

34. Install or connect the following:
- Transmission support
- 3 torque converter bolts and tighten to 44 ft. lbs. (60 Nm)
- Torque converter bolt cover
- AIR pipes to the AIR pump
- Fuel tank shield, if equipped
- Engine protection shield
- Propeller shaft to the front axle pinion yoke
- Exhaust pipe to the manifold and tighten the bolts to 37 ft. lbs. (50 Nm)
- Oil level switch and oil pressure sender electrical connectors
- Oil pan drain plug and tighten to 19 ft. lbs. (26 Nm)
- Lower radiator hose
- Left and right wheel driveshafts

35. Lower the vehicle.
- 4 grounds on the left side of the block
- CMP, CKP and knock sensor electrical connectors
- Alternator and starter electrical connectors and battery leads. Torque the nuts to 80 inch lbs. (9 Nm).
- Fuel lines at the fuel pressure regulator

- Engine harness bracket and bolt. Torque the bolt to 37 ft. lbs. (50 Nm).
- Front differential vent hose, to the engine harness bracket
- PCM
- Power brake hose to the booster
- Harness retainer to its original location
- Ignition coil harness connectors
- Transmission cooler lines to clips on right side of engine block
- Camshaft phaser actuator valve electrical connector
- Front axle actuator electrical connector
- A/C line at the accumulator

36. Remove the lift hook.
37. Install or connect the following:
- AIR adapter and secure with the studs. Tighten to 18 ft. lbs. (25 Nm).
- Transmission filler tube bracket to AIR adapter stud and secure the bracket with the nut. Torque the nut to 89 inch lbs. (10 Nm).
- Heater hoses to the heater core
- Power steering pump and tighten the bolts to 18 ft. lbs. (25 Nm).

38. The remainder of installation is the reverse of removal, but please note the following important steps:
39. Connect the negative battery cable
40. Check all powertrain fluid levels and add, as necessary.
41. Refill the engine crankcase.
42. Refill the engine cooling system.
43. Perform the CKP System Variation Learn Procedure, as follows:

a. Install a suitable scan tool and check for Diagnostic Trouble Codes (DTCs). If any DTCs, other than P1336 are set, resolve those codes first, before proceeding with this procedure.

b. With the scan tool, select the crankshaft position variation learn procedure.

c. Observe the fuel cut-off for the 4.2L engine.

d. The scan tool will instruct you to perform certain steps, make sure you follow all directions given by the scan tool exactly.

e. Enable the crankshaft position system variation learn procedure.

➡While the learn procedure is in progress, release the throttle immediately when the engine started to decelerate. The engine control is returned to the operator and the engine responds to throttle position after the learn procedure is complete.

f. Slowly increase the engine speed to the RPM that you observed.

g. Immediately release the throttle when fuel cut-out is reached.

h. The scan tool displays: Learn Status: Learned this ignition. If the scan tool does NOT display this message and not other DTCs set, you must perform further troubleshooting.

i. Turn the ignition **OFF** for 30 seconds after the learn procedure has been completed successfully.

44. Start and run the engine, then check for leaks.

5.3L Engine

1. Drain the engine cooling system
2. Drain the engine oil.
3. Remove and recover the refrigerant, if equipped with A/C.
4. Remove or disconnect the following:
- Negative battery cable
- Hood
- Radiator
- Radiator support brace
- Front axle, if 4WD
- Drive shafts
- Intake manifold
- Oil pressure sensor connector
- Oxygen (O_2) sensor connector

- Camshaft Position (CMP) sensor connector
- A/C compressor hose
- Rear auxiliary A/C compressor pipe fitting
- Rear auxiliary A/C compressor pipe nut and bolt. Tie the pipe out of the way.
- Engine Coolant Temperature (ECT) sensor
- Ground terminal bolt
- Retaining clips from the brackets
- A/C pressure switch electrical connector
- Retaining clip from the cylinder head
- Ground terminal bolts
- Starter
- Battery cable channel bolt
- Battery cable channel from the oil pan
- A/C compressor electrical connector

5. Collect all branches of the engine wiring harness, then position the harness out of the way.

- Alternator cable from the alternator
- Alternator bracket bolts, then position the bracket and alternator assembly aside
- Inlet and outlet hoses from the

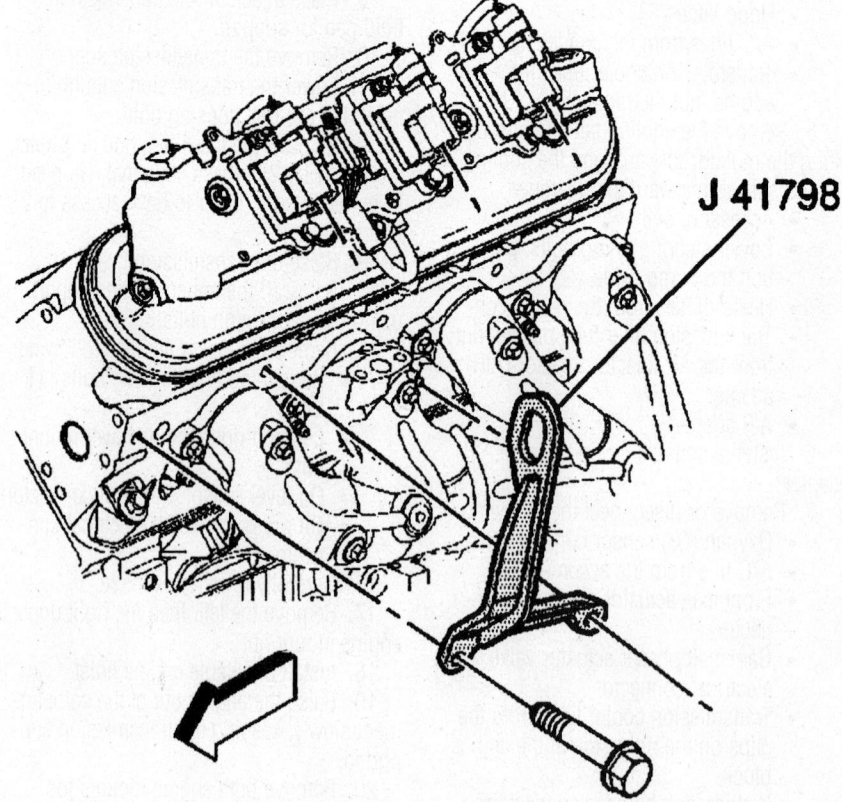

J 41798

42372-BLAZ-G03

If necessary, remove ignition coil(s) to install the engine lifting brackets—5.3L engine

water outlet, using J 38185 to move the hose clamps
- Auxiliary heater inlet and outlet hose/pipe assembly from the heater water shutoff valve pipes
- Auxiliary heater inlet and outlet hoses/pipes from the water pump, using Hose Clamp Pliers J 38185
- Remove ignition coils, if necessary, to install Engine Lifting Brackets J 41798 to the cylinder heads

6. Install Engine Lifting Brackets J 41798 to the cylinder heads. Tighten the M8 bolts to 18 ft. lbs. (25 Nm) and the M10 bolts to 37 ft. lbs. (50 Nm).
- Catalytic converter
- 3 frame engine mount bracket bolts from the right and left sides
- Torque converter bolts
- Transmission oil level dipstick tube nut and tube
- Transmission bolt and stud on the right side
- Lower transmission bolt/studs
- 3 upper transmission bolts/studs

7. Install a suitable engine hoist to the engine lifting brackets.

8. Place a floor jack under the transmission for support.

9. Separate the engine from the transmission.

10. Remove the engine from the vehicle and place on a suitable engine stand.

11. Install Converter Holding Strap J 21366 to the transmission to hold the torque converter.

To install:

12. Remove Converter Holding Strap J 21366 from the transmission.

13. Attach the engine to a hoist and remove it from the engine stand

14. Install or connect the following:
- Engine into the vehicle. Match the transmission up to the engine, then remove the floor jack.
- 3 upper transmission bolts/studs and tighten to 37 ft. lbs. (50 Nm)
- Lower transmission bolts/studs and tighten to 37 ft. lbs. (50 Nm)
- Transmission bolt and stud on the right side and tighten to 37 ft. lbs. (50 Nm)
- Transmission oil level dipstick tube and nut. Torque to 89 inch lbs. (10 Nm).
- Torque converter bolts and tighten to 44 ft. lbs. (60 Nm)
- 3 frame engine mount bracket bolts to both the right and left sides. Torque the bolts to 37 ft. lbs. (50 Nm).
- Catalytic converter

15. Remove the engine lifting brackets from the cylinder heads
- Ignition coils, if removed, and tighten the bolts to 71 inch lbs. (8 Nm)
- Auxiliary heater inlet and outlet hoses
- Auxiliary heater inlet and outlet hose/pipe assembly to the heater water shutoff valve pipes
- Outlet and inlet hoses to the water outlet
- Bracket and alternator assembly. Tighten the bolts to 37 ft. lbs. (50 Nm).
- Cable to the alternator
- Position the engine wiring harness back over the engine
- A/C compressor electrical connector
- Battery cable channel to the oil pan and secure with the bolt. Torque to 106 inch lbs. (12 Nm).
- Starter
- Ground terminal bolts and tighten to 18 ft. lbs. (25 Nm)
- Retaining clip to the cylinder head
- A/C pressure switch electrical connector
- Retaining clips to the brackets
- Ground terminal bolt and tighten to 18 ft. lbs. (25 Nm)
- ECT sensor connector
- Rear auxiliary A/C compressor pipe nut and bolt. Torque to 15 ft. lbs. (20 Nm).
- A/C compressor hose
- Oil pressure sensor connector
- O_2 sensor connector
- CMP sensor connector
- Intake manifold
- Drive shafts
- Front axle, if removed
- Radiator support brace
- Radiator

16. Recharge the A/C system
- Negative battery cable
- Hood

17. Check all powertrain fluid levels and add, as necessary.

18. Refill the engine crankcase.

19. Refill the engine cooling system.

20. Start and run the engine, then check for leaks.

Water Pump

REMOVAL & INSTALLATION

1. Disconnect the negative battery cable.

2. Drain the engine cooling system.

3. On 5.3L engines, loosen the air cleaner outlet duct clamps at the throttle body and Mass Airflow/Intake Air Temperature (MAP/IAT) sensor. Remove the bolt and air cleaner outlet duct.

4. Relieve the belt tension and remove the accessory drive belts or the serpentine drive belt, as applicable.

5. Remove or disconnect the following:
- Upper fan shroud
- Fan or fan and clutch assembly, as applicable
- Water pump pulley; use a suitable tool to hold the pulley while removing the bolts
- Coolant hose(s) from the water pump

➡ **For the hoses on some engines, removal may be easier if the hose is left attached until the pump is free from the block. Once the pump is removed from the engine, the pump may be pulled (giving a better grip and greater leverage) from the tight hose connection.**

- Water pump retainers
- Water pump from the engine

❈❈ WARNING

Note the positions of all retainers as some engines will utilize different length fasteners in different locations and/or bolts and studs in different locations.

To install:

6. Clean the gasket mounting surfaces.

➡ **The water pumps on some of the engines covered may have been installed using sealer only, no gasket, at the factory. If a gasket is supplied with the replacement part, it should be used. Otherwise, a ⅛ in. (3mm) bead of RTV sealer should be used around the sealing surface of the pump.**

7. Apply sealant to the water pump retainer threads.

8. Install or connect the following:
- Water pump using a new gasket. Tighten the water pump retainers to 89 inch lbs. (10 Nm) for 4.2L engines, or on 5.3L engines to 11 ft. lbs. (15 Nm), then to 22 ft. lbs. (30 Nm).
- Coolant hose(s)
- Water pump pulley. Tighten the pulley bolts to 18 ft. lbs. (25 Nm).
- Fan or fan and clutch assembly
- Serpentine drive belt (if equipped) by

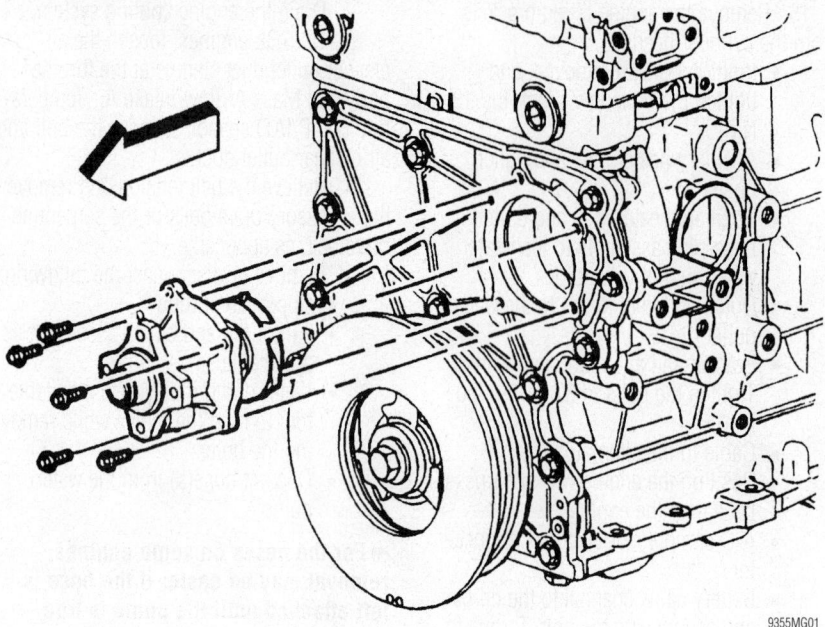

Exploded view of the water pump assembly mounting—4.2L engine

9355MG01

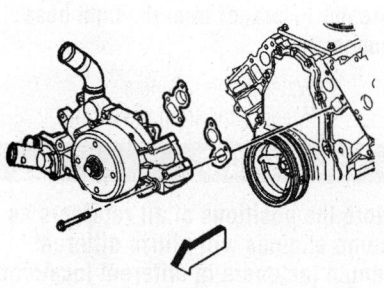

Exploded view of the water pump assembly mounting—5.3L engine

42372-BLAZ-G04

positioning the belt over the pulleys and carefully allow the tensioner back into contact with the belt.
- V-belts (if equipped) and adjust the tension
- Upper fan shroud
- Negative battery cable
9. Refill the engine cooling system.
10. Run the engine and check for leaks.

Heater Core

REMOVAL & INSTALLATION

1. Drain the engine cooling system.
2. Remove or disconnect the following:
   - Negative battery cable
   - Heater hoses from the heater core
3. Remove the instrument panel as follows:
   a. Disable the air bag system.
   b. Set the parking brake and block the wheels.

c. Disconnect the parking brake release cable from the parking brake lever.

d. Unfasten the screws that retain the DLC instrument panel left side sound insulator. Feed the DLC through the hole in the sound insulator.

e. Unfasten the right side sound insulator panel screws and remove the panel.

f. Unfasten the screws that attach the instrument panel left side sound insulator to the knee bolster and cowl panel.

g. Unfasten the nut that attaches the left side sound insulator to the accelerator pedal bracket.

h. Unplug the remote control door lock receiver module electrical connector.

i. Remove the door lock receiver module from the left side sound insulator. Remove the left side sound insulator.

j. Unfasten the screws that attach the instrument panel center sound insulator to the knee bolster, instrument panel, heater assembly and floor duct.

k. Remove the center sound insulator.

l. Unfasten the screws that attach the courtesy lamp to the knee bolster.

m. Unfasten the screws that attach the knee bolster to the instrument panel.

n. Disconnect the lap cooler duct from the knee bolster.

o. Unplug the lighter electrical connection and remove the knee bolster.

p. Unfasten the steering column-to-instrument panel nuts and lower the column.

q. Unfasten the screws that attach the instrument panel accessory trim plate to the instrument panel.

r. Remove the trim plate and unplug all necessary electrical connection.

s. Remove the heater and/or air conditioning control assembly.

t. Remove the radio and the storage compartment assembly (if equipped).

u. If necessary, remove the instrument cluster.

v. Unfasten the left and right instrument panel pivot bolts and the panel lower support bolt.

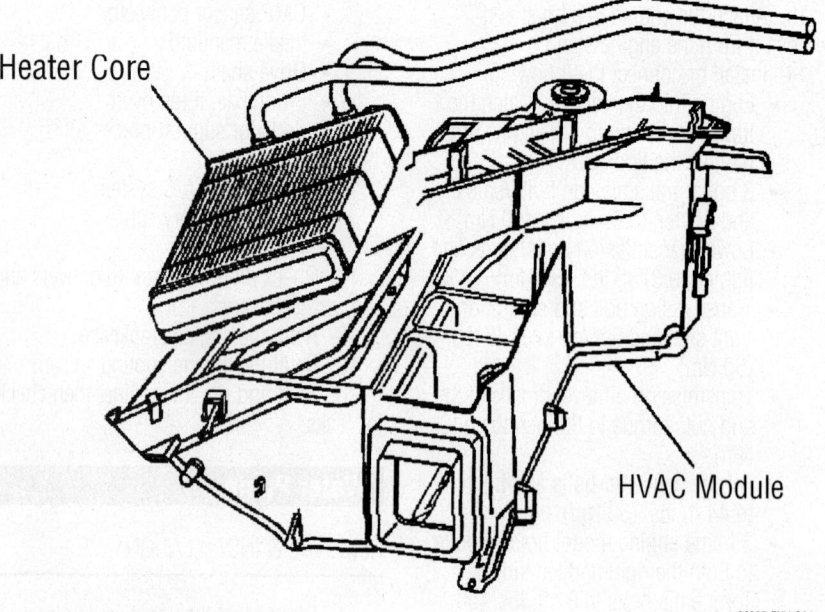

Heater Core

HVAC Module

Heater core mounting

06025-ENV-G04

w. Unfasten the speaker grilles retaining screws and remove the speaker grilles.

x. Remove the windshield defroster grille using a flat-bladed prytool. Start at one end of the grille and work your way down the grille.

y. Unfasten the 4 instrument panel upper support screws.

z. Tag and unplug all necessary electrical connections.

aa. Remove the instrument panel from the vehicle.

4. Remove or disconnect the following:
- Air inlet assembly, if equipped
- Vacuum hoses
- Heater assembly studs, from inside the engine compartment
- Blower motor resistor
- Stud from inside the heater case assembly; the stud is located behind the blower motor resistor
- Heater assembly-to-chassis screws
- Heater assembly from the vehicle
- Access cover screws and cover from the heater assembly
- Heater core from the heater case assembly

To install:

5. Install or connect the following:
- Heater core to the heater case assembly
- Access cover to the heater assembly and the cover screws
- Heater assembly to the vehicle
- Heater assembly-to-chassis screws and torque them to 40 inch lbs. (4.5 Nm)
- Stud, working from inside the heater case assembly; the stud is located behind the blower motor resistor
- Blower motor resistor
- Heater assembly studs, working inside the engine compartment, torque them to 17 inch lbs. (1.9 Nm)
- Vacuum hoses
- Air inlet assembly, if equipped

6. Install the instrument panel as follows:

a. Rest the instrument panel on the lower pivot studs.

b. Attach the electrical connections.

c. Install but do not tighten the 4 upper instrument panel support screws.

d. Install the left and right panel pivot bolts. Tighten the bolts to 102 inch lbs. (11.5 Nm).

e. Install the panel lower support bolt. Tighten the bolt to 102 inch lbs. (11.5 Nm).

f. Tighten the upper support screws to 17 inch lbs. (1.9 Nm).

g. Install the windshield defroster grille and the speaker grilles.

h. Install the radio and storage compartment assembly (if equipped).

i. If removed, install the instrument cluster.

j. Install the heater and/or air conditioning control assembly.

k. Attach the electrical connections to the instrument panel accessory trim plate.

l. Place the trim plate in position and install its retaining screws. Tighten the screws to 17 inch lbs. (1.9 Nm).

m. Place the steering column into position and install its retaining nuts. Tighten the nuts to 22 ft. lbs. (30 Nm).

n. Attach the lighter electrical connection and the lap cooler duct to the knee bolster.

o. Place the knee bolster into position and install its retaining screws. Tighten the Torx® head screws to 80 inch lbs. (9 Nm) and the hex head screws to 17 inch lbs. (1.9 Nm).

p. Place the courtesy lamp in position and install its screws. Tighten the screws to 17 inch lbs. (1.9 Nm).

q. Place the instrument panel center sound insulator in position. Install the screws that attach the center sound insulator to the knee bolster, instrument panel and the floor duct. Tighten the screws to 17 inch lbs. (1.9 Nm).

r. Install the screw that attaches the center sound insulator to the heater assembly. Tighten the screw to 13 inch lbs. (1.5 Nm).

s. Install the remote control door lock receiver module to the instrument panel left side sound insulator.

t. Attach the door lock receiver electrical connection.

u. Install the nut that attaches the left side sound insulator to the accelerator pedal bracket. Tighten the nut to 35 inch lbs. (4 Nm).

v. Install the screw that attaches left side sound insulator to cowl panel. Tighten the screw to 13 inch lbs. (1.5 Nm).

w. Install the screws that attach the left side sound insulator to knee bolster. Tighten the screw to 17 inch lbs. (1.9 Nm).

x. Feed the DLC through the hole in the sound insulator, place the DLC in position and install its retaining screws. Tighten the screws to 21 inch lbs. (2.4 Nm).

y. Install the right side sound insulator and tighten the screws

z. Connect the parking brake release cable to the lever.

aa. Enable the air bag system.

7. Install the heater hoses to the heater core.

8. Refill the cooling system.

9. Connect the negative battery cable.

10. Run the engine to normal operating temperatures; then, check the climate control operation and check for leaks.

Cylinder Head

REMOVAL & INSTALLATION

4.2L Engine

1. Disconnect the negative battery cable.

2. Drain the engine cooling system.

3. Remove or disconnect the following:
- Camshaft cover
- Exhaust manifold
- Front cover
- Cylinder head access hole plugs
- Timing chain tensioner shoe bolt and shoe
- Timing chain tensioner guide bolts and guide
- Timing chain and sprockets

4. Unfasten the cylinder head bolts by loosening them in the reverse of the torque sequence, then carefully remove the cylinder head.

5. Remove the cylinder head gasket.

To install:

6. Carefully clean and inspect the cylinder head and the gasket mounting surfaces.

➡The gasket surfaces on both the head and block must be clean of any foreign matter and free of nicks or heavy scratches. The cylinder bolt threads in the block and thread on the bolts must be cleaned (dirt will affect the bolt torque).

➡DO NOT apply sealer to composition steel-asbestos gaskets.

✳✳ WARNING

Make sure the number 1 cylinder is at Top Dead Center (TDC).

7. If using a steel only gasket, apply a thin and even coat of sealer to both sides of the gaskets.

8. Place a new gasket over the dowel pins with the bead or the words "This Side Up" facing upwards (as applicable), then

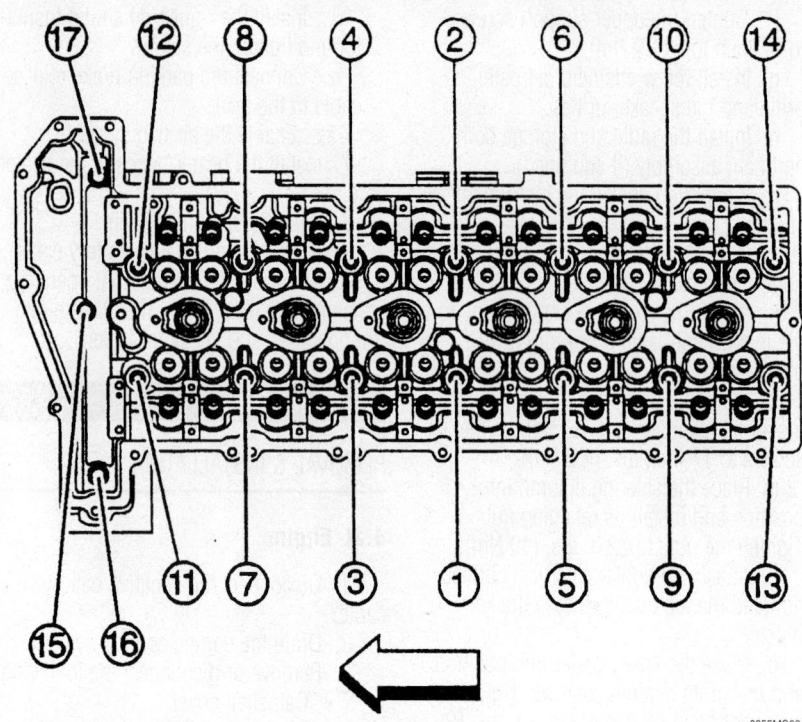

Cylinder head bolt tightening sequence—4.2L engine

9355MG02

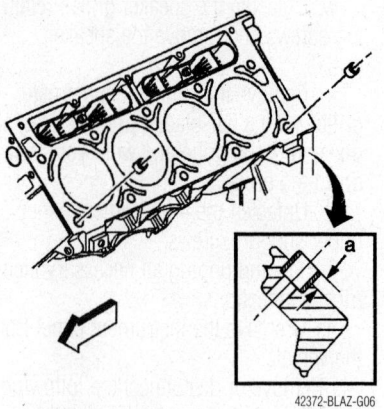

42372-BLAZ-G06

Make sure the cylinder head locating pins are properly installed, see dimension (a)

carefully lower the cylinder head into position over the gasket and dowels.

9. Apply a coating of 12345493 or equivalent sealer to the threads of the cylinder head bolts, then thread the bolts into position until finger-tight.

10. Tighten the cylinder head bolts in sequence as follows:

a. Tighten bolts 1-14 in sequence to 22 ft. lbs. (30 Nm), plus an additional 155 degrees.

b. Tighten the 2 short end bolts to 62 inch lbs. (7 Nm), plus an additional 60 degrees.

c. Tighten the 1 long end bolt to 62 inch lbs. (7 Nm) plus an additional 120 degrees.

11. Install or connect the following:
- Cylinder head access hole plugs and tighten to 44 inch lbs. (5 Nm)
- Timing chain and sprockets
- Front cover
- Camshaft cover
- Exhaust manifold
- Negative battery cable

12. Properly refill the engine cooling system.

13. Run the engine to check for leaks.

5.3L Engine

LEFT SIDE

1. Drain the engine cooling system.
2. Remove or disconnect the following:

- Negative battery cable
- Alternator bracket
- Coolant air bleed pipe
- Left exhaust manifold
- Pushrods
- Auxiliary A/C bracket bolt, if equipped
- Cylinder head bolts and discard the bolts
- Cylinder head
- Cylinder head gasket and discard

To install:

3. Carefully clean and inspect the cylinder head and the gasket mounting surfaces.

➡**The gasket surfaces on both the head and block must be clean of any foreign matter and free of nicks or heavy scratches. The cylinder bolt threads in the block and thread on the bolts must be cleaned (dirt will affect the bolt torque).**

➡**DO NOT apply any type sealer to the cylinder head gasket, unless otherwise specified.**

4. Check the cylinder head locating pins for proper installation, location (a) 0.327 in. (8.3mm) on 2003–04 engines, or 0.236 in. (6.0mm) on 2005 engines, as shown.

5. Place a new gasket over the dowel pins. When installed properly, the word "FRONT" on the left side, the tab on the gasket should be left of center or closer to the front of the engine.

6. Install or connect the following:
- Cylinder head

➡**You must use new cylinder head bolts during reassembly. Do NOT reuse the old head bolts.**

- NEW cylinder head bolts 1, 2 and 3.

7. Tighten the cylinder head bolts in sequence as follows on 2003–04 engines:

a. Tighten the M11 bolts to 22 ft. lbs. (30 Nm).

b. Tighten the M11 an additional 90 degrees.

c. Tighten M11 bolts 1–8, an additional 90 degrees and M11 bolts 9 and 10 an additional 50 degrees.

d. Tighten the M8 bolts (11–15) to 22 ft. lbs. (30 Nm). Tighten all the bolts beginning with the center bolt and working outward, alternating sides

8. Tighten the cylinder head bolts in sequence as follows on 2005 engines:

a. Tighten the M11 bolts to 22 ft. lbs. (30 Nm).

b. Tighten the M11 bolts an additional 90 degrees.

c. Tighten M11 bolts an additional 70 degrees.

d. Tighten the M8 bolts (11–15) to 22 ft. lbs. (30 Nm). Tighten all the bolts beginning with the center bolt and working outward, alternating sides

9. Install or connect the following:
- Auxiliary A/C bracket, if equipped. Torque the bolt to 15 ft. lbs. (20 Nm).
- Pushrods
- Left exhaust manifold
- Coolant air bleed pipe
- Alternator bracket

10. Properly refill the engine cooling system.

11. Run the engine to check for leaks.

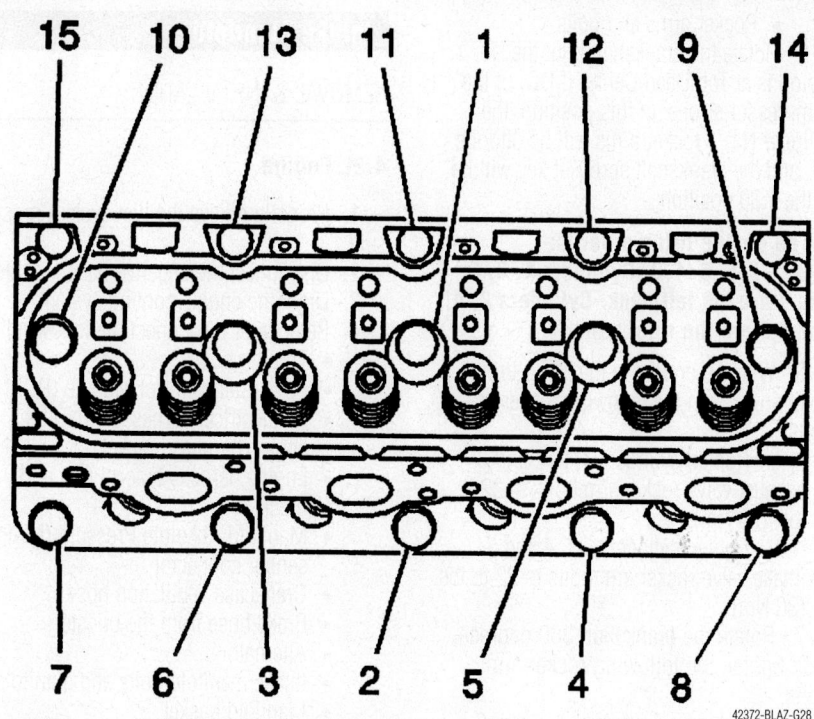

Cylinder head bolt torque sequence—5.3L engine

RIGHT SIDE

1. Drain the engine cooling system.
2. Remove or disconnect the following:
   - Negative battery cable
   - Oil level dipstick
   - Coolant air bleed pipe
   - Right exhaust manifold
   - Pushrods
   - Auxiliary A/C bracket nut, if equipped
   - Cylinder head bolts 1, 2 and 3. Discard the bolts
   - Cylinder head
   - Cylinder head gasket and discard

To install:

3. Carefully clean and inspect the cylinder head and the gasket mounting surfaces.

➥**The gasket surfaces on both the head and block must be clean of any foreign matter and free of nicks or heavy scratches. The cylinder bolt threads in the block and thread on the bolts must be cleaned (dirt will affect the bolt torque).**

➥**DO NOT apply any type sealer to the cylinder head gasket, unless otherwise specified.**

4. Check the cylinder head locating pins for proper installation, location (a) 0.327 in. (8.3mm) on 2003–04 engines, or 0.236 in. (6.0mm) on 2005 engines, as shown.
5. Place a new gasket over the dowel pins. When installed properly, the word "FRONT" on the right side, the tab on the gasket should be right of center or closer.
6. Install or connect the following:
   - Cylinder head

➥**You must use new cylinder head bolts during reassembly. Do NOT reuse the old head bolts.**

   - NEW cylinder head bolts 1, 2 and 3.

7. Tighten the cylinder head bolts in sequence as follows on 2003–04 engines:
   a. Tighten the M11 bolts to 22 ft. lbs. (30 Nm).
   b. Tighten the M11 an additional 90 degrees.
   c. Tighten M11 bolts 1–8, an additional 90 degrees and M11 bolts 9 and 10 an additional 50 degrees.
   d. Tighten the M8 bolts (11–15) to 22 ft. lbs. (30 Nm). Tighten all the bolts beginning with the center bolt and working outward, alternating sides
8. Tighten the cylinder head bolts in sequence as follows on 2005 engines:
   a. Tighten the M11 bolts to 22 ft. lbs. (30 Nm).
   b. Tighten the M11 bolts an additional 90 degrees.
   c. Tighten M11 bolts an additional 70 degrees.
   d. Tighten the M8 bolts (11–15) to 22 ft. lbs. (30 Nm). Tighten all the bolts beginning with the center bolt and working outward, alternating sides

9. Install or connect the following:
   - Auxiliary A/C bracket, if equipped. Torque the nut to 15 ft. lbs. (20 Nm).
   - Pushrods
   - Right exhaust manifold
   - Coolant air bleed pipe
   - Oil level dipstick
10. Properly refill the engine cooling system.
11. Run the engine to check for leaks.

Rocker Arms/Shafts

REMOVAL & INSTALLATION

4.2L Engine

1. Remove or disconnect the following:
   - Camshaft cover
   - Exhaust and intake camshaft sprocket bolts
2. Install Camshaft Sprocker Holding tool J-4422 onto the cylinder head and adjust the horizontal bolts into the camshaft sprockets in order to maintain chain tension and keep from disturbing the timing chain components.
3. Carefully move the sprockets with the timing chain off of the camshafts.

➥**Make sure to place the camshaft caps in a rack to keep them in order, so they may be installed in their original locations.**

   - Camshaft cap bolts and caps
   - Camshafts

➥**If valve train components, such as the rocker arms or lash adjusters, are to be reused, they must be tagged or arranged to insure installation in their original locations.**

   - Rocker arms
   - Valve lash adjusters

To install:

4. Lubricate and fill the valve lash adjusters and the rocker arm roller with engine oil.
5. Install or connect the following
   - Valve lash adjusters, in their original locations
   - Rocker arm rollers in their original positions
   - Camshafts
6. Carefully move the camshaft sprockets back onto the camshafts and remove the holding tool.
7. Install the intake camshaft sprocket washer, and the bolt, and the exhaust camshaft actuator bolt.

8. Tighten the intake camshaft sprocket bolt to 15 ft. lbs. (20 Nm), plus an additional 100 degrees.

9. Tighten the exhaust camshaft sprocket bolt to 18 ft. lbs. (25 Nm), plus an additional 135 degrees.
- Camshaft cap bolts and tighten to 106 inch lbs. (12 Nm).
- Camshaft cover

5.3L Engine

1. Remove or disconnect the following:
- Rocker arm cover(s)

➡ **If valve train components, such as the rocker arms, pushrods or pivot supports, are to be reused, they must be tagged or arranged to insure installation in their original locations.**

- Rocker arm bolts
- Rocker arms
- Rocker arm pivot support
- Pushrods

To install:

2. Inspect and replace components if worn or damaged.

3. Coat the bearing surfaces of the rocker arms, pushrods and the flange of the rocker arm bolts with clean engine oil.

4. Install or connect the following:
- Rocker arm pivot support
- Pushrods making sure they seat properly in the lifter sockets

➡ **Make sure the pushrods are seated properly to the ends of the rocker arms, but do not tighten the bolts yet.**

- Rocker arms and bolts

5. Rotate the crankshaft until the No. 1 piston is at Top Dead Center (TDC) of the compressor stroke. In this position, the cylinder No. 1 rocker arms will be off lobe lift, and the crankshaft sprocket key will be at the 1:30 position.

➡ **The engine firing order is: 1–8–7–2–6–5–4–3. Cylinders 1, 3, 5 and 7 are the left bank. Cylinders 2, 4, 6 and 8 are the right bank.**

6. With the engine in the No. 1 firing position, tighten the following rocker arm bolts:

a. Tighten cylinders 1, 2, 7 and 8 exhaust valve rocker arm bolts to 22 ft. lbs. (30 Nm).

b. Tighten cylinder 1, 3, 4 and 5 intake valve rocker arm bolts to 22 ft. lbs. (30 Nm).

7. Rotate the crankshaft 360 degrees, then tighten the following rocker arm bolts:

a. Tighten cylinders 3, 4, 5 and 6 exhaust valve rocker arm bolts to 22 ft. lbs. (30 Nm).

b. Tighten cylinder 2, 6, 7 and 8 intake valve rocker arm bolts to 22 ft. lbs. (30 Nm).

8. Install the rocker arm cover(s).

Intake Manifold

REMOVAL & INSTALLATION

4.2L Engine

1. Properly relieve the fuel system pressure.

2. Disconnect the negative battery cable.

3. Drain the engine cooling system.

4. Remove or disconnect the following:
- Throttle body
- Powertrain Control Module (PCM)
- All electrical harnesses from the engine harness bracket
- Engine harness bracket bolt and bracket
- Manifold Absolute Pressure (MAP) sensor connector
- Crankcase ventilation hose
- Brake hose from the booster
- Alternator
- Intake manifold bolts and manifold.
- Manifold gasket

To install:

5. Clean the gasket mounting surfaces. Be sure to inspect the manifold for warpage and/or cracks. If necessary, replace it.

6. Properly position a new intake manifold gasket.

7. Install or connect the following:

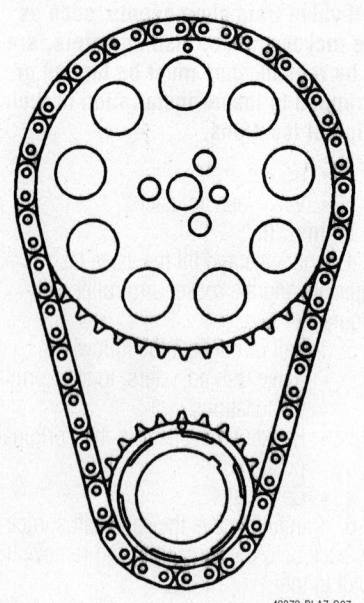

42372-BLAZ-G07

View of the crankshaft key with the No. 1 piston at TDC—5.3L engine

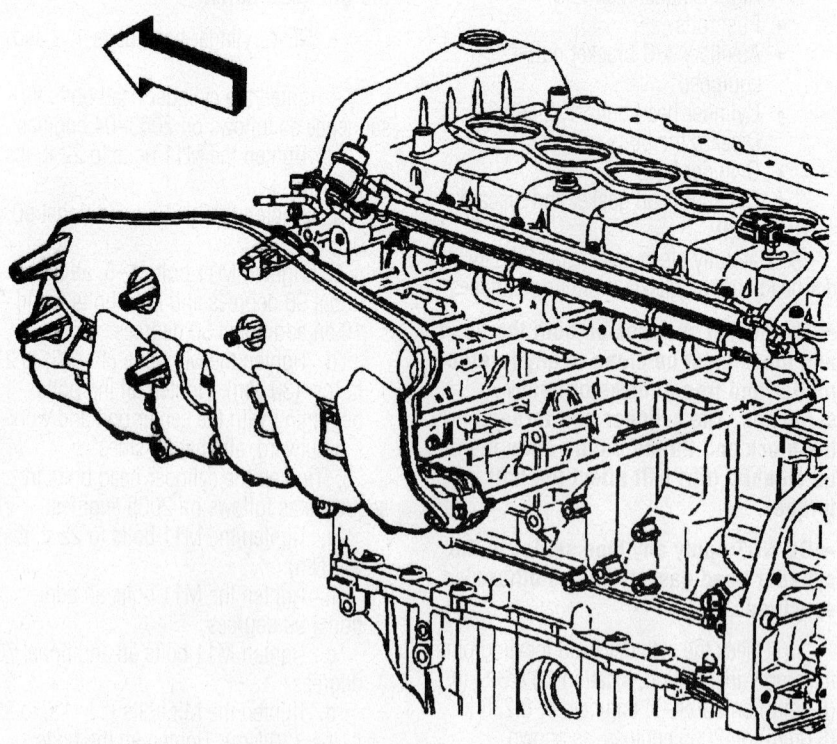

06025-ENV-G05

Exploded view of the intake manifold—4.2L engine

- Intake manifold and bolts. Torque the bolts to 89 inch lbs. (10 Nm).
- Alternator
- Brake hose to the booster
- Crankcase ventilation hose, lubricating the inner diameter first with 12345884, or equivalent lubricant
- MAP sensor electrical connector
- Engine harness bracket. Tighten the retaining bolt to 37 ft. lbs. (50 Nm).
- All harnesses to their original locations onto the engine harness bracket
- PCM
- Throttle body
- Negative battery cable

8. Refill the engine cooling system.

5.3L Engine

➡The intake manifold, throttle body, fuel rail and injectors can be removed as an assembly. If you are not servicing these components individually, remove the intake manifold as a complete assembly.

1. Properly relieve the fuel system pressure.
2. Disconnect the negative battery cable.
3. Drain the engine cooling system.
4. Remove or disconnect the following:
- Air cleaner outlet duct
- A/C compressor pressure switch electrical connector
- Harness clip from the cylinder head and fuel raid
- Mass Airflow/Intake Air Temperature sensor connector
5. Disconnect the electrical connectors from the following:
a. Main coil
b. Electronic Throttle Control (ETC)
c. Fuel injectors. Matchmark the connectors, pull the Connector Position Assurance (CPA) retainer up 1 click. Push the tab on the connector in, then detach the injector connector.
- Alternator connector
- Evaporative emission (EVAP) purge solenoid electrical connector
- Knock Sensor (KS) electrical connector
- Main coil
- Fuel injector electrical connector
- Electrical harness clips from the fuel rail
- KS harness electrical connector from the intake manifold
- Positive Crankcase Ventilation (PCV) valve hose and valve
- Heater water shutoff valve actuator

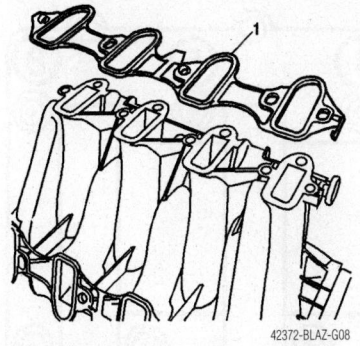

Make sure to use NEW intake manifold gaskets (1)—5.3L engine

inlet hose from the intake manifold
- EVAP purge solenoid vent tube
- Vacuum brake booster hose from the rear of the intake manifold
- Upper engine wire harness retainer nut. Position the wire harness aside.
- Intake manifold bolts
- Intake manifold and gaskets. Discard the gaskets.

To install:

6. Clean the gasket mounting surfaces. Be sure to inspect the manifold for warpage and/or cracks. If necessary, replace it.

7. Properly position a new intake manifold gasket.
8. Apply a 0.20 in. (5mm) band of a suitable threadlocking material to the intake manifold bolt threads.
9. Install or connect the following:
- Intake manifold and bolts. Torque the bolts, in sequence to 44 inch lbs. (5 Nm), then to 89 inch lbs. (10 Nm).
- Route the electrical harness into position over the engine.
- Engine harness bracket nut and tighten to 89 inch lbs. (10 Nm)
- Vacuum brake booster hose to the rear of the intake manifold
- EVAP purge solenoid valve
- Heater water shutoff valve actuator inlet hose to the intake manifold
- PCV valve and hose
- EVAP purge solenoid, KS, MAP sensor, main coil & fuel injector electrical connectors
- Harness clips to the fuel rail
- Alternator electrical connector
- Main coil, ETC, fuel injector electrical connectors
- Electrical harness clips to the fuel rail
- A/C compressor pressure switch electrical connector
- Harness clip to the cylinder head

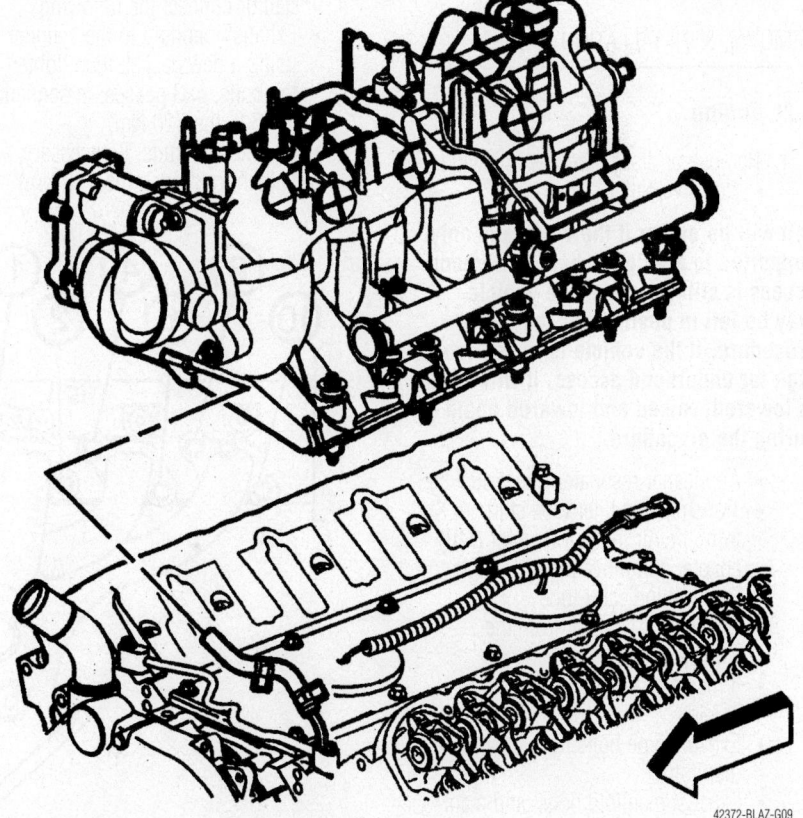

Exploded view of the intake manifold—5.3L engine

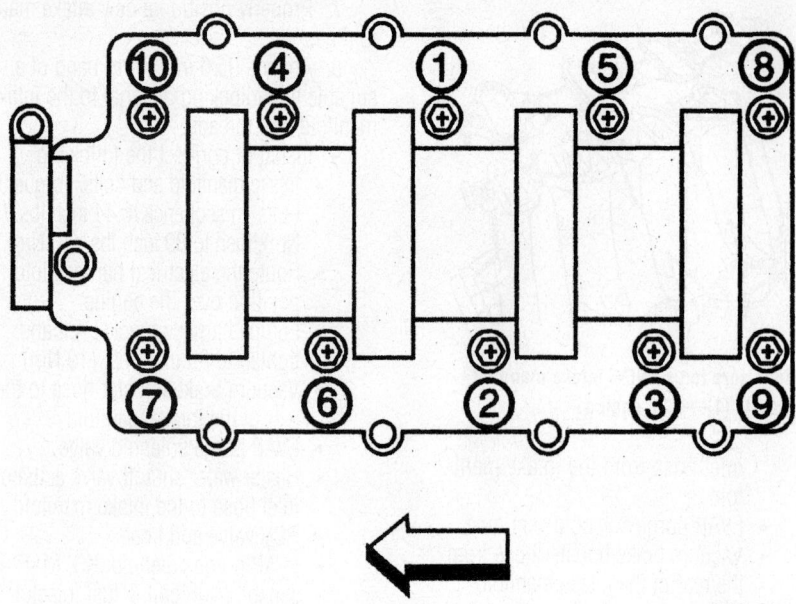

Intake manifold bolt tightening sequence—5.3L engine

- Mass Airflow/Intake Air Temperature sensor connector
- Air cleaner outlet duct
- Fuel fill cap
- Negative battery cable
10. Refill the engine cooling system.

Exhaust Manifold

REMOVAL & INSTALLATION

4.2L Engine

1. Remove or disconnect the following:
   - Negative battery cable

➡It will be easier if the vehicle is only supported to a height where underhood access is still possible, the vehicle may be left in position for the entire procedure. If the vehicle is raised too high for underhood access, it will have to lowered, raised and lowered again during the procedure.

- Air cleaner resonator outlet duct
- Transmission filler tube stud nut from the Air Injector Reactor (AIR) adapter and move the tube aside
- Oil level indicator tube
- Oxygen (O_2) sensor from the exhaust manifold
- 4 manifold heat shield nuts and shield
- Exhaust pipe bolts from the exhaust manifold
- Exhaust manifold bolts, and manifold
- Old gaskets and discard

To install:

2. Using a putty knife, clean the gasket mounting surfaces. Inspect the exhaust manifold for distortion, cracks or damage; replace if necessary.

3. Apply a threadlock such as GM 12345493 to the threads of the manifold retainers prior to installation.

4. Install or connect the following:
   - Exhaust manifold to the cylinder using a new gasket, then tighten the bolts, in 3 passes, in sequence, to 15 ft. lbs. (20 Nm)
   - Heat shield studs, if necessary, and tighten to 89 inch lbs. (10 Nm)

- O_2 sensor
- Exhaust manifold heat shield

➡Apply a suitable anti-seize compound to the exhaust manifold heat shield nuts prior to installation.

- Heat shield nuts and tighten to 44 inch lbs. (5 Nm)
- Exhaust pipe to the manifold with seal and retaining nuts. Tighten the nuts to 37 ft. lbs. (50 Nm).
- Oil level indicator tube
- Transmission filler tube back onto the AIR adapter block stud and secure with the nut. Tighten the bracket nut to 89 inch lbs. (10 Nm).
- Air cleaner resonator outlet duct
- Negative battery cable.

5.3L Engine

1. Remove or disconnect the following:
   - Negative battery cable
   - Spark plug wires from the spark plugs. Don't disconnect the wires from the ignition coil unless necessary for clearance.
   - Exhaust manifold bolts, manifold and gasket. Discard the gasket.
   - Heat shield bolt and shield, if necessary

To install:

2. Apply a 0.2 inch (5mm) bead of threadlock GM P/N 12345493, or equivalent to the threads of the exhaust manifold bolts. Do NOT apply sealer to the first 3 threads of the bolts.

3. Install or connect the following:

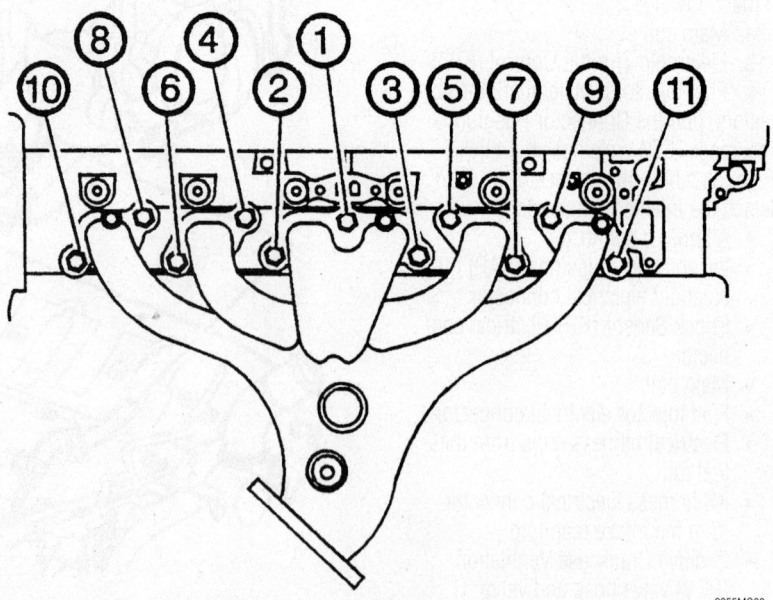

Exhaust manifold bolt tightening sequence—4.2L engine

9355MG03

- New exhaust manifold gasket
- Exhaust manifold
- Exhaust manifold bolts. Tighten in two passes. First to 11 ft. lbs. (15 Nm), then to 15 ft. lbs. (20 Nm) starting with the center bolts and working outward.

4. Bend over the exposed edge of the gasket at the rear of the cylinder head using a flat punch or equivalent tool.

- Heat shield and bolts, if removed. Torque the bolts to 80 inch lbs. (9 Nm).
- Spark plug wires to the spark plugs
- Negative battery cable

Camshaft and Valve Lifters

REMOVAL & INSTALLATION

4.2L Engine

1. Disconnect the negative battery cable.
2. Discharge and recover the refrigerant from the air conditioning system, using the proper equipment.
3. Remove or disconnect the following:

- Intake manifold
- A/C line from the oil level indicator tube
- A/C line from the accumulator
- A/C bracket bolt from the engine lift hook
- Engine lift bracket
- Ignition control module electrical connectors
- Ignition control module bolts and module

✳✳ WARNING

Be careful not to damage the clips that hold the harness housing in place.

- Engine electrical harness housing from the camshaft cover
- Fuel injection harness electrical connector
- Camshaft cover bolts and cover
- Exhaust and intake sprocket bolts

4. Install a suitable sprocket holding tool onto the cylinder head and adjust the horizontal bolts into the camshaft sprockets to maintain timing chain tension and avoid disturbing the timing chain components.

5. Carefully move the sprockets with the timing chain off of the camshafts.

➡**Make sure to place the camshaft caps in a rack to keep them in order, so they may be installed in their original locations.**

6. Remove or disconnect the following:

- Camshaft cap bolts and caps
- Camshafts

To install:

7. Coat the camshaft journals with engine oil.

- Camshafts, in their original position
- Camshaft caps, in their original locations. Tighten the bolts to 106 inch lbs. (12 Nm).

8. Carefully place the camshaft sprockets back onto the camshafts and remove the holding tool.

9. Install or connect the following:

- Intake camshaft sprocket washer and bolt and the exhaust camshaft actuator bolt. Tighten the intake camshaft sprocket bolt to 15 ft. lbs. (20 Nm), plus an additional 100 degrees and the exhaust camshaft actuator bolt to 18 ft. lbs. (25 Nm), plus an additional 135 degrees.
- New camshaft cover seal
- New rubber ignition control module seals
- Camshaft cover and bolts. Tighten the bolts to 89 inch lbs. (10 Nm).
- Ignition control module. Tighten the bolts to 89 inch lbs. (10 Nm).
- Ignition control module electrical connectors
- Fuel injector electrical connectors
- Engine electrical harness housing
- A/C line bracket to the oil level indicator tube stud and secure with the nut. Tighten the nut to 62 inch lbs. (7 Nm).
- Engine lift bracket and secure the lift hook with the bolts. Tighten the bolts to 37 ft. lbs. (50 Nm).
- A/C line bracket to the engine lift bracket. Tighten the bolt to 89 inch lbs. (10 Nm).
- Intake manifold

10. Using the proper equipment, recharge the A/C system.

5.3L Engine

1. Disconnect the negative battery cable.
2. Discharge and recover the refrigerant from the air conditioning system, using the proper equipment.
3. Remove or disconnect the following:

- Condenser
- Cylinder head and gasket
- Valve lifter guide bolts
- Valve lifters and guide

➡**If the lifters are stuck in the bores due to built up deposits, use Valve Lifter Remover tool No. J 3049-A or equivalent to remove the lifters**

- Valve lifters from the guide

➡**Make sure to keep the lifters in order as you are removing them. They must be installed in their original locations.**

4. Clean and inspect the lifters for damage.

- Camshaft sensor bolt and sensor

5. Rotate the crankshaft until the timing marks on the crankshaft and camshaft sprockets are aligned.

- Camshaft sprocket bolts

✳✳ WARNING

Do NOT turn the crankshaft after the timing chain has been removed to avoid damaging the pistons or valves!

- Camshaft sprocket and reposition the timing chain
- Camshaft retaining bolts and retainer
- Camshaft by installing three M8-1.25 x 4.0 in. (M8-1.25 x 1.00mm)

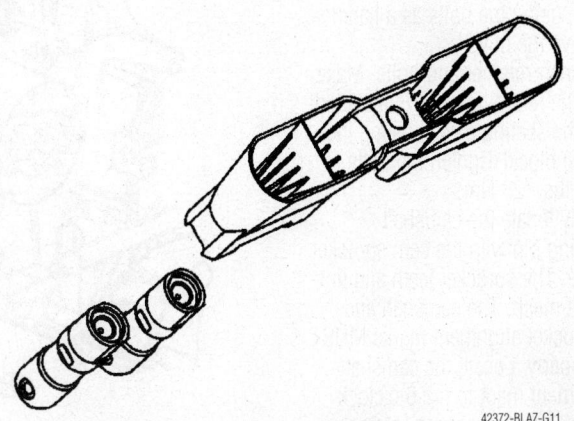

42372-BLAZ-G11

Remove the lifters from the guides, making sure to keep them in order—5.3L engine

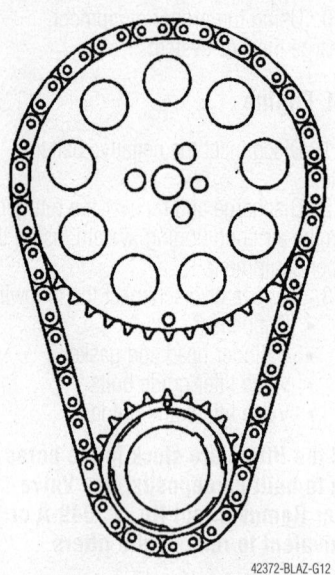

Make sure the crankshaft and camshaft timing marks are aligned

bolts in the front of the camshaft to act as a handle; then, remove the camshaft while turning slightly from side to side, as necessary. Remove the bolts from the camshaft.

➡ **Take care not to damage the camshaft bearings when removing the camshaft.**

6. Clean and inspect the camshaft and bearings.

To install:

➡ **If the camshaft must be replaced, you must also replace the lifters.**

7. Lubricate the camshaft journals with clean engine oil.
8. Install or connect the following:
   - Three bolts used during removal into the bolt hold in the front of the camshaft
   - Camshaft carefully into the engine block, using the bolts as a handle. Remove the bolts.
   - Camshaft retainer and bolts. Make sure the retaining plate is installed with the sealing gasket facing the engine block. Tighten the bolts to 18 ft. lbs. (25 Nm).
9. Properly locate the camshaft sprocket locating pin with the cam sprocket alignment hole. The sprocket teeth and timing chain must mesh. The camshaft and crankshaft sprocket alignment marks MUST be aligned properly. Locate the camshaft sprocket alignment mark in the 6 o'clock position. It may be necessary to rotate the camshaft or crankshaft to align the marks.

- Camshaft sprocket and timing chain
- Camshaft sprocket bolts and tighten to 26 ft. lbs. (35 Nm)
- Camshaft sensor O-ring, after making sure it is not damaged and lubricating it with clean engine oil
- Camshaft sensor and bolt. Torque the bolt to 18 ft. lbs. (25 Nm).

10. Lubricate the valve lifters and engine block lifter bores with clean engine oil.
11. Install or connect the following:
   - Lifters into the lifter guides. Align the area on top of the lifter with the flat area in the lifter guide bore. Push the lifter completely into the guide bore.
   - Valve lifters and guide to the engine block
   - Valve lifter guide bolt and tighten to 106 inch lbs. (12 Nm)
   - Cylinder head and gasket
   - Condenser

12. Using the proper equipment, recharge the A/C system.

Valve Lash

ADJUSTMENT

The 4.2L and 5.3L engines do not require a periodic valve lash adjustment.

Starter Motor

REMOVAL & INSTALLATION

4.2L Engine

1. Disconnect the negative battery cable
2. Disconnect the brake vacuum booster hose.
3. Disconnect the positive battery lead from the solenoid.
4. Remove or disconnect the following:
   - Starter mount bolt and nut
   - Starter motor

To install:

5. Install or connect the following:
   - Starter motor
   - Starter mounting bolt and nut. Tighten to 37 ft. lbs. (50 Nm).
   - Positive battery cable to the starter. Tighten the nut to 80 inch lbs. (9 Nm).
   - Brake booster vacuum hose

6. Connect the negative battery cable.

5.3L Engine

1. Remove or disconnect the following:
   - Negative battery cable
   - Rear steering gear crossmember
   - Wire harness from the wire harness retaining clips on the transmission oil cooler line bracket

Starter mounting—4.2L engine

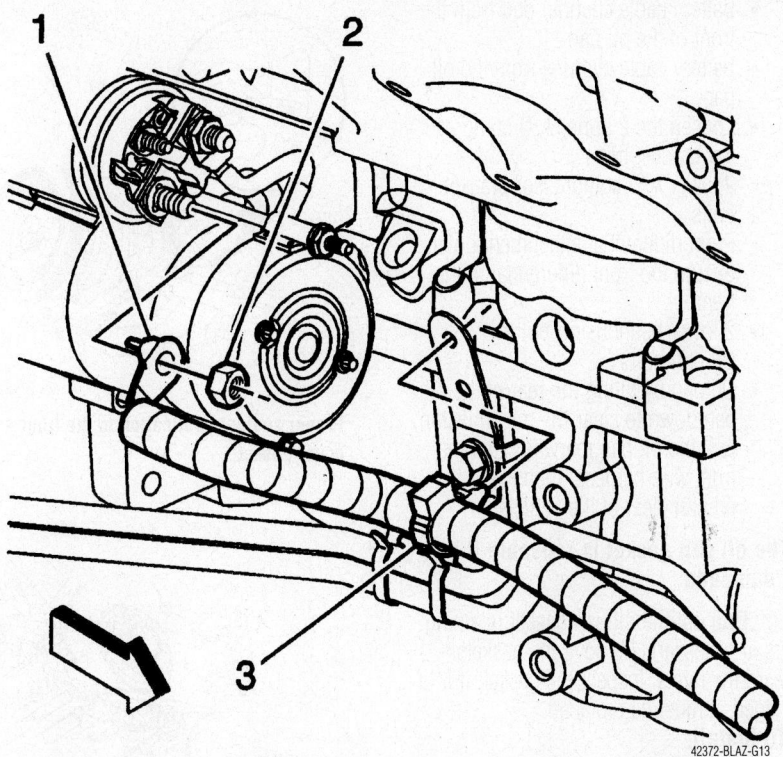

View of the starter, positive cable (1) and starter lead nut —5.3L engine

- Transmission oil cooler line bracket bolt
- Right transmission cover bolt
- Starter bolts
- Transmission cover and shield, after repositioning the starter

2. Position the starter down, with the terminals facing toward the front of the vehicle.

- Starter solenoid nut
- Starter lead from the solenoid stud
- Starter lead nut
- Positive cable from the starter stud
- Starter

To install:

3. Install or connect the following:
- Starter in the vehicle. Position the starter down, with the terminals facing toward the front of the vehicle.
- Positive cable to the starter stud.
- Starter lead nut and tighten to 80 inch lbs. (9 Nm)
- Starter solenoid lead to the stud
- Starter solenoid nut and tighten to 30 inch lbs. (3.4 Nm)
- Install the shield and transmission cover, after repositioning the starter

4. Slide the starter rearward.
- Starter bolts and tighten to 37 ft. lbs. (50 Nm)
- Right transmission cover bolt and tighten to 80 inch lbs. (9 Nm)

- Transmission oil cooler line bracket bolt
- Wire harness to the wire harness retaining clips on the transmission oil cooler line bracket
- Rear steering gear crossmember
- Negative battery cable

5. Start the vehicle to check for proper operation.

Oil Pan

REMOVAL & INSTALLATION

4.2L Engine

1. Disconnect the negative battery cable.
2. Remove or disconnect the following:
- A/C compressor bottom bolts and loosen the top bolts
- Oil dipstick and tube
3. Raise and safely support the vehicle.
4. Drain the engine crankcase oil.
5. Remove or disconnect the following:
- Left and right front tire and wheel assemblies
- Engine protection shield mounting bolts and shield
- Front steering gear crossmember
- Left and right driveshafts
- Front drive axle clutch fork assembly

- Prop shaft from the front axle pinion yoke
- Unclip the transmission cooler lines from the engine block
- Front differential bolts and position the differential aside
- 4 transmission bell housing-to-oil pan bolts
- Remaining oil pan bolts
- Oil pan, by placing 2 oil pan bolts in the jack screws on the oil pan and tighten evenly to release the oil pan from the engine

To install:

6. Clean the gasket mounting surfaces.

➡**The alignment between the rear of the oil pan and the rear of the block is critical. When the oil pan is installed it could be inadvertently shifted front or back a small amount which could cause a transmission alignment problem. The back to the oil pan needs to be flush with the engine block.**

7. Apply a 0.12 in. (3mm) bead of sealant to engine block, rather than the oil pan.

➡**The oil pan MUST be installed within 10 minutes of applying the sealant to the engine block.**

8. Install or connect the following:
- Oil pan, maneuvering it to clear the oil pump and screen assembly

➡**After the bolts are installed, before tightening them to specifications, check the oil pan alignment. Use a straight edge on the back to the block and the oil pan transmission mounting surface.**

- Oil pan bolts; tighten the side bolts to 18 ft. lbs. (25 Nm) and the end bolts to 89 inch lbs. (10 Nm)
- Transmission bell housing-to-oil pan bolts and tighten to 35 ft. lbs. (47 Nm)
- A/C compressor bottom bolts. Tighten to 37 ft. lbs. (50 Nm)
- Front differential bolts and tighten to 63 ft. lbs. (85 Nm)
- Front drive axle and clutch fork assembly
- Transmission cooler lines to block
- Prop shaft to front differential
- Steering gear crossmember
- Left and right driveshaft
- Oil pan drain plug. Tighten to 19 ft. lbs. (26 Nm)
- Engine protection shield. Tighten the bolts to 18 ft. lbs. (25 Nm)

- Left and right front wheel and tire assemblies

9. Carefully lower the vehicle.

10. Refill the crankcase with fresh oil. Start the engine, establish normal operating temperatures and check for leaks.

5.3L Engine

1. Disconnect the negative battery cable.

2. Drain the engine crankcase oil and differential oil.

3. Remove or disconnect the following:
- Oil level dipstick
- Front shock upper retaining nuts
- Tires and wheels
- Engine shield bolts and shield
- Power steering gear
- Left and right Antilock Brake System (ABS) wiring harnesses from the retainers
- Wheel Speed Sensor (WSS) electrical connectors
- Brake hose retaining bolts from the frame
- Sway bar link pins from the lower control arm on both sides

4. Place an adjustable jackstand under the lower control arm.
- Upper ball joint pinch bolt and nut
- Upper control arm from the upper ball joint

5. Lower and remove the jackstand, letting the suspension hang.
- Left driveshaft
- Right driveshaft from the intermediate shaft bearing only. Do not remove the driveshaft from the steering knuckle. Position the driveshaft aside.

6. Using wire or hooks, secure the front shock modules to the frame. Do NOT let the shocks and steering knuckle hang without being supported.

7. Matchmark the position of the propeller shaft to the front axle pinion yoke.

8. Remove or disconnect the following:
- Yoke retainer bolt and yoke retainers from the front axle pinion yoke. Wrap the bearing caps with tap to avoid losing the bearing rollers. Secure the propeller shaft to the frame.
- Transmission oil cooler lines from the retainer
- Transmission oil cooler line retaining bracket bolt and bracket
- Inner axle shaft
- Starter
- Flywheel inspection cover from the left side of the transmission

- Battery cable channel bolt from the front of the oil pan
- Battery cable channel from the oil pan
- Loosen the 2 upper A/C compressor bracket bolts
- 2 lower A/C compressor bracket bolts
- Front differential attachment bolts. Secure the front differential to the frame.
- 2 lower bellhousing bolts
- Oil pan bolts
- Oil pan by tilting the rear of the oil pan down to clear the transmission, pull the oil pan rearward past the front wire harness, then lower the oil pan clear of the vehicle

➥The oil pan gasket is reusable if it is not damaged.

9. Drill out the oil pan gasket retaining rivets, if necessary. Remove the gaskets. Discard the rivets. Inspect the gasket, if it is damaged, it must be replaced.

To install:

➥The proper alignment of the oil pan is very important. The rear bolt hold location of the oil pan provide mounting points for the transmission bellhousing. To ensure the rigidity of the powertrain and correct transmission alignment, make sure that the rear of the block and rear of the oil pan NEVER protrude beyond the engine block and transmission bellhousing plane.

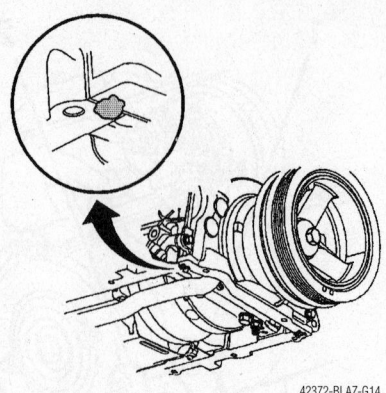

42372-BLAZ-G14

Proper sealant application to the front cover gasket

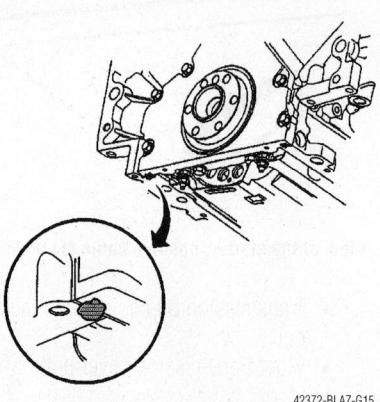

42372-BLAZ-G15

Proper sealant application to the rear cover gasket

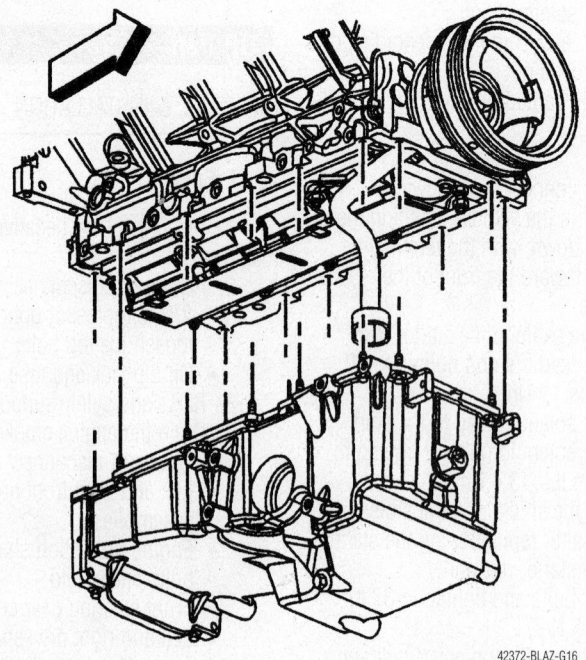

42372-BLAZ-G16

Oil pan mounting—5.3 engine

➡️**If replacing the oil pan gasket, it is not necessary to rivet the NEW gasket to the pan.**

10. Apply a 0.20 in. (5mm) bead of sealant 0.80 in. (20mm) long to the engine block. Apply the sealant directly onto the tabs of the front cover gasket that protrudes into the oil pan surface.

11. Apply a 0.20 in. (5mm) bead of sealant 0.80 in. (20mm) long to the engine block. Apply the sealant directly onto the tabs of the rear cover gasket that protrudes into the oil pan surface.

12. Pre-assemble the oil pan gasket and bolts to the pan. Install the gasket onto the pan. Install the oil pan bolts to the pan and through the gasket.

13. Install the oil pan, oil pan gasket and bolts to the engine block as an assembly.

14. Hand-start the bolts into the engine block snug-tight. Do not fully tighten yet.

15. Install the 2 lower bellhousing bolts and tighten to 37 ft. lbs. (50 Nm).

16. Tighten the 2 rear oil pan-to-rear cover bolts to 106 inch lbs. (12 Nm) and the remaining oil pan bolts to 18 ft. lbs. (25 Nm).

17. Release the differential from the frame and install to the oil pan. Install and tighten the bolts to 63 ft. lbs. (85 Nm).

18. Install or connect the following:
- 2 lower A/C compressor bracket bolts. Tighten the lower and upper compressor bolts to 37 ft. lbs. (50 Nm).
- Battery cable channel to the oil pan
- Battery cable channel bolt and tighten to 106 inch lbs. (12 Nm)
- Flywheel inspection cover to the left side of the transmission
- Starter
- Inner axle shaft
- Transmission oil cooler line retaining bracket and bolt. Torque the bolt to 80 inch lbs. (9 Nm).
- Transmission oil cooler lines to the retainer

19. Unhook the right driveshaft from the frame.
- Left and right driveshafts

20. Unsecure the shocks from the frame. Put adjustable jackstand under the lower control arm. Using the jackstand, raise the lower control arm and knuckle assembly in order to connect the upper ball joint to the upper control arm.
- Upper ball joint pinch nut and bolt and tighten to 30 ft. lbs. (40 Nm). Remove the jackstand.

- Sway bar link pins to the lower control arm on both sides
- Steering gear

21. Unsecure the prop shaft from the frame. Align the matchmarks on the prop shaft to the marks on the front axle pinion yoke.
- Propeller shaft to the front axle pinion yoke
- Yoke retainers and yoke retainer bolts to the front axle pinion yoke. Torque the bolts to 15 ft. lbs. (20 Nm).
- Brake hose retaining bolts to the frame and tighten to 18 ft. lbs. (25 Nm).
- WSS electrical connectors
- Left and right ABS wiring harnesses to the retainers
- Differential with oil
- Engine shield and bolts. Tighten the bolts to 18 ft. lbs. (25 Nm).
- Tires and wheels

22. Fill the engine with oil. Fill the power steering system with fluid.
- Upper shock nuts and tighten to 74 ft. lbs. (100 Nm).
- Oil dipstick
- Negative battery cable

Oil Pump

REMOVAL & INSTALLATION

4.2L Engine

1. Remove or disconnect the following:
- Engine front cover
- Oil pump cover bolts
- Oil pump cover. Mark the inner and outer gears in relation to the pump housing.
- Inner and outer pump gears
- Oil pump pressure relief valve plug
- Oil pump pressure relief valve and spring

To install:

2. Install or connect the following:
- Oil pump pressure relief valve and spring
- Oil pump pressure relief valve plug. Tighten to 10 ft. lbs. (14 Nm).
- Oil pump outer and inner gears, as marked during removal
- Oil pump cover and bolts. Tighten the bolts to 89 inch lbs. (10 Nm).
- Front cover

5.3L Engine

1. Remove or disconnect the following:
- Oil pan

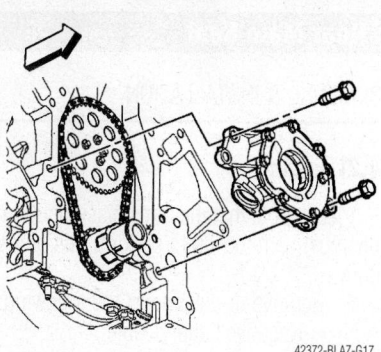

42372-BLAZ-G17

Exploded view of the oil pump mounting— 5.3L engine

- Engine front cover
- Oil pump screen bolt and nuts
- Oil pump screen with O-ring seal
- O-ring seal from the pump screen. Discard the O-ring seal.
- Remaining crankshaft oil deflector nuts
- Crankshaft oil deflector
- Oil pump bolts
- Oil pump

➡️**Do not let any dirt or debris into the oil pump or cap end.**

- Clean and inspect the oil pump.

To install:

2. Align the splined surfaces of the crankshaft sprocket and the oil pump drive gear and install the oil pump.

3. Install or connect the following:
- Oil pump onto the crankshaft sprocket until the pump housing contacts the face of the engine block
- Oil pump bolts and tighten to 18 ft. lbs. (25 Nm)
- Crankshaft oil deflector and nuts until snug
- New oil pump screen O-ring seal into the oil pump screen, after lubricating with clean engine oil

➡️**Push the oil pump screen tube completely into the oil pump prior to tightening the bolt. Do not let the bolt pull the tube into the pump.**

4. Align the oil pump screen mounting brackets with the correct crankshaft bearing cap studs.
- Oil pump screen
- Oil pump screen bolts and nuts. Tighten the bolts to 106 inch lbs. (12 Nm) and the nuts to 18 ft. lbs. (25 Nm).
- Engine front cover
- Oil pan

Rear Main Seal

REMOVAL & INSTALLATION

4.2L Engine

Please note that the transmission assembly must be removed to perform this procedure.

1. Remove or disconnect the following:
   - Negative battery cable
   - Transmission
   - Flywheel
   - Crankshaft rear main seal housing bolts. Install 2 bolts into the jackscrew holes to release the cover from the block
   - Crankshaft and rear main seal housing
   - Rear main seal from the crankshaft snout

To install:

2. Install or connect the following:
   - Rear main seal, using a suitable seal installation tool, then remove the tool
   - Apply a 0.12 in. (3mm) bead of 12378521, or equivalent sealant to the rear mail seal housing
   - Suitable cover alignment pins into the block

➡ When you install a new seal, make sure to use the plastic installation sleeve supplies with the new seal. The sleeve should come off and be discarded after the seal is installed.

3. Slide the crankshaft rear main seal housing over the alignment pins and crankshaft.

4. Install the crankshaft rear main seal housing bolts, except the 2 in place of the guide pins.

5. Remove the guide pins.

6. Install or connect the following:
   - Remaining 2 crankshaft rear main seal housing bolts and tighten to 89 inch lbs. (10 Nm). Wipe off any excess sealant.
   - Flywheel and secure with the mounting bolts. Tighten, in sequence, to 18 ft. lbs. (25 Nm), plus an additional 50 degrees.
   - Transmission

5.3L Engine

Please note that the transmission assembly must be removed to perform this procedure.

1. Remove or disconnect the following:
   - Negative battery cable

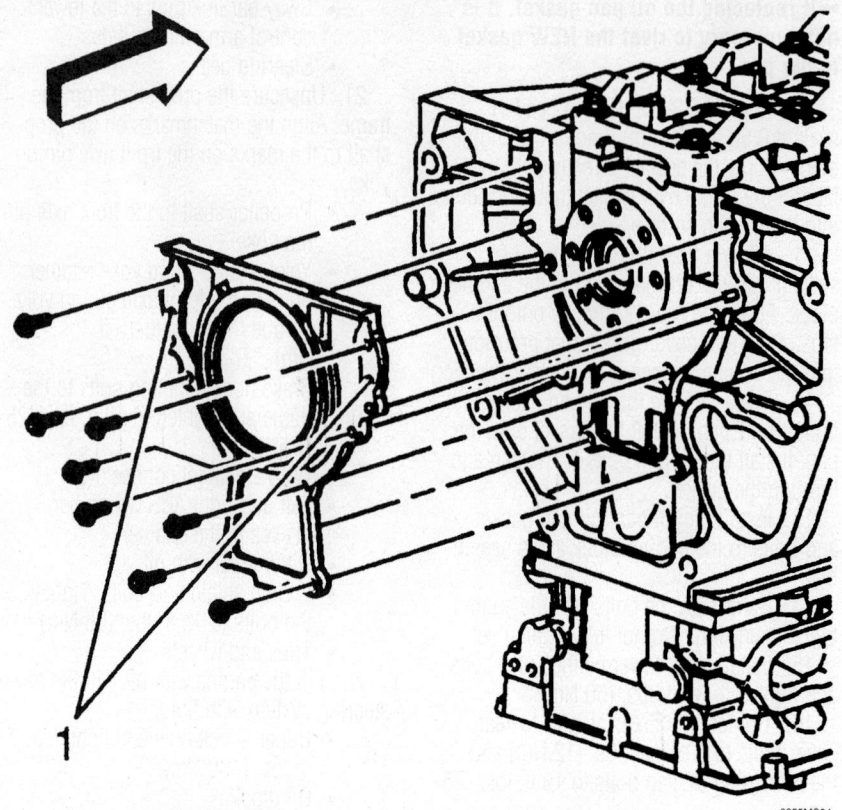

Install 2 bolts into the jackscrew holes (1) to push the cover off of the block

9355MG04

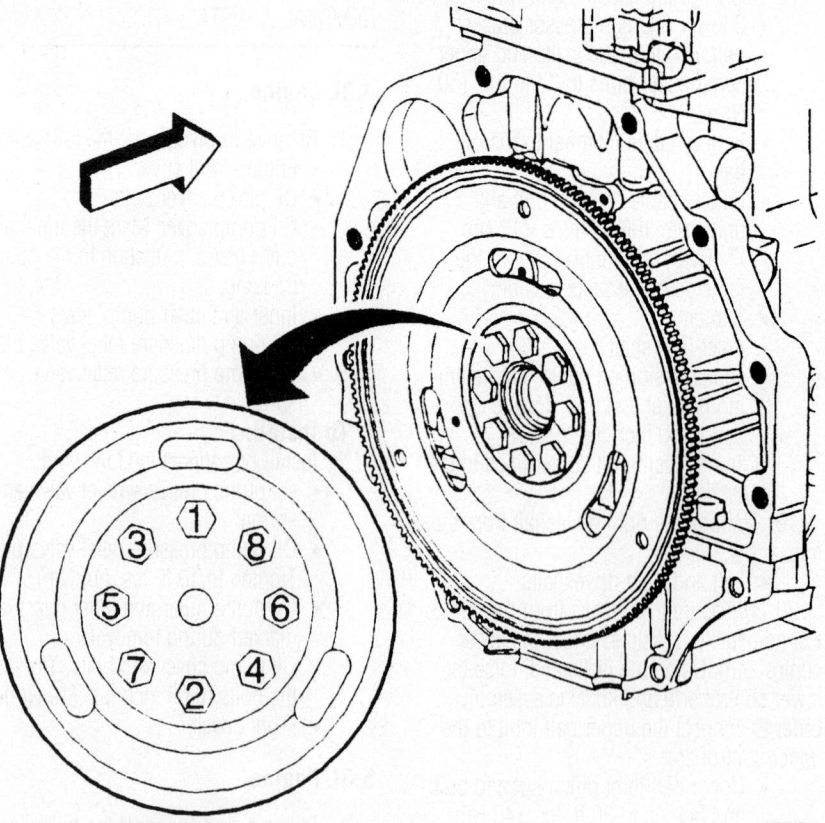

Flywheel bolt tightening sequence—4.2L engine

9355MG05

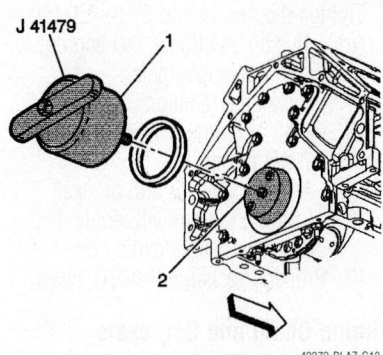

J 41479

Installing the rear main seal—5.3L engine

- Transmission
- Flywheel
- Crankshaft rear main oil seal from the rear cover

To install:

➡The flywheel spacer (if applicable) must be removed prior to oil seal installation. Do not lubricate the oil seal Inside Diameter (ID) or crankshaft surface. Never reuse the rear main seal. Once it is removed, it must be replaced with a new seal.

2. Lubricate the Outside Diameter (OD) of the rear main seal and the rear cover oil seal bore with clean engine oil. Do NOT let oil contact the seal surface or the crankshaft surface.

3. Install or connect the following:
- Crankshaft Rear Oil Seal Installer Tool No. J 41479 tapered cone and bolts onto the rear of the crankshaft. Tighten the bolts until just snug, being careful not to overtighten.
- Rear oil seal onto the tapered cone until the tool contacts the oil seal

4. Align the oil seal into the tool, Rotate the handle of the tool clockwise until the seal enters the rear cover and bottoms into the cover bore. Remove the tool.

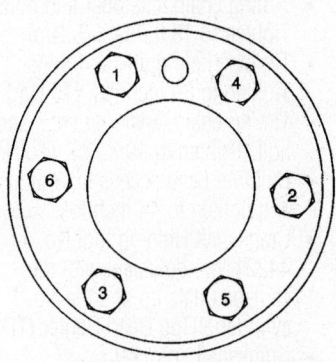

Flywheel bolt tightening sequence—5.3L engine

- Flywheel and secure with the mounting bolts.

5. Tighten the flywheel mounting bolts, in sequence, as follows:
- a. 1st pass: 15 ft. lbs. (20 Nm)
- b. 2nd pass: 37 ft. lbs. (50 Nm)
- c. Final pass: 74 ft. lbs. (100 Nm)
- Transmission
- Negative battery cable

6. Start the engine and verify no oil leaks.

Timing Chain, Sprockets, Front Cover and Seal

REMOVAL & INSTALLATION

Front Cover and Seal

4.2L ENGINE

1. Remove or disconnect the following:
- Negative battery cable
- Drain the engine cooling system.
- Cooling fan and shroud
- Accessory belt
- Water pump
- Crankshaft balancer

✷✷ WARNING

When removing the seal, be careful not to damage the front cover or crankshaft.

- Seal from the front cover, using a suitable prytool in the slots provided
- Power steering pump

2. Raise and safely support the vehicle.
- Oil pan, then carefully lower the vehicle
- 7mm center bolt
- Remaining front cover bolts. Place two of the front cover bolts in the jackscrew holes on the front cover and tighten the bolts evenly to release the front cover from the engine.
- 2 bolts from the front cover
- Oil pump

To install:

3. Clean the gasket mating surfaces of the engine and cover of all remaining gasket or sealer material. Be careful not to score or damage the surfaces.

4. Install or connect the following:
- Suitable cover alignment pins, onto the engine

➡The front cover MUST be installed within 10 minutes of applying the sealant.

- Apply a 0.12 in. (3mm) beat of 12378521 or equivalent sealant to the trace grooves on the back side of the engine front cover. Apply sealant on the inside 3 bolt hole bosses on the cover also.

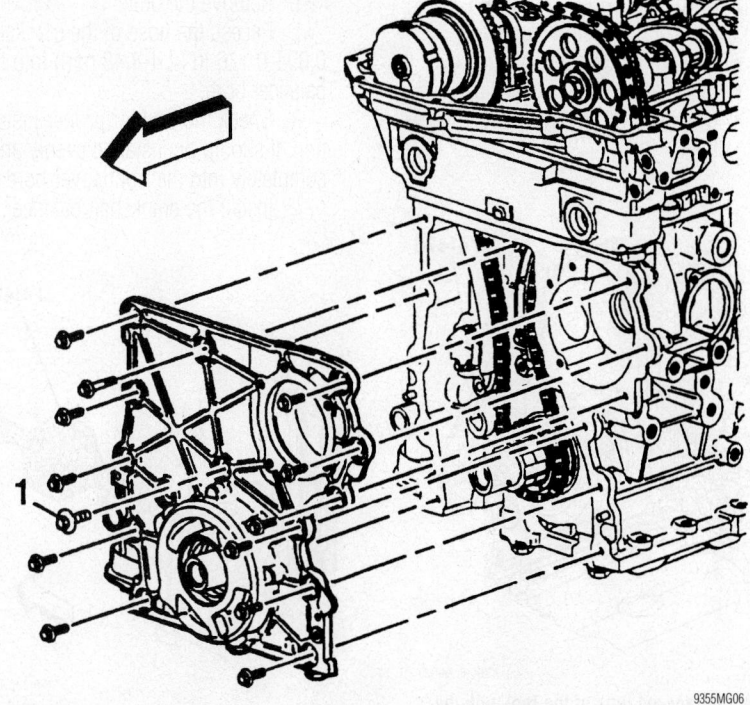

Place 2 front cover bolts in the jackscrew holes on the cover and tighten to push the cover off of the engine—4.2L engine

- Oil pump to the crankshaft splines
- Front cover and bolts, tighten the center bolt last. Tighten to 89 inch lbs. (10 Nm).

5. Remove the alignment pins and raise and safely support the vehicle. Install the oil pan, then lower the vehicle.
- Power steering pump
- Crankshaft balancer
- Water pump
- Accessory belt
- Cooling fan and shroud
- Negative battery cable

6. Properly refill the engine cooling system.

7. Run the engine until normal operating temperature has been reached, then check for leaks.

5.3L ENGINE

1. Properly discharge the A/C system.
2. Drain the engine cooling system.
3. Remove or disconnect the following:
- Negative battery cable
- A/C compressor and bracket
- Water pump
- Crankshaft balancer
- Oil pan-to-front cover bolts
- Front cover bolts
- Front cover and gasket. Discard the gasket.

4. Clean and inspect the front cover.

To install:

5. Apply a 0.20 in. (5mm) bead of sealant 0.80 in. (20mm) long to the oil pan-to-engine block junction.

6. Install or connect the following:
- New front cover gasket and cover
- Front cover bolts, finger-tight
- Oil pan-to-front cover bolts, finger-tight

- Front and Rear Cover Alignment Tool No. J 41476 to the front cover. Align the tapered legs of the tool with the machined alignment surfaces on the front cover
- Crankshaft balancer bolt, finger-tight
- Oil pan-to-front cover bolts to 18 ft. lbs. (25 Nm)
- Front cover bolts to 18 ft. lbs. (25 Nm)

7. Remove the tool.

8. Install a NEW crankshaft front oil seal as follows:

a. Remove the radiator for access.

b. Remove the crankshaft balancer.

c. Remove the crankshaft oil seal.

d. Lubricate the outer edge ONLY of the NEW crankshaft oil seal with clean engine oil.

e. Install the crankshaft front oil seal into the Crankshaft Front Seal Installation Tool No. J 41478 guide.

f. Install the J 41478 threaded rod (with nut, washer, guide and oil seal) into the end of the crankshaft.

g. Use J 41478 to install the oil seal into the cover bore. Use a wrench and hold the hex on the installer bolt. Use a second wrench to rotate the installer nut clockwise until the seal bottoms in the cover bore. Remove the tool.

h. Install the used crankshaft balancer bolt and tighten to 240 ft. lbs. (220 Nm) to seat the balancer.

i. Remove the bolts.

j. Recess the nose of the crankshaft 0.094-0.176 in (2.4-4.48 mm) into the balancer bore.

k. Check the seal for proper installation. It should be installed evenly and completely into the front cover bore.

l. Install the crankshaft balancer.

Tighten the new bolt to 37 ft. lbs. (50 Nm), plus an additional 140 degrees using a torque angle meter.

m. Install the radiator.

9. Install or connect the following:
- Water pump
- A/C compressor and bracket
- Cooling system with coolant
- Negative battery cable

10. Properly recharge the A/C system

Timing Chain and Sprockets

4.2L ENGINE

➡ **The following procedure requires the use of the Crankshaft Holding tool No. J-44221 and a suitable torque angle meter.**

1. Remove or disconnect the following:
- Camshaft cover
- Timing chain (front) cover
- Tension on the timing chain by moving the tensioner shoe in. Place a tee into the tension to hold the shoe in place.
- Top chain guide bolts and guide
- Exhaust camshaft position actuator bolt and actuator
- Intake camshaft sprocket bolt and sprocket
- Timing chain
- Crankshaft sprocket
- Cylinder head access hole plugs
- Timing chain tensioner shoe bolt and shoe
- Timing chain tensioner guide bolts and guide
- Timing chain tensioner bolts and tensioner

To install:

➡ **Every seventh link of the timing chain is darkened to help in aligning the timing marks.**

2. Install or connect the following:
- Timing chain tensioner and bolts. Tighten to 18 ft. lbs. (25 Nm).
- Timing chain guide and bolts. Tighten to 89 inch lbs. (10 Nm).
- Timing chain tensioner shoe and bolt. Tighten to 19 ft. lbs. (26 Nm).
- Cylinder head access hole plugs and tighten to 44 inch lbs. (5 Nm)
- Crankshaft Holding tool No. J-44221, or equivalent with the camshaft flats up and the No. 1 cylinder at Top Dead Center (TDC)
- Crankshaft sprocket
- Intake camshaft sprocket into the timing chain

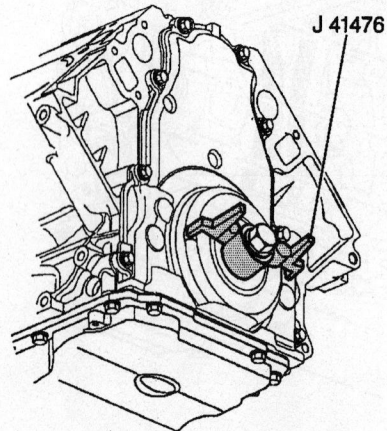

42372-BLAZ-G20

Align the tapered legs of the tool with the machined alignment surfaces on the front cover—5.3L engine

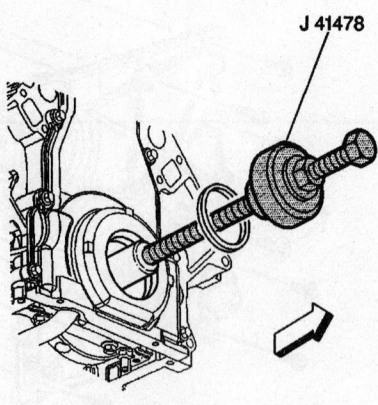

42372-BLAZ-G21

Front cover seal installation using the proper tool—5.3L engine

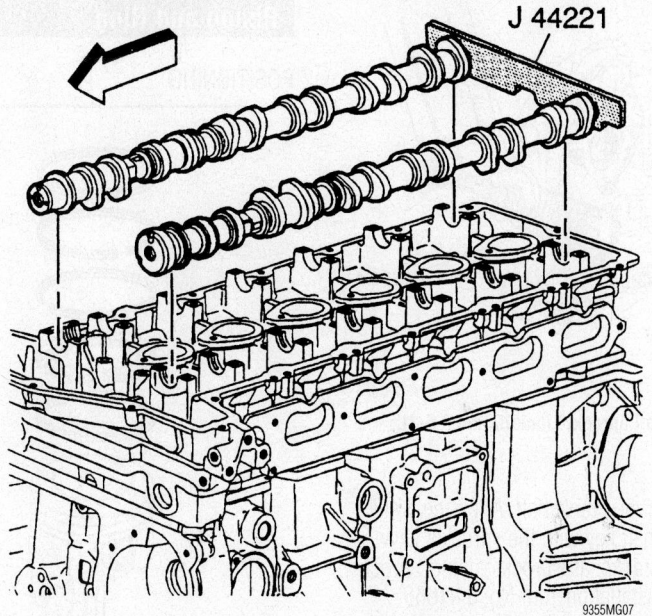

J 44221

9355MG07

Proper installation of the crankshaft holding tool with the No. 1 cylinder at TDC

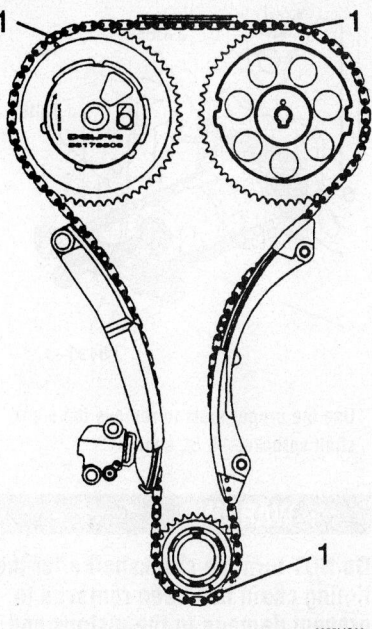

9355MG09

3. Align the dark link of the timing chain with the timing mark on the intake camshaft sprocket. Feed the timing chain down through the opening in the head.

- Timing chain onto the crankshaft sprocket. Align the dark link of the timing chain with the timing mark on the crankshaft sprocket.

➡️**It may be necessary to remove the crankshaft holding tool to rotate and hold the camshaft hex to align the pin to the camshaft sprocket**

- Intake camshaft sprocket onto the intake camshaft
- Intake camshaft washer and bolt
- Exhaust camshaft actuator into the timing chain. Align the dark link of the timing chain with the timing mark on the exhaust camshaft actuator.

➡️**It may be necessary to remove the crankshaft holding tool to rotate and hold the camshaft hex to align the pin to the camshaft sprocket**

- Exhaust camshaft actuator onto the exhaust camshaft

➡️**Rotate the camshaft actuator clockwise relative to the camshaft prior to tightening the bolt.**

4. Rotate the camshaft actuator clockwise (as seen from the front of the vehicle).

❋❋ WARNING

The camshaft actuator must be fully advanced during installation. Engine

damage may occur if the camshaft actuator is not fully advanced.

5. Install the exhaust camshaft actuator bolt and tighten to 18 ft. lbs. (25 Nm), plus an additional 135 degrees, using a torque angle meter.

6. Tighten the intake camshaft sprocket bolt to 15 ft. lbs. (20 Nm), plus an additional 100 degrees, using a torque angle meter.

7. Remove the tee from the timing chain tensioner to regain tension on the timing chain.

8. Remove the crankshaft holding tool. The dark lines on the timing chain should be aligned with the marks on the sprockets.

The dark lines on the timing chain should be aligned with the marks on the sprockets—4.2L engine

9. Install or connect the following:
- Top chain guide
- Suitable threadlock to the top chain guide bolt threads, then install and tighten to 89 inch lbs. (10 Nm)
- Engine front cover
- Camshaft cover

5.3L ENGINE

1. Remove the engine front cover.
2. Remove the oil pump.
3. Rotate the crankshaft until the timing marks on the crankshaft and the camshaft sprockets are aligned.

SEE MANUAL

DELPHI
25178506

9355MG08

Rotate the camshaft actuator clockwise—4.2L engine

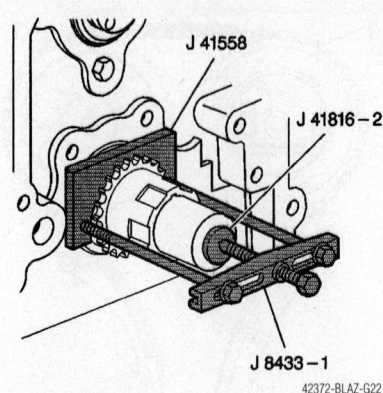

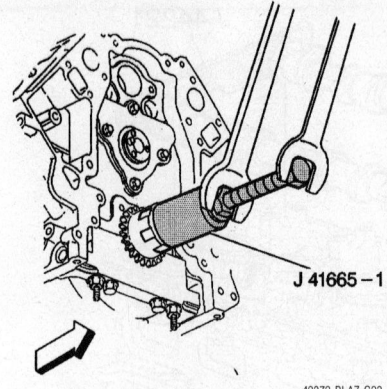

Piston and Ring

POSITIONING

Use the proper tools to remove the crank-shaft sprocket—5.3L engine

Crankshaft sprocket installation—5.3L engine

Piston ring positioning—4.2L engine

✳✳ WARNING

Do NOT turn the crankshaft after the timing chain has been removed to prevent damage to the pistons and valves.

4. Remove or disconnect the following:
   • Camshaft sprocket bolts
   • Camshaft sprocket and timing chain
   • Crankshaft sprocket using Pulley Puller No. J 8433, Crankshaft End Protector Tool No. J 41816-2 and Crankshaft Sprocket Removal Tool No. J 41558
   • Crankshaft sprocket key, if necessary
5. Clean and inspect the timing chain and sprockets.

To install:
6. Install or connect the following:
   • Key into the crankshaft keyway, if removed. Tap the key into the keyway until both ends of the key bottom into the crankshaft.
   • Crankshaft sprocket onto the front

of the crankshaft. Align the crankshaft key with the sprocket keyway.
   • Crankshaft sprocket using Sprocket Installation Tool No. J 41665. Install the sprocket onto the crankshaft until fully seated against the crankshaft flange. Rotate the crankshaft sprocket until the alignment mark is in the 12 o'clock position.

➡**Properly locate the camshaft sprocket locating pin with the cam sprocket alignment hole. The sprocket teeth and timing chain must mesh. The camshaft and crankshaft sprocket alignment marks MUST be aligned properly. Locate the camshaft sprocket alignment mark in the 6 o'clock position. It may be necessary to rotate the camshaft or crankshaft to align the marks.**

   • Camshaft sprocket and timing chain
   • Camshaft sprocket bolts and tighten to 26 ft. lbs. (35 Nm)
   • Oil pump
   • Engine front cover

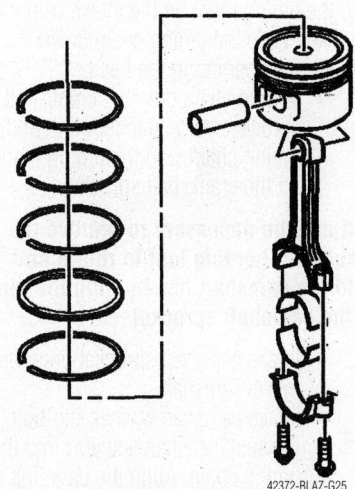

Piston ring positioning—5.3L engine

FUEL SYSTEM

Fuel System Service Precautions

Safety is the most important factor when performing not only fuel system maintenance but also any type of maintenance. Failure to conduct maintenance and repairs in a safe manner may result in serious personal injury or death. Maintenance and testing of the vehicle's fuel system components can be accomplished safely and effectively by adhering to the following rules and guidelines.

• To avoid the possibility of fire and personal injury, always disconnect the negative battery cable unless the repair or test procedure requires that battery voltage be applied.

• Always relieve the fuel system pressure prior to disconnecting any fuel system component (injector, fuel rail, pressure regulator, etc.), fitting or fuel line connection. Exercise extreme caution whenever relieving fuel system pressure, to avoid exposing skin, face and eyes to fuel spray. Please be advised that fuel under pressure may penetrate the skin or any part of the body that it contacts.

• Always place a shop towel or cloth around the fitting or connection prior to loosening to absorb any excess fuel due to spillage. Ensure that all fuel spillage (should it occur) is quickly removed from engine surfaces. Ensure that all fuel soaked

cloths or towels are deposited into a suitable waste container.

• Always keep a dry chemical (Class B) fire extinguisher near the work area.

• Do not allow fuel spray or fuel vapors to come into contact with a spark or open flame.

• Always use a back-up wrench when loosening and tightening fuel line connection fittings. This will prevent unnecessary stress and torsion to fuel line piping. Always follow the proper torque specifications.

• Always replace worn fuel fitting O-rings with new. Do not substitute fuel hose or equivalent where fuel pipe is installed.

Fuel System Pressure

RELIEVING

The fuel systems operate under high fuel pressures. It is very important that the pressure be properly relieved prior to servicing the system or any of its components.

4.2L Engine

1. Before servicing the vehicle, refer to the precautions in the beginning of this section.

✳✳ WARNING

Do not perform this procedure for more than 2 minutes to avoid damaging the catalytic converter.

2. Loosen the fuel filler cap to release the fuel tank pressure.
3. Remove the fuel pump relay from the junction block.
4. Crank the engine, allowing it to start and stall.
5. Crank the engine for an additional 3 seconds to relieve any remaining fuel pressure.
6. Disconnect the negative battery cable to avoid repressurizing the fuel system.
7. Install the fuel pump relay in the junction block.
8. Tighten the fuel filler cap.
9. After you are finished working on the fuel system, connect the negative battery cable.

5.3L Engine

1. Disconnect the negative battery cable.
2. Install Fuel Pressure Gauge J 34730-1A or equivalent to the fuel pressure connection.
3. Loosen the fuel fill cap to relieve the fuel tank vapor pressure.
4. Open the valve on the fuel pressure gauge to bleed the system pressure. The fuel connections are now safe for servicing. Drain any fuel remaining in the gauge into an approved container. Once the system pressure is completely relieved, remove the fuel pressure gauge.

Fuel Filter

REMOVAL & INSTALLATION

1. Before servicing the vehicle, refer to the precautions in the beginning of this section.

2. Properly relieve the fuel system pressure.
3. Remove or disconnect the following:
   - Negative battery cable and fuel filler cap, if not already done
4. Raise and support the vehicle.
   - Fuel tank shield, if equipped
   - Quick connect fittings from the filter
   - Filter feed nut and the clamp bolt
   - Filter and the clamp from the vehicle

To install:
5. Install or connect the following:
   - Filter and clamp with the directional arrow facing away from the fuel tank, toward the throttle body

➡ **The filter has an arrow (fuel flow direction) on the side of the case, be sure to install it correctly in the system, the with arrow facing away from the fuel tank.**

   - Tighten the fuel feed nut
   - Tighten the filter clamp assembly bolt
   - Fuel quick disconnect fittings to the filter
   - Fuel tank shield, if equipped
   - Fuel filler cap
   - Negative battery cable
6. Start the engine and check for leaks.

Fuel Pump

REMOVAL & INSTALLATION

1. Before servicing the vehicle, refer to the precautions in the beginning of this section.
2. Properly relieve the fuel system pressure.
3. Drain the fuel tank.
4. Support the fuel tank.
5. Remove or disconnect the following:
   - Negative battery cable
   - Filler neck from the tank
   - Shield from tank and tank straps
   - Fuel lines and vapor hose from pump
   - Electrical connection from fuel pump
   - Fuel tank
   - Fuel pump/sending unit assembly by turning the locking ring (located on top of the fuel tank) counterclockwise using a spanner wrench
   - Fuel pump from the fuel level sending device

To install:
6. Install or connect the following:

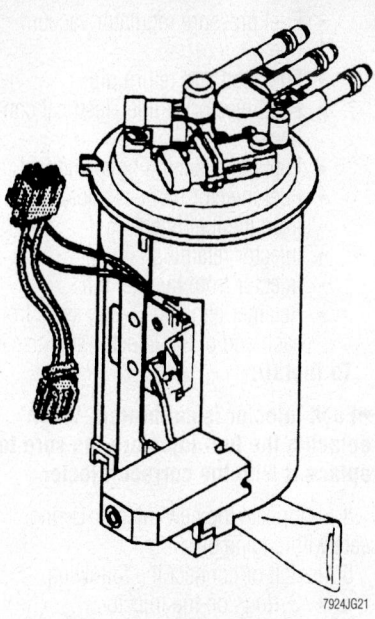

View of the in-tank fuel pump assembly

   - Fuel pump in tank with new seal around opening

➡ **The fuel pump strainer must be in a horizontal position when the fuel sender is installed in the tank. When installing the sender assembly, make sure that the fuel pump strainer does not block full travel of the float arm.**

   - Tank and connect fuel lines and vapor hose
   - Tank to the frame. Torque the fasteners to 33 ft. lbs. (45 nm).
   - Shield
   - Fuel filler neck and clamp
   - Negative battery cable
7. Refill the tank.
8. Run the engine and check for leaks.

Fuel Injector

REMOVAL & INSTALLATION

4.2L Engine

1. Before servicing the vehicle, refer to the precautions in the beginning of this section.
2. Relieve the fuel system pressure. Refer to the fuel system relief procedure in this section.
3. Remove or disconnect the following:
   - Negative battery cable, if not done already
   - Intake manifold

➡ **Clean the fuel rail assembly with a suitable spray cleaner before proceeding. Never soak the fuel rail in a cleaning solvent.**

- Fuel pressure regulator vacuum line
- Fuel feed and return pipes
- Fuel injector in-line electrical connector
- Fuel rail attaching bolts and fuel rail
- Fuel injector harness connector from the fuel injectors
- Injector retaining clip
- Injector from the fuel rail
- Retainer clip and O-ring seals from each end of the injector and discard

To install:

➡**Each injector is calibrated. When replacing the fuel injectors, be sure to replace it with the correct injector.**

4. Lubricate the new injector O-ring seats with engine oil.

5. Install or connect the following:
- O-rings on the injector
- New retainer clip on the injector

6. Push the fuel injector into the fuel rail socket, making sure the connector faces outward. The retainer clip locks to a flange on the fuel rail injector socket.
- Fuel rail assembly. Tighten the bolts to 89 inch lbs. (10 Nm).
- Fuel feed and return lines to the rail
- Fuel injector electrical connectors
- Fuel pressure regulator vacuum line
- Intake manifold
- Negative battery cable

7. Turn the ignition **ON** for 2 seconds and then turn it **OFF** for 10 seconds. Again turn the ignition **ON** and check for leaks.

5.3L Engine

1. Before servicing the vehicle, refer to the precautions in the beginning of this section.

2. Relieve the fuel system pressure. Refer to the fuel system relief procedure in this section.

3. Remove or disconnect the following:
- Negative battery cable, if not done already
- Air cleaner outlet duct
- PCV foul air hose
- A/C compressor pressure switch electrical connector
- Wire harness from the clip on the cylinder head
- Mass Airflow/Intake Air Temperature (MAF/IAT) sensor connector
- Alternator electrical connector
- Right side electrical connectors from the coil main electrical harness, Electronic Throttle Control (ETC) and fuel injectors.

4. To detach the injector connector: Matchmark the connectors, pull the Connector Position Assurance (CPA) retainer up 1 click. Push the tab on the connector in, then detach the injector connector.
- Electrical harness from the clips on the ignition coil bracket
- Evaporative emission (EVAP) purge solenoid electrical connector
- Knock Sensor (KS) electrical connector
- Manifold Absolute Pressure (MAP) electrical connector
- Main coil
- Fuel injector electrical connector (right side)
- Electrical harness from the clips on the ignition coil bracket
- Upper engine wire harness retainer nut. Position the wire harness aside.
- Fuel feed and return pipes from the rail
- Fuel pressure regulator vacuum line
- Fuel rail bolts
- Fuel rail, after cleaning with a spray-type cleaner

✳✳ WARNING

Be very careful when removing the fuel rail and injectors not to damage the connector terminals or injector spray tips

- Fuel injector from the fuel rail
- Fuel injector retainer clip and discard
- Fuel injector lower O-ring seals and discard

To install:

5. Install or connect the following:
- New O-ring seals on the injectors, after lubricating with clean engine oil
- New retainer clip on the injector
- Fuel injector by pushing it into the fuel rail socket
- Fuel rail
- Apply 0.20 (5mm) band of threadlock to the threads of the fuel rail bolts
- Fuel rail bolts and tighten to 89 inch lbs. (10 Nm)
- Fuel pressure regulator vacuum line
- Fuel feel and return pipes
- Route the upper electrical harness into position over the engine.
- Engine harness bracket nut and tighten to 89 inch lbs. (10 Nm)
- PCV valve and hose
- EVAP purge solenoid, KS, MAP

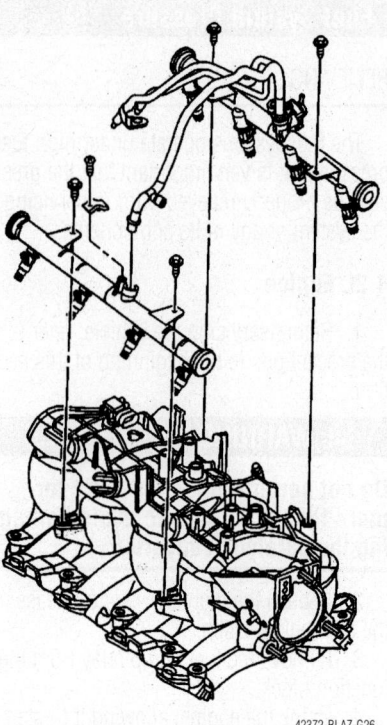

Exploded view of the fuel rail mounting— 5.3L engine

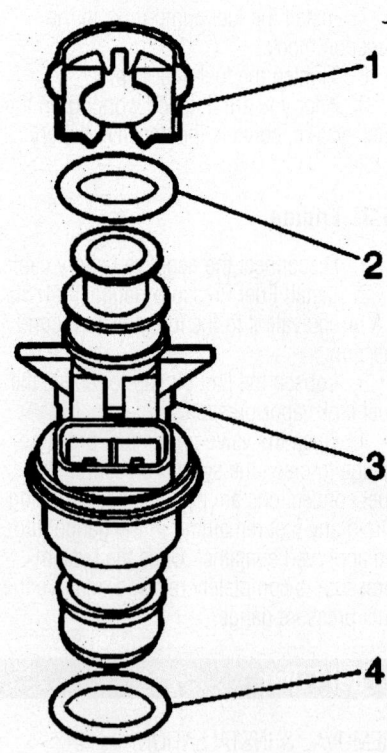

Exploded view of the fuel injector (3), retainer (1) and O-ring seals (2, 4)—5.3L engine

sensor, main coil & fuel injector electrical connectors
- Harness to the clips on the ignition coil bracket
- Main coil, ETC, fuel injector electrical connectors

- Harness to the clips on the ignition coil bracket
- Alternator electrical connector
- MAF/IAT sensor connector
- Wire harness to the clip on the cylinder head

- A/C compressor switch electrical connector
- Air cleaner outlet duct
- Fuel fill cap
- Negative battery cable

6. Refill the engine cooling system.

DRIVE TRAIN

Automatic Transmission Assembly

REMOVAL & INSTALLATION

4.2L Engine

➡This procedure requires the use of a Converter Holding Strap tool No. J 21366 to secure the torque converter to the transmission during removal and installation.

1. Disconnect the negative battery.
2. Drain the transmission fluid.
3. Remove the filler tube nut and stud located on the right side of the engine.
4. Raise the vehicle.
5. If equipped with 2 wheel drive (2WD), remove the rear propeller shaft.
6. If equipped with 4 wheel drive (4WD), remove the transfer case.
7. Support the transmission with a transmission jack.
8. Remove the fuel tank shield if equipped.
9. Remove the transmission support.
10. Remove the transmission mount bolts and mount.
11. Remove the front exhaust pipe assembly.
12. Lower the transmission for access to the top and sides of the transmission.
13. Remove the range selector cable end from the transmission range selector lever ball stud and bracket.
14. Remove the transmission heat shield, transmission vent hose park/neutral position switch connector, and main connector from the transmission.
15. Remove the bolt that secures the fuel line bracket to the left side of the transmission.
16. Remove the flywheel-to-torque converter bolts. Be careful not to drop the bolts into the bell housing.
17. Disconnect the transmission oil cooler lines from the transmission. Plug the transmission oil cooler lines connectors in the transmission case.
18. Install a safety chain around the transmission.

19. Remove the bolt that secures the fuel line bracket to the bell housing.
20. Remove the bolts that secure the coolant pipe to the bell housing.
21. Remove the remaining nuts, studs and/or bolts that secure the transmission to the engine.
22. Install Converter Holding Strap tool No. J 21366 onto the transmission bell housing to hold the torque converter.
23. Pull the transmission straight back and remove it from the vehicle.

To install:

Installation is the reverse of removal, but please note the following important steps.
24. Make sure the torque converter is fully seated in the pump drive. If not, the transmission will not fit tightly to the rear of the engine block.
25. Raise the transmission into position and remove the torque converter holding strap. Carefully slide the transmission forward until the dowel pins are engaged while lining up the marks on the flywheel made during removal.
26. The torque converter should be flush with the flywheel and turn freely by hand.
27. Tighten the torque converter-to-flywheel bolts to 44 ft. lbs. (66 Nm).
28. Install the transmission-to-engine nuts, studs and or bolts. Tighten the studs and/or bolts to 37 ft. lbs. (50 Nm).
29. Tighten the bolts securing the heat shield to the transmission to 13 ft. lbs. (17 Nm).
30. Tighten the bolts and washers securing the transmission mount to 18 ft. lbs. (25 Nm).
31. Tighten the nut and washer securing the transmission mount to the transmission support to 35 ft. lbs. (46 Nm).
32. Refill the transmission with the proper amount and type of fluid.
33. Connect the negative battery cable. Start the vehicle and allow to warm while checking for leaks. Road test the vehicle to check for shift quality.

5.3L Engine

➡This procedure requires the use of a Converter Holding Strap tool No. J 21366 to secure the torque converter to

the transmission during removal and installation.

1. Disconnect the negative battery.
2. Drain the transmission fluid.
3. Raise and support the vehicle.
4. Remove the rear propeller shaft.
5. Support the transmission with a jack.
6. Remove the nuts securing the transmission mount to the transmission support.
7. Remove the transmission support from the vehicle.
8. Remove the transmission mount.
9. Remove the front exhaust pipe assembly.
10. Lower the transmission to gain access to the top and sides of the transmission.
11. Remove the transfer case, if equipped.
12. Remove the range selector cable end from the transmission range selector lever ball stud and the bracket.
13. Remove the transmission heat shield.
14. Disconnect the transmission vent hose, the park/neutral position switch connectors, and the main electrical connector from the transmission.
15. Remove the transmission harness from the retainers.
16. Remove the bolt that secures the fuel line bracket to the left side of the transmission.
17. Remove the torque converter access plug.
18. Mark the flywheel and torque converter orientation for reassembly.
19. Remove the flywheel to torque converter bolts. Use care not to drop the bolts into the bell housing.
20. Disconnect the transmission oil cooler lines from the transmission.
21. Plug the transmission oil cooler line connectors in the transmission case.
22. Install a safety chain around the transmission.
23. Remove the nut that secures the filler tube to the bell housing.
24. Remove the transmission filler tube.
25. Remove the remaining nuts, studs and/or bolts that secure the transmission to the engine.

26. Install the J-21366 onto the transmission bell housing to retain the torque converter.

27. Pull the transmission straight back.

28. Remove the transmission from the vehicle.

To install:

29. Raise the transmission into place and remove the torque converter holding tool.

30. Slide the transmission straight onto the locating pins while lining up the marks on the flywheel and the torque converter made during removal. The torque converter must be flush onto the flywheel and rotate freely by hand.

31. Install nuts, studs and/or bolts securing the transmission to the engine and tighten to 37 ft. lbs. (50 Nm).

32. Install the fuel line retaining bracket to the transmission.

33. Install the flywheel-to-torque converter bolts and tighten to 44 ft. lbs. (66 Nm).

34. Install the torque converter access plug.

35. Remove the safety chain from the transmission.

36. Install the transmission filler tube.

37. Install the filler tube nut.

38. Install the transmission vent hose, fuel lines, and the wiring harness to the transmission.

39. Install the transmission harness to the retainers.

40. Install the heat shield to the transmission.

41. Install the bolts securing the heat shield to the transmission and tighten to 13 ft. lbs. (17 Nm).

42. Install the shift cable end to the transmission shift lever ball stud and bracket.

43. Install the transfer case, if equipped.

44. Install the front exhaust pipe assembly.

45. Install the transmission mount to the vehicle.

46. Install the bolts securing the transmission mount to the transmission and tighten to 18 ft. lbs. (25 Nm).

47. Install the transmission support to the vehicle.

48. Lower the transmission and remove the transmission jack.

49. Install the nuts securing the transmission mount to the transmission support and tighten to 35 ft. lbs. (46 Nm).

50. Install the rear propeller shaft.

51. Flush the transmission oil cooler and cooling lines at this time, if necessary.

52. Connect the transmission oil cooler lines to the transmission.

53. Lower the vehicle.

54. Connect the battery cable.

55. Fill the transmission to the proper level with DEXRON® III transmission fluid and check for leaks.

56. Road test the vehicle and check for proper operation.

Transfer Case Assembly

REMOVAL & INSTALLATION

1. Disconnect the negative battery cable.

2. Raise and support the vehicle. Drain the transfer case.

3. Remove or disconnect the following:
   • Fuel tank shield mounting bolts and shield
   • Rear propeller shaft. Matchmark the shafts prior to removal.
   • Fuel lines from the retainer
   • Electrical harness from the retainers on the right and left sides
   • Speed sensor electrical connectors
   • Motor/encoder electrical connector
   • Transfer case wiring harness
   • Vent hose

4. Install a transmission jack to support the transfer case.
   • Transfer case mounting bolts
   • Transfer case from the vehicle
   • Transfer case gasket and discard if damaged

To install:

5. Install or connect the following:

➡**You must replace the transfer case gasket if it is damaged. Never use silicone sealant in place of, or with the transfer case gasket.**

   • Transfer case, using a new gasket if necessary
   • Transfer case mounting bolts and tighten to 35 ft. lbs. (47 Nm)

6. Remove the transmission jack.

7. The remainder of installation is the reverse of removal.

8. Refill the transfer case.

Halfshaft

REMOVAL & INSTALLATION

1. Remove or disconnect the following:
   • Front wheel

➡**Place a drift through the caliper into the edge of the rotor to keep the rotor from turning when the nut is removed**

   • Engine protection shield
   • Wheel speed sensor from the harness

   • Wheel center cap, if equipped
   • Halfshaft nut and discard. A new nut must be used for installation.
   • Drift from the rotor
   • Brake caliper and support it with a piece of wire to avoid damaging the brake hose
   • Brake rotor

2. To remove the steering knuckle, remove or disconnect the following:
   • Front stabilizer bar link from the lower control arm
   • Upper shock module retaining from the shock tower
   • Outer tie rod retaining nut
   • Outer tie rod end from the steering knuckle using a puller
   • Brake hose bracket retaining bolts
   • Brake hose bracket
   • Upper control arm-to-steering knuckle pinch bolt and nut
   • Upper control arm from the steering knuckle
   • Shock module from the shock tower
   • Lower ball joint retaining nut
   • Steering knuckle from the control arm using a puller
   • Support the front shock module/steering knuckle to the frame

3. Remove the left side halfshaft from differential carrier, or right halfshaft from the clutch fork housing as follows:
   a. Place a brass drift against the tripot housing.
   b. Use a hammer to strike the drift outward from the case, striking hard enough to overcome the snapring tension holding the halfshaft.

4. Pull the halfshaft straight out of the differential carrier or clutch fork housing.

To install:

5. Install the halfshaft as follows:
   a. With both hands on the tripot housing, align the splines on the shaft with the differential carrier assembly (left) or clutch fork housing (right).
   b. Center the halfshaft into the differential carrier or clutch fork housing assembly seal.
   c. Firmly push the shaft straight into the differential carrier or clutch fork housing assembly until the snapring is properly seated.

6. To install the steering knuckle, install or connect the following:
   • Steering knuckle to the lower control arm
   • Upper control arm to the steering knuckle
   • Upper control arm pinch bolt and

nut and tighten to 30 ft. lbs. (40 Nm)
- Shock module to the shock tower
- Brake hose bracket. Tighten the bolts to 7 ft. lbs. (10 Nm).
- Outer tie rod to the steering knuckle and tighten the nut to 33 ft. lbs. (45 Nm)
- Wheel speed sensor
- Stabilizer bar link
- Engine protection shield

7. Install or connect the following:
- New halfshaft nut and tighten to 103 ft. lbs. (140 Nm)
- Wheel

8. Lower the vehicle. Adjust the front toe.

CV-Joints

OVERHAUL

Outer CV-Joint

1. Remove or disconnect the following:

- Front wheel
- Halfshaft and position it in a vise
- Large CV-joint boot clamp and discard it
- Small CV-joint boot clamp and discard it
- CV-joint boot and slide it back on the shaft
- Outer race from the halfshaft, by spreading the outer race-to-halfshaft retaining ring, using Snapring Pliers J-8059
- Retaining ring from the halfshaft and discard it
- CV-joint boot from the halfshaft and discard it, if damaged

2. Disassemble the chrome alloy balls from the CV-joint cage as follows:
a. Position a brass drift against the CV-joint cage and tap it with a hammer to tilt the cage.
b. Remove the 1st chrome alloy ball from the cage.
c. Tilt the cage in the opposite direction.
d. Remove the opposite chrome alloy ball.
e. Repeat the procedure until all 6 balls are removed.
3. Disassemble the CV-joint cage and inner race as follows:
a. Pivot the cage and race 90 degrees to the center line of the outer race.
b. Align the cage windows with outer race lands.

c. Remove the cage from the outer race.
d. Rotate the inner race upward and remove it from the cage.
4. Thoroughly clean and inspect all parts.

To install:

5. Lubricate the parts with a light coat of grease.
6. Assemble the CV-joint cage and inner race, as follows:
a. Rotate the inner race 90 degrees to the cage centerline.
b. Align the cage windows with inner race lands.
c. Insert the inner race into the cage by rotating the inner race downward.
d. Insert the cage/inner race into the outer race.
7. Assemble the chrome alloy balls into the CV-joint cage, as follows:
a. Position a brass drift against the CV-joint cage and tap it with a hammer to tilt the cage.
b. Insert the 1st chrome alloy ball into the cage.
c. Tilt the cage in the opposite direction.
d. Insert the opposite chrome alloy ball.
e. Repeat the procedure until all 6 balls are inserted.
8. Install ½ kit grease into the CV-joint.
9. Install or connect the following:
- Small ring clamp on the CV boot
- New retaining ring on the halfshaft
- Large ring clamp on the CV boot
- Outer race assembly onto the halfshaft until the ring engages the halfshaft groove
10. Slide the small end of the CV-joint boot/clamp into place, with the seal lip in the halfshaft groove

➡ **Make sure the boot lies flat against the halfshaft.**

11. Using the Crimp tool J-35910, a torque wrench and a breaker bar, crimp the small CV-joint boot clamp to 100 ft. lbs. (136 Nm).
12. Check the clamp gap dimension; if it is not 0.085 in. (2.15mm), continue tightening the clamp until it is.
13. Install ½ kit grease into the CV-joint boot.
14. Measure approximately 0.687 in. (17.5mm) up from the bottom edge of the outer CV-joint assembly.
15. Slide the large end of the CV boot/clamp into place, with the seal lip in place over the outer race.

➡ **Make sure the boot lies flat against the outer race.**

16. Using the Crimp tool J-35910, a torque wrench and a breaker bar, crimp the large CV-joint boot clamp to 130 ft. lbs. (176 Nm).
17. Check the clamp gap dimension; if it is not 0.102 in. (2.60mm), continue tightening the clamp until it is.
18. Install the halfshaft and the front wheel.

Inner (Tri-Pot) Joint

1. Remove or disconnect the following:
- Front wheel
- Halfshaft and place it in a vise
- Snapring from the stub shaft and discard it
- Small CV-joint boot clamp, cut and discard it
- Large CV-joint boot clamp, cut and discard it
- CV-joint boot by sliding it away from the tri-pot joint
2. Install a Stub Shaft Removal tool J-38868-A to the stub shaft snapring groove.
3. Using a slide hammer puller, press the stub shaft from the tri-pot housing.
4. Remove or disconnect the following:
- Tri-pot housing from the tri-pot spider
- Inboard spacer ring slide it rearward on the shaft using Snapring Pliers tool J-8059
- Outboard retaining ring using Snapring Pliers tool J-8059 and discard it
- Tri-pot joint spider assembly
- Inboard spacer ring and discard it
- CV-joint boot
- Trilobal tri-pot bushing from the housing
5. Thoroughly clean and inspect all parts.

To install:

6. Install or connect the following:
- New snapring onto the stub shaft
- Small boot clamp
- CV-joint boot
7. Using the Crimp tool J-35910, a torque wrench and a breaker bar, crimp the small CV-joint boot clamp to 100 ft. lbs. (136 Nm).
8. Install or connect the following:
- Inboard spacer ring slide it rearward on the shaft using Snapring Pliers tool J-8059, past the 2nd groove
- Tri-pot joint spider assembly onto the shaft until it passes the 2nd groove

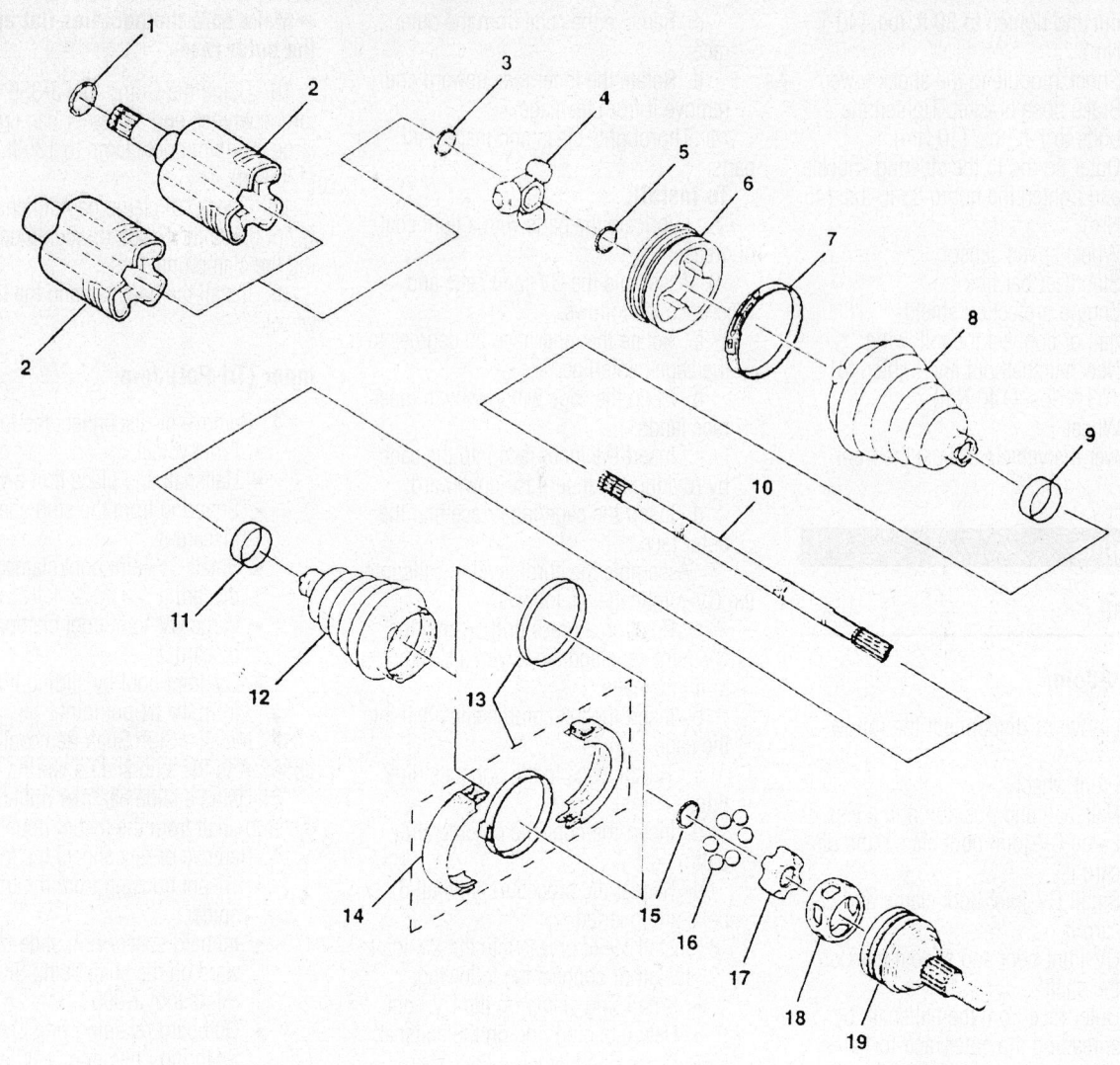

(1) Differential Shaft Ring
(2) Tripot Housing Assembly
(3) Spacer Ring
(4) Tripot Joint Spider Assembly
(5) Spacer Ring
(6) Tripot Bushing
(7) Boot Retaining Clamp
(8) Tripot Joint Boot
(9) Halfshaft Swage Ring
(10) Halfshaft Bar

(11) Halfshaft Swage Ring
(12) CV Joint Boot
(13) Swage Ring
(14) Clamp Protector
(15) Race Retaining Ring
(16) Ball
(17) CV Joint Inner Race
(18) CV Joint Cage
(19) CV Joint Outer Race

9308JG09

Exploded view of the CV-Joint Assembly

- Outboard retaining ring into the axle shaft groove using Snapring Pliers tool J-8059
- Tri-pot joint spider assembly, slide it against the outboard retaining ring
- Inboard spacer ring, seat it in the groove
- ½ kit grease into the boot
- ½ kit grease into the tri-pot housing

- Trilobal tip-pot bushing flush with the tri-pot housing face
- New large seal clamp onto the CV-joint boot
- Tri-pot housing, slide it over the tri-pot joint spider assembly
- CV-joint boot/clamp, slide it into place, over the trilobal tri-pot bushing with the seal lip in the groove

➡**Make sure the boot lies flat against the trilobal bushing.**

9. Position the CV-joint boot so it measures 4.9 in. (125mm).

10. Using the Crimp tool J-35566, latch the large CV-joint boot clamp.

11. Install the halfshaft and the front wheel.

Axle Shaft, Bearing and Seal

REMOVAL & INSTALLATION

Front

For the Axle Shaft, Bearing and Seal, Removal and Installation, please refer to Front Wheel Bearing procedure located in this section.

Rear

1. Raise the vehicle.
2. Remove the tire and wheel assembly.
3. Remove the brake caliper.
4. Remove the rear axle housing cover and the gasket.
5. Remove the pinion shaft locking bolt.
6. On axles without a locking differential, remove the pinion shaft.
7. On axles with a locking differential, remove the shaft part way. Rotate the case until the pinion shaft touches the housing.
8. On axles with a locking differential, use a screwdriver, or a similar tool, in order to enter the differential case and rotate the C-lock until the C-lock aligns with the thrust block.
9. Push the flange of the axle shaft toward the differential.
10. Remove the C-lock from the button end of the axle shaft.

➡**When removing the axle shaft, do not rotate the shaft. Rotating the shaft will misalign the gears. Misaligning the gears will make the installing of the axle shaft difficult.**

11. Remove the axle shaft from the housing.
12. Remove the axle shaft seal and/or the bearing from the axle housing using a slide hammer and bearing remover.

To install:
13. Install the axle shaft bearing using a universal driver and bearing installer.
14. Drive the axle shaft bearing into the axle housing until the tool bottoms against the tube.
15. Install the axle shaft seal using a bearing installer.
16. Drive the tool into the bore until the axle shaft seal bottoms flush with the tube.
17. Install the axle shaft into the rear axle housing.
18. Slide the axle shaft into place allowing the splines to engage the differential side gear.

19. On axles without a locking differential, place the C-lock on the button end of the axle shaft.
20. On axles with a locking differential, keep the pinion shaft partially withdrawn.
21. On axles with a locking differential, place the C-lock on the axle shaft so that the ends are flush with the thrust block.
22. Pull the shaft flange outward in order to seat the C-lock in the differential gear.
23. Align the hole in the pinion shaft with the bolt hole in the differential case.
24. Install the new pinion shaft locking bolt and tighten to 27 ft. lbs. (36 Nm).
25. Install the rear axle housing cover and the gasket. Tighten the bolts in a cross-wise pattern to 20 ft. lbs. (30 m) on 8.0 in. axles, or 18 ft. lbs. (25 Nm) on 8.6 in. axles.
26. Install the brake caliper.
27. Install the tire and wheel.
28. Fill the rear axle with axle lubricant.
29. Lower the vehicle.

Pinion Seal

REMOVAL & INSTALLATION

➡**The following procedure requires the use of the Pinion Holding tool J-8614-10, the Pinion Flange Removal tool J-8614-1, J-8614-2, J-8614-3 and the Pinion Seal Installation tool J-23911 or J-33782.**

1. Remove or disconnect the following:

- Engine protection shield
- Driveshaft from the pinion flange. Matchmark the driveshaft prior to removal.

- Driveshaft from the rear axle pinion flange and support the shaft up in body tunnel by wiring it to the exhaust pipe.
- Rear steering gear crossmember

➡**If the U-joint bearings are not retained by a retainer strap, use a piece of tape to hold bearings on their journals.**

2. Mark the position of the pinion stem, flange and nut for reference.
3. Use an inch lbs. torque wrench to measure the amount of torque necessary to turn the pinion, then note this measurement as it is the combined pinion bearing, seal, carrier bearing, axle bearing and seal pre-load.
4. Remove or disconnect the following:
- Pinion flange nut and washer, using a Pinion Holding tool J-8614-10 and a Pinion Flange Removal tool J-8614-1, J-8614-2, J-8614-3, as applicable
- Pinion flange
- Pinion oil seal by driving it out of the differential with a blunt chisel; DO NOT damage the carrier
- Dust deflector

To install:
5. Examine the seal surface of pinion flange for tool marks, nicks or damage, such as a groove worn by the seal. If damaged, replace flange.
6. Examine the carrier bore and remove any burrs that might cause leaks around the O.D. of the seal.
7. Apply GM seal lubricant 12346004 to the outside diameter of the pinion flange and sealing lip of new seal.

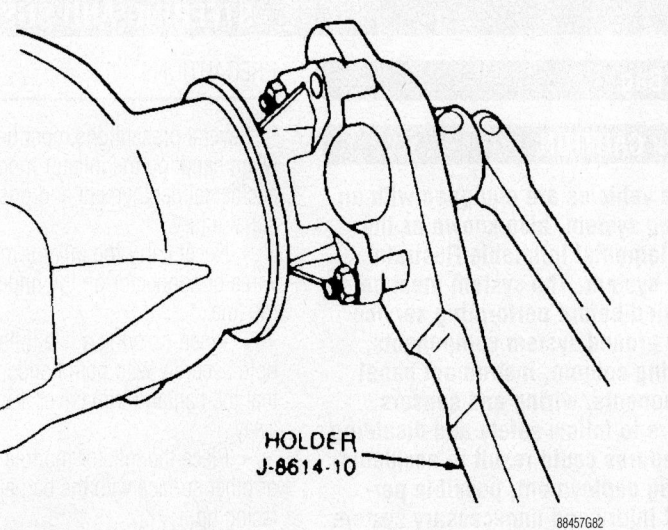

HOLDER
J-8614-10

88457G82

Removing the pinion nut using a pinion holding fixture tool

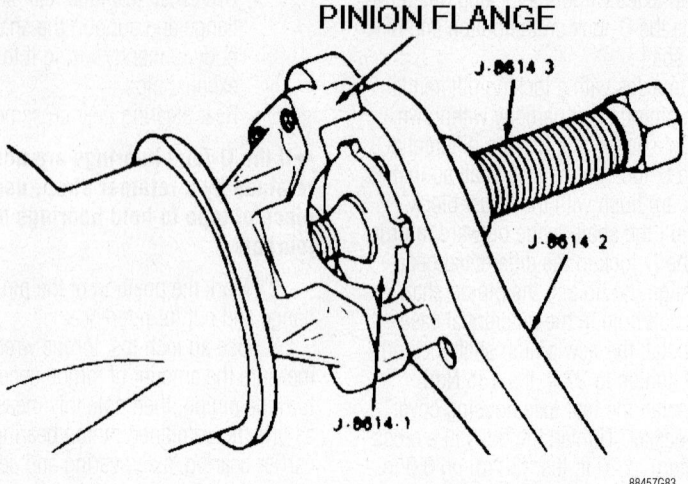

PINION FLANGE

J-8814 3

J-8814 2

J-8814 1

88457G83

A puller and adapter should be used to withdraw the pinion from the housing

8. Install or connect the following:
- Dust deflector
- New pinion oil seal using a seal installer tool
- Pinion flange and tighten nut to the same position as marked earlier. Tighten the nut a little at a time and turn the pinion flange several times after each tightening in order to set the rollers.

9. Measure the torque necessary to turn the pinion and compare this to the reading taken during removal. The rotating torque of the pinion nut should be 3–5 inch lbs. (0.40-0.57 Nm) greater than the torque recorded during removal.

➡**If fluid was lost from the differential housing during this procedure, be sure to check and add additional fluid, as necessary.**

10. Remove the support then align and secure the driveshaft assembly to the pinion flange.

➡**The original matchmarks MUST be aligned to assure proper shaft balance and prevent vibration.**

11. Install the rear steering gear cross-member.

12. Install the engine protection shield.

Axle Housing

REMOVAL & INSTALLATION

1. Raise and support the vehicle.
2. Place safety stands at the front-end of the vehicle.
3. Support the rear axle with safety stands.
4. Remove the rear tires and wheels.

5. Disconnect the rear axle vent tube.
6. Remove the rear propeller shaft and support with mechanics wire.
7. Disconnect the left and right rear cables of the parking brake from the rear axle.
8. Remove the caliper assemblies from the rear axle.
9. Remove the stabilizer shaft from the rear axle.
10. Remove the coil springs.
11. Disconnect the rear axle tie rod from the rear axle.
12. Disconnect the lower control arms from the rear axle.
13. Disconnect the upper control arms from the rear axle.
14. Remove the rear-axle assembly from the vehicle.

To install:

15. Install the rear-axle assembly to the vehicle.
16. Connect the upper control arm to the rear axle.
17. Connect the lower control arm to the rear axle.
18. Connect the rear axle tie rod to the rear axle.
19. Install the coil springs.
20. Install the stabilizer shaft to the rear axle.
21. Install the caliper assemblies to the rear axle.
22. Connect the right and left rear cables of the parking brake to the rear axle.
23. Install the propeller shaft.
24. Connect the rear axle vent tube.
25. Install the rear tires and the rear wheels.
26. Fill the axle with lubricant.
27. Remove the safety stands.
28. Lower the vehicle.

STEERING AND SUSPENSION

Air Bag

✳✳ CAUTION

Some vehicles are equipped with an air bag system, also known as the Supplemental Inflatable Restraint (SIR) system. The system must be disabled before performing service on or around system components, steering column, instrument panel components, wiring and sensors. Failure to follow safety and disabling procedures could result in accidental air bag deployment, possible personal injury and unnecessary system repairs.

PRECAUTIONS

Several precautions must be observed when handling the inflator module to avoid accidental deployment and possible personal injury.

- Never carry the inflator module by the wires or connector on the underside of the module.
- When carrying a live inflator module, hold securely with both hands, and ensure that the bag and trim cover are pointed away.
- Place the inflator module on a bench or other surface with the bag and trim cover facing up.
- With the inflator module on the bench,

never place anything on or close to the module, that may be thrown in the event of an accidental deployment.

DISARMING

2002 Models

1. Turn the steering wheel so that the vehicle's wheels are pointing straight ahead.
2. Turn the ignition switch to **LOCK**, remove the key, then disconnect the negative battery cable.
3. Remove the AIR BAG or SIR fuse from the fuse block.
4. Remove the steering column filler panel or knee bolster.
5. Unplug the Connector Position

Assurance (CPA) and yellow two way connector at the base of the steering column.

6. Open the glove compartment door, lift the stop and let the door fully open.

7. Remove the Connector Position Assurance (CPA) from the passenger yellow two way connector located behind the glove box.

8. Unplug the yellow two way connector located behind the glove box.

9. Connect the negative battery cable.

➡**With the AIR BAG fuse removed, the battery cable connected and the ignition in the ON position, the AIR BAG warning lamp will be ON. This is normal and does not indicate a system malfunction.**

2003–05 Models

1. Turn the steering wheel so that the vehicle's wheels are pointing straight ahead.

2. Turn the ignition switch to **LOCK**, remove the key, then disconnect the negative battery cable.

3. Remove the SIR fuse from the fuse block located in the underhood fuse block.

4. Remove the grille.

5. Remove Front End Sensor bracket from the bumper.

➡**These vehicles are equipped with two inflatable restraint Electronic Frontal Sensors (EFS). When performing this procedure be sure to include both EFS's.**

6. Remove the Connector Position Assurance (CPA) from both electronic frontal sensor (EFS) connectors.

7. Disconnect both EFS connectors.

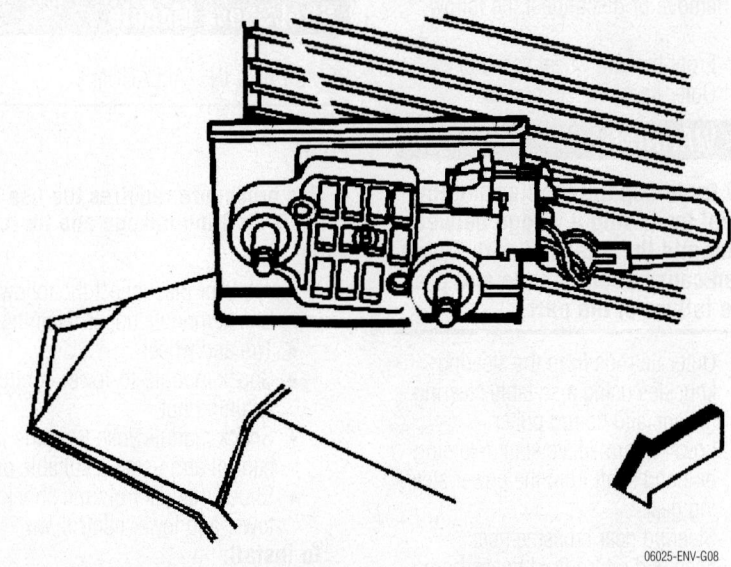

Electronic Frontal Sensor location on bumper– 2003–05 vehicles

06025-ENV-G08

ARMING

2002 Models

1. Disconnect the negative battery cable.

2. Attach the yellow two way connector located behind the glove box.

3. Install the Connector Position Assurance (CPA) to the passenger yellow two way connector located behind the glove box.

4. Close the glove compartment door.

5. Turn the ignition switch to **LOCK**, then remove the key.

6. Attach the two way connector at the base of the steering column and the Connector Position Assurance (CPA).

7. Install the steering column filler panel or knee bolster.

8. Install the AIR BAG fuse to the fuse block.

9. Connect the negative battery cable.

10. From the passenger seat, turn the ignition switch to **RUN** and make sure that the AIR BAG warning lamp flashes seven times and then shuts off. If the warning lamp does not shut off, make sure that the wiring is properly connected. If the light remains on, take the vehicle to a reputable repair facility for service.

2003–05 Models

1. Connect the EFS connectors to both EFS's.

2. Install the CPAs to both EFS connectors.

3. Install sensor bracket to bumper.

4. Install the grille.

5. Install the SIR fuse into the fuse block.

6. Staying well away from all air bags, turn ON the ignition, with the engine OFF.

7. The AIR BAG indicator will flash 7 times. The AIR BAG indicator will then turn OFF.

8. Perform the Diagnostic System Check-SIR if the AIR BAG indicator does not operate as described.

Power Steering Gear

REMOVAL & INSTALLATION

1. Raise and support the vehicle.

2. Position a fluid catch pan under the power steering gear.

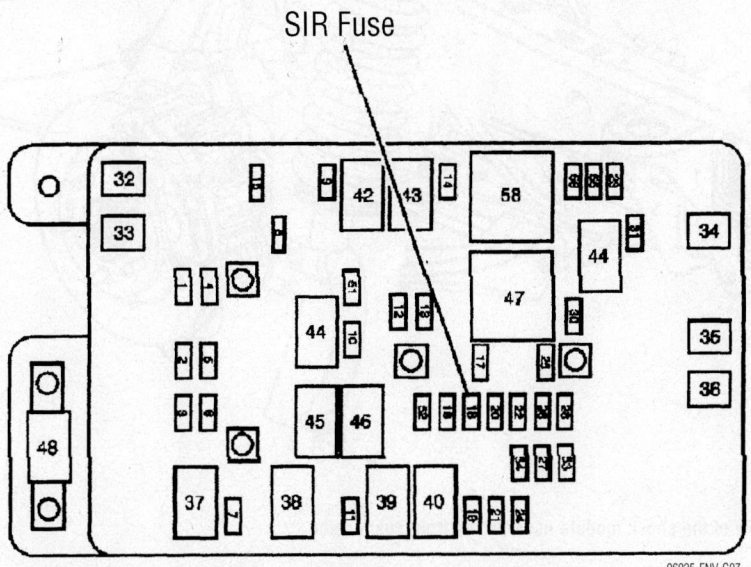

SIR fuse location in underhood fuse block– 2003–05 vehicles

06025-ENV-G07

3. Remove or disconnect the following:

- Front tire and wheel assemblies
- Outer tie rod retaining nuts

✳✳ WARNING

Do not try to separate a steering linkage joint by driving a wedge between the joint and the attached part. Doing this can cause seal damage and premature failure of the part.

- Outer tie rods from the steering knuckles using a suitable steering linkage and tie rod puller
- Lower intermediate shaft retaining bolt and shaft from the power steering gear
- Steering gear crossmember
- Feed and return fluid hoses from the steering gear. Immediately cap or plug all openings to prevent system contamination or excessive fluid loss.

4. Support the power steering gear.

- Power steering gear mounting bolts, then remove the gear from the vehicle

5. Loosen the outer tie rod jam nuts, then remove the outer tie rods from the inner tie rods and discard the jam nut.

To install:

6. Lubricate the inner tie rod threads with a suitable lubricant before installing the outer tie rod.

7. Install or connect the following:

- New jam nuts to the outer tie rods
- Outer tie rods to the inner tie rods
- Power steering gear to the vehicle. Tighten the retaining bolts to 81 ft. lbs. (110 Nm).

8. Remove the support from the power steering gear.

- Power steering hose(s) to the gear. Tighten the retaining bolt to 9 ft. lbs. (12 Nm).
- Steering gear crossmember
- Lower intermediate shaft to the power steering gear. Tighten the retaining bolt to 30 ft. lbs. (40 Nm).
- Outer tie rod ends to the steering knuckles. Tighten the retaining nuts to 33 ft. lbs. (45 Nm).
- Front tire and wheel assemblies

9. Remove the drain pan, then lower the vehicle.

10. Bleed the power steering system and adjust the front toe as necessary.

Strut/Shock Module

REMOVAL & INSTALLATION

Front

➡ **This procedure requires the use of a suitable steering linkage and tie rod puller.**

1. Remove or disconnect the following:

- Shock module upper retaining nuts
- Tire and wheel
- Shock module-to-lower control arm retaining nut
- Shock module yoke from the lower control arm using a suitable puller
- Shock module from the shock tower and lower control arm

To install:

2. Install or connect the following:

- Shock module to the shock tower and lower control arm
- Shock module yoke to the lower control arm
- Shock module upper retaining nuts and tighten to 33 ft. lbs. (45 Nm)
- Shock module-to-lower control arm retaining nut and tighten to 81 ft. lbs. (110 Nm)
- Tire and wheel

Shock Absorbers

REMOVAL & INSTALLATION

Rear

1. Properly support the rear axle assembly.

2. Remove or disconnect the following:

- Automatic level control air lines from the shock absorber, if equipped
- Shock absorber-to-frame retainer(s) at the top of the shock
- Shock-to-axle retainer(s) at the bottom of the shock
- Shock absorber

To install:

3. Install the shock in the vehicle and loosely install the upper mounting fasteners to retain it

4. Align the lower-end of the shock absorber with the axle mounting, then loosely install the retainers.

5. Tighten the upper and lower shock retainers to 59 ft. lbs. (80 Nm).

6. If equipped, attach the automatic level control air lines to the shock absorber.

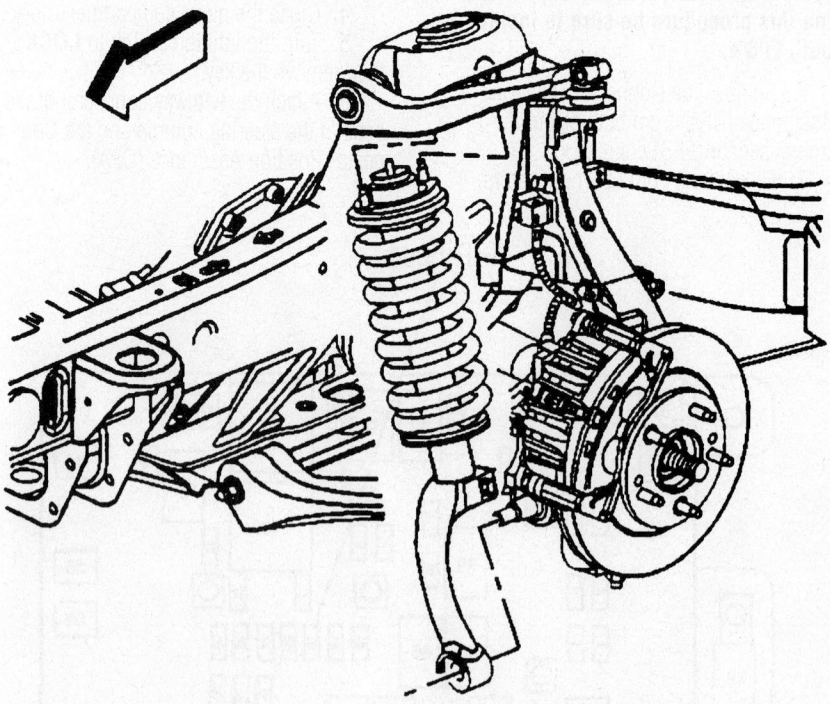

View of the shock module used on the front suspension

9355MG11

Coil Springs

REMOVAL & INSTALLATION

Front

➡**This procedure requires the use of a suitable spring compressor.**

1. Remove or disconnect the following:
   - Wheel
   - Shock module
   - Shock module yoke-to-shock absorber pinch bolt and nut
2. Spread the shock module yoke at the pinch bolt using a suitable flat-bladed tool.
   - Shock module yoke from the shock absorber
3. Install pieces of heater hose or equivalent material to the shock module spring where the spring compressor contacts the lower part of the spring.
4. Install the shock module into the spring compressor.

➡**The spring is compressed when the shock absorber moves freely.**

5. Turn the spring compressor forcing screw until the coil spring is compressed.

6. Remove or disconnect the following:
   - Shock absorber upper retaining nut
   - Shock absorber from the shock module
7. Loosen the compressor forcing screw until the upper mounting plate and coil spring can be removed.
   - Upper mounting plate and coil spring from the spring compressor

To install:

8. Install or connect the following:
   - Coil spring and upper mounting plate to the spring compressor
9. Turn the compressor forcing screw until the coil spring is compressed.
   - Shock absorber to the shock module. Tighten the retaining nut to 33 ft. lbs. (45 Nm)
10. Remove the shock module from the spring compressor. Remove the pieces of heater hose from the spring.
   - Shock module yoke to the shock absorber
   - Shock module yoke-to-shock pinch bolt and nut and tighten to 52 ft. lbs. (70 Nm)
   - Shock module to the vehicle
   - Tire and wheel
11. Lower the vehicle

Rear

1. Raise and support the vehicle.
2. Support the rear axle.
3. Remove the shock absorber lower mounting bolts.

➡**Do not lower the rear axle so the upper control arms contact the frame. This will damage the upper control arms.**

4. Lower the rear axle, then remove the coil springs.

➡**Be careful not to chip or scratch the coating of the coil springs when removing and installing the springs. Damaging the coating will cause premature failure of the coil springs.**

To install:

5. Install the coil springs, then raise the rear axle.
6. Install the shock absorber lower and upper mounting bolts and tighten to 59 ft. lbs. (80 Nm).
7. Remove the rear axle support.
8. Lower the vehicle.

Stabilizer Bar

REMOVAL & INSTALLATION

Front and Rear

1. Raise and support the vehicle.
2. Remove the stabilizer shaft links to the stabilizer shaft retaining nuts.
3. Remove the stabilizer shaft insulator clamp mounting bolts.
4. Remove the stabilizer shaft insulator clamp from the stabilizer shaft insulator.
5. Remove the stabilizer shaft insulators from the stabilizer shaft.
6. Remove the stabilizer shaft from the vehicle.

To install:

7. Install the stabilizer shaft to the vehicle.
8. Install the stabilizer shaft insulators to the stabilizer shaft.
9. Install the stabilizer shaft insulator clamp to the stabilizer shaft insulator.
10. Install the stabilizer shaft insulator clamp mounting bolts and tighten to 41 ft. lbs. (55 Nm) on the front or 52 ft. lbs. (70 Nm) on the rear.
11. Install the stabilizer shaft links to the stabilizer shaft and tighten to 74 ft. lbs. (100 Nm).
12. Lower the vehicle.

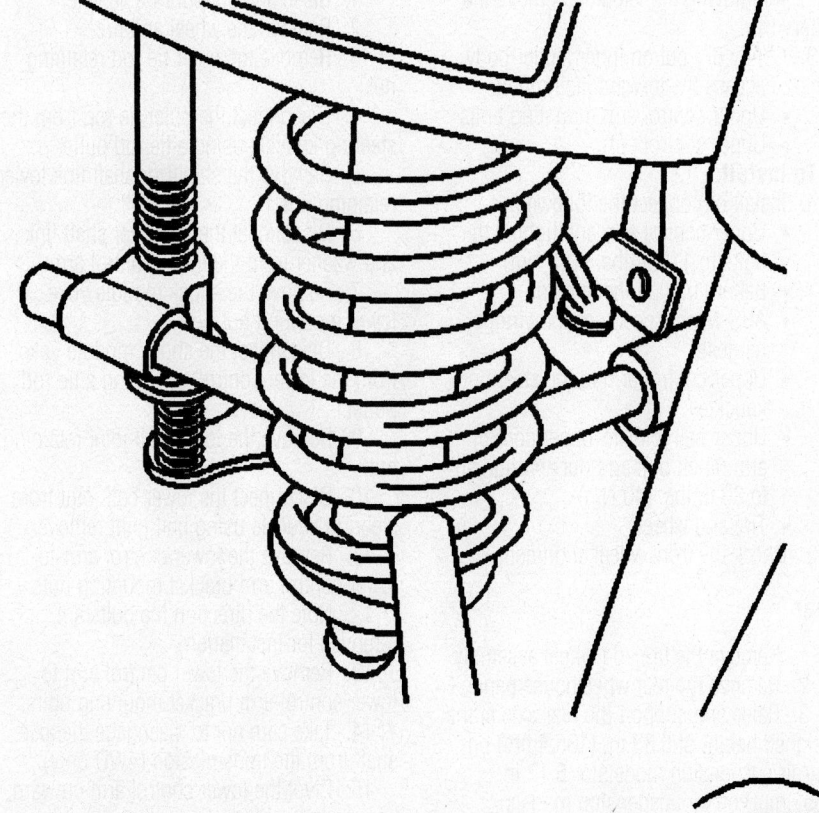

71461-BLAZ-G01

Place pieces of heater hose to the spring where the compressor contacts the lower part of the spring

Upper Ball Joints

REMOVAL & INSTALLATION

Front

➡**This procedure requires the use of the following special tools: J 9519-E Lower Ball Joint Remover and Installer, J 21474-01 Control Arm Bushing Set and J 45117 Ball Joint Installation Spacer.**

1. Raise and support the vehicle.
2. Remove or disconnect the following:
   • Tire and wheel
   • Wheel hub and bearing with the steering knuckle
   • Upper ball joint retaining clip
   • Upper ball joint from the steering knuckle using Lower Ball Joint Removal and Installer tool No. J 9519-E

To install:

3. Install or connect the following:
   • Upper ball joint to the steering knuckle using J 9519-E, J 21474-01 and J 45117
   • Upper ball joint retaining clip
   • Lower ball joint retaining nut and tighten to 81 ft. lbs. (110 Nm)
   • Wheel hub and bearing with steering knuckle
   • Tire and wheel
4. Lower the vehicle
5. Check the front wheel alignment.

Lower Ball Joints

REMOVAL & INSTALLATION

➡**This procedure requires the use of the following special tools: J 9519-E Lower Ball Joint Remover and Installer, J 34874 Booster Seal Remover/ Installer, J 41435 Ball Joint Installer, J 45105-1 Ball Joint Flaring Adapter and J 45105-2 Receiver.**

1. Raise and support the vehicle.
2. Remove or disconnect the following:
   • Tire and wheel
   • Wheel hub and bearing with the steering knuckle
   • Lower ball joint flange with a chisel
3. Install tools J 9519-E and J 34874 to the lower ball joint, then use those tools to remove the lower ball joint from the lower control arm.

To install:

4. Install or connect the following:
   • Lower ball joint to the lower control

arm, using tools J 9519-E, J 41435 and J 45105-2
5. Remove the tools from the lower control arm.
   • Tools J 9519-E and J 45105-1 to the lower ball joint
6. Flare the lower ball joint flange with J 9519-E and J 45105-1, then remove the tools from the lower ball joint.
   • Lower ball joint retaining nut and tighten to 81 ft. lbs. (110 Nm)
   • Wheel hub and bearing with steering knuckle
   • Tire and wheel
7. Lower the vehicle
8. Check the front wheel alignment.

Upper Control Arm

REMOVAL & INSTALLATION

Front

1. Remove or disconnect the following:
   • Tire and wheel assembly
   • Upper ball joint-to-upper control arm pinch bolt and nut
   • Upper control arm from the knuckle
   • Anti-lock Brake System (ABS) wheel speed sensor wiring harness
2. If removing the left side, remove the battery tray.
3. Gently pry out on inner fender body panel to access the forward facing bolt.
   • Upper control arm mounting bolts
   • Upper control arm

To install:

4. Install or connect the following:
   • Upper control arm and tighten the bolts to 111 ft. lbs. (150 Nm)
   • Battery tray on the left side
   • ABS wheel speed sensor wiring harness
   • Upper control arm to the steering knuckle
   • Upper ball joint-to-upper control arm pinch bolt and nut and tighten to 30 ft. lbs. (40 Nm)
   • Tire and wheel
5. Check the front wheel alignment.

Rear

1. Remove the tire and wheel assembly.
2. Remove the rear wheelhouse panel.
3. Raise and support the rear axle at the designed height of 5.33 in. (135.4mm) on non-air suspension models or 6.12 in. (155.4mm) on air suspension models.
4. If equipped with air suspension, depressurize the air suspension system.
5. Disconnect the air suspension level-

ing sensor link from the rear axle upper control arm.
6. Remove the rear axle upper control arm to axle mounting bolt and nut.
7. Remove the rear axle upper control arm to frame mounting bolt.
8. Remove the rear axle upper control arm.

To install:

9. Install the rear axle upper control arm.
10. Install the rear axle upper control arm to frame mounting bolt.
11. Install the rear axle upper control arm to axle mounting nut and bolt and tighten to 97 ft. lbs. (131 Nm).
12. If equipped with air suspension, connect the air suspension leveling sensor link to the rear axle upper control arm.
13. Remove the rear axle support.
14. Install the wheelhouse panel.
15. Install the tire and wheel.
16. Lower the vehicle.

Lower Control Arm

REMOVAL & INSTALLATION

Front

1. Raise and support the vehicle.
2. Remove the wheel and tire.
3. Remove the outer tie rod retaining nut.
4. Disconnect the outer tie rod from the steering knuckle using a tie rod puller.
5. Remove the stabilizer shaft link lower retaining nut.
6. Disconnect the stabilizer shaft link and washer from the lower control arm.
7. Remove the shock module yoke lower mounting nut.
8. Disconnect the shock module yoke from the lower control arm using a tie rod puller.
9. Remove the lower ball joint retaining nut.
10. Disconnect the lower ball joint from steering knuckle using ball joint remover.
11. Remove the lower control arm-to-lower control arm bracket mounting nuts.
12. Note the direction the bolts are removed for installation.
13. Remove the lower control arm to lower control arm bracket mounting bolts.
14. Take care not to disengage the axle shaft from the transmission (4WD only).
15. Pivot the lower control arm outward and downward in order to disconnect the lower control arm from the lower control arm bracket.

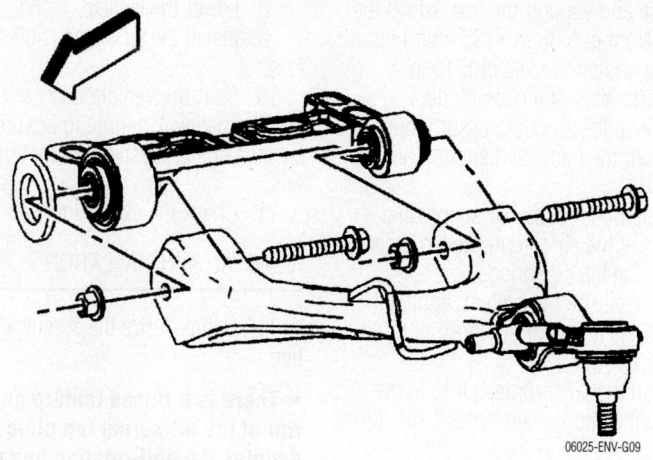

06025-ENV-G09

Front lower control arm mounting

16. Ensure that the spacer stays in position on the front control arm bracket front bushing.

17. Remove the lower control arm from the vehicle.

To install:

18. Position the lower control arm ball joint stud to the steering knuckle.

19. Ensure that the spacer stays in position on the front control arm bracket front bushing.

20. Pivot the lower control arm outward and upward in order to connect the lower control arm to the lower control arm bracket.

21. Install the lower control arm to lower control arm bracket mounting bolts.

➡**Ensure that the lower control arm is parallel to the lower control arm bracket during the installation and tightening of the lower control arm mounting bolts and nuts. This will ensure correct alignment of the lower control arm bushings.**

22. Install the lower control arm to lower control arm bracket mounting nuts and tighten to 96 ft. lbs. (130 Nm).

23. Connect the shock module yoke to the lower control arm.

24. Install the shock module yoke lower mounting nut and tighten to 81 ft. lbs. (110 Nm).

25. Install the lower ball joint retaining nut and tighten to 81 ft. lbs. (110 Nm).

26. Install the stabilizer shaft link and washer to the lower control arm.

27. Install the stabilizer shaft link retaining nut and tighten to 114 ft. lbs. (155 Nm).

28. Install the outer tie rod to the steering knuckle.

29. Install the outer tie rod retaining nut and tighten to 33 ft. lbs. (45 Nm).

30. Install the tire and wheel.

31. Lower the vehicle.

32. Check the front wheel alignment.

Rear

1. Raise and support the vehicle.

2. Remove the wheel and tire.

3. Raise and support the rear axle at the designed height of 5.33 in. (135.4mm) on non-air suspension models or 6.12 in. (155.4mm) on air suspension models.

4. If equipped with air suspension, depressurize the air suspension system.

5. Remove the rear axle lower control arm to the axle mounting nut and bolt.

6. Remove the rear axle lower control arm to the frame mounting nut and bolt.

7. Remove the lower control arm.

To install:

8. Install the lower control arm.

9. Install the rear axle lower control arm to the frame mounting nut and bolt.

10. Install the rear axle lower control arm to the axle mounting bolt and nut and tighten to 74 ft. lbs. (100 Nm).

11. Remove the rear axle support.

12. Lower the vehicle.

Wheel Bearings and Hub

ADJUSTMENT

The wheel bearings on these vehicles are not adjustable. If the bearings become loose or make noise, the wheel hub and bearing must be replaced.

REMOVAL & INSTALLATION

Front

1. On 4WD vehicles, remove wheel center cap, if equipped, and the drive axle nut and washer

2. Raise and support the vehicle.

3. Remove or disconnect the following:
   - Tire and wheel
   - Caliper, leaving the fluid lines connected
   - Brake rotor
   - Halfshaft from the hub and bearing on 4WD vehicles. Place a brass drift against the outer edge of the halfshaft to protect the shaft threads. Use a hammer to sharply strike the brass drift, but to do not remove the halfshaft at this time.
   - Wheel speed sensor
   - Wheel hub and bearing-to-steering knuckle bolts and hub and bearing

➡**Lay the hub and bearing on the wheel studs on the outboard side. This will avoid damaging the bearing seal.**

   - Splash shield from the steering knuckle

To install:

4. Install or connect the following:
   - Wheel hub and bearing seal
   - Splash shield to the steering knuckle, making sure it's properly aligned
   - Hub and bearing to the steering knuckle, aligning the threaded holes
   - Hub and bearing bolts and tighten to 77 ft. lbs. (105 Nm)
   - Wheel speed sensor. Tighten the bolt to 13 ft. lbs. (18 Nm).
   - Rotor and brake caliper
   - Tire and wheel

5. Lower the vehicle

6. On 4WD vehicles, install the drive axle nut and tighten to 103 ft. lbs. (140 Nm), then install the center cap.

Steering Knuckle

REMOVAL & INSTALLATION

Front

1. Raise and support the vehicle.

2. Remove the tire and wheel.

3. On 4WD vehicles, remove wheel center cap, if equipped, and the drive axle nut and washer

4. Remove the wheel hub and bearing.

5. Remove the outer tie rod retaining nut.

6. Disconnect the outer tie rod from the steering knuckle using a tie rod puller.

7. Remove the brake hose bracket retaining bolts.

8. Remove the brake hose bracket from the steering knuckle.

9. Disconnect the ABS wheel speed sensor wiring harness bracket from the steering knuckle.

10. Remove the upper control arm to the steering knuckle pinch bolt and nut.

11. Disconnect the upper control arm from the steering knuckle.

12. Remove the lower ball joint retaining nut.

13. Remove the steering knuckle from the lower control arm.

14. Remove the steering knuckle from the vehicle.

To install:

15. Install the steering knuckle to the lower control arm.

16. Install the lower ball joint retaining nut and tighten to 81 ft. lbs. (110 Nm).

17. Connect the upper control arm to the steering knuckle.

18. Install upper control arm pinch bolt and nut and tighten to 30 ft. lbs. (40 Nm).

19. Connect the ABS wheel speed sensor wiring harness bracket to the steering knuckle.

20. Install the brake hose bracket to the steering knuckle.

21. Install the brake hose bracket retaining bolts and tighten to 89 inch lbs. (10 Nm).

22. Install the outer tie rod to the steering knuckle.

23. Install the new outer tie rod retaining nut and tighten to 33 ft. lbs. (45 Nm) on 2WD models, or 44 ft. lbs. (60 Nm) on 4WD models.

24. Install the wheel hub and bearing.

25. On 4WD vehicles, install the drive axle nut and tighten to 103 ft. lbs. (140 Nm), then install the center cap.

26. Install the tire and wheel.

27. Lower the vehicle.

28. Adjust the front toe.

Air Suspension

DEPRESSURIZING

❊❊ CAUTION

A sudden release of pressure may cause personal injury or damage to the vehicle. The air suspension system is under pressure until the air supply lines are disconnected. Wear gloves, ear protection, and eye protection. Wrap a clean cloth around the air supply lines.

1. Remove the air suspension system fuse.

2. Raise and support the vehicle.

3. Raise and support the rear axle at the designed height of 5.33 in. (135.4mm) on non-air suspension models or 6.12 in. (155.4mm) on air suspension models.

4. Remove the air compressor mounting bolts from the frame and support air compressor.

5. Loosen both of the air supply line connections at the air compressor in order to depressurize the air springs.

6. To pressurize the system, tighten the air supply lines to the air compressor to 20 inch lbs. (2.25 Nm).

7. Install the air compressor to frame mounting bolts and tighten to 15 ft. lbs. (20 Nm).

8. Lower the vehicle.

9. Install the air suspension system fuse.

10. Start the vehicle and run for approximately 1 minute to ensure that the air suspension system is functioning properly.

11. Check the axle height.

REMOVAL & INSTALLATION

1. Depressurize the air suspension system.

➡**There is a raised feature on the outer rim of the air spring top plate that denotes the anti-rotation peg position.**

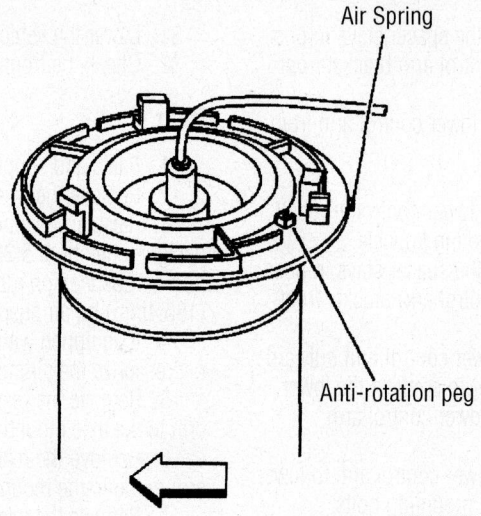

Air Spring

Anti-rotation peg

06025-ENV-G10

Locating anti-rotation peg in air spring

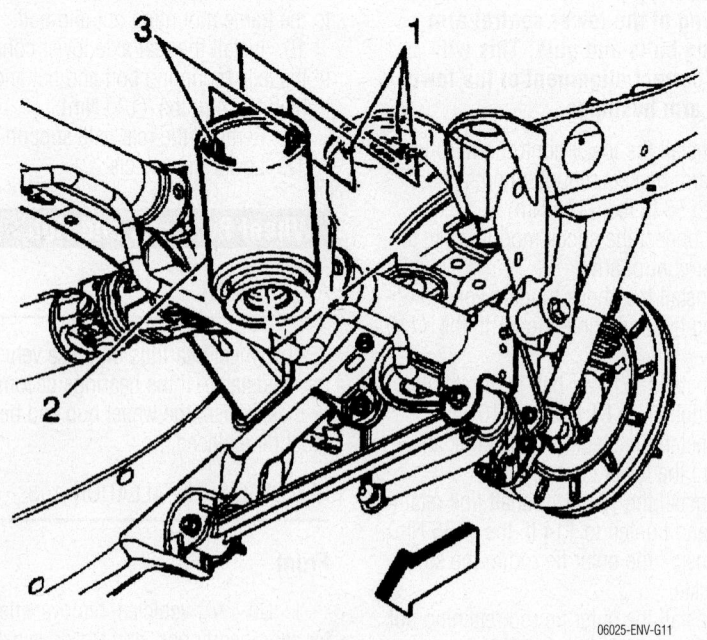

06025-ENV-G11

Air spring mounting

2. Depress the anti-rotation peg (2) in the air spring top plate located in the upper spring seat.

3. With the anti rotation peg (2) depressed, rotate the air spring counter-clockwise and remove the air spring from the upper spring seat.

4. Disconnect the air supply line from the air spring in the following way:

a. Push the air supply line into the air spring connection and hold in place.

b. Depress and hold the air supply line collet down.

c. Remove the air supply line®from the air spring.

5. Remove the air spring.

To install:

6. Install the air supply line to the air spring. Ensure the air supply line is fully seated.

✳✳ CAUTION

Ensure that the air spring is fully seated and properly positioned on the axle pilot. Failure to properly position the air spring may cause the air spring to break apart, possibly resulting in personal injury or damage to the vehicle.

7. Install the air spring (2) to the frame by aligning the mounting tabs (3) with the keyhole slots (1) in the upper spring seat.

8. Apply upward pressure to the air spring and rotate clockwise until the anti-rotation peg snaps into place.

9. Pressurize the system.

10. Lower the vehicle.

11. Install the air suspension system fuse.

12. Start the vehicle and run for approximately 1 minute to ensure that the air suspension system is functioning properly.

13. Check the axle height.

BRAKES

Brake Caliper

REMOVAL & INSTALLATION

FRONT

1. Remove or disconnect the following:
   - ⅔ of the brake fluid from the master cylinder reservoir
   - Tire and wheel assembly
   - Caliper fluid line, then plug
   - Bolts retaining the caliper to the rotor
   - Caliper from the rotor
   - Disc brake pads from the caliper
   - Disc brake pad retaining clips from inside the caliper

To install:

2. Clean and lubricate the guide pins and bushings with silicon grease.

3. Install or connect the following:
   - Pads in the caliper
   - Caliper in position over the rotor
   - Mounting bolts and tighten to 31 ft. lbs. (42 Nm)
   - Fluid lines to the caliper using new copper washers and tighten to 30 ft. lbs. (40 Nm)
   - Wheel and tire assembly

4. Refill the master cylinder to the correct level. Bleed the brake system if the fluid lines were disconnected from the caliper.

REAR

1. Raise and safely support the vehicle.

2. Remove or disconnect the following:
   - Rear wheels
   - Brake hose and cap line
   - Retainers from caliper and remove caliper

To install:

3. Clean and lubricate the guide pins and bushings with silicon grease.

4. Install or connect the following:
   - Brake pads if removed
   - Caliper over rotor, and onto mounts
   - Retainers, and tighten to 23 ft. lbs. or (31 Nm)
   - Brake hose using new copper washers, and tighten to 30 ft. lbs. (40 Nm)

5. Bleed brake system.

6. Install tires.

7. Refill the master cylinder and pump pedal to attain full brake pedal before Road-testing the vehicle.

Disc Brake Pads

REMOVAL & INSTALLATION

FRONT

1. Remove or disconnect the following:
   - ⅔ of the brake fluid from the master cylinder

2. Place a C-clamp around the outer pad and caliper; tighten the C-clamp until the piston is fully compressed in the caliper.

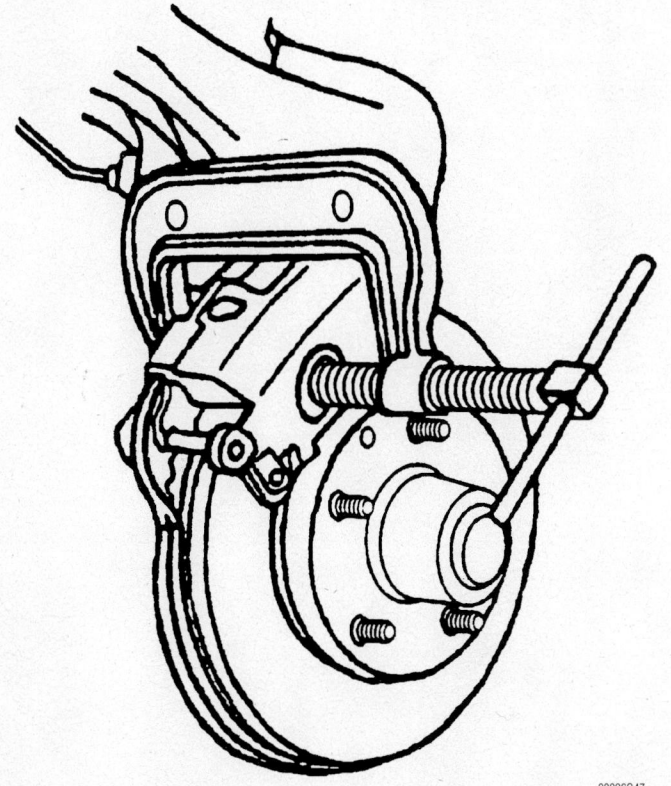

93026G47

Compressing the caliper piston with a C-clamp

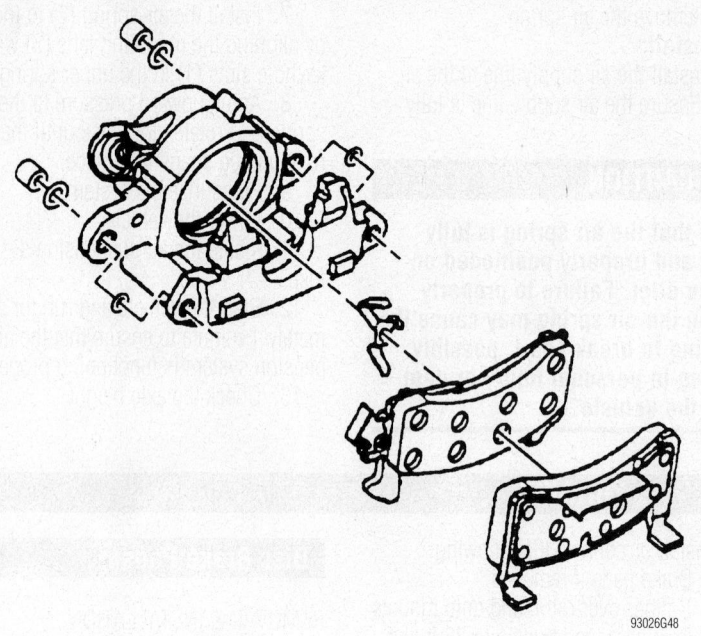

93026G48

Exploded view of the disc brake assembly

- Brake pads
- Inboard pad and retaining spring from the caliper
- Outboard pad from the caliper
- Sleeves and bushings

To install:

3. Clean and lubricate the sleeves and bushing with silicone lubricant and install them in the caliper.

4. Clip the retaining spring onto the inboard pad and install the pad in the caliper.

5. Install or connect the following:
- Outboard pad into the caliper
- Caliper in position over the rotor and install the mounting bolts.
- Wheel and tire assemblies

6. Refill the master cylinder and pump pedal to attain full brake pedal before Road-testing the vehicle.

REAR

1. Remove or disconnect the following:
- ⅔ of the brake fluid from the master cylinder
- Wheels

2. Place a C-clamp around the outer pad and caliper; tighten the C-clamp until the piston is fully compressed in the caliper.
- Top caliper retainer, and rotate caliper away from rotor
- Inboard pad and retaining spring from the caliper
- Outboard pad from the caliper

To install:

3. Clean and lubricate the sleeves and bushing with silicone lubricant

4. Install or connect the following:
- Sleeves and bushings into the caliper
- Clip the retaining spring onto the inboard pad and install the pad in the caliper
- Outboard pad into the caliper
- Caliper in position over the rotor and install the mounting bolts
- Wheel and tire assemblies

5. Refill the master cylinder and pump pedal to attain full brake pedal before Road-testing the vehicle.

CHEVROLET AND GMC

Canyon • Colorado

BRAKES 10-32
DRIVE TRAIN 10-22
ENGINE REPAIR..............10-11
FUEL SYSTEM 10-20
SPECIFICATIONS AND
  MAINTENANCE CHARTS......10-2
Engine and Vehicle Identification ..10-2
General Engine Specifications 10-2
Engine Tune-Up Specifications......10-2
Accessory Drive Belt Routing 10-3
Capacities.................................10-3
Valve Specifications.....................10-4
Crankshaft and Connecting
  Rod Specifications....................10-4
Piston and Ring Specifications 10-5
Torque Specifications 10-5
Wheel Alignment 10-6
Tire, Wheel and Ball Joint
  Specifications..........................10-7
Brake Specifications 10-8
Scheduled Maintenance Intervals ..10-9
STEERING AND
  SUSPENSION 10-26

A
Air Bag....................................10-26
  Arming.................................10-26
  Disarming.............................10-26
Alternator.................................10-11
  Removal & Installation............10-11
Automatic Transmission
  Assembly..............................10-22
  Removal & Installation............10-22
Axle Shaft, Bearing and Seal........10-24
  Removal & Installation............10-24

B
Balance Shafts 10-19
  Removal & Installation............10-19
Ball Joints................................10-28
  Removal & Installation............10-28
Brake Caliper 10-32
  Removal & Installation............10-32
Brake Drums.............................10-32
  Removal & Installation............10-32
Brake Shoes..............................10-32
  Removal & Installation............10-32

C
Camshafts................................10-15
  Removal & Installation............10-15

Clutch.....................................10-22
  Removal & Installation............10-22
Coil Springs 10-27
  Removal & Installation............10-27
CV-Joints.................................10-23
  Overhaul 10-23
Cylinder Head 10-13
  Removal & Installation............10-13

D
Differential Carrier 10-25
  Removal & Installation............10-25
Disc Brake Pads........................10-32
  Removal & Installation............10-32

E
Engine Assembly 10-11
  Removal & Installation............10-11
Exhaust Manifold.......................10-15
  Removal & Installation............10-15

F
Fuel Filter 10-20
  Removal & Installation............10-20
Fuel Pump 10-20
  Removal & Installation............10-20
Fuel Rail and Injectors 10-21
  Removal & Installation............10-21
Fuel System Pressure 10-20
  Relieving 10-20

H
Halfshaft.................................10-23
  Removal & Installation............10-23
Heater Core..............................10-13
  Removal & Installation............10-13
Hydraulic Clutch System 10-23
  Bleeding...............................10-23

I
Ignition Timing 10-11
  Adjustment...........................10-11
Intake Manifold.........................10-14
  Removal & Installation............10-14

L
Leaf Springs 10-27
  Removal & Installation............10-27
Lower Control Arm 10-29
  Control Arm Bushing
    Replacement.........................10-30
  Removal & Installation............10-29

M
Manual Transmission Assembly ..10-22
  Removal & Installation............10-22

O
Oil Pan....................................10-16
  Removal & Installation............10-16
Oil Pump 10-17
  Removal & Installation............10-17

P
Pinion Seal 10-25
  Removal & Installation............10-25
Piston and Ring 10-20
  Positioning 10-20
Power Steering Gear 10-26
  Removal & Installation............10-26

R
Rear Main Seal 10-17
  Removal & Installation............10-17
Rocker Arms 10-14
  Removal & Installation............10-14

S
Shock Absorbers 10-27
  Removal & Installation............10-27
Stabilizer Bar 10-31
  Removal & Installation............10-31
Starter Motor 10-16
  Removal & Installation............10-16
Steering Knuckle........................10-31
  Removal & Installation............10-31

T
Timing Chain, Sprockets, Front
  Cover and Seal 10-17
  Removal & Installation............10-17
Torsion Bar 10-27
  Removal & Installation............10-27
Transfer Case Assembly...............10-23
  Removal & Installation............10-23

U
Upper Control Arm 10-28
  Removal & Installation............10-28

W
Water Pump..............................10-12
  Removal & Installation............10-12
Wheel Bearings 10-30
  Adjustment...........................10-30
Wheel Hub/Bearing Assembly 10-30
  Removal & Installation............10-30

SPECIFICATIONS AND MAINTENANCE CHARTS

ENGINE AND VEHICLE IDENTIFICATION

Engine							Model Year	
Code ①	Liters (cc)	Cu. In.	Cyl.	Fuel Sys.	Engine Type	Eng. Mfg.	Code ②	Year
8	2.8 (2786)	170	4	SFI	DOHC	CPC	4	2004
6	3.5 (3474)	212	5	SFI	DOHC	CPC	5	2005
							6	2006

CPC: Chevrolet/Pontiac/Canada

SFI: Sequential Fuel Injection

① 8th position of VIN

② 10th position of VIN

06025-COLO-C01

GENERAL ENGINE SPECIFICATIONS
All measurements are given in inches.

Year	Model	Engine Displacement Liters	Engine Series VIN	Net Horsepower @ rpm	Net Torque @ rpm (ft. lbs.)	Bore x Stroke (in.)	Com-pression Ratio	Oil Pressure @ rpm
2004	Canyon	2.8	8	175@5600	185@4400	3.66x4.02	10.0:1	12@1200
		3.5	6	225@2800	225@4000	3.66x4.02	10.0:1	12@1200
	Colorado	2.8	8	175@5600	185@4400	3.66x4.02	10.0:1	12@1200
		3.5	6	220@5600	225@4000	3.66x4.02	10.0:1	12@1200
2005	Canyon	2.8	8	175@5600	185@4400	3.66x4.02	10.0:1	12@1200
		3.5	6	220@5600	225@4000	3.66x4.02	10.0:1	12@1200
	Colorado	2.8	8	175@5600	185@4400	3.66x4.02	10.0:1	12@1200
		3.5	6	220@5600	225@4000	3.66x4.02	10.0:1	12@1200
2006	Canyon	2.8	8	175@5600	185@4400	3.66x4.02	10.0:1	12@1200
		3.5	6	220@5600	225@4000	3.66x4.02	10.0:1	12@1200
	Colorado	2.8	8	175@5600	185@4400	3.66x4.02	10.0:1	12@1200
		3.5	6	220@5600	225@4000	3.66x4.02	10.0:1	12@1200

06025-COLO-C02

GASOLINE ENGINE TUNE-UP SPECIFICATIONS

Year	Engine Displacement Liters	Engine VIN	Spark Plug Gap (in.)	Ignition Timing (deg.) MT	Ignition Timing (deg.) AT	Fuel Pump (psi)	Idle Speed (rpm) MT	Idle Speed (rpm) AT	Valve Clearance In.	Valve Clearance Ex.
2004	2.8	8	0.042	①	①	48-54	②	②	HYD	HYD
	3.5	6	0.042	①	①	48-54	②	②	HYD	HYD
2005	2.8	8	0.042	①	①	50-57	②	②	HYD	HYD
	3.5	6	0.042	①	①	50-57	②	②	HYD	HYD
2006	2.8	8	0.042	①	①	50-57	②	②	HYD	HYD
	3.5	6	0.042	①	①	50-57	②	②	HYD	HYD

NOTE: The Vehicle Emission Control Information label often reflects specification changes made during production.

The label figures must be used if they differ from those in this chart.

HYD: Hydraulic

① Ignition timing is preset and cannot be adjusted

② Idle speed is maintained by the PCM

06025-COLO-C03

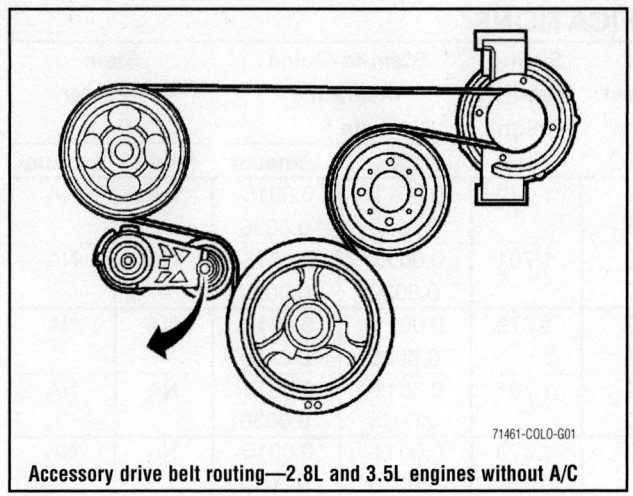

Accessory drive belt routing—2.8L and 3.5L engines without A/C

71461-COLO-G01

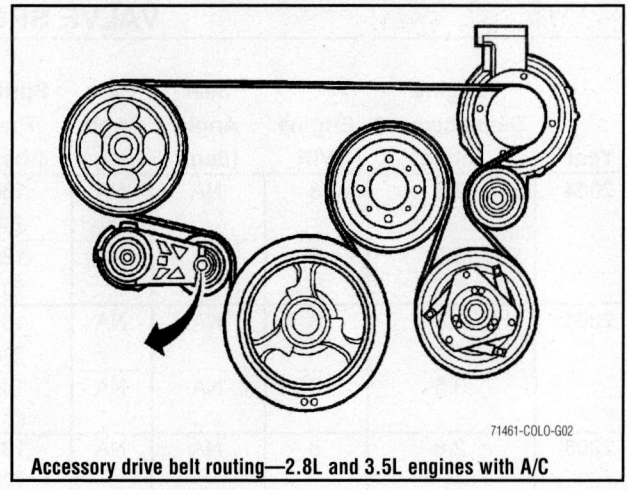

Accessory drive belt routing—2.8L and 3.5L engines with A/C

71461-COLO-G02

CAPACITIES

Year	Model	Engine Displacement Liters	Engine VIN	Engine Oil with Filter (qts.)	Transmission (pts.)		Transfer Case (pts.)	Drive Axle		Fuel Tank (gal.)	Cooling System (qts.)
					5-Spd	Auto.		Front (pts.)	Rear (pts.)		
2004	Canyon	2.8	8	5	①	②	2.7	—	3.4-3.8	19.0	10.4
		3.5	6	6	①	②	2.7	3.2	3.4-3.8	19.0	10.6
	Colorado	2.8	8	5	①	②	2.7	—	3.4-3.8	19.0	10.4
		3.5	6	6	①	②	2.7	3.2	3.4-3.8	19.0	10.6
2005	Canyon	2.8	8	5	①	②	2.7	3.2	3.4-3.8	19.0	10.4
		3.5	6	6	①	②	2.7	3.2	3.4-3.8	19.0	10.6
	Colorado	2.8	8	5	①	②	2.7	3.2	3.4-3.8	19.0	10.4
		3.5	6	6	①	②	2.7	3.2	3.4-3.8	19.0	10.6
2006	Canyon	2.8	8	5	①	②	2.7	3.2	3.4-3.8	19.0	10.4
		3.5	6	6	①	②	2.7	3.2	3.4-3.8	19.0	10.6
	Colorado	2.8	8	5	①	②	2.7	3.2	3.4-3.8	19.0	10.4
		3.5	6	6	①	②	2.7	3.2	3.4-3.8	19.0	10.6

NOTE: All capacities are approximate. Add fluid gradually and check to be sure a proper fluid level is obtained.

① RWD: 4.6
   4WD: 4.8

② w/245mm Torque Converter (Dry Fill) 19.8 qts./39.6 pts.

   w/258mm Torque Converter (Dry Fill) 20.3 qts./40.6 pts.

06025-COLO-C04

VALVE SPECIFICATIONS

Year	Engine Displacement Liters	Engine VIN	Seat Angle (deg.)	Face Angle (deg.)	Spring Test Pressure (lbs. @ in.)	Spring Installed Height (in.)	Stem-to-Guide Clearance (in.)		Stem Diameter (in.)	
							Intake	Exhaust	Intake	Exhaust
2004	2.8	8	NA	NA	130-142 @0.965	1.379	0.0011-0.0020	0.0015-0.0030	NA	NA
	3.5	6	NA	NA	130-142 @1.260	1.701	0.0011-0.0020	0.0016-0.0030	NA	NA
2005	2.8	8	NA	NA	130-142 @0.965	1.379	0.0011-0.0025	0.0015-0.0030	NA	NA
	3.5	6	NA	NA	130-142 @1.260	1.701	0.0011-0.0025	0.0015-0.0030	NA	NA
2006	2.8	8	NA	NA	130-142 @0.965	1.379	0.0011-0.0025	0.0015-0.0030	NA	NA
	3.5	6	NA	NA	130-142 @0.965	1.701	0.0011-0.0025	0.0015-0.0030	NA	NA

NA: Not Available

06025-COLO-C05

CRANKSHAFT AND CONNECTING ROD SPECIFICATIONS
All measurements are given in inches.

Year	Engine Displacement Liters	Engine VIN	Crankshaft				Connecting Rod		
			Main Brg. Journal Dia.	Main Brg. Oil Clearance	Shaft End-play	Thrust on No.	Journal Diameter	Oil Clearance	Side Clearance
2004	2.8	8	2.7567-2.7574	0.0004-0.0025	0.0044-0.0153	3	2.3749-2.3755	0.0008-0.0025	0.0019-0.0137
	3.5	6	2.7567-2.7574	0.0004-0.0025	0.0044-0.0153	4	2.3749-2.3755	0.0008-0.0025	0.0019-0.0137
2005	2.8	8	2.7567-2.7574	0.0004-0.0025	0.0044-0.0153	3	2.3749-2.3755	0.0008-0.0025	0.0019-0.0137
	3.5	6	2.7567-2.7574	0.0004-0.0025	0.0044-0.0153	4	2.3749-2.3755	0.0008-0.0025	0.0019-0.0137
2006	2.8	8	2.7567-2.7574	0.0004-0.0025	0.0044-0.0153	3	2.3749-2.3755	0.0008-0.0025	0.0019-0.0137
	3.5	6	2.7567-2.7574	0.0004-0.0025	0.0044-0.0153	4	2.3749-2.3755	0.0008-0.0025	0.0019-0.0137

06025-COLO-C06

PISTON AND RING SPECIFICATIONS

All measurements are given in inches.

Year	Engine Displ. Liters	Engine VIN	Piston Clearance	Ring Gap			Ring Side Clearance		
				Top Compression	Bottom Compression	Oil Control	Top Compression	Bottom Compression	Oil Control
2004	2.8	8	0.0004-0.0017	0.00787-0.0157	0.0142-0.0201	0.010-0.029	0.0017-0.0037	0.0021-0.0037	0.0023-0.0085
	3.5	6	0.0004-0.0017	0.0079-0.0157	0.0142-0.0201	0.010-0.029	0.0017-0.0037	0.0021-0.0037	0.0023-0.0085
2005	2.8	8	0.0006-0.0014	0.00787-0.0157	0.0142-0.0201	0.0098-0.0299	0.0017-0.0037	0.0021-0.0037	0.0023-0.0085
	3.5	6	0.0004-0.0017	0.0079-0.0157	0.0142-0.0201	0.0098-0.0299	0.0017-0.0037	0.0021-0.0037	0.0023-0.0085
2006	2.8	8	0.0006-0.0014	0.00787-0.0157	0.0142-0.0201	0.0098-0.0299	0.0017-0.0037	0.0021-0.0037	0.0023-0.0085
	3.5	6	0.0004-0.0017	0.0079-0.0157	0.0142-0.0201	0.0098-0.0299	0.0017-0.0037	0.0021-0.0037	0.0023-0.0085

06025-COLO-C07

TORQUE SPECIFICATIONS

All readings in ft. lbs.

Year	Engine Displacement Liters	Engine VIN	Cylinder Head Bolts	Main Bearing Bolts	Rod Bearing Bolts	Crankshaft Damper Bolts	Flywheel Bolts	Manifold		Spark Plugs	Oil Pan Drain Plug
								Intake	Exhaust		
2004	2.8	8	①	②	③	④	⑤	⑥	⑦	13	19
	3.5	6	①	②	③	④	⑤	⑥	⑦	13	19
2005	2.8	8	①	②	③	④	⑤	⑥	⑦	13	19
	3.5	6	①	②	③	④	⑤	⑥	⑦	13	19
2006	2.8	8	①	②	③	④	⑤	⑥	⑦	13	19
	3.5	6	①	②	③	④	⑤	⑥	⑦	13	19

① 1st pass: 22 ft. lbs.
2nd pass: Plus 155 degrees
Short end bolt
1st pass: 62 inch lbs.
2nd pass: plus 60 degrees
Long end bolt
1st pass: 62 inch lbs.
2nd pass: plus 120 degrees

② 1st pass: 18 ft. lbs.
2nd pass: plus 180 degrees

③ 1st pass: 18 ft. lbs.
2nd pass: plus 110 degrees

④ 1st pass: 110 ft. lbs.
2nd pass: plus 180 degrees

⑤ 1st pass: 30 ft. lbs.
2nd pass: plus 45 degrees

⑥ 89 inch lbs.

⑦ 1st pass: 15 ft. lbs.
Repeat twice more in sequence

06025-COLO-C08

WHEEL ALIGNMENT

Year	Model		Caster Range (+/-Deg.)	Caster Preferred Setting (Deg.)	Camber Range (+/-Deg.)	Camber Preferred Setting (Deg.)	Toe-in (Deg.)
2004	Sport	Left	1.0	+4.7	0.50	0	0+/-0.20
		Right	1.0	+4.7	0.50	0	0+/-0.20
	2WD & 4WD	Left	1.0	+3.6	0.50	0	0+/-0.20
	w/RPO Z71	Right	1.0	+4.0	0.50	0	0+/-0.20
	2WD	Left	1.0	+4.3	0.50	0	0+/-0.20
	w/RPO Z85	Right	1.0	+4.5	0.50	0	0+/-0.20
	4WD	Left	1.0	+3.8	0.50	0	0+/-0.20
	w/RPO Z85	Right	1.0	+4.0	0.50	0	0+/-0.20
2005	2WD	Left	1.0	+4.7	0.50	0	0+/-0.20
	w/RPO ZQ8	Right	1.0	+4.7	0.50	0	0+/-0.20
	2WD & 4WD	Left	1.0	+3.6	0.50	0	0+/-0.20
	w/RPO Z71	Right	1.0	+4.0	0.50	0	0+/-0.20
	2WD	Left	1.0	+4.3	0.50	0	0+/-0.20
	w/RPO Z85	Right	1.0	+4.5	0.50	0	0+/-0.20
	4WD	Left	1.0	+3.8	0.50	0	0+/-0.20
	w/RPO Z85	Right	1.0	+4.0	0.50	0	0+/-0.20
2006	2WD	Left	1.0	+4.3	0.60	0	0+/-0.20
	wo/RPO Z71 & ZQ8	Right	1.0	+4.5	0.60	0	0+/-0.20
	2WD	Left	1.0	+4.7	0.60	0	0+/-0.20
	w/RPO ZQ8	Right	1.0	+4.7	0.60	0	0+/-0.20
	2WD	Left	1.0	+3.6	0.60	0	0+/-0.20
	w/RPO Z71	Right	1.0	+4.0	0.60	0	0+/-0.20
	4WD	Left	1.0	+3.8	0.60	0	0+/-0.20
		Right	1.0	+4.0	0.60	0	0+/-0.20

06025-COLO-C09

TIRE, WHEEL AND BALL JOINT SPECIFICATIONS

Year	Model	OEM Tires		Tire Pressures (psi)		Wheel Size	Ball Joint Inspection	Lug Nut Torque (Ft. Lbs.)
		Standard	Optional	Front	Rear			
2004	Z-71	P265/75R15	None	36	36	NA	NA	103
	2WD Sport	P235/50R17	None	36	36	NA	NA	103
	Z85 2WD	P205/75R15	P225/75R15	36	36	NA	NA	103
	Z85 4WD	P235/75R15	None	36	36	NA	NA	103
2005	Z-71	P265/75R15	None	④	④	15x7	NA	103
	2WD Sport ①	P235/50R17	P235/50R18	④	④	17x8 18x8	NA	103
	Z85 2WD ②	P205/75R15	P225/75R15	④	④	15x6 15x6.5	NA	103
	Z85 4WD ③	P235/75R15	None	④	④	15x6 15x6.5	NA	103
2006	Z-71	P265/75R15	None	④	④	15x7	NA	103
	2WD Sport ①	P235/50R17	P235/50R18	④	④	17x8 18x8	NA	103
	Z85 2WD ②	P205/75R15	P225/75R15	④	④	15x6 15x6.5	NA	103
	Z85 4WD ③	P235/75R15	None	④	④	15x6 15x6.5	NA	103

NA: Not Available

OEM: Original Equipment Manufacturer

OPT: Optional

PSI: Pounds Per Square Inch

STD: Standard

① P205/75R15 Spare Tire

② Compact Spare Standard, P225/75R15 Optional

③ Compact Spare Standard, P235/75R15 Optional

④ Refer to placard on vehicle for proper inflation pressure

BRAKE SPECIFICATIONS
All measurements in inches unless noted

Year	Model		Brake Disc Original Thickness	Brake Disc Minimum Thickness	Brake Disc Maximum Runout	Brake Drum Diameter Original Inside Diameter	Brake Drum Diameter Max. Wear Limit	Brake Drum Diameter Maximum Machine Diameter	Minimum Lining Thickness	Brake Caliper Bracket Bolts (ft. lbs.)	Brake Caliper Mounting Bolts (ft. lbs.)
2004	Canyon	F	1.060	1.000	0.002	—	—	—	0.070	129	29
		R	—	—	—	—	—	11.673	0.030	—	—
	Colorado	F	▢▢▢▢▢	1.000	0.002	—	—	—	0.070	129	29
		R	—	—	—	—	—	11.673	0.030	—	—
2005	Canyon	F	1.060	1.000	0.002	—	—	—	0.070	129	29
		R	—	—	—	—	—	11.673	0.030	—	—
	Colorado	F	1.060	1.000	0.002	—	—	—	0.070	129	29
		R	—	—	—	—	—	11.673	0.030	—	—
2006	Canyon	F	1.060	1.000	0.002	—	—	—	0.070	129	29
		R	—	—	—	—	—	11.673	0.030	—	—
	Colorado	F	1.060	1.000	0.002	—	—	—	0.070	129	29
		R	—	—	—	—	—	11.673	0.030	—	—

06025-COLO-C11

SCHEDULED MAINTENANCE INTERVALS
2004-2006 CHEVROLET COLORADO & GMC CANYON

When the CHANGE ENGINE OIL light appears, certain services and inspections are required.
Required services are described as Maintenance I and Maintenance II.
The first service on a vehicle should be Maintenance I, and the second service should be Maintenance II.
Alternate between the 2 thereafter. However, in some cases, Maintenance II may be required more often.
Maintenance I: Use Maintenance I if the CHANGE ENGINE OIL light comes on within 10 months since
vehicle was purchased or, if Maintenance II was performed.
Maintenance II: Use Maintenance II if the previous service performed was Maintenance I.
Always use Maintenance II whenever the CHANGE ENGINE OIL light comes on 10 months or more
since the last service, or, if the CHANGE ENGINEOIL light has not come on at all for one year.

Service	Maintenance I	Maintenance II
Change oil and oil filter, then Reset Oil Life Monitor ①	✓	✓
Visually inspect vehicle for any leaks or damage	✓	✓
Inspect engine air filter and replace as necessary		✓
Rotate tires, check for unusual wear and reset tire pressures ②	✓	✓
Inspect brake system	✓	✓
Check engine coolant and add fluid as needed	✓	✓
See additional required services	✓	✓
Inspect suspension and steering components		✓
Inspect engine cooling system		✓
Inspect wiper blades and windshield washer fluid level		✓
Inspect restraint system components		✓
Lubricate body components		✓
Check transmission and transfer case fluid level and add fluid as required		✓

① Mileage interval varies based on your driving habits.

② See Placard on Vehicle for Proper Inflation Pressure

Engine Oil Life System Reset Instructions

1). Begin with the engine off and the key in the "Lock" position.

2). Turn key to the "On" position

3). Press and release the stem located at the lower center of the instrument cluster until
the "Oil Life" message is displayed.

4). Wait for the "Oil Life" and "Reset" message appear, then press and hold the stem
down until several beeps are heard.

5). Turn the key to the"Lock" position.

06025-COLO-C12

ADDITIONAL SERVICE REQUIREMENTS
2004 CHEVROLET COLORADO & GMC CANYON

TO BE SERVICED	TYPE OF SERVICE	VEHICLE MILEAGE INTERVAL (x1000)					
		25	50	75	100	125	150
Fuel system	I	✓	✓	✓	✓	✓	✓
Exhaust system	I	✓	✓	✓	✓	✓	✓
Fuel filter	R	✓	✓	✓	✓	✓	✓
Air filter	R	✓	✓	✓	✓	✓	✓
Automatic transmission fluid and filter (Severe Service)	S		✓		✓		✓
Automatic transmission fluid and filter (Normal Service)	S				✓		
Spark plugs	R				✓		
Cooling systerm	S						✓
Accessory drive belt	I						✓

06025-COLO-C13

ADDITIONAL SERVICE REQUIREMENTS
2005 CHEVROLET COLORADO & GMC CANYON

TO BE SERVICED	TYPE OF SERVICE	VEHICLE MILEAGE INTERVAL (x1000)					
		25	50	75	100	125	150
Fuel system	I	✓	✓	✓	✓	✓	✓
Exhaust system	I	✓	✓	✓	✓	✓	✓
Fuel filter	R	✓	✓	✓	✓	✓	✓
Air filter	R		✓		✓		✓
Automatic transmission fluid and filter (Severe Service)	S		✓		✓		✓
Automatic transmission fluid and filter (Normal Service)	S				✓		
Spark plugs	R				✓		
Cooling systerm	S						✓
Accessory drive belt	I						✓

06025-COLO-C14

ADDITIONAL SERVICE REQUIREMENTS
2006 CHEVROLET COLORADO & GMC CANYON

TO BE SERVICED	TYPE OF SERVICE	VEHICLE MILEAGE INTERVAL (x1000)					
		25	50	75	100	125	150
Fuel system	I	✓	✓	✓	✓	✓	✓
Exhaust system	I	✓	✓	✓	✓	✓	✓
Air filter	R		✓		✓		✓
Automatic transmission fluid and filter (Severe Service)	S		✓		✓		✓
Automatic transmission fluid and filter (Normal Service)	S				✓		
Spark plugs	R				✓		
Cooling system	S						✓
Accessory drive belt	I						✓

06025-COLO-C15

ENGINE REPAIR

Alternator

REMOVAL & INSTALLATION

1. Before servicing the vehicle, refer to the Precautions Section.
2. Remove or disconnect the following:
   - Negative battery cable
   - Accessory drive belt
   - Left front wheel
   - Left front inner fender liner
   - A/C compressor electrical connector

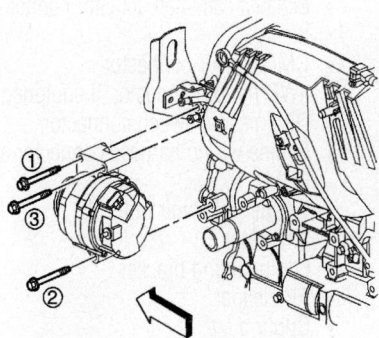

Alternator mounting bolt tightening sequence—2.8L and 3.5L engines

71461-COLO-G03

- Lower A/C compressor mounting bolts

➡**The lower mounting bolts are removed to allow the engine lift bracket to be removed. Do not remove the upper A/C compressor mounting bolt.**

- Alternator wiring
- Alternator mounting bolts
- A/C compressor hose bracket from the engine lift bracket
- Engine lift bracket
- Alternator

To install:

3. Install the engine lift bracket. Tighten the bolts in sequence as follows:
   a. Step 1: 44 inch lbs. (5 Nm)
   b. Step 2: 37 ft. lbs. (50 Nm)
4. Install or connect the following:
   - Alternator. Torque the bolts in sequence to 37 ft. lbs. (50 Nm).
   - A/C compressor hose bracket to the engine lift bracket. Tighten the bolt to 80 inch lbs. (9 Nm).
   - Alternator electrical connectors. Torque the battery feed wire nut to 15 ft. lbs. (20 Nm).
   - A/C compress lower mounting bolts. Tighten the bolts to 37 ft. lbs. (50 Nm).
   - Left front inner fender liner

- Left front wheel
- Accessory drive belt
- Negative battery cable

Ignition Timing

ADJUSTMENT

The ignition timing is preset and cannot be adjusted.

Engine Assembly

REMOVAL & INSTALLATION

1. Disconnect the battery cables and properly relieve the fuel system pressure.
2. Drain the engine cooling system.
3. Drain the engine oil.
4. Remove or disconnect the following:

- Hood
- Battery box
- Radiator hoses
- Cooling fan
- Air cleaner assembly
- Engine lifting bracket
- Alternator

➡**Reinstall the engine lift bracket after removing the alternator. Tighten the bolts to 44 inch lbs. (5 Nm) and then retighten to 37 ft. lbs. (50 Nm).**

- Washer reservoir bolts
- Engine wiring harness connector at PCM
- Wiring harness retainers at fender, power steering pump, throttle body and camshaft cover
- Oil pressure switch connector
- 4WD motor connector, if equipped
- Camshaft Position (CMP) sensor connector
- Exhaust camshaft actuator connector
- Coolant temperature sensor connector
- Injector harness connector
- Ignition coil connectors
- Oxygen Sensor (O₂S) connector
- Wiring harness conduit at camshaft cover
- Automatic transmission filler tube, if equipped
- Air injection pipe cover
- Install engine lifting eye in air pipe cover location
- Heater hoses from heater core

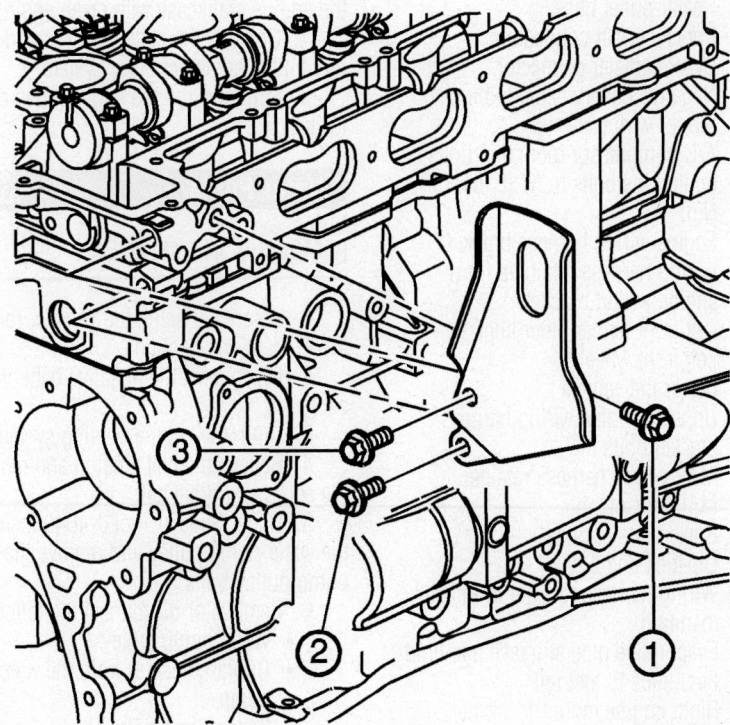

Engine lift bracket torque sequence

06025-COLO-G01

- A/C suction hose bracket
- Power steering pump bolts and position pump aside
- Right engine mount-to-frame bracket bolt
- Wiring harness retainer from intake manifold and position aside
- Fuel lines from fuel rail
- Evaporative pipe at intake manifold
- Dipstick and tube
- Brake booster hose
- Manifold Absolute Pressure (MAP) sensor
- MAP wiring harness retainer
- Upper 2 engine wiring harness bracket bolts
- Raise and support vehicle
- Left front wheel
- Left front fender inner liner
- Wiring harness retainers from engine wiring bracket
- Engine wiring harness bracket
- A/C compressor mounting bolts and position compressor aside
- Starter wiring
- Negative battery cable from block
- EVAP canister connector
- Knock sensor connectors
- Heater outlet hose
- Crankcase Position (CKP) sensor connector
- Engine wiring ground leads from block
- Engine wiring harness retainer at oil pan rail and position harness aside
- Catalytic converter
- Exhaust donut gasket
- Automatic transmission oil cooler and fuel line brackets, if equipped
- Left engine mount-to-frame bracket bolt
- On 2WD drive models, front cross-member
- On 4WD models, differential carrier
- On automatic transmission, torque converter bolts after marking torque converter-to-flexplate location
- Leave 2 upper transmission-to-engine bolts, but remove all other bolts

5. Lower the vehicle and place a transmission jack under the transmission

6. Remove the remaining transmission mounting bolts

7. Install a suitable lifting device to the engine.

8. Remove the engine mount bolts and carefully lift the engine from the vehicle. Pause several times while lifting the engine to make sure no wires or hoses have become snagged.

To install:

9. Carefully lower the engine into the vehicle and align the engine dowels with the transmission.

10. Install the engine mount bolts and tighten the bolts to 37 ft. lbs. (50 Nm).

11. Lower the engine onto the engine mounts.

12. Remove the engine lifting device.

13. Raise and support the vehicle.

14. Install the transmission-to-engine bolts and tighten the bolts to 37 ft. lbs. (50 Nm).

15. Remove the transmission jack.

16. Aligning the torque converter to the flexplate, install the bolts and tighten to 44 ft. lbs. (60 Nm).

17. On 2WD, install the crossmember and tighten the bolts to 44 ft. lbs. (60 Nm).

18. On 4WD, install the differential carrier.

19. Install or connect the following:
- Automatic transmission oil cooler and fuel line brackets, if equipped
- Left engine mount-to-frame bracket bolt and tighten to 63 ft. lbs. (85 Nm).
- New exhaust donut gasket
- Catalytic converter and tighten the bolts to 37 ft. lbs. (50 Nm).
- Engine wiring harness retainers at oil pan rail
- Engine wiring ground leads to block
- CKP sensor connector
- Heater outlet hose
- Knock sensor connectors
- EVAP canister connector
- Negative battery cable to block
- Starter wiring
- A/C compressor mounting bolts tighten the bolts to 37 ft. lbs. (50 Nm)
- Engine wiring harness bracket
- Wiring harness retainers from engine wiring bracket
- Left front fender inner liner
- Left front wheel
- Lower the vehicle
- Upper 2 engine wiring harness bracket bolts
- MAP wiring harness retainer
- MAP sensor
- Brake booster hose
- Dipstick and tube
- Wiring harness retainer to intake manifold
- Evaporative pipe at intake manifold
- Fuel lines to fuel rail
- Right engine mount-to-frame bracket bolt and tighten bolt to 63 ft. lbs. (85 Nm)

- Power steering pump
- A/C suction hose bracket
- Heater hoses from heater core
- Remove engine lifting eye in air pipe cover location
- Air injection pipe cover using new gasket
- Automatic transmission filler tube
- Wiring harness conduit at camshaft cover
- Wiring harness retainers at fender, power steering pump, throttle body and camshaft cover
- Coolant temperature sensor connector
- Injector harness connector
- Ignition coil connectors
- Oxygen Sensor (O_2S) connector
- Exhaust camshaft actuator connector
- CMP sensor connector
- 4WD motor connector, if equipped
- Oil pressure switch connector
- Engine wiring harness connector at PCM
- Washer reservoir bolts
- Alternator
- Engine lifting bracket
- Air cleaner
- Cooling fan
- Radiator hoses
- Battery box
- Hood

20. Check all powertrain fluid levels and add, as necessary. Be sure to properly fill the engine crankcase with clean engine oil.

21. Connect the battery cables and properly fill the engine cooling system.

22. Start and run the engine, then check for leaks.

Water Pump

REMOVAL & INSTALLATION

1. Before servicing the vehicle, refer to the Precautions Section.

2. Disconnect the negative battery cable.

3. Drain the engine cooling system.

4. Relieve the belt tension and remove the accessory drive belt.

5. Using Special Tool J-46406, secure the water pump pulley and remove the water pump pulley bolts.

6. Remove or disconnect the following:
- Water pump pulley
- Coolant hose(s) from the water pump
- Water pump bolts
- Water pump

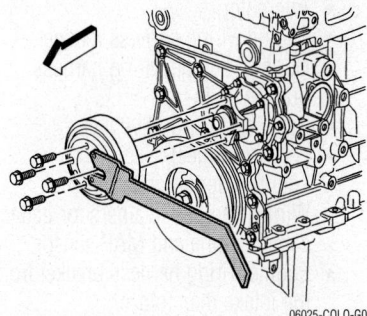

06025-COLO-G02

Use Special Tool J-46406 to secure the pump pulley—2.8L and 3.5L engines.

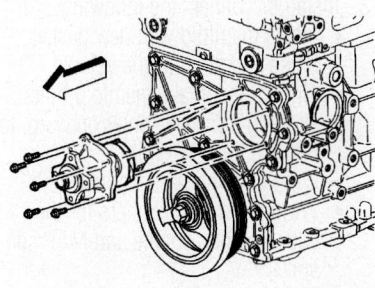

71461-COLO-G04

Exploded view of the water pump mounting—Canyon & Colorado

To install:

7. Clean the gasket mounting surfaces.
8. Install or connect the following:
   - Water pump using a new gasket. Tighten the water pump bolts to 89 inch lbs. (10 Nm).
   - Coolant hose(s)
   - Water pump pulley and tighten the bolts to 18 ft. lbs. (25 Nm).
   - Drive belt and adjust the tension
   - Negative battery cable
9. Refill the engine cooling system.
10. Run the engine and check for leaks.

Heater Core

REMOVAL & INSTALLATION

1. Before servicing the vehicle, refer to the Precautions Section.
2. Disable the air bag system.
3. Disconnect the negative battery cable.
4. Drain the engine cooling system.
5. Discharge and recovery the A/C refrigerant using approved recycling equipment.
6. Remove or disconnect the following:
   - Glove box door
   - A pillar trim panels
   - Door sill plates

- Hinge pillar trim panels
- Lower center instrument panel extension
- Accessory trim plate panel
- Radio
- A/C-heater control panel
- Center air outlets
- Left and right air outlets
- Knee bolster trim panel and bolster
- Instrument luster bezel
- Instrument cluster
- Headlight switch
- Daytime running light sensor
- Upper 3 instrument panel nuts
- Upper HVAC module screws
- Instrument panel screws at instrument cluster and glove box openings
- Hazard warning light connector
- Six passenger air bag fasteners and passenger air bag module
- Two screws in passenger air bag opening
- Open the left side compartment and remove the screw
- Screw at the back of the center storage compartment
- Grasp the lower edge of the center storage compartment and the right lower edge of the instrument panel and pull out to disengage the clips
- Partially pull out the instrument panel from the carrier
- Release all the wiring harness clips from the instrument panel

7. With the aid of an assistant, carefully remove the instrument panel.
8. Disconnect the heater hoses from the HVAC module.
9. Disconnect the A/C lines from the thermal expansion valve.
10. Disconnect all HVAC module electrical connectors.
11. Remove the HVAC module retaining screws from the firewall
12. Remove the HVAC module.
13. Remove the core pipe clamp screws and clamp.
14. Remove the heater core from the HVAC module.

To install:

15. Install or connect the following:
   - Heater core to the HVAC module
   - Heater core pipe clamp screws and clamp
   - HVAC module
   - HVAC module retaining screws at the firewall
   - HVAC module electrical connectors
   - A/C lines at the thermal expansion valve
   - Heater hoses to HVAC module

- With the aid of an assistant, carefully install the instrument panel
- Wiring harness clips to instrument panel
- Push the lower edge of the center storage compartment and the right lower edge of the instrument panel in to engage the clips
- Screw at the back of the center storage compartment
- Screw inside the left side compartment
- Two screws in passenger air bag opening
- Hazard warning light connector
- Passenger air bag fasteners and passenger air bag module. Tighten the fasteners to 80 inch lbs. (9 Nm).
- Instrument panel screws at instrument cluster and glove box openings
- Upper HVAC module screws
- Upper 3 instrument panel nuts
- Daytime running light sensor
- Headlight switch
- Instrument cluster
- Instrument luster bezel
- Knee bolster trim panel and bolster
- Left and right air outlets
- Center air outlets
- A/C-heater control panel
- Radio
- Accessory trim plate panel
- Lower center instrument panel extension
- Hinge pillar trim panels
- Door sill plates
- A pillar trim panels
- Glove box door

16. Charge the A/C refrigerant using approved equipment.
17. Refill the cooling system.
   a. Enable the air bag system.
18. Connect the negative battery cable.
19. Run the engine to normal operating temperatures; then, check the climate control operation and check for leaks.

Cylinder Head

REMOVAL & INSTALLATION

1. Before servicing the vehicle, refer to the Precautions Section.
2. Relieve the fuel system pressure.
3. Disconnect the negative battery cable.
4. Drain the engine cooling system.
5. Remove or disconnect the following:
   - Exhaust manifold
   - Timing chain and camshaft sprockets

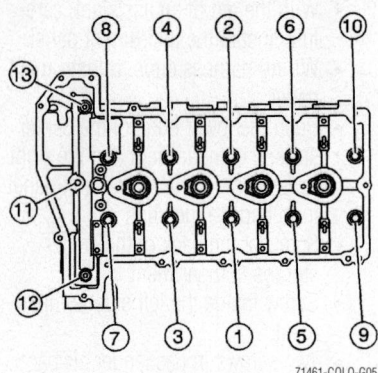

Cylinder head bolt torque sequence—2.8L engine

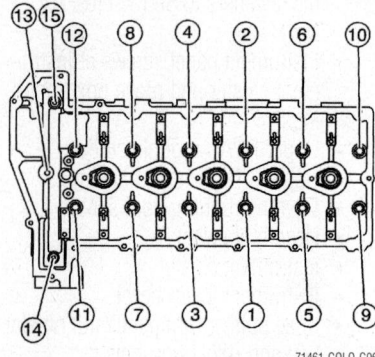

Cylinder head bolt torque sequence—3.5L engine

- Cylinder head bolts
- Cylinder head from the engine

To install:

6. Clean and inspect the gasket mounting surfaces.

7. Install the cylinder head using a new gasket. Tighten the new bolts as follows:

a. Standard bolts in sequence to 22 ft. lbs. Plus an additional 155°.

b. Tighten the 2 short end bolts to 62 inch lbs. (7 Nm), plus an additional 60°.

c. Tighten the 2 long end bolts to 62 inch lbs. (7 Nm), plus an additional 120°.

8. Install or connect the following:
- Timing chain and camshaft sprockets
- Exhaust manifold
- Negative battery cable

9. Refill the engine cooling system.

10. Start the engine to check for leaks.

Rocker Arms

REMOVAL & INSTALLATION

1. Before servicing the vehicle, refer to the Precautions Section.

2. Remove or disconnect the following:
- Camshafts
- Rocker arms
- Lash adjusters

➡**Rocker arms and adjusters, being reused, must be installed in their original positions. Be sure to tag or arrange all rocker arms.**

To install:

3. Inspect the rocker arms and lash adjusters for damage or wear and replace as necessary.

4. Fill the lash adjusters with clean engine oil.

5. Install or connect the following:
- Lash adjusters in their original locations
- Rocker arms after lubricating them
- Camshafts

6. Start and run the engine to check for leaks.

Intake Manifold

REMOVAL & INSTALLATION

1. Before servicing the vehicle, refer to the Precautions Section.

2. Remove or disconnect the following:
- Negative battery cable
- Air intake assembly
- Throttle body
- Battery and battery box
- Dipstick and tube
- Brake booster hose
- Manifold Absolute Pressure (MAP) sensor connector and harness retainer
- PCV tube

- Alternator
- Engine wiring harness retainer
- Upper 2 engine wiring harness bracket bolts
- Raise and support vehicle
- Left front wheel
- Left front fender inner liner
- Wiring harness retainers for battery cable, engine and MAP sensor
- Engine wiring harness bracket from the intake manifold
- Intake manifold bolts
- Lower the vehicle
- Intake manifold and gasket

To install:

3. Install or connect the following:
- Intake manifold with new gasket
- Raise and support the vehicle
- Tighten the intake manifold bolts, working from the center outward, to 89 inch lbs. (10 Nm)
- Engine wiring harness bracket
- Wiring harness retainers for the battery cable, engine and MAP sensor
- Left front fender inner liner
- Left front wheel
- Lower the vehicle
- Upper 2 engine wiring harness bracket bolts
- Engine wiring harness retainer
- Alternator
- PCV tube
- MAP sensor connector and harness retainer
- Brake booster hose
- Dipstick and tube
- Throttle body
- Negative battery cable

4. Start the engine and check for leaks.

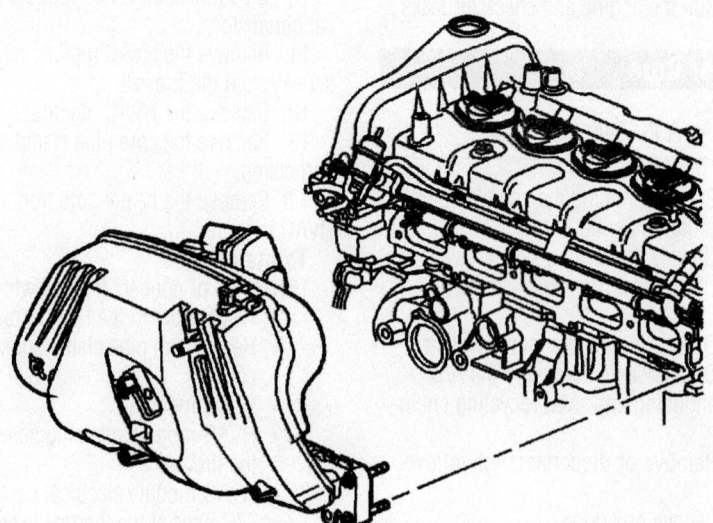

Intake manifold mounting—Canyon & Colorado

Exhaust Manifold

REMOVAL & INSTALLATION

1. Before servicing the vehicle, refer to the Precautions Section.
2. Remove or disconnect the following:
   - Negative battery cable
   - Catalytic converter from the exhaust manifold
   - Exhaust seal
   - Air intake assembly
   - Transmission fill tube bracket nut, if equipped with automatic transmission
   - Heated Oxygen Sensor (HO2S) from exhaust manifold
   - Exhaust manifold heat shield
   - Exhaust manifold bolts
   - Exhaust manifold and gasket

To install:

3. Clean the exhaust manifold retainer threads and the gasket mating surfaces.
4. Coat the bolt threads with a suitable threadlock.

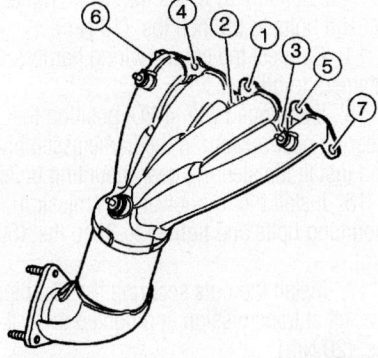

Exhaust manifold tightening sequence—
2.8L engine

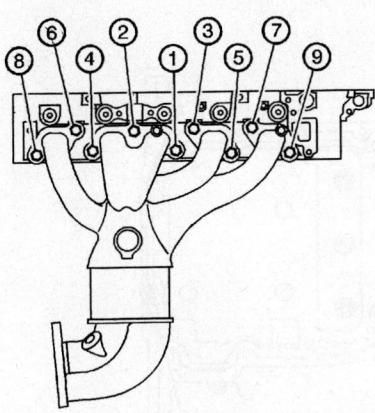

Exhaust manifold tightening sequence—
3.5L engine

5. Install the exhaust manifold and tighten the bolts in sequence to 15 ft. lbs. (20 Nm), tighten the bolts again to 15 ft. lbs. (20 Nm), then tighten the bolts again to 15 ft. lbs. (20 Nm).
6. Install or connect the following:
   - Heat shield and tighten the nuts to 89 inch lbs. (10 Nm)
   - HO2S and tighten to 31 ft. lbs. (42 Nm)
   - Transmission fill tube bracket nut, if equipped with automatic transmission.
   - Air intake assembly
   - New exhaust seal to the exhaust manifold flange
   - Catalytic converter to the manifold. Tighten the nuts to 37 ft. lbs. (50 Nm).
   - Negative battery cable

Camshafts

REMOVAL & INSTALLATION

1. Before servicing the vehicle, refer to the Precautions Section.
2. Properly relieve the fuel system pressure.
3. Disconnect the negative battery cable.
4. Drain the engine cooling system and the engine oil.
5. Remove or disconnect the following:
   - Intake manifold
   - Ignition coils
   - Coolant temperature sensor connector
   - Injector harness connector
   - Ignition coil connectors

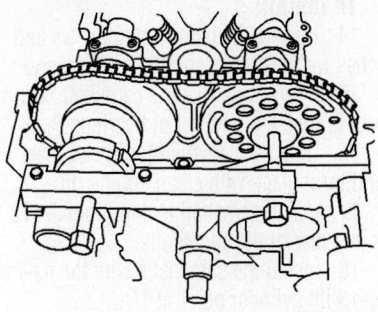

Use Special Tool J-44222 camshaft sprocket holding tool to prevent the timing chain and sprockets from turning.

   - Oxygen Sensor (O2S) connector
   - Fuel pressure regulator screw
   - Camshaft cover
   - Both Camshaft Position (CMP) sensors

6. Rotate the crankshaft clockwise until the no. 1 cylinder is at TDC on the compression stroke. The word DELPHI on the camshaft position actuator will be parallel with the cylinder head surface.
7. Install camshaft locking tool J-44221 to the rear of the camshafts.
8. Remove the camshaft sprocket bolts and discard them.
9. Install tension tool J-44222 on the cylinder head and install the holding bolts in the camshaft sprocket bolt holes to lock the timing chain and sprockets in position.
10. Carefully slide the sprockets and timing chain onto the tension tool.
11. Remove the camshaft bearing caps.
12. Remove the Camshaft Locking Tool from the camshafts.
13. Remove the camshafts.

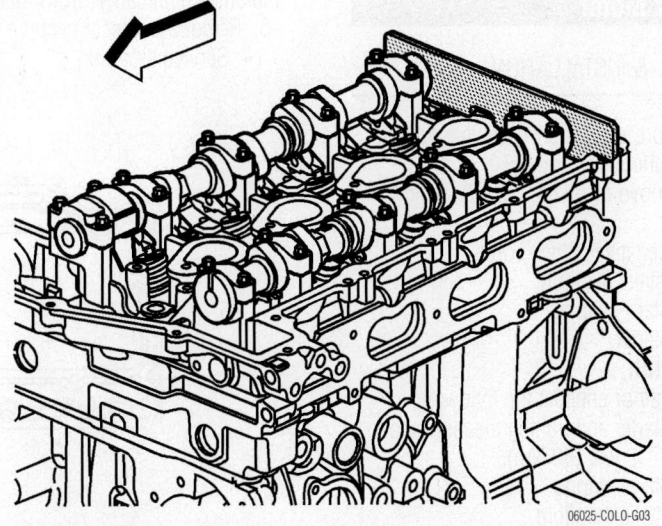

Install the camshaft locking tool J-44221 to the rear of the camshafts.

To install:

14. Inspect the camshaft, journals and lobes for wear and replace, if necessary.

15. If removed, use the camshaft bearing tool to install a new set of bearings.

16. Coat the camshaft lobes, journals and thrust face with clean engine oil.

17. Install camshaft locking tool J-44221 to the rear of the camshafts.

18. Install the camshafts with the flats up and with cylinder no. 1 at TDC.

19. Install the bearing caps in their original position and tighten the bolts to 106 inch lbs. (12 Nm).

20. Carefully slide the sprockets and timing chain onto the camshafts, ensuring the alignment pins are engaged between the camshafts and sprockets.

21. Install new camshaft sprocket bolts and washers. Tighten the intake sprocket bolt to 15 ft. lbs. (20 Nm), plus it additional 100°. Tighten the exhaust sprocket bolt to 18 ft. lbs. (25 Nm), plus an additional 135°.

22. Remove the camshaft locking plate tool.

23. Install or connect the following:
- Both Camshaft Position (CMP) sensors
- Camshaft cover
- Fuel pressure regulator screw
- Oxygen Sensor (O$_2$S) connector
- Ignition coil connectors
- Injector harness connector
- Coolant temperature sensor connector
- Ignition coils
- Intake manifold
- Negative battery cable

24. Refill the engine cooling system and engine oil.

Starter Motor

REMOVAL & INSTALLATION

1. Before servicing the vehicle, refer to the Precautions Section.

2. Remove or disconnect the following:
- Negative battery cable
- Intake manifold
- Starter wiring
- Starter

To install:

3. Install or connect the following:
- Starter and tighten the fasteners to 37 ft. lbs. (50 Nm)
- Starter wiring
- Intake manifold
- Negative battery cable

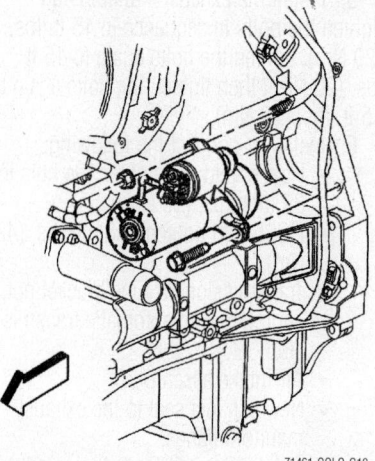

Starter motor mounting—Canyon & Colorado

Oil Pan

REMOVAL & INSTALLATION

1. Before servicing the vehicle, refer to the Precautions Section.

2. Drain the engine oil.

3. Remove or disconnect the following:
- Oil dipstick tube
- Engine splash guard
- Right front halfshaft, if equipped with 3.5L Engine
- Power steering gear, if equipped with RWD

4. If equipped with 4WD:
   a. Remove the front driveshaft.
   b. Remove the differential carrier assembly bushing to frame bolts only.
   c. Pull the differential carrier assembly downward.
   d. Secure the pinion yoke to prevent the differential carrier from rotating.

5. Remove or disconnect the following:
- Service slot plug

- Fuel pipe bracket at transmission and position aside
- Four lower transmission mounting bolts attached to oil pan

6. If equipped with 4WD, remove the power steering gear mounting bolts only and pull gear down far enough to access oil pan.

7. Disconnect the engine wiring harness retainers from oil pan

8. Remove the oil pan mounting bolts.

9. Install 2 bolts in the threaded holes at the rear of the oil pan to act as jack screws. Tighten evenly to release the oil pan from the engine block.

10. Remove the oil pan and 2 bolts from the jack screw holes.

To install:

11. Apply a bead of sealant around the oil pan as shown.

➡ **Install the oil pan within 10 minutes of applying the sealant.**

12. Install the oil pan, making sure that the pan if positioned fully rearward against the transmission mounting surface.

13. Install the oil pan bolts and tighten the side bolts to 18 ft. lbs. (25 Nm). Tighten the end bolts to 89 inch lbs. (10 Nm).

14. Connect the engine wiring harness retainers to oil pan.

15. If equipped with 4WD, position the steering gear upward to the frame assembly and install the steering gear mounting bolts.

16. Install the four lower transmission mounting bolts and tighten to 37 ft. lbs. (50 Nm).

17. Install the nuts securing the fuel pipe bracket at transmission and tighten to 15 ft. lbs. (20 Nm).

18. Install the service slot plug.

19. Right front halfshaft, if equipped with 3.5L Engine

20. If equipped with RWD, install the power steering gear.

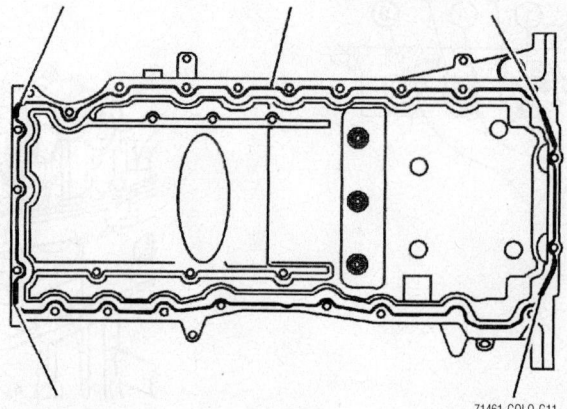

Oil pan sealant application areas—Canyon & Colorado

21. If equipped with 4WD:

a. Position the differential carrier assembly to the frame.

b. Install the differential carrier assembly bushing to frame bolts. Tighten to 112 ft. lbs. (152 Nm).

c. Install the front driveshaft.

22. Install the engine splash shield.

23. Install the oil dipstick tube. Tighten the bolt to 89 inch lbs. (10 Nm).

24. Fill the engine with oil to the correct level.

25. Start the engine and check for leaks.

Oil Pump

REMOVAL & INSTALLATION

1. Before servicing the vehicle, refer to the Precautions Section.

2. Remove or disconnect the following:
   - Engine front cover
   - Oil pump cover bolts
   - Oil pump cover

3. Matchmark the inner and outer gears in relation to the oil pump housing.

4. Remove or disconnect the following:
   - Inner and outer oil pump gears
   - Oil pump pressure relief valve plug
   - Oil pump pressure relief valve and spring

To install:

5. Install or connect the following:
   - Oil pump pressure relief valve and spring
   - Oil pump pressure relief valve plug and tighten to 124 inch lbs. (14 Nm)
   - Oil pump outer and inner gears
   - Oil pump cover and tighten the bolts to 89 inch lbs. (10 Nm).
   - Engine front cover

6. Start the engine and check for leaks.

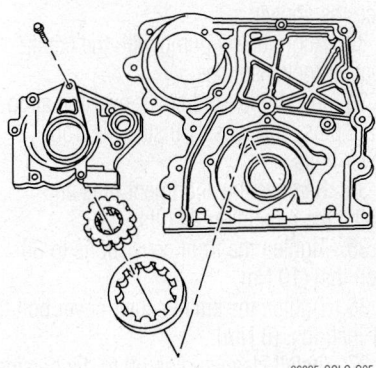

06025-COLO-G05

Exploded view of the oil pump and related components—2.8L and 3.5L engines

Rear Main Seal

REMOVAL & INSTALLATION

1. Before servicing the vehicle, refer to the Precautions Section.

2. Remove or disconnect the following:
   - Negative battery cable
   - Transmission assembly and transfer case, if equipped
   - Clutch assembly, if equipped
   - Flywheel
   - Crankshaft seal by prying it from out oil seal housing

➡ **Be careful not to damage the crankshaft seal surface with the prying tool.**

To install:

3. Install the new rear seal by lubricating it with engine oil and using a seal installer Special Tool J–44215. The spring side goes toward the engine and the seal will bottom out when installed fully.

4. Install or connect the following:
   - Flywheel/clutch assembly or flexplate
   - Transmission assembly and transfer case, if equipped
   - Negative battery cable

5. Start the engine and check for leaks.

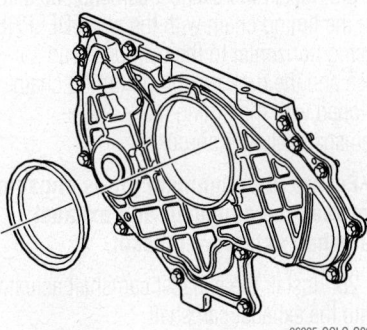

06025-COLO-G06

Remove the seal from the oil seal housing.

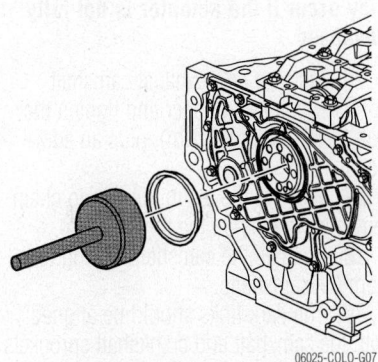

06025-COLO-G07

Use Special Tool J-44215 to install the rear oil seal–2.8L and 3.5L engines.

Timing Chain, Sprockets, Front Cover and Seal

REMOVAL & INSTALLATION

1. Before servicing the vehicle, refer to the Precautions Section.

2. Drain the cooling system.

3. Drain the engine oil.

4. Remove or disconnect the following:
   - Negative battery cable
   - Number 1 cylinder spark plug
   - Intake manifold
   - Ignition coils
   - Engine coolant temperature (ECT) sensor electrical connector from camshaft cover
   - Fuel injector connector from camshaft cover
   - Heated oxygen sensor (HO2S) connector from camshaft cover
   - Fuel pressure regulator screw
   - Camshaft cover
   - Camshaft position (CMP) sensor
   - Water pump

5. Remove the service slot plug and install Flywheel Holding Tool EN-46547 into the flywheel teeth.

6. Remove the crankshaft balancer bolt and discard.

7. Install Crankshaft end protector J–41816-2 into the end of the crankshaft and remove the crankshaft balancer using a 3-jaw puller.

8. Remove or disconnect the following:
   - Drive belt tensioner
   - Power steering pump
   - Oil pan
   - Oil pump pipe and screen assembly
   - 7mm center front cover bolt
   - Remaining engine front cover bolts

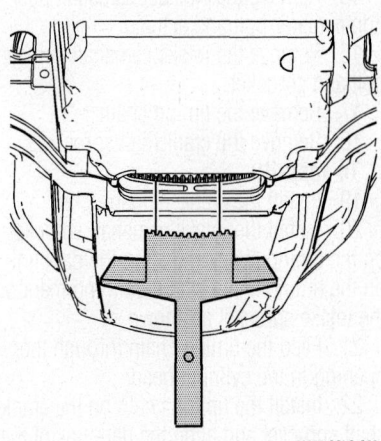

06025-COLO-G08

Use Special tool EN-46547 to prevent the flywheel from turning.

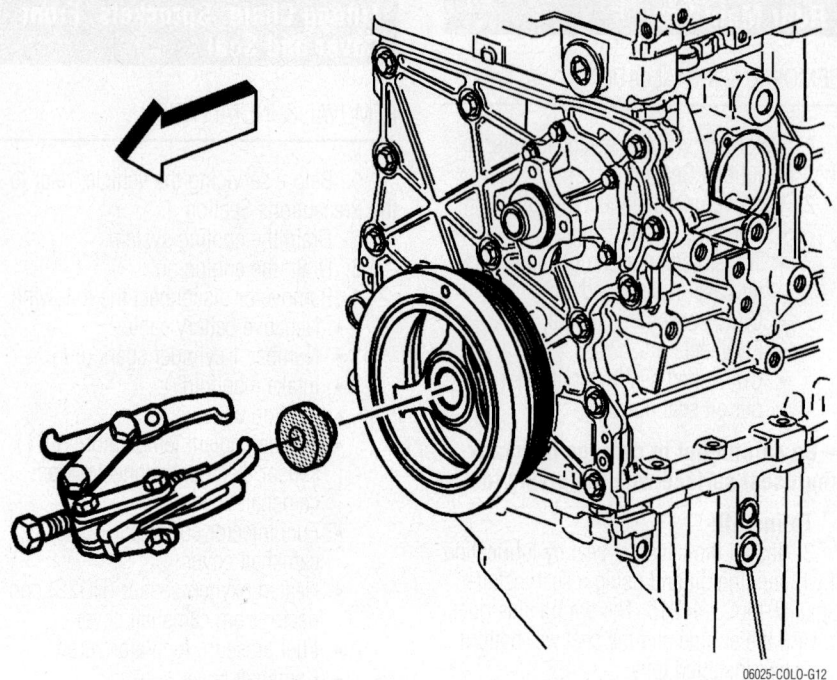

06025-COLO-G12

Remove the crankshaft balancer using a suitable puller after installing End Protector J-41816-2.

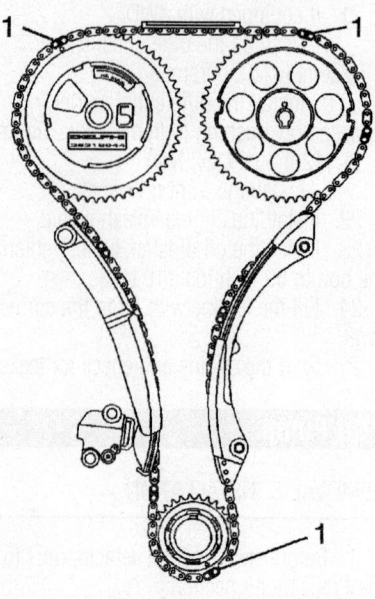

1. Timing chain dark link locations

71461-COLO-G12

Aligning timing chain dark links with camshaft and crankshaft sprockets— Canyon & Colorado

9. Install 2 bolts into the threaded holes to act as jack screws and tighten evenly to release the front cover.

10. Rotate the crankshaft clockwise until the no. 1 cylinder is at TDC on the compression stroke. The word DELPHI on the camshaft position actuator will be parallel with the cylinder head surface.

11. Install camshaft locking tool J-44221 to the rear of the camshafts.

12. Release the tension on the timing chain by moving the tensioner shoe in by hand.

13. Place a tee in the tensioner to hold the shoe in place.

14. Remove the top timing chain guide.

15. Remove the exhaust camshaft position actuator bolt and actuator.

16. Remove the intake camshaft sprocket bolt and sprocket.

17. Remove the timing chain.

18. Remove the crankshaft sprocket.

To install:

19. Install the crankshaft sprocket.

20. Install the intake camshaft sprocket on the timing chain and align the dark link on the timing chain with the timing mark on the intake sprocket as shown.

21. Feed the timing chain through the opening in the cylinder head.

22. Install the timing chain on the crankshaft sprocket and align the dark link of the timing chain with the timing mark on the crankshaft sprocket.

23. Install a new intake camshaft sprocket bolt and washer and tighten the bolt to 15 ft. lbs. (20 Nm), plus an additional 100°.

24. Install the exhaust camshaft actuator on the timing chain with the word DELPHI facing horizontal to the cylinder head surface and the dark link of the timing chain aligned with the timing mark on the camshaft actuator sprocket.

➡**Ensure the alignment pin is engaged between the camshaft and exhaust camshaft actuator sprocket.**

25. Install the exhaust camshaft actuator onto the exhaust camshaft.

➡**Rotate the camshaft actuator clockwise until it stops. This will fully advance the actuator. Engine damage may occur if the actuator is not fully advanced.**

26. Install a new exhaust camshaft sprocket bolt and washer and tighten the bolt to 18 ft. lbs. (25 Nm), plus an additional 135°.

27. Remove the tee in the timing chain tensioner to tension the timing chain.

28. Remove the camshaft locking tool from the camshafts.

29. The dark links should be aligned with the camshaft and crankshaft sprockets as shown.

30. Thread alignment pins into the engine block to aid front cover installation.

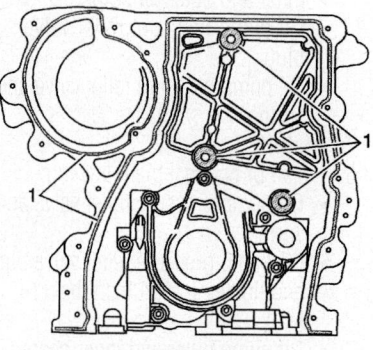

1. Sealant application areas

71461-COLO-G13

Front cover sealant application—Canyon & Colorado

31. Apply sealant to the front cover surfaces as shown.

32. Align the oil pump with the crankshaft sprocket splines.

33. Install the front cover over the alignment pins, and loosely install the front cover bolts.

34. Remove the alignment pins and install the remaining 2 bolts.

35. Tighten the front cover bolts to 89 inch lbs. (10 Nm).

36. Tighten the small center cover bolt to 71 inch lbs. (8 Nm).

37. Install clean engine oil to the outside diameter of the new front crankshaft oil seal.

38. Using a seal installer, install the front oil seal.

39. Install a new o-ring to the oil pump pipe screen assembly and install the oil pump pipe screen.

40. Apply sealant to the oil pump pipe bolt threads and tighten the bolts to 89 inch lbs. (10 Nm).

41. Install or connect the following:
- Oil pan
- Power steering pump
- Drive belt tensioner
- Crankshaft damper and tighten new bolt and tighten to 111 ft. lbs. (150 Nm), plus an additional 180°.

42. Remove the flywheel locking tool.

43. Install or connect the following:
- Flywheel access service plug
- Water pump
- Accessory drive belt
- Cooling fan
- Negative battery cable

44. Fill the engine with coolant and oil.

45. Start the engine and check for leaks.

Balance Shafts

REMOVAL & INSTALLATION

1. Before servicing the vehicle, refer to the Precautions Section.

2. Remove or disconnect the following:
- Engine
- Crankshaft rear oil seal housing

3. On 2.8L engines the left hand balance shaft sprocket timing mark is at the 1:00 o'clock position, the right side

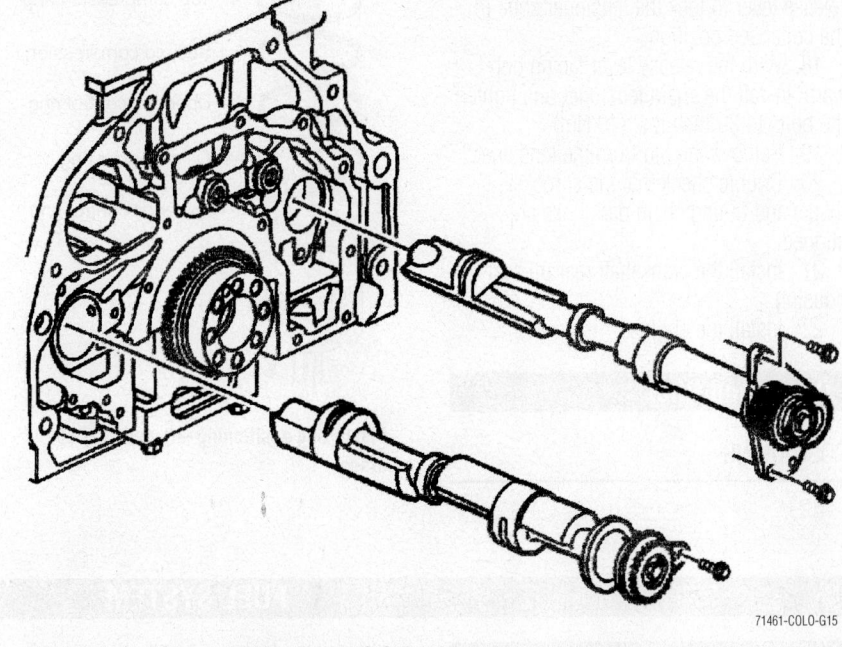

Removing the balance shafts—Canyon & Colorado

sprocket is at the 1:30 position and the crankshaft sprocket is at the 5:30 position. The timing marks should line up with the dark links on the timing chain as shown.

4. On 3.5L engines the left hand balance shaft sprocket timing mark is at the 12:00 o'clock position, the right side sprocket is at the 2:30 position and the crankshaft sprocket is at the 4:30 position. The timing marks should line up with the dark links on the timing chain as shown.

5. Remove the balance shaft chain tensioner.

6. Remove the balance shaft timing chain.

➡ It may be necessary to remove the right balance shaft bolts and rotate the retainer plate counterclockwise to relieve the tension on the chain.

7. Rotate the balance shafts to check for free rotation. If the shafts do not turn freely inspect the balance shaft bearings and bearing surface for damage.

8. Remove the balance shaft chain guide.

9. Remove the balance shaft bolts and remove the balance shafts.

To install:

10. Lubricate the balance shaft bearing journals with clean engine oil.

11. Install the balance shafts with the counterweight down to prevent damage to the shaft bearings.

12. Install new bolts and tighten the bolts to 106 inch lbs. (12 Nm).

13. Install the chain guide and tighten the bolts to 89 inch lbs. (10 Nm).

14. Install the balance shaft timing chain and ensure the dark links of the chain are aligned with the sprocket timing marks.

15. Using both hands, rotate the timing chain tensioner ratchet release lever clockwise and hold, compress the tensioner shoe and hold, then release the ratchet lever.

16. Slowly release the pressure on the shoe until the ratchet lever moves to the first detent and a click is heard.

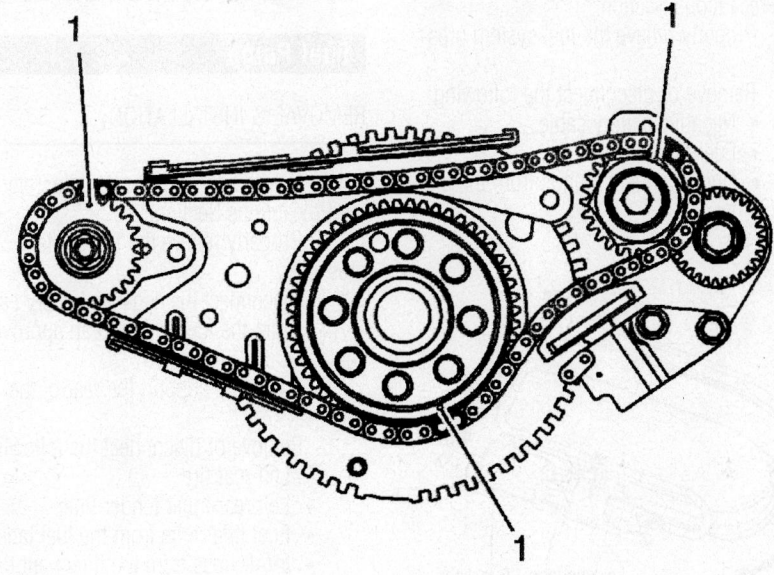

1. Sprocket timing marks

Balance shaft sprocket timing mark locations—Canyon & Colorado

17. Insert a pin into the hole of the release lever to lock the tensioner shoe in the collapsed position.

18. With the release lever facing outward, install the chain tensioner and tighten the bolts to 89 inch lbs. (10 Nm).

19. Remove the pin from the tensioner.

20. Double check that the sprocket marks and timing chain dark links are aligned.

21. Install the crankshaft rear oil seal housing.

22. Install the engine.

Piston and Ring

POSITIONING

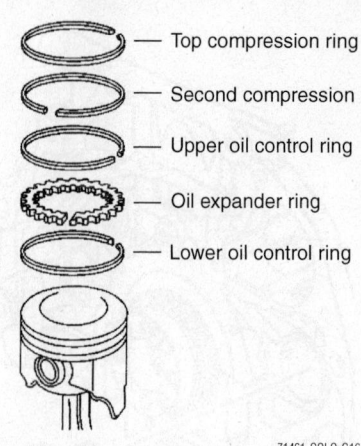

Top compression ring
Second compression r
Upper oil control ring
Oil expander ring
Lower oil control ring

71461-COLO-G16

Piston ring positioning—Canyon & Colorado

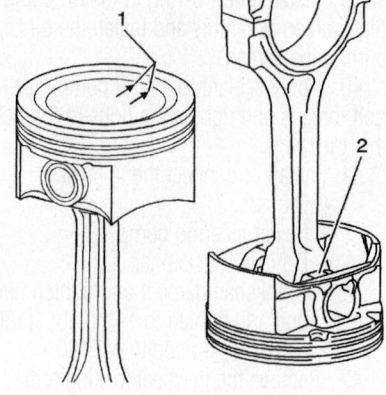

1 - Piston arrow face forward in block
2 - Flat casting surface faces forward

71461-COLO-G17

Piston and connecting rod orientation— Canyon & Colorado

FUEL SYSTEM

Fuel System Pressure

RELIEVING

The fuel systems operate under high fuel pressures. It is very important that the pressure be properly relieved prior to servicing the system or any of its components.

A Schrader valve is provided on these fuel systems to conveniently test or release the system pressure. A fuel pressure gauge and adapter will be necessary to connect the gauge to the fitting. This system utilizes a service valve on one end of the fuel rail assembly.

1. Before servicing the vehicle, refer to the Precautions Section.

2. Disconnect the negative battery cable to assure the prevention of fuel spillage if the ignition switch is accidentally turned **ON** while a fitting is still detached.

3. Loosen the fuel filler cap to release the fuel tank pressure.

4. Be sure the release valve on the fuel gauge is closed, then connect the fuel gauge to the pressure fitting located on the inlet fuel pipe fitting.

✷✷ CAUTION

When connecting the gauge to the fitting, be sure to wrap a rag around the fitting to avoid spillage. After repairs, place the rag in an approved container.

5. Install the bleed hose portion of the fuel gauge assembly into an approved container, then open the gauge release valve and bleed the fuel pressure from the system.

6. When the gauge is removed, be sure to open the bleed valve and drain all fuel from the gauge assembly.

7. When fuel service is finished, tighten the fuel filler cap and connect the negative battery cable.

Fuel Filter

REMOVAL & INSTALLATION

1. Before servicing the vehicle, refer to the Precautions Section.

2. Properly relieve the fuel system pressure.

3. Remove or disconnect the following:
   • Negative battery cable
   • Fuel filler cap
   • Quick connect fittings from the filter

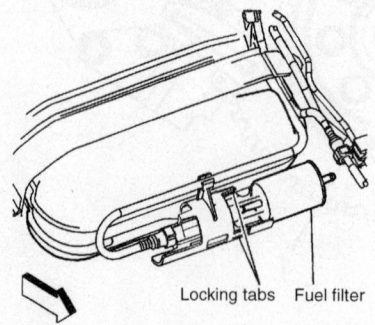

Locking tabs   Fuel filter

71461-COLO-G18

Fuel filter mounting at the fuel tank— Canyon & Colorado

4. Pry open the locking tabs of the mounting bracket enough to remove the fuel filter and remove the filter.

To install:

5. Slide the filter into the mounting bracket until the lacking tabs are fully engaged.

6. Install or connect the following:
   • Fuel quick disconnect fittings to the filter
   • Fuel filler cap
   • Negative battery cable

7. Turn the ignition **ON** for 10 seconds and then turn it **OFF** for 10 seconds. Again turn the ignition **ON** and check for leaks.

Fuel Pump

REMOVAL & INSTALLATION

1. Before servicing the vehicle, refer to the Precautions Section.

2. Properly relieve the fuel system pressure.

3. Disconnect the negative battery cable.

4. Drain the fuel tank into an approved container.

5. Raise and support the rear of the vehicle.

6. Remove or disconnect the following:
   • Left rear tire
   • Left rear inner fender liner
   • Fuel filler tube from the fuel tank
   • EVAP hose from the filler vent tube
   • Electrical connectors and wiring harness retainers from the fuel tank
   • Fuel return line and fuel filter
   • Upper fuel tank retaining strap

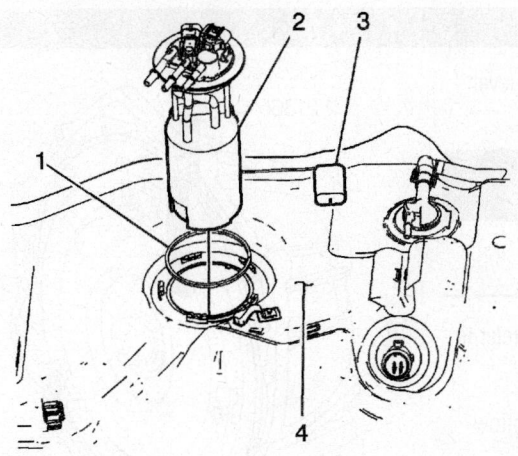

1. Seal
2. Fuel pump
3. Fuel level sensor and float
4. Fuel tank

71461-COLO-G19

Fuel pump mounting in fuel tank—Canyon & Colorado

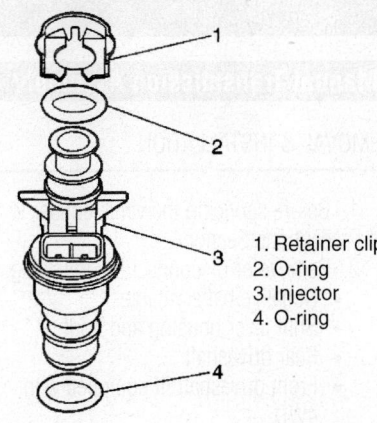

1. Retainer clip
2. O-ring
3. Injector
4. O-ring

71461-COLO-G20

Exploded view of fuel injector—Canyon & Colorado

7. Support the fuel tank.

8. Remove the lower tank retaining strap and lower the fuel tank.

9. Disconnect the fuel lines from the fuel pump module.

10. Using tool J-39765, rotate the fuel pump cam locking ring counterclockwise and remove the ring.

11. Raise the fuel pump and tilt it back to allow the fuel level sensor and float to clear the opening.

12. Remove and discard the fuel pump seal.

To install:

13. Install a new seal on the fuel pump.

14. Tilt the fuel pump until the fuel level sensor and float can enter the opening.

15. Lower the fuel pump and align the tang on the pump with the notch in the opening.

16. Using tool J-39765, rotate the fuel pump cam locking ring clockwise until fully seated.

17. Reconnect the fuel lines to the fuel pump module.

18. Raise the fuel tank and install the lower tank retaining strap.

19. Install or connect the following:
- Upper fuel tank retaining strap and tighten both strap bolts to 24 ft. lbs. (32 Nm)

- Fuel return line and fuel filter
- Electrical connectors and wiring harness retainers to the fuel tank
- EVAP hose to the vent tube
- Fuel filler tube to the fuel tank
- Left rear inner fender liner
- Left rear tire

20. Lower the vehicle.

21. Fill the tank with gasoline.

22. Turn the ignition **ON** for 10 seconds and then turn it **OFF** for 10 seconds. Again turn the ignition **ON** and check for leaks.

Fuel Rail and Injectors

REMOVAL & INSTALLATION

1. Before servicing the vehicle, refer to the Precautions Section.

2. Relieve the fuel system pressure.

3. Remove or disconnect the following:
- Negative battery cable
- Fuel feed and return lines from fuel rail.
- Vent hoses from the air cleaner resonator and fuel pressure regulator
- EVAP purge hose
- Intake manifold
- Fuel injector harness from the engine wiring harness

- Fuel rail mounting bolts
- Fuel rail
- Fuel injector connector from the injector
- Injector retaining clip
- Fuel injector from the fuel rail

4. Discard the retainer clip and remove and discard the 2 O-rings from the injector.

To install:

5. Lubricate the new injector O-ring seats with engine oil.

6. Install or connect the following:
- O-rings and retainer clip on the injector
- Fuel injector into the fuel rail socket
- Fuel rail. Torque the bolts to 89 inch lbs. (10 Nm).
- Fuel injector harness
- Intake manifold
- EVAP purge hose
- Vent hoses to air cleaner resonator and fuel pressure regulator
- Fuel feed and return lines
- Negative battery cable

7. Turn the ignition **ON** for 10 seconds and then turn it **OFF** for 10 seconds. Again turn the ignition **ON** and check for leaks.

DRIVE TRAIN

Manual Transmission Assembly

REMOVAL & INSTALLATION

1. Before servicing the vehicle, refer to the Precautions Section.
2. Remove or disconnect the following:
   - Negative battery cable
   - Shift lever housing and boot
   - Rear driveshaft
   - Front driveshaft, if equipped with 4WD
   - Transfer case and shift lever, if equipped with 4WD
   - All wiring harness that would interfere with transmission removal
3. Disconnect the hydraulic clutch quick-connect from the slave cylinder using special tool J–42371 to depress the white plastic sleeve on the quick connect to separate the clutch line end from the slave cylinder quick connect.
   - Engine wiring harness and fuel line retainers from transmission
4. Support the transmission with a transmission jack or equivalent.
5. Remove the transmission crossmember.
6. Remove the transmission mounting bolts. Pull the transmission straight back on the clutch hub splines.
7. Lower the transmission using the transmission jack.

To install:

Installation is the reverse of removal, but please note the following important steps.

8. Place a THIN coat of high-temperature grease on the main drive gear (input shaft) splines.
9. Secure the transmission to the floor jack and raise the transmission into position.
10. Slowly insert the input shaft through the clutch. Rotate the output shaft slowly to engage the splines of the input shaft into the clutch while pushing the transmission forward into place. Do not force the transmission into position, the transmission should easily fall into place once everything is properly aligned.
11. Tighten the transmission mounting bolts to 37 ft. lbs. (50 Nm).
12. Tighten the transmission crossmember horizontal nuts to 37 ft. lbs. (50 Nm).
13. Tighten the transmission crossmember vertical bolts to 74 ft. lbs. (100 Nm).
14. Do not remove the transmission jack until the crossmember has been installed.

15. Check the transmission fluid level and replenish as necessary.

Automatic Transmission Assembly

REMOVAL & INSTALLATION

1. Before servicing the vehicle, refer to the Precautions Section.
2. Drain the transmission fluid.
3. Remove or disconnect the following:
   - Negative battery cable
   - Dipstick and filler tube
   - Rear driveshaft
   - Front driveshaft, if equipped with 4WD
   - Transfer case, if equipped with 4WD
   - Range selector cable from selector lever
   - Transmission main harness connector
   - Engine wiring harness retainers
   - Park/neutral switch connector
   - Vent hose retainer
   - Fuel line bracket retainers and position fuel line aside
   - Transmission service access plug
   - 3 bolts securing the torque converter to the flywheel
   - Transmission cooler lines from the transmission. Plug the lines and the ports in the transmission.
4. Place a transmission jack under the transmission.
5. Remove the transmission crossmember.
6. Inspect for any other wiring, brackets etc. which may interfere with the removal of the transmission.
7. Remove the transmission from the engine by pulling the transmission rearward to disengage it from the locator dowel pins on the back of the block. Carefully lower transmission from the vehicle. Use care that the torque converter does not fall out of the front of the transmission.

➡**Use converter holding strap tool No. J-21366, to secure the torque converter to the transmission during removal and installation procedures.**

To install:

Installation is the reverse of removal, but please note the following important steps.

8. Make sure the torque converter is

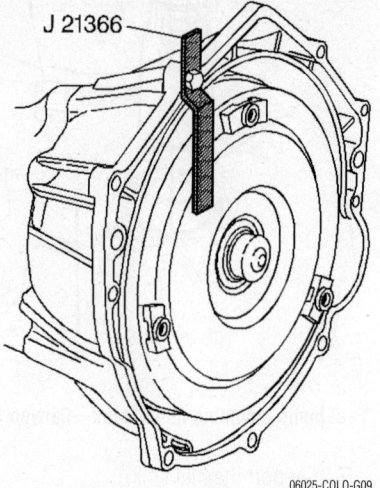

J 21366

06025-COLO-G09

Install Special tool 21366 to secure the torque converter to the transmission— Automatic transmission

fully seated in the pump drive. If not, the transmission will not fit tightly to the rear of the engine block.

9. Raise the transmission into position and remove the torque converter holding strap and carefully. Slide the transmission forward until the dowel pins are engaged.
10. The torque converter should be flush with the flywheel and turn freely by hand.
11. Install the transmission–to–engine bolts. Tighten the bolts to 37 ft. lbs. (50 Nm).
12. Tighten the torque converter-to-flywheel bolts to 44 ft. lbs. (60 Nm).
13. Refill the transmission with the proper amount and type of fluid.
14. Connect the negative battery cable. Start the vehicle and allow to warm while checking for leaks. Road test the vehicle to check for shift quality.

Clutch

REMOVAL & INSTALLATION

1. Before servicing the vehicle, refer to the Precautions Section.
2. Remove or disconnect the following:
   - Negative battery cable
   - Transmission
3. Install a clutch alignment tool or a used transmission input shaft to support the clutch.
4. If the clutch assembly is going to be reused, mark the flywheel, clutch cover and a pressure plate lug for alignment when installing.

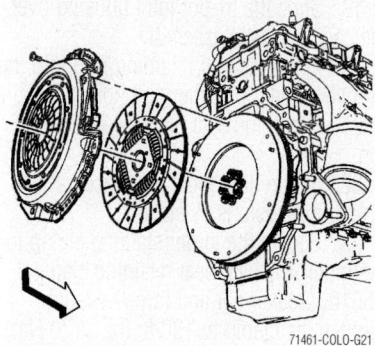

Exploded view of the clutch disc and related components—Canyon & Colorado

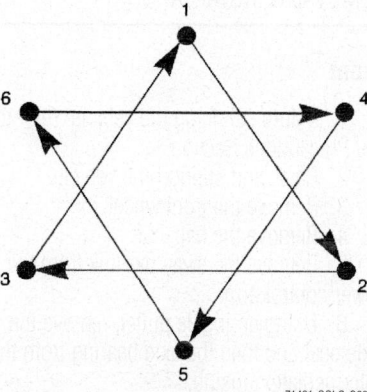

Clutch pressure plate bolt tightening sequence—Canyon & Colorado

5. Remove or disconnect the following:
   - Clutch cover bolts and washers
   - Clutch cover assembly and the clutch plate
   - Clutch alignment tool
6. Clean all parts and inspect for damage.

To install:

7. Install or connect the following:
   - Clutch alignment tool, to support the clutch
   - Clutch plate/clutch cover assembly to the flywheel. Tighten the bolts in sequence shown to 15 ft. lbs. (20 Nm).
8. Remove the clutch alignment tool.
9. Install or connect the following:
   - Transmission
   - Negative battery cable

Hydraulic Clutch System

Bleeding air from the hydraulic clutch system is necessary whenever any part of the system has been disconnected or the fluid level (in the reservoir) has been allowed to fall so low, that air has been drawn into the master cylinder.

BLEEDING

1. Before servicing the vehicle, refer to the Precautions Section.
2. Fill master cylinder reservoir with new brake fluid conforming to DOT 3 specifications.

✳✳ CAUTION

Always use new fluid from a sealed container. Never, under any circumstances, use fluid that has been bled from a system to fill the reservoir as it may be aerated, have too much moisture content and possibly be contaminated.

3. Pump the clutch pedal up and down at least 15 times.
4. Have an assistant fully depress and hold the clutch pedal, then open the bleeder screw.
5. Close the bleeder screw and have your assistant release the clutch pedal.
6. Repeat the procedure until all of the air is evacuated from the system. Check and refill master cylinder reservoir as required to prevent air from being drawn through the master cylinder.

➡**Never release a depressed clutch pedal with the bleeder screw open or air will be drawn into the system.**

7. Test the clutch for proper operation.

Transfer Case Assembly

REMOVAL & INSTALLATION

1. Before servicing the vehicle, refer to the Precautions Section.
2. Disconnect the negative battery cable.
3. Drain the transfer case fluid.
4. Support the transfer case.
5. Remove or disconnect the following:
   - Front and rear driveshafts from the transfer case. Matchmark the shafts prior to removal.
   - Vacuum lines and/or the electrical connectors, as equipped
   - Transfer case encoder motor
   - Engine wiring harness and position aside
   - Transfer case
6. Remove all traces of old gasket material from the mating surfaces.

To install:

7. Install or connect the following:
   - New gasket using sealer to hold it in position

- Transfer case. Torque the bolts to 37 ft. lbs. (50 Nm).
   - Engine wiring harness
   - Encoder motor
   - Vacuum lines and/or electrical connections, as necessary
   - Front and rear driveshafts by aligning the matchmarks
8. Refill the transfer case.
   - Negative battery cable

Halfshaft

REMOVAL & INSTALLATION

1. Before servicing the vehicle, refer to the Precautions Section.
2. Unlock the steering column so the steering linkage is free to move.
3. Remove or disconnect the following:
   - Negative battery cable
   - Front wheel
   - Halfshaft nut and washer
   - Steering knuckle assembly
4. Place a drift against the tripot housing and hammer the drift to release the shaft.
5. Remove the halfshaft from the vehicle.

To install:

6. Install the halfshaft. A snap or pop should be heard and felt when the shaft properly seats in the differential housing.
7. Install the steering knuckle assembly.
8. Install a new halfshaft nut and tighten the nut to 191 ft. lbs. (260 Nm).
9. Install the front wheel.
10. Lower the vehicle.

CV-Joints

OVERHAUL

Outer CV-Joint

1. Before servicing the vehicle, refer to the Precautions Section.

➡**When replacing the outer CV-joint, you must replace the inner CV-joint.**

2. Remove or disconnect the following:
   - Front wheel
   - Halfshaft and position it in a vise
   - Inner CV-joint
   - Large CV-joint boot clamp and discard it
   - Small CV-joint boot clamp and discard it
3. Slide the outer seal away from the CV-joint, remove from the halfshaft and discard.

➡**Do not remove remove the CV-joint from the halfshaft or you must replace the entire halfshaft assembly.**

4. Clean the old grease from the CV-joint and allow to dry.

To install:

5. Pack the CV-joint assembly with the kit supplied grease.

6. Place new retaining clamps onto the new outer seal.

7. Position the small end of the outer seal into the CV-joint outer seal groove on the halfshaft.

8. Secure the small to the outer seal with seal retaining clamp J-35910, breaker bar and torque wrench. Tighten the clamp to 100 ft. lbs. (136 Nm).

9. Slide the large end of the outer seal with large retaining ring in place over the outside edge of the CV-joint. Position the lip of the outer seal into the groove on the CV-joint.

10. Remove any excess air from the outer seal.

11. Secure the large retaining clamp to the outer seal with seal retaining clamp J-35910, breaker bar and torque wrench. Tighten the clamp to 130 ft. lbs. (176 Nm).

12. Check the clamp gap dimension; if it is not 0.085 in. (2.15mm), continue tightening the clamp until it is.

13. Install or connect the following:
- Inner CV-joint
- Halfshaft
- Front wheel

Inner (Tri-Pot) Joint

1. Before servicing the vehicle, refer to the Precautions Section.

2. Remove or disconnect the following:
- Front wheel
- Halfshaft and place it in a vise
- Large and small boot clamp, cut and discard
- Ball retaining ring using a screwdriver or equivalent
- Tri-pot housing and bushing from the shaft

3. Using Snapring Pliers tool J-8059, spread the small retaining ring located in the cage and inner race assembly.

4. Remove the cage and inner race assembly from the halfshaft

5. Remove the balls from the cage.

6. Remove the inner race from the cage.

7. Remove the seal and discard.

8. Thoroughly clean and inspect all parts.

To install:

9. Install a new small retaining clamp onto the neck of the seal.

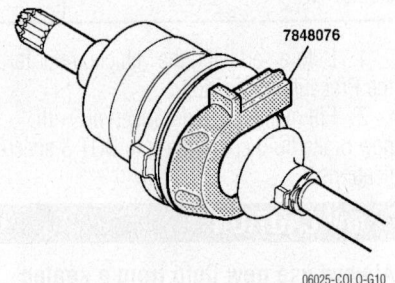

Install retainer 7848076 over the boot—Inner Joint

10. Slide the inner seal with the retaining clamp to the proper position on the halfshaft.

11. Secure the small retaining clamp retaining clamp J-35910, breaker bar and torque wrench. Tighten the clamp to 100 ft. lbs. (136 Nm).

12. Place the inner race with the retaining ring side up into the cage.

13. Place the six balls in the cage windows.

14. Slide the cage and inner race assembly, small diameter first, onto the halfshaft.

15. Using Snapring pliers J-8059, install the small retaining ring into the groove of the halfshaft.

16. Install convolute retainer 7848076 over the boot engaging 4 boot convolutions.

17. Pack the joint assembly with the kit-supplied grease.

18. Install a new large retaining clamp on the seal.

19. Slide the tri-pot joint housing over the cage and race assembly.

20. Insert the ball retaining ring into the groove at the top of the joint housing.

21. Install the seal onto the joint housing.

22. Ensure the boots are placed at the proper dimensions as shown.

23. Secure the large retaining clamp to the inner seal with seal retaining clamp J-35910, breaker bar and torque wrench. Tighten the clamp to 130 ft. lbs. (176 Nm).

24. Remove the convolute retainer tool.

Axle Shaft, Bearing and Seal

REMOVAL & INSTALLATION

Front

1. Before servicing the vehicle, refer to the Precautions Section.

2. Raise and support the vehicle.

3. Remove the front wheel.

4. Remove the halfshaft.

5. Remove the sway bar link from the lower control arm.

6. Using an inside puller, remove the axle seal and then the axle bearing from the intermediate housing.

To install:

7. Position the new bearing in the housing and use a bearing installer to press it in.

8. Install a new axle seal using a seal installer.

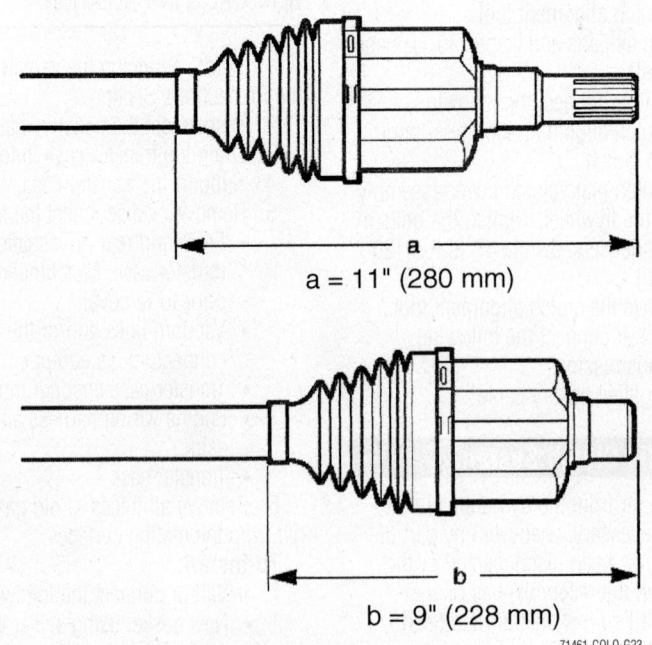

a = 11" (280 mm)

b = 9" (228 mm)

Inner (tri pot) joint boot dimensions—Canyon & Colorado

9. Install the halfshaft.

10. Connect the sway bar link to the control arm.

11. Install the wheel and tighten the lug nuts to 103 ft. lbs. (140 Nm).

Rear

1. Before servicing the vehicle, refer to the Precautions Section.

2. Raise and support the vehicle.

3. Remove the rear wheel.

4. Remove the brake drum.

5. Remove the rear axle housing cover.

6. Remove the pinion shaft lock bolt.

7. Remove the pinion shaft.

8. Remove the c-lock from the rear axle.

9. Pull the axle shaft out of the housing.

10. Remove the axle shaft seal and bearing.

To install:

11. Press on a new bearing and install the seal.

12. Install the axle shaft.

13. Install the c-lock to the rear axle.

14. Install the pinion shaft.

15. Install the pinion shaft lock bolt and tighten to 18 ft. lbs. (25 Nm).

16. Install a new gasket and the rear axle housing cover.

17. Install the brake drum.

18. Fill the rear axle with the correct lubricant.

19. Install the wheel and tighten the lug nuts to 103 ft. lbs. (140 Nm).

20. Lower the vehicle.

Pinion Seal

REMOVAL & INSTALLATION

1. Before servicing the vehicle, refer to the Precautions Section.

➡**The following procedure requires the use of the Pinion Holding tool J-8614-01 and the Pinion Seal Installation tool J-33782.**

2. Raise and support the vehicle.

3. Remove the front wheels.

4. Remove or disconnect the following:
- Engine undercover
- Driveshaft from the pinion flange. Matchmark the driveshaft prior to removal.
- Brake calipers and wire out of the way.

5. Mark the position of the pinion stem, flange and nut for reference.

6. Use an inch lbs. torque wrench to measure the amount of torque necessary to turn the pinion, then note this measurement as it is the combined pinion bearing, seal, carrier bearing, axle bearing and seal preload.

7. Remove or disconnect the following:
- Pinion flange nut and washer, using a Pinion Holding tool J-8614-01 and a Pinion Flange Removal tool J-8614-1, J-8614-2, J-8614-3, as applicable
- Pinion flange
- Pinion oil seal by driving it out of the differential with a blunt chisel; DO NOT damage the carrier

To install:

8. Examine the seal surface of pinion flange for tool marks, nicks or damage, such as a groove worn by the seal. If damaged, replace flange.

9. Examine the carrier bore and remove any burrs that might cause leaks around the O.D. of the seal.

10. Install a new pinion oil seal using a seal installer tool

11. Apply GM seal lubricant 12346004 to the splines of the pinion yoke.

12. Install or connect the following:
- Pinion flange aligning the reference marks made earlier.

13. Seat the flange on the shaft by tapping in with a soft-faced hammer.

14. Install the washer and new nut.

15. Install Pinion Holding tool J-8614-01 and tighten the nut while holding the tool until the end play is removed.

16. Continue tightening the nut until the preload torque is 3–5 inch lbs. (0.40–0.57 Nm) greater than the previously measured torque.

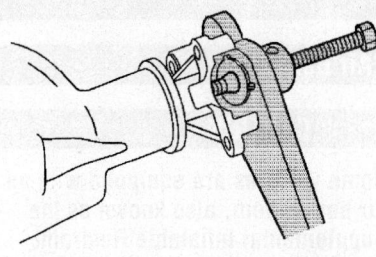

71461-COLO-G24

Removing the pinion nut using a pinion holding fixture tool

17. Rotate the pinion several times to ensure the bearing is seated.

18. Install the brake calipers.

19. Install the driveshaft to the pinion flange using the matchmarks made earlier.

20. Install the engine undercover.

➡**If fluid was lost from the differential housing during this procedure, be sure to check and add additional fluid, as necessary.**

Differential Carrier

REMOVAL & INSTALLATION

Front

1. Before servicing the vehicle, refer to the Precautions Section.

2. Remove the front wheel.

3. Drain the differential fluid.

4. Remove the front driveshaft. Matchmark the driveshaft prior to removal.

5. Remove the left sway bar link from the lower control arm.

6. Remove both halfshafts.

7. Remove the electrical connectors and vent hose.

8. Place a transmission jack under the differential.

9. Remove the differential mounting bracket.

10. Remove the differential.

11. To install, reverse the removal procedure.

STEERING AND SUSPENSION

Air Bag

⁂ CAUTION

Some vehicles are equipped with an air bag system, also known as the Supplemental Inflatable Restraint (SIR) system. The system must be disabled before performing service on or around system components, steering column, instrument panel components, wiring and sensors. Failure to follow safety and disabling procedures could result in accidental air bag deployment, possible personal injury and unnecessary system repairs.

DISARMING

1. Turn the steering wheel so that the vehicle's wheels are pointing straight ahead.
2. Turn the ignition switch to **LOCK**, remove the key, then disconnect the negative battery cable.
3. Remove the SIR fuse from the fuse block.
4. Remove the steering column filler panel or knee bolster.
5. Unplug the Connector Position Assurance (CPA) and yellow four way connector at the base of the steering column.
6. Remove the Connector Position Assurance (CPA) from the passenger yellow four way connector located behind the glove box.
7. Unplug the yellow four way connector located behind the glove box.
8. Connect the negative battery cable.

➡ With the AIR BAG fuse removed, the battery cable connected and the ignition in the ON position, the AIR BAG warning lamp will be ON. This is normal and does not indicate a system malfunction.

ARMING

1. Disconnect the negative battery cable.
2. Attach the yellow four way connector located behind the glove box.
3. Install the Connector Position Assurance (CPA) to the passenger yellow four way connector located behind the glove box.
4. Turn the ignition switch to **LOCK**, then remove the key.
5. Attach the four way connector at the base of the steering column and the Connector Position Assurance (CPA).
6. Install the steering column filler panel or knee bolster.
7. Install the AIR BAG fuse to the fuse block.
8. Connect the negative battery cable.
9. Staying away from the air bags, turn the ignition switch to **RUN** and make sure that the AIR BAG warning lamp flashes seven times and then shuts off. If the warning lamp does not shut off, make sure that the wiring is properly connected. If the light remains on, take the vehicle to a reputable repair facility for service.

Power Steering Gear

REMOVAL & INSTALLATION

1. Before servicing the vehicle, refer to the Precautions Section.
2. Position a fluid catch pan under the power steering gear.
3. Place the front wheels in the straight ahead position.
4. Turn the ignition key to the LOCK position and remove the key.
5. Insert Steering Column Locking Pin J-42640 into the access hole in the lower steering column trim cover.
6. Raise and support the vehicle.
7. Remove or disconnect the following:
   - Front wheels
   - Engine skid plate, if equipped
   - Front axle housing, if equipped
   - Outer tie rod ends from steering knuckle
   - Feed and return fluid hoses from the steering gear. Immediately cap or plug all openings to prevent system contamination or excessive fluid loss.
   - Intermediate shaft-to-lower steering column shaft pinch bolt
   - Lower shaft-to-steering gear pinch bolt
   - Power steering gear-to-frame bolts and washers
   - Front crossmember, if equipped with RWD only
   - Power steering gear

To install:

8. Install or connect the following:
   - Steering gear to the vehicle. Loosely install the mounting nuts, washers and bolts.
   - Crossmember, if equipped with RWD only. Tighten the mounting bolts to 44 ft. lbs. (60 Nm).
   - Steering gear mounting bolts. Tighten the vertical bolts to 96 ft. lbs. (130 Nm) and the horizontal bolts to 74 ft. lbs. (100 Nm).
   - Lower shaft-to-steering gear pinch bolt and tighten the bolt to 33 ft. lbs. (45 Nm).
   - Intermediate shaft-to-lower shaft pinch bolt, making sure the matchmarks line up. Tighten the bolt to 17 ft. lbs. (23 Nm).
   - Pressure and return hoses to the power steering gear. Tighten the hoses to 106 inch lbs. (12 Nm).
   - Outer tie rod ends
   - Front axle housing, if equipped

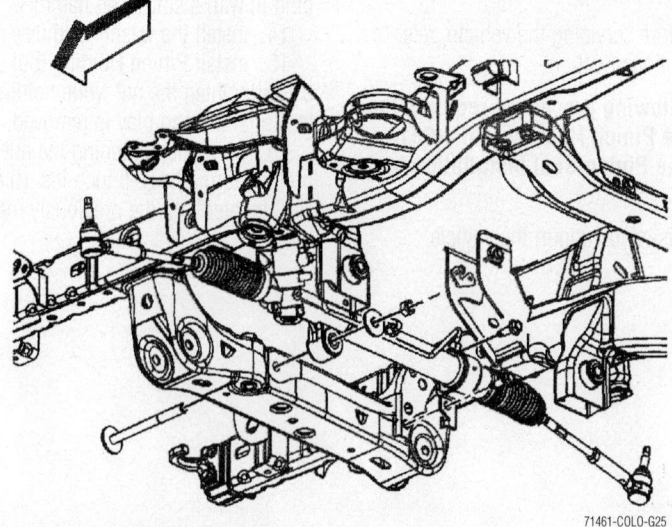

Power steering gear mounting—Canyon & Colorado

71461-COLO-G25

- Engine undercover
- Front wheels

9. Remove the steering column lock pin
10. Bleed the power steering system

Shock Absorbers

REMOVAL & INSTALLATION

Front

2WD MODELS

1. Before servicing the vehicle, refer to the Precautions Section.
2. Remove or disconnect the following:
   - Upper shock mounting nuts
   - Wheel
   - Lower mounting bolt and nut
   - Shock absorber/coil spring assembly
3. Place the shock absorber/coil spring assembly into a coil spring compressor and compress the spring.
4. Remove the upper shock retaining nut, bushings and washer.
5. Remove the shock absorber from the spring compressor.
6. If the coil spring is being replaced, remove the spring and mounting plate from the compressor, noting the mounting plate-to-spring orientation.

To install:

7. If the coil spring was replaced, align the spring to the mounting plate.
8. Align the centerline of the upper mounting studs with the centerline of the lower shock mount.

Install the mounting plate and spring into the compressor and compress the spring.

9. Install the shock absorber into the coil spring and compressor.
10. Install the washers, bushings and upper retaining nut. Tighten the retaining nut to 15 ft. lbs. (20 Nm).
11. Relax the tension on the spring compressor and remove the compressor.
12. Install the shock absorber/coil spring assembly to the lower mount. Tighten the lower mounting bolt and nut to 81 ft. lbs. (110 Nm).
13. Install the wheel and lower the vehicle.
14. Install the upper mounting nuts and tighten to 20 ft. lbs. (27 Nm).

4WD MODELS

1. Before servicing the vehicle, refer to the Precautions Section.
2. Place a jack or stand under the lower control arm.
3. Remove or disconnect the following:

- Shock absorber upper nut and isolator
- Lower nut/bolt
- Shock absorber through the control arm

To install:

4. Install the shock through the lower control arm and insert it through the mounting hole in the upper spring pocket.
5. Install or connect the following:
   - Shock absorber to the lower control arm. Tighten the nuts/bolts to 52 ft. lbs. (70 Nm).
   - Upper isolator to the shock
6. Tighten the upper mounting nut to 18 ft. lbs. (25 Nm).

Rear

1. Before servicing the vehicle, refer to the Precautions Section.
2. Support the rear axle assembly at ride height.
3. Remove or disconnect the following:
   - Shock absorber-to-frame retainers at the top of the shock
   - Shock-to-axle retainers at the bottom of the shock
   - Shock absorber

To install:

4. Install the shock in the vehicle and loosely install the upper mounting fasteners to retain it.
5. Align the lower-end of the shock absorber with the axle mounting, then loosely install the retainers.
6. Tighten the upper shock retainers to 26 ft. lbs. (35 Nm). Tighten the lower shock retainers to 70 ft. lbs. (95 Nm).

Coil Springs

REMOVAL & INSTALLATION

➡**On 2WD models, the coil springs are removed with the front shock absorber/coil spring assembly. 4WD models do not use coil springs. Coil springs are not used on the rear suspension.**

Leaf Springs

REMOVAL & INSTALLATION

1. Before servicing the vehicle, refer to the Precautions Section.

➡**The following procedure requires the use of two sets of jackstands.**

2. Support the rear axle with jackstands, support the axle and the body sep-

arately in order to relieve the load on the rear spring.

3. Remove or disconnect the following:
   - Wheel
   - Parking brake cable
   - Trailer hitch, if equipped
   - Shock absorber lower mounting nut and bolt
   - U-bolt nuts, washers, anchor plate and bolts
   - Rear spring hanger bracket nut and bolt
   - Front spring bracket bolt
   - Leaf spring

To install:

4. Install or connect the following:
   - Spring to the front bracket using the bolt, washers and nut, but do not fully tighten at this time.
   - Rear bracket U-bolts, anchor plate, washers and U-bolt nuts. Torque the nuts to 56 ft. lbs. (76 Nm).
   - Tighten the front nuts to 59 ft. lbs. (80 Nm), plus an additional 80°.
   - Tighten the rear shackle nut to 63 ft. lbs. (85 Nm).
   - Shock absorber lower mounting nut and bolt
   - Trailer hitch, if equipped
   - Parking brake cable
   - Wheel

Torsion Bar

Instead of the coil spring used on the front suspension of 2WD vehicles, the 4WD vehicles are equipped with a torsion bar.

REMOVAL & INSTALLATION

1. Before servicing the vehicle, refer to the Precautions Section.
2. Raise and support the vehicle allowing the front suspension to hang in the rebound position.
3. Remove the front wheels.
4. Mark the rear torsion bar adjuster bolt location and count the number of turns required to remove the bolt.
5. Remove the adjuster bolt spacer and adjuster nut.
6. Remove the torsion bar and adjustment arms as a unit, moving it rearward to disengage it from the lower control arm.

➡**Note the direction of the forward end and side of the torsion bar being removed**

To install:

7. Install the adjustment arms and torsion bars in the same position as where they were removed.

8. Install the adjustment arms to the torsion bar and slide the bar forward until it engages the lower control arm fully.

9. Install the adjuster bolt, spacer and adjuster nut and tighten it the same number of turns as counted on removal.

10. Install the wheels and lower the vehicle.

Ball Joints

REMOVAL & INSTALLATION

2WD Vehicles

UPPER

1. Before servicing the vehicle, refer to the Precautions Section.
2. Raise and support the vehicle.
3. Remove the front wheels.
4. Support the lower control arm with a suitable jack.
5. Disconnect the brake lines from the upper control arm.
6. Remove the wheel speed sensor bracket bolt and disconnect the sensor brackets.
7. Remove the upper ball joint nut.
8. Separate the ball joint from the knuckle using a separator.
9. Remove the 4 ball joint nuts and bolts from the control arm and discard them.
10. Remove the ball joint from the arm.

To install:

11. Install or connect the following:
   - Ball joint in the upper control arm
   - New ball joint retaining nuts and bolts. Tighten the ball joint retainers to 12 ft. lbs. (16 Nm).
   - Ball joint to steering knuckle. Tighten the new nut to 55 ft. lbs. (75 Nm).
   - Speed sensor bracket and tighten the nuts to 15 ft. lbs. (20 Nm).

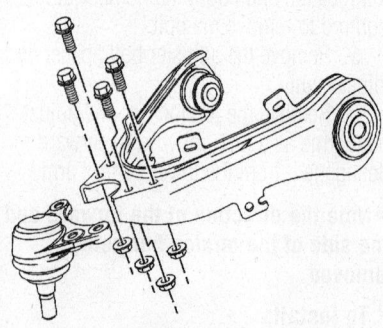

Upper ball joint mounting— Canyon & Colorado 2WD

- Brake hose to the control arm
- Remove the lower control arm jack
- Tire and wheel assembly

12. Check and adjust the front end alignment, as necessary.

LOWER

1. Before servicing the vehicle, refer to the Precautions Section.
2. Remove the steering knuckle.
3. Remove the ball joint nuts and bolts.
4. Remove the ball joint.

To install:

5. Install or connect the following:
   - Ball joint nuts and bolts and tighten to 44 ft. lbs. (60 Nm)
   - Steering knuckle

6. Check and adjust the front end alignment, as necessary.

4WD Vehicles

UPPER

1. Before servicing the vehicle, refer to the Precautions Section.
2. Raise and support the vehicle.
3. Remove the front wheels.
4. Support the lower control arm with a suitable jack.
5. Disconnect the brake lines from the upper control arm.
6. Remove the wheel speed sensor bracket bolt and disconnect the sensor brackets.
7. Remove the upper ball joint nut.
8. Disconnect the upper control arm from the ball stud by removing the retaining nuts.
9. Remove and discard the upper ball joint retaining nut.

10. Separate the ball joint from the knuckle using a separator.
11. Remove the ball joint from the arm.

To install:

12. Install or connect the following:
   - Ball joint in the steering knuckle
   - Ball joint retaining nut and tighten the nut to 55 ft. lbs. (75 Nm).
   - Upper control arm to the ball stud and tighten the nut to 35 ft. lbs. (47 Nm).
   - Wheel speed sensor bracket bolt and tighten to 15 ft. lbs. (20 Nm).
   - Brake line to the upper control arm
   - Front tires

13. Check and adjust the front end alignment, as necessary.

LOWER

1. Before servicing the vehicle, refer to the Precautions Section.
2. Remove the steering knuckle.
3. Remove the ball joint nuts and bolts.
4. Remove the ball joint.

To install:

5. Install or connect the following:
   - Ball joint nuts and bolts and tighten to 47 ft. lbs. (64 Nm)
   - Steering knuckle

6. Check and adjust the front end alignment, as necessary.

Upper Control Arm

REMOVAL & INSTALLATION

2WD Vehicles

1. Before servicing the vehicle, refer to the Precautions Section.

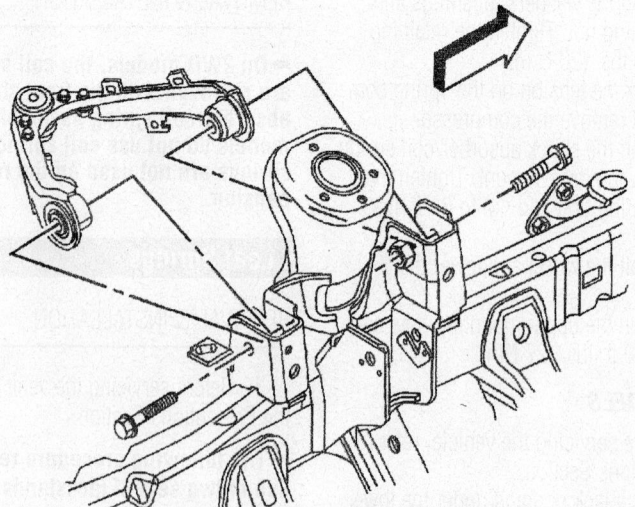

Upper control arm mounting—2WD Canyon & Colorado

2. Disconnect the negative battery cable.

3. Raise and support the vehicle.

4. Place a jack stand under the lower control arm to support it at ride height.

5. Remove or disconnect the following:
- Wheel
- Wheel speed sensor harness bracket retaining bolt and nut, if equipped
- Brake hose from upper arm
- Upper ball joint nut
- Upper arm from steering knuckle
- Mounting bolts
- Upper control arm

To install:

6. Install or connect the following:
- Upper control arm and tighten the bolts to 118 ft. lbs. (160 Nm).
- Steering knuckle to upper control arm ball joint
- New ball joint stud nut and tighten to 55 ft. lbs. (75 Nm)
- Brake hose to upper arm
- Wheel speed sensor harness bracket retaining bolt and nut, if equipped
- Wheel

7. Check the front wheel alignment.

4WD Vehicles

1. Before servicing the vehicle, refer to the Precautions Section.

2. Disconnect the negative battery cable.

3. Raise and support the vehicle.

4. Place a jack stand under the lower control arm to support it at ride height.

5. Remove or disconnect the following:
- Wheel
- Wheel speed sensor brackets
- Brake hose from upper arm
- Upper arm from ball joint stud nuts

➡️**The 4WD vehicle upper control arm bolts are equipped with cams, which are rotated to achieve caster and camber adjustments. In order to preserve adjustment and ease installation, matchmark the cams to the control arm before removal. If the control arm is being replaced, transfer the alignment marks to the new component before installation.**

- Nuts and adjustments cams
- Upper arm bolts and upper arm

To install:

6. Install or connect the following:
- Upper arm and mounting bolts. Tighten the bolts to 114 ft. lbs. (155 Nm).

7. Align the cams to the reference marks made earlier, then tighten the cam nuts to 114 ft. lbs. (155 Nm).

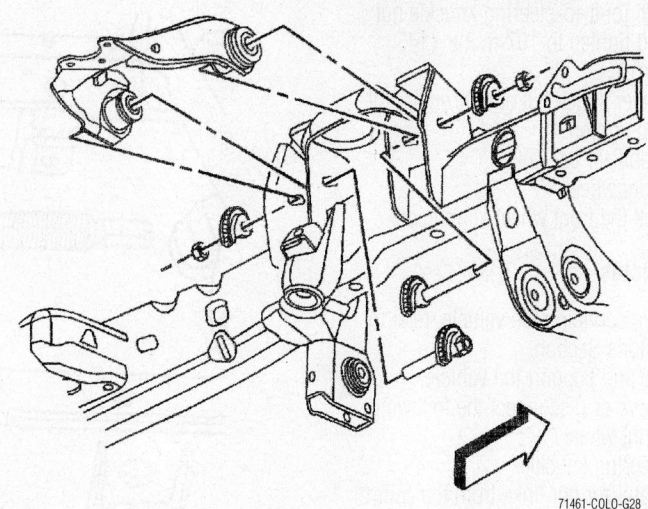

Upper control arm mounting—4WD Canyon & Colorado

- Ball joint stud nuts and tighten to 47 ft. lbs. (64 Nm).
- Brake hose to upper arm
- Wheel speed sensor brackets
- Wheel

Lower Control Arm

REMOVAL & INSTALLATION

2WD Vehicles

1. Before servicing the vehicle, refer to the Precautions Section.

2. Raise and support the vehicle. Remove or disconnect the following:
- Front wheel
- Stabilizer bar links from control arm
- Lower shock nut and through bolt
- Lower ball joint stud from steering knuckle

➡️**The 2WD vehicle lower control arm bolts are equipped with cams, which are rotated to achieve caster and camber adjustments. In order to preserve adjustment and ease installation, matchmark the cams to the control arm before removal. If the control arm is being replaced, transfer the alignment marks to the new component before installation.**

- Adjustment cam nuts and cams
- Lower control arm bolts and control arm

To install:

3. Install or connect the following:

➡️**The nuts must be tightened in sequence.**

- Lower control arm. Tighten the rear nuts first, then the front nuts to 114 ft. lbs. (155 Nm).
- Lower ball joint stud into the steering knuckle

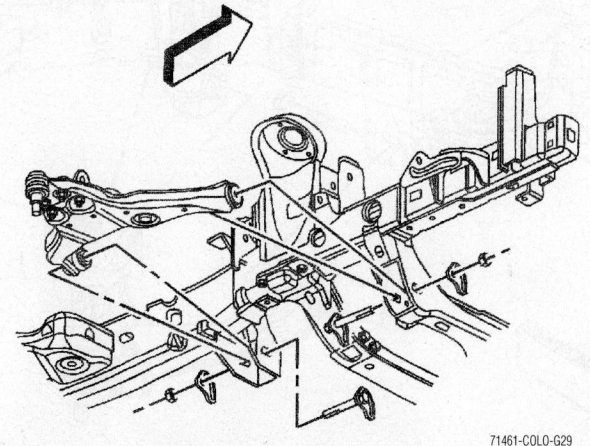

Lower control arm mounting—2WD Canyon & Colorado

- Ball joint-to-steering knuckle nut and tighten to 107 ft. lbs. (145 Nm).
- Lower shock nut and through bolt and tighten
- Stabilizer bar links
- Front wheel

4. Check the front wheel alignment.

4WD Vehicles

1. Before servicing the vehicle, refer to the Precautions Section.
2. Raise and support the vehicle.
3. Remove or disconnect the following:
   - Front wheels
   - Steering knuckle
   - Stabilizer bar links from the control arm
   - Lower shock bolt and nut
   - Torsion bar, if necessary
   - Lower control arm

To install:

4. Install the control arm, bolts and nuts. Install the washer on the front bolt with the shoulder facing the control arm.

➡**The nuts must be tightened in sequence.**

5. Tighten the rear nut first to 107 ft. lbs. (145 Nm), then tighten the front nut to 122 ft. lbs. (165 Nm).
6. Install or connect the following:
   - Steering knuckle
   - Torsion bar, if removed
   - Lower shock bolt and nut
   - Stabilizer bar links
   - Front wheels

7. Check and adjust the front end alignment, as necessary.

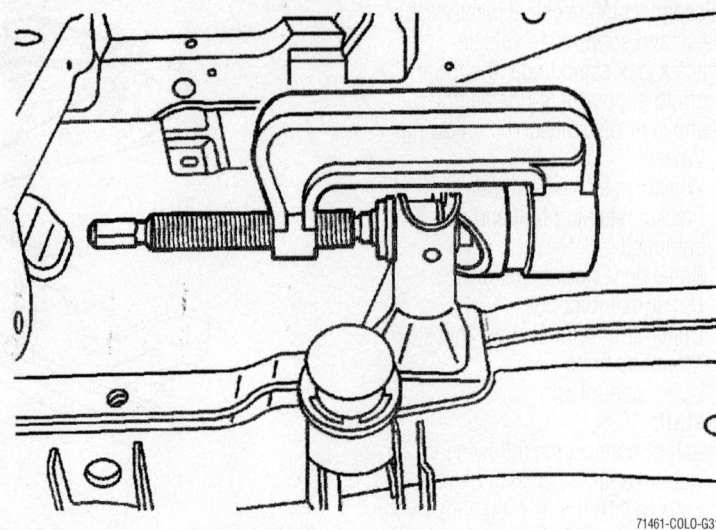

Removing and installing the lower control arm rear bushing using tool J 41805—Canyon & Colorado

CONTROL ARM BUSHING REPLACEMENT

1. Before servicing the vehicle, refer to the Precautions Section.
2. Remove the lower control arm.
3. Measure and record the distance from the bushing flange to the bracket.
4. Install tool J 41805 on the rear bushing and tighten until the bushing is removed.

To install:

5. Install the rear bushing into the control arm
6. Install tools J 41805. Tighten until the bushing is fully seated.
7. Install the lower control arm.

Wheel Bearings

ADJUSTMENT

➡**All models use sealed wheel bearings that are pre-adjusted. If the bearing need replacing, replace the front wheel hub/bearing assembly.**

Wheel Hub/Bearing Assembly

REMOVAL & INSTALLATION

➡**To replace the rear bearing see the Axle Shaft, Bearing and Seal procedure in DRIVETRAIN.**

Front

1. Before servicing the vehicle, refer to the Precautions Section.

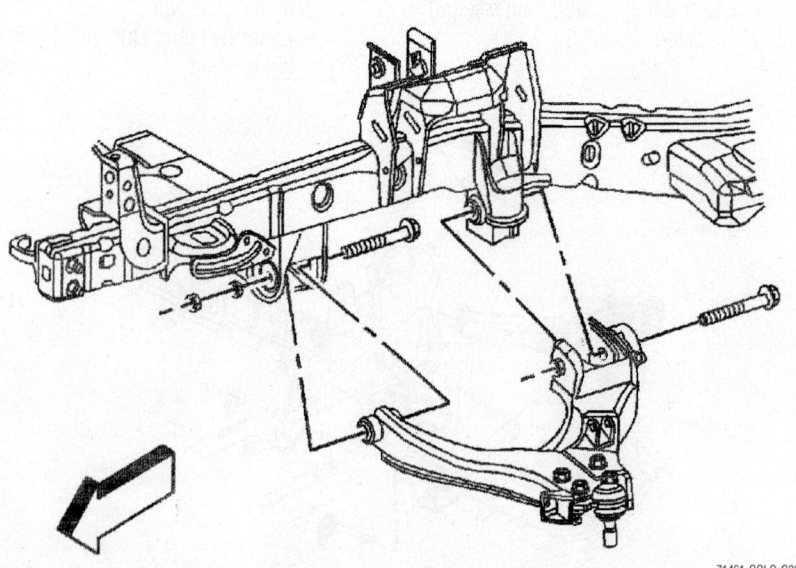

71461-COLO-G30

Lower control arm mounting—4WD Canyon & Colorado

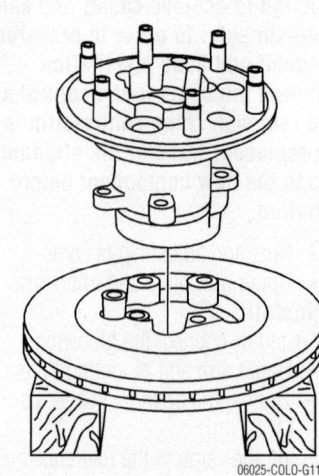

06025-COLO-G11

Separating the wheel hub from the brake rotor.

2. Remove or disconnect the following:
- Wheel
- Brake caliper with the pads without disconnecting the brake line
- Wheel speed sensor harness from upper control arm
- Speed sensor bracket from control arm
- Speed sensor electrical connector from body

3. Using white paint, mark the location of the speed sensor wiring harness to the steering knuckle for installation reference. Coil up the sensor wiring so it is out of the way of suspension components.

4. On 4WD models, remove the steering knuckle.

5. On all models, remove the wheel hub/brake rotor rear mounting bolts and remove the hub/rotor assembly. The backing plate will come off when the assembly is removed.

6. Separate the brake rotor from the wheel hub by removing the 6 mounting bolts.

To install:

7. Clean the contact surface between the wheel hub and brake rotor.

8. Position the new hub/bearing assembly onto the brake rotor and tighten the bolts to 15 ft. lbs. (20 Nm) in a crisscross pattern.

9. Install or connect the following:
- Wheel hub/rotor assembly to backing plate
- On 4WD models, the steering knuckle
- Wheel hub/rotor to steering knuckle. Tighten the bolts to 92 ft. lbs. (125 Nm).
- Speed sensor electrical connector to body

10. While holding the rotor from turning,

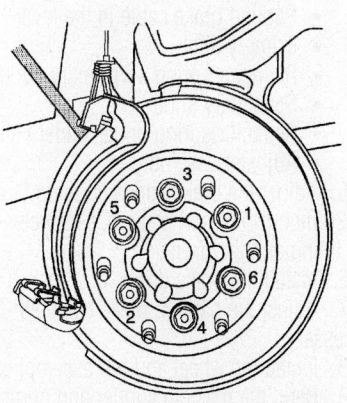

Brake rotor/wheel hub mounting bolt tightening sequence—Canyon & Colorado

tighten the rotor mounting bolts in the sequence shown to 88 ft. lbs. (120 Nm).

11. Install or connect the following:
- Brake caliper
- Speed sensor bracket to control arm
- Wheel speed sensor harness to upper control arm
- Wheel

Steering Knuckle

REMOVAL & INSTALLATION

2WD Vehicles

1. Before servicing the vehicle, refer to the Precautions Section.
2. Raise and support the vehicle.
3. Place a jack stand under the lower control arm.
4. Remove or disconnect the following:
- Wheel
- Wheel hub/bearing assembly
- Outer tie rod end
- Upper and lower ball joint nuts and discard
- Separate the ball joints from the steering knuckle
- Steering knuckle

To install:

5. Clean the ball joints and tie rod ends. Inspect the tapered holes and mounting surfaces of the steering knuckle for damage or being out of round. Replace the knuckle if the holes are damaged.

6. Install the steering knuckle and connect the lower ball joint. Tighten the nut to 107 ft. lbs. (145 Nm).

7. Connect the upper ball joint to the knuckle and tighten the nut to 55 ft. lbs. (75 Nm).

8. Install or connect the following:
- Outer tie rod end
- Wheel hub/bearing assembly
- Wheel

9. Check the front wheel alignment.

4WD Vehicles

1. Before servicing the vehicle, refer to the Precautions Section.
2. Raise and support the vehicle.
3. Place a jack stand under the lower control arm.
4. Remove or disconnect the following:
- Wheel
- Wheel hub nut and discard
- Brake caliper bracket
- Speed sensor harness from chassis harness and fender panel
- Outer tie rod end from steering knuckle
- Upper ball joint stud nuts
- Lower ball joint nut
- Separate ball joint from knuckle
- Steering knuckle

5. If the knuckle is being replaced, remove the wheel hub/bearing assembly.

To install:

6. Clean the ball joints and tie rod ends. Inspect the tapered holes and mounting surfaces of the steering knuckle for damage or being out of round. Replace the knuckle if the holes are damaged.

7. If removed, install the wheel hub/bearing assembly.

8. Install the steering knuckle and connect the lower ball joint. Tighten the nut to 107 ft. lbs. (145 Nm).

9. Connect the upper ball joint to the knuckle and tighten the nut to 55 ft. lbs. (75 Nm).

10. Install or connect the following:
- Outer tie rod end
- Speed sensor harness to chassis harness and fender panel
- Brake caliper bracket
- New wheel hub nut and tighten to 191 ft. lbs. (260 Nm).
- Wheel

11. Check the front wheel alignment.

Stabilizer Bar

REMOVAL & INSTALLATION

1. Before servicing the vehicle, refer to the Precautions Section.
2. Raise and support the vehicle.
3. Remove or disconnect the following:
- Front wheels
- Stabilizer link nuts from control arm
- Stabilizer links
- Stabilizer bar insulator clamps
- Stabilizer bar
- Insulators

To install:

4. Install new insulators on the stabilizer bar.

5. Install or connect the following:
- Stabilizer bar and insulator clamps. Tighten the bolts to 37 ft. lbs. (50 Nm).

6. Support the lower control arms at ride height.

7. Install the stabilizer links and tighten the nuts to 32 ft. lbs. (44 Nm).

8. Install the front wheels and lower the vehicle.

BRAKES

Brake Caliper

REMOVAL & INSTALLATION

1. Before servicing the vehicle, refer to the Precautions Section.
2. If brake fluid level is midway between MAX and MIN level in the reservoir, no fluid needs to be removed. If brake fluid level is higher than midway, remove the brake fluid to the midway point.
3. Remove or disconnect the following:
   - Tire and wheel assembly
   - Brake caliper fluid line, then plug it
   - Caliper slide pin bolts
   - Caliper from the mounting bracket

To install:
4. Clean and lubricate the sleeves and bushings with silicon grease.
5. Install or connect the following:
   - Caliper in mounting bracket
   - Slide pin bolts. Tighten to 29 ft. lbs. (40 Nm).
   - Fluid lines to the caliper using new Copper washers, and tighten to 29 ft. lbs. (40 Nm)
   - Wheel and tire assembly
6. Refill the master cylinder to the correct level. Bleed the brake system.

Disc Brake Pads

REMOVAL & INSTALLATION

Front

1. Before servicing the vehicle, refer to the Precautions Section.
2. If brake fluid level is midway between MAX and MIN level in the reservoir, no fluid needs to be removed. If brake fluid level is higher than midway, remove the brake fluid to the midway point.
3. Remove or disconnect the following:
4. Place a C-clamp around the outer pad and caliper; tighten the C-clamp until the piston is fully compressed in the caliper.
   - Remove top caliper retainer, and rotate caliper away from rotor
   - Inboard pad and retaining clips
   - Outboard pad from the caliper
   - Shims

To install:
5. Install or connect the following:
   - New retaining clips onto the inboard pad
   - Shims
   - Outboard pad into the caliper

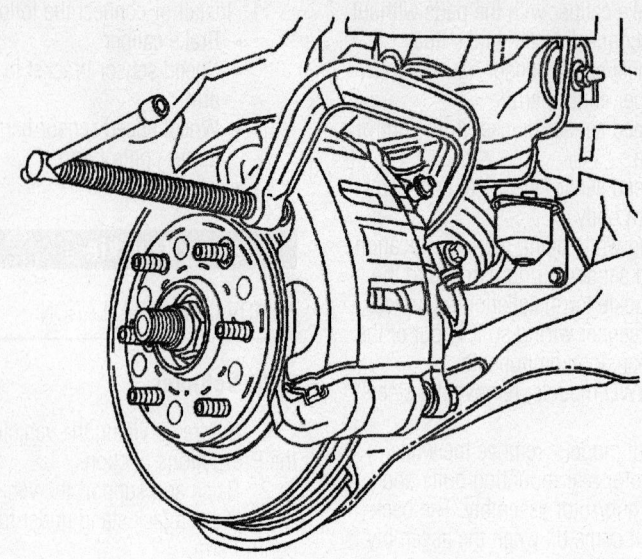

Compressing the caliper piston with a C-clamp—Canyon & Colorado

- Inboard pad in the caliper
- Caliper in position over the rotor
- Caliper bolts and tighten to 29 ft. lbs. (40 Nm).
- Wheel and tire

6. Refill the master cylinder and pump pedal to attain full brake pedal before road-testing the vehicle.

Brake Drums

REMOVAL & INSTALLATION

1. Before servicing the vehicle, refer to the Precautions Section.
2. Remove or disconnect the following:
   - Wheel and tire assembly
   - Retaining clip
   - Brake drum. If the drum will not pull of the axle, use a rubber mallet and tap it around the edge.

To install:
3. Install or connect the following:
   - Drum on the axle
   - Retaining clip
   - Wheel and tire assembly
4. Refill the master cylinder and pump pedal to attain full brake pedal before road-testing the vehicle.

Brake Shoes

REMOVAL & INSTALLATION

1. Before servicing the vehicle, refer to the Precautions Section.

2. Remove or disconnect the following:
   - Wheel and tire assembly
   - Brake drum
   - Adjuster assembly
   - Retractor spring from secondary shoe
   - Secondary shoe
   - Retractor spring from primary shoe
   - Primary shoe
   - Return spring
   - Depress lock tab on parking brake cable
   - Hold lock tab and push parking brake cable forward
   - Parking brake cable from lever

To install:
3. Lubricate the contact points on the backing plate with high temperature silicone grease.
4. Install or connect the following:
   - Parking brake cable in the lever
   - Primary shoe
   - Retractor spring on primary shoe
   - Secondary shoe
   - Retractor spring on secondary shoe
   - Adjuster assembly
5. Adjust the brake shoes so there is 0.030 inch (0.76 mm) clearance between the lining and the drum.
6. Install the brake drum.
7. Adjust the parking brake cable as necessary.
8. Install the wheel and tire assemblies.
9. Refill the master cylinder and pump pedal to attain full brake pedal before Road-testing the vehicle.

CHEVROLET AND PONTIAC

Cavalier • Sunfire

BRAKES **11-55**
DRIVE TRAIN **11-42**
ENGINE REPAIR **11-9**
FUEL SYSTEM **11-39**
SPECIFICATION AND
  MAINTENANCE CHARTS......**11-2**
Engine and Vehicle
  Identification..................11-2
General Engine Specifications 11-2
Engine Tune-Up Specifications 11-2
Firing Order 11-3
Accessory Drive Belt Routing ...11-3
Capacities 11-4
Valve Specifications......................11-5
Crankshaft and Connecting
  Rod Specifications..............11-5
Piston and Ring Specifications 11-5
Torque Specifications 11-6
Wheel Alignment 11-6
Tire, Wheel and Ball Joint
  Specifications 11-7
Brake Specifications 11-7
Scheduled Maintenance
  Intervals..............................11-8
STEERING AND
  SUSPENSION **11-47**
A
Air Bag..............................11-47
  Arming.................................11-48
  Disarming 11-48
  Precautions 11-48
Alternator 11-9
  Installation 11-9
  Removal 11-9
B
Brake Caliper 11-55
  Removal & Installation............11-55
Brake Drums.........................11-55
  Removal & Installation............11-55
Brake Shoes.........................11-56
  Removal & Installation............11-56
C
Camshaft and Valve Lifters 11-28
  Removal & Installation............11-28

Clutch 11-43
  Removal & Installation............11-43
Coil Spring 11-51
  Removal & Installation............11-51
CV-Joints............................11-44
  Overhaul 11-44
Cylinder Head 11-23
  Removal & Installation............11-23
D
Disc Brake Pads......................11-55
  Removal & Installation............11-55
E
Engine Assembly 11-9
  Removal & Installation............11-9
Exhaust Manifold 11-27
  Removal & Installation............11-27
F
Fuel Filter 11-39
  Removal & Installation............11-39
Fuel Injector.............................11-40
  Removal & Installation............11-40
Fuel Pump 11-39
  Removal & Installation............11-39
Fuel System Pressure 11-39
  Relieving 11-39
Fuel System Service
  Precautions..............................11-39
H
Halfshaft..............................11-44
  Removal & Installation............11-44
Heater Core...........................11-18
  Removal & Installation............11-18
Hydraulic Clutch System 11-43
  Bleeding..............................11-43
I
Ignition Timing 11-9
  Adjustment.................................11-9
Intake Manifold 11-26
  Removal & Installation............11-26
L
Lower Ball Joint 11-52
  Removal & Installation............11-52

Lower Control Arm 11-52
  Control Arm Bushing
    Replacement......................11-53
  Removal & Installation............11-52
O
Oil Pan...............................11-31
  Removal & Installation............11-31
Oil Pump 11-32
  Removal & Installation............11-32
P
Piston and Ring 11-38
  Positioning 11-38
R
Rack and Pinion Steering Gear 11-49
  Removal & Installation............11-49
Rear Main Seal 11-33
  Removal & Installation............11-33
Rocker Arms...........................11-26
  Removal & Installation............11-26
S
Shock Absorber 11-50
  Removal & Installation............11-50
Starter Motor 11-31
  Removal & Installation............11-31
Strut.................................11-49
  Removal & Installation............11-49
T
Timing Chain, Sprockets,
  Front Cover and Seal................11-34
  Removal & Installation............11-34
Transmission Assembly 11-42
  Removal & Installation............11-42
V
Valve Lash 11-31
  Adjustment...............................11-31
W
Water Pump 11-16
  Removal & Installation............11-16
Wheel Bearings.........................11-53
  Adjustment...............................11-53
  Removal & Installation............11-53

SPECIFICATION AND MAINTENANCE CHARTS

ENGINE AND VEHICLE IDENTIFICATION CHART

Code ①	Liters (cc)	Cu. In.	Cyl.	Fuel Sys.	Engine Type	Eng. Mfg.
4	2.2 (2180)	133	4	MFI	OHV	CUS
F	2.2 (2180)	133	4	MFI	OHV	CUS
T	2.4 (2392)	146	4	MFI	DOHC	CUS

Code ②	Year
2	2002
3	2003
4	2004
5	2005
6	2006

CUS: Chevrolet/United States

MFI: Multi-point Fuel Injection

OHV: Overhead Valves

DOHC: Double Overhead Camshafts

① 8th position of VIN

② 10th position of VIN

06025-JBOD-C01

GENERAL ENGINE SPECIFICATIONS

Year	Model	Engine Displacement Liters (VIN)	Net Horsepower @ rpm	Net Torque @ rpm (ft. lbs.)	Bore x Stroke (in.)	Com-pression Ratio	Oil Pressure @ rpm
2002	Cavalier	2.2 (4)	115@5000	136@3600	3.50x3.46	9.0:1	56@3000
		2.4 (T)	150@5600	155@4400	3.54x3.70	9.5:1	30@3000
	Sunfire	2.2 (4)	115@5000	136@3600	3.50x3.46	9.0:1	56@3000
		2.4 (T)	150@5600	155@4400	3.54x3.70	9.5:1	30@3000
2003	Cavalier	2.2 (F)	140@5600	150@4000	3.50x3.46	10.0:1	50-80@1000
	Sunfire	2.2 (F)	140@5600	150@4000	3.50x3.46	10.0:1	50-80@1000
2004	Cavalier	2.2 (F)	140@5600	150@4000	3.50x3.46	10.0:1	50-80@1000
	Sunfire	2.2 (F)	140@5600	150@4000	3.50x3.46	10.0:1	50-80@1000
2005	Cavalier	2.2 (F)	140@5600	150@4000	3.38x3.72	10.0:1	50-80@1000
	Sunfire	2.2 (F)	140@5600	150@4000	3.38x3.72	10.0:1	50-80@1000

06025-JBOD-C02

ENGINE TUNE-UP SPECIFICATIONS

Year	Engine Displacement Liters (VIN)	Spark Plug Gap (in.)	Ignition Timing (deg.)	Fuel Pump (psi)	Idle Speed (rpm)	Valve Clearance Intake	Valve Clearance Exhaust
2002	2.2 (4)	0.040	①	50-60	①	HYD	HYD
	2.4 (T)	0.050	①	53-59	①	HYD	HYD
2003	2.2 (F)	0.042	①	50-60	①	HYD	HYD
2004	2.2 (F)	0.042	①	50-60	①	HYD	HYD
2005	2.2 (F)	0.043	①	50-60	①	HYD	HYD

NOTE: The Vehicle Emission Control Information label often reflects specification changes made during production.

The label figures must be used if they differ from those in this chart.

HYD: Hydraulic

① Refer to Vehicle Emission Control Information label

06025-JBOD-C03

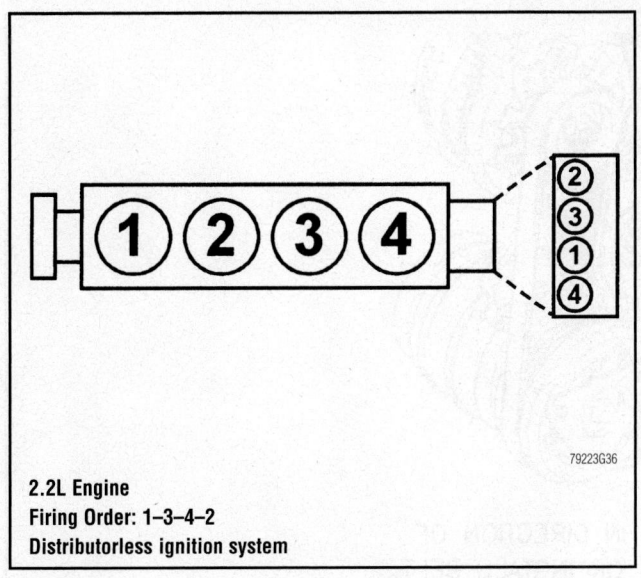

2.2L Engine
Firing Order: 1–3–4–2
Distributorless ignition system

79223G36

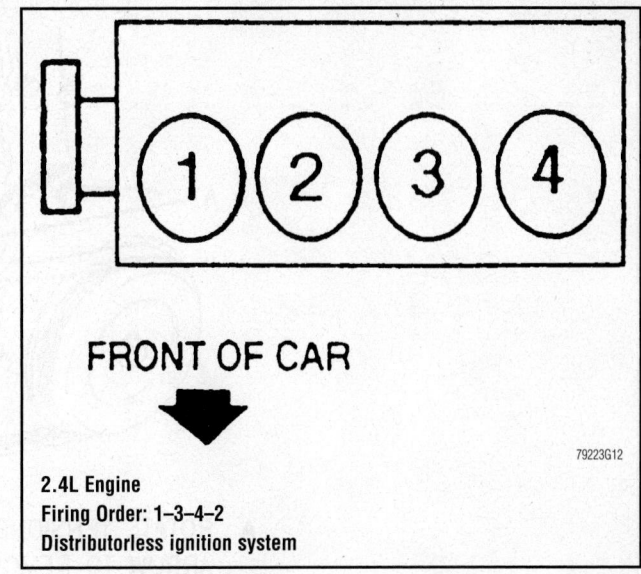

FRONT OF CAR

2.4L Engine
Firing Order: 1–3–4–2
Distributorless ignition system

79223G12

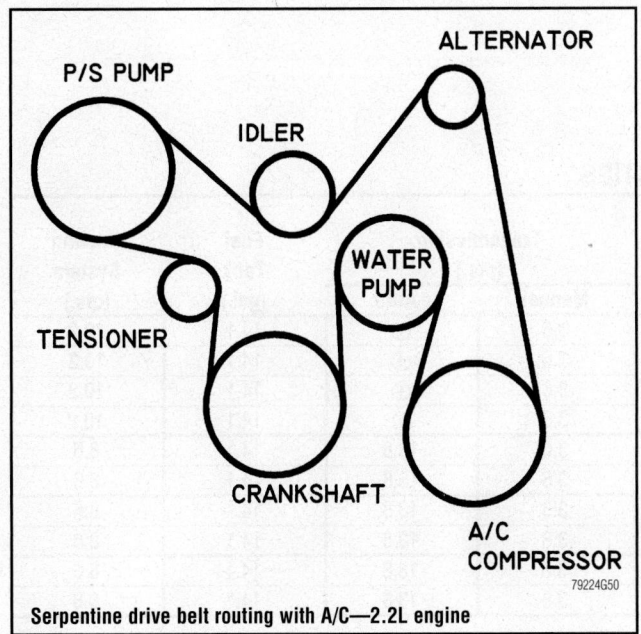

79224G50

Serpentine drive belt routing with A/C—2.2L engine

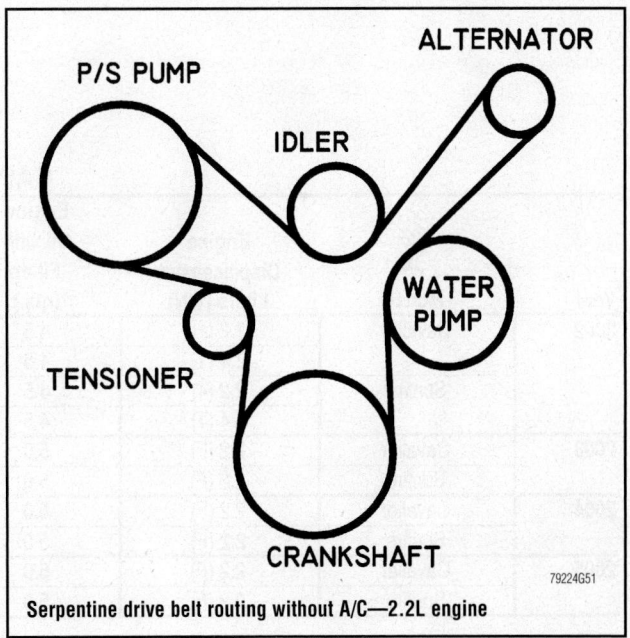

79224G51

Serpentine drive belt routing without A/C—2.2L engine

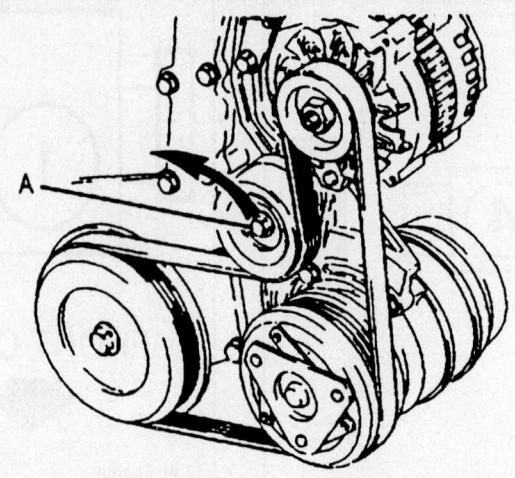

A ROTATE TENSIONER IN DIRECTION OF
ARROW TO REMOVE OR INSTALL BELT.

79224G52

Serpentine drive belt routing—2.4L engine

CAPACITIES

Year	Model	Engine Displacement Liters (VIN)	Engine Oil with Filter (qts.)	Transmission (pts.)		Fuel Tank (gal.)	Cooling System (qts.)
				Manual	Auto.		
2002	Cavalier	2.2 (4)	4.5	3.6	①	14.1	10.2
		2.4 (T)	4.5	3.6	①	14.1	10.2
	Sunfire	2.2 (4)	4.5	3.6	①	14.1	10.2
		2.4 (T)	4.5	3.6	①	14.1	10.2
2003	Cavalier	2.2 (F)	5.0	3.6	13.8	14.1	8.6
	Sunfire	2.2 (F)	5.0	3.6	13.8	14.1	8.6
2004	Cavalier	2.2 (F)	5.0	3.6	13.8	14.1	8.6
	Sunfire	2.2 (F)	5.0	3.6	13.8	14.1	8.6
2005	Cavalier	2.2 (F)	5.0	3.6	13.8	14.1	8.6
	Sunfire	2.2 (F)	5.0	3.6	13.8	14.1	8.6

NOTE: All capacities are approximate. Add fluid gradually and ensure a proper fluid level is obtained.

① 3 Speed: 8.0 pts.
  4 Speed: 13.8 pts.

06025-JB0D-C04

VALVE SPECIFICATIONS

Year	Engine Displacement Liters (VIN)	Seat Angle (deg.)	Face Angle (deg.)	Spring Test Pressure (lbs. @ in.)	Spring Installed Height (in.)	Stem-to-Guide Clearance (in.)		Stem Diameter (in.)	
						Intake	Exhaust	Intake	Exhaust
2002	2.2 (4)	46	45	72-81@1.60	1.600	0.0007-0.0020	0.0014-0.0029	0.2740-0.2743	0.2731-0.2736
	2.4 (T)	45	46	50-55@1.44	1.437	0.0009-0.0025	0.0016-0.0032	0.2331-0.2339	0.2326-0.2334
2003	2.2 (F)	46	45	NA	NA	0.2362-0.2367	0.2362-0.2367	0.2344-0.2355	0.2337-0.2343
2004	2.2 (F)	46	45	NA	NA	0.2362-0.2367	0.2362-0.2367	0.2344-0.2355	0.2337-0.2343
2005	2.2 (F)	46	45	NA	NA	0.0012-0.0022	0.0020-0.0026	0.2344-0.2355	0.2337-0.2343

NA: Not available

06025-JBOD-C05

CRANKSHAFT AND CONNECTING ROD SPECIFICATIONS
All measurements are given in inches.

Year	Engine Displacement Liters (VIN)	Crankshaft				Connecting Rod		
		Main Brg. Journal Dia.	Main Brg. Oil Clearance	Shaft End-play	Thrust on No.	Journal Diameter	Oil Clearance	Side Clearance
2002	2.2 (4)	2.4945-2.4954	0.0006-0.0019	0.0020-0.0070	4	1.9983-1.9994	0.0010-0.0030	0.0039-0.0149
	2.4 (T)	2.3612-2.3631	0.0004-0.0023	0.0034-0.0095	3	1.8887-1.8897	0.0004-0.0026	0.0059-0.0177
2003	2.2 (F)	2.2045-2.2050	0.0012-0.0026	0.0012-0.0150	3	2.0519-2.0525	0.004-0.0026	0.0028-0.0146
2004	2.2 (F)	2.2045-2.2050	0.0012-0.0026	0.0012-0.0150	3	2.0519-2.0525	0.004-0.0026	0.0028-0.0146
2005	2.2 (F)	2.2045-2.2050	0.0012-0.0026	0.0012-0.0150	3	1.9291-1.9297	0.0011-0.0027	0.0028-0.0146

06025-JBOD-C06

PISTON AND RING SPECIFICATIONS
All measurements are given in inches.

Year	Engine Displacement Liters (VIN)	Piston Clearance	Ring Gap			Ring Side Clearance		
			Top Compression	Bottom Compression	Oil Control	Top Compression	Bottom Compression	Oil Control
2002	2.2 (4)	0.0006-① 0.0018	0.0100-0.0200	0.0120-0.0177	0.0100-0.0300	0.0020-0.0035	0.0016-0.0031	0.0005-0.0087
	2.4 (T)	0.0006-② 0.0009	0.0060-0.0120	0.0098-0.0157	0.0098-0.0299	0.0016-0.0031	0.0012-0.0028	NA
2003	2.2 (F)	0.0004-0.0016	0.008-0.0160	0.014-0.0220	0.0100-0.0300	0.0015-0.0031	0.0012-0.0028	0.0035-0.0052
2004	2.2 (F)	0.0004-0.0016	0.008-0.0160	0.014-0.0220	0.0100-0.0300	0.0015-0.0031	0.0012-0.0028	0.0035-0.0052
2005	2.2 (F)	0.0004-0.0016	0.008-0.0160	0.014-0.0220	0.0100-0.0300	0.0015-0.0031	0.0012-0.0027	0.0035-0.0052

NA: Not Available

① 42.7mm from top of piston
② 42.7mm from top of piston

06025-JBOD-C07

TORQUE SPECIFICATIONS
All readings in ft. lbs.

| Year | Engine Displacement Liters (VIN) | Cylinder Head Bolts | Main Bearing Bolts | Rod Bearing Bolts | Crankshaft Damper Bolts | Flywheel Bolts | Manifold | | Spark Plugs | Oil Pan Drain Plug |
							Intake	Exhaust		
2002	2.2 (4)	①	70	38	77 ②	55	③	10	13	18
	2.4 (T)	④	⑤	⑥	⑦	⑧	⑨	⑩	13	18
2003	2.2 (F)	⑪	⑫	⑬	⑭	⑮	⑯	⑰	15	18
2004	2.2 (F)	⑪	⑫	⑬	⑭	⑮	⑯	⑰	15	18
2005	2.2 (F)	⑪	⑫	⑬	⑭	⑮	⑯	⑰	15	18

NA: Not Available

① Step 1: Long bolts: 46 ft. lbs.
Step 2: Short bolts: 43 ft. lbs.
Step 3: Long bolts an additional 90 degree turn
Step 4: Short bolts an additional 90 degree turn

② Center bolt spec shown; Pulley-to-hub bolts: 37 ft. lbs.

③ Bolts: 17 ft. lbs.
Nuts: 17 ft. lbs.
Studs: 9 ft. lbs.

④ Step 1: bolts 1-8; 40 ft. lbs.
Step 2: bolts 9-10: 30 ft. lbs.
Step 3: An additional 90 degree turn

⑤ 15 ft. lbs. plus 90 degrees
⑥ 18 ft. lbs. plus 80 degrees
⑦ 129 ft. lbs. plus 90 degrees
⑧ 15 ft. lbs. plus 45 degrees
⑨ Nuts: 18 ft. lbs.
Studs: 97 inch lbs.
⑩ Nuts: 31 ft. lbs.
Studs: 97 inch lbs.
⑪ Step 1: 22 ft. lbs.
Step 2: plus an additional 155 degrees
⑫ 15 ft. lbs. plus 70 degrees

⑬ 18 ft. lbs. plus 100 degrees
⑭ 74 ft. lbs. plus 125 degrees
⑮ 39 ft. lbs. plus 25 degrees
⑯ Bolts: 89 inch lbs.
Nuts: 89 inch lbs.
Studs: 53 inch lbs.
⑰ Bolts: 89 inch lbs.
Studs: 12 ft. lbs.

06025-JBOD-C08

WHEEL ALIGNMENT

| Year | Model | | Caster | | Camber | | Toe-in (in.) |
			Range (+/-Deg.)	Preferred Setting (Deg.)	Range (+/-Deg.)	Preferred Setting (Deg.)	
2002	All	F	+1.00	+4.30	+1.00	0	0 +/- 0.25
		R	—	—	+0.75	-0.40	0.20 +/- 0.30
2003	All	F	4.10° ± 0.75°	+4.30	-0.20° ± 0.70°	0	0.10° ± 0.20°
		R	—	—	0.25° ± 0.25°	—	0.20 +/- 0.30
2004	All	F	4.10° ± 0.75°	+4.30	-0.20° ± 0.70°	0	0.10° ± 0.20°
		R	—	—	0.25° ± 0.25°	—	0.20 +/- 0.30
2005	All	F	4.10° ± 0.75°	+4.30	-0.20° ± 0.70°	0	0.10° ± 0.20°
		R	—	—	0.25° ± 0.25°	—	0.20 +/- 0.30

06025-JBOD-C09

TIRE, WHEEL AND BALL JOINT SPECIFICATIONS

Year	Model	OEM Tires		Tire Pressures (psi)		Wheel Size	Ball Joint Inspection	Lug Nuts
		Standard	Optional	Front	Rear			
2002	Cavalier base	P195/70R14	None	30	30	6-JJ	①	100
	Cavalier LS, RS	P195/65R15	None	30	30	6-JJ	①	100
	Cavalier Z24	P205/55R16	None	30	30	6-JJ	①	100
	Sunfire SE	P195/70R14	P195/65R15	30	30	6-JJ	0.125 in.	100
	Sunfire GT	P205/55R16	None	30	30	6-JJ	0.125 in.	100
2003	Cavalier base	P195/70R14	P195/65R15	30	30	6-JJ	①	100
	Cavalier Z24, LS	P195/65R15	P205/55R16	30	30	6-JJ	①	100
	Cavalier Z24	P205/55R16	None	30	30	6-JJ	①	100
	Sunfire SE	P195/70R14	P195/65R15	30	30	6-JJ	0.125 in.	100
	Sunfire GT	P205/55R16	None	30	30	6-JJ	0.125 in.	100
2004	Cavalier base	P195/70R14	P195/65R15	30	30	6-JJ	①	100
	Cavalier Z24, LS	P195/65R15	P205/55R16	30	30	6-JJ	①	100
	Cavalier Z24	P205/55R16	None	30	30	6-JJ	①	100
	Sunfire SE	P195/70R14	P195/65R15	30	30	6-JJ	0.125 in.	100
	Sunfire GT	P205/55R16	None	30	30	6-JJ	0.125 in.	100
2005	Cavalier base	P195/70R14	P195/65R15	30	30	6-JJ	①	100
	Cavalier Z24, LS	P195/65R15	P205/55R16	30	30	6-JJ	①	100
	Cavalier Z24	P205/55R16	None	30	30	6-JJ	①	100
	Sunfire SE	P195/70R14	P195/65R15	30	30	6-JJ	0.125 in.	100
	Sunfire GT	P205/55R16	None	30	30	6-JJ	0.125 in.	100

OEM: Original Equipment Manufacturer

PSI: Pounds Per Square Inch

① Replace if any measurable movement is found

06025-JBOD-C10

BRAKE SPECIFICATIONS
All measurements in inches unless noted

Year	Model	Brake Disc			Brake Drum			Minimum Lining Thickness		Caliper Mounting Bolts (ft. lbs.)
		Original Thickness	Minimum Thickness	Maximum Run-out	Original Inside Diameter	Max. Wear Limit	Maximum Machine Diameter	Front	Rear	
2002	Cavalier	0.786	0.751	0.002	NA	8.909	8.879	0.030	0.030	38
	Sunfire	0.786	0.751	0.002	NA	8.909	8.879	0.030	0.030	38
2003	Cavalier	0.786	0.751	0.002	9.060	9.094	9.075	0.030	0.030	38
	Sunfire	0.786	0.751	0.002	9.060	9.094	9.075	0.030	0.030	38
2004	Cavalier	0.786	0.736	0.002	9.060	9.094	9.075	0.030	0.030	38
	Sunfire	0.786	0.751	0.002	9.060	9.094	9.075	0.030	0.030	38
2005	Cavalier	0.806-0.796	0.736	0.002	9.060	9.094	9.075	0.030	0.030	38
	Sunfire	0.806-0.796	0.736	0.002	9.060	9.094	9.075	0.030	0.030	38

NA: Not Available

06025-JBOD-C11

SCHEDULED MAINTENANCE INTERVALS
GM J BODY—CHEVROLET CAVALIER & PONTIAC SUNFIRE

TO BE SERVICED	TYPE OF SERVICE	3	6	9	12	15	18	21	24	27	30	33	36	39
Engine oil & filter	R	✓	✓	✓	✓	✓	✓	✓	✓	✓	✓	✓	✓	✓
Coolant level, hoses & clamps	S/I	✓	✓	✓	✓	✓	✓	✓	✓	✓	✓	✓	✓	✓
Exhaust system & brake hoses	S/I	✓	✓	✓	✓	✓	✓	✓	✓	✓	✓	✓	✓	✓
Lubricate chassis and suspension	S/I	✓	✓	✓	✓	✓	✓	✓	✓	✓	✓	✓	✓	✓
Lubricate steering linkage and transaxle linkage	S/I	✓	✓	✓	✓	✓	✓	✓	✓	✓	✓	✓	✓	✓
Throttle linkage	S/I	✓	✓	✓	✓	✓	✓	✓	✓	✓	✓	✓	✓	✓
Brake linings	S/I	✓		✓		✓		✓		✓		✓		✓
Rotate tires	S/I	✓		✓		✓		✓		✓		✓		✓
Air filter element	R				✓				✓				✓	
Engine coolant ①	R													
Spark plugs ②	R				✓				✓				✓	
Accessory drive belt(s)	S/I				✓				✓				✓	
Automatic transaxle fluid & filter ③	S/I													
Fuel system	S/I				✓				✓				✓	
Ignition cables	R				✓				✓				✓	
Inspect throttle body bore & throttle plate for deposits	S/I				✓				✓				✓	
Supercharger oil	S/I				✓				✓				✓	

R: Replace

S/I: Service or Inspect

① Engine coolant: replace every 100,000 miles. Use O.E. specified (DEX-COOL™) coolant only. If any silicate coolant is used, the service interval is every 30,000 miles.

② Platinum tip spark plugs: replace every 100,000 miles.

③ Replace fuid every 50, 000 miles if driven in etxreme traffic or in places where the temperature exceeds 90 degrees F

FREQUENT OPERATION MAINTENANCE (SEVERE SERVICE)

If a vehicle is operated under any of the following conditions it is considered severe service:

- Extremely dusty areas.

- 50% or more of the vehicle operation is in 32°C (90°F) or higher temperatures, or constant operation in temperatures below 0°C (32°F).

- Prolonged idling (vehicle operation in stop and go traffic).

- Frequent short running periods (engine does not warm to normal operating temperatures).

- Police, taxi, delivery usage or trailer towing usage.

CV joints & front suspension components: service or inspect every 3000 miles.

Engine oil & filter change: change every 3000 miles.

Brake linings: check every 6000 miles.

Chassis lubrication: lubricate every 6000 miles.

Suspension, steering linkage, transaxle shift linkage, parking cable guides, underbody contact points: lubricate every 6000 miles.

Throttle body mount bolt torque: tighten at 6000 miles.

Air filter element: service or inspect every 15,000 miles.

Inspect throttle body bore & throttle plate for deposits: clean as required every 15,000 miles.

Rotate tires at 6000 miles, then every 15,000 miles.

06025-JBOD-C12

ENGINE REPAIR

Alternator

REMOVAL

2.2L (VIN F) Engines

1. Before servicing the vehicle, refer to the precautions section.
2. Remove or disconnect the following:
   - Negative battery cable
   - Accessory drive belt
   - Alternator mounting bolts
   - Alternator electrical connectors
   - Alternator

2.2L (VIN 4) Engines

1. Before servicing the vehicle, refer to the precautions section.
2. Remove or disconnect the following:
   - Negative battery cable
   - Accessory drive belt
   - Alternator mounting bolts
   - Alternator electrical connectors
   - Alternator

2.4L Engines

1. Before servicing the vehicle, refer to the precautions section.
2. Remove or disconnect the following:
   - Negative battery cable
   - Accessory drive belt
   - Alternator electrical connectors
   - Power steering line clip
   - Rear alternator brace
   - Alternator mounting bolts
   - Alternator

INSTALLATION

2.2L (VIN F) Engines

1. Install or connect the following:
   - Alternator. Torque the bolts to 16 ft. lbs. (22 Nm).
   - Alternator electrical connectors
   - Accessory drive belt
   - Negative battery cable

2.2L (VIN 4) Engines

1. Before servicing the vehicle, refer to the precautions section.
   - Alternator. Torque the upper bolts to 37 ft. lbs. (50 Nm) and the lower bolt to 22 ft. lbs. (30 Nm).
   - Alternator electrical connectors
   - Accessory drive belt
   - Negative battery cable

2.4L Engines

   - Alternator
   - Alternator mounting bolts. Torque the bolts to 37 ft. lbs. (50 Nm).
   - Rear alternator brace
   - Power steering line clip
   - Alternator electrical connectors
   - Accessory drive belt
   - Negative battery cable

Ignition Timing

ADJUSTMENT

Ignition timing is controlled by the Powertrain Control Module (PCM). No adjustment is necessary or possible.

Engine Assembly

REMOVAL & INSTALLATION

With A Manual Transmission

2002 2.2L ENGINE

1. Before servicing the vehicle, refer to the precautions section.
2. Relieve the fuel system pressure.
3. Drain the cooling system.
4. Remove or disconnect the following:
   - Both battery cables
   - Throttle body air inlet duct
   - Upper radiator hose
   - Vacuum hose from the power brake booster

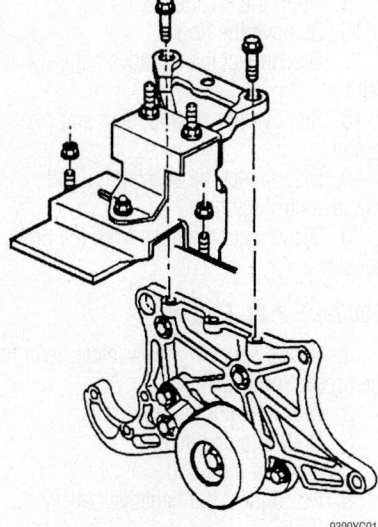

Exploded view of the upper engine mount—2002 2.2L engine

9300YG01

   - Idle Air Control (IAC) valve electrical connector
   - Alternator electrical connectors
   - Throttle Position (TPS) sensor electrical connector
   - Manifold Absolute Pressure (MAP) sensor electrical connector
   - EVAP emission solenoid electrical connection
   - Fuel injector electrical connectors
   - Exhaust Gas Recirculation (EGR) valve electrical connector
   - Engine Coolant Temperature (ECT) sender electrical connector
   - Oxygen (O_2S) sensor electrical connector
   - Engine grounds electrical connection
   - Accessory drive belt
   - Shift control cable
   - Coolant surge tank
   - Vacuum line near the master cylinder

5. Install an engine support fixture.
   - Lower radiator hose
   - Wheels
   - Wheel well splash shields
   - Exhaust pipe from the manifold
   - Engine mount strut
   - Wheel Speed Sensor (WSS) electrical connection
   - Ball joints from the steering knuckles
   - Tie rod ends from the steering knuckles
   - Brake lines from the suspension support
   - Air conditioning compressor. DO NOT disconnect the lines.
   - Ignition Control Module (ICM) electrical connector
   - Vehicle Speed Sensor (VSS) electrical connector
   - Cooling fan electrical connector
   - Starter electrical connector
   - Oil pressure sensor electrical connector
   - Power steering lines from the rack and pinion
   - Intermediate shaft
   - Accelerator and cruise control cables
   - Rack and pinion bolts
   - Suspension support extension brace
   - Stabilizer bar
   - Suspension support
   - Heater hoses
   - Halfshafts from the transmission

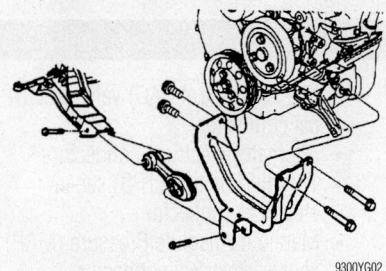

Exploded view of the lower engine mount and strut—2002 2.2L engine

9300YG02

- Fuel lines
- Transmission mount
- Engine mount
6. Remove the engine support fixture.
7. Remove or disconnect the following:
- Engine and transmission assembly from the vehicle
- Engine from the transmission

To install:

8. Install or connect the following:
- Transaxle to the engine. Torque the bolts to 55 ft. lbs. (75 Nm).
- Engine and transmission assembly to the vehicle
- Engine support fixture
- Engine mount. Torque the bolts to 49 ft. lbs. (66 Nm).
- Transmission mount. Torque the bolts to 49 ft. lbs. (66 Nm).
- Halfshafts
- Heater hoses
- Fuel lines
- Suspension support. Torque the bolts to 71 ft. lbs. (110 Nm) plus an additional 90 degree turn.
- Stabilizer bar. Torque the bolts to 49 ft. lbs. (66 Nm).
- Suspension support extension brace. Torque the bolts to 53 ft. lbs. (72 Nm).
- Rack and pinion bolts. Torque the bolts to 89 ft. lbs. (120 Nm).
- Accelerator and cruise control cables
- Intermediate shaft. Torque the pinch bolt to 30 ft. lbs. (41 Nm).
- Power steering lines
- Ignition module electrical connector
- VSS electrical connector
- Cooling fan electrical connector
- Starter electrical connector
- Oil pressure sensor electrical connector
- A/C compressor. Torque the front bolts to 33 ft. lbs. (45 Nm) and the rear bolt to 48 ft. lbs. (65 Nm).
- Brake lines to the suspension support

- Tie rod ends to the steering knuckles. Torque the nuts to 7 ft. lbs. (10 Nm) plus an additional 120 degree turn.
- Ball joints to the steering knuckles. Torque the nuts to 46 ft. lbs. (62 Nm).
- WSS electrical connector
- Engine mount strut. Torque the bolts to 74 ft. lbs. (100 Nm) plus an additional turn of 90 degrees.
- Exhaust pipe to the manifold. Torque the bolts to 26 ft. lbs. (35 Nm).
- Wheel well splash shields
- Wheels
- Lower radiator hose
9. Remove the engine support fixture.
10. Install or connect the following:
- Coolant surge tank. Torque the bolts to 89 inch lbs. (10 Nm).
- Shift control cable
- Accessory drive belt
- O_2S electrical connector
- ECT sender electrical connector
- EGR valve electrical connector
- Fuel injector electrical connectors
- MAP sensor electrical connector
- TPS sensor electrical connector
- Alternator electrical connector
- IAC electrical connector
- Brake booster vacuum hose
- Upper radiator hose
- Throttle body air inlet duct
- Both battery cables
11. Refill the cooling system.
12. Start the engine and check for proper operation.
13. Before servicing the vehicle, refer to the precautions section.
14. Drain the cooling system.
15. Drain the engine oil.
16. Remove the hood.
17. Disconnect the negative battery cable.
18. Remove the air inlet duct and resonator
19. Disconnect the accelerator and cruise control cable.
20. Disconnect the hose from the brake booster.

2003–05 2.2L ENGINE

1. Before servicing the vehicle, refer to the precautions section.
2. Drain the cooling system.
3. Drain the engine oil.
4. Remove the hood.
5. Disconnect the negative battery cable.
6. Remove the air inlet duct and resonator

7. Disconnect the accelerator and cruise control cable.
8. Disconnect the hose from the brake booster.
9. Remove the power steering pump bolts and move the pump aside.
10. Disconnect the fuel lines.
11. Disconnect the transmission shift control cables.
12. Disconnect the clutch actuator cylinder from the transmission.
13. Remove the radiator inlet hose.
14. Disconnect the hose from the surge tank to the cylinder head.
15. Disconnect the outlet hose from the surge tank to the radiator.
16. Remove the bolt attaching the cooling system fill hose to the intake manifold.
17. Remove the radiator outlet hose.
18. Disconnect the heater hoses.
19. Disconnect the following electrical connectors:
- Idle Air Control (IAC) motor
- Throttle Position (TPS) sensor
- Manifold Absolute Pressure (MAP) sensor
- Crankshaft position (CKP) sensor
- Camshaft (CMP) sensor
- Oil pressure sensor
- Purge solenoid
- Ignition coil and module assembly
- Oxygen (O_2S) sensor
- Vehicle Speed (VSS) Sensor
- Engine Coolant Temperature (ECT) sensor
- Back-up lamp switch
20. Disconnect the electrical harness from the engine and set the harness aside.
21. Remove the front suspension crossmember .
22. Remove the front wheel drive shafts.
23. Remove the accessory drive belt.
24. Remove the A/C compressor bolts and set the compressor aside.
25. Disconnect the alternator electrical connectors.
26. Disconnect the starter electrical connectors.
27. Disconnect the front exhaust pipe from the manifold.
28. Use a block of wood to support the front of the engine at the front of the oil pan.
29. Lower the vehicle onto an engine support table.
30. Remove the bolts which attach the bracket to the engine mount.
31. Remove the bolts which attach the transmission mount to the frame.
32. Raise the vehicle away from the engine and transmission assembly.
33. Install engine lift bracket J 4251 to the right rear of the cylinder head.

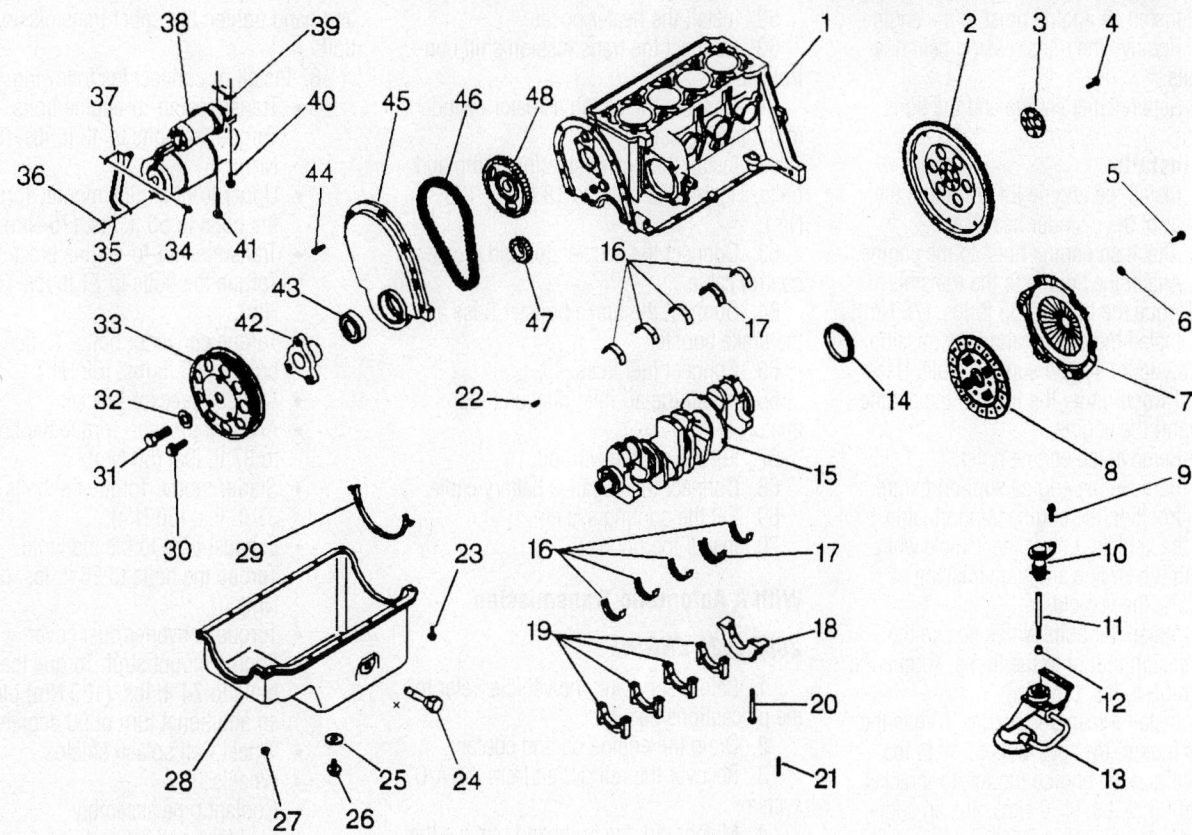

1 Cylinder Block
2 Flywheel
3 Flywheel Retainer
4 Flywheel Bolt
5 Clutch Pressure Plate Bolt
6 Clutch Pressure Plate Washer
7 Clutch Pressure Plate
8 Clutch Disc
9 Oil Pump Drive Bolt
10 Oil Pump Drive
11 Oil Pump Drive Shaft
12 Oil Pump Drive Shaft Retainer
13 Oil Pump
14 Crankshaft Rear Oil Seal
15 Crankshaft
16 Crankshaft Bearing
17 Crankshaft Thrust Bearing
18 Main Bearing Cap
19 Main Bearing Cap
20 Crankshaft Bearing Cap Bolt
21 Oil Pan Stud
22 Connecting Rod Nut
23 Oil Pan Bolt
24 Connecting Rod Nut

25 Oil Pan Drain Plug Gasket
26 Oil Pan Drain Plug
27 Oil Pan Nut
28 Oil Pan
29 Oil Pan Rear Seal
30 Crankshaft Pulley Bolt
31 Crankshaft Pulley Hub Bolt
32 Crankshaft Pulley Hub Bolt Washer
33 Crankshaft Pulley
34 Bracket Bolt
35 Starter Motor Bracket Nut
36 Starter Motor Bracket Nut
37 Starter Motor Bracket
38 Starter Motor
39 Starter Motor Shim
40 Starter Motor Bolt
41 Starter Motor Bolt
42 Crankshaft Pulley Hub
43 Crankshaft Front Oil Seal
44 Crankcase Front Cover Bolt
45 Crankcase Front Cover
46 Timing Chain
47 Crankshaft Sprocket
48 Camshaft Sprocket

9300YG06

Exploded view of the crankshaft and related components—2002 2.2L engine

34. Install an engine hoist to the engine.
35. Remove the transmission bellhousing bolts.
36. Separate the engine and the transmission.

To install:

37. Install the engine lift bracket to the right rear of the cylinder head.
38. Attach an engine hoist to the engine.
39. Attach the engine to the transmission. Torque the bolts to 55 ft. lbs. (75 Nm).
40. Install the engine and transmission assembly to an engine support table. Use a block of wood under the front of the engine to support the engine.
41. Remove the engine hoist.
42. Remove the engine support fixture.
43. Position the engine support table under the vehicle. Lower the vehicle while aligning the engine and transmission mounts to the vehicle.
44. Install the bolts which secure the transmission mount to the frame. Torque the bolts to 55 ft. lbs. (75 Nm).
45. Install the engine mount. Torque the engine mount–to–body nuts to 49 ft. lbs. (66 Nm) and the engine mount–to–bracket bolts to 96 ft. lbs. (130 Nm) plus an additional 45 degrees
46. Connect the front exhaust pipe to the manifold.
47. Connect the alternator electrical connectors.
48. Connect the starter electrical connectors.
49. Install the A/C compressor to the engine.
50. Install the engine drive belt.
51. Install the front wheel drive shafts.
52. Install the front suspension cross member.
53. Connect the following electrical connectors:
   - IAC motor
   - TPS
   - MAP sensor
   - CKP sensor
   - CMP sensor
   - Oil pressure sensor
   - Purge solenoid
   - Ignition coil and module assembly
   - O_2 sensor
   - VSS
   - ECT
54. Install the radiator inlet hose.
55. Connect the hose from the surge tank to the cylinder head.
56. Connect the outlet hose from the surge tank to the radiator.
57. Install the bolt attaching the cooling system fill hose to the intake manifold.
58. Install the radiator outlet hose.

59. Install the heater hoses.
60. Connect the transmission shift control cables.
61. Connect the clutch actuator cylinder to the transmission.
62. Install the power steering pump and bolts. Tighten the bolts to 18 ft. lbs. (25 Nm).
63. Connect the accelerator and cruise control cable.
64. Connect the brake booster hose at the brake booster.
65. Connect fuel lines.
66. Install the air inlet duct and resonator.
67. Fill the engine with oil.
68. Connect the negative battery cable.
69. Fill the cooling system.
70. Install the hood.

With A Automatic Transmission

2002 2.2L ENGINE

1. Before servicing the vehicle, refer to the precautions section.
2. Drain the engine oil and coolant.
3. Recover the refrigerant from the A/C system.
4. Matchmark the bolts and remove the hood.
5. Remove or disconnect the following:
   - Both battery cables
   - Air cleaner assembly
   - Brake booster vacuum line
   - Fuel injector sight shield
   - Accelerator and cruise control cables
   - Engine wiring harness
   - Cruise control module
   - A/C line from the accumulator
   - Accessory drive belt
   - Coolant surge tank
   - Upper and lower radiator hoses
   - Fuel feed and return lines from the engine
   - Coolant pipe assembly
   - Wheels
   - Wheel well splash shields
   - Engine mount strut
   - Torque converter dust cover
   - Exhaust pipe from the manifold
   - Starter motor
   - A/C compressor and bracket
   - Torque converter bolts
   - Transmission-to-engine support brace
   - Upper front engine mount
   - Transmission-to-engine bolts
   - Transmission from the engine
6. Lift the engine from the vehicle.

To install:

7. Position the engine in the vehicle,

and hand tighten the upper transmission bolts.
8. Install or connect the following:
   - Transmission-to-engine bolts. Torque the bolts to 43 ft. lbs. (58 Nm).
   - Upper front engine mount. Torque the bolts to 55 ft. lbs. (75 Nm).
   - Transmission-to-engine brace. Torque the bolts to 71 ft. lbs. (96 Nm).
   - Torque converter bolts. Torque the bolts to 46 ft. lbs. (62 Nm).
   - A/C compressor bracket
   - A/C compressor. Torque the bolts to 37 ft. lbs. (50 Nm).
   - Starter motor. Torque the bolts to 37 ft. lbs. (50 Nm).
   - Exhaust pipe to the manifold. Torque the bolts to 26 ft. lbs. (35 Nm).
   - Torque converter dust cover
   - Engine mount strut. Torque the bolts to 74 ft. lbs. (100 Nm) plus an additional turn of 90 degrees.
   - Wheel well splash shields
   - Wheels
   - Coolant pipe assembly
   - Fuel feed and return lines
   - Upper and lower radiator hoses
   - Coolant surge tank. Torque the bolts to 89 inch lbs. (10 Nm).
   - Accessory drive belt
   - A/C line to the accumulator
   - Cruise control module
   - Engine wiring harness
   - Accelerator and cruise control cables
   - Brake booster vacuum line
   - Air cleaner assembly
   - Hood. Torque the bolts to 15 ft. lbs. (20 Nm).
   - Battery cables
9. Evacuate and recharge the A/C system.
10. Refill the engine oil and coolant.
11. Start the engine and check for leaks.
12. Install the fuel injector sight shield.

2003–05 2.2L ENGINE

1. Before servicing the vehicle, refer to the precautions section.
2. Drain the cooling system.
3. Drain the engine oil.
4. Remove the hood.
5. Disconnect the negative battery cable.
6. Remove the air inlet duct and resonator
7. Disconnect the accelerator and cruise control cable.
8. Disconnect the hose from the brake booster.

9. Remove the power steering pump bolts and move the pump aside.

10. Disconnect the fuel lines.

11. Disconnect the transmission shift control cables.

12. Disconnect the clutch actuator cylinder from the transmission.

13. Remove the radiator inlet hose.

14. Disconnect the hose from the surge tank to the cylinder head.

15. Disconnect the outlet hose from the surge tank to the radiator.

16. Remove the bolt attaching the cooling system fill hose to the intake manifold.

17. Remove the radiator outlet hose.

18. Disconnect the heater hoses.

19. Disconnect the following electrical connectors:

- Idle Air Control (IAC) motor
- Throttle Position (TPS) sensor
- Manifold Absolute Pressure (MAP) sensor
- Crankshaft position (CKP) sensor
- Camshaft (CMP) sensor
- Oil pressure sensor
- Purge solenoid
- Ignition coil and module assembly
- Oxygen (O_2S) sensor
- Vehicle Speed (VSS) Sensor
- Engine Coolant Temperature (ECT) sensor
- Back-up lamp switch

20. Disconnect the electrical harness from the engine and set the harness aside.

21. Remove the accessory drive belt.

22. Remove the A/C compressor bolts and set the compressor aside.

23. Remove the crankshaft balancer.

24. Disconnect the alternator electrical connectors.

25. Disconnect the starter electrical connectors.

26. Disconnect the front exhaust pipe from the manifold.

27. Remove the front wheel drive shafts.

28. Remove the starter.

29. Remove the flywheel to torque converter bolts.

30. Remove the lower transmission bellhousing bolts.

31. Remove the transmission–to–engine brace.

32. Install engine lift bracket J 4251 to the right rear of the cylinder head.

33. Install an engine hoist to the engine.

34. Remove the bolts which attach the bracket to the engine mount.

35. Remove the bolts which attach the transmission mount to the frame.

36. Raise the vehicle away from the engine and transmission assembly.

37. Remove the transmission bellhousing bolts.

38. Separate the engine and the transmission.

39. Remove the engine from the vehicle. Heated Oxygen (HO$_2$S) sensor

To install:

40. Install the engine lift bracket to the right rear of the cylinder head.

41. Attach an engine hoist to the engine.

42. Install the upper engine to the transmission bolts. Torque the bolts to 66 ft. lbs. (90 Nm).

43. Install the engine and transmission assembly.

44. Remove the engine hoist.

45. Remove the engine support fixture.

46. Install the lower engine to the transmission bolts. Torque the bolts to 66 ft. lbs. (90 Nm).

47. Install the torque converter bolts and tighten to 46 ft. lbs. (62 Nm).

48. Install the starter.

49. Install the brace that attaches the transmission to the engine. Tighten the brace bolts to 53 ft. lbs. (72 Nm).

50. Connect the front exhaust pipe to the manifold.

51. Connect the alternator electrical connectors.

52. Connect the starter electrical connectors.

53. Install the A/C compressor to the engine.

54. Install the engine drive belt.

55. Install the front wheel drive shafts.

56. Connect the following electrical connectors:

- IAC motor
- TPS
- MAP sensor
- CKP sensor
- CMP sensor
- Oil pressure sensor
- Purge solenoid
- Ignition coil and module assembly
- O$_2$ sensor
- VSS
- ECT

57. Install the radiator inlet hose.

58. Connect the hose from the surge tank to the cylinder head.

59. Connect the outlet hose from the surge tank to the radiator.

60. Install the bolt attaching the cooling system fill hose to the intake manifold.

61. Install the radiator outlet hose.

62. Install the heater hoses.

63. Connect the transmission shift control cables.

64. Connect the clutch actuator cylinder to the transmission.

65. Install the power steering pump and bolts. Tighten the bolts to 18 ft. lbs. (25 Nm).

66. Connect the accelerator and cruise control cable.

67. Connect the brake booster hose at the brake booster.

68. Connect fuel lines.

69. Install the air inlet duct and resonator.

70. Fill the engine with oil.

71. Connect the negative battery cable .

72. Fill the cooling system.

73. Install the hood.

2.4L Engine

WITH MANUAL TRANSMISSION

1. Before servicing the vehicle, refer to the precautions section.

2. Discharge and recover the air conditioning refrigerant.

3. Properly drain the cooling system and engine oil.

4. Relieve the fuel system pressure.

5. Remove or disconnect the following:

- Both battery cables
- Clutch pushrod from the pedal assembly
- Heater hose at the thermostat housing
- Upper radiator hose
- Air cleaner assembly
- Coolant fan
- Refrigerant hose assembly at the compressor
- Both vacuum hoses from the front of the engine
- Alternator electrical connector
- Air conditioning compressor electrical connector
- Fuel injector harness
- Idle Air Control (IAC) electrical connector
- Throttle Position (TPS) sensor electrical connector
- Manifold Absolute Pressure (MAP) sensor electrical connector

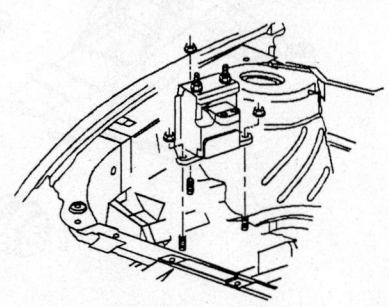

9300YG03

Exploded view of the upper engine mount—2.4L engine

- Intake Air Temperature (IAT) sensor electrical connector
- Evaporative Emissions (EVAP) canister purge solenoid electrical connector
- Negative battery cable from the transaxle
- Electronic ignition module electrical connector
- Engine Coolant Temperature (ECT) sensors electrical connectors
- Oil pressure sensor/switch electrical connector
- Oxygen (O2S) sensor electrical connector
- Crankshaft Position (CKP) sensor electrical connector
- Back-up lamp switch electrical connector
- Power brake vacuum hose from the throttle body
- Throttle cable and bracket
- Power steering pump and move it aside with the lines attached
- Fuel lines
- Shift cables
- Clutch actuator line
- Exhaust manifold and heat shield
- Lower radiator hose

6. Install an engine support fixture.
7. Remove or disconnect the following:
- Coolant recovery/surge tank, move it aside with the hoses attached
- Engine strut mount
- Front wheels
- Right splash shield
- Vehicle Speed Sensor (VSS) electrical connector

- Knock Sensor (KS) electrical connector
- Starter solenoid electrical connector
- Both front Anti-lock Brake System (ABS) wheel speed sensor electrical connectors
- Engine mount strut
- Transmission mount
- Ball joints from the steering knuckles
- Suspension supports, crossmember and stabilizer shaft as an assembly
- Heater outlet hose from the radiator outlet pipe
- Halfshaft from the transmission and intermediate shaft
- Flywheel housing cover

8. Position a suitable support below the engine, then carefully lower the vehicle onto the support.
9. Raise the vehicle slowly off the engine/transaxle assembly.

➡**It may be necessary to move the engine/transaxle assembly rearward to clear the intake manifold.**

10. Remove the engine from the transaxle.
To install:

❄❄ WARNING

Be sure the retaining bolts are in their correct locations. If not, engine damage may occur.

11. Install or connect the following:
- Engine to the transaxle. Torque the mounting bolts to 55 ft. lbs. (75 Nm).
- Engine/transaxle assembly under the vehicle, then lower the vehicle over the assembly
- Engine support fixture, making sure to adjust it to the previous setting
- Halfshafts to the transaxle
- Heater outlet hose to the radiator outlet pipe
- Suspension supports, crossmember and stabilizer shaft assembly. Torque the bolts to 53 ft. lbs. (72 Nm).
- Ball joints to the steering knuckles. Torque the nuts to 44 ft. lbs. (60 Nm) plus an additional turn of 180 degrees.
- Engine mount assembly. Torque the bolts to 46 ft. lbs. (62 Nm).
- Transmission mount. Torque the bolts to 66 ft. lbs. (90 Nm).
- Both front ABS wheel speed sensor electrical connectors
- Starter solenoid electrical connector
- KS electrical connector
- VSS electrical connector
- Right splash shield
- Front wheels

12. Carefully raise the vehicle off the support.
13. Install or connect the following:
- Engine strut mount. Torque the bolts to 74 ft. lbs. (100 Nm).
- Coolant recovery/surge tank
- Lower radiator hose
- Exhaust manifold. Torque the nuts to 31 ft. lbs. (42 Nm).
- Exhaust manifold heat shield. Torque the bolts to 19 ft. lbs. (26 Nm).
- Clutch actuator line
- Shift cables
- Fuel lines
- Power steering pump. Torque the bolts to 22 ft. lbs. (30 Nm).
- Throttle cable and bracket
- Vacuum hoses to the intake manifold and the brake booster
- Back-up lamp switch electrical connector
- CKP sensor electrical connector
- O2S electrical connector
- Oil pressure sensor/switch electrical connector
- ECT sensors electrical connectors
- Electronic ignition module electrical connector
- Negative battery cable to the transaxle

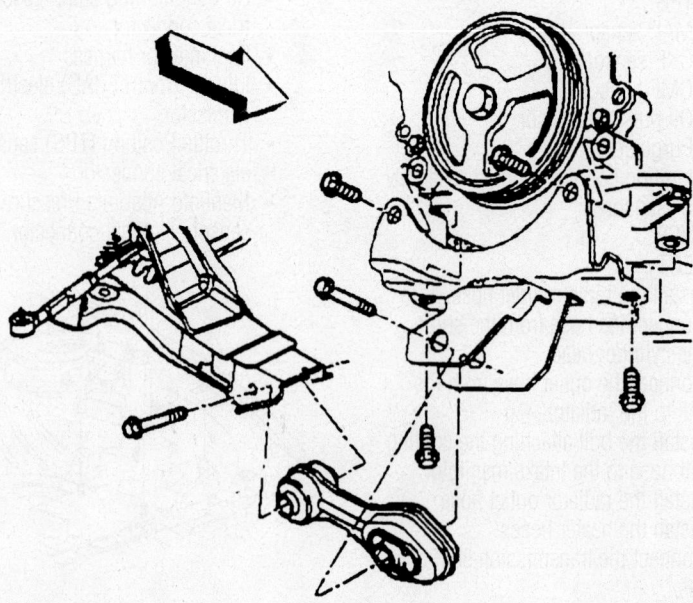

9300YG04

Exploded view of the lower engine mount and strut—2.4L engine

- EVAP canister purge solenoid electrical connector
- IAT sensor electrical connector
- MAP sensor electrical connector
- TPS sensor electrical connector
- IAC sensor electrical connector
- Fuel injector harness
- Air conditioning compressor electrical connector
- Alternator electrical connector

- Refrigerant hose assembly to the compressor
- Coolant fan
- Air cleaner assembly
- Upper radiator hose
- Heater hose at the thermostat housing
- Clutch pushrod to the pedal assembly
- Both battery cables

14. Refill the transaxle and the crankcase.
15. Evacuate and recharge the air conditioning system.
16. Refill the cooling system.
17. Start the engine and check for leaks.

WITH AUTOMATIC TRANSMISSION

1. Before servicing the vehicle, refer to the precautions section.

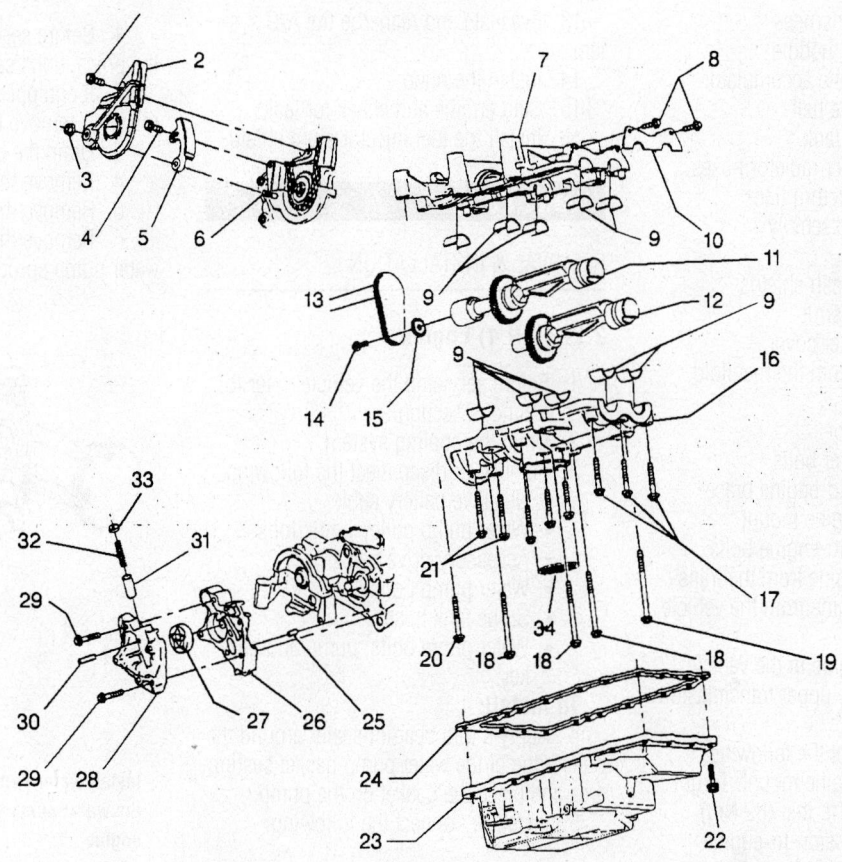

1 Balance Shaft Chain Cover Bolt
2 Balance Shaft Chain Cover
3 Balance Shaft Guide Adjuster Nut
4 Balance Shaft Chain Guide Adjuster Bolt
5 Balance Shaft Chain Adjuster Guide
6 Stud
7 Balance Shaft Upper Housing
8 Balance Shaft Thrust Plate Retainer Bolt
9 Balance Shaft Bearings
10 Balance Shaft Thrust Plate Retainer
11 Balance Shaft Drive
12 Balance Shaft Driven
13 Balance Shaft Drive Chain
14 Driven Sprocket Bolt-Note: Left Hand Thread
15 Balance Shaft Driven Sprocket
16 Balance Shaft Lower Housing

17 Balance Shaft Housing Bolt
18 Balance Shaft Housing to Block Bolt
19 Balance Shaft Housing to Block Bolt
20 Balance Shaft Housing to Block Bolt
21 Balance Shaft Housing Bolt
22 Oil Pan Bolt
23 Oil Pan
24 Oil Pan Gasket
25 Dowel Pin
26 Oil Pump Body
27 Gerotor
28 Oil Pump Cover
29 Oil Pump Cover to Body Bolt
30 Relief Valve Pin
31 Relief Valve Spring Guide
32 Relief Valve Spring
33 Relief Valve
34 Pickup Screen

9300YG09

Exploded view of the balance shafts and related components—2.4L engine

2. Matchmark the bolts and remove the hood.

3. Drain the engine coolant.

4. Recover the refrigerant.

5. Drain the engine oil.

6. Relieve the fuel system pressure.

7. Remove or disconnect the following:
- Air cleaner assembly
- Brake booster vacuum line
- Fuel injector sight shield
- Accelerator and cruise control cables
- Engine wiring harness
- Cruise control module
- A/C line from the accumulator
- Accessory drive belt
- Coolant surge tank
- Upper and lower radiator hoses
- Fuel feed and return lines
- Coolant pipe assembly
- Wheels
- Wheel well splash shields
- Engine mount strut
- Torque converter cover
- Exhaust pipe from the manifold
- Starter motor
- A/C compressor
- Torque converter bolts
- Transmission-to-engine brace
- Upper front engine mount
- Transmission-to-engine bolts

8. Separate the engine from the transmission and lift the engine from the vehicle.

To install:

9. Position the engine in the vehicle.

10. Hand tighten the upper transmission bolts.

11. Install or connect the following:
- Upper front engine mount. Torque the bolts to 46 ft. lbs. (62 Nm).
- Lower transmission-to-engine bolts. Torque all bolts to 55 ft. lbs. (75 Nm).
- Transmission-to-engine brace. Torque the bolts to 71 ft. lbs. (96 Nm).
- Torque converter bolts. Torque the bolts to 46 ft. lbs. (62 Nm).
- A/C compressor. Torque the bolts to 37 ft. lbs. (50 Nm).
- Starter motor. Torque the bolts to 37 ft. lbs. (50 Nm).
- Exhaust pipe to the manifold. Torque the bolts to 26 ft. lbs. (35 Nm).
- Torque converter cover
- Engine mount strut. Torque the bolts to 74 ft. lbs. (100 Nm).
- Wheel well splash shields
- Wheels
- Coolant pipe assembly
- Fuel feed and return lines

- Upper and lower radiator hoses
- Coolant surge tank
- Accessory drive belt
- A/C line to the accumulator
- Cruise control module
- Engine wiring harness
- Accelerator and cruise control cables
- Brake booster vacuum line
- Air cleaner assembly
- Both battery cables

12. Refill the crankcase and cooling system.

13. Evacuate and recharge the A/C system.

14. Install the hood.

15. Start engine and check for leaks.

16. Install the fuel injector sight shield.

Water Pump

REMOVAL & INSTALLATION

2.2L (VIN 4) Engine

1. Before servicing the vehicle, refer to the precautions section.

2. Drain the cooling system.

3. Remove or disconnect the following:
- Negative battery cable
- Water pump pulley bolts, loosen
- Accessory drive belt
- Water pump pulley
- Surge tank hose
- Water pump bolts, pump and gasket

To install:

4. Apply a thin bead of sealer around the outer edge of the water pump gasket seating area and place the gasket on the pump.

5. Install or connect the following:

- Water pump. Torque the bolts to 18 ft. lbs. (25 Nm).
- Water pump pulley and hand-tighten the bolts
- Surge tank hose
- Accessory drive belt
- Water pump pulley bolts. Torque the bolts to 22 ft. lbs. (30 Nm).

6. Connect the negative battery cable.

7. Refill and bleed the cooling system.

2.2L (VIN F) Engine

1. Before servicing the vehicle, refer to the precautions section.

2. If equipped with an automatic transmission, remove the exhaust manifold.

3. Drain the cooling system.

4. Remove the right front tire.

5. Remove the front fender liner.

6. Remove the access plate on the water pump sprocket from the timing cover.

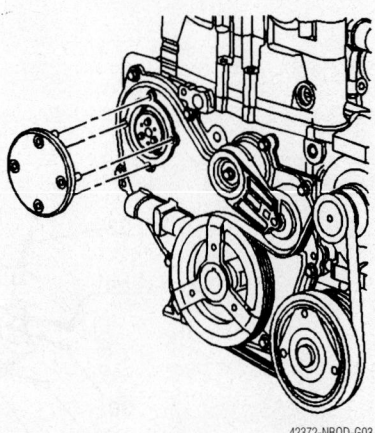

42372-NBOD-G03

Install water pump holding tool J 43651 to the water pump sprocket—2.2L (VIN F) engine

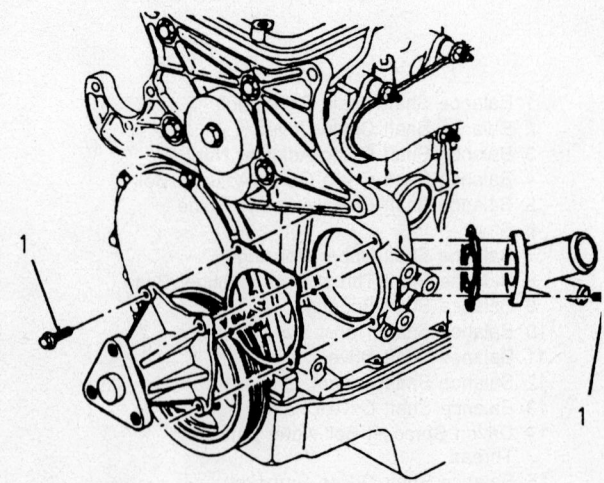

1 BOLT - 25 N·m (18 LBS. FT.)

7922YG01

Exploded view of the water pump mounting—2.2L (VIN 4) engine

7. Install water pump holding tool J 43651 to the water pump sprocket.

8. Use the access plate bolts to secure the pump holding tool to the engine front cover.

9. Remove the bolts that secure the sprocket to the water pump.

10. Remove the bolt attaching the engine block to the water pump.

11. Remove the bolt attaching the engine front cover to the water pump.

12. Remove the feed pipe that joins the thermostat housing to the water pump.

13. Remove the 2 bolts that secure the water pump to the engine block.

14. Remove the water pump.

To install:

15. Use a threaded stud in the hub to align the hub to the water pump sprocket.

16. Install the water pump.

➡**Make sure to use the correct fastener in the correct location. Replacement fasteners must be the correct part number.**

17. Install the 2 bolts that secure the water pump to the engine block and tighten to 15 ft. lbs. (20 Nm).

18. Install the feed pipe that joins the thermostat housing to the water pump.

19. Install the engine front cover bolt and the bolt that attaches the engine block to the water pump. Torque the bolts to 15 ft. lbs. (20 Nm)

20. Install 2 of the bolts that secure the water pump sprocket to the water pump.

21. Remove the threaded stud.

22. Install the last bolt and tighten to 89 inch lbs. (10 Nm).

23. Remove the holding tool.

24. Install the access plate on the water pump sprocket to the timing cover.

25. Install the bolts that secure the access plate to the timing cover. Torque the bolts to 89 inch lbs. (10 Nm) on 2002–04

models or 62 inch lbs. (7 Nm) on 2005 models.

26. Install the front fender liner and wheel.

27. If equipped with an automatic transmission, install the exhaust manifold.

28. Fill the cooling system.

2.4L Engine

1. Before servicing the vehicle, refer to the precautions section.

2. Drain the cooling system.

3. Remove or disconnect the following:

- Negative battery cable
- Oxygen (O_2S) sensor electrical connector
- Upper exhaust manifold heat shield
- Exhaust manifold brace-to-manifold bolt
- Manifold-to-exhaust pipe spring loaded bolts
- Both radiator outlet pipe-to-water pump cover bolts
- Exhaust manifold pipe

❊❊ WARNING

DO NOT rotate the flex coupling more than 4 degrees or damage may occur.

- Radiator outlet pipe from the oil pan
- Brake vacuum pipe from the camshaft housing
- Exhaust manifold from the cylinder head
- Front cover
- Timing chain tensioner
- Water pump cover and water pump as an assembly

4. Separate the water pump from the water pump cover.

To install:

5. Install or connect the following:

- Water pump with a new gasket to the cover and tighten the bolts finger-tight

➡**Lubricate the splines with clean grease**

- Water pump and cover assembly using new gaskets and tighten the fasteners finger-tight

6. Torque the bolts in the following sequence;

a. Pump assembly-to-chain housing nuts: 19 ft. lbs. (26 Nm).

b. Pump cover-to-pump assembly: 124 inch lbs. (14 Nm).

c. Cover-to-block, bottom bolt first: 19 ft. lbs. (26 Nm).

7. Install or connect the following:

- Timing chain tensioner
- Front cover. Torque the bolts to 97 inch lbs. (11 Nm).
- Exhaust manifold with a new gasket. Torque the bolts to 31 ft. lbs. (42 Nm).
- Brake vacuum pipe to the camshaft housing
- Radiator outlet pipe to the water pump cover. Torque the bolts to 125 inch lbs. (14 Nm).
- Radiator outlet pipe to the oil pan. Torque the bolts to 18 ft. lbs. (25 Nm).
- Exhaust pipe to the manifold. Tighten the exhaust pipe flange bolts evenly and gradually to avoid binding, until fully seated. Torque the bolts to 26 ft. lbs. (35 Nm).
- Exhaust manifold brace to the manifold. Torque the bolts to 41 ft. lbs. (56 Nm).
- Upper heat shield. Torque the bolts to 125 inch lbs. (14 Nm).
- O_2S electrical connector
- Negative battery cable

8. Refill the cooling system until it comes out of the heater hose outlet; then connect the heater hose.

9. Start the engine. Run the vehicle until the thermostat opens, refill the radiator and recovery tank, then turn the engine **OFF**.

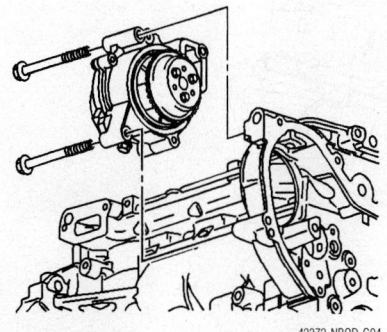

42372-NB0D-G04

Exploded view of the water pump—2.2L (VIN F) engine

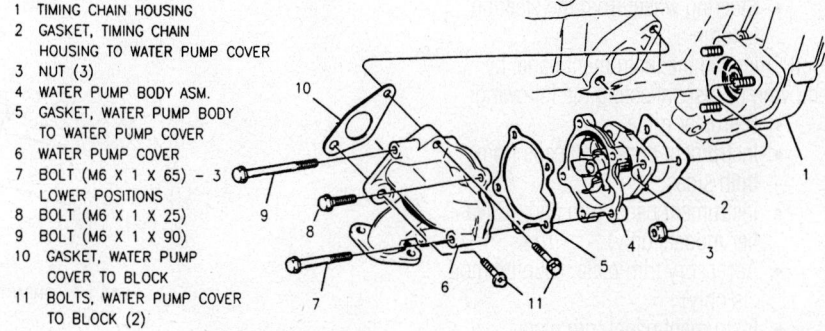

1 TIMING CHAIN HOUSING
2 GASKET, TIMING CHAIN HOUSING TO WATER PUMP COVER
3 NUT (3)
4 WATER PUMP BODY ASM.
5 GASKET, WATER PUMP BODY TO WATER PUMP COVER
6 WATER PUMP COVER
7 BOLT (M6 X 1 X 65) – 3 LOWER POSITIONS
8 BOLT (M6 X 1 X 25)
9 BOLT (M6 X 1 X 90)
10 GASKET, WATER PUMP COVER TO BLOCK
11 BOLTS, WATER PUMP COVER TO BLOCK (2)

7922YG02

Exploded view of the water pump mounting and related components—2.4L engine

10. Once the vehicle has cooled, recheck the coolant level.

Heater Core

REMOVAL & INSTALLATION

2002 Models

1. Before servicing the vehicle, refer to the precautions section.

2. Disable the SIR system by performing the following procedure:

a. Point the wheel in the straight-ahead position.

b. Turn the ignition switch to the LOCK position.

c. Remove the AIR BAG fuse from the fuse block.

d. At the base of the steering column, remove the left sound insulator.

e. At the base of the steering column, disconnect the Connector Position Assurance (CPA), the yellow 2-way electrical connectors and the passenger's side module electrical connector.

3. Disconnect the negative battery cable.

4. Drain the cooling system into a clean container for reuse.

5. Disconnect the heater hoses from the heater core.

6. If equipped, remove the Diagnostic Energy Reserve Module (DERM) with attaching brackets.

7. Remove the steering wheel by removing or disconnecting the following:

- SIR module-to-steering wheel screws
- SIR module and disconnect the electrical connector

❄❄ CAUTION

Place the SIR module in a safe place with the front facing upward.

- Steering wheel-to-steering column nut
- Steering wheel from the steering column

8. Remove the instrument panel by removing or disconnecting the following:

- Defroster grille
- Instrument panel end caps from both sides
- Instrument panel trim pad, (Cavalier models only)
- Accessory trim plate, (Sunfire models only)
- Instrument panel trim plate

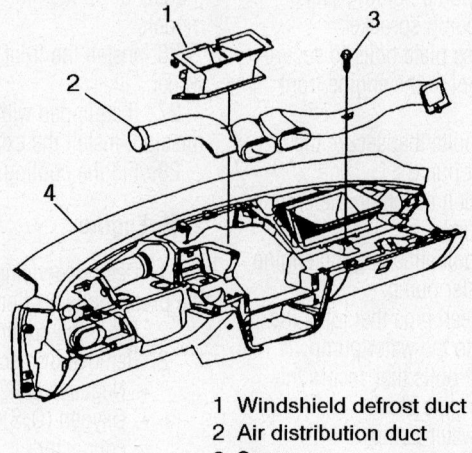

1 Windshield defrost duct
2 Air distribution duct
3 Screw
4 Lower I/P

87950083

Air distribution duct mounting—2002 Cavalier shown

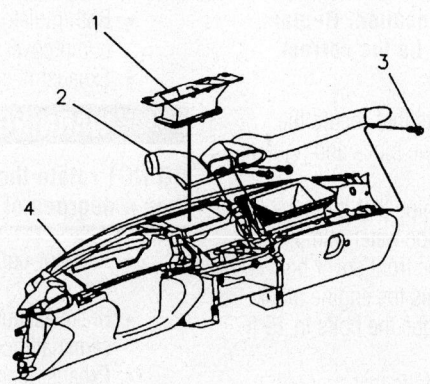

1 DUCT, WINDSHIELD DEFROST
2 DUCT, AIR DISTRIBUTION
3 SCREW
4 LOWER I/P

87950084

Location of the air distribution duct mounting—2002 Sunfire shown

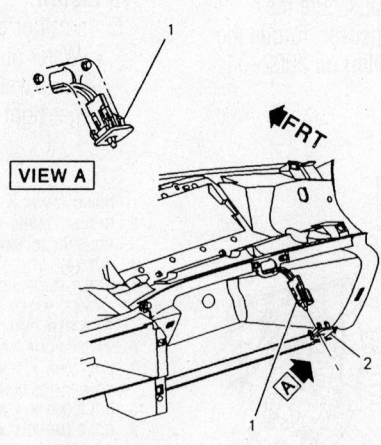

VIEW A

FRT

1 LAMP, I/P COMPARTMENT
2 RETAINER

87950085

Detach the instrument panel lamp connector—2002 Cavalier and Sunfire

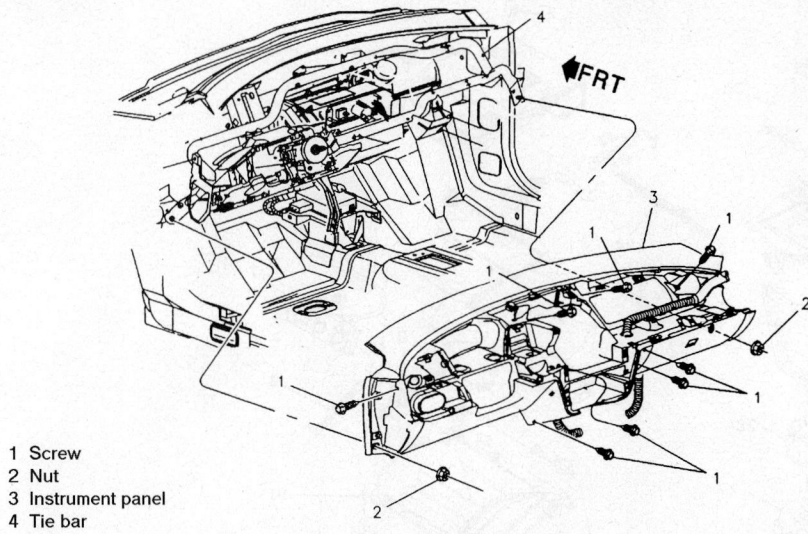

1 Screw
2 Nut
3 Instrument panel
4 Tie bar

87950086

Instrument panel-to-tie bar attachments—2002 Cavalier

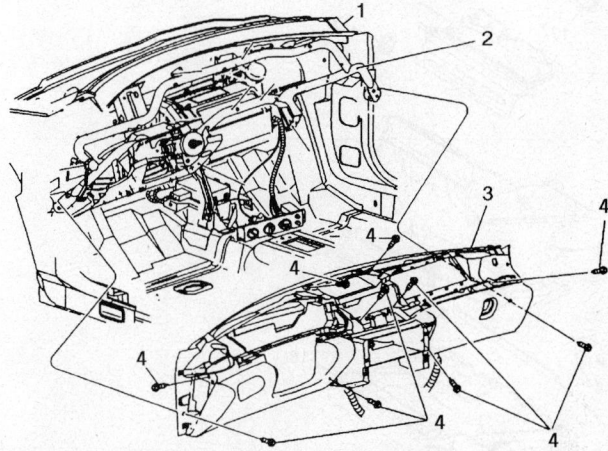

1 Front of dash
2 Tie bar
3 Instrument panel
4 Screw

87950087

Instrument panel-to-tie bar mounting—2002 Sunfire

- Heater/air conditioning control assembly
- Radio
- Air ventilation ducts from the heater housing
- Instrument panel light electrical connector
- Tie bar screws
- Instrument panel from the tie bar
- Heater core outlet screws and the outlet
- Heater core cover-to-heater case screws and the cover.

➡The heater core mounting screw is located in the recess in the center of the heater core cover.

- Heater core clamp from the heater assembly
- Heater core from the heater assembly

To install:

9. Install or connect the following:
   - Heater core to the heater assembly
   - Heater core clamp to the heater assembly and torque the clamp screws to 9 inch lbs. (1 Nm)
   - Heater core cover and the cover-to-heater case screw and torque the screws to 9 inch lbs. (1 Nm)
   - Heater core outlet and torque the screws to 9 inch lbs. (1 Nm)
   - Diagnostic Energy Reserve Module (DERM) with attaching bracket, if equipped
10. Install the instrument panel by installing or connecting the following:
    - Instrument panel to the tie bar
    - Tie bar screws
    - Instrument panel light electrical connector

- Air ventilation ducts to the heater housing
- Radio
- Heater/air conditioning control assembly
- Instrument panel trim plate
- Instrument panel trim pad, (Cavalier models only)
- Accessory trim plate, (Sunfire models only)
- Instrument panel end caps on both sides
- Defroster grill
- Heater hoses to the heater core

11. Install the steering wheel by installing or connecting the following:
    - Steering wheel to the steering column
    - Steering wheel-to-steering column nut and torque the nut to 30 ft. lbs. (41 Nm)
    - Electrical connector and install the SIR module
    - SIR module-to-steering wheel screws and torque the screws to 89 inch lbs. (10 Nm)
12. Refill the cooling system.
13. Connect the negative battery cable.
14. Enable the SIR system by performing the following procedure:
    a. Turn the ignition switch to the LOCK position.
    b. At the base of the steering column, connect the Connector Position Assurance (CPA), the yellow 2-way electrical connectors and the passenger's side module electrical connector.
    c. At the base of the steering column, install the left sound insulator.
    d. Install the AIR BAG fuse to the fuse block.
    e. Turn the ignition switch to the RUN position; the INFL REST warning light should flash 7–9 times then turn OFF.
15. Operate the engine to normal operating temperatures; then, check the climate control operation and check for leaks.

2003–05 Models

CAVALIER

1. Drain the cooling system.
2. Recover the refrigerant.
3. Remove the bolt connecting the evaporator lines to the evaporator.
4. Remove the evaporator lines from the evaporator.
5. Use hose clamp pliers to reposition the hose clamps.
6. Disconnect the heater hoses from the heater core.

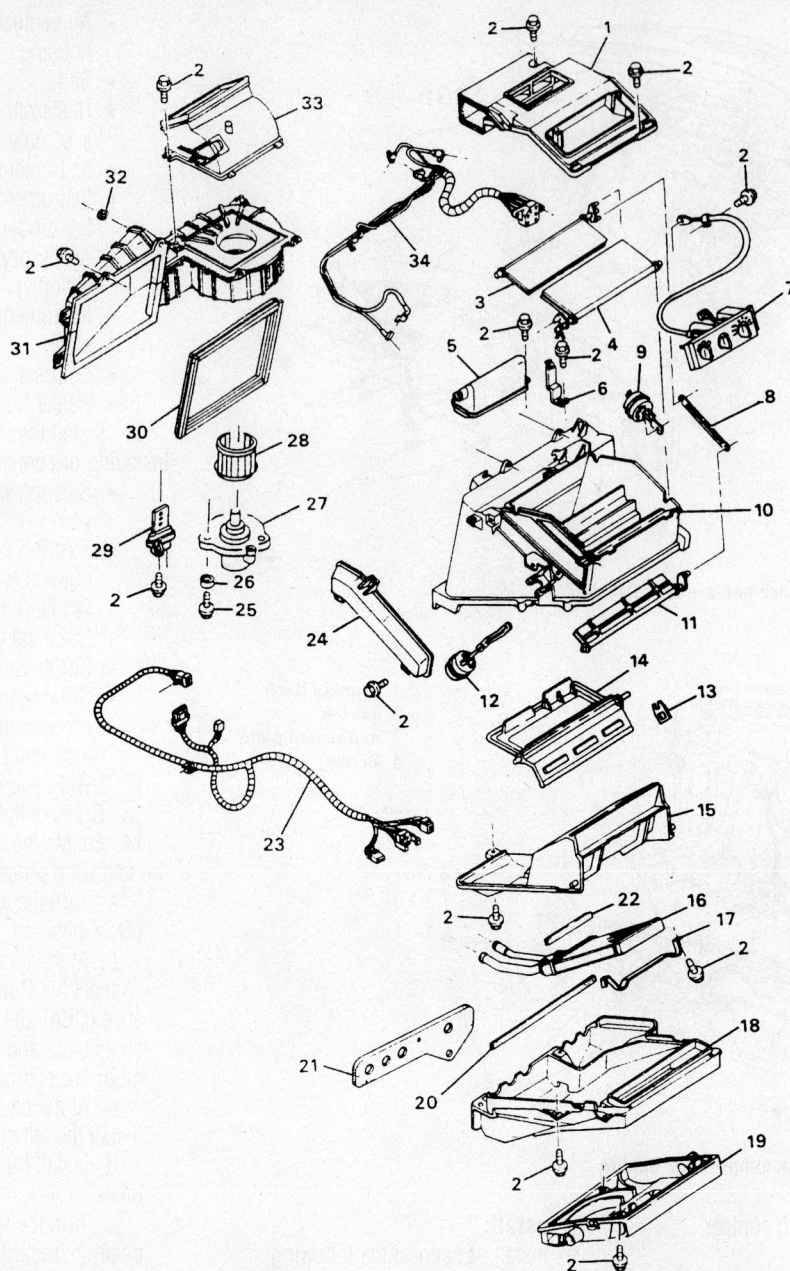

1 VALVE HOUSING COVER	18 HEATER COVER
2 HEATER/BLOWER MODULE BOLT	19 HEATER OUTLET
3 DEFROSTER VALVE	20 HEATER CORE SHROUD SEAL
4 MODE VALVE	21 HEATER CORE TUBE AND MOUNT SEAL
5 HEATER–VACUUM TANK	22 HEATER CORE SEAL
6 HEATER MODULE MOUNTING BRACKET	23 HEATER AND A/C CONTROL SWITCH HARNESS
7 HEATER–CONTROL	24 DEFROSTER DUCT
8 HEATER VALVE LEVER LINK	25 BLOWER MOTOR BOLTS
9 DEFROSTER VALVE ACTUATOR	26 BLOWER MOTOR ISOLATOR
10 HEATER CASE	27 BLOWER MOTOR
11 HEATER VALVE	28 BLOWER FAN
12 MODE VALVE ACTUATOR	29 BLOWER RESISTOR
13 TEMPERATURE VALVE CLIP	30 HEATER CASE SEAL
14 TEMPERATURE VALVE	31 BLOWER AND AIR INLET CASE
15 HEATER CORE SHROUD	32 MOUNTING SEAL
16 HEATER CORE	33 AIR INLET HOUSING
17 HEATER CORE STRAP	34 VACUUM HARNESS

93111GB4

Exploded view of the heater/evaporator housing assembly—2002 Cavalier and Sunfire

7. Remove the evaporator case drain tube.

8. Turn the steering wheel so that the vehicle wheels are pointing straight ahead.

9. Turn the ignition switch to the OFF position.

10. Remove the key from the ignition switch.

11. Remove the defroster grille screw.

12. With a flat-bladed tool disengage the defroster grille from the I/P trim pad.

13. With a flat-bladed tool disengage the outer trim covers from the I/P carrier. Pull the cover rearward to disengage the locating tabs from I/P carrier.

14. Open the I/P compartment.

15. Remove the I/P trim pad screws.

16. Remove the I/P trim pad from the I/P carrier.

17. Remove the AIR BAG fuse from the left hand I/P wiring harness junction block.

➡**With the AIR BAG fuse removed and the ignition switch in the RUN position, the AIR BAG warning lamp illuminates. This is a normal operation and does not indicate an SIR system malfunction.**

18. Remove the Connector Position Assurance (CPA) from the inflatable restraint steering wheel module coil connector, located under the instrument panel, left of the steering column.

19. Disconnect the steering wheel module coil connector from the vehicle harness connector.

20. Remove the screws from the top of the I/P cluster.

21. Remove the I/P cluster pulling rearward.

22. Disconnect the electrical connector from the I/P cluster.

23. Remove the I/P cluster.

24. Remove the air distribution duct screws.

25. Remove the air distribution duct.

26. Remove the right side air distribution duct.

27. Remove the CPA from the inflatable restraint I/P module yellow 2-way pigtail connector.

28. Disconnect the yellow 2-way pigtail connector from the inflatable restraint I/P module.

29. Remove the mounting bolts from the inflatable restraint I/P module.

30. Remove the inflatable restraint module from the I/P

31. Remove the screw from the HVAC temperature control cable.

32. Remove the control cable end from the door lever.

33. Remove the HVAC control assembly from the I/P carrier. Leave the harness connected to the HVAC control and position aside.

34. Remove the radio from the I/P carrier

35. Remove the steering wheel.

36. Remove the tilt lever.

37. Remove the washer switch.

38. Remove the screws from the I/P carrier.

39. Remove the nuts from the I/P carrier.

40. Remove the I/P carrier from the lower tie bar.

41. Remove the I/P carrier from the vehicle.

42. Disconnect the wiring harness from the cross vehicle beam.

43. Remove the attaching bolts from the right and left side of the HVAC module.

44. Remove the bolts from the cross vehicle beam.

➡**Place the steering column in the straight tilt position using the tilt lever.**

45. Remove the cross vehicle beam from the front of the vehicle.

46. Remove the floor air outlet.

47. Disconnect the wiring harness from the HVAC module assembly.

48. Disconnect the electrical connections at the blower motor and blower resistor.

49. Remove the HVAC module attaching bolts near the heater hoses and the A/C evaporator lines.

50. Remove the HVAC module assembly screw from the mounting bracket.

➡**Do not tilt the HVAC module assembly, or coolant may leak out.**

51. Remove the HVAC module from the vehicle.

52. Remove the heat stakes and the screws from the heater cover case.

53. Remove the heater cover case.

54. Remove the heater core from the HVAC module assembly.

To install:

55. Install the heater core to the HVAC module assembly.

56. Install the heater core cover case.

➡**Use the correct fastener in the correct location. Replacement fasteners must be the correct part number for that application. Fasteners requiring replacement or fasteners requiring the use of thread locking compound or sealant are identified in the service procedure. Do not use paints, lubricants, or corrosion inhibitors on fasteners or fastener joint surfaces unless specified. These coatings affect fas-**

tener torque and joint clamping force and may damage the fastener. Use the correct tightening sequence and specifications when installing fasteners in order to avoid damage to parts and systems.

57. Install the heater core cover case screws. Tighten to 9 inch lbs. (1 Nm).

58. Install the HVAC module assembly. Align the mounting bracket to the front of dash slot and mounting bolt hole.

Install the bolt to the mounting bracket.

59. Install the HVAC module attaching bolts near the heater hoses and the A/C evaporator lines.

60. Connect the electrical connections at the blower motor and resistor.

61. Connect the wiring harness to the HVAC module assembly.

62. Install the floor air outlet.

63. Install the cross vehicle beam to the front of the dash.

64. Install the bolts and the studs to the cross vehicle beam. Tighten to 89 inch lbs. (10 Nm).

65. Install the support bolts on the right and left side of the HVAC module.

66. Connect the wiring harness to the cross vehicle beam.

67. Place the I/P carrier into the vehicle.

68. Install the I/P carrier to the lower tie bar.

69. Install the screws to the I/P carrier. Tighten the screws to 20 inch lbs. (2.3 Nm) and the nuts to 44 inch lbs. (5 Nm).

70. Install the washer switch.

71. Install the tilt lever.

72. Install the steering wheel.

73. Install the air distribution duct.

74. Install the radio.

75. Install the HVAC control assembly to the I/P carrier.

76. Install the end of the HVAC temperature control cable to the door lever.

77. Install the control cable screw and tighten to 18 inch lbs. (2 Nm).

78. Install the inflatable restraint module to the I/P

79. Install the mounting bolts to the inflatable restraint module and tighten to 44 inch lbs. (5 Nm).

80. Connect the yellow 2-way pigtail connector to the inflatable restraint module.

81. Install the CPA to the yellow 2-way pigtail connector on the inflatable restraint module.

82. Install the air distributor duct.

83. Connect the electrical connector to the I/P cluster.

84. Position the I/P cluster retainer tabs into the I/P.

85. Install the instrument panel cluster to the I/P.

86. Install the I/P cluster screws.

87. Install the I/P trim pad to the I/P carrier.

88. Install the I/P trim pad screws.

89. Close the I/P compartment door.

90. Align the outer trim covers locating tabs into the I/P carrier. Press the cover into the I/P carrier until fully engaged.

91. Install the defroster grille to the I/P trim pad.

92. Remove the key from the ignition switch.

93. Connect the steering wheel module coil connector to the vehicle harness connector.

94. Install the CPA to the steering wheel module coil connector.

95. Install the AIR BAG fuse to the left hand I/P wiring harness junction block.

96. Install the left hand I/P outer trim cover.

97. Use caution while reaching in and turn the ignition switch to the ON position. The AIR BAG indicator will flash then turn OFF.

98. Connect the negative battery cable

99. Install the evaporator case drain tube.

100. Connect the heater hoses to the heater core.

101. Reposition the hose clamps.

102. Connect the evaporator tube to the evaporator. Tighten the fittings to 18 ft. lbs. (24 Nm).

103. Evacuate and recharge the A/C system.

104. Leak test the fittings of the component.

105. Fill the cooling system.

SUNFIRE

1. Drain the cooling system.

2. Recover the refrigerant.

3. Remove the bolt connecting the evaporator lines to the evaporator.

4. Remove the evaporator lines from the evaporator.

5. Use hose clamp pliers to reposition the hose clamps.

6. Disconnect the heater hoses from the heater core.

7. Remove the evaporator case drain tube.

8. Turn the steering wheel so that the vehicle wheels are pointing straight ahead.

9. Turn the ignition switch to the OFF position.

10. Remove the key from the ignition switch.

11. Remove the left hand Instrument Panel (I/P) outer trim cover.

➡With the AIR BAG fuse removed and the ignition switch in the RUN position, the AIR BAG warning lamp illuminates. This is a normal operation and does not indicate an SIR system malfunction.

12. Remove the AIR BAG fuse from the left hand I/P wiring harness junction block.

13. Remove the Connector Position Assurance (CPA) from the inflatable restraint steering wheel module coil connector, located under the instrument panel, left of the steering column.

14. Disconnect the steering wheel module coil connector from the vehicle harness connector

15. With a flat-bladed tool, remove the I/P defroster grille.

16. Remove the I/P outer trim cover screws.

17. Remove the outer trim covers from the I/P carrier.

18. Remove the radio from the I/P carrier.

19. Remove the right I/P trim panel screws.

20. Remove the right I/P trim panel from the I/P carrier (3).

21. Remove the I/P upper trim panel by lifting upward to release the retainers.

22. Remove the screws from the I/P outer trim cover.

23. Remove the I/P outer trim covers.

24. Remove the I/P accessory trim plate.

25. Remove the screws from the right I/P trim panel.

26. Remove the right I/P trim panel.

27. Remove the screws attaching the I/P trim pad.

28. Remove the I/P trim pad from the I/P carrier.

29. Remove the air distributor duct.

30. Remove the CPA from the inflatable restraint I/P module yellow 2-way pigtail connector.

31. Disconnect the yellow 2-way pigtail connector from the inflatable restraint I/P module.

32. Remove the mounting bolts from the inflatable restraint I/P module.

33. Remove the inflatable restraint module from the I/P

34. Remove the screw from the HVAC temperature control cable.

35. Remove the control cable end from the door lever.

36. Remove the HVAC control assembly from the I/P carrier. Leave the harnesses connected to the HVAC control and position aside.

37. Remove the air distribution duct.

38. Remove the steering wheel.

39. Remove the tilt lever.

40. Remove the washer lever.

41. Remove the screws from the I/P carrier.

42. Remove the I/P carrier from the lower tie bar.

43. Remove the I/P carrier from the vehicle.

44. Disconnect the wiring harness from the cross vehicle beam.

45. Remove the attaching bolts from the right and left side of the HVAC module.

46. Remove the bolts from the cross vehicle beam.

➡Place the steering column in the straight tilt position using the tilt lever.

47. Remove the cross vehicle beam from the front of the vehicle.

48. Remove the floor air outlet.

49. Disconnect the wiring harness from the HVAC module assembly.

50. Disconnect the electrical connections at the blower motor and blower resistor.

51. Remove the HVAC module attaching bolts near the heater hoses and the A/C evaporator lines.

52. Remove the HVAC module assembly screw from the mounting bracket.

➡Do not tilt the HVAC module assembly, or coolant may leak out.

53. Remove the HVAC module from the vehicle.

54. Remove the heat stakes and the screws from the heater cover case.

55. Remove the heater cover case.

56. Remove the heater core from the HVAC module assembly.

To install:

57. Install the heater core to the HVAC module assembly.

58. Install the heater core cover case.

➡Use the correct fastener in the correct location. Replacement fasteners must be the correct part number for that application. Fasteners requiring replacement or fasteners requiring the use of thread locking compound or sealant are identified in the service procedure. Do not use paints, lubricants, or corrosion inhibitors on fasteners or fastener joint surfaces unless specified. These coatings affect fastener torque and joint clamping force and may damage the fastener. Use the correct tightening sequence and specifications when installing fasteners in order to avoid damage to parts and systems.

59. Install the heater core cover case screws. Tighten to 9 inch lbs. (1 Nm).

60. Install the HVAC module assembly. Align the mounting bracket to the front of dash slot and mounting bolt hole.

Install the bolt to the mounting bracket.

61. Install the HVAC module attaching bolts near the heater hoses and the A/C evaporator lines.

62. Connect the electrical connections at the blower motor and resistor.

63. Connect the wiring harness to the HVAC module assembly.

64. Install the floor air outlet.

65. Install the cross vehicle beam to the front of the dash.

66. Install the bolts and the studs to the cross vehicle beam. Tighten to 89 inch lbs. (10 Nm).

67. Install the support bolts on the right and left side of the HVAC module.

68. Connect the wiring harness to the cross vehicle beam.

69. Place the I/P carrier into the vehicle.

70. Place the I/P carrier into the vehicle.

71. Install the I/P carrier to the tie bar.

72. Install the washer switch.

73. Install the tilt lever.

74. Install the steering wheel.

75. Install the air distribution duct.

76. Install the HVAC control assembly to the I/P carrier.

77. Install the end of the HVAC temperature control cable to the door lever.

78. Install the control cable screw and tighten to 18 inch lbs. (2 Nm).

79. Install the inflatable restraint module to the I/P. Tighten the bolts to 44 inch lbs. (5 Nm).

80. Connect the yellow 2-way pigtail connector to the inflatable restraint module.

81. Install the CPA to the yellow 2-way pigtail connector on the inflatable restraint module.

82. Install the air distributor duct.

83. Install the I/P trim pad.

84. Install the right I/P trim panel to the I/P carrier.

85. Install the radio to the I/P carrier.

86. Install the I/P outer trim covers.

87. Install the I/P defroster grille to the I/P trim pad. Press into place until fully seated.

88. Remove the key from the ignition switch.

89. Connect the I/P module connector to the vehicle harness connector.

90. Install the CPA to the I/P module connector.

91. Install the AIR BAG fuse to the left hand I/P wiring harness junction block.

92. Install the left hand I/P outer trim cover.

93. Use caution while reaching in and turn the ignition switch to the ON position. The AIR BAG indicator will flash then turn OFF.

94. Connect the negative battery cable

95. Install the evaporator case drain tube.

96. Connect the heater hoses to the heater core.

97. Reposition the hose clamps.

98. Connect the evaporator tube to the evaporator. Tighten the fittings to 18 ft. lbs. (24 Nm).

99. Evacuate and recharge the A/C system.

100. Leak test the fittings of the component.

101. Fill the cooling system.

Cylinder Head

REMOVAL & INSTALLATION

2.2L (VIN 4) Engine

1. Before servicing the vehicle, refer to the precautions section.

2. Relieve the fuel system pressure.

3. Drain the cooling system.

4. Remove or disconnect the following:
- Accessory drive belt
- Air cleaner outlet duct
- Engine mount
- Intake manifold
- Exhaust manifold
- Engine Coolant Temperature (ECT) sensor electrical connector

- Power steering pump
- Radiator inlet pipe
- Spark plug wires
- Rocker arm cover

➡**Whenever valve train components are removed, keep them in order for installation purposes.**

- Rocker arms and pushrods
- Lifters
- Alternator rear brace and the alternator
- Power steering pump
- Radiator inlet pipe
- Ignition coil assembly
- Accessory bracket
- Spark plug wires
- Cylinder head bolts

✺✺ WARNING

Two sizes of bolts are used; note the location of each.

- Cylinder head

5. Inspect the cylinder head and block surface for cracks, nicks, heavy scratches and flatness.

To install:

6. Install or connect the following:
- Cylinder head with a new gasket and new cylinder head bolts. Torque the long bolts, in sequence to 46 ft. lbs. (63 Nm) and the short bolts to 43 ft. lbs. (58 Nm) plus an additional 90 degree turn on all the bolts.
- Lifters
- Pushrods and rocker arms. Torque nuts to 22 ft. lbs. (30 Nm).

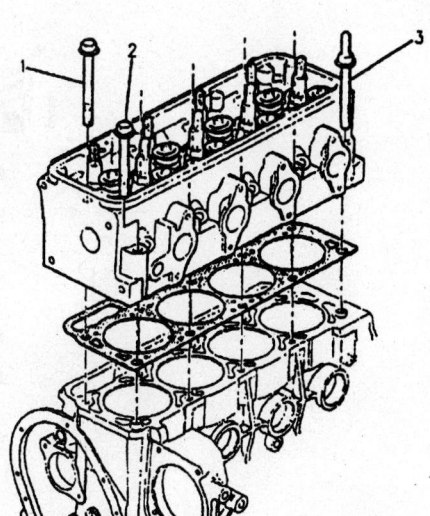

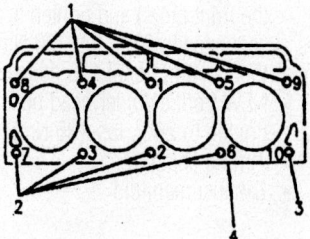

1. Long bolts
2. Short bolts
3. Stud
4. Numbers on gasket indicate torque sequence

Cylinder head torquing sequence—2.2L (VIN 4) engine

7922YG03

- Rocker arm cover. Torque the bolts to 89 inch lbs. (10 Nm).
- Spark plug wires
- Accessory bracket
- Ignition coil assembly
- Radiator inlet pipe
- Power steering pump. Torque the bolts to 22 ft. lbs. (30 Nm).
- ECT sensor electrical connector
- Exhaust manifold. Torque the nuts to 118 inch lbs. (13 Nm).
- Alternator. Torque the bolts to 37 ft. lbs. (50 Nm).
- Intake manifold. Torque the bolts to 17 ft. lbs. (24 Nm).
- Engine mount. Torque the bolts to 55 ft. lbs. (75 Nm).
- Air cleaner outlet duct
- Accessory drive belt
- Negative battery cable

7. Refill and bleed the cooling system.
8. Start the vehicle and inspect for leaks.

2.2L (VIN F) Engine

1. Before servicing the vehicle, refer to the precautions section.
2. Remove or disconnect the following:
   - Negative battery cable
   - Intake manifold
   - Power steering pump bolts and set pump aside
   - Exhaust manifold
   - Timing chain
3. Drain the cooling system.
   - Cylinder head bolts in sequence and discard
   - Cylinder head and gasket

To install:
4. Install or connect the following:
   - Cylinder head and gasket
   - NEW cylinder head bolts (except the front bolts) and tighten in sequence to 22 ft. lbs. (30 Nm) and then an additional 155 degrees
   - NEW front cylinder head bolts and tighten to 26 ft. lbs. (35 Nm)
   - Timing chain
   - Exhaust manifold

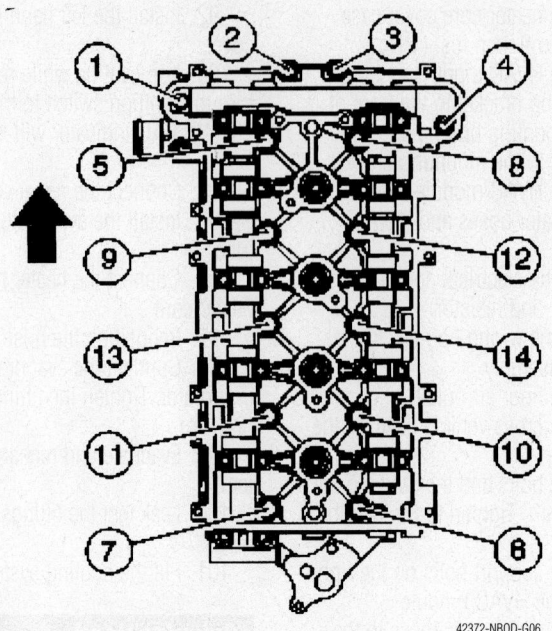

42372-NBOD-G06

Remove the cylinder head bolts in sequence and discard —2.2L (VIN F) engine

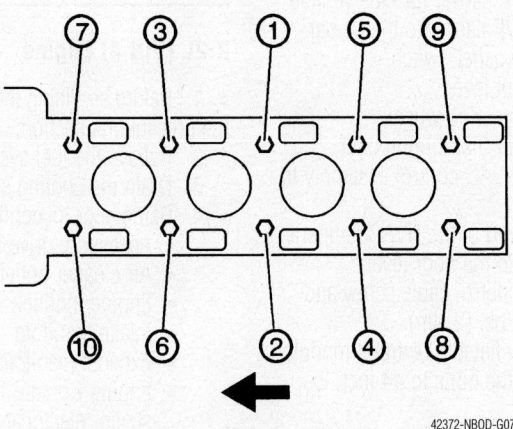

42372-NBOD-G07

Install the cylinder head bolts in sequence (except the front bolts)—2.2L (VIN F) engine

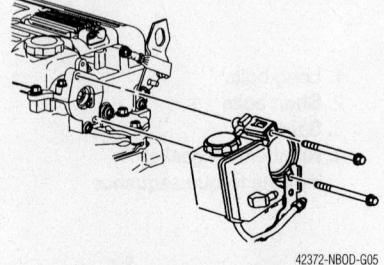

42372-NBOD-G05

Remove the power steering pump bolts and set pump aside—2.2L (VIN F) engine

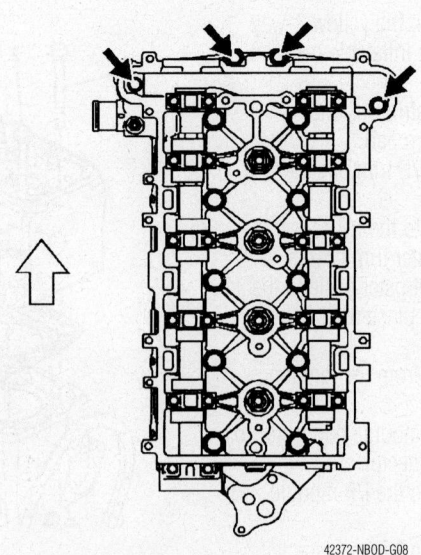

42372-NBOD-G08

Install the front cylinder head bolts—2.2L (VIN F) engine

- Intake manifold
- Power steering pump and tighten the bolts to 19 ft. lbs. (25 Nm).

5. Fill the cooling system.
6. Connect the negative battery cable.

2.4L Engine

1. Before servicing the vehicle, refer to the precautions section.
2. Relieve the fuel system pressure.
3. Drain the cooling system.
4. Remove or disconnect the following:

- Throttle body-to-air cleaner duct
- Heater inlet hose
- Throttle body heater hose
- Power brake vacuum hose from throttle body
- Manifold Absolute Pressure (MAP) sensor electrical connector
- Intake Air Temperature (IAT) sensor electrical connector
- Evaporative Emission (EVAP) canister purge solenoid
- Camshaft Position (CMP) sensor electrical connector
- Intake manifold
- Exhaust manifold
- Ignition coil assembly
- CMP sensor
- Power steering pump
- Vacuum line from the fuel pressure regulator
- Fuel injector harness connector
- Fuel rail
- Timing chain housing from the intake camshaft housing. DO NOT remove it from the vehicle.
- Oil pressure switch electrical connector
- Transmission fluid level indicator tube
- Intake and exhaust camshaft housing
- Upper radiator hose
- Coolant temperature sensor connector
- Cylinder head bolts
- Cylinder head and gasket

To install:

➡ Do not use abrasive pads to clean the cylinder head or block surfaces. An abrasive pad may damage the cylinder head and/or block.

5. Install or connect the following:
- Cylinder head with a new gasket. Torque the new bolts 1-8 to 40 ft. lbs. (54 Nm) and bolts 9-10 to 30 ft. lbs. (40 Nm); then, turn all bolts an additional 90 degrees (¼ turn) in sequence.

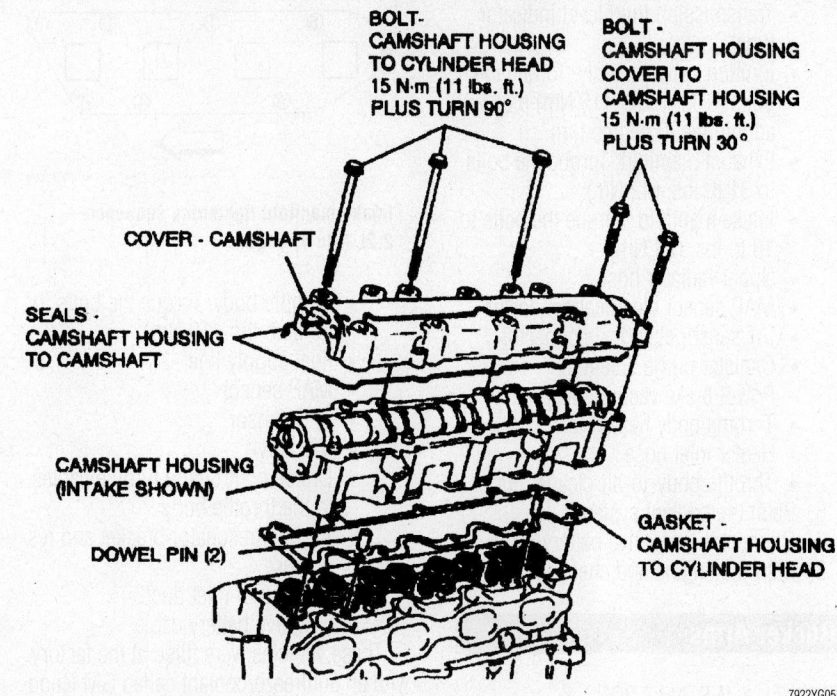

BOLT - CAMSHAFT HOUSING TO CYLINDER HEAD 15 N·m (11 lbs. ft.) PLUS TURN 90°

BOLT - CAMSHAFT HOUSING COVER TO CAMSHAFT HOUSING 15 N·m (11 lbs. ft.) PLUS TURN 30°

COVER - CAMSHAFT

SEALS - CAMSHAFT HOUSING TO CAMSHAFT

CAMSHAFT HOUSING (INTAKE SHOWN)

DOWEL PIN (2)

GASKET - CAMSHAFT HOUSING TO CYLINDER HEAD

7922YG05

Exploded view of the camshaft housing cover mounting—2.4L engine

- Upper radiator hose to coolant outlet
- Intake and exhaust camshaft housings
- Timing chain housing. Torque the bolts to 19 ft. lbs. (26 Nm).
- Fuel rail with new o-rings. Torque the bolts to 19 ft. lbs. (26 Nm).

- Vacuum line to the fuel pressure regulator
- Fuel injector electrical connectors
- CMP
- Power steering pump. Torque the bolts to 19 ft. lbs. (26 Nm).
- Oil pressure switch electrical connector

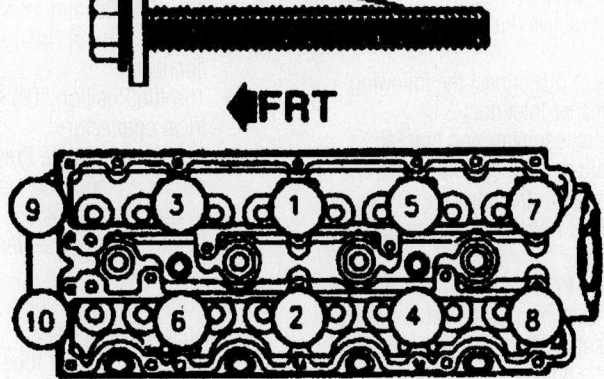

SPARINGLY APPLY CLEAN ENGINE OIL HERE

◀FRT

A. Tighten the bolts to the following N·m (lb. ft.) specification in sequence:
bolts 1 through 8: 65 N·m (40 ft. lb.)
bolts 9 and 10: 40 N·m (30 ft. lb.)
B. Then turn all 10 bolts an additional 90 degrees in sequence

7922YG04

Head bolt torque sequence—2.4L engine

- Transmission fluid level indicator tube
- Ignition coil assembly. Torque the bolts to 11 ft. lbs. (15 Nm) plus an additional 30 degree turn.
- Exhaust manifold. Torque the bolts to 31 ft. lbs. (42 Nm).
- Intake manifold. Torque the bolts to 18 ft. lbs. (24 Nm).
- Upper radiator hose
- MAP sensor electrical connector
- IAT sensor electrical connector
- Canister purge solenoid
- Power brake vacuum hose
- Throttle body heater hose
- Heater inlet hose
- Throttle body-to-air cleaner duct

6. Refill the coolant system.
7. Connect the negative battery cable.
8. Start the engine and check for leaks.

Rocker Arms

REMOVAL & INSTALLATION

2.2L Engines

These engines are not equipped with rocker arms. The camshafts directly actuate the valves

Intake Manifold

REMOVAL & INSTALLATION

2.2L (VIN 4) Engine

1. Before servicing the vehicle, refer to the precautions section.
2. Properly relieve the fuel system pressure.
3. Remove or disconnect the following:
   - Air cleaner inlet duct
   - Air inlet resonator and bracket
   - Throttle and cruise control cable from the throttle body
   - Manifold Absolute Pressure (MAP) sensor
   - Throttle Position (TPS) sensor
   - Idle Air Control (IAC) valve
   - Fuel supply line
   - Throttle body
   - Fuel inlet pipe
   - Intake manifold

To install:
4. Install or connect the following:
   - Intake manifold with a new gasket. Torque the bolts/nuts, in sequence, to 18 ft. lbs. (24 Nm).
   - Fuel inlet pipe. Torque the bolts to 18 ft. lbs. (24 Nm).

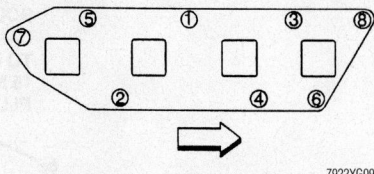

7922YG09

Intake manifold tightening sequence—2.2L (Vin 4) engine

   - Throttle body. Torque the bolts to 89 inch lbs. (10 Nm).
   - Fuel supply line
   - MAP sensor
   - TPS sensor
   - IAC valve
   - Cruise control and throttle cables to the throttle body
   - Air inlet resonator bracket and resonator
   - Air cleaner inlet duct
   - Negative battery cable

These vehicles were filled at the factory with an antifreeze/coolant called GM Goodwrench DEX-COOL®. When adding coolant to vehicles, it is important that you use GM Goodwrench DEX-COOL (orange-colored, silicate-free) coolant. **Propylene glycol is not recommended for use in GM vehicles**. A 50/50 mixture of DEX-COOL and clean water will provide all the recommended protection. **DO NOT mix DEX-COOL with any other type of antifreeze**.

2.2L (VIN F) Engine

1. Before servicing the vehicle, refer to the precautions section.
2. Remove or disconnect the following:
   - Air inlet duct and resonator
   - Idle Air Control (IAC) electrical connector
   - Throttle Position (TP) sensor electrical connector
   - Manifold Absolute Pressure (MAP) sensor electrical connector
   - Evaporative Emissions (EVAP) hose
   - Positive Crankcase Ventilation (PCV) hose
   - Purge solenoid tube
   - Brake booster hose
   - Oil level indicator tube bolt
   - Accelerator and cruise control cables
   - Throttle body
   - Fuel rail
   - Knock Sensor (KS) connector from the intake manifold
   - Intake manifold nuts and bolts
   - Intake manifold and gasket. The gasket is reusable if the gasket is not damaged.

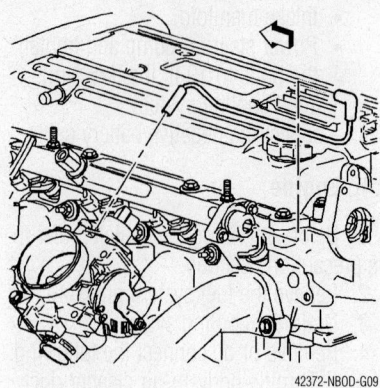

42372-NBOD-G09

Remove the throttle body—2.2L (VIN F) engine

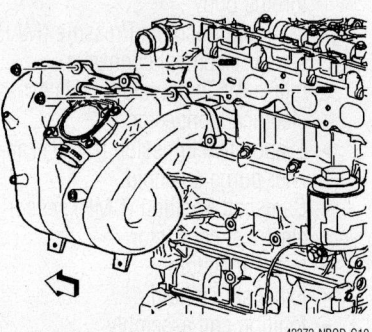

42372-NBOD-G10

Remove the intake manifold—2.2L (VIN F) engine

To install:
3. Install or connect the following:
   - Intake manifold gasket
   - Intake manifold
   - Intake manifold nuts and bolts and tighten to 89 inch lbs. (10 Nm)
   - KS connector
   - Fuel rail
   - Throttle body
   - IAC, TP, and MAP electrical connectors
   - EVAP hose
   - PCV hose
   - Purge solenoid tube
   - Brake booster hose
   - Oil level indicator tube bolt
   - Accelerator and the cruise control cables
   - Air inlet duct and resonator

2.4L Engine

1. Before servicing the vehicle, refer to the precautions section.
2. Properly relieve the fuel system pressure.
3. Drain the cooling system.
4. Remove or disconnect the following:
   - Manifold Absolute Pressure (MAP) sensor electrical connector

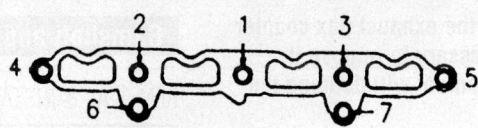

TIGHTENING SEQUENCE

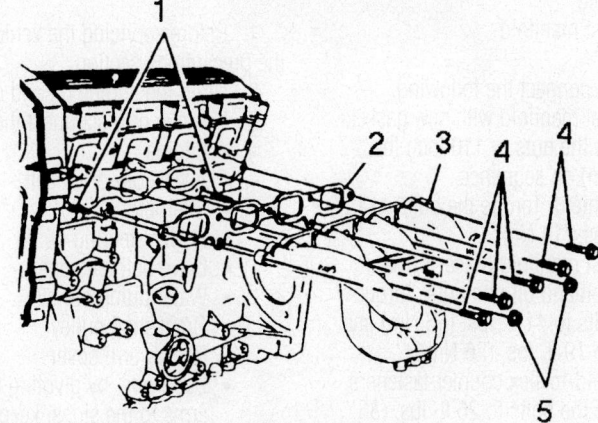

1. STUD – 12 N•M (100 LB. IN.)
2. INTAKE MANIFOLD GASKET
3. INTAKE MANIFOLD
4. BOLT – 24 N•M (17 LB. FT.)
5. NUT – 24 N•M (17 LB. FT.)

7922YG08

Intake manifold torque sequence—2.4L engine

- Intake Air Temperature (IAT) sensor electrical connector
- Evaporative Emission (EVAP) canister purge solenoid electrical connector
- Fuel injector harness
- Fuel regulator vacuum hose
- EVAP canister purge solenoid to canister vacuum hose
- Air cleaner duct
- Accelerator control cable bracket
- Stud-ended alternator mount bolt
- Exhaust Gas Recirculation (EGR) pipe from the EGR adapter
- Intake manifold support brace
- Intake manifold

To install:
5. Install or connect the following:
- Intake manifold with a new gasket. Torque the bolts/nuts, in sequence, to 18 ft. lbs. (24 Nm).
- EGR pipe to the adapter. Torque the fasteners to 19 ft. lbs. (26 Nm).
- Stud-ended alternator bolt
- Accelerator control cable bracket
- Fuel injector electrical connectors
- Vacuum hoses to the fuel regulator and EVAP canister purge solenoid
- MAP sensor electrical connector
- IAT sensor electrical connector
- EVAP canister purge solenoid electrical connector

- Air cleaner duct
- Negative battery cable
6. Refill the cooling system.
7. Start the engine and inspect for leaks.

Exhaust Manifold

REMOVAL & INSTALLATION

2.2L (VIN 4) Engine

1. Before servicing the vehicle, refer to the precautions section.
2. Drain the cooling system.
3. Remove or disconnect the following:
- Negative battery cable
- Oxygen (O2S) sensor electrical connector
- Accessory drive belt
- Alternator
- Radiator inlet pipe
- Exhaust pipe-to-exhaust manifold bolts
- Oil filler tube
- Heater outlet hose assembly-to-exhaust manifold nut
- Exhaust manifold

To install:
4. Install or connect the following:
- Exhaust manifold with new gaskets.

Torque the nuts to 115 inch lbs. (13 Nm).
- Heater outlet hose assembly-to-exhaust manifold nut. Torque the nut to 18 ft. lbs. (25 Nm).
- Oil filler tube
- Exhaust pipe-to-exhaust manifold bolts. Torque the bolts to 26 ft. lbs. (35 Nm).
- Radiator inlet pipe
- Alternator. Torque the bolts to 37 ft. lbs. (50 Nm).
- Accessory drive belt
- O2S sensor electrical connector
- Negative battery cable
5. Refill the cooling system.
6. Start the engine and check for exhaust leaks.

2.2L (VIN F) Engine

1. Before servicing the vehicle, refer to the precautions section.
2. Remove or disconnect the following:
- Negative battery cable
- Exhaust manifold heat shield
- Oxygen (O2S) sensor
- Manifold to the exhaust flex decoupler retainers
3. Pull down and back on the exhaust pipe in order to disconnect the pipe from the exhaust manifold.
- Exhaust manifold to cylinder head retaining nuts
- Exhaust manifold and gasket

To install:
4. Install or connect the following:
- New exhaust manifold gasket
- Exhaust manifold and tighten the retainers in sequence to 115 inch lbs. (13 Nm)
- New exhaust manifold to flex coupler gasket
5. Push the flex coupler into position on the exhaust manifold.
- Retaining nuts and tighten to 22 ft. lbs. (30 Nm)

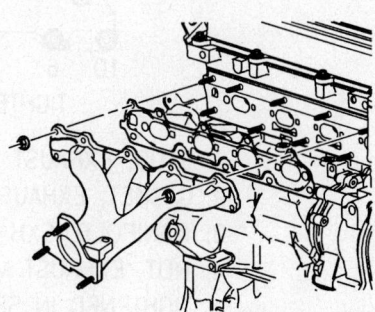

Remove the exhaust manifold—2.2L engine

42372-NBOD-G11

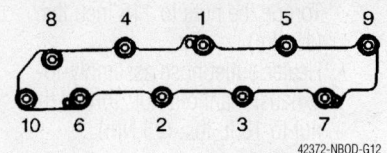

Tighten the exhaust manifold retainers in this sequence—2.2L engine

42372-NBOD-G12

- O$_2$S sensor
- Exhaust manifold heat shield and tighten the bolts to 18 ft. lbs. (25 Nm)
- Negative battery cable

2.4L Engine

1. Before servicing the vehicle, refer to the precautions section.
2. Remove or disconnect the following:
- Negative battery cable
- Oxygen (O$_2$S) sensor electrical connector
- Upper heat shield
- Exhaust manifold brace-to-manifold bolt
- Exhaust manifold-to-oil pan nuts

➡Do not bend the exhaust flex coupler more than necessary to remove it. Excessive movement will damage the flex coupler.

- Exhaust pipe from the exhaust manifold
- Exhaust manifold

To install:

3. Install or connect the following:
- Exhaust manifold with new gaskets. Torque the nuts to 110 inch lbs. (12 Nm), in sequence.
- Heat shield. Torque the bolts to 124 inch lbs. (14 Nm).
- Exhaust manifold brace-to-manifold bolt and oil pan nuts. Torque the bolts to 41 ft. lbs. (56 Nm) and nuts to 19 ft. lbs. (26 Nm).
- Manifold-to-flex coupler fasteners. Torque the bolts to 26 ft. lbs. (35 Nm).
- O$_2$S sensor electrical connector
- Negative battery cable

4. Start the engine and check for exhaust leaks.

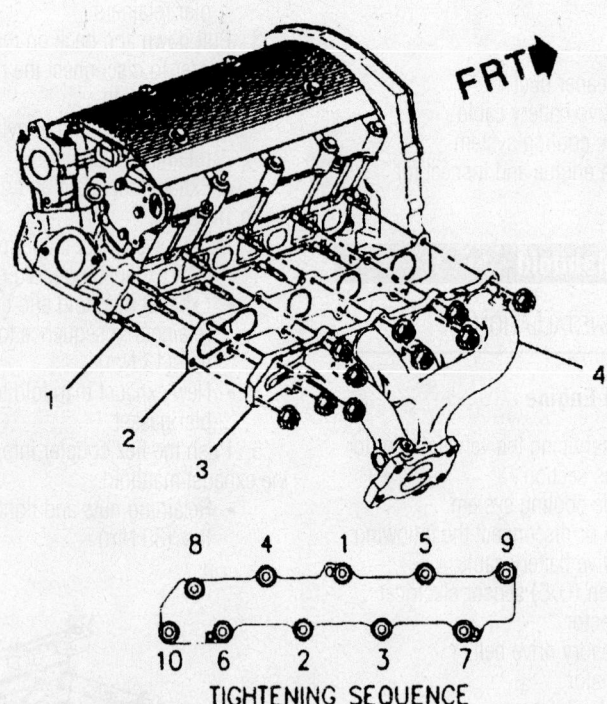

TIGHTENING SEQUENCE

1. STUD, EXHAUST MANIFOLD
2. GASKET, EXHAUST MANIFOLD
3. MANIFOLD, EXHAUST
4. NUT, EXHAUST MANIFOLD, MUST BE TIGHTENED IN SEQUENCE SHOWN TO 12.5 N•m (110 LB. IN.)

7922YG11

Exploded view of the exhaust manifold, showing the torque sequence—2.4L engine

Camshaft and Valve Lifters

REMOVAL & INSTALLATION

2.2L (Vin 4) Engine

1. Before servicing the vehicle, refer to the precautions section.
2. Drain the engine oil and coolant.
3. Remove or disconnect the following:
- Engine
- Accessory drive belt
- Alternator
- Power steering pump
- Drive belt tensioner
- Water pump pulley
- Crankshaft pulley
- Rocker arm cover
- Pushrods, by pivoting the rocker arms to the sides, keeping them in order
- Cylinder head
- Valve lifters, keeping them in order
- Front cover
- Camshaft timing sprocket
- Timing chain
- Timing chain tensioner
- Camshaft thrust plate
- Camshaft Position (CMP) sensor
- Camshaft from the block, being sure the lobes do not contact the bearings

To install:

✳✳ WARNING

The camshaft lobes and journals must be adequately lubricated or engine damage could occur upon start up.

4. Install or connect the following:
- Camshaft, being careful not to contact the bearings with the cam lobes
- CMP sensor
- Camshaft thrust plate. Torque the bolts to 106 inch lbs. (12 Nm).
- Timing chain and sprocket. Torque the bolts to 96 ft. lbs. (130 Nm).
- Timing chain tensioner. Torque the bolts to 18 ft. lbs. (24 Nm).
- Front cover. Torque the bolts to 97 inch lbs. (11 Nm).
- Valve lifters
- Cylinder head. Torque the long bolts to 46 ft. lbs. (63 Nm) and the short bolts to 43 ft. lbs. (58 Nm), plus an additional 90 degree turn on all bolts.
- Pushrods
- Rocker arms

5. Adjust the valve lash after installing the engine.

6. Install or connect the following:
- Rocker arm cover. Torque the bolts to 89 inch lbs. (10 Nm).
- Crankshaft pulley. Torque the bolts to 37 ft. lbs. (50 Nm).
- Water pump pulley. Torque the bolts to 22 ft. lbs. (30 Nm).
- Drive belt tensioner. Torque the bolts to 37 ft. lbs. (50 Nm).
- Power steering pump. Torque the bolts to 22 ft. lbs. (30 Nm).
- Alternator. Torque the bolts to 37 ft. lbs. (50 Nm).
- Accessory drive belt
- Engine

7. Replace the oil filter and refill engine with new oil.

8. Refill the cooling system.

2.2L (VIN F) Engine

1. Before servicing the vehicle, refer to the precautions section.

2. Remove or disconnect the following:
- Camshaft cover
- Upper timing chain guide

3. Install Camshaft Sprocket Holding Tool J 43665.
- Both the intake and exhaust camshaft sprocket bolts and discard

4. Slide the camshaft sprockets forward.
- Power steering pump bolts and set pump aside

5. Mark bearing caps to ensure they are installed in the original position.

➡**Remove each bolt on each cap one turn at a time until there is no spring tension on the camshaft.**

- Bearing caps
- Camshaft
- Camshaft roller followers
- Hydraulic lash adjusters

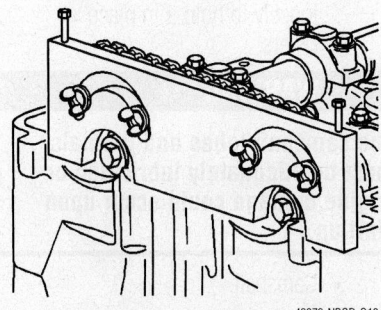

Install Camshaft Sprocket Holding Tool J 43665—2.2L (VIN F) engine

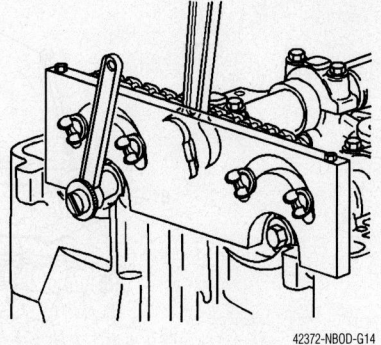

42372-NBOD-G14

Remove both the intake and exhaust camshaft sprocket bolts and discard—2.2L (VIN F) engine

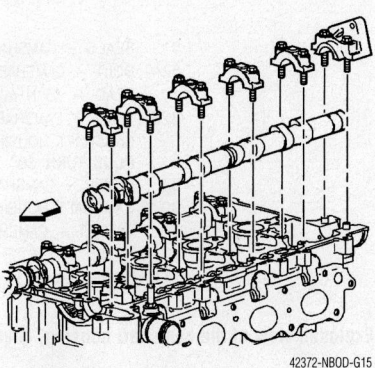

42372-NBOD-G15

Remove the bearing caps and camshaft—2.2L (VIN F) engine

To install:

6. Lubricate the valve tips.

7. Install or connect the following:
- Hydraulic lash adjusters
- Camshaft roller followers

8. Make sure that the alignment notches are aligned with the camshaft sprocket.
- Camshaft
- Camshaft bearing caps

9. Tighten the camshaft bearing cap bolts in 3 steps to 89 inch lbs. (10 Nm).

10. Apply a 0.197 inch (5mm) bead of anaerobic sealer97 in) bead to the rear intake camshaft bearing cap.

11. Install or connect the following:
- Rear intake camshaft bearing cap bolts and tighten to 18 ft. lbs. (25 Nm)
- Camshaft sprockets onto the camshafts and hand-tighten the NEW camshaft sprocket bolts

12. Remove the sprocket holding tool.

13. Tighten the camshaft sprocket bolts to 63 ft. lbs. (85 Nm), then an additional 30 degrees.

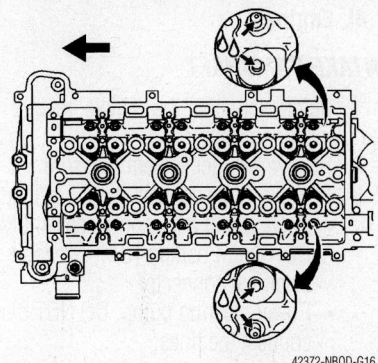

42372-NBOD-G16

Lubricate the valve tips—2.2L (VIN F) engine

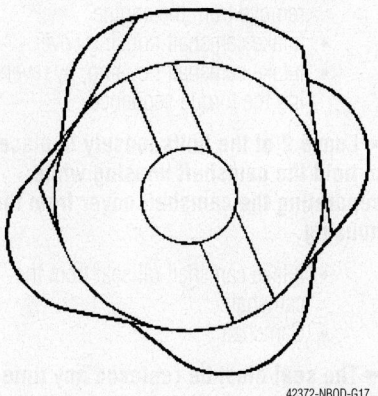

42372-NBOD-G17

Make sure that the alignment notches are aligned with the camshaft sprocket—2.2L (VIN F) engine

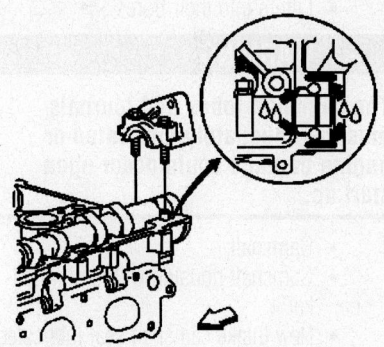

Apply a 0.197 inch (5mm) bead of anaerobic sealer to the rear intake camshaft bearing cap—2.2L (VIN F) engine

42372-NBOD-G18

- Upper timing chain guide and tighten to 89 in. lbs. (10 Nm).
- Power steering pump and tighten the bolts to 19 ft. lbs. (25 Nm).
- Camshaft cover

2.4L Engine

INTAKE CAMSHAFT

1. Before servicing the vehicle, refer to the precautions section.
2. Relieve the fuel system pressure.
3. Remove or disconnect the following:
   - Ignition coil and module assembly
   - Camshaft Position (CMP) sensor electrical connector
   - Power steering pump. DO NOT disconnect the lines.
   - Fuel pressure regulator vacuum line
   - Fuel rail
   - Timing chain housing, DO NOT remove from the engine
   - Intake camshaft housing cover
   - Intake camshaft housing, by reversing the torque sequence

➡ **Leave 2 of the bolts loosely in place to hold the camshaft housing while separating the camshaft cover from the housing.**

   - Intake camshaft oil seal from the camshaft
   - Camshaft

➡ **The seal must be replaced any time the housing and cover are separated.**

   - Camshaft housing from the cylinder head

To install:

4. Install or connect the following:
   - Lifters into their bores

✳✳ WARNING

The camshaft lobes and journals must be adequately lubricated or engine damage could occur upon start up.

   - Camshaft
   - Camshaft housing with a new gasket
   - New intake camshaft seal lubricated with engine oil

➡ **The timing chain sprocket dowel pin should be straight up and align with the centerline of the lifter bores.**

   - Camshaft housing cover with a new gasket. Torque the bolts, in sequence, to 11 ft. lbs. (15 Nm) plus an additional 90 degree turn (except for the 2 rear fuel pipe-to-camshaft housing bolts). Torque the 2 rear bolts to 11 ft. lbs. (15 Nm), plus an additional 30 degree turn.
   - Timing chain housing and the timing chain

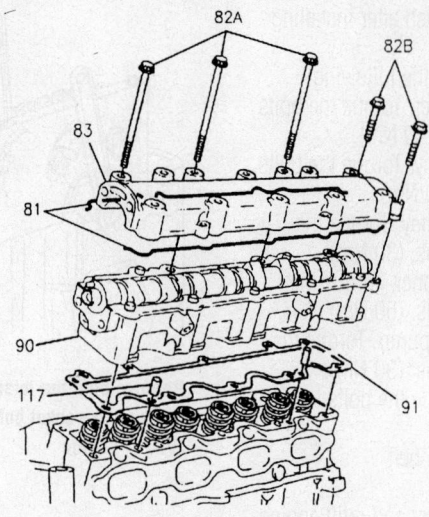

81   SEALS – CAMSHAFT HOUSING TO CAMSHAFT
82A BOLT – CAMSHAFT HOUSING TO CYLINDER HEAD – 15 N·m (11 LBS. FT.) PLUS TURN 90°
82B BOLT – CAMSHAFT HOUSING COVER TO CAMSHAFT HOUSING – 15 N·m (11 LBS. FT.) PLUS TURN 30°
83   COVER – CAMSHAFT
90   CAMSHAFT HOUSING (INTAKE SHOWN)
91   GASKET – CAMSHAFT HOUSING TO CYLINDER HEAD
117 DOWEL PIN (2)

7922YG31

Exploded view of the camshaft housing, cover and gaskets—2.4L engine

   - Fuel rail with new injector o-rings
   - Vacuum line to the fuel pressure regulator
   - Fuel injector harness connectors
   - Power steering pump. Torque the bolts to 22 ft. lbs. (30 Nm).
   - Camshaft position sensor
   - Ignition coil/module assembly. Torque the bolts to 11 ft. lbs. (15 Nm) plus 30 degrees.
   - Negative battery cable

5. Start the engine and check for leaks.

EXHAUST CAMSHAFT

1. Before servicing the vehicle, refer to the precautions section.
2. Relieve the fuel system pressure.
3. Remove or disconnect the following:
   - Ignition coil/module assembly
   - Oil pressure switch electrical connector
   - Transaxle fluid level indicator tube assembly, (if equipped) from exhaust camshaft cover
   - Timing chain housing; do not remove from the engine
   - Exhaust camshaft cover
   - Exhaust camshaft housing by reversing of the torque sequence

4. Press the cover off the housing by threading 4 housing-to-head bolts into the camshaft housing cover tapped holes.

Tighten the bolts evenly so the cover does not bind on the dowel pins.

5. Remove the camshaft housing cover.
6. Loosely install a camshaft housing-to-cylinder head bolt to retain the housing during camshaft and lifter removal.

➡ **Note the position of the chain sprocket dowel pin for reassembly.**

7. Remove or disconnect the following:
   - Lifters
   - Camshaft
   - Camshaft housing from the cylinder head

To install:

8. Install or connect the following:
   - Camshaft housing with a new gasket on the cylinder head with 1 bolt loosely to hold it in place
   - Lifters

✳✳ WARNING

The camshaft lobes and journals must be adequately lubricated or engine damage could occur upon start up.

   - Camshaft

➡ **The timing chain sprocket dowel pin should be straight up and align with the centerline of the lifter bores.**

➡ **Apply thread locking compound to the camshaft housing cover bolt threads.**

- Camshaft housing cover with a new gasket. Torque the bolts, in sequence, to 11 ft. lbs. (15 Nm) plus an additional 90 degree turn (except for the 2 rear fuel pipe-to-camshaft housing bolts). Torque the 2 rear bolts to 11 ft. lbs. (15 Nm), plus an additional 30 degree turn.
- Timing chain housing
- Transaxle fluid level indicator tube assembly to the exhaust camshaft cover
- Oil pressure switch electrical connector
- Ignition coil/module assembly. Torque the bolts to 11 ft. lbs. (15 Nm) plus an additional 30 degrees.
- Negative battery cable
9. Start the engine and check for leaks.

Valve Lash

ADJUSTMENT

All of the engines are equipped with hydraulic valve lifters. There is no adjustment.

Starter Motor

REMOVAL & INSTALLATION

1. Before servicing the vehicle, refer to the precautions section.
2. Remove or disconnect the following:
- Negative battery cable
- Air inlet duct
- Top starter bolt
- Lower starter bolt
- Starter electrical connectors
- Starter motor

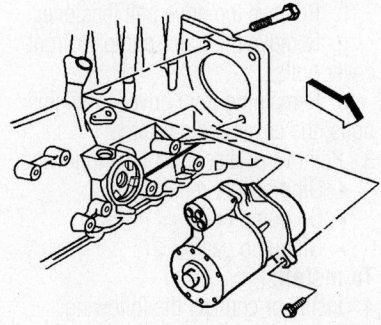

Exploded view of the starter motor—2.4L engine

9306YG02

To install:
3. Install or connect the following:
- Starter motor. Torque the bolts to 66 ft. lbs. (90 Nm) for 2.4L engine or 37 ft. lbs. (50 Nm) for the 2.2L (VIN 4) engine or 30 ft. lbs. (40 Nm) for the 2.2L (VIN F) engine.
- Starter electrical connectors
- Air inlet duct to the throttle body
- Negative battery cable

Oil Pan

REMOVAL & INSTALLATION

2.2L (VIN 4) Engine

1. Before servicing the vehicle, refer to the precautions section.
2. Drain the engine oil.
3. Remove or disconnect the following:
- Negative battery cable
- Right front wheel
- Right inner fender splash shield
- Starter motor and bracket
- Engine mount strut bracket
- Oil pan

To install:
4. Place a 2mm bead of RTV sealer to the oil pan sealing surface except at the rear seal mounting surface. Using a new oil pan rear seal, apply a thin coat of RTV sealer on the end down to the ears.
5. Install or connect the following:
- Oil pan with a new gasket. Torque the nuts and bolts to 89 inch lbs. (10 Nm).

Exploded view of the oil pan diagram:

1 SEAL
2 BOLT, 10 N•m (89 LB. IN.)
3 OIL PAN
4 NUT, OIL PAN 10 N•m (89 LB. IN.)

END SEALER
EAR

VIEW A

7922YG32

Exploded view of the oil pan mounting and related components—2.2L (VIN 4) engine

- Engine mount strut bracket. Torque the bolts to 49 ft. lbs. (66 Nm).
- Starter motor and bracket. Torque the bolts to 37 ft. lbs. (50 Nm).
- Right fender splash shield
- Right front wheel. Torque the nuts to 100 ft. lbs. (140 Nm).
- Negative battery cable
6. Refill the crankcase.
7. Start the vehicle and verify no leaks.

2.2L (VIN F) Engine

1. Before servicing the vehicle, refer to the precautions section.
2. Drain the engine oil.
3. Remove or disconnect the following:
- Engine mount strut bracket
- Drive belt
- Lower then upper AC compressor bolts
- Oil pan bolts
- Oil pan

To install:
4. Apply a 2 mm bead of RTV sealant around the perimeter of the oil pan and the oil suction port opening. Do not over apply the RTV. More than a 2 mm bead is not required.
5. Install or connect the following:
- Oil pan and tighten the bolts to 18 ft. lbs. (25 Nm)
- AC compressor bolts
- Engine mount bracket
- Drive belt
6. Refill the crankcase.
7. Start the vehicle and verify no leaks.

2.4L Engine

1. Before servicing the vehicle, refer to the precautions section.
2. Drain the engine oil.
3. Drain the cooling system.
4. Remove or disconnect the following:
- Negative battery cable
- Flywheel/converter cover
- Right wheel

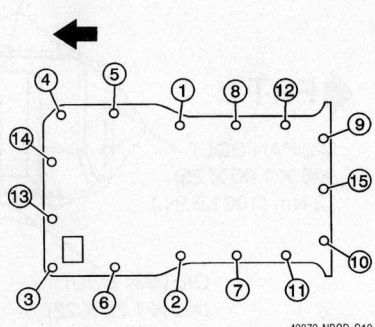

42372-NBOD-G19

Remove/Install the oil pan bolts in this sequence—2.2L (VIN F) engine

- Right wheel well splash shield
- Accessory drive belt
- Air conditioning compressor lower bolts
- Transmission-to-engine brace
- Engine mount strut bracket
- Radiator outlet pipe bolts
- Radiator outlet pipe from the oil pan
- Oil pan to the flywheel cover bolt and nut
- Flywheel cover stud for clearance
- Radiator outlet pipe from the lower radiator hose and oil pan
- Oil level sensor connector
- Oil pan

To install:

5. Inspect the oil pan gasket; it is reusable if not damaged.
6. Install or connect the following:
   - Oil pan with the gasket. Torque the M8 bolts to 18 ft. lbs. (24 Nm) and the M6 bolts to 106 inch lbs. (12 Nm).
   - Oil pan to the transmission nut
   - Oil level sensor connector
   - Radiator outlet pipe to the lower radiator hose and oil pan
   - Exhaust manifold brace
   - Radiator outlet pipe. Torque the bolts to 124 inch lbs. (14 Nm).
   - Engine mount strut bracket. Torque the bolts to 55 ft. lbs. (75 Nm).
   - Transmission to the engine brace
   - Air conditioning compressor lower bolts. Torque the bolts to 37 ft. lbs. (50 Nm).
   - Accessory drive belt
   - Right splash shield
   - Right front wheel
   - Flywheel/converter cover
   - Negative battery cable
7. Refill the crankcase.

8. Refill the cooling system.
9. Start the vehicle and verify no leaks.

Oil Pump

REMOVAL & INSTALLATION

2.2L (VIN4) Engine

1. Before servicing the vehicle, refer to the precautions section.
2. Drain the crankcase.
3. Remove or disconnect the following:
   - Negative battery cable
   - Oil pan
   - Oil pump and extension shaft

To install:

❄❄ WARNING

A plastic extension shaft retainer connects the oil pump driveshaft to the oil pump. Heat the extension shaft retainer in hot water prior to assembly. Be sure the retainer does not crack upon installation.

4. Fill the oil pump cavities with petroleum jelly before installing the gears into the pump body.
5. Install or connect the following:
   - Extension shaft and oil pump. Torque the oil pump-to-bearing cap bolt to 32 ft. lbs. (43 Nm) and the upper oil pump drive bolt to 18 ft. lbs. (25 Nm).
   - Oil pan
   - Negative battery cable
6. Refill the crankcase.
7. Start the engine and check oil pressure and check for leaks.

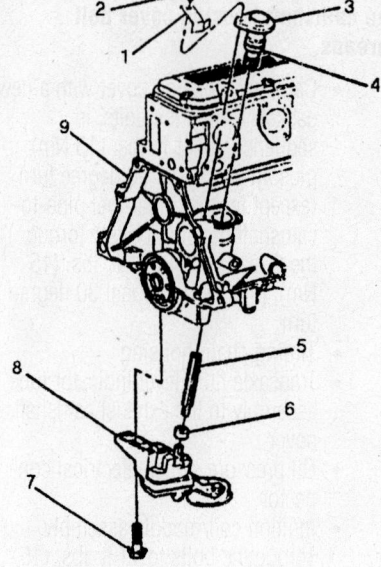

1. Bracket
2. Bolt
3. Oil pump drive assembly
4. O-ring
5. Shaft
6. Retainer; Heat and water soak prior to installation
7. Bolt
8. Oil pump
9. Cylinder block

Exploded view of the oil pump mounting to engine block—2.2L (VIN 4) engine

2.2L (VIN F) Engine

1. Before servicing the vehicle, refer to the precautions section.
2. Remove the engine front cover as follows:
   a. Remove the front wheel.
   b. Remove the engine splash shield.
   c. Remove the drive belt.
   d. Use crankshaft holding tool J 38122-A to prevent the crankshaft from rotating while loosening the crankshaft balancer bolt.
   e. Remove the crankshaft balancer bolt and discard. Remove the balancer.
   f. Remove the drive belt tensioner.
   g. Remove the water pump-to-front cover bolts.
   h. Remaining front cover-to-engine bolts, the cover and gasket.
3. Remove or disconnect the following:
   - Oil pressure relief valve
   - Oil pump cover
   - Oil pump gears
To install:
4. Install or connect the following:
   - Oil pump gears
   - Oil pump cover and tighten the bolts to 53 inch lbs. (6 Nm)

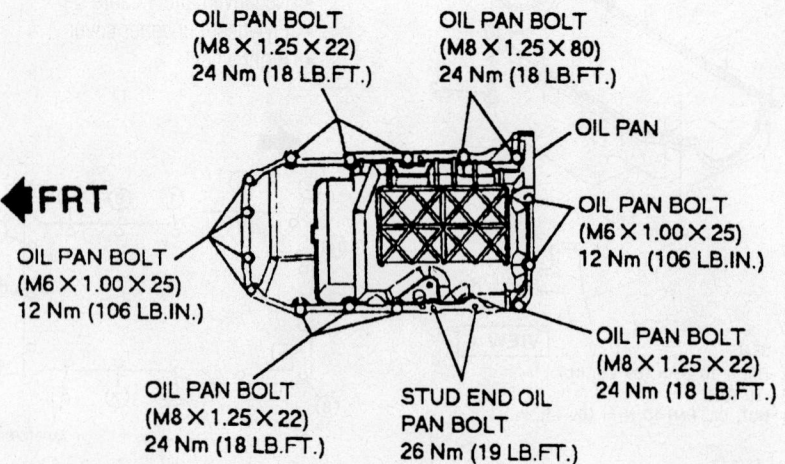

OIL PAN BOLT
(M8 × 1.25 × 22)
24 Nm (18 LB.FT.)

OIL PAN BOLT
(M8 × 1.25 × 80)
24 Nm (18 LB.FT.)

OIL PAN

FRT

OIL PAN BOLT
(M6 × 1.00 × 25)
12 Nm (106 LB.IN.)

OIL PAN BOLT
(M6 × 1.00 × 25)
12 Nm (106 LB.IN.)

OIL PAN BOLT
(M8 × 1.25 × 22)
24 Nm (18 LB.FT.)

OIL PAN BOLT
(M8 × 1.25 × 22)
24 Nm (18 LB.FT.)

STUD END OIL
PAN BOLT
26 Nm (19 LB.FT.)

Oil pan fastener torque specifications—2.4L engine

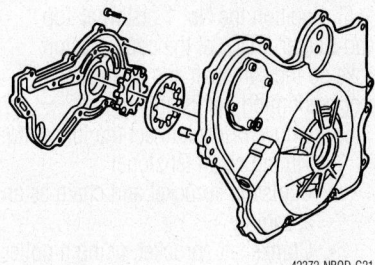

42372-NBOD-G21

Oil pump assembly—2.2L (VIN F) engine

- Oil pressure relief valve and tighten the plug to 30 ft. lbs. (40 Nm).

5. Install the engine front cover as follows:

a. Install front cover gasket, cover and cover-to-engine bolts.

b. Install the water pump-to-front cover bolts. Tighten all the cover bolts to 15 ft. lbs. (20 Nm).

c. Install the drive belt tensioner and tighten the bolt to 33 ft. lbs. (45 Nm).

d. Install the crankshaft balancer and bolt.

e. Use crankshaft holding tool J 38122-A to prevent the crankshaft from rotating while tightening the crankshaft balancer bolt. Tighten the bolt to 74 ft. lbs. (100 Nm) plus an additional 75 degrees.

f. Install the engine splash shield.

g. Install the drive belt.

h. Install the front wheel.

2.4L Engine

1. Before servicing the vehicle, refer to the precautions section.

2. Disconnect the negative battery cable.

3. Install an engine support fixture.

4. Drain the crankcase.

5. Remove or disconnect the following:

- Oil pan

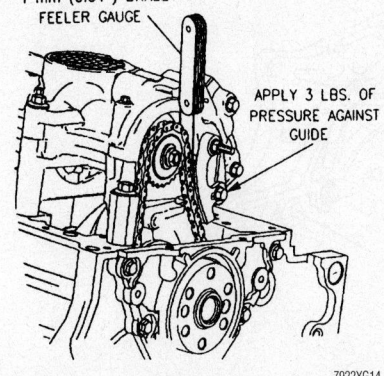

1 mm (0.04") BRASS FEELER GAUGE

APPLY 3 LBS. OF PRESSURE AGAINST GUIDE

7922YG14

Using a feeler gauge to check the chain tension—2.4L engine

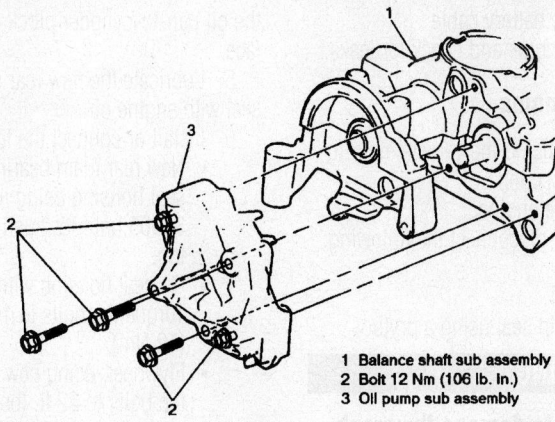

1	Balance shaft sub assembly
2	Bolt 12 Nm (106 lb. In.)
3	Oil pump sub assembly

7922YG15

Exploded view of the oil pump assembly mounting—2.4L engine

- Balance shaft chain cover and tensioner
- Oil pump

6. Disassemble the oil pump as follows:

a. Remove the oil pump cover.

b. Remove the pump gear.

c. Remove the sub-assembly from the balance shaft assembly

d. Remove the gerotor from the oil pump housing.

e. Remove the oil pump from the balance shaft housing.

f. Remove the pressure relief valve.

To install:

7. Lubricate the gears with clean engine oil.

8. Assemble the oil pump as follows:

a. Place the gerotor gear into the housing.

b. Fill the oil pump cavities with petroleum jelly.

c. Install the pressure relief valve using a 9⁄16 in. deep well socket to seat the valve.

d. Connect the pump housing to the balance shaft assembly.

e. Install the pump cover.

9. Install or connect the following:

- Oil pump. Torque the bolts to 40 ft. lbs. (54 Nm).
- Balance shaft chain tensioner and chain

➡**A brass feeler gauge must be used to ensure that correct measurements are obtained. If a steel gauge is used, it will not bend to conform to the guide and will allow for incorrect measurements.**

10. Adjust the chain tension as follows:

a. Insert a 0.40 in. (1mm) brass feeler between the chain guide and the chain.

b. Press the guide against the chain using about 3 pounds of force.

c. Torque the chain tensioner fastener to 115 inch lbs. (13 Nm).

11. Install or connect the following:

- Balance shaft chain cover. Torque the nut and bolt to 115 inch lbs. (13 Nm).
- Oil pan
- Negative battery cable.

12. Refill the crankcase.

13. Start the vehicle and verify oil pressure and no leaks.

Rear Main Seal

REMOVAL & INSTALLATION

2.2L (VIN 4) Engine

1. Before servicing the vehicle, refer to the precautions section.

2. Remove or disconnect the following:

- Negative battery cable
- Transaxle
- Clutch/pressure plate assembly, if equipped with a manual transmission
- Flywheel
- Rear main bearing seal by prying it from the engine

➡**Be careful not to damage or scratch the seal mounting surfaces.**

To install:

3. Lubricate the new rear main bearing seal with engine oil.

4. Install or connect the following:

- New rear main bearing seal using seal installer J-34686 until it's flush with the block
- Flywheel
- Clutch/pressure plate assembly, if equipped with a manual transmission
- Transaxle

- Negative battery cable
5. Start the engine and check for leaks.

2.2L (VIN F) Engine

1. Before servicing the vehicle, refer to the precautions section.
2. Support the engine.
3. Remove or disconnect the following:
   - Transaxle
   - Flywheel
   - Rear main seal using a prytool

✳✳ WARNING

Be careful not to damage the crankshaft sealing surface.

To install:

4. Install or connect the following:
   - Seal using tool J 42067
   - Flywheel. Tighten the flywheel-to-crankshaft bolts to 39 ft. lbs. (53 Nm) plus an additional 25 degrees.
   - Transaxle
5. Start the engine and check for leaks.

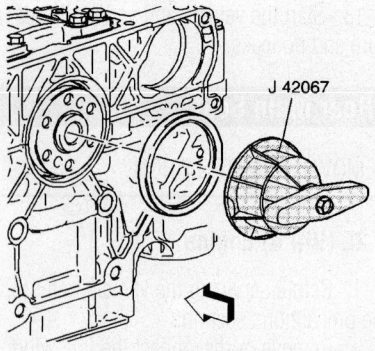

J 42067

42372-NB0D-G22

Install rear main seal using tool J 42067—2.2L (VIN F) engine

2.4L Engine

1. Before servicing the vehicle, refer to the precautions section.
2. Remove or disconnect the following:
   - Negative battery cable
   - Transaxle
   - Clutch/pressure plate assembly, if equipped with a manual transmission
   - Flywheel
   - Seal housing and discard the gasket
   - Oil seal from the transaxle side of the seal housing

To install:

3. Clean and inspect the gasket mounting surfaces.
4. If necessary, add silicone sealer along the oil pan-to-cylinder block mating surface.
5. Lubricate the new rear main bearing seal with engine oil.
6. Install or connect the following:
   - New rear main bearing seal into the seal housing using installer tool J-36005 until it's flush with the housing.
   - Oil seal housing with a new gasket. Torque the bolts to 106 inch lbs. (12 Nm).
   - Flywheel, using new bolts. Torque the bolts to 22 ft. lbs. (30 Nm) plus an additional 45 degree turn.
   - Clutch/pressure plate assembly, if equipped with a manual transmission
   - Transaxle
   - Negative battery cable
7. Start the engine and check for leaks.

Timing Chain, Sprockets, Front Cover and Seal

REMOVAL & INSTALLATION

2.2L (VIN 4) Engine

1. Before servicing the vehicle, refer to the precautions section.
2. Remove or disconnect the following:
   - Negative battery cable
   - Accessory drive belt
   - Drive belt tensioner
3. Install an engine support fixture.
4. Remove or disconnect the following:
   - Engine mount assembly
   - Alternator
   - Power steering pump and move it aside with the lines attached
   - Oil pan
   - Crankshaft pulley/hub
   - Front cover
5. Position the No. 1 piston at Top Dead Center (TDC) of the compression stroke so the camshaft and crankshaft sprockets timing marks are aligned.
6. Remove or disconnect the following:
   - Timing chain tensioner
   - Camshaft sprocket and chain as an assembly
   - Crankshaft sprocket, using a puller

To install:

7. Lubricate the timing chain with clean engine oil.
8. Install the crankshaft sprocket, press it onto the crankshaft.
9. Install the timing chain over the camshaft sprocket, then around the crankshaft sprocket. Be sure the marks on the 2 sprockets are in alignment. Lubricate the thrust surface with Molykote® or equivalent.
10. Install or connect the following:
    - Camshaft sprocket. Torque the bolts to 96 ft. lbs. (130 Nm).
11. Compress the timing chain tensioner spring. Insert a cotter pin or a nail into the hole in the tensioner to retain the timing chain tensioner shoe.
    - Chain tensioner. Torque the bolts to 18 ft. lbs. (24 Nm).
12. Remove the cotter pin or nail from the hole in the tensioner.
    - Front cover with a new gasket. Torque the bolts to 97 inch lbs. (11 Nm).
    - Crankshaft pulley/hub. Torque the pulley bolts to 37 ft. lbs. (50 Nm) and the hub bolt to 77 ft. lbs. (105 Nm).
    - Oil pan with new gaskets. Torque the nuts and bolts to 89 inch lbs. (10 Nm).
    - Power steering pump. Torque the bolts to 22 ft. lbs. (30 Nm).
    - Alternator. Torque the bolts to 37 ft. lbs. (50 Nm).

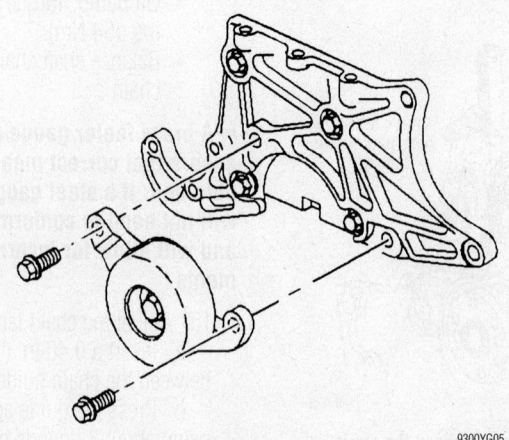

9300YG05

Exploded view of the accessory drive belt tensioner mounting—2.2L (VIN4) engine

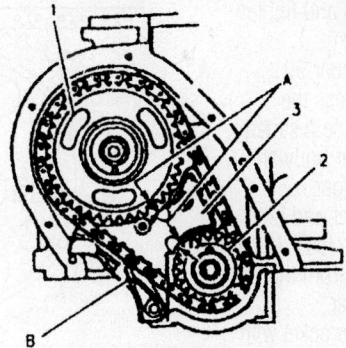

1. Camshaft sprocket
2. Crankshaft sprocket
3. Timing chain tensioner
A. Line up timing marks on sprockets with tabs on timing chain tensioner
B. Remove pin after timing chain is installed

7922YG16

Align the sprocket timing marks with the alignment tabs on the tensioner during timing chain installation—2.2L (VIN4) engine

- Engine mount assembly. Torque the bolts to 55 ft. lbs. (75 Nm).
13. Remove the engine support fixture.
14. Install or connect the following:
    - Accessory drive belt tensioner. Torque the bolts to 37 ft. lbs. (50 Nm).
    - Accessory drive belt
    - Negative battery cable

2.2L (VIN F) Engine

1. Before servicing the vehicle, refer to the precautions section.
2. Remove the valve cover.
3. Remove the engine front cover as follows:
    a. Remove the front wheel.
    b. Remove the engine splash shield.
    c. Remove the drive belt.
    d. Use crankshaft holding tool J 38122-A to prevent the crankshaft from rotating while loosening the crankshaft balancer bolt.

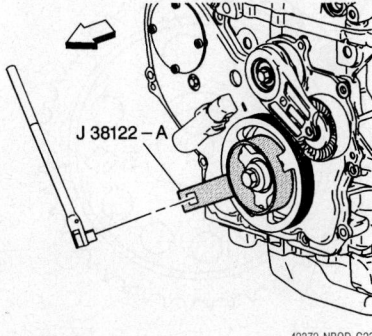

J 38122 – A

42372-NBOD-G23

Use crankshaft holding tool J 38122-A to prevent the crankshaft from rotating while loosening the crankshaft balancer bolt— 2.2L (VIN F) engine

e. Remove the crankshaft balancer bolt and discard. Remove the balancer.
f. Remove the drive belt tensioner.
g. Remove the water pump-to-front cover bolt.
h. Remaining front cover-to-engine bolts, the cover and gasket.
4. Rotate the engine until the crankshaft sprocket mark aligns with the second silver link (2) at the 5 o'clock position. Refer to the illustration.
5. Make sure the INT diamond on the intake camshaft sprocket is aligned with the copper link at (1) at the 2 o'clock position. Refer to the illustration.
6. Make sure the EXH triangle on the exhaust camshaft sprocket is aligned with the silver link (3). Refer to the illustration.
7. Remove or disconnect the following:
    - Timing chain tensioner
    - Timing chain tensioner guide
    - Fixed timing chain guide access plug
    - Fixed timing chain guide

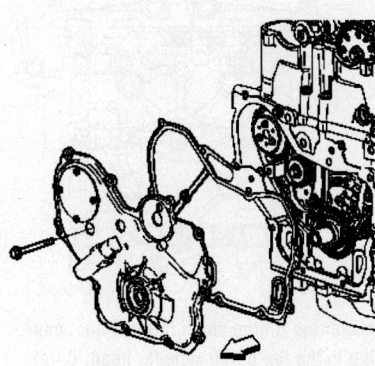

42372-NBOD-G24

Remove the water pump-to-front cover bolt—2.2L (VIN F) engine

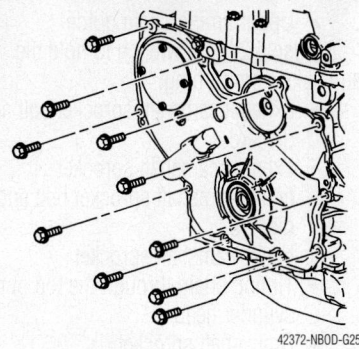

42372-NBOD-G25

Remove the front cover bolts and cover— 2.2L (VIN F) engine

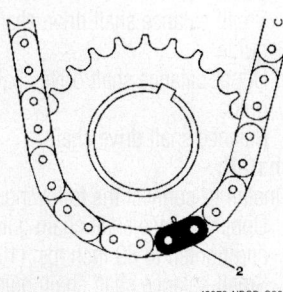

42372-NBOD-G26

Rotate the engine until the crankshaft sprocket mark aligns with the second silver link (2) at the 5 o'clock position—2.2L (VIN F) engine

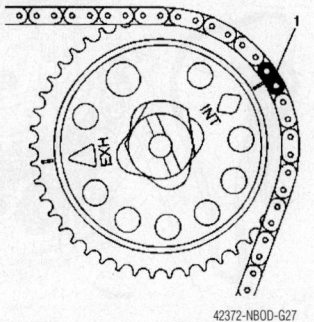

42372-NBOD-G27

Make sure the INT diamond on the intake camshaft sprocket is aligned with the copper link at (1) at the 2 o'clock position— 2.2L (VIN F) engine

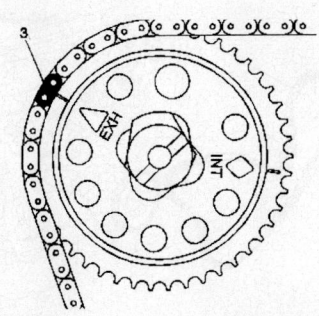

42372-NBOD-G28

Make sure the EXH triangle on the exhaust camshaft sprocket is aligned with the silver link (3)—2.2L (VIN F) engine

- Upper timing chain guide

8. Use a 24 mm wrench to hold the camshafts from turning.
- Exhaust camshaft sprocket bolt and discard
- Exhaust camshaft sprocket
- Intake camshaft sprocket bolt and discard
- Intake camshaft sprocket
- Timing chain through the top of the cylinder head
- Crankshaft sprocket
- Balance shaft drive chain tensioner
- Adjustable balance shaft chain guide
- Small balance shaft drive chain guide
- Upper balance shaft drive chain guide
- Balance shaft drive chain.

To install:

9. Install or connect the following:
- Upper balance shaft chain guide and tighten to 89 inch lbs. (10 Nm)
- Small balance shaft chain guide and bolts and tighten to 89 inch lbs. (10 Nm)
- Adjustable balance shaft drive

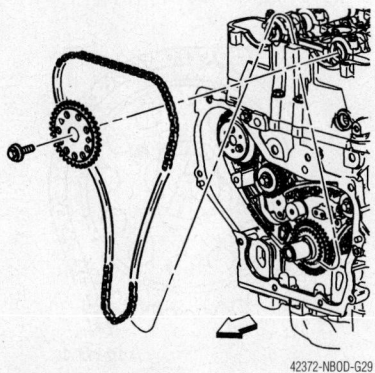

Remove the intake camshaft sprocket, then the chain through the top of the head—2.2L (VIN F) engine

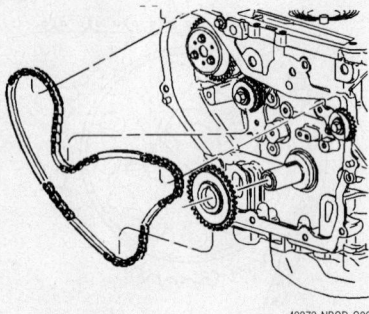

Remove the upper balance shaft chain guide, then the chain—2.2L (VIN F) engine

chain guide and bolts and tighten to 89 inch lbs. (10 Nm)

10. Turn the tensioner plunger 90 degrees in its bore and compress the plunger until a paper clip can be inserted through the hole in the plunger body and into hole in the tensioner plunger.
- Timing chain tensioner and bolts and tighten to 89 inch lbs. (10 Nm)

11. Remove the paper clip from the balance shaft drive chain tensioner.

12. Install the crankshaft sprocket with timing mark at the 5 o'clock position.

13. Lower the timing chain through the opening in the top of the cylinder head. Carefully ensure that the chain goes around both sides of the cylinder block bosses.

14. Install or connect the following:
- Intake camshaft sprocket with the INT diamond at the 2 o'clock position

15. Hand tighten a NEW intake camshaft sprocket bolt.

16. Route the timing chain around the

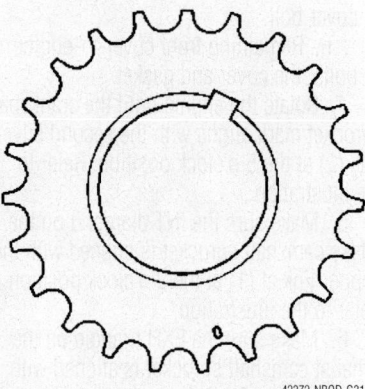

Install the crankshaft sprocket with timing mark at the 5 o'clock position—2.2L (VIN F) engine

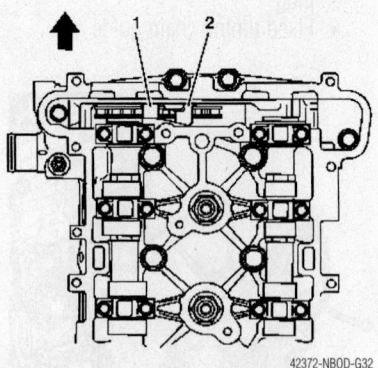

Lower the timing chain through the opening in the top of the cylinder head. Carefully ensure that the chain goes around both sides of the cylinder block bosses (1 & 2)—2.2L (VIN F) engine

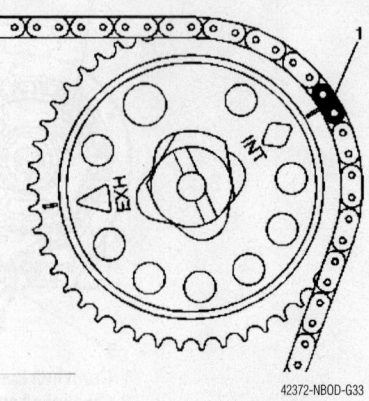

Install the intake camshaft sprocket with the INT diamond at the 2 o'clock position—2.2L (VIN F) engine

crankshaft sprocket with the second silver link aligning with the timing mark.

17. Route the timing chain around the intake camshaft sprocket with the copper colored link aligning with the INT diamond.
- Timing chain tensioner guide through the opening in the top of the cylinder head and tighten the bolts to 89 inch lbs. (10 Nm).
- Exhaust camshaft sprocket with the timing chain silver link at EXH triangle aligned at the 10 o'clock position.

18. Use a 24 mm wrench to rotate the camshaft slightly, until exhaust sprocket aligns with the camshaft.

19. Hand tighten the NEW exhaust camshaft sprocket bolt.

20. Install the fixed timing chain guide and tighten the bolts to 89 inch lbs. (10 Nm).

21. Apply sealant, GM P/N 12345382 compound to thread and install the timing chain guide bolt access hole plug. Tighten the access hole plug to 30 ft. lbs. (40 Nm).

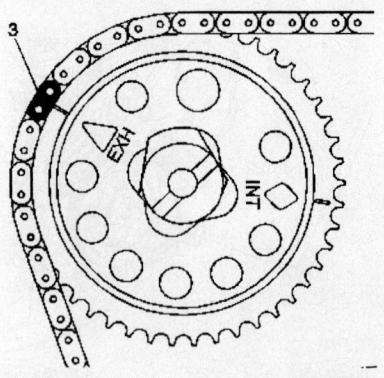

Install the exhaust camshaft sprocket with the timing chain silver link at EXH triangle aligned at the 10 o'clock position—2.2L (VIN F) engine

- Timing chain upper guide and bolts and tighten to 89 inch lbs. (10 Nm)

22. Measure the timing chain tensioner. In a fully compressed, non-active state, the tensioner will measure 2.83 in (72mm). A tensioner in the active state will measure 3.35 in (85mm). To put the tensioner in a non-active state, hold the flat end of the tensioner with a wrench and rotate the piston clockwise for slightly less than one full turn.

- Timing chain tensioner and tighten the tensioner to 55 ft. lbs. (75 Nm)

23. Use a suitable tool with a rubber tip on the end. Feed the tool down through the camshaft drive chant to rest on the timing chain. Then give a sharp jolt diagonally downwards to release the tensioner.

24. Use a 24 mm wrench to hold the camshaft and tighten the new bolts to 63 ft. lbs. (85 Nm) plus an additional 30 degrees.

25. Install the valve cover.

26. Install the engine front cover as follows:

   a. Install front cover gasket, cover and cover-to-engine bolts.

   b. Install the water pump-to-front cover bolts. Tighten all the cover bolts to 15 ft. lbs. (20 Nm).

   c. Install the drive belt tensioner and tighten the bolt to 33 ft. lbs. (45 Nm).

   d. Install the crankshaft balancer and bolt.

   e. Use crankshaft holding tool J 38122-A to prevent the crankshaft from rotating while tightening the crankshaft balancer bolt. Tighten the bolt to 74 ft. lbs. (100 Nm) plus an additional 75 degrees.

   f. Install the engine splash shield.

   g. Install the drive belt.

   h. Install the front wheel.

27. Connect negative battery cable.

2.4L Engine

➡**It is recommended that the entire procedure be reviewed before attempting to service the timing chain.**

1. Before servicing the vehicle, refer to the precautions section.

2. Disconnect the negative battery cable.

3. Drain the cooling system.

4. Install an engine support.

5. Remove or disconnect the following:

- Coolant surge tank
- Accessory drive belt
- Upper fasteners from the front cover
- Right engine mount and bracket
- Right front wheel
- Splash shield from the right wheel well
- Crankshaft balancer
- Front cover

6. Rotate the crankshaft clockwise, as

viewed from front of engine (normal rotation), until the camshaft sprocket's timing dowel pin holes align with the holes in the timing chain housing. The mark on the crankshaft sprocket should align with the mark on the cylinder block. The crankshaft sprocket keyway should point upwards and align with the center line of the cylinder bores. This is the normal timed position.

7. Remove or disconnect the following:

- Timing chain guides

➡**Make sure that all of the slack in the timing chain is above the tensioner assembly.**

The timing chain must be disengaged from any wear grooves in the tensioner shoe in order to remove the shoe. Slide a screwdriver blade under the timing chain while pulling the shoe outward.

- Tensioner assembly

8. Mark the crankshaft sprocket and the timing chain outer surface.

- Timing chain

To install:

9. Install the intake camshaft sprocket. Torque the bolt to 52 ft. lbs. (70 Nm).

➡**Install the Special tool J 36008-A through the holes in the camshaft sprockets and into the holes in the timing chain housing. This positions the camshafts for correct timing.**

10. If the camshafts are out of position and must be rotated more than ⅛ turn in order to install the alignment dowel pins, perform the following:

   a. Rotate the crankshaft 90 degrees clockwise off Top Dead Center (TDC) in order to give the valves adequate clearance to open.

   b. Once the camshafts are in position and the dowels installed, rotate the crankshaft counterclockwise back to TDC.

✳✳ WARNING

Do not rotate the crankshaft clockwise to TDC or valve and piston damage may occur.

11. Install the timing chain over the exhaust camshaft sprocket, around the coolant pump sprocket and around the crankshaft sprocket.

12. Remove the alignment dowel pin from the intake camshaft. Using tool J 39579, rotate the intake camshaft sprocket counterclockwise enough to slide the timing chain over the intake camshaft sprocket. Release the camshaft sprocket wrench. The length of chain between the 2 camshaft sprockets will tighten. If properly timed, the

intake camshaft alignment dowel pin should slide in easily. If the dowel pin does not fully index, the camshafts are not timed correctly and the procedure must be repeated.

13. Leave the alignment dowel pins installed.

14. With slack removed from the chain between the intake camshaft sprocket and the crankshaft sprocket, the crankshaft keyway and the cylinder block mark should be aligned. If not aligned, move the chain 1 tooth forward or rearward. Remove the slack and recheck marks.

15. Reload timing chain tensioner assembly to its **0** position as follows:

   a. Insert the tensioner plunger assembly into the tensioner housing.

   b. With the tensioner plunger fully extended, turn the complete assembly upside down on a flat surface.

   c. Press the bottom of the tensioner housing to compress the plunger into the housing until it is seated.

   d. Make sure that the plunger does not extend out of the tensioner housing more than 0.07 in. (1.7mm).

   e. Loosely install the tensioner assembly to the timing chain housing.

16. Install or connect the following:

- Tensioner shoe on the stud.
- Tensioner assembly by applying hand pressure on the timing chain tensioner shoe until the locking tab seats in the stud groove. Tighten the bolts to 89 inch lbs. (10 Nm).

✳✳ WARNING

If the timing chain tensioner plunger is not released from the installation position, engine damage will occur upon start up.

17. Release the tensioner plunger by firmly pressing a flat blade tool against the plunger face.

18. Remove tool J 36008-A from the camshaft sprockets.

   a. Rotate crankshaft clockwise 2 full rotations. Align crankshaft keyway with mark on cylinder block and reinstall alignment dowel pins. Alignment dowel pins will slide in easily if engine is timed correctly.

19. Install or connect the following:

- Timing chain guides
- New seal lubricated with engine oil
- Front cover using new gaskets. Tighten the nuts and bolts to 115 inch lbs. (13 Nm).
- Crankshaft balancer. Torque the bolt to 129 ft. lbs. (175 Nm) plus a 90 degree turn.

- Right front wheel well splash shield
- Wheel
- Right engine mount bracket. Torque the bolts to 44 ft. lbs. (60 Nm) plus an additional 60 degree turn.
- Right engine mount. Torque the bolts to 55 ft. lbs. (75 Nm).

20. Remove the engine support.
21. Install or connect the following:
- Accessory drive belt
- Coolant surge tank
- Negative battery cable

22. Refill the cooling system and check for leaks.

Piston and Ring

POSITIONING

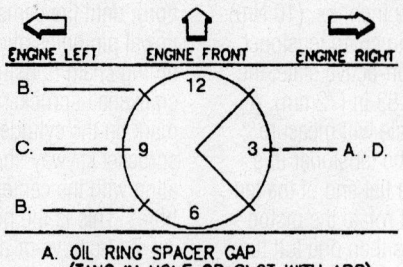

A. OIL RING SPACER GAP
   (TANG IN HOLE OR SLOT WITH ARC)
B. OIL RING RAIL GAPS
C. 2ND COMPRESSION RING GAP
D. TOP COMPRESSION RING GAP

7922AG46

Piston ring end-gap spacing—2.2L engine

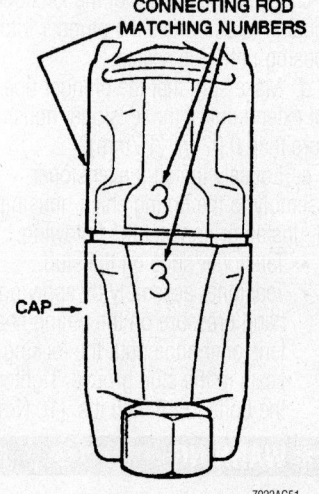

Connecting rod and cap installation. Be sure to matchmark the cap and rod prior to disassembly—2.2L and 2.4L engines

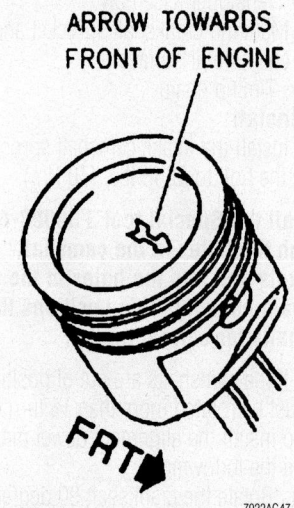

Piston positioning. Often the arrow is replaced by a notch, which also must face toward the front of the engine—2.2L engine

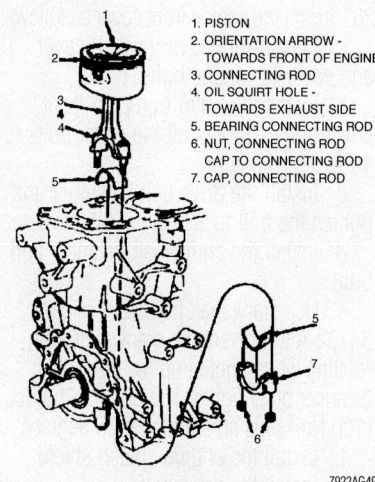

1. PISTON
2. ORIENTATION ARROW - TOWARDS FRONT OF ENGINE
3. CONNECTING ROD
4. OIL SQUIRT HOLE - TOWARDS EXHAUST SIDE
5. BEARING CONNECTING ROD
6. NUT, CONNECTING ROD CAP TO CONNECTING ROD
7. CAP, CONNECTING ROD

7922AG49

Piston and connecting rod assembly positioning—2.4L engine

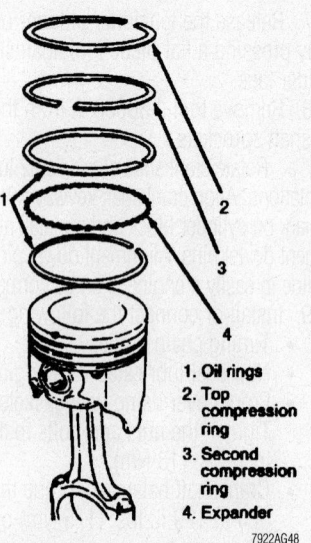

1. Oil rings
2. Top compression ring
3. Second compression ring
4. Expander

7922AG48

Piston ring positioning—2.2L engine

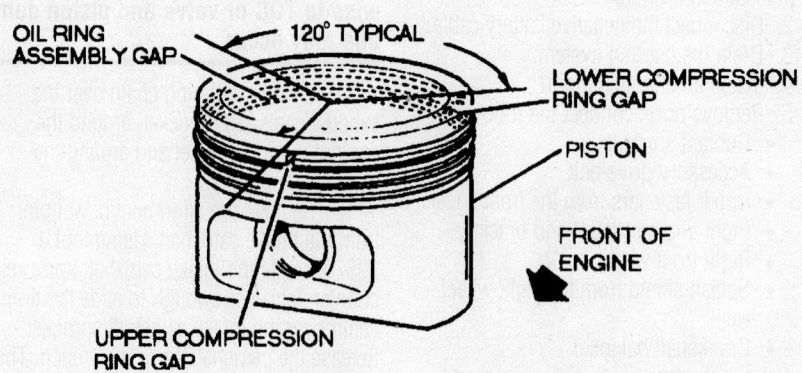

Piston ring end-gap spacing—2.4L engine

7922AG50

FUEL SYSTEM

Fuel System Service Precautions

Safety is the most important factor when performing not only fuel system maintenance but any type of maintenance. Failure to conduct maintenance and repairs in a safe manner may result in serious personal injury or death. Maintenance and testing of the vehicle's fuel system components can be accomplished safely and effectively by adhering to the following rules and guidelines.

• To avoid the possibility of fire and personal injury, always disconnect the negative battery cable unless the repair or test procedure requires that battery voltage be applied.

• Always relieve the fuel system pressure prior to disconnecting any fuel system component (injector, fuel rail, pressure regulator, etc.), fitting or fuel line connection. Exercise extreme caution whenever relieving fuel system pressure, to avoid exposing skin, face and eyes to fuel spray. Please be advised that fuel under pressure may penetrate the skin or any part of the body that it contacts.

• Always place a shop towel or cloth around the fitting or connection prior to loosening to absorb any excess fuel due to spillage. Ensure that all fuel spillage (should it occur) is quickly removed from engine surfaces. Ensure that all fuel soaked cloths or towels are deposited into a suitable waste container.

• Always keep a dry chemical (Class B) fire extinguisher near the work area.

• Do not allow fuel spray or fuel vapors to come into contact with a spark or open flame.

• Always use a back-up wrench when loosening and tightening fuel line connection fittings. This will prevent unnecessary stress and torsion to fuel line piping.

• Always replace worn fuel fitting O-rings with new. Do not substitute fuel hose or equivalent, where fuel pipe is installed.

Fuel System Pressure

RELIEVING

2.4L Engine

1. Before servicing the vehicle, refer to the precautions section.
2. Loosen the fuel filler cap in order to relieve the pressure in the tank (do not tighten at this time).

3. Detach the fuel pump electrical connector.
4. Start and run the vehicle until it stalls, then engage the starter for an additional 3 seconds to ensure the relief of any remaining pressure.
5. Disconnect the negative battery cable.
6. Once the tests or repairs are completed, reattach the fuel pump electrical connector.
7. Connect the negative battery cable.
8. Tighten the fuel filler cap.
9. Prime the fuel system by cycling the ignition switch **ON** for 2 seconds, **OFF** for 10 seconds, then **ON** again. Repeat, if necessary to build system pressure.

2.2L Engines

1. Before servicing the vehicle, refer to the precautions section.
2. Disconnect the negative battery cable in order to avoid possible fuel discharge if an accidental attempt is made to start the engine.
3. Loosen the fuel tank filler cap in order to relieve fuel tank pressure.
4. Connect a fuel pressure gauge (with bleed hose) to the fuel pressure test port connection. Wrap a towel around the fuel pressure connection when installing the fuel pressure gauge in order to avoid fuel spillage.
5. Install the bleed hose into an approved container and open the valve in order to bleed the fuel system pressure. The fuel pipe connections are now safe for servicing.
6. Drain any fuel remaining in the fuel pressure gauge into an approved container.

Fuel Filter

REMOVAL & INSTALLATION

1. Before servicing the vehicle, refer to the precautions section.
2. Relieve the fuel system pressure.
3. Remove the fuel filter from the fuel line using a back-up wrench on the filter.
4. Grasp the filter and nylon connection line fitting. Twist the quick-connect fitting ¼ turn in each direction to loosen any dirt within the fitting. Disconnect the quick-connect fitting from the fuel filter.
5. Remove the fuel filter from the bracket.

To install:

6. Before installing a new filter, always apply a few drops of clean engine oil to the

male tube end of the filter and to the fuel sending unit assembly connection. This will help ensure proper connection and prevent possible fuel leaks. During normal operation, the O-rings located in the female connector will swell and may prevent proper connection if not lubricated.
7. Install the fuel filter in the mounting bracket.
8. Connect the quick-connect fitting to the fuel filter using the following procedure:
   a. Apply a few drops of clean engine oil to the male ends of the filter and the fuel sender assembly.
   b. Push the connectors together to cause the retaining tabs/fingers to snap into place.
   c. Once installed, pull on both ends of each connection to be sure they are secure.
9. Install or connect the following:
   • New O-ring
   • Fuel filter. Torque the fitting to 20 ft. lbs. (27 Nm).
   • Negative battery cable
10. Pressurize the fuel system and verify no leaks.

Fuel Pump

REMOVAL & INSTALLATION

1. Before servicing the vehicle, refer to the precautions section.
2. Relieve the fuel system pressure.
3. Drain the fuel tank.
4. Remove or disconnect the following:
5. While holding the modular fuel sender assembly down, remove the snapring, if equipped or if equipped with a cam lock ring use Fuel Sender Lock Nut Tool J-39765 to press down and rotate the cam lock ring to remove it.

✳✳ WARNING

If the modular fuel sender assembly is retained by a snapring, it may spring up from its position. When removing the modular fuel sender from the tank, be aware that the reservoir bucket is full of fuel. It must be tipped slightly during removal to avoid damage to the float.

• Fuel sender assembly
• External fuel strainer
• Connector retainer from the wiring harness and the fuel pump

6. Gently release the tabs on the sides of the fuel sender at the cover assembly. Begin by squeezing the sides of the reservoir and releasing the tab opposite the fuel level sensor. Move clockwise to release the second and third tab in the same manner.

7. Remove or disconnect the following:

- Fuel pump electrical connection by lifting the cover assembly
- Baffle and pump assembly from the retainer by rotating the fuel pump baffle counterclockwise
- Fuel pump outlet by sliding it out of slot
- Fuel pump outlet seal

To install:

8. Install or connect the following:

- Fuel pump outlet with a new seal by sliding it in the reservoir cover slots
- Fuel pump and baffle assembly onto the reservoir retainer by rotating it clockwise until seated
- Lower retainer assembly partially into the reservoir by aligning all 3 sleeve tabs and pressing the retainer onto the reservoir making sure all 3 tabs are firmly seated

➥**Gently pull on the fuel pump reservoir to assure it is secure. If not secure, replace the entire fuel sender.**

- Connector retainer to the wiring harness and the fuel pump
- External fuel strainer
- Snapring, if equipped to the retainer slots while holding the modular fuel sender assembly down or use Fuel Sender Lock Nut Tool J-39765 in order to install the cam lock ring
- Fuel tank
- Negative battery cable

9. Pressurize the fuel system and verify no leaks.

Fuel Injector

REMOVAL & INSTALLATION

2.2L Engine

1. Before servicing the vehicle, refer to the precautions section.
2. Relieve the fuel system pressure.
3. Remove or disconnect the following:

- Air cleaner outlet resonator
- Vacuum pipe from the fuel pressure regulator
- Fuel pressure regulator, on 2004–05 models

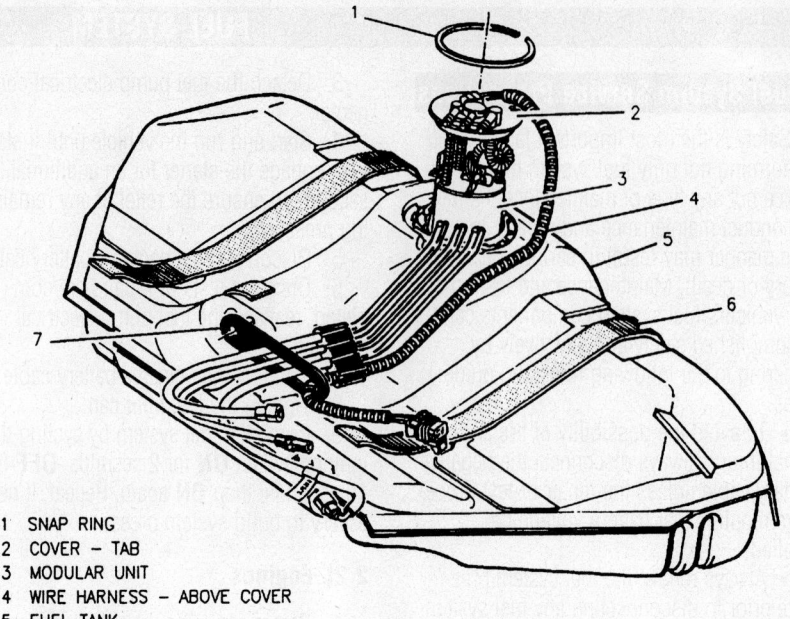

1 SNAP RING
2 COVER – TAB
3 MODULAR UNIT
4 WIRE HARNESS – ABOVE COVER
5 FUEL TANK
6 TANK ISOLATION STRIPS (3)
7 RUBBER ISOLATOR

7922Z325

Exploded view of the fuel sender assembly mounting to the tank

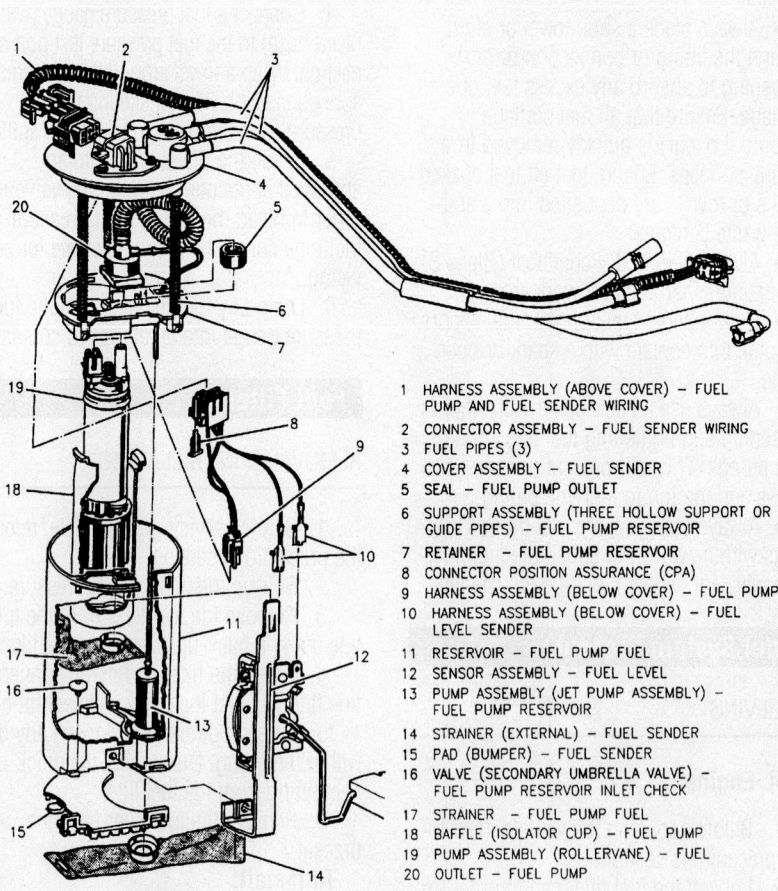

1 HARNESS ASSEMBLY (ABOVE COVER) – FUEL PUMP AND FUEL SENDER WIRING
2 CONNECTOR ASSEMBLY – FUEL SENDER WIRING
3 FUEL PIPES (3)
4 COVER ASSEMBLY – FUEL SENDER
5 SEAL – FUEL PUMP OUTLET
6 SUPPORT ASSEMBLY (THREE HOLLOW SUPPORT OR GUIDE PIPES) – FUEL PUMP RESERVOIR
7 RETAINER – FUEL PUMP RESERVOIR
8 CONNECTOR POSITION ASSURANCE (CPA)
9 HARNESS ASSEMBLY (BELOW COVER) – FUEL PUMP
10 HARNESS ASSEMBLY (BELOW COVER) – FUEL LEVEL SENDER
11 RESERVOIR – FUEL PUMP FUEL
12 SENSOR ASSEMBLY – FUEL LEVEL
13 PUMP ASSEMBLY (JET PUMP ASSEMBLY) – FUEL PUMP RESERVOIR
14 STRAINER (EXTERNAL) – FUEL SENDER
15 PAD (BUMPER) – FUEL SENDER
16 VALVE (SECONDARY UMBRELLA VALVE) – FUEL PUMP RESERVOIR INLET CHECK
17 STRAINER – FUEL PUMP FUEL
18 BAFFLE (ISOLATOR CUP) – FUEL PUMP
19 PUMP ASSEMBLY (ROLLERVANE) – FUEL
20 OUTLET – FUEL PUMP

7922Z326

Exploded view of the fuel pump assembly

- Engine fuel supply and return pipes
- Fuel injector harness connectors
- Fuel rail attaching studs.

➡**Be careful when removing the fuel rail so that you do not damage the injector tips or electrical connections.**

4. Remove the fuel rail as follows:

a. Pull the fuel rail back and upward to remove the fuel injectors from the ports.

b. Rotate the fuel rail to position the injectors downward.

c. Remove the fuel rail.

5. Remove or disconnect the following:
- Fuel injector retainer clip
- Fuel injector from the fuel rail

6. Inspect the fuel injector in order to determine if the upper O-ring was also removed. If the upper O-ring in not removed, remove the O-ring from the fuel rail assembly.
- Fuel injector O-rings and discard

To install:

7. Install or connect the following:
- O-rings on the fuel injector
- Fuel injector clip on the fuel injector

➡**The fuel injector will click when the injector is installed correctly.**

- Fuel injector to the fuel rail with the connector facing upward

8. Install the fuel rail as follows:

a. With the fuel injectors positioned downward, lower the fuel injectors into the ports.

b. Align the injectors by rotating the fuel rail forward.

c. Carefully push the fuel injectors into the cylinder head ports.

d. Install and tighten the fuel rail retainers to 89 inch lbs. (10 Nm).
- Fuel injector harness connectors. Gently pull on the connectors to make sure they are fully engaged.
- Fuel supply and return pipes
- Fuel pressure regulator, on 2004 models and tighten the bolt to 89 inch lbs. (10 Nm)
- Vacuum pipe to the fuel pressure regulator
- Air cleaner outlet resonator
- Negative battery cable

9. Inspect for fuel leaks using the following procedure:

a. Turn ON the ignition, with the engine OFF for 2 seconds.

b. Turn OFF the ignition for 10 seconds.

c. Turn ON the ignition.

d. Inspect for fuel leaks.

2.4L Engine

1. Before servicing the vehicle, refer to the precautions section.

2. Relieve the fuel system pressure.

3. Remove or disconnect the following:
- Air cleaner resonator
- Fuel feed and return lines
- Camshaft Position (CMP) sensor electrical connector
- Fuel injector electrical connectors
- Vacuum hose from the fuel pressure regulator
- Fuel rail
- Fuel pressure regulator from the fuel rail, twist it back and forth
- Fuel injector-to-fuel rail retaining clip
- Fuel injector

To install:

4. Lubricate the O-rings with engine oil prior to installation.

5. Install or connect the following:
- Fuel injector(s) with new O-rings
- Fuel rail. Torque the bolts to 18 ft. lbs. (24 Nm).
- Fuel pressure regulator with a new O-ring to the fuel rail. Torque the bolt to 53 inch lbs. (6 Nm).
- Fuel feed and return lines
- Fuel injector electrical connectors

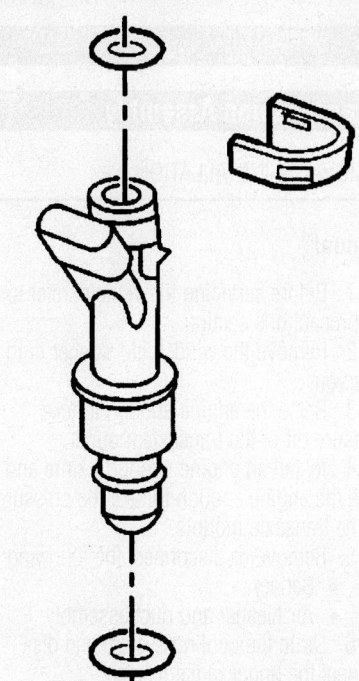

9306YG06

Exploded view of the fuel injector—2.4L engine

- Vacuum hose to the fuel pressure regulator
- CMP sensor electrical connector
- Air cleaner resonator
- Negative battery cable

6. Pressurize the fuel system.

7. Inspect for fuel leaks.

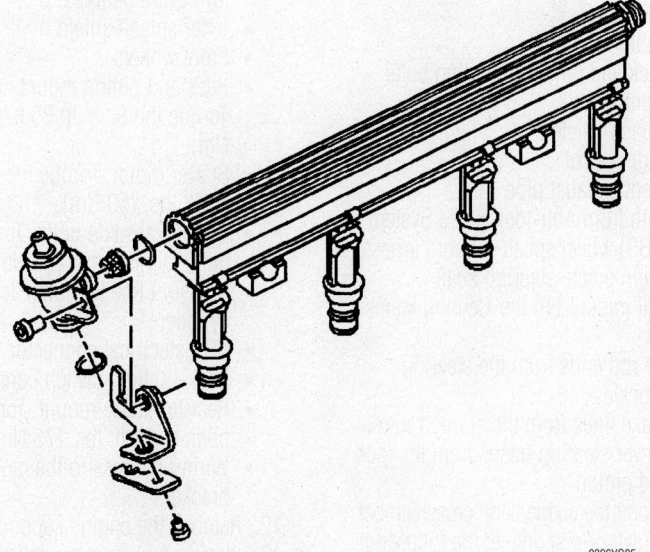

9306YG05

Exploded view of the fuel rail assembly—2.4L engine

DRIVE TRAIN

Transmission Assembly

REMOVAL & INSTALLATION

Manual

1. Before servicing the vehicle, refer to the precautions section.
2. Remove the windshield washer fluid reservoir.
3. Raise the engine enough to take pressure off of the transaxle mounts.
4. Install an engine support fixture and raise the engine enough to take the pressure off the transaxle mounts.
5. Remove or disconnect the following:
   - Battery
   - Air cleaner and duct assembly
6. Drain the cooling system and disconnect the upper radiator hose.
   - Wiring harness from the upper transmission mount bracket
   - Upper transaxle mount-to-transaxle bolts
   - Negative battery cable from the transmission
   - Starter motor
   - Pressure line from the clutch actuator cylinder
   - Back-up light switch connector
   - Rear transmission mount bolts
   - Vehicle Speed Sensor (VSS) electrical connector
   - Upper transmission mounting bolts
   - Shift cables
   - Cable bracket
   - Rack and pinion mounting bolts
   - Front wheels
   - Splash shields
   - Engine strut
   - Front exhaust pipe
   - Both front Anti-lock Brake System (ABS) wheel speed sensor harness
   - Intermediate steering shaft
   - Ball joints from the steering knuckles
   - Tie rod ends from the steering knuckles
   - Brake lines from the support frame
   - Power steering hoses from the rack and pinion
7. Support the suspension crossmember.
8. Remove or disconnect the following:
   - Left side suspension support bolts
   - Flywheel cover
   - Halfshafts from the transaxle
   - Front lower transaxle mount
9. Position a suitable jack under the transaxle.

10. Remove or disconnect the following:
   - Transaxle-to-engine mounting bolts (noting their location)
   - Transaxle from the engine

To install:

11. Install or connect the following:
   - Transaxle
   - Transaxle-to-engine mounting bolts. Torque the bolts to 55 ft. lbs. (75 Nm).
   - Front transaxle mount. Torque the bolt to 44 ft. lbs. (60 Nm).
   - Flywheel cover
   - Halfshafts into the transaxle
   - Suspension support. Torque the bolts to 81 ft. lbs. (110 Nm).
   - Power steering hoses to the rack and pinion
   - Brake lines to the support frame
   - Ball joints to the steering knuckle. Torque the nuts to 48 ft. lbs. (65 Nm).
   - Tie rod ends to the steering knuckles. Torque the nuts to 33 ft. lbs. (45 Nm).
   - Intermediate steering shaft. Torque the pinch bolt to 12 ft. lbs. (17 Nm).
   - ABS wheel speed sensor wiring harness connectors
   - Front exhaust pipe. Torque the bolts to 26 ft. lbs. (35 Nm).
   - Engine strut. Torque the bolts to 74 ft. lbs. (100 Nm) plus an additional turn of 90 degrees.
   - Inner splash shield
   - Front wheels
   - Rack and pinion mounting bolts. Torque the bolts to 89 ft. lbs. (120 Nm).
   - Starter motor. Torque the bolts to 37 ft. lbs. (50 Nm).
   - Upper transaxle bolts. Torque the bolts to 71 ft. lbs. (96 Nm).
   - Pressure line to clutch actuator cylinder
   - VSS electrical connector
   - Back-up light switch connector
   - Rear transaxle mount. Torque the bolts to 55 ft. lbs. (75 Nm).
   - Wiring harness to the mount bracket
12. Remove the engine support fixture.
13. Install or connect the following:
   - Shift cable clamp. Torque the nut to 89 inch lbs. (10 Nm).
   - Shift cables
   - Negative battery cable to the transmission

   - Air cleaner and duct assembly to the throttle body
   - Upper radiator hose
   - Battery
14. Refill the transaxle and cooling system.
15. Refill and bleed the clutch hydraulic system.
16. Road test the vehicle and verify proper operation.

Automatic

1. Before servicing the vehicle, refer to the precautions section.
2. Disconnect the negative battery cable
3. Disconnect the air duct hose from the intake plenum.
4. Install an engine support fixture.
5. Disconnect the wiring harness from the transaxle and the PNP switch.
6. Remove the upper transmission–to–engine bolts and the stud.
7. Remove the front wheels.
8. Remove the left front splash shield.
9. Disconnect the engine strut from the lower engine mount strut bracket.
10. Remove the wiring harness from the retaining clips.
11. Remove the steering gear mounting bolts and secure the steering gear using mechanics wire.
12. Disconnect the wheel speed sensor wires from the both front wheels and from the frame.
13. Separate the ball joints from the steering knuckles
14. Remove the front exhaust pipe.
15. Disconnect both front Anti-lock Brake System (ABS) wheel speed sensors and harness from left suspension support
16. Disconnect both lower ball joints.
17. Remove the tie rods from the steering knuckles.
18. On vehicles with a brake line running along the support frame, remove the brake pipe from the retainers.
19. Remove the front suspension support brace.
20. Lower the vehicle until the front suspension crossmember rests on jackstands.
21. Remove the front suspension crossmember mounting bolts.
22. Raise the vehicle off of the front suspension crossmember.
23. Remove the lower control arms
24. Remove the 2 bolts from the transmission brace.
25. Disconnect the shift cable from the shift linkage.

26. Disconnect the cable from the bracket.

27. Remove the flywheel inspection cover.

28. Remove the starter.

29. Matchmark the position of the flywheel in relation to the torque converter for assembly purposes.

30. Remove the torque converter–to–flywheel bolts.

31. Remove the transmission cooler lines by first removing the nut holding the bracket to the transaxle case.

32. Disconnect the VSS wiring harness from the sensor.

33. Disconnect the halfshafts from the transaxle.

34. Remove the body–to–transmission mount bolts.

35. Lower the transmission and engine with the engine support fixture enough to remove the transmission.

36. Raise the vehicle.

37. Remove the transmission oil pan, drain the fluid from the transmission and reinstall the pan.

38. Support the transmission with a suitable jack.

39. Remove the transaxle–to–engine nut.

40. Separate the engine and the transaxle and remove the transaxle from the vehicle.

To install:

41. Install the transaxle in the vehicle.

42. Install the lower transmission–to–engine bolts and nuts. Tighten to 66 ft. lbs. (90 Nm).

43. Lubricate the transaxle cooler pipes before inserting into seals. Connect the transaxle cooler pipes to the transaxle and tighten the pipes to 71 inch lbs. (8 Nm).

44. Install the torque converter–to–flywheel bolts and tighten to 46 ft. lbs. (62 Nm)

45. Install the halfshafts.

46. Connect the wiring harness to the VSS.

47. Install the starter.

48. Install the flywheel inspection cover bolts and tighten to 89 inch lbs. (10 Nm).

49. Use the engine support fixture to raise the engine and transmission assembly.

50. Install the transaxle mount–to–body bolts and tighten to 66 ft. lbs. (90 Nm).

51. Attach the wiring harness to the clips on the rail.

52. Install the lower control arms.

53. Lower the vehicle on to the front crossmember support.

54. Install the front suspension cross-member mounting bolts and tighten the fas-teners as follows:

- Left rear outboard bolt to 71 ft. lbs. (110 Nm)
- Right rear outboard bolt to 71 ft. lbs. (110 Nm)
- Tighten the front upper bolt to 71 ft. lbs. (110 Nm)
- Tighten the rear inboard bolts to 71 ft. lbs. (110 Nm)

55. Install the front suspension support brace.

56. Install the power steering gear mounting bolts and tighten to 81 ft. lbs. (110 Nm).

57. On vehicles with the brake line run-ning along the frame, install the brake pipe into the retainers.

58. Install the tie rods to the knuckle.

59. Install the ball joints to the knuckle.

60. Connect the ABS harness to the lower control arms.

61. Install the front exhaust pipe.

62. Install the engine strut to the lower engine mount and frame.

63. Connect the wheel speed sensor wiring into the clips.

64. Install the steering gear to the front suspension crossmember.

65. Install the transmission brace bolts and tighten the bolts to 53 ft. lbs. (72 Nm).

66. Install the left front splash shield.

67. Install the front wheels.

68. Install the upper transmission–to–engine bolts and stud and tighten 66 ft. lbs. (90 Nm).

69. Connect the engine wiring harness.

70. Remove the engine support fixture.

71. Connect the shift linkage to the transmission.

72. Connect the electrical connectors to the PNP switch and transaxle.

73. Connect the air duct hose to the intake plenum.

74. Connect the negative battery cable.

75. Refill the transmission with correct type and amount of fluid.

Clutch

REMOVAL & INSTALLATION

1. Before servicing the vehicle, refer to the precautions section.

2. Remove or disconnect the following:
- Negative battery cable
- Clutch master cylinder pushrod from the clutch pedal, if necessary
- Transaxle

3. If any of the parts are to be reused, mark the pressure plate assembly and the flywheel so they can be assembled in the same position.

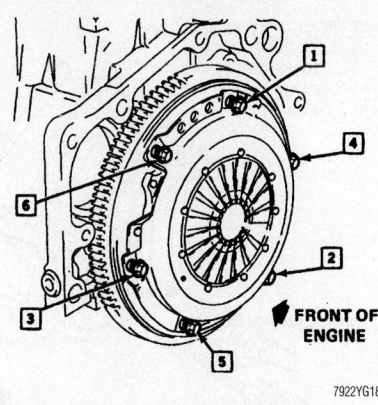

Clutch cover bolt tightening sequence

4. Loosen the attaching bolts 1 turn at a time until spring tension is relieved.

5. Support the pressure plate and remove the bolts.

6. Remove the pressure plate and clutch disc as an assembly.

➡**Do not disassemble the pressure plate assembly. Replace it if defective.**

To install:

7. Install the pressure plate and clutch disc assembly, supporting it with a clutch aligning tool.

8. Torque the pressure plate-to-flywheel bolts, gradually, in a cross pattern, as fol-lows:
   a. Step 1: Lightly seat all bolts.
   b. Step 2: Tighten the bolts to 18 ft. lbs. (24 Nm).

9. Lubricate the outside groove and the inside recess of the release bearing with high temperature grease.

10. Install or connect the following:
- Release bearing
- Transaxle
- Clutch master cylinder pushrod to the clutch pedal and secure with the retaining clip, if removed
- Negative battery cable

11. Bleed clutch system as necessary and road test vehicle.

Hydraulic Clutch System

BLEEDING

1. Before servicing the vehicle, refer to the precautions section.

2. Attach a hose to the bleeder screw on the clutch actuator assembly and submerge the other end of the hose in a container of hydraulic clutch fluid.

3. Depress the clutch pedal slowly and hold.

4. Loosen the bleeder screw to purge air.

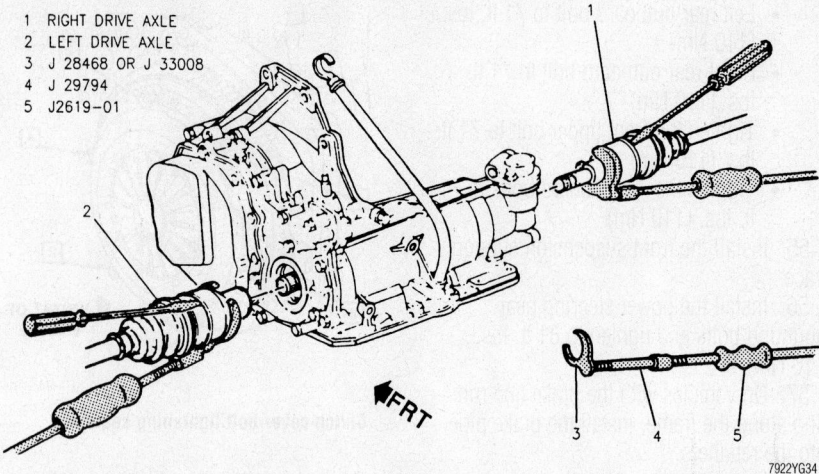

1 RIGHT DRIVE AXLE
2 LEFT DRIVE AXLE
3 J 28468 OR J 33008
4 J 29794
5 J2619-01

FRT

7922YG34

To prevent damaging the transaxle or halfshaft, use the tools as shown to remove the half-shafts

5. Tighten the bleeder screw to 18 inch lbs. (2 Nm).

6. Release the clutch pedal.

7. Repeat Steps 3 through 6 until all air is purged from the system.

8. Refill the reservoir with hydraulic clutch fluid.

9. Repeat this bleeding procedure if there is a grinding noise during the clutch spin down procedure.

Halfshaft

REMOVAL & INSTALLATION

1. Before servicing the vehicle, refer to the precautions section.

2. Remove or disconnect the following:
   • Negative battery cable
   • Front wheel
   • Hub nut and washer
   • Lower ball joint
   • Wheel Speed Sensor (WSS) electrical connector
   • Stabilizer bar link
   • Halfshaft, press it from wheel bearing/hub assembly
   • Halfshaft from the transaxle

✳✳ WARNING

Do not pull the halfshaft by the CV-joint boot or on the joint itself.

To install:

3. Install the halfshaft into the transaxle (or intermediate shaft, if equipped) by placing a brass drift pin into the groove on the joint housing and tapping until seated. Be careful not to damage the axle seal or dislodge the seal garter spring when installing the axle.

➡**Be sure the halfshaft is fully engaged in the transaxle. Verify that the half-shaft is seated by grasping the inner joint housing and pulling outward. Do not pull on the shaft or the boot, but on the inner joint housing only.**

4. Install or connect the following:
   • Halfshaft into the hub/bearing assembly
   • Stabilizer link. Torque the bolts to 13 ft. lbs. (17 Nm).
   • WSS electrical connector
   • Lower ball joint to the steering knuckle. Torque the nut to 41–48 ft. lbs. (55–65 Nm).
   • Washer and a new hub nut. Torque the nut to 144 ft. lbs. (200 Nm).
   • Front wheel
   • Negative battery cable

CV-Joints

OVERHAUL

Outer CV-Joint

1. Before servicing the vehicle, refer to the precautions section.

2. Remove or disconnect the following:
   • Front wheel
   • Halfshaft and position it in a vise
   • Large CV-joint boot clamp
   • Small CV-joint boot clamp
   • CV-joint boot and slide it back on the shaft

3. Perform the following:
   a. Choose a reference mark on the halfshaft.
   b. Measure the distance between the reference mark and the CV-joint inner race face; retain this measurement.

4. Attach CV Puller tool J-41398 to the outer race threaded area.

5. Remove or disconnect the following:
   • CV-joint outer race from the halfshaft using a slide hammer puller
   • CV Puller tool J-41398 and slide hammer puller from the outer race
   • Retaining ring from the halfshaft
   • CV-joint boot from the halfshaft

6. Disassemble the chrome alloy balls from the CV-joint cage as follows:
   a. Position a brass drift against the CV-joint cage and tap it with a hammer to tilt the cage.
   b. Remove the 1st chrome alloy ball from the cage.
   c. Tilt the cage in the opposite direction.
   d. Remove the opposite chrome alloy ball.
   e. Repeat the procedure until all 6 balls are removed.

7. Disassemble the CV-joint cage and inner race as follows:
   a. Pivot the cage and race 90 degrees to the center line of the outer race.
   b. Align the cage windows with outer race lands.
   c. Remove the cage from the outer race.
   d. Rotate the inner race upward and remove it from the cage.

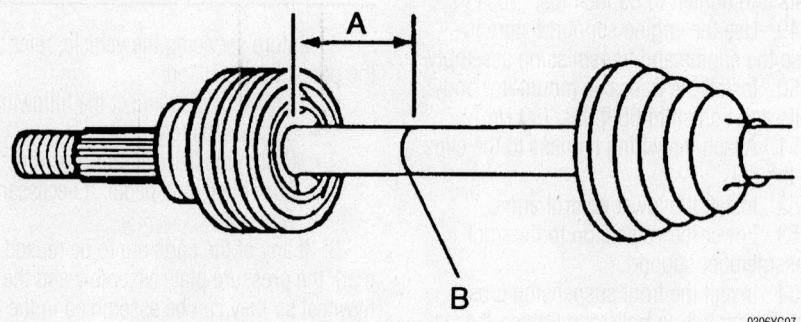

A

B

9306YG07

Measuring the halfshaft-to-inner race reference distance—Outer CV-joint

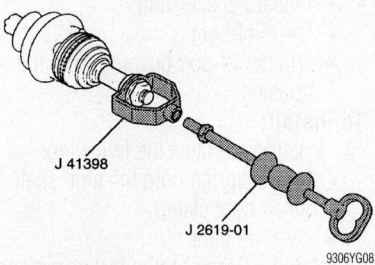

Removing the outer race from the half-shaft—Outer CV-joint

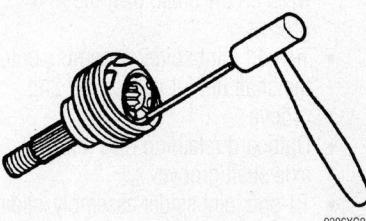

Tilting the cage—Outer CV-joint

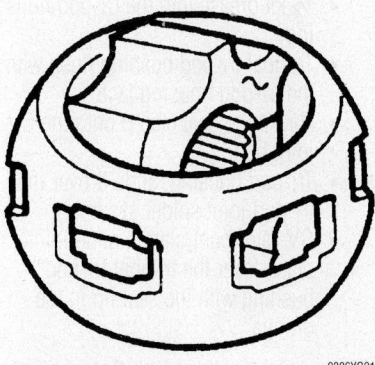

View the cage and inner race—Outer CV-joint

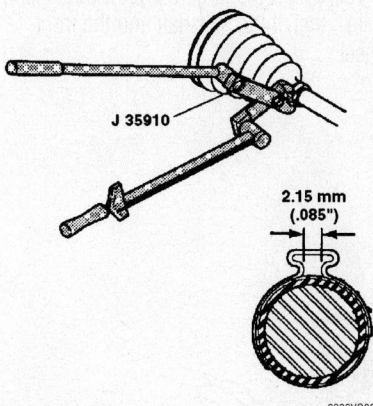

Crimping the small boot clamp—Outer CV-joint

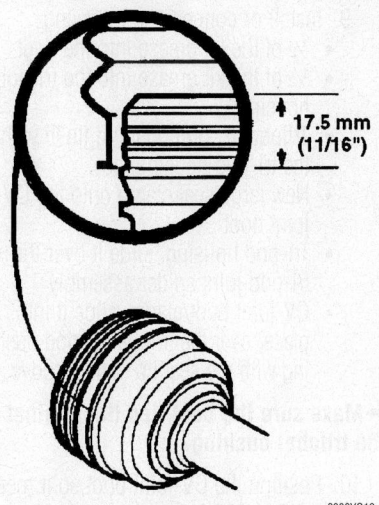

Positioning the CV boot onto the outer race—Outer CV-joint

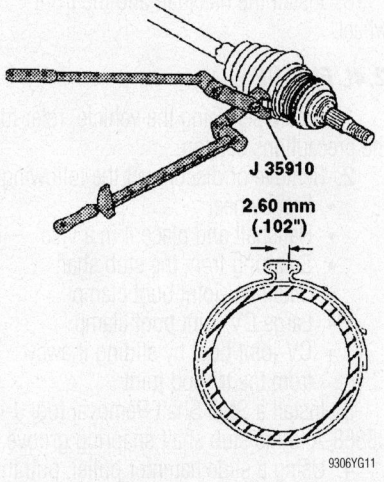

Crimping the large boot clamp—Outer CV-joint

To install:

8. Lubricate the parts with a light coat of grease.

9. Assemble the CV-joint cage and inner race, as follows:

    a. Rotate the inner race 90 degrees to the cage centerline.

    b. Align the cage windows with inner race lands.

    c. Insert the inner race into the cage by rotating the inner race downward.

    d. Insert the cage/inner race into the outer race.

10. Assemble the chrome alloy balls into the CV-joint cage, as follows:

    a. Position a brass drift against the CV-joint cage and tap it with a hammer to tilt the cage.

    b. Insert the 1st chrome alloy ball into the cage.

    c. Tilt the cage in the opposite direction.

    d. Insert the opposite chrome alloy ball.

    e. Repeat the procedure until all 6 balls are inserted.

11. Install ½ of the kit grease into the CV-joint.

12. Install or connect the following:

- Small ring clamp on the CV boot, do not crimp at this time
- CV-joint boot and slide it up the shaft to expose the reference mark
- New retaining ring on the halfshaft
- Large ring clamp on the CV boot
- Outer race assembly by pressing it onto the halfshaft until the ring engages

✳✳ WARNING

When installing the outer race assembly onto the halfshaft, do not exceed 4,000 lbs. pressure.

13. Measure the distance between the reference mark and the CV-joint inner race face; the distance must be +/- 0.039 in. (1mm) of the original measurement. If the measurement is not correct, repress the outer race assembly and recheck it.

14. Slide the small end of the CV-joint boot/clamp into place, with the seal lip in the halfshaft groove.

➡**Make sure the boot lies flat against the halfshaft.**

15. Using a Crimp tool, a torque wrench and a breaker bar, crimp the small CV-joint boot clamp to 100 ft. lbs. (136 Nm).

16. Check the clamp gap dimension; if it is not 0.085 in. (2.15mm), continue tightening the clamp until it is.

17. Install ½ of the kit grease into the CV-joint boot.

18. Slide the large end of the CV boot/clamp into place, with the seal lip in place over the outer race.

➡**Make sure the boot lies flat against the outer race.**

19. Using a Crimp tool, a torque wrench and a breaker bar, crimp the large CV-joint boot clamp to 130 ft. lbs. (176 Nm).

20. Check the clamp gap dimension; if it is not 0.102 in. (2.60mm), continue tightening the clamp until it is.

21. Install the halfshaft and the front wheel.

Inner (Tri-Pod) Joint

2.2L ENGINE

1. Before servicing the vehicle, refer to the precautions section.

2. Remove or disconnect the following:
- Front wheel
- Halfshaft
- Small CV-joint boot clamp
- Large CV-joint boot clamp
- CV-joint boot by sliding it away from the tri-pod joint
- Inboard spacer ring
- Outboard retaining ring
- Tri-pod joint spider assembly by tapping it from the halfshaft with a brass drift
- Tri-pod spider retaining ring
- Trilobal tri-pod bushing from the housing
- CV-joint boot

To install:

3. Install or connect the following:
- Small boot clamp
- CV-joint boot
- New inboard spacer ring slide it rearward on the shaft past the 2nd groove
- Tri-pod joint spider assembly onto the shaft

4. Assemble the tri-pod spider assembly onto the halfshaft as follows:

a. Position the tri-pod spider assembly onto the shop press plate.

b. Position the halfshaft onto the tri-pod spider assembly, in the shop press.

c. Press the halfshaft into the tri-pod spider assembly until the spider assembly passes the 2nd groove.

❊❊ WARNING

When assembling the tri-pod assembly onto the halfshaft, do not exceed 4,000 lbs. pressure.

5. Remove the halfshaft from the shop press and place it in vise.

6. Install or connect the following:
- New outboard retaining ring into the axle shaft groove
- Tri-pod joint spider assembly, slide it against the outboard retaining ring using a brass drift

7. Seat the inboard spacer ring in the proper groove.

8. Slide the CV-joint boot into place and crimp the small boot clamp.

9. Install or connect the following:
- ½ of the kit grease into the boot
- ½ of the kit grease into the tri-pod housing
- Trilobal tri-pod bushing flush with the tri-pod housing face
- New large seal clamp onto the CV-joint boot
- Tri-pod housing, slide it over the tri-pod joint spider assembly
- CV-joint boot/clamp, slide it into place, over the trilobal tri-pod bushing with the seal lip in the groove

➡ **Make sure the boot lies flat against the trilobal bushing.**

10. Position the CV-joint boot so it measures or 3.7 in. (95.7).

11. Using a Crimp tool, a torque wrench and a breaker bar, crimp the small CV-joint boot clamp to 100 ft. lbs. (136 Nm).

12. Crimp the large CV-joint boot clamp.

13. Install the halfshaft and the front wheel.

2.4L ENGINE

1. Before servicing the vehicle, refer to the precautions section.

2. Remove or disconnect the following:
- Front wheel
- Halfshaft and place it in a vise
- Snapring from the stub shaft
- Small CV-joint boot clamp
- Large CV-joint boot clamp
- CV-joint boot by sliding it away from the tri-pod joint

3. Install a Stub Shaft Removal tool J-38868-A to the stub shaft snapring groove.

4. Using a slide hammer puller, pull the stub shaft from the tri-pod housing.

5. Remove or disconnect the following:
- Tri-pod housing from the tri-pod spider
- Inboard spacer ring slide it rearward on the shaft
- Outboard retaining ring
- Tri-pod joint spider assembly

- Inboard spacer ring
- CV-joint boot
- Trilobal tri-pod bushing from the housing

To install:

6. Install or connect the following:
- New snapring onto the stub shaft
- Small boot clamp
- CV-joint boot

7. Using a Crimp tool, a torque wrench and a breaker bar, crimp the small CV-joint boot clamp to 100 ft. lbs. (136 Nm).

8. Install or connect the following:
- Inboard spacer ring slide it rearward on the shaft, past the 2nd groove
- Tri-pod joint spider assembly onto the shaft until it passes the 2nd groove
- Outboard retaining ring into the axle shaft groove
- Tri-pod joint spider assembly, slide it against the outboard retaining ring
- Inboard spacer ring, seat it in the groove
- ½ kit grease into the boot
- ½ kit grease into the tri-pod housing
- Trilobal tri-pod bushing flush with the tri-pod housing face
- New large seal clamp onto the CV-joint boot
- Tri-pod housing, slide it over the tri-pod joint spider assembly
- CV-joint boot/clamp, slide it into place, over the trilobal tri-pod bushing with the seal lip in the groove

➡ **Make sure the boot lies flat against the trilobal bushing.**

9. Position the CV-joint boot so it measures 4.9 in. (125mm).

10. Crimp the large CV-joint boot clamp.

11. Install the halfshaft and the front wheel.

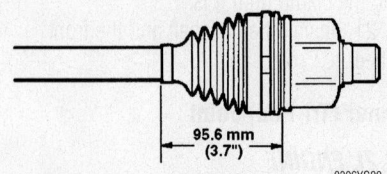

CV-joint boot measurement—Inner (tri-pod) joint—2.2L engine

95.6 mm (3.7")

9306YG20

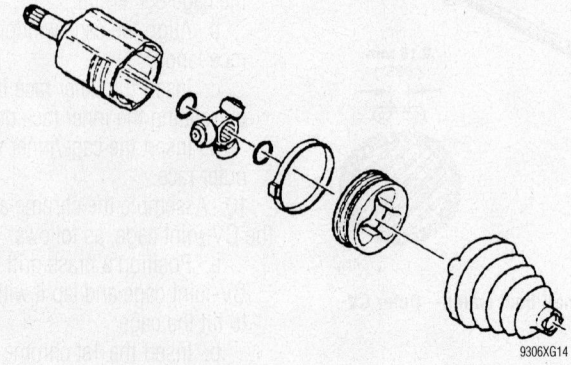

Exploded view of the inner (tri-pod) joint

9306XG14

STEERING AND SUSPENSION

Air Bag

❊❊ CAUTION

Some vehicles are equipped with an air bag system. The system must be disabled before performing service on or around system components, steering column, instrument panel components, wiring and sensors. Failure to follow safety and disabling procedures could result in accidental air bag deployment, possible personal injury and unnecessary system repairs.

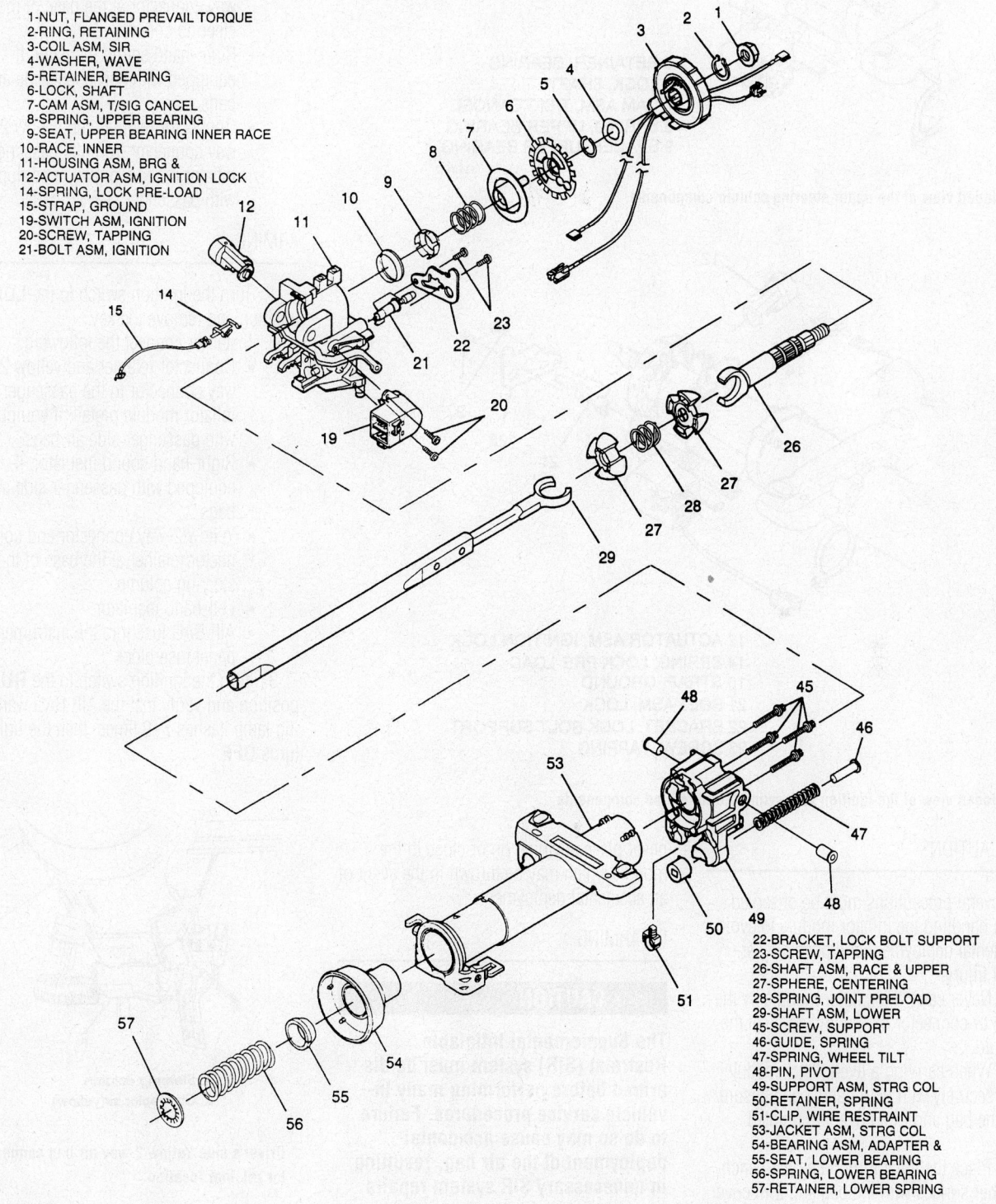

1-NUT, FLANGED PREVAIL TORQUE
2-RING, RETAINING
3-COIL ASM, SIR
4-WASHER, WAVE
5-RETAINER, BEARING
6-LOCK, SHAFT
7-CAM ASM, T/SIG CANCEL
8-SPRING, UPPER BEARING
9-SEAT, UPPER BEARING INNER RACE
10-RACE, INNER
11-HOUSING ASM, BRG &
12-ACTUATOR ASM, IGNITION LOCK
14-SPRING, LOCK PRE-LOAD
15-STRAP, GROUND
19-SWITCH ASM, IGNITION
20-SCREW, TAPPING
21-BOLT ASM, IGNITION

22-BRACKET, LOCK BOLT SUPPORT
23-SCREW, TAPPING
26-SHAFT ASM, RACE & UPPER
27-SPHERE, CENTERING
28-SPRING, JOINT PRELOAD
29-SHAFT ASM, LOWER
45-SCREW, SUPPORT
46-GUIDE, SPRING
47-SPRING, WHEEL TILT
48-PIN, PIVOT
49-SUPPORT ASM, STRG COL
50-RETAINER, SPRING
51-CLIP, WIRE RESTRAINT
53-JACKET ASM, STRG COL
54-BEARING ASM, ADAPTER &
55-SEAT, LOWER BEARING
56-SPRING, LOWER BEARING
57-RETAINER, LOWER SPRING

9300YG10

Exploded view of the steering column with tilt wheel

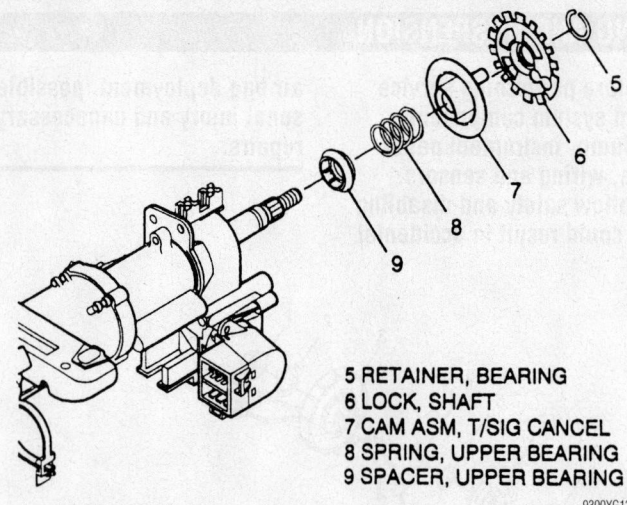

5 RETAINER, BEARING
6 LOCK, SHAFT
7 CAM ASM, T/SIG CANCEL
8 SPRING, UPPER BEARING
9 SPACER, UPPER BEARING

9300YG12

Exploded view of the upper steering column components

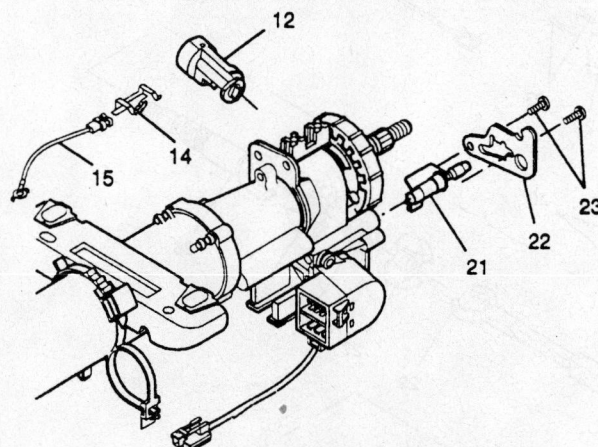

12 ACTUATOR ASM, IGNITION LOCK
14 SPRING, LOCK PRE-LOAD
15 STRAP, GROUND
21 BOLT ASM, LOCK
22 BRACKET, LOCK BOLT SUPPORT
23 SCREW, TAPPING

9300YG11

Exploded view of the ignition lock cylinder and related components

PRECAUTIONS

Several precautions must be observed when handling the inflator module to avoid accidental deployment and possible personal injury.

• Never carry the inflator module by the wires or connector on the underside of the module.

• When carrying a live inflator module, hold securely with both hands, and ensure that the bag and trim cover are pointed away.

• Place the inflator module on a bench or other surface with the bag and trim cover facing up.

• With the inflator module on the bench, never place anything on or close to the module which may be thrown in the event of an accidental deployment.

DISARMING

✳✳ CAUTION

The Supplemental Inflatable Restraint (SIR) system must be disarmed before performing many in-vehicle service procedures. Failure to do so may cause accidental deployment of the air bag, resulting in unnecessary SIR system repairs and/or personal injury.

1. Turn the steering wheel so the vehicle's wheels are pointing straight-ahead.
2. Turn the ignition switch to the **LOCK** position.
3. Remove or disconnect the following:
- Ignition key
- AIR BAG fuse from the instrument panel fuse block
- Left-hand sound insulator
- Connector retainer and yellow 2-way connector at the base of the steering column.
- Right-hand sound insulator, if equipped with passenger side air bags
- Connector retainer and yellow 2-way connector from the passenger inflator module pigtail, if equipped with passenger side air bags

ARMING

1. Turn the ignition switch to the **LOCK** position and remove the key.
2. Install or connect the following:
- Connector retainer and yellow 2-way connector to the passenger inflator module pigtail, if equipped with passenger side air bags
- Right-hand sound insulator, if equipped with passenger side air bags
- Yellow 2-way connector and connector retainer at the base of the steering column
- Left-hand insulator
- AIR BAG fuse into the instrument panel fuse block

3. Turn the ignition switch to the **RUN** position and verify that the AIR BAG warning lamp flashes 7–9 times, then the light turns **OFF**.

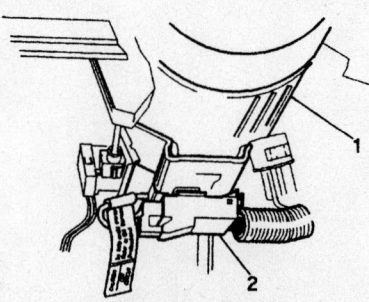

1. Steering column
2. Connector, sir (yellow)

7922YG19

Driver's side Yellow 2-way air bag connector retainer location

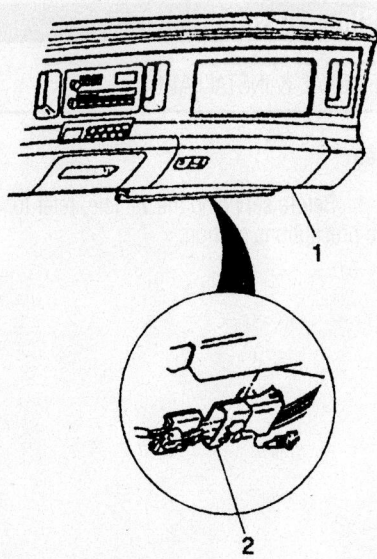

1. I/P compartment
2. Connector,sir(yellow)

7922YG20

Passenger's side Yellow 2-way air bag connector location

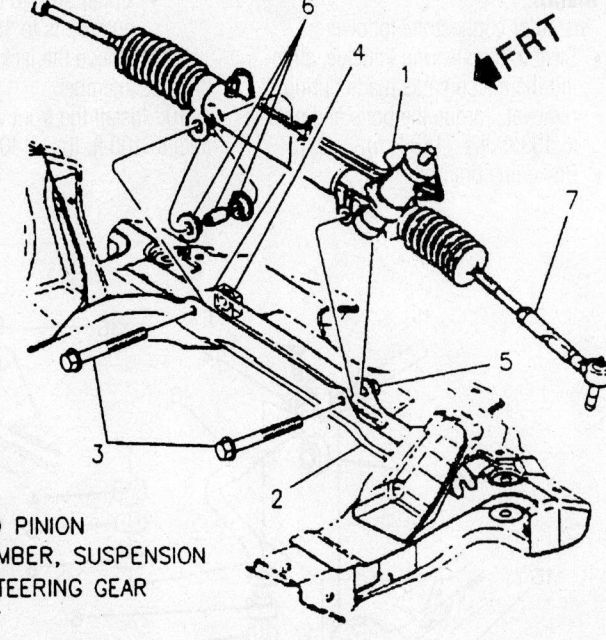

1 RACK AND PINION
2 CROSSMEMBER, SUSPENSION
3 BOLTS, STEERING GEAR
4 CAGE NUT
5 WELD NUTS
6 BUSHING AND SLEEVE
7 TIE ROD

7922YG21

Exploded view of the rack and pinion steering gear mounting

Rack and Pinion Steering Gear

REMOVAL & INSTALLATION

1. Before servicing the vehicle, refer to the precautions section.
2. Remove or disconnect the following:
   - Negative battery cable
   - Front wheels
   - Inlet and outlet hose assemblies from the rack and pinion
   - Tie rod ends from the steering knuckles
   - Intermediate shaft
   - Brake pipes from the retainers on the front suspension support
   - Crossmember bolts, loosen them to gain additional clearance
   - Rack and pinion mounting bolts
   - Rack and pinion

To install:

3. Install the rack and pinion and hand tighten the bolts.
4. Install the crossmember bolts in the following order:
   a. Step 1: Left rear outboard to 71 ft. lbs. (110 Nm).
   b. Step 2: Right rear outboard 2nd to 71 ft. lbs. (110 Nm).
   c. Step 3: Front upper bolts to 71 ft. lbs. (110 Nm).
   d. Step 4: Rear inboard bolts to 71 ft. lbs. (110 Nm).
   e. Tighten the rack and pinion

gear–to–crossmember bolts to 89 ft. lbs. (120 Nm).
5. Install or connect the following:
   - Intermediate shaft. Torque the pinch bolt to 30 ft. lbs. (41 Nm).
   - Brake pipes to the retainers on the front suspension crossmember
   - Tie rod ends. Torque the nuts to 44 ft. lbs. (60 Nm), use new cotter pins.
   - Inlet and outlet hoses to the rack and pinion. Torque the fittings to 20 ft. lbs. (27 Nm).
   - Wheels
   - Negative battery cable
6. Refill and bleed the power steering system.
7. Check and/or adjust the toe setting.
8. Road test vehicle and verify no leaks.

Strut

REMOVAL & INSTALLATION

Front

1. Before servicing the vehicle, refer to the precautions section.
2. Remove the upper strut-to-body nuts and/or bolts.
3. Support the front crossmember with jack stands.

4. Lower the vehicle so the weight of the vehicle rests on the jack stands and NOT the control arms.
5. Before removing front suspension components, their positions should be marked so they may be assembled correctly.

✳✳ WARNING

Whenever working near the halfshaft, use care to prevent damage from over-extension of the halfshaft joints. When either end of the shaft is disconnected, over-extension of the joint could result in separation of the internal components and possible joint failure.

6. Install a drive axle joint protective cover.
7. Remove or disconnect the following:
   - Front wheel
   - Brake line bracket, if necessary
   - Strut from the steering knuckle. Matchmark the assembly before removal.

✳✳ WARNING

The steering knuckle MUST be supported to prevent axle joint over-extension.

To install:

8. Install or connect the following:
- Strut to the steering knuckle, aligning the matchmarks made during removal. Torque the bolts and nuts to 133 ft. lbs. (180 Nm).
- Brake line bracket

- Upper strut to the body, Torque the nuts/bolts to 18 ft. lbs. (25 Nm).

9. Remove the jack stands from under the crossmember.

10. Install the front wheel. Torque the lug nuts to 100 ft. lbs. (140 Nm).

Shock Absorber

REMOVAL & INSTALLATION

Rear

1. Before servicing the vehicle, refer to the precautions section.

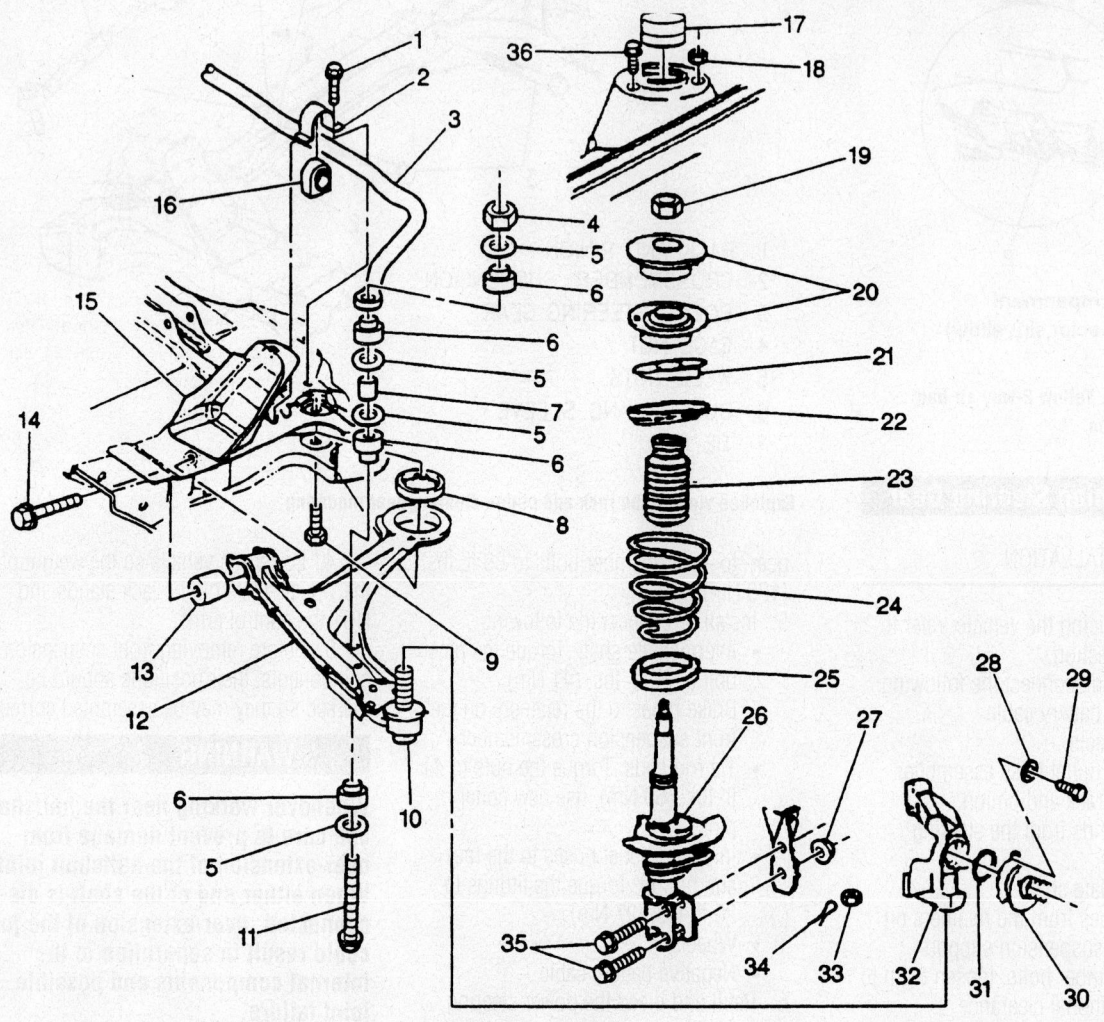

Legend
(1) Bolt
(2) Clamp, Stabilizer Shaft
(3) Stabilizer Shaft
(4) Nut, Stabilizer Link
(5) Washer
(6) Insulator, Stabilizer Link
(7) Spacer Stabilizer Link
(8) Bushing, Vertical
(9) Bolt, Vertical Bushing
(10) Ball Joint
(11) Bolt, Stabilizer Link

(12) Control Arm
(13) Bushing, Control Arm
(14) Lower Spring Insulator
(15) Suspension Support
(16) Insulator, Stabilizer Shaft
(17) Cover, Strut Mount
(18) Nut
(19) Nut, Strut Dampener Shaft
(20) Strut Mount and Rate Washer Assembly
(21) Spring Seat
(22) Upper Spring Insulator
(23) Strut Bumper and Shield

(24) Spring
(25) Lower Spring Insulator
(26) Strut
(27) Nut
(28) Washer
(29) Bolt
(30) Hub and Bearing Assembly
(31) Seal (Part of 5)
(32) Steering Knuckle
(33) Nut, Ball Joint
(34) Cotter Pin
(35) Bolt
(36) Bolt

Exploded view of the front suspension

7922YG22

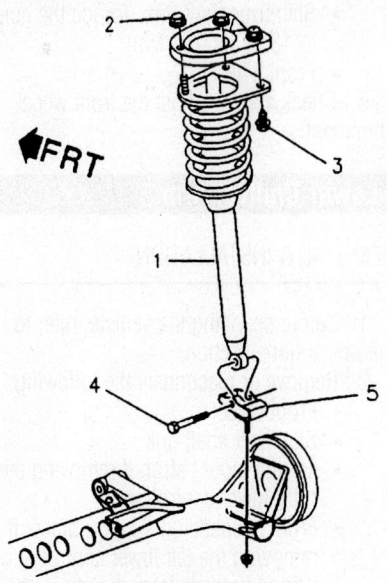

1 SHOCK, COIL-OVER
2 NUT
3 BOLT, UPPER STRUT (2)
4 BOLT, AXLE

7922YG23

Exploded view of the rear shock mounting

2. Open the deck lid and position the carpet out of the way.
3. Raise and safely support the rear axle with jack stands.
4. Remove or disconnect the following:
   • Wheel
   • Upper mounting bolts
   • Lower mounting bolt
   • Shock absorber

To install:
5. Install or connect the following:
   • Shock
   • Lower mounting bolt. Torque it to 51 ft. lbs. (70 Nm).
   • Upper mounting bolts. Torque them to 18 ft. lbs. (25 Nm).
   • Wheel
   • Rear compartment carpet
6. Road test the vehicle.

Coil Spring

REMOVAL & INSTALLATION

Front

1. Before servicing the vehicle, refer to the precautions section.
2. Remove the strut from the vehicle.
3. Mount the strut assembly into a strut compressor. Note that the strut compressor has strut mounting holes drilled for specific vehicle lines.

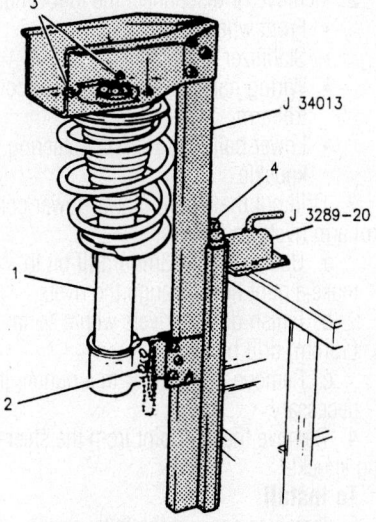

1 STRUT ASSEMBLY
2 INSTALL LOCKING PINS THROUGH STRUT ASSEMBLY
3 TIGHTEN NUTS UNTIL FLUSH WITH STRUT COMPRESSOR
4 COMPRESSOR FORCING SCREW

7922YG24

View of a typical strut assembly mounted in a compressor

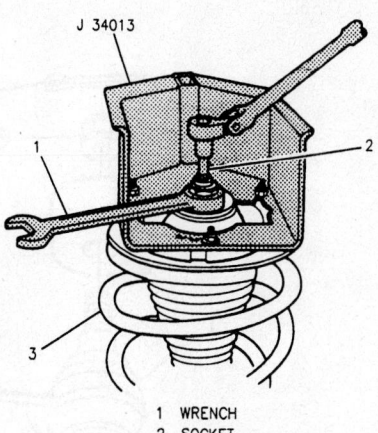

1 WRENCH
2 SOCKET
3 STRUT ASSEMBLY

7922YG25

Use a socket and a wrench to remove the damper shaft nut spring cap while compressing the spring

4. Compress the strut approximately ½ its height after initial contact with the top cap.

✳✳ WARNING

Never bottom the spring or damper rod!

5. Remove the nut from the strut damper shaft and place an alignment/guiding rod on top of the damper shaft. Use the rod to guide the damper shaft straight down

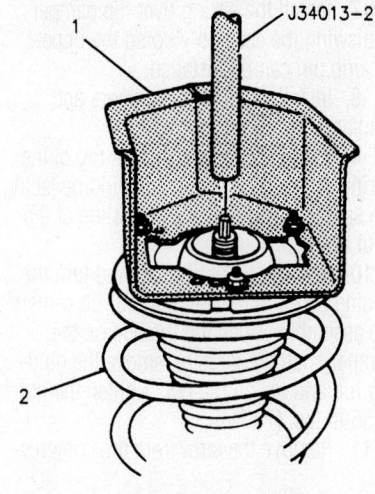

1 STRUT COMPRESSOR
2 STRUT ASSEMBLY

7922YG26

Install the rod to guide the damper shaft straight down through the spring cap while compressing the spring

through the spring cap while compressing the spring. Remove the components.

To install:
6. Mount the strut assembly in the strut compressor, using the bottom locking pin only.

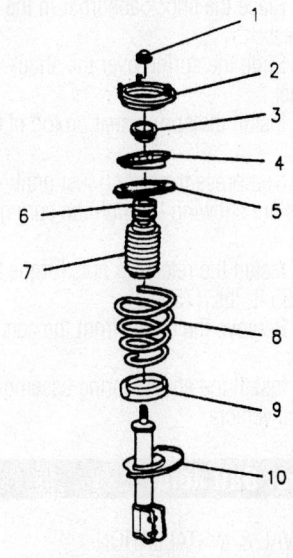

1 STRUT MOUNT NUT
2 STRUT MOUNT
3 RATE WASHER
4 SPRING SEAT
5 SPRING UPPER INSULATOR
6 JOUNCE BUMPER
7 STRUT DUST SHIELD
8 SPRING
9 SPRING LOWER INSULATOR
10 STRUT

7922YG27

Exploded view of the front strut assembly

7. Install the spring over the damper and swing the assembly up so the upper locking pin can be installed.

8. Install all shields, bumpers and insulators on the spring seat.

9. Install the spring seat on top of the spring. The spring seat flat should be facing the same direction as the centerline of the strut assembly spindle.

10. Install the guiding rod and turn the forcing screw while the guiding rod centers the assembly. When the threads on the damper shaft are visible, remove the guiding rod and install the nut. Tighten the nut to 55 ft. lbs. (75 Nm).

11. Remove the strut from the compressor.

12. Install the strut to the vehicle.

Rear

1. Remove the shock absorber from the vehicle.

2. Place the shock absorber in a strut compressor tool using a shock adapter.

3. Compress the spring.

4. Remove the nut.

5. Slowly release the tension on the spring.

6. Remove the spring and shock absorber from the compressor.

To install:

7. Place the shock absorber in the strut compressor.

8. Slide the spring over the shock absorber.

9. Install the spring seat on top of spring.

10. Compress the spring just until threads are showing through the spring seat.

11. Install the retaining nut. Torque the nut to 55 ft. lbs. (75 Nm).

12. Remove the shock from the compressor.

13. Install the shock/spring assembly onto the vehicle.

Lower Ball Joint

REMOVAL & INSTALLATION

1. Before servicing the vehicle, refer to the precautions section.

※ WARNING

Care must be exercised to prevent the axle shaft joints from being over-extended.

2. Remove or disconnect the following:
- Front wheel
- Stabilizer shaft link
- Wiring harness from the lower control arm
- Lower ball joint from the steering knuckle

3. Drill out the 3 ball joint-to-lower control arm rivets as follows:

   a. Use an ⅛ in. (3mm) drill bit to make a pilot hole through the rivets.

   b. Finish drilling rivets with a ½ in. (13mm) drill bit.

   c. Remove the rivets with a punch, if necessary.

4. Remove the ball joint from the steering knuckle.

To install:

5. Install or connect the following:
- Ball joint into the control arm
- 3 new nuts/bolts. Tighten the nuts/bolts to the specifications supplied with new ball joint
- Ball joint to the steering knuckle. Torque the nuts to 41 ft. lbs. (55 Nm) plus an additional 180 degree turn, use a new cotter pin.
- Wiring harness to the lower control arm

- Stabilizer shaft link. Torque the nuts to 13 ft. lbs. (17 Nm).
- Front wheel

6. Check and/or adjust the front wheel alignment.

Lower Control Arm

REMOVAL & INSTALLATION

1. Before servicing the vehicle, refer to the precautions section.

2. Remove or disconnect the following:
- Front wheel
- Stabilizer shaft link
- Engine mount strut, if removing the right lower control arm
- Front suspension support brace, if removing the left lower control arm
- Wiring harness from the lower control arm
- Ball joint from the steering knuckle
- Lower control arm

To install:

3. Install or connect the following:
- Lower control arm to the cross-member. DO NOT torque the bolts at this time.

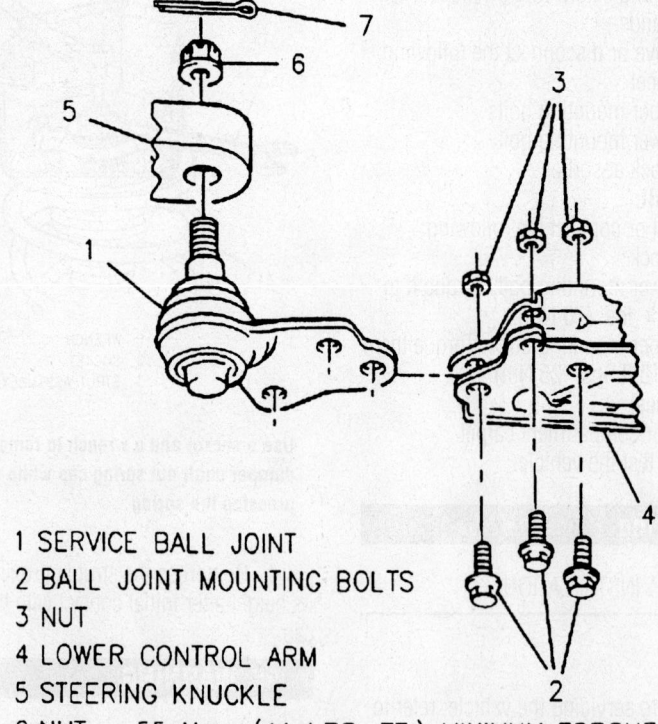

1 SERVICE BALL JOINT
2 BALL JOINT MOUNTING BOLTS
3 NUT
4 LOWER CONTROL ARM
5 STEERING KNUCKLE
6 NUT — 55 N·m (41 LBS. FT.) MINIMUM TORQUE
        65 N·m (48 LBS. FT.) MAXIMUM TORQUE
        TO INSTALL PIN
7 PIN

7922YG28

Exploded view of the ball joint mounting

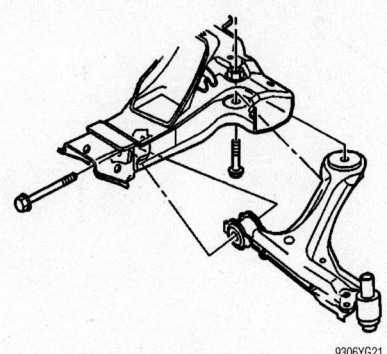

View of the lower control arm

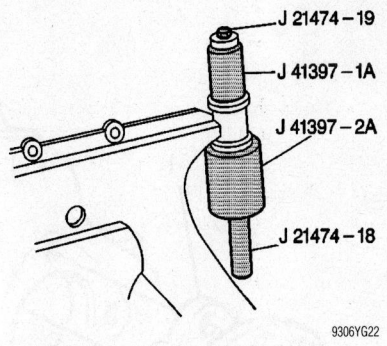

Removing the lower control arm front bushing

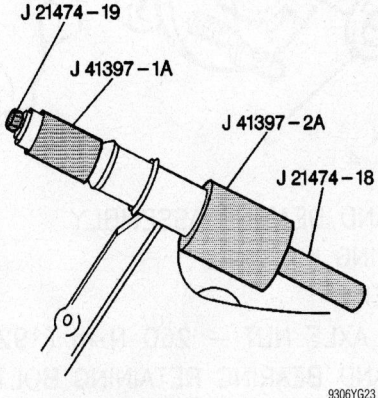

Installing the lower control arm front bushing

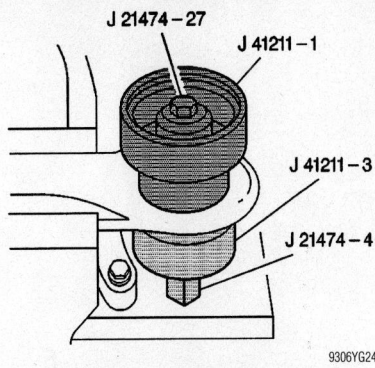

Removing/Installing the lower control arm rear bushing

- Wiring harness to the lower control arm
- Engine mount strut, if removed
- Front suspension support brace ,if removing removed
- Ball joint to the steering knuckle. Torque the nut to 41 ft. lbs. (55 Nm) plus an additional 180 degree turn, use a new cotter pin.
- Stabilizer shaft link. Torque the nut to 13 ft. lbs. (17 Nm).
- Front wheel

4. Lower the vehicle and allow it to assume proper trim height.

5. Torque the front control arm bolt to 74 ft. lbs. (100 Nm) plus an additional 90 degree turn and the rear bolt to 125 ft. lbs. (170 Nm).

6. Check and/or adjust the front wheel alignment.

CONTROL ARM BUSHING REPLACEMENT

Front Bushing

1. Before servicing the vehicle, refer to the precautions section.
2. Remove or disconnect the following:
   - Front wheel
   - Lower control arm and place it in a vise
3. Lubricate the threads of tool J-21474-19 with High Pressure Lubricant J-23444-A.
4. Assemble ⅜ in. Bolt J-21474-19, Remover/Installer tool J-41397-1A, Remover/Installer tool J-41397-2A and ⅜ in. Nut J-21474-18 onto the lower control arm bushing.
5. Tighten the ⅜ in. Nut J-21474-18 to press the front bushing from the lower control arm.
6. Disassemble the tools.

To install:

7. Lubricate the outer case of the new front bushing.

8. Insert the new front bushing into the lower control arm.
9. Assemble ⅜ in. Bolt J-21474-19, Remover/Installer tool J-41397-1A, Remover/Installer tool J-41397-2A and ⅜ in. Nut J-21474-18 onto the lower control arm.
10. Tighten the ⅜ in. Nut J-21474-18 to press the bushing into the lower control arm.
11. Disassemble the tools.

Rear Bushing

1. Before servicing the vehicle, refer to the precautions section.
2. Remove or disconnect the following:
   - Front wheel
   - Lower control arm and place it in a vise
3. Lubricate the threads of tool J-21474-27 with High Pressure Lubricant J-23444-A.
4. Assemble bolt J-21474-27, Remover/Installer tool J-41211-1, Remover/Installer tool J-41211-3 and ½ in. Nut J-21474-4.
5. Tighten the Bolt J-21474-27 to press the rear bushing from the lower control arm.
6. Disassemble the tools.

To install:

7. Insert the rear lower control arm bushing.
8. Assemble bolt J-21474-27, Remover/Installer tool J-41211-1, Remover/Installer tool J-41211-3 and ½ in. Nut J-21474-4.
9. Tighten the Nut J-21474-4 to press the rear bushing into the lower control arm.
10. Disassemble the tools.

Wheel Bearings

ADJUSTMENT

These vehicles are equipped with sealed hub and bearing assemblies. The hub and bearing assemblies are non-serviceable. If the assembly is damaged, the complete unit must be replaced.

REMOVAL & INSTALLATION

Front

1. Before servicing the vehicle, refer to the precautions section.
2. Remove or disconnect the following:
   - Front wheel
   - Halfshaft nut
   - Brake caliper without disconnecting the fluid line
   - Brake rotor
   - 3 hub/bearing assembly-to-steering knuckle bolts
   - Hub/bearing assembly from the steering knuckle

To install:

3. Install or connect the following:
   - Hub/bearing assembly to the steering knuckle. Torque the bolts to 70 ft. lbs. (95 Nm).
   - Brake rotor
   - Brake caliper
   - Halfshaft through hub/knuckle

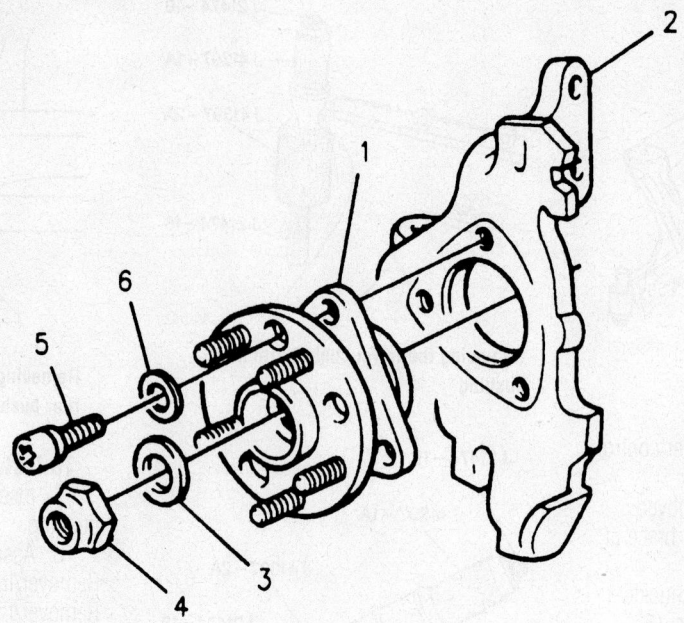

1 HUB AND BEARING ASSEMBLY
2 STEERING KNUCKLE
3 WASHER
4 DRIVE AXLE NUT – 260 N·m (192 LBS. FT.)
5 HUB AND BEARING RETAINING BOLT
6 WASHER

7922YG29

Exploded view of the front hub and bearing assembly

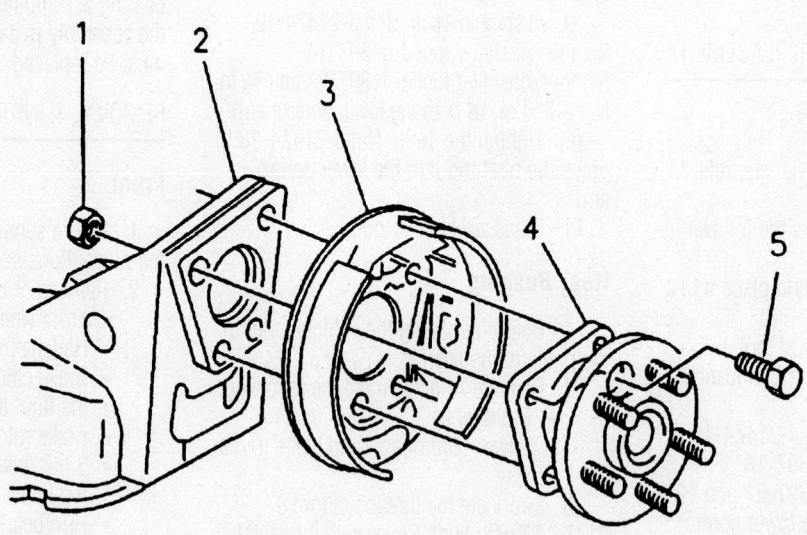

1 NUT
2 REAR AXLE ASSEMBLY
3 BACKING PLATE
4 HUB AND BEARING ASSEMBLY
5 BOLT

7922YG30

Exploded view of the rear hub and bearing assembly

assembly. Torque the nut to 144 ft. lbs. (200 Nm).
- Front wheel

4. Road test the vehicle and verify proper operation.

Rear

A single-unit hub/bearing assembly is bolted to both ends of the rear axle assembly or rear knuckle assembly. The hub/bearing assembly is a sealed unit.

1. Before servicing the vehicle, refer to the precautions section.
2. Remove or disconnect the following:
- Rear wheel
- Brake drum
- Anti-lock Brake System (ABS) wheel speed sensor electrical connector
- Hub/bearing assembly-to-rear axle assembly nuts/bolts

➡**The top rear mounting bolt will not clear the brake shoe when removing the hub and bearing assembly. Partially remove the hub and bearing assembly prior to removing this bolt.**

- Hub/bearing assembly from the rear axle

To install:

3. Install or connect the following:
- Hub/bearing assembly. Torque the nuts/bolts to 44 ft. lbs. (60 Nm).

➡**Position the top rear mounting bolt in the hub and bearing assembly prior to installation to the axle housing.**

- Rear ABS wheel speed sensor wire connector
- Brake drum
- Rear wheel

4. Road test the vehicle.

BRAKES

Brake Caliper

REMOVAL & INSTALLATION

1. Before servicing the vehicle, refer to the precautions section.
2. Siphon ⅔ of the brake fluid out of the master cylinder.
3. Remove the tire and wheel assembly.
4. Compress the caliper piston back into the caliper bore using a large pair of pliers, C-clamp or special piston retracting tool.
5. If the caliper is to be completely removed from the vehicle for bench service, disconnect and cap the brake line from the caliper. Discard the old washers.
6. Remove the caliper mounting bolts and sleeves.
7. Remove the caliper from the knuckle.
8. Remove the brake pads from the caliper, if caliper is being replaced.

To install:

9. Install the brake pads in the caliper.
10. Install the caliper on the steering knuckle.
11. Install the mounting bolts and sleeves and torque to 38 ft. lbs. (51 Nm).
12. If the caliper brake line was disconnected, uncap and connect the brake hose to the caliper using new copper washers. Torque the bolt to 37 ft. lbs. (50 Nm).
13. Refill the master cylinder and bleed the brake system.
14. Install the tire and wheel assembly and tighten to specification.
15. Verify correct brake operation.

Disc Brake Pads

REMOVAL & INSTALLATION

1. Before servicing the vehicle, refer to the precautions section.
2. Siphon ⅔ of the brake fluid out of the master cylinder reservoir.
3. Remove the tire and wheel assembly.
4. Compress the caliper piston back into the caliper bore using a large pair of pliers, C-clamp or special caliper piston retracting tool.
5. Remove the caliper mounting bolts and sleeves.
6. Remove the caliper from the steering knuckle without disconnecting the brake hose. Do not allow the caliper to hang from the brake hose. Support the caliper with a piece of wire.
7. Remove the outboard pad by pushing in on the outside edge of the pad to release the mounting dowel from the hole in the caliper. When both dowels are unseated, push the pad out the bottom of the caliper.
8. Remove the inboard pad from the caliper by pulling it out of the caliper.

To install:

9. Inspect the caliper for any signs of leakage. If the caliper is leaking, new brake pads will be damaged. Also inspect the condition of the caliper support for rust and corrosion which will hinder the travel on the caliper.
10. Inspect the caliper mounting hardware. New bolts are usually recommended. Clean all parts well. Lubricate the mounting bushings and sleeves with silicone grease as required.
11. Install the inboard pad in the caliper so the spring clip on the pad back engages in the caliper piston.
12. Install the outboard pad over the caliper end until the mounting dowels snap into the mounting holes in the caliper.
13. Install the caliper over the rotor onto the steering knuckle.
14. Install the mounting bolts and sleeves and torque to 38 ft. lbs. (51 Nm).
15. Install the tire and wheel assembly.
16. Pump the brake pedal several times to seat the pads against the rotor before attempting to move the vehicle.
17. Check the master cylinder level and add fluid as necessary.

Brake Drums

REMOVAL & INSTALLATION

1. Before servicing the vehicle, refer to the precautions section.
2. Remove the wheel and tire assembly.
3. Pull the brake drum off. It may be necessary to gently tap the rear edge of the drum to start it off the studs. If extreme resistance to removal is encountered, it will be necessary to retract the brake shoe self-adjuster screw, sometimes called the star-wheel. Knock out the access hole in the brake drum and turn the adjuster to retract the linings from the drum. Install a replacement hole cover before reinstalling the drum.

➡**Do not hammer on the brake drum to remove it.**

To install:

4. Inspect the inside of the brake drum. If worn, heavily grooved or if the opening is

distorted, the drum should be refinished or replaced. If refinishing, observe the maximum drum diameter specification.

5. Inspect the wheel cylinder for signs of brake fluid leakage. Inspect the brake shoe springs and self-adjuster mechanism. The adjuster should usually be disassembled, cleaned and lubricated when the drum is removed for brake service.

6. Install the drum over the brake shoes.

7. Install the wheel and tire assembly.

8. Adjust the brakes.

9. Check brake operation.

Brake Shoes

REMOVAL & INSTALLATION

2002 Models

1. Before servicing the vehicle, refer to the precautions section.

2. Remove the tire and wheel assembly.

3. Remove the brake drum. It may be necessary to gently tap the rear edge of the drum to start it off the studs. If extreme resistance to removal is encountered, it will be necessary to retract the brake shoe self-adjuster screw, sometime called the starwheel. Turn the adjuster to retract the linings from the drum.

➡ **Do not hammer on the brake drum to remove it.**

4. Remove the upper return springs from the shoes using tool J-8057 brake spring tool.

5. Remove the hold-down springs using J-8049 brake spring tool.

6. Remove the shoe hold-down pins from behind the brake backing plate.

7. Lift up the actuator lever for the self-adjusting mechanism and remove the actuating link. Remove the actuator lever, pivot, and the pivot return spring.

8. Spread the shoes apart to clear the wheel cylinder pistons, then remove the parking brake strut and spring.

9. Disconnect the parking brake cable from the lever. Remove the shoes, still connected by their adjusting screw spring.

10. With the shoes removed, note the position of the adjusting spring and remove the spring and adjusting screw.

11. Remove the C-clip from the parking brake lever and remove the lever from the secondary shoe.

12. Use a damp cloth to remove all dirt and dust from the backing plate and brake parts.

To install:

13. Check the backing plate attaching bolts to make sure they are tight. Use fine emery cloth to clean all rust and dirt from the shoe contact surfaces on the plate and lubricate with brake grease. Check the wheel cylinder for signs of leakage.

14. Clean all parts completely in brake solvent and air dry.

15. Clean the backing plate shoe contact points.

16. Inspect the inside of the brake drum. If worn, heavily grooved or if the opening is distorted, the drum should be refinished or replaced.

17. Inspect the brake shoe springs and self-adjuster mechanism. Disassemble the adjuster mechanism and clean the threads and coat with grease. Make sure the adjuster assembly turns freely before installing in the vehicle.

18. Install the parking brake lever on the secondary shoe and secure with C-clip.

19. Install the adjusting screw and spring on the shoes, connecting them together. The coils of the spring must not be over the starwheel on the adjuster. The left and right hand springs are not interchangeable.

20. Spread the shoe assemblies apart and connect the parking brake cable. Install the shoes on the backing plate, engaging the shoes at the top temporarily with the wheel cylinder pistons. Make sure the starwheel on the adjuster is lined up with the adjusting hole in the backing plate, if equipped.

21. Spread the shoes apart slightly and install the parking brake strut and spring. Make sure the end of the strut without the spring engages the parking brake lever. The end with the spring engages the primary shoe (the one with the shorter lining).

22. Install the actuator pivot, lever and return spring. Install the actuating link in the shoe retainer. Lift up the actuator lever and hook the link into the lever.

23. Install the hold-down pins through the back of the plate using J-8057. Install the lever pivots and hold-down springs. Install the shoe return springs using J-8049. Be careful not to stretch or distort the springs.

24. Make sure the linings are in the right place, the self-adjusting mechanism is correctly installed, and the parking brake parts are hooked up.

25. Measure the distance from the edge of the primary lining to the edge secondary lining, then measure the inside width of the

drum. Adjust the linings by means of the adjuster so the drum will fit onto the linings.

26. Install the hub and bearing assembly onto the axle if removed. Torque the retaining bolts to 35 ft. lbs. (55 Nm.).

27. Install the drum.

28. Install the wheel and tire assembly.

29. Adjust the brakes. Install a rubber hole cover in the adjustment knock-out hole after the adjustment is complete. Adjust the parking brake.

30. Road test the vehicle and verify proper brake operation.

2003–05 Models

1. Before servicing the vehicle, refer to the precautions section.

➡ **Brake shoe spanner and spring removal tool J 38400 or its equivalent is required to remove the brake components**

2. Remove the wheel assembly.

3. Remove the brake drum. It may be necessary to gently tap the rear edge of the drum to start it off the studs. If extreme resistance to removal is encountered, it will be necessary to retract the brake shoe self-adjuster screw, sometime called the starwheel. Turn the adjuster to retract the linings from the drum.

➡ **Do not hammer on the brake drum to remove it.**

4. Remove the adjuster spring.

5. Disengage the adjuster spring hook end from the tab on the adjuster actuator lever, then release the spring from the brake shoe web hole.

6. Remove the adjuster actuator lever from the pivot.

7. Spread the top of the brake shoes apart and remove the adjuster assembly from the brake shoes.

8. Position the hook end of the removal/installation tool under the universal spring and lightly pull the universal spring end out of the shoe web hole.

9. Hold the universal spring with the tool while removing the secondary brake shoe.

10. Release the park brake cable from the park brake lever on the secondary shoe.

11. Position the hook end of the removal/installation tool under the universal spring and lightly pull the universal spring end out of the shoe web hole.

12. Hold the universal spring with the tool while removing the primary brake shoe.

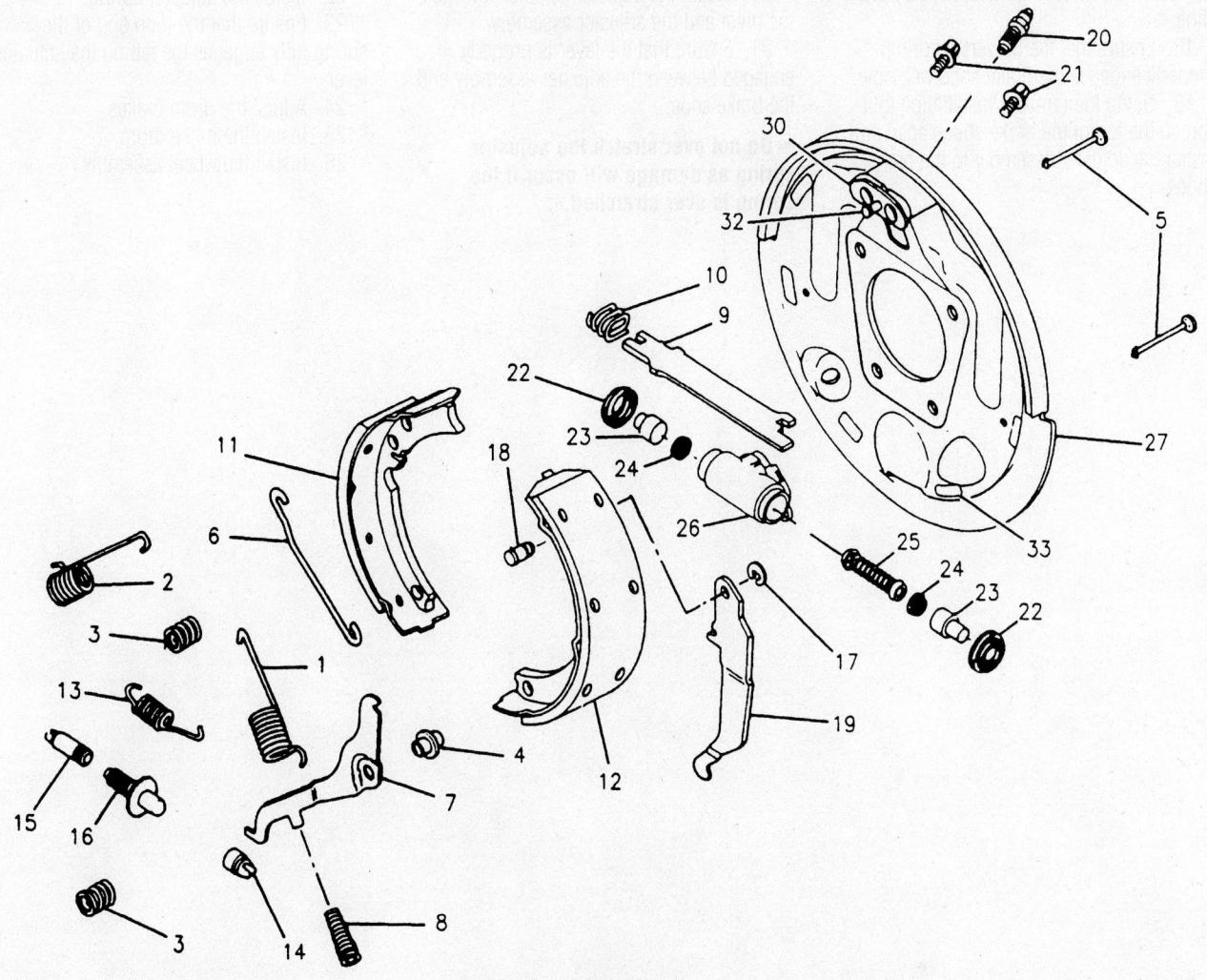

1	RETURN SPRING	11	PRIMARY SHOE AND LINING	21	BOLT
2	RETURN SPRING	12	SECONDARY SHOE AND LINING	22	BOOT
3	HOLD DOWN SPRING	13	ADJUSTING SCREW SPRING	23	PISTON
4	BEARING SLEEVE	14	SOCKET	24	SEAL
5	HOLD–DOWN PIN	15	PIVOT NUT	25	SPRING ASSEMBLY
6	ACTUATOR LINK	16	ADJUSTING SCREW	26	WHEEL CYLINDER
7	ACTUATOR LEVER	17	RETAINING RING	27	BACKING PLATE
8	LEVER RETURN SPRING	18	PIN	30	SHOE RETAINER
9	PARKING BRAKE STRUT	19	PARKING BRAKE LEVER	32	ANCHOR PIN
10	STRUT SPRING	20	BLEEDER VALVE	33	SHOE PADS

93006G97

Exploded view of drum brake components—2002 models

To install:

13. Apply a thin, light coat of high temperature silicone brake lubricant to the brake shoe contact surfaces of the brake backing plate.

14. Position the hook end of the removal/installation tool under the universal spring and lightly pull the universal spring end out while installing the primary brake shoe.

15. Ensure that the universal spring engages the brake shoe web hole.

16. Install the park brake cable to the park brake lever on the secondary brake shoe.

17. Position the hook end of the removal/installation tool under the universal spring and lightly pull the universal spring

end out while installing the secondary brake shoe.

18. Ensure that the universal spring properly engages the brake shoe web hole.

19. Using the removal/installation tool spread the top of the brake shoes apart and install the adjuster assembly to the brake shoes.

20. Install the adjuster actuator lever to the pivot and the adjuster assembly.

21. Ensure that the lever is properly engaged between the adjuster assembly and the brake shoe.

➡ Do not over stretch the adjuster spring as damage will occur if the spring is over stretched.

22. Install the adjuster spring.

23. Ensure that the loop end of the spring fully engages the tab on the actuator lever.

24. Adjust the drum brakes.

25. Install the brake drum.

26. Install the wheel assembly.

BRAKES **12-111**
DRIVE TRAIN **12-91**
ENGINE REPAIR **12-31**
FUEL SYSTEM **12-88**
SPECIFICATIONS AND
  MAINTENANCE CHARTS......**12-2**
Engine and Vehicle
  Identification 12-2
General Engine Specifications 12-3
Engine Tune-Up Specifications 12-4
Firing Orders 12-5
Accessory Drive Belt Routing 12-6
Capacities 12-8
Valve Specifications 12-9
Crankshaft and Connecting
  Rod Specifications 12-11
Piston and Ring Specifications 12-13
Torque Specifications 12-14
Wheel Alignment 12-16
Tire, Wheel and Ball Joint
  Specifications 12-19
Brake Specifications 12-22
Scheduled Maintenance
  Intervals.............................12-24
STEERING AND
  SUSPENSION **12-94**

A
Air Bag...................................12-94
  Disarming And Arming 12-94
  Precautions 12-94
Alternator 12-31
  Removal & Installation.............12-31
B
Brake Caliper 12-111
  Removal & Installation..........12-111
Brake Drums...........................12-117
  Removal & Installation.............12-117
Brake Shoes...........................12-117
  Removal & Installation.............12-117
C
Camshaft and Valve Lifters 12-64
  Removal & Installation.............12-64

Coil Spring 12-107
  Removal & Installation...........12-107
CV-Joint 12-92
  Overhaul 12-92
Cylinder Head...........................12-44
  Removal & Installation.............12-44
D
Disc Brake Pads.......................12-115
  Removal & Installation...........12-115
Distributor.................................12-31
E
Engine Assembly 12-32
  Removal & Installation.............12-32
Exhaust Manifold 12-56
  Removal & Installation.............12-56
F
Front Cover and Seal,
  Timing Chain, and Sprockets 12-78
  Removal & Installation.............12-78
Front Crankshaft Seal 12-61
  Removal & Installation.............12-61
Fuel Filter 12-88
  Removal & Installation.............12-88
Fuel Injector.............................12-89
  Removal & Installation.............12-89
Fuel Pump 12-88
  Removal & Installation.............12-88
Fuel System Pressure 12-88
  Relieving 12-88
Fuel System Service
  Precautions.........................12-88
H
Halfshaft..................................12-92
  Removal & Installation.............12-92
Heater Core.............................12-39
  Removal & Installation.............12-39
I
Ignition Timing 12-32
  Adjustment...........................12-32
Intake Manifold.........................12-52
  Removal & Installation.............12-52
L
Lower Ball Joint.........................12-108
  Removal & Installation...........12-108

Lower Control Arm 12-108
  Control Arm Bushing
    Replacement.......................12-109
  Removal & Installation...........12-108
O
Oil Pan...................................12-69
  Removal & Installation.............12-69
Oil Pump 12-74
  Removal & Installation.............12-74
P
Piston and Ring 12-86
  Positioning 12-86
Power Rack and Pinion
  Steering Gear.......................12-105
  Removal & Installation...........12-105
R
Rear Main Seal 12-76
  Removal & Installation.............12-76
Rocker Arms 12-48
  Removal & Installation.............12-48
S
Starter Motor 12-68
  Removal & Installation.............12-68
Strut......................................12-107
  Removal & Installation...........12-107
Supercharger 12-51
  Removal & Installation.............12-51
T
Transaxle Assembly 12-91
  Removal & Installation.............12-91
V
Valve Lash 12-68
  Adjustment...........................12-68
W
Water Pump.............................12-38
  Removal & Installation.............12-38
Wheel Bearings.........................12-109
  Adjustment...........................12-109
  Removal & Installation.............12-109

SPECIFICATIONS AND MAINTENANCE CHARTS

ENGINE AND VEHICLE IDENTIFICATION

		Engine Code						Model Year	
Code ①	Liters (cc)	Cu. In.	Cyl.	Fuel Sys.	Engine Type	Eng. Mfg.		Code ②	Year
1 ③	3.8 (3785)	231	6	MFI	OHV	GM		2	2002
2	3.8 (3785)	231	6	SFI	OHV	GM		3	2003
4	3.8 (3785)	231	6	SFI	OHV	GM		4	2004
C	5.3 (5326)	325	8	SFI	OHV	GM		5	2005
E	3.4 (3393)	207	6	MFI	OHV	GM		6	2006
H	3.5 (3475)	212	6	MFI	DOHC	GM			
J	3.1 (3130)	191	6	SFI	OHV	GM			
K	3.8 (3785)	231	6	MFI	OHV	GM			
M	3.1 (3130)	191	6	MFI	OHV	GM			

MFI: Multi-point Fuel Injection

DOHC: Dual Overhead Camshafts

OHV: Overhead Valves

① 8th position of VIN

② 10th position of VIN

③ Supercharged

06025-GMWB-C01

GENERAL ENGINE SPECIFICATIONS

Year	Model	Engine Displacement Liters	Engine Series VIN	Net Horsepower @ rpm	Net Torque @ rpm (ft. lbs.)	Bore x Stroke (in.)	Compression Ratio	Oil Pressure @ rpm
2002	Century	3.1	M	170@5200	185@4000	3.50x3.31	9.5:1	15@1100
	Century	3.8	1	240@5200	280@3200	3.80x3.40	9.0:1	60@1850
	Century	3.8	K	205@5200	230@4000	3.80x3.40	9.4:1	60@1850
	Grand Prix	3.8	1	240@5200	280@3200	3.80x3.40	9.0:1	60@1850
	Grand Prix	3.8	K	205@5200	230@4000	3.80x3.40	9.5:1	60@1850
	Grand Prix	3.1	J	170@5200	185@4000	3.50x3.31	9.5:1	15@1100
	Impala	3.4	E	210@5200	215@4000	3.62x3.31	9.5:1	15@1100
	Impala	3.8	K	205@5200	230@4000	3.80x3.40	9.4:1	60@1850
	Intrigue	3.5	H	215@5600	230@4400	3.52x3.62	9.3:1	29@2000
	Monte Carlo	3.8	K	205@5200	230@4000	3.80x3.40	9.4:1	60@1850
	Monte Carlo	3.4	E	210@5200	215@4000	3.62x3.31	9.5:1	15@1100
	Regal	3.8	1	240@5200	280@3200	3.80x3.40	9.0:1	60@1850
	Regal	3.8	K	205@5200	230@4000	3.80x3.40	9.4:1	60@1850
	Regal	3.1	M	170@5200	185@4000	3.50x3.31	9.5:1	15@1100
2003	Century	3.1	M	170@5200	185@4000	3.50x3.31	9.5:1	15@1100
	Century	3.8	1	240@5200	280@3200	3.80x3.40	9.0:1	60@1850
	Century	3.8	K	205@5200	230@4000	3.80x3.40	9.4:1	60@1850
	Grand Prix	3.8	1	240@5200	280@3200	3.80x3.40	9.0:1	60@1850
	Grand Prix	3.8	K	205@5200	230@4000	3.80x3.40	9.5:1	60@1850
	Grand Prix	3.1	J	170@5200	185@4000	3.50x3.31	9.5:1	15@1100
	Impala	3.4	E	210@5200	215@4000	3.62x3.31	9.5:1	15@1100
	Impala	3.8	K	205@5200	230@4000	3.80x3.40	9.4:1	60@1850
	Monte Carlo	3.8	K	205@5200	230@4000	3.80x3.40	9.4:1	60@1850
	Monte Carlo	3.4	E	210@5200	215@4000	3.62x3.31	9.5:1	15@1100
	Regal	3.8	1	240@5200	280@3200	3.80x3.40	9.0:1	60@1850
	Regal	3.8	K	205@5200	230@4000	3.80x3.40	9.4:1	60@1850
	Regal	3.1	M	170@5200	185@4000	3.50x3.31	9.5:1	15@1100
2004	Century	3.1	J	170@5200	185@4000	3.50x3.31	9.6:1	60@1850
	Century	3.8	1	240@5200	280@3200	3.80x3.40	8.5:1	60@1850
	Century	3.8	K	205@5200	230@4000	3.80x3.40	9.4:1	60@1850
	Grand Prix	3.8	2	240@5200	280@3200	3.80x3.40	9.4:1	60@1850
	Grand Prix	3.8	4	205@5200	230@4000	3.80x3.40	8.5:1	60@1850
	Impala	3.4	E	210@5200	215@4000	3.62x3.31	9.6:1	60@1850
	Impala	3.8	1	240@5200	280@3200	3.80x3.40	8.5:1	60@1850
	Impala	3.8	K	205@5200	230@4000	3.80x3.40	9.4:1	60@1850
	Monte Carlo	3.8	K	205@5200	230@4000	3.80x3.40	9.4:1	60@1850
	Monte Carlo	3.8	1	240@5200	280@3200	3.80x3.40	8.5:1	60@1850
	Monte Carlo	3.4	E	210@5200	215@4000	3.62x3.31	9.6:1	60@1850
	Regal	3.8	1	240@5200	280@3200	3.80x3.40	8.5:1	60@1850
	Regal	3.8	K	205@5200	230@4000	3.80x3.40	9.4:1	60@1850
	Regal	3.1	J	170@5200	185@4000	3.50x3.31	9.6:1	60@1850

06025-GMWB-C02

GENERAL ENGINE SPECIFICATIONS

Year	Model	Engine Displacement Liters	Engine Series VIN	Net Horsepower @ rpm	Net Torque @ rpm (ft. lbs.)	Bore x Stroke (in.)	Com-pression Ratio	Oil Pressure @ rpm
2005	Century	3.1	J	175@5200	195@4000	3.50x3.31	9.6:1	60@1850
	Grand Prix	3.8	2	240@5200	280@3200	3.80x3.40	9.4:1	60@1850
	Grand Prix	3.8	4	205@5200	230@4000	3.80x3.40	8.5:1	60@1850
	Grand Prix	5.3	C	303@5600	323@4400	3.78x3.62	9.9:1	18@2000
	Impala	3.4	E	180@5200	205@4000	3.62x3.31	9.6:1	60@1850
	Impala	3.8	1	240@5200	280@3200	3.80x3.40	8.5:1	60@1850
	Impala	3.8	K	205@5200	230@4000	3.80x3.40	9.4:1	60@1850
	Monte Carlo	3.8	K	205@5200	230@4000	3.80x3.40	9.4:1	60@1850
	Monte Carlo	3.8	1	240@5200	280@3200	3.80x3.40	8.5:1	60@1850
	Monte Carlo	3.4	E	210@5200	215@4000	3.62x3.31	9.6:1	60@1850

06025-GMWB-C03

ENGINE TUNE-UP SPECIFICATIONS

Year	Engine Displacement Liters	Engine VIN	Spark Plug Gap (in.)	Ignition Timing (deg.)	Fuel Pump (psi)	Idle Speed (rpm)	Valve Clearance In.	Valve Clearance Ex.
2002	3.1	M	0.060	①	41-47	②	HYD	HYD
	3.1	J	0.060	①	41-47	②	HYD	HYD
	3.4	E	0.045	①	41-47	②	HYD	HYD
	3.5	H	0.050	①	41-47	②	HYD	HYD
	3.8	1	0.060	①	41-47	②	HYD	HYD
	3.8	K	0.060	①	41-47	②	HYD	HYD
2003	3.1	M	0.060	①	41-47	②	HYD	HYD
	3.1	J	0.060	①	41-47	②	HYD	HYD
	3.4	E	0.045	①	41-47	②	HYD	HYD
	3.8	1	0.060	①	41-47	②	HYD	HYD
	3.8	K	0.060	①	41-47	②	HYD	HYD
2004	3.1	J	0.060	①	41-47	②	HYD	HYD
	3.4	E	0.060	①	41-47	②	HYD	HYD
	3.8	1	0.060	①	41-47	②	HYD	HYD
	3.8	2	0.060	①	41-47	②	HYD	HYD
	3.8	4	0.060	①	41-47	②	HYD	HYD
	3.8	K	0.060	①	41-47	②	HYD	HYD
2005	3.1	J	0.060	①	52-59	②	HYD	HYD
	3.4	E	0.060	①	52-59	②	HYD	HYD
	3.8	1	0.060	①	53-59	②	HYD	HYD
	3.8	2	0.060	①	56-62	②	HYD	HYD
	3.8	4	0.060	①	56-62	②	HYD	HYD
	3.8	K	0.060	①	53-59	②	HYD	HYD
	5.3	C	0.04	①	55-62	②	HYD	HYD

NOTE: The Vehicle Emission Control Information label often reflects specification changes made during production.
The label figures must be used if they differ from those in this chart.

HYD: Hydraulic

① Distributorless Ignition System (DIS) timing is not adjustable

② Idle speed is maintained by the Engine Control Module (ECM). There is no recommended adjustment procedure.

06025-GMWB-C04

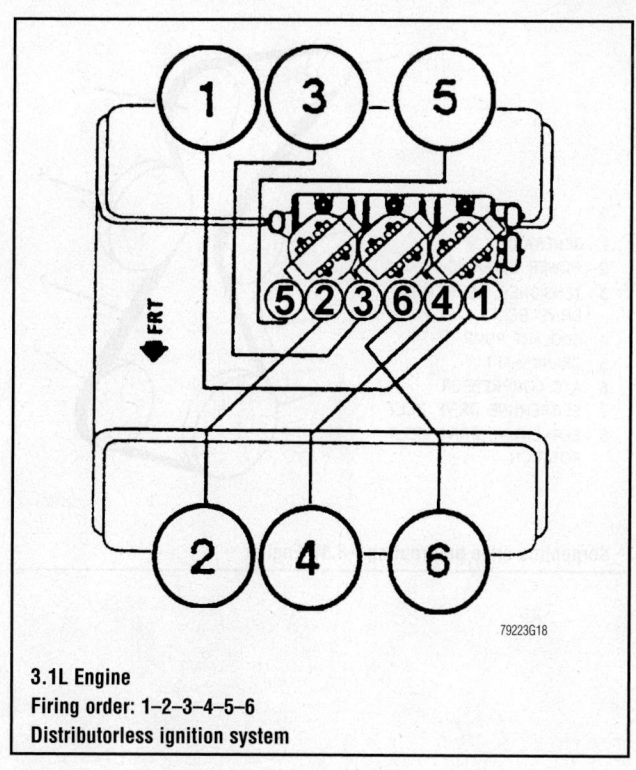

3.1L Engine
Firing order: 1–2–3–4–5–6
Distributorless ignition system

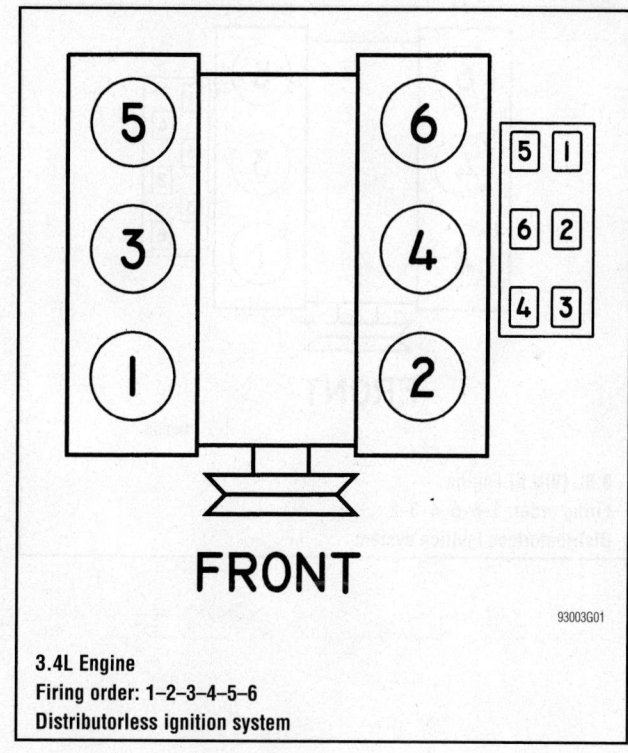

3.4L Engine
Firing order: 1–2–3–4–5–6
Distributorless ignition system

3.5L Engine
Firing order: 1–2–3–4–5–6
Distributorless ignition system

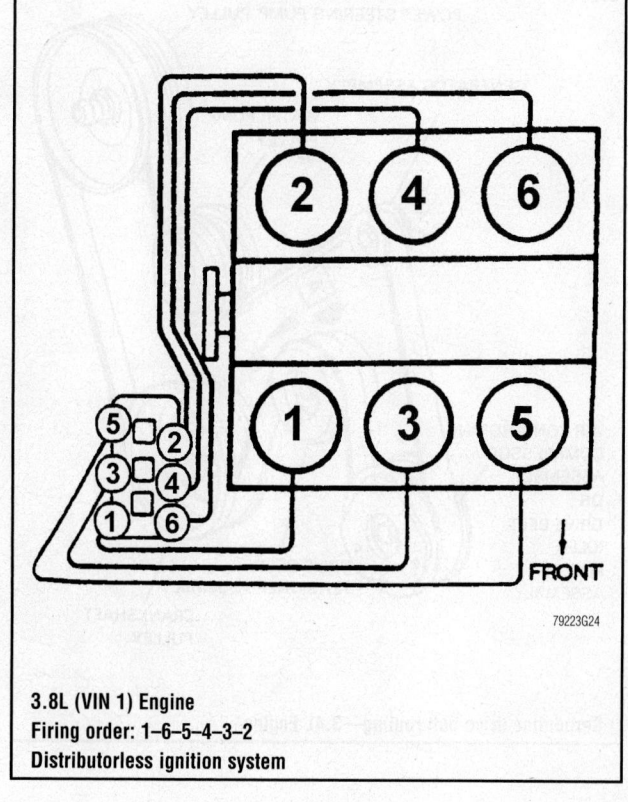

3.8L (VIN 1) Engine
Firing order: 1–6–5–4–3–2
Distributorless ignition system

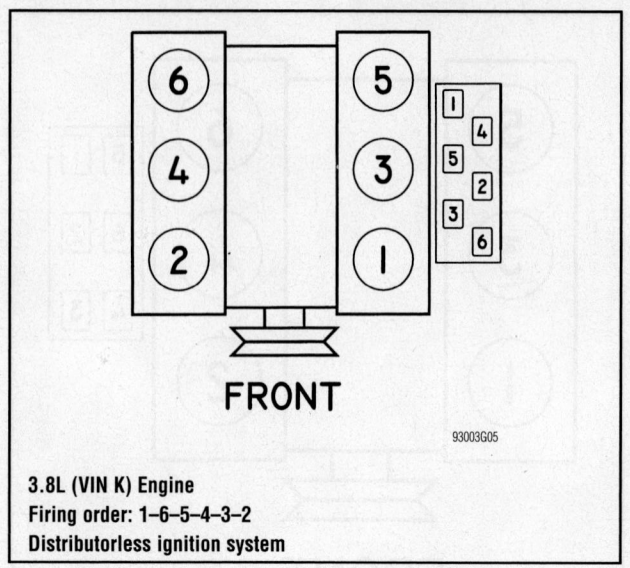

FRONT

93003G05

3.8L (VIN K) Engine
Firing order: 1–6–5–4–3–2
Distributorless ignition system

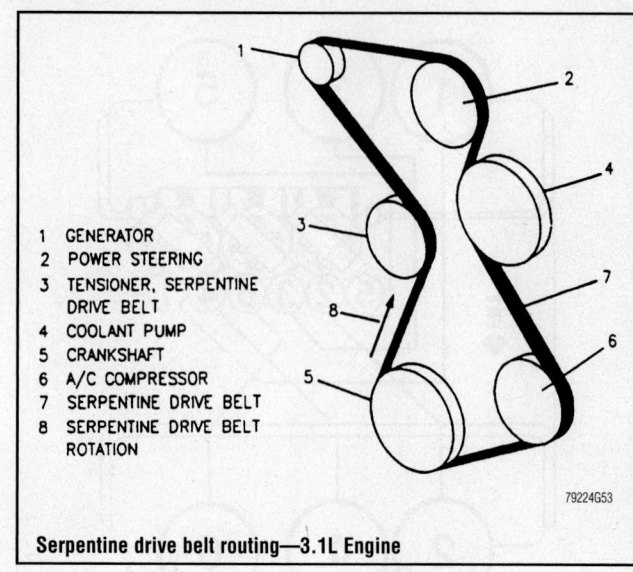

1 GENERATOR
2 POWER STEERING
3 TENSIONER, SERPENTINE
   DRIVE BELT
4 COOLANT PUMP
5 CRANKSHAFT
6 A/C COMPRESSOR
7 SERPENTINE DRIVE BELT
8 SERPENTINE DRIVE BELT
   ROTATION

79224G53

Serpentine drive belt routing—3.1L Engine

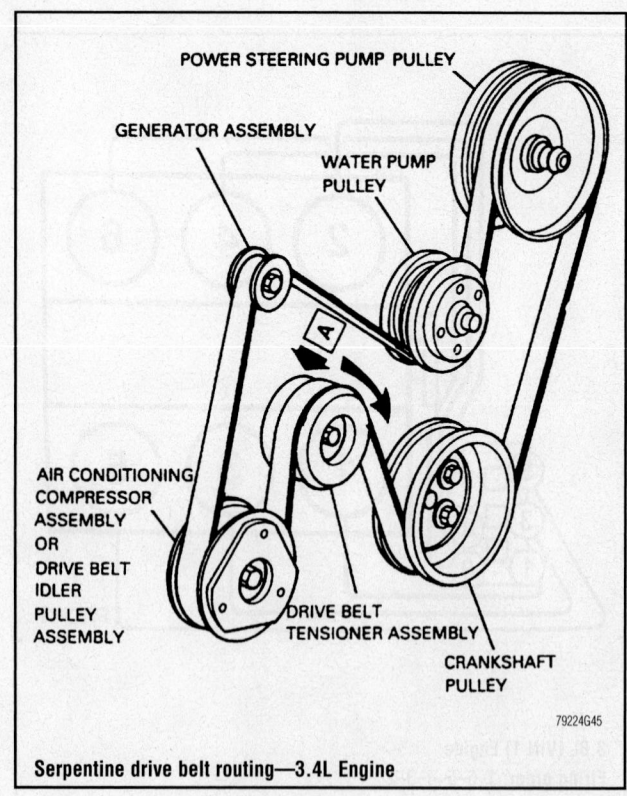

POWER STEERING PUMP PULLEY

GENERATOR ASSEMBLY

WATER PUMP PULLEY

AIR CONDITIONING
COMPRESSOR
ASSEMBLY
OR
DRIVE BELT
IDLER
PULLEY
ASSEMBLY

DRIVE BELT
TENSIONER ASSEMBLY

CRANKSHAFT
PULLEY

79224G45

Serpentine drive belt routing—3.4L Engine

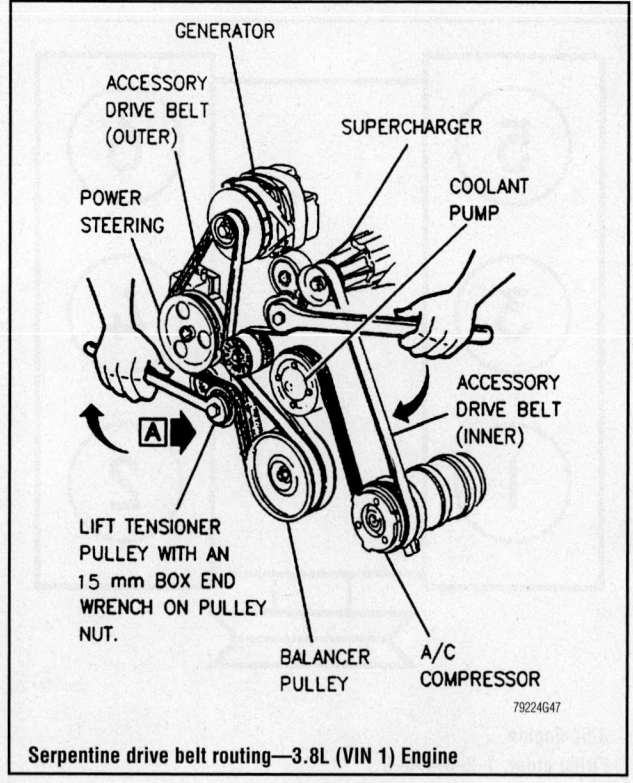

GENERATOR

ACCESSORY
DRIVE BELT
(OUTER)

SUPERCHARGER

POWER
STEERING

COOLANT
PUMP

ACCESSORY
DRIVE BELT
(INNER)

LIFT TENSIONER
PULLEY WITH AN
15 mm BOX END
WRENCH ON PULLEY
NUT.

BALANCER
PULLEY

A/C
COMPRESSOR

79224G47

Serpentine drive belt routing—3.8L (VIN 1) Engine

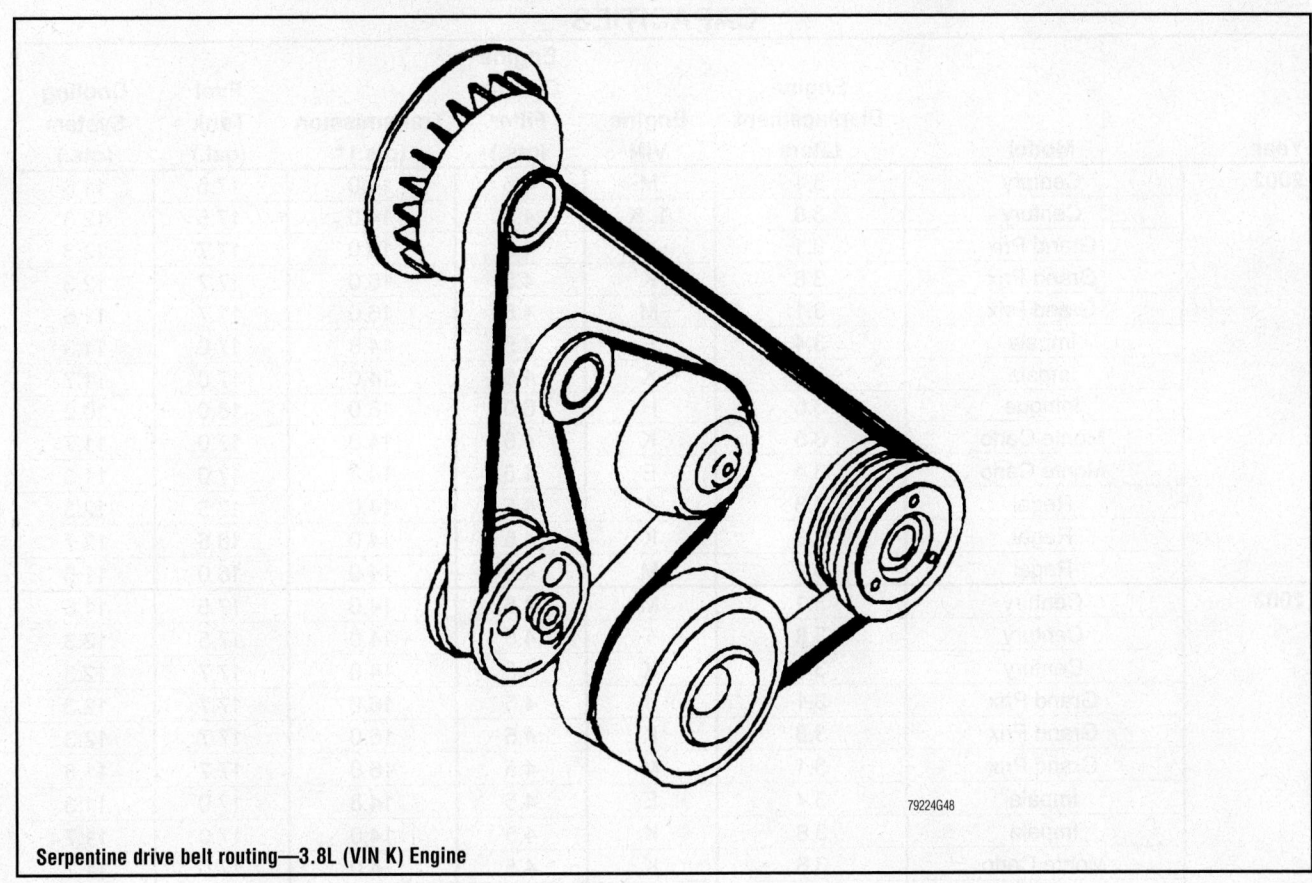

79224G48

Serpentine drive belt routing—3.8L (VIN K) Engine

CAPACITIES

Year	Model	Engine Displacement Liters	Engine VIN	Engine Oil with Filter (qts.)	Transmission (pts.) *	Fuel Tank (gal.)	Cooling System (qts.)
2002	Century	3.1	M	4.5	14.0	17.5	11.6
	Century	3.8	1, K	4.5	14.0	17.5	12.3
	Grand Prix	3.1	J	4.5	16.0	17.7	12.3
	Grand Prix	3.8	K	4.5	16.0	17.7	12.3
	Grand Prix	3.1	M	4.5	16.0	17.7	11.6
	Impala	3.4	E	4.5	14.8	17.0	11.3
	Impala	3.8	K	4.5	14.0	17.0	11.7
	Intrigue	3.5	H	6.0	16.0	18.0	10.2
	Monte Carlo	3.8	K	4.5	14.0	17.0	11.7
	Monte Carlo	3.4	E	4.5	14.8	17.0	11.3
	Regal	3.8	1	4.5	14.0	17.5	12.3
	Regal	3.8	K	4.5	14.0	16.6	12.7
	Regal	3.1	M	4.5	14.0	16.0	11.6
2003	Century	3.1	M	4.5	14.0	17.5	11.6
	Century	3.8	1	4.5	14.0	17.5	12.3
	Century	3.8	K	4.5	14.0	17.7	12.3
	Grand Prix	3.1	J	4.5	16.0	17.7	12.3
	Grand Prix	3.8	K	4.5	16.0	17.7	12.3
	Grand Prix	3.1	M	4.5	16.0	17.7	11.6
	Impala	3.4	E	4.5	14.8	17.0	11.3
	Impala	3.8	K	4.5	14.0	17.0	11.7
	Monte Carlo	3.8	K	4.5	14.0	17.0	11.7
	Monte Carlo	3.4	E	4.5	14.8	17.0	11.3
	Regal	3.8	1	4.5	14.0	17.5	12.3
	Regal	3.8	K	4.5	14.0	16.6	12.7
	Regal	3.1	M	4.5	14.0	16.0	11.6
2004	Century	3.1	J	4.5	14.8	17.0	11.7
	Grand Prix	3.8	2	4.5	14.8	17.0	11.2
	Grand Prix	3.8	4	4.5	14.8	17.0	11.2
	Impala	3.4	E	4.5	14.8	17.0	11.3
	Impala	3.8	1	4.5	14.8	17.0	11.7
	Impala	3.8	K	4.5	14.8	17.0	11.7
	Monte Carlo	3.8	K	4.5	14.8	17.0	11.7
	Monte Carlo	3.8	1	4.5	14.8	17.0	11.7
	Monte Carlo	3.4	E	4.5	14.8	17.0	11.3
2005	Century	3.1	J	4.5	14.8	17.0	11.7
	Grand Prix	3.8	2	4.5	14.8	17.0	11.2
	Grand Prix	3.8	4	4.5	14.8	17.0	11.2
	Grand Prix	5.3	C	6.0	14.8	17.5	13.0
	Impala	3.4	E	4.5	14.8	17.0	11.3
	Impala	3.8	1	4.5	14.8	17.0	11.7
	Impala	3.8	K	4.5	14.8	17.0	11.7
	Monte Carlo	3.8	K	4.5	14.8	17.0	11.7
	Monte Carlo	3.8	1	4.5	14.8	17.0	11.7
	Monte Carlo	3.4	E	4.5	14.8	17.0	11.3

NOTE: All capacities are approximate. Add fluid gradually and ensure a proper fluid is obtained.

* Drain and refill

06025-GMWB-C05

VALVE SPECIFICATIONS

Year	Engine Displacement Liters	Engine VIN	Seat Angle (deg.)	Face Angle (deg.)	Spring Test Pressure (lbs. @ in.)	Spring Installed Height (in.)	Stem-to-Guide Clearance (in.) Intake	Exhaust	Stem Diameter (in.) Intake	Exhaust
2002	3.1	M	45	45	250@1.239	1.710	0.0001-0.0027	0.0010-0.0027	0.0010-0.0027	0.0010-0.0027
	3.4	E	46	45	75@1.40	1.400	0.0011-0.0026	0.0014-0.0031	0.0010-0.0027	0.0010-0.0027
	3.5	H	45.75	45	130-142@0.964	1.377	0.0010-0.0030	0.0020-0.0040	0.233-0.234	0.233-0.234
	3.8	J	45	45	80@1.750	1.690-1.720	0.0015-0.0032	0.0015-0.0032	0.0012-0.0028	0.0014-0.0029
	3.8	1	45	45	80@1.750	1.690-1.720	0.0015-0.0032	0.0015-0.0032	0.0012-0.0028	0.0014-0.0029
	3.8	K	45	45	210@1.32	1.690-1.720	0.0015-0.0035	0.0015-0.0032	0.0012-0.0028	0.0014-0.0029
2003	3.1	M	45	45	250@1.239	1.710	0.0001-0.0027	0.0010-0.0027	0.0010-0.0027	0.0010-0.0027
	3.4	E	46	45	75@1.40	1.400	0.0011-0.0026	0.0014-0.0031	0.0010-0.0027	0.0010-0.0027
	3.8	J	45	45	80@1.750	1.690-1.720	0.0015-0.0032	0.0015-0.0032	0.0012-0.0028	0.0014-0.0029
	3.8	1	45	45	80@1.750	1.690-1.720	0.0015-0.0032	0.0015-0.0032	0.0012-0.0028	0.0014-0.0029
	3.8	K	45	45	210@1.32	1.690-1.720	0.0015-0.0035	0.0015-0.0032	0.0012-0.0028	0.0014-0.0029
2004	3.1	J	45	45	250@1.239	1.710	0.0001-0.0027	0.0010-0.0027	0.0010-0.0027	0.0010-0.0027
	3.4	E	46	45	75@1.40	1.400	0.0011-0.0026	0.0014-0.0031	0.0010-0.0027	0.0010-0.0027
	3.8	1	45	45	80@1.750	1.690-1.720	0.0015-0.0032	0.0015-0.0032	0.0012-0.0028	0.0014-0.0029
	3.8	2	45	45	80@1.750	1.690-1.720	0.0015-0.0032	0.0015-0.0032	0.0012-0.0028	0.0014-0.0029
	3.8	4	45	45	80@1.750	1.690-1.720	0.0015-0.0032	0.0015-0.0032	0.0012-0.0028	0.0014-0.0029
	3.8	K	45	45	210@1.32	1.690-1.720	0.0015-0.0035	0.0015-0.0032	0.0012-0.0028	0.0014-0.0029
2005	3.1	J	45	45	250@1.239	1.710	0.0001-0.0027	0.0010-0.0027	0.0010-0.0027	0.0010-0.0027
	3.4	E	46	45	75@1.40	1.400	0.0011-0.0026	0.0014-0.0031	0.0010-0.0027	0.0010-0.0027
	3.8	1	45	45	80@1.750	1.690-1.720	0.0015-0.0032	0.0015-0.0032	0.0012-0.0028	0.0014-0.0029
	3.8	2	45	45	80@1.750	1.690-1.720	0.0015-0.0032	0.0015-0.0032	0.0012-0.0028	0.0014-0.0029
	3.8	4	45	45	80@1.750	1.690-1.720	0.0015-0.0032	0.0015-0.0032	0.0012-0.0028	0.0014-0.0029

VALVE SPECIFICATIONS

Year	Engine Displacement Liters	Engine VIN	Seat Angle (deg.)	Face Angle (deg.)	Spring Test Pressure (lbs. @ in.)	Spring Installed Height (in.)	Stem-to-Guide Clearance (in.)		Stem Diameter (in.)	
							Intake	Exhaust	Intake	Exhaust
2005 cont.	3.8	K	45	45	210@1.32	1.690-1.720	0.0015-0.0035	0.0015-0.0032	0.0012-0.0028	0.0014-0.0029
	5.3	C	46	45	220@1.32	1.801	0.0010-0.0026	0.0010-0.0026	0.3130-0.3140	0.3130-0.3140

NA: Not Available

06025-GMWB-C07

CRANKSHAFT AND CONNECTING ROD SPECIFICATIONS

All measurements are given in inches.

Year	Engine Displacement Liters	Engine VIN	Crankshaft				Connecting Rod		
			Main Brg. Journal Dia.	Main Brg. Oil Clearance	Shaft End-play	Thrust on No.	Journal Diameter	Oil Clearance	Side Clearance
2002	3.1	M	2.6473-2.6383	0.0008- ① 0.0025	0.0024-0.0083	3	1.9987-1.9994	0.0007-0.0024	0.0070-0.0170
	3.4	E	2.6472-2.6479	0.0008-0.0025	0.0024-0.0083	3	1.9987-1.9994	0.0007-0.0024	0.0070-0.0170
	3.5	H	2.7550-2.7560	0.0006-0.0021	0.0050-0.0200	3	2.1829-2.1835	0.0009-0.0025	0.0040-0.0130
	3.8	J	2.4988-2.4998	0.0008-0.0022	0.0030-0.0110	2	2.3738-2.3745	0.0005-0.0026	0.0030-0.0150
	3.8	1	2.4988-2.4998	0.0008-0.0022	0.0030-0.0110	2	2.3738-2.3745	0.0005-0.0026	0.0030-0.0150
	3.8	K	2.4988-2.4998	0.0008-0.0022	0.0030-0.0110	2	2.3738-2.3745	0.0005-0.0026	0.0030-0.0150
2003	3.1	M	2.6473-2.6383	0.0008- ① 0.0025	0.0024-0.0083	3	1.9987-1.9994	0.0007-0.0024	0.0070-0.0170
	3.4	E	2.6472-2.6479	0.0008-0.0025	0.0024-0.0083	3	1.9987-1.9994	0.0007-0.0024	0.0070-0.0170
	3.8	J	2.4988-2.4998	0.0008-0.0022	0.0030-0.0110	2	2.3738-2.3745	0.0005-0.0026	0.0030-0.0150
	3.8	1	2.4988-2.4998	0.0008-0.0022	0.0030-0.0110	2	2.3738-2.3745	0.0005-0.0026	0.0030-0.0150
	3.8	K	2.4988-2.4998	0.0008-0.0022	0.0030-0.0110	2	2.3738-2.3745	0.0005-0.0026	0.0030-0.0150
2004	3.1	J	2.6473-2.6483	0.0008- ① 0.0025	0.0024-0.0083	3	1.9987-1.9994	0.0007-0.017	0.010-0.015
	3.4	E	2.6473-2.6483	0.0008-0.0025	0.0024-0.0083	3	1.9987-1.9994	0.0007-0.017	0.010-0.015
	3.8	1	2.4988-2.4998	②	0.0030-0.0110	2	2.2487-2.2499	0.0005-0.0026	0.0040-0.0200
	3.8	2	2.4988-2.4998	②	0.0030-0.0110	2	2.2487-2.2499	0.0005-0.0026	0.0040-0.0200
	3.8	4	2.4988-2.4998	②	0.0030-0.0110	2	2.2487-2.2499	0.0005-0.0026	0.0040-0.0200
	3.8	K	2.4988-2.4998	②	0.0030-0.0110	2	2.2487-2.2499	0.0005-0.0026	0.0040-0.0200
2005	3.1	J	2.6473-2.6483	0.0008- ① 0.0025	0.0024-0.0083	3	1.9987-1.9994	0.0007-0.017	0.010-0.015
	3.4	E	2.6473-2.6483	0.0008-0.0025	0.0024-0.0083	3	1.9987-1.9994	0.0007-0.017	0.010-0.015
	3.8	1	2.4988-2.4998	②	0.0030-0.0110	2	2.2487-2.2499	0.0005-0.0026	0.0040-0.0200
	3.8	2	2.4988-2.4998	②	0.0030-0.0110	2	2.2487-2.2499	0.0005-0.0026	0.0040-0.0200
	3.8	4	2.4988-2.4998	②	0.0030-0.0110	2	2.2487-2.2499	0.0005-0.0026	0.0040-0.0200

06025-GMWB-C08

CRANKSHAFT AND CONNECTING ROD SPECIFICATIONS

All measurements are given in inches.

| Year | Engine Displacement Liters | Engine VIN | Crankshaft | | | | Connecting Rod | | |
			Main Brg. Journal Dia.	Main Brg. Oil Clearance	Shaft End-play	Thrust on No.	Journal Diameter	Oil Clearance	Side Clearance
2005 cont.	3.8	K	2.4988-2.4998	②	0.0030-0.0110	2	2.2487-2.2499	0.0005-0.0026	0.0040-0.0200
	5.3	C	2.5580-2.5590	0.0008-0.0021	0.0015-0.0078	3	2.0991-2.0999	0.0009-0.0025	0.0043-0.0200

① Thrust bearing: 0.0012 - 0.0030
② Bearing No. 1: 0.0007 - 0.0016
   Bearing Nos. 2, 3, and 4: 0.0009 - 0.0018

06025-GMWB-C09

PISTON AND RING SPECIFICATIONS

All measurements are given in inches.

Year	Engine Displacement Liters	Engine VIN	Piston Clearance	Ring Gap			Ring Side Clearance		
				Top Compression	Bottom Compression	Oil Control	Top Compression	Bottom Compression	Oil Control
2002	3.1	M	0.0010-0.0018	0.0118-0.0196	0.0118-0.0196	0.0157-0.0551	0.0008-0.0015	0.0008-0.0015	0.0004-0.0012
	3.4	E	0.0008-0.0020	0.0080-0.0180	0.0220-0.0320	0.0098-0.0229	0.0013-0.0031	0.0013-0.0031	0.0011-0.0081
	3.5	H	0.0010-0.0025	0.0080-0.0180	0.0140-0.0200	0.0100-0.0300	0.0016-0.0037	0.0016-0.0037	side-sealing
	3.8	J	0.0004-0.0020	0.0100-0.0160	0.0300-0.0400	0.0100-0.0300	0.0013-0.0031	0.0013-0.0031	0.0009-0.0079
	3.8	1	0.0004-0.0020	0.0100-0.0160	0.0300-0.0400	0.0100-0.0300	0.0013-0.0031	0.0013-0.0031	0.0009-0.0079
	3.8	K	0.0004-0.0020	0.0120-0.0220	0.0300-0.0400	0.0100-0.0300	0.0013-0.0031	0.0013-0.0031	0.0009-0.0079
2003	3.1	M	0.0010-0.0018	0.0118-0.0196	0.0118-0.0196	0.0157-0.0551	0.0008-0.0015	0.0008-0.0015	0.0004-0.0012
	3.4	E	0.0008-0.0020	0.0080-0.0180	0.0220-0.0320	0.0098-0.0229	0.0013-0.0031	0.0013-0.0031	0.0011-0.0081
	3.8	J	0.0004-0.0020	0.0100-0.0160	0.0300-0.0400	0.0100-0.0300	0.0013-0.0031	0.0013-0.0031	0.0009-0.0079
	3.8	1	0.0004-0.0020	0.0100-0.0160	0.0300-0.0400	0.0100-0.0300	0.0013-0.0031	0.0013-0.0031	0.0009-0.0079
	3.8	K	0.0004-0.0020	0.0120-0.0220	0.0300-0.0400	0.0100-0.0300	0.0013-0.0031	0.0013-0.0031	0.0009-0.0079
2004	3.1	J	0.0010-0.0018	0.0118-0.0196	0.0118-0.0196	0.0157-0.0551	0.0008-0.0015	0.0008-0.0015	0.0004-0.0012
	3.4	E	0.0008-0.0020	0.0080-0.0180	0.0220-0.0320	0.0098-0.0229	0.0013-0.0031	0.0013-0.0031	0.0011-0.0081
	3.8	1	0.0004-0.0020	0.0100-0.0160	0.0300-0.0400	0.0100-0.0300	0.0013-0.0031	0.0013-0.0031	0.0009-0.0079
	3.8	2	0.0004-0.0020	0.0100-0.0160	0.0300-0.0400	0.0100-0.0300	0.0013-0.0031	0.0013-0.0031	0.0009-0.0079
	3.8	4	0.0004-0.0020	0.0100-0.0160	0.0300-0.0400	0.0100-0.0300	0.0013-0.0031	0.0013-0.0031	0.0009-0.0079
	3.8	K	0.0004-0.0020	0.0120-0.0220	0.0300-0.0400	0.0100-0.0300	0.0013-0.0031	0.0013-0.0031	0.0009-0.0079
2005	3.1	J	0.0010-0.0018	0.0118-0.0196	0.0118-0.0196	0.0157-0.0551	0.0008-0.0015	0.0008-0.0015	0.0004-0.0012
	3.4	E	0.0008-0.0020	0.0080-0.0180	0.0220-0.0320	0.0098-0.0229	0.0013-0.0031	0.0013-0.0031	0.0011-0.0081
	3.8	1	0.0004-0.0020	0.0100-0.0160	0.0300-0.0400	0.0100-0.0300	0.0013-0.0031	0.0013-0.0031	0.0009-0.0079
	3.8	2	0.0004-0.0020	0.0100-0.0160	0.0300-0.0400	0.0100-0.0300	0.0013-0.0031	0.0013-0.0031	0.0009-0.0079
	3.8	4	0.0004-0.0020	0.0100-0.0160	0.0300-0.0400	0.0100-0.0300	0.0013-0.0031	0.0013-0.0031	0.0009-0.0079
	3.8	K	0.0004-0.0020	0.0120-0.0220	0.0300-0.0400	0.0100-0.0300	0.0013-0.0031	0.0013-0.0031	0.0009-0.0079
	5.3	C	0-0.0006	0.0090-0.0170	0.0170-0.0270	0.0070-0.0290	0.0016-0.0034	0.0016-0.0031	0.0005-0.0078

TORQUE SPECIFICATIONS
All readings in ft. lbs.

Year	Engine Displacement Liters	Engine VIN	Cylinder Head Bolts	Main Bearing Bolts	Rod Bearing Bolts	Crankshaft Damper Bolts	Flywheel Bolts	Manifold Intake	Manifold Exhaust	Spark Plug	Oil Pan Drain Plug
2002	3.1	M	①	②	③	76	61	④	10	11	18
	3.1	J	①	②	③	76	61	④	10	11	18
	3.4	E	⑥	②	39	78	61	18	⑦	11	18
	3.5	H	⑧	⑨	⑩	⑪	⑫	5	18	15	15
	3.8	1	⑬	⑭	⑮	⑯	⑰	⑱	22	11	22
	3.8	K	⑬	⑭	⑮	⑯	⑰	⑲	38	11	22
2003	3.1	M	①	②	③	76	61	④	10	13	22
	3.1	J	①	②	③	76	61	④	10	13	22
	3.4	E	⑥	②	39	78	61	18	⑦	11	22
	3.8	1	⑬	⑭	⑮	⑯	⑰	⑱	22	11	22
	3.8	K	⑬	⑭	⑮	⑯	⑰	⑲	38	11	22
2004	3.1	J	①	②	③	76	61	④	10	13	22
	3.4	E	⑥	②	39	78	61	18	⑦	11	18
	3.8	1	⑬	⑭	⑮	⑯	⑰	⑱	22	11	22
	3.8	2	⑬	⑭	⑮	⑯	⑰	⑱	22	11	22
	3.8	4	⑬	⑭	⑮	⑯	⑰	⑱	22	11	22
	3.8	K	⑬	⑭	⑮	⑯	⑰	⑲	38	11	22
2005	3.1	J	⑳	②	③	㉑	52	㉒	12	㉓	22
	3.4	E	⑳	②	③	㉑	52	㉒	12	㉓	18
	3.8	1	㉔	⑭	⑮	⑯	⑰	㉕	22	㉖	22
	3.8	2	㉔	⑭	⑮	⑯	⑰	㉕	22	11	22
	3.8	4	⑬	⑭	⑮	⑯	⑰	⑱	22	11	22
	3.8	K	㉔	⑭	⑮	⑯	⑰	⑲	38	㉖	22
	5.3	C	㉗	㉘	③	㉙	㉚	㉛	㉜	11	18

* Lower manifold, unless otherwise noted

① Coat threads with sealer torque to 33 ft. lbs., then turn 1/4 turn (90 degrees)

② 37 ft. lbs. plus 77 degrees

③ 15 ft. lbs. plus 75 degrees

④ Torque all bolts to 15 ft. lbs. Retorque to 24 ft. lbs.

⑤ 20 ft. lbs. plus 50 degrees

⑥ 37 ft. lbs. plus 90 degrees

⑦ 115 inch lbs.

⑧ Torque the M11 bolts in sequence to:
Step 1: 22 ft. lbs.
Step 2: 100 degree turn.
Step 3: 100 degree turn.
Torque the long M6 bolt to 106 inch lbs.
Torque both shorter M6 bolts to 106 inch lbs.

⑨ Cap bolts
Step 1: 15 ft. lbs.
Step 2: plus 70 degrees
Perimeter bolts: 22 ft. lbs.

㉒ Lower intake center:
Step 1: 62 inch lbs.
Step 2: 115 inch lbs.
Lower intake corners:
Step 1: 62 inch lbs.
Step 2: 18 ft. lbs.
Upper manifold: 18 ft. lbs.

㉓ New installation: 15 ft. lbs.
Re-installation: 13 ft. lbs.

㉔ Step 1: 37 ft. lbs.
Step 2: plus 120 degrees

㉕ Upper manifold: 89 inch lbs.
Lower manifold: 11 ft. lbs.

㉖ New installation: 20 ft. lbs.
Re-installation: 11 ft. lbs.

㉗ M11 bolts
Step 1: 22 ft. lbs.
Step 2: plus 90 degrees
Step 3: plus 70 degrees
M8 bolts: 22 ft. lbs.

06025-GMWB-C11

TORQUE SPECIFICATIONS
FOOTNOTES CONTINUED

⑩ Step 1: 22 ft. lbs.

    Step 2: Loosen completely

    Step 3: 18 ft. lbs.

    Step 4: plus 110 degrees

⑪ Step 1: 37 ft. lbs.

    Step 2: plus 150 degrees

⑫ Step 1: 11 ft. lbs.

    Step 2: plus 50 degrees

⑬ Step 1: 35 ft. lbs.

    Step 2: plus 130 degrees

⑭ 30 ft. lbs. plus 110 degrees

    Side bolts: 11 ft. lbs. plus 45 degrees

⑮ 20 ft. lbs. plus 50 degrees

⑯ 110 ft. lbs. plus 76 degrees

⑰ 11 ft. lbs. plus 50 degrees

⑱ Upper manifold: 8 ft. lbs.

    Lower manifold: 11 ft. lbs.

⑲ Upper manifold: 18 ft. lbs.

    Lower manifold bolt/nut: 22 ft. lbs.

    Upper manifold studs: 89 inch lbs.

⑳ Step 1: 44 ft. lbs.

    Step 2: plus 95 degrees

㉑ Step 1: 52 ft. lbs.

    Step 2: plus 72 degrees

㉘ M10 bolts

    Step 1: 15 ft. lbs

    Step 2: plus 80 degrees

  M10 studs

    Step 1: 15 ft. lbs

    Step 2: plus 51 degrees

  M8 bolts: 18 ft. lbs.

㉙ Step 1: install the old bolt and tighten to 240 ft. lbs.

    Step 2: install a NEW bolt and tighten to 37 ft. lbs.

    Step 3: plus 140 degrees

㉚ Step 1: 15 ft. lbs.

    Step 2: 37 ft. lbs.

    Step 3: 74 ft. lbs.

㉛ Step 1: 44 inch lbs.

    Step 2: 89 inch lbs.

㉜ Step 1: 11 ft. lbs.

    Step 2: 15 ft. lbs.

06025-GMWB-C12

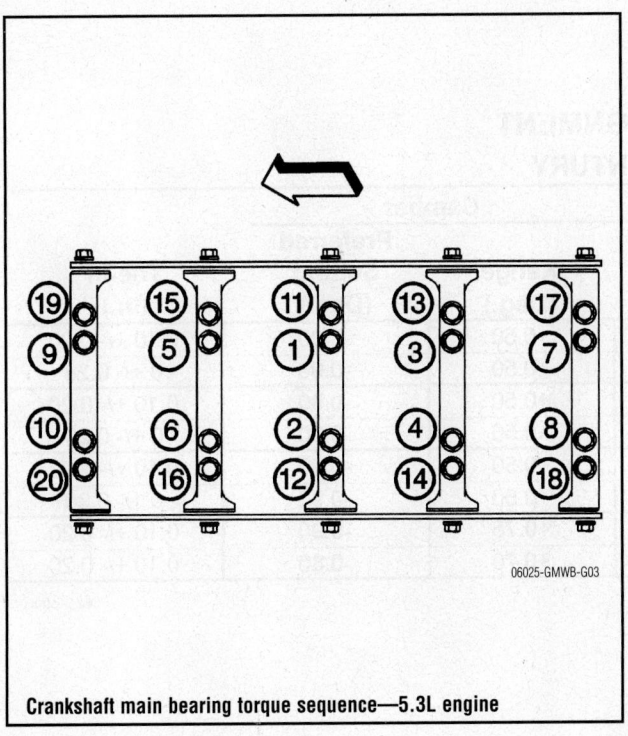

Crankshaft main bearing torque sequence—5.3L engine

06025-GMWB-G03

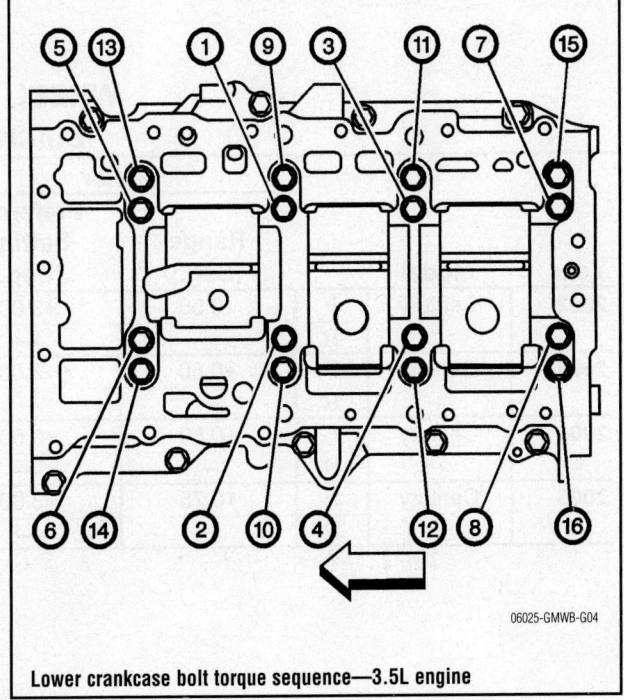

Lower crankcase bolt torque sequence—3.5L engine

06025-GMWB-G04

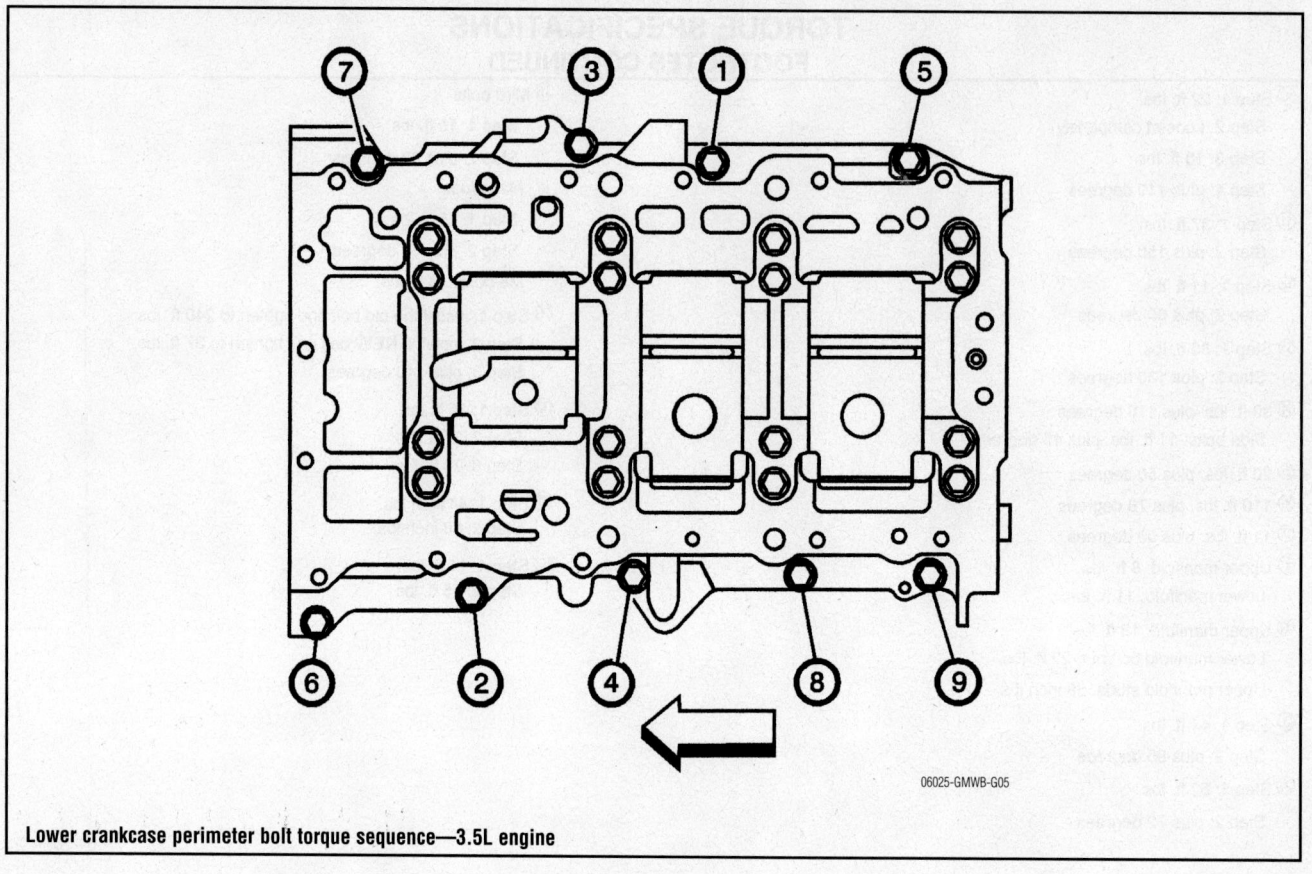

06025-GMWB-G05

Lower crankcase perimeter bolt torque sequence—3.5L engine

WHEEL ALIGNMENT
BUICK CENTURY

Year	Model		Caster Range (Deg.)	Caster Preferred Setting (Deg.)	Camber Range (Deg.)	Camber Preferred Setting (Deg.)	Toe-in (in.)
2002	Century	F	+0.50	+3.00	+0.50	-0.90	0.10 +/- 0.20
		R	—	—	+0.50	-0.90	0 +/- 0.20
2003	Century	F	+0.50	+3.00	+0.50	-0.90	0.10 +/- 0.20
		R	—	—	+0.50	-0.90	0 +/- 0.20
2004	Century	F	+0.50	+3.00	+0.50	-0.90	0.10 +/- 0.20
		R	—	—	+0.50	-0.90	0 +/- 0.20
2005	Century	F	+0.75	+3.00	+0.75	-0.90	0.10 +/- 0.20
		R	—	—	+0.70	-0.80	0.10 +/- 0.20

06025-GMWB-C13

WHEEL ALIGNMENT

BUICK REGAL

Year	Model		Caster Range (Deg.)	Caster Preferred Setting (Deg.)	Camber Range (Deg.)	Camber Preferred Setting (Deg.)	Toe-in (in.)
2002	Regal	F	+0.50	+3.00	+0.50	-0.90	0.10 +/- 0.20
		R	—	—	+0.50	-0.90	0 +/- 0.20
2003	Regal	F	+0.50	+3.00	+0.50	-0.90	0.10 +/- 0.20
		R	—	—	+0.50	-0.90	0 +/- 0.20
2004	Regal	F	+0.50	+3.00	+0.50	-0.90	0.10 +/- 0.20
		R	—	—	+0.50	-0.90	0 +/- 0.20

06025-GMWB-C14

WHEEL ALIGNMENT

CHEVROLET IMPALA

Year	Model		Caster Range (Deg.)	Caster Preferred Setting (Deg.)	Camber Range (Deg.)	Camber Preferred Setting (Deg.)	Toe-in (in.)
2002	Impala	F	+/-0.50	+3.20	+/-0.50	+0.85	0.10 +/- 0.20
		R	—	—	+/-0.50	0	0.10 +/- 0.20
2003	Impala	F	+/-0.50	+3.20	+/-0.50	+0.85	0.10 +/- 0.20
		R	—	—	+/-0.50	0	0.10 +/- 0.20
2004	Impala	F	+/-0.50	+3.20	+/-0.50	+0.85	0.10 +/- 0.20
		R	—	—	+/-0.50	0	0.10 +/- 0.20
2005	RPO FE1 & FE2	F	+/-0.75	+3.00	+/-0.75	-0.75	0.10 +/- 0.20
		R	—	—	+/-0.70	-0.65	-0.10 +/- 0.20
	RPO FE4	F	+/-0.75	+3.00	+/-0.75	-0.75	0.10 +/- 0.20
		R	—	—	+/-0.70	-0.65	-0.10 +/- 0.20
	Police	F	+/-0.75	+3.50	+/-0.75	-0.60	0.10 +/- 0.20
		R	—	—	+/-0.70	-0.70	0.10 +/- 0.20

06025-GMWB-C15

WHEEL ALIGNMENT

CHEVROLET MONTE CARLO

Year	Model		Caster Range (Deg.)	Caster Preferred Setting (Deg.)	Camber Range (Deg.)	Camber Preferred Setting (Deg.)	Toe-in (in.)
2002	Monte Carlo	F	0.50	+1.80	+0.50	+0.70	0 +/- 0.20
		R	—	—	+0.50	-0.15	0.10 +/- 0.10
2003	Monte Carlo	F	0.50	+1.80	+0.50	+0.70	0 +/- 0.20
		R	—	—	+0.50	-0.15	0.10 +/- 0.10
2004	Monte Carlo	F	0.50	+1.80	+0.50	+0.70	0 +/- 0.20
		R	—	—	+0.50	-0.15	0.10 +/- 0.10
2005	Monte Carlo	F	0.75	+3.40	+/-0.75	-0.85	0.10 +/- 0.20
		R	—	—	+/-0.50	-0.85	-0.10 +/- 0.10

06025-GMWB-C16

WHEEL ALIGNMENT
OLDSMOBILE INTRIGUE

Year	Model		Caster Range (Deg.)	Caster Preferred Setting (Deg.)	Camber Range (Deg.)	Camber Preferred Setting (Deg.)	Toe-in (in.)
2002	Intrigue	F	0.50	3.00	0.50	-0.90	0.05 +/- 0.10
		R	—	—	0.50	-0.90	0 +/- 0.10

06025-GMWB-C17

WHEEL ALIGNMENT
PONTIAC GRAND PRIX

Year	Model		Caster Range (Deg.)	Caster Preferred Setting (Deg.)	Camber Range (Deg.)	Camber Preferred Setting (Deg.)	Toe-in (in.)
2002	Grand Prix	F	0.50	3.00	0.50	-0.90	0.05 +/- 0.10
		R	—	—	0.50	-0.90	0 +/- 0.10
2003	Grand Prix	F	0.50	3.00	0.50	-0.90	0.05 +/- 0.10
		R	—	—	0.50	-0.90	0 +/- 0.10
2004	Grand Prix	F	0.50	3.00	0.50	-0.90	0.05 +/- 0.10
		R	—	—	0.50	-0.90	0 +/- 0.10
2005	Except GXP	F	0.75	3.15	0.75	-0.80	0.10 +/- 0.20
		R	—	—	0.50	-0.95	0.10 +/- 0.20
	GXP	F	0.75	3.20	0.75	-1.00	0.10 +/- 0.20
		R	—	—	0.70	-1.15	0.10 +/- 0.20

06025-GMWB-C18

TIRE, WHEEL AND BALL JOINT SPECIFICATIONS
Buick Century and Regal

Year	Model	OEM Tires Standard	OEM Tires Optional	Tire Pressures (psi) Front	Tire Pressures (psi) Rear	Wheel Size	Ball Joint Inspection	Lug Nut Torque (ft. lbs.)
2002	Regal, exc GS	P205/70R15	P215/70R15 P225/60R16	①	①	6-JJ	②	100
	Regal GS	P225/60R16	None	①	①	6.5-JJ	②	100
	Century 2dr	P225/60R16	None	①	①	6.5-JJ	②	100
	Century 4dr	P205/70R15	P225/60R16	①	①	Std: 6-JJ Opt: 6.5-JJ	②	100
2003	Regal, exc GS	P205/70R15	P215/70R15 P225/60R16	①	①	6-JJ	②	100
	Regal GS	P225/60R16	None	①	①	6.5-JJ	②	100
	Century 2dr	P225/60R16	None	①	①	6.5-JJ	②	100
	Century 4dr	P205/70R15	P225/60R16	①	①	Std: 6-JJ Opt: 6.5-JJ	②	100
2004	Regal LS	P215/70R15	P225/70R16R	①	①	6-JJ	②	100
	Regal GS	P225/70R16R	None	①	①	6.5-JJ	②	100
	Century SE	P205/70R16	None	①	①	6.5-JJ	②	100
	Century, exc SE	P205/70R15	None	①	①	6-JJ	②	100
2005	Century SE	P225/70R16	None	①	①	6.5-JJ	③	100
	Century, exc SE	P205/70R15	None	①	①	6-JJ	③	100

OEM: Original Equipment Manufacturer

PSI: Pounds Per Square Inch

STD: Standard

OPT: Optional

L: Lower

U: Upper

① See placard on vehicle

② Replace if any measurable movement is found.

③ Remove all load from the joint. Vertical and horizontal movement is 0.125 in. max.

06025-GMWB-C19

TIRE, WHEEL AND BALL JOINT SPECIFICATIONS
Chevrolet Impala and Monte Carlo

Year	Model	OEM Tires Standard	OEM Tires Optional	Tire Pressures (psi) Front	Tire Pressures (psi) Rear	Wheel Size	Ball Joint Inspection	Lug Nut Torque (ft. lbs.)
2002	Monte Carlo	P195/70R14	None	①	①	6-JJ	②	100
			P225/60R16N	①	①	6-JJ		
	Impala	P225/60R16	P225/60R16	①	①	6-JJ	②	100
			P225/60R16N	①	①	6-JJ		
2003	Monte Carlo	P195/70R14	None	①	①	6-JJ	②	100
			P225/60R16N	①	①	6-JJ		
	Impala	P225/60R16	P225/60R16	①	①	6-JJ	②	100
			P225/60R16N	①	①	6-JJ		
2004	Monte Carlo	P225/60R16	P225/60R16	①	①	6-JJ	②	100
			P235/55R17	①	①	6-JJ		
	Impala	P225/60R16	P225/60R16	①	①	6-JJ	②	100
			P235/55R17	①	①	6-JJ		
2005	Monte Carlo	P225/60R16	P225/60R16	①	①	6-JJ	③	100
			P235/55R17	①	①	6-JJ		
	Impala	P225/60R16	P225/60R16	①	①	6-JJ	③	100
			P235/55WR17	①	①	6-JJ		

OEM: Original Equipment Manufacturer

PSI: Pounds Per Square Inch

① See placard on vehicle

② Replace if any measurable movement is found

③ Remove all load from the joint. Vertical and horizontal movement is 0.125 in. max.

06025-GMWB-C20

TIRE, WHEEL AND BALL JOINT SPECIFICATIONS
Oldsmobile Intrigue

Year	Model	OEM Tires Standard	OEM Tires Optional	Tire Pressures (psi) Front	Tire Pressures (psi) Rear	Wheel Size	Ball Joint Inspection	Lug Nut Torque (ft. lbs.)
2002	Intrigue	P225/60R16	None	①	①	6-JJ	②	100

OEM: Original Equipment Manufacturer

PSI: Pounds Per Square Inch

STD: Standard

OPT: Optional

① See placard on vehicle

② Replace if any measurable movement is found

06025-GMWB-C21

TIRE, WHEEL AND BALL JOINT SPECIFICATIONS
Pontiac Grand Prix

Year	Model	OEM Tires		Tire Pressures (psi)		Wheel Size	Ball Joint Inspection	Lug Nut Torque (ft. lbs.)
		Standard	Optional	Front	Rear			
2002	Grand Prix	P205/70R15	P225/60R16	①	①	7-JJ	0.125 in.	100
2003	Grand Prix	P205/70R15	P225/60R16	①	①	7-JJ	0.125 in.	100
2004	Grand Prix	P225/60R16	P225/55VR17	①	①	7-JJ	②	100
2005	Base & GT	P225/60R16	none	①	①	7-JJ	④	100
	GTP	P225/55HR17	none	①	①	7-JJ	④	100
	GXP	P225/50WR18	P225/45WR18 ③	①	①	7-JJ	④	100

OEM: Original Equipment Manufacturer

PSI: Pounds Per Square Inch

① See placard on vehicle

② Replace if any measurable movement is found

③ Front only

④ Remove all load from the joint. Vertical and horizontal movement is 0.125 in. max.

06025-GMWB-C22

BRAKE SPECIFICATIONS
All measurements in inches unless noted

Year	Model		Brake Disc			Brake Drum Diameter			Minimum Lining Thickness		Brake Caliper	
			Original Thickness	Minimum Thickness	Maximum Runout	Original Inside Diameter	Max. Wear Limit	Maximum Machine Diameter	Front	Rear	Bracket Bolts (ft. lbs.)	Mounting Bolts (ft. lbs.)
2002	Century	F	1.270	1.210	0.002	—	—	—	NA	NA	133	70
		R	0.430	0.350	0.002	8.863	8.920	8.909	NA	NA	85	32
	Grand Prix	F	1.270	1.210	0.002	—	—	—	NA	NA	133	70
		R	0.430	0.350	0.002	—	—	—	NA	NA	85	32
	Impala	F	1.270	1.210	0.002	—	—	—	NA	NA	133	70
		R	0.430	0.350	0.002	—	—	—	NA	NA	85	32
	Intrigue	F	1.270	1.210	0.002	—	—	—	NA	NA	133	70
		R	0.430	0.350	0.002	—	—	—	NA	NA	85	32
	Monte Carlo	F	1.270	1.210	0.002	—	—	—	NA	NA	133	70
		R	0.430	0.350	0.002	—	—	—	NA	NA	85	32
	Regal	F	1.270	1.210	0.002	—	—	—	NA	NA	133	70
		R	0.430	0.350	0.002	8.863	8.920	8.909	NA	NA	85	32
2003	Century	F	1.270	1.210	0.002	—	—	—	NA	NA	133	70
		R	0.430	0.350	0.002	8.863	8.920	8.909	NA	NA	85	32
	Grand Prix	F	1.270	1.210	0.002	—	—	—	NA	NA	133	70
		R	0.430	0.350	0.002	—	—	—	NA	NA	85	32
	Impala	F	1.270	1.210	0.002	—	—	—	NA	NA	133	70
		R	0.430	0.350	0.002	—	—	—	NA	NA	85	32
	Monte Carlo	F	1.270	1.210	0.002	—	—	—	NA	NA	133	70
		R	0.430	0.350	0.002	—	—	—	NA	NA	85	32
	Regal	F	1.270	1.210	0.002	—	—	—	NA	NA	133	70
		R	0.430	0.350	0.002	8.863	8.920	8.909	NA	NA	85	32
2004	Century	F	1.270	1.210	0.002	—	—	—	NA	NA	133	70
		R	0.430	0.350	0.002	—	—	—	NA	NA	85	32
	Grand Prix	F	1.270	1.210	0.002	—	—	—	NA	NA	133	70
		R	0.430	0.350	0.002	—	—	—	NA	NA	85	32
	Impala	F	1.270	1.210	0.002	—	—	—	NA	NA	133	70
		R	0.430	0.350	0.002	—	—	—	NA	NA	85	32
	Monte Carlo	F	1.270	1.210	0.002	—	—	—	NA	NA	133	70
		R	0.430	0.350	0.002	—	—	—	NA	NA	85	32
	Regal	F	1.270	1.210	0.002	—	—	—	NA	NA	133	70
		R	0.430	0.350	0.002	—	—	—	NA	NA	85	32

06025-GMWB-C23

BRAKE SPECIFICATIONS

All measurements in inches unless noted

Year	Model		Brake Disc Original Thickness	Brake Disc Minimum Thickness	Brake Disc Maximum Runout	Brake Drum Diameter Original Inside Diameter	Brake Drum Diameter Max. Wear Limit	Brake Drum Diameter Maximum Machine Diameter	Minimum Lining Thickness Front	Minimum Lining Thickness Rear	Brake Caliper Bracket Bolts (ft. lbs.)	Brake Caliper Mounting Bolts (ft. lbs.)
2005	Century	F	1.270	1.210	0.002	—	—	—	NA	NA	133	70
		R	0.430	0.350	0.002	—	—	—	NA	NA	85	32
	Grand Prix	F	①	②	0.002	—	—	—	NA	NA	③	④
		R	⑤	⑥	0.002	—	—	—	NA	NA	⑦	⑧
	Impala	F	⑨	⑩	0.002	—	—	—	NA	NA	133	70
		R	0.430	0.350	0.002	—	—	—	NA	NA	85	32
	Monte Carlo	F	⑨	⑩	0.002	—	—	—	NA	NA	133	70
		R	0.430	0.350	0.002	—	—	—	NA	NA	85	32

F: Front

R: Rear

NA: Information not available

① RPO Codes L 26, L 32: 1.27 in.
  RPO Code LS 4: 1.26 in.

② RPO Codes L 26, L 32: 1.21 in.
  RPO code LS 4: 1.14 in.

③ RPO Codes L 26, L 32: 133 ft. lbs.
  RPO code LS 4: 136 ft. lbs.

④ RPO Codes L 26, L 32: 70 ft. lbs.
  RPO code LS 4: 44 ft. lbs.

⑤ RPO Codes L 26, L 32: 0.55 in.
  RPO Code LS 4: 1.024 in.

⑥ RPO Codes L 26, L 32: 0.49 in.
  RPO code LS 4: 0.91 in.

⑦ RPO Codes L 26, L 32: 89 ft. lbs.
  RPO code LS 4: 96 ft. lbs.

⑧ RPO Codes L 26, L 32: 25 ft. lbs.
  RPO code LS 4: 44 ft. lbs.

⑨ 1st design, before VIN break 59143076 (RPO JL9), or before VIN break 59148924 (RPO J65): 1.27 in.
  2nd design, after above VIN breaks: 1.18 in.

⑩ First design, before VIN break 59143076 (RPO JL9), or before VIN break 59148924 (RPO J65): 1.21 in
  2nd design, after above VIN breaks: 1.13 in.

06025-GMWB-C24

SCHEDULED MAINTENANCE INTERVALS
2002-03 BUICK CENTURY, REGAL, CHEVROLET IMPALA, MONTE CARLO, OLDSMOBILE INTRIGUE & PONTIAC GRAND PRIX

TO BE SERVICED	SERVICE	VEHICLE MILEAGE INTERVAL (x1000)												
		7.5	15	22.5	30	37.5	45	52.5	60	67.5	75	82.5	90	97.5
Engine oil & filter	R	✓	✓	✓	✓	✓	✓	✓	✓	✓	✓	✓	✓	✓
Automatic transaxle fluid & filter ①	S/I	✓	✓	✓	✓	✓	✓	✓	✓	✓	✓	✓	✓	✓
Brake hoses	S/I	✓	✓	✓	✓	✓	✓	✓	✓	✓	✓	✓	✓	✓
Coolant level, hoses & clamps	S/I	✓	✓	✓	✓	✓	✓	✓	✓	✓	✓	✓	✓	✓
Drive shaft boots & front suspension components	S/I	✓	✓	✓	✓	✓	✓	✓	✓	✓	✓	✓	✓	✓
Exhaust system & throttle linkage	S/I	✓	✓	✓	✓	✓	✓	✓	✓	✓	✓	✓	✓	✓
Lubricate chassis, suspension, steering linkage, transaxle shift linkage, parking brake cable guides, underbody contact points & linkage	S/I	✓	✓	✓	✓	✓	✓	✓	✓	✓	✓	✓	✓	✓
Rotate tires	S/I	✓		✓		✓		✓		✓		✓		✓
Air filter element	R				✓				✓				✓	
Engine coolant ②	R				✓				✓				✓	
PCV filter	R				✓				✓				✓	
Spark plugs ③	R				✓				✓				✓	
Accessory drive belt(s)	S/I				✓				✓				✓	
Ignition cables, EGR & fuel systems	S/I				✓				✓				✓	
Camshaft timing belt	R								✓					

R: Replace   S/I: Service or Inspect

① Automatic transaxle fluid & filter: replace at 100,000 miles (if not changed previously).

② Engine coolant: replace every 100,000 miles. Use DEX-COOL™ coolant only. If any silicate coolant is used, the service interval is every 30,000 miles

③ Platinum tip spark plugs: replace every 100,000 miles.

FREQUENT OPERATION MAINTENANCE (SEVERE SERVICE)
If a vehicle is operated under any of the following conditions it is considered severe service:
- Extremely dusty areas.
- 50% or more of the vehicle operation is in 32°C (90°F) or higher temperatures, or constant operation in temperatures below 0°C (32°F).
- Prolonged idling (vehicle operation in stop and go traffic).
- Frequent short running periods (engine does not warm to normal operating temperatures).
- Police, taxi, delivery usage or trailer towing usage.

Oil & oil filter: change every 3000 miles

Chassis lubrication: lubricate every 6000 miles.

Rotate tires at 6000 miles, then every 12,000 miles.

Air filter element: service or inspect every 15,000 miles.

Camshaft timing belt: change every 60,000 miles.

06025-GMWB-C25

MAINTENANCE I AND II SERVICE SCHEDULES
2004-05 Buick Century & Regal

When the CHANGE ENGINE OIL light appears, certain services and inspections are required.
Required services are described as Maintenance I and Maintenance II.
The first service on a vehicle should be Maintenance I, and the second service should be Maintenance II.
Alternate between the 2 thereafter. However, in some cases, Maintenance II may be required more often.
Maintenance I: Use Maintenance I if the CHANGE ENGINE OIL light comes on within 10 months since vehicle was purchased or, if Maintenance II was performed.
Maintenance II: Use Maintenance II if the previous service performed was Maintenance I. Always use Maintenance II whenever the CHANGE ENGINE OIL light comes on 10 months or more since the last service, or, if the CHANGE ENGINE OIL light has not come on at all for one year.

Service	Maintenance I	Maintenance II
Change the engine oil and filter. Reset the oil life system.	✓	✓
Visually inspect the vehicle for leaks or damage. A fluid loss in the vehicle system could indicate a problem. Inspected, repair and add fluid to the system if necessary.	✓	✓
Inspect the engine air cleaner filter. If necessary, replace the filter.		✓
Rotate the tires. Inspect the tire inflation pressures and the tire wear.	✓	✓
Visually inspect the brake lines and hoses for proper hook-up, binding, leaks, cracks, chafing, etc. Inspect the disc brake pads for wear and the rotors for surface condition. Inspect the drum brake linings for wear or cracks. Inspect other brake parts, including drums, wheel cylinders, calipers, parking brake, etc. Inspect the parking brake adjustment.	✓	✓
Inspect the engine coolant and the windshield washer fluid levels. Add fluid as needed.	✓	✓
Inspect the suspension and steering components. Inspect the front and rear suspension and the steering system for damaged, loose or missing parts, or signs of wear. Inspect the power steering lines and the hoses for proper hook-up, binding, leaks, cracks, chafing, etc.	--	✓
Visually inspect the coolant hoses and replace the hoses if they are cracked, swollen or deteriorated. Inspect all pipes, fittings and clamps; replace with GM parts as needed. To help ensure proper operation, a pressure test of the cooling system and pressure cap and cleaning the outside of the radiator and air conditioning condenser is recommended at least once a year.	--	✓
Inspect the front and rear suspension and the steering system for damaged, loose or missing parts, or signs of wear. Inspect power steering lines and hoses for proper hook-up, binding, leaks, cracks, chafing, etc.	--	✓
Inspect the throttle system for interference or binding and for damaged or missing parts. Replace the parts as needed. Replace any components that have high effort or excessive wear. Do not lubricate the accelerator or the cruise control cables.	--	✓
Replace the passenger compartment air filter.	--	✓

To reset the CHANGE ENGINE OIL LIGHT:
1. Turn the ignition key to RUN with the engine off.
2. Fully press and release the accelerator pedal three times within five seconds. The change engine oil light will flash while the system is resetting
3. Turn the key to OFF. The oil life will change to 100%.

If the change engine oil light comes back on and stays on when you start your vehicle, the engine oil life system has not reset repeat the procedure

06025-GMWB-C26

ADDITIONAL MAINTENANCE SERVICES
2004-05 Buick Century & Regal

TO BE SERVICED	TYPE OF SERVICE	VEHICLE MILEAGE INTERVAL (x1000)					
		25	50	75	100	125	150
Air cleaner filter	R		✓		✓		✓
Accessory drive belt	I						✓
Auto. Trans. Fluid and Filter①	R		✓		✓		✓
Cooling system hoses and clamps	S/I						✓
Engine coolant	R						✓
Fuel system	I	✓	✓	✓	✓	✓	✓
Exhaust system & heat shields	S/I	✓	✓	✓	✓	✓	✓
Spark plugs	R				✓		

R: Replace   S/I: Inspect and service, if necessary

① Replace if any of the following conditions are met:

Heavy city traffic where the outside temperature regularly reaches 32°C (90°F) or higher

Hilly or mountainous terrain

Frequent trailer towing

Taxi, police or delivery service

Otherwise, change every 100,000 miles

06025-GMWB-C27

MAINTENANCE I AND II SERVICE SCHEDULES
2004-05 Chevrolet Impala & Monte Carlo

When the CHANGE ENGINE OIL light appears, certain services and inspections are required.

Required services are described as Maintenance I and Maintenance II.

The first service on a vehicle should be Maintenance I, and the second service should be Maintenance II.

Alternate between the 2 thereafter. However, in some cases, Maintenance II may be required more often.

Maintenance I: Use Maintenance I if the CHANGE ENGINE OIL light comes on within 10 months since vehicle was purchased or, if Maintenance II was performed.

Maintenance II: Use Maintenance II if the previous service performed was Maintenance I. Always use Maintenance II whenever the CHANGE ENGINE OIL light comes on 10 months or more since the last service, or, if the CHANGE ENGINE OIL light has not come on at all for one year.

Service	Maintenance I	Maintenance II
Change the engine oil and filter. Reset the oil life system.	✓	✓
Visually inspect the vehicle for leaks or damage. A fluid loss in the vehicle system could indicate a problem. Inspected, repair and add fluid to the system if necessary.	✓	✓
Inspect the engine air cleaner filter. If necessary, replace the filter.		✓
Rotate the tires. Inspect the tire inflation pressures and the tire wear.	✓	✓
Visually inspect the brake lines and hoses for proper hook-up, binding, leaks, cracks, chafing, etc. Inspect the disc brake pads for wear and the rotors for surface condition. Inspect the drum brake linings for wear or cracks. Inspect other brake parts, including drums, wheel cylinders, calipers, parking brake, etc. Inspect the parking brake adjustment.	✓	✓
Inspect the engine coolant and the windshield washer fluid levels. Add fluid as needed.	✓	✓
Inspect the suspension and steering components. Inspect the front and rear suspension and the steering system for damaged, loose or missing parts, or signs of wear. Inspect the power steering lines and the hoses for proper hook-up, binding, leaks, cracks, chafing, etc.	--	✓
Visually inspect the coolant hoses and replace the hoses if they are cracked, swollen or deteriorated. Inspect all pipes, fittings and clamps; replace with GM parts as needed. To help ensure proper operation, a pressure test of the cooling system and pressure cap and cleaning the outside of the radiator and air conditioning condenser is recommended at least once a year.	--	✓
Inspect the transaxle fluid level and add fluid as needed.	--	✓
Inspect the wiper blades and replace as necessary	✓	✓
Inspect the restraint system components.		✓
Inspect the throttle system.	--	✓
Replace the passenger compartment air filter.		✓

To reset the CHANGE ENGINE OIL LIGHT:

1. Turn the ignition key to RUN with the engine off.

2. Fully press and release the accelerator pedal three times within five seconds. The change engine oil light will flash while the system is resetting

3. Turn the key to OFF. The oil life will change to 100%.

If the change engine oil light comes back on and stays on when you start your vehicle, the engine oil life system has not reset repeat the procedure

06025-GMWB-C28

ADDITIONAL MAINTENANCE SERVICES
2004-05 Chevrolet Impala & Monte Carlo

TO BE SERVICED	TYPE OF SERVICE	VEHICLE MILEAGE INTERVAL (x1000)					
		25	50	75	100	125	150
Air cleaner filter	R		✓		✓		✓
Accessory drive belt	I						✓
Auto. Trans. Fluid and Filter①	R		✓		✓		✓
Cooling system hoses and clamps	S/I						✓
Engine coolant	R						✓
Fuel system	I	✓	✓	✓	✓	✓	✓
Exhaust system & heat shields	S/I	✓	✓	✓	✓	✓	✓
Supercharger oil level	S/I	✓	✓	✓	✓	✓	✓
Spark plugs and wires	R				✓		

R: Replace   S/I: Inspect and service, if necessary

① Replace if any of the following conditions are met:

    Heavy city traffic where the outside temperature regularly reaches 32°C (90°F) or higher

    Hilly or mountainous terrain

    Frequent trailer towing

    Taxi, police or delivery service

    Otherwise, change every 100,000 miles

06025-GMWB-C29

MAINTENANCE I AND II SERVICE SCHEDULES
2004-05 Pontiac Grand Prix

When the CHANGE ENGINE OIL light appears, certain services and inspections are required.
Required services are described as Maintenance I and Maintenance II.
The first service on a vehicle should be Maintenance I, and the second service should be Maintenance II.
Alternate between the 2 thereafter. However, in some cases, Maintenance II may be required more often.
Maintenance I: Use Maintenance I if the CHANGE ENGINE OIL light comes on within 10 months since vehicle was purchased or, if Maintenance II was performed.
Maintenance II: Use Maintenance II if the previous service performed was Maintenance I. Always use Maintenance II whenever the CHANGE ENGINE OIL light comes on 10 months or more since the last service, or, if the CHANGE ENGINE OIL light has not come on at all for one year.

Service	Maintenance I	Maintenance II
Change the engine oil and filter. Reset the oil life system.	✓	✓
Visually inspect the vehicle for leaks or damage. A fluid loss in the vehicle system could indicate a problem. Inspected, repair and add fluid to the system if necessary.	✓	✓
Inspect the engine air cleaner filter. If necessary, replace the filter.		✓
Rotate the tires. Inspect the tire inflation pressures and the tire wear.	✓	✓
Visually inspect the brake lines and hoses for proper hook-up, binding, leaks, cracks, chafing, etc. Inspect the disc brake pads for wear and the rotors for surface condition. Inspect the drum brake linings for wear or cracks. Inspect other brake parts, including drums, wheel cylinders, calipers, parking brake, etc. Inspect the parking brake adjustment.	✓	
Inspect the engine coolant and the windshield washer fluid levels. Add fluid as needed.	✓	✓
Inspect the suspension and steering components. Inspect the front and rear suspension and the steering system for damaged, loose or missing parts, or signs of wear. Inspect the power steering lines and the hoses for proper hook-up, binding, leaks, cracks, chafing, etc.	--	✓
Visually inspect the coolant hoses and replace the hoses if they are cracked, swollen or deteriorated. Inspect all pipes, fittings and clamps; replace with GM parts as needed. To help ensure proper operation, a pressure test of the cooling system and pressure cap and cleaning the outside of the radiator and air conditioning condenser is recommended at least once a year.	--	✓
Inspect the transaxle fluid level and add fluid as needed.	--	✓
Inspect the wiper blades and replace as necessary	✓	✓
Inspect the restraint system components.		✓
Inspect the throttle system.	--	✓
Replace the passenger compartment air filter.		✓

To reset the CHANGE ENGINE OIL LIGHT:
1. Turn the ignition key to RUN with the engine off.
2. Fully press and release the accelerator pedal three times within five seconds. The change engine oil light will flash while the system is resetting
3. Turn the key to OFF. The oil life will change to 100%.

If the change engine oil light comes back on and stays on when you start your vehicle, the engine oil life system has not reset repeat the procedure

06025-GMWB-C30

ADDITIONAL MAINTENANCE SERVICES
2004-05 Pontiac Grand Prix

TO BE SERVICED	TYPE OF SERVICE	VEHICLE MILEAGE INTERVAL (x1000)					
		25	50	75	100	125	150
Air cleaner filter	R		✓		✓		✓
Accessory drive belt	I						✓
Auto. Trans. Fluid and Filter①	R		✓		✓		✓
Cooling system hoses and clamps	S/I						✓
Engine coolant	R						✓
Fuel system	I	✓	✓	✓	✓	✓	✓
Exhaust system & heat shields	S/I	✓	✓	✓	✓	✓	✓
Supercharger oil level	S/I	✓	✓	✓	✓	✓	✓
Spark plugs and wires	R				✓		

R: Replace   S/I: Inspect and service, if necessary

① Replace if any of the following conditions are met:

  Heavy city traffic where the outside temperature regularly reaches 32°C (90°F) or higher

  Hilly or mountainous terrain

  Frequent trailer towing

  Taxi, police or delivery service

  Otherwise, change every 100,000 miles

06025-GMWB-C31

ENGINE REPAIR

Distributor

The engines in this section all utilize a Distributorless Ignition System (DIS).

Alternator

REMOVAL & INSTALLATION

3.1L Engine

1. Before servicing the vehicle, refer to the Precautions Section.
2. Remove or disconnect the following:
   - Negative battery cable
   - Engine compartment cross vehicle brace, if equipped
   - Drive belt from the alternator
   - Coolant recovery reservoir and place it away from the alternator
   - Alternator electrical connector
   - Alternator bolts and the alternator

To install:

3. Install or connect the following:
   - Alternator
   - Alternator output BAT terminal and torque the nut to 15 ft. lbs. (20 Nm)
   - Protective boot from the output terminal
   - Electrical connector
   - Alternator bolts and torque them to 37 ft. lbs. (50 Nm)
   - Coolant recovery reservoir
   - Drive belt
   - Engine compartment cross vehicle brace, if removed
   - Negative battery cable

3.4L Engine

1. Before servicing the vehicle, refer to the Precautions Section.
2. Remove or disconnect the following:
   - Negative battery cable
   - Engine compartment cross brace
   - Accessory drive belt
   - Coolant recovery reservoir and place it aside
   - Alternator bolts
   - Alternator electrical connector
   - Alternator

To install:

3. Install or connect the following:
   - Alternator
   - Alternator output BAT wire and torque the nut to 15 ft. lbs. (20 Nm)
   - Alternator electrical connector
   - Alternator bolts and torque them to 37 ft. lbs. (50 Nm)

- Coolant recovery reservoir
- Drive belt
- Negative battery cable

3.5L Engine

1. Before servicing the vehicle, refer to the Precautions Section.
2. Drain the cooling system.
3. Remove or disconnect the following:
   - Battery and tray
   - Drive belt
   - Engine cooling fan assembly
   - Thermostat housing and radiator hose
   - Outboard/inboard alternator bolts
   - Idler pulley bolt and pulley
   - Alternator electrical connectors
   - Alternator bolts and the alternator

To install:

4. Install or connect the following:
   - Alternator
   - Alternator electrical connectors. Torque the positive battery terminal to 15 ft. lbs. (20 Nm).
   - Idler pulley and torque the bolt to 37 ft. lbs. (50 Nm)
   - Alternator bolts and torque them to 37 ft. lbs. (50 Nm)
   - Thermostat housing with radiator hose and torque the bolts to 80 inch lbs. (9 Nm)
   - Engine cooling fan assembly
   - Drive belt
   - Battery tray and torque the bolts to 44 inch lbs. (5 Nm)
   - Battery
5. Refill the cooling system.

3.8L Engine

1. Before servicing the vehicle, refer to the Precautions Section.
2. Remove or disconnect the following:
   - Negative battery cable
   - Drive belt
   - Alternator electrical connectors
   - Rear alternator brace
   - Alternator bolts and the alternator

To install:

3. Install or connect the following:
   - Alternator
   - Alternator bolts and torque them to 37 ft. lbs. (50 Nm)
   - Alternator electrical connectors and torque the positive battery terminal to 15 ft. lbs. (20 Nm)
   - Rear alternator brace and torque the bolts to 37 ft. lbs. (50 Nm)
   - Accessory drive belt
   - Negative battery cable

5.3L Engine

1. Before servicing the vehicle, refer to the Precautions Section.
2. Disconnect the negative battery cable. .
3. Remove the engine sight shield.
4. Remove the drive belt.
5. Disconnect the generator electrical connector (2).
6. Position aside the protective boot (1) from the generator output BAT terminal for access.
7. Remove the generator output BAT

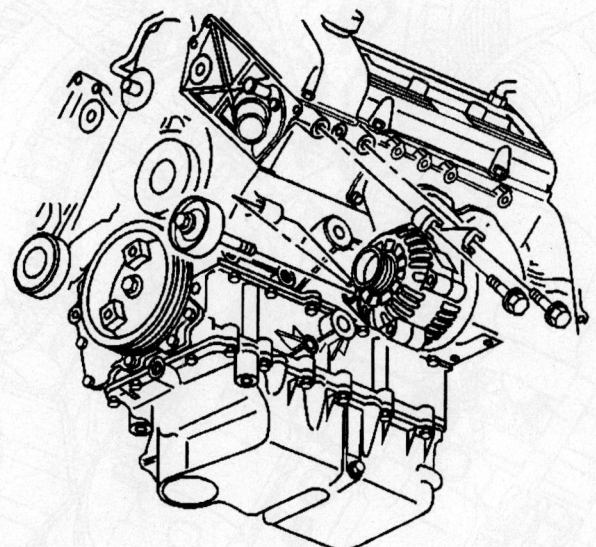

9306VG01

View of the alternator—3.5L engine

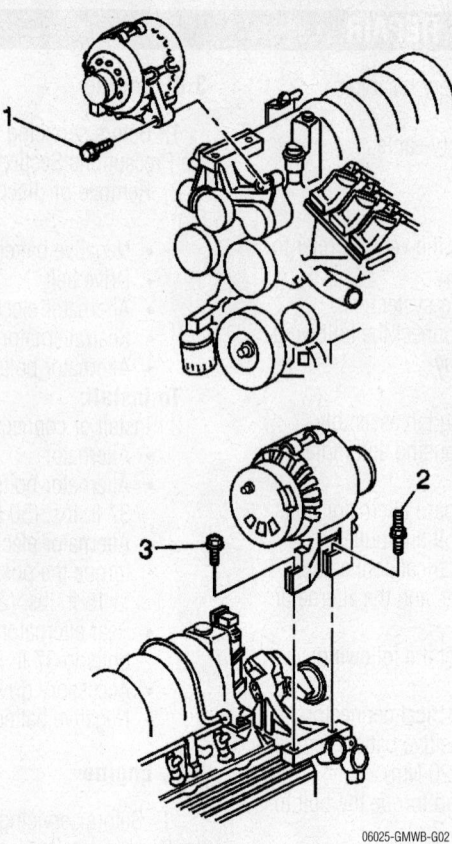

06025-GMWB-G02

Alternator mounting—3.8L engine

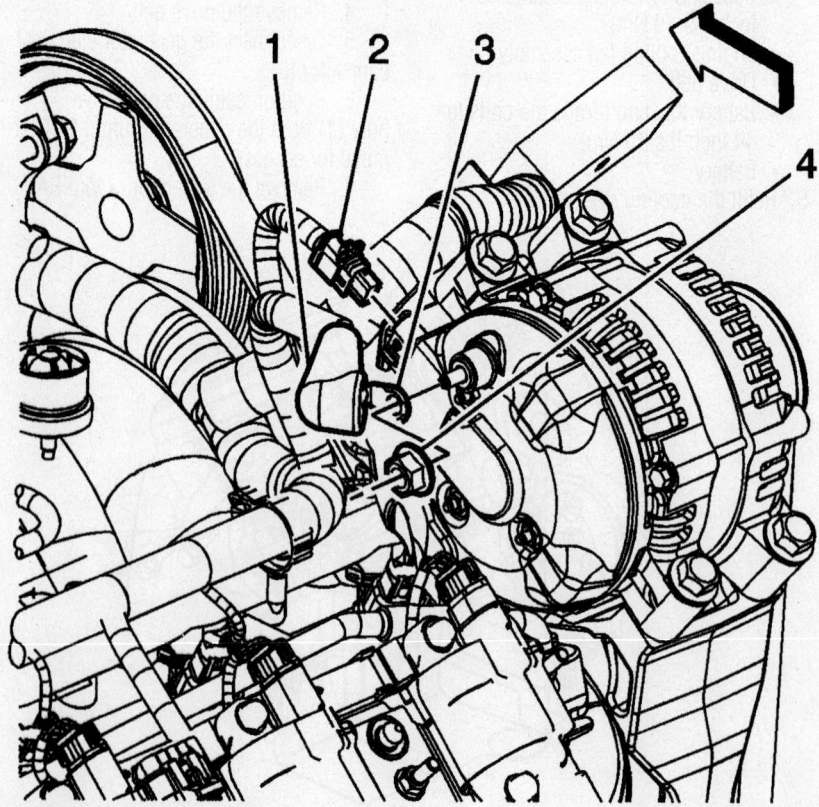

06025-GMWB-G01

Alternator mounting—5.3L engine

terminal nut (4) and remove the positive battery lead (3) from the generator.

8. Remove the generator bolts and generator.

To install:

9. Install the generator and bolts. Tighten the bolts to 50 Nm (37 ft. lbs.).

10. Install the positive battery lead (3) and generator output BAT terminal nut (4) to the generator. Tighten the nut to 20 Nm (15 ft. lbs.).

11. Position the protective boot (1) to the generator output BAT terminal.

12. Connect the generator electrical connector (2).

13. Install the drive belt.

14. Install the engine sight shield.

15. Connect the negative battery cable.

Ignition Timing

ADJUSTMENT

The ignition timing is not adjustable.

Engine Assembly

REMOVAL & INSTALLATION

3.1L Engine

1. Before servicing the vehicle, refer to the Precautions Section.

2. Drain the engine coolant.

3. Drain the engine oil.

4. Relieve the fuel system pressure.

5. Remove or disconnect the following:
- Negative battery cable
- Hood
- Throttle body air inlet duct
- Air cleaner
- Engine mount struts
- Drive belt
- Knock Sensor (KS) electrical connector
- Heated Oxygen Sensor (HO2S) electrical connector
- Camshaft Position (CMP) sensor electrical connector
- Crankshaft Position (CKP) sensor electrical connector
- Manifold Air Pressure (MAP) sensor electrical connector
- Exhaust Gas Recirculation (EGR) valve electrical connector
- Evaporative (EVAP) emissions canister purge solenoid valve
- Throttle Position Sensor (TPS) electrical connector
- Idle Air Control (IAC) valve electrical connector

- Starter motor electrical connector
- Alternator electrical connector
- Ignition coil electrical connector
- Wiring harness ground connections
- Two body wiring harness-to-engine harness connectors
- Oil filter
- Catalytic converter pipe from the rear exhaust manifold
- Engine mount lower nuts
- Torque converter cover
- A/C compressor
- Right side splash shield
- Torque converter bolts
- Transaxle brace
- Radiator outlet hose from the engine
- Accelerator and cruise control cables
- Heater inlet and outlet hoses
- Vacuum hoses from the upper intake manifold
- Fuel feed and return hoses
- Power steering pump and reposition
- Radiator inlet and outlet hoses

6. Attach an engine lifting device.
7. Remove the transmission-to-engine bolts.
8. Remove the engine from the vehicle.

To install:

9. Install or connect the following:
   - Engine to the vehicle
   - Transmission-to-engine bolts and torque them to 55 ft. lbs. (75 Nm)
10. Remove the engine lifting device.
   - Power steering pump and torque the bolts to 25 ft. lbs. (34 Nm)
   - Power steering lines to the pump and torque the fasteners to 20 ft. lbs. (27 Nm)
   - Radiator inlet hose
   - Heater inlet and outlet hoses
   - Fuel feed and return hoses
   - Vacuum hoses to the upper intake manifold

➡ **In order to avoid possible injury or vehicle damage, always replace the accelerator control cable with a NEW cable whenever you remove the engine from the vehicle.**

11. Remove the trim panel under the left instrument panel and detach the throttle cable from the top of the pedal then squeeze the retainer and push the cable through the bulkhead to remove it.
12. Install or connect the following:
   - New accelerator cable
   - Cruise control cable
   - Radiator outlet hose
   - Transmission brace and torque the

transmission bolts to 32 ft. lbs. (43 Nm) and the engine bolts to 46 ft. lbs. (63 Nm)
   - Right side splash shield and torque the fasteners to 44 inch lbs. (5 Nm)
   - Torque converter bolts and torque them to 47 ft. lbs. (63 Nm)
   - A/C compressor and torque the bolts to 37 ft. lbs. (50 Nm)
   - Starter motor and torque the bolts to 32 ft. lbs. (43 Nm)
   - Torque converter cover and torque the bolts to 89 inch lbs. (10 Nm)
   - Engine mount lower nuts and torque them to 32 ft. lbs. (43 Nm)
   - Catalytic converter pipe to the rear exhaust manifold and torque the fasteners to 26 ft. lbs. (35 Nm)
   - Oil Filter
   - Two body wiring to engine harness
   - Wiring harness grounds
   - Ignition coil electrical connector
   - Alternator electrical connector
   - A/C compressor electrical connector
   - Starter motor electrical connector
   - IAC valve electrical connector
   - TPS electrical connector
   - EVAP canister purge solenoid valve electrical connector
   - EGR valve electrical connector
   - MAP sensor electrical connector
   - CKP sensor electrical connector
   - CMP sensor electrical connector
   - HO2 sensor electrical connector
   - KS electrical connector
   - Drive belt
   - Engine mount struts and torque the fasteners to 35 ft. lbs. (48 Nm)
   - Hood and torque the bolts to 18 ft. lbs. (25 Nm)
   - Throttle body air inlet duct
   - Air cleaner
   - Negative battery cable
13. Fill the engine with clean oil.
14. Fill the coolant system.
15. Start the vehicle and check for leaks, repair if necessary.

3.4L Engine

1. Before servicing the vehicle, refer to the Precautions Section.
2. Drain the cooling system.
3. Drain the engine oil.
4. Relieve the fuel system pressure.
5. Remove or disconnect the following:
   - Hood
   - Cross vehicle brace
   - Engine mount struts
   - Drive belt
   - Brake booster vacuum hose

- Heated Oxygen Sensor (HO2S) electrical connector
- Secondary Air Injection (AIR) check valve electrical connector
- Exhaust Gas Recirculation (EGR) valve electrical connector
- Evaporative Emissions (EVAP) canister purge solenoid valve electrical connector
- Throttle Position Sensor (TPS) electrical connector
- Idle Air Control (IAC) valve electrical connector
- Alternator electrical connector
- Ignition coil electrical connector
- Wiring harness grounds
- Two engine wiring harness connectors
- Lower radiator air baffle
- Engine splash shields
- Oil filter
- Vehicle Speed Sensor (VSS) electrical connector
- Oil level sensor electrical connector
- Oil pressure switch electrical connector
- Engine block heater (if equipped)
- Knock Sensor (KS) electrical connector
- Crankshaft Position (CKP) sensor
- A/C compressor electrical connector
- Catalytic converter
- Engine mount lower nuts
- Torque converter cover
- Starter motor
- A/C compressor without disconnecting the lines
- Torque converter bolts
- Transaxle brace
- Radiator outlet hose
- Accelerator and cruise control cables
- Vacuum hoses from the upper intake manifold
- Fuel feed and return hoses
- Power steering pump without disconnecting the hoses
- Heater inlet and outlet hoses
- Radiator inlet hose

6. Attach an engine lifting device.
7. Remove the transmission bolts.
8. Remove the engine from the vehicle.

To install:

9. Install or connect the following:
   - Engine to the transaxle and torque the bolts to 55 ft. lbs. (75 Nm)
   - Radiator inlet hose
   - Heater inlet and outlet hoses
   - Power steering pump and torque the bolts to 25 ft. lbs. (34 Nm)
   - Fuel feed and return hoses

- Vacuum hoses to the upper intake manifold

➡In order to avoid possible injury or vehicle damage, always replace the accelerator control cable with a NEW cable whenever you remove the engine from the vehicle.

10. Remove the trim panel under the left instrument panel and detach the throttle cable from the top of the pedal then squeeze the retainer and push the cable through the bulkhead to remove it.

11. Install or connect the following:
- New accelerator cable
- Cruise control cable
- Radiator outlet hose
- Transmission brace and torque the transmission bolts to 32 ft. lbs. (43 Nm) and the engine bolts to 46 ft. lbs. (63 Nm)
- Torque converter-to-flywheel bolts and torque them to 47 ft. lbs. (63 Nm)
- A/C compressor and torque the bolts to 37 ft. lbs. (50 Nm)
- Starter motor and torque the bolts to 32 ft. lbs. (43 Nm)
- Torque converter cover and torque the bolts to 89 inch lbs. (10 Nm)
- Engine mount lower nuts and torque them to 32 ft. lbs. (43 Nm)
- Catalytic converter and torque the nuts to 24 ft. lbs. (32 Nm)
- New oil filter
- VSS electrical connector
- Oil level sensor electrical connector
- Oil pressure switch electrical connector
- Engine block heater electrical connector
- KS electrical connector
- HO2S electrical connector
- CKP sensor electrical connector
- A/C compressor electrical connector
- Wiring harness grounds
- Lower radiator air baffle and torque the bolts to 15 ft. lbs. (20 Nm)
- Engine splash shield and torque the fasteners to 18 inch lbs. (2 Nm)
- AIR check valve solenoid electrical connector
- EGR valve electrical connector
- EVAP canister purge solenoid valve electrical connector
- TPS electrical connector
- IAC valve electrical connector
- Alternator electrical connector
- Ignition coil electrical connector
- Wiring harness grounds
- Engine wiring harness connectors

- Drive belt
- Engine mount strut and torque the bolt to 35 ft. lbs. (48 Nm)
- Throttle body air inlet duct
- Cross vehicle brace and torque the nuts to 29 ft. lbs. (40 Nm)
- Hood and torque the bolts to 18 ft. lbs. (25 Nm)
- Negative battery cable

12. Fill the engine oil.
13. Fill the engine coolant.
14. Inspect the transmission fluid level and top off if necessary.
15. Turn the ignition to the **ON** position several times to pressurize the fuel system.
16. Start the engine and inspect for any leaks, repair if necessary.
17. Check and top off the fluid levels if required.

3.5L Engine

1. Before servicing the vehicle, refer to the Precautions Section.
2. Relieve the fuel system pressure.
3. Drain the engine oil.
4. Drain the engine cooling system.
5. Remove or disconnect the following:

- Front wheels
- Lower air deflector
- Fuel injector sight shield
- Left diagonal brace
- Air cleaner assembly
- Radiator inlet hose
- Alternator
- Fuel lines from the rail and position
- Fuel vapor line
- Throttle and cruise cables with the mounting bracket from the throttle body
- Automatic range selector cable from the Park/Neutral Position (PNP) switch
- Vacuum brake booster hose from the booster
- Transaxle oil cooler lines from the radiator
- Heater hoses from the engine
- Right side AIR control valve, if equipped
- Upper engine electrical connectors, including grounds
- Lower AIR hose from the elbow, if equipped and discard the clamp
- Lower electrical connectors, including grounds and the harness retainer
- Fog lamp electrical connectors, if equipped
- Torque converter cover
- Starter

- Bolts attaching the engine flywheel to the torque converter
- Engine splash shields from the inner fender
- A/C compressor and position aside without disconnecting the lines
- Catalytic converter pipe from the rear exhaust manifold
- Bolts attaching the lower transaxle to the engine
- Connectors from the Wheel Speed Sensors, and unclip the harnesses from the lower control arms
- Drive axles from the steering knuckles

✳✳ CAUTION

Failure to disconnect the intermediate shaft from the rack and pinion stub shaft can result in damage to the steering gear and/or damage to the intermediate shaft. This damage may cause loss of steering control which could result in personal injury.

- Pinch bolt from the intermediate steering shaft
- Intermediate shaft from the steering gear
- Upper and lower engine wiring harnesses position them aside

6. Place a frame table under the engine/transaxle/frame. With an assistant, lower the vehicle to the frame table in order to support the engine/transaxle/frame.
- Bolts attaching the frame to the body
- Frame table with the engine/transaxle/frame from under the vehicle
- Power steering pump from the engine and lay the pump aside
- Secondary air injection check valve/pipe, if equipped
- Dipstick tube
- Exhaust manifolds and crossover pipe

7. Install an engine lifting device to the engine.
- Bolts attaching the engine mount bracket to the engine
- Transaxle brace
- Bolts attaching the upper transmission to the engine
- Engine from the transaxle/frame

To install:

8. Installation is the reverse of removal, please note the following torques:
9. Install or connect the following:
- Bolts attaching the transmission to the engine to 55 ft. lbs. (75 Nm)

- Frame bolts to 133 ft. lbs. (180 Nm)
10. Center the steering angle sensor.
11. Fill the engine with new oil.

➡ **A filter change is recommended.**

12. Fill the cooling system.
13. Perform the Crankshaft Position (CKP) system variation learn procedure.
14. Start the engine and check for leaks, repair if necessary.

3.8L Engine

1. Before servicing the vehicle, refer to the Precautions Section.
2. Relieve the fuel system pressure.
3. Matchmark the hood hinges and remove the hood.
4. Drain the cooling system.
5. Drain the engine oil.
6. Remove or disconnect the following:
- Negative battery cable
- Hood
- Fuel injector sight shield
- Torque converter cover
- Starter motor
- Engine flywheel-to-torque converter bolts
- Transaxle bracket
- Engine mount lower nuts
- Catalytic converter pipe from the right exhaust manifold
- Oil level harness retainer from the engine
7. Unplug and reposition the following electrical connections:
- Vehicle Speed Sensor (VSS)
- Oil pressure sensor
- Oil level sensor
- Knock sensors
- Heated Oxygen Sensor (HO$_2$S)
8. Remove or disconnect the following:
- Oil filter adapter housing
- Engine ground nut and engine ground from the transaxle stud
- Bolt and the stud which attach the lower transaxle to the engine
9. Lower the vehicle while supporting the transaxle.
- Air cleaner assembly
- Fuel lines from the fuel rail
- Fuel vapor line
- Throttle and cruise control cables and bracket from the throttle body
- Water pump pulley bolts, loosen only
- Drive belt
- Water pump pulley bolts and the pulley
- Right and left engine mount struts
- Engine mount brackets from the radiator support

- Positive battery cable
- Engine cooling fans
- Vacuum booster hose from the engine
- A/C vacuum hose from the engine
- Power steering pump and lay aside with the lines attached
10. Unplug the upper engine electrical connectors from the following components:
- Fuel injectors
- Throttle Position (TP) sensor
- Idle Air Control (IAC) valve
- Evaporative Emissions (EVAP) purge solenoid
- Exhaust Gas Recirculation (EGR) valve
- Manifold Absolute Pressure (MAP) sensor
- Upper engine wiring harness from retaining clips and reposition the engine wiring harness for engine removal
- Radiator inlet and outlet hoses from the engine
- Heater hoses from the drive belt tensioner
- A/C compressor from the engine and set aside without disconnecting the lines
11. Install an engine lifting device.
- Bolts that attach the upper transaxle to the engine
- Engine from the vehicle

To install:
12. Installation is the reverse of removal, please note the following torques:
- Bolts attaching the upper transaxle to the engine to 55 ft. lbs. (75 Nm)
- A/C compressor-to-bracket nuts to 22 ft. lbs. (30 Nm)
- Power steering bolts to 25 ft. lbs. (34 Nm)
- Water pump pulley bolts to 116 inch lbs. (13 Nm)
- Bolt and the stud attaching the lower transaxle to the engine to 55 ft. lbs. (75 Nm)
- Engine ground nut to 26 ft. lbs. (35 Nm)
- Engine mount lower nuts to 58 ft. lbs. (78 Nm)
13. Fill the engine with new oil.

➡ **A filter change is recommended.**

14. Fill the cooling system.
15. Perform the Crankshaft Position (CKP) system variation learn procedure.
16. Start the engine and check for leaks, repair if necessary.

5.3L Engine

1. Before servicing the vehicle, refer to the Precautions Section.
2. Disconnect the negative battery cable.
3. Remove the engine sight shield.
4. Evacuate the air conditioning (A/C) system and recover the refrigerant.
5. Remove the front tires and wheels.
6. Drain the cooling system.
7. Drain the engine oil.
8. Lower the vehicle.
9. Reposition the inlet hose clamp at the radiator.
10. Remove the inlet hose from the radiator.
11. Reposition the outlet hose clamp at the radiator.
12. Disengage the outlet hose clips from the fan shroud.
13. Remove the outlet hose from the radiator.
14. Remove the A/C compressor hose nut at the A/C receiver/dehydrator tube.
15. Remove the A/C compressor hose.
16. Remove the A/C compressor hose nut at the A/C condenser.
17. Remove the A/C compressor hose.
18. Remove the brake booster vacuum hose from the brake booster.
19. Remove the brake booster vacuum hose from the intake manifold.
20. Disconnect the engine harness electrical connectors from the instrument panel (I/P) harness electrical connectors.
21. Disconnect the brake fluid level switch electrical connector from the master cylinder.
22. Remove brake fluid from the master cylinder.
23. Disconnect the from brake pipe fittings from the master cylinder.
24. Disconnect the brake pipe fittings from the antilock brake (ABS) module.
25. Remove the master cylinder nuts.
26. Position the master cylinder to the engine. Hold the master cylinder in place using mechanic's wire.
27. Relieve the fuel system pressure.
28. Disconnect the fuel feed line from the fuel rail.
29. Disconnect the evaporative emission (EVAP) line from the purge solenoid.
30. Remove the right front fender diagonal brace.
31. Remove the underhood electrical center cover.
32. Loosen the 4 integral bolts attaching the fuse block. Reposition the fuse block.

33. Loosen the engine harness connector bolt. Remove the engine harness connector from the bracket.

34. Disconnect the camshaft position sensor lead.

35. Disconnect the engine harness electrical connector from the ABS module.

36. Disconnect the engine harness electrical connector from the electronic brake control module (EBCM).

37. Remove the air cleaner assembly.

38. Disconnect the following electrical connectors:
- Transmission control module (TCM)
- Engine control module (ECM)
- A/C pressure sensor

39. Disconnect the engine harness electrical connector from the crankshaft position (CKP) sensor harness.

40. Disconnect the engine harness electrical connector from the power steering gear harness.

41. Remove the clip attaching the power steering gear harness to the bracket.

42. Remove the shift cable clip.

43. Disconnect the shift cable from the transaxle selector lever stud.

44. Remove the shift cable from the bracket.

45. Remove the vehicle speed sensor (VSS) shield nut and bolt. Remove the shield.

46. Disconnect the engine harness electrical connector from the VSS.

47. Set all branches of the engine wiring harness on top of the engine.

48. Reposition the intermediate shaft lower boot.

49. Remove the intermediate shaft to steering gear bolt.

50. Separate the intermediate shaft from the steering gear.

51. Disconnect both front wheel speed sensors.

52. Unclip the ABS wire harness from the lower control arm.

53. Reposition the heater inlet and outlet hose clamps.

54. Remove the heater inlet and outlet hoses from the heater inlet/outlet pipe.

55. Raise and support the vehicle.

56. Disconnect the stabilizer links from the stabilizer shaft.

57. Disconnect the ball joints from the steering knuckles.

58. Disconnect the power steering pressure hose from the steering gear.

59. Disconnect the power steering pressure line clips from the frame.

60. Disconnect the outer tie rod ends from the steering knuckles.

61. Loosen the power steering hose clamp at the inlet pipe.

62. Remove the power steering hose from the inlet pipe.

63. Remove the left and right halfshafts from the transaxle.

64. Support the halfshafts using mechanic's wire.

65. Remove the transaxle oil cooler line bracket bolt/stud.

66. Disconnect the transaxle oil cooler lines from the transaxle.

67. Remove the positive battery cable nut from the starter.

68. Remove the cable terminal from the starter.

69. Remove the battery cable ground nut.

70. Remove the cable ground terminal from the stud.

71. Remove the battery cable retainers from the engine frame.

72. Disconnect the exhaust system.

73. Disconnect the O$_2$sensor harness pigtail.

74. Remove the transaxle converter cover bolt/stud.

75. Remove the converter cover.

76. Remove the flywheel bolts.

77. Remove the front air deflectors.

78. Raise the vehicle enough to place engine holding fixture J39580 under the engine, frame, and front suspension.

79. Support the rear of the vehicle with suitable jackstand.

80. Strap the front of the vehicle to the hoist.

81. Raise the engine holding fixture, or lower the vehicle to preload the weight of the engine, frame, and front suspension.

82. Remove the radiator to frame brackets.

83. Remove the engine frame front bolts.

84. Remove the engine frame rear bolts.

85. With the aid of an assistant, lower the engine holding fixture and/or raise the vehicle to remove the engine and the frame from the vehicle.

86. Ensure that all hoses, wires, and pipes clear the vehicle during the removal process.

87. Use a suitable engine lift to support the engine.

88. Remove the front engine mount to frame nuts.

89. Remove the rear engine mount to frame nuts.

90. Remove the transaxle to engine bolts and stud.

91. Separate the engine from the transaxle.

92. Using a suitable engine lift, remove the engine from the frame.

93. Install the engine onto a suitable engine stand.

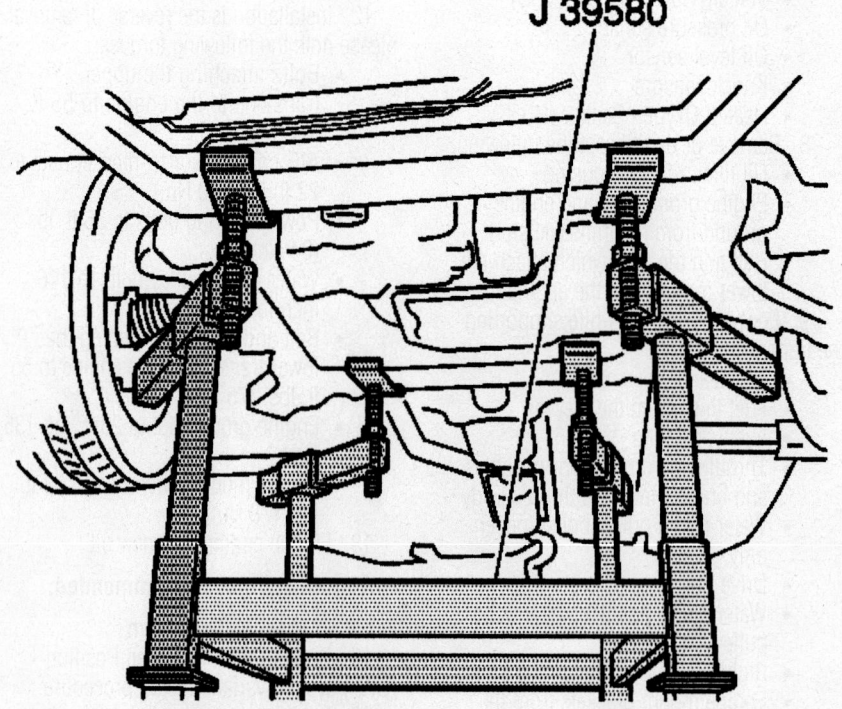

Engine holding fixture J39580—5.3L engine

06025-GMWB-G06

To install:

94. Using a suitable engine lift, remove the engine from the engine stand.

95. Position and install the engine to the frame and the transaxle.

96. Install the transaxle to engine bolts and stud. Tighten the bolts/stud to 75 Nm (55 ft. lbs.).

97. Install the rear engine mount to frame nuts. Tighten the nuts to 50 Nm (37 ft. lbs.).

98. Install the front engine mount to frame nuts. Tighten the nuts to 50 Nm (37 ft. lbs.).

99. Ensure that all hoses, wires, and pipes clear the vehicle during the installation process.

100. With the aid of an assistant, raise the engine holding fixture and/or lower the vehicle to install the engine and frame to the vehicle.

101. Install the engine frame rear bolts. Tighten the bolts to 160 Nm (118 ft. lbs.).

102. Install the engine frame front bolts. Tighten the bolts to 145 Nm (107 ft. lbs.).

103. Install radiator to front frame brackets.

104. Lower the engine holding fixture or raise the vehicle to preload the weight of the engine, frame, and front suspension.

105. Remove the support from under the rear of the vehicle.

106. Raise the vehicle enough to remove the engine holding fixture from under the engine, frame, and front suspension.

107. Connect the exhaust system and connect the O_2 sensor harness pigtail.

108. Install the flywheel bolts. Tighten the bolts to 63 Nm (47 ft. lbs.).

109. Install the converter cover.

110. Install the transaxle converter cover bolt/stud. Tighten the bolt/stud to 10 Nm (89 inch lbs.).

111. Install the left and right halfshafts into the transaxle.

112. Install the battery cable retainers to the engine frame.

113. Install the front air deflector.

114. Install the cable ground terminal to the stud.

115. Install the battery cable ground nut. Tighten the nut to 30 Nm (22 ft. lbs.).

116. Install the cable terminal to the starter.

117. Install the positive battery cable nut to the starter. Tighten the nut to 10 Nm (89 inch lbs.).

118. Connect the transaxle oil cooler lines to the transaxle.

119. Install the transaxle oil cooler line bracket bolt/stud. Tighten the bolt/stud to 25 Nm (18 ft. lbs.).

120. Install the power steering hose to the inlet pipe.

121. Tighten the power steering hose clamp at the inlet pipe. Tighten the clamp to 6 Nm (53 inch lbs.).

122. Connect the power steering pipe to the steering gear.

123. Connect the power steering pipe to the frame clips.

124. Lower the vehicle.

125. Position and install the ball joint nuts.

126. Install the sway links to the stabilizer shaft.

127. Install the tie rod ends to the steering knuckles.

128. Connect the wheel speed sensor electrical connectors.

129. Connect the wheel speed sensor wiring harness to the lower control arm.

130. Install the heater inlet and outlet hoses to the heater inlet/outlet pipe.

131. Reposition the heater inlet and outlet hose clamps.

132. Install the intermediate shaft to the steering gear.

133. Install the intermediate shaft to steering gear bolt. Tighten the bolt to 48 Nm (35 ft. lbs.).

134. Position the intermediate shaft lower boot.

135. Route all branches of the engine wiring harness to their correct locations.

136. Connect the engine harness electrical connector to the VSS.

137. Install the VSS shield. Install the VSS shield nut and bolt. Tighten the bolt/nut to 25 Nm (18 ft. lbs.).

138. Install the shift cable to the bracket.

139. Connect the shift cable to the transaxle selector lever stud.

140. Install the shift cable clip.

141. Install the clip attaching the power steering gear harness to the bracket.

142. Connect the engine harness electrical connector to the CKP sensor harness.

143. Connect the engine harness electrical connector to the power steering gear harness.

144. Connect the camshaft position sensor lead to the sensor.

145. Connect the following electrical connectors:
   • ECM
   • TCM
   • A/C pressure sensor

146. Install the air cleaner assembly.

147. Connect the engine harness electrical connector to the EBCM.

148. Connect the engine harness electrical connector to the ABS module.

149. Install the engine harness connector to the bracket. Tighten the engine harness connector bolt. Tighten the bolt to 10 Nm (89 inch lbs.).

150. Position the fuse block. Tighten the 4 integral bolts attaching the fuse block. Tighten the bolts to 10 Nm (89 inch lbs.).

151. Install the electrical center cover.

152. Connect the fuel feed line to the fuel rail.

153. Connect the EVAP line to the purge solenoid.

154. Remove the mechanic's wire holding the master cylinder. Position the master cylinder to the brake booster.

155. Install the master cylinder nuts. Tighten the nuts to 33 Nm (24 ft. lbs.).

156. Connect the brake pipe fittings to the ABS module. Tighten the fittings to 15 Nm (11 ft. lbs.).

157. Connect the from brake pipe fittings to the master cylinder. Tighten the fittings to 30 Nm (22 ft. lbs.).

158. Connect the brake fluid level switch electrical connector to the master cylinder.

159. Connect the engine harness electrical connectors to the I/P harness electrical connectors.

160. Install the brake booster vacuum hose to the intake manifold.

161. Install the brake booster vacuum hose to the brake booster.

162. Install the A/C compressor hose.

163. Install the A/C compressor hose nut at the A/C condenser. Tighten the nut to 16 Nm (12 ft. lbs.).

164. Install the A/C compressor hose.

165. Install the A/C compressor hose nut at the A/C receiver/dehydrator tube. Tighten the nut to 16 Nm (12 ft. lbs.).

166. Install the outlet hose to the radiator.

167. Connect the outlet hose clips to the fan shroud.

168. Reposition the outlet hose clamp at the radiator.

169. Install the inlet hose to the radiator.

170. Reposition the inlet hose clamp at the radiator.

171. Install the front tires and wheels.

172. Connect the negative battery cable.

173. Install the right fender diagonal brace.

174. Refill the engine with oil.

175. Recharge the A/C system.

176. Refill the cooling system.

177. Bleed the brake system.

178. Check the transaxle fluid level, add fluid if necessary.

179. Install the engine sight shield.

180. Run the engine and inspect for leaks.

Water Pump

REMOVAL & INSTALLATION

3.1L and 3.4L Engines

1. Before servicing the vehicle, refer to the Precautions Section.
2. Drain the cooling system.
3. Remove or disconnect the following:
   - Negative battery cable
   - Accessory drive belt guard
   - Water pump pulley bolts, loosen
   - Accessory drive belt
   - Water pump pulley
   - Water pump

To install:

4. Install or connect the following:
   - Water pump with a new gasket and torque the bolts to 89 inch lbs. (10 Nm)
   - Water pump pulley and torque the bolts to 18 ft. lbs. (25 Nm)
   - Accessory drive belt
   - Drive belt guard
5. Fill the cooling system.
6. Start the engine and check for leaks, repair if necessary.

3.5L Engine

1. Before servicing the vehicle, refer to the Precautions Section.
2. Partially drain the cooling system.
3. Remove or disconnect the following:
   - Surge tank outlet hose
   - Drive belt
   - Idler pulley
   - Water pump pulley

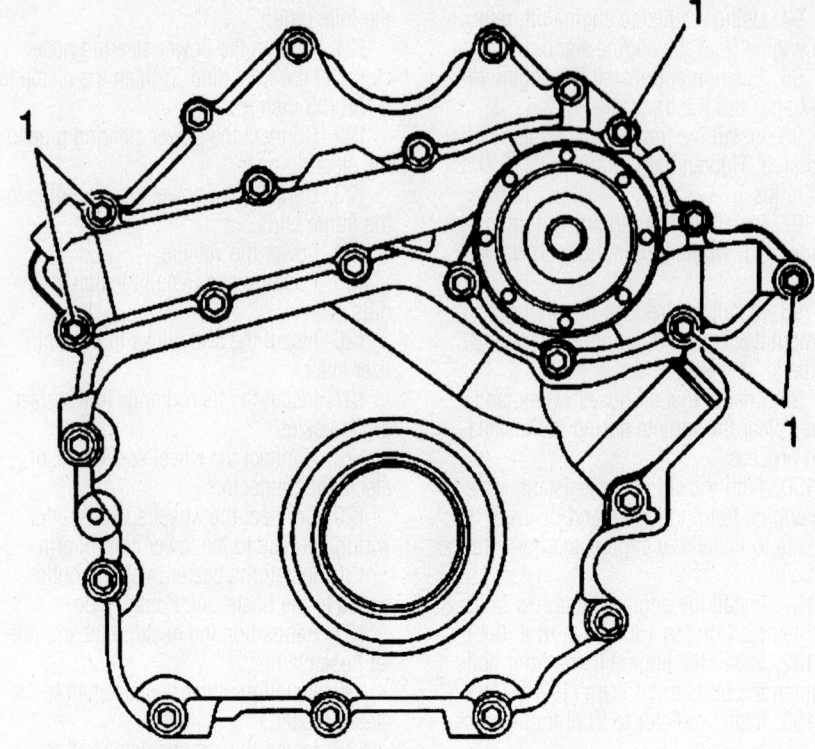

Be sure to install the 5 long water pump bolts in the correct locations—3.5L engine

➡ **The water pump is attached to the engine with both long and short bolts, be sure to note their locations.**

   - Water pump

To install:

4. Install or connect the following:
   - Water pump using a new gasket and torque the bolts to 124 inch lbs. (14 Nm)
   - Water pump pulley and torque the bolts to 106 inch lbs. (12 Nm)
   - Idler pulley and torque the bolt to 37 ft. lbs. (50 Nm)
   - Drive belt
   - Surge tank hose
5. Refill the cooling system.
6. Start the engine and check for leaks.

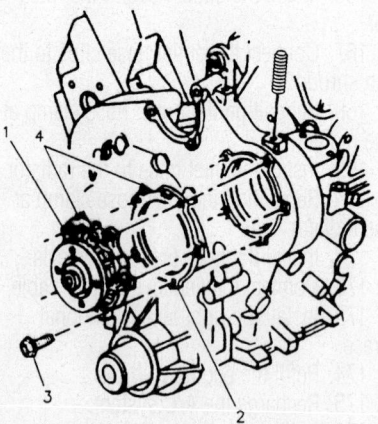

```
1  WATER PUMP
2  GASKET
3  10 N·m (89 LB. IN.)
4  LOCATOR — MUST BE VERTICAL
```
7924LG01

Water pump assembly mounting—3.1L and 3.4L engines

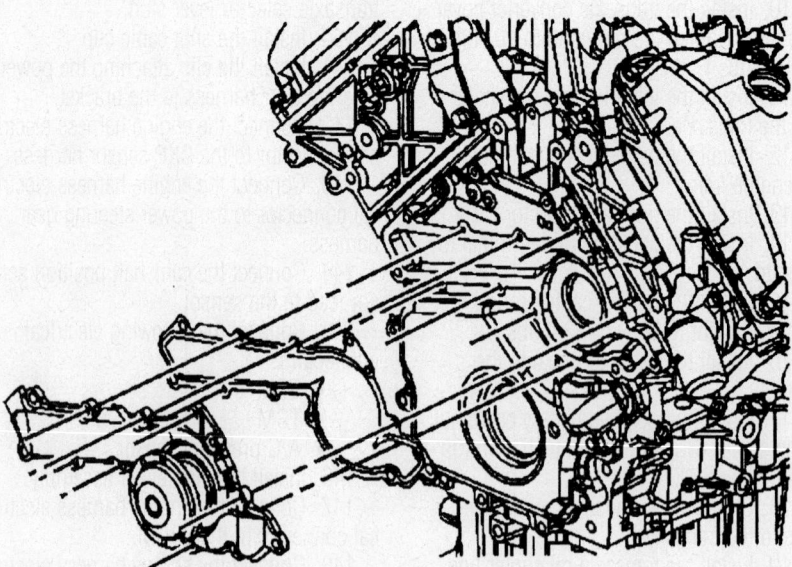

Water pump mounting—3.5L engine

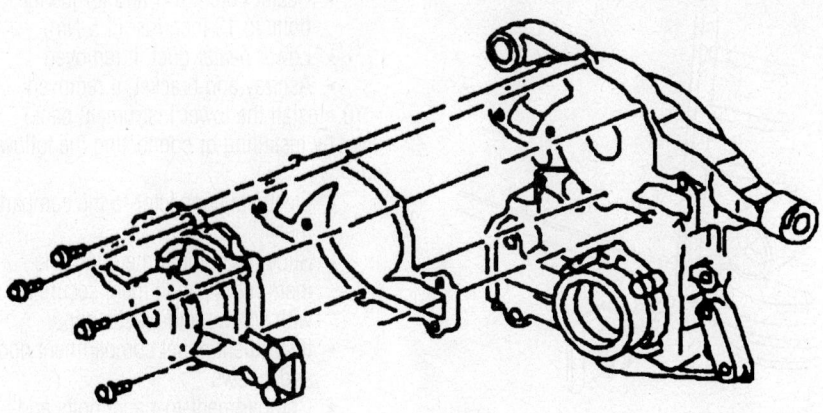

Exploded view of the water pump assembly mounting—3.8L engine

79222Z504

3.8L Engines

1. Before servicing the vehicle, refer to the Precautions Section.
2. Drain the engine coolant.
3. Remove or disconnect the following:

- Negative battery cable.
- Drive belt
- Water pump pulley
- Power steering pump and move it aside
- Water pump

To install:

4. Install or connect the following:
- Water pump using a new gasket and torque the long bolts to 25 ft. lbs. (34 Nm) and the short bolts to 16 ft. lbs. (22 Nm)
- Water pump pulley and torque the bolts to 115 inch lbs. (13 Nm)
- Power steering pump and torque the bolts to 25 ft. lbs. (34 Nm)
- Drive belt
- Negative battery cable
5. Fill the cooling system.

6. Start the vehicle and check for leaks, repair if necessary.

5.3L Engine

1. Before servicing the vehicle, refer to the Precautions Section.
2. Drain the cooling system.
3. Remove the battery and battery tray.
4. Remove the drive belt.
5. Remove the water pump bolts (3).
6. Remove the water pump (1) and gasket (2). Discard the gasket.
7. Clean and inspect the water pump gasket mating surfaces.

To install:

8. Install the water pump (1) and a NEW gasket (2).
9. Tighten the water pump bolts (3). Tighten the bolts to 10 Nm (89 inch lbs.).
10. Install the drive belt.
11. Install the battery tray and battery.
12. Fill the cooling system.

Heater Core

REMOVAL & INSTALLATION

Century and Regal

1. Before servicing the vehicle, refer to the Precautions Section.
2. Disconnect the negative battery cable.
3. If equipped with a 3.1L engine, remove the air cleaner and duct assembly.
4. If equipped with a 3.8L engine, remove the fuel injector sight shield by performing the following procedures:
    a. Clean the area around the oil filler cap/tube assembly location.
    b. Rotate the oil filler cap/tube assembly counterclockwise from the valve cover.
    c. Lift the fuel injector sight shield up at the front and slide it from the rear engine bracket.
    d. Reinstall the oil filler cap/tube assembly in the valve cover.

✳✳ CAUTION

Before draining the cooling system, allow the engine to cool to relieve the systems internal pressure and to avoid scalding coolant.

5. Drain the cooling system by performing the following procedure:
    a. Raise and safely support the front of the vehicle.
    b. Remove and clean the coolant recovery tank.

Water pump mounting—5.3L engine

06025-GMWB-G07

c. Place a 2 gallon pan under the radiator to catch the coolant.

d. At the bottom of the radiator, open the drain valve and drain the coolant to a level lower than the heater core.

e. Remove the radiator cap and open the air bleed screw (2–3 turns) located on top of the thermostat housing.

f. After sufficient coolant has been drained from the system, close the drain valve.

✳✳ CAUTION

Engine coolant is a hazardous waste; it should be stored for reuse or submitted for recycling. NEVER dispose of it by dumping it into the environment.

6. Disconnect and plug the heater hoses at the heater core.

7. If equipped, remove the lower center console by removing or disconnecting the following:
- Cigarette lighter
- Automatic transaxle shift handle
- Upper console trim plate from the front floor console by unsnapping it
- Electrical connectors from the upper console trim plate and remove the trim plate
- CD storage compartment from the front floor console by unsnapping it
- Raise the front floor console armrest and remove the compartment mat
- Console-to-chassis bolts and screws
- Electrical connectors from the console and remove the console from the vehicle

8. Remove the lower instrument panel lower compartment by removing or disconnecting the following:
- Lower right instrument panel insulator
- Compartment-to-panel bolts and screws, from under the instrument panel compartment
- Instrument panel compartment door screws and the door
- Instrument panel compartment screws and plastic clips; then slide the compartment from the instrument panel
- Electrical connector from the compartment
- Ashtray and bracket, if necessary
- Lower heater duct

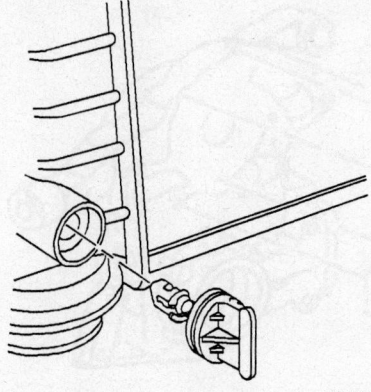

93111G01

View of the radiator drain valve—Century, Grand Prix, Intrigue, Monte Carlo and Regal

- Heater core cover and discard the cover seals
- Heater core mounting clip and bracket
- Heater core

To install:

9. Install and connect the following:
- Heater core in the vehicle
- Heater core mounting clip and bracket
- New seals on the heater core cover

- Heater core cover and torque the bolts to 13 inch lbs. (1.5 Nm)
- Lower heater duct, if removed
- Ashtray and bracket, if removed

10. Install the lower instrument panel lower by installing or connecting the following:
- Electrical connector to the compartment
- Slide the compartment into the instrument panel; then, secure it with screws and plastic clips
- Instrument panel compartment door and screws
- Compartment-to-panel bolts and screws, located under the instrument panel compartment
- Lower right instrument panel insulator

11. If equipped, install the lower center console install or connect the following:

- Console in the vehicle and connect the electrical connectors
- Console-to-chassis bolts and screws, tighten in sequence beginning at the front right and continue in a clockwise order to 106 inch lbs. (12 Nm)
- Armrest compartment mat and

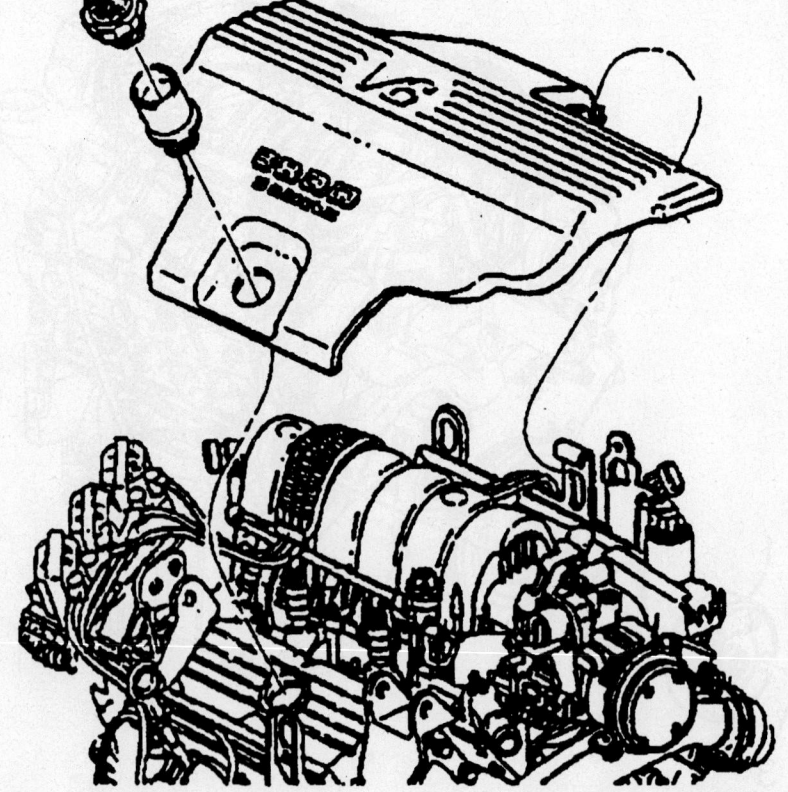

93111G02

View of the fuel injector sight shield—Century, Grand Prix, Intrigue, Monte Carlo and Regal

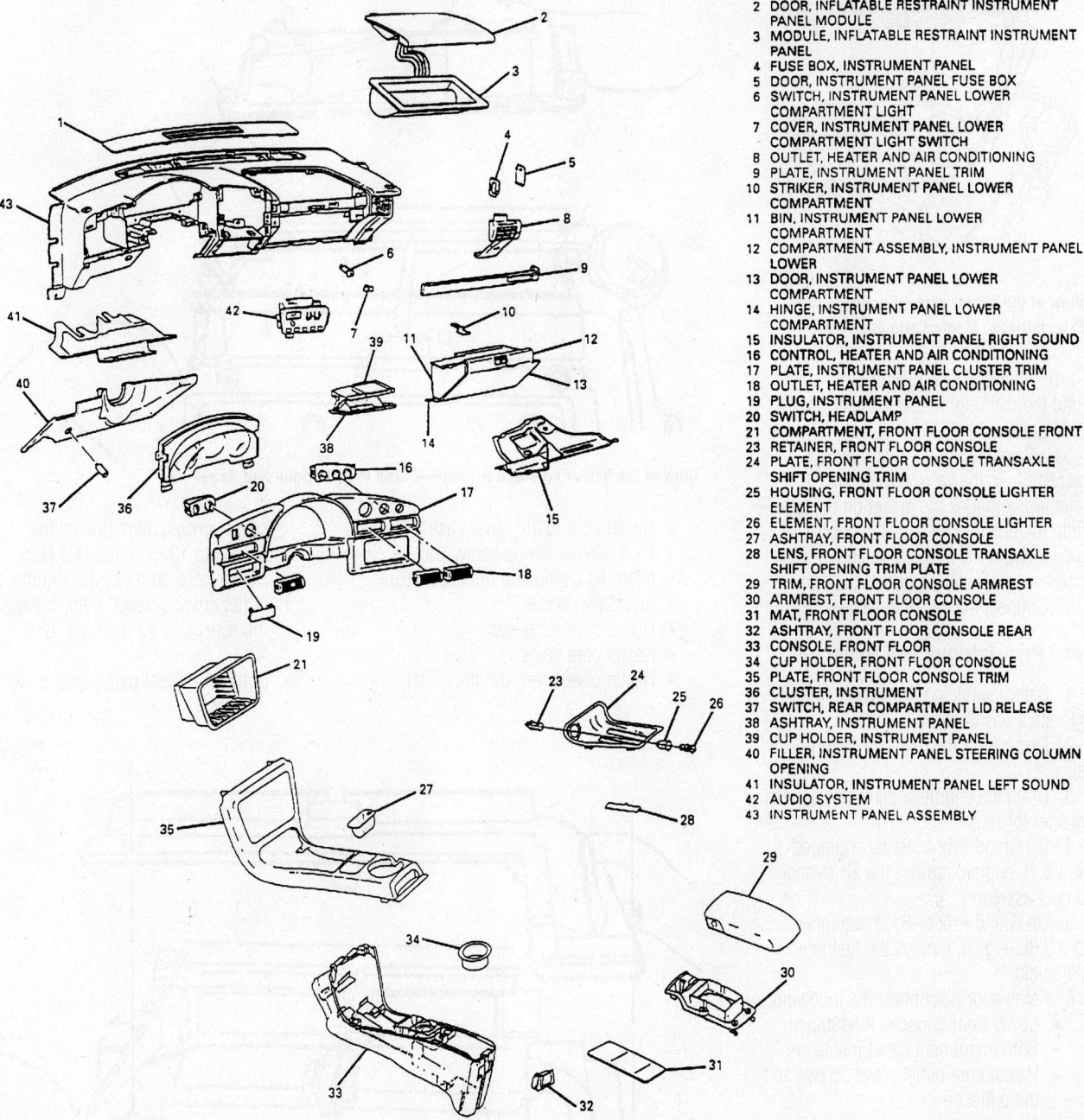

1 PANEL, INSTRUMENT PANEL UPPER TRIM
2 DOOR, INFLATABLE RESTRAINT INSTRUMENT PANEL MODULE
3 MODULE, INFLATABLE RESTRAINT INSTRUMENT PANEL
4 FUSE BOX, INSTRUMENT PANEL
5 DOOR, INSTRUMENT PANEL FUSE BOX
6 SWITCH, INSTRUMENT PANEL LOWER COMPARTMENT LIGHT
7 COVER, INSTRUMENT PANEL LOWER COMPARTMENT LIGHT SWITCH
8 OUTLET, HEATER AND AIR CONDITIONING
9 PLATE, INSTRUMENT PANEL TRIM
10 STRIKER, INSTRUMENT PANEL LOWER COMPARTMENT
11 BIN, INSTRUMENT PANEL LOWER COMPARTMENT
12 COMPARTMENT ASSEMBLY, INSTRUMENT PANEL LOWER
13 DOOR, INSTRUMENT PANEL LOWER COMPARTMENT
14 HINGE, INSTRUMENT PANEL LOWER COMPARTMENT
15 INSULATOR, INSTRUMENT PANEL RIGHT SOUND
16 CONTROL, HEATER AND AIR CONDITIONING
17 PLATE, INSTRUMENT PANEL CLUSTER TRIM
18 OUTLET, HEATER AND AIR CONDITIONING
19 PLUG, INSTRUMENT PANEL
20 SWITCH, HEADLAMP
21 COMPARTMENT, FRONT FLOOR CONSOLE FRONT
23 RETAINER, FRONT FLOOR CONSOLE
24 PLATE, FRONT FLOOR CONSOLE TRANSAXLE SHIFT OPENING TRIM
25 HOUSING, FRONT FLOOR CONSOLE LIGHTER ELEMENT
26 ELEMENT, FRONT FLOOR CONSOLE LIGHTER
27 ASHTRAY, FRONT FLOOR CONSOLE
28 LENS, FRONT FLOOR CONSOLE TRANSAXLE SHIFT OPENING TRIM PLATE
29 TRIM, FRONT FLOOR CONSOLE ARMREST
30 ARMREST, FRONT FLOOR CONSOLE
31 MAT, FRONT FLOOR CONSOLE
32 ASHTRAY, FRONT FLOOR CONSOLE REAR
33 CONSOLE, FRONT FLOOR
34 CUP HOLDER, FRONT FLOOR CONSOLE
35 PLATE, FRONT FLOOR CONSOLE TRIM
36 CLUSTER, INSTRUMENT
37 SWITCH, REAR COMPARTMENT LID RELEASE
38 ASHTRAY, INSTRUMENT PANEL
39 CUP HOLDER, INSTRUMENT PANEL
40 FILLER, INSTRUMENT PANEL STEERING COLUMN OPENING
41 INSULATOR, INSTRUMENT PANEL LEFT SOUND
42 AUDIO SYSTEM
43 INSTRUMENT PANEL ASSEMBLY

93111G03

Exploded view of the instrument panel—Century, Grand Prix, Intrigue, Monte Carlo and Regal

lower the front floor console armrest
- C/D storage compartment into the front floor console
- Electrical connectors to the upper console trim plate and snap the trim plate onto the console
- Automatic transaxle shift handle and the cigarette lighter
- Heater hoses to the heater core and secure with the clamps

12. Refill the cooling system by performing the following procedure:

a. Close the radiator drain valve.

b. If the air bleed screws (located at the top of the thermostat housing) is closed, open it by turning it 2–3 turns.

c. Slowly add coolant until it reaches the radiator neck.

d. Wait for 2 minutes and recheck the coolant level; then, add more coolant if necessary.

e. Fill the coolant reservoir to the COLD mark.

f. Close the air bleed screw.

❄❄ WARNING

Do not over-tighten the air bleed screw for it is made of brass.

g. Install the radiator cap and make sure that the arrows align with the overflow tube.

13. If equipped with a 3.8L engine, install the fuel injector sight shield by performing the following procedures:

a. Remove the oil filler cap/tube assembly from the valve cover.

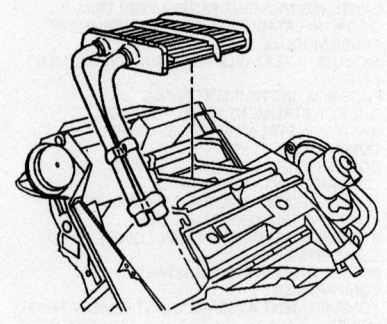

View of the heater core—Century, Grand Prix, Intrigue, Monte Carlo and Regal

93111G04

b. Slide the fuel injector sight shield into the rear engine bracket and lower it into place.

c. Reinstall the oil filler cap/tube assembly in the valve cover. Twist it clockwise to lock the detent on the tube into the notch in the valve cover.

14. If equipped with a 3.1L engine, install the air cleaner and duct assembly.

15. Connect the negative battery cable.

Grand Prix, Intrigue and Regal

1. Before servicing the vehicle, refer to the Precautions Section.

2. Disconnect the negative battery cable.

3. Drain the engine coolant into a clean container for reuse.

4. On Grand Prix or Regal equipped with a 3.1L engine, remove the air cleaner and duct assembly.

5. On Grand Prix or Regal equipped with a 3.8L engine, remove the fuel injector sight shield.

6. Remove or disconnect the following:
- Lower floor console, if equipped
- Both instrument panel insulators
- Heater core outlet cover screws and the outlet cover
- Heater core cover screws and the cover
- Heater core seals. Discard the seals
- Heater core outer seal. Discard the seal
- Heater core line clamp screw, the retaining clamp and the pipe retainer clamp screw
- Heater core from the lower case
- Heater core lower, center, upper and side seals from the lower heater core case. Discard the seals

To install:

7. Install or connect the following:
- New heater core lower, center, upper and side seals to the lower heater core case

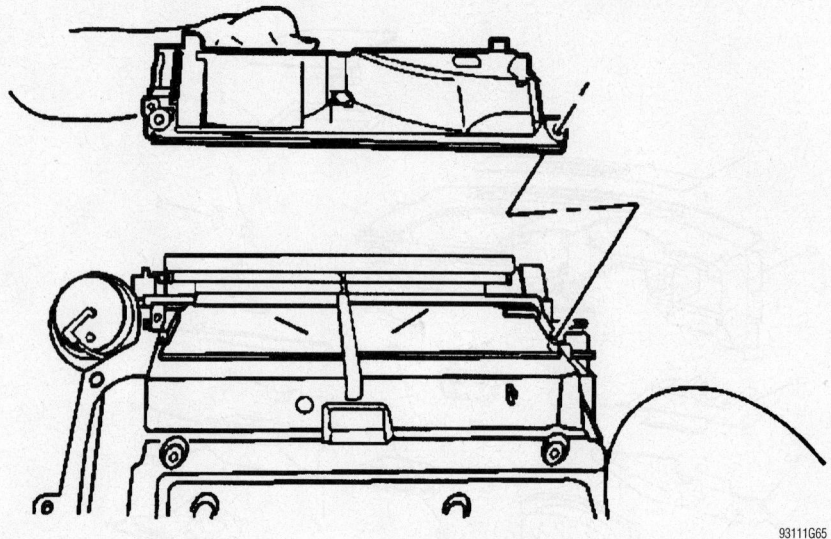

View of the heater core outlet cover—Grand Prix, Intrigue and Regal

93111G65

- Heater core to the lower case
- Pipe retainer clamp screw, the retaining clamp and the heater core line clamp screw
- Heater core outer seal
- Heater core seals
- Heater core cover and the outlet cover screws, then, tighten the screws to 13 inch lbs. (1.5 Nm)
- Heater core outlet cover and the outlet cover screws, then, tighten the screws to 13 inch lbs. (1.5 Nm)
- Both instrument panel insulators

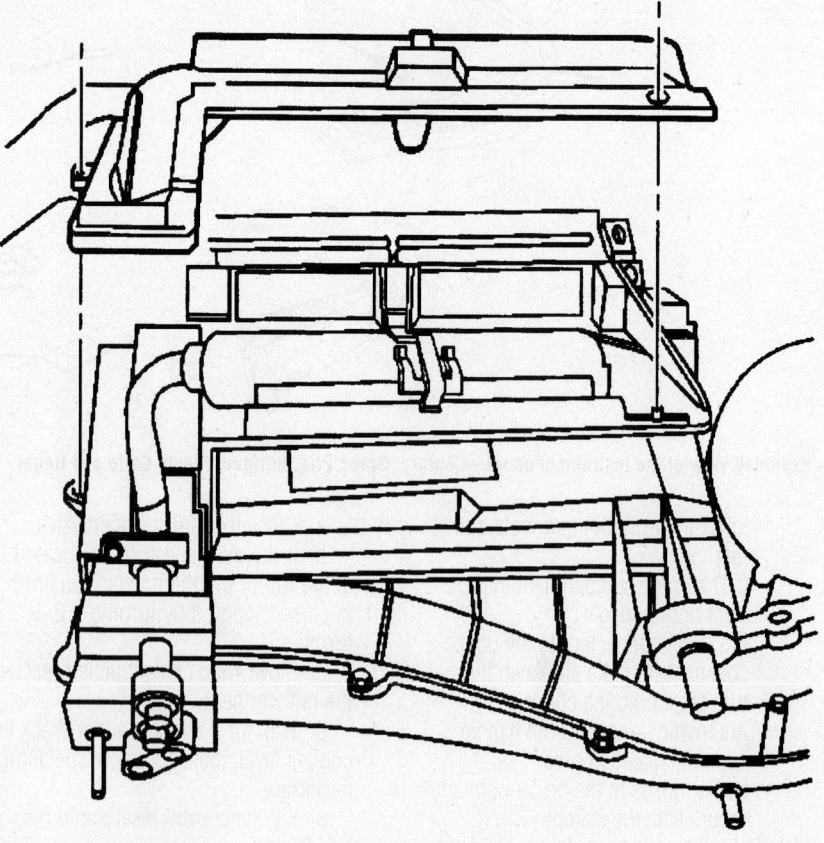

View of the heater core cover—Grand Prix, Intrigue and Regal

93111G66

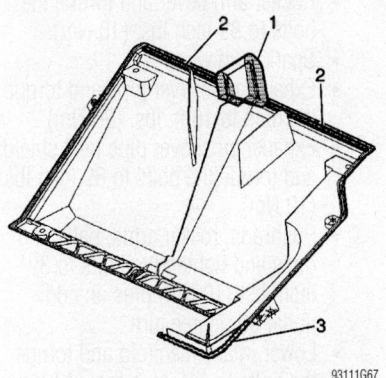

Location of the heater core cover seals—Grand Prix, Intrigue and Regal

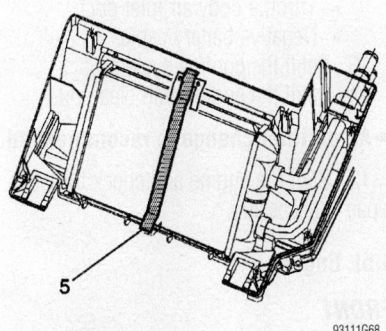

View of the heater core outer seal—Grand Prix, Intrigue and Regal

- Lower floor console, if equipped
8. On Grand Prix or Regal equipped with a 3.8L engine, install the fuel injector sight shield.
9. On Grand Prix or Regal equipped with a 3.1L engine, install the air cleaner and duct assembly.
10. Refill the engine cooling system.
11. Connect the negative battery cable.
12. Operate the engine to normal operating temperatures; then, check the climate control operation and check for leaks.

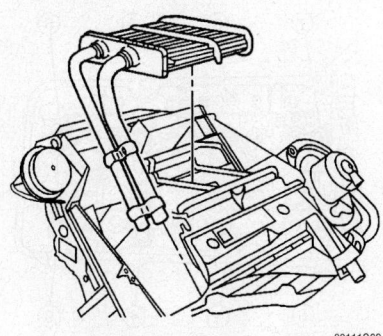

View of the heater core—Grand Prix, Intrigue and Regal

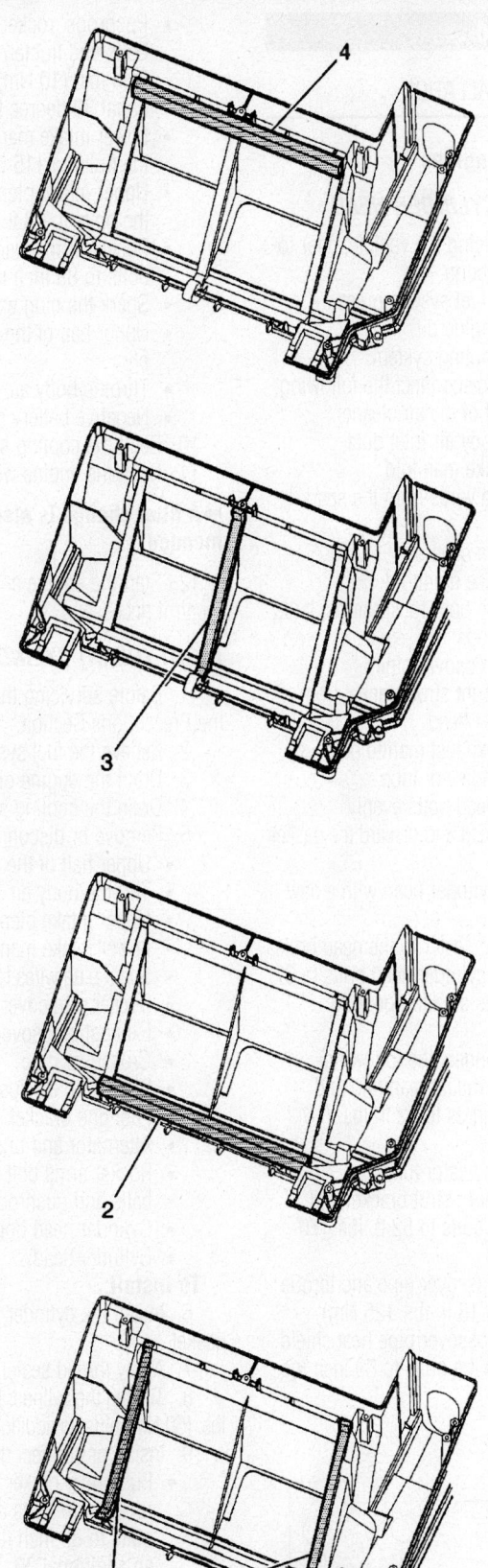

View of the heater core lower, center, upper and side seals of the lower heater core case—Grand Prix, Intrigue and Regal

Cylinder Head

REMOVAL & INSTALLATION

3.1L and 3.4L Engines

LEFT (FRONT) CYLINDER HEAD

1. Before servicing the vehicle, refer to the Precautions Section.
2. Relieve the fuel system pressure.
3. Drain the engine oil.
4. Drain the cooling system.
5. Remove or disconnect the following:
   - Upper half of the air cleaner
   - Throttle body air inlet duct
   - Upper intake manifold
   - Spark plug wires from the spark plugs
   - Rocker arm covers
   - Lower intake manifold
   - Rocker arm bolt, rocker arms, balls and pushrods
   - Exhaust crossover pipe
   - Engine mount strut bracket from the cylinder head
   - Left side exhaust manifold
   - Oil level indicator tube
   - Cylinder head bolts evenly
   - Cylinder head and discard the gasket

To install:

6. Install the cylinder head with a new gasket.
7. Apply thread sealer to the head bolts.
8. Tighten the cylinder head bolts to 37 ft. lbs. (50 Nm) plus an additional 90 degree turn.
9. Install or connect the following:
   - Left side exhaust manifold and torque the nuts to 12 ft. lbs. (16 Nm)
   - Oil level indicator tube
   - Engine mount strut bracket and torque the bolts to 52 ft. lbs. (70 Nm)
   - Exhaust crossover pipe and torque the nuts to 18 ft. lbs. (25 Nm)
   - Exhaust crossover pipe heat shield and torque the bolts to 89 inch lbs. (10 Nm)

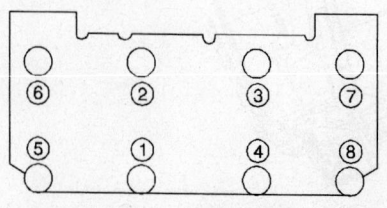

Tighten the cylinder head bolts using the following sequence—31L and 3.4L

9346ZGA9

   - Pushrods, rocker arms, balls and bolts and tighten the bolts to 89 inch lbs. (10 Nm) plus an additional 30 degree turn
   - Lower intake manifold and torque the bolts to 115 inch lbs. (13 Nm)
   - Upper intake plenum and torque the bolts to 18 ft. lbs. (25 Nm)
   - Rocker arm cover and torque the bolts to 89 inch lbs. (10 Nm)
   - Spark the plug wires
   - Upper half of the air cleaner assembly
   - Throttle body air inlet duct
   - Negative battery cable
10. Refill the cooling system.
11. Refill the engine with clean oil.

➡**A filter change is also recommended.**

12. Start the engine and check for leaks, repair if necessary.

RIGHT (REAR) CYLINDER HEAD

1. Before servicing the vehicle, refer to the Precautions Section.
2. Relieve the fuel system pressure.
3. Drain the engine oil.
4. Drain the cooling system.
5. Remove or disconnect the following:
   - Upper half of the air cleaner
   - Throttle body air inlet duct
   - Upper intake plenum
   - Lower intake manifold
   - Spark plug wires from the spark plugs
   - Rocker arm covers
   - Exhaust crossover pipe heat shield
   - Crossover pipe
   - Right side exhaust manifold
   - Fuel line bracket
   - Alternator and bracket
   - Rocker arms bolt, rocker arms, balls and pushrods
   - Cylinder head bolts evenly
   - Cylinder head

To install:

6. Install the cylinder head with a new gasket.
7. Apply thread sealer to the head bolts.
8. Tighten the cylinder head bolts to 37 ft. lbs. (50 Nm) plus an additional 90 degree turn.
9. Install or connect the following:
   - Pushrods, rocker arms, balls and rocker arm bolts and tighten the bolts to 89 inch lbs. (10 Nm) plus an additional 30 degrees
   - Exhaust manifold and torque the nuts to 12 ft. lbs. (16 Nm)
   - Alternator bracket and torque the bolts to 37 ft. lbs. (50 Nm)
   - Alternator and torque the bolts to 37 ft. lbs. (50 Nm)

   - Rocker arm cover and torque the bolts to 89 inch lbs. (10 Nm)
   - Spark plug wires
   - Exhaust crossover pipe and torque the nuts to 18 ft. lbs. (25 Nm)
   - Exhaust crossover pipe heat shield and torque the bolts to 89 inch lbs. (10 Nm)
   - Pushrods, rocker arms, balls and bolts and tighten the bolts to 89 inch lbs. (10 Nm) plus an additional 30 degree turn
   - Lower intake manifold and torque the bolts to 115 inch lbs. (13 Nm)
   - Upper intake plenum and torque the bolts to 18 ft. lbs. (25 Nm)
   - Upper half of the air cleaner assembly
   - Throttle body air inlet duct
   - Negative battery cable
10. Refill the cooling system.
11. Refill the engine with clean oil.

➡**An oil filter change is recommended.**

12. Start the engine and check for leaks, repair if necessary.

3.5L Engine

FRONT

1. Before servicing the vehicle, refer to the Precautions Section.
2. Drain the engine oil.
3. Drain the cooling system.
4. Remove or disconnect the following:
   - Negative battery cable
   - Intake manifold
   - Water outlet housing
   - Engine mount strut bracket
   - Coolant crossover pipe
   - Front exhaust manifold
   - Camshaft cover
5. Install a holding tool on the camshafts to hold them in position.
6. Remove or disconnect the following:
   - Camshaft primary chain

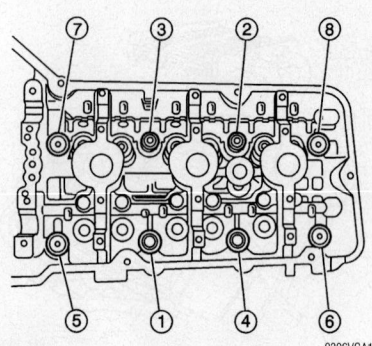

Front cylinder head bolt torque sequence—3.5L engine

9306VGA1

- Camshafts from the front cylinder head
- Rocker arms and valve lifters

➡**Be sure to keep the arms and lifters in order so they can be installed the their original locations.**

- M6 bolts from the front of the cylinder head
- M11 cylinder head bolts and discard
- Cylinder head

To install:

7. Be sure the dowels are securely mounted in the engine block.

8. Install or connect the following:
- New gasket
- Cylinder head
- New M11 bolts
- M6 bolts in the front of the cylinder head

9. Torque the M11 bolts in sequence to:
  a. Step 1: 22 ft. lbs. (30 Nm).
  b. Step 2: 100 degree turn.
  c. Step 3: 100 degree turn.

10. Torque the long M6 bolt to 106 inch lbs. (12 Nm).

11. Torque both shorter M6 bolts to 106 inch lbs. (12 Nm).

12. Install or connect the following:
- Lifters and rocker arms in their original positions
- Camshafts and torque the bearing cap bolts to 71 inch lbs. (8 Nm) plus an additional 22 degree turn
- Primary camshaft drive chain
- Camshaft cover and torque the bolts to 80 inch lbs. (9 Nm)
- Exhaust manifold and torque the bolts to 18 ft. lbs. (25 Nm)
- Coolant crossover pipe and torque the bolts to 18 ft. lbs. (25 Nm)
- Engine mount strut bracket and torque the bolts to 37 ft. lbs. (50 Nm)

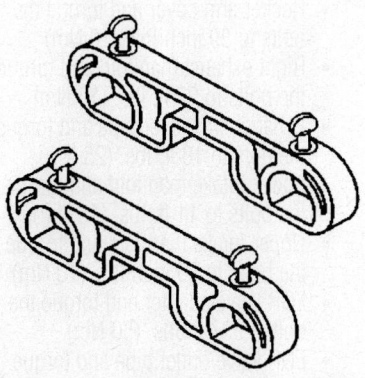

Camshaft holding fixture J-42038—3.5L engine

9300Z505

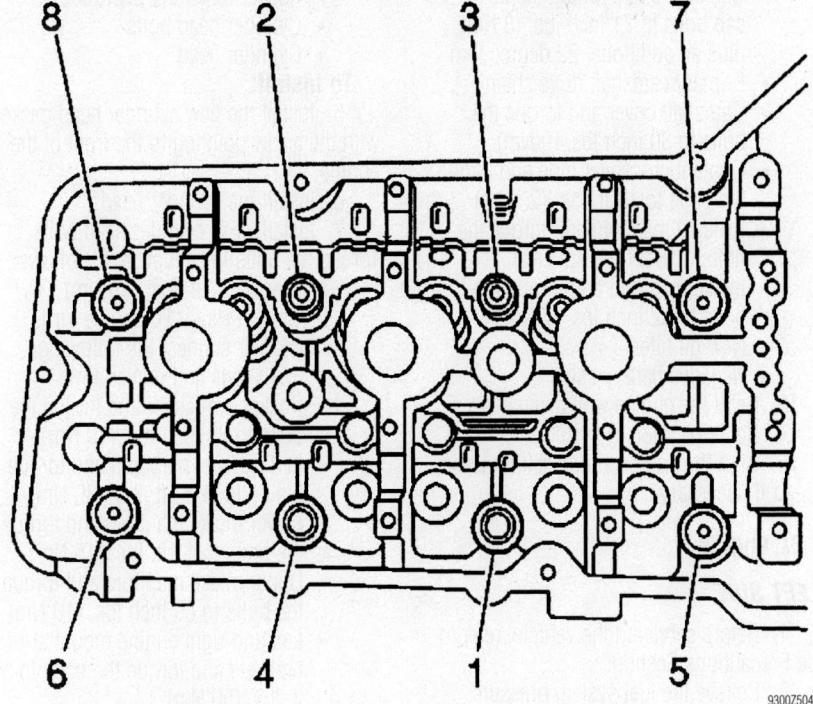

9300Z504

Rear cylinder head bolt torque sequence—3.5L engine

- Water outlet housing and torque the bolts to 80 inch lbs. (9 Nm)
- Intake manifold and torque the bolts to 62 inch lbs. (7 Nm)
- New oil filter
- Negative battery cable

13. Refill the engine with clean oil.

14. Refill the cooling system.

15. Start the engine and check for leaks, repair if necessary.

REAR

1. Before servicing the vehicle, refer to the Precautions Section.

2. Drain the engine oil.

3. Drain the cooling system.

4. Remove or disconnect the following:
- Negative battery cable
- Intake manifold

➡**Do not remove the rear exhaust manifold. Detach it from the cylinder head and the connection from the front manifold; then, move it aside.**

- Exhaust manifold
- Coolant crossover pipe
- Camshaft cover

5. Install a holding tool on the camshafts to hold them in position.

6. Remove or disconnect the following:

- Primary camshaft drive chain
- Camshafts
- Rocker arms and valve lifters

➡**Keep the valve train parts in order so they can be installed in their original positions.**

- Engine Coolant Temperature (ECT) sensor from the cylinder head
- M6 bolts from the front of the cylinder head, note the location of the longer bolt
- M11 cylinder head bolts and discard the bolts
- Cylinder head

To install:

7. Be sure the dowels are securely mounted in the engine block.

8. Install or connect the following:
- New gasket
- Cylinder head
- New M11 bolts
- M6 bolts in the front of the cylinder head

9. Torque the M11 bolts in sequence to:
  a. Step 1: 22 ft. lbs. (30 Nm).
  b. Step 2: 100 degree turn.
  c. Step 3: 100 degree turn.

10. Torque the long M6 bolt to 106 inch lbs. (12 Nm).

11. Torque both shorter M6 bolts to 106 inch lbs. (12 Nm).

12. Install or connect the following:
- ECT sensor and torque the fastener to 15 ft. lbs. (20 Nm)
- Lifters and rocker arms in their original positions

- Camshafts and torque the bearing cap bolts to 71 inch lbs. (8 Nm) plus an additional 22 degree turn
- Primary camshaft drive chain
- Camshaft cover and torque the bolts to 80 inch lbs. (9 Nm)
- Coolant crossover pipe and torque the bolts to 18 ft. lbs. (25 Nm)
- Exhaust manifold and torque the bolts to 18 ft. lbs. (25 Nm)
- Intake manifold and torque the bolts to 62 inch lbs. (7 Nm)
- New oil filter
- Negative battery cable

13. Refill the engine with clean oil.
14. Refill the cooling system.
15. Start the engine and check for leaks, repair if necessary.

3.8L Engine

LEFT SIDE

1. Before servicing the vehicle, refer to the Precautions Section.
2. Relieve the fuel system pressure.
3. Drain the cooling system.
4. Remove or disconnect the following:
- Fuel injector sight shield
- Throttle body air inlet duct
- Right and left engine mount strut brackets
- Fuel lines from the fuel rail
- Upper intake manifold
- Lower intake manifold
- Left exhaust manifold
- Rocker arm cover

- Rocker arms and pushrods
- Cylinder head bolts
- Cylinder head

To install:

5. Install the new cylinder head gasket with the arrow pointing to the front of the engine.
6. Install the cylinder head.
7. Install NEW cylinder head bolts. Torque the bolts in sequence, as follows:
   a. Step 1: 37 ft. lbs. (50 Nm).
   b. Step 2: Plus 120 degree turn.
8. Install or connect the following:
- Push rods and rocker arms
- Rocker arm cover and torque the bolts to 89 inch lbs. (10 Nm)
- Left exhaust manifold and torque the bolts to 22 ft. lbs. (30 Nm)
- Lower intake manifold and torque the bolts to 11 ft. lbs. (15 Nm)
- Upper intake manifold and torque the bolts to 89 inch lbs. (10 Nm)
- Left and right engine mount strut brackets and torque the bolts to 37 ft. lbs. (50 Nm)
- Throttle body air inlet duct
- Fuel injector sight shield and torque the bolts to 27 inch lbs. (3 Nm)
- Negative battery cable

9. Refill the cooling system.
10. Start the engine and check for leaks.

RIGHT SIDE

1. Before servicing the vehicle, refer to the Precautions Section.

2. Relieve the fuel system pressure.
3. Drain the cooling system.
4. Remove or disconnect the following:
- Throttle body air inlet duct
- Fuel injector sight shield
- Accessory drive belt
- Alternator
- Catalytic converter from the exhaust manifold
- Engine mount struts

5. Place the transmission in neutral and rotate the engine forward for access.
6. Remove or disconnect the following:
- Drive belt tensioner
- Power steering pump without disconnecting the lines
- Heater hoses from the engine
- Spark plug wires
- Oxygen Sensor (O_2S) electrical connector
- Throttle and cruise control cables
- Fuel lines from the fuel rail
- Fuel rail
- Exhaust Gas Recirculation (EGR) valve heat shield
- EGR valve
- EGR valve outlet pipe
- EGR valve adapter
- Upper intake manifold
- Lower intake manifold
- Exhaust crossover pipe
- Right exhaust manifold
- Rocker arm cover
- Rocker arms and pushrods
- Cylinder head

To install:

7. Install a new cylinder head gasket with the arrow pointing to the front of the engine.
8. Install the cylinder head and secure with NEW head bolts. Torque the bolts, in sequence, as follows:
   a. Step 1: 37 ft. lbs. (50 Nm).
   b. Step 2: Plus a 120 degree turn.
9. Install or connect the following:
- Push rods and rocker arms
- Rocker arm cover and torque the bolts to 89 inch lbs. (10 Nm)
- Right exhaust manifold and torque the bolts to 22 ft. lbs. (30 Nm)
- Exhaust crossover pipe and torque the nuts to 18 ft. lbs. (25 Nm)
- Lower intake manifold and torque the bolts to 11 ft. lbs. (15 Nm)
- Upper intake manifold and torque the bolts to 89 inch lbs. (10 Nm)
- EGR valve adapter and torque the bolts to 37 ft. lbs. (50 Nm)
- EGR valve outlet pipe and torque the bolts to 21 ft. lbs. (29 Nm)
- EGR valve and torque the nuts to 21 ft. lbs. (29 Nm)

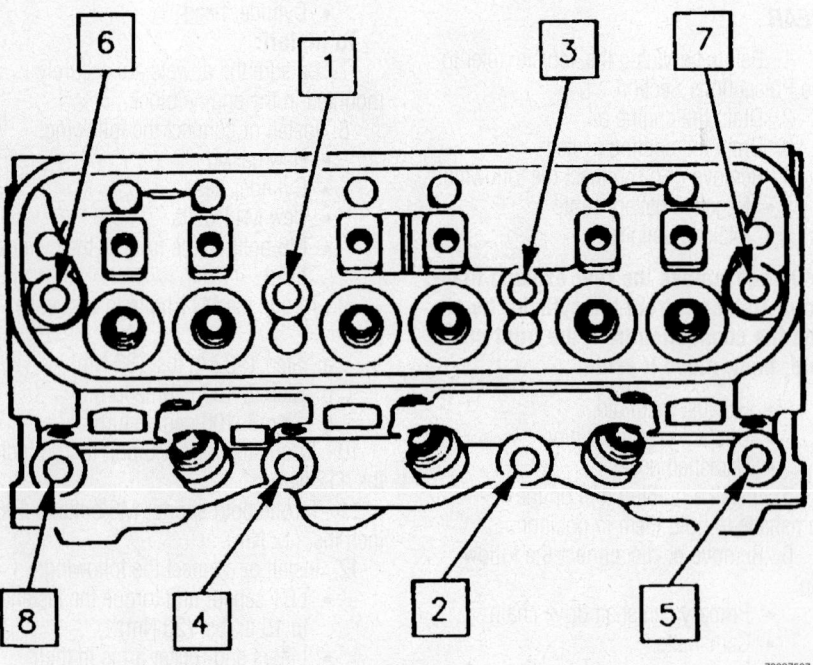

79222507

Cylinder head bolt torque sequence—3.8L engine

- EGR valve heat shield and torque the bolts to 89 inch lbs. (10 Nm)
- Fuel rail
- Fuel lines to the fuel rail
- Throttle and cruise control cables
- O2S electrical connector
- Spark plugs and torque them to 20 ft. lbs. (27 Nm)
- Spark plug wires
- Heater hoses
- Power steering pump and torque the bolts to 25 ft. lbs. (34 Nm)
- Drive belt tensioner and torque the bolts to 37 ft. lbs. (50 Nm)

10. Carefully return the engine to its proper position.
11. Place the transmission in park.
12. Install or connect the following:
- Throttle body air inlet duct
- Engine mount struts and torque the bolts to 35 ft. lbs. (48 Nm)
- Catalytic converter and torque the bolts to 26 ft. lbs. (35 Nm)
- Alternator and torque the bolts to 37 ft. lbs. (50 Nm)
- Accessory drive belt
- Fuel injector sight shield
- Negative battery cable

13. Refill the cooling system.
14. Start the engine and check for leaks.

5.3L Engine

LEFT SIDE

1. Remove the intake manifold.
2. Before servicing the vehicle, refer to the Precautions Section.
3. Disconnect the hose clamp and hose from the coolant fill neck.
4. Remove the coolant air bleed pipe bolts.
5. Remove the coolant air bleed pipe (1) with hose and seals.
6. Remove the hose (2) and clamps (3) from the coolant air bleed pipe as required.
7. Remove the engine coolant air bleed cover bolts.
8. Remove the covers with seals.
9. Remove the seals from the pipe and covers. Discard the seals.
10. Remove the left exhaust manifold.
11. Remove the pushrods.
12. Remove the power steering pump pulley.
13. Remove the 3 coolant manifold bolts to the cylinder head.
14. Remove the forward engine mount strut bolt and nut.
15. Remove the rear engine mount strut bolt and nut.
16. Remove the engine mount strut.

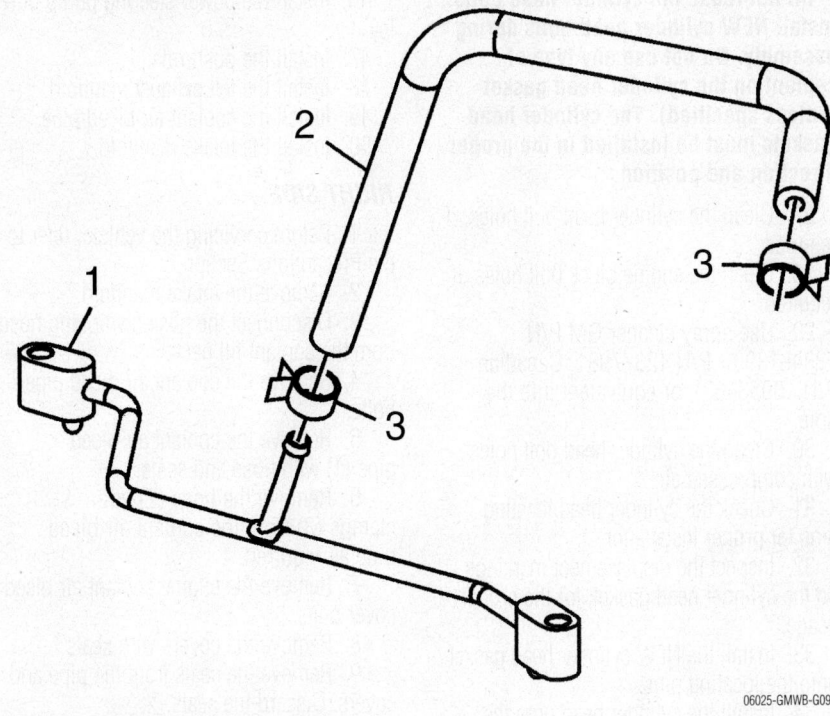

Coolant air bleed pipe—5.3L engine

17. Inspect the rubber in the engine mount strut for the following conditions:
18. If necessary, remove the engine mount strut bracket bolts (body side).
19. If necessary, remove the engine mount strut bracket (body side).
20. Remove the heater pipe bracket bolt from the engine mount.
21. If necessary, remove the engine mount strut bracket bolts (engine side).
22. If necessary, remove the engine mount strut bracket (engine side).

➡The cylinder head bolts are of a torque to yield design and are NOT to be used again.

23. Remove and discard the cylinder head bolts.

✳✳ WARNING

After removal, place the cylinder head on 2 wood blocks in order to prevent damage to the sealing surfaces.

24. Remove the cylinder head.
25. Remove and discard the cylinder head gasket.
26. Clean and inspect the cylinder head.
To install:

✳✳ CAUTION

Wear safety glasses in order to avoid eye damage.

✳✳ WARNING

Clean all dirt, debris, and coolant from the engine block cylinder head bolt holes. Failure to remove all foreign material may result in damaged threads, improperly tightened fasteners or damage to components.

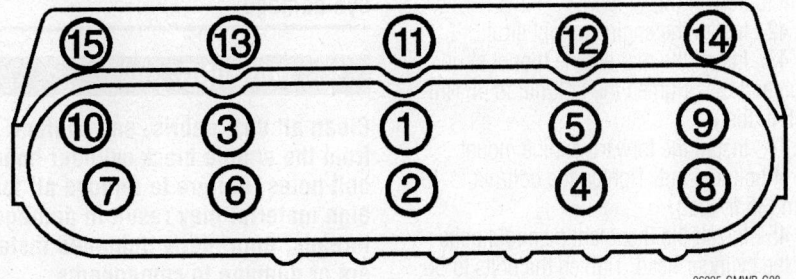

Cylinder head bolt torque sequence—5.3L engine

➡ **Do not reuse the cylinder head bolts. Install NEW cylinder head bolts during assembly. Do not use any type of sealant on the cylinder head gasket (unless specified). The cylinder head gaskets must be installed in the proper direction and position.**

27. Clean the cylinder head bolt holes, if required.

28. Clean the engine block bolt holes, if required.

29. Use spray cleaner GM P/N 12346139 or P/N 12377981, Canadian P/N 10953463), or equivalent into the hole.

30. Clean the cylinder head bolt holes with compressed air.

31. Check the cylinder head locating pins for proper installation.

32. Inspect the displacement markings on the cylinder head gasket, for the proper usage.

33. Install the NEW cylinder head gasket onto the locating pins.

34. Install the cylinder head onto the locating pins.

35. Install the NEW cylinder head bolts.

36. Tighten the cylinder head bolts. Tighten the M11 bolts (1-10) a first pass in sequence to 30 Nm (22 ft. lbs.).

  a. Tighten the M11 bolts (1-10) a second pass in sequence to 90 degrees.

  b. Tighten the M11 bolts (1-10) a final pass in sequence to 70 degrees.

  c. Tighten the M8 bolts (11-15) to 30 Nm (22 ft. lbs.). Begin with the center bolt (11) and alternating side-to-side, work outward tightening all of the bolts.

37. If necessary, position the engine mount strut bracket (engine side) to the cylinder head.

38. If necessary, install the engine mount strut bracket bolts (engine side). Tighten the bolts to 48 Nm (35 ft. lbs.).

39. Install the heater pipe bracket bolt. Tighten the bolt to 16 Nm (12 ft. lbs.).

40. If necessary, position the engine mount strut bracket (body side) to the vehicle.

41. If necessary, install the engine mount strut bracket bolts (body side). Tighten the bolt to 48 Nm (35 ft. lbs.).

42. Install the engine mount strut.

43. Install the rear engine mount strut bolt and nut. Tighten the bolt/nut to 48 Nm (35 ft. lbs.).

44. Install the forward engine mount strut bolt and nut. Tighten the bolt/nut to 48 Nm (35 ft. lbs.).

45. Install the 3 coolant manifold bolts to the cylinder head. Tighten the bolts to 50 Nm (37 ft. lbs.).

46. Install the power steering pump pulley.

47. Install the pushrods.

48. Install the left exhaust manifold.

49. Install the coolant air bleed pipe.

50. Install the intake manifold.

RIGHT SIDE

1. Before servicing the vehicle, refer to the Precautions Section.

2. Remove the intake manifold.

3. Disconnect the hose clamp and hose from the coolant fill neck.

4. Remove the coolant air bleed pipe bolts.

5. Remove the coolant air bleed pipe (1) with hose and seals.

6. Remove the hose (2) and clamps (3) from the coolant air bleed pipe as required.

7. Remove the engine coolant air bleed cover bolts.

8. Remove the covers with seals.

9. Remove the seals from the pipe and covers. Discard the seals.

10. Remove the generator bracket.

11. Remove the right exhaust manifold.

12. Remove the pushrods.

➡ **The cylinder head bolts are of a torque to yield design and are NOT to be used again.**

13. Remove and discard the cylinder head bolts.

✳✳ WARNING

After removal, place the cylinder head on 2 wood blocks in order to prevent damage to the sealing surfaces.

14. Remove the cylinder head.

15. Remove and discard the cylinder head gasket.

16. Clean and inspect the cylinder head.

To install:

✳✳ CAUTION

Wear safety glasses in order to avoid eye damage.

✳✳ WARNING

Clean all dirt, debris, and coolant from the engine block cylinder head bolt holes. Failure to remove all foreign material may result in damaged threads, improperly tightened fasteners or damage to components.

➡ **Do not reuse the cylinder head bolts. Install NEW cylinder head bolts during assembly. Do not use any type of sealant on the cylinder head gasket (unless specified). The cylinder head gaskets must be installed in the proper direction and position.**

17. Clean the cylinder head bolt holes, if required.

18. Clean the engine block bolt holes, if required.

19. Use spray cleaner GM P/N 12346139 or P/N 12377981, Canadian P/N 10953463), or equivalent into the hole.

20. Clean the cylinder head bolt holes with compressed air.

21. Check the cylinder head locating pins for proper installation.

22. Inspect the displacement markings on the cylinder head gasket, for the proper usage.

23. Install the NEW cylinder head gasket onto the locating pins.

24. Install the cylinder head onto the locating pins.

25. Install the NEW cylinder head bolts.

26. Tighten the cylinder head bolts.

  a. Tighten the M11 bolts (1-10) a first pass in sequence to 30 Nm (22 ft. lbs.).

  b. Tighten the M11 bolts (1-10) a second pass in sequence to 90 degrees.

  c. Tighten the M11 bolts (1-10) a final pass in sequence to 70 degrees.

  d. Tighten the M8 bolts (11-15) to 30 Nm (22 ft. lbs.). Begin with the center bolt (11) and alternating side-to-side, work outward tightening all of the bolts.

27. Install the pushrods.

28. Install the right exhaust manifold.

29. Install the generator bracket.

30. Install the coolant air bleed pipe.

31. Install the intake manifold.

Rocker Arms

REMOVAL & INSTALLATION

3.1L and 3.4L Engines

LEFT SIDE

1. Before servicing the vehicle, refer to the Precautions Section.

2. Disconnect the negative battery cable.

3. Drain the cooling system.

4. Remove or disconnect the following:

- Spark plug wires
- Engine mount strut
- Secondary Air Injection (AIR) check valve pipe, if equipped

- Thermostat bypass hose and pipe
- Positive Crankcase Ventilation (PCV) valve from the rocker arm cover
- Rocker arm cover

➡**Keep the pushrods in order. Intake pushrods are 5¾ inches long and exhaust pushrods are 6 inches long.**

- Rocker arm bolts
- Rocker arms
- Pushrods

To install:

5. Lubricate all the valvetrain components with engine oil.

6. Install or connect the following:
- Pushrods and the rocker arms and tighten the bolts to 14 ft. lbs. (19 Nm) plus 30 degrees
- Rocker arm cover using a new gasket and tighten the rocker cover bolts to 89 inch lbs. (10 Nm)
- PCV valve to the rocker arm cover
- Coolant tube with the thermostat bypass hose and tighten the bolt to 98 inch lbs. (11 Nm) and the nut to 18 ft. lbs. (25 Nm)
- AIR check valve pipe and torque the adapters to 22 ft. lbs. (30 Nm)
- Engine mount strut and torque the bolt to 35 ft. lbs. (48 Nm)
- Spark plug wires
- Negative battery cable

7. Refill the cooling system.

8. Start the vehicle and verify no leaks.

RIGHT SIDE

1. Before servicing the vehicle, refer to the Precautions Section.

2. Remove or disconnect the following:
- Negative battery cable
- Accessory drive belt
- Alternator
- Alternator Bracket
- Spark plug wires
- Ignition coil and Evaporative (EVAP) emissions canister purge solenoid as an assembly
- Power brake booster vacuum pipe from the intake plenum
- Rocker arm cover
- Rocker arm bolts, balls, rocker arms and pushrods

To install:

3. Lubricate all the valve train components with engine oil.

4. Install or connect the following:
- Pushrods and the rocker arms and tighten the bolts to 14 ft. lbs. (19 Nm) plus 30 degrees

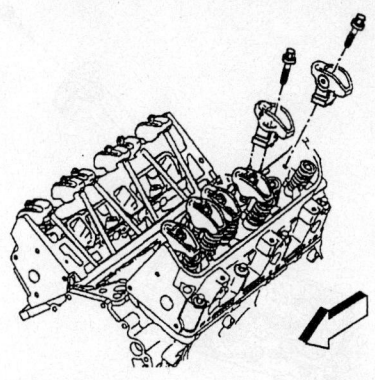

93062G09

Rocker arm components—3.1L and 3.4L engines

- Rocker arm cover using a new gasket and tighten the rocker cover bolts to 89 inch lbs. (10 Nm)
- Power brake booster vacuum pipe to the plenum
- EVAP solenoid and ignition coil assembly
- Spark plug wires
- Alternator bracket and torque the bolts to 37 ft. lbs. (50 Nm)
- Alternator and torque the bolts to 37 ft. lbs. (50 Nm)
- Accessory drive belt
- Negative battery cable

5. Start the vehicle and verify no leaks.

3.5L Engine

Refer to the camshaft removal and installation procedure for rocker arm service.

3.8L Engine

> ✳✳ **WARNING**

The rocker arm bolts have been permanently stretched during installation. New bolts must be used each time the rocker arm assemblies are removed.

LEFT SIDE (FRONT) ROCKER ARMS

1. Before servicing the vehicle, refer to the Precautions Section.

2. Remove or disconnect the following:
- Negative battery cable
- Engine lift bracket from the exhaust manifold studs
- Fuel injector sight shield
- Engine mount strut bracket
- Spark plug wires
- Spark plug wire cover from the valve rocker arm cover
- Rocker arm cover

➡**The rocker arms, pushrods, pedestals and bolts must be kept in order for installation in the same locations they were removed from.**

- Rocker arm bolts, pedestals, rocker arms and pushrods

To install:

3. Apply thread locking compound to the rocker arm bolts.

4. Install or connect the following:
- Pushrod in the lifter
- Rocker arm, pedestal and mounting bolt and tighten the mounting bolt

79222508

Rocker arm mounting—3.1L engine

to 11 ft. lbs. (15 Nm) plus an additional 90 degrees

- Rocker arm cover with a new gasket and torque the mounting bolts to 89 inch lbs. (10 Nm)
- Spark plug wire cover to the rocker arm cover
- Spark plug wires
- Engine mount strut bracket and torque the bolt to 75 ft. lbs. (102 Nm)
- Fuel injector sight shield
- Engine lift bracket to the exhaust manifold and torque the bolt to 22 ft. lbs. (30 Nm)
- Negative battery cable

5. Start the vehicle and check for leaks, repair if necessary.

RIGHT SIDE (REAR) ROCKER ARMS

1. Before servicing the vehicle, refer to the Precautions Section.
2. Remove or disconnect the following:
- Negative battery cable
- Fuel injector sight shield
- Accessory drive belt
- Alternator
- Drive belt tensioner
- Engine mount struts
- Throttle body air inlet duct
3. Rotate the engine forward for proper access.
- Evaporative (EVAP) emissions purge solenoid and brace
- Right side spark plug wires
- Fuel injector sight shield bracket
- Rocker arm cover

➡ **The rocker arms, pushrods, pedestals and bolts must be kept in order for installation in the same locations they were removed from.**

- Rocker arm bolts, pedestals, rocker arms and pushrods

To install:

4. Seat the pushrod in the lifter and install the rocker arm, pedestal and mounting bolt. Tighten the mounting bolt to 11 ft. lbs. (15 Nm) plus an additional 90 degrees.
5. Install or connect the following:
- Rocker arm cover using a new gasket and torque the bolts to 89 inch lbs. (10 Nm)
- Fuel injector sight shield bracket and torque the bolts to 22 ft. lbs. (30 Nm)
- EVAP purge solenoid and brace
- Right side spark plug wires
6. Return the engine back to its original position.
- Throttle body air inlet duct
- Engine mount struts and torque the bolt to 35 ft. lbs. (48 Nm)

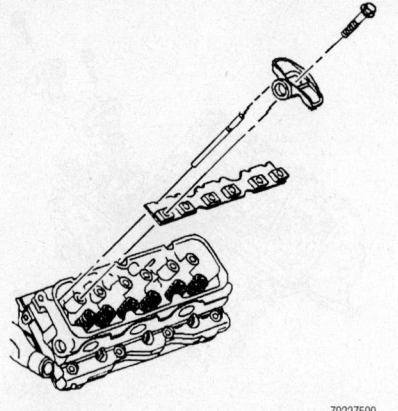

Exploded view of the rocker arm, pushrod and pushrod guide plate assembly—3.8L engine

79222509

- Drive belt tensioner and torque the bolts to 37 ft. lbs. (50 Nm)
- Alternator and torque the bolt to 37 ft. lbs. (50 Nm)
- Fuel injector sight shield
- Negative battery cable

7. Start the vehicle and check for leaks, repair if necessary.

5.3L Engine

LEFT SIDE

1. Before servicing the vehicle, refer to the Precautions Section.
2. Remove the engine sight shield, if required.
3. Remove the connector position assurance (CPA) retainer.
4. Disconnect the ignition coil main electrical connector.
5. Reposition the engine harness, as necessary.
6. Remove the spark plug wires from the ignition coils.
7. Remove the ignition coil bracket bolts.
8. Remove the ignition coil and bracket assembly.
9. Remove the positive crankcase ventilation (PCV) clean air tube from the valve rocker arm cover.
10. Remove the valve rocker arm cover bolts and cover.
11. Remove and discard the old the gasket from the valve rocker arm cover.
12. Remove the valve rocker arm cover bolt grommets, if cut or damaged.

➡ **Do not remove the oil fill tube from the rocker cover unless service is required. If the oil fill tube has been removed, install a NEW tube during assembly.**

13. Remove the oil fill cap from the oil fill tube, if necessary.
14. Remove the oil fill tube, if necessary.
15. Discard the oil fill tube, if necessary.

➡ **Place the valve rocker arm, pushrod, and pivot support, in a rack so that the can be installed in the same location from which they were removed.**

16. Remove the valve rocker arm bolt.
17. Remove the valve rocker arm.
18. Remove the valve rocker arm pivot support.
19. Remove the pushrod.
20. Clean and inspect the valve rocker arm and pushrod.

To install:

➡ **When reusing the valve train components, always install the components to their original location and position. Valve lash is net build, no valve adjustment is required.**

21. Lubricate the valve rocker arm and pushrod with clean engine oil.
22. Lubricate the flange of the valve rocker arm bolt with clean engine oil.
23. Install the valve rocker arm pivot support.

➡ **Make sure that the pushrod seats properly to the valve lifter socket.**

24. Install the pushrod.

➡ **Ensure that the pushrod seat properly to the ends of the rocker arm. DO NOT tighten the rocker arm bolt at this time.**

25. Install the rocker arm and bolt.
26. Rotate the crankshaft until the number one piston is at top dead center (TDC) of the compression stroke. In this position, cylinder number one rocker arms will be off lobe lift, and the crankshaft sprocket key will be at the 1:30 position. The engine firing order is 1, 8, 7, 2, 6, 5, 4, 3. Cylinders 1, 3, 5, and 7 are the left bank. Cylinder 2, 4, 6, and 8 are the right bank.
27. With the engine in the number one firing position, tighten the following valve rocker arm bolts:
- Tighten cylinders 1, 2, 7, and 8 exhaust valve rocker arm bolts to 30 Nm (22 ft. lbs.).
- Tighten cylinders 1, 3, 4, and 5 intake valve rocker arm bolts to 30 Nm (22 ft. lbs.).
28. Rotate the crankshaft 360 degrees.
29. Tighten the following valve rocker arm bolts:
- Tighten cylinders 3, 4, 5, and 6 exhaust valve rocker arm bolts to 30 Nm (22 ft. lbs.).

- Tighten cylinders 2, 6, 7, and 8 intake valve rocker arm bolts to 30 Nm (22 ft. lbs.).

➡**All gasket surfaces should be free of oil an/or other foreign material during assembly. DO NOT reuse the valve rocker arm cover gasket. The valve rocker arm cover bolt grommets may be reused.**

30. Install a NEW gasket (504) into the groove of the valve rocker arm cover.

31. Install a NEW oil fill tube to the valve rocker arm cover, if necessary.

32. Install NEW valve rocker arm cover bolt grommets, if necessary.

33. Install the valve rocker arm cover and bolts. Tighten the bolts to 12 Nm (106 inch lbs.).

34. Remove the positive crankcase ventilation (PCV) clean air tube from the valve rocker arm cover.

35. Apply threadlock GM P/N 12345382 (Canadian P/N 10953489) or equivalent to the threads of the ignition coil bracket bolts.

36. Install the ignition coil and bracket assembly.

37. Install the ignition coil bracket bolts. Tighten the bolts to 12 Nm (106 inch lbs.).

38. Install the spark plug wires to the ignition coils.

39. Position the engine harness, as necessary.

40. Connect the ignition coil main electrical connector.

41. Install the CPA retainer.

42. Install the engine sight shield, if required.

RIGHT SIDE

1. Before servicing the vehicle, refer to the Precautions Section.

2. Remove the engine sight shield, if required.

3. Remove the connector position assurance (CPA) retainer.

4. Disconnect the ignition coil main electrical connector.

5. Reposition the engine harness, as necessary.

6. Remove the spark plug wires from the ignition coils.

7. Remove the ignition coil bracket bolts.

8. Remove the ignition coil and bracket assembly.

9. Remove the positive crankcase ventilation (PCV) clean air tube from the valve rocker arm cover.

10. Remove the valve rocker arm cover bolts and cover.

11. Remove and discard the old the gasket from the valve rocker arm cover.

12. Remove the valve rocker arm cover bolt grommets, if cut or damaged.

➡**Do not remove the oil fill tube from the rocker cover unless service is required. If the oil fill tube has been removed, install a NEW tube during assembly.**

13. Remove the oil fill cap from the oil fill tube, if necessary.

14. Remove the oil fill tube, if necessary.

15. Discard the oil fill tube, if necessary.

➡**Place the valve rocker arm, pushrod, and pivot support, in a rack so that the can be installed in the same location from which they were removed.**

16. Remove the valve rocker arm bolt.

17. Remove the valve rocker arm.

18. Remove the valve rocker arm pivot support.

19. Remove the pushrod.

20. Clean and inspect the valve rocker arm and pushrod.

To install:

➡**When reusing the valve train components, always install the components to their original location and position. Valve lash is net build, no valve adjustment is required.**

21. Lubricate the valve rocker arm and pushrod with clean engine oil.

22. Lubricate the flange of the valve rocker arm bolt with clean engine oil.

23. Install the valve rocker arm pivot support.

➡**Make sure that the pushrod seats properly to the valve lifter socket.**

24. Install the pushrod.

➡**Ensure that the pushrod seat properly to the ends of the rocker arm. DO NOT tighten the rocker arm bolt at this time.**

25. Install the rocker arm and bolt.

26. Rotate the crankshaft until the number one piston is at top dead center (TDC) of the compression stroke. In this position, cylinder number one rocker arms will be off lobe lift, and the crankshaft sprocket key will be at the 1:30 position. The engine firing order is 1, 8, 7, 2, 6, 5, 4, 3. Cylinders 1, 3, 5, and 7 are the left bank. Cylinder 2, 4, 6, and 8 are the right bank.

27. With the engine in the number one firing position, tighten the following valve rocker arm bolts:

- Tighten cylinders 1, 2, 7, and 8

exhaust valve rocker arm bolts to 30 Nm (22 ft. lbs.).

- Tighten cylinders 1, 3, 4, and 5 intake valve rocker arm bolts to 30 Nm (22 ft. lbs.).

28. Rotate the crankshaft 360 degrees.

29. Tighten the following valve rocker arm bolts:

- Tighten cylinders 3, 4, 5, and 6 exhaust valve rocker arm bolts to 30 Nm (22 ft. lbs.).
- Tighten cylinders 2, 6, 7, and 8 intake valve rocker arm bolts to 30 Nm (22 ft. lbs.).

➡**All gasket surfaces should be free of oil an/or other foreign material during assembly. DO NOT reuse the valve rocker arm cover gasket. The valve rocker arm cover bolt grommets may be reused.**

30. Install a NEW gasket (504) into the groove of the valve rocker arm cover.

31. Install a NEW oil fill tube to the valve rocker arm cover, if necessary.

32. Install NEW valve rocker arm cover bolt grommets, if necessary.

33. Install the valve rocker arm cover and bolts. Tighten the bolts to 12 Nm (106 inch lbs.).

34. Remove the positive crankcase ventilation (PCV) clean air tube from the valve rocker arm cover.

35. Apply threadlock GM P/N 12345382 (Canadian P/N 10953489) or equivalent to the threads of the ignition coil bracket bolts.

36. Install the ignition coil and bracket assembly.

37. Install the ignition coil bracket bolts. Tighten the bolts to 12 Nm (106 inch lbs.).

38. Install the spark plug wires to the ignition coils.

39. Position the engine harness, as necessary.

40. Connect the ignition coil main electrical connector.

41. Install the CPA retainer.

42. Install the engine sight shield, if required.

Supercharger

REMOVAL & INSTALLATION

3.8L (VIN 1) Engine

1. Before servicing the vehicle, refer to the Precautions Section.

2. Relieve the fuel system pressure.

3. Remove or disconnect the following:

- Fuel injector sight shield
- Exhaust Gas Recirculation (EGR) valve heat shield
- Supercharger drive belt
- Right side spark plug wires from the ignition module
- Alternator brace
- Fuel injector electrical connectors
- Manifold Absolute Pressure (MAP) sensor bracket
- Fuel lines
- Fuel rail with the injectors
- Boost control solenoid
- Electrical and vacuum connections as necessary
- Throttle and cruise control cables with the bracket
- Throttle body air inlet duct
- Supercharger

To install:

4. Install or connect the following:
- Supercharger with new a new gasket and O-rings and torque the bolts to 17 ft. lbs. (23 Nm)
- Throttle body air inlet duct
- Throttle and cruise control cables with the bracket and torque the bolts to 142 inch lbs. (16 Nm)
- Electrical and vacuum connectors
- Boost control solenoid and torque the nut to 72 inch lbs. (8 Nm)
- Fuel rail with fuel injectors and torque the hold-down bolts to 7 ft. lbs. (10 Nm) and the stud to 18 ft. lbs. (25 Nm)
- Fuel lines
- MAP sensor bracket
- Fuel injector electrical connectors
- Alternator brace and torque the bolt and nut to 37 ft. lbs. (50 Nm)
- Spark plug wires
- Supercharger drive belt
- EGR valve heat shield
- Fuel injector sight shield
- Negative battery cable

5. Start the engine and ensure proper operation.

Intake Manifold

REMOVAL & INSTALLATION

3.1L and 3.4L Engines

1. Before servicing the vehicle, refer to the Precautions Section.
2. Relieve the fuel system pressure.
3. Drain the cooling system.
4. Remove or disconnect the following:
- Intake Air Temperature (IAT) sensor electrical connector

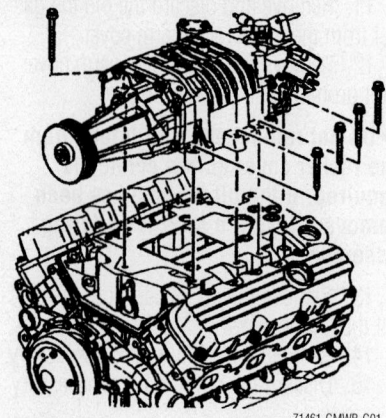

71461-GMWB-G01

Supercharger bolt locations–3.8L engine

- Throttle body air inlet duct
- Throttle and cruise control cables with bracket
- Throttle Position Sensor (TPS) electrical connector
- Idle Air Control (IAC) valve electrical connector
- Camshaft Position (CMP) sensor wiring harness from the intake
- Spark plug wires from the intake
- Thermostat bypass hoses from the throttle body
- Ignition coil bracket with the coils, the purge solenoid and the vacuum canister solenoid
- Vacuum lines from the upper intake manifold
- Manifold Absolute Pressure (MAP) sensor
- Alternator brace
- Exhaust Gas Recirculation (EGR) valve
- Upper intake manifold
- Both rocker arm covers
- Engine Coolant Temperature (ECT) sensor electrical connector
- Fuel injector and MAP wiring harness
- Fuel injector electrical connectors
- Fuel lines from the fuel rail and fuel line bracket
- Fuel rail with the injectors
- Power steering pump and move it aside without disconnecting the lines
- Coolant inlet pipe from the coolant outlet housing
- Coolant bypass hose
- Upper radiator hose at thermostat housing
- Lower intake manifold bolts

➡**When removing the valvetrain components, keep them in order for installation purposes.**

- Rocker arm bolts, rocker arms and pushrods
- Intake manifold

To install:

5. Place a 8–12mm bead of RTV, on each ridge, where the front and rear of the intake manifold contact the block.
6. Install or connect the following:
- Intake manifold gasket
- Pushrods and rocker arms
- Rocker arm bolts and torque the bolts to 89 inch lbs. (10 Nm) plus an additional 30 degree turn

✳✳ WARNING

In order to prevent oil leaks, tighten the vertical bolts before the diagonal bolts.

- Lower intake manifold and torque the bolts to 115 inch lbs. (13 Nm) and if equipped, tighten the vertical and diagonal lower intake manifold bolts to 62 inch lbs. (7 Nm)
- Both rocker arm covers and torque the bolts to 89 inch lbs. (10 Nm)
- Thermostat bypass hose
- Upper radiator hose
- Heater inlet pipe and heater hose to the lower intake
- Power steering pump and torque the bolts to 25 ft. lbs. (34 Nm)
- Fuel fail with the injectors
- Fuel feed and return lines
- Fuel injector electrical connectors
- ECT sensor electrical connector
- Upper intake manifold using a new gasket and torque the bolts to 18 ft. lbs. (25 Nm)
- EGR valve
- Alternator brace
- Vacuum lines
- MAP sensor
- Ignition coil bracket with the coils, the purge solenoid and the vacuum canister solenoid
- Thermostat bypass pipe coolant hoses to the throttle body
- CMP sensor wiring harness to the intake
- Spark plug wires
- TPS electrical connector
- IAC valve electrical connector
- Throttle and cruise control cables with the bracket and torque the bolts to 115 inch lbs. (13 Nm) and the nuts to 89 inch lbs. (10 Nm)
- Throttle body air inlet duct
- IAT sensor electrical connector
- Negative battery cable

7. Refill the cooling system.

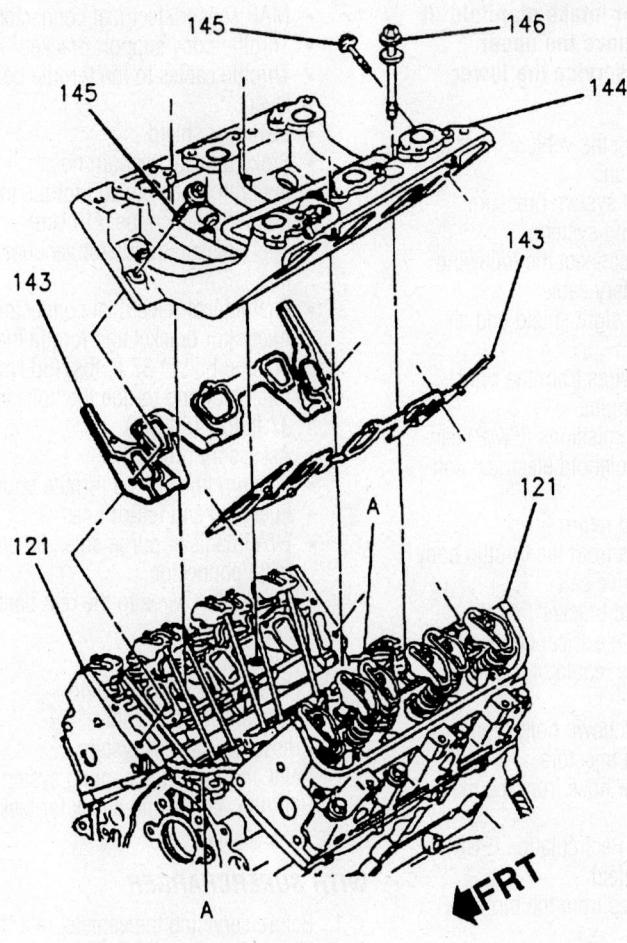

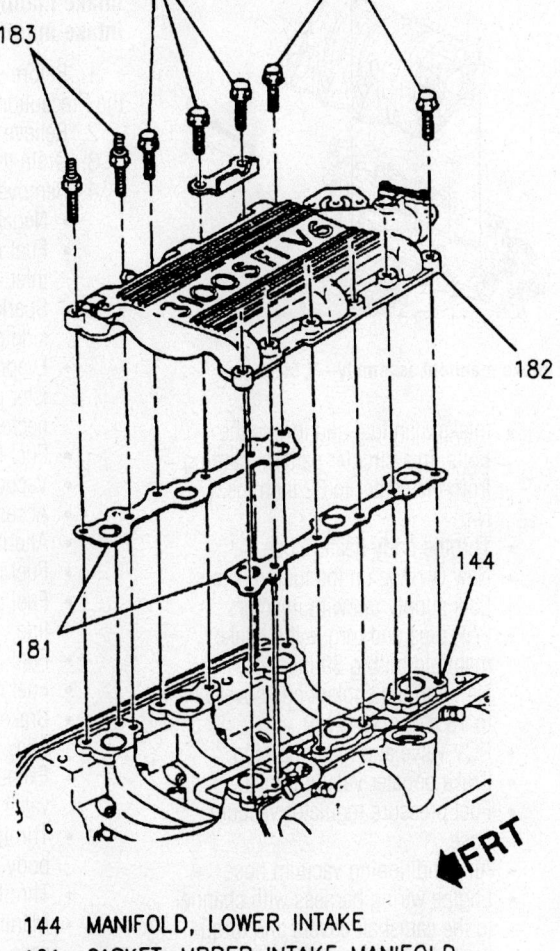

A   APPLY SEALANT
121 HEAD ASSEMBLY, CYLINDER
143 GASKET, LOWER INTAKE MANIFOLD
144 BOLT, LOWER INTAKE
145 BOLT, LOWER INTAKE MANIFOLD
146 BOLT, LOWER INTAKE MANIFOLD

144 MANIFOLD, LOWER INTAKE
181 GASKET, UPPER INTAKE MANIFOLD
182 MANIFOLD, UPPER INTAKE
183 STUD, UPPER INTAKE MANIFOLD
184 BOLT, UPPER INTAKE MANIFOLD

79227310

Lower intake manifold torque sequence—3.1L and 3.4L engines

➡**An engine oil and filter change is recommended.**

8. Start the vehicle and check for leaks, repair if necessary.

3.5L Engine

1. Before servicing the vehicle, refer to the Precautions Section.
2. Partially drain the engine coolant.
3. Remove or disconnect the following:
   - Negative battery cable
   - Throttle body air inlet duct
   - Fuel injector sight shield
   - Throttle and cruise control cables from the throttle body with the bracket
   - Coolant hoses from the throttle body
   - Fuel lines from the fuel supply rail

   - Fuel vapor line from the Evaporative Emission (EVAP) canister purge solenoid
   - Brake booster vacuum hose
   - Air conditioning vacuum hose from the engine
   - Surge tank inlet pipe retainer from the fuel supply rail
   - Fuel injector electrical connectors
   - Throttle Position Sensor (TPS) electrical connectors
   - Idle Air Control (IAC) valve electrical connector
   - EVAP canister purge solenoid connector
   - Manifold Absolute Pressure (MAP) sensor connector
   - Wiring harness channels from the camshaft covers

   - Vacuum tube from the fuel pressure regulator
   - Positive Crankcase Ventilation (PCV) valve and both feed tubes
   - Exhaust Gas Recirculation (EGR) valve outlet pipe
   - Fuel supply rail with injectors

➡**Disengage the snap-lock retainers by pushing toward the camshaft covers and lifting.**

   - Throttle body coolant hose
   - Intake manifold

➡**The manifold-to-cylinder head seals are reusable unless cut or damaged.**

To install:
4. Install or connect the following:
   - New intake manifold-to-cylinder head seals

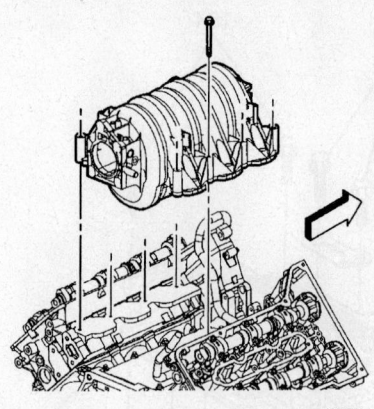

Intake manifold assembly—3.5L engine

- Intake manifold and torque the bolts, in a circular pattern, starting from the center to 62 inch lbs. (7 Nm)
- Throttle body heater hose
- New O-rings on the fuel injectors
- Fuel supply rail with injectors
- EGR pipe and torque the intake manifold bolt to 89 inch lbs. (10 Nm) and the coolant crossover bolt to 18 ft. lbs. (24 Nm)
- PCV valve and related tubing
- Brake booster vacuum hose
- Fuel pressure regulator vacuum hose
- Air conditioning vacuum hose
- Engine wiring harness with channel to the camshaft covers and torque the bolts to 89 inch lbs. (10 Nm)
- Surge tank pipe retainer to the fuel supply rail
- TPS electrical connector
- IAC electrical connector
- EVAP solenoid electrical connector
- MAP sensor electrical connector
- Fuel injector electrical connectors
- Coolant hoses to the throttle body
- Throttle and cruise control cables to the throttle body
- Fuel lines to the fuel supply rail
- Fuel vapor line to the purge solenoid
- Fuel injector sight shield
- Throttle body air inlet duct
- Negative battery cable

5. Refill the cooling system.

6. Start the engine and check for leaks, repair if necessary.

3.8L Engine

WITHOUT SUPERCHARGER

➡There are 2 bolts hidden beneath the upper intake manifold. These bolts are located in the right front and left rear corners of the lower intake manifold. It is necessary to remove the upper intake manifold to service the lower intake manifold.

1. Before servicing the vehicle, refer to the Precautions Section.
2. Relieve the fuel system pressure.
3. Drain the cooling system.
4. Remove or disconnect the following:
   - Negative battery cable
   - Fuel injector sight shield and air inlet duct
   - Spark plug wires from the right side of the engine
   - Evaporative emissions (EVAP) canister purge solenoid electrical connector
   - Fuel feed and return lines
   - Vacuum lines from the throttle body
   - Accessory drive belt
   - Alternator and bracket
   - Fuel injector electrical connectors
   - Fuel pressure regulator vacuum line
   - Fuel rail hold down bolts
   - Fuel rail with injectors
   - Brake booster hose from the manifold
   - Exhaust Gas Recirculation (EGR) valve heat shield
   - Throttle cables from the throttle body lever
   - Throttle body support bracket
   - Manifold Absolute Pressure (MAP) sensor electrical connector
   - Upper intake plenum
   - Drive belt tensioner
   - EGR valve outlet pipe
   - Upper radiator hose
   - Engine Coolant Temperature (ECT) sensor electrical connector
   - Evaporative emissions canister purge solenoid valve
   - Alternator brace
   - Lower intake manifold

To install:

5. Install or connect the following:
   - New intake manifold gaskets
   - Intake manifold and torque the bolts, in sequence, to 11 ft. lbs. (15 Nm)
   - Evaporative emissions canister purge solenoid valve
   - Alternator brace
   - ECT sensor electrical connector
   - Upper radiator hose
   - EGR valve outlet pipe
   - Drive belt tensioner and torque the bolts to 37 ft. lbs. (50 Nm)
   - Intake plenum using a new gasket and torque the bolts, in sequence, to 89 inch lbs. (10 Nm)
   - MAP sensor electrical connector
   - Throttle body support bracket
   - Throttle cables to the throttle body lever
   - EGR heat shield
   - Brake booster vacuum hose
   - Fuel rail assembly and torque the bolts to 84 inch lbs. (10 Nm)
   - Fuel pressure regulator vacuum line
   - Fuel injector electrical connectors
   - Alternator bracket and torque the nut and bolt to 37 ft. lbs. (50 Nm)
   - Alternator and torque the bolts to 37 ft. lbs. (50 Nm)
   - Accessory drive belt
   - Vacuum lines to the throttle body
   - Fuel feed and return lines
   - EVAP canister purge solenoid electrical connector
   - Spark plug wires to the rear bank spark plugs
   - Air inlet duct
   - Fuel injector sight shield
   - Negative battery cable

6. Pressurize the fuel system.
7. Refill and bleed the cooling system.
8. Start the vehicle and check for leaks, repair if necessary.

WITH SUPERCHARGER

1. Before servicing the vehicle, refer to the Precautions Section.
2. Relieve the fuel system pressure.
3. Drain the cooling system.
4. Remove or disconnect the following:
   - Negative battery cable
   - Exhaust Gas Recirculation (EGR) valve heat shield
   - Supercharger drive belt
   - Right side spark plug wires
   - Alternator rear brace
   - Fuel injector electrical connectors
   - Manifold Absolute Pressure (MAP) sensor and bracket
   - Fuel feed and return lines
   - Fuel rail with the injectors
   - Boost control solenoid
   - Cruise control and throttle cables
   - Air inlet duct
   - Supercharger
   - Upper radiator hose
   - Thermostat housing
   - EGR tube
   - Temperature sensor electrical connector
   - Intake manifold

To install:

5. Install or connect the following:
   - Intake manifold with new gaskets and torque the bolts to 11 ft. lbs. (15 Nm)

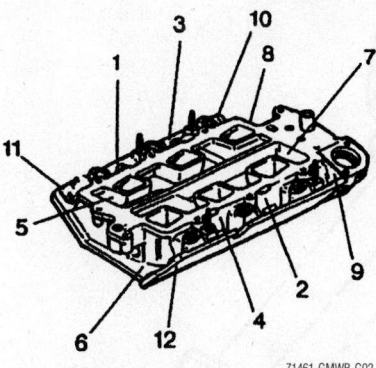

Always tighten the lower intake manifold bolts in sequence—3.8L engines

71461-GMWB-G02

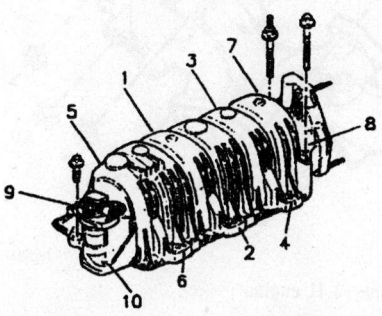

Upper intake manifold bolt tightening sequence—3.8L engines

71461-GMWB-G03

- Temperature sensor electrical connector
- EGR tube
- Thermostat housing with a new gasket and torque the bolts to 20 ft. lbs. (27 Nm)
- Upper radiator hose
- Supercharger with new gaskets and torque the bolts to 17 ft. lbs. (23 Nm)
- Air inlet duct
- Cruise control and throttle cables
- Boost control solenoid and torque the bolts to 72 inch lbs. (8 Nm)
- Fuel rail and torque the hold down bolts to 7 ft. lbs. (10 Nm)
- Fuel feed and return lines
- MAP sensor and bracket
- Fuel injector electrical connectors
- Alternator rear brace and torque the bolts to 37 ft. lbs. (50 Nm)
- Spark plug wires
- Supercharger drive belt
- EGR valve heat shield
- Fuel injector sight shield
- Negative battery cable

6. Fill the cooling system.
7. Start the engine and check for leaks.

5.3L Engine

1. Before servicing the vehicle, refer to the Precautions Section.

➡**The intake manifold, throttle body, fuel rail, and injectors may be removed as an assembly. If not servicing the individual components, remove the manifold as a complete assembly.**

2. Remove the throttle body.
3. Remove the fuel injectors.
4. Remove the brake booster vacuum hose from the intake manifold and the booster check valve.
5. Remove the positive crankcase ventilation (PCV) clean air tube from the air intake duct and valve rocker arm cover.
6. Remove the PCV foul air tube from the intake manifold and valve rocker arm cover.
7. Remove the evaporative emission (EVAP) purge solenoid tube (2) from the solenoid and reposition.
8. Disconnect the following electrical connectors:
- EVAP purge solenoid
- Manifold absolute pressure (MAP) sensor
- Oil pressure sensor
- Valve lifter oil manifold

9. Remove the MAP sensor from the intake manifold.
10. Remove the EVAP purge solenoid valve and bracket.
11. Remove the intake manifold bolts.
12. Remove the intake manifold.

13. Remove and discard the intake manifold gaskets.

To install:
14. Install the NEW intake manifold gaskets.
15. Install the intake manifold.
16. Apply a band of threadlock GM P/N 12345382 (Canadian P/N 10953489) or equivalent to the threads of the intake manifold bolts.
17. Install the intake manifold bolts. Tighten the bolts a first pass in sequence to 5 Nm (44 inch lbs.). Tighten the bolts a second pass in sequence to 10 Nm (89 inch lbs.).
18. Install the EVAP purge solenoid bracket and solenoid.
19. Install the MAP sensor to the intake manifold.
20. Connect the following electrical connectors:
- EVAP purge solenoid
- MAP sensor
- Oil pressure sensor
- Valve lifter oil manifold

21. Position and install the EVAP purge solenoid tube to the solenoid.
22. Install the PCV foul air tube to the intake manifold and valve rocker arm cover.
23. Install the PCV clean air tube to the air intake duct and valve rocker arm cover.
24. Install the brake booster vacuum hose to the intake manifold and the booster check valve.
25. Install the fuel injectors.
26. Install the throttle body.

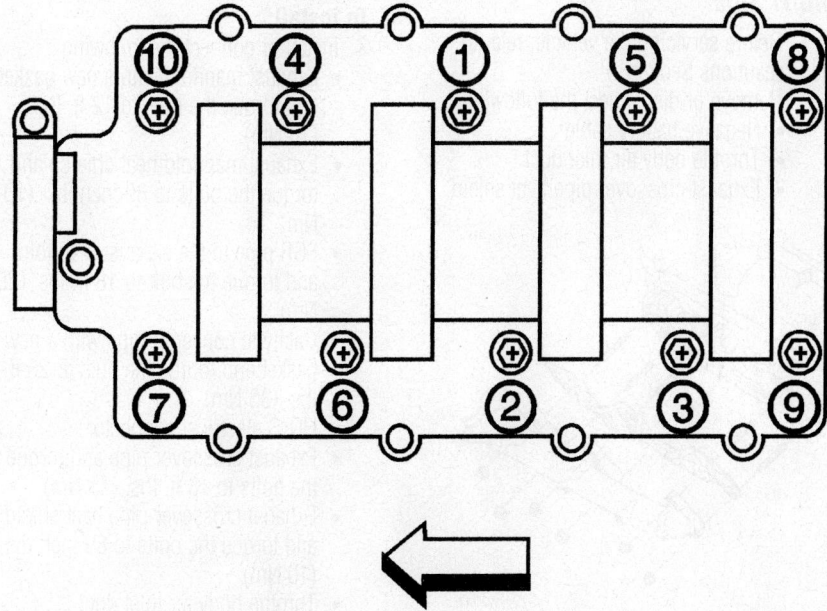

Intake manifold bolt torque sequence—5.3L engine

06025-GMWB-G10

Exhaust Manifold

REMOVAL & INSTALLATION

3.1L Engine

LEFT SIDE

1. Before servicing the vehicle, refer to the Precautions Section.
2. Remove or disconnect the following:
   - Negative battery cable
   - Throttle body air inlet duct
   - Right engine mount strut bracket
   - Crossover pipe heat shield
   - Crossover pipe nuts to the left exhaust manifold
   - Exhaust manifold heat shield
   - Exhaust manifold

To install:

3. Install or connect the following:
   - Exhaust manifold with a new gasket and torque the nuts to 12 ft. lbs. (16 Nm)
   - Exhaust manifold heat shield and torque the nuts to 89 inch lbs. (10 Nm)
   - Exhaust crossover pipe to the manifold and torque the bolt to 18 ft. lbs. (25 Nm)
   - Crossover pipe heat shield and torque the bolt to 89 inch lbs. (10 Nm)
   - Left side engine mount strut bracket and torque the bolt to 37 ft. lbs. (50 Nm)
   - Throttle body air inlet duct
   - Negative battery cable

RIGHT SIDE

1. Before servicing the vehicle, refer to the Precautions Section.
2. Remove or disconnect the following:
   - Negative battery cable
   - Throttle body air inlet duct
   - Exhaust crossover pipe heat shield

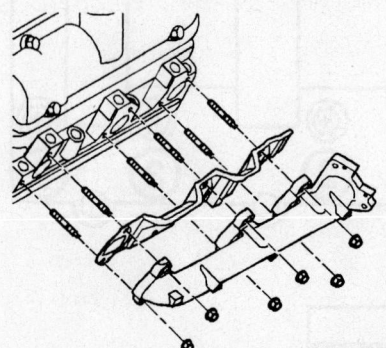

Exploded view of the left exhaust manifold mounting—3.1L engine

79222Z513

- Exhaust crossover pipe-to-right exhaust manifold nuts
- Heated Oxygen Sensor (HO2S) electrical connector
- Catalytic converter pipe from the exhaust manifold
- Exhaust Gas Recirculation (EGR) pipe from the exhaust manifold
- Exhaust manifold heat shields
- Exhaust manifold

To install:

3. Install or connect the following:
   - Exhaust manifold with a new gasket and torque the nuts to 12 ft. lbs. (16 Nm)
   - Exhaust manifold heat shields and torque the bolts to 89 inch lbs. (10 Nm)
   - EGR pipe to the exhaust manifold and torque the bolt to 18 ft. lbs. (25 Nm)
   - Catalytic converter pipe with a new gasket and torque the nuts to 26 ft. lbs. (35 Nm).
   - HO2S electrical connector
   - Exhaust crossover pipe and torque the bolts to 18 ft. lbs. (25 Nm)
   - Exhaust crossover pipe heat shield and torque the bolts to 89 inch lbs. (10 Nm)
   - Throttle body air inlet duct
   - Negative battery cable

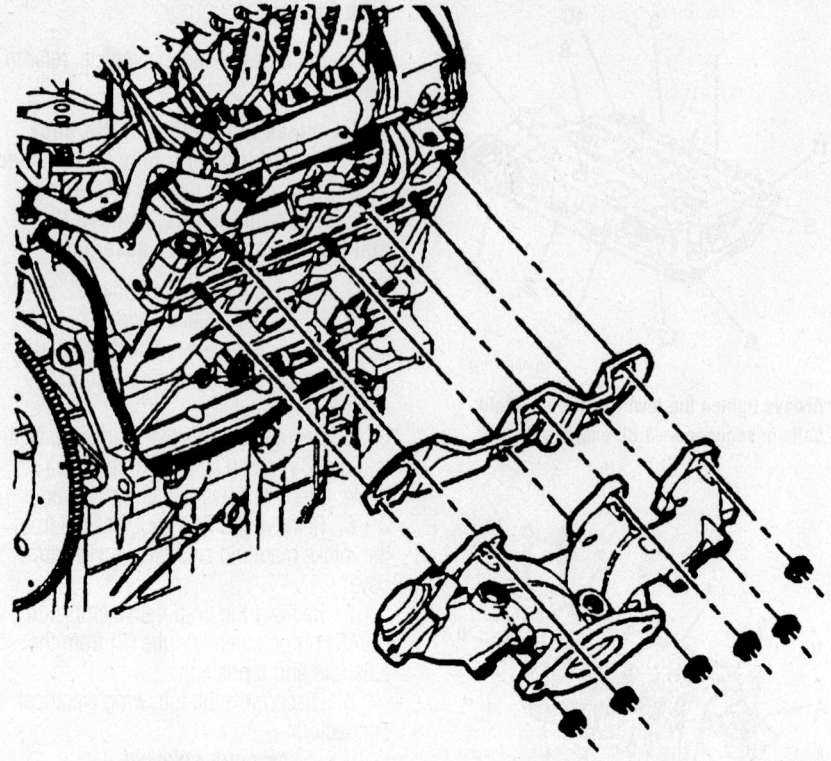

Exploded view of the right exhaust manifold mounting—3.1L engine

79222Z514

3.4L Engine

LEFT SIDE

1. Before servicing the vehicle, refer to the Precautions Section.
2. Remove or disconnect the following:
   - Negative battery cable
   - Air cleaner
   - Throttle body air inlet duct
   - Right engine mount strut bracket
   - Exhaust crossover pipe heat shield
   - Exhaust crossover pipe nuts to the left exhaust manifold
   - Exhaust manifold heat shield
   - Exhaust manifold

To install:

3. Install or connect the following:
   - Exhaust manifold with a new gasket and torque the bolts to 12 ft. lbs. (16 Nm)
   - Exhaust manifold heat shield and torque the bolts to 89 inch lbs. (10 Nm)
   - Crossover pipe and torque the bolts to 18 ft. lbs. (25 Nm)
   - Crossover pipe heat shield and torque the bolts to 89 inch lbs. (10 Nm)
   - Right engine mount strut bracket
   - Negative battery cable
   - Throttle body air inlet duct
   - Air cleaner

4. Start the vehicle and check for leaks, repair if necessary.

RIGHT SIDE

1. Before servicing the vehicle, refer to the Precautions Section.
2. Remove or disconnect the following:
   - Negative battery cable
   - Throttle body air inlet duct
   - Accelerator cable bracket from the throttle body
3. Rotate the engine for service access.
   - Manifold Air Pressure (MAP) sensor
   - Exhaust Gas Recirculation (EGR) pipe
   - Ignition module and the ignition coils with the bracket
   - Spark plug wires from the plugs
   - Heated Oxygen Sensor (HO$_2$S) electrical connector
   - Evaporative Emissions (EVAP) solenoid bracket
   - Exhaust crossover heat shield
   - Exhaust crossover pipe from the manifold
   - Catalytic converter from the manifold
   - Exhaust manifold heat shields
   - Exhaust manifold

To install:

4. Install or connect the following:
   - Exhaust manifold with a new gasket and torque the nuts to 12 ft. lbs. (16 Nm)
   - Exhaust manifold heat shields and torque the bolts to 89 inch lbs. (10 Nm)
   - Catalytic converter and torque the nuts to 24 ft. lbs. (32 Nm)
   - Crossover pipe to the manifold and torque the bolts to 18 ft. lbs. (25 Nm)
   - Crossover pipe heat shield and torque the bolts to 89 inch lbs. (10 Nm)
   - EVAP solenoid bracket
   - HO$_2$S electrical connector
   - Spark plug wires from the plugs
   - Ignition module, ignition coils and bracket
   - EGR pipe
   - MAP sensor
5. Rotate the engine back to the original position.
   - Accelerator cable bracket to the throttle body
   - Throttle body air inlet duct
   - Negative battery cable
6. Start the vehicle and check for leaks, repair if necessary.

3.5L Engine

LEFT SIDE

1. Before servicing the vehicle, refer to the Precautions Section.
2. Drain the cooling system.
3. Remove or disconnect the following:
   - Negative battery cable
   - Fuel injector sight shield
   - Engine mount strut and bracket
   - Cooling fans
   - Upper and lower radiator hoses
   - Transaxle oil cooler lines from the radiator
   - Radiator
   - Alternator
   - Exhaust manifold heat shield
   - Oil level indicator tube
   - Secondary Air Injection (AIR) control valve assembly from the engine mount strut bracket
   - Exhaust manifold-to-crossover pipe studs
   - Exhaust manifold

To install:

4. Install or connect the following:
   - Manifold to the crossover pipe using a new gasket and torque the studs to 18 ft. lbs. (25 Nm)
   - Exhaust manifold with a new gasket and torque the bolts to 18 ft. lbs. (25 Nm)
   - AIR valve and torque the pipe nut to 44 ft. lbs. (60 Nm) and the bolt to 80 inch lbs. (9 Nm)
   - Oil level indicator tube and torque the bolt to 80 inch lbs. (9 Nm)
   - Exhaust manifold heat shield and torque the bolts to 80 inch lbs. (9 Nm)
   - Alternator and torque the bolts to 37 ft. lbs. (50 Nm)
   - Radiator and torque the bolts to 89 inch lbs. (10 Nm)
   - Transaxle oil cooler lines and torque the fitting to 17 ft. lbs. (23 Nm)
   - Upper and lower radiator hoses
   - Cooling fans and torque the bolts to 89 inch lbs. (10 Nm)
   - Engine mount strut bracket and torque the bolts to 37 ft. lbs. (50 Nm)
   - Engine mount strut and torque the bolts to 35 ft. lbs. (48 Nm)
   - Fuel injector sight shield and torque the nuts to 27 inch lbs. (3 Nm)
   - Negative battery cable
5. Fill the cooling system.
6. Start the vehicle and check for leaks, repair if necessary.

RIGHT SIDE

1. Before servicing the vehicle, refer to the Precautions Section.
2. Relieve the fuel system pressure.
3. Drain the cooling system.
4. Remove or disconnect the following:
   - Negative battery cable
   - Fuel injector cover
   - Air intake duct
   - Engine mount strut
   - Fuel lines from the supply rail
   - Cruise control and accelerator cables from the throttle body
   - Transaxle selector range cable and cable brackets
   - Transaxle shift cable from the shift module
   - Brake booster vacuum hose from the engine
   - Wiring harness connectors from the engine and transaxle
   - Upper radiator hose
   - Transaxle oil cooler lines from the radiator
   - Surge tank inlet hose
   - Heater hoses from the engine
   - Lower radiator air deflector
   - Battery cables from the retainers
   - Lower radiator hose
   - A/C compressor without disconnecting the lines and move it aside
   - Starter wiring
   - Catalytic converter from the manifold
   - Front wheels and splash shields
   - Wheel Speed Sensor (WSS) wiring from the lower control arms
   - Tie rod ends from the steering knuckles
   - Lower ball joints from the knuckles
   - Halfshafts
   - Intermediate shaft from the steering rack
5. Secure the vehicle to the lift in preparation for engine removal.
6. Position an engine/frame support

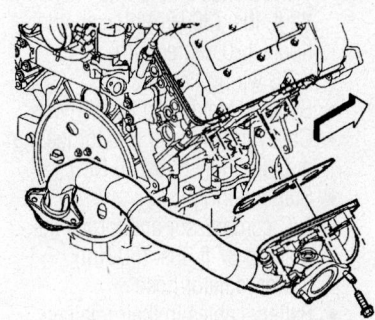

9300Z507

Exploded view of the right exhaust manifold—3.5L engine

table under the vehicle and lower the vehicle to meet the table.

7. Remove or disconnect the following:
- Frame-to-body bolts
- Crossover pipe from the front manifold
- Exhaust Gas Recirculation (EGR) pipe from the crossover pipe
- Right exhaust manifold from the engine

To install:

8. Install or connect the following:
- Right exhaust manifold with a new gasket and torque the bolts to 18 ft. lbs. (25 Nm)
- Crossover pipe to the front exhaust manifold and torque the bolts to 18 ft. lbs. (25 Nm)
- EGR pipe to the crossover pipe and torque the pipe nut to 44 ft. lbs. (60 Nm)

9. Position the engine/transaxle assembly under the vehicle.

10. Coat the sub-frame bushings with rubber lubricant.

11. Lower the vehicle onto the assembly. Align the sub-frame on the vehicle using 2 bolts or drill bits, ¾ inches thick by 8 inches long through the alignment holes on the right side of the frame.

12. Install new frame-to-body bolts. Torque the bolts to 133 ft. lbs. (180 Nm) starting with the rear bolts and then the front bolts.

13. Raise the vehicle and remove the frame table.

14. Install or connect the following:
- Intermediate shaft to the steering rack and torque the bolts to 35 ft. lbs. (48 Nm)

➡**Be sure the shaft is fully seated on the stub before installing the pinch bolt.**

- Halfshafts
- Ball joints and torque the nuts to 40 ft. lbs. (55 Nm)
- Tie rod ends and torque the nuts to 22 ft. lbs. (30 Nm) plus an additional 120 degree turn
- WSS wiring harness
- Splash shields and front wheels
- Catalytic converter and torque the nuts to 53 inch lbs. (6 Nm)
- Starter wiring
- A/C compressor and torque the bolts to 37 ft. lbs. (50 Nm)
- Lower radiator hose
- Battery cables in their retainers
- Lower radiator air deflector and torque the bolts to 15 ft. lbs. (20 Nm)

15. Remove the straps securing the vehicle to the lift.
- Heater hoses
- Surge tank hose
- Transaxle oil cooler lines to the radiator and torque the fittings to 17 ft. lbs. (23 Nm)
- Upper radiator hose
- Wiring harness connectors to the engine and transaxle
- Transaxle shift cable to the shift module and bracket
- Transaxle range selector cable
- Cruise control and accelerator cables to the throttle body
- Fuel lines to the supply rail
- Engine mount strut and torque the bolts to 35 ft. lbs. (48 Nm)
- Air inlet duct
- Fuel injector sight shield
- Negative battery cable

16. Refill the cooling system.

17. Start the engine and check for leaks, repair if necessary.

3.8L Engines

LEFT SIDE

1. Before servicing the vehicle, refer to the Precautions Section.

2. Remove or disconnect the following:
- Negative battery cable
- Fuel injector sight shield
- Engine mount struts

- Exhaust crossover from the manifold
- Spark plug wires from the plugs
- Oil level indicator tube
- Exhaust manifold heat shield
- Engine lift hook
- Exhaust manifold

To install:

3. Install or connect the following:
- Exhaust manifold with a new gasket and torque the bolts to 22 ft. lbs. (30 Nm)
- Engine lift hook and torque the bolts to 22 ft. lbs. (30 Nm)
- Exhaust manifold heat shield and torque the bolts to 15 ft. lbs. (20 Nm).
- Oil level indicator tube and torque the nut to 14 ft. lbs. (19 Nm)
- Spark plug wires
- Exhaust crossover and torque the bolts to 15 ft. lbs. (20 Nm)
- Engine mount struts and torque the bolts to 35 ft. lbs. (48 Nm)
- Fuel injector sight shield
- Negative battery cable

4. Start the vehicle and check for leaks, repair if necessary.

RIGHT SIDE

1. Before servicing the vehicle, refer to the Precautions Section.

2. Remove or disconnect the following:
- Negative battery cable

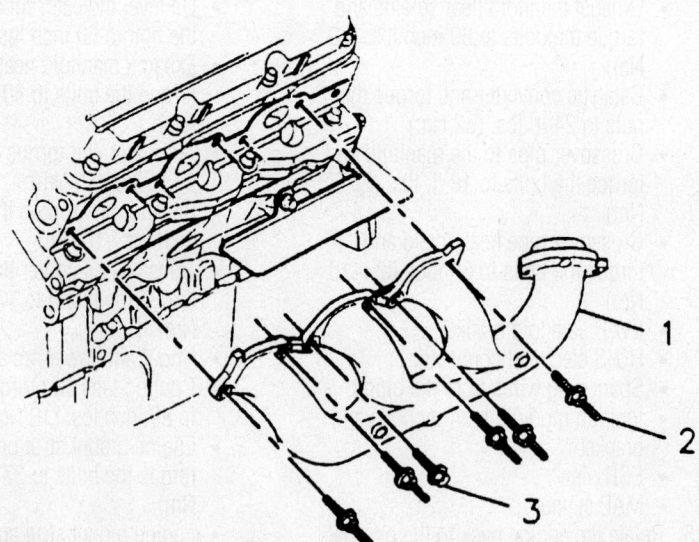

1   LEFT (FRONT) EXHAUST MANIFOLD
2   STUD 30 N•m (22 LB. FT.)
3   BOLT 30 N•m (22 LB. FT.)

7922XG30

Exploded view of the left exhaust manifold mounting—3.8L engine

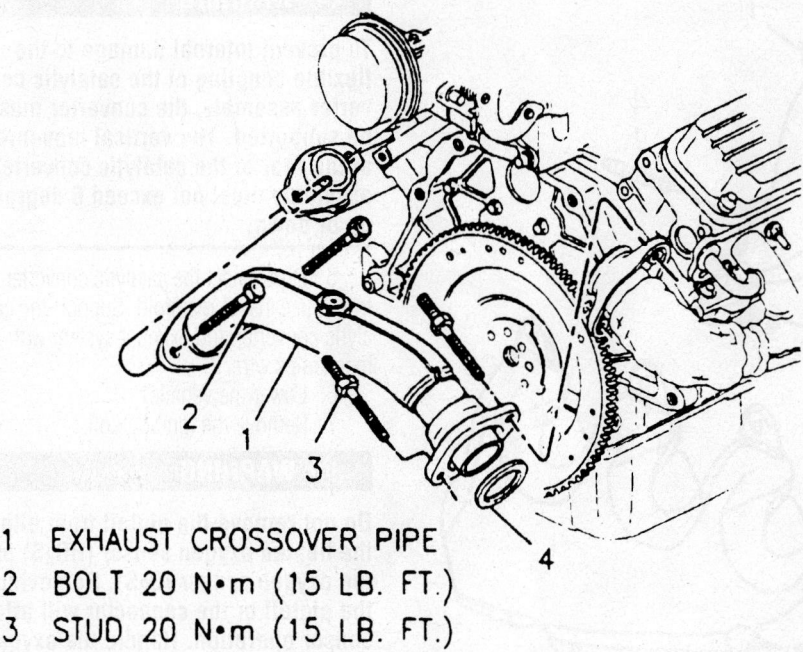

1   EXHAUST CROSSOVER PIPE
2   BOLT 20 N•m (15 LB. FT.)
3   STUD 20 N•m (15 LB. FT.)
4   SEAL

7922XG31

Exploded view of the crossover pipe mounting—3.8L engine

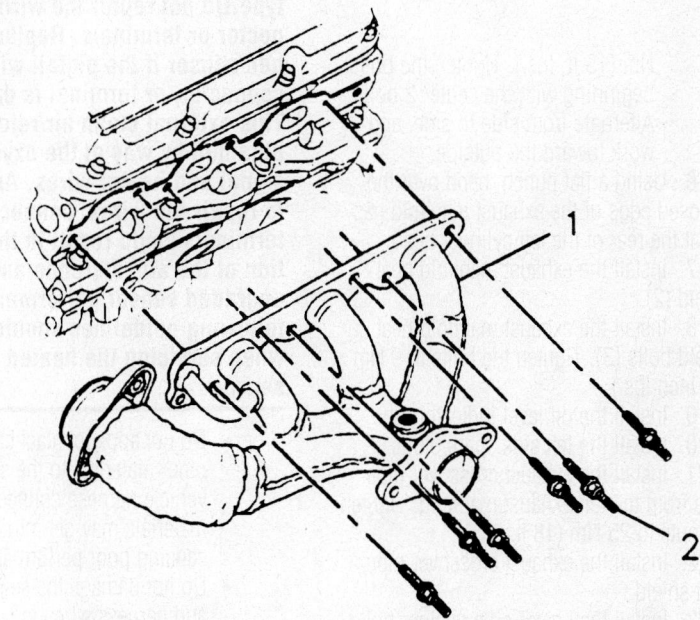

1   RIGHT (REAR) EXHAUST MANIFOLD
2   STUD 30 N•m (22 LB. FT.)

7922XG29

Exploded view of the right exhaust manifold mounting—3.8L engine

- Fuel injector sight shield
- Cross vehicle brace
- Exhaust crossover pipe from the right exhaust manifold
- AIR pipe, if equipped
- Exhaust manifold pipe stud nuts and the catalytic converter pipe from the right exhaust manifold
- Exhaust Gas Recirculation (EGR) adapter pipe from the manifold

3. Rotate the engine to access the right side of the engine.
- Heated Oxygen Sensor (HO2S) electrical connector
- Spark plug wires from the plugs
- Right engine lift bracket
- Fuel injector sight shield bracket
- Transaxle fluid filler tube
- Exhaust manifold

To install:
4. Install or connect the following:
- Exhaust manifold with a new gasket and torque the bolts to 22 ft. lbs. (30 Nm)
- Engine lift bracket and torque the bolts to 22 ft. lbs. (30 Nm)
- Transaxle fluid filler tube
- Fuel injector sight shield bracket and torque the bolts to 22 ft. lbs. (30 Nm)
- Spark plug wires
- Heated Oxygen Sensor (HO2S) electrical connector

5. Rotate the engine back to the original position.
- EGR adapter pipe and torque the bolt to 21 ft. lbs. (29 Nm)
- Catalytic converter and torque the bolts to 26 ft. lbs. (35 Nm)
- Exhaust crossover pipe and torque the bolts to 13 ft. lbs. (18 Nm)
- AIR pipe, if equipped
- Cross vehicle brace
- Fuel injector sight shield
- Negative battery cable

6. Start the vehicle and check for leaks, repair if necessary.

5.3L Engine

LEFT SIDE

1. Before servicing the vehicle, refer to the Precautions Section.
2. Remove the engine sight shield.
3. Remove the exhaust crossover pipe heat shield in order to access the cross pipe to the exhaust manifold nuts.
4. Remove the exhaust crossover pipe nuts from the left exhaust manifold.
5. Remove the left side spark plugs.
6. Remove the oil level indicator tube.
7. Remove the exhaust manifold heat shield bolts (3).
8. Remove the exhaust manifold heat shield (2).
9. Remove the heater hose retainer bolts in order to be able to remove the manifold.
10. Remove the exhaust manifold bolts.
11. Remove the exhaust manifold (1).
12. Remove and discard the exhaust manifold gasket.

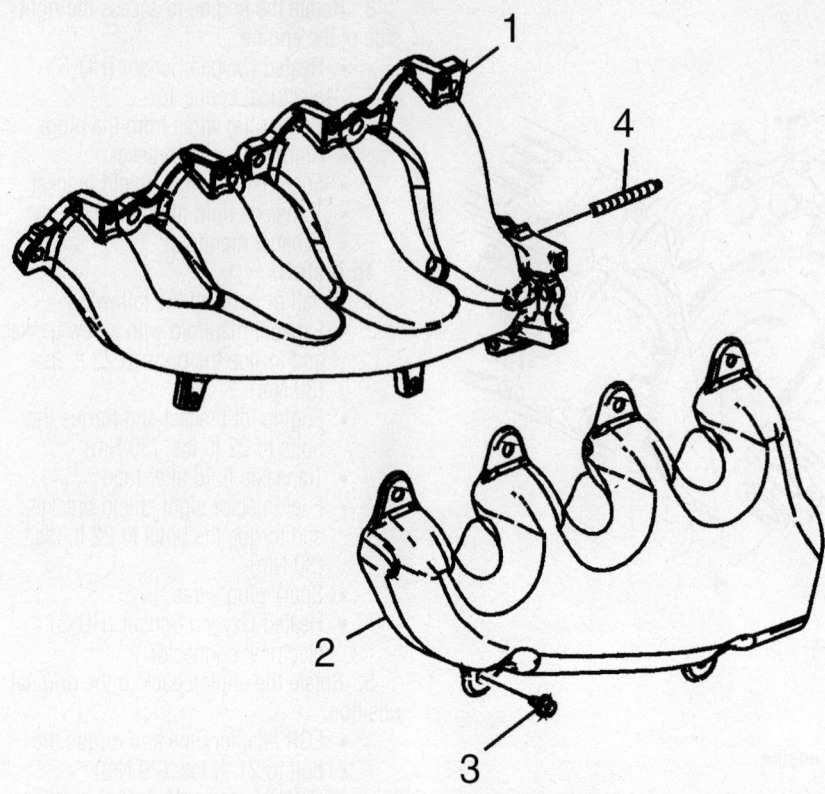

06025-GMWB-G11

Left side exhaust manifold—5.3L engine

13. Clean and inspect the left exhaust manifold.

To install:

> ❊❊ **WARNING**
>
> Tighten the exhaust manifold bolts as specified in the service procedure. Improperly installed and/or leaking exhaust manifold gaskets my affect vehicle emissions and/or On Board Diagnostic (OBD) II system performance. The cylinder head exhaust manifold bolt hole threads must be clean and free of debris or threadlocking material. **DO NOT** apply threadlock to the first three thread of the bolts.

14. Apply a 5 mm (0.2 in.) wide band of threadlock GM P/N 12345493 (Canadian P/N 10953488) or equivalent to the threads of the exhaust manifold bolts.

15. Install a NEW exhaust manifold gasket, the exhaust manifold and bolts. Tighten as follows:

- Tighten the bolts a first pass to 15 Nm (11 ft. lbs.). Tighten the bolts beginning with the center 2 bolts. Alternate from side to side, and work toward the outside.
- Tighten the bolts a final pass to 20 Nm (15 ft. lbs.). Tighten the bolts beginning with the center 2 bolts. Alternate from side to side, and work toward the outside.

16. Using a flat punch, bend over the exposed edge of the exhaust manifold gasket at the rear of the left cylinder head.

17. Install the exhaust manifold heat shield (2).

18. Install the exhaust manifold heat shield bolts (3). Tighten the bolts to 9 Nm (80 inch lbs.).

19. Install the oil level indicator tube.

20. Install the left side spark plugs.

21. Install the exhaust crossover pipe nuts from the left exhaust manifold. Tighten the nuts to 25 Nm (18 ft. lbs.).

22. Install the exhaust crossover pipe heat shield.

23. Install the heater pipe retainer bolts. Tighten the bolts to 16 Nm (12 ft. lbs.).

24. Install the engine sight shield.

RIGHT SIDE

1. Before servicing the vehicle, refer to the Precautions Section.

2. Remove the engine sight shield.

3. Remove the spark plugs.

4. Remove the exhaust crossover pipe nuts from the right exhaust manifold.

> ❊❊ **WARNING**
>
> To prevent internal damage to the flexible coupling of the catalytic converter assembly, the converter must be supported. The vertical movement at the rear of the catalytic converter assembly must not exceed 6 degrees up or down.

5. Disconnect the catalytic converter from the exhaust manifold. Support the catalytic converter and exhaust system with the mechanics wire.

6. Lower the vehicle.

7. Remove the ignition coil.

> ❊❊ **WARNING**
>
> Do not remove the pigtail from either the heated oxygen sensor (HO$_2$S) or the oxygen sensor (O$_2$S). Removing the pigtail or the connector will affect sensor operation. Handle the oxygen sensor carefully. Do not drop the HO$_2$S. Keep the in-line electrical connector and the louvered end free of grease, dirt, or other contaminants. Do not use cleaning solvents of any type.Do not repair the wiring, connector or terminals. Replace the oxygen sensor if the pigtail wiring, connector, or terminal is damaged. This external clean air reference is obtained by way of the oxygen sensor signal and heater wires. Any attempt to repair the wires, connectors, or terminals could result in the obstruction of the air reference and degraded sensor performance. The following guidelines should be used when servicing the heated oxygen sensor:

- Do not apply contact cleaner or other materials to the sensor or vehicle harness connectors. These materials may get into the sensor causing poor performance.
- Do not damage the sensor pigtail and harness wires in such a way that the wires inside are exposed. This could provide a path for foreign materials to enter the sensor and cause performance problems.
- Ensure the sensor or vehicle lead wires should not be bent sharply or kinked. Sharp bends or kinks could block the reference air path through the lead wire.
- Do not remove or defeat the oxygen sensor ground wire (where applica-

ble). Vehicles that utilize the ground wired sensor may rely on this ground as the only ground contact to the sensor. Removal of the ground wire will cause poor engine performance.

- Ensure that the peripheral seal remains intact on the vehicle harness connector in order to prevent damage due to water intrusion. The engine harness may be repaired using Packard's Crimp and Splice Seals Terminal Repair Kit. Under no circumstances should repairs be soldered since this could result in the air reference being obstructed.

8. Remove the O₂ sensors.

9. Remove the exhaust manifold heat shield bolts.

10. Remove the exhaust manifold heat shield.

11. Remove the exhaust manifold bolts.

12. Remove the exhaust manifold.

13. Remove and discard the exhaust manifold gasket.

14. Clean and inspect the left exhaust manifold.

To install:

✳✳ WARNING

Tighten the exhaust manifold bolts as specified in the service procedure. Improperly installed and/or leaking exhaust manifold gaskets my affect vehicle emissions and/or On Board Diagnostic (OBD) II system performance. The cylinder head exhaust manifold bolt hole threads must be clean and free of debris or threadlocking material. DO NOT apply threadlock to the first three threads of the bolts.

15. Apply a 5 mm (0.2 in.) wide band of threadlock GM P/N 12345493 (Canadian P/N 10953488) or equivalent to the threads of the exhaust manifold bolts.

16. Install a NEW exhaust manifold gasket, the exhaust manifold and bolts. Tighten as follows:

- Tighten the bolts a first pass to 15 Nm (11 ft. lbs.). Tighten the bolts beginning with the center 2 bolts. Alternate from side to side, and work toward the outside.
- Tighten the bolts a final pass to 20 Nm (15 ft. lbs.). Tighten the bolts beginning with the center 2 bolts. Alternate from side to side, and work toward the outside.

17. Using a flat punch, bend over the exposed edge of the exhaust manifold gasket at the rear of the right cylinder head.

18. Install the exhaust manifold heat shield.

19. Install the exhaust manifold heat shield bolts. Tighten the bolts to 9 Nm (80 inch lbs.).

➡**A special anti-seize compound is used on the HO₂S threads. The compound consists of liquid graphite and glass beads. The graphite tends to burn away, but the glass beads remain, making the sensor easier to remove. New or service replacement sensors already have the compound applied to the threads. If the sensor is removed from an exhaust component and if for any reason the sensor is to be reinstalled, the threads must have anti-seize compound applied before reinstallation.**

20. If re-installing the old sensor, coat the threads with anti-seize compound P/N 12377953, or equivalent.

21. Install the bank 1 sensor 1 to the exhaust manifold. Tighten the sensor to 42 Nm (31 ft. lbs.).

22. Connect the bank 1 sensor 1 electrical connector.

23. Install the CPA retainer.

24. Raise the vehicle.

25. Install NEW gaskets to the exhaust manifold and muffler studs.

26. Position the catalytic converter into place.

27. Install the catalytic converter hangers to the converter.

28. Install the catalytic converter pipe stud nuts. Tighten the nuts to 35 Nm (26 ft. lbs.).

29. Install the catalytic converter nuts. Tighten the nuts to 60 Nm (44 ft. lbs.).

30. Remove the exhaust system support.

31. Install the exhaust crossover pipe nuts to the right exhaust manifold. Tighten the nuts to 25 Nm (18 ft. lbs.).

32. Apply threadlock GM P/N 12345382 (Canadian P/N 10953489) or equivalent to the threads of the ignition coil bolts.

33. Install the ignition coil(s), as necessary.

34. Install the ignition coil bolts, as necessary. Tighten the bolts to 8 Nm (71 inch lbs.).

35. Install the spark plug wire(s) to the ignition coil(s), as necessary.

36. Connect the left ignition coil main electrical connector.

37. Install the CPA retainer.

38. Install the spark plugs.

39. Install the fuel injector sight shield.

Front Crankshaft Seal

REMOVAL & INSTALLATION

3.1L, 3.4L and 3.8L Engines

1. Before servicing the vehicle, refer to the Precautions Section.

2. Remove the crankshaft balancer using an appropriate puller.

3. Pry the front crankshaft seal out.

To install:

4. Coat the new seal with engine oil to ease installation.

5. Install the front crankshaft seal using seal installer J-35468 on 3.1L and 3.4L engines. On 3.8L engines use tool J 35354. Make sure that the crankshaft front oil seal lip faces the engine.

6. Make sure the seal is flush with the engine when installed.

- Crankshaft balancer and torque the bolt to 76 ft. lbs. (103 Nm)

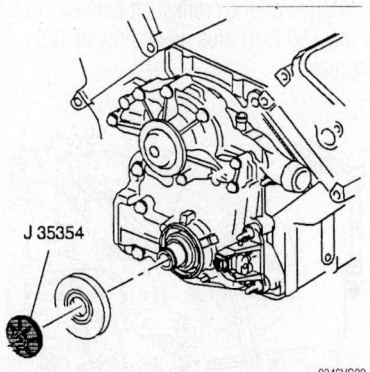

Installing the front crankshaft seal—3.1L, 3.4L and 3.8L engines

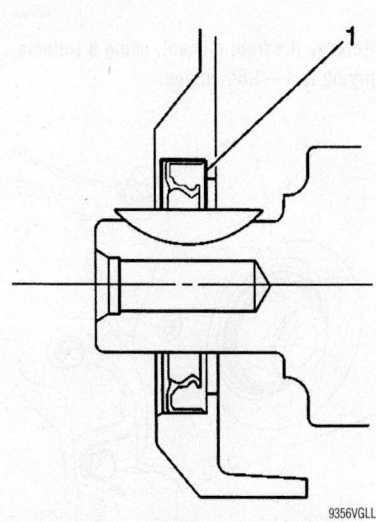

Make sure the crankshaft front oil seal is installed flush to the engine front cover (1)—3.1L, 3.4L and 3.8L engines

3.5L Engines

1. Before servicing the vehicle, refer to the Precautions Section.

2. Remove the crankshaft balancer using an appropriate puller.

3. Pry the front crankshaft seal out.

To install:

4. Coat the new seal with engine oil to ease installation.

5. Place the front seal onto the mandrel of the installation tool J-42041.

6. Slide the mandrel onto the screw of the installation tool.

7. Thread the installation tool in the crankshaft. Ensure you engage at least ten threads of the tool before pressing the seal in place.

8. Push the seal into position by tightening the nut on the installation tool until the mandrel bottoms out on the crankshaft sprocket.

9. Check the installed depth of the seal, which should protrude 1–2 mm (0.039–0.078 in.), from the front cover.

10. Install the crankshaft balancer to 37 ft. lbs. (50 Nm) plus an additional 125 degrees.

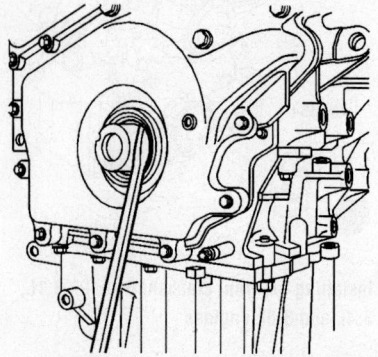

Remove the front oil seal, using a suitable prying tool—3.5L engine

9346VG04

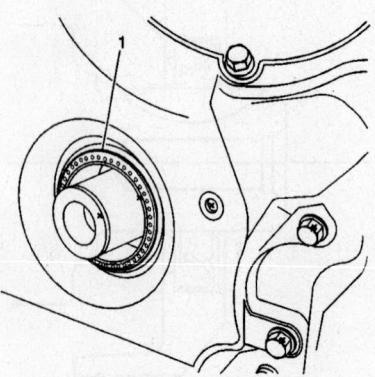

The installed depth of the seal (1) should protrude 1-2 mm (0.039-0.078 in.) from the front cover—3.5L engine

9346GV05

5.3L Engine

1. Before servicing the vehicle, refer to the Precautions Section.

2. Install a J 42640, Steering Column Anti-Rotation Pin.

3. Remove the accessory drive belt.

4. Remove the air cleaner upper housing.

5. Remove the engine mount strut. Refer to Engine Removal and Installation.

6. Remove the starter motor.

7. Remove the front fender splash shield.

8. Remove the transmission bellhousing bolt located at approximately the 10 o'clock position when looking from the rear of the engine.

9. Disconnect the transaxle cooler lines at the transaxle.

10. Remove the stabilizer shaft link lower nuts.

11. Remove the intermediate steering shaft pinch bolt and separate the shaft from the steering gear.

12. Remove the front lower air deflector braces and the deflector.

13. Remove the radiator to frame braces.

14. Install the engine support fixture. Refer to Engine Removal & Installation.

15. Raise and support the vehicle.

16. Remove the frame to body bolts.

17. Install a flywheel locking tool to the block and flywheel.

18. Remove the right front tire and wheel assembly.

19. Lower the engine approximately 100 mm (4 in.).

20. Remove the crankshaft balancer bolt. Do not discard the crankshaft balancer bolt. The balancer bolt will be used during the balancer installation procedure.

21. Install a puller to the crankshaft balancer.

22. Remove the crankshaft balancer.

23. Using a punch and pick, remove and discard the crankshaft oil seal.

To install:

➡**Do not lubricate the oil seal sealing surface. Do not reuse the crankshaft oil seal.**

24. Lubricate the outer edge of the oil seal (1) with clean engine oil.

25. Lubricate the front cover oil seal bore with clean engine oil.

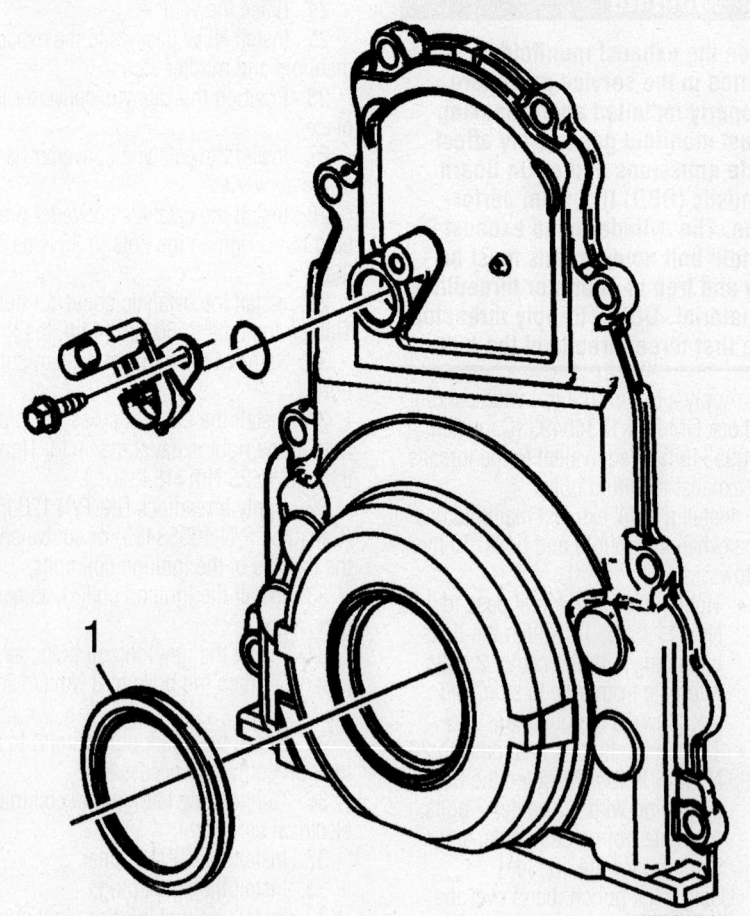

Crankshaft front oil seal removal—5.3L engine

06025-GMWB-G12

26. Install the crankshaft front oil seal (1) onto a driver such as J 41478.

27. Install the J 41478 threaded rod with nut, washer, guide, and oil seal into the end of the crankshaft.

28. Using the J 41478, install the oil seal into the cover bore.

   a. Use a wrench and hold the hex on the installer bolt.

   b. Use a second wrench and rotate the installer nut clockwise until the seal bottoms in the cover bore.

   c. Remove the J 41478.

   d. Inspect the oil seal for proper installation. The oil seal should be installed evenly and completely into the front cover bore.

➡**The used crankshaft balancer bolt will be used only during the first pass of the balancer installation procedure. Install a NEW bolt and tighten as described in the second, third and forth passes of the balancer bolt tightening procedure. The crankshaft balancer installation and bolt tightening involves a four stage tightening process. The first pass ensures that the balancer is installed completely onto the crankshaft. The second, third, and forth passes tighten the new bolt to the proper torque.**

➡**The balancer should be positioned onto the end of the crankshaft as straight as possible prior to tool installation.**

29. Position the crankshaft balancer onto the end of the crankshaft.

30. Using tool J 41665, install the crankshaft balancer.

   a. Assemble the threaded rod, nut, washer and installer. Insert the smaller end of the installer into the front of the balancer.

   b. Use a wrench and hold the hex end of the threaded rod.

   c. Use a second wrench and rotate the installation tool nut clockwise until the balancer is started onto the crankshaft.

   d. Remove the tool and reverse the installation tool. Position the larger end of the installer against the front of the balancer.

   e. Use a wrench and hold the hex end of the threaded rod.

   f. Use a second wrench and rotate the installation tool nut clockwise until the balancer is installed onto the crankshaft.

   g. Remove the balancer installation tool.

31. Install the USED crankshaft balancer bolt. Tighten the USED bolt to 330 Nm (240 lb ft).

32. Remove the USED crankshaft balancer bolt.

➡**The nose of the crankshaft should be recessed 2.4-4.48 mm (0.094-0.176 in) into the balancer bore.**

33. Measure for a correctly installed balancer. If the balancer is not installed to the proper dimensions, install the J 41665 and repeat the installation procedure.

34. Install a NEW crankshaft balancer bolt.

   • Tighten the bolt a first pass to 50 Nm (37 lb ft).

   • Tighten the bolt a second pass to 140 degrees.

35. Remove the holding tool and bolt from the block and flywheel.

36. Install the transmission bellhousing bolt located at approximately the 10 o'clock position when looking from the rear of the engine. Tighten the bolt to 75 Nm (55 lb ft).

37. Raise and properly position the frame and install the frame to body bolts. Refer to Engine Removal & Installation.

38. Install the frame to radiator braces.

39. Install the front lower air deflector.

40. Connect the intermediate steering shaft to the steering gear.

41. Install the stabilizer shaft link lower nuts.

42. Connect the transaxle cooler lines to the transaxle.

43. Remove the engine support fixture.

44. Install the front fender splash shield.

45. Install the starter motor.

46. Install the air cleaner upper housing.

47. Install the accessory drive belt.

48. Install the engine mount strut. Refer to Engine Removal & Installation.

49. Install the right front tire and wheel assembly.

50. Perform the crankshaft position (CKP) system variation learn procedure.

   a. Install a scan tool.

   b. Monitor the engine control module (ECM) for DTCs with a scan tool.

   c. Select the crankshaft position (CKP) variation learn procedure with a scan tool.

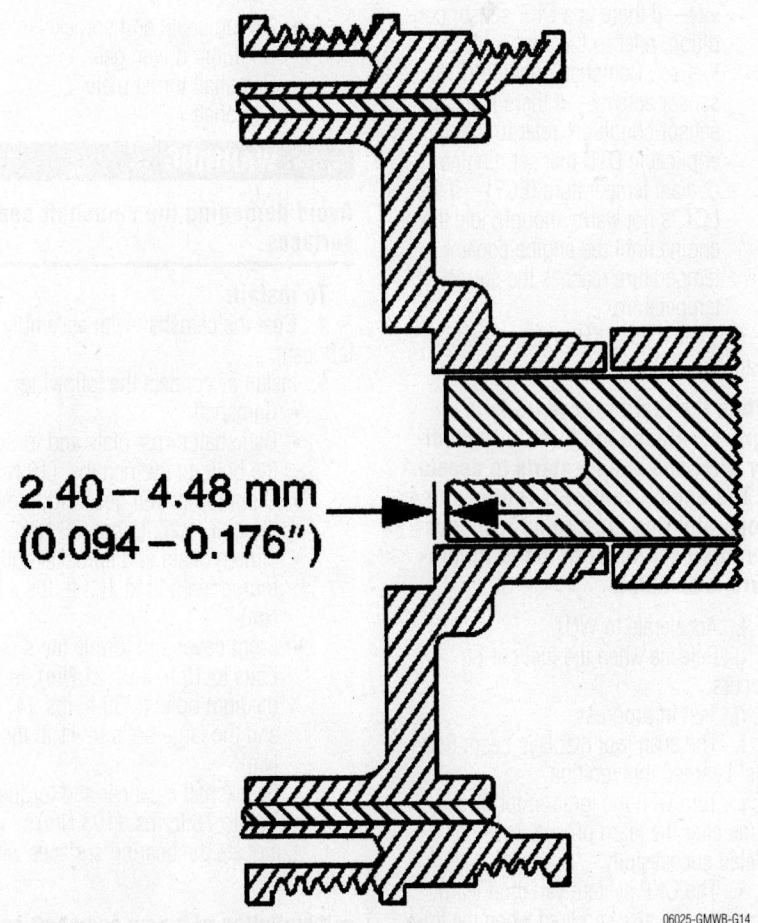

2.40 – 4.48 mm (0.094 – 0.176")

06025-GMWB-G14

Balancer installation—5.3L engine

d. The scan tool instructs you to perform the following:
- Accelerate to wide open throttle (WOT).
- Release throttle when fuel cut-off occurs.
- Observe fuel cut-off for applicable engine.
- Engine should not accelerate beyond calibrated RPM value.
- Release throttle immediately if value is exceeded.
- Block drive wheels.
- Set parking brake.
- DO NOT apply brake pedal.
- Cycle ignition from OFF to ON.
- Apply and hold brake pedal.
- Start and idle engine.
- Turn A/C OFF.
- Vehicle must remain in Park or Neutral.
- The scan tool monitors certain component signals to determine if all the conditions are met to continue with the procedure. The scan tool only displays the condition that inhibits the procedure. The scan tool monitors the following components: CKP sensors activity—If there is a CKP sensor condition, refer to the applicable DTC that set. Camshaft position (CMP) sensor activity—If there is a CMP sensor condition, refer to the applicable DTC that set. Engine coolant temperature (ECT)—If the ECT is not warm enough, idle the engine until the engine coolant temperature reaches the correct temperature.

e. Enable the CKP system variation learn procedure with a scan tool.

➥While the learn procedure is in progress, release the throttle immediately when the engine starts to decelerate. The engine control is returned to the operator and the engine responds to throttle position after the learn procedure is complete.

f. Accelerate to WOT.
g. Release when the fuel cut-off occurs.
h. Test in progress
i. The scan tool displays Learn Status: Learned this ignition.
j. Turn OFF the ignition for 30 seconds after the learn procedure is completed successfully.
k. The CKP system variation learn procedure is also required when the following service procedures have been

performed, regardless of whether DTC P0315 is set:
- A CKP sensor replacement
- An engine replacement
- An ECM replacement
- A harmonic balancer replacement
- A crankshaft replacement
- Any engine repairs which disturb the CKP sensor relationship

Camshaft and Valve Lifters

REMOVAL & INSTALLATION

3.1L and 3.4L Engines

1. Before servicing the vehicle, refer to the Precautions Section.
2. Relieve the fuel system pressure.
3. Remove or disconnect the following:
- Engine assembly
- Rocker arm covers
- Intake manifold
- Rocker arm bolts, balls, rocker arms and pushrods
- Lifter guide bolts and the guide
- Valve lifters from the bores
- Crankshaft balancer
- Front cover
- Timing chain and sprockets
- Oil pump driven gear
- Camshaft thrust plate
- Camshaft

✳✳ WARNING

Avoid damaging the camshaft bearing surfaces.

To install:
4. Coat the camshaft with assembly lubricant.
5. Install or connect the following:
- Camshaft
- Camshaft thrust plate and torque the bolts to 89 inch lbs. (10 Nm)
- Oil pump driven gear and torque the bolt to 27 ft. lbs. (36 Nm)
- Timing chain and sprocket and torque the bolt to 103 ft. lbs. (140 Nm)
- Front cover and torque the small bolts to 15 ft. lbs. (21 Nm), the medium bolts to 35 ft. lbs. (47 Nm) and the large bolts to 41 ft. lbs. (55 Nm)
- Crankshaft balancer and torque the bolt to 76 ft. lbs. (103 Nm)
6. Lubricate the bearing surfaces with Molykote ®.

➥Installation of a new camshaft or a wear pattern on the old valve lifter will

require the replacement of the camshaft and lifters together. If camshaft replacement is not necessary, be sure to install the used valve lifters in their original position.

7. Install or connect the following:
- Lifters in their original locations
- Lifter guide and torque guide bolts to 89 inch lbs. (10 Nm)
- Pushrods, rocker arms, balls and bolts and torque the nuts to 89 inch lbs. (10 Nm) plus an additional 30 degree turn
- Intake manifold and torque the lower manifold bolts to 115 inch lbs. (13 Nm) and the upper manifold bolts to 18 ft. lbs. (25 Nm)
- Rocker arm covers and torque the bolts to 89 inch lbs. (10 Nm)
- Engine assembly
- Negative battery cable

➥The only time valve adjustment in needed is if there was a valve job performed or the rocker studs have been replaced with an adjustable rocker arm stud. The rocker arm stud installed from the factory should be shouldered and not need any adjustment.

8. Start the engine and check for leaks, repair if necessary.

3.5L Engine

1. Before servicing the vehicle, refer to the Precautions Section.
2. Drain the cooling system.
3. Remove or disconnect the following:
- Negative battery cable
- Fuel injector sight shield
- Thermostat housing for clearance when installing the camshaft Holding Fixture
- Camshaft cover
4. Rotate the crankshaft so the camshaft flats are parallel to the camshaft's sealing surface, then install a camshaft holding fixture.
5. Remove the camshaft sprocket bolts.
6. Install a timing chain/sprocket holding fixture J 42042.
7. Evenly slide the camshaft sprocket and chain from the camshafts onto the holding tool.

➥The camshaft bearing caps are marked. Be sure the raised portion of the cap faces the outside of the engine. They must always be installed in their original positions.

8. Remove or disconnect the following:
- Camshaft bearing caps

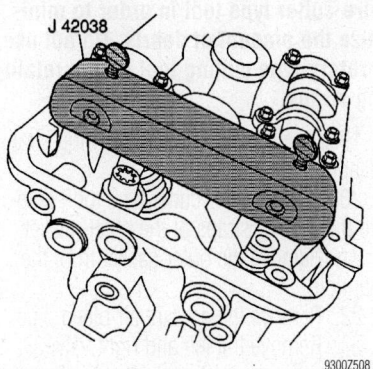

Camshaft holding fixture installed on the camshafts—3.5L engine

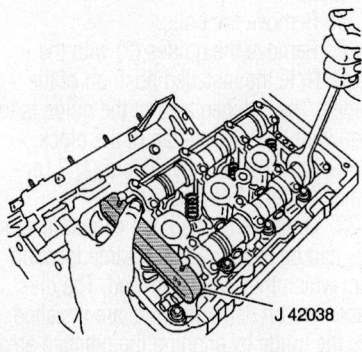

Timing chain/sprocket holding fixture installed on the cylinder head; use the flats on the camshaft if rotation is necessary for installation of the holding tool—3.5L engine

- Camshaft holding fixture
- Camshafts

To install:

9. Coat the rocker arms with engine oil and place them in their original positions.

➡**Be sure to install the rounded end on the lash adjuster and the flat end on the tip of the valve.**

1. Left intake
2. Left exhaust
3. Right intake
4. Right exhaust

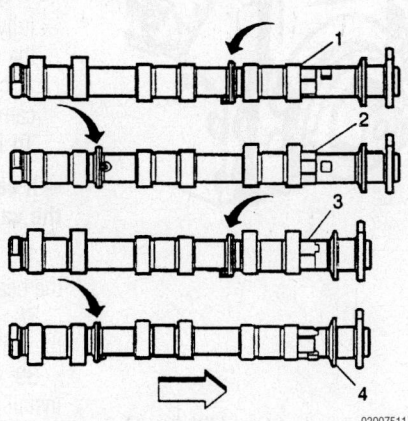

Camshaft identification—3.5L engine

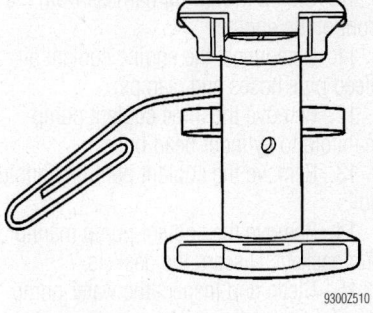

Before installation, compress the tensioner and lock it in place with a piece of wire—3.5L engine

10. Install or connect the following:
- Camshafts, lubricated with engine oil, with the sprocket drive pin notch located at the top
- Bearing caps and torque the bolts to 71 inch lbs. (8 Nm) plus an additional 22 degree turn
- Camshaft Holding Fixture tool onto the camshaft(s) at the rear of the cylinder head

➡**Use the camshaft flats to turn the camshaft.**

11. Compress the secondary timing chain tensioner by hand and insert a wire into the access hole to lock it in place.

12. Slide the camshaft sprockets/timing chain off the tool and onto the camshafts. Be sure to align the drive pins.

13. Remove the timing chain/sprocket holder from the front of the cylinder head. Torque the sprocket bolts to 18 ft. lbs. (25 Nm); then, an additional 45 degree turn.

14. Remove the wire from the chain tensioner and allow the tensioner to apply pressure to the chain.

15. Remove the camshaft holding fixture.

16. Install or connect the following:
- Camshaft cover and torque the bolts to 80 inch lbs. (9 Nm)

- Thermostat housing, if removed
- Fuel injector sight shield and torque the nuts to 27 inch lbs. (3 Nm)
- Negative battery cable

17. Refill the cooling system.

18. Start the vehicle and check for leaks, repair if necessary.

3.8L Engine

1. Before servicing the vehicle, refer to the Precautions Section.

2. Discharge the A/C system.

3. Drain the cooling system.

4. Remove or disconnect the following:
- Negative battery cable
- Radiator with the air conditioning condenser assembly
- Rocker arm cover
- Valve lifters
- Front cover
- Camshaft sprocket and timing chain
- Camshaft thrust plate

5. Remove the camshaft assembly, as follows:

a. Step 1: Install 1⁄2–20 x 6 inch bolt in the camshaft front bolt hole

b. Step 2: Carefully rotate and pull the camshaft assembly out of the bearings.

6. Inspect the camshaft for damage and replace if necessary.

To install:

7. Install or connect the following:
- Camshaft lubricated with assembly lubricant

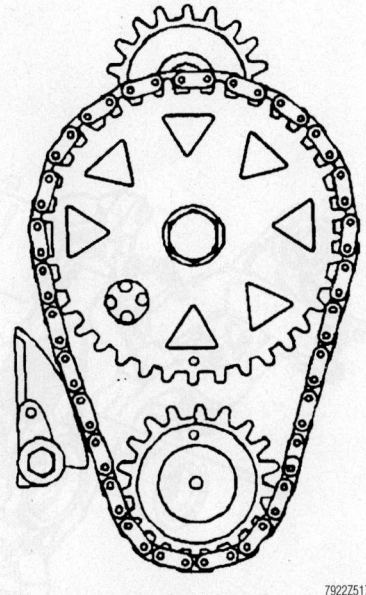

The timing marks should face each other if the chain and gears are installed properly—3.8L engine

- Thrust plate and torque the bolts to 11 ft. lbs. (15 Nm)
- Front cover and torque the bolts to 15 ft. lbs. (20 Nm) plus an additional 40 degree turn
- Valve lifters
- Rocker arm cover and torque the bolts to 89 inch lbs. (10 Nm)
- Radiator with the air conditioning condenser and torque the bolts to 18 ft. lbs. (25 Nm)
- Negative battery cable

8. Evacuate and recharge the A/C system.

9. Refill the cooling system.

10. Start the engine and check for leaks, repair if necessary.

5.3L Engine

1. Before servicing the vehicle, refer to the Precautions Section.

2. Remove the engine.

3. Remove the crankshaft balancer. Refer to Crankshaft Front Seal.

4. Remove the oil level indicator.

5. Remove the left and right exhaust manifolds.

6. Remove the water pump.

7. Remove the water pump bolts.

8. Remove the water pump and gasket. Discard the gasket.

9. Remove the camshaft position sensor.

10. Unclip the wiring harness from the front of the engine.

11. Disconnect the engine coolant air bleed pipe hoses and clamps.

12. Remove the three coolant pump manifold to cylinder head bolts.

13. Remove the coolant pump manifold bolts.

14. Remove the coolant pump manifold and gaskets. Discard the gaskets.

15. Clean and inspect the water pump manifold mounting surfaces.

16. Remove the intake manifold.

✱✱ WARNING

Do not lift the manifold assembly by the electrical lead frame.

17. Remove the valve lifter oil manifold bolts.

→Do not allow dirt or debris to enter the oil passages of the manifold. Plug, as required.

18. Remove the valve lifter oil manifold.

→Remove only the outer gasket from the manifold. Do not disassemble any of the internal components of the manifold in an attempt to remove the 8 inner sealing gaskets. If the inner gaskets are cut or damaged, replace the manifold as an assembly. Only use a wire-cutter type tool in order to minimize the amount of debris. Do not use a rotary-type cutting tool on the retaining straps.

19. Identify the 8 gasket retaining strap locations.

20. Using a wire-cutter type tool, snip the 8 retaining straps of the outer gasket.

21. Remove the outer gasket from the manifold.

22. Remove the coolant air bleed pipe.

23. Remove the left and right valve rocker arm covers. Refer to Rocker Arm Removal & Installation.

24. Remove the valve rocker arms and pushrods. Refer to Rocker Arm Removal & Installation.

25. Remove the bolts.

26. Remove the guides (2) with the lifters. Note the installed position of the guides. The notched area of the guide is to align with the locating tab of the block.

27. Remove the valve lifters (1, 3 [on-demand lifter]) from the guide.

28. Organize or mark the components so they can be installed in the same location from which they were removed. The displacement on demand lifters are installed into the guide by aligning the notched area of the guide with the raised surface on the side of the lifter.

29. Remove the oil pan-to-front cover bolts.

30. Remove the front cover bolts.

31. Remove the front cover and gasket.

32. Rotate the crankshaft in order to align the timing marks (1, 2).

33. Remove the camshaft sprocket bolts.

34. Remove the camshaft sprocket and timing chain.

35. Remove the camshaft retainer bolts and retainer.

36. Remove the camshaft.

  a. Install 3 M8-125 x 100 mm bolts in the camshaft front bolt holes.

  b. Using the bolts as a handle, carefully rotate and pull the camshaft out of the engine block.

  c. Remove the bolts from the camshaft.

To install:

→If camshaft replacement is required, the valve lifters must also be replaced.

37. Lubricate the camshaft journals and the bearings with clean engine oil.

38. Install 3 M8-125 x 100 mm bolts in the NEW camshaft front bolt holes.

39. Using the bolts as a handle, carefully install the NEW camshaft into the engine block.

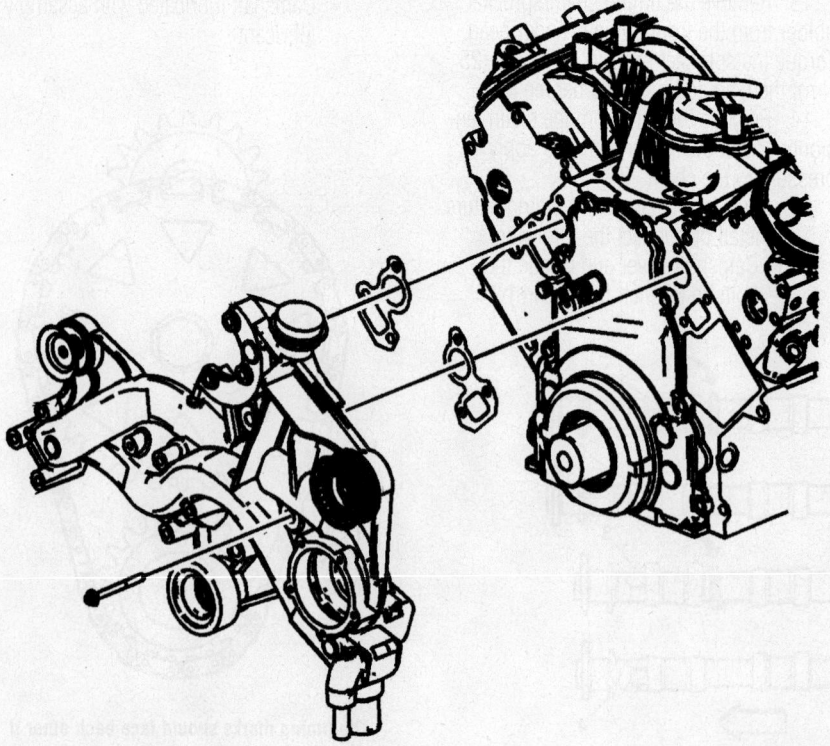

06025-GMWB-G19

Coolant pump manifold—5.3L engine

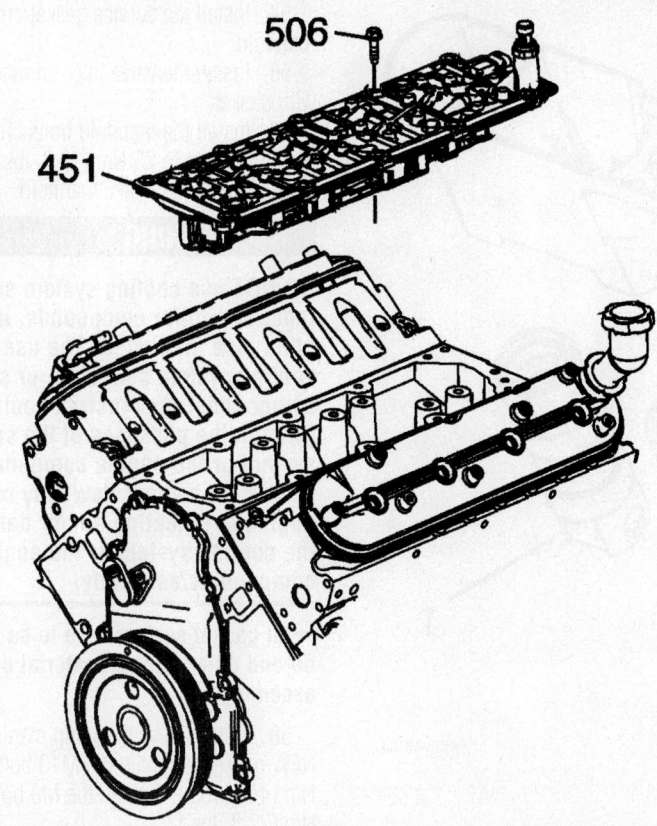

06025-GMWB-G16

Valve lifter oil manifold—5.3L engine

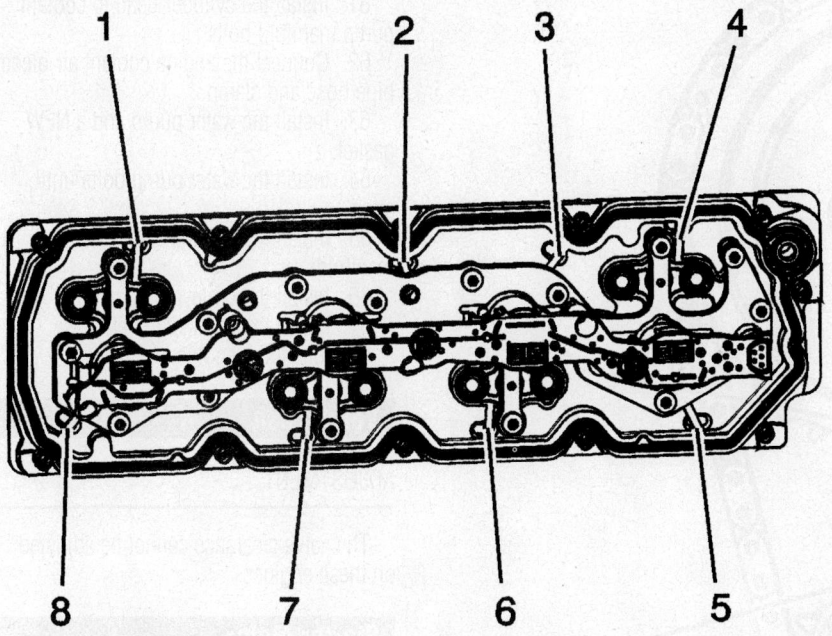

06025-GMWB-G17

The 8 gasket retaining strap locations—5.3L engine

40. Remove the 3 bolts from the camshaft.

➡**Install the retainer plate with the sealing gasket facing the engine block. The gasket surface on the engine block should be clean and free of dirt or debris.**

41. Install the camshaft retainer and bolts. Tighten the bolts to 25 Nm (18 ft. lbs.).

➡**Properly locate the camshaft sprocket onto the locating pin of the camshaft. The sprocket teeth and timing chain teeth must mesh. The camshaft and crankshaft sprocket alignment marks MUST be aligned properly. Position the camshaft sprocket alignment mark in the 6 o'clock position.**

42. Install the camshaft sprocket and timing chain. If necessary, rotate the camshaft or crankshaft sprocket in order to align the timing marks.

43. Install the camshaft sprocket bolts. Tighten the bolts to 25 Nm (18 ft. lbs.).

44. Inspect the camshaft and crankshaft sprockets for proper timing mark alignment.

45. Install the front cover.

46. Install the oil pan-to-front cover bolts. Tighten the bolts to 25 Nm (18 ft. lbs.).

➡**When using the valve lifters again, install the lifters to their original locations. If camshaft replacement is required, the valve lifters must also be replaced. Each of the 4 valve guide assemblies will contain 2 displacement on demand valve lifters and 2 non-displacement on demand valve lifters. With the lifters and guides properly installed, cylinders 1, 4, 6, and 7 lifter bores will each contain 2 displacement on demand valve lifters.**

47. Lubricate the valve lifters and engine block valve lifter bores with clean engine oil.

48. Insert the valve lifters into the lifter guides. Align the flat area on the top of the non displacement on demand lifter with the flat area in the lifter guide bore. Push the lifter completely into the guide bore. The displacement on demand lifters are to be installed into the guide, with the notch in the guide aligned with the raised area of the lifter.

49. Install the valve lifters and guide assembly to the engine block.

50. Install the valve lifter guide bolts. Tighten the valve lifter guide bolts to 12 Nm (106 inch lbs.).

51. Install the valve rocker arms and pushrods.

52. Install the left and right valve rocker arm covers.

53. Install the coolant air bleed pipe.

➡**All gasket surfaces should be free of oil or other foreign material during assembly. Do not allow dirt or debris to enter the oil passages of the manifold. Plug as required.**

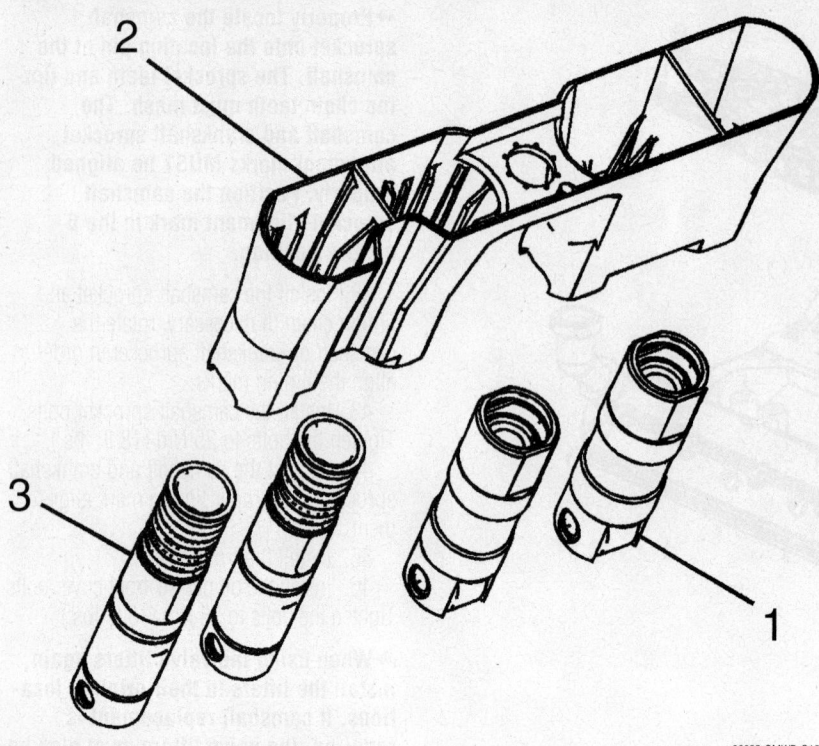

06025-GMWB-G18

Valve lifters and guide—5.3L engine

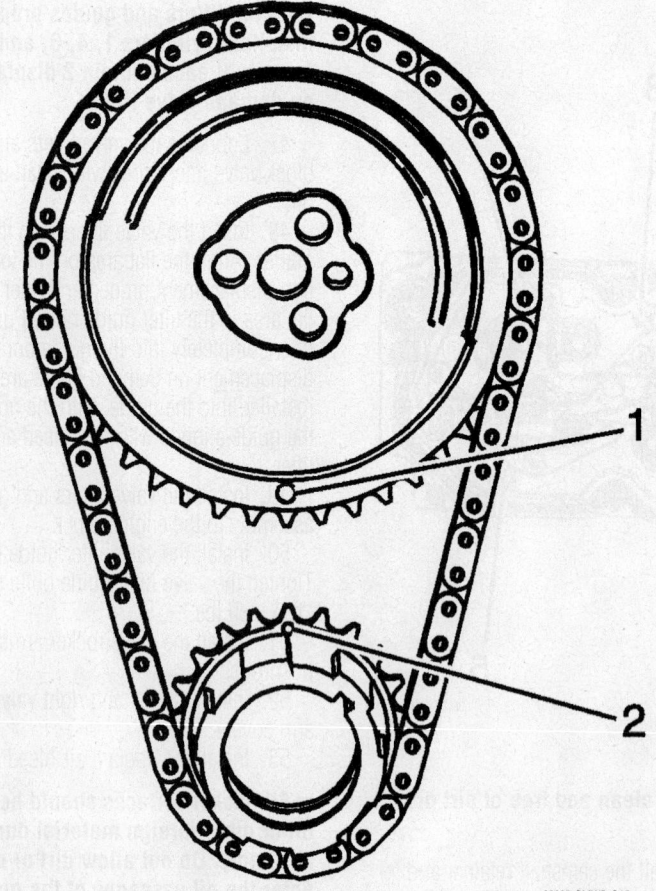

06025-GMWB-G15

Timing mark alignment—5.3L engine

54. Install the service gasket onto the manifold.

55. Install the valve lifter oil manifold with gasket.

56. Install the manifold bolts. Tighten the manifold bolts to 25 Nm (18 ft. lbs.).

57. Install the intake manifold.

✳✳ WARNING

DO NOT use cooling system seal tabs, or similar compounds, unless otherwise instructed. The use of cooling system seal tabs, or similar compounds, may restrict coolant flow through the passages of the cooling system or the engine components. Restricted coolant flow may cause engine overheating and/or damage to the cooling system or the engine components/assembly.

➡**All gasket surfaces are to be free of oil and other foreign material during assembly.**

58. Install the water pump manifold and NEW gaskets. Tighten the M10 bolts to 60 Nm (44 ft. lbs.). Tighten the M8 bolts to 30 Nm (22 ft. lbs.).

59. Install the water pump manifold bolts.

60. Install the camshaft position sensor.

61. Install the cylinder head to coolant pump manifold bolts.

62. Connect the engine coolant air bleed pipe hose and clamp.

63. Install the water pump and a NEW gasket.

64. Install the water pump bolts until sung.

65. Install the left and right exhaust manifolds.

66. Install the oil level indicator.

67. Install the crankshaft balancer.

68. Install the engine.

Valve Lash

ADJUSTMENT

The valve clearance cannot be adjusted on these engines.

Starter Motor

REMOVAL & INSTALLATION

Except 5.3L Engine

1. Before servicing the vehicle, refer to the Precautions Section.

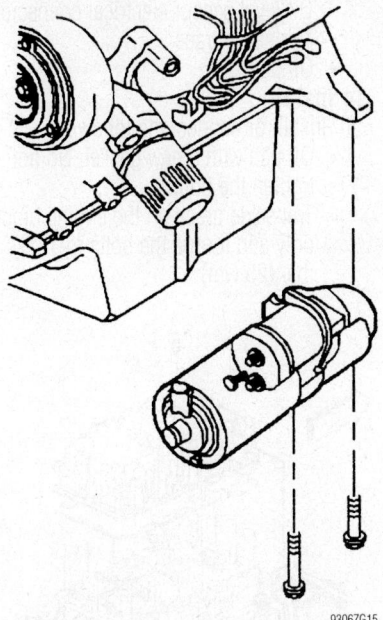

Exploded view of the starter—3.1L and 3.4L engines

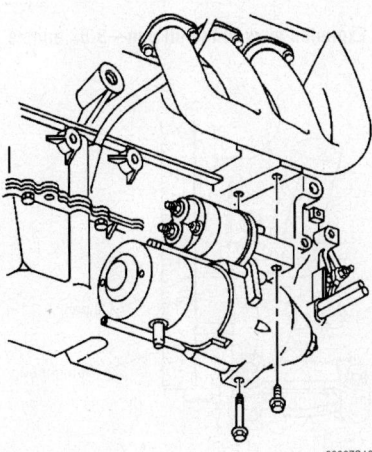

Exploded view of the starter—3.5L engine

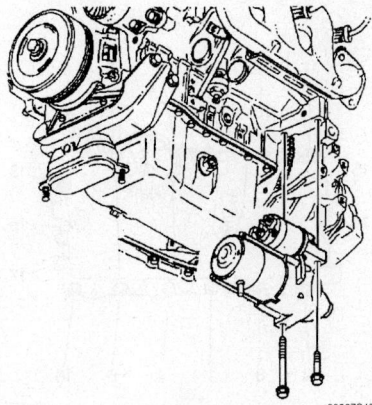

Exploded view of the starter—3.8L engine

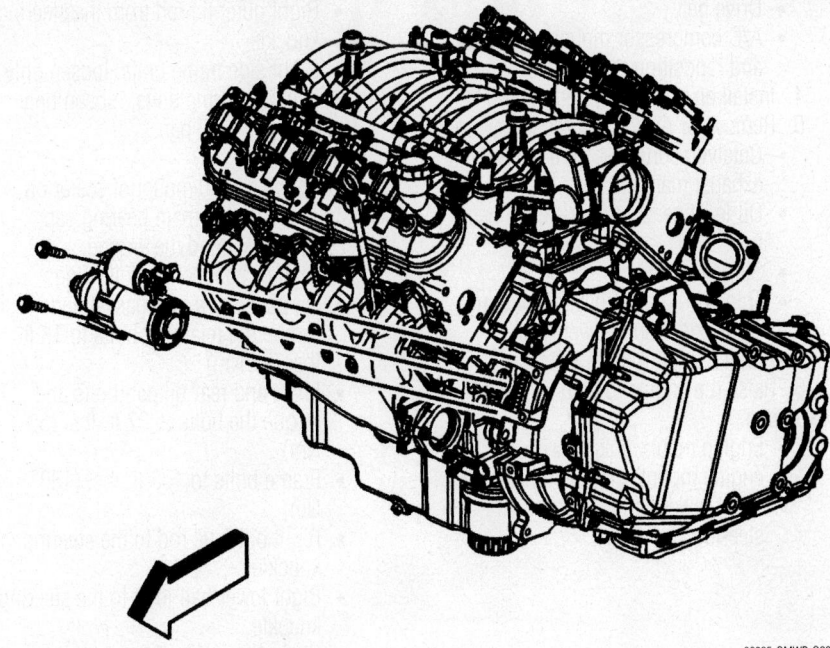

Starter mounting—5.3L engine

➡**The starter motors used in these vehicles are of varying lengths, make sure to use the correct size starter motor.**

2. Remove or disconnect the following:
   - Negative battery cable
   - Radiator lower air deflector
   - Torque converter cover
   - Electrical connections from the starter
   - Starter motor

To install:

3. Install or connect the following:
   - Starter motor and torque the bolts to 32 ft. lbs. (43 Nm) on all models except Intrigue. On Intrigue models, tighten the bolts to 37 ft. lbs. (50 Nm).
   - Starter solenoid BAT terminal
   - Starter solenoid S terminal
   - Torque converter cover and torque the bolts to 89 inch lbs. (10 Nm)
   - Radiator lower air deflector and torque the bolts to 15 ft. lbs. (20 Nm)
   - Negative battery cable

5.3L Engine

1. Before servicing the vehicle, refer to the Precautions Section.
2. Disconnect the negative battery cable.
3. Remove the starter solenoid BAT terminal nut and remove the positive battery cable from the starter motor.

4. Remove the engine harness terminal from the starter motor.
5. Disconnect the starter motor electrical connector.
6. Remove the air cleaner assembly.
7. Remove the starter motor bolts and starter motor.

To install:

8. Position the starter motor to the engine.
9. Install the starter bolts. Tighten the bolts to 50 Nm (37 ft. lbs).
10. Connect the starter motor electrical connector.
11. Install the engine harness terminal to the starter motor.
12. Install the positive battery cable and the starter solenoid BAT terminal nut to the starter motor. Tighten the solenoid BAT terminal nut to 10 Nm (89 inch lbs.).
13. Install the air cleaner assembly.
14. Connect the negative battery cable.

Oil Pan

REMOVAL & INSTALLATION

3.1L and 3.4L Engines

1. Before servicing the vehicle, refer to the Precautions Section.
2. Drain the engine oil.
3. Remove or disconnect the following:
   - Negative battery cable
   - Engine mount struts

- Drive belt
- A/C compressor mounting bolts and reposition the compressor

4. Install an engine support fixture.

5. Remove or disconnect the following:
- Catalytic converter from the right exhaust manifold
- Oil level sensor electrical connection
- Starter motor
- Transaxle brace from the oil pan
- Transaxle mount lower nuts
- Engine mount lower nuts

6. Raise the engine to gain access to the oil pan.
- Engine mount bracket with the engine mount from the oil pan
- Right lower ball joint from the steering knuckle

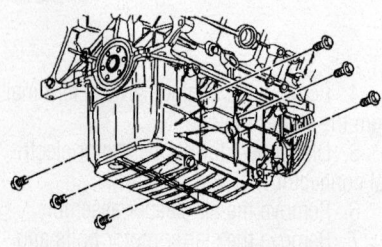

Exploded view of the oil pan side bolts—3.1L and 3.4L engines

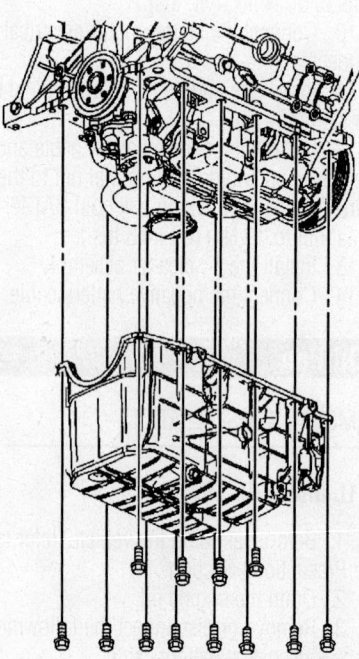

Exploded view of the oil pan retaining bolts—3.1L and 3.4L engines

- Right outer tie rod from the steering knuckle
- Right side frame bolts, loosen only
- Left side frame bolts, loosen only

7. Remove the oil pan.

To install:

8. Apply a small amount of sealer on both sides of the rear main bearing cap between the gasket and the engine.

9. Install or connect the following:
- Oil pan with a new gasket and torque the retaining bolts to 18 ft. lbs. (25 Nm)
- Front and rear oil pan bolts and torque the bolts to 37 ft. lbs. (50 Nm)
- Frame bolts to 133 ft. lbs. (180 Nm)
- Right outer tie rod to the steering knuckle
- Right lower ball joint to the steering knuckle
- Engine mount and bracket to the oil pan and torque the bolts to 43 ft. lbs. (58 Nm)

10. Lower the engine into position.

11. Install or connect the following:
- Engine mount lower nuts and torque the nuts to 32 ft. lbs. (43 Nm)
- Transaxle mount lower nuts and torque the nuts to 35 ft. lbs. (47 Nm)
- Transaxle brace to the oil pan and torque the bolts to 46 ft. lbs. (63 Nm)
- Starter motor and torque the bolts to 32 ft. lbs. (43 Nm)
- Oil level sensor electrical connection
- Catalytic converter to the rear manifold and torque the nuts to 26 ft. lbs. (35 Nm)

12. Remove the engine support fixture.
- A/C compressor and torque the bolts to 37 ft. lbs. (50 Nm)
- Drive belt
- Engine mount struts and torque the bolts to 35 ft. lbs. (48 Nm)
- Negative battery cable

13. Fill the engine with new oil.

14. Start the vehicle and check for leaks, repair if necessary.

3.5L Engine

1. Before servicing the vehicle, refer to the Precautions Section.

2. Drain the engine oil.

3. Remove or disconnect the following:
- Negative battery cable
- Oil filter cap and filter
- Oil level sensor electrical connector
- Transaxle brace
- Oil pan

To install:

4. Install or connect the following:
- Oil pan with a new gasket, Do not tighten the bolts
- Transaxle brace on the engine block only and torque the bolts to 18 ft. lbs. (25 Nm)

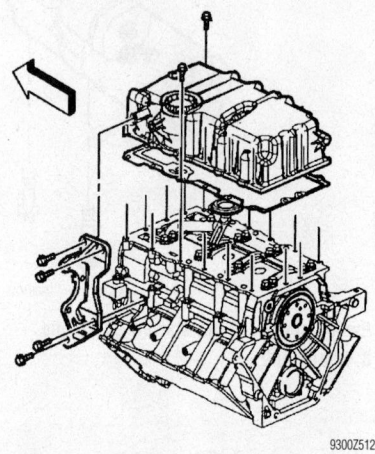

Exploded view of the oil pan—3.5L engine

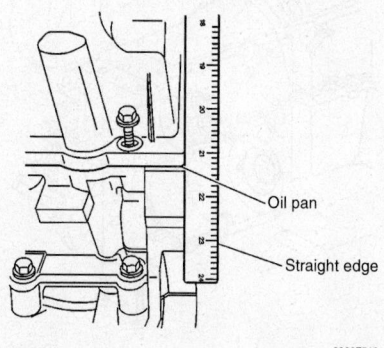

Oil pan

Straight edge

Use a straight-edge to align the rear of the oil pan to the rear of the engine—3.5L engine

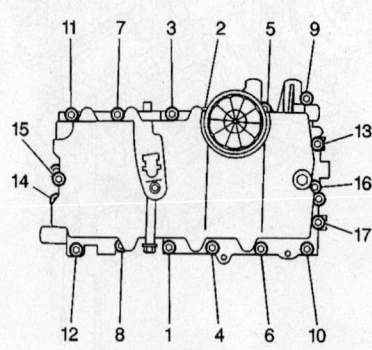

Oil pan mounting bolt tightening sequence—3.5L engine

- Brace-to-oil pan bolts, loosely install them

5. Align the rear of the oil pan flush with the rear of the engine block. Use a straight edge for reference.

6. Press the front of the oil pan against the transaxle brace; then, torque the brace-to-oil pan bolts to 18 ft. lbs. (25 Nm). Be sure to keep the rear of the pan flush with the rear of the engine.

7. Torque the oil pan bolts in sequence to 18 ft. lbs. (25 Nm).

8. Torque the brace-to-transaxle bolts to 32 ft. lbs. (43 Nm).

9. Install or connect the following:
- Oil level sensor electrical connection
- Drain plug and torque it to 15 ft. lbs. (20 m)
- New oil filter and torque the cap to 18 ft. lbs. (25 Nm)
- Negative battery cable

10. Refill the engine with clean oil.

11. Start the vehicle and check for leaks, repair if necessary.

3.8L Engine

1. Before servicing the vehicle, refer to the Precautions Section.

2. Drain the engine oil.

3. Remove or disconnect the following:
- Negative battery cable
- Throttle body air inlet duct
- Engine mount struts
- Accessory drive belt

4. Install an engine support fixture.
- Catalytic converter from the right exhaust manifold
- Right front wheel
- Right side splash shield
- Oil filter and discard
- A/C compressor bracket and reposition the compressor
- Power steering oil cooler pipe brackets from the frame
- Engine mount bracket bolts
- Lower engine mount nuts
- Torque converter cover
- Oil level sensor electrical connector
- Oil level sensor

5. Raise the vehicle and place a support under the frame.
- Right side frame bolts and loosen the left side bolts
- Engine mount and bracket
- Oil pan
- Oil pump pipe and screen

To install:

6. Install or connect the following:
- New oil pan gasket and the oil pump pipe and screen assembly

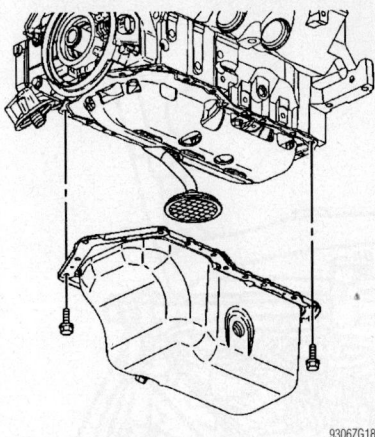

Exploded view of the oil pan mounting— 3.8L engine

and torque the bolts to 11 ft. lbs. (15 Nm)

➡**Do not over-tighten the oil pan bolts or damage to the pan may occur.**

- Oil pan and torque the bolts to 125 inch lbs. (14 Nm)
- Engine mount and bracket and torque the bolts to 50 ft. lbs. (68 Nm)
- New right side frame bolts and torque all the bolts to 133 ft. lbs. (180 Nm)
- Lower engine mount and torque the nuts to 35 ft. lbs. (47 Nm)
- Oil level sensor and torque the fastener to 15 ft. lbs. (20 Nm)
- Oil level wire connector
- Torque converter cover and torque the bolts to 89 inch lbs. (10 Nm)
- Power steering oil cooler pipe brackets to the frame
- A/C compressor and torque the upper bolt to 37 ft. lbs. (50 Nm) and the lower bolt to 59 ft. lbs. (80 Nm)
- New oil filter
- Drain plug and torque the plug to 22 ft. lbs. (30 Nm)
- Splash shield
- Right front wheel
- Catalytic converter pipe to the right exhaust manifold and torque the bolts to 26 ft. lbs. (35 Nm)

7. Remove the engine support fixture.
- Drive belt
- Engine mount struts to the engine and torque the bolts to 41 ft. lbs. (56 Nm)
- Throttle body air inlet duct
- Negative battery cable

8. Fill the engine with new oil.

9. Start the vehicle and check for leaks, repair if necessary.

10. Road test the vehicle, check the front end alignment and adjust if necessary.

5.3L Engine

1. Before servicing the vehicle, refer to the Precautions Section.

2. Disconnect the negative battery cable.

3. Remove the left engine mount strut and the bracket from the upper radiator support.

4. Assemble the J-28467-501 (2) to the J 28467-B cross bar (1).

5. Install the J 28467-B (1) and the J-28467-501 (2) to the fender rails.

6. Install the J 36462-A (1) to the cross bar (2).

7. Install the support lift hook (2) to the cross bar (1).

8. Install the support hook (1) to the right engine lift hook (3).

9. Install the support hook (1) to the J 36462-A (2).

10. Install the support hook (1) to the left engine lift hook (3).

11. Raise the engine to release the pressure off of the engine mounts.

12. Raise and support the vehicle.

13. Remove the front tires and wheels.

14. Remove and discard the 2 plastic braces from the front of the radiator lower air deflector. The plastic braces are directly below the front cradle bolts.

15. Remove the positive battery cable and the retainers from the frame and position aside.

16. Disconnect the power steering return hose from the frame.

17. Secure the power steering return hose.

18. Remove the stabilizer shaft links and rotate the stabilizer shaft upward to gain access to the mounting bolts in the power steering gear.

19. Remove the mounting bolts from the power steering gear.

20. Secure the power steering gear.

21. Remove the nuts that secure the engine mount to the frame.

22. Remove the nuts which secure the transaxle mount to the frame.

23. If applicable, disconnect the front wheel speed sensor harness connectors.

24. If applicable, disconnect the wheel speed sensor harness from the frame and lower control arms.

25. If applicable, remove the retainers at the front wheel speed harness from the frame and from the lower control arms.

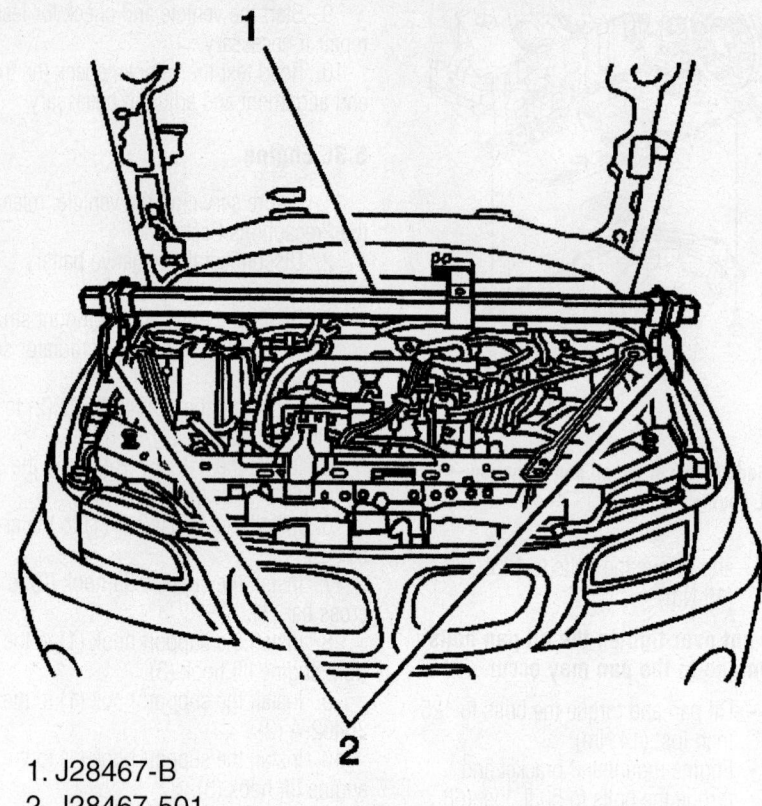

1. J28467-B
2. J28467-501

06025-GMWB-G21

Assemble the J-28467-501 (2) to the J 28467-B cross bar (1)—5.3L engine

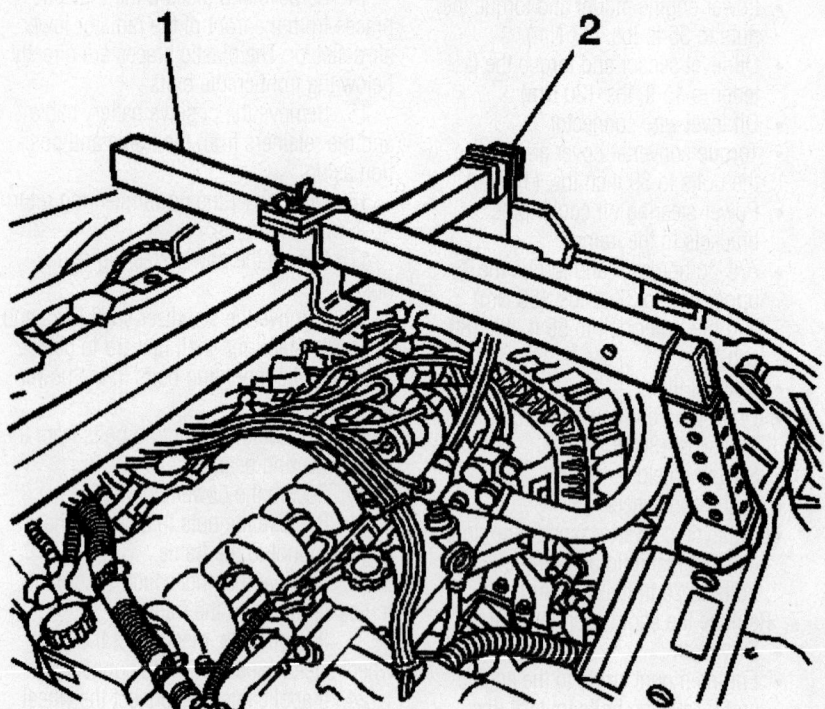

1. Crossbar
2. J28467-501

06025-GMWB-G22

Install the J 28467-B (1) and the J-28467-501 (2) to the fender rails—5.3L engine

26. Separate both of the lower ball joints from the steering knuckle.

27. Remove both front drivetrain reinforcements using the following procedure.

    a. Remove the drivetrain reinforcement to support brace bolts.

    b. Remove the drivetrain reinforcement to front frame mounting stud nut.

    c. Remove the drivetrain reinforcement from the vehicle.

28. Lower the vehicle until the frame contacts the J 39580.

29. Remove the radiator to front frame brackets.

30. Remove the bolts which secure the front frame to the body.

31. Remove the bolts which secure the rear frame to the body.

32. Raise the vehicle in order to separate the frame from the body.

33. Drain the engine oil and remove the engine oil filter.

34. Reinstall the drain plug and oil filter until snug.

35. Remove the transaxle converter cover bolt/stud and cover.

36. Disconnect the oil level sensor electrical connector.

37. Remove engine harness retainer from the front of oil pan.

38. Remove the oil pan bolts.

39. Remove the oil pan.

➡ **DO NOT allow foreign material to enter the oil passages of the oil pan, cap or cover the openings as required.**

40. Drill out the oil pan gasket retaining rivets, if required.

41. Remove the gasket from the pan.

42. Discard the gasket and rivets.

43. Clean and inspect the engine oil pan.

To install:

➡ **The alignment of the structural oil pan is critical. The rear bolt hole locations of the oil pan provide mounting points for the transmission housing. To ensure the rigidity of the powertrain and correct transmission alignment, it is important that the rear of the block and the rear of the oil pan are flush, or even. The rear of the oil pan must NEVER protrude beyond the engine block and transmission housing plane. Do NOT reuse the oil pan gasket. It is NOT necessary to rivet the NEW gasket to the oil pan.**

44. Apply a 5 mm (0.20 in.) bead of sealant GM P/N 12346141 or 12378577 (Canadian P/N 89022195), or equivalent 20 mm (0.80 in.) long to the engine block. Apply the sealant directly onto the tabs of

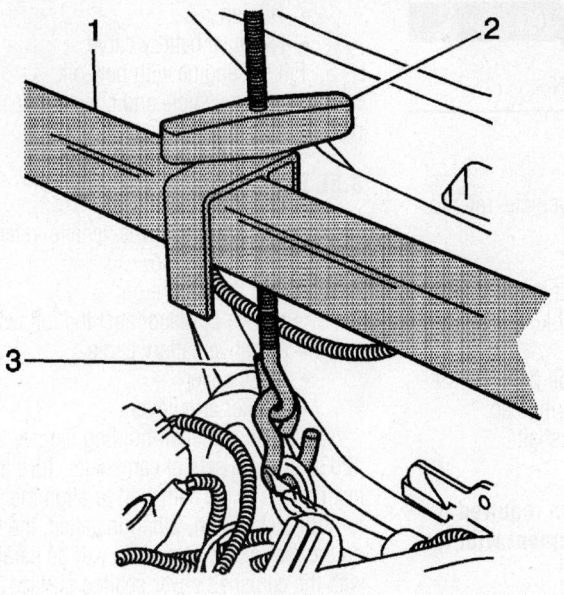

1. Crossbar
2. Lift hook t-nut
3. Lift hook

06025-GMWB-G23

Assembling lift hook—5.3L engine

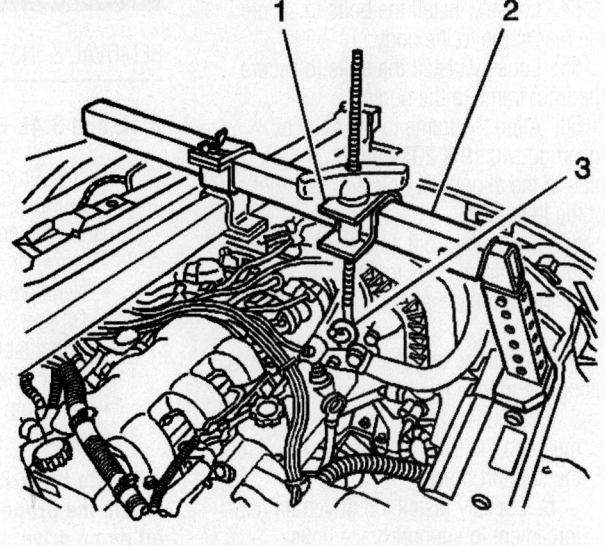

1. Support bracket
2. Crossbar
3. Lift hook

06025-GMWB-G24

Final assembly of support fixture—5.3L engine

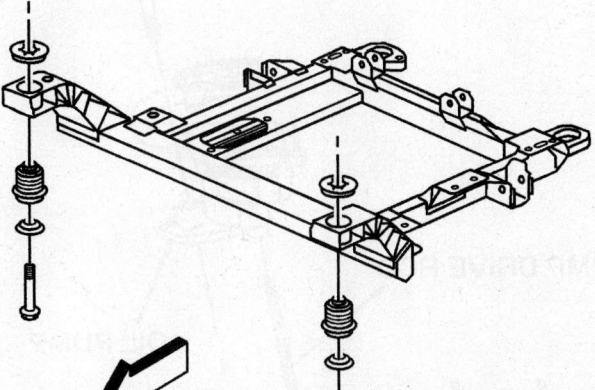

06025-GMWB-G25

Remove the bolts which secure the front frame to the body—5.3L engine

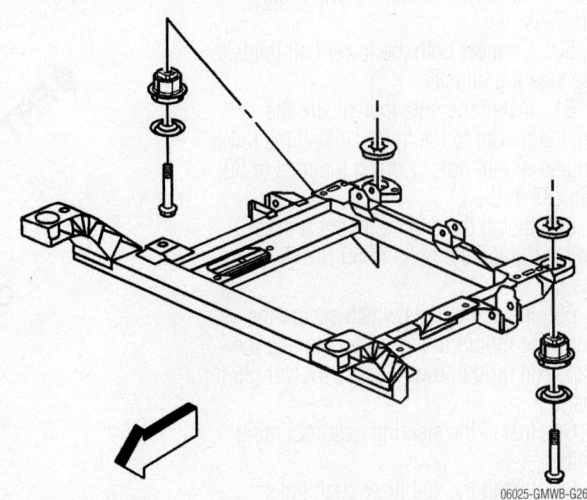

06025-GMWB-G26

Remove the bolts which secure the rear frame to the body—5.3L engine

the front/rear cover gasket that protrudes into the oil pan surface.

➡**Be sure to align the oil gallery passages in the oil pan and engine block properly with the oil pan gasket.**

45. Install the gasket onto the pan.

46. Install the oil pan bolts to the pan and through the gasket.

47. Install the oil pan, gasket, and bolts to the engine block. Tighten the oil pan and oil pan-to-front cover bolts to 25 Nm (18 ft. lbs.). Tighten the oil pan-to-rear cover bolts to 12 Nm (106 inch lbs.). Tighten the trans-

mission housing, converter cover, and transmission bolts/stud to 50 Nm (37 ft. lbs.).

48. Install engine harness to front of oil pan.

49. Connect the oil level sensor electrical connector.

50. Install the transaxle converter cover and bolt/stud. Tighten the bolt/stud to 12 Nm (106 inch lbs.).

51. Install new engine oil and a new oil filter.

52. Position the engine support table with the frame under the vehicle.

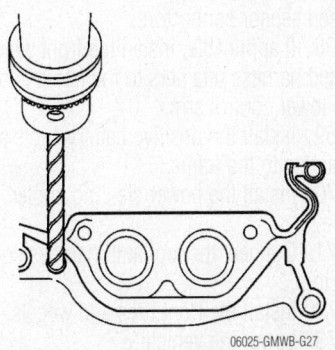

06025-GMWB-G27

Drill out the oil pan gasket retaining rivets—5.3L engine

53. Lower the vehicle to the frame.
54. Loosely install the bolts to secure the rear frame to the body.
55. Loosely install the bolts to secure the front frame to the body.
56. Align the frame to the body by inserting two 19 X 203 mm (0.74 X 8 in) pins in the alignment holes on the right side of the frame.
57. Install the front and rear frame bolts. Tighten the front bolts to 145 Nm (107 ft. lbs.). Tighten the rear bolts to 160 Nm (118 ft. lbs.).
58. Install the drivetrain reinforcements using the following procedure:
   a. Position the drivetrain reinforcements to the front frame mount stud to the support brace.
   b. Loosely install the drivetrain reinforcement to support brace bolts.
   c. Install the drivetrain reinforcement to cradle mount nut. Tighten the drivetrain reinforcement brace nut to 50 Nm (37 ft. lbs.). Tighten the drivetrain reinforcement brace bolts to 25 Nm (18 ft. lbs.).
59. Install the radiator to front frame brackets.
60. Connect both the lower ball joints to the steering knuckle.
61. Install the nuts that secure the engine mount to the frame. Install the lower engine mount nuts. Tighten the nuts to 50 Nm (37 ft. lbs.).
62. Install the engine mount bracket bolts. Tighten the bolts to 50 Nm (37 ft. lbs.).
63. Install the nuts which secure the transaxle mount to the frame. Tighten the transaxle mount lower nuts to 47 Nm (35 ft. lbs.).
64. Install the steering gear mounting bolts.
65. Install the stabilizer shaft links.
66. If applicable, connect the wheel speed sensor wiring harness to the frame and lower control arm.
67. If applicable, connect the front wheel speed sensor connectors.
68. If applicable, install the front wheel speed harness retainers to the frame and to the lower control arm.
69. Install the positive battery cable and retainers to the frame.
70. Install the power steering cooler pipe.
71. Connect the fog lamp harness connectors.
72. Install the front tires and wheels.
73. Lower the vehicle.
74. Remove the engine support fixture.
75. Inspect the front wheel alignment.

Oil Pump

REMOVAL & INSTALLATION

3.1L and 3.4L Engines

1. Before servicing the vehicle, refer to the Precautions Section.
2. Drain the engine oil.
3. Remove or disconnect the following:
   - Negative battery cable
   - Oil pan
   - Bolt attaching the oil pump to the rear crankshaft bearing cap
   - Oil pump and driveshaft

To install:

➡**Rotate the driveshaft as required to obtain the proper engagement with the oil pump drive unit.**

4. Install or connect the following:
   - Oil pump and driveshaft and torque the bolt to 30 ft. lbs. (41 Nm)
   - Oil pan
   - Negative battery cable
5. Fill the engine with new oil.
6. Start the vehicle and check for leaks, repair if necessary.

3.5L Engine

1. Before servicing the vehicle, refer to the Precautions Section.
2. Drain the engine oil.
3. Remove or disconnect the following:
   - Negative battery cable
   - Front cover
   - Rocker arm covers
4. Install camshaft holding fixtures J 42038 on both sets of camshafts. Turn the hex portion of the camshaft to align them for tool installation. When installed, the flats on the rear of the camshafts will be parallel with the camshaft cover sealing surface.
5. Remove or disconnect the following:
   - Oil pan
   - Oil pump pipe and screen

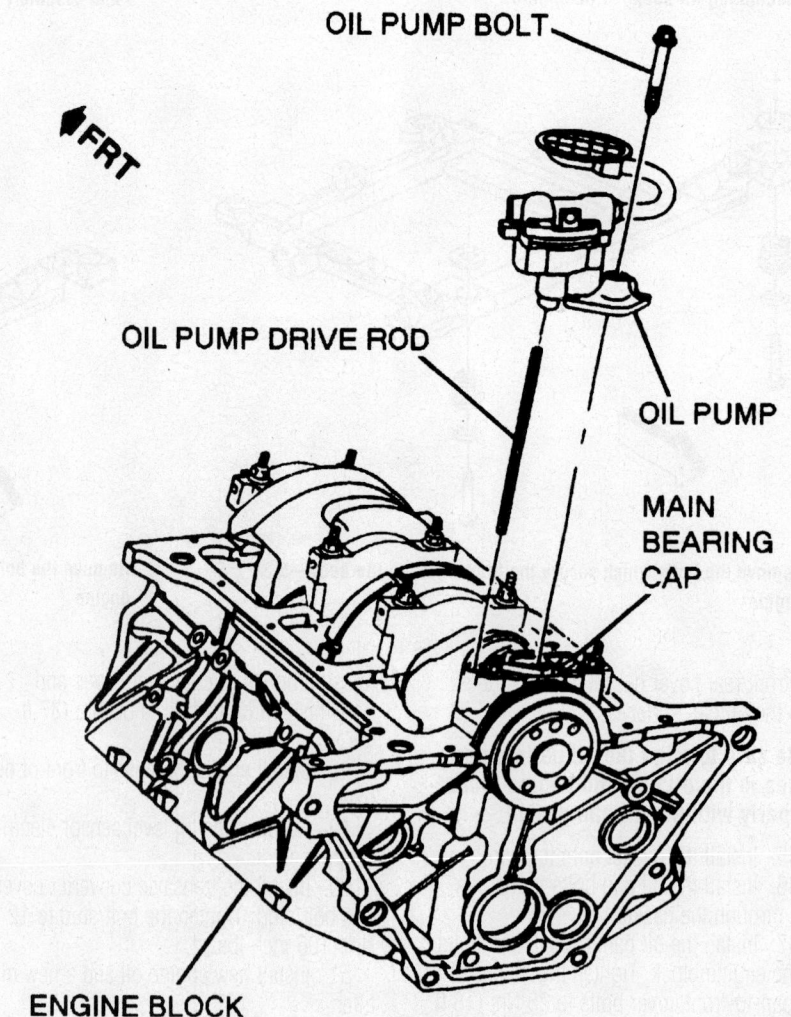

Oil pump and driveshaft components—3.1L and 3.4L engines

79222519

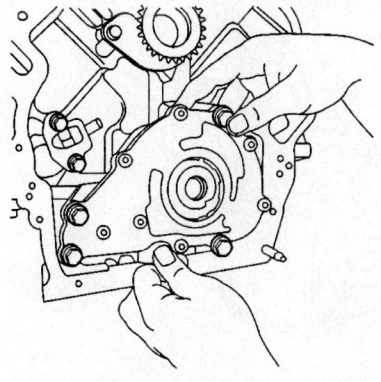

The oil pump is mounted on the front of the engine and driven by the crankshaft—3.5L engine

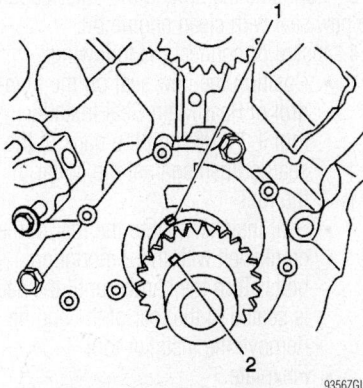

Align the crankshaft sprocket splines with the oil pump gear and install the sprocket in the oil pump–3.5L engine

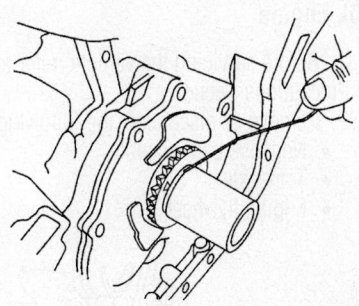

Correct position of the crankshaft sprocket when the oil pump is installed correctly—3.5L engine

- Primary chain tensioner
- Primary chain from the drive sprocket
- Oil pump by sliding it off the crankshaft

To install:

6. Pack the oil pump housing with white petroleum jelly to insure priming.
7. Align the oil pump sprocket with the

1. 97 inch lbs. (11 Nm)
2. Oil pump cover
3. Pump outer gear
4. Pump inner gear
5. Front cover

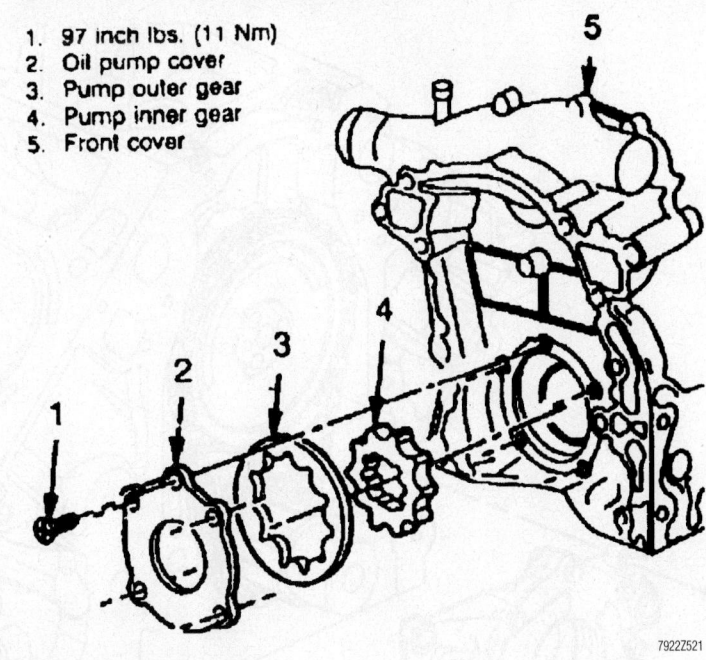

Oil pump assembly—3.8L engine

crankshaft and install the pump on the engine until a positive stop is felt. When installed properly, the sprocket will protrude slightly from the oil pump and the face of the sprocket will be behind the machined step in the crankshaft.

8. Install or connect the following:
- Oil pump and torque the bolts to 18 ft. lbs. (25 Nm)
- Primary chain on the sprocket

➡ **Be sure to maintain correct timing.**

- Chain tensioner
- Oil pump pipe with the screen and torque the nut and bolt to 89 inch lbs. (10 Nm)
- Oil pan and torque the bolts to 18 ft. lbs. (25 Nm)
9. Remove the camshaft holding tools.
10. Install or connect the following:
- Camshaft covers and torque the bolts to 80 inch lbs. (99 Nm)
- Engine front cover and torque the bolts to 124 inch lbs. (14 Nm)
- Negative battery cable
11. Fill the engine with new oil.
12. Start the vehicle and check for leaks, repair if necessary.

3.8L Engines

1. Before servicing the vehicle, refer to the Precautions Section.
2. Drain the engine oil.
3. Remove or disconnect the following:
- Negative battery cable
- Front cover. Refer to the timing

chain removal and installation procedure for front cover removal.
- Oil pump cover
- Oil pump gear set

To install:

4. Lubricate the oil pump gears with petroleum jelly and install the gears into the housing.
5. Pack the gear cavity with petroleum jelly after the gears have been installed.
6. Install or connect the following:
- Oil pump cover and torque the screws to 98 inch lbs. (11 Nm)
- Front cover. Refer to the timing chain removal and installation procedure for front cover installation.
- Negative battery cable
7. Fill the engine oil. A new oil filter is recommended.
8. Start the vehicle and check for leaks, repair if necessary.

5.3L Engine

1. Before servicing the vehicle, refer to the Precautions Section.
2. Remove the oil pan.
3. Remove the engine front cover.
4. Remove the oil pump screen bolt and nuts.
5. Remove the oil pump screen with O-ring seal.
6. Remove the O-ring seal from the pump screen.
7. Discard the O-ring seal.
8. Remove the remaining crankshaft oil deflector nuts.

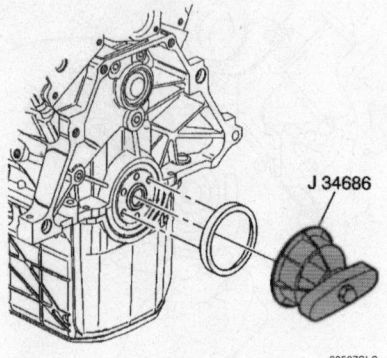

9356ZGLO

Installing the rear main seal—except 3.5L engines

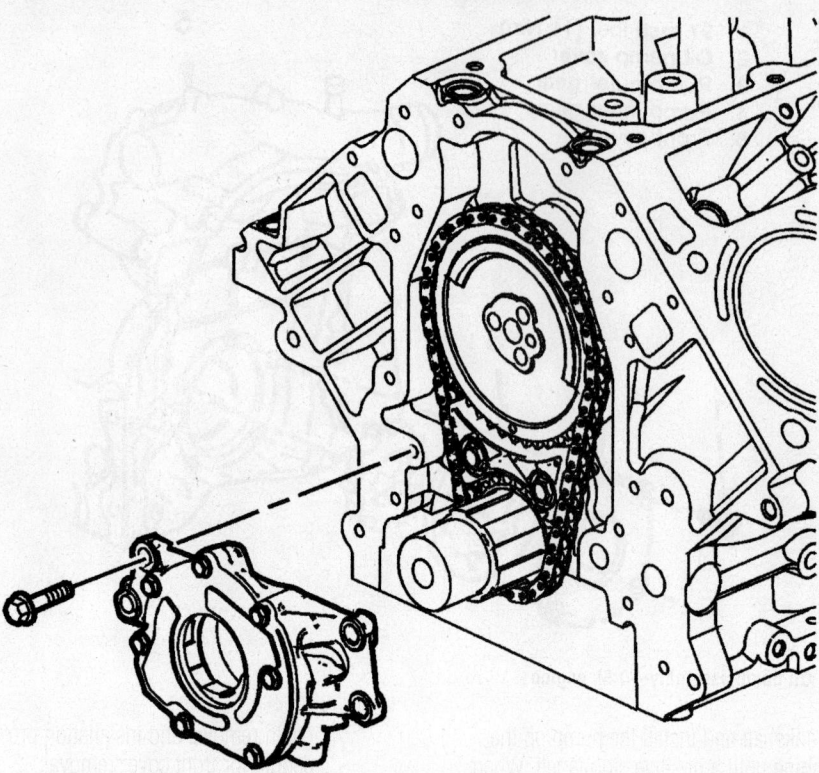

06025-GMWB-G28

Oil pump—5.3L engine

9. Remove the crankshaft oil deflector.
10. Remove the oil pump bolts.

➡**Do not allow dirt or debris to enter the oil pump assembly, cap end as necessary.**

11. Remove the oil pump.
12. Clean and inspect the oil pump.

To install:

13. Align the splined surfaces of the crankshaft sprocket and the oil pump drive gear and install the oil pump.

14. Install the oil pump onto the crankshaft sprocket until the pump housing contacts the face of the engine block.

15. Install the oil pump bolts. Tighten the bolts to 25 Nm (18 ft. lbs.).

16. Install the crankshaft oil deflector and nuts until snug.

17. Lubricate a NEW oil pump screen O-ring seal with clean engine oil.

18. Install the NEW O-ring seal onto the oil pump screen.

➡**Push the oil pump screen tube completely into the oil pump prior to tightening the bolt. Do not allow the bolt to pull the tube into the pump.**

19. Align the oil pump screen mounting brackets with the correct crankshaft bearing cap studs.

20. Install the oil pump screen.

21. Install the oil pump screen bolt and nuts. Tighten the bolt to 12 Nm (106 lb in). Tighten the nuts to 25 Nm (18 ft. lbs.).

22. Install the engine front cover.
23. Install the oil pan.

Rear Main Seal

REMOVAL & INSTALLATION

Except 3.5L and 5.3L Engines

1. Before servicing the vehicle, refer to the Precautions Section.

2. Remove or disconnect the following:
   - Negative battery cable
   - Transmission
   - Flexplate

7922Z522

Carefully pry the seal from the bore without damaging the crankshaft seal surface

   - Rear main seal by prying it out

To install:

3. Lubricate the lip and the outer edge of the new seal with clean engine oil.

4. Install or connect the following:
   - Position the new seal on the mandrel of Rear Main Seal Installer tool J 34686 until the back of the seal is flush against the collar of the tool
   - Seal installer tool to the rear of the crankshaft with the 2 mounting bolts. Turn the handle until the seal is seated in the rear of the engine. Remove the installer tool
   - Flexplate
   - Transmission
   - Negative battery cable

5. Start the vehicle and check for leaks, repair if necessary.

3.5L Engine

1. Before servicing the vehicle, refer to the Precautions Section.

2. Remove or disconnect the following:
   - Negative battery cable
   - Transaxle
   - Engine flywheel

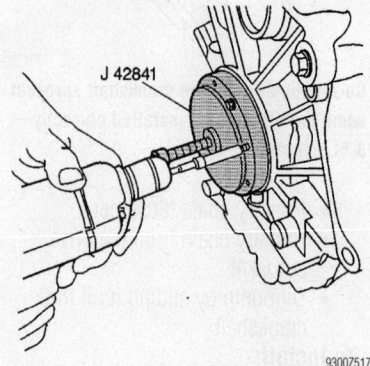

9300Z517

Use the guide holes in Tool J 42841 to install the screws in the seal—3.5L engine

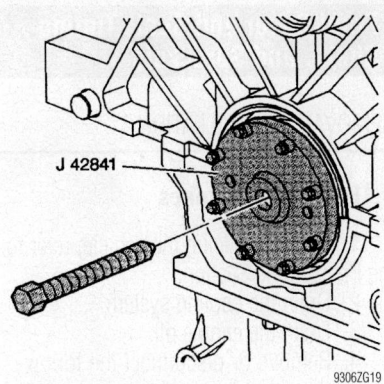

Install a center forcing screw into the removal tool—3.5L engine

3. Place Rear Seal Remover Tool J 42841 on the crankshaft with retaining bolts.

4. Install eight one-inch self starting screws through the guide holes of the tool. Tighten the screws.

5. Install the two retaining bolts.

6. Install a center forcing screw into the removal tool and pull the seal off the end of the crankshaft.

To install:

7. Clean debris from the crankshaft rear seal drain. The seal may leak if the drain is not properly cleaned.

8. Place a small amount of gasket maker to the crankcase split line across the end of the upper and lower crankcase seal.

9. Coat the outer diameter of the block with clean engine oil.

10. Clean the outer diameter of the flywheel flange with a lint-free cloth.

✳✳ CAUTION

Do not apply any oil on the green coating of the new seal.

11. Loosen the center bolt of the seal installer tool until the hub protrudes past the outer plate (approximately ½ inch).

12. Install or connect the following:
- Three mounting bolts into the crankshaft flange until the tool is fully seated on the crankshaft
- New seal by tightening the center bolt until the tool bottoms out against the crankshaft

13. Remove the removal/installer tool and make certain the seal is installed properly

14. Install or connect the following:
- Flywheel. Torque the bolts to 11 ft. lbs. (15 Nm) plus an additional 50 degrees with a torque angle meter.
- Transaxle

- Negative battery cable

15. Top off the engine oil if needed.

16. Start the vehicle and check for leaks, repair if necessary.

5.3L Engine

1. Before servicing the vehicle, refer to the Precautions Section.

2. Remove the automatic transmission.

3. Remove the flywheel bolts and flywheel.

4. Remove and discard the crankshaft rear oil seal.

To install:

➡**Do not lubricate the oil seal inside diameter (ID) or the crankshaft surface.**

5. Lubricate the outside diameter (OD) of the oil seal with clean engine oil. DO NOT allow oil or other lubricants to contact the seal surface.

6. Lubricate the rear cover oil seal bore with clean engine oil. DO NOT allow oil or other lubricants to contact the crankshaft surface.

7. Install the J 41479 tapered cone (2) and bolts onto the rear of the crankshaft.

8. Tighten the bolts until snug. Do not overtighten.

9. Install the rear oil seal onto the

tapered cone (2) and push the seal to the rear cover bore.

10. Thread the J 41479 threaded rod into the tapered cone until the tool (1) contacts the oil seal.

11. Align the oil seal into the tool (1).

12. Rotate the handle of the tool (1) clockwise until the seal enters the rear cover and bottoms into the cover bore.

13. Remove the J 41479.

➡**The flywheel does not use a locating pin for alignment and will not initially seat against the crankshaft flange or spacer if applicable, but will be pulled onto the crankshaft by the engine flywheel bolts. This procedure requires a 3-stage tightening process.**

14. Install the flywheel to the crankshaft.

15. Apply threadlock GM P/N 12345382 (Canadian P/N 10953489), or equivalent to the threads of the flywheel bolts.

16. Install the engine flywheel bolts:
- First pass in sequence: 20 Nm (15 ft. lbs.).
- Second pass in sequence: 50 Nm (37 ft. lbs.).
- Third pass in sequence: 100 Nm (74 ft. lbs.).

17. Install the automatic transmission.

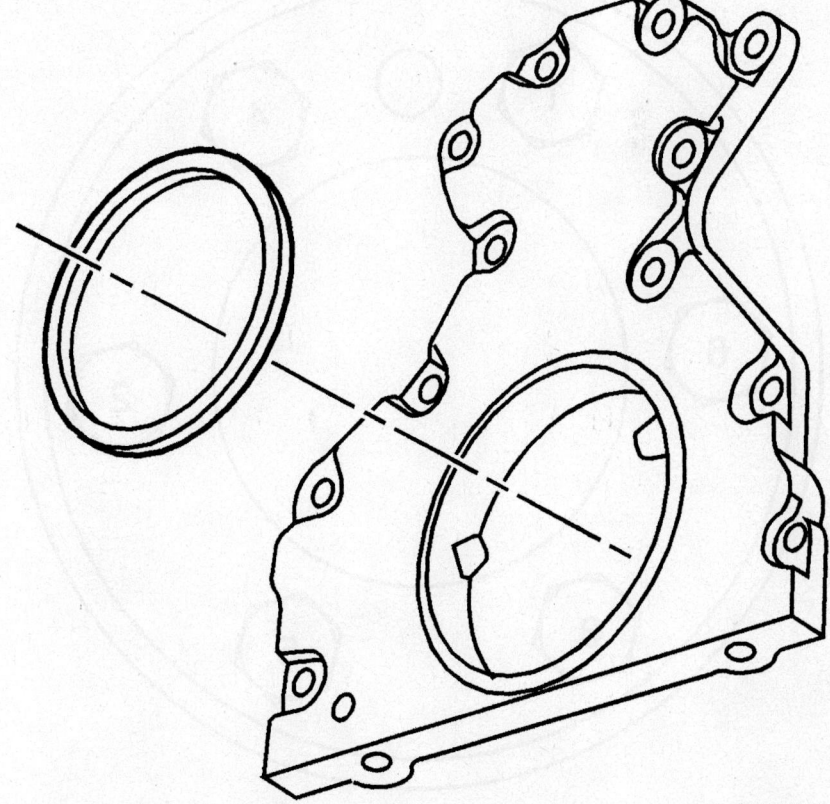

Rear main seal—5.3L engine

J 41479

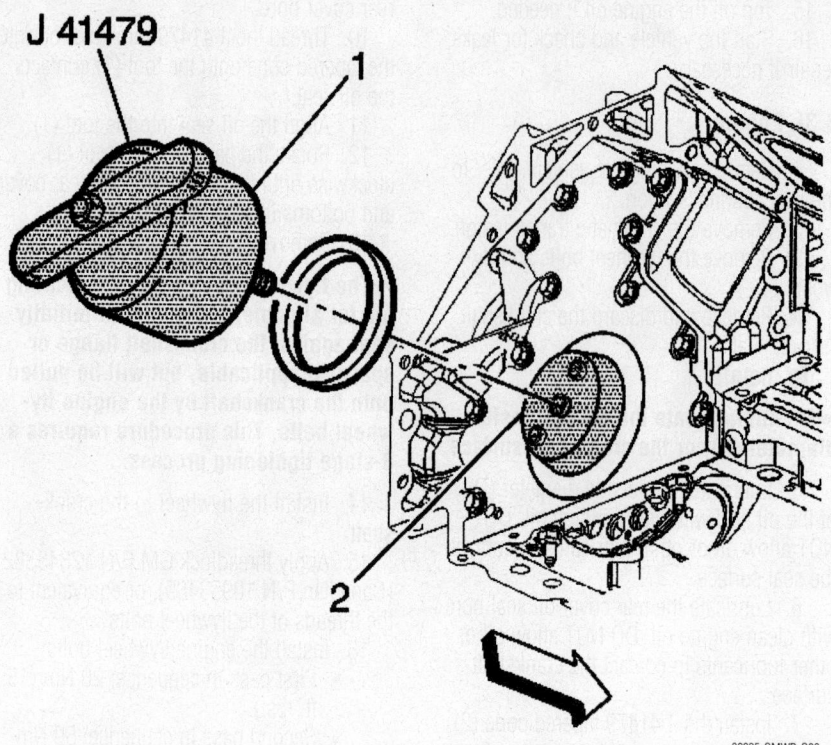

06025-GMWB-G30

Rear main seal installation—5.3L engine

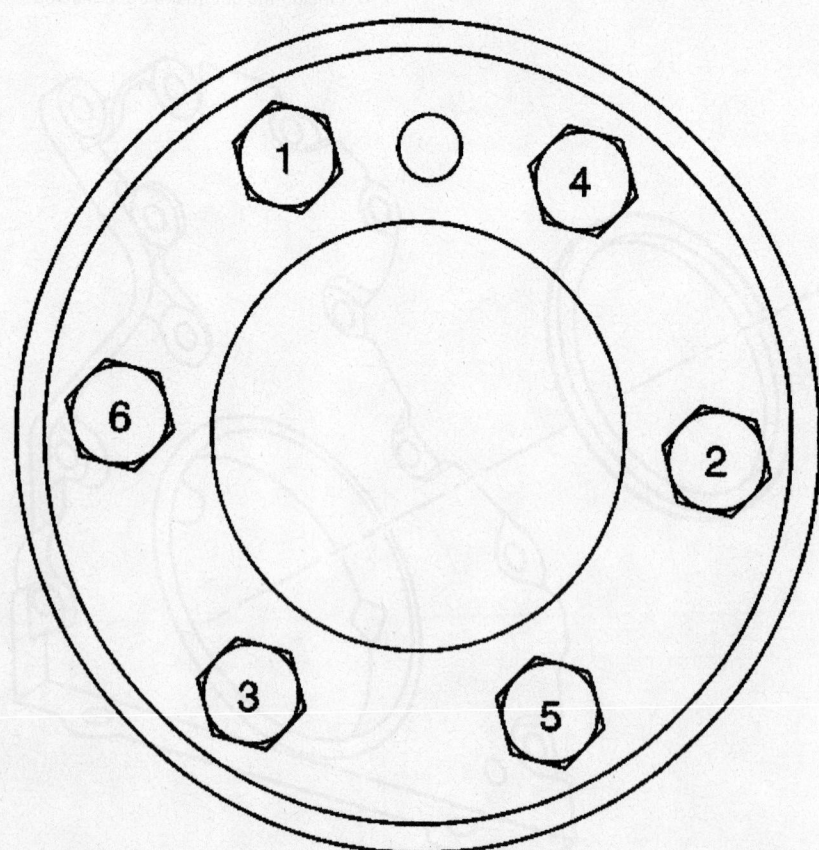

06025-GMWB-G31

Flywheel torque sequence—5.3L engine

Front Cover and Seal, Timing Chain, and Sprockets

REMOVAL & INSTALLATION

3.1L and 3.4L Engines

1. Before servicing the vehicle, refer to the Precautions Section.
2. Drain the cooling system.
3. Drain the engine oil.
4. Remove or disconnect the following:

- Negative battery cable
- Coolant reservoir
- Accessory drive belt
- Crankshaft balancer
- Drive belt tensioner
- Power steering pump, DO NOT disconnect the lines
- Thermostat bypass pipe from the front cover
- Upper radiator hose
- Water pump pulley
- Lower Crankshaft Position (CKP) sensor
- Front cover

5. Rotate the crankshaft until the timing marks on the camshaft and crankshaft sprockets are in alignment (facing each other).

- Camshaft sprocket bolt, sprocket and timing chain
- Crankshaft sprocket
- Timing chain damper bolts and damper, if necessary

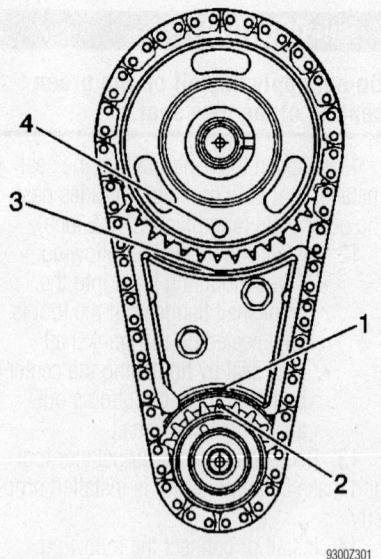

9300Z301

Be sure to align the damper mark (1) with the crankshaft mark (2) and the damper mark (3) with the camshaft sprocket mark (4)—3.1L and 3.4L engines

To install:

6. Install or connect the following:
   - Timing chain damper (if removed) and torque the bolts to 15 ft. lbs. (21 Nm)
   - Crankshaft sprocket

➡ **Be sure the timing mark on the crankshaft sprocket is pointing toward the mark on the chain damper.**

   - Timing chain over the camshaft sprocket

7. Loop the timing chain under the crankshaft sprocket and install the camshaft sprocket on the camshaft.

8. Verify that the marks are aligned; the camshaft sprocket will be at the 6 o'clock position and the crankshaft sprocket at the 12 o'clock position.

➡ **The No. 1 piston will be at Top Dead Center (TDC) and the No. 4 piston will also be at TDC but on the compression stroke.**

9. Tighten the camshaft sprocket bolt to 103 ft. lbs. (140 Nm).

10. Lubricate the timing chain components with engine oil.

11. Install or connect the following:
    - New front cover seal
    - Front cover using a new gasket and torque the small bolts to 20 ft. lbs. (27 Nm) and the large bolts to 41 ft. lbs. (55 Nm)
    - Water pump pulley and torque the bolts to 18 ft. lbs. (25 Nm)
    - Upper radiator hose
    - Thermostat bypass pipe
    - Power steering pump and torque the bolts to 25 ft. lbs. (34 Nm)
    - Drive belt tensioner and torque the bolt to 37 ft. lbs. (50 Nm)
    - CKP sensor
    - Crankshaft balancer and torque the bolt to 76 ft. lbs. (103 Nm)
    - Accessory drive belt
    - Coolant reservoir
    - Negative battery cable

12. Refill the fluids.

13. Start the engine and check for leaks, repair if necessary.

3.5L Engine

PRIMARY CHAIN

1. Before servicing the vehicle, refer to the Precautions Section.

2. Drain the engine oil.

3. Drain the cooling system.

4. Disconnect the negative battery cable.

5. Remove the camshaft covers.

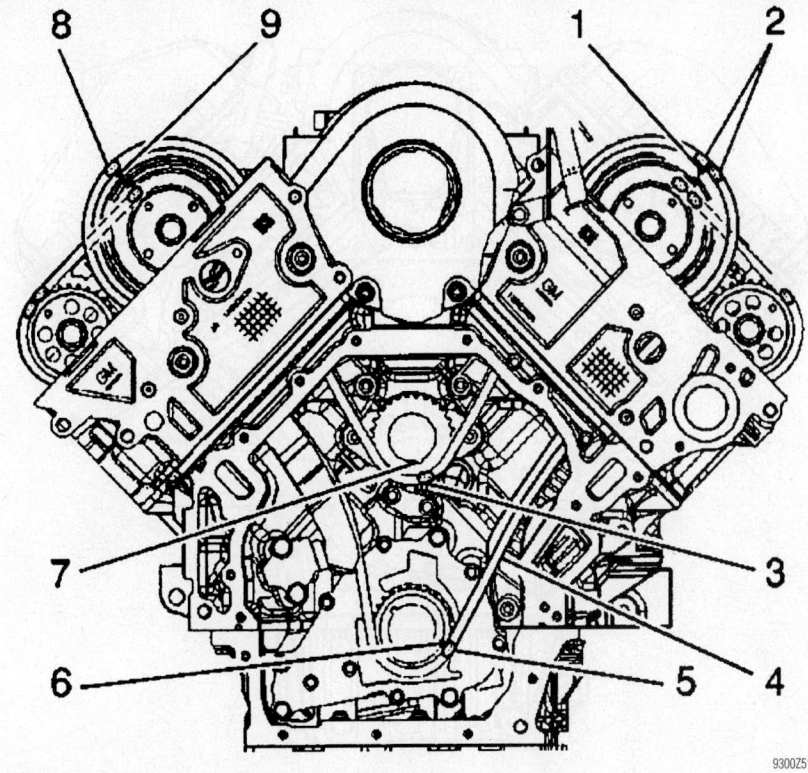

Primary timing chain alignment marks—3.5L engine

6. Rotate the crankshaft so the No. 1 piston is at Top Dead Center (TDC) and the flats on the rear of the camshafts are parallel with the camshaft cover sealing surface.

7. Install camshaft holding fixtures on both sets of camshafts. Turn the hex portion of the camshaft to align them for tool installation.

➡ **When installed, the flats on the rear of the camshafts will be parallel with the camshaft cover sealing surface.**

8. Remove or disconnect the following:
   - Right diagonal brace
   - Battery and tray
   - Coolant reservoir
   - Underhood accessory wiring junction block, move it aside
   - Drive belt
   - Power steering pump pulley
   - Idler pulley and belt tensioner
   - Water pump pulley
   - Water pump drive belt shield
   - Water pump

9. Support the engine cradle.

10. Remove the right side engine cradle bolts.

11. Lower the cradle.

12. Remove or disconnect the following:
    - Crankshaft balancer
    - Front cover
    - Engine lift bracket from the front of the engine

   - Camshaft Position (CMP) sensor
   - Sprocket bolt from the exhaust camshaft on the right cylinder head to allow for clearance of the chain guide
   - Four chain guide access plugs from the cylinder heads

➡ **Note that each plug has an O-ring.**

   - Primary chain tensioner

➡ **Remove the lower bolt allowing the tensioner to swing down and expand.**

   - Primary chain tensioner shoe, by removing the bolt, pushing the guide downward slightly and pulling it up through the cylinder head

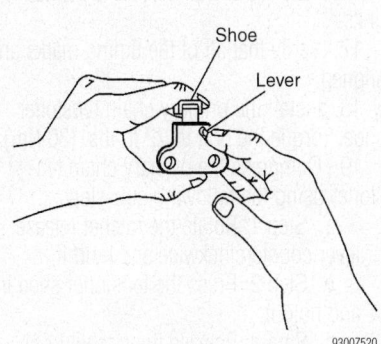

Compressing the primary chain tensioner—3.5L engine

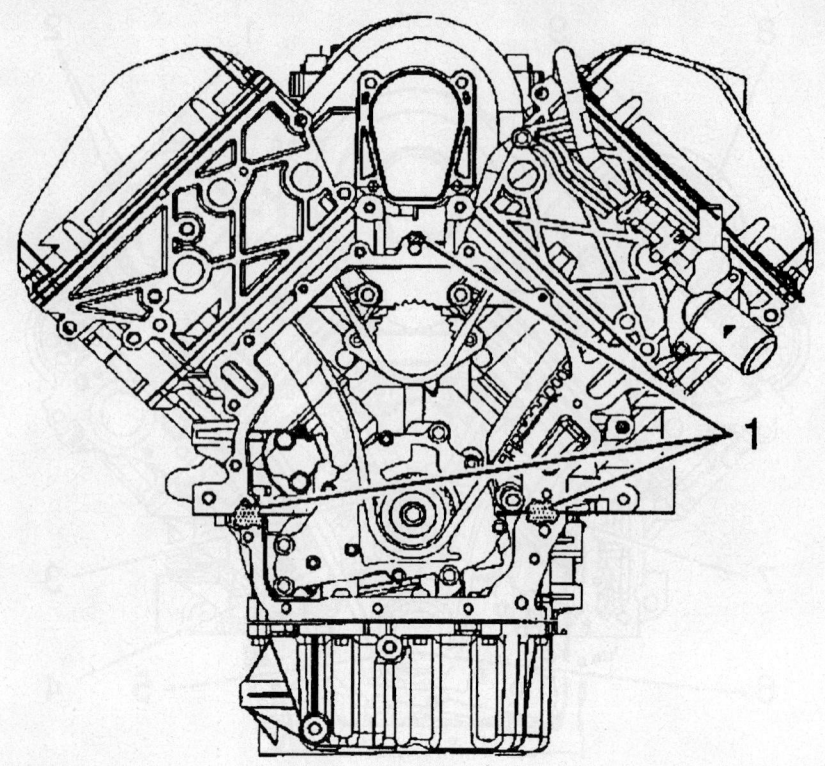

9300Z521

Apply RTV sealant to the 3 areas indicated before installing the front cover and gasket—3.5L engine

- Primary chain from the right camshaft, allowing it to fall into the oil pump area
- Primary chain

To install:

13. Rotate the crankshaft so the No. 1 piston is at Top Dead Center (TDC) and the mark on the crankshaft is at the 4 o'clock position.

14. Rotate the balance shaft so the timing mark is at the 5 o'clock position.

➡**Be sure the painted links are facing the front of the engine.**

15. Install the timing chain on the sprockets.

16. Center the mark on the left intake camshaft sprocket between the 2 painted links.

17. Verify that all of the timing marks are aligned.

18. Install the primary chain tensioner shoe. Torque the bolt to 22 ft. lbs. (30 Nm).

19. Compress the primary chain tensioner using the following sub-steps:

    a. Step 1: Rotate the ratchet release lever counterclockwise and hold it.

    b. Step 2: Press the tensioner shoe in and hold it.

    c. Step 3: Release the ratchet lever and slowly release the pressure on the shoe.

    d. Step 4: Insert a pin through the hole in the lever as the lever moves to the first click. The ratchet should hold the shoe in the compressed position.

➡**Be sure the lever on the tensioner is facing you when installed.**

20. Install or connect the following:
- Primary chain tensioner and torque the bolts to 18 ft. lbs. (25 Nm); then, remove the chain tensioner pin
- Four chain guide access plugs and torque the plugs to 44 inch lbs. (5 Nm)
- Front engine lift bracket and torque the hex head bolt to 37 ft. lbs. (50 Nm) and the internal drive bolt to 18 ft. lbs. (25 Nm)
- CMP sensor and torque the bolts to 80 inch lbs. (9 Nm)

21. Remove the camshaft holding tools.

22. Install the rocker arm covers. Torque the bolts to 80 inch lbs. (9 Nm).

23. Place a small bead of RTV sealant on the 3 areas indicated in the diagram.

24. Install or connect the following:
- Front cover with a new gasket and torque the bolts to 124 inch lbs. (14 Nm) and the coolant drain plug to 89 inch lbs. (10 Nm)
- Crankshaft balancer and torque the

bolt to 37 ft. lbs. (50 Nm) plus an additional 120 degree turn

25. Raise the engine cradle and install new bolts loosely.

26. Coat the sub-frame bushings with rubber lubricant.

27. Lower the vehicle onto the assembly. Align the sub-frame on the vehicle using 2 bolts or drill bits, ¾ inches thick by 8 inches long through the alignment holes on the right side of the frame.

28. Install or connect the following:
- New frame-to-body bolts and torque the bolts to 133 ft. lbs. (180 Nm)
- Water pump with a new gasket and torque the bolts to 124 inch lbs. (14 Nm)
- Water pump pulley and torque the bolts to 106 inch lbs. (12 Nm)
- Belt tensioner and torque the bolt to 37 ft. lbs. (50 Nm)
- Power steering pump pulley
- Drive belt
- Underhood accessory wiring junction
- Coolant reservoir and torque the nuts to 30 inch lbs. (3 Nm)
- Battery and tray
- Right diagonal brace and torque the bolts to 35 ft. lbs. (47 Nm)
- Camshaft covers and torque the bolts to 80 inch lbs. (9 Nm)
- Negative battery cable

29. Refill the engine cooling system.

30. Refill the engine with new oil.

31. Start the vehicle and check for leaks, repair if necessary.

SECONDARY TIMING CHAIN

1. Before servicing the vehicle, refer to the Precautions Section.

2. Drain the cooling system.

3. Remove or disconnect the following:
- Negative battery cable
- Thermostat housing for clearance

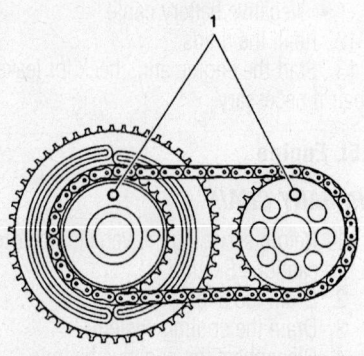

9300Z522

Correct sprocket alignment for secondary timing chain—3.5L engine

(when working on the front cylinder head)

- Rocker arm cover and install a Camshaft Holding Fixture
- Camshaft Position (CMP) sensor
- Camshaft sprocket bolts

4. Install the timing chain/sprocket holding fixture on the cylinder head.

5. Evenly slide the secondary drive chain and sprockets off the camshafts.

To install:

6. Install the secondary timing chain on the sprockets, with the drive pins at the 12 o'clock positions.

7. Install the sprockets/chain assembly onto the camshafts, with the chain properly aligned on the tensioner.

8. Remove the sprocket holding fixture from the cylinder head.

9. Install the sprocket bolts and torque the bolts to 18 ft. lbs. (25 Nm) plus an additional 45 degree turn.

10. Remove the camshaft holding fixture.

11. Install or connect the following:
- Rocker arm cover and torque the bolts to 80 inch lbs. (9 Nm)
- Thermostat housing (if removed)

and torque the bolts to 80 inch lbs. (9 Nm)
- CMP sensor and torque the bolts to 80 inch lbs. (9 Nm)
- Negative battery cable

12. Refill the cooling system.

13. Start the vehicle and check for leaks, repair if necessary.

3.8L Engines

1. Before servicing the vehicle, refer to the Precautions Section.
2. Drain the coolant system.
3. Drain the engine oil.

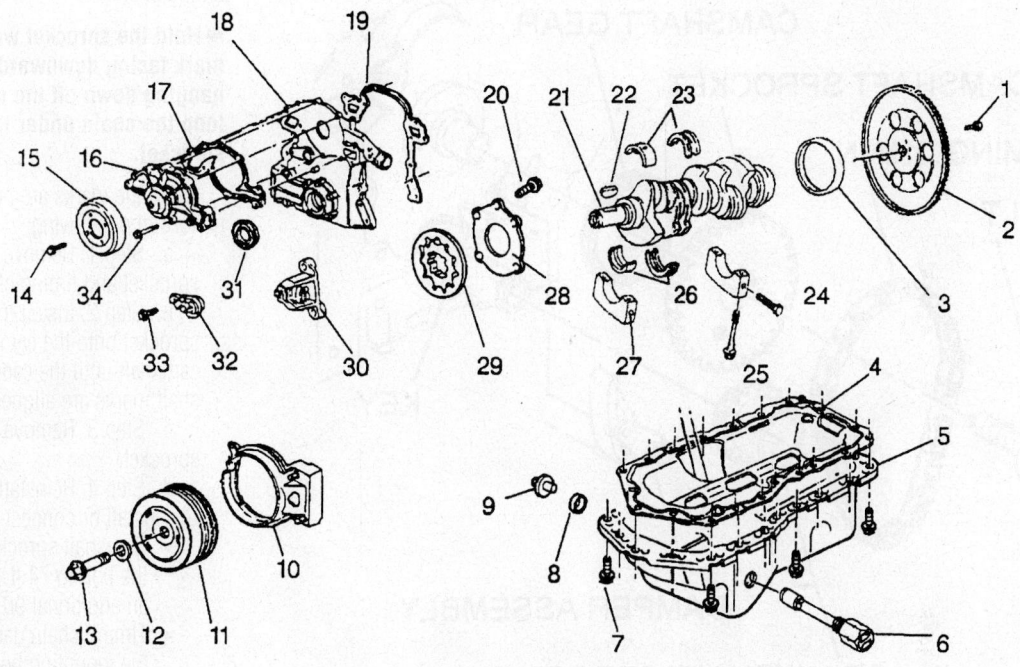

Legend

(1) Flywheel Bolt	(18) Engine Front Cover
(2) Flywheel	(19) Engine Front Cover Gasket
(3) Crankshaft Rear Oil Seal	(20) Oil Pump Cover Bolt
(4) Oil Pan Gasket (Includes Baffle)	(21) Engine Crankshaft
(5) Engine Oil Pan	(22) Crankshaft Balancer Key
(6) Oil Level Sensor	(23) Crankshaft Upper Bearing
(7) Oil Pan Bolt	(24) Side Main Bolt
(8) Oil Pan Drain Plug	(25) Crankshaft Main Bearing Cap Bolt
(9) Oil Pan Drain Gasket	(26) Crankshaft Lower Bearing
(10) Crankshaft Position Sensor Shield	(27) Crankshaft Main Bearing Cap
(11) Crankshaft Balancer	(28) Oil Pump Cover
(12) Crankshaft Balancer Washer	(29) Oil Pump Gear Set
(13) Crankshaft Balancer Bolt	(30) Crankshaft Position Sensor
(14) Water Pump Pulley Bolt	(31) Crankshaft Front Oil Seal
(15) Water Pump Pulley	(32) Camshaft Position Sensor
(16) Water Pump	(33) Camshaft Position Sensor Bolt
(17) Water Pump Gasket	(34) Water Pump Bolt

9300XG04

Exploded view of lower engine components—3.8L engine

4. Install an Engine Support Fixture.
5. Raise the engine so that the weight is removed from the engine mount.
6. Remove or disconnect the following:
- Negative battery cable
- Water pump pulley bolts
- Accessory drive belt(s)
- Drive belt tensioner
- Water pump pulley
- Crankshaft balancer
- Crankshaft Position (CKP) and

Camshaft Position (CMP) sensor connections
- Oil pressure sensor connection
- CKP sensor shield
- Engine mount bracket
- Front oil pan-to-front cover bolts
- Oil filter
- Lower radiator hose
- CKP sensor
- Front cover

7. Rotate the crankshaft until the timing mark on the camshaft sprocket is aligned with the crankshaft sprocket timing mark.
- Timing chain damper assembly
- Camshaft sprocket bolt
- Camshaft sprocket with the timing chain
- Crankshaft sprocket

To install:
8. Install the crankshaft sprocket.

➡It may be necessary to use a gear installer to fully seat the gear. Be sure the timing mark on the crankshaft gear is pointing straight up.

9. Install the camshaft gear with the timing chain.

➡Hold the sprocket with the timing mark facing downward and the chain hanging down off the sprocket; then, loop the chain under the crankshaft sprocket.

10. If the marks are not in alignment perform the following:
  a. Step 1: Remove the camshaft sprocket and timing chain.
  b. Step 2: Install the camshaft sprocket onto the camshaft and rotate the camshaft until the camshaft and crankshaft marks are aligned.
  c. Step 3: Remove the camshaft sprocket.
  d. Step 4: Reinstall the assembly.
11. Install or connect the following:
- Camshaft sprocket bolt and torque the bolt to 74 ft. lbs. (100 Nm) plus an additional 90 degree turn
- Timing chain damper and torque the mounting bolts to 16 ft. lbs. (22 Nm)
- Front cover seal lubricated with engine oil, using the appropriate seal driver
- New front cover gasket
- Front cover and torque the bolts to 15 ft. lbs. (20 Nm) plus an additional 40 degree turn
- Oil pan-to-front cover bolts and torque the bolts to 125 inch lbs. (14 Nm)
- CKP sensor and torque the bolts to 21 ft. lbs. (28 Nm)
- CKP sensor shield
- Engine mount bracket and torque the bolts to 65 ft. lbs. (87 Nm)
- Water pump pulley and torque the bolts to 115 inch lbs. (13 Nm)
- Lower radiator hose
- Oil filter
- Crankshaft balancer and torque the bolt to 111 ft. lbs. (150 Nm) plus an additional 75 degree turn

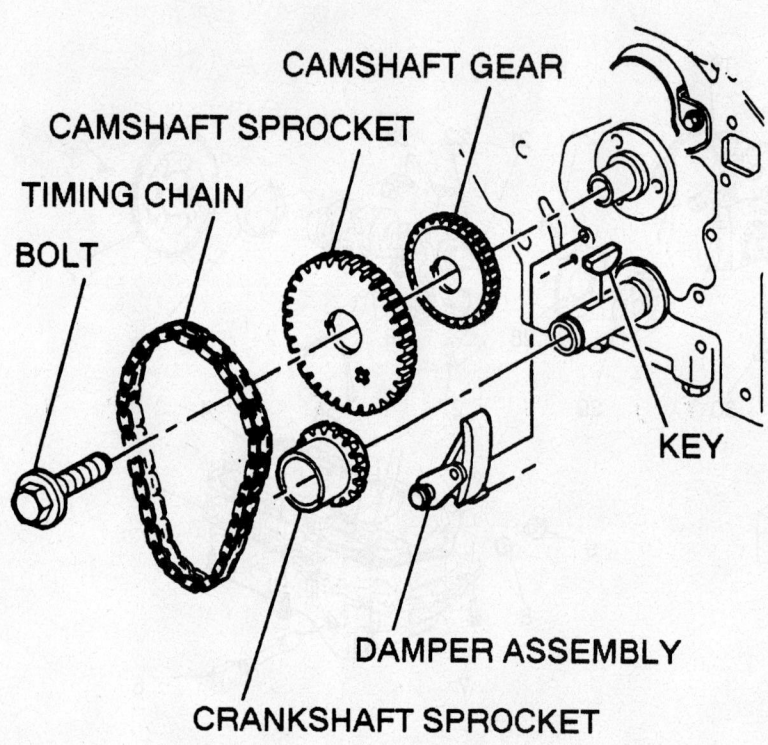

CAMSHAFT GEAR
CAMSHAFT SPROCKET
TIMING CHAIN
BOLT
KEY
DAMPER ASSEMBLY
CRANKSHAFT SPROCKET

7922XG16

Exploded view of the timing chain and sprockets—3.8L engines

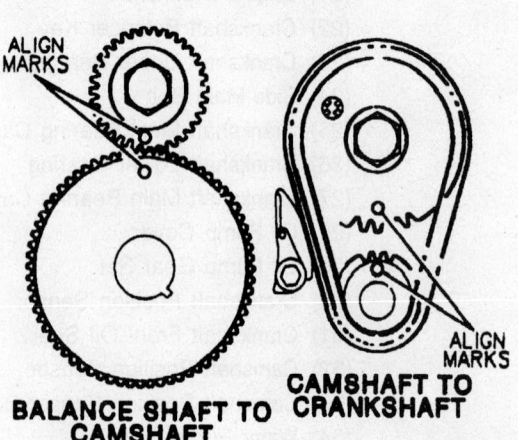

ALIGN MARKS
ALIGN MARKS
BALANCE SHAFT TO CAMSHAFT
CAMSHAFT TO CRANKSHAFT

7922XG17

Balance shaft-to-camshaft and camshaft-to-crankshaft timing mark alignment—3.8L engine

- Belt tensioner and torque the bolts to 37 ft. lbs. (50 Nm)
- Accessory drive belt
- Negative battery cable

12. Remove the engine support fixture.

13. Refill the cooling system.

14. Refill the engine oil.

15. Start the engine and check for leaks, repair if necessary.

5.3L Engine

1. Before servicing the vehicle, refer to the Precautions Section.

2. Remove the generator.

3. Remove the power steering pump.

4. Remove the coolant pump manifold.

5. Install a J 42640 Steering Column Anti-Rotation Pin.

6. Remove the accessory drive belt.

7. Remove the air cleaner upper housing.

8. Remove the engine mount strut. Refer to Engine Removal & Installation.

9. Remove the starter motor.

10. Remove the front fender splash shield.

11. Remove the transmission bellhousing bolt located at approximately the 10 o'clock position when looking from the rear of the engine.

12. Disconnect the transaxle cooler lines at the transaxle.

13. Remove the stabilizer shaft link lower nuts.

14. Remove the intermediate steering shaft pinch bolt and separate the shaft from the steering gear.

15. Remove the front lower air deflector braces and the deflector.

16. Remove the radiator to frame braces.

17. Install the engine support fixture. Refer to Engine Removal & Installation.

18. Raise and support the vehicle.

19. Remove the frame to body bolts.

20. Install a flywheel holding tool to the block and flywheel.

21. Remove the right front tire and wheel assembly.

22. Lower the engine approximately 100 mm (4 in.).

23. Remove the crankshaft balancer bolt. Do not discard the crankshaft balancer bolt. The balancer bolt will be used during the balancer installation procedure.

24. Install a puller to the crankshaft balancer.

25. Remove the crankshaft balancer.

26. Remove the belt tensioner bolt and reposition the tensioner (which blocks the front cover bolt).

27. Remove the oil pan-to-front cover bolts.

28. Remove the front cover bolts (501).

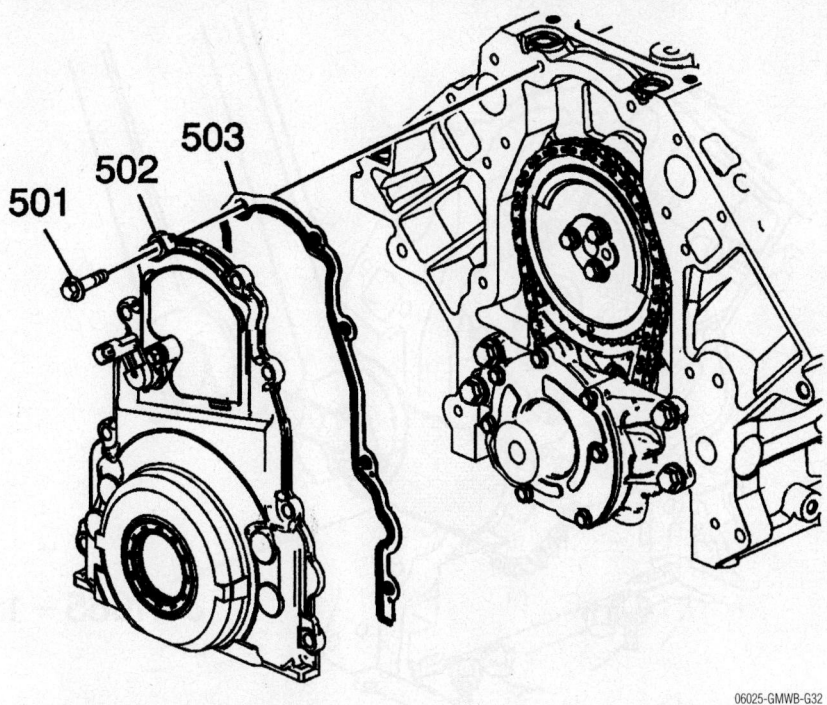

06025-GMWB-G32

Front cover and gasket—5.3L engine

29. Remove the front cover (502) and gasket (503).

30. Discard the front cover gasket.

31. Remove the oil seal, if necessary.

32. Remove the camshaft position (CMP) sensor bolt and sensor, if necessary.

33. Remove the O-ring seal from the sensor, if necessary.

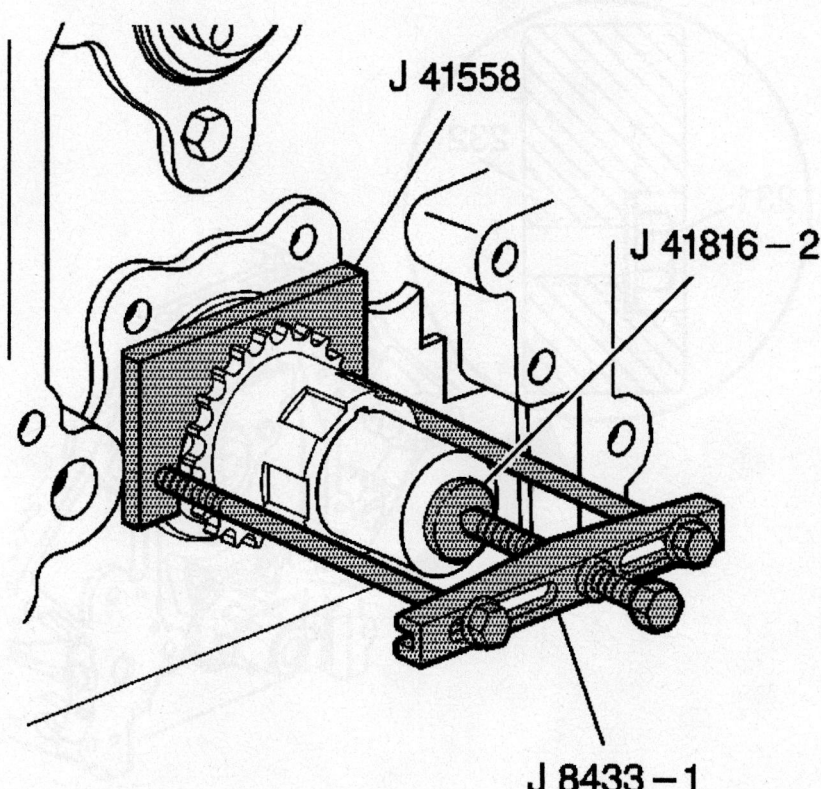

06025-GMWB-G34

Crankshaft sprocket removal—5.3L engine

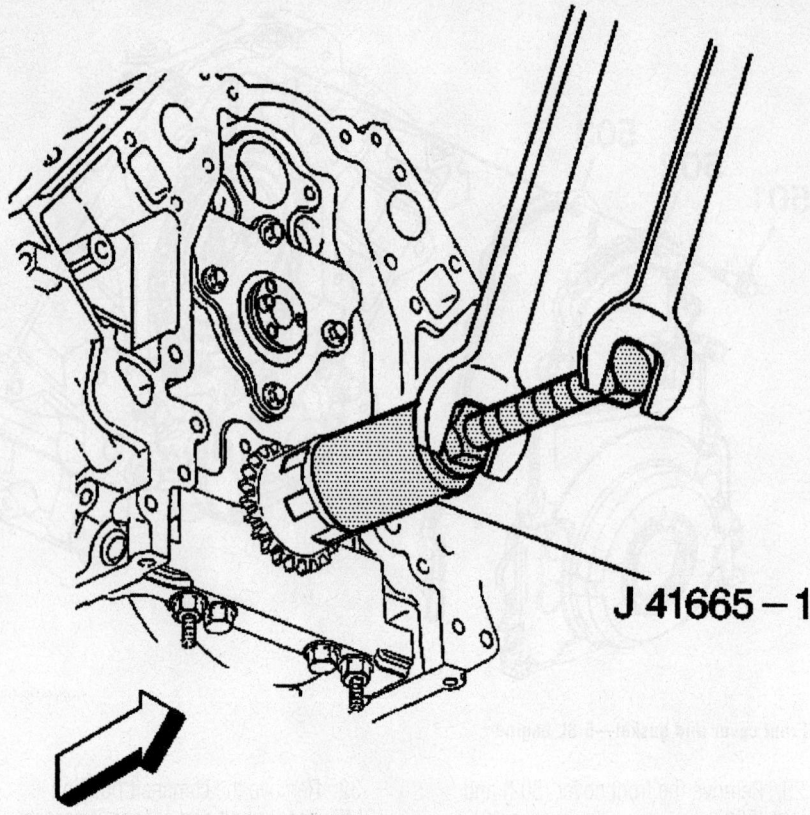

Using the J 41665, install the crankshaft sprocket—5.3L engine

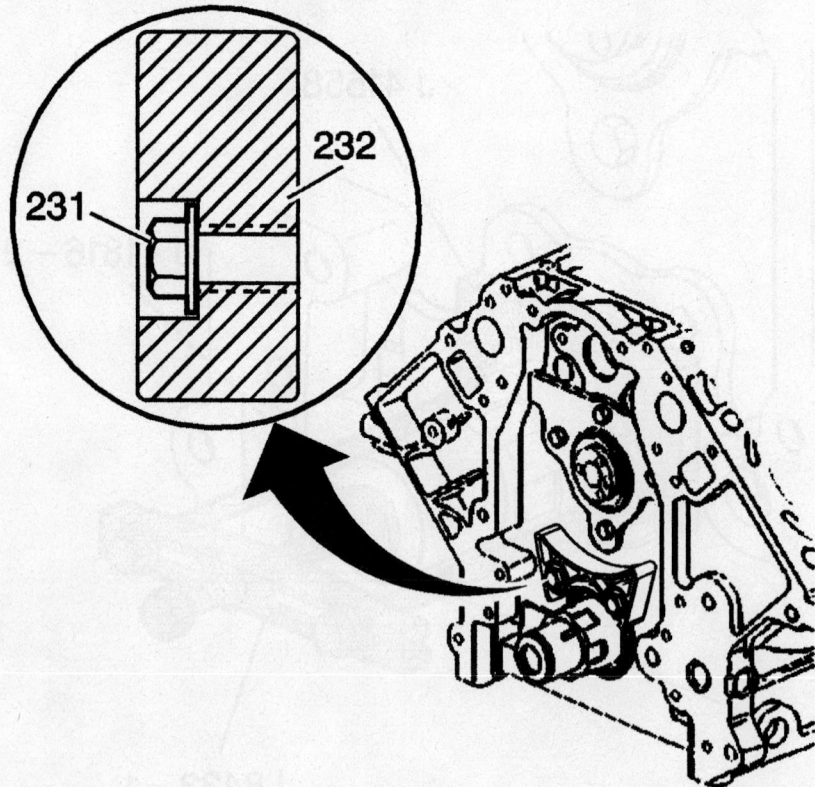

With the chain dampener properly installed, the heads of the bolts (231) should install to flush or below the face of the guide (232)—5.3L engine

34. Clean and inspect the engine front cover.

35. Rotate the crankshaft until the timing marks on the crankshaft and the camshaft sprockets are aligned.

❈❈ WARNING

Do not turn the crankshaft assembly after the timing chain has been removed in order to prevent damage to the piston assemblies or the valves.

36. Remove the camshaft sprocket bolts.

37. Remove the camshaft sprocket and timing chain.

38. Remove the timing chain dampener and bolts.

39. Using a puller, remove the crankshaft sprocket.

40. Remove the crankshaft sprocket.

41. Remove the crankshaft sprocket key, if required.

42. Clean and inspect the timing chain and sprockets.

To install:

43. Install the key into the crankshaft keyway, if previously removed.

44. Tap the key into the keyway until both ends of the key bottom onto the crankshaft.

45. Install the crankshaft sprocket onto the front of the crankshaft. Align the crankshaft key with the crankshaft sprocket keyway.

46. Using the J 41665, install the crankshaft sprocket onto the crankshaft until fully seated against the crankshaft flange.

47. Rotate the crankshaft sprocket until the alignment mark is in the 12 o'clock position.

48. Install the timing change guide and bolts. Tighten the bolts to 25 Nm (18 ft. lbs.).

➡Properly locate the camshaft sprocket locating pin with the camshaft sprocket alignment hole. The sprocket teeth and timing chain must mesh. The camshaft and the crankshaft sprocket alignment marks MUST be aligned properly. Position the camshaft sprocket alignment mark in the 6 o'clock position.

49. Install the camshaft sprocket and timing chain. If necessary, rotate the camshaft or crankshaft sprockets in order to align the timing marks.

50. Install the camshaft sprocket bolts. Tighten the bolts to 25 Nm (18 ft. lbs.).

51. Inspect the camshaft and crankshaft sprockets for proper timing mark alignment.

52. Install the oil pump.

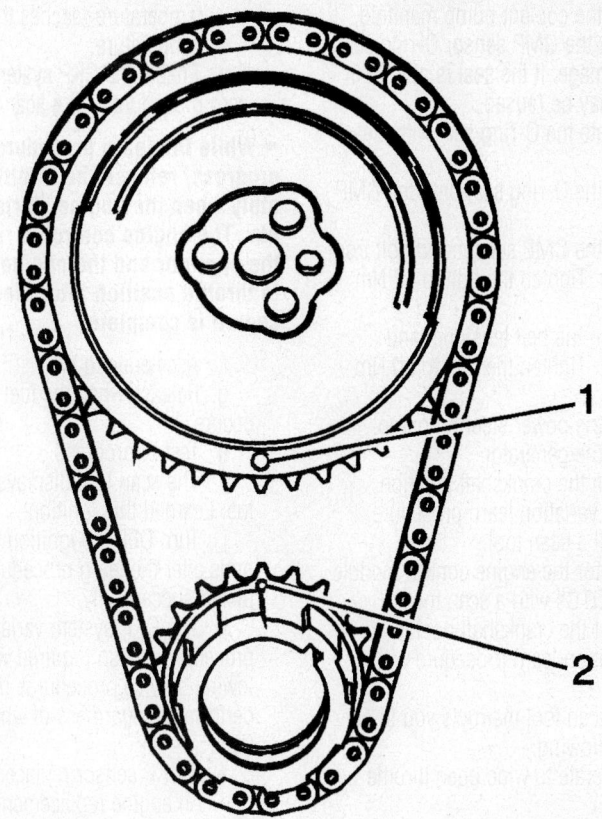

Timing mark alignment—5.3L engine

06025-GMWB-G36

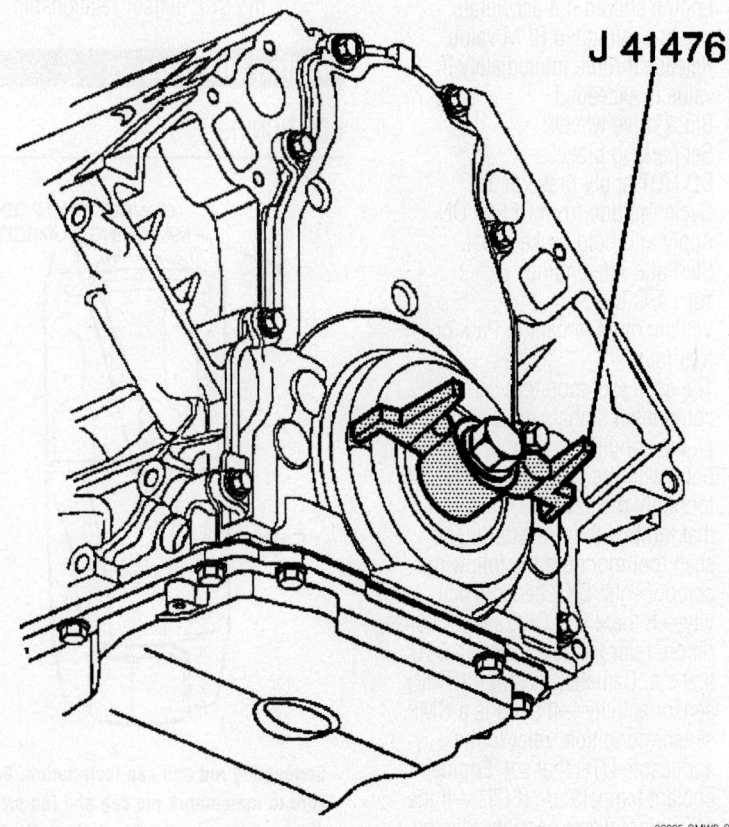

J 41476

Install the J 41476 to the front cover—5.3L engine

06025-GMWB-G37

➡ Do not reuse the crankshaft oil seal or front cover gasket. Do not apply any type of sealant to the front cover gasket (unless specified). The special tool in this procedure is used to properly center the front crankshaft front oil seal. All gasket surfaces should be free of oil or other foreign material during assembly. The crankshaft front oil seal MUST be centered in relation to the crankshaft. An improperly aligned front cover may cause premature front oil seal wear and/or engine oil leaks.

53. Apply a 5 mm (0.20 in.) bead of sealant GM P/N 12378577 or 12346141 (Canadian P/N 89022195) or equivalent 20 mm (0.80 in.) long to the oil pan to engine block junction.

54. Install the front cover gasket and cover.

55. Install the front cover bolts until snug. Do not overtighten.

56. Install the oil pan-to-front cover bolts until snug. Do not over tighten.

57. Install the J 41476 to the front cover.

58. Align the tapered legs of the J 41476 with the machined alignment surfaces on the front cover.

59. Install the crankshaft balancer bolt until snug. Do not overtighten. Tighten the oil pan to front cover bolts to 25 Nm (18 ft. lbs.). Tighten the engine front cover bolts to 25 Nm (18 ft. lbs.).

60. Remove the J 41476.

61. Install a NEW crankshaft front oil seal.

➡ The used crankshaft balancer bolt will be used only during the first pass of the balancer installation procedure. Install a NEW bolt and tighten as described in the second, third and forth passes of the balancer bolt tightening procedure. The crankshaft balancer installation and bolt tightening involves a four stage tightening process. The first pass ensures that the balancer is installed completely onto the crankshaft. The second, third, and forth passes tighten the new bolt to the proper torque.

➡ The balancer should be positioned onto the end of the crankshaft as straight as possible prior to tool installation.

62. Position the crankshaft balancer onto the end of the crankshaft.

63. Using tool J 41665, install the crankshaft balancer.

　a. Assemble the threaded rod, nut, washer and installer. Insert the smaller end of the installer into the front of the balancer.

b. Use a wrench and hold the hex end of the threaded rod.

c. Use a second wrench and rotate the installation tool nut clockwise until the balancer is started onto the crankshaft.

d. Remove the tool and reverse the installation tool. Position the larger end of the installer against the front of the balancer.

e. Use a wrench and hold the hex end of the threaded rod.

f. Use a second wrench and rotate the installation tool nut clockwise until the balancer is installed onto the crankshaft.

g. Remove the balancer installation tool.

64. Install the USED crankshaft balancer bolt. Tighten the USED bolt to 330 Nm (240 ft. lbs.).

65. Remove the USED crankshaft balancer bolt.

➡ **The nose of the crankshaft should be recessed 2.4-4.48 mm (0.094-0.176 in.) into the balancer bore.**

66. Measure for a correctly installed balancer. If the balancer is not installed to the proper dimensions, install the J 41665 and repeat the installation procedure.

67. Install a NEW crankshaft balancer bolt. Tighten the bolt a first pass to 50 Nm (37 ft. lbs.). Tighten the bolt a second pass to 140 degrees.

68. Remove the flywheel holding tool from the block and flywheel.

69. Install the transmission bellhousing bolt located at approximately the 10 o'clock position when looking from the rear of the engine. Tighten the bolt to 75 Nm (55 ft. lbs.).

70. Raise and properly position the frame and install the frame to body bolts. Refer to Engine Removal & Installation.

71. Install the frame to radiator braces.

72. Install the front lower air deflector.

73. Connect the intermediate steering shaft to the steering gear.

74. Install the stabilizer shaft link lower nuts.

75. Connect the transaxle cooler lines to the transaxle.

76. Remove the engine support fixture.

77. Install the front fender splash shield.

78. Install the starter motor.

79. Install the air cleaner upper housing.

80. Install the accessory drive belt.

81. Install the engine mount strut. Refer to Engine Removal & Installation.

82. Install the right front tire and wheel assembly.

83. Install the coolant pump manifold.

84. Inspect the CMP sensor O-ring seal for cuts or damage. If the seal is not cut or damaged, it may be reused.

85. Lubricate the O-ring seal with clean engine oil.

86. Install the O-ring seal onto the CMP sensor.

87. Install the CMP sensor and bolt from the front cover. Tighten the bolt to 12 Nm (106 inch lbs.).

88. Position the belt tensioner and install the bolt. Tighten the bolt to 50 Nm (37 ft. lbs.).

89. Install the power steering pump.

90. Install the generator.

91. Perform the crankshaft position (CKP) system variation learn procedure.

a. Install a scan tool.

b. Monitor the engine control module (ECM) for DTCs with a scan tool.

c. Select the crankshaft position (CKP) variation learn procedure with a scan tool.

d. The scan tool instructs you to perform the following:

• Accelerate to wide open throttle (WOT).
• Release throttle when fuel cut-off occurs.
• Observe fuel cut-off for applicable engine.
• Engine should not accelerate beyond calibrated RPM value.
• Release throttle immediately if value is exceeded.
• Block drive wheels.
• Set parking brake.
• DO NOT apply brake pedal.
• Cycle ignition from OFF to ON.
• Apply and hold brake pedal.
• Start and idle engine.
• Turn A/C OFF.
• Vehicle must remain in Park or Neutral.
• The scan tool monitors certain component signals to determine if all the conditions are met to continue with the procedure. The scan tool only displays the condition that inhibits the procedure. The scan tool monitors the following components: CKP sensors activity—If there is a CKP sensor condition, refer to the applicable DTC that set. Camshaft position (CMP) sensor activity—If there is a CMP sensor condition, refer to the applicable DTC that set. Engine coolant temperature (ECT)—If the ECT is not warm enough, idle the engine until the engine coolant

temperature reaches the correct temperature.

e. Enable the CKP system variation learn procedure with a scan tool.

➡ **While the learn procedure is in progress, release the throttle immediately when the engine starts to decelerate. The engine control is returned to the operator and the engine responds to throttle position after the learn procedure is complete.**

f. Accelerate to WOT.

g. Release when the fuel cut-off occurs.

h. Test in progress

i. The scan tool displays Learn Status: Learned this ignition.

j. Turn OFF the ignition for 30 seconds after the learn procedure is completed successfully.

k. The CKP system variation learn procedure is also required when the following service procedures have been performed, regardless of whether DTC P0315 is set:

• A CKP sensor replacement
• An engine replacement
• An ECM replacement
• A harmonic balancer replacement
• A crankshaft replacement
• Any engine repairs which disturb the CKP sensor relationship

Piston and Ring

POSITIONING

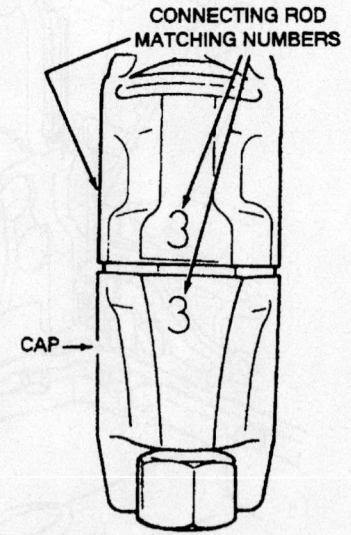

Connecting rod and cap installation. Be sure to matchmark the cap and rod prior to disassembly, as shown—3.1L, 3.4L, 3.8L and 5.3L engines

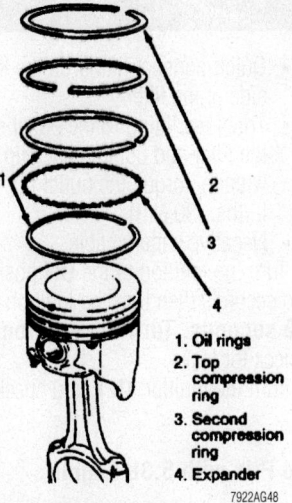

1. Oil rings
2. Top compression ring
3. Second compression ring
4. Expander

7922AG48

Piston ring positioning—3.1L, 3.4L, 3.8L and 5.3L engines

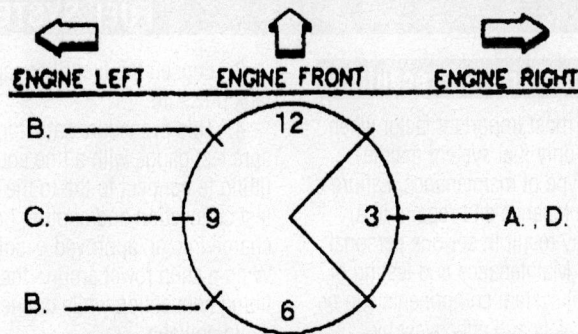

A. OIL RING SPACER GAP (TANG IN HOLE OR SLOT WITH ARC)
B. OIL RING RAIL GAPS
C. 2ND COMPRESSION RING GAP
D. TOP COMPRESSION RING GAP

7922AG46

Piston ring end-gap spacing—3.1L, 3.4L, 3.8L and 5.3L engines

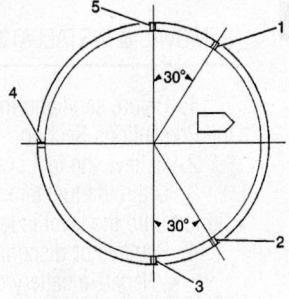

1. Lower oil control ring
2. Upper oil control ring
3. Top Ring
4. Oil control ring expander
5. Second ring

9306XG05

Piston ring end-gap positioning—3.5L engine

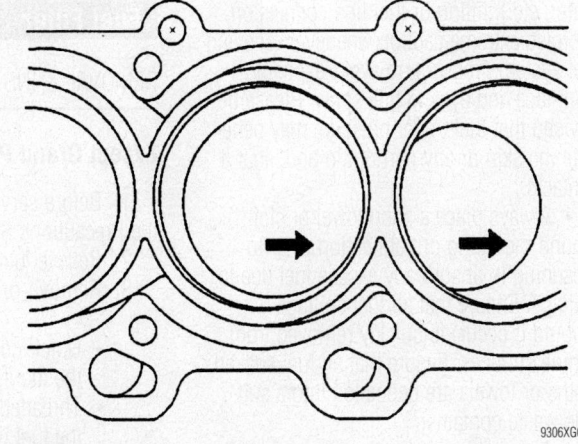

9306XG06

Piston positioning—3.5L engine

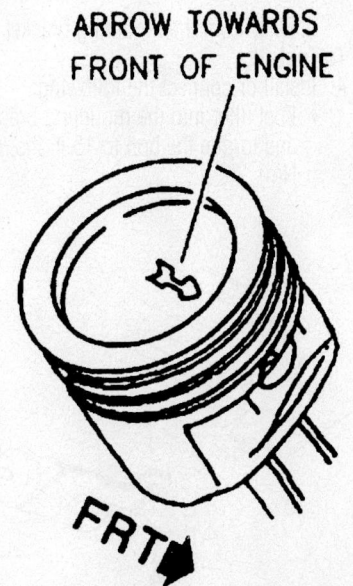

ARROW TOWARDS FRONT OF ENGINE

FRT

7922AG47

Piston positioning. Often the arrow is replaced by a notch, which also must face toward the front of the engine—3.1L, 3.4L, 3.8L and 5.3L engines

FUEL SYSTEM

Fuel System Service Precautions

Safety is the most important factor when performing not only fuel system maintenance but any type of maintenance. Failure to conduct maintenance and repairs in a safe manner may result in serious personal injury or death. Maintenance and testing of the vehicle's fuel system components can be accomplished safely and effectively by adhering to the following rules and guidelines.

• To avoid the possibility of fire and personal injury, always disconnect the negative battery cable unless the repair or test procedure requires that battery voltage be applied.

• Always relieve the fuel system pressure prior to disconnecting any fuel system component (injector, fuel rail, pressure regulator, etc.), fitting or fuel line connection. Exercise extreme caution whenever relieving fuel system pressure, to avoid exposing skin, face and eyes to fuel spray. Please be advised that fuel under pressure may penetrate the skin or any part of the body that it contacts.

• Always place a shop towel or cloth around the fitting or connection prior to loosening to absorb any excess fuel due to spillage. Ensure that all fuel spillage (should it occur) is quickly removed from engine surfaces. Ensure that all fuel soaked cloths or towels are deposited into a suitable waste container.

• Always keep a dry chemical (Class B) fire extinguisher near the work area.

• Do not allow fuel spray or fuel vapors to come into contact with a spark or open flame.

• Always use a backup wrench when loosening and tightening fuel line connection fittings. This will prevent unnecessary stress and torsion to fuel line piping. Always follow the proper torque specifications.

• Always replace worn fuel fitting O-rings with new. Do not substitute fuel hose or equivalent, where fuel pipe is installed.

Fuel System Pressure

RELIEVING

1. Before servicing the vehicle, refer to the Precautions Section.
2. Disconnect the negative battery cable to prevent possible discharge of fuel if an accidental attempt is made to start the engine.

3. Loosen the fuel filler cap to relieve tank pressure.
4. This procedure calls for a fuel pressure test gauge with a line equipped with a fitting to connect to the to the fuel pressure test connection and another hose to discharge into an approved gasoline container. Wrap a shop towel around the pressure test fitting connection while connecting gauge to avoid spillage.
5. Install the bleed hose into an approved container and open the valve to bleed fuel system pressure. The fuel connections are now safe for servicing.
6. Drain any fuel remaining in the gauge into an approved container.
7. Reconnect the negative battery cable unless additional service work is being performed.

Fuel Filter

REMOVAL & INSTALLATION

Except Grand Prix w/5.3L Engine

1. Before servicing the vehicle, refer to the Precautions Section.
2. Relieve fuel system pressure.
3. Remove or disconnect the following:

• Quick-connect fitting at the inlet of the fuel filter
• Threaded fitting at the outlet side of the fuel filter.

➡**Have a container available to retrieve any fuel remaining in the filter.**

• Filter from the mounting bracket
To install:
4. Install or connect the following:
• Fuel filter into the mounting bracket and torque the bolt to 15 ft. lbs. (20 Nm)

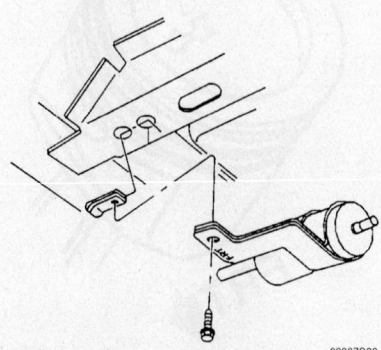

Common fuel filter mounting

9306ZG20

• Quick-connect fitting on the inlet side of the filter
• Threaded fitting to the outlet side of the filter and using a back up wrench, torque the outlet nut to 22 ft. lbs. (30 Nm)
• Negative battery cable
5. Turn the ignition to the **ON** position for two seconds then turn the ignition **OFF- for 10 seconds. Turn the ignition ON** and check for leaks.
6. Turn the ignition **OFF** and check for leaks.

Grand Prix with 5.3L Engine

The fuel filter is located inside the fuel tank and is not serviceable.

Fuel Pump

REMOVAL & INSTALLATION

1. Before servicing the vehicle, refer to the Precautions Section.
2. Relieve the fuel system pressure.
3. Drain the fuel tank with a hand held siphon until the level is less than ¼ full.
4. Remove or disconnect the following:
• Negative battery cable
• Spare tire and jack
• Floor trunk liner by pulling it back
• Fuel sender access panel
• Fuel tank pressure sensor electrical connector
• Fuel sender electrical connector
• Fuel sender assembly quick connect fittings
• Retaining lockring from the fuel sender

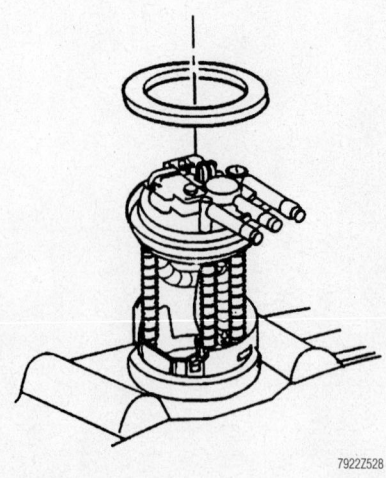

7922Z528

Remove the fuel pump from the tank after removing the locking ring

⁕⁕ WARNING

When the lockring is removed from the fuel sender, the sender assembly will spring up. Downward pressure should be kept on the assembly and slowly released to ensure the sender assembly does not get damaged.

- Modular fuel sender assembly

To install:

5. Install or connect the following:
- New O-ring on the fuel tank
- Fuel sender
- Lockring on the fuel sender
- Fuel sender electrical connector
- Fuel tank pressure sensor electrical connector
- Quick connect fittings at the fuel sender
- Negative battery cable

6. Add a small amount of fuel to the fuel tank.

7. Turn the ignition to the **ON** position for 2 seconds then turn the ignition **OFF** for 10 seconds. Turn the ignition **ON** and check for leaks.

8. Turn the ignition **OFF** and check for leaks.

9. Install or connect the following:
- Fuel sender access panel and torque the nuts to 89 inch lbs. (10 Nm)

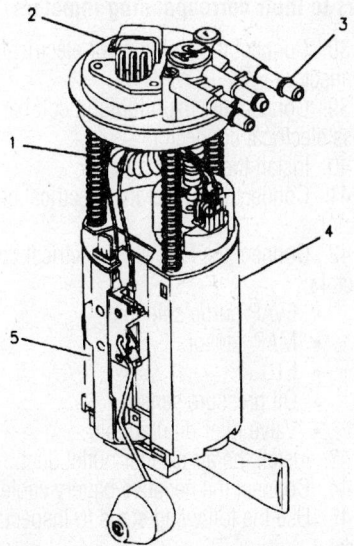

1 SUPPORT ASSEMBLY - FUEL SENDER
2 COVER ASSEMBLY - FUEL SENDER
3 FUEL PIPES (ABOVE COVER)
4 RESERVOIR - FUEL PUMP FUEL
5 SENSOR ASSEMBLY - FUEL LEVEL

7922XG19

Fuel pump and sending unit module assembly

- Trunk liner
- Spare tire, jack and spare tire cover
10. Refill the fuel tank.

Fuel Injector

REMOVAL & INSTALLATION

3.1L and 3.4L Engines

1. Before servicing the vehicle, refer to the Precautions Section.

2. Relieve the fuel system pressure.

3. Remove or disconnect the following:
- Upper intake manifold
- Fuel inlet and return lines
- Fuel injector electrical connectors
- Coolant temperature sensor electrical connector
- Fuel rail
- Fuel injector retaining clips
- Fuel injectors

To install:

4. Coat the new fuel injector O-rings with clean engine oil.

5. Install or connect the following:
- Lower backup O-ring
- Upper O-ring
- Lower O-ring
- Fuel injector
- Fuel injector retaining clips
- Fuel rail and torque the bolts to 7 ft. lbs. (10 Nm)
- Coolant temperature sensor electrical connector
- Fuel injector electrical connectors

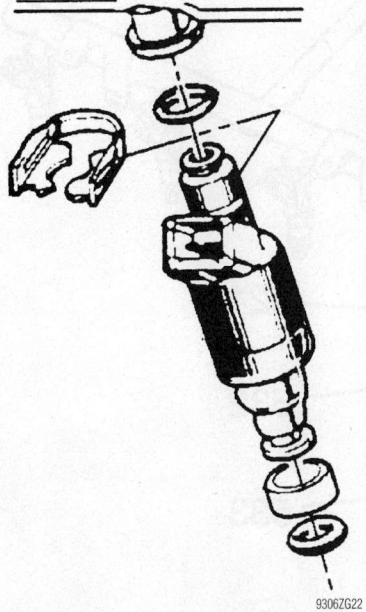

9306ZG22

Exploded view of the fuel injector—3.1L and 3.4L engines

- Fuel inlet and return lines with new O-rings and torque the fittings to 13 ft. lbs. (17 Nm)
- Upper intake manifold and torque the bolts to 18 ft. lbs. (25 Nm)
- Negative battery cable

6. Start the vehicle and check for leaks, repair if necessary.

3.5L and 3.8L Engines

1. Before servicing the vehicle, refer to the Precautions Section.

2. Relieve the fuel system pressure.

3. Remove or disconnect the following:
- Fuel injector sight shield
- Fuel feed and return lines
- Fuel pressure regulator vacuum connection
- Ignition coil wires from the coil
- Fuel injector electrical connectors
- Fuel rail
- Fuel injector retaining clips
- Fuel injectors

To install:

4. Coat the new fuel injector O-rings with clean engine oil.

5. Install or connect the following:
- Lower backup O-ring
- Upper O-ring
- Lower O-ring
- Fuel injector
- Fuel injector retaining clips
- Fuel rail and torque the bolts to 7 ft. lbs. (10 Nm) if equipped with bolt retainers. If equipped with snap-lock tab retainers, push down until the retainers snap into place.
- Fuel injector electrical connectors
- Ignition coil wires to the coil
- Fuel pressure regulator vacuum connection
- Fuel feed and return lines
- Negative battery cable

6. Start the vehicle and check for leaks; repair if necessary.

7. Install the fuel injector sight shield and torque the nuts to 27 inch lbs. (3 Nm).

5.3L Engine

1. Before servicing the vehicle, refer to the Precautions Section.

➡ **Clean the fuel and evaporative emission (EVAP) connections and surrounding areas prior to disconnecting the lines in order to avoid possible system contamination.**

2. Relieve the fuel system pressure.

3. Remove the air cleaner outlet duct.

4. Disconnect the following electrical connectors:

- EVAP purge solenoid
- Manifold absolute pressure (MAP) sensor
- Electronic throttle control (ETC)
- Oil pressure sensor
- Valve lifter oil manifold

5. Disconnect the generator electrical connector.

6. Remove the connector position assurance (CPA) retainer.

7. Disconnect the main ignition coil harness electrical connector.

➡**Mark the injector connectors to their corresponding injectors to ensure correct reassembly.**

8. Disconnect the fuel injector electrical connectors.

9. Remove the CPA retainer.

10. Disconnect the main ignition coil harness electrical connector.

➡**Mark the injector connectors to their corresponding injectors to ensure correct reassembly.**

11. Disconnect the fuel injector electrical connectors.

12. Remove the engine wiring harness retainers from the tabs on the fuel rail.

13. Reposition the harness out of the way.

14. Disconnect the fuel feed and EVAP lines.

15. Note the location of the fuel rail ground strap.

16. Remove the intake manifold bolt and ground strap.

17. Remove the fuel rail bolts.

18. Remove the fuel rail with injectors. Lift evenly on both sides of the fuel rail until all injectors have been removed from their bores.

➡**Do not separate the fuel injectors from the fuel rail unless component service is required.**

19. Remove the fuel injector retaining clip, as required.

20. Remove the fuel injector, as required

21. Remove and discard the fuel injector O-ring seals, as required.

To install

➡**Do not reuse the fuel injector O-ring seals. Install NEW O-ring seals during assembly.**

22. Note the installed location of the fuel rail ground strap.

23. Lubricate the NEW O-ring seals with clean engine oil.

24. Install the NEW fuel injector O-ring seals (532, 534) onto the injector, as required.

25. Install the fuel injector (533), as required

26. Install the fuel injector retaining clip (521), as required.

27. If necessary, lubricate the NEW O-ring seals with clean engine oil.

28. If necessary, install NEW O-ring seals to the fuel injectors.

29. Install the fuel rail (510) with injectors. Push firmly on both sides of the rail until all the injectors have been seated into their bores.

30. Apply a 5 mm (0.2 in.) band of threadlock GM P/N 12345382 (Canadian P/N 10953489) or equivalent to the threads of the fuel rail bolts.

31. Install the fuel rail bolts. Tighten the bolts to 10 Nm (89 inch lbs.).

32. Install the ground strap and intake manifold bolt. Tighten the bolt to 10 Nm (89 inch lbs.).

33. Connect the fuel feed and EVAP lines.

34. Position the harness to the engine. Install the engine wiring harness retainers to the tabs on the fuel rail.

➡**Install the marked injector connectors to their corresponding injectors.**

35. Connect the fuel injector electrical connectors.

36. Connect the main ignition coil harness electrical connector.

37. Install the CPA retainer.

➡**Install the marked injector connectors to their corresponding injectors.**

38. Connect the fuel injector electrical connectors.

39. Connect the main ignition coil harness electrical connector.

40. Install the CPA retainer.

41. Connect the generator electrical connector.

42. Connect the following electrical connectors:

- EVAP purge solenoid
- MAP sensor
- ETC
- Oil pressure sensor
- Valve lifter oil manifold

43. Install the air cleaner outlet duct.

44. Connect the negative battery cable.

45. Use the following steps to inspect for leaks:

   a. Turn the ignition ON, with the engine OFF, for 2 seconds.

   b. Turn the ignition OFF for 10 seconds.

   c. Turn the ignition ON with the engine OFF.

   d. Inspect for leaks.

46. Install the engine sight shield.

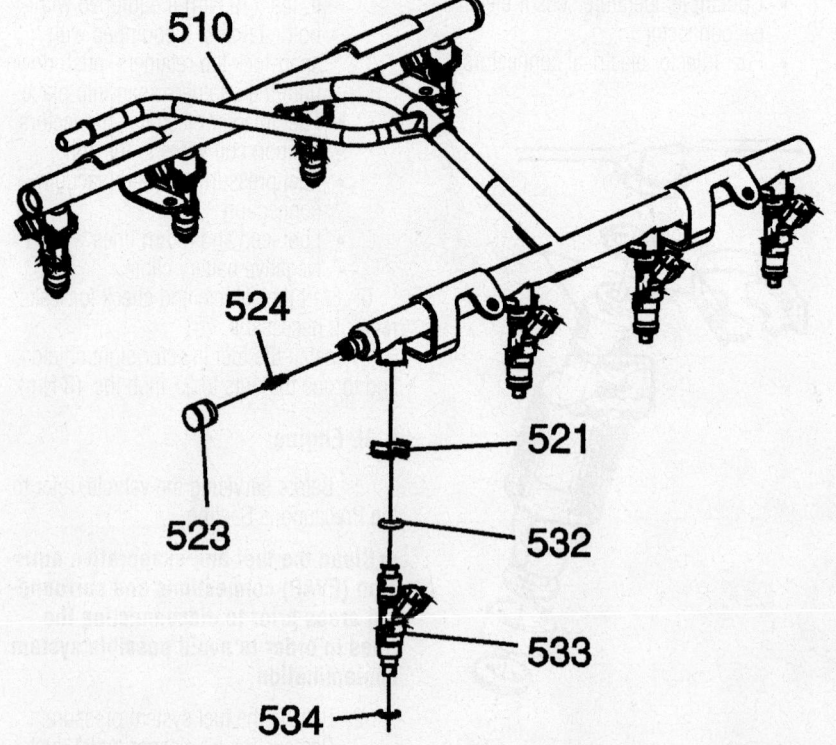

06025-GMWB-G38

Fuel rail and injectors—5.3L engine

DRIVE TRAIN

Transaxle Assembly

REMOVAL & INSTALLATION

➡ **These transaxles were used in a variety of General Motor's vehicles. Due to model year, vehicle model and installed options, the removal and installation procedures may vary slightly. The procedures given here should suffice for most all vehicles using these transaxles.**

4T60-E Transmission

1. Before servicing the vehicle, refer to the Precautions Section.
2. Drain the transmission fluid.
3. Install an engine support fixture.
4. Remove or disconnect the following:
   - Negative battery cable
   - Throttle body air inlet duct
   - Engine mount struts
   - Wire harness connectors from the transmission
   - Vacuum hose and pipe from the modulator
   - Range selector cable from the Park/Neutral Position (PNP) switch
   - PNP switch
   - Fluid filler tube
   - Upper transmission bolts
   - Wire harness grounds
   - Both front wheels
   - Engine splash shields
   - Both tie rod ends from the steering knuckles
   - Power steering gear from the frame and secure it to the body of the vehicle
   - Power steering cooler line clamps
   - Engine mount lower nuts
   - Lower ball joints from the steering knuckles
   - Torque converter cover
   - Starter motor
   - Torque converter bolts
   - Oil cooler hoses
   - Drive axles and secure them to the steering knuckles and struts
   - Wheel Speed Sensor (WSS) connectors
   - Vehicle Speed Sensor (VSS) connectors
5. Use a transmission table to support the transmission.
   - Transmission brace
   - Lower transmission bolts
   - Frame-to-body bolts
6. Lower the frame from the vehicle.

7. Remove the transmission from the frame.

To install:

8. Install or connect the following:
   - Transmission
   - Lower transmission bolts and torque them to 55 ft. lbs. (75 Nm)
   - New frame to body bolts and torque them to 125 ft. lbs. (170 Nm)
   - Transmission brace and torque the bolts to 35 ft. lbs. (47 Nm)
   - VSS electrical connector
   - WSS electrical connectors
   - Drive axles to the transmission
   - Oil cooler hoses
   - Torque converter bolts and torque them to 46 ft. lbs. (63 Nm)
   - Starter motor and torque the bolts to 32 ft. lbs. (43 Nm)
   - Torque converter cover and torque the bolts to 89 inch lbs. (10 Nm)
   - Lower ball joints to the steering knuckle and torque the nuts to 40 ft. lbs. (55 Nm)
   - Engine mount lower nuts and torque them to 32 ft. lbs. (43 Nm)
   - Power steering cooler line clamps to the frame
   - Power steering gear to the frame and torque the bolts to 59 ft. lbs. (80 Nm)
   - Tie rod ends to the steering knuckles and torque the nuts to 63 ft. lbs. (85 Nm)
   - Engine splash shields
   - Front wheels
   - Fluid filler tube
   - Upper transaxle bolts and torque them to 55 ft. lbs. (75 Nm)
   - PNP switch and torque the bolts to 18 ft. lbs. (25 Nm)
   - Range selector cable with the bracket and torque the nut to 15 ft. lbs. (20 Nm)
   - Range selector cable to the PNP switch
   - Wire harness connectors to the transmission
   - Vacuum hose and pipe to the modulator
   - Engine mount struts and torque the bolts to 37 ft. lbs. (50 Nm)
9. Remove the engine support fixture.
   - Throttle body air inlet duct
   - Negative battery cable
10. Fill the transmission with fluid.
11. Start the engine and check the engine and transaxle oil level. Add oil if necessary.

4T65-E Transmission

1. Before servicing the vehicle, refer to the Precautions Section.
2. Drain the transmission fluid.
3. Install an engine support fixture.
4. Remove or disconnect the following:
   - Negative battery cable
   - Throttle body air inlet duct
   - Engine mount struts
   - Wire harness connectors from the transmission
   - Range selector cable from the Park Neutral Position (PNP) switch
   - Range selector cable and bracket
   - PNP switch
   - Power steering gear to frame retaining bolts
   - Fluid filler tube
   - Upper transmission bolts
   - Wire harness grounds
   - Both front wheels
   - Engine splash shields
   - Both tie rod ends from the steering knuckles
   - Power steering gear from the frame and secure it to the body of the vehicle
   - Power steering cooler line clamps
   - Engine mount lower nuts
   - Lower ball joints from the steering knuckles
   - Torque converter cover
   - Starter motor
   - Torque converter bolts
   - Oil cooler hoses
   - Drive axles and secure them to the steering knuckles
   - Wheel Speed Sensor (WSS) electrical connectors
   - Vehicle Speed Sensor (VSS) electrical connectors
5. Use a transmission table to support the transmission.
   - Engine mount lower nuts
   - Transmission brace
   - Lower transmission bolts
   - Frame-to-body bolts
6. Separate the transmission from the engine.
7. Lower the transmission and frame from the vehicle.
8. Remove the transmission.

To install:

9. Install or connect the following:
   - Transmission to the frame and raise the assembly into position
   - New frame-to-body bolts and torque them to 133 ft. lbs. (180 Nm)

- Lower transmission-to-engine bolts and torque them to 55 ft. lbs. (75 Nm)
- Transmission brace and torque the transmission bolts to 32 ft. lbs. (43 Nm) and the engine bolts to 46 ft. lbs. (63 Nm)
- VSS electrical connector
- WSS electrical connectors
- Drive axles to the transmission
- Oil cooler hoses
- Torque converter bolts and torque them to 46 ft. lbs. (63 Nm)
- Starter motor and torque the bolts to 32 ft. lbs. (43 Nm)
- Torque converter cover and torque the bolts to 89 inch lbs. (10 Nm).
- Lower ball joints to the steering knuckle and torque the nuts to 40 ft. lbs. (55 Nm)
- Engine mount lower nuts and torque them to 35 ft. lbs. (47 Nm)
- Power steering cooler line clamps to the frame
- Power steering gear to the frame and torque the bolts to 59 ft. lbs. (80 Nm)
- Tie rod ends to the steering knuckles and torque the nuts to 63 ft. lbs. (85 Nm)
- Engine splash shields
- Front wheels
- Fluid filler tube
- Upper transmission bolts and torque them to 55 ft. lbs. (75 Nm)
- PNP switch and torque the bolts to 18 ft. lbs. (25 Nm)
- Range selector cable with the bracket and torque the nut to 15 ft. lbs. (20 Nm)
- Range selector cable to the PNP switch
- Wire harness connectors to the transmission
- Engine mount strut and torque the bolts to 37 ft. lbs. (50 Nm)

10. Remove the engine support fixture.
- Throttle body air inlet duct
- Negative battery cable

11. Fill the transmission with fluid.
12. Adjust the wheel alignment.
13. Start engine and check the engine and transaxle oil levels. Add oil if necessary.

Halfshaft

REMOVAL & INSTALLATION

1. Before servicing the vehicle, refer to the Precautions Section.
2. Remove or disconnect the following:

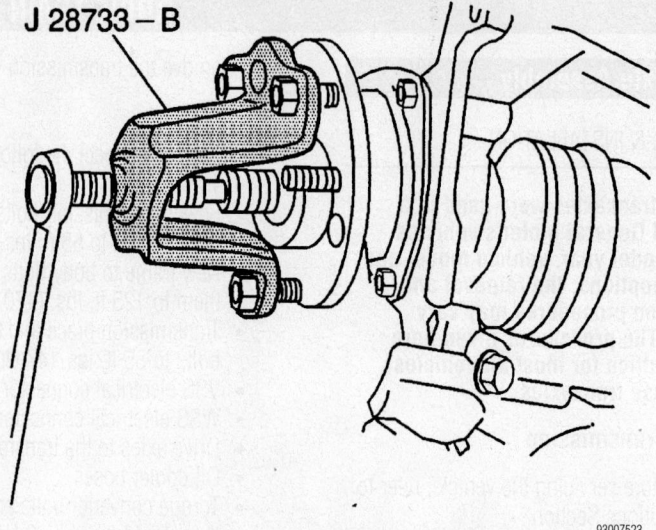

J 28733 – B

9300Z523

Use a puller to press the axle shaft through the hub/bearing assembly

- Front wheel
- Stabilizer shaft link
- Drive shaft nut
- Outer tie rod end from the steering knuckle
- Ball joint from the steering knuckle

3. Press the axle shaft through the hub.

❊❊ WARNING

To prevent damage to the inner CV-joint, do not pull on the axle shaft to remove it from the transaxle.

4. Place a drain pan under the transaxle to catch any transaxle fluid that leaks out when the axle shaft is removed.
5. Remove the axle shaft from the transaxle by prying between the transaxle and the inner CV-joint housing.

To install:

6. Install or connect the following:
- Axle shaft in the transaxle. Verify that it is seated by pulling on the housing
- Axle shaft through the hub/bearing assembly
- Ball joint and torque the nut to 15 ft. lbs. (22 Nm) plus 120 degrees
- Tie rod end and torque the nut to 22 ft. lbs. (30 Nm) plus an additional 120 degree turn
- New drive shaft nut and torque it to 118 ft. lbs. (160 Nm)
- Stabilizer shaft link and torque the nut to 17 ft. lbs. (23 Nm)
- Front wheel

7. Check the transaxle fluid level.
8. Check the front alignment and adjust, if necessary.

CV-Joint

OVERHAUL

Inner (Tri-Pod) Joint

1. Before servicing the vehicle, refer to the Precautions Section.
2. Remove or disconnect the following:
- Front wheel
- Halfshaft
- Swage ring using a hand grinder
- Large CV-joint boot clamp, cut and discard it
- CV-joint boot by sliding it away from the tri-pod joint
- Tri-pod housing from the tri-pod spider
- Trilobal tri-pod bushing from the housing
- Inboard spacer ring slide it rearward on the shaft using Snapring Pliers Tool J-8059
- Outboard retaining ring using Snapring Pliers Tool J-8059
- Tri-pod joint spider assembly
- Inboard spacer ring and CV-joint boot

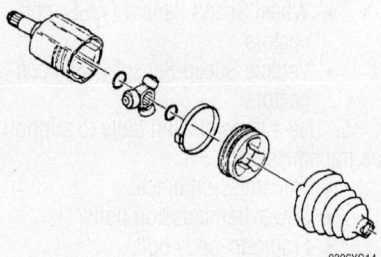

9306XG14

Exploded view of the inner (tri-pod) joint

3. Thoroughly clean and inspect all parts.

To install:

4. Install or connect the following:
- Swage ring clamp

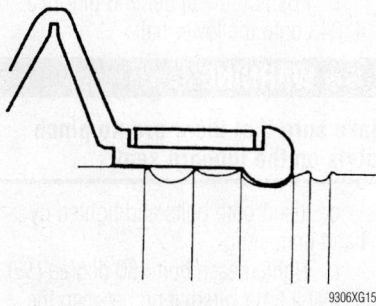

9306XG15

Positioning the inner CV-joint boot seal and swage ring—Inner (tri-pod) joint

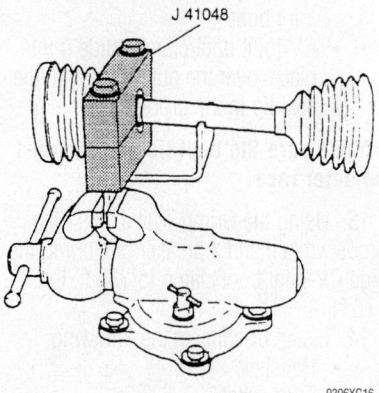

J 41048

9306XG16

View of the swage ring crimping tool—Inner (tri-pod) joint

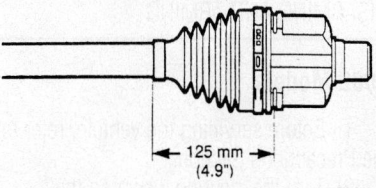

← 125 mm →
(4.9")

9306XG17

Boot measurement—Inner (tri-pod) joint

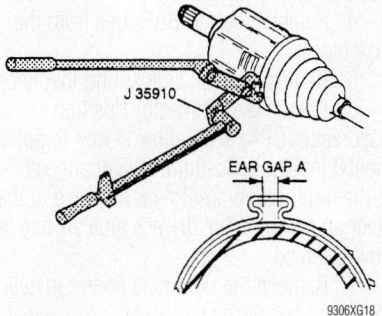

J 35910

EAR GAP A

9306XG18

Crimping the large CV-joint boot ring—Inner (tri-pod) joint

- CV-joint boot

5. Position the CV-joint boot seal into the axle shaft's joint seal groove and align the swage ring clamp on the boot.

6. Secure the swage ring clamp as follows:

   a. Mount the lower half of Tool J-41048 in a vise.

   b. Position the outboard of the half-shaft in the tool.

   c. Position the upper end of Tool J-41048 onto the lower half.

✳✳ WARNING

Make sure that there are no pinch points on the inboard seal.

   d. Insert both bolts and tighten by hand until snug.

   e. Tighten each bolt 180 degree (½ turn) at a time, alternating between the bolts, until both sides are bottomed.

   f. Remove the tool.

7. Install or connect the following:
- Inboard spacer ring, slide it rearward on the shaft using Snapring Pliers Tool J-8059
- Tri-pod joint spider assembly onto the shaft
- Outboard retaining ring into the axle shaft groove using Snapring Pliers Tool J-8059
- Tri-pod joint spider assembly, slide it against the outboard retaining ring
- Inboard spacer ring, seat it in the groove
- ½ kit grease into the boot
- ½ kit grease into the tri-pod housing
- Trilobal tip-pot bushing flush with the tri-pod housing face
- New large seal clamp onto the CV-joint boot
- Tri-pod housing, slide it over the tri-pod joint spider assembly
- CV-joint boot/clamp, slide it into place, over the trilobal tri-pod bushing with the seal lip in the groove

➡ **Make sure the boot lies flat against the trilobal bushing.**

8. Position the CV-joint boot so it measures 4.9 in. (125mm).

9. Using the Crimp Tool J-35910, a torque wrench and a breaker bar, crimp the large CV-joint boot clamp to 130 ft. lbs. (176 Nm).

10. Install or connect the following:
- Halfshaft
- Front wheel

Outer Joint

1. Before servicing the vehicle, refer to the Precautions Section.

2. Remove or disconnect the following:
- Front wheel
- Halfshaft
- Swage ring using a hand grinder
- Large boot clamp, cut and discard it
- CV-joint boot, slide it away from the CV-joint
- CV-joint assembly by spreading the inner race-to-axle shaft retaining ring ears using Snapring Pliers Tool J-8059
- CV-joint boot from the axle shaft and discard it

3. Disassemble the chrome alloy balls from the CV-joint cage as follows:

   a. Position a brass drift against the CV-joint cage and tap it with a hammer to tilt the cage.

   b. Chrome alloy ball from the cage.

   c. Tilt the cage in the opposite direction.

   d. Remove the opposite chrome alloy ball.

   e. Repeat the procedure until all 6 balls are removed.

4. Disassemble the CV-joint cage and inner race as follows:

   a. Pivot the cage and race 90 degrees to the center line of the outer race.

   b. Align the cage windows with outer race lands.

   c. Remove the cage from the outer race.

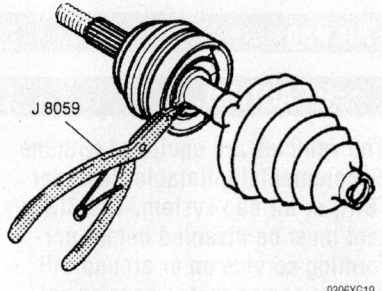

J 8059

9306XG19

Disconnecting the outer CV-joint from the axle shaft

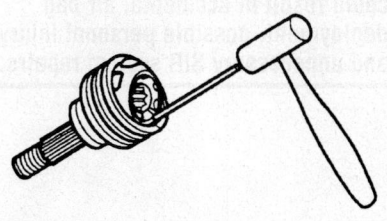

9306XG20

Tilting the cage—Outer CV-joint

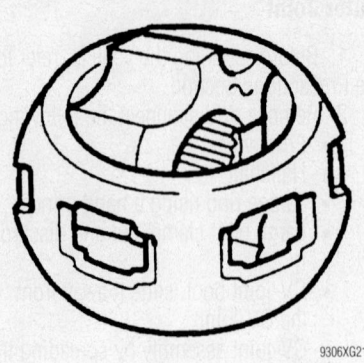

View the cage and inner race—Outer CV-joint

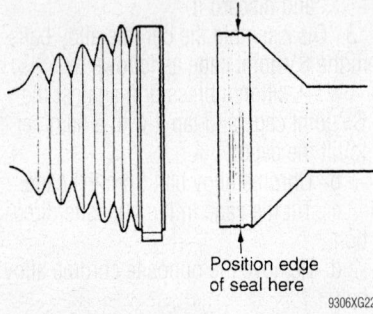

Position edge of seal here

9306XG22

Positioning the boot—Outer CV-joint

d. Rotate the inner race upward and remove it from the cage.

5. Thoroughly clean and inspect all parts.

To install:

6. Lubricate the parts with a light coat of grease.

7. Assemble the CV-joint cage and inner race, as follows:

a. Rotate the inner race 90 degrees to the cage centerline.

b. Align the cage windows with inner race lands.

c. Insert the inner race into the cage by rotating the inner race downward.

d. Insert the cage/inner race into the outer race.

8. Assemble the chrome alloy balls into the CV-joint cage, as follows:

a. Position a brass drift against the CV-joint cage and tap it with a hammer to tilt the cage.

b. Insert the 1st chrome alloy ball into the cage.

c. Tilt the cage in the opposite direction.

d. Insert the opposite chrome alloy ball.

e. Repeat the procedure until all 6 balls are inserted.

9. Install or connect the following:

- ½ kit grease into the CV-joint boot
- ½ kit grease into the CV-joint
- Swage ring clamp
- CV-joint boot
- CV-joint onto the axle shaft until the retaining ring seats into the groove

10. Position the CV-joint boot seal into the axle shaft's joint seal groove and align the swage ring clamp on the boot.

11. Secure the swage ring clamp as follows:

a. Mount the lower half of tool J-41048 in a vise.

b. Position the outboard of the half-shaft in the tool.

c. Position the upper end of tool J-41048 onto the lower half.

❊❊ WARNING

Make sure that there are no pinch points on the inboard seal.

d. Insert both bolts and tighten by hand until snug.

e. Tighten each bolt 180 degree (½) turn at a time, alternating between the bolts, until both sides are bottomed.

f. Remove the tool.

12. Install or connect the following:

- New large seal clamp onto the CV-joint boot
- CV-joint boot/clamp, slide it into place, over the outer race with the seal lip in the groove

➡Make sure the boot lies flat against the outer race.

13. Using the Crimp Tool J-35910, a torque wrench and a breaker bar, crimp the large CV-joint boot clamp to 130 ft. lbs. (176 Nm).

14. Install or connect the following:

- Halfshaft
- Front wheel

STEERING AND SUSPENSION

Air Bag

❊❊ CAUTION

The vehicles are equipped with the Supplemental Inflatable Restraint (SIR) or air bag system. The SIR system must be disabled before performing service on or around SIR system components, steering column, instrument panel components, wiring and sensors. Failure to follow safety and disabling procedures could result in accidental air bag deployment, possible personal injury and unnecessary SIR system repairs.

PRECAUTIONS

Several precautions must be observed when handling the inflator module to avoid accidental deployment and possible personal injury.

- Never carry the inflator module by the wires or connector on the underside of the module.
- When carrying a live inflator module, hold securely with both hands, and ensure that the bag and trim cover are pointed away.
- Place the inflator module on a bench or other surface with the bag and trim cover facing up.
- With the inflator module on the bench, never place anything on or close to the module which may be thrown in the event of an accidental deployment.

DISARMING AND ARMING

2002 Models

1. Before servicing the vehicle, refer to the Precautions Section.

2. Turn the steering wheel so the vehicle wheels are pointing straight-ahead.

3. Turn the ignition key to the **LOCK** position and remove the key.

4. Remove the AIR BAG fuse from the fuse block.

5. Remove the left side sound insulator.

6. Detach the Connector Position Assurance (CPA) and yellow 2-way Supplemental Inflatable Restraint (SIR) connector at the multi-use bracket near the base of the steering column. The driver's side air bag is now disabled.

7. Remove the right side sound insulator.

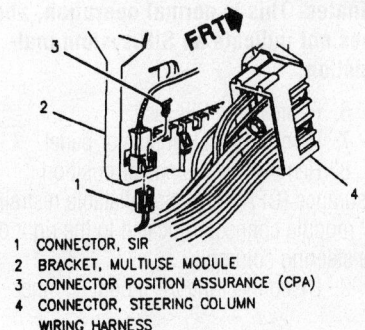

1 CONNECTOR, SIR
2 BRACKET, MULTIUSE MODULE
3 CONNECTOR POSITION ASSURANCE (CPA)
4 CONNECTOR, STEERING COLUMN
  WIRING HARNESS

7922XG21

SIR 2-way connector location—driver's side

8. Detach the CPA and yellow 2-way SIR connector which is located at the base of the steering wheel. The passenger's side air bag is now disabled.

9. On models equipped with side air bags, disconnect the CPA from the inflatable restraint side impact module located under the driver seat.

To arm:

After necessary repairs are made, re-enable the air bag system as follows:

10. Be sure the ignition is locked and the key is removed.

11. On models equipped with side air bags, attach the Connector Position Assurance CPA from the inflatable restraint side impact module located under the driver seat.

12. Attach the yellow 2-way Supplemental Inflatable Restraint (SIR) connector and CPA at the base of the steering wheel.

13. Install the right side sound insulator.

14. Attach the yellow 2-way SIR connector and CPA at the multi-use bracket at the base of the column.

15. Install the left side sound insulator.

16. Install the AIR BAG fuse.

17. Turn the ignition switch to the **RUN** position and verify the AIR BAG light flashes 7 times, then shuts off.

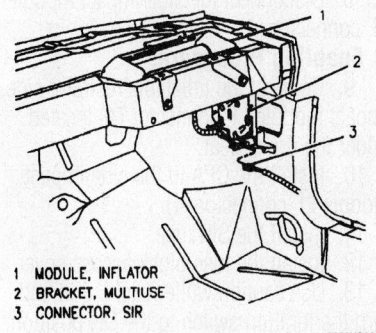

1 MODULE, INFLATOR
2 BRACKET, MULTIUSE
3 CONNECTOR, SIR

7922XG22

SIR 2-way connector location—passenger's side

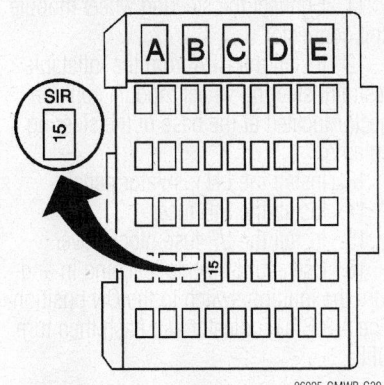

06025-GMWB-G39

SIR fuse location—2003–05 Buick Century; 2003–04 Buick Regal; 2003 Pontiac Grand Prix

2003–04 Buick Regal and 2003–05 Buick Century

ZONE 2

1. Before servicing the vehicle, refer to the Precautions Section.

2. Turn the steering wheel so that the vehicle's wheels are pointing straight ahead.

3. Turn OFF the ignition.

4. Remove the key from the ignition switch.

5. Remove the I/P fuse block cover.

➡With the SIR Fuse removed and the ignition ON, the AIR BAG indicator illuminates. This is normal operation, and does not indicate an SIR system malfunction.

6. Remove the SIR Fuse.

7. Remove the side impact sensor (SIS)-LF from the center pillar (3).

8. Remove the connector position assurance (CPA) from the SIS-LF connector (2).

9. Disconnect the SIS-LF connector (2) from the SIS-LF (1).

Enabling Procedure

10. Connect the SIS-LF connector (2) to the SIS-LF (1).

11. Install the CPA to the SIS-LF connector (2).

12. Install the side impact sensor (SIS)-LF into the center pillar (3).

13. Install the SIR Fuse.

14. Install the fuse block access cover.

15. Staying well away from all inflator modules and pre-tensioners, turn ON the ignition.

   a. The AIR BAG indicator will flash 7 times.

   b. The AIR BAG indicator will then turn OFF.

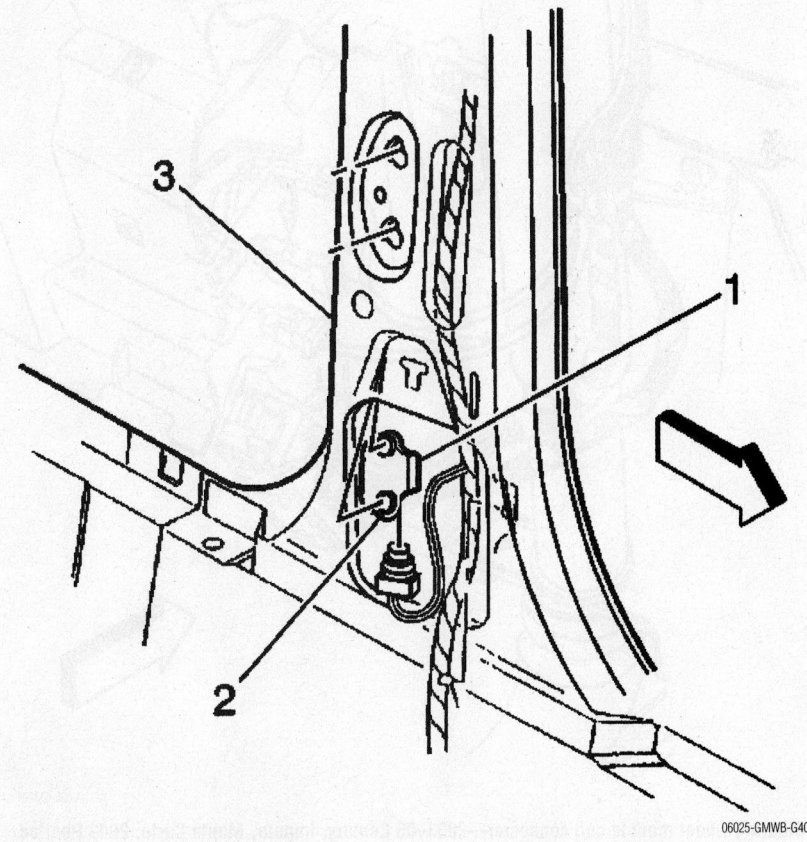

06025-GMWB-G40

Zone 2 sensor—Century and Impala

ZONE 3

1. Before servicing the vehicle, refer to the Precautions Section.

2. Turn the steering wheel so that the vehicle's wheels are pointing straight ahead.

3. Turn OFF the ignition.

4. Remove the key from the ignition switch.

5. Remove the instrument panel (I/P) fuse block cover.

➡**With the SIR fuse removed and the ignition ON, the AIR BAG indicator illuminates. This is normal operation, and does not indicate an SIR system malfunction.**

6. Remove the SIR fuse.

7. Remove the LH insulator panel.

8. Remove the connector position assurance (CPA) from the inflatable restraint steering wheel module coil connector located at the base of the steering column.

9. Disconnect the steering wheel module coil connector.

Enabling Procedure:

10. Remove the key from the ignition.

11. Connect the steering wheel module coil connector.

12. Install the CPA from the inflatable restraint steering wheel module coil connector located at the base of the steering column.

13. Install the LH insulator panel.

14. Install the SIR fuse.

15. Install the I/P fuse block cover.

16. Use caution while reaching in and turn the ignition switch to the ON position. The AIR BAG indicator will flash then turn OFF.

ZONE 5

1. Before servicing the vehicle, refer to the Precautions Section.

2. Turn the steering wheel so that the vehicle's wheels are pointing straight ahead.

3. Turn OFF the ignition.

4. Remove the key from the ignition switch.

5. Remove the instrument panel (I/P) fuse block cover.

➡**With the SIR fuse removed and the ignition ON, the AIR BAG indicator illu-**

minates. **This is normal operation, and does not indicate an SIR system malfunction.**

6. Remove the SIR fuse.

7. Remove the LH insulator panel.

8. Remove the connector position assurance (CPA) from the inflatable restraint I/P module connector located to the right of the steering column.

9. Disconnect the I/P module connector.

Enabling Procedure:

10. Remove the key from the ignition.

11. Connect the I/P module connector.

12. Install the CPA into the I/P module connector located to the right of the steering column.

13. Install the LH insulator panel.

14. Install the SIR fuse.

15. Install the I/P fuse block cover.

16. Use caution while reaching in and turn the ignition switch to the ON position. The AIR BAG indicator will flash then turn OFF.

ZONE 7

1. Before servicing the vehicle, refer to the Precautions Section.

2. Turn the steering wheel so that the vehicle's wheels are pointing straight ahead.

3. Turn OFF the ignition.

4. Remove the key from the ignition switch.

5. Remove the fuse block access cover.

➡**With the SIR fuse removed and the ignition ON, the AIR BAG indicator illuminates. This is normal operation, and does not indicate an SIR system malfunction.**

6. Remove the SIR fuse.

7. Remove the connector position assurance (CPA) from the side impact module LF connector (4) located under the driver seat.

8. Disconnect the side impact module LF connector (4).

Enabling Procedure:

9. Connect the inflatable restraint side impact module LF connector (4) located under the driver seat.

10. Install the CPA to the side impact module LF connector (4).

11. Install the SIR fuse.

12. Install the fuse block access cover.

13. Use caution while reaching in and turn the ignition switch to the ON position. The AIR BAG indicator will flash then turn OFF.

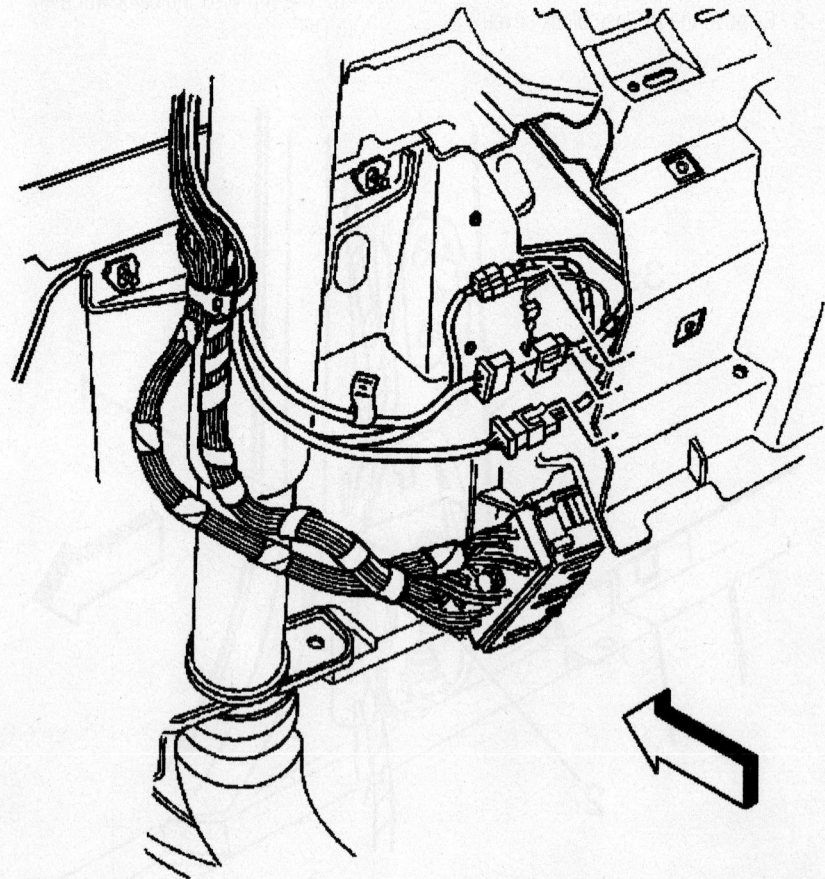

06025-GMWB-G41

Steering wheel module coil connector—2003–05 Century, Impala, Monte Carlo; 2003 Pontiac Grand Prix

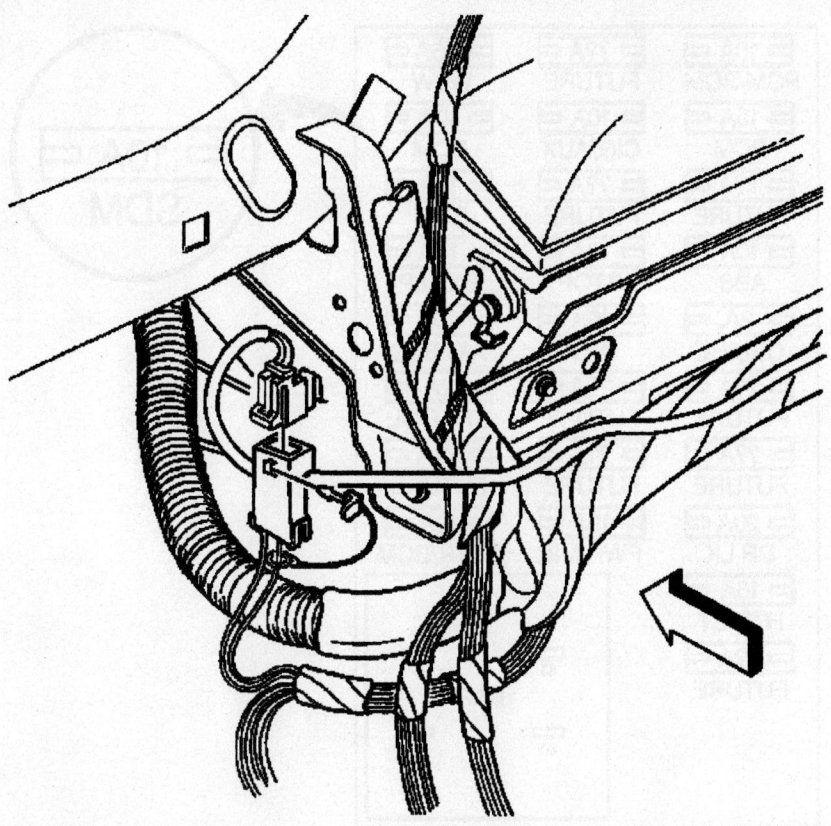

06025-GMWB-G42

I/P module connector—2003–05 Buick Century

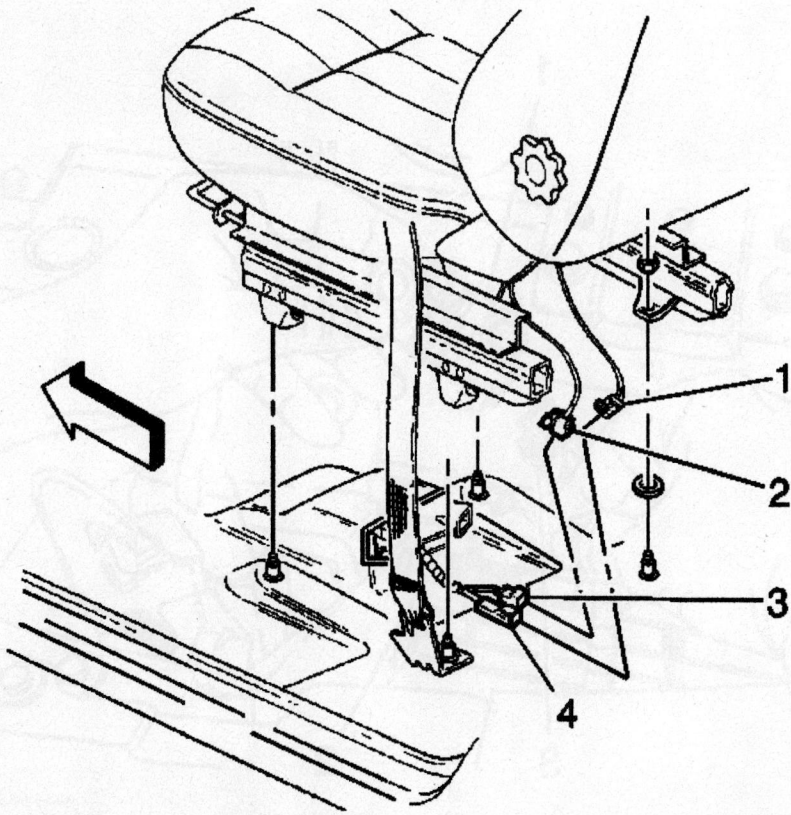

06025-GMWB-G43

Inflatable restraint side impact module LF connector (4)—2003–05 Buick Century

ZONE 9

1. Before servicing the vehicle, refer to the Precautions Section.

2. Turn the steering wheel so that the vehicle wheels are pointing straight ahead.

3. Turn OFF the ignition.

4. Remove the key from the ignition switch.

5. Remove the instrument panel (I/P) fuse block cover.

➡ **With the SIR fuse removed and the ignition ON, the AIR BAG indicator illuminates. This is normal operation, and does not indicate an SIR system malfunction.**

6. Remove the SIR fuse.

7. Remove the LH insulator panel.

8. Remove the connector position assurance (CPA) from the inflatable restraint steering wheel module coil connector located at the base of the steering column.

9. Disconnect the steering wheel module coil connector.

10. Remove the LH insulator panel.

11. Remove the CPA from the inflatable restraint I/P module connector located to the right of the steering column.

12. Disconnect the I/P module connector.

13. Remove the side impact sensor (SIS) LF from the center pillar.

14. Remove the connector position assurance (CPA) from the SIS LF connector.

15. Disconnect the SIS LF connector from the SIS LF.

Enabling Procedure:

16. Remove the key from the ignition.

17. Connect the SIS—LF connector to the SIS LF.

18. Install the CPA to the SIS LF connector.

19. Install the side impact sensor (SIS) LF into the center pillar.

20. Connect the I/P module connector.

21. Install the CPA into the I/P module connector located to the right of the steering column.

22. Install the LH insulator panel.

23. Connect the steering wheel module coil connector.

24. Install the CPA from the inflatable restraint steering wheel module coil connector located at the base of the steering column.

25. Install the LH insulator panel.

26. Install the SIR fuse.

27. Install the I/P fuse block cover.

28. Use caution while reaching in and turn the ignition switch to the ON position. The AIR BAG indicator will flash then turn OFF.

2003–05 Chevrolet Impala and Monte Carlo

ZONE 1

1. Before servicing the vehicle, refer to the Precautions Section.

2. Turn the steering wheel so that the vehicle wheels are pointing straight ahead.

3. Turn OFF the ignition.

4. Remove the key from the ignition switch.

5. Remove the instrument panel (I/P) fuse block cover.

➡**With the SIR fuse removed and the ignition ON, the AIR BAG indicator illuminates. This is normal operation, and does not indicate an SIR system malfunction.**

6. Remove the SDM fuse.

7. Remove the radiator upper air baffle and deflector.

8. Remove the connector position assurance (CPA) (2) from the front end sensor harness connector (3).

9. Disconnect the front end sensor harness connector (3) from the front end sensor (1).

Enabling Procedure:

10. Remove the key from the ignition.

11. Connect the front end sensor harness connector (3) to the front end sensor (1).

12. Install the CPA (2) into the front end sensor harness connector (3).

13. Install the radiator upper air baffle and deflector.

14. Install the SDM fuse.

15. Install the I/P fuse block cover.

16. Use caution while reaching in and turn the ignition switch to the ON position. The AIR BAG indicator will flash then turn OFF.

ZONE 2— IMPALA

1. Before servicing the vehicle, refer to the Precautions Section.

2. Turn the steering wheel so that the vehicle wheels are pointing straight ahead.

3. Turn OFF the ignition.

4. Remove the key from the ignition switch.

5. Remove the instrument panel (I/P) fuse block cover.

➡**With the SIR fuse removed and the ignition ON, the AIR BAG indicator illuminates. This is normal operation, and does not indicate an SIR system malfunction.**

6. Remove the SDM fuse.

7. Remove the connector position

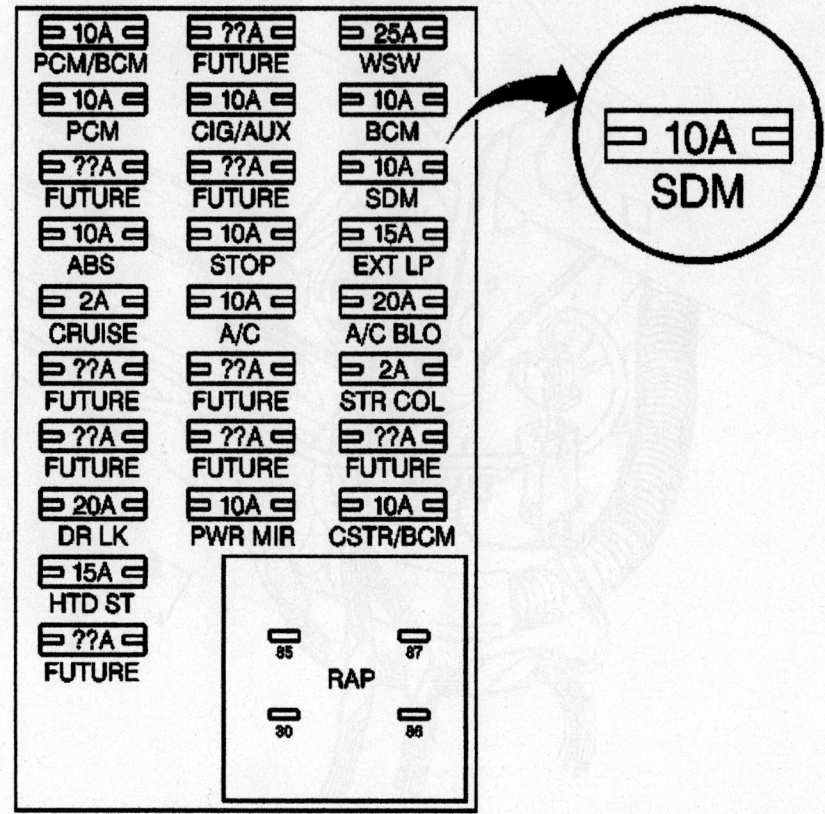

06025-GMWB-G44

SDM fuse—2003–05 Chevrolet Impala and Monte Carlo

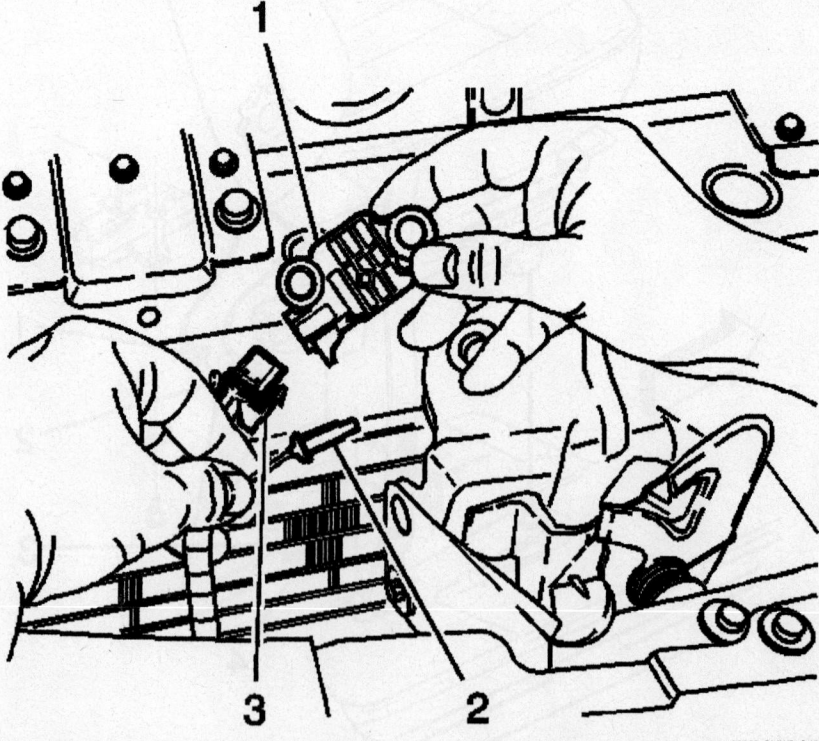

06025-GMWB-G45

Front end sensor harness—2003–05 Chevrolet Impala and Monte Carlo; 2004–05 Pontiac Grand Prix assurance (CPA) from the left side impact sensor (SIS) connector.

8. Disconnect the left SIS connector from the left SIS.

Enabling Procedure:

9. Remove the key from the ignition.

10. Connect the left SIS connector to the left SIS.

11. Install the CPA to the left SIS connector.

12. Install the SDM fuse.

13. Install the fuse block access cover.

14. Use caution while reaching in and turn the ignition switch to the ON position. The AIR BAG indicator will flash then turn OFF.

ZONE 2 —MONTE CARLO

1. Before servicing the vehicle, refer to the Precautions Section.

2. Turn the steering wheel so that the vehicle wheels are pointing straight ahead.

3. Turn OFF the ignition.

4. Remove the key from the ignition switch.

5. Remove the instrument panel (I/P) fuse block cover.

➡ **With the SIR fuse removed and the ignition ON, the AIR BAG indicator illuminates. This is normal operation, and does not indicate an SIR system malfunction.**

6. Remove the SDM fuse.

7. Remove the connector position assurance (CPA) from the left side impact sensor (SIS) connector.

8. Disconnect the left SIS connector from the left SIS (2).

Enabling Procedure:

9. Remove the key from the ignition.

10. Connect the left SIS connector to the left SIS (2).

11. Install the CPA to the left SIS connector.

12. Install the SDM fuse.

13. Install the fuse block access cover.

14. Use caution while reaching in and turn the ignition switch to the ON position. The AIR BAG indicator will flash then turn OFF.

ZONE 3

1. Before servicing the vehicle, refer to the Precautions Section.

2. Turn the steering wheel so that the vehicle wheels are pointing straight ahead.

3. Turn OFF the ignition.

4. Remove the key from the ignition switch.

5. Remove the instrument panel (I/P) fuse block cover.

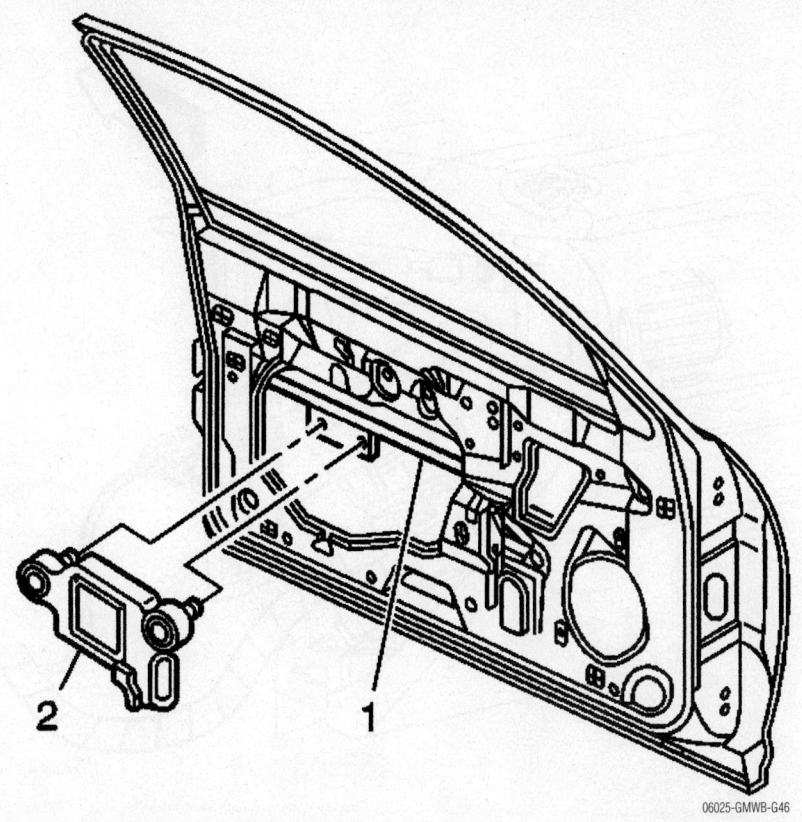

06025-GMWB-G46

Left side impact sensor—Monte Carlo

➡ **With the SIR fuse removed and the ignition ON, the AIR BAG indicator illuminates. This is normal operation, and does not indicate an SIR system malfunction.**

6. Remove the SDM fuse.

7. Remove the LH insulator panel.

8. Remove the connector position assurance (CPA) from the steering wheel module coil connector located at the base of the steering column.

9. Disconnect the steering wheel module coil connector.

Enabling Procedure:

10. Remove the key from the ignition.

11. Connect the steering wheel module coil connector.

12. Install the CPA to the steering wheel module coil connector located at the base of the steering column.

13. Install the LH insulator panel.

14. Install the SDM fuse.

15. Install the I/P fuse block cover.

16. Use caution while reaching in and turn the ignition switch to the ON position. The AIR BAG indicator will flash then turn OFF.

ZONE 5

1. Before servicing the vehicle, refer to the Precautions Section.

2. Turn the steering wheel so that the vehicle wheels are pointing straight ahead.

3. Turn OFF the ignition.

4. Remove the key from the ignition switch.

5. Remove the instrument panel (I/P) fuse block cover.

➡ **With the SIR fuse removed and the ignition ON, the AIR BAG indicator illuminates. This is normal operation, and does not indicate an SIR system malfunction.**

6. Remove the SDM fuse.

7. Remove the RH I/P access hole cover.

8. Remove the connector position assurance (CPA) from the I/P module connector located to the right of the steering column.

9. Disconnect the I/P module connector.

Enabling Procedure:

10. Remove the key from the ignition.

11. Connect the I/P module connector.

12. Install the CPA into the I/P module connector located to the right of the steering column.

13. Install the LH insulator panel.

14. Install the SDM fuse.

15. Install the I/P fuse block cover.

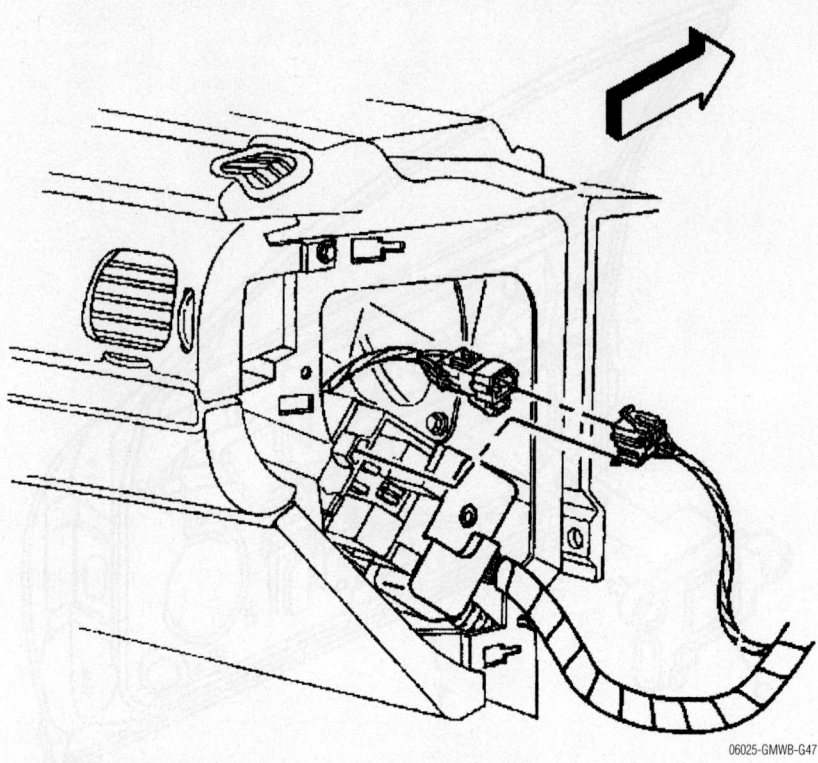

06025-GMWB-G47

Right I/P module connector—Impala and Monte Carlo; 2004–05 Pontiac Grand Prix

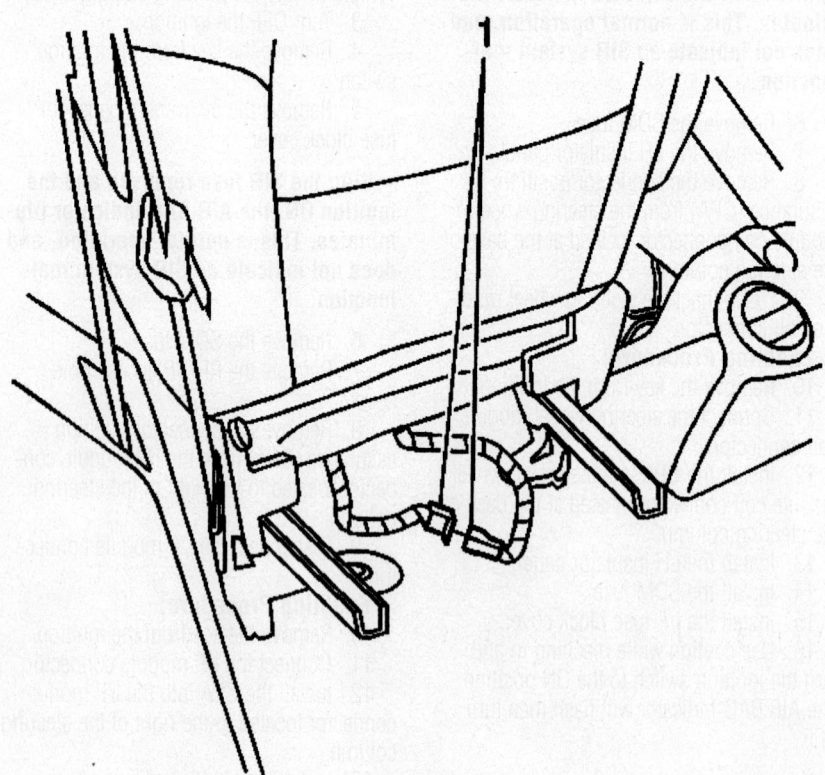

06025-GMWB-G48

Side impact module—Impala and Monte Carlo

16. Use caution while reaching in and turn the ignition switch to the ON position. The AIR BAG indicator will flash then turn OFF.

ZONE 7

1. Before servicing the vehicle, refer to the Precautions Section.
2. Turn the steering wheel so that the vehicle's wheels are pointing straight ahead.
3. Turn OFF the ignition.
4. Remove the key from the ignition switch.
5. Remove the fuse block access cover.

➡ **With the SIR fuse removed and the ignition ON, the AIR BAG indicator illuminates. This is normal operation, and does not indicate an SIR system malfunction.**

6. Remove the SDM fuse.
7. Remove the connector position assurance (CPA) from the side impact module LF connector (1) located under the driver seat.
8. Disconnect the side impact module LF connector (1).

Enabling Procedure:

9. Remove the key from the ignition.
10. Connect the side impact module LF connector (1) located under the driver seat.
11. Install the CPA to the side impact module LF connector (1).
12. Remove the SDM fuse.
13. Install the fuse block access cover.
14. Use caution while reaching in and turn the ignition switch to the ON position. The AIR BAG indicator will flash then turn OFF.

ZONE 9

1. Before servicing the vehicle, refer to the Precautions Section.
2. Turn the steering wheel so that the vehicle wheels are pointing straight ahead.
3. Turn OFF the ignition.
4. Remove the key from the ignition switch.
5. Remove the instrument panel (I/P) fuse block cover.

➡ **With the SIR fuse removed and the ignition ON, the AIR BAG indicator illuminates. This is normal operation, and does not indicate an SIR system malfunction.**

6. Remove the SDM fuse.
7. Remove the RH I/P access hole cover.
8. Remove the connector position assurance (CPA) from the I/P module connector, located at the RH side of the I/P.

9. Disconnect the I/P module connector.

10. Remove the LH insulator panel.

11. Remove the CPA from the steering wheel module coil connector located at the base of the steering column.

12. Disconnect the steering wheel module coil connector.

13. Remove the CPA from the side impact module LF connector (1), located under the driver seat.

14. Disconnect the side impact module LF connector.

Enabling Procedure:

15. Remove the key from the ignition.

16. Connect the side impact module LF connector (1), located under the driver seat.

17. Install the CPA to the side impact module LF connector.

18. Connect the steering wheel module coil connector.

19. Install the CPA to the steering wheel module coil connector located at the base of the steering column.

20. Install the LH insulator panel.

21. Connect the I/P module connector, located at the RH side of the I/P.

22. Install the CPA to the I/P module connector.

23. Install the RH I/P access hole cover.

24. Install the SDM fuse.

25. Install the LH I/P access hole cover.

26. Use caution while reaching in and turn the ignition switch to the ON position. The AIR BAG indicator will flash then turn OFF.

2003 Pontiac Grand Prix

ZONE 3

1. Before servicing the vehicle, refer to the Precautions Section.

2. Turn the steering wheel so that the vehicle's wheels are pointing straight ahead.

3. Turn OFF the ignition.

4. Remove the key from the ignition switch.

5. Remove the IP fuse block cover.

➡ **With the SIR Fuse removed and the ignition ON, the AIR BAG indicator illuminates. This is normal operation, and does not indicate an SIR system malfunction.**

6. Remove the SIR Fuse.

7. Remove the LH insulator panel.

8. Remove the connector position assurance (CPA) from the inflatable restraint steering wheel module coil connector located at the base of the steering column.

9. Disconnect the steering wheel module coil connector.

Enabling Procedure:

10. Remove the key from the ignition.

11. Connect the steering wheel module coil connector.

12. Install the connector position assurance (CPA) from the inflatable restraint steering wheel module coil connector located at the base of the steering column.

13. Install the LH insulator panel.

14. Install the SIR Fuse.

15. Install the IP fuse block cover.

16. Staying well away from all inflator modules and pretensioner, turn ON the ignition.

   a. The AIR BAG indicator will flash 7 times.

   b. The AIR BAG indicator will then turn OFF.

ZONE 5

1. Before servicing the vehicle, refer to the Precautions Section.

2. Turn the steering wheel so that the vehicle's wheels are pointing straight ahead.

3. Turn OFF the ignition.

4. Remove the key from the ignition switch.

5. Remove the IP fuse block cover.

➡ **With the SIR Fuse removed and the ignition ON, the AIR BAG indicator illuminates. This is normal operation, and does not indicate an SIR system malfunction.**

6. Remove the SIR Fuse.

7. Remove the LH insulator panel.

8. Remove the (CPA) from the inflatable restraint IP module connector located to the right of the steering column.

9. Disconnect the IP module connector.

Enabling Procedure:

10. Remove the key from the ignition.

11. Connect the IP module connector.

12. Install the connector position assurance (CPA) from the inflatable restraint CPA from the inflatable restraint IP module connector located to the right of the steering column.

13. Install the LH insulator panel.

14. Install the SIR Fuse.

15. Install the IP fuse block cover.

16. Staying well away from all inflator modules and pretensioner, turn ON the ignition.

   a. The AIR BAG indicator will flash 7 times.

   b. The AIR BAG indicator will then turn OFF.

ZONE 9

1. Before servicing the vehicle, refer to the Precautions Section.

2. Turn the steering wheel so that the vehicle's wheels are pointing straight ahead.

3. Turn OFF the ignition.

4. Remove the key from the ignition switch.

5. Remove the IP fuse block cover.

➡ **With the SIR Fuse removed and the ignition ON, the AIR BAG indicator illuminates. This is normal operation, and does not indicate an SIR system malfunction.**

6. Remove the SIR Fuse.

7. Remove the LH insulator panel.

8. Remove the connector position assurance (CPA) from the inflatable restraint steering wheel module coil connector located at the base of the steering column.

9. Disconnect the steering wheel module coil connector.

10. Remove the CPA from the inflatable restraint IP module connector located to the right of the steering column.

11. Disconnect the IP module connector.

12. Remove the CPA from the inflatable restraint sensing and diagnostic module (SDM) wiring harness connector.

13. Disconnect the SDM wiring harness connector from the SDM.

Enabling Procedure:

14. Remove the key from the ignition.

15. Connect the SDM wiring harness connector from the SDM.

16. Install the CPA from the inflatable restraint sensing and diagnostic module (SDM) wiring harness connector.

17. Connect the IP module connector.

18. Install the connector position assurance CPA from the inflatable restraint CPA from the inflatable restraint IP module connector located to the right of the steering column.

19. Connect the steering wheel module coil connector.

20. Install the connector position assurance (CPA) from the inflatable restraint steering wheel module coil connector located at the base of the steering column.

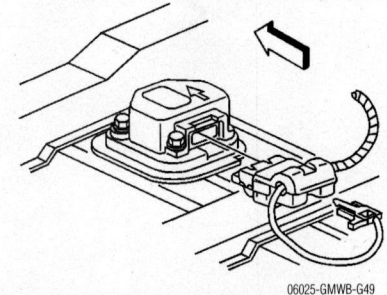

06025-GMWB-G49

Inflatable restraint sensing and diagnostic module (SDM)—2003 Pontiac Grand Prix

21. Install the LH insulator panel.
22. Install the SIR Fuse.
23. Install the IP fuse block cover.
24. Staying well away from all inflator modules and pretensioner, turn ON the ignition.

   a. The AIR BAG indicator will flash 7 times.

   b. The AIR BAG indicator will then turn OFF.

2004–05 Pontiac Grand Prix

ZONE 1

1. Before servicing the vehicle, refer to the Precautions Section.
2. Turn the steering wheel so that the vehicles wheels are pointing straight ahead.
3. Turn the ignition switch to the OFF position.
4. Remove the key from the ignition switch.
5. Open the hood and locate the underhood fuse center on right/passenger shock tower.

➡**With the SIR Fuse removed and the ignition ON, the AIR BAG indicator illu-minates. This is normal operation, and does not indicate an SIR system mal-function.**

6. Lift the cover for the underhood fuse center.
7. Locate and remove the SIR fuse from the underhood fuse center.
8. Remove the radiator upper air baffle and deflector and locate the front end sensor also known as electronic frontal sensor (EFS).
9. Remove the connector position assurance (CPA) from the front end sensor connector.
10. Remove the front end sensor connector from the front end sensor.

Enabling Procedure:

11. Remove the key from the ignition switch.
12. Connect the front end sensor connector to the front end sensor.
13. Install the CPA into the front end sensor connector.
14. Install the radiator upper air baffle and deflector.
15. Install the SIR Fuse.
16. Close the underhood fuse center cover.
17. Use caution while reaching in and turn the ignition switch to the ON position. The AIR BAG indicator will flash then turn OFF.

ZONE 2

1. Before servicing the vehicle, refer to the Precautions Section.
2. Turn the steering wheel so that the vehicles wheels are pointing straight ahead.
3. Turn the ignition switch to the OFF position.
4. Remove the key from the ignition switch.
5. Open the hood and locate the underhood fuse center on right/passenger shock tower.

➡**With the SIR Fuse removed and the ignition ON, the AIR BAG indicator illu-minates. This is normal operation, and does not indicate an SIR system mal-function.**

6. Lift the cover for the underhood fuse center.
7. Locate and remove the SIR fuse from the underhood fuse center.
8. When disabling the roof rail module

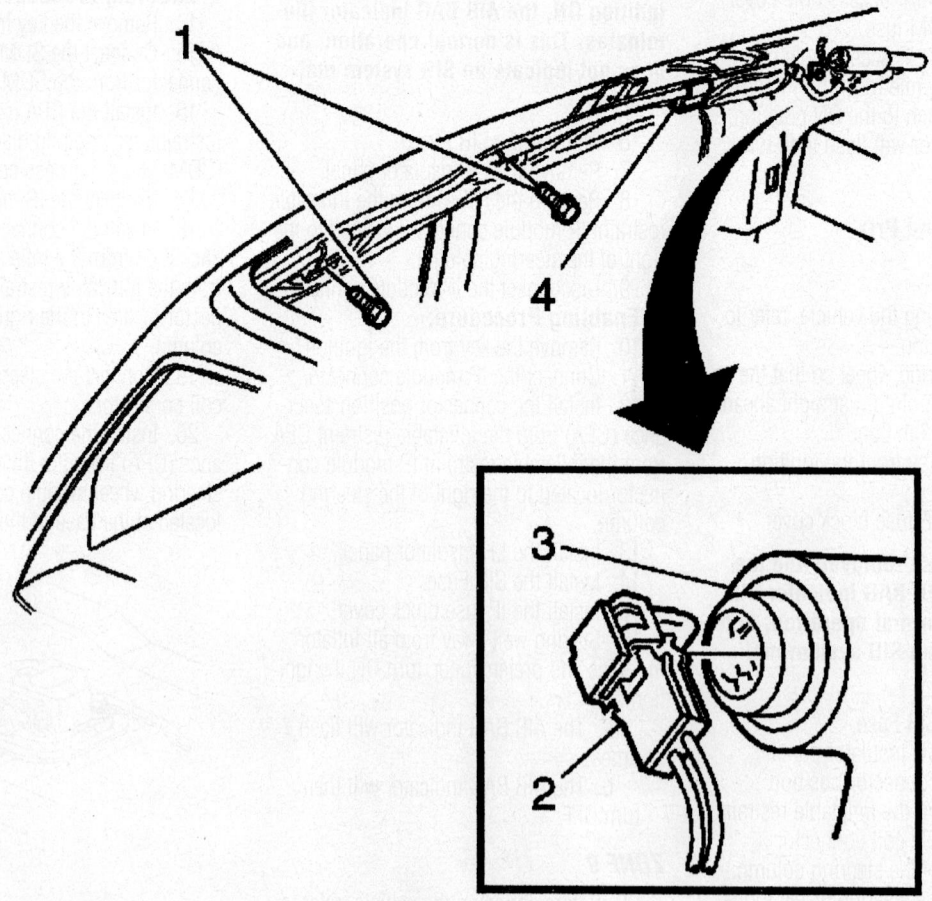

06025-GMWB-G51

Left roof rail module—2004–05 Pontiac Grand Prix

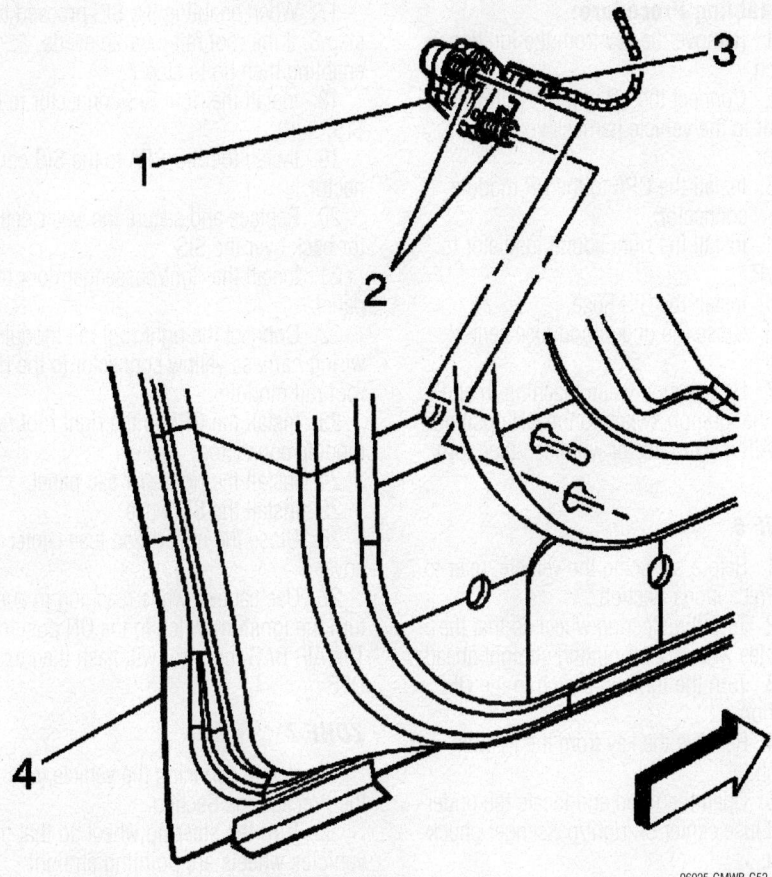

Side impact sensor—2004–05 Pontiac Grand Prix

06025-GMWB-G52

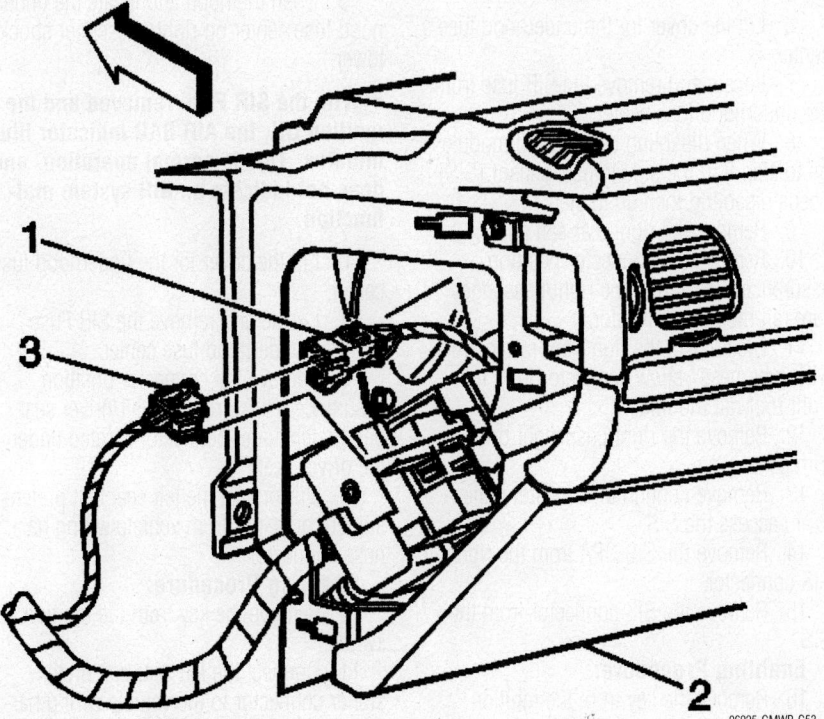

Steering wheel module coil yellow connector—2004–05 Pontiac Grand Prix

06025-GMWB-G53

go to step 8, if the side impact sensor (SIS) needs disabling then go to step 11.

9. Remove the left rear sail panel.

10. Remove the connector position assurance (CPA) from the left/driver roof rail module connector (2).

11. Disconnect the left roof rail module wiring harness yellow connector (2) from the left roof rail module (3).

12. Remove the left/driver door trim panel.

13. Remove enough of the water deflector to access the SIS.

14. Remove the SIS CPA from the left SIS connector (3).

15. Remove the SIS connector (3) from the SIS (1).

Enabling Procedure:

16. Remove the key from the ignition switch.

17. When enabling the SIS proceed to step 3, if the roof rail module needs enabling then go to step 7.

18. Install the left SIS connector (3) to the SIS (1).

19. Install the SIS CPA to the SIS connector (3).

20. Replace and secure the water deflector back over the SIS.

21. Install the left/driver door trim panel.

22. Connect the left roof rail module wiring harness yellow connector (2) to the left roof rail module (3).

23. Install the CPA to the left roof rail module connector (2).

24. Install the left rear sail panel.

25. Close the underhood fuse center cover.

26. Use caution while reaching in and turn the ignition switch to the ON position. The AIR BAG indicator will flash then turn OFF.

ZONE 3

1. Before servicing the vehicle, refer to the Precautions Section.

2. Turn the steering wheel so that the vehicles wheels are pointing straight ahead.

3. Turn the ignition switch to the OFF position.

4. Remove the key from the ignition switch.

5. Open the hood and locate the underhood fuse center on right/passenger shock tower.

➡**With the SIR Fuse removed and the ignition ON, the AIR BAG indicator illuminates. This is normal operation, and does not indicate an SIR system malfunction.**

6. Lift the cover for the underhood fuse center.

7. Locate and remove the SIR fuse from the underhood fuse center.

8. Remove the left/driver sound insulator from the instrument panel (I/P) (2).

9. Remove the connector position assurance (CPA) from the steering wheel module coil yellow connector (1).

10. Disconnect the steering wheel module coil yellow connector (1) from the vehicle harness yellow connector (3).

Enabling Procedure:

11. Remove the key from the ignition switch.

12. Connect the steering wheel module coil yellow connector (3) to the vehicle harness yellow connector (1).

13. Install the CPA to the steering wheel module coil yellow connector (1).

14. Install the left sound insulator to the I/P (2).

15. Install the SIR Fuse.

16. Close the underhood fuse center cover.

17. Use caution while reaching in and turn the ignition switch to the ON position. The AIR BAG indicator will flash then turn OFF.

ZONE 5

1. Before servicing the vehicle, refer to the Precautions Section.

2. Turn the steering wheel so that the vehicle wheels are pointing straight ahead.

3. Turn the ignition switch to the OFF position.

4. Remove the key from the ignition switch.

5. Open the hood and locate the underhood fuse center on right/passenger shock tower.

➡**With the SIR Fuse removed and the ignition ON, the AIR BAG indicator illuminates. This is normal operation, and does not indicate an SIR system malfunction.**

6. Lift the cover for the underhood fuse center.

7. Locate and remove the SIR Fuse from the underhood fuse center.

8. Remove the right/passenger sound insulator from the instrument panel (I/P).

9. Remove the connector position assurance (CPA) from the I/P module yellow connector.

10. Disconnect the I/P module yellow connector from the vehicle harness yellow connector.

Enabling Procedure:

11. Remove the key from the ignition switch.

12. Connect the I/P module yellow connector to the vehicle harness yellow connector.

13. Install the CPA to the I/P module yellow connector.

14. Install the right sound insulator to the I/P.

15. Install the SIR Fuse.

16. Close the underhood fuse center cover.

17. Use caution while reaching in and turn the ignition switch to the ON position. The AIR BAG indicator will flash then turn OFF.

ZONE 6

1. Before servicing the vehicle, refer to the Precautions Section.

2. Turn the steering wheel so that the vehicles wheels are pointing straight ahead.

3. Turn the ignition switch to the OFF position.

4. Remove the key from the ignition switch.

5. Open the hood and locate the underhood fuse center on right/passenger shock tower.

➡**With the SIR Fuse removed and the ignition ON, the AIR BAG indicator illuminates. This is normal operation, and does not indicate an SIR system malfunction.**

6. Lift the cover for the underhood fuse center.

7. Locate and remove the SIR fuse from the underhood fuse center.

8. When disabling the roof rail module go to step 8, if the side impact sensor (SIS) needs disabling then go to step 11.

9. Remove the right rear sail panel.

10. Remove the connector position assurance (CPA) from the right/passenger roof rail module connector.

11. Disconnect the right roof rail module wiring harness yellow connector from the right roof rail module.

12. Remove the right/passenger door trim panel.

13. Remove enough of the water deflector to access the SIS.

14. Remove the SIS CPA from the right SIS connector.

15. Remove the SIS connector from the SIS.

Enabling Procedure:

16. Remove the key from the ignition switch.

17. When enabling the SIS proceed to step 3, if the roof rail module needs enabling then go to step 7.

18. Install the right SIS connector to the SIS.

19. Install the SIS CPA to the SIS connector.

20. Replace and secure the water deflector back over the SIS.

21. Install the right/passenger door trim panel.

22. Connect the right roof rail module wiring harness yellow connector to the right roof rail module.

23. Install the CPA to the right roof rail module connector.

24. Install the right rear sail panel.

25. Install the SIR Fuse.

26. Close the underhood fuse center cover.

27. Use caution while reaching in and turn the ignition switch to the ON position. The AIR BAG indicator will flash then turn OFF.

ZONE 7

1. Before servicing the vehicle, refer to the Precautions Section.

2. Turn the steering wheel so that the vehicles wheels are pointing straight ahead.

3. Turn the ignition switch to the OFF position.

4. Remove the key from the ignition switch.

5. Open the hood and locate the underhood fuse center on right/passenger shock tower.

➡**With the SIR Fuse removed and the ignition ON, the AIR BAG indicator illuminates. This is normal operation, and does not indicate an SIR system malfunction.**

6. Lift the cover for the underhood fuse center.

7. Locate and remove the SIR Fuse from the underhood fuse center.

8. Remove the connector position assurance (CPA) from the left/driver seat belt pretensioner connector located under the driver seat.

9. Disconnect the left seat belt pretensioner connector from vehicle wiring harness connector.

Enabling Procedure:

10. Remove the key from the ignition switch.

11. Connect the left seat belt pretensioner connector to the vehicle wiring harness connector.

12. Install the CPA to the seat belt pre-tensioner connector.

13. Install the SIR Fuse.

14. Close the underhood fuse center cover.

15. Use caution while reaching in and turn the ignition switch to the ON position. The AIR BAG indicator will flash then turn OFF.

ZONE 9

1. Before servicing the vehicle, refer to the Precautions Section.

2. Turn the steering wheel so that the vehicles wheels are pointing straight ahead.

3. Turn the ignition switch to the OFF position.

4. Remove the key from the ignition switch.

5. Open the hood and locate the under-hood fuse center on right/passenger shock tower.

➡With the SIR Fuse removed and the ignition ON, the AIR BAG indicator illuminates. This is normal operation, and does not indicate an SIR system malfunction.

6. Lift the cover for the underhood fuse center.

7. Locate and remove the SIR fuse from the underhood fuse center.

8. When disabling the SDM use the entire procedure, if the right/passenger seat belt pretensioner needs disabling then proceed to step 22.

9. Remove the right rear sail panel.

10. Remove the CPA from the right/passenger roof rail module connector.

11. Disconnect the right roof rail module wiring harness yellow connector from the right roof rail module.

12. Remove the right/passenger sound insulator from the instrument panel (I/P).

13. Remove the CPA from the I/P module yellow connector.

14. Disconnect the I/P module yellow connector from the vehicle harness yellow connector.

15. Remove the left/driver sound insulator from the instrument panel (I/P).

16. Remove the CPA from the steering wheel module coil yellow connector.

17. Disconnect the steering wheel module coil yellow connector from the vehicle harness yellow connector.

18. Remove the CPA from the left/driver seat belt pretensioner connector located under the driver seat.

19. Disconnect the left seat belt pretensioner connector from vehicle wiring harness connector.

20. Remove the left rear sail panel.

21. Remove the CPA from the left/driver roof rail module connector.

22. Disconnect the left roof rail module wiring harness yellow connector from the left roof rail module.

23. Remove the connector position assurance (CPA) from the right/passenger seat belt pretensioner connector located under the passenger seat.

24. Disconnect the right seat belt pretensioner connector from vehicle wiring harness connector.

Enabling Procedure:

25. Remove the key from the ignition switch.

26. When enabling the SDM use the entire procedure, if the right/passenger seat belt pretensioner needs enabling then proceed to step 17.

27. Connect the right roof rail module wiring harness yellow connector to the right roof rail module.

28. Install the CPA to the right roof rail module connector.

29. Install the right rear sail panel.

30. Connect the I/P module yellow connector to the vehicle harness yellow connector.

31. Install the CPA to the I/P module yellow connector.

32. Install the right sound insulator to the I/P (3).

33. Connect the steering wheel module coil yellow connector to the vehicle harness yellow connector.

34. Install the CPA to the steering wheel module coil yellow connector.

35. Install the left sound insulator to the I/P.

36. Connect the left seat belt pretensioner connector to the vehicle wiring harness connector.

37. Install the CPA to the seat belt pretensioner connector.

38. Connect the left roof rail module wiring harness yellow connector to the left roof rail module.

39. Install the CPA to the left roof rail module connector.

40. Install the left rear sail panel.

41. Connect the right seat belt pretensioner connector to the vehicle wiring harness connector.

42. Install the CPA to the seat belt pretensioner connector.

43. Install the SIR Fuse.

44. Close the underhood fuse center cover.

45. Use caution while reaching in and turn the ignition switch to the ON position. The AIR BAG indicator will flash then turn OFF.

Power Rack and Pinion Steering Gear

REMOVAL & INSTALLATION

1. Before servicing the vehicle, refer to the Precautions Section.

2. Remove or disconnect the following:
   • Negative battery cable
   • Front wheels

✱✱ CAUTION

Failure to disconnect the intermediate shaft from the rack and pinion stub shaft may result in damage to the steering gear. This damage may cause a loss of steering control and may cause personal injury.

➡Set the steering shaft so that the block tooth on the upper steering shaft is at the 12 o'clock position. The wheels should be straight ahead. Set the ignition key lock to the LOCK position. Failure to follow these procedures could result in damage to the SIR coil assembly.

3. Remove or disconnect the following:
   • Intermediate steering shaft lower pinch bolt from the steering gear stub shaft
   • Intermediate steering shaft
   • Both tie rod ends from the steering knuckles

4. Support the frame at the center rear.

➡DO NOT lower the frame too far. Engine components near the firewall may be damaged.

5. Remove or disconnect the following:
   • Frame bolts from the rear of the frame. Lower the frame slightly
   • Power steering pressure line from the steering gear
   • Power steering return hose
   • Magnasteer Variable Assist electrical connector from the power steering gear assembly, if equipped
   • Steering gear mounting bolts
   • Power steering gear through the left wheel opening

To install:

6. Install or connect the following:
   • Power steering gear through the left wheel opening
   • Mounting bolts and torque them to 66 ft. lbs. (90 Nm)
   • Magnasteer Variable Assist electrical connector to the power steering gear assembly, if equipped

- Power steering lines with new O-rings to the steering gear. Torque the fasteners to 20 ft. lbs. (27 Nm).
- Power steering return hose

7. Raise the rear frame to its original position.

8. Install or connect the following:
- New rear frame bolts and torque them to 133 ft. lbs. (180 Nm) on all models except Impala and Monte Carlo models. On Impala and Monte

Carlo models, align the frame to the body by inserting two 19 x 203 mm (0.74 x 8.0 in) pins in the alignment holes on the right side of the frame. Tighten the front bolts to 118 ft. lbs. (160 Nm) and the rear bolts to 122 ft. lbs. (165 Nm)
- Tie rod ends to the steering knuckles and torque the nuts to 22 ft. lbs. (30 Nm)
- Intermediate steering shaft to the

steering gear stub shaft and torque the lower pinch bolt to 35 ft. lbs. (48 Nm)
- Front wheels
- Negative battery cable

9. Fill and bleed the power steering system.

10. Start the vehicle and check for leaks, repair if necessary.

11. Check the front end alignment and adjust as needed.

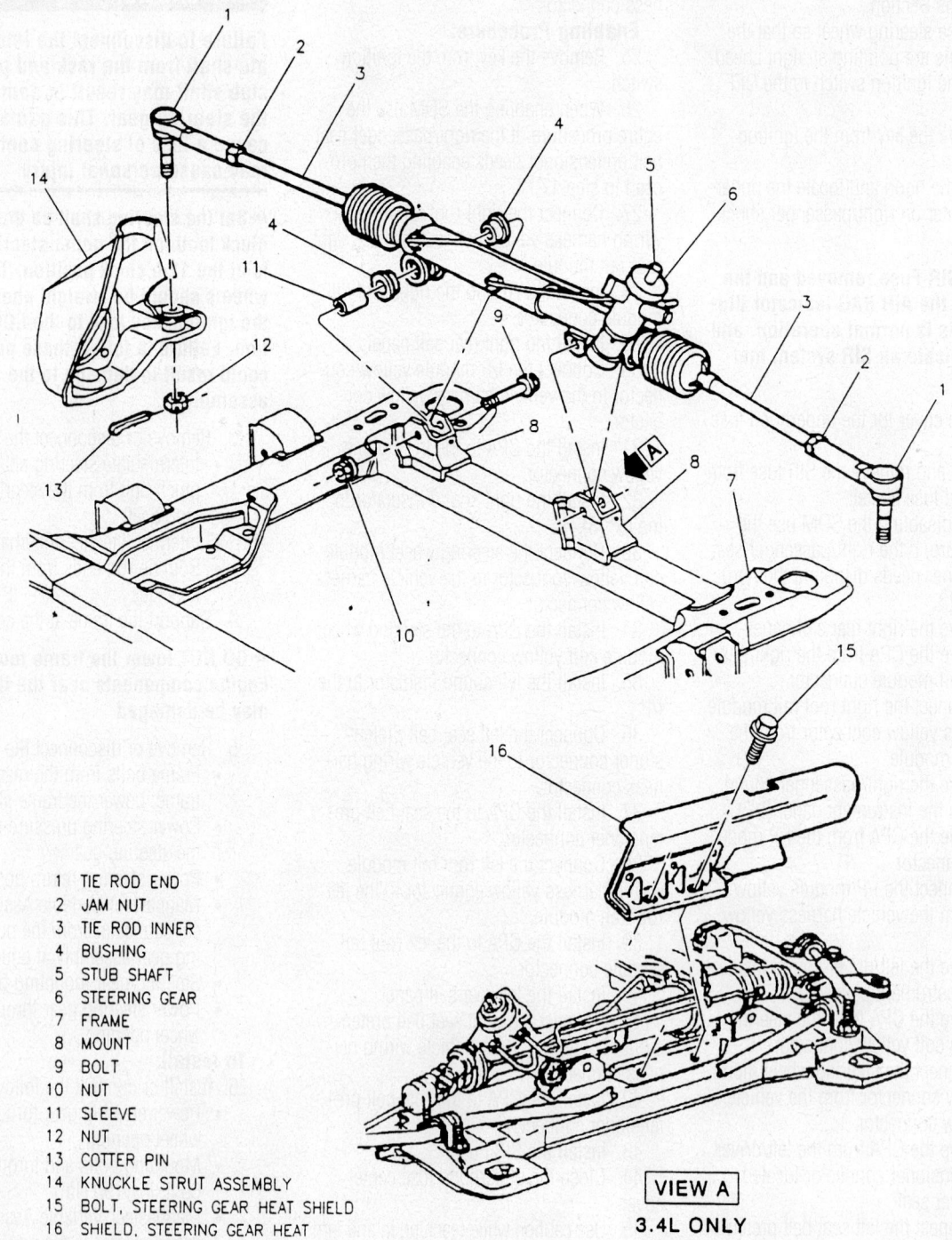

1 TIE ROD END
2 JAM NUT
3 TIE ROD INNER
4 BUSHING
5 STUB SHAFT
6 STEERING GEAR
7 FRAME
8 MOUNT
9 BOLT
10 NUT
11 SLEEVE
12 NUT
13 COTTER PIN
14 KNUCKLE STRUT ASSEMBLY
15 BOLT, STEERING GEAR HEAT SHIELD
16 SHIELD, STEERING GEAR HEAT

VIEW A
3.4L ONLY

79222529

Common rack and pinion steering gear mounting

Strut

REMOVAL & INSTALLATION

Front Strut

1. Before servicing the vehicle, refer to the Precautions Section.
2. Scribe the strut to the steering knuckle for proper installation.
3. Remove or disconnect the following:

- Front wheel
- Three upper strut nuts
- Lower strut bolts
- Strut

To install:

4. Install or connect the following:
- Strut
- Three upper strut nuts and torque them to 24 ft. lbs. (33 Nm)
- Lower strut bolts

5. Align the strut to the scribe mark on the steering knuckle. Torque the lower bolts to 90 ft. lbs. (123 Nm).
6. Install the front wheel.
7. Road test the vehicle and check the front end alignment, adjust if necessary.

Rear Strut

1. Before servicing the vehicle, refer to the Precautions Section.
2. Remove or disconnect the following:

- Negative battery cable
- Three strut-to-body nuts
- Rear tire and wheel
- Stabilizer shaft link from the strut

3. Scribe the strut to the knuckle.

➡**The knuckle must be retained after the strut to knuckle bolts have been removed. Damage may occur to the ball joint or drive axle if the knuckle is not retained.**

4. Remove or disconnect the following:

- Strut to knuckle bolts
- Strut

To install:

5. Install or connect the following:
- Strut
- Strut to knuckle bolts and torque the bolts to 90 ft. lbs. (122 Nm)
- Stabilizer shaft link to the strut
- Rear tire and wheel
- Three strut to body mount nuts and torque them to 30 ft. lbs. (41 Nm)

6. Road test the vehicle and adjust the rear wheel alignment if needed.

Coil Spring

REMOVAL & INSTALLATION

1. Before servicing the vehicle, refer to the Precautions Section.
2. Remove the strut from the vehicle.
3. Mount the strut assembly into a strut compressor. Note that the strut compressor has strut mounting holes drilled for specific vehicle lines.
4. Compress the strut approximately ½ its height after initial contact with the top cap.

✳✳ WARNING

Never bottom the spring or damper rod!

5. Remove the nut from the strut damper shaft and place an alignment/guid-ing rod on top of the damper shaft. Use the rod to guide the damper shaft straight down through the spring cap while compressing the spring. Remove the components.

To install:

6. Mount the strut assembly in the strut compressor, using the bottom locking pin only.
7. Install the spring over the damper and swing the assembly up so the upper locking pin can be installed.
8. Install all shields, bumpers and insulators on the spring seat.
9. Install the spring seat on top of the spring. The spring seat flat should be facing the same direction as the centerline of the strut assembly spindle.
10. Install the guiding rod and turn the forcing screw while the guiding rod centers the assembly. When the threads on the damper shaft are visible, remove the guiding rod and install the nut. tighten the front

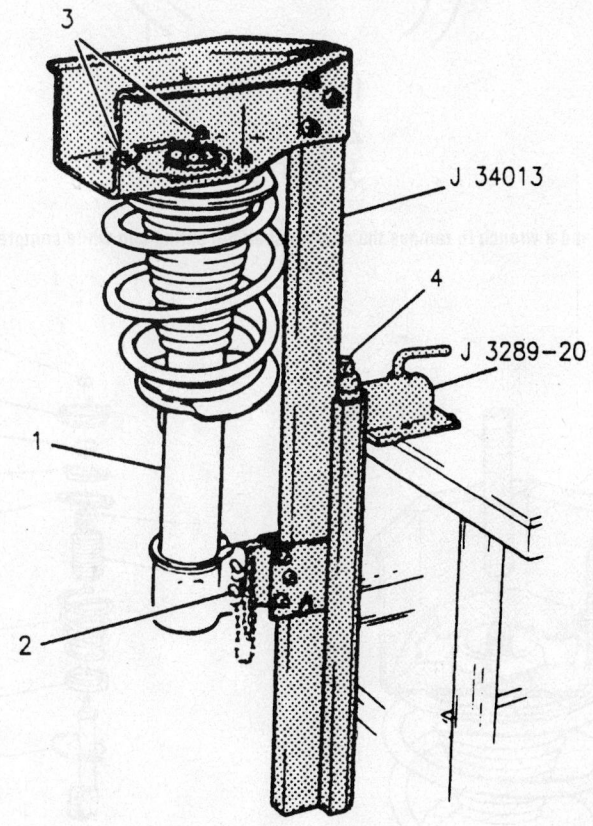

1 STRUT ASSEMBLY
2 INSTALL LOCKING PINS THROUGH
   STRUT ASSEMBLY
3 TIGHTEN NUTS UNTIL FLUSH WITH
   STRUT COMPRESSOR
4 COMPRESSOR FORCING SCREW

7922YG24

View of a typical strut assembly mounted in a compressor

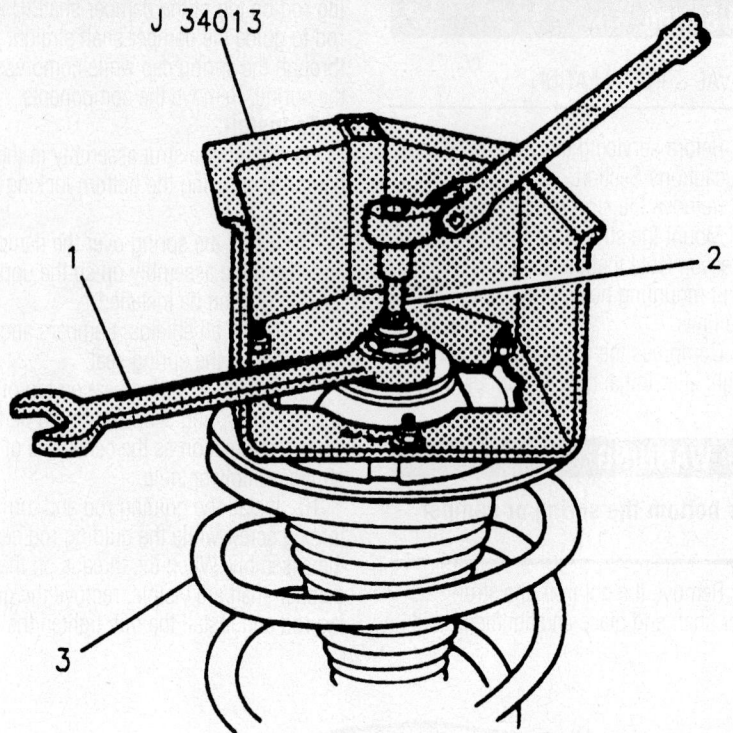

J 34013

1 WRENCH
2 SOCKET
3 STRUT ASSEMBLY

7922YG25

Use a socket and a wrench to remove the damper shaft nut spring cap while compressing the spring

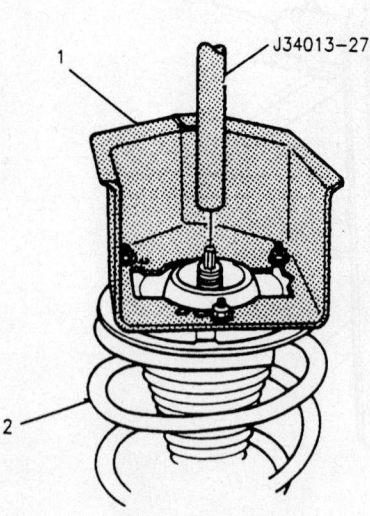

J34013–27

1 STRUT COMPRESSOR
2 STRUT ASSEMBLY

7922YG26

Install the rod to guide the damper shaft straight down through the spring cap while compressing the spring

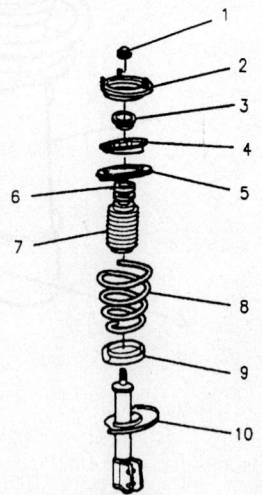

1 STRUT MOUNT NUT
2 STRUT MOUNT
3 RATE WASHER
4 SPRING SEAT
5 SPRING UPPER INSULATOR
6 JOUNCE BUMPER
7 STRUT DUST SHIELD
8 SPRING
9 SPRING LOWER INSULATOR
10 STRUT

7922YG27

Exploded view of the front strut assembly

strut nut to 63 ft. lbs. (85 Nm). Tighten the rear strut nut to 55 ft. lbs. (75 Nm).

11. Remove the strut from the compressor.

12. Install the strut to the vehicle.

Lower Ball Joint

REMOVAL & INSTALLATION

1. Before servicing the vehicle, refer to the Precautions Section.

2. Remove the wheel.

3. Remove the lower control arm from the vehicle.

4. Remove the ball joint from the lower control arm by drilling out the 4 rivets retaining the ball joint to the control arm. Use an ⅛ in. drill bit to make a pilot hole through the rivets. Finish drilling the rivets using a ½ in. drill bit.

5. Remove the ball joint.

To install:

6. Install or connect the following:
   - Ball joint to the control arm
   - Bolts with the heads facing down and torque them to 50 ft. lbs. (68 Nm)
   - Lower control arm to the vehicle
   - Wheel

➡A 4-wheel alignment is recommended after any steering/suspension repairs are performed.

Lower Control Arm

REMOVAL & INSTALLATION

1. Before servicing the vehicle, refer to the Precautions Section.

2. Remove or disconnect the following:
   - Front wheel
   - Antilock Brake System (ABS) wheel speed sensor connector and jumper harness from the retainer
   - Stabilizer shaft link
   - Cotter pin from the ball stud and loosen the nut

3. Install a ball joint removal tool over the ball joint and lower control arm. Rotate the ball stud nut counterclockwise to separate the ball stud from the steering knuckle.

4. Remove the lower control arm.

To install:

5. Install the lower control arm and bolts. Do not tighten the nuts at this time.

➡Align the ball stud cotter pin hole parallel to the knuckle to ease the cotter pin installation.

6. Install or connect the following:
- Ball stud to the knuckle and torque the nut to 15 ft. lbs. (20 Nm) plus an additional 120 degrees on Impala models.
- New cotter pin and bend the ends. Make certain the ends do not make contact with the ABS wheel speed sensor
- Stabilizer shaft link
- ABS wheel speed sensor wire harness to the retainer clips
- ABS wheel speed sensor connector
- Lower control arm nuts and torque them to 92 ft. lbs. (125 Nm).
- Front wheel

CONTROL ARM BUSHING REPLACEMENT

1. Before servicing the vehicle, refer to the Precautions Section.
2. Remove or disconnect the following:
- Front wheel
- Lower control arm
3. Mark the lower control arm along the flat edge of the bushing flange.
4. Coat the threads of tool J 21474-27 with a high pressure lubricant.
5. Assemble the following bushing removal tools as illustrated:
- J 21474-27
- J 21474-13
- J 34126
- J 35379
- J21474-2
- J 21474-4
6. Tighten J 21474-4 to remove the bushing.

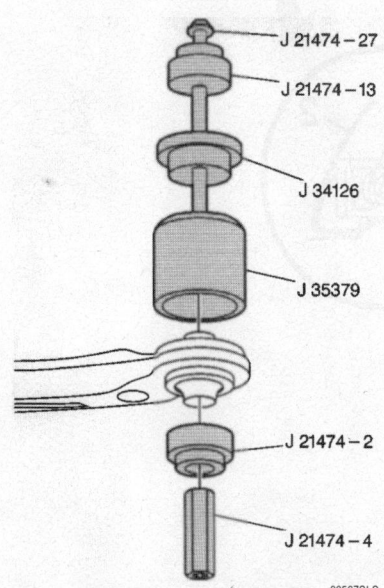

Removing the control arm bushing

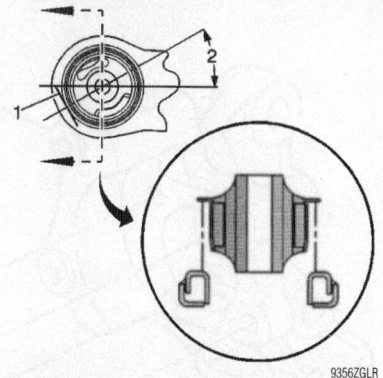

Positioning the control arm bushing

To install:

➡ You MUST install the lower control arm vertical bushing in the same position in order to maintain the original vehicle ride, handling, and road feel.

7. Align the flat edge of the bushing flange to the mark in the control arm (1). Ensure that the flat edge of the bushing flange is 30 degrees (2) from the centerline of the lower control arm. Ensure that the thin slot in the bushing is facing outboard. Insert the bushing into the control arm. Refer to the accompanying illustration for clarification of positions (1) and (2).
8. Coat the threads of tool J 21474-27 with a high pressure lubricant.

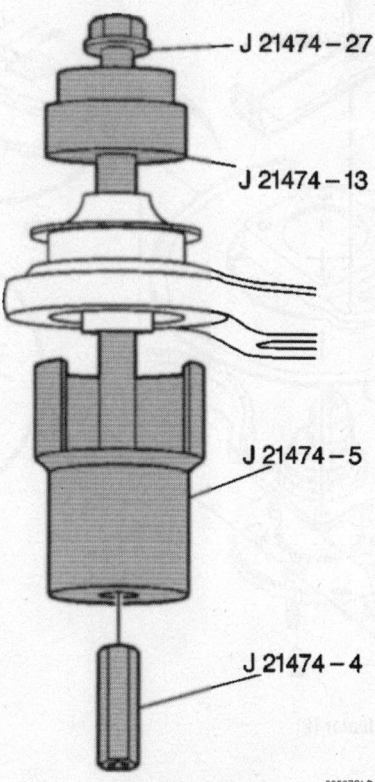

Installing the control arm bushing

9. Assemble the following bushing installation tools as illustrated:
- J 21474-27
- J 21474-13
- J 21474-5
- J 21474-4
10. Tighten J 21474-4 to remove the bushing.
11. Install the lower control arm.
- Front wheel

Wheel Bearings

ADJUSTMENT

The wheel bearings are not adjustable. If a wheel bearing is out of specification, it must be replaced. Using a dial indicator, check for looseness. If it exceeds 0.005 in. (0.127mm) on drum or disc brakes the bearing wear is excessive and the hub and bearing should be replaced.

REMOVAL & INSTALLATION

Front

1. Before servicing the vehicle, refer to the Precautions Section.
2. Remove or disconnect the following:
- Front wheel
- Wheel Speed Sensor (WSS) electrical connector
- Brake caliper and bracket
- Rotor
- Driveshaft nut
3. Install a front hub removal tool to the wheel bearing/hub assembly with three wheel nuts. Use the tool to push the driveshaft out of the wheel bearing/hub.
4. Remove the wheel bearing/hub assembly and discard the bolts.

To install:

5. Install or connect the following:
- Wheel bearing/hub assembly using new bolts and torque them to 96 ft. lbs. (130 Nm)
- New drive shaft nut and torque it to 118 ft. lbs. (160 Nm)
- Brake rotor
- Front caliper with the bracket
- WSS electrical connector
- Front wheel

Rear

The rear wheel bearing/hub is integrated into one unit. The unit is non-serviceable. If the hub or bearing is damaged, the complete hub and bearing unit must be replaced.

1. Before servicing the vehicle, refer to the Precautions Section.
2. Remove or disconnect the following:
   - Rear wheel
   - Brake drum, if equipped
   - Rear caliper and bracket, if equipped
   - Brake rotor, if equipped
   - Antilock Brake System (ABS) Wheel Speed Sensor (WSS) electrical connector
   - Rear wheel hub to knuckle bolts
   - Parking brake lever bracket and parking brake actuator
   - Rear wheel hub from the knuckle

To install:

3. Install or connect the following:
   - Parking brake lever bracket and actuator
   - Hub and bearing assembly and torque the bolts to 55 ft. lbs. (75 Nm)
   - WSS electrical connector
   - Brake rotor, if equipped
   - Brake caliper with the bracket
   - Brake drum, if equipped
   - Rear wheel
4. A 4-wheel alignment is recommended after any steering/suspension repairs have been performed.

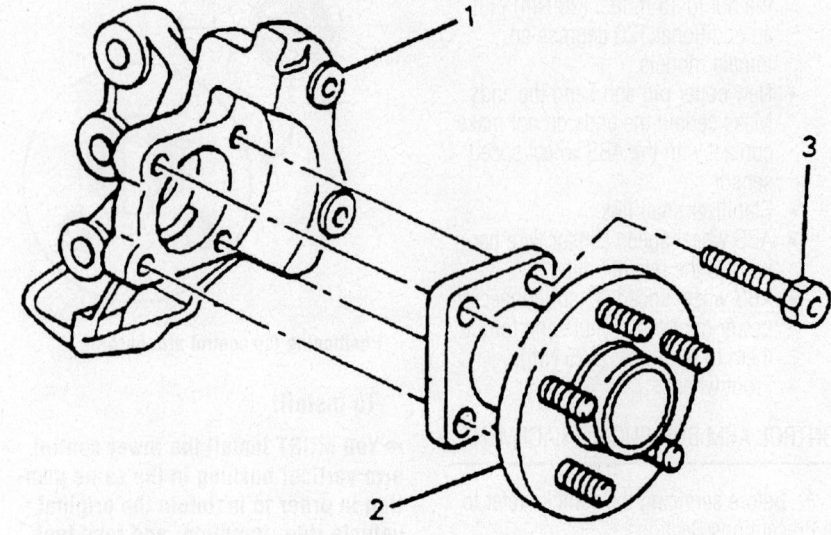

1 KNUCKLE ASSEMBLY, REAR SUSPENSION
2 HUB AND BEARING ASSEMBLY
3 BOLT/SCREW, WHEEL

79222535

The rear hub/bearing assembly is bolted to the knuckle

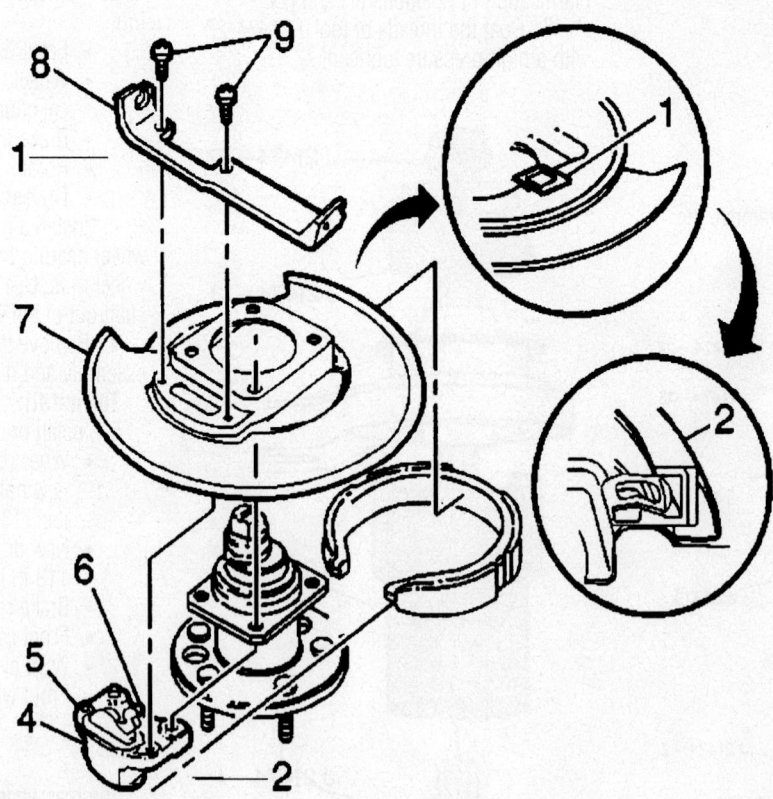

9306ZG23

Parking brake lever bracket (1) and parking brake actuator (2)

BRAKES

Brake Caliper

REMOVAL & INSTALLATION

Front

EXCEPT 2005 GRAND PRIX

1. Remove ⅔ of the brake fluid from the master cylinder assembly.

2. Remove the tire and wheel assembly. Mark a relationship between the wheel and the wheel stud for re-installation purposes.

3. Install 2 wheel nuts to retain the rotor on the vehicle.

4. Install a large C-clamp over top of the caliper housing and against the back of the outboard shoe. Slowly tighten the C-clamp until the piston(s) are pushed into the caliper bore.

5. Disconnect the bolt attaching the brake hose fitting to the brake caliper.

6. Plug the opening in the caliper housing and brake line to prevent brake fluid loss and contamination.

7. Remove the caliper mounting bolts.

8. Remove the brake caliper housing from the rotor and mounting bracket.

9. If the caliper is to be replaced or repaired remove the brake pads from the caliper or mounting bracket.

To install:

10. Fully inspect the brake caliper bushing assemblies for cuts, tears or deterioration and replace parts as needed.

11. Inspect the slide bolts for corrosion. If corrosion is found, replace the slide bolts and bushings before installing the brake caliper assembly.

12. Before installing the caliper, make sure the piston(s) are seated in the bore, and that the brake pads are correctly installed.

13. Lubricate the caliper slide bolts with silicone grease. Do not lubricate the bolt threads.

14. Install brake pads onto caliper or mounting bracket.

15. Install the caliper bolts and torque the bolts to 70 ft. lbs. (95 Nm).

16. Install the brake hose inlet fitting to the caliper and torque the bolt to 24 ft. lbs. (32 Nm).

17. Remove wheel nuts securing the brake rotor to the hub.

18. Install the tire and wheel assembly.

19. Fill the master cylinder to the proper level.

20. Bleed the brake system.

21. Verify correct brake operation.

2005 GRAND PRIX RPO J65 AND JL9

1. Before servicing the vehicle, refer to the Precautions Section.

2. Inspect the fluid level in the brake master cylinder reservoir.

3. If the brake fluid level is midway between the maximum-full point and the minimum allowable level, then no brake fluid needs to be removed from the reservoir before proceeding. If the brake fluid level is higher than midway between the maximum-full point and the minimum allowable level, then remove brake fluid to the midway point before proceeding.

4. Raise and suitably support the vehicle.

5. Remove the front tire and the wheel assembly.

6. Hand tighten 2 wheel lug nuts to retain the rotor to the hub.

7. Install a large C-clamp over the top of the brake caliper and against the back of the outboard brake pad.

8. Tighten the C-clamp until the caliper piston is pushed into the caliper bore enough to slide the caliper off the rotor.

9. Remove the C-clamp from the caliper.

10. Remove the brake hose-to-caliper bolt from the caliper. Discard the 2 copper brake hose gaskets. These gaskets may be stuck to the brake caliper and/or the brake hose end.

11. Plug the opening in the front brake hose to prevent excessive brake fluid loss and contamination.

➡ Note the location of the caliper pin bolts. The leading caliper pin bolt, or top bolt, has a bushing as part of the assembly. The trailing caliper pin bolt, or bottom bolt, is a solid design.

12. Remove the caliper pin bolts. Note the location of the caliper pin bolts. The leading caliper pin, or top bolt, has a bushing as part of the assembly. The trailing caliper pin, or bottom bolt, is a solid design.

13. Remove the caliper from the rotor and the caliper bracket.

14. Inspect the caliper bolt boots in the caliper bracket for damage. Replace any damaged caliper bolt boots.

15. Inspect the caliper bolts for corrosion or damage. If corrosion or damage is found, use new caliper pin bolts when installing the caliper.

To install:

16. If reusing the brake caliper pin bolts, wipe away any debris and old lubricant with a with a clean shop cloth.

17. Apply lubricant, GM P/N 18047666, or equivalent, to the brake caliper pin bolts. Apply a thin layer to the pin bushing and to the caliper pin bolt shank. Ensure that there is not a buildup of excess lubricant at the end of the leading caliper pin, in front of the bushing.

18. Install the caliper over the rotor and onto the caliper bracket.

➡ The leading caliper pin, or top bolt, has a bushing as part of the assembly. The trailing caliper pin, or bottom bolt, is a solid design.

19. Install the caliper pin bolts. The leading caliper pin bolt, or top bolt, has a

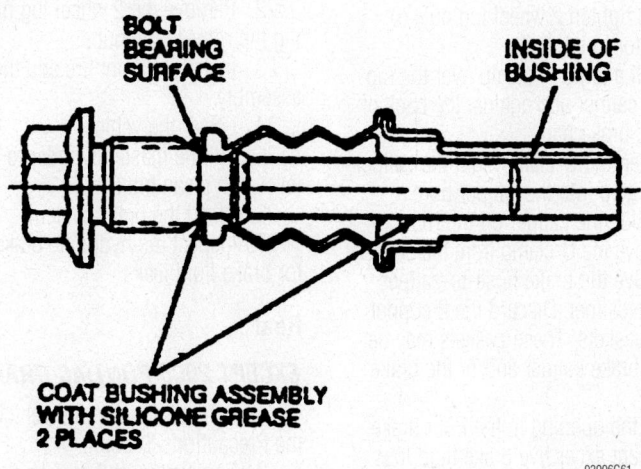

BOLT BEARING SURFACE

INSIDE OF BUSHING

COAT BUSHING ASSEMBLY WITH SILICONE GREASE 2 PLACES

93006G61

Slide bolts lubrication points

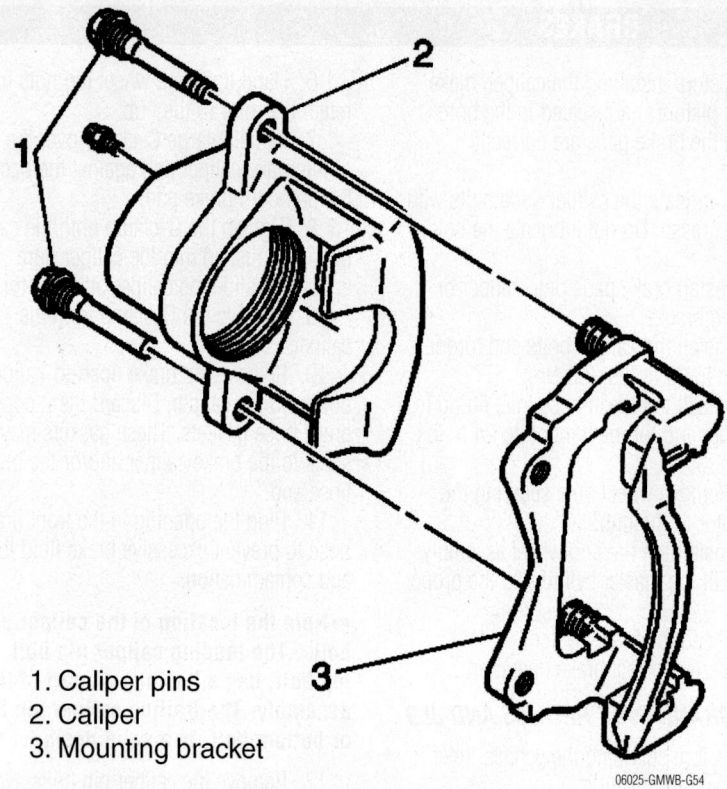

1. Caliper pins
2. Caliper
3. Mounting bracket

06025-GMWB-G54

Front caliper mounting—Except 2005 Grand Prix RPO Z7U

bushing as part of the assembly. The trailing caliper pin bolt, or bottom bolt, is a solid design. Ensure that the bolt boots fit securely in the groove of the caliper pin bolts. If the boots are damaged, they must be replaced. Tighten the bolts to 95 Nm (70 ft. lbs.).

❊❊ WARNING

Install NEW copper brake hose gaskets.

20. Assemble the brake hose bolt and the NEW copper brake hose gaskets to the brake hose.
21. Install the brake hose-to-caliper bolt to the brake caliper. Tighten the bolt to 54 Nm (40 ft. lbs.).
22. Remove the 2 wheel lug nuts retaining the rotor to the hub.
23. Install the front tire and the wheel assembly.
24. Lower the vehicle.
25. Fill the master cylinder to the proper level with clean brake fluid.
26. Bleed the brake system.
27. Inspect the hydraulic brake system for brake fluid leaks.

2005 GRAND PRIX RPO Z7U

1. Before servicing the vehicle, refer to the Precautions Section.

2. Inspect the fluid level in the brake master cylinder reservoir.
3. If the brake fluid level is midway between the maximum-full point and the minimum allowable level, then no brake fluid needs to be removed from the reservoir before proceeding. If the brake fluid level is higher than midway between the maximum-full point and the minimum allowable level, then remove brake fluid to the midway point before proceeding.
4. Raise and suitably support the vehicle.
5. Remove the front tire and the wheel assembly.
6. Hand tighten 2 wheel lug nuts to retain the rotor to the hub.
7. Install a large C-clamp over the top of the brake caliper and against the back of the outboard brake pad.
8. Tighten the C-clamp until the caliper piston is pushed into the caliper bore enough to slide the caliper off the rotor.
9. Remove the C-clamp from the caliper.
10. Remove the brake hose-to-caliper bolt from the caliper. Discard the 2 copper brake hose gaskets. These gaskets may be stuck to the brake caliper and/or the brake hose end.
11. Plug the opening in the front brake hose to prevent excessive brake fluid loss and contamination.

12. Remove the caliper pin bolts.
13. Remove the caliper from the rotor and the caliper bracket.
14. Inspect the caliper bolt boots in the caliper bracket for damage. Replace any damaged caliper bolt boots.
15. Inspect the caliper bolts for corrosion or damage. If corrosion or damage is found, use new caliper pin bolts when installing the caliper.

To install:

16. If reusing the brake caliper pin bolts, wipe away any debris and old lubricant with a with a clean shop cloth.
17. Apply lubricant, GM P/N 18047666, or equivalent, to the brake caliper pin bolts. Apply a thin layer to the pin bushing and to the caliper pin bolt shank. Ensure that there is not a buildup of excess lubricant at the end of the leading caliper pin, in front of the bushing.
18. Install the caliper over the rotor and onto the caliper bracket.

➡**The leading caliper pin, or top bolt, has a bushing as part of the assembly. The trailing caliper pin, or bottom bolt, is a solid design.**

19. Install the caliper pin bolts. Ensure that the bolt boots fit securely in the groove of the caliper pin bolts. If the boots are damaged, they must be replaced. Tighten the bolts to 44 Nm (60 ft. lbs.).

❊❊ WARNING

Install NEW copper brake hose gaskets.

20. Assemble the brake hose bolt and the NEW copper brake hose gaskets to the brake hose.
21. Install the brake hose-to-caliper bolt to the brake caliper. Tighten the bolt to 54 Nm (40 ft. lbs.).
22. Remove the 2 wheel lug nuts retaining the rotor to the hub.
23. Install the front tire and the wheel assembly.
24. Lower the vehicle.
25. Fill the master cylinder to the proper level with clean brake fluid.
26. Bleed the brake system.
27. Inspect the hydraulic brake system for brake fluid leaks.

Rear

EXCEPT 2005 PONTIAC GRAND PRIX

1. Before servicing the vehicle, refer to the Precautions Section.
2. Remove ⅔ of the brake fluid from the master cylinder assembly.

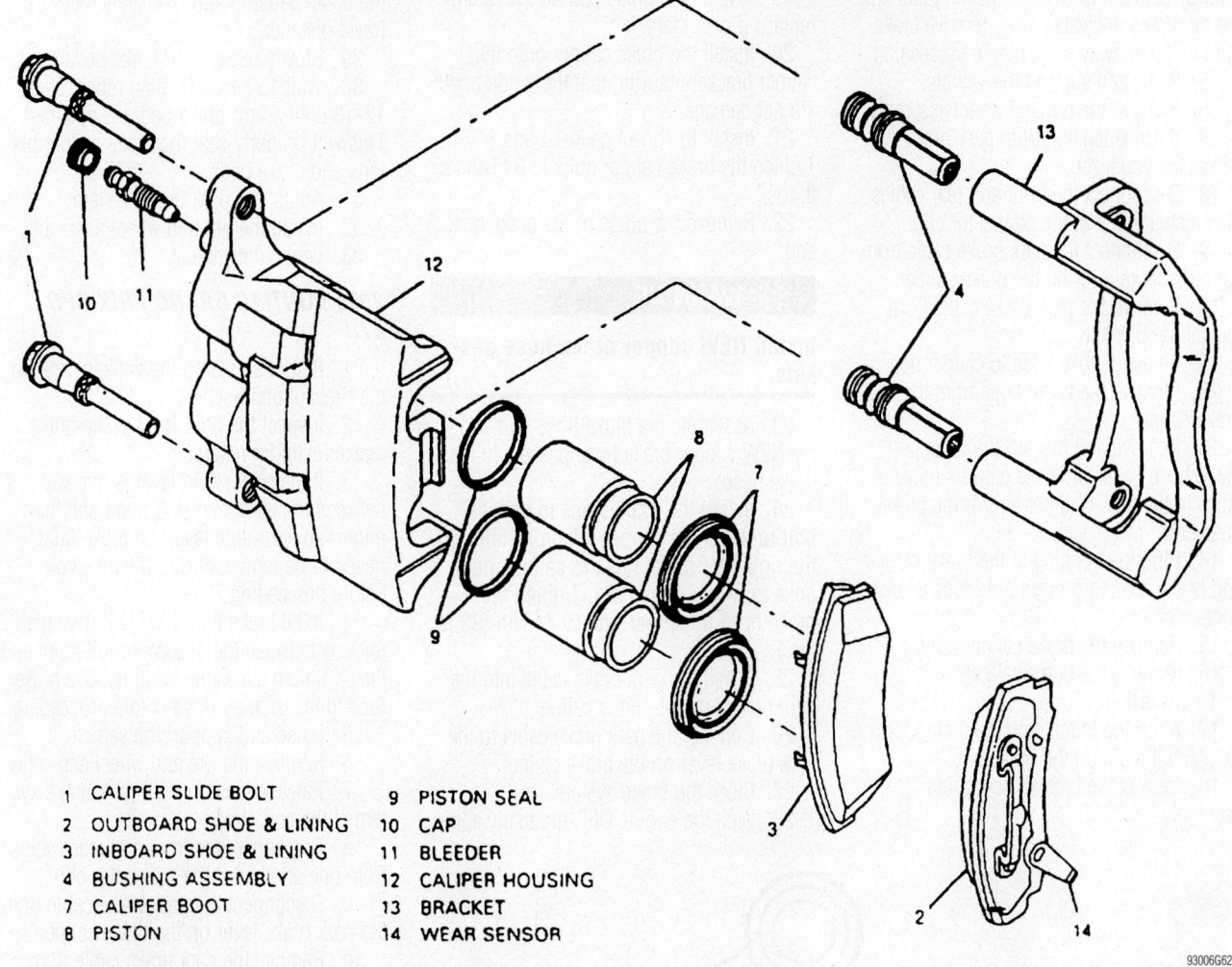

1	CALIPER SLIDE BOLT	9	PISTON SEAL
2	OUTBOARD SHOE & LINING	10	CAP
3	INBOARD SHOE & LINING	11	BLEEDER
4	BUSHING ASSEMBLY	12	CALIPER HOUSING
7	CALIPER BOOT	13	BRACKET
8	PISTON	14	WEAR SENSOR

Caliper attachments—except 2005 Pontiac Grand Prix RPO Z7U

93006G62

3. Remove the tire and wheel assembly. Mark a relationship between the wheel and the wheel stud for re-installation purposes.

4. Install 2 wheel nuts to retain the rotor.

5. Remove the brake hose from the caliper and discard the copper washers.

6. Plug the openings in the caliper and the brake hose to prevent brake fluid loss and contamination.

7. Remove the two caliper mounting bolts.

8. Remove the caliper body assembly from the vehicle. Pivot the caliper assembly up to clear the rotor and then slide it inboard off the pin sleeve.

To install:

9. Inspect the caliper bolt boots, pins and sleeve bolt for cuts, tears or deterioration. Replace as necessary.

10. Lubricate the mounting surfaces and the mounting sleeves.

11. Hold the caliper body assembly in the position from which it was removed, and start it over the end of the pin sleeve.

12. As the caliper body assembly approaches the pin boot, work the large end of the pin boot in the caliper body groove. Push the caliper body fully onto the pin.

13. Pivot the caliper body assembly down, using care not to damage the piston boot on the inboard shoe. Compress the sleeve boot by hand as the caliper body moves into position to prevent boot damage.

14. After installing the caliper assembly into position, recheck the installation of the pad clips. If necessary, use a small prying tool to reset or center the pad clips.

15. Torque caliper mounting bolts to 32 ft. lbs. (43 Nm)

16. Connect the brake hose to the brake caliper and torque the bolt to 40 ft. lbs. (54 Nm).

17. Remove the wheel nuts securing the rotor to the hub and bearing assembly.

18. Install the tire and wheel assembly.

19. Fill the master cylinder to the proper level with clean brake fluid.

20. Bleed the brake system using the recommended procedure.

21. Apply approximately 175 lbs. (79 kg) of force, 3 times, to properly seat the brake shoe and linings against the rotor.

22. Adjust the parking brake cable as necessary (some models).

2005 PONTIAC GRAND PRIX RPO J65 AND JL9

1. Before servicing the vehicle, refer to the Precautions Section.

2. Inspect the fluid level in the brake master cylinder reservoir.

3. If the brake fluid level is midway between the maximum-full point and the minimum allowable level, no brake fluid needs to be removed from the reservoir before proceeding.

4. If the brake fluid level is higher than

midway between the maximum-full point and the minimum allowable level, remove brake fluid to the midway point before proceeding.

5. Raise and support the vehicle.

6. Remove the tire and wheel assembly.

7. Release tension from park brake system at the equalizer.

8. Disconnect the front and rear cables from one another at the connector clip.

9. Disconnect the park brake cable from the park brake lever on the brake caliper.

10. Remove the park brake cable from the caliper bracket.

11. Remove brake hose to caliper bolt.

12. Remove the brake hose from the brake caliper.

13. Remove and discard the 2 copper brake hose gaskets. These gaskets may be stuck to the brake caliper and/or the brake hose end.

14. Plug the opening in the brake caliper and brake hose to prevent fluid loss and/or contamination.

15. Remove the brake caliper bolts.

16. Remove the brake caliper.

To install:

17. Align the indents on the piston face to match the pin on the brake pad.

18. Inspect the bracket bolt guide assembly.

19. Inspect the brake pad hardware and replace if necessary.

20. Install the brake caliper onto the caliper bracket insuring that the guide boots are not damaged.

21. Install the brake caliper bolts. Tighten the brake caliper bolts to 34 Nm (25 ft. lbs.).

22. Remove the plugs in the brake hose end.

✳✳ WARNING

Install NEW copper brake hose gaskets.

23. Assemble the brake hose bolt and the NEW copper brake hose gaskets to the brake hose.

24. Install the brake hose to caliper bolt to the brake caliper. When installing the right rear brake hose to caliper, hold hose up while tightening. Tighten the brake hose to caliper bolt to 44 Nm (33 ft. lbs.).

25. Install the park brake cable into the park brake bracket on the caliper.

26. Connect the park brake cable to the park brake lever on the brake caliper.

27. Bleed the brake system.

28. With the engine OFF, gradually apply the brake pedal to approximately ⅔ of its travel distance.

29. Slowly release the brake pedal.

30. Wait 15 seconds, then repeat steps 12-13 until a firm brake pedal is obtained. This will properly seat the brake caliper pistons and brake pads.

31. Adjust the park brake system.

32. Install the tire and wheel assembly.

33. Lower the vehicle.

2005 PONTIAC GRAND PRIX RPO Z7U

1. Before servicing the vehicle, refer to the Precautions Section.

2. Inspect the fluid level in the brake master cylinder reservoir.

3. If the brake fluid level is midway between the maximum-full point and the minimum allowable level, no brake fluid needs to be removed from the reservoir before proceeding.

4. If the brake fluid level is higher than midway between the maximum-full point and the minimum allowable level, remove brake fluid to the midway point before proceeding.

5. Raise and support the vehicle.

6. Remove the tire and wheel assembly.

7. Release tension from park brake system at the equalizer.

8. Disconnect the front and rear cables from one another at the connector clip.

9. Disconnect the park brake cable from the park brake lever on the brake caliper.

10. Remove the park brake cable from the caliper bracket.

11. Remove brake hose to caliper bolt.

12. Remove the brake hose from the brake caliper.

13. Remove and discard the 2 copper brake hose gaskets. These gaskets may be stuck to the brake caliper and/or the brake hose end.

14. Plug the opening in the brake caliper and brake hose to prevent fluid loss and/or contamination.

15. Remove the brake caliper bolts.

16. Remove the brake caliper.

To install:

17. Inspect the bracket bolt guide assembly.

18. Inspect the brake pad hardware and replace if necessary.

19. Install the brake caliper onto the caliper bracket insuring that the guide boots are not damaged.

20. Install the brake caliper bolts. Tighten the brake caliper bolts to 60 Nm (44 ft. lbs.).

21. Remove the plugs in the brake hose end.

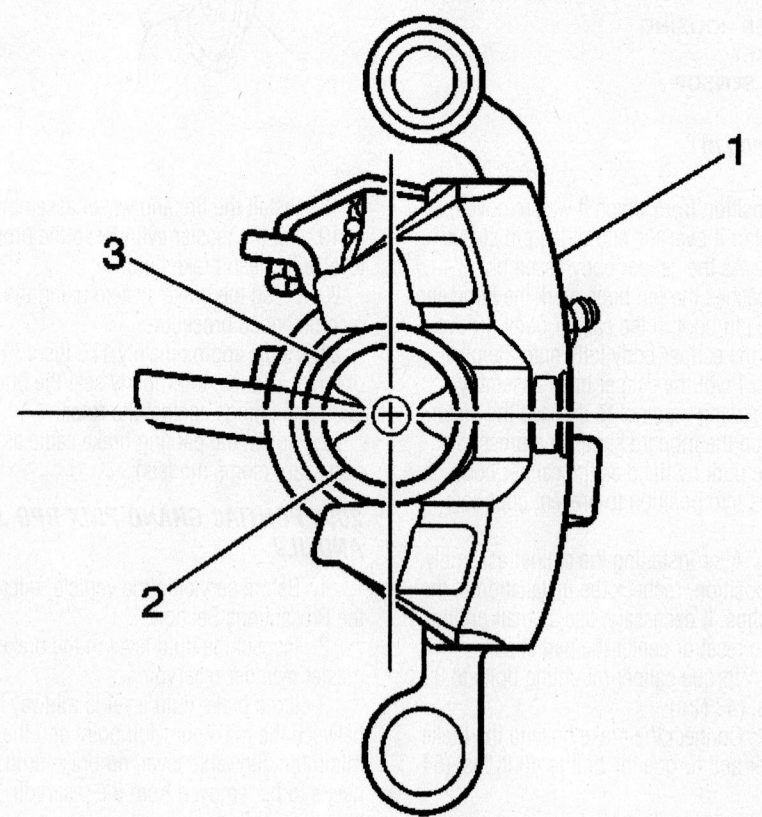

06025-GMWB-G55

Align the indents on the piston face to match the pin on the brake pad

✳✳ WARNING

Install NEW copper brake hose gaskets.

22. Assemble the brake hose bolt and the NEW copper brake hose gaskets to the brake hose.

23. Install the brake hose to caliper bolt to the brake caliper. When installing the right rear brake hose to caliper, hold hose up while tightening. Tighten the brake hose to caliper bolt to 44 Nm (33 ft. lbs.).

24. Install the park brake cable into the park brake bracket on the caliper.

25. Connect the park brake cable to the park brake lever on the brake caliper.

26. Bleed the brake system.

27. With the engine OFF, gradually apply the brake pedal to approximately ⅔ of its travel distance.

28. Slowly release the brake pedal.

29. Wait 15 seconds, then repeat steps 12-13 until a firm brake pedal is obtained. This will properly seat the brake caliper pistons and brake pads.

30. Adjust the park brake system.

31. Install the tire and wheel assembly.

32. Lower the vehicle.

Disc Brake Pads

REMOVAL & INSTALLATION

Front

EXCEPT 2005 GRAND PRIX RPO Z7U

1. Before servicing the vehicle, refer to the Precautions Section.

2. Siphon ⅔ of the brake fluid out of the master cylinder.

3. Mark the relationship of the wheel to the wheel stud for re-installation purposes. Remove the tire and wheel assembly.

4. Install 2 lug nuts to secure the rotor in place when the caliper is removed.

5. Install a large C-clamp over the top of the caliper housing and against the back of the outboard shoe. Slowly tighten the C-clamp until the caliper pistons are pushed into the caliper bore enough to slide the caliper assembly off the rotor. Use care not to tighten the C-clamp too far or the outboard shoe retaining spring will be deformed and require replacement.

6. Remove the caliper mounting bolts and remove the brake caliper from the mounting bracket.

7. DO NOT disconnect the brake hose from the caliper or allow the brake hose to support the weight of the caliper. Support the caliper on a piece of wire out of the way.

8. Remove the outer brake pad from the caliper using a suitable prying tool to lift the outboard shoe retaining spring so that it will clear the caliper center lug and pull the brake pad out of the caliper.

9. Remove the inner brake pad by unsnapping the shoe springs from the piston.

To install:

10. Clean all parts well. If the brake pads were worn so badly that the brake rotor is damaged, it must be replaced. Light scoring of the rotor surfaces not exceeding 0.060 inch (1.5mm) in depth is not harmful to brake operation and may result from normal use. Brake rotors may be refinished. Do not use a rotor that, after refinishing, will not meet the thickness specification cast in the rotor. Always replace with a new rotor.

11. If not done at removal, now use a C-clamp and clamp both pistons at the same time with a metal plate or wooden block across the face of both pistons. Take care not to damage the pistons or caliper boots.

➡**After bottoming the pistons into the caliper bore, lift the inner edge of each caliper boot next to the piston and press out any trapped air. Make sure each boot convolution is tucked back into place. Boots must lay flat.**

12. Inspect the caliper bushings for wear. Replace as necessary. Carefully inspect the slide bolts for corrosion. If corrosion if found, use new parts including the bushing assemblies when installing the caliper. Do not attempt to polish away corrosion. Lubricate caliper slide bolts with silicone grease.

13. Install the new inner disc brake pad in the caliper by snapping the shoe retainer springs into the piston making sure both sets of locking tabs are seated in the caliper pistons. The pad must seat flat against the pistons.

14. Install the outer pad into the caliper by snapping the outboard shoe retaining spring over the caliper center lug and into the housing slot. The pad will slide up onto the caliper and the retaining ring will lock into place on the groove in the caliper.

15. The outer pad wear sensor should be at the trailing edge of the shoe during forward wheel rotation.

16. Install the caliper mounting bolts.

17. Remove the 2 nuts temporarily securing the rotor.

18. Install the tire and wheel assembly and tighten to specification.

19. Pump the brake pedal several times to seat the pads against the rotor.

20. Check the brake fluid level and top off as necessary.

21. Road test the vehicle to ensure the proper brake performance.

2005 GRAND PRIX RPO Z7U

1. Before servicing the vehicle, refer to the Precautions Section.

2. Siphon ⅔ of the brake fluid out of the master cylinder.

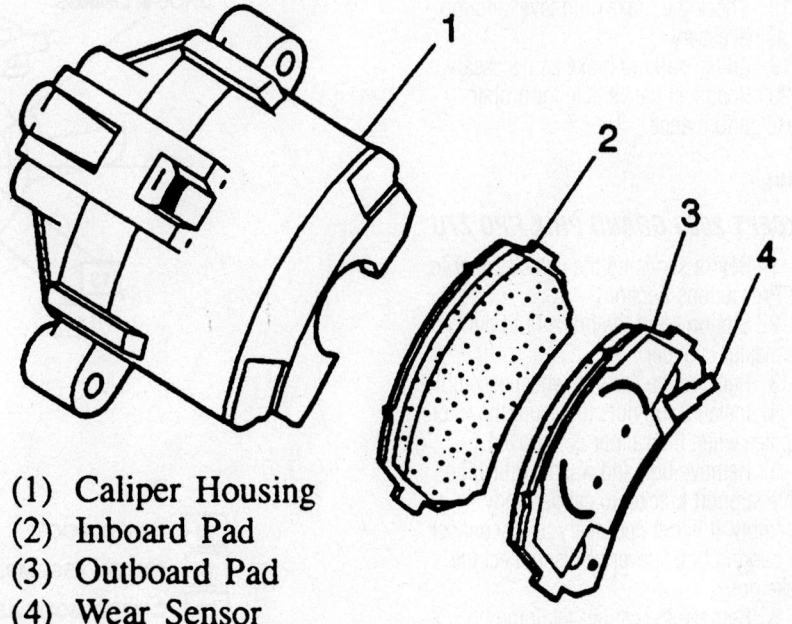

(1) Caliper Housing
(2) Inboard Pad
(3) Outboard Pad
(4) Wear Sensor

93006G71

Front brake pads and caliper—Except 2005 Grand Prix RPO Z7U

3. Remove the tire and wheel assembly.

4. Install 2 lug nuts to secure the rotor in place when the caliper is removed.

5. Remove bolt and washer attaching cable support bracket to caliper body assembly. It is not necessary to disconnect the parking brake lever or disconnect the brake hose.

6. Remove the caliper retaining bolt.

7. Pivot the caliper body assembly up from the rotor and remove from the bracket. Do not completely remove the caliper assembly body.

8. Remove the outboard and inboard shoe and linings from the caliper support assembly.

9. Remove 2 brake lining clips from the caliper support.

To install:

10. Install 2 pad clips in the caliper support.

11. Lubricate the inner pad where it contacts the piston and mounting surfaces.

12. Install outboard and inboard shoe and linings in caliper support. Do not re-use the shims. Install new shims.

13. Pivot the caliper body assembly down over the brake pad. Compress the sleeve boot by hand as the caliper body moves into position to prevent boot damage.

14. Torque the bolt to 44 ft. lbs. (60 Nm).

15. Remove the 2 lug nuts securing the rotor.

16. Install the tire and wheel assembly and tighten to specification.

17. Pump the brake pedal several times to seat the pads against the rotor.

18. Check the brake fluid level and top off as necessary.

19. Adjust parking brake as necessary.

20. Road test the vehicle for proper brake performance.

Rear

EXCEPT 2005 GRAND PRIX RPO Z7U

1. Before servicing the vehicle, refer to the Precautions Section.

2. Siphon ⅔ of the brake fluid out of the master cylinder.

3. Remove the tire and wheel assembly.

4. Install 2 lug nuts to secure the rotor in place when the caliper is removed.

5. Remove bolt and washer attaching cable support bracket to caliper body assembly. It is not necessary to disconnect the parking brake lever or disconnect the brake hose.

6. Remove the caliper retaining bolt.

7. Pivot the caliper body assembly up from the rotor and remove from the bracket. Do not completely remove the caliper assembly body.

8. Remove the outboard and inboard shoe and linings from the caliper support assembly.

9. Remove 2 brake lining clips from the caliper support.

To install:

➡ In order for the rear pads to seat in the caliper properly, the cut outs in the caliper piston must be at the 6 and 12 o'clock positions. Failure to align the piston correctly can lead to brake drag, premature brake wear and possible brake failure.

10. Using a suitable type spanner tool turn the piston in to bottom the piston fully into the caliper bore. Once the caliper is fully seated make sure the cutouts in the piston are at the 6 and 12 o'clock positions.

11. After bottoming the piston into the caliper bore, lift the inner edge of boot next to the piston assembly and press out any trapped air.

12. Install 2 pad clips in the caliper support.

13. Lubricate the inner pad where it contacts the piston and mounting surfaces.

14. Install outboard and inboard shoe and linings in caliper support. Position the wear sensors downward at the leading edge of the rotor during forward wheel rotation.

15. Hold the metal shoe edge against the spring end of clips in the caliper support. Push brake pad in towards the hub, bending spring ends slightly and engage shoe notches with support abutments.

16. Pivot the caliper body assembly down over the brake pad.

➡ After the caliper body assembly is in position, recheck installation of the brake pad clips. If necessary, use a small prying tool to reseat or center the pad clip on the support abutments.

17. Install the bolts.

18. Install the cable support bracket with the cable attached and bolt washer. Torque bolt to 32 ft. lbs. (43 Nm).

19. Remove the 2 lug nuts securing the rotor.

20. Install the tire and wheel assembly and tighten to specification.

21. Pump the brake pedal several times to seat the pads against the rotor.

22. Check the brake fluid level and top off as necessary.

23. Adjust parking brake as necessary.

24. Road test the vehicle for proper brake performance.

2005 GRAND PRIX RPO Z7U

1. Before servicing the vehicle, refer to the Precautions Section.

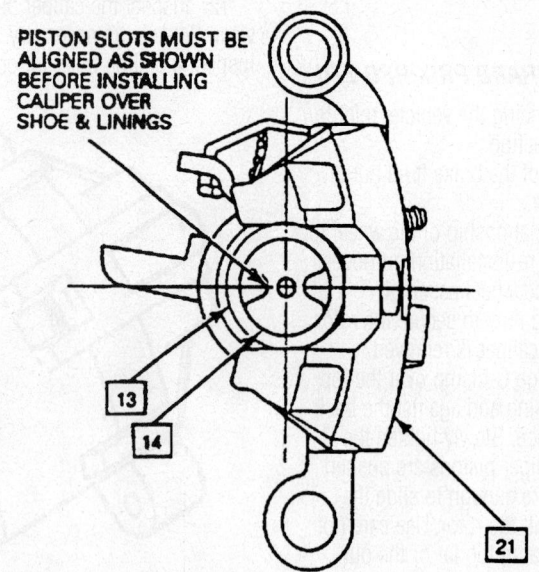

PISTON SLOTS MUST BE ALIGNED AS SHOWN BEFORE INSTALLING CALIPER OVER SHOE & LININGS

13	PISTON BOOT
14	PISTON ASSEMBLY
21	CALIPER BODY ASSEMBLY

93006G72

Positioning rear caliper piston slots—Except 2005 Grand Prix RPO Z7U

2. Inspect the fluid level in the brake master cylinder reservoir.

3. If the brake fluid level is midway between the maximum-full point and the minimum allowable level, no brake fluid needs to be removed from the reservoir before proceeding.

4. If the brake fluid level is higher than midway between the maximum-full point and the minimum allowable level, remove brake fluid to the midway point before proceeding.

5. Raise and support the vehicle.

6. Remove the tire and wheel assembly.

7. Release tension from park brake system at the equalizer.

8. Remove the brake caliper bolts.

9. Remove the brake caliper. Support it out of the way.

10. Remove the pads and discard the shims.

To install:

11. Inspect the brake pad hardware and replace if necessary.

12. Install the pads and new shims.

13. Install the brake caliper onto the caliper bracket insuring that the guide boots are not damaged.

14. Install the brake caliper bolts. Tighten the brake caliper bolts to 60 Nm (44 ft. lbs.).

15. With the engine OFF, gradually apply the brake pedal to approximately ⅔ of its travel distance.

16. Slowly release the brake pedal.

17. Wait 15 seconds, then repeat steps 12-13 until a firm brake pedal is obtained. This will properly seat the brake caliper pistons and brake pads.

18. Adjust the park brake system.

19. Install the tire and wheel assembly.

20. Lower the vehicle.

Brake Drums

REMOVAL & INSTALLATION

1. Before servicing the vehicle, refer to the Precautions Section.

2. Mark the relationship of the wheel to the axle flange to help maintain wheel balance after assembly.

3. Remove the tire and wheel assembly.

4. Mark the relationship of the brake drum to the axle flange.

5. If difficulty is encountered in removing the brake drum, the following steps may be of assistance.

   a. Make sure the parking brake is released.

   b. Back off the parking brake cable adjustment.

   c. Remove the access hole plug from the backing plate.

   d. Using a screwdriver, back off the adjusting screw.

   e. Install the access hole plug to prevent dirt or contamination from entering the drum brake assembly.

   f. Use a small amount of penetrating oil applied around the brake drum pilot hole.

   g. Carefully remove the brake drum from the vehicle.

6. After removing the brake drum it should be checked for the following:

   a. Inspecting for cracks and deep grooves.

   b. Inspect for out of round and taper.

   c. Inspecting for hot spots (black in color).

To install:

7. Install the brake drum onto the vehicle aligning the reference marks on the axle flange.

8. Install the tire and wheel assembly, torquing it to specifications.

9. Road test for proper brake operation.

Brake Shoes

REMOVAL & INSTALLATION

1. Before servicing the vehicle, refer to the Precautions Section.

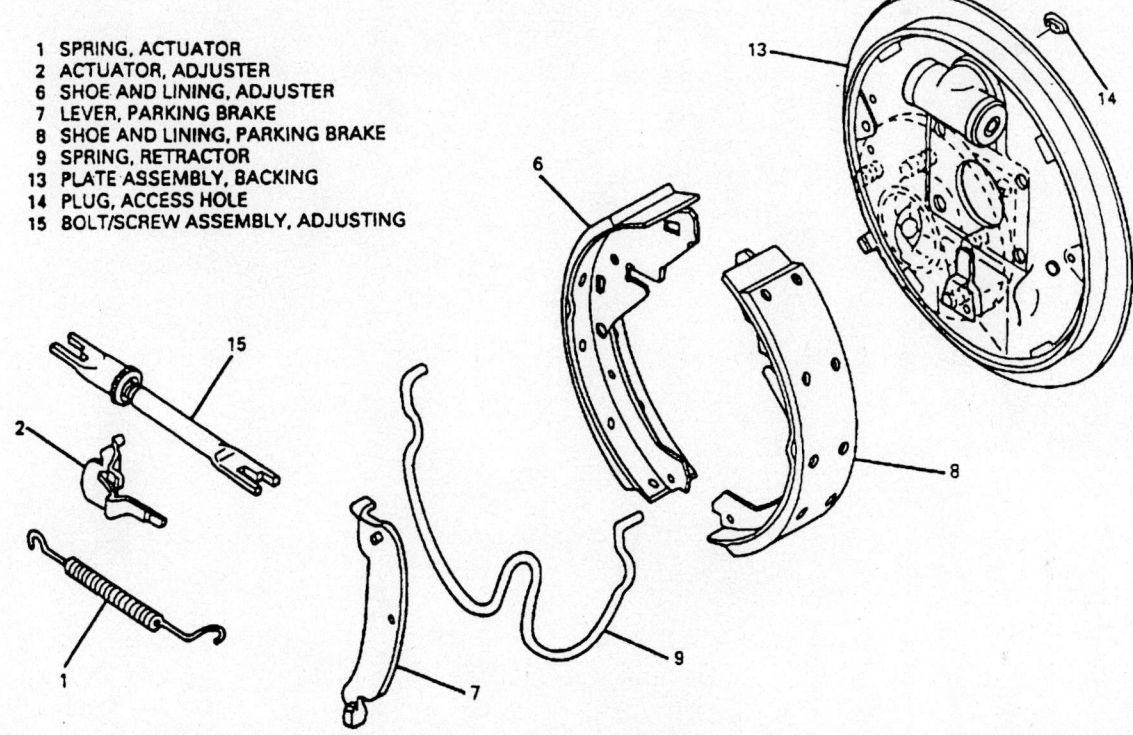

1 SPRING, ACTUATOR
2 ACTUATOR, ADJUSTER
6 SHOE AND LINING, ADJUSTER
7 LEVER, PARKING BRAKE
8 SHOE AND LINING, PARKING BRAKE
9 SPRING, RETRACTOR
13 PLATE ASSEMBLY, BACKING
14 PLUG, ACCESS HOLE
15 BOLT/SCREW ASSEMBLY, ADJUSTING

93006G79

Rear brake assembly component alignment

2. Remove the tire and wheel assembly.

3. Remove the brake drum.

4. Using tool J-38400 brake spanner and remover, remove the actuator spring from the adjuster lever. Use care not to distort the spring when removing it.

✳✳ CAUTION

During the following steps when removing the retractor spring from either shoe and lining assembly, do not over stretch the spring. This will reduce its effectiveness. Keep fingers away from retractor spring to prevent fingers from being pinched between the spring and shoe web or spring and the backing plate.

5. Lift the end of the retractor spring from the adjuster shoe assembly. Insert the hook end of the J-38400, between the retractor spring and the shoe. Pry slightly to remove the spring end from the hole in the shoe.

6. Pry the end of the retractor spring toward the axle with the flat end of the tool until the spring snaps down off the shoe web onto the backing plate.

7. Remove the one brake shoe and remove the adjuster assembly.

8. Disconnect the parking brake lever from the shoe. DO NOT remove the parking brake lever from the cable end unless it is being replaced.

9. Using J-38400, lift the end of the retractor spring from the adjuster shoe assembly. Insert the hook end of the J-38400, between the retractor spring and the shoe. Pry slightly to remove the spring end from the hole in the shoe. Pry the end of the retractor spring toward the axle with the flat end of the tool until the spring snaps down off the shoe web onto the backing plate.

10. Remove the brake shoe.

To install:

11. Clean all the brake spring completely with brake solvent and allow to air dry.

12. Disassemble, clean and lubricate the adjuster screw. Once lubricated, reassemble.

13. Clean the backing plate and after it is dry apply a thin coat of brake grease to the brake shoe contact points on the backing plate.

14. Position the brake shoe that connects to the parking brake lever, on the backing plate. Using J-38400, pull the end of the retractor spring up to rest on the web of the shoe. Pull the end of the retractor spring up until it snaps into the slot in the brake shoe.

15. Connect the parking brake lever.

16. Install the remaining shoe and the adjuster screw assembly.

17. Position the brake shoe, using J-38400, pull the end of the retractor spring up to rest on the web of the shoe. Pull the end of the retractor spring up until it snaps into the slot in the brake shoe.

18. Using J-38400, spread the brake shoes and work the adjuster screw into position.

19. Install the actuator spring with the U-shaped end going through the web.

20. Install the brake drum.

21. Install the tire and wheel assembly.

22. Adjust the brakes.

23. Road test for proper brake operation.

CHEVROLET

Cobalt

13

**BRAKE HYDRAULIC
  SYSTEM** **13-78**
DISC BRAKES **13-67**
DRIVE TRAIN **13-40**
DRUM BRAKES **13-75**
ENGINE REPAIR **13-10**
FRONT SUSPENSION **13-59**
FUEL SYSTEM **13-33**
PARKING BRAKE **13-77**
REAR SUSPENSION **13-63**
**SPECIFICATIONS AND
  MAINTENANCE CHARTS**......**13-3**
Engine and Vehicle Identification...13-3
General Engine Specifications 13-3
Engine Tune-Up Specifications 13-3
Accessory Drive Belt Routing 13-4
Capacities 13-5
Valve Specifications......................13-5
Crankshaft and Connecting
  Rod Specifications 13-5
Piston and Ring Specifications 13-6
Torque Specifications 13-6
Wheel Alignment 13-6
Tire, Wheel and Ball Joint
  Specifications 13-7
Brake Specifications 13-7
Scheduled Maintenance
  Intervals..............................13-8
Additional Maintenance Services...13-9
STEERING **13-55**

A
Air Bag..13-55
  Disabling And Enabling
    The System............................13-55
  Precautions..............................13-55
Alternator 13-10
  Removal..................................13-10
Automatic Transaxle.....................13-40
  Removal & Installation............13-40
B
Balance Shaft..............................13-32
  Removal & Installation............13-32
Brake Bleeding............................13-78
  Manual Bleeding......................13-79
  Pressure Bleeding...................13-78

Brake Caliper 13-67
  Removal & Installation............13-67
Brake Shoes................................13-75
  Adjustment............................13-76
  Removal & Installation............13-75
C
Caliper Bracket 13-73
  Removal & Installation............13-73
Camshaft Followers and
  Adjusters 13-24
  Removal & Installation............13-24
Charge Air Cooler Coolant
  Pump.....................................13-18
  Removal & Installation............13-18
Clutch 13-45
  Adjustment............................13-46
  Removal & Installation............13-45
Coil Spring 13-63
  Removal & Installation............13-63
Crankshaft Balancer 13-27
  Removal & Installation............13-27
CV-Joints....................................13-49
  Overhaul 13-49
Cylinder Head..............................13-21
  Removal & Installation............13-21
D
Disc Brake Hardware....................13-72
  Removal & Installation............13-72
Disc Brake Pads...........................13-69
  Removal & Installation............13-69
E
Engine Assembly 13-11
  Removal & Installation............13-11
Exhaust Manifold.........................13-23
  Removal & Installation............13-23
F
Front Cover and Seal 13-27
  Removal & Installation............13-27
Fuel Pump 13-36
  Removal & Installation............13-36
Fuel Rail and Injectors 13-37
  Removal & Installation............13-37
Fuel System Service
  Precautions............................13-33
H
Halfshaft......................................13-47
  Removal & Installation............13-47

Heater Core.................................13-19
  Removal & Installation............13-19
Hydraulic Clutch System 13-47
  Bleeding................................13-47
I
Intake Manifold...........................13-22
  Removal & Installation............13-22
Intermediate Shaft.......................13-48
  Removal & Installation............13-48
L
Lower Ball Joint...........................13-60
  Removal & Installation............13-60
Lower Control Arm 13-60
  Lower Control Arm Bushing
    Replacement.........................13-61
  Removal And & Installation 13-60
M
Manual Transaxle.........................13-42
  Removal & Installation............13-42
O
Oil Pan.......................................13-25
  Removal & Installation............13-25
Oil Pump 13-26
  Removal & Installation............13-26
P
Piston and Ring 13-32
  Positioning 13-32
Power Steering Gear 13-58
  Removal & Installation............13-58
Q
Quick-Connect Fittings...............13-34
  Opening And Closing 13-34
R
Rear Axle Beam............................13-63
  Control Arm Bushing
    Replacement.........................13-65
  Removal & Installation............13-63
Rear Main Seal 13-27
  Removal & Installation............13-27
Relieving Fuel System
  Pressure 13-33
S
Shock Absorber 13-63
  Removal & Installation............13-63
Stabilizer Shaft and Links 13-61
  Removal & Installation............13-61

Steering Knuckle...........................13-62
    Removal & Installation............13-62
Strut..13-59
    Disassembly And Assembly....13-60
    Removal & Installation............13-59
Supercharger 13-22
    Removal & Installation............13-22

T

Timing Chains and Sprockets......13-28
    Removal & Installation............13-28

W

Water Pump 13-17
    Removal & Installation............13-17

Wheel Bearings............................13-63
    Removal & Installation............13-63
Wheel Bearings (Rear) 13-66
    Removal & Installation............13-66
Wheel Cylinder 13-76
    Removal & Installation............13-76

SPECIFICATIONS AND MAINTENANCE CHARTS

ENGINE AND VEHICLE IDENTIFICATION

Engine							Model Year	
Code ①	Liters	Cu. In.	Cyl.	Fuel Sys.	Engine Type	Eng. Mfg.	Code ②	Year
P	2.0	122	4	MFI	DOHC SC	GM	5	2005
F	2.2	134	4	MFI	DOHC	GM	6	2006

MFI: Multi-port Fuel Injection

DOHC: Double Overhead Camshafts

SC: Supercharged

① 8th digit of VIN

② 10th digit of VIN

06025-COBALT-C01

GENERAL ENGINE SPECIFICATIONS

Engine Displacement Liters	Engine VIN	Year	Net Horsepower @ rpm	Net Torque @ rpm (ft. lbs.)	Bore x Stroke (in.)	Com-pression Ratio	Oil Pressure @ rpm
2.0	P	2005	205@5600	200@4400	NA	9.5:1	50-80@1000
2.2	F	2005	145@5600	150@4000	3.386x3.727	10:01	50-80@1000

NA: Information not available

06025-COBALT-C02

ENGINE TUNE-UP SPECIFICATIONS

Engine Displacement Liters	Engine VIN	Year	Spark Plug Gap (in.)	Ignition Timing (deg.) MT	Ignition Timing (deg.) AT	Fuel Pump (psi)	Idle Speed (rpm) MT	Idle Speed (rpm) AT	Valve Clearance Intake	Valve Clearance Exhaust
2.0	P	2005	0.040	NA	NA	50-60	NA	NA	HYD	HYD
2.2	F	2005	0.043	NA	NA	50-60	NA	NA	HYD	HYD

HYD: Hydraulic lash adjusters

NA: Information not available

06025-COBALT-C03

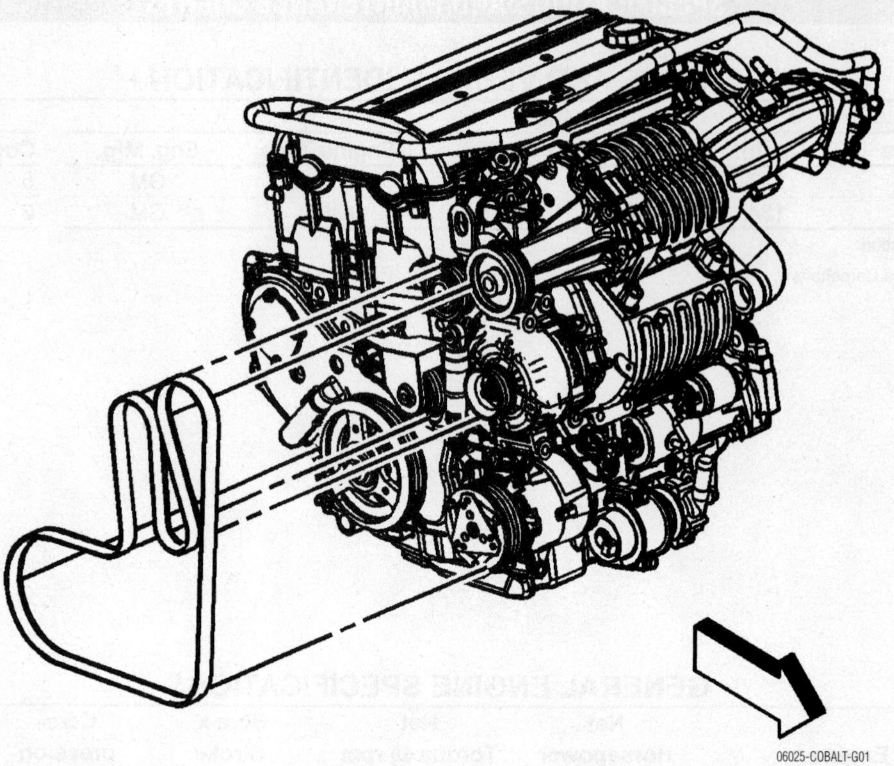

06025-COBALT-G01

Accessory drive belt routing—2.0L engine

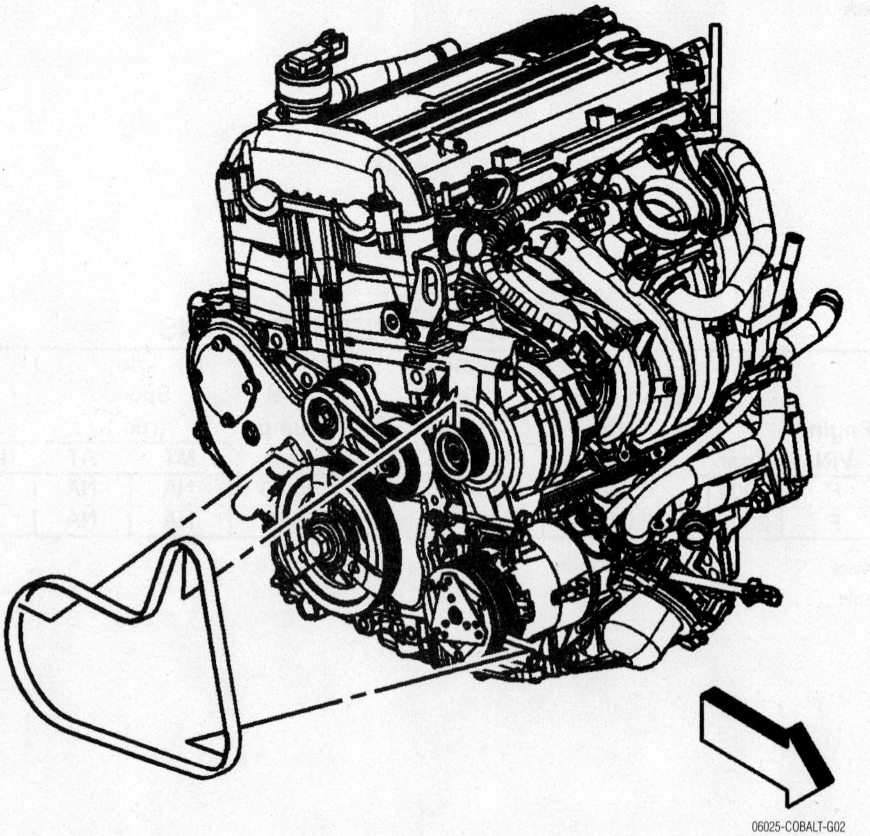

06025-COBALT-G02

Accessory drive belt routing—2.2L engine

CAPACITIES

Engine Displacement Liters	Engine VIN	Model	Year	Engine Oil with Filter (qts.)	Transmission (pts.)		Fuel Tank (gal.)	Cooling System (qts.)
					Manual	Auto.*		
2.0	P	SS	2005	5.0	4.0	13.8	13.5	7.4 ①
2.2	F	All others	2005	5.0	4.0	13.8	13.5	6.8

NOTE: All capacities are approximate. Add fluid gradually and check to be sure a proper fluid level is obtained.

* Drain and refill

① Intercooler system: 2.0 qts.

06025-COBALT-C04

VALVE SPECIFICATIONS

Engine Displacement Liters	Engine VIN	Year	Seat Angle (deg.)	Face Angle (deg.)	Spring Test Pressure (lbs. @ in.)	Spring Installed Height (in.)	Stem-to-Guide Clearance (in.)		Stem Diameter (in.)	
							Intake	Exhaust	Intake	Exhaust
2.0	P	2005	NA	NA	118-129@0.905	1.279	0.0012-0.0022	0.0020-0.0026	0.2344-0.2355	0.2337-0.2343
2.2	F	2005	NA	NA	118-129@0.905	1.279	0.0012-0.0022	0.0020-0.0026	0.2344-0.2355	0.2337-0.2343

NA: Information not available

06025-COBALT-C05

CRANKSHAFT AND CONNECTING ROD SPECIFICATIONS

All measurements in inches

Engine Displacement Liters)	Engine VIN	Year	Crankshaft				Connecting Rod		
			Main Brg. Journal Dia.	Main Brg. Oil Clearance	Shaft End-play	Thrust on No.	Journal Diameter	Oil Clearance	Side Clearance
2.0	P	2005	2.2045-2.2050	0.0012-0.0026	0.0012-0.0150	2	1.9291-1.9297	0.0011-0.0027	0.0028-0.0146
2.2	F	2005	2.2045-2.2050	0.0012-0.0026	0.0012-0.0150	2	1.9291-1.9297	0.0011-0.0027	0.0028-0.0146

06025-COBALT-C06

PISTON AND RING SPECIFICATIONS

All measurements in inches

Engine Displacement Liters	Engine VIN	Year	Piston Clearance	Ring Gap			Ring Side Clearance		
				Top Compression	Bottom Compression	Oil Control	Top Compression	Bottom Compression	Oil Control
2.0	P	2005	0.0004-0.0016	0.0080-0.0160	0.0140-0.0220	0.0100-0.0300	0.0015-0.0031	0.0012-0.0027	0.0035-0.0042
2.2	F	2005	0.0004-0.0016	0.0080-0.0160	0.0140-0.0220	0.0100-0.0300	0.0015-0.0031	0.0012-0.0027	0.0035-0.0042

06025-COBALT-C07

TORQUE SPECIFICATIONS

All values in ft. lbs. unless otherwise noted

Engine Displacement Liters	Engine VIN	Year	Cylinder Head Bolts	Main Bearing Bolts	Rod Bearing Bolts	Crankshaft Damper Bolts	Flywheel Bolts	Manifold		Spark Plugs	Oil Pan Drain Plug
								Intake	Exhaust		
2.0	P	2005	①	②	③	④	⑤	16	9	15	18
2.2	F	2005	①	②	③	④	⑤	⑥	10	15	18

① Step 1: 22 ft. lbs.
  Step 2: plus 155 degrees

② Step 1: 15 ft. lbs.
  Step 2: plus 70 degrees

③ Step 1: 18 ft. lbs.
  Step 2: plus 90 degrees

④ Step 1: 74 ft. lbs.
  Step 2: plus 125 degrees

⑤ Step 1: 39 ft. lbs.
  Step 2: plus 25 degrees

⑥ 89 inch lbs.

06025-COBALT-C08

WHEEL ALIGNMENT SPECIFICATIONS

Model	Year	RPO		Caster		Camber		Toe-in (in.)
				Range (+/-Deg.)	Preferred Setting (Deg.)	Range (+/-Deg.)	Preferred Setting (Deg.)	
Cobalt	2005	FE1	F	0.75	+3.00	0.75	-1.00	0.20+/-0.20
			R	—	—	0.75	-0.80	0.25+/-0.30
		FE3 & 5	F	0.75	+3.65	0.75	-1.05	0.20+/-0.20
			R	—	—	0.75	-0.80	0.25+/-0.30

06025-COBALT-C09

TIRE, WHEEL AND BALL JOINT SPECIFICATIONS

Model	Year	OEM Tires Standard	OEM Tires Optional	Tire Pressures (psi) Front	Tire Pressures (psi) Rear	Wheel Size	Ball Joint Inspection	Lug Nuts (ft. lbs.)
SS	2005	P215/45R18	none	①	①	NA	0.125 in. max.	100
2dr &4dr exc. LT	2005	P195/60R15	none	①	①	NA	0.125 in. max.	100
LT 4dr	2005	P205/55R16	none	①	①	NA	0.125 in. max.	100

OEM: Original Equipment Manufacturer
PSI: Pounds Per Square Inch
STD: Standard
OPT: Optional
① See placard on vehicle

06025-COBALT-C10

BRAKE SPECIFICATIONS
All measurements in inches, unless otherwise noted

Model	Year		Brake Disc Original Thickness	Brake Disc Minimum Thickness	Brake Disc Maximum Run-out	Brake Drum Original Inside Diameter	Brake Drum Maximum Machine Diameter	Minimum Lining Thickness	Brake Caliper Bracket Bolts (ft. lbs.)	Brake Caliper Mounting Bolts (ft. lbs.)
Exc. SS	2005		0.933	0.870	0.002	9.06	9.075	①	85	25
SS	2005	F	1.023	0.898	0.002	—	—	NA	85	25
		R	0.551	0.465	0.002	—	—	NA	85	25

NA: Information not available
① Front: 0.039 in.; Rear: 0.020 in.

06025-COBALT-C11

MAINTENANCE I AND II SERVICE SCHEDULES
2005 Chevrolet Cobalt

When the CHANGE ENGINE OIL light appears, certain services and inspections are required.

Required services are described as Maintenance I and Maintenance II.

The first service on a vehicle should be Maintenance I, and the second service should be Maintenance II.

Alternate between the 2 thereafter. However, in some cases, Maintenance II may be required more often.

Maintenance I: Use Maintenance I if the CHANGE ENGINE OIL light comes on within 10 months since vehicle was purchased or, if Maintenance II was performed.

Maintenance II: Use Maintenance II if the previous service performed was Maintenance I. Always use Maintenance II whenever the CHANGE ENGINE OIL light comes on 10 months or more since the last service, or, if the CHANGE ENGINE OIL light has not come on at all for one year.

Service	Maintenance I	Maintenance II
Change the engine oil and filter. Reset the oil life system.	✓	✓
Visually inspect the vehicle for leaks or damage. A fluid loss in the vehicle system could indicate a problem. Inspected, repair and add fluid to the system if necessary.	✓	✓
Inspect the engine air cleaner filter. If necessary, replace the filter.	--	✓
Rotate the tires. Inspect the tire inflation pressures and the tire wear.	✓	✓
Visually inspect the brake lines and hoses for proper hook-up, binding, leaks, cracks, chafing, etc. Inspect the disc brake pads for wear and the rotors for surface condition. Inspect the drum brake linings for wear or cracks. Inspect other brake parts, including drums, wheel cylinders, calipers, parking brake, etc. Inspect the parking brake adjustment.	✓	✓
Inspect the engine coolant and the windshield washer fluid levels. Add fluid as needed.	✓	✓
With a 2.0L engine, check the intercooler fluid level. Add as necessary.	✓	✓
Inspect the suspension and steering components. Inspect the front and rear suspension and the steering system for damaged, loose or missing parts, or signs of wear. Inspect the power steering lines and the hoses for proper hook-up, binding, leaks, cracks, chafing, etc.	--	✓
Visually inspect the coolant hoses and replace the hoses if they are cracked, swollen or deteriorated. Inspect all pipes, fittings and clamps; replace with GM parts as needed. To help ensure proper operation, a pressure test of the cooling system and pressure cap and cleaning the outside of the radiator and air conditioning condenser is recommended at least once a year.	--	✓
Inspect the front and rear suspension and the steering system for damaged, loose or missing parts, or signs of wear. Inspect power steering lines and hoses for proper hook-up, binding, leaks, cracks, chafing, etc.	--	✓
Inspect the throttle system for interference or binding and for damaged or missing parts. Replace the parts as needed. Replace any components that have high effort or excessive wear. Do not lubricate the accelerator or the cruise control cables.	--	✓
Replace the passenger compartment air filter.	--	✓

To reset the CHANGE ENGINE OIL LIGHT:

1. Turn the ignition switch to RUN with the engine OFF.

2. Press the Information and Reset buttons on the DIC at the same time to enter the personalization menu.

3. Press the Information button to scroll through the menu to Oil Life Reset.

4. Press and hold the Reset button until the display shows Acknowledged.

5. Turn the key to OFF.

06025-COBALT-C12

ADDITONAL MAINTENANCE SERVICES
2005 Chevrolet Cobalt

TO BE SERVICED	TYPE OF SERVICE	VEHICLE MILEAGE INTERVAL (x1000)					
		25	50	75	100	125	150
Air cleaner filter	R	✓	✓	✓	✓	✓	✓
Accessory drive belt	I						✓
Auto. Trans. Fluid ①	R		✓		✓		✓
Cooling system hoses and clamps	S/I						✓
Engine coolant	R						✓
Fuel system	I	✓	✓	✓	✓	✓	✓
Exhaust system & heat shields	S/I	✓	✓	✓	✓	✓	✓
Spark plugs	R				✓		

R: Replace   S/I: Inspect and service, if necessary

① Replace if any of the following conditions are met:

    Heavy city traffic where the outside temperature regularly reaches 32°C (90°F) or higher

    Hilly or mountainous terrain

    Frequent trailer towing

    Taxi, police or delivery service

    Otherwise, change every 100,000 miles

06025-COBALT-C13

ENGINE REPAIR

Alternator

REMOVAL

2.0L Engine

1. Before servicing the vehicle, refer to the Precautions Section.
2. Disconnect negative battery cable.
3. Remove the accessory drive belt.
4. Remove the supercharger.
5. Remove the oil level indicator tube bolt and reposition the tube slightly for clearance.
6. Disconnect the alternator connectors.
7. Remove the alternator bolts.
8. Remove the alternator from the vehicle.

To install:

9. Position the alternator on the engine.
10. Install the alternator bolts. Tighten the alternator bolts to 25 Nm (18 ft. lbs.).

11. Connect the positive battery harness to the alternator battery terminal. Tighten the alternator terminal nut to 20 Nm (15 ft. lbs.).
12. Connect the alternator harness connectors.
13. Position the oil level indicator tube into the correct installed position and install the retaining bolt. Tighten the bolt to 10 Nm (89 inch lbs.).
14. Install the supercharger.
15. Install the drive belt.
16. Connect the battery negative cable.

2.2L Engine

1. Before servicing the vehicle, refer to the Precautions Section.
2. Disconnect negative battery cable.
3. Remove the drive belt.

4. Remove the air cleaner outlet resonator.
5. Disconnect the alternator connectors.
6. Remove the alternator bolts.
7. Remove the alternator from the vehicle.

To install:

8. Position the alternator on the engine.
9. Install the alternator bolts. Tighten the alternator bolts to 22 Nm (16 ft. lbs.).
10. Connect the positive battery harness to the alternator battery terminal. Tighten the alternator terminal nut to 20 Nm (15 ft. lbs.).
11. Connect the alternator harness connectors.
12. Install the air cleaner outlet resonator.
13. Install the drive belt.
14. Connect the battery negative cable.

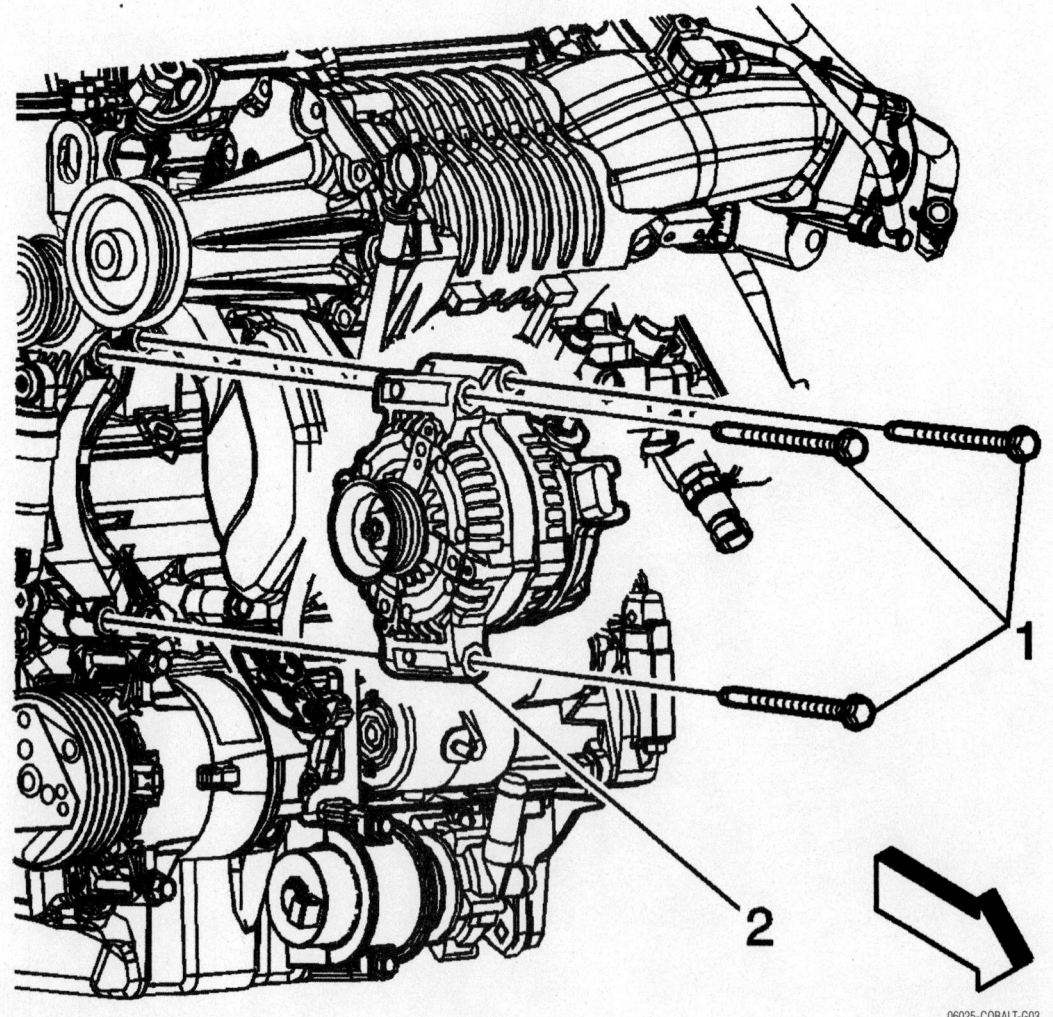

Alternator mounting—2.0L engine

06025-COBALT-G03

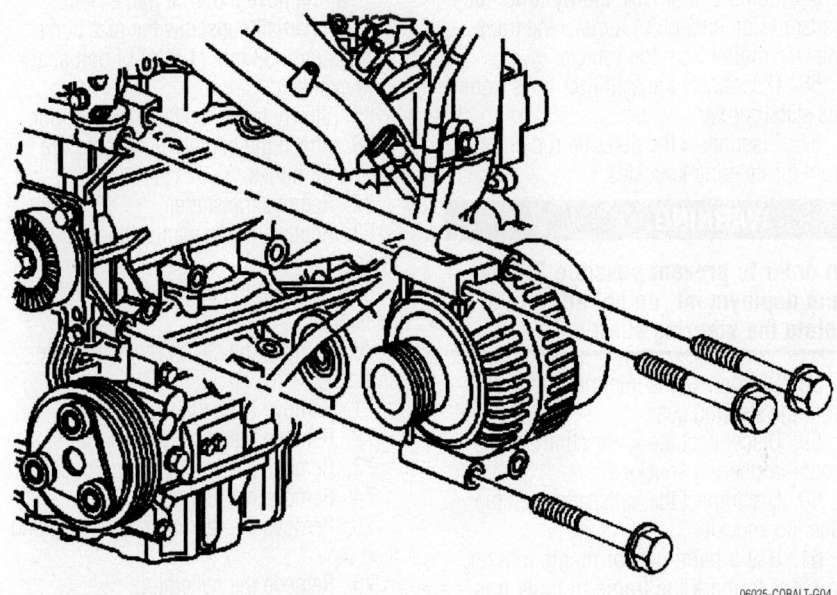

Alternator mounting—2.2L engine

06025-COBALT-G04

Engine Assembly

REMOVAL & INSTALLATION

2.0L Engine

1. Before servicing the vehicle, refer to the Precautions Section.

2. With the tires in the straight forward position, remove the key from the ignition.

3. Disconnect the negative battery cable.

4. Remove the air outlet duct.

5. Secure the cooling module to the upper body structure.

6. Relieve the fuel system pressure.

7. Disconnect the fuel line from the fuel rail.

8. Drain the cooling system.

9. Remove the radiator inlet hose.

10. Remove the surge tank to cylinder head pipe.

11. Remove the radiator outlet hose.

12. Remove the inlet and outlet heater hoses.

13. Disconnect the following harness connectors:
   - TMAP sensor
   - Electronic temperature control (ETC)
   - Manifold absolute pressure (MAP) sensor
   - Barometric pressure (BARO) sensor
   - Crankshaft sensor
   - Oil pressure sensor
   - Purge solenoid
   - Ignition coil modules
   - Oxygen (O_2) sensor
   - Vehicle speed sensor
   - Engine temperature sensor
   - Boost solenoid
   - Back-up lamp switch

14. Remove the accessory drive belt.

15. Raise and suitably support the vehicle.

16. Recover the refrigerant.

17. Remove the right front fender liner.

18. Disconnect the electrical connector from the compressor.

19. Remove the compressor hose from the evaporator hose.

20. Remove and discard the seal washer.

21. Remove the lower right radiator mount.

22. Remove the cap from the charge air cooler reservoir.

23. Place a drain pan under the charge air cooler radiator.

24. Reposition the inlet hose clamp at the auxiliary water pump.

25. Remove the inlet hose from the auxiliary water pump and allow the cooling system to drain.

26. Remove the auxiliary intercooler pump.

27. Remove the mounting cover from the charge air cooler pump and position the pump forward.

28. Disconnect the electrical connector from the pressure transducer.

29. Remove the compressor and condenser hose assembly bolt from the compressor.

30. Remove the compressor and condenser hose assembly from the compressor.

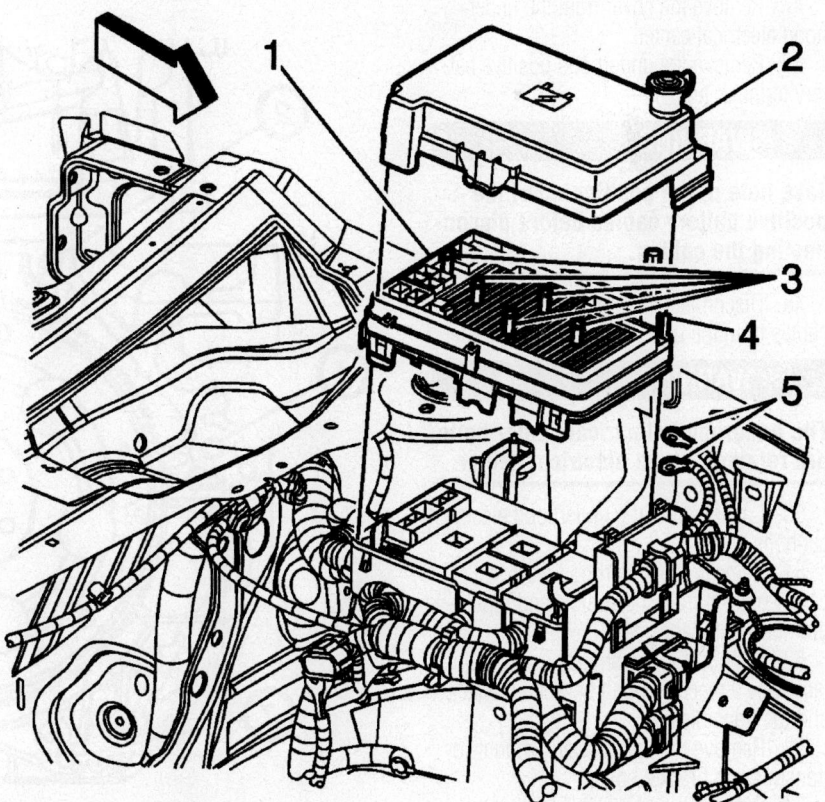

Underhood electrical center

06025-COBALT-G07

31. Remove and discard the sealing washers.

32. Remove the compressor hose from the condenser.

33. Remove and discard the seal washer.

34. Remove the auxiliary intercooler pump bolts and position the pump toward the driver's side of the vehicle.

35. Remove the compressor mounting bolts.

36. Remove the compressor.

37. Disconnect the starter harness connectors.

38. Disconnect the generator harness connectors.

39. Drain the engine oil.

40. Disconnect the front exhaust pipe from the exhaust manifold.

41. Disconnect the transmission shift cable from the transmission.

42. Use blocks of wood to support the powertrain assembly between the frame and the powertrain.

43. Support the engine with a hydraulic floor jack. Use a piece of wood between the jack and the oil pan.

44. Remove the engine mount to intermediate bracket bolts.

45. Remove the engine mount to mid-rail nuts.

46. Remove the engine mount from the engine compartment.

47. Remove the cover from the underhood electrical center.

48. Remove the underhood positive battery terminal lug.

✹✹ WARNING

Take note of the positioning of the positive battery cables before disconnecting the cables.

49. Disconnect the positive battery cables from the underhood electrical center.

✹✹ WARNING

The underhood electrical center bolts are retained in the electrical center.

50. Loosen all of the underhood electrical bolts.

51. Remove the underhood electrical center bracket from the vehicle and reposition the electrical center.

52. Support the transmission with a floor jack. Use a piece of wood between the jack and the transmission.

53. Remove the transmission mount-to-transmission bracket bolts.

54. Remove the transmission mount to mid-rail bolts.

55. Using a floor jack, slowly lower the transmission enough to remove the transmission mount from the vehicle.

56. Disconnect the stabilizer links from the stabilizer bar.

57. Disconnect the outer tie rod ends from the steering knuckles.

✹✹ WARNING

In order to prevent possible SIR system deployment, do not attempt to rotate the steering shaft.

58. Disconnect the intermediate shaft from the steering gear.

59. Disconnect the lower control arms from the steering knuckles.

60. Disconnect the halfshafts from the steering knuckle.

61. Use a paint pen or magic marker in order to mark the frame to body position.

62. Lower the vehicle to about 1 meter (3 feet) off the ground in order to position the lift table under the frame.

63. Use wood blocks as necessary between the lift table and the frame to support the assembly.

64. Slowly remove the frame bolts using the following sequence:

a. Remove the front frame bolts.

b. Partially unscrew the rear frame bolts until 38 mm (1.5 in) of bolt shank is exposed.

65. Slowly lower the table to the floor.

66. Attach the engine lift hoist to the engine lift hooks.

67. Remove the starter.

68. Remove the transmission to engine bolts.

69. Separate the engine from the transmission.

70. Remove the clutch pressure plate and disk.

71. Remove the exhaust manifold

72. Remove the exhaust manifold studs

73. Remove the engine mount bracket

74. Remove the fuel rail

75. Remove the thermostat housing and feed pipe

76. Remove the generator

77. Remove the engine from the engine lift.

To install:

78. Attach the engine lift hoist to the engine lift hooks.

79. Install the exhaust manifold

80. Install the fuel rail

81. Install the idler pulley

82. Install the drive belt tensioner

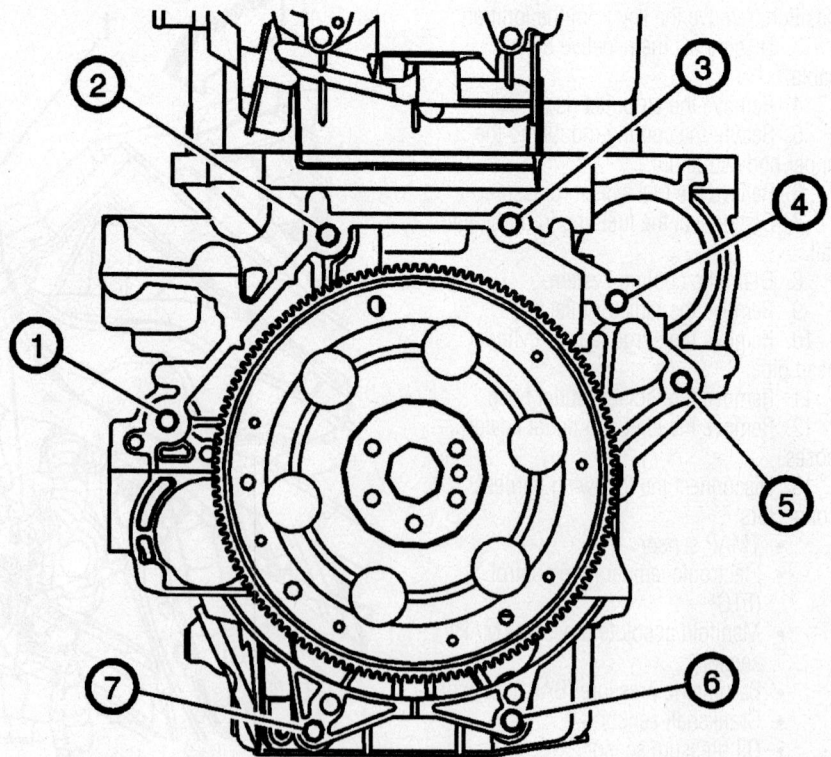

06025-COBALT-G06

Transmission to engine bolts. The number 3 bolt location is not used

83. Install the thermostat housing and feed pipe. Install the thermostat housing cap bolts and cap. Tighten the bolt to 10 Nm (18 inch lbs.).

84. Install the generator

85. Install the flywheel. Install the flywheel bolts. Tighten the flywheel bolts to 53 Nm (39 ft. lbs.) plus 25 degrees.

86. Install the clutch pressure plate and disk.

✳✳ WARNING

NEW CVC compressor assemblies are shipped with a partial poly-alkylene glycol (PAG) refrigerant oil charge. Use of the incorrect PAG oil can result in compressor failure.

87. Install the compressor and the bolts to the engine. Tighten the bolts to 25 Nm (18 ft. lbs.).

88. Install new sealing washers on the hose fittings.

89. Install the hose assembly to the compressor.

90. Install the assembly bolt. Tighten the bolt to 20 Nm (15 ft. lbs.).

91. Connect the compressor electrical connector.

92. Install the auxiliary intercooler pump. Tighten the pump bolt to 10 Nm (88 inch lbs.).

93. Install the front fender liner.

94. Align the engine to the transmission.

✳✳ WARNING

The number 3 bolt location is not used.

95. Secure the engine to the transmission. Tighten the transmission to engine bolts to 75 Nm (55 ft. lbs.).

96. Install the starter.

97. Remove the engine lift from the engine.

98. Raise and position the frame and powertrain assembly to the vehicle.

99. Hand start all the frame bolts while aligning the frame to the paint marks.

100. Tighten the frame bolts to 100 Nm (74 ft. lbs.) plus 180 degrees.

101. Remove the lift table.

102. Connect the halfshafts to the steering knuckles.

103. Connect the lower control arm to the steering knuckle.

104. Connect the intermediate steering shaft to the steering gear.

105. Connect the outer tie rod ends to the steering knuckles.

106. Connect the stabilizer links to the stabilizer bar.

107. Install the transmission mount to the mid-rail.

108. Install the transmission mount to mid-rail bolts. Tighten the bolts to 27 Nm (20 ft. lbs.).

109. Using a floor jack, raise the transmission until it contacts the transmission mount.

✳✳ WARNING

The transmission mount to transmission bolts must be hand started. Do not pry the transmission or mount to align the holes.

110. Hand start the transmission mount to bracket bolts using the following sequence:
   a. Rear bolt
   b. Middle bolt
   c. Front bolt

111. Using the previous sequence, tighten the transmission mount bolts. Tighten the bolts to 50 Nm (37 ft. lbs.).

112. Install the underhood electrical center bracket to the vehicle and install the electrical center into position on the bracket.

113. Place the engine mount onto the mid-rail and hand start the nuts.

114. Tighten the engine mount to mid-rail nuts. Tighten the nuts to 100 Nm (74 ft. lbs.).

✳✳ WARNING

The engine mount to intermediate bracket bolts must be hand started. Do not pry the engine mount to align the holes.

115. Hand start the engine mount to intermediate bracket bolts.

✳✳ WARNING

The engine mount bracket bolts must be torqued in a mandatory torque sequence as shown.

116. Tighten the engine mount to intermediate bracket bolts. Tighten the bolts to 50 Nm (37 ft. lbs.).

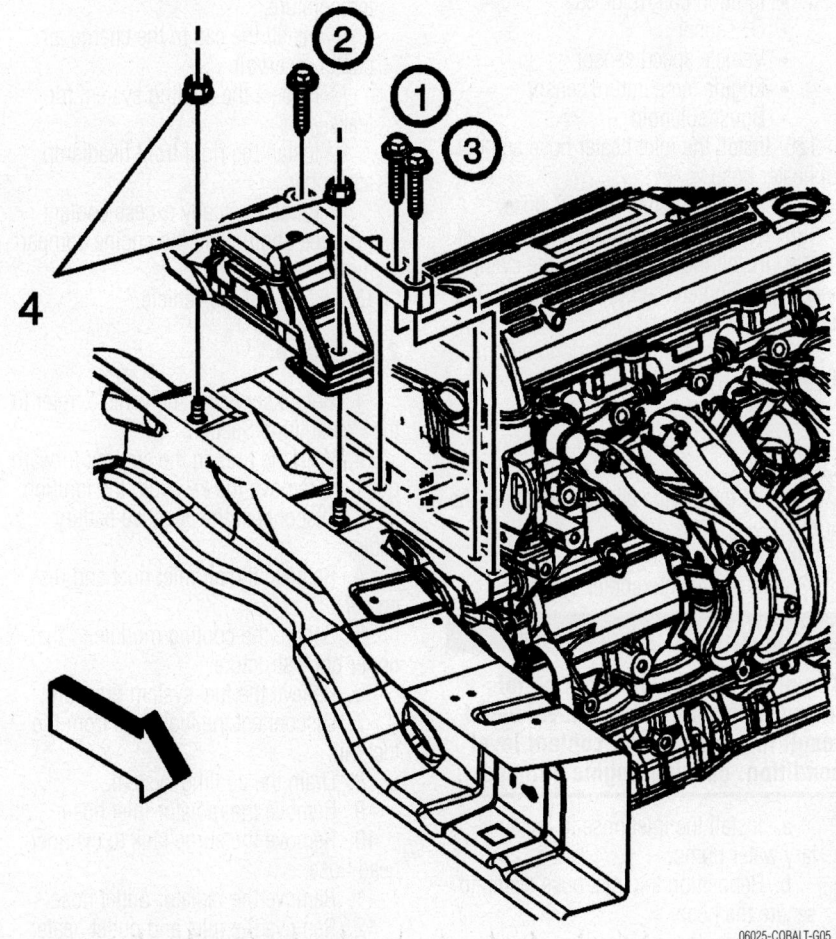

Engine mount bracket bolt torque sequence

06025-COBALT-G05

117. Remove the wood blocks between the powertrain and frame.

118. Connect the transmission shift cable to the transmission.

119. Connect the exhaust takedown pipe to the exhaust manifold. Tighten the nuts to 30 Nm (22 ft. lbs.).

120. Connect the generator harness connectors. Tighten the generator terminal nut to 20 Nm (15 ft. lbs.).

121. Connect the starter harness connectors. Tighten the battery terminal nut to 17 Nm (13 ft. lbs.). Tighten the S-terminal nut to 3 Nm (27 inch lbs.).

122. Install the compressor and condenser hose assembly to the compressor.

123. Evacuate and charge the refrigerant system.

124. Install the accessory drive belt.

125. Connect the following harness connectors:

- TMAP sensor
- ETC
- MAP sensor
- Crankshaft sensor
- Oil pressure sensor
- Purge solenoid
- BARO sensor
- Ignition coil modules
- O$_2$ sensor
- Vehicle speed sensor
- Engine temperature sensor
- Boost solenoid

126. Install the inlet heater hose and outlet heater hose.

127. Install the radiator outlet hose.

128. Connect the fuel line to the fuel rail.

129. Install the surge tank to the cylinder head pipe. Tighten the bolt to 10 Nm (89 inch lbs.).

130. Release the cooling module from the upper body structure.

131. Install the air outlet duct.

132. Connect the negative battery cable.

133. Fill the engine with engine oil to the proper level.

134. Fill the cooling system.

135. To fill the intercooler system:

☀ WARNING

The procedure below must be followed. Improper coolant level could result in a low or high coolant level condition, causing engine damage.

a. Install the inlet hose to the auxiliary water pump.

b. Reposition the inlet hose clamp to secure the hose.

c. Lower the vehicle.

d. Remove the right front headlamp assembly.

e. Remove the bleeder screw from the radiator on the passenger top side of the charge air cooler.

☀ WARNING

Use a 50/50 mixture of DEX-COOL antifreeze and clean, drinkable water. It is necessary to maintain the coolant level near the cold fill line on the surge tank to ensure all the air has been purged from the cooling system.

f. Slowly add a mixture of 50/50 DEX-COOL antifreeze and clean, drinkable water to the cooling system until the coolant level reaches the top of the bleeder screw opening.

g. Install the bleeder screw.

h. Slowly add a mixture of 50/50 DEX-COOL antifreeze and clean, drinkable water to the cooling system until the coolant level is just visible in the bottom of the charge air cooler reservoir. The coolant level will reach the hot fill line once the system has reached operating temperature.

i. Install the cap to the charge air cooler reservoir.

j. Inspect the cooling system for leaks.

k. Install the right front headlamp assembly.

l. Rinse away any excess coolant from the engine and the engine compartment.

136. Road test the vehicle.

2.2L Engine

1. Before servicing the vehicle, refer to the Precautions Section.

2. With the tires in the straight forward position, remove the key from the ignition.

3. Disconnect the negative battery cable.

4. Remove the air inlet duct and resonator.

5. Secure the cooling module to the upper body structure.

6. Relieve the fuel system pressure.

7. Disconnect the fuel lines from the fuel rail.

8. Drain the cooling system.

9. Remove the radiator inlet hose.

10. Remove the surge tank to cylinder head hose.

11. Remove the radiator outlet hose.

12. Remove the inlet and outlet heater hoses.

13. Disconnect the following harness connectors:

- Idle air control (IAC) motor
- TPS
- Manifold absolute pressure (MAP) sensor
- Crankshaft sensor
- Oil pressure sensor
- Purge solenoid
- Ignition coil and module assembly
- Oxygen (O$_2$) sensor
- Vehicle speed sensor
- Engine temperature sensor
- Backup lamp switch

14. Raise and suitably support the vehicle.

15. Remove the engine accessory drive belt.

16. Remove the front fender liner.

17. Rotate the drive belt tensioner counterclockwise to release the spring tension.

18. Remove the drive belt.

19. Disconnect the electrical connector from the compressor.

20. Remove the AC compressor bolts and set the compressor aside.

21. Disconnect the starter harness connectors.

22. Disconnect the generator harness connectors.

23. Drain the engine oil.

24. Disconnect the front exhaust pipe from the exhaust manifold.

25. Disconnect the transmission harness connectors.

26. Disconnect the transmission shift cable from the transmission.

27. Use blocks of wood to support the powertrain assembly between the frame and the powertrain.

28. Support the engine with a hydraulic floor jack. Use a piece of wood between the jack and the oil pan.

29. Remove the engine mount to intermediate bracket bolts.

30. Remove the engine mount to mid-rail nuts.

31. Remove the engine mount from the engine compartment.

32. With an automatic transaxle:

a. Remove the front transmission mount thru bolt.

b. Remove the rear transmission mount thru bolt.

c. Lower the vehicle.

d. Remove the under hood electrical center cover.

e. Disconnect the engine control module harness connector.

f. Disconnect the positive battery cables from the under hood electrical center.

g. Disconnect the surge tank inlet hose from the surge tank.

h. Remove the under hood electrical center tray bracket nuts and bolt.

i. Disconnect the wiring harness retainer from the tray bracket.

j. Lift the electrical center up and swing it back and out of the way.

k. Support the transmission with a hydraulic floor jack. Use a block of wood between the jack and the transmission.

l. Remove the transmission mount to transmission bolts.

m. Remove the transmission mount to mid-rail bolts.

n. Using the floor jack, slowly lower the transmission just enough to remove the transmission mount from the vehicle.

33. With an MU3 manual transmission:

a. Remove the cover from the under-hood electrical center.

b. Remove the underhood positive battery terminal lug.

✳✳ WARNING

Take note of the positioning of the positive battery cables before disconnecting the cables.

c. Disconnect the positive battery cables from the underhood electrical center.

✳✳ WARNING

The underhood electrical center bolts are retained in the electrical center.

d. Loosen all of the underhood electrical bolts.

e. Remove the underhood electrical center bracket from the vehicle and reposition the electrical center.

f. Support the transmission with a floor jack. Use a piece of wood between the jack and the transmission.

g. Remove the transmission mount-to-transmission bracket bolts.

h. Remove the transmission mount to mid-rail bolts.

i. Using a floor jack, slowly lower the transmission enough to remove the transmission mount from the vehicle.

34. With a Getrag 5-speed manual transmission:

35. Remove the underhood electrical center cover.

36. Disconnect the engine control module (ECM) harness connector.

37. Disconnect the positive battery cables from the underhood electrical center.

38. Disconnect the surge tank inlet hose from the surge tank.

39. Remove the underhood electrical center tray bracket nuts and bolt.

40. Disconnect the wiring harness retainer from the tray bracket.

41. Lift the electrical center up and swing it back and out of the way.

42. Support the transmission with a floor jack. Use a piece of wood between the jack and the transmission.

43. Remove the transmission mount to transmission bolts.

44. Remove the transmission mount to mid-rail bolts.

45. Using a floor jack, slowly lower the transmission enough to remove the transmission mount from the vehicle.

46. Disconnect the stabilizer links from the stabilizer bar.

47. Disconnect the outer tie rod ends from the steering knuckles.

✳✳ WARNING

In order to prevent possible SIR system deployment, do not attempt to rotate the steering shaft.

48. Disconnect the intermediate shaft from the steering gear.

49. Disconnect the lower control arms from the steering knuckles.

50. Disconnect the drive axles from the steering knuckle.

51. Use a paint pen or magic marker in order to mark the frame to body position.

52. Lower the vehicle to about 3 feet off the ground in order to position the lift table under the frame.

53. Use wood blocks as necessary between the lift table and the frame to support the assembly.

54. Slowly remove the frame bolts using the following sequence:

a. Remove the front frame bolts.

b. Partially unscrew the rear frame bolts until 1.5 inches of bolt shank is exposed.

55. Slowly lower the table to the floor with the cradle and powertrain assembly.

56. Attach the engine lift hoist to the engine lift hooks.

57. Remove the starter.

58. If applicable, remove the torque converter to flywheel bolts.

59. Remove the transmission to engine bolts.

60. Separate the engine from the transmission.

61. If applicable, remove the clutch pressure plate and disk.

62. Remove the following components:

63. Remove the exhaust manifold

64. Remove the exhaust manifold studs

65. Remove the engine mount bracket

66. Remove the engine block heater

67. Remove the thermostat housing and feed pipe

68. Remove the generator

69. Remove the engine from the engine lift.

To install:

70. Attach the engine lift hoist to the engine lift hooks.

71. Install the exhaust manifold.

72. Install the intermediate bracket to the engine.

73. Hand tighten the engine mount intermediate bracket bolts in the following locations:

- The long bolts in the forward and front lower holes
- The short bolt in the rear upper hole

74. Tighten the intermediate bracket bolts to 100 Nm (74 ft. lbs.).

75. Install the fuel rail.

76. Install the engine block heater, if equipped.

77. Install the drive belt tensioner.

78. Install the thermostat and retaining sleeve with the dimple placed into the housing slot.

✳✳ WARNING

Lubricate the O-ring with soapy water or coolant before installing the O-ring in the water pump.

79. Install the feed pipe that connects the thermostat housing to the water pump.

80. Install the bolt that secures the water pump feed pipe. Tighten the bolt to 10 Nm (18 inch lbs.).

81. Install the generator.

82. Install the flywheel. Tighten the flywheel bolts to 53 Nm (39 ft. lbs.) plus 25 degrees

83. If applicable, install the clutch pressure plate and disk.

84. Align the engine to the transmission.

✳✳ WARNING

The number 3 bolt location is not used.

85. Secure the engine to the transmission. Tighten the transmission to engine bolts to 75 Nm (55 ft. lbs.).

86. If applicable, install the torque converter bolts. Tighten the bolts to 60 Nm (44 ft. lbs.).

87. Install the starter.

88. Remove the engine lift from the engine.

89. Raise and position the frame and powertrain assembly to the vehicle.

90. Hand start all the frame bolts while aligning the frame to the paint marks.

91. Tighten the frame bolts. Tighten the frame bolts to 100 Nm (74 ft. lbs. +180 degrees).

92. Remove the lift table.

93. Connect the drive axles to the steering knuckles.

94. Connect the lower control arm to the steering knuckle.

95. Connect the intermediate steering shaft to the steering gear.

96. Connect the outer tie rod ends to the steering knuckles.

97. Connect the stabilizer links to the stabilizer bar.

98. With an automatic transaxle:

a. Install the transmission mount to the mid-rail.

b. Hand start the transmission mount to mid-rail bolts. Tighten the bolts to 34 Nm (25 ft. lbs.).

c. Using a hydraulic jack, raise the transmission until it contacts the transmission mount.

⁕⁕ WARNING

The transmission mount to transmission bolts must be hand started. Do not pry the transmission or mount to align the holes.

d. Hand start the transmission mount to transmission bolts using the following sequence:

- Rear Bolt
- Middle Bolt
- Front Bolt

e. Using the previous sequence, tighten the transmission mount bolts. Tighten the bolts to 50 Nm (37 ft. lbs.).

f. Reposition the under hood electrical center.

g. Connect the wiring harness retainer to the tray bracket.

h. Install the electrical center nuts and bolts. Tighten the nut to 10 Nm (89 inch lbs.). Tighten the bolt to 25 Nm (18 ft. lbs.).

i. Connect the surge tank inlet hose to the surge tank.

j. Install the positive battery cables to the under hood electrical center. Tighten the positive cable nut to 15 Nm (11 ft. lbs.).

k. Connect the engine control module harness connectors.

l. Install the under hood electrical center cover.

m. Raise the vehicle.

n. Hand tighten the front transmission mount thru bolt.

o. Hand tighten the rear transmission mount thru bolt.

p. Tighten the front transmission mount thru bolt. Tighten the bolt to 100 Nm (74 ft. lbs.).

q. Tighten the rear transmission mount thru bolt. Tighten the bolt to 100 Nm (74 ft. lbs.).

99. With an MU3 manual transmission:

a. Install the transmission mount to the mid-rail.

b. Install the transmission mount to mid-rail bolts. Tighten the bolts to 27 Nm (20 ft. lbs.).

c. Using a floor jack, raise the transmission until it contacts the transmission mount.

⁕⁕ WARNING

The transmission mount to transmission bolts must be hand started. Do not pry the transmission or mount to align the holes.

d. Hand start the transmission mount to bracket bolts using the following sequence:

- Rear bolt
- Middle bolt
- Front bolt

e. Using the previous sequence, tighten the transmission mount bolts. Tighten the bolts to 50 Nm (37 ft. lbs.).

f. Install the underhood electrical center bracket to the vehicle and install the electrical center into position on the bracket.

100. With a Getrag 5-speed:

a. Install the transmission mount to the mid-rail.

b. Hand start the transmission mount to mid-rail bolts. Tighten the bolts to 34 Nm (25 ft. lbs.).

c. Using a floor jack, raise the transmission until it contacts the transmission mount.

⁕⁕ WARNING

The transmission mount to transmission bolts must be hand started. Do not pry the transmission or mount to align the holes.

d. Hand start the transmission mount to transmission bolts using the following sequence:

- Rear bolt
- Middle bolt
- Front bolt

e. Using the previous sequence, tighten the transmission mount bolts. Tighten the bolts to 45 Nm (33 ft. lbs.).

101. Reposition the underhood electrical center.

102. Connect the wiring harness to the tray bracket.

103. Install the electrical center nuts and bolt. Tighten the nuts to 10 Nm (89 inch lbs.). Tighten the bolt to 25 Nm (18 ft. lbs.).

104. Connect the surge tank inlet hose to the surge tank.

105. Install the positive battery cables to the underhood electrical center. Tighten the positive cable nut to 15 Nm (11 ft. lbs.).

106. Connect the engine control module harness connectors.

107. Install the underhood electrical center cover.

108. Place the engine mount onto the mid rail and hand start the nuts.

109. Tighten the engine mount to mid-rail nuts. Tighten the nuts to 100 Nm (74 ft. lbs.).

⁕⁕ WARNING

The engine mount to intermediate bracket bolts must be hand started. Do not pry the engine mount to align the holes.

110. Hand start the engine mount to intermediate bracket bolts.

111. Tighten the engine mount to intermediate bracket bolts. Tighten the bolts to 50 Nm (37 ft. lbs.).

112. Remove the hydraulic floor jack.

113. Install the air cleaner assembly.

114. Remove the wood blocks between the powertrain and frame.

115. Connect the transmission shift cable to the transmission.

116. Connect the transmission harness connector.

117. Connect the exhaust takedown pipe to the exhaust manifold. Tighten the nuts to 30 Nm (22 ft. lbs.).

118. Connect the generator harness connectors. Tighten the generator terminal nut to 20 Nm (15 ft. lbs.).

119. Connect the starter harness connectors. Tighten the battery terminal nut to 17 Nm (13 ft. lbs.). Tighten the S-terminal nut to 3 Nm (27 inch lbs.).

120. Install the AC compressor to the engine. Tighten the bolts to 25 Nm (18 ft. lbs.).

121. Install the engine drive belt.

122. Connect the following harness connectors:
- IAC motor
- TPS
- MAP sensor
- Crankshaft sensor
- Oil pressure sensor
- Purge solenoid
- Ignition coil and module assembly
- O$_2$ sensor
- Vehicle speed sensor
- Engine temperature sensor

123. Install the inlet heater hose and outlet heater hose.

124. Install the radiator outlet hose.

125. Connect the fuel line to the fuel rail.

126. Connect the brake booster hose at the brake booster.

127. Release the cooling module from the upper body structure.

128. Install the air inlet duct and resonator.

129. Connect the negative battery cable.

130. Fill the engine with engine oil to the proper level.

131. Fill the cooling system.

132. Road test the vehicle.

Water Pump

REMOVAL & INSTALLATION

2.0L Engine

1. Before servicing the vehicle, refer to the Precautions Section.

2. Remove the thermostat housing cap and bolts.

3. Remove the thermostat.

4. Remove the oxygen sensor clip.

5. Remove the thermostat housing and water feed pipe retaining bolts.

➡**Twist the water feed pipe while pulling to remove it from the water pump cover.**

6. Remove the thermostat housing and water feed pipe from the water pump cover.

7. Remove the water pump retaining bolts. Be sure to remove the bolt that goes through the front of the engine block.

8. Remove the water pump assembly.

To install:

9. Install the water pump assembly.

10. Install the water pump bolts. Finger tighten the bolts.

11. Tighten the water pump bolts to 25 Nm (18 ft. lbs.).

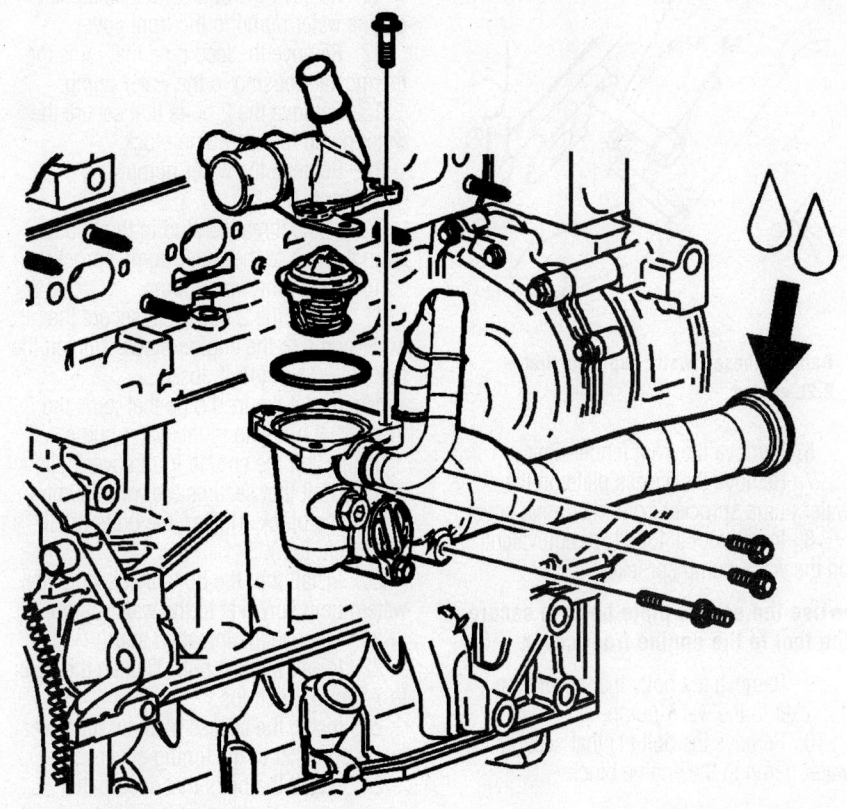

06025-COBALT-G13

Thermostat housing and feed pipe—2.0L engine

12. Apply sealant P/N 12378521 (Canadian P/N 88901148) to the water pump drain plug.

13. Install the water pump drain plug, if necessary. Tighten the water pump drain plug to 20 Nm (15 ft. lbs.).

14. Install the water feed tube.

15. Lubricate the feed tube O-ring with anti-freeze.

16. Install the water feed tube by twisting and pushing toward the water pump. Take care not to tear or damage the O-ring.

17. Install the thermostat housing to block bolts. Tighten the bolts to 10 Nm (89 inch lbs.).

18. Install the thermostat.

19. Install the thermostat housing cap and bolts.

2.2L Engine

1. Before servicing the vehicle, refer to the Precautions Section.

2. If equipped with an automatic transmission, remove the exhaust manifold.

3. Drain the cooling system.

4. Raise and suitably support the vehicle.

5. Remove the right front tire and wheel.

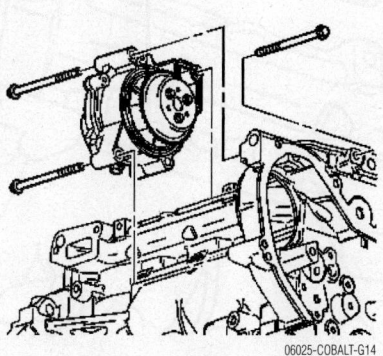

06025-COBALT-G14

Water pump removal—2.0L engine

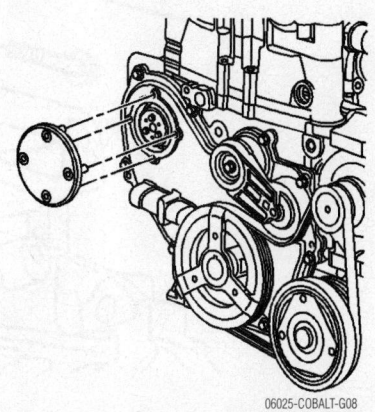

06025-COBALT-G08

Water pump access cover—2.2L engine

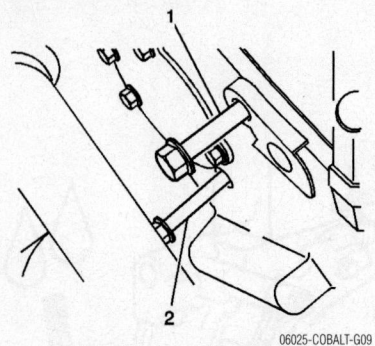

Remove these 2 water pup bolts first— 2.2L engine

06025-COBALT-G09

6. Remove the front fender liner.

7. Remove the access plate on the water pump sprocket from the timing cover.

8. Install tool J 43651, or equivalent, on the water pump sprocket.

➡**Use the access plate bolts to secure the tool to the engine front cover.**

9. Remove the bolts that secure the sprocket to the water pump.

10. Remove the bolt (1) that secures the water pump to the engine block.

11. Remove the bolt (2) that secures the engine water pump to the front cover.

12. Remove the feed pipe that joins the thermostat housing to the water pump.

13. Remove the 2 bolts that secure the water pump to the engine block.

14. Remove the water pump.

To install:

15. Use a threaded stud in the hub to align the hub to the water pump sprocket.

16. Install the water pump.

17. Install the 2 bolts that secure the water pump to the engine block. Tighten the bolts to 20 Nm (15 ft. lbs.).

18. Install the feed pipe that joins the thermostat housing to the water pump.

19. Install the engine front cover bolt and the bolt that secures the water pump to the engine block. Tighten the bolts to 20 Nm (15 ft. lbs.).

20. Install 2 of the bolts that secure the water pump sprocket to the water pump.

21. Remove the threaded stud.

22. Install the last bolt. Tighten the bolts to 7 Nm (62 inch lbs.).

23. Install the access plate on the water pump sprocket to the timing cover.

24. Install the bolts that secure the

access plate to the timing cover. Tighten the bolts to 10 Nm (18 inch lbs.).

25. Install the front fender liner.

26. Install the right front tire and wheel.

27. If equipped with an automatic transmission, install the exhaust manifold.

28. Fill the cooling system.

Charge Air Cooler Coolant Pump

REMOVAL & INSTALLATION

2.0L Engine

1. Before servicing the vehicle, refer to the Precautions Section.

2. Drain the charge air cooling system.

3. Compress the auxiliary water pump inlet hose clamp at the charge air cooler pump.

4. Remove the auxiliary water pump inlet hose from the charge air cooler pump.

5. Compress the auxiliary water pump outlet hose clamp at the charge air cooler pump.

6. Remove the auxiliary water pump outlet hose from the charge air cooler pump.

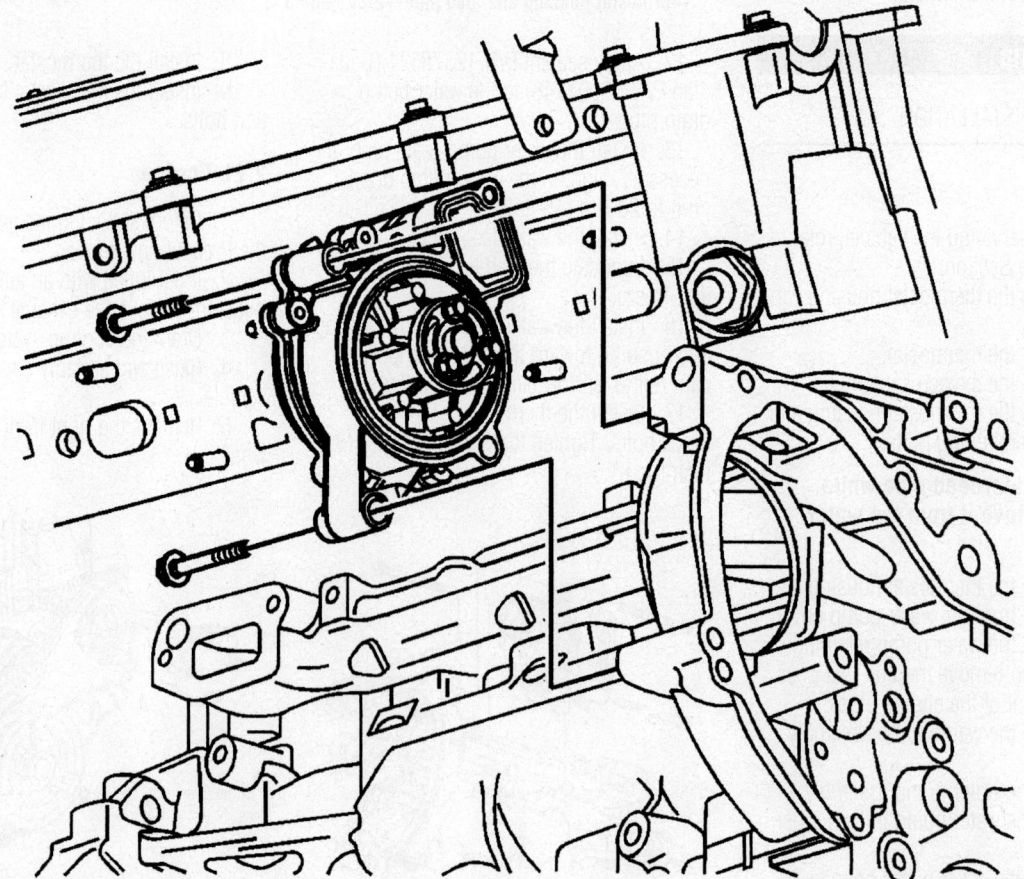

06025-COBALT-G10

Water pump removal—2.2L engine

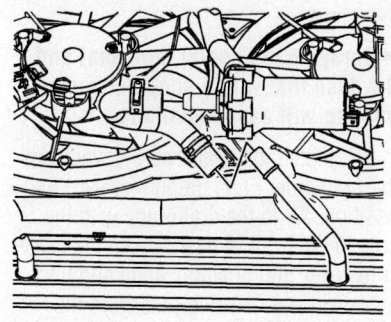

06025-COBALT-G11

Auxiliary coolant pump—2.0L engine

7. Disconnect the charge air cooler pump electrical connector.

8. Remove the charge air cooler pump mounting clamp bolts.

9. Remove the charge air cooler pump.

To install:

10. Install the charge air cooler pump.

11. Install the mounting clamp bolts to the charge air cooler pump. Tighten the bolts to 10 Nm (88 inch lbs.).

12. Connect the charge air cooler pump electrical connector.

13. Compress the auxiliary water pump inlet hose clamp to secure the hose.

14. Install the auxiliary water pump inlet hose to the charge air cooler pump.

15. Compress the auxiliary water pump outlet hose clamp to secure the hose.

16. Install the auxiliary water pump outlet hose to the charge air cooler pump.

17. Lower the vehicle.

18. Fill the cooling system.

Heater Core

REMOVAL & INSTALLATION

✳✳ CAUTION

With a pressurized cooling system, the coolant temperature in the radiator can be considerably higher than the boiling point of the solution at atmospheric pressure. Removal of the surge tank cap, while the cooling system is hot and under high pressure, causes the solution to boil instantaneously with explosive force. This will cause the solution to spew out over the engine, the fenders, and the person removing the cap. Serious bodily injury may result.

1. Before servicing the vehicle, refer to the Precautions Section.

2. Drain the cooling system.

3. Raise and support the vehicle.

4. Place a drain pan under the water pump drain port.

5. Loosen the water pump drain bolt and drain the coolant from the water pump.

6. Close and tighten the water pump drain bolt. Tighten the water pump drain bolt to 10 Nm (88 inch lbs.).

7. Lower the vehicle.

8. Reposition the heater outlet hose clamp at the heater core.

9. Remove the heater outlet hose from the heater core.

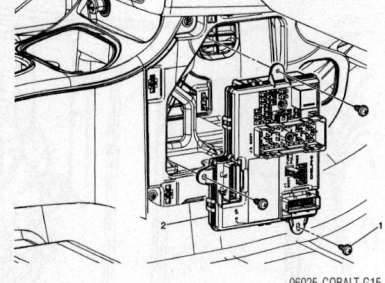

06025-COBALT-G15

Body control module

10. Reposition the heater inlet hose clamp at the heater core.

11. Remove the heater inlet hose from the heater core.

12. Remove the right console extension panel.

13. Remove the body control module (BCM) from the vehicle.

14. Remove the front floor console left side extension panel.

15. Pull back the carpet at the bottom of the left instrument panel (I/P) center support bracket and remove the left I/P center support bracket nuts.

16. Remove the left I/P center support bracket.

17. Remove the accelerator control pedal from the front of the dash and position out of the way.

18. Raise the center floor outlet duct while pushing the floor ducts down to disengage the ducts.

19. Rotate the center floor outlet duct forward in the vehicle and pull down to disengage it from the HVAC module.

20. Remove the heater core cover heat stakes with a small chisel.

21. Loosen the nut that is behind the fuel line bracket and remove the stud from the dash panel at the heater hoses.

➡**Make certain that all of the heater core cover screws are removed before attempting to remove the heater core cover.**

22. Pull the heater core cover down just enough to clear the locating pins from the HVAC module. Slide the heater core cover rearward until the drain tube clears the front of dash. Slide the heater core cover down, rearward, and to the right to remove.

23. Remove the heater core.

To install:

24. Inspect the foam heater core seal on the lower HVAC case. If damaged, replace using Kent Industries adhesive black foam tape P/N 46480, or equivalent.

25. Install the heater core into the HVAC module.

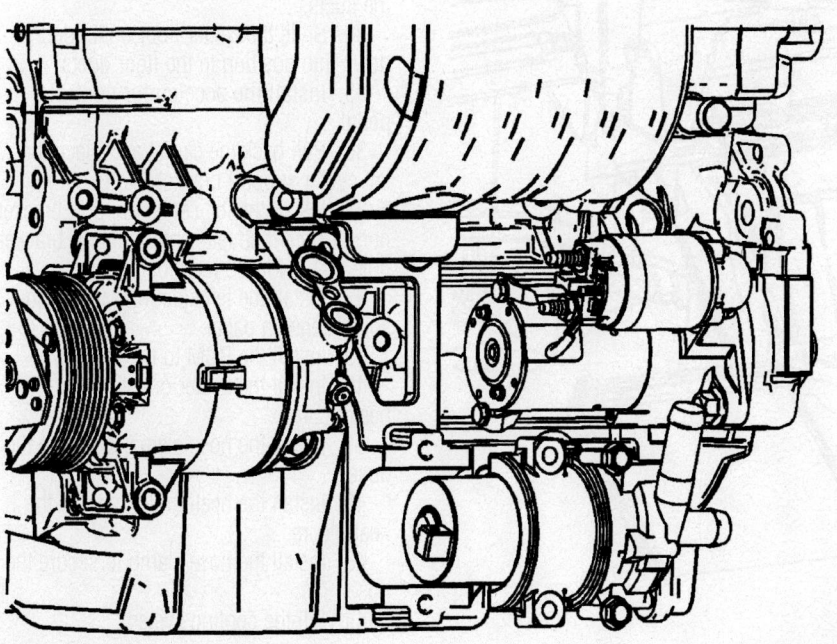

06025-COBALT-G12

Charge air coolant pump—2.0L engine

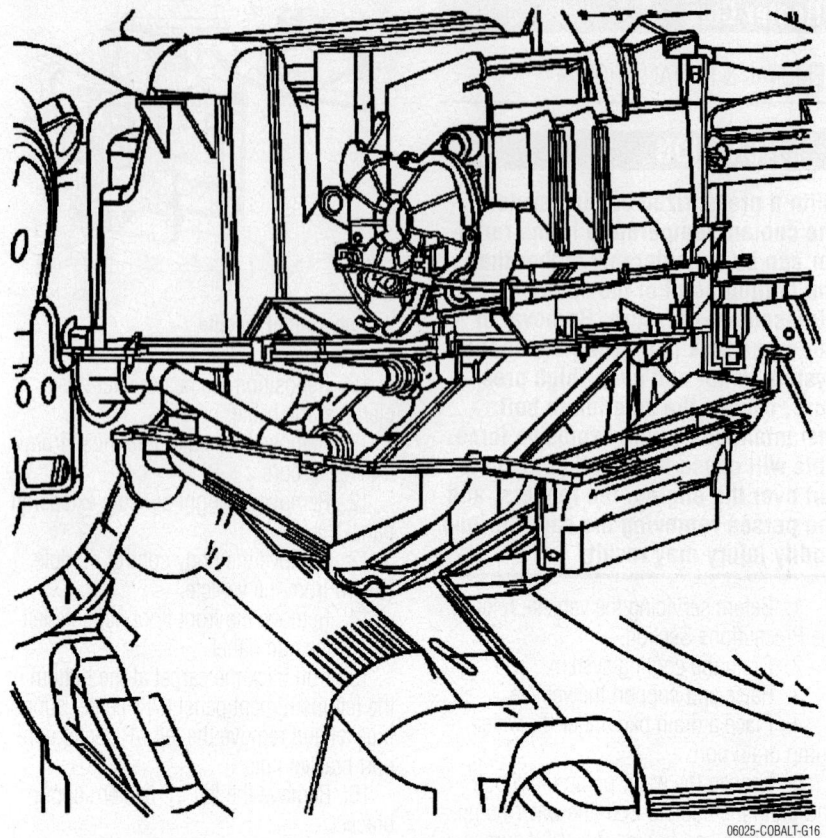

Heater core cover removal

06025-COBALT-G16

Heater core removal

06025-COBALT-G17

26. Install a new drain tube seal onto the drain tube.

➡Spraying the heater core seal and the dash mat with a soap and water mixture will ease installation.

27. Install the heater core cover from the right side. Slide up and forward into position. Align the drain tube with the hole in the front of dash. Raise the heater core cover into position while aligning holes with the locating pins from the HVAC module.

28. Cut the sound insulator at the cowl near the center screw approximately 76 mm (3 in.) and fold the sound insulator back to ease in installation of the screw. Ensure the sound insulator is positioned back after the screws are tightened.

➡Make certain that the heater core cover is properly positioned and is fully seated on the HVAC module before installing the screws. Be sure to install all heater core cover screws.

29. Install the heater core cover screws. Tighten the heater core cover screws to 1.8 Nm (15 inch lbs.).

30. Install the stud into the dash panel at the heater hoses and tighten the nut that is behind the fuel line bracket.

31. Align the center floor duct with the HVAC module.

32. Push the center floor duct up while rotating rearward in the vehicle to install on the HVAC module.

33. Push down on the floor duct while rotating the center floor outlet ducts to align the ducts.

34. Slide the center floor outlet duct down into position in the floor ducts.

35. Install the accelerator control pedal.

36. Pull back the carpet and place the left center support bracket into position.

37. Install the left center support bracket nuts. Tighten the left center support bracket nuts to 10 Nm (88 inch lbs.).

38. Install the left side front floor console extension panel.

39. Install the BCM to the vehicle.

40. Install the heater outlet hose to the heater core.

41. Install the hose clamp to secure the hose.

42. Install the heater inlet hose to the heater core.

43. Install the hose clamp to secure the hose.

44. Fill the cooling system.

Cylinder Head

REMOVAL & INSTALLATION

2.0L Engine

1. Before servicing the vehicle, refer to the Precautions Section.
2. Remove the supercharger.
3. Remove the intake manifold.
4. Remove the exhaust manifold.
5. Remove the timing chain.
6. Drain the cooling system.
7. Remove the cylinder head bolts in sequence and discard.
8. Remove the cylinder head and gasket.
9. Clean all gasket surfaces.

To install:

10. Install the cylinder head and gasket.

➡**Install NEW cylinder head bolts.**

11. Tighten the cylinder head bolts in sequence to 30 Nm (22 ft. lbs.) plus 155 degrees.
12. Install the NEW front cylinder head bolts. Tighten the front cylinder head bolts to 35 Nm (26 ft. lbs.).
13. Install the timing chain.
14. Install the exhaust manifold.
15. Lower the vehicle.
16. Install the intake manifold.
17. Install the supercharger.
18. Fill the cooling system.

2.2L Engine

1. Before servicing the vehicle, refer to the Precautions Section.
2. Remove the intake manifold.
3. Remove the exhaust manifold.
4. Remove the timing chain.
5. Drain the cooling system.
6. Remove the cylinder head bolts in sequence and discard.
7. Remove the cylinder head and gasket.
8. Clean all gasket surfaces.

To install:

9. Install the cylinder head and gasket.

➡**Install NEW cylinder head bolts.**

10. Tighten the cylinder head bolts in sequence to 30 Nm (22 ft. lbs.) plus 155 degrees.
11. Install the NEW front cylinder head bolts. Tighten the front cylinder head bolts to 35 Nm (26 ft. lbs.).
12. Install the timing chain.
13. Install the exhaust manifold.
14. Lower the vehicle.
15. Install the intake manifold.
16. Fill the cooling system.

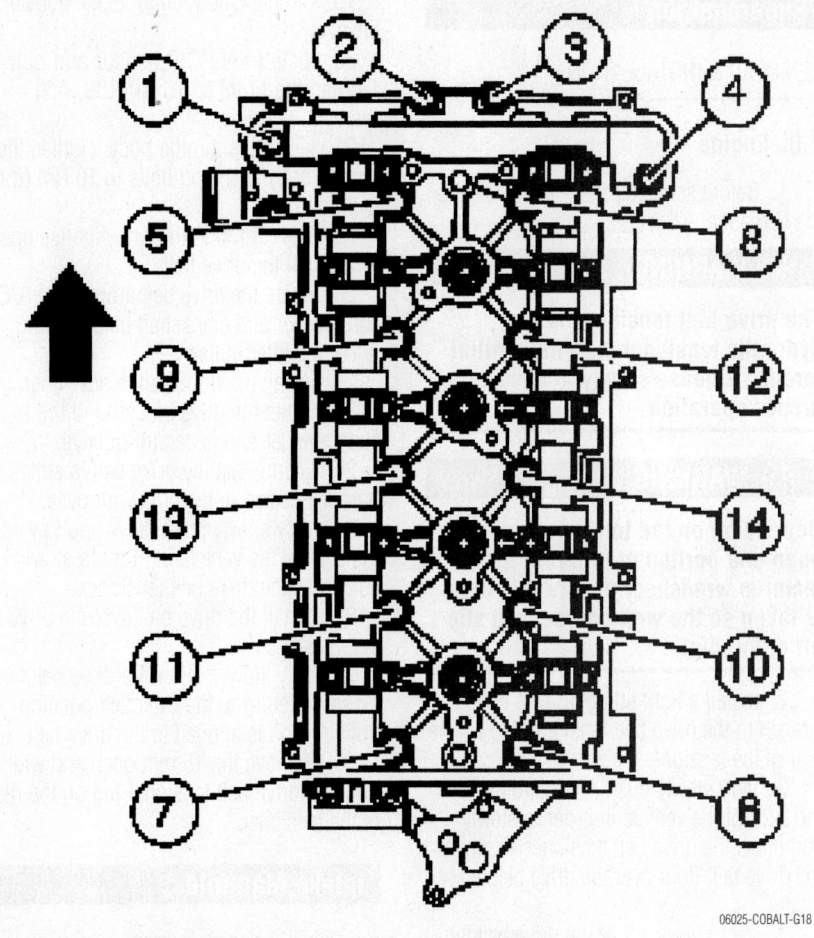

06025-COBALT-G18

Cylinder head bolt removal sequence—2.0L and 2.2L engines

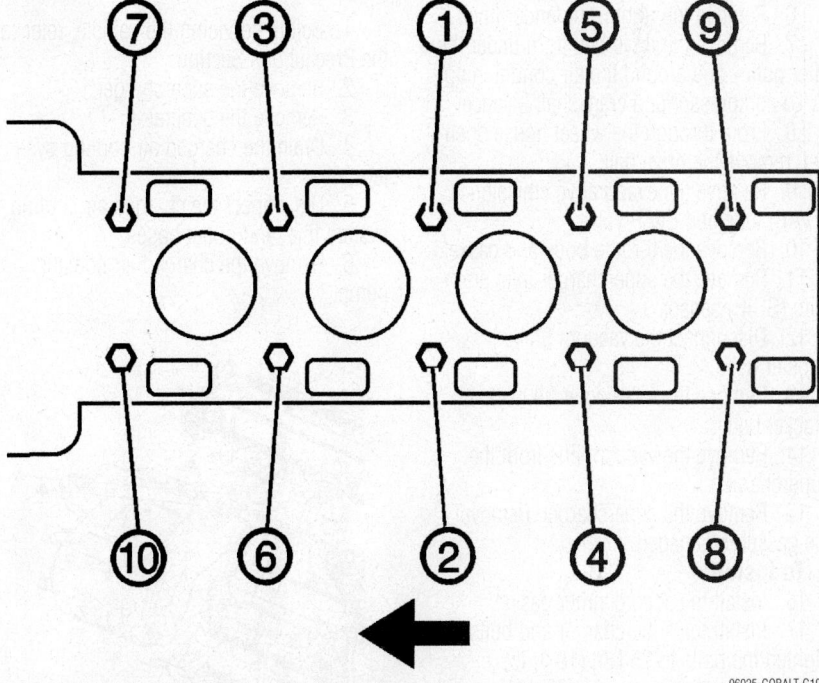

06025-COBALT-G19

Cylinder head tightening sequence—2.0L and 2.2L engines

Supercharger

REMOVAL & INSTALLATION

2.0L Engine

1. Before servicing the vehicle, refer to the Precautions Section.

> ※※ **WARNING**
>
> **The drive belt tensioner is a hydraulic tensioner with high initial torque. Release slowly to ensure proper operation.**

> ※※ **WARNING**
>
> **Depending on the tolerances of the open end portion of different manufactures wrenches, ensure that care is taken so the wrench does not slip off of the lug.**

2. Install a tight fitting 15 mm open end wrench to the drive belt tensioner lug on the rear of the tensioner.

3. Very slowly push down and towards the back of the vehicle in order to slowly compress the drive belt tensioner. Remove the drive belt from over the drive belt tensioner.

4. Very slowly, allow the drive belt tensioner to return to the extended position.

5. Remove the drive belt from around the supercharger and generator pulleys.

6. Remove the right front fender liner.

7. Remove the drive belt from under the idler pulley and around the air conditioning (A/C) compressor and crankshaft balancer.

8. From through the wheel house opening, remove the drive belt.

9. Remove the evaporative emission (EVAP) tube and EVAP valve.

10. Remove the throttle body and gasket.

11. Remove the supercharger inlet pressure (SCIP) sensor.

12. Disconnect the vacuum brake booster hose.

13. Remove the intercooler fill neck bracket bolts.

14. Remove the vacuum line from the supercharger.

15. Remove the supercharger. Remove the gasket if damaged.

To install:

16. Install the supercharger gasket.

17. Install the supercharger and bolts. Tighten the bolts to 25 Nm (18 ft. lbs.).

18. Install the intercooler fill neck bracket bolts. Tighten the bolts to 10 Nm (89 inch lbs.).

19. Connect the vacuum brake booster hose.

20. Install the SCIP sensor and bolt. Tighten the bolts to 10 Nm (89 inch lbs.).

21. Install the throttle body. Tighten the throttle body attaching bolts to 10 Nm (89 inch lbs.).

22. From through the wheel house opening, install the drive belt.

23. Route the drive belt around the A/C compressor and crankshaft balancer and under the idler pulley.

24. Install the right front fender liner.

25. Route the drive belt around the supercharger and generator pulleys.

26. Ensure that the drive belt is still properly seated in the pulley grooves.

27. Very slowly push down and towards the back of the vehicle in order to slowly compress the drive belt tensioner.

28. Install the drive belt over the drive belt tensioner.

29. Very slowly, allow the drive belt tensioner to return to the extended position until tension is applied to the drive belt.

30. Remove the 15 mm open end wrench from the drive belt tensioner lug on the rear of the tensioner.

Intake Manifold

REMOVAL & INSTALLATION

2.0L Engine

1. Before servicing the vehicle, refer to the Precautions Section.

2. Remove the supercharger.

3. Remove the generator.

4. Drain the charged air cooling system.

5. Disconnect the charged air cooling system inlet and outlet hoses.

6. Remove the charged air coolant pump.

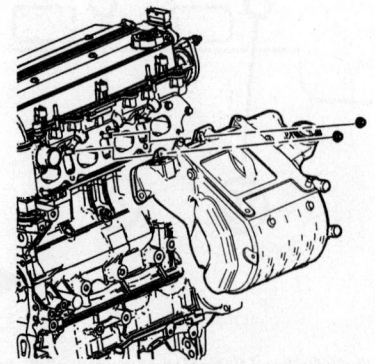

Intake manifold—2.0L engine

06025-COBALT-G20

7. Remove the cooling fan assembly.

8. Remove the oil level indicator tube bolt.

9. Remove the electrical connector from the intake manifold.

10. Remove the coolant hoses from the intake manifold.

➡ **Be sure to remove all fasteners before attempting to remove the intake manifold.**

11. Remove the intake manifold nuts and bolts.

12. Remove the intake manifold.

➡ **The intake manifold gasket is reusable. Only replace the gasket if damage has occurred.**

13. Remove the intake manifold gasket.

To install:

14. Install the intake manifold gasket.

15. Install the intake manifold.

16. Install the intake manifold nuts and bolts. Tighten the intake manifold nuts and bolts to 10 Nm (89 inch lbs.).

17. Install the coolant hoses to the intake manifold. Tighten the hose clamps to 17 Nm (13 ft. lbs.).

18. Connect the electrical connector to the intake manifold.

19. Install the oil level indicator tube and bolt. Tighten the oil level indicator tube bolt to 10 Nm (89 inch lbs.).

20. Reposition the AC compressor and install the bolts. Tighten the bolts to 22 Nm (16 ft. lbs.).

21. Install the cooling fan assembly.

22. Install the charged air coolant pump.

23. Install the generator.

24. Connect the charged air cooling system inlet and outlet hoses.

25. Fill the charged air cooling system.

26. Install the supercharger.

2.2L Engine

1. Before servicing the vehicle, refer to the Precautions Section.

2. Remove the air cleaner outlet resonator.

3. Remove the throttle body.

4. Disconnect the positive crankcase ventilation (PCV) hose.

5. Disconnect the purge solenoid tube.

6. Disconnect the brake booster hose.

7. Remove the oil level indicator tube bolt.

8. Remove the fuel rail.

9. Disconnect the knock sensor electrical connectcr.

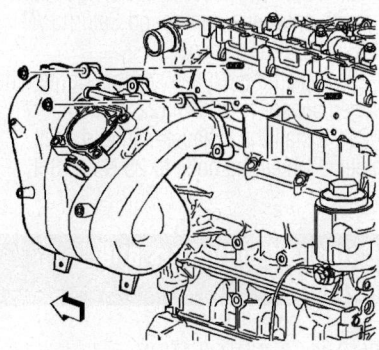

Intake manifold—2.2L engine

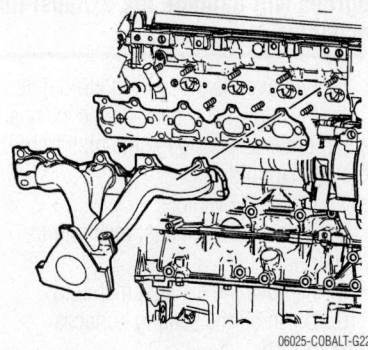

Exhaust manifold—2.0L engine; 2.2L similar

10. Remove the knock sensor harness connector from the intake manifold.

11. Remove the intake manifold nuts and bolts.

12. Remove the intake manifold.

➡The intake manifold gasket is reusable, only replace the gasket if damage has occurred.

13. If applicable, remove the intake manifold gasket.

To install:

14. If applicable, install the intake manifold gasket.

15. Install the intake manifold.

16. Install the intake manifold nuts and bolts. Tighten the intake manifold nuts and bolts to 10 Nm (89 inch lbs.).

17. Install the knock sensor harness connector to the intake manifold.

18. Connect the knock sensor electrical connector.

19. Install the fuel rail.

20. Install the throttle body. Tighten the throttle body attaching bolts to 10 Nm (89 inch lbs.).

21. Install the oil level indicator tube bolt. Tighten the oil level indicator tube bolt to 10 Nm (89 inch lbs.).

22. Connect the brake booster hose.

23. Connect the purge solenoid tube.

24. Connect the PCV hose.

25. Install the throttle body.

26. Install the air cleaner outlet resonator.

Exhaust Manifold

REMOVAL & INSTALLATION

2.0L Engine

1. Before servicing the vehicle, refer to the Precautions Section.

2. Remove the exhaust manifold heat shield.

✴✴ WARNING

The oxygen sensor uses a permanently attached pigtail and connector. Do not remove the pigtail from the oxygen sensor. Damage to or removal of the pigtail connector could affect proper operation of the oxygen sensor.

✴✴ WARNING

The use of excessive force may damage the threads in the exhaust manifold/pipe.

➡The in-line connector and louvered end must be kept clear of grease, dirt or other contaminants. Avoid using cleaning solvents of any type. DO NOT drop or roughly handle the oxygen sensor.

➡The oxygen sensor may be difficult to remove when the engine temperature is less than 48°C (120°F).

3. Remove the oxygen sensor.

✴✴ WARNING

Do not bend the exhaust flex decoupler more than 3 degrees in any direction. Movement of more than 3 degrees will damage the exhaust flex decoupler.

4. Disconnect the exhaust pipe from the manifold.

5. Remove and discard the exhaust manifold to cylinder head retaining nuts.

6. Remove the exhaust manifold.

7. Clean all of the sealing surfaces.

To install:

8. Install new exhaust manifold studs. Tighten the studs to 10 Nm (89 inch lbs.).

9. Install the exhaust manifold gasket.

10. Install the exhaust manifold to the cylinder head.

11. Install NEW exhaust manifold to cylinder head retaining nuts finger tight.

12. Tighten the NEW exhaust manifold to cylinder head retaining nuts in sequence. Tighten the nuts to 14 Nm (124 inch lbs.).

13. Connect the exhaust pipe. Tighten the nuts to 50 Nm (37 ft. lbs.).

14. Coat the threads of the oxygen sensor with anti-seize compound.

➡A special anti-seize compound is used on the oxygen sensor threads. The compound consists of a liquid graphite and glass beads. The graphite will burn away, but the glass beads will remain, making the sensor easier to remove. New or service sensors will have the compound applied to the threads. If a sensor is removed and is to be reinstalled, the threads must have an anti-seize compound applied before installation.

15. Coat the threads of the oxygen sensor with anti-seize compound Saturn P/N 21485279, if necessary.

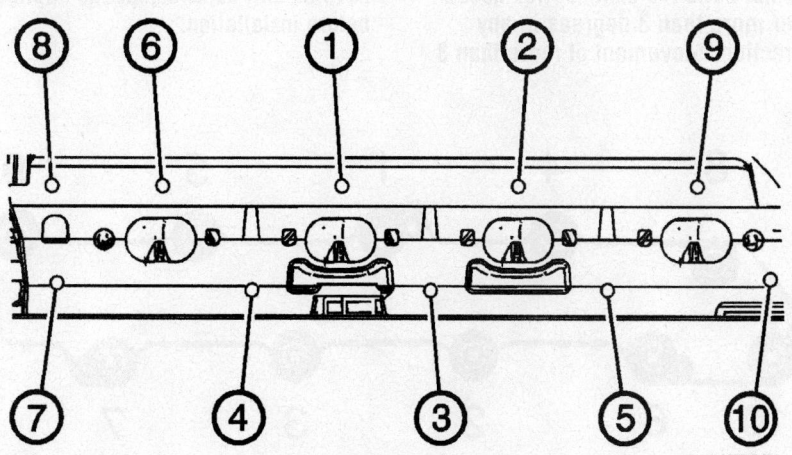

Exhaust manifold torque sequence—2.0L engine

16. Install the oxygen sensor. Tighten the oxygen sensor to 30 Nm (22 ft. lbs.).

17. Install the exhaust manifold heat shield.

18. Install the exhaust manifold heat shield bolts. Tighten the bolts to 23 Nm (17 ft. lbs.).

2.2L Engine

1. Before servicing the vehicle, refer to the Precautions Section.

2. Remove the exhaust manifold heat shield.

> ❊❊ **WARNING**
>
> **The oxygen sensor uses a permanently attached pigtail and connector. Do not remove the pigtail from the oxygen sensor. Damage to or removal of the pigtail connector could affect proper operation of the oxygen sensor.**

> ❊❊ **WARNING**
>
> **The use of excessive force may damage the threads in the exhaust manifold/pipe.**

➡ **The in-line connector and louvered end must be kept clear of grease, dirt or other contaminants. Avoid using cleaning solvents of any type. DO NOT drop or roughly handle the oxygen sensor.**

➡ **The oxygen sensor may be difficult to remove when the engine temperature is less than 48°C (120°F).**

3. Remove the oxygen sensor.

4. Raise and support the vehicle.

> ❊❊ **WARNING**
>
> **Do not bend the exhaust flex decoupler more than 3 degrees in any direction. Movement of more than 3 degrees will damage the exhaust flex decoupler.**

5. Remove the pipe to manifold nuts.

6. Pull down and back on the exhaust pipe in order to disengage the pipe from the exhaust manifold.

7. Lower the vehicle.

8. Remove the exhaust manifold to cylinder head nuts.

9. Remove the exhaust manifold.

10. Clean all the sealing surfaces.

To install:

11. Install a new exhaust manifold gasket.

12. Install the exhaust manifold to the cylinder head.

➡ **Install the new exhaust nuts.**

13. Install the new exhaust manifold to cylinder head retaining nuts. Follow the tightening sequence. Tighten the nuts to 13 Nm (115 inch lbs.).

14. Raise and support the vehicle.

15. Install a new exhaust manifold to flex coupler gasket.

16. Push the flex coupler into position on the exhaust manifold.

➡ **Install the new exhaust nuts.**

17. Install the retaining nuts which secure the manifold to the flex decoupler. Tighten the nuts to 50 Nm (37 ft. lbs.).

18. Lower the vehicle.

➡ **A special anti-seize compound is used on the oxygen sensor threads. The compound consists of a liquid graphite and glass beads. The graphite will burn away, but the glass beads will remain, making the sensor easier to remove. New or service sensors will have the compound applied to the threads. If a sensor is removed and is to be reinstalled, the threads must have an anti-seize compound applied before installation.**

19. Coat the threads of the oxygen sensor with anti-seize compound Saturn P/N 21485279, if necessary.

20. Install the oxygen sensor. Tighten the oxygen sensor to 30 Nm (22 ft. lbs.)

21. Install the exhaust manifold heat shield. Tighten the bolts to 25 Nm (18 ft. lbs.).

Camshaft Followers and Adjusters

REMOVAL & INSTALLATION

2.0L and 2.2L Engines

1. Before servicing the vehicle, refer to the Precautions Section.

2. Remove the ignition coils.

3. Remove the ground strap and stud.

4. Disconnect the positive crankcase ventilation (PCV) hose from the cam cover.

5. Disconnect the fuel feed pipe from the fuel rail.

6. Remove the camshaft cover bolts.

7. Remove the camshaft cover.

8. Remove the upper timing chain guide. Install camshaft sprocket holding tool J 43655, or equivalent.

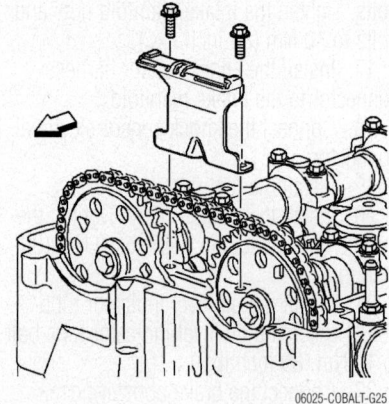

06025-COBALT-G25

Remove the upper timing chain guide—2.0L engine

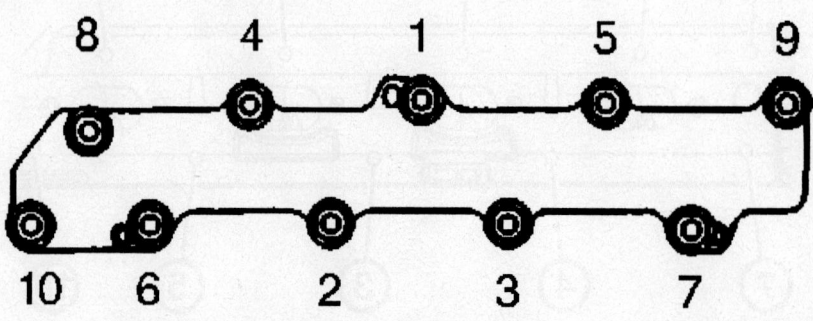

06025-COBALT-G24

Exhaust manifold torque sequence—2.2L engine

06025-COBALT-G26

Tool J43655 installed—2.0L engine

9. Remove both the intake and exhaust camshaft sprocket bolts and discard.

10. Slide the camshaft sprockets forward.

11. Mark the caps to ensure they are installed in the original position.

➡ **Remove each bolt on each cap one turn at a time until there is no spring tension on the camshaft.**

12. Remove the caps.

13. Remove the camshaft.

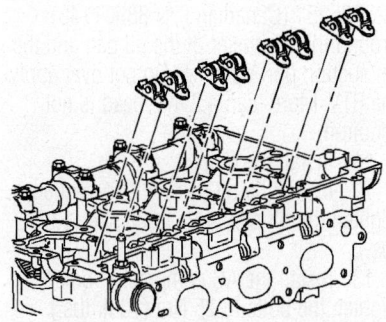

06025-COBALT-G27

Camshaft roller followers—2.0L engine

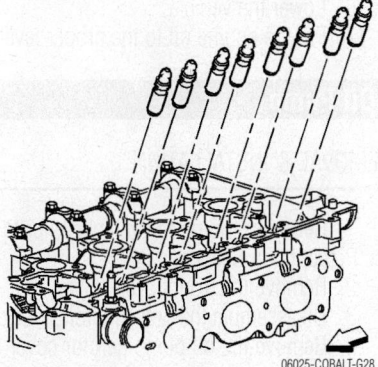

06025-COBALT-G28

Hydraulic lash adjusters—2.0L engine

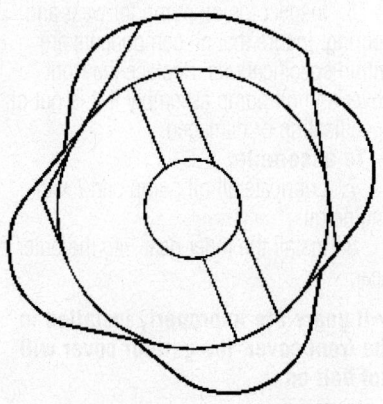

06025-COBALT-G29

Ensure that the alignment notches are aligned with the camshaft sprocket—2.0L engine

14. Remove the camshaft roller followers.

15. Remove the hydraulic lash adjusters.

To install:

16. Lubricate the valve tips.

17. Install the hydraulic lash adjusters.

18. Install the camshaft roller followers.

19. Ensure that the alignment notches are aligned with the camshaft sprocket.

20. Install the camshaft.

21. Install the caps.

22. Tighten the camshaft bearing cap bolts in increments of 3 turns until they are seated. Tighten the camshaft bearing cap bolts to 10 Nm (89 inch lbs.).

23. Apply a 3.5 mm (0.138 in) bead of GM P/N 12378521 (Canadian P/N 88901148) to the rear cap.

24. Install the rear intake camshaft bearing cap bolts. Tighten the bolts to 25 Nm (18 ft. lbs.).

25. Install camshaft sprockets onto the camshafts.

26. Hand tighten NEW camshaft sprocket bolts.

27. Remove the tool.

28. Tighten the camshaft sprocket bolts. Tighten the bolts to 85 Nm (63 ft. lbs.) plus 30 degrees.

29. Install the upper timing chain guide.

Tighten the upper timing chain guide to 10 Nm (89 inch lbs.).

30. Install the camshaft cover and bolts. Tighten the camshaft cover bolts to 10 Nm (89 inch lbs.).

31. Install the ground strap to the camshaft cover. Tighten the ground strap stud to 10 Nm (89 inch lbs.).

➡ **Make sure that the ignition coil seals are properly seated to the valve cover.**

32. Install the ignition coils. Tighten the ignition coil retaining bolts to 10 Nm (89 inch lbs.).

33. Connect the fuel feed pipe to the fuel rail.

34. Install the fuel pipe bracket. Tighten the fuel pipe bracket bolt to 10 Nm (89 inch lbs.).

35. Connect the PCV hose to the cam cover.

Oil Pan

REMOVAL & INSTALLATION

2.0L and 2.2L Engines

1. Before servicing the vehicle, refer to the Precautions Section.

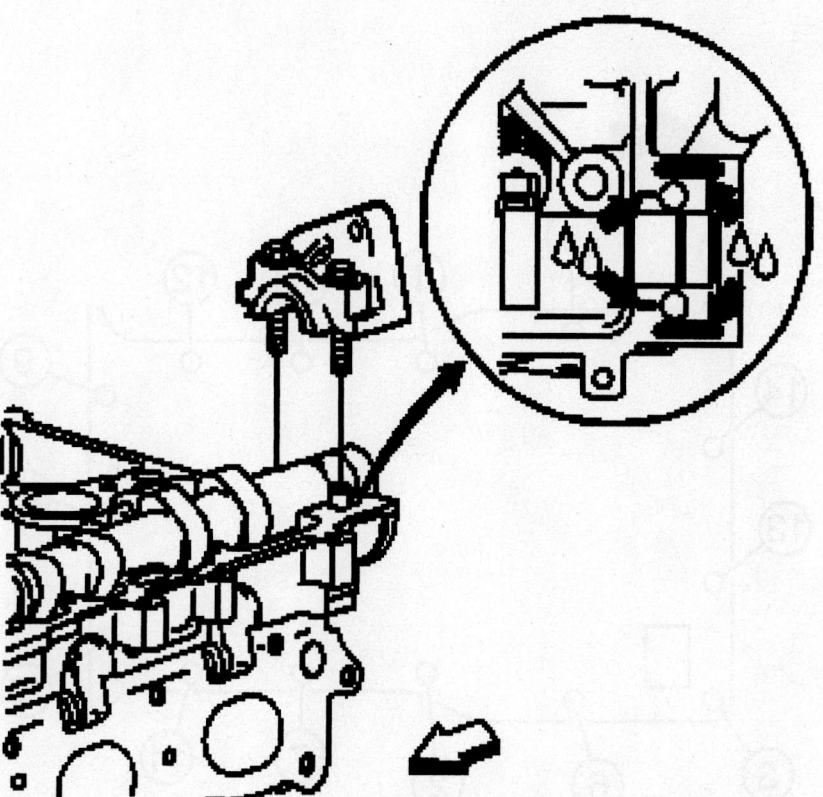

06025-COBALT-G30

Apply a 3.5 mm (0.138 in) bead of GM P/N 12378521 (Canadian P/N 88901148) to the rear cap—2.0L engine

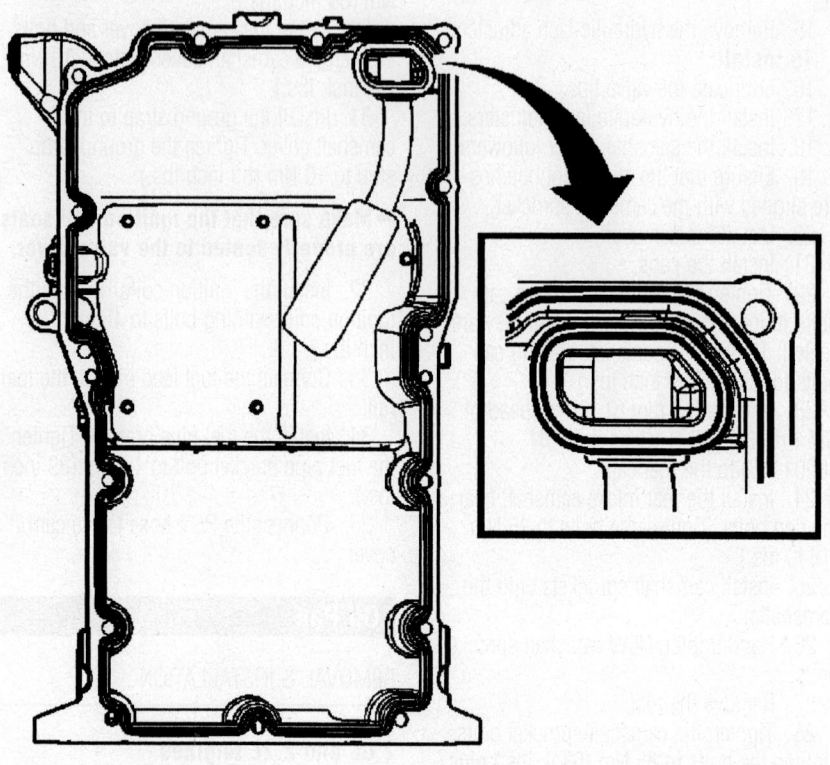

Apply a 2 mm bead of GM P/N 123785251 (Canadian P/N 88901148) around the perimeter of the oil pan and the oil suction port opening—2.0L and 2.2L engines

06025-COBALT-G31

2. Raise and support the vehicle.

3. Drain the engine oil.

4. Remove the engine drive belt.

5. Remove the intercooler pump bracket bolts from the oil pan.

6. Remove the lower AC compressor bolt from the oil pan.

7. Remove the oil pan bolts.

8. Remove the oil pan.

To install:

9. Make sure that the oil pan and mounting surface on the lower crankcase are free of all oil and debris.

10. Apply a 2 mm bead of GM P/N 123785251 (Canadian P/N 88901148) around the perimeter of the oil pan and the oil suction port opening. Do not over apply the RTV. More than a 2 mm bead is not required.

11. Install the oil pan.

12. Install the oil pan bolts in sequence. Tighten the oil pan bolts to 25 Nm (18 ft. lbs.).

13. Install the AC compressor bolts. Tighten the bolts to 25 Nm (18 ft. lbs.).

14. Install the intercooler pump bracket bolts. Tighten the bolts to 25 Nm (18 ft. lbs.).

15. Install the engine drive belt.

16. Lower the vehicle.

17. Fill the engine oil to the proper level.

Oil Pump

REMOVAL & INSTALLATION

1. Before servicing the vehicle, refer to the Precautions Section.

2. Remove the front cover.

3. Disassemble the pressure relief valve.

4. Remove the oil pump gerotor cover and bolts.

5. Clean all of the parts in cleaning solvent. Remove varnish, sludge and dirt.

6. Inspect the oil pump for wear and scoring. Insure that all components are within specifications. Replace the front cover and oil pump assembly if it is out of specification or damaged.

To assemble:

7. Lubricate all oil pump parts with engine oil.

8. Install the inner gear into the outer gear.

➡**If gears are improperly installed in the front cover, the gerotor cover will not bolt on.**

9. Install the gears together into the front cover with the hub of the center gear facing the front cover.

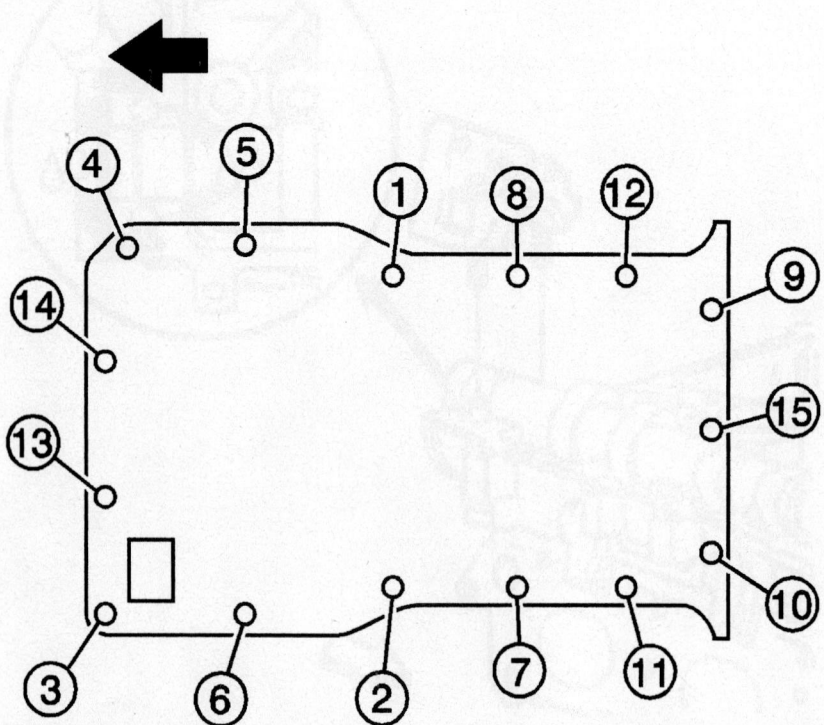

06025-COBALT-G32

Oil pan bolt torque sequence—2.0L and 2.2L engines

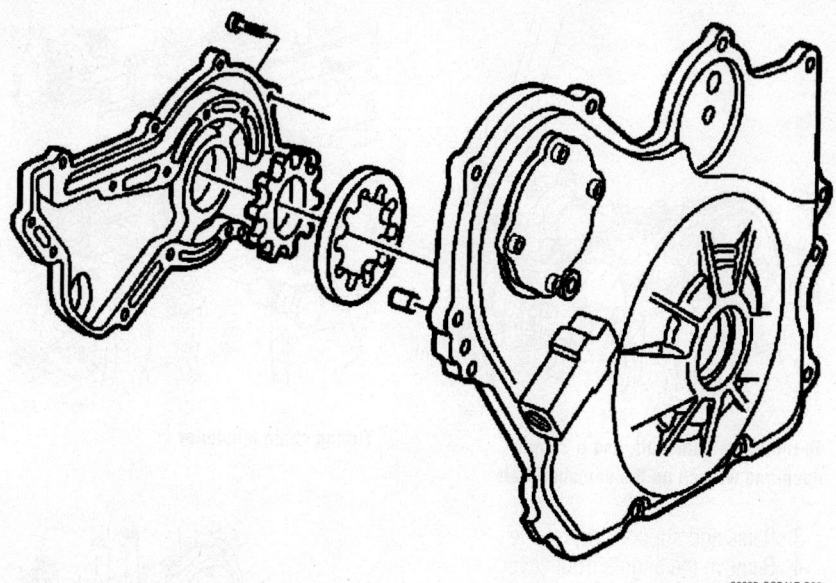

Oil pump disassembled—2.0L and 2.2L engines

06025-COBALT-G33

10. Install the oil pump gerotor cover and bolts. Tighten the oil pump gerotor bolts to 6 Nm (53 inch lbs.).

11. Install the pressure relief valve piston.

12. Install the pressure relief valve spring. Tighten the pressure relief valve plug to 40 Nm (30 ft. lbs.).

13. Install the front cover.

Rear Main Seal

REMOVAL & INSTALLATION

1. Before servicing the vehicle, refer to the Precautions Section.

2. Remove the transmission.

3. Remove the flywheel.

➡**Do not damage the outside diameter of the crankshaft or chamber with any tool.**

4. Pry the crankshaft rear oil seal with a flat-bladed tool.

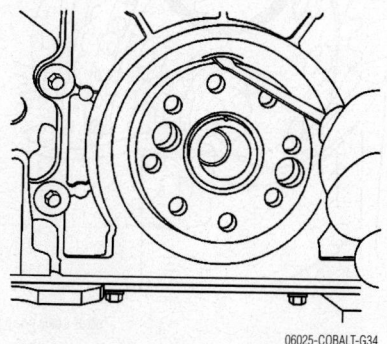

06025-COBALT-G34

Rear main seal removal—2.0L and 2.2L engines

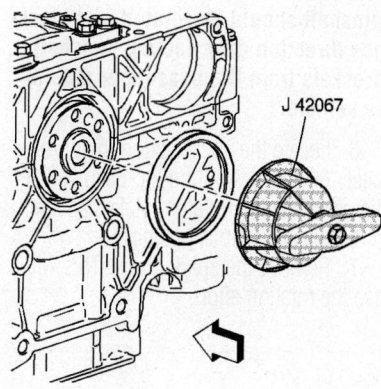

06025-COBALT-G35

Rear main seal installation—2.0L and 2.2L engines

To install:

5. Use a seal driver such as J 42067 and install the seal.

6. Install the flywheel. Tighten the flywheel bolts to 53 Nm (39 ft. lbs.) plus 25 degrees.

7. Install the transmission.

Crankshaft Balancer

REMOVAL & INSTALLATION

1. Before servicing the vehicle, refer to the Precautions Section.

2. Remove the engine drive belt.

3. Use tool J 38122-A, or equivalent, to prevent the crankshaft from rotating while loosening the crankshaft balancer bolt.

4. Remove the crankshaft balancer bolt. Discard the bolt.

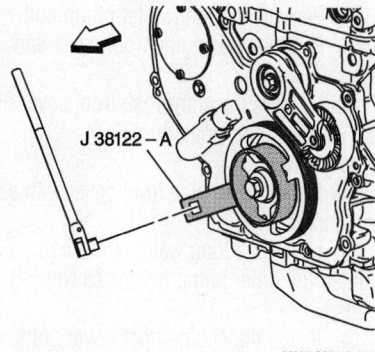

06025-COBALT-G37

Balancer holding tool installed

5. Remove the crankshaft balancer.

To install:

6. Install the crankshaft balancer.

7. Install a NEW crankshaft balancer bolt.

8. Use J 38122-A to prevent the crankshaft from rotating while tightening the crankshaft balancer bolt. Tighten the bolt to 100 Nm (74 ft. lbs.) plus 75 degrees.

9. Install the engine drive belt.

Front Cover and Seal

REMOVAL & INSTALLATION

2.0L Engine

1. Before servicing the vehicle, refer to the Precautions Section.

2. Remove the accessory drive belt tensioner bolts.

3. Remove the accessory drive belt tensioner.

4. Remove the idler pulley bolts.

5. Remove the idler pulley.

6. Remove the alternator bracket and lift hook assembly bolts.

7. Remove the alternator bracket and lift hook assembly.

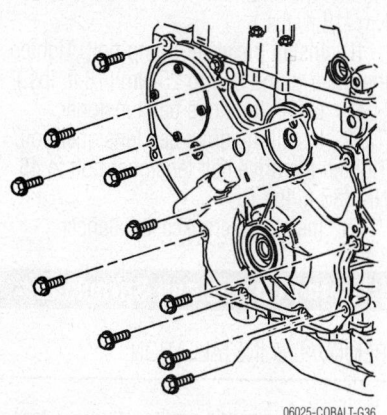

06025-COBALT-G36

Front cover bolt positions—2.0L and 2.2L engines

8. Remove the engine front cover bolts.

9. Remove the long water pump bolt.

10. Remove the engine front cover and gaskets.

11. Remove the crankshaft front cover oil seal with an appropriate tool.

To install:

12. Install the engine front cover with a new gasket.

13. Install the long water pump bolt. Tighten the water pump bolt to 25 Nm (18 ft. lbs.).

14. Install the engine front cover bolts. Tighten the engine front cover bolts to 25 Nm (18 ft. lbs.).

15. Install the generator bracket and lift hook assembly.

16. Install the generator bracket and lift hook assembly bolts. Tighten the bolts to 42 Nm (31 ft. lbs.).

17. Install the idler pulley.

18. Install the idler pulley bolts. Tighten the pulley bolts to 22 Nm (16 ft. lbs.).

19. Install the accessory drive belt tensioner.

20. Install the accessory drive belt tensioner bolts. Tighten the accessory drive belt tensioner bolts to 32 Nm (24 ft. lbs.).

2.2L Engine

1. Before servicing the vehicle, refer to the Precautions Section.

2. Remove the crankcase balancer.

3. Remove the drive belt tensioner.

4. Remove the engine front cover to water pump bolt.

5. Remove the remaining engine front cover bolts.

6. Remove the engine front cover.

To install:

7. If removed install a new engine front cover gasket.

8. Install the engine front cover.

9. Install the engine front cover bolts. Tighten the engine front cover bolts to 25 Nm (18 ft. lbs.).

10. Install the water pump bolt. Tighten the water pump bolt to 25 Nm (18 ft. lbs.).

11. Install the drive belt tensioner.

12. Install the drive belt tensioner bolt. Tighten the drive belt tensioner bolt to 45 Nm (33 ft. lbs.).

13. Install the crankcase balancer.

Timing Chains and Sprockets

REMOVAL & INSTALLATION

1. Before servicing the vehicle, refer to the Precautions Section.

2. Remove the camshaft cover.

06025-COBALT-G41

To rotate the camshaft, use a 24 mm open-end wrench on the camshaft flats

06025-COBALT-G39

Timing chain tensioner

3. Raise and support the vehicle.

4. Remove the engine front cover.

5. Lower the vehicle.

➡ **To rotate the camshaft, use a 24 mm open-end wrench on the camshaft flats. Camshaft should be rotated in a clockwise direction only, facing camshaft sprockets from the passenger side of the vehicle.**

6. Locate the No. 1 piston to approximately 60 degrees before top dead center (diamond shaped hole on intake camshaft sprocket at 12 o'clock position).

7. Remove the spark plugs. This will ease the rotation effort.

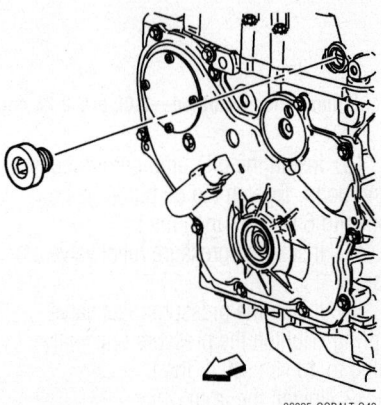

06025-COBALT-G40

Fixed timing chain guide access plug

06025-COBALT-G38

Locate the No. 1 piston to approximately 60 degrees before top dead center (diamond shaped hole on intake camshaft sprocket at 12 o'clock position)

8. Remove the timing chain tensioner.

9. Remove the fixed timing chain guide access plug.

10. Remove the fixed timing chain guide.

11. Remove the upper timing chain guide.

12. Use a 24 mm wrench to hold the camshafts from turning.

13. Remove the exhaust camshaft sprocket bolt and discard.

14. Remove the exhaust camshaft sprocket.

15. Remove the timing chain tensioner guide.

16. Remove the intake camshaft sprocket bolt and discard.

17. Remove the intake camshaft sprocket.

18. Remove the timing chain through the top of the cylinder head.

19. Remove the crankshaft sprocket.

20. Remove the oil nozzle and bolt.

21. Remove the balance shaft drive chain tensioner.

22. Remove the adjustable balance shaft chain guide.

23. Remove the small balance shaft drive chain guide.

24. Remove the upper balance shaft drive chain guide.

➥It may ease removal of the balance shaft drive chain to get all of the slack in the chain between the crankshaft and water pump sprockets.

25. Remove the balance shaft drive chain.

To install:

✳ WARNING

If the balance shafts are not properly timed to the engine, the engine may vibrate and make noise.

26. Install the upper balance shaft chain guide. Tighten the upper balance shaft chain guide bolts to 15 Nm (11 ft. lbs.).

27. Install the balance shaft drive chain with the colored links lined up on with the marks on the balance shaft drive sprockets and the crankshaft sprocket. Use the following procedure to line up the links with the sprockets: Orient the chain so that the copper colored and chrome links are visible.

　a. Place the uniquely colored link (1) so that it lines up with the timing mark on the intake side balance shaft sprocket.

　b. Working clockwise around the chain, place the first matching colored link (2) in line with the timing mark on the crankshaft drive sprocket. (approxi-

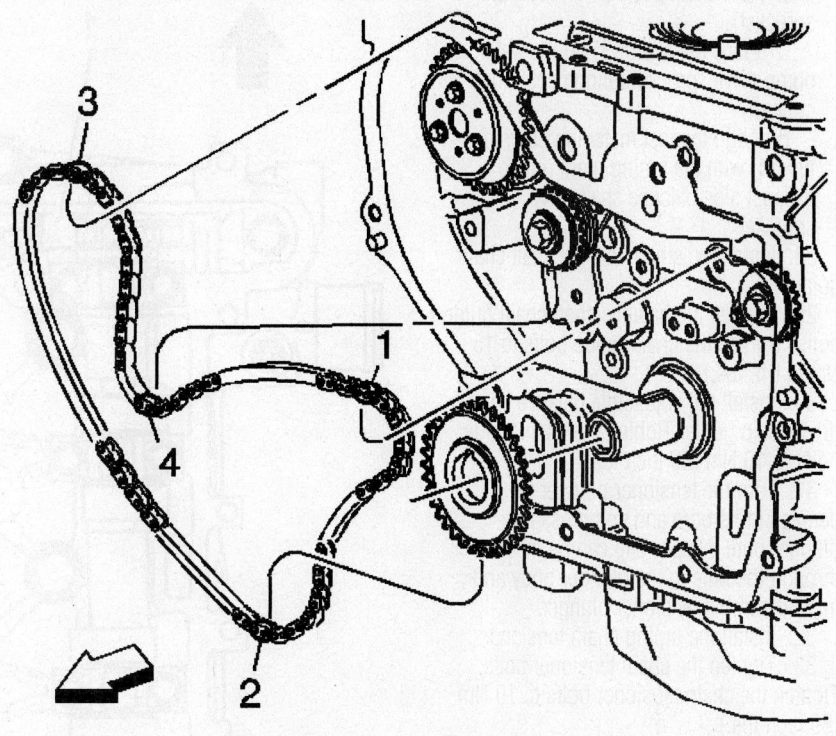

Balance shaft chain installation

06025-COBALT-G42

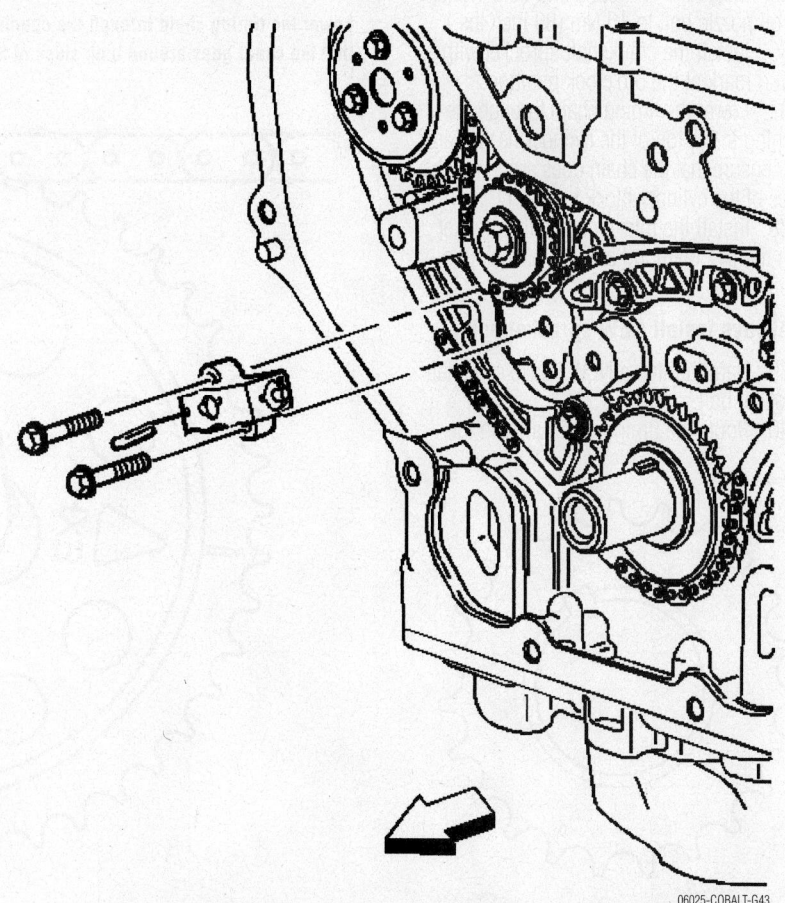

06025-COBALT-G43

Turn the tensioner plunger 90 degrees in its bore and compress the plunger until a paper clip can be inserted through the hole in the plunger body and into hole in the tensioner plunger

mately 5 o'clock position on the crank sprocket).

 c. Place the chain (3) on the water pump drive sprocket (alignment is not critical).

 d. Align the last matching colored link (4) with the timing mark on the exhaust side balance shaft drive sprocket.

28. Install the small balance shaft chain guide.

29. Tighten the balance shaft chain guide bolts. Tighten the chain guide bolts to 15 Nm (11 ft. lbs.).

30. Install the adjustable balance shaft drive chain guide. Tighten the chain guide bolts to 10 Nm (89 inch lbs.).

31. Turn the tensioner plunger 90 degrees in its bore and compress the plunger until a paper clip can be inserted through the hole in the plunger body and into hole in the tensioner plunger.

32. Install the timing chain tensioner.

33. Tighten the chain tensioner bolts. Tighten the chain tensioner bolts to 10 Nm (89 inch lbs.).

34. Remove the paper clip from the balance shaft drive chain tensioner.

35. Install the oil nozzle and bolt. Tighten the oil nozzle bolt to 10 Nm (89 inch lbs.).

36. Install the crankshaft sprocket with timing mark at the 5 o'clock position.

37. Lower the timing chain through the opening in the top of the cylinder head. Carefully ensure that the chain goes around both sides of the cylinder block bosses (1, 2).

38. Install the intake camshaft sprocket with the INT diamond at the 2 o'clock position.

➡ **Always install NEW sprocket bolts.**

39. Hand tighten a NEW intake camshaft sprocket bolt.

40. Route the timing chain around the

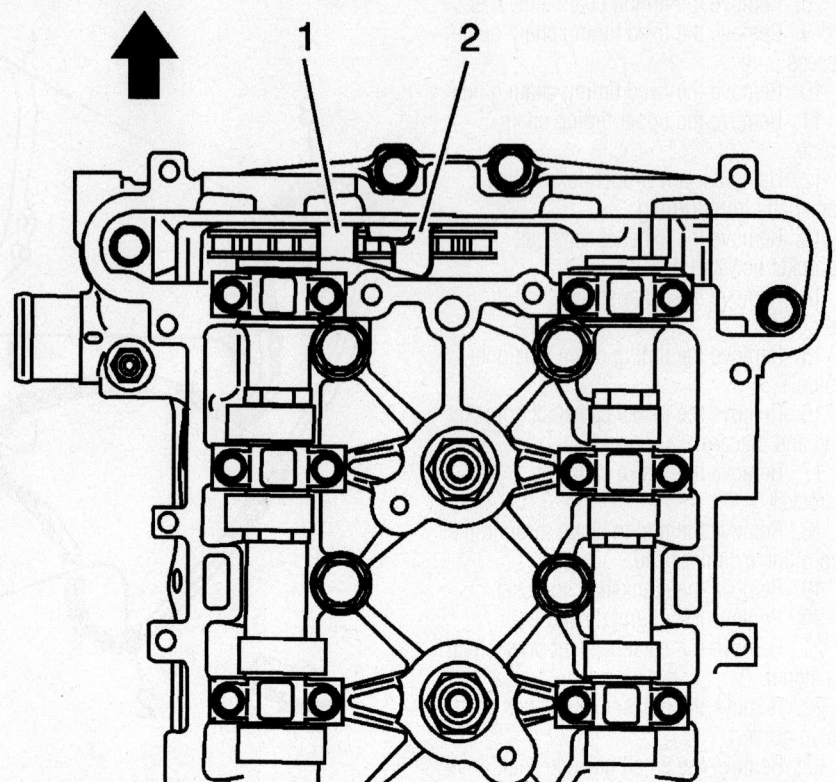

06025-COBALT-G45

Lower the timing chain through the opening in the top of the cylinder head. Carefully ensure that the chain goes around both sides of the cylinder block bosses (1, 2)

06025-COBALT-G44

Install the crankshaft sprocket with timing mark at the 5 o'clock position

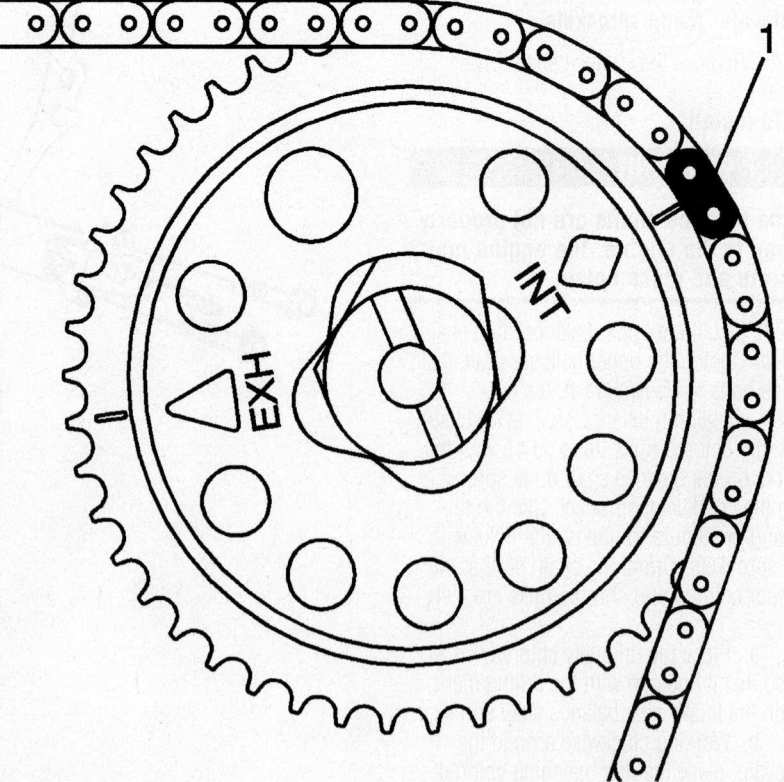

06025-COBALT-G46

Install the intake camshaft sprocket with the INT diamond at the 2 o'clock position

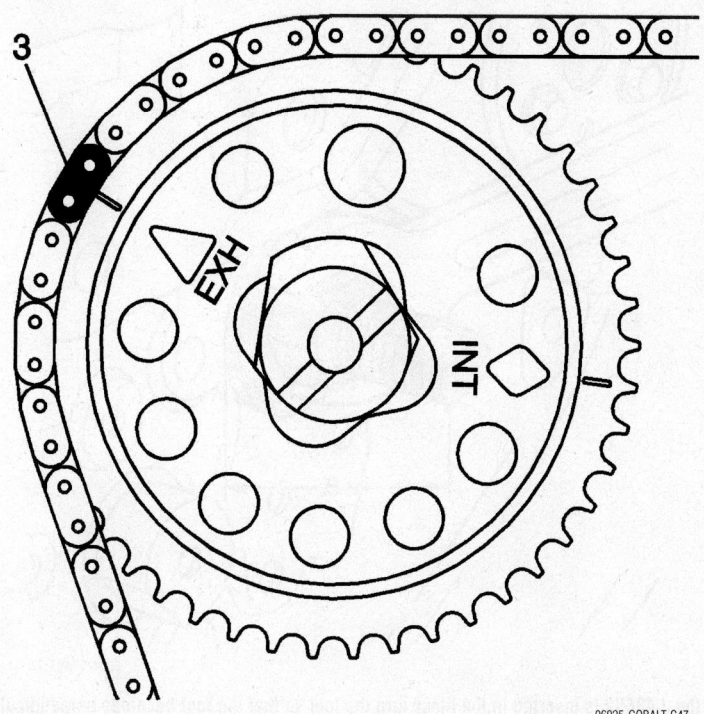

06025-COBALT-G47

Install the exhaust camshaft sprocket with the timing chain matching colored link (3) at EXH triangle aligned at the 10 o'clock position

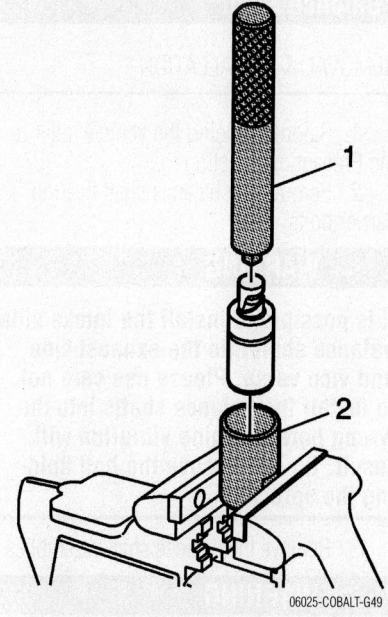

06025-COBALT-G49

Compressing the timing chain tensioner

crankshaft sprocket with the matching colored link aligning with the timing mark.

41. Route the timing chain around the intake camshaft sprocket with the uniquely colored link (1) aligning with the INT diamond.

42. Install the timing chain tensioner guide through the opening in the top of the cylinder head. Tighten the timing chain tensioner guide bolt to 10 Nm (89 inch lbs.).

43. Install the exhaust camshaft sprocket with the timing chain matching colored link (3) at EXH triangle aligned at the 10 o'clock position.

44. Use a 24 mm wrench to rotate the camshaft slightly, until exhaust sprocket aligns with the camshaft.

➡ **Always install NEW sprocket bolts.**

45. Hand-tighten the NEW exhaust camshaft sprocket bolt.

46. Install the fixed timing chain guide. Tighten the fixed timing chain bolts to 10 Nm (89 inch lbs.).

47. Apply sealant, GM P/N 12378521 (Canadian P/N 88901148) compound to thread and install the timing chain guide bolt access hole plug. Tighten the chain guide plug to 90 Nm (59 ft. lbs.).

48. Install the timing chain upper guide. Tighten the timing chain upper guide bolts to 10 Nm (89 inch lbs.).

49. Inspect the timing chain tensioner. If

the timing chain tensioner, O-ring seal, or washer is damaged, replace the timing chain tensioner.

50. Measure the timing chain tensioner assembly from end to end. A new tensioner should be supplied in the fully compressed non-active state. A tensioner in the compressed state will measure 72 mm (2.83 in) (a) from end to end. A tensioner in the active state will measure 85 mm (3.35 in) (a) from end to end.

51. If the timing chain tensioner is not in the compressed state, perform the following steps:

    a. Remove the piston assembly from the body of the timing chain tensioner by pulling it out.

    b. Install the J 45027-2 (2) into a vise.

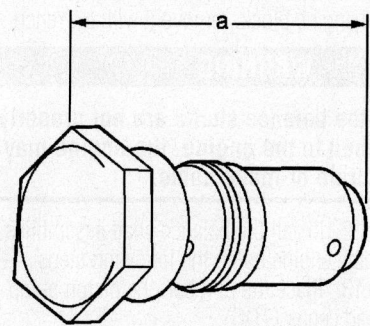

06025-COBALT-G48

Measure the timing chain tensioner assembly from end to end

    c. Install the notch end of the piston assembly into the J 45027-2 (2).

    d. Using the J 45027-1 (1), turn the ratchet cylinder into the piston.

52. Inspect the bore of the tensioner body for dirt, debris, and damage. If any damage appears, replace the tensioner. Clean dirt or debris out with a lint-free cloth.

53. Install the compressed piston assembly back into the timing chain tensioner body until it stops at the bottom of the bore. Do not compress the piston assembly against the bottom of the bore. If the piston assembly is compressed against the bottom of the bore, it will activate the tensioner, which will then need to be reset again.

54. At this point the tensioner should measure approximately 72 mm (2.83 in) (a) from end to end. If the tensioner does not read 72 mm (2.83 in) (a) from end to end repeat steps 26.1 through 26.4.

55. Install the timing chain tensioner. Tighten the timing chain tensioner to 75 Nm (55 ft. lbs.).

56. Use a suitable tool with a rubber tip on the end. Feed the tool down through the camshaft drive chain to rest on the timing chain. Then give a sharp jolt diagonally downwards to release the tensioner.

57. Use a 24 mm wrench to hold the camshaft. Tighten the NEW camshaft bolts to 85 Nm (63 ft. lbs.) plus 30 degrees.

58. Install the camshaft cover.

59. Raise the vehicle.

60. Install the engine front cover.

61. Lower the vehicle.

Balance Shaft

REMOVAL & INSTALLATION

1. Before servicing the vehicle, refer to the Precautions Section.
2. Remove the balance shaft bearing carrier bolts.

✳✳ WARNING

It is possible to install the intake side balance shaft into the exhaust side and vice versa. Please use care not to install the balance shafts into the wrong bores. Engine vibration will result. Do not remove the bolt holding the sprocket.

3. Remove the balance shaft assemblies.

✳✳ WARNING

Proper centering of the tool is required on the balance shaft bushing. If the tool is not properly centered then damage to the bearing bore and block will occur.

4. Install tool J 43650 into the balance shaft hole. Insert the tool with the foot parallel to the shaft.
5. When the J 43650 is inserted in the block turn the tool so that the foot becomes perpendicular to the shaft.

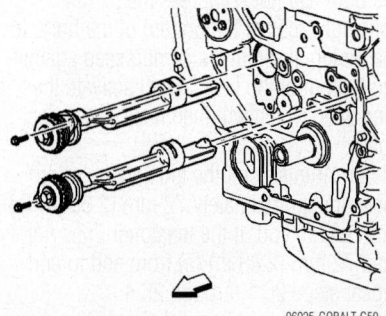

Balance shaft assemblies

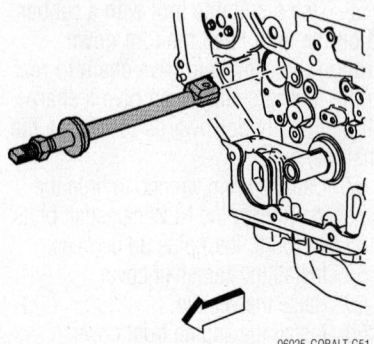

06025-COBALT-G51

Tool J43650 installed

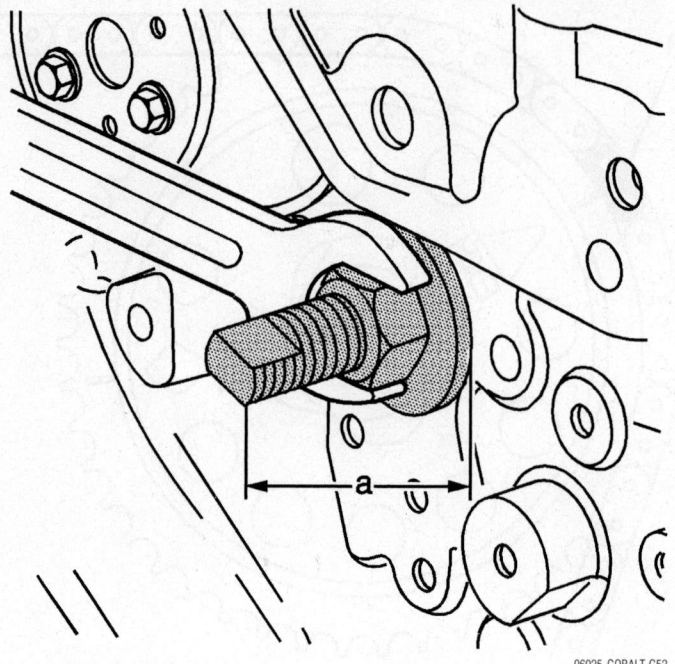

06025-COBALT-G52

When the J 43650 is inserted in the block turn the tool so that the foot becomes perpendicular to the shaft

6. Center the foot of the tool on the balance shaft bushing.
7. Once the tool is centered on the balance shaft bushing, then insert the centering guide into the front balance shaft bore and tighten the nut with an appropriate wrench. When the tool is properly installed, before removing the bushing, the end of the tool should be 116 mm (4.6 in) (a) from the block face. If the tool is less than approximately 114 mm (4.5 in) (a), recheck the tool alignment.
8. Tighten the nut on the tool until the tension releases. When the tension releases, remove the tool and the balance shaft bushing.
   To install:
9. Install the balance shaft bushing using the J 43650.
10. Seat the balance shaft bushing into the bore using the J 43650 and a wrench.
11. When the J 43650 is fully seated in the engine block, remove it with a wrench.

✳✳ WARNING

If the balance shafts are not properly timed to the engine, the engine may vibrate or make noise.

12. Install the balance shaft assemblies to the engine using the following steps:
13. Place the number one piston at top dead center (TDC).
14. Lubricate the balance shaft lobes with engine oil.

15. Install the balance shafts into their bores.
16. Install the balance shaft retaining bolts. Tighten the balance shaft retaining bolts to 10 Nm (89 inch lbs.).

Piston and Ring

POSITIONING

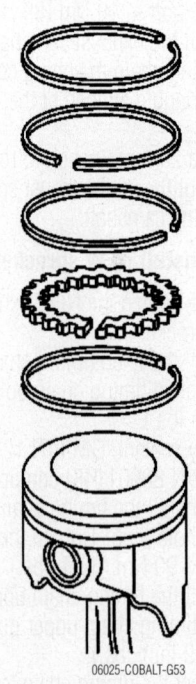

06025-COBALT-G53

Piston and ring assembly

FUEL SYSTEM

Fuel System Service Precautions

Safety is the most important factor when performing not only fuel system maintenance but any type of maintenance. Failure to conduct maintenance and repairs in a safe manner may result in serious personal injury or death. Maintenance and testing of the vehicle's fuel system components can be accomplished safely and effectively by adhering to the following rules and guidelines.

• To avoid the possibility of fire and personal injury, always disconnect the negative battery cable unless the repair or test procedure requires that battery voltage be applied.

• Always relieve the fuel system pressure prior to disconnecting any fuel system component (injector, fuel rail, pressure regulator, etc.), fitting or fuel line connection. Exercise extreme caution whenever relieving fuel system pressure, to avoid exposing skin, face and eyes to fuel spray. Please be advised that fuel under pressure may penetrate the skin or any part of the body that it contacts.

• Always place a shop towel or cloth around the fitting or connection prior to loosening to absorb any excess fuel due to spillage. Ensure that all fuel spillage (should it occur) is quickly removed from engine surfaces. Ensure that all fuel soaked cloths or towels are deposited into a suitable waste container.

• Always keep a dry chemical (Class B) fire extinguisher near the work area.

• Do not allow fuel spray or fuel vapors to come into contact with a spark or open flame.

• Always use a back-up wrench when loosening and tightening fuel line connection fittings. This will prevent unnecessary stress and torsion to fuel line piping. Always follow the proper torque specifications.

• Always replace worn fuel fitting O-rings with new ones. Do not substitute fuel hose, or equivalent, where fuel pipe is installed.

✲✲ CAUTION

In order to reduce the risk of fire and personal injury observe the following items:

• Replace all nylon fuel pipes that are nicked, scratched or damaged during installation, do not attempt to repair the sections of the nylon fuel pipes

• Do not hammer directly on the fuel harness body clips when installing new fuel pipes. Damage to the nylon pipes may result in a fuel leak.

• Always cover nylon vapor pipes with a wet towel before using a torch near them. Also, never expose the vehicle to temperatures higher than 115°C (239°F) for more than one hour, or more than 90°C (194°F) for any extended period.

• Apply a few drops of clean engine oil to the male pipe ends before connecting fuel pipe fittings. This will ensure proper reconnection and prevent a possible fuel leak. (During normal operation, the O-rings located in the female connector will swell and may prevent proper reconnection if not lubricated.)

Relieving Fuel System Pressure

2.0L Engine

1. Before servicing the vehicle, refer to the Precautions Section.

✲✲ CAUTION

Remove the fuel tank cap and relieve the fuel system pressure before servicing the fuel system in order to reduce the risk of personal injury. After you relieve the fuel system pressure, a small amount of fuel may be released when servicing the fuel lines, the fuel injection pump, or the connections. In order to reduce the risk of personal injury, cover the fuel system components with a shop towel before disconnection. This will catch any fuel that may leak out. Place the towel in an approved container when the disconnection is complete.

2. Turn the ignition OFF.
3. Disconnect the battery negative cable in order to avoid possible fuel discharge if an accidental attempt is made to start the engine.
4. Loosen the fuel filler cap to relieve the fuel tank vapor pressure.
5. Remove the cap from the fuel pressure service port.

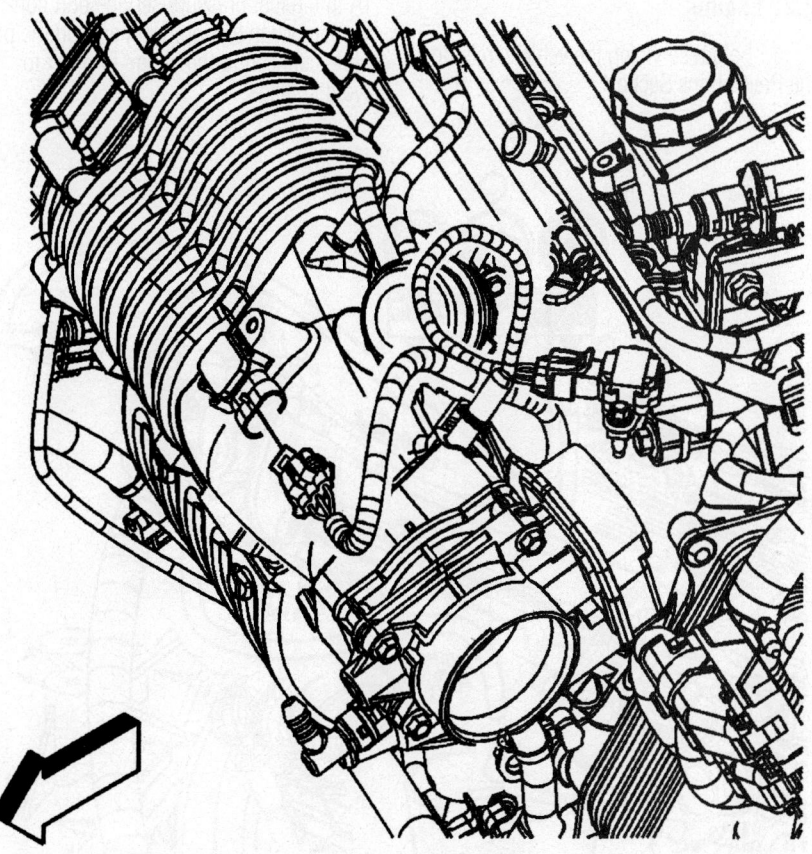

06025-COBALT-G54

Fuel pressure service port connection—2.0L engine

6. Remove the engine identification cover nuts and cover.

7. Connect fuel pressure gauge J 34730-1A, or equivalent, to the fuel pressure service port connection. Wrap a shop towel around the port while connecting the gage in order to avoid spillage.

❋❋ CAUTION

Do not drain the fuel into an open container. Never store the fuel in an open container due to the possibility of a fire or an explosion.

8. Install the bleed hose of the gauge into an approved fuel container.

9. Open the bleed valve on the gauge in order to bleed the fuel system pressure. The fuel connections are now safe for servicing.

10. Place a shop towel under the fuel pressure service port to catch any remaining fuel spillage.

11. Disconnect the gauge from the fuel pressure service port connection.

12. Drain any fuel remaining in the gage into an approved fuel container.

13. Install the cap to the fuel pressure service port.

2.2L Engine

1. Before servicing the vehicle, refer to the Precautions Section.

❋❋ CAUTION

Remove the fuel tank cap and relieve the fuel system pressure before servicing the fuel system in order to reduce the risk of personal injury. After you relieve the fuel system pressure, a small amount of fuel may be released when servicing the fuel lines, the fuel injection pump, or the connections. In order to reduce the risk of personal injury, cover the fuel system components with a shop towel before disconnection. This will catch any fuel that may leak out. Place the towel in an approved container when the disconnection is complete.

2. Turn the ignition OFF.

3. Disconnect the battery negative cable in order to avoid possible fuel discharge if an accidental attempt is made to start the engine.

4. Loosen the fuel filler cap to relieve the fuel tank vapor pressure.

5. Remove the cap from the fuel pressure service port.

6. Connect tool SA9127E or J 34730-1A to the fuel pressure service port connection. Wrap a shop towel around the port while connecting the gauge in order to avoid spillage.

❋❋ CAUTION

Do not drain the fuel into an open container. Never store the fuel in an open container due to the possibility of a fire or an explosion.

7. Install the bleed hose into an approved fuel container.

8. Open the bleed valve on the in order to bleed the fuel system pressure. The fuel connections are now safe for servicing.

9. Place a shop towel under the fuel pressure service port to catch any remaining fuel spillage.

10. Disconnect the gauge from the fuel pressure service port connection.

11. Drain any fuel remaining in the gauge into an approved fuel container.

12. Install the cap to the fuel pressure service port.

Quick-Connect Fittings

OPENING AND CLOSING

Metal Collar

1. Before servicing the vehicle, refer to the Precautions Section.

2. Before servicing the vehicle, refer to the Precautions Section.

➡**Tool Required: J 37088-A Fuel Line Disconnect Tool Set, or equivalent.**

3. Relieve the fuel system pressure before servicing any fuel system connection.

4. Remove the retainer from the quick-connect fitting.

❋❋ CAUTION

Wear safety glasses when using compressed air, as flying dirt particles may cause eye injury.

5. Blow dirt out of the fitting using compressed air.

6. Choose the correct tool from the set for the size of the fitting. Insert the tool into

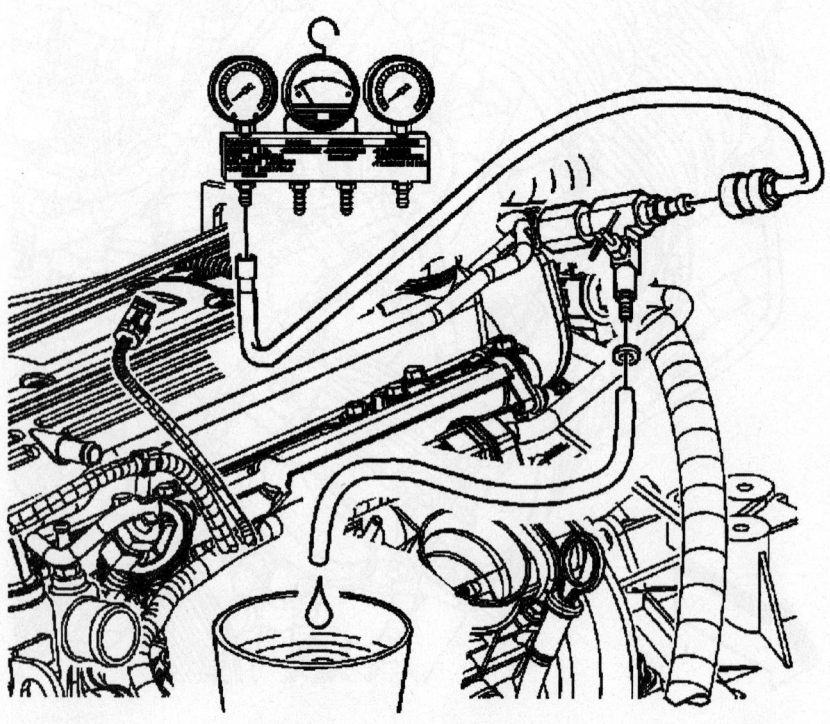

06025-COBALT-G55

Fuel bleed procedure—2.2L engine

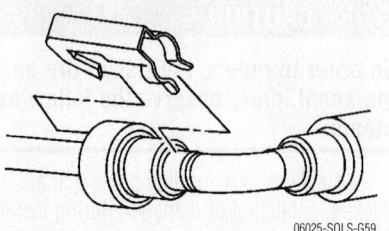

06025-SOLS-G59

Remove the retainer from the quick-connect fitting—metal collar type

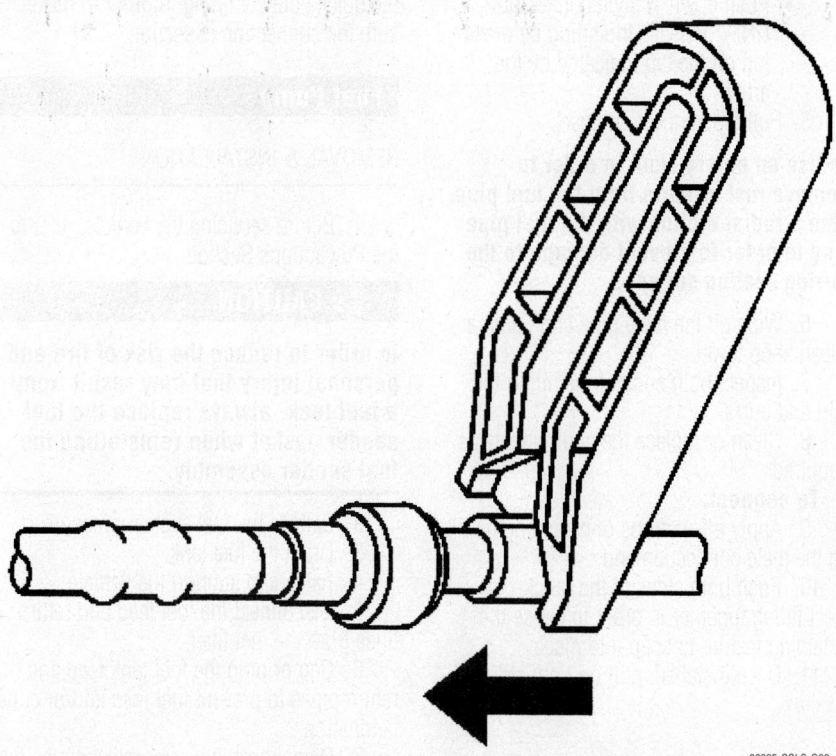

Insert the tool into the female connector, then push inward in order to release the locking

06025-SOLS-G60

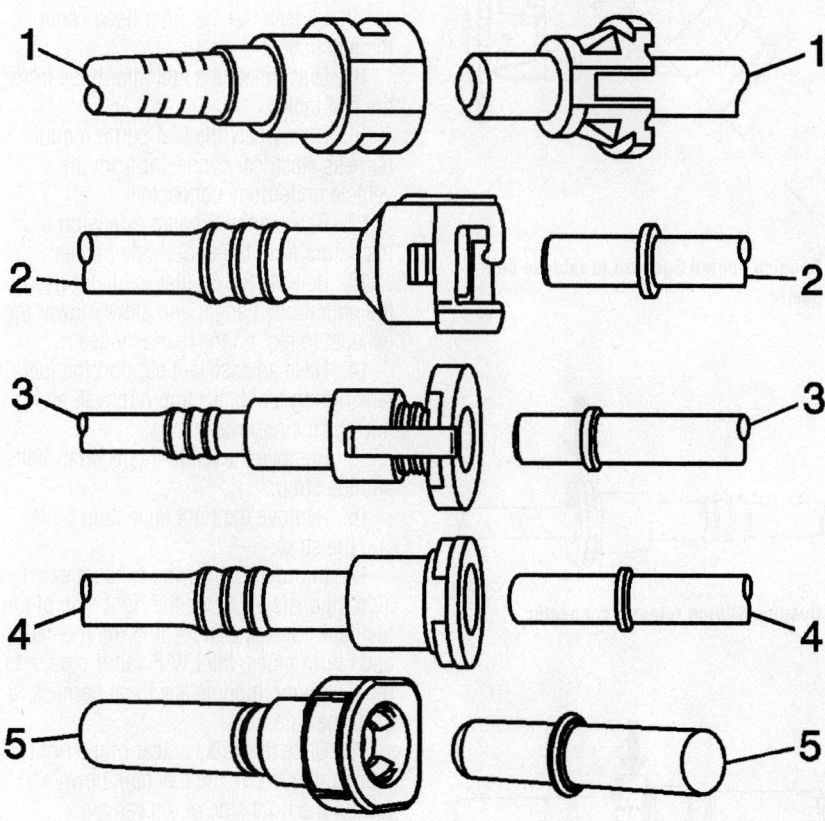

06025-SOLS-G61

Plastic collar type fittings

the female connector, then push inward in order to release the locking tabs.

7. Pull the connection apart.

➡ **If necessary, remove rust or burrs from the fuel pipes with an emery cloth. Use a radial motion with the fuel pipe end in order to prevent damage to the O-ring sealing surface. Use a clean shop towel in order to wipe off the male tube ends. Inspect all the connections for dirt and burrs. Clean or replace the components and assemblies as required.**

8. Use a clean shop towel in order to wipe off the male pipe end.

9. Inspect both ends of the fitting for dirt and burrs. Clean or replace the components as required.

To connect

✷✷ CAUTION

In order to reduce the risk of fire and personal injury, before connecting fuel pipe fittings, always apply a few drops of clean engine oil to the male pipe ends. This will ensure proper reconnection and prevent a possible fuel leak. During normal operation, the O-rings located in the female connector will swell and may prevent proper reconnection if not lubricated.

10. Apply a few drops of clean engine oil to the male pipe end.

11. Push both sides of the fitting together in order to snap the retaining tabs into place.

12. Once installed, pull on both sides of the fitting in order to make sure the connection is secure.

13. Install the retainer to the quick-connect fitting.

Plastic Collar

1. Before servicing the vehicle, refer to the Precautions Section.

2. Before servicing the vehicle, refer to the Precautions Section.

➡ **There are several types of plastic collar fuel and evaporative emission quick connect fittings that may be used on this vehicle.**

- Bartholomew (1)
- Q release (2)
- Squeeze to release (3)
- Sliding retainer (4)
- Push down TI (5)

The following instructions apply to all of these types of fittings except where indicated.

3. Relieve the fuel system pressure.

Wear safety glasses when using compressed air in order to prevent eye injury.

4. Using compressed air, blow any dirt out of the quick-connect fitting.

- Bartholomew style connectors ONLY: Squeeze the plastic quick-connect fitting release tabs.
- Q release style connector ONLY: Release the fitting by pushing the tab toward the other side of the slot in the fitting.
- Squeeze to release style connectors ONLY: Squeeze where indicated by arrows on both sides of the plastic ring surrounding the quick connect fitting.
- Sliding retainer style connectors ONLY: Release the fitting by pressing on one side of the release tab causing it to push in slightly. If the tab does not move, try pressing the tab in from the opposite side. The tab will only move in one direction.

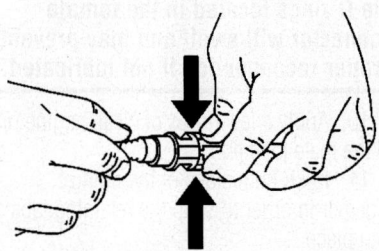

Opening Bartholomew connector

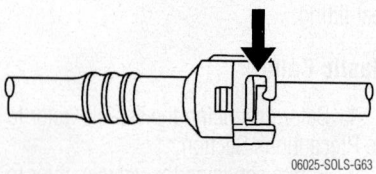

Opening Q release connector

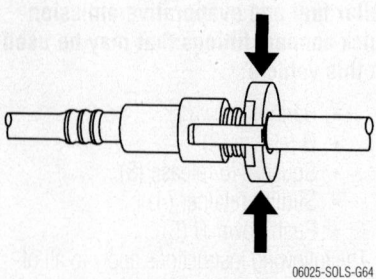

Opening straight Squeeze to release connector

- Push down TI style connectors ONLY: Release the fitting by pressing on the tab indicated by the arrow.

5. Pull the connection apart.

➡**Use an emery cloth in order to remove rust or burrs from the fuel pipe. Use a radial motion with the fuel pipe end in order to prevent damage to the O-ring sealing surface.**

6. Wipe off the male pipe end using a clean shop towel.

7. Inspect both ends of the fitting for dirt and burrs.

8. Clean or replace the components as required.

To connect:

9. Apply a few drops of clean engine oil to the male connection end.

10. Push both sides of the quick-connect fitting together in order to cause the retaining feature to snap into place.

11. Once installed, pull on both sides of

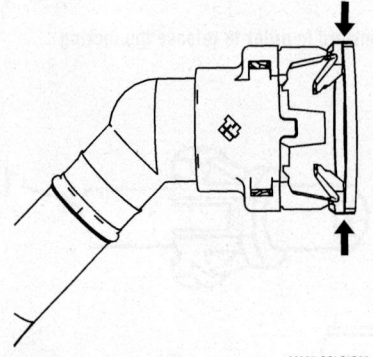

Opening angled Squeeze to release connector

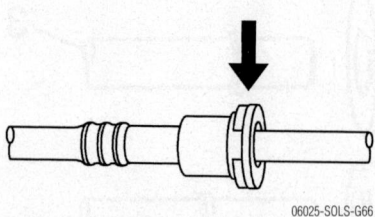

Opening Sliding retainer connector

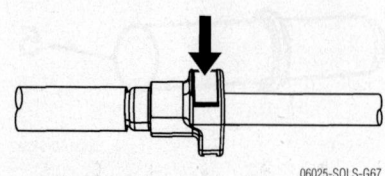

Opening TI connector

the quick connect fitting in order to make sure the connection is secure.

REMOVAL & INSTALLATION

1. Before servicing the vehicle, refer to the Precautions Section.

In order to reduce the risk of fire and personal injury that may result from a fuel leak, always replace the fuel sender gasket when reinstalling the fuel sender assembly.

2. Relieve the fuel system pressure.

3. Drain the fuel tank.

4. Raise and support the vehicle.

5. Disconnect the fuel feed and return lines from the fuel filter.

6. Cap or plug the fuel tank feed and return pipes to prevent fuel loss and/or contamination.

7. Disconnect the evaporative emission (EVAP) pipe fittings, for access to disconnect the fuel filler hose from the tank.

8. Cap or plug the EVAP purge and vapor pipes to prevent contamination.

9. Loosen the fuel filler hose clamp at the fuel tank.

10. Disconnect the fuel filler hose from the fuel tank.

11. Disconnect the fuel pump module harness electrical connector from the vehicle underbody connector.

12. Release the exhaust extension pipe insulators from the underbody hangers.

13. Release the muffler insulator from the underbody hanger and slowly lower the exhaust to rest on the rear axle beam.

14. Have an assistant support the fuel tank during fuel tank strap removal, and during tank removal.

15. Remove the left fuel tank strap bolts and the strap.

16. Remove the right tank strap bolts and the strap.

17. In order to clear the exhaust extension pipe, slowly lower the right side of the fuel tank. Use care in feeding the fuel feed and return pipes, the EVAP vapor pipe, and the fuel pump module electrical harness to clear the axle.

18. Once the tank is clear of the right frame rail, remove the fuel tank down and toward the right side of the vehicle.

➡**Take note of the positioning of the fuel tank rear shield prior to releasing the pump module pipe retainer.**

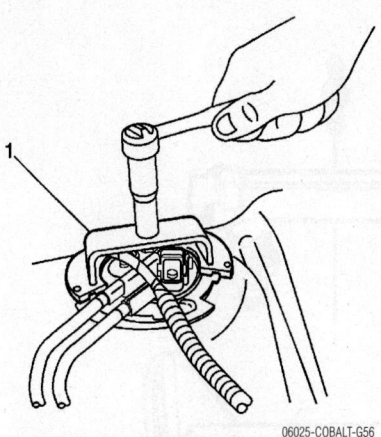

Using tool J 39765 (1), or equivalent, carefully rotate to release the fuel pump module retaining lock ring

19. Release the retaining tab on the fuel tank retainer used to secure the fuel pump module pipes in position on the tank.

20. Release the fuel pump module electrical harness from the retaining slot on the tank.

21. Disconnect the fuel pump module harness electrical connector from the fuel tank pressure sensor.

22. Using tool J 39765, or equivalent, carefully rotate to release the fuel pump module retaining lock ring.

23. Remove the fuel pump module retaining lock ring, by sliding the ring over the module pipes and electrical harness.

24. Slowly raise the fuel pump module assembly until the fuel level sensor float arm is just visible. Ensure that the fuel level sensor harness connector clears the tank opening.

✶✶ CAUTION

When removing the fuel pump module assembly from the fuel tank, be aware that the pump module reservoir bowl is full of fuel. The reservoir must be tipped slightly during removal to avoid bending the fuel level sensor float arm.

25. Tilt the pump module toward the rear of the fuel tank to enable the level sensor float arm to clear the tank opening. Remove the pump module from the tank.

26. Carefully discard the fuel in the pump module reservoir bowl into an approved fuel container.

✶✶ WARNING

Do NOT reuse the old fuel pump module-to-fuel tank seal.

27. Remove and discard the fuel pump module-to-fuel tank seal.

To install:

28. Install a NEW fuel pump module-to-fuel tank seal.

➡**The reservoir must be tipped slightly during installation to avoid bending the fuel level sensor float arm.**

29. Tilt the pump module toward the rear of the fuel tank to enable the level sensor float arm to clear the tank opening. Install the fuel pump module to the fuel tank.

30. Slowly lower the fuel pump module assembly into the tank. Ensure that the fuel level sensor harness connector is positioned properly.

31. Install the fuel pump module retaining lock ring over the module pipes and electrical harness, and into position on the top of the module.

➡**Ensure that the fuel pump module retaining lock ring is fully seated within the fuel tank retaining tab slots.**

32. Carefully rotate to fully secure the fuel pump module retaining lock ring.

33. Connect the fuel pump module harness electrical connector to the fuel tank pressure sensor.

34. Secure the fuel pump module electrical harness into the retaining slot on the tank.

➡**Ensure that the fuel tank rear shield is positioned properly on the pump module pipe retainer.**

35. Secure the retaining tab on the fuel tank retainer used to secure the fuel pump module pipes in position on the tank.

36. Have an assistant support the fuel tank during fuel tank and fuel tank strap installation.

37. Begin to install the left side of the fuel tank over the exhaust pipe.

38. Raise the right side of the fuel tank into position inboard of the right frame rail. Use care in feeding the fuel feed and return pipes, the EVAP vapor pipe, and the fuel pump module electrical harness over the rear axle.

39. Install the right fuel tank strap and strap bolts.

40. Install the left fuel tank strap and strap bolts.

41. Tighten the fuel tank strap bolts. Tighten the bolts to 25 Nm (18 ft. lbs.).

42. Raise the exhaust into position and install the muffler insulators to the underbody hanger.

43. Install the exhaust extension pipe insulators to the underbody hangers.

44. Connect the fuel pump module harness electrical connector to the vehicle underbody connector.

45. Connect the fuel filler hose to the fuel tank.

46. Tighten the fuel filler hose clamp at the fuel tank. Tighten the clamp to 4.5 Nm (40 inch lbs.).

47. Remove the caps or plugs from the EVAP purge and vapor pipes.

48. Connect the EVAP purge pipe and vapor pipe to the EVAP canister.

49. Remove the caps or plugs from the fuel tank feed and return pipes.

50. Connect the fuel feed and return lines to the fuel filter.

51. Lower the vehicle.

52. Refill the fuel tank.

53. Connect the negative battery cable.

54. Inspect for fuel leaks using the following procedure:

   a. Turn ON the ignition, with the engine OFF for 2 seconds.

   b. Turn OFF the ignition for 10 seconds.

   c. Turn ON the ignition, with the engine OFF.

   d. Inspect for fuel leaks.

Fuel Rail and Injectors

REMOVAL & INSTALLATION

2.0L Engine

1. Before servicing the vehicle, refer to the Precautions Section.

2. Remove the coolant overflow pipe.

3. Relieve the fuel system pressure.

4. Use a back up wrench on the fuel rail and disconnect the fuel supply pipe.

5. Remove the fuel rail attaching studs.

✶✶ WARNING

Use care when removing the fuel rail assembly in order to prevent damage to the fuel injectors electrical connector terminals and spray tips.

6. Remove the fuel rail using the following procedure:

   a. Pull the fuel rail back and upward to remove the fuel injectors from

   b. the cylinder head ports.

   c. Rotate the fuel rail in order to position the injectors downward.

   d. Remove the fuel rail.

7. Disconnect the fuel injector harness connectors.

8. Remove the fuel injector retainer clip.

9. Remove the fuel injectors from the fuel rail.

➡**Visually inspect the fuel injector in order to determine if the upper O-ring was also removed. If the upper O-ring is not removed, remove the O-ring from the fuel rail assembly.**

10. Remove and discard the fuel injector O-rings.

To install:

➡**Always install new injector O-rings when servicing the fuel injectors. Lubricate the new injector O-rings with clean engine oil**

11. Install the O-rings on the fuel injector.

12. Install the fuel injector clip on the fuel injector.

➡**The fuel injector will click when the injector is installed correctly.**

13. Install the fuel injector in the fuel rail with the connector facing upward.

14. Connect the fuel injector harness connectors. Pull back to insure the connectors are locked in place.

➡**Install new lower O-rings when reusing fuel injectors. Lubricate the injector tip O-rings prior to installing the injectors into the intake manifold.**

15. Install the fuel rail using the following procedure:
   a. With the fuel injectors positioned downward, lower the fuel injectors into the cylinder head ports.
   b. Align the injectors by rotating the fuel rail forward.
   c. Carefully push the fuel injectors into the cylinder head ports.

16. Install the fuel rail attaching studs. Tighten the fuel rail studs to 10 Nm (89 inch lbs.).

17. Install the fuel supply pipe. Using a backup wrench on the fuel rail tighten the fuel supply pipe to 14 Nm (10 ft. lbs.).

18. Install the coolant overflow pipe. Tighten the pipe bolt to 8 Nm (71 inch lbs.).

19. Connect the negative battery cable.

20. Inspect for fuel leaks using the following procedure:
   a. Turn ON the ignition, with the engine OFF for 2 seconds.
   b. Turn OFF the ignition for 10 seconds.
   c. Turn ON the ignition.
   d. Inspect for fuel leaks.

2.2L Engine

1. Before servicing the vehicle, refer to the Precautions Section.

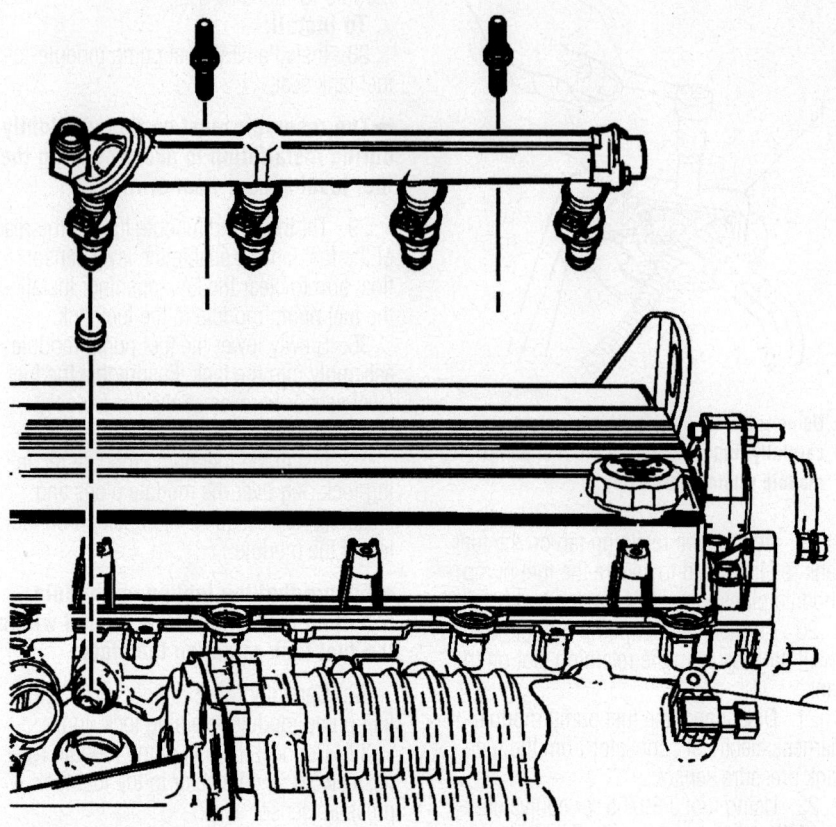

06025-COBALT-G57

Fuel rail and injectors—2.0L engine

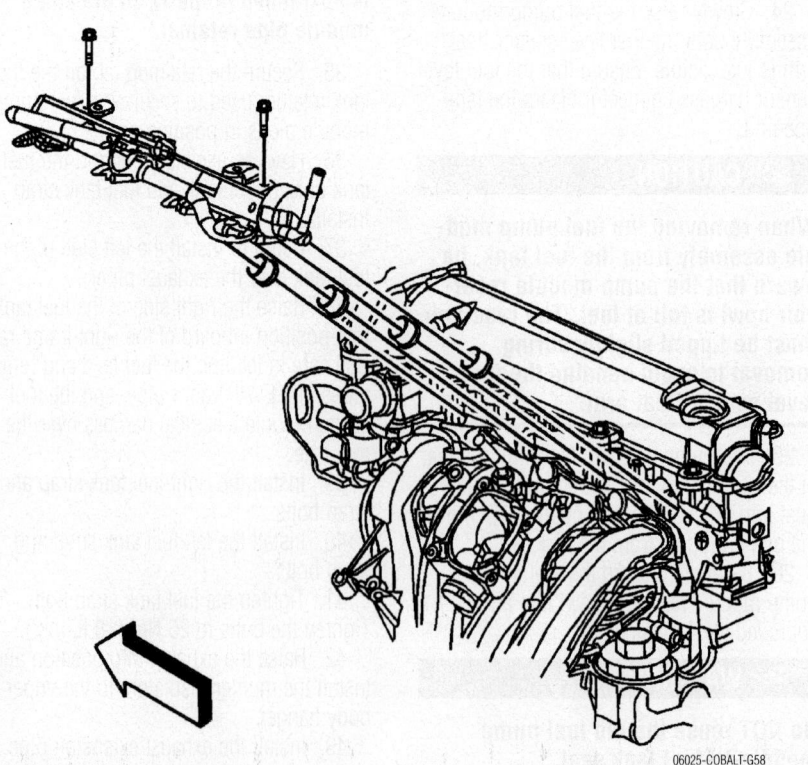

06025-COBALT-G58

Fuel rail removal—2.2L engine

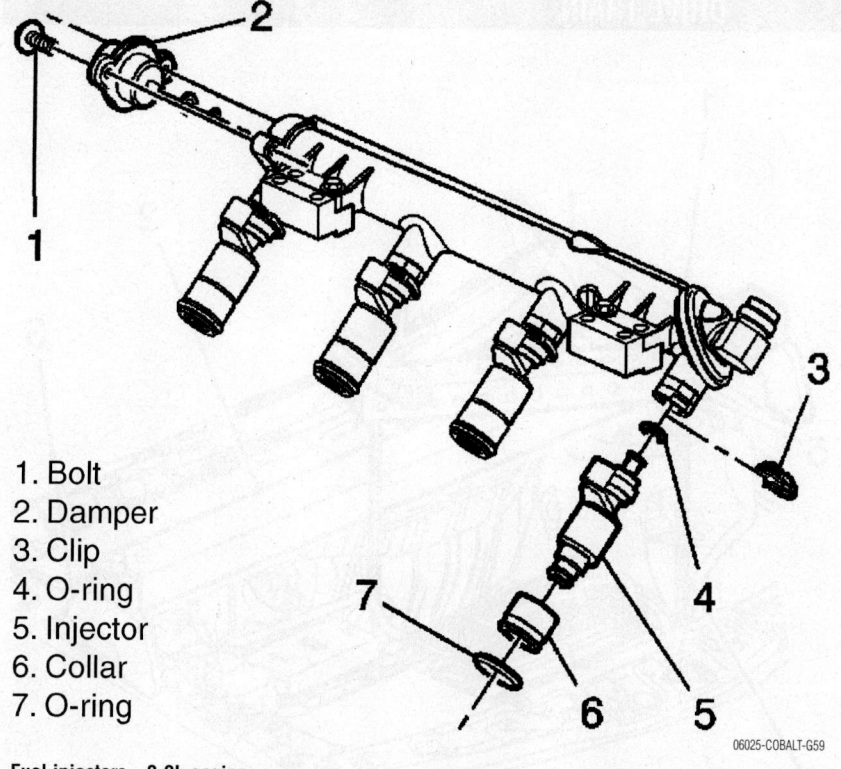

1. Bolt
2. Damper
3. Clip
4. O-ring
5. Injector
6. Collar
7. O-ring

Fuel injectors—2.2L engine

06025-COBALT-G59

2. Relieve the fuel system pressure.

3. Remove the air cleaner outlet resonator.

4. Disconnect the fuel line fitting.

5. Disconnect the fuel injector harness connectors.

6. Remove the fuel rail attaching studs.

➡**Use care when removing the fuel rail assembly in order to prevent damage to the fuel injectors electrical connector terminals and spray tips.**

7. Remove the fuel rail using the following procedure:

a. Pull the fuel rail back and upward to remove the fuel injectors from the cylinder head ports.

b. Rotate the fuel rail in order to position the injectors downward.

c. Remove the fuel rail.

8. Remove the fuel injector retainer clip.

9. Remove the fuel injectors from the fuel rail.

➡**Visually inspect the fuel injector in order to determine if the upper O-ring was also removed. If the upper O-ring is not removed, remove the O-ring from the fuel rail assembly.**

10. Remove and discard the fuel injector O-rings.

To install:

➡**Always install new injector O-rings when servicing the fuel injectors.**

Lubricate the new injector O-rings with clean engine oil.

11. Install the O-rings on the fuel injector.

12. Install the fuel injector clip on the fuel injector.

➡**The fuel injector will click when the injector is installed correctly.**

13. Install the fuel injector in the fuel rail with the connector facing upward.

➡**Install new lower O-rings when reusing fuel injectors. Lubricate the injector tip O-rings prior to installing the injectors into the intake manifold.**

14. Install the fuel rail using the following procedure:

a. With the fuel injectors positioned downward, lower the fuel injectors into the cylinder head ports.

b. Align the injectors by rotating the fuel rail forward.

c. Carefully push the fuel injectors into the cylinder head ports.

15. Install the fuel rail attaching studs. Tighten the fuel rail studs to 10 Nm (89 inch lbs.).

16. Connect the fuel injector harness connectors. Pull back to insure the connectors are locked in place.

17. Connect the fuel line fitting.

18. Install the air cleaner outlet resonator.

19. Connect the negative battery cable..

20. Inspect for fuel leaks using the following procedure:

a. Turn ON the ignition, with the engine OFF for 2 seconds.

b. Turn OFF the ignition for 10 seconds.

c. Turn ON the ignition.

d. Inspect for fuel leaks.

DRIVE TRAIN

Automatic Transaxle

REMOVAL & INSTALLATION

1. Before servicing the vehicle, refer to the Precautions Section.

2. Disconnect the negative battery cable.

3. Disconnect the air inlet duct hose from the intake plenum.

4. Disconnect the transaxle wiring harness from the transaxle and the PNP switch.

5. Remove the upper transmission to engine bolts and stud.

6. Install the engine support fixture.

    a. Place the engine support fixture long bar (2), from kit SA9105E across the engine compartment.

    b. Install the engine support fixture legs (3) from the kit on the engine support fixture long bar and center above the engine.

    c. Install the engine support fixture hooks (4) and engine support fixture handle to the engine support cross bar.

    d. Place the engine support cross bar over the engine support long bar and connect the hooks to the engine lift brackets.

7. Raise and support the vehicle.

8. Remove the front wheel and tire assemblies.

9. Remove the both front fender liners.

10. Remove the steering gear mounting bolts and secure the steering gear with mechanic's wire.

11. Disconnect the wheel speed sensor wires from the both front wheels and unclip from the frame.

12. Separate the ball joints from the steering knuckles.

13. With the wheels in the straight ahead position, remove the key from the ignition switch.

14. Secure the cooling module to the upper body structure.

15. Remove the left and right splash shields.

16. Remove the front transaxle mount to cradle through bolt.

17. Remove the rear transaxle mount to frame bolts.

18. Remove both stabilizer link to stabilizer shaft nuts.

19. Remove both tie rod to steering knuckle nuts.

20. Separate the outer tie rods from the steering knuckles.

21. Remove the intermediate steering

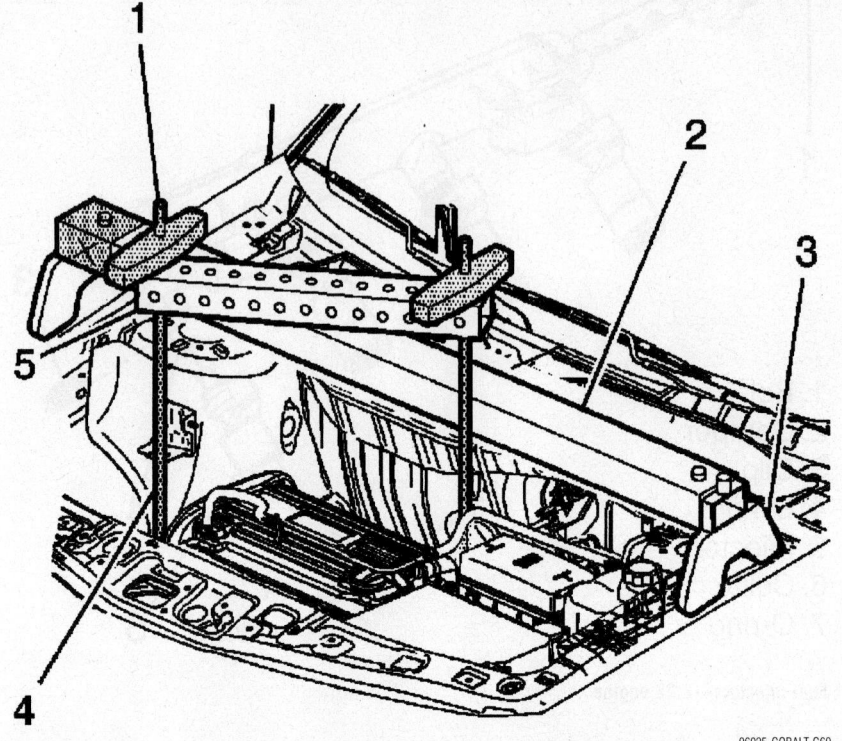

Engine support fixture installed—2.2L engine

06025-COBALT-G60

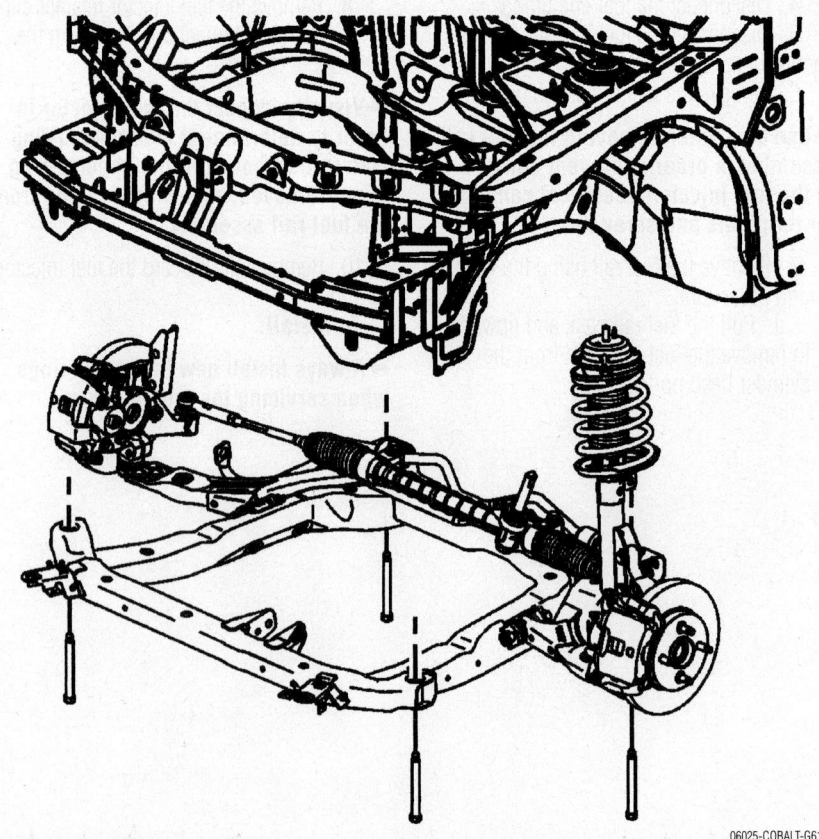

Frame assembly removal

06025-COBALT-G61

shaft to steering gear pinch bolt and discard.

DO NOT rotate the intermediate shaft once separated from the gear. Possible damage or a malfunction could occur.

22. Disconnect the intermediate steering shaft from the steering gear.
23. Remove both lower control arm ball stud to steering knuckle pinch bolts.

Do not free the ball stud by using a pickle fork or a wedge-type tool. Damage to the seal or bushing may result.

24. Lower the lower control arms in order to disengage the steering knuckle.
25. Mark the frame to body position with a paint pen or permanent marker.
26. Lower the vehicle to about 3 feet off the ground in order to place a hydraulic lift table under the frame.
27. Use two 2 x 4's between the lift table and the frame and lift the table to the frame.
28. Slowly remove the frame bolts using the following sequence:
    a. Remove the front frame bolts.
    b. Partially unscrew the rear frame bolts exposing 25.4 mm (1.5 in.) of bolt shank.
29. Slowly lower the lift table to the floor.
30. Remove the 2 bolts from the transmission brace.
31. Disconnect the shift cable from the shift linkage.
32. Disconnect the cable from the bracket.
33. Remove the flywheel inspection cover.
34. Remove the starter.
35. Mark the relationship of the flywheel to the torque converter for reassembly.
36. Prevent the crankshaft from rotating.
37. Remove the torque converter to flywheel bolts.
38. Remove the transmission cooler lines by removing the nut holding the bracket to the transaxle case.
39. Disconnect the VSS wiring harness from the sensor.
40. Disconnect the halfshafts from the transaxle.
41. Lower the vehicle.
42. Remove the body to transmission mount bolts.
43. Lower the transmission with the

engine support fixture enough to remove the transmission.
44. Raise the vehicle.
45. Loosen the oil pan bolts.
46. Drain the oil. Use a suitable container to catch the transmission fluid.
47. Remove the oil pan attaching bolts.
48. Remove the oil pan to drain the transmission.
49. Install the oil pan gasket. Use a new gasket if the sealing ribs are damaged.
50. Install the oil pan.
51. Install the oil pan attaching bolts. Tighten the bolts to 10 Nm (89 inch lbs.).
52. Support the transmission with a suitable jack.
53. Remove the transaxle to engine nut.
54. Separate the engine and the transaxle.
55. Remove the transaxle from the vehicle.
56. Remove the PNP switch.
57. Remove the shifter cable bracket.
58. Remove the lower transmission to engine stud.
59. Remove the transmission mount.
60. Flush the transmission cooler and lines.

To install:
61. Install the PNP switch.
62. Install the shifter cable bracket.
63. Install the lower transmission to engine stud.
64. Install the transmission mount.
65. Position the transaxle in the vehicle.
66. Install the lower transmission to engine bolts and nuts. Tighten the bolts and nuts to 90 Nm (66 ft. lbs.).
67. Connect the transaxle cooler pipes to the transaxle. Tighten the cooler pipes to 8 Nm (71 inch lbs.).
68. Prevent the crankshaft from rotating.
69. Install the torque converter to flywheel bolts. Tighten the torque converter bolts to 62 Nm (46 ft. lbs.).
70. Install the halfshafts to the transaxle.
71. Connect the wiring harness to the VSS.
72. Install the starter.
73. Install the flywheel inspection cover bolts. Tighten the bolts to 10 Nm (89 inch lbs.).
74. Lower the vehicle.
75. Use the engine support fixture to raise the engine and transmission assembly.
76. Install the transaxle mount to body bolts. Tighten the bolts to 90 Nm (66 ft. lbs.).
77. Raise the vehicle.
78. With the frame on the lift table, raise the frame to the vehicle.

79. Hand-start all the frame bolts while aligning the frame to the paint marks.
80. Tighten the frame bolts. Tighten the bolts to 100 Nm plus 180 degrees (74 ft. lbs.) plus 180 degrees.
81. Lower and remove the hydraulic table.
82. Connect the lower control arm to the steering knuckle.

➡**The torque sequence must be followed in the order that is listed.**

83. Install the ball joint pinch bolt and nut.
    • First Pass: Tighten the nut to 50 Nm (37 ft. lbs.). Reverse nut ¾ turn
    • Second Pass: Tighten the nut to 50 Nm (37 ft. lbs.) plus 30 degrees.

➡**The front and rear transmission mounts must be allowed to settle with the through bolts loosened.**

84. Hand-start the front transaxle mount through bolt.
85. Loosen the rear transmission mount through bolt.
86. Tighten the rear transaxle mount to frame bolts. Tighten the rear bolts to 50 Nm (37 ft. lbs.).
87. Tighten the front and rear transaxle mount through bolts in the following order. Tighten the rear bolt to 100 Nm (74 ft. lbs.). Tighten the front bolt to 100 Nm (74 ft. lbs.).
88. Install the outer tie rods to the steering knuckles.
89. Install the new outer tie rod to the knuckle nuts. Tighten the nuts to 20 Nm (15 ft. lbs.) plus 180 degrees.
90. Connect the stabilizer links to the stabilizer shaft.
91. Connect the intermediate shaft to the steering gear.
92. Install a new intermediate shaft pinch bolt. Tighten the bolt to 34 Nm (25 ft. lbs.).
93. Install the left and right splash shields.
94. Install the front wheels.
95. Lower the vehicle.
96. Connect the ball joints to the steering knuckles.
97. Route and clip the wheel speed sensor wiring into the proper position on both sides.
98. Install the steering gear to the front suspension crossmember.
99. Install the steering gear bolts to the front suspension crossmember. Tighten the bolts to 110 Nm (81 ft. lbs.).
100. Install the transmission to engine brace bolts. Tighten the bolts to 72 Nm (53 ft. lbs.).

101. Install the both front fender liners.

102. Install the front wheel and tire assemblies.

103. Lower the vehicle.

104. Install the upper transmission to engine bolts and stud. Tighten all of the bolts to 90 Nm (66 ft. lbs.).

105. Install the engine wiring harness grounds to the transaxle to engine mount stud and nut. Tighten the nut to 8 Nm (71 inch lbs.).

106. Remove the engine support fixture.

107. Install the shift linkage to the transmission.

108. Connect the electrical connectors to the PNP switch and transaxle.

109. Connect the air duct hose to the intake plenum.

110. Connect the negative battery cable.

111. Inspect the transmission fluid level.

Manual Transaxle

REMOVAL & INSTALLATION

MU3 Transaxle

1. Before servicing the vehicle, refer to the Precautions Section.

2. Disconnect the negative battery cable.

3. Remove the cover from the underhood electrical center.

4. Release the forward lamp harness retainer from the ABS modulator bracket, or the proportioning valve bracket, to allow the underhood electrical center to be repositioned adequately.

5. Remove the underhood electrical center bracket from the vehicle and reposition the electrical center to access the bracket.

6. Release the wiring harness retainers above the brake booster, to allow the electrical center to be repositioned adequately.

7. Disconnect the hydraulic clutch hose from the clutch actuator cylinder and the clutch master cylinder.

8. Install the engine support fixture.

   a. Place the engine support fixture long bar (2), from the J 28467-B across the engine compartment.

   b. Install the engine support fixture legs (3) from the J 43405 on the engine support fixture long bar and center above the engine.

   c. Install the engine support fixture hooks (4) and engine support fixture handle to the engine support cross bar.

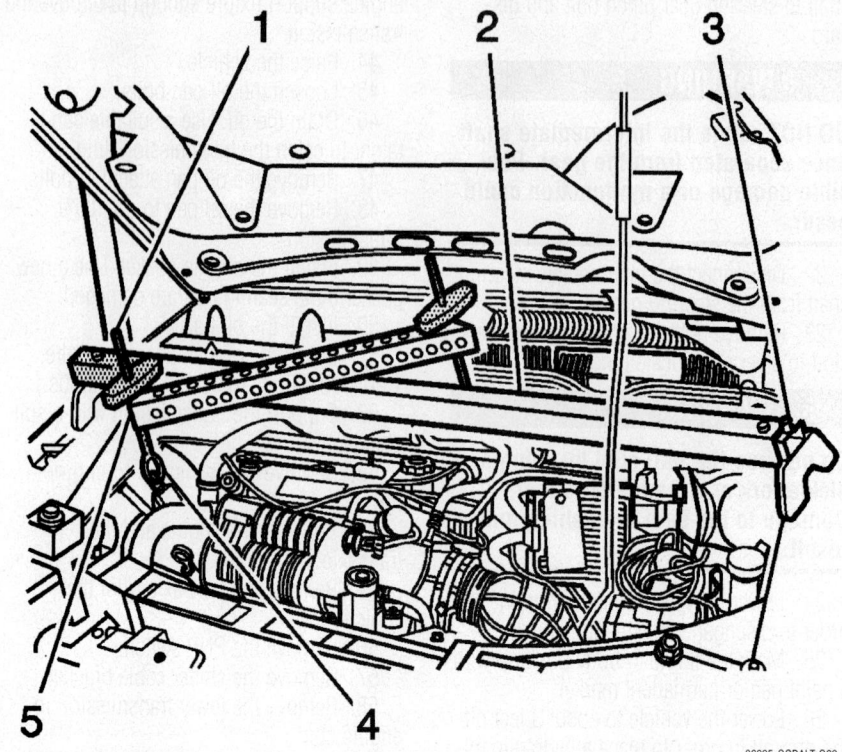

Engine support fixture installed—2.0L engine

06025-COBALT-G62

   d. Place the engine support cross bar over the engine support long bar and connect the hooks to the engine lift brackets.

9. Secure the cooling module to the upper body structure.

10. Remove the upper transmission to mount bolts.

11. Disconnect the wiring harness retainer from the transmission stud.

12. Remove the upper transmission to engine stud and bolt.

13. Remove the front wheel and tire assemblies.

14. Remove the both front fender liners.

15. Remove the steering gear mounting bolts and secure the steering gear with mechanic's wire.

16. Disconnect the wheel speed sensor wires from the both front wheels and unclip from the frame.

17. Separate the ball joints from the steering knuckles.

18. With the wheels in the straight ahead position, remove the key from the ignition switch.

19. Secure the cooling module to the upper body structure.

20. Remove the left and right splash shields.

21. Remove the front transaxle mount to cradle through bolt.

22. Remove the rear transaxle mount to frame bolts.

23. Remove both stabilizer link to stabilizer shaft nuts.

24. Remove both tie rod to steering knuckle nuts.

25. Separate the outer tie rods from the steering knuckles.

26. Remove the intermediate steering shaft to steering gear pinch bolt and discard.

✴✴ WARNING

DO NOT rotate the intermediate shaft once separated from the gear. Possible damage or a malfunction could occur.

27. Disconnect the intermediate steering shaft from the steering gear.

28. Remove both lower control arm ball stud to steering knuckle pinch bolts.

✴✴ WARNING

Do not free the ball stud by using a pickle fork or a wedge-type tool. Damage to the seal or bushing may result.

29. Lower the lower control arms in order to disengage the steering knuckle.

30. Mark the frame to body position with a paint pen or permanent marker.

31. Lower the vehicle to about 3 feet off the ground in order to place a hydraulic lift table under the frame.

32. Use two 2 x 4's between the lift table and the frame and lift the table to the frame.

33. Slowly remove the frame bolts using the following sequence:

a. Remove the front frame bolts.

b. Partially unscrew the rear frame bolts exposing 25.4 mm (1.5 in.) of bolt shank.

34. Slowly lower the lift table to the floor.

35. Drain the transaxle.

36. Disconnect the drive axle and intermediate shaft from the transmission and secure out of the way.

37. Remove the starter.

38. Disconnect the shift cables from the transmission. Disconnect the backup lamp switch harness connector.

39. Lower the vehicle.

40. Use the engine support fixture rear hook to lower the powertrain enough to allow clearance between the side rail and powertrain.

41. Raise the vehicle.

42. Use a transmission jack to secure the transmission, and remove the transmission to engine bolts.

43. Remove the transmission from the vehicle.

44. Remove the front transmission mount from the transmission.

45. Remove the rear transmission mount and bracket from the transmission.

To install:

46. Install the rear transmission mount to the transmission. Tighten the thru bolt to 100 Nm (74 ft. lbs.).).

47. Install the front transmission mount to the transmission. Tighten the thru bolt to 100 Nm (74 ft. lbs.).

48. Use a transmission jack to position the transmission to the vehicle.

49. Secure the transmission to the engine. Tighten the bolts to 75 Nm (55 ft. lbs.).

50. Connect the backup lamp switch harness connector.

51. Connect the shift cable to the transmission.

52. Install the starter.

53. Connect the drive axle and intermediate shaft to the transmission.

54. Lower the vehicle.

55. Use the engine support fixture in order to raise the powertrain assembly.

56. Install the left transmission mount:

a. Install the transmission mount to the mid-rail.

b. Install the transmission mount to mid-rail bolts. Tighten the bolts to 27 Nm (20 ft. lbs.).

c. Using a floor jack, raise the transmission until it contacts the transmission mount.

➡**The transmission mount to transmission bolts must be hand started. Do not pry the transmission or mount to align the holes.**

d. Hand start the transmission mount to bracket bolts using the following sequence:

• Rear bolt
• Middle bolt
• Front bolt

e. Using the previous sequence, tighten the transmission mount bolts. Tighten the bolts to 50 Nm (37 ft. lbs.).

f. Install the underhood electrical center bracket to the vehicle and install the electrical center into position on the bracket.

57. With the frame on the lift table, raise the frame to the vehicle.

58. Hand-start all the frame bolts while aligning the frame to the paint marks.

59. Tighten the frame bolts. Tighten the bolts to 100 Nm plus 180 degrees (74 ft. lbs.) plus 180 degrees.

60. Lower and remove the hydraulic table.

61. Connect the lower control arm to the steering knuckle.

➡**The torque sequence must be followed in the order that is listed.**

62. Install the ball joint pinch bolt and nut.

• First Pass: Tighten the nut to 50 Nm (37 ft. lbs.). Reverse nut ¾ turn
• Second Pass: Tighten the nut to 50 Nm (37 ft. lbs.) plus 30 degrees.

➡**The front and rear transmission mounts must be allowed to settle with the through bolts loosened.**

63. Hand-start the front transaxle mount through bolt.

64. Loosen the rear transmission mount through bolt.

65. Tighten the rear transaxle mount to frame bolts. Tighten the rear bolts to 50 Nm (37 ft. lbs.).

66. Tighten the front and rear transaxle mount through bolts in the following order. Tighten the rear bolt to 100 Nm (74 ft. lbs.). Tighten the front bolt to 100 Nm (74 ft. lbs.).

67. Install the outer tie rods to the steering knuckles.

68. Install the new outer tie rod to the knuckle nuts. Tighten the nuts to 20 Nm (15 ft. lbs.) plus 180 degrees.

69. Connect the stabilizer links to the stabilizer shaft.

70. Connect the intermediate shaft to the steering gear.

71. Install a new intermediate shaft pinch bolt. Tighten the bolt to 34 Nm (25 ft. lbs.).

72. Install the left and right splash shields.

73. Remove the engine support fixture.

74. Install the top engine to transmission bolt. Tighten the bolt to 75 Nm (55 ft. lbs.).

75. Install the top engine to transmission stud. Tighten the stud to 75 Nm (55 ft. lbs.).

76. Connect the wiring harness retainer to the transmission stud.

77. Connect the hydraulic clutch hose to the clutch actuator cylinder.

Bleed the clutch hydraulic system.

78. Install the underhood electrical center bracket to the vehicle and install the electrical center into position on the bracket.

79. Secure the forward lamp harness retainer to the ABS modulator bracket, or the proportioning valve bracket.

80. Connect the electrical connector to the brake fluid level sensor, then press forward on the connector position assurance (CPA) tab of the connector to secure.

81. Install the cover to the underhood electrical center.

82. Release the cooling module from the upper body structure.

83. Connect the negative battery cable.

84. Fill the transmission to the proper level.

Getrag Transaxle

1. Before servicing the vehicle, refer to the Precautions Section.

2. Disconnect the negative battery cable.

3. Remove the positive battery post from the underhood junction block.

4. Disconnect the positive cables from the underhood junction block.

5. Disconnect the surge tank inlet hose from the surge tank.

6. Remove the underhood junction block bracket nuts.

7. Loosen the underhood junction block bracket bolt.

8. Disconnect the front wiring harness from the underhood junction block bracket.

9. Reposition the underhood junction block bracket aside.

10. Disconnect the hydraulic clutch hose from the clutch actuator cylinder.

11. Remove the front wheel and tire assemblies.

12. Remove the both front fender liners.

13. Remove the steering gear mounting bolts and secure the steering gear with mechanic's wire.

14. Disconnect the wheel speed sensor wires from the both front wheels and unclip from the frame.

15. Separate the ball joints from the steering knuckles.

16. With the wheels in the straight ahead position, remove the key from the ignition switch.

17. Secure the cooling module to the upper body structure.

18. Remove the left and right splash shields.

19. Remove the front transaxle mount to cradle through bolt.

20. Remove the rear transaxle mount to frame bolts.

21. Remove both stabilizer link to stabilizer shaft nuts.

22. Remove both tie rod to steering knuckle nuts.

23. Separate the outer tie rods from the steering knuckles.

24. Remove the intermediate steering shaft to steering gear pinch bolt and discard.

✳✳ WARNING

DO NOT rotate the intermediate shaft once separated from the gear. Possible damage or a malfunction could occur.

25. Disconnect the intermediate steering shaft from the steering gear.

26. Remove both lower control arm ball stud to steering knuckle pinch bolts.

✳✳ WARNING

Do not free the ball stud by using a pickle fork or a wedge-type tool. Damage to the seal or bushing may result.

27. Lower the lower control arms in order to disengage the steering knuckle.

28. Mark the frame to body position with a paint pen or permanent marker.

29. Lower the vehicle to about 3 feet off the ground in order to place a hydraulic lift table under the frame.

30. Use two 2 x 4's between the lift table and the frame and lift the table to the frame.

31. Slowly remove the frame bolts using the following sequence:

   a. Remove the front frame bolts.

   b. Partially unscrew the rear frame bolts exposing 25.4 mm (1.5 in.) of bolt shank.

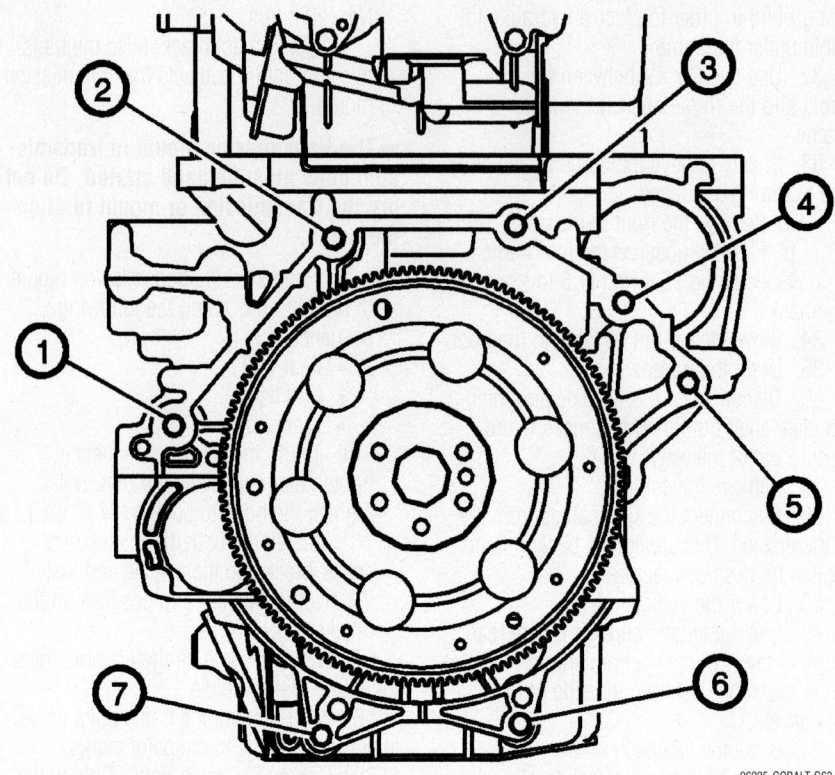

Engine-to-transmission bolts—Getrag transmission

06025-COBALT-G63

32. Slowly lower the lift table to the floor. Secure the cooling module to the upper body structure.

33. Remove the upper transmission to mount bolts.

34. Remove the upper transmission to engine bolt.

35. Drain the transaxle.

36. Disconnect the drive axles from the transmission and secure out of the way.

37. Remove the starter.

38. Disconnect the shift cables from the transmission.

39. Disconnect the backup lamp switch harness connector.

40. Disconnect the vehicle speed sensor.

41. Lower the vehicle.

42. Use the engine support fixture rear hook to lower the powertrain enough to allow clearance between the side rail and powertrain.

43. Raise the vehicle.

44. Use a transmission jack to secure the transmission, and remove the transmission to engine bolts.

45. Remove the transmission from the vehicle.

46. Remove the front transmission mount from the transmission.

47. Remove the rear transmission mount and bracket from the transmission.

To install:

48. Install the rear transmission mount to the transmission.

49. Install the front transmission mount to the transmission.

50. Tighten the transmission mount through bolts in the following order: Tighten the rear bolt to 100 Nm (74 ft. lbs.). Tighten the front bolt to 100 Nm (74 ft. lbs.).

51. Use a transmission jack to position the transmission to the vehicle.

➡**The number 3 position does not require a bolt.**

52. Secure the transmission to the engine. Tighten the bolts to 75 Nm (55 ft. lbs.).

53. Connect the vehicle speed sensor.

54. Connect the backup lamp switch harness connector.

55. Connect the shift cable to the transmission.

56. Install the starter.

57. Connect the drive axles to the transmission.

58. Lower the vehicle.

59. Use the engine support fixture in order to raise the powertrain assembly.

60. Install the transmission mount to the mid-rail.

61. Hand start the transmission mount to mid-rail bolts. Tighten the bolts to 34 Nm (25 ft. lbs.).

62. Using a floor jack, raise the transmission until it contacts the transmission mount.

➡**The transmission mount to transmission bolts must be hand started. Do not pry the transmission or mount to align the holes.**

63. Hand start the transmission mount to transmission bolts using the following sequence:
   a. Rear bolt
   b. Middle bolt
   c. Front bolt

64. Using the previous sequence, tighten the transmission mount bolts. Tighten the bolts to 45 Nm (33 ft. lbs.).

65. Reposition the underhood electrical center.

66. Connect the wiring harness to the tray bracket.

67. Install the electrical center nuts and bolt. Tighten the nuts to 10 Nm (89 inch lbs.). Tighten the bolt to 25 Nm (18 ft. lbs.).

68. Position the underhood junction block to the original position.

69. Install the surge tank inlet hose to the surge tank.

70. Install the front wiring harness to the junction block bracket.

71. Connect the positive battery cables to the junction block bracket.

72. Install the positive battery post to the junction block bracket.

73. Install the junction block bracket, bolt and nuts. Tighten the nuts to 10 Nm (89 inch lbs.). Tighten the bolt to 25 Nm (18 ft. lbs.).

74. With the frame on the lift table, raise the frame to the vehicle.

75. Hand-start all the frame bolts while aligning the frame to the paint marks.

76. Tighten the frame bolts. Tighten the bolts to 100 Nm plus 180 degrees (74 ft. lbs.) plus 180 degrees.

77. Lower and remove the hydraulic table.

78. Connect the lower control arm to the steering knuckle.

➡**The torque sequence must be followed in the order that is listed.**

79. Install the ball joint pinch bolt and nut.
   • First Pass: Tighten the nut to 50 Nm (37 ft. lbs.). Reverse nut ¾ turn
   • Second Pass: Tighten the nut to 50 Nm (37 ft. lbs.) plus 30 degrees.

➡**The front and rear transmission mounts must be allowed to settle with the through bolts loosened.**

80. Hand-start the front transaxle mount through bolt.

81. Loosen the rear transmission mount through bolt.

82. Tighten the rear transaxle mount to frame bolts. Tighten the rear bolts to 50 Nm (37 ft. lbs.).

83. Tighten the front and rear transaxle mount through bolts in the following order. Tighten the rear bolt to 100 Nm (74 ft. lbs.). Tighten the front bolt to 100 Nm (74 ft. lbs.).

84. Install the outer tie rods to the steering knuckles.

85. Install the new outer tie rod to the knuckle nuts. Tighten the nuts to 20 Nm (15 ft. lbs.) plus 180 degrees.

86. Connect the stabilizer links to the stabilizer shaft.

87. Connect the intermediate shaft to the steering gear.

88. Install a new intermediate shaft pinch bolt. Tighten the bolt to 34 Nm (25 ft. lbs.).

89. Install the left and right splash shields.

90. Remove the engine support fixture.

91. Install the top engine to transmission bolt. Tighten the bolt to 75 Nm (55 ft. lbs.).

92. Connect the hydraulic clutch hose to the clutch actuator cylinder.

93. Bleed the clutch hydraulic system.

94. Release the cooling module from the upper body structure.

95. Connect the negative battery cable.

96. Fill the transmission to the proper level.

Clutch

REMOVAL & INSTALLATION

With MU3 Transaxle

1. Before servicing the vehicle, refer to the Precautions Section.

2. Remove the transmission.

3. Remove the clutch cover bolts one turn at a time, until spring pressure is relieved.

4. Remove the clutch cover and the clutch disc.

To install:

5. Install the clutch disc and the clutch cover.

6. Hand start the clutch cover to flywheel bolts, leaving the clutch cover loose enough to reposition for alignment.

7. Install a clutch alignment tool.

8. Tighten the clutch cover to flywheel bolts in the sequence shown. Tighten the bolts to 30 Nm (22 ft. lbs.).

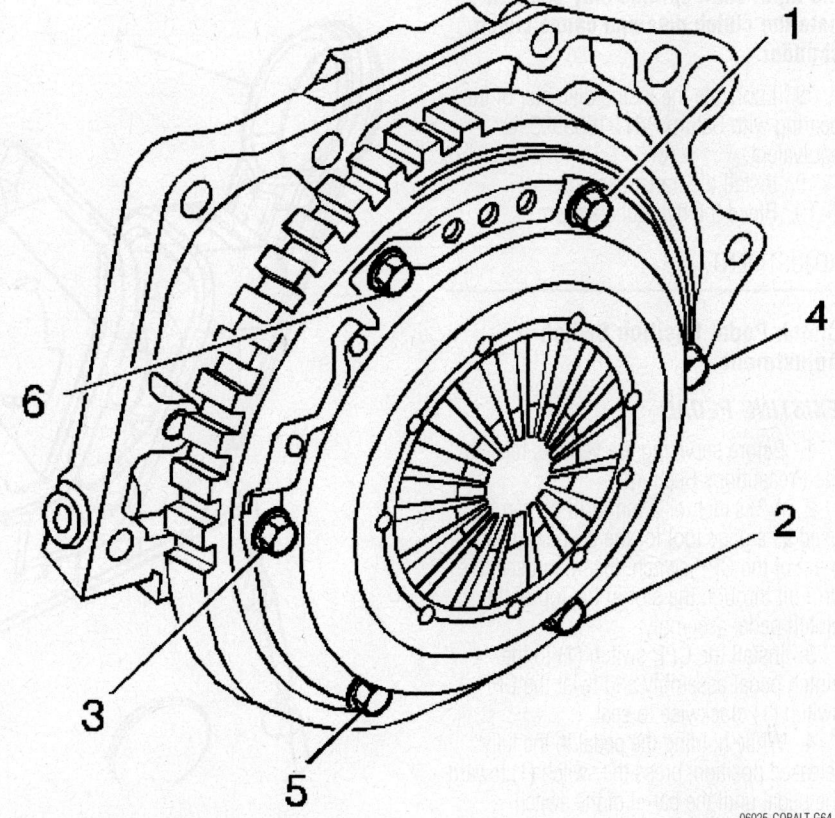

Clutch pressure plate torque sequence

06025-COBALT-G64

9. Recheck each bolt torque using the tightening sequence.

➡️**Excessive amounts of lubricant on the input shaft splines may contaminate the clutch disc and cause clutch shudder.**

10. Lubricate the inside diameter of the bearing with Saturn P/N 21005995, or equivalent.
11. Install the transmission.
12. Bleed the hydraulic system.
13. Connect the negative battery cable.

With Getrag Transaxle

1. Before servicing the vehicle, refer to the Precautions Section.
2. Remove the transmission.
3. Remove the clutch cover bolts one turn at a time, until spring pressure is relieved.
4. Remove the clutch cover and the clutch disc.

To install:
5. Align the Heavy Side of the flywheel assembly, with the clutch cover Light Side.
6. Install an alignment tool.
7. Install the clutch cover to flywheel bolts. Follow the tightening sequence. Tighten the bolts to 24 Nm (18 ft. lbs.).

➡️**Excessive amounts of lubricant on the input shaft splines may contaminate the clutch disc and cause clutch shudder.**

8. Lubricate the inside diameter of the bearing with Saturn P/N 21005995, or equivalent.
9. Install the transmission.
10. Bleed the hydraulic system.

ADJUSTMENT

Clutch Pedal Position Switch Adjustment

EXISTING PEDAL

1. Before servicing the vehicle, refer to the Precautions Section.
2. A ³⁄₃₂ inch or 2 mm drill bit can be used as a gage tool for the proper adjustment of the CPP switch. Insert and seat the drill bit through the slot in the top of the clutch pedal assembly.
3. Install the CPP switch (1) to the clutch pedal assembly and twist the CPP switch (1) clockwise to seat.
4. While holding the pedal in the fully released position, press the switch (1) toward the pedal until the barrel of the switch touches the drill bit on the pedal assembly.

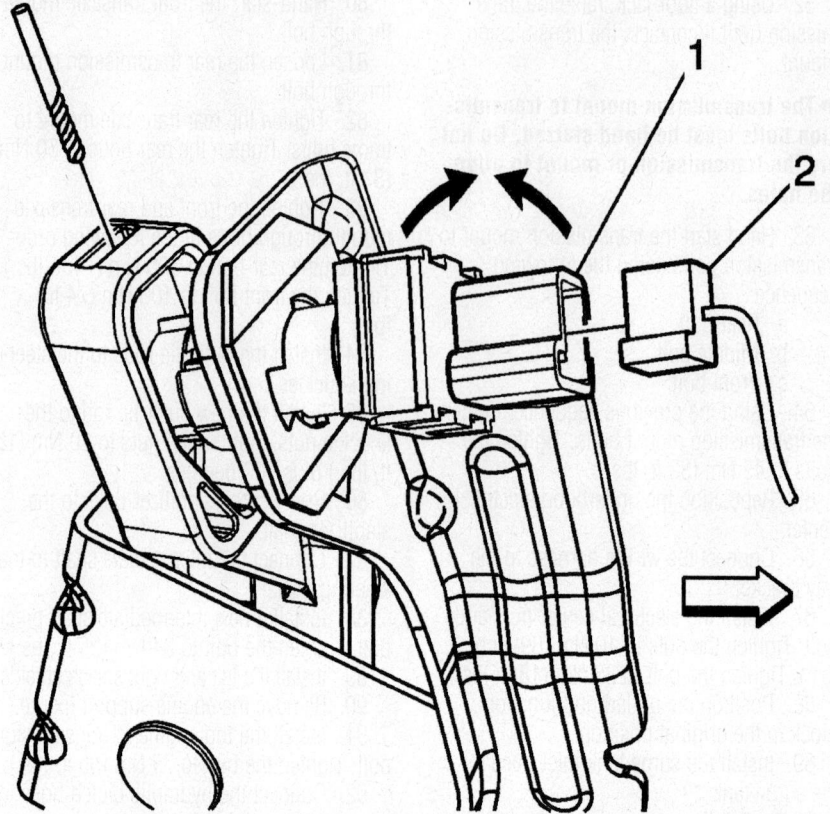

Clutch pedal position switch adjustment—existing pedal

06025-COBALT-G65

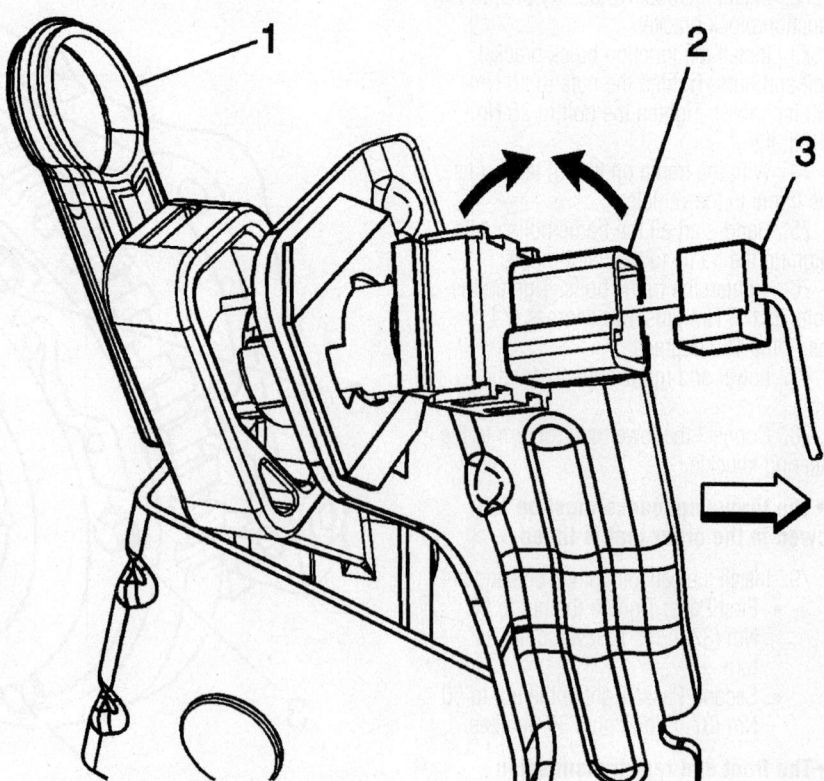

Clutch pedal position switch—new pedal

06025-COBALT-G66

5. Remove the drill bit.
6. Connect the electrical connector (2) to the CPP switch (1).

NEW PEDAL

1. While holding the pedal in the fully released position, press the switch (2) toward the pedal until the barrel of the switch (2) touches the shipping clip (1).
2. Remove the shipping clip (1).
3. Connect the electrical connector (3).

Hydraulic Clutch System

BLEEDING

Vacuum Bleeding

1. Before servicing the vehicle, refer to the Precautions Section.
2. Verify that all the hydraulic lines are dry and secure.
3. Clean dirt and grease from the reservoir cap in order to ensure that no foreign substances enter the system.
4. Remove the reservoir cap.
5. Fill the reservoir using DOT 3 hydraulic fluid.

❊❊ WARNING

Brake fluid will deteriorate the rubber on the adapter, use a clean shop towel to wipe away all fluid after each use.

6. Install a vacuum pump such as adapter J 43485 and pump J 35555 to the reservoir.
7. Hold the adapter to position while applying 51–68 kPa (15–20 hg) of vacuum.
8. Remove the adapter and refill the reservoir
9. Fully depress the clutch pedal cycling the pedal for 30 seconds.
10. Lift the clutch pedal to the up stop position and hold for 30 seconds.
11. Repeat steps 4-9 until all air is removed from the clutch system.

❊❊ CAUTION

Do not start the engine while the transaxle is engaged, only while in the neutral position. This vehicle is equipped with a concentric slave cylinder (CSC) and will move if started in gear.

12. Place the vehicle into the neutral position and start the engine.
13. Pump the clutch pedal until firm.

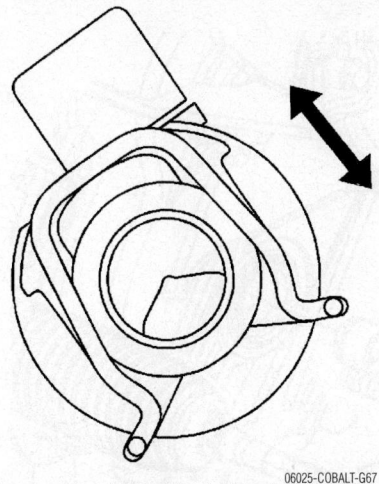

06025-COBALT-G67

Push the clip in order to move the clutch line into the bleed position

❊❊ CAUTION

The clutch and braking systems are integrated into one reservoir. The brake may be soft when first applying.

14. Pump the brake pedal until firm.
15. If needed, add additional hydraulic fluid.
16. Road test the vehicle to ensure proper operation.

Manual Bleeding

1. Before servicing the vehicle, refer to the Precautions Section.

❊❊ WARNING

Do not reuse the fluid that has been bled from a system in order to fill the clutch master cylinder reservoir.

➡**Maintain the fluid level in the clutch reservoir to the top step with DOT 3 hydraulic fluid.**

2. Clean dirt and grease from the cap in order to ensure that no foreign substances enter the system.
3. Attach a hose to the bleeder port on the clutch actuator assembly. Submerge the other end of the hose in a container of DOT 3 hydraulic fluid.
4. Depress the clutch pedal quickly to the full depressed position.
5. Push the clip in order to move the clutch line into the bleed position.
6. Move the clutch line into the normal position ensuring that the clip returns.
7. Lift the clutch pedal to the up stop position and hold for 5 seconds.

8. Repeat steps 3-6 until air is purged from the clutch system.

Halfshaft

REMOVAL & INSTALLATION

1. Before servicing the vehicle, refer to the Precautions Section.
2. Raise and support the vehicle.
3. Remove the tire and wheel assembly.
4. Remove the halfshaft nut. Insert a drift or a flat-bladed tool into the caliper and the rotor to prevent the rotor from turning.
5. Carefully loosen the halfshaft splines from the wheel bearing/hub assembly using a wood block and a hammer. The nut can be partially reinstalled to protect the threads.

➡**Be sure that the wheel speed sensor wiring harness is repositioned away from the ball joint after disconnecting the electrical connector from the sensor.**

6. Disconnect the ball joint from the steering knuckle.
7. Separate the halfshaft from the wheel bearing/hub assembly, then, support the shaft assembly.
8. Assemble tools J 45341 and J-2619-A, or equivalents, to the halfshaft inner tripot housing assembly.
9. Separate the halfshaft from the transaxle, then, remove the shaft assembly.
10. Inspect the transaxle output shaft seal for damage and/or contamination and replace if necessary.

To install:
11. Install tool J 44394 into the transaxle output shaft seal.
12. Install the halfshaft into the transaxle until the drive shaft splines are past the seal, remove the tool, then fully install the drive shaft.
13. Verify that the halfshaft is properly engaged:
    a. Grasp the inner tripod housing and pull the inner housing outward. Do NOT pull on the wheel drive axle shaft.
    b. The halfshaft will remain firmly in place when properly engaged.
14. Install the halfshaft to the wheel bearing/hub assembly.
15. Connect the ball joint to the steering knuckle.
16. Install the halfshaft nut to the halfshaft assembly. Tighten the nut to 210 Nm (155 ft. lbs.).
17. Install the tire and wheel assembly.
18. Lower the vehicle.
19. Inspect the transaxle fluid level.

06025-COBALT-G68

Assemble tools J 45341 and J-2619-A, or equivalents, to the halfshaft inner tripot housing assembly

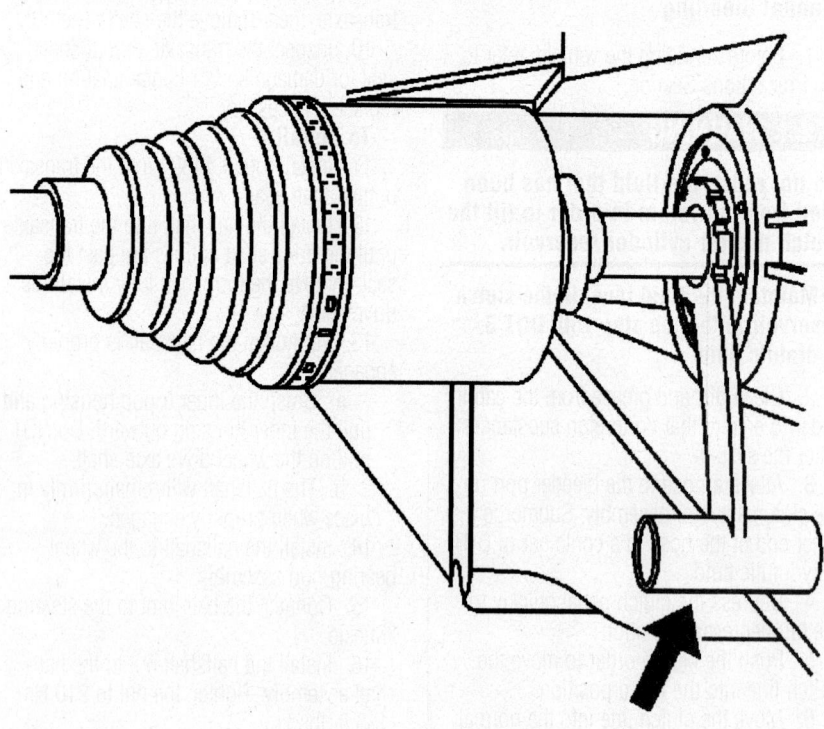

06025-COBALT-G69

Install tool J 44394 into the transaxle output shaft seal

Intermediate Shaft

REMOVAL & INSTALLATION

1. Before servicing the vehicle, refer to the Precautions Section.

2. Raise and support the vehicle.

3. Remove the RH tire and wheel assembly.

4. Disconnect the ball joint pinch bolt.

5. Rotate the steering knuckle to access the halfshaft inner joint.

6. Separate the halfshaft from the intermediate drive shaft.

7. Reposition and support the halfshaft from the intermediate drive shaft.

8. Inspect the halfshaft-to-intermediate drive shaft seal for excessive wear, damage, and/or contamination and replace if necessary.

9. Remove the rear, or LH intermediate drive shaft bracket-to-engine block bolts.

10. Remove the remaining intermediate shaft bracket-to-engine block bolts.

11. Using care to not damage the transaxle output shaft seal, remove the intermediate drive shaft assembly.

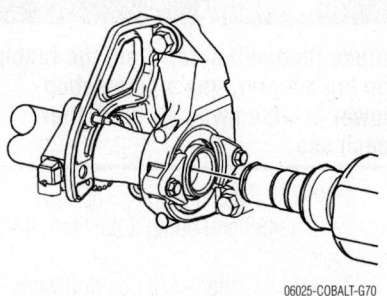

06025-COBALT-G70

Separate the halfshaft from the intermediate drive shaft

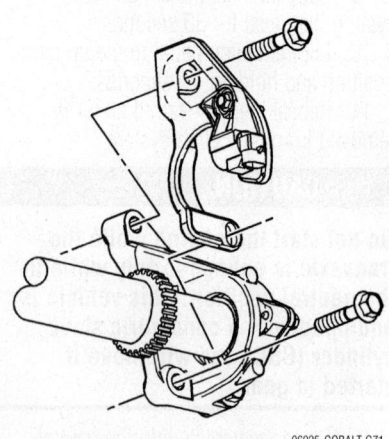

06025-COBALT-G71

Remove the remaining intermediate shaft bracket-to-engine block bolts

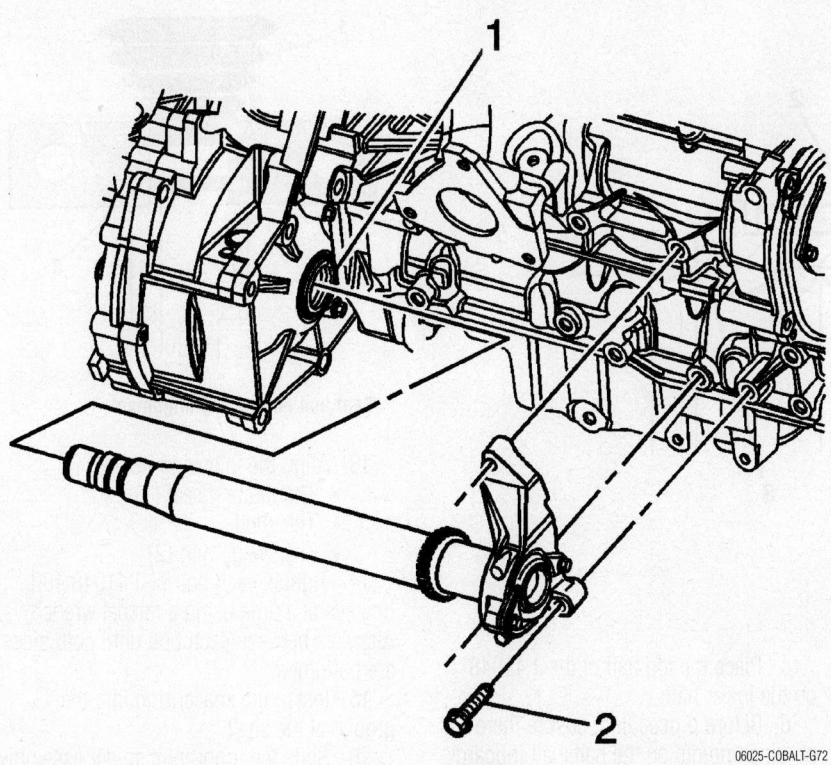

06025-COBALT-G72

Intermediate shaft removal

12. Inspect the transaxle output shaft seal for damage and/or contamination and replace if necessary.

To install:

13. Install tool J 44394 into the transaxle output shaft seal.

14. Install the intermediate drive shaft into the transaxle until the drive shaft splines are past the seal, remove the tool, then fully install the drive shaft.

15. Install, but do NOT tighten the intermediate drive shaft bracket-to-engine block forward, or RH bolt.

16. Tighten the intermediate drive shaft bracket-to-engine block bolts, beginning with the upper bolt. Tighten the bolts to 50 Nm (37 ft. lbs.).

17. Apply a very small amount of grease, GM P/N 1051344 (Canadian P/N 993037), or equivalent to the splines of the halfshaft inner joint.

18. Install the halfshaft into the intermediate drive shaft.

19. Verify that the halfshaft is properly engaged:

a. Grasp the inner tripod housing and pull the inner housing outward. Do NOT pull on the wheel drive axle shaft.

b. The halfshaft will remain firmly in place when properly engaged.

20. Install the tire and wheel assembly.

21. Lower the vehicle.

22. Inspect the transaxle fluid level.

CV-Joints

OVERHAUL

Inner Joint

GKN TYPE

✳✳ WARNING

Do not cut through the halfshaft inboard seal during service. Cutting

through the seal may damage the sealing surface of the housing and the tripot bushing. Damage to the sealing surface may lead to water and dirt intrusion and premature wear of the constant velocity joint.

1. Before servicing the vehicle, refer to the Precautions Section.

2. Disconnect the swage ring from the shaft using a hand grinder to cut through the ring, taking care not to damage the shaft.

3. Remove the large seal retaining clamp (2) from the tripot joint with side cutters. Discard the large seal retaining clamp.

4. Separate the inboard seal from the trilobal tripot bushing (3) at the large diameter.

5. Slide the seal away from the joint along the shaft.

6. Remove the housing (1) from the tripot joint spider and the shaft (2).

7. Remove the spacer ring, spider assembly, spacer ring, and tripot boot. Discard the boot and rings.

8. Clean the shaft. Use a wire brush in order to remove any rust in the boot mounting area (grooves).

9. Inspect the needle rollers, needle bearings, and trunion. Check the tripot housing for unusual wear, cracks, or other damage. Replace any damaged parts with the appropriate kit.

To assemble:

10. Place the new small swage ring (2) onto the small end of the joint seal (1). Slide the joint seal (1) and the small swage ring (2) onto the shaft.

11. Position the small end of the joint

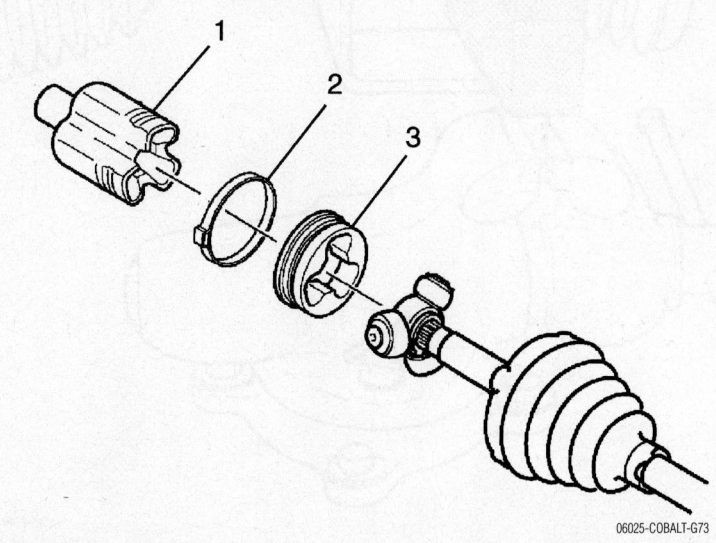

06025-COBALT-G73

GKN inner joint

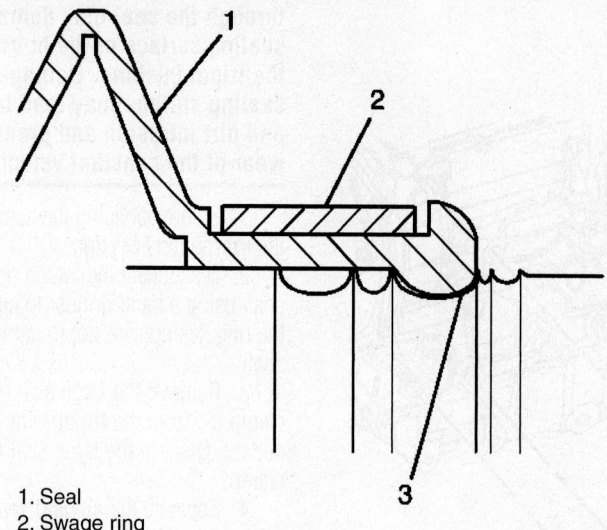

1. Seal
2. Swage ring
3. Seal groove

06025-COBALT-G74

Swage ring installation—GKN inner joint

seal (1) into the joint seal groove (3) on the shaft.

12. Mount tool J 41048 in a vise and proceed as follows:

a. Position the inboard end (1) of the halfshaft assembly in tool J 41048.

b. Align the top of seal neck on the bottom die using the indicator.

c. Place the top half of the J 41048 on the lower half.

d. Before proceeding, ensure there are no pinch points on the halfshaft inboard seal. This could cause damage to the inboard seal.

e. Insert the bolts (2).

f. Tighten the bolts by hand until snug.

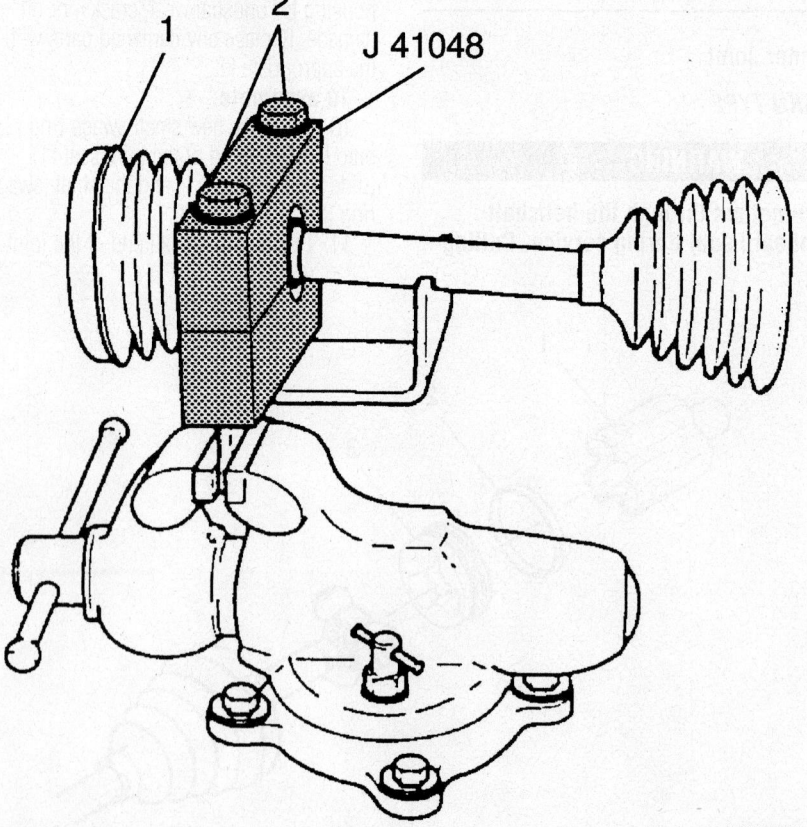

J 41048

06025-COBALT-G75

Using tool J41048 or DT-47732

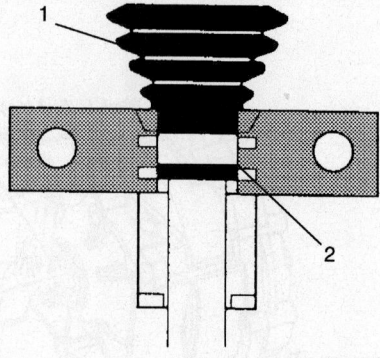

06025-COBALT-G76

Seal and swage ring alignment

13. Align the following items:
- The inboard seal (1)
- The shaft
- The swage ring (2)

14. Tighten each bolt of J 41048 180 degrees at a time using a ratchet wrench. Alternate between each bolt until both sides are bottomed.

15. Install the spacer ring into the groove of the shaft.

16. Slide the tripot joint spider assembly toward the spacer ring as far as it will go on the shaft.

17. Install the spacer ring into the groove of the shaft.

18. Place approximately half of the grease from the service kit in the inboard seal. Use the remainder of the grease to repack the housing.

➡**Ensure the trilobal tripot bushing is flush with the face of the housing.**

19. Install the new trilobal tripot bushing to housing.

20. Position the larger new seal retaining clamp on the inboard seal.

21. Slide the housing over the tripot joint spider assembly on the shaft.

22. Slide the large diameter of the inboard seal, with the larger clamp in place, over the outside of the trilobal tripot bushing and locate the lip of the seal in the groove.

➡**The inboard seal must not be dimpled, stretched out or out of shape in any way. If the inboard seal is not shaped correctly, carefully insert a thin flat blunt tool, no sharp edges, between the large seal opening and the trilobal tripot bushing in order to equalize the pressure. Shape the inboard seal properly by hand.**

23. Remove the tool.

24. Position the joint assembly at the proper vehicle dimension: 95 mm (3.75 in).

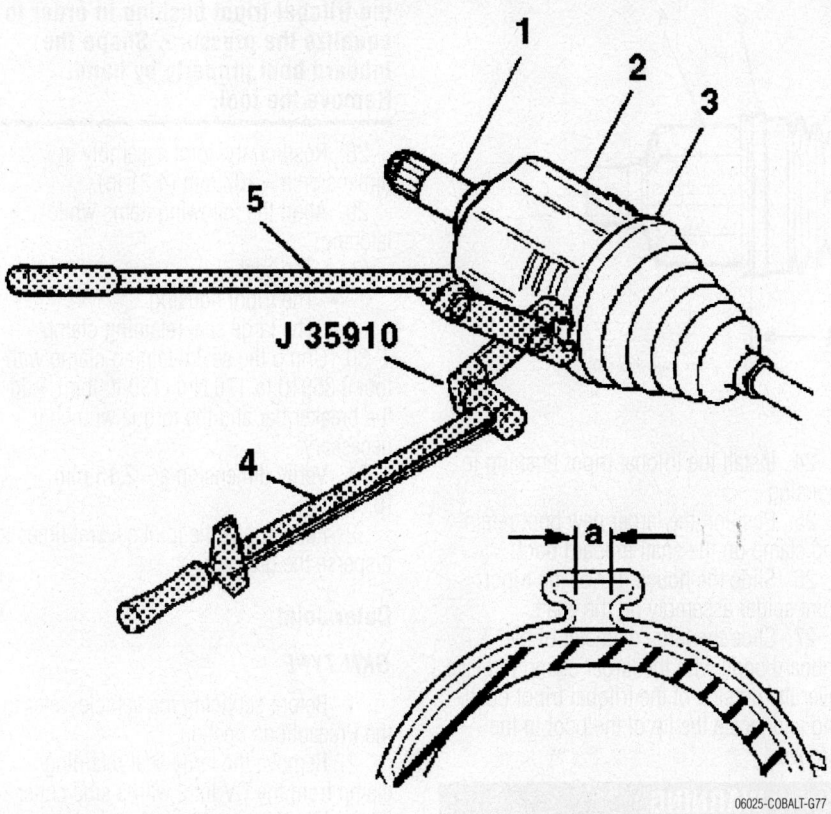

J 35910

Crimping the retaining clamp

25. Align the following items while latching:

- The inboard seal
- The tripot housing
- The large seal retaining clamp

26. Crimp the seal retaining clamp with J 35910, or equivalent, to 176 Nm (130 ft. lbs.). Add the breaker bar (5) and the torque wrench (4) to the tool, if necessary.

27. Check the gap dimension (a) on the clamp ear. If gap dimension is larger than shown, continue tightening until gap dimension of 2.60 mm (0.102 in.) is reached.

28. Fully stroke the joint several times to disperse the grease.

DELPHI TYPE

❄❄ WARNING

With removal of the wheel driveshaft for any reason, the transmission sealing surface (the tripot male/female shank of the wheel driveshaft) should be inspected for corrosion. If corrosion is evident, the surface should be cleaned with 320 grit cloth or equivalent in a rotation motion. Do not clean with an oscillating motion. Transmission fluid may be used to clean off any remaining debris. The surface should be wiped

06025-COBALT-G77

dry and the wheel driveshaft reinstalled free of any buildup.

1. Before servicing the vehicle, refer to the Precautions Section.

2. Use a hand grinder in order to cut through the swage ring. Do not damage the tripot housing (1).

3. Remove the large boot retaining clamp from the tripot joint with side cutter.

4. Dispose of the large boot retaining clamp.

❄❄ WARNING

Do not cut through the halfshaft inboard seal during service. Cutting through the seal may damage the sealing surface of the housing and the tripot bushing. Damage to the sealing surface may lead to water and dirt intrusion and premature wear of the constant velocity joint.

5. Separate the wheel driveshaft inboard boot from the trilobal tripot bushing (3) at the large diameter.

6. Slide the boot (4) away from the joint along the shaft.

7. Remove the housing (1) from the tripot joint spider (2) and the shaft.

8. Remove and discard the trilobal tripot bushing (3) from the housing (1).

9. Remove and discard the lower spacer ring from the groove on the shaft.

10. Slide tripot joint spider assembly off of the shaft.

11. If present, remove and discard the second spacer ring from the shaft.

12. Clean the following items with cleaning solvent:

- The tripot balls
- The needle rollers
- The housing

13. Remove all traces of old grease and any contaminates. Dry all parts.

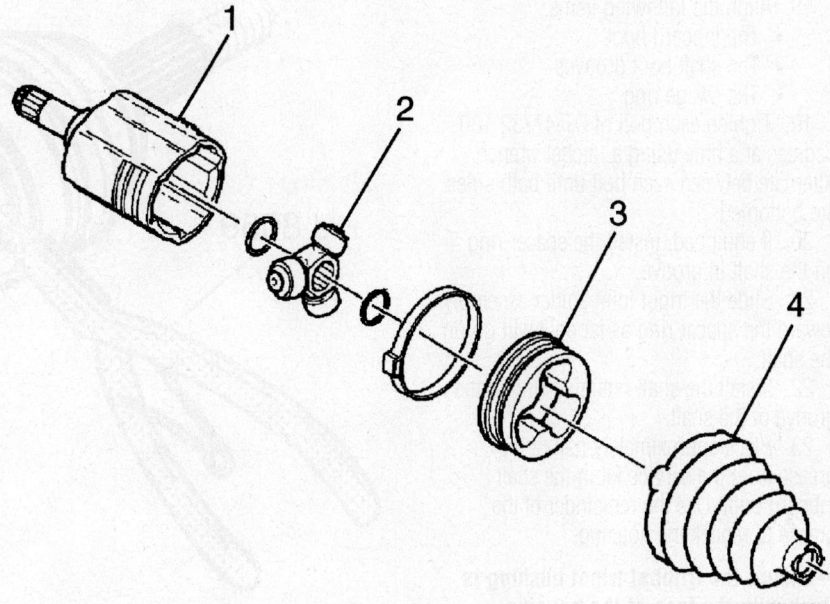

Delphi type inner joint

06025-COBALT-G78

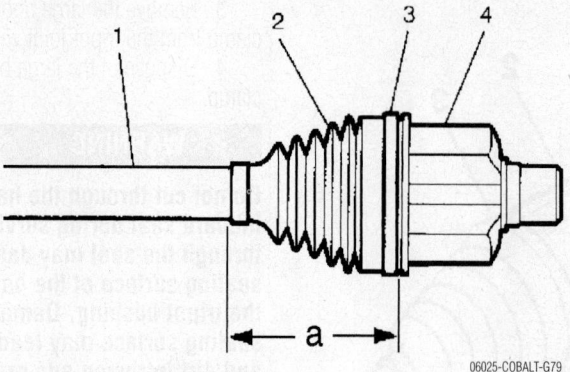

Joint assembly measurement

14. Inspect the parts for damage or wear.

To assemble:

15. Place the new small boot clamp onto the small end of the joint boot. Slide the joint boot and the small boot clamp onto the shaft.

16. Position the small end of the joint boot into the joint boot groove on the shaft.

17. Mount tool DT-47732 in a vise and proceed as follows:

a. Position the inboard end of the wheel driveshaft assembly in tool DT-47732.

b. Align the top of boot neck on the bottom die using the indicator.

c. Place the top half of tool DT-47732 on the lower half of tool.

d. Before proceeding, ensure there are no pinch points on the wheel driveshaft inboard boot. This could cause damage to the inboard boot.

e. Insert the bolts.

f. Tighten the bolts by hand until snug.

18. Align the following items:

- The inboard boot
- The shaft boot grooves
- The swage ring

19. Tighten each bolt of DT-47732 180 degrees at a time using a ratchet wrench. Alternate between each bolt until both sides are bottomed.

20. If equipped, install the spacer ring on the shaft in groove.

21. Slide the tripot joint spider assembly toward the spacer ring as far as it will go on the shaft.

22. Install the shaft retaining ring in the groove of the shaft.

23. Place approximately half of the grease from the service kit in the shaft inboard boot. Use the remainder of the grease to repack the housing.

➡ Ensure the trilobal tripot bushing is flush with the face of the housing.

24. Install the trilobal tripot bushing to housing.

25. Position the larger new boot retaining clamp on the shaft inboard boot.

26. Slide the housing over the tripot joint spider assembly on the shaft.

27. Slide the large diameter of the inboard boot, with the larger clamp in place, over the outside of the trilobal tripot bushing and locate the lip of the boot in the groove.

✷✷ WARNING

The inboard boot must not be dimpled, stretched out or out of shape in any way. If the inboard boot is not shaped correctly, carefully insert a thin flat blunt tool (no sharp edges) between the large boot opening and

---

the trilobal tripot bushing in order to equalize the pressure. Shape the inboard boot properly by hand. Remove the tool.

28. Position the joint assembly at dimension a = 107 mm (4.21 in).

29. Align the following items while latching:

- The seal
- The tripot housing
- The large seal retaining clamp

30. Crimp the seal retaining clamp with tool J 35910 to 176 Nm (130 ft. lbs.). Add the breaker bar and the torque wrench if necessary.

31. Verify dimension a = 2.15 mm (0.085 in).

32. Fully stroke the joint several times to disperse the grease.

Outer Joint

GKN TYPE

1. Before servicing the vehicle, refer to the Precautions Section.

2. Remove the large seal retaining clamp from the CV joint with a side cutter. Discard the seal retaining clamp.

3. Use a hand grinder to cut through the swage ring in order to remove the swage ring.

4. Separate the outboard seal from CV joint outer race (1) at large diameter.

5. Slide the seal (5) away from joint along shaft (4).

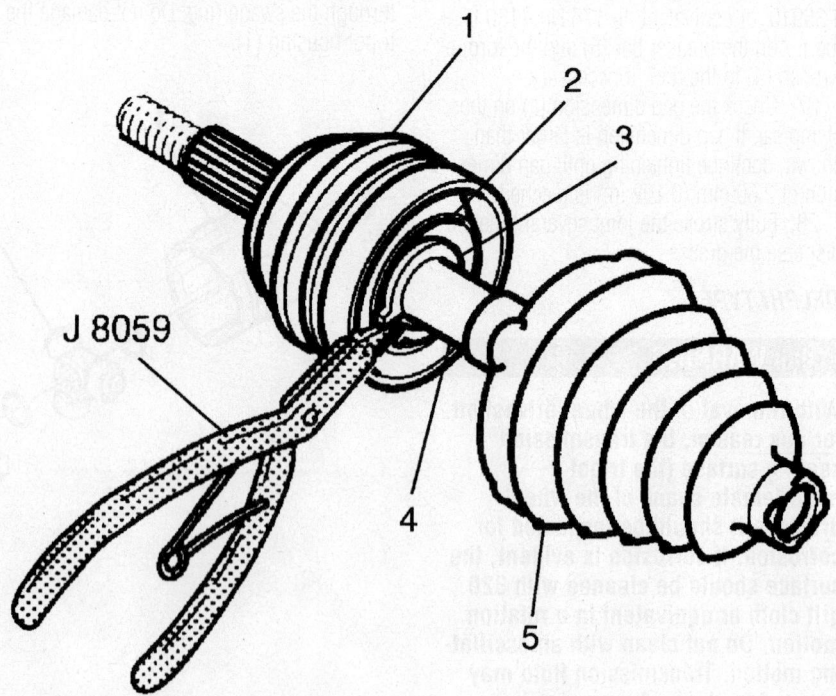

GKN outer joint

6. Wipe the grease from the face of the CV joint inner race.

7. Spread the ears on the race retaining ring.

8. Remove the CV joint assembly from the shaft.

9. Remove the outboard seal (5) from the shaft.

10. Discard the old outboard seal.

11. Place a brass drift against the CV joint cage.

12. Tap gently on the brass drift with a hammer in order to tilt the cage.

13. Remove the first chrome alloy ball when the CV joint cage tilts.

14. Tilt the CV joint cage in the opposite direction to remove the opposing chrome alloy ball.

15. Repeat this process to remove all 6 of the balls.

16. Pivot the CV joint cage and the inner race 90 degrees to the center line of the outer race. At the same time, align the cage windows with the lands of the outer race.

17. Lift out the cage and the inner race.

18. Remove the inner race from the cage by rotating the inner race upward.

19. Clean the all items thoroughly with cleaning solvent. Remove all traces of old grease and any contaminates.

20. Dry all the parts.

21. Check the CV joint assembly for the following items wear and damage.

22. Replace any damaged parts.

23. Clean the shaft. Use a wire brush to remove any rust in the grooves of the seal mounting area.

To assemble:

24. Install the new swage ring on the neck of the outboard seal. Do not swage.

25. Slide the outboard seal onto the shaft and position the neck of the outboard seal in the seal groove on the shaft. The largest groove below the sight groove on the shaft is the seal groove.

26. The swage ring is swaged using tool J 41048 by the following method:

    a. Position the outboard end of the wheel drive shaft assembly in J 41048.

    b. Align the swage ring.

    c. Place the top half of J 41048 on the lower half of J 41048.

➡**Align the following items during this procedure:**

- The outboard seal
- The shaft
- The swage ring

    d. Insert the bolts and tighten by hand until snug. Tighten each bolt 180 degrees at a time using a ratchet wrench. Alternate between each bolt until both sides are bottomed.

27. Loosen the bolts and separate the dies.

28. Check swaged ring for any lip deformities. If present, place the ring back into the J 41048 making sure the ring covers the whole swaging area. Then re-swage the ring.

29. Put a light coat of grease from the service kit on the ball grooves of the inner race and the outer race.

30. Hold the inner race 90 degrees to centerline of cage with the lands of the inner race aligned with the windows of the cage and insert the inner race into the cage.

31. Hold the cage and the inner race 90 degrees to centerline of the outer race and align the cage windows with the lands of the outer race.

➡**Verify that the retaining ring side of the inner race faces the shaft.**

32. Place the cage and the inner race into the outer race.

33. Insert the first chrome ball then tilt the cage in the opposite direction to insert the opposing ball.

34. Repeat this process until all 6 balls are in place.

35. Place approximately half the grease from the service kit inside the outboard seal and pack the CV joint with the remaining grease.

36. Push the CV joint onto the shaft until the retaining ring is seated in the groove on the shaft.

➡**The outboard seal must not be dimpled, stretched or out of shape in any way. If the outboard seal is not shaped correctly, equalize the pressure in the outboard seal and shape the seal properly by hand.**

37. Slide large diameter of the outboard seal with the large seal retaining clamp in place over the outside of the CV joint outer race and locate the seal lip in the groove on the CV joint outer race.

38. Crimp the seal retaining clamp using J 35910. Crimp the clamp to 174 Nm (130 ft. lbs.).

39. Check the gap dimension on the clamp ear. Continue tightening until the gap dimension is reached. Dimension a = 2.3 mm (³⁄₃₂ in.)

DELPHI TYPE

➡**Due to the helical splines on the shaft, disassembly and assembly of the joints will be difficult.**

1. Before servicing the vehicle, refer to the Precautions Section.

2. Remove the large seal retaining clamp from the CV joint with a side cutter. Discard the large seal retaining clamp.

3. Remove the small seal retaining clamp from the shaft with a side cutter and discard.

4. Separate the CV joint seal from the CV joint race at the large diameter.

5. Slide the seal away from the joint along the shaft.

6. Wipe the grease from the face of the CV joint inner race.

7. Before removing the CV joint assembly from the half shaft bar, perform the following procedure:

    a. Choose a reference mark (B) on the shaft.

    b. Measure the distance between the reference mark (B) and the face of the CV joint inner race. Make a note of this measurement (A).

    c. Mark as a reference the inboard side, toward the center of the half shaft bar, of the CV joint inner race.

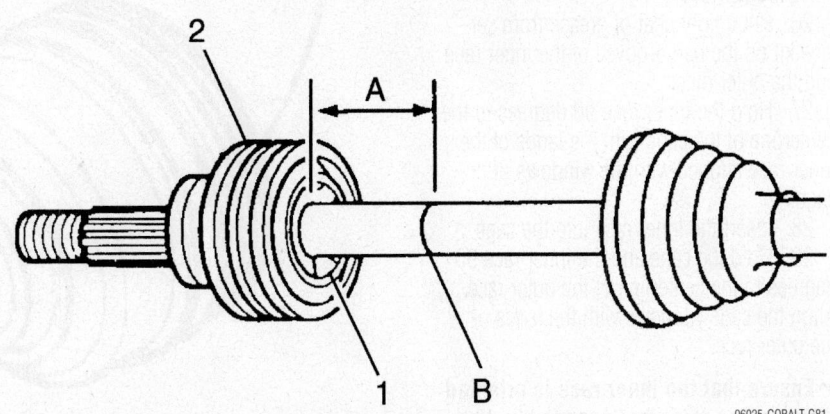

Reference marking the shaft—Delphi type outer joint

d. Inspect the CV joint inner race/shaft interface for the presence of a retaining ring. If present, remove the retaining ring using.

8. Clamp the shaft into a vise.

9. Attach a slide hammer to the threaded area of the outer race.

10. Remove the CV joint from the shaft.

➡**Never reuse the retaining ring.**

11. Remove the retaining ring from the shaft. Discard the retaining ring.

12. Remove the CV joint seal from the shaft.

13. Place a brass drift against the CV joint cage.

14. Tap gently on the brass drift with a hammer in order to tilt the cage.

15. Remove the first chrome alloy ball when the CV joint cage tilts.

16. Tilt the CV joint cage in the opposite direction to remove the opposing chrome alloy ball.

17. Repeat this process to remove all 6 balls.

18. Pivot the CV joint cage and the inner race 90 degrees to the centerline of the outer race. At the same time, align the cage windows with the lands of the outer race.

19. Lift out the cage and the inner race.

20. Remove the inner race from the cage by rotating the inner race upward.

21. Clean the inner and outer race assemblies, the CV joint cage and the chrome alloy balls thoroughly with cleaning solvent. Remove all traces of old grease and any contaminates.

22. Dry all the parts.

23. Check the CV joint assembly for unusual wear, cracks, or other damage.

24. Replace any damaged parts.

25. Clean the shaft. Use a wire brush to remove any rust in the seal mounting area (grooves).

To assemble:

26. Put a light coat of grease from service kit on the ball grooves of the inner race and the outer race.

27. Hold the inner race 90 degrees to the centerline of the cage with the lands of the inner race aligned with the windows of cage.

28. Insert the inner race into the cage.

29. Hold the cage and the inner race 90 degrees to the centerline of the outer race. Align the cage windows with the lands of the outer race.

➡**Ensure that the inner race is oriented the same as prior to disassembly. Use the reference mark placed earlier.**

30. Install the cage and the inner race into outer race.

31. Place a brass drift against the CV joint cage.

32. Tap gently on the brass drift with a hammer in order to tilt the cage.

33. Install the first chrome alloy ball when the CV joint cage tilts.

34. Repeat this process to install all 6 balls.

35. Pack the CV joint with half of the grease supplied in the service kit.

36. Install the new swage ring on the neck of the seal. Do not crimp.

37. Clean the shaft. Use a wire brush to remove any rust in the seal mounting area (grooves).

38. Slide the CV joint seal onto the shaft. Expose the reference mark by sliding the CV joint seal up the shaft toward the tripot end.

39. Position the large seal retaining clamp around the joint seal.

40. Place the new retaining ring onto the shaft.

41. While supporting the tripot assembly, place the wheel drive shaft assembly onto the arbor press with the CV assembly under the press head.

42. Lower the arbor press head onto the CV joint assembly until the press cannot move any further. This ensures that the retaining ring engages in the inner race. Do not exceed 1 779 N (4,000 lb) press load during assembly.

43. Remove the wheel drive shaft assembly from the arbor press.

44. Measure the distance between the reference mark on the shaft (B) and the face of the CV joint inner race.

45. Compare the mark with the measurement made before disassembly.

46. Repeat the previous 4 steps if the

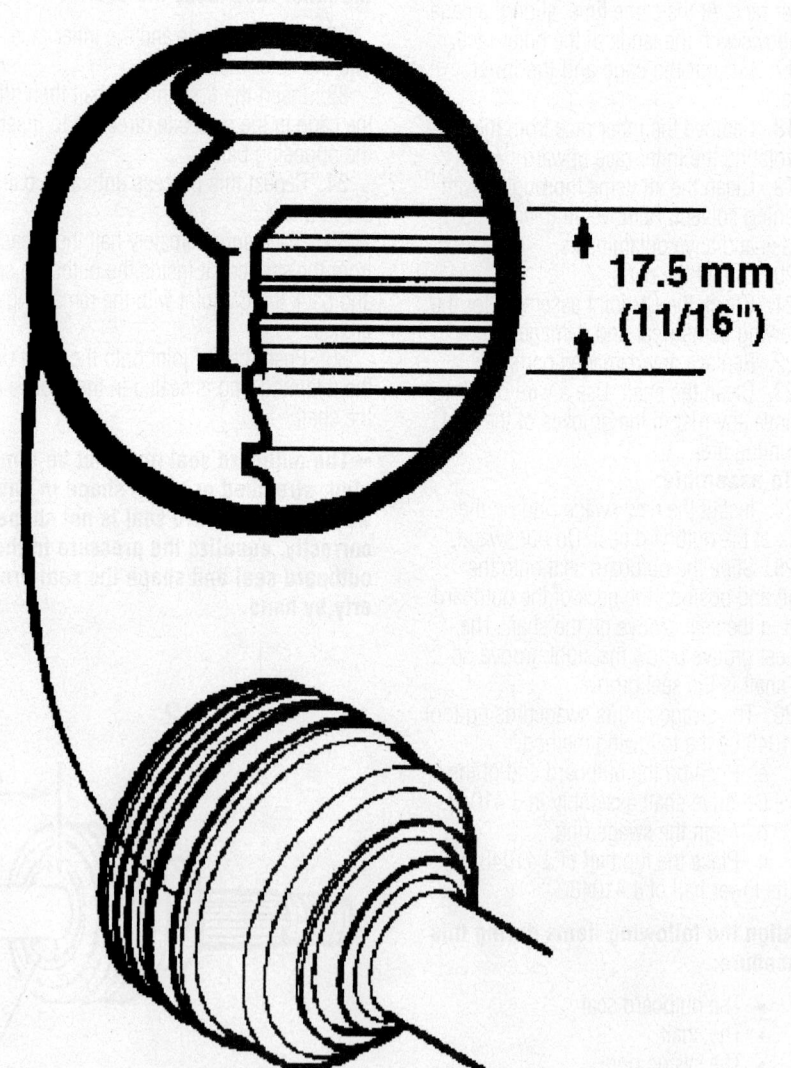

Measure approximately 17.5 mm (11/16 in.) up from the bottom edge of the CV outer joint assembly—Delphi type outer joint

06025-COBALT-G82

current measurement does not match the measurement made before disassembly.

47. Place the new swage ring onto the small end of the joint boot. Slide the boot and the swage ring onto the half shaft bar.

48. Position the small end of the boot into the boot groove on the half shaft bar.

49. Mount tool DT-47732 in a vise and proceed as follows:

    a. Position the outboard end of the wheel drive shaft assembly in tool.

    b. Align the top of boot neck on the bottom die using the indicator.

    c. Place the top half of the DT-47732 on the lower half of the DT-47732.

    d. Before proceeding, ensure there are no pinch points on the boot. This could cause damage to the boot.

    e. Insert the bolts.

    f. Tighten the bolts by hand until snug.

50. Align the following items:
- The shaft outboard boot
- The shaft boot grooves
- The swage ring

51. Tighten each bolt of DT-47732 180 degrees at a time using a ratchet wrench. Alternate between each bolt until both sides are bottomed.

52. Place the remaining grease from the service kit inside the seal.

53. Mount DT-47732 in a vise and proceed as follows: Measure approximately 17.5 mm (11/16 in.) up from the bottom edge of the CV outer joint assembly.

54. Slide the large diameter of the seal with the large seal retaining clamp in place over the outside of CV joint.

✳✳ WARNING

The CV joint seal must not be dimpled, stretched or out of shape in any way. If seal is not shaped correctly, equalize pressure in the seal and shape the seal properly by hand.

55. Locate the seal lip to the ridge of the CV outer joint assembly as measured in the previous step.

56. Crimp seal retaining clamp with tool J 35910, a breaker bar, and a torque wrench. Tighten seal retaining clamp to 176 Nm (130 ft. lbs.).

57. Check the gap dimension. Continue tightening until the correct gap dimension is reached. Dimension a = 2.3 mm (0.091 in.).

STEERING

Air Bag

PRECAUTIONS

✳✳ CAUTION

When performing service on or near the SIR components or the SIR wiring, the SIR system must be disabled. Refer to SIR Disabling and Enabling Zones. Failure to observe the correct procedure could cause deployment of the SIR components, personal injury, or unnecessary SIR system repairs.

The inflatable restraint sensing and diagnostic module (SDM) maintains a reserved energy supply. The reserved energy supply provides deployment power for the air bags. Deployment power is available for as much as 1 minute after disconnecting the vehicle power. Disabling the SIR system prevents deployment of the air bags from the reserved energy supply.

DISABLING AND ENABLING THE SYSTEM

Zone 1

1. Before servicing the vehicle, refer to the Precautions Section.

2. Turn the steering wheel so that the vehicles wheels are pointing straight ahead.

3. Turn the ignition switch to the OFF position.

4. Remove the key from the ignition switch.

5. Locate the body control module (BCM) fuse center then remove fuse center cover.

✳✳ WARNING

This inflatable restraint sensing and diagnostic module (SDM) has two fused power inputs. To ensure there is no unwanted SIR deployment, personal injury, or unnecessary SIR system repairs, remove both AIR BAG

(IGN) and AIR BAG (BATT) fuses from the BCM fuse center. With the AIR BAG fuses removed and the ignition switch in the ON position, the AIR BAG warning indicator illuminates. This is normal operation, and does not indicate an SIR system malfunction.

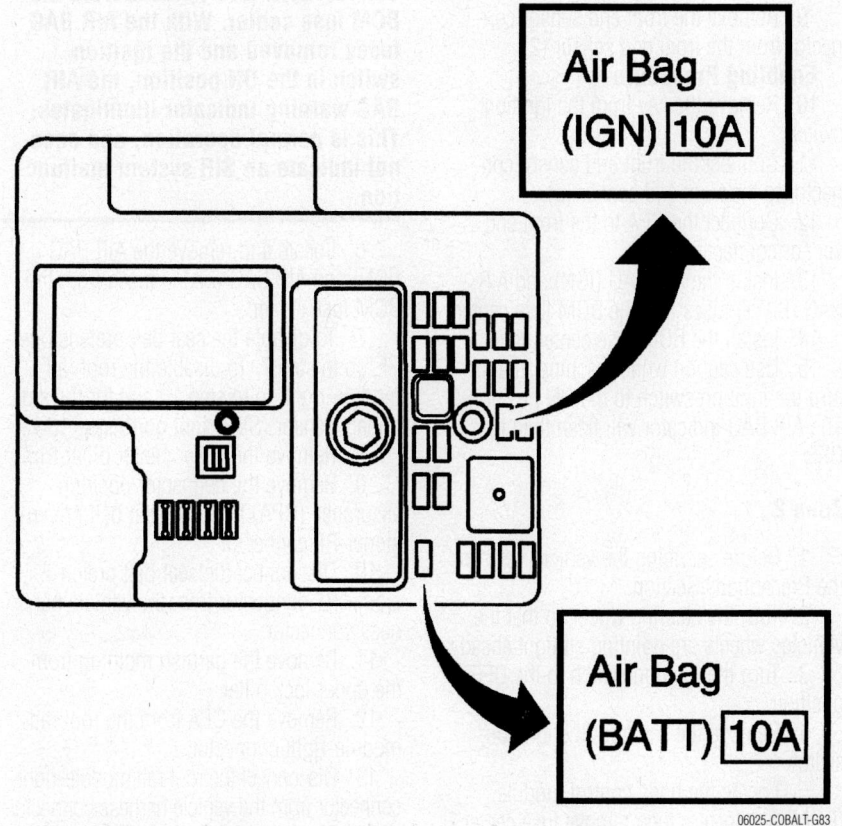

BCM fuse center

06025-COBALT-G83

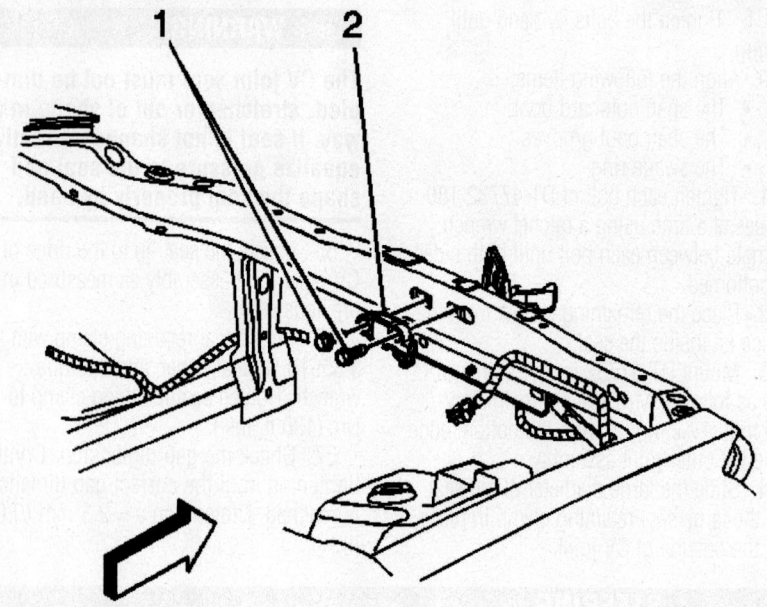

Front end sensor

06025-COBALT-G84

6. Locate and remove the AIR BAG (IGN) and AIR BAG (BATT) fuses from the BCM fuse center.

7. Open front hood, and locate the front end sensor also known as the electronic frontal sensor (EFS) (2).

8. Remove the connector position assurance (CPA) from the front end sensor connector.

9. Remove the front end sensor connector from the front end sensor (2).

Enabling Procedure

10. Remove the key from the ignition switch.

11. Connect the front end sensor connector to the front end sensor (2).

12. Connect the CPA to the front end sensor connector.

13. Install the AIR BAG (IGN) and AIR BAG (BATT) fuses into the BCM fuse center.

14. Install the BCM fuse center cover.

15. Use caution while reaching in and turn the ignition switch to the ON position. The AIR BAG indicator will flash then turn OFF.

Zone 2

1. Before servicing the vehicle, refer to the Precautions Section.

2. Turn the steering wheel so that the vehicles wheels are pointing straight ahead.

3. Turn the ignition switch to the OFF position.

4. Remove the key from the ignition switch.

5. Locate the body control module (BCM) fuse center then remove fuse center cover.

⁑ WARNING

This inflatable restraint sensing and diagnostic module (SDM) has 2 fused power inputs. To ensure there is no unwanted SIR deployment, personal injury, or unnecessary SIR system repairs, remove both AIR BAG (IGN) and AIR BAG (BATT) fuses from the BCM fuse center. With the AIR BAG fuses removed and the ignition switch in the ON position, the AIR BAG warning indicator illuminates. This is normal operation, and does not indicate an SIR system malfunction.

6. Locate and remove the AIR BAG (IGN) and AIR BAG (BATT) fuses from the BCM fuse center.

7. To disable the seat belt pretensioner-RF, go to step 7. To disable the roof rail module-right go to step 10 and for the side impact sensor (SIS)-right go to step 13.

8. Remove the lower center pillar trim.

9. Remove the connector position assurance (CPA) from the seat belt pretensioner-RF connector.

10. Disconnect the seat belt pretensioner-RF connector from the vehicle harness connector.

11. Remove the garnish molding from the upper lock pillar.

12. Remove the CPA from the roof rail module-right connector.

13. Disconnect the roof rail module-right connector from the vehicle harness connector.

14. Remove the RH door trim panel.

---

15. Peel back the door water deflector far enough to access the SIS-right.

16. Remove the CPA from the SIS connector.

17. Disconnect the SIS connector from the SIS-right.

Enabling Procedure

18. Remove the key from the ignition switch.

19. To enable the SIS-right go to step 3. To enable the roof rail module-right go to step 7, and to enable the seat belt pretensioner-RF go to step 10.

20. Install the SIS-right connector to the SIS-right.

21. Install the CPA to the SIS connector.

22. Reinstall door water deflector.

23. Install the RH door trim panel.

24. Connect the roof rail module-right connector to the vehicle harness connector.

25. Install the CPA to the roof rail module-right connector.

26. Install the garnish molding to the upper lock pillar.

27. Connect the seat belt pretensioner-LF and install the CPA.

28. Install the lower center pillar.

29. Install the AIR BAG (IGN) and AIR BAG (BATT) fuses into the BCM fuse center.

30. Install the BCM fuse center cover.

31. Use caution while reaching in and turn the ignition switch to the ON position. The AIR BAG indicator will flash then turn OFF.

Zone 3

1. Before servicing the vehicle, refer to the Precautions Section.

2. Turn the steering wheel so that the vehicles wheels are pointing straight ahead.

3. Turn the ignition switch to the OFF position.

4. Remove the key from the ignition switch.

Locate the body control module (BCM) fuse center then remove the fuse center cover.

⁑ WARNING

This inflatable restraint sensing and diagnostic module (SDM) has two fused power inputs. To ensure there is no unwanted SIR deployment, personal injury, or unnecessary SIR system repairs, remove both AIR BAG (IGN) and AIR BAG (BATT) fuses from the BCM fuse center. With the AIR BAG fuses removed and the ignition switch in the ON position, the AIR BAG warning indicator illuminates. This is normal operation, and does not indicate an SIR system malfunction.

5. Locate and remove the AIR BAG (IGN) and AIR BAG (BATT) fuses from the BCM fuse center.

6. Remove the left/driver outer trim cover from the instrument panel (I/P).

7. Remove the connector position assurance (CPA) from the steering wheel module coil connector.

8. Disconnect the steering wheel module coil connector from the vehicle harness connector.

Enabling Procedure

9. Remove the key from the ignition switch.

10. Connect the steering wheel module coil connector to the vehicle harness connector.

11. Install the CPA to the steering wheel module coil connector.

12. Install the left outer trim cover to the I/P.

13. Install the AIR BAG (IGN) and AIR BAG (BATT) fuses into the BCM fuse center.

14. Install the BCM fuse center cover.

15. Use caution while reaching in and turn the ignition switch to the ON position. The AIR BAG indicator will flash then turn OFF.

Zone 5

1. Before servicing the vehicle, refer to the Precautions Section.

2. Turn the steering wheel so that the vehicle's wheels are pointing straight ahead.

3. Turn the ignition switch to the OFF position.

4. Remove the key from the ignition switch.

5. Locate the body control module (BCM) fuse center, then, remove the fuse center cover.

✳✳ WARNING

This inflatable restraint sensing and diagnostic module (SDM) has two fused power inputs. To ensure there is no unwanted SIR deployment, personal injury, or unnecessary SIR system repairs, remove both AIR BAG (IGN) and AIR BAG (BATT) fuses from the BCM fuse center. With the AIR BAG fuses removed and the ignition switch in the ON position, the AIR BAG warning indicator illuminates. This is normal operation, and does not indicate an SIR system malfunction.

6. Locate and remove the AIR BAG (IGN) and AIR BAG (BATT) fuses from the BCM fuse center.

7. Remove the right/passenger outer trim cover from the instrument panel (I/P).

8. Remove the connector position assurance (CPA) from the I/P module connector.

9. Disconnect the I/P module connector from the vehicle harness connector.

Enabling Procedure

10. Remove the key from the ignition switch.

11. Connect the I/P module connector to the vehicle harness connector.

12. Install the CPA to the I/P module connector.

13. Install the right outer trim cover to the I/P.

14. Install the AIR BAG (IGN) and AIR BAG (BATT) fuses into the BCM fuse center.

15. Install the BCM fuse center cover.

16. Use caution while reaching in and turn the ignition switch to the ON position. The AIR BAG indicator will flash then turn OFF.

Zone 6

1. Before servicing the vehicle, refer to the Precautions Section.

2. Turn the steering wheel so that the vehicles wheels are pointing straight ahead.

3. Turn the ignition switch to the OFF position.

4. Remove the key from the ignition switch.

5. Locate the body control module (BCM) fuse center, then, remove fuse center cover.

✳✳ WARNING

This inflatable restraint sensing and diagnostic module (SDM) has 2 fused power inputs. To ensure there is no unwanted SIR deployment, personal injury, or unnecessary SIR system repairs, remove both AIR BAG (IGN) and AIR BAG (BATT) fuses from the BCM fuse center. With the AIR BAG fuses removed and the ignition switch in the ON position, the AIR BAG warning indicator illuminates. This is normal operation, and does not indicate an SIR system malfunction.

6. Locate and remove the AIR BAG (IGN) and AIR BAG (BATT) fuses from the BCM fuse center.

7. To disable the seat belt pretensioner-LF go to step 7. To disable the roof rail module-left go to step 10 and for the side impact sensor (SIS)-left go to step 13.

8. Remove the lower center pillar trim.

9. Remove the connector position assurance (CPA) from the seat belt pretensioner-LF connector.

10. Disconnect the seat belt pretensioner-LF connector from the vehicle harness connector.

11. Remove the garnish molding from the upper lock pillar.

12. Remove the CPA from the roof rail module-left connector.

13. Disconnect the roof rail module-left connector from the vehicle harness connector.

14. Remove the LH door trim panel.

15. Peel back the door water deflector far enough to access the SIS-left.

16. Remove the CPA from the SIS connector.

17. Disconnect the SIS connector from the SIS-left.

Enabling Procedure

18. Remove the key from the ignition switch.

19. To enable the SIS-left go to step 3. To enable the roof rail module-left go to step 7, and to enable the seat belt pretensioner-LF go to step 10.

20. Install the SIS-left connector to the SIS-left.

21. Install the CPA to the SIS connector.

22. Reinstall door water deflector.

23. Install the LH door trim panel.

24. Connect the roof rail module-left connector to the vehicle harness connector.

25. Install the CPA to the roof rail module-left connector.

26. Install the garnish molding to the upper lock pillar.

27. Connect the seat belt pretensioner-LF and install the CPA.

28. Install the lower center pillar.

29. Install the AIR BAG (IGN) and AIR BAG (BATT) fuses into the BCM fuse center.

30. Install the BCM fuse center cover.

31. Use caution while reaching in and turn the ignition switch to the ON position. The AIR BAG indicator will flash, then, turn OFF.

Zone 8

1. Before servicing the vehicle, refer to the Precautions Section.

2. Turn the steering wheel so that the vehicles wheels are pointing straight ahead.

3. Turn the ignition switch to the OFF position.

4. Remove the key from the ignition switch.

5. Locate the body control module (BCM) fuse center then remove fuse center cover.

✳✳ WARNING

This inflatable restraint sensing and diagnostic module (SDM) has two fused power inputs. To ensure there is no unwanted SIR deployment, personal injury, or unnecessary SIR system repairs, remove both AIR BAG (IGN) and AIR BAG (BATT) fuses from the BCM fuse center. With the AIR BAG fuses removed and the ignition switch in the ON position, the AIR BAG warning indicator illuminates. This is normal operation, and does not indicate an SIR system malfunction.

6. Locate and remove the AIR BAG (IGN) and AIR BAG (BATT) fuses from the BCM fuse center.

7. Remove the garnish molding from the right upper lock pillar.

8. Remove the connector position assurance (CPA) from the roof rail module-right connector.

9. Disconnect the roof rail module-right connector from the vehicle harness connector.

10. Remove the lower right center pillar trim.

11. Remove the CPA from seat belt pretensioner-RF connector.

12. Disconnect the seat belt pretensioner-RF connector from the vehicle harness connector.

13. Remove the passenger/right outer trim cover from the instrument panel (I/P). Remove the CPA from the I/P module connector.

14. Disconnect the I/P module connector from the vehicle harness connector.

15. Remove the driver/left outer trim cover from the I/P.

16. Remove the CPA from the steering wheel module coil connector.

17. Disconnect the steering wheel module coil connector from the vehicle harness connector.

18. Remove the lower left center pillar trim.

19. Remove the CPA from seat belt pretensioner-LF connector.

20. Disconnect the seat belt pretensioner-LF connector from the vehicle harness connector.

21. Remove the garnish molding from the upper lock pillar.

22. Remove the CPA from the roof rail module-left connector.

23. Disconnect the roof rail module-left connector from the vehicle harness connector.

Enabling Procedure

24. Remove the key from the ignition switch.

25. Connect the roof rail module-left connector to the vehicle harness connector.

26. Install the CPA to the roof rail module-left connector.

27. Install the garnish molding to the left upper lock pillar.

28. Connect the seat belt pretensioner-LF connector.

29. Install the CPA to the seat belt pretensioner-LF connector.

30. Install the left/driver lower center pillar trim.

31. Connect the steering wheel module coil connector to the vehicle harness connector.

32. Install the CPA to the steering wheel module coil connector.

33. Install the driver/left outer trim cover to the I/P.

34. Connect the I/P module connector to the vehicle harness connector.

35. Install the CPA to the I/P module connector.

36. Install the passenger/right outer trim cover to the I/P.

37. Connect the seat belt pretensioner-RF connector.

38. Install the CPA to the seat belt pretensioner-RF connector.

39. Install the right lower center pillar trim.

40. Connect the roof rail module-right connector to the vehicle harness connector.

41. Install the CPA to the roof rail module-right connector.

42. Install the garnish molding to the right upper lock pillar.

43. Install the AIR BAG (IGN) and AIR BAG (BATT) fuses into the BCM fuse center.

44. Install the BCM fuse center cover.

45. Use caution while reaching in and turn the ignition switch to the ON position. The AIR BAG indicator will flash then turn OFF.

Power Steering Gear

REMOVAL & INSTALLATION

1. Before servicing the vehicle, refer to the Precautions Section.

2. Turn the steering wheel to the

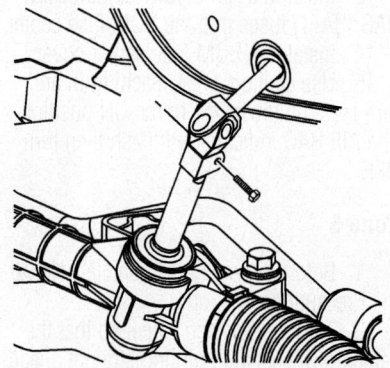

Steering shaft pinch bolt

06025-COBALT-G85

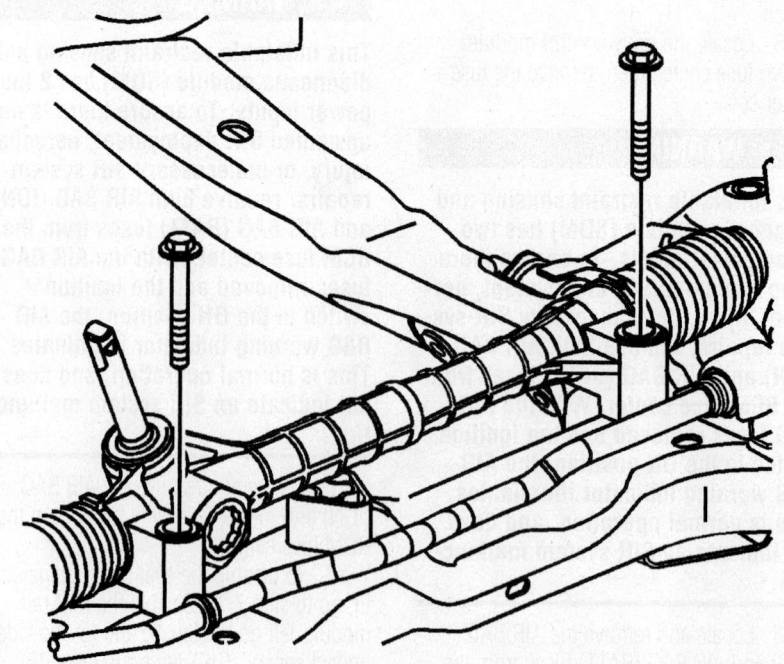

Steering gear mounting

06025-COBALT-G86

straight ahead position and remove the key from the ignition.

3. Turn the wheel counterclockwise in order to lock the steering column in place.

4. Raise and support the vehicle.

5. Remove the front wheels.

6. Remove the steering gear to intermediate shaft pinch bolt and discard.

7. Disconnect the intermediate shaft from the steering gear.

8. Remove both steering gear outer tie rod to knuckle nuts and discard the nuts.

9. Separate the outer tie rods from the steering knuckles.

10. Remove the steering gear bolts.

11. Remove the rear transmission mount.

12. Carefully remove the steering gear from the frame and the vehicle through the LH wheel opening.

To install:

13. Install the steering gear to the frame through the LH wheel opening.

14. Install the steering gear bolts. Tighten the bolts to 110 Nm (81 ft. lbs.).

15. Install the rear transmission mount. Tighten the rear transaxle mount to frame bolts. Tighten the bolts to 60 Nm (44 ft. lbs.).

16. Tighten the rear transmission mount thru bolt. Tighten the thru bolt to 100 Nm (74 ft. lbs.).

17. Connect the intermediate shaft to the steering gear.

18. Install a new intermediate shaft pinch bolt. Tighten the bolts to 34 Nm (25 ft. lbs.).

19. Install the outer tie rods to the steering knuckles.

20. Install new outer tie rod nuts. Tighten the bolts to 60 Nm (44 ft. lbs.).

21. Install the front wheels.

22. Lower the vehicle.

23. Adjust the toe if necessary.

FRONT SUSPENSION

Strut

REMOVAL & INSTALLATION

1. Before servicing the vehicle, refer to the Precautions Section.

2. Remove the strut upper mounting nuts.

➥**Lift the vehicle using ONLY a frame-contact vehicle lift. Do NOT lift the vehicle using a suspension-contact vehicle lift.**

3. Raise and support the vehicle.

4. Remove the tire and wheel.

5. Disconnect the stabilizer shaft link from the strut assembly.

6. Remove the strut lower bolts, nuts and antilock brake system (ABS) wiring bracket, if equipped.

7. Remove the strut.

To install:

8. Install the strut.

9. Install the strut upper mounting nuts. Tighten the nuts to 20 Nm (15 ft. lbs.).

➥**This is a prevailing torque type fastener. This fastener may be reused ONLY if: The fastener and its counterpart are clean and free from rust. The fastener develops 3 Nm (27 inch lbs.)**

of torque/drag against its counterpart prior to the fastener seating. If the fastener does not meet these criteria, **REPLACE the fastener.**

10. Install the strut lower bolts, nuts and ABS wiring bracket, if equipped. Tighten the strut lower nuts to 120 Nm (89 ft. lbs.).

11. Connect the stabilizer shaft link to the strut assembly. Tighten the stabilizer shaft link nut to 65 Nm (48 ft. lbs.).

12. Install the tire and wheel.

13. Lower the vehicle.

14. Align the front wheels.

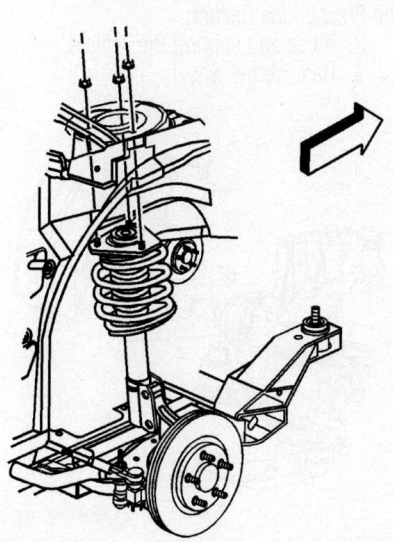

06025-COBALT-G87

Upper strut mounting nuts

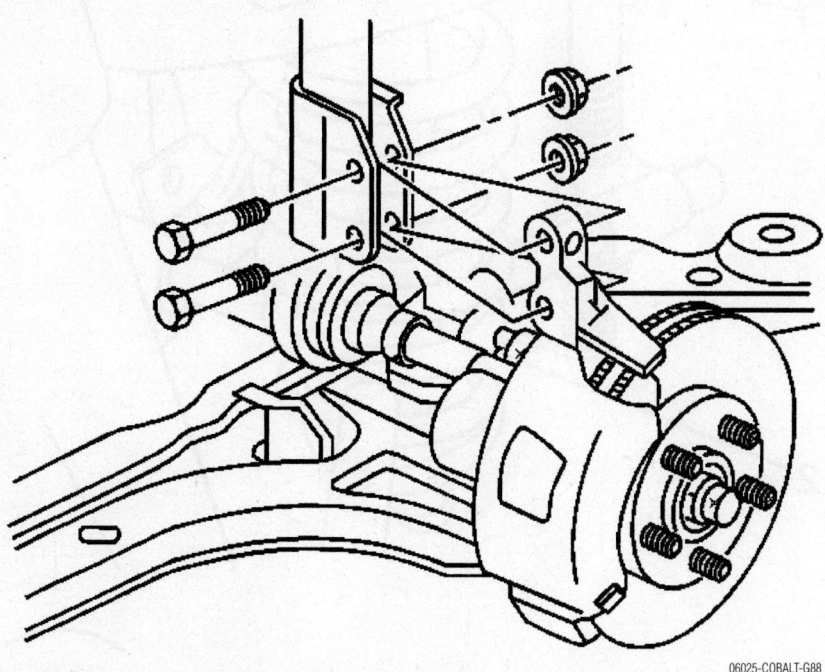

06025-COBALT-G88

Lower strut mount bolts/nuts

DISASSEMBLY AND ASSEMBLY

1. Before servicing the vehicle, refer to the Precautions Section.

2. Remove the strut from the vehicle.

3. Install the strut (2) in the J 45400 (1).

➡ **The spring is compressed when the strut moves freely.**

4. Turn the spring compressor forcing screw until the coil spring is compressed.

5. Loosen the compressor forcing screw until the upper strut mount and coil spring may be removed.

6. Use a 45 TORX® socket in order to hold the strut shaft. Use a strut rod socket or equivalent to remove the upper strut mount nut.

7. Remove the upper strut mount and the coil spring from the tool.

8. Remove the strut from the tool.

To assemble:

9. Install the coil spring and upper strut mount to the tool.

10. Turn the spring compressor forcing screw until the coil spring is compressed.

11. Install the strut to the coil spring and upper strut mount.

12. Loosely install the strut retaining nut.

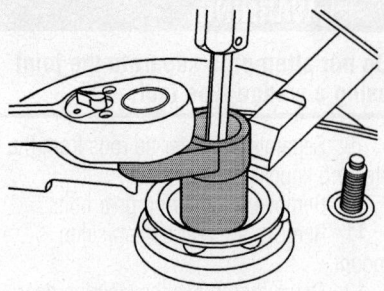

06025-COBALT-G90

Use a 45 TORX socket in order to hold the strut shaft. Use a strut rod socket or equivalent to remove the upper strut mount nut

13. Use a 45 TORX® socket in order to hold the strut shaft. Use a strut rod socket or equivalent to install the upper strut mount nut. Tighten the strut mount nut to 75 Nm (55 ft. lbs.).

14. Remove the strut from the tool.

15. Install the strut to the vehicle.

Lower Ball Joint

REMOVAL & INSTALLATION

1. Before servicing the vehicle, refer to the Precautions Section.

2. Raise and support the vehicle.

3. Remove the lower control arm.

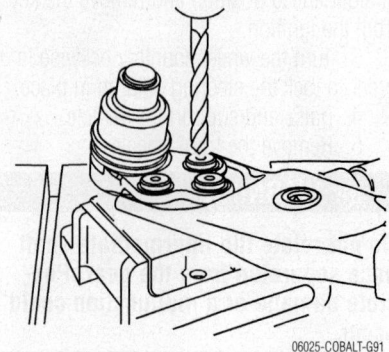

06025-COBALT-G91

Drilling out the lower ball joint rivets

4. Place the lower control arm in a vise.

5. Remove the ball joint to control arm rivets using the following procedure:

   a. Use a 3 mm (⅛ in.) drill bit in order to make a pilot hole through the rivets.

   b. Use a 13 mm (³¹⁄₆₄ in.) drill bit to complete drilling the rivets.

6. Remove the ball joint from the lower control arm.

To install:

7. Install the ball joint to the lower control arm.

8. Install the ball joint bolts and the nuts. Follow the instructions in the ball joint kit.

9. Tighten the ball joint bolts to 68 Nm (50 ft. lbs.).

10. Install the lower control arm.

11. Lower the vehicle.

Lower Control Arm

REMOVAL & INSTALLATION

1. Before servicing the vehicle, refer to the Precautions Section.

2. Raise and support the vehicle.

3. Remove the wheel.

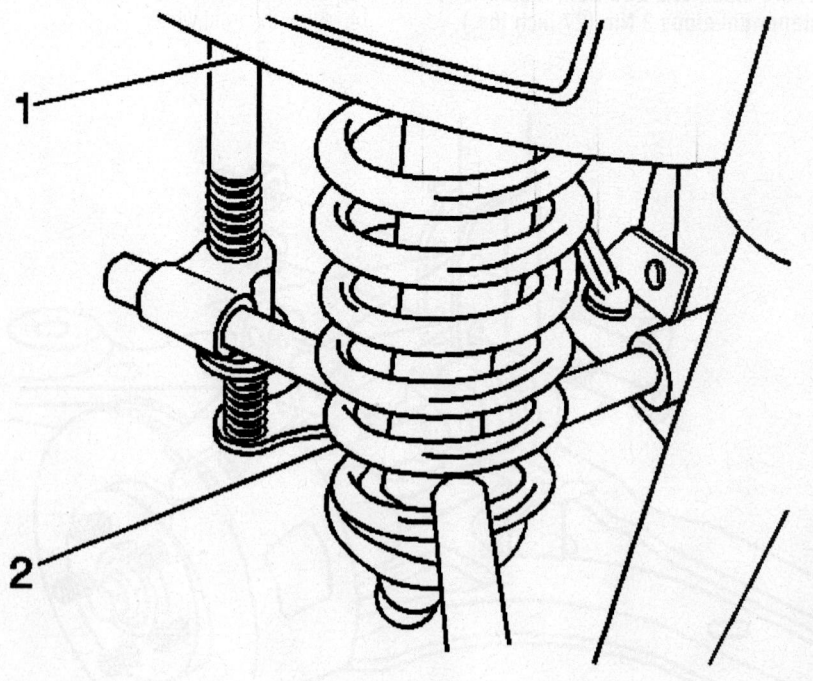

06025-COBALT-G89

Install the strut in tool J 45400

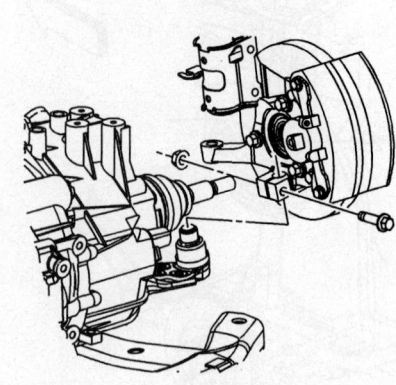

06025-COBALT-G92

Ball stud to steering knuckle pinch bolt and nut

Do not free the ball stud by using a pickle fork or a wedge-type tool. Damage to the seal or bushing may result.

4. Remove the ball stud to steering knuckle pinch bolt and nut.

5. Separate the ball stud from the steering knuckle.

6. Remove the rear frame bolt.

7. Remove the control arm to frame bolts.

8. Remove the control arm from the frame.

To install:

9. Insert the rear portion of the control arm into the frame.

10. Loosely install the rear frame bolt.

11. Lower the control arm and insert the ball stud into the steering knuckle.

❄❄ WARNING

The control arm contains 2 fore/aft movement limiting brackets. Failure to install these brackets will result in abnormal handling characteristics.

12. Install the fore/aft movement limiting brackets onto the control arm forward bushing.

13. Install both control arm to frame

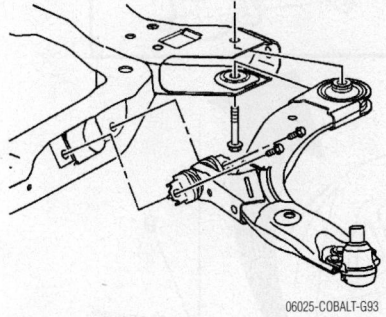

06025-COBALT-G93

Control arm to frame bolts

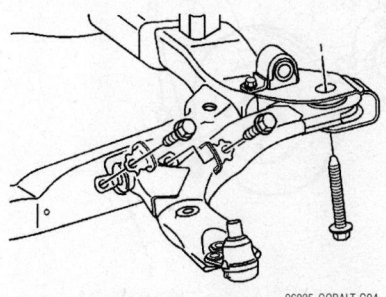

06025-COBALT-G94

Fore/aft movement limiting brackets

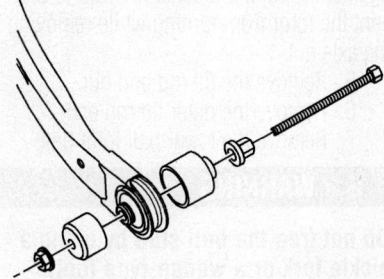

06025-COBALT-G95

Control arm bushing replacement

bolts. Tighten the bolts to 55 Nm (41 ft. lbs.).

14. Tighten the rear frame bolt. Tighten the bolt to 100 Nm plus 180 degree (74 ft. lbs.) plus 180 degrees.

➡ **The torque sequence must be followed in the order that is listed.**

15. Install the ball stud pinch bolt and nut.

   a. First Pass: Tighten the nut to 50 Nm (37 ft. lbs.).

   b. Back off nut ¾ turn

   c. Second Pass: Tighten the nut to 50 Nm (37 ft. lbs.) + 30 degrees.

16. Install the wheel.

17. Lower the vehicle.

18. Road test the vehicle in order to test for leads or pulls. Align as needed.

LOWER CONTROL ARM BUSHING REPLACEMENT

1. Before servicing the vehicle, refer to the Precautions Section.

2. Raise and support the vehicle.

3. Remove the lower control arm.

4. Wrap the control arm with a shop towel and place it in a vise.

➡ **Note the depth and orientation of the old bushing before removal.**

5. Using tool J 41211, pull the control arm bushing through the control arm.

6. Disassemble the tools and remove the bushing.

To install:

7. Place the NEW bushing to the tapered side of the control arm.

8. Using the tool, pull the control arm bushing through the opposite direction of the control arm.

9. Install the bushing to the same depth and orientation as noted during removal.

10. Remove the tool from the control arm.

11. Install the control arm.

REMOVAL & INSTALLATION

1. Before servicing the vehicle, refer to the Precautions Section.

2. Raise and support the vehicle.

3. Remove the front wheels.

4. Remove the rear transaxle mount.

5. Remove the steering gear.

6. Disconnect the stabilizer links from the stabilizer shaft.

7. Remove the stabilizer bar mounting clamp bolts and clamps from both sides of the vehicle.

8. Remove the bushings from the stabilizer bar.

9. Lift and rotate the stabilizer bar up and to the right.

10. Carefully remove the stabilizer bar from the right side of the vehicle.

11. Disconnect the stabilizer link from the strut assembly and remove from the vehicle.

To install:

12. Connect the stabilizer link to the strut assembly. Tighten the stabilizer link nut to 65 Nm (48 ft. lbs.).

13. Move the stabilizer bar into position from the right side of the vehicle.

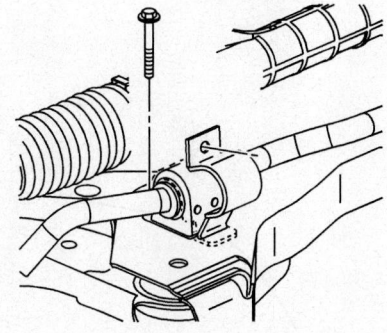

06025-COBALT-G97

Front stabilizer bar mounting clamps

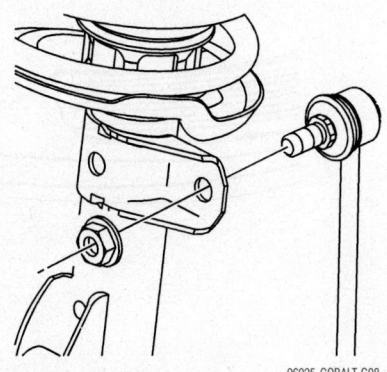

06025-COBALT-G98

End link-to-strut mounting

14. Install the stabilizer bushings on the stabilizer bar with the cut line facing rearward.

15. Install the stabilizer bar clamps and bolts. Tighten the bolts to 50 Nm (37 ft. lbs.).

16. Connect the stabilizer links to the stabilizer shaft. Tighten the stabilizer link nut to 80 Nm (59 ft. lbs.).

17. Install the steering gear.

18. Install the rear transaxle mount.

19. Install the front wheels.

20. Lower the vehicle.

Steering Knuckle

REMOVAL & INSTALLATION

1. Before servicing the vehicle, refer to the Precautions Section.

2. Raise and support the vehicle.

3. Remove the wheel.

4. Remove the axle shaft nut. Insert a drift or flat-bladed tool into the rotor and against the caliper bracket in order to prevent the rotor from turning while removing the axle nut.

5. Remove the tie rod end nut.

6. Remove the outer tie rod end.

7. Remove the lower ball joint bolt.

✳✳ WARNING

Do not free the ball stud by using a pickle fork or a wedge-type tool. Damage to the seal or bushing may result.

8. Remove the lower ball joint nut.

9. Remove the lower ball joint.

10. Remove the caliper bracket bolts.

✳✳ WARNING

Support the brake caliper with heavy mechanic's wire, or equivalent, whenever it is separated from its mount and the hydraulic flexible brake hose is still connected. Failure to support the caliper in this manner will cause the flexible brake hose to bear the weight of the caliper, which may cause damage to the brake hose and in turn may cause a brake fluid leak.

➡ **Do not disconnect the hydraulic brake flexible hose from the caliper.**

11. Remove the front brake rotor.

12. Remove the strut bolts.

13. Remove the wheel bearing bolt.

14. Remove the wheel bearing.

15. Remove the wheel bearing spacer.

16. Remove the steering knuckle.

17. Remove the road test the vehicle to verify alignment.

18. Installation is the reverse of removal. Observe the following torques:

- Hub: 115 Nm (85 ft. lbs.)
- Strut: 120 Nm (89 ft. lbs.)
- Caliper bracket: 115 Nm (85 ft. lbs.)
- Lower ball joint: 50 Nm (37 ft. lbs.), back off nut ¾ turn; 50 Nm (37 ft. lbs.) + 30 degrees

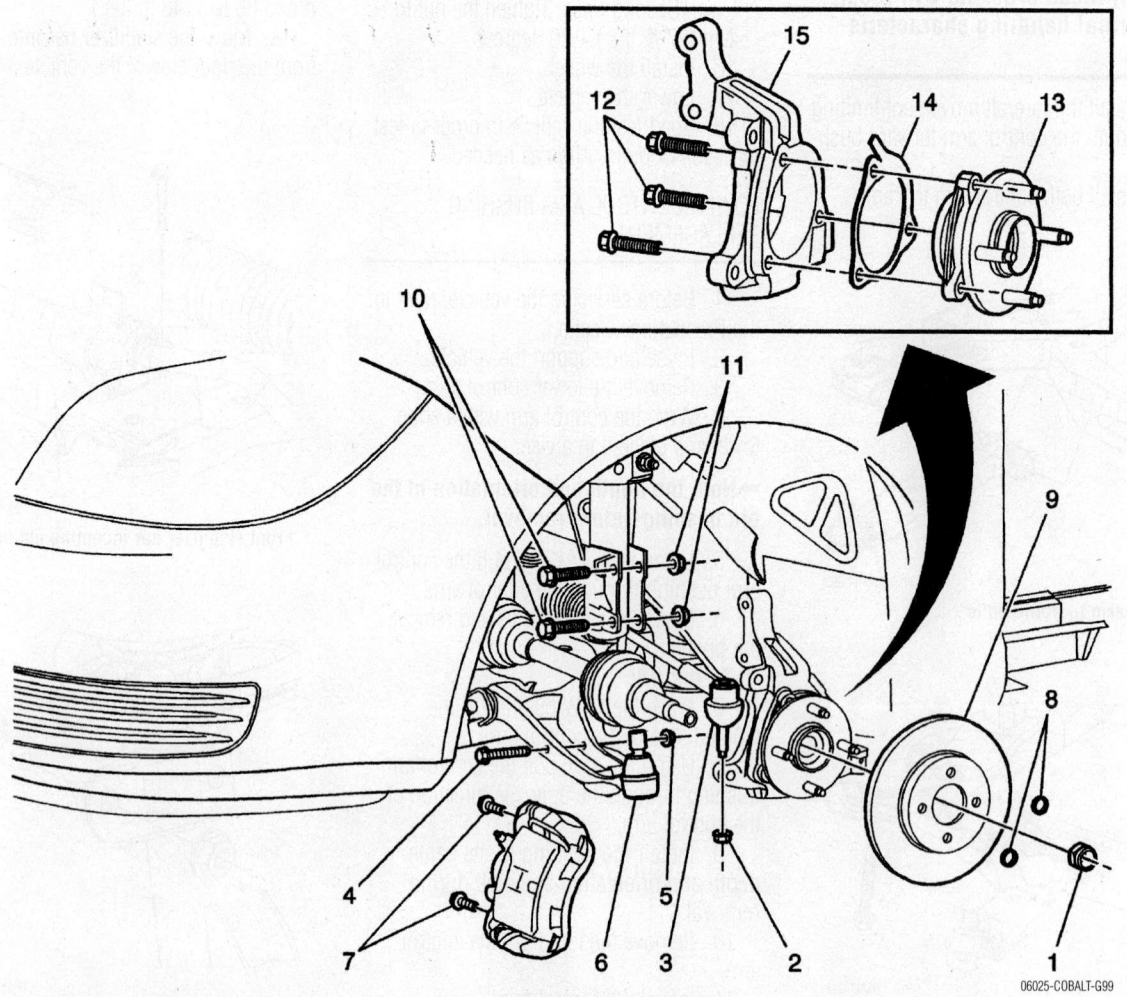

06025-COBALT-G99

Steering knuckle and related parts

Wheel Bearings

REMOVAL & INSTALLATION

1. Before servicing the vehicle, refer to the Precautions Section.
2. Raise and support the vehicle.
3. Remove the wheel.
4. Remove the rotor.
5. Remove the axle shaft nut.
6. Remove the wheel speed sensor connector.
7. Remove the wheel bearing bearing/hub assembly bolt.
8. Remove the wheel bearing/hub assembly bearing.
9. Remove the wheel bearing/hub assembly spacer.
10. Installation is the reverse of removal. Torque the axle shaft nut to 110 Nm (81 ft. lbs.); the hub bolts to 115 Nm (85 ft. lbs.).

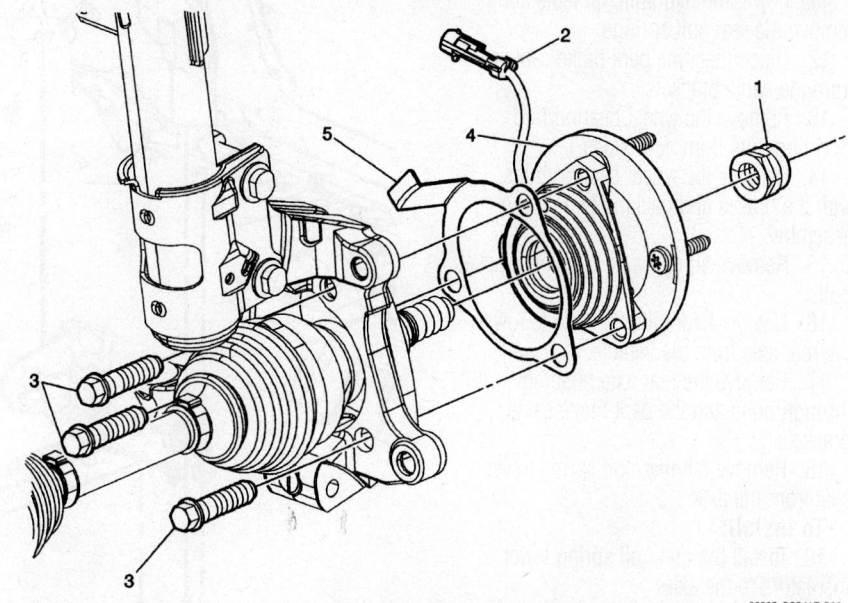

06025-COBALT-G96

Front hub/bearing removal

REAR SUSPENSION

Shock Absorber

REMOVAL & INSTALLATION

1. Before servicing the vehicle, refer to the Precautions Section.
2. Raise and support the vehicle.
3. Remove the wheel.
4. Support the rear axle with a tall jack-stand near the shock absorber.
5. Remove the upper and lower shock bolts.
6. Remove the shock from the vehicle.

To install:

7. Position the shock absorber to the vehicle.
8. Install NEW upper and lower shock bolts. Tighten the upper bolt to 90 Nm (66

06025-COBALT-G100

Support the rear axle with a tall jackstand near the shock absorber

ft. lbs.). Tighten the lower bolt to 110 Nm (81 ft. lbs.).

9. Remove the jackstand.
10. Install the wheel.
11. Lower the vehicle.

Coil Spring

REMOVAL & INSTALLATION

1. Before servicing the vehicle, refer to the Precautions Section.
2. Raise and support the vehicle.
3. Support the rear axle with tall jack stands near each rear shock absorber.
4. Remove the U-clips from the rear brake hose brackets at the rear axle.
5. Remove the lower shock bolts.
6. Using the tall jack stands, slowly lower the rear axle in order to remove tension from the rear springs.
7. Remove the spring.
8. Remove the upper spring seat/jounce bumper from the spring, while leaving the lower spring seat on the axle.

To install:

➡**The rear springs are indexed with the colored tag toward the rear of the vehicle. No up/down or side to side orientation is required.**

9. Install the upper spring seat/jounce bumper on the spring.
10. Install the spring with the spring tag toward the rear of the vehicle, making sure

the lower coil is seated into the lower spring seat.

11. Using the jack stands, raise the rear axle in order to compress the rear springs.
12. Install the lower shock absorber bolts.
13. Reposition the rear brake hoses in the axle brackets.
14. Install the U-clips to secure the brake hoses.
15. Lower the vehicle.

Rear Axle Beam

REMOVAL & INSTALLATION

1. Before servicing the vehicle, refer to the Precautions Section.
2. Raise and support the vehicle.
3. Remove the rear wheels.
4. Disconnect the left and right rear brake pipes from the rear brake hoses at the axle.
5. Disconnect the brake hoses from the axle brake hose bracket.
6. Plug the brake pipes and hoses in order to prevent additional brake fluid loss.
7. Disconnect both rear parking brake cables at the rear brake.
8. If applicable, disconnect the antilock brake system (ABS) harness connectors and disconnect from the axle.
9. Support the rear axle with a hydraulic lift table.
10. Remove the lower shock bolts.

11. Lower the hydraulic lift table and remove the rear coil springs.

12. Disconnect the park brake cables from the cable brackets.

13. Remove the wheel bearing/hub retaining nuts from both sides.

14. Remove the wheel bearing/hubs, with the brakes and backing plate as an assembly.

15. Remove all rear axle bushing bracket bolts.

16. Use the hydraulic lift table to lower the rear axle from the vehicle.

17. Remove the rear axle bushing through bolts and the park brake cable brackets.

18. Remove the rear coil spring lower seat from the axle.

To install:

19. Install the rear coil spring lower insulators to the axle.

20. Install the axle brackets to the axle bushings, with the alignment slot on the outboard side.

➡**The axle bushing through bolts must be installed with the bolt head facing inboard.**

21. Loosely install the bushing bolts and nuts.

22. Place the axle on the hydraulic lift table.

23. Raise the axle into position.

24. Hand tighten the axle bracket to body bolts just enough to hold the brackets flush to the body.

➡**The axle through bolts must be tightened with the axle at the correct trim height and prior to torquing the axle bracket to body bolts.**

25. Using the lift table, raise the axle to the proper trim height specification by measuring the vertical distance between the bottom edge of the upper spring seat and the bottom of the notch in the lower spring seat.

26. Before setting the D height measurement:

a. Set the tire pressure to the specifications shown on the certification label.

b. Check the fuel level. Add additional weight if necessary to simulate a full tank.

c. Make sure the passenger and rear compartments are empty, except for the spare tire.

d. Make sure the vehicle is on a flat and level surface, such as an alignment rack.

e. Check that all the vehicle doors are securely closed.

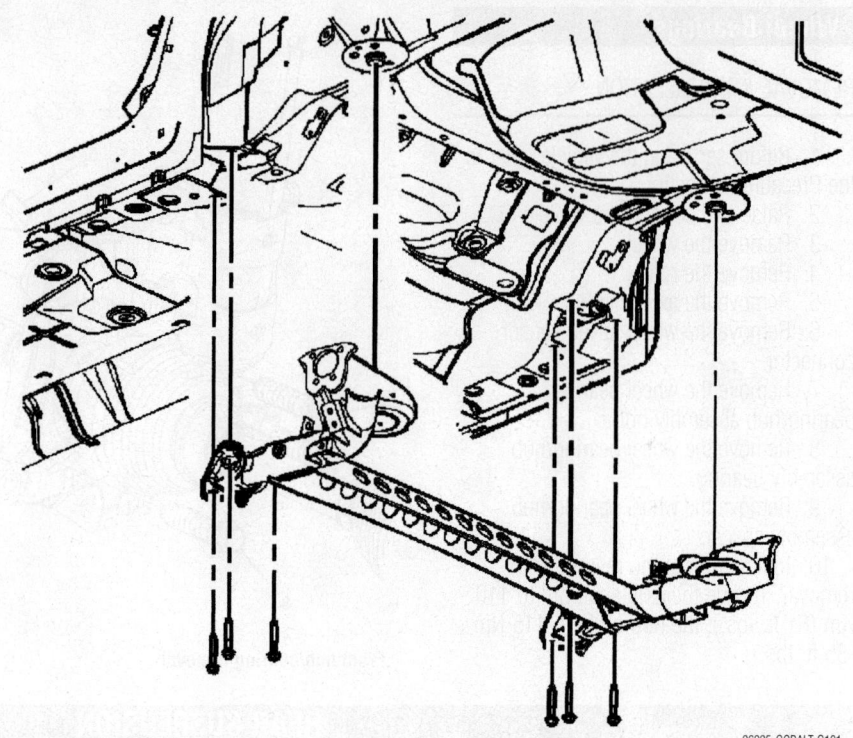

06025-COBALT-G101

Rear axle beam removal

f. Check that the vehicle hood and rear deck lids are securely closed.

g. Check for installed after market accessories or modifications that could affect trim height measurement.

➡**All dimensions are measured vertical to the ground. Trim height should be within +/-10 mm (+/-0.39 in.) to be considered correct.**

The D height dimension measurement determines the proper rear end ride height.

There is no adjustment procedure. Repair may require replacement of suspension components.

27. Use the following procedure to check the D dimension:

a. With the vehicle on a flat level surface, lift upward on the rear bumper 38 mm (1.59 in).

b. Gently remove your hands and allow the vehicle to settle.

c. Repeat the jouncing operation 2 more times.

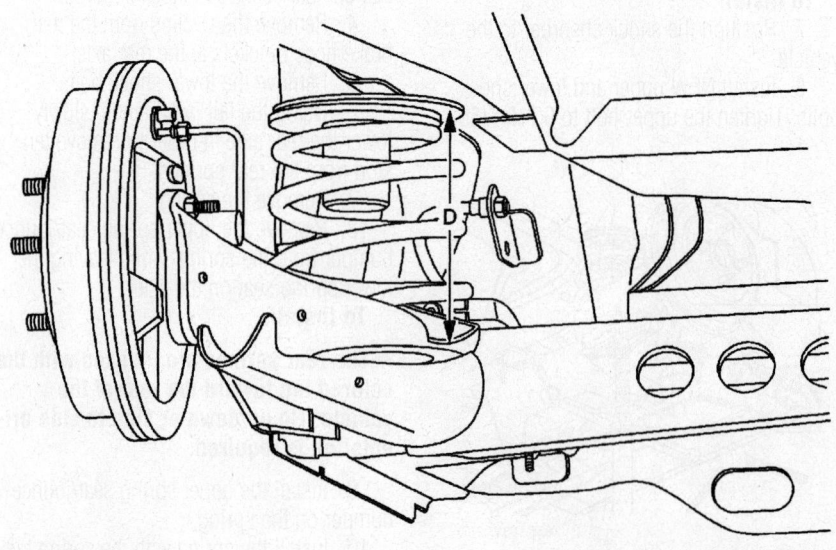

06025-COBALT-G102

D height measurement

d. Measure the D height for the left and right side of the vehicle. Measure the vertical distance between the bottom edge of the upper spring seat to the bottom of the notch in the lower spring seat.

e. Using your hands, jounce the front of the vehicle downward approximately 38 mm (1.59 in).

f. Gently remove your hands and allow the vehicle to settle .

g. Repeat the jouncing operation 2 more times.

h. Measure the D height dimension.

i. The true D height dimension number is the average of the high and the low measurements:

- 2.0L engine: 213mm (8.39 in.)
- 2.2L engine with P195/60R15 tires: 238mm (9.37 in.)
- 2.2L engine with P205/50R16 tires: 228mm (8.98 in.)

If these measurements are out of specifications, inspect for worn or damaged suspension components and/or collision damage.

28. Tighten the axle bushing through bolts. Tighten the bolts to 90 Nm (66 ft. lbs.) +60 degrees.

29. Insert two 12 mm diameter pins through the axle brackets into the underbody.

30. Align the left side axle bracket and snug down the bolts.

31. Align the right side axle bracket and snug down the bolts.

32. Tighten all the bracket to body bolts. Tighten the bolts to 90 Nm (66 ft. lbs.) + 45 degrees.

33. Install the wheel bearing/hubs, with the brakes and backing plate assemblies.

34. Install the bearing/hub nuts. Tighten the nuts to 45 Nm + 30 degrees (33 ft. lbs. + 30 degrees).

35. Connect the brake hoses to the rear axle brackets.

36. Connect the brake pipes to the brake hoses at the axle. Tighten the brake pipe fittings to 19 Nm (14 ft. lbs.).

37. Install the rear coil springs.

38. Install the lower shock bolts.

39. Lower and remove the hydraulic lift table.

40. If applicable, connect the ABS sensor harness connector and harness to axle retainer.

41. Connect the park brake cables to the axle brackets and rear brakes.

42. Bleed the brake system.

43. Install the rear wheels.

44. Lower the vehicle.

CONTROL ARM BUSHING REPLACEMENT

1. Before servicing the vehicle, refer to the Precautions Section.

2. Raise and support the vehicle.

3. Remove the rear wheels.

4. Place 2 screw type jack stands under both ends of the rear axle.

5. Remove the rear brake hose bracket attaching nuts from the body.

6. Detach the rear brake hose brackets

from the body allowing the lines to hang free.

7. Remove the lower shock bolts.

※※ WARNING

Do not kink the brake pipes while lowering the axle.

8. Lower the jacks in order to remove the coil springs.

9. Temporarily re-install the lower shock bolts to support the axle.

10. Remove the bushing bracket to body bolts from both ends of the rear axle.

11. Using the jackstands, raise the rear of the axle until the bushing brackets pivot away from the body.

12. Remove the axle bushing through bolts and remove the bushing brackets.

➡**Note the depth and orientation of the old bushing before removal.**

13. Using tool J 44570 , install tool J 44570-1 with the lip between the axle sleeve and bushing flange. It may require tapping with a hammer to fully seat the tool.

14. Insert J 44570-3 through tool J 44570-1 and the axle bushing.

15. Install the washer and nut by hand, tightening until the tool is snug.

16. Using a hammer, drive the bushing from the axle sleeve.

17. Disassemble the tool and remove the bushing.

To install:

➡**The axle bushings must be installed in the correct orientation as shown.**

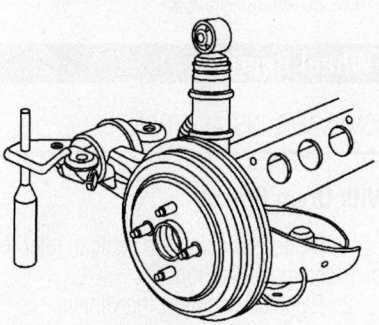

06025-COBALT-G103

Aligning the left side axle bracket

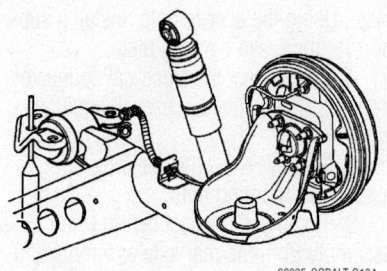

06025-COBALT-G104

Aligning the right side axle bracket

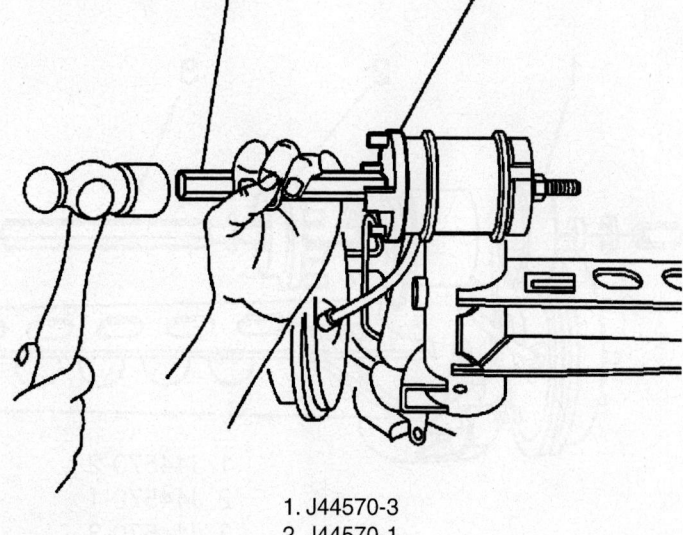

1. J44570-3
2. J44570-1

06025-COBALT-G105

Driving out the old bushings with tool set J4450

18. Slide the new bushings into the axle sleeve in the same orientation noted during removal. Make sure the rubber end is facing inboard and the largest void is in line with the wheel hub center.

19. Place the J 44570-1 onto the bushing. Make sure the bushing is still oriented correctly.

20. Insert the J 44570-3 through the J 44570-1 and the axle bushing.

21. Install the J 44570-2 bearing, washer, and nut.

22. Pull the bushing into the axle sleeve by holding the hex end of the threaded shaft while turning the nut.

23. Disassemble and remove the bushing installation tool from the axle.

24. Install the axle brackets to the axle

bushings with the alignment slot on the outboard side.

➡ The axle bushing through bolts must be installed with the bolt head facing inboard.

25. Loosely install the bushing bolts, park brake cable brackets and nuts.

26. Using the jack stands, lower the rear of the axle until the bushing brackets contact the body.

27. Hand tighten the axle bracket to body bolts just enough to hold the brackets flush to the body.

➡ The axle through bolts must be tightened with the axle at the correct trim height and prior to torquing the axle bracket to body bolts.

28. Using the jack stands, raise the axle to the proper trim height specification by measuring the vertical distance between the bottom edge of the upper spring seat and the bottom of the notch in the lower spring seat. Refer to Rear Axle Beam Removal & Installation for D Height Measurement.

29. Tighten the axle bushing through bolts. Tighten the bolts to 90 Nm (66 ft. lbs.) +60 degrees

30. Insert two 12 mm diameter pins through the axle brackets into the underbody.

31. Align the left side axle bracket and snug down the bolts.

32. Align the right side axle bracket and snug down the bolts.

33. Tighten all of the bracket-to-body bolts. Tighten the bolts to 90 Nm (66 ft. lbs.) +30 degrees.

34. With the axle supported by the jack stands, remove the lower shock bolts.

35. Lower the jacks in order to install the coil springs.

36. Install the coil springs, making sure the colored tag is facing the rear of the vehicle.

37. Raise the jacks until the springs are slightly compressed in order to install the lower shock bolts. Tighten the lower bolts to 110 Nm (81 ft. lbs.).

38. Remove the jack stands.

39. Reposition the rear brake hose brackets to the body.

40. Install the brake hose bracket attaching nuts.

41. Install the rear wheels.

42. Lower the vehicle.

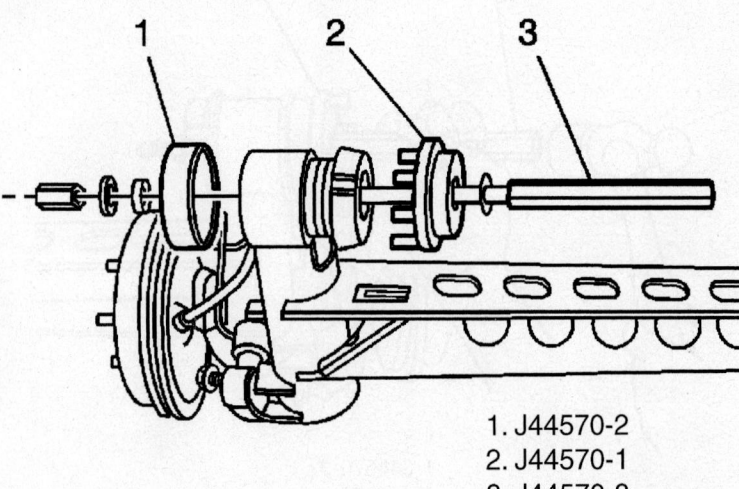

1. Rubber end
2. Largest void
3. Wheel hub center

06025-COBALT-G106

Correct axle bushing orientation

1. J44570-2
2. J44570-1
3. J44570-3

06025-COBALT-G107

Axle bushing installation

Wheel Bearings

REMOVAL & INSTALLATION

With Drum Brakes

1. Before servicing the vehicle, refer to the Precautions Section.

2. Raise and support the vehicle.

3. Remove the tire and wheel assembly.

4. Remove the brake drum.

5. Remove the plug from the drum brake actuator access hole in the backing plate. Using the access hole, install a support for the brake backing plate.

6. Disconnect the electrical connector from the wheel speed sensor, if equipped with ABS.

7. Remove the wheel bearing/hub assembly mounting nuts.

8. Remove the wheel bearing/hub assembly from the rear axle assembly and brake backing plate.

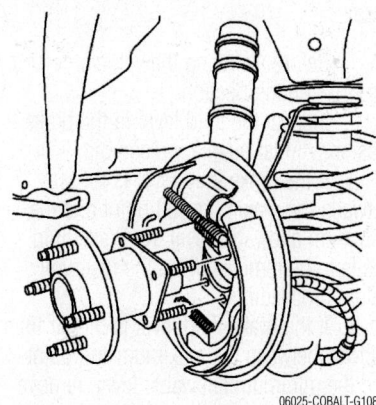

Rear wheel bearing removal with drum brakes

To install:

9. Install the wheel bearing/hub assembly to the brake backing plate and the rear axle assembly.

10. Install the wheel bearing/hub assembly mounting nuts to the axle assembly. Tighten the nuts evenly, in a cross-pattern. Tighten the nuts to 45 Nm (33 ft. lbs.) plus 30 degrees.

11. Connect the electrical connector to the wheel speed sensor, if equipped with ABS.

---

12. Remove the support from the brake backing plate.

13. Install the plug to the drum brake actuator access hole in the backing plate.

14. Install the brake drum.

15. Install the tire and wheel assembly.

16. Lower the vehicle.

With Disc Brakes

1. Before servicing the vehicle, refer to the Precautions Section.

2. Raise and support the vehicle.

3. Remove the tire and wheel assembly.

4. Without disconnecting the hydraulic brake flex hose, remove and support the rear brake caliper and bracket as an assembly, and remove the rear brake rotor.

5. Disconnect the electrical connector from the wheel speed sensor.

6. Remove the wheel bearing/hub assembly mounting nuts.

7. Remove the wheel bearing/hub assembly and the disc brake backing plate from the rear axle assembly.

To install:

8. Install the wheel bearing/hub assembly and the brake backing plate to the rear axle assembly.

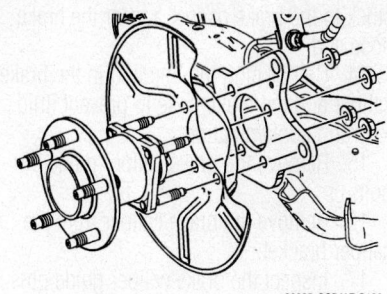

Rear wheel bearing removal with disc brakes

9. Install the wheel bearing/hub assembly mounting nuts to the axle assembly. Tighten the nuts evenly, in a cross-pattern. Tighten the nuts to 45 Nm (33 ft. lbs.) plus 30 degrees.

10. Connect the electrical connector to the wheel speed sensor.

11. Install the brake rotor, and install the brake caliper and bracket as an assembly.

12. Install the tire and wheel assembly.

13. Lower the vehicle.

DISC BRAKES

Brake Caliper

REMOVAL & INSTALLATION

Front

1. Before servicing the vehicle, refer to the Precautions Section.

2. Inspect the fluid level in the brake master cylinder reservoir.

3. If the brake fluid level is midway between the maximum-full point and the minimum allowable level, no brake fluid needs to be removed from the reservoir before proceeding.

4. If the brake fluid level is higher than midway between the maximum-full point and the minimum allowable level, remove brake fluid to the midway point before proceeding.

5. Raise and support the vehicle.

6. Remove the tire and wheel assembly.

7. Install and firmly hand tighten 2 wheel nuts to opposite wheel studs in order to retain the rotor to the hub.

8. Install a large C-clamp over the body of the brake caliper with the C-clamp ends against the rear of the caliper body and against the outer brake pad.

---

9. Tighten the C-clamp until the caliper piston is compressed into the caliper bore enough to allow the caliper to slide past the brake rotor.

10. Remove the C-clamp from the caliper.

---

11. Remove the brake hose-to-caliper bolt from the brake caliper.

12. Remove the brake hose from the brake caliper.

13. Remove and discard the 2 copper brake hose gaskets. These gaskets may be

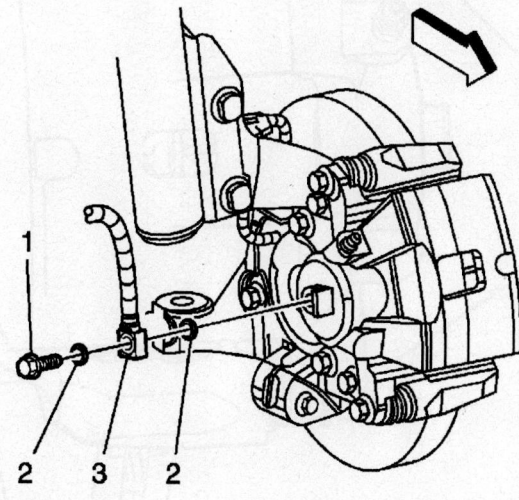

1. Bolt
2. Washer
3. Brake hose

Front brake hose

stuck to the brake caliper and/or the brake hose end.

14. Cap or plug the opening in the brake caliper and the brake hose to prevent fluid loss and contamination.

15. Remove the brake caliper guide pin bolts.

16. Remove the brake caliper from the caliper bracket.

17. Inspect the brake caliper guide pins for freedom of movement, and inspect the condition of the guide pin boots. Move the guide pins inboard and outboard within the bracket bores, without disengaging the slides from the boots, and observe for the following:

- Restricted caliper guide pin movement
- Looseness in the brake caliper mounting bracket
- Seized or binding caliper guide pins
- Split or torn boots

18. If any of the conditions listed are found, the brake caliper guide pins and/or boots require replacement.

To install:

19. Install the brake caliper to the brake caliper bracket.

20. Install the brake caliper guide pin

bolts. Tighten the bolts to 34 Nm (25 ft. lbs.).

21. Remove the caps or plugs from the brake caliper opening and the brake hose.

✳✳ WARNING

Do not reuse the copper brake hose gaskets.

22. Install NEW copper brake hose gaskets to the brake hose-to-caliper bolt and to the brake hose.

23. Install the brake hose and the brake hose-to-brake caliper bolt to the brake caliper. Tighten the bolt to 48 Nm (35 ft. lbs.).

24. Bleed the hydraulic brake system.

25. Remove the wheel nuts retaining the brake rotor to the wheel hub.

26. Install the tire and wheel assembly.

27. Lower the vehicle.

28. With the engine OFF, gradually apply the brake pedal to approximately ⅔ of its travel distance.

29. Slowly release the brake pedal.

30. Wait 15 seconds, then gradually apply the brake pedal approximately ⅔ of its travel distance again until a firm brake pedal apply is obtained. This will properly seat the brake caliper pistons and brake pads.

Rear

1. Before servicing the vehicle, refer to the Precautions Section.

2. Inspect the fluid level in the brake master cylinder auxiliary reservoir.

3. If the brake fluid level is midway between the maximum-full point and the minimum allowable level, no brake fluid needs to be removed from the reservoir before proceeding.

4. If the brake fluid level is higher than midway between the maximum-full point and the minimum allowable level, remove brake fluid to the midway point before proceeding.

5. Release the park brake lever boot from the floor console by applying light pressure inward on the sides of the boot retainer, and pull the boot back.

6. Release the tension from the park brake cables. With the park brake lever in the released position, using ONLY HAND TOOLS, loosen the adjusting nut completely to the end of the front cable threaded rod.

7. Raise and support the vehicle.

8. Remove the tire and wheel assembly.

9. Install and firmly hand tighten 2 wheel nuts to opposite wheel studs in order to retain the rotor to the hub.

10. Release the park brake cable end from the lever on the caliper.

11. Release the retaining tabs securing the park brake cable to the bracket on the caliper.

12. Install a large C-clamp, over the body of the brake caliper with the C-clamp ends against the rear of the caliper body and against the outer brake pad.

✳✳ WARNING

When using a large C-clamp to compress a caliper piston into a caliper bore of a caliper equipped with an integral park brake mechanism, do not exceed more than 1 mm (0.039 in.) of piston travel. Exceeding this amount of piston travel will cause damage to the internal adjusting mechanism and/or the integral park brake mechanism.

13. Tighten the C-clamp just enough to compress the caliper piston 1 mm (0.039 in.) of travel only.

14. Remove the C-clamp from the caliper.

15. Remove the brake hose-to-caliper bolt from the brake caliper.

16. Remove the brake hose from the brake caliper.

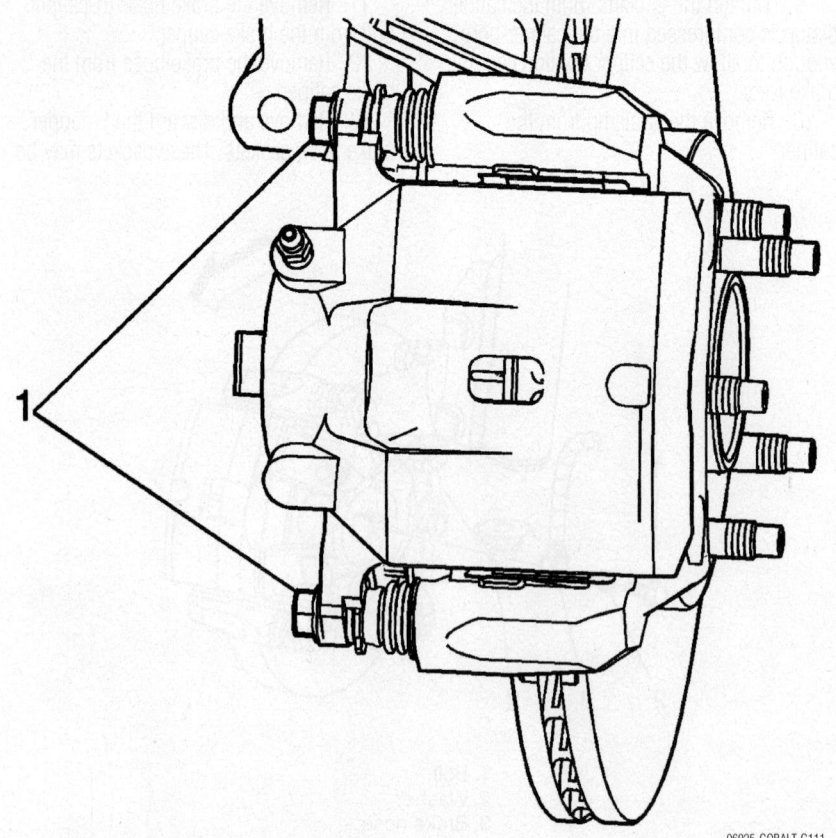

Front caliper pins

06025-COBALT-G111

17. Remove and discard the 2 copper brake hose gaskets. These gaskets may be stuck to the brake caliper and/or the brake hose end.

18. Cap or plug the opening in the brake caliper and the brake hose to prevent fluid loss and contamination.

19. While using a wrench on the flats of the caliper guide pins, remove the brake caliper guide pin bolts.

20. Remove the brake caliper from the caliper bracket.

21. Inspect the brake caliper guide pins for freedom of movement, and inspect the condition of the guide pin boots. Move the guide pins inboard and outboard within the bracket bores, without disengaging the slides from the boots, and observe for the following:

- Restricted caliper guide pin movement
- Looseness in the brake caliper mounting bracket
- Seized or binding caliper guide pins
- Split or torn boots

22. If any of the conditions listed are found, the brake caliper guide pins and/or boots require replacement.

To install:

23. Install the brake caliper to the caliper bracket.

24. While using a wrench on the flats of the caliper guide pins, install the brake caliper guide pin bolts. Tighten the bolts to 34 Nm (25 ft. lbs.).

25. Press the park brake cable end fitting into the bracket on the caliper to secure the retaining tabs.

26. Secure the park brake cable end to the lever on the caliper.

27. Remove the caps or plugs from the brake caliper opening and the brake hose.

✳✳ WARNING

Do not reuse the copper brake hose gaskets.

28. Install NEW copper brake hose gaskets to the brake hose-to-caliper bolt and to the brake hose.

29. Install the brake hose and the brake hose-to-brake caliper bolt to the caliper. Tighten the bolt to 48 Nm (35 ft. lbs.).

30. Bleed the hydraulic brake system.

31. Remove the wheel nuts retaining the brake rotor to the wheel hub.

32. Install the tire and wheel assembly.

33. Lower the vehicle.

34. With the engine OFF, gradually apply the brake pedal to approximately ⅔ of its travel distance.

35. Slowly release the brake pedal.

36. Wait 15 seconds, then gradually apply the brake pedal approximately ⅔ of its travel distance again until a firm brake pedal apply is obtained. This will properly seat the brake caliper pistons and brake pads.

37. Adjust the park brake cable tension.

38. Position the park brake lever boot to the floor console and press the boot retainer into place to secure.

Disc Brake Pads

REMOVAL AND INSTALLATION

Front

1. Before servicing the vehicle, refer to the Precautions Section.

2. Inspect the fluid level in the brake master cylinder auxiliary reservoir.

3. If the brake fluid level is midway between the maximum-full point and the minimum allowable level, no brake fluid needs to be removed from the reservoir before proceeding.

4. If the brake fluid level is higher than midway between the maximum-full point and the minimum allowable level, remove brake fluid to the midway point before proceeding.

5. Raise and support the vehicle.

6. Remove the tire and wheel assembly.

7. Install and firmly hand tighten 2 wheel nuts to opposite wheel studs in order to retain the rotor to the hub.

8. Using a piston compressing tool, the caliper piston is compressed into the caliper bore enough to allow the caliper to slide past the brake rotor. Or, install a large C-clamp over the body of the brake caliper with the C-clamp ends against the rear of the caliper body and against the outboard brake pad.

9. Tighten the C-clamp evenly until the caliper piston is compressed into the caliper bore enough to allow the caliper to slide past the brake rotor.

10. Remove the C-clamp from the caliper.

11. Remove the brake caliper lower guide pin bolt.

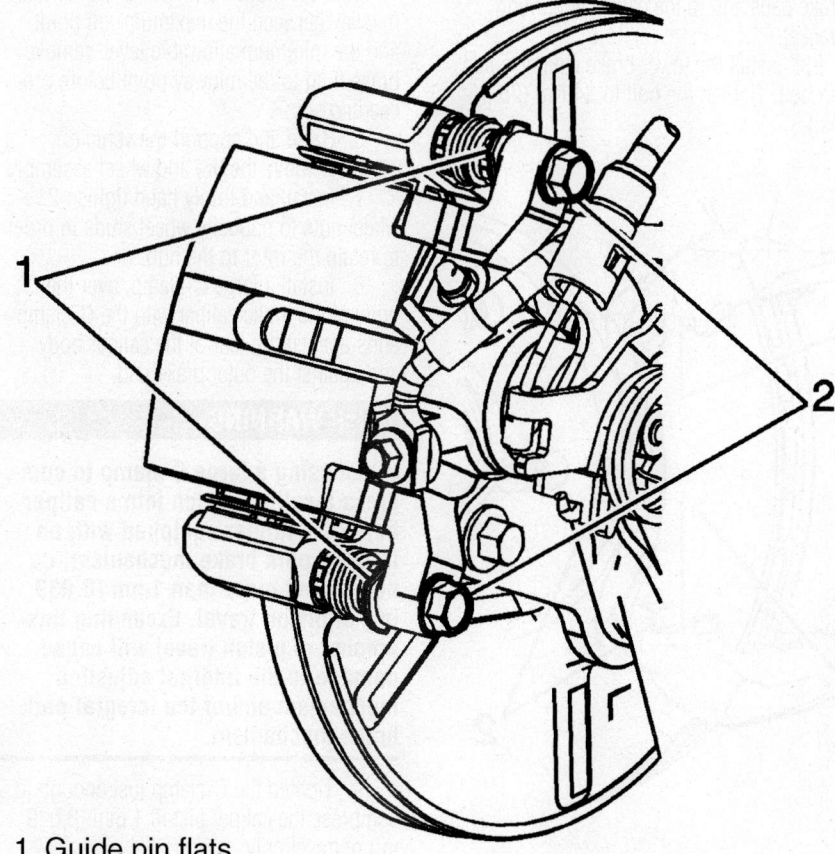

1. Guide pin flats
2. Guide pin bolts

06025-COBALT-G112

Rear caliper

Support the brake caliper with heavy mechanic's wire, or equivalent, whenever it is separated from its mount and the hydraulic flexible brake hose is still connected. Failure to support the caliper in this manner will cause the flexible brake hose to bear the weight of the caliper, which may cause damage to the brake hose and in turn may cause a brake fluid leak.

12. Without disconnecting the hydraulic brake flexible hose, pivot the caliper upward and secure the caliper with heavy mechanics wire, or equivalent.

13. Remove the brake pads from the caliper mounting bracket.

14. Remove the brake pad retainers (1) from the caliper bracket.

15. Fully compress the piston in its bore.

16. Thoroughly clean the brake pad hardware mating surfaces of the caliper bracket (2), of any debris and corrosion.

17. Inspect the brake caliper guide pins for freedom of movement, and inspect the condition of the guide pin boots. Move the guide pins inboard and outboard within the bracket bores, without disengaging the slides from the boots, and observe for the following:

- Restricted caliper guide pin movement
- Looseness in the brake caliper mounting bracket
- Seized or binding caliper guide pins
- Split or torn boots

18. If any of the conditions listed are found, the brake caliper guide pins and/or boots require replacement.

To install:

19. Apply a very thin coating of high temperature silicone brake lubricant to the pad hardware mating surfaces of the caliper bracket (2) only.

20. Install the brake pad retainers to the brake caliper bracket.

➡**The wear sensor equipped disc brake pad must be mounted inboard of the rotor with the leading edge of the sensor facing the brake rotor during forward wheel rotation, or at the top of the pad when installed in vehicle position.**

21. Install the brake pads to the caliper bracket.

22. Remove the support, and rotate the brake caliper into position over the disc brake pads and to the caliper mounting bracket.

23. Install the lower brake caliper guide pin bolt. Tighten the bolt to 34 Nm (25 ft. lbs.).

24. Remove the wheel nuts retaining the brake rotor to the hub.

25. Install the tire and wheel assembly.

26. Lower the vehicle.

27. With the engine OFF, gradually apply the brake pedal approximately ⅔ of its travel distance.

28. Slowly release the brake pedal.

29. Wait 15 seconds, then gradually apply the brake pedal approximately ⅔ of its travel distance again until a firm brake pedal apply is obtained. This will properly seat the brake caliper pistons and brake pads.

30. Fill the master cylinder auxiliary reservoir to the proper level.

31. Burnish the pads and rotors.

Rear

1. Before servicing the vehicle, refer to the Precautions Section.

2. Inspect the fluid level in the brake master cylinder auxiliary reservoir.

3. If the brake fluid level is midway between the maximum-full point and the minimum allowable level, no brake fluid needs to be removed from the reservoir before proceeding.

4. If the brake fluid level is higher than midway between the maximum-full point and the minimum allowable level, remove brake fluid to the midway point before proceeding.

5. Raise and support the vehicle.

6. Remove the tire and wheel assembly.

7. Install and firmly hand tighten 2 wheel nuts to opposite wheel studs in order to retain the rotor to the hub.

8. Install a large C-clamp, over the body of the brake caliper with the C-clamp ends against the rear of the caliper body and against the outer brake pad.

When using a large C-clamp to compress a caliper piston into a caliper bore of a caliper equipped with an integral park brake mechanism, do not exceed more than 1mm (0.039 in.) of piston travel. Exceeding this amount of piston travel will cause damage to the internal adjusting mechanism and/or the integral park brake mechanism.

9. Tighten the C-clamp just enough to compress the caliper piston 1 mm (0.039 in.) of travel only.

10. Remove the C-clamp from the caliper.

11. While using a wrench on the flats of

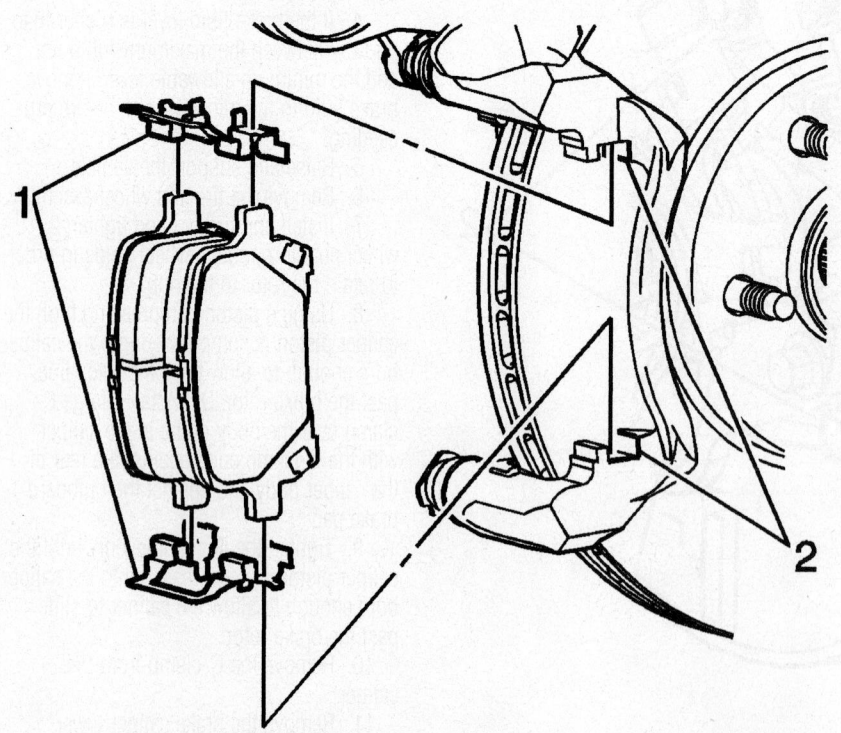

Front brake pads and retainers

06025-COBALT-G113

the caliper guide pins, remove the brake caliper guide pin bolts.

WARNING

Support the brake caliper with heavy mechanic's wire, or equivalent, whenever it is separated from its mount and the hydraulic flexible brake hose is still connected. Failure to support the caliper in this manner will cause the flexible brake hose to bear the weight of the caliper, which may cause damage to the brake hose and in turn may cause a brake fluid leak.

12. Without disconnecting the hydraulic brake flexible hose, remove the caliper from the mounting bracket and secure the caliper with heavy mechanics wire, or equivalent.
13. Remove the brake pads from the caliper mounting bracket.
14. Remove the brake pad retainers (1) from the caliper bracket.
15. Thoroughly clean the brake pad hardware mating surfaces of the caliper bracket (2), of any debris and corrosion.
16. Inspect the brake caliper guide pins for freedom of movement, and inspect the condition of the guide pin boots. Move the guide pins inboard and outboard within the bracket bores, without disengaging the slides from the boots, and observe for the following:
    - Restricted caliper guide pin movement
    - Looseness in the brake caliper mounting bracket
    - Seized or binding caliper guide pins
    - Split or torn boots
17. If any of the conditions listed are found, the brake caliper guide pins and/or boots require replacement.
18. Using a spanner wrench type caliper piston installer, fully retract the piston into the caliper bore.

To install:

19. Apply a very thin coating of high temperature silicone brake lubricant to the pad hardware mating surfaces of the caliper bracket only.
20. Install the brake pad retainers to the brake caliper bracket.

➡The wear sensor equipped disc brake pad must be mounted inboard of the rotor with the leading edge of the sensor facing the brake rotor during forward wheel rotation, or at the bottom of the pad when installed in vehicle position.

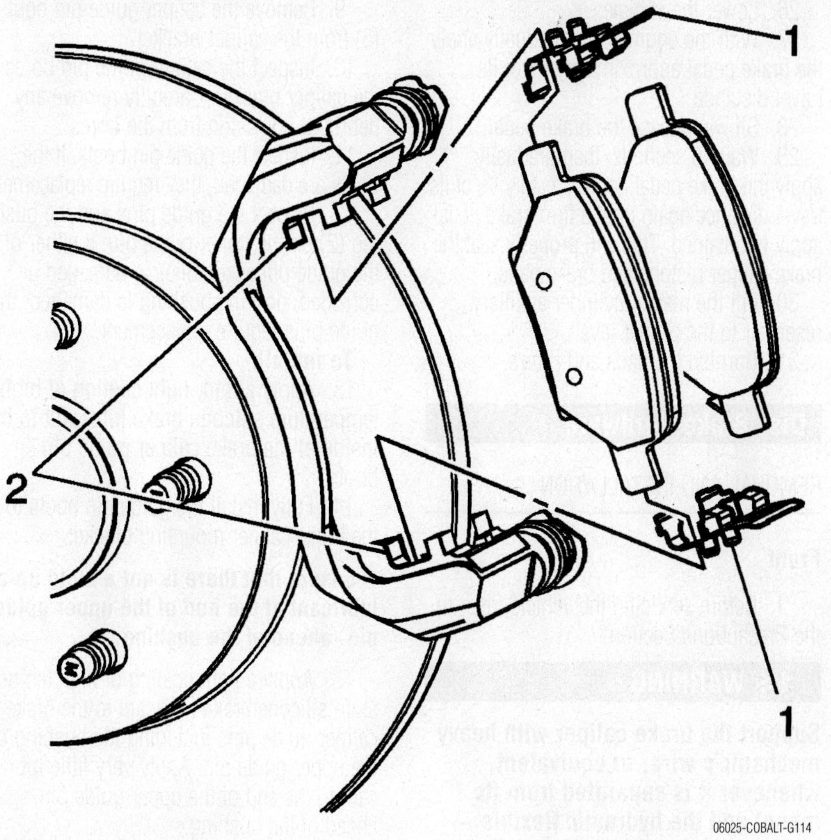
1
2
1
06025-COBALT-G114

Rear brake pads and retainers

21. Install the brake pads to the caliper bracket.
22. Remove the support, and install the caliper into position over the disc brake pads and to the caliper mounting bracket.
23. While using a wrench on the flats of the caliper guide pins, install the brake caliper guide pin bolts. Tighten the bolts to 34 Nm (25 ft. lbs.).
24. Remove the wheel nuts retaining the brake rotor to the hub.
25. Install the tire and wheel assembly.

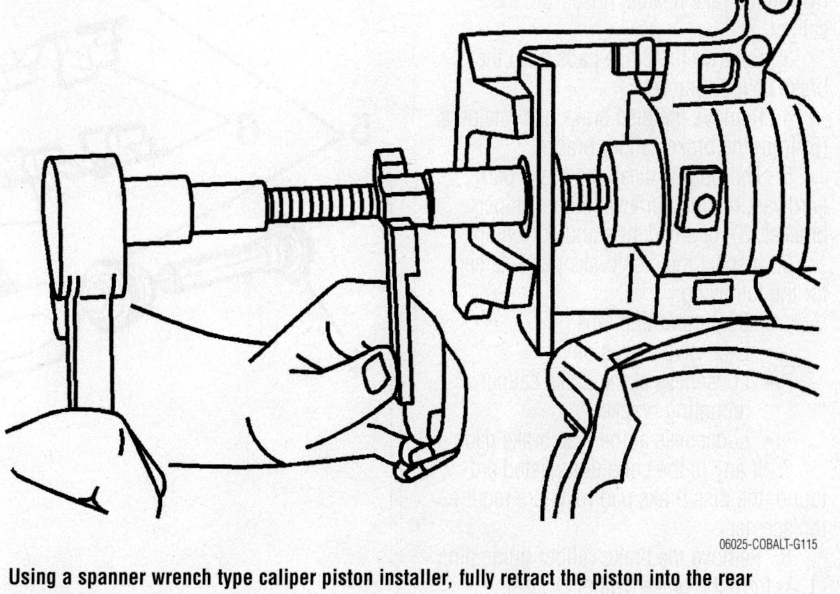
06025-COBALT-G115

Using a spanner wrench type caliper piston installer, fully retract the piston into the rear caliper bore

26. Lower the vehicle.

27. With the engine OFF, gradually apply the brake pedal approximately ⅔ of its travel distance.

28. Slowly release the brake pedal.

29. Wait 15 seconds, then gradually apply the brake pedal approximately ⅔ of its travel distance again until a firm brake pedal apply is obtained. This will properly seat the brake caliper pistons and brake pads.

30. Fill the master cylinder auxiliary reservoir to the proper level.

31. Burnish the pads and rotors.

Disc Brake Hardware

REMOVAL AND INSTALLATION

Front

1. Before servicing the vehicle, refer to the Precautions Section.

✳✳ WARNING

Support the brake caliper with heavy mechanic's wire, or equivalent, whenever it is separated from its mount and the hydraulic flexible brake hose is still connected. Failure to support the caliper in this manner will cause the flexible brake hose to bear the weight of the caliper, which may cause damage to the brake hose and in turn may cause a brake fluid leak.

2. Remove the brake caliper from the brake caliper mounting bracket and support the brake caliper with heavy mechanic's wire, or equivalent. Do NOT disconnect the hydraulic brake flexible hose from the caliper.

3. Remove the brake pads from the brake caliper bracket.

4. Remove the disc brake pad retainers (6) from the brake caliper bracket.

5. Thoroughly clean the brake pad hardware mating surfaces of the caliper bracket (3), of any debris and corrosion.

6. Inspect the disc brake pad retainers for the following:
- Bent mounting tabs
- Excessive corrosion
- Looseness at the brake caliper mounting bracket
- Looseness at the disc brake pads

7. If any of the conditions listed are found, the disc brake pad retainers require replacement.

8. Remove the brake caliper guide pins (1, 4) from the brake caliper bracket.

9. Remove the caliper guide pin boots (5) from the caliper bracket.

10. Inspect the caliper guide pin bores in the caliper bracket. Carefully remove any debris or corrosion from the bores.

11. Inspect the guide pin boots. If the boots are damaged, they require replacement.

12. Inspect the guide pins and the bushing (2) on the upper guide pin. If either of the guide pin assemblies is damaged or corroded, or if the bushing is damaged, the guide pins require replacement.

To install:

13. Apply a thin, light coating of high temperature silicone brake lubricant to the inside of the brake caliper guide pin boots.

14. Fully install the guide pin boots to the brake caliper mounting bracket.

➡**Ensure that there is not a build up of lubricant at the end of the upper guide pin, ahead of the bushing.**

15. Apply a light coating of high temperature silicone brake lubricant to the brake caliper guide pins including the bushing on the upper guide pin. Apply very little lubricant to the end of the upper guide pin, ahead of the bushing.

16. Install the brake caliper guide pins to the caliper mounting bracket. Ensure that

the rim of the guide pin boots is fully seated in the groove on the guide pins.

17. Apply a very thin coating of high temperature silicone brake lubricant to the pad hardware mating surfaces of the caliper bracket only.

18. If reusing the brake pad retainers, clean the brake pad mating surfaces of the brake pad retainers.

19. Install the brake pad retainers to the brake caliper bracket.

✳✳ WARNING

The wear sensor equipped disc brake pad must be mounted inboard of the rotor with the leading edge of the sensor facing the brake rotor during forward wheel rotation, or at the top of the pad when installed in vehicle position.

20. Install the brake pads to the brake caliper bracket.

21. Remove the support and reposition the brake caliper over the brake pads and to the mounting bracket.

Rear

1. Before servicing the vehicle, refer to the Precautions Section.

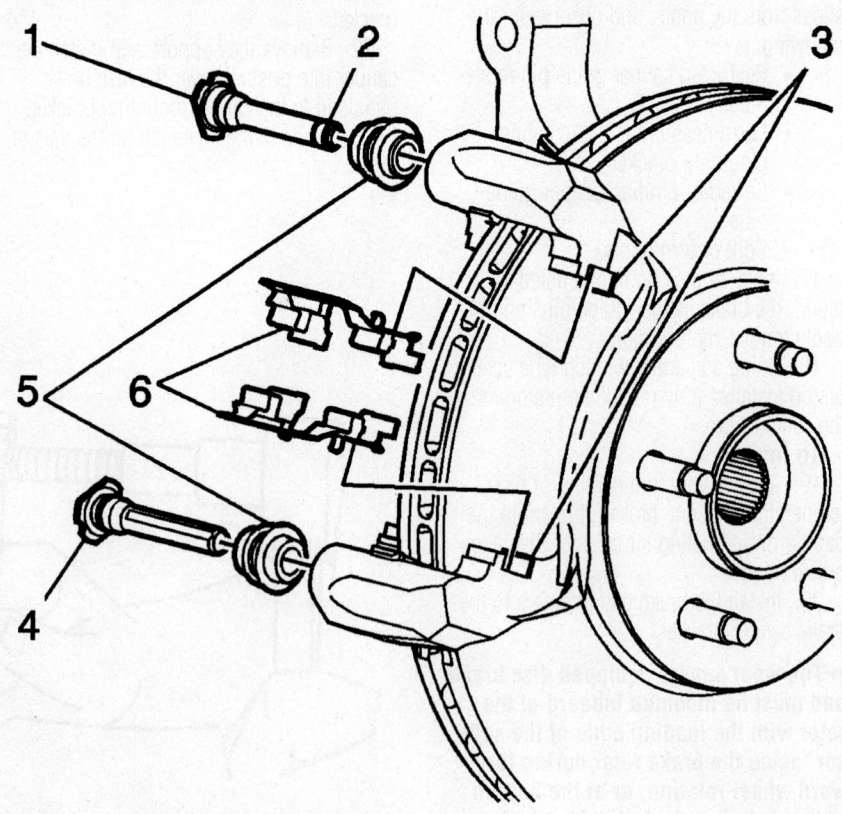

Front disc brake hardware

06025-COBALT-G116

Support the brake caliper with heavy mechanic's wire, or equivalent, whenever it is separated from its mount and the hydraulic flexible brake hose is still connected. Failure to support the caliper in this manner will cause the flexible brake hose to bear the weight of the caliper, which may cause damage to the brake hose and in turn may cause a brake fluid leak.

2. Remove the brake caliper from the brake caliper mounting bracket and support the brake caliper with heavy mechanic's wire, or equivalent. Do NOT disconnect the hydraulic brake flexible hose from the caliper.

3. Remove the brake pads from the brake caliper bracket.

4. Remove the disc brake pad retainers from the brake caliper bracket.

5. Thoroughly clean the brake pad hardware mating surfaces of the caliper bracket, of any debris and corrosion.

6. Inspect the disc brake pad retainers for the following:
- Bent mounting tabs
- Excessive corrosion
- Looseness at the brake caliper mounting bracket

- Looseness at the disc brake pads

7. If any of the conditions listed are found, the disc brake pad retainers require replacement.

8. Remove the brake caliper guide pins (1, 2) from the brake caliper bracket.

9. Remove the caliper guide pin boots (4) from the caliper bracket.

10. Inspect the caliper guide pin bores in the caliper bracket. Carefully remove any debris or corrosion from the bores.

11. Inspect the guide pin boots. If the boots are damaged, they require replacement.

12. Inspect the guide pins and the bushing (3) on the lower guide pin. If either of the guide pin assemblies is damaged or corroded, or if the bushing is damaged, the guide pins require replacement.

To install:

13. Apply a thin, light coating of high temperature silicone brake lubricant to the inside of the brake caliper guide pin boots.

14. Fully install the guide pin boots to the brake caliper mounting bracket.

➡**Ensure that there is not a build up of lubricant at the end of the lower guide pin, ahead of the bushing.**

15. Apply a light coating of high temperature silicone brake lubricant to the brake caliper guide pins, including the bushing on the lower guide pin. Apply very little lubri-

cant to the end of the lower guide pin, ahead of the bushing.

16. Install the brake caliper guide pins to the caliper mounting bracket. Ensure that the rim of the guide pin boots is fully seated in the groove on the guide pins.

17. Apply a very thin coating of high temperature silicone brake lubricant to the pad hardware mating surfaces of the caliper bracket only.

18. If reusing the brake pad retainers, clean the brake pad mating surfaces of the brake pad retainers.

19. Install the brake pad retainers to the brake caliper bracket.

The wear sensor equipped disc brake pad must be mounted inboard of the rotor with the leading edge of the sensor facing the brake rotor during forward wheel rotation, or at the bottom of the pad when installed in vehicle position.

20. Install the brake pads to the brake caliper bracket.

21. Remove the support and reposition the brake caliper over the brake pads and to the mounting bracket.

Caliper Bracket

REMOVAL & INSTALLATION

Front

1. Before servicing the vehicle, refer to the Precautions Section.

Support the brake caliper with heavy mechanic's wire, or equivalent, whenever it is separated from its mount and the hydraulic flexible brake hose is still connected. Failure to support the caliper in this manner will cause the flexible brake hose to bear the weight of the caliper, which may cause damage to the brake hose and in turn may cause a brake fluid leak.

2. Remove the brake caliper from the brake caliper mounting bracket and support the brake caliper with heavy mechanic's wire, or equivalent. Do NOT disconnect the hydraulic brake flexible hose from the caliper.

3. Remove the brake pads from the brake caliper bracket.

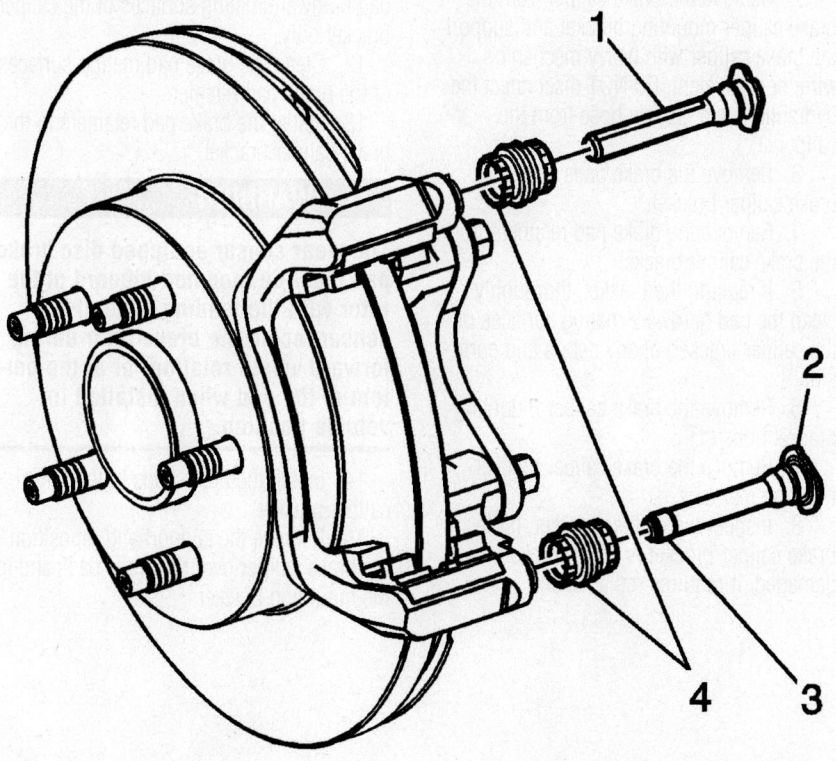

06025-COBALT-G117

Rear disc brake hardware

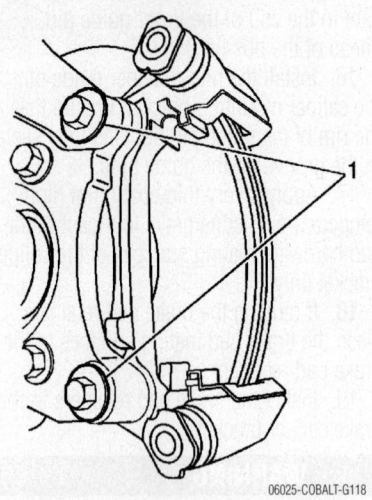

Front caliper bracket mounting bolts

06025-COBALT-G118

4. Remove the brake pad retainers from the brake caliper bracket.

5. If reusing the bracket, thoroughly clean the pad hardware mating surfaces of the caliper bracket, of any debris and corrosion.

6. Remove the brake caliper mounting bracket bolts (1).

7. Remove the brake caliper bracket from the steering knuckle.

8. Inspect the caliper bracket. If the brake caliper bracket is bent, cracked, or damaged, it requires replacement.

To install:

9. Install the brake caliper mounting bracket to the steering knuckle.

✳✳ WARNING

This is a self-retaining fastener joint that does not require thread locking compounds. Do not attempt to clean the threads with a standard tap. If a standard tap is used, damage to the joint threads will occur.

10. Install the brake caliper mounting bracket bolts. Tighten the bolts to 115 Nm (85 ft. lbs.).

11. Apply a very thin coating of high temperature silicone brake lubricant to the pad hardware mating surfaces of the caliper bracket only.

12. Clean the brake pad mating surfaces of the brake pad retainers.

13. Install the brake pad retainers to the brake caliper bracket.

✳✳ WARNING

The wear sensor equipped disc brake pad must be mounted inboard of the rotor with the leading edge of the sensor facing the brake rotor during forward wheel rotation, or at the top of the pad when installed in vehicle position.

14. Install the brake pads to the brake caliper bracket.

15. Remove the support and reposition the brake caliper over the brake pads and to the mounting bracket.

Rear

1. Before servicing the vehicle, refer to the Precautions Section.

✳✳ WARNING

Support the brake caliper with heavy mechanic's wire, or equivalent, whenever it is separated from its mount and the hydraulic flexible brake hose is still connected. Failure to support the caliper in this manner will cause the flexible brake hose to bear the weight of the caliper, which may cause damage to the brake hose and in turn may cause a brake fluid leak.

2. Remove the brake caliper from the brake caliper mounting bracket and support the brake caliper with heavy mechanic's wire, or equivalent. Do NOT disconnect the hydraulic brake flexible hose from the caliper.

3. Remove the brake pads from the brake caliper bracket.

4. Remove the brake pad retainers from the brake caliper bracket.

5. If reusing the bracket, thoroughly clean the pad hardware mating surfaces of the caliper bracket, of any debris and corrosion.

6. Remove the brake caliper mounting bracket bolts (1).

7. Remove the brake caliper bracket from the rear axle.

8. Inspect the caliper bracket. If the brake caliper bracket is bent, cracked, or damaged, it requires replacement.

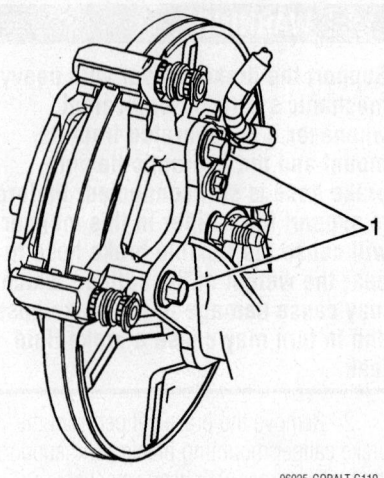

Rear caliper anchor bolts

06025-COBALT-G119

To install:

9. Install the brake caliper mounting bracket to the rear axle.

➡**This is a self-retaining fastener joint that does not require thread locking compounds. Do not attempt to clean the threads with a standard tap. If a standard tap is used, damage to the joint threads will occur.**

10. Install the brake caliper mounting bracket bolts. Tighten the bolts to 115 Nm (85 ft. lbs.).

11. Apply a very thin coating of high temperature silicone brake lubricant to the pad hardware mating surfaces of the caliper bracket only.

12. Clean the brake pad mating surfaces of the brake pad retainers.

13. Install the brake pad retainers to the brake caliper bracket.

✳✳ WARNING

The wear sensor equipped disc brake pad must be mounted inboard of the rotor with the leading edge of the sensor facing the brake rotor during forward wheel rotation, or at the bottom of the pad when installed in vehicle position.

14. Install the brake pads to the brake caliper bracket.

15. Remove the support and reposition the brake caliper over the brake pads and to the mounting bracket.

DRUM BRAKES

Brake Shoes

REMOVAL & INSTALLATION

1. Before servicing the vehicle, refer to the Precautions Section.
2. Raise and support the vehicle.
3. Remove the tire and wheel assembly.
4. Remove the brake drum.

※ WARNING

Do not over stretch the adjuster spring. Damage can occur if the spring is over stretched.

5. Remove the adjuster spring. Disengage the adjuster spring hook end from the tab on the adjuster actuator lever, then release the spring from the brake shoe web hole.
6. Remove the adjuster actuator lever (1) from the pivot.
7. Using brake tool J 38400 (1), or equivalent, spread the top of the brake shoes apart.
8. Remove the adjuster assembly (2) from the brake shoes.

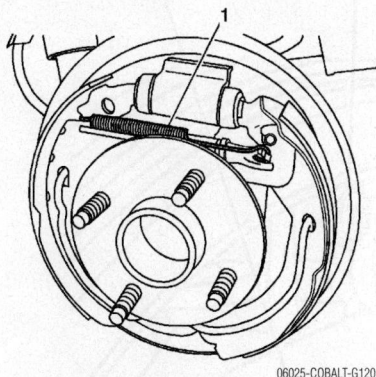

06025-COBALT-G120

Adjuster spring (1)

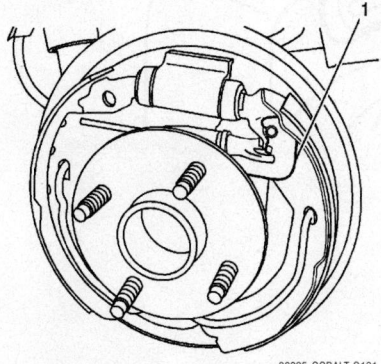

06025-COBALT-G121

Adjuster lever (1)

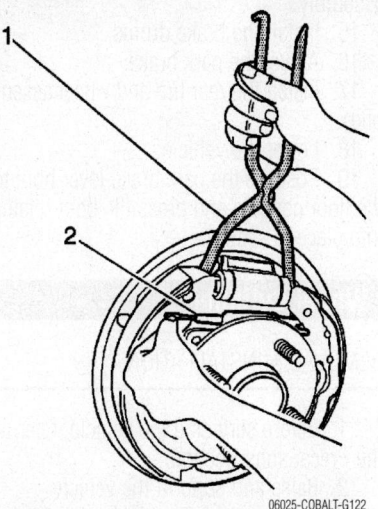

06025-COBALT-G122

Using brake tool J 38400 (1), or equivalent, spread the top of the brake shoes apart

9. Position the hook end of the tool under the universal spring and lightly pull the universal spring end out of the shoe web hole. Hold the universal spring while removing the trailing brake shoe.
10. Release the park brake cable from the park brake lever on the trailing shoe.
11. Position the hook end of the tool under the universal spring and lightly pull the universal spring end out of the shoe web hole. Hold the universal spring while removing the leading brake shoe.

To install:

12. Measure the brake shoe lining thickness. Brake shoe lining minimum thickness is 0.5 mm (0.020 in.).
13. If the brake shoe lining thickness is at or below the minimum specification, replace the brake shoes.
14. Apply a thin, light coat of high tem-

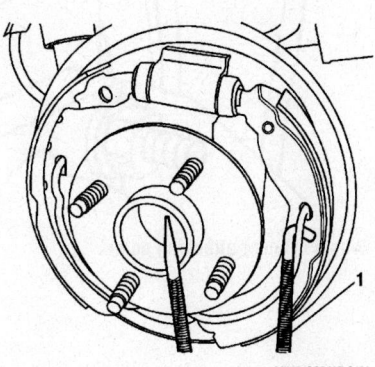

06025-COBALT-G123

Position the hook end of the tool under the universal spring and lightly pull the universal spring end out of the shoe web hole

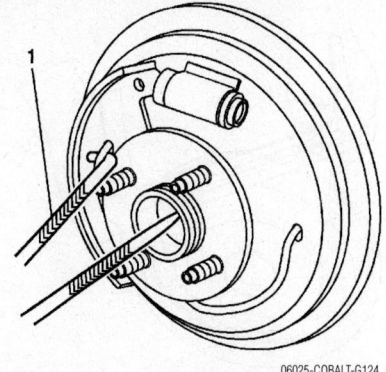

06025-COBALT-G124

Position the hook end of the tool under the universal spring and lightly pull the universal spring end out of the shoe web hole

perature silicone brake lubricant to the brake shoe contact surfaces of the brake backing plate.

15. Position the hook end of the tool under the universal spring and lightly pull the universal spring end out while installing the leading brake shoe. Ensure that the universal spring engages the brake shoe web hole.
16. Install the park brake cable to the park brake lever on the trailing brake shoe.
17. Position the hook end of the tool under the universal spring and lightly pull the universal spring end out while installing the trailing brake shoe. Ensure that the universal spring properly engages the brake shoe web hole.
18. Using the tool, spread the top of the brake shoes apart.
19. Install the adjuster assembly to the brake shoes.
20. Install the adjuster actuator lever to the brake shoe and the adjuster assembly. Ensure that the lever is properly engaged between the adjuster assembly and the brake shoe.

※ WARNING

Do not over stretch the adjuster spring. Damage can occur if the spring is over stretched.

21. Install the adjuster spring. Ensure that the loop end of the spring fully engages the tab on the actuator lever.
22. Adjust the drum brakes.
23. Install the brake drum.
24. Install the tire and wheel assembly.
25. Lower the vehicle.

ADJUSTMENT

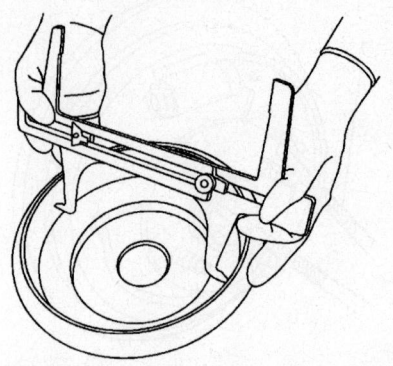

06025-COBALT-G125

Measuring the drum inner diameter

1. Before servicing the vehicle, refer to the Precautions Section.

2. Ensure that the park brake lever is in the fully released position.

3. Release the park brake lever boot from the floor console by applying light pressure inward on the sides of the boot retainer.

4. Pull the boot away from the console to expose the front park brake cable adjusting nut.

5. Release the tension from the park brake cable system at the front cable adjusting nut. Using ONLY HAND TOOLS, loosen the adjusting nut completely to the end of the front cable threaded rod.

6. Raise and support the vehicle.

7. Remove the rear tire and wheel assemblies.

8. Remove the brake drums.

9. Measure the inner diameter of the drum with a caliper such as tool J 21177-A at its widest point.

10. Firmly hand tighten the set screw on the tool.

11. Remove the tool from the brake drum and position it over the corresponding brake shoe assembly at its widest point.

12. While holding the tool in position, insert a 0.635 mm (0.025 in.) feeler gage between one side of the tool, and the corresponding brake shoe lining.

13. Rotate the brake shoe adjuster screw until the brake shoe linings contact the tool, and the feeler gage.

14. Repeat the above steps for the opposite brake drum and brake shoe assembly.

15. Install the brake drums.

16. Adjust the park brake.

17. Install the rear tire and wheel assemblies.

18. Lower the vehicle.

19. Position the park brake lever boot to the floor console and press the boot retainer into place to secure.

Wheel Cylinder

REMOVAL & INSTALLATION

1. Before servicing the vehicle, refer to the Precautions Section.

2. Raise and support the vehicle.

3. Remove the tire and wheel assembly.

4. Remove the brake drum.

5. Clean any debris and contaminants from around the wheel cylinder.

6. Remove the wheel cylinder bleeder cap and valve (2).

7. Disconnect the brake pipe fitting (1) from the wheel cylinder. Cap the exposed brake pipe end to prevent fluid loss and contamination.

8. Remove the wheel cylinder mounting bolts (3).

9. Spread the top of the brakes shoes apart, then remove the wheel cylinder from the brake backing plate.

To install:

10. Spread the top of the brakes shoes apart, then, install the wheel cylinder to the brake backing plate.

11. Install the wheel cylinder mounting bolts. Tighten the bolts to 16 Nm (12 ft. lbs.).

12. Remove the cap from the brake pipe end.

13. Connect the brake pipe fitting at the wheel cylinder. Tighten the fitting to 19 Nm (14 ft. lbs.).

14. Install the wheel cylinder bleeder valve. Tighten the valve to 8 Nm (71 inch lbs.).

15. Install the brake drum.

16. Bleed the hydraulic brake system.

17. Install the bleeder valve cap.

18. Adjust the drum brakes.

19. Install the tire and wheel assembly.

20. Lower the vehicle.

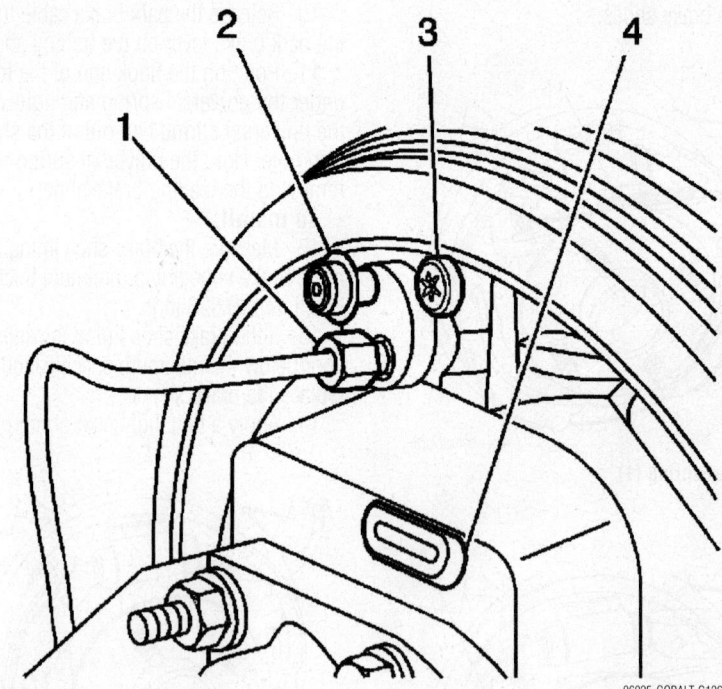

06025-COBALT-G126

Wheel cylinder attaching points

PARKING BRAKE

ADJUSTMENT

Disc Brakes

1. Before servicing the vehicle, refer to the Precautions Section.

✳✳ WARNING

The park brake cable adjusting nut is a nylon lock type. Use ONLY HAND TOOLS whenever tightening or loosening the adjusting nut.

2. Apply and fully release the park brake several times. Verify that the park brake lever releases completely.

3. Turn ON the ignition. Verify the red BRAKE warning lamp is not illuminated.

4. If the red BRAKE warning lamp is illuminated, verify the following:
  • The park brake lever is in the fully released position and against the stop
  • There is no slack in the park brake cables

5. Turn OFF the ignition.

6. Release the park brake lever boot from the floor console by applying light pressure inward on the sides of the boot retainer, and pull the boot back.

7. With the park brake lever in the released position, loosen the adjusting nut (1) enough to completely relieve tension on the front cable.

8. Raise and support the vehicle. Raise the vehicle just enough to observe the rear calipers and rotate the rear tire and wheel assemblies.

9. With all tension relieved from the park brake cables, rotate the rear tire and wheel assemblies, or the rear brake rotors if the wheels have been removed. Observe the amount of effort required for rotation, and the amount of drag if present.

10. Tighten the park brake cable adjusting nut until all slack is taken out of the front cable.

11. Further tighten the adjusting nut until one of the park brake levers on the rear calipers is just lifted off the stop on the caliper housing.

12. Slowly back off the adjusting nut until the park brake lever just rests on the stop.

13. Back off the adjusting nut one full turn.

14. Fully apply and release the park brake lever 3–5 times.

15. Raise the park brake lever 3 detent positions and attempt to rotate the rear tire and wheel assemblies, or the rear brake rotors. If rotating the tire and wheel assemblies, they should be difficult to rotate, but should not be locked. If rotating the brake rotors, they should be locked.

16. Raise the park brake lever one additional detent position and attempt to rotate the rear tire and wheel assemblies, or the rear brake rotors. The tire and wheel assemblies, or the rear brake rotors should be locked.

17. Fully release the park brake lever.

18. Verify the park brake is released by rotating the rear tire and wheel assemblies, or the rear brake rotors. The rotors should rotate freely and exhibit no brake shoe drag from the park brake system.

19. With the lever released, if the rotors required more effort to rotate, or exhibited more drag than noted previously when all cable tension was relieved, check the park brake levers on the rear calipers. The levers should be on the stops.

20. If the levers are not against the stops, loosen the adjusting nut just until the levers rest against the stops, then repeat steps 14–18.

21. If the rotors still do not rotate freely, with the park lever fully released, park brake adjustment is not the cause of any drag in the brake system.

22. Lower the vehicle.

23. Position the park brake lever boot to the floor console and press the boot retainer into place to secure.

24. Release the park brake lever.

Drum Brakes

1. Before servicing the vehicle, refer to the Precautions Section.

2. Apply and fully release the park brake several times. Verify that the park brake lever releases completely.

3. Turn ON the ignition. Verify the red BRAKE warning lamp is not illuminated.

4. If the red BRAKE warning lamp is illuminated, verify the following:
  • The park brake lever is in the fully released position and against the stop
  • There is no slack in the park brake cables

5. Turn OFF the ignition.

6. Release the park brake lever boot from the floor console by applying light pressure inward on the sides of the boot retainer, and pull the boot back.

7. With the park brake lever in the released position, loosen the adjusting nut enough to completely relieve tension on the front cable.

8. Raise and support the vehicle. Raise the vehicle just enough to allow rear tire and wheel assembly removal and rear drum adjustment.

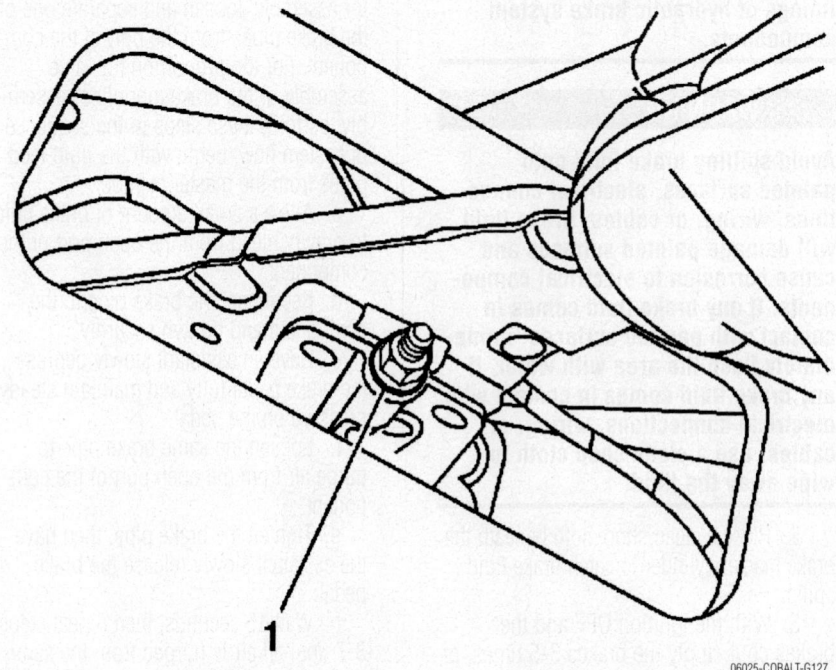

1

06025-COBALT-G127

Parking brake adjusting nut

9. Remove the rear tire and wheel assemblies.

10. Adjust the rear drum brakes.

11. Ensure there is no brake shoe drag after adjustment by rotating the brake drums. If drag exists, re-center the brake shoes and perform the brake shoe adjustment again.

12. Install 2 wheel nuts to the wheel studs and firmly hand-tighten in order to retain the brake drums.

13. Raise the park brake lever 6 detent positions.

14. Tighten the park brake cable adjusting nut. Tighten the nut to 3.9 Nm (35 inch lbs.).

15. Attempt to rotate the rear brake drums. There should be no rotation forward or rearward.

16. Fully release the park brake lever.

17. Verify the park brake is released by rotating the rear brake drums. The drums should rotate freely and exhibit no brake shoe drag.

18. If the drums do not rotate freely, repeat the park brake cable adjustment procedure.

19. Raise the park brake lever 3 detent positions and attempt to rotate the rear brake drums. One of the brake drums should not rotate forward or rearward. The other brake drum should not rotate forward

or rearward, or should require substantial effort to rotate.

20. Raise the park brake lever one additional detent position and attempt to rotate the rear brake drums.

21. Verify that the left and right brake drums cannot be rotated.

22. Remove the wheel nuts retaining the brake drums.

23. Install the rear tire and wheel assemblies.

24. Lower the vehicle.

25. Position the park brake lever boot to the floor console and press the boot retainer into place to secure.

26. Release the park brake lever.

BRAKE HYDRAULIC SYSTEM

Brake Bleeding

PRESSURE BLEEDING

1. Before servicing the vehicle, refer to the Precautions Section.

✳✳ WARNING

When adding fluid to the brake master cylinder reservoir, use only GM approved or equivalent DOT-3 brake fluid from a clean, sealed brake fluid container. The use of any type of fluid other than the recommended type of brake fluid may cause contamination which could result in damage to the internal rubber seals and/or rubber linings of hydraulic brake system components.

✳✳ WARNING

Avoid spilling brake fluid onto painted surfaces, electrical connections, wiring, or cables. Brake fluid will damage painted surfaces and cause corrosion to electrical components. If any brake fluid comes in contact with painted surfaces, immediately flush the area with water. If any brake fluid comes in contact with electrical connections, wiring, or cables, use a clean shop cloth to wipe away the fluid.

2. Place a clean shop cloth beneath the brake master cylinder to catch brake fluid spills.

3. With the ignition OFF and the brakes cool, apply the brakes 3-5 times, or until the brake pedal becomes firm, in

order to deplete the brake booster power reserve.

4. If you have performed a brake master cylinder bench bleeding on this vehicle, or if you disconnected the brake pipes from the master cylinder, or if you have disconnected the brake pipes from the proportioning valve assembly or the brake modulator assembly, you must perform the following steps to bleed air at the ports of the hydraulic component:

   a. If removal of the reservoir cap and diaphragm is necessary, clean the outside of the reservoir on and around the cap prior to removal.

   b. With the brake pipes installed securely to the master cylinder, proportioning valve assembly, or brake modulator assembly, loosen and separate one of the brake pipes from the port of the component. For the proportioning valve assembly or the brake modulator assembly, perform these steps in the sequence of system flow; begin with the fluid feed pipes from the master cylinder.

   c. Allow a small amount of brake fluid to gravity bleed from the open port of the component.

   d. Reconnect the brake pipe to the component and tighten securely.

   e. Have an assistant slowly depress the brake pedal fully and maintain steady pressure on the pedal.

   f. Loosen the same brake pipe to purge air from the open port of the component.

   g. Tighten the brake pipe, then have the assistant slowly release the brake pedal.

   h. Wait 15 seconds, then repeat steps 3-7 until all air is purged from the same port of the component.

   i. With the brake pipe installed

securely to the master cylinder, proportioning valve assembly, or brake modulator assembly, and after all air has been purged from the first port of the component that was bled, loosen and separate the next brake pipe from the component, then repeat steps 3.3-3.8 until each of the ports on the component has been bled.

   j. After completing the final component port bleeding procedure, ensure that each of the brake pipe-to-component fittings is properly tightened.

5. Clean the outside of the reservoir on and around the reservoir cap prior to removing the cap and diaphragm.

6. Install a pressure bleeder such as J 44894-A to the brake master cylinder reservoir.

7. Connect the J 29532, or equivalent, to the J 44894-A.

8. Charge the J 29532, or equivalent, air tank to 175-205 kPa (25–30 psi).

9. Open the J 29532, or equivalent, fluid tank valve to allow pressurized brake fluid to enter the brake system.

10. Wait approximately 30 seconds, then, inspect the entire hydraulic brake system in order to ensure that there are no existing external brake fluid leaks. Any brake fluid leaks identified require repair prior to completing this procedure.

11. Install a proper box-end wrench onto the RIGHT REAR wheel hydraulic circuit bleeder valve.

12. Install a transparent hose over the end of the bleeder valve.

13. Loosen the bleeder valve to purge air from the wheel hydraulic circuit. Allow fluid to flow until air bubbles stop flowing from the bleeder, then tighten the bleeder valve.

14. With the right rear wheel hydraulic circuit bleeder valve tightened securely, and

after all air has been purged from the right rear hydraulic circuit, install a proper box-end wrench onto the LEFT FRONT wheel hydraulic circuit bleeder valve.

15. Install a transparent hose over the end of the bleeder valve, then repeat steps 13-14.

16. With the left front wheel hydraulic circuit bleeder valve tightened securely, and after all air has been purged from the left front hydraulic circuit, install a proper box-end wrench onto the LEFT REAR wheel hydraulic circuit bleeder valve.

17. Install a transparent hose over the end of the bleeder valve, then, repeat steps 13-14.

18. With the left rear wheel hydraulic circuit bleeder valve tightened securely, and after all air has been purged from the left rear hydraulic circuit, install a proper box-end wrench onto the RIGHT FRONT wheel hydraulic circuit bleeder valve.

19. Install a transparent hose over the end of the bleeder valve, then, repeat steps 13-14.

20. After completing the final wheel hydraulic circuit bleeding procedure, ensure that each of the 4 wheel hydraulic circuit bleeder valves is properly tightened.

21. Close the J 29532, or equivalent, fluid tank valve, then disconnect the J 29532, or equivalent, from the J 44894-A.

22. Remove the J 44894-A from the brake master cylinder reservoir.

23. Slowly depress and release the brake pedal. Observe the feel of the brake pedal.

24. If the brake pedal feels spongy perform the following steps:

a. Inspect the brake system for external leaks.

b. If equipped with anti-lock brakes, using a scan tool, perform the antilock brake system automated bleeding procedure to remove any air that may have been trapped in the brake pressure modulator valve (BPMV).

Performing the Automated Bleed Procedure

1. Before servicing the vehicle, refer to the Precautions Section.

✱✱ WARNING

The Auto Bleed Procedure may be terminated at any time during the process by pressing the EXIT button. No further Scan Tool prompts pertaining to the Auto Bleed procedure will be given. After exiting the bleed procedure, relieve bleed pressure and disconnect bleed equipment per

manufacturers instructions. Failure to properly relieve pressure may result in spilled brake fluid causing damage to components and painted surfaces.

2. Raise and support the vehicle.

3. Remove all four tire and wheel assemblies.

4. Inspect the brake system for leaks and visual damage. Repair or replace components as needed.

5. Lower the vehicle.

6. Inspect the battery state of charge.

7. Install a scan tool.

8. Turn the ignition ON, with the engine OFF.

9. With the scan tool, establish communications with the ABS system. Select Special Functions. Select Automated Bleed from the Special Functions menu.

10. Raise and support the vehicle.

11. Following the directions given on the scan tool, pressure bleed the base brake system.

12. Follow the scan tool directions until the desired brake pedal height is achieved.

13. If the bleed procedure is aborted, a malfunction exists. Perform the following steps before resuming the bleed procedure:

a. If a DTC is detected, diagnose the appropriate DTC.

b. If the brake pedal feels spongy, perform the conventional brake bleed procedure again.

14. When the desired pedal height is achieved, press the brake pedal to inspect for firmness.

15. Lower the vehicle.

16. Remove the scan tool.

17. Install the tire and wheel assemblies.

18. Inspect the brake fluid level.

19. Road test the vehicle while inspecting that the pedal remains high and firm.

20. Turn the ignition key ON, with the engine OFF. Check to see if the brake system warning lamp remains illuminated.

✱✱ WARNING

DO NOT allow the vehicle to be driven until it is diagnosed and repaired.

MANUAL BLEEDING

1. Before servicing the vehicle, refer to the Precautions Section.

✱✱ WARNING

When adding fluid to the brake master cylinder reservoir, use only GM approved or equivalent DOT-3 brake

fluid from a clean, sealed brake fluid container. The use of any type of fluid other than the recommended type of brake fluid may cause contamination which could result in damage to the internal rubber seals and/or rubber linings of hydraulic brake system components.

✱✱ WARNING

Avoid spilling brake fluid onto painted surfaces, electrical connections, wiring, or cables. Brake fluid will damage painted surfaces and cause corrosion to electrical components. If any brake fluid comes in contact with painted surfaces, immediately flush the area with water. If any brake fluid comes in contact with electrical connections, wiring, or cables, use a clean shop cloth to wipe away the fluid.

2. Place a clean shop cloth beneath the brake master cylinder to catch brake fluid spills.

3. With the ignition OFF and the brakes cool, apply the brakes 3-5 times, or until the brake pedal effort increases significantly, in order to deplete the brake booster power reserve.

4. If you have performed a brake master cylinder bench bleeding on this vehicle, or if you disconnected the brake pipes from the master cylinder, or if you have disconnected the brake pipes from the proportioning valve assembly or the brake modulator assembly, you must perform the following steps to bleed air at the ports of the hydraulic component:

a. If removal of the reservoir cap and diaphragm is necessary, clean the outside of the reservoir on and around the cap prior to removal.

b. With the brake pipes installed securely to the master cylinder, proportioning valve assembly, or brake modulator assembly, loosen and separate one of the brake pipes from the port of the component. For the proportioning valve assembly or the brake modulator assembly, perform these steps in the sequence of system flow; begin with the fluid feed pipes from the master cylinder.

c. Allow a small amount of brake fluid to gravity bleed from the open port of the component.

d. Reconnect the brake pipe to the component and tighten securely.

e. Have an assistant slowly depress

the brake pedal fully and maintain steady pressure on the pedal.

f. Loosen the same brake pipe to purge air from the open port of the component.

g. Tighten the brake pipe, then have the assistant slowly release the brake pedal.

h. Wait 15 seconds, then repeat steps 3-7 until all air is purged from the same port of the component.

i. With the brake pipe installed securely to the master cylinder, proportioning valve assembly, or brake modulator assembly, and after all air has been purged from the first port of the component that was bled, loosen and separate the next brake pipe from the component, then repeat steps 3-8 until each of the ports on the component has been bled.

j. After completing the final component port bleeding procedure, ensure that each of the brake pipe-to-component fittings is properly tightened.

5. Ensure the brake master cylinder reservoir remains at least half-full during this bleeding procedure. Add fluid as needed to maintain the proper level. Clean the outside of the reservoir on and around the reservoir cap prior to removing the cap and diaphragm.

6. Install a proper box-end wrench onto the RIGHT REAR wheel hydraulic circuit bleeder valve.

7. Install a transparent hose over the end of the bleeder valve.

8. Have an assistant slowly depress the brake pedal fully and maintain steady pressure on the pedal.

9. Loosen the bleeder valve to purge air from the wheel hydraulic circuit.

10. Tighten the bleeder valve, then have the assistant slowly release the brake pedal.

11. Wait 15 seconds, then repeat steps 8-10 until all air is purged from the same wheel hydraulic circuit.

12. With the right rear wheel hydraulic circuit bleeder valve tightened securely, and after all air has been purged from the right rear hydraulic circuit, install a proper box-end wrench onto the LEFT FRONT wheel hydraulic circuit bleeder valve.

13. Install a transparent hose over the end of the bleeder valve, then repeat steps 7-11.

14. With the left front wheel hydraulic circuit bleeder valve tightened securely, and after all air has been purged from the left front hydraulic circuit, install a proper box-end wrench onto the LEFT REAR wheel hydraulic circuit bleeder valve.

15. Install a transparent hose over the end of the bleeder valve, then repeat steps 7-11.

16. With the left rear wheel hydraulic circuit bleeder valve tightened securely, and after all air has been purged from the left rear hydraulic circuit, install a proper box-end wrench onto the RIGHT FRONT wheel hydraulic circuit bleeder valve.

17. Install a transparent hose over the end of the bleeder valve, then repeat steps 7-11.

18. After completing the final wheel hydraulic circuit bleeding procedure, ensure that each of the 4 wheel hydraulic circuit bleeder valves is properly tightened.

19. Slowly depress and release the brake pedal. Observe the feel of the brake pedal.

20. If the brake pedal feels spongy, repeat the bleeding procedure again. If the brake pedal still feels spongy after repeating the bleeding procedure, perform the following steps:

a. Inspect the brake system for external leaks.

b. Pressure bleed the hydraulic brake system in order to purge any air that may still be trapped in the system.

21. Turn the ignition key ON, with the engine OFF. Check to see if the brake system warning lamp remains illuminated.

✴✴ WARNING

DO NOT allow the vehicle to be driven until it is diagnosed and repaired.

CHEVROLET

14

Corvette

BRAKES **14-47**
DRIVE TRAIN **14-28**
ENGINE REPAIR **14-9**
FUEL SYSTEM **14-26**
SPECIFICATIONS AND
  MAINTENANCE CHARTS......**14-2**
Engine and Vehicle
  Identification 14-2
General Engine Specifications 14-2
Engine Tune-Up Specifications 14-2
Firing Order 14-3
Accessory Drive Belt Routing 14-3
Capacities 14-3
Valve Specifications.....................14-4
Crankshaft and Connecting
  Rod Specifications..................14-4
Piston and Ring Specifications 14-5
Torque Specifications 14-5
Wheel Alignment 14-6
Tire, Wheel and Ball Joint
  Specifications 14-6
Brake Specifications 14-6
Scheduled Maintenance
  Intervals.........................14-7
STEERING AND
  SUSPENSION.................**14-37**

A
Air Bag..............................14-37
  Disabling & Enabling...............14-37
Alternator 14-9
  Removal...........................14-9

B
Brake Caliper 14-47
  Removal & Installation.............14-47

C
Camshaft 14-19
  Removal & Installation............14-19
Clutch..............................14-32
  Adjustment.......................14-32
  Removal & Installation............14-32
CV-Joints...........................14-33
  Overhaul 14-33

Cylinder Head.....................14-15
  Removal & Installation............14-15

D
Disc Brake Pad 14-48
  Removal & Installation............14-48
Distributor..........................14-9

E
Engine Assembly 14-9
  Removal & Installation.............14-9
Exhaust Manifold 14-18
  Removal & Installation.............14-18

F
Fuel Filter 14-26
  Removal & Installation.............14-26
Fuel Injector 14-27
  Removal & Installation.............14-27
Fuel Pump 14-26
  Removal & Installation.............14-26
Fuel System Pressure 14-26
  Relieving.........................14-26

H
Halfshaft............................14-32
  Removal & Installation.............14-32
Heater Core 14-12
  Removal & Installation.............14-12
Hydraulic Clutch System 14-32
  Bleeding..........................14-32

I
Intake Manifold 14-16
  Removal & Installation.............14-16

L
Lower Ball Joint......................14-45
  Removal & Installation.............14-45
Lower Control Arm 14-45
  Control Arm Bushing
    Replacement.....................14-46
  Removal & Installation.............14-45

O
Oil Pan.............................14-20
  Removal & Installation.............14-20
Oil Pump 14-22
  Removal & Installation.............14-22

P
Piston and Ring 14-25
  Positioning 14-25
Power Rack and Pinion
  Steering Gear.....................14-41
  Removal & Installation.............14-41

R
Rear Main Seal 14-23
  Removal & Installation.............14-23
Rocker Arms 14-16
  Removal & Installation.............14-16

S
Shock Absorber 14-42
  Removal & Installation.............14-42
Starter Motor 14-20
  Removal & Installation.............14-20

T
Timing Chain, Sprockets, Front
  Cover and Seal 14-24
  Removal & Installation.............14-24
Transmission Assembly 14-28
  Removal & Installation.............14-28
Transverse Spring 14-43
  Removal & Installation.............14-43

U
Upper Ball Joint.....................14-44
  Removal & Installation.............14-44
Upper Control Arm 14-45
  Control Arm Bushing
    Replacement.....................14-45
  Removal & Installation.............14-45

V
Valve Lash 14-20
  Adjustment.......................14-20

W
Water Pump 14-12
  Removal & Installation.............14-12
Wheel Bearings.....................14-46
  Adjustment.......................14-46
  Removal & Installation.............14-46

SPECIFICATION AND MAINTENANCE CHARTS

ENGINE AND VEHICLE IDENTIFICATION

Code ①	Liters (cc)	Cu. In.	Cyl.	Fuel Sys.	Engine Type	Eng. Mfg.
G	5.7 (5665)	350	8	SFI	OHV	CPC
S ③	5.7 (5665)	350	8	SFI	OHV	CPC
U	6.0 (5967)	364	8	SFI	OHV	CPC

Code ②	Year
2	2002
3	2003
4	2004
5	2005

CPC: Chevrolet/Pontiac/Canada

MFI: Multi-point Fuel Injection

OHV: Over Head Valves

① 8th position of VIN

② 10th position of VIN

③ High output engine

06025-CORV-C01

GENERAL ENGINE SPECIFICATIONS

Year	Model	Engine Displacement Liters (VIN)	Net Horsepower @ rpm	Net Torque @ rpm (ft. lbs.)	Bore x Stroke (in.)	Compression Ratio	Oil Pressure @ rpm
2002	Corvette	5.7 (G)	345@5600	350@4400	3.89x3.62	10.1:1	18@2000
		5.7 (S)	405@6000	400@4800	3.89x3.62	10.5:1	18@2000
2003	Corvette	5.7 (G)	345@5600	350@4400	3.89x3.62	10.1:1	18@2000
		5.7 (S)	405@6000	400@4800	3.89x3.62	10.5:1	18@2000
2004	Corvette	5.7 (G)	345@5600	350@4400	3.89x3.62	10.1:1	18@2000
		5.7 (S)	405@6000	400@4800	3.89x3.62	10.5:1	18@2000
2005	Corvette	6.0 (U)	400@6000	400@4400	4.00x3.62	10.9:1	18@2000

06025-CORV-C02

ENGINE TUNE-UP SPECIFICATIONS

Year	Engine Displacement Liters (VIN)	Spark Plug Gap (in.)	Ignition Timing (deg.) MT	Ignition Timing (deg.) AT	Fuel Pump (psi)	Idle Speed (rpm) MT	Idle Speed (rpm) AT	Valve Clearance In.	Valve Clearance Ex.
2002	5.7 (G)	0.060	①	①	48-55	①	①	HYD	HYD
	5.7 (S)	0.060	①	①	48-55	①	①	HYD	HYD
2003	5.7 (G)	0.060	①	①	48-55	①	①	HYD	HYD
	5.7 (S)	0.060	①	①	48-55	①	①	HYD	HYD
2004	5.7 (G)	0.060	①	①	48-55	①	①	HYD	HYD
	5.7 (S)	0.060	①	①	48-55	①	①	HYD	HYD
2005	6.0 (U)	0.040	①	①	55-62	①	①	HYD	HYD

NOTE: The Vehicle Emission Control Information label often reflects specification changes made during production. The label figures must be used if they differ from those in this chart.

HYD: Hydraulic

① Refer to Vehicle Emission Control Information label

06025-CORV-C03

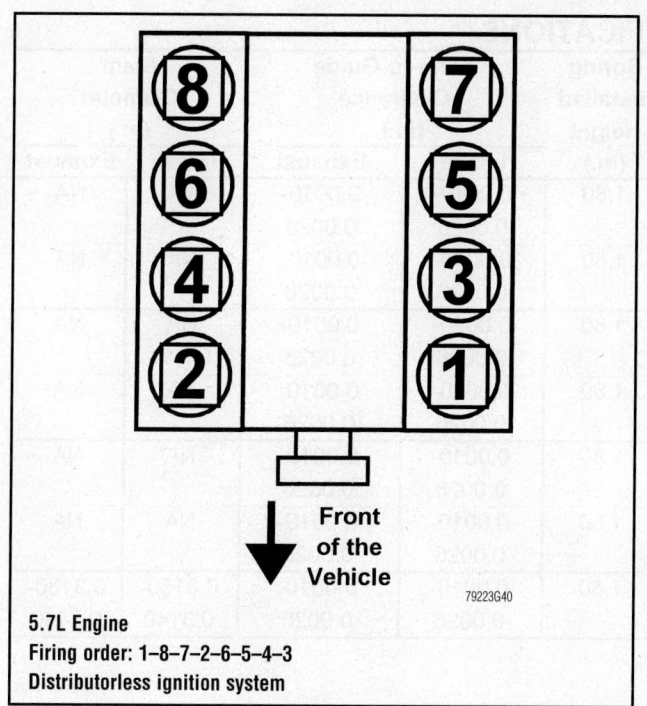

5.7L Engine
Firing order: 1–8–7–2–6–5–4–3
Distributorless ignition system

79223G40

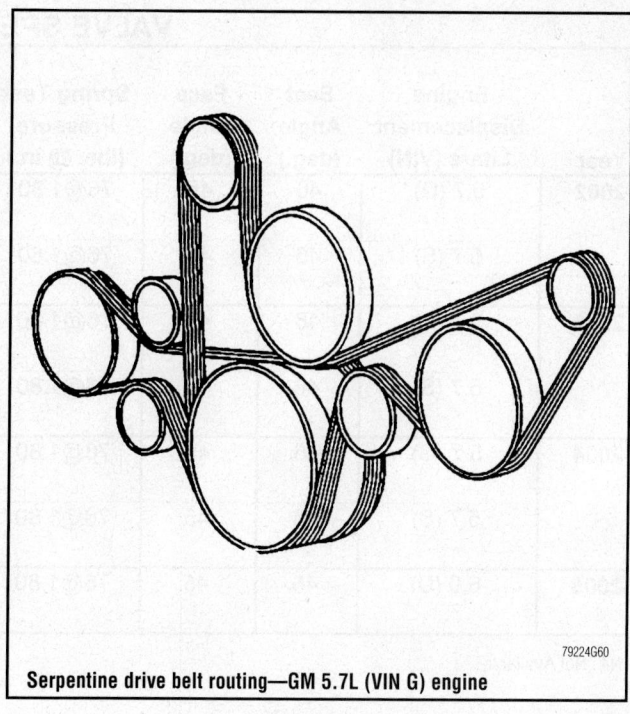

Serpentine drive belt routing—GM 5.7L (VIN G) engine

79224G60

CAPACITIES

Year	Model	Engine Displacement Liters (VIN)	Engine Oil with Filter (qts.)	Transmission (pts.) 6-Spd	Auto.	Drive Axle (pts.)	Fuel Tank (gal.)	Cooling System (qts.)
2002	Corvette	5.7 (G)	6.5	①	②	3.8	18.5	12.6
		5.7 (S)	6.5	①	②	3.8	18.5	12.6
2003	Corvette	5.7 (G)	6.5	①	②	3.8	18.5	12.6
		5.7 (S)	6.5	①	②	3.8	18.5	12.6
2004	Corvette	5.7 (G)	6.5	①	②	3.8	18.5	12.6
		5.7 (S)	6.5	①	②	3.8	18.5	12.6
2005	Corvette	6.0 (U)	③	8.2	11.4	3.6	18.0	12.6

NOTE: All capacities are approximate. Add fluid gradually and ensure a proper fluid level is obtained.

① MM 6 speed trans: 8.2 pts.

② 4L60E trans: 10.0 pts.

③ 6.0L - 5.5 qts.
　6.0L with Z51 Package - 6.0 qts.

06025-CORV-C04

VALVE SPECIFICATIONS

Year	Engine Displacement Liters (VIN)	Seat Angle (deg.)	Face Angle (deg.)	Spring Test Pressure (lbs. @ in.)	Spring Installed Height (in.)	Stem-to-Guide Clearance (in.)		Stem Diameter (in.)	
						Intake	Exhaust	Intake	Exhaust
2002	5.7 (G)	46	45	76@1.80	1.80	0.0010-0.0026	0.0010-0.0026	NA	NA
	5.7 (S)	46	45	76@1.80	1.80	0.0010-0.0026	0.0010-0.0026	NA	NA
2003	5.7 (G)	46	45	76@1.80	1.80	0.0010-0.0026	0.0010-0.0026	NA	NA
	5.7 (S)	46	45	76@1.80	1.80	0.0010-0.0026	0.0010-0.0026	NA	NA
2004	5.7 (G)	46	45	76@1.80	1.80	0.0010-0.0026	0.0010-0.0026	NA	NA
	5.7 (S)	46	45	76@1.80	1.80	0.0010-0.0026	0.0010-0.0026	NA	NA
2005	6.0 (U)	46	45	76@1.80	1.80	0.0010-0.0026	0.0010-0.0026	0.3130-0.3140	0.3130-0.3140

NA: Not Available

06025-CORV-C05

CRANKSHAFT AND CONNECTING ROD SPECIFICATIONS

All measurements are given in inches.

Year	Engine Displacement Liters (VIN)	Crankshaft				Connecting Rod		
		Main Brg. Journal Dia.	Main Brg. Oil Clearance	Shaft End-play	Thrust on No.	Journal Diameter	Oil Clearance	Side Clearance
2002	5.7 (G)	2.558-2.559	0.0007-0.0021	0.0015-0.0078	5	2.0987	0.0006-0.0025	0.0043-0.0200
	5.7 (S)	2.558-2.559	0.0007-0.0021	0.0015-0.0078	5	2.0987	0.0006-0.0025	0.0043-0.0200
2003	5.7 (G)	2.558-2.559	0.0007-0.0021	0.0015-0.0078	5	2.0987	0.0006-0.0025	0.0043-0.0200
	5.7 (S)	2.558-2.559	0.0007-0.0021	0.0015-0.0078	5	2.0987	0.0006-0.0025	0.0043-0.0200
2004	5.7 (G)	2.558-2.559	0.0007-0.0021	0.0015-0.0078	5	2.0987	0.0006-0.0025	0.0043-0.0200
	5.7 (S)	2.558-2.559	0.0007-0.0021	0.0015-0.0078	5	2.0987	0.0006-0.0025	0.0043-0.0200
2005	6.0 (U)	2.558-2.559	0.0008-0.0021	0.0015-0.0078	5	2.0991-2.0999	0.0009-0.0025	0.0043-0.0200

06025-CORV-C06

PISTON AND RING SPECIFICATIONS
All measurements are given in inches.

Year	Engine Displacement Liters (VIN)	Piston Clearance	Ring Gap			Ring Side Clearance		
			Top Compression	Bottom Compression	Oil Control	Top Compression	Bottom Compression	Oil Control
2002	5.7 (G)	0.0007-0.0021	0.009-0.015	0.017-0.025	0.007-0.027	0.0016-0.0033	0.0016-0.0031	0.0004-0.0087
	5.7 (S)	0.0007-0.0021	0.009-0.015	0.017-0.025	0.007-0.027	0.0016-0.0033	0.0016-0.0031	0.0004-0.0087
2003	5.7 (G)	0.0007-0.0021	0.009-0.015	0.017-0.025	0.007-0.027	0.0016-0.0033	0.0016-0.0031	0.0004-0.0087
	5.7 (S)	0.0007-0.0021	0.009-0.015	0.017-0.025	0.007-0.027	0.0016-0.0033	0.0016-0.0031	0.0004-0.0087
2004	5.7 (G)	0.0007-0.0021	0.009-0.015	0.017-0.025	0.007-0.027	0.0016-0.0033	0.0016-0.0031	0.0004-0.0087
	5.7 (S)	0.0007-0.0021	0.009-0.015	0.017-0.025	0.007-0.027	0.0016-0.0033	0.0016-0.0031	0.0004-0.0087
2005	6.0 (U)	0.0009-0.0012	0.0080-0.0160	0.0150-0.0270	0.0090-0.0310	0.0012-0.0040	0.0014-0.0031	0.0005-0.0079

06025-CORV-C07

TORQUE SPECIFICATIONS
All readings in ft. lbs.

Year	Engine Displacement Liters (VIN)	Cylinder Head Bolts	Main Bearing Bolts	Rod Bearing Bolts	Crankshaft Damper Bolts	Flywheel Bolts	Manifold		Spark Plugs	Oil Pan Drain Plug
							Intake	Exhaust		
2002	5.7 (G)	①	②	③	④	⑤	⑥	⑦	11	18
	5.7 (S)	①	②	③	④	⑤	⑥	⑦	11	18
2003	5.7 (G)	①	②	③	④	⑤	⑥	⑦	11	18
	5.7 (S)	①	②	③	④	⑤	⑥	⑦	11	18
2004	5.7 (G)	①	②	③	④	⑤	⑥	⑦	11	18
	5.7 (S)	①	②	③	④	⑤	⑥	⑦	11	18
2005	6.0 (U)	⑧	②	③	④	⑤	⑥	⑨	11	18

① M11 bolts: 22 ft. lbs.
  M11 bolts: plus 90 degrees
  M11 bolts 1-8: plus 90 degrees
  M11 bolts 9-10: plus 50 degrees
  M8 bolts (11-15): 22 ft. lbs.
② Inner Bolts:
  Step 1: 15 ft. lbs.
  Step 2: 80 degrees
  Side bolts: 18 ft. lbs.
  Outer studs:
  Step 1: 15 ft. lbs.
  Step 2: 53 degrees

③ Step 1: 15 ft. lbs.
  Step 2: Rotate 75 degrees
④ Step 1: 240 ft. lbs.
  Step 2: Install a new bolt and tighten to 37 ft. lbs.
  Step 3: Rotate 140 degrees
⑤ Step 1: 15 ft. lbs.
  Step 2: 37 ft. lbs.
  Step 3: 74 ft. lbs.
⑥ Step 1: 44 inch lbs.
  Step 2: 89 inch lbs.
⑦ Step 1: 11 ft. lbs.
  Step 2: 18 ft. lbs.

⑧ M11 bolts: 22 ft. lbs.
  M11 bolts: plus 90 degrees
  M11 bolts: plus 70 degrees
  M8 bolts: 22 ft. lbs.
⑨ Step 1: 11 ft. lbs.
  Step 2: 15 ft. lbs.

06025-CORV-C08

WHEEL ALIGNMENT

Year	Model		Caster Range (+/-Deg.)	Caster Preferred Setting (Deg.)	Camber Range (+/-Deg.)	Camber Preferred Setting (Deg.)	Toe-in (in.)
2002	Corvette	F	0.50	+6.90	0.50	-0.20	0.05 +/- 0.10
		R	—	—	0.50	-0.18	0 +/- 0.10
2003	Corvette	F	0.50	+6.90	0.50	-0.20	0.05 +/- 0.10
		R	—	—	0.50	-0.18	0 +/- 0.10
2004	Corvette	F	0.50	+6.90	0.50	-0.20	0.05 +/- 0.10
		R	—	—	0.50	-0.18	0 +/- 0.10
2005	Corvette	F	0.60	+7.90	0.60	-0.45	0.05 +/- 0.10
		R	—	—	0.50	-0.45	0 +/- 0.10

06025-CORV-C09

TIRE, WHEEL AND BALL JOINT SPECIFICATIONS

Year	Model	OEM Tires Standard	OEM Tires Optional	Tire Pressures (psi) Front	Tire Pressures (psi) Rear	Wheel Size	Ball Joint Inspection	Lug Nut
2002	Corvette	Fr: P245/45ZR17 Rr: P275/40ZR18	None	30	30	8.5	U: 0.125 in. L: 0.047 in.	①
2003	Corvette	Fr: P245/45ZR17 Rr: P275/40ZR18	None	30	30	8.5	U: 0.125 in. L: 0.047 in.	①
2004	Corvette	Fr: P245/45ZR17 Rr: P275/40ZR18	None	30	30	8.5	U: 0.125 in. L: 0.047 in.	①
2005	Corvette	Fr: P245/40ZR18 Rr: P285/35ZR19	None	30	30	②	NA	①

NA: Not available

OEM: Original Equipment Manufacturer

PSI: Pounds Per Square Inch

L: Lower

U: Upper

Fr: Front

Rr: Rear

① Wheel lug nut torque: 100 ft. lbs.

② 8.5 Front
10.0 Rear

06025-CORV-C10

BRAKE SPECIFICATIONS
All measurements in inches unless noted

Year	Model	Brake Disc Original Thickness	Brake Disc Minimum Thickness	Brake Disc Maximum Runout	Minimum Lining Thickness Front	Minimum Lining Thickness Rear	Brake Caliper Bracket Bolts (ft. lbs.)	Brake Caliper Mounting Bolts (ft. lbs.)
2002	Corvette	1.26	1.19	0.002	0.030	0.030	125	23
2003	Corvette	1.26	1.19	0.002	0.030	0.030	125	23
2004	Corvette	1.26	1.19	0.002	0.030	0.030	125	23
2005	Corvette	1.26	1.19	0.002	0.030	0.030	125	23

① Heavy duty: 1.110; Std.: 0.795

② Heavy duty: 1.059; Std.: 0.744

③ Front: not available, Rear: upper 26 ft. lbs., lower 16 ft. lbs.

06025-CORV-C11

MAINTENANCE I AND II SERVICE SCHEDULES
Corvette

When the CHANGE ENGINE OIL light appears, certain services and inspections are required.
Required services are described as Maintenance I and Maintenance II.
The first service on a vehicle should be Maintenance I, and the second service should be Maintenance II.
Alternate between the 2 thereafter. However, in some cases, Maintenance II may be required more often.
Maintenance I: Use Maintenance I if the CHANGE ENGINE OIL light comes on within 10 months since vehicle was purchased or, if Maintenance II was performed.
Maintenance II: Use Maintenance II if the previous service performed was Maintenance I. Always use Maintenance II whenever the CHANGE ENGINE OIL light comes on 10 months or more since the last service, or, if the CHANGE ENGINE OIL light has not come on at all for one year.

Service	Maintenance I	Maintenance II
Change the engine oil and filter. Reset the oil life system.	✓	✓
Visually inspect the vehicle for leaks or damage. A fluid loss in the vehicle system could indicate a problem. Inspected, repair and add fluid to the system if necessary.	✓	✓
Inspect the engine air cleaner filter. If necessary, replace the filter.	✓	✓
Rotate the tires. Inspect the tire inflation pressures and the tire wear.	✓	✓
Visually inspect the brake lines and hoses for proper hook-up, binding, leaks, cracks, chafing, etc. Inspect the disc brake pads for wear and the rotors for surface condition. Inspect the drum brake linings for wear or cracks. Inspect other brake parts, including drums, wheel cylinders, calipers, parking brake, etc. Inspect the parking brake adjustment.	✓	✓
Inspect the engine coolant and the windshield washer fluid levels. Add fluid as needed.	✓	✓
Inspect the suspension and steering components. Inspect the front and rear suspension and the steering system for damaged, loose or missing parts, or signs of wear. Inspect the power steering lines and the hoses for proper hook-up, binding, leaks, cracks,	--	✓
Visually inspect the coolant hoses and replace the hoses if they are cracked, swollen or deteriorated. Inspect all pipes, fittings and clamps; replace with GM parts as needed. To help ensure proper operation, a pressure test of the cooling system and pressure cap and cleaning the outside of the radiator and air conditioning condenser is recommended at least once a year.	--	✓
Inspect the front and rear suspension and the steering system for damaged, loose or missing parts, or signs of wear. Inspect power steering lines and hoses for proper hook-up, binding, leaks, cracks, chafing, etc.	--	✓
Inspect the throttle system for interference or binding and for damaged or missing parts. Replace the parts as needed. Replace any components that have high effort or excessive wear. Do not lubricate the accelerator or the cruise control cables.	--	✓
Replace the passenger compartment air filter.	--	✓

To reset the CHANGE OIL SOON message after an oil change, do the following:

1. Turn the ignition to ON, with the engine off.
2. Press the TRIP button so the OIL LIFE percentage is displayed.
3. Press RESET and hold for two seconds. OIL LIFE REMAIN 100% will appear.

06025-CORV-C12

ADDITIONAL MAINTENANCE SERVICES
Corvette

TO BE SERVICED	TYPE OF SERVICE	VEHICLE MILEAGE INTERVAL (x1000)					
		25	50	75	100	125	150
Air cleaner filter	R		✓		✓		✓
Accessory drive belt	I						✓
Auto. Trans. Fluid and Filter①	R		✓		✓		✓
Cooling system hoses and clamps	S/I						✓
Engine coolant	R						✓
Fuel system	I	✓	✓	✓	✓	✓	✓
Exhaust system & heat shields	S/I	✓	✓	✓	✓	✓	✓
Spark plugs and wires	R				✓		

R: Replace   S/I: Inspect and service, if necessary

① Replace if any of the following conditions are met:

   Heavy city traffic where the outside temperature regularly reaches 32°C (90°F) or higher

   Hilly or mountainous terrain

   Frequent trailer towing

   Taxi, police or delivery service

   Otherwise, change every 100,000 miles

06025-CORV-C13

ENGINE REPAIR

Distributor

All 5.7L and 6.0L engines use a Direct Ignition System.

Alternator

REMOVAL

1. Before servicing the vehicle, refer to the Precautions Section.
2. Remove or disconnect the following:
   - Negative battery cable
   - Regulator connector and battery-to-alternator terminal from the rear of the alternator
   - Accessory drive belt
   - Alternator mounting bolts
   - Alternator

To install:
3. Install or connect the following:
   - Alternator
   - Alternator mounting bolts and torque them to 37 ft. lbs. (50 Nm)
   - Regulator connector and battery-to-alternator terminals and torque the alternator terminal nut to 10 inch lbs. (13 Nm)
   - Drive belt
   - Negative battery cable

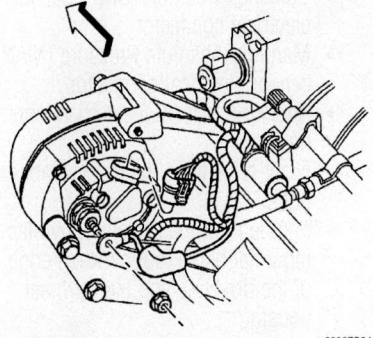

9306ZG24

Exploded view of the alternator—5.7L engine, 6.0L engine similar

Engine Assembly

REMOVAL & INSTALLATION

1. Before servicing the vehicle, refer to the Precautions Section.
   The following tools will be required in addition to the basic hand tools:
   - Transverse spring compressor and adapters
   - Ball joint separator
   - Engine support table
   - Driveshaft support strap
   - Fuel pressure gauge
2. Relieve the fuel system pressure.
3. Drain the cooling system.
4. Drain the engine oil.
5. Evacuate the A/C system.
6. Remove or disconnect the following:
   - Intake Air Temperature (IAT) sensor electrical connector
   - Mass Air Flow (MAF) sensor electrical connector
   - Air intake duct and air cleaner
   - Upper radiator support
   - Radiator
   - Electronic Brake Control Traction Module/Brake Pressure Modulator Valve (EBTCM/BPMV) and bracket
   - Brake pipes
   - Accessory drive belt
   - Fuel line from the connector at the front of the dash

➡**Cap and plug all openings for the fuel system to prevent contaminants from entering the system.**

   - Fuel rail covers
   - Fuel line at the fuel rail
   - Radiator hoses from the water pump
   - Heater hoses from the water pump
   - Fuel injector electrical connectors
   - Ignition coil main harness connector
   - Evaporative Emission (EVAP) solenoid electrical connector
   - Electric throttle motor electrical connector
   - Throttle Position Sensor (TPS) electrical connector
   - Engine Coolant Temperature (ECT) sensor electrical connector
   - A/C compressor electrical connector
   - Alternator electrical connectors
   - Alternator
   - Brake booster vacuum hose
   - Intermediate shaft to steering gear bolt
   - Intermediate shaft from the steering gear
   - Secondary Air Injection (AIR) hose from the left exhaust manifold
   - Both front wheels
   - Heated Oxygen Sensor (HO$_2$S) connectors from the intermediate exhaust pipes
   - Intermediate exhaust pipes
   - Catalytic converter
   - Close-out panel
   - Starter motor electrical connectors
   - Starter motor
   - Crankshaft Position (CKP) sensor electrical connector
   - Oil level sensor connector
   - Right Heated HO$_2$S
   - A/C compressor hose assembly
   - Engine oil temperature sensor electrical connector
   - Left Heated HO$_2$S
   - Ground straps
   - Front stabilizer shaft
   - Connectors from the cooling fans
   - Cooling fan assembly
   - Tie rod ends from the steering knuckles
   - Antilock Brake System (ABS) electrical connectors from the crossmember, if equipped
   - Electronic Variable Orifice (EVO) connector clips from the crossmember, if equipped
   - Real Time Damping (RTD) connector clips from the crossmember, if equipped
   - Shock absorber lower mounting bolts
   - Front transverse leaf spring
   - Automatic transmission cooler pipes at the flywheel housing junction
   - Automatic transmission cooler pipe clamps from the engine oil pan
   - Automatic transmission cooler pipe from the radiator

❋❋ CAUTION

Failure to use the minimum fastener length specified will prevent the proper retention of the propeller shaft during assembly.

7. For vehicles with an automatic transmission, perform the following steps:
   a. Remove the two plug bolts in the driveline support assembly.
   b. Install an M10 x 55 bolt, or longer, in each plug location. Torque the bearing support bolts to 26 ft. lbs. (35 Nm).
   c. Remove the bell housing inspection cover, then turn the flywheel hub collar to access the bolt and loosen it.
   d. Remove the engine flywheel housing access plug.
   e. Rotate the automatic transmission collar so that the bolt is facing down.
   f. Loosen the bolt and unclip the wire

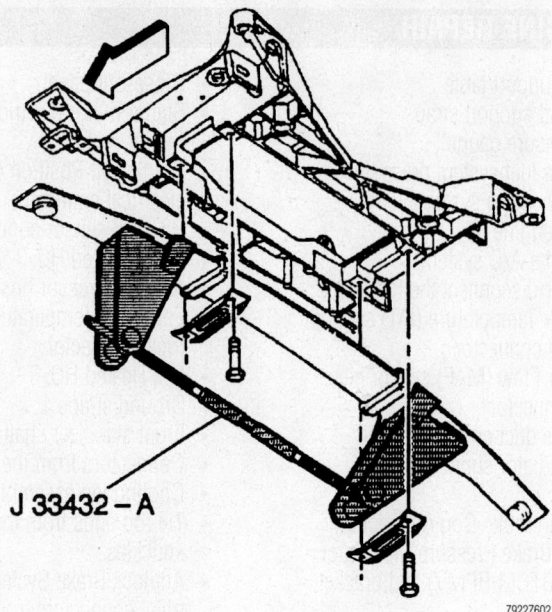

Use a compressor such as J 33432-A to compress the front transverse spring

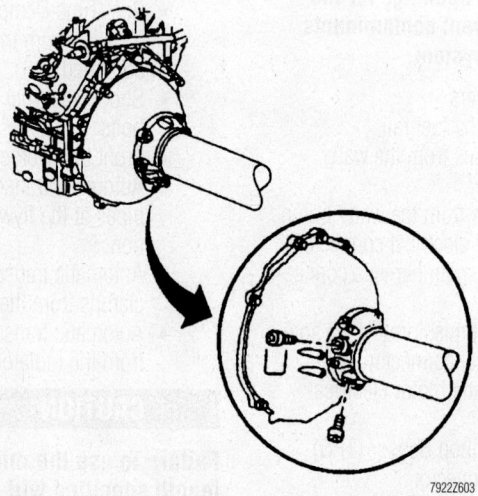

On automatic transmission vehicles, remove the plugs from the driveline support assembly and install 2 M10 x 1.5 bolts into the plug holes

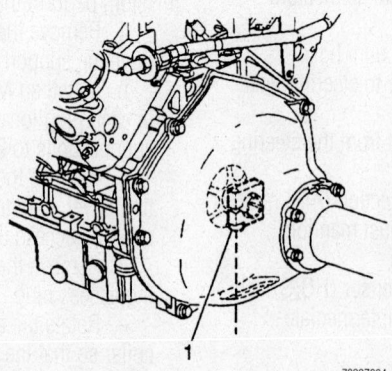

Loosen the bolt on the flywheel hub collar after turning it for access

harness from the engine and reposition it to the driveline.

g. Install a driveline support tool to the close out panel flange.

✳✳ CAUTION

Never use a driveline support tool to support the weight of the engine assembly.

8. If equipped with a manual transmission, perform the following steps:

a. Unclip the clutch actuator hose from the engine flywheel housing clip.

b. Using a hydraulic clutch line separator tool, depress the white release ring on the actuator hose and pull lightly on the master cylinder hose.

c. Remove the flywheel housing bolts from the driveline support.

9. Support the engine and crossmember assembly.

10. Remove or disconnect the following:
- Front crossmember nuts BY HAND and lower the engine assembly slightly
- Secondary Air Injection (AIR) tube bracket bolt and reposition the bracket to gain access to the ground strap
- Ground strap from the rear of the left cylinder head
- Engine oil pressure sensor electrical connector
- Camshaft Position (CMP) sensor electrical connector
- Manifold Absolute Pressure (MAP) sensor electrical connector
- Knock Sensor (KS) electrical connector
- Front driveline support assembly bolts
- Engine from the driveline, by placing a flat blade tool between edge of the driveline and the flywheel housing

11. Pull the engine away from the propeller shaft.

12. Raise the vehicle off of the engine and crossmember.
- Power steering pump (with reservoir) from the engine and reposition them to the crossmember
- A/C, alternator and power steering pump brackets
- AIR pipe and gasket
- Spark plugs

13. Remove the engine mount-to-crossmember nuts.

14. Remove the engine from the crossmember.

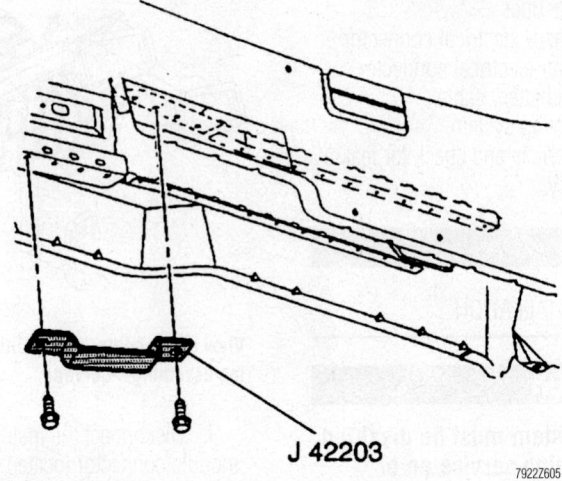

J 42203

79222Z605

Install a Driveline Support tool J 42203 to the under cover flange

To install:

15. Install the engine on the crossmember and torque the nuts to 48 ft. lbs. (65 Nm).

16. Lower the vehicle onto the engine/crossmember assembly.

17. Install or connect the following:
- Spark plugs
- AIR pipe and gasket
- A/C, alternator and power steering pump brackets
- Power steering pump and reservoir on the engine and torque the bolts to 18 ft. lbs. (25 Nm)
- Ground strap to the rear of the left cylinder head and torque the bolt to 24 ft. lbs. (32 Nm)
- Engine oil pressure sensor electrical connector
- CMP sensor electrical connector
- MAP sensor electrical connector
- KS electrical connector
- Install or connect the following:
- Front driveline support bolts and torque them to 37 ft. lbs. (50 Nm)
- Air tube and bracket and torque the bolt to 15 ft. lbs. (20 Nm)
- New crossmember nuts and torque them to 81 ft. lbs. (110 Nm)
- Flywheel housing bolts and torque them to 37 ft. lbs. (50 Nm)
- Clutch actuator hose to the flywheel housing (if equipped)
- Master cylinder hose to the clutch actuator hose (if equipped)
- Wire harness to the rear of the engine

18. Remove the driveline support tool.

19. Remove the M10 x 55mm bolts from the driveline support assembly.
- Two plugs in the driveline support assembly after removing the M10 x

55 bolts and torque them to 37 ft. lbs. (50 Nm)
- Automatic transmission oil cooler pipes at the flywheel housing junction and torque them to 20 ft. lbs. (27 Nm), if equipped
- Automatic transmission cooler pipes to the front and rear of the oil pan
- Front transverse spring and torque the bolts to 46 ft. lbs. (62 Nm)
- Shock absorber lower mounting nuts and torque them to 21 ft. lbs. (28 Nm)
- Front stabilizer shaft and torque the bolts to 53 ft. lbs. (72 Nm)
- ABS electrical connector
- EVO electrical connector
- RTD electrical connector
- Tie rods to the steering knuckle and torque the nuts to 33 ft. lbs. (45 Nm).
- Radiator
- Cooling fan assembly to the radiator and attach the electrical connectors to the fans
- Ground wires to the left side of the engine
- Left HO$_2$S connector
- Engine oil temperature sensor connector
- A/C compressor hoses to the compressor
- A/C compressor line retaining bolt and torque it to 26 ft. lbs. (35 Nm)
- Right HO$_2$S connector
- CKP sensor electrical connector
- Oil level sensor electrical connector
- Wire harness ground wires to the right side of the engine
- Starter motor and torque the bolts to 37 ft. lbs. (50 Nm)

- Driveline close out panel and torque the bolts to 106 inch lbs. (12 Nm)
- Catalytic converters and torque the mounting nuts to 15 ft. lbs. (20 Nm)
- Intermediate exhaust pipe and hangers and torque the bolts to 37 ft. lbs. (50 Nm) and the nuts to 15 ft. lbs. (20 Nm)
- Front wheels
- Automatic transmission cooler pipes to the radiator (if equipped) and torque the fittings to 26 ft. lbs. (35 Nm)
- AIR hose to the left exhaust manifold
- Intermediate steering shaft to the steering gear and torque the bolt to 35 ft. lbs. (48 Nm)
- Brake booster vacuum hose
- Alternator bracket and torque the bolts to 37 ft. lbs. (50 Nm)
- Alternator and torque the bolts to 37 ft. lbs. (50 Nm)
- Alternator connectors and torque the nuts to 10 ft. lbs. (13 Nm)
- Fuel injector electrical connectors
- Ignition coil main electrical connectors
- EVAP solenoid electrical connectors
- Throttle motor electrical connector
- ECT sensor electrical connector
- TPS electrical connector
- A/C compressor electrical connector
- Heater hoses to the water pump
- Radiator hoses to the water pump
- Fuel line to the fuel rail
- Fuel rail covers
- Drive belt
- EBTCM/BPMV bracket and torque the bolts to 20 ft. lbs. (27 Nm)
- EBTCM/BPMV to the bracket and torque the nuts to 89 inch lbs. (10 Nm)
- Upper radiator hose
- Upper radiator support
- Air cleaner and air intake duct
- IAT sensor electrical connector
- MAF sensor electrical connector
- Negative battery cable
- Engine flywheel housing inspection plug

20. Fill the engine with new oil.

21. Fill the cooling system.

22. Evacuate and recharge the A/C system.

23. On vehicles with an automatic transmission, start the engine and allow it to idle for 10 minutes. Torque the hub collar bolt to

92 ft. lbs. (125 Nm) and install the flywheel inspection plug.

24. A wheel alignment is recommended when the engine and crossmember have been removed from the vehicle.

25. Check the vehicle for leaks and repair if necessary.

Water Pump

REMOVAL & INSTALLATION

1. Before servicing the vehicle, refer to the Precautions Section.
2. Drain the cooling system.
3. Remove or disconnect the following:
   - Negative battery cable
   - Intake Air Temperature (IAT) sensor electrical connector
   - Mass Air Flow (MAF) sensor electrical connector
   - Air intake duct
   - Accessory drive belt
   - Inlet and outlet hoses from the water pump
   - Heater hoses from the water pump
   - Water pump pulley bolts and pulley (if equipped)
   - Six water pump retaining bolts
   - Water pump

To install:

4. Install or connect the following:
   - New gaskets with the tabs facing up
   - Water pump. Torque the bolts to 11 ft. lbs. (15 Nm) and then to 22 ft. lbs. (30 Nm).
   - Water pump pulley (if equipped). Torque the bolts to 89 inch lbs. (10 Nm) and then to 18 ft. lbs. (25 Nm).
   - Heater hoses to the water pump
   - Radiator inlet and outlet hoses to the water pump
   - Accessory drive belt

- Air intake duct
- MAF sensor electrical connector
- IAT sensor electrical connector
- Negative battery cable

5. Fill the cooling system.
6. Start the vehicle and check for leaks, repair if necessary.

Heater Core

REMOVAL & INSTALLATION

✳✳ CAUTION

The air bag system must be disabled before performing service on or around the air bag, instrument panel components, wiring and sensors. Failure to follow safety and disabling procedures could result in accidental air bag deployment, possible personal injury and unnecessary air bag system repairs.

1. Before servicing the vehicle, refer to the Precautions Section.
2. If equipped, disable the SIR system by using the following procedure:
   a. Remove the SIR fuse from the fuse panel.
   b. Remove the left side sound insulator.
   c. Disconnect the Connector Positive Assurance (CPA) from the yellow 2-way SIR harness connector at the base of the steering column and separate the connector.
   d. Disconnect the steering wheel module coil connector located at the base of the steering column.
   e. Disconnect the CPA from the inflatable restraint instrument panel module connector located at the near the base of the steering column.

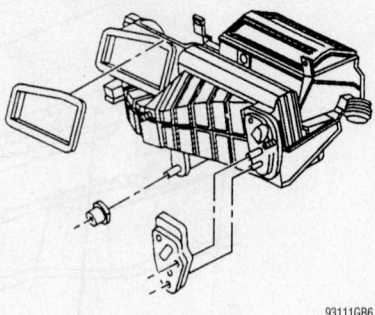

View of the heater/air conditioning housing assembly—Corvette

   f. Disconnect the instrument panel module connector located near the base of the steering column.
   g. Disconnect the negative battery cable.
3. Discharge and recover the air conditioning system refrigerant.
4. Drain the cooling system into a clean container for reuse.
5. Remove or disconnect the following:
   - Battery heat shield
   - Intake manifold
   - Hose from the right hand air injection check valve
   - Bolt attaching the right hand air injection check valve to the cylinder head(s), if equipped
   - Nut attaching the heater pipe bracket to the cowl and set the bracket aside
   - Heater pipe-to-heater core bolt
   - Heater pipe assembly from the core, allow any remaining coolant to drain and cap the lines. Discard the sealing washers.
   - Accumulator hose-to-evaporator bolt
   - Accumlator hose and evaporator tube from the evaporator. Discard the O–ring seals and cap hose and tube ends and also the evaporator.
   - HVAC module drain tube from the module
6. Remove the following components to access the heater core assembly as follows:
   a. Remove the folding top stowage compartment lid extension by opening the compartment, removing the screws from the lower sides and the top of the panel. Lift the panel up off the bracket.
   b. Open the console door.
   c. Pull up on the rear of the traction control ride switch to release it from the retaining clips. If the switch will not release, carefully insert a suitable prytool into the recess at the rear of the switch and carefully pull up at the rear of the switch.

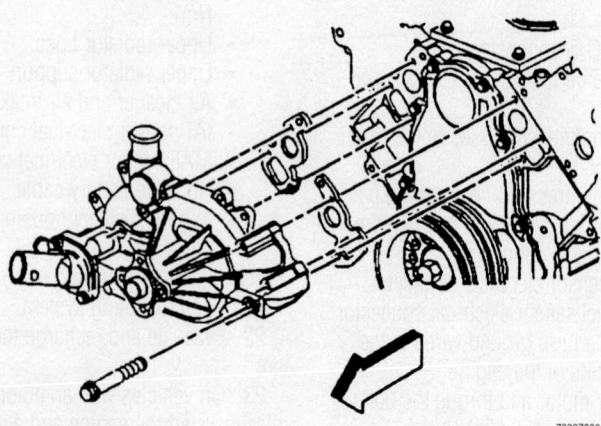

79222608

Exploded view of the water pump mounting assembly—5.7L engine, 6.0L engine similar

d. Disconnect the connector from the traction switch.

e. Disconnect the LED connector from the harness connector and remove the switch.

f. Remove the console retaining nut covers using a suitable prytool.

g. Remove the console retaining nuts.

h. Lift the rear of the console slightly and pull back gently to release the console from under the accessory trim plate.

i. Disconnect the accessory power plug connector, unscrew the plug housing retainer and remove the plug housing.

j. Disconnect the fuel door release electrical connector and if equipped, the rear lift window switch. Remove the switch.

k. Remove the console.

l. Apply the parking brake for additional clearance.

m. Place the transmission in second gear on models with an Automatic Transmission (AT) or fourth on models with a Manual Transmission (MT).

n. On models equipped with MT, grasp the shift control boot and apply light pressure in towards the shift lever to release the tabs from the IP accessory trim plate.

o. Release the remaining tabs using light pressure and remove the boot from the trim plate.

p. Remove the ashtray.

q. Remove the IP accessory trim plate grille.

r. Remove the accessory trim plate screws next to the cigar lighter and behind the ashtray.

s. Remove the accessory trim plate retaining screws in the grille opening.

t. Grasp the sides of the trim plate near the curve at the base.

u. Pull the trim plate rearwards to release the tabs, making sure to lift the rear of the trim plate to clear the driveline tunnel studs.

v. Disconnect the cigar lighter electrical connector.

w. Rotate the shift control boot and reposition one end down into the shifter opening in the trim plate, on models with MT.

x. Lift the trim plate over the shifter and boot and remove the trim plate.

y. Remove the fog lamp, rear compartment lid release switch.

z. Remove the driver knee bolster trim panel screw from behind the switch.

aa. Remove the lower driver knee bolster screws.

bb. Grasp the knee bolster trim panel and pull firmly rearward to disengage the locking tabs. Disconnect the connector from the inside air temperature sensor and remove the panel.

cc. Open the glove box and disconnect the light switch.

dd. Remove the trim plugs from the bottom of the door by reaching behind the door and pushing the plugs out while using a prytool on the front.

ee. Remove the lower bolts and the upper and side screws.

ff. Slide the glove box forward and unplug the wiring harness connector at the passenger side air bag.

gg. Remove the glove box.

hh. Remove the inboard side window glass defroster duct from the windshield defroster duct on both sides

ii. Remove the DRL electrical connector and the sun load temperature sensor from the windshield defroster duct, if equipped

jj. Remove the windshield side garnish moldings.

kk. Remove the screws attaching the upper trim pad to the defroster duct.

ll. Remove the screws attaching the upper trim pad to the left and right hand hinge pillars.

mm. Remove the screws attaching the IP cluster bezel to the upper trim pad.

nn. Remove the screws attaching the upper trim panel to the driver knee bolster outer bracket and center support bracket.

oo. Remove the screw attaching the upper trim panel to the passenger SIR bracket.

pp. Tilt the steering wheel to its lowest position.

qq. Lift the rearward edge of the trim panel approximately 2 inches (5 CM) to provide clearance for the air distribution duct.

rr. Slowly pull the pad away from the windshield while guiding the tabs past the hinge pillars.

ss. Disconnect the remaining electrical connectors to the upper trim panel and remove the panel.

tt. Remove the left hand side window defogger outlet duct.

uu. Remove the inside air temperature sensor duct.

vv. Disconnect the DRL sensor, if equipped.

ww. Disconnect the left hand temperature sensor connector, if equipped.

xx. Turn the key to the ON position.

yy. Using a suitable prytool, depress

the park/lock cable retaining tab located on the underside of the ignition switch and pull to release the cable from the ignition switch.

zz. Using a suitable prytool, release the tab retaining the park/lock cable to the slot in the shift control cable.

aaa. Lift the park lock cable from the slot.

bbb. Grasp the cable end and pull rearwards to unlock the cable from the shifter pivot arm stud.

ccc. Remove the park/lock cable.

ddd. Remove the CPA from the electrical connector on the inflatable restraint and Sensing and Diagnostic Module (SDM).

eee. Disconnect the SDM.

fff. Remove the ignition switch bolts and position the switch aside.

ggg. Mark the position of the IP center support bracket.

hhh. Remove the right bolt that attaches the bracket to the driveline tunnel.

iii. Remove the bolts that attach the bracket to the passenger knee bolster bracket.

jjj. Remove the bolt that attaches the bracket to the IP lower support beam.

kkk. Remove the left bolt that attaches the bracket to the driveline tunnel.

lll. Mark the location of the ignition switch bracket.

mmm. Remove the bolt that attaches the ignition switch housing bracket to the steering column bracket.

nnn. Remove the bolt that attaches the c enter support bracket to the IP lower support beam.

ooo. Pull the bracket away from the IP to access the radio control and HVAC connectors.

ppp. Disconnect connectors from the radio.

qqq. Disconnect the electrical connectors and vacuum harness connectors from the HVAC control.

rrr. Disconnect the IP wiring harness and parking brake connector clip from the bracket.

sss. Remove the bolts attaching the ignition switch housing bracket to the center support bracket.

ttt. Note the routing of the IP wiring harness and remove the harness.

uuu. Remove the center bracket.

vvv. Remove the left hand floor outlet duct.

www. Remove the right hand side window defogger outlet duct.

xxx. Remove the SIR bracket and the passenger knee bolster bracket.

yyy. Disconnect the sunload electrical connector.

zzz. Disconnect the right hand temperature actuator electrical connector.

aaaa. Remove the right hand floor air duct.

bbbb. Move the carpet away from the right hand side of the driveline tunnel to remove the carpet air duct (rear).

cccc. Remove the right hand floor air duct (rear).

dddd. Disconnect the blower motor connector.

eeee. Remove the blower motor.

ffff. Disconnect the vacuum supply from the HVAC module.

gggg. Disconnect the vacuum solenoid connector.

hhhh. Remove the windshield defroster duct.

iiii. Remove the HVAC module and discard seals.

jjjj. Remove the heater core outlet cover screws and the cover.

kkkk. Remove the heater core cover screws and the cover. Discard the heater core cover seals.

llll. Remove and discard the heater core outer seal.

mmmm. Remove the core clamp screw and clamp.

nnnn. Remove the core pipe retaining clamp screw and remove the core from the case.

oooo. Remove and discard the core lower, center, upper and side seals.

pppp. Remove the core pipe clamp using a suitable tool.

To install:

7. Install the core pipe clamp.

8. Install new core lower, center, upper and side seals.

9. Install the core in the case.

10. Install the core pipe retaining clamp and screw.

11. Install the core clamp and screw.

12. Install a new heater core outer seal.

13. Install new cavity seals.

14. Install the heater core cover and screws.

15. Install the heater core outlet cover and screws.

16. Install the HVAC module using new seals. Make sure the dashmat is aligned. The opening in the dashmat for the HVAC drain should be aligned so the drain opening in the cowl is approximately centered in the dashmat opening allowing room from the module drain seal to fully align against the cowl. Make sure the heater core joint fitting, condenser fitting module drain and module studs correspond with the openings

in the cowl. Insert the left hand stud into the cowl first, then the module drain, then the right hand stud. Install the right hand then the left hand bolts and tighten to 89 inch lbs. (10 Nm).

17. Install the windshield defroster duct.

18. Connect the vacuum solenoid connector.

19. Connect the vacuum supply to the HVAC module.

20. Install the blower motor.

21. Connect the blower motor connector.

22. Install the right hand floor air duct (rear).

23. Install the carpet air duct (rear) and replace carpet.

24. Install the right hand floor air duct.

25. Connect the right hand temperature actuator electrical connector.

26. Connect the sunload electrical connector.

27. Install the SIR bracket and the passenger knee bolster bracket and tighten the bolts to 106 ft. lbs. (12 Nm).

28. Install the right hand side window defogger outlet duct.

29. Install the left hand floor outlet duct.

30. Position the ignition switch housing bracket to the center bracket aligning the marks made during removal, tighten the bolts to 106 inch lbs. (12 Nm).

31. Install the center bracket.

32. Rout the IP wiring harness. Connect the IP wiring harness and parking brake connector clip to the bracket.

33. Connect the electrical connectors and vacuum harness connectors from the HVAC control.

34. Connect connectors to the radio.

35. Install the center support bracket aligning the marks made during removal and tighten the left bolt that attaches the bracket to the IP lower support beam to 106 inch lbs. (12 Nm).

36. Install the ignition switch housing aligning the marks made during removal and tighten the bolt to 17 inch lbs. (2 Nm).

37. Install the left bolt that attaches the bracket to the driveline tunnel and tighten to 106 ft. lbs. (12 Nm).

38. Install the bolt that attaches the bracket to the IP lower support beam and tighten to 106 ft. lbs. (12 Nm).

39. Install the bolts that attach the bracket to the passenger knee bolster bracket and tighten to 106 ft. lbs. (12 Nm).

40. Install the right bolt that attaches the bracket to the driveline tunnel and tighten to 106 ft. lbs. (12 Nm).

41. Install the ignition switch.

42. Connect the SDM.

43. Install the CPA electrical.

44. Route the park/lock cable through the IP center support bracket. Make sure to route the cable under the wiring harness and over the floor rear outlet.

45. Engage the park/lock cable to the ignition switch.

46. Turn the ignition switch OFF.

47. Engage the park/lock cable to the shift control pivot arm stud once aligned pull back to lock.

48. Insert the cable into the slot on the shift control making sure it locks in place.

49. Turn the ignition ON.

50. Place the shift lever in each position but park and try to turn the key to the OFF position.

51. If the key turns to the OFF position except in park the cable is working properly.

52. Place the shift lever in PARK. Turn the key OFF and remove the key.

53. Install the trim panel and attach the hazard warning switch connector.

54. Install the upper trim panel.

55. Install the screws attaching the upper trim pad to the defroster duct.

56. Install the screws attaching the upper trim pad to the left and right hand hinge pillars.

57. Install the screws attaching the upper trim panel to the driver knee bolster outer bracket and center support bracket.

58. Install the screw attaching the upper trim panel to the passenger SIR bracket.

59. Install the screws attaching the IP cluster bezel to the upper trim pad.

60. Install the windshield side garnish moldings.

61. Install the DRL electrical connector and the sun load temperature sensor to the windshield defroster duct, if equipped

62. Install the inboard side window glass defroster duct to the windshield defroster duct on both sides

63. Connect the passenger side air bag connector and install the glove box.

64. Install the upper and side screws and tighten to 7 inch lbs. (2 Nm).

65. Install the lower retaining bolts and tighten to 106 inch lbs. (12 Nm). Install the trim plugs.

66. Connect the light switch and close the door.

67. Connect the inside air temperature switch connector and install the drivers knee bolster trim in position. Insert the connector for the fog lamp, rear compartment lid release switch through the opening in the panel.

68. Install the trim panel and tighten the screws to 16 inch lbs. (1.8 Nm).

69. Install the fog lamp, rear compartment lid release switch.

70. Lower the trim plate over the shifter and under the parking brake lever.

71. Place the shifter boot in position and insert a tool up through the shifter opening in the trim plate.

72. Connect the cigar lighter electrical connector.

73. Install the trim plate into position, install the upper locator tabs and upper locking tabs, then work downwards to engage the remaining tabs.

74. Install the trim plate screws and tighten to 17 inch lbs. (2 Nm).

75. Install trim plate grille.

76. Install the ashtray.

77. Install the shift boot in position and press to engage the tabs.

78. Place the transmission in Park on models with an Automatic Transmission (AT) or reverse on models with a Manual Transmission (MT).

79. Release the parking brake.

80. Install the console.

81. Install the fuel door release electrical switch and if equipped, the rear lift window switch and attach the connections.

82. Install the accessory power plug and attach connector.

83. Slide the console under the accessory trim plate.

84. Install the console retaining nuts.

85. Install the console retaining nut covers.

86. Install the traction switch.

87. Connect the LED connector to the harness connector.

88. Connect the traction switch connector.

89. Engage traction switch to the retaining clips.

90. Close the console door.

91. Install the stowage compartment and tighten the screws to 35 inch lbs. (4 Nm).

92. Install or connect the following:

- Battery heat shield
- Intake manifold
- Hose to the right hand air injection check valve
- Bolt attaching the right hand air injection check valve to the cylinder head(s), if equipped
- Heater pipe bracket to the cowl and tighten the bolt
- Heater pipe-to-heater core bolt
- Heater pipe assembly to the core.
- Accumulator hose-to-evaporator bolt
- Accumlator hose and evaporator tube to the evaporator using new O–ring seals.
- HVAC module drain tube to the module

93. Refill the cooling system.

94. Evacuate, charge and leak test the air conditioning system refrigerant.

95. Connect the negative battery cable.

96. If equipped, enable the SIR system by performing the following procedure:

    a. Connect the Connector Positive Assurance (CPA) to the yellow 2-way SIR harness connector at the base of the steering column.

    b. Connect the steering wheel module coil connector located at the base of the steering column.

    c. Connect the CPA from the inflatable restraint instrument panel module connector located at the near the base of the steering column.

    d. Connect the instrument panel module connector located near the base of the steering column.

    e. Install the left side sound insulator.

    f. Install the SIR fuse to the fuse panel.

97. Operate the engine to normal operating temperatures; then, check the climate control operation and check for leaks.

Cylinder Head

REMOVAL & INSTALLATION

2002-2004 Models

1. Before servicing the vehicle, refer to the Precautions Section.

2. Drain the engine coolant.

3. Relieve the fuel system pressure.

4. Remove or disconnect the following:

- Negative battery cable
- Rocker arm cover
- Alternator bracket, if removing the left side

- Exhaust manifold from the cylinder head
- Oil level indicator tube, if removing the right side
- Wiring harness from the clip at the rear of the cylinder head, if removing the right side
- Intake manifold
- Vapor vent pipe
- Power steering pump pulley, if removing the left side
- Power steering pump and reposition it, if removing the left side
- Power steering pump bracket, if removing the left side
- Ground wire bolt from the rear of the cylinder head, if removing the left side
- Cylinder head

To install:

➡**New M11 cylinder head bolts must be used when installing the cylinder head assembly.**

5. Install the new cylinder head gasket onto the locating pins.

6. Place the cylinder head on the engine and tighten the bolts in sequence to:

    a. M11 bolts: 22 ft. lbs. (30 Nm).

    b. M11 bolts: plus 90 degrees.

    c. M11 bolts 1–8: plus 90 degrees.

    d. M11 bolts 9–10: plus 50 degrees.

    e. M8 bolts (11–15): 22 ft. lbs. (30 Nm).

7. Install or connect the following:

- Ground wire to the rear of the cylinder head and torque the bolt to 24 ft. lbs. (32 Nm), if removed
- Power steering pump bracket and torque the bolts to 37 ft. lbs. (50 Nm) , if removed
- Power steering pump and torque

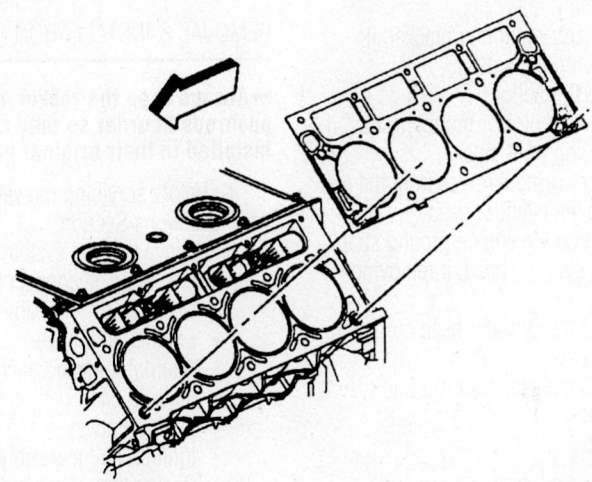

79222611

Be sure the tab on the edge of the gasket is closer to the front of the engine when installed

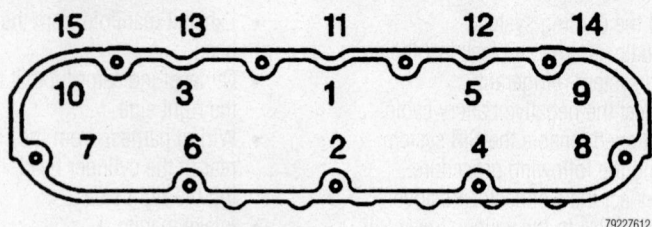

15 13 11 12 14
10 3 1 5 9
7 6 2 4 8

79222612

Cylinder head bolt torque sequence—All engines

the bolts to 18 ft. lbs. (25 Nm) , if removed
- Power steering pump pulley, if removed
- Exhaust manifold and torque the bolts, in sequence, to 11 ft. lbs. (15 Nm), then to 18 ft. lbs. (25 Nm)
- Oil level indicator tube, if removed
- Wiring harness from the clip to the rear of the cylinder head, if removed
- Valve rocker arm cover and torque the bolts to 106 inch lbs. (12 Nm)
- Intake manifold and torque the bolts in sequence to 44 inch lbs. (5 Nm) then to 89 inch lbs. (10 Nm)
- Vapor vent pipe and torque the bolt to 106 inch lbs. (12 Nm)
- Alternator bracket, if removed
- Negative battery cable

8. Fill the cooling system.
9. Start the vehicle and check for leaks, repair if necessary.

2005 Models

1. Before servicing the vehicle, refer to the Precautions Section.
2. Drain the cooling system.
3. Remove or disconnect the following:
- Negative battery cable
- Rocker arms
- Pushrods
- Engine coolant air bleed pipe assembly
- Alternator mounting bracket, if removing left side
- Exhaust manifold
- Oil level indicator mounting bolt, if removing right side
- Engine wiring harness from the rear of the left cylinder head

4. Reposition the engine ground strap away from the cylinder head, if removing left side.
5. Remove the cylinder head mounting bolts and discard.
6. Remove the cylinder head and discard the gasket.

To install:

➡**New cylinder head bolts must be used.**

7. Install the new cylinder head gasket onto the locating pins.
8. Install the cylinder head onto the locating pins and gasket and tighten the bolts in sequence as follows:
   a. M11 bolts (1–10): 22 ft. lbs. (30 Nm)
   b. M11 bolts (1–10): plus 90 degrees
   c. M11 bolts (1–10): plus 70 degrees.
   d. M8 bolts (11–15): 22 ft. lbs. (30 Nm)
9. Position the engine ground strap against the cylinder head, if installing left side.
10. Install or connect the following:
- Engine wiring harness ground bolt and tighten to 24 ft. lbs. (32 Nm)
- Oil level indicator mounting bolt, if installing right side. Tighten to 18 ft. lbs. (25 Nm).
- Exhaust manifold
- Alternator mounting bracket, if installing left side
- Engine coolant air bleed pipe assembly
- Rocker arms
- Pushrods
- Negative battery cable
11. Refill the cooling system.
12. Start the engine and check for leaks.

Rocker Arms

REMOVAL & INSTALLATION

➡**Always keep the rocker arms and pushrods in order so they can be installed in their original positions.**

1. Before servicing the vehicle, refer to the Precautions Section.
2. Relieve the fuel system pressure.
3. Remove or disconnect the following:
- Negative battery cable
- Engine sight cover
- Alternator electrical connector, left side
- Fuel rail cover
- Ignition coil assembly
- Rocker arm cover and gasket
- Rocker arms

- Push rods
- Rocker arm pivot support, if necessary

To install:

4. Lubricate the rocker arms and pushrods with clean engine oil.
5. Lubricate the flange and washer surface of the rocker arm mounting bolts.
6. Install the rocker arm pivot support, if removed.
7. Install the pushrods, if removed.
8. Install the rocker arms but do not tighten the bolts at this time.
9. Rotate the crankshaft so the No. 1 piston is at Top Dead Center (TDC).
10. With the engine in this position, tighten the exhaust valve rocker arm bolts on cylinders No. 1, 2, 7 and 8 to 22 ft. lbs. (30 Nm), then, tighten the intake valve rocker arm bolts on cylinders No. 1, 3, 4 and 5 to 22 ft. lbs. (30 Nm).
11. Rotate the crankshaft 1 revolution (360 degrees).
12. With the engine in this position, tighten the exhaust valve rocker arm bolts on cylinders No. 3, 4, 5 and 6 to 22 ft. lbs. (30 Nm), then, tighten the intake valve rocker arm bolts on cylinders No. 2, 6, 7 and 8 to 22 ft. lbs. (30 Nm).
13. Install or connect the following:
- Rocker arm cover using a new gasket and torque the bolts to 106 inch lbs. (12 Nm)
- Ignition coil assembly
- Fuel rail cover
- Alternator electrical connectors, left side
- Engine sight cover
- Negative battery cable
14. Start the vehicle and check for leaks.

Intake Manifold

REMOVAL & INSTALLATION

5.7L Engines

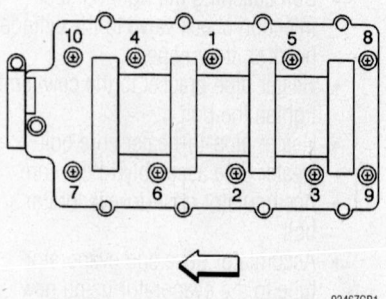

10 4 1 5 8
7 6 2 3 9

9346ZGB1

Intake manifold torque sequence—All engines

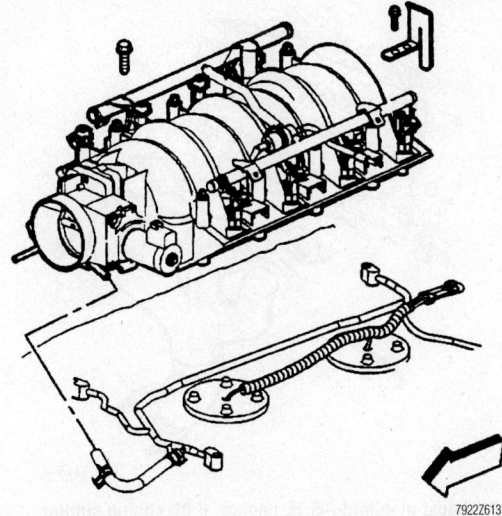

79222613

Be sure to reinstall the fuel stop bracket when installing the intake manifold

➡**The intake manifold, throttle body, fuel injectors and rail may be removed from the engine as an assembly.**

1. Before servicing the vehicle, refer to the Precautions Section.
2. Drain the cooling system.
3. Relieve the fuel system pressure.
4. Remove or disconnect the following:
- Engine sight covers
- Fuel feed hose
- Evaporative Emission (EVAP) canister purge tube from the intake manifold
- EVAP canister purge tube from the EVAP canister purge solenoid valve
- EVAP canister purge tube
- EVAP canister purge tube from the EVAP canister purge solenoid valve
- Disconnect the EVAP canister purge tube from the fuel feed pipe
- EVAP canister purge tube
- Throttle Position (TP) sensor connector
- Coolant air bleed hose
- Throttle body heater outlet hose
- Fuel injector electrical connectors
- EVAP canister purge solenoid valve connector
- Electronic Throttle Control (ETC) connector
- EVAP canister purge solenoid valve from the bracket
- Harness clips at the fuel rails
- Power brake booster vacuum hose at the booster
- Knock Sensor (KS) wire harness clip from the fuel rail stop bracket
- TP sensor harness clip from the Positive Crankcase Ventilation (PCV) tube

- PCV tube from the right rocker arm cover and throttle body
- PCV valve pipe from the left rocker arm cover, if equipped with the VIN G engine
- PCV valve pipe strap nut, if equipped with the VIN G engine
- PCV valve pipe from the right rocker arm cover and intake manifold
- PCV valve hose from the valley cover and intake manifold, if equipped with the VIN S engine
- Intake manifold bolts and fuel rail stop bracket
5. Position the intake manifold forward.
- Manifold Absolute Pressure (MAP) sensor vacuum hose
- MAP sensor connector
- Intake manifold
- Intake manifold gaskets and discard

To install:
6. Install or connect the following:
- New intake manifold gaskets
- Intake manifold
7. Position the intake manifold forward.
- MAP sensor vacuum hose
- MAP sensor connector
8. Position the intake manifold into place.
9. Apply threadlock GM P/N 12345382 to the threads of the intake manifold bolts.

✳✳ CAUTION

The fuel rail stop bracket must be installed onto the engine assembly. The stop bracket serves as a protective shield for the fuel rail in the event of a vehicle frontal crash. If the fuel rail stop bracket is not installed and the vehicle is involved in a frontal crash, fuel could be sprayed

possibly causing a fire and personal injury from burns.

10. Install or connect the following:
- Fuel rail stop bracket
- Intake manifold bolts
11. Tighten the intake manifold bolts as follows:
    a. Step 1: in sequence to 44 inch lbs. (5 Nm).
    b. Step 2: in sequence to 89 inch lbs. (10 Nm).
12. Install or connect the following:
- PCV valve hose to the valley cover and intake manifold, if equipped
- PCV valve pipe to the right valve rocker arm cover and intake manifold, if equipped
- PCV pipe strap nut and tighten to 106 inch lbs. (12 Nm)
- PCV valve pipe to the left rocker arm cover, if equipped
- PCV tube to the right rocker arm cover
- TP sensor harness clip to the PCV tube
- KS wire harness to the fuel rail stop bracket
- Power brake booster vacuum hose to the booster
- Intake manifold branches of the wiring harness
- Harness clips at the fuel rails
- EVAP canister purge solenoid valve to the bracket
- ETC connector
- EVAP canister purge solenoid valve connector
- Fuel injector connectors
- Throttle body heater outlet hose
- Throttle body heater outlet hose to the throttle body
- Throttle body heater outlet hose clamp at the throttle body
- Coolant air bleed hose
- TP sensor connector
- EVAP canister purge tube
- EVAP canister purge tube to the fuel feed pipe
- EVAP canister purge tube to the EVAP canister purge solenoid valve
- EVAP canister purge tube
- EVAP canister purge tube to the EVAP canister purge solenoid valve
- EVAP canister purge tube to the intake manifold
- Fuel feed hose
- Fuel rail covers
- Negative battery cable
13. Fill the cooling system.
14. Start the vehicle and check for leaks, repair if necessary.

6.0L Engines

1. Before servicing the vehicle, refer to the Precautions Section.
2. Relieve the fuel system pressure.
3. Drain the cooling system.
4. Remove or disconnect the following:
   - Negative battery cable
   - Engine sight covers
   - Fuel injector electrical connectors
   - Throttle body electrical connectors
   - Fuel supply line for the fuel injectors
   - Fuel rail
   - Brake booster vacuum hose
   - Manifold absolute pressure (MAP) sensor
   - Evaporative emission (EVAP) valve and tube assembly from the intake manifold assembly
   - Intake manifold mounting bolts
   - Fuel rail stop bracket
   - Intake manifold and gaskets

To install:

✳✳ WARNING

Do not reuse the intake manifold gaskets.

5. Install new intake manifold gaskets.
6. Install the intake manifold.
7. Install the fuel rail stop bracket.
8. Install the intake manifold bolts and tighten in sequence as follows:
   a. Step 1: Tighten to 44 inch lbs.
   b. Step 2: Tighten to 89 inch lbs.
9. Install or connect the following:
   - MAP sensor
   - EVAP valve and tube assembly to the intake assembly. Tighten the mounting bolt to 37 ft. lbs. (50 Nm).
   - Fuel rail and fuel supply hose
   - Brake booster vacuum hose
   - Throttle body electrical connectors
   - Fuel injector electrical connectors
   - Engine sight covers
   - Negative battery cable
10. Refill the cooling system to the correct level.
11. Start the engine and check for leaks.

Exhaust Manifold

REMOVAL & INSTALLATION

5.7L Engines

LEFT SIDE

1. Before servicing the vehicle, refer to the Precautions Section.

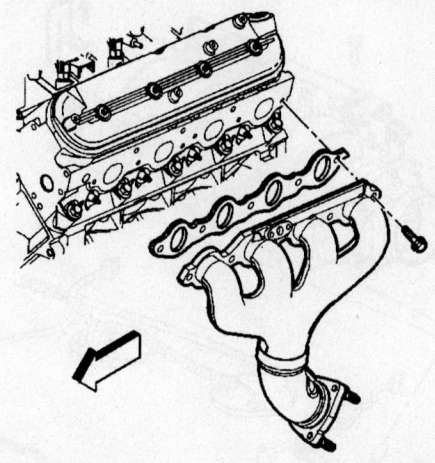

Exploded view of the exhaust manifold—5.7L engine, 6.0L engine similar

2. Relieve the fuel system pressure.
3. Remove or disconnect the following:
   - Negative battery cable
   - Fuel rail cover
   - Alternator
   - Exhaust manifold nuts
   - Connector Position Assurance (CPA) lock
   - Oxygen Sensor (O$_2$S)
   - O$_2$S connector clip at the body
   - Secondary Air Injection (AIR) hose clamp
   - AIR hose from the pipe
   - Hose clamps at the right and left check valves
   - AIR injection pipe hose from the right check valve
   - AIR injection pipe hose from the left check valve
   - AIR pipe
   - AIR pipe and gasket, and discard the old gasket
   - Brake booster vacuum hose from the booster
   - Ignition coil bracket
   - Spark plugs
   - Exhaust manifold

To install:

4. Apply threadlock to the threads of the bolts.
5. Install the exhaust manifold using a new gasket.
6. Beginning with the center 2 bolts, torque the bolts to 11 ft. lbs. (15 Nm) alternating from side-to-side until all the bolts are tight. Torque the bolts again in the same sequence to 18 ft. lbs. (25 Nm).
7. Bend over the exposed portion of the gasket at the rear of the cylinder head.
8. Install or connect the following:
   - Spark plugs and torque them to 11 ft. lbs. (15 Nm)

- Ignition coil bracket and tighten to 106 inch lbs. (12 Nm)
- Brake booster vacuum hose
- AIR pipe and gasket
- AIR pipe bolts and tighten to 15 ft. lbs. (20 Nm)
- AIR injection pipe hose to the left check valve
- AIR injection pipe hose to the right check valve
- Hose clamps at the right and left check valves
- AIR hose to the pipe
- AIR hose clamp
- O$_2$S and tighten to 30 ft. lbs. (40 Nm)
- O$_2$S connector clip at the body
- CPA lock
- Exhaust manifold nuts and tighten to 15 ft. lbs. (20 Nm)
- Alternator
- Fuel rail cover
- Negative battery cable

RIGHT SIDE

1. Before servicing the vehicle, refer to the Precautions Section.
2. Relieve the fuel system pressure.
3. Remove or disconnect the following:
   - Negative battery cable
   - Fuel rail cover
   - Exhaust manifold nuts
   - Connector Position Assurance (CPA) lock
   - Oxygen (O$_2$S) sensor
   - Secondary Air Injection (AIR) hose from the right check valve
4. Loosen, DO NOT remove the AIR pipe bolt at the rear of the left cylinder head.
   - AIR pipe and gasket, and discard the old gasket
   - Oil level indicator tube

- Ignition coil bracket
- Spark plugs
- Exhaust manifold

To install:

5. Apply threadlock to the threads of the bolts.

6. Install the exhaust manifold using a new gasket.

7. Beginning with the center 2 bolts, torque the bolts to 11 ft. lbs. (15 Nm) alternating from side-to-side until all the bolts are tight. Torque the bolts again in the same sequence to 18 ft. lbs. (25 Nm).

8. Bend over the exposed portion of the gasket at the rear of the cylinder head.

9. Install or connect the following:

- Spark plugs and torque them to 11 ft. lbs. (15 Nm)
- Ignition coil bracket and tighten to 106 inch lbs. (12 Nm)
- Oil level indicator tube
- AIR pipe and gasket
- AIR pipe bolts and tighten to 15 ft. lbs. (20 Nm)
- AIR injection pipe to the right check valve
- Hose clamps at the right check valve
- O_2S and tighten to 30 ft. lbs. (40 Nm)
- CPA lock
- Exhaust manifold nuts and tighten to 15 ft. lbs. (20 Nm)
- Alternator
- Fuel rail cover
- Negative battery cable

6.0L Engines

LEFT SIDE

1. Before servicing the vehicle, refer to the Precautions Section.

2. Remove or disconnect the following:

- Negative battery cable
- Engine sight cover
- Bank 1, sensor 1 Oxygen (O_2) sensor from exhaust manifold
- Spark plugs
- Alternator
- Intermediate pipe
- Catalytic converter O_2 sensor electrical connectors
- Left catalytic converter
- Exhaust manifold mounting bolts
- Exhaust manifold

To install:

3. Install the intermediate pipe.

4. Position the catalytic converter on the intermediate pipe but do not tighten the clamp at this time.

5. Connect the catalytic converter O_2 sensor electrical connectors.

6. Position the exhaust manifold with a new gasket into place.

7. Tighten the exhaust manifold mounting bolts as follows:

    a. Step 1: Tighten to 11 ft. lbs. (15 Nm). Start with the center 2 bolts and alternate side-to-side working toward the outside bolts.

    b. Step 2: Tighten to 18 ft. lbs. (25 Nm). Start with the center 2 bolts and alternate side-to-side working toward the outside bolts.

8. Install or connect the following:

- Bank 1, sensor 1 O_2 sensor to the exhaust manifold
- Spark plugs
- Alternator
- Catalytic converter to exhaust manifold nuts and tighten to 15 ft. lbs. (20 Nm)
- Tighten the catalytic converter exhaust clamp to 15 ft. lbs. (20 Nm)
- Engine sight cover
- Negative battery cable

9. Start the engine and check for leaks.

RIGHT SIDE

1. Before servicing the vehicle, refer to the Precautions Section.

2. Remove or disconnect the following:

- Negative battery cable
- Engine sight cover
- Spark plugs
- Bank 2, sensor 1 Oxygen (O_2) sensor from the exhaust manifold
- Oil level indicator tube assembly
- Intermediate pipe
- Catalytic converter O_2 electrical connectors
- Right catalytic converter
- Exhaust manifold mounting bolts
- Exhaust manifold

To install:

3. Install the intermediate pipe.

4. Position the catalytic converter on the intermediate pipe but do not tighten the clamp at this time.

5. Connect the catalytic converter O_2 electrical connectors.

6. Install the exhaust manifold and gasket.

7. Tighten the exhaust manifold mounting bolts as follows:

    a. Step 1: Tighten to 11 ft. lbs. (15 Nm). Start with the center 2 bolts and alternate side-to-side working toward the outside bolts.

    b. Step 2: Tighten to 18 ft. lbs. (25 Nm). Start with the center 2 bolts and alternate side-to-side working toward the outside bolts.

8. Install or connect the following:

- Oil level indicator tube assembly. Tighten to the mounting bolt to 18 ft. lbs. (25 Nm).
- Spark plugs
- Bank 2, sensor 1 O_2 sensor
- Catalytic converter to exhaust manifold nuts and tighten to 15 ft. lbs. (20 Nm)
- Tighten the catalytic converter exhaust clamp to 15 ft. lbs. (20 Nm)
- Engine sight cover
- Negative battery cable

9. Start the engine and check for leaks.

Camshaft

REMOVAL & INSTALLATION

1. Before servicing the vehicle, refer to the Precautions Section.

2. Drain the cooling system.

3. Relieve the fuel system pressure.

4. Remove or disconnect the following:

- Engine
- Crankshaft balancer
- Oil level indicator tube
- Left and right exhaust manifolds
- Water pump
- Intake manifold
- Coolant air bleed pipe
- Left and right valve rocker arm covers
- Valve rocker arms and push rods
- Left and right cylinder heads
- Valve lifters
- Oil pan-to-front cover bolts
- Front cover and gasket. Discard the old gasket.

5. Rotate the engine to Top Dead Center (TDC).

- Camshaft sprocket bolts
- Timing chain from the camshaft sprocket, and allow the timing chain to rest on the crankshaft sprocket
- Camshaft Position (CMP) sensor bolt and sensor
- Camshaft retainer plate

❈❈ WARNING

All camshaft bearings are the same diameter, extreme care must be taken during camshaft removal or installation to avoid damaging them.

6. Install 3 M8 x 100mm long bolts in the front of the camshaft to use as a handle.

7. Carefully remove the camshaft from the engine, be sure not to damage the camshaft bearings.

8. Clean and inspect the camshaft and bearings.

To install:

9. Install or connect the following:

10. Lubricate the camshaft with clean engine oil.

- Three M8–1.25 x 100mm long bolts in the front of the camshaft to use as a handle
- Camshaft into the engine and remove the 3 bolts
- Camshaft retainer using a new gasket and torque the bolts to 18 ft. lbs. (25 Nm)
- CMP sensor with a new O-ring and torque the bolt to 18 ft. lbs. (25 Nm)

11. Align the camshaft sprocket alignment mark in the 6 o'clock position.

- Camshaft sprocket and timing chain
- Camshaft sprocket bolts and tighten to 26 ft. lbs. (35 Nm)

12. Apply a 5 mm bead of sealant GM P/N 12378190, 0.8 inch (20mm) long to the oil pan-to-engine block junction.

- Front cover with a new gasket
- Front cover bolts until snug
- Oil pan-to-front cover bolts until snug

13. Install tool J 41476 and crankshaft balancer bolt to the front cover. Align the tapered legs of the tool with the machined alignment surfaces on the front cover.

14. Install the crankshaft balancer bolt until snug. Tighten the oil pan-to-front cover bolts to 18 ft. lbs. (25 Nm) and the engine front cover bolts to 18 ft. lbs. (25 Nm).

15. Remove the tool.

16. Install or connect the following:

- New crankshaft front oil seal
- Valve lifters
- Right and left cylinder heads
- Valve rocker arms and push rods
- Right and left valve rocker arm covers
- Coolant air bleed pipe
- Intake manifold
- Water pump
- Exhaust manifolds
- Oil level indicator tube
- Crankshaft balancer
- Engine assembly

Valve Lash

ADJUSTMENT

The rocker arms on the 5.7L (VIN G) and 6.0L (VIN U) engines are bolted into position so that no adjustment is possible. If the valves are noisy, suspect low oil pressure or worn valve train components.

Starter Motor

REMOVAL & INSTALLATION

1. Before servicing the vehicle, refer to the Precautions Section.

2. Remove or disconnect the following:

- Negative battery cable
- Intermediate exhaust pipe
- Starter electrical connectors
- Starter bolts
- Starter motor

To install:

3. Install or connect the following:

- Starter motor and torque the bolts to 37 ft. lbs. (50 Nm)

- Starter electrical connectors
- Intermediate exhaust pipe and torque intermediate pipe-to-manifold nuts to 15 ft. lbs. (20 Nm) and all other bolts to 37 ft. lbs. (50 Nm)
- Negative battery cable

Oil Pan

REMOVAL & INSTALLATION

5.7L Engines

1. Before servicing the vehicle, refer to the Precautions Section.

2. Drain the engine oil.

3. Remove or disconnect the following:

- Negative battery cable
- Alternator
- Washer reservoir
- Engine Coolant Temperature (ECT) sensor electrical connector
- Headlamp electrical connector
- Tie rods from the steering linkage
- Real Time Damping (RTD) sensor links, if equipped
- Stabilizer shaft
- Intermediate shaft from the steering gear
- Electronic Brake Control Module/Brake Pressure Modulator Valve (EBCM/BPMV) bracket and reposition
- Power steering gear
- Lower shock absorber mounting bolts
- Transverse leaf spring using a suitable compressor
- Lower control arm bolts from the suspension crossmember

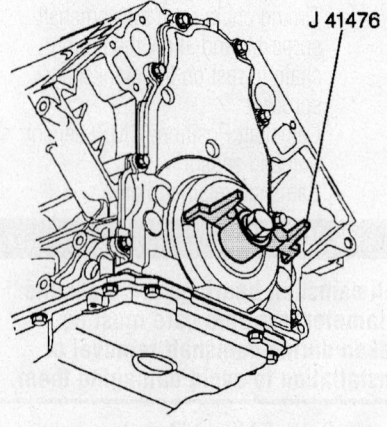

J 41476

9356ZGZZ

Install tool J 41476 and crankshaft balancer bolt to the front cover. Align the tapered legs of the tool with the machined alignment surfaces on the front cover

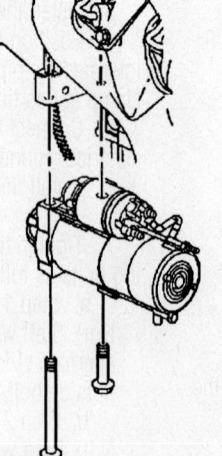

9306ZG26

Exploded view of the starter assembly

- Engine mount lower nuts
- Wheel Speed Sensor (WSS) wiring harness from the crossmember
- Brake line from the crossmember

4. Support the suspension crossmember.

- Crossmember mounting nuts
- Front crossmember
- Oil filter
- Automatic transmission cooler line front and rear retaining clamp bolts (if equipped)
- Engine flywheel housing-to-oil pan bolts
- Engine flywheel close out bolts
- Engine oil level sensor
- Engine oil temperature sensor electrical connector
- Oil pan assembly
- Lower oil pan bolts and separate the pans (if necessary)

➡**To remove the oil pan gasket it is necessary to drill out 3 rivets securing it to the pan.**

To install:

5. If separated, connect the upper and lower oil pans with a new gasket and torque the bolts to 106 inch lbs. (12 Nm).

6. Apply a 0.20 in. (5mm) bead of RTV sealant directly on the tabs of the front and rear cover gaskets that extend onto the oil pan mounting surface.

7. Install or connect the following:

- New gasket on the oil pan. It is not necessary to rivet the new gasket on the oil pan.
- Oil pan and hand-tighten the bolts
- Oil pan-to-engine bolts and torque them to 18 ft. lbs. (25 Nm)
- Oil pan-to-front cover bolts and torque them to 18 ft. lbs. (25 Nm)
- Oil pan-to-rear cover bolts and torque them to 106 inch lbs. (12 Nm)
- Flywheel housing bolts and torque them to 37 ft. lbs. (50 Nm)
- Left and right close outs and bolts and torque the bolts to 106 inch lbs. (12 Nm)
- Engine oil temperature sensor electrical connector
- Engine oil level sensor and torque it to 26 ft. lbs. (35 Nm)
- Automatic transmission cooler pipe front retaining clamp bolt and torque it to 106 inch lbs. (12 Nm)
- Automatic transmission cooler pipe rear retaining clamp bolt and torque it to 22 inch lbs. (3 Nm)
- Oil filter and torque it to 22 ft. lbs. (30 Nm)

- Oil drain plug and torque it to 18 ft. lbs. (25 Nm)

8. Raise the suspension crossmember into position.

- Crossmember mounting nuts and torque them to 81 ft. lbs. (110 Nm)
- Engine mount lower nuts and torque them to 48 ft. lbs. (65 Nm)
- WSS wiring harness to the crossmember
- Brake line to the crossmember
- Transverse leaf spring
- Lower control arm to the crossmember, DO NOT torque the bolts at this time

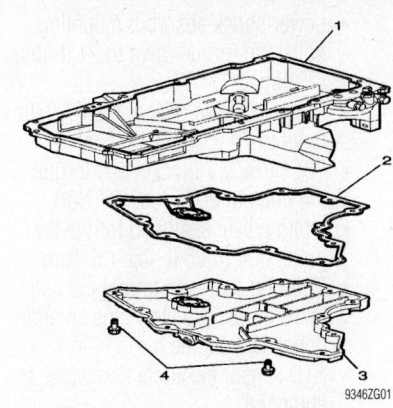

Separate the upper (1) and lower (3) oil pans

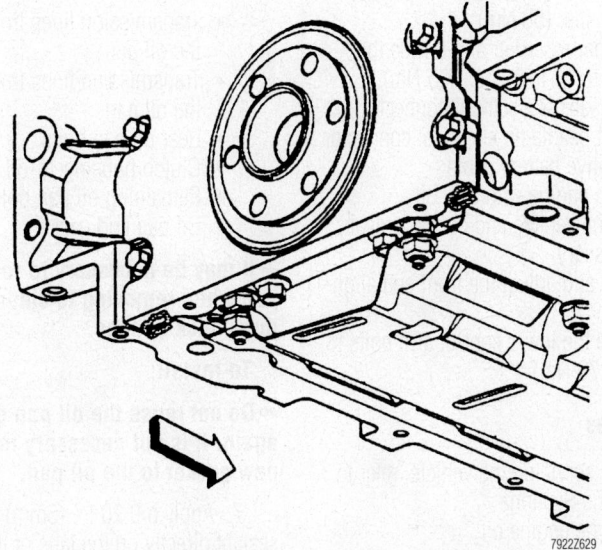

Apply sealant to the areas where the front and rear covers attach to the engine block

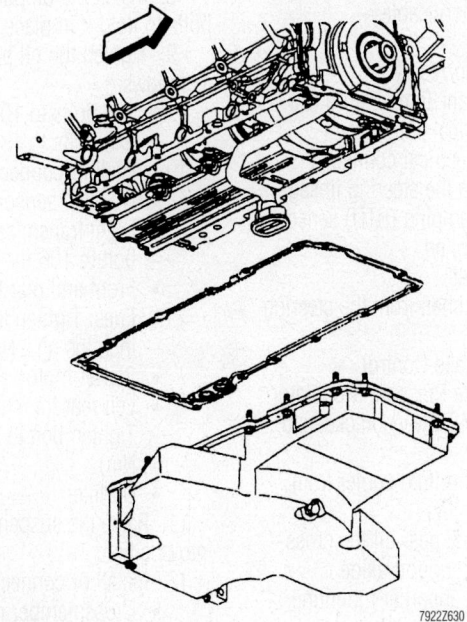

Exploded view of the oil pan mounting

- Lower shock absorber mounting bolts and torque them to 21 ft. lbs. (28 Nm)
- Power steering gear and torque the bolts to 74 ft. lbs. (100 Nm)
- EBCM/BPMV bracket and torque the bolts to 20 ft. lbs. (27 Nm)
- Intermediate shaft and torque the pinch bolt to 25 ft. lbs. (35 Nm)
- Outer tie rods and torque the nuts to 18 ft. lbs. (25 Nm) plus an additional 180 degree turn
- RTD sensor electrical connector, if equipped
- Stabilizer shaft and torque the link nuts to 53 ft. lbs. (72 Nm) and the clamp bolts to 43 ft. lbs. (58 Nm)
- Alternator and torque the bolts to 37 ft. lbs. (50 Nm)
- Washer reservoir and torque the nuts to 66 inch lbs. (7.5 Nm)
- ECT sensor electrical connector
- Front headlamp electrical connector
- Negative battery cable

9. Fill the engine with new oil.
10. Start the vehicle and check for leaks, repair if necessary.
11. Check and adjust the front end alignment.
12. Torque the lower control arm bolts to 125 ft. lbs. (170 Nm).

6.0L Engines

1. Before servicing the vehicle, refer to the Precautions Section.
2. Drain the engine oil.
3. Support the engine with a suitable engine support fixture.
4. Remove or disconnect the following:
- Negative battery cable
- Alternator
- Washer pump/reservoir
- Engine Coolant Temperature (ECT) sensor electrical connector
- Headlamp electrical connector
- Tie rods from the steering linkage
- Real Time Damping (RTD) sensor links, if equipped
- Stabilizer shaft
- Intermediate shaft from the steering gear
- Electronic Brake Control Module/Brake Pressure Modulator Valve (EBCM/BPMV) bracket and reposition
- Power steering fluid cooler from the crossmember
- Power steering gear off the crossmember and support aside
- Lower shock absorber mounting bolts

- Transverse leaf spring using a suitable compressor
- Lower control arm bolts from the suspension crossmember
- Engine mount lower nuts
- Wheel Speed Sensor (WSS) wiring harness from the crossmember
- Brake line from the crossmember

5. Remove the crossmember mounting nuts and lower the crossmember out of the vehicle.
6. Remove or disconnect the following:
- Oil filter
- Left rear transmission cover
- Starter motor
- Right transmission cover
- Engine oil level sensor electrical connector
- Transmission lines from the rear of the oil pan
- Transmission lines from the front of the oil pan
- Rear oil pan bolts
- Clutch housing-to-oil pan bolts
- Remaining oil pan bolts
- Oil pan and gasket

➡**It may be necessary to rotate the oil pan when removing to clear the oil pump pick up tube.**

To install:

➡**Do not reuse the oil pan gasket again. It is not necessary to rivet the new gasket to the oil pan.**

7. Apply a 0.20 in. (5mm) bead of RTV sealant directly on the tabs of the front and rear cover gaskets that extend onto the oil pan mounting surface.
8. Install the oil pan and hand tighten 2 bolts to hold it in place.
9. Tighten the oil pan mounting bolts as follows:
   a. M6 bolts to 106 inch lbs. (12 Nm)
   b. M8 bolts to 18 ft. lbs. (25 Nm)
10. Install or connect the following:
- Oil level sensor
- Right transmission cover. Tighten bolt to 106 inch lbs. (12 Nm).
- Front and rear transmission cooler lines. Tighten the retainers to 106 inch lbs. (12 Nm).
- Starter motor
- Left rear transmission cover. Tighten bolt to 106 inch lbs. (12 Nm).
- Oil filter
11. Raise the suspension crossmember into position.
12. Install or connect the following:
- Crossmember mounting nuts. Tighten to 81 ft. lbs. (110 Nm).

- Engine mount lower nuts. Tighten to 48 ft. lbs. (65 Nm).
- WSS wiring harness to the crossmember
- Brake line to the crossmember
- Transverse leaf spring
- Lower control arm to the crossmember, Do not tighten the bolts until the front end alignment is correct.
- Lower shock absorber mounting bolts and torque them to 21 ft. lbs. (28 Nm)
- Power steering gear and torque the bolts to 74 ft. lbs. (100 Nm)
- EBCM/BPMV bracket and torque the bolts to 20 ft. lbs. (27 Nm)
- Intermediate shaft and torque the pinch bolt to 25 ft. lbs. (35 Nm)
- Outer tie rods and torque the nuts to 18 ft. lbs. (25 Nm) plus an additional 180 degree turn
- RTD sensor electrical connector, if equipped
- Stabilizer shaft and torque the link nuts to 53 ft. lbs. (72 Nm) and the clamp bolts to 43 ft. lbs. (58 Nm)
- Alternator and torque the bolts to 37 ft. lbs. (50 Nm)
- Washer reservoir and torque the nuts to 66 inch lbs. (7.5 Nm)
- ECT sensor electrical connector
- Front headlamp electrical connector
- Negative battery cable

13. Fill the engine with oil to the correct level.
14. Start the vehicle and check for leaks.
15. Check and adjust the front end alignment. Tighten the lower control arm bolts to 125 ft. lbs. (170 Nm).

Oil Pump

REMOVAL & INSTALLATION

1. Before servicing the vehicle, refer to the Precautions Section.
2. Drain the engine oil.
3. Drain the cooling system.
4. Remove or disconnect the following:
- Negative battery cable
- Engine front cover
- Oil pan
- Oil pump screen and O-ring seal
- Oil pump

To install:
5. Install or connect the following:
- Oil pump and torque the bolts to 18 ft. lbs. (25 Nm)

➡**Be sure the crankshaft sprocket and oil pump drive gear are aligned properly.**

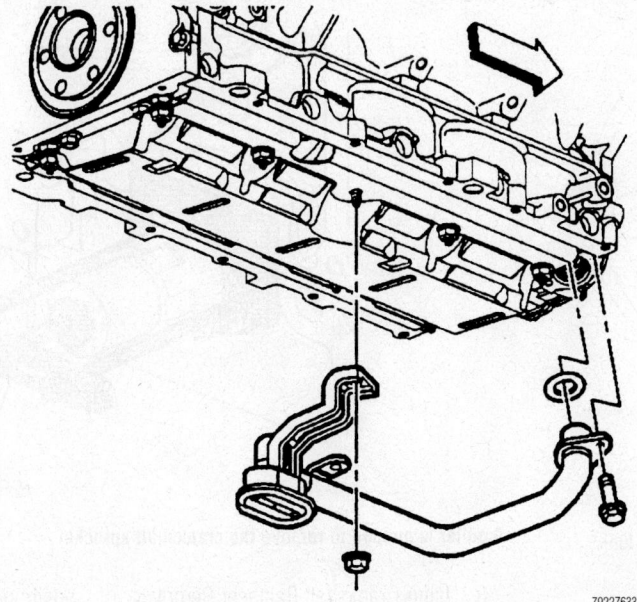

7922Z633

Be sure to seat the tube into the pump before installing the bolt

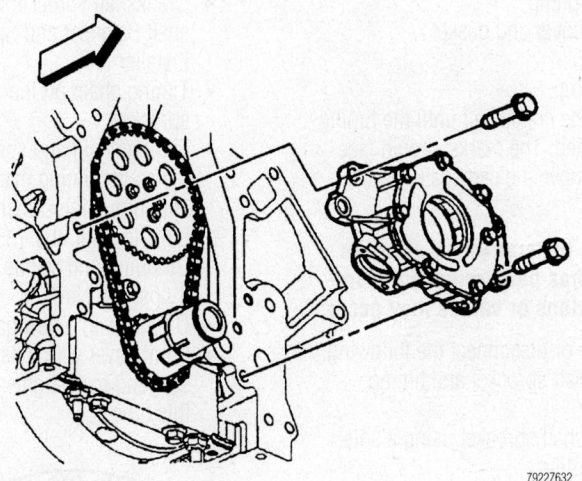

7922Z632

Oil pump assembly mounting

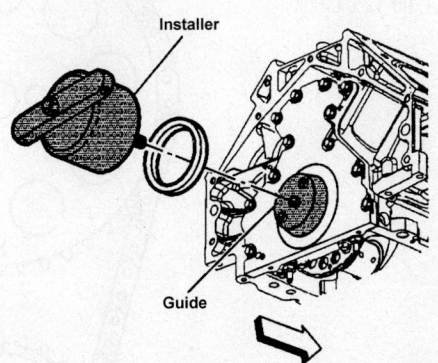

7922Z634

Use a seal installer to press the seal on the crankshaft

- Oil pump screen with a new O-ring and torque the bolt to 106 inch lbs. (12 Nm) and the nut to 18 ft. lbs. (25 Nm)

- Oil pan
- Engine front cover
- Negative battery cable
6. Fill the engine with new oil.

7. Fill the cooling system.
8. Start the vehicle and check for leaks, repair if necessary.

Rear Main Seal

REMOVAL & INSTALLATION

➡**The rear main seal is a 1-piece unit. It can be removed or installed without removing the oil pan or crankshaft.**

1. Before servicing the vehicle, refer to the Precautions Section.
2. Remove or disconnect the following:
   - Transmission
   - Clutch and pressure plate, if equipped with a manual transmission
   - Flywheel inspection cover
   - Flywheel
3. Using a prytool, carefully pry the old seal out.
   To install:
4. Inspect the crankshaft for nicks or burrs, correct as required.
5. Clean the area and coat the seal and crankshaft with engine oil.
6. Install or connect the following:
   - Seal, using a seal installer
   - Flywheel and torque the bolts in sequence to 15 ft. lbs. (20 Nm) then to 37 ft. lbs. (50 Nm) and finally to 74 ft. lbs. (100 Nm)
   - Flywheel inspection cover and torque the bolts 18 ft. lbs. (25 Nm)
   - Clutch and pressure plate if equipped with a manual transmission

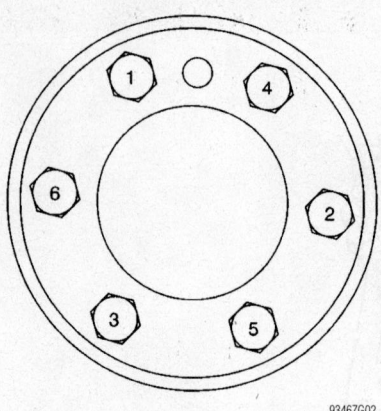

Flywheel torque sequence

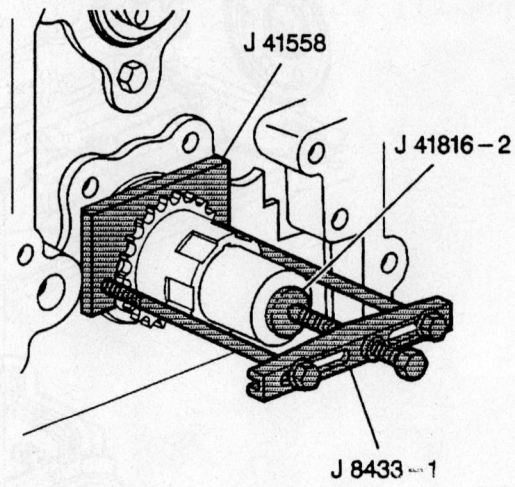

A puller is needed to remove the crankshaft sprocket

- Transmission

7. Start the vehicle and check for leaks, repair if necessary.

Timing Chain, Sprockets, Front Cover and Seal

REMOVAL & INSTALLATION

1. Before servicing the vehicle, refer to the Precautions Section.
2. Drain the cooling system.
3. Drain the engine oil.
4. Remove or disconnect the following:
   - Negative battery cable
   - Accessory drive belt
   - Power steering gear
   - Starter motor
   - Right transmission cover and bolt
5. Install the flywheel holding tool J-42386-A using one M10-1.5x120mm and one M10-1.5x45mm bolt.
6. Matchmark the position of the crankshaft balancer.
7. Remove the crankshaft balancer bolt.

8. Using Crankshaft Balancer Remover J-41816, remove the crankshaft balancer.
9. Remove or disconnect the following:
   - Water pump
   - Front cover and gasket
   - Oil pan
   - Oil pump
10. Rotate the crankshaft until the timing marks are aligned. The marks should face each other. Remove the camshaft sprocket bolts.

➡**Do not turn the crankshaft after the timing chain has been removed. Damage to the pistons or valves may occur.**

11. Remove or disconnect the following:
   - Camshaft sprocket and timing chain
   - Crankshaft sprocket using a suitable puller

To install:

12. Be sure the timing mark on the crankshaft sprocket is at the 12 o'clock

position. If not rotate the crankshaft to the correct position.

13. Install or connect the following:
   - Crankshaft sprocket with a Crankshaft Balancer and Sprocket Installer tool
   - Timing chain on the camshaft sprocket.
   - Timing chain and sprocket assembly so the timing marks are aligned and torque the camshaft sprocket bolts to 26 ft. lbs. (35 Nm)
   - Oil pump and torque the bolts to 18 ft. lbs. (25 Nm)
   - Oil pan with a new gasket
   - Front cover with a new gasket and seal and hand-tighten the bolts at this time

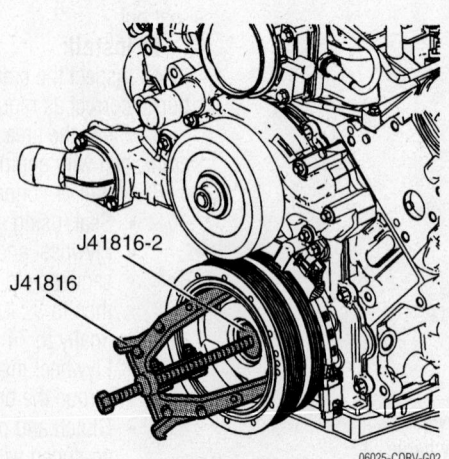

Using Special Tool J-41816 to remove the crankshaft balancer.

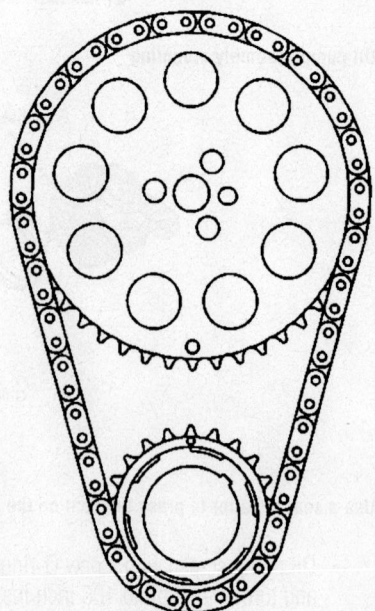

Timing chain alignment

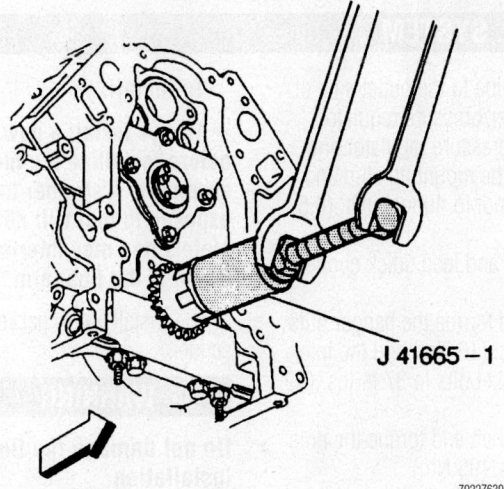

Be sure to install the crankshaft sprocket properly

14. Install the oil pan to cover bolts and hand-tighten them at this time. Install Front Cover alignment plate J-41480 and Front Cover Alignment Tool J-41476 to ensure front cover is properly aligned. Torque the oil pan-to-front cover bolts to 18 ft. lbs. (25 Nm) then the remaining bolts to 18 ft. lbs. (25 Nm).

15. Remove the alignment tools.
16. Install the water pump.
17. Position the balancer on the end of the crankshaft, using the matchmarks as a reference.
18. Using Crankshaft Balancer and Sprocket Installer J-41665, install the crankshaft balancer.
19. Install the old crankshaft balancer bolt and tighten to 240 ft. lbs. (330 Nm) to completely seat the balancer.

20. Remove the old bolt and replace with a new bolt. Tighten the bolt as follows:
   a. Step 1: Tighten to 37 ft. lbs. (50 Nm).
   b. Step 2: Tighten an additional 140 degrees.
21. Remove the flywheel holding tool.
22. Install or connect the following:
   • Starter motor
   • Power steering gear
   • Accessory drive belt
   • Negative battery cable
23. Fill the cooling system to the correct level.
24. Fill the engine with oil to the correct level.
25. Start the engine and check for leaks.

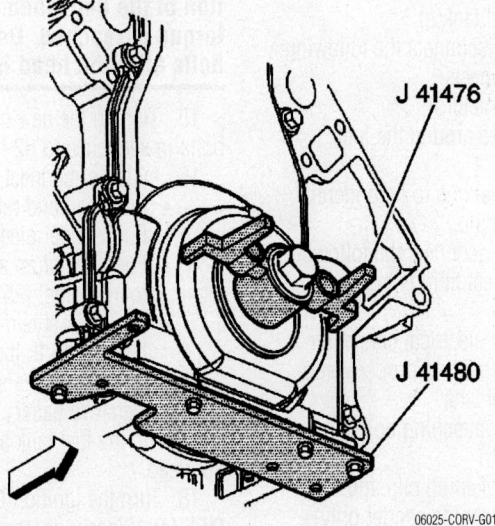

Front cover alignment tools J-41476 for crankshaft oil seal area and J-41480 for oil pan area.

Piston and Ring

POSITIONING

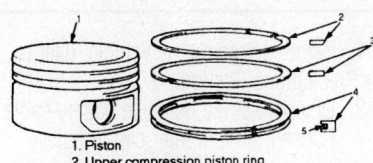

1. Piston
2. Upper compression piston ring
3. Lower compression piston ring
4. Oil control piston ring
5. Oil control ring spring w/spacer

Piston ring positioning—5.7L engine

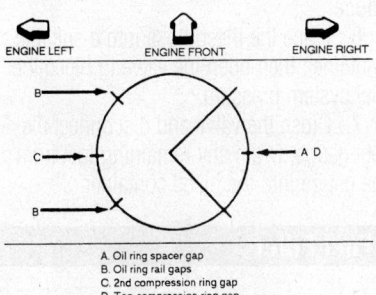

A. Oil ring spacer gap
B. Oil ring rail gaps
C. 2nd compression ring gap
D. Top compression ring gap

Piston ring end-gap spacing—5.7L engine

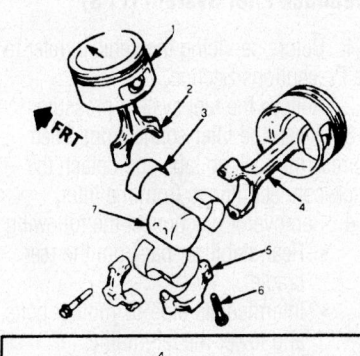

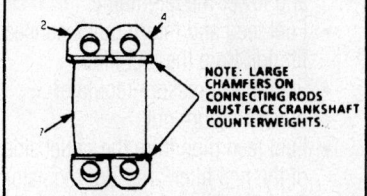

1. Piston
2. Connecting rod LH
3. Connecting rod bearing
4. Connecting rod RH
5. Connecting rod bearing cap
6. Connecting rod bearing cap bolt
7. Crankshaft

Piston and connecting rod positioning—5.7L engine

FUEL SYSTEM

Fuel System Pressure

RELIEVING

1. Before servicing the vehicle, refer to the Precautions Section.
2. Disconnect the negative battery cable.
3. Loosen the fuel filler cap to relieve the tank pressure.
4. Remove the left fuel rail cover.
5. Wrap a shop towel around the fuel pressure valve fitting (located on the side or end of the fuel rail assembly) to catch any fuel spray and connect a fuel pressure gauge.
6. Place the bleed hose into a suitable container, then open the valve to bleed the fuel system pressure.
7. Close the valve and disconnect the fuel gauge. Drain any remaining fuel from the gauge into the bleed container.

Fuel Filter

REMOVAL & INSTALLATION

2002 and 2003 Models without Feedback Fuel System (FFS)

1. Before servicing the vehicle, refer to the Precautions Section.
2. Relieve the fuel system pressure.
3. Clean the filter connections, then depress the locking tabs and detach the quick-connect fittings from the filter.
4. Remove or disconnect the following:
   • Rear stabilizer bar from the rear cradle
   • Intermediate pipe-to-muffler bolts and lower the left muffler
   • Fuel feed and return quick connect fittings from the fuel filter
   • Fuel filter/pressure regulator bracket mount nut
   • Fuel feed pipe from the outlet side of the fuel filter/pressure regulator
   • Fuel filter/pressure regulator and bracket
   • Fuel system ground strap
   • Fuel filter/pressure regulator from the bracket

To install:
5. Install or connect the following:
   • New plastic quick connector retainers on the inlet and outlet tubes
   • Fuel filter/pressure regulator into the bracket
   • Fuel system ground strap

• Fuel feed pipe to the outlet side of the fuel filter/pressure regulator
• Fuel filter/pressure regulator and bracket to the mounting stud and torque the nut to 40 inch lbs. (4.5 Nm)
• Fuel return and feed quick connect fittings
• Muffler and torque the hanger nuts to 12 ft. lbs. (16 Nm) and the intermediate pipe bolts to 37 ft. lbs. (50 Nm)
• Stabilizer shaft and torque the nuts to 70 ft. lbs. (95 Nm)
6. Connect the negative battery cable.
7. Turn the ignition switch **ON** for 2 seconds, then **OFF** for 10 seconds. Turn the ignition switch back **ON** and inspect for leaks.

2003 with Feedback Fuel System (FFS) and 2004-2005 Models

The fuel filter on these models is contained in the fuel sender assembly inside the left fuel tank. Refer to the Fuel Pump section for removal.

Fuel Pump

REMOVAL & INSTALLATION

2002 and 2003 Models without Feedback Fuel System (FFS)

1. Before servicing the vehicle, refer to the Precautions Section.
2. Properly relieve the fuel system pressure.
3. Drain the fuel tank(s).
4. Remove or disconnect the following:
   • Both rear wheels
   • Fuel tank shield(s)
5. Clean the area around the fuel sender assembly.
6. Mark each fuel line to help identify them during installation.
7. Remove or disconnect the following:
   • Quick connect fittings from the fuel sender
   • Fuel sender electrical connector
   • Fuel tank strap and properly support the fuel tank
   • Fuel sender attaching bolts and discard them
   • Float arm retaining clip and the float arm (left side sender only)
   • Fuel sender and gasket

To install:

→**Always install a new fuel strainer before installing the fuel sender assembly. A strainer that has been exposed to fuel will not unfold completely and may interfere with the full travel of the float arm.**

8. Install a new gasket on the fuel sender.

✳✳ WARNING

Do not damage the float arm during installation.

9. Fold the long strainer over itself and hold the strainer in this position.
10. Pinch both strainers upward toward each other.
11. Install the float arm through the fuel tank opening with the folded strainers.

→**It may be necessary to rotate the sender for proper installation.**

12. Look into the fuel tank opening and make certain that the long strainer is visible. If it is not visible, rotate the fuel sender until the strainer is free
13. Align the fuel sender gasket tab with the fuel sender cover mark and align the cover mark with the fuel tank mark.
14. Install the new attaching bolts to the fuel sender and hand-tighten them.

✳✳ CAUTION

The upper hex head portion of the fuel sender attaching bolts is designed to shear off the lower section of the bolt when the proper torque is reached. Do not tighten the bolts after the head is sheared off.

15. Tighten the new break away attaching bolts in sequence to 62 inch lbs. (7 Nm).
16. Install or connect the following:
   • Fuel feed and return pipes
   • Fuel sender electrical connector
   • Fuel tank strap and torque the bolts to 18 ft. lbs. (25 Nm)
   • Fuel tank shield and torque the bolts to 18 ft. lbs. (25 Nm)
   • Both rear wheels
   • Negative battery cable
17. Fill the fuel tank and install the fuel filler cap.
18. Turn the ignition **ON** for 2 seconds, **OFF** for 10 seconds, then **ON** again and inspect the system for leaks.

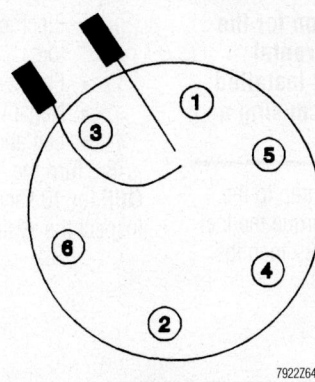

Tighten the fuel pump mounting bolts in the sequence shown—5.7L engine

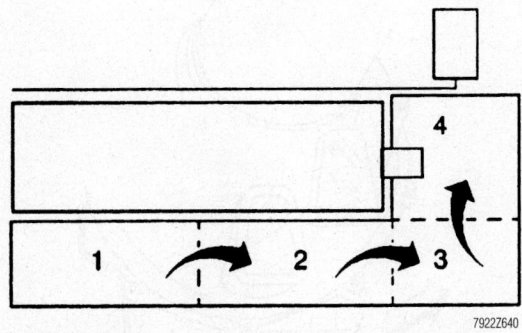

Fold the strainer on itself 3 times so it will fit into the opening in the fuel tank

2003 with Feedback Fuel System (FFS) and 2004-2005 Models

These models have 2 fuel tanks (right and left). The fuel pump is part of the fuel sender assembly in the left fuel tank. The right fuel tank contains a fuel sender assembly with a siphon jet pump which supplies fuel to the left tank through the fuel sender feed pipe. Although there are 2 sending units, the removal and installation procedure is the same for both.

1. Before servicing the vehicle, refer to the Precautions Section.
2. Relieve the fuel system pressure.
3. Drain the fuel tank.

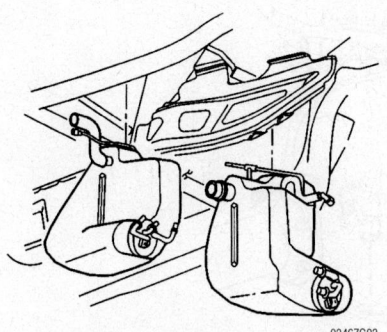

These models have 2 fuel tanks connected by a fuel sender feed pipe

4. Remove or disconnect the following:
   • Negative battery cable
   • Rear tire
   • Rear wheelhouse panel
   • Mufflers
   • Driveline support assembly
   • Fuel filler hose
   • Fuel pump electrical connector
   • Crossover tube
   • Evaporative emission (EVAP) crossover pipe quick connect fitting
   • Fuel tank strap
   • Fuel tank
   • Jet line insert connector from crossover tube to fuel tank opening
   • Fuel supply hose from clip on side of the tank
5. Using Fuel Sender Lock Ring Tool J-39765-A, remove the fuel pump module locking ring.
6. Carefully remove the fuel pump module from the tank.
   To install:
7. Install a new fuel pump module O-ring to the fuel tank opening.
8. Place tape around the jet line with the connector. This will permit line access once the pump module is inserted into the tank.
9. Install the pump module into the tank halfway.
10. Using the tape as a guide, gently

pull the jet line up through the fuel pump module opening.
11. Secure the line into the module retaining clip and remove the tape.
12. Compress and align the fuel pump module into the fuel tank and fully lock the ring in place using Fuel Sender Lock Ring Tool J-39765-A.
13. Install or connect the following:
   • Fuel supply hose
   • Jet line insert connector into the crossover tube to fuel tank opening
   • Fuel tank
   • Fuel tank strap
   • EVAP crossover pipe
   • Crossover tube

➡ **Lubricate the O-ring mating surfaces with the rubber lubricant.**

   • Fuel filler hose. Tighten the clamp to 35 inch lbs. (4 Nm).
   • Fuel pump electrical connector
   • Driveline support assembly
   • Mufflers
   • Rear wheelhouse panel
   • Rear tire
   • Negative battery cable
14. Turn the ignition **ON** for 2 seconds, **OFF** for 10 seconds, then **ON** again and inspect the system for leaks.

Fuel Injector

REMOVAL & INSTALLATION

1. Before servicing the vehicle, refer to the Precautions Section.
2. Relieve the fuel system pressure.
3. Remove or disconnect the following:
   • Both fuel rail covers
   • Fuel feed hose from the fuel rail

➡ **Matchmark the fuel injector to the electrical connector before removal.**

   • Fuel injector electrical connectors
   • Fuel rail mounting bolts
   • Fuel rail
4. Spread the injector clip to release the injector from the fuel rail.
5. Fuel injector by spreading the retaining clip.
   To install:

➡ **The fuel injector is stamped with a part number identification, manufacturing date, week code and plant number. Make certain the correct injector is ordered when replacing them.**

6. Lubricate the new injector seals with clean oil.

7. Install or connect the following:
- New O-ring seals on the injectors
- New retainer clip on the injector
- Fuel injector into the fuel rail socket

✷✷ CAUTION

The fuel rail stop bracket must be installed onto the engine. The bracket serves as protection for the fuel rail in the event of a frontal crash. If the bracket is not installed, fuel could spray possibly causing a fire and personal injury.

- Fuel rail and ground strap to the intake manifold and torque the fuel rail attaching bolts to 89 inch lbs. (10 Nm)

- Electrical connectors to the injectors
- Fuel feed hose to the fuel rail
- Negative battery cable
- Left and right fuel rail covers

8. Turn the ignition **ON** for 2 seconds, **OFF** for 10 seconds, then **ON** again and inspect the system for leaks.

DRIVE TRAIN

Transmission Assembly

REMOVAL & INSTALLATION

Automatic

1. Before servicing the vehicle, refer to the Precautions Section.
2. Disconnect the negative battery cable.
3. Shift the transmission into **N**.
4. Remove or disconnect the following:
- Rear wheels
- Intermediate exhaust pipe
- Right side muffler and tie the left side muffler to the underbody
- Driveline tunnel closeout panel
- Rear bell housing access plug and matchmark the flexplate to the torque converter
- Flexplate-to-torque converter bolts
- 2 plug bolts from the front of the driveline support assembly

5. Install two M10 x 1.5 x 55mm or longer bolts into the bolt holes. Torque the bolts to 26 ft. lbs. (35 Nm). These bolts must remain installed until instructed to remove them in order to maintain the position of the input shaft bearing.

6. Remove or disconnect the following:
- Engine flywheel housing access plug
- Propeller shaft hub clamp bolt
- Shift cable bracket nuts
- Shift control cable from the shift lever
- Rear transverse spring using Spring Compressor J-33432-A and support the lower control arms
- Outer tie rod ends from the knuckle
- Shock absorber lower mounting bolts
- Lower ball joints from the knuckles

7. Install a transmission support fixture to the transmission.

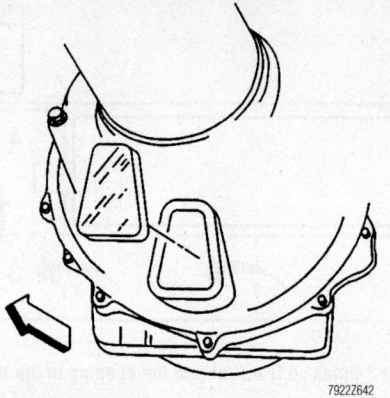

Remove the inspection plug, then remove the flexplate-to-torque converter bolts

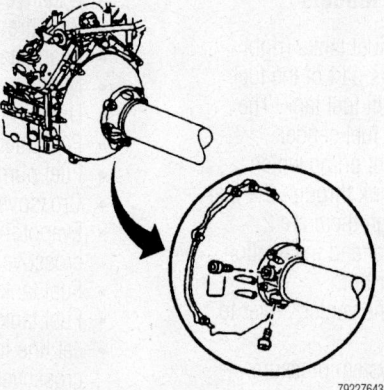

Remove the plugs and install two M10 x 1.5 x 55mm or longer bolts into the bolt holes to secure the bearing

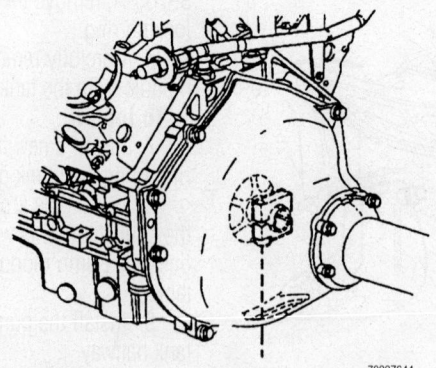

Rotate the flywheel to gain access to the clamp bolt, then loosen it

8. Remove or disconnect the following:
- Wiring harness and brake pipe clip retainers from the rear crossmember
- Lower transmission-to-differential nut
- Transaxle mount-to-rear crossmember nut
- Rear crossmember retaining nuts while supporting the crossmember
- Crossmember
- Transaxle mount bracket

9. Separate the axle shafts from the differential and tie them to the underbody.

10. Release the retainer securing the wiring harness from the "L" shaped brackets along the driveline support and move the harness out of the way.

11. Lower the driveline slightly and tilt it to gain access to the electrical connectors.

12. Remove or disconnect the following:
- Vehicle Speed Sensor (VSS) electrical connector
- Wire harness retainer from the differential rear cover stud
- Wire harness retainer clip from the top of the differential
- Transmission harness 20 way connector
- Park Neutral Position (PNP) switch electrical connectors
- Bolt securing the wire harness to the left side of the case

13. Lower the driveline and angle it while observing the top rear of the differential and the lowest part of the rear compartment floor panel. The engine Positive Crankcase Ventilation (PCV) pipes will most likely contact the dash panel. Make certain not to damage the dash.

14. Remove or disconnect the following:
- Wire harness from the retainer along the top of transmission
- Transmission rear oil cooler pipes from the junction fittings at the flywheel housing
- 5 driveline to flywheel housing bolts after supporting the rear of the engine

15. Bend the wiring harness bracket away from the driveline, toward the tunnel wall to have access to remove the driveline.

16. Separate the driveline from the engine by prying them apart with a flat blade tool.

17. Lower the driveline and tilt it away from the engine until the input shaft clears the flywheel housing.

18. Remove or disconnect the following:
- Driveline
- Transmission oil cooler rear pipes from the fittings

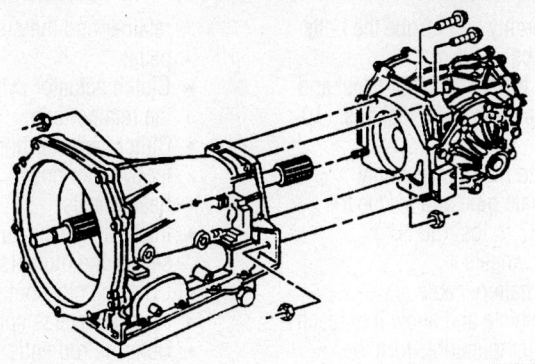

79222645

Transmission-to-differential mounting bolt locations

- Transmission to driveline bolts
- Separate the transmission from the driveline with a flat blade tool

19. Using a lifting device, place the assembly on a bench.
- Rear transmission oil cooler fittings from the transmission
- Transmission-to-driveline assembly support bolts
- Driveline from the transmission while supporting the torque converter
- Differential-to-transmission bolts
- Differential from the transmission

To install:

20. Install or connect the following:
- Differential to the transmission and torque the bolts to 37 ft. lbs. (50 Nm)
- Transmission to the driveline support and torque the bolts to 37 ft. lbs. (50 Nm)
- Rear transmission oil cooler fittings to the transmission and torque the fittings to 30 ft. lbs. (40 Nm)

21. Using a chain hoist, place the assembly on a transmission jack.

22. Carefully raise the assembly into the vehicle while placing the wiring harness loosely into the harness retaining slots.

23. Align the assembly for installation into the engine. The driveline will slide into the rear of the engine as long as the angles are the same.

24. Install or connect the following:
- Driveline assembly in the vehicle. Reposition the wiring harness bracket to align with the appropriate hole in the driveline assembly.
- Driveline support-to-flywheel housing bolts and torque them to 37 ft. lbs. (50 Nm)
- Oil cooler lines and torque the fittings to 20 ft. lbs. (27 Nm)
- Wiring harness to the left side of the transmission and torque the bolt to 22 inch lbs. (2.5 Nm)

- PNP switch electrical connector
- Transmission harness 20 way connector
- Wiring harness clip to the top of the differential
- VSS electrical connector

25. Raise the transmission to installation height.
- Axle shafts
- Transmission mount to the crossmember and torque the nuts to 37 ft. lbs. (50 Nm)
- Suspension crossmember and torque the nuts to 81 ft. lbs. (110 Nm)
- Lower transmission-to-differential nut and torque it to 37 ft. lbs. (50 Nm)
- Wiring harness and brake pipe clip retainers to the crossmember
- Lower ball joints to the suspension knuckles and torque the nuts to 52 ft. lbs. (70 Nm)
- Shock absorber lower mounting bolts and torque them to 162 ft. lbs. (220 Nm)
- Outer tie rod ends to the knuckles and torque the nuts to 15 ft. lbs. plus an additional 160 degrees
- Rear transverse spring and torque the bolts to 46 ft. lbs. (62 Nm)
- Wiring harness into the "L" shaped brackets
- Transmission cable and bracket and torque the nuts to 15 ft. lbs. (20 Nm)
- Transmission flexplate to the torque converter using the matchmarks made during removal and torque the bolts to 47 ft. lbs. (63 Nm)
- Rear bell housing access plug
- Propeller shaft hub bolt until it is finger-tight

26. Remove the two M10 x 55mm bolts from the input shaft front bearing.

27. Install or connect the following:
- 2 plug bolts to the driveline sup-

port assembly and torque the bolts to 37 ft. lbs. (50 Nm)
- Driveline tunnel closeout panel and torque the bolts to 89 inch lbs. (10 Nm)
- Right hand muffler assembly
- Intermediate pipe and torque the bolts to 37 ft. lbs. (50 Nm)
- Both rear wheels
- Negative battery cable

28. Start the vehicle and allow it to reach normal operating temperature. Turn the engine **OFF**. Allow the engine to cool to ambient temperature, then tighten the flywheel hub collar bolt to 96 ft. lbs. (130 Nm).

29. Install the bell housing inspection plug.

30. Flush the automatic transmission oil cooler.

31. A front end alignment is recommended when the crossmember is removed.

Manual

1. Before servicing the vehicle, refer to the Precautions Section.
2. Remove or disconnect the following:
- Negative battery cable
- Both rear wheels
- Folding top stowage compartment lid extension panel (on convertible models)
- Traction control/ride control switch
- Console retaining nut covers
- Console retaining nuts
- Accessory plug electrical connector
- Fuel door release electrical connector
- Console
- Shift control knob button
- Shift control knob retainer
- Shift control knob
- Shift control boot by grabbing both sides and pulling it in toward the lever
- Ashtray
- Trim plate grill
- Retaining screws located behind the grill and behind the ashtray
- Instrument panel accessory trim plate
- Shift control closeout boot retaining nuts and boot
- Shift rod clamp bolt
- Shift control mounting bolts
- Shift control assembly
- Courtesy lamp
- Left side lower closeout panel while guiding the courtesy lamp through the hole
- Clutch master cylinder pushrod

retainer and the pushrod from the pedal
- Clutch actuator cylinder hose from the retainer clip
- Clutch actuator hose from the master cylinder hose
- Rear wheels
- Intermediate exhaust pipe and secure the mufflers out of the way
- Driveline closeout panel
- Rear transverse spring
- Outer tie rod ends
- Lower shock absorber bolts
- Lower ball joints

3. Support the transmission
- Wiring harness and brake lines from the suspension crossmember
- Lower transmission-to-differential nut
- Transmission mount-to-crossmember nuts
- Rear suspension crossmember nuts
- Transmission mount and bracket from the differential

- Axle shafts from the differential and position them out of the way
- Wiring harness from the driveline support assembly
- Harness clip from the top of the differential

Do not lower the top rear portion of the differential past the bottom of the storage compartment or the PCV pipe will hit the dash panel possibly causing damage.

4. Lower the transmission enough to remove the wiring harness from the top of the assembly.

5. Remove or disconnect the following:
- Vehicle Speed Sensor (VSS) electrical connector
- Backup lamp switch electrical connector
- Reverse lockout solenoid electrical connector

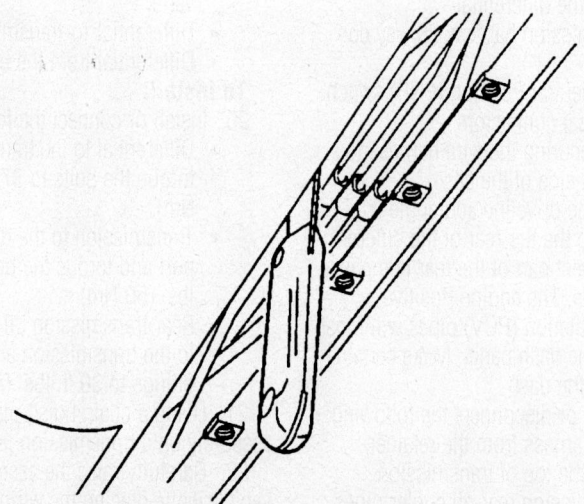

Insert a flat-bladed tool between the shifter bracket and brake liner retainer before lowering the transmission assembly

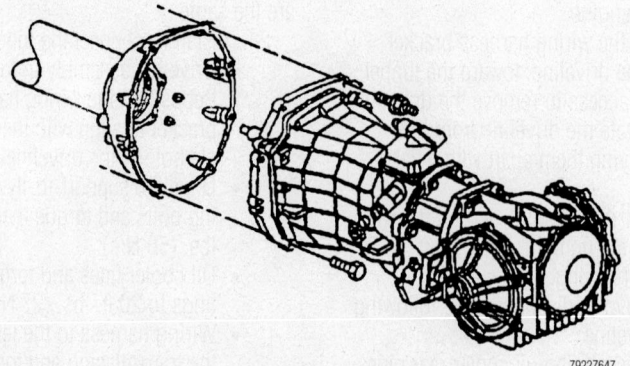

Driveline support-to-transmission mounting

- Gear select (skip shift) solenoid electrical connector
- Transmission fluid temperature sensor electrical connector (if equipped)

6. Support the rear of the engine
- Driveline support-to-bell housing bolts
- Driveline support assembly out of the bell housing while moving the assembly rearward

7. Carefully lower the assembly away from the vehicle while simultaneously adjusting the angle.

8. Attach a chain hoist to the assembly, then remove it from the jack and place it on a workbench.

9. Remove or disconnect the following:
- Driveline support-to-transmission mounting bolts. Carefully pry the driveline assembly away from the transmission while guiding the shift rod through the opening in the driveline support
- Transmission shift rod from the transmission
- Transmission-to-differential mounting bolts and separate the differential from the transmission

To install:

10. Install or connect the following:
- Differential to the transmission and torque the bolts to 37 ft. lbs. (50 Nm)
- Transmission shift rod
- Driveline support to the transmission and torque the bolts to 37 ft. lbs. (50 Nm)

11. To aid in later installation place a rubber band around the shift rod then tape the shift rod to the driveline support assembly.

12. Place the assembly on the transmission jack with a chain hoist.

13. Begin to raise the assembly into the vehicle while loosely installing the wiring harness along the driveline assembly retaining slots.

14. Have an assistant guide the front of the driveline support to the bell housing.

15. Be sure the driveline assembly is at the same angle as the engine before trying to install it.

16. Install or connect the following:
- Input propeller shaft into the clutch disc
- Wiring harness bracket to align with the appropriate hole in the driveline assembly
- Driveline support-to-flywheel housing bolts and torque them to 37 ft. lbs. (50 Nm)
- Wiring harness to the retainer on the top of the transmission and attach the connectors
- Transmission fluid temperature sensor electrical connector (if equipped)
- Gear select/skip shift solenoid electrical connector
- Reverse lockout solenoid electrical connector
- Backup lamp switch electrical connector
- Wiring harness to the top of the differential
- VSS electrical connector
- Axle shafts in the differential
- Transmission mount on the differential and torque the bolts to 37 ft. lbs. (50 Nm)
- Rear suspension crossmember and torque the nuts to 81 ft. lbs. (110 Nm)
- Transmission mount-to-crossmember nuts and torque them to 37 ft. lbs. (50 Nm)
- Lower transmission-to-differential

nut and torque it to 37 ft. lbs. (50 Nm)
- Wiring harness and brake line retainers to the crossmember
- Lower ball joints and torque the nuts to 52 ft. lbs. (70 Nm)
- Shock absorber lower mounting bolts and torque them to 162 ft. lbs. (220 Nm)
- Outer tie rod ends and torque the nuts to 15 ft. lbs. (20 Nm) plus an additional 160 degrees
- Transverse spring and torque the mounting bolts to 46 ft. lbs. (62 Nm)
- Clutch actuator hose to the master cylinder hose
- Clutch actuator hose to the retaining clip
- Driveline tunnel closeout panel and torque the bolts to 89 inch lbs. (10 Nm)
- Muffler assemblies
- Intermediate exhaust pipe and torque the bolts to 37 ft. lbs. (50 Nm)
- Rear wheels
- Clutch master cylinder pushrod to the clutch pedal
- Clutch master cylinder pushrod retainer
- Left lower closeout panel
- Courtesy lamp

17. Pull up the shift rod to break the tape and hook the rubber band on the rear stud on the top of the driveline tunnel.
- Shift control assembly and torque the bolts to 22 ft. lbs. (30 Nm)
- Shift control closeout boot and torque the nuts to 106 inch lbs. (12 Nm)
- Instrument panel accessory trim plate
- Shift control boot
- Shift control knob
- Shift control knob retainer and button

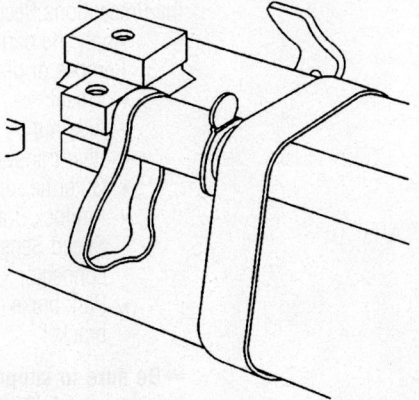

79222648

Place a rubber band on the shift rod, then tape the rod to the driveline support assembly

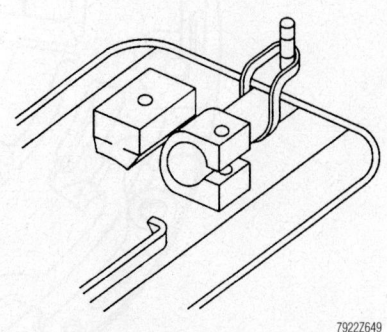

79222649

Pull the shift rod up to break the tape and hook the rubber band on the rear stud

- Console
- Fuel door release electrical connector
- Accessory plug electrical connector
- Console retaining nuts and torque them to 89 inch lbs. (10 Nm)
- Console retaining nut covers
- Traction control/ride control switch
- Folding top stowage compartment lid extension panel
- Rear wheels
- Negative battery cable

18. Bleed the clutch system.

Clutch

ADJUSTMENT

1. Before servicing the vehicle, refer to the Precautions Section.
2. Remove the flywheel inspection cover.
3. Have an assistant depress the clutch pedal until the tension is released from the stepped adjusting ring.
4. Using 2 screwdrivers, rotate the stepped adjusting ring counterclockwise until fully adjusted out.
5. Continue to hold them in this position and have the assistant release the clutch pedal.
6. Remove the screwdrivers.
7. Install the inspection cover. Torque the bolts to 18 ft. lbs. (25 Nm).

REMOVAL & INSTALLATION

1. Before servicing the vehicle, refer to the Precautions Section.
2. Remove or disconnect the following:
   - Negative battery cable
   - Exhaust system

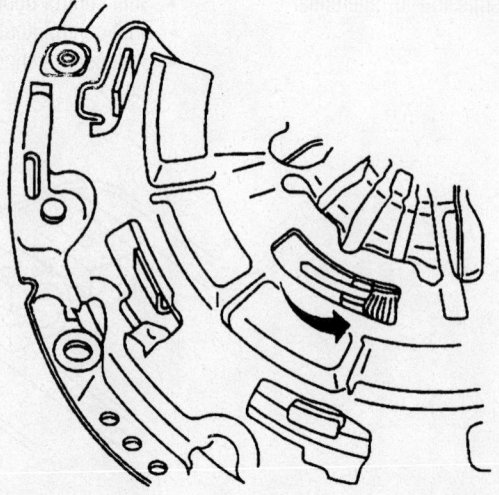

Rotate the stepped adjusting ring counterclockwise to compress the springs

- Driveline support and transmission as an assembly
- Flywheel inspection cover
- Pressure plate bolts. It will be necessary to rotate the flywheel to access all the bolts
- Pressure plate and disc

To install:

3. Install or connect the following:
   - Clutch disc and pressure plate on the flywheel
   - Clutch disc alignment tool through the disc to keep it in place
   - Pressure plate bolts finger-tight. Turn the flywheel to access all the bolt holes
4. Torque the flywheel bolts in the sequence shown, using 3 steps to 52 ft. lbs. (70 Nm).
5. Install the inspection cover and tighten the bolts as follows:
   a. 2002–2004 Models: 18 ft. lbs. (25 Nm)
   b. 2005 Models: 37 ft. lbs. (50 Nm)

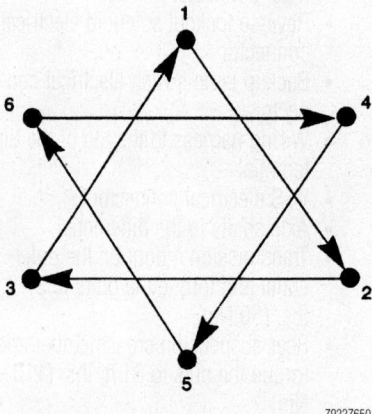

Clutch assembly torque sequence

6. Install or connect the following:
   - Driveline support and transmission assembly
   - Exhaust system
   - Negative battery cable
7. Bleed the clutch system.

Hydraulic Clutch System

BLEEDING

1. Before servicing the vehicle, refer to the Precautions Section.
2. Fill the clutch master cylinder with clean clutch hydraulic fluid.
3. Raise and safely support the vehicle with an assistant in it.
4. Remove the intermediate exhaust pipe.
5. Remove the driveline tunnel cover.
6. Have the assistant depress and hold the clutch pedal down.
7. Loosen the bleeder screw on the actuator cylinder to release the air, then tighten the screw.

✳✳ WARNING

Do not allow the clutch pedal to be released until the bleeder screw is closed or air will be drawn into the system.

8. Repeat the prior two steps until the air has been purged from the system. Check the master cylinder fluid level and refill as needed.
9. Install the driveline tunnel cover.
10. Install the intermediate exhaust pipe.
11. Lower the vehicle.

Halfshaft

REMOVAL & INSTALLATION

1. Before servicing the vehicle, refer to the Precautions Section.
2. Apply the parking brake.
3. Remove or disconnect the following:
   - Wheel
   - Axle nut
   - Rear transverse spring
   - Outer tie rod end from the knuckle
   - Antilock Brake System (ABS) Wheel Speed Sensor (WSS) electrical connector
   - Park brake cable from the lever and bracket

➡**Be sure to support the halfshaft until it is removed. Do not let it hang by the CV-joint.**

4. Attach a puller on the wheel studs and start to push the axle shaft into the hub assembly. This will provide clearance for the ball joint to be separated from the knuckle assembly.

5. Remove or disconnect the following:
- Ball joint from the knuckle
- Axle shaft from the hub
- Halfshaft assembly from the differential by inserting a suitable tool between the CV-joint and differential and prying them apart

To install:

6. Install or connect the following:
- Halfshaft on the differential output shaft. Use light force to be sure it is fully seated
- Halfshaft through the hub assembly but do not install completely. This

will provide clearance for installing the ball joint

- Ball joint to the knuckle and torque the nut to 41 ft. lbs. (55 Nm)
- Halfshaft through the hub assembly completely
- Park brake cable to the bracket and the lever
- WSS electrical connector
- Tie rod end to the knuckle and torque the nut to 15 ft. lbs. (20 Nm) plus an additional 160 degrees
- Rear transverse spring and torque the bolts to 46 ft. lbs. (62 Nm)
- Halfshaft retaining nut and torque the nut to 118 ft. lbs. (160 Nm)
- Wheel

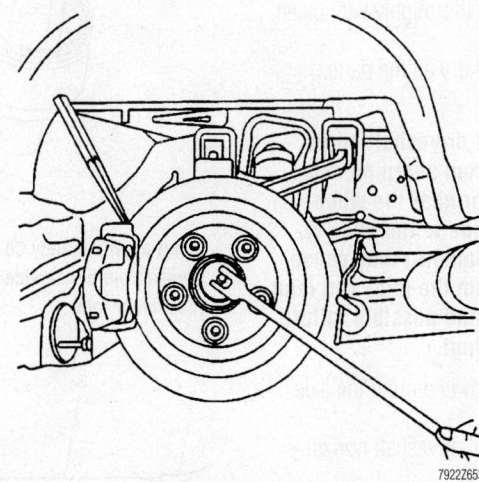

Insert a large drift through the cooling fins to keep the hub assembly from turning while removing the retaining nut

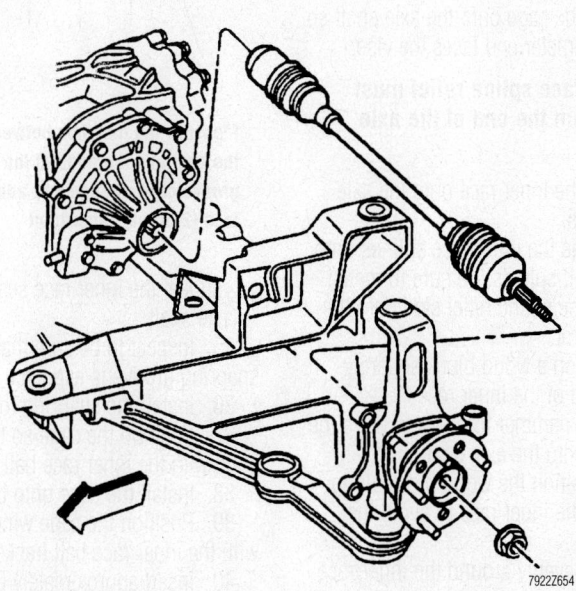

Exploded view of the halfshaft mounting

OVERHAUL

Inner Joint

1. Before servicing the vehicle, refer to the Precautions Section.

2. Remove the halfshaft.

3. Wrap a shop towel around the axle shaft.

4. Place the wheel driveshaft horizontally in a bench vise.

5. Remove the large seal retaining clamp from the CV joint seal.

6. Use a side cutter or other suitable tool and discard the clamp.

7. Remove the small seal retaining clamp from the joint seal.

8. Use a side cutter or other suitable tool and discard the clamp.

9. Separate the seal from the joint outer race at the large diameter end.

10. Position the seal behind the joint face.

11. Position the wheel driveshaft vertically in the bench vise so the inner joint is up.

12. Slide the joint outer race down toward the vise.

13. Disengage the outer race retaining ring as follows:

a. Insert a small flat-bladed screwdriver between the retaining ring and the outer race.

b. Pry the retaining ring from the outer race.

c. Position the retaining ring along the axle shaft away from the outer race.

➡ **The balls may fall out of the cage and inner race when the outer race is removed.**

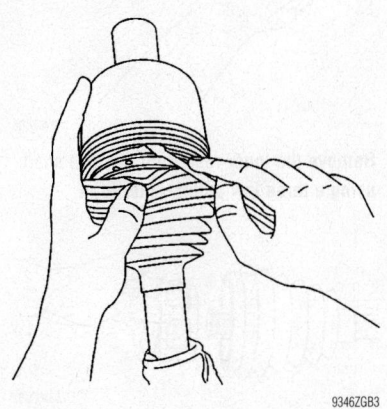

Separate the seal from the joint outer race at the large diameter end–inner joint

14. Remove the outer race from the axle shaft as follows:

    a. Use the seal to catch any balls which are not retained by grease.

    b. Lift the outer race off the axle shaft.

15. Remove any remaining balls from the cage and inner race.

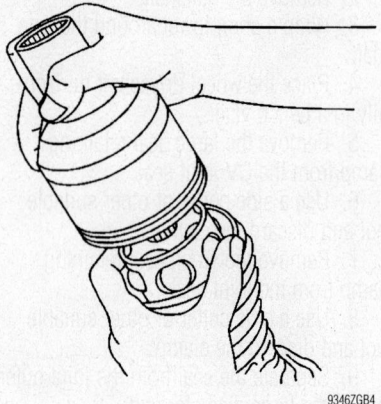

Remove the outer race from the axle shaft–inner joint

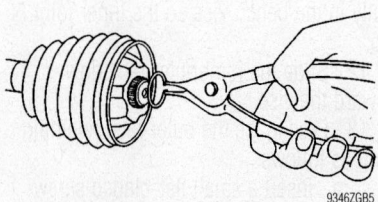

Remove the snapring from the axle shaft–inner joint

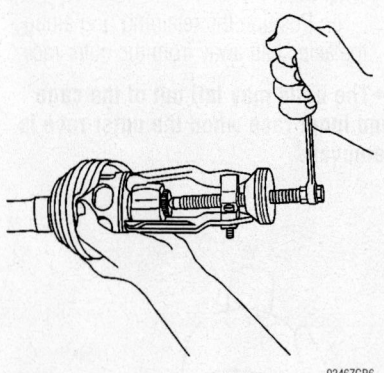

Remove the inner race from the axle shaft using a three jaw puller–inner joint

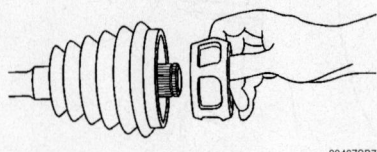

Remove the cage from the axle shaft–inner joint

16. Remove any balls caught by the seal.

17. Position the wheel driveshaft horizontally in the bench vise.

18. Remove the outer race retaining ring from the axle shaft.

19. Remove the snapring from the axle shaft.

20. Align the cage lands with the inner race ball tracks.

21. Reposition the cage along the axle shaft away from the inner race.

22. Wipe the grease from the inner race.

23. Remove the inner race from the axle shaft using a three jaw puller.

24. Remove the cage from the axle shaft.

25. Remove the seal from the axle shaft.

26. Remove the wheel driveshaft from the bench vise.

➡**All traces of old grease and any contaminates must be removed.**

27. Clean all parts thoroughly with clean solvent.

28. Thoroughly air dry all the parts.

To install:

➡**Protect the wheel driveshaft boots, seals and clamps from sharp objects when servicing on or near the wheel driveshaft(s). Damage to the boot(s), the seal(s) or the clamp(s) may cause lubricant to leak from the joint and lead to increased noise and possible failure of the wheel driveshaft.**

29. Wrap a shop towel around the axle shaft.

30. Place the wheel driveshaft horizontally in a bench vise.

31. Install a new small seal retaining clamp onto the wheel driveshaft.

32. Install the seal onto the axle shaft.

33. Install the cage onto the axle shaft so the smaller diameter end faces the vise.

➡**The inner race spline relief must face away from the end of the axle shaft.**

34. Install the inner race onto the axle shaft as follows:

    a. Engage the inner race splines onto the axle shaft splines. Be sure to install the inner race spline relief side onto the axle shaft first.

    b. Position a wood block squarely over the end of the inner race.

    c. Use a hammer to begin to drive the inner race onto the axle shaft.

    d. Reposition the wood block along the face of the inner race to avoid the axle shaft.

    e. Work evenly around the inner race and continue to drive the inner race, until

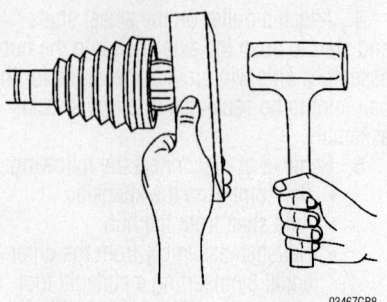

Use a hammer and a block of wood to drive the inner race onto the axle shaft–inner joint

Insert approximately 60 percent of the grease from the service kit into the outer race–inner joint

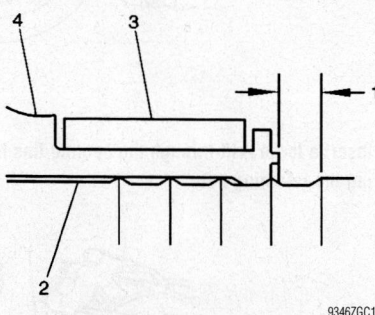

Measure the distance between the edge of the seal and the edge of the last axle shaft groove closing edge and adjust fit to 0.10 inch (2.5mm)–inner joint

you feel the inner race seat fully onto the axle shaft.

35. Inspect to be sure that the axle shaft snapring groove is exposed.

36. Install the snapring to the axle shaft.

37. Position the cage so the cage lands align with the inner race ball tracks.

38. Install the cage onto the inner race.

39. Position the cage windows to align with the inner race ball tracks.

40. Insert approximately 60 percent of

the grease from the service kit into the outer race.

41. Position the wheel driveshaft vertically in the bench vise so the inner joint end is up.

42. Apply a small amount of the grease from the service kit to the cage windows and inner race ball tracks.

43. Insert the remaining grease from the service kit into the seal.

44. Install the outer race retaining ring onto the axle shaft.

45. Position the retaining ring below the cage, toward the vise.

46. Install the balls through the cage windows to the inner race ball tracks.

47. Use the seal to keep the balls in position if necessary.

48. Install the outer race onto the axle shaft as follows:

    a. Be careful not to allow the grease in the outer race to leak out.

    b. Align the outer race ball tracks to the balls.

    c. Slide the outer race down over the balls.

49. Position the wheel driveshaft horizontally in the bench vise. Engage the outer race retaining ring as follows:

    a. Slide the outer race toward the vise.

    b. Insert the outer race retaining ring into the groove along the outer edge of the outer race.

    c. Position the outer race retaining ring so the opening in the ring aligns with an outer race land (not a ball track).

50. Position the large diameter end of the seal onto the outer race.

51. Position the small seal retaining clamp onto the neck of the seal.

52. Position the seal and small retaining clamp to the axle shaft.

53. Measure the distance between the edge of the seal and the edge of the last axle shaft groove closing edge and adjust fit to 0.10 inch (2.5mm).

➡The seal retaining clamp must not be over-tightened or under-tightened.

54. Crimp the small seal retaining clamp using tool J 42572. Tighten the small seal retaining clamp until the base of the omega shape has a gap width between 0.079–0.118 inch (2–3mm), with a difference in the gap width from side to side no greater than 0.016 inch (0.4mm). The clamping hold time must be no less than 2 seconds.

55. Measure the distance between the end of the seal and the end of the joint outer

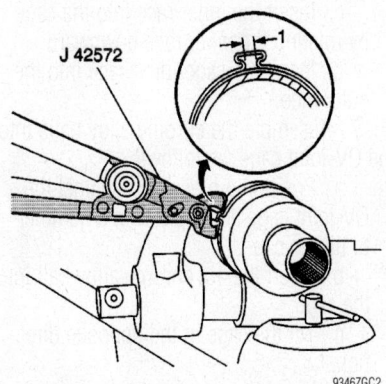

Tighten the small seal retaining clamp until the base of the omega shape has a gap width between 0.079–0.118 inch (2–3mm), with a difference in the gap width from side to side no greater than 0.016 inch (0.4mm)–inner joint

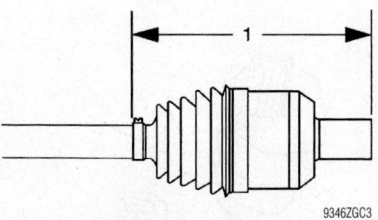

Measure the distance between the end of the seal and the end of the joint outer race and adjust the plunging motion of the joint to 8.90 inch within 0.16 inch (226mm within 4mm–inner joint

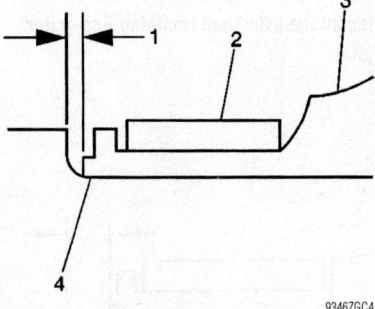

Measure the distance between the edge of the seal and the edge of the joint outer race last groove closing edge and adjust fit to 0.03 inch (0.8mm)–inner joint

race and adjust the plunging motion of the joint to 8.90 inch within 0.16 inch (226mm within 4mm).

56. Position the large seal retaining clamp onto the seal.

57. Position the seal and large retaining

clamp to the joint outer race. Measure the distance between the edge of the seal and the edge of the joint outer race last groove closing edge and adjust fit to 0.03 inch (0.8mm).

➡The seal must not be dimpled, stretched or out of shape in any way.

58. Inspect the seal for proper shape. If the seal is NOT shaped correctly, equalize the pressure in the seal and shape the seal properly by hand. Inspect the seal for damage. If the seal has been cut or punctured during assembly, you must discard and replace the seal.

➡The seal retaining clamp must not be over-tightened or under-tightened.

59. Crimp the large seal retaining clamp using tool J 42572. Tighten the small seal retaining clamp until the base of the omega shape has a gap width between 0.079–0.118 inch (2–3mm), with a difference in the gap width from side to side no greater than 0.016 inch (0.4mm). The clamping hold time must be no less than 2 seconds.

60. Remove the wheel driveshaft from the bench vise.

61. Distribute the grease within the inner CV joint.

62. Plunge the joint back and forth four or five times.

63. Inspect the inner CV joint and wheel driveshaft for smooth operation as follows:

    a. Hold the wheel driveshaft vertically, with the outer joint at the bottom.

    b. Rotate the wheel driveshaft four or five times in a circular motion.

64. Install the axle shaft.

Outer Joint

1. Before servicing the vehicle, refer to the Precautions Section.

2. Remove or disconnect the following:
- Axle shaft from the vehicle
- Large CV boot retaining clamp
- Small CV boot retaining clamp
- CV boot from the joint
- Outer joint from the axle shaft using a hammer and wood block
- Axle shaft retaining ring
- CV boot

3. Disassemble the chrome alloy balls from the CV-joint cage as follows:

    a. Position a brass drift against the CV-joint cage and tap it with a hammer to tilt the cage.

    b. Remove the 1st chrome alloy ball from the cage.

    c. Tilt the cage in the opposite direction.

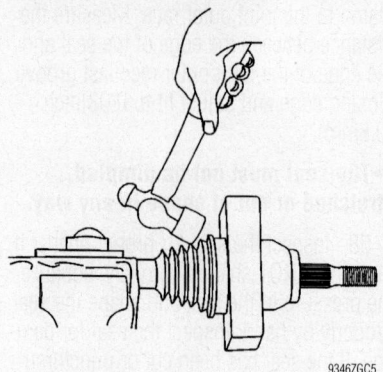

Separate the outer joint from the axle shaft using a hammer and wood block–outer joint

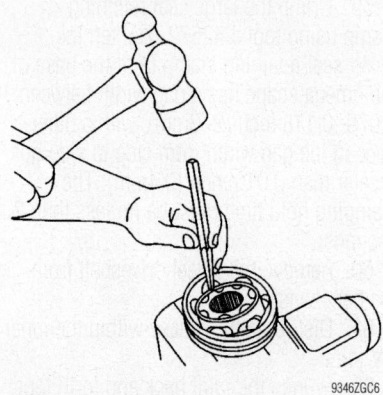

Position a brass drift against the CV-joint cage and tap it with a hammer to tilt the cage–outer joint

d. Remove the opposite chrome alloy ball.

e. Repeat the procedure until all 6 balls are removed.

4. Disassemble the CV-joint cage and inner race as follows:

a. Pivot the cage and race 90 degrees to the center line of the outer race.

b. Align the cage windows with outer race lands.

c. Remove the cage from the outer race.

d. Rotate the inner race upward and remove it from the cage.

To install:

5. Lubricate the parts with a light coat of grease.

6. Assemble the CV-joint cage and inner race, as follows:

a. Rotate the inner race 90 degrees to the cage centerline.

b. Align the cage windows with inner race lands.

c. Insert the inner race into the cage by rotating the inner race downward.

d. Insert the cage/inner race into the outer race.

7. Assemble the chrome alloy balls into the CV-joint cage, as follows:

a. Position a brass drift against the CV-joint cage and tap it with a hammer to tilt the cage.

b. Insert the 1st chrome alloy ball into the cage.

c. Tilt the cage in the opposite direction.

d. Insert the opposite chrome alloy ball.

e. Repeat the procedure until all 6 balls are inserted.

8. Install ½ of the grease provided, into the CV-joint.

9. Install or connect the following:
- Small CV boot retaining ring
- CV boot on the halfshaft
- New retaining ring on the halfshaft

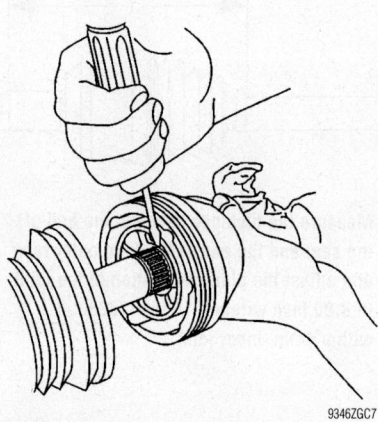

Install the axle shaft retaining ring–outer joint

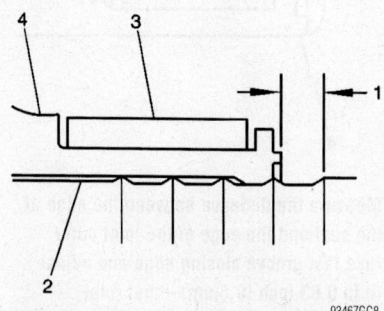

Measure the distance between the edge of the seal and the edge of the last axle shaft groove closing edge and adjust fit to 0.10 inch (2.5mm)

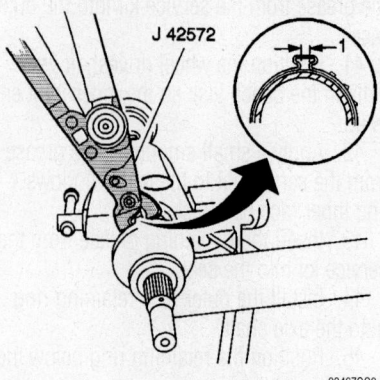

Tighten the small seal retaining clamp until the base of the omega shape has a gap width between 0.079–0.118 inch (2–3mm), with a difference in the gap width from side to side no greater than 0.016 inch (0.4mm)–outer joint

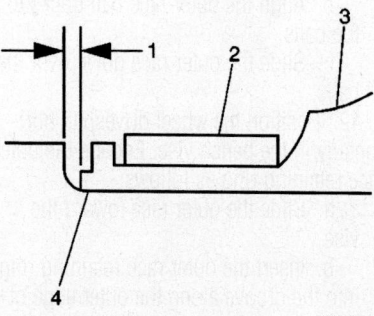

Measure the distance between the edge of the seal and the edge of the last axle shaft groove closing edge and adjust fit to 0.3 inch (8mm)—outer joint

- Large ring clamp on the CV boot
- Outer joint onto the axle shaft

10. Compress the retaining ring using a small flat-bladed tool while pushing the outer joint onto the axle shaft.

11. Seat the outer joint on the shaft using a hammer and block of wood.

12. Install the remaining grease into the CV boot.

13. Position the CV boot and the small boot clamp.

14. Crimp the small boot clamp.

15. Position and crimp in place the large boot clamp.

16. Install the halfshaft in the vehicle.

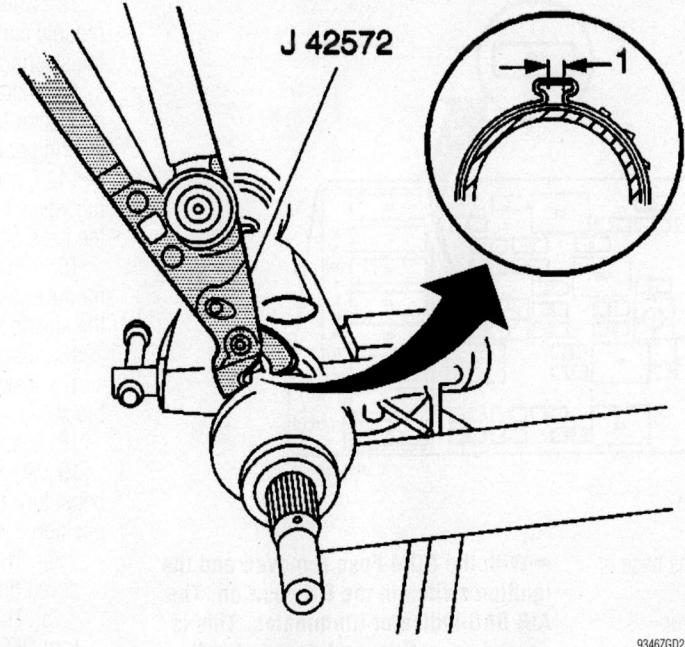

J 42572

Tighten the small seal retaining clamp until the base of the omega shape has a gap width between 0.079–0.118 inch (2–3mm), with a difference in the gap width from side to side no greater than 0.016 inch (0.4mm)–outer joint

9346ZGD2

STEERING AND SUSPENSION

Air Bag

❄❄ CAUTION

All vehicles are equipped with an air bag system. The system must be disabled before performing service on or around system components, steering column, instrument panel components, wiring and sensors. Failure to follow safety and disabling procedures could result in accidental air bag deployment, possible personal injury and unnecessary system repairs.

DISABLING & ENABLING

2002 Models

1. Before servicing the vehicle, refer to the Precautions Section.
2. Turn the steering wheel to align the wheels in the straight-ahead position.
3. Turn the ignition switch to the **LOCK** position.
   a. Remove the SIR fuse from the fuse panel.
   b. Remove the left side sound insulator.
   c. Disconnect the Connector Positive Assurance (CPA) from the yellow 2-way

SIR harness connector at the base of the steering column and separate the connector.
   d. Disconnect the steering wheel module coil connector located at the base of the steering column.
   e. Disconnect the CPA from the inflatable restraint instrument panel module connector located at the near the base of the steering column.

To enable:
After the necessary repairs have been made, re-enable the air bag system as follows:
4. Turn the ignition switch to the **LOCK** position.
   a. Connect the Connector Positive Assurance (CPA) to the yellow 2-way SIR harness connector at the base of the steering column.
   b. Connect the steering wheel module coil connector located at the base of the steering column.
   c. Connect the CPA from the inflatable restraint instrument panel module connector located at the near the base of the steering column.
   d. Connect the instrument panel module connector located near the base of the steering column.
   e. Install the left side sound insulator.
   f. Install the SIR fuse to the fuse panel.

2003–04 Models

ZONE 3

1. Turn the steering wheel so that the vehicle's wheels are pointing straight ahead.
2. Turn the ignition switch to the OFF position.
3. Remove the key from the ignition.
4. Remove the front floor kick-up panel.

➡**With the SDM Fuse removed and the ignition switch in the ON position, the AIR BAG indicator illuminates. This is normal operation and does not indicate an SIR system malfunction.**

5. Remove the SDM Fuse from the I/P fuse block.
6. Remove the left sound insulator.
7. Remove the connector position assurance (CPA) from the inflatable restraint steering wheel module coil connector located at the base of the steering column.
8. Disconnect the steering wheel module coil connector located at the base of the steering column.

Enabling Procedure:

9. Remove the key from the ignition switch.
10. Connect the inflatable restraint steering wheel module coil connector located at the base of the steering.
11. Install the CPA to the steering wheel

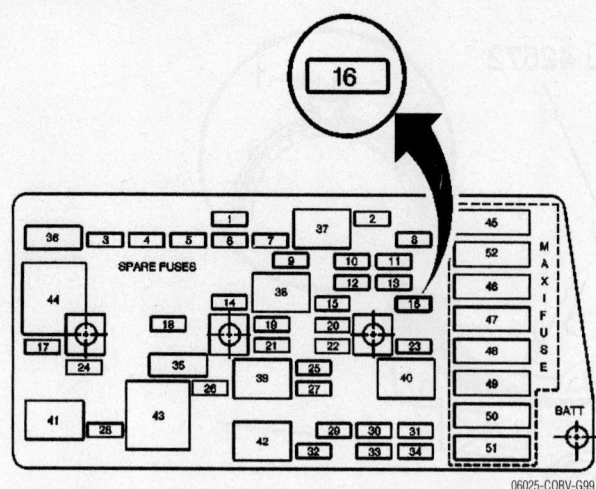

SDM fuse location—2003–05 models

module coil connector located at the base of the steering column.

12. Install the left sound insulator.

13. Install the SDM Fuse to the I/P fuse block.

14. Install the front floor kick-up panel.

15. Staying well away from both air bags, turn the ignition switch to the ON position.

   a. The AIR BAG indicator will flash seven times.

   b. The AIR BAG indicator will then turn OFF.

ZONE 4

1. Turn the steering wheel so that the vehicle's wheels are pointing straight ahead.

2. Turn the ignition switch to the OFF position.

3. Remove the key from the ignition switch.

4. Remove the front floor kick-up panel.

➡**With the SDM Fuse removed and the ignition switch in the ON position, The AIR BAG indicator illuminates. This is normal operation, and does not indicate an SIR system malfunction.**

5. Remove the SDM Fuse from the I/P fuse block.

6. Remove the left sound insulator.

7. Remove the connector position assurance (CPA) from the inflatable restraint steering wheel module coil connector located at the base of the steering column.

8. Disconnect the steering wheel module coil connector located at the base of the steering column.

9. Remove the CPA from the inflatable restraint I/P module connector located near the base of the steering column.

10. Disconnect the I/P module connector located near the base of the steering column.

Enabling Procedure:

11. Remove the key from the ignition switch.

12. Connect the inflatable restraint I/P module connector located near the base of the steering column.

13. Install the CPA to the I/P module connector located near the base of the steering column.

14. Connect the inflatable restraint steering wheel module coil connector located at the base of the steering column.

15. Install the CPA to the steering wheel module coil connector located at the base of the steering column.

16. Install the left sound insulator.

17. Install the SDM Fuse to the I/P fuse block.

18. Install the front floor kick-up panel.

19. Staying well away from both air bags, turn the ignition switch to the ON position.

   a. The AIR BAG indicator will flash seven times.

   b. The AIR BAG indicator will then turn OFF.

ZONE 5

1. Turn the steering wheel so that the vehicle's wheels are pointing straight ahead.

2. Turn the ignition switch to the OFF position.

3. Remove the key from the ignition.

4. Remove the front floor kick-up panel.

➡**With the SDM Fuse removed and the ignition switch in the ON position, the AIR BAG indicator illuminates. This is normal operation and does not indicate an SIR system malfunction.**

5. Remove the SDM Fuse from the I/P fuse block.

6. Remove the left sound insulator.

7. Remove the connector position assurance (CPA) from the inflatable restraint I/P module connector located at the base of the steering column.

8. Disconnect the I/P module connector located at the base of the steering column.

Enabling Procedure:

9. Remove the key from the ignition switch.

10. Connect the inflatable restraint I/P module connector located at the base of the steering column.

11. Install the CPA to the I/P module coil connector located at the base of the steering column.

12. Install the left sound insulator.

13. Install the SDM Fuse to the I/P fuse block.

14. Install the front floor kick-up panel.

15. Staying well away from both air bags, turn the ignition switch to the ON position.

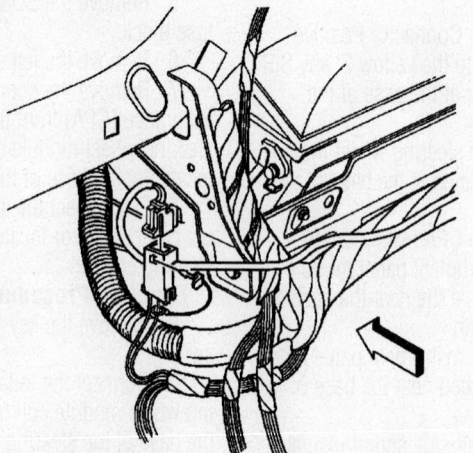

I/P module connector—2003–04 models

a. The AIR BAG indicator will flash seven times.

b. The AIR BAG indicator will then turn OFF.

2005 Models

ZONE 1

1. Turn the steering wheel so the vehicle wheels are pointing straight ahead.

2. Place the ignition in the OFF position.

3. Remove the kick-up panel (2) to expose the instrument panel (I/P) fuse block.

4. Remove the I/P fuse block cover (1).

➡**With the sensing and diagnostic module (SDM) fuse removed and the ignition in the ON position, the AIR BAG warning indicator illuminates. This is normal operation and does not indicate a SIR system malfunction.**

5. Locate the SDM fuse within the I/P fuse block.

6. Remove the SDM fuse from the I/P fuse block.

7. Open the front hood, and remove the air cleaner assembly.

8. Locate both the right and left front end sensors, also known as the electronic frontal sensor (EFS) (1, 4).

9. Remove both the right and left connector position assurance (CPA) (3, 6) from the right and left front end sensor connector (2, 5).

10. Remove both the right and left front end sensor connector (2, 5) from the right and left front end sensor (1, 4).

Enabling Procedure:

11. Place the ignition in the OFF position.

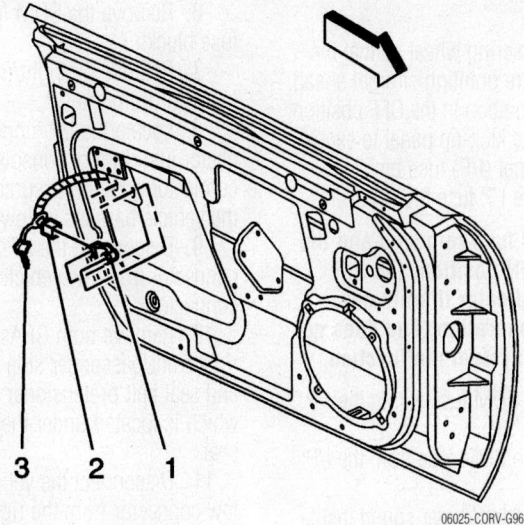

06025-CORV-G96

Left side impact sensor—2005 models

12. Connect both the right and left front end sensor connector (2, 5) to the right and left front end sensor (1, 4).

13. Connect both the right and left CPA (3, 6) to the right and left front end sensor connector (2, 5).

14. Install the air cleaner assembly.

15. Install the SDM fuse into the I/P fuse block.

16. Install the I/P fuse block cover (1).

17. Install the kick-up panel (2) to cover the I/P fuse block (1).

18. Use caution while reaching in and place the ignition to the ON position. The AIR BAG indicator will flash then turn OFF.

ZONE 2

1. Turn the steering wheel so that the vehicle's wheels are pointing straight ahead.

2. Place the ignition in the OFF position.

3. Remove the kick-up panel (2) to expose the instrument panel (I/P) fuse block.

4. Remove the I/P fuse block cover (1).

➡**With the SDM fuse removed and the ignition in the ON position, the AIR BAG warning indicator illuminates. This is normal operation, and does not indicate an SIR system malfunction.**

5. Locate the SDM fuse within the I/P fuse block.

6. Remove the SDM fuse from the I/P fuse block.

7. Open left/driver door and remove the door trim panel.

8. Locate left side impact sensor (SIS) (1).

9. Remove the connector position assurance (CPA) (3) from SIS connector (2).

10. Remove the SIS connector (2) from the SIS (1).

Enabling Procedure:

11. Place the ignition in the OFF position.

12. Connect the SIS connector (2) to the SIS (1).

13. Connect the CPA (3) to the SIS connector (2).

14. Install the door trim panel.

15. Install the SDM fuse into the I/P fuse block.

16. Install the I/P fuse block cover (1).

17. Install the kick-up panel (2) to cover the I/P fuse block (1).

18. Use caution while reaching in and place the ignition to the ON position. The AIR BAG indicator will flash then turn OFF.

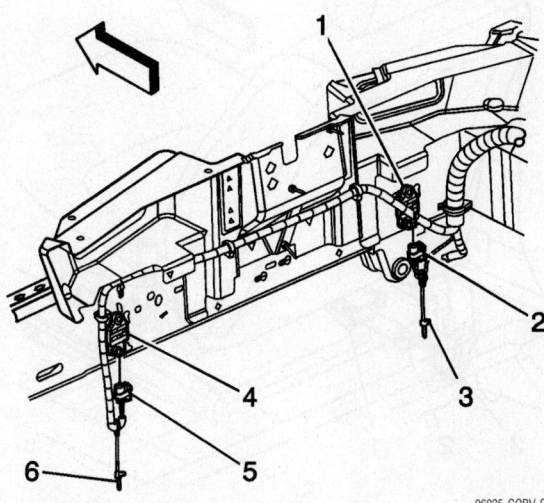

06025-CORV-G97

Front end sensor locations—2005 models

ZONE 3

1. Turn the steering wheel so that the vehicle's wheels are pointing straight ahead.

2. Place the ignition in the OFF position.

3. Remove the kick-up panel to expose the instrument panel (I/P) fuse block.

4. Remove the I/P fuse block cover.

➡ **With the SDM fuse removed and the ignition in the ON position, the AIR BAG warning indicator illuminates. This is normal operation, and does not indicate an SIR system malfunction.**

5. Locate the SDM fuse within the I/P fuse block.

6. Remove the SDM fuse from the I/P fuse block.

7. Remove the left/driver sound insulator from the I/P.

8. Remove the connector position assurance (CPA) from the vehicle harness yellow connector.

9. Disconnect the steering wheel module coil yellow connector from the vehicle harness yellow connector.

Enabling Procedure:

10. Place the ignition in the OFF position.

11. Connect the steering wheel module coil yellow connector to the vehicle harness yellow connector.

12. Install the CPA to the vehicle harness yellow connector.

13. Install the left sound insulator to the I/P.

14. Install the SDM fuse into the I/P fuse block.

15. Install the I/P fuse block cover.

16. Install the kick-up panel to cover the I/P fuse block.

17. Use caution while reaching in and place the ignition to the ON position. The AIR BAG indicator will flash then turn OFF.

ZONE 4

1. Turn the steering wheel so that the vehicle's wheels are pointing straight ahead.

2. Place the ignition in the OFF position.

3. Remove the kick-up panel to expose the instrument panel (I/P) fuse block.

4. Remove the I/P fuse block cover.

➡ **With the sensing and diagnostic module (SDM) fuse removed and the ignition in the ON position, the AIR BAG warning indicator illuminates. This is normal operation and does not indicate a SIR system malfunction.**

5. Locate the SDM fuse within the I/P fuse block.

6. Remove the SDM fuse from the I/P fuse block.

7. Remove the right/passenger sound insulator from the I/P.

8. Locate the I/P module yellow connector just right of console and remove the connector position assurance (CPA) from the vehicle harness yellow connector.

9. Disconnect the I/P module yellow connector from the vehicle harness yellow connector.

10. Remove both CPAs (1) from the right front/passenger side impact module and seat belt pretensioner yellow connector which is located under the passenger seat.

11. Disconnect the vehicle harness yellow connector from the right front side impact module and pretensioner yellow connector.

12. Remove the left/driver sound insulator from the I/P.

13. Remove the CPA from the vehicle harness yellow connector.

14. Disconnect the steering wheel module coil yellow connector from the vehicle harness yellow connector.

15. Remove both CPAs (3) from the left front/driver's side impact module and pretensioner yellow connector (1) which is located under the driver seat.

16. Disconnect the vehicle harness yellow connector from the left front side impact module and pretensioner yellow connector.

Enabling Procedure:

17. Place the ignition in the OFF position.

18. Connect the vehicle harness yellow connector to the left front side impact module and pretensioner yellow connector.

19. Install both CPAs to the left front side impact module and pretensioner yellow connector.

20. Connect the steering wheel module coil yellow connector to the vehicle harness yellow connector.

21. Install the CPA to the vehicle harness yellow connector.

22. Install the left sound insulator to the I/P.

23. Connect the vehicle harness yellow connector to the left front side impact module and pretensioner yellow connector.

24. Install both CPAs to the left front side impact module and pretensioner yellow connector.

25. Connect the I/P module yellow connector to the vehicle harness yellow connector.

26. Install the CPA to the vehicle harness yellow connector.

27. Install the right sound insulator to the I/P.

28. Install the SDM fuse into the I/P fuse block.

29. Install the I/P fuse block cover.

30. Install the kick-up panel to cover the I/P fuse block.

31. Use caution while reaching in and place the ignition to the ON position. The AIR BAG indicator will flash then turn OFF.

ZONE 5

1. Turn the steering wheel so that the vehicle's wheels are pointing straight ahead.

2. Place the ignition in the OFF position.

3. Remove the kick-up panel to expose the instrument panel (I/P) fuse block.

4. Remove the I/P fuse block cover.

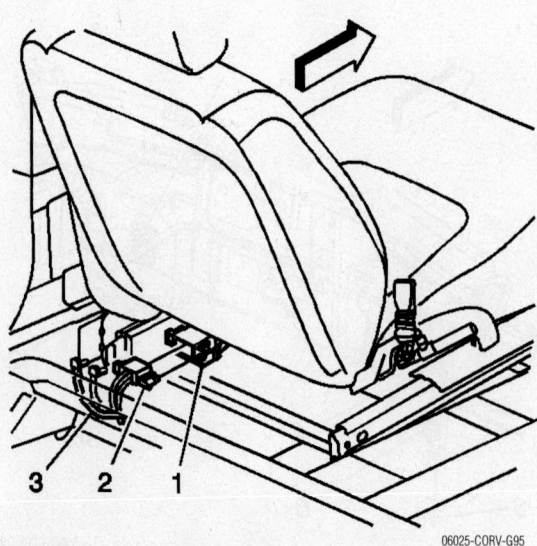

06025-CORV-G95

Left front/driver side impact module and pretensioner—2005 models

➡**With the sensing and diagnostic module (SDM) fuse removed and the ignition in the ON position, the AIR BAG warning indicator illuminates. This is normal operation and does not indicate a SIR system malfunction.**

5. Locate the SDM fuse within the I/P fuse block.

6. Remove the SDM fuse from the I/P fuse block.

7. Remove the right/passenger sound insulator from the I/P.

8. Locate the I/P module yellow connector just right of console and remove the connector position assurance (CPA) from the vehicle harness yellow connector.

9. Disconnect the I/P module yellow connector from the vehicle harness yellow connector.

Enabling Procedure:

10. Place the ignition in the OFF position.

11. Connect the I/P module yellow connector to the vehicle harness yellow connector.

12. Install the CPA to the vehicle harness yellow connector.

13. Install the right sound insulator to the I/P.

14. Install the SDM fuse into the I/P fuse block.

15. Install the I/P fuse block cover.

16. Install the kick-up panel to cover the I/P fuse block.

17. Use caution while reaching in and place the ignition to the ON position. The AIR BAG indicator will flash then turn OFF.

ZONE 6

1. Turn the steering wheel so that the vehicle's wheels are pointing straight ahead.

2. Place the ignition in the OFF position.

3. Remove the kick-up panel to expose the instrument panel (I/P) fuse block.

4. Remove the I/P fuse block cover.

➡**With the SDM fuse removed and the ignition in the ON position, the AIR BAG warning indicator illuminates. This is normal operation, and does not indicate an SIR system malfunction.**

5. Locate the SDM fuse within the I/P fuse block.

6. Remove the SDM fuse from the I/P fuse block.

7. Open right/passenger door and remove the door trim panel.

8. Locate right side impact sensor (SIS).

9. Remove the connector position assurance (CPA) from SIS connector.

10. Remove the SIS connector from the SIS.

Enabling Procedure:

11. Place the ignition in the OFF position.

12. Connect the SIS connector to the SIS.

13. Connect the CPA to the SIS connector.

14. Install the door trim panel.

15. Install the SDM fuse into the I/P fuse block.

16. Install the I/P fuse block cover.

17. Install the kick-up panel to cover the I/P fuse block.

18. Use caution while reaching in and place the ignition to the ON position. The AIR BAG indicator will flash then turn OFF.

ZONE 7

1. Turn the steering wheel so that the vehicle's wheels are pointing straight ahead.

2. Place the ignition in the OFF position.

3. Remove the kick-up panel to expose the instrument panel (I/P) fuse block.

4. Remove the I/P fuse block cover.

➡**With the SDM fuse removed and the ignition in the ON position, the AIR BAG warning indicator illuminates. This is normal operation, and does not indicate an SIR system malfunction.**

5. Locate the SDM fuse within the I/P fuse block.

6. Remove the SDM fuse from the I/P fuse block.

7. Remove both connector position assurance (CPA) from the left front/driver side impact module and pretensioner yellow connector which is located under the driver seat.

8. Disconnect the vehicle harness yellow connector from the left front side impact module and pretensioner yellow connector.

Enabling Procedure:

9. Place the ignition in the OFF position.

10. Connect the vehicle harness yellow connector to the left front side impact module and pretensioner yellow connector.

11. Install both CPA's to the left front side impact module and pretensioner yellow connector.

12. Install the SDM fuse into the I/P fuse block.

13. Install the I/P fuse block cover.

14. Install the kick-up panel to cover the I/P fuse block.

15. Use caution while reaching in and place the ignition to the ON position. The AIR BAG indicator will flash then turn OFF.

ZONE 9

1. Turn the steering wheel so that the vehicle's wheels are pointing straight ahead.

2. Place the ignition in the OFF position.

3. Remove the kick-up panel to expose the instrument panel (I/P) fuse block.

4. Remove the I/P fuse block cover.

➡**With the SDM fuse removed and the ignition in the ON position, the AIR BAG warning indicator illuminates. This is normal operation, and does not indicate an SIR system malfunction.**

5. Locate the SDM fuse within the I/P fuse block.

6. Remove the SDM fuse from the I/P fuse block.

7. Remove both Connector Position Assurance (CPA) from the right front/passenger side impact module and seat belt pretensioner yellow connector which is located under the passenger seat.

8. Disconnect the vehicle harness yellow connector from the right front side impact module and pretensioner yellow connector.

Enabling Procedure:

9. Place the ignition in the OFF position.

10. Connect the vehicle harness yellow connector to the right front side impact module and pretensioner yellow connector.

11. Install both CPA's to the right front side impact module and pretensioner yellow connector.

12. Install the SDM fuse into the I/P fuse block.

13. Install the I/P fuse block cover.

14. Install the kick-up panel to cover the I/P fuse block.

15. Use caution while reaching in and place the ignition to the ON position. The AIR BAG indicator will flash then turn OFF.

Power Rack and Pinion Steering Gear

REMOVAL & INSTALLATION

1. Before servicing the vehicle, refer to the Precautions Section.

2. Drain the power steering fluid.

3. Remove or disconnect the following:

- Negative battery cable
- Brake Pressure Modulator Valve (BPMV) bracket, if equipped
- Both front wheels
- Intermediate shaft shield
- Intermediate shaft lower coupling from the power steering gear

- Power steering inlet hose from the steering gear
- Steering cooler pipe from the steering gear
- Power steering cooler
- Both outer tie rod ends from the knuckles
- Magnasteer electrical connector
- Stabilizer shaft from the crossmember
- Wire harness clips from the crossmember
- Brake pipe from the crossmember
- Steering gear mounting bolts and loosen the crossmember mounting nuts
- Power steering gear through the left wheel opening

To install:

4. Install or connect the following:
- Steering gear to the crossmember and torque the nuts to 74 ft. lbs. (100 Nm)

5. Torque the crossmember mounting nuts 81 ft. lbs. (110 Nm)
- Brake pipe to the crossmember
- Stabilizer shaft
- Both outer tie rod ends to the steering knuckle and torque the nuts to 15 ft. lbs. (20 Nm) plus and additional 160 degrees
- Intermediate shaft lower steering coupling to the steering gear and torque the bolt to 25 ft. lbs. (34 Nm)
- Magnasteer electrical connector
- Wire harness clips to the crossmember
- Cooler to the crossmember
- Cooler pipe to the steering gear and torque the fitting to 20 ft. lbs. (27 Nm)

- Inlet hose to the steering gear and torque the fitting to 20 ft. lbs. (27 Nm)
- Intermediate shaft shield and torque the clamp to 32 inch lbs. (3.5 Nm)
- Both front wheels
- Negative battery cable

6. Refill and bleed the power steering system.

7. Check and adjust the front end alignment.

Shock Absorber

REMOVAL & INSTALLATION

Front

1. Before servicing the vehicle, refer to the Precautions Section.
2. Remove or disconnect the following:
- Negative battery cable
- Wheel

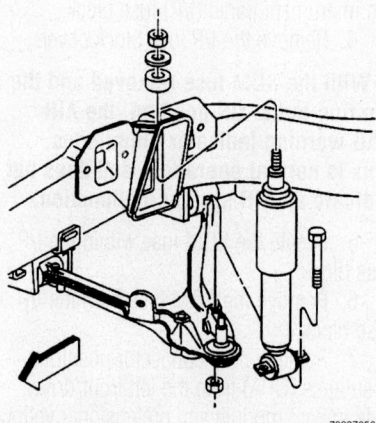

79222656

Exploded view of the front shock absorber mounting

- Real Time Damper (RTD) electrical connector, if equipped
- Upper mounting nut retainer and insulator
- Shock absorber lower mounting bolts
- Shock absorber from the upper tower

3. If equipped with heavy duty shocks (FE3) or Magnetic Selective Ride Control (F55) perform the following steps:
   a. Compress the shock absorber from the bottom upward.
   b. Install a shock support tool to the shock while it is compressed.
   c. Remove the shock from the vehicle.
   d. Remove the tool from the shock.
4. Remove the shock absorber.

To install:

5. Install or connect the following:
- Retainer and insulator to the shock absorber
- Shock absorber to the upper shock tower
- Upper insulator, retainer and nut and torque the nut to 19 ft. lbs. (26 Nm)
- Lower mounting bolts and torque them to 21 ft. lbs. (28 Nm)
- RTD electrical connector

6. If equipped with heavy duty shocks (FE3) or Magnetic Selective Ride Control (F55) perform the following steps:
   a. Install a shock support tool to the shock absorber.
   b. Install the shock to the vehicle.
   c. Install the upper insulator, retainer and nut and torque the nut to 19 ft. lbs. (26 Nm).
   d. Remove the support tool from the shock and install a spring compressor to the spring.
7. Raise the lower control arm and install the shock absorber lower mounting bolts and torque them to 21 ft. lbs. (28 Nm).
8. Remove the spring compressor tool.
9. Install or connect the following:
- Wheel
- Negative battery cable

Rear

1. Before servicing the vehicle, refer to the Precautions Section.
2. Remove or disconnect the following:
- Negative battery cable
- Wheel
- Rear position sensor electrical connector, if equipped
- Lower mounting bolt
- Upper mounting bolts

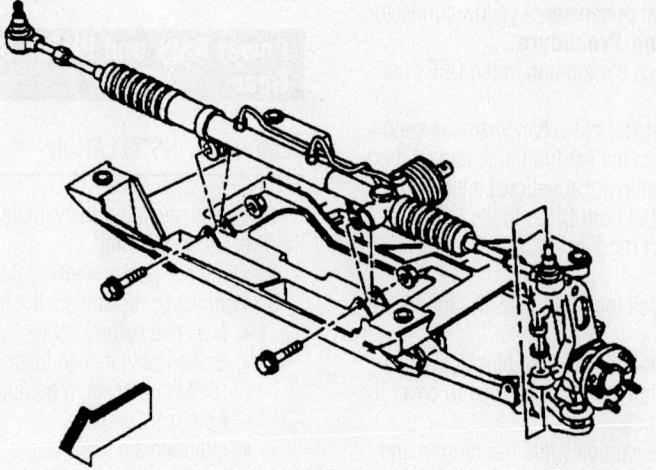

79222655

Exploded view of the power steering gear mounting

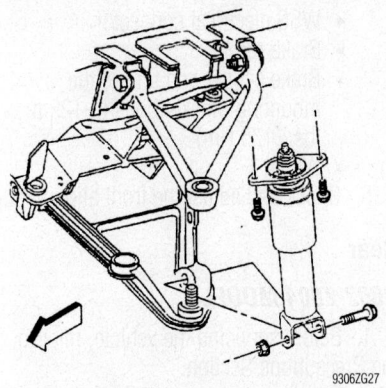

Exploded view of the rear shock absorber

- Tie rod from the control arm
- Shock absorber from the lower control arm and shock tower
- Upper insulator and retainer from the shock absorber

To install:
3. Install or connect the following:
- Upper insulator and retainer to the shock absorber, if removed
- Shock absorber to the shock tower and lower control arm
- Tie rod to the control arm
- Upper mounting bolts and torque them to 22 ft. lbs. (30 Nm)
- Lower shock absorber mounting bolt and torque it to 162 ft. lbs. (220 Nm)
- Rear position sensor electrical connector
- Wheel
- Negative battery cable

Transverse Spring

REMOVAL & INSTALLATION

Front

1. Before servicing the vehicle, refer to the Precautions Section.
2. Remove or disconnect the following:
- Negative battery cable
- Front wheels
3. Measure the front spring adjuster bolt gap to ease the installation procedure and setting the proper vehicle trim height.
4. Install a spring compressor tool J 33432-A to the spring and compress it.
5. Remove or disconnect the following:
- Lower shock absorber mounting bolts from one of the lower control arms
- Stabilizer shaft link from the lower control arm
- Lower ball joint from the steering knuckle

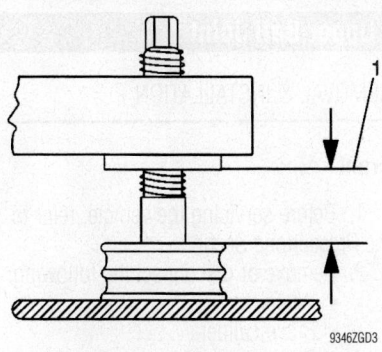

Measure the front spring adjuster bolt gap to ease the installation procedure and setting the proper vehicle trim height—front assembly

- Cam bolts from the lower control arm after matchmarking them
- Lower control arm
- Transverse spring bolts and retainers
- Transverse spring and the compressor tool

To install:
6. Install or connect the following:
- Spring to the crossmember
- New spring retainers and bolts to the crossmember and torque them to 46 ft. lbs. (62 Nm)
- Lower control arm to the crossmember
- Cam bolts to the matchmarks made during the removal procedure. Hand-tighten the cam bolts at this time
- Lower control arm ball joint stud to the steering knuckle and torque the

nut to 20 ft. lbs. (30 Nm) plus an additional 180 degrees
- Shock absorber
- Shock absorber lower mounting bolts and torque them to 21 ft. lbs. (28 Nm)
- Stabilizer shaft link to the lower control arm and torque the nut to 53 ft. lbs. (72 Nm)
7. Remove the spring compressor tool from the transverse spring and move the lower control arm supports.
8. Install or connect the following:
- Front wheels
- Negative battery cable
9. Adjust the front trim height.
10. Perform a front end alignment.
11. Torque the control arm cam bolts to 125 ft. lbs. (170 Nm)

Rear

1. Before servicing the vehicle, refer to the Precautions Section.
2. Remove or disconnect the following:
- Negative battery cable
- Rear wheels
3. Measure the transverse spring stud height to ease the installation procedure and setting the proper vehicle trim height.
4. Install a spring compressor tool J 33432-A to the spring and compress it.
5. Remove the spring mounting bolts, spacers and insulators from the crossmember and control arm.
6. Remove the spring from the vehicle and remove the compressor tool.

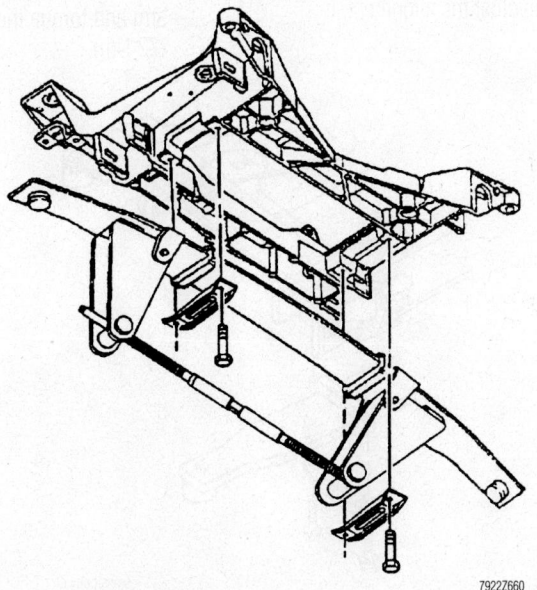

Exploded view of the front transverse spring

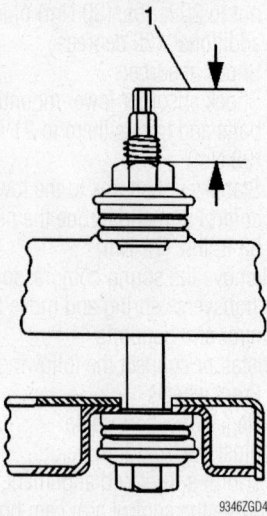

Measure the transverse spring stud height to ease the installation procedure and setting the proper vehicle trim height—rear assembly

To install:

7. Install or connect the following:
   - Spring compressor to the transverse spring
   - Spring to the vehicle
   - Spring spacers, insulators and NEW mounting bolts to the crossmember and torque the bolts to 46 ft. lbs. (62 Nm)
   - Spring to the lower control arm
   - Lower control arm to spring bolts and insulators and release the compressor tool
   - Rear wheel
   - Negative battery cable

8. Adjust the trim height and place the retainers on the control arm bolts.

9. Check and adjust the alignment, if needed.

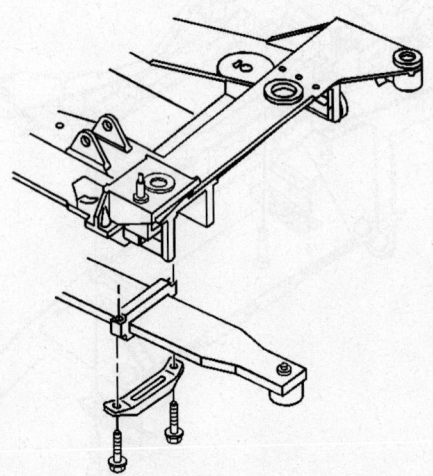

Removing the spring mounting bolts, spacers and insulators.

Upper Ball Joint

REMOVAL & INSTALLATION

Front

1. Before servicing the vehicle, refer to the Precautions Section.
2. Remove or disconnect the following:
   - Wheel
   - Brake caliper
   - Brake rotor
   - Stabilizer shaft link from the lower control arm
   - Wheel Speed Sensor (WSS) electrical connector
   - Outer tie rod from the steering knuckle
   - Upper ball joint from the control arm using Ball Joint Separator J-42188 or equivalent
   - Steering knuckle from the vehicle
3. Press the ball joint from the steering knuckle.

To install:

4. Install or connect the following:
   - Ball joint to the steering knuckle
   - Upper ball joint to the control arm and torque the nut to 20 ft. lbs. (30 Nm) plus an additional 225 degree turn
   - Lower ball joint to the steering knuckle and torque the nut to 20 ft. lbs. (30 Nm) plus an additional 180 degree turn
   - Outer tie rod to the steering knuckle and torque the nut to 15 ft. lbs. (20 Nm) plus an additional 160 degree turn
   - Stabilizer shaft link to the control arm and torque the nut to 53 ft. lbs. (72 Nm)

- WSS electrical connector
- Brake rotor
- Brake caliper and torque the mounting bracket bolts to 125 ft. lbs. (175 Nm)
- Wheel

5. Check and adjust the front alignment.

Rear

2002-2004 MODELS

1. Before servicing the vehicle, refer to the Precautions Section.
2. Remove or disconnect the following:
   - Wheel
   - Wheel Speed Sensor (WSS) electrical connector
   - Real Time Damping (RTD) position sensor, if equipped
   - Brake caliper
   - Brake rotor
   - Shock absorber solenoid electrical connector, if equipped
   - Outer tie rod from the suspension knuckle
   - Axle shaft nut
   - Upper control arm from the knuckle
   - Lower ball joint from the knuckle
   - Suspension knuckle from the vehicle
3. Press the upper ball joint from the knuckle.

To install:

4. Press the ball joint into the suspension knuckle.

5. Install or connect the following:
   - Knuckle to the vehicle
   - Lower ball joint to the knuckle and torque the nut to 41 ft. lbs. (55 Nm)
   - Upper control arm to the knuckle and torque the nut to 15 ft. lbs. (20 Nm) plus an additional 250 degree turn
   - Axle shaft nut and torque it to 118 ft. lbs. (160 Nm)
   - Outer tie rod to the knuckle and torque the nut to 15 ft. lbs. (20 Nm) plus an additional 160 degree turn
   - Brake rotor
   - Brake caliper and torque the mounting bracket bolts to 125 ft. lbs. (175 Nm)
   - WSS electrical connector
   - Shock absorber solenoid electrical connector, if equipped
   - RTD position sensor, if equipped
   - Wheel

6. Check and adjust the rear wheel alignment.

Lower Ball Joint

REMOVAL & INSTALLATION

Front and Rear

1. Before servicing the vehicle, refer to the Precautions Section.
2. Remove the lower control arm from the vehicle.
3. Press the ball joint from the control arm.

To install:

4. Press the ball joint into the control arm.
5. Install the control arm in the vehicle.

Upper Control Arm

REMOVAL & INSTALLATION

Front

1. Before servicing the vehicle, refer to the Precautions Section.
2. Remove or disconnect the following:
   - Negative battery cable
   - Front wheel
   - Real Time Damping (RTD) sensor link and support the lower control arm

➡**Loosen the ball joint stud, but do not remove the nut.**

   - Upper ball joint from the upper control arm using Ball Joint Separate J-42188 or equivalent
   - Upper control arm bolts and shims

➡**Note the number and position of the shims.**

   - Upper control arm

To install:

3. Install or connect the following:
   - Upper control arm
   - Upper control arm shims and bolts and torque the bolts to 48 ft. lbs. (65 Nm)
   - Ball joint stud into the upper control arm and torque the nut to 20 ft. lbs. (30 Nm) plus an additional 225 degrees
   - RTD sensor link
   - Front wheel
   - Negative battery cable
4. Check and adjust the front wheel alignment.

Rear

1. Before servicing the vehicle, refer to the Precautions Section.

2. Remove or disconnect the following:
   - Negative battery cable
   - Rear wheel
   - Wheel Speed Sensor (WSS) electrical connector
   - Real Time Damping (RTD) position sensor link, if equipped
3. Support the lower control arm with a jack stand.
4. Remove or disconnect the following:
   - Ball joint stud from the control arm
   - Upper control arm bolts
   - Upper control arm

To install:

5. Install or connect the following:
   - Upper control arm and torque the bolts to 81 ft. lbs. (110 Nm)
   - Suspension knuckle ball joint stud nut into the upper control arm. It may be necessary to use an Allen wrench to keep the ball joint stud from spinning while tightening the ball joint stud nut. Torque the nut to 22 ft. lbs. (30 Nm) plus an additional 195 degrees
   - WSS electrical connector
   - RTD position sensor link, if equipped
   - Rear wheel
   - Negative battery cable
6. Check and adjust the rear wheel alignment.

CONTROL ARM BUSHING REPLACEMENT

The manufacturer does not provide a bushing replacement procedure. The bushing is replaced when a new control arm is installed.

Lower Control Arm

REMOVAL & INSTALLATION

Front

1. Before servicing the vehicle, refer to the Precautions Section.
2. Remove or disconnect the following:
   - Front wheel
   - Front transverse spring
   - Wheel Speed Sensor (WSS) electrical connector
   - Real Time Damping (RTD) electrical connector, if equipped
3. Support the lower control arm with a jack stand
   - Shock absorber from the lower control arm
   - Stabilizer shaft link from the lower control arm

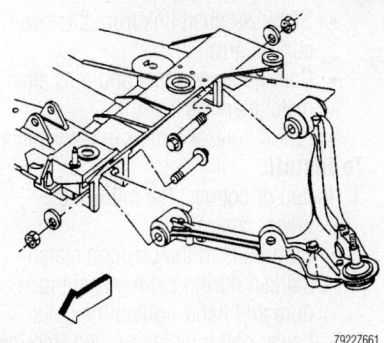

Lower control arm mounting bolts—

   - Ball joint stud from the knuckle
   - Cam bolts, washers and nuts after matchmarking them
   - Lower control arm from the vehicle

To install:

4. Install or connect the following:
   - Lower control arm
5. Support the lower control arm with a jack stand
   - Cam bolts to the position matchmarked during the removal procedure and hand-tighten the bolts
   - Lower control arm ball joint stud to the steering knuckle and torque the nut to 20 ft. lbs. (30 Nm) plus an additional 180 degrees
   - Stabilizer shaft link to the lower control arm and torque the nut to 53 ft. lbs. (72 Nm)
   - Transverse spring and remove
   - Shock absorber lower mounting bolts and torque them to 21 ft. lbs. (28 Nm)
   - RTD electrical connector, if equipped
   - WSS electrical connector
   - Front wheel
   - Negative battery cable
6. Perform a front wheel alignment.
7. Tighten the lower control arm nuts to 125 ft. lbs. (170 Nm).

Rear

1. Before servicing the vehicle, refer to the Precautions Section.
2. Remove or disconnect the following:
   - Negative battery cable
   - Rear wheel
   - Suspension position link, if equipped
   - Transverse spring from the lower control arm
   - Shock absorber from the lower control arm
   - Lower ball joint stud nut from the knuckle using Ball Joint Separator J-42188 or equivalent

- Stabilizer shaft link from the lower control arm
- Cam bolts, washers and nuts after matchmarking them
- Lower control arm from the vehicle

To install:

3. Install or connect the following:
   - Lower control arm
   - Cam bolts to the position match-marked during the removal procedure and hand-tighten the bolts
   - Lower ball joint stud to the steering knuckle and torque the nut to 22 ft. lbs. (30 Nm) plus an additional 180 degrees
   - Stabilizer shaft link to the lower control arm and torque the link nut to 53 ft. lbs. (72 Nm)
   - Shock absorber lower mounting bolts and torque the bolts to 107 ft. lbs. (145 Nm)
   - Transverse spring to the lower control arm
   - Rear wheel
   - Negative battery cable
4. Perform a rear wheel alignment.
5. Torque the lower control arm front cam bolt to 107 ft. lbs. (145 Nm) and the rear cam bolt to 70 ft. lbs. (95 Nm).

CONTROL ARM BUSHING REPLACEMENT

The manufacturer does not provide a bushing replacement procedure. The bushing is replaced when a new control arm is installed.

Wheel Bearings

ADJUSTMENT

No periodic wheel bearing adjustment is necessary. The wheel bearings are a sealed unit that must be replaced if loose or noisy.

REMOVAL & INSTALLATION

Front

1. Before servicing the vehicle, refer to the Precautions Section.
2. Remove or disconnect the following:
   - Negative battery cable
   - Front wheel
   - Wheel Speed Sensor (WSS) electrical connector
   - Brake caliper
   - Brake rotor
   - Stabilizer shaft link from the lower control arm and support the lower control arm

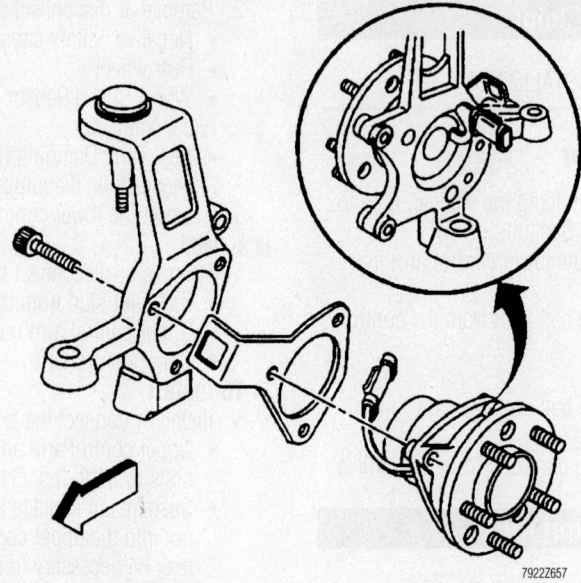

79222657

Exploded view of the front hub/wheel bearing and knuckle assembly

- Separate the outer tie rod ball stud from the steering knuckle using Ball Joint Separator J-42188 or equivalent
- Lower ball joint stud from the steering knuckle using Ball Joint Separator J-42188 or equivalent
- Wheel hub mounting bolts
- Wheel hub/bearing assembly from the knuckle

To install:

3. Install or connect the following:
   - Hub/bearing assembly to the steering knuckle. Make certain that the speed sensor cable connection is facing rearward and torque the mounting bolts to 96 ft. lbs. (130 Nm)
   - Lower control arm ball stud to the steering knuckle and torque the nut to 20 ft. lbs. (30 Nm) plus 180 degrees
   - Outer tie rod to the steering knuckle and torque the nut to 33 ft. lbs. (45 Nm)
   - WSS electrical connector
   - Stabilizer shaft link to the lower control arm and torque the nut to 53 ft. lbs. (72 Nm)
   - Brake rotor
   - Brake caliper and torque the bracket mounting bolts to 125 ft. lbs. (175 Nm)
   - Front wheel
   - Negative battery cable

Rear

1. Before servicing the vehicle, refer to the Precautions Section.

2. Remove or disconnect the following:
   - Negative battery cable
   - Rear wheel
   - Wheel Speed Sensor (WSS) electrical connector
   - Real Time Damping (RTD) position sensor link, if equipped
   - Brake caliper
   - Brake rotor
   - Shock absorber solenoid electrical connector, if equipped
   - Outer tie rod end from the suspension knuckle
   - Axle nut
   - Separate the upper control arm from the suspension knuckle
   - Steering knuckle from the lower control arm ball joint stud
   - Suspension knuckle
   - Hub/bearing bolts
   - Hub/bearing from the knuckle

To install:

3. Install or connect the following:
   - Hub/bearing assembly to the steering knuckle. Make certain that the speed sensor cable connection is facing rearward and torque the hub mounting bolts to 96 ft. lbs. (130 Nm)
   - Upper control arm to the steering knuckle and torque the bolt to 22 ft. lbs. (30Nm) plus 195 degrees
   - Lower control arm ball stud to the suspension knuckle and torque the nut to 22 ft. lbs. (30 Nm) plus 180 degrees
   - Axle nut and torque it to 118 ft. lbs. (160 Nm)
   - Outer tie rod ball stud to the steer-

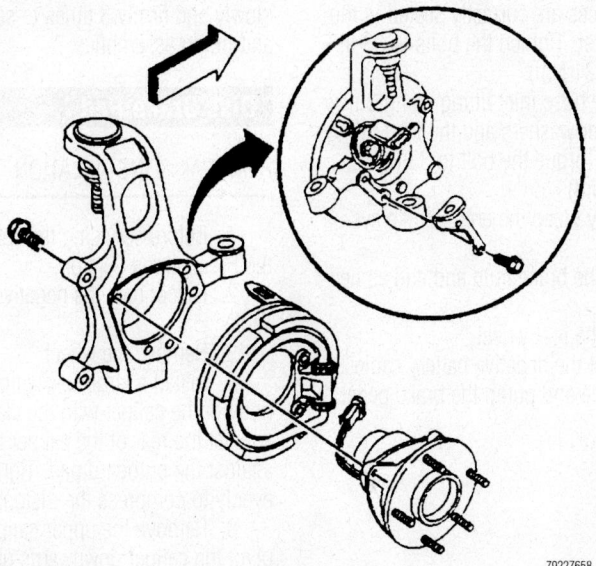

ing knuckle and torque the nut to 45 ft. lbs. (60 Nm)
- Brake rotor
- Brake caliper and torque the mounting bracket bolts to 125 ft. lbs. (175 Nm)
- WSS electrical connector
- Shock absorber solenoid electrical connector
- RTD position sensor link, if equipped
- Rear wheel
- Negative battery cable

79222658

Exploded view of the rear hub/wheel bearing and knuckle assembly

BRAKES

Brake Caliper

REMOVAL & INSTALLATION

Front

1. Before servicing the vehicle, refer to the Precautions Section.
2. Disconnect the negative battery cable and remove ⅔ of the brake fluid from the master cylinder reservoir.
3. Remove or disconnect the following:
- Wheel assembly
- Brake hose fitting at the caliper by removing the bolt. Discard the 2 copper washers and plug the hose to prevent fluid contamination or loss.

➡ **Do not allow the fluid to come into contact with the front transverse spring, as damage to the spring may occur.**

- Two caliper mounting bolts
- Caliper housing from the rotor and the caliper mounting bracket.

To install:

➡ **Inspect the caliper slide boots for damage and replace if necessary.**

4. Compress the pistons using a suitable tool, if necessary.
5. Install or connect the following:
- Caliper over the brake rotor and into the caliper mounting bracket. Make sure the shoe lining guiding

surfaces are correctly seated in the bracket. Tighten the bolts to 23 ft. lbs. (31 Nm).
- Brake hose inlet fitting using two NEW copper washers and the inlet fitting bolt. Torque the bolt to 33 ft. lbs. (45 Nm).
6. Properly bleed the entire brake system.
7. Check the brake fluid and add as necessary.
8. Install the wheel.
9. Connect the negative battery cable, start the engine and pump the brake pedal slowly and firmly 3 times to seat the shoe and lining assemblies.

Rear

1. Before servicing the vehicle, refer to the Precautions Section.
2. Disconnect the negative battery cable and remove ⅔ of the brake fluid from the master cylinder reservoir.
3. Remove or disconnect the following:
- Wheel assembly
- Brake hose fitting at the caliper by removing the bolt. Discard the 2 copper washers and plug the hose to prevent fluid contamination or loss.

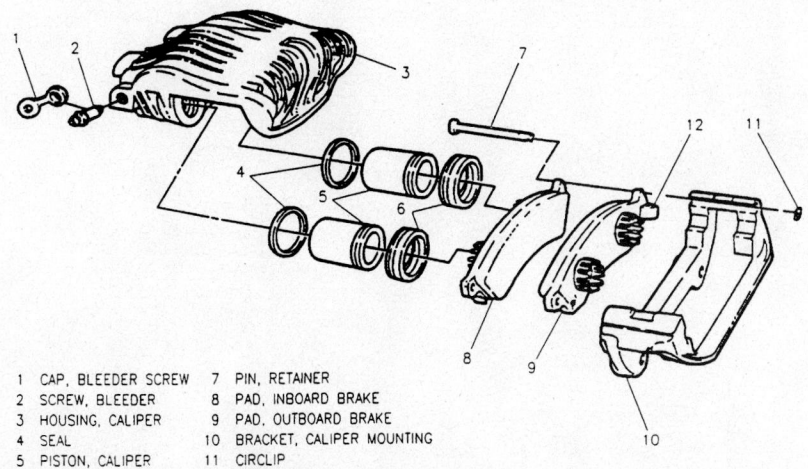

1	CAP, BLEEDER SCREW	7	PIN, RETAINER
2	SCREW, BLEEDER	8	PAD, INBOARD BRAKE
3	HOUSING, CALIPER	9	PAD, OUTBOARD BRAKE
4	SEAL	10	BRACKET, CALIPER MOUNTING
5	PISTON, CALIPER	11	CIRCLIP
6	BOOT	12	SPRING, BIAS

93006G63

Exploded view of the front caliper assembly

➡**Do not allow the fluid to come into contact with the front transverse spring, as damage to the spring may occur.**

- Caliper mounting bolts
- Caliper housing from the rotor and the caliper mounting bracket.

To install:

4. Compress the pistons using a suitable tool, if necessary.

5. Install or connect the following:
- Caliper over the brake rotor and into the caliper mounting bracket. Make sure the shoe lining guiding surfaces are correctly seated in the bracket. Tighten the bolts to 23 ft. lbs. (31 Nm).
- Brake hose inlet fitting using 2 new copper washers and the inlet fitting bolt. Torque the bolt to 33 ft. lbs. (45 Nm).

6. Properly bleed the entire brake system.

7. Check the brake fluid and add as necessary.

8. Install the rear wheel.

9. Connect the negative battery cable, start the engine and pump the brake pedal slowly and firmly 3 times to seat the shoe and lining assemblies.

Disc Brake Pad

REMOVAL & INSTALLATION

1. Before servicing the vehicle, refer to the Precautions Section.

2. Disconnect the negative battery cable.

3. Remove the wheel.

4. Install a large C–clamp over the body of the caliper with the clamp ends against the rear of the caliper body and against the outboard pad. Tighten the clamp evenly to compress the pistons.

5. Remove the upper caliper bolt, then pivot the caliper downwards until enough clearance is achieved and support the caliper with wire.

6. Remove the pad and lining assemblies from the caliper.

To install:

7. Clean all residue from the pad and lining assembly guiding surfaces on the caliper housing and the mounting bracket.

8. Ensure the caliper pistons are fully compressed in their bores.

9. Install the outboard pad with the insulator to the caliper housing and the inboard pad with the wear sensor into the caliper pistons. Press the pads firmly until they are they are fully seated.

10. Install the caliper in position, insert and tighten the upper bolt to 23 ft. lbs. (31 Nm).

11. Install the wheel.

12. Connect the negative battery cable.

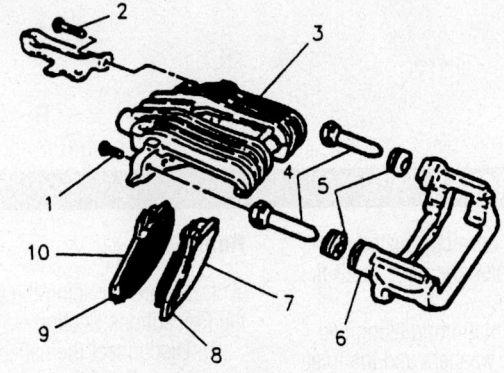

1	BOLT, UPPER GUIDE PIN
2	BOLT, LOWER GUIDE PIN
3	HOUSING, CALIPER
4	PIN, GUIDE
5	BOOT
6	BRACKET, MOUNTING
7	INSULATOR
8	PAD, OUTBOARD BRAKE
9	SENSOR, WEAR
10	PAD, INBOARD BRAKE

93006G64

Exploded view of the rear caliper assembly

CADILLAC

CTS • CTS-V

15

BRAKES 15-59
DRIVE TRAIN 15-45
ENGINE REPAIR 15-9
FUEL SYSTEM 15-43
SPECIFICATIONS AND
  MAINTENANCE CHARTS......15-2
Engine and Vehicle Identification
  Chart.................................15-2
General Engine Specifications 15-2
Engine Tune-Up Specifications......15-2
Firing Order 15-3
Accessory Drive Belt Routing 15-3
Capacities 15-3
Valve Specifications...................15-4
Crankshaft and Connecting Rod
  Specifications 15-4
Piston and Ring Specifications 15-5
Torque Specifications 15-5
Wheel Alignment 15-7
Tire, Wheel and Ball Joint
  Specifications 15-7
Brake Specifications 15-7
Scheduled Maintenance Intervals ..15-8
STEERING AND
  SUSPENSION 15-51
A
Air Bag...................................15-51
  Disabling And Enabling 15-51
  Precautions.......................15-51
Alternator.................................15-9
  Removal & Installation...........15-9
B
Brake Caliper...........................15-59
  Removal & Installation...........15-59
C
Camshaft 15-25
  Removal & Installation............15-25
Clutch 15-47
  Adjustments........................15-47
  Removal & Installation............15-47
Coil Spring 15-55
  Removal & Installation............15-55

CV-Joints.................................15-48
  Removal & Installation............15-48
Cylinder Head 15-16
  Removal & Installation............15-16
D
Disc Brake Pads..........................15-60
  Removal & Installation............15-60
E
Engine Assembly 15-10
  Removal & Installation............15-10
Exhaust Manifold.........................15-22
  Removal & Installation............15-22
F
Front Crankshaft Seal 15-24
  Removal And Installation 15-24
Fuel Filter 15-43
  Removal & Installation............15-43
Fuel Injector 15-44
  Removal & Installation............15-44
Fuel Pump 15-43
  Removal & Installation............15-43
Fuel System Pressure 15-43
  Relieving 15-43
Fuel System Service
  Precautions............................15-43
H
Halfshaft...................................15-47
  Removal & Installation............15-47
Heater Core 15-14
Hydraulic Clutch System 15-47
  Bleeding............................15-47
I
Ignition Timing 15-10
  Adjustment.........................15-10
Intake Manifold Assembly..............15-20
  Removal & Installation............15-20
L
Lower Ball Joint..........................15-56
  Removal & Installation............15-56
Lower Control Arm 15-56
  Control Arm Bushing
    Replacement 15-57
  Removal & Installation............15-56

O
Oil Pan....................................15-29
  Removal & Installation............15-29
Oil Pump 15-30
  Removal & Installation............15-30
P
Pinion Seal 15-50
  Removal & Installation............15-50
Piston and Ring 15-42
  Positioning 15-42
R
Rack and Pinion Power Steering
  Gear...................................15-54
  Removal & Installation............15-54
Rear Main Seal 15-32
  Removal & Installation............15-32
S
Shock Absorber 15-55
  Removal & Installation............15-55
Starter Motor 15-28
  Removal & Installation............15-28
T
Timing Belt 15-39
  Removal & Installation............15-39
Timing Chain, Sprockets, Front
  Cover and Seal 15-34
  Removal & Installation............15-34
Transmission Assembly 15-45
  Removal & Installation............15-45
U
Upper Control Arm 15-56
  Removal & Installation............15-56
V
Valve Lash 15-28
  Adjustment..........................15-28
W
Water Pump...............................15-13
  Removal & Installation............15-13
Wheel Bearings..........................15-58
  Adjustment..........................15-58
  Removal & Installation............15-58

SPECIFICATION AND MAINTENANCE CHARTS

ENGINE AND VEHICLE IDENTIFICATION

	Engine						Model Year	
Code ①	Liters (cc)	Cu. In.	Cyl.	Fuel Sys.	Engine Type	Eng. Mfg.	Code ②	Year
T	2.8 (2792)	171	6	SMFI	DOHC	GM	3	2003
N	3.2 (3173)	195	6	SMFI	DOHC	Opel	4	2004
7	3.6 (3564)	217	6	SMFI	DOHC	GM	5	2005
S	5.7 (5665)	346	8	MFI	OHV	GM	6	2006

SMFI: Sequential Multi-port Fuel Injection

DOHC: Double Overhead Camshaft

① 8th position of VIN

② 10th position of VIN

06025-CTSD-C01

GENERAL ENGINE SPECIFICATIONS

Year	Model	Engine Displacement Liters	Engine Series (VIN)	Fuel System	Net Horsepower @ rpm	Net Torque @ rpm (ft. lbs.)	Bore x Stroke (in.)	Compression Ratio	Oil Pressure @ rpm
2003	CTS	3.2	N	MFI	220@6000	220@3400	3.45x3.47	10.0:1	22@900
2004	CTS	3.2	N	MFI	220@6000	220@3400	3.45x3.47	10.0:1	22@900
	3.6 CTS	3.6	7	MFI	255@6500	252@2800	3.70x3.37	10.2:1	20@2000
	CTS-V	5.7	S	MFI	400@6000	385@4800	3.90x3.60	10.5:1	①
2005	2.8 CTS	2.8	T	MFI	210@6500	195@3200	3.50x2.94	10.0:1	20@2000
	3.6 CTS	3.6	7	MFI	255@6200	252@3200	3.70x3.37	10.2:1	20@2000
	CTS-V	5.7	S	MFI	400@6000	395@4800	3.90x3.60	10.5:1	①

MFI: Multi-point Fuel Injection

① 6psi @ 1000RPM
   18psi @ 2000RPM
   24psi @ 4000RPM

06025-CTSD-C02

ENGINE TUNE-UP SPECIFICATIONS

Year	Engine Displacement Liters	Engine VIN	Spark Plug Gap (in.)	Ignition Timing (deg.)	Fuel Pump (psi)	Idle Speed (rpm)	Valve Clearance In.	Valve Clearance Ex.
2003	3.2	N	0.060	①	41-47	①	HYD	HYD
2004	3.2	N	0.060	①	41-47	①	HYD	HYD
	3.6	7	0.043	①	NA	①	HYD	HYD
	5.7	S	0.040	①	NA	①	HYD	HYD
2005	2.8	T	0.043	①	NA	①	HYD	HYD
	3.6	7	0.043	①	NA	①	HYD	HYD
	5.7	S	0.040	①	NA	①	HYD	HYD

NOTE: The Vehicle Emission Control Information label often reflects specification changes made during production.

The label figures must be used if they differ from those in this chart.

HYD: Hydraulic

NA: Information not available

① Refer to Vehicle Emission Control Information label

06025-CTSD-C03

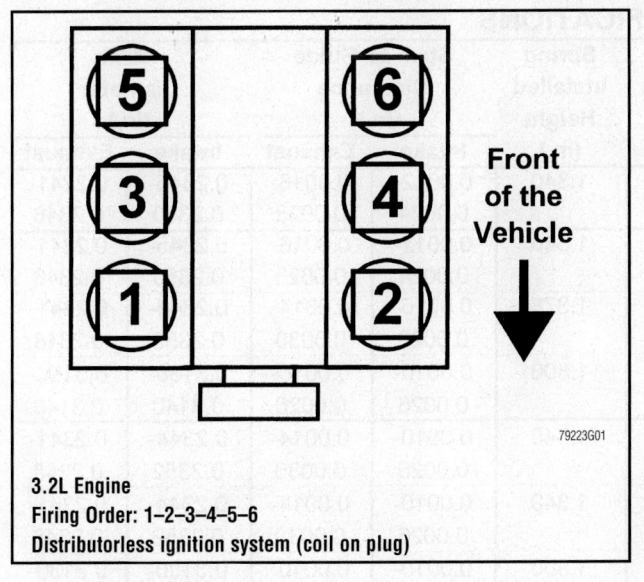

3.2L Engine
Firing Order: 1–2–3–4–5–6
Distributorless ignition system (coil on plug)

79223G01

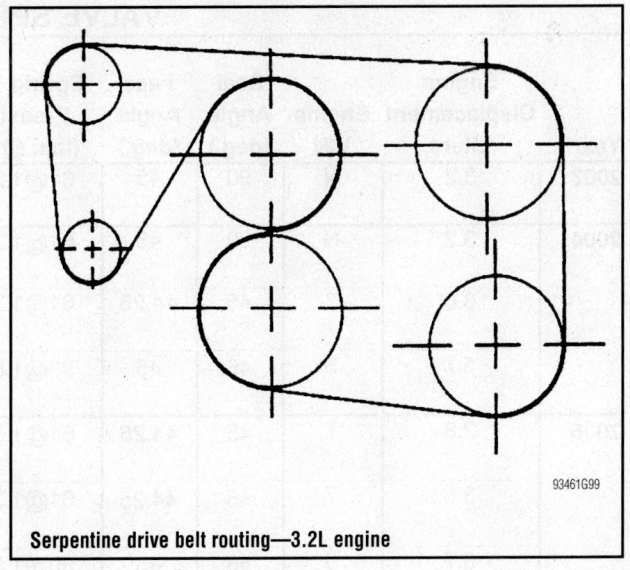

Serpentine drive belt routing—3.2L engine

93461G99

CAPACITIES

Year	Model	Engine Displacement Liters	Engine VIN	Engine Oil with Filter (qts.)	Transmission (pts.)		Drive Axle Rear (pts.)	Fuel Tank (gal.)	Cooling System (qts.)
					Auto.	Man.			
2003	CTS	3.2	N	5.0	18	2.6	3.5	17.0	10.4
2004	CTS	3.2	N	5.0	18	2.6	2.75	17.5	10.4
	3.6 CTS	3.6	7	6.0	18	N/A	2.75	17.5	9.7
	CTS-V	5.7	S	6.0	N/A	7.4	2.75	17.5	13.2
2005	2.8 CTS	2.8	T	6.0	18	3.8	2.75	17.5	10.6
	3.6 CTS	3.6	7	6.0	18	N/A	2.75	17.5	12.0
	CTS-V	5.7	S	6.0	N/A	7.4	2.75	17.5	13.4

NOTE: All capacities are approximate. Add fluid gradually and ensure a proper fluid level is obtained.

06025-CTSD-C04

VALVE SPECIFICATIONS

Year	Engine Displacement Liters	Engine VIN	Seat Angle (deg.)	Face Angle (deg.)	Spring Test Pressure (lbs. @ in.)	Spring Installed Height (in.)	Stem-to-Guide Clearance (in.)		Stem Diameter (in.)	
							Intake	Exhaust	Intake	Exhaust
2003	3.2	N	90	45	61@1.340	1.340	0.0012-0.0024	0.0016-0.0028	0.2345-0.2350	0.2341-0.2346
2004	3.2	N	90	45	61@1.340	1.340	0.0012-0.0024	0.0016-0.0028	0.2345-0.2350	0.2341-0.2346
	3.6	7	45	44.25	61@1.378	1.378	0.0010-0.0026	0.0014-0.0030	0.2344-0.2352	0.2341-0.2348
	5.7	S	46	45	90@1.800	1.800	0.0010-0.0026	0.0010-0.0026	0.3130-0.3140	0.3130-0.3140
2005	2.8	T	45	44.25	61@1.340	1.340	0.0010-0.0026	0.0014-0.0030	0.2344-0.2352	0.2341-0.2348
	3.6	7	45	44.25	61@1.340	1.340	0.0010-0.0026	0.0014-0.0030	0.2344-0.2352	0.2341-0.2348
	5.7	S	46	45	90@1.800	1.800	0.0010-0.0026	0.0010-0.0026	0.3130-0.3140	0.3130-0.3140

06025-CTSD-C05

CRANKSHAFT AND CONNECTING ROD SPECIFICATIONS

All measurements are given in inches.

Year	Engine Displacement Liters	Engine VIN	Crankshaft				Connecting Rod		
			Main Brg. Journal Dia.	Main Brg. Oil Clearance	Shaft End-play	Thrust on No.	Journal Diameter	Oil Clearance	Side Clearance
2003	3.2	N	2.6670-2.6760	0.0006-0.0017	0.0039-0.0079	2	2.6764-2.6770	0.0008-0.0024	0.0027-0.0110
2004	3.2	N	2.6670-2.6760	0.0006-0.0017	0.0039-0.0079	2	2.6764-2.6770	0.0008-0.0024	0.0027-0.0110
	3.6	7	2.6768-2.6775	0.0004-0.0024	0.0039-0.0130	2	2.2044-2.2050	0.0004-0.0028	0.0140-0.0374
	5.7	S	2.5580-2.5590	0.0008-0.0021	0.0015-0.0078	3	2.0991-2.0999	0.0009-0.0025	0.0043-0.0200
2005	2.8	T	2.6772	0.0004-0.0024	0.0039-0.0130	2	2.2044-2.2050	0.0004-0.0028	0.0037-0.0140
	3.6	7	2.6768-2.6775	0.0004-0.0024	0.0039-0.0130	2	2.2044-2.2050	0.0004-0.0028	0.0140-0.0374
	5.7	S	2.5580-2.5590	0.0008-0.0021	0.0015-0.0078	3	2.0991-2.0999	0.0009-0.0025	0.0043-0.0200

06025-CTSD-C06

PISTON AND RING SPECIFICATIONS

All measurements are given in inches.

Year	Engine Displacement Liters	Engine VIN	Piston Clearance	Ring Gap			Ring Side Clearance		
				Top Compression	Bottom Compression	Oil Control	Top Compression	Bottom Compression	Oil Control
2003	3.2	N	0.0010-0.0018	0.0118-0.0196	0.0118-0.0196	0.0157-0.0551	0.0008-0.0024	0.0008-0.0024	0.0008-0.0024
2004	3.2	N	0.0010-0.0018	0.0118-0.0196	0.0118-0.0196	0.0157-0.0551	0.0008-0.0024	0.0008-0.0024	0.0008-0.0024
	3.6	7	0.0010-0.0021	0.0059-0.0118	0.0110-0.0189	0.0059-0.0236	0.0012-0.0026	0.0006-0.0024	0.0012-0.0067
	5.7	S	0.0005-0.0019	0.0090-0.0170	0.0170-0.0270	0.0070-0.0290	0.0016-0.0033	0.0020-0.0034	(-)0.0003-+0.0069
2005	2.8	T	0.0008-0.0013	0.0059-0.0118	0.0110-0.0189	0.0059-0.0236	0.0012-0.0026	0.0006-0.0024	0.0012-0.0669
	3.6	7	0.0010-0.0021	0.0059-0.0118	0.0110-0.0189	0.0059-0.0236	0.0012-0.0026	0.0006-0.0024	0.0012-0.0067
	5.7	S	0.0001-0.0011	0.0090-0.0170	0.0170-0.0270	0.0070-0.0290	0.0016-0.0033	0.0020-0.0034	(-)0.0003-+0.0069

06025-CTSD-C07

TORQUE SPECIFICATIONS

All readings in ft. lbs.

Year	Engine Displacement Liters	Engine VIN	Cylinder Head Bolts	Main Bearing Bolts	Rod Bearing Bolts	Crankshaft Damper Bolts	Flywheel Bolts	Manifold		Spark Plugs	Oil Pan Drain Plug
								Intake	Exhaust		
2003	3.2	N	①	②	③	15	④	15	15	18	8
2004	3.2	N	①	②	③	15	④	15	15	18	8
	3.6	7	⑤	⑥	⑦	⑧	⑨	17	15	13	18
	5.7	S	⑩	⑪	⑫	⑬	⑭	⑮	⑯	11	18
2005	2.8	T	⑤	⑥	⑦	⑧	⑨	17	15	13	18
	3.6	7	⑤	⑥	⑦	⑧	⑨	17	15	13	18
	5.7	S	⑰	⑪	⑫	⑬	⑭	⑮	⑯	11	18

① Step 1: 18 ft. lbs.
  Step 2: Rotate 90 degrees
  Step 3: Rotate 90 degrees
  Step 4: Rotate 90 degrees
  Step 5: Rotate 15 degrees
② Step 1: 37 ft. lbs.
  Step 2: Rotate 60 degrees
  Step 3: Rotate 15 degrees
③ Step 1: 26 ft. lbs.
  Step 2: Rotate 45 degrees
  Step 3: Rotate 15 degrees
④ Step 1: 48 ft. lbs.
  Step 2: Rotate 30 degrees
  Step 3: Rotate 15 degrees
⑤ M8 bolts:
  Step 1: 10 ft. lbs.
  Step 2: Rotate 60 degrees
  M11 bolts:
  Step 1: 33 ft. lbs.
  Step 2: Rotate 120 degrees

⑥ Inner:
  Step 1: 15 ft. lbs.
  Step 2: Rotate 80 degrees
  Outer:
  Step 1: 10 ft. lbs.
  Step 2: Rotate 110 degrees
  Side:
  Step 1: 22 ft. lbs.
  Step 2: Rotate 60 degrees
⑦ Step 1: 22 ft. lbs.
  Step 2: Back off to zero
  Step 3: 18 ft. lbs.
  Step 4: Rotate to 110 degrees
⑧ Step 1: 74 ft. lbs.
  Step 2: Rotate 150 degrees
⑨ Step 1: 22 ft. lbs.
  Step 2: Rotate 45 degrees

⑩ First Design:
  Step 1: M11 bolts to 22 ft. lbs.
  Step 2: M11 bolts plus 90 degrees
  Step 3: M11 bolts 1-8 plus 90 degrees
  Step 4: M11 bolts 9-10 plus 50 degrees
  Step 5: M8 bolts to 22 ft. lbs.
  Second Design:
  Step 1: M11 bolts to 22 ft. lbs.
  Step 2: M11 bolts plus 90 degrees
  Step 3: M11 bolts plus 70 degrees
  Step 4: M8 bolts to 22 ft. lbs.
⑪ Inner:
  Step 1: 15 ft. lbs.
  Step 2: Rotate 80 degrees
  Outer:
  Step 1: 15 ft. lbs.
  Step 2: Rotate 51 degrees
  Side:
  Step 1: 18 ft. lbs.

⑫ Step 1: 15 ft. lbs.
  Step 2: Rotate 75 degrees
⑬ Step 1: 37 ft. lbs.
  Step 2: Rotate 140 degrees
⑭ Step 1: 15 ft. lbs.
  Step 2: 37 ft. lbs
  Step 2: 74 ft. lbs
⑮ Step 1: 44 ft. lbs.
  Step 2: 89 inch lbs.
⑯ Step 1: 11 ft. lbs.
  Step 2: 18 ft. lbs.
⑰ Step 1: M11 bolts to 22 ft. lbs.
  Step 2: M11 bolts plus 90 degrees
  Step 3: M11 bolts plus 70 degrees
  Step 4: M8 bolts to 22 ft. lbs.

06025-CTSD-C08

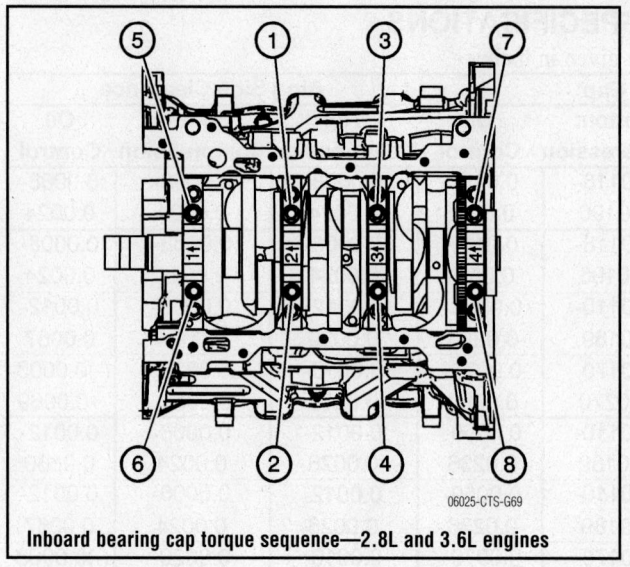

Inboard bearing cap torque sequence—2.8L and 3.6L engines

06025-CTS-G69

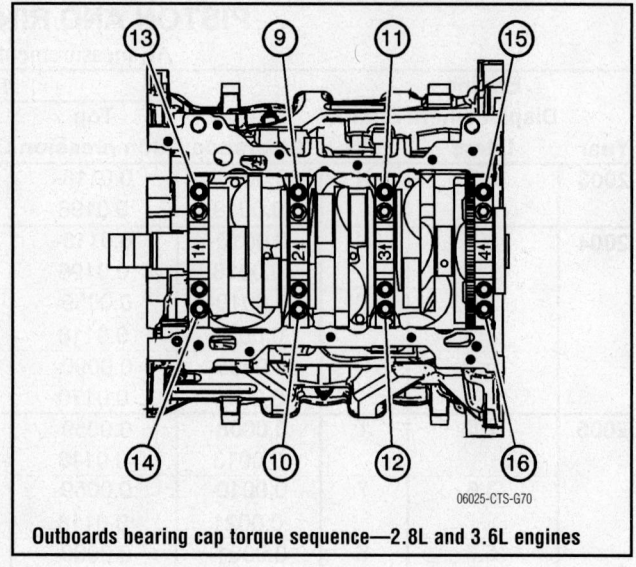

Outboards bearing cap torque sequence—2.8L and 3.6L engines

06025-CTS-G70

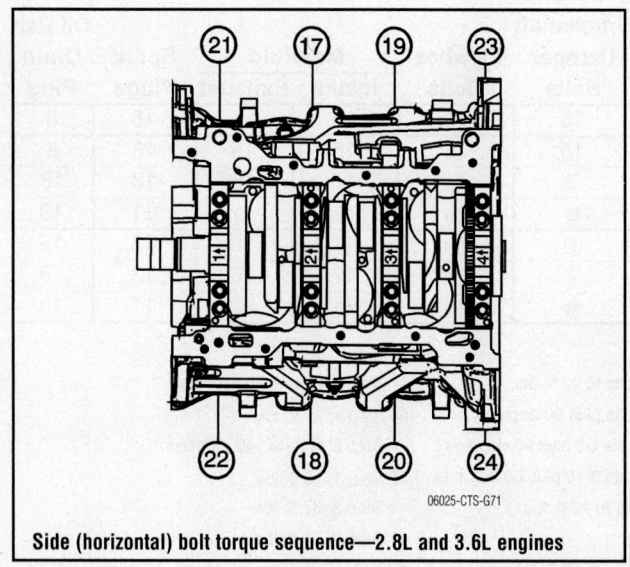

Side (horizontal) bolt torque sequence—2.8L and 3.6L engines

06025-CTS-G71

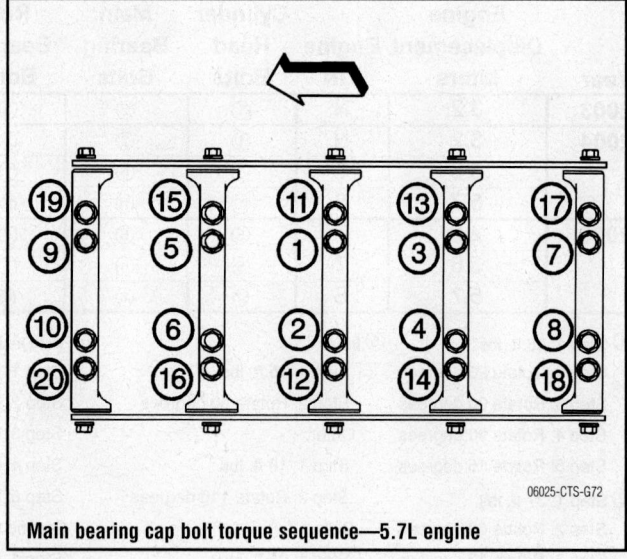

Main bearing cap bolt torque sequence—5.7L engine

06025-CTS-G72

WHEEL ALIGNMENT

Year	Model		Caster Range (+/-Deg.)	Caster Preferred Setting (Deg.)	Camber Range (+/-Deg.)	Camber Preferred Setting (Deg.)	Toe-in (in.)
2003	CTS	F	1.00	+5.10	0.50	-0.50	0.02 +/- 0.02
		R	—	—	0.40	-1.00	0.2 +/-0.2
2004	CTS	F	1.00	+5.10	0.50	-0.50	0.02 +/- 0.02
		R	—	—	0.40	-1.00	0.2 +/-0.2
2005	CTS	F	0.60	+5.10	0.60	-0.50	0.02 +/- 0.02
		R	—	—	0.50	-1.50	0.2 +/-0.2

06025-CTSD-C09

TIRE, WHEEL AND BALL JOINT SPECIFICATIONS

Year	Model	OEM Tires Standard	OEM Tires Optional	Tire Pressures (psi) Front	Tire Pressures (psi) Rear	Wheel Size	Ball Joint Inspection	Wheel Lug Torque (ft. lbs.)
2003	CTS	P225/55R16	P225/50R17	32	32	7-JJ	0.125 in. ①	100
2004	CTS	P225/55R16	P225/50R17	33	33	7-JJ	0.125 in. ①	100
	CTS-V	P245/45R18	NA	33	33	NA	NA	100
2005	CTS	P225/55R16	P225/50R17	32	32	7-JJ	0.125 in. ①	100
	CTS-V	P245/45R18	NA	②	②	NA	NA	100

NA: Information not available

OEM: Original Equipment Manufacturer

PSI: Pounds Per Square Inch

① Replace if vertical or horizontal movement exceeds specification

② See placard on vehicle

06025-CTSD-C10

BRAKE SPECIFICATIONS

All measurements in inches unless noted

Year	Model		Brake Disc Original Thickness	Brake Disc Minimum Thickness	Brake Disc Maximum Runout	Minimum Lining Thickness	Brake Caliper Bracket Bolts (ft. lbs.)	Brake Caliper Mounting Bolts (ft. lbs.)
2003	CTS	F	1.267	1.209	0.002	0.039	136	46
		R	1.020	0.944	0.002	0.039	88	44
2004	CTS	F	1.267	1.209	0.002	0.039	96	46
		R	1.020	0.944	0.002	0.039	88	44
	CTS-V	F	1.259	1.181	0.059	0.039	96	NA
		R	1.102	1.062	0.059	0.039	88	NA
2005	CTS	F	1.267	1.209	0.002	0.039	96	46
		R	1.023	0.944	0.059	0.039	88	44
	CTS-V	F	1.259	1.181	0.002	0.039	96	NA
		R	1.102	1.062	0.059	0.039	88	NA

F: Front

R: Rear

NA: Information not available

06025-CTSD-C11

MAINTENANCE I AND II SERVICE SCHEDULES
Cadillac CTS

When the CHANGE ENGINE OIL light appears, certain services and inspections are required.
Required services are described as Maintenance I and Maintenance II.
The first service on a vehicle should be Maintenance I, and the second service should be Maintenance II.
Alternate between the 2 thereafter. However, in some cases, Maintenance II may be required more often.
Maintenance I: Use Maintenance I if the CHANGE ENGINE OIL light comes on within 10 months since vehicle was purchased or, if Maintenance II was performed.
Maintenance II: Use Maintenance II if the previous service performed was Maintenance I. Always use Maintenance II when the CHANGE ENGINE OIL light comes on 10 months or more since the last service, or, if the CHANGE ENGINE OIL light has not come on at all for one year.

Service	Maintenance I	Maintenance II
Change the engine oil and filter. Reset the oil life system.	✓	✓
Visually inspect the vehicle for leaks or damage. A fluid loss in the vehicle system could indicate a problem. Inspected, repair and add fluid to the system if necessary.	✓	✓
Inspect the engine air cleaner filter. If necessary, replace the filter.	✓	✓
Rotate the tires. Inspect the tire inflation pressures and the tire wear.	✓	✓
Visually inspect the brake lines and hoses for proper hook-up, binding, leaks, cracks, chafing, etc. Inspect the disc brake pads for wear and the rotors for surface condition. Inspect the drum brake linings for wear or cracks. Inspect other brake parts, including drums, wheel cylinders, calipers, parking brake, etc. Inspect the parking brake adjustment.	✓	✓
Inspect the engine coolant and the windshield washer fluid levels. Add fluid as needed.	✓	✓
Inspect the suspension and steering components. Inspect the front and rear suspension and the steering system for damaged, loose or missing parts, or signs of wear. Inspect the power steering lines and the hoses for proper hook-up, binding, leaks, cracks,	--	✓
Visually inspect the coolant hoses and replace the hoses if they are cracked, swollen or deteriorated. Inspect all pipes, fittings and clamps; replace with GM parts as needed. To help ensure proper operation, a pressure test of the cooling system and pressure cap and cleaning the outside of the radiator and air conditioning condenser is recommended at least once a year.	--	✓
Inspect the front and rear suspension and the steering system for damaged, loose or missing parts, or signs of wear. Inspect power steering lines and hoses for proper hook-up, binding, leaks, cracks, chafing, etc.	--	✓
Inspect the throttle system for interference or binding and for damaged or missing parts. Replace the parts as needed. Replace any components that have high effort or excessive wear. Do not lubricate the accelerator or the cruise control cables.	--	✓
Replace the passenger compartment air filter.	--	✓

06025-CTSD-C12

ENGINE REPAIR

➡**Disconnecting the negative battery cable on some vehicles may interfere with the operation of the on board computer system. The computer may undergo a relearning process once the negative battery cable is reconnected.**

Alternator

REMOVAL & INSTALLATION

3.2L Engine

1. Before servicing the vehicle, refer to the Precautions Section.
2. Remove or disconnect the following:
   - Negative battery cable
   - Intake air resonator
   - Drive belt
   - Coolant heater, if equipped
   - Alternator cooling duct
   - Alternator electrical connectors
   - Alternator

To install:
3. Install or connect the following:
   - Alternator. Tighten the bolts to 26 ft. lbs. (35 Nm).
   - Cooling duct
   - Alternator electrical connectors
   - Coolant heater, if equipped
   - Drive belt
   - Intake air resonator
   - Negative battery cable
4. Perform a charging system test and verify that the system is operating properly.

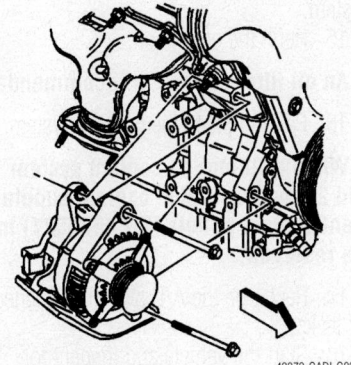

Alternator mounting–3.2L

2.8L and 3.6L Engines

1. Before servicing the vehicle, refer to the Precautions Section.
2. Install the engine support fixture.
3. Remove or disconnect the following:

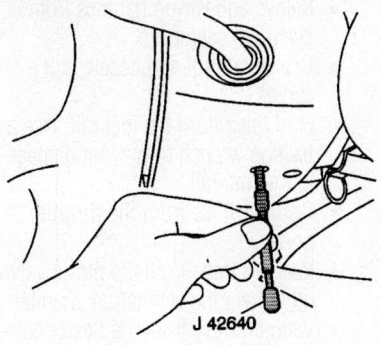

J 42640

06025-CTS-G01

Alternator mounting–2.8L and 3.6L

- Negative battery cable
- Accessory drive belt
- Alternator electrical connections
- Electronic brake control module (EBCM) electrical connector
- Engine mount lower nuts
- Lower alternator mounting bolt

4. Raise the engine using the support fixture.
5. Remove the alternator upper mounting bolt.
6. Remove the alternator.

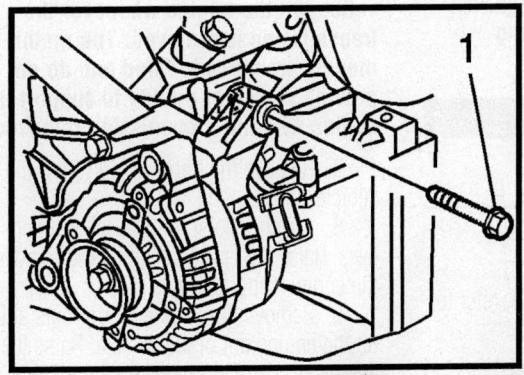

To install:
7. Install the alternator.
8. Lower the engine using the support fixture.
9. Install or connect the following:
   - Engine mount lower nuts and tighten to 59 ft. lbs. (80 Nm)
   - Lower alternator mounting bolt and tighten to 37 ft. lbs. (50 Nm)
   - Alternator electrical connections
   - EBCM electrical connector
   - Upper alternator mounting bolt and tighten to 37 ft. lbs. (50 Nm)
10. Remove the engine support fixture.
11. Install the accessory drive belt.
12. Connect the negative battery cable.

5.7L Engine

1. Before servicing the vehicle, refer to the Precautions Section.
2. Remove or disconnect the following:
   - Negative battery cable
   - Air intake assembly
   - Radiator fan assembly
   - Accessory drive belt
   - Power steering pressure hose bracket mounting bolt

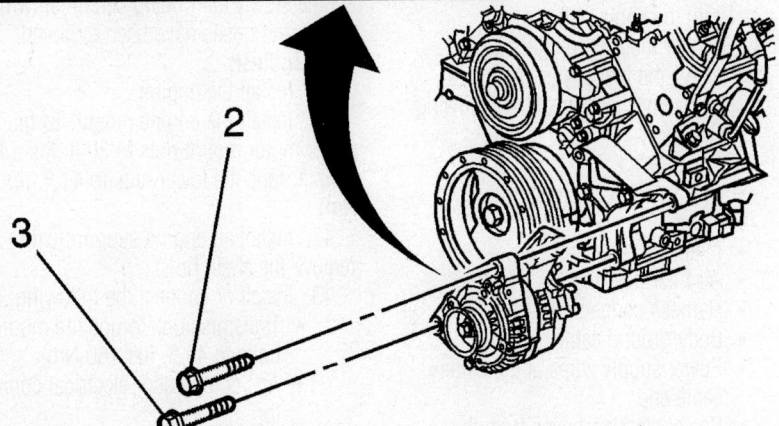

06025-CTS-G02

Alternator mounting bolt tightening sequence–5.7L

- Alternator mounting bolts

3. Lift the alternator off of the mounting bracket to gain access to the wiring electrical connections.

4. Disconnect the alternator electrical connections.

5. Remove the alternator.

To install:

6. Connect the alternator electrical connections. Tighten the battery terminal nut to 115 inch lbs. (13 Nm).

7. Install the alternator and tighten the mounting bolts to 37 ft. lbs. (50 Nm) in sequence as follows:

   a. Front mounting bolt (3)
   b. Front mounting bolt (2)
   c. Rear mounting bolt (1)

8. Install or connect the following:

   - Power steering pressure hose bracket mounting bolt. Tighten to 80 ft. lbs. (9 Nm).
   - Accessory drive belt
   - Radiator fan assembly
   - Air intake assembly
   - Negative battery cable

Ignition Timing

ADJUSTMENT

➡**The 3.2L DOHC engine used in the CTS utilizes a Distributorless Ignition System (DIS). No ignition timing adjustment is possible.**

Engine Assembly

REMOVAL & INSTALLATION

3.2L Engine

1. Before servicing the vehicle, refer to the Precautions Section.

2. Drain the cooling system.

3. Drain the engine oil.

4. Recover the A/C refrigerant.

5. Relieve the fuel system pressure.

6. Remove or disconnect the following:

   - Battery
   - Wiper arms
   - Left and right air inlet grilles
   - Hood
   - Intake air resonator
   - Air filter housing
   - Harness connectors
   - Body ground cable
   - Power supply wires at the battery cable end
   - Power steering hoses from the power steering pump
   - Brake booster vacuum lines

- Electronic Control Module (ECM) and harness from the electrical center
- Relays and wiring harness from electrical center box
- Cruise control and accelerator cables
- Fuel lines from the fuel rail. Use a backup wrench to prevent damage to the fuel rail.
- Coolant hose from the throttle body
- Vacuum hose from the purge valve on the engine ventilation chamber
- Vacuum hose from the heater control valve
- Coolant reservoir hose from the coolant inlet pipe
- Electric water pump with hoses attached
- Coolant hoses from the heater core
- Radiator
- Refrigerant line from the A/C compressor and compressor bracket
- Condenser line from the A/C condenser
- A/C quick-connect hose near the low pressure service valve
- Splash shield
- A/C compressor electrical connector

➡**Support the engine whenever the transmission is removed. The motor mounts are silicone filled and do not provide sufficient rigidity to support the engine with the transmission removed.**

7. Remove the transmission from the vehicle.

8. Attach a hoist to the engine lifting eyes. Raise the engine slightly and remove the engine support fixture.

9. Remove the engine mount nuts and lift the engine out of the vehicle. Raise the engine slowly after being certain all wiring, cables and hoses have been removed.

To install:

10. Install the engine.

11. Install the engine mount. Torque the upper motor mount nuts to 30 ft. lbs. (40 Nm). Torque the lower nuts to 41 ft. lbs. (55 Nm).

12. Install an engine support fixture and remove the chain hoist.

13. Install or connect the following:

   - Transmission. Torque the mounting bolts to 44 ft. lbs. (60 Nm).
   - A/C compressor electrical connector
   - Splash shield and tighten securely
   - Quick connects for the high and low pressure fittings

- A/C condenser line to the condenser
- A/C compressor line
- Radiator
- Coolant hoses to the heater core
- Electric water pump
- Coolant reservoir hose to the coolant inlet pipe
- Vacuum hose to the heater control valve
- Vacuum hose for the purge valve on the engine ventilation chamber
- Coolant hose to the throttle body
- Fuel return and supply lines. Torque the lines to 11 ft. lbs. (15 Nm).
- Cruise control and accelerator cables to the throttle body
- Relays and wiring harness to the electrical center box
- ECM
- Brake booster vacuum lines
- Power steering hoses to the steering pump. Torque the discharge hose fasteners to 21 ft. lbs. (28 Nm).
- Power supply wires at the battery cable ends
- Body ground cable
- Red, white and black wiring harness connectors
- Air filter housing
- Intake air resonator
- Hood
- Left and right air inlet grilles
- Wiper arms
- Battery
- Negative battery cable

14. Fill and bleed the power steering system.

15. Refill the engine oil.

➡**An oil filter change is recommended.**

16. Fill and bleed the coolant system.

➡**When refilling the coolant system add 2 crushed engine coolant supplement sealant pellets (PN 3634621) into the reservoir.**

17. Recharge the A/C system and check for leaks.

18. Start the vehicle and inspect for leaks.

2.8L and 3.6L Engines

1. Before servicing the vehicle, refer to the Precautions Section.

2. Drain the cooling system.

3. Drain the engine oil.

4. Recover the A/C refrigerant.

5. Relieve the fuel system pressure.

6. Center the steering wheel and install Special Tool J-42640 Steering Column Anti-Rotation Pin to the steering column.

7. Remove or disconnect the following:
- Negative battery cable
- Engine shroud
- Battery
- Air intake assembly
- Cooling fan wiring harnesses
- Surge tank hoses
- Heater hoses
- Vacuum hose from purge solenoid
- Fuel hose
- Wiper module
- A/C suction hose from the evaporator

✳✳ WARNING

Do NOT disconnect suction hose from A/C compressor.

- A/C pressure switch electrical connector and liquid line

➡**Do NOT disconnect the liquid line from the condenser.**

- Radiator support brackets
- Brake booster vacuum hoses
- Brake fluid level switch electrical connector
- Mass air flow sensor electrical connector
- Engine wiring harnesses from the underhood electrical center
- Transmission control module (TCM) wiring harness
- Ground wire
- Master cylinder and secure to the engine

➡**Do NOT disconnect the brake lines from the master cylinder.**

- Muffler assembly
- Driveshaft
- Air deflector
- Washer bottle bracket

➡**Do NOT remove the washer bottle.**

- Radiator side baffles
- Left front brake line retainer
- Right front brake line from bundle
- Rear brake lines from the brake pressure modulator valve (BPVM)

➡**Cap the brake lines to minimize fluid loss.**

- Front wheels
- Intermediate steering shaft
- Lower engine mount nuts
- Transmission shift linkage
- Low oil level sensor electrical connector. Secure to the engine mount bracket.

- Headlight leveling sensors
- Left and right shock module upper mounting bolts

➡**Secure the shock modules to the lower control arm with a strap to prevent damage to the front brake hoses.**

8. Place a suitable lift table under the engine, transmission, front frame and front suspension assembly.

9. Position a suitable powertrain lift table below the frame, engine and transmission. Raise the lift table to support the frame, engine and transmission.

10. Remove the transmission brace and front frame mounting bolts.

11. Lower the lift table to remove the engine, transmission front frame and front suspension assembly.

12. Remove or disconnect the following:
- Engine control module (ECM) harness
- Catalytic converters
- Thermostat housing with heater hoses, heater lines and surge tank hoses
- Starter motor
- Flywheel bolts
- Heater Oxygen sensor (HO2S) connector bracket from left cylinder head
- Remaining engine wiring harnesses
- Transmission oil cooler (TOC) lines
- Radiator hoses from water outlet and coolant inlet housing
- Power steering cooler hoses

➡**Cap the power steering lines to prevent fluid loss and contamination.**

- Condenser, radiator and fan module (CRFM) with radiator houses attached

13. Install engine lift brackets EN-46114 to the engine.

14. Attach an engine hoist to the lift brackets and partially support the engine. Position a second powertrain lift table below the transmission.

15. Remove or disconnect the following:
- Transmission
- Accessory drive belt
- Left and right drive belt tensioners
- Idler pulley
- Alternator and mounting bracket
- A/C compressor
- Power steering pump
- Power steering reservoir
- Oil level indicator tube
- Left exhaust manifold
- Oil filter adapter
- Right exhaust manifold
- Crankshaft pulley

- Flywheel
- Intake manifold
- Water outlet
- Block heater
- ECM and mounting bracket

16. Using the engine hoist, remove the engine assembly from the frame.

To install:

17. Using an engine hoist, lower the engine assembly to the frame.

18. Install or connect the following:
- ECM with the mounting bracket
- Water outlet
- Block heater
- Flywheel
- Intake manifold
- Crankshaft pulley. Tighten to 74 ft. lbs. (100 Nm) plus 150 degrees.
- Right exhaust manifold
- Oil filter adapter
- Left exhaust manifold
- Oil level indicator tube
- Alternator with mounting bracket
- Power steering pump
- Power steering reservoir
- A/C compressor
- Idler pulley. Tighten to 37 ft. lbs. (50 Nm).
- Left and right drive belt tensioners
- Accessory drive belt
- Transmission
- Flywheel-to-torque converter bolts
- Starter

19. Remove the engine lifting brackets from the engine.

20. Install or connect the following:
- CRFM with radiator hoses to the frame
- Power steering cooler hoses to the CRFM
- Radiator hoses to the water outlet and coolant inlet housing
- TOC lines
- Engine wiring harnesses
- HO2S connector bracket to the left cylinder head
- Thermostat housing with heater hoses, heater lines and surge tank hoses
- Catalytic converters
- Fuel lines
- ECM upper electrical connector

21. Raise the lift table to install the engine, transmission, front frame and front suspension assembly into the vehicle.

22. Install the front frame and transmission support mounting bolts. Tighten as follows:

a. Front frame bolts: 141 ft. lbs. (191 Nm)

b. Transmission support bolts: 44 ft. lbs. (60 Nm)

23. Remove the powertrain lift table.
24. Install or connect the following:
- Left and right shock module upper mounting bolts
- Headlight leveling sensors
- Transmission shift linkage
- Low oil level sensor electrical connector
- Lower engine mount nuts. Tighten to 59 ft. lbs. (80 Nm).
- Intermediate steering shaft
- Front wheels
- Brake lines
- Radiator side air baffles to the radiator
- Washer bottle bracket
- Air deflector
- Drive shaft
- Muffler assembly
- Brake master cylinder
- Ground wire
- TCM wiring harness
- Engine wiring harnesses to the underhood electrical connector
- Mass air flow sensor electrical connector
- Brake fluid level switch electrical sensor
- Brake booster vacuum hoses
- Radiator support brackets
- Vacuum hose to the purge solenoid
- Heater hoses
- Air intake assembly
- Surge tank lines
- A/C pressure switch electrical connector and liquid line to the evaporator
- A/C suction hose to the evaporator
- Cooling fan wiring harnesses
- Wiper module
- Battery and battery cables
- Engine shroud

25. Remove the steering column anti-rotation pin.
26. Fill and bleed the power steering system.
27. Refill the engine oil.
28. Fill and bleed the coolant system.
29. Recharge the A/C system and check for leaks.
30. Start the vehicle and inspect for leaks.

5.7L Engine

1. Before servicing the vehicle, refer to the Precautions Section.
2. Drain the cooling system.
3. Drain the engine oil.
4. Recover the A/C refrigerant.
5. Relieve the fuel system pressure.
6. Center the steering wheel and install

Special Tool J-42640 Steering Column Anti-Rotation Pin to the steering column.

7. Remove or disconnect the following:
- Negative battery cable
- Shifter assembly mounting bolts
- Engine cross brace
- Engine shroud
- Air intake assembly
- Throttle body heater outlet hose from throttle body
- A/C suction hose from the evaporator
- A/C liquid line
- Brake booster vacuum hose from the brake booster
- Brake fluid level switch electrical connector
- Brake master cylinder mounting nuts and secure master cylinder to the engine with wire
- Electrical connectors from the top of right and left shock tower
- Chassis fuel line at the right rear of the engine
- Evaporative emissions (EVAP) hose at the right rear of the engine
- Engine wiring harness connector at the front of the dash
- Underhood fuse block wiring harness

- Battery
- Throttle actuator control (TAC) module
- Ground strap
- Electric cooling fan assembly
- A/C condenser assembly
- Power steering oil cooler mounting nuts and secure unit to the engine

➡**Do NOT disconnect the lines from the cooler.**

- Front wheels
- Right front fender splash guard
- Windshield washer bottle-to-frame brace
- Separate the upper intermediate steering shaft from the center intermediate shaft
- Heater hoses
- Rear brake lines from the brake pressure modulator valve (BPMV)
- Transmission
- Clutch slave cylinder
- Rear brake lines from bundle clips
- Powertrain control module (PCM)
- Right and left shock module upper mounting bolts

➡**Secure the shock modules to the front frame to avoid stretching the front brake hoses.**

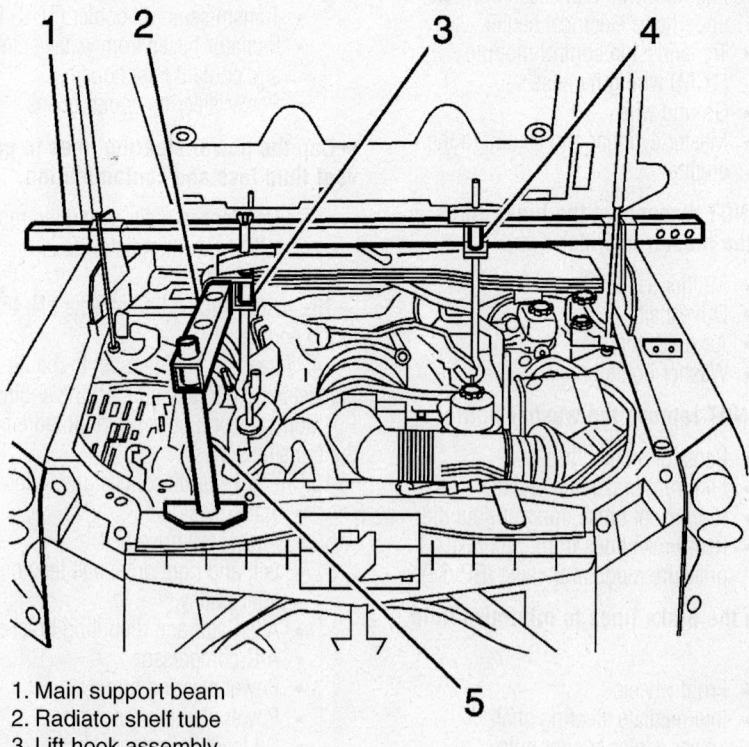

1. Main support beam
2. Radiator shelf tube
3. Lift hook assembly
4. Lift hook assembly
5. Support assembly

Engine support fixture–5.7L

06025-CTS-G60

8. Please a suitable lift table under the engine, front frame and front suspension assembly.

9. Raise the lift table to preload the weight of the engine, front frame and suspension assembly.

10. Remove the front frame bolts.

11. Lower the lift table and remove the engine, front frame and suspension assembly from the vehicle.

12. Using a suitable engine hoist, remove the engine from the front frame assembly.

To install:

13. Lower the engine onto the front frame assembly.

14. Raise the engine, front frame and suspension assembly into the vehicle.

15. Install the frame bolts and tighten to 141 ft. lbs. (191 Nm).

16. Install the right and left shock module upper mounting bolts. Tighten to 83 ft. lbs. (112 Nm).

17. Remove the engine lift table.

18. Install or connect the following:
- PCM and wiring harness
- Rear brake lines into the bundle clips
- Clutch slave cylinder
- Transmission
- Rear brake lines to the BPMV
- Heater hoses
- Upper intermediate steering shaft to the center intermediate shaft
- Windshield washer bottle-to-frame brace
- Right fender splash guard
- Power steering hose and A/C suction hose to proper positions
- Power steering oil cooler. Tighten the bolts to 18 ft. lbs. (25 Nm).
- A/C condenser
- Radiator assembly
- Electric cooling fan assembly
- Ground strap
- Underhood fuse block wiring harness
- TAC module
- Electrical connectors to the top of the right and left shock towers
- Battery
- Engine wiring harness connect at the front of the dash
- Chassis-to-EVAP hose at the right rear of the engine
- Chassis fuel line at the right rear of the engine
- Brake master cylinder. Tighten the mounting nuts to 18 ft. lbs. (25 Nm).
- Brake fluid level switch electrical connector
- Brake booster vacuum line

- A/C liquid line
- A/C suction hose to the evaporator
- Throttle body heater outlet hose to the throttle body
- Air intake assembly
- Engine shroud
- Engine cross brace
- Shifter assembly mounting bolts. Tighten to 13 ft. lbs. (18 Nm).
- Battery cables

19. Remove the steering column anti-rotation pin.

20. Fill and bleed the power steering system.

21. Fill the transmission with fluid to the correct level.

22. Refill the engine oil.

23. Fill and bleed the coolant system.

24. Recharge the A/C system and check for leaks.

25. Start the vehicle and inspect for leaks.

Water Pump

REMOVAL & INSTALLATION

3.2L Engine

1. Before servicing the vehicle, refer to the Precautions Section.

2. Drain the coolant.

3. Remove or disconnect the following:
- Negative battery cable
- Intake air resonator
- Water pump pulley bolts, loosen only
- Front timing belt cover
- Water pump pulley
- Water pump

To install:

4. Install or connect the following:
- Water pump with a new O-ring.

Torque the bolts to 18 ft. lbs. (25 Nm).
- Water pump pulley
- Front timing belt cover
- Intake air resonator
- Negative battery cable

5. Fill the cooling system through the reservoir tank.

➡**When refilling the cooling system, add 2 crushed engine coolant supplement sealant pellets (PN 3634621) into the coolant reservoir.**

6. Start the vehicle and inspect the coolant systems for leaks.

2.8L and 3.6L Engines

1. Before servicing the vehicle, refer to the Precautions Section.

2. Drain the cooling system.

3. Disconnect the negative battery cable.

4. Remove the alternator and water pump drive belt.

5. Install Water Pump holding tool EN-46104 to retain the water pump pulley.

6. Remove or disconnect the following:
- Water pump pulley bolts
- Water pump pulley
- Water pump mounting bolts
- Water pump

To install:

7. Install or connect the following:
- Water pump with a NEW gasket. Tighten mounting bolts to 89 inch lbs. (10 Nm).
- Water pump pulley. Tighten pulley bolts to 106 inch lbs. (12 Nm).
- Drive belt
- Negative battery cable

8. Fill the cooling system to the correct level.

9. Start the engine and check for leaks.

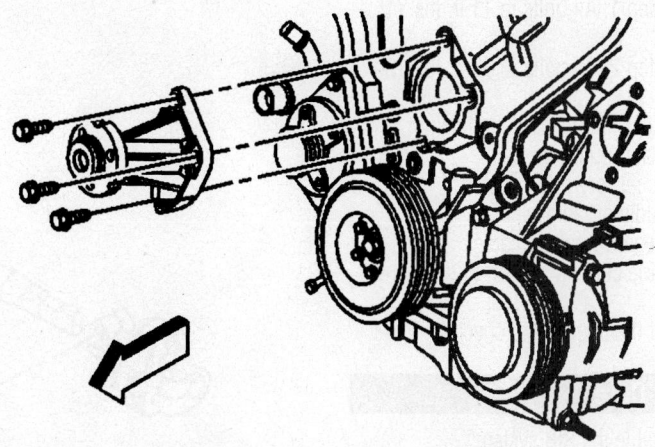

Exploded view of the water pump mounting

42372-CADI-G03

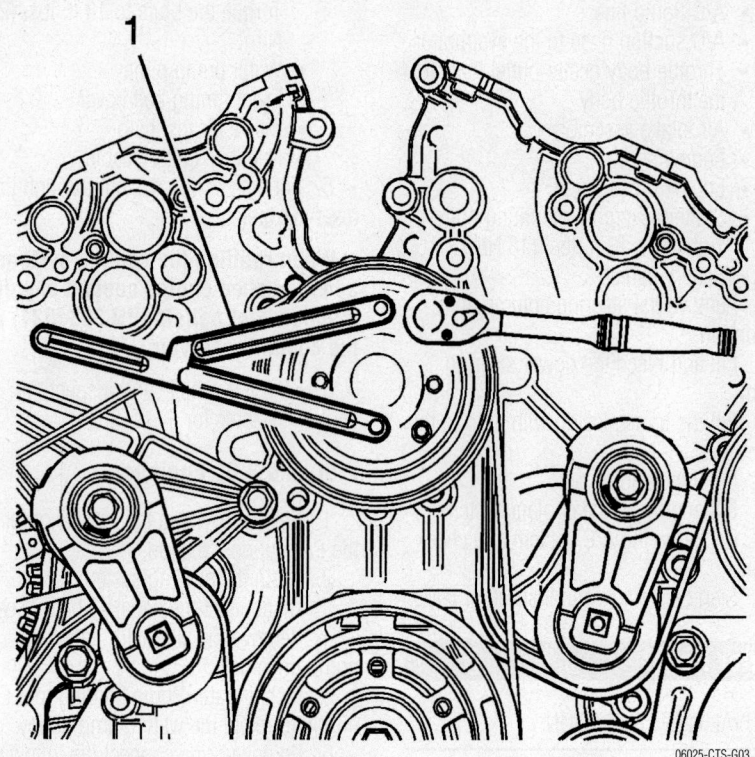

Using special tool EN-46104 to retain the pulley when installing water pump pulley bolts–2.8L and 3.6L

5.7L Engine

1. Before servicing the vehicle, refer to the Precautions Section.
2. Drain the cooling system.
3. Remove or disconnect the following:
   - Negative battery cable
   - Radiator fan assembly
   - Accessory drive belt
   - Water pump hoses
   - Water pump mounting bolts
   - Water pump and gaskets

To install:

4. Install the water pump and gaskets. Tighten the bolts as follows:
   a. Step 1: All bolts to 11 ft. lbs. (15 Nm).
   b. Step 2: All bolts to 22 ft. lbs. (30 Nm).
5. Install or connect the following:
   - Water pump hoses
   - Accessory drive belt
   - Radiator fan assembly
   - Negative battery cable
6. Fill the cooling system to the correct level.
7. Start the engine and check for leaks.

Heater Core

1. Disable the SIR system.
2. Disconnect the negative battery cable.

3. Drain the cooling system.
4. Remove the heater hose quick connects from the heater core pipes by performing the following procedure:
   a. At the passenger side, raise the air inlet screen and open the access door near the pollen filter.
   b. Unlock the quick connect collars by squeezing the tabs and carefully pulling back on the tabs to disconnect the sleeve.
   c. If green assembly marks are attached, discard them.

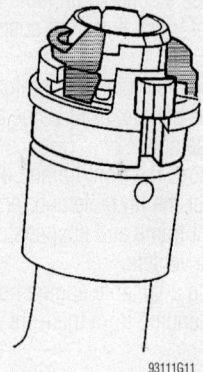

View of a heater hose connector—CTS

✳✳ WARNING

The front wheels must be maintained in the straight-ahead position and the steering column must be in the LOCK position. Failure to do so will cause improper alignment of some components during installation and may result to damage to the SIR coil assembly.

5. Remove the steering column by removing or disconnecting the following:
   - Instrument panel driver knee bolster energy absorber and sound insulator
   - Steering column electrical connector(s)
   - Coupler bolt from the lower steering column connection and slightly separate the coupler to aid in the shaft removal
   - Using a chisel and a hammer, rotate the forward support strap shear nut and bolt (located under the steering column) counterclockwise in order to remove them

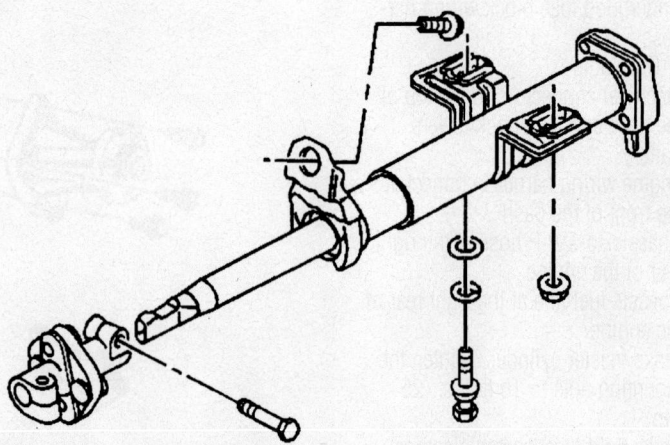

View of the steering column assembly—CTS

- Steering column-to-rear support bracket bolt
- Pull the steering column straight back and through the dash panel and remove it from the vehicle.

✳ WARNING

Handle the steering column with care for it is very susceptible to damage. Dropping, leaning or hammering on it could cause damage to its collapsible design.

6. Remove the instrument panel carrier by removing or disconnecting the following:
- Windshield pillar moldings
- Access panel, the air deflector outlet screw, the air deflector outlet and the air outlet duct, located at the right-side of the instrument panel
- Instrument panel SIR (air bag) module cover, the instrument panel compartment and the SIR module
- Upper center console and the lower center console
- Center console air duct screw and the center console air duct
- Radio tape player bezel and the radio
- Climate control head
- Center air outlet deflector, the outlet screws and the outlet housing
- Driver's side access panel
- Driver's side air outlet deflector, the outlet screw and the outlet housing
- Driver's side lower outlet duct
- Instrument cluster
- Fuse and relay panel screws and the panels
- Instrument panel carrier bolts
- Electrical connectors from the instrument panel and/or the wiring harness clips from the instrument panel carrier, if necessary
- Instrument panel carrier

➡ **It is not necessary to physically remove the instrument panel but it is necessary to pull the carrier rearward to enable the heater core to be removed from its housing.**

- Blower motor housing screws and the housing with the motor
- Heater core pipe bracket-to-chassis screw and the bracket
- Heater core inlet/outlet pipe bracket-to-heater core screw and the inlet/outlet pipe bracket
- Instrument panel support brace bolts from the instrument panel and the transmission well, then remove the brace

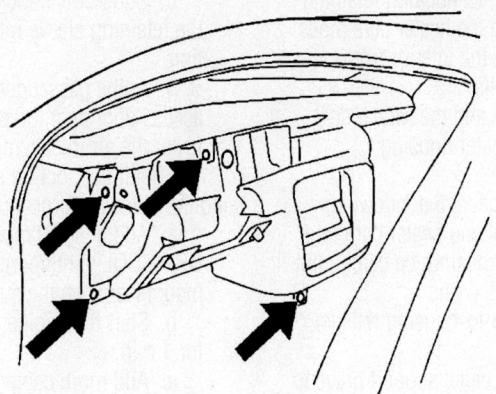

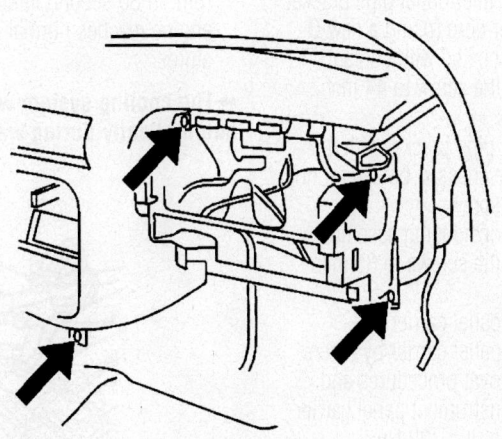

93111G09

View of the instrument panel carrier bolt locations—CTS

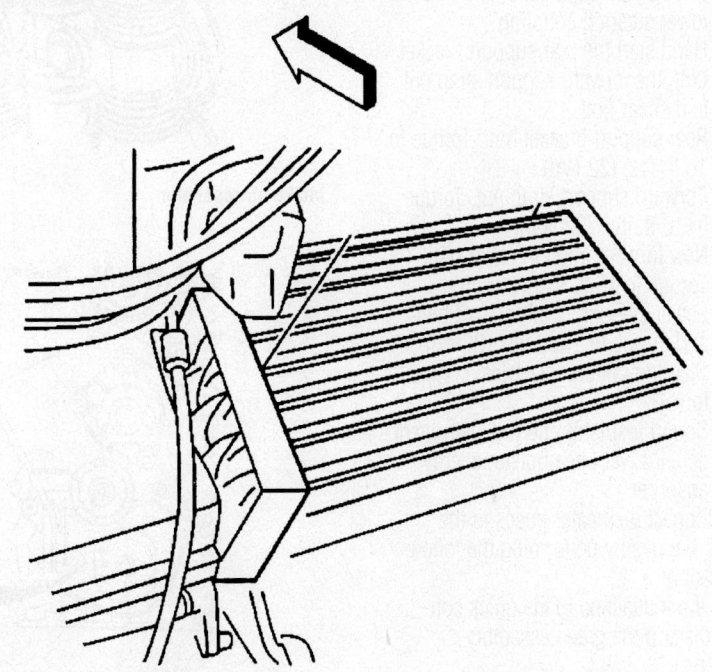

93111G12

View of a heater core and seal—CTS

- Heater core-to-housing retaining screw. Plug the heater core pipes and protect the interior from coolant spills
- Heater core and the rubber seal from the heater housing

To install:

7. Install or connect the following:

- Rubber seal and heater core into the heater housing; be careful not to damage the fins
- Heater core-to-housing retainer screw
- Instrument panel support brace to the transmission well and instrument panel, then torque the bolts to 16 ft. lbs. (22 Nm)
- Heater core inlet/outlet pipe bracket to the heater core (using a new O-ring lightly coated with coolant), and torque the screw to 44 inch lbs. (5 Nm)
- Heater core pipe bracket with the screw to the chassis, be careful not to strip the screw
- Blower motor/housing assembly and torque the screws to 35 inch lbs. (4 Nm)
- Instrument panel carrier
- Instrument panel carrier by reversing the removal procedures and torque the instrument panel carrier bolts to 16 ft. lbs. (22 Nm)

8. Remove the steering column by installing or connecting the following:

- Steering column into the vehicle, through the dash panel and into the lower steering coupling
- Hand start the rear support bracket bolt, the forward support strap nut and shear bolt
- Rear support bracket bolt. Torque to 16 ft. lbs. (22 Nm)
- Forward support strap nut. Torque to 16 ft. lbs. (22 Nm)
- New forward support shear bolt. Torque to 15 ft. lbs. (21 Nm)
- Lower steering column shaft bolt and torque to 16 ft. lbs. (22 Nm)
- Steering column electrical connector(s)
- Sound insulator and the instrument panel driver knee bolster energy absorber

9. Connect the heater hoses to the heater core pipes by performing the following procedure:

a. If not attached to the quick connect, discard the green assembly marker(s).

b. Push the quick connects into the pipes until they are fully seated.

c. Squeeze the locking tabs and press the retaining sleeve into the locked position.

d. At the passenger side, close the access door near the pollen filter and lower the air inlet screen.

10. Refill the cooling system by performing the following procedure:

a. Add a 50/50 mixture of water and DEX-COOL® antifreeze to the KALT/COLD mark (seam) on the surge tank.

b. Start the engine and allow it to idle for 1 min.

c. Add more coolant to the surge tank as necessary.

d. Install the radiator sure tank cap.

e. Cycle the engine, from idle to 3000 rpm, in 30 second intervals, until the engine reaches normal operating temperatures.

➡**The cooling system will bleed itself automatically during warm-up.**

f. Turn the engine OFF and recheck the coolant level when the engine is cool

11. Connect the negative battery cable.

12. Enable the SIR system.

13. Reprogram the necessary accessories.

Cylinder Head

REMOVAL & INSTALLATION

3.2L Engine

1. Before servicing the vehicle, refer to the Precautions Section.

2. Drain the engine coolant.

3. Relieve the fuel system pressure.

4. Remove or disconnect the following:

- Intake plenum
- Intake air resonator
- Intake manifold and spacer
- Both Engine Coolant Temperature (ECT) sensor electrical connectors

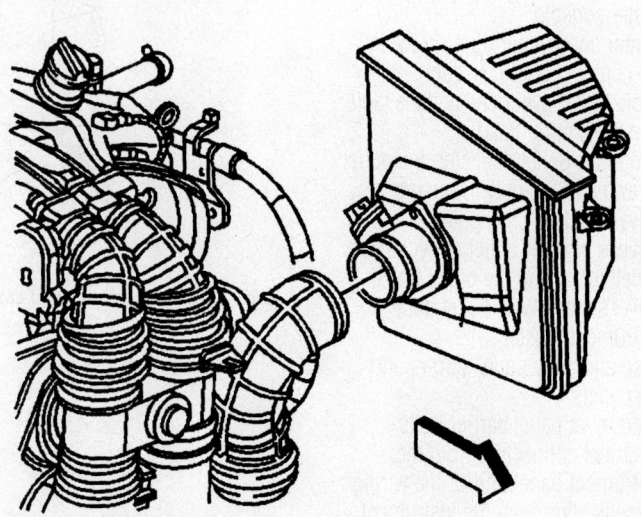

42372-CADI-G04

Intake air resonator

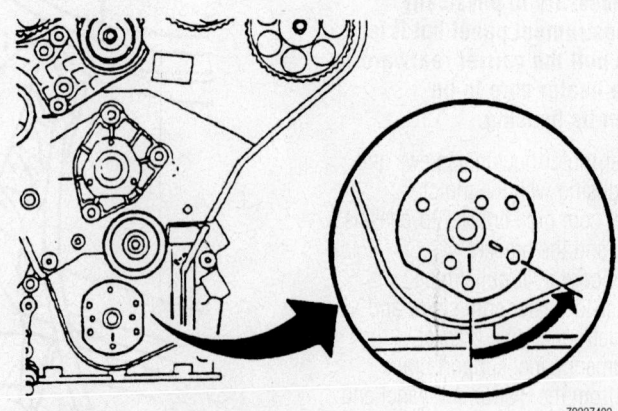

79222402

Before removing the timing belt, be sure to turn the crankshaft 60 degrees BTDC to avoid valve-to-piston contact and subsequent engine damage

- Coolant crossover
- Ignition coil assembly
- Camshaft cover
- Front timing belt cover

5. Position the crankshaft 60 degrees Before Top Dead Center (BTDC) to avoid contact between the valves and the pistons.

6. Remove or disconnect the following:

- Timing belt
- Timing belt tensioner bracket
- Four camshaft gears
- Water pump
- Rear timing belt cover
- Camshaft Position (CMP) sensor electrical connector
- Exhaust camshaft
- Coolant pipe/engine lift bracket from the cylinder head
- Oil level indicator tube
- Upper radiator hose from the coolant pipe
- Coolant pipe
- Exhaust manifold from the cylinder head
- Cylinder head

To install:

➡**Be sure to use new cylinder head bolts when installing the cylinder head. The old bolts have been stretched and are not reusable.**

7. Install a new cylinder head gasket.
8. Install the cylinder head.
9. Install new cylinder head bolts and tighten the bolts as follows:

  a. Step 1: Torque the bolts to 18 ft. lbs. (25 Nm).

  b. Step 2: An additional 90 degrees.

  c. Step 3: An additional 90 degrees.

  d. Step 4: An additional 90 degrees.

  e. Step 5: An additional 15 degrees.

  f. Step 6: An additional 15 degrees.

10. Install or connect the following:

- Exhaust manifold using a new gasket. Torque the nuts to 15 ft. lbs. (20 Nm).
- Coolant pipe with new O-rings
- Oil level indicator tube
- Coolant pipe/engine lift bracket. Torque the bolt to 15 ft. lbs. (20 Nm).
- Upper radiator hose to the coolant pipe
- Exhaust camshaft. Torque the bearing cap bolts to 71 inch lbs. (8 Nm).
- CMP electrical connector
- Rear timing belt cover. Torque the bolts to 71 inch lbs. (8 Nm).
- Water pump. Torque the bolts to 18 ft. lbs. (25 Nm).
- Camshaft gears. Torque the bolts to

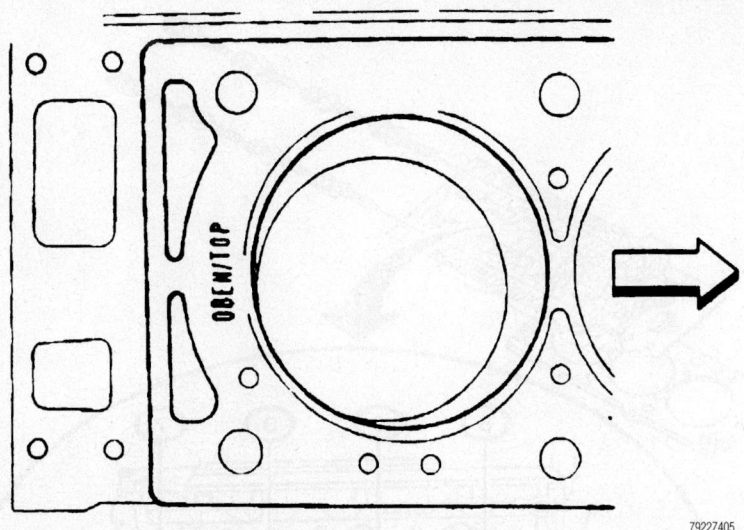

Proper head gasket installation position—right cylinder head

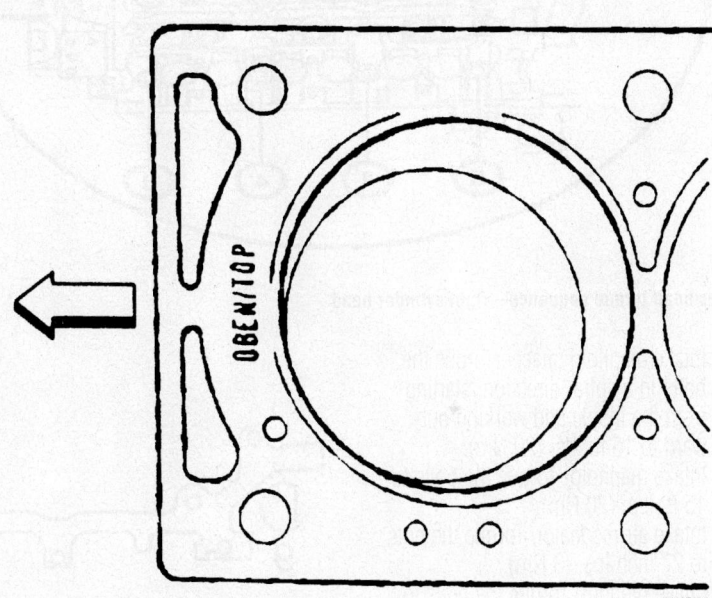

Proper head gasket installation position—left cylinder head

37 ft. lbs. (50 Nm) plus a 60 degree turn, plus another 15 degree turn.

- Timing belt tensioner bracket. Torque the bolts to 30 ft. lbs. (40 Nm).
- Timing belt
- Front timing belt cover. Torque the bolts to 71 inch lbs. (8 Nm).
- Left camshaft cover with new O-rings and gaskets. Torque the fasteners to 71 inch lbs. (8 Nm).
- Ignition coil assembly
- Coolant crossover. Torque the bolts to 22 ft. lbs. (30 Nm).
- Both ECT sensor electrical connectors

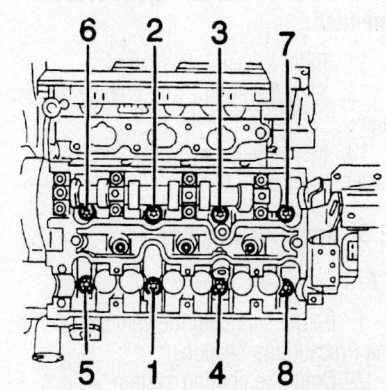

Cylinder head torque sequence—left cylinder head

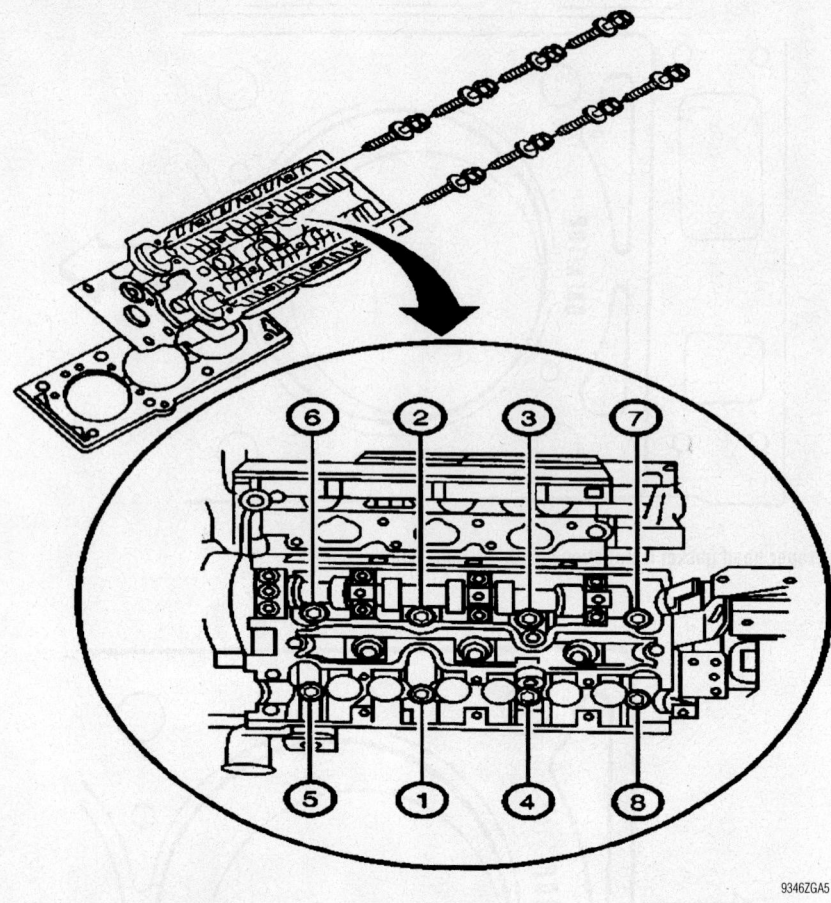

Cylinder head torque sequence—right cylinder head

- Intake manifold spacer. Torque the bolts in a spiral direction, starting from the inside and working outward to 15 ft. lbs. (20 Nm).
- Intake manifold. Torque the bolts to 15 ft. lbs. (20 Nm).
- Intake air resonator. Torque the nuts to 27 inch lbs. (3 Nm).
- Intake plenum. Torque the bolts to 71 inch lbs. (8 Nm).
- Negative battery cable

➡ **An oil and filter change is recommended.**

11. Refill the cooling system.
12. Start the engine and inspect for leakage.
13. Inspect all fluid levels and top off, if necessary.

2.8L and 3.6L Engines

LEFT

1. Before servicing the vehicle, refer to the Precautions Section.
2. Drain the cooling system.
3. Relieve the fuel system pressure.
4. Remove or disconnect the following:

- Negative battery cable
- Spark plugs
- Front cover
- Left bank secondary camshaft timing chain
- Coolant temperature sensor
- Wiring harnesses from the cylinder head
- Power steering pump mounting bolts
- Surge tank hose
- Wiring harness bracket
- Catalytic converter
- Oil level indicator tube
- Oil filter adapter upper bolt

➡ **Do NOT remove the oil filter adapter.**

- Exhaust manifold
- Cylinder head

To install:

5. Install the exhaust manifold to the cylinder head. Tighten the bolts to 18 ft. lbs. (25 Nm).

6. Install the cylinder head with a new gasket. Tighten using new bolts as follows:

    a. Step 1: Bolts 1-8 to 33 ft. lbs. (45 Nm).

    b. Step 2: Bolts 1-8 plus 120 degrees.

    c. Step 3: Bolts 9-10 to 11 ft. lbs. (15 Nm).

    d. Step 4: Bolts 9-10 plus 60 degrees.

7. Install or connect the following:

- Oil filter adapter upper bolt. Tighten to 17 ft. lbs. (23 Nm).

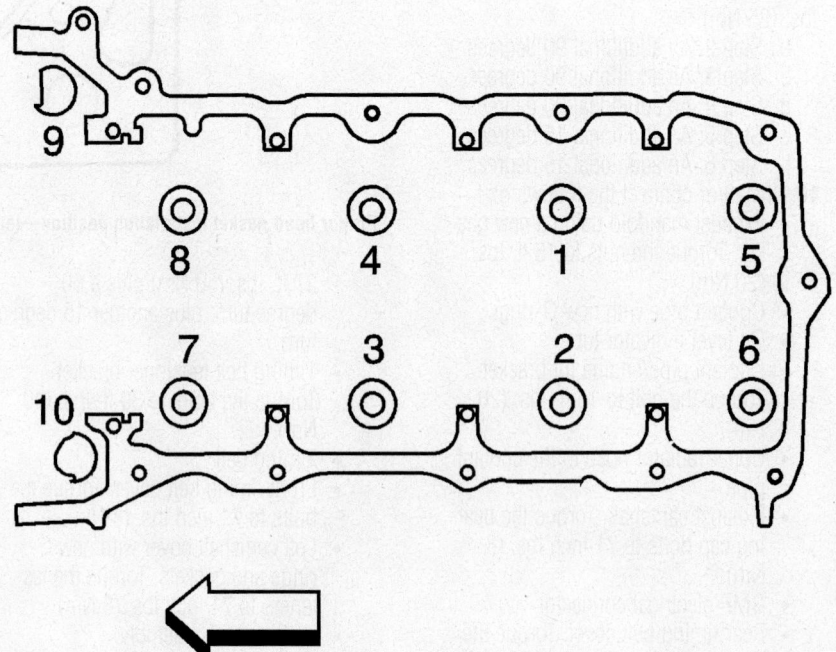

Left cylinder head torque sequence–2.8L and 3.6L engines

06025-CTS-G04

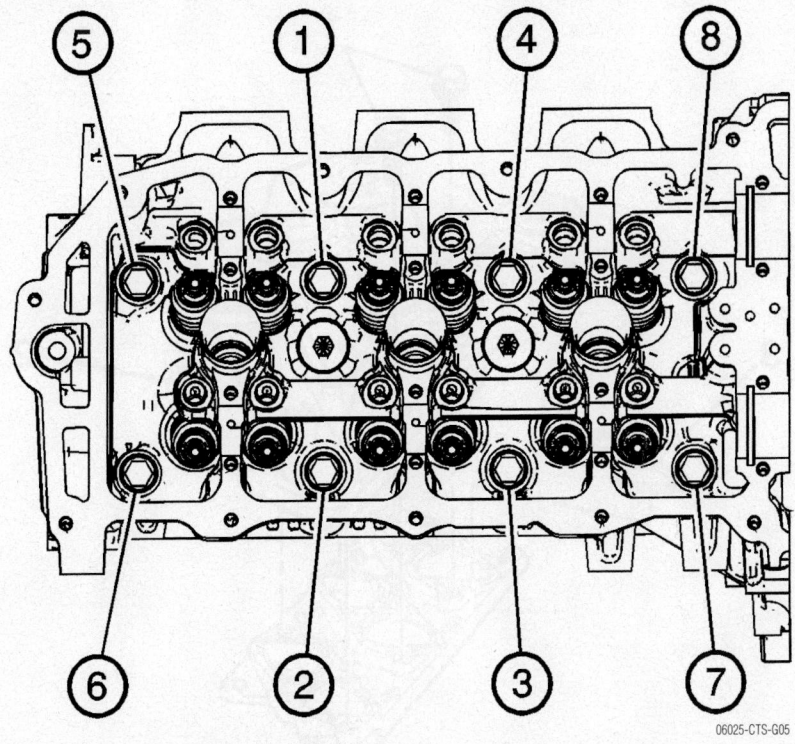

Right cylinder head torque sequence–2.8L and 3.6L

06025-CTS-G05

- Oil level indicator tube
- Catalytic converter
- Wiring harnesses bracket
- Surge tank hose
- Power steering pump bolts. Tighten bolts to 37 ft. lbs. (50 Nm).
- Wiring harnesses
- Coolant temperature sensor
- Left bank secondary camshaft timing chain
- Spark plugs
- Front cover
- Negative battery cable

8. Fill the cooling system to the correct level.

9. Start the engine and check for leaks.

RIGHT

1. Before servicing the vehicle, refer to the Precautions Section.

2. Drain the cooling system.

3. Relieve the fuel system pressure.

4. Remove or disconnect the following:
- Negative battery cable
- Spark plugs
- Front cover
- Right bank secondary camshaft timing chain
- Coolant hoses
- Catalytic converter
- Wiring harnesses from the cylinder head
- Exhaust manifold
- Cylinder head

To install:

5. Install the exhaust manifold to the cylinder head. Tighten the bolts to 18 ft. lbs. (25 Nm).

6. Install the cylinder head with a new gasket. Tighten the bolts as follows:
   a. Step 1: Tighten to 33 ft. lbs. (45 Nm).
   b. Step 2: Tighten an additional 120 degrees.

7. Install or connect the following:
- Wiring harnesses to the cylinder head
- Catalytic converter
- Coolant hoses
- Right bank secondary camshaft timing chain
- Spark plugs
- Front cover
- Negative battery cable

8. Fill the cooling system to the correct level.

9. Start the engine and check for leaks.

5.7L Engine

1. Start the engine and check for leaks.

2. Drain the cooling system.

3. Relieve the fuel system pressure.

4. Remove or disconnect the following:
- Valve cover
- Rocker arms
- Pushrods
- Intake manifold
- Coolant air bleed pipe
- Power steering pump and mounting bracket
- Exhaust manifold
- Engine wiring harness
- Ground strap
- Cylinder head bolts
- Cylinder head

To install:

5. Install the cylinder head with a new gasket. Tighten bolts in sequence as follows:
   a. Step 1: Tighten bolts 1-10 in to 22 ft. lbs. (30 Nm).
   b. Step 2: Tighten bolts 1-10 in an additional 90 degrees.
   c. Step 3: Tighten bolts 1-10 an additional 70 degrees.
   d. Step 4: Tighten bolts 11-15 to 22 ft. lbs. (30 Nm).

6. Install or connect the following:
- Engine wiring harness
- Ground strap. Tighten bolt to 24 ft. lbs. (32 Nm).
- Exhaust manifold
- Power steering pump. Tighten mounting bolts to 18 ft. lbs. (25 Nm).
- Coolant air bleed pipe. Tighten bolts to 106 inch lbs. (12 Nm).
- Intake manifold
- Pushrods
- Rocker arms
- Valve cover
- Negative battery cable

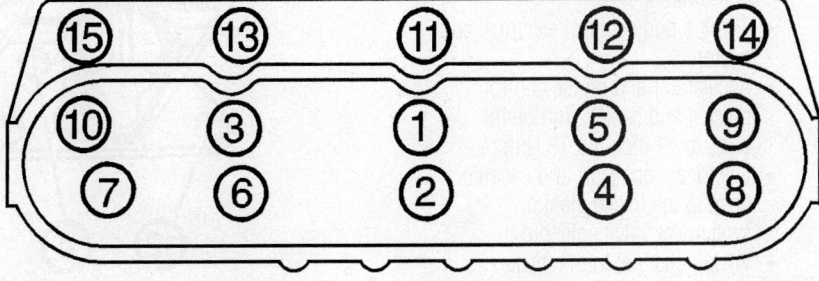

06025-CTS-G59

Cylinder head bolt sequence–5.7L

7. Refill the cooling system.
8. Start the engine and check for leaks.

Intake Manifold Assembly

REMOVAL & INSTALLATION

3.2L Engine

INTAKE PLENUM

1. Before servicing the vehicle, refer to the Precautions Section.
2. Remove or disconnect the following:
   - Negative battery cable
   - Intake plenum air inlet hoses from the throttle body
   - Brake booster vacuum hose from the intake plenum
   - Wiring harness channel from the plenum
   - Switch over valve electrical connector and vacuum line
   - Accelerator, cruise control cables and the bracket from the throttle body
   - Throttle body control electrical connector
   - Idle Air Control (IAC) inlet hose and electrical connector
   - Throttle Position Sensor (TPS) electrical connection
   - Throttle body from the intake plenum
   - Crankcase vent tube adapter and cover from the intake plenum
   - Intake plenum and O-rings

To install:
3. Install or connect the following:
   - Intake plenum with new O-rings. Torque the bolts to 71 inch lbs. (8 Nm).
   - Crankcase vent tube adapter to the intake plenum. Torque the fastener to 71 inch lbs. (8 Nm).
   - Throttle body to the intake plenum. Torque the fasteners to 106 inch lbs. (12 Nm).
   - TPS electrical connector
   - IAC electrical connector and vacuum line
   - Throttle body control electrical connector
   - Accelerator and cruise control bracket and cables. Torque the bolts to 71 inch lbs. (8 Nm).
   - Electrical connector and vacuum hose to the intake plenum switchover valve solenoid
   - Wiring channel to the intake plenum. Torque the bolts to 71 inch lbs. (8 Nm).

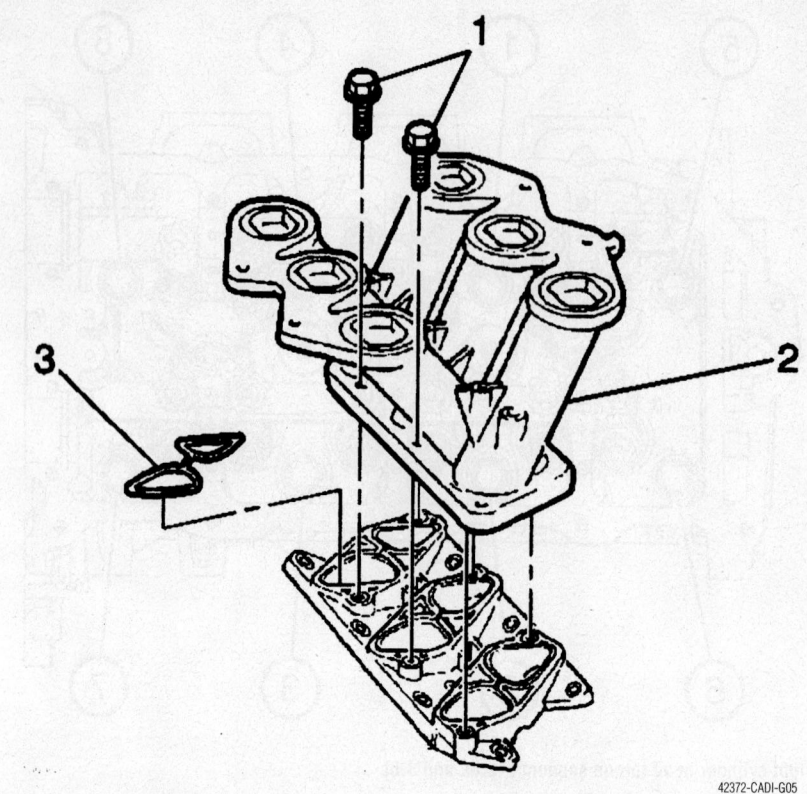

Upper intake exploded view

   - Threaded brake booster vacuum hose to the intake plenum

➡**Be certain not to strip the threads on the brake booster vacuum hose.**

   - Intake plenum air inlet hoses to the throttle body
   - Negative battery cable

4. Start the engine and check for proper performance.

INTAKE MANIFOLD

1. Before servicing the vehicle, refer to the Precautions Section.
2. Relieve the fuel system pressure.
3. Remove or disconnect the following:
   - Intake plenum
   - Fuel supply/return lines from the fuel rail by loosening the fittings
   - Fuel injector electrical harness connector

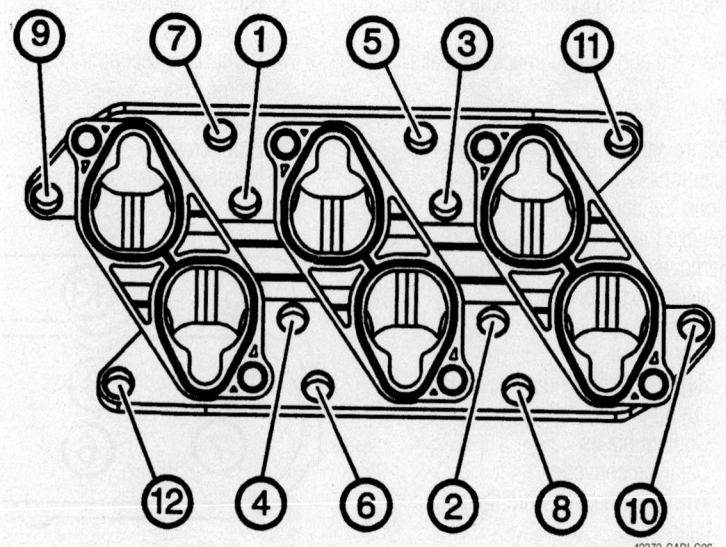

Lower intake torque sequence

- Fuel pressure regulator vacuum line
- Intake manifold bolts

✳✳ WARNING

Cover the open ports to the intake manifold spacer. Severe damage may occur if foreign material enters the engine.

- Intake manifold

To install:

4. Install or connect the following:
- Intake manifold with new gaskets. Torque the bolts to 15 ft. lbs. (20 Nm).
- Fuel pressure regulator vacuum line

➡ **The fuel injector connectors are numbered. Be sure to attach the correct connector to the proper fuel injector.**

- Fuel injector electrical harness connectors
- Fuel supply/return hoses. Torque the fittings to 11 ft. lbs. (15 Nm).
- Intake plenum
- Negative battery cable
5. Start the vehicle and inspect for leaks.

2.8L and 3.6L Engines

1. Before servicing the vehicle, refer to the Precautions Section.
2. Remove or disconnect the following:
- Engine shroud
- Air intake assembly
- Brake booster vacuum hose from manifold
- Purge solenoid valve electrical connector
- Throttle body electrical connector
- Upper intake manifold brace
- Positive crankcase valve (PCV) hose from the camshaft cover
- Barometric pressure (BARO) sensor electrical connector
- Intake manifold runner control solenoid electrical connector
- Left bank ignition coil harness
- Fuel injector harness bracket
- Intake manifold bolts
- Intake manifold assembly
3. To disassemble the intake manifold:
   a. Remove the upper-to-lower intake manifold bolts.
   b. Remove the fuel injector wiring harness bracket bolt from the upper intake manifold
   c. Remove the upper intake manifold from the lower intake manifold.

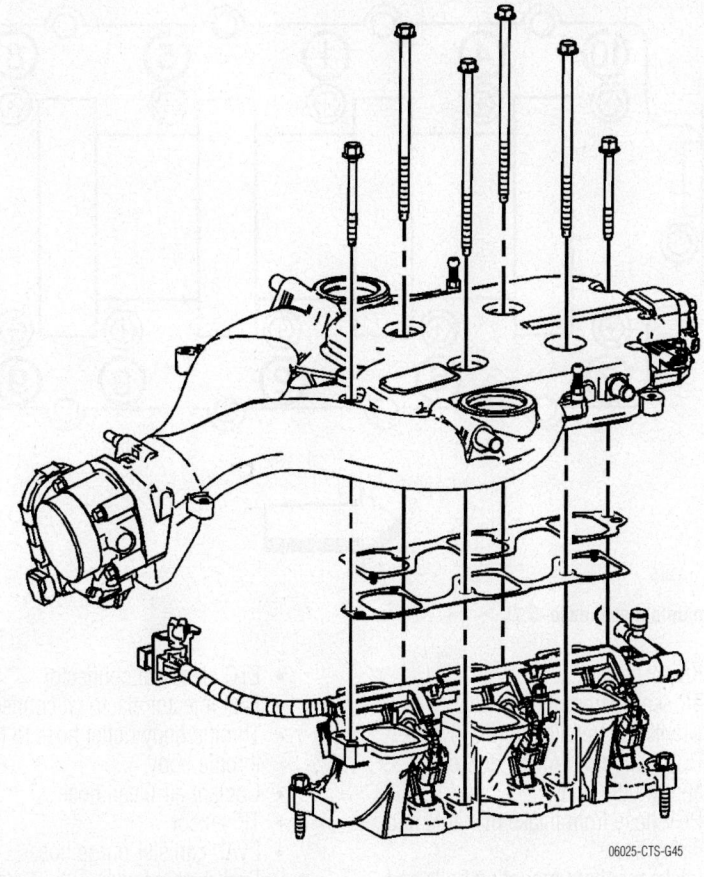

View of the intake manifold—2.8L and 3.6L

06025-CTS-G45

To install:

4. If necessary, assembly the intake manifold:
   a. Install the upper manifold to the lower manifold using a new gasket.
   b. Install the fuel injector wiring harness bracket bolt. Tighten to 89 inch lbs. (10 Nm).
   c. Tighten the upper-to-lower intake manifold bolts to 17 ft. lbs. (23 Nm).
5. Install the intake manifold with a new gasket. Tighten the bolts to 17 ft. lbs. (23 Nm).
6. Install or connect the following:
- Fuel injector harness bracket. Tighten bolt to 89 inch lbs. (10 Nm).
- Left bank ignition coil harness
- BARO sensor electrical connector
- Intake manifold runner control solenoid electrical connector
- PCV hose to the camshaft cover
- Upper intake manifold brace. Tighten bolts to 48 ft. lbs. (65 Nm).
- Throttle body electrical connector
- Purge solenoid valve electrical connector
- Brake booster vacuum hose to the manifold

- Air intake assembly
- Engine shroud
7. Start the engine and check for leaks.

5.7L ENGINE

➡ **The intake manifold, throttle body, fuel rail and injectors may be removed as an assembly.**

1. Before servicing the vehicle, refer to the Precautions Section.
2. Drain the cooling system.
3. Relieve the fuel system pressure.
4. Remove or disconnect the following:
- Negative battery cable
- Engine shroud
- Air intake assembly
- Fuel supply hose
- Evaporative emissions (EVAP) canister purge hose
- Throttle position (TP) sensor
- Coolant air bleed hose
- Throttle body outlet hose from the throttle body
- Fuel injector electrical connectors
- Electronic throttle control (ETC) electrical connector
- EVAP canister purge solenoid valve
- Power brake booster vacuum hose at the booster

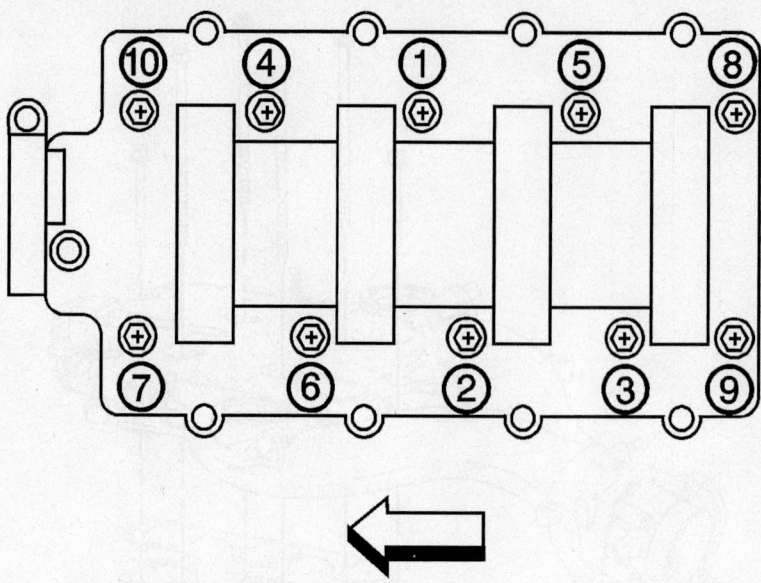

06025-CTS-G64

Intake manifold sequence–5.7L

- Knock sensor
- TP sensor from the Positive Crankcase Ventilation (PCV) hose
- PCV hose from rocker arm cover and throttle body
- PCV hose from intake manifold and valley cover
- Intake manifold mounting bolts and fuel rail stop bracket
5. Slide the intake manifold forward.
6. Remove or disconnect the following:
   - Manifold absolute pressure (MAP) sensor vacuum hose
   - MAP sensor
   - Intake manifold

To install:

7. Install the intake manifold with new gaskets. Slide the intake manifold forward.
8. Connect the MAP sensor vacuum hose
9. Connect the MAP sensor
10. Position the intake manifold into place.
11. Apply a thread locking compound to the intake manifold bolts. Tighten the bolts as follows:
   a. Step 1: Tighten the bolts in sequence to 44 inch lbs. (5 Nm).
   b. Step 2: Tighten the bolts in sequence to 89 inch lbs. (10 Nm).
12. Install or connect the following:
   - PCV hose to the valley cover and intake manifold
   - PCV hose to the throttle body and right rocker arm cover
   - TP sensor to the PCV hose
   - Knock sensor
   - Power brake booster vacuum hose
   - EVAP canister purge solenoid valve

- ETC electrical connector
- Fuel injector electrical connectors
- Throttle body outlet hose to the throttle body
- Coolant air bleed hose
- TP sensor
- EVAP canister purge hose
- Fuel supply hose
- Air intake assembly
- Engine shroud
- Negative battery cable
13. Start the engine and check for leaks.

Exhaust Manifold

REMOVAL & INSTALLATION

3.2L Engine

LEFT SIDE

1. Before servicing the vehicle, refer to the Precautions Section.
2. Drain the engine coolant.
3. Remove or disconnect the following:
   - Engine assembly
   - Exhaust manifold lower/upper heat shields
   - Secondary Air Injection (AIR) pipe from the exhaust manifold
   - Coolant pipe and engine lift bracket from the cylinder head
   - Oil level indicator tube
   - Exhaust manifold

To install:

4. Install or connect the following:
   - Exhaust manifold with a new gasket. Torque the nuts to 15 ft. lbs. (20 Nm).
   - Oil level indicator tube. Torque the bolt to 15 ft. lbs. (20 Nm).
   - Engine lifting bracket and coolant pipe. Torque the bolt to 15 ft. lbs. (20 Nm).
   - Air injection pipe to the exhaust manifold. Torque the bolts to 15 ft. lbs. (20 Nm).
   - Exhaust manifold upper/lower heat

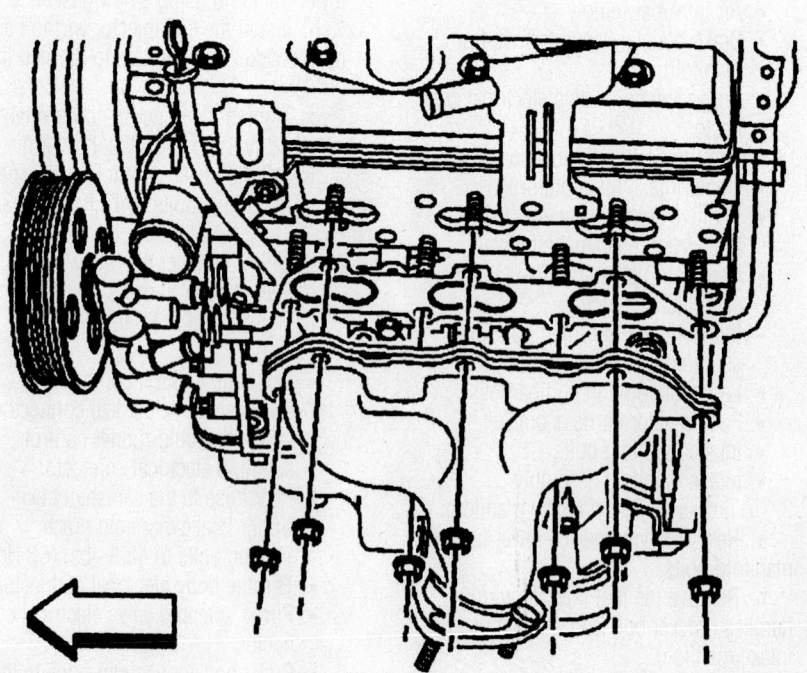

42372-CADI-G07

Left exhaust manifold

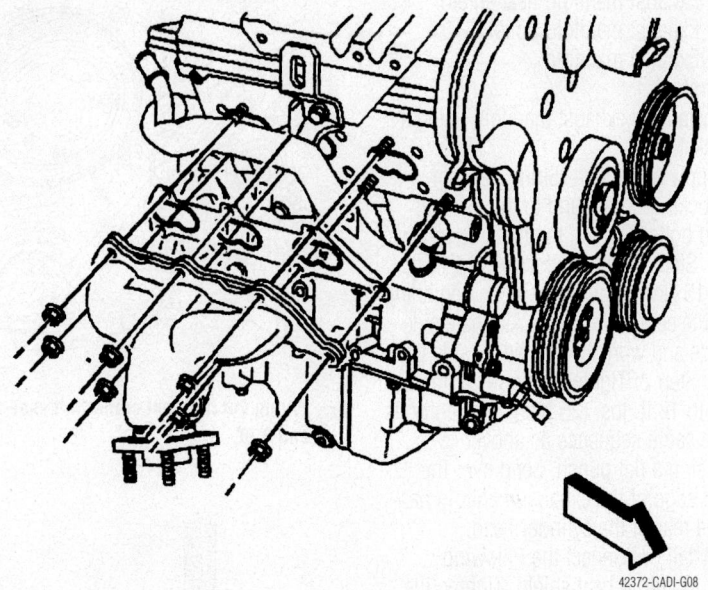

42372-CADI-G08

Right exhaust manifold

shields. Torque the bolts to 71 inch lbs. (8 Nm).
• Engine assembly
5. Refill the engine coolant.
6. Start the vehicle and inspect for any leaks.

RIGHT SIDE

1. Before servicing the vehicle, refer to the Precautions Section.
2. Drain the engine coolant.
3. Remove or disconnect the following:
• Transmission assembly
• Coolant intake pipe
4. Remove or disconnect the following:
• Exhaust manifold lower heat shield
• Exhaust manifold lower rear nuts
• Exhaust manifold upper rear heat shield bolt
• Catalytic converter nuts from the exhaust manifold
• Drive belt tensioner
• Exhaust manifold front two nuts
• Exhaust manifold upper heat shield bolts

➡ **It is not necessary to remove the upper heat shield.**

• Secondary Air Injection (AIR) injection pipe from the exhaust manifold
• Exhaust manifold
To install:
5. Install or connect the following:
• Exhaust manifold with a new gasket. Torque the 2 upper nuts to 15 ft. lbs. (20 Nm).
• AIR injection pipe to the exhaust manifold. Torque the bolts to 15 ft. lbs. (20 Nm).

• Upper heat shield bolts. Torque the bolts to 71 inch lbs. (8 Nm).
• Exhaust manifold upper heat shield bolts. Torque the bolts to 71 inch lbs. (8 Nm).
• Exhaust manifold lower front nuts. Torque the nuts to 15 ft. lbs. (20 Nm).
• Drive belt tensioner. Torque the bolts to 30 ft. lbs. (40 Nm).
• Catalytic converter nuts. Torque the nuts to 15 ft. lbs. (20 Nm).

• Exhaust manifold lower rear nuts. Torque the nuts to 15 ft. lbs. (20 Nm).
• Exhaust manifold lower heat shield. Torque the bolts to 71 inch lbs. (8 Nm).
• Coolant intake pipe. Torque the bolts to 15 ft. lbs. (20 Nm).
• Transmission assembly
6. Refill the engine coolant.
7. Start the vehicle and inspect for leaks.

2.8L and 3.6L Engines

1. Before servicing the vehicle, refer to the Precautions Section.
2. Remove or disconnect the following:
• Exhaust manifold heat shield
• Oil level indicator, when removing the left exhaust manifold
• Catalytic converter nuts
• Exhaust manifold bolts
• Exhaust manifold
To install:
3. Install a new gasket to the exhaust manifold.
4. Install the exhaust manifold to the catalytic converter and cylinder head.
5. Tighten the exhaust manifold bolts to 15 ft. lbs. (20 Nm).
6. Tighten the catalytic converter nuts to 37 ft. lbs. (50 Nm).
7. Install the oil level indicator, when install the left exhaust manifold.
8. Install the exhaust manifold heat

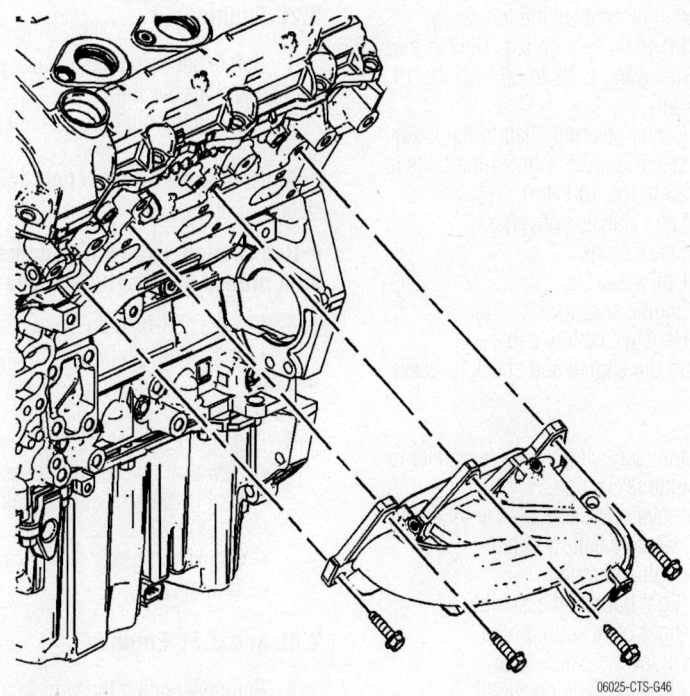

06025-CTS-G46
View of the exhaust manifold–2.8L and 3.6L

shield. Tighten the bolts to 89 inch lbs. (10 Nm).

9. Start the engine and check for leaks.

5.7L Engine

LEFT

1. Before servicing the vehicle, refer to the Precautions Section.
2. Remove or disconnect the following:
   - Negative battery cable
   - Engine shroud
   - Left bank coil assembly
   - Left bank spark plugs
   - Left catalytic converter
   - Center steering shaft from the lower steering shaft.
   - Exhaust manifold heat shield
   - Exhaust manifold bolts
   - Exhaust manifold

To install:

3. Install the exhaust manifold using a new gasket.
4. Apply a 0.2 inch (5mm) band of thread locking compound to the exhaust manifold bolts. Tighten the bolts as follows:
   a. Step 1: Tighten the bolts to 12 ft. lbs. (15 Nm). Tighten the bolts beginning with the center two bolts. Alternate side-to-side and work toward the outside bolts.
   b. Step 2: Tighten the bolts a final pass to 18 ft. lbs. (25 Nm). Tighten bolts in the same sequence as above.
5. Using a flat punch, bend over the exposed edge of the exhaust manifold gasket at the rear of the cylinder head.
6. Install or connect the following:
   - Manifold heat shield. Tighten the mounting bolts to 80 inch lbs. (9 Nm).
   - Center steering shaft to the lower steering shaft. Tighten the bolts to 23 ft. lbs. (30 Nm).
   - Left catalytic converter
   - Spark plugs
   - Coil assembly
   - Engine shroud
   - Negative battery cable
7. Start the engine and check for leaks.

RIGHT

1. Before servicing the vehicle, refer to the Precautions Section.
2. Remove or disconnect the following:
   - Negative battery cable
   - Engine shroud
   - Right bank coil assembly
   - Right bank spark plugs
   - Oil level dipstick tube
   - Right catalytic converter
   - Starter motor

- Exhaust manifold heat shield
- Exhaust manifold bolts
- Exhaust manifold

To install:

3. Install the exhaust manifold using a new gasket.
4. Apply a 0.2 inch (5mm) band of thread locking compound to the exhaust manifold bolts. Tighten the bolts as follows:
   a. Step 1: Tighten the bolts to 12 ft. lbs. (15 Nm). Tighten the bolts beginning with the center two bolts. Alternate side-to-side and work toward the outside bolts.
   b. Step 2: Tighten the bolts a final pass to 18 ft. lbs. (25 Nm). Tighten bolts in the same sequence as above.
5. Using a flat punch, bend over the exposed edge of the exhaust manifold gasket at the rear of the cylinder head.
6. Install or connect the following:
   - Manifold heat shield. Tighten the mounting bolts to 80 inch lbs. (9 Nm).
   - Starter motor
   - Right catalytic converter
   - Spark plugs
   - Coil assembly
   - Oil level dipstick tube
   - Engine shroud
   - Negative battery cable
7. Start the engine and check for leaks.

Front Crankshaft Seal

REMOVAL AND INSTALLATION

3.2L Engine

1. Before servicing the vehicle, refer to the Precautions Section.
2. Remove the timing belt and crankshaft gear.
3. Drill a small shallow hole into the steel ring of the seal.

➡**Use caution so as not to damage the area around and behind the seal.**

4. Insert a self-tapping screw.
5. Using pliers, pull the front crankshaft seal out.

To install:

6. Install or connect the following:
   - New seal coated with grease using Seal Installer Tool (such as J35268-A)
   - Crankshaft gear
   - Timing belt

2.8L and 3.6L Engines

1. Before servicing the vehicle, refer to the Precautions Section.

Prying out the front crankshaft seal—2.8L and 3.6L

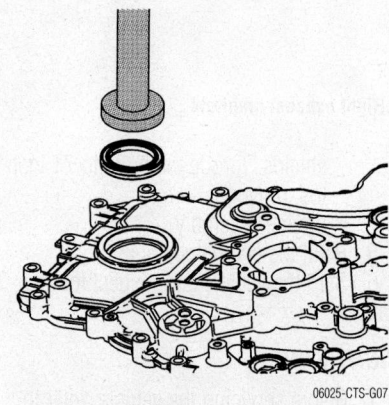

Installing front crankshaft seal using Special Tool J-29184—2.8L and 3.6L

2. Remove the accessory drive belts.
3. Remove the crankshaft pulley.
4. Using a flat-bladed tool to remove the crankshaft seal.

To install:

5. Using Special Tool J-29184 or equivalent, install the crankshaft front seal.
6. Install the crankshaft pulley. Tighten the bolt to 74 ft. lbs. (100 Nm) plus 150 degrees.
7. Install the accessory drive belts.
8. Start the engine and check for leaks.

5.7L Engine

1. Before servicing the vehicle, refer to the Precautions Section.
2. Remove or disconnect the following:
   - Negative battery cable
   - Radiator assembly
   - A/C condenser
   - A/C drive belt
   - Starter motor
   - Crankshaft pulley
3. Gently pry the crankshaft seal from the front cover.

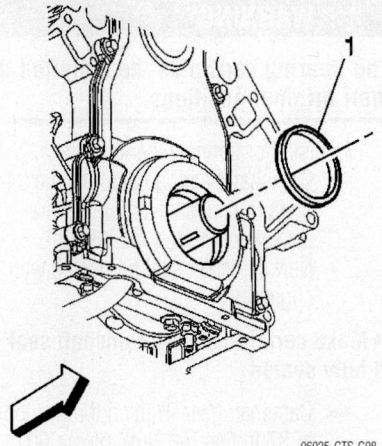

Removing the front crankshaft seal–5.7L

06025-CTS-G08

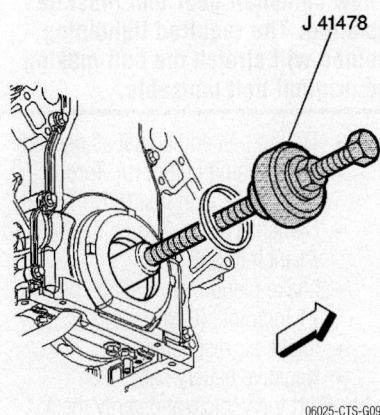

J 41478

Install the front crankshaft seal using Special Tool J-41478–5.7L

06025-CTS-G09

To install:

4. Lubricate the outer edge of the oil seal and front cover oil seal bore with clean engine oil.

5. Install the front oil seal onto Special Tool J-41478 Installer guide.

6. Install the J-41478 threaded rod into the end of the crankshaft.

7. Use a wrench to hold the hex on the installer tool.

8. Use a second wrench to rotate the installer nut clockwise until the seal bottoms out in the front cover oil seal bore.

9. Install or connect the following:
- Crankshaft pulley. Tighten the bolt to 240 ft. lbs. (330 Nm).
- Starter motor
- A/C drive belt
- A/C condenser
- Radiator assembly
- Negative battery cable

10. Start the engine and check for leaks.

Camshaft

REMOVAL & INSTALLATION

3.2L Engine

1. Before servicing the vehicle, refer to the Precautions Section.

2. Remove or disconnect the following:
- Negative battery cable
- Intake plenum
- Intake air resonator
- Camshaft cover
- Front timing belt cover
- Timing belt

3. Rotate the crankshaft 60 degrees counterclockwise Before Top Dead Center (BTDC) to prevent valve/piston contact.

4. Use a Camshaft Locking tool to hold the camshaft gears in place and loosen the camshaft gear bolt.

5. Remove the locking tool from the gears.

6. Remove the camshaft gear(s).

➡**Be certain that the camshaft is not under load from the lifters.**

7. Gradually loosen the camshaft bearing cap bolts sequentially, starting in the center and working outward in a spiral. Note the identification marks on the caps.

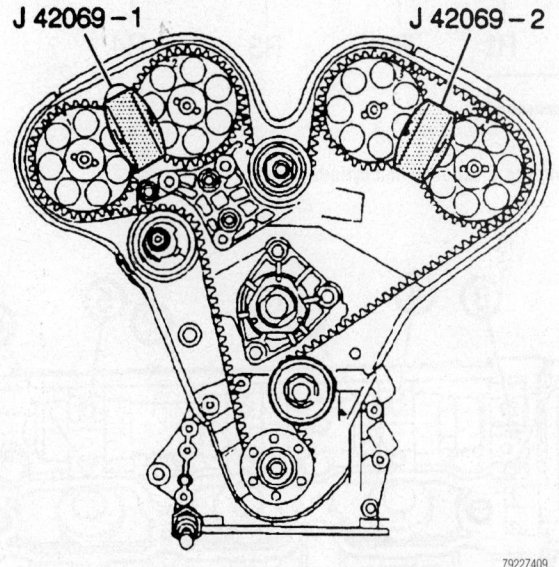

J 42069 – 1   J 42069 – 2

79222409

Camshaft gear holding tools must be installed before attempting to remove the gears from the camshafts

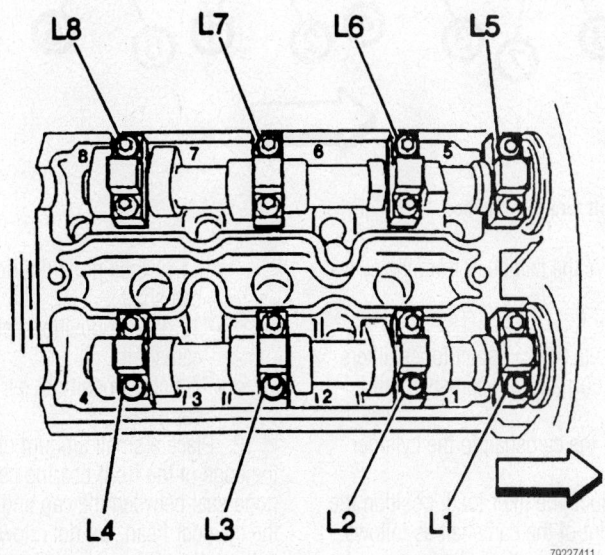

79222411

Camshaft bearing cap locations-right cylinder head

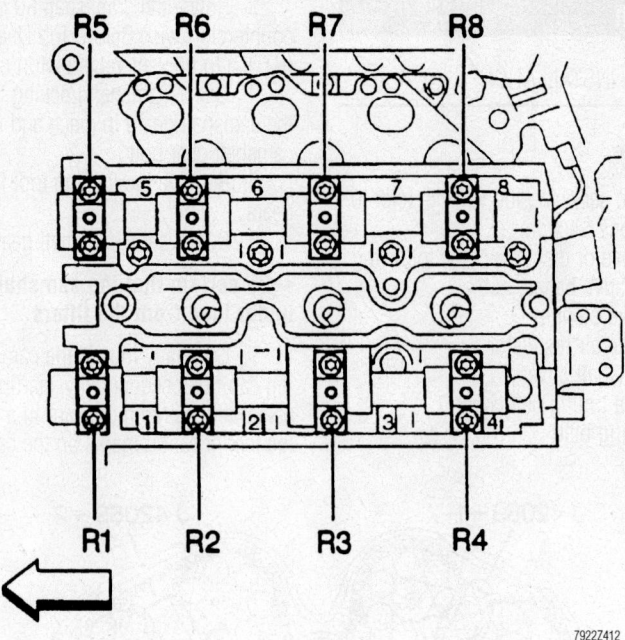

Camshaft bearing cap locations—left cylinder head

7922Z412

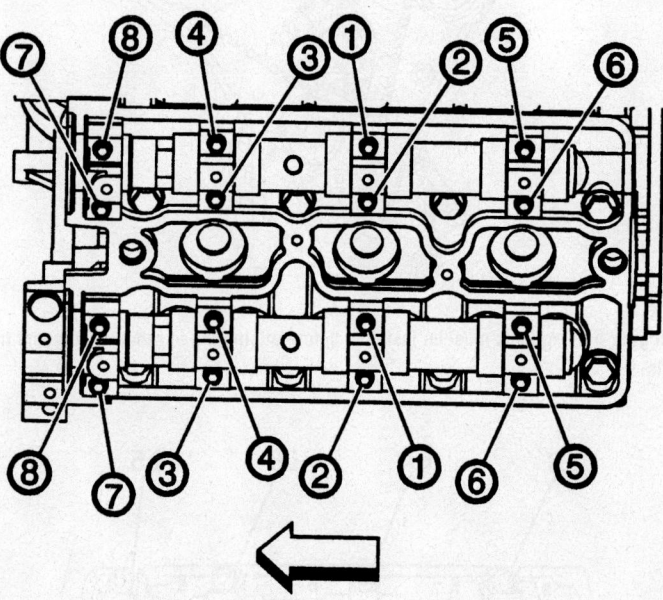

42372-CADI-G09

Camshaft bolt torque sequence

8. Remove the camshaft from the cylinder head.

To install:

9. Lubricate the camshaft lobes, lifters and bearing journals with camshaft lubricant.

10. Install the camshaft to the cylinder head.

11. To reduce the lifter load, position the pin on the front of the camshaft as follows:
- 1 o'clock position: right exhaust camshaft
- 11 o'clock position: right intake camshaft
- 12 o'clock position: left exhaust camshaft
- 7 o'clock position: left intake camshaft

12. Place a small amount of Loctite® on the edge of the front bearing cap to ensure a good seal between the cap and surface of the cylinder head. Do not allow the sealer to get into the oil journal of the cap.

❊❊ WARNING

The bearing caps must be installed in their original positions.

13. Install or connect the following:
- Camshaft bearing caps. Torque the bolts in sequence and in several passes to 71 inch lbs. (8 Nm).
- New camshaft seal lubricated with engine oil

➡**Make certain that the camshaft seal is fully seated.**

- Camshaft gear. Tighten the new bolt to 37 ft. lbs. (50 Nm), plus a 60 degree turn, plus a 15 degree turn.

❊❊ WARNING

A new camshaft gear bolt must be installed. The required tightening method will stretch the bolt making the original bolt unusable.

- Timing belt and adjust as needed
- Front timing belt cover. Torque the bolts to 71 inch lbs. (8 Nm).
- Camshaft cover. Torque the bolts to 71 inch lbs. (8 Nm).
- Intake plenum. Torque the bolts to 71 inch lbs. (8 Nm).
- Intake air resonator
- Negative battery cable

14. Start the vehicle and verify the engine is running properly.

2.8L and 3.6L Engines

1. Before servicing the vehicle, refer to the Precautions Section.

2. Remove or disconnect the following:
- Negative battery cable
- Intake manifold assembly
- Ignition coil electrical connectors
- Wiring harnesses and wiring conduit retainers from the camshaft cover
- Ignition coils
- Camshaft cover
- Camshaft position (CMP) sensors
- CMP actuator solenoid
- Crankshaft pulley

3. Rotate the crankshaft until the camshafts are in the neutral (low) position. The camshaft flats will be parallel with the camshaft cover rail.

4. Loosen the camshaft position actuator bolt.

5. Remove the front cover.

➡**Do not remove the bolt at this time.**

6. Install the Timing Chain Retention Tool EN-46108 to retain the timing chain.

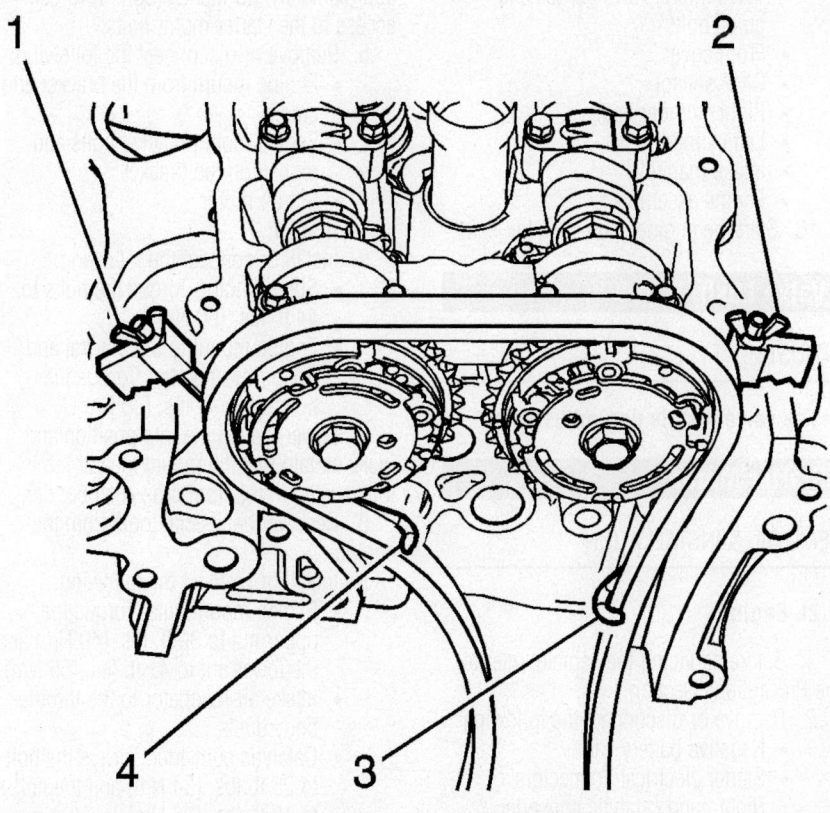

Installing camshaft retention tool EN-46108 and ensure the tips are fully engaged on the timing chain (3 and 4)–2.8L and 3.6L

06025-CTS-G10

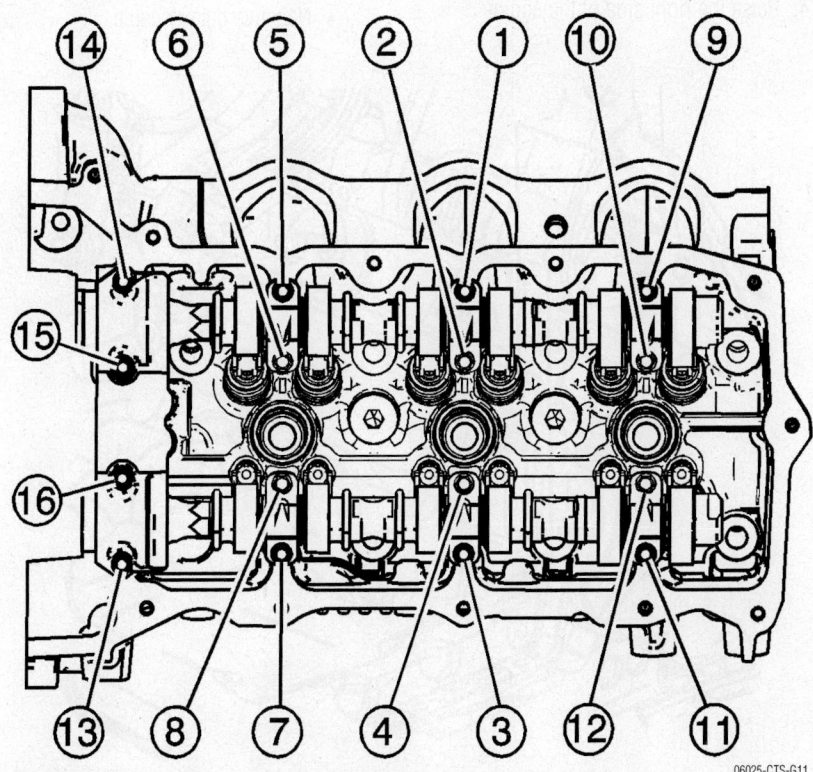

Camshaft bearing cap torque sequence–2.8L and 3.6L

06025-CTS-G11

7. Matchmark the timing chain on the camshaft position actuators.

8. Remove or disconnect the following:
- CMP actuator bolt
- Camshaft bearing caps
- Camshafts

To install:

9. Lubricate the camshaft lobes, lifters and bearing journals with camshaft lubricant.

10. Position the camshaft lobes in a neutral position with the flats on the back of the camshafts up and parallel with the left cylinder head camshaft cover rail.

✳✳ WARNING

The bearing caps must be installed in their original positions.

11. Install the camshaft bearing thrust cap in the first journal of the cylinder head.

12. Install the remaining bearing caps with their orientation mark toward the center of the cylinder head.

13. Hand start all the camshaft bearing cap bolts.

14. Tighten the bearing cap bolts in sequence as follows:

a. Step 1: Tighten all bolts to 89 inch lbs. (10 Nm).

b. Step 2: Loosen the center intake camshaft bearing cap bolts (1,2) and center exhaust camshaft bearing cap (3.4).

c. Step 3: Retighten center camshaft bearing cap bolts to 89 inch lbs. (10 Nm).

15. Remove the Timing chain retention tool EN-46108.

16. Install or connect the following:
- CMP actuators. Tighten the bolts to 43 ft. lbs. (58 Nm).
- Front cover
- CMP actuator solenoid
- CMP sensors
- Crankshaft pulley
- Camshaft cover
- Ignition coils
- Wiring harnesses and conduit retainers to the camshaft cover
- Ignition coil electrical connectors
- Intake manifold
- Negative battery cable

17. Start the engine and verify proper operation.

5.7L Engine

1. Before servicing the vehicle, refer to the precautions in the beginning of this engine.

2. Remove or disconnect the following:

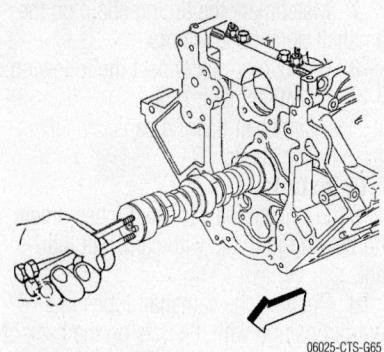

06025-CTS-G65

Removing the camshaft using the bolts as a handle–5.7L

- Engine assembly
- Intake manifold
- Left cylinder head
- Right cylinder head
- Camshaft position (CMP) sensor
- Front cover
- Valve lifter guide bolts
- Valve lifters and guides

3. Rotate the crankshaft in order to align the timing marks.

4. Remove the camshaft sprocket bolts.

5. Remove the timing chain from the camshaft sprocket, and allow the chain to rest on the crankshaft sprocket.

6. Remove the camshaft retainer bolts and retainer.

7. Install 3 M8—4 inch (100 mm) in the camshaft front bolt holes.

8. Using the bolts as a handle, rotate and pull the camshaft out of the engine block.

To install:

9. Lubricate the camshaft journals and the bearings with clean engine oil.

10. Install 3 M8—4 inch (100 mm) in the camshaft front bolt holes.

11. Using the bolts as a handle, carefully install the camshaft into the engine block.

12. Remove the 3 bolts from the front of the camshaft.

13. Install the camshaft retainer. Tighten the bolts as follows:

   a. Hex head bolts to 18 ft. lbs (25 Nm)

   b. TORX® head bolts to 11 ft. lbs. (15 Nm)

14. CMP sensor. Tighten the bolt to 18 ft. lbs (25 Nm).

➡**Lubricate the sensor O-ring seal with clean engine oil.**

15. Install or connect the following:
- Timing chain to the camshaft sprocket
- Camshaft sprocket and bolts to the camshaft. Tighten the bolts to 26 ft. lbs. (35 Nm).

- Valve lifters, valve guides and guide bolts
- Front cover
- CMP sensor
- Right cylinder head
- Left cylinder head
- Intake manifold
- Engine assembly

16. Start the engine and check for leaks.

Valve Lash

ADJUSTMENT

➡**The valve lash is non-adjustable.**

Starter Motor

REMOVAL & INSTALLATION

3.2L Engine

1. Before servicing the vehicle, refer to the Precautions Section.

2. Remove or disconnect the following:
- Negative battery cable
- Starter electrical connectors
- Right hand catalytic converter
- Right side engine mount nuts
- Intake air resonator to the throttle body ducts

3. Install an Engine Support Fixture to the right side of the engine.

4. Raise the right side of the engine approximately 1.5 inches (38mm) to gain access to the starter motor bolts.

5. Remove or disconnect the following:
- Engine mount from the bracket and cradle
- Engine mount bracket bolts and reposition the bracket
- Starter motor

To install:

6. Install or connect the following:
- Starter motor. Torque the bolts to 44 ft. lbs. (60 Nm).
- Engine mount to the bracket and front crossmember. Torque the bolts to 30 ft. lbs. (40 Nm).

7. Lower the engine into position and make certain that the mount locator tab is fully seated in the front crossmember slot.

8. Remove the special tools from the engine.

9. Install or connect the following:
- Engine mount nuts. Torque the upper nut to 30 ft. lbs. (40 Nm) and the lower nut to 41 ft. lbs. (55 Nm).
- Intake air resonator to the throttle body ducts
- Catalytic converter. Torque the bolt to 25 ft. lbs. (34 Nm) and the nuts to 18 ft. lbs. (25 Nm).
- Starter electrical connectors. Torque the battery cable nut to 115 inch lbs. (13 Nm) and the starter solenoid nut to 35 inch lbs. (4 Nm).
- Negative battery cable

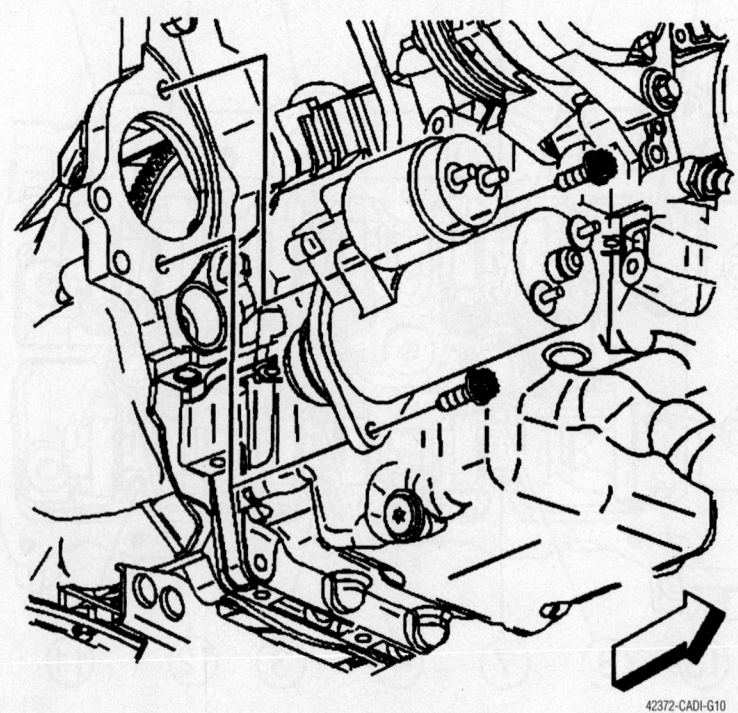

42372-CADI-G10

Remove the starter motor and bolts

2.8L and 3.6L Engine

1. Before servicing the vehicle, refer to the Precautions Section.
2. Remove or disconnect the following:
   - Negative battery cable
   - Starter electrical connections
   - Starter motor mounting bolts
   - Starter motor

To install:

3. Install or connect the following:
   - Starter motor. Tighten the bolts to 37 ft. lbs. (50 Nm).
   - Starter motor electrical connections
   - Negative battery cable

5.7L Engine

1. Before servicing the vehicle, refer to the Precautions Section.
2. Remove or disconnect the following:
   - Negative battery cable
   - Right side catalytic converter
   - Positive battery cable
   - Starter electrical connections
   - Starter motor mounting bolts
   - Starter motor

To install:

3. Install or connect the following:
   - Starter motor. Tighten the mounting bolts to 37 ft. lbs. (50 Nm).
   - Starter motor electrical connections
   - Positive battery cable
   - Right side catalytic converter
   - Negative battery cable

Oil Pan

REMOVAL & INSTALLATION

3.2L Engine

LOWER

1. Before servicing the vehicle, refer to the Precautions Section.
2. Drain the engine oil.
3. Remove or disconnect the following:
   - Negative battery cable
   - Splash shield
   - Oil level sensor wiring C-clip from the oil pan housing
   - Oil level sensor electrical connector
   - Oil pan from the oil pan housing

To install:

4. Clean the oil pan and housing sealing surfaces with a non-abrasive cleaner.
5. Install or connect the following:
   - Oil level sensor wire connector to the oil pan housing C-clip
   - Oil pan with a new gasket. Torque the bolts to 71 inch lbs. (8 Nm).
   - Oil level sensor electrical connector

   - Splash shield. Torque the bolts until they are fully seated.
   - Negative battery cable
6. Refill the engine oil.
7. Start the vehicle and check for leaks.

UPPER

1. Before servicing the vehicle, refer to the Precautions Section.
2. Remove or disconnect the following:
   - Negative battery cable
   - Lower oil pan
   - Engine mount lower nuts
   - A/C compressor hose strap from the oil pan
   - Transmission bolts from the oil pan
   - All but the 4 corner bolts from the oil pan
   - Propeller shaft
3. Support the propeller shaft out of the way.
4. Remove or disconnect the following:
   - Catalytic converter
   - Idler arm bolts and lower the relay rod out of the way
5. Install an engine support fixture and raise the engine slightly to allow removal of the oil pan.
6. Remove or disconnect the following:
   - Remaining oil pan bolts
   - Oil intake pipe
   - Oil pan from the vehicle

To install:

7. Apply a bead of silicone sealer in the bottom of the upper oil pan groove.
8. Install a new rubber seal into the groove in the pan.
9. Apply a 3mm (0.12 in) bead of silicone sealant on the OUTSIDE edge of the seal, at the front of the upper oil pan and apply an identical bead on the INSIDE edge of the seal at the rear of the upper oil pan. The beads will overlap at the middle of the upper oil pan.
10. Install or connect the following:
    - Upper oil pan
    - Oil intake pipe. Torque the bolt to 71 inch lbs. (8 Nm).
    - Four oil pan corner bolts finger tight
11. Lower the engine into place.
12. Install or connect the following:
    - Remaining oil pan bolts. Torque the bolts to 11 ft. lbs. (15 Nm).
    - Transmission-to-oil pan bolts. Torque the bolts to 30 ft. lbs. (40 Nm).
    - Idler arm bolts. Torque the bolts to 44 ft. lbs. (60 Nm).
    - Catalytic converter. Torque the nuts to 18 ft. lbs. (25 Nm).
    - Propeller shaft. Torque the bolts to 70 ft. lbs. (95 Nm).

   - A/C compressor hose strap. Torque the bolt to 71 inch lbs. (8 Nm).
   - Engine mount lower nuts. Torque the nuts to 41 ft. lbs. (55 Nm).
   - Lower oil pan
13. Refill the engine oil.

2.8L and 3.6L Engines

1. Before servicing the vehicle, refer to the Precautions Section.
2. Drain the engine oil.
3. Remove or disconnect the following:
   - Front cover
   - Power steering hose retainer from the A/C compressor bracket
   - Intermediate steering shaft
   - Engine mount lower nuts
   - A/C compress bracket bolts and reposition aside

➡ **Do not disconnect the A/C lines.**

   - Transmission oil cooler hose retainer
4. Install an engine support fixture. Tighten the support fixture wing nuts in order to provide clearance for the oil pan.
5. Remove the oil pan.

To install:

6. Apply a 0.12 in (3mm) bead of sealant on the block pan rail and the crankshaft rear oil seal housing.
7. Position the oil pan on the loosely install the oil pan bolts.
8. Tighten the oil pan bolts in sequence as follows:
   a. Tighten the 0.30 in (8mm) bolts (1-11) to 17 ft. lbs. (23 Nm).
   b. Tighten the 0.23 in (6mm) bolts (12-13) to 89 inch lbs. (10 Nm).
9. Loosen the engine support fixture wing nuts in order to lower the engine and engage the engine mounts to the frame.
10. Remove the engine support fixture.

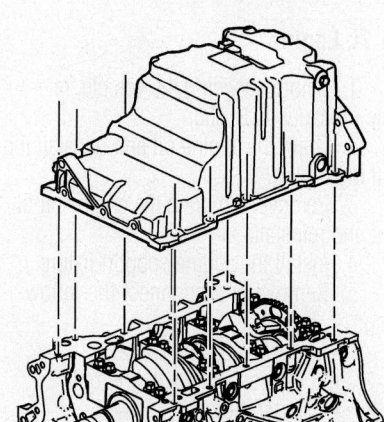

Oil pan removal–2.8L and 3.6L

06025-CTS-G12

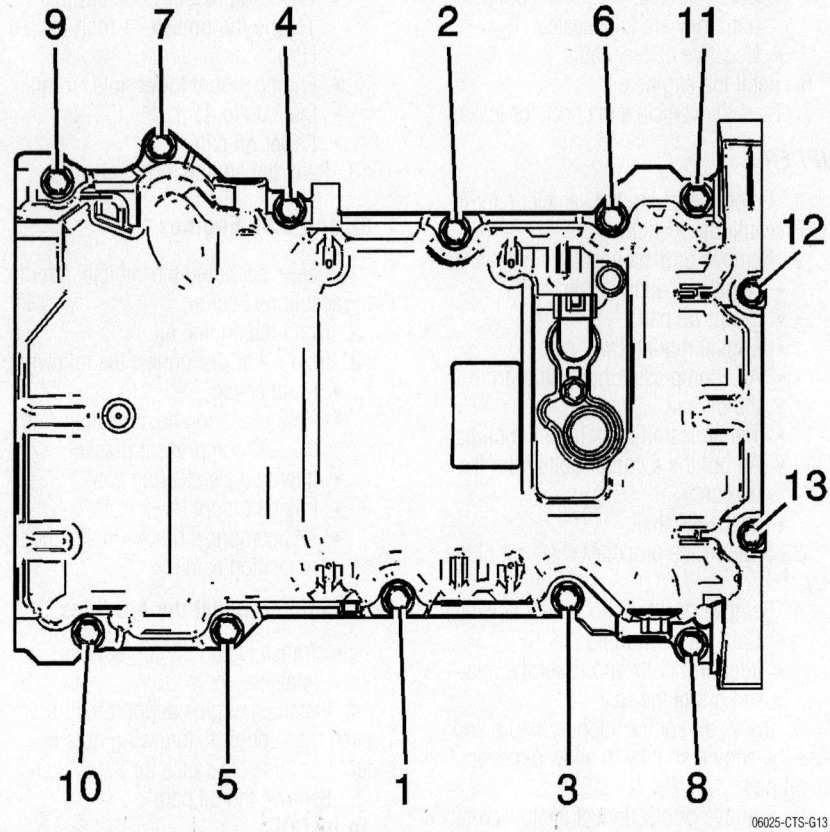

Oil pan torque sequence—2.8L and 3.6L

06025-CTS-G13

11. Install or connect the following:
- A/C compressor bracket. Tighten the bolts to 37 ft. lbs. (50 Nm).
- Transmission oil cooler hose retainer
- Engine mount lower nuts. Tighten nuts to 59 ft. lbs. (80 Nm).
- Intermediate steering shaft
- Power steering hose retainer
- Front cover

12. Fill the engine with oil to the correct level.

13. Start the engine and check for leaks.

5.7L Engine

1. Before servicing the vehicle, refer to the Precautions Section.

2. Drain the engine oil and reinstall the oil drain plug.

3. Remove and drain the engine oil filter and reinstall.

4. Install the engine support fixture.

5. Remove or disconnect the following:
- Negative battery cable
- Left closeout cover and bolt
- Starter motor
- Right transmission closeout cover and bolt

- Bottom two transmission housing-to-oil pan bolts
- Engine oil temperature sensor
- Front frame assembly
- Power steering and air conditioning line retainers from the front of the oil pan
- Oil level sensor
- Oil pan mounting bolts
- Oil pan

To install:

6. Apply a 0.2 in (5mm) bead of sealant directly to the tabs of the front and rear cover gasket that protrudes into the into oil pan surface.

7. Install the oil pan gasket to the pan and install the oil pan bolts to the pan and through the gasket.

➡**Be sure to align the oil gallery passages in the oil pan and engine block properly with the oil pan gasket.**

8. Install the oil pan to the engine block and tighten the bolts finger tight.

9. Place a straight edge across the rear of the engine block and the rear of the oil pan at the transmission housing mounting surfaces. Align the oil pan until the rear of the engine block and rear of the oil pan are flush.

⁂ **WARNING**

The rear of the oil pan must never protrude beyond the edge of the engine block and transmission housing mounting surfaces.

10. Tighten the oil pan mounting bolts as follows:

a. Tighten the oil pan-to-block and oil pan-to-front cover bolts to 18 ft. lbs. (25 Nm).

b. Tighten the oil pan-to-rear cover bolts to 106 inch lbs. (12 Nm).

11. Install or connect the following:
- Oil level sensor
- Power steering and air conditioning line retainers to the front of the oil pan
- Front frame assembly
- Engine oil temperature sensor
- Transmission housing-to-oil pan bolts. Tighten the bolts to 37 ft. lbs. (50 Nm).
- Right transmission closeout cover and bolt. Tighten the bolt to 106 inch lbs. (12 Nm).
- Starter motor
- Left closeout cover and bolt. Tighten the bolt to 106 inch lbs. (12 Nm).
- Negative battery cable

12. Remove the engine support fixture.

13. Fill the engine with oil to the correct level using a NEW engine oil filter.

14. Start the engine and check for leaks.

Oil Pump

REMOVAL & INSTALLATION

3.2L Engine

1. Before servicing the vehicle, refer to the Precautions Section.

2. Drain the engine oil.

3. Drain the engine coolant.

4. Discharge and recover the A/C system.

5. Remove or disconnect the following:
- Negative battery cable
- Intake air resonator
- Front timing belt cover
- Intake plenum
- Timing belt
- Rear timing belt cover
- A/C compressor
- A/C compressor and power steering pump bracket and position it out of the way of the oil pump housing

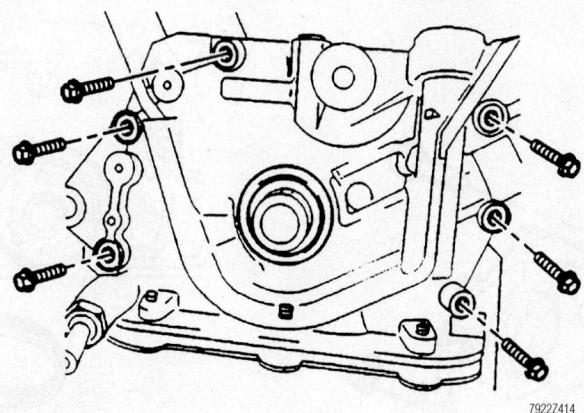

Oil pump mounting bolt locations

- Lower alternator mounting bolt and loosen the upper mounting bolt then, position the alternator aside
- Lower and upper oil pans
- Crankshaft drive gear
- Oil pump
- Front main oil seal and collar

To install:

6. Coat the pump side of the oil pump gasket with a thin layer of sealant. Do not cover or restrict the gasket openings with sealant.

7. Install or connect the following:
- Oil pump collar
- New front main oil seal
- Oil pump with a new gasket by aligning the guide pins. Torque the bolts to 9 ft. lbs. (12 Nm).
- Crankshaft gear. Torque the bolt to 184 ft. lbs. (250 Nm) plus an additional 60 degree turn.
- Upper oil pan. Torque the bolts to 11 ft. lbs. (15 Nm).
- Lower oil pan. Torque the bolts to 71 inch lbs. (8 Nm).
- Alternator. Torque the upper and lower bolts to 30 ft. lbs. (40 Nm).

8. Re-tighten the oil pump bolts to 15 ft. lbs. (20 Nm).

9. Install or connect the following:
- A/C compressor and power steering pump bracket. Torque the fasteners to 30 ft. lbs. (40 Nm).
- A/C compressor. Torque the bolts to 30 ft. lbs. (40 Nm).
- Rear timing belt cover. Torque the bolts to 71 inch lbs. (8 Nm).
- Timing belt
- Front timing belt cover. Torque the bolts to 71 inch lbs. (8 Nm).
- Intake plenum. Torque the bolts to 71 inch lbs. (8 Nm).
- Intake air resonator
- Negative battery cable

10. Fill the engine oil.
11. Fill the coolant.
12. Recharge the A/C system.
13. Start the vehicle and checks for leaks.

2.8L and 3.6L Engines

1. Before servicing the vehicle, refer to the Precautions Section.
2. Remove or disconnect the following:
- Negative battery cable
- Timing chain
- Crankshaft sprocket
- Oil pump mounting bolts
- Oil pump

To install:

3. Install or connect the following:
- Oil pump. Tighten the mounting bolts to 17 ft. lbs. (23 Nm).
- Crankshaft sprocket
- Timing chain
- Negative battery cable
4. Start the engine and check for leaks.

5.7L Engine

1. Before servicing the vehicle, refer to the Precautions Section.
2. Remove or disconnect the following:
- Negative battery cable
- Front cover
- Oil pan
- Oil pump screen
- O-ring seal from the pump screen
- Crankshaft oil deflector nuts
- Crankshaft oil deflector
- Oil pump bolts
- Oil pump

To install:

3. Align the splined surfaces of the crankshaft sprocket and the oil pump drive gear and install the oil pump. Tighten the oil pump bolts to 18 ft. lbs. (25 Nm).

4. Install or connect the following:
- Crankshaft oil deflector. Tighten the nuts to 18 ft. lbs. (25 Nm).
- New O-ring seal onto the oil pump screen

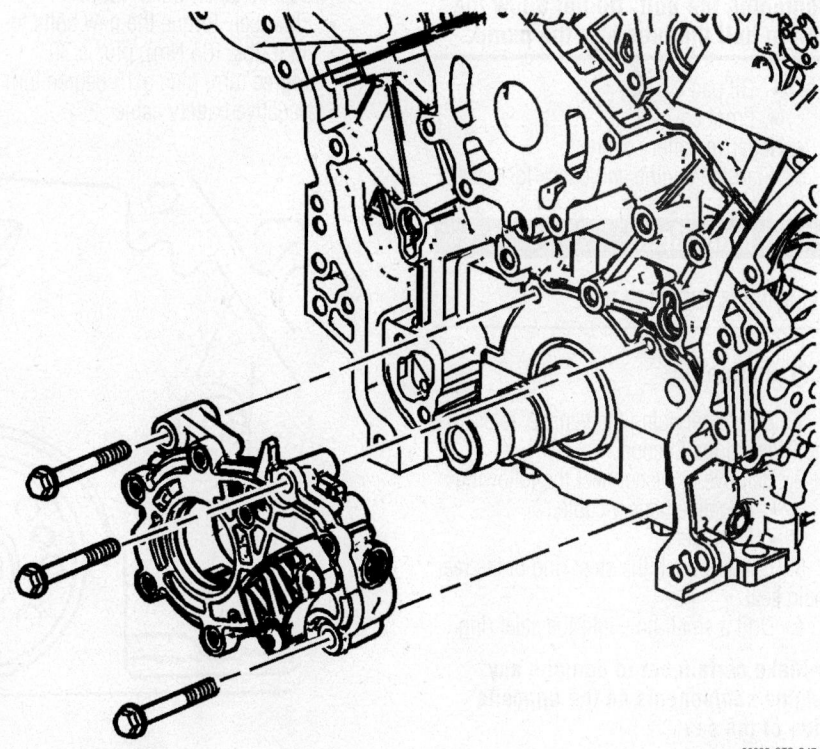

Oil pump mounting bolt–2.8L and 3.6L

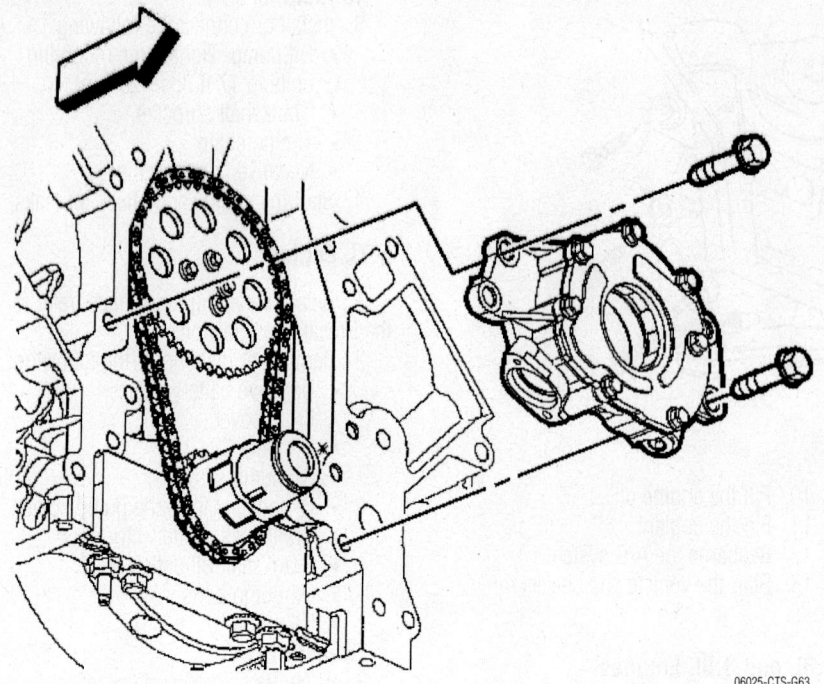

Exploded view of the oil pump—5.7L

- Oil pump screen. Tighten the bolt to 106 inch lbs. (12 Nm). Tighten the nut to 18 ft. lbs. (25 Nm).

✳✳ WARNING

Push the oil pump screen tube completely into the oil pump prior to tightening the bolt. Do not allow the bolt to pull the tube into the pump.

- Oil pan
- Front cover
- Negative battery cable
5. Start the engine and check for leaks.

Rear Main Seal

REMOVAL & INSTALLATION

3.2L Engine

1. Before servicing the vehicle, refer to the Precautions Section.
2. Remove or disconnect the following:
   - Negative battery cable.
   - Flywheel
3. Center punch the steel ring of the rear main seal.
4. Drill a small hole into the steel ring.

➡**Make certain not to damage any engine components on the opposite side of the seal.**

5. Install a small self tapping screw.
6. Remove the rear main oil seal.

To install:
7. Clean all sealing surfaces with a non-abrasive cleaner.
8. Install or connect the following:
   - New seal lubricated with chassis grease
   - Rear main oil seal using an Oil Seal Installer tool J 42067
   - Flywheel. Torque the new bolts to 48 ft. lbs. (65 Nm), plus a 30 degree turn, plus a 15 degree turn.
   - Negative battery cable

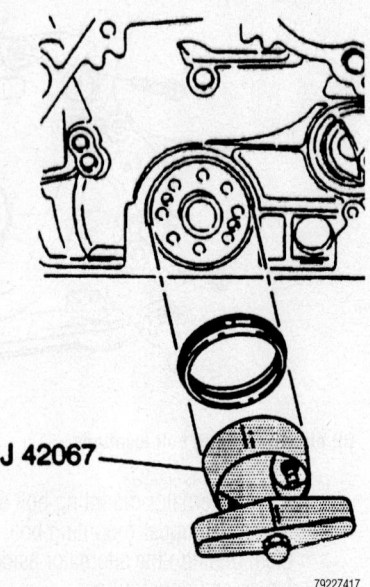

J 42067

Press the rear main seal into position using a threaded seal installer like J 42067

2.8L and 3.6L Engines

1. Before servicing the vehicle, refer to the Precautions Section.
2. Remove or disconnect the following:
   - Negative battery cable
   - Flywheel
   - Oil pan
   - Rear oil seal housing bolts
3. Using the pry point located at the edge of the crankshaft rear main seal housing, shear the RTV sealant.
4. Remove the rear main oil seal.

Thread a self-tapping screw into the metal part of the seal and remove the seal

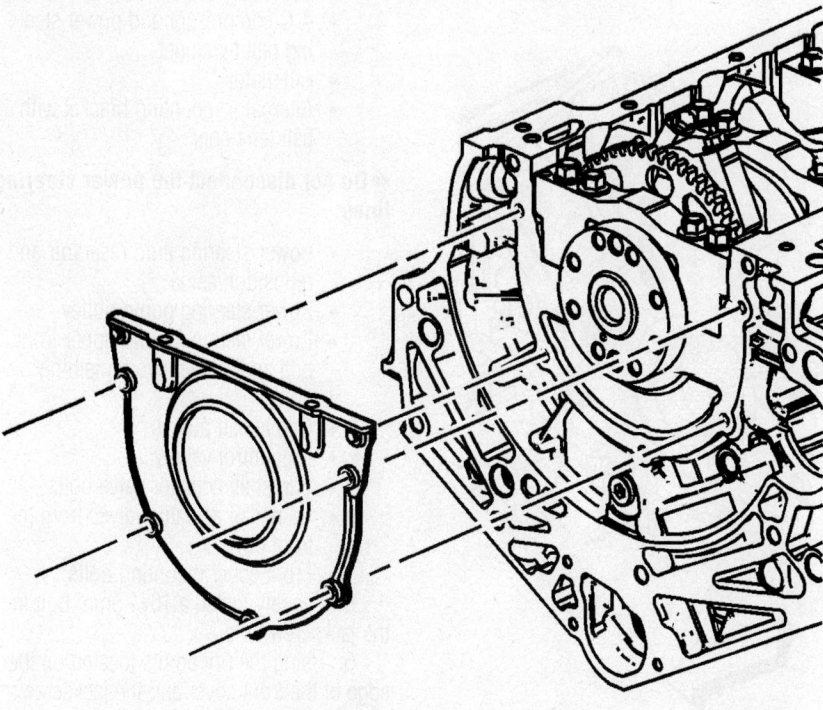

06025-CTS-G50

Exploded view of the rear oil seal—2.8L and 3.6L

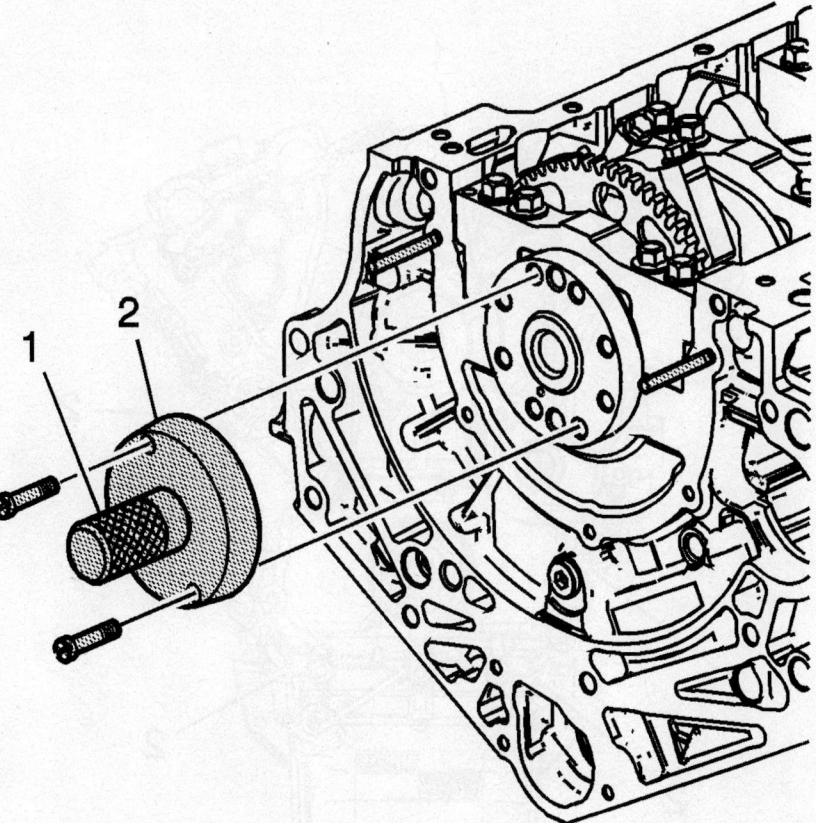

06025-CTS-G51

Using the Crankshaft Rear Seal Installation Tool—2.8L and 3.6L

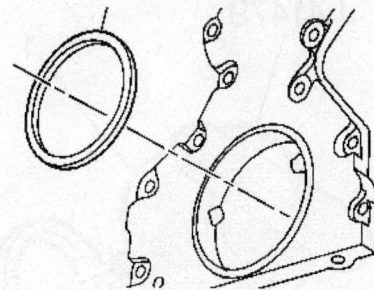

06025-CTS-G48

View of the rear oil seal—5.7L

To install:

5. Install guide pins into the two crankshaft rear oil seal housing corner bolt hoses of the engine block.

6. Install the crankshaft rear seal installation tool onto the rear of the crankshaft flange.

7. Apply a 0.2 inch (3mm) bead of silicone sealant to the rear oil seal housing.

8. Install the rear seal housing to the engine block.

9. Remove the guide pins and install the housing bolts. Tighten the bolts to 89 inch lbs. (10 Nm).

10. Remove the rear seal installation tool from the crankshaft flange.

11. Install or connect the following:
   • Oil pan
   • Flywheel
   • Negative battery cable

12. Start the engine and check for leaks.

5.7L Engine

1. Before servicing the vehicle, refer to the Precautions Section.

2. Remove the flywheel.

3. Gently pry the rear oil seal from the rear cover.

To install:

4. Lubricate the outside diameter of the new rear oil seal and rear cover oil seal bore with clean engine oil.

✷✷ WARNING

Do not allow oil or other lubricants to contact the crankshaft surface or surface or the seal.

5. Install the Rear Oil Seal Installer cone onto the rear of the crankshaft.

6. Install the rear oil seal onto the tapered cone of the installer tool and push the seal to the rear cover bore.

7. Install the threaded rod Installation

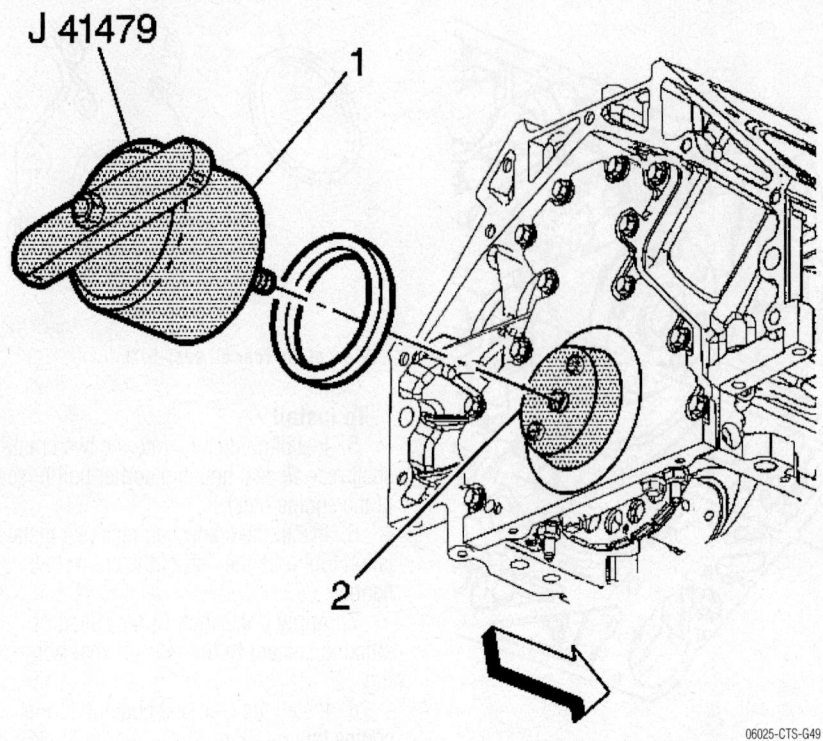

J 41479

06025-CTS-G49

Install the oil seal installer cone (2) onto the crankshaft. Install the threaded rod of the Oil Seal Installer (1) into the cone.

Tool into the tapered cone until the tool contacts the seal.

8. Rotate the handle of the tool clockwise until the seal enters the rear cover and bottoms into the cover bore.

9. Remove the Rear Oil Seal Installer Tool.

10. Install the flywheel.

11. Start the engine and check for leaks.

Timing Chain, Sprockets, Front Cover and Seal

REMOVAL & INSTALLATION

2.8L and 3.6L Engines

1. Before servicing the vehicle, refer to the Precautions Section.

2. Drain the engine oil.

3. Drain the cooling system.

4. Remove or disconnect the following:

- Negative battery cable
- Engine shroud
- Intake manifold
- Camshaft covers
- Purge vent hose from the water outlet
- Water outlet housing

- Accessory drive belts
- A/C compressor and power steering belt tensioner
- Alternator
- Alternator mounting bracket with belt tensioner

➡**Do not disconnect the power steering lines.**

- Power steering fluid reservoir and reposition aside
- Power steering pump pulley
- Power steering pump upper front bolt and loosen two remaining bolts
- Crankshaft pulley
- Oil control valves
- Camshaft actuator valve bolts
- Camshaft actuator valves from the front cover
- Front cover mounting bolts

5. Loosely install a 10x1.5mm bolt in the jackscrew hole.

6. Using the pry points located on the edge of the front cover and the jackscrew, shear the RTV sealant.

7. Remove the front cover.

8. Remove or disconnect the following:

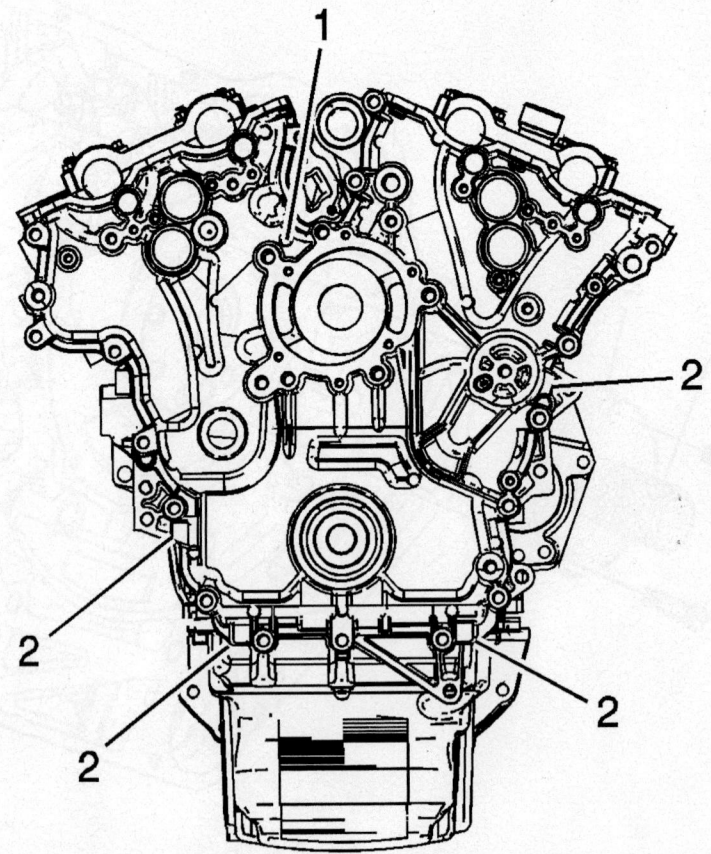

06025-CTS-G54

Front cover pry locations—2.8L and 3.6L

- Right bank secondary camshaft drive chain tensioner
- Right bank secondary camshaft drive chain shoe
- Right bank secondary camshaft drive chain guide
- Right bank secondary camshaft drive chain
- Primary camshaft drive chain tensioner
- Primary camshaft drive chain upper guide
- Primary camshaft drive chain timing chain

To install:

9. Install the primary timing chain. Ensure all the timing marks (2,3,6) are properly aligned with the timing camshaft drive chain links (1,4,5).

10. Install the primary camshaft drive chain upper guide. Tighten the bolts to 17 ft. lbs. (23 Nm).

11. Install the primary camshaft drive chain tensioner as follows:

 a. Use the Special Tensioner Tool J-45027 to reset the primary camshaft drive chain tensioner plunger.

 b. Install the plunger into the tensioner body

 c. Compress the plunger into the body and lock the tensioner by inserting the Special Retraction Tool EN-46112 into the access hole in the side of the tensioner body.

 d. Install a new gasket to the drive chain tensioner.

 e. Place the primary camshaft drive chain tensioner into position and loosely install the bolts to the block.

 f. Tighten the tensioner bolts to 44 inch lbs. (5 Nm) and then retighten to 17 ft. lbs. (23 Nm).

 g. Release the tensioner by pulling out the Retraction Tool EN-46112.

12. Install the right bank secondary camshaft drive chain.

13. Install the right secondary camshaft drive chain guide. Tighten the bolts to 17 ft. lbs. (23 Nm).

14. Install the right secondary camshaft drive chain shoe. Tighten the bolt to 17 ft. lbs. (23 Nm).

15. Install the right secondary camshaft drive chain tensioner as follows:

 a. Use the Special Tensioner Tool J-45027 to reset the primary camshaft drive chain tensioner plunger.

 b. Install the plunger into the tensioner body

 c. Compress the plunger into the body and lock the tensioner by inserting

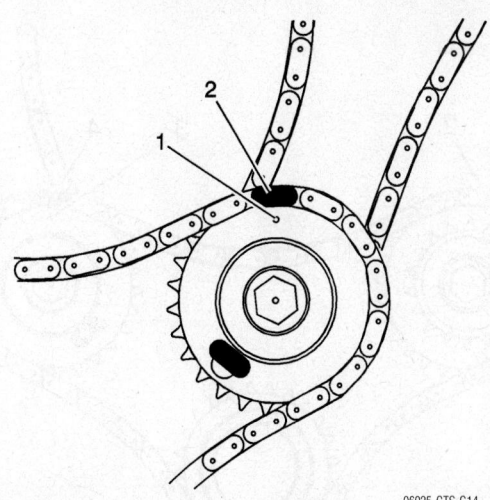

06025-CTS-G14

The left camshaft intermediate drive chain idler timing mark (1) will align with a timing camshaft drive chain link (2)

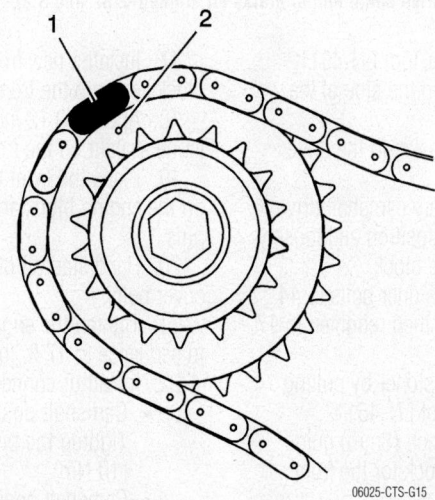

06025-CTS-G15

The right camshaft intermediate drive chain idler timing mark (2) will align with a timing camshaft drive chain link (1).

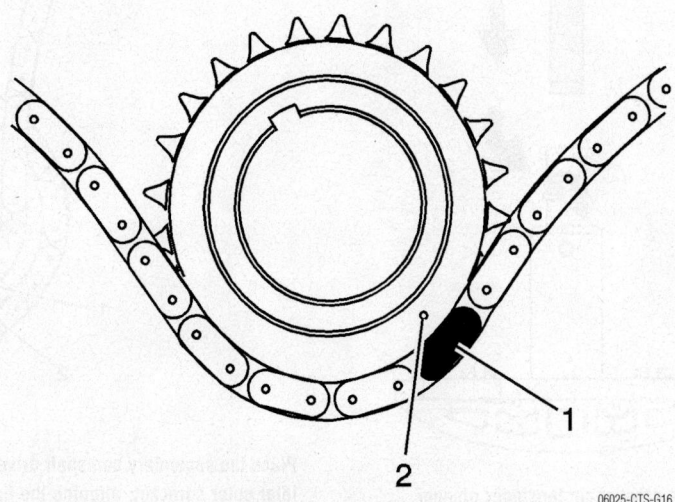

06025-CTS-G16

The crankshaft sprocket timing mark (2) will align with a timing camshaft drive chain link (1).

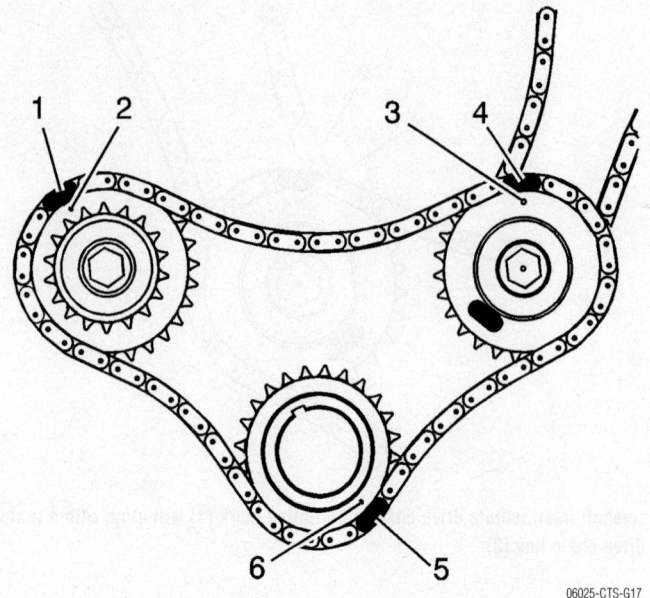

06025-CTS-G17

Make sure all primary drive chain timing marks all aligned—2.8L and 3.6L

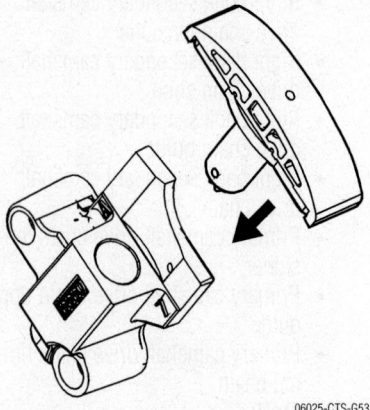

06025-CTS-G53

Lock the tensioner with Special Tool EN-46112—2.8L and 3.6L

the Special Retraction Tool EN-46112 into the access hole in the side of the tensioner body.

d. Install a new gasket to the drive chain tensioner.

e. Place the primary camshaft drive chain tensioner into position and loosely install the bolts to the block.

f. Tighten the tensioner bolts to 44 inch lbs. (5 Nm) and then retighten to 17 ft. lbs. (23 Nm).

g. Release the tensioner by pulling out the Retraction Tool EN-46112.

16. Install the 0.32 inch (8mm) guide pins into the cylinder block for the front cover.

17. Install a new front cover-to-cylinder block seal into the front cover.

18. Apply a 0.12 inch (3mm) bead of silicone sealant on the front cover as shown.

19. Place the front cover into position on the engine block and remove the guide pins.

20. Hand start all of the engine front cover bolts.

21. Tighten the engine front cover bolts in sequence to 17 ft. lbs. (23 Nm).

22. Install or connect the following:
- Camshaft position sensors. Tighten the bolts to 89 inch lbs. (10 Nm).
- Camshaft position actuator valves.

Tighten the bolts to 89 inch lbs. (10 Nm).
- Oil control valves
- Crankshaft pulley. Tighten the bolt to 74 ft. lbs. (100 Nm) plus 150 degrees.
- Power steering pump. Tighten the bracket bolts to 37 ft. lbs. (50 Nm).
- Power steering pump pulley
- Power steering fluid reservoir. Tighten the upper bolts to 80 inch lbs. (9 Nm) and lower bolt to 19 ft. lbs. (25 Nm).
- Alternator mounting bracket with belt tensioner. Tighten front bracket bolts to 37 ft. lbs. (50 Nm) and side bolt to 17 ft. lbs. (23 Nm).
- Alternator
- A/C compressor and power steering belt tensioner. Tighten the bolt to 37 ft. lbs. (50 Nm).

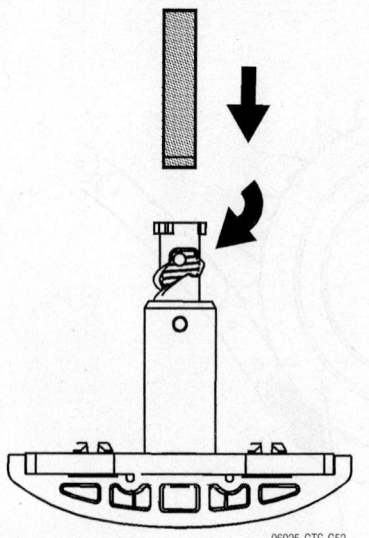

06025-CTS-G52

Reset the drive chain tensioner plunger with Special Tool J-45027.

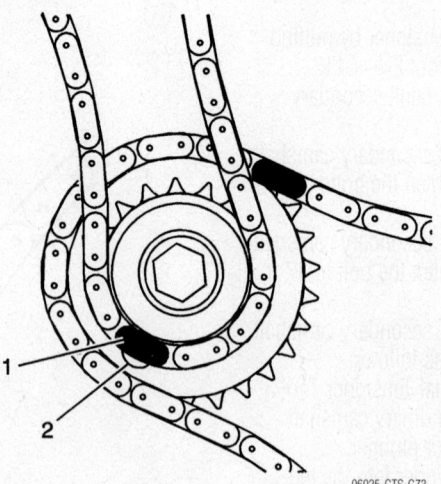

06025-CTS-G73

Place the secondary camshaft drive chain around the right camshaft intermediate drive chain idler outer sprocket, aligning the timing camshaft drive chain link (1) with the alignment access hole (2) made in the right camshaft intermediate drive chain idler inner sprocket.

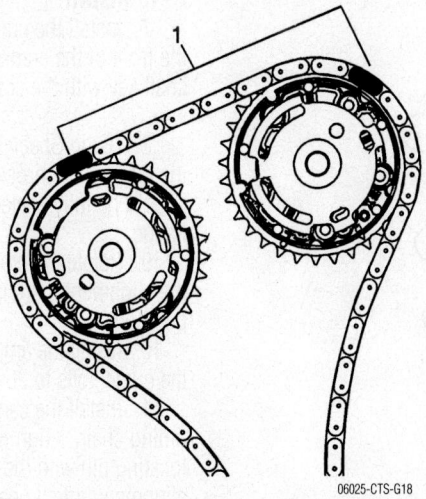

06025-CTS-G18

Ensure there are 7 links (1) between the timing camshaft drive chain links for the camshaft position actuator sprockets

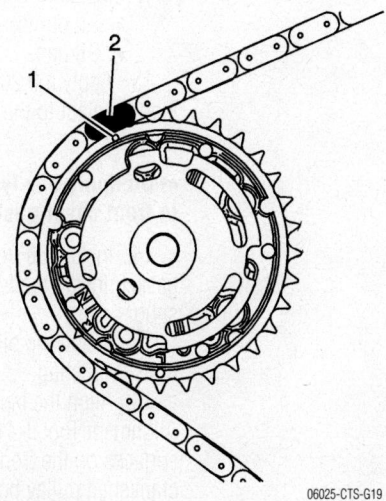

06025-CTS-G19

Align the right exhaust camshaft position actuator sprocket alignment triangle mark (1) with the timing camshaft drive chain link (2).

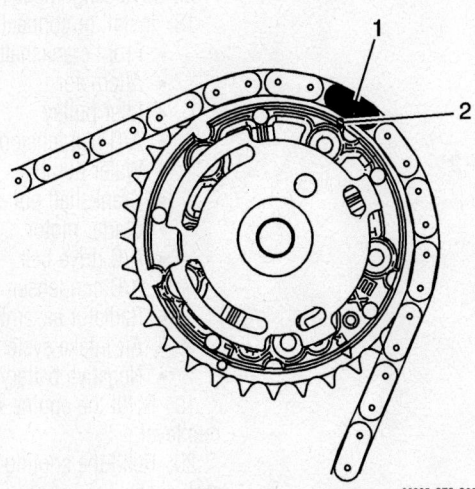

06025-CTS-G20

Align the right intake camshaft position actuator sprocket alignment triangle mark (2) with the timing camshaft drive chain link (1).

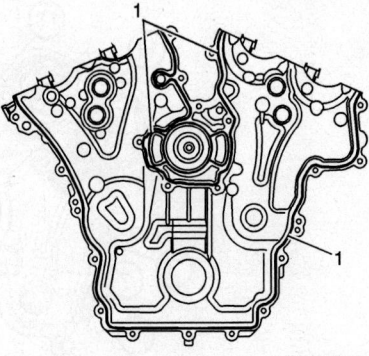

06025-CTS-G21

Front cover sealant locations–2.8L and 3.6L

- Accessory drive belts
- Water outlet housing. Tighten the bolts to 89 inch lbs. (10 Nm).
- Purge vent hose to the water outlet housing
- Camshaft covers
- Intake manifold
- Engine shroud
- Negative battery cable

23. Fill the engine with oil to the correct level.

24. Fill the cooling system to the correct level.

25. Start the engine and check for leaks.

5.7L Engine

1. Before servicing the vehicle, refer to the Precautions Section.

2. Drain the engine oil.

3. Drain the cooling system.

4. Remove or disconnect the following:
- Negative battery cable
- Air intake system
- Radiator assembly
- A/C condenser
- A/C drive belt
- Starter motor
- Crankshaft pulley
- Water pump
- A/C belt tensioner
- Idler pulley
- Alternator and mounting bracket
- Oil pan-to-front cover bolts
- Front cover mounting bolts
- Front cover
- Oil pan
- Oil pump
- Camshaft sprocket bolts
- Camshaft sprocket
- Timing chain
- Timing chain guide

5. Install Crankshaft Sprocket Removal Tool J-41558, J-41816-2 and J-8433 onto the crankshaft sprocket.

6. Remove the crankshaft sprocket.

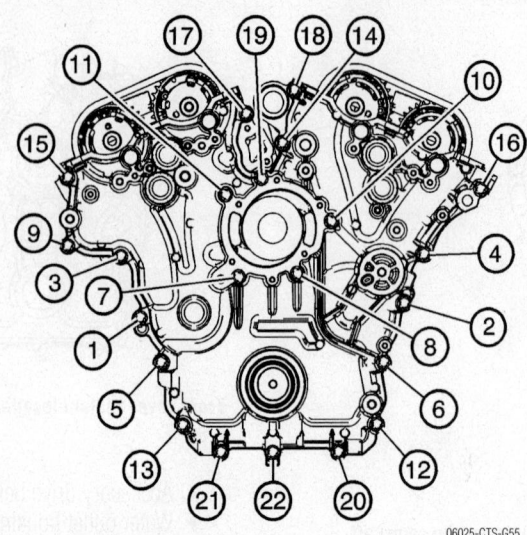

Front cover torque sequence–2.8L and 3.6L

06025-CTS-G55

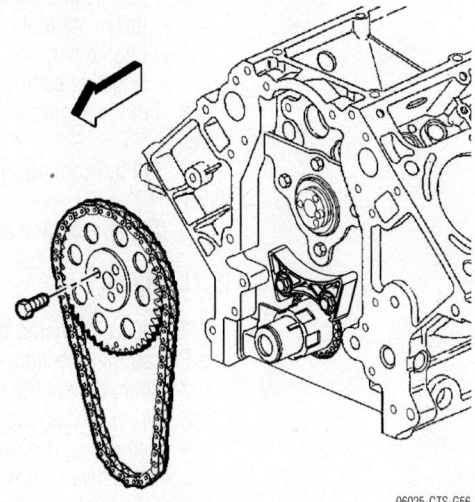

Removing the camshaft sprocket and timing chain–5.7L

06025-CTS-G56

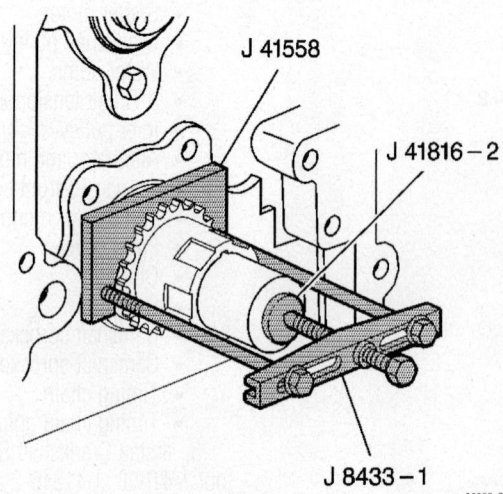

J 41558

J 41816 – 2

J 8433 – 1

06025-CTS-G57

Using the Crankshaft Sprocket Removal Tools–5.7L

To install:

7. Install the crankshaft sprocket onto the front of the crankshaft. Align the crankshaft key with the crankshaft sprocket keyway.

8. Using Special Tool J-41665, install the crankshaft sprocket. The crankshaft should be fully seated against the crankshaft flange.

9. Rotate the crankshaft sprocket until the alignment mark is in the 12 o'clock position.

10. Install the timing chain guide. Tighten the guide bolts to 26 ft. lbs. (35 Nm).

11. Install the camshaft sprocket and timing chain. Align the camshaft sprocket locating pin with the camshaft sprocket alignment hole. Locate the camshaft sprocket alignment mark in the 6 o'clock position. Tighten the camshaft sprocket bolts to 26 ft. lbs. (35 Nm).

12. Install or connect the following:
• Oil pump
• Oil pan

13. Apply a 0.20 inch (5mm) bead of silicone sealant to the oil pan-to-engine block junction.

➡**Do apply any type of sealant directly to front cover gasket.**

14. Install the front cover with a new gasket. Install the front cover bolts until snug.

15. Install the oil pan-to-front cover bolts until snug.

16. Align the tapered legs of Front Cover Alignment Tool J-41476 with the alignment surfaces on the front cover. Install the crankshaft pulley bolt through the alignment tool and tighten until snug.

17. Tighten the front cover mounting bolts to 18 ft. lbs. (25 Nm) and remove the Front Cover Alignment Tool.

18. Install or connect the following:
• Front crankshaft seal
• Alternator
• Idler pulley
• A/C belt tensioner
• Water pump
• Crankshaft pulley
• Starter motor
• A/C drive belt
• A/C condenser
• Radiator assembly
• Air intake system
• Negative battery cable

19. Refill the engine with oil to the correct level.

20. Refill the cooling system to the correct level.

21. Start the engine and check for leaks.

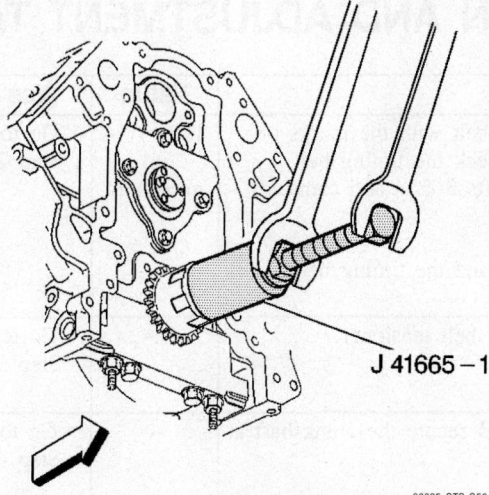

Install the crankshaft sprocket using Special Tool J-41665–5.7L

06025-CTS-G58

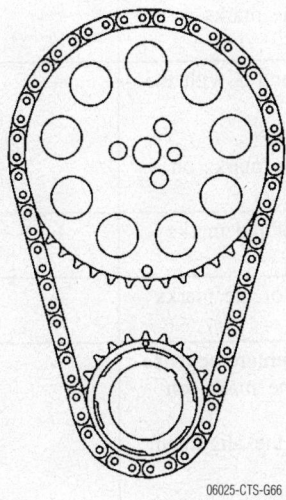

Timing chain alignment–5.7L

06025-CTS-G66

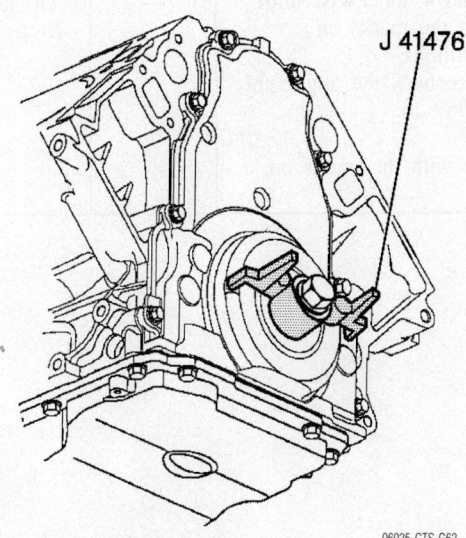

J 41476

Front cover alignment tool J-41476 installed–5.7L

06025-CTS-G62

Timing Belt

REMOVAL & INSTALLATION

3.2L Engine

1. Before servicing the vehicle, refer to the Precautions Section.
2. Remove or disconnect the following:
   - Negative battery cable
   - Intake air resonator
   - Intake plenum
   - Splash shield
   - Secondary air injection (AIR) pipe
   - Accessory drive belt
   - Water pump and power steering pump pulleys
   - Drive belt tensioner
   - Front timing belt cover
   - Crankshaft balancer
3. Rotate the crankshaft clockwise to 60 degrees before top dead center (BTDC).
4. Loosen the timing belt tensioner and idler pulleys.
5. Remove the timing belt.

To install:

6. Start at the crankshaft sprocket and install the timing belt with the double dash (TDC) aligned with the marks on the oil pump and on the belt drive gear.
7. Set the initial timing belt tension as follows:

   a. Turn the tensioner nut COUNTER-CLOCKWISE to full stop, then, turn the nut back until the reference mark is 1 mm (0.003 in) over the flange.

   b. Tighten the timing belt tensioner locking nut until snug, the locking nut will be tightened to specifications after all final adjustments are made.
8. Rotate the engine in the clockwise direction two revolutions stopping at 60 degrees BTDC.
9. Inspect the alignment of the reference marks on the camshaft gears with the notches on the rear timing belt cover, as well as, the mark on the crankshaft sprocket and oil pump housing.
10. If timing belt adjustment IS NOT required, set the final timing belt tension:

    a. Loosen the timing belt tensioner locking nut.

    b. Turn the locking nut counterclockwise to full stop, then back until the reference mark is 2-4 mm (0.078-0.157 in) ABOVE the reference mark on the flange.
11. Tighten the timing belt tensioner locking nut 15 ft. lbs. (20 Nm).
12. Tighten the idler pulley bolts to 30 ft. lbs. (40 Nm).
13. Install or connect the following:

TIMING BELT INSTALLATION AND ADJUSTMENT TABLE

Step	Action	Value	Yes	No
1	Install the timing belt and align marks on the belt with the marks on the camshaft gears and the crankshaft gear. Check the timing belt deflection between the idler pulley for camshafts 3 & 4 and camshaft number 4. Is the timing belt installed, the marks aligned and the timing belt deflection adjusted?	1 cm (0.4 in) maximum	Go to Step 2	—
2	Set the initial timing belt tension at the timing belt tensioner. Is the initial timing belt tension set?	—	Go to Step 3	—
3	Rotate the engine two complete revolutions and secure the crankshaft at Top Dead Center (TDC) with the J 42069-10. Has the engine been rotated and the crankshaft secured to TDC?	—	Go to Step 4	—
4	Starting with camshafts 3 and 4, check the alignment of the marks on the camshaft gears with the marks on the J 42069-20 checking gauge. Do the marks on the camshaft gears align exactly with the marks on J 42069-20?	—	Go to Step 5	Go to Step 6
5	Check the alignment of the marks on camshafts gears 1 and 2 with the marks on the J 42069-20 checking gauge. Do the marks on the camshaft gears align exactly with the marks on J 42069-20?	—	Go to Step 14	Go to Step 10
6	Do the camshaft gear marks line up to the left (BTDC) of the marks on the J 42069-20 checking gauge?	—	Go to Step 8	Go to Step 7
7	Do the camshaft gear marks line up to the right (ATDC) of the marks on the J 42069-20 checking gauge?	—	Go to Step 9	—
8	Turn the idler pulley eccentric, for camshafts 3 and 4, counterclockwise until the marks on the camshaft gear align exactly with the marks on J 42069-20. Rotate the engine two complete revolutions, lock the crankshaft at TDC with J 42069-10 and recheck the alignment of the camshaft gear marks to the marks on J 42069-20. Do the marks on the camshaft gears align exactly with the marks on J 42069-20?	—	Go to Step 5	Go to Step 6
9	Turn the idler pulley eccentric, for camshafts 3 and 4, clockwise until the marks on the camshaft gear align exactly with the marks on J 42069-20. Rotate the engine two complete revolutions, lock the crankshaft at TDC with J 42069-10 and recheck the alignment of the camshaft gear marks to the marks on J 42069-20. Do the marks on the camshaft gears align exactly with the marks on J 42069-20?	—	Go to Step 5	Go to Step 6

79225G35

Timing belt installation and adjustment table—GM 3.2L (VIN N) engine

Step	Action	Value	Yes	No
10	Do the camshaft gear marks line up to the left (BTDC) of the marks on the J 42069-20 checking gauge?	—	Go to Step 12	Go to Step 11
11	Do the camshaft gear marks line up to the right (ATDC) of the marks on the J 42069-20 checking gauge?	—	Go to Step 13	—
12	Turn the idler pulley eccentric, for camshafts 1 and 2, counterclockwise until the marks on the camshaft gear align exactly with the marks on J 42069-20. Rotate the engine two complete revolutions, lock the crankshaft at TDC with J 42069-10 and recheck the alignment of the camshaft gear marks to the marks on J 42069-20. Do the marks on the camshaft gears align exactly with the marks on J 42069-20?	—	Go to Step 14	Go to Step 10
13	Turn the idler pulley eccentric, for camshafts 1 and 2, clockwise until the marks on the camshaft gear align exactly with the marks on J 42069-20. Rotate the engine two complete revolutions, lock the crankshaft at TDC with J 42069-10 and recheck the alignment of the camshaft gear marks to the marks on J 42069-20. Do the marks on the camshaft gears align exactly with the marks on J 42069-20?	—	Go to Step 14	Go to Step 10
14	Set the final timing belt tension at the timing belt tensioner. Is the final timing belt tension set?	—	Go to Step 15	—
15	Again, rotate the engine two complete revolutions and lock the crankshaft at TDC. Do a final inspection of the camshaft gear marks' relationship to the J 42069-20 marks. The marks must align exactly. Do the marks on the camshaft gears align exactly with the marks on the J 42069-20?	—	Go to Step 16	Go to Step 2
16	Remove all checking tools and ensure all idler pulleys and the tensioner locking nut are tightened to specifications. Continue with re-assembly of the engine.			

79225G36

Timing belt installation and adjustment table (continued)—GM 3.2L (VIN N) engine

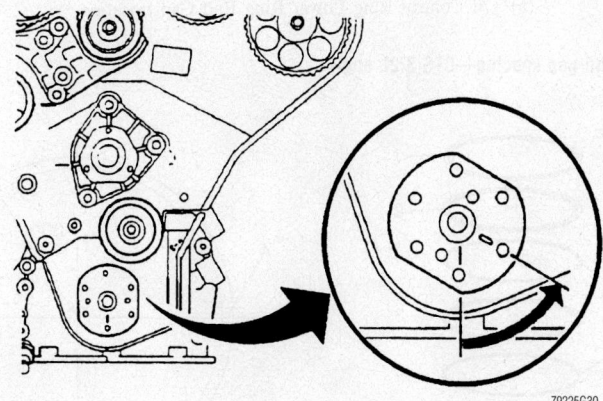

79225G30

Crankshaft alignment to 60 degrees BTDC—GM 3.2L (VIN N) engine

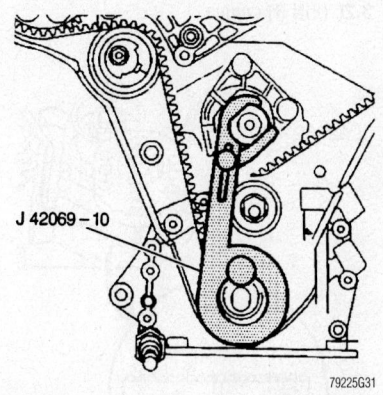

J 42069-10

79225G31

Securing the crankshaft—GM 3.2L (VIN N) engine

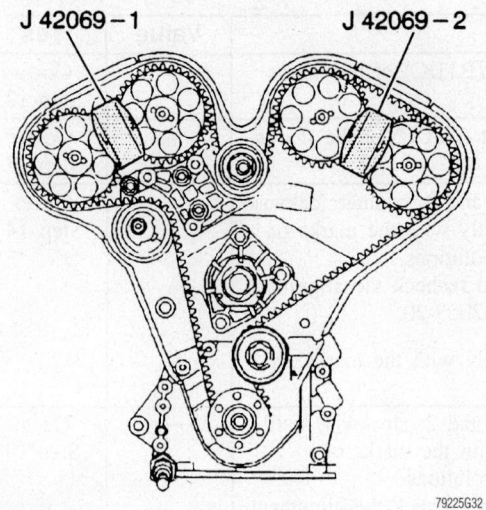

Locking the camshaft—GM 3.2L (VIN N) engine

79225G32

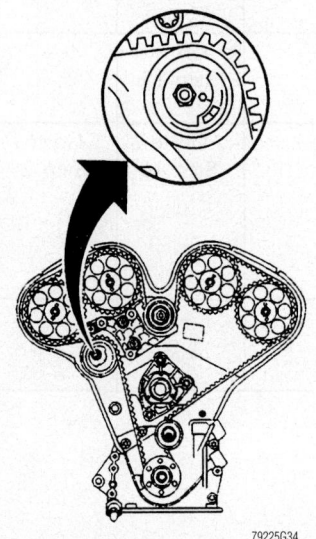

Initial timing belt tension adjustment—GM 3.2L (VIN N) engine

79225G34

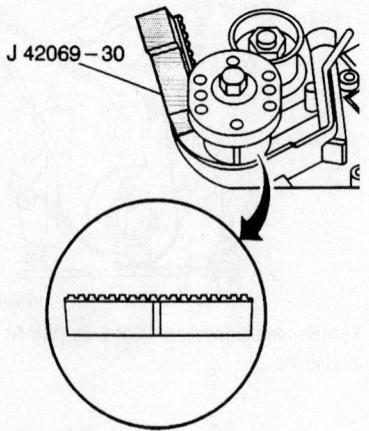

Using the tool to pin the timing belt—GM 3.2L (VIN N) engine

79225G33

- Crankshaft balancer. Torque the bolts to 15 ft. lbs. (20 Nm).
- Front timing belt cover. Torque the bolts to 71 inch lbs. (8 Nm).
- Drive belt tensioner. Torque the bolts to 30 ft. lbs. (40 Nm).
- Water pump pulley. Torque the bolts to 71 inch lbs. (8 Nm).
- Power steering pump pulley. Torque the bolts to 15 ft. lbs. (20 Nm).
- Accessory drive belt
- AIR injection pipe
- Splash shield
- Intake plenum
- Intake air resonator

Piston and Ring

POSITIONING

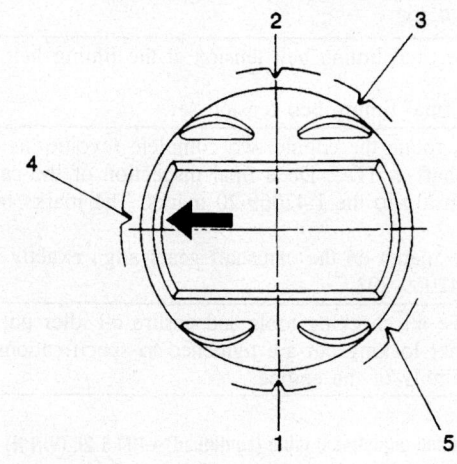

(1) 1st Compression Ring End Gap Location
(2) 2nd Compression Ring End Gap Location
(3) Oil Control Ring Upper Ring End Gap Location
(4) Oil Control Ring Spacer End Gap Location
(5) Oil Control Ring Lower Ring End Gap Location

7922AG55

Piston ring end-gap spacing—CTS 3.2L engine

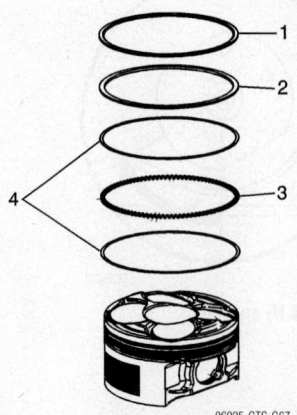

06025-CTS-G67

Piston ring installation order—2.8L and 3.6L Engine

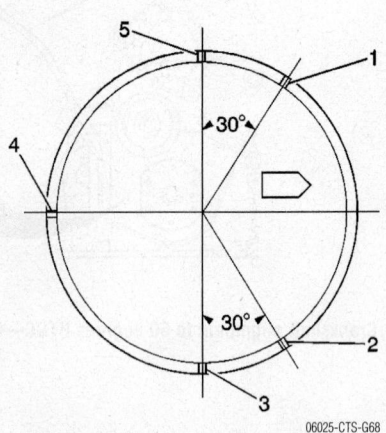

06025-CTS-G68

Piston ring end-gap spacing–2.8L and 3.6L Engine

FUEL SYSTEM

Fuel System Service Precautions

Safety is the most important factor when performing not only fuel system maintenance but any type of maintenance. Failure to conduct maintenance and repairs in a safe manner may result in serious personal injury or death. Maintenance and testing of the vehicle's fuel system components can be accomplished safely and effectively by adhering to the following rules and guidelines.

• To avoid the possibility of fire and personal injury, always disconnect the negative battery cable unless the repair or test procedure requires that battery voltage be applied.

• Always relieve the fuel system pressure prior to disconnecting any fuel system component (injector, fuel rail, pressure regulator, etc.), fitting or fuel line connection. Exercise extreme caution whenever relieving fuel system pressure, to avoid exposing skin, face and eyes to fuel spray. Please be advised that fuel under pressure may penetrate the skin or any part of the body that it contacts.

• Always place a shop towel or cloth around the fitting or connection prior to loosening to absorb any excess fuel due to spillage. Ensure that all fuel spillage (should it occur) is quickly removed from engine surfaces. Ensure that all fuel soaked cloths or towels are deposited into a suitable waste container.

• Always keep a dry chemical (Class B) fire extinguisher near the work area.

• Do not allow fuel spray or fuel vapors to come into contact with a spark or open flame.

• Always use a backup wrench when loosening and tightening fuel line connection fittings. This will prevent unnecessary stress and torsion to fuel line piping. Always follow the proper torque specifications.

• Always replace worn fuel fitting O-rings with new. Do not substitute fuel hose or equivalent, where fuel pipe is installed.

Fuel System Pressure

RELIEVING

1. Before servicing the vehicle, refer to the Precautions Section.
2. Loosen the fuel filler cap to relieve the tank pressure.
3. Remove or disconnect the following:

• Negative battery cable
• Intake manifold top cover

4. Install a Fuel Pressure Gauge to the fuel pressure fitting. Wrap a shop towel around the fitting while installing the gauge.
5. Connect a bleed hose into an approved container and open the valve to bleed the system.
6. Close the valve and disconnect the gauge.
7. Drain any remaining fuel from the gauge into the approved container.

Fuel Filter

REMOVAL & INSTALLATION

1. Before servicing the vehicle, refer to the Precautions Section.
2. Loosen the fuel filler cap to relieve pressure in the tank.
3. Relieve the fuel system pressure.
4. Remove or disconnect the following:

• Fuel feed and return lines
• Fuel filter from the bracket

To install:
5. Clean the bolts and fittings on both fuel lines.

6. Install or connect the following:

• Filter in the retaining strap making certain that the flow is in the proper direction
• Both fuel lines. Torque the fittings to 18 ft. lbs. (25 Nm).
• Fuel filter bracket mounting bolt. Torque the bolt to 13 ft. lbs. (18 Nm).
• Fuel filler cap

7. Crank the engine for a few seconds and check for leakage.

Fuel Pump

REMOVAL & INSTALLATION

1. Before servicing the vehicle, refer to the Precautions Section.
2. Relieve the fuel system pressure.
3. Drain the fuel tank.
4. Remove or disconnect the following:

• Fuel tank
• Fuel tank pressure sensor
• Fuel tank connection cover electrical connector
• Spring loaded clamp around the tank boot
• Tank boot from the housing

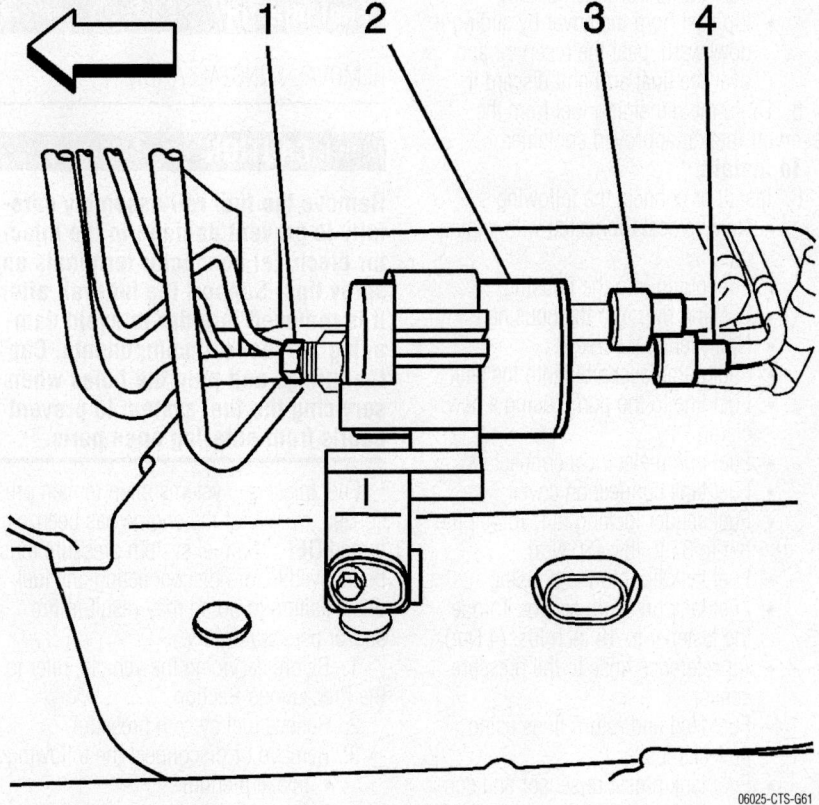

Fuel filter–5.7L

06025-CTS-G61

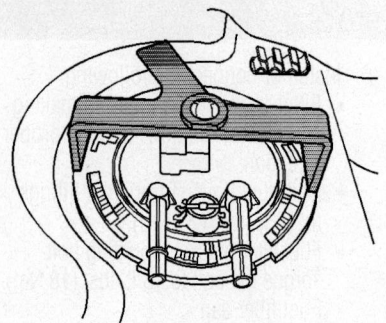

06025-CTS-G22

Remove the locking nut using J-45747 fuel tank wrench

➡️**The fuel pump assembly may spring up from its position. The reservoir bucket on the assembly is full of fuel. Tip the assembly slightly so the float is not damaged during removal.**

- Fuel sender locking nut from the tank using a Fuel Tank Sender Wrench (J 45747)
- Fuel pump electrical connectors
- Fuel tank connection cover
- Hose from the pump
- Fuel pump from the tank by pushing the tabs inward
- Locking ring by pushing the tabs in and pushing the locking ring from the housing simultaneously
- Fuel pump from the housing
- Lip seal from the cover by sliding it downward, past the reservoir and over the float arm and discard it

5. Drain the remaining fuel from the reservoir into an approved container.

To install:

6. Install or connect the following:
- New lip seal lubricated with engine oil
- Fuel pump into the housing
- Locking ring into the housing
- New fuel inlet screen
- Fuel pump assembly into the tank
- Fuel line to the pump using a new clamp
- Fuel pump electrical connectors
- Fuel tank connection cover
- Fuel sender locking nut. Torque the nut to 37 ft. lbs. (50 Nm).
- Fuel tank boot to the housing
- Fuel tank pressure sensor. Torque the fastener to 18 inch lbs. (4 Nm).
- Air reference hose to the pressure sensor
- Fuel feed and return lines using new clamps
- Fuel tank pressure sensor and connection cover electrical connectors

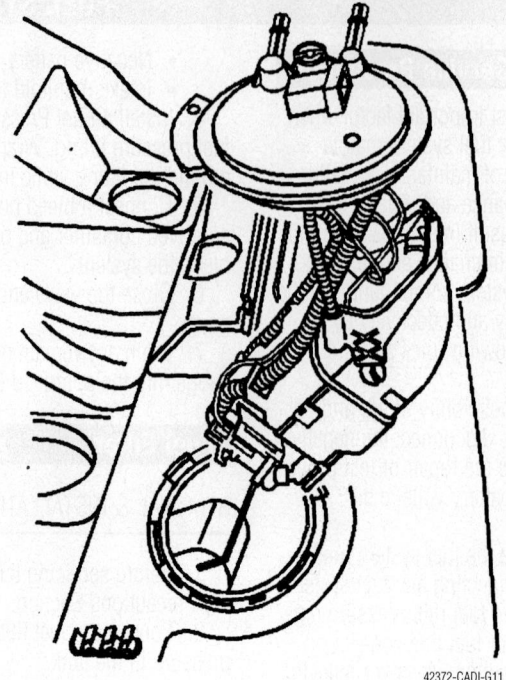

42372-CADI-G11

Fuel pump module

- Fuel tank to the vehicle. Torque the bolts to 22 ft. lbs. (30 Nm).
- Negative battery cable

7. Refill the fuel tank.

8. Start the vehicle and inspect for leaks.

Fuel Injector

REMOVAL & INSTALLATION

❋❋ CAUTION

Remove the fuel rail assembly carefully to prevent damage to the injector electrical connector terminals and spray tips. Support the fuel rail after it is removed in order to avoid damaging the fuel rail components. Cap the fittings and plug the holes when servicing the fuel system to prevent debris from entering open ports.

Fuel injection systems often remain pressurized, even after the engine has been turned **OFF**. The fuel system pressure must be relieved before disconnecting any fuel lines. Failure to do so may result in fire and/or personal injury.

1. Before servicing the vehicle, refer to the Precautions Section.

2. Relieve fuel system pressure.

3. Remove or disconnect the following:
- Intake plenum
- Fuel rail supply and return lines from the fuel rail
- Fuel injector electrical connectors
- Harness from the fuel rail
- Fuel rail attaching bolts from the intake manifold
- Fuel pressure regulator vacuum lines
- Fuel rail assembly
- Injectors from the fuel rail and discard the O-rings

To install:

4. Install or connect the following:
- New O-rings lubricated with clean engine oil onto the fuel injectors
- Fuel injectors with new retaining clips
- Fuel rail to the intake manifold. Torque the bolt to 71 inch lbs. (8 Nm).
- Fuel pressure regulator vacuum line
- Fuel injector electrical connectors
- Harness to the fuel rail
- Fuel supply/return lines to the fuel rail. Torque the fasteners to 11 ft. lbs. (15 Nm).
- Fuel line bracket. Torque the bolts to 37 ft. lbs. (50 Nm).
- Intake plenum. Torque the bolts to 71 inch lbs. (8 Nm).
- Negative battery cable

5. Crank the engine several times to pressurize the system.

6. Check for leaks and repair, if necessary.

DRIVE TRAIN

Transmission Assembly

REMOVAL & INSTALLATION

Automatic Transmission

1. Before servicing the vehicle, refer to the Precautions Section.

2. Disconnect the shift linkage from the transmission.

3. Place the transmission in neutral by rotating the shift shaft clockwise 2 clicks.

4. Remove or disconnect the following:
   - Exhaust system
   - Driveshaft and secure to the shift control level with mechanics wire or equivalent
   - Catalytic converter hanger bracket
   - Transmission wiring harness and harness clips
   - Transmission close out plug

➡**Matchmark the torque converter to flywheel orientation to ensure proper realignment.**

   - Torque converter bolt close out plug
   - Front air deflector
   - Torque converter bolts
   - Transmission cooler pipes

5. Support the transmission with a suitable transmission jack.

6. Remove the transmission mount.

7. Remove the 3 lower transmission mounting bolts.

8. Lower the rear of the transmission enough to gain access to the upper mounting bolts.

9. Disconnect the engine wiring harness clips from the transmission mounting bolts.

10. Remove the 3 upper transmission mounting bolts.

11. Slide the transmission free from the engine dowels and lower the transmission from the vehicle.

To install:

12. Raise the transmission into the vehicle and align the transmission on the engine dowels.

13. Install the 3 lower transmission mounting bolts. Tighten as follows:
   a. Transmission bolts (2) to 55 ft. lbs. (75 Nm)
   b. Transmission bolts (1) to 37 ft. lbs. (50 Nm)

14. Install the 3 upper transmission mounting bolts to 55 ft. lbs. (75 Nm).

15. Install or connect the following:

   - Engine wiring harness clips to the transmission mounting bolts
   - Transmission mount. Tighten the mounting bolts to 44 ft. lbs. (60 Nm).
   - Transmission cooler pipes
   - Torque converter bolts. Tighten to 46 ft. lbs. (63 Nm).
   - Torque converter bolt plug

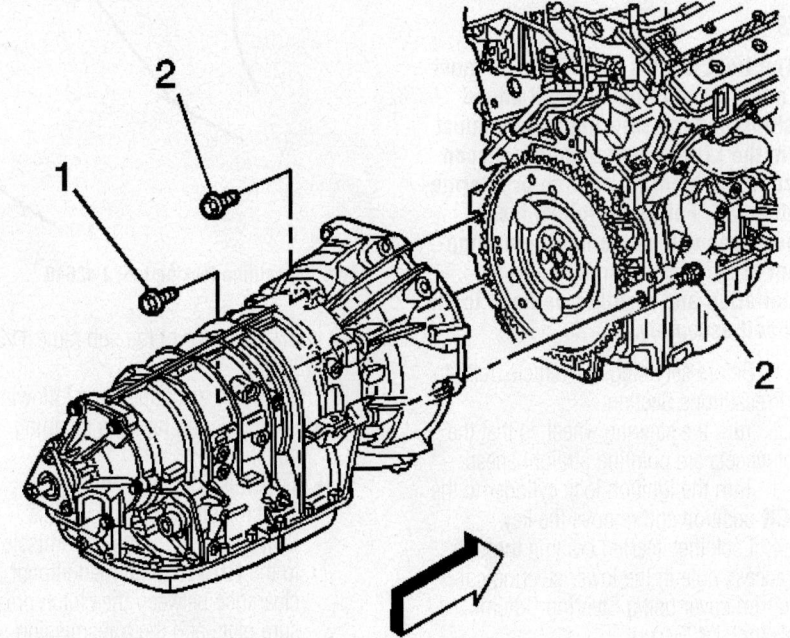

Location of automatic transmission lower mounting bolts.

06025-CTS-G23

   - Transmission close out plug
   - Front air deflector
   - Transmission wiring harness clips and wiring harnesses
   - Catalytic converter hanger bracket
   - Driveshaft. Tighten the shaft coupler-to-transmission flange bolts to 63 ft. lbs. (85 Nm).

16. Place the transmission in park by

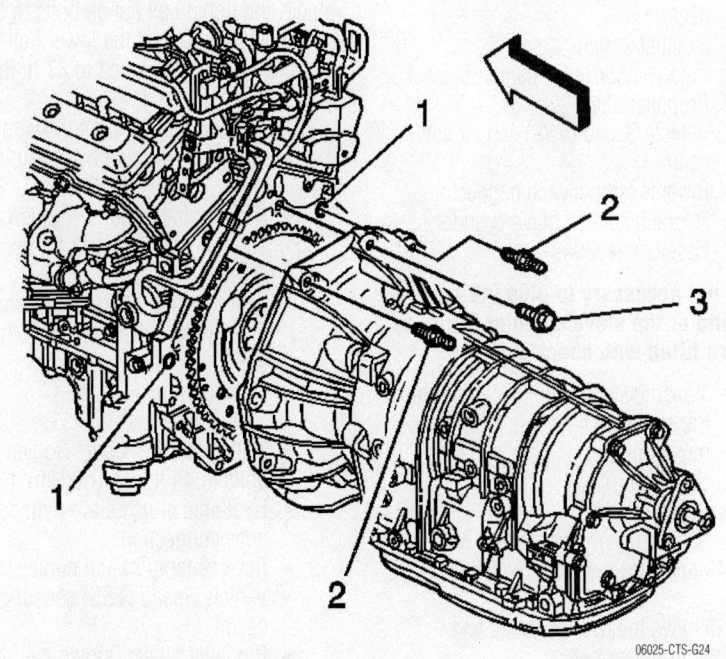

Location of automatic transmission upper mounting bolts.

06025-CTS-G24

rotating the shift shaft fully counter clock-wise.

17. Connect the shift linkage to the transmission. Tighten the shift shaft nut to 11 ft. lbs. (15 Nm).

18. Install the exhaust system.

19. Check the transmission fluid level.

20. Start the engine and check for leaks.

21. Road test the vehicle for proper operation.

Manual Transmission

CTS

➥The front wheels of the vehicle must be maintained in the straight ahead position and the steering column must be in the LOCK position before disconnecting the steering column or intermediate shaft. Failure to follow these procedures will cause improper alignment of some components during installation and result in damage to the SIR coil assembly.

1. Before servicing the vehicle, refer to the Precautions Section.

2. Turn the steering wheel so that the front wheels are pointing straight ahead.

3. Turn the ignition lock cylinder to the LOCK position and remove the key.

4. Lock the steering column through the access hole in the lower steering column trim cover using Steering Column Lock Tool J 42640.

5. Remove or disconnect the following:
- Negative battery cable
- Heated Oxygen (HO2S) sensor connectors
- Exhaust system
- Catalytic converter hanger bracket
- Propeller shaft
- Vehicle Speed (VSS) sensor connector
- Reverse lamp switch connector
- Hydraulic clutch slave cylinder hose connection

➥It is not necessary to plug the lower hose end or the slave cylinder fitting as they are fitted with check valves.

- Transmission mount. Support the transmission with a suitable transmission jack.
- Shift control rod
- Reaction arm
- Steering gear intermediate shaft

6. Support the subframe with a jack stand.

7. Remove the two rear bolts and loosen the front two bolts.

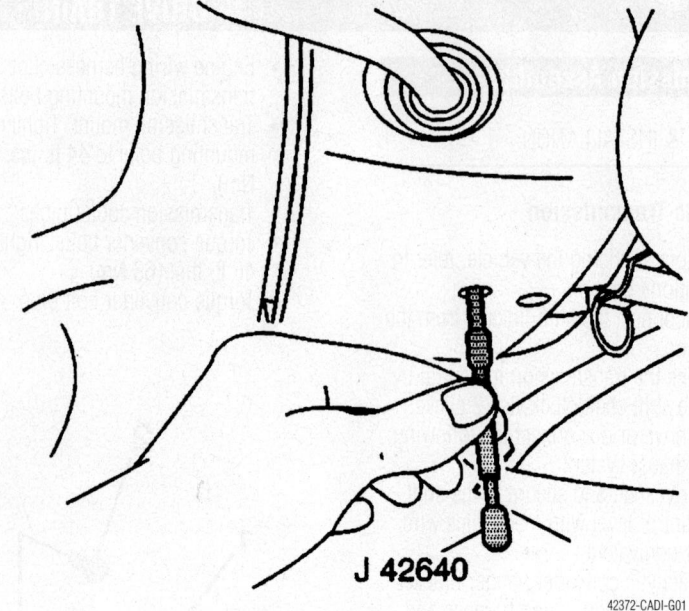

J 42640

42372-CADI-G01

Steering column locking tool J 42640

8. Lower the rear of the subframe 1½ inches (38mm).

9. Remove or disconnect the following:
- Engine wiring harness retaining clips
- Transmission flange bolts
- Transmission from the vehicle. Swing the rear of the transmission to the RIGHT to gain additional clearance between the clutch pressure plate and the transmission input shaft.

To install:

10. Install the transmission to the vehicle and tighten all flange bolts to 55 ft. lbs. (75 Nm) except for the lower bolt on the left side, which is tightened to 37 ft. lbs. (50 Nm).

11. Install or connect the following:
- Engine wiring harness retaining clips
- Rear subframe bolts. Tighten all subframe bolts to 141 ft. lbs. (191 Nm).
- Steering gear intermediate shaft. Tighten the pinch bolt to 35 ft. lbs. (48 Nm).
- Reaction arm
- Shift control rod
- Transmission mount. Tighten the bolts to 44 ft. lbs. (60 Nm).
- Hydraulic clutch slave cylinder hose connection
- Reverse lamp switch connector
- Vehicle Speed (VSS) sensor connector
- Propeller shaft. Tighten the bolts to 63 ft. lbs. (85 Nm).

- Catalytic converter hanger bracket. Tighten the bolts to 37 ft. lbs. (50 Nm).
- Exhaust system
- Heated Oxygen (HO2S) sensor connectors
- Negative battery cable

12. Remove Steering Column Lock Tool J 42640 from the steering column.

13. Bleed the clutch hydraulic system.

14. Check the transmission fluid level.

15. Road test the vehicle for proper operation.

CTS-V

1. Before servicing the vehicle, refer to the Precautions Section.

2. Remove the shift knob and boot assembly.

3. Remove the shift control adapter plate.

4. Remove the exhaust system.

5. Remove the driveshaft.

6. Support the transmission with a suitable jack.

7. Remove the transmission support and lower the transmission assembly to gain access to the top of the transmission.

8. Remove or disconnect the following:
- Shift control assembly
- Transmission fluid temperature (TFT) sensor
- Backup lamp switch
- Vehicle speed sensor (VSS)
- Reverse lockout solenoid
- Gear select solenoid
- Bolts securing the transmission to the clutch housing

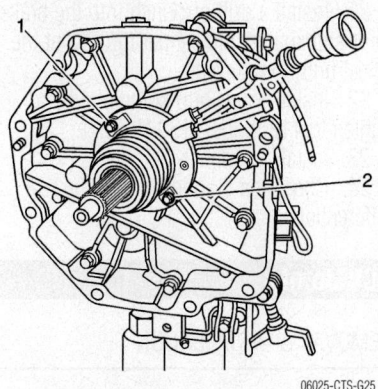

Bolts securing the concentric slave cylinder to the transmission–5.7L

06025-CTS-G25

9. Carefully separate the transmission approximately 1.5 inches (3.8cm) from the clutch housing.

10. Remove the bolts securing the concentric slave cylinder to the transmission.

11. Remove the transmission.

➡**Ensure the concentric slave cylinder does not bind on the input shaft during removal.**

To install:

12. Partially install the transmission until the concentric slave cylinder bolts can be installed. Tighten the bolts to 106 inch lbs. (12 Nm).

13. Install the transmission until flush with the clutch housing. Tighten the housing bolts to 35 ft. lbs. (48 Nm).

14. Install or connect the following:
- TFT sensor
- Backup lamp switch
- VSS
- Reverse lockout solenoid
- Gear select solenoid
- Shift control assembly
- Transmission support. Tighten the bolts to 44 ft. lbs. (60 Nm).
- Driveshaft
- Exhaust system
- Shift control adapter plate. Tighten the bolts to 13 ft. lbs. (18 Nm).
- Shift knob and boot assembly
- Negative battery cable

15. Check the transmission fluid level.
16. Bleed the clutch hydraulic system.
17. Road test the vehicle for proper operation.

Clutch

ADJUSTMENTS

The CTS uses a hydraulic clutch system. No adjustments are necessary.

REMOVAL & INSTALLATION

CTS

1. Before servicing the vehicle, refer to the Precautions Section.
2. Remove or disconnect the following:
- Negative battery cable
- Transmission from the vehicle
- Clutch pressure plate
- Clutch driven plate

To install:

3. Install or connect the following:
- Clutch driven plate
- Clutch pressure plate. Tighten the bolts in a star pattern and in several passes to 21 ft. lbs. (28 Nm).
- Transmission to the vehicle
- Negative battery cable

CTS-V

1. Before servicing the vehicle, refer to the Precautions Section.
2. Remove or disconnect the following:
- Negative battery cable
- Transmission from the vehicle
- Concentric slave cylinder
- Clutch housing
- Clutch pressure plate
- Clutch driven plate

To install:

3. Install or connect the following:
- Clutch driven plate
- Clutch pressure plate. Tighten the clutch pressure plate bolts in sequence evenly over 3 increments with the fourth increment to 52 ft. lbs. (70 Nm).
- Clutch housing. Tighten the bolts to 37 ft. lbs. (50 Nm).
- Concentric slave cylinder
- Transmission to the vehicle
- Negative battery cable

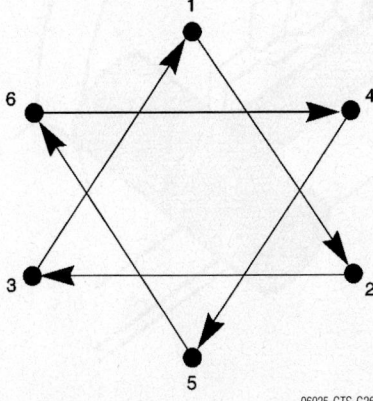

06025-CTS-G26
Clutch tightening sequence

Hydraulic Clutch System

BLEEDING

1. Before servicing the vehicle, refer to the Precautions Section.
2. Fill the clutch cylinder reservoir with new hydraulic fluid.
3. Stroke the clutch pedal from the up stop to the down stop position at least 15 times.
4. With the pedal in the down stop position, open the bleeder valve at the slave cylinder to release trapped air.
5. Close the bleeder valve and slowly return the clutch pedal to the up stop position.
6. Open the bleeder valve and slowly depress the clutch pedal from the up stop position to the down stop position until fluid escapes from the bleeder valve.
7. Close the bleeder valve and slowly return the clutch pedal to the up stop position.
8. Depress the clutch pedal from the up stop to the down stop position.
9. Open the bleeder valve and allow fluid with air bubbles to escape.
10. Close the bleeder valve.
11. Repeat until fluid without air bubbles escapes from the bleeder valve.

Halfshaft

REMOVAL & INSTALLATION

1. Before servicing the vehicle, refer to the Precautions Section.
2. Remove the wheel.
3. Insert a drift or punch into the brake rotor and against the caliper to prevent the wheel hub from turning.
4. Remove and discard the hub spindle nut.
5. Remove or disconnect the following:
- Brake rotor
- ABS sensor
- Parking brake cable bracket
- Upper ball joint nut
6. Separate the upper ball stud from the knuckle.
7. Support the lower control arm with a suitable jack.
8. Remove or disconnect the following:
- Lower shock mounting bolt
- Lower control arm-to-knuckle mounting bolt
- Trailing arm-to-knuckle mounting bolt
- Adjustment link-to-knuckle mounting bolt

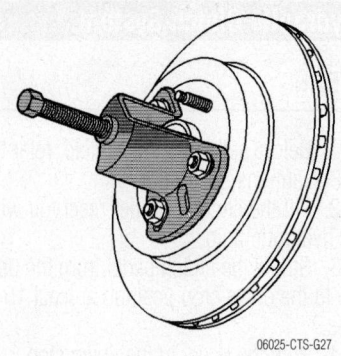

06025-CTS-G27

Using wheel hub removal tool J-42129–CTS

9. Install Wheel Hub Removal Tool J-42129 to the wheel hub.

10. Use the Wheel Hub Removal Tool to disengage the halfshaft from the wheel hub.

11. Remove the knuckle assembly from the vehicle.

12. Using a suitable tool, carefully release the halfshaft from the rear differential enough to install Seal Protector J-44394 over the halfshaft into the differential output shaft seal.

13. Remove the halfshaft from the vehicle and discard the halfshaft retaining ring.

To install:

14. Install a new halfshaft retaining ring.

15. If previously removed, carefully install Seal Protector J-44394 into the differential output shaft seal.

➡**Failure to install J-44394 as indicated may cause the splines of the wheel drive shaft to cut the differential output seal.**

16. Install the halfshaft until the splines are past Seal Protector J-44394. Ensure the retaining ring is installed in the upright position.

17. Remove Seal Protector J-44394.

18. Install or connect the following:

➡**Loosely install all fasteners before tightening.**

- Knuckle assembly
- Adjustment link-to-knuckle mounting bolt. Tighten to 129 ft. lbs. (175 Nm).
- Trailing arm-to-knuckle mounting bolt. Tighten to 129 ft. lbs. (175 Nm).
- Lower shock mounting bolt. Tighten to 111 ft. lbs. (150 Nm).
- Lower control arm-to-knuckle mounting bolt. Tighten to 129 ft. lbs. (175 Nm).
- Upper ball joint mounting nut. Tighten to 15 ft. lbs. (20 Nm) plus 210 degrees.
- Parking brake cable bracket. Tighten to 44 ft. lbs. (60 Nm).

19. Remove the jack from the lower control arm.

20. Install or connect the following:
- ABS sensor
- Brake rotor

21. Install a drift or punch into the brake rotor and against the caliper to prevent the wheel hub from turning.

22. Install a new hub spindle nut and tighten to 118 ft. lbs. (160 Nm).

23. Install the wheel.

24. Check the fluid level in the rear differential.

CV-Joints

REMOVAL & INSTALLATION

Outer

1. Before servicing the vehicle, refer to the Precautions Section.

2. Remove the halfshaft from the vehicle.

3. Remove the large boot retaining clamp.

4. Use a hand grinder or suitable equivalent to cut through and remove the swage ring.

5. Separate the halfshaft outboard seal from the CV joint outer race.

6. Slide the boot away from the joint.

7. Spread the ears on the race retaining ring with Special Tool J-8059.

8. Remove the CV joint assembly from the halfshaft.

9. Using a brass drift and hammer, gently tap against the CV joint cage in order to tilt the cage.

10. Remove the first chrome alloy ball with the cage tilts.

11. Tilt the cage in the opposite direction to remove the opposing alloy ball.

12. Repeat this process to remove all 6 balls.

13. Pivot the CV joint cage and the inner race 90 degrees to the center line of the outer race and align the cage windows with the lands of the outer race.

14. Lift out the cage and the inner race.

15. Remove the inner race from the cage by rotating the inner race upward.

To install:

16. Clean and inspect all parts.

17. Install the new swage ring on the neck of the outer boot.

18. Slide the outer boot onto the halfshaft and position the neck of the boot in the seal groove on the halfshaft.

19. Position the outboard end of the halfshaft assembly in the Drive Axle Swage Ring Clamp J-41048. Insert the bolts and tighten by hand until snug.

20. Tighten each bolt 180 degrees at a time until both sides are bottomed. Loosen the bolts and separate the dies.

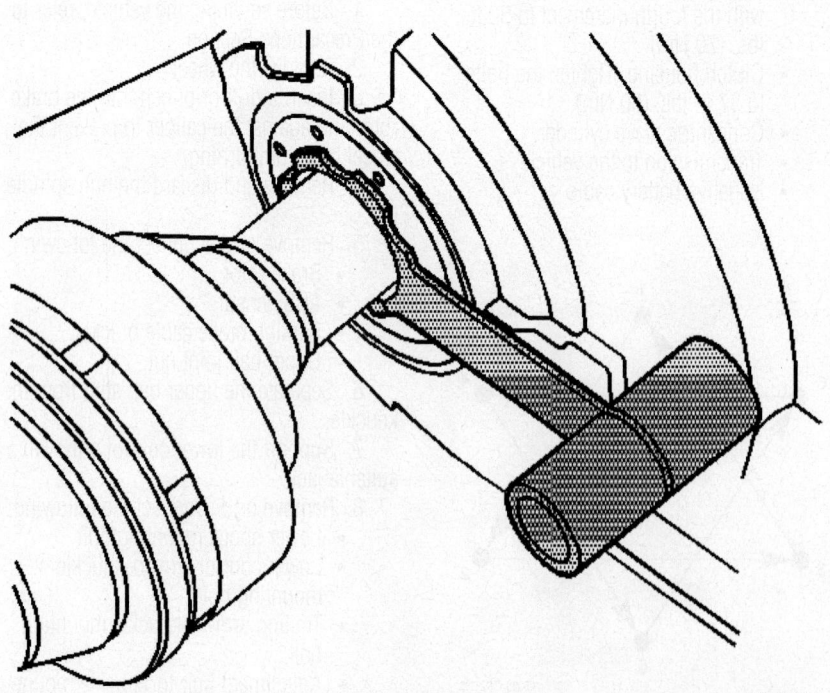

06025-CTS-G28

Install seal protector J-44394 to prevent the splines of the halfshaft to cut the output shaft seal.

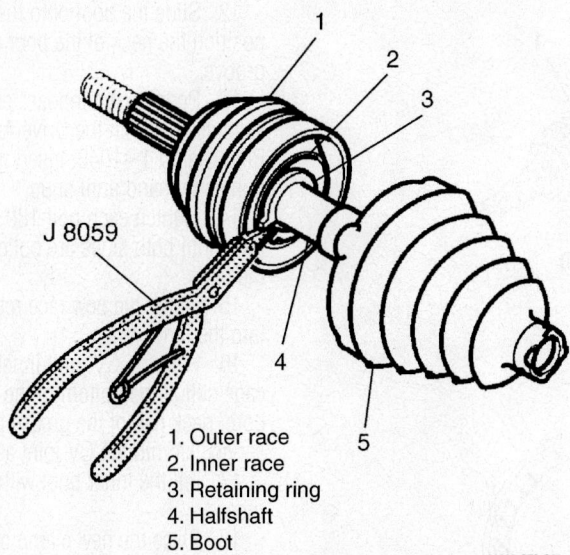

1. Outer race
2. Inner race
3. Retaining ring
4. Halfshaft
5. Boot

06025-CTS-G29

Spread the ears of the retaining ring using Special Tool J-8059.

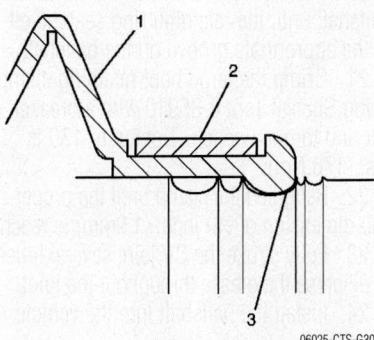

06025-CTS-G30

The largest groove below the sight groove on the halfshaft is the seal groove (3).

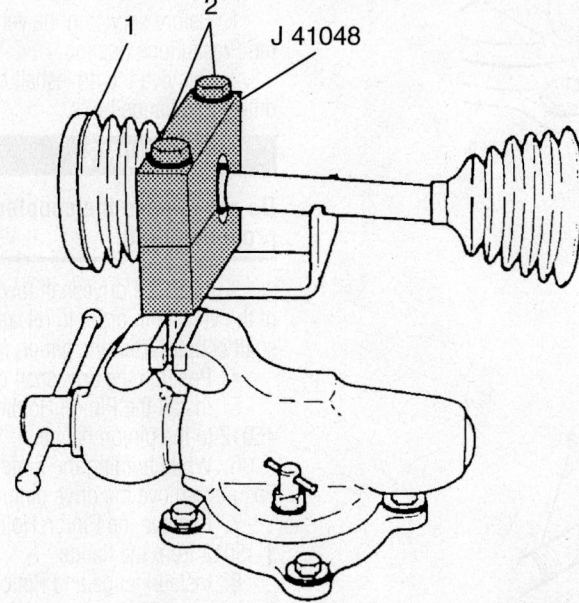

06025-CTS-G31

Install the Halfshaft in the Ring Clamp Tool J-41048.

21. Check the swaged ring for any lip deformities. Re-swage the ring if necessary.

22. Put a light coat of grease from the service kit on the ball grooves of the inner and outer race.

23. Hold the inner race 90 degrees to the centerline of the cage and align the cage windows the lands of the inner race and install the inner race into the cage.

24. Hold the cage and inner race assembly at 90 degrees to the centerline of the outer race and align the cage windows with the lands of the outer race.

25. Install the cage and inner race assembly into the outer race.

26. Insert the first chrome ball and then tilt the cage in the opposite direction to insert the opposing ball.

27. Repeat this process until all 6 balls are in place.

28. Place approximately half of the grease from the service kit inside the outer boot.

29. Pack the CV joint with the remaining grease.

30. Push the CV joint onto the halfshaft until the retaining ring is seated in the groove on the halfshaft.

31. Slide the outer boot, with the boot retaining clamp, in place over the outside of the CV joint outer face and locate the seal lip in the groove on the CV joint outer race.

32. Crimp the boot clamp using Special J-35910 or equivalent clamp pliers to 130 ft. lbs. (174 Nm).

33. Check the gap dimension on the clamp ear and continue tightening until the gap dimension is 5⁄64 inch (1.9mm).

34. Install the halfshaft in the vehicle.

Inner

1. Before servicing the vehicle, refer to the Precautions Section.

2. Wrap a towel around the halfshaft and place it in a vise.

3. Cut the large boot retaining clamp and discard.

4. Using a hand grinder, cut through the halfshaft swage ring.

5. Slide the boot away from the CV joint and wipe away any grease from the CV joint assembly.

6. Spread the ears of the CV joint race retaining ring using Special Tool J-8059 or equivalent snap ring pliers.

✳✳ WARNING

The CV joint shank must be parallel to the halfshaft prior to removal. Never allow the inner race or cage to rotate within the outer race when the CV joint is separated from the halfshaft.

7. Remove the CV joint assembly from the halfshaft.

8. Remove the transmission retaining ring from the CV joint assembly using Special Tool J-8059 or equivalent snap ring pliers.

9. Remove the inner boot and swage ring from the halfshaft and discard.

To install:

10. Inspect the inner CV joint for visible damage or wear. The assembly is not serviceable and must be replaced if necessary.

11. Install a new swage ring on the neck of a new inner boot.

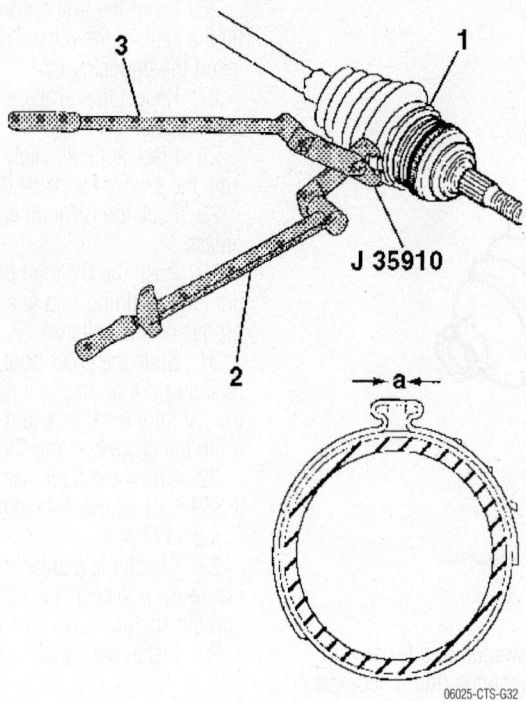

J 35910

Use Special Tool J-35910 to clamp the ear to the proper gap dimension.

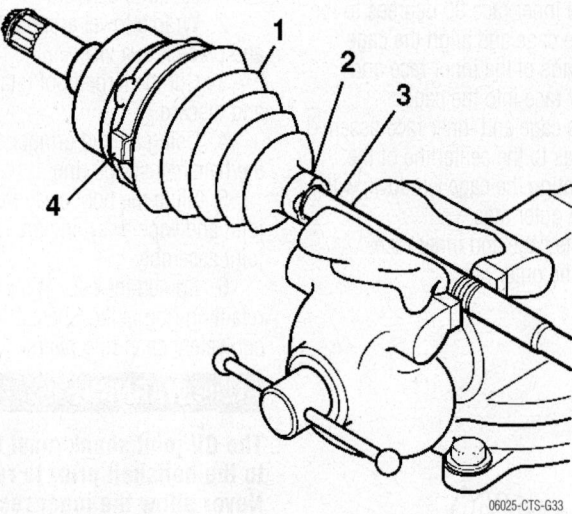

06025-CTS-G33

The halfshaft mounted in vise.

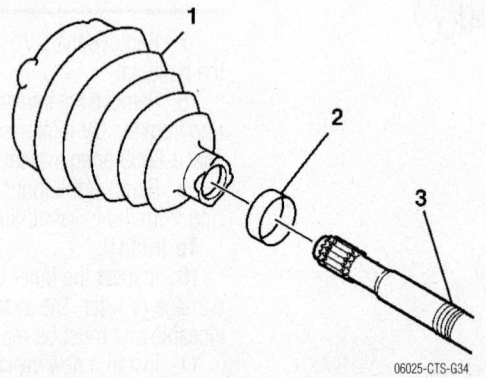

06025-CTS-G34

The largest groove below the sight groove is the boot groove.

12. Slide the boot onto the halfshaft and position the neck of the boot onto the boot groove.

13. Position the inboard end of the half-shaft assembly in the Drive Axle Swage Ring Clamp J-41048. Insert the bolts and tighten by hand until snug.

14. Tighten each bolt 180 degrees at a time until both sides are bottomed. Loosen the bolts.

15. Install the new race retaining ring into the inner race.

16. With the CV joint inner race and cage still at the bottom of the outer race bore, pack half of the grease provided in the service kit into the CV joint assembly.

17. Fill the inner boot with the remaining grease.

18. Place the new clamp over the large end of the inner boot.

19. Install a new transmission retaining ring onto the CV joint assembly.

20. Push the CV joint assembly onto the halfshaft until the retaining ring seats itself in the appropriate groove on the halfshaft.

21. Crimp the large boot retaining clamp using Special Tool J-35910 with a breaker bar and torque wrench. Tighten to 130 ft. lbs. (176 Nm).

22. Retighten the clamp until the proper gap dimension of 5/64 inch. (1.9mm) is reach.

23. Fully stroke the CV joint several times to disperse the grease throughout the joint.

24. Install the halfshaft into the vehicle.

Pinion Seal

REMOVAL & INSTALLATION

1. Before servicing the vehicle, refer to the Precautions Section.

2. Remove the driveshaft coupler-to-differential flange bolts.

❄❄ WARNING

Do not remove the coupler from the propeller shaft.

3. Push the driveshaft toward the front of the vehicle in order to release the drive-shaft coupler from the pinion flange.

4. Position the driveshaft out of the way.

5. Install the Pinion Holding Fixture J-45012 to the pinion flange.

6. While holding the Pinion Holding Fixture, remove the drive pinion nut.

7. Remove the Pinion Holding Fixture J-45012 from the flange.

8. Install Flange and Pinion Cage Remover J-45019 and remove the pinion flange.

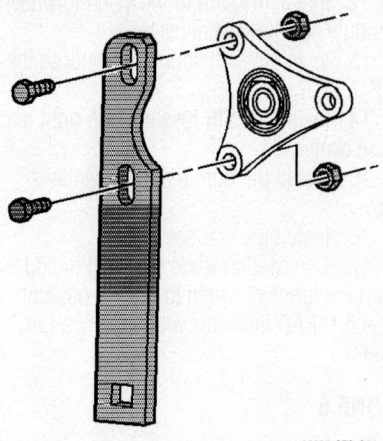

06025-CTS-G35

Installing Special Tool J-45012 Pinion Holding fixture to the pinion flange.

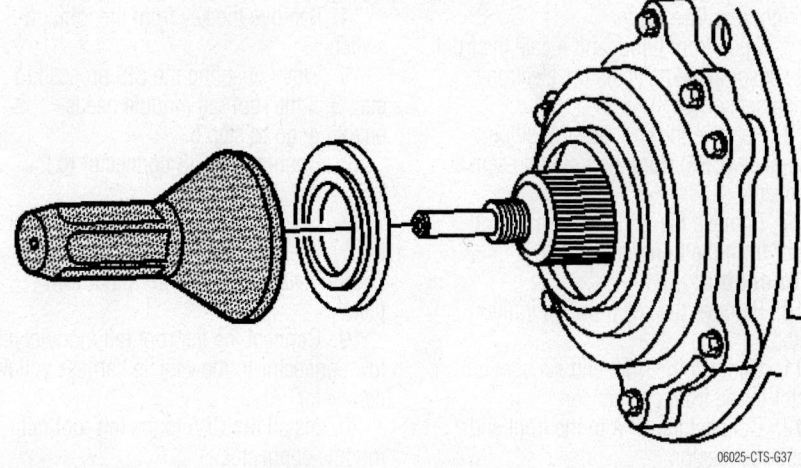

06025-CTS-G37

Use Special Tool J-45005 Seal installer to install the new pinion seal.

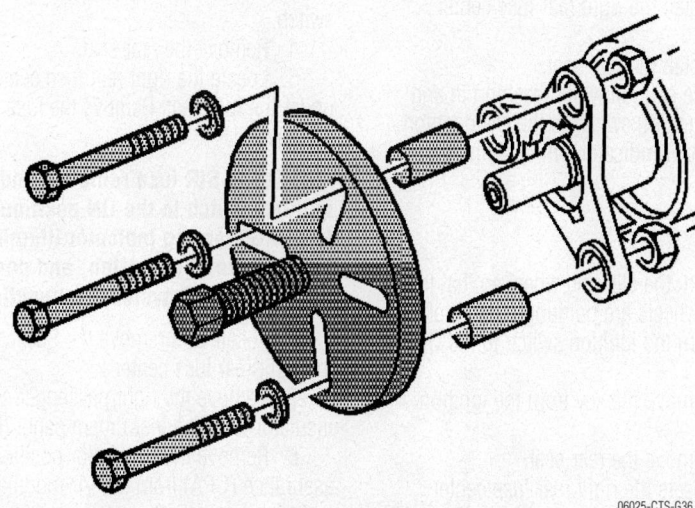

06025-CTS-G36

Remove the pinion flange using Special Tool J-45019 Flange Remover.

9. Remove the pinion seal using a flat bladed tool to pry it out.

To install:

10. Lubricate the pinion flange sealing surface of the new pinion seal with synthetic gear oil.

11. Install the new seal to Special Tool J-45005 Seal Installer.

12. Using Special Tool J-45005, install the new pinion seal to the differential.

13. Install Special Tool J-45012 to the pinion flange.

14. Install the pinion flange to drive pinion shaft. Apply threadlocker to ⅔ of the threaded area of the pinion shaft threads. Tighten the pinion nut to 210 ft. lbs. (285 Nm).

15. Install the driveshaft coupler-to-differential flange bolts. Tighten to 63 ft. lbs. (85 Nm).

16. Check the fluid level of the differential.

STEERING AND SUSPENSION

Air Bag

☀☀ CAUTION

These vehicles are equipped with an air bag system, known as a Supplemental Inflatable Restraint (SIR) system. The system must be disabled before performing service on or around system components, steering column, instrument panel components, wiring and sensors. Failure to follow safety and disabling procedures could result in accidental air bag deployment, possible personal injury and unnecessary system repairs.

PRECAUTIONS

Several precautions must be observed when handling the inflator module to avoid accidental deployment and possible personal injury.

• Never carry the inflator module by the wires or connector on the underside of the module.

• When carrying a live inflator module, hold securely with both hands, and ensure that the bag and trim cover are pointed away.

• Place the inflator module on a bench or other surface with the bag and trim cover facing up.

• With the inflator module on the bench, never place anything on or close to the module which may be thrown in the event of an accidental deployment.

DISABLING AND ENABLING

ZONE 1

1. Turn the steering wheel so that the vehicle's wheels are pointing straight ahead.

2. Turn the ignition switch to the OFF position.

3. Remove the key from the ignition switch.

4. Remove the rear seat.

5. Locate the right rear fuse center under the rear seat. Remove the fuse center top cover.

➡**With the SIR fuse removed and the ignition switch in the ON position, the AIR BAG warning indicator illuminates. This is normal operation, and does not indicate an SIR system malfunction.**

6. Locate and remove the SIR fuse from the right rear fuse center.

7. Open front hood, and locate the front end sensor also known as the electronic frontal sensor (EFS).

8. Remove the connector position assurance (CPA) from the front end sensor connector.

9. Remove the front end sensor connector from the front end sensor.

To enable:

10. Remove the key from the ignition switch.

11. Connect the front end sensor connector to the front end sensor

12. Connect the CPA to the front end sensor connector.

13. Install the SIR fuse into the right rear fuse center.

14. Install the right rear fuse center cover.

15. Install the rear seat.

16. Use caution while reaching in and turn the ignition switch to the ON position. The AIR BAG indicator will flash then turn OFF.

ZONE 2

1. Turn the steering wheel so that the vehicle's wheels are pointing straight ahead.

2. Turn the ignition switch to the OFF position.

3. Remove the key from the ignition switch.

4. Remove the rear seat.

5. Locate the right rear fuse center under the rear seat. Remove the fuse center top cover.

➡**With the SIR fuse removed and the ignition switch in the ON position, the AIR BAG warning indicator illuminates. This is normal operation, and does not indicate an SIR system malfunction.**

6. Locate and remove the SIR fuse from the right rear fuse center.

7. When disabling the roof rail module, go to step 8. If the side impact sensor (SIS) needs disabling, go to step 11.

8. Remove the left rear sail panel.

9. Remove the connector position assurance (CPA) from the left/driver roof rail module connector.

10. Disconnect the left roof rail module yellow connector from the vehicle harness yellow connector.

11. Remove the left center pillar trim panel.

12. Remove the SIS CPA from the left SIS connector.

13. Remove the SIS connector from the SIS.

To enable:

14. Remove the key from the ignition switch.

15. When enabling the SIS proceed to step 3. If the roof rail module needs enabling, go to step 6.

16. Connect the SIS connector to the SIS.

17. Connect the CPA to the SIS connector.

18. Install the left center pillar trim panel.

19. Connect the left roof rail module yellow connector to the vehicle harness yellow connector.

20. Install the CPA to the left roof rail module connector.

21. Install the left rear sail panel.

22. Install the SIR fuse into the right rear fuse center.

23. Install the right rear fuse center cover.

24. Install the rear seat.

25. Use caution while reaching in and turn the ignition switch to the ON position. The AIR BAG indicator will flash then turn OFF.

ZONE 3

1. Turn the steering wheel so that the vehicle's wheels are pointing straight ahead.

2. Turn the ignition switch to the OFF position.

3. Remove the key from the ignition switch.

4. Remove the rear seat.

5. Locate the right rear fuse center under the rear seat. Remove the fuse center top cover.

➡**With the SIR fuse removed and the ignition switch in the ON position, the AIR BAG warning indicator illuminates. This is normal operation, and does not indicate an SIR system malfunction.**

6. Locate and remove the SIR fuse from the right rear fuse center.

7. Remove the left/driver sound insulator from the instrument panel (I/P).

8. Remove the connector position assurance (CPA) from the steering wheel module coil yellow connector.

9. Disconnect the steering wheel module coil yellow connector from the vehicle harness yellow connector.

To enable:

10. Remove the key from the ignition switch.

11. Connect the steering wheel module coil yellow connector to the vehicle harness yellow connector.

12. Install the CPA to the steering wheel module coil yellow connector.

13. Install the left sound insulator to the I/P.

14. Install the SIR fuse into the right rear fuse center.

15. Install the right rear fuse center cover.

16. Install the rear seat.

17. Use caution while reaching in and turn the ignition switch to the ON position. The AIR BAG indicator will flash then turn OFF.

ZONE 5

1. Turn the steering wheel so that the vehicle's wheels are pointing straight ahead.

2. Turn the ignition switch to the OFF position.

3. Remove the key from the ignition switch.

4. Remove the rear seat.

5. Locate the right rear fuse center under the rear seat. Remove the fuse center top cover.

➡**With the SIR fuse removed and the ignition switch in the ON position, the AIR BAG warning indicator illuminates. This is normal operation, and does not indicate an SIR system malfunction.**

6. Locate and remove the SIR fuse from the right rear fuse center.

7. Remove the right/passenger sound insulator from the instrument panel (I/P).

8. Remove the connector position assurance (CPA) from the I/P module yellow connector.

9. Disconnect the I/P module yellow connector from the vehicle harness yellow connector.

To enable:

10. Remove the key from the ignition switch.

11. Connect the I/P module yellow connector to the vehicle harness yellow connector.

12. Install the CPA to the I/P module yellow connector.

13. Install the right sound insulator to the I/P.

14. Install the SIR fuse into the right rear fuse center.

15. Install the right rear fuse center cover.

16. Install the rear seat.

17. Use caution while reaching in and turn the ignition switch to the ON position. The AIR BAG indicator will flash then turn OFF.

ZONE 6

1. Turn the steering wheel so that the vehicle's wheels are pointing straight ahead.

2. Turn the ignition switch to the OFF position.

3. Remove the key from the ignition switch.

4. Remove the rear seat.

5. Locate the right rear fuse center under the rear seat. Remove the fuse center top cover.

➡ **With the SIR fuse removed and the ignition switch in the ON position, the AIR BAG warning indicator illuminates. This is normal operation, and does not indicate an SIR system malfunction.**

6. Locate and remove the SIR fuse from the right rear fuse center.

7. When disabling the roof rail module, go to step 8. If the side impact sensor (SIS) needs disabling, go to step 11.

8. Remove the right rear sail panel.

9. Remove the connector position assurance (CPA) from the right/passenger roof rail module connector.

10. Disconnect the right roof rail module yellow connector from the vehicle harness yellow connector.

11. Remove the right center pillar trim panel.

12. Remove the SIS CPA from the right SIS connector.

13. Remove the SIS connector from the SIS.

To enable:

14. Remove the key from the ignition switch.

15. When enabling the SIS, proceed to step 3. If the roof rail module needs enabling, go to step 6.

16. Connect the SIS connector to the SIS.

17. Connect the CPA to the SIS connector.

18. Install the right center pillar trim panel.

19. Connect the right roof rail module yellow connector to the vehicle harness yellow connector.

20. Install the CPA to the right roof rail module connector.

21. Install the right rear sail panel.

22. Install the SIR fuse into the right rear fuse center.

23. Install the right rear fuse center cover.

24. Install the rear seat.

25. Use caution while reaching in and turn the ignition switch to the ON position. The AIR BAG indicator will flash then turn OFF.

ZONE 7

1. Turn the steering wheel so that the vehicle's wheels are pointing straight ahead.

2. Turn the ignition switch to the OFF position.

3. Remove the key from the ignition switch.

4. Remove the rear seat.

5. Locate the right rear fuse center under the rear seat. Remove the fuse center top cover.

➡ **With the SIR fuse removed and the ignition switch in the ON position, the AIR BAG warning indicator illuminates. This is normal operation, and does not indicate an SIR system malfunction.**

6. Locate and remove the SIR fuse from the right rear fuse center.

7. Remove both connector position assurance (CPA) from the LF/driver side impact module and seat belt pretensioner yellow connector located under the front of driver seat.

8. Disconnect the LF side impact module and pretensioner yellow connector from the vehicle harness yellow connector.

To enable:

9. Remove the key from the ignition switch.

10. Connect the LF side impact module and pretensioner yellow connector to the vehicle harness yellow connector.

11. Install both CPA locks to the LF side impact module and pretensioner yellow connector.

12. Install the SIR fuse into the right rear fuse center.

13. Install the right rear fuse center cover.

14. Install the rear seat.

15. Use caution while reaching in and turn the ignition switch to the ON position. The AIR BAG indicator will flash then turn OFF.

ZONE 8

1. Turn the steering wheel so that the vehicle's wheels are pointing straight ahead.

2. Turn the ignition switch to the OFF position.

3. Remove the key from the ignition switch.

4. Remove the rear seat.

5. Locate the right rear fuse center under the rear seat. Remove the fuse center top cover.

➡ **With the SIR fuse removed and the ignition switch in the ON position, the AIR BAG warning indicator illuminates.**

This is normal operation, and does not indicate an SIR system malfunction.

6. Locate and remove the SIR fuse from the right rear fuse center.

7. Remove the right rear sail panel.

8. Remove the connector position assurance (CPA) from the right/passenger roof rail module yellow connector.

9. Disconnect the right roof rail module yellow connector from the vehicle harness yellow connector.

10. Remove the passenger/right sound insulator from the instrument Panel (I/P).

11. Remove the CPA from the I/P module yellow connector.

12. Disconnect the I/P module yellow connector from the vehicle harness yellow connector.

13. Remove both CPA locks from the passenger/RF side impact module and seat belt pretensioner yellow connector located under the front of passenger seat.

14. Disconnect the RF side impact module and pretensioner yellow connector from the vehicle harness yellow connector.

15. Remove the driver/left sound insulator from the I/P.

16. Remove the CPA from the steering wheel module coil yellow connector.

17. Disconnect the steering wheel module coil yellow connector from the vehicle harness yellow connector.

18. Remove both CPA locks from the driver/LF side impact module and seat belt pretensioner yellow connector located under the front of driver seat.

19. Disconnect the LF side impact module and pretensioner yellow connector from the vehicle harness yellow connector.

20. Remove the left rear sail panel.

21. Remove the CPA from the left/driver roof rail module yellow connector.

22. Disconnect the left roof rail module yellow connector from the vehicle harness yellow connector.

To enable:

23. Remove the key from the ignition switch.

24. Connect the steering wheel module coil yellow connector to the vehicle harness yellow connector.

25. Install the CPA to the steering wheel module coil yellow connector.

26. Install the driver/left sound insulator to the I/P.

27. Connect the driver/LF side impact module and seat belt pretensioner yellow connector to the vehicle harness yellow connector located under the front of driver seat.

28. Install both CPA locks to the LF side

impact module and pretensioner yellow connector.

29. Connect the left/driver roof rail module yellow connector to the vehicle harness yellow connector.

30. Install the CPA to the left roof rail module yellow connector.

31. Install the left rear sail panel.

32. Connect the passenger/I/P module yellow connector to the vehicle harness yellow connector.

33. Install the CPA to the I/P module yellow connector.

34. Install the passenger/right sound insulator to the I/P.

35. Connect the passenger/RF side impact module and seat belt pretensioner yellow connector to the vehicle harness yellow connector located under the front of passenger seat.

36. Install both CPA locks to the RF side impact module and pretensioner yellow connector.

37. Connect the passenger/right roof rail module yellow connector to the vehicle harness yellow connector.

38. Install the CPA to the right roof rail module yellow connector.

39. Install the right rear sail panel.

40. Install the SIR fuse into the right rear fuse center.

41. Install the right rear fuse center cover.

42. Install the rear seat.

43. Use caution while reaching in and turn the ignition switch to the ON position. The AIR BAG indicator will flash then turn OFF.

ZONE 9

1. Turn the steering wheel so that the vehicle's wheels are pointing straight ahead.

2. Turn the ignition switch to the OFF position.

3. Remove the key from the ignition switch.

4. Remove the rear seat.

5. Locate the right rear fuse center under the rear seat. Remove the fuse center top cover.

➡ **With the SIR fuse removed and the ignition switch in the ON position, the AIR BAG warning indicator illuminates. This is normal operation, and does not indicate an SIR system malfunction.**

6. Locate and remove the SIR fuse from the right rear fuse center.

7. Remove both connector position assurance (CPA) from the RF/passenger side impact module and seat belt preten-

sioner yellow connector located under the front of passenger seat.

8. Disconnect the RF side impact module and pretensioner yellow connector from the vehicle harness yellow connector.

To enable:

9. Remove the key from the ignition switch.

10. Connect the RF side impact module and pretensioner yellow connector to the vehicle harness yellow connector.

11. Install both CPA locks to the RF side impact module and pretensioner yellow connector.

12. Install the SIR fuse into the right rear fuse center.

13. Install the right rear fuse center cover.

14. Install the rear seat.

15. Use caution while reaching in and turn the ignition switch to the ON position. The AIR BAG indicator will flash then turn OFF.

Rack and Pinion Power Steering Gear

REMOVAL & INSTALLATION

➡**The front wheels of the vehicle must be maintained in the straight ahead position and the steering column must be in the LOCK position before disconnecting the steering column or intermediate shaft. Failure to follow these procedures will cause improper alignment of some components during installation and result in damage to the SIR coil assembly.**

1. Before servicing the vehicle, refer to the Precautions Section.

2. Center the front wheels, then turn the ignition key to the **LOCK**position.

3. Lock the steering column through the access hole in the lower steering column trim cover using Steering Column Lock Tool J 42640.

4. Remove or disconnect the following:
- Front wheels
- Front air deflector
- Intermediate shaft
- Outer tie rod ends
- Variable effort steering electrical connector, if equipped
- Stabilizer shaft
- Power steering pressure and return hoses
- Power steering pressure hose retaining bolt
- Steering gear mounting bolts

5. Remove the steering gear through the left wheel opening.

To install:

6. Install or connect the following:
- Steering gear to the vehicle. Tighten the mounting bolts to 70 ft. lbs. (95 Nm).
- Power steering pressure hose retaining bolt. Tighten the bolt to 80 inch lbs. (9 Nm).
- Power steering pressure and return hoses. Use new seals and tighten the hoses to 22 ft. lbs. (30 Nm).
- Stabilizer shaft
- Variable effort steering electrical connector, if equipped
- Outer tie rod ends. Tighten the nuts to 52 ft. lbs. (70 Nm).

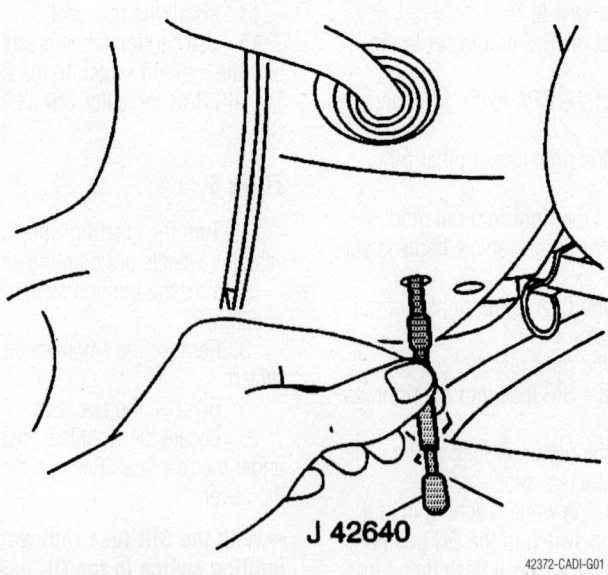

J 42640

42372-CADI-G01

Steering column locking tool J 42640

- Intermediate shaft. Tighten the pinch bolt to 37 ft. lbs. (50 Nm).
- Front air deflector
- Front wheels

7. Remove the Steering Column Lock Tool J 42640.

8. Bleed the power steering hydraulic system.

9. Align the front wheels as necessary.

Shock Absorber

REMOVAL & INSTALLATION

Front

1. Before servicing the vehicle, refer to the Precautions Section.

2. Remove or disconnect the following:
- Front wheel
- Upper control arm-to-steering knuckle pinch bolt
- Upper ball joint from steering knuckle
- Lower shock mounting bolts
- A/C line bracket from shock tower and position aside
- Upper shock mounting bolts
- Shock module assembly

3. Install the shock assembly into a spring compressor.

4. Compress the coil spring and remove the shock absorber upper retaining nut.

5. Remove the shock absorber from the shock module assembly.

To install:

6. Install the shock absorber into the shock module assembly. Tighten the upper retaining nut to 18 ft. lbs. (25 Nm).

7. Remove the shock module assembly from the spring compressor.

8. Install or connect the following:
- Shock module assembly into the vehicle. Tighten the upper mounting bolts to 83 ft. lbs. (112 Nm).
- A/C line bracket. Tighten to 80 inch lbs. (9 Nm).
- Lower shock mounting bolts. Tighten to 18 ft. lbs. (25 Nm).
- Upper ball joint to the steering knuckle
- Upper control arm-to-steering knuckle pinch bolt. Tighten to 44 ft. lbs. (60 Nm).
- Front wheel

Rear

1. Before servicing the vehicle, refer to the Precautions Section.

2. Remove the rear seat back to allow access to the upper shock absorber mounting.

3. Remove or disconnect the following:
- Sound insulator from the shock tower
- Upper mounting nut, washer and grommet
- Lower shock mounting bolt
- Shock absorber

To install:

4. Install or connect the following:
- Lower rubber mounting grommet and washer
- Shock into the tower
- Protective cap to the shock tower
- Lower shock mounting bolt. Torque the bolt 111 ft. lbs. (150 Nm).
- Upper shock mounting grommet, washer and nut. Torque the nut 18 ft. lbs. (25 Nm).
- Rear seat back

Coil Spring

REMOVAL & INSTALLATION

Front

1. Before servicing the vehicle, refer to the Precautions Section.

2. Remove or disconnect the following:
- Front wheel
- Upper control arm-to-steering knuckle pinch bolt

- Upper ball joint from steering knuckle
- Lower shock mounting bolts
- A/C line bracket from shock tower and position aside
- Upper shock mounting bolts
- Shock module assembly

3. Install the shock assembly into a spring compressor.

4. Matchmark the locations of the upper control arm bracket and insulator to the coil spring.

5. Compress the coil spring and remove the shock absorber upper retaining nut.

6. Remove the shock absorber from the coil spring assembly.

7. Remove the upper control arm bracket assembly, insulator and coil spring from the spring compressor.

To install:

8. Install the coil spring, insulator, upper control arm bracket assembly, and shock absorber to the spring compressor aligning all marks made during disassembly.

9. Compress the coil spring and install the shock absorber retaining nut. Tighten to 18 ft. lbs. (25 Nm).

10. Remove the coil spring compressor from the shock module assembly.

11. Install or connect the following:

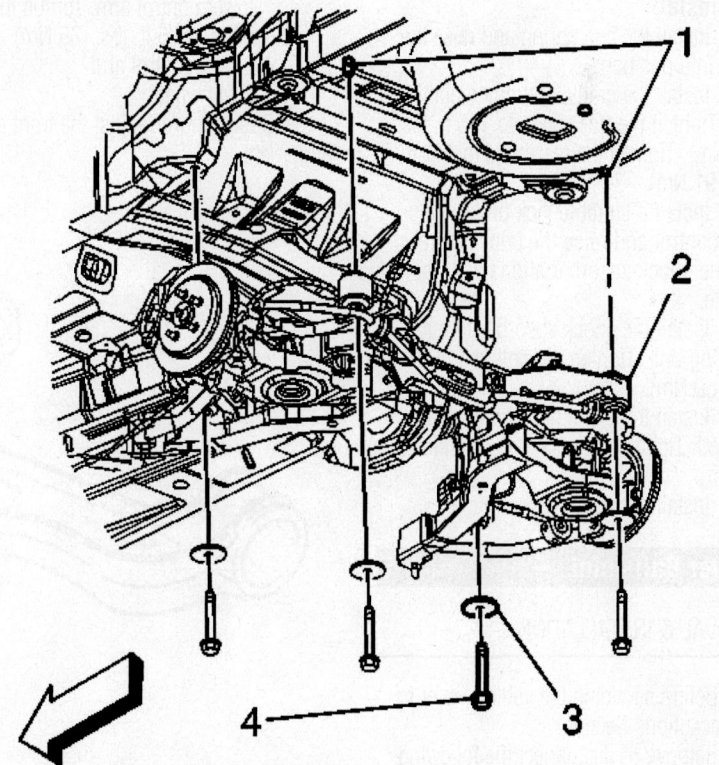

View of the bolt locations of the rear cradle–CTS.

06025-CTS-G38

- Shock module assembly into the vehicle. Tighten the upper mounting bolts to 83 ft. lbs. (112 Nm).
- A/C line bracket. Tighten to 80 inch lbs. (9 Nm).
- Lower shock mounting bolts. Tighten to 18 ft. lbs. (25 Nm).
- Upper ball joint to the steering knuckle
- Upper control arm-to-steering knuckle pinch bolt. Tighten to 44 ft. lbs. (60 Nm).
- Front wheel

Rear

1. Before servicing the vehicle, refer to the Precautions Section.
2. Remove the wheel.
3. Remove the brake hose bracket from the mounting studs when removing the right coil spring.
4. Support the lower control arm using a suitable jack.
5. Remove the lower shock mounting bolt.
6. Lower the lower control arm and remove the jack.
7. Support the cradle with a suitable jack.
8. Remove the side cradle-to-body mounting bolts.
9. Lower the side of the cradle and remove the coil spring.
   To install:
10. Install the coil spring and raise the cradle into position.
11. Install the cradle-to-body mounting bolts. Tighten the front bolts to 195 ft. lbs. (265 Nm). Tighten the rear bolts to 141 ft. lbs. (191 Nm).
12. Install a suitable jack under the lower control and raise the control arm until the shock absorber aligns with the knuckle.
13. Install the shock absorber lower mounting bolt. Tighten the bolt to 111 ft. lbs. (150 Nm).
14. Install the brake hose bracket, if removed. Tighten the nuts to 89 inch lbs. (10 Nm).
15. Install the wheel.

Lower Ball Joint

REMOVAL & INSTALLATION

1. Before servicing the vehicle, refer to the Precautions Section.
2. Remove or disconnect the following:
   - Front wheel
   - Lower control arm

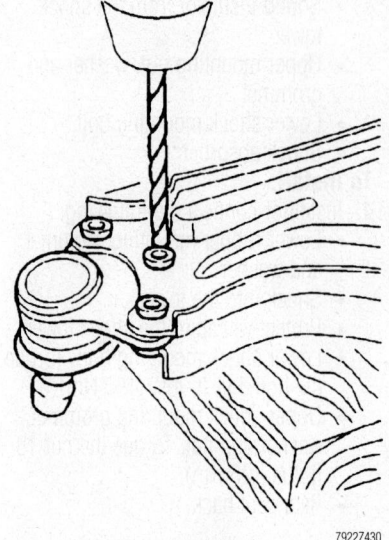

7922Z430

Drill out the rivets to remove the ball joint from the lower control arm

3. Using a 0.5 inch (13mm) drill, drill out the 3 rivet heads that attach the ball stud to the lower control arm.
4. Remove the ball stud from the lower control arm.
   To install:
5. Install or connect the following:
   - Ball stud to the lower control arm
   - Bolts through the upper side of the lower control arm. Torque the new bolts to 26 ft. lbs. (35 Nm).
   - Lower control arm
   - Front wheel
6. Check and/or adjust the front end alignment.

Upper Control Arm

REMOVAL & INSTALLATION

Rear

1. Before servicing the vehicle, refer to the Precautions Section.
2. Remove or disconnect the following:
   - Rear wheel
   - Shock absorber
   - Brake rotor
   - Upper control arm mounting bolts
   - Upper ball joint from knuckle
   - Upper control arm
   To install:
3. Install or connect the following:
   - Upper control arm
   - Upper ball joint to the knuckle. Tighten the nut to 15 ft. lbs. (20 Nm) plus 210 degrees.
   - Upper control arm mounting bolts. Tighten to 129 ft. lbs. (175 Nm).
   - Brake rotor
   - Shock absorber
   - Rear wheel

Lower Control Arm

REMOVAL & INSTALLATION

Front

1. Before servicing the vehicle, refer to the Precautions Section.
2. Remove or disconnect the following:
   - Front wheel
   - Stabilizer shaft link
   - Shock assembly lower mounting bolts

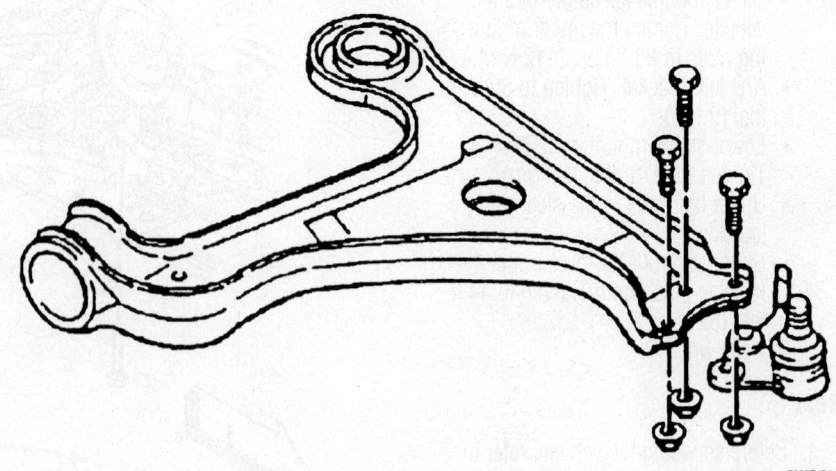

7922Z431

The replacement ball joint will be bolted to the lower control arm with the bolts supplied in the kit

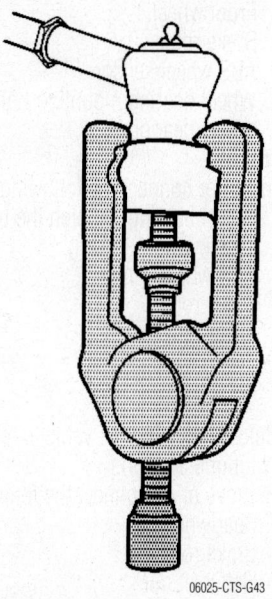

Separate the outer tie rod from the knuckle using Special Tool J-24319-B–Lower control arm.

- Outer tie rod from the steering knuckle using Puller J-24319-B
- ABS wire harness from the lower control arm
- Lower control arm mounting bolts

3. Lower and support the lower control arm to gain access to the lower ball joint.

4. Separate the lower ball joint from the steering knuckle using Special Tool J-43631 or equivalent Ball Joint Remover.

5. Remove the lower control arm.

To install:

6. Install or connect the following:
- Lower control arm. Tighten the mounting bolts to 100 ft. lbs. (135 Nm).

- ABS wiring harness
- Lower ball joint to the steering knuckle. Tighten the nut to 15 ft. lbs. (20 Nm) plus 210 degrees.
- Outer tie rod to the steering knuckle. Tighten the retaining nut to 55 ft. lbs. (75 Nm).
- Shock assembly lower mounting bolts. Tighten to 18 ft. lbs. (25 Nm).
- Stabilizer shaft link to the lower control arm. Tighten the retaining nut to 37 ft. lbs. (50 Nm).
- Front wheel.

7. Check and/or adjust the front wheel alignment.

Rear

1. Before servicing the vehicle, refer to the Precautions Section.

2. Remove or disconnect the following:
- Wheel
- Stabilizer shaft link connection at the lower control arm
- Halfshaft from the rear wheel hub flange
- Lower shock mounting bolts
- Lower control arm-to-knuckle mounting bolt
- Lower control arm and coil spring

To install:

3. Install the rear coil spring and the lower control arm to a suitable jack.

4. Raise the jack until the lower control arm and spring is in position in the frame.

5. Install or connect the following:
- Lower control arm. Torque the control arm-to-frame mounting bolts to 111 ft. lbs. (150 Nm).
- Lower control arm-to-knuckle

mounting bolt. Tighten the bolt to 129 ft. lbs. (175 Nm).
- Lower shocking mounting bolt. Tighten to 111 ft. lbs. (150 Nm).
- Halfshaft
- Stabilizer shaft link to the lower control arm. Tighten the nut to 37 ft. lbs. (50 Nm).
- Wheel

6. Check and/or adjust the rear toe.

CONTROL ARM BUSHING REPLACEMENT

Front

1. Before servicing the vehicle, refer to the Precautions Section.

2. Remove the control arm from the vehicle.

3. Press the bushings from the control arm.

To install:

4. Press the new bushings into the control arm.

5. Install the control arm in the vehicle.

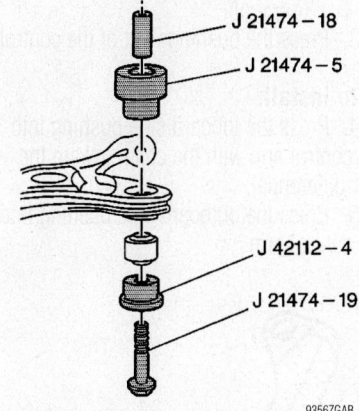

Use the following tools to install the front control arm vertical bushing

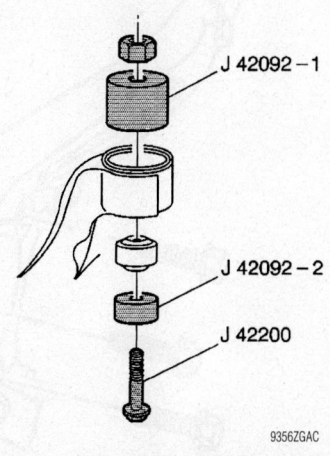

Use the following tools to install the front control arm horizontal bushing

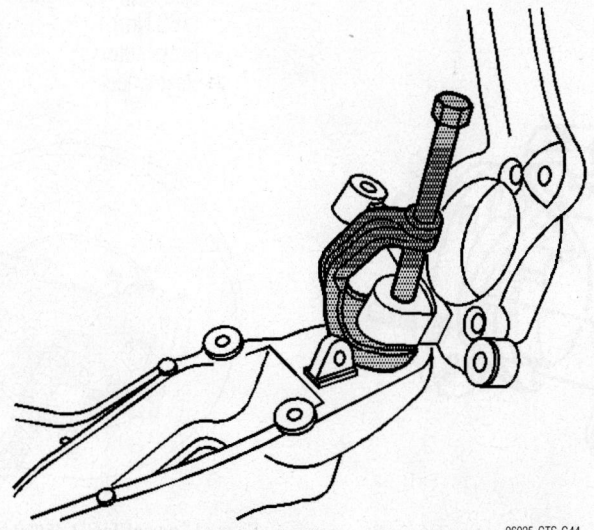

Separate the lower ball joint from the knuckle using Special Tool J-43631–Lower control arm.

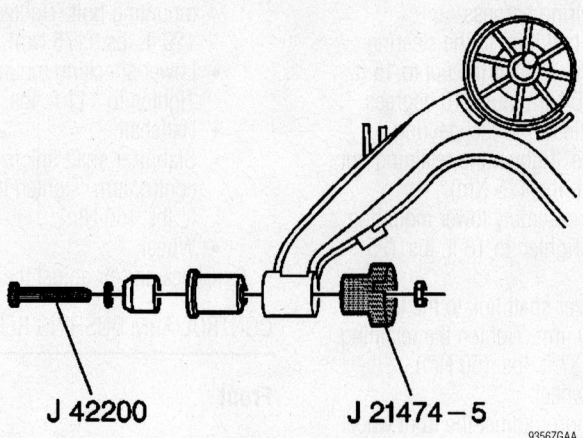

J 42200   **J 21474−5**

9356ZGAA

Use the following tools to remove and install the rear control arm bushing

Rear

1. Before servicing the vehicle, refer to the Precautions Section.
2. Remove or disconnect the following:
   - Wheel
   - Rear axle lower control arm
   - Collar for the inboard and outboard lower control arm bushings by cutting it off
3. Press the bushings out of the control arm.

To install:

4. Press the inboard side bushing into the control arm with the collar toward the rear differential.
5. Press the outboard side bushing into the control arm.

6. Install or connect the following:
   - Lower control arm
   - Wheel

Wheel Bearings

ADJUSTMENT

The wheel bearings are not adjustable.

REMOVAL & INSTALLATION

Front

1. Before servicing the vehicle, refer to the Precautions Section.
2. Remove or disconnect the following:
   - Front wheel
   - Brake rotor
   - ABS wheel sensor
   - Wheel bearing mounting bolts
   - Wheel bearing

To install:

3. Install or connect the following:
   - Wheel bearing. Tighten the bolts to 100 ft. lbs. (135 Nm).
   - ABS wheel sensor
   - Brake rotor
   - Front wheel

Rear

1. Before servicing the vehicle, refer to the Precautions Section.
2. Remove or disconnect the following:
   - Rear wheel
   - Brake rotor
   - Halfshaft nut
   - ABS wheel sensor
   - Parking brake cable bracket from the knuckle
3. Install Special Tool J-45859 Axle Remover to the wheel bearing.
4. Disengage the halfshaft from the wheel bearing using Special Tool J-45859.
5. Remove or disconnect the following:
   - Wheel bearing mounting bolts
   - Wheel bearing and backing plate from the knuckle
   - Special Tool J-45859 from the wheel bearing

To install:

6. Install or connect the following:
   - Wheel bearing and backing plate to the knuckle. Tighten to 92 ft. lbs. (125 Nm).
   - Parking brake cable bracket. Tighten the bolts to 44 ft. lbs. (60 Nm).
   - ABS wheel sensor
   - Halfshaft nut. Tighten to 118 ft. lbs. (160 Nm).
   - Brake rotor
   - Rear wheel

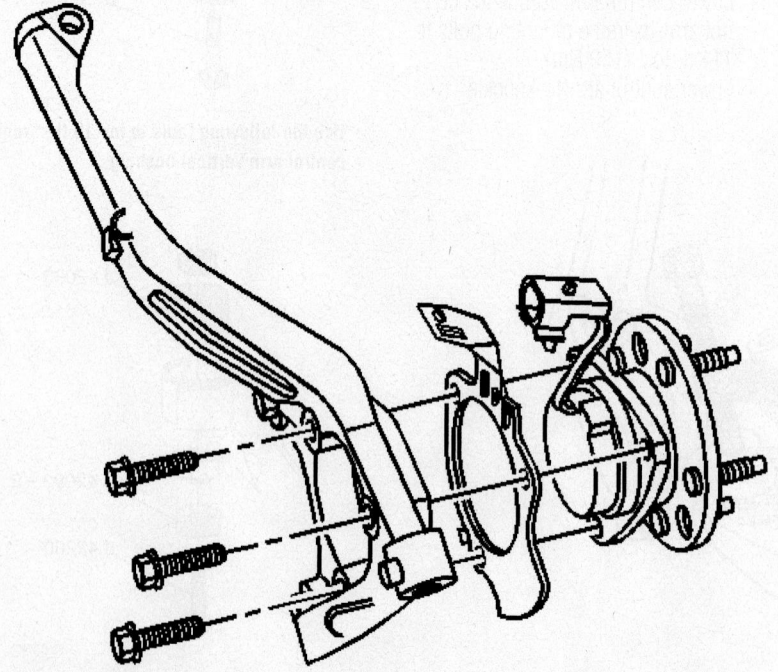

06025-CTS-G42

Location of the bearing hub mounting bolts–Front

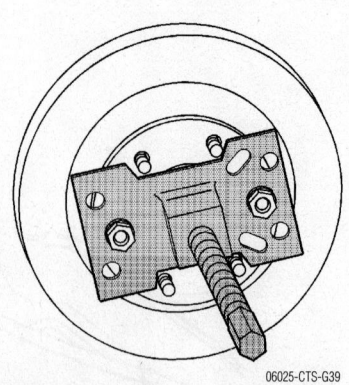

06025-CTS-G39

View of Special Tool J-45859 installed on the wheel bearing.

Brake Caliper

REMOVAL & INSTALLATION

Front

CTS

1. Disconnect the negative battery cable.
2. Remove ⅔ of the brake fluid from the master cylinder.
3. Remove the front wheel. Mark the relationship between the wheel and the wheel stud for re-installation purposes.
4. Install 2 wheel nuts to keep the rotor in place.
5. Using a large C-clamp against the inboard pad, compress the caliper piston into the caliper to provide clearance during removal.
6. Place a catch pan under the caliper.
7. Disconnect the brake hose from the caliper. Cap the line to prevent excessive fluid loss or contamination.
8. Remove the caliper mounting bolts and remove the caliper from the vehicle.
9. Inspect the mounting bolts; sleeves and boots for wear and/or damage. Replace parts as necessary.

To install:

10. Before installing the caliper, make sure the piston is fully seated in the bore and the brake pads are properly seated.
11. Lubricate the mounting bolt shafts and inner diameter of the sleeves with silicone grease.
12. Install the caliper in the caliper mounting bracket and install the mounting bolts. Torque the mounting bolts to 46 ft. lbs. (63 Nm).
13. Connect the brake hose with the bolt and new gaskets. Torque the brake hose bolt to 37 ft. lbs. (50 Nm).
14. Refill the master cylinder and bleed the brake system.
15. Remove the 2 wheel nuts securing the rotor.
16. Install the wheel and tire assembly.
17. Connect the negative battery cable.
18. Road test the vehicle for proper brake system operation.

CTS-V

1. If the brake fluid level is midway between the maximum-full point and the minimum allowable level, no brake fluid needs to be removed from the reservoir before proceeding.

2. If the brake fluid level is higher than midway between the maximum-full point and the minimum allowable level, remove brake fluid to the midway point before proceeding.
3. Remove the front wheel.
4. Carefully insert a plastic flat-bladed trim tool between the rotor and inboard brake pad.
5. Carefully apply pressure to the inboard brake pad until both caliper inner pistons are compressed into the caliper piston bores.
6. Carefully insert a plastic flat-bladed trim tool between the rotor and outboard brake pad.
7. Carefully apply pressure to the outboard brake pad until both caliper outer pistons are compressed into the bores.
8. Remove the brake hose and copper gaskets from the brake caliper. Cap the line to prevent excessive fluid loss or contamination.
9. Remove the caliper-to-knuckle mounting bolts.
10. Remove the brake caliper.

To install:

11. Apply threadlocker to two-thirds of the threaded length of the caliper to knuckle mounting bolts.
12. Apply a thin coat of high temperature silicone brake lubricant to the brake caliper pin.
13. Install the brake caliper. Tighten the mounting bolts to 96 ft. lbs. (130 Nm).
14. Connect the brake hose with the bolt and new gaskets. Tighten the bolt to 25 ft. lbs. (34 Nm).
15. Refill the master cylinder and bleed the brake system.
16. Install the wheel.
17. Road test the vehicle for proper brake system operation.

Rear

CTS

1. Remove ⅔ of the brake fluid from the master cylinder.
2. Remove the rear wheel.
3. Install 2 wheel nuts to keep the rotor in place.
4. Using a large C-clamp against the inboard pad, compress the caliper piston into the caliper to provide clearance during removal.
5. Place a catch pan under the caliper.
6. Disconnect the brake hose from the caliper. Cap the line to prevent excessive fluid loss or contamination.

7. Remove the caliper mounting bolts and remove the caliper from the vehicle.
8. Inspect the mounting bolts; sleeves and boots for wear and/or damage. Replace parts as necessary.

To install:

9. Before installing the caliper, make sure the piston is fully seated in the bore and the brake pads are properly seated.
10. Lubricate the mounting bolt shafts and inner diameter of the sleeves with silicone grease.
11. Install the caliper in the caliper mounting bracket and install the mounting bolts. Torque the mounting bolts to 46 ft. lbs. (63 Nm).
12. Connect the brake hose with the bolt and new gaskets. Torque the brake hose bolt to 37 ft. lbs. (50 Nm).
13. Refill the master cylinder and bleed the brake system.
14. Remove the 2 wheel nuts securing the rotor.
15. Install the rear wheel.
16. Road test the vehicle for proper brake system operation.

CTS-V

1. If the brake fluid level is midway between the maximum-full point and the minimum allowable level, no brake fluid needs to be removed from the reservoir before proceeding.
2. If the brake fluid level is higher than midway between the maximum-full point and the minimum allowable level, remove brake fluid to the midway point before proceeding.
3. Remove the front wheel.
4. Carefully insert a plastic flat-bladed trim tool between the rotor and inboard brake pad.
5. Carefully apply pressure to the inboard brake pad until both caliper inner pistons are compressed into the caliper piston bores.
6. Carefully insert a plastic flat-bladed trim tool between the rotor and outboard brake pad.
7. Carefully apply pressure to the outboard brake pad until both caliper outer pistons are compressed into the bores.
8. Remove the brake hose and copper gaskets from the brake caliper. Cap the line to prevent excessive fluid loss or contamination.
9. Remove the caliper-to-knuckle mounting bolts.
10. Remove the brake caliper.

To install:

11. Apply threadlocker to two-thirds of the threaded length of the caliper to knuckle mounting bolts.

12. Apply a thin coat of high temperature silicone brake lubricant to the brake caliper pin.

13. Install the brake caliper. Tighten the mounting bolts to 88 ft. lbs. (120 Nm).

14. Connect the brake hose with the bolt and new gaskets. Tighten the bolt to 25 ft. lbs. (34 Nm).

15. Refill the master cylinder and bleed the brake system.

16. Install the wheel.

17. Road test the vehicle for proper brake system operation.

Disc Brake Pads

REMOVAL & INSTALLATION

Front

CTS

1. Remove ⅔ of the brake fluid from the master cylinder reservoir.

2. Remove the front wheel.

3. Remove the caliper mounting bolts.

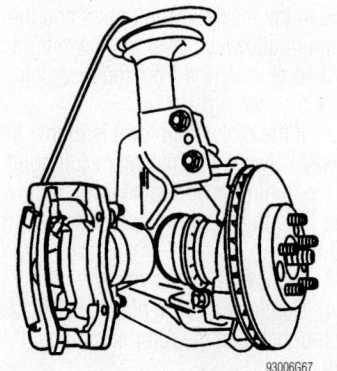

Supporting the front caliper—CTS

93006G67

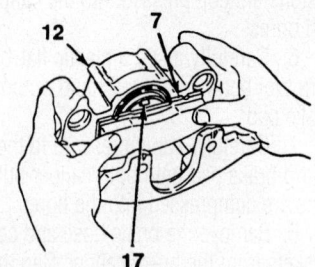

7. Inboard Shoe & Lining
12. Caliper Housing
17. Shoe Retainer Spring

93006G68

Installing the inboard disc brake pad into the front caliper—CTS

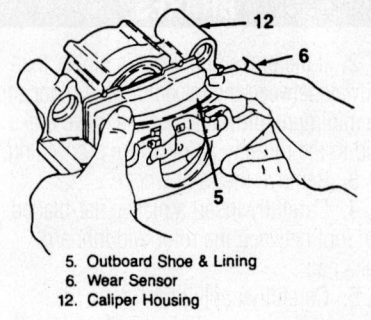

5. Outboard Shoe & Lining
6. Wear Sensor
12. Caliper Housing

93006G69

Installing the outboard disc brake pad in the front caliper—CTS

4. Remove the caliper from the steering knuckle without disconnecting the brake hose.

5. Suspend the caliper from the coil spring with wire. Do not let the caliper hang from the brake hose.

6. Remove brake pads from anchor bracket.

To install:

7. Fully seat the caliper piston into the bore using a large C-clamp.

8. Lubricate the brake caliper mounting surfaces.

9. Install the brake pads onto the anchor bracket using new shims and clips.

10. Place the caliper onto the anchor bracket and install the mounting bolts.

11. Torque the mounting bolts to 63 ft. lbs. (85 Nm).

12. Install the wheel and tire assembly and torque to specification.

13. Lower the vehicle.

14. Pump the brake pedal several times to seat the brake pads.

15. Check the fluid level in the master cylinder and fill as necessary.

16. Road test the vehicle for proper brake operation.

CTS-V

1. If the brake fluid level is midway between the maximum-full point and the minimum allowable level, no brake fluid needs to be removed from the reservoir before proceeding.

2. If the brake fluid level is higher than midway between the maximum-full point and the minimum allowable level, remove brake fluid to the midway point before proceeding.

3. Remove the front wheel.

4. Holding the lower end of the retainer (1) down and using a hammer and punch carefully tap the lower caliper guide pin (2) inward out of the caliper.

Caliper guide pin–CTS-V

06025-CTS-G40

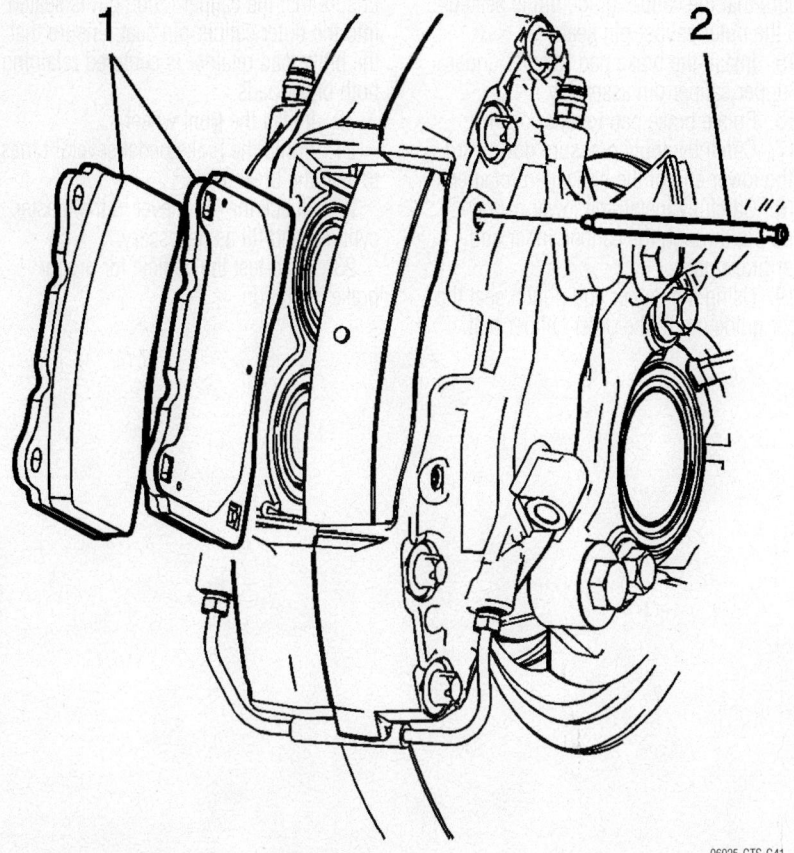

Upper caliper pin-to-brake pad mounting pin–CTS-V

4. Remove the caliper from the steering knuckle without disconnecting the brake hose.

5. Suspend the caliper from the coil spring with wire. Do not let the caliper hang from the brake hose.

6. Remove brake pads from anchor bracket.

To install:

7. Fully seat the caliper piston into the bore using a large C-clamp.

8. Lubricate the brake caliper mounting surfaces.

9. Install the brake pads onto the anchor bracket using new shims and clips.

10. Place the caliper onto the anchor bracket and install the mounting bolts.

11. Torque the mounting bolts to 44 ft. lbs. (60 Nm).

12. Install the wheel and tire assembly and torque to specification.

13. Lower the vehicle.

14. Pump the brake pedal several times to seat the brake pads.

15. Check the fluid level in the master cylinder and fill as necessary.

16. Road test the vehicle for proper brake operation.

CTS-V

1. If the brake fluid level is midway between the maximum-full point and the minimum allowable level, no brake fluid needs to be removed from the reservoir before proceeding.

2. If the brake fluid level is higher than midway between the maximum-full point and the minimum allowable level, remove brake fluid to the midway point before proceeding.

3. Remove the front wheel.

4. Holding the lower end of the retainer down and using a hammer and punch carefully tap the lower caliper guide pin inward out of the caliper.

5. Rotate the brake pad retainer upward and remove the retainer.

6. Using a hammer and punch tap the upper caliper to brake pad mounting pin inward out of the caliper.

7. Carefully insert a plastic flat-bladed trim tool between the rotor and inboard brake pad.

8. Carefully apply pressure to the inboard brake pad until both caliper inner pistons are fully compressed into the caliper piston bores.

9. Carefully insert a plastic flat-bladed trim tool between the rotor and outboard brake pad.

5. Rotate the brake pad retainer (1) upward and remove the retainer.

6. Using a hammer and punch tap the upper caliper to brake pad mounting pin (2) inward out of the caliper.

7. Carefully insert a plastic flat-bladed trim tool between the rotor and inboard brake pad.

8. Carefully apply pressure to the inboard brake pad until both caliper inner pistons are fully compressed into the caliper piston bores.

9. Carefully insert a plastic flat-bladed trim tool between the rotor and outboard brake pad.

10. Carefully apply pressure to the outboard brake pad until both caliper outer pistons are fully compressed into the bores.

11. Remove the brake pads from the caliper.

To install:

12. Install the brake pads to the caliper.

13. Install the upper caliper guide pin (2) through the caliper, inner and outer brake pads.

14. Using a hammer and punch seat the upper guide pin (2) to the outer caliper half. Ensure that the caliper guide pin is seated into the outer caliper pin seat.

15. Install the brake pad retainer (1) under the upper caliper pin assembly.

16. Rotate brake pad retainer (1) down.

17. Carefully apply pressure downward on the lower end of the brake pad retainer.

18. Carefully install the lower caliper guide pin (2) through the caliper, inner and outer brake pads.

19. Using a hammer and punch seat the upper guide pin (2) to the outer caliper half. Ensure that the caliper guide pin is seated into the outer caliper pin seat. Ensure that the brake pad retainer is centered retaining both brake pads.

20. Install the front wheel.

21. Pump the brake pedal several times to seat the brake pads.

22. Check the fluid level in the master cylinder and fill as necessary.

23. Road test the vehicle for proper brake operation.

Rear

CTS

1. Remove ⅔ of the brake fluid from the master cylinder reservoir.

2. Remove the front wheel.

3. Remove the caliper mounting bolts.

10. Carefully apply pressure to the outboard brake pad until both caliper outer pistons are fully compressed into the bores.

11. Remove the brake pads from the caliper.

To install:

12. Install the brake pads to the caliper.

13. Install the upper caliper guide pin through the caliper, inner and outer brake pads.

14. Using a hammer and punch seat the upper guide pin to the outer caliper half.

Ensure that the caliper guide pin is seated into the outer caliper pin seat.

15. Install the brake pad retainer under the upper caliper pin assembly.

16. Rotate brake pad retainer down.

17. Carefully apply pressure downward on the lower end of the brake pad retainer.

18. Carefully install the lower caliper guide pin through the caliper, inner and outer brake pads.

19. Using a hammer and punch seat the upper guide pin to the outer caliper half.

Ensure that the caliper guide pin is seated into the outer caliper pin seat. Ensure that the brake pad retainer is centered retaining both brake pads.

20. Install the front wheel.

21. Pump the brake pedal several times to seat the brake pads.

22. Check the fluid level in the master cylinder and fill as necessary.

23. Road test the vehicle for proper brake operation.

BRAKES 16-40
DRIVE TRAIN 16-27
ENGINE REPAIR 16-12
FUEL SYSTEM 16-25
SPECIFICATION AND
  MAINTENANCE CHARTS......16-2
Engine and Vehicle
  Identification 16-2
General Engine Specifications 16-2
Engine Tune-Up Specifications 16-3
Firing Order 16-3
Accessory Drive Belt Routing 16-3
Capacities 16-4
Valve Specifications 16-4
Crankshaft and Connecting
  Rod Specifications 16-5
Piston and Ring Specifications 16-5
Torque Specifications 16-6
Tire, Wheel and Ball Joint
  Specifications 16-6
Wheel Alignment 16-7
Brake Specifications 16-8
Scheduled Maintenance
  Intervals 16-10
STEERING AND
  SUSPENSION 16-31
A
Air Bag..............................16-31
  Arming.............................16-32
  Disarming.........................16-31
  Precautions 16-31
Alternator 16-12
  Installation 16-12
  Removal............................16-12
B
Brake Caliper 16-40
  Removal & Installation............16-40
C
Camshaft and Valve Lifters 16-19
  Removal & Installation............16-19

Coil Spring.............................16-36
  Removal & Installation............16-36
CV-Joints..............................16-30
  Overhaul 16-30
Cylinder Head 16-18
  Removal & Installation............16-18
D
Disc Brake Pads......................16-41
  Removal & Installation............16-41
E
Engine Assembly 16-12
  Removal & Installation............16-12
Exhaust Manifold 16-19
  Removal & Installation............16-19
F
Fuel Filter 16-26
  Removal & Installation............16-26
Fuel Injectors 16-26
  Removal & Installation............16-26
Fuel Pump 16-26
  Removal & Installation............16-26
Fuel System Pressure 16-25
  Relieving 16-25
Fuel System Service
  Precautions.........................16-25
H
Halfshaft..............................16-29
  Removal & Installation............16-29
Heater Core...........................16-16
  Removal And Installation 16-16
I
Ignition Timing 16-12
  Adjustment.........................16-12
Intake Manifold 16-18
  Removal & Installation............16-18
L
Lower Ball Joint........................16-37
  Removal & Installation............16-37
Lower Control Arm 16-37

Control Arm Bushing
  Replacement........................16-38
  Removal & Installation............16-37
O
Oil Pan................................16-20
  Removal & Installation............16-20
Oil Pump 16-20
  Removal & Installation............16-20
P
Piston and Ring 16-24
  Positioning 16-24
Power Rack and Pinion
  Steering Gear.......................16-32
  Removal & Installation............16-32
R
Rear Main Seal 16-21
  Removal & Installation............16-21
S
Shock Absorber 16-35
  Removal & Installation............16-35
Starter Motor 16-20
  Removal & Installation............16-20
Strut.................................16-34
  Removal & Installation............16-34
T
Timing Chain, Sprockets,
  Front Cover and Seal.................16-22
  Removal & Installation............16-22
Transaxle Assembly 16-27
  Removal & Installation............16-27
V
Valve Lash 16-20
  Adjustment.........................16-20
W
Water Pump 16-16
  Removal & Installation............16-16
Wheel Bearings 16-38
  Adjustment.........................16-38
  Removal & Installation............16-38

SPECIFICATION AND MAINTENANCE CHARTS

ENGINE AND VEHICLE IDENTIFICATION

	Engine							Model Year	
Code ①	Liters (cc)	Cu. In.	Cyl.	Fuel Sys.	Engine Type	Eng. Mfg.		Code ②	Year
9	4.6 (4565)	279	8	MFI	DOHC	Cadillac		2	2002
Y	4.6 (4565)	279	8	MFI	DOHC	Cadillac		3	2003
								4	2004
								5	2005
								6	2006

DOHC: Double Overhead Camshafts

MFI: Multi-point Fuel Injection

① 8th position of VIN

② 10th position of VIN

06025-CADI-C01

GENERAL ENGINE SPECIFICATIONS

Year	Model	Engine Displacement Liters	Engine Series (ID/VIN)	Net Horsepower @ rpm	Net Torque @ rpm (ft. lbs.)	Bore x Stroke (in.)	Com- pression Ratio	Oil Pressure @ rpm
2002	DeVille	4.6	Y	275@5600	300@4000	3.66x3.31	10.0:1	35@2000
	DeVille DTS	4.6	9	300@6000	295@4400	3.66x3.31	10.0:1	35@2000
	DeVille DHS	4.6	Y	275@5600	300@4000	3.66x3.31	10.0:1	35@2000
	Eldorado	4.6	Y	275@5600	300@4000	3.66x3.31	10.0:1	35@2000
	Eldorado ETC	4.6	9	300@6000	295@4400	3.66x3.31	10.0:1	35@2000
	Seville SLS	4.6	Y	275@5600	300@4000	3.66x3.31	10.0:1	35@2000
	Seville STS	4.6	9	300@6000	295@4400	3.66x3.31	10.0:1	35@2000
2003	DeVille	4.6	Y	275@5600	300@4000	3.66x3.31	10.0:1	35@2000
	DeVille DTS	4.6	9	300@6000	295@4400	3.66x3.31	10.0:1	35@2000
	DeVille DHS	4.6	Y	275@5600	300@4000	3.66x3.31	10.0:1	35@2000
	Seville SLS	4.6	Y	275@5600	300@4000	3.66x3.31	10.0:1	35@2000
	Seville STS	4.6	9	300@6000	295@4400	3.66x3.31	10.0:1	35@2000
2004	DeVille	4.6	Y	275@5600	300@4000	3.66x3.31	10.0:1	35@2000
	DeVille DTS	4.6	9	300@6000	295@4400	3.66x3.31	10.0:1	35@2000
	DeVille DHS	4.6	Y	275@5600	300@4000	3.66x3.31	10.0:1	35@2000
	Seville SLS	4.6	Y	275@5600	300@4000	3.66x3.31	10.0:1	35@2000
	Seville STS	4.6	9	300@6000	295@4400	3.66x3.31	10.0:1	35@2000
2005	DeVille	4.6	Y	275@5600	300@4000	3.66x3.31	10.0:1	35@2000
	DeVille DTS	4.6	9	300@6000	295@4400	3.66x3.31	10.0:1	35@2000
	DeVille DHS	4.6	Y	275@5600	300@4000	3.66x3.31	10.0:1	35@2000

MFI: Multi-point Fuel Injection

06025-CADI-C02

ENGINE TUNE-UP SPECIFICATIONS

Year	Engine Displacement Liters	Engine ID/VIN	Spark Plug Gap (in.)	Ignition Timing (deg.)	Fuel Pump (psi)	Idle Speed (rpm)	Valve Clearance	
							Intake	Exhaust
2002	4.6	9	0.050	①	40-50	①	HYD	HYD
	4.6	Y	0.050	①	40-50	①	HYD	HYD
2003	4.6	9	0.050	①	40-50	①	HYD	HYD
	4.6	Y	0.050	①	40-50	①	HYD	HYD
2004	4.6	9	0.050	①	40-50	①	HYD	HYD
	4.6	Y	0.050	①	40-50	①	HYD	HYD
2005	4.6	9	0.050	①	40-50	①	HYD	HYD
	4.6	Y	0.050	①	40-50	①	HYD	HYD

NOTE: The Vehicle Emission Control Information label often reflects specification changes made during production. The label figures must be used if they differ from those in this chart.

HYD: Hydraulic

① Refer to Vehicle Emission Control Information label

06025-CADI-C03

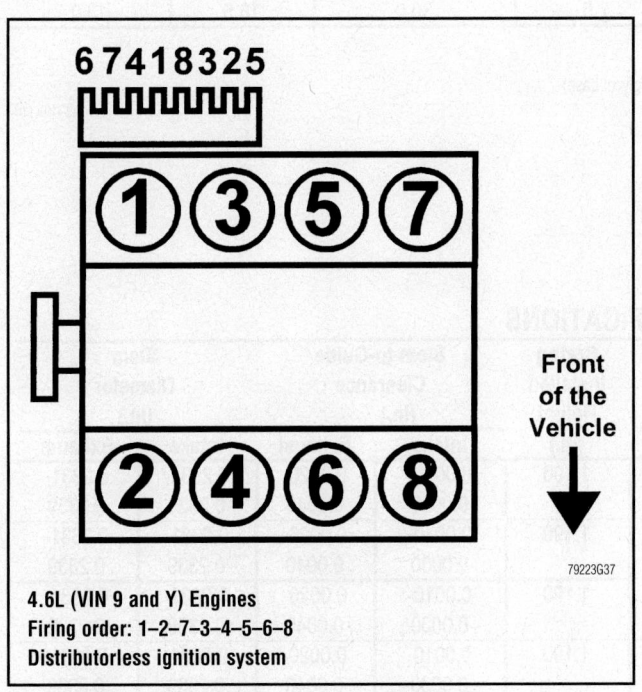

4.6L (VIN 9 and Y) Engines
Firing order: 1–2–7–3–4–5–6–8
Distributorless ignition system

79223G37

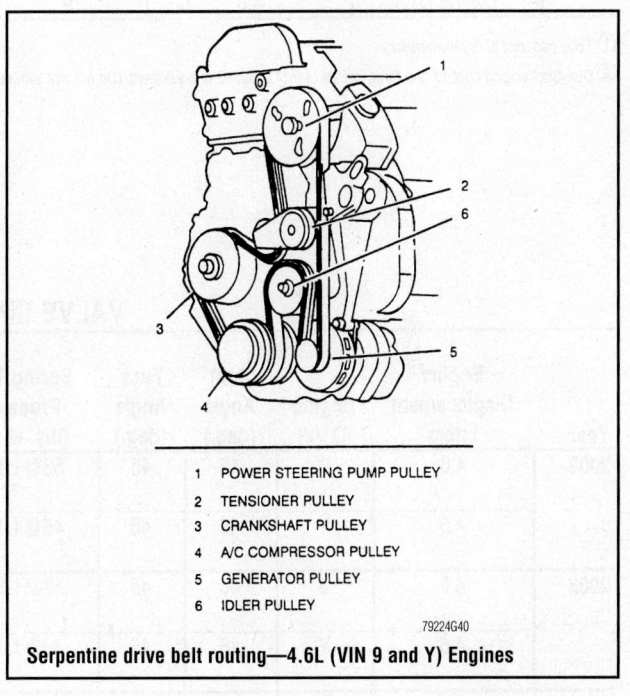

1 POWER STEERING PUMP PULLEY
2 TENSIONER PULLEY
3 CRANKSHAFT PULLEY
4 A/C COMPRESSOR PULLEY
5 GENERATOR PULLEY
6 IDLER PULLEY

79224G40

Serpentine drive belt routing—4.6L (VIN 9 and Y) Engines

CAPACITIES

Year	Model	Engine Displacement Liters	Engine ID/VIN	Engine Oil with Filter (qts.)	Transmission (pts.) ①	Fuel Tank (gal.)	Cooling System (qts.) ②
2002	DeVille	4.6	Y	7.5	30.0	18.5	12.5
	DeVille DHS	4.6	9	7.5	30.0	18.5	12.5
	DeVille DTS	4.6	9	7.5	30.0	18.5	12.5
	Eldorado	4.6	Y	7.0	30.0	19.0	12.5
	Eldorado ETC	4.6	9	7.0	30.0	19.0	12.5
	Seville SLS	4.6	Y	7.0	30.0	18.5	12.5
	Seville STS	4.6	9	7.0	30.0	18.5	12.5
2003	DeVille	4.6	Y	7.5	30.0	18.5	12.5
	DeVille DHS	4.6	9	7.5	30.0	18.5	12.5
	DeVille DTS	4.6	9	7.5	30.0	18.5	12.5
	Seville SLS	4.6	Y	7.0	30.0	18.5	12.5
	Seville STS	4.6	9	7.0	30.0	18.5	12.5
2004	DeVille	4.6	Y	7.5	30.0	18.5	12.5
	DeVille DHS	4.6	9	7.5	30.0	18.5	12.5
	DeVille DTS	4.6	9	7.5	30.0	18.5	12.5
	Seville SLS	4.6	Y	7.0	30.0	18.5	12.5
	Seville STS	4.6	9	7.0	30.0	18.5	12.5
2005	DeVille	4.6	Y	7.5	30.0	18.5	13.0
	DeVille DHS	4.6	9	7.5	30.0	18.5	13.0
	DeVille DTS	4.6	9	7.5	30.0	18.5	13.0

① Total capacity of dry transmission

② Dex-Cool engine coolant and three pellets of P/N 1052753 or equivalent. (Do not mix with ethylene glycol base).

06025-CADI-C04

VALVE SPECIFICATIONS

Year	Engine Displacement Liters	Engine ID/VIN	Seat Angle (deg.)	Face Angle (deg.)	Spring Test Pressure (lbs. @ in.)	Spring Installed Height (in.)	Stem-to-Guide Clearance (in.)		Stem Diameter (in.)	
							Intake	Exhaust	Intake	Exhaust
2002	4.6	9	46	45	53@1.190	1.190	0.0010-0.0030	0.0020-0.0040	0.2331-0.2339	0.2331-0.2339
	4.6	Y	46	45	46@1.190	1.190	0.0010-0.0030	0.0020-0.0040	0.2331-0.2339	0.2331-0.2339
2003	4.6	9	46	45	53@1.190	1.190	0.0010-0.0030	0.0020-0.0040	0.2331-0.2339	0.2331-0.2339
	4.6	Y	46	45	46@1.190	1.190	0.0010-0.0030	0.0020-0.0040	0.2331-0.2339	0.2331-0.2339
2004	4.6	9	46	45	53@1.190	1.190	0.0010-0.0030	0.0020-0.0040	0.2331-0.2339	0.2331-0.2339
	4.6	Y	46	45	46@1.190	1.190	0.0010-0.0030	0.0020-0.0040	0.2331-0.2339	0.2331-0.2339
2005	4.6	9	46	45	53@1.190	1.190	0.0010-0.0030	0.0020-0.0040	0.2331-0.2339	0.2331-0.2339
	4.6	Y	46	45	46@1.190	1.190	0.0010-0.0030	0.0020-0.0040	0.2331-0.2339	0.2331-0.2339

06025-CADI-C05

CRANKSHAFT AND CONNECTING ROD SPECIFICATIONS
All measurements are given in inches.

| Year | Engine Displacement Liters | Engine ID/VIN | Crankshaft | | | | Connecting Rod | | |
			Main Brg. Journal Dia.	Main Brg. Oil Clearance	Shaft End-play	Thrust on No.	Journal Diameter	Oil Clearance	Side Clearance
2002	4.6	9	2.5335-2.5337	0.0006-0.0025	0.0020-0.0200	3	2.1239-2.1235	0.001-0.003	0.0080-0.0200
	4.6	Y	2.5335-2.5337	0.0006-0.0025	0.0020-0.0200	3	2.1239-2.1235	0.001-0.003	0.0080-0.0200
2003	4.6	9	2.5335-2.5337	0.0006-0.0025	0.0020-0.0200	3	2.1239-2.1235	0.001-0.003	0.0080-0.0200
	4.6	Y	2.5335-2.5337	0.0006-0.0025	0.0020-0.0200	3	2.1239-2.1235	0.001-0.003	0.0080-0.0200
2004	4.6	9	2.5335-2.5337	0.0006-0.0025	0.0020-0.0200	3	2.1239-2.1235	0.001-0.003	0.0080-0.0200
	4.6	Y	2.5335-2.5337	0.0006-0.0025	0.0020-0.0200	3	2.1239-2.1235	0.001-0.003	0.0080-0.0200
2005	4.6	9	2.5335-2.5341	0.0006-0.0025	0.0020-0.0200	3	2.1239-2.1245	0.001-0.003	0.0080-0.0200
	4.6	Y	2.5335-2.5341	0.0006-0.0025	0.0020-0.0200	3	2.1239-2.1245	0.001-0.003	0.0080-0.0200

06025-CADI-C06

PISTON AND RING SPECIFICATIONS
All measurements are given in inches.

| Year | Engine Displacement Liters | Engine ID/VIN | Piston Clearance | Ring Gap | | | Ring Side Clearance | | |
				Top Compression	Bottom Compression	Oil Control	Top Compression	Bottom Compression	Oil Control
2002	4.6	9	0.0008-0.0020	0.010-0.016	0.014-0.020	0.010-0.030	0.0016-0.0037	0.0016-0.0037	①
	4.6	Y	0.0008-0.0020	0.010-0.016	0.014-0.020	0.010-0.030	0.0016-0.0037	0.0016-0.0037	①
2003	4.6	9	0.0008-0.0020	0.010-0.016	0.014-0.020	0.010-0.030	0.0016-0.0037	0.0016-0.0037	①
	4.6	Y	0.0008-0.0020	0.010-0.016	0.014-0.020	0.010-0.030	0.0016-0.0037	0.0016-0.0037	①
2004	4.6	9	0.0008-0.0020	0.010-0.016	0.014-0.020	0.010-0.030	0.0016-0.0037	0.0016-0.0037	①
	4.6	Y	0.0008-0.0020	0.010-0.016	0.014-0.020	0.010-0.030	0.0016-0.0037	0.0016-0.0037	①
2005	4.6	9	0.0008-0.0020	0.010-0.016	0.014-0.020	0.010-0.030	0.0016-0.0037	0.0016-0.0037	①
	4.6	Y	0.0008-0.0020	0.010-0.016	0.014-0.020	0.010-0.030	0.0016-0.0037	0.0016-0.0037	①

① Side sealing

06025-CADI-C07

TORQUE SPECIFICATIONS
All readings in ft. lbs.

Year	Engine Displacement Liters	Engine ID/VIN	Cylinder Head Bolts	Main Bearing Bolts	Rod Bearing Bolts	Crankshaft Damper Bolts	Flywheel Bolts	Manifold Intake	Manifold Exhaust	Spark Plugs	Oil Pan Drain Plug
2002	4.6	9	①	②	③	④	⑤	⑥	18	11	15
	4.6	Y	①	②	③	④	⑤	⑥	18	11	15
2003	4.6	9	①	②	③	④	⑤	⑥	18	11	15
	4.6	Y	①	②	③	④	⑤	⑥	18	11	15
2004	4.6	9	①	②	③	④	⑤	⑥	18	11	15
	4.6	Y	①	②	③	④	⑤	⑥	18	11	15
2005	4.6	9	⑦	②	③	④	⑤	⑥	18	11	15
	4.6	Y	⑦	②	③	④	⑤	⑥	18	11	15

① M11 bolts:
  Step 1: 30 ft. lbs. plus an additional 70 degrees
  Step 2: Plus 70 degrees
  Step 3: Plus 60 degrees
  Step 4: Plus 60 degrees for a total of 190 degrees

② M10 bolts:
  Step 1: 15 ft. lbs.
  Step 2: Plus 65 degrees
  M8 bolts:
  22 ft. lbs.

③ Step 1: 22 ft. lbs.
  Step 2: Loosen completely
  Step 3: 18 ft. lbs.
  Step 4: Plus 110 degrees

④ Step 1: 37 ft. lbs.
  Step 2: Plus 120 degrees

⑤ Step 1: 11 ft. lbs.
  Step 2: Plus 50 degrees

⑥ 89 inch lbs.

⑦ M11 bolts:
  Step 1: 22 ft. lbs.
  Step 2: Plus 60 degrees
  Step 3: Plus 60 degrees
  Step 4: Plus 60 degrees for a total of 180 degrees

06025-CADI-C08

TIRE, WHEEL AND BALL JOINT SPECIFICATIONS
Cadillac

Year	Model	OEM Tires Standard	OEM Tires Optional	Tire Pressures (psi) Front	Tire Pressures (psi) Rear	Wheel Size	Ball Joint Inspection	Wheel Lug Torque (ft. lbs.)
2002	DeVille	P225/60R16	None	30	30	7-JJ	①	100
	Eldorado	P225/60R16	P225/60ZR16	Std: 28 Opt: 29	Std: 26 Opt: 29	7-JJ	①	100
	Seville	P225/60R16	P225/60ZR16	Std: 28 Opt: 29	Std: 26 Opt: 29	7-JJ	①	100
2003	DeVille	P225/60R16	None	30	30	7-JJ	①	100
	Seville	P225/60R16	P225/60ZR16	Std: 28 Opt: 29	Std: 26 Opt: 29	7-JJ	①	100
2004	DeVille	P225/60R16	None	30	30	7-JJ	①	100
	Seville	P225/60R16	P225/60ZR16	Std: 28 Opt: 29	Std: 26 Opt: 29	7-JJ	①	100
2005	DeVille	P225/60R16	None	30	30	7-JJ	①	100

OEM: Original Equipment Manufacturer

PSI: Pounds Per Square Inch

STD: Standard

OPT: Optional

L: Lower

U: Upper

① Replace if any measurable movement is found.

06025-CADI-C09

WHEEL ALIGNMENT
CADILLAC SEVILLE

Year	Model		Caster Range (Deg.)	Caster Preferred Setting (Deg.)	Camber Range (Deg.)	Camber Preferred Setting (Deg.)	Toe-in (in.)
2002	Seville	F	+0.50	+6.00	+0.50	-0.20	0.09 +/- 0.09
		R	—	—	+0.50	-0.30	0.09 +/- 0.09
2003	Seville	F	+0.50	+6.00	+0.50	-0.20	0.09 +/- 0.09
		R	—	—	+0.50	-0.30	0.09 +/- 0.09
2004	Seville	F	+0.50	+6.00	+0.50	-0.20	0.09 +/- 0.09
		R	—	—	+0.50	-0.30	0.09 +/- 0.09

06025-CADI-C10

WHEEL ALIGNMENT
CADILLAC DeVille

Year	Model		Caster Range (Deg.)	Caster Preferred Setting (Deg.)	Camber Range (Deg.)	Camber Preferred Setting (Deg.)	Toe-in (in.)
2002	DeVille	F	+0.50	+6.00	+0.50	-0.20	0.09 +/- 0.09
		R	—	—	+0.50	-0.30	0.09 +/- 0.09
2003	DeVille	F	+0.50	+6.00	+0.50	-0.20	0.09 +/- 0.09
		R	—	—	+0.50	-0.30	0.09 +/- 0.09
2004	DeVille	F	+0.50	+6.00	+0.50	-0.20	0.09 +/- 0.09
		R	—	—	+0.50	-0.30	0.09 +/- 0.09
2005	DeVille	F	+0.50	+6.00	+0.50	-0.20	0.09 +/- 0.09
		R	—	—	+0.50	-0.30	0.09 +/- 0.09

06025-CADI-C11

WHEEL ALIGNMENT
CADILLAC ELDORADO

Year	Model		Caster Range (Deg.)	Caster Preferred Setting (Deg.)	Camber Range (Deg.)	Camber Preferred Setting (Deg.)	Toe-in (in.)
2002	Eldorado	F	+1.00	+2.30	+0.50	0	0.20 +/- 0.20
		R	—	—	+0.50	0	0.20 +/- 0.20

06025-CADI-C12

BRAKE SPECIFICATIONS
All measurements in inches unless noted

| Year | Model | | Brake Disc | | | Minimum Lining Thickness | Brake Caliper Mounting Bolts (ft. lbs.) |
			Original Thickness	Minimum Thickness	Maximum Runout		
2002	DeVille	F	1.268	1.209	0.002	0.030	38
		R	0.433	0.374	0.002	0.030	20
	DeVille DHS	F	1.268	1.209	0.002	0.030	38
		R	0.433	0.374	0.002	0.030	20
	DeVille DTS	F	1.268	1.209	0.002	0.030	38
		R	0.433	0.374	0.002	0.030	20
	Eldorado	F	1.268	1.209	0.002	0.030	38
		R	0.433	0.374	0.002	0.030	20
	Eldorado ETC	F	1.268	1.209	0.002	0.030	38
		R	0.433	0.374	0.002	0.030	20
	Seville SLS	F	1.268	1.209	0.002	0.030	38
		R	0.433	0.374	0.002	0.030	20
	Seville STS	F	1.268	1.209	0.002	0.030	38
		R	0.433	0.374	0.002	0.030	20
2003	DeVille	F	1.268	1.209	0.002	0.030	38
		R	0.433	0.374	0.002	0.030	20
	DeVille DHS	F	1.268	1.209	0.002	0.030	38
		R	0.433	0.374	0.002	0.030	20
	DeVille DTS	F	1.268	1.209	0.002	0.030	38
		R	0.433	0.374	0.002	0.030	20
	Seville SLS	F	1.268	1.209	0.002	0.030	38
		R	0.433	0.374	0.002	0.030	20
	Seville STS	F	1.268	1.209	0.002	0.030	38
		R	0.433	0.374	0.002	0.030	20
2004	DeVille	F	1.268	1.209	0.002	0.030	38
		R	0.433	0.374	0.002	0.030	20
	DeVille DHS	F	1.268	1.209	0.002	0.030	38
		R	0.433	0.374	0.002	0.030	20
	DeVille DTS	F	1.268	1.209	0.002	0.030	38
		R	0.433	0.374	0.002	0.030	20
	Seville SLS	F	1.268	1.209	0.002	0.030	38
		R	0.433	0.374	0.002	0.030	20
	Seville STS	F	1.268	1.209	0.002	0.030	38
		R	0.433	0.374	0.002	0.030	20

06025-CADI-C13

BRAKE SPECIFICATIONS
All measurements in inches unless noted

| Year | Model | | Brake Disc | | | Minimum Lining Thickness | Brake Caliper Mounting Bolts (ft. lbs.) |
			Original Thickness	Minimum Thickness	Maximum Runout		
2005	DeVille	F	①	②	0.001	0.030	③
		R	④	⑤	0.002	0.030	20
	DeVille DHS	F	①	②	0.001	0.030	③
		R	④	⑤	0.002	0.030	20
	DeVille DTS	F	①	②	0.001	0.030	③
		R	④	⑤	0.002	0.030	20

All rotors have a discard thickness stamped on them. Discard any rotor that does not meet minimum thickeness specification

Caution: two different designs were used on the front brake rotors and pads. Do not interchange the components before the first and second designs or loss of braking can occur.

① First design before VIN 5U129029: 1.267 inch

Second design after VIN 5U129029: 1.181 inch

J55: 1.496 inch

② First design before VIN 5U129029: 1.224 inch

Second design after VIN 5U129029: 1.125 inch

J55: 1.457 inch

③ Except J55 63 ft. lbs.

J55: 83 ft. lbs.

④ Except J55 0.433 inch

J55: 1.142 inch

⑤ Except J55 0.354 inch

J55: 1.083 inch

06025-CADI-C14

SCHEDULED MAINTENANCE INTERVALS
GM E & K BODIES—2002-04 CADILLAC DEVILLE, ELDORADO & SEVILLE

TO BE SERVICED	OF SERVI	VEHICLE MILEAGE INTERVAL (x1000)												
		7.5	15	22.5	30	37.5	45	52.5	60	67.5	75	82.5	90	97.5
Engine oil & filter	R	✓	✓	✓	✓	✓	✓	✓	✓	✓	✓	✓	✓	✓
Coolant level, hoses & clamps	S/I	✓	✓	✓	✓	✓	✓	✓	✓	✓	✓	✓	✓	✓
Drive shaft boots & front suspension components	S/I	✓	✓	✓	✓	✓	✓	✓	✓	✓	✓	✓	✓	✓
Exhaust system, brake hoses & throttle linkage	S/I	✓	✓	✓	✓	✓	✓	✓	✓	✓	✓	✓	✓	✓
Lubricate chassis, suspension, steering linkage, transaxle shift linkage, parking brake cable guides, underbody contact points & linkage	S/I	✓	✓	✓	✓	✓	✓	✓	✓	✓	✓	✓	✓	✓
Brake linings	S/I	✓		✓		✓		✓		✓		✓		✓
Rotate tires	S/I	✓		✓		✓		✓		✓		✓		✓
Inspect throttle body bore & throttle plate for deposits	S/I		✓				✓				✓		✓	
Air filter element	R				✓				✓				✓	
Engine coolant ①	R													
PCV valve	R				✓				✓				✓	
Spark plugs ②	R				✓				✓				✓	
Accessory drive belt(s)	S/I				✓				✓				✓	
Automatic transaxle fluid & filter	S/I				✓				✓				✓	
EGR & fuel systems	S/I				✓				✓				✓	
Ignition cables	S/I				✓				✓				✓	

R: Replace   S/I: Service or Inspect

① Engine coolant: replace every 100,000 miles. Use O.E. specified (DEX-COOL™) coolant only. If any silicate coolant is used, the service interval is every 30,000 miles.

② Platinum tip spark plugs: replace every 100,000 miles.

FREQUENT OPERATION MAINTENANCE (SEVERE SERVICE)

If a vehicle is operated under any of the following conditions it is considered severe service:

- Extremely dusty areas.
- 50% or more of the vehicle operation is in 32°C (90°F) or higher temperatures, or constant operation in temperatures below 0°C (32°F).
- Prolonged idling (vehicle operation in stop and go traffic).
- Frequent short running periods (engine does not warm to normal operating temperatures).
- Police, taxi, delivery usage or trailer towing usage.

CV joints & front suspension components: service or inspect every 3000 miles.

Engine oil & filter change: change every 3000 miles.

Brake linings: check every 6000 miles.

Chassis lubrication: lubricate every 6000 miles.

Suspension, steering linkage, transaxle shift linkage, parking cable guides, underbody contact points: lubricate every 6000 miles.

Air filter element: service or inspect every 15,000 miles.

Automatic transaxle fluid: change every 50,000 miles (1997 only).

Inspect throttle body bore & throttle plate for deposits: clean as required every 15,000 miles.

Rotate tires at 6000 miles, then every 15,000 miles.

SCHEDULED MAINTENANCE INTERVALS
GM E & K BODIES—2005 CADILLAC DEVILLE

TO BE SERVICED	OF SERVI	VEHICLE MILEAGE INTERVAL (x1000)												
		3	6	9	12	15	18	21	24	27	30	33	36	39
Engine oil & filter ①	R	✓	✓	✓	✓	✓	✓	✓	✓	✓	✓	✓	✓	✓
Coolant level, hoses & clamps	S/I	✓	✓	✓	✓	✓	✓	✓	✓	✓	✓	✓	✓	✓
Drive shaft boots & front suspension components	S/I	✓	✓	✓	✓	✓	✓	✓	✓	✓	✓	✓	✓	✓
Exhaust system, brake hoses & throttle linkage	S/I	✓	✓	✓	✓	✓	✓	✓	✓	✓	✓	✓	✓	✓
Lubricate chassis, suspension, steering linkage, transaxle shift linkage, parking brake cable guides, underbody contact points & linkage	S/I	✓	✓	✓	✓	✓	✓	✓	✓	✓	✓	✓	✓	✓
Brake linings	S/I	✓		✓		✓		✓		✓		✓		✓
Rotate tires	S/I	✓		✓		✓		✓		✓		✓		✓
Inspect throttle body bore & throttle plate for deposits	S/I		✓				✓				✓		✓	
Air filter element	R				✓				✓				✓	
Engine coolant ②	R													
PCV valve	R				✓				✓				✓	
Spark plugs ③	R				✓				✓				✓	
Accessory drive belt(s)	S/I				✓				✓				✓	
Automatic transaxle fluid & filter	S/I				✓				✓				✓	
EGR & fuel systems	S/I				✓				✓				✓	
Ignition cables	S/I				✓				✓				✓	

R: Replace   S/I: Service or Inspect

① When the change engine oil message in the Driver Information Center (DIC) comes on, it means that service is required for your vehicle. Have your vehicle serviced as soon as possible within the next 600 miles (1 000 km). It is possible that, if you are driving under the best conditions, the engine oil life system may not indicate that vehicle service is necessary for over a year. However, your engine oil and filter must be changed at least once a year and at this time the system must be reset.

After the oil has been changed, the CHANGE ENGINE OIL message must be reset. To reset the message, do the following:

Turn the key to the ON position without starting the engine.

Press the INFO button on the Driver Information Center (DIC) until ENGINE OIL LIFE is displayed.

Press and hold the INFO RESET button until 100% ENGINE OIL LIFE is displayed. This resets the oil life indicator.

Turn the key to OFF.

If the CHANGE ENGINE OIL message comes back on when you start your vehicle, the engine oil life system has not reset. Repeat the procedure

② Engine coolant: replace every 100,000 miles. Use O.E. specified (DEX-COOL™) coolant only. If any silicate coolant is used, the service interval is every 30,000 miles.

③ Platinum tip spark plugs: replace every 100,000 miles.

FREQUENT OPERATION MAINTENANCE (SEVERE SERVICE)

If a vehicle is operated under any of the following conditions it is considered severe service:

- Extremely dusty areas.

- 50% or more of the vehicle operation is in 32°C (90°F) or higher temperatures, or constant operation in temperatures below 0°C (32°F).

- Prolonged idling (vehicle operation in stop and go traffic).

- Frequent short running periods (engine does not warm to normal operating temperatures).

- Police, taxi, delivery usage or trailer towing usage.

CV joints & front suspension components: service or inspect every 3000 miles.

Engine oil & filter change: change every 3000 miles.

Brake linings: check every 6000 miles.

Chassis lubrication: lubricate every 6000 miles.

Suspension, steering linkage, transaxle shift linkage, parking cable guides, underbody contact points: lubricate every 6000 miles.

Air filter element: service or inspect every 15,000 miles.

Automatic transaxle fluid: change every 50,000 miles (1997 only).

Inspect throttle body bore & throttle plate for deposits: clean as required every 15,000 miles.

Rotate tires at 6000 miles, then every 15,000 miles.

06025-CADI-C16

ENGINE REPAIR

Alternator

REMOVAL

Eldorado

1. Before servicing the vehicle, refer to the precautions section.
2. Drain the cooling system.
3. Remove or disconnect the following:
   - Negative battery cable
   - Accessory drive belt
   - Alternator upper mounting bolt
   - Engine splash shield
   - Radiator support access panel
   - Rear alternator bracket from the engine
   - Alternator mounting bolts
   - Duct on the back of the alternator
   - Electrical connectors
   - Front alternator-to-A/C bracket
   - Alternator

DeVille

1. Before servicing the vehicle, refer to the precautions section.
2. Drain the cooling system.
3. Remove or disconnect the following:
   - Negative battery cable
   - Accessory drive belt
   - Radiator
   - Electrical connectors
   - Alternator mounting bolts
   - Alternator

Seville

1. Before servicing the vehicle, refer to the precautions section.
2. Drain the cooling system.
3. Remove or disconnect the following:
   - Negative battery cable

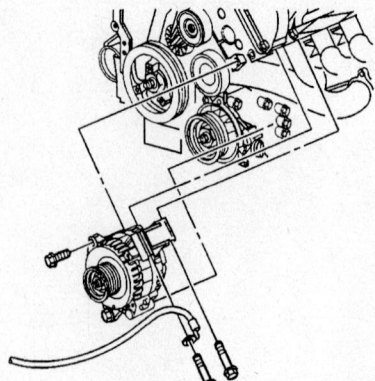

9306AG01
Alternator mounting and wires

- Accessory drive belt
- Cooling fans
- Alternator cooler outlet hose, if equipped
- Alternator cooler inlet hose, if equipped
- Electrical connectors
- Alternator mounting bolts
- Alternator

INSTALLATION

Eldorado

1. Install or connect the following:
   - Alternator
   - Front alternator-to-A/C bracket. Torque the bolt to 37 ft. lbs. (50 Nm).
   - Electrical connectors. Torque the battery connection nut to 115 inch lbs. (13 Nm).
   - Duct on the back of the alternator
   - Alternator mounting bolts. Torque the bolts to 37 ft. lbs. (50 Nm).
   - Upper front bolt. Torque the bolt to 35 ft. lbs. (47 Nm).
   - Rear bracket to the engine. Torque the bolt to 37 ft. lbs. (50 Nm).
   - Radiator support access panel
   - Engine splash shield
   - Accessory drive belt
   - Negative battery cable
2. Refill the cooling system.

DeVille

1. Install or connect the following:
   - Alternator
   - Alternator mounting bolts. Torque the bolts to 37 ft. lbs. (50 Nm).
   - Electrical connectors. Torque the battery connection nut to 111 inch lbs. (12 Nm).
   - Radiator
   - Accessory drive belt
   - Negative battery cable
2. Refill the cooling system.

Seville

1. Install or connect the following:
   - Alternator. Torque the bolt to 37 ft. lbs. (50 Nm).
   - Electrical connectors. Torque the battery connection nut to 115 inch lbs. (13 Nm).
   - Alternator cooler inlet hose, if equipped
   - Alternator cooler outlet hose, if equipped. Torque the bolt to 80 inch lbs. (9 Nm).

- Cooling fans
- Accessory drive belt
- Negative battery cable
2. Refill the cooling system.

Ignition Timing

ADJUSTMENT

The 4.6L Northstar engine is equipped with a Distributorless Ignition System (DIS). The ignition timing is controlled by the Powertrain Control Module (PCM). Therefore no adjustments are necessary.

Engine Assembly

REMOVAL & INSTALLATION

DeVille And Seville

1. Before servicing the vehicle, refer to the precautions section.
2. Disable the air bag system.
3. Recover the air-conditioning system.
4. Drain the cooling system.
5. Recover the refrigerant from the air conditioning system.
6. Relieve the fuel pressure.
7. Drain the crankcase.
8. Remove or disconnect the following:
   - Negative battery cable
   - Vacuum brake booster hose from the vacuum connection
   - Fuel inlet and return quick-connect fittings at the fuel
   - Hose from the evaporative emission canister purge valve
   - Upper filler panel
   - Air cleaner assembly
   - 2 nuts from the intake manifold sight shield
   - Sight shield from the engine
   - Nut securing the positive battery cable to the remote positive terminal
   - Secondary AIR relay from the relay bracket
9. Disconnect the following electrical connections
   - Powertrain Control Module (PCM)
   - Engine electrical harness
10. Remove or disconnect the following:
   - Bolt attaching the engine ground cable to the right side body frame rail

❋❋ **CAUTION**

Always replace the accelerator control cable with a NEW cable whenever you remove the engine from the vehicle.

❋❋ **CAUTION**

Position the cruise control cable out of the way while you remove or install the engine. Do not pry or lean against the cable and do not kink the cable. A damaged cable must be replaced.

- Cruise control cable from the throttle body bracket and lever
- Throttle cable from the throttle body
- Shift cable from the bracket and manual shift lever
- Radiator inlet hose from the water housing crossover
- Radiator outlet hose from the thermostat housing
- Surge tank inlet hose from the tank
- Heater hoses from the heater pipes

➡Mark the location of the brake pipes to the Brake Pressure Modulator Valve (BPMV) to aid during installation.

- 2 master cylinder brake pipes from the BPMV and plug the outlet ports to prevent fluid loss and contamination
- Upper transaxle oil cooler pipe retaining bolt from the fan shroud
- Plastic cap off the upper transaxle oil cooler pipe quick connect fitting
- Upper transaxle oil cooler pipe from the radiator
- Lower transaxle oil cooler pipe fitting from the radiator

➡Make sure the vehicle wheels are in the straight position and the steering column in the LOCK position before disconnecting the steering column or intermediate shaft from the steering gear. Failure to do so will cause the SIR coil assembly to become uncentered, which may cause damage to the coil assembly.

11. Lock the steering column by installing the J 42640 into the underside of the steering column.

- Right and left side strut tower bolts
- Rear exhaust manifold pipe
- Front wheels
- Front Wheel Speed Sensor (WSS) electrical leads from the frame rail

- Road sensing suspension electrical connector at the frame rail, if equipped
- Electronic suspension position sensor links from the lower control arms, if equipped
- Front air deflector
- Front fascia extensions
- Secondary AIR inlet hose from the secondary AIR pump
- 2 nuts attaching the front brake pipe frame brackets to the frame rails
- Front brake pipes from the retainers at the frame rails
- Front brake pipes away from the frame rails by gently pulling
- 2 rear brake pipes at the rear of the engine frame and plug the open outlet ports
- A/C pressure sensor
- A/C discharge hose from the compressor and it secure to the cooling fan assembly
- A/C suction hose from the compressor and secure it to the cooling fan assembly

❋❋ **CAUTION**

Failure to disconnect the intermediate shaft from the rack and pinion stub shaft could result in damage to the steering gear and/or damage to the intermediate shaft. This damage may cause loss of steering control which could result in personal injury.

- Intermediate shaft pinch bolt
- Steering gear from the intermediate shaft
- Post Heated Oxygen (HO₂S) sensor connection
- Engine oil cooler quick connect fittings from the engine oil filter adapter, with the oil pipes still attached, and position aside, if equipped
- Brace between the engine oil pan and the transaxle case
- Torque converter cover
- Torque converter-to-flywheel bolts

12. Position J 39580 powertrain support dolly under the engine frame and lower the vehicle onto the dolly.

13. If a support dolly is unavailable, support the powertrain with four suitable jackstands.

14. Place a 2 in x 4 in block of wood between the front of the engine oil pan and the engine frame.

15. Remove or disconnect the following:

- Nut securing the right engine mount to the right engine mount bracket
- Nut securing the left transaxle mount to the left transaxle mount bracket

❋❋ **CAUTION**

To avoid any vehicle damage, serious personal injury or death when major components are removed from the vehicle and the vehicle is supported by a hoist, support the vehicle with jack stands at the opposite end from which the components are being removed.

16. Secure the front hoist pads to the vehicle

- 6 frame-to-body mounting bolts

17. Make sure clearance is maintained between the powertrain assembly and the following components:

- A/C accumulator hose
- A/C compressor hose
- Brake pipes
- Heater hoses
- Radiator hoses
- Wheel Speed Sensor (WSS) cables
- Wiring harnesses

18. Carefully raise the vehicle in order to clear the supported engine/transaxle assembly.

19. Drain the engine oil

20. Remove or disconnect the following:

- Heater pipes
- Intermediate hose from the secondary AIR valve at bank 2
- Nut securing the intermediate hose to the secondary AIR valve at bank 1
- Nut securing the coil cassette ground wire to the right cylinder head
- Engine wiring harness from the engine
- Power steering hose from the power steering pump reservoir
- Power steering return hose retaining bolt from the cylinder head
- Power steering pressure hose from the power steering pump
- Nut securing the power steering pressure hose to the right engine mount bracket
- 4 bolts securing the right engine mount bracket to the engine
- Right engine mount bracket
- Bolt securing the rear transaxle brace to the transaxle
- Nuts securing the rear transaxle

brace to the stud located on the right cylinder head
- Rear transaxle brace
- Bolts securing the front transaxle brace to the transaxle and right cylinder head
- Nuts securing the Vehicle Speed Sensor (VSS) heat shield to the transaxle
- Bolts securing the transaxle brace to the engine and transaxle

21. Attach an engine lift bracket to the cylinder head
- Engine lift chain to the engine lift brackets and attach to an engine lift devise
- Nut securing the front engine mount to the engine frame
- Bolts attaching the engine to the transaxle

22. Raise the engine from the supported frame and transaxle assembly.
- Front engine mount bracket

To install:

23. Installation is the reverse of removal, please note the following torque specifications:
- Bolts attaching the engine to the transaxle to 55 ft. lbs. (75 Nm)
- Nut securing the front engine mount to the front engine frame to 52 ft. lbs. (70 Nm)
- 4 bolts securing the center transaxle brace to the engine and transaxle to 37 ft. lbs. (50 Nm)
- Retaining nuts attaching the VSS heat shield to the transaxle to 37 ft. lbs. (50 Nm)
- Bolts attaching the front transaxle brace to the transaxle and right cylinder head to 37 ft. lbs. (50 Nm)
- Rear transaxle brace bolt and nuts to 37 ft. lbs. (50 Nm)
- 4 bolts securing the right engine mount bracket to the engine to 37 ft. lbs. (50 Nm)
- Nut securing the power steering pressure hose to the right engine mount bracket to 80 inch lbs. (9 Nm)
- Power steering pressure hose to the power steering pump to 20 ft. lbs. (27 Nm)
- Power steering return hose retaining bolt to the cylinder head to 37 ft. lbs. (50 Nm)
- Nut securing the coil cassette ground wire to the cylinder head to 13 ft. lbs. (17 Nm)
- Nut securing the intermediate hose to the secondary AIR valve at bank 1 to 80 inch lbs. (9 Nm)

- 6 frame mounting bolts to 141 ft. lbs. (191 Nm)
- Nut securing the right engine mount to the right engine mount bracket to 59 ft. lbs. (80 Nm)
- Nut securing the left transaxle mount to the left transaxle mount bracket to 59 ft. lbs. (80 Nm)
- Flywheel-to-torque converter bolts to 44 ft. lbs. (60 Nm)
- Oil pan-to-transaxle brace bolts to 37 ft. lbs. (50 Nm)
- Pinch bolt to 33 ft. lbs. (45 Nm)
- A/C suction hose nut to 15 ft. lbs. (20 Nm)
- A/C discharge hose nut to 15 ft. lbs. (20 Nm)
- 2 rear brake pipes at the rear of the engine frame to 11 ft. lbs. (15 Nm)
- Brake pipe frame bracket nuts to 11 ft. lbs. (15 Nm)
- Transaxle oil cooler pipe fitting to 26 ft. lbs. (35 Nm)
- 2 master cylinder brake pipes to the BPMV to 11 ft. lbs. (15 Nm)
- Brake pipes to 11 ft. lbs. (15 Nm)

24. Fill the engine with oil.
25. Fill the cooling system.
26. Bleed the brake system.
27. Recharge the A/C refrigerant system.
28. Bleed the power steering system.
29. Check the wheel alignment.
30. Complete the following steps after the engine is installed in the vehicle:

a. With the ignition **OFF** or disconnected, crank the engine several times. Listen for any unusual noises or evidence that any parts are binding.

b. Start the engine and listen for abnormal conditions.

c. Check the vehicle oil pressure gauge or light and confirm that the engine has acceptable oil pressure.

d. Run the engine at approximately 1000 RPM until the engine reaches normal operating temperature.

e. While the engine continues to idle, raise and support the vehicle

f. Inspect for oil, coolant and exhaust leaks while the engine is idling.

g. Lower the vehicle.

h. Perform the Crankshaft Position (CKP) system variation learn procedure.

i. Perform a final inspection for the proper engine oil and coolant levels.

31. Road test the vehicle.

Eldorado

1. Before servicing the vehicle, refer to the precautions section.
2. Disable the air bag system.

3. Recover the air-conditioning system.
4. Drain the cooling system.
5. Recover the refrigerant from the air conditioning system.
6. Relieve the fuel pressure.
7. Drain the crankcase.
8. Remove or disconnect the following:
- Negative battery cable
- Vacuum brake booster hose from the vacuum connection
- Air cleaner assembly
- Fuel inlet and return quick-connect fittings at the fuel
- Hose from the evaporative emission canister purge valve
- Radiator support sight shield
- Sight shield from the engine

9. Disconnect the following electrical connections:
- Powertrain Control Module (PCM)
- Engine electrical harness
- Nut securing the positive battery cable to engine

✳✳ CAUTION

Always replace the accelerator control cable with a NEW cable whenever you remove the engine from the vehicle.

✳✳ CAUTION

Position the cruise control cable out of the way while you remove or install the engine. Do not pry or lean against the cable and do not kink the cable. A damaged cable must be replaced.

- Cruise control cable from the throttle body bracket and lever
- Throttle cable from the throttle body
- Shift cable from the bracket and manual shift lever
- Radiator inlet hose from the water housing crossover
- Radiator outlet hose from the thermostat housing
- Cooling fans
- Surge tank inlet hose from the tank
- Accumulator tube, and position the accumulator aside
- Heater hoses from the heater pipes
- Upper transaxle oil cooler pipe from the radiator
- Lower transaxle oil cooler pipe fitting from the radiator
- Right and left side strut tower nuts
- Engine mount struts
- Rear exhaust manifold pipe

- Front wheels
- Front Wheel Speed Sensor (WSS) electrical leads from the frame rail
- Road sensing suspension electrical connector at the frame rail, if equipped
- Electronic suspension position sensor links from the lower control arms, if equipped
- Splash shield
- Front fascia extensions
- Secondary AIR inlet hose from the secondary AIR pump
- Front left brake pipe fitting from the brake hose
- Front left brake hose retainer from the brake hose retaining bracket
- Bolt attaching the front brake pipes to the frame rails
- Front brake pipes away from the frame rails by gently pulling

➡**Mark the location of the brake pipes to the brake pressure modulator valve (BPMV) to aid during installation.**

- Master cylinder, left and rear brake pipes from the BPMV and plug the outlet ports to prevent fluid loss and contamination
- A/C pressure sensor
- A/C discharge hose from the compressor and it secure to the cooling fan assembly
- A/C suction hose from the compressor and secure it to the cooling fan assembly
- Intermediate shaft pinch bolt
- Steering gear from the intermediate shaft
- Post Heated Oxygen (HO2S) sensor connection
- Engine oil cooler quick connect fittings from the engine oil filter adapter, with the oil pipes still attached, and position aside, if equipped
- Wiring from the alternator
- Positive battery cable from the front of the frame
- Brace between the engine oil pan and the transaxle case
- Torque converter cover
- Torque converter-to-flywheel bolts

10. Position J 39580 powertrain support dolly under the engine frame and lower the vehicle onto the dolly.

11. If a support dolly is unavailable, support the powertrain with four suitable jackstands.

12. Secure the front hoist pads to the vehicle
- 6 frame-to-body mounting bolts

13. Make sure clearance is maintained between the powertrain assembly and the following components:
- A/C accumulator hose
- A/C compressor hose
- Brake pipes
- Heater hoses
- Radiator hoses
- Wheel Speed Sensor (WSS) cables
- Wiring harnesses

14. Carefully raise the vehicle in order to clear the supported engine/transaxle assembly.

15. Drain the engine oil

16. Remove or disconnect the following:
- Intermediate hose from the AIR valve at bank 2
- Nut securing the intermediate hose to the AIR valve at bank 1
- Bolt securing the AIR pipe to the left engine mount strut bracket
- Exhaust Gas Recirculation (EGR) pipes
- Heater pipes
- Nut securing the coil cassette ground wire to the right cylinder head
- Engine wiring harness from the engine
- Power steering hose from the power steering pump reservoir
- Power steering return hose bolt to the right engine mount bracket
- Power steering pressure hose from the power steering pump
- Bolts securing the right engine mount bracket to the engine
- Bolt securing the top of the generator to the right engine mount strut bracket
- Right engine strut bracket

17. Attach an engine lift bracket to the cylinder head
- Power steering pump
- Engine lift chain to the engine lift brackets and attach to an engine lift devise
- Nuts securing the right transaxle mount to the engine frame
- 3 bolts securing the right transaxle mount to the engine and transaxle
- Right transaxle mount
- Bolt securing the rear transaxle brace to the transaxle
- Nuts securing the rear transaxle brace to the stud located on the right cylinder head
- Rear transaxle brace
- Nuts securing the Vehicle Speed Sensor (VSS) heat shield to the transaxle
- Bolts securing the front transaxle brace to the engine and transaxle

- 2 nuts securing the front engine mount to the engine frame
- Bolts attaching the engine to the transaxle
- Engine

To install:

18. Installation is the reverse of removal, please note the following torque specifications:
- Bolts attaching the engine to the transaxle to 55 ft. lbs. (75 Nm)
- Nuts securing the front engine mount to the engine frame to 37 ft. lbs. (55 Nm)
- Bolts securing the front transaxle brace to the engine and transaxle to 37 ft. lbs. (50 Nm)
- Nuts securing the VSS heat shield to the transaxle to 37 ft. lbs. (50 Nm)
- Bolts securing the rear transaxle brace to the transaxle to 37 ft. lbs. (50 Nm)
- Bolts securing the right transaxle mount to the engine and transaxle to 37 ft. lbs. (50 Nm)
- Nuts securing the right transaxle mount to the engine frame to 37 ft. lbs. (50 Nm)
- Bolts securing the right engine mount strut bracket to the cylinder head to 37 ft. lbs. (50 Nm)
- Bolt securing the top of the generator to the right engine mount strut bracket to 37 ft. lbs. (50 Nm)
- Power steering pressure hose to the power steering pump to 22 ft. lbs. (30 Nm)
- 6 frame mounting bolts retaining the frame to the vehicle to 72 ft. lbs. (98 Nm)
- Bolts securing the flywheel to the torque converter to 44 ft. lbs. (60 Nm)
- Oil pan-to-transaxle brace to 37 ft. lbs. (50 Nm)
- A/C discharge/suction hose to the compressor to 20 ft. lbs. (30 Nm)
- Lower transaxle oil cooler pipe fitting to the radiator to 26 ft. lbs. (35 Nm)
- Upper transaxle oil cooler pipe fitting to the radiator to 26 ft. lbs. (35 Nm)

19. Fill the engine with oil.
20. Fill the cooling system.
21. Bleed the brake system.
22. Recharge the A/C refrigerant system.
23. Bleed the power steering system.
24. Check the wheel alignment.
25. Complete the following steps after the engine is installed in the vehicle:

a. With the ignition **OFF** or disconnected, crank the engine several times. Listen for any unusual noises or evidence that any parts are binding.

b. Start the engine and listen for abnormal conditions.

c. Check the vehicle oil pressure gauge or light and confirm that the engine has acceptable oil pressure.

d. Run the engine at approximately 1000 RPM until the engine reaches normal operating temperature.

e. While the engine continues to idle raise and support the vehicle

f. Inspect for oil, coolant and exhaust leaks while the engine is idling.

g. Lower the vehicle.

h. Perform the Crankshaft Position (CKP) system variation learn procedure.

i. Perform a final inspection for the proper engine oil and coolant levels.

26. Road test the vehicle.

Water Pump

REMOVAL & INSTALLATION

1. Before servicing the vehicle, refer to the precautions section.
2. Drain the coolant.
3. Remove or disconnect the following:
   - Negative battery cable
   - Upper fill panel, if equipped
   - AIR injection control valve, if equipped

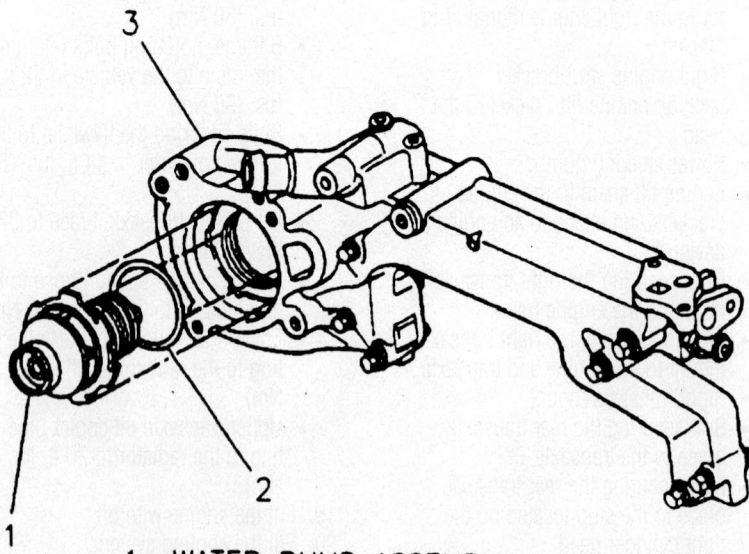

1 WATER PUMP ASSEMBLY
2 O-RING SEAL
3 WATER PUMP HOUSING ASSEMBLY

7922VG01

To ensure proper operation, be sure to install a new O-ring

- Air cleaner
- Oil level indicator tube nut, if necessary
- Water pump belt cover
- Water pump drive belt
- Lower radiator hose
- Thermostat bypass hose from the coolant inlet housing
- Water pump cover

4. Remove the water pump by rotating it clockwise using special Tool J 38816-1A (water pump remover/installer).

To install:

5. Install the water pump with a new O-ring, by turning it counterclockwise using the special tool, until it stops. Torque the water pump to 73 ft. lbs. (100 Nm).

6. Install or connect the following:
   - Water pump cover. Torque the bolts to 89 inch lbs. (10 Nm).
   - Lower radiator hose
   - Thermostat bypass hose to the coolant inlet housing
   - Water pump drive belt
   - Water pump belt cover
   - Oil level indicator tube nut, if necessary
   - Air cleaner
   - AIR injection control valve, if equipped
   - Upper fill panel, if equipped
   - Negative battery cable

7. Refill and bleed the cooling system.
8. Run the engine and check for leaks.

Heater Core

REMOVAL AND INSTALLATION

Deville and Eldorado

1. Disconnect the negative battery cable.
2. Drain the cooling system into a clean container for reuse.
3. Remove or disconnect the following:
   - Instrument panel compartment
   - Sound insulator from the right side
   - Electrical connector and remove the heater/air conditioning programmer
   - Driver's air mix actuator from the bottom of the heater assembly
   - Passenger's air mix actuator from the bottom of the heater assembly
   - Heater core cover-to-heater assembly screws and the cover
   - Heater hoses from the heater core, located in the engine compartment
   - Heater core-to-heater assembly screws
   - Heater core

To install:

4. Install or connect the following:
   - Heater core
   - Heater core-to-heater assembly screws, then, tighten to 18 inch lbs. (2 Nm)
   - Heater hoses to the heater core
   - Heater core cover and the cover-to-heater assembly screws, then, tighten to 18 inch lbs. (2 Nm)
   - Passenger's air mix actuator at the top of the heater assembly
   - Driver's air mix actuator at the bottom of the heater assembly
   - Heater/air conditioning programmer and connect the electrical connector
   - Sound insulator on the right side
   - Instrument panel compartment

5. Refill the cooling system.
6. Connect the negative battery cable.

Seville

1. Disconnect the negative battery cable.
2. Drain the cooling system into a clean container for reuse.
3. Disconnect the heater hoses from the heater core.
4. Disable the SIR system.
5. Remove the instrument panel by removing or disconnecting the following:
   - Console
   - Right insulator panel
   - Instrument panel end caps from both sides

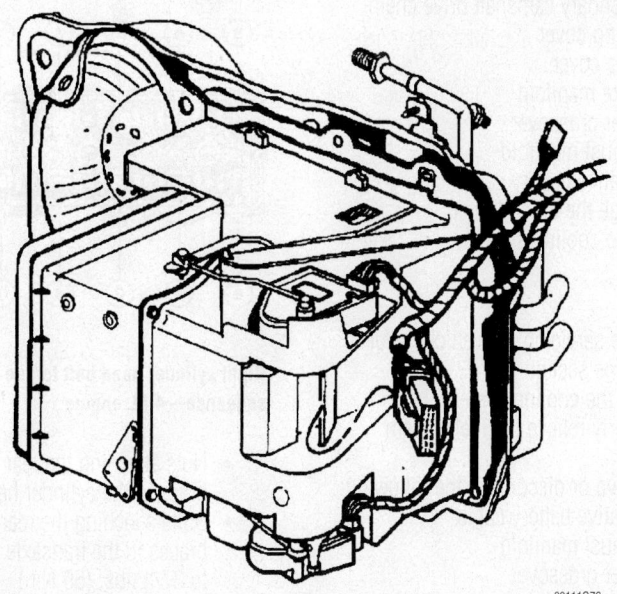

93111G76

View of the heater assembly with the driver's and passenger's air mix actuators—DeVille and Eldorado

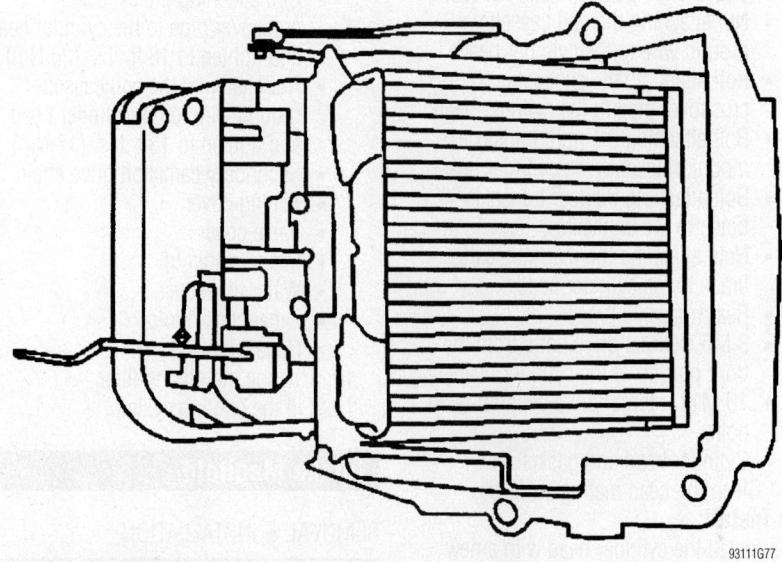

93111G77

View of the heater core—DeVille and Eldorado

- Side windows air outlets
- Upper instrument panel trim pad
- Passenger's side SIR module
- Instrument panel trim plate from the right side
- Instrument panel cluster
- Center instrument panel trim plate
- Radio
- Lap cooler duct
- Knee bolster.
- Steering column
- Headlight switch.
- Fuel door/rear compartment release switch.

- Instrument panel storage compartment
- 10 instrument panel retainer-to-instrument panel carrier fasteners

➡**One fastener is located in the fuel/rear compartment release switch opening.**

- Instrument panel retainer from the carrier and feed the wiring through the openings, as necessary
- Inside air temperature sensor electrical connector
- Instrument panel

- Heater core-to-heater housing cover screws
- Heater core-to-heater housing screws
- Heater core retaining straps
- Heater core
- Heater core seals

To install:

6. Install or connect the following:
   - Heater core seals
   - Heater core.
   - Heater core retaining straps
   - Heater core-to-heater housing screws.
   - Heater core-to-heater housing cover screws.

7. Install the instrument panel by installing or connecting the following:
   - Instrument panel
   - Inside air temperature sensor electrical connector
   - Instrument panel retainer to the carrier and feed the wiring through the openings, as necessary
   - 10 instrument panel retainer-to-instrument panel carrier fasteners

➡**One fastener is located in the fuel/rear compartment release switch opening.**

   - Instrument panel storage compartment
   - Fuel door/rear compartment release switch
   - Headlight switch

8. Install or connect the following:
   - Steering column
   - Knee bolster
   - Lap cooler duct
   - Radio
   - Center instrument panel trim plate
   - Instrument panel cluster
   - Install the instrument panel trim plate on the right side
   - Passenger's side SIR module
   - Upper instrument panel trim pad
   - Side windows air outlets
   - Instrument panel end caps on both sides
   - Right insulator panel
   - Console

9. Enable the SIR system.
10. Connect the heater hoses to the heater core.
11. Refill the cooling system.
12. Connect the negative battery cable.
13. Operate the engine to normal operating temperatures; then, check the climate control operation and check for leaks.

Cylinder Head

REMOVAL & INSTALLATION

Left Side

1. Before servicing the vehicle, refer to the precautions section.
2. Drain the cooling system.
3. Properly relieve the fuel system pressure.
4. Remove or disconnect the following:
   - Negative battery cable
   - Exhaust manifold
   - Alternator
   - Water crossover
   - Intake manifold
   - Valve cover
   - Timing cover
   - Secondary camshaft drive chain
   - Power steering return hose retaining bolt from the cylinder head
   - 3 M6 external drive bolts from the front portion of the cylinder head
   - 10 M11 internal drive cylinder head bolts
   - Cylinder head and gasket.
5. Clean the head mating surfaces.

To install:

6. Install the cylinder head with a new gasket. Lubricate the bolts with engine oil prior to installation.
7. Torque the M11 bolts in sequence as follows:
   a. Step 1: 30 ft. lbs. (40 Nm), plus an additional 70 degrees
   b. Step 2: Turn an additional 60 degrees
   c. Step 3: Turn an additional 60 degrees (total 190 degrees)
8. Torque the M6 bolts to 106 inch lbs. (12 Nm)
9. Install or connect the following:
   - Power steering return hose retaining bolt to the cylinder head and tighten to 37 ft. lbs. (50 Nm)

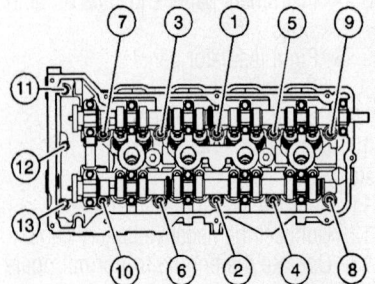

Left cylinder head bolt torque sequence— 4.6L engine

- Secondary camshaft drive chain
- Timing cover
- Valve cover
- Intake manifold
- Water crossover
- Exhaust manifold
- Negative battery cable
10. Change the oil and filter.
11. Fill the cooling system.

Right Side

1. Before servicing the vehicle, refer to the precautions section.
2. Drain the cooling system.
3. Properly relieve the fuel system pressure.
4. Remove or disconnect the following:
   - Negative battery cable
   - Exhaust manifold
   - Water crossover
   - Intake manifold
   - Valve cover
   - Timing cover
   - Secondary camshaft drive chain
   - Electrical connector from the Engine Coolant Temperature (ECT) sensor
   - Nut attaching the coil cassette ground wire to the cylinder head
   - Bolt attaching the exhaust crossover pipe to the cylinder head
   - Bolt attaching the right transaxle mount bracket to the cylinder head
   - Bolt attaching the rear transaxle brace to the transaxle
   - Nuts attaching the rear transaxle brace to the cylinder head
   - Rear transaxle brace
   - 3 M6 external drive bolts from the front portion of the cylinder head
   - 10 M11 internal drive cylinder head bolts
   - Cylinder head and gasket.
5. Clean the head mating surfaces.

To install:

6. Install the cylinder head with a new gasket. Lubricate the bolts with engine oil prior to installation.
7. Torque the M11 bolts in sequence as follows:
   a. Step 1: 30 ft. lbs. (40 Nm), plus an additional 70 degrees.
   b. Step 2: Turn an additional 60 degrees.
   c. Step 3: Turn an additional 60 degrees (total 190 degrees).
8. Torque the M6 bolts to 106 inch lbs. (12 Nm).
9. Install or connect the following:
   - Rear transaxle brace over the studs located at the rear of the right cylinder head

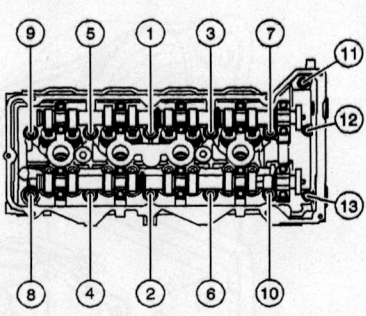

Right cylinder head bolt torque sequence—4.6L engine

- Nuts attaching the rear transaxle brace to the cylinder head, loosely
- Bolts attaching the rear transaxle braces to the transaxle and tighten to 37 ft. lbs. (50 Nm)
- Nuts attaching the rear transaxle brace to the cylinder headand tighten 37 ft. lbs. (50 Nm)
- Electrical connector to the ECT sensor
- Bolt attaching the exhaust crossover pipe to the cylinder head and tighten to 18 ft. lbs. (25 Nm)
- Nut attaching the coil cassette ground wire to the cylinder head and tighten to 13ft. lbs. (17 Nm)
- Secondary camshaft drive chain
- Timing cover
- Valve cover
- Intake manifold
- Water crossover
- Exhaust manifold
- Negative battery cable
10. Change the oil and filter.
11. Fill the cooling system.

Intake Manifold

REMOVAL & INSTALLATION

1. Before servicing the vehicle, refer to the precautions section.
2. Relieve the fuel system pressure.
3. Drain the cooling system.
4. Remove or disconnect the following:
   - Negative battery cable
   - Intake manifold heat shield
   - Sight shield
   - Coil module connectors from the coil modules located on the valve covers
   - Positive Crankcase Ventilation (PCV) hose and valve from the valve cover
   - Fuel regulator vacuum tube

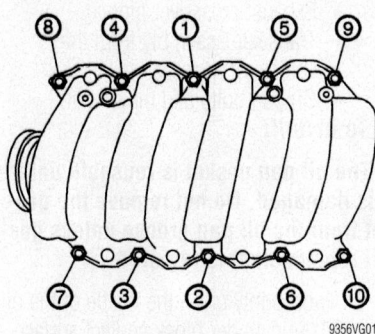

Intake manifold torque sequence—4.6L engines

9356VG01

- Vacuum tubes from the AIR solenoid
- Fuel inlet and return lines
- Fuel rail bracket retaining nut at the rear lift bracket
- 2 pushnuts attaching the engine coolant heater wire and set it aside, if equipped
- Fuel injector electrical connections
- Fuel rail and injectors
- Plenum duct clamp at the rear of the intake manifold, loosen
- Intake manifold bolts and the manifold

5. Clean the manifold mating surfaces.

To install:

6. Grease the inside edge of the rubber plenum duct.

7. Position the intake manifold by performing the following sub-steps:

   a. Place the rear of the intake manifold into the plenum duct.

   b. Place the front of the intake manifold downward on to the cylinder heads.

8. Install the intake manifold bolts and torque to 89 inch lbs. (10 Nm) in the sequence illustrated. DO NOT torque the intake manifold bolts when the engine is HOT or at operating temperature.

9. Make sure the plenum duct is fully attached to the rear of the intake manifold and tighten the plenum duct clamp to 20 inch lbs. (2.25 Nm).

10. Install or connect the following:

- Fuel rail and injectors and tighten the rail retainers to 89 inch lbs. (10 Nm)
- Fuel injector electrical connections
- Alternator coolant pipe/surge tank pipe and position onto the fuel rail studs
- Engine coolant heater wire onto the studs and install the pushnuts, if equipped
- Fuel rail bracket retaining nut at the rear lift bracket and tighten to 89 inch lbs. (10 Nm)
- Fuel inlet and return lines
- Vacuum tubes to the AIR solenoid

- Fuel regulator vacuum tube
- Positive Crankcase Ventilation (PCV) valve and hose to the valve cover
- Coil module connectors to the coil modules located on the valve covers
- Sight shield
- Intake manifold heat shield
- Negative battery cable

11. Fill and bleed the cooling system.

Exhaust Manifold

REMOVAL & INSTALLATION

Left Side

1. Before servicing the vehicle, refer to the precautions section.

2. Remove or disconnect the following:

- Negative battery cable
- 2 nuts attaching the AIR tube to the exhaust manifold
- Engine mount and bracket
- 2 bolts at the manifold outlet flange
- Oxygen Sensor (O_2S), if necessary
- Exhaust manifold bolts
- Manifold and the gasket. Discard the gasket.

To install:

3. Install or connect the following:

- Exhaust manifold by inserting the outlet pipe partially into the exhaust crossover pipe
- Exhaust manifold with a new gasket. Torque the nuts to 18 ft. lbs. (25 Nm).
- O_2S. Torque the sensor to 30 ft. lbs. (40 Nm).
- Engine mount and bracket
- 2 nuts attaching the AIR tube to the exhaust manifold and tighten to 106 inch lbs. (12 Nm)
- Negative battery cable

Right Side

1. Before servicing the vehicle, refer to the precautions section.

2. Remove or disconnect the following:

- Negative battery cable
- Oxygen Sensor (O_2S)
- 2 nuts attaching the AIR tube to the exhaust manifold
- Exhaust front pipe
- Real time dampening sensor links from the lower control arms
- Intermediate shaft from the steering gear

3. Support the rear cross member of the engine cradle with a tall screw jack.

- 4 rearward cradle-to-body bolts, then lower the rear of the engine cradle
- Right side cylinder head to the transaxle brace
- Exhaust manifold nuts
- Manifold and the gasket

To install:

4. Install or connect the following:

- Exhaust manifold with a new gasket. Torque the nuts to 18 ft. lbs. (25 Nm).
- Right side cylinder head to the transaxle brace and tighten the bolt and nut to 35 ft. lbs. (47 Nm)

5. Raise the engine cradle into position.

- 4 rearward cradle-to-body bolts and tighten to 141 ft. lbs. (191 Nm)

6. Remove the screw jack.

- Intermediate shaft to the steering gear and tighten the pinch bolt to 35 ft. lbs. (47 Nm)
- Real time dampening sensor links to the lower control arms
- Exhaust front pipe
- New gasket to the AIR tube
- 2 nuts attaching the AIR tube to the exhaust manifold and tighten to 106 inch lbs. (12 Nm)
- O_2S sensor
- Negative battery cable

Camshaft and Valve Lifters

REMOVAL & INSTALLATION

1. Before servicing the vehicle, refer to the precautions section.

2. Remove both camshaft covers.

3. Secure the camshaft sprockets to the timing chain by installing tie wraps through the camshaft sprocket holes.

4. Working behind the sprockets, install camshaft Chain Holder J38822 so that it is positioned between the chain tensioner and guide. Apply tension to the tool by tightening the tension adjusting screw.

5. Remove or disconnect the following:

- Both camshaft sprocket bolts
- Both camshaft sprockets

6. Alternately loosen the camshaft bearing cap screws a few turns at a time until all valve spring pressure has been released.

7. Remove the bolts and bearing caps.

8. Remove the camshaft.

To install:

9. Lubricate the camshaft lobes and camshaft journals with clean engine oil.

10. Position the camshaft bearing caps in the cylinder head and loosely install the bolts.

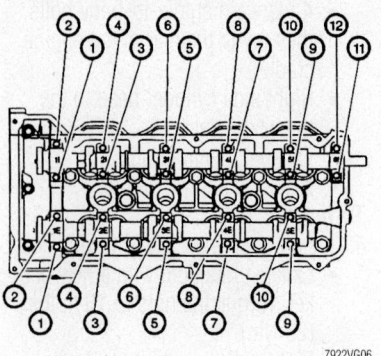

Left cylinder head camshaft bearing cap torque sequence—4.6L engine

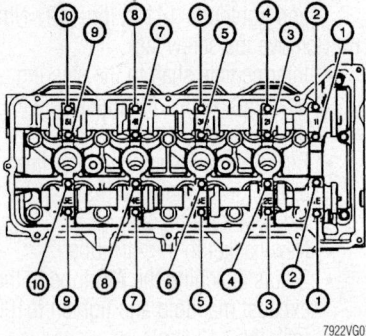

Right cylinder head camshaft bearing cap torque sequence—4.6L engine

11. Tighten the camshaft bearing cap bolts in the sequence shown, to 44 inch lbs. (5 Nm), plus 30 degrees.

12. Using the hex cast into the camshaft, position the camshafts to accept the camshaft sprockets.

13. Install the camshaft sprockets onto the camshafts. Torque the bolts to 89 ft. lbs. (120 Nm).

14. Remove the Chain Holder J38822 and the tie wraps.

15. Install the left and right camshaft covers.

Valve Lash

ADJUSTMENT

The valve clearance cannot be adjusted. The engines in this section are equipped with hydraulic lifters and no adjustment is possible.

Starter Motor

REMOVAL & INSTALLATION

1. Before servicing the vehicle, refer to the precautions section.

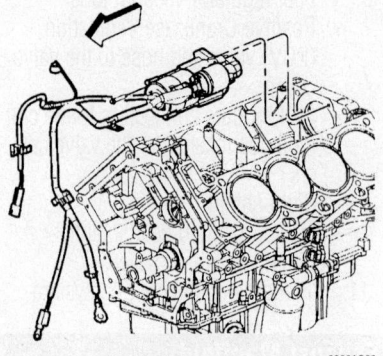

View of starter motor removal and wires

2. Remove or disconnect the following:
   - Negative battery cable
   - Intake manifold
   - Starter electrical connectors
   - Starter motor

To install:
3. Install or connect the following:
   - "S" terminal wire. Torque the nut to 35 inch lbs. (4 Nm).
   - Starter electrical connectors. Torque the nut to 89 inch lbs. (10 Nm).
   - Starter motor
   - Mounting bolts. Torque the bolts to 22 ft. lbs. (30 Nm).
   - Intake manifold
   - Negative battery cable

Oil Pan

REMOVAL & INSTALLATION

1. Before servicing the vehicle, refer to the precautions section.
2. Drain the crankcase.
3. Remove or disconnect the following:
   - Negative battery cable
   - Oil level indicator harness connector, if equipped

   - Exhaust crossover pipe
   - Transaxle assembly from the vehicle, if necessary
   - Oil pan bolts and the oil pan

To install:

➡ The oil pan gasket is reusable unless it is damaged. Do not remove the gasket from the oil pan groove unless gasket replacement is required.

4. Thoroughly clean the inside of the oil pan and the cylinder block contact surface. If the oil pan gasket is being reused, be careful not to damage it. Do not expose the gasket to cleaning solvents.

5. If a new gasket is being installed, start the gasket into the oil pan groove and work the gasket into the groove in both directions. Once the gasket is exposed to oil, it will expand and no longer stay in the groove without wrinkles. If this condition exists, replace the gasket.

6. Install or connect the following:
   - Oil pan. Torque the bolts in sequence to 89 inch lbs. (10 Nm).
   - Oil level indicator connector, if equipped
   - Exhaust crossover pipe
   - Transaxle assembly, if necessary
   - Oil pan drain plug. Torque it to 15 ft. lbs. (20 Nm).
   - Negative battery cable

7. Refill the crankcase.
8. Run the engine and check for leaks.

Oil Pump

REMOVAL & INSTALLATION

1. Before servicing the vehicle, refer to the precautions section.
2. Remove or disconnect the following:
   - Negative battery cable
   - Drive belt tensioner
   - Drive belt idler pulley

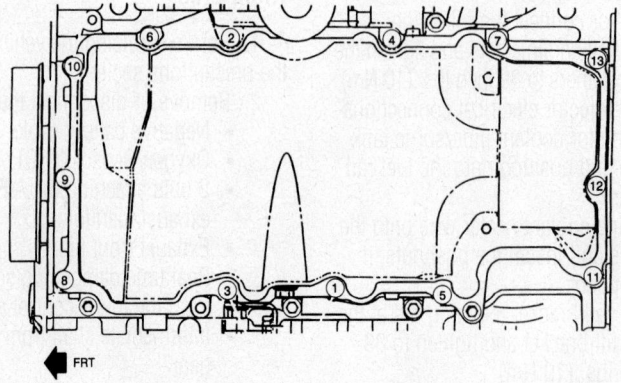

Oil pan bolt torque sequence

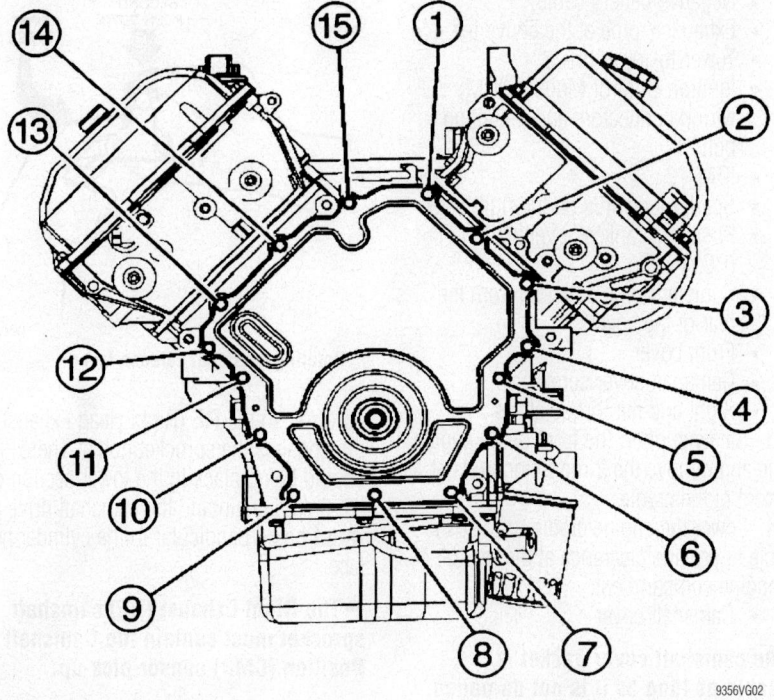

Front cover torque sequence

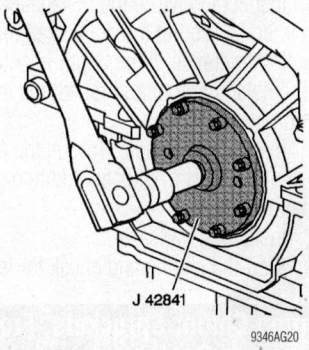

Removing the rear main seal—4.6L engines

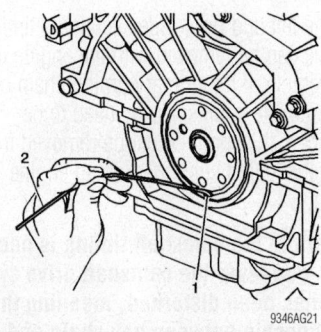

Clean any debris from the crankshaft rear oil seal drain using wire or an unbound plastic tie-wrap

- Crankshaft balancer
- Right engine mount bracket
- Engine front cover and gasket. The gasket is reusable. Do not discard unless it is damaged.
- 3 oil pump assembly retaining bolts identified the bolts by the larger head size
- Oil pump assembly off the nose of the crankshaft with the drive collar in place

3. Clean and inspect the oil pump.

To install:

4. Install or connect the following:
- Oil pump drive spacer into the oil pump so that the drive flat engages the pump rotor
- Oil pump on the crankshaft
- Pump retaining bolts and tighten to 89 inch lbs. (10 Nm) plus an additional 35 degrees

5. Place a small amount of sealant at the split line of the upper and lower crankcases.

6. Install or connect the following:
- Engine front cover gasket over the crankcase dowel pins
- Engine front cover in position on the crankcase
- Engine front cover retaining bolts in the sequence illustrated to 89 inch lbs. (10 Nm)
- Crankshaft balancer
- Engine mount bracket
- Drive belt idler pulley and tighten the bolt to 37 ft. lbs. (50 Nm)

- Drive belt tensioner
- Negative battery cable

Rear Main Seal

REMOVAL & INSTALLATION

1. Before servicing the vehicle, refer to the precautions section.

2. Remove or disconnect the following:
- Transaxle
- Flexplate

3. Place oil seal removal Tool J 42841 on to the crankshaft.

4. Install the tool retaining bolts.

5. Using a drill with a socket adapter, install eight one-inch self-drilling crews into the seal using the guide holes in the removal tool. When drilling, make sure you reduce the drill speed when the screw begins threading into the seal.

6. With all eight removal screws installed, remove the removal tool retaining bolts.

7. Install the removal tool center screw.

8. Tighten the screw until the seal is removed.

To install:

9. Clean any debris from the crankshaft rear oil seal drain using wire or an unbound plastic tie-wrap.

10. Coat the outer diameter of the cylinder block crankshaft rear oil seal area with clean engine oil

➡**DO NOT allow any engine oil on the area where the crankshaft rear oil seal is to be pressed onto the crankshaft. The green coating pre-applied to the inner diameter of the crankshaft rear oil seal must not be contaminated.**

11. Wipe the outer diameter of the flywheel flange clean with a lint-free cloth.

12. Lubricate the outer rubber surface of the crankshaft rear oil seal with clean engine oil.

13. Loosen the center bolt of the removal tool until the center hub protrudes approximately one-half inch beyond the outer plate. It is not necessary to completely unthread the center bolt and separate the two pieces of the installation Tool J 42842. Place the installation tool on to the crankshaft.

14. Thread the three mounting bolts into the crankshaft flange.

15. Tighten the bolts until the installation tool is firmly mounted on the crankshaft.

16. Install the crankshaft rear oil seal by tightening the center bolt until the installation tool bottoms against the crankcase.

17. Loosen the center bolt to release pressure on the crankcase.

18. Loosen the three mounting bolts.

19. Remove the installation tool from the crankshaft flange.

20. Inspect to ensure the installation depth is equal around the crankshaft rear oil seal's circumference. If the depth is not equal reinstall the tool and repeat the installation procedures.

21. Install the flexplate. Torque the bolts to 11 ft. lbs. (15 Nm) plus an additional 50 degree turn.

22. Install the transaxle.

23. Start the engine and check for leaks.

Timing Chain, Sprockets, Front Cover and Seal

REMOVAL & INSTALLATION

The left and right-side secondary timing chains can be removed with the engine in the vehicle. If the primary timing chain or intermediate shaft sprocket need to be replaced, the engine must be removed from the vehicle and supported on an engine stand.

➡ **Setting the camshaft timing is necessary whenever the camshaft drive system has been disturbed, meaning the relationship between any chain and sprocket has been lost. Correct timing exists when the crankshaft and intermediate shaft sprocket timing marks are in alignment and all 4 camshaft drive pins are perpendicular (90 degrees) to the cylinder head surface.**

Right Side Secondary Chain

1. Before servicing the vehicle, refer to the precautions section.

2. Remove or disconnect the following:

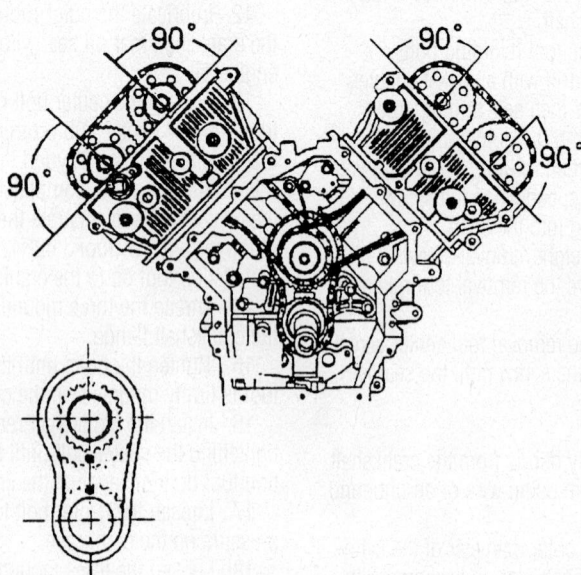

Correct timing chain alignment

- Negative battery cable
- Exhaust Y-pipe at the converter
- Tower-to-tower brace
- Ignition Control Module (ICM) wiring connectors and mounting bolts
- ICM
- Spark plug wires on the right bank
- Positive Crankcase Ventilation (PCV) valve
- Canister purge solenoid from the rear of the cover
- Front cover
- Camshaft cover screws
- Right and left torque struts

3. Safely support the front of the engine cradle and remove the 2 mounting bolts at the front of the cradle.

4. Lower the engine cradle or raise the vehicle to provide clearance at the rear of the engine compartment.
- Camshaft cover

➡ **The camshaft cover gasket is reusable as long as it is not damaged.**

- Right side secondary chain tensioner
- Right side chain guide. Access the upper chain guide mounting bolt through the hole in the cylinder head capped with the plastic plug.
- Right side camshaft sprocket bolts and camshaft sprockets
- Secondary drive chain

To install:

5. Install the secondary timing chain over the outer row of teeth on the intermediate shaft sprocket. Route the chain over the chain guide and install the exhaust camshaft

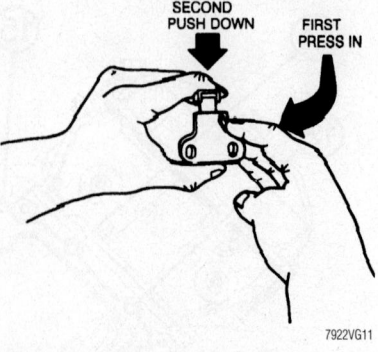

Rotating tensioner release lever

sprocket so the **RE** (Right Head Exhaust) pin engages the sprocket notch. There should be no slack in the lower section of the timing chain and the camshaft drive pin **must** be perpendicular to the cylinder head face.

➡ **The Right Exhaust (RE) camshaft sprocket must contain the Camshaft Position (CMP) sensor pick-up.**

6. Install the intake camshaft sprocket into the chain so the sprocket notch **RI** (Right Head Intake) engages the camshaft and the camshaft drive pin remains perpendicular to the cylinder head face. A hex is cast into the camshafts behind the lobes for cylinder No. 1, so an open end wrench may be used to provide minor repositioning of the camshafts.

7. Install or connect the following:
- Both camshaft sprocket bolts loosely
- Chain tensioner. Torque the mounting bolts to 20 ft. lbs. (27 Nm).

8. Torque the camshaft sprocket bolts to 90 ft. lbs. (120 Nm).
- Spark plug and camshaft cover seals
- Camshaft cover. Torque the screws to 84 inch lbs. (10 Nm).
- Engine cradle. Torque both bolts to 75 ft. lbs. (100 Nm).

9. Install the right and left torque struts. Torque the retaining bolts to 45 ft. lbs. (60 Nm).

➡ **It is important during installation that the engine torque struts are not preloaded in their installed position. Adjustment is provided at the point the strut fastens to the core support bracket. Be sure this bolt is loose during assembly.**

- Front cover. Torque the bolts to 89 inch lbs. (10 Nm).
- Strut-to-core support bracket bolt. Torque it to 45 ft. lbs. (60 Nm).

- Canister purge solenoid to the rear of the cover
- PCV valve
- ICM and the wiring connectors
- Spark plug wires on the right-bank
- Tower-to-tower brace
- Exhaust Y-pipe to the converter. Torque the bolts to 20 ft. lbs. (27 Nm).
- Negative battery cable
10. Start the engine and check for leaks.

Left Side Secondary Chain

1. Before servicing the vehicle, refer to the precautions section.
2. Drain the cooling system.
3. Remove or disconnect the following:
   - Negative battery cable
   - Accessory drive belt
   - Front cover bolts
   - Front cover and gasket
   - Power steering hose
   - Right front wheel
   - Splash shields from the wheel well
   - Flexplate cover
   - Crankshaft balancer bolt
4. Support the engine cradle.
   - 3 right-side engine cradle bolts
   - Road Sensing Suspension (RSS) sensor from the right control arm
5. Lower the cradle to gain access for the crankshaft balancer puller.
   - Crankshaft balancer
   - Drive belt tensioner
   - Drive belt idler pulley
   - Upper radiator hose at the water crossover
   - Spark plug wires
   - Right side cooling fan
   - Battery cable at the alternator
   - Battery cable harness at the camshaft cover
   - Positive Crankcase Ventilation (PCV) fresh air tube from the camshaft cover
   - Water pump pulley
   - Camshaft seal retainer screws and the seal
   - Battery cable retainer at the front of the camshaft cover
   - Camshaft cover by pivoting the entire cover around the water pump driveshaft. Continue moving the cover upward and pivoting so that the edge of the cover closely follows the left edge of the intake manifold cover.

➡**The camshaft cover gasket is reusable as long as it is not damaged.**

- Right side secondary chain
- Left side secondary chain tensioner

- Left side chain guide. Access the upper chain guide mounting bolt through the hole in the cylinder head capped with the plastic plug.
- Left side camshaft sprocket bolts and sprockets
- Secondary drive chain

To install:
6. Route the secondary timing chain for the left side over the inner row of intermediate sprocket teeth.
7. Install or connect the following:
   - Secondary timing chain over the inner row of teeth on the intermediate shaft sprocket. Route the chain over the chain guide and install the exhaust camshaft sprocket so the **LE** (Left Head Exhaust) pin engages the sprocket notch. There should be no slack in the lower section of the timing chain and the camshaft drive pin **must** be perpendicular to the cylinder head face.
   - Intake camshaft sprocket into the chain so the sprocket notch **LI** (Left Head Intake) engages the camshaft and the camshaft drive pin remains perpendicular to the cylinder head face. A hex is cast into the camshafts behind the lobes for cylinder No. 2, so an open-end wrench may be used to provide minor repositioning of the camshafts.
   - Both camshaft sprocket bolts loosely
   - Chain tensioner. Torque the mounting bolts to 20 ft. lbs. (27 Nm).
8. Torque the camshaft sprocket bolts to 90 ft. lbs. (120 Nm).
   - Front cover gasket on the dowel pins on the block
   - Front cover. Torque the bolts to 89 inch lbs. (10 Nm). Apply a dab of RTV to the split line between the upper and lower crankcase assemblies.
   - Drive belt idler pulley. Torque the bolt to 35 ft. lbs. (47 Nm).
   - Drive belt tensioner. Torque the nut to 35 ft. lbs. (47 Nm).
   - Crankshaft balancer. Torque the bolt to 44 ft. lbs. (60 Nm) plus an additional 120 degree turn.
   - Engine cradle. Torque the bolts to 75 ft. lbs. (102 Nm).
   - RSS sensor
   - Wheel well splash shields
   - Flexplate cover
   - Spark plug and camshaft cover seals
   - Intake camshaft through the hole in the camshaft cover

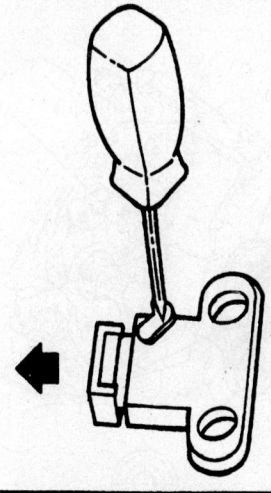

1 RELEASE TO FIRST CLICK
2 INSTALL LOCK PIN

7922VG12

Locking the tensioner in the collapsed position

✳✳ WARNING

Use care to prevent the exposed section of the camshaft cover seal from being damaged by the edge of the cylinder head casting.

- Camshaft cover
- Camshaft cover screws. Torque the screws to 84 inch lbs. (10 Nm).
- Battery cable retainer to the front of the camshaft cover
- Battery cable at the alternator
- Camshaft seal to the end of the intake camshaft. Seal the screw threads with sealer.
- Water pump pulley
- PCV fresh air tube to the camshaft cover
- Right side fan
- Spark plug wires
- Upper radiator hose to the water crossover
9. Refill the cooling system.

Primary Chain/Intermediate Sprocket

1. Before servicing vehicle, refer to the precautions in the beginning of this section.
2. Remove or disconnect the following:
   - Engine
   - Accessory drive belt
   - Drive belt pulley
   - Dive belt tensioner
   - Front cover
   - Right and left camshaft covers

➡**Align all timing marks prior to removal of the timing chains.**

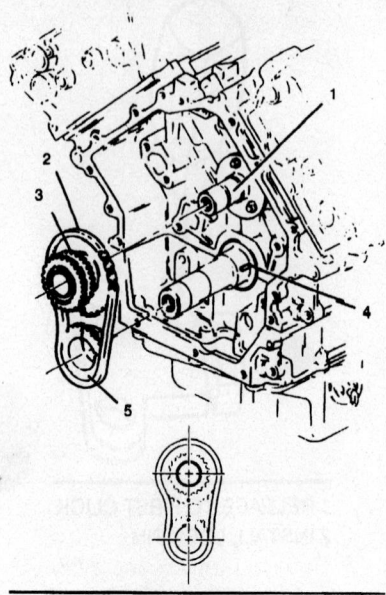

1 INTERMEDIATE SHAFT
2 PRIMARY CHAIN
3 INTERMEDIATE SHAFT SPROCKET
4 CRANKSHAFT SPROCKET KEY
5 SPROCKET

7922VG10

Primary drive chain components

- Timing chain tensioners
- Camshaft sprocket bolts
- Camshaft sprockets (4)
- Right and left secondary chains
- Intermediate shaft sprocket bolt and the sprocket
- Primary timing sprockets and primary chain off the engine

To install:

➡ **The following procedure must be followed to set the camshaft timing on the vehicle.**

3. Install the primary and secondary chain guides.

4. Rotate the crankshaft until the sprocket drive key is at the 1 o'clock position.

5. Install or connect the following:
- Crankshaft sprocket and intermediate shaft sprocket in the primary timing chain so the timing marks are aligned
- Primary timing chain assembly

➡ **The crankshaft sprocket keyway will have to slide over the key on the crankshaft. If it is necessary to turn the crankshaft sprocket, the intermediate shaft sprocket will also have to be turned so the timing mark remains aligned with the crankshaft sprocket.**

- Intermediate shaft sprocket bolt. Torque the bolt to 45 ft. lbs. (61 Nm).
- Primary timing chain tensioner. Torque the tensioner mounting bolts to 20 ft. lbs. (27 Nm).
- Right and left secondary chains with the timing marks aligned
- Camshaft sprockets (4). Torque the bolts to 90 ft. lbs. (120 Nm).
- Right and left camshaft covers. Torque the bolts to 89 inch lbs. (10 Nm).

- Front engine cover. Torque the bolts to 89 inch lbs. (10 Nm).
- Accessory drive belt
- Drive belt pulley
- Drive belt tensioner
- Engine
6. Refill the cooling system.
7. Start the engine and check for leaks.

Piston and Ring

POSITIONING

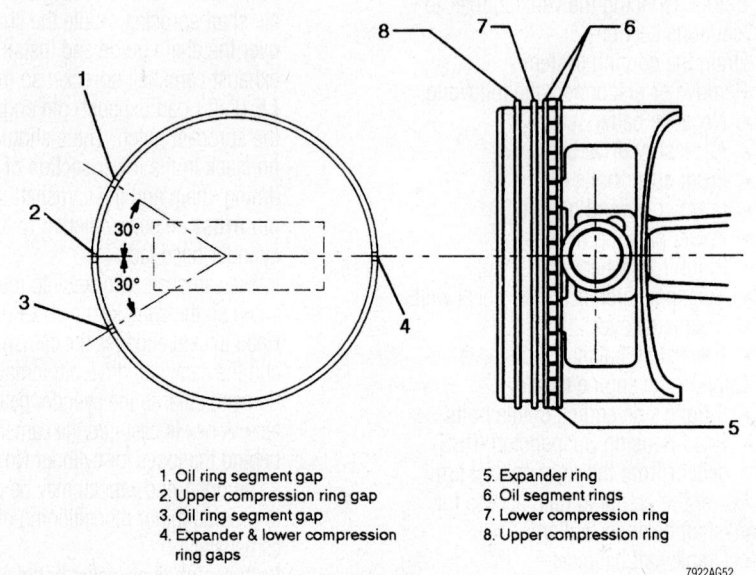

1. Oil ring segment gap
2. Upper compression ring gap
3. Oil ring segment gap
4. Expander & lower compression ring gaps
5. Expander ring
6. Oil segment rings
7. Lower compression ring
8. Upper compression ring

7922AG52

Piston ring and ring end-gap positioning—4.6L engine

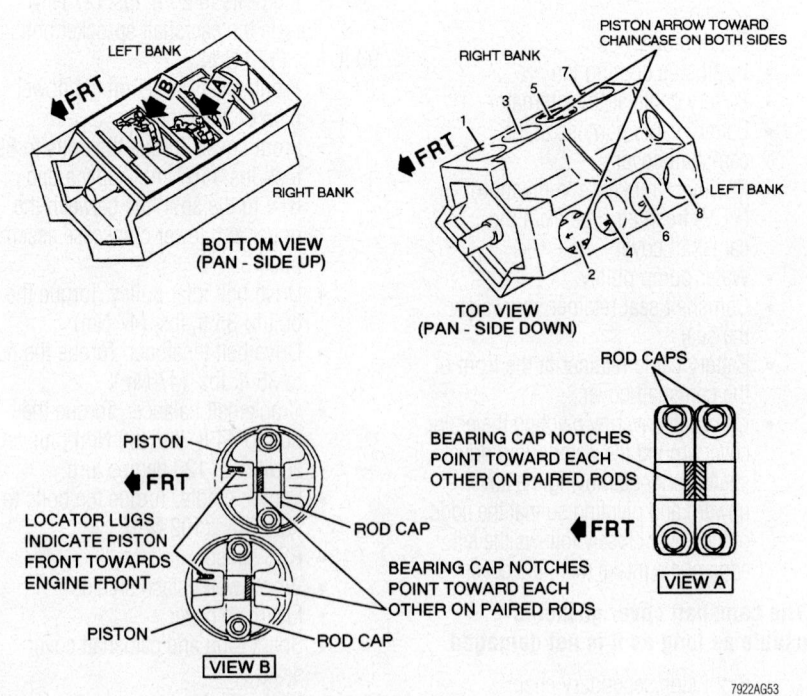

7922AG53

Piston and connecting rod assembly positioning—4.6L engine

FUEL SYSTEM

Fuel System Service Precautions

Safety is the most important factor when performing not only fuel system maintenance but also any type of maintenance. Failure to conduct maintenance and repairs in a safe manner may result in serious personal injury or death. Maintenance and testing of the vehicle's fuel system components can be accomplished safely and effectively by adhering to the following rules and guidelines.

• To avoid the possibility of fire and personal injury, always disconnect the negative battery cable unless the repair or test procedure requires that battery voltage be applied.

• Always relieve the fuel system pressure prior to disconnecting any fuel system component (injector, fuel rail, pressure regulator, etc.), fitting or fuel line connection. Exercise extreme caution whenever relieving fuel system pressure, to avoid exposing skin, face and eyes to fuel spray. Please be advised that fuel under pressure may penetrate the skin or any part of the body that it contacts.

• Always place a shop towel or cloth around the fitting or connection prior to

loosening to absorb any excess fuel due to spillage. Ensure that all fuel spillage (should it occur) is quickly removed from engine surfaces. Ensure that all fuel soaked cloths or towels are deposited into a suitable waste container.

• Always keep a dry chemical (Class B) fire extinguisher near the work area.

• Do not allow fuel spray or fuel vapors to come into contact with a spark or open flame.

• Always use a back-up wrench when loosening. Torque the fuel line connection fittings. This will prevent unnecessary stress and torsion to fuel line piping.

• Always replace worn fuel fitting O-rings with new. Do not substitute fuel hose or equivalent, where fuel pipe is installed.

Fuel System Pressure

RELIEVING

✳✳ CAUTION

The fuel injection system remains under pressure, even when the engine has been turned OFF. The fuel

system pressure must be relieved before disconnecting any fuel lines. Failure to do so may result in fire and/or personal injury.

1. Before servicing vehicle, refer to the precautions at the beginning of this section.
2. Loosen the fuel filler cap to relieve tank vapor pressure.

✳✳ CAUTION

Observe all applicable safety precautions when working around fuel. Whenever servicing the fuel system, always work in a well-ventilated area. Do not allow fuel spray or vapors to come in contact with a spark or open flame. Keep a dry chemical fire extinguisher near the work area. Always keep fuel in a container specifically designed for fuel storage; also, always properly seal fuel containers to avoid the possibility of fire or explosion.

3. Be sure the ignition switch is in the **OFF** position.
4. Disconnect the negative battery cable.
5. Remove the engine cover.

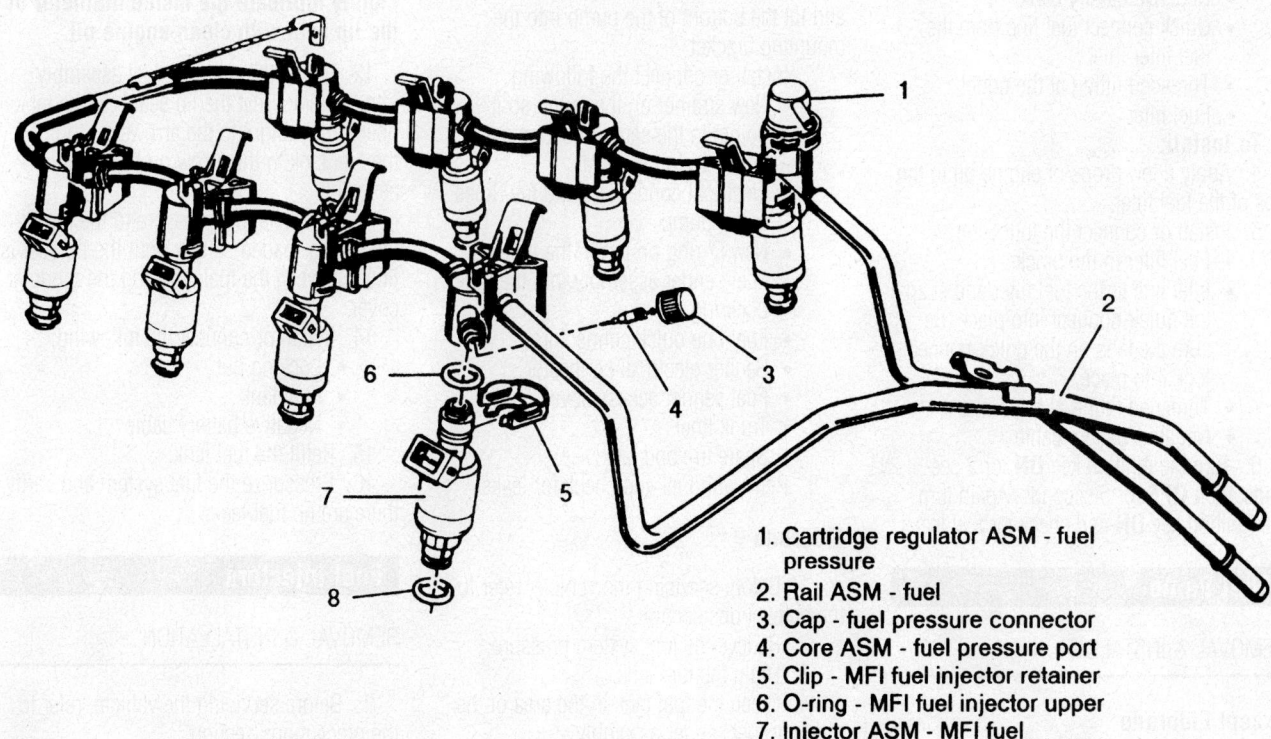

1. Cartridge regulator ASM - fuel pressure
2. Rail ASM - fuel
3. Cap - fuel pressure connector
4. Core ASM - fuel pressure port
5. Clip - MFI fuel injector retainer
6. O-ring - MFI fuel injector upper
7. Injector ASM - MFI fuel
8. O-ring - MFI fuel injector lower

View of the fuel rail assembly, showing the fuel system service port location

There may still be residual fuel in the system, and a small amount of fuel may be released when servicing fuel lines or connections. In order to reduce the chance of personal injury, cover the fuel line fittings with a shop towel before disconnecting to catch any fuel that may leak out.

6. Install a fuel pressure gauge with a drain hose attached, J-34730–1, or equivalent. Wrap a shop towel around the fitting while connecting the gauge to avoid spillage.

7. Install the drain hose into an approved container and open the valve to drain the system pressure. Fuel connections are now safe for servicing.

8. Drain any remaining fuel from inside the gauge into the approved container.

Fuel Filter

REMOVAL & INSTALLATION

1. Before servicing the vehicle, refer to the precautions section.

2. Properly relieve the fuel system pressure.

3. Remove or disconnect the following:
- Negative battery cable
- Quick connect fuel line from the fuel filter inlet
- Threaded fitting at the outlet
- Fuel filter

To install:

4. Apply a few drops of engine oil to the tips of the fuel filter.

5. Install or connect the following:
- Fuel filter in the bracket
- Inlet line to the fuel filter and snap the quick-connect into place. Be sure the tabs on the quick-connects lock into place.
- Threaded fitting at the outlet
- Negative battery cable

6. Turn the ignition key **ON** for 2 seconds, then **OFF** for 5 seconds. Again turn the ignition key **ON** and check for fuel leaks.

Fuel Pump

REMOVAL & INSTALLATION

Except Eldorado

1. Before servicing the vehicle, refer to the precautions section.

2. Relieve the fuel system pressure.

3. Drain the fuel tank.

4. Remove or disconnect the following:
- Negative battery cable
- Spare tire and jack
- Trunk lining
- Fuel sender access panel
- Sender and quick connect fittings from the sender
- Electrical connector from the sender and position harness and hoses aside

When removing the fuel sender from the tank, the reservoir bucket is full of fuel. Use caution in containing the fuel.

- Sender retaining ring
- Sender and take note of its position
- Fuel sender O-ring and discard it

➡ Note the direction the strainer is pointing.

- Strainer from the pump by pulling it down and twisting
- Pump electrical wires and hoses
- Pump assembly out of the rubber connectors

To install:

5. Transfer any insulators and grommets from the old pump to the new one.

6. Connect the pump to the fuel hose and tilt the bottom of the pump into the mounting bracket.

7. Install or connect the following:
- New strainer on the pump so it points in the same direction as noted during removal
- Electrical connectors and fuel lines to the pump
- New O-ring on top of the fuel tank
- Fuel sender assembly into the tank
- Lockring
- Fuel line quick connectors
- Sender electrical connector
- Fuel sender access cover
- Trunk liner
- Spare tire and jack

8. Refill with fuel and check for leaks.

Eldorado

1. Before servicing the vehicle, refer to the precautions section.

2. Relieve the fuel system pressure.

3. Drain the fuel tank.

4. Clean the fuel tank in the area of the modular fuel sender assembly.

5. Remove or disconnect the following:
- Fuel tank from the vehicle
- Locking nut
- Modular fuel sender assembly from the fuel tank

The modular fuel sender assembly may spring up from its position. When removing the assembly, be aware that the reservoir bucket is full of fuel. Tip the assembly slightly during removal to avoid damaging the float. Have a shop towel ready to absorb any leakage.

6. Remove modular fuel sender assembly from the fuel tank.

7. Remove fuel sender seal.

8. Slide the fuel sender lip seal downward, past the reservoir and carefully over the float arm assembly.

9. Discard the fuel sender lip seal.

10. Carefully discard the reservoir fuel into an approved container.

To install:

➡ The lip seal should be carefully positioned over the float arm assembly, moved up over the reservoir and half way up the guide posts.

11. Install a new lip seal on the modular fuel sender assembly.

➡ Always use a new seal when servicing the modular fuel sender assembly. Lightly lubricate the inside diameter of the lip seal with clean engine oil.

12. Install the modular fuel assembly into the tank. Seat the lip seal into the tank opening by aligning the arrows on top of the fuel tank to the arrow on the modular assembly.

13. Slowly apply pressure to the top of the spring-loaded sender until the lip seal is flush between the fuel tank and the modular cover.

14. Install or connect the following:
- Locking nut
- Fuel tank
- Negative battery cable

15. Refill the fuel tank.

16. Pressurize the fuel system and verify there are no fuel leaks.

Fuel Injectors

REMOVAL & INSTALLATION

1. Before servicing the vehicle, refer to the precautions section.

2. Relieve the fuel system pressure.

3. Remove or disconnect the following:
- Intake manifold cover

- Intake Air Temperature (IAT) sensor electrical connector
- Mass Airflow (MAF) sensor electrical connector
- Air cleaner intake duct
- Quick-connect fittings from the fuel rail
- Fuel pressure regulator vacuum hose
- Positive Crankcase Ventilation (PCV) air tube
- Fuel rail end-point bracket retainer nut
- Fuel injector electrical connectors
- Fuel rail attaching bolts
- Fuel injectors from the intake manifold with injector removal tool J 43013
- Fuel rail

- Injector from the rail and discard the O-rings

To install:

➡**Each fuel injector is calibrated for a specific flow rate. Be sure to use the correct part number when ordering replacement fuel injectors.**

4. Install or connect the following:
- New O-rings lubricated with clean engine oil
- Fuel injector into the fuel rail
- Fuel rail into the intake manifold and tighten the attaching bolts to 89 inch lbs. (10 Nm)
- Fuel injector electrical connectors
- Fuel rail end-point bracket retainer nut and tighten to 89 inch lbs. (10 Nm)

- PCV air tube
- Fuel pressure regulator vacuum hose
- Quick-connect fittings from the fuel rail
- Air cleaner intake duct
- MAF sensor electrical connector
- IAT sensor electrical connector
- Negative battery cable

5. Inspect for fuel leaks as follows:
   a. Step 1: Turn ignition switch to the **ON** position for 2 seconds.
   b. Step 2: Turn ignition switch **OFF** for 10 seconds.
   c. Step 3: Turn ignition switch **ON**.
   d. Step 4: Check for leaks.

6. Install the intake manifold cover. Torque the nuts to 27 inch lbs. (3 Nm).

DRIVE TRAIN

Transaxle Assembly

REMOVAL & INSTALLATION

2002–05 DeVille and Seville

1. Before servicing the vehicle, refer to the precautions section.
2. Drain the cooling system.
3. Remove or disconnect the following:
- Negative battery cable
- Front air deflector
- Starter motor
- Upper filler panel
- Upper and lower air cleaner assemblies
- Electrical connector from the Engine Coolant temperature (ECT) sensor.
- Nut securing the coil cassette ground wire to the right cylinder head
- Range selector cable from the shift lever
- Nut that retains the heater inlet pipe bracket
- Locking tabs and pull the heater inlet pipe out of the adapter
- Nut that retains the heater outlet pipe
- Heater outlet pipe from the thermostat housing
- Range selector cable bracket from the studs
- Nuts securing the front transaxle brace to the right cylinder head.

4. Attach an engine lift bracket to the left cylinder head.
- Air inlet grille panel

5. Install an engine support fixture and preload the engine support fixture with the weight of the engine and transaxle.
6. Reposition the main wiring harness to gain access to the upper transaxle to engine bolts.
- Transaxle-to-engine bolts. The remaining bolts will hold the transaxle until it is removed and will be removed from the underside of the vehicle.
- Upper transaxle oil cooler pipe retaining bolt from the radiator fan shroud
- Upper transaxle oil cooler pipe from the radiator plug the lines
- Lower transaxle oil cooler pipe from the transaxle and plug the lines

7. The following are exceptions while following the frame removal procedure:
   a. Do not remove the front stabilizer shaft from the frame. Remove the stabilizer shaft links only.
   b. Do not remove the A.I.R pump assembly from the vehicle. Retain the pump to the body using mechanics wire.
   c. Do not remove the insulators from the frame.
   d. Do not remove the control arms from the frame.

8. Remove or disconnect the following:
- Front frame assembly from the vehicle and position frame aside
- Right and left drive shafts from the transaxle end only. Support the shafts using wire
- Engine-to-transaxle bracket lower bolt and bracket
- Torque converter cover assembly

- Flywheel-to-converter bolts
- Main electrical connector and the ground wires from the front of the transaxle
- Vehicle Speed Sensor (VSS) wiring harness

9. Support the transaxle with a suitable transmission jack.
- Nut securing the transaxle mount to the transaxle mount bracket

10. Slowly lower the transmission jack and allow the engine support fixture to hold the weight of the engine and transaxle assembly.

11. Position the transmission jack aside and lower the vehicle.

12. Using the engine support fixture, lower the engine and transaxle assembly until the left mount bracket clears the mount stud.

13. Raise the vehicle.

14. Support the transaxle with a suitable transmission jack.

15. Remove the engine-to-transaxle brace and heat shield.

16. Remove the remaining transaxle to engine bolts.

17. Separate the transaxle from the engine.

18. Carefully lower the transaxle from the vehicle.

To install:

19. Installation is the reverse of removal, please note the following torques:
- Engine-to-transaxle bolts to 55 ft. lbs. (75 Nm)
- Nut securing the transaxle mount to the transaxle mount bracket to 55 ft. lbs. (75 Nm)

- Flywheel-to-converter bolts to 44 ft. lbs. (60 Nm)
- Engine-to-transaxle bracket lower bolt to 37 ft. lbs. (50 Nm)
- Front transaxle brace nuts to 37 ft. lbs. (50 Nm)

20. Fill the transmission with fluid.
21. Fill the cooling system.
22. Check the front suspension alignment and as necessary.

➡The PCM maintains 3 types of transaxle adapt parameters which are used to modify transaxle line pressure. The line pressure is modified to maintain shift quality regardless of wear or tolerance variations within the transaxle. Whenever the transaxle is replaced, the transaxle adapts must be reset as follows:

a. Step 1: Turn the ignition key **ON**. Enter the self-diagnostic system.

b. Step 2: Select PCM override PS13 (TPS SENSOR LEARN).

c. Step 3: Press the WARMER button. The Driver Information Center (DIC) should display 09, indicating that the Garage Shift Adapt value has been reset.

d. Step 4: Select PCM override PS14 (TRAN ADAPT).

e. Step 5: Press the COOLER button. The DIC should display 90, indicating the Upshift Adapt (UA) value has been reset.

f. Step 6: Press the WARMER button. The DIC should display 09, indicating the Steady State Adapt (SSA) value has been reset.

➡The PCM maintains a value for transaxle oil life. This value indicates the percentage of oil life remaining and is calculated based on transaxle temperature and speed. When the vehicle is new, the transaxle oil life value is 100. As the vehicle operates, the percentage will decrease. Whenever the transaxle is replaced, the transaxle oil life indicator should be reset to 100 as follows:

g. Step 1: Turn the ignition key **ON**, but leave the engine OFF.

h. Step 2: Press and hold the OFF and REAR DEFOG buttons on the DIC until the message TRANSAXLE OIL LIFE RESET is displayed on the DIC.

Eldorado

1. Before servicing the vehicle, refer to the precautions section.
2. Drain the cooling system.
3. Remove or disconnect the following:

- Negative battery cable
- Headlamp housing upper filler panel
- Cross vehicle brace
- Air cleaner assembly
- Shift control cable from the shift lever
- Shift control cable from the bracket
- Engine mount struts
- Oil cooler lines at the cooler and the oil sending line at the transmission
- Heater bypass pipe
- Power steering return hose at the auxiliary cooler and plug all lines to avoid contamination
- Both heater tube retainers from the upper case side cover studs
- Electrical connector from the Coolant Temperature Sensor (CTS)
- Ground wire from the stud near the rear transmission-to-engine brace
- Upper nuts from the rear engine-to-transmission brace
- Transmission vent hose at transmission

4. Carefully pull the wiring harness up from beneath the vehicle and set aside.

5. Move the engine harness at the top of the transmission to access the upper transmission bolts.

- Upper transmission bolts

6. Install an engine support fixture.

7. Raise the left side of the powertrain (the transmission side) 1 inch (25.4 mm) above the powertrain resting position with the adjusting screws.

8. Remove or disconnect the following:
- Both drive axles
- Both front wheel opening splash shields
- Engine splash shield
- Vehicle Speed Sensor (VSS) connector
- Both front suspension sensor rods from lower control arm
- ABS modulator from the bracket and the support
- Torque converter cover
- Engine the transmission brace
- Front engine-to-transmission pencil brace
- Three ground connections at the front of the transmission
- Transmission main harness
- Flywheel to converter bolts
- Powertrain mount nuts from the frame
- Clamp bolt (Rotate the steering intermediate shaft so that the steering gear stub shaft clamp bolt is accessible from the left wheel opening)

Failure to disconnect the intermediate shaft from the rack and pinion steering gear stub shaft can result in damage to the steering gear or to the intermediate shaft. This damage may cause loss of steering control, which could result in an accident and possible personal injury.

➡**Never rotate the steering wheel or move the position of the steering gear once the intermediate shaft has been disconnected. This will uncenter the Inflatable Restraint coil in the steering column. If the Inflatable Restraint coil becomes uncentered, it may become damaged during vehicle operation.**

- Steering intermediate shaft from the steering gear
- Electrical harness and connector from the front of the frame

9. Support the rear of frame with a jack stand.

- Four rear frame bolts

10. Lower the jack stand a few inches to gain access to the power steering gear heat shield and the return line fitting.

- Power steering gear heat shield and the power steering return line at the gear and plug the lines to avoid contamination
- Power steering electrical connector

11. Raise the jack stand and reinstall one rear frame bolt on each side finger-tight in order to support the frame.

- Jack stand
- Support frame
- Six frame mount bolts.
- Lower
- Left transmission mount and bracket from the transmission
- Engine-to-transmission bolts
- Engine from the transmission

12. Lower the transmission down and to the left on a slight angle so the transmission case can clear the end of the starter.

13. Lower the transmission.

To install:

14. Installation is the reverse of removal, please note the following torques:

- Engine-to-transaxle bolts to 55 ft. lbs. (75 Nm)
- Left transmission bracket and mount to the transmission bolts to 35 ft. lbs. (47 Nm)
- Power steering pressure hose fitting to 20 ft. lbs. (27 Nm)

- Steering intermediate shaft to the steering gear clamp bolt to 35 ft. lbs. (47 Nm)
- Left and right transmission mount nuts and the right engine mount nuts to 35 ft. lbs. (47 Nm)
- Manual shaft nut to 15 ft. lbs. (20 Nm)
- Upper transaxle case-to-engine bolts to 55 ft. lbs. (75 Nm)

15. Fill the transmission with fluid.
16. Fill the cooling system.
17. Check the front suspension alignment and as necessary.

➡The PCM maintains 3 types of transaxle adapt parameters which are used to modify transaxle line pressure. The line pressure is modified to maintain shift quality regardless of wear or tolerance variations within the transaxle. Whenever the transaxle is replaced, the transaxle adapts must be reset as follows:

   a. Step 1: Turn the ignition key **ON**. Enter the self-diagnostic system.

   b. Step 2: Select PCM override PS13 (TPS SENSOR LEARN).

   c. Step 3: Press the WARMER button. The Driver Information Center (DIC) should display 09, indicating that the Garage Shift Adapt value has been reset.

   d. Step 4: Select PCM override PS14 (TRAN ADAPT).

   e. Step 5: Press the COOLER button. The DIC should display 90, indicating the Upshift Adapt (UA) value has been reset.

   f. Step 6: Press the WARMER button. The DIC should display 09, indicating the Steady State Adapt (SSA) value has been reset.

➡The PCM maintains a value for transaxle oil life. This value indicates the percentage of oil life remaining and is calculated based on transaxle temperature and speed. When the vehicle is new, the transaxle oil life value is 100. As the vehicle operates, the percentage will decrease. Whenever the transaxle is replaced, the transaxle oil life indicator should be reset to 100 as follows:

   a. Step 1: Turn the ignition key **ON**, but leave the engine OFF.

   b. Step 2: Press and hold the OFF and REAR DEFOG buttons on the DIC until the message TRANSAXLE OIL LIFE RESET is displayed on the DIC.

Halfshaft

REMOVAL & INSTALLATION

❈❈ WARNING

Use care when removing the halfshaft to prevent the inner CV-joint from becoming over-extended. Over-extension of the joint could result in separation of internal components and possible joint failure.

1. Before servicing the vehicle, refer to the precautions section.
2. Remove the front wheel.
3. Install a boot protector on the outer CV-joint boot to protect it from damage.
4. Remove or disconnect the following:
   - Hub nut
   - Stabilizer link
   - Ball joint cotter pin and nut
   - Ball joint from the steering knuckle

➡**Partially install the hub nut to protect the threads when removing the halfshaft from the hub.**

5. Remove the halfshaft from the transaxle.

➡**If equipped with anti-lock brakes, care must be used to prevent damage to the toothed sensor ring on the halfshaft and the wheel speed sensor on the steering knuckle.**

To install:
6. Install or connect the following:
   - Halfshaft into the transaxle

➡**To verify the halfshaft is properly seated, grasp the inner CV-joint hous-**

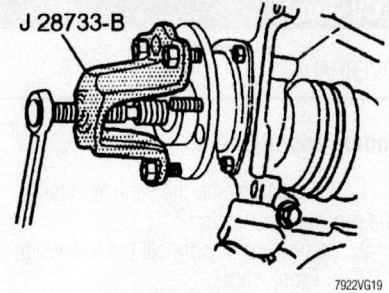

J 28733-B

Removing the halfshaft from the hub

ing and pull it outward. DO NOT pull on the halfshaft. If the CV-joint is properly seated, the halfshaft will not pull back out.

- Halfshaft into the hub/bearing assembly
- New hub nut loosely
- Ball joint into the steering knuckle
- Ball joint castle nut. Torque the nut to 88 inch lbs. (10 Nm), plus an additional 150 degrees. A minimum of 41 ft. lbs. (51 Nm) of torque must be attained.

➡**If necessary, the nut can be tightened up to 20 degrees additional. NEVER loosen the castle nut to install the cotter pin.**

7. Torque the hub nut to 118 ft. lbs. (160 Nm).
8. Install the stabilizer link. Torque the nut to 13 ft. lbs. (17 Nm).
9. If a halfshaft protector was used, remove it now.
10. Install the wheel. Torque the lug nuts to 100 ft. lbs. (140 Nm).
11. Road test and check vehicle operation.

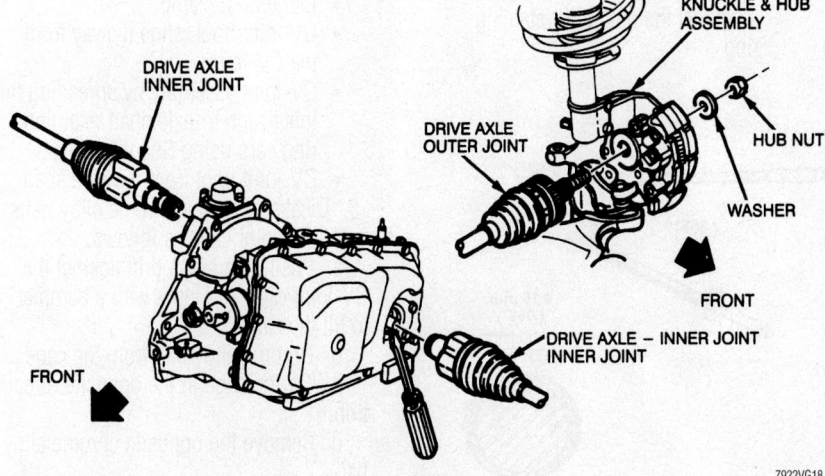

Removing the halfshaft from the transaxle

CV-Joints

OVERHAUL

Inner (Tripod) Joint

1. Before servicing the vehicle, refer to the precautions section.
2. Remove or disconnect the following:
   - Front wheel
   - Halfshaft and place it in a vise
   - Small CV-joint boot clamp
   - Large CV-joint boot clamp
   - CV-joint boot by sliding it away from the tripod joint
   - Tripod housing from the tripod spider
   - Inboard spacer ring, slide it rearward on the shaft
   - Outboard retaining ring
   - Tripod joint spider assembly
   - Inboard spacer ring
   - CV-joint boot
   - Trilobal tripod bushing from the housing

To install:

3. Install or connect the following:
   - New snapring onto the stub shaft
   - Small boot clamp
   - CV-joint boot

4. Using a crimp tool, a torque wrench and a breaker bar, crimp the small CV-joint boot clamp to 100 ft. lbs. (136 Nm) until the crimped gap measures .085 inch (2.15mm).
   - Inboard spacer ring, slide it rearward on the shaft past the 2nd groove
   - Tripod joint spider assembly onto the shaft until it passes the 2nd groove
   - Outboard retaining ring into the axle shaft groove
   - Tripod joint spider assembly, slide it against the outboard retaining ring

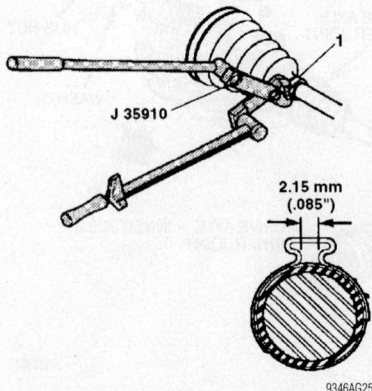

Crimp the small CV-boot clamp as shown

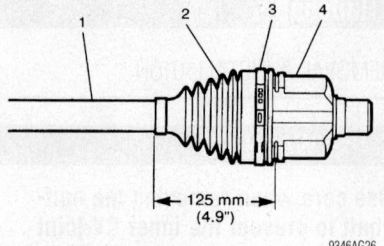

Position the boot as illustrated before clamping the large clamp

   - Inboard spacer ring, seat it in the groove

5. Place ½ of the grease provided in the service kit into the boot and the other ½ of the grease into the tripod housing.
   - Trilobal tripod bushing flush with the tripod housing face
   - New large seal clamp onto the CV-joint boot
   - Tripod housing, slide it over the tripod joint spider assembly
   - CV-joint-boot clamp, slide it into place, over the trilobal tripod bushing with the seal lip in the groove

➡**Make sure the boot lies flat against the trilobal bushing.**

6. Make sure the boot is positioned as illustrated, then using a crimp tool, latch the large CV-joint boot clamp.
7. Install the halfshaft and the front wheel. Torque the lug nuts to 100 ft. lbs. (140 Nm).

Outer CV-Joint

1. Before servicing the vehicle, refer to the precautions section.
2. Remove or disconnect the following:
   - Front wheel
   - Halfshaft
   - Swage ring using a hand grinder
   - Large boot clamp
   - CV-joint boot, slide it away from the CV-joint
   - CV-joint assembly by spreading the inner race-to-axle shaft retaining ring ears using Snapring Pliers
   - CV-joint boot from the axle shaft
3. Disassemble the chrome alloy balls from the CV-joint cage as follows:
   a. Position a brass drift against the CV-joint cage and tap it with a hammer to tilt the cage.
   b. Chrome alloy ball from the cage.
   c. Tilt the cage in the opposite direction.
   d. Remove the opposite chrome alloy ball.
   e. Repeat the procedure until all 6 balls are removed.

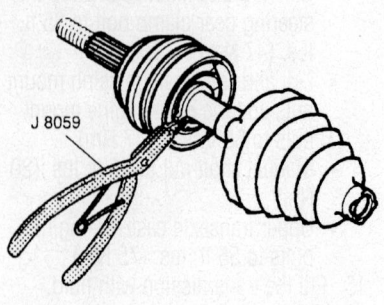

Disconnecting the outer CV-joint from the axle shaft

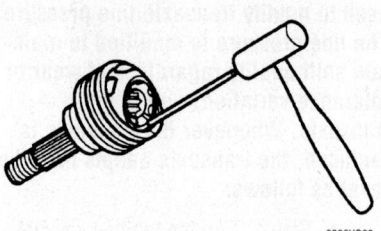

Tilting the cage—Outer CV-joint

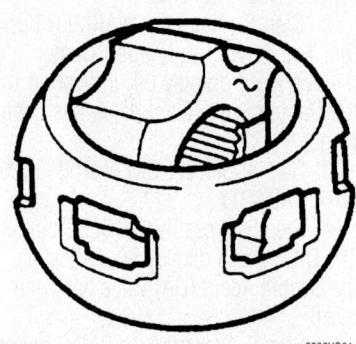

View the cage and inner race—Outer CV-joint

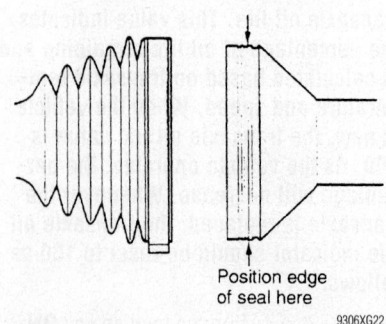

Positioning the boot—Outer CV-joint

4. Disassemble the CV-joint cage and inner race as follows:
   a. Pivot the cage and race 90 degrees to the center line of the outer race.

b. Align the cage windows with outer race lands.

c. Remove the cage from the outer race.

d. Rotate the inner race upward and remove it from the cage.

5. Thoroughly clean and inspect all parts.

To install:

6. Lubricate the parts with a light coat of grease.

7. Assemble the CV-joint cage and inner race, as follows:

a. Rotate the inner race 90 degrees to the cage centerline.

b. Align the cage windows with inner race lands.

c. Insert the inner race into the cage by rotating the inner race downward.

d. Insert the cage/inner race into the outer race.

8. Assemble the chrome alloy balls into the CV-joint cage, as follows:

a. Position a brass drift against the CV-joint cage and tap it with a hammer to tilt the cage.

b. Insert the 1st chrome alloy ball into the cage.

c. Tilt the cage in the opposite direction.

d. Insert the opposite chrome alloy ball.

e. Repeat the procedure until all 6 balls are inserted.

9. Install or connect the following:
- Swage ring clamp
- CV-joint boot
- CV-joint onto the axle shaft until the retaining ring seats into the groove

10. Position the CV-joint boot seal into the axle shaft's joint seal groove and align the swage ring clamp on the boot.

11. Secure the swage ring clamp using appropriate crimping tool.

✳✳ WARNING

Make sure that there are no pinch points on the inboard seal.

12. Install or connect the following:
- ½ kit grease into the CV-joint boot
- ½ kit grease into the CV-joint
- New large seal clamp onto the CV-joint boot
- CV-joint boot/clamp, slide it into place, over the outer race with the seal lip in the groove

➡ **Make sure the boot lies flat against the outer race.**

13. Using a Crimp tool, a torque wrench and a breaker bar, crimp the large CV-joint boot clamp to 130 ft. lbs. (176 Nm).

14. Install the halfshaft and the front wheel.

STEERING AND SUSPENSION

Air Bag

✳✳ CAUTION

Some vehicles are equipped with an air bag system. The system must be disabled before performing service on or around system components, steering column, instrument panel components, wiring and sensors. Failure to follow safety and disabling procedures could result in accidental air bag deployment, possible personal injury and unnecessary system repairs.

PRECAUTIONS

Several precautions must be observed when handling the inflator module to avoid accidental deployment and possible personal injury.

- Never carry the inflator module by the wires or connector on the underside of the module.

- When carrying a live inflator module, hold securely with both hands, and ensure that the bag and trim cover are pointed away.

- Place the inflator module on a bench or other surface with the bag and trim cover facing up.

- With the inflator module on the bench, never place anything on or close to the module, which may be thrown in the event of an accidental deployment.

DISARMING

Eldorado

✳✳ CAUTION

The air bag system must be disarmed before performing service procedures around the air bag or wiring. Failure to do so may cause accidental deployment, resulting in unnecessary repairs and/or personal injury.

1. Before servicing the vehicle, refer to the precautions section.

2. Turn the steering wheel so that the vehicle's wheels are pointing straight-ahead.

3. Turn the ignition switch to the **LOCK** position and remove the key.

4. Remove or disconnect the following:
- Negative battery cable
- **AIR BAG** fuse from the fuse block. On some models the fuse compartment is in the trunk, on other models the fuse panel is located under the rear seat.
- Left sound insulator

5. Disconnect the connector retainer clip from the yellow 2-way connector at the base of the steering column, and detach the harness.

6. Remove the connector retainer clip and detach the yellow 2-way connector from the passenger air bag lead located behind the instrument panel.

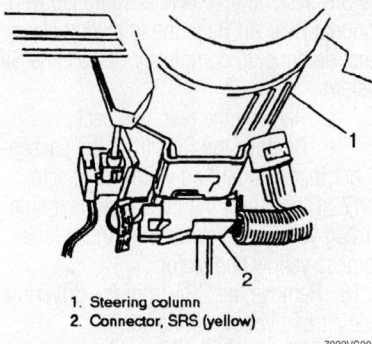

1. Steering column
2. Connector, SRS (yellow)

7922VG20

Detach the SRS yellow 2-way connector

Except Eldorado

1. Turn the steering wheel so that the vehicle's wheels are pointing straight ahead.

2. Turn the ignition switch to the **OFF** position.

3. Remove the key from the ignition switch.

➡ **With the SIR fuse removed and the ignition switch in the ONposition, The AIR BAG warning lamp illuminates. This is normal operation, and does not indicate an SIR system malfunction.**

4. Remove the rear seat.

5. Remove the SIR fuse from the rear fuse block located under the rear seat.

6. Remove the driver sound insulator.

7. Remove the Connector Position Assurance (CPA) from the driver yellow connector located next to steering column.

8. Disconnect the driver frontal air bag yellow connector from the vehicle harness yellow connector.

9. Remove the passenger sound insulator.

10. Remove the CPA from the passenger yellow connector located above the passenger sound insulator.

11. Disconnect the passenger (IP) frontal air bag yellow connector from the vehicle harness yellow connector.

12. Remove both CPA locks from the driver side (seat) air bag and pretensioner yellow connector located under the driver seat.

13. Disconnect the driver side air bag and pretensioner yellow connector from the vehicle harness yellow connector.

14. Remove both CPA locks from the passenger side (seat) air bag and pretensioner yellow connector located under the passenger seat.

15. Disconnect the passenger side air bag and pretensioner yellow connector from the vehicle harness yellow connector.

16. The above procedures will disable the SIR system. If vehicle is equipped with optional Rear Air Bags the following steps must be done, to completely disable the SIR system:

a. Remove the rear seat back.

b. Remove the CPA from the passenger rear side air bag yellow connector.

17. Disconnect the passenger rear side air bag yellow connector from the vehicle harness yellow connector.

18. Remove the CPA from the driver rear side air bag yellow connector.

19. Disconnect the driver rear side air bag yellow connector from the vehicle harness yellow connector.

ARMING

Eldorado

After the applicable service is concluded, enable the air bag system as follows:

1. Turn the ignition switch to the **LOCK** position and remove the key.

2. Install or connect the following:

• Yellow 2-way connector at the base of steering column and secure it with the connector retainer clip. Attach the yellow 2-way connector at the passenger air bag lead and secure it with the connector retainer clip.

• Left sound insulator

• **AIR BAG** fuse in the fuse block

3. Connect the negative battery cable.

4. Turn the ignition switch to the **RUN** position and verify that the **AIR BAG** warning lamp flashes 7 times, then turns **OFF**.

Except Eldorado

➡**If vehicle is equipped with optional Rear Air Bags (AW9) the following steps must be done, to completely enable the SIR system.**

1. Remove the key from the ignition switch.

2. Connect the passenger rear side air bag yellow connector to the vehicle harness yellow connector.

3. Install the (CPA) to the passenger rear side air bag yellow connector.

4. Connect the driver rear side air bag yellow connector to the vehicle harness yellow connector.

5. Install the CPA to the driver rear side air bag yellow connector.

6. Install the rear seat back.

7. Connect the driver side (seat) air bag and pretensioner yellow connector to the vehicle harness yellow connector.

8. Install both CPA locks to the driver side (seat) air bag and pretensioner yellow connector located under the driver seat.

9. Connect the passenger side (seat) air bag and pretensioner yellow connector to the vehicle harness yellow connector.

10. Install both CPA locks to the passenger side (seat) air bag and pretensioner yellow connector located under the passenger seat.

11. Connect the passenger (IP) frontal air bag yellow connector to the vehicle harness yellow connector located above the passenger sound insulator.

12. Install the CPA to the passenger yellow connector.

13. Install the passenger sound insulator.

14. Connect the driver frontal air bag yellow connector to the vehicle harness yellow connector located next to steering column.

15. Install the CPA to the driver yellow connector.

16. Install the driver sound insulator.

17. Install the SIR fuse to the rear fuse block.

18. Install the rear seat.

19. Turn the ignition switch to the **RUN** position and verify that the **AIR BAG** warning lamp flashes 7 times, then turns **OFF**.

Power Rack and Pinion Steering Gear

REMOVAL & INSTALLATION

✳✳ CAUTION

Failure to disconnect the intermediate shaft from the rack and pinion stub shaft can result in damage to the steering gear and/or intermediate shaft. This damage can cause loss of steering control, which could result in personal injury.

✳✳ WARNING

The wheels of the vehicle must be straight-ahead and the steering column in the LOCK position before disconnecting the steering column or intermediate shaft from the steering gear. Failure to do so will cause the coil assembly in the steering column to become off-centered, which will cause damage to the coil assembly.

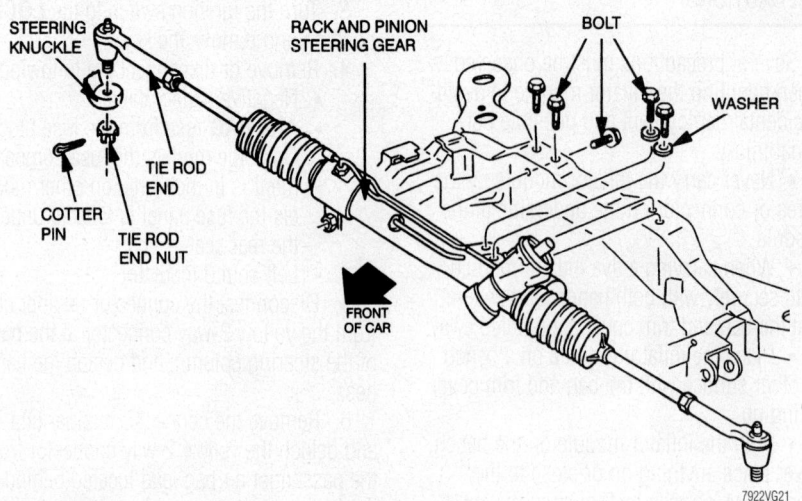

7922VG21

Exploded view of the power steering rack components

Eldorado

1. Before servicing the vehicle, refer to the precautions section.

2. Remove or disconnect the following:
- Wheel
- Road sensing suspension position sensor
- Outer tie rod ends from the steering knuckles
- Intermediate shaft lower pinch bolt
- Intermediate shaft from the power steering gear
- Power steering gear heat shield
- Variable effort steering electrical connector
- Electrical harness clips from the power steering gear
- Power steering pressure hose retaining clip at the power steering gear

3. Support the rear of the frame using a screw jack.
- Power steering pressure and return hoses from the power steering gear
- Frame rear body mount bolts

4. Slowly lower the rear of the frame.
- Power steering gear mounting bolts
- Power steering gear out the left side wheel opening
- Outer tie rod ends when replacing the power steering gear

To install:

5. Install or connect the following:
- Outer tie rod ends when replacing the power steering gear
- Power steering gear through the left wheel opening
- Power steering gear mounting bolts and tighten in the sequence illustrated to 50 ft. lbs. (68 Nm)
- Power steering pressure and return hoses to the power steering gear and tighten to 20 ft. lbs. (27 Nm)

6. Raise the rear of the frame.
- Frame rear body mounting bolts and tighten to 141 ft. lbs. (191 Nm)
- Power steering pressure hose retaining clip to the power steering gear
- Electrical harness clips to the power steering gear
- Variable effort steering electrical connector
- Plastic line retainer.
- Power steering gear heat shield
- Outer tie rod ends to the steering knuckles
- Intermediate shaft to the power steering gear
- Intermediate shaft lower pinch bolt and tighten to 35 ft. lbs. (47 Nm)

- Road sensing suspension position sensor
- Wheels

7. Bleed the power steering system.

8. Adjust the front toe.

DeVille

2002–04 MODELS

1. Before servicing the vehicle, refer to the precautions section.

2. Lock the steering column by placing the wheels in the straight ahead position and remove the keys.

3. Remove or disconnect the following:
- Intermediate shaft pinch bolt and
- Intermediate shaft lower coupling
- Heat shield
- Rear transaxle mount
- Pressure and return lines from steering gear
- Electrical connector
- Steering gear attaching bolts

4. Install a jack under the rear portion of the frame.
- Rear mounting bolts from the frame

✳✳ CAUTION

The frame must be properly supported before partially lowering. The frame should not be lowered any further than needed to gain access to the steering gear.

5. Lower the rear portion of the frame, if necessary.
- Steering gear assembly

To install

6. Install or connect the following:
- Steering gear into vehicle.

7. Raise the rear portion of the frame and tighten the bolts to 142 ft. lbs. (192 Nm), if removed.
- Rear transaxle mount

8. Tighten the bolts to 75 ft. lbs. (90 Nm).

- Electrical connector
- Pressure and return lines. Torque the lines to 22 ft. lbs. (30 Nm).
- Heat shield
- Intermediate shaft to steering gear. Torque the pinch bolt to 33 ft. lbs. (45 Nm).

9. Fill and bleed the power steering system.

2005 MODELS

1. Before servicing the vehicle, refer to the precautions section.

2. Lock the steering column by placing the wheels in the straight ahead position and remove the keys.

3. Install holding tool J 42640 into the steering column access hole in order to lock the steering column. This will maintain the correct steering orientation.

4. Remove the tires and wheels.

5. Remove the outer tie rods retaining nuts.

6. Remove the outer tie rod from the steering knuckle.

7. Remove the intermediate shaft pinch bolt.

8. Disconnect the intermediate shaft from the power steering gear.

9. Remove the steering gear heat shield.

10. Remove the rear transmission mount upper mounting nuts.

11. Disconnect the variable effort steering electrical connector, if equipped.

12. Install a drain pan under the vehicle.

13. Remove the power steering pressure and return hoses from the power steering gear.

14. Remove the steering gear mounting bolts.

15. Install a jack under the rear portion of the frame.

16. Remove the rear mounting bolts from the frame.

✳✳ CAUTION

The frame must be properly supported before partially lowering. The frame should not be lowered any further than needed to gain access to the steering gear.

17. Lower the rear portion of the frame.

18. Remove the rack and pinion from the drivers side of the vehicle.

19. Transfer necessary components.

To install:

20. Install the rack and pinion to the vehicle.

21. Raise the rear portion of the frame.

➡**Use the correct fastener in the correct location. Replacement fasteners must be the correct part number for that application. Fasteners requiring replacement or fasteners requiring the use of thread locking compound or sealant are identified in the service procedure. Do not use paints, lubricants, or corrosion inhibitors on fasteners or fastener joint surfaces unless specified. These coatings affect fastener torque and joint clamping force and may damage the fastener. Use the correct tightening sequence and specifications when installing fasteners in order to avoid damage to parts and systems.**

22. Install the frame mounting bolts. Tighten to 142 ft. lbs. (192 Nm).

23. Install the transmission mount mounting nuts. Tighten to 30 ft. lbs. (40 Nm).

24. Install the rack and pinion mounting bolts. Tighten to 70 ft. lbs. (95 Nm).

25. Install the power steering pressure and return hoses to the power steering gear. Tighten to 20 ft. lbs. (27 Nm).

26. Remove the drain pan from under the vehicle.

27. Connect the variable effort steering electrical connector.

28. Install the power steering gear heat shield.

29. Install the outer tie rods to the steering knuckles. Tighten the nuts 35 ft. lbs (47 Nm) plus an additional ⅛ turn to a maximum torque of 52 ft. lbs. (70 Nm).

30. Install the intermediate shaft to the steering gear.

31. Install the intermediate shaft pinch bolt. Tighten to 33 ft. lbs. (45 Nm).

32. Install the tires and wheels.

33. Remove the steering wheel holding tool J42640 from the steering column access hole.

34. Bleed the power steering system..

35. Adjust front toe.

Seville

1. Before servicing the vehicle, refer to the precautions section.

2. Lock the steering column by installing the Steering Column Anti Rotation Pin tool J 42640 into the underside of the steering column.

3. Remove or disconnect the following:
- Outer tie rod ends from the steering knuckles
- Intermediate shaft lower pinch bolt
- Heat shield
- Rear transaxle mount
- Electrical connector
- Pressure and return lines from steering gear
- Steering gear attaching bolts

4. Install a jack under the rear portion of the engine frame.
- Rear mounting bolts from the engine frame

❋❋ CAUTION

The frame must be properly supported before partially lowering. The frame should not be lowered any further than needed to gain access to the steering gear.

5. Lower the rear portion of the engine frame.

- Steering gear assembly

To install

6. Install or connect the following:
- Steering gear into vehicle

7. Raise the rear portion of the engine frame and tighten the bolts to mounting bolts to 142 ft. lbs. (192 Nm)
- Transaxle mount. Torque the nuts to 55 ft. lbs. (75 Nm).
- Steering gear bolts to 89 ft. lbs. (120 Nm).
- Pressure and return lines. Torque the lines to 22 ft. lbs. (30 Nm).
- Electrical connector
- Heat shield
- Outer tie rod ends to the steering knuckles and tighten the nuts to 35 ft. lbs. (47 Nm)
- Intermediate shaft to steering gear. Torque the pinch bolt to 33 ft. lbs. (45 Nm).

8. Remove the steering column anti rotation pin from the steering column.

9. Fill and bleed the power steering system.

Strut

REMOVAL & INSTALLATION

Front

ELDORADO

❋❋ WARNING

When working near the halfshafts, use care to prevent the inner Tripod CV-joint from being overextended. If the joint is overextended, the internal joint components could separate, resulting in CV-joint failure.

1. Before servicing the vehicle, refer to the precautions section.

2. Remove or disconnect the following:
- Negative battery cable
- Front wheel
- Electrical connector from the top of the strut, if equipped
- Upper strut mounting nuts
- Road Sensing Suspension (RSS) position sensor from the lower control arm, if equipped
- Anti-lock Brake System (ABS) wheel speed sensor, if equipped
- Wheel speed sensor from the strut bracket, if equipped
- Brake line bracket from the strut
- Stabilizer link from the strut

3. Scribe a mark on the strut referencing the lower strut bracket to the steering knuckle.

4. Remove the strut.

To install:

5. Install or connect the following:
- Strut assembly into the vehicle
- Strut-to-knuckle bolts and nuts, but do not torque yet
- Stabilizer link to the strut assembly, but do not torque the nuts yet
- Brake line bracket to the strut
- Speed sensor bracket on the strut, if equipped
- ABS sensor, if equipped
- RSS sensor to the lower control arm, if equipped
- Electrical connector to the top of the strut, if equipped

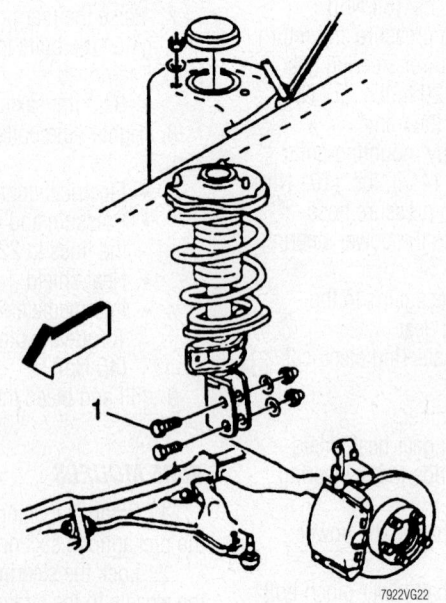

7922VG22

Exploded view of the strut mounting

- Upper strut mounting nuts. Torque the nuts to 15 ft. lbs. (21 Nm).

6. Torque the stabilizer link nuts to 41 ft. lbs. (55 Nm).

7. Torque the strut-to-knuckle bolt to 140 ft. lbs. (190 Nm).

8. Install the front wheel. Torque the lug nuts to 100 ft. lbs. (140 Nm).

DEVILLE

✷✷ WARNING

When working near the halfshafts, use care to prevent the inner Tripod CV-joint from being overextended. If the joint is overextended, the internal joint components could separate, resulting in CV-joint failure.

1. Before servicing the vehicle, refer to the precautions section.

2. Remove or disconnect the following:
- Negative battery cable
- 3 strut mount bolts

3. Raise the vehicle and suitably support by the frame allowing the control arms to hang free.
- Front wheel
- Wheel speed sensor connector
- Wheel speed sensor bracket from the strut
- Brake line bracket from the strut

4. Scribe a mark on the strut referencing the lower strut bracket to the steering knuckle.
- Strut-to-knuckle bolts
- Strut from the vehicle

To install:
- Strut
- 3 upper strut mount bolts and washers. Tighten the bolts to 49 ft. lbs. (66 Nm) on models with the FE7 system or 30 ft. lbs. (40 Nm) on FE1 & 3 systems.

5. Align the scribe marks made on the steering knuckle with the strut.
- Strut-to-knuckle bolts and nuts and hand tighten
- Brake line bracket to strut
- Wheel speed sensor bracket to strut
- Wheel speed sensor connector

6. Tighten the strut-to-knuckle bolts to 131 ft. lbs. (177 Nm) on all models with the FE7 system or 108 ft. lbs. (147 Nm) on models with the FE1 & 3 system

7. Tighten the brake line and wheel speed sensor bracket to 17 ft. lbs. (23 Nm).

8. Install the front wheel.

SEVILLE

1. Before servicing the vehicle, refer to the precautions section.

2. Remove or disconnect the following:
- Tire
- Wheel speed sensor connector
- Wheel speed sensor bracket from the strut
- Brake line bracket from the strut

➡ **Care should be taken to avoid chipping or scratching the coating when handling the suspension coil spring. Damage to the coating can cause premature failure.**

- Strut-to-knuckle bolts
- Three strut mount bolts
- Strut

To install:
3. Install or connect the following:
- Strut
- Three upper strut mount bolts and washers. Tighten the bolts to 33 ft. lbs. (45 Nm).
- New strut-to-knuckle bolts and nuts
- Brake line bracket to strut
- Wheel speed sensor bracket to strut
- Wheel speed sensor connector
- Tighten the strut-to-knuckle bolts to 108 ft. lbs. (147 Nm)
- Tighten the brake line and wheel speed sensor bracket to 13 ft. lbs. (17 Nm)
- Tire

Shock Absorber

REMOVAL & INSTALLATION

Rear

ELDORADO

1. Before servicing the vehicle, refer to the precautions section.

2. Remove or disconnect the following:
- Rear wheel
- Negative battery cable
- Air line from the top of the strut, if equipped
- Shock absorber electrical connector or air line from the rear suspension support, if equipped

3. Support the lower control arm with a jack to relieve the tension on the shock absorber.
- Lower shock absorber mounting bolt and nut
- Upper mounting nut, retainer and insulator
- Shock absorber, compress by hand and remove through the upper control arm

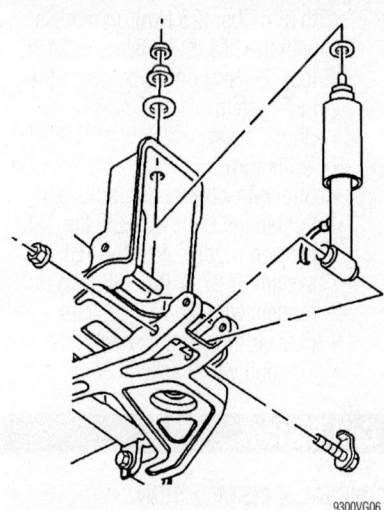

9300VG06

Exploded view of the rear shock absorber mounting

To install:
4. Position the top of the shock absorber with the insulator attached into the suspension support.

5. Install or connect the following:
- Upper shock insulator, retainer and nut. Torque the nut to 55 ft. lbs. (74 Nm).
- Shock absorber lower mounting nut and bolt. Torque the nut to 75 ft. lbs. (102 Nm).
- Shock absorber electrical connector or air line to the rear suspension support, if equipped
- Air line from the top of the strut, if equipped
- Rear wheel. Torque the lug nuts to 100 ft. lbs. (140 Nm).
- Negative battery cable

SEVILLE AND DEVILLE

1. Before servicing the vehicle, refer to the precautions section.

2. Remove the tire and wheel assembly.

3. Support the control arm with a jack stand.

4. Remove or disconnect the following:
- Electronic Level Control (ELC) air tube from the shock
- Two bolts securing the shock to the control arm
- Trunk trim to gain access to the shock upper mounting nuts.
- Cover, the two nuts, and the reinforcement from the top of the shock
- Shock from the vehicle

To install:
5. Install or connect the following:
- Shock, reinforcement, and the two nuts. Tighten the mounting nuts

to18 ft. lbs. (25 Nm) on models with the FE1 & 3 system or 21 ft. lbs. (29 Nm) on models with the FE7 system.
- Shock cover
- Trunk trim
- Shock-to-control arm bolts and tighten the bolts to 18 ft. lbs. (25 Nm) on models with the FE1 & 3 system or 27 ft. lbs. (36 Nm) on models with the FE7 system
- ELC air tube to the shock.
- Tire and wheel assembly

Coil Spring

REMOVAL & INSTALLATION

Front

1. Before servicing the vehicle, refer to the precautions section.
2. Remove the strut from the vehicle.
3. Disassemble the strut as follows:
   a. Step 1: Place the strut assembly into compressor tool, to compress the coil spring.

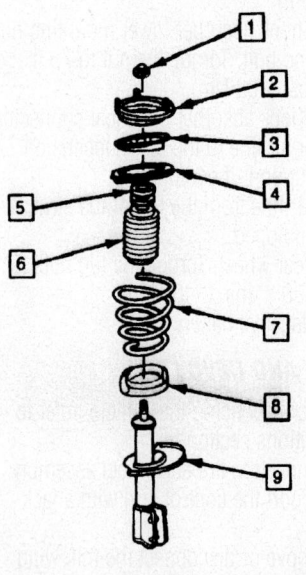

1	NUT, STRUT TO MOUNT
2	STRUT MOUNT
3	FRONT SPRING SEAT
4	FRONT SPRING UPPER INSULATOR
5	JOUNCE BUMPER
6	DUST SHIELD
7	SPRING
8	FRONT SPRING LOWER INSULATOR
9	FRONT STRUT

7922VG23

Disassembled view of strut

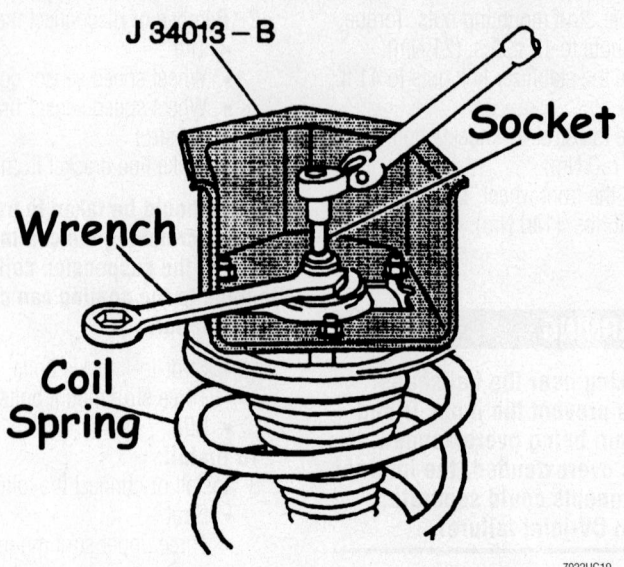

7922UG19

Use a Torx® socket to keep the piston rod from turning while removing the upper nut—front strut shown

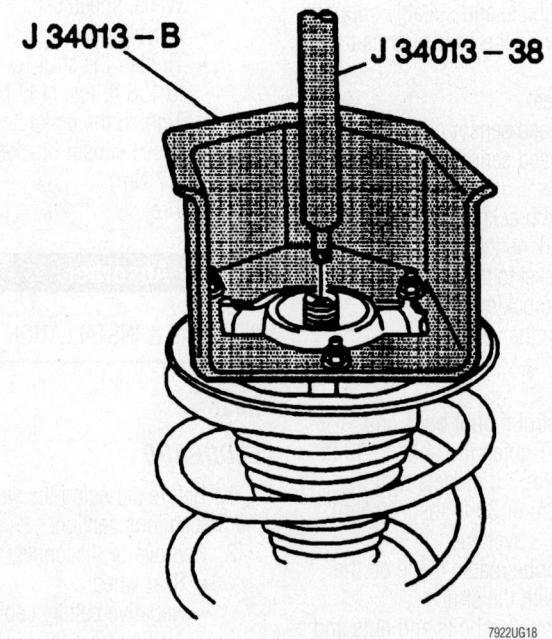

7922UG18

Install Rod J 34013-38 to help guide the strut shaft from the upper mount assembly—front strut shown

b. Step 2: Compress the spring slightly.
c. Step 3: Hold the strut shaft from turning using a No. 50 Torx® socket and remove the 24mm nut on the top end of the strut.
d. Step 4: Install Rod Tool J 34013-38 to help guide the strut shaft from the upper mount assembly.
e. Step 5: Loosen the spring compressor tool until the coil spring and mount can be removed as an assembly.

Remove the lower spring insulator, if equipped.
To install:
4. Assemble the strut as follows:
   a. Step 1: Place the strut in compressor tool.
   b. Step 2: Install the coil spring over the strut.
   c. Step 3: Compress the coil spring while guiding strut shaft through the top of the strut assembly.
   d. Step 4: Install the top strut nut.

Torque the nut to 55 ft. lbs. (75 Nm) on 2002–04 models or 63 ft. lbs. (85 Nm) on 2005 models.

e. Step 5: Remove the strut from the compressor.

Rear

ELDORADO

1. Before servicing the vehicle, refer to the precautions section.
2. Support the outboard end of the lower control arm.
3. Remove or disconnect the following:
   • Wheel
   • Stabilizer link lower mounting bolt
   • Shock absorber lower mounting bolt
   • Height sensor link attachment from the lower control arm
4. Separate the ball joint from the control arm.
5. Slowly lower the control arm until spring pressure is released.
6. Remove the spring.
   To install:
7. Place spring into lower control arm.
8. Raise the lower control arm until the spring is properly seated. Install the bolts finger-tight, do not torque the bolts until proper trim height is obtained.
9. Insert the ball joint into the knuckle. Torque the nut to 129 ft. lbs. (175 Nm).
10. Install the shock absorber lower mounting bolt. Torque the bolt to 80 ft. lbs. (108 Nm).
11. Connect the stabilizer link lower mounting bolt. Do not torque the bolt at this time.
12. Install the wheel. Torque the lug nuts to 100 ft. lbs. (140 Nm).
13. Lower the vehicle to allow the suspension to obtain proper trim height.
14. Torque the lower control arm nuts to 80 ft. lbs. (108 Nm).
15. Torque the stabilizer link bolt to 38 ft. lbs. (52 Nm).

DEVILLE AND SEVILLE

1. Before servicing the vehicle, refer to the precautions section.
2. Support the outboard end of the lower control arm.
3. Remove or disconnect the following:
   • Wheel
   • Electronic level control tube from the shock, if equipped
   • Adjustment link bolt and separate it from the knuckle
   • Shock absorber lower mounting bolt
4. Separate the ball joint from the control arm.

5. Slowly lower the control arm until spring pressure is released.
6. Remove the spring.
   To install:
7. Place spring into lower control arm making sure the insulator is properly seated.
8. Raise the lower control arm until the spring is properly seated.
9. Tighten the shock lower bolts.
10. Connect the adjustment link bolt and tighten to 22 ft. lbs. (30 Nm) plus an additional 180 degrees
11. Connect the electronic level control tube to the shock, if equipped.
12. Install the wheel.
13. Check the alignment.

Lower Ball Joint

REMOVAL & INSTALLATION

Commercial Chassis

FRONT

1. Before servicing the vehicle, refer to the precautions section.

❋❋ CAUTION

Be careful when working in the area of the CV-boot. Damage to the boot could result in eventual joint failure.

2. Remove or disconnect the following:
   • Wheel
   • Ball joint cotter pin and nut
   • Ball joint from steering knuckle
3. Press the ball joint from the lower control arm.
   To install:
4. Press the ball joint into the control arm.
5. Install or connect the following:
   • Hex nut. Torque the nut to 129 ft. lbs. (175 Nm).
   • New cotter pin

➡**If the cotter pin cannot be installed because the hole in the stud does not align with a nut slot, torque the nut an additional 60 degrees to allow for installation. NEVER loosen the nut to provide for cotter pin installation.**

6. Install the wheel and torque the lug nuts to 100 ft. lbs. (140 Nm).

REAR

1. Before servicing the vehicle, refer to the precautions section.
2. Remove the wheel.
3. Support the lower control arm.

➡**The tension of the coil spring must be supported with a suitable jack stand.**

4. Remove the cotter pin and ball stud nut.
5. Separate the ball joint from the knuckle.
6. Press the ball joint out of the control arm.
   To install:
7. Press the ball joint into the control arm.
8. Install the ball joint into the knuckle.
9. Install the ball stud nut. Torque the nut to 129 ft. lbs. (175 Nm) and install a new cotter pin.
10. Remove the support from the lower control arm.
11. Install the wheel. Torque the lug nuts to 100 ft. lbs. (140 Nm).

All Others

The ball joint is an integral part of the lower control arm and if found to be defective the control arm should be replaced.

Lower Control Arm

REMOVAL & INSTALLATION

Front

ELDORADO

1. Before servicing the vehicle, refer to the precautions section.
2. Remove or disconnect the following:
   • Wheel
   • Road Sensing Suspension (RSS) sensor
   • Lower ball joint from the steering knuckle
3. On the left control arm, support the transaxle and remove two transaxle mount nuts. Then raise the transaxle to access the control arm bolt.
4. Remove or disconnect the following:
   • Control arm nuts and bolts
   • Control arm
   To install:
5. Install or connect the following:
   • Control arm
   • Front and rear bolts, do not tighten yet
   • Transaxle mount and lower the transaxle
   • Lower ball joint. Torque nut to 84 inch lbs. (10 Nm) then an additional 120 degrees, reaching a minimum of 37 ft. lbs. (50 Nm).
   • RSS position sensor

- Wheel. Torque the lug nuts to 100 ft. lbs. (140 Nm).

6. Lower vehicle and allow the vehicle to assume proper trim heights.

7. Torque the control arm front bolt to 93 ft. lbs. (126 Nm).

8. Torque the control arm rear bolt to 116 ft. lbs. (157 Nm).

DEVILLE AND SEVILLE

1. Before servicing the vehicle, refer to the precautions section.

2. Remove or disconnect the following:
- Wheel
- Stabilizer link bolt from the control arm
- Ball joint from the lower control arm
- Control arm mounting nuts and bolts
- Control arm

To install:

3. Install or connect the following:
- Control arm
- Front and rear bolts, do not tighten yet
- Stabilizer link to control arm
- Ball joint stud in the lower control arm

4. Tighten the stabilizer shaft link assembly nut to 17 ft. lbs. (23 Nm).

5. On models with the FE7 system, tighten the ball joint stud nut to 22 ft. lbs. (30 Nm), then tighten the nut an additional 190 degrees.

6. On models with the FE1 & 3 system, tighten the ball joint stud nut to 88 inch lbs. (10 Nm), then tighten the nut an additional 150 degrees.
- Cotter pin
- Wheel

7. Lower the vehicle.

8. Bounce the vehicle and tighten the front lower control arm nut to 120 ft. lbs. (162 Nm) and the rear nut to 108 ft. lbs. (146 Nm) on models with the FE7 system or on the FE1 & 3 systems to 116 ft. lbs. (157 Nm) on the front nut and the rear to 116 ft. lbs. (157 Nm).

Rear

ELDORADO

1. Before servicing the vehicle, refer to the precautions section.

2. Support the outboard end of the lower control arm.

3. Remove or disconnect the following:
- Wheel
- Electronic level control tube from the shock, if equipped
- Adjustment link bolt and separate it from the knuckle

- Shock absorber lower mounting bolt

4. Separate the ball joint from the control arm.

5. Slowly lower the control arm until spring pressure is released.

6. Remove the spring.

7. Remove the control arm bolt and the arm

To install:

8. Lower control arm and inboard control arm bolt and nut and tighten to 75 ft. lbs. (102 Nm).

9. Install the spring and insulators.

10. Raise the lower control arm until the spring is properly seated.

11. Install the stabilizer link lower attachment.

12. Remove the transmission jack. Place the jack under the outboard end of the lower control arm in order to bring the suspension to the suspension's design position.

13. The inner control arm nuts must be tightened in the design position in order to reduce wind up in the bushings as follows:

   a. Stabilizer link lower nut to 44 ft. lbs. (60 Nm).

   b. Tighten the shock absorber lower nut to 75 ft. lbs. (102 Nm).

   c. Tighten the lower control arm inner nuts to 75 ft. lbs. (102 Nm).

14. Install the wheel.

DEVILLE AND SEVILLE

1. Before servicing the vehicle, refer to the precautions section.

2. Remove or disconnect the following:
- Wheels
- Exhaust system
- Coil springs
- Brake calipers
- Wheel speed sensor electrical connectors
- Antilock Brake System (ABS) electrical connectors
- Electronic level control electrical connectors

3. Support the rear suspension support assembly.

4. Remove the bolts and lower the support assembly from the vehicle.
- Stabilizer link bolt
- Hub and bearing
- Lower control arm

To install

5. Install or connect the following:
- Lower control arm. Do not torque the bolts at this time.
- Hub and bearing. Torque the bolts to 52 ft. lbs. (70 Nm).
- Stabilizer link. Torque the bolt to 11 ft. lbs. (15 Nm).

6. Raise the rear suspension support assembly into place. Torque the bolts to 141 ft. lbs. (191 Nm).
- Electronic level control electrical connectors
- ABS electrical connectors
- Wheel speed sensor electrical connectors
- Brake calipers. Torque the bolts to 63 ft. lbs. (85 Nm).
- Coil springs
- Exhaust system
- Wheels

7. To obtain proper trim height, raise the control arm until the spindle face is parallel to the ground and torque the nuts to 78 ft. lbs. (106 Nm).

CONTROL ARM BUSHING REPLACEMENT

Front

1. Before servicing the vehicle, refer to the precautions section.

2. Remove or disconnect the following:
- Wheel
- Lower control arm

3. Press the bushing out of the lower control arm.

To install

4. Lubricate the outer case of the new bushing.

5. Press the new bushing into the control arm.

6. Install or connect the following:
- Lower control arm
- Wheel. Torque the lug nuts to 100 ft. lbs. (140 Nm).

Wheel Bearings

ADJUSTMENT

The wheel bearings are not adjustable. If a wheel bearing is out of specifications, it must be replaced. Using a dial indicator, check for looseness. If play exceeds 0.005 inches (0.127mm), the bearing wear is excessive and the hub and bearing should be replaced.

REMOVAL & INSTALLATION

Front

1. Before servicing the vehicle, refer to the precautions section.

2. Remove or disconnect the following:
- Front wheel
- Halfshaft nut and washer
- Caliper from the steering knuckle and support it on a wire

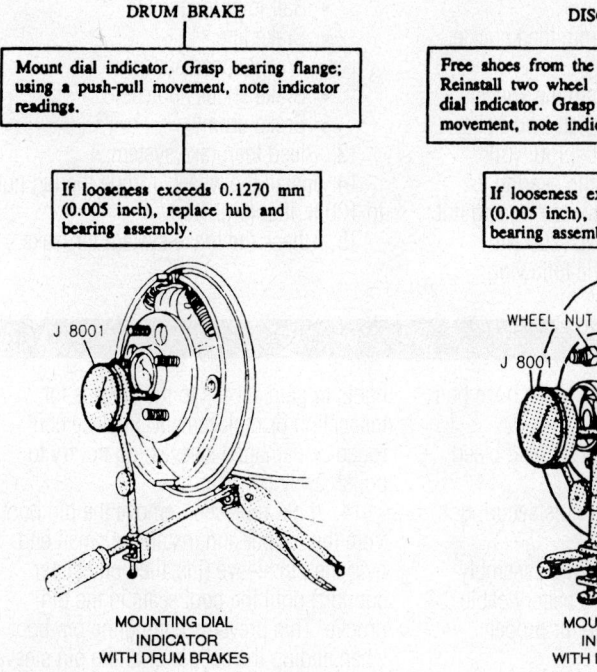

WHEEL BEARING LOOSENESS DIAGNOSIS

DRUM BRAKE

Mount dial indicator. Grasp bearing flange; using a push-pull movement, note indicator readings.

If looseness exceeds 0.1270 mm (0.005 inch), replace hub and bearing assembly.

DISC BRAKE

Free shoes from the disc, or remove calipers. Reinstall two wheel nuts to secure disc. Mount dial indicator. Grasp disc; using a push-pull movement, note indicator readings.

If looseness exceeds 0.1270 mm (0.005 inch), replace hub and bearing assembly.

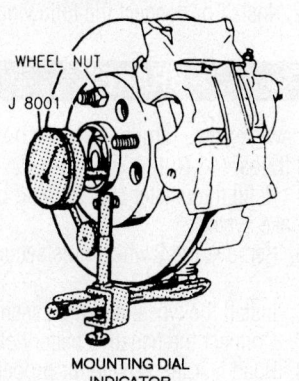

J 8001

WHEEL NUT
J 8001

MOUNTING DIAL INDICATOR WITH DRUM BRAKES

MOUNTING DIAL INDICATOR WITH DISK BRAKES

7922VG33

Inspect the wheel bearings for play with a dial indicator

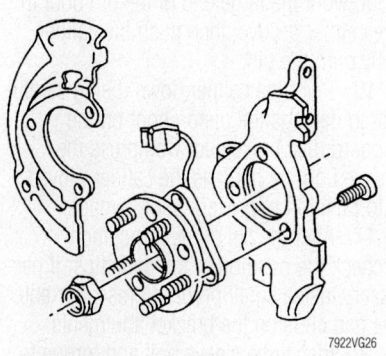

7922VG26

Exploded view of the front wheel bearing mounting

- Brake rotor
- Anti-lock Brake System (ABS) speed sensor electrical connector
- Hub/bearing assembly to steering knuckle bolts
- Dust shield
- Hub and bearing assembly from the halfshaft
- Hub and bearing assembly from the steering knuckle

To install:

3. Install the hub and bearing assembly over the halfshaft splines.

➡**Be sure the splines engage smoothly.**

4. Apply a light coating of grease to the steering knuckle bore.

5. Install the hub/bearing assembly onto the halfshaft as far as possible. If the hub will not bottom out on the halfshaft, install the hub bolts and use the hub nut to draw the hub onto the halfshaft.

6. Once the hub is flush with the steering knuckle, remove the mounting bolts and install the dust shield. Reinstall the mounting bolts and torque to 70 ft. lbs. (95 Nm) on Eldorado. Tighten to 95 ft. lbs. (130 Nm) on all Seville and all models with the FE1 & 3 system or on FE7 systems to 112 ft. lbs. (152 Nm).

7. Install or connect the following:

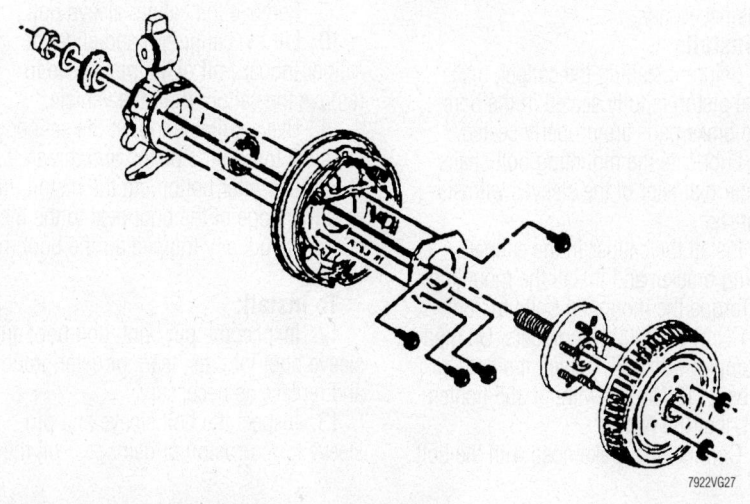

7922VG27

Exploded view of the rear wheel bearing hub mounting

- ABS speed sensor electrical connector
- Brake rotor
- Caliper. Torque the bolts to 38 ft. lbs. (51 Nm).
- Halfshaft nut. Torque it to 107 ft. lbs. (145 Nm).
- Wheel. Torque the lug nuts to 100 ft. lbs. (140 Nm).

8. Road test the vehicle for proper operation.

Rear

EXCEPT COMMERCIAL CHASSIS

➡**The wheel bearing and hub are serviced as an assembly. The individual components are not serviceable separately.**

1. Before servicing the vehicle, refer to the precautions section.
2. Remove or disconnect the following:
- Rear wheel
- Wheel speed sensor, if equipped
- Caliper bracket from the knuckle
- Brake rotor
- Hub and bearing assembly bolts
- Hub and bearing assembly

To install:

3. Install or connect the following:
- Hub and bearing assembly. Torque the bolts to 52 ft. lbs. (70 Nm).
- Brake rotor
- Caliper
- Wheel speed sensor, if equipped
- Wheel. Torque the lug nuts to 100 ft. lbs. (140 Nm).

4. Road test the vehicle.

COMMERCIAL CHASSIS

1. Before servicing the vehicle, refer to the precautions section.

2. Remove or disconnect the following:
- Wheel
- Brake drum
- Brake shoes and cable
- Brake line
- Wheel speed sensor

3. Support the lower control arm.

4. Separate the ball joint from the knuckle.
- Upper knuckle bolt
- Bearing retainer
- Nut from backside of knuckle

5. Press the hub from the knuckle.

6. Remove the brake backing plate from the knuckle.

7. Press the bearing from the knuckle.

To install

8. Press the bearing into the knuckle.

9. Install the backing plate and bolts. Torque the bolts to 37 ft. lbs. (50 Nm).

10. Press the hub into the bearing.

11. Install the speed sensor ring and nut. Torque the nut to 145 ft. lbs. (200 Nm).

12. Install or connect the following:

- Upper knuckle bolt and nut. Torque the nut to 80 ft. lbs. (108 Nm).
- Ball joint to knuckle
- Brake line
- Wheel speed sensor
- Brake shoes and cable
- Brake drum

13. Bleed the brake system.

14. Install the wheel. Torque the lug nuts to 100 ft. lbs. (140 Nm).

15. Check the brake system for leaks.

BRAKES

Brake Caliper

REMOVAL & INSTALLATION

Front

1. Disconnect the negative battery cable.

2. Remove ⅔ of the brake fluid from the master cylinder.

3. Remove the front wheel. Mark the relationship between the wheel and the wheel stud for re-installation purposes.

4. Install 2 wheel nuts to keep the rotor in place.

5. Using a large C-clamp against the inboard pad, compress the caliper piston into the caliper to provide clearance during removal.

6. Place a catch pan under the caliper.

7. Disconnect the brake hose from the caliper. Cap the line to prevent excessive fluid loss or contamination.

8. Remove the caliper mounting bolts and remove the caliper from the vehicle.

9. Inspect the mounting bolts; sleeves and boots for wear and/or damage. Replace parts as necessary.

To install:

10. Before installing the caliper, make sure the piston is fully seated in the bore and the brake pads are properly seated.

11. Lubricate the mounting bolt shafts and inner diameter of the sleeves with silicone grease.

12. Install the caliper in the caliper mounting bracket and install the mounting bolts. Torque the mounting bolts to 38 ft. lbs. (51 Nm) on 2002–04 models. On models equipped with J55, tighten the bolts to 83 ft. lbs. (113 Nm) or without J55 tighten to 63 ft. lbs. (85 Nm).

13. Connect the brake hose with the bolt

and new gaskets. Torque the brake hose bolt to 33 ft. lbs. (45 Nm).

14. Refill the master cylinder and bleed the brake system.

15. Remove the 2 wheel nuts securing the rotor.

16. Install the wheel and tire assembly.

17. Connect the negative battery cable.

18. Road test the vehicle for proper brake system operation.

Rear

1. Disconnect the negative battery cable.

2. Remove ⅔ of the brake fluid from the master cylinder.

3. Remove the rear wheel.

4. Install 2 wheel nuts to keep the rotor in place.

5. Place a catch pan under the caliper.

6. Disconnect the brake hose from the caliper. Cap the line to prevent fluid loss or contamination.

7. Loosen the tension on the parking brake at the equalizer.

8. Remove the parking brake cable mounting lever, and remove the cable end by lifting up and disengaging the end.

9. Remove the caliper sleeve bolt.

10. Lift the caliper up and slide the caliper inboard off of the pin sleeve to remove the caliper from the vehicle.

11. Use a suitable tool in the caliper piston slots to turn the piston and thread it into the caliper. After bottoming the piston, lift the inner edge of the boot next to the piston and press out any trapped air the boot must lay flat.

To install:

12. Inspect the pin boot, bolt boot and sleeve boot for cuts, tears or deterioration and replace as necessary.

13. Inspect the bolt sleeve and pin sleeve for corrosion or damage. Pull the

boots to gain access to the sleeves for inspection or replacement. Replace corroded or damaged sleeves; do not try to polish away corrosion.

14. If not replaced, remove the pin boot from the caliper and install the small end over the pin sleeve (installed on caliper support) until the boot seats in the pin groove. This prevents cutting the pin boot when sliding the caliper onto the pin sleeve.

15. Hold the caliper in the position as removed and start it over the end of the pin sleeve. As the caliper approaches the pin boot, work the large end of the pin boot in the caliper groove, then push the caliper fully onto the pin.

16. Pivot the caliper down, being careful not to damage the piston boot on the inboard disc brake pad. Compress the sleeve boot by hand as the caliper moves into position to prevent boot damage.

17. After the caliper is in position, recheck the position of the pad clips. If necessary, use a small prybar to reseat or enter the pad clips on the bracket abutments.

18. Install the sleeve bolt and torque to 20 ft. lbs. (27 Nm) .

19. Install the parking brake cable bracket, with the cable attached, and torque the bolt to 32 ft. lbs. (43 Nm).

20. Install the parking brake cable onto the parking brake lever and the retaining clip onto the parking brake cable.

21. Connect the brake hose with the bolt and new gaskets and torque the bolt to 32 ft. lbs. (43 Nm).

22. Adjust the parking brake cable.

23. Refill the master cylinder and bleed the brake system.

24. Remove the wheel nuts retaining the rotor and install the wheel.

25. Connect the negative battery cable.

26. Road test the vehicle for proper brake system operation.

Disc Brake Pads

REMOVAL & INSTALLATION

Front

1. Remove ⅔ of the brake fluid from the master cylinder reservoir.
2. Remove the front wheel.
3. Remove the caliper mounting bolts.
4. Remove the caliper from the steering knuckle without disconnecting the brake hose.

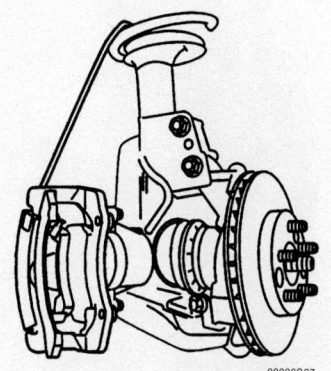

Supporting the front caliper

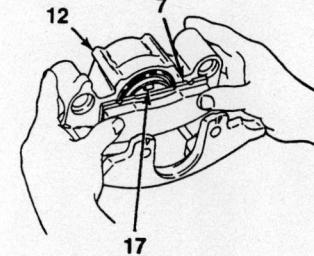

7. Inboard Shoe & Lining
12. Caliper Housing
17. Shoe Retainer Spring

93006G68

Installing the inboard disc brake pad into the front caliper

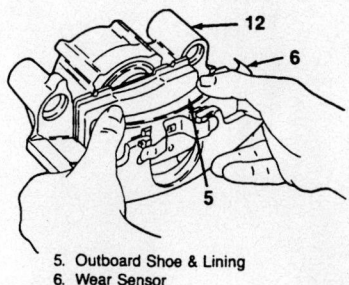

5. Outboard Shoe & Lining
6. Wear Sensor
12. Caliper Housing

93006G69

Installing the outboard disc brake pad in the front caliper

5. Suspend the caliper from the coil spring with wire. Do not let the caliper hang from the brake hose.
6. Remove brake pads from anchor bracket.

To install:

7. Fully seat the caliper piston into the bore using a large C-clamp.
8. Lubricate the brake caliper mounting surfaces.
9. Install the brake pads onto the anchor bracket using new shims and clips.
10. Place the caliper onto the anchor bracket and install the mounting bolts.
11. Torque the mounting bolts to 38 ft. lbs. (51 Nm) on 2002–04 models. On models equipped with J55, tighten the bolts to 83 ft. lbs. (113 Nm) or without J55 tighten to 63 ft. lbs. (85 Nm).
12. Install the wheel and tire assembly and torque to specification.
13. Lower the vehicle.
14. Pump the brake pedal several times to seat the brake pads.
15. Check the fluid level in the master cylinder and fill as necessary.
16. Road test the vehicle for proper brake operation.

Rear

1. Remove ⅔ of the brake fluid from the master cylinder reservoir.
2. Remove the rear wheel.
3. Remove the park brake cable from the caliper.

➡ **Some models use a brake pad wear sensor. Disconnect the sensor connector from the vehicle harness.**

4. Remove the caliper mounting bolts and position and suspend the caliper from the strut with a wire. Do not let the caliper hang from the brake hose.
5. Remove the inboard and the outboard brake pads from the caliper mounting bracket.

To install:

6. Lubricate mounting surfaces and install new brake pad clips.
7. The caliper piston must be fully seated in the bore. A spanner type tool can be used to bottom the piston into its bore by turning it in. Make sure the slots in the piston are straight across from each other so the notches on the brake pad will seat.
8. Install the inboard pad by inserting the pad into the straight tabs on the retainer, then pressing down and snapping the pad

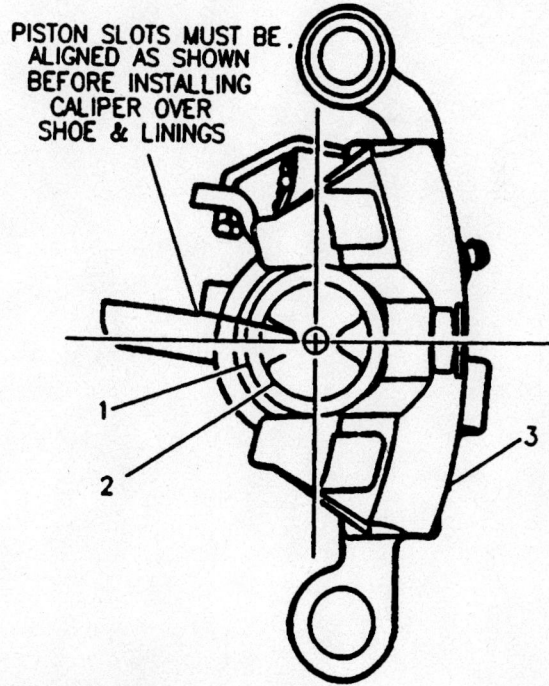

PISTON SLOTS MUST BE ALIGNED AS SHOWN BEFORE INSTALLING CALIPER OVER SHOE & LININGS

1 **PISTON BOOT**
2 **PISTON ASSEMBLY**
3 **CALIPER BODY ASSEMBLY**

93006G70

Aligning the slots on the rear caliper

under the S-shaped tabs. The pad should lay flat against the rotor. Make sure the D-shaped notches are in line with the buttons on the back of the pad lining. If they are not in alignment, rotate the piston until the D-shaped notches face the caliper mounting bolt holes.

9. Install the outer pad into the brake pad clips. Make sure the wear indicator is at the leading edge of the pad during forward wheel rotation.

10. Install the caliper on the mounting bracket and install the bolts. Torque the mounting bolts to 63 ft. lbs. (85 Nm).

11. Install the parking brake cable onto the caliper.

12. Install the wheel and tire assembly and torque to specification.

13. Fill the master cylinder reservoir to the FULL mark and pump the brake pedal several times to seat the brake pads.

14. Check the fluid level in the master cylinder again and fill as necessary.

15. Road test the vehicle for proper brake operation.

DISC BRAKES **17-61**
DRIVE TRAIN **17-38**
DRUM BRAKES **17-65**
ENGINE REPAIR **17-9**
FRONT SUSPENSION **17-52**
FUEL SYSTEM **17-34**
REAR SUSPENSION **17-56**
SPECIFICATIONS AND
  MAINTENANCE CHARTS......**17-3**
Engine and Vehicle
  Identification 17-3
General Engine Specifications 17-3
Engine Tune-Up Specifications......17-3
Accessory Drive Belt Routing 17-4
Capacities.............................17-4
Valve Specifications....................17-4
Crankshaft and Connecting Rod
  Specifications.......................17-5
Piston and Ring Specifications......17-5
Torque Specifications 17-5
Wheel Alignment 17-6
Tire, Wheel and Ball Joint
  Specifications.......................17-6
Brake Specifications 17-6
Maintenance I and II Service
  Schedules............................17-7
Additional Maintenance Services...17-8
STEERING AND
  SUSPENSION **17-46**
A
Air Bag...................................17-46
  Disabling And Enabling The
    System 17-48
  Precautions............................17-46
Alternator 17-9
  Removal & Installation..............17-9
B
Brake Caliper 17-61
  Removal & Installation............17-61
Brake Drums.............................17-65
  Removal And Installation 17-65
Brake Shoes.............................17-66
  Removal And Installation 17-66
C
Caliper Bracket 17-64
  Removal & Installation............17-64
Camshaft and Bearings................17-25
  Removal & Installation............17-25

Coil Spring.............................17-57
  Removal & Installation............17-57
Crankshaft Damper 17-29
  Removal & Installation............17-29
CV-Joints.................................17-43
  Overhaul 17-43
Cylinder Head 17-18
  Removal & Installation............17-18
D
Disc Brake Hardware...................17-64
  Removal & Installation............17-64
Disc Brake Pads.........................17-62
  Removal And Installation 17-62
Driveshaft 17-45
  Removal & Installation............17-45
E
Engine Assembly 17-9
  Removal & Installation..............17-9
Exhaust Manifold........................17-24
  Removal & Installation............17-24
F
Front Cover 17-30
  Removal & Installation............17-30
Front Crankshaft Seal 17-30
  Removal & Installation............17-30
Front Halfshaft 17-40
  Removal & Installation............17-40
Front Hub and Bearing
  Assembly..............................17-56
  Removal & Installation............17-56
Fuel Filter 17-35
  Removal & Installation............17-35
Fuel Injectors 17-37
  Removal & Installation............17-37
Fuel Pump 17-35
  Removal & Installation............17-35
Fuel System Service
  Precautions............................17-34
Fuel Tank 17-35
  Removal & Installation............17-35
H
Heater Core..............................17-13
  Removal & Installation............17-13
Hydraulic System........................17-68
  Bleeding................................17-68
I
Ignition Timing 17-9
  Adjustment............................17-9

Intake Manifold.........................17-22
  Removal & Installation............17-22
Intermediate shaft 17-42
  Removal & Installation............17-42
K
Knuckle...................................17-55
  Removal & Installation............17-55
Knuckle...................................17-60
  Removal & Installation............17-60
L
Lower Ball Joint.........................17-54
  Removal & Installation............17-54
Lower Control Arm 17-54
  Lower Control Arm Bushing
    Replacement.......................17-55
  Removal And & Installation 17-54
Lower Control Arm 17-58
  Bushing Replacement 17-59
  Removal & Installation............17-58
O
Oil Pan....................................17-27
  Removal & Installation............17-27
Oil Pump 17-27
  Removal & Installation............17-27
P
Parking Brake 17-67
  Adjustment............................17-67
Power Steering Gear 17-51
  Removal & Installation............17-51
Primary Fuel Tank Module............17-35
  Removal & Installation............17-35
R
Rear Differential Pinion Seal 17-44
  Removal & Installation............17-44
Rear Halfshaft 17-41
  Removal & Installation............17-41
Rear Hub and Bearing
  Assembly..............................17-61
  Removal & Installation............17-61
Rear Main Seal 17-28
  Removal & Installation............17-28
Relieving Fuel System
  Pressure 17-34
Rocker Arm Covers.....................17-20
  Removal & Installation............17-20
Rocker Arms, Pushrods and
  Lifters 17-21
  Removal & Installation............17-21

Rotor...17-63
    Removal & Installation............17-63

S

Secondary Fuel Tank Module 17-36
    Removal & Installation............17-36
Shock Absorber 17-57
    Removal & Installation............17-57
Stabilizer Bar and Links..............17-52
    Removal & Installation............17-52
Stabilizer Bar and Links..............17-56
    Removal & Installation............17-56

Strut..17-53
    Assembly 17-54
    Disassembly 17-53
    Removal & Installation............17-53

T

Timing Chain and Sprockets........17-33
    Removal & Installation............17-33
Toe Link......................................17-60
    Removal & Installation............17-60
Trailing Arm 17-59
    Removal & Installation............17-59

Transaxle.....................................17-38
    Removal & Installation............17-38
Transfer Case Assembly...............17-39
    Removal & Installation............17-39

U

Upper Control Arm 17-57
    Removal & Installation............17-57

W

Water Pump................................17-12
    Removal & Installation............17-12
Wheel Cylinder 17-67
    Removal & Installation............17-67

SPECIFICATIONS AND MAINTENANCE CHARTS

ENGINE AND VEHICLE IDENTIFICATION

Engine							Model Year	
Code ①	Liters	Cu. In.	Cyl.	Fuel Sys.	Engine Type	Eng. Mfg.	Code ②	Year
F	3.4	204	6	MFI	OHV	Chev.	5	2005
							6	2006

MFI: Multi-port Fuel Injection

OHV: Overhead valves

① 8th digit of VIN

② 10th digit of VIN

06025-EQUI-C01

GENERAL ENGINE SPECIFICATIONS

Engine Displacement Liters	Engine VIN	Year	Net Horsepower @ rpm	Net Torque @ rpm (ft. lbs.)	Bore x Stroke (in.)	Com- pression Ratio	Oil Pressure @ rpm
3.4	F	2005	210@3800	185@5200	3.62x3.31	9.6:1	30-35@1850

06025-EQUI-C02

ENGINE TUNE-UP SPECIFICATIONS

Engine Displacement Liters	Engine VIN	Year	Spark Plug Gap (in.)	Ignition Timing	Fuel Pump (psi)	Idle Speed (rpm)	Valve Clearance	
							Intake	Exhaust
3.4	F	2005	0.060	①	50-60	①	Hyd	Hyd

Hyd: Hydraulic

NA: Information not available

① Ignition timing and idle speed are not adjustable

06025-EQUI-C04

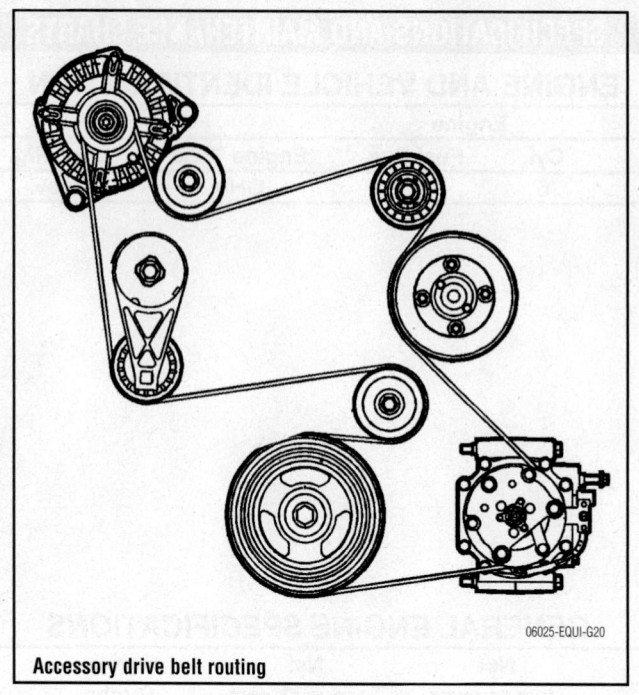

06025-EQUI-G20

Accessory drive belt routing

CAPACITIES

Model	Engine Displacement Liters	Engine VIN	Year	Engine Oil with Filter (qts.)	Transmission (pts.)	Transfer Case (pts.)	Rear Drive Axle (pts.)	Fuel Tank (gal.)	Cooling System (qts.)
Equinox	3.4	F	2005	4.5	15 ①	②	1.58	16.7	10.6

NA: Information not available

NOTE: All capacities are approximate. Add fluid gradually and check to be sure a proper fluid level is obtained.

① Drain and refill

② Included in transaxle capacity

06025-EQUI-C03

VALVE SPECIFICATIONS

Engine Displacement Liters	Engine VIN	Year	Seat Angle (deg.)	Face Angle (deg.)	Spring Test Pressure (lbs. @ in.)	Spring Installed Height (in.)	Stem-to-Guide Clearance (in.) Intake	Stem-to-Guide Clearance (in.) Exhaust	Stem Diameter (in.) Intake	Stem Diameter (in.) Exhaust
3.4	F	2005	46	45	230@1.260	1.701	0.0010-0.0027	0.0010-0.0027	NA	NA

NA: Information not available

06025-EQUI-C05

CRANKSHAFT AND CONNECTING ROD SPECIFICATIONS
All measurements are given in inches.

Engine Displacement Liters	Engine VIN	Year	Crankshaft				Connecting Rod		
			Main Brg. Journal Dia.	Main Brg. Oil Clearance	Shaft End-play	Thrust on No.	Journal Diameter	Oil Clearance	Side Clearance
3.4	F	2005	2.6473-2.6483	①	0.0024-0.0083	3	1.9987-1.9994	0.0007-0.0017	0.0100-0.0150

① Except No.3: 0.0008-0.0025 in.
   No.3: 0.0012-0.0030 in.

06025-EQUI-C06

PISTON AND RING SPECIFICATIONS
All measurements given in inches

Engine Displacement Liters	Engine VIN	Year	Piston Clearance	Ring Gap			Ring Side Clearance		
				Top Compression	Bottom Compression	Oil Control	Top Compression	Bottom Compression	Oil Control
3.4	F	2005	0.0036	0.006-0.014	0.019-0.029	0.0098-0.0303	0.002-0.003	0.002-0.003	0.003-0.004

06025-EQUI-C07

TORQUE SPECIFICATIONS
All readings are ft. lbs. unless otherwise noted

Engine Displacement Liters	Engine VIN	Year	Cylinder Head Bolts	Main Bearing Bolts	Rod Bearing Bolts	Crankshaft Damper Bolt	Flywheel Bolts	Manifold		Spark Plugs	Oil Pan Drain Plug
								Intake	Exhaust		
3.4	F	2005	①	②	③	④	52	⑤	12	⑥	18

① Step 1: 44 ft. lbs.
   Step 2: plus 95 degrees
② Step 1: 37 ft. lbs.
   Step 2: plus 77 degrees
③ Step 1: 15 ft. lbs.
   Step 2: plus 75 degrees
④ Step 1: 52 ft. lbs.
   Step 2: plus 72 degrees

⑤ Lower intake center
   Step 1: 62 inch lbs.
   Step 2: 115 inch lbs.
   Lower intake corber
   Step 1: 115 inch lbs.
   Step 2: 18 ft. lbs.
⑥ Initial installation: 15 ft. lbs.
   Re-installation: 13 ft. lbs.

06025-EQUI-C08

WHEEL ALIGNMENT SPECIFICATIONS

Model	Year		Caster		Camber		Toe-in (in.)
			Range (+/-Deg.)	Preferred Setting (Deg.)	Range (+/-Deg.)	Preferred Setting (Deg.)	
Equinox	2005	F	0.75	+3.00	0.75	-0.60	+0.15+/-0.20
		R	—	—	0.75	-0.50	+0.20+/-0.20

06025-EQUI-C09

TIRE WHEEL AND BALL JOINT SPECIFICATIONS

Model	Year	OEM Tires		Tire Pressures (psi)		Wheel Size	Ball Joint Inspection	Lug Nuts (ft. lbs.)
		Standard	Optional	Front	Rear			
Equinox	2005	P235/65R16	none	①	①	6.5	②	92

OEM: Original Equipment Manufacturer

PSI: Pounds Per Square Inch

STD: Standard

OPT: Optional

① See placard on vehicle

② Vertical and horizontal play: 0.125 in.

06025-EQUI-C10

BRAKE SPECIFICATIONS

All measurements in inches unless otherwise noted

Model	Year	Brake Disc			Brake Drum		Minimum Lining Thickness	Brake Caliper	
		Original Thickness	Minimum Thickness	Maximum Run-out	Original Inside Diameter	Maximum Machine Diameter		Bracket Bolts (ft. lbs.)	Mounting Bolts (ft. lbs.)
Equinox	2005	1.024	0.960	0.002	9.84	9.90	①	136	32

① Front: 0.08 in.

  Rear: 0.24 in.

06025-EQUI-C11

MAINTENANCE I AND II SERVICE SCHEDULES
2005 Chevrolet Equinox

When the CHANGE ENGINE OIL light appears, certain services and inspections are required.
Required services are described as Maintenance I and Maintenance II.
The first service on a vehicle should be Maintenance I, and the second service should be Maintenance II.
Alternate between the 2 thereafter. However, in some cases, Maintenance II may be required more often.
Maintenance I: Use Maintenance I if the CHANGE ENGINE OIL light comes on within 10 months since vehicle was purchased or, if Maintenance II was performed.
Maintenance II: Use Maintenance II if the previous service performed was Maintenance I. Always use Maintenance II whenever the CHANGE ENGINE OIL light comes on 10 months or more since the last service, or, if the CHANGE ENGINE OIL light has not come on at all for one year.

Service	Maintenance I	Maintenance II
Change the engine oil and filter. Reset the oil life system.	✓	✓
Visually inspect the vehicle for leaks or damage. A fluid loss in the vehicle system could indicate a problem. Inspected, repair and add fluid to the system if necessary.	✓	✓
Inspect the engine air cleaner filter. If necessary, replace the filter.	✓	✓
Rotate the tires. Inspect the tire inflation pressures and the tire wear.	✓	✓
Visually inspect the brake lines and hoses for proper hook-up, binding, leaks, cracks, chafing, etc. Inspect the disc brake pads for wear and the rotors for surface condition. Inspect the drum brake linings for wear or cracks. Inspect other brake parts, including drums, wheel cylinders, calipers, parking brake, etc. Inspect the parking brake adjustment.	✓	✓
Inspect the engine coolant and the windshield washer fluid levels. Add fluid as needed.	✓	✓
Inspect the suspension and steering components. Inspect the front and rear suspension and the steering system for damaged, loose or missing parts, or signs of wear. Inspect the power steering lines and the hoses for proper hook-up, binding, leaks, cracks, chafing, etc.	--	✓
Visually inspect the coolant hoses and replace the hoses if they are cracked, swollen or deteriorated. Inspect all pipes, fittings and clamps; replace with GM parts as needed. To help ensure proper operation, a pressure test of the cooling system and pressure cap and cleaning the outside of the radiator and air conditioning condenser is recommended at least once a year.	--	✓
Inspect the front and rear suspension and the steering system for damaged, loose or missing parts, or signs of wear. Inspect power steering lines and hoses for proper hook-up, binding, leaks, cracks, chafing, etc.	--	✓
Inspect the throttle system for interference or binding and for damaged or missing parts. Replace the parts as needed. Replace any components that have high effort or excessive wear. Do not lubricate the accelerator or the cruise control cables.	--	✓
Replace the passenger compartment air filter.	--	✓

To reset the CHANGE ENGINE OIL LIGHT:
1. Turn the ignition key to RUN with the engine off.
2. Fully press and release the accelerator pedal three times within five seconds. The change engine oil light will flash while the system is resetting
3. Turn the key to LOCK.

If the change engine oil light comes back on and stays on when you start your vehicle, the engine oil life system has not reset repeat the procedure

06025-EQUI-C12

ADDITONAL MAINTENANCE SERVICES
2005 Chevrolet Equinox

TO BE SERVICED	TYPE OF SERVICE	VEHICLE MILEAGE INTERVAL (x1000)					
		25	50	75	100	125	150
Air cleaner filter	R	✓	✓	✓	✓	✓	✓
Accessory drive belt	I						✓
Auto. Trans. Fluid ①	R		✓		✓		✓
Cooling system hoses and clamps	S/I						✓
Engine coolant	R						✓
Fuel system	I	✓	✓	✓	✓	✓	✓
Exhaust system & heat shields	S/I	✓	✓	✓	✓	✓	✓
Spark plugs	R				✓		

R: Replace   S/I: Inspect and service, if necessary

① Replace if any of the following conditions are met:

    Heavy city traffic where the outside temperature regularly reaches 32°C (90°F) or higher

    Hilly or mountainous terrain

    Frequent trailer towing

    Taxi, police or delivery service

    Otherwise, change every 100,000 miles

06025-EQUI-C13

ENGINE REPAIR

Alternator

REMOVAL & INSTALLATION

1. Before servicing the vehicle, refer to the Precautions Section.
2. Remove the battery ground (negative) cable from the battery.
3. Remove the alternator B+ terminal nut.
4. Remove the alternator B+ lead.
5. Remove the alternator electrical connector.
6. Remove the drive belt from the alternator.
7. Remove the alternator bolts.
8. Remove the alternator from the vehicle.
   To install:
9. Install the alternator to the alternator bracket.
10. Install the alternator bolts. Tighten the alternator bolts to 50 Nm (37 ft. lbs.).
11. Install the drive belt.
12. Install the alternator electrical connector.
13. Install the alternator B+ lead.

14. Install the alternator B+ terminal nut. Tighten the alternator B+ terminal nut to 20 Nm (15 ft. lbs.).
15. Install the battery ground (negative) cable to the battery.

Ignition Timing

ADJUSTMENT

Ignition timing is not adjustable.

Engine Assembly

REMOVAL & INSTALLATION

1. Before servicing the vehicle, refer to the Precautions Section.
2. Remove the battery box and carefully set the engine control module (ECM) on top of the engine.
3. Remove the battery cable retainers from the battery tray.
4. Remove the negative battery cable nut from the inner fender body ground stud.
5. Remove the negative battery cable from the inner fender body ground stud.

6. Remove the fuel injector sight shield.
7. Release the clamp from the brake booster vacuum hose connection.
8. Disconnect the brake booster vacuum hose from the intake manifold.
9. Remove the transaxle control module (TCM) from the bracket and set the TCM on top of the engine.
10. Remove the air cleaner assembly and air intake duct.
11. Discharge the fuel system.
12. Disconnect the evaporative emission (EVAP) hose/pipe from the EVAP canister purge solenoid valve.
13. Disconnect the engine fuel hose/pipe from the chassis fuel hose/pipe.
14. Properly discharge the air conditioning (A/C) system.
15. Remove the A/C compressor hose assembly from the compressor. Cap or plug the hoses and compressor to prevent contamination.
16. Disconnect the transaxle shift control cable from the transaxle.
17. Disconnect engine to body inline connector C102.
18. Drain the engine coolant from the cooling system.
19. Tie the radiator, A/C condenser, and fan module assembly to the upper radiator support to keep the assembly with the vehicle when the frame and drivetrain is removed.
20. Remove the coolant surge tank.
21. Disconnect the heater hoses from the engine.
22. Remove the radiator inlet hose.
23. Raise and support the vehicle.
24. Remove the radiator outlet hose.
25. Disconnect the transaxle oil cooler lines from the transaxle and remove the seals.
26. Cap the transaxle oil cooler lines and plug the transaxle oil cooler line fittings to prevent loss of transmission fluid.
27. Remove the front bumper fascia air deflector.
28. Disconnect the heated oxygen sensor (HO2S) wiring harness.
29. Remove the HO2S wiring harness retainers from the vehicle underbody.
30. Remove the catalytic converter and secure the rear half of the exhaust system to the vehicle underbody.
31. If equipped with all wheel drive (AWD), disconnect the transfer case vent hose from the transfer case.
32. Remove the front tires.
33. Remove the right and left engine splash shields.

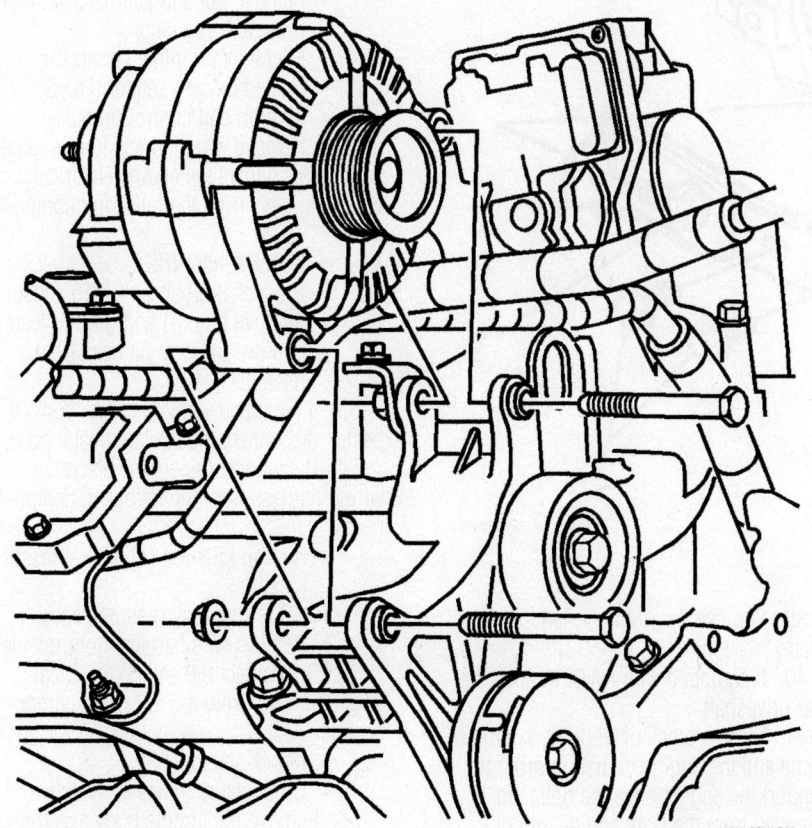

06025-EQUI-G01

Alternator mounting

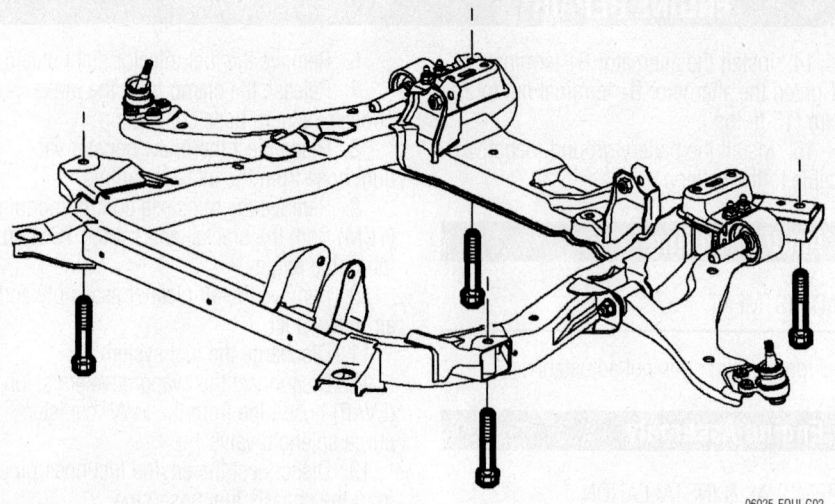

06025-EQUI-G02

Frame-to-body bolts

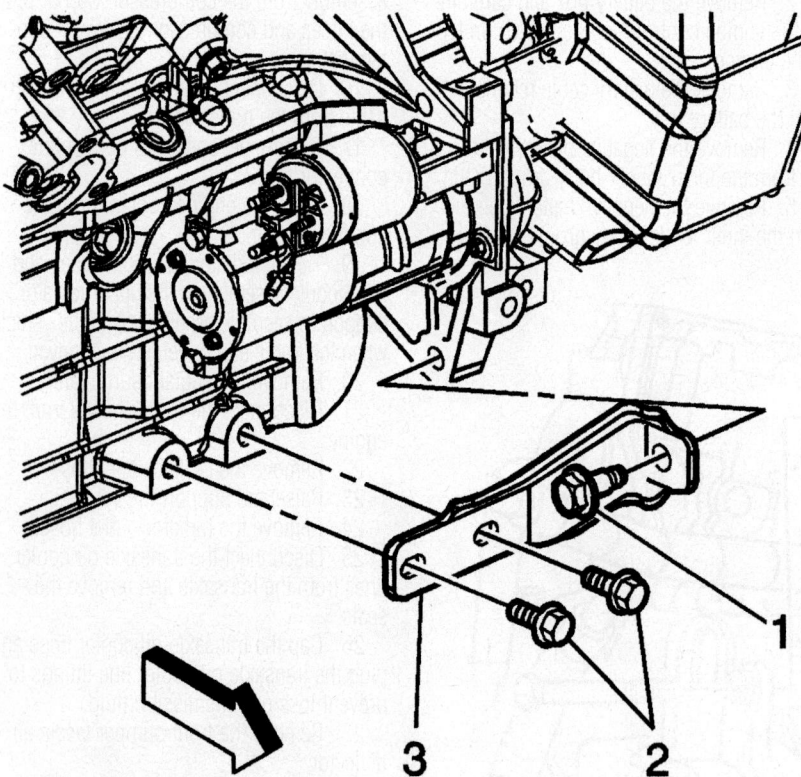

06025-EQUI-G03

Engine-to-transaxle brace

34. Remove the steering intermediate shaft pinch bolt and discard the bolt.

35. Disconnect the steering intermediate shaft from the steering gear.

36. Remove the right and left outer tie rod ends from the steering knuckles.

37. Remove the right and left stabilizer shaft links from the stabilizer shaft.

38. Remove the right and left lower ball joints from the steering knuckles.

39. Remove the right and left front half-shafts.

40. If equipped with AWD, remove the rear driveshaft.

41. Place a block of wood between the frame and the engine oil pan in order to support the engine once the bolts are removed from the right engine mount.

42. Place a block of wood between the frame and the transaxle in order to support the transaxle once the bolts are removed from the left transaxle mount.

43. Lower the vehicle.

44. Remove the bolts that secure the right engine mount to the engine.

45. Remove the bolts that secure the left transaxle mount to the transaxle.

➥**Insure the vehicle body is secured to the hoist.**

46. Raise the vehicle.

47. Place a universal frame support fixture or jackstands under the frame.

48. Lower the vehicle until the frame contacts the frame support fixture or jackstands.

49. Remove the frame-to-body bolts. Discard the bolts.

➥**Inspect for areas of body to powertrain contact or entanglement of wires and hoses while separating the vehicle body and powertrain.**

50. Carefully raise the vehicle body up away from the powertrain.

51. Disconnect the engine electrical wiring harness from the following components:

- Exhaust gas recirculation (EGR) valve
- Throttle body EVAP purge solenoid
- Remove nut and alternator B+ lead
- Generator regulator
- Fuel injector inline connector
- Heated oxygen sensor (HO$_2$S)
- Ignition coil/control module. Remove wire harness from retainers

52. Disconnect the engine electrical wiring harness from the following components:

- Knock sensor (KS)
- Crankshaft position (CKP) sensor
- Remove bolt (4) and ground lead
- Remove the wire harness from retainers

53. If equipped with an engine coolant heater, disconnect the coolant heater cord.

54. Disconnect the engine electrical wiring harness from the following components:

- Air conditioning (AC) compressor clutch
- AC refrigerant pressure sensor
- Remove wire harness from retainer

55. Disconnect the engine electrical wiring harness from the following components:

- Knock sensor (KS)
- Oil pressure indicator switch

56. Remove the throttle body assembly.

57. Remove the nut securing the fuel pipe to the transaxle.

58. Remove the fuel pipe retainer from the threaded stud on the transaxle.

59. If equipped with AWD, remove the transfer case mounting bracket bolts and bracket.

60. If equipped with front wheel drive (FWD), remove the intermediate shaft.

61. Remove the negative battery cable-to-transaxle nut from the transaxle stud

62. Remove the negative battery cable from the transaxle stud.

63. Remove the engine-to-transaxle brace bolts and brace.

64. Remove the starter motor.

65. Remove the torque converter bolts.

66. Install an engine lift chain to the engine lift brackets.

67. Support the engine weight with an engine hoist.

68. Remove the automatic transaxle bolts.

69. Separate the automatic transaxle from the engine.

70. Lift the engine away from the frame and the automatic transaxle.

71. Secure the engine to an engine stand.

72. Remove any additional engine components as necessary. Refer to appropriate component sections in manual if needed.

To install:

73. Remove the engine from the engine stand.

74. Align the engine to the frame and automatic transaxle.

75. Install the automatic transaxle bolts. Tighten the bolts to 75 Nm (55 ft. lbs.).

76. Place a block of wood between the frame and the engine oil pan in order to support the engine on the frame once the engine hoist is removed.

77. Remove the engine hoist and lift chain.

78. Install the torque converter bolts. Tighten the bolts to 60 Nm (44 ft. lbs.).

79. Install the starter motor.

80. Install the engine to transaxle brace and bolts. Tighten the bolts to 50 Nm (37 ft. lbs.).

81. Install the negative battery cable to the transaxle stud.

82. Install the negative battery cable-to-transaxle nut to the transaxle stud. Tighten the nut to 45 Nm (33 ft. lbs.).

83. If equipped with AWD, install the transfer case mounting bracket and bolts. Tighten the bolts to 60 Nm (44 ft. lbs.).

84. If equipped with FWD, install the intermediate shaft.

85. Install the fuel pipe retainer to the threaded stud on the transaxle.

86. Install the nut securing the fuel pipe

to the transaxle. Tighten the nut to 28 Nm (21 ft. lbs.).

87. Install the throttle body assembly.

88. Connect the engine electrical wiring harness to the following components:
- Oil pressure indicator switch
- Knock sensor (KS)

89. Connect the engine electrical wiring harness to the following components:
- Install wire harness to retainer
- AC refrigerant pressure sensor
- Air conditioning (AC) compressor clutch

90. If equipped with an engine coolant heater, connect the coolant heater cord.

91. Connect the engine electrical wiring harness to the following components:
- Install the wire harness to retainers
- Install ground lead and bolt. Tighten the bolt to 25 Nm (18 ft. lbs.)
- CKP sensor
- KS

92. Connect the engine electrical wiring harness to the following components:
- Install wire harness to retainers
- Ignition coil/control module HO2S 1
- Fuel injector inline connector
- Generator regulator
- Install alternator B+ lead and nut. Tighten the nut to 13 Nm (115 inch lbs.)
- EVAP purge solenoid
- Throttle body EGR valve

➡**Inspect for areas of body to power-train contact or entanglement of wires and hoses while joining the vehicle body to the powertrain.**

93. Carefully lower the vehicle body down to the powertrain.

94. Install NEW frame-to-body bolts. Tighten the bolts to 155 Nm (114 ft. lbs.).

95. Raise the vehicle up away from the frame support fixture or jackstands and remove the support fixture or jackstands from under the vehicle.

96. Lower the vehicle.

97. Install the bolts that secure the left transaxle mount to the transaxle. Tighten the bolts to 50 Nm (37 ft. lbs.).

98. Install the bolts that secure the right engine mount to the engine. Tighten the bolts to 50 Nm (37 ft. lbs.).

99. Raise the vehicle.

100. Remove the block of wood between the frame and the transaxle used to support the transaxle while the bolts were removed from the left transaxle mount.

101. Remove the block of wood between the frame and the engine oil pan used to

support the engine while the bolts were removed from the right engine mount.

102. If equipped with AWD, install the rear driveshaft.

103. Install the right and left front half-shafts.

104. Install the right and left lower ball joints to the steering knuckles.

105. Install the right and left stabilizer shaft links to the stabilizer shaft.

106. Install the right and left tie rod ends to the steering knuckles.

107. Connect the steering intermediate shaft to the steering gear.

108. Install a NEW pinch bolt to the steering intermediate shaft. Tighten the bolt to 34 Nm (25 ft. lbs.).

109. Install the right and left engine splash shields.

110. Install the front tires.

111. If equipped with AWD, connect the transfer case vent hose to the transfer case.

112. Install the catalytic converter.

113. Install the HO2S 2 wiring harness retainers to the vehicle underbody.

114. Connect the HO2S 2 wiring harness.

115. Install the front bumper fascia air deflector.

116. Install new seals and connect the transaxle oil cooler lines to the transaxle.

117. Install the radiator outlet hose.

118. Lower the vehicle.

119. Install the radiator inlet hose.

120. Connect the heater hoses to the engine.

121. Install the coolant surge tank.

122. Untie the radiator, AC condenser, and fan module assembly from the upper radiator support.

123. Connect engine to body inline connector C102.

124. Connect the transaxle shift control cable to the transaxle.

125. Install the AC compressor hose assembly to the compressor.

126. Connect the engine fuel hose/pipe to the chassis fuel hose/pipe.

127. Connect the EVAP hose/pipe to the EVAP canister purge solenoid valve.

128. Install the air cleaner assembly and air intake duct.

129. Install the TCM to the TCM bracket.

130. Connect the brake booster vacuum hose to the intake manifold.

131. Position the clamp on the brake booster vacuum hose connection.

132. Install the fuel injector sight shield.

133. Install the negative battery cable from the inner fender body ground stud.

134. Install the negative battery cable nut to the inner fender body ground stud. Tighten the nut to 12 Nm (106 inch lbs.).

135. Install the battery cable retainers to the battery tray.

136. Install the battery box, battery and ECM.

137. Fill the engine with engine oil.

138. Fill the engine with coolant.

139. Check the transaxle fluid level.

140. Charge the AC system.

141. Prime the fuel system.

   a. Cycle the ignition ON for 5 seconds then OFF for 10 seconds. Repeat cycling twice.

   b. Crank the engine until it starts. The maximum starter motor cranking time is 20 seconds.

   c. If the engine does not start, repeat the steps.

142. Install a scan tool.

143. Monitor the powertrain control module (PCM) for DTCs with a scan tool. If other DTCs are set, except DTC P0315, refer to Diagnostic Trouble Code (DTC) List.

144. Select the crankshaft position variation learn procedure with a scan tool.

145. The scan tool instructs you to perform the following:

   a. Accelerate to wide open throttle (WOT).

   b. Observe fuel cut-off for applicable engine.

   c. Release throttle when fuel cut-off occurs.

   d. Engine should not accelerate beyond calibrated RPM value.

   e. Release throttle immediately if value is exceeded.

   f. Block drive wheels.

   g. Set parking brake.

   h. DO NOT apply brake pedal.

   i. Cycle ignition from OFF to ON.

   j. Apply and hold brake pedal.

   k. Start and idle engine.

146. Turn the A/C OFF. Vehicle must remain in Park or Neutral. The scan tool monitors certain component signals to determine if all the conditions are met to continue with the procedure. The scan tool only displays the condition that inhibits the procedure. The scan tool monitors the following components:

- Crankshaft position (CKP) sensors activity—If there is a CKP sensor condition, refer to the applicable DTC
- Camshaft position (CMP) signal activity—If there is a CMP signal condition, refer to the applicable DTC
- Engine coolant temperature (ECT)—If the engine coolant temperature is not warm enough, idle the engine until the engine coolant temperature reaches the correct temperature

➡ **While the learn procedure is in progress, release the throttle immediately when the engine starts to decelerate. The engine control is returned to the operator and the engine responds to throttle position after the learn procedure is complete.**

147. Enable the CKP system variation learn procedure with the scan tool and perform the following:

- Accelerate to WOT
- Release throttle when fuel cut-off occurs

148. The scan tool displays Learn Status: Learned this ignition. If the scan tool indicates that DTC P0315 ran and passed, the CKP variation learn procedure is complete. If the scan tool indicates DTC P0315 failed or did not run, refer to DTC P0315. If any other DTCs set, refer to Diagnostic Trouble Code (DTC) List.

149. Turn OFF the ignition for 30 seconds after the learn procedure is completed successfully.

150. The CKP system variation learn procedure is also required when the following service procedures have been performed, regardless of whether or not DTC P0315 is set:

- An engine replacement
- A PCM replacement
- A harmonic balancer replacement
- A crankshaft replacement
- A CKP sensor replacement
- Any engine repairs which disturb the crankshaft to CKP sensor relationship

Water Pump

REMOVAL & INSTALLATION

1. Before servicing the vehicle, refer to the Precautions Section.

2. Drain the cooling system until the coolant is below the level of the water pump.

3. Loosen the water pump pulley bolts.

4. Rotate the drive belt tensioner to release the tension on the drive belt.

5. Remove the drive belt from the right idler pulley.

6. Carefully release the drive belt tensioner spring tension.

7. Remove the water pump pulley bolts.

8. Remove the water pump pulley.

9. Remove the water pump bolts.

10. Remove the water pump.

11. Remove the water pump gasket.

12. Clean the water pump mating surfaces.

To install:

13. Install the water pump gasket (2).

14. Install the water pump (1).

15. Install the water pump bolts (3). Tighten the bolts to 10 Nm (89 inch lbs.).

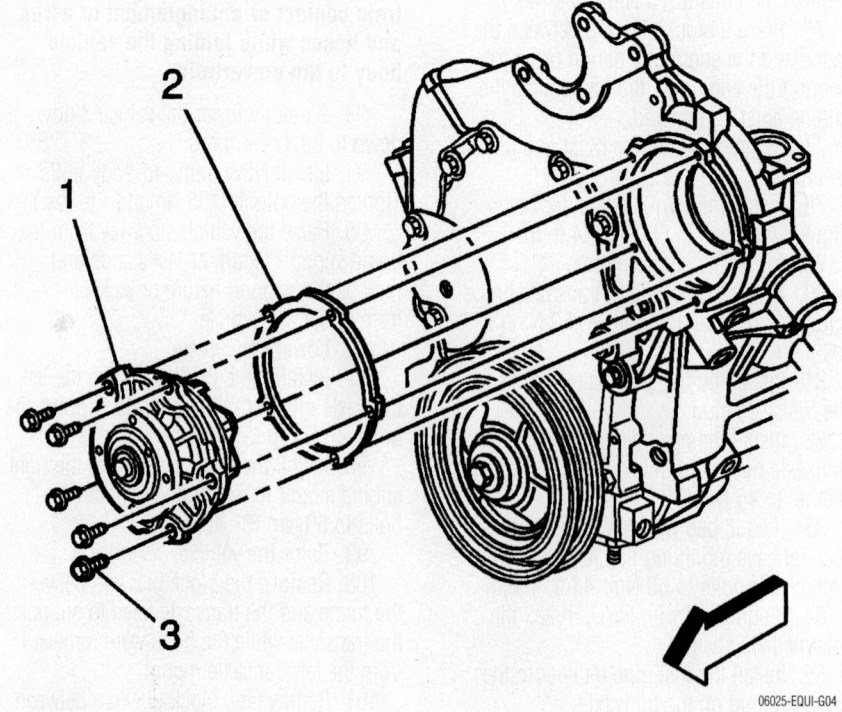

06025-EQUI-G04

Water pump installation

16. Install the water pump pulley. Loosely install the pulley bolts.

17. Insure the drive belt is properly centered on all the pulleys except the right idler puller.

18. Rotate the drive belt tensioner away from the drive belt.

19. Install the drive belt to the right idler pulley.

20. Release the tensioner allowing the drive belt tensioner to come in contact with the drive belt.

21. Inspect the drive belt to insure the belt is properly centered on all the pulleys.

22. Tighten the water pump pulley bolts. Tighten the bolts to 25 Nm (18 ft. lbs.).

23. Fill the cooling system.

24. Inspect the cooling system for leaks.

Heater Core

REMOVAL & INSTALLATION

1. Before servicing the vehicle, refer to the Precautions Section.

2. Drain the engine coolant.

3. Remove the heater outlet hose clamp from the heater core.

4. Remove the heater outlet hose from the heater core.

5. Remove the heater inlet hose clamp from the heater core.

6. Remove the heater inlet hose from the heater core.

7. Disconnect the negative battery cable.

8. Disable the SIR system Zone 5.

➡**Use care when working around the head curtain inflator module. Sharp tools may puncture the curtain airbag. If the head curtain inflator module is damaged in any way, it must be replaced.**

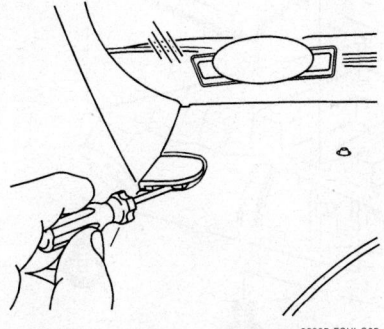

06025-EQUI-G05

Using a small flat-bladed tool, remove the fastener cover from the windshield upper garnish molding, by prying at the bottom edge

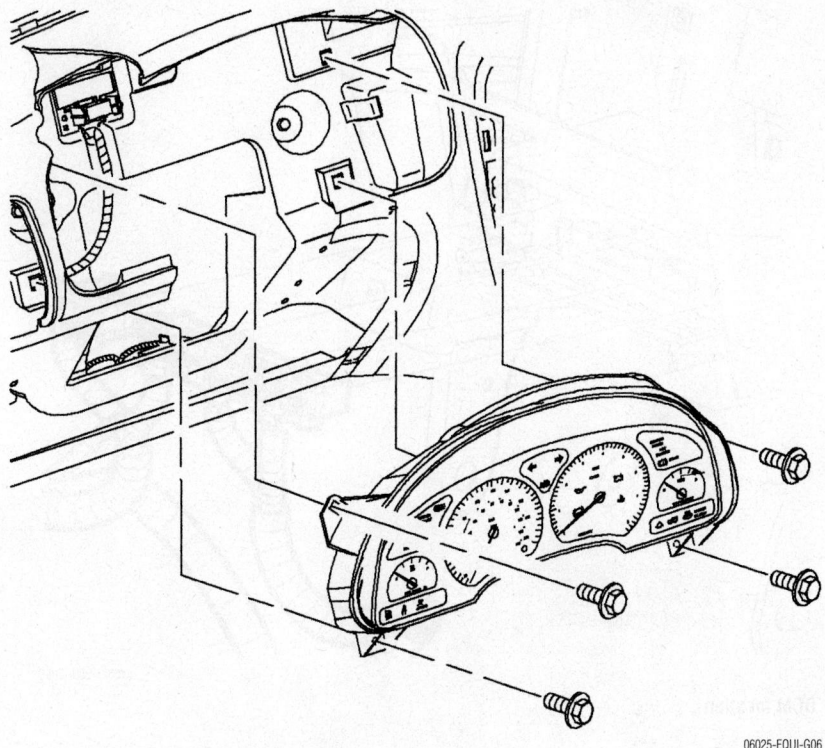

06025-EQUI-G06

Instrument cluster removal

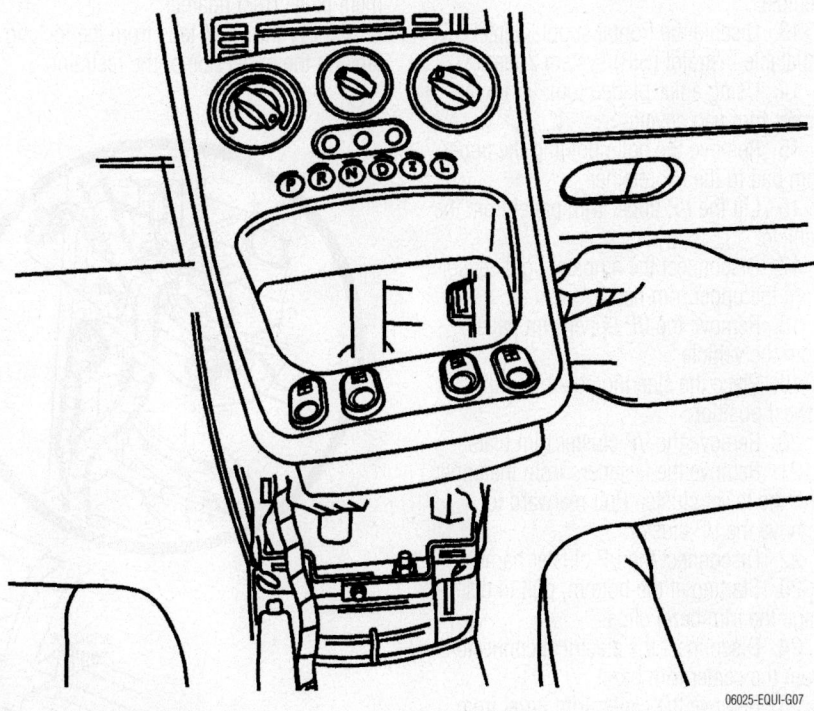

06025-EQUI-G07

Center trim bezel removal

➡**If a vehicle is equipped with a head curtain inflator module ensure that the inflator module and tether are undamaged. If tether or curtain airbag are damaged in any way, they must be replaced.**

9. Using a small flat-bladed tool, remove the fastener cover from the windshield upper garnish molding, by prying at the bottom edge.

10. Remove the fastener from the garnish molding.

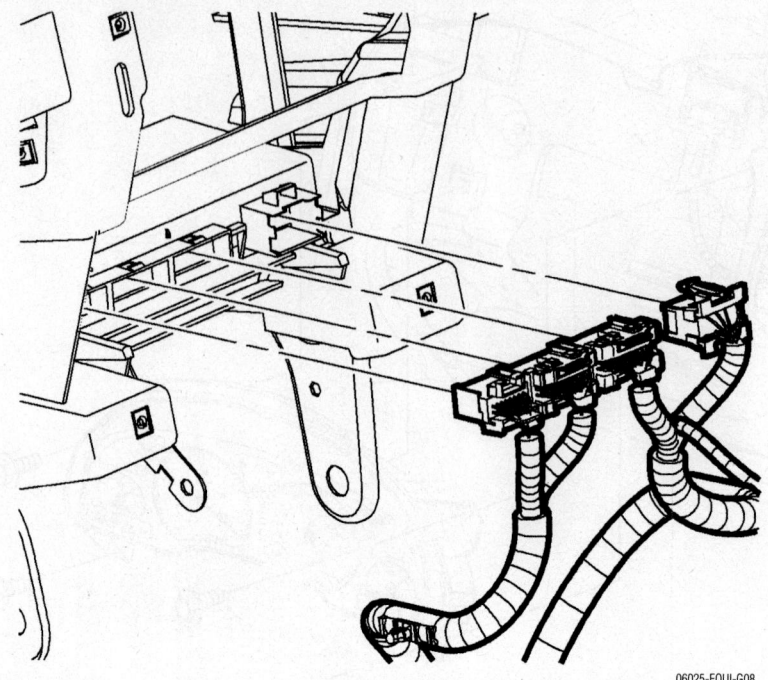

BCM location

11. Starting at the top, gently pull on the garnish molding to disengage the clips.

12. Remove the molding from the vehicle.

13. Disable the frontal supplemental inflatable restraint (SIR) system Zone 3.

14. Using a flat-bladed tool, lift up the upper trim pad covers.

15. Remove the bolts holding the upper trim pad to the I/P retainer.

16. Lift the I/P upper trim panel from the vehicle.

17. Disconnect the ambient light sensor from the upper trim pad.

18. Remove the I/P upper trim panel from the vehicle.

19. Place the steering column in the lowest position.

20. Remove the I/P cluster trim plate.

21. Remove the fasteners from the upper and the lower cluster. Pull rearward to remove the I/P cluster.

22. Disconnect the I/P cluster harness.

23. Starting at the bottom, pull to disengage the trim bezel clips.

24. Disconnect the electrical connectors from the center trim bezel.

25. Remove the center trim bezel from the vehicle.

26. Remove the radio screws. Disconnect the radio electrical connectors and the antenna.

27. Disconnect the ground strap.

28. Remove the radio.

29. Rotate the ignition switch to the OFF position.

30. Disconnect the body control module (BCM) electrical harnesses.

31. Remove the BCM from the instrument panel (I/P) retainer:

32. Pry the BCM tabs from the locking tabs on the underside of the restraint.

33. Slide the BCM out of the retainer.

34. Position the front seats rearward.

35. Remove the center trim panel by pulling outward.

36. Remove the screws from the right console trim panel.

37. Remove the push-in fastener from the trim panel.

38. Remove the trim panel from the vehicle.

39. Remove the screws from the left console trim panel. Remove the push-in fastener from the trim panel.

40. Remove the trim panel from the vehicle.

41. Open the compartment door.

42. Remove the compartment door pivot pins.

43. Rotate the compartment door stops and remove the stops.

44. Remove the I/P compartment.

45. Remove the screws from the steering column filler panel (the housing between the steering wheel and instrument cluster.

46. Pull on the trim panel to disengage the trim clips.

47. Remove the filler panel from the vehicle.

48. Remove the right side I/P trim panel.

49. Remove the panel which contains the power mirror switch and the dimmer switch.

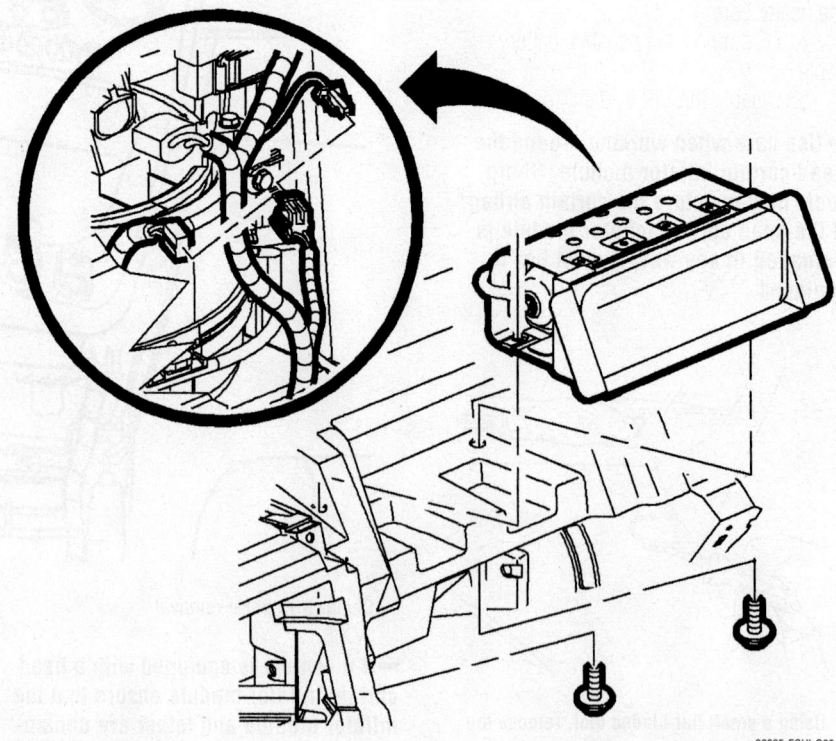

Reach through the radio and the air outlet duct areas and remove the fasteners securing the SIR to the I/P beam

A deployed dual stage inflator module will look the same whether one or both stages were used. Always assume a deployed dual stage inflator module has an active stage 2. Improper handling or servicing can activate the inflator module and cause personal injury.

50. Remove the I/P top cover.

51. Remove the fastener from the passenger side air outlet duct.

52. Remove the air outlet duct.

53. Reach through the radio and the air outlet duct areas to access the fasteners. Remove the fasteners securing the SIR to the I/P beam.

54. Unsnap the module from the I/P carrier retainers.

55. Remove the module from the vehicle.

56. Starting at the rear of the door sill plate, pull upward to disengage the attaching clips.

57. Reach under the instrument panel and pull inward to disengage the additional attaching clips.

58. Remove the door sill plate from the vehicle.

59. Remove the screws and the bolts which hold the I/P retainer to the cross beam. Remove the I/P retainer from the vehicle.

60. Disable the SIR system Zone 8.

61. Remove the console.

62. Disconnect the shift control cable from the shift control range select lever pin.

63. Squeeze the tabs on the shift control cable retainer clip and remove the cable from the shift control.

64. Squeeze the tabs on the park lock cable clip and pull upward to remove the cable from the shift control.

65. Disconnect the park lock cable from the shift control park lock lever.

66. Disconnect the shift control electrical connector.

67. Remove the nuts from the shift control.

68. Remove the shift control from the shift control mounting bracket.

69. Remove the heater duct screws.

70. Remove the heater duct.

71. Remove the heater core cover screws.

72. Remove the heater core cover.

73. Remove the heater core pipe cover screw.

74. Remove the heater core pipe cover.

75. Remove the heater core pipe foam seal.

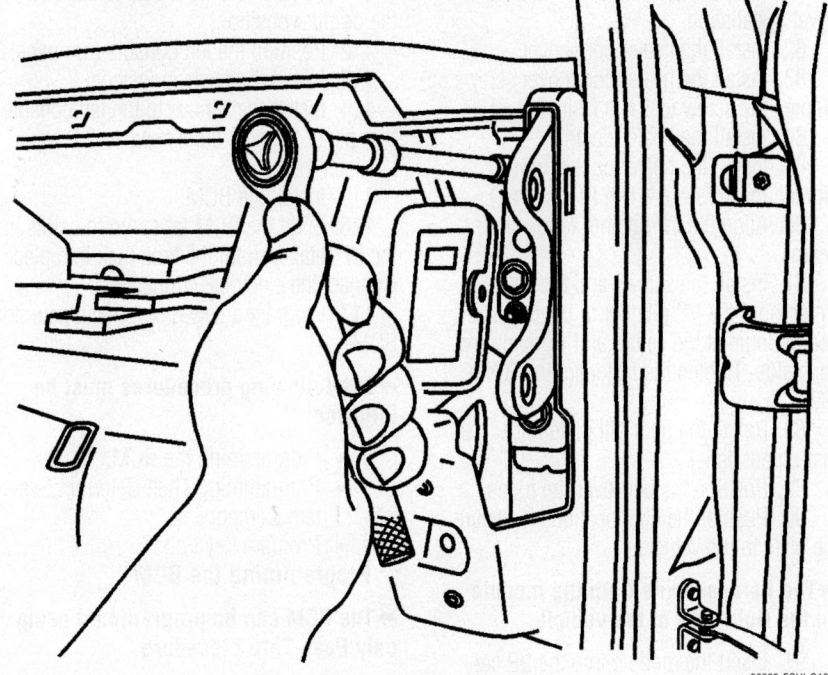

06025-EQUI-G10

Remove the screws and the bolts which hold the I/P retainer to the cross beam

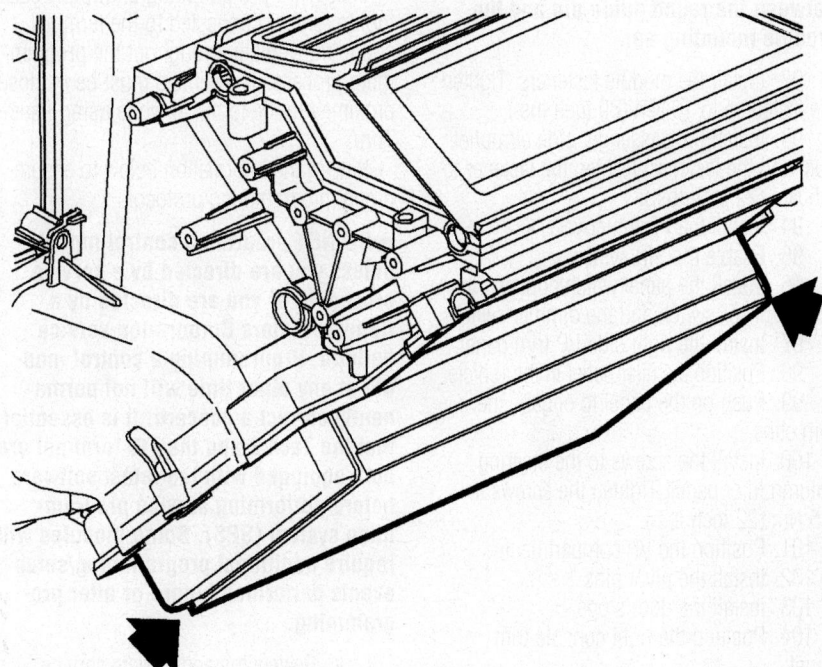

06025-EQUI-G11

Grasp the heater core at end tanks

76. Grasp the heater core at end tanks and remove heater core. Spray the perimeter of the heater core seal and the heater core pipes at the front of dash with a soap and water mixture to ease removal.

To install:

77. Install the heater core. Spray the perimeter of the heater core seal and the heater core pipes at the front of dash with a soap and water mixture to ease installation.

78. Install the heater core pipe seal.

79. Install the heater core pipe cover.

80. Install the heater core pipe cover screw. Tighten the screw to 1 Nm (9 inch lbs.).

81. Spray the front of dash seal at the

drain tube opening to aid in heater core cover installation.

82. Install the heater core cover.

83. Install the heater core cover screws. Tighten the screw to 1 Nm (9 inch lbs.).

84. Install the heater duct.

85. Install the heater duct screws. Tighten the screw to 1 Nm (9 inch lbs.).

86. Align the I/P retainer to the cross beam.

87. Install the screws and the bolts which hold the I/P retainer to the cross beam. Begin at the center and work toward each side. Tighten the fasteners to 10 Nm (89 inch lbs.).

88. Rotate the door sill plate under the instrument panel.

89. Position the part over the frame.

90. Push at the clip locations to secure the plate to the vehicle.

➡ **The harness exits from the module on the right side of the vehicle.**

91. Insert the module into the I/P carrier. Snap the module into place with 2 retainers.

➡ **Route the harness to the module between the round guide pin and the module mounting ear.**

92. Install the module fasteners. Tighten the fasteners to 10 Nm (89 inch lbs.).

93. Install the passenger side air outlet duct and the fastener. Tighten the fastener to 2.5 Nm (22 inch lbs.).

94. Install the I/P top cover.

95. Enable the SIR system.

96. Install the panel which contains the power mirror switch and the dimmer switch.

97. Install the right side I/P trim panel.

98. Position the filler panel in the vehicle.

99. Push on the panel to engage the trim clips.

100. Install the screws to the steering column filler panel. Tighten the screws to 2.5 Nm (22 inch lbs.).

101. Position the I/P compartment.

102. Install the pivot pins.

103. Install the door stops.

104. Position the right console trim panel.

105. Install the push-in fastener.

106. Install the screws to the right console trim panel. Tighten the screws to 2.5 Nm (22 inch lbs.).

107. Position the center trim plate to the front console.

108. Push until the trim clips engage.

109. Connect the electrical connectors to the center trim panel.

110. Snap the center trim panel into place.

111. Position the front seat forward to the original position.

112. Position the left console trim panel.

113. Install the push-in fastener.

114. Install the screws to the left console trim panel. Tighten the screws to 2.5 Nm (22 inch lbs.).

115. Install the BCM.

116. Slide the BCM tabs into the slots in the I/P retainer until the tabs lock into place. Connect the electrical connectors.

117. Program a new or a remanufactured BCM.

➡ **The following procedures must be followed:**

- Programming the BCM.
- Programming Theft Deterrent System Components.
- Program Key Fobs

Programming the BCM

➡ **The BCM can be programmed using only Pass-Thru procedure.**

After the procedure is completed, the personalization settings of the BCM are set to a default setting.

Pass-Thru programming allows the scan tool to remain connected to the terminal and to the vehicle throughout the programming process. The vehicle must be in close proximity to the terminal while using Pass-Thru.

Review the information below to ensure proper programming protocol.

➡ **DO NOT program a control module unless you are directed by a service procedure or you are directed by a General Motors Corporation service bulletin. Programming a control module at any other time will not permanently correct a concern. It is essential that the Tech 2 and the TIS terminal are both equipped with the latest software before performing service programming system (SPS). Some modules will require additional programming/setup events performed before or after programming.**

a. Review the appropriate service information for these procedures:
- Ensure the following conditions are met before programming a control module:
- There is not a charging system concern. All charging system concerns must be repaired before programming a control module
- Battery voltage is greater than 12 volts but less than 16 volts. The battery must be charged before

programming the control module if the battery voltage is low
- A battery charger is NOT connected to the vehicle's battery. Incorrect system voltage or voltage fluctuations from a battery charger, may cause programming failure or control module damage
- Turn OFF or disable any system that may put a load on the vehicle's battery, such as the following components
- The ignition switch must be in the proper position. The Tech 2 prompts you to turn ON the ignition, with the engine OFF. DO NOT change the position of the ignition switch during the programming procedure, unless instructed to do so
- Make certain all tool connections are secure

b. DO NOT disturb the tool harnesses while programming. If an interruption occurs during the programming procedure, programming failure or control module damage may occur

c. DO NOT turn OFF the ignition if the programming procedure is interrupted or unsuccessful. Ensure that all control module and DLC connections are secure and the TIS terminal operating software is up to date. Attempt to reprogram the control module. If the control module cannot be programmed, replace the control module

Programming Theft Deterrent System Components

- The body control module (BCM) must be programmed with the proper regular production option (RPO) configurations before performing learn procedures.
- If replacing the BCM with a GM Service Parts Operations (SPO) replacement part, the module will learn Passlock® sensor data code immediately. The existing PCM however, must learn the new fuel continue password when the BCM is replaced.
- If replacing a PCM with a GM SPO replacement part, after programming, these modules will learn the incoming fuel continue password immediately upon receipt of a password message. Once a password message is received, and a password is learned, a learn procedure must be performed to change this password again. A PCM which has been previously installed in another

vehicle will have learned the other vehicle's fuel continue password, and will require a learn procedure after programming to learn the current vehicle's password.

a. Conditions

Use this procedure after replacing:

- Passlock® Sensor
- BCM
- PCM

a. **Learn Procedures**

There are 2 available methods to perform the programming procedure:

- A 10-minute procedure which requires a Tech 2 and a techline terminal.
- A 30-minute procedure which does not require the use of any tools.

10-Minute Learn Procedure

Tools Required:

- Tech 2
- Techline terminal with current Service Programming System (SPS) software

a. Connect the Tech 2 to the vehicle.

b. Select "Request Information" under "Service Programming".

c. Disconnect the Tech 2 from the vehicle and connect it to a techline terminal.

d. On the techline terminal, select "Theft Module Re-Learn" under "Service Programming".

e. Disconnect the Tech 2 from the techline terminal and connect it to the vehicle.

f. Turn ON the ignition, with the engine OFF.

g. Select "VTD Re-Learn" under "Service Programming".

h. Attempt to start the engine, then release the key to ON. The vehicle will not start.

i. Observe the SECURITY telltale. After approximately 10 minutes, the telltale will turn OFF. The vehicle is now ready to relearn the Passlock® Sensor Data Code and/or password on the next ignition switch transition from OFF to CRANK.

j. Turn OFF the ignition, and wait 5 seconds.

k. Start the engine. The vehicle has now learned the password.

l. With the Tech 2 (scan tool), clear any DTCs.

30-Minute Learn Procedure

a. Turn ON the ignition, with the engine OFF.

b. Attempt to start the engine, then release the key to ON. The vehicle will not start.

c. Observe the SECURITY telltale. After approximately 10 minutes, the telltale will turn OFF.

d. Turn OFF the ignition, and wait 5 seconds.

e. Repeat the first 4 two more times for a total of 3 cycles/30 minutes. The vehicle is now ready to relearn the Passlock® Sensor Data Code and/or passwords on the next ignition switch transition from OFF to CRANK.

➡The vehicle learns the Passlock® Sensor Data Code and/or password on the next ignition switch transition from OFF to CRANK. You must turn the ignition OFF before attempting to start the vehicle.

f. Start the engine. The vehicle has now learned the Passlock® Sensor Data Code and/or password.

g. With a scan tool, clear any DTCs if needed. History DTCs will self clear after 100 ignition cycles.

If the Programming Theft Deterrent System Components in the Theft Deterrent procedure is not performed after a BCM replacement, one of the following conditions will occur:

- The vehicle will not be protected against theft by the PASSLOCK system.
- The engine will not crank or start.

Program Key Fobs

➡Up to 4 transmitters can be programmed. Do not operate or program the transmitters in the vicinity of other vehicles that are in the keyless entry program mode. This prevents the programming of the transmitters to the incorrect vehicle. The order in which the transmitters are programmed is important. The first transmitter programmed will be transmitter #1, and the second transmitter programmed will be transmitter #2. Use care to program the transmitters correctly.

a. Install a scan tool.

b. Turn ON the ignition with the engine OFF.

c. Select Body Control Module.

d. Select Special Functions.

e. Select Program Key Fobs.

f. If you are adding a transmitter to the system, or are replacing a transmitter, choose the Add/Replace Key Fobs option. This allows new transmitters to be added without deleting those currently programmed to the module. If you are replacing all transmitters or none of the transmitters are working, choose the

Clear memory and Program All Fobs option. This prevents lost or stolen transmitters from being used to access the vehicle.

g. Follow the scan tool on-screen directions.

h. After programming the last transmitter, remove the scan tool.

i. Exit the vehicle. Attempt to lock/unlock the doors to verify the programming.

j. Operate the transmitter functions in order to verify correct system operation.

➡After programming, perform the following to avoid future misdiagnosis:

k. Turn the ignition OFF for 10 seconds.

l. Connect the scan tool to the data link connector.

m. Turn the ignition ON, with the engine OFF.

n. Use the scan tool in order to retrieve history DTCs from all modules.

o. Clear all history DTCs.

118. Position the radio in the vehicle.

119. Connect the ground strap.

120. Connect the electrical connectors and the antenna.

121. Install the radio screws. Tighten the screws to 2.5 Nm (22 inch lbs.).

122. Install the I/P center trim bezel.

123. Set up the radio.

124. Connect the electrical connectors from the center trim bezel.

125. Position the center trim bezel in the vehicle.

126. Snap the center trim bezel in place.

127. Align the I/P retainer to the cross beam.

128. Install the screws and the bolts which hold the I/P retainer to the cross beam. Begin at the center and work toward each side. Tighten the fasteners to 10 Nm (89 inch lbs.).

129. Connect the I/P cluster harness.

130. Install the I/P cluster in the I/P housing. Install and tighten the fasteners. Tighten the fasteners to 2 Nm (18 inch lbs.).

131. Install the I/P cluster trim plate.

132. Enable the SIR system.

133. Connect the ambient light sensor to the upper trim pad.

134. Position the I/P upper trim panel in the vehicle. Snap the trim panel into place.

135. Install the bolts to the I/P upper trim panel. Tighten the bolts to 10 Nm (89 inch lbs.). Insert the bolt caps.

136. Install the windshield garnish moldings.

137. Connect the ambient light sensor to the upper trim pad.

138. Install the windshield garnish moldings.

139. Position the windshield upper garnish molding to the top of the instrument panel. Align the molding to the body structure.

➡**If the vehicle is equipped with a side inflator module, position the module tether rearward of the molding attaching clips.**

140. Gently push at the clip locations to install the garnish molding. Install the fastener. Tighten the fastener to 2 Nm (18 inch lbs.).

141. Install the fastener plug to the garnish molding.

142. Replace the door seal over the edge of the garnish molding.

143. Install the shift control to the shift control mounting bracket.

144. Install the shift control nuts. Tighten the nuts to 25 Nm (18 ft. lbs.).

145. Connect the shift control electrical connector.

146. Install the park lock cable onto the shift control.

147. Install the park lock cable end terminal onto the park lock lever by aligning the cable fitting to the park lock lever pin and pushing forward. When installing, be sure not to install the cable end fitting on an angle or push down on the cable portion as this may cause cable to misadjust.

148. Install the shift control cable onto the shift control.

149. Connect the shift control cable onto the shift control range select lever pin.

150. Install the console.

151. Install the heater inlet hose to the heater core.

152. Install the heater inlet hose clamp to the heater core.

153. Install the heater outlet hose to the heater core.

154. Install the heater outlet hose clamp to the heater core.

155. Fill the cooling system.

Cylinder Head

REMOVAL & INSTALLATION

Left Side

1. Before servicing the vehicle, refer to the Precautions Section.

2. Drain the engine coolant from the cooling system.

3. Remove the engine left side spark plug wires from the spark plugs.

4. Remove the spark plug wire retainer support bolt and support.

5. Remove the engine left side spark plugs.

6. Remove the foul air tube bolt and the clip.

7. Remove the foul air tube.

➡**Valve rocker arm cover gasket and sealant must be carefully trimmed away from lower intake manifold gasket. Failure to do so will damage the lower intake manifold gasket, causing a severe oil leak.**

➡**When removing the valve rocker arm cover make sure the gasket stays in place attached to the cylinder head.**

8. Remove the valve rocker arm cover bolts.

9. Remove the valve rocker arm cover.

10. Trim valve cover gasket and sealant away from lower intake manifold gasket at the cylinder head to lower intake manifold joints.

11. Remove the valve rocker arm cover gasket.

12. Remove the lower intake manifold.

13. Remove the oil level indicator tube.

14. Remove the battery box.

15. Remove the exhaust crossover pipe nuts.

16. Remove the exhaust crossover pipe.

17. Remove the exhaust manifold studs, if replacement of the stud is necessary.

18. Remove the engine left side exhaust manifold.

19. Remove the cylinder head bolts and discard.

20. Remove the cylinder head.

21. Remove the cylinder head gasket.

➡**All gasket mating surfaces must remain free of oil and foreign material. Use GM P/N 12346139 (Canadian P/N 10953463) or equivalent to clean surfaces.**

22. Clean the following areas:
   - The gasket sealing surfaces on the cylinder head, cylinder block, intake manifold, and exhaust manifold
   - The cylinder block bolt threads

To install:

23. Install the cylinder head gasket.

24. Install the cylinder head.

➡**This component uses torque-to-yield bolts. When servicing this component do not reuse the bolts, New torque-to-yield bolts must be installed. Reusing used torque-to-yield bolts will not provide proper bolt torque and clamp load. Failure to install NEW torque-to-yield bolts may lead to engine damage.**

25. Install the NEW cylinder head bolts. Tighten the bolts in sequence to 60 Nm (44 ft. lbs.). Turn the bolts an additional 95 degrees.

26. Install the engine left side exhaust manifold.

27. Install any previously removed exhaust manifold studs. Tighten the studs to 25 Nm (18 ft. lbs.).

28. Install the exhaust crossover pipe.

➡**Maintain approximately 6.35 mm (0.25 in) between the thermostat housing and the exhaust crossover pipe.**

29. Install the exhaust crossover pipe nuts. Tighten the nuts to 25 Nm (18 ft. lbs.).

30. Install the battery box.

31. Install the oil level indicator tube.

32. Install the lower intake manifold.

➡**All gasket mating surfaces need to be free of oil and foreign material. Use GM P/N 12346139 (Canadian P/N 10953463) or equivalent to clean surfaces.**

➡**Apply sealant GM P/N 12378521 (Canadian P/N 88901148) or equivalent, at the cylinder head to lower intake manifold joint.**

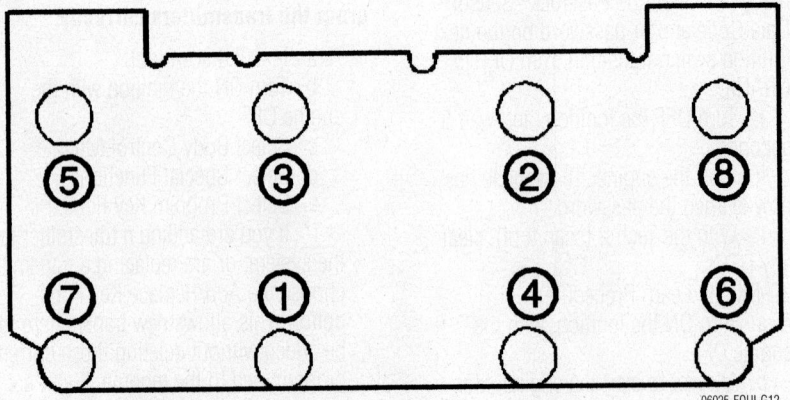

06025-EQUI-G12

Cylinder head bolt torque sequence

33. Apply sealant at the cylinder head to lower intake manifold joints.

34. Install a new gasket to the valve rocker arm cover. Ensure the gasket is properly seated in the groove of the valve rocker arm cover.

35. Install the valve rocker arm cover. Tighten the bolts to 10 Nm (89 inch lbs.).

36. Install the oil filler cap to the valve rocker arm cover.

37. Connect the PCV foul air pipe to the PCV valve.

38. Install the spark plug wire support and bolt. Tighten the bolt to 25 Nm (18 ft. lbs.).

39. Install the engine left side spark plugs.

40. Install the spark plug wire retainer support and bolt. Tighten the bolt to 25 Nm (18 ft. lbs.).

41. Install the engine left side spark plug wires to the spark plugs.

42. Fill the cooling system with engine coolant.

Right Side

1. Before servicing the vehicle, refer to the Precautions Section.

2. Drain the engine coolant from the cooling system.

3. Remove the injector sight shield.

4. Drain the cooling system.

5. Remove the air cleaner air intake duct.

6. Remove the manifold absolute pressure (MAP) sensor bolt and bracket.

7. Rotate the MAP sensor out of the way of the heater outlet pipe.

8. Disconnect the evaporative emissions (EVAP) pipe from the EVAP canister purge solenoid.

9. Disconnect the electrical connector from the EVAP canister purge solenoid.

10. Disconnect the exhaust gas recirculation (EGR) valve electrical connector.

11. Release and slide the heater outlet hose clamp away from the heater outlet pipe connection.

12. Disconnect the heater outlet hose from the heater outlet pipe.

13. Release and slide the heater core outlet hose clamp away from the heater outlet pipe connection.

14. Disconnect the heater core outlet hose from the heater outlet pipe.

15. Remove the heater outlet pipe nut securing the heater outlet pipe to the intake manifold.

16. Remove the two nuts and bolt securing the heater outlet pipe to the throttle body.

17. Remove the heater outlet pipe from the engine.

18. Remove the nut securing the hose/pipe retainer to the right cylinder head.

19. Remove the engine coolant temperature sensor.

20. Remove the ignition control module and bracket.

21. Remove the ignition control module bracket studs.

22. Remove the engine right side spark plug wires from the spark plugs.

23. Remove the engine right side spark plugs.

24. Remove the manifold absolute pressure (MAP) sensor bolt, the clip, and the sensor from the upper intake manifold.

25. Remove the heater outlet pipe hose clamp and hose from the pipe.

26. Remove the heater outlet pipe nuts and bolt from the upper intake manifold.

27. Remove the heater outlet pipe.

28. Remove the fresh air tube and the wire harness retaining bolt.

29. Remove the ignition module nuts and bolts.

30. Remove the ignition module with the ignition wires attached.

31. Remove the ignition module studs.

32. Remove the PCV fresh air tube.

➡**Valve rocker arm cover gasket and sealant must be carefully trimmed away from lower intake manifold gasket. Failure to do so will damage the lower intake manifold gasket, causing a severe oil leak.**

➡**When removing the valve rocker arm cover make sure the gasket stays in place attached to the cylinder head.**

33. Remove the valve rocker arm cover bolts.

34. Remove the valve rocker arm cover.

35. Trim valve cover gasket and sealant away from lower intake manifold gasket at the cylinder head to lower intake manifold joints.

36. Remove the valve rocker arm cover gasket.

37. Remove the lower intake manifold.

38. Remove the battery box.

39. Remove the exhaust crossover pipe nuts.

40. Remove the exhaust crossover pipe.

41. Remove the exhaust manifold studs, if replacement of the stud is necessary.

42. Remove the engine right side exhaust manifold.

43. Remove the alternator bracket and engine lift bracket.

44. Remove the cylinder head bolts and discard.

45. Remove the cylinder head.

46. Remove the cylinder head gasket.

➡**All gasket mating surfaces must remain free of oil and foreign material. Use GM P/N 12346139 (Canadian P/N 10953463) or equivalent to clean surfaces.**

47. Clean the following areas:
- The gasket sealing surfaces on the cylinder head, cylinder block, intake manifold, and exhaust manifold
- The cylinder block bolt threads

To install:

48. Install the cylinder head gasket.

49. Install the cylinder head.

➡**This component uses torque-to-yield bolts. When servicing this component do not reuse the bolts, New torque-to-yield bolts must be installed. Reusing used torque-to-yield bolts will not provide proper bolt torque and clamp load. Failure to install NEW torque-to-yield bolts may lead to engine damage.**

50. Install the NEW cylinder head bolts. Tighten the bolts in sequence to 60 Nm (44 ft. lbs.). Turn the bolts an additional 95 degrees.

51. Install the alternator bracket and engine lift bracket. Torque to 37 ft. lbs. (50 Nm). Install the engine right side exhaust manifold.

52. Install any previously removed exhaust manifold studs. Tighten the studs to 25 Nm (18 ft. lbs.).

53. Install the exhaust crossover pipe.

➡**Maintain approximately 6.35 mm (0.25 in) between the thermostat housing and the exhaust crossover pipe.**

54. Install the exhaust crossover pipe nuts. Tighten the nuts to 25 Nm (18 ft. lbs.).

55. Install the battery box.

56. Install the lower intake manifold.

➡**All gasket mating surfaces need to be free of oil and foreign material. Use GM P/N 12346139 (Canadian P/N 10953463) or equivalent to clean surfaces.**

➡**Apply sealer GM P/N 12378521 (Canadian P/N 88901148) or equivalent, at the cylinder head to lower intake manifold joint.**

57. Apply sealant at the cylinder head to lower intake manifold joints.

58. Install a new gasket to the valve

rocker arm cover. Ensure the gasket is properly seated in the groove of the valve rocker arm cover.

59. Install the valve rocker arm cover. Tighten the bolts to 10 Nm (89 inch lbs.).

60. Connect the PCV fresh air pipe to the right valve rocker arm cover.

61. Install the ignition control module bracket with the ignition control module and spark plug wired still attached.

62. Install the alternator bracket. Torque to 37 ft. lbs. (50 Nm).

63. Install the engine left side spark plugs.

64. Install the engine left side spark plug wires to the spark plugs.

65. Install the ignition module bracket studs. Tighten the studs to 25 Nm (18 ft. lbs.).

66. Install the ignition control module and bracket.

67. Install the engine coolant temperature sensor.

68. Install the nut securing the hose/pipe retainer to the right cylinder head. Tighten the nut to 25 Nm (18 ft. lbs.).

69. Install the heater outlet pipe to the engine.

70. Install the heater outlet pipe nut securing the heater outlet pipe to the intake manifold. Tighten the nut to 25 Nm (18 ft. lbs.).

71. Install the heater outlet pipe to the throttle body nuts and bolt. Tighten the nuts and bolt to 10 Nm (89 inch lbs.).

72. Connect the heater core outlet hose to the heater outlet pipe.

73. Position the heater core outlet hose clamp over the heater outlet pipe connection.

74. Connect the heater outlet hose to the heater outlet pipe.

75. Position the heater outlet hose clamp over the heater outlet pipe connection.

76. Connect the electrical connector to the EGR valve.

77. Connect the electrical connector to the EVAP canister purge solenoid.

78. Connect the EVAP pipe to the EVAP canister purge solenoid.

79. Reposition the MAP sensor.

80. Install the MAP sensor bracket and bolt. Tighten the bolt to 10 Nm (89 inch lbs.).

81. Install the air cleaner air intake duct.

82. Fill the cooling system.

83. Install the injector sight shield.

84. Fill the cooling system with engine coolant.

Rocker Arm Covers

REMOVAL & INSTALLATION

Left Side

1. Before servicing the vehicle, refer to the Precautions Section.

2. Remove the fuel injector sight shield.

3. Disconnect the left side spark plug wires from the retainers.

4. Remove the bolt and the spark plug wire support.

5. Disconnect the positive crankcase ventilation (PCV) foul air pipe from the PCV valve.

6. Remove the oil filler cap from the valve rocker arm cover.

➡**Valve rocker arm cover gasket and sealant must be carefully trimmed away from lower intake manifold gasket. Failure to do so will damage the lower intake manifold gasket, causing a severe oil leak.**

7. Loosen the valve rocker arm cover bolts.

➡**When removing the valve rocker arm cover, ensure the gasket stays in place attached to the cylinder head.**

8. Remove the valve rocker arm cover. Bump the end of the valve rocker arm cover with the palm of your hand or a soft rubber mallet if the cover adheres to the cylinder head.

9. Trim the valve cover gasket and sealant away from the lower intake manifold gasket at the cylinder head to the lower intake manifold joints.

10. Remove the rocker arm cover gasket.

11. Clean the sealing surface on the cylinder head with degreaser.

12. Clean the valve rocker arm cover.

To install:

➡**All gasket mating surfaces need to be free of oil and foreign material. Use GM P/N 12346139 (Canadian P/N 10953463) or equivalent to clean surfaces.**

➡**Apply sealant GM P/N 12378521 (Canadian P/N 88901148) or equivalent, at the cylinder head to lower intake manifold joint.**

13. Apply sealant at the cylinder head to lower intake manifold joints.

14. Install a new gasket to the valve rocker arm cover. Ensure the gasket is properly seated in the groove of the valve rocker arm cover.

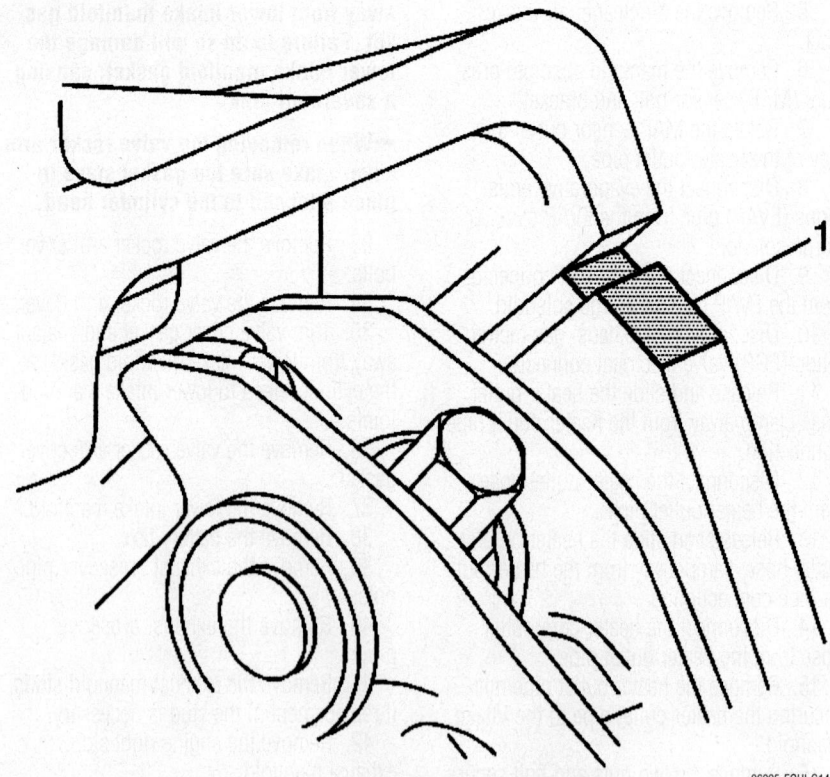

06025-EQUI-G14

Apply sealant at the cylinder head to lower intake manifold joints

15. Install the valve rocker arm cover. Tighten the bolts to 10 Nm (89 inch lbs.).

16. Install the oil filler cap to the valve rocker arm cover.

17. Connect the PCV foul air pipe to the PCV valve.

18. Install the spark plug wire support and bolt. Tighten the bolt to 25 Nm (18 ft. lbs.).

19. Connect the left side spark plug wires to the spark plug wire retainers.

20. Install the fuel injector sight shield.

Right Side

1. Before servicing the vehicle, refer to the Precautions Section.

2. Remove the fuel injector sight shield.

3. Remove the alternator bracket.

4. Remove the ignition control module bracket from the engine with the ignition control module and spark plug wires still attached. Position out of the way.

5. Disconnect the positive crankcase ventilation (PCV) fresh air pipe from the right valve rocker arm cover.

➥ **Valve rocker arm cover gasket and sealant must be carefully trimmed away from lower intake manifold gasket. Failure to do so will damage the lower intake manifold gasket, causing a severe oil leak.**

6. Loosen the valve rocker arm cover bolts.

➥ **When removing the valve rocker arm cover, ensure the gasket stays in place attached to the cylinder head.**

7. Remove the valve rocker arm cover. Bump the end of the valve rocker arm cover with the palm of your hand or a soft rubber mallet if the cover adheres to the cylinder head.

8. Trim the valve cover gasket and sealant away from the lower intake manifold gasket at the cylinder head to lower intake manifold joints.

9. Remove the valve rocker arm cover gasket.

10. Clean the sealing surface on the cylinder head with degreaser.

11. Clean the valve rocker arm cover.
To install:

➥ **All gasket mating surfaces need to be free of oil and foreign material. Use GM P/N 12346139 (Canadian P/N 10953463) or equivalent to clean surfaces.**

➥ **Apply sealer GM P/N 12378521 (Canadian P/N 88901148) or equiva-**

lent, at the cylinder head to lower intake manifold joint.

12. Apply sealant at the cylinder head to lower intake manifold joints.

13. Install a new gasket to the valve rocker arm cover. Ensure the gasket is properly seated in the groove of the valve rocker arm cover.

14. Install the valve rocker arm cover. Tighten the bolts to 10 Nm (89 lb in).

15. Connect the PCV fresh air pipe to the right valve rocker arm cover.

16. Install the ignition control module bracket with the ignition control module and spark plug wired still attached.

17. Install the alternator bracket. Torque to 37 ft. lbs. (50 Nm).

18. Install the fuel injector sight shield.

Rocker Arms, Pushrods and Lifters

REMOVAL & INSTALLATION

1. Before servicing the vehicle, refer to the Precautions Section.

2. Remove the valve rocker arm cover(s).

3. Remove the rocker arm bolt(s).

4. Remove the valve rocker arm(s) and the pushrod(s).

5. Remove the lifter guide bolts.

6. Remove the lifter guide.

➥ **Place the valve lifters in an organized order to ensure that they are installed in the same location from which they were removed.**

7. Remove the lifters.
To install:

8. Coat the valve lifters using prelube GM P/N 1052367 (Canadian P/N 992869) or the equivalent.

9. Install the valve lifters in their original locations.

10. Install the lifter guide.

11. Apply threadlock GM P/N 12345382 (Canadian P/N 10953489) or the equivalent to the lifter guide bolt threads and install the bolts. Tighten the bolts to 10 Nm (89 inch lbs.).

12. Coat the ends of the push rods using prelube GM P/N 1052367 (Canadian P/N 992869) or the equivalent.

➥ **The intake valve push rods measure 146.0 mm (5.75 in) in length. The exhaust valve push rods measure 152.5 mm (6.0 in) in length.**

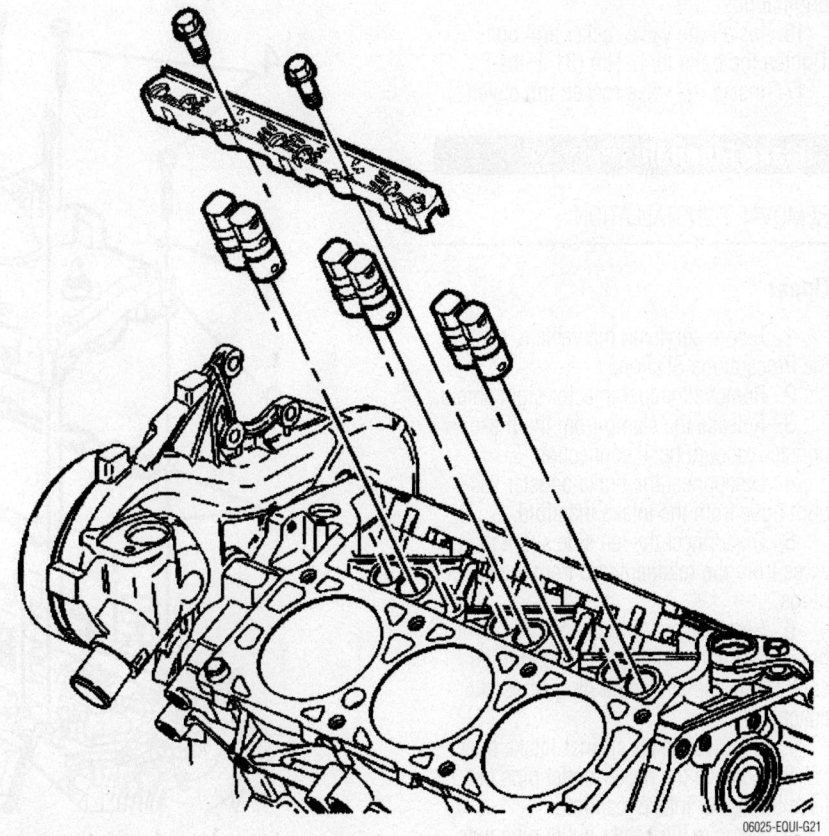

Lifter removal

06025-EQUI-G21

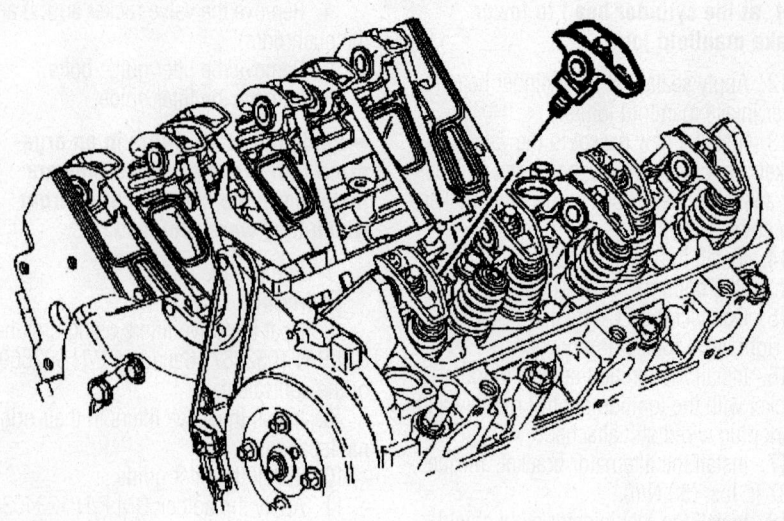

06025-EQUI-G13

Rocker arm removal

13. Install the push rods in their original location.

14. Coat the rocker arm friction surfaces using prelube GM P/N 1052367 (Canadian P/N 992869) or the equivalent.

➡ **Shims may be required under the valve rocker arm pedestals if reconditioning has been performed on the cylinder head or its components.**

15. Install the valve rocker arms in their original positions.

16. Install the valve rocker arm bolts. Tighten the bolts to 42 Nm (31 ft. lbs.) .

17. Install the valve rocker arm cover(s).

Intake Manifold

REMOVAL & INSTALLATION

Upper

1. Before servicing the vehicle, refer to the Precautions Section.

2. Remove the fuel injector sight shield.

3. Release the clamp from the brake booster vacuum hose connection.

4. Disconnect the brake booster vacuum hose from the intake manifold.

5. Disconnect the left side spark plug wires from the retainers and from the spark plugs.

6. Remove the ignition control module bracket from the engine with the ignition control module and spark plug wires still attached.

7. Remove the air cleaner intake duct.

8. Remove the heater outlet pipe nut from the upper intake manifold.

9. Remove the heater outlet pipe nuts and bolt from the throttle body.

10. Position the heater outlet pipe out of the way without disconnecting the heater hoses.

11. Remove the exhaust gas recirculation (EGR) pipe.

12. Remove the positive crankcase ventilation (PCV) foul air hose.

13. Loosen but do not completely remove the alternator attachment bolt most near the intake manifold.

14. Remove the alternator brace nut.

15. Remove the alternator brace.

16. Remove the upper intake manifold bolts.

17. Remove the spark plug wire retainer.

18. Remove the upper intake manifold.

19. Remove the upper intake manifold gaskets.

To install:

20. Install the upper intake manifold gaskets (1) to the lower intake manifold and install the fir tree retainers to retain the upper intake manifold gasket position.

21. Install the upper intake manifold (2).

22. Install the spark plug wire retainer (3).

23. Apply threadlock GM P/N 12345382 (Canadian P/N 10953489) to the bolt threads. Install the upper intake manifold bolts (4, 5). Tighten the bolts to 25 Nm (18 ft. lbs.).

24. Install the alternator brace.

25. Install the alternator brace nut. Tighten the nut to 25 Nm (18 ft. lbs.).

26. Fully insert the alternator attachment bolt most near the intake manifold. Tighten the bolt to 25 Nm (18 ft. lbs.).

27. Install the PCV foul air hose.

28. Install the EGR pipe.

29. Position the heater outlet pipe to the throttle body and the upper intake manifold.

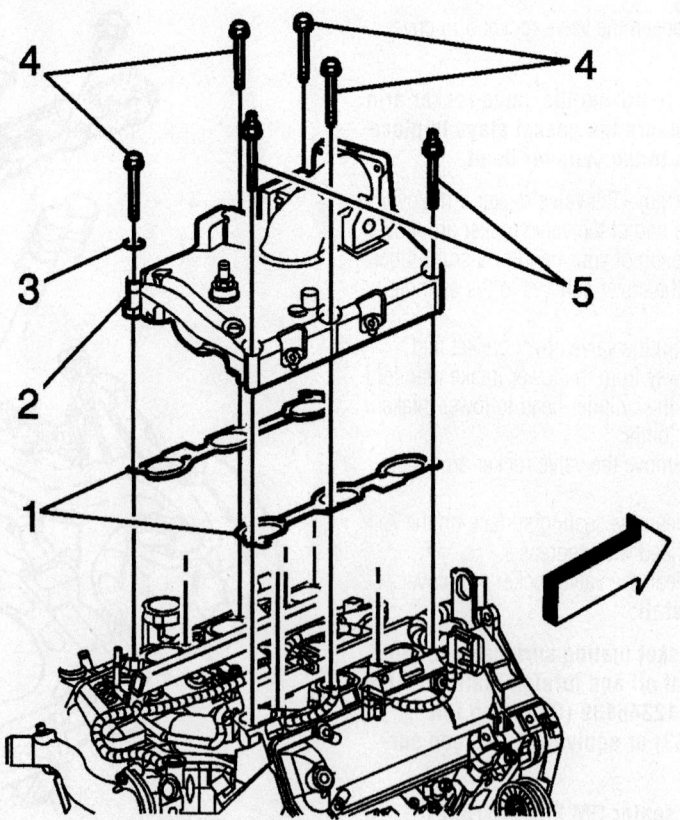

06025-EQUI-G15

Upper intake manifold removal

30. Install the heater outlet pipe nuts to the throttle body. Tighten the nut to 10 Nm (89 inch lbs.).

31. Install the heater outlet pipe bolt to the throttle body. Tighten the bolt to 10 Nm (89 inch lbs.).

32. Install the heater outlet pipe nut to the upper intake manifold. Tighten the nut to 25 Nm (18 ft. lbs.).

33. Install the air cleaner intake duct.

34. Install the ignition control module bracket.

35. Connect the left side spark plug wires to the spark plugs and to the spark plug wire retainers.

36. Connect the brake booster vacuum hose to the intake manifold.

37. Install the clamp to the brake booster vacuum hose connection.

38. Install the fuel injector sight shield.

Lower

1. Before servicing the vehicle, refer to the Precautions Section.

2. Drain the cooling system.

3. Remove the upper intake manifold.

4. Remove the heater inlet pipe.

5. Disconnect the radiator inlet hose from the thermostat housing.

6. Remove the fuel rail assembly.

7. Remove the lower intake manifold bolts.

8. Remove the lower intake manifold from the engine.

9. Remove the rocker arms and push rods.

10. Remove the lower intake manifold gaskets.

To install:

➡ **All gasket mating surfaces must remain free of oil and foreign material. Use GM P/N 12346139 (Canadian P/N 10953463) or equivalent to clean surfaces.**

➡ **Do not apply room temperature vulcanizing (RTV) sealer to the engine block prior to the installation of the manifold gaskets. RTV sealer is not to be placed under the lower intake manifold gaskets.**

11. Install the lower intake manifold gaskets.

12. Install the rocker arms and push rods.

13. Install the lower intake manifold seals.

14. With the seals in place, apply a small drop 8–10 mm (0.31–0.39 in.) of RTV sealer GM P/N 12346141 (Canadian P/N 10953433).

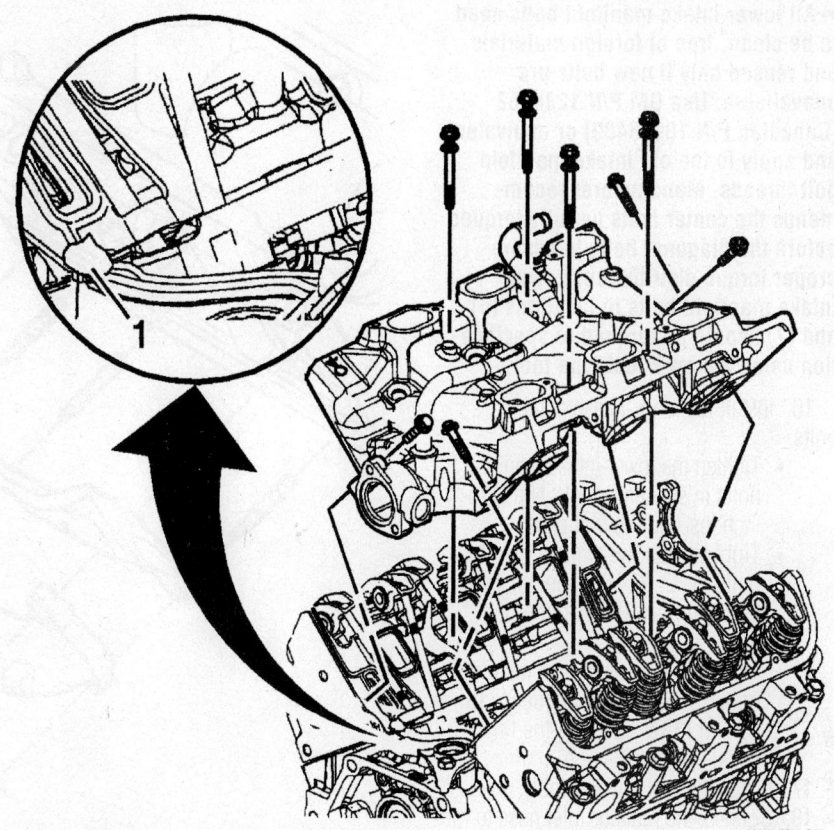

06025-EQUI-G16

With the seals in place, apply a small drop 8–10 mm (0.31–0.39 in.) of RTV sealer as shown

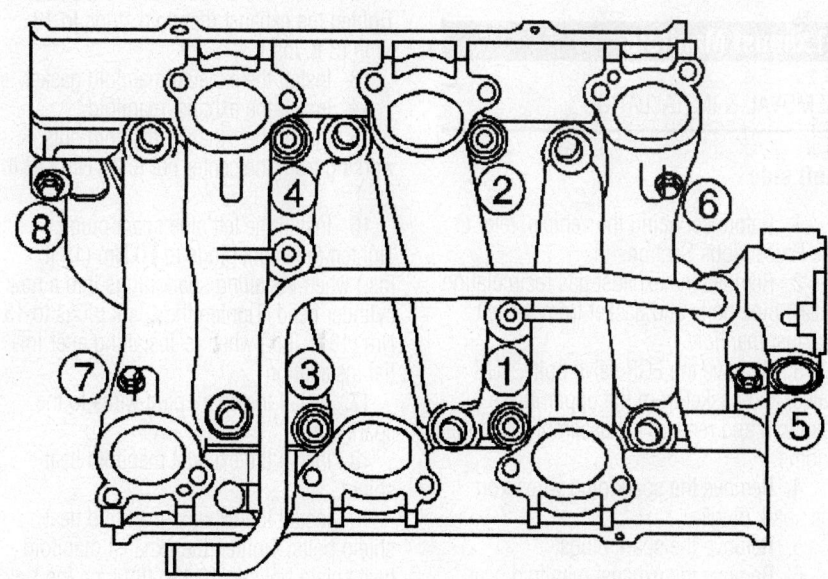

06025-EQUI-G17

Lower intake manifold torque sequence

15. Install the lower intake manifold to the engine.

➡ **Maximum gasket performance is achieved when using new fasteners, which contain a thread-locking patch. If** the fasteners are not replaced, a thread locking chemical must be applied to the fastener threads. Failure to replace the fasteners or apply a thread-locking chemical MAY reduce gasket sealing capability.

➡ All lower intake manifold bolts need to be clean, free of foreign materials, and reused only if new bolts are unavailable. Use GM P/N 1234382 (Canadian P/N 10953489) or equivalent and apply to the old intake manifold bolt threads. Manufacturer recommends the center bolts be fully torqued before the diagonal bolts to assure proper torque distribution. Lower intake manifold bolts in locations (6) and (7) should be torqued to specification using a crow's foot type tool.

16. Install the lower intake manifold bolts.

- Tighten the lower intake manifold bolts in sequence to 13 Nm (115 inch lbs.) on the first pass.
- Tighten the lower intake manifold bolts (1, 2, 3, 4) in sequence to 20 Nm (15 ft. lbs.) on the final pass.
- Tighten the lower intake manifold bolts (5, 6, 7, 8) in sequence to 25 Nm (18 ft. lbs.) on the final pass.

17. Install the fuel rail assembly.
18. Connect the radiator inlet hose to the thermostat housing.
19. Install the heater inlet pipe.
20. Install the upper intake manifold.
21. Fill the cooling system.

Exhaust Manifold

REMOVAL & INSTALLATION

Left side

1. Before servicing the vehicle, refer to the Precautions Section.
2. Remove the exhaust gas recirculation (EGR) pipe bolts and gasket from the left exhaust manifold.
3. Remove the EGR valve bolts, EGR valve, and gasket from the upper intake manifold and remove the assembly from the engine.
4. Remove the spark plug wires from the spark plugs.
5. Remove the spark plugs.
6. Remove the exhaust manifold heat shield bolts.
7. Remove the exhaust manifold heat shield.
8. Remove the exhaust manifold nuts.
9. Remove the exhaust manifold.
10. Remove the exhaust manifold gasket.
11. Remove the exhaust studs, if required.

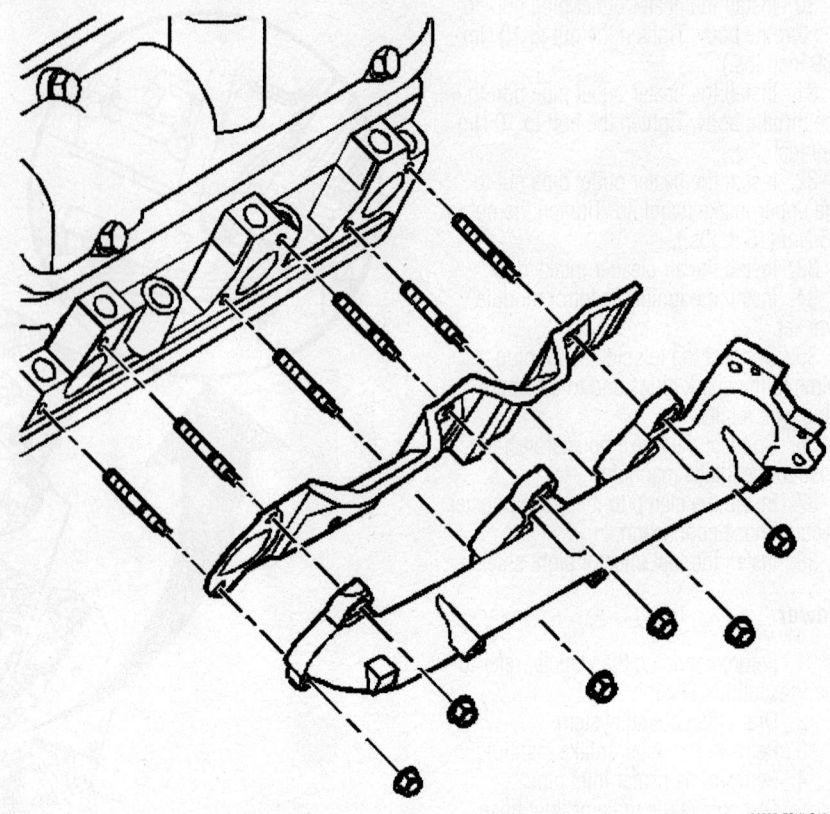

06025-EQUI-G18

Left side exhaust manifold

To install:

12. Install the exhaust manifold studs. Tighten the exhaust manifold studs to 18 Nm (13 ft. lbs.).
13. Install the exhaust manifold gasket.
14. Install the exhaust manifold.
15. Install the exhaust manifold nuts, working from the center out to 16 Nm (12 ft. lbs.).
16. Install the left side spark plugs. Tighten the spark plugs to 20 Nm (15 ft. lbs.) when installing spark plugs into a new cylinder head. Tighten the spark plugs to 15 Nm (13 ft. lbs.) when re-installing after initial installation
17. Install the spark plug wires to the spark plugs.
18. Install the exhaust manifold heat shield.
19. Install the exhaust manifold heat shield bolts. Tighten the exhaust manifold heat shield bolts to 10 Nm (89 inch lbs.).
20. Install the exhaust gas recirculation (EGR) valve gasket and the EGR assembly to the upper intake manifold.
21. Install the EGR valve bolts. Tighten the EGR valve bolts to 30 Nm (22 ft. lbs.).
22. Install the EGR pipe gasket and pipe to the left exhaust manifold.
23. Install the EGR pipe bolts. Tighten the EGR pipe bolts to 30 Nm (22 ft. lbs.).

Right Side

1. Before servicing the vehicle, refer to the Precautions Section.
2. Remove the spark plug wires from the spark plugs.
3. Remove the spark plugs.
4. Remove the heated oxygen sensor.
5. Remove the exhaust manifold heat shield bolts.
6. Remove the exhaust manifold heat shields.
7. Remove the exhaust manifold nuts.
8. Remove the exhaust manifold.
9. Remove the exhaust manifold gasket.
10. Remove the exhaust studs, if required.

To install:

11. Install the exhaust manifold studs. Tighten the exhaust manifold studs to 18 Nm (13 ft. lbs.).
12. Install the exhaust manifold gasket.
13. Install the exhaust manifold.
14. Install the exhaust manifold nuts. Tighten the exhaust manifold nuts working from the center out to 16 Nm (12 ft. lbs.).
15. Install the right side spark plugs. Tighten the spark plugs to 20 Nm (15 ft. lbs.) when installing spark plugs into a new cylinder head Tighten the spark plugs to 15 Nm (13 ft. lbs.) when re-installing after initial installation

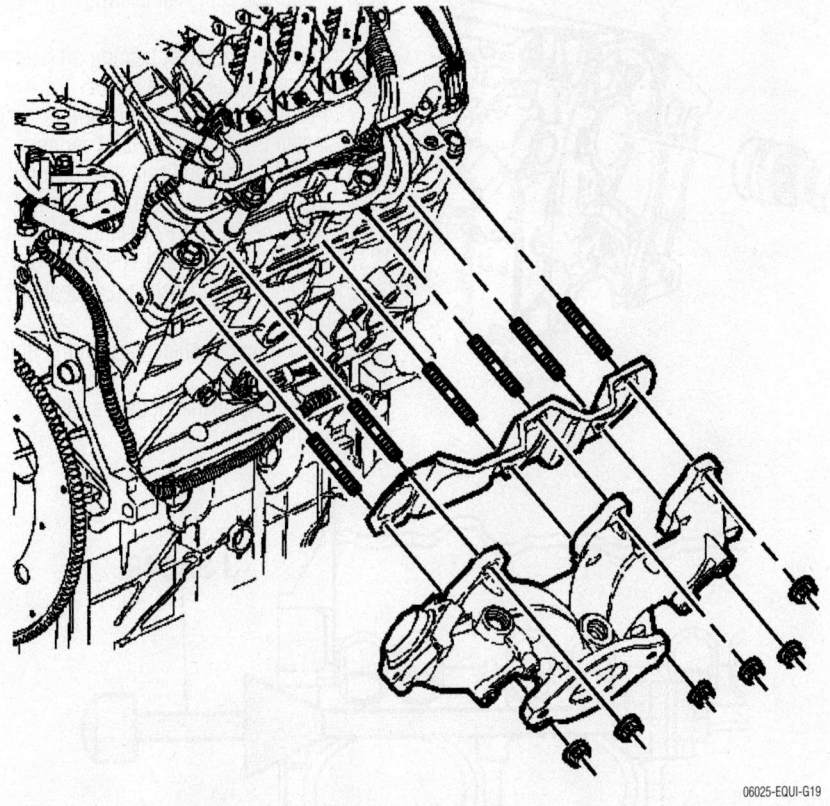

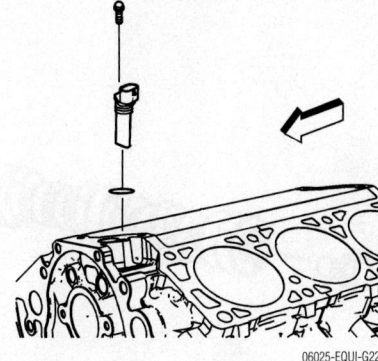

Camshaft position sensor bolt

06025-EQUI-G22

8. Remove the camshaft thrust plate screws.

9. Remove the camshaft thrust plate.

➡**All camshaft journals are the same diameter, so care must be used in removing or installing the camshaft to avoid damage to the camshaft bearings.**

10. Complete the following steps in order to remove the camshaft.

    a. Install the camshaft sprocket bolt into the camshaft. Tighten finger tight only.

06025-EQUI-G19

Right side exhaust manifold

16. Install the spark plug wires on to the spark plugs.

17. Install the lower exhaust manifold heat shield.

18. Install the upper exhaust manifold heat shield.

19. Install the exhaust manifold heat shield bolts. Tighten the exhaust manifold heat shield bolts to 10 Nm (89 inch lbs.).

20. Coat the threads of the heated oxygen sensor with anti seize compound.

21. Install the heated oxygen sensor. Tighten the heated oxygen sensor to 42 Nm (31 ft. lbs.).

22. Install the spark plug wires.

Camshaft and Bearings

REMOVAL & INSTALLATION

1. Before servicing the vehicle, refer to the Precautions Section.

2. Remove the engine.

3. Remove the cylinder heads.

4. Remove the lifters.

5. Remove the oil pan.

6. Remove the camshaft position sensor bolt.

7. Remove the camshaft position sensor.

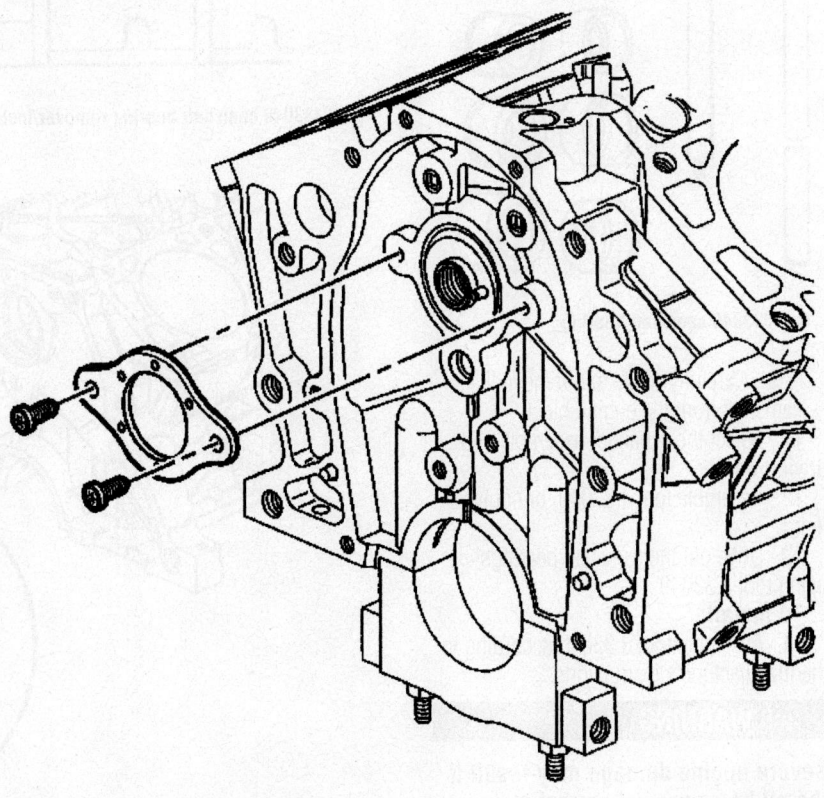

06025-EQUI-G23

Camshaft thrust plate

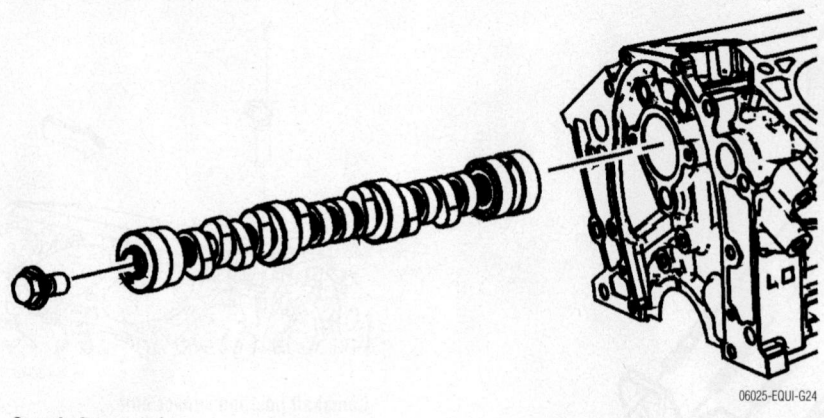

Camshaft removal

15. Install the camshaft bearings in the following order:

16. Index the camshaft bearing oil holes with the engine block oil passages.

17. Place the bearing on the tool.

18. Install the third camshaft bearing.

19. Install the second camshaft bearing.

20. Install the outer camshaft bearings.

21. Apply sealer GM P/N United States 12377901, GM P/N Canada 10953504 or the equivalent to the camshaft rear bearing hole plug.

22. Install the camshaft rear bearing hole plug.

23. Coat the camshaft journals with clean engine oil.

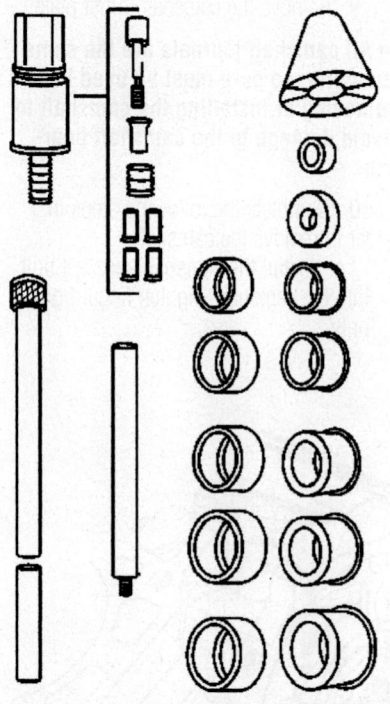

Tool J33049 camshaft bearing

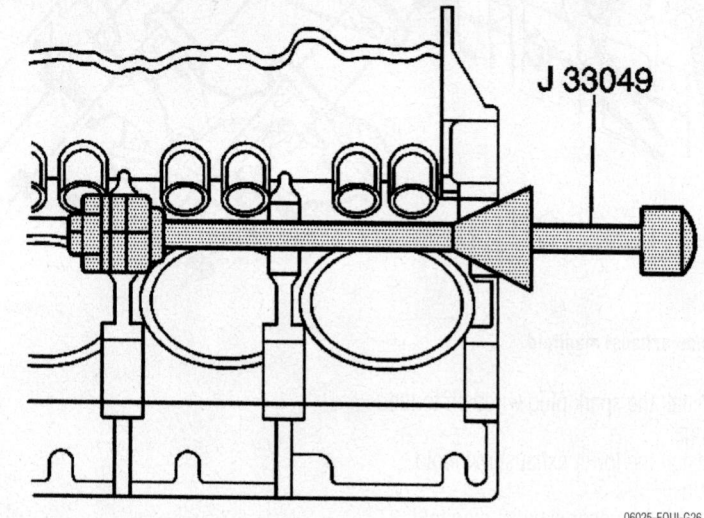

Tool J33049 camshaft bearing remover/installer installed

b. Carefully rotate and remove the camshaft from the engine block.

11. Select the expander assembly and driving washer.

12. Assemble tool J 33049, or equivalent.

13. Drive out the camshaft bearings using tool J 33049.

To install:

14. Assemble tool J 33049 according to the manufacturer's instructions.

✳✳ WARNING

Severe engine damage may result if the oil holes are not correctly aligned.

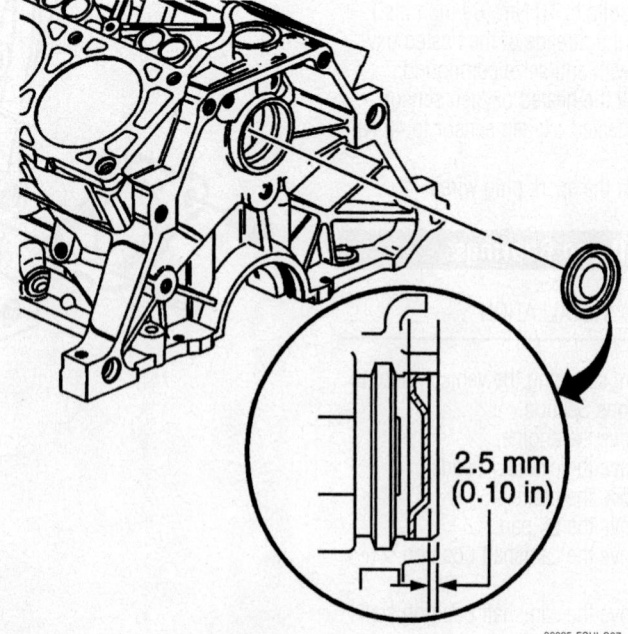

2.5 mm
(0.10 in)

Camshaft rear bearing hole plug installation

24. Coat the camshaft lobes with prelube GM P/N 12345501 (Canadian P/N 992704) or the equivalent.

25. Install the camshaft using the following procedure:

   a. Install the camshaft sprocket bolt into the camshaft. Tighten finger tight only.

   b. Carefully rotate the camshaft while installing the camshaft into the camshaft bearings.

26. Install the camshaft thrust plate.

27. Install the camshaft thrust plate screws. Tighten the camshaft thrust plate screws to 10 Nm (89 inch lbs.).

28. Install the camshaft position sensor.

29. Install the camshaft position sensor bolt. Tighten the camshaft position sensor bolt to 10 Nm (89 inch lbs.).

Oil Pan

REMOVAL & INSTALLATION

1. Before servicing the vehicle, refer to the Precautions Section.

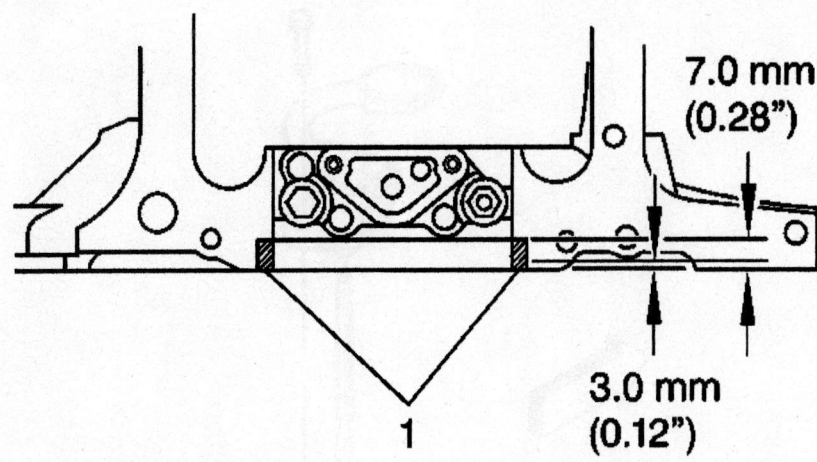

7.0 mm (0.28")

3.0 mm (0.12")

1

06025-EQUI-G29

Oil pan sealer application

Oil pan removal and installation is best done with the engine out of the vehicle.

2. Remove the oil pan side bolts.
3. Remove the oil pan flange bolts.
4. Remove the oil pan.
5. Remove the oil pan gasket.

To install:

6. Apply sealer GM P/N 12378521, (Canadian P/N 88901148) or the equivalent to both sides of the crankshaft rear main bearing cap (1).

7. Install the oil pan gasket.

8. Install the oil pan.

9. Install the oil pan flange bolts. Tighten the oil pan bolts to 25 Nm (18 ft. lbs.).

10. Install the oil pan side bolts. Tighten the oil pan side bolts to 50 Nm (37 ft. lbs.).

11. Install the oil pan drain plug. Tighten the oil pan drain to 25 Nm (18 ft. lbs.).

Oil Pump

REMOVAL & INSTALLATION

1. Before servicing the vehicle, refer to the Precautions Section.

2. Remove the oil pump bolt.

3. Remove the oil pump and oil pump drive shaft.

4. Remove the crankshaft oil deflector nuts.

5. Remove the crankshaft oil deflector.

To install:

6. Install the crankshaft oil deflector.

7. Install the crankshaft oil deflector nuts. Tighten the crankshaft oil deflector nuts to 25 Nm (18 ft. lbs.).

➡**Do not reuse the oil pump driveshaft retainer. During assembly, install a NEW oil pump driveshaft retainer.**

8. Install the oil pump.

9. Position the oil pump onto the pins.

10. Install the oil pump bolt attaching the oil pump to the rear crankshaft bearing cap. Tighten the oil pump bolt to 41 Nm (30 ft. lbs.).

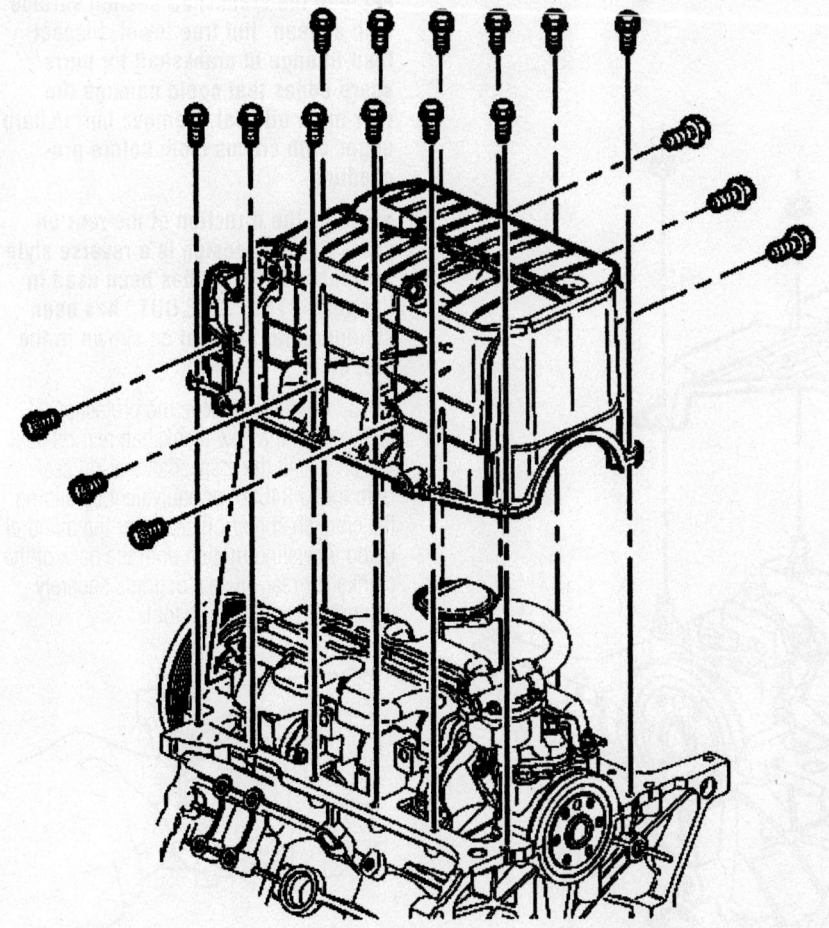

06025-EQUI-G28

Oil pan bolts

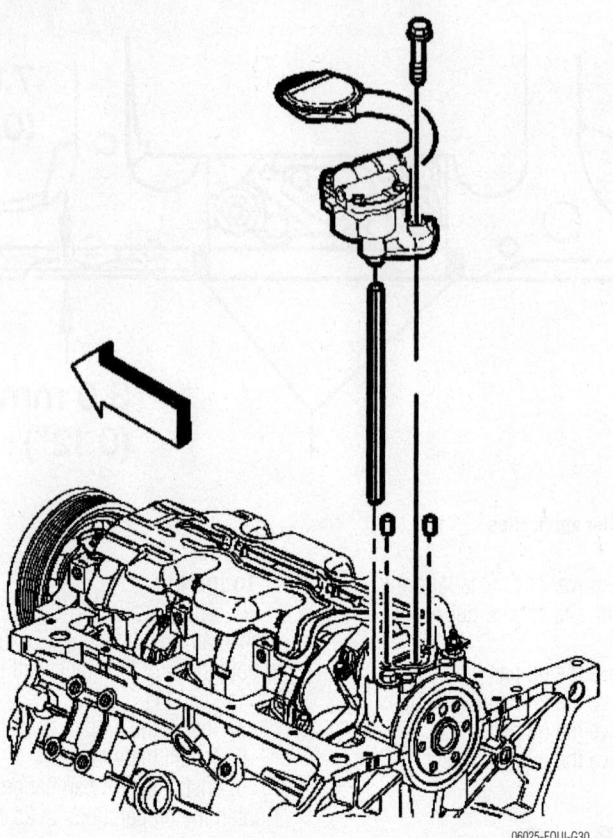

Oil pump installation

06025-EQUI-G30

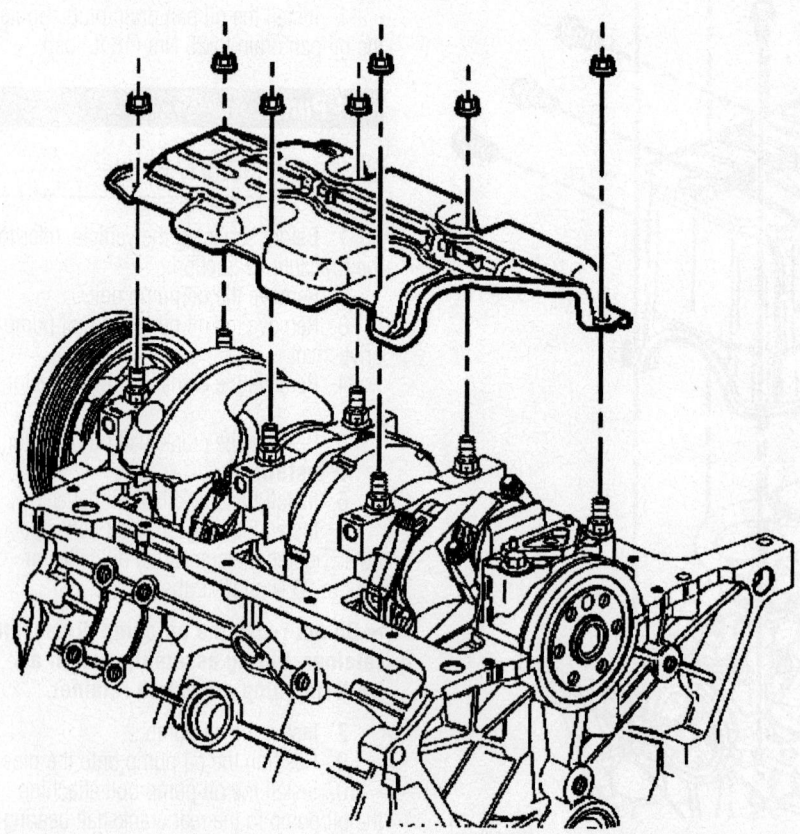

Oil deflector installation

06025-EQUI-G31

Rear Main Seal

REMOVAL & INSTALLATION

1. Before servicing the vehicle, refer to the Precautions Section.

2. Remove the automatic transaxle.

3. Remove the engine flywheel bolts and flywheel.

4. Clean the engine flywheel bolt threads and bolt holes.

5. Clean and inspect the engine flywheel.

➡**Do not damage the crankshaft or seal bore.**

6. Remove the engine flywheel.

7. Remove the crankshaft rear oil seal. Pry the crankshaft rear oil seal out using a suitable tool.

   To install:

➡**Do not apply or use any oil lubrication on the crankshaft rear oil seal, or the seal installer. Do not touch the sealing lip of the oil seal once the protective sleeve is removed. Doing so will damage/deform the seal.**

➡**Clean the crankshaft sealing surface with a clean, lint free towel. Inspect lead-in edge of crankshaft for burrs/sharp edges that could damage the rear main oil seal. Remove burrs/sharp edges with crocus cloth before proceeding.**

➡**Notice the direction of the rear oil seal. The new design is a reverse style as opposed to what has been used in the past. "THIS SIDE OUT" has been stamped into the seal as shown in the graphic.**

8. Carefully remove the protection sleeve from the new crankshaft rear oil seal.

9. Install the crankshaft rear oil seal onto tool J 34686, or equivalent by sliding the crankshaft rear oil seal over the mandrel using a twisting motion until the back of the crankshaft rear oil seal bottoms squarely against the collar of the tool.

06025-EQUI-G32

Prying out the rear main seal

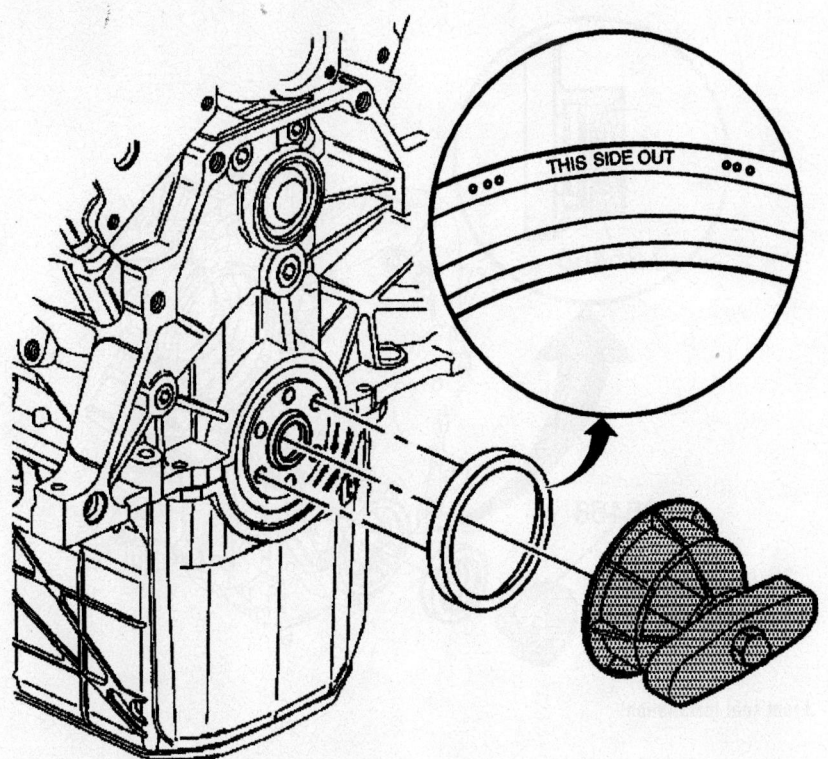

Rear main seal installation

06025-EQUI-G33

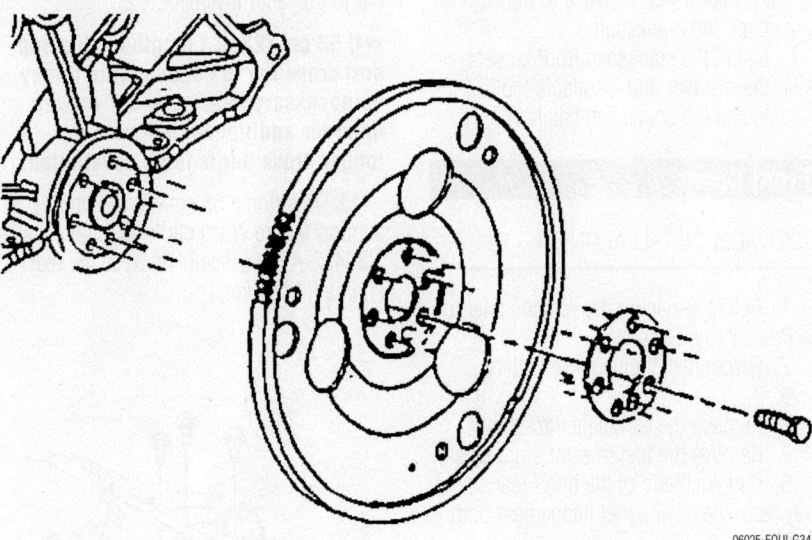

Flywheel installation

06025-EQUI-G34

10. Perform the following steps in order to install the crankshaft rear oil seal:

a. Align the dowel pin of the tool with the dowel pin in the crankshaft.

b. Attach the tool to the crankshaft by hand, or tighten the attaching screws to 5 Nm (45 inch lbs.).

c. Turn the T-handle of the tool in order to engage allow the collar to push the seal into the bore. Turn the handle until the collar is tight against the engine block. Ensure that the seal is seated properly.

d. Loosen the T-handle until the handle comes to a stop.

e. Remove the attaching screws.

11. Install the engine flywheel and bolts Tighten bolts to 71 Nm (52 ft. lbs.).

12. Install the automatic transaxle.

Crankshaft Damper

REMOVAL & INSTALLATION

1. Before servicing the vehicle, refer to the Precautions Section.

2. Rotate the drive belt tensioner to release the tension on the drive belt.

3. Remove the drive belt from the right idler pulley.

4. Carefully release the drive belt tensioner spring tension.

5. Raise and support the vehicle.

6. Remove the right front wheel.

7. Remove the wheelhouse liner.

8. Remove the crankshaft damper bolt and washer.

※ WARNING

The inertial weight section of the crankshaft damper is assembled to the hub with a rubber type material. The correct installation procedures (with the proper tool) must be followed or movement of the inertial weight section of the hub will destroy the tuning of the crankshaft damper.

➡ **Do NOT use a power-assisted tool with the special tool in order to remove or install this component. You cannot properly control the alignment of this component using a power-assisted tool, and this can damage the component.**

9. Remove the crankshaft damper using tool J 41816-A along with EN 46359, or equivalent.

To install:

10. Apply sealer GM P/N 12378521 (Canadian P/N 88901148) or the equivalent, to the keyway of the crankshaft damper.

11. Place the crankshaft damper into position over the key in the crankshaft.

➡ **Do NOT use a power-assisted tool with the special tool in order to remove or install this component.**

12. You cannot properly control the alignment of this component using a power-assisted tool, and this can damage the component.

13. Install tool J 29113, or equivalent onto the crankshaft.

14. Rotate the hex nut on the tool to install the crankshaft damper onto the crankshaft.

15. Remove the tool from the crankshaft.

16. Install the crankshaft damper bolt. Tighten the bolt to 70 Nm (52 ft. lbs.). Turn the bolt an additional 70 degrees.

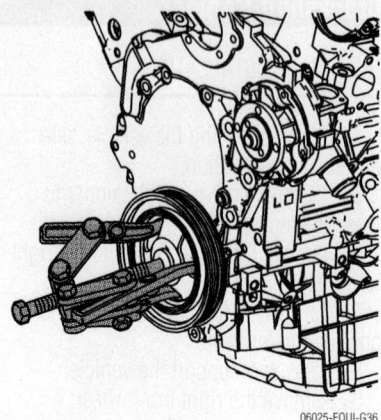

Damper removal

06025-EQUI-G36

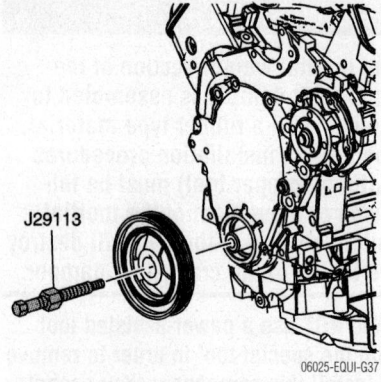

J29113

Damper installation

06025-EQUI-G37

17. Install the wheelhouse liner.

18. Install the right front wheel.

19. Lower the vehicle.

20. Insure the drive belt is properly centered on all the pulleys except the right idler pulley.

21. Rotate the drive belt tensioner away from the drive belt.

22. Install the drive belt to the right idler pulley.

23. Carefully release the drive belt tensioner to come in contact with the drive belt.

24. Inspect the drive belt to insure the belt is properly centered on all the pulleys.

Front Crankshaft Seal

REMOVAL & INSTALLATION

1. Before servicing the vehicle, refer to the Precautions Section.

2. Remove the crankshaft balancer.

3. Pry out the crankshaft front oil seal using a suitable tool. Use care not to damage the engine front cover or the crankshaft.

4. Inspect the crankshaft, the crankshaft balancer and the engine front cover for wear and/or damage.

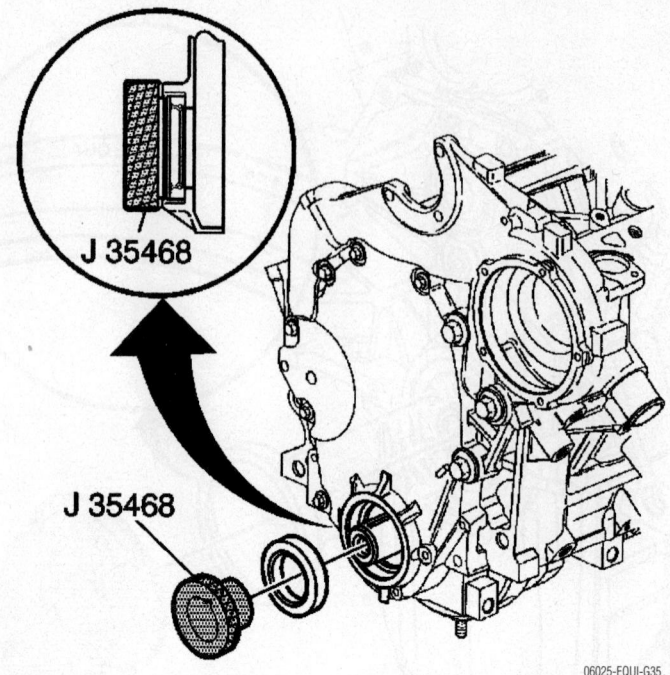

J 35468

J 35468

Front seal installation

06025-EQUI-G35

5. Replace the components as necessary.

To install:

6. Align tool J 35468, or equivalent and the crankshaft front oil seal with the engine front cover and crankshaft.

7. Install the crankshaft front oil seal using the installer and a suitable tool.

8. Install the crankshaft balancer.

Front Cover

REMOVAL & INSTALLATION

1. Before servicing the vehicle, refer to the Precautions Section.

2. Disconnect the negative battery cable.

3. Remove the air cleaner assembly.

4. Remove the fuel injector sight shield.

5. Pull each end of the hood rear seal away from the cowl panel flange near both strut towers.

6. Replace 2 strut bolts with studs GM P/N 11519137 on right and left sides of the vehicle for installation of support fixture. Tighten the studs to 25 Nm (18 ft. lbs.).

7. Install tool J-28467-13 and J 28467-5 strut tower adapter to the top of the right strut tower.

8. Install tool J-28467-13 and J 28467-5 strut tower adapter to the top of the left strut tower.

9. Install a 127 cm (50 in.) engine support fixture cross bar (2) transversely across

the vehicle between both J 28467-5 strut tower adapters.

10. Insert safety pins through the J 28467-5 strut tower adapters and the cross bar to prevent movement.

➡ **If 58 cm (23 in.) length engine support cross bar is not available it may be necessary to remove the vehicle hood for additional clearance if a longer cross bar is to be substituted.**

11. Position a 58 cm (23 in.) engine support fixture cross bar longitudinally with J 36462-A leg assembly next to the rear engine lift bracket.

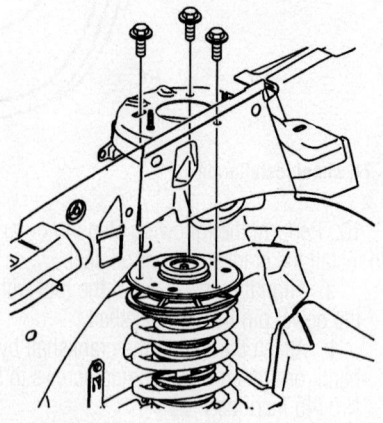

06025-EQUI-G38

Strut tower bolts

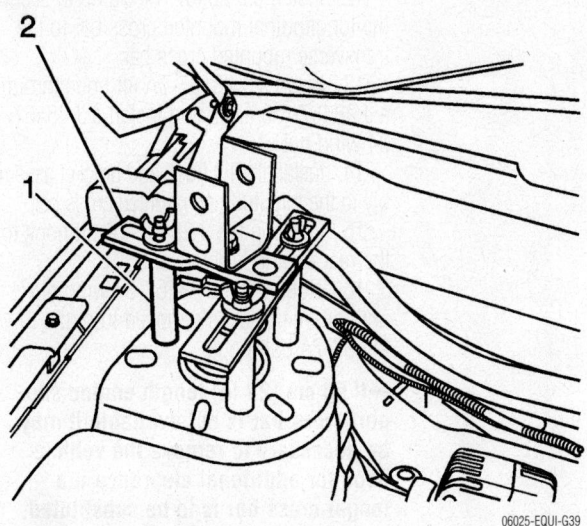

Install tool J-28467-13 (1) and J 28467-5 strut tower adapter (2) to the top of the right strut tower

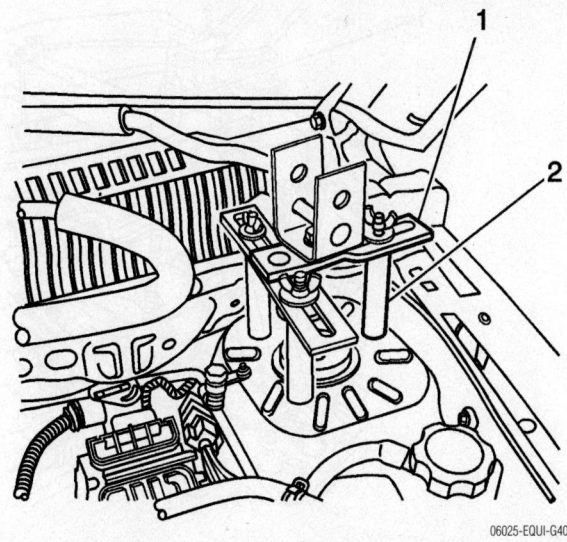

Install tool J-28467-13 (2) and J 28467-5 strut tower adapter (1) to the top of the left strut tower

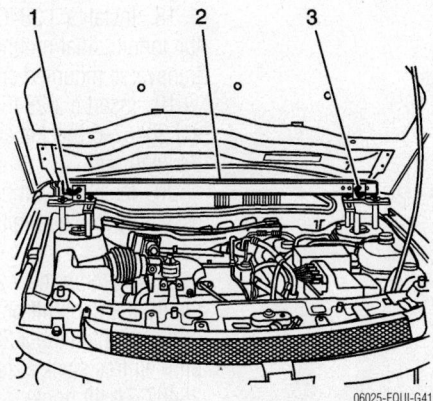

Install a 127 cm (50 in) engine support fixture cross bar (2) transversely across the vehicle between both J 28467-5 strut tower adapters. Insert safety pins (1, 3) through the J 28467-5 strut tower adapters and the cross bar (2) to prevent movement

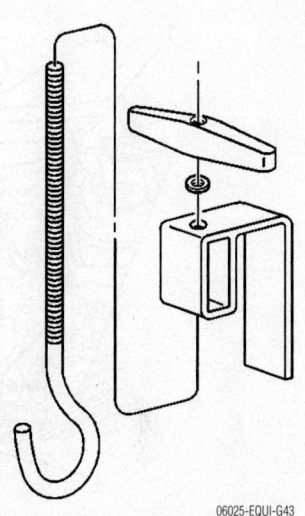

Insert a J 28467-7A lift hook through a J 28467-6 bracket and install a J 28467-34 wing nut

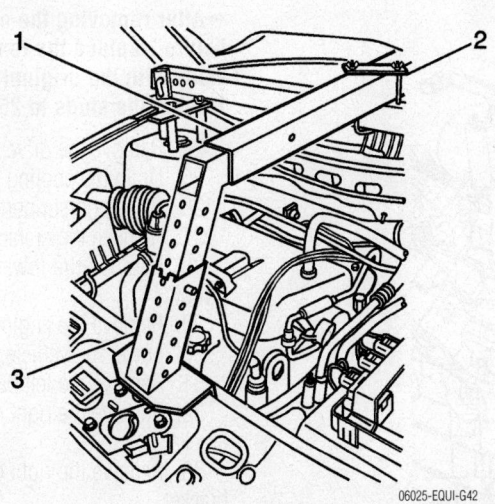

Position a 58 cm (23 in.) engine support fixture cross bar (1) longitudinally with J 36462-A leg assembly (3) next to the rear engine lift bracket. Install a J 28467-1A clamp (2) to secure the longitudinal mounted cross bar to the transverse mounted cross bar

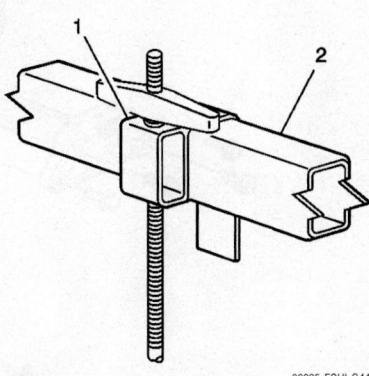

Install the lift hook and bracket assembly (1) to the longitudinal mounted cross bar (2)

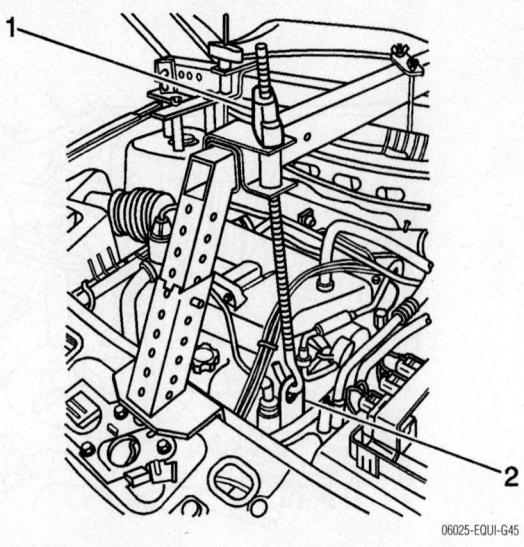

Position the J 28467-7A lift hook to the rear engine lift bracket (2). Tighten the J 28467-34 wing nut (1) until all free slack is removed from the J 28467-7A bolt hook

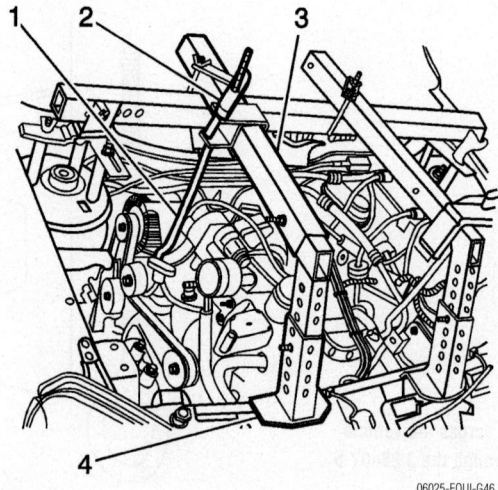

06025-EQUI-G46

Engine support fixture installed

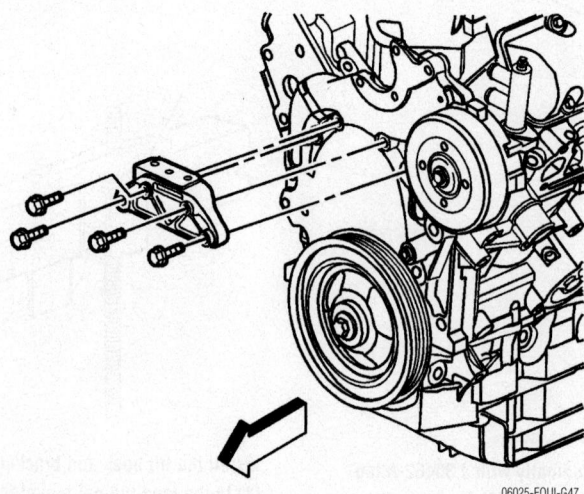

06025-EQUI-G47

Right engine mount bracket

12. Install a J 28467-1A clamp to secure the longitudinal mounted cross bar to the transverse mounted cross bar.

13. Insert a J 28467-7A lift hook through a J 28467-6 bracket and install a J 28467-34 wing nut.

14. Install the lift hook and bracket assembly to the longitudinal mounted cross bar.

15. Position the J 28467-7A lift hook to the rear engine lift bracket.

16. Tighten the J 28467-34 wing nut until all free slack is removed from the J 28467-7A bolt hook.

➡ **If 58 cm (23 in) length engine support cross bar is not available it may be necessary to remove the vehicle hood for additional clearance if a longer cross bar is to be substituted.**

17. Position a 58 cm (23 in) engine support fixture cross bar longitudinally with J 36462-A leg assembly next to the front engine lift bracket.

18. Install a J 28467-1A clamp to secure the longitudinal mounted cross bar to the transverse mounted cross bar.

19. Insert a J 28467-7A lift hook through a J 28467-6 bracket and install a J 28467-34 wing nut.

20. Install the lift hook and bracket assembly to the longitudinal mounted cross bar.

21. Position the J 28467-7A bolt hook to the front engine lift bracket.

22. Tighten the J 28467-34 wing nut until all free slack is removed from the J 28467-7A lift hook.

23. Evenly tighten both wing nuts until the engine weight is supported by the engine support fixture and no longer carried by the engine mounts.

➡ **After removing the engine support fixture, replace the temporary strut studs with the original strut bolts. Tighten the studs to 25 Nm (18 ft. lbs.).**

24. Remove the drive belt.
25. Drain the cooling system.
26. Raise and support the vehicle.
27. Remove the crankshaft balancer.
28. Remove the lower belt idler pulley.
29. Remove the engine oil pan.
30. Lower the vehicle.
31. Remove the left belt idler pulley.
32. Remove the right engine mount bracket bolts.
33. Remove the right engine mount bracket.
34. Remove the water pump.
35. Remove the thermostat bypass hose adapter.

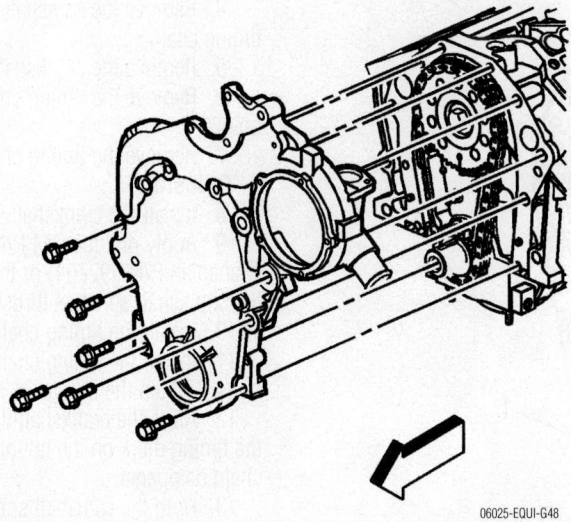

Front cover

06025-EQUI-G48

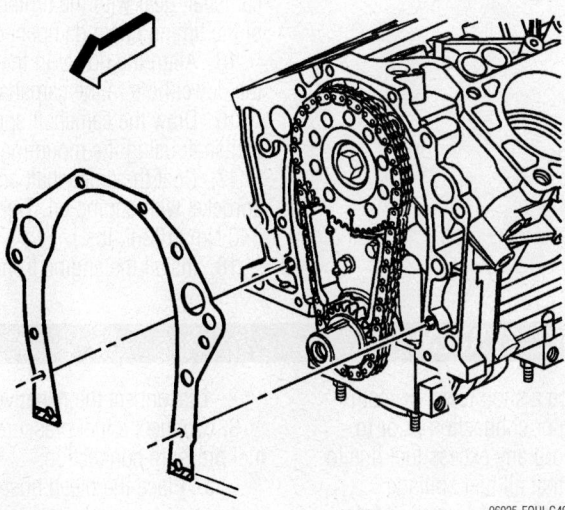

Front cover gasket

06025-EQUI-G49

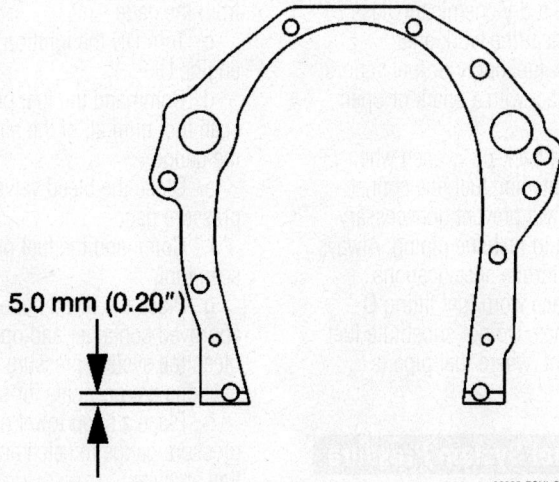

5.0 mm (0.20")

06025-EQUI-G50

Apply sealer GM P/N 12346004 (Canadian P/N 10953480) or the equivalent, to the lower tabs of the engine front cover gasket

36. Remove the radiator outlet hose from the engine front cover.

37. Remove the engine front cover bolts.

38. Remove the engine front cover.

39. Remove the engine front cover gasket.

40. Clean the engine block and front cover gasket sealing surfaces.

To install:

41. Apply sealer GM P/N 12346004 (Canadian P/N 10953480) or the equivalent, to the lower tabs of the engine front cover gasket. Uniformly apply the sealer to both sides of the gasket lower tabs with the sealant bead being no less than 5.0 mm (0.20 in) wide as shown. Torque front cover immediately after installation of sealer coated gasket.

42. Install the gasket to the engine block positioning the gasket on the locating pins.

43. Install the engine front cover to the engine block aligning the cover with the locating pins.

44. Install the engine front cover bolts. Tighten the bolts to 55 Nm (41 ft. lbs.).

45. Install the engine front cover bolts. Tighten the bolts to 27 Nm (20 ft. lbs.).

46. Install the radiator outlet hose to the engine front cover.

47. Install the thermostat bypass hose adapter.

48. Install the water pump.

49. Install the right engine mount bracket.

50. Install the right engine mount bracket bolts. Tighten the bolts to 55 Nm (41 ft. lbs.).

51. Install the engine mount bracket bolts. Tighten the bolts to 25 Nm (18 ft. lbs.).

52. Install the left belt idler pulley.

53. Raise and support the vehicle.

54. Install the engine oil pan.

55. Install the lower belt idler pulley.

56. Install the crankshaft balancer.

57. Lower the vehicle.

58. Install the drive belt.

59. Remove the engine support fixture.

60. Install the air cleaner assembly.

61. Connect the negative battery cable.

62. Fill the cooling system.

Timing Chain and Sprockets

REMOVAL & INSTALLATION

1. Before servicing the vehicle, refer to the Precautions Section.

2. Remove the engine front cover.

3. Remove the camshaft sprocket bolt.

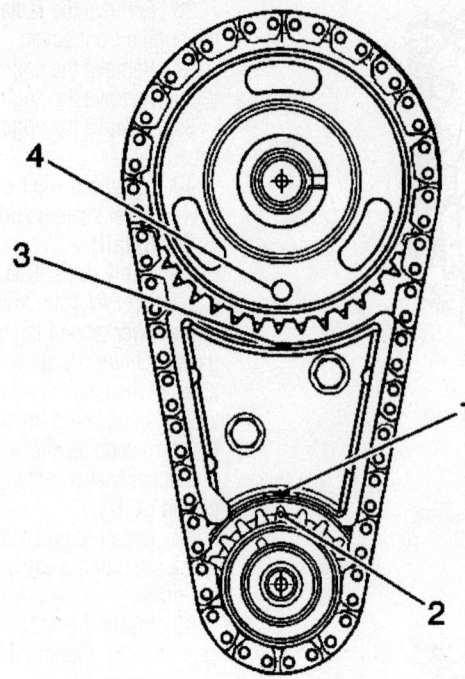

1. Dampener timing mark
2. Crankshaft sprocket timing mark
3. Dampener upper mark
4. Camshaft sprocket timing mark

06025-EQUI-G51

Timing mark alignment

4. Remove the camshaft sprocket and timing chain.

5. Remove the crankshaft sprocket.

6. Remove the timing chain dampener bolts.

7. Remove the timing chain dampener.

To install:

8. Install the crankshaft sprocket.

9. Apply prelube GM P/N 12345501 (Canadian P/N 992704) or the equivalent to the crankshaft sprocket thrust surface.

10. Install the timing chain dampener.

11. Install the timing chain dampener bolts. Tighten the bolt to 21 Nm (15 ft. lbs.).

12. Align the crankshaft timing mark to the timing mark on the bottom of the timing chain dampener.

13. Hold the camshaft sprocket with the timing chain hanging down and install the timing chain to the crankshaft gear.

14. Align the timing mark on the camshaft gear with the timing mark on top of the timing chain dampener.

15. Align the dowel in the camshaft with the dowel hole in the camshaft sprocket.

16. Draw the camshaft sprocket onto the camshaft using the mounting bolt.

17. Coat the crankshaft and camshaft sprocket with engine oil. Tighten the bolt to 140 Nm (103 ft. lbs.).

18. Install the engine front cover.

FUEL SYSTEM

Fuel System Service Precautions

Safety is the most important factor when performing not only fuel system maintenance but any type of maintenance. Failure to conduct maintenance and repairs in a safe manner may result in serious personal injury or death. Maintenance and testing of the vehicle's fuel system components can be accomplished safely and effectively by adhering to the following rules and guidelines.

• To avoid the possibility of fire and personal injury, always disconnect the negative battery cable unless the repair or test procedure requires that battery voltage be applied.

• Always relieve the fuel system pressure prior to disconnecting any fuel system component (injector, fuel rail, pressure regulator, etc.), fitting or fuel line connection. Exercise extreme caution whenever relieving fuel system pressure, to avoid exposing skin, face and eyes to fuel spray. Please be advised that fuel under pressure may penetrate the skin or any part of the body that it contacts.

• Always place a shop towel or cloth around the fitting or connection prior to loosening to absorb any excess fuel due to spillage. Ensure that all fuel spillage (should it occur) is quickly removed from engine surfaces. Ensure that all fuel soaked cloths or towels are deposited into a suitable waste container.

• Always keep a dry chemical (Class B) fire extinguisher near the work area.

• Do not allow fuel spray or fuel vapors to come into contact with a spark or open flame.

• Always use a back-up wrench when loosening and tightening fuel line connection fittings. This will prevent unnecessary stress and torsion to fuel line piping. Always follow the proper torque specifications.

• Always replace worn fuel fitting O-rings with new ones. Do not substitute fuel hose, or equivalent, where fuel pipe is installed.

Relieving Fuel System Pressure

1. Before servicing the vehicle, refer to the Precautions Section.

2. Disconnect the negative battery cable.

3. Connect a fuel pressure gauge to the fuel pressure connection:

a. Place the bleed hose of the fuel pressure gage into an approved gasoline container.

b. Open the bleed valve on the fuel pressure gage in order to bleed the air from the gage.

c. Turn ON the ignition, with the engine OFF.

d. Command the fuel pump ON with a scan tool until all of the air is bled out of the gauge.

e. Close the bleed valve on the fuel pressure gage.

f. Command the fuel pump ON with a scan tool.

g. Place the bleed hose into an approved container and open the valve to bleed the system pressure. The fuel connections are now safe for servicing.

h. Place a shop towel under the fuel pressure gauge to catch any remaining fuel spillage.

i. Remove the gauge from fuel pressure connection.

j. Drain any fuel remaining in the fuel pressure gage into an approved container.

k. Install the cap on the fuel pressure connection.

l. Place the shop towel in an approved container.

Fuel Filter

REMOVAL & INSTALLATION

The fuel filter is part of the Primary Fuel Tank Module, and is not normally replaced.

Fuel Pump

REMOVAL & INSTALLATION

The fuel pump is part of the Primary Fuel Tank Module.

Fuel Tank

REMOVAL & INSTALLATION

1. Before servicing the vehicle, refer to the Precautions Section.

✳✳ CAUTION

Do not allow smoking or the use of open flames in the area where work on the fuel or EVAP system is taking place. Anytime work is being done on the fuel system, disconnect the negative battery cable, except for those tests where battery voltage is required.

2. Ensure that the fuel level in the tank is less than ¼ full. If necessary, drain the fuel tank to at least this level.

✳✳ CAUTION

Fuel supply lines will remain pressurized for long periods of time after the engine is shutdown. This pressure must be relieved before servicing the fuel system.

3. Relieve the fuel system pressure.
4. Disconnect the negative battery cable.
5. Raise and support the vehicle.
6. Remove the catalytic converter pipe flange to exhaust system pipe flange nuts.
7. Separate the pipes and discard the gasket.
8. Separate the rubber isolators from the hangers.

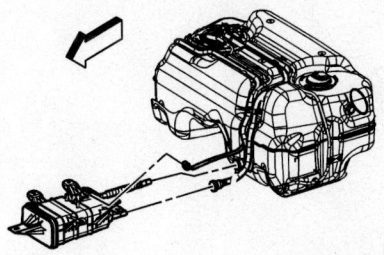

06025-EQUI-G54

EVAP canister

9. Remove the exhaust system.
10. With All Wheel Drive, remove the driveshaft and driveshaft guard.
11. Remove the evaporative emission (EVAP) canister.

➡**Clean all fuel pipe connections and surrounding areas before disconnecting the fuel pipes to avoid contamination of the fuel system.**

✳✳ CAUTION

Whenever fuel lines are removed, catch fuel in an approved container. Container opening must be a minimum of 300 mm (12 in.) diameter to adequately catch the fluid.

12. Disconnect the chassis fuel supply line from the fuel tank.
13. Disconnect the fuel filler tube, EVAP vent hose, and fresh air hose from the fuel tank.
14. Disconnect the fuel tank electrical connector and remove the electrical connector retainer from the rear frame.

➡**Do not bend the fuel tank straps. Bending the fuel tank straps may cause damage to the straps.**

✳✳ WARNING

Do not lower the rear frame. It is not necessary to lower the rear frame for fuel tank removal.

15. Support the fuel tank.
16. Remove the fuel tank strap bolts and fuel tank straps.
17. Lower the fuel tank from the underbody of the vehicle.
To install:
18. Install the fuel tank heat shield and fuel tank assembly to the vehicle.
19. Install the fuel tank straps and the fuel tank strap-to-body bolts. Tighten the bolts to 25 Nm (18 ft. lbs.).
20. Connect the fuel tank electrical connector and install the electrical connector retainer to the rear frame.

21. Connect the EVAP vent, and fresh air hoses to the fuel tank.
22. Connect the fuel filler tube to the fuel tank. Tighten the fuel filler tube clamp to 5 Nm (44 inch lbs.).
23. Connect the chassis fuel supply line to the fuel tank.
24. Install the evaporative emission (EVAP) canister.
25. Install the exhaust system.
26. With All Wheel Drive, install the driveshaft and driveshaft guard.
27. Lower the vehicle.
28. Fill the fuel tank with gasoline.
29. Connect the negative battery cable.
30. Prime the fuel system:
a. Cycle the ignition ON for 5 seconds and then OFF for 10 seconds.
b. Repeat the previous step twice.
c. Crank the engine until it starts. The maximum starter motor cranking time is 20 seconds.
d. If the engine does not start, repeat the procedure.

Primary Fuel Tank Module

REMOVAL & INSTALLATION

1. Before servicing the vehicle, refer to the Precautions Section.

✳✳ WARNING

NEW fuel tank module seals are necessary each time the fuel tank module is serviced. Obtain NEW seals for both the primary and secondary modules prior to beginning this service procedure.

✳✳ CAUTION

Whenever fuel line fittings are loosened or removed, wrap a shop cloth around the fitting and have an approved container available to collect any fuel.

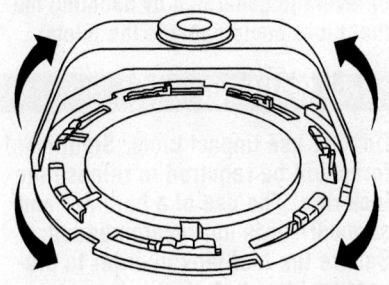

06025-EQUI-G52

Tool J-45722

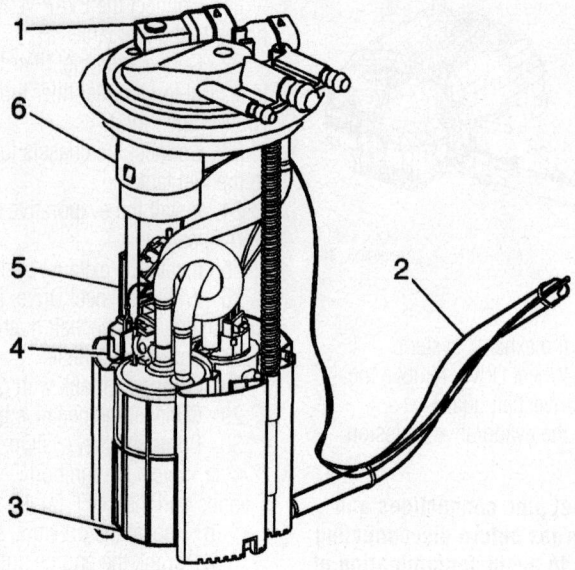

1. Fuel pressure sensor
2. Fuel transfer pipe
3. Fuel filter
4. Fuel level sensor
5. Fuel pressure regulator
6. Fill limiter vent valve

06025-EQUI-G53

Primary fuel tank module

➡Clean all fuel pipe and hose connections and surrounding areas before disassembling to avoid possible contamination of the fuel system. Spray the fuel pump module cam-lock ring tang with penetrating oil prior to attempting removal.

2. Remove the fuel tank.
3. Remove the secondary fuel pump module.
4. Disconnect the electrical connectors from the primary fuel pump module and fuel tank pressure sensor.

➡Avoid damaging the lock ring. Use only tool J-45722 to prevent damage to the lock ring.

➡Do Not handle the fuel sender assembly by the fuel pipes. The amount of leverage generated by handling the fuel pipes could damage the joints.

❋❋ WARNING

Do NOT use impact tools. Significant force will be required to release the lock ring. The use of a hammer and screwdriver is not recommended. Secure the fuel tank in order to prevent fuel tank rotation.

5. Use tool J 45722, or equivalent and a long breaker-bar in order to unlock the fuel sender lock ring. Turn the fuel sender lock ring in a counterclockwise direction.
6. Disconnect the fuel feed and vent lines from the fuel tank.

➡To prevent bending of the sending unit float arm during removal, lift the pump module up slightly to disengage the orientation tabs in the tank and rotate the module 45 degrees.

7. Remove the primary fuel pump module assembly.

❋❋ WARNING

Always replace the fuel pump module-to-tank seal, O-ring, when the fuel pump module is removed.

8. Discard the fuel pump module-to-tank seal.

❋❋ WARNING

Some lock ring were manufactured with DO NOT REUSE stamped into them. These lock rings may be reused if they are not damaged or warped.

❋❋ WARNING

Inspect the lock ring for damage due to improper removal or installation procedures. If damage is found, install a NEW lock ring.

❋❋ WARNING

Check the lock ring for flatness.

9. Place the lock ring on a flat surface. Measure the clearance between to lock ring and the flat surface using a feeler gage at 7 points. If the warpage is less than 0.41 mm (0.016 in.), the lock ring does not require replacement. If the warpage is greater than 0.41 mm (0.016 in.), the lock ring must be replaced.
   To install:
10. Insert the new primary fuel pump module assembly with the level sender and the new fuel pump-to-tank seal. Ensure the orientation tabs are aligned.

❋❋ WARNING

Always replace the fuel sender seal when installing the fuel sender assembly. Replace the lock ring if necessary. Do not apply any type of lubrication in the seal groove.

11. Ensure the lock ring is installed with the correct side facing upward. A correctly installed lock ring will only turn in a clockwise direction.
12. Use tool J 45722 in order to install the fuel sender lock ring. Turn the fuel sender lock ring in a clockwise direction.
13. Connect the wiring harness to the primary fuel pump module and fuel tank pressure sensor.
14. Install the secondary fuel pump module.
15. Install the fuel tank.

Secondary Fuel Tank Module

REMOVAL & INSTALLATION

1. Before servicing the vehicle, refer to the Precautions Section.

❋❋ WARNING

A NEW fuel tank module seal is necessary each time the fuel tank module is serviced. Obtain a NEW seal prior to beginning this service procedure.

❋❋ CAUTION

Whenever fuel line fittings are loosened or removed, wrap a shop cloth around the fitting and have an approved container available to collect any fuel.

➡ Clean all fuel pipe and hose connections and surrounding areas before disassembling to avoid possible contamination of the fuel system. Spray the fuel pump module cam-lock ring tang with penetrating oil prior to attempting removal.

2. Remove the fuel tank.
3. Disconnect the EVAP vent line quick connect.

➡ To prevent retainer damage, do not attempt to remove the retainer with a 12 in. or shorter ratchet/breaker bar.

4. Use tool J39765-A and remove the fuel pump module retaining ring.
5. Disconnect the secondary level sensor electrical connector.
6. Disconnect the suction port attaching tube by pressing down on the tab (1).

➡ To prevent bending of the sending unit float arm during removal, lift the pump module up slightly to disengage the orientation tabs in the tank and rotate the module 45 degrees.

7. Remove the secondary fuel pump module.

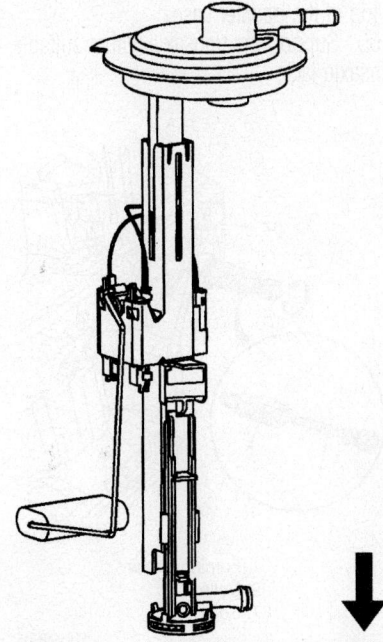

Secondary fuel tank module

06025-EQUI-G55

❋❋ WARNING

Always replace the fuel pump module-to-tank seal, O-ring, when the fuel pump module is removed.

8. Discard the fuel pump module-to-tank seal.
   To install:
9. Connect the suction port
10. Insert the new secondary fuel pump module with the level sender and new fuel pump-to-tank seal.
11. Ensure the orientation tabs are aligned.
12. Use the tool to install the fuel pump lock ring.
13. Connect the EVAP line quick connect.
14. Install the fuel tank.

Fuel Injectors

REMOVAL & INSTALLATION

1. Before servicing the vehicle, refer to the Precautions Section.

❋❋ CAUTION

In order to reduce the risk of fire and personal injury that may result from a fuel leak, always install the fuel injector O-rings in the proper position. If the upper and lower O-rings are different colors (black and brown), be sure to install the black O-ring in the upper position and the brown O-ring in the lower position on the fuel injector. The O-rings are the same size but are made of different materials.

❋❋ WARNING

When servicing the fuel rail assembly, precautions must be taken to prevent dirt and other contaminants from entering the fuel passages. It is recommended that the fittings be capped, and the holes be plugged during servicing.

2. Relieve the fuel system pressure.
3. Remove the upper intake manifold.
4. Disconnect the engine fuel feed pipe at the fuel rail.
5. Disconnect the main fuel injector harness electrical connector.
6. Disconnect the coolant temperature sensor and camshaft position sensor electrical connectors

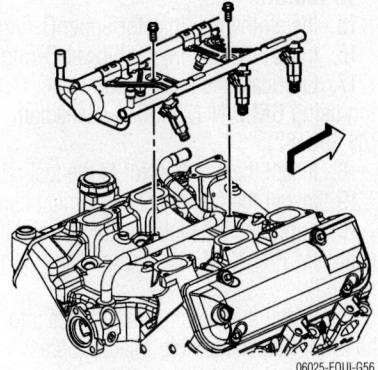

Fuel rail assembly

06025-EQUI-G56

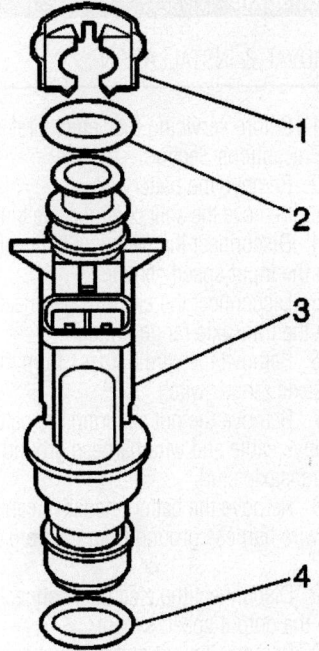

1. Retainer
2. Upper o-ring
3. Injector
4. Lower o-ring

06025-EQUI-G57

Fuel injector parts

7. Disconnect the electrical connectors from the fuel injectors.
8. Remove the fuel rail retaining bolts.
9. Remove the fuel rail assembly.
10. Remove the injector O-ring seal from the spray tip end of each injector.
11. Remove the fuel injector retaining clip.
12. Remove the fuel injector from the fuel rail.
13. Remove the fuel injector upper O-ring.
14. Remove the fuel injector lower O-ring.

To install:

15. Install the fuel injector upper O-ring.

16. Install the fuel injector lower O-ring.

17. Lubricate the fuel injector upper O-ring using GM P/N 12345616 (Canadian P/N 993182).

18. Install the fuel injector to the fuel rail.

19. Install the fuel injector retaining clip.

20. Lubricate the fuel injector O-rings using GM P/N 12345616, (Canadian P/N 993182)

21. Install the fuel injector nozzles into the lower intake manifold injector bores.

22. Press on the fuel rail using the palms of both hands until the fuel injectors are fully seated.

23. Install the fuel rail attaching bolts. Tighten the bolt to 10 Nm (89 inch lbs.).

24. Install the injector electrical harness to the fuel rail.

25. Apply lubricant to the fuel injector electrical connectors; GM P/N 12377900, (Canadian P/N 10953529).

26. Connect the fuel injector electrical connectors.

27. Connect the coolant temperature sensor and camshaft position sensor electrical connectors

28. Connect the main fuel injector electrical harness connector.

29. Connect the fuel feed pipe at the fuel rail.

30. Install the upper intake manifold.

31. Connect the negative battery cable.

32. Inspect for fuel leaks.

    a. Turn ON the ignition for 2 seconds.

    b. Turn OFF the ignition for 10 seconds.

    c. Turn ON the ignition.

    d. Inspect for fuel leaks.

DRIVE TRAIN

Transaxle

REMOVAL & INSTALLATION

1. Before servicing the vehicle, refer to the Precautions Section.

2. Remove the battery tray.

3. Remove the shift control cable bracket.

4. Disconnect the electrical connector from the input speed sensor.

5. Disconnect the electrical connectors from the transaxle range switch.

6. Remove the wire harness from the transaxle range switch.

7. Remove the nut securing the battery negative cable and wire harness ground to the transaxle stud.

8. Remove the battery negative cable and wire harness ground from the transaxle stud.

9. Disconnect the electrical connector from the output speed sensor.

10. Remove the nut and fuel line retaining clip from the transaxle.

11. Disconnect the transaxle vent tube.

12. Secure the wire harness, vent hose, and shift cable up away from the transaxle.

13. Remove the upper 4 transaxle-to-engine bolts.

14. Tie the radiator, air conditioning condenser and fan module assembly to the upper radiator support to keep the assembly with the vehicle when the frame and drivetrain are removed.

15. Install the engine support fixture. See the procedure as described under Front Cover Removal & Installation.

16. Remove the left hand transaxle mount-to-transaxle bolts.

17. Remove the left hand transaxle mount-to-side rail bolts.

18. Remove the left hand transaxle mount from the vehicle.

19. Raise and support the vehicle.

20. Remove both front wheels.

21. Remove the left and right side engine splash shields.

22. Remove the steering intermediate shaft pinch bolt and discard the bolt.

23. Disconnect the steering intermediate shaft from the steering gear.

24. Remove the right and left outer tie rod ends from the steering knuckles.

25. Remove the right and left stabilizer shaft links from the stabilizer shaft.

26. Remove the right and left lower ball joints from the steering knuckles.

27. Remove the front bumper fascia air deflector.

28. Drain the transaxle fluid.

29. Remove the transaxle oil cooler lines from the transaxle.

➡**Ensure that the J 45201, or equivalent is fully seated into the transaxle seal bore.**

30. Insert the collet piece of tool J 45201 into the cooler line seal.

31. Insert the forcing screw piece of the J 45201 into the collet.

32. Tighten the forcing screw until snug.

33. Thread the collar piece of J 45201 onto the collet until snug.

34. Turn the collar clockwise in order to remove the cooler line seal.

35. Discard the seal.

36. Clean the case bores for the cooler line seals.

37. Remove the engine-to-transaxle brace bolts and brace.

38. Remove the starter motor.

39. Turn the crankshaft balancer bolt clockwise to gain access to the torque converter-to-flywheel bolts through the starter motor hole.

➡**Mark the relation of the flywheel to torque converter for reassembly.**

40. Remove the torque converter-to-flywheel bolts.

41. Remove the front engine mount.

42. Remove the through bolt from the rear transaxle mount and bracket.

43. Place a universal frame support fixture under the frame.

44. Lower the vehicle until the frame contacts the frame support fixture.

45. Remove the frame-to-body bolts. Discard the bolts.

46. Raise the vehicle up away from the frame and remove the frame from under the vehicle.

47. Disconnect the right and left half-shafts from the intermediate shaft and transaxle. Secure both out of the way.

48. Remove the intermediate shaft.

49. If vehicle is equipped with all wheel drive (AWD) complete the following steps:

    a. Remove the rear driveshaft.

    b. Remove the transfer case mounting bracket.

    c. Disconnect the vent hose from the top of the transfer case.

50. Support the transaxle with a suitable transaxle jack.

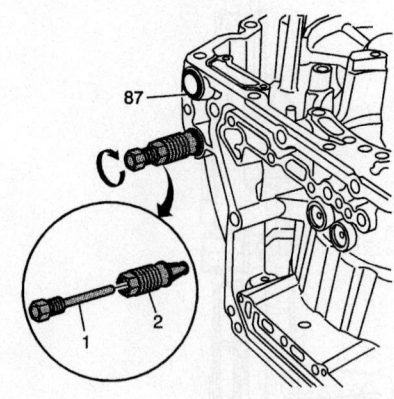

1. Forcing screw
2. Collet
87. Seal

06025-EQUI-G58

Collet and forcing screw

51. Remove the 4 lower transaxle-to-engine bolts.

52. Slide the transaxle away from the engine until the transaxle torque-converter clears the flywheel.

53. If equipped with a drive axle seal dust cover, discard it and do not replace it.

54. Lower the transaxle away from the vehicle.

55. If the vehicle is equipped with AWD complete the following steps:

　a. Remove the retaining ring from the stub shaft for tool installation. Discard the retainer ring.

　b. Remove the stub shaft from the transfer case using a slidehammer and adapter.

　c. Remove the bolts securing the transfer case to the transaxle.

　d. Remove the transfer case from the transaxle.

56. If the vehicle is equipped with front wheel drive (FWD) complete the following steps:

　a. Remove the bolts securing the rear transaxle mount bracket to the transaxle.

　b. Remove the rear transaxle mount bracket from the transaxle.

To install:

57. If the vehicle is equipped with FWD complete the following steps:

　a. Install the rear transaxle mount bracket to the transaxle.

　b. Install the bolts securing the rear transaxle mount bracket to the transaxle. Tighten the bolts to 55 Nm (41 ft. lbs.).

58. If the vehicle is equipped with AWD complete the following steps:

　a. Install the transfer case to the transaxle.

　b. Install the bolts securing the transfer case to the transaxle. Tighten the bolts to 60 Nm (44 ft. lbs.).

　c. Install the stub shaft.

　d. Install a NEW retaining ring on the stub shaft

59. Raise the transaxle up into the vehicle engine compartment.

60. Align and install the transaxle to the engine.

61. Install the 4 lower transaxle-to-engine bolts. Tighten the bolts to 75 Nm (55 ft. lbs.).

62. If the vehicle is equipped with AWD complete the following steps:

　a. Connect the vent hose to the top of the transfer case.

　b. Install the transfer case mounting bracket.

　c. Install the rear driveshaft.

63. Install the intermediate shaft.

64. Install the right and left halfshafts to the intermediate shaft and transaxle.

65. Install the frame to the vehicle.

66. Install NEW frame-to-body bolts. Tighten the bolts to 155 Nm (114 ft. lbs.).

67. Install the bolt through the rear transaxle mount and transaxle mount bracket. Tighten the bolt to 110 Nm (80 ft. lbs.).

68. Install the front engine mount. Torque the mount-to-frame bolt to 110 Nm (81 ft. lbs.). Torque the mount-to-transmission bolts to 50 Nm (37 ft. lbs.).

69. Turn the crankshaft balancer bolt clockwise to gain access to the torque converter-to-flywheel bolts through the starter motor hole.

➡**Align the reference marks on the flywheel and torque converter.**

70. Install the torque converter to flywheel bolts. Tighten the bolt to 60 Nm (44 ft. lbs.).

71. Install the starter motor.

72. Install the engine-to-transaxle brace.

73. Install the engine-to-transaxle brace bolts. Tighten the bolts to 50 Nm (37 ft. lbs).

74. Insert a new transaxle cooler line seal into the case bore.

75. Remove the nub from tool J 41239-1.

76. Install the nub of tool J 41239-1 on the transaxle cooler line seal.

77. Tap the new transaxle cooler line seal into the case bore.

78. Install the transaxle oil cooler line assembly to the transaxle.

79. Install the transaxle oil cooler line assembly nut. Tighten the transaxle cooler line retaining nut to 7 Nm (62 inch lbs.).

80. Install the front bumper fascia air deflector.

81. Install the right and left lower ball joints to the steering knuckles.

82. Install the right and left stabilizer shaft links to the stabilizer shaft.

83. Install the right and left outer tie rod ends to the steering knuckles.

84. Connect the steering intermediate shaft to the steering gear.

85. Install a NEW pinch bolt to the steering intermediate shaft. Tighten the bolt to 34 Nm (25 ft. lbs.).

86. Install the right and left side engine splash shields.

87. Install both front wheels.

88. Lower the vehicle.

89. Install the left hand transaxle mount to the vehicle.

90. Install the left hand transaxle mount-to-side rail bolts. Tighten the bolts to 37 Nm (27 ft. lbs.).

91. Install the left hand transaxle mount to transaxle bolts. Tighten the bolts to 50 Nm (37 ft. lbs.).

92. Remove the engine support fixture.

93. Untie the radiator, air conditioning condenser and fan module assembly from the upper radiator support.

94. Install the upper 4 transaxle-to-engine bolts. Tighten the bolts to 75 Nm (55 ft. lbs.).

95. Connect the transaxle vent tube.

96. Connect the electrical connector to the output speed sensor.

97. Install the nut and fuel line retaining clip to the transaxle. Tighten the nut to 25 Nm (18 inch lbs.).

98. Install the battery negative cable and wire harness ground to the transaxle stud.

99. Install the nut securing the battery negative cable and wire harness ground to the transaxle stud. Tighten the nut to 45 Nm (33 ft. lbs.).

100. Connect the electrical connectors to the transaxle range switch.

101. Connect the electrical connector from the input speed sensor.

102. Install the shift control cable bracket.

103. Install the battery tray.

104. Fill the transaxle with fluid.

105. Perform the transmission adaptive learn procedure.

Transfer Case Assembly

REMOVAL & INSTALLATION

1. Before servicing the vehicle, refer to the Precautions Section.

2. Raise and support the vehicle.

3. Drain the transfer case fluid.

4. Remove the driveshaft.

5. Remove the right wheel halfshaft.

6. Remove the intermediate shaft.

7. Remove the retainer ring from the stub shaft for tool installation. Discard the used retainer ring.

8. Remove the stub shaft using a slidehammer and adapter.

9. Remove the transfer case mounting bracket.

10. Disconnect the transfer case vent hose.

11. Support the transaxle with a jackstand.

12. Remove the 4 bolts securing the transfer case to the transaxle.

➡**Remove the rear transaxle mount from the transfer case after the transfer case has been removed from the vehicle.**

13. Remove the 3 bolts securing the rear transaxle mount to the vehicle frame.

14. Slide the transfer case away from the transaxle.

15. Rotate the transfer case so that the driveshaft drive flange faces the transaxle.

16. Lift and rotate the transfer case so that the driveshaft drive flange is pointing down toward the floor.

17. Lower the transfer case through the opening between the engine oil pan and the vehicle frame.

18. Remove the 3 bolts and rear transaxle mount from the transfer case.

To install:

19. Install the rear transaxle mount to the transfer case.

20. Install the 3 bolts securing the rear transaxle mount to the transfer case. Tighten the bolts to 110 Nm (81 ft. lbs.).

21. Ensure the torque converter cover is in the proper location.

22. With the transfer case driveshaft drive flange pointing down toward the floor, lift the transfer case up between the engine oil pan and the vehicle frame.

23. Rotate and align the transfer case with the transaxle.

24. Install the 4 bolts securing the transfer case to the transaxle. Tighten the bolts to 60 Nm (44 ft. lbs.).

25. Install the 3 bolts securing the rear transaxle mount to the vehicle frame. Tighten the bolts to 50 Nm (37 ft. lbs.).

26. Remove the jackstand supporting the transaxle.

27. Connect the transfer case vent hose.

28. Install the transfer case mounting bracket. Tighten the bolts to 50 Nm (37 ft. lbs.).

29. Install the stub shaft.

30. Install a NEW retainer ring on the stub shaft.

31. Install the intermediate shaft.

32. Install the right wheel halfshaft.

33. Fill the transfer case with fluid.

34. Lower the vehicle.

Front Halfshaft

REMOVAL & INSTALLATION

1. Before servicing the vehicle, refer to the Precautions Section.

2. Raise and support the vehicle.

3. Remove the tire and wheel assembly.

4. Remove and discard the halfshaft spindle nut.

➡**Hold the ball stud from turning when removing/installing the nut. The boot**

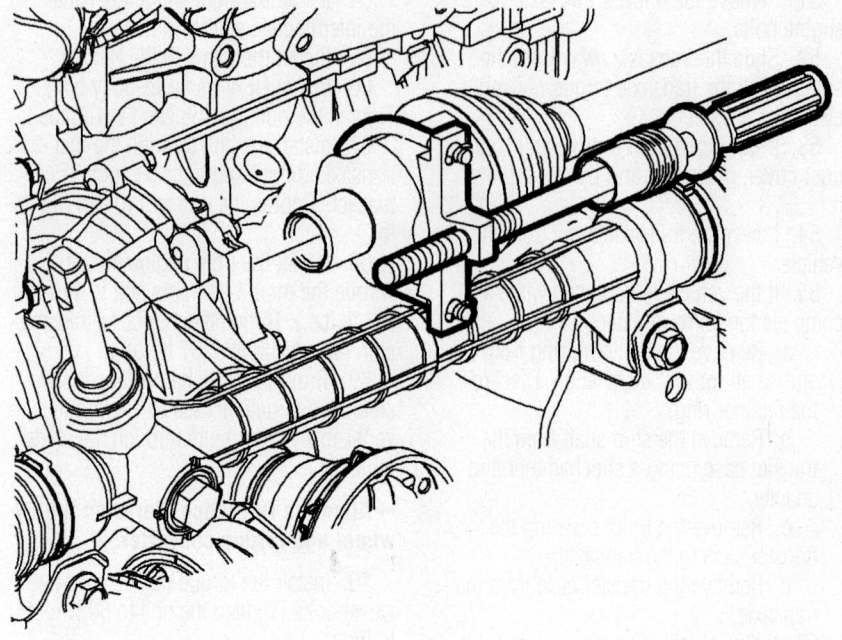

J45341 and J2619-01 assembled on the halfshaft.

06025-EQUI-G59

can become torn and damaged if the ball stud turns.

5. Remove the outer tie rod end-to-steering knuckle nut. Do not loosen the tie rod end jam nut.

❋❋ WARNING

Do not use a wedge type tool to separate the tie rod end from the steering knuckle.

6. Using a 2-jawed puller, separate the tie rod end from the steering knuckle.

7. Remove and discard the cotter pin from the lower ball joint stud.

8. Remove the ball joint stud nut.

9. Using a ball joint separator, separate the lower ball joint stud from the steering knuckle.

10. Using a backup wrench on the stud, remove the nut securing the lower stabilizer bar link and disengage the link.

11. Disengage the halfshaft spindle from the wheel hub assembly. If necessary, place a wood block against the end of the halfshaft spindle and tap with a hammer to aid removal.

❋❋ WARNING

Use care not to damage the joint seal when removing the halfshaft.

12. Assemble tools J 45341 and J 2619-01, or equivalent to the halfshaft inner tripot joint.

❋❋ WARNING

On vehicles equipped with all-wheel drive (AWD), the stub shaft may disengage from the power takeoff unit (PTU). If necessary, cap the opening in the PTU to prevent fluid loss.

13. Disengage the halfshaft from the transmission or power takeoff unit (PTU), if equipped.

14. Remove the halfshaft from the vehicle.

To install:

15. Install a new halfshaft retaining ring to the output shaft.

16. Install tool J 44394, or equivalent to the halfshaft oil seal.

17. Install the halfshaft to the output shaft:

   a. Guide the halfshaft tripot joint squarely onto the output shaft.

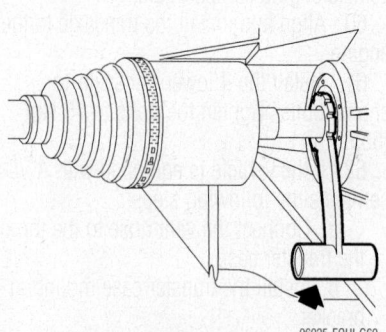

06025-EQUI-G60

J44394 installed on the seal

b. After the splined end of the half-shaft passes the oil seal, remove the tool from the oil seal.

c. Firmly engage the halfshaft to the output shaft.

d. Ensure that the tripot joint is fully seated on the output shaft by grasping the tripot joint and attempting to pull free of the output shaft.

18. Insert the constant velocity (CV) joint spindle to the wheel hub/bearing assembly of the steering knuckle.

19. Hand install a new halfshaft spindle nut.

20. Install the lower ball joint stud to the steering knuckle.

21. Install the lower ball joint castle nut to the stud. Tighten the nut to 10 Nm (89 lb in). Tighten the nut an additional 150 degrees.

22. Install the cotter pin to the ball joint stud.

23. If necessary, tighten the nut one additional flat at a time until the castle nut aligns with the hole in the ball joint stud.

24. Secure the cotter pin to the ball joint stud by folding one tine over the end of the ball joint stud. Cut off any excess length of the cotter pin tines.

25. Install the lower link to the stabilizer bar.

26. Install a new nut to the stabilizer bar link stud.

✵✵ WARNING

In order to prevent damaging the stabilizer bar link stud seal, do not allow the stud to rotate while tightening the nut.

27. Use a back up wrench on the stud and tighten the nut. Tighten the nut to 65 Nm (48 ft. lbs.).

28. Install the tie rod end to the steering knuckle. Install a new nut to the tie rod end stud. Tighten the nut to 50 Nm (37 ft. lbs.).

29. Tighten the halfshaft spindle nut. Tighten the nut to 205 Nm (151 ft. lbs.).

30. Install the tire and wheel assembly.
31. Lower the vehicle.
32. Inspect the transmission fluid level.

Rear Halfshaft

REMOVAL & INSTALLATION

1. Before servicing the vehicle, refer to the Precautions Section.

2. Raise and support the vehicle.

3. Remove the tire and wheel assembly.

4. Remove and discard the halfshaft spindle nut.

5. While holding the stabilizer link with a wrench, remove the stabilizer link-to-lower control arm nut.

6. Disconnect the link from the control arm.

7. Place a stand under the lower control arm and support the control arm.

8. Remove the lower shock absorber mounting bolt and nut.

9. Remove the toe link nut, bolt, and washer.

10. Loosen, but do not remove, the lower suspension jounce bumper nut.

11. Remove the lower control arm-to-suspension knuckle bolt and nut.

➡**Relieve spring tension slowly in order to avoid sudden release of the coil spring.**

12. Slowly lower support stand until coil spring tension is relieved and remove coil spring.

13. Loosen, but do not remove, the upper control arm-to-suspension knuckle nut.

➡**Support the halfshaft while it is disengaged from the wheel hub and bearing assembly in order to avoid damaging the halfshaft seals.**

14. Place a block of wood against the halfshaft spindle and tap with a hammer to release the spindle from the wheel hub and bearing assembly.

15. Rotate the suspension knuckle upward and secure with heavy mechanics wire, or equivalent.

16. Assemble tools J 45341 and J-2619-A to the halfshaft inner tripot joint.

17. Disengage the tripot joint from the rear drive module (RDM).

18. Remove the halfshaft from the vehicle.

19. Remove and discard the halfshaft retaining ring.

To install:

20. Install a new halfshaft retaining ring to the inner tripot joint.

21. Install tool J 44394, or equivalent to the halfshaft oil seal.

22. Align the splines of the inner tripot joint to the output shaft of the RDM.

23. Install the halfshaft to the output shaft:

a. Guide the halfshaft tripot joint squarely onto the output shaft.

b. After the splined end of the half-shaft passes the oil seal, remove the tool from the oil seal.

c. Firmly engage the halfshaft to the output shaft.

d. Ensure that the tripot joint is fully

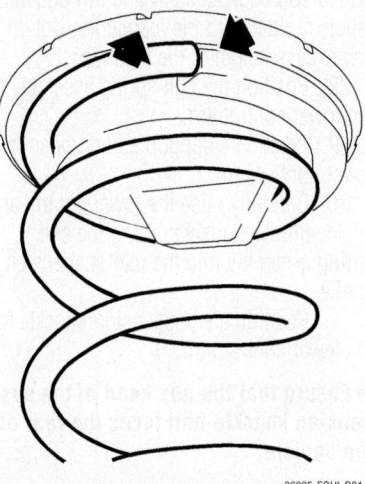

06025-EQUI-G61

Position the insulators to the coil spring and align the ends of the coil spring with the abutments of the insulators

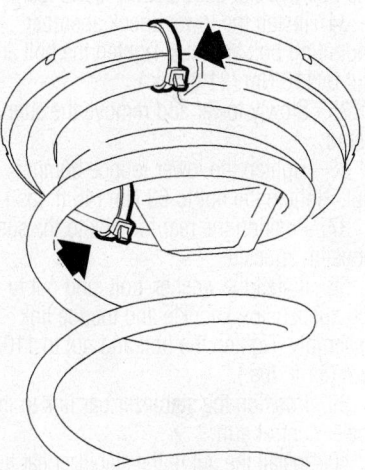

06025-EQUI-G62

Secure each of the insulators to the coil spring using 2 plastic tie straps positioned 180 degrees apart and through the reliefs molded into the insulators

seated on the output shaft by grasping the tripot joint and attempting to pull free of the output shaft.

e. Ensure that the tripot slinger does not become damaged.

24. Rotate the suspension knuckle downward while simultaneously guiding the constant velocity (CV) joint spindle to the wheel hub and bearing assembly of the suspension knuckle.

25. Hand install a new halfshaft spindle nut.

26. Position the insulators to the coil spring and align the ends of the coil spring with the abutments of the insulators.

27. Secure each of the insulators to the coil spring using 2 plastic tie straps posi-

tioned 180 degrees apart and through the reliefs molded into the insulators. Cut off any excess length of the tie straps.

28. Position the coil spring assembly to the lower control arm.

29. Position a support stand under the lower control arm.

30. Carefully raise the lower control arm while simultaneously guiding the coil spring assembly into the rear suspension cradle.

31. Position the suspension knuckle to the lower control arm.

➡**Ensure that the hex head of the suspension knuckle bolt faces the rear of the vehicle.**

32. Install the lower control arm-to-suspension knuckle bolt and nut. Tighten the bolt and nut to 160 Nm (118 ft. lbs.).

33. Tighten the upper control arm-to-suspension knuckle bolt and nut. Tighten the bolt and nut to 135 Nm (100 ft. lbs.).

34. Install the lower shock absorber mounting bolt and nut. Tighten the bolt and nut to 110 Nm (81 ft. lbs.).

35. Slowly lower and remove the support stand.

36. Tighten the lower jounce bumper nut. Tighten the nut to 63 Nm (46 ft. lbs.).

37. Position the rear toe link to the suspension knuckle.

38. Install the washer, bolt, and nut to the suspension knuckle and the toe link assembly. Tighten the bolt and nut to 110 Nm (81 ft. lbs.).

39. Position the stabilizer bar link to the lower control arm.

40. Install the nut to the stabilizer bar link.

41. While holding the stabilizer bar link stationary with a wrench, tighten the nut. Tighten the nut to 15 Nm (11 ft. lbs.).

42. Tighten the halfshaft spindle nut. Tighten the nut to 110 Nm (81 ft. lbs.).

43. Install the tire and wheel assembly.

44. Lower the vehicle.

Intermediate shaft

REMOVAL & INSTALLATION

RPO L61

1. Before servicing the vehicle, refer to the Precautions Section.

2. Raise and support the vehicle.

3. Remove the right tire and wheel assembly.

4. Remove the right front halfshaft.

5. Remove and discard the halfshaft retaining ring.

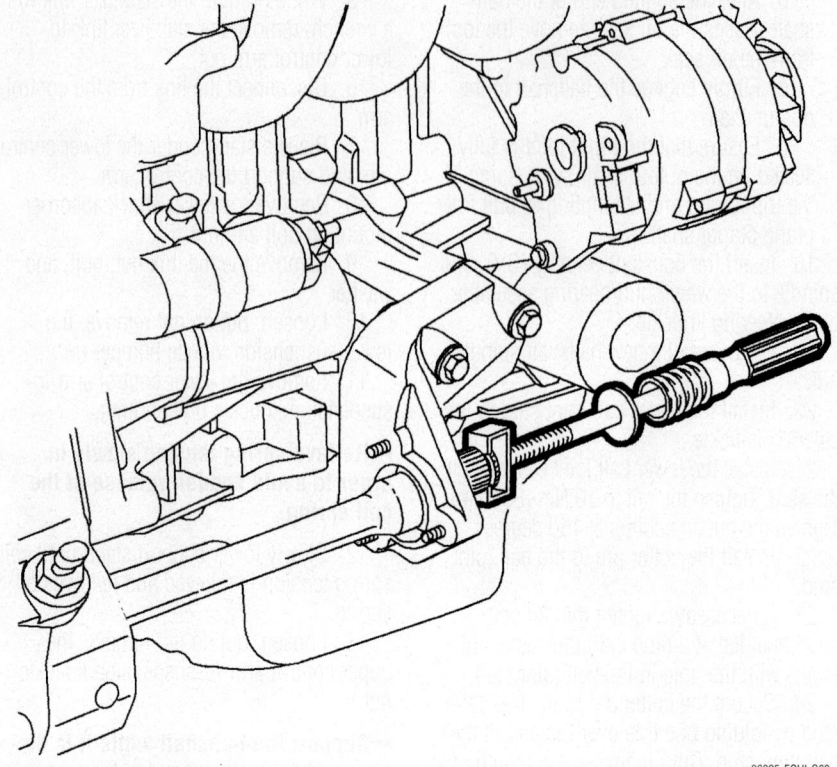

06025-EQUI-G63

Intermediate shaft removal

6. Remove the O-ring seal from the intermediate drive shaft at the bearing retainer.

7. Remove the bearing retainer-to-support bracket mounting bolts.

8. Remove the bearing retainer.

9. Assemble tools J 44467 and the J 2619-01, or equivalent to the intermediate drive shaft retainer ring groove.

10. Disengage the intermediate drive shaft from the transaxle.

11. Remove the intermediate drive shaft from the vehicle.

12. Remove the intermediate drive shaft support bracket mounting bolts.

13. Remove the support bracket from the engine.

To install:

14. Install the intermediate drive shaft support bracket to the engine.

15. Install the support bracket mounting bolts. Tighten the bolts to 35 Nm (26 ft. lbs.).

16. Install tool J 44394, or equivalent to the transaxle oil seal.

17. Install the intermediate drive shaft to the bearing retainer bracket and the transaxle.

18. Remove the tool when the intermediate shaft splines pass the oil seal.

19. Ensure that the intermediate drive shaft is fully seated by grasping the inter-

mediate drive shaft and attempting to pull free of the transaxle.

20. Position the bearing retainer to the support bracket.

21. Install the bearing retainer bolts. Tighten the bolts to 30 Nm (22 ft. lbs.).

22. Install the O-ring seal to the intermediate drive shaft.

23. Install a new halfshaft retaining ring to the intermediate driveshaft.

24. Install the right halfshaft.

25. Install the right tire and wheel assembly.

26. Lower the vehicle.

RPO L66

1. Before servicing the vehicle, refer to the Precautions Section.

2. Raise and support the vehicle.

3. Remove the right tire and wheel assembly.

4. Remove the right front halfshaft.

5. Remove and discard the halfshaft retaining ring.

6. Remove the O-ring seal from the intermediate drive shaft at the bearing retainer.

7. Remove the intermediate shaft support bracket bolts at the engine.

8. Assemble tools J 44467 and the J 2619-01, or equivalent to the intermediate drive shaft retainer ring groove.

9. Disengage the intermediate shaft from the transaxle.

10. Remove the intermediate shaft from the vehicle.

11. Remove tools J 44467 and J 2619-01 from the intermediate drive shaft.

To install:

12. Install tool J 44394, or equivalent to the transaxle oil seal.

13. Install the intermediate drive shaft to the transaxle.

14. Remove the tool when the intermediate shaft splines pass the oil seal.

15. Ensure that the intermediate drive shaft is fully seated by grasping the intermediate drive shaft and attempting to pull free of the transaxle.

16. Position the support bracket to the engine.

17. Install the support bracket bolts. Tighten the bolts to 50 Nm (37 ft. lbs.).

18. Install the O-ring seal to the intermediate drive shaft.

19. Install a new halfshaft retaining ring to the intermediate drive shaft.

20. Install the right halfshaft.

21. Install the right tire and wheel assembly.

22. Lower the vehicle.

CV-Joints

OVERHAUL

Front or Rear Inner Joint

1. Before servicing the vehicle, refer to the Precautions Section.

2. Position halfshaft bar in a soft jawed vise and clamp securely.

3. Using side cutters, remove and discard the small seal clamp.

4. Remove large seal retaining clamp using a flat-bladed tool and discard the clamp.

5. Separate the seal from the tripot housing at the large diameter and slide the seal away from the joint along the axle shaft.

6. Wipe the excess grease from the face of the tripot spider and the inside of the tripot housing.

7. Remove the tripot housing from the spider and shaft.

8. Remove the retaining ring from the groove on the halfshaft bar and remove the spider assembly.

9. Remove the seal from the halfshaft bar.

10. Thoroughly clean all parts with a suitable solvent, removing all traces of grease and contaminants.

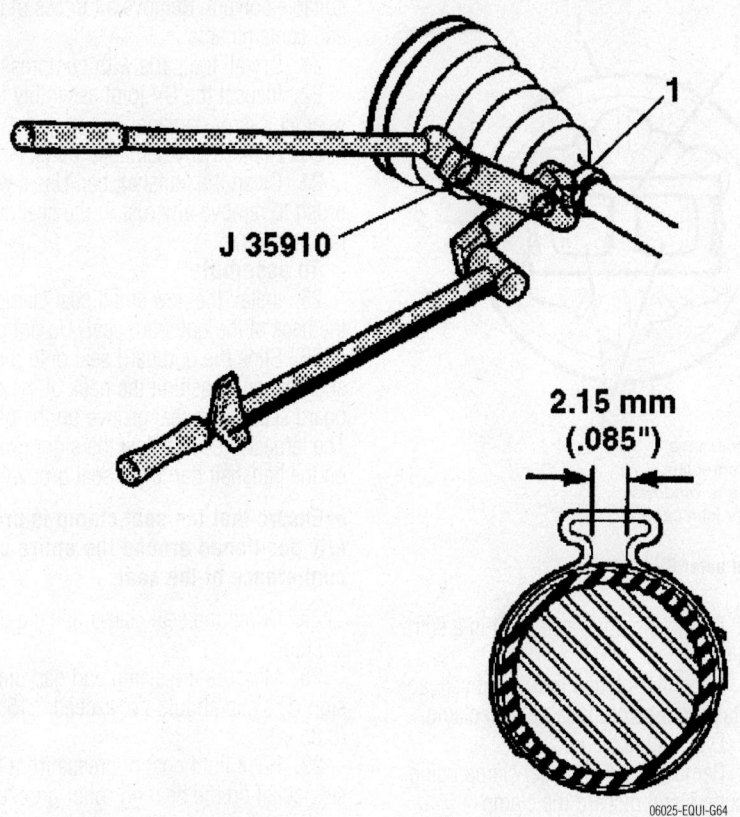

J 35910

2.15 mm (.085")

06025-EQUI-G64

Inner boot clamp gap width

11. Dry all parts with compressed air.

12. Inspect the tripot joint components for unusual wear, cracks, and other damage. Replace any damaged components.

To assemble:

13. Install the small seal clamp to the seal. Do not crimp the clamp.

14. Slide the inner seal onto the halfshaft bar and locate the lip of the seal groove on the halfshaft bar.

➡**Ensure the seal clamp is positioned correctly in the seal groove.**

15. Using a crimping tool, crimp the small seal clamp.

16. Measure the clamp gap width. Clamp gap width should not exceed 2.15 mm (0.85 in.).

17. Install the tripot spider assembly to the halfshaft bar, until seated against shoulder.

18. Install the retaining ring in the groove of the halfshaft bar with suitable pliers.

19. Place approximately half of the grease in the kit to the seal and place the remainder in the tripot housing.

20. Install the large clamp over the large diameter of the seal.

21. Install the tripot housing to the tripot spider assembly on the halfshaft bar.

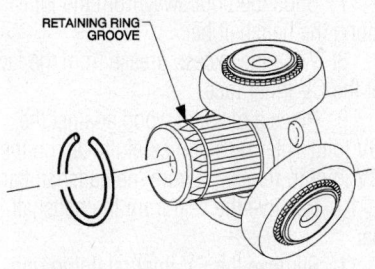

RETAINING RING GROOVE

06025-EQUI-G65

Retaining ring installation

22. Slide the large diameter of the seal over the outside of the tripot housing and position the lip of the seal in the housing groove.

23. Place the large seal retaining clamp around the seal and close using the tool.

24. Inspect the gap dimension on the clamp ear. Continue tightening until the gap dimension is reached. Gap should be 1.9 mm (5/64 in).

25. Rotate the housing in a circular motion to distribute the grease in the tripot joint.

Front or Rear Outer Joint

1. Before servicing the vehicle, refer to the Precautions Section.

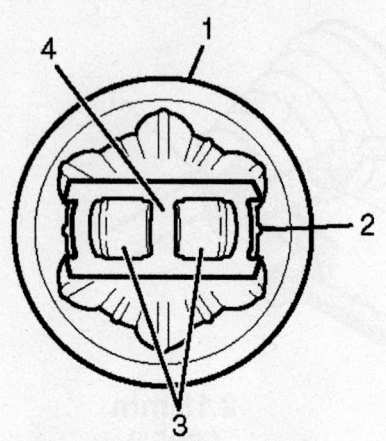

1. Inner race
2. Center line
3. Cage windows
4. CV joint cage

06025-EQUI-G66

Front outer CV joint

2. Clamp the drive axle shaft in a soft jawed vice.

3. Use a flat-bladed tool and disengage the retaining tabs of the large seal clamp.

4. Discard the clamp.

5. Remove the small seal clamp using side cutters and discard the clamp.

6. Separate the constant velocity (CV) joint boot from the CV joint race at the large diameter.

7. Slide the boot away from the joint along the halfshaft bar.

8. Wipe the excess grease from the face of the CV inner race.

9. Place a block of wood against the CV joint outer race and carefully tap on the CV joint to remove it from the halfshaft bar.

10. Remove the seal from the halfshaft bar.

11. Remove the CV joint retaining ring from the halfshaft bar.

12. Place a brass drift against the CV joint inner race.

13. Tap gently on the brass drift with a hammer in order to tilt the joint race.

14. Remove the first bearing roller when the CV race tilts.

15. Tilt the CV joint inner race in the opposite direction to remove the opposing bearing roller.

16. Repeat the process to remove all 6 of the bearing rollers.

17. Pivot the CV joint cage and the inner race 90 degrees to the centerline of the outer race. At the same time, align the cage windows with the lands of the outer race.

18. Lift out the cage and the inner race.

19. Remove the inner race from the cage by rotating the inner race upward.

20. Clean the all items thoroughly with a suitable solvent. Remove all traces of grease and contaminants.

21. Dry all the parts with compressed air.

22. Inspect the CV joint assembly for wear, cracks or damage.

23. Replace any damaged parts.

24. Clean the halfshaft bar. Use a wire brush to remove any rust in the seal mounting grooves.

To assemble:

25. Install the new small seal clamp on the neck of the outboard seal. Do not clamp.

26. Slide the outboard seal onto the halfshaft bar and position the neck of the outboard seal in the seal groove on the bar. The largest groove below the sight groove on the halfshaft bar is the seal groove seal.

➡**Ensure that the seal clamp is properly positioned around the entire circumference of the seal.**

27. Crimp the seal clamp using a crimping tool.

28. Measure the clamp end gap dimension. The gap should not exceed 2.15 mm (0.85 in).

29. Put a light coat of grease from the service kit on the bearing roller grooves of the inner race and outer race.

30. Hold the inner race 90 degrees to centerline of cage with the lands of the inner race aligned with the windows of the cage and insert the inner race into the cage.

31. Hold the cage and inner race 90 degrees to the center line of the outer race and align the cage windows with the lands of the outer race.

➡**Be sure that the retaining ring side of the inner race faces the halfshaft bar.**

32. Place the cage and the inner race into the outer race.

33. Insert the first bearing roller, then, tilt the cage in the opposite direction to insert the opposing bearing roller.

34. Repeat this process until all 6 bearing rollers are in place.

35. Install the CV joint retaining ring to the halfshaft bar.

36. Place approximately half the grease from the service kit inside the outboard seal and pack the CV joint with the remaining grease.

37. Place a block of wood against the CV joint spindle and tap on the block of wood until the CV joint inner race engages the retaining ring.

38. Slide the large diameter of the seal over the outside of the CV race and locate the lip of the seal in the housing groove.

39. Install the large seal retaining clamp over the seal and close.

40. Inspect the gap dimension on the clamp ear. Continue tightening until the gap 1.9 mm ($5/64$ in) is reached.

41. Remove the halfshaft from the bench vise.

42. Distribute the grease within the outer CV joint by rotating the joint in a circular motion four to five times.

Rear Differential Pinion Seal

REMOVAL & INSTALLATION

1. Before servicing the vehicle, refer to the Precautions Section.

2. Raise and support the vehicle.

3. Index mark the driveshaft at the power take-off unit (PTU) output flange and at the rear drive module input flange.

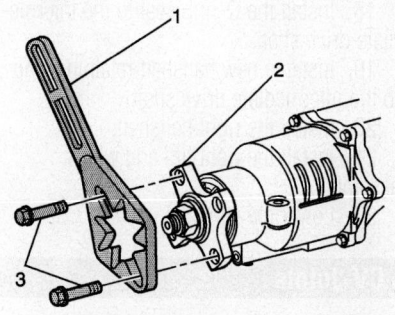

06025-EQUI-G67

Pinion flange holding tool

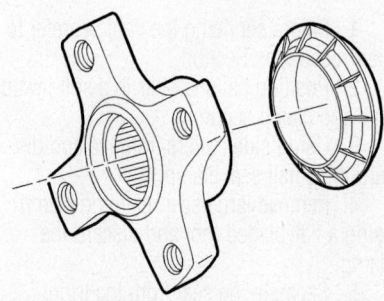

06025-EQUI-G68

Pinion flange and deflector

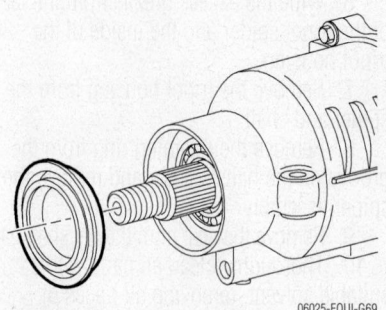

06025-EQUI-G69

Pinion seal

4. Place an adjustable support at the front and the rear of the driveshaft.

5. Remove the rear drive module (RDM) driveshaft guard mounting bolts.

6. Remove the bolts securing the driveshaft to the RDM flange.

7. Remove the bolts securing the driveshaft to the PTU.

8. Remove the bolts securing the support bearing to the vehicle underbody.

9. Remove the driveshaft from the vehicle.

10. Install tool J-08614-A (1) to the pinion flange (2) using J 44873-2 shoulder bolts (3), or equivalent.

11. Loosen the pinion flange nut.

12. Remove and discard the pinion flange nut.

13. Remove the flange.

14. Remove the dust deflector from the pinion flange.

15. Remove and discard the pinion oil seal.

To install:

16. Thoroughly clean the pinion oil seal mounting surface of the RDM housing.

17. Using a seal driver, install a new pinion oil seal to the RDM.

18. Ensure that the seal flange seats squarely against the face of the RDM.

19. Install the dust deflector to the drive pinion flange.

20. Install the pinion flange and nut to the RDM.

21. Install the holding tool to the pinion flange.

22. Tighten the pinion flange nut to 203 Nm (150 ft. lbs.).

23. Remove the tool from pinion flange.

24. Install the driveshaft to the vehicle using support stands to assist in positioning the driveshaft.

25. Hand-install the support bearing retainer bolts to the vehicle underbody.

26. Align the reference marks on the driveshaft and the PTU output flange.

27. Install the bolts to the driveshaft and the PTU flange. Tighten the bolts to 25 Nm (18 ft. lbs.).

28. Align the reference marks on the driveshaft and the RDM pinion flange.

29. Thoroughly clean and apply threadlocker, GM P/N 89021297 (Canadian P/N 10953488), to the driveshaft flange mounting bolt threads.

30. Install the bolts to the driveshaft and the pinion flange. Tighten the bolts to 50 Nm (37 ft. lbs.).

31. Remove the support stands.

32. Install the driveshaft guard to the RDM.

33. Install the bolts to the driveshaft guard. Tighten the bolts to 25 Nm (18 ft. lbs.).

34. Inspect the RDM fluid level.

35. Lower the vehicle.

Driveshaft

REMOVAL & INSTALLATION

1. Before servicing the vehicle, refer to the Precautions Section.

2. Place the transmission in neutral.

3. Raise and support the vehicle.

4. Index mark the relationship of the driveshaft to the rear drive module flange.

5. Remove the bolts securing the underbody guard loop.

6. Remove the underbody guard loop. Place a support under the driveshaft at the rear drive module.

7. Remove the bolts securing the driveshaft yoke flange to the rear drive module flange.

8. Index mark the relationship of the driveshaft to the power take-off unit (PTU) flange.

9. Place a support under the driveshaft at the PTU.

10. Remove the bolts securing the driveshaft to the PTU flange.

11. Place a support under the driveshaft at the support bearing.

12. Remove the bolts securing the drive-

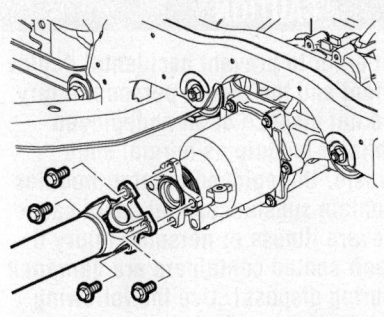

06025-EQUI-G87

Bolts securing the driveshaft yoke flange to the rear drive module flange

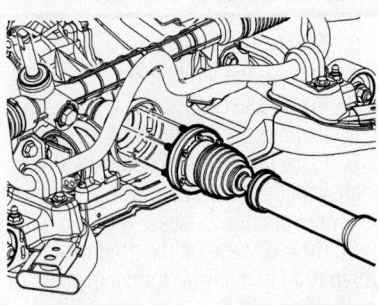

06025-EQUI-G88

Bolts securing the driveshaft to the PTU flange

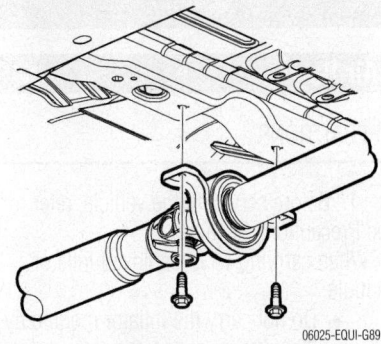

06025-EQUI-G89

Bolts securing the driveshaft support bearing to the vehicle underbody

shaft support bearing to the vehicle underbody.

13. While supporting the driveshaft, move the driveshaft rearward to disengage the constant velocity joint from the PTU flange.

14. Remove the driveshaft from the vehicle.

To install

15. While supporting the front, center, and rear of the driveshaft, install the driveshaft to the vehicle.

16. Install, but do not tighten, the support bearing mounting bolts.

17. Pull the forward section of the driveshaft rearward and install the driveshaft to the PTU flange.

18. Align the index marks on the driveshaft constant velocity joint and the PTU flange.

19. Thoroughly clean the driveshaft flange mounting bolts and apply threadlocker, GM P/N 89021297 (Canadian P/N 10953488), to the bolt threads.

20. Install the front driveshaft mounting bolts. Tighten the bolts to 25 Nm (19 ft. lbs.).

21. Align the index marks on the driveshaft yoke flange and the rear drive module flange and install the propeller .

22. Thoroughly clean the yoke mounting bolts and apply threadlocker, GM P/N 89021297 (Canadian P/N 10953488), to the bolt threads.

23. Install the bolts to the driveshaft yoke and rear drive module flange. Tighten the bolts to 50 Nm (37 ft. lbs.).

24. Tighten the support bearing mounting bolts. Tighten the bolts to 25 Nm (19 ft. lbs.).

25. Remove the support stands from the driveshaft.

26. Install the guard loop to the vehicle underbody.

27. Install the bolts to the guard loop. Tighten the bolts to 25 Nm (19 ft. lbs.).

28. Lower the veuhicle.

STEERING AND SUSPENSION

Air Bag

PRECAUTIONS

1. Before servicing the vehicle, refer to the Precautions Section.

When carrying an undeployed inflator module:

- Do not carry the inflator module by the wires or connector
- Make sure the air bag opening points away from you

When storing an undeployed inflator module:

- Make sure the air bag opening points away from the surface on which the inflator module rests
- Provide free space for the air bag to expand in case of an accidental deployment
- When storing a steering column, do not rest the column with the air bag opening facing down and the column vertical. Lay the column on its side

Failure to observe these guidelines may result in personal injury.

➡**Dual stage inflator modules have two deployment stages. If stage 1 was used to deploy a dual stage inflator module, stage 2 may still be active. Therefore, a deployed dual stage inflator module must be treated as an active module. If disposal of a dual stage module is required, both deployment loops must be energized to deploy the air bag.**

Scrapping Procedure

During the course of a vehicles useful life, certain situations may arise which will require the disposal of a live (undeployed) inflator module. Do NOT dispose a live (undeployed) inflator module through normal disposal channels until the inflator module has been deployed.

Do not deploy the inflator module in the following situations:

- After replacement of an inflator module under warranty. The inflator module may need to be returned undeployed to the manufacturer
- If the vehicle is the subject of a Product Liability report, GM-1241, related to the SIR system and is subject to a preliminary investigation. Do NOT alter the SIR system in any manner
- If the vehicle is involved in a campaign affecting the inflator modules

Deployment Procedures

You can deploy the inflator module either inside or outside of the vehicle. The method used depends upon the final disposition of the vehicle. Review the following procedures in order to determine which will work best in a given situation:

Deployment Outside Vehicle (Steering Wheel Module, I/P Module, and Roof Rail Module)

Deploy the inflator module outside of the vehicle when the vehicle will be returned to service. Situations that require deployment outside of the vehicle include the following:

- Using the SIR diagnostics, you determine that the inflator module is malfunctioning
- The inflator module is cosmetically damaged, scratched, or ripped
- The inflator module pigtail is damaged
- The inflator module connector is damaged
- The inflator module connector terminals are damaged
- Deployment and disposal of a malfunctioning inflator module is subject to any required retention period

✳✳ CAUTION

In order to prevent accidental deployment and the risk of personal injury, do not dispose of an undeployed inflator module as normal shop waste. Undeployed inflator modules contain substances that could cause severe illness or personal injury if their sealed containers are damaged during disposal. Use the following deployment procedures to safely dispose of an undeployed inflator module. Failure to observe the following disposal methods may be a violation of federal, state, or local laws.

2. Turn OFF the ignition.
3. Remove the ignition key.
4. Put on safety glasses.
5. Remove the inflator module.
6. Place the inflator module on a work bench, with the vinyl trim cover facing up and away from the surface.
7. Clear a space on the ground about 1.85 m (6 ft) in diameter for deployment of the inflator module or deployment fixture. If possible, use a paved, outdoor location free of activity. Otherwise, use a space free of

activity on the shop floor. Ensure you have sufficient ventilation.

8. Clear the area of loose or flammable objects.

➡**Dual stage deployments are only used in steering wheel and I/P inflator modules. If stage 1 was used to deploy a dual stage inflator module, stage 2 may still be active. If disposal of a dual stage module is required, both deployment loops must be energized to deploy the air bag.**

9. If you are deploying a steering wheel inflator module, place the inflator module in the center of the space.

10. When deploying an I/P inflator module, perform the following instructions:

a. Place tool J 39401-B, or equivalent in the center of the cleared area.

b. Fill the deployment fixture with water or sand.

c. Using the proper nuts and bolts, mount the I/P module to the deployment fixture, with the vinyl trim facing up.

d. Securely tighten all fasteners that hold the I/P module to the deployment fixture.

When deploying a roof rail module, perform the following instructions:

a. Place the SIR deployment fixture in the center of the cleared area.

b. Fill the deployment fixture with water or sand to provide sufficient stabilization of fixture during deployment.

c. Adjust and secure the fixture arms to the deployment fixture, using the proper nuts and bolts.

d. Attach the roof rail module in the deployment fixture and securely tighten all fasteners.

Inspect the J 38826 and the appropriate pigtail adapter for damage. Replace as needed.

11. Short the 2 SIR deployment harness leads together using one banana plug seated into the other.

12. Connect the appropriate pigtail adapter to the SIR deployment harness.

13. Extend the SIR deployment harness and adapter to the full length from the deployment fixture or area.

Connect the inflator module to the adapter on the SIR deployment harness.

➡**The rapid expansion of gas involved with deploying an inflator module is very loud. Notify all the people in the immediate area that you intend to deploy the inflator module. When the** inflator module deploys, the deployment fixture may jump about 30 cm (1 ft) vertically. This is a normal reaction of the inflator module due to the force of the rapid expansion of gas inside the inflator module. If you are deploying a dual stage inflator module with stage 1 already deployed, the fixture may not move and the noise may have been reduced.

14. Clear the area of people.

15. Place a 12-volt minimum/2-amp minimum power source, such as a vehicle battery, near the shorted end of the harness.

16. Separate the 2 banana plugs on the SIR deployment harness that were shorted together earlier in the procedure.

17. Connect the SIR deployment harness wires to the power source. Deployment of the Inflator module will occur when contact is made.

18. Disconnect the SIR deployment harness from the power source after the inflator module deploys.

19. Seat one banana plug into the other, in order to short the deployment harness leads.

20. If the inflator module did not deploy, disconnect the adapter and discontinue the procedure and contact the Technical Assistance Group. If deployment was successful, proceed to the following steps.

➡**Put on a pair of shop gloves.**

21. Disconnect the pigtail adapter from the inflator module as soon as possible.

22. Inspect the pigtail adapter and the SIR deployment harness. Replace as needed.

23. Dispose of the deployed inflator module through normal refuse channels.

24. Wash your hands with a mild soap.

Deployment Inside Vehicle (Vehicle Scrapping Procedure)

Deploy the inflator modules inside of the vehicle when destroying the vehicle or when salvaging the vehicle for parts. This includes, but is not limited to, the following situations:

- The vehicle has completed all useful life
- Irreparable damage occurred to the vehicle in a non-deployment type accident
- Irreparable damage occurred to the vehicle during a theft
- The vehicle is being salvaged for parts to be used on a vehicle with a different VIN, as opposed to rebuilding as the same VIN

✳✳ CAUTION

After deployment, the metal surfaces of the SIR component may be very hot. To help avoid a fire or personal injury: Allow sufficient time for cooling before touching any metal surface of the SIR component. Do not place the deployed SIR component near any flammable objects.

---

25. Lower the driver and passenger windows.

26. Turn the ignition switch to the OFF position and remove the ignition key.

27. Check that all inflator modules which will be deployed are mounted securely.

28. Put on safety glasses.

29. Remove all loose objects from the front seats.

30. Disable the SIR system.

✳✳ CAUTION

A deployed dual stage inflator module will look the same whether one or both stages were used, always assume a deployed dual stage inflator module has an active stage 2. Improper handling or servicing can activate the inflator module and cause personal injury.

Disconnect the steering wheel module yellow connector from vehicle harness yellow connector.

➡**If the vehicle is equipped with dual stage air bags the steering wheel module and I/P module will each have 4 wires.**

31. Cut the yellow harness connector out of the vehicle, leaving at least 16 cm (6 in) of wire at the connector.

32. Strip 13 mm (0.5 in) of insulation from each of the connector wire leads.

33. Cut two 6.1 m (20 ft) deployment wires from a 0.8 mm (18 gage) or thicker multi-strand wire. Use these wires to fabricate the driver deployment harness.

34. Strip 13 mm (0.5 in) of insulation from both ends of the wires.

35. Twist together one end from each of the wires in order to short the wires. Deployment wires shall remain shorted, and not connected to a power source until you are ready to deploy the inflator module.

36. Twist together the 2 connector wire leads from the high circuits from both stages of the steering wheel module, to one set of deployment wires.

37. Inspect that the 3-wire connection is secure.

---

38. Bend flat the twisted connection.

39. Secure and insulate the 3-wire connection to the deployment harness using electrical tape.

40. Twist together the 2 connector wire leads from the low circuits from both stages of the steering wheel module, to one set of deployment wires.

41. Inspect that the 3-wire connection is secure.

42. Bend flat the twisted connection.

43. Secure and insulate the 3-wire connection to the deployment harness using electrical tape.

44. Connect the deployment harness to the steering wheel module in-line connector.

45. Route the deployment harness out of the driver side of the vehicle.

46. Disconnect the left/driver roof rail yellow harness connector from the vehicle harness connector.

47. Cut the harness connector out of the vehicle, leaving at least 16 cm (6 in) of wire at the connector.

48. Strip 13 mm (0.5 in) of insulation from each of the connector wire leads.

49. Cut two 6.1 m (20 ft) deployment wires from a 0.8 mm (18 gage) or thicker multi-strand wire. These wires will be used to fabricate the roof rail air bag deployment harness.

50. Strip 13 mm (0.5 in) of insulation from both ends of the wires.

51. Twist together one end from each of the wires in order to short the wires.

52. Twist together one connector wire lead to one deployment wire.

53. Bend flat the twisted connection.

54. Secure and insulate the connection using electrical tape.

55. Twist together, bend, and tape the remaining connector wire lead to the remaining deployment wire.

56. Connect the deployment harness to the yellow connector of the left/driver roof rail module.

57. Route the deployment harness out of the driver side of the vehicle.

58. Disconnect the I/P module yellow harness connector from the vehicle harness connector.

➡**If the vehicle is equipped with dual stage air bags the steering wheel module and I/P module will each have 4 wires.**

59. Cut the yellow harness connector out of the vehicle, leaving at least 16 cm (6 in) of wire at the connector.

60. Strip 13 mm (0.5 in) of insulation from each of the connector wire leads.

61. Cut two 6.1 m (20 ft) deployment

wires from a 0.8 mm (18 gage) or thicker multi-strand wire. These wires will be used to fabricate the passenger deployment harness.

62. Strip 13 mm (0.5 in) of insulation from both ends of the wires.

63. Twist together one end from each of the wires in order to short the wires.

64. Twist together the 2 connector wire leads from the high circuits from both stages of the I/P module to one set of deployment wires.

65. Inspect that the 3-wire connection is secure.

66. Bend flat the twisted connection.

67. Secure and insulate the 3-wire connection to the deployment harness using electrical tape.

68. Twist together the 2 connector wire leads from the low circuits from both stages of the I/P module to one set of deployment wires.

69. Inspect that the 3-wire connection is secure.

70. Bend flat the twisted connection.

71. Secure and insulate the 3-wire connection to the deployment harness using electrical tape.

72. Connect the deployment harness to the I/P module in-line connector.

73. Route the deployment harness out of the passenger side of the vehicle.

74. Disconnect the yellow harness connector to the right/passenger roof rail air bag from the vehicle harness connector.

75. Cut the harness connector out of the vehicle, leaving at least 16 cm (6 in) of wire at the connector.

76. Strip 13 mm (0.5 in) of insulation from each of the connector wire leads.

77. Cut two 6.1 m (20 ft) deployment wires from a 0.8 mm (18 gage) or thicker multi-strand wire. These wires will be used to fabricate the roof rail air bag deployment harness.

78. Strip 13 mm (0.5 in) of insulation from both ends of the wires.

79. Twist together one end from each of the wires in order to short the wires.

80. Twist together one connector wire lead to one deployment wire.

81. Bend flat the twisted connection.

82. Secure and insulate the connection using electrical tape.

83. Twist together, bend, and tape the remaining connector wire lead to the remaining deployment wire.

84. Connect the deployment harness to the right/passenger roof rail module yellow connector.

85. Route the deployment harness out of the passenger side of the vehicle.

86. Completely cover the windshield and the front door window openings with a drop cloth.

87. Stretch to the full length all of the deployment harness wires on the passenger side of the vehicle.

88. Deploy each deployment loop one at a time.

89. Place a power source, 12-volt minimum/2-amp minimum, such as a vehicle battery, near the shorted end of the harnesses.

90. Separate one set of wires and touch the wire ends to the power source in order to deploy the selected inflator module.

91. Disconnect the deployment harness from the power source and twist the wire ends together.

92. Continue the same process with the remaining deployment harnesses.

93. Stretch to the full length all of the deployment harness wires on the driver side of the vehicle.

94. Deploy each deployment loop one at a time.

95. Place a power source, 12-volt minimum/2-amp minimum, such as a vehicle battery, near the shorted end of the harnesses.

96. Separate one set of wires and touch the wires ends to the power source in order to deploy the selected inflator modules.

97. Disconnect the deployment harness from the power source and twist the wire ends together.

98. Continue the same process with the remaining deployment harnesses.

99. Remove the drop cloth from the vehicle.

100. Disconnect all harnesses from the vehicle.

101. Discard the harnesses.

102. Scrap the vehicle in the same manner as a non-SIR equipped vehicle.

DISABLING AND ENABLING THE SYSTEM

Zone 1

DISABLING PROCEDURE

1. Before servicing the vehicle, refer to the Precautions Section.

2. Turn the steering wheel so that the vehicles wheels are pointing straight ahead.

3. Turn the ignition switch to the OFF position.

4. Remove the key from the ignition switch.

5. Locate the body control module fuse center. Remove the fuse cover.

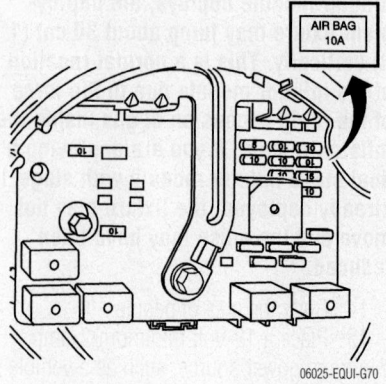

Air Bag fuse

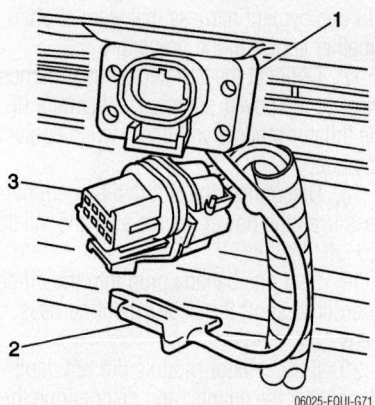

EFS connector

→With the SIR Fuse removed and the ignition ON, the AIR BAG indicator illuminates. This is normal operation, and does not indicate an SIR system malfunction.

6. Locate and remove the AIR BAG fuse from the body control module fuse center.

7. Open the front hood. Locate the front end sensor, also known as electronic frontal sensor (EFS) (1).

8. Remove the connector position assurance (CPA) (2) from the EFS connector (3).

9. Remove the EFS connector (3) from the EFS (1).

ENABLING PROCEDURE

1. Before servicing the vehicle, refer to the Precautions Section.

2. Remove the key from the ignition switch.

3. Connect the EFS connector (3) to the EFS (1).

4. Install the CPA (2) into the EFS connector (3).

5. Install the AIR BAG fuse and fuse center cover.

6. Use caution while reaching in and turn the ignition switch to the ON position. The AIR BAG indicator will flash then turn OFF.

Zone 2

DISABLING PROCEDURE

1. Before servicing the vehicle, refer to the Precautions Section.

2. Turn the steering wheel so that the vehicles wheels are pointing straight ahead.

3. Turn the ignition switch to the OFF position.

4. Remove the key from the ignition switch.

5. Locate the body control module fuse center. Remove fuse center cover.

➡ **With the AIR BAG fuse removed and the ignition switch in the ON position, the AIR BAG warning indicator illuminates. This is normal operation, and does not indicate an SIR system malfunction.**

6. Locate and remove the AIR BAG fuse from the body control module fuse center.

7. To disable the roof rail module—left, perform steps 7-13. To disable both the side impact sensor (SIS)—LF and the seat belt pretensioner—LF, perform steps 14-18.

8. Remove the upper rear window molding.

9. Remove both the left and right rear corner garnish moldings.

10. Remove the rear headliner push-in retainers.

11. Remove the rear coat hooks.

12. Gently pull down the left rear corner of the headliner to access the roof rail module—left connector.

13. Remove the connector position assurance (CPA) from the roof rail module—left connector.

14. Disconnect the roof rail module—left connector from the vehicle harness connector.

15. Remove the lower center pillar trim.

16. Remove the CPA from the seat belt pretensioner.

17. Disconnect the seat belt pretensioner.

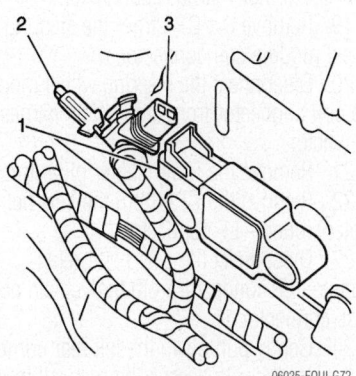

06025-EQUI-G72

LF connector; RF similar

18. Remove the CPA (2) from the SIS—LF connector.

19. Disconnect the SIS—LF (1) from the vehicle harness connector (3).

ENABLING PROCEDURE

1. Before servicing the vehicle, refer to the Precautions Section.

2. Remove the key from the ignition switch.

3. To enable the roof rail module—left, continue to step 3. To enable both the side impact sensor (SIS)—LF and the seat belt pretensioner—LF, continue to step 9.

4. Connect the roof rail module—left connector to the vehicle harness connector.

5. Install the CPA to the roof rail module—left connector.

6. Install the rear coat hooks.

7. Install the rear headliner push-in retainers.

8. Install both the left and right rear corner garnish moldings.

9. Install the upper rear window molding.

10. Connect the seat belt pretensioner.

11. Connect the SIS—LF (1) to the vehicle harness connector (3) and install the CPA (2).

12. Install the lower center pillar trim.

13. Install the AIR BAG fuse into the body control module fuse center.

14. Install the body control module fuse center cover.

15. Use caution while reaching in and turn the ignition switch to the ON position. The AIR BAG indicator will flash then turn OFF.

Zone 3

DISABLING PROCEDURE

1. Before servicing the vehicle, refer to the Precautions Section.

2. Turn the steering wheel so that the vehicles wheels are pointing straight ahead.

3. Turn the ignition switch to the OFF position.

4. Remove the key from the ignition switch.

5. Locate the body control module fuse center then remove the fuse center cover.

➡ **With the AIR BAG fuse removed and the ignition switch in the ON position, the AIR BAG warning indicator illuminates. This is normal operation, and does not indicate an SIR system malfunction.**

6. Locate and remove the AIR BAG fuse from the body control module fuse center.

7. Remove the connector position assurance (CPA) from the steering wheel module coil connector.

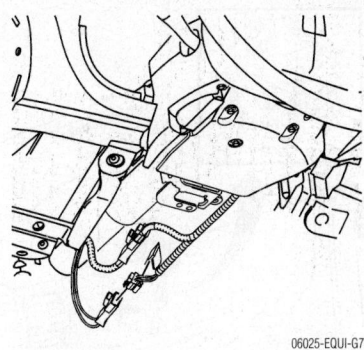

06025-EQUI-G73

Connector position assurance (CPA) at the steering column

8. Disconnect the steering wheel module coil connector from the vehicle harness connector.

ENABLING PROCEDURE

1. Before servicing the vehicle, refer to the Precautions Section.

2. Remove the key from the ignition switch.

3. Connect the steering wheel module coil connector to the vehicle harness connector.

4. Install the CPA to the steering wheel module coil connector.

5. Install the AIR BAG fuse into the body control module fuse center.

6. Install the body control module fuse center cover.

7. Use caution while reaching in and turn the ignition switch to the ON position. The AIR BAG indicator will flash then turn OFF.

Zone 5

DISABLING PROCEDURE

1. Before servicing the vehicle, refer to the Precautions Section.

2. Turn the steering wheel so that the vehicles wheels are pointing straight ahead.

3. Turn the ignition switch to the OFF position.

4. Remove the key from the ignition switch.

5. Locate the body control module fuse center, then remove the fuse center cover.

➡ **With the AIR BAG fuse removed and the ignition switch in the ON position, the AIR BAG warning indicator illuminates. This is normal operation, and does not indicate an SIR system malfunction.**

6. Locate and remove the AIR BAG fuse from the body control module fuse center.

7. Remove the connector position assurance (CPA) from the I/P module connector.

Connector position assurance (CPA) at the I/P module

8. Disconnect the I/P module connector from the vehicle harness connector.

ENABLING PROCEDURE

1. Before servicing the vehicle, refer to the Precautions Section.
2. Remove the key from the ignition switch.
3. Connect the I/P module connector to the vehicle harness connector.
4. Install the CPA to the I/P module connector.
5. Install the AIR BAG fuse into the body control module fuse center.
6. Install the body control module fuse center cover.
7. Use caution while reaching in and turn the ignition switch to the ON position. The AIR BAG indicator will flash then turn OFF.

Zone 6

DISABLING PROCEDURE

1. Before servicing the vehicle, refer to the Precautions Section.
2. Turn the steering wheel so that the vehicles wheels are pointing straight ahead.
3. Turn the ignition switch to the OFF position.
4. Remove the key from the ignition switch.
5. Locate the body control module fuse center. Remove the fuse center cover.

➡ **With the AIR BAG fuse removed and the ignition switch in the ON position, the AIR BAG warning indicator illuminates. This is normal operation, and does not indicate an SIR system malfunction.**

6. Locate and remove the AIR BAG fuse from the body control module fuse center.
7. To disable the roof rail module—right, perform steps 7-13. To disable both the side impact sensor (SIS)—RF and the seat belt pretensioner—RF, perform steps 14-18.

8. Remove the upper rear window molding.
9. Remove both the left and right rear corner garnish moldings.
10. Remove the rear headliner push-in retainers.
11. Remove the rear coat hooks.
12. Gently pull down the right rear corner of the headliner to access the roof rail module—right connector.
13. Remove the connector position assurance (CPA) from the roof rail module—right connector.
14. Disconnect the roof rail module—right connector from the vehicle harness connector.
15. Remove the lower center pillar trim.
16. Remove the CPA from the seat belt pretensioner—RF connector.
17. Disconnect the seat belt pretensioner—RF connector from the vehicle harness connector.
18. Remove the CPA (2) from the SIS—RF connector.
19. Disconnect the SIS—RF (1) from the vehicle harness connector (3).

ENABLING PROCEDURE

1. Before servicing the vehicle, refer to the Precautions Section.
2. Remove the key from the ignition switch.
3. To enable the roof rail module—right, continue to step 3. To enable both the side impact sensor (SIS)—RF and the seat belt pretensioner—RF, continue to step 9.
4. Connect the roof rail module—right connector to the vehicle harness connector.
5. Install the CPA to the roof rail module—right connector.
6. Install the rear coat hooks.
7. Install the rear headliner push-in retainers.
8. Install both the left and right rear corner garnish moldings.
9. Install the upper rear window molding.
10. Connect the seat belt pretensioner—RF connector to the vehicle harness connector and install the CPA.
11. Connect the SIS—RF (1) to the vehicle harness connector (3) and install the CPA (2).
12. Install the lower center pillar trim.
13. Install the AIR BAG fuse into the body control module fuse center.
14. Install the body control module fuse center cover.
15. Use caution while reaching in and turn the ignition switch to the ON position. The AIR BAG indicator will flash then turn OFF.

Zone 8

DISABLING PROCEDURE

1. Before servicing the vehicle, refer to the Precautions Section.
2. Turn the steering wheel so that the vehicles wheels are pointing straight ahead.
3. Turn the ignition switch to the OFF position.
4. Remove the key from the ignition switch.
5. Locate the body control module fuse center. Remove the fuse center cover.

➡ **With the AIR BAG fuse removed and the ignition switch in the ON position, the AIR BAG warning indicator illuminates. This is normal operation, and does not indicate an SIR system malfunction.**

6. Locate and remove the AIR BAG fuse from the body control module fuse center.
7. Remove the upper rear window molding.
8. Remove both the left and right rear corner garnish moldings.
9. Remove the rear headliner push-in retainers.
10. Remove the rear coat hooks.
11. Gently pull down the right rear corner to access the roof rail module—right connector.
12. Remove the connector position assurance (CPA) from the roof rail module—right connector.
13. Disconnect the roof rail module—right connector from the vehicle harness connector.
14. Remove the lower center pillar trim.
15. Remove the CPA from the seat belt pretensioner—RF connector.
16. Disconnect the seat belt pretensioner—RF connector from the vehicle harness connector.
17. Remove the CPA from the I/P module connector.
18. Disconnect the I/P module connector from the vehicle harness connector.
19. Remove the CPA from the steering wheel module coil connector.
20. Disconnect the steering wheel module coil connector from the vehicle harness connector.
21. Remove the lower center pillar trim.
22. Remove the CPA from the seat belt pretensioner—LF connector.
23. Disconnect the seat belt pretensioner—LF connector from the vehicle harness connector.
24. Gently pull down the left rear corner of the headliner to access the roof rail module—left connector.

25. Remove the CPA from the roof rail module—left connector.

26. Disconnect the roof rail module—left connector from the vehicle harness connector.

ENABLING PROCEDURE

1. Before servicing the vehicle, refer to the Precautions Section.

2. Remove the key from the ignition switch.

3. Connect the steering wheel module coil connector to the vehicle harness connector.

4. Install the CPA to the steering wheel module coil connector.

5. Connect the seat belt pretensioner—LF connector to the vehicle harness connector and install the CPA.

6. Install the lower center pillar trim.

7. Connect the roof rail module—left connector to the vehicle harness connector.

8. Install the CPA to the roof rail module—left connector.

9. Connect the I/P module connector to the vehicle harness connector.

10. Install the CPA to the I/P module connector.

11. Connect the seat belt pretensioner—RF connector to the vehicle harness connector and install the CPA.

12. Install the lower center pillar trim.

13. Connect the roof rail module—right connector to the vehicle harness connector.

14. Install the CPA to the roof rail module—right connector.

15. Install the rear coat hooks.

16. Install the rear headliner push-in retainers.

17. Install both the left and right rear corner garnish moldings.

18. Install the upper rear window molding.

19. Install the AIR BAG fuse into the body control module fuse center.

20. Install the body control module fuse center cover.

21. Use caution while reaching in and turn the ignition switch to the ON position. The AIR BAG indicator will flash then turn OFF.

Power Steering Gear

REMOVAL & INSTALLATION

1. Before servicing the vehicle, refer to the Precautions Section.

2. Raise and support the vehicle.

3. Remove the front tires.

4. Remove both outer tie rod to steering knuckle nuts. Discard the nuts.

✵✵ WARNING

Do not free the ball stud by using a pickle fork or a wedge-type tool. Damage to the seal or bushing may result.

➡**Hold the ball stud to prevent turning during removal of the nut.**

5. Separate the tie rods from the steering knuckles.

6. Rotate the intermediate steering shaft in order to gain access to the intermediate shaft pinch bolt.

7. Remove the intermediate to steering gear pinch bolt. Discard the bolt.

8. Disconnect the intermediate shaft from the steering gear.

9. Disconnect the stabilizer links from the stabilizer bar.

10. Remove the steering gear to cradle mounting bolts.

11. Remove the steering gear through the right side of the vehicle.

12. With heat shield equipped steering gears, remove the heat shield. Save for installation.

To install:

13. If applicable, install the heat shield.

➡**Ensure the stabilizer is swung in the uppermost position for gear clearance.**

14. Install the steering gear from the right side of the vehicle.

15. Center the gear mounting bushings into the cradle supports.

16. Hand start both steering gear to cradle mounting bolts. Tighten the bolts to 110 Nm (81 ft. lbs.).

17. Connect the intermediate shaft to the steering gear and install a new pinch bolt. Tighten the intermediate pinch bolt to 34 Nm (25 ft. lbs.).

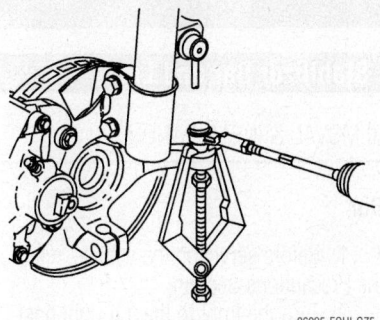

06025-EQUI-G75

Separate the tie rods from the steering knuckles

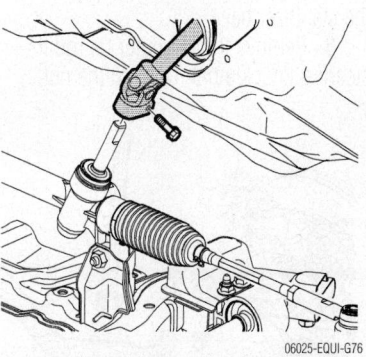

06025-EQUI-G76

Remove the intermediate to steering gear pinch bolt

18. Connect the stabilizer links to the stabilizer bar.

➡**Hold the ball stud to prevent turning during installation of the nut.**

19. Connect the tie rod to the knuckle and install a new nut. Tighten the nut to 60 Nm (44 ft. lbs.).

20. Install the front tires and wheels.

21. Check the front wheel alignment and align as necessary.

22. Lower the vehicle.

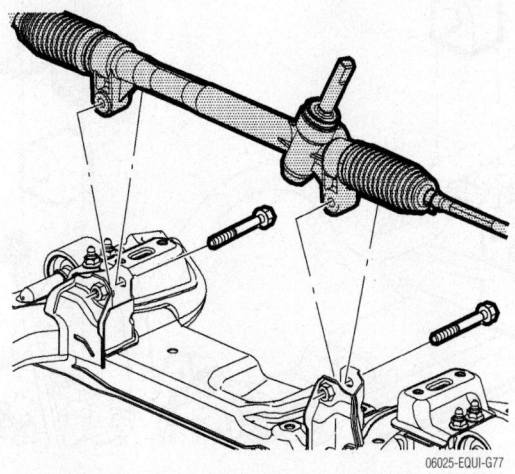

06025-EQUI-G77

Steering gear mounting bolts

FRONT SUSPENSION

Stabilizer Bar and Links

REMOVAL & INSTALLATION

Bar

1. Before servicing the vehicle, refer to the Precautions Section.

2. Turn the front to the full right position.

3. Raise and support the vehicle.

4. Remove the front tire and wheels.

5. Disconnect the stabilizer link from the stabilizer bar.

6. Remove the left outer tie rod to steering knuckle nut. Discard the nut.

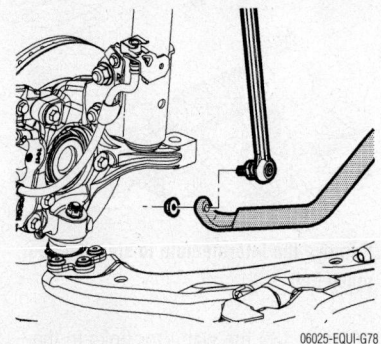

06025-EQUI-G78

Disconnect the front stabilizer link from the stabilizer bar

7. Separate the outer tie rod from the steering knuckle.

8. Remove the stabilizer bar clamp to cradle bolts.

9. Remove the stabilizer bar clamps and bushings from the stabilizer bar.

➡**Take care not to catch the transmission shift cable or left wheel house plastic trim when removing the stabilizer bar.**

10. Remove the stabilizer bar from the vehicle through the left wheel opening.

To install:

➡**Take care not to catch the transmission shift cable or left wheel house plastic trim when installing the stabilizer bar.**

11. Install the stabilizer bar to the vehicle through the left wheel opening.

12. Install the stabilizer bar clamps and bushings to the stabilizer bar.

13. Install the stabilizer bar clamp bolts. Tighten the bolts to 50 Nm (37 ft. lbs.).

14. Inspect the stabilizer link boots for damage and replace the stabilizer link if needed.

➡**Hold the ball stud when tightening the nut.**

15. Connect the stabilizer links to the stabilizer bar. Do not allow the boot to twist.

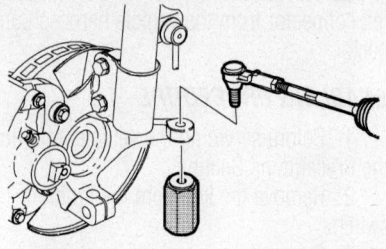

06025-EQUI-G80

Use tool J 44015, or equivalent to seat the ball stud taper

Tighten the bar to link nut to 65 Nm (48 ft. lbs.).

16. Connect the left outer tie rod to the steering knuckle.

17. Use tool J 44015 to seat the ball stud taper to 40 Nm (30 ft. lbs.).

18. Install a new tie rod retention nut. Tighten the nut to 50 Nm (37 ft. lbs.).

19. Install the front tire and wheels.

20. Lower the vehicle.

Links

1. Before servicing the vehicle, refer to the Precautions Section.

2. Raise and support the vehicle.

3. Remove the tire and wheel.

➡**Do not allow the stabilizer link ball stud to rotate while removing the link nut.**

4. Remove the stabilizer link to strut nut.

➡**Do not allow the stabilizer link ball stud to rotate while removing the link nut.**

5. Remove the stabilizer link to stabilizer bar nut.

6. Remove the stabilizer link.

To install:

➡**Do not allow the stabilizer link ball stud to rotate while installing the link nut.**

7. Connect the stabilizer link to the strut. Tighten the nut to 65 Nm (48 ft. lbs.).

➡**Do not allow the stabilizer link ball stud to rotate while installing the link nut.**

8. Connect the stabilizer link to the stabilizer bar. Tighten the nut to 65 Nm (48 ft. lbs.).

9. Install the tire and wheel.

10. Lower the vehicle.

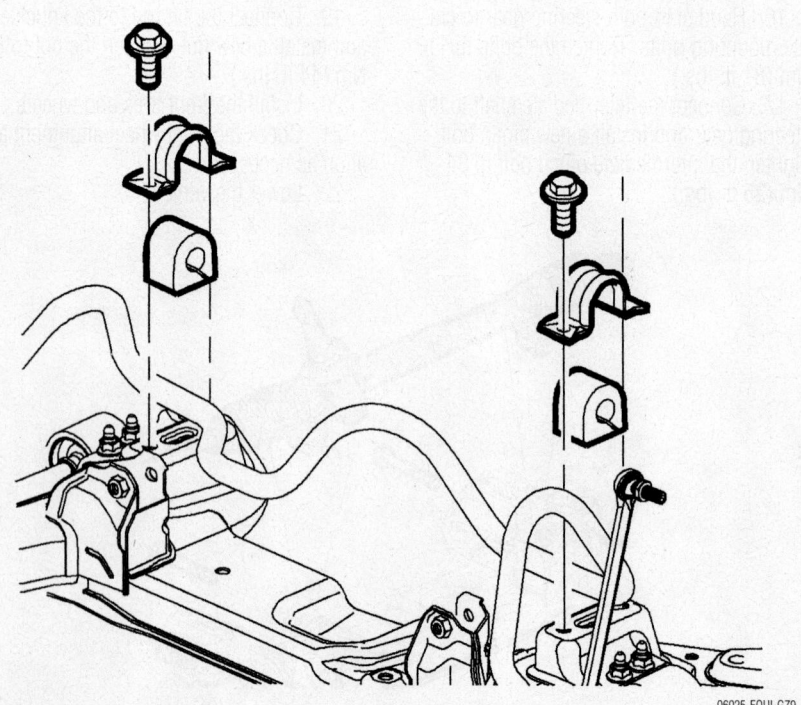

06025-EQUI-G79

Front stabilizer bar clamp to cradle bolts

Strut

REMOVAL & INSTALLATION

1. Before servicing the vehicle, refer to the Precautions Section.

2. Raise and support the vehicle.

3. Remove the strut assembly to body fasteners.

4. Remove the wheel and tire.

5. Remove the brake hose bracket from the strut assembly.

6. Loosen, do not remove the strut to knuckle bolts and nuts.

7. Disconnect the stabilizer link from the strut assembly.

8. Remove the strut to knuckle bolts and nuts. Discard the bolts and nuts.

9. Remove the strut assembly from the vehicle.

To install:

10. Install the strut assembly to the vehicle. Tighten the strut to body bolts to 25 Nm (18 lb ft).

11. Attach the strut to the steering knuckle using new bolts and nuts. Tighten the bolts and nuts to 180 Nm (133 lb ft).

12. Inspect the stabilizer link seals for damage and replace the link as necessary.

➡**Do not allow the stabilizer link ball stud to rotate while installing the link nut.**

13. Connect the stabilizer link to the strut. Tighten the nut to 65 Nm (48 lb ft).

14. Install the brake hose bracket to the strut assembly. Tighten the brake bracket bolt to 15 Nm (11 lb ft).

15. Install the wheel and tire.

16. Lower the vehicle.

17. Perform a wheel alignment.

DISASSEMBLY

1. Before servicing the vehicle, refer to the Precautions Section.

2. Install the strut assembly in tool J 45400 using the following procedure:

a. Adjust the lower legs of the tool to the lowest possible coil of the spring.

b. Adjust the upper legs of the tool to the highest possible coil of the spring.

c. Inspect the strut assembly to insure hooks on the strut compress legs are properly installed on the spring coils.

d. Verify the strut assembly is parallel with the tool.

3. Compress the spring enough to unload the upper strut mount.

✳✳ WARNING

Do not allow the front strut stud to rotate during disassembly/reassembly. Use hand tools to keep the strut stud from rotating. If air tools are used, and the stud is allowed to rotate, damage to the strut may occur.

4. Remove the strut shaft nut.

➡**Leave the spring in the spring compressor.**

5. Lower the strut from the spring assembly.

✳✳ WARNING

Do not handle the top mount assembly by the plastic portion. Handle the top mount assembly by the metal portion when removing/installing the top mount from/to the strut assembly. Holding the top mount assembly by the plastic portion may loosen the snap fit of the bearing components and cause the bearing to fall apart.

6. Remove the upper mount assembly, inspect for damage and deterioration. Replace as necessary.

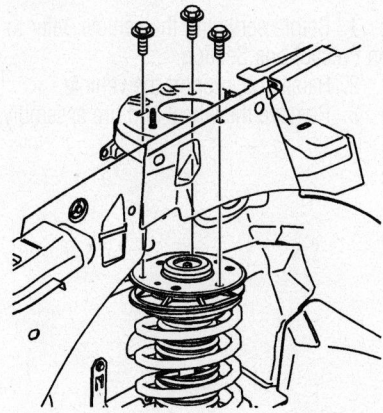

06025-EQUI-G83

Upper strut attaching bolts

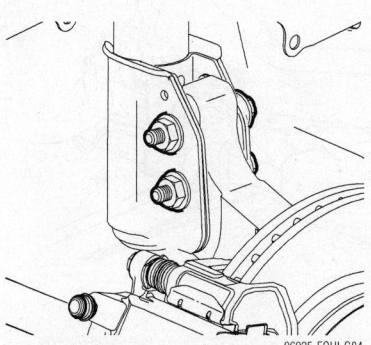

06025-EQUI-G84

Lower strut attaching bolts

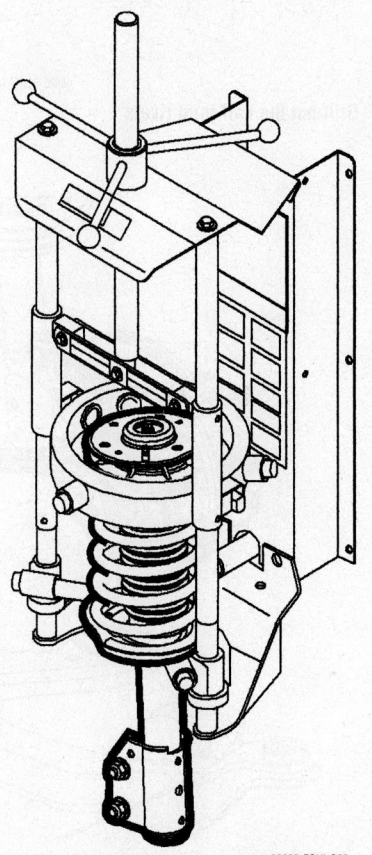

06025-EQUI-G85

Strut mounted in the J45400 fixture

06025-EQUI-G86

Strut components

7. Remove the strut dust shield and inspect for damage and deterioration. Replace as necessary.

8. Remove the hollow bumper from the strut shaft and inspect for damage and deterioration. Replace as necessary.

9. Inspect the spring for damage. Replace as necessary.

ASSEMBLY

1. Before servicing the vehicle, refer to the Precautions Section.

2. Extend the strut to its limit of travel.

3. Install the hollow bumper and dust boot to the strut shaft.

➡The tag identifying the spring will be closer to the bottom of the spring. The end of the cold sits up against the tab on the spring seat.

4. With the spring in the compressor, install the strut into the spring.

➡The anti-rotation tab on the spring seat must face 180 degrees from the direction that the knuckle bracket points.

5. Assemble the upper spring seat onto the strut shaft and align the flat with the strut to knuckle mounting bracket.

✳✳ WARNING

Do not handle the top mount assembly by the plastic portion. Handle the top mount assembly by the metal portion when removing/installing the top mount from/to the strut assembly. Holding the top mount assembly by the plastic portion may loosen the snap fit of the bearing components and cause the bearing to fall apart.

➡The flat on the metal plate of the top mount assembly must face the same direction of the anti-rotation tab on the spring seat.

6. Assemble the top mount onto the strut shaft and align the flat 180 degrees from flat on the upper spring seat.

7. Loosely install the strut shaft nut.

✳✳ WARNING

Do not allow the front strut stud to rotate during disassembly/reassembly. Use hand tools to keep the strut stud from rotating. If air tools are used, and the stud is allowed to rotate, damage to the strut may occur.

8. Hold the strut shaft and tighten the shaft while verifying that the upper spring seat flats align with the top mount. Tighten the strut shaft to 75 Nm (55 ft. lbs).

9. Release the tension on the tool.

10. Remove the strut assembly from the tool.

Lower Ball Joint

REMOVAL & INSTALLATION

1. Before servicing the vehicle, refer to the Precautions Section.

2. Remove the lower control arm.

3. Place the control arm in a vise or suitable holding device.

4. Remove the ball joint rivets using the following procedure:

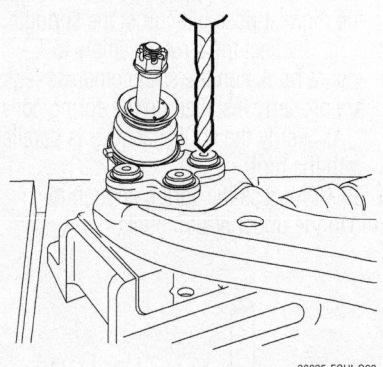

06025-EQUI-G93

Drill out the ball joint rivets

a. Drill through the rivets using an 8 mm (5/16 in.) drill bit.

b. Enlarge the hole using a 12 mm (31/64 in.) drill bit.

c. Remove any remaining burs from the control arm.

5. Remove the ball joint from the control arm. Note the position of the ball joint for reassembly.

To install:

➡**The control arm must be clean and free of debris.**

6. Install the ball joint to the control arm as previously noted.

✳✳ WARNING

Only use hardware provided with the new ball joint. The bolts must be installed with the bolt head on top of the ball joint.

7. Install the ball joint to control arm bolts. Tighten the bolts and nuts to 68 Nm (50 ft. lbs.).

8. Install the lower control arm.

Lower Control Arm

REMOVAL AND & INSTALLATION

1. Before servicing the vehicle, refer to the Precautions Section.

2. Raise and support the vehicle.

3. Remove the wheel and tire assembly.

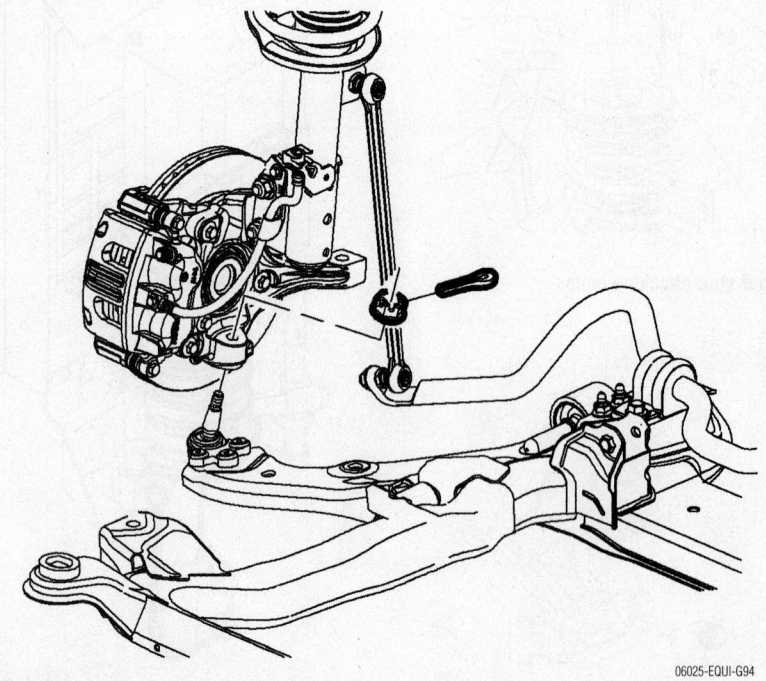

06025-EQUI-G94

Remove the control arm ball stud cotter pin

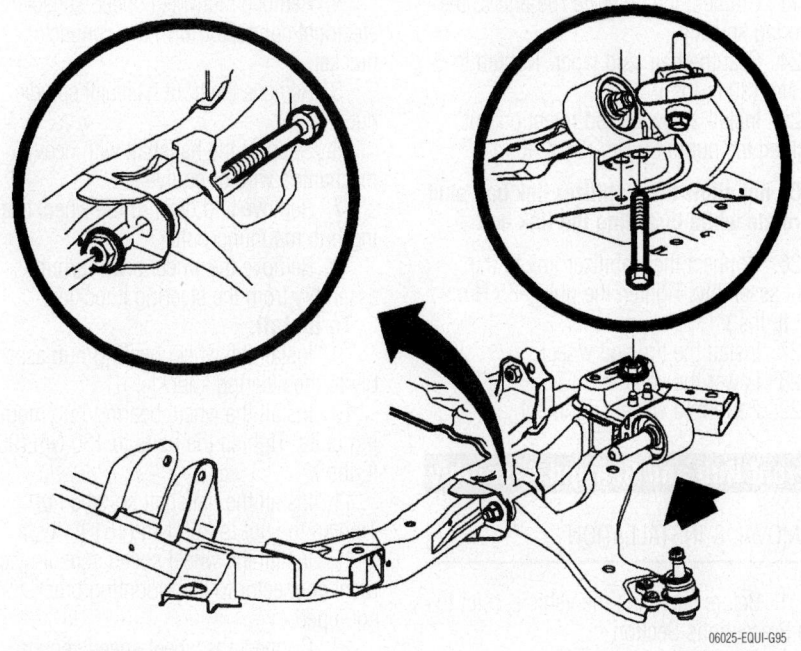

Control arm removal

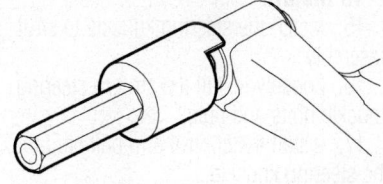

Press out the front control arm bushing

4. Remove the control arm ball stud cotter pin. Discard the cotter pin.

5. Loosen the ball stud nut until the nut is level with the top of the ball stud.

6. Separate the lower control arm from the steering knuckle.

7. Remove the ball stud nut.

8. Remove the control arm-to-frame front bolt and nut. Discard the bolt and nut.

9. Remove the control arm-to-frame rear bolts and nuts. Discard the bolts and nuts.

10. Remove the control arm.

To install:

11. Install the control arm to the frame.

12. Install new control arm-to-frame rear bolts and nuts. Tighten the nuts to 70 Nm (52 ft. lbs.).

13. Install a new arm-to-frame front bolt and nut. Install new control arm-to-frame rear bolts and nuts. Tighten the bolt to 200 Nm (148 ft. lbs.).

✳✳ WARNING

There are 2 different ball studs used on this vehicle. The ball stud type must be identified in order to use the correct torque value when tightening the ball stud nut. If the wrong torque is used, damage could occur to the ball stud.

14. Install the control arm ball stud into the steering knuckle.

15. If the bottom of the ball stud has a cup and is silver, tighten the nut to 60 Nm (44 ft. lbs.). If the bottom of the ball stud is

flat and black, tighten the nut to 40 Nm (30 ft. lbs.).

✳✳ WARNING

Do not loosen the castle nut, only tighten to align the ball stud slot. Ensure that the cotter pin ends do not contact the antilock brake system (ABS) sensor harness or drive axle.

16. Continue to tighten the nut only enough to align the castle nut slots with the ball stud, install the cotter pin.

17. Install the wheel and tire assembly.

18. Lower the vehicle.

LOWER CONTROL ARM BUSHING REPLACEMENT

1. Before servicing the vehicle, refer to the Precautions Section.

2. Raise and support the vehicle.

3. Remove the front tire and wheel assembly.

4. Remove the lower control arm.

5. Place the control arm in a vise or suitable holding device.

6. Press out the front control arm bushing.

To install:

7. Press in the front control arm bushing.

8. Install the lower control arm.

9. Install the front tire and wheel assembly.

10. Lower the vehicle.

Knuckle

REMOVAL & INSTALLATION

1. Before servicing the vehicle, refer to the Precautions Section.

2. Raise and support the vehicle.

3. Remove the tire and wheel.

➡**Do not allow the stabilizer link ball stud to rotate while removing the link nut.**

4. Disconnect the stabilizer link from the strut assembly.

5. Loosen the steering knuckle to strut bolts and nuts.

6. Remove the wheel bearing/hub assembly.

7. Remove and discard the lower ball joint cotter pin.

8. Loosen the ball stud nut, until level with the top of the ball stud.

9. Separate the lower control arm from the steering knuckle.

10. Remove the lower control arm and nut from the steering knuckle.

11. Remove the outer tie rod end to knuckle nut.

12. Separate the outer tie rod from the steering knuckle.

13. Remove the steering knuckle to strut bolts and nuts. Discard the bolts and nuts.

14. Remove the steering knuckle from the vehicle.

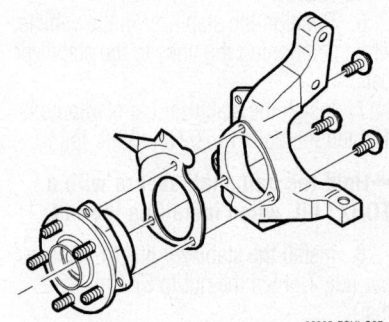

Hub and knuckle assembly

To install:

15. Install the steering knuckle to strut assembly.

16. Loosely install the strut to steering knuckle bolts and nuts.

17. Install the control arm ball stud into the steering knuckle.

18. Install the ball stud nut. If the bottom of the ball stud has a cup and is silver, tighten the nut to 60 Nm (44 ft. lbs.). If the bottom of the ball stud is flat and black, tighten the nut to 40 Nm (30 ft. lbs.).

19. Tighten the strut to steering knuckle bolts and nuts. Tighten the bolts and nuts to 180 Nm (133 ft. lbs.).

➡**Do not loosen the castle nut for cotter pin installation.**

20. Tighten the castle nut enough to allow for cotter pin installation.

➡**The cotter pin must not contact the wheel speed sensor or drive axle.**

21. Install the cotter pin.

22. Install the wheel bearing/hub assembly.

23. Connect the outer tie rod end to the steering knuckle.

24. Seat the ball stud taper. Tighten to 40 Nm (30 ft. lbs.).

25. Install a new tie rod retention nut. Tighten the nut to 50 Nm (37 ft. lbs.).

➡**Do not allow the stabilizer link ball stud to rotate while installing the link nut.**

26. Connect the stabilizer link to the strut assembly. Tighten the nut to 65 Nm (48 ft. lbs.).

27. Install the tire and wheel.

28. Lower the vehicle.

29. Perform a wheel alignment.

Front Hub and Bearing Assembly

REMOVAL & INSTALLATION

1. Before servicing the vehicle, refer to the Precautions Section.

2. Remove the front brake rotor.

3. Disconnect the wheel speed sensor electrical connector, if equipped.

4. Remove the wheel speed sensor electrical connector form the connector bracket.

5. Remove the front halfshaft spindle nut.

6. Support the halfshaft with heavy mechanic's wire or equivalent.

7. Remove and discard the wheel bearing/hub mounting bolts.

8. Remove the wheel bearing/hub assembly from the steering knuckle.

To install:

9. Install the wheel bearing/hub assembly to the steering knuckle.

10. Install the wheel bearing/hub mounting bolts. Tighten the bolts to 130 Nm (96 ft. lbs.).

11. Install the halfshaft spindle nut. Tighten the nut to 205 Nm (151 ft. lbs.).

12. Install the wheel speed sensor electrical connector to the mounting bracket, if equipped.

13. Connect the wheel speed sensor electrical connector.

14. Install the front brake rotor.

REAR SUSPENSION

Stabilizer Bar and Links

REMOVAL & INSTALLATION

Bar

1. Before servicing the vehicle, refer to the Precautions Section.

2. Raise and support the vehicle.

➡**Hold the ball shaft secure with a TORX® bit, when removing the nut.**

3. Remove the stabilizer link to stabilizer bar nut.

4. Remove the stabilizer bar clamp bolts.

5. Disengage the stabilizer bar from the stabilizer link ball studs, while removing the stabilizer bar from the vehicle.

To install

6. Position the stabilizer in the vehicle, while positioning the links to the stabilizer bar.

7. Install the stabilizer bar clamp bolts. Tighten the bolts to 70 Nm (52 ft. lbs.).

➡**Hold the ball shaft secure with a TORX® bit, when installing the nut.**

8. Install the stabilizer link to stabilizer bar nut. Tighten the nut to 57 Nm (42 ft. lbs.).

9. Lower the vehicle.

Links

1. Before servicing the vehicle, refer to the Precautions Section.

2. Raise and support the vehicle.

3. Loosen the stabilizer bar clamp bolts.

➡**Hold the ball shaft secure with a TORX® bit, when removing the nut.**

4. Remove the stabilizer link to stabilizer bar nut.

➡**When disconnecting the stabilizer link, hold the link with a wrench to prevent turning.**

5. Remove the stabilizer link to control arm nut.

6. Remove the stabilizer link from the vehicle.

To install:

7. Position the stabilizer link through the control arm.

➡**When connecting the stabilizer link, hold the link with a wrench to prevent turning.**

8. Install the stabilizer link to control arm nut. Tighten the nut to 15 Nm (11 ft. lbs.).

➡**Hold the ball shaft secure with a TORX® bit, when installing the nut.**

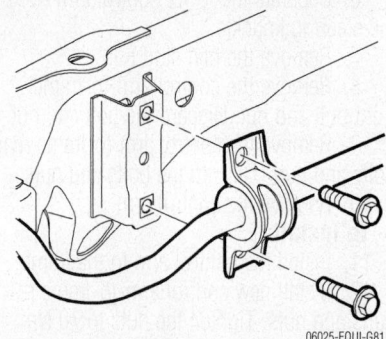

Rear stabilizer bar clamp bolts

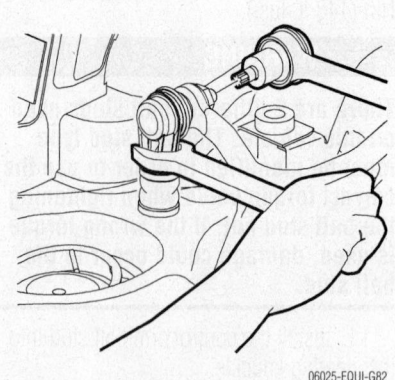

Hold the ball shaft secure with a TORX® bit, when removing the nut

9. Install the stabilizer link to stabilizer bar nut. Tighten the nut to 57 Nm (42 ft. lbs.).

10. Install the stabilizer bar clamp bolts. Tighten the bolts to 70 Nm (52 ft. lbs.).

11. Lower the vehicle.

Shock Absorber

REMOVAL & INSTALLATION

1. Before servicing the vehicle, refer to the Precautions Section.

2. Raise and support the vehicle.

3. Remove the lower shock bolt.

4. For right side only, remove splash shield.

5. Remove the upper shock bolt.

6. Remove the shock from the vehicle.

To install:

7. Install the shock to the vehicle.

8. Install the upper shock bolt. Tighten the bolt to 110 Nm (81 ft. lbs.).

9. For right side only, install splash shield

10. Install the lower shock bolt. Tighten the bolt to 110 Nm (81 ft. lbs.).

11. Lower the vehicle.

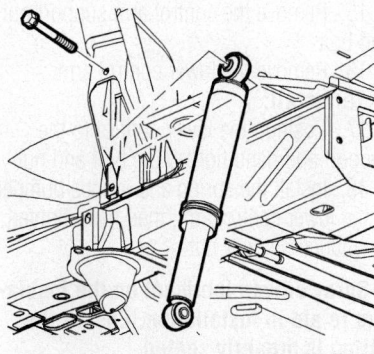

06025-EQUI-G90

Shock absorber upper bolt

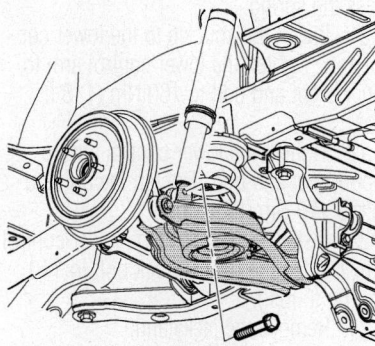

06025-EQUI-G91

Shock absorber lower bolt

Coil Spring

REMOVAL & INSTALLATION

1. Before servicing the vehicle, refer to the Precautions Section.

2. Raise and support the vehicle.

3. Remove the tire and wheel.

➡ **Hold the link with a wrench during nut removal.**

4. Remove the stabilizer link to lower control arm nut.

5. Remove the trailing arm bracket to underbody bolts.

6. Place a screw-type jackstand under the lower control arm.

7. Using the jackstand, compress the coil spring.

8. Remove the lower shock bolt.

9. Loosen the lower control arm to support frame bolt.

10. Remove the lower control arm to knuckle nut and bolt.

11. Lower the control arm in order to unload the coil spring.

12. Remove the coil spring.

To install:

13. Fully seat the top and bottom coil spring insulators to the spring.

➡ **Spray silicon lubricant on the insulators to aid in installation. Ensure that part number identification tape located on the coil spring is oriented outboard of the vehicle and at the top of the spring.**

14. Position the spring with the rubber insulators into the vehicle.

15. Use a screw-type jackstand to compress the spring.

16. Install the knuckle to the lower control arm. Tighten the lower control arm to knuckle bolt to 160 Nm (118 ft. lbs.).

17. Tighten the lower control arm to

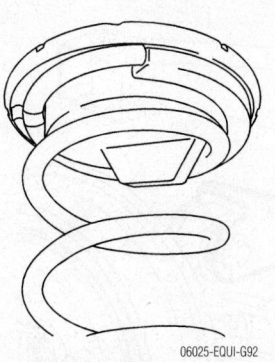

06025-EQUI-G92

Fully seat the top and bottom coil spring insulators to the spring

support nut and bolt. Tighten the bolt to 110 Nm (81 ft. lbs.).

18. Install the shock to the lower control arm. Tighten the lower shock bolt to 110 Nm (81 ft. lbs.).

19. Remove the jackstand.

➡ **Hold the link with a wrench during nut installation.**

20. Install the stabilizer link to the lower control arm. Tighten the nut to 15 Nm (11 ft. lbs.).

21. Push the trailing arm upward to align the front bracket to body bolt.

22. Use a drift to aid in bracket alignment and install the remaining bolts. Tighten the bracket to body bolts to 110 Nm (81 ft. lbs.).

23. Install the tire and wheel.

24. Lower the vehicle.

25. Check the rear alignment.

Upper Control Arm

REMOVAL & INSTALLATION

1. Before servicing the vehicle, refer to the Precautions Section.

2. Raise and support the vehicle.

3. Remove the trailing arm bracket to body bolts.

4. If applicable, remove the antilock brake system (ABS) brake harness from the upper control arm.

5. Remove the upper control arm to knuckle nut and bolt.

6. Remove the upper control to support nut and bolt.

7. Remove the upper control arm.

To install

8. Install the upper control arm to the knuckle.

9. Loosely install the upper control arm to knuckle nut and bolt.

10. Install the upper control to support bolt and cam nut.

11. Tighten the upper control arm to knuckle nut and bolt. Tighten the nut and bolt to 160 Nm (118 ft. lbs.).

12. Tighten the upper control arm to support bolt. Tighten the upper control arm to support bolt to 160 Nm (118 ft. lbs.).

13. If applicable, install the ABS harness to the upper control arm.

14. Push upward on the trailing arm and loosely install the front bolt.

15. Use a drift to align the remaining bolts. Tighten the trailing arm bracket to body bolts to 110 Nm (81 ft. lbs.).

16. Lower the vehicle.

17. Check the rear alignment.

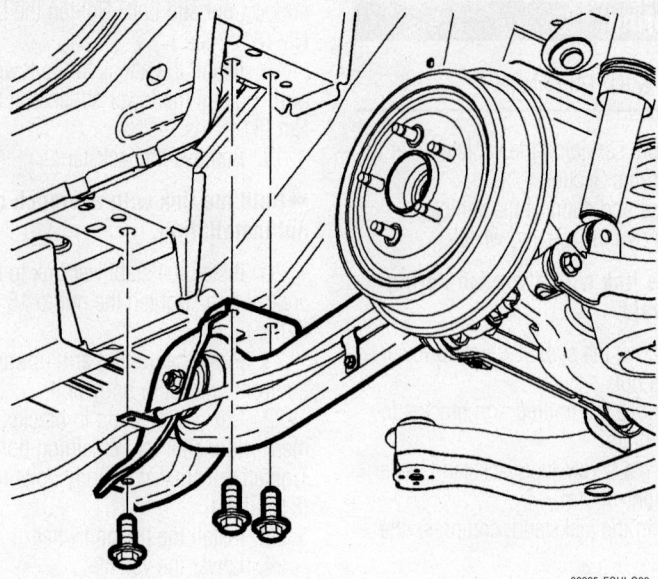

Trailing arm bracket to body bolts

06025-EQUI-G98

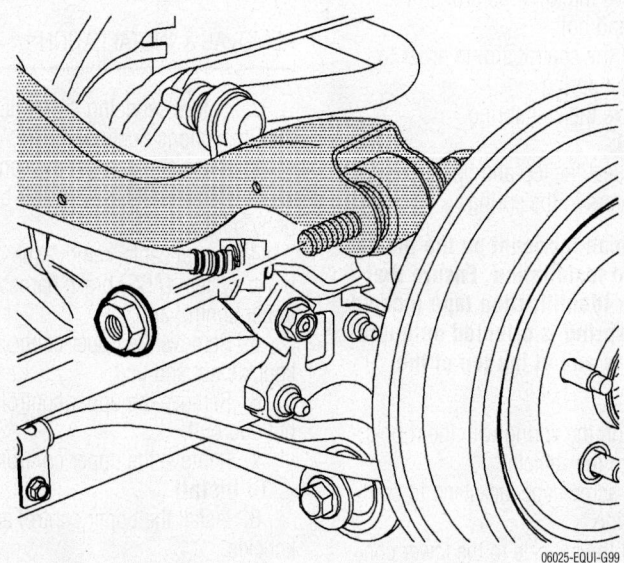

Upper control arm to knuckle nut and bolt

06025-EQUI-G99

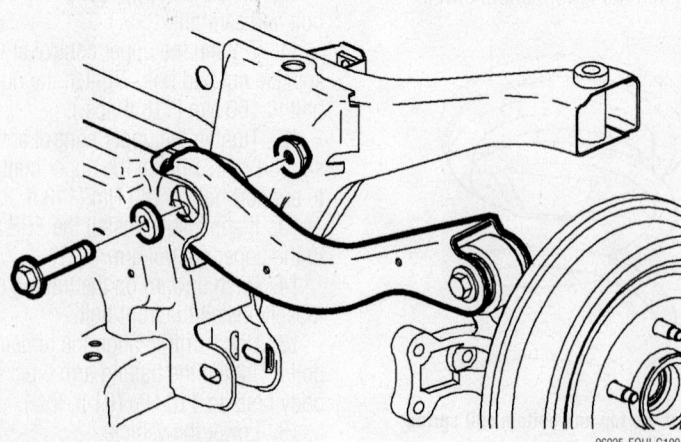

Upper control to support nut and bolt

06025-EQUI-G100

Lower Control Arm

REMOVAL & INSTALLATION

1. Before servicing the vehicle, refer to the Precautions Section.
2. Raise and support the vehicle.
3. Remove the tire and wheel.

➡**Hold the link with a wrench during nut removal.**

4. Remove the stabilizer link to lower control arm nut.
5. Remove the trailing arm bracket to underbody bolts.
6. Place a screw type jackstand under the lower control arm.
7. Using the jackstand, compress the coil spring.
8. Remove the lower shock bolt.
9. Remove the jounce bumper nut at the lower control arm.
10. Loosen the lower control arm to support the frame bolt.
11. Remove the lower control arm to knuckle nut and bolt.
12. Lower the control arm in order to unload the coil spring.
13. Remove the coil spring.
14. Remove the jounce bumper.
15. Remove the control arm support nut and bolt.
16. Remove the lower control arm.

To install:

17. Position the lower control to the support and hand tighten the bolt and nut.
18. Install the spring and jounce bumper to the lower control arm; then hand tighten the jounce bumper nut.

➡**Spray silicon lubricant on the insulators to aid in installation. Ensure the spring is properly seated.**

19. Position the spring with the rubber insulators into the vehicle.
20. Use a screw type jack stand to compress the spring.
21. Install the knuckle to the lower control arm. Tighten the lower control arm to knuckle nut and bolt to 160 Nm (118 ft. lbs.).
22. Tighten the lower control arm to support nut and bolt. Tighten the nut and bolt to 110 Nm (81 ft. lbs.).
23. Install the shock to the lower control arm. Tighten the lower shock bolt to 110 Nm (81 ft. lbs.).
24. Remove the jackstand.
25. Tighten the jounce bumper to the lower control arm nut. Tighten the nut to 63 Nm (46 ft. lbs.).

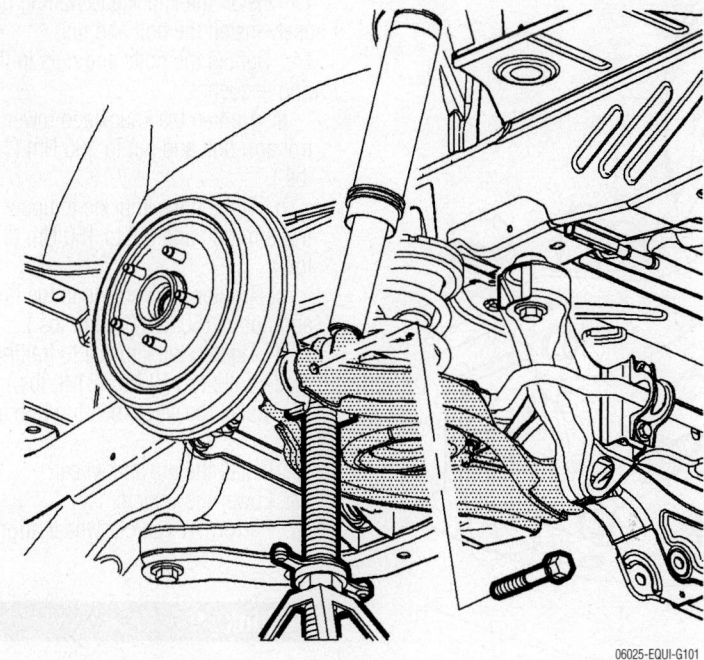

Remove the lower control arm to knuckle nut and bolt

06025-EQUI-G101

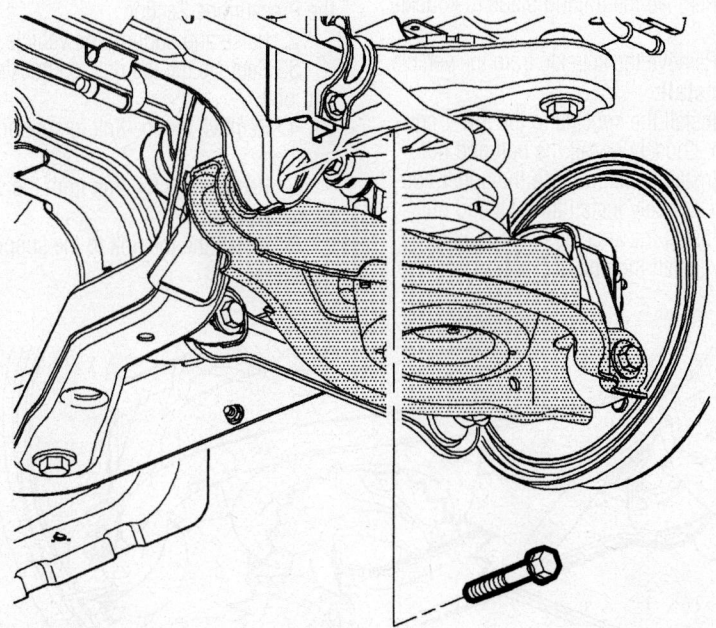

Remove the control arm support nut and bolt

06025-EQUI-G102

➡**Hold the link with a wrench during nut installation.**

26. Install the stabilizer link to the lower control arm. Tighten the nut to 15 Nm (11 ft. lbs.).

27. Push the trailing arm upward to align the front bracket to the body bolt.

28. Use a drift to aid in bracket alignment and install the remaining bolts. Tighten the bracket to body bolts to 110 Nm (81 ft. lbs.).

29. Install the tire and wheel.
30. Lower the vehicle.
31. Check the rear alignment.

BUSHING REPLACEMENT

1. Before servicing the vehicle, refer to the Precautions Section.
2. Remove the lower control arm.
3. Using tool J 45097, or equivalent,

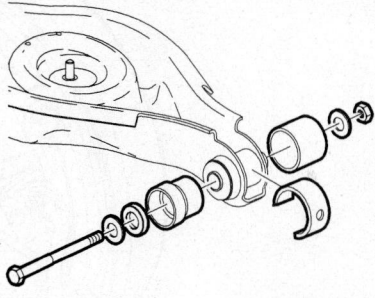

06025-EQUI-G103

Removing the rear lower control arm bushing

remove the bushing in the direction of the arrow.

4. Install the push out socket against the bushing from the flanged side of the control arm.

5. Install the through bolt with the washer and bearing against the push out socket.

6. Install a backing socket against the control arm, opposite of the flange.

7. Install the stabilizer between the control arm ears.

8. Install flat washer and nut.

➡**Apply high pressure lube to the threads of the tool.**

9. Tighten the nut until the bushing is removed.

To install:

➡**Apply high pressure lube to the threads of the tool.**

10. Install a new bushing from the opposite direction.

11. Using tool J 45097, reverse the tool layout from the removal procedure.

12. Tighten the nut until the bushing is fully installed.

13. Install the lower control arm.

Trailing Arm

REMOVAL & INSTALLATION

1. Before servicing the vehicle, refer to the Precautions Section.
2. Raise and support the vehicle.
3. Remove the trailing arm bracket to body bolts.
4. Remove the trailing arm bushing to bracket nut and bolt.
5. Remove the park brake cable clip from the trailing arm.
6. Remove the trailing arm to knuckle bolts.
7. Remove the trailing arm.

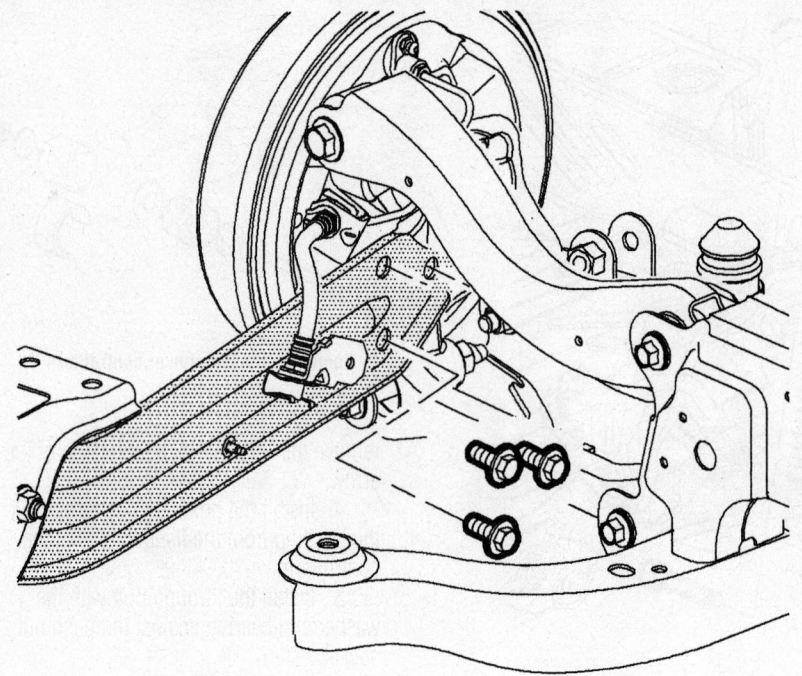

06025-EQUI-G104

Trailing arm-to-knuckle bolts

To install:

8. Install the trailing arm to the knuckle. Tighten the bolts to 110 Nm (81 ft. lbs.).

9. Position the trailing arm bracket to the trailing arm.

10. Loosely install the trailing arm bushing to bracket nut and bolt.

11. Push upward on the trailing arm and loosely install the front bolt.

12. Use a drift to align the remaining bolts. Tighten the trailing arm bracket to body bolts to 110 Nm (81 ft. lbs.).

13. Tighten the trailing arm bushing to bracket nut and bolt. Tighten the bolt to 160 Nm (118 ft. lbs.).

14. Install the park brake cable clip. Tighten the bolt to 25 Nm (18 ft. lbs.).

15. Lower the vehicle.

Knuckle

REMOVAL & INSTALLATION

1. Before servicing the vehicle, refer to the Precautions Section.

2. Raise and support the vehicle.

3. Remove the tire and wheel.

4. Remove the wheel bearing/hub assembly.

5. Remove the upper control arm to knuckle bolt and nut.

6. Remove the lower control arm to knuckle bolt and nut.

7. Remove the toe link to knuckle bolt and nut.

8. Remove the trailing blade to knuckle bolts.

9. Remove the knuckle from the vehicle.

To install:

10. Install the knuckle to the lower control arm. Loosely install the bolt and nut.

11. Install the knuckle to the upper control arm. Loosely install the bolt and nut.

12. Install the knuckle to the toe link. Loosely install the bolt and nut.

13. Install the knuckle to trailing blade. Loosely install the bolt and nut.

14. Tighten the bolts and nuts in the following sequence:

    a. Tighten the knuckle to lower control arm bolt and nut to 160 Nm (118 ft. lbs.).

    b. Tighten the knuckle to upper control arm bolt and nut to 160 Nm (118 ft. lbs.).

    c. Tighten the knuckle to toe link bolt and nut to 160 Nm (118 ft. lbs.).

    d. Tighten the knuckle to trailing blade bolts to 110 Nm (81 ft. lbs.).

15. Install the wheel bearing/hub assembly.

16. Install the tire and wheel.

17. Lower the vehicle.

18. Perform a vehicle wheel alignment.

Toe Link

REMOVAL & INSTALLATION

1. Before servicing the vehicle, refer to the Precautions Section.

2. Raise and support the vehicle.

3. Remove the toe link to knuckle nut and bolt.

4. Remove the toe link to support nut and bolt.

5. Remove the toe link from the vehicle.

To install:

6. Install the toe link to the support assembly.

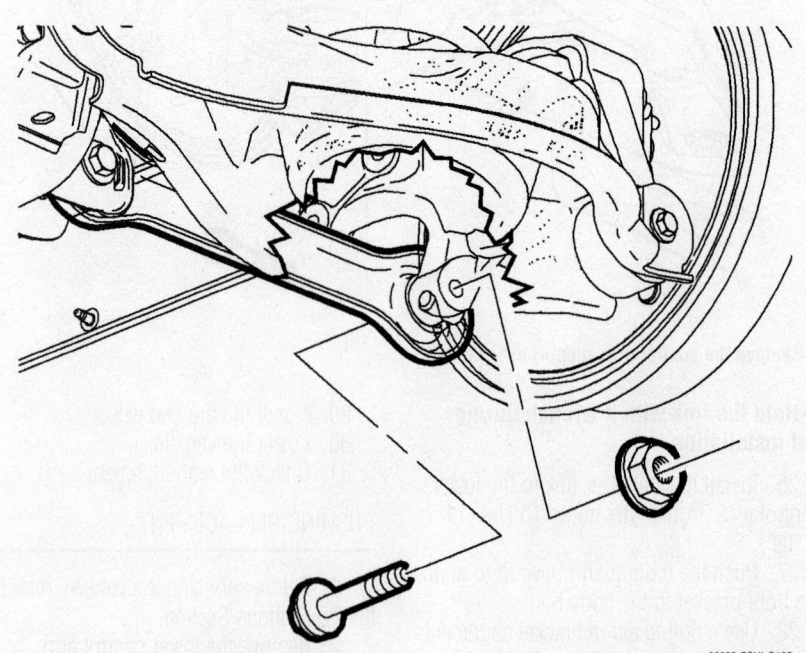

06025-EQUI-G105

Toe link-to-knuckle bolt/nut

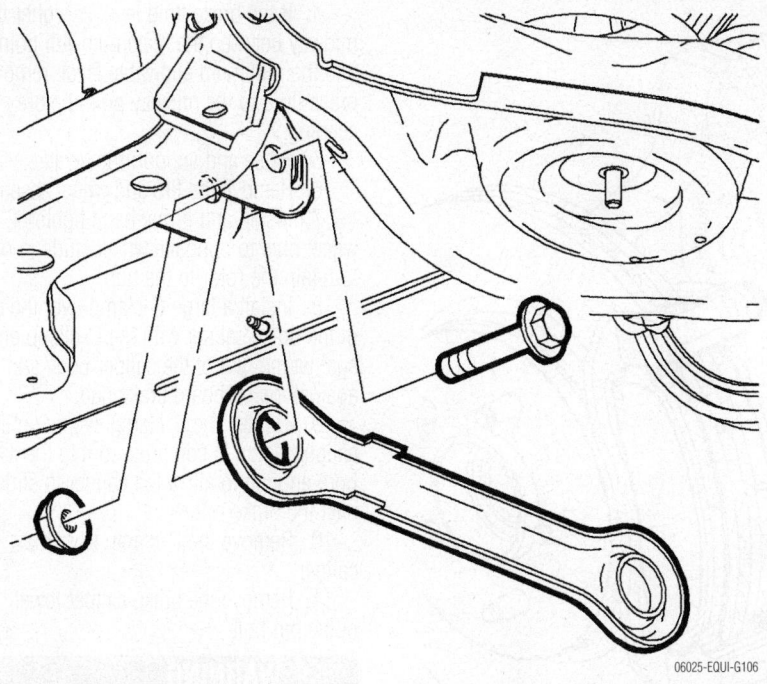

06025-EQUI-G106

Toe link-to-support bolt/nut

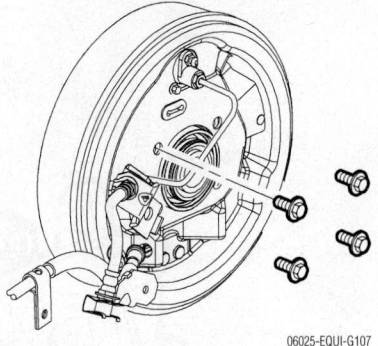

06025-EQUI-G107

Rear hub/bearing bolts

➡**Install the bolt with the head towards the front of the vehicle. Position the cam nut in same position as in the upper control arm.**

7. Install the toe link to support nut and bolt.

8. Install the toe link to the knuckle. Tighten the bolt to 160 Nm (118 ft. lbs.).

9. Tighten the toe link to support bolt. Tighten the bolt to 160 Nm (118 ft. lbs.).

10. Lower the vehicle.

11. Check the rear alignment.

Rear Hub and Bearing Assembly

REMOVAL & INSTALLATION

1. Before servicing the vehicle, refer to the Precautions Section.

2. Remove the rear brake drum.

3. On vehicles with all-wheel drive, remove the halfshaft spindle nut.

4. Disconnect the wheel speed sensor electrical connector, if equipped.

➡**Do not damage the halfshaft joint seal.**

5. Support the halfshaft with heavy mechanic's wire, or equivalent.

6. Remove the wheel bearing/hub mounting bolts.

7. Remove the wheel bearing/hub assembly from the suspension knuckle.

To install:

8. Install the wheel bearing/hub assembly to the steering knuckle.

9. Install the wheel bearing/hub mounting bolts. Tighten the bolts to 84 Nm (62 ft. lbs.).

10. On all-wheel drive vehicles, install the halfshaft spindle nut. Tighten the nut to 125 Nm (92 ft. lbs.).

11. If equipped, route the wheel speed sensor electrical harness through the backing plate and seat the grommet.

12. Connect the wheel speed sensor electrical connector.

13. Install the rear brake drum.

DISC BRAKES

※※ CAUTION

Do not move the vehicle until a firm brake pedal is obtained. Failure to obtain a firm pedal before moving vehicle may result in personal injury.

Brake Caliper

REMOVAL & INSTALLATION

1. Before servicing the vehicle, refer to the Precautions Section.

2. Inspect the fluid level in the brake master cylinder reservoir.

3. If the brake fluid level is midway between the maximum-full point and the minimum allowable level, no brake fluid needs to be removed from the reservoir before proceeding.

4. If the brake fluid level is higher than midway between the maximum-full point and the minimum allowable level, remove brake fluid to the midway point before proceeding.

5. Raise and support the vehicle.

6. Remove the tire and wheel assembly.

7. Install and firmly hand tighten 2 wheel nuts to opposite wheel studs in order to retain the rotor to the hub.

8. Install a large C-clamp over the body of the brake caliper with the C-clamp ends against the rear of the caliper body and against the outer brake pad.

9. Tighten the C-clamp until the caliper piston is compressed into the caliper bore enough to allow the caliper to slide past the brake rotor.

10. Remove the C-clamp from the caliper.

11. Remove the brake hose-to-caliper bolt from the brake caliper.

12. Remove the brake hose from the brake caliper.

13. Remove and discard the 2 copper brake hose gaskets. These gaskets may be stuck to the brake caliper and/or the brake hose end.

14. Cap or plug the opening in the brake caliper and the brake hose to prevent fluid loss and contamination.

15. Remove the brake caliper guide pin bolts.

16. Remove the brake caliper from the caliper bracket.

17. Inspect the brake caliper guide pins for freedom of movement, and inspect the condition of the guide pin boots. Move the guide pins inboard and outboard within the bracket bores, without disengaging the

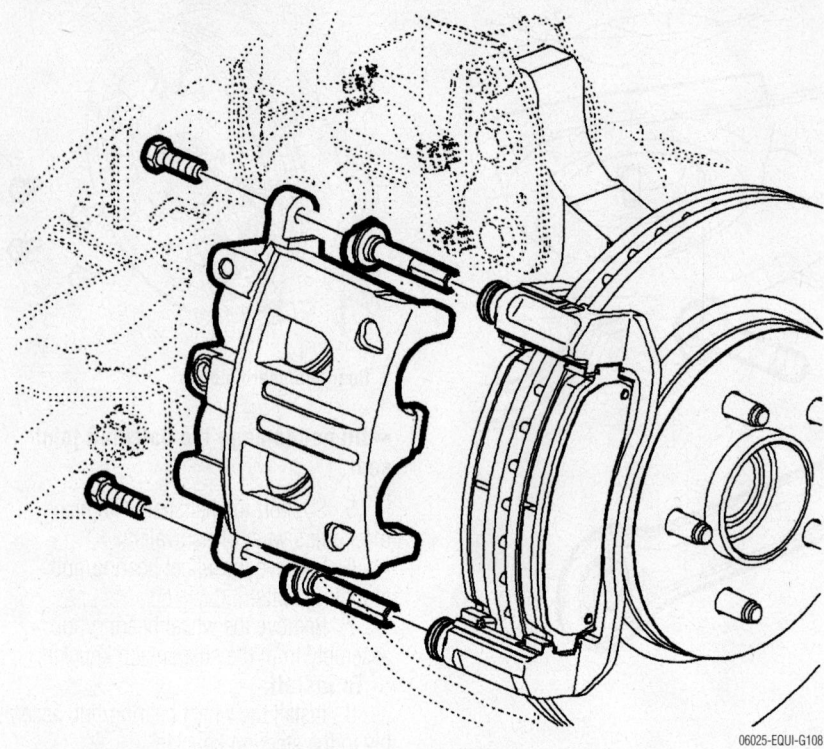

06025-EQUI-G108

Caliper mounting

slides from the boots, and observe for the following:

- Restricted caliper guide pin movement
- Looseness in the brake caliper mounting bracket
- Seized or binding caliper guide pins
- Split or torn boots

If any of the conditions listed are found, the brake caliper guide pins and/or boots require replacement.

To install:

18. Apply a light, thin coat of high temperature silicone brake lubricant to the caliper guide pins.

19. Install the guide pins to the brake caliper bracket.

20. Install the brake caliper to the brake caliper bracket.

21. Install the brake caliper guide pin bolts. Tighten the bolts to 44 Nm (32 ft. lbs.).

22. Remove the caps or plugs from the brake caliper opening and the brake hose.

✳✳ WARNING

Do not reuse the copper brake hose gaskets.

23. Install NEW copper brake hose gaskets to the brake hose-to-caliper bolt and to the brake hose.

24. Install the brake hose and the brake hose-to-brake caliper bolt to the brake caliper. Tighten the bolt to 44 Nm (32 ft. lbs.).

25. Bleed the hydraulic brake system.

26. Remove the wheel nuts retaining the brake rotor to the wheel hub.

27. Install the tire and wheel assembly.

28. Lower the vehicle.

29. With the engine OFF, gradually apply the brake pedal to approximately ⅔ of its travel distance.

30. Slowly release the brake pedal.

31. Wait 15 seconds, then gradually apply the brake pedal approximately ⅔ of its travel distance again until a firm brake pedal apply is obtained. This will properly seat the brake caliper pistons and brake pads.

Disc Brake Pads

REMOVAL AND INSTALLATION

1. Before servicing the vehicle, refer to the Precautions Section.

2. Inspect the fluid level in the brake master cylinder auxiliary reservoir.

3. If the brake fluid level is midway between the maximum-full point and the minimum allowable level, no brake fluid needs to be removed from the reservoir before proceeding.

4. If the brake fluid level is higher than midway between the maximum-full point and the minimum allowable level, remove brake fluid to the midway point before proceeding.

5. Raise and support the vehicle.

6. Remove the tire and wheel assembly.

7. Install and firmly hand tighten 2 wheel nuts to opposite wheel studs in order to retain the rotor to the hub.

8. Install a large C-clamp over the body of the brake caliper with the C-clamp ends against the rear of the caliper body and against the outboard brake pad.

9. Tighten the C-clamp evenly until the caliper piston is compressed into the caliper bore enough to allow the caliper to slide past the brake rotor.

10. Remove the C-clamp from the caliper.

11. Remove the brake caliper lower guide pin bolt.

✳✳ WARNING

Support the brake caliper with heavy mechanic's wire, or equivalent, whenever it is separated from its mount and the hydraulic flexible brake hose is still connected. Failure to support the caliper in this manner will cause the flexible brake hose to bear the weight of the caliper, which may cause damage to the brake hose and in turn may cause a brake fluid leak.

12. Without disconnecting the hydraulic brake flexible hose, pivot the caliper upward and secure the caliper with heavy mechanics wire, or equivalent.

13. Remove the brake pads from the caliper mounting bracket.

14. Remove the brake pad retainers from the caliper bracket.

15. Thoroughly clean the brake pad hardware mating surfaces of the caliper bracket, of any debris and corrosion.

16. Inspect the brake caliper guide pins for freedom of movement, and inspect the condition of the guide pin boots. Move the guide pins inboard and outboard within the bracket bores, without disengaging the slides from the boots, and observe for the following:

- Restricted caliper guide pin movement
- Looseness in the brake caliper mounting bracket
- Seized or binding caliper guide pins
- Split or torn boots

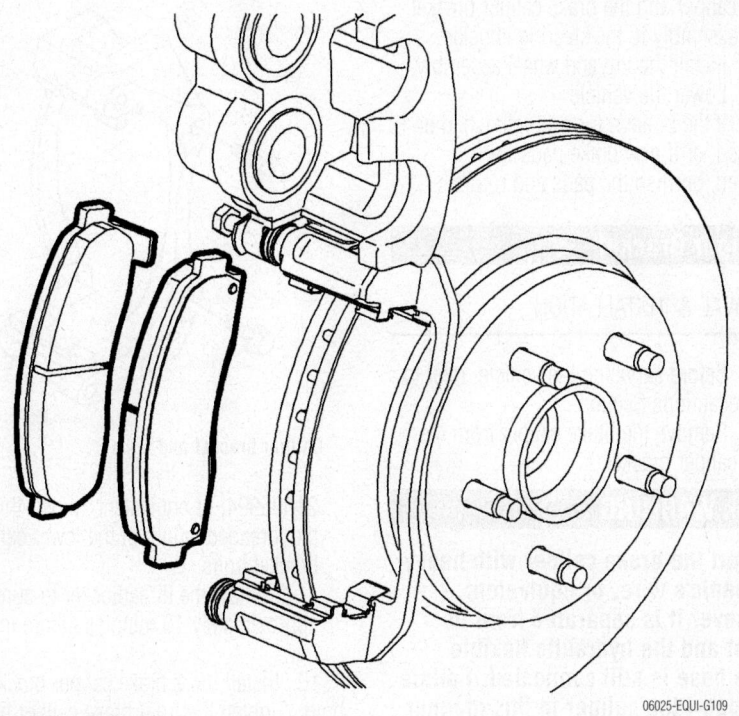

Disc brake pad removal

06025-EQUI-G109

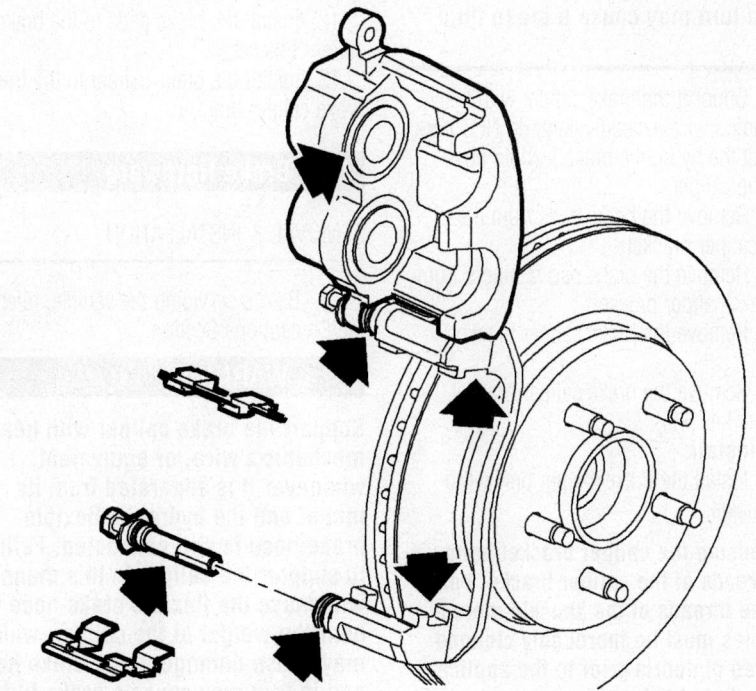

Brake pad retainers

06025-EQUI-G110

If any of the conditions listed are found, the brake caliper guide pins and/or boots require replacement.

To install:

17. Install a large C-clamp over the body of the brake caliper, with the C-clamp ends against the rear of the caliper body and against an old inboard brake pad or a wood block installed against the caliper piston.

18. Tighten the C-clamp evenly until the caliper piston is compressed completely into the caliper bore.

19. Remove the C-clamp and the old brake pad or wood block from the caliper.

20. Apply a very thin coating of high temperature silicone brake lubricant to the pad hardware mating surfaces of the caliper bracket only.

21. Install the brake pad retainers to the brake caliper bracket.

⁂ WARNING

The wear sensor equipped disc brake pad must be mounted inboard of the rotor with the leading edge of the sensor facing the brake rotor during forward wheel rotation, or at the top of the pad when installed in vehicle position.

22. Install the brake pads to the caliper bracket.

23. Remove the support, and rotate the brake caliper into position over the disc brake pads and to the caliper mounting bracket.

24. Install the lower brake caliper guide pin bolt. Tighten the bolt to 44 Nm (32 ft. lbs.).

25. Remove the wheel nuts retaining the brake rotor to the hub.

26. Install the tire and wheel assembly.

27. Lower the vehicle.

28. With the engine OFF, gradually apply the brake pedal approximately ⅔ of its travel distance.

29. Slowly release the brake pedal.

30. Wait 15 seconds, then gradually apply the brake pedal approximately ⅔ of its travel distance again until a firm brake pedal apply is obtained. This will properly seat the brake caliper pistons and brake pads.

31. Fill the master cylinder auxiliary reservoir to the proper level.

32. Burnish the pads and rotors.

Rotor

REMOVAL & INSTALLATION

1. Before servicing the vehicle, refer to the Precautions Section.

2. Raise and support the vehicle.

3. Remove the tire and wheel assembly.

4. Install a C-clamp over the body of the brake caliper, with the C-clamp ends against the rear of the caliper body and the outboard disc brake pad.

5. Tighten the C-clamp until the caliper piston is compressed into the caliper bore enough to allow the caliper to slide past the brake rotor.

6. Remove the C-clamp.

✳✳ WARNING

Support the brake caliper with heavy mechanic's wire, or equivalent, whenever it is separated from its mount and the hydraulic flexible brake hose is still connected. Failure to support the caliper in this manner will cause the flexible brake hose to bear the weight of the caliper, which may cause damage to the brake hose and in turn may cause a brake fluid leak.

➡ **Do NOT disconnect the hydraulic brake flexible hose from the caliper.**

7. Remove the brake caliper and the caliper mounting bracket as an assembly from the steering knuckle and support the assembly with heavy mechanic's wire, or equivalent. Ensure that there is no tension on the hydraulic brake flexible hose.

8. Matchmark the position of the brake rotor to the wheel studs.

9. Remove the brake rotor.

To install:

✳✳ WARNING

Whenever the brake rotor has been separated from the hub/axle flange, any rust or contaminants should be cleaned from the hub/axle flange and the brake rotor mating surfaces. Failure to do this may result in excessive assembled lateral runout (LRO) of the brake rotor, which could lead to brake pulsation.

10. Thoroughly clean any rust or corrosion from the mating surface of the hub/axle flange.

11. Thoroughly clean any rust or corrosion from the mating surface and mounting surface of the brake rotor.

12. Inspect the mating surfaces of the hub/axle flange and the rotor to ensure that there are no foreign particles or debris remaining.

13. Install the brake rotor to the hub/axle flange. Use the matchmark made prior to removal for proper orientation to the flange.

14. If the brake rotor was removed and installed as part of a brake system repair, measure the assembled LRO of the brake rotor to ensure optimum performance of the disc brakes.

15. If the brake rotor assembled LRO measurement exceeds the specification, bring the LRO to within specifications.

16. Remove the support, and install the brake caliper and the brake caliper bracket as an assembly to the steering knuckle.

17. Install the tire and wheel assembly.

18. Lower the vehicle.

19. If the brake rotor was refinished or replaced, or if new brake pads were installed, burnish the pads and rotors.

Caliper Bracket

REMOVAL & INSTALLATION

1. Before servicing the vehicle, refer to the Precautions Section.

2. Remove the brake caliper from the brake caliper bracket.

✳✳ WARNING

Support the brake caliper with heavy mechanic's wire, or equivalent, whenever it is separated from its mount and the hydraulic flexible brake hose is still connected. Failure to support the caliper in this manner will cause the flexible brake hose to bear the weight of the caliper, which may cause damage to the brake hose and in turn may cause a brake fluid leak.

3. Support the brake caliper with heavy mechanic's wire, or equivalent; do NOT disconnect the hydraulic brake flexible hose from the caliper.

4. Remove the brake pads from the brake caliper bracket.

5. Remove the brake pad retainers from the brake caliper bracket.

6. Remove the brake caliper bracket bolts.

7. Remove the brake caliper bracket from the knuckle.

To install:

8. Install the brake caliper bracket to the knuckle.

➡ **If reusing the caliper bracket bolts, the threads of the caliper bracket bolts and the threads of the knuckle mounting holes must be thoroughly cleaned and free of debris prior to the application of threadlocker.**

9. Prepare the bolts and the threaded holes for assembly:

   a. Thoroughly clean the residue from the bolt threads by using denatured alcohol or equivalent and allow to dry.

   b. Thoroughly clean the residue from the threaded holes by using denatured alcohol or equivalent and allow to dry.

   c. Apply threadlocker Saturn P/N

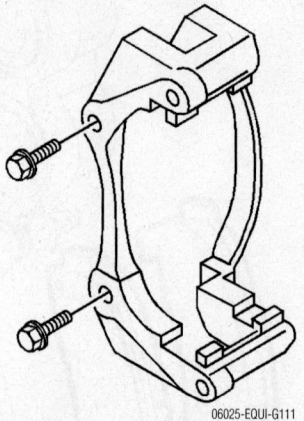

06025-EQUI-G111

Caliper bracket and bolts

21005994, or equivalent, to two-thirds of the threaded length of the lower caliper bracket bolts.

   d. Allow the threadlocker to cure approximately 10 minutes before installation.

10. Install the 2 brake caliper bracket bolts. Tighten the front brake caliper bracket bolts to 185 Nm (136 ft. lbs.).

11. Install the brake pad retainers to the brake caliper bracket.

12. Install the brake pads to the brake caliper bracket.

13. Install the brake caliper to the front brake caliper bracket.

Disc Brake Hardware

REMOVAL & INSTALLATION

1. Before servicing the vehicle, refer to the Precautions Section.

✳✳ WARNING

Support the brake caliper with heavy mechanic's wire, or equivalent, whenever it is separated from its mount and the hydraulic flexible brake hose is still connected. Failure to support the caliper in this manner will cause the flexible brake hose to bear the weight of the caliper, which may cause damage to the brake hose and in turn may cause a brake fluid leak.

2. Remove the brake caliper from the brake caliper mounting bracket and support the brake caliper with heavy mechanic's wire, or equivalent. Do NOT disconnect the hydraulic brake flexible hose from the caliper.

3. Remove the brake pads from the brake caliper bracket.

4. Remove the disc brake pad retainers from the brake caliper bracket.

5. Thoroughly clean the brake pad hardware mating surfaces of the caliper bracket, of any debris and corrosion.

6. Inspect the disc brake pad retainers for the following:

7. Bent mounting tabs

8. Excessive corrosion

9. Looseness at the brake caliper mounting bracket

10. Looseness at the disc brake pads

11. If any of the conditions listed are found, the disc brake pad retainers require replacement.

12. Remove the brake caliper guide pins from the brake caliper bracket.

13. Remove the caliper guide pin boots from the caliper bracket.

14. Inspect the caliper guide pin bores in the caliper bracket. Carefully remove any debris or corrosion from the bores.

15. Inspect the guide pin boots. If the boots are damaged, they require replacement.

16. Inspect the caliper guide pins. If either of the guide pin assemblies is damaged or corroded, the guide pins require replacement.

To install:

17. Apply a thin, light coating of high temperature silicone brake lubricant to the brake caliper guide pin boots.

18. Fully install the guide pin boots to the brake caliper mounting bracket.

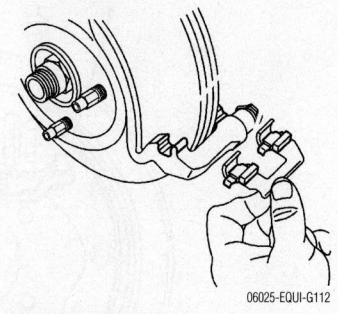

Brake pad retainers

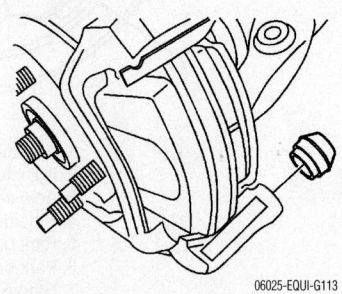

Guide pin boot

➡**Ensure that there is not a build up of lubricant at the end of the upper guide pin, ahead of the bushing.**

19. Apply a light coating of high temperature silicone brake lubricant to the brake caliper guide pins.

20. Install the brake caliper guide pins to the caliper mounting bracket. Ensure that the rim of the guide pin boots is fully seated in the groove on the guide pins.

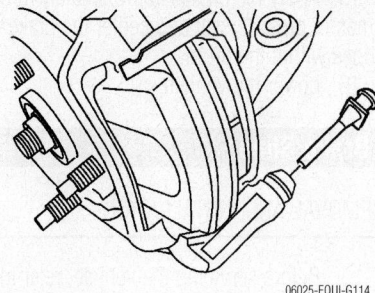

Guide pin

21. Apply a very thin coating of high temperature silicone brake lubricant to the pad hardware mating surfaces of the caliper bracket only.

22. If reusing the brake pad retainers, clean the brake pad mating surfaces of the brake pad retainers.

23. Install the brake pad retainers to the brake caliper bracket.

➡**The wear sensor equipped disc brake pad must be mounted inboard of the rotor with the leading edge of the sensor facing the brake rotor during forward wheel rotation, or at the top of the pad when installed in vehicle position.**

24. Install the brake pads to the brake caliper bracket.

25. Remove the support and reposition the brake caliper over the brake pads and to the mounting bracket.

DRUM BRAKES

❊❊ CAUTION

Do not move the vehicle until a firm brake pedal is obtained. Failure to obtain a firm pedal before moving vehicle may result in personal injury.

Brake Drums

REMOVAL AND INSTALLATION

1. Before servicing the vehicle, refer to the Precautions Section.

2. Check to ensure that the park brake is fully released.

3. Raise and support the vehicle.

4. Remove the tire and wheel assembly.

5. Remove the brake drum.

6. If the brake drum is to be reinstalled to the vehicle, clean any rust or corrosion from the hub/flange mating surface of the brake drum. If necessary, carefully remove

any corrosion from the edge of the drum braking surface in order to ease installation.

7. Clean the wheel hub flange.

To install:

8. Adjust the brake shoe diameter:

9. Relieve cable tension from the park brake system at the equalizer. There should be no tension on the park brake cables, so

Measuring drum diameter

that the brake shoes are positioned only by the adjuster strut.

10. Set a caliper so that it contacts the inside diameter of the drum at the widest point.

11. Position the caliper over the shoes at the widest point.

12. Turn the adjuster nut until the shoes just contact the caliper.

13. Install the brake drum.

14. Install the tire and wheel assembly.

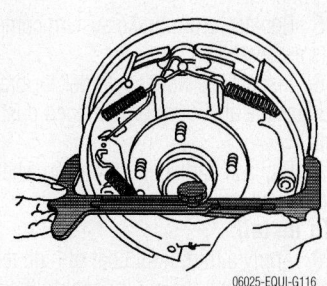

Measuring shoe width

15. Apply the brakes approximately three times in order to seat and center the brake shoes within the drum.

16. Lower the vehicle.

Brake Shoes

REMOVAL AND INSTALLATION

1. Before servicing the vehicle, refer to the Precautions Section.

✷✷ CAUTION

Keep fingers away from rear brake shoe springs to prevent fingers from being pinched between spring and shoe web or spring and backing plate.

2. Raise and support the vehicle.
3. Remove the brake drum.

✷✷ WARNING

Do not over stretch the adjuster spring. Damage can occur if the spring is over stretched.

4. Disengage the adjuster spring hook end from the tab on the adjuster actuator.
5. Remove the straight end of the adjuster spring from the brake shoe.
6. Remove the adjuster actuator from the brake shoe.
7. Remove the return spring from the brake shoes.
8. Remove the park brake cable from the park brake actuator lever.
9. Remove the brake shoe hold-down springs and retainers from the brake shoes.
10. Remove the adjuster from the brake shoes and the park brake actuator lever.
11. Remove the horseshoe clip retaining the park brake actuator lever to the brake shoe.
12. Remove the park brake actuator lever and wave washer from the brake shoe.
13. Clean all of the drum brake system components with denatured alcohol.
14. Inspect all of the drum brake system components.
15. Replace drum brake system components as necessary.
16. Inspect the wheel cylinder for brake fluid leakage and worn or damaged dust boots.
17. Replace damaged or leaking wheel cylinders as necessary.

To install:

18. Apply a thin, light coat of high temperature, silicone brake lubricant to the following areas:

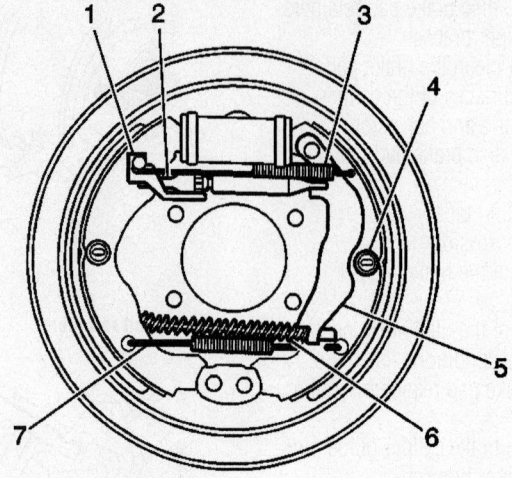

1. Adjuster actuator
2. Adjuster
3. Adjuster spring
4. Retainer
5. Park brake actuator lever
6. Park brake cable
7. Return spring

06025-EQUI-G117

Drum brake components

• The brake shoe contact points on the backing plate
• The adjuster screw threads
• The inside diameter of the adjuster socket

19. Install the park brake actuator lever to the lever pivot pin.
20. Install the horseshoe clip to the park brake actuator lever pivot pin.
21. Install the brake shoes to the brake backing plate.
22. Install the brake shoe hold-down pins, springs and retainers to the brake shoes.
23. Install the park brake cable to the park brake actuator lever.

➡**Ensure that the adjuster (2) engages the brake shoe (4) and the park brake actuator (3) properly.**

24. Install the adjuster screw to the brake shoe and the park brake actuator.
25. Apply a thin, light coat of high temperature, silicone brake lubricant to the adjuster actuator/brake shoe interface.
26. Install the adjuster actuator to the brake shoe.

✷✷ WARNING

Do not over stretch the adjuster spring. Damage can occur if the spring is over stretched.

27. Install the straight end of the adjuster spring to the brake shoe.

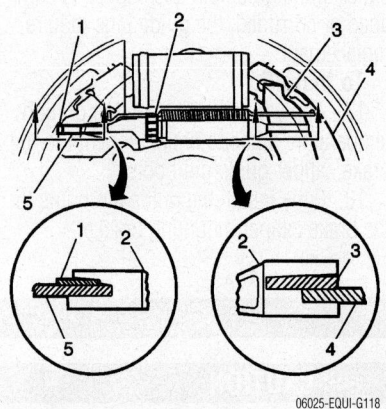

06025-EQUI-G118

Proper adjuster installation

28. Install the adjuster spring hook end to the tab on the adjuster actuator.
29. Install the return spring to the brake shoes.

➡**Ensure that the adjuster operates properly.**

30. Move the park brake actuator lever in order to spread the brake shoes apart. The adjuster actuator lever should move downward, then upward as the park brake actuator lever is released, forcing the adjuster wheel to rotate. If the adjuster does not operate properly, remove then reinstall the adjuster.
31. Adjust the brake shoes.
32. Adjust the park brake cable.
33. Install the brake drum.
34. Lower the vehicle.

Wheel Cylinder

REMOVAL & INSTALLATION

1. Before servicing the vehicle, refer to the Precautions Section.
2. Raise and support the vehicle.
3. Remove the tire and wheel.
4. Clean all dirt and foreign material from the wheel cylinder brake pipe fitting and the bleeder valve.
5. Remove the brake drum.
6. Remove the brake pipe nut.
7. Install a cap over the end of the brake pipe in order to prevent brake fluid loss and/or brake fluid contamination.
8. Remove the wheel cylinder bolts.
9. Spread the brake shoes apart.
10. Remove the wheel cylinder.
11. Clean the old sealant from the backing plate where the wheel cylinder was installed. Clean the sealant from the wheel cylinder if you are reusing the wheel cylinder.

To install:
12. Spread the brake shoes apart.
13. Install the wheel cylinder .
14. Install the wheel cylinder bolts. Tighten the bolts to 15 Nm (11 ft. lbs.).
15. Remove the cap from the brake pipe and install the brake pipe to the wheel cylinder. Tighten the nut to 19 Nm (14 ft. lbs.).
16. Install the brake drum.
17. Bleed the hydraulic brake system at the wheel cylinder.
18. Install the tire and wheel.
19. Lower the vehicle.

Parking Brake

ADJUSTMENT

1. Before servicing the vehicle, refer to the Precautions Section.
2. Apply and fully release the park brake several times. Verify that the park brake lever releases completely.
3. Turn ON the ignition. Verify the red BRAKE warning lamp is not illuminated.
4. If the red BRAKE warning lamp is illuminated, verify the following conditions:
   - The park brake lever is in the fully released position and against the stop
   - There is no slack in the park brake cable

5. Turn OFF the ignition.
6. Disable the supplemental inflatable restraint (SIR) system Zone 8. See the procedure under Steering.
7. Remove the front floor console.
8. With the park brake lever in the released position, using ONLY hand tools, loosen the adjusting nut completely to the end of the front cable threaded rod.
9. Raise and support the vehicle.
10. Remove the rear tire and wheel assemblies.
11. Adjust the rear drum brakes.
12. Ensure there is no brake shoe drag after adjustment by rotating the brake drums. If drag exists, re-center the brake shoes and perform the brake shoe adjustment again.

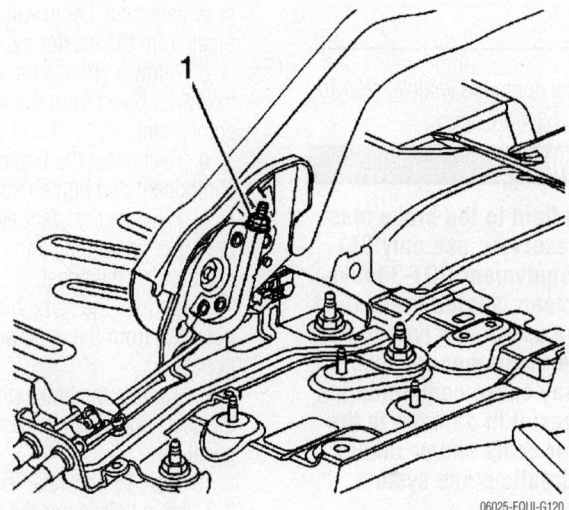

Parking brake cable adjusting nut

06025-EQUI-G120

13. Install 2 wheel nuts to the wheel studs and firmly hand tighten in order to retain the brake drums.
14. Lower the vehicle to permit access to the park brake lever.
15. Raise the park brake lever 1 detent position.
16. Using ONLY hand tools, tighten the park brake cable adjusting nut (1) until light to moderate drag is exhibited while rotating the rear brake drums.
17. Attempt to rotate the rear brake drums. There should be no rotation forward or rearward.
18. Fully release the park brake lever.
19. Verify the park brake is released by rotating the rear brake drums. The drums should rotate freely and exhibit no brake shoe drag.
20. If the drums do not rotate freely, repeat the park brake cable adjustment procedure.
21. Raise the park brake lever 3 detent

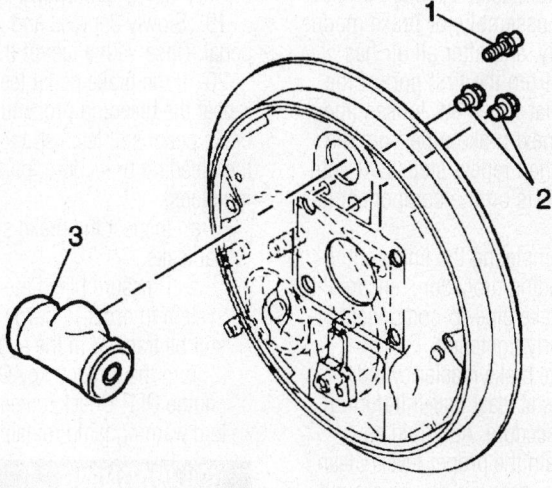

1. Bleed valve
2. Bolts
3. Wheel cylinder

06025-EQUI-G119

Wheel cylinder

positions and attempt to rotate the rear brake drums. One of the brake drums should not rotate forward or rearward. The other brake drum should not rotate forward or rearward, or should require substantial effort to rotate.

22. Raise the vehicle.

23. Remove the wheel nuts retaining the brake drums.

24. Install the rear tire and wheel assemblies.

25. Lower the vehicle.

26. Install the front floor console.

27. Release the park brake lever.

28. Enable the SIR system Zone 8. See the procedure under Steering.

Hydraulic System

BLEEDING

1. Before servicing the vehicle, refer to the Precautions Section.

✳✳ WARNING

When adding fluid to the brake master cylinder reservoir, use only GM approved or equivalent DOT-3 brake fluid from a clean, sealed brake fluid container. The use of any type of fluid other than the recommended type of brake fluid may cause contamination which could result in damage to the internal rubber seals and/or rubber linings of hydraulic brake system components.

✳✳ WARNING

Avoid spilling brake fluid onto painted surfaces, electrical connections, wiring, or cables. Brake fluid will damage painted surfaces and cause corrosion to electrical components. If any brake fluid comes in contact with painted surfaces, immediately flush the area with water. If any brake fluid comes in contact with electrical connections, wiring, or cables, use a clean shop cloth to wipe away the fluid.

2. Place a clean shop cloth beneath the brake master cylinder to catch brake fluid spills.

3. With the ignition OFF and the brakes cool, apply the brakes 3-5 times, or until the brake pedal effort increases significantly, in order to deplete the brake booster power reserve.

4. If you have performed a brake master

cylinder bench bleeding on this vehicle, or if you disconnected the brake pipes from the master cylinder, or if you have disconnected the brake pipes from the proportioning valve assembly or the brake modulator assembly, you must perform the following steps to bleed air at the ports of the hydraulic component:

a. If removal of the reservoir cap and diaphragm is necessary, clean the outside of the reservoir on and around the cap prior to removal.

b. With the brake pipes installed securely to the master cylinder, proportioning valve assembly, or brake modulator assembly, loosen and separate one of the brake pipes from the port of the component. For the proportioning valve assembly or the brake modulator assembly, perform these steps in the sequence of system flow; begin with the fluid feed pipes from the master cylinder.

c. Allow a small amount of brake fluid to gravity bleed from the open port of the component.

d. Reconnect the brake pipe to the component and tighten securely.

e. Have an assistant slowly depress the brake pedal fully and maintain steady pressure on the pedal.

f. Loosen the same brake pipe to purge air from the open port of the component.

g. Tighten the brake pipe, then have the assistant slowly release the brake pedal.

h. Wait 15 seconds, then repeat steps 3-7 until all air is purged from the same port of the component.

i. With the brake pipe installed securely to the master cylinder, proportioning valve assembly, or brake modulator assembly, and after all air has been purged from the first port of the component that was bled, loosen and separate the next brake pipe from the component, then repeat steps 3-8 until each of the ports on the component has been bled.

j. After completing the final component port bleeding procedure, ensure that each of the brake pipe-to-component fittings is properly tightened.

5. Ensure the brake master cylinder reservoir remains at least half-full during this bleeding procedure. Add fluid as needed to maintain the proper level. Clean the outside of the reservoir on and around the reservoir cap prior to removing the cap and diaphragm.

6. Install a proper box-end wrench onto the RIGHT REAR wheel hydraulic circuit bleeder valve.

7. Install a transparent hose over the end of the bleeder valve.

8. Have an assistant slowly depress the brake pedal fully and maintain steady pressure on the pedal.

9. Loosen the bleeder valve to purge air from the wheel hydraulic circuit.

10. Tighten the bleeder valve, then have the assistant slowly release the brake pedal.

11. Wait 15 seconds, then repeat steps 8-10 until all air is purged from the same wheel hydraulic circuit.

12. With the right rear wheel hydraulic circuit bleeder valve tightened securely, and after all air has been purged from the right rear hydraulic circuit, install a proper box-end wrench onto the LEFT FRONT wheel hydraulic circuit bleeder valve.

13. Install a transparent hose over the end of the bleeder valve, then repeat steps 7-11.

14. With the left front wheel hydraulic circuit bleeder valve tightened securely, and after all air has been purged from the left front hydraulic circuit, install a proper box-end wrench onto the LEFT REAR wheel hydraulic circuit bleeder valve.

15. Install a transparent hose over the end of the bleeder valve, then repeat steps 7-11.

16. With the left rear wheel hydraulic circuit bleeder valve tightened securely, and after all air has been purged from the left rear hydraulic circuit, install a proper box-end wrench onto the RIGHT FRONT wheel hydraulic circuit bleeder valve.

17. Install a transparent hose over the end of the bleeder valve, then repeat steps 7-11.

18. After completing the final wheel hydraulic circuit bleeding procedure, ensure that each of the 4 wheel hydraulic circuit bleeder valves is properly tightened.

19. Slowly depress and release the brake pedal. Observe the feel of the brake pedal.

20. If the brake pedal feels spongy, repeat the bleeding procedure again. If the brake pedal still feels spongy after repeating the bleeding procedure, perform the following steps:

a. Inspect the brake system for external leaks.

b. Pressure bleed the hydraulic brake system in order to purge any air that may still be trapped in the system.

Turn the ignition key ON, with the engine OFF. Check to see if the brake system warning lamp remains illuminated.

✳✳ WARNING

DO NOT allow the vehicle to be driven until it is diagnosed and repaired.

CADILLAC, CHEVROLET AND GMC

18

Escalade • Suburban • Tahoe • Yukon • Yukon Denali • Yukon XL

BRAKES **18-71**
DRIVE TRAIN **18-48**
ENGINE REPAIR **18-23**
FUEL SYSTEM **18-45**
SPECIFICATIONS AND
  MAINTENANCE CHARTS......**18-3**
Engine and Vehicle
  Identification 18-3
General Engine Specifications 18-3
Engine Tune-Up Specifications 18-4
Firing Order 18-5
Accessory Drive Belt Routing 18-5
Capacities 18-6
Valve Specifications 18-9
Crankshaft and Connecting
  Rod Specifications....................18-11
Piston and Ring Specifications 18-13
Torque Specifications 18-15
Wheel Alignment 18-16
Tire, Wheel and Ball Joint
  Specifications 18-17
Brake Specifications 18-18
Scheduled Maintenance
  Intervals..................................18-20
STEERING AND
  SUSPENSION **18-60**

A
Air Bag.......................................18-60
  Disarming & Arming...............18-60
  Precautions 18-60
Alternator 18-23
  Removal & Installation............18-23

B
Brake Caliper 18-71
  Removal & Installation............18-71

C
Camshaft and Valve Lifters 18-35
  Removal & Installation............18-35
Coil Springs 18-63
  Removal & Installation............18-63
CV-Joints....................................18-52
  Overhaul 18-52
Cylinder Head 18-28
  Removal & Installation............18-28

D
Disc Brake Pads..........................18-71
  Removal & Installation............18-71

Distributor...................................18-23
  Removal 18-23

E
Engine Assembly 18-23
  Removal & Installation............18-23
Exhaust Manifold 18-34
  Removal & Installation............18-34

F
Front Axle Shaft, Bearing and
  Seal..18-54
  Removal & Installation............18-54
Front Drive Axle Differential
  Carrier 18-59
  Removal & Installation............18-59
Front Drive Axle Pinion Seal........18-58
  Removal & Installation............18-58
Fuel Filter...................................18-45
  Removal & Installation............18-45
Fuel Injector...............................18-46
  Removal & Installation............18-46
Fuel Pump..................................18-46
  Removal & Installation............18-46
Fuel System Pressure 18-45
  Relieving 18-45
Fuel System Service
  Precautions............................18-45

H
Halfshaft.....................................18-51
  Removal & Installation............18-51
Heater Core................................18-27
  Removal & Installation............18-27

I
Ignition Timing 18-23
  Adjustment.............................18-23
Intake Manifold 18-32
  Removal & Installation............18-32

L
Leaf Springs 18-65
  Removal & Installation............18-65
Lower Ball Joint..........................18-67
  Removal & Installation............18-67
Lower Control Arm and
  Bushing..................................18-68
    Control Arm Bushing
    Replacement 18-69
  Removal & Installation............18-68

O
Oil Pan.......................................18-37
  Removal & Installation............18-37
Oil Pump 18-39
  Removal & Installation............18-39

P
Piston and Ring 18-44
  Positioning 18-44
Power Steering Pump 18-61
  Removal & Installation............18-61

R
Rack & Pinion Steering Gear 18-61
  Removal & Installation............18-61
Rear Axle Shaft, Bearing and
  Seal..18-56
  Removal & Installation............18-56
Rear Drive Axle Housing............18-59
  Removal & Installation............18-59
Rear Drive Axle Pinion Seal........18-58
  Removal & Installation............18-58
Rear Main Seal 18-41
  Removal & Installation............18-41
Rear Wheel Steering Gear
  Actuator 18-62
  Removal & Installation............18-62
Rear Wheel Steering Gear
  Motor.....................................18-61
  Removal & Installation............18-61
Rocker Arms 18-31
  Removal & Installation............18-31

S
Shock Absorber 18-62
  Removal & Installation............18-62
Stabilizer Bar 18-70
  Removal & Installation............18-70
Starter Motor 18-37
  Removal & Installation............18-37
Steering Knuckle.........................18-71
  Removal & Installation............18-71

T
Timing Chain, Sprockets, Front
  Cover and Seal 18-41
  Removal & Installation............18-42
Torsion Bars 18-66
  Removal & Installation............18-66
Transfer Case Assembly.............18-50
  Removal & Installation............18-50

Transfer Case Encoder Motor.......18-53
   Removal & Installation............18-53
Transmission Assembly 18-48
   Removal & Installation............18-48

U
Upper Ball Joint.........................18-67
   Removal & Installation............18-67

Upper Control Arm 18-67
   Control Arm Bushing
     Replacement.........................18-68
   Removal & Installation............18-67

V
Valve Lash 18-37
   Adjustment............................18-37

W
Water Pump 18-26
   Removal & Installation............18-26
Wheel Hub, Bearings and Seal 18-69
   Removal & Installation............18-69

SPECIFICATIONS AND MAINTENANCE CHARTS

ENGINE AND VEHICLE IDENTIFICATION

Code ①	Engine						Model Year	
	Liters (cc)	Cu. In.	Cyl.	Fuel Sys.	Engine Type	Eng. Mfg.	Code ②	Year
B	5.3 (5327)	325	8	MFI	OHV	CPC	2	2002
G	8.1 (8128)	496	8	MFI	OHV	CPC	3	2003
N	6.0 (5966)	364	8	MFI	OHV	CPC	4	2004
T	5.3 (5327)	325	8	MFI	OHV	CPC	5	2005
U	6.0 (5966)	364	8	MFI	OHV	CPC	6	2006
V	4.8 (4802)	293	8	MFI	OHV	CPC		
Z	5.3 (5327)	325	8	MFI	OHV	CPC		

CPC: Chevrolet/Pontiac/Canada

MFI: Multi-port Fuel Injection

① 8th position of VIN

② 10th position of VIN

06025-YUKO-C01

GENERAL ENGINE SPECIFICATIONS

Engine Displ. Liters	Engine VIN	Models	Years	Net Horsepower @ rpm	Net Torque @ rpm (ft. lbs.)	Bore x Stroke (in.)	Compression Ratio	Oil Pressure @ rpm
4.8	V	Escalade	2005-06	270@5200	285@4000	3.78x3.27	9.5:1	18@2000
		Suburban	2005-06					
		Tahoe/Yukon	2002-06					
5.3	B	Escalade	2006	285@4000	360@4000	3.78x3.62	9.9:1	18@2000
		Suburban	2006					
		Tahoe/Yukon	2006					
5.3	T	Escalade	2002-06	285@4000	360@4000	3.78x3.62	9.5:1	18@2000
		Suburban	2002-06					
		Tahoe/Yukon	2002-06					
5.3	Z	Escalade	2004-06	285@4000	360@4000	3.78x3.62	9.5:1	18@2000
		Suburban	2004-06					
		Tahoe/Yukon	2004-06					
6.0	N	Escalalade	2003-06	300@4400	360@4000	4.00x3.62	9.4:1	18@2000
		Suburban/Yukon ①	2003-06					
6.0	U	Escalalade	2002-06	300@4400	360@4000	4.00x3.62	9.4:1	18@2000
		Suburban/Yukon ①	2002-06					
8.1	G	Escalade	2006	340@4200	455@3200	4.25x4.37	9.1:1	10@2000
		Suburban/Yukon	2002-06					
		Tahoe	2006					

① Yukon Denali, Denali XL and XL

06025-YUKO-C02

GASOLINE ENGINE TUNE-UP SPECIFICATIONS

Year	Engine Displacement Liters	Engine ID/VIN	Spark Plugs Gap (in.)	Ignition Timing (deg.) MT	AT	Fuel Pump (psi)	Idle Speed (rpm) MT	AT	Valve Clearance In.	Ex.
2002	4.8	V	0.060	—	①	55-62 ②	—	③	HYD	HYD
	5.3	T	0.060	—	①	55-62 ②	—	③	HYD	HYD
	6.0	U	0.060	—	①	55-62 ②	—	③	HYD	HYD
	8.1	G	0.060	—	①	55-62 ②	—	③	HYD	HYD
2003	4.8	V	0.060	—	①	55-62 ②	—	③	HYD	HYD
	5.3	T	0.060	—	①	55-62 ②	—	③	HYD	HYD
	5.3	Z	0.060	—	①	55-62 ②	—	③	HYD	HYD
	6.0	N	0.060	—	①	55-62 ②	—	③	HYD	HYD
	6.0	U	0.060	—	①	55-62 ②	—	③	HYD	HYD
	8.1	G	0.060	—	①	55-62 ②	—	③	HYD	HYD
2004	4.8	V	0.060	—	①	55-62 ②	—	③	HYD	HYD
	5.3	T	0.060	—	①	55-62 ②	—	③	HYD	HYD
	5.3	Z	0.060	—	①	55-62 ②	—	③	HYD	HYD
	6.0	N	0.060	—	①	55-62 ②	—	③	HYD	HYD
	6.0	U	0.060	—	①	55-62 ②	—	③	HYD	HYD
	8.1	G	0.060	—	①	55-62 ②	—	③	HYD	HYD
2005	4.8	V	0.060	—	①	55-62 ②	—	③	HYD	HYD
	5.3	T	0.060	—	①	55-62 ②	—	③	HYD	HYD
	5.3	Z	0.060	—	①	55-62 ②	—	③	HYD	HYD
	6.0	N	0.060	—	①	55-62 ②	—	③	HYD	HYD
	6.0	U	0.060	—	①	55-62 ②	—	③	HYD	HYD
	8.1	G	0.060	—	①	55-62 ②	—	③	HYD	HYD
2006	4.8	V	0.060	—	①	55-62 ②	—	③	HYD	HYD
	5.3	B	0.060	—	①	55-62 ②	—	③	HYD	HYD
	5.3	T	0.060	—	①	55-62 ②	—	③	HYD	HYD
	5.3	Z	0.060	—	①	55-62 ②	—	③	HYD	HYD
	6.0	N	0.060	—	①	55-62 ②	—	③	HYD	HYD
	6.0	U	0.060	—	①	55-62 ②	—	③	HYD	HYD
	8.1	G	0.060	—	①	55-62 ②	—	③	HYD	HYD

NOTE: The Vehicle Emission Control Information label often reflects specification changes made during production.

The label figures must be used if they differ from those in this chart.

HYD: Hydraulic

① Ignition timing is preset and cannot be adjusted

② With key ON and engine OFF

③ Idle speed is maintained by the Powertrain Control Module (PCM)

06025-YUKO-C03

8
7
6
5
4
3
2
1
FRONT
93023G01

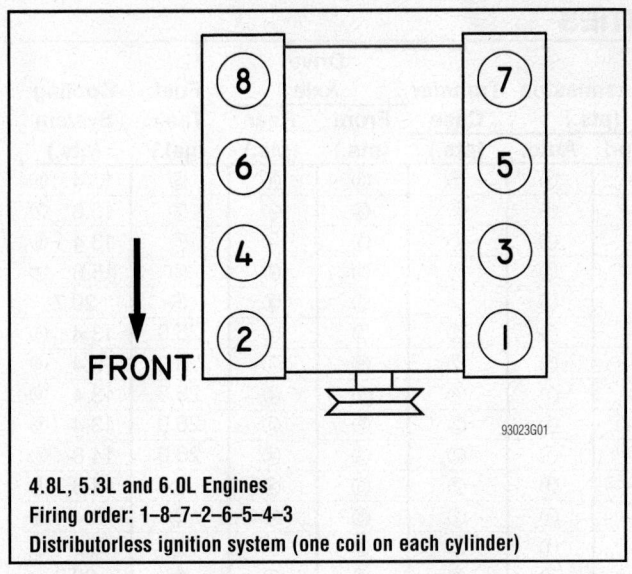

4.8L, 5.3L and 6.0L Engines
Firing order: 1–8–7–2–6–5–4–3
Distributorless ignition system (one coil on each cylinder)

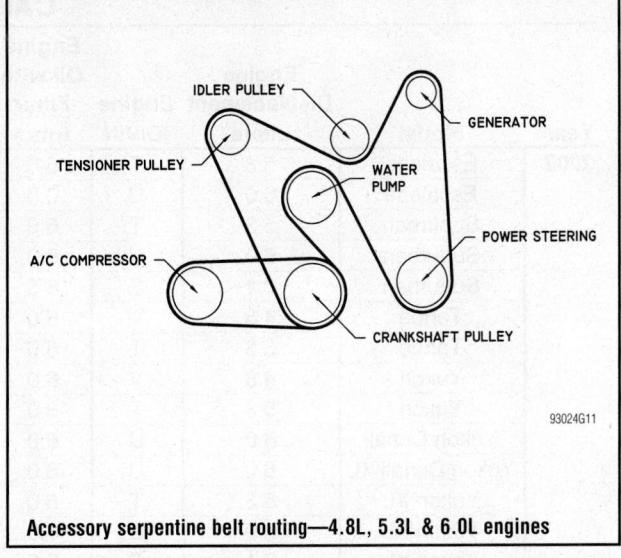

IDLER PULLEY
GENERATOR
TENSIONER PULLEY
WATER PUMP
POWER STEERING
A/C COMPRESSOR
CRANKSHAFT PULLEY
93024G11

Accessory serpentine belt routing—4.8L, 5.3L & 6.0L engines

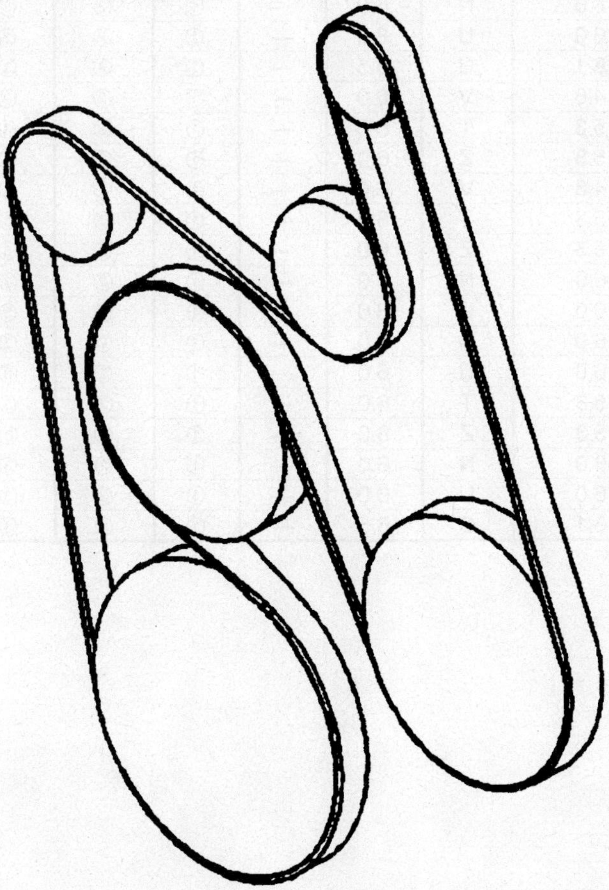

93551GMM

Accessory serpentine belt routing—8.1L engines

CAPACITIES

Year	Model	Engine Displacement Liters	Engine ID/VIN	Engine Oil with Filter (qts.)	Transmission (pts.) 5-Spd	Transmission (pts.) Auto.	Transfer Case (pts.)	Drive Axle Front (pts.)	Drive Axle Rear (pts.)	Fuel Tank (gal.)	Cooling System (qts.)
2002	Escalade	5.3	T	6.0	—	①	②	③	④	⑤	13.4 ⑥
	Escalade	6.0	U	6.0	—	①	②	③	④	⑤	15.8 ⑦
	Suburban	5.3	T	6.0	—	①	②	③	④	⑤	13.4 ⑥
	Suburban	6.0	U	6.0	—	①	②	③	④	⑤	15.8 ⑦
	Suburban	8.1	G	6.5	—	①	②	③	④	⑤	20.7
	Tahoe	4.8	V	6.0	—	①	②	③	④	26.0	13.4 ⑥
	Tahoe	5.3	T	6.0	—	①	②	③	④	26.0	13.4 ⑥
	Yukon	4.8	V	6.0	—	①	②	③	④	26.0	13.4 ⑥
	Yukon	5.3	T	6.0	—	①	②	③	④	26.0	13.4 ⑥
	Yukon Denali	6.0	U	6.0	—	①	②	③	④	26.0	14.8 ⑥
	Yukon Denali XL	6.0	U	6.0	—	①	②	③	④	⑤	14.8 ⑥
	Yukon XL	5.3	T	6.0	—	①	②	③	④	⑤	13.4 ⑥
	Yukon XL	6.0	U	6.0	—	①	②	③	④	⑤	15.8 ⑦
	Yukon XL	8.1	G	6.5	—	①	②	③	④	⑤	20.7
2003	Escalade	5.3	Z	6.0	—	①	②	③	④	⑤	13.4 ⑥
	Escalade	5.3	T	6.0	—	①	②	③	④	⑤	13.4 ⑥
	Escalade	6.0	N	6.0	—	①	②	③	④	⑤	15.8 ⑦
	Escalade	6.0	U	6.0	—	①	②	③	④	⑤	15.8 ⑦
	Suburban	5.3	T	6.0	—	①	②	③	④	⑤	13.4 ⑥
	Suburban	5.3	Z	6.0	—	①	②	③	④	⑤	13.4 ⑥
	Suburban	6.0	N	6.0	—	①	②	③	④	⑤	15.8 ⑦
	Suburban	6.0	U	6.0	—	①	②	③	④	⑤	15.8 ⑦
	Suburban	8.1	G	6.5	—	①	②	③	④	⑤	20.7
	Tahoe	4.8	V	6.0	—	①	②	③	④	26.0	13.4 ⑥
	Tahoe	5.3	T	6.0	—	①	②	③	④	26.0	13.4 ⑥
	Tahoe	5.3	Z	6.0	—	①	②	③	④	26.0	13.4 ⑥
	Yukon	4.8	V	6.0	—	①	②	③	④	26.0	13.4 ⑥
	Yukon	5.3	T	6.0	—	①	②	③	④	26.0	13.4 ⑥
	Yukon	5.3	Z	6.0	—	①	②	③	④	26.0	13.4 ⑥
	Yulkon Denali	6.0	N	6.0	—	①	②	③	④	26.0	14.8 ⑥
	Yukon Denali	6.0	U	6.0	—	①	②	③	④	⑤	15.8 ⑦
	Yukon Denali XL	6.0	N	6.0	—	①	②	③	④	⑤	14.8 ⑥
	Yukon Denali XL	6.0	U	6.0	—	①	②	③	④	⑤	15.8 ⑦
	Yukon XL	5.3	T	6.0	—	①	②	③	④	⑤	13.4 ⑥
	Yukon XL	5.3	Z	6.0	—	①	②	③	④	⑤	13.4 ⑥
	Yukon XL	6.0	N	6.0	—	①	②	③	④	⑤	15.8 ⑦
	Yukon XL	6.0	U	6.0	—	①	②	③	④	⑤	15.8 ⑦
	Yukon XL	8.1	G	6.5	—	①	②	③	④	⑤	20.7

06025-YUKO-C04

CAPACITIES

Year	Model	Engine Displacement Liters	Engine ID/VIN	Engine Oil with Filter (qts.)	Transmission (pts.) 5-Spd	Transmission (pts.) Auto.	Transfer Case (pts.)	Drive Axle Front (pts.)	Drive Axle Rear (pts.)	Fuel Tank (gal.)	Cooling System (qts.)
2004	Escalade	5.3	Z	6.0	—	①	②	③	④	⑤	13.4 ⑥
	Escalade	5.3	T	6.0	—	①	②	③	④	⑤	13.4 ⑥
	Escalade	6.0	N	6.0	—	①	②	③	④	⑤	15.8 ⑦
	Escalade	6.0	U	6.0	—	①	②	③	④	⑤	15.8 ⑦
	Suburban	5.3	T	6.0	—	①	②	③	④	⑤	13.4 ⑥
	Suburban	5.3	Z	6.0	—	①	②	③	④	⑤	13.4 ⑥
	Suburban	6.0	N	6.0	—	①	②	③	④	⑤	15.8 ⑦
	Suburban	6.0	U	6.0	—	①	②	③	④	⑤	15.8 ⑦
	Suburban	8.1	G	6.5	—	①	②	③	④	⑤	20.7
	Tahoe	4.8	V	6.0	—	①	②	③	④	26.0	13.4 ⑥
	Tahoe	5.3	T	6.0	—	①	②	③	④	26.0	13.4 ⑥
	Tahoe	5.3	Z	6.0	—	①	②	③	④	26.0	13.4 ⑥
	Yukon	4.8	V	6.0	—	①	②	③	④	26.0	13.4 ⑥
	Yukon	5.3	T	6.0	—	①	②	③	④	26.0	13.4 ⑥
	Yukon	5.3	Z	6.0	—	①	②	③	④	26.0	13.4 ⑥
	Yukon Denali	6.0	N	6.0	—	①	②	③	④	26.0	14.8 ⑥
	Yukon Denali	6.0	U	6.0	—	①	②	③	④	⑤	15.8 ⑦
	Yukon Denali XL	6.0	N	6.0	—	①	②	③	④	⑤	14.8 ⑥
	Yukon Denali XL	6.0	U	6.0	—	①	②	③	④	⑤	15.8 ⑦
	Yukon XL	5.3	T	6.0	—	①	②	③	④	⑤	13.4 ⑥
	Yukon XL	5.3	Z	6.0	—	①	②	③	④	⑤	13.4 ⑥
	Yukon XL	6.0	N	6.0	—	①	②	③	④	⑤	15.8 ⑦
	Yukon XL	6.0	U	6.0	—	①	②	③	④	⑤	15.8 ⑦
	Yukon XL	8.1	G	6.5	—	①	②	③	④	⑤	20.7
2005	Escalade	5.3	B	6.0	—	①	②	③	④	⑤	16.7 ⑧
	Escalade	5.3	T	6.0	—	①	②	③	④	⑤	16.7 ⑧
	Escalade	5.3	Z	6.0	—	①	②	③	④	⑤	16.7 ⑧
	Escalade	6.0	N	6.0	—	①	②	③	④	⑤	16.2 ⑧
	Escalade	6.0	U	6.0	—	①	②	③	④	⑤	16.2 ⑧
	Suburban	5.3	B	6.0	—	①	②	③	④	⑤	16.7 ⑧
	Suburban	5.3	T	6.0	—	①	②	③	④	⑤	16.7 ⑧
	Suburban	5.3	Z	6.0	—	①	②	③	④	⑤	16.7 ⑧
	Suburban	6.0	N	6.0	—	①	②	③	④	⑤	16.2 ⑧
	Suburban	6.0	U	6.0	—	①	②	③	④	⑤	16.2 ⑧
	Suburban	8.1	G	6.5	—	①	②	③	④	⑤	26.9 ⑧
	Tahoe	4.8	V	6.0	—	①	②	③	④	26.0	16.7 ⑧
	Tahoe	5.3	B	6.0	—	①	②	③	④	26.0	16.7 ⑧
	Tahoe	5.3	T	6.0	—	①	②	③	④	26.0	16.7 ⑧
	Tahoe	5.3	Z	6.0	—	①	②	③	④	26.0	16.7 ⑧
	Yukon	4.8	V	6.0	—	①	②	③	④	26.0	16.7 ⑧
	Yukon	5.3	B	6.0	—	①	②	③	④	26.0	16.7 ⑧
	Yukon	5.3	T	6.0	—	①	②	③	④	26.0	16.7 ⑧
	Yukon	5.3	Z	6.0	—	①	②	③	④	26.0	16.7 ⑧
	Yukon Denali	6.0	N	6.0	—	①	②	③	④	26.0	16.2 ⑧
	Yukon Denali	6.0	U	6.0	—	①	②	③	④	⑤	16.2 ⑧

06025-YUKO-C05

CAPACITIES

Year	Model	Engine Displacement Liters	Engine ID/VIN	Engine Oil with Filter (qts.)	Transmission (pts.) 5-Spd	Transmission (pts.) Auto.	Transfer Case (pts.)	Drive Axle Front (pts.)	Drive Axle Rear (pts.)	Fuel Tank (gal.)	Cooling System (qts.)
2005 cont.	Yukon Denali XL	6.0	N	6.0	—	①	②	③	④	⑤	16.2 ⑧
	Yukon Denali XL	6.0	U	6.0	—	①	②	③	④	⑤	16.2 ⑧
	Yukon XL	5.3	B	6.0	—	①	②	③	④	26.0	16.7 ⑧
	Yukon XL	5.3	T	6.0	—	①	②	③	④	26.0	16.7 ⑧
	Yukon XL	5.3	Z	6.0	—	①	②	③	④	⑤	16.7 ⑧
	Yukon XL	6.0	N	6.0	—	①	②	③	④	⑤	16.2 ⑧
	Yukon XL	6.0	U	6.0	—	①	②	③	④	⑤	16.2 ⑧
	Yukon XL	8.1	G	6.5	—	①	②	③	④	⑤	26.9 ⑧
2006	Escalade	5.3	B	6.0	—	①	②	③	④	⑤	16.7 ⑧
	Escalade	5.3	T	6.0	—	①	②	③	④	⑤	16.7 ⑧
	Escalade	5.3	Z	6.0	—	①	②	③	④	⑤	16.7 ⑧
	Escalade	6.0	N	6.0	—	①	②	③	④	⑤	16.2 ⑧
	Escalade	6.0	U	6.0	—	①	②	③	④	⑤	16.2 ⑧
	Suburban	5.3	B	6.0	—	①	②	③	④	⑤	16.7 ⑧
	Suburban	5.3	T	6.0	—	①	②	③	④	⑤	16.7 ⑧
	Suburban	5.3	Z	6.0	—	①	②	③	④	⑤	16.7 ⑧
	Suburban	6.0	N	6.0	—	①	②	③	④	⑤	16.2 ⑧
	Suburban	6.0	U	6.0	—	①	②	③	④	⑤	16.2 ⑧
	Suburban	8.1	G	6.5	—	①	②	③	④	⑤	26.9 ⑧
	Tahoe	4.8	V	6.0	—	①	②	③	④	26.0	16.7 ⑧
	Tahoe	5.3	B	6.0	—	①	②	③	④	26.0	16.7 ⑧
	Tahoe	5.3	T	6.0	—	①	②	③	④	26.0	16.7 ⑧
	Tahoe	5.3	Z	6.0	—	①	②	③	④	26.0	16.7 ⑧
	Yukon	4.8	V	6.0	—	①	②	③	④	26.0	16.7 ⑧
	Yukon	5.3	B	6.0	—	①	②	③	④	26.0	16.7 ⑧
	Yukon	5.3	T	6.0	—	①	②	③	④	26.0	16.7 ⑧
	Yukon	5.3	Z	6.0	—	①	②	③	④	26.0	16.7 ⑧
	Yukon Denali	6.0	N	6.0	—	①	②	③	④	26.0	16.2 ⑧
	Yukon Denali	6.0	U	6.0	—	①	②	③	④	⑤	16.2 ⑧
	Yukon Denali XL	6.0	N	6.0	—	①	②	③	④	⑤	16.2 ⑧
	Yukon Denali XL	6.0	U	6.0	—	①	②	③	④	⑤	16.2 ⑧
	Yukon XL	5.3	B	6.0	—	①	②	③	④	⑤	16.7 ⑧
	Yukon XL	5.3	T	6.0	—	①	②	③	④	⑤	16.7 ⑧
	Yukon XL	5.3	Z	6.0	—	①	②	③	④	⑤	16.7 ⑧
	Yukon XL	6.0	N	6.0	—	①	②	③	④	⑤	16.2 ⑧
	Yukon XL	6.0	U	6.0	—	①	②	③	④	⑤	16.2 ⑧
	Yukon XL	8.1	G	6.5	—	①	②	③	④	⑤	26.9 ⑧

NOTE: All capacities are approximate. Add fluid gradually and check to be sure a proper fluid level is obtained.

① 4L60E trans.: 10.0 pts.
   4L80E trans.: 15.4 pts.

② NV241 and NV243: 4.5 pts.
   4401 and 4470: 6.6 pts.
   NV246: 4.0 pts.
   BW4480/81: 3.0 pts.

③ 8.25 in ring gear: 3.02 pts.
   9.25 ring gear: 3.7 pts.

④ 8.6 in. ring gear: 4.3 pts.
   9.5 & 10.5 in. ring gear: 5.5 pts.
   9.75 in. ring gear: 6.0 pts.
   11.5 in. ring gear: 7.7 pts.

⑤ 1500: 31 gallon tank
   2500: 37 gallon tank

⑥ With optional A/C: 14.9 qts.
   With front A/C: 14.4 qts.
   With front and rear A/C: 15.8 qts.

⑦ With optional oil cooler: 15.4 qts.

⑧ With rear heater add 2.1 qts.

06025-YUKO-C06

VALVE SPECIFICATIONS

Year	Engine Displacement Liters	Engine ID/VIN	Seat Angle (deg.)	Face Angle (deg.)	Spring Test Pressure (lbs. @ in.)	Spring Installed Height (in.)	Stem-to-Guide Clearance (in.)		Stem Diameter (in.)	
							Intake	Exhaust	Intake	Exhaust
2002	4.8	V	46	45	220@1.32	1.80	0.0010-0.0026	0.0010-0.0026	0.3132-0.3140	0.3132-0.3140
	5.3	T	46	45	220@1.32	1.80	0.0010-0.0026	0.0010-0.0026	0.3132-0.3140	0.3132-0.3140
	6.0	U	46	45	220@1.32	1.80	0.0010-0.0026	0.0010-0.0026	0.3132-0.3140	0.3132-0.3140
	8.1	G	46	45	216-236@1.34	1.81-1.84	0.0010-0.0029	0.0012-0.0031	0.3715-0.3722	0.3713-0.3720
2003	4.8	V	46	45	220@1.32	1.80	0.0010-0.0026	0.0010-0.0026	0.3132-0.3140	0.3132-0.3140
	5.3	T	46	45	220@1.32	1.80	0.0010-0.0026	0.0010-0.0026	0.3132-0.3140	0.3132-0.3140
	5.3	Z	46	45	220@1.32	1.80	0.0010-0.0026	0.0010-0.0026	0.3132-0.3140	0.3132-0.3140
	6.0	N	46	45	220@1.32	1.80	0.0010-0.0026	0.0010-0.0026	0.3132-0.3140	0.3132-0.3140
	6.0	U	46	45	220@1.32	1.80	0.0010-0.0026	0.0010-0.0026	0.3132-0.3140	0.3132-0.3140
	8.1	G	46	45	216-236@1.34	1.81-1.84	0.0010-0.0029	0.0012-0.0031	0.3715-0.3722	0.3713-0.3720
2004	4.8	V	46	45	220@1.32	1.80	0.0010-0.0026	0.0010-0.0026	0.3132-0.3140	0.3132-0.3140
	5.3	T	46	45	220@1.32	1.80	0.0010-0.0026	0.0010-0.0026	0.3132-0.3140	0.3132-0.3140
	5.3	Z	46	45	220@1.32	1.80	0.0010-0.0026	0.0010-0.0026	0.3132-0.3140	0.3132-0.3140
	6.0	N	46	45	220@1.32	1.80	0.0010-0.0026	0.0010-0.0026	0.3132-0.3140	0.3132-0.3140
	6.0	U	46	45	220@1.32	1.80	0.0010-0.0026	0.0010-0.0026	0.3132-0.3140	0.3132-0.3140
	8.1	G	46	45	216-236@1.34	1.81-1.84	0.0010-0.0029	0.0012-0.0031	0.3715-0.3722	0.3713-0.3720
2005	4.8	V	46	45	220@1.32	1.80	0.0010-0.0026	0.0010-0.0026	0.3132-0.3140	0.3132-0.3140
	5.3	B	46	45	220@1.32	1.80	0.0010-0.0026	0.0010-0.0026	0.3132-0.3140	0.3132-0.3140
	5.3	T	46	45	220@1.32	1.80	0.0010-0.0026	0.0010-0.0026	0.3132-0.3140	0.3132-0.3140
	5.3	Z	46	45	220@1.32	1.80	0.0010-0.0026	0.0010-0.0026	0.3132-0.3140	0.3132-0.3140
	6.0	N	46	45	220@1.32	1.80	0.0010-0.0026	0.0010-0.0026	0.3132-0.3140	0.3132-0.3140
	6.0	U	46	45	220@1.32	1.80	0.0010-0.0026	0.0010-0.0026	0.3132-0.3140	0.3132-0.3140
	8.1	G	46	45	216-236@1.34	1.81-1.84	0.0010-0.0029	0.0012-0.0031	0.3715-0.3722	0.3713-0.3720

06025-YUKO-C07

VALVE SPECIFICATIONS

Year	Engine Displacement Liters	Engine ID/VIN	Seat Angle (deg.)	Face Angle (deg.)	Spring Test Pressure (lbs. @ in.)	Spring Installed Height (in.)	Stem-to-Guide Clearance (in.)		Stem Diameter (in.)	
							Intake	Exhaust	Intake	Exhaust
2006	4.8	V	46	45	220@1.32	1.80	0.0010-0.0026	0.0010-0.0026	0.3132-0.3140	0.3132-0.3140
	5.3	B	46	45	220@1.32	1.80	0.0010-0.0026	0.0010-0.0026	0.3132-0.3140	0.3132-0.3140
	5.3	T	46	45	220@1.32	1.80	0.0010-0.0026	0.0010-0.0026	0.3132-0.3140	0.3132-0.3140
	5.3	Z	46	45	220@1.32	1.80	0.0010-0.0026	0.0010-0.0026	0.3132-0.3140	0.3132-0.3140
	6.0	N	46	45	220@1.32	1.80	0.0010-0.0026	0.0010-0.0026	0.3132-0.3140	0.3132-0.3140
	6.0	U	46	45	220@1.32	1.80	0.0010-0.0026	0.0010-0.0026	0.3132-0.3140	0.3132-0.3140
	8.1	G	46	45	216-236@1.34	1.81-1.84	0.0010-0.0029	0.0012-0.0031	0.3715-0.3722	0.3713-0.3720

06025-YUKO-C08

CRANKSHAFT AND CONNECTING ROD SPECIFICATIONS

All measurements are given in inches.

Year	Engine Displacement Liters	Engine ID/VIN	Crankshaft				Connecting Rod		
			Main Brg. Journal Dia.	Main Brg. Oil Clearance	Shaft End-play	Thrust on No.	Journal Diameter	Oil Clearance	Side Clearance
2002	4.8	V	2.5580-2.5593	0.0007-0.0021	0.0015-0.0078	5	2.0990-2.1000	0.0006-0.0030	0.0043-0.0200
	5.3	T	2.5580-2.5593	0.0007-0.0021	0.0015-0.0078	5	2.0990-2.1000	0.0006-0.0030	0.0043-0.0200
	6.0	U	2.5580-2.5593	0.0007-0.0021	0.0015-0.0078	5	2.0990-2.1000	0.0006-0.0030	0.0043-0.0200
	8.1	G	2.7482-2.7489	①	0.0050-0.0110	NA	2.1990-2.1996	0.0008-0.0025	0.0151-0.0270
2003	4.8	V	2.5580-2.5593	0.0007-0.0021	0.0015-0.0078	5	2.0990-2.1000	0.0006-0.0030	0.0043-0.0200
	5.3	T	2.5580-2.5593	0.0007-0.0021	0.0015-0.0078	5	2.0990-2.1000	0.0006-0.0030	0.0043-0.0200
	5.3	Z	2.5580-2.5593	0.0008-0.0021	0.0015-0.0078	5	2.0991-2.0999	0.0009-0.0025	0.0043-0.0200
	6.0	N	2.5580-2.5593	0.0007-0.0021	0.0015-0.0078	5	2.0990-2.1000	0.0006-0.0030	0.0043-0.0200
	6.0	U	2.5580-2.5593	0.0007-0.0021	0.0015-0.0078	5	2.0990-2.1000	0.0006-0.0030	0.0043-0.0200
	8.1	G	2.7482-2.7489	①	0.0050-0.0110	NA	2.1990-2.1996	0.0008-0.0025	0.0151-0.0270
2004	4.8	V	2.5580-2.5593	0.0007-0.0021	0.0015-0.0078	5	2.0990-2.1000	0.0006-0.0030	0.0043-0.0200
	5.3	T	2.5580-2.5593	0.0007-0.0021	0.0015-0.0078	5	2.0990-2.1000	0.0006-0.0030	0.0043-0.0200
	5.3	Z	2.5580-2.5593	0.0008-0.0021	0.0015-0.0078	5	2.0991-2.0999	0.0009-0.0025	0.0043-0.0200
	6.0	N	2.5580-2.5593	0.0007-0.0021	0.0015-0.0078	5	2.0990-2.1000	0.0006-0.0030	0.0043-0.0200
	6.0	U	2.5580-2.5593	0.0007-0.0021	0.0015-0.0078	5	2.0990-2.1000	0.0006-0.0030	0.0043-0.0200
	8.1	G	2.7482-2.7489	①	0.0050-0.0110	NA	2.1990-2.1996	0.0008-0.0025	0.0151-0.0270
2005	4.8	V	2.5580-2.5593	0.0008-0.0021	0.0015-0.0078	5	2.0990-2.1000	0.0009-0.0025	0.0043-0.0200
	5.3	B	2.5580-2.5593	0.0008-0.0021	0.0015-0.0078	5	2.0991-2.1000	0.0009-0.0025	0.0043-0.0200
	5.3	T	2.5580-2.5593	0.0008-0.0021	0.0015-0.0078	5	2.0991-2.1000	0.0009-0.0025	0.0043-0.0200
	5.3	Z	2.5580-2.5593	0.0008-0.0021	0.0015-0.0078	5	2.0991-2.0999	0.0009-0.0025	0.0043-0.0200
	6.0	N	2.5580-2.5593	0.0008-0.0021	0.0015-0.0078	5	2.0991-2.1000	0.0009-0.0025	0.0043-0.0200
	6.0	U	2.5580-2.5593	0.0008-0.0021	0.0015-0.0078	5	2.0991-2.1000	0.0009-0.0025	0.0043-0.0200
	8.1	G	2.7482-2.7489	②	0.0050-0.0138	NA	2.1990-2.1996	0.0013-0.0027	0.0151-0.0270

06025-YUKO-C09

CRANKSHAFT AND CONNECTING ROD SPECIFICATIONS

All measurements are given in inches.

Year	Engine Displacement Liters	Engine ID/VIN	Crankshaft				Connecting Rod		
			Main Brg. Journal Dia.	Main Brg. Oil Clearance	Shaft End-play	Thrust on No.	Journal Diameter	Oil Clearance	Side Clearance
2006	4.8	V	2.5580-2.5593	0.0008-0.0021	0.0015-0.0078	5	2.0990-2.1000	0.0009-0.0025	0.0043-0.0200
	5.3	B	2.5580-2.5593	0.0008-0.0021	0.0015-0.0078	5	2.0991-2.1000	0.0009-0.0025	0.0043-0.0200
	5.3	T	2.5580-2.5593	0.0008-0.0021	0.0015-0.0078	5	2.0991-2.1000	0.0009-0.0025	0.0043-0.0200
	5.3	Z	2.5580-2.5593	0.0008-0.0021	0.0015-0.0078	5	2.0991-2.0999	0.0009-0.0025	0.0043-0.0200
	6.0	N	2.5580-2.5593	0.0008-0.0021	0.0015-0.0078	5	2.0991-2.1000	0.0009-0.0025	0.0043-0.0200
	6.0	U	2.5580-2.5593	0.0008-0.0021	0.0015-0.0078	5	2.0991-2.1000	0.0009-0.0025	0.0043-0.0200
	8.1	G	2.7482-2.7489	②	0.0050-0.0138	NA	2.1990-2.1996	0.0013-0.0027	0.0151-0.0270

NA: Not Available

① No. 1, 2, 3, 4: 0.0008-0.0020 in.
   No. 5: 0.0014-0.0026 in.

② No. 1, 2, 3, 4: 0.0008-0.0022 in.
   No. 5: 0.0013-0.0027 in.

06025-YUKO-C10

PISTON AND RING SPECIFICATIONS

All measurements are given in inches.

Year	Engine Displacement Liters	Engine ID/VIN	Piston Clearance	Ring Gap			Ring Side Clearance		
				Top Compression	Bottom Compression	Oil Control	Top Compression	Bottom Compression	Oil Control
2002	4.8	V	0.0010-0.0024	0.009-0.015	0.017-0.025	0.007-0.027	0.0016-0.0033	0.0016-0.0031	0.0004-0.0087
	5.3	T	0.0010-0.0024	0.009-0.015	0.017-0.025	0.007-0.027	0.0016-0.0033	0.0016-0.0031	0.0004-0.0087
	6.0	U	0.0010-0.0024	0.009-0.015	0.017-0.025	0.007-0.027	0.0016-0.0033	0.0016-0.0031	0.0004-0.0087
	8.1	G	①	0.012-0.018	0.017-0.025	0.010-0.030	0.0012-0.0029	0.0012-0.0029	0.002-0.008
2003	4.8	V	0.0010-0.0024	0.009-0.015	0.017-0.025	0.007-0.027	0.0016-0.0033	0.0016-0.0031	0.0004-0.0087
	5.3	T	0.0010-0.0024	0.009-0.015	0.017-0.025	0.007-0.027	0.0016-0.0033	0.0016-0.0031	0.0004-0.0087
	5.3	Z	-0.0014 0.0006	0.009-0.017	0.017-0.027	0.007-0.029	0.0016-0.0033	0.0016-0.0031	0.0005-0.0078
	6.0	N	0.0010-0.0024	0.009-0.015	0.017-0.025	0.007-0.027	0.0016-0.0033	0.0016-0.0031	0.0004-0.0087
	6.0	U	0.0010-0.0024	0.009-0.015	0.017-0.025	0.007-0.027	0.0016-0.0033	0.0016-0.0031	0.0004-0.0087
	8.1	G	①	0.012-0.018	0.017-0.025	0.010-0.030	0.0012-0.0029	0.0012-0.0029	0.002-0.008
2004	4.8	V	0.0010-0.0024	0.009-0.015	0.017-0.025	0.007-0.027	0.0016-0.0033	0.0016-0.0031	0.0004-0.0087
	5.3	T	0.0010-0.0024	0.009-0.015	0.017-0.025	0.007-0.027	0.0016-0.0033	0.0016-0.0031	0.0004-0.0087
	5.3	Z	-0.0014 0.0006	0.009-0.017	0.017-0.027	0.007-0.029	0.0016-0.0033	0.0016-0.0031	0.0005-0.0078
	6.0	N	0.0010-0.0024	0.009-0.015	0.017-0.025	0.007-0.027	0.0016-0.0033	0.0016-0.0031	0.0004-0.0087
	6.0	U	0.0010-0.0024	0.009-0.015	0.017-0.025	0.007-0.027	0.0016-0.0033	0.0016-0.0031	0.0004-0.0087
	8.1	G	①	0.012-0.018	0.017-0.025	0.010-0.030	0.0012-0.0029	0.0012-0.0029	0.002-0.008
2005	4.8	V	-0.0014 0.0006	0.009-0.017	0.017-0.027	0.007-0.029	0.0016-0.0033	0.0016-0.0031	0.0005-0.0078
	5.3	B	-0.0014 0.0006	0.009-0.017	0.017-0.027	0.007-0.029	0.0016-0.0033	0.0016-0.0031	0.0005-0.0078
	5.3	T	-0.0014 0.0006	0.009-0.017	0.017-0.027	0.007-0.029	0.0016-0.0033	0.0016-0.0031	0.0005-0.0078
	5.3	Z	-0.0014 0.0006	0.009-0.017	0.017-0.027	0.007-0.029	0.0016-0.0033	0.0016-0.0031	0.0005-0.0078
	6.0	N	-0.0009 0.0012	0.012-0.020	0.020-0.030	0.012-0.034	0.0014-0.0031	0.0013-0.0031	0.0004-0.0078
	6.0	U	-0.0009 0.0012	0.012-0.020	0.020-0.030	0.012-0.034	0.0015-0.0031	0.0013-0.0031	0.0006-0.0078
	8.1	G	①	0.012-0.018	0.017-0.025	0.010-0.030	0.0012-0.0029	0.0012-0.0029	0.002-0.008

06025-YUKO-C11

PISTON AND RING SPECIFICATIONS
All measurements are given in inches.

Year	Engine Displacement Liters	Engine ID/VIN	Piston Clearance	Ring Gap			Ring Side Clearance		
				Top Compression	Bottom Compression	Oil Control	Top Compression	Bottom Compression	Oil Control
2006	4.8	V	-0.0014 0.0006	0.009- 0.017	0.017- 0.027	0.007- 0.029	0.0016- 0.0033	0.0016- 0.0031	0.0005- 0.0078
	5.3	B	-0.0014 0.0006	0.009- 0.017	0.017- 0.027	0.007- 0.029	0.0016- 0.0033	0.0016- 0.0031	0.0005- 0.0078
	5.3	T	-0.0014 0.0006	0.009- 0.017	0.017- 0.027	0.007- 0.029	0.0016- 0.0033	0.0016- 0.0031	0.0005- 0.0078
	5.3	Z	-0.0014 0.0006	0.009- 0.017	0.017- 0.027	0.007- 0.029	0.0016- 0.0033	0.0016- 0.0031	0.0005- 0.0078
	6.0	N	-0.0009 0.0012	0.012- 0.020	0.020- 0.030	0.012- 0.034	0.0014- 0.0031	0.0013- 0.0031	0.0004- 0.0078
	6.0	U	-0.0009 0.0012	0.012- 0.020	0.020- 0.030	0.012- 0.034	0.0015- 0.0031	0.0013- 0.0031	0.0006- 0.0078
	8.1	G	①	0.012- 0.018	0.017- 0.025	0.010- 0.030	0.0012- 0.0029	0.0012- 0.0029	0.002- 0.008

① Interference fit (coated piston)

06025-YUKO-C12

TORQUE SPECIFICATIONS
All readings in ft. lbs.

Year	Engine Displacement Liters	Engine ID/VIN	Cylinder Head Bolts	Main Bearing Bolts	Rod Bearing Bolts	Crankshaft Damper Bolts	Flywheel Bolts	Manifold Intake *	Exhaust	Spark Plugs	Oil Pan Drain Plug
2002	4.8	V	①	②	③	④	⑤	⑥	⑦	11	18
	5.3	T	①	②	③	④	⑤	⑥	⑦	11	18
	6.0	U	①	②	③	④	⑤	⑥	⑦	11	18
	8.1	G	⑧	⑨	⑨	189	⑩	⑪	⑫	22	21
2003	4.8	V	①	②	③	④	⑤	⑥	⑦	11	18
	5.3	T	①	②	③	④	⑤	⑥	⑦	11	18
	5.3	Z	①	②	③	④	⑤	⑥	⑦	11	18
	6.0	N	①	②	③	④	⑤	⑥	⑦	11	18
	6.0	U	①	②	③	④	⑤	⑥	⑦	11	18
	8.1	G	⑧	⑨	⑨	189	⑩	⑬	⑫	22	21
2004	4.8	V	①	②	③	④	⑤	⑥	⑦	11	18
	5.3	T	①	②	③	④	⑤	⑥	⑦	11	18
	5.3	Z	①	②	③	④	⑤	⑥	⑦	11	18
	6.0	N	①	②	③	④	⑤	⑥	⑦	11	18
	6.0	U	①	②	③	④	⑤	⑥	⑦	11	18
	8.1	G	⑧	⑨	⑨	189	⑩	⑬	⑫	22	21
2005	4.8	V	⑭	②	③	④	⑤	⑥	⑦	11	18
	5.3	B	⑭	②	③	④	⑤	⑥	⑦	11	18
	5.3	T	⑭	②	③	④	⑤	⑥	⑦	11	18
	5.3	Z	⑭	②	③	④	⑤	⑥	⑦	11	18
	6.0	N	⑭	②	③	④	⑤	⑥	⑦	11	18
	6.0	U	⑭	②	③	④	⑤	⑥	⑦	11	18
	8.1	G	⑭	⑨	⑨	189	⑩	⑬	⑫	22	21
2006	4.8	V	⑭	②	③	④	⑤	⑥	⑦	11	18
	5.3	B	⑭	②	③	④	⑤	⑥	⑦	11	18
	5.3	T	⑭	②	③	④	⑤	⑥	⑦	11	18
	5.3	Z	⑭	②	③	④	⑤	⑥	⑦	11	18
	6.0	N	⑭	②	③	④	⑤	⑥	⑦	11	18
	6.0	U	⑭	②	③	④	⑤	⑥	⑦	11	18
	8.1	G	⑭	⑨	⑨	189	⑩	⑬	⑫	22	21

* NOTE: Applies to Lower Manifold only.

① Step 1: 22 ft. lbs.
Step 2: 90 degrees
Step 3: 90 degrees
(except medium length bolts at front and rear)
Step 4: Tighten medium length bolts,
at front and rear an additional 50 degrees

② Inner bolts: 15 ft. lbs., plus 80 degrees
Outer bolts: 15 ft. lbs. , plus 51 degrees

③ Step 1: 15 ft. lbs.
Step 2 (2002): 60 degrees
Step 2 (2003-06): 75 degrees

④ Installation pass: 240 ft. lbs. (to be sure damper is seated)
Step 1: Remove bolt and replace with a new bolt
Step 2: 37 ft. lbs.
Step 3: 140 degrees

⑤ Step 1: 15 ft. lbs.
Step 2: 37 ft. lbs.
Step 3: 74 ft. lbs.

⑥ Step 1: 44 inch lbs.
Step 2: 89 inch lbs.

⑦ Step 1: 11 ft. lbs.
Step 2: 18 ft. lbs.

⑧ 1st pass: 22 ft. lbs.
2nd pass: 22 ft. lbs.,
plus 120 degrees
Final pass:
Short bolt: Plus 60 degrees
Medium bolt: Plus 45 degrees
Long bolt: Plus 30 degrees

⑨ 22 ft. lbs., plus 90 degrees

⑩ 1st pass: 30 ft. lbs.
2nd pass: 59 ft. lbs.
3rd pass: 74 ft. lbs.

⑪ 1st & 2nd pass: 44 inch lbs.
3rd pass: 89 inch lbs.
4th pass: 106 inch lbs.

⑫ Center bolt: 26 ft. lbs.
Nut: 12 ft. lbs.
Stud: 15 ft. lbs.

⑬ 1st pass: 44 inch lbs.
2nd pass: 71 inch lbs.
3rd pass: 106 inch lbs.
Final pass: 12 ft. lbs.

⑭ M8 bolts: 22 ft. lbs.
M11 bolts 1st pass: 22 ft. lbs.
M11 bolts 2nd pass: 90 degrees
M11 bolts final pass: 70 degrees

WHEEL ALIGNMENT

Year	Model	Caster Range (+/-Deg.)	Caster Preferred Setting (Deg.)	Camber Range (+/-Deg.)	Camber Preferred Setting (Deg.)	Toe-in (in.)
2001	2WD	1.00	①	0.50	+0.25	0.10+/-0.20
	2WD HD	1.00	②	0.50	+0.25	0.10+/-0.12
	4WD	1.00	③	0.50	+0.25	0.10+/-0.20
	4WD HD	1.00	②	0.50	+0.25	0.10+/-0.20
2002	2WD	1.00	①	0.50	+0.25	0.10+/-0.20
	2WD HD	1.00	②	0.50	+0.25	0.10+/-0.12
	4WD	1.00	③	0.50	+0.25	0.10+/-0.20
	4WD HD	1.00	②	0.50	+0.25	0.10+/-0.20
2003	2WD	1.00	①	0.50	+0.25	0.10+/-0.20
	2WD HD	1.00	②	0.50	+0.25	0.10+/-0.12
	4WD	1.00	③	0.50	+0.25	0.10+/-0.20
	4WD HD	1.00	②	0.50	+0.25	0.10+/-0.20
2004	2WD	1.00	①	0.50	+0.25	0.10+/-0.20
	2WD HD	1.00	②	0.50	+0.25	0.10+/-0.12
	4WD	1.00	③	0.50	+0.25	0.10+/-0.20
	4WD HD	1.00	②	0.50	+0.25	0.10+/-0.20
2005	2WD	1.00	①	0.50	+0.25	0.10+/-0.20
	2WD Police	1.00	④	0.50	-0.10	0.10+/-0.20
	2WD HD	1.00	②	0.50	+0.25	0.10+/-0.20
	4WD	1.00	⑤	0.50	+0.25	0.10+/-0.20
	4WD HD	1.00	②	0.50	+0.25	0.10+/-0.20
2006	2WD	1.00	①	0.50	+0.25	0.10+/-0.20
	2WD Police	1.00	④	0.50	-0.10	0.10+/-0.20
	2WD HD	1.00	②	0.50	+0.25	0.10+/-0.20
	4WD	1.00	⑤	0.50	+0.25	0.10+/-0.20
	4WD HD	1.00	②	0.50	+0.25	0.10+/-0.20

① Left side: 3.90
   Right side: 4.70

② Left side: 4.50
   Right side: 4.75

③ Left side: 3.50
   Right side: 4.50

④ Left side: 4.35
   Right side: 4.85

⑤ With 16" tires Left side: 3.60
   Right side: 4.40
   With 17" tires Left side: 3.80
   Right side: 4.50
   With 20" tires Left side: 4.10
   Right side: 5.10

06025-YUKO-C14

TIRE, WHEEL AND BALL JOINT SPECIFICATIONS

| Year | Model | OEM Tires | | Tire Pressures (psi) | | Wheel Size | Ball Joint Inspection | Lug Nut (ft. lbs.) |
		Standard	Optional	Front	Rear			
2002	Escalade	P265/70R16	None	36	36	7-JJ	U ① L: 0.090 in.	140
	Tahoe/Yukon, 2WD	P235/75R15	None	36	36	6.5-JJ	U ② L: 0.090 in.	140
	Tahoe/Yukon, 4WD	P245/75R16	P265/75R16	36	36	7-JJ	①	140
	1500 Suburban 2WD	P235/75R15XL	LT245/75R16E	36	36	7-JJ	L ③	140
	1500 Suburban 4WD	P245/75R16C	LT245/75R16E	36	36	6.5-JJ	L ③	140
	2500 Suburban	LT245/75R16E	None	36	36	6.5-JJ	U: 0.125 in. L ③	140
2003	Escalade	P265/70R16	None	36	36	7-JJ	U ① L: 0.090 in.	140
	Tahoe/Yukon, 2WD	P235/75R15	None	36	36	6.5-JJ	U ② L: 0.090 in.	140
	Tahoe/Yukon, 4WD	P245/75R16	P265/75R16	36	36	7-JJ	①	140
	1500 Suburban 2WD	P235/75R15XL	LT245/75R16E	36	36	7-JJ	L ③	140
	1500 Suburban 4WD	P245/75R16C	LT245/75R16E	36	36	6.5-JJ	L ③	140
	2500 Suburban	LT245/75R16E	None	36	36	6.5-JJ	U: 0.125 in. L ③	140
2004	Escalade	P265/70R16	None	36	36	7-JJ	U ① L: 0.090 in.	140
	Tahoe/Yukon, 2WD	P235/75R15	None	36	36	6.5-JJ	U ② L: 0.090 in.	140
	Tahoe/Yukon, 4WD	P245/75R16	P265/75R16	36	36	7-JJ	①	140
	1500 Suburban 2WD	P235/75R15XL	LT245/75R16E	36	36	7-JJ	L ③	140
	1500 Suburban 4WD	P245/75R16C	LT245/75R16E	36	36	6.5-JJ	L ③	140
	2500 Suburban	LT245/75R16E	None	36	36	6.5-JJ	U: 0.125 in. L ③	140
2005	Escalade	P265/70R16	④	36	36	7-JJ	U ① L: 0.090 in.	140
	Tahoe/Yukon, 2WD	P235/75R15	None	36	36	6.5-JJ	U ② L: 0.090 in.	140
	Tahoe/Yukon, 4WD	P245/75R16	P265/75R16		36	7-JJ	①	140
	1500 Suburban 2WD	P235/75R15XL	LT245/75R16E	36	36	7-JJ	L ③	140
	1500 Suburban 4WD	P245/75R16C	LT245/75R16E	36	36	6.5-JJ	L ③	140
	2500 Suburban	LT245/75R16E	None	36	36	6.5-JJ	U: 0.125 in. L ③	140
2006	Escalade	P265/70R16	④	36	36	7-JJ	U ① L: 0.090 in.	140
	Tahoe/Yukon, 2WD	P235/75R15	None	36	36	6.5-JJ	U ② L: 0.090 in.	140
	Tahoe/Yukon, 4WD	P245/75R16	P265/75R16	36	36	7-JJ	①	140
	1500 Suburban 2WD	P235/75R15XL	LT245/75R16E	36	36	7-JJ	L ③	140
	1500 Suburban 4WD	P245/75R16C	LT245/75R16E	36	36	6.5-JJ	L ③	140
	2500 Suburban	LT245/75R16E	None	36	36	6.5-JJ	U: 0.125 in. L ③	140

OEM: Original Equipment Manufacturer

PSI: Pounds Per Square Inch

STD: Standard

OPT: Optional

L: Lower

U: Upper

① Ball joint is adjustable, refer to manual for procedure

② Replace if any movement is noted or if stud can be moved by hand

③ Do not lift truck. Inspect the boss into which the grease fitting is threaded. Replace if the boss is flush or receded below the surface of the ball joint

④ 17" and 20" tires are optional but sizes not available.

BRAKE SPECIFICATIONS
All measurements in inches unless noted

Year	Model		Brake Disc — Original Thickness	Brake Disc — Minimum Thickness	Max. Runout	Brake Drum Diameter — Original Inside Diameter	Brake Drum Diameter — Max. Wear Limit	Brake Drum Diameter — Maximum Machine Diameter	Minimum Lining Thickness	Brake Caliper — Bracket Bolts (ft. lbs.)	Brake Caliper — Mounting Bolts (ft. lbs.)
2002	Escalade	F	①	②	0.005	—	—	—	0.030	③	80
		R	④	⑤	0.005	—	—	—	0.030	⑥	⑦
	Suburban	F	①	②	0.005	—	—	—	0.030	③	80
		R	④	⑤	0.005	—	—	—	0.030	⑥	⑦
	Tahoe	F	①	②	0.005	—	—	—	0.030	③	80
		R	④	⑤	0.005	—	—	—	0.030	⑥	⑦
	Yukon	F	①	②	0.005	—	—	—	0.030	③	80
		R	④	⑤	0.005	—	—	—	0.030	⑥	⑦
	Yukon Denali	F	①	②	0.005	—	—	—	0.030	③	80
		R	④	⑤	0.005	—	—	—	0.030	⑥	⑦
	Yukon Denali XL	F	①	②	0.005	—	—	—	0.030	③	80
		R	④	⑤	0.005	—	—	—	0.030	⑥	⑦
	Yukon XL	F	①	②	0.005	—	—	—	0.030	③	80
		R	④	⑤	0.005	—	—	—	0.030	⑥	⑦
2003	Escalade	F	①	②	0.005	—	—	—	0.030	③	80
		R	④	⑤	0.005	—	—	—	0.030	⑥	⑦
	Suburban	F	①	②	0.005	—	—	—	0.030	③	80
		R	④	⑤	0.005	—	—	—	0.030	⑥	⑦
	Tahoe	F	①	②	0.005	—	—	—	0.030	③	80
		R	④	⑤	0.005	—	—	—	0.030	⑥	⑦
	Yukon	F	①	②	0.005	—	—	—	0.030	③	80
		R	④	⑤	0.005	—	—	—	0.030	⑥	⑦
	Yukon Denali	F	①	②	0.005	—	—	—	0.030	③	80
		R	④	⑤	0.005	—	—	—	0.030	⑥	⑦
	Yukon Denali XL	F	①	②	0.005	—	—	—	0.030	③	80
		R	④	⑤	0.005	—	—	—	0.030	⑥	⑦
	Yukon XL	F	①	②	0.005	—	—	—	0.030	③	80
		R	④	⑤	0.005	—	—	—	0.030	⑥	⑦
2004	Escalade	F	①	②	0.005	—	—	—	0.030	③	80
		R	④	⑤	0.005	—	—	—	0.030	⑥	⑦
	Suburban	F	①	②	0.005	—	—	—	0.030	③	80
		R	④	⑤	0.005	—	—	—	0.030	⑥	⑦
	Tahoe	F	①	②	0.005	—	—	—	0.030	③	80
		R	④	⑤	0.005	—	—	—	0.030	⑥	⑦
	Yukon	F	①	②	0.005	—	—	—	0.030	③	80
		R	④	⑤	0.005	—	—	—	0.030	⑥	⑦
	Yukon Denali	F	①	②	0.005	—	—	—	0.030	③	80
		R	④	⑤	0.005	—	—	—	0.030	⑥	⑦
	Yukon Denali XL	F	①	②	0.005	—	—	—	0.030	③	80
		R	④	⑤	0.005	—	—	—	0.030	⑥	⑦
	Yukon XL	F	①	②	0.005	—	—	—	0.030	③	80
		R	④	⑤	0.005	—	—	—	0.030	⑥	⑦

06025-YUKO-C16

BRAKE SPECIFICATIONS
All measurements in inches unless noted

Year	Model		Brake Disc Original Thickness	Brake Disc Minimum Thickness	Brake Disc Max. Runout	Brake Drum Diameter Original Inside Diameter	Brake Drum Diameter Max. Wear Limit	Brake Drum Diameter Maximum Machine Diameter	Minimum Lining Thickness	Brake Caliper Bracket Bolts (ft. lbs.)	Brake Caliper Mounting Bolts (ft. lbs.)
2005	Escalade	F	⑧	⑨	0.005	—	—	—	0.030	③	80
		R	⑩	⑪	0.005	—	—	—	0.030	⑥	⑦
	Suburban	F	⑧	⑨	0.005	—	—	—	0.030	③	80
		R	⑩	⑪	0.005	—	—	—	0.030	⑥	⑦
	Tahoe	F	⑧	⑨	0.005	—	—	—	0.030	③	80
		R	⑩	⑪	0.005	—	—	—	0.030	⑥	⑦
	Yukon	F	⑧	⑨	0.005	—	—	—	0.030	③	80
		R	⑩	⑪	0.005	—	—	—	0.030	⑥	⑦
	Yukon Denali	F	⑧	⑨	0.005	—	—	—	0.030	③	80
		R	⑩	⑪	0.005	—	—	—	0.030	⑥	⑦
	Yukon Denali XL	F	⑧	⑨	0.005	—	—	—	0.030	③	80
		R	⑩	⑪	0.005	—	—	—	0.030	⑥	⑦
	Yukon XL	F	⑧	⑨	0.005	—	—	—	0.030	③	80
		R	⑩	⑪	0.005	—	—	—	0.030	⑥	⑦
2006	Escalade	F	⑧	⑨	0.005	—	—	—	0.030	③	80
		R	⑩	⑪	0.005	—	—	—	0.030	⑥	⑦
	Suburban	F	⑧	⑨	0.005	—	—	—	0.030	③	80
		R	⑩	⑪	0.005	—	—	—	0.030	⑥	⑦
	Tahoe	F	⑧	⑨	0.005	—	—	—	0.030	③	80
		R	⑩	⑪	0.005	—	—	—	0.030	⑥	⑦
	Yukon	F	⑧	⑨	0.005	—	—	—	0.030	③	80
		R	⑩	⑪	0.005	—	—	—	0.030	⑥	⑦
	Yukon Denali	F	⑧	⑨	0.005	—	—	—	0.030	③	80
		R	⑩	⑪	0.005	—	—	—	0.030	⑥	⑦
	Yukon Denali XL	F	⑧	⑨	0.005	—	—	—	0.030	③	80
		R	⑩	⑪	0.005	—	—	—	0.030	⑥	⑦
	Yukon XL	F	⑧	⑨	0.005	—	—	—	0.030	③	80
		R	⑩	⑪	0.005	—	—	—	0.030	⑥	⑦

NA: Not Available

① Vacuum: 1.14 in.
Hydraulic: 1.50 in.

② Vacuum: 1.10 in.
Hydraulic: 1.46 in.

③ 15 series: 129 ft. lbs.
25 series: 221 ft. lbs.

④ Vacuum: 0.787 in.
Hydraulic: 1.14 in.

⑤ Vacuum: 0.748 in.
Hydraulic: 1.10 in.

⑥ Vacuum: 148 ft. lbs.
Hydraulic (9000 lbs.): 122 ft. lbs.
Hydraulic (12,000 lbs.): 221 ft. lbs.

⑦ 15 series: 31 ft. lbs.
25 series: 80 ft. lbs.

⑧ 6400/7000/7200 GVW: 1.142
9900 GVW: 1.496.

⑨ 6400/7000/7200 GVW: 1.102
9900 GVW: 1.457

⑩ 6400/7000/7200 GVW: 1.181
9900 GVW: 1.142

⑪ 6400/7000/7200 GVW: 1.142
9900 GVW: 1.102

06025-YUKO-C17

SCHEDULED MAINTENANCE INTERVALS
2002-04 ESCALADE, SUBURBAN, TAHOE, YUKON, YUKON DENALI, & YUKON XL

TO BE SERVICED	TYPE OF SERVICE	VEHICLE MILEAGE INTERVAL (x1000)															
		7.5	15	22.5	30	37.5	45	52.5	60	67.5	75	82.5	90	97.5	105	112.5	120
Accessory drive belt	S/I								✓								✓
Automatic transmission fluid ①	R	Every 50,000 miles															
Brake system	S/I	✓	✓	✓	✓	✓	✓	✓	✓	✓	✓	✓	✓	✓	✓	✓	✓
Chassis & suspension grease points	L	✓	✓	✓	✓	✓	✓	✓	✓	✓	✓	✓	✓	✓	✓	✓	✓
Cooling fan operation	S/I		✓		✓		✓		✓		✓		✓		✓		✓
CV-joint boots & axle seals	S/I	✓	✓	✓	✓	✓	✓	✓	✓	✓	✓	✓	✓	✓	✓	✓	✓
EGR system	S/I								✓								✓
Engine coolant	R	Every 150,000 miles															
Engine oil & filter	R	✓	✓	✓	✓	✓	✓	✓	✓	✓	✓	✓	✓	✓	✓	✓	✓
EVAP system	S/I								✓								✓
Front wheel bearings ①	S/I & L				✓				✓				✓				✓
Fuel filter	R								✓								✓
Fuel system	S/I								✓								✓
Rear/front axle fluid level	S/I	✓	✓	✓	✓	✓	✓	✓	✓	✓	✓	✓	✓	✓	✓	✓	✓
Rotate tires	S/I	✓	✓	✓	✓	✓	✓	✓	✓	✓	✓	✓	✓	✓	✓	✓	✓
Shields & underhood insulation ②	S/I		✓		✓		✓		✓		✓		✓		✓		✓
Spark plugs	R	Every 100,000 miles															
Spark plug wires	S/I	Every 100,000 miles															

R: Replace   S/I: Inspect and service, if necessary   L: Lubricate

① 2-wheel drive models only.

② Vehicles with a GVWR or 8500 lbs. or more only.

FREQUENT OPERATION MAINTENANCE (SEVERE SERVICE)

If a vehicle is operated under any of the following conditions it is considered severe service:

- Towing a trailer or using a camper or car-top carrier.
- Repeated short trips of less than 5 miles in temperatures below freezing, or trips of less than 10 miles in any temperature.
- Extensive idling or low-speed driving for long distances as in heavy commercial use, such as delivery, taxi or police cars.
- Operating on rough, muddy or salt-covered roads.
- Operating on unpaved or dusty roads.
- Driving in extremely hot (over 90°) conditions.

Engine oil & filter: replace every 3000 miles or 3 months, whichever occurs first.

Chassis and suspension grease points: lubricate every 3000 miles.

Rear/front axle fluid level: inspect every 3000 miles.

Rotate the tires ever 6000 miles.

Brake system components: inspect ever 6000 miles.

Front wheel bearings (2-wheel drive only): clean, inspect and repack every 15,000 miles.

Shields & underhood insulation (vehicles w/GVWR over 8500 lbs. only): inspect every 15,000 miles.

Cooling fan system hoses & connections: inspect every 15,000 miles.

Fuel filter: replace every 30,000 miles.

Air cleaner filter: inspect every 45,000 miles.

Automatic transmission fluid & filter: replace every 50,000 miles.

Accessory drive belt: inspect every 60,000 miles.

Fuel system tank, cap and lines: inspect every 60,000 miles.

EVAP system: inspect every 60,000 miles.

EGR system: inspect every 60,000 miles.

PCV system: inspect every 100,000 miles.

Engine cooling system components: inspect and clean every 150,000 miles.

06025-YUKO-C18

MAINTENANCE I AND II SERVICE SCHEDULES
2005-06 GENERAL MOTORS—ESCALADE, SUBURBAN, TAHOE, YUKON, YUKON DENALI, YUKON XL

When the CHANGE ENGINE OIL light appears, certain services and inspections are required.
Required services are described as Maintenance I and Maintenance II.
The first service on a vehicle should be Maintenance I, and the second service should be Maintenance II.
Alternate between the 2 thereafter. However, in some cases, Maintenance II may be required more often.
Maintenance I: Use Maintenance I if the CHANGE ENGINE OIL light comes on within 10 months
since vehicle was purchased or, if Maintenance II was performed.
Maintenance II: Use Maintenance II if the previous service performed was Maintenance I.
Always use Maintenance II whenever the CHANGE ENGINE OIL light comes on 10 months or more since the last
service, or, if the CHANGE ENGINE OIL light has not come on at all for one year.

Service	Maintenance I	Maintenance II
Change the engine oil and filter. Reset the oil life system.	✓	✓
Visually inspect the vehicle for leaks or damage. A fluid loss in the vehicle system could indicate a problem. Inspected, repair and add fluid to the system if necessary.	✓	✓
Inspect the engine air cleaner filter. If necessary, replace the filter.	✓	✓
Rotate the tires. Inspect the tire inflation pressures and the tire wear.	✓	✓
Visually inspect the brake lines and hoses for proper hook-up, binding, leaks, cracks, chafing, etc. Inspect the disc brake pads for wear and the rotors for surface condition. Inspect the drum brake linings for wear or cracks. Inspect other brake parts, including drums, wheel cylinders, calipers, parking brake, etc. Inspect the parking brake adjustment.	✓	✓
Inspect the engine coolant and the windshield washer fluid levels. Add fluid as needed.	✓	✓
Inspect the suspension and steering components. Inspect the front and rear suspension and the steering system for damaged, loose or missing parts, or signs of wear. Inspect the power steering lines and the hoses for proper hook-up, binding, leaks, cracks, chafing, etc.	--	✓
Visually inspect the coolant hoses and replace the hoses if they are cracked, swollen or deteriorated. Inspect all pipes, fittings and clamps; replace with GM parts as needed. To help ensure proper operation, a pressure test of the cooling system and pressure cap and cleaning the outside of the radiator and air conditioning condenser is recommended at least once a year.		✓
Inspect the wiper blades for wear or cracking.	--	✓
Inspect the restraint system components.Ensure the safety belt reminder light and all the belts, buckles, latch plates, retractors and anchorages are working properly. Look for any other loose or damaged safety belt system parts. If you see anything that might keep a safety belt system from working correctly, repair or replaced the damaged part. Replace torn or frayed safety belts, refer to Operational and Functional Checks in Seat Belts. Inspect for any opened or broken air bag coverings, and repair or replace as needed. The air bag system does require regular maintenance.	--	✓

06025-YUKO-C19

MAINTENANCE I AND II SERVICE SCHEDULES
2005-06 GENERAL MOTORS—ESCALADE, SUBURBAN, TAHOE, YUKON, YUKON DENALI, YUKON XL

Lubricate the body components.Lubricate all key lock cylinders, hood latch assemblies, secondary latches, pivots, spring anchor and release pawl, hood and door hinges, rear folding seats and liftgate hinges. Frequent lubrication may be required when exposed to a corrosive environment, refer to Fluid and Lubricant Recommendations . Applying dielectric silicone grease GM P/N 12345579 (Canadian P/N 1974984) or equivalent on the weatherstrips with a clean cloth.	--	✓
Inspect the transaxle fluid level and add fluid as needed.	--	✓
Inspect the suspension and steering components.Inspect the front and rear suspension and the steering system for damaged, loose or missing parts, or signs of wear. Inspect power steering lines and hoses for proper hook-up, binding, leaks, cracks, chafing, etc.	--	✓
Inspect the throttle system for interference or binding and for damaged or missing parts. Replace the parts as needed. Replace any components that have high effort or excessive wear. Do not lubricate the accelerator or the cruise control cables.	--	✓
Replace the passenger compartment air filter.	--	✓

06025-YUKO-C20

ENGINE REPAIR

Distributor

REMOVAL

All engines use a distributorless ignition system.

Alternator

REMOVAL & INSTALLATION

4.8L, 5.3L and 6.0L Engines

1. Disconnect the negative battery cable.
2. Remove or disconnect the following:
   - Accessory drive belt
   - Engine sight shield, if necessary
   - Electrical connections from the generator
   - Mounting bolts
   - Generator

To install:

3. Install or connect the following:
   - Generator
   - Generator mounting bolts and tighten to 37 ft. lbs. (50 Nm).
   - Electrical connections to the generator. Tighten the B+ nut to 13 ft. lbs. (18 Nm).
   - Engine sight shield, if removed
   - Accessory drive belt
   - Negative battery cable

8.1L Engine

1. Disconnect the negative battery cable.
2. Remove or disconnect the following:
   - Accessory drive belt
   - Electrical connections from the generator
3. Remove the cable from the alternator as follows:
   a. Slide the boot down, to reveal the terminal stud.
   b. Unfasten the cable nut from the stud, then remove the cable.
4. Remove the mounting bolts and the alternator.

To install:

5. Install or connect the following:
   - Alternator
   - Alternator mounting bolts and tighten to 37 ft. lbs. (50 Nm)
   - Accessory drive belt
   - Cable, secure with the nut and tighten to 80 inch lbs. (9 Nm)
   - Boot back over the terminal stud.
   - Electrical connections
   - Negative battery cable

Ignition Timing

ADJUSTMENT

Ignition timing is controlled by the PCM. Always refer to the Vehicle Emissions Control Information label in the engine compartment for base ignition timing specification.

Engine Assembly

REMOVAL & INSTALLATION

4.8L, 5.3L and 6.0L Engines

✻✻ CAUTION

Before servicing any electrical component, the ignition key must be in the OFF or LOCK position and all electrical loads must be OFF, unless instructed otherwise in these procedures.

1. Remove or disconnect the following:
   - Negative battery cable
   - Coolant
   - A/C refrigerant
2. Raise the hood to the servicing position. Move the hood hinge bolt to hold the hood in the servicing position.
   - Upper and the lower radiator hoses from the engine
   - Air cleaner duct from the engine
   - A/C condenser mounting bolts
   - Radiator support from the vehicle
   - A/C compressor
   - Coolant hose from the throttle body
   - Heater hoses from the engine and the cowl
   - Engine sight shield from the intake manifold
   - Accelerator control cable mounting bracket from the intake manifold

✻✻ CAUTION

In order to avoid possible injury or vehicle damage, always replace the accelerator control cable with a NEW cable whenever you remove the engine from the vehicle. In order to avoid cruise control cable damage, position the cable out of the way while you remove or install the engine.

   - Accelerator control cable and the cruise control cable, if equipped, from the throttle shaft
3. Open the large electrical harness retainer. Remove one 10mm nut in order to release the engine harness from the intake manifold.
4. Disconnect the electrical connectors from the following:

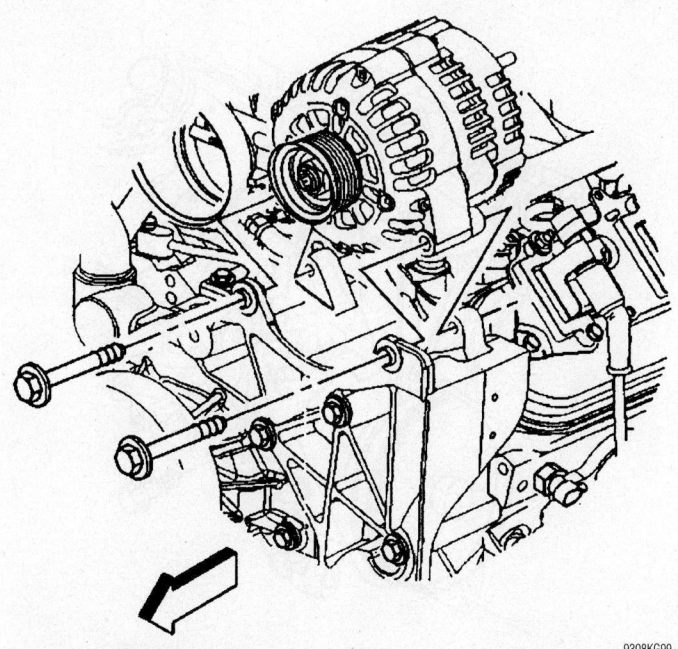

9308KG99

Alternator mounting—4.8L, 5.3L and 6.0L engines

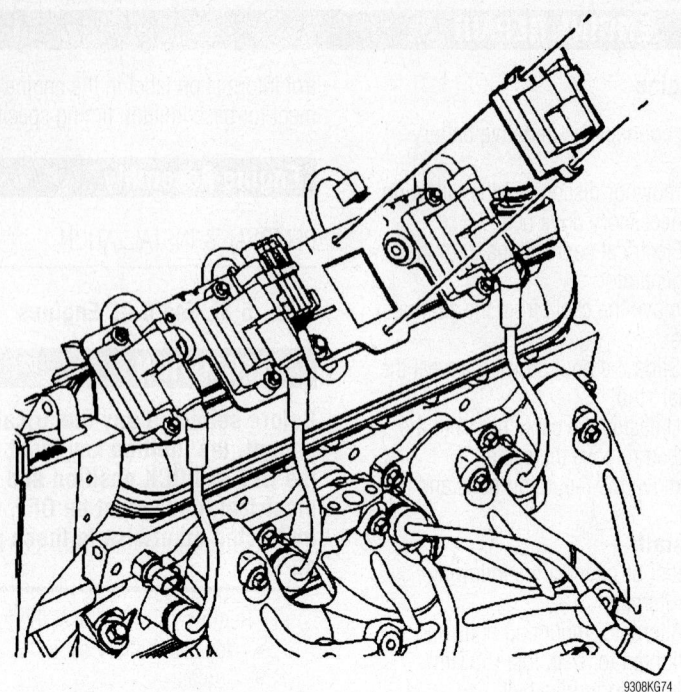

Ignition coil removal—4.8L, 5.3L and 6.0L engines

- Eight injector connectors
- Idle Air Control (IAC) motor
- Throttle Position (TP) sensor
- Evaporative Emissions (EVAP) canister purge solenoid
- Manifold Absolute Pressure (MAP) sensor
- Camshaft Position (CMP) sensor
- Ground splice at the rear of the right side of the block
- Ground splice and the ground strap at the rear of the left side of the block
- Coolant Temperature (CTS) sensor
- Oil pressure sensor/switch
- Intake electrical and disconnect from harness
- Junction block bracket from alternator bracket

5. Set the electrical harness aside.
6. Remove or disconnect the following:
   - EVAP canister purge solenoid vent tube from the solenoid by squeezing the retainer, then release the tube from the solenoid
   - Battery negative cable from the engine block
   - Drive belt
   - Bolts holding the alternator mounting bracket to the cylinder head and block
   - Bolt behind the power steering pump to engine block
   - Alternator mounting bracket and position the bracket aside
   - Fuel pipes from the engine

7. Raise the vehicle.
   - Steering linkage under body shield, if equipped
   - Engine oil pan under body shield, if equipped
   - Engine oil
   - Starter motor
8. Disconnect the engine wiring harness from the following components:
   - Crankshaft Position (CKP) sensor
   - Engine oil level sensor
   - Block heater, if equipped

- Wiring harness from the oil pan
- Reposition wiring from lower engine area

9. Remove or disconnect the following:
   - Exhaust pipes from the exhaust manifolds
   - Transmission cooler pipe retainer from the right side of the engine block, if equipped
   - Torque converter shield from the engine
   - Torque converter bolts
   - Nut and the transmission oil level indicator tube from the bellhousing stud
   - Lower bellhousing studs from the engine
10. Lower the vehicle.
    - Remaining bellhousing bolts
    - Engine electrical harness; position aside
    - Ignition coil(s)
11. Install an engine crane.
12. Install a floor jack or stands to transmission for support.
13. Remove the engine mount bolts.

➡️**Use care while moving the engine assembly in order to avoid breaking the MAP sensor locating tabs. Broken MAP sensor tabs may result in decreased engine performance.**

14. Remove the engine from the vehicle.
To install:
15. Install or connect the following:
    - Engine to the vehicle
    - Engine mount bolts
    - Upper bellhousing bolts

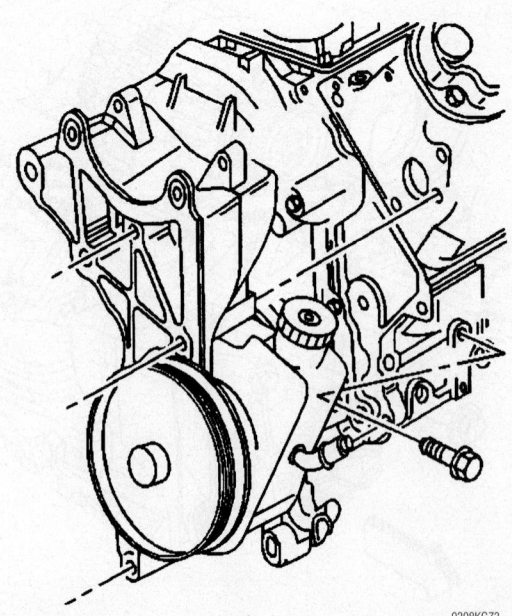

Power steering pump removal—4.8L, 5.3L and 6.0L engines

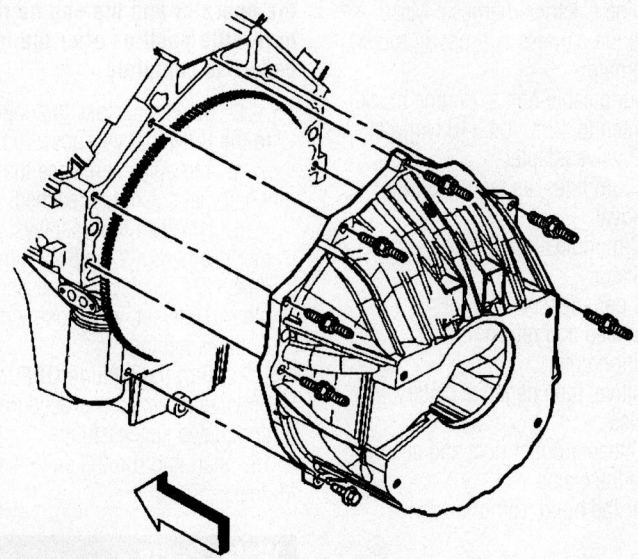

9308KG75

Bellhousing bolt removal—4.8L, 5.3L and 6.0L engines

16. Remove transmission support apparatus.

17. Remove the lifting device.

18. Remove the lift brackets from both cylinder heads.

19. Install the ignition coil(s) and the spark plug wire(s).

20. Route the engine wiring harness to the lower right hand side of the engine.

21. Raise the vehicle.

22. Install or connect the following:
- Remaining bellhousing bolts
- Torque converter bolts
- Torque converter shield
- Transmission oil level indicator tube and nut to bellhousing stud
- A/C compressor
- Transmission cooler pipe retainer to right side of engine block
- Engine exhaust pipes to the exhaust manifolds

23. Reroute wiring to lower engine area and install bolt to oil pan.

24. Install or connect the following:
- CKP sensor electrical connector
- Engine oil level sensor and the block heater electrical connectors, if equipped.
- Starter motor
- Engine oil pan under body shield, if equipped
- Steering linkage under body shield

25. Lower the vehicle.
- Fuel pipes to the engine
- Alternator mounting bracket to the cylinder head using the nuts and the bolts. Tighten the bolts to 37 ft. lbs. (50 Nm).
- Bolt at the rear of the power steering pump to the engine block and tighten to 37 ft. lbs. (50 Nm).
- Alternator
- Drive belt
- Negative battery cable to the engine block
- EVAP canister purge solenoid to the intake manifold

26. Route the engine harness over the top of the engine. Attach the connectors to the following components:
- Eight injector connectors
- IAC motor
- TP sensor
- EVAP canister purge solenoid.
- MAP sensor
- CMP sensor
- Ground splice at the rear of the right side of engine block
- Ground splice and the ground strap at the rear of the left side of engine block
- CTS sensor
- Oil pressure sensor/switch

27. Install or connect the following:
- Nut to the engine wiring harness bracket and tighten to 89 inch lbs. (10 Nm)

❊❊ CAUTION

In order to avoid possible injury or vehicle damage, always replace the accelerator control cable with a NEW cable whenever you remove the engine from the vehicle. In order to avoid cruise control cable damage, position the cable out of the way while you remove or install the engine.

- NEW accelerator control cable
- Cruise control cable, if equipped, to the throttle shaft
- Bolts for the accelerator control cable mounting bracket and tighten to 89 inch lbs. (10 Nm)
- Engine sight shield to the intake manifold
- Heater hoses to the cowl and the engine
- Coolant hose to the throttle body
- Radiator support in the vehicle
- A/C condenser mounting bolts
- Air cleaner duct
- Lower radiator hoses to the engine

28. Lower the hood.

29. Fill the engine with oil.

30. Fill the engine with coolant.

31. Connect the negative battery cable.

8.1L Engine

1. Raise the hood to the servicing position. Move the hood hinge bolt to hold the hood in the servicing position.

2. Release the fuel system pressure.

3. Remove or disconnect the following:
- Negative, then positive battery cables
- Coolant
- A/C refrigerant
- Engine oil cooler lines from the engine block
- Transmission-to-engine bolts
- Clutch pressure plate bolts, if equipped
- Torque converter bolts, if equipped
- Catalytic converter
- Exhaust manifold pipe

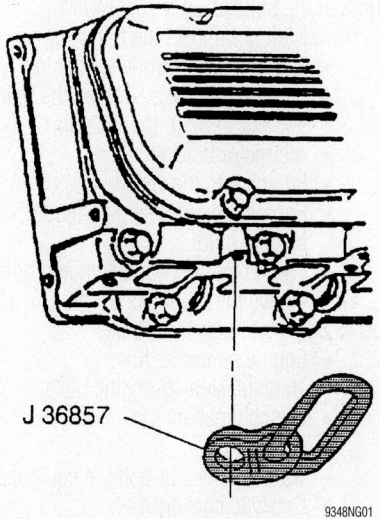

J 36857

9348NG01

Install suitable lift brackets to the rear of the right head and the front of the left head

- Hoses from power steering pump, then plug the lines and ports
- Starter motor

4. Raise the vehicle.
- Engine electrical harness and tie aside
- Alternator
- Ground cable bolt from engine block
- Exhaust Gas Recirculation (EGR) valve adapter
- Vacuum lines (tag before removal)
- Throttle Actuator Control (TAC) module electrical connector

5. Install Engine Lift Brackets part No. J 36857, or equivalent, to the rear of the right cylinder head and the front of the left cylinder head.

6. Install the attaching bolt and washer. Use part No. 9428217 with 1560963. Tighten the bolts to 30 ft. lbs. (40 Nm).

7. Remove or disconnect the following:
- Engine mount heat shield bolt and shields
- Engine mount-to-engine mount bracket bolts
- Engine from the vehicle, using a suitable lifting device. Place on a suitable stand.
- A/C compressor/power steering pump bracket from the cylinder head
- Lift brackets from the cylinder head

To install:

8. Install Engine Lift Brackets part No. J 36857, or equivalent, to the rear of the right cylinder head and the front of the left cylinder head.

9. Install the attaching bolt and washer. Use part No. 9428217 with 1560963. Tighten the bolts to 30 ft. lbs. (40 Nm).

10. Install or connect the following:
- A/C compressor/power steering mounting bracket. Tighten the bolts and nut to 37 ft. lbs. (50 Nm)
- Alternator bracket
- Engine into the vehicle
- Engine mount-to-engine mount bracket bolts
- Engine mount heat shield and bolts

11. Remove the lift hooks from the cylinder heads, then raise the vehicle.
- Engine oil cooler lines
- Transmission-to-engine bolts
- Clutch pressure plate bolts, if equipped
- Torque converter bolts, if equipped
- Catalytic converter
- Exhaust manifold pipe
- Hoses to the power steering pump
- Starter motor

12. Lower the vehicle.

- Engine electrical harness. Make sure the harness is properly routed.
- Alternator
- Ground cable bolt to engine block. Tighten to 12 ft. lbs. (16 Nm).
- EGR valve adapter
- Vacuum lines, as tagged during removal
- TAC module electrical connector
- Radiator
- A/C compressor
- Fuel feed and return lines
- Ignition coils
- Positive, then negative battery cables
- Air cleaner outlet duct and secure with the clamp

13. Lower the hood from the service position.

14. Properly recharge the A/C system.

15. Fill the engine with oil.

16. Fill the engine with coolant.

17. Perform the Crankshaft Position (CKP) sensor variation learn procedure:

   a. Install a suitable scan tool and check for Diagnostic Trouble Codes (DTCs). If any DTCs, other than P1336 are set, resolve those codes first, before proceeding with this procedure.

   b. With the scan tool, select the crankshaft position variation learn procedure.

   c. Observe the fuel cut-off for the 8.1L engine.

   d. The scan tool will instruct you to perform certain steps, make sure you follow all directions given by the scan tool exactly.

   e. Enable the crankshaft position system variation learn procedure.

➡ **While the learn procedure is in progress, release the throttle immediately when the engine started to decelerate. The engine control is returned to**

the operator and the engine responds to throttle position after the learn procedure is complete.

   f. Slowly increase the engine speed to the RPM that you observed.

   g. Immediately release the throttle when fuel cut-out is reached.

   h. The scan tool displays: Learn Status: Learned this ignition. If the scan tool does NOT display this message and not other DTCs set, you must perform further troubleshooting.

   i. Turn the ignition **OFF** for 30 seconds after the learn procedure has been completed successfully.

18. Start and run the engine, then check for leaks.

Water Pump

REMOVAL & INSTALLATION

4.8L, 5.3L and 6.0L Engines

1. Remove or disconnect the following:
- Air inlet and outlet duct
- Coolant
- Inlet radiator hose from the water pump
- Upper fan shroud
- Cooling fan and clutch assembly
- Drive belt
- Radiator outlet hose from the coolant pump
- Surge tank hose
- Heater hose
- Water pump

To install:

2. Install or connect the following:
- Water pump. Install the water pump bolts. Tighten the bolts to 11 ft. lbs. (15 Nm) for the first pass; then tighten to 22 ft. lbs. (30 Nm) for the final pass.

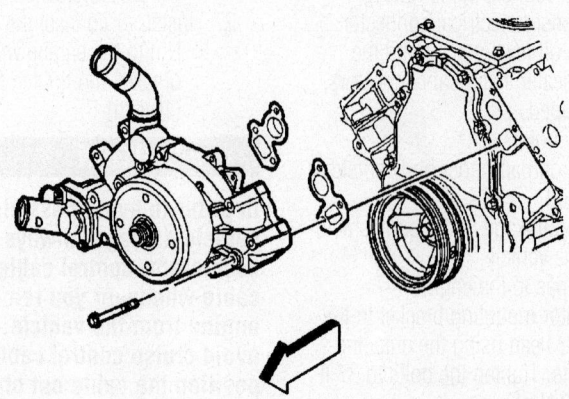

9302KG01

Exploded view of the water pump assembly—4.8L, 5.3L and 6.0L engines

- Water pump drive belt pulley and bolts (if applicable). Tighten the bolts to 89 inch lbs. (10 Nm) for the first pass; then tighten to 18 ft. lbs. (25 Nm) for the final pass.
- Surge tank hose
- Heater hose
- Outlet radiator hose to the coolant pump
- Drive belt
- Cooling fan and clutch assembly
- Upper fan shroud
- Inlet radiator hose to the water pump
- Air inlet and outlet duct
- Coolant

8.1L Engines

1. Remove or disconnect the following:
   - Coolant
   - Drive belt
   - Fan clutch
   - Outlet hose clamp and hose
2. Reposition the bypass hose clamps at the water pump and water crossover.
   - Bypass hose
   - Water pump bolt and pump. Discard the water pump gaskets.

To install:
3. Install or connect the following:
   - New water pump gaskets
   - Water pump and bolts. Tighten the water pump bolts 37 ft. lbs. (50 Nm).
   - Bypass hose and clamps
   - Outlet hose and clamp
   - Fan clutch
   - Drive belt
   - Surge tank hose
   - Heater hose

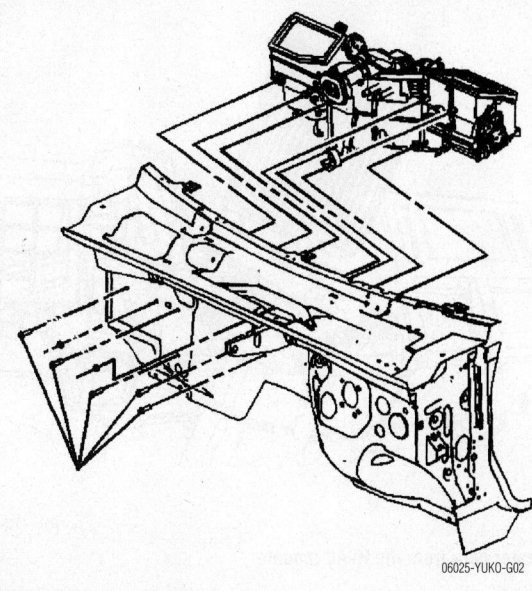

View of the HVAC assembly mounting

06025-YUKO-G02

- Outlet radiator hose to the coolant pump
- Drive belt
- Cooling fan and clutch assembly
- Upper fan shroud
- Inlet radiator hose to the water pump
- Air inlet duct
- Coolant

Heater Core

REMOVAL & INSTALLATION

Front Heater

1. Drain the engine cooling system into a clean container for reuse.

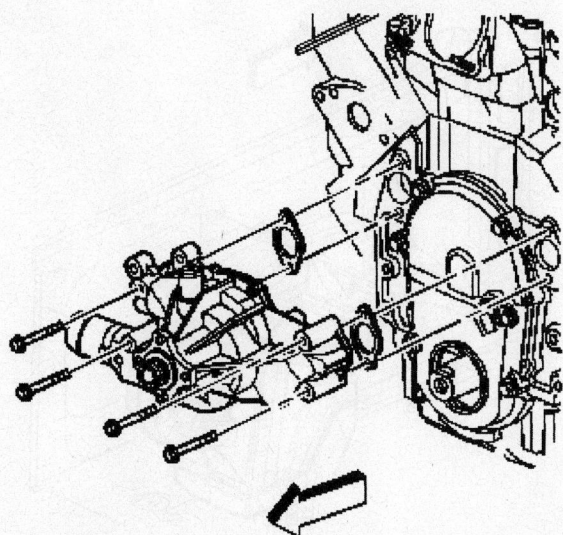

Exploded view of the water pump assembly—8.1L engines

06025-YUKO-G01

2. Remove or disconnect the following:
   - Negative battery cable
   - Heater hoses from the heater core
   - Instrument panel storage compartment
   - Electrical connectors, as necessary, that may be in the way
   - Center floor air distribution duct
   - Hinge pillar trim kick panels
   - Blower motor cover screws and the cover
   - Blower motor screws and the blower motor
   - Steering wheel and the steering column
   - Instrument panel fasteners and pull the instrument panel back far enough to gain access to the heater assembly
   - Screw located on the interior side near the evaporator pipe, if equipped, while holding the heater assembly against the firewall
   - 4 heater assembly-to-chassis screws and the 2 heater assembly-to-chassis nuts, working in the engine compartment

➡**Removal of the heater assembly may require the help of an assistant.**

   - 7 heater cover-to-heater assembly screws and the cover
   - Heater core from the heater assembly

To install:
3. Install or connect the following:
   - Heater core to the heater assembly
   - Heater cover and the 7 heater cover-to-heater assembly screws

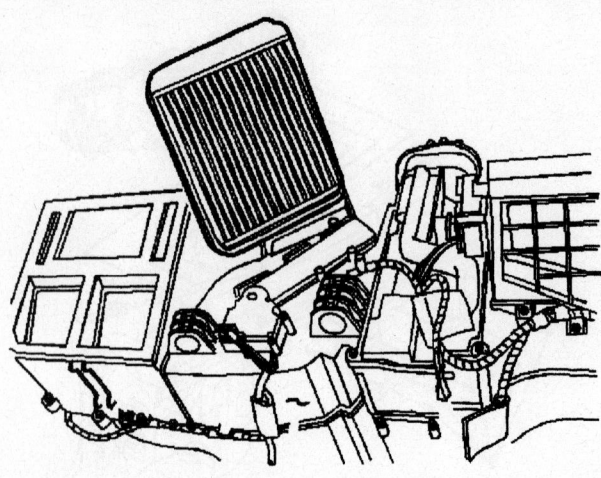

06025-YUKO-G03

Removing the heater core from the HVAC module

➡**Installation of the heater assembly may require the help of an assistant.**

- 4 heater assembly-to-chassis screws and the 2 heater assembly-to-chassis nuts. Torque the screws to 17 inch lbs. (1.9 Nm) and the nuts to 25 inch lbs. (2.8 Nm).
- Screw located on the interior side near the evaporator pipe, if equipped. Torque the screw to 97 inch lbs. (11 Nm).
- Instrument panel and the instrument panel fasteners
- Steering column and the steering wheel
- Blower motor and the blower motor screws
- Blower motor the cover and the cover screws
- Hinge pillar trim kick panels
- Center floor air distribution duct
- Electrical connectors that were disconnected
- Instrument panel storage compartment
- Heater hoses to the heater core
- Negative battery cable
4. Refill the engine cooling system.
5. Run the engine to normal operating temperatures; then, check the climate control operation and check for leaks.

Rear Auxiliary Heater

1. Drain the engine cooling system into a clean container for reuse.
2. Remove or disconnect the following:
   - Negative battery cable
   - Rear quarter trim panel, as necessary
   - Right rear quarter trim panel

- Right rear wheelhouse
- Heater hoses from the rear auxiliary heater core
- Electrical connectors, as necessary
- Drain valve
- Rear auxiliary heater assembly-to-chassis nuts and bolts
- Rear auxiliary heater assembly
- Blower motor from the heater assembly, if necessary
- Rear auxiliary heater assembly cover
- Heater core from the rear auxiliary assembly

To install:
3. Install or connect the following:
   - Heater core to the rear auxiliary assembly
   - Rear auxiliary heater assembly cover
   - Blower motor to the heater assembly, if removed

- Rear auxiliary heater assembly
- Rear auxiliary heater assembly-to-chassis nuts and bolts. Torque the bolts to 13 inch lbs. (1.5 Nm) and the nuts to 89 inch lbs. (10 Nm).
- Drain valve
- Electrical connectors, as necessary
- Heater hoses to the rear auxiliary heater core
- Right rear wheelhouse
- Right rear quarter trim panel
- Rear quarter trim panel, as necessary
4. Refill the engine cooling system.
5. Connect the negative battery cable.
6. Run the engine to normal operating temperatures; then, check the climate control operation and check for leaks.

Cylinder Head

REMOVAL & INSTALLATION

4.8L, 5.3L and 6.0L Engines

RIGHT SIDE

1. Remove or disconnect the following:
   - Negative battery cable
   - Oil level indicator tube
   - Intake manifold
   - Coolant air bleed pipe
   - Exhaust manifold
   - Push rods
   - Cylinder head bolts and discard
   - Cylinder head from the engine

➡**After removal, place the cylinder head on two wood blocks to prevent damage.**

2. Remove and discard the gasket.

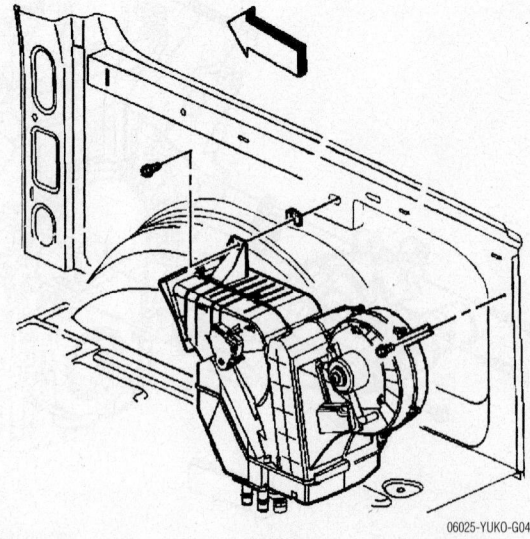

06025-YUKO-G04

View of the rear auxiliary HVAC assembly—Suburban shown; other models similar

To install:

➡️**Do not use any type of sealant on the cylinder head gasket (unless specified). The cylinder head gaskets must be installed in the proper direction and position.**

3. Clean the engine block cylinder head bolt holes (if required). Thread repair tool J 42385-107 may be used to clean the threads of old threadlocking material.

4. Spray cleaner GM P/N 12346139, P/N 12377981, or equivalent into the hole.

5. Clean the cylinder head bolt holes with compressed air.

6. Check the cylinder head locating pins for proper installation.

➡️**When properly installed, with FRONT on the right side, the tab on the cylinder head gasket should be located right of center or closer to the front of the engine.**

7. Install or connect the following:
   - NEW right cylinder head gasket onto the locating pins
   - Cylinder head onto the locating pins and the gasket
   - NEW M11 cylinder head bolts. Apply a 0.20 in. (5mm) band of threadlock GM P/N 12345382 or equivalent to the threads of the M8 cylinder head bolts.
   - M8 cylinder head bolts.

8. On 2002–04 models, tighten the cylinder head bolts, as follows:
   a. M11 bolts, first pass: in sequence to 22 ft. lbs. (30 Nm).
   b. M11 bolts, second pass: in sequence + 90 degrees.
   c. M11 bolts (1,2,3,4,5,6,7,8): + 90 degrees.
   d. M11 bolts (9 and 10): + 50 degrees.
   e. M8 bolts (11,12,13,14,15): to 22 ft. lbs. (30 Nm). Begin with the center bolt (11) and alternating side-to-side, work outward tightening all of the bolts.

9. On 2005–06 models, tighten the cylinder head bolts, as follows:
   a. Tighten the M11 cylinder head bolts (1-10) a first pass in sequence to 22 ft. lbs. (30 Nm).
   b. Tighten the M11 cylinder head bolts (1-10) a second pass in sequence to 90 degrees.
   c. Tighten the M11 cylinder head bolts (1-10) a final pass to 70 degrees in sequence.
   d. Tighten the M8 cylinder head bolts (11-15) to 22 ft. lbs. (30 Nm). Begin with the center bolt (11) and alternating side-

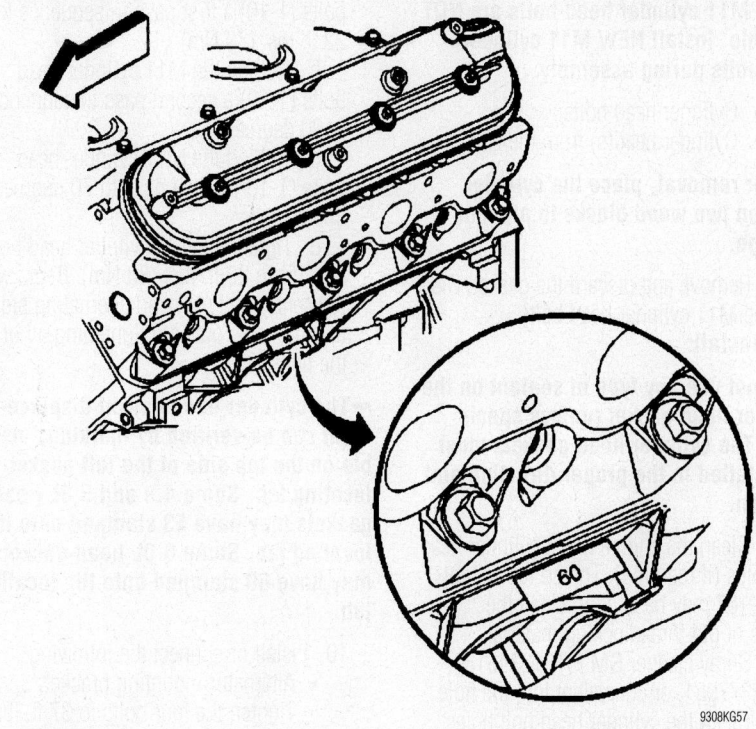

Locating tab—4.8L, 5.3L and 6.0L engines

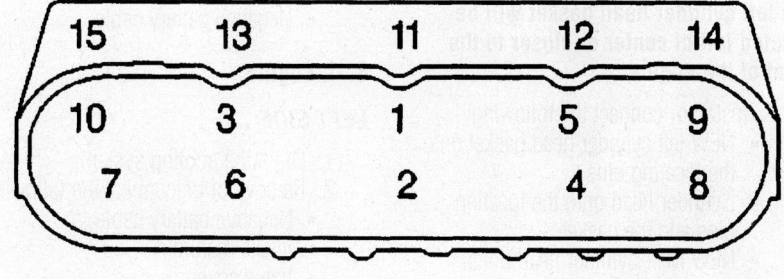

Cylinder head bolt tightening sequence—4.8L, 5.3L and 6.0L engines

to-side, work outward tightening all of the bolts.

➡️**The cylinder head gasket displacement can be verified by markings visible on the underside of the right gasket locating tab. Some 4.8 and 5.3L head gaskets may have 53 stamped onto the locating tab. Some 6.0L head gaskets may have 60 stamped onto the locating tab.**

10. Install or connect the following:
   - Alternator
   - Exhaust manifold(s)
   - Pushrods
   - Intake manifold
   - Coolant air bleed pipe
   - Oil level indicator tube
   - Negative battery cable

LEFT SIDE

❈❈ CAUTION

Before servicing any electrical component, the ignition key must be in the OFF or LOCK position and all electrical loads must be OFF, unless instructed otherwise in these procedures.

1. Remove or disconnect the following:
   - Negative battery cable
   - Alternator bracket
   - Coolant air bleed pipe
   - Intake manifold
   - Pushrods
   - Exhaust manifold
   - Spark plugs from the cylinder head

➡The M11 cylinder head bolts are NOT reusable. Install NEW M11 cylinder head bolts during assembly.

- Cylinder head bolts
- Cylinder head(s) from the engine

➡After removal, place the cylinder head on two wood blocks to prevent damage.

2. Remove and discard the gasket. Discard the M11 cylinder head bolts.

To install:

➡Do not use any type of sealant on the cylinder head gasket (unless specified). The cylinder head gaskets must be installed in the proper direction and position.

3. Clean the engine block cylinder head bolt holes (if required). Thread repair tool J 42385-107 may be used to clean the threads of old threadlocking material.

4. Spray cleaner GM P/N 12346139, P/N 12377981, or equivalent into the hole.

5. Clean the cylinder head bolt holes with compressed air.

6. Check the cylinder head locating pins for proper installation.

➡When properly installed, the tab on the left cylinder head gasket will be located left of center or closer to the front of the engine.

7. Install or connect the following:
- NEW left cylinder head gasket onto the locating pins
- Cylinder head onto the locating pins and the gasket
- NEW M11 cylinder head bolts. Apply a 0.20 in. (5mm) band of threadlock GM P/N 12345382 or equivalent to the threads of the M8 cylinder head bolts.
- M8 cylinder head bolts

8. On 2002–04 models, tighten the cylinder head bolts, as follows:
a. M11 bolts, first pass: in sequence to 22 ft. lbs. (30 Nm).
b. M11 bolts, second pass: in sequence + 90 degrees.
c. M11 bolts (1,2,3,4,5,6,7,8): + 90 degrees.
d. M11 bolts (9 and 10): + 50 degrees.
e. M8 bolts (11,12,13,14,15): to 22 ft. lbs. (30 Nm). Begin with the center bolt (11) and alternating side-to-side, work outward tightening all of the bolts.

9. On 2005–06 models, tighten the cylinder head bolts, as follows:
a. Tighten the M11 cylinder head

bolts (1-10) a first pass in sequence to 22 ft. lbs. (30 Nm).
b. Tighten the M11 cylinder head bolts (1-10) a second pass in sequence to 90 degrees.
c. Tighten the M11 cylinder head bolts (1-10) a final pass to 70 degrees in sequence.
d. Tighten the M8 cylinder head bolts (11-15) to 22 ft. lbs. (30 Nm). Begin with the center bolt (11) and alternating side-to-side, work outward tightening all of the bolts.

➡The cylinder head gasket displacement can be verified by markings visible on the top side of the left gasket locating tab. Some 4.8 and 5.3L head gaskets may have 53 stamped onto the locating tab. Some 6.0L head gaskets may have 60 stamped onto the locating tab.

10. Install or connect the following:
- Alternator mounting bracket. Tighten the four bolts to 37 ft. lbs. (50 Nm).
- Exhaust manifold
- Pushrods
- Intake manifold
- Coolant air bleed pipe
- Negative battery cable

8.1L Engine

LEFT SIDE

1. Drain the cooling system.
2. Remove or disconnect the following:
- Negative battery cable
- Intake manifold
- Valve cover
- Rocker arms and pushrods, keeping them in order for installation
- Engine harness ground bolts
3. Reposition the engine harness grounds and ground straps from the cylinder head.
- Water crossover
- Exhaust manifold
- Cylinder head bolts, then discard

➡The cylinder head bolts must be replaced for installation.

- Cylinder head. Place the head on 2 wood blocks to protect the sealing surfaces while it is removed.

To install:

➡The cylinder head should be cleaned and inspected for warpage or damage before installation.

4. Thoroughly clean the mating surfaces of the head and block. Clean the bolt holes thoroughly.

➡If a composition gasket is used, do not use sealer.

5. Align the cylinder head gasket locating marks to face up. Make sure that the gasket tabs are located of the no. 1 and 2 cylinder for proper installation.

6. Install or connect the following:
- New cylinder head gasket
- Cylinder head
- Sealer to the threads of new cylinder head bolts, if not pre-applied
- Cylinder head bolts and hand-tighten

7. On 2002–04 models, tighten the head bolts a little at a time, in the proper sequence, in 3 stages:
a. Step 1: Torque to 30 ft. lbs. (40 Nm).
b. Step 2: Tighten an additional 120 degrees using a torque angle meter.
c. Step 3: Torque bolt numbers. 1, 2, 3, 6, 7, 8, 9, 10, 11, 14, 16 and 17 an additional 60 degrees. Tighten bolts 15 and 18 an additional 45 degrees, and bolt numbers 4, 5, 12 and 13 an additional 30 degrees.

8. On 2005–06 models, tighten the head bolts a little at a time, in the proper sequence, in 3 stages:
a. Tighten the bolts a first pass in sequence to 22 ft. lbs. (30 Nm).
b. Tighten the bolts a second pass in sequence to 22 ft. lbs. (30 Nm) then an additional 120 degrees,

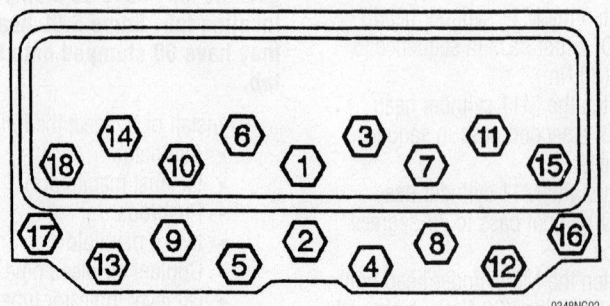

Cylinder head bolt tightening sequence—8.1L engine

9348NG02

c. Tighten the bolts (1, 2, 3, 6, 7, 8, 9, 10, 11, 14, 16, 17) an additional 60 degrees, bolts (15, and 18) an additional 45 degrees, and bolts (4, 5, 12, and 13) an additional 30 degrees a final pass in sequence.

9. Install or connect the following:
- Exhaust manifold
- Water crossover
- Engine harness grounds and ground strap
- Rocker arms and pushrods
- Valve cover
- Intake manifold

10. Connect the battery cable and refill the cooling system.

RIGHT SIDE

1. Drain the cooling system.
2. Remove or disconnect the following:
- Negative battery cable
- Intake manifold
- Valve cover
- Rocker arms and pushrods, keeping them in order for installation
- Engine Coolant Temperature (ECT) sensor clip from the bracket
- ECT sensor
- ECT sensor bracket bolt and bracket
- Heater inlet and outlet hoses from the hose bracket
- Oil level indicator tube
- Water crossover
- Exhaust manifold
- Cylinder head bolts, then discard

➥The cylinder head bolts must be replaced for installation.

- Cylinder head. Place the head on 2 wood blocks to protect the sealing surfaces while it is removed.

To install:

➥The cylinder head should be cleaned and inspected for warpage or damage before installation.

3. Thoroughly clean the mating surfaces of the head and block. Clean the bolt holes thoroughly.

➥If a composition gasket is used, do not use sealer.

4. Align the cylinder head gasket locating marks to face up. Make sure that the gasket tabs are located of the no. 1 and 2 cylinder for proper installation.

5. Install or connect the following:
- New cylinder head gasket
- Cylinder head
- Sealer to the threads of new cylinder head bolts, if not pre-applied

- Cylinder head bolts and hand-tighten

6. On 2002–04 models, tighten the head bolts a little at a time, in the proper sequence, in 3 stages:
   a. Step 1: Torque to 30 ft. lbs. (40 Nm).
   b. Step 2: Tighten an additional 120 degrees using a torque angle meter.
   c. Step 3: Torque bolt numbers. 1, 2, 3, 6, 7, 8, 9, 10, 11, 14, 16 and 17 an additional 60 degrees. Tighten bolts 15 and 18 an additional 45 degrees, and bolt numbers 4, 5, 12 and 13 an additional 30 degrees.

7. On 2005–06 models, tighten the head bolts a little at a time, in the proper sequence, in 3 stages:
   a. Tighten the bolts a first pass in sequence to 22 ft. lbs. (30 Nm).
   b. Tighten the bolts a second pass in sequence to 22 ft. lbs. (30 Nm) then an additional 120 degrees,
   c. Tighten the bolts (1, 2, 3, 6, 7, 8, 9, 10, 11, 14, 16, 17) an additional 60 degrees, bolts (15, and 18) an additional 45 degrees, and bolts (4, 5, 12, and 13) an additional 30 degrees a final pass in sequence.

8. Install or connect the following:
- Exhaust manifold
- Water crossover
- Oil level indicator tube
- Heater hose bracket and bolts. Tighten the bolts to 37 ft. lbs. (50 Nm).
- ECT sensor bracket and bolt. Tighten to 37 ft. lbs. (50 Nm).
- ECT sensor
- ECT sensor clip
- Rocker arms and pushrods
- Valve cover
- Intake manifold

9. Connect the battery cable and refill the cooling system.

Rocker Arms

REMOVAL & INSTALLATION

4.8L, 5.3L and 6.0L Engines

➥Do not remove the ignition coils from the valve rocker arm cover unless required. Do not remove the oil fill tube from the cover unless service is required. If the oil fill tube has been removed from the cover, install a NEW tube during assembly.

On the right side:
1. Remove or disconnect the following:
- Ignition coil bracket bolts from the rocker arm cover (if required)
- Ignition coil and bracket assembly from the cover
- Valve rocker arm cover bolts
- Valve rocker arm cover
- Gasket from the cover. Discard the gasket. The bolt grommets may be reused if not damaged.
- Oil fill cap from the oil fill tube
- Oil fill tube (if required). Discard the oil fill tube.

On the left side:

➥Do not remove the Positive Crankcase Ventilation (PCV) valve grommet from the cover unless service is required.

2. Remove or disconnect the following:
- Ignition coil bracket bolts from the rocker arm cover (if required)
- Ignition coil and bracket assembly from the cover
- Valve rocker arm cover bolts

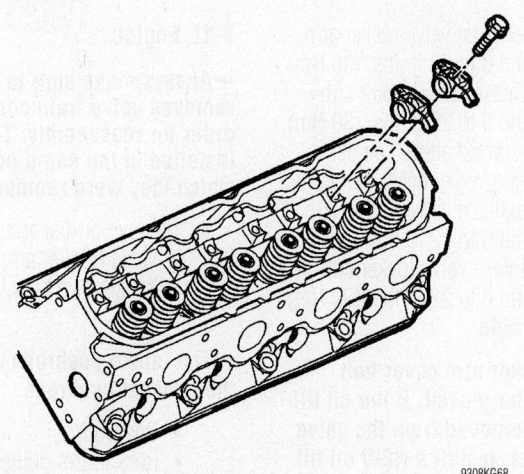

9308KG68

Rocker arm removal—4.8L, 5.3L and 6.0L engines

- Valve rocker arm cover
- Gasket from the cover. Discard the gasket. The bolt grommets may be reused if not damaged.
- Valve rocker arm bolts
- Valve rocker arms
- Valve rocker arm pivot support
- Pushrods

To install:

➡**Valve lash is built in. No valve adjustment is required.**

3. Lubricate the valve rocker arms and pushrods with clean engine oil.

4. Lubricate the flange of the valve rocker arm bolts with clean engine oil.

5. Lubricate the flange or washer surface of the bolt that will contact the valve rocker arm.

6. Install or connect the following:
- Valve rocker arm pivot support

➡**Make sure that the pushrods seat properly to the valve lifter sockets.**

- Pushrods

➡**Make sure that the pushrods seat properly to the ends of the rocker arms.**

- Rocker arms and bolts. DO NOT tighten the rocker arm bolts at this time

7. Rotate the crankshaft until number one piston is at top dead center of compression stroke. In this position, cylinder number one rocker arms will be off lobe lift, and the crankshaft sprocket key will be at the 1:30 position. If viewing from the rear of the engine, the additional crankshaft pilot hole (non-threaded) will be in the 10:30 position. The engine firing order is 1, 8, 7, 2, 6, 5, 4, 3. Cylinders 1, 3, 5 and 7 are left bank. Cylinders 2, 4, 6, and 8 are right bank.

8. With the engine in the number one firing position, tighten the following valve rocker arm bolts:
   a. Tighten exhaust valve rocker arm bolts 1, 2, 7, and 8 to 22 ft. lbs. (30 Nm).
   b. Tighten intake valve rocker arm bolts 1, 3, 4, and 5 to 22 ft. lbs. (30 Nm).

9. Rotate the crankshaft 360 degrees. Tighten the following valve rocker arm bolts:
   a. Tighten exhaust valve rocker arm bolts 3, 4, 5, and 6 to 22 ft. lbs. (30 Nm).
   b. Tighten intake valve rocker arm bolts 2, 6, 7, and 8 to 22 ft. lbs. (30 Nm).

On the right side:

➡**The valve rocker arm cover bolt grommets may be reused. If the oil fill tube has been removed from the valve rocker arm cover, install a NEW oil fill tube during assembly.**

10. Lubricate the O-ring seal of the NEW oil fill tube with clean engine oil.

11. Install or connect the following:
- NEW oil fill tube into the rocker arm cover and rotate the tube clockwise until locked in the proper position
- Oil fill cap into the tube and rotate clockwise until locked in the proper position
- NEW cover gasket into the valve rocker arm cover
- Valve rocker arm cover onto the cylinder head
- Cover bolts with grommets and tighten to 106 inch lbs. (12 Nm)

12. Apply threadlock GM P/N 12345382 or equivalent to the threads of the bracket bolts.
- Ignition coil and bracket assembly and bolts. Tighten to 106 inch lbs. (12 Nm).

On the left side:

➡**DO NOT reuse the valve rocker arm cover gasket. The valve rocker arm cover bolt grommets may be reused. If the vapor vent grommet has been removed from the valve rocker arm cover, install a NEW vapor vent gourmet during assembly.**

13. Install or connect the following:
- NEW cover gasket (1) into the valve rocker arm cover
- Valve rocker arm cover onto the cylinder head
- Cover bolts with grommets and tighten to 106 inch lbs. (12 Nm)

14. Apply threadlock GM P/N 12345382 or equivalent to the threads of the bracket bolts.
- Ignition coils and bracket assembly and bolts. Tighten the bolts to 106 inch lbs. (12 Nm).

8.1L Engine

➡**Always make sure to keep all removed valve train components in order for reassembly. They must be installed in the same position from which they were removed.**

1. Remove or disconnect the following:
- Valve (rocker arm) cover
- Rocker arm nuts, balls and rocker arms

➡**The intake pushrods are shorter than the exhaust pushrods.**

- Pushrods
- Rocker arm guides and pushrod guides

2. Clean and inspect all components for damage.

To install:

3. Apply a suitable sealer to the rocker arm stud-to-cylinder head threads.

4. Install or connect the following:
- Pushrod guides and rocker arm studs. Tighten to 37 ft. lbs. (50 Nm).
- Pushrods

5. Coat the rocker arm and ball bearing surfaces with a suitable prelube.
- Rocker arms, balls and nuts. Tighten the nuts slowly to 18 ft. lbs. (25 Nm) while guiding the tips of the rocker arms over the tips of the valves.
- Valve (rocker arm) cover

Intake Manifold

REMOVAL & INSTALLATION

4.8L, 5.3L and 6.0L Engines

➡**The intake manifold, throttle body, fuel injection rail, and fuel injectors may be removed as an assembly. If not servicing the individual components, remove the manifold as a complete assembly.**

1. Partially drain the engine coolant.

2. Remove or disconnect the following:
- Air cleaner outlet duct
- Engine sight shield
- Throttle actuator motor electrical connector
- Coolant air bleed hose from the throttle body
- Throttle body
- Fuel rail with injectors
- Manifold Absolute Pressure (MAP) sensor, if required
- Knock sensor connector
- Vacuum brake booster hose from the rear of the intake manifold
- Positive Crankcase Ventilation (PCV) hose and valve
- EVAP solenoid, bolt, and isolator
- Intake manifold bolts
- Intake manifold with gaskets
- Intake manifold-to-cylinder head gaskets from the manifold. Discard the intake manifold gaskets.
- Throttle body and gasket

3. Clean the intake manifold in solvent.

4. Dry the intake manifold with compressed air.

5. Inspect all components for damage, and replace the necessary parts.

6. Locate a straight edge across the

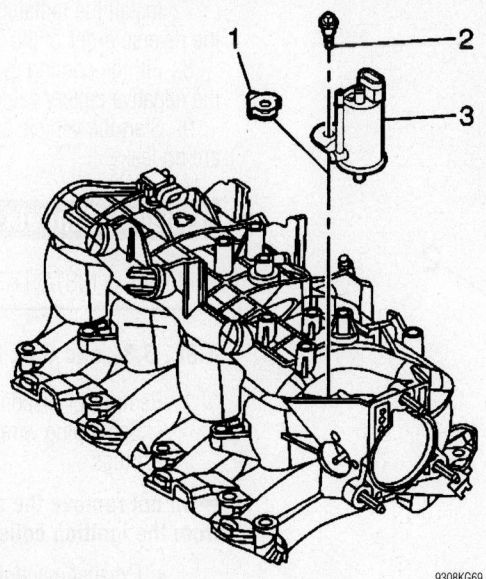

EVAP solenoid removal—4.8L, 5.3L and 6.0L engines

9308KG69

intake manifold cylinder head deck surface. Position the straight edge across a minimum of two runner port openings.

7. Insert a feeler gauge between the intake manifold and the straight edge. A intake manifold with warpage in excess of 0.118 in. (3mm) over a 7.87 in. (200mm) area is warped and should be replaced.

To install:

8. Install or connect the following:
- MAP sensor
- EVAP solenoid, bolt, and isolator. Tighten the bolt to 89 inch lbs. (10 Nm).
- NEW intake manifold-to-cylinder head gaskets
- Intake manifold. Apply a 0.20 in. (5mm) band of threadlock GM P/N 12345382 or equivalent to the threads of the intake manifold bolts.
- Intake manifold bolts. Tighten intake manifold bolts, in sequence, to 44 inch lbs. (5 Nm). Then,

tighten the bolts, in sequence, to 89 inch lbs. (10 Nm).
- Vacuum brake booster hose from the rear of the intake manifold
- Knock sensor connector
- Manifold Absolute Pressure (MAP) sensor, if required
- Air cleaner outlet duct
- Fuel rail with injectors
- Throttle body
- Coolant air bleed hose to the throttle body
- Throttle actuator motor electrical connector
- Engine sight shield

8.1L Engine

➡The intake manifold, throttle body, fuel rail and injectors can be removed as an assembly. If you do not need to service these components individually, remove the manifold as a complete assembly.

1. Relieve the fuel system pressure and drain the cooling system.
2. Remove or disconnect the following:
- Air cleaner outlet duct
- Intake manifold sight shield
- Fuel feed and return pipes
- Engine harness clips from the studs on the front of the dash
- Engine harness clip from the wheelhouse splash shield
- Pressure cycling switch, surge tank switch and Mass Air Flow (MAF) electrical connectors
- Reposition the engine harness to the top of the engine
- Connector Position Assurance (CPA) retainer from the ignition coil harness
- Manifold Absolute Pressure (MAP) sensor electrical connector
- Ignition coil electrical connector
- Engine Coolant Temperature (ECT) sensor electrical connector
- Engine harness bolt and studs
- CPA retainer from the ignition coil harness
- Alternator, injector harness and ignition coil harness connectors
- Throttle Position (TP) sensor connector
- Electronic Throttle Control (ETC) and purge valve solenoid connectors

3. Reposition the engine harness to the drivers side of the engine compartment.
- Bypass valve vacuum hose from the intake manifold
- Unclip the EVAP tube from the fuel rail
- EVAP tube from the vent pipe at rear of engine
- EVAP tube
- Fuel rail studs and fuel rail, ONLY if replacing the manifold
- Intake manifold bolts

✲✲ WARNING

Do NOT try to remove the intake manifold by prying under the sealing surfaces.

- Intake manifold
- Intake manifold side gaskets and end seals and discard

➡The splash shield is reusable and secured using a snap-in fit. Do not distort the shield during removal.

- Splash shield

To install:

4. Clean all gasket surfaces completely.

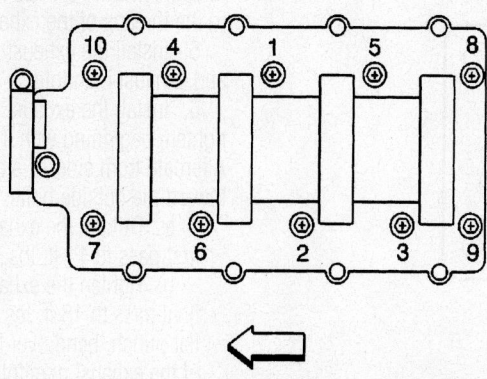

Lower intake manifold bolt tightening sequence—4.8L, 5.3L and 6.0L engines

9302KG03

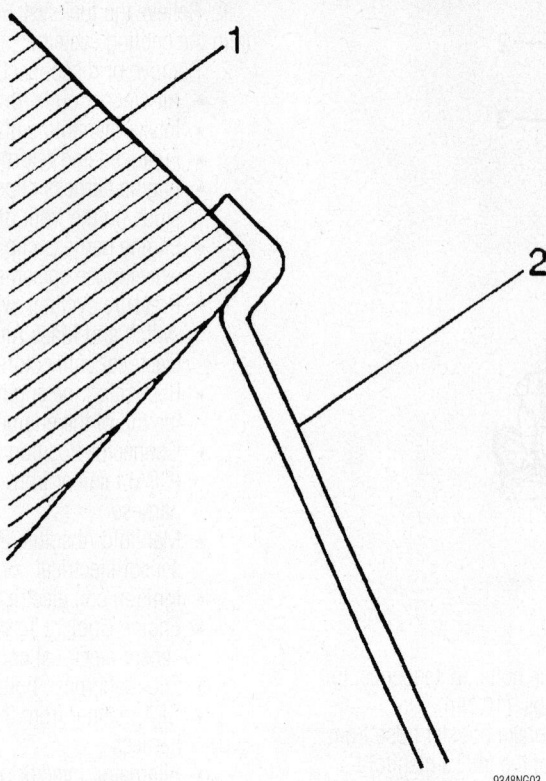

9348NG03

Make sure that the splash shield snap fits between the cylinder heads

5. Install or connect the following:
- Splash shield. Make sure the shield fits properly between the cylinder head.

➡**Make sure the manifold gasket tabs align with the hole in the head gasket.**

- New intake manifold end seals
- New intake manifold side gaskets onto the heads. Make sure the stamped THIS SIDE UP is showing.
- Intake manifold to the block
- Apply a suitable thread locking material to at least 8 threads of the intake manifold bolts

6. Install the intake manifold bolts
- On 2002–04 models, tighten the bolts in the sequence shown, in 4 passes:
  a. 1st pass: 44 inch lbs. (5 Nm).
  b. 2nd pass: 44 inch lbs. (5 Nm).
  Check the manifold joints for shifting and fix as necessary.
  c. 3rd pass: 89 inch lbs. (10 Nm).
  d. 4th pass: 106 inch lbs. (12 Nm).
- On 2005–06 models, tighten the bolts in the sequence shown, in 4 passes:
  e. 1st pass: 44 inch lbs. (5 Nm)
  f. 2nd pass: 71 inch lbs. (8 Nm).
  Check the manifold joints for shifting and fix as necessary.
  g. 3rd pass: 106 inch lbs. (12 Nm).
  h. 4th pass: 11 ft. lbs. (15 Nm).

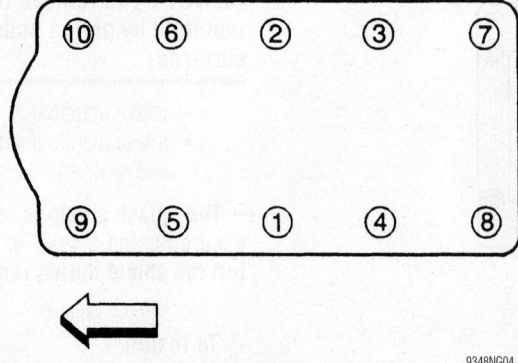

9348NG04

Intake manifold bolt tightening sequence—8.1L engine

7. Install the remaining components in the reverse order of the removal procedure.
8. Fill the cooling system, then connect the negative battery cable
9. Start the vehicle and verify that there are no leaks.

Exhaust Manifold

REMOVAL & INSTALLATION

4.8L, 5.3L and 6.0L Engines

1. Remove or disconnect the following:
- Spark plug wires from the spark plugs

➡**Do not remove the spark plug wires from the ignition coils unless required.**

- Exhaust manifold, bolts, and gasket. Discard the gasket.
- Heat shield and bolts from the manifold, if required

To install:

➡**Do not reuse the exhaust manifold-to-cylinder head gaskets. Upon installation of the exhaust manifold, install a NEW gasket. A improperly installed gasket or leaking exhaust system may effect On-Board Diagnostics (OBD) II system performance.**

2. Clean the exhaust manifold and heat shield in solvent. Dry the exhaust manifold with compressed air.
3. Use a straight edge and a feeler gauge and measure the exhaust manifold cylinder head deck for warpage. An exhaust manifold deck with warpage in excess of 0.01 in. (0.25mm) within the two front or two rear runners or 0.02 in. (0.5mm) overall, may cause an exhaust leak and may effect OBD II system performance. Exhaust manifolds not within specifications must be replaced.
4. Apply a 0.2 in. (5mm) wide band of threadlock GM P/N 12345493 or equivalent to the threads of the exhaust manifold bolts.
5. Install the exhaust manifold gasket and exhaust manifold
6. Install the exhaust manifold bolts and tighten, beginning with the center two bolts. Alternate from side-to-side, and work toward the outside bolts.
   a. Tighten the exhaust manifold bolts first pass to 11 ft. lbs. (15 Nm).
   b. Tighten the exhaust manifold bolts final pass to 18 ft. lbs. (25 Nm). Using a flat punch, bend over the exposed edge of the exhaust manifold gasket at the front of the right cylinder head.

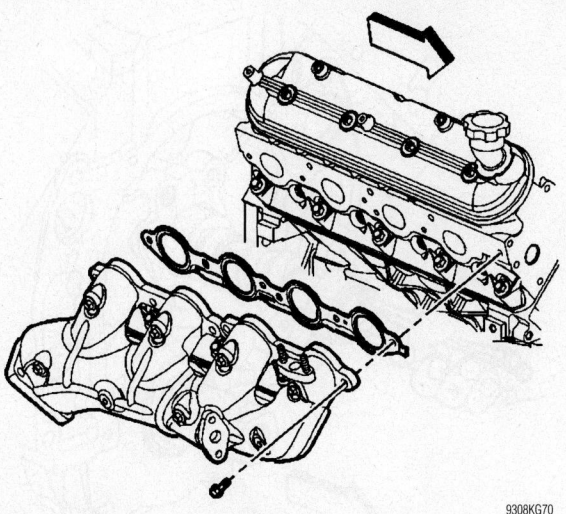

Right exhaust manifold removal—4.8L, 5.3L and 6.0L; left side similar

9308KG70

7. Install or connect the following:
   - Heat shield and bolts and tighten to 80 inch lbs. (9 Nm)
   - Spark plug wires

8.1L Engines

1. Remove or disconnect the following:
   - Spark plug wires
   - Spark plugs
   - Exhaust manifold heat shield bolts and shield
   - Exhaust manifold bolt and nuts
   - Exhaust manifold
   - Exhaust manifold gasket and discard

To install:

2. Clean the mating surfaces and the retainer threads.

3. Install or connect the following:
   - New exhaust manifold gasket
   - Exhaust manifold
   - Exhaust manifold bolt and nuts. Tighten the bolt to 26 ft. lbs. (35 Nm) and the nuts to 12 ft. lbs. (16 Nm).
   - If removed, tighten the studs to 15 ft. lbs. (20 Nm).
   - Heat shield. Tighten the retaining bolts and nuts to 18 ft. lbs. (25 Nm).
   - Spark plugs and plug wires

Camshaft and Valve Lifters

REMOVAL & INSTALLATION

4.8L, 5.3L and 6.0L Engines

1. Raise the hood to the servicing position and secure it. Move the hood hinge bolt to hold the hood in the servicing position.

2. Remove or disconnect the following:
   - Battery negative cable
   - Coolant
   - Upper and lower radiator hoses from the engine
   - Air cleaner duct from the engine
   - A/C condenser mounting bolts, if equipped
   - Radiator support and radiator
   - Engine cooling fan
   - Drive belt
   - A/C drive belt, if equipped
   - Engine sight shield
   - Electrical wiring harness from the thermostat housing
   - Water pump
3. Raise the vehicle.
   - Starter motor
   - Right side closeout cover and bolt

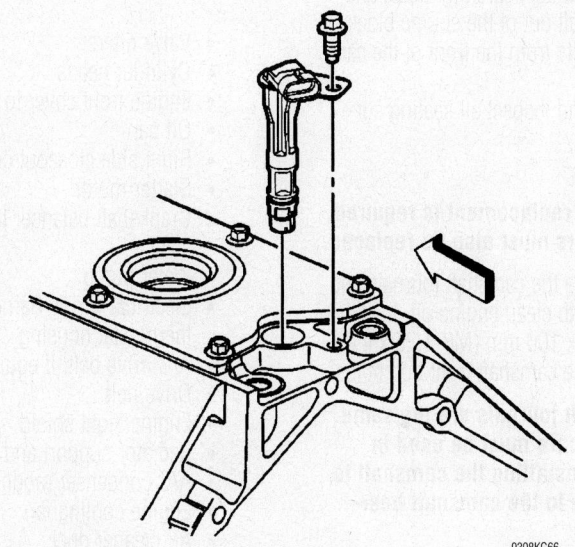

Camshaft sensor removal—4.8L, 5.3L and 6.0L engines

9308KG66

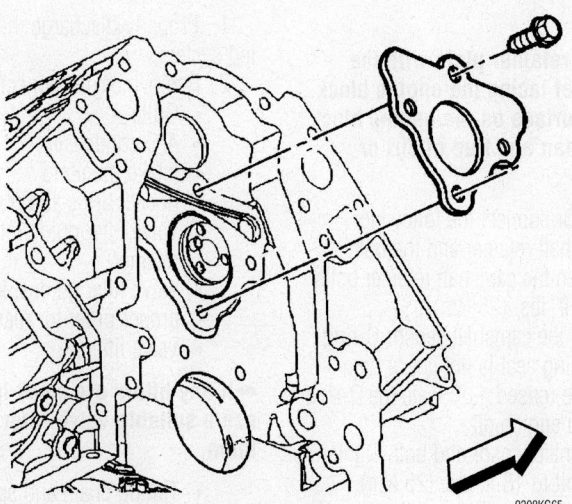

Camshaft retainer removal—4.8L, 5.3L and 6.0L engines

9308KG65

- Crankshaft balancer
- Engine oil pan
- Engine front cover
- Cylinder heads from the engine
- Valve lifters from the engine

4. Align the timing marks on the camshaft and crankshaft sprockets. Make sure that the number 1 piston is in the firing position.

- Camshaft sprocket
- Camshaft sensor bolt and sensor
- Camshaft retainer bolts and retainer

➡ **All camshaft journals are the same diameter, so care must be used in removing or installing the camshaft to avoid damage to the camshaft bearings.**

5. Install the three M8-1.25 x 100 mm bolts in the camshaft front bolt holes. Using the bolts as a handle, carefully rotate and pull the camshaft out of the engine block. Remove the bolts from the front of the camshaft.

6. Clean and inspect all sealing surfaces.

To install:

➡ **If camshaft replacement is required, the valve lifters must also be replaced.**

7. Lubricate the camshaft journals and the bearings with clean engine oil. Install three M8-1.25 x 100 mm (M8-1.25 x 4.0 in) bolts into the camshaft front bolt holes.

➡ **All camshaft journals are the same diameter, so care must be used in removing or installing the camshaft to avoid damage to the camshaft bearings.**

8. Using the bolts as a handle, carefully install the camshaft into the engine block. Remove the three bolts from the front of the camshaft.

➡ **Install the retainer plate with the sealing gasket facing the engine block. The gasket surface on the engine block should be clean and free of dirt or debris.**

9. Install or connect the following:
- Camshaft retainer and the bolts. Tighten the camshaft retainer bolts to 18 ft. lbs. (25 Nm).

10. Inspect the camshaft sensor O-ring seal. If the O-ring seal is not cut or damaged, it may be reused. Lubricate the O-ring seal with clean engine oil.
- Camshaft sensor and bolt. Tighten the bolt to 18 ft. lbs. (25 Nm).
- Camshaft sprocket and timing chain

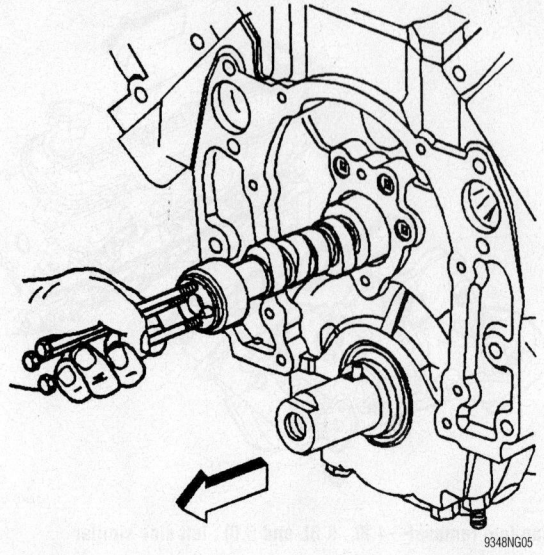

Use the 3 bolts as a handle to carefully remove and install the camshaft

- Valve lifters
- Cylinder heads
- Engine front cover to the engine
- Oil pan
- Right side closeout cover
- Starter motor
- Crankshaft balancer to the crankshaft
- Water pump
- Electrical wiring harness to the thermostat housing
- A/C drive belt, if equipped
- Drive belt
- Engine sight shield
- Radiator support and radiator
- A/C condenser mounting bolts
- Engine cooling fan
- Air cleaner duct
- Negative battery cable

8.1L Engines

1. Properly discharge the air conditioning system.

2. Remove or disconnect the following:
- Grille
- A/C condenser
- Intake manifold
- Rocker arms and pushrods
- Valve lifter guide retainer bolts and retainer
- Valve lifter guides, keeping them in proper order for reassembly
- Valve lifters

➡ **If any lifters are stuck in their bores, use a suitable valve lifter to remove them.**

- Timing chain and sprocket
- Camshaft retaining bolts
- Camshaft retainer

✹✹ WARNING

All of the cam journals are the same size so be very careful when removing and installing the camshaft that you do not damage the bearings.

3. Install three 8-1.25 x 100mm bolts in the holes in the front of the camshaft and carefully pull the camshaft from the block.

4. Remove the bolts from the front of the camshaft.

5. Clean and inspect the camshaft for damage.

To install:

6. Liberally coat camshaft and bearings with heavy engine oil or engine assembly lubricant.

7. Install the camshaft, using the 3 bolts threaded into the camshaft bolt holes as a handle, then remove the bolts.

8. Install or connect the following:
- Camshaft retainer and bolts. Tighten to 106 inch lbs. (12 Nm).

➡ **If a new camshaft is installed, you MUST install new valve lifters.**

- Timing chain and sprocket
- Valve lifters
- Valve lifter guides over the flats on the lifters. Make sure the rollers of the lifters are properly aligned with the cam lobes.
- Valve lifter guide retainer and tighten the bolts to 18 ft. lbs. (25 Nm)
- Rocker arms and pushrods
- Intake manifold
- A/C condenser
- Grille

9. Recharge the A/C system.

Valve Lash

ADJUSTMENT

All engines use hydraulic lifters, which require no periodic adjustment.

Starter Motor

REMOVAL & INSTALLATION

4.8L, 5.3L and 6.0L Engines

❊❊ CAUTION

Before servicing any electrical component, the ignition key must be in the OFF or LOCK position and all electrical loads must be OFF, unless instructed otherwise in these procedures.

1. Disconnect the negative battery cable.
2. Raise and support the vehicle.
3. Remove or disconnect the following:
   - Protective shields, as necessary
   - Starter solenoid shield
   - Starter-to-transmission close out cover bolt
   - Engine oil level sensor connection
   - On 4WD, front axle mounting bracket through bolt nut
4. Reposition the front axle mounting bracket through bolt until the bolt tip is

flush with the support bushing. Do not remove the bolt.
   - Mounting bolts from the engine block. Slide the starter forward until the starter clears the transmission.
   - Starter transmission close out cover
   - Positive battery cable and wiring harness from the starter
   - Starter from the vehicle

To install:
5. Install or connect the following:
   - Starter
   - Positive battery cable to the starter. Tighten the nut to 12 ft. lbs. (16 Nm).
   - Starter transmission close out cover
   - Mounting bolts to the engine block and tighten to 37 ft. lbs. (50 Nm)
6. Reposition the front axle mounting bracket through bolt until the bolt is fully seated.
   - Front axle mounting bracket through bolt nut and tighten to 70 ft. lbs. (95 Nm)
   - Engine oil level sensor connection
   - Starter-to-transmission close out cover bolt
   - Starter solenoid shield
   - Protective shields as necessary
7. Remove the safety stands.
8. Lower the vehicle.
9. Connect the negative battery cable.

8.1L Engine

1. Remove or disconnect the following:
   - Negative battery cable
   - On 4WD, front axle mounting bracket through bolt nut
2. Reposition the front axle mounting bracket through bolt until the bolt tip is

flush with the support bushing. Do not remove the bolt.
   - Positive battery cable nut
   - Positive cable from the solenoid
   - Engine harness ground nut and ground from the solenoid
   - Mounting bolts and starter
   - Heat shield bolts, nut and shield, if necessary

To install:
3. Install or connect the following:
   - Heat shield, bolts and nut if removed. Tighten the bolts to 35 inch lbs. (3 Nm) and the nut to 44 inch lbs. (5 Nm).
   - Starter and bolts. Tighten to 37 ft. lbs. (50 Nm).
   - Ground wire and nut. Tighten to 30 inch lbs. (3.4 Nm).
   - Positive cable and nut. Tighten to 80 inch lbs. (9 Nm).
4. Reposition the front axle mounting bracket through bolt until the bolt is fully seated.
   - Front axle mounting bracket through bolt nut and tighten to 70 ft. lbs. (95 Nm)
   - Negative battery cable

Oil Pan

REMOVAL & INSTALLATION

4.8L, 5.3L and 6.0L Engines

➡ **The original oil pan gasket is retained and aligned to the oil pan by rivets. When installing a new gasket, it is not necessary to install new rivets. DO NOT reuse the oil pan gasket. When installing the oil pan, install a NEW oil pan gasket.**

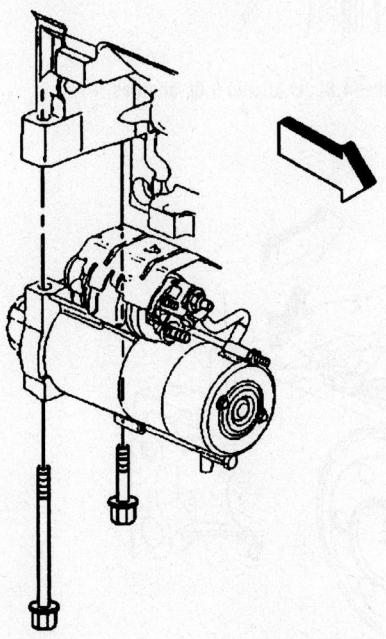

9308KG00

Starter removal—4.8L, 5.3L and 6.0L engines

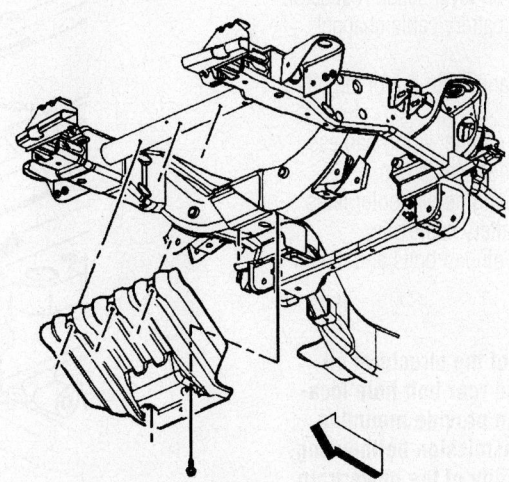

9308KG81

Oil pan shield—4.8L, 5.3L and 6.0L engines

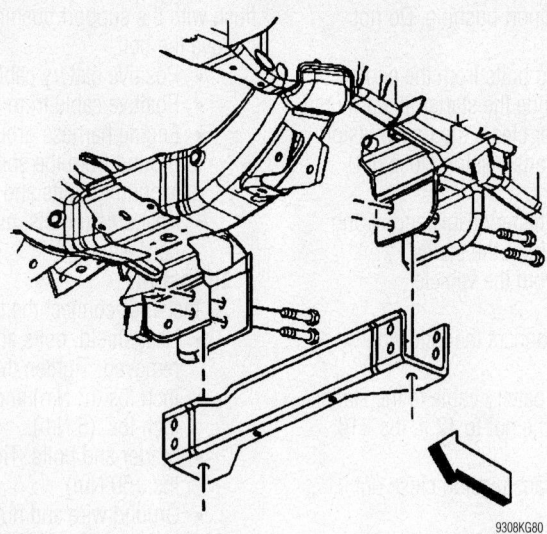

Cross brace—4.8L, 5.3L and 6.0L engines

1. Remove or disconnect the following:
   - Negative battery cable
   - On 4WD, the inner axle housing nuts and washers from the bracket

2. Support the front drive axle with a suitable jack.

3. On 4WD, remove the differential carrier lower and upper mounting bolts and nut.

4. Lower the front drive axle.

5. Remove the transmission cover bolt.

6. Remove the oil pan shield.
   - Cross brace if equipped
   - Engine oil and filter

7. If equipped with the 4L60-E automatic transmission, remove the transmission bolt and stud on the right side and the bottom bolt on the left side.

8. If equipped with the 4L80-E automatic transmission, remove the transmission converter cover bolts.

9. Remove the oil level sensor conector.

10. Remove the battery cable channel bolt.

11. Slide the channel pin out of the oil pan tab.

12. Remove the engine wiring harness and the positive battery cable clip.

13. Remove the engine oil cooler lines from the positive battery cable clip.

14. Remove the oil pan bolts and the oil pan.

To install:

➡The alignment of the structural oil pan is critical. The rear bolt hole locations of the oil pan provide mounting points for the transmission bellhousing. To ensure the rigidity of the powertrain and correct transmission alignment, it is important that the rear of the block

and the rear of the oil pan must NEVER protrude beyond the engine block and transmission bellhousing plane.

15. Apply a 0.20 in. (5mm) bead of sealant GM P/N 12378190 or equivalent 0.8 in. (20mm) long to the engine block. Apply the sealant directly onto the tabs of the front cover gasket that protrudes into the oil pan surface.

16. Apply a 0.20 in. (5mm) bead of sealant GM P/N 12378190 or equivalent 0.8 in. (20mm) long to the engine block. Apply the sealant directly onto the tabs of the rear cover gasket that protrudes into the oil pan surface.

➡Be sure to align the oil gallery passages in the oil pan and engine block properly with the oil pan gasket.

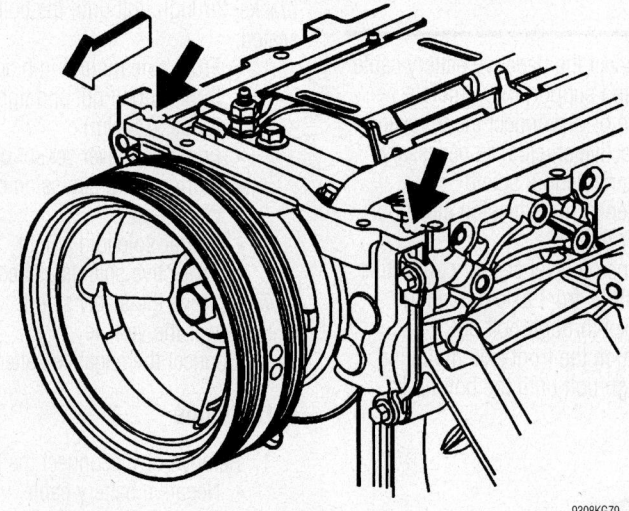

Apply sealant at these points at the front of the block—4.8L, 5.3L and 6.0L engines

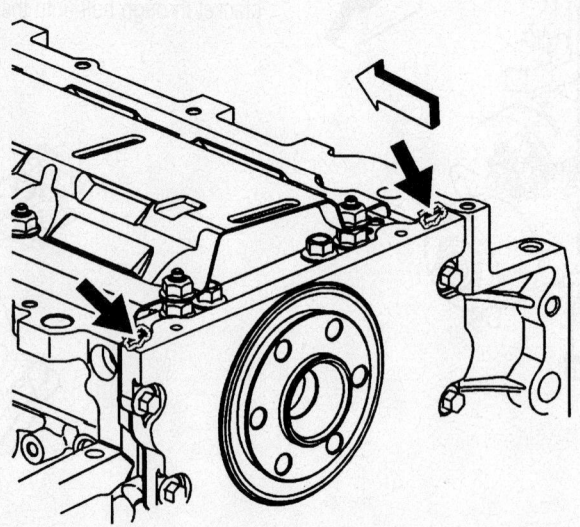

Apply sealant at these points at the rear of the block—4.8L, 5.3L and 6.0L engines

17. Pre-assemble the oil pan gasket to the pan. Install the oil pan bolts to the pan through the gasket.

18. Install or connect the following:
- Oil pan, gasket and bolts to the engine block. Snug the oil pan bolts finger-tight. Do not overtighten.

19. Install the transmission converter cover bolts until snug, if equipped with the 4L80-E automatic transmission.

20. Install the transmission bolt and stud on the right side and bottom side until snug, if equipped with the 4L60-E automatic transmission.

21. Snug the lower bellhousing bolt finger-tight. Do not overtighten. Tighten the oil pan-to-block and oil pan-to-oil pan front cover bolts to 18 ft. lbs. (25 Nm). Tighten the oil pan-to-rear cover bolts to 106 inch lbs. (12 Nm). Tighten the bellhousing bolts to 37 ft. lbs. (50 Nm).

22. Install the positive battery cable clip and bolt to the oil pan.

23. Install the engine oil cooler lines to the positive battery cable clip.

24. Install the engine wiring harness and the positive battery cable clip.

25. Slide the channel pin into the oil pan tab.

26. Install the battery cable channel bolt.

27. Install the oil level sensor conector.

28. Install the transmission cover.

29. Install the crossbar and tighten the bolts to 75 ft. lbs. (100 Nm).

30. If equipped, install the oil pan skid plate and tighten to 15 ft. lbs. (20 Nm).

31. Raise the front drive axle into position.

32. If equipped with 4WD, install the differential carrier upper mounting bolt and nut until snug. Do not tighten at this time.

33. If equipped with 4WD, instal the differential carrier lower mounting bolt and nut and tighten to 75 ft. lbs. (100 Nm).

34. If equipped with 4WD, install the inner axle housing washers and nuts to the bracket and tighten to 75 ft. lbs. (100 Nm).

35. Remove the jack from the front drive axle.

36. Lower the vehicle. Fill the engine with oil and install the engine oil filter.

37. Connect the negative battery cable.

8.1L Engine

1. Disconnect the negative battery cable and drain the engine oil.

2. Remove or disconnect the following:
- Front differential, if equipped with 4WD
- Starter motor
- Oil pan skid plate bolts and plate

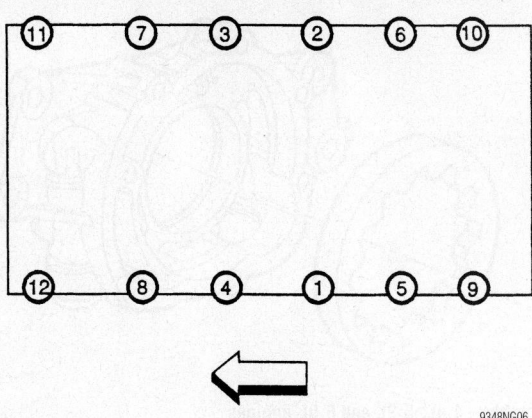

Oil pan bolt tightening sequence—8.1L engine

- Crossbar bolt(s) and crossbar
- Oil level dipstick
- Oil level sensor electrical connector
- Engine harness clip from the oil pan
- Battery cable channel bolt
- Battery cable channel and reposition
- Oil pan bolts, oil pan and gasket

➡ **You can reuse the oil pan gasket, if it is not damaged**

To install:

➡ **You must install the oil pan within 5 minutes of applying the sealer.**

3. Apply sealant to the sides of the front and rear crankshaft bearing caps on the left and right sides.

4. Install or connect the following:
- Oil pan gasket into the oil pan groove
- Oil pan and bolts

5. Tighten the oil pan bolts, in sequence, as follows:
  a. 1st pass: 89 inch lbs. (10 Nm).
  b. 2nd pass: 18 ft. lbs. (25 Nm).

6. Install or connect the following:
- Battery cable channel and bolt. Tighten to 80 inch lbs. (9 Nm).
- Oil level sensor and tighten to 15 ft. lbs. (20 Nm).
- Engine harness clip
- Oil level sensor connector
- Oil level dipstick
- Crossbar and bolt(s). Tighten to 74 ft. lbs. (100 Nm).
- Skid plate. Tighten the bolts to 15 ft. lbs. (20 Nm).
- Starter motor
- Front differential
- Negative battery cable

7. Fill the crankcase with oil.

Oil Pump

REMOVAL & INSTALLATION

4.8L, 5.3L and 6.0L Engines

1. Remove or disconnect the following:
- Engine front cover
- Oil pan
- Oil pump screen bolt and nuts

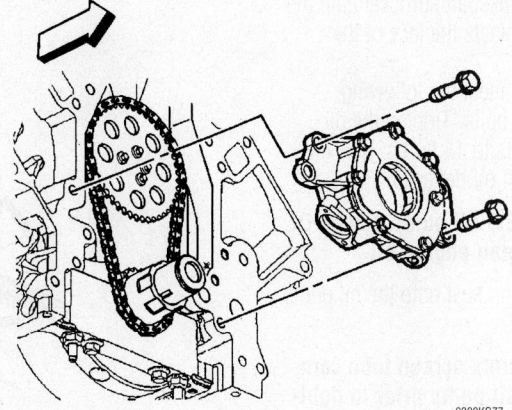

Oil pump removal—4.8L, 5.3L and 6.0L engines

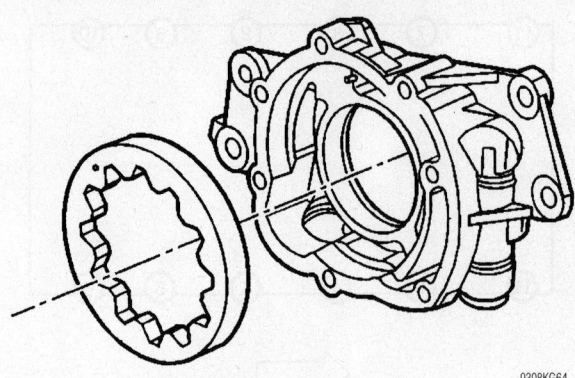

9308KG64

Oil pump disassembly—4.8L, 5.3L and 6.0L engines

- Oil pump screen with O-ring seal.
- O-ring seal from the pump screen. Discard the O-ring seal.
- Remaining crankshaft oil deflector nuts
- Crankshaft oil deflector
- Oil pump bolts

➡**Do not allow dirt or debris to enter the oil pump assembly, cap ends as necessary.**

- Oil pump

➡**The internal parts of the oil pump assembly are not serviced separately (excluding the spring). If the oil pump components are worn or damaged, replace the oil pump as an assembly. Do not attempt to repair the wire mesh portion of the pump and screen assembly.**

To install:

➡**Inspect the oil pump and engine block oil gallery passages. These surfaces must be clear and free of debris or restrictions.**

2. Align the splined surfaces of the crankshaft sprocket and the oil pump drive gear and install the oil pump. Install the oil pump onto the crankshaft sprocket until the pump housing contacts the face of the engine block.

3. Install or connect the following:
- Oil pump bolts. Tighten the oil pump bolts to 18 ft. lbs. (25 Nm).
- Crankshaft oil deflector.

➡**Lubricate a NEW oil pump screen O-ring seal with clean engine oil.**

- NEW O-ring seal onto the oil pump screen

➡**Push the oil pump screen tube completely into the oil pump prior to tightening the bolt. Do not allow the bolt to pull the tube into the pump.**

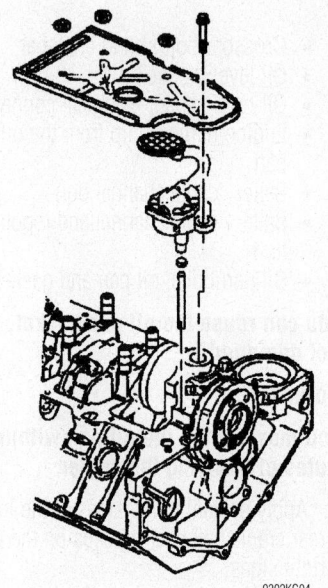

9302KG04

Exploded view of the oil pump mounting—4.8L, 5.3L and 6.0L engines

4. Align the oil pump screen mounting brackets with the correct crankshaft bearing cap studs.
- Oil pump screen, screen bolt and reflector nuts. Tighten the bolt (4) to 106 inch lbs. (12 Nm) and the oil deflector nuts (2) to 18 ft. lbs. (25 Nm).
- Oil pan
- Engine front cover

8.1L Engine

1. Remove or disconnect the following:
- Oil pan
- Oil pump screen bolt
- Oil pump, retainer and driveshaft. Discard the driveshaft retainer.
- Crankshaft oil deflector nuts
- Crankshaft oil deflector
- Oil pump bolts
- Oil pump

2. Clean and inspect the oil pump
To install:

3. Install the crankshaft oil deflector. Tighten the nuts to 37 ft. lbs. (50 Nm).

➡**Always replace the retainer between the oil pump and the shaft, when installing the oil pump. During assembly, install a new oil pump driveshaft retainer. To ease installation, slightly heat the retainer to above room temperature.**

4. Assemble the oil pump, driveshaft and a new retainer.

5. Install or connect the following:
- Oil pump, positioning it on the locating pins

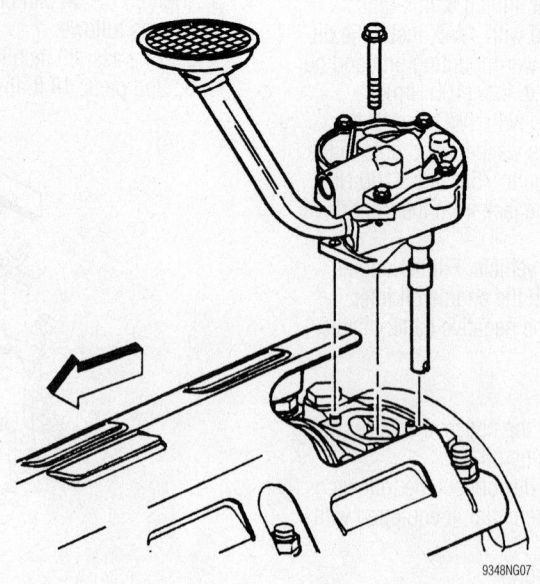

9348NG07

Oil pump removal—8.1L engine

- Oil pump bolt and tighten to 56 ft. lbs. (75 Nm)
- Oil pan

6. Refill the engine crankcase

7. Disable the ignition system; crank engine for approximately 10 seconds to aid in priming the oil pump and reducing the risk of engine damage.

➡ **If the oil pump does not build up oil pressure almost immediately, remove the pan and check for a loose oil pump-to-pick-up tube attachment. If necessary dismantle the pump and pack the pump cavity with petroleum jelly. Running the engine without measurable oil pressure will cause extensive damage.**

Rear Main Seal

REMOVAL & INSTALLATION

Except 8.1L Engine

➡**Please note that the entire transmission assembly and flywheel/flexplate must be removed to perform this procedure.**

1. Remove or disconnect the following:
- Negative battery cable
- Transfer case, if equipped
- Transmission assembly
- Flywheel
- Crankshaft spacer
- Crankshaft rear main oil seal by inserting a suitable prying tool and prying the seal out. Take care not to damage the crankshaft sealing surface.

To install:

2. Clean the oil seal bore in the block thoroughly before installation of the new seal.

3. Inspect the crankshaft for grit, rust or burrs and correct as necessary. Also inspect the portion of the crankshaft where the oil seal makes contact, for wear due to the rubbing action of the oil seal.

4. Clean the seal running surface of the crankshaft with a non-abrasive cleaner.

5. Lubricate the inner diameter of the new seal and the outer diameter of the crankshaft with engine oil.

6. Install or connect the following:
- Rear main oil seal, using installation tool J 38841, until the tool bottoms against the block and crankshaft rear main bearing cap.
- Flywheel

7. Tighten the flywheel bolts in a crisscross pattern as follows:
   a. Tighten the bolts a first pass to 15 ft. lbs. (20 Nm).

b. Tighten the bolts a second pass to 37 ft. lbs. (50 Nm).
   c. Tighten the bolts a final pass to 75 ft. lbs. (100 Nm).
- Transmission assembly
- Transfer case, if equipped
- Negative battery cable

8. Start the engine and verify no oil leaks.

8.1L Engine

Please note that the entire transmission assembly and flywheel/flexplate must be removed to perform this procedure. This procedure requires the use of the following tools: Crankshaft Rear Seal Puller tool No. J 43320 and Crankshaft Rear Seal Installer tool No. J 42849.

1. Remove or disconnect the following:
- Negative battery cable
- Transfer case, if equipped
- Transmission assembly
- Flywheel

2. Install the guide pins from the Crankshaft Rear Sear Puller into the crankshaft.

3. Install the Rear Seal Puller over the guide pins.

4. Using a drill, insert 8 of the self-drilling sheet metal screws into the rear crankshaft seal, using a crisscross pattern as shown. The self tapping screws are included with the Crankshaft Rear Seal Puller.

5. Thread the center bolt of the Crankshaft Rear Seal Puller into the crankshaft to remove the seal.

6. Remove the guide pins from the crankshaft.

To install:

7. Make sure there is no dirt, rust or loose burrs on the crankshaft.

8. Apply a light coating of engine oil to the crankshaft sealing surface. Do NOT get oil on the sealing surface of the engine block.

9. Install the new rear main seal onto the Crankshaft Rear Seal Installation Tool.

10. Position the Rear Seal Installation Tool against the crankshaft. Thread the attaching screws into the tapped holes in the crankshaft.

11. Use a screwdriver to tighten the screws securely to make sure the seal is squarely installed against the crankshaft.

12. Rotate the center nut until the installation tool bottoms, then remove the seal installation tool.

13. Install the flywheel bolts and tighten as follows:
   a. Tighten the bolts a first pass to 30 ft. lbs. (40 Nm).
   b. Tighten the bolts a second pass to 59 ft. lbs. (80 Nm).
   c. Tighten the bolts a final pass to 75 ft. lbs. (100 Nm).

14. Install or connect the following:
- Transmission assembly
- Transfer case, if equipped
- Negative battery cable

Timing Chain, Sprockets, Front Cover and Seal

➡**The manufacturer recommends that the front cover oil seal be replaced whenever the cover is removed.**

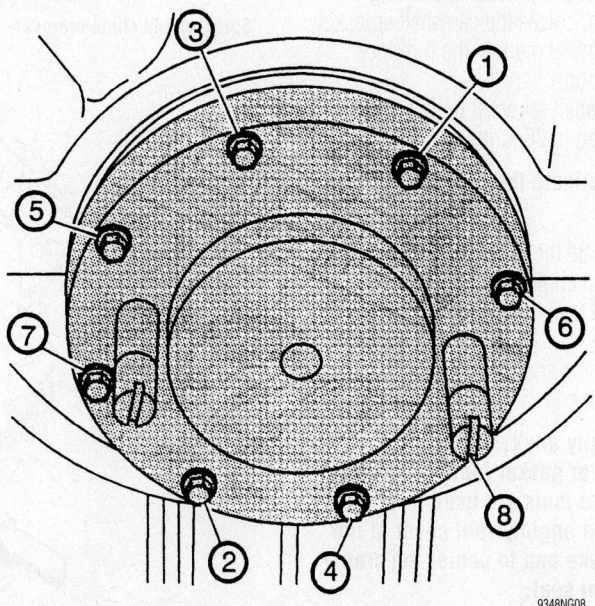

Drill the screws into the rear main seal using a crisscross pattern—8.1L engine

9348NG08

REMOVAL & INSTALLATION

4.8L, 5.3L and 6.0L Engine

1. Drain the cooling system.
2. Remove or disconnect the following:
   - Negative battery cable
   - Water pump
   - Crankshaft balancer from the crankshaft
   - Front cover bolts
   - Front cover and gasket. Discard the front cover gasket.
   - Crankshaft front oil seal from the cover
   - Oil pump
3. Rotate the crankshaft until the timing marks on the crankshaft and the camshaft sprockets are aligned.

➡ **Do not turn the crankshaft assembly after the timing chain has been removed in order to prevent damage to the piston assemblies or the valves.**

4. Remove or disconnect the following:
   - Camshaft sprocket bolts
   - Camshaft sprocket and timing chain
   - Crankshaft sprocket
   - Crankshaft sprocket key

To install:

5. Install or connect the following:
   - Key into the crankshaft keyway
   - Crankshaft sprocket onto the front of the crankshaft. Align the crankshaft key with the crankshaft sprocket keyway. Rotate the crankshaft sprocket until the alignment mark is in the 12 o'clock position.
   - Camshaft sprocket and timing chain. Locate the camshaft sprocket alignment mark in the 6 o'clock position.
   - Camshaft sprocket bolts and tighten to 26 ft. lbs. (35 Nm)

➡ **Do not lubricate the oil seal sealing surface.**

6. Lubricate the outer edge of the oil seal with clean engine oil. Lubricate the front cover oil seal bore with clean engine oil.

7. Install the crankshaft front oil seal with an installer.

➡ **Do not apply any type of sealant to the front cover gasket (unless specified). Special tools are used to properly align the engine front cover at the oil pan surface and to center the crankshaft front oil seal.**

8. Install the front cover gasket, cover,

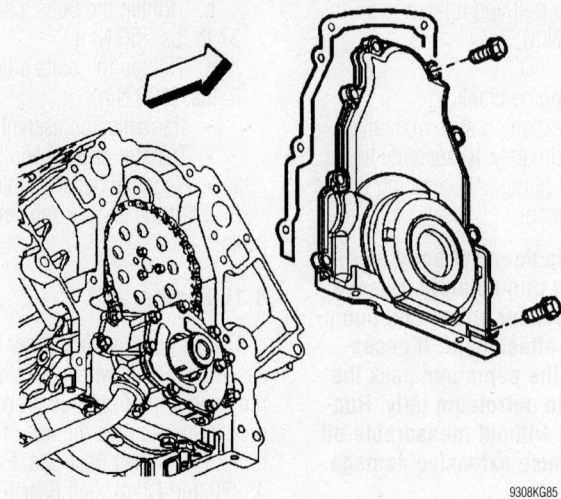

Front cover and gasket—4.8L, 5.3L and 6.0L engines

9308KG85

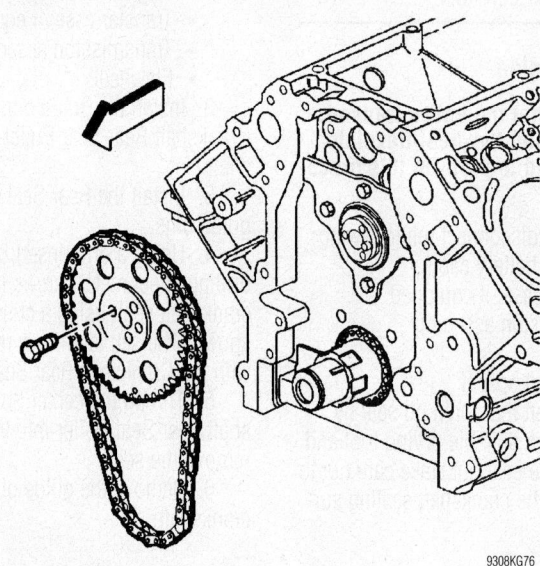

Sprocket and chain removal—4.8L, 5.3L and 6.0L engines

9308KG76

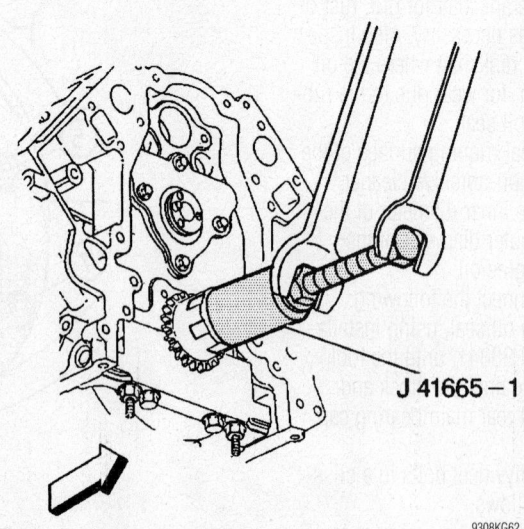

J 41665 – 1

Crankshaft sprocket installation—4.8L, 5.3L and 6.0L engines

9308KG62

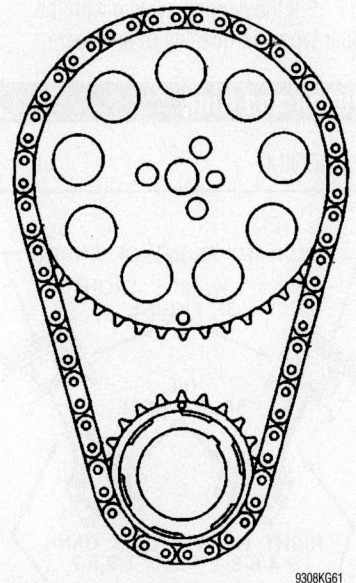

Timing mark alignment—4.8L, 5.3L and 6.0L engines

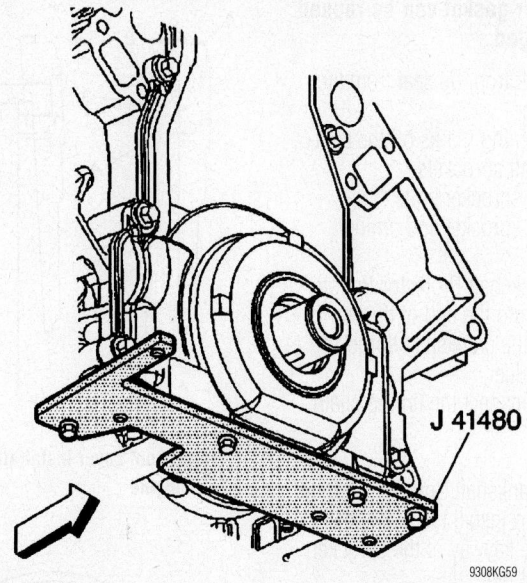

J41480 installation—4.8L, 5.3L and 6.0L engines

and bolts onto the engine. Tighten the cover bolts finger-tight. Do not overtighten.

9. Start the J41480 tool-to-front cover bolts. Don't tighten the bolts yet.

➡**Align the tapered legs of the tool with the machined alignment surfaces on the front cover.**

10. Install tool J41476 . Install the crankshaft balancer bolt. Tighten the crankshaft balancer bolt by hand until snug. Do not overtighten. Tighten the J41480 bolts and front cover bolts to 18 ft. lbs. (25 Nm).

11. Remove the tools.

12. Install the used crankshaft balancer bolt and tighten to 240 ft. lbs. (330 Nm).

13. Remove the used bolt.

➡**The nose of the crankshaft should be recessed 2.4-4.48 mm (0.094-0.176 in) into the balancer bore.**

14. Install a NEW crankshaft balancer bolt and tighten to 37 ft. lbs. (50 Nm), then tighten an additional 140 degrees.

15. Place a straight edge across the engine block and front cover oil pan sealing surfaces. Avoid contact with the portion of the gasket that protrudes into the oil pan surface. Insert a feeler gauge between the front cover and the straight edge tool. The cover must be flush with the oil pan surface or no more than 0.02 in. (0.5mm) below flush. If the front cover-to-engine block oil pan surface alignment is not within specifications, repeat the cover alignment procedure. If the correct front cover-to-engine block alignment cannot be obtained, replace the front cover.

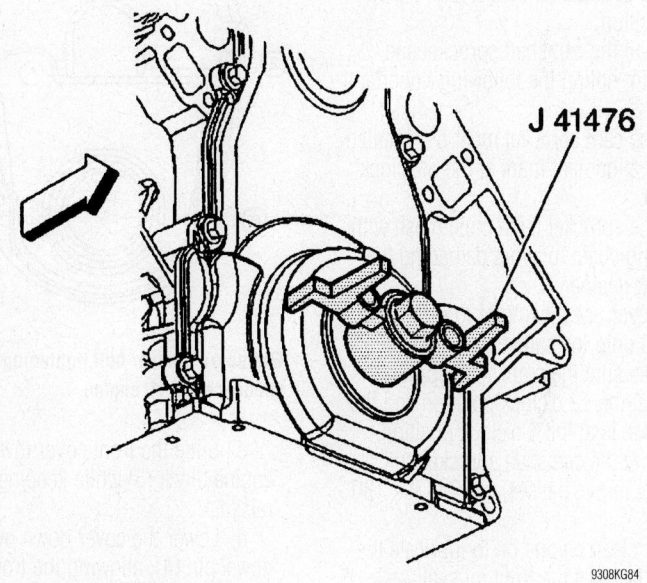

Seal alignment tool installation—4.8L, 5.3L and 6.0L engines

16. Install the crankshaft balancer bolt. Tighten the crankshaft balancer bolt by hand until snug. Do not overtighten the bolt.

17. Snug the oil pan-to-cover bolts in order to position the cover at the pan rail.

18. Tighten the oil pan-to-front cover bolts to 18 ft. lbs. (25 Nm).

19. Tighten the front cover bolts to 18 ft. lbs. (25 Nm).

20. Install the water pump.

8.1L Engine

➡**This procedure requires the use of Crankshaft Sprocket Installer tool No. J 22102 and Crankshaft Protector Button tool No. J 42846.**

1. Drain the cooling system.

2. Remove or disconnect the following:
   - Negative battery cable
   - Water pump
   - Crankshaft balancer from the crankshaft
   - Camshaft Position (CMP) sensor connector
   - Engine harness clips from the battery cable channel
   - CMP sensor bolt and sensor
   - Battery cable channel bolt
   - Battery cable channel and reposition
   - Front cover bolts, front cover and gasket

➡ **The front cover gasket can be reused if it is not damaged.**

- Crankshaft front oil seal from the front cover

3. Align the timing marks on the camshaft and crankshaft sprockets.

- Camshaft sprocket bolts
- Camshaft sprocket and timing chain

4. Install Crankshaft Protector Button tool No. J 42846 into the end of the crankshaft and remove the crankshaft sprocket using a 3-jawed puller.

5. Clean and inspect the timing chain and sprockets.

To install:

6. Use the Crankshaft Sprocket Installer tool No. J 22102 to install the crankshaft sprocket. Align the keyway of the sprocket with the crankshaft pin.

7. Remove the installation tool.

8. Rotate the crankshaft until the crankshaft sprocket alignment mark is in the 12 o'clock position.

9. Install the camshaft sprocket and timing chain, noting the following important points:

a. The cam sprocket must be installed with the alignment mark at the 6 o'clock position.

b. The sprocket teeth must mesh with the timing chain to avoid damaging the camshaft retainer.

c. Never use a hammer to install the sprocket onto the camshaft.

10. Make sure the crankshaft sprocket is alignment at the 12 o'clock position and the cam sprocket is at the 6 o'clock position.

11. Install the camshaft sprocket bolts and tighten, in two passes, to 22 ft. lbs. (30 Nm)

12. Use clean engine oil to lubricate the sealing surfaces of the front oil seal.

13. Install or connect the following:

- New seal into the front cover, using a suitable seal installation tool

➡ **The front cover must be installed while the sealant is still wet to the touch.**

- Sealant to the 2 places on the engine block where the front cover meets the oil pan
- Front cover gasket into the cover

14. Install the front cover, referring to the accompanying figure and using the following steps only:

a. Hold the front cover (1) up to the crankshaft (2).

b. Lift the cover (1) while sliding the cover over the crankshaft (2).

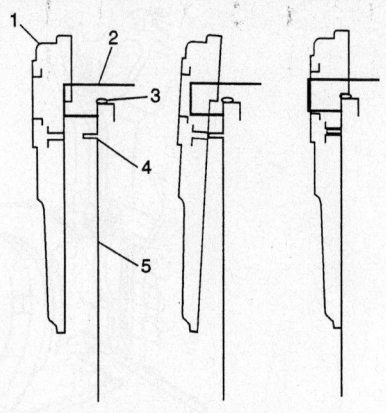

9348NG09

Proper front cover installation sequence— 8.1L engine

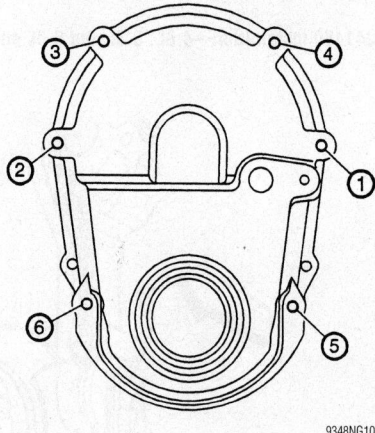

9348NG10

Engine front cover bolt tightening sequence—8.1L engine

c. Slide the front cover toward the engine block (5) while keeping the cover raised.

d. Lower the cover down over the dowel pin (4), allowing the front cover to rest on the sealant (3).

15. Install the front cover bolts and tighten, in sequence, as follows:

a. 1st pass: 53 inch lbs. (6 Nm)

b. 2nd pass: 106 inch lbs. (12 Nm)

16. Install or connect the following:

- Battery cable channel and bolt. Tighten to 80 inch lbs. (9 Nm).
- CMP sensor. Inspect the O-ring first, replace if necessary and coat with oil before installation
- CMP sensor bolt to 106 inch lbs. (12 Nm)
- Engine harness clips to the battery cable channel
- CMP sensor electrical connector
- Crankshaft balancer
- Water pump
- Negative battery cable.

17. Fill the cooling system with the proper type and quantity of antifreeze.

Piston and Ring

POSITIONING

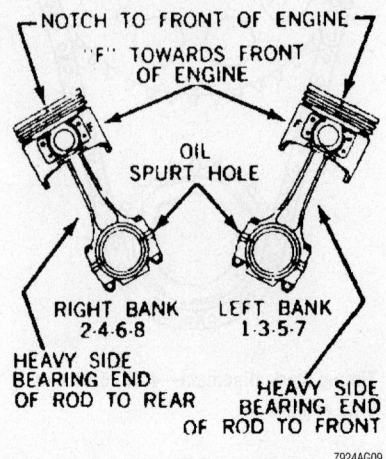

7924AG09

Piston and connecting rod assembly positioning —4.8L, 5.3L, and 6.0L engines

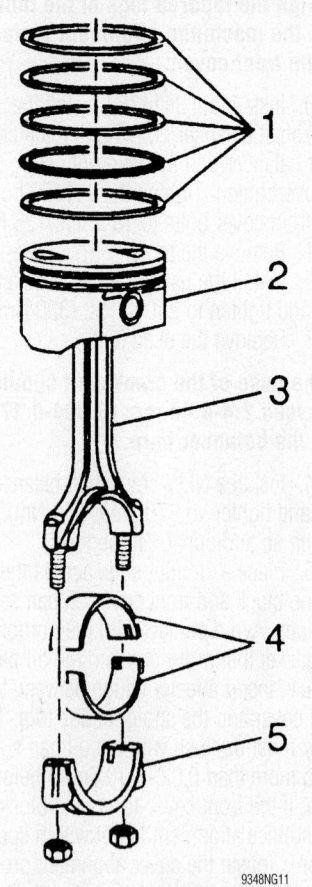

9348NG11

Piston rings (1), piston (2), connecting rod (3) and related components—8.1L engine

FUEL SYSTEM

Fuel System Service Precautions

Safety is the most important factor when performing not only fuel system maintenance but any type of maintenance. Failure to conduct maintenance and repairs in a safe manner may result in serious personal injury or death. Maintenance and testing of the vehicle's fuel system components can be accomplished safely and effectively by adhering to the following rules and guidelines.

• To avoid the possibility of fire and personal injury, always disconnect the negative battery cable unless the repair or test procedure requires that battery voltage be applied.

• Always relieve the fuel system pressure prior to disconnecting any fuel system component (injector, fuel rail, pressure regulator, etc.), fitting or fuel line connection. Exercise extreme caution whenever relieving fuel system pressure, to avoid exposing skin, face and eyes to fuel spray. Please be advised that fuel under pressure may penetrate the skin or any part of the body that it contacts.

• Always place a shop towel or cloth around the fitting or connection prior to loosening to absorb any excess fuel due to spillage. Ensure that all fuel spillage (should it occur) is quickly removed from engine surfaces. Ensure that all fuel soaked cloths or towels are deposited into a suitable waste container.

• Always keep a dry chemical (Class B) fire extinguisher near the work area.

• Do not allow fuel spray or fuel vapors to come into contact with a spark or open flame.

• Always use a back-up wrench when loosening and tightening fuel line connection fittings. This will prevent unnecessary stress and torsion to fuel line piping. Always follow the proper torque specifications.

• Always replace worn fuel fitting O-rings with new. Do not substitute fuel hose or equivalent where fuel pipe is installed.

Fuel System Pressure

RELIEVING

A Schrader valve is provided on these fuel systems, in order to conveniently test or release the system pressure. A fuel pressure gauge and adapter will be necessary to connect the gauge to the fitting. Most of the

MFI systems utilize a service valve on one end of the fuel rail assembly. The CMFI system covered here uses a valve located on the inlet pipe fitting, immediately before it enters the CMFI assembly (toward the rear of the engine)

1. Before servicing the vehicle, refer to the precautions in the beginning of this section.
2. Turn the ignition **OFF**.
3. Disconnect the negative battery cable.
4. Loosen the fuel filler cap in order to relieve the fuel tank vapor pressure.
5. Connect a fuel pressure gauge to the fuel pressure valve/fitting.
6. Wrap a shop towel around the fitting while connecting the gauge in order to avoid spillage.
7. Install the bleed hose of the gauge into an approved container.
8. Open the valve on the gauge to bleed the system pressure.

The fuel connections are now safe for servicing. Drain any fuel remaining in the gauge into an approved container.

Fuel Filter

REMOVAL & INSTALLATION

➡ **On 2005-06 gasoline engines except 5.3L Flex Fuel the fuel filter is integral with the fuel pump/sender assembly in the fuel tank.**

1. Before servicing the vehicle, refer to the precautions in the beginning of this section.
2. Disconnect the negative battery cable.
3. Relieve the fuel system pressure.
4. Raise the vehicle.
5. Clean all the fuel filter connections and the surrounding areas before disconnecting the fuel pipes in order to avoid possible contamination of the fuel system.
6. Disconnect the threaded fittings from the fuel filter.
7. Cap the fuel pipes in order to prevent possible fuel system contamination.
8. Slide the fuel filter from the bracket.
9. Inspect the fuel pipe O-rings for cuts, nicks, swelling, or distortion. Replace the O-rings if necessary.

To install:

10. Slide the fuel filter into the bracket. Remove the caps from the fuel pipes.
11. Connect the threaded fittings to the fuel filter. Tighten the fittings to 18 ft. lbs. (25 Nm).
12. Lower the vehicle.
13. Tighten the fuel filler cap.
14. Connect the negative battery cable.
15. Turn the ignition **ON** for 2 seconds.
16. Turn the ignition **OFF** for 10 seconds.
17. Turn the ignition **ON**.
18. Inspect for fuel leaks.

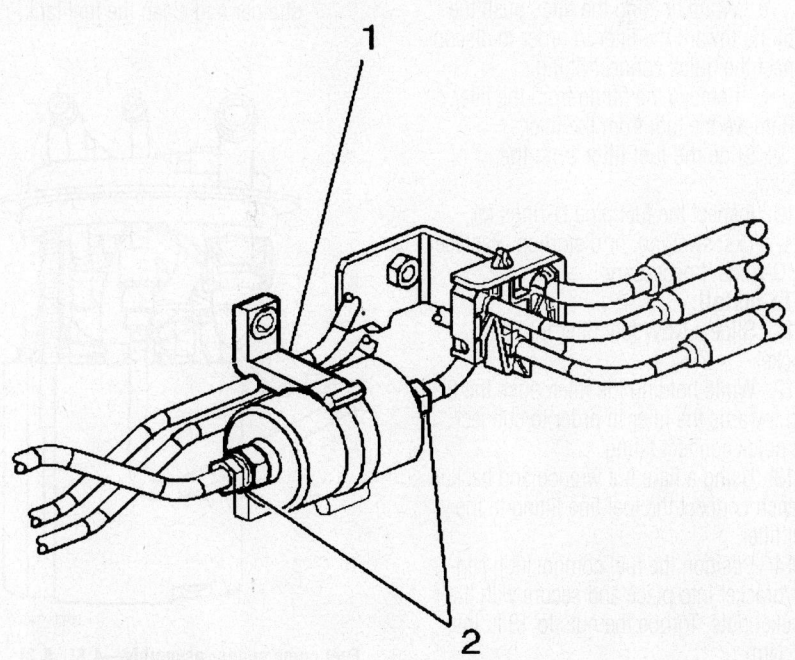

9308KG55

Fuel filter—2002-04 4.8L, 5.3L and 6.0L engines

5.3L FLEX FUEL ENGINE

➡**This procedure requires the use of tool J-46363, Fuel Line Release Tool, or equivalent.**

1. Before servicing the vehicle, refer to the precautions in the beginning of this section.
2. Disconnect the negative battery cable.
3. Relieve the fuel system pressure.
4. Raise the vehicle.
5. Remove the fuel composition sensor bracket nuts and reposition it aside.
6. Clean all the fuel filter connections and the surrounding areas before disconnecting the fuel pipes in order to avoid possible contamination of the fuel system.
7. Using a flare nut wrench and backup wrench disconnect the fuel line fitting from the fuel filter.
8. Use the Fuel Line Release Tool, J-46363, or equivalent to disconnect the quick connect fittings as follows:

    a. Pull the quick connect fitting back until the internal retainers hit the filter stop.

    b. Twist one end of J-46363 and insert it between the fitting and the filter

    c. Twist the opposite end of the tool and insert it between the fitting and filter

    d. Once tool J-46363 is inserted, make sure it is parallel to the filter and fitting.

    e. While holding the filter, push the fitting toward the filter in order to disconnect the quick connect fitting.

    f. Remove the fitting from the filter. Remove the tool from the filter.

9. Slide the fuel filter from the bracket.
10. Inspect the fuel pipe O-rings for cuts, nicks, swelling, or distortion. Replace the O-rings if necessary.

To install:

11. Slide a NEW fuel filter into the bracket.
12. While holding the filter, push the fitting towards the filter in order to connect the quick connect fitting.
13. Using a flare nut wrench and backup wrench connect the fuel line fitting to the fuel filter.
14. Position the fuel composition sensor/bracket into place and secure with the bracket nuts. Torque the nuts to 13 ft. lbs. (17 Nm).
15. Lower the vehicle

Fuel Pump

REMOVAL & INSTALLATION

1. Before servicing the vehicle, refer to the precautions in the beginning of this section.
2. Remove or disconnect the following:
    • Negative battery cable
3. Relieve the fuel system pressure.
4. Drain the fuel tank.
5. Remove or disconnect the following:

    • Fuel tank

✲✲ WARNING

Do not handle the fuel sender assembly by the fuel pipes. The amount of leverage generated by handling the fuel pipes could damage the joints.

6. Disconnect the fuel and evaporative emission (EVAP) lines from the fuel tank module.
    • Fuel sender assembly retaining ring using a fuel tank sending unit wrench. Remove the fuel sender assembly and the seal. Discard the seal.
    • Note the position of the fuel strainer on the fuel sender. Support the fuel sender assembly with one hand and grasp the strainer with the other hand. Pull the strainer off the fuel sender. Discard the strainer after inspection. Inspect the strainer. Replace a contaminated strainer and clean the fuel tank.

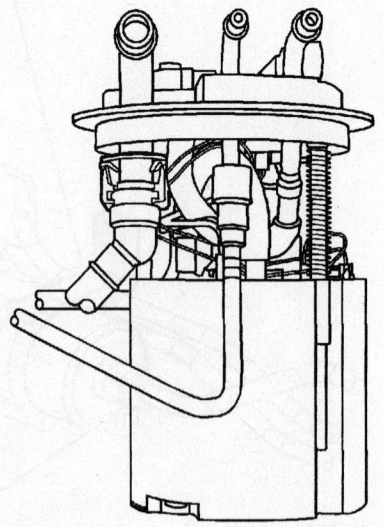

Fuel pump/sender assembly—4.8L, 5.3L and 6.0L engines

06025-YUKO-G05

• Fuel pump electrical connector
• Electrical connector retaining clip from the fuel level sensor
• Sensor electrical connector from under the fuel sender cover
• Fuel level sensor retaining clip

7. Squeeze the locking tangs and remove the fuel level sensor.
8. Remove the fuel pressure sensor.

To install:

9. Install or connect the following:
    • Fuel pressure sensor
    • Fuel level sensor
    • Sensor retaining clip
    • Electrical connector to the fuel level sensor
    • Electrical connector retaining clip to the fuel level sensor
    • Fuel pump electrical connector

➡**Always install a new fuel strainer when replacing the fuel tank fuel pump module.**

    • New fuel strainer in the same position as noted during disassembly. Push the strainer on the bottom of the fuel sender until the strainer is fully seated.
    • New seal on the fuel tank

➡**The fuel pump strainer must be in a horizontal position when the fuel sender is installed in the tank. When installing the fuel sender assembly, assure that the fuel pump strainer does not block full travel of the float arm.**

    • Fuel sender assembly into the fuel tank
    • Fuel sender assembly retaining ring
    • Fuel tank. Install the fuel tank strap attaching bolts. Tighten the bolts to 30 ft. lbs. (40 Nm).

10. Refill the fuel tank. Install the fuel filler cap. Connect the negative battery cable.
11. Turn the ignition **ON** for 2 seconds.
12. Turn the ignition **OFF** for 10 seconds.
13. Turn the ignition **ON**.
14. Inspect for fuel leaks.

Fuel Injector

REMOVAL & INSTALLATION

4.8L, 5.3L and 6.0L Engines

1. Before servicing the vehicle, refer to the precautions in the beginning of this section.

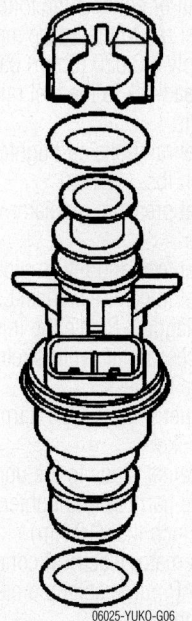

06025-YUKO-G06

Fuel injector—4.8L, 5.3L and 6.0L engines

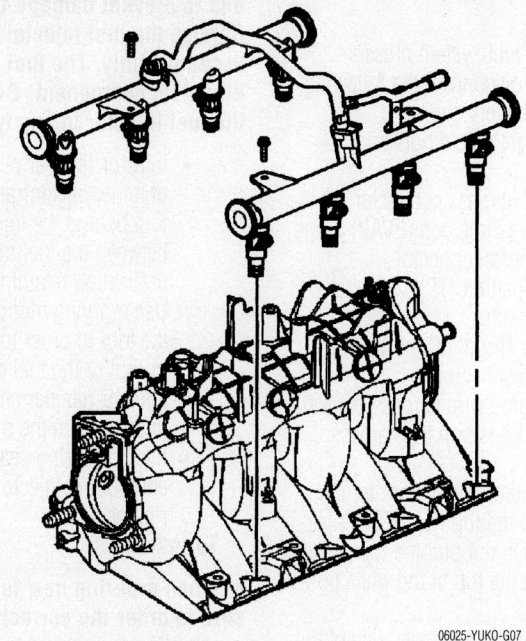

06025-YUKO-G07

Fuel rail assembly—4.8L, 5.3L and 6.0L engines

2. Relieve the fuel system pressure.

3. Remove the wire harness bracket nut.

4. Disconnect the evaporative emission (EVAP) purge solenoid electrical connector.

5. Disconnect the generator electrical connector.

6. Disconnect the MAP and knock sensor connectors.

7. Remove the knock sensor harness connector from the intake manifold.

8. Remove the connector position assurance (CPA) lock.

9. Disconnect the main coil and fuel injector connectors from both sides of the engine.

10. Remove the harness clips from the fuel rail.

11. Reposition the engine wire harness aside.

12. Perform the following steps in order to disconnect the fuel injector electrical connectors.

    a. Mark the connectors to their corresponding injectors to ensure correct reassembly.

    b. Pull the connector position assurance (CPA) retainer on the connector up 1 click.

    c. Push the tab on the connector in.

    d. Disconnect the fuel injector electrical connector.

    e. Repeat the steps for each injector electrical connector.

13. Disconnect the fuel feed pipe from the fuel rail.

14. Loosen the crossover tube-to-right fuel rail bolt.

15. Remove the fuel rail bolts.

16. Remove the fuel rail.

17. Remove the fuel injector lower O-ring seal from each injector, if necessary.

18. Discard the O-ring seal.

19. Remove and discard the fuel injector retainer clip.

20. Remove the fuel injector.

21. Remove and discard the fuel injector O-ring seals.

To install:

➡**When ordering new fuel injectors, be sure to order the correct injector for the application being serviced. The fuel injector assembly is stamped with a part number identification.**

22. Lubricate the new injector O-ring seals with clean engine oil.

23. Install or connect the following:
• New injector O-ring seals on the injector
• New retainer clip on the injector

24. Push the fuel injector into the fuel rail injector socket with the electrical connector facing outward. The retainer clip locks on to a flange on the fuel rail injector socket.

25. Install the fuel rail.

26. Apply a 5 mm (0.2 in) band of threadlocker, or equivalent to the threads of the fuel rail bolts.

27. Install the fuel rail bolts and tighten to 89 inch lbs. (10 Nm).

28. Connect the fuel feed pipe to the fuel rail.

29. Install the PCV hose.

30. Perform the following steps in order to connect the fuel injector electrical connectors:

    a. Install the connectors to their corresponding injectors to ensure correct reassembly.

    b. Connect the fuel injector electrical connector.

    c. Push the CPA retainer on the connector in 1 click.

    d. Repeat the steps for each injector electrical connector.

31. Position the engine wire harness.

32. Connect the main coil and fuel injector connectors to both sides of the engine.

33. Install the harness clips to the fuel rail.

34. Install the CPA retainers.

35. Connect the MAP and knock sensor connectors.

36. Install the knock sensor harness connector to the intake manifold.

37. Connect the EVAP purge solenoid electrical connector.

38. Connect the generator electrical connector.

39. Install the wire harness bracket nut.

40. Tighten the fuel cap.

41. Connect the negative battery cable.

42. Turn the ignition **ON** for 2 seconds.

43. Turn the ignition **OFF** for 10 seconds.

44. Turn the ignition **ON**.

45. Inspect for fuel leaks.

46. Install the engine sight shield.

Tighten the engine sight shield bolts to 89 inch lbs. (10 Nm)

8.1L Engine

1. Relieve the fuel system pressure.
2. Remove or disconnect the following:
   - Negative battery cable
   - Engine sight shield nuts and bracket
   - Alternator harness connector
   - Evaporative Emission (EVAP) purge valve harness connector
   - Throttle Position (TP) sensor electrical connector
   - Electronic Throttle Control (ETC) electrical connector
   - Upper engine wire harness bracket studs, and position the harness aside
3. Tag the injector connectors for identification, then pull the top part of the injector connector up. Do not pull the top part of the connector past the top of the white portion.
4. Push the tab on the lower side of the injector connector to release the connect from the injector. Perform these steps on each injector connector.
5. Remove or disconnect the following:
   - Fuel feed and return pipes from the fuel rail
   - Fuel pressure regulator vacuum line
   - Fuel rail attaching bolts and fuel rail

➡Use care in removing the fuel injectors in order to prevent damage to the electrical connector pins on the injector and to prevent damage to the nozzle. Service the fuel injector as a complete assembly only. The fuel injector is an electrical component. DO NOT immerse the fuel injector in any type of cleaner.

   - Injector retainer clip. Insert the fork of a fuel injector assembly removal tool behind the injector connector between the fuel rail pod and the 3 protruding retaining clip ledges. Use a prying motion while inserting the tool in order to force the injector out of the fuel rail pod.
   - Injector retainer clip
   - Injector from the fuel rail pod
   - Injector O-ring seals from both ends of the injector. Discard the O-ring seals.

To install:

➡When ordering new fuel injectors, be sure to order the correct injector for the application being serviced. The fuel injector assembly is stamped with a part number identification.

6. Lubricate the new injector O-ring seals with clean engine oil.
7. Install or connect the following:
   - New injector O-ring seals on the injector
   - New retainer clip on the injector
8. Push the fuel injector into the fuel rail injector socket with the electrical connector facing outward. The retainer clip locks on to a flange on the fuel rail injector socket.

9. Install or connect the following:
   - Fuel rail to the intake manifold
   - Apply a 0.020 (5mm) band of threadlock to the fuel rail retaining bolts
   - Fuel rail bolts and tighten to 106 inch lbs. (12 Nm)
   - Fuel pressure regulator vacuum line
   - Fuel feed and return pipes
   - Fuel injector electrical connectors, as tagged. Rotate the injectors as necessary to avoid stretching the wire harness
   - Upper engine wire harness bracket
   - Retainer studs to the upper engine wire harness and tighten the nut to 89 inch lbs. (10 Nm)
   - Alternator electrical connector
   - EVAP purge solenoid electrical connector
   - TP and ETC sensor connectors
   - Engine sight shield mounting bracket and bolts
10. Tighten the fuel cap.
11. Connect the negative battery cable.
12. Turn the ignition **ON** for 2 seconds.
13. Turn the ignition **OFF** for 10 seconds.
14. Turn the ignition **ON**.
15. Inspect for fuel leaks.
16. Install the engine sight shield. Tighten the engine sight shield bolts to 89 inch lbs. (10 Nm).

DRIVE TRAIN

Transmission Assembly

REMOVAL & INSTALLATION

Automatic Transmission

4L60-E/4L65-E

1. Drain the transmission fluid.
2. Remove the propeller shaft.
3. Support the transmission with a transmission jack.
4. Remove the transmission mount nuts.
5. Remove the transmission support bolts/nuts.
6. Raise the transmission slightly and remove the transmission support from the vehicle.
7. Remove the exhaust pipe.
8. Remove the starter motor.

9. Remove the transfer case, if equipped.
10. Remove the transmission mount bolts and mount.
11. Ensure the transmission manual shaft is positioned in mechanical park.
12. Remove the retainer that secures the cable to the bracket.
13. Remove the range selector cable end from the transmission range selector lever ball stud.
14. Depress the tangs and remove the cable from the bracket.
15. Remove the exhaust hanger bracket bolts and bracket.
16. Remove the harness clip from the fuel line bracket.
17. Disconnect the park/neutral position (PNP) switch electrical connectors.
18. Disconnect the main electrical connector.

19. Disconnect the vehicle speed sensor (VSS) electrical connector.
20. Remove the bolt that secures the fuel line bracket to the left side of the transmission.
21. Remove the torque converter bolts.
22. Disconnect the transmission oil cooler lines.
23. Plug the transmission oil cooler line fittings.
24. Remove the fuel feed/return and evaporative emission (EVAP) return pipe bracket nut.
25. Remove the lower right transmission bolt and stud.
26. Remove the remaining transmission studs and bolt.
27. Install the torque converter holding tool J-21366 onto the transmission in order to retain the torque converter.
28. Pull the transmission straight back.

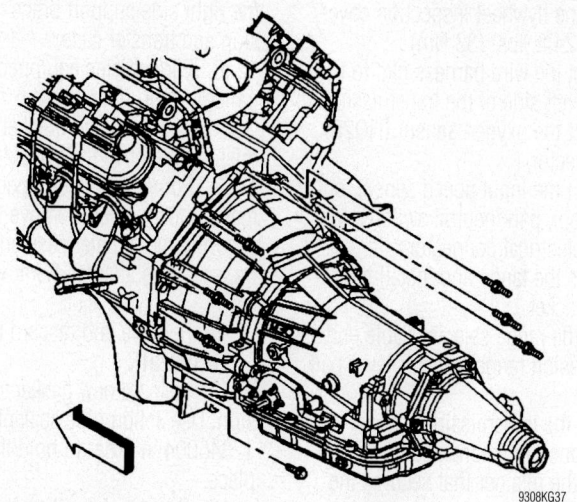

4L60-E/4L65-E transmission removal

29. Remove the transmission from the vehicle while simultaneously removing the fluid level indicator tube.

30. Flush the transmission oil cooler and cooling lines when you remove the transmission.

To install:

31. Install Tool J21366 onto the transmission bell housing to retain the torque converter.

32. Raise the transmission into place while simultaneously installing the fluid indicator tube.

33. Remove the holding tool.

34. Slide the transmission straight onto the locating pins while lining up the marks on the flywheel and the torque converter. The torque converter must rotate freely by hand.

35. Install the transmission studs and bolt and tighten to 37 ft. lbs. (50 Nm).

36. Install the lower right transmission stud and bolt and tighten to 37 ft. lbs. (50 Nm).

37. Install the fuel feed/return and EVAP return pipe bracket nut.

38. Remove the plugs from the transmission oil cooler line fittings.

39. Connect the transmission oil cooler lines.

40. If reusing the torque converter bolts, clean the bolt threads and apply Loctite 242, or equivalent to the threads prior to installation.

41. Install the torque converter bolts and tighten to 46 ft. lbs. (63 Nm).

42. Install the bolt that secures the fuel line bracket to the left side of the transmission.

43. Connect the VSS and tha main electrical connectors.

44. Connect the PNP switch electrical connectors.

45. Install the harness clip to the fuel line bracket.

46. Install the exhaust hanger bracket and bolts and tighten to 13 ft. lbs. (17 Nm).

47. Install the cable to the bracket.

48. Install the range selector cable end to the transmission range selector lever ball stud.

49. Ensure the transmission manual shaft is positioned in mechanical park.

50. Install the retainer that secures cable to the bracket.

51. Install the transmission mount and bolts and tighten to 18 ft. lbs. (25 Nm).

52. Install the transfer case, if equipped,

53. Install the starter.

54. Install the exhaust pipe

55. Raise the transmission slightly and install the transmission support.

56. Install the transmission support bolts/nuts and tighten to 70 ft. lbs. (95 Nm).

57. Install the transmission mount nuts and tighten to 37 ft. lbs. (50 Nm).

58. Remove the support from the transmission.

59. Install the propeller shaft.

60. Connect the negative battery cable.

61. Fill the transmission to the proper level with DEXRON VI transmission fluid.

4L80-E/4L85-E

1. Drain the transmission fluid.

2. Remove the propeller shaft.

3. Support the transmission with a suitable jack.

4. Remove the transmission mount nuts.

5. Remove the transmission support bolts/nuts.

6. Remove the transmission support.

7. Raise the transmission slightly and remove the transmission support from the vehicle.

8. Remove the exhaust manifold pipe.

9. Remove the transfer case, if equipped.

10. Remove the transmission mount bolts and mount.

11. Ensure the transmission manual shaft is positioned in mechanical park.

12. Remove the retainer that secures the cable to the bracket.

13. Remove the range selector cable end from the transmission range selector lever ball stud.

14. Depress the tangs and remove the cable from the bracket.

15. Disconnect the input speed sensor, output speed sensor, park/neutral switch and transmission electrical connectors.

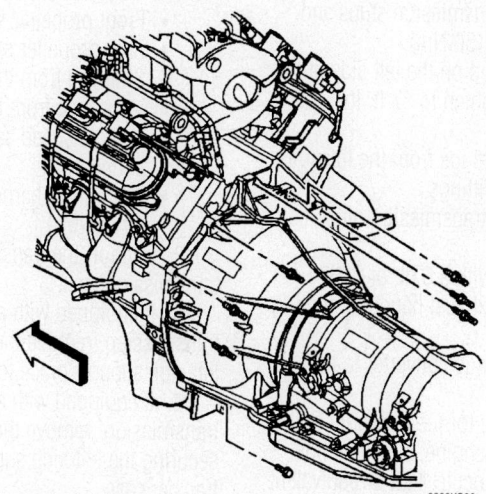

4L80-E/4L85-E transmission removal

16. Disconnect the oxygen sensor (HO2S) electrical connector.

17. Disconnect the wire harness clip from the bracket on the left side of the transmission.

18. Cut the tie strap to separate the vent hose from the wiring harness.

19. Remove the flywheel inspection cover.

20. Mark the torque converter to flywheel orientation.

21. Remove the torque converter bolts.

22. If equipped with a 6.0L or 8.1L engine, remove the exhaust hanger mounting bracket bolts and bracket.

23. Disconnect and plug the transmission oil cooler lines.

24. Remove the stud on the left side of the transmission.

25. Remove the remaining transmission studs.

26. Install torque converter holding tool J-21366 onto the transmission bell housing to retain the torque converter.

27. Pull the transmission straight back.

28. Remove the transmission from the vehicle.

29. Flush the transmission oil cooler and lines at this time.

To install:

30. Install Tool J21366 onto the transmission bell housing to retain the torque converter.

31. Install the transmission to the vehicle.

32. Raise the transmission into place and remove the holding tool.

33. Slide the transmission straight onto the locating pins while lining up the marks on the flywheel and the torque converter. The torque converter must rotate freely by hand.

34. Install the transmission studs and tighten to 37 ft. lbs. (50 Nm).

35. Install the stud on the left side of the transmission and tighten to 37 ft. lbs. (50 Nm).

36. Remove the plugs from the transmission oil cooler line fittings.

37. Connect the transmission oil cooler lines.

38. If equipped with a 6.0L or 8.1L engine, install the exhaust hanger mounting bracket and bolts.

39. Align the torque converter to the flywheel.

40. If reusing the torque converter bolts with a 6.0L or 8.1L engine, clean the bolt threads and apply Loctite 242, or equivalent to the threads prior to installation.

41. Install the torque converter bolts and tighten to 44 ft. lbs. (60 Nm).

42. Install the flywheel inspection cover and tighten to 24 ft. lbs. (33 Nm).

43. Connect the wire harness clip to the bracket on the left side of the transmission.

44. Connect the oxygen sensor (HO2S) electrical connector.

45. Connect the input speed sensor, output speed sensor, park/neutral switch and transmission electrical connectors.

46. Depress the tangs and install the cable to the bracket.

47. Install the range selector cable end to the transmission range selector lever ball stud.

48. Ensure the transmission manual shaft is positioned in mechanical park.

49. Install the retainer that secures the cable to the bracket.

50. Install the transmission mount and bolts and tighten to 18 ft. lbs. (25 Nm).

51. Install the transfer case, if equipped.

52. Install the exhaust manifold pipe.

53. Install the transmission support.

54. Install the transmission support bolts/nuts and tighten to 75 ft. lbs. (95 Nm).

55. Install the transmission mount nuts and tighten to 37 ft. lbs. (50 Nm).

56. Remove the support from the transmission.

57. Install the propeller shaft.

58. Connect the negative battery cable.

59. Fill the transmission to the proper level with Dexron VI transmission fluid.

Transfer Case Assembly

REMOVAL & INSTALLATION

NVG261-NP2 Transfer Case

1. Remove or disconnect the following:
- Transfer case shields
- Front propeller shaft
- Rear propeller shaft
- Shift rod from the transfer case
- Vent hose from the transfer case
- Vehicle speed sensor electrical connectors
- Any wiring harness from the transfer case

2. Support the transfer case with a transmission jack.

3. If equipped with a NV3500 manual transmission, remove the bolt securing the left side support brace to the transmission.

4. If equipped with a NV3500 manual transmission, remove the bolt and stud securing the left side support brace to the transfer case.

5. If equipped with a NV3500 manual transmission, remove the 2 bolts securing the right side support brace to the transmission and transfer case.

6. For vehicles equipped with a manual transmission, remove the 6 nuts securing the transfer case and bracket to the transmission. Remove the transfer case.

7. For vehicles equipped with a automatic transmission, remove the 6 nuts securing the transfer case and bracket to the transmission adapter. Remove the transfer case.

8. Remove and discard the gasket.

To install:

9. Install a new gasket to the transmission. Use Teflon pipe sealant GM P/N 12346004 in order to hold the gasket in place.

10. Raise and position the transfer case to the vehicle

11. For vehicles equipped with a automatic transmission, install the 6 nuts securing the transfer case and bracket to the transmission adapter. Tighten the nuts to 37 ft. lbs. (50 Nm).

12. For vehicles equipped with a manual transmission, install the 6 nuts securing the transfer case and bracket to the transmission. Tighten the nuts to 37 ft. lbs. (50 Nm).

13. If equipped with a NVG 261 manual transmission, install the bolt securing the left side support brace to the transmission. Tighten the bolt to 37 ft. lbs. (50 Nm).

14. If equipped with a NVG 261 manual transmission, Install the bolt and stud securing the left side support brace to the transfer case. Tighten the bolt and stud to 37 ft. lbs. (50 Nm).

15. If equipped with a NVG 261 manual transmission, install the 2 bolts securing the right side support brace to the transmission and transfer case. Tighten the bolts to 37 ft. lbs. (50 Nm).

16. Install or connect the following:
- Vent hose to the transfer case

17. Check the transfer case oil level.
- Speed sensor electrical connectors
- Wiring harness to the transfer case
- Shift rod to the transfer case
- Rear propeller shaft
- Front propeller shaft
- Transfer case shields

18. Lower the vehicle.

NVG246-NP8 Transfer Case

1. Remove or disconnect the following:
- Transfer case shields
- Front propeller shaft
- Rear propeller shaft
- Vent hose from the transfer case
- Vehicle speed sensor electrical connectors

- Electrical connectors from the transfer case motor/encoder
- Any wiring harness from the transfer case

2. Support the transfer case with a transmission jack.

3. If equipped with a NV3500 manual transmission, remove the bolt securing the left side support brace to the transmission.

4. If equipped with a NV3500 manual transmission, remove the bolt and stud securing the left side support brace to the transfer case.

5. If equipped with a NV3500 manual transmission, remove the two bolts securing the right side support brace to the transmission and transfer case.

6. For vehicles equipped with a manual transmission, remove the six nuts securing the transfer case and bracket to the transmission. Remove the transfer case.

7. For vehicles equipped with a automatic transmission, remove the six nuts securing the transfer case and bracket to the transmission adapter. Remove the transfer case.

8. Remove and discard the gasket.

To install:

9. Install a new gasket to the transmission. Use Teflon Pipe Sealant GM P/N 12346004 in order to hold the gasket in place.

10. Raise and position the transfer case to the vehicle

11. For vehicles equipped with a automatic transmission, install the six nuts securing the transfer case and bracket to the transmission adapter. Tighten the nuts to 37 ft. lbs. (50 Nm).

12. For vehicles equipped with a manual transmission, install the six nuts securing the transfer case and bracket to the transmission. Tighten the nuts to 37 ft. lbs. (50 Nm).

13. If equipped with a NV3500 manual transmission, install the bolt securing the left side support brace to the transmission. Tighten the bolt to 37 ft. lbs. (50 Nm).

14. If equipped with a NV3500 manual transmission, Install the bolt and stud securing the left side support brace to the transfer case. Tighten the bolt and stud to 37 ft. lbs. (50 Nm).

15. If equipped with a NV3500 manual transmission, install the two bolts securing the right side support brace to the transmission and transfer case. Tighten the bolts to 37 ft. lbs. (50 Nm).

16. Install or connect the following:
- Vent hose to the transfer case
- Oil
- Speed sensor electrical connectors

- Electrical connectors to the transfer case motor/encoder
- Any wiring harness to the transfer case
- Rear propeller shaft
- Front propeller shaft
- Transfer case shields

BW4481-NR3 and 4482-NR4 Transfer Case

1. Raise and support the vehicle.
2. Remove the transfer case shield, if equipped.
3. Remove the front and rear rear propeller shafts.
4. Drain the fluid from the transfer case.
5. Disconnect the speed sensor electrical connector.
6. Remove the clip from the transfer case bracket.
7. Reposition the speed sensor wire.
8. Remove the clips from the fuel pipe bracket.
9. Remove the fuel pipe clip from the fuel pipe bracket.
10. Remove the transmission mount nuts.
11. Remove the crossmember bolts.
12. Remove the crossmember.
13. Install a suitable transmission jack to the transfer case.
14. Remove the transfer case adapter nuts.
15. Remove the fuel pipe bracket from the studs.
16. Remove the transfer case from the adapter.
17. Rotate the transfer case so that it is perpendicular to the torsion bar mounting bracket.
18. Lower the transfer case.
19. Remove the gasket from the transfer case.
20. Remove the transfer case from the transmission jack.

To install:

21. Install the transfer case onto a suitable transmission jack.
22. When installing a new transfer case gasket, the gasket must be installed with the tab oriented up, and the yellow printing toward the front of the vehicle. Install the gasket without the use of any type of sealant or of lubricant.
23. Install a new transfer case gasket.
24. Rotate the transfer case so that it is parallel to the torsion bar mounting bracket.
25. Raise the transfer case into position.
26. Rotate the transfer case so that it is aligned with the adapter.

27. Install the transfer case to the adapter.
28. Install the fuel pipe bracket onto the studs.
29. Install the transfer case adapter nuts and tighten to 37 ft. lbs. (50 Nm).
30. Position the crossmember.
31. Install the crossmember bolts and tighten to 52 ft. lbs. (70 Nm).
32. Install the transmission mount nuts and tighten to 30 ft. lbs. (40 Nm).
33. Remove the jack stand from the transmission.
34. Install the fuel pipe clip to the fuel pipe bracket.
35. Install the clips to the fuel pipe bracket.
36. Position the speed sensor wire.
37. Install the clip to the transfer case bracket.
38. Connect the speed sensor electrical connector.
39. Install the propeller shafts
40. Fill the transfer case with fluid.
41. Install the transfer case shield, if equipped.

Halfshaft

REMOVAL & INSTALLATION

1. Remove or disconnect the following:
- Wheels

2. Insert a drift or a large screwdriver through the brake caliper into one of the brake rotor vanes in order to prevent the drive axle wheel drive shaft from turning.

3. Remove or disconnect the following:

- Nut and the washer from the hub

➡**Do not reuse the hub nut. A new nut must be used when installing the wheel drive shaft.**

- Bolts (6) securing the wheel drive shaft inboard flange to the output shaft flange
- Drift from the rotor
- Stabilizer shaft link from the lower control arm

4. Wrap shop towels around both the inner and the outer wheel drive shaft boots in order to avoid damage to the boots during removal and installation.

5. Pull the wheel drive shaft through the lower control arm opening.

To install:

6. Wrap shop towels around both the inner and the outer wheel drive shaft boots in order to avoid damage to the boots during removal and installation.

➡Clean the steering knuckle and the wheel drive shaft splines and threads. These areas must be dry and free of grease, dirt, and contamination.

7. Insert the wheel drive shaft splined shank into the knuckle hub.

➡Use only a genuine GM front wheel drive shaft nut. Installation of anything but an OEM front wheel drive shaft nut could cause damage to the vehicle.

8. Install or connect the following:
- Washer and the new hub nut to the wheel driveshaft. Do not tighten.
- The wheel drive shaft inboard flange to the output shaft flange using the inboard flange bolts

9. Insert a drift or a large screwdriver through the brake caliper into 1 of the brake rotor vanes in order to prevent the wheel drive shaft from turning. Tighten the inboard flange bolts to 58 ft. lbs. (78 Nm). Tighten the hub nut to 177 ft. lbs. (240 Nm).

10. Remove the drift from the rotor.
11. Install the stabilizer shaft link.
12. Install the wheel and tire assembly.

CV-Joints

OVERHAUL

Inner Joint

➡With removal of the halfshaft for any reason, the transmission sealing surface (the tripot male/female shank of the halfshaft) should be inspected for corrosion. If corrosion is evident, the surface should be cleaned with 320 grit cloth or equivalent. Transmission fluid may be used to clean off any remaining debris. The surface should be wiped dry and the halfshaft reinstalled free of any buildup.

1. Use a hand grinder in order to cut through the swage ring.

2. Remove the tripot housing from the halfshaft. Wipe the grease off of the tripot assembly roller bearings and the tripot housing. Thoroughly degrease the tripot housing. Allow the tripot housing to dry prior to assembly.

➡Handle the tripot spider assembly with care. Tripot balls and needle rollers may separate from the spider trunnion if the tripot balls and needle rollers are not handled carefully.

3. Before servicing the vehicle, refer to the precautions in the beginning of this section.

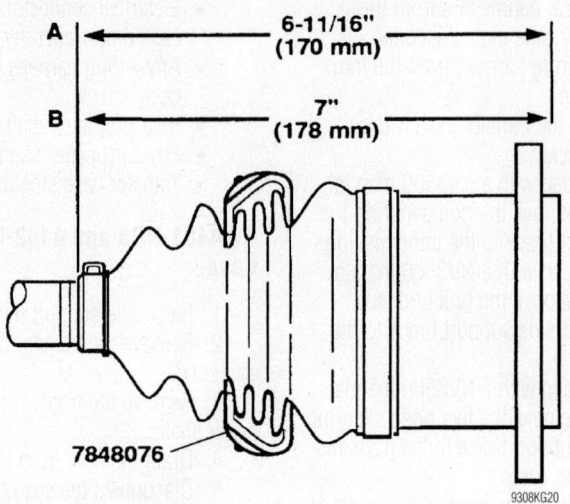

7848076

9308KG20

Assembled inner CV-Joint measurement

4. Use side cutters to cut away the small boot clamp.

5. Compress the tripot boot up the halfshaft away from the tripot spider assembly toward the outboard (CV joint assembly) end of the halfshaft.

6. Spread the spider spacer ring with tool J8059, or equivalent.

7. Remove the following items from the halfshaft bar:
   a. The spacer ring.
   b. The spider assembly.
   c. The tripot boot.

8. Clean the halfshaft bar. Use a wire brush in order to remove any rust in the boot mounting area (grooves).

9. Inspect the needle rollers, needle bearings, and trunnion. Check the tripot housing for unusual wear, cracks, or other damage. Replace any damaged parts.

To assemble:

10. Place the new small boot clamp onto the small end of the joint boot.

11. Compress the joint boot and small boot clamp onto the halfshaft bar.

12. Position the small end of the joint boot into the joint boot groove on the halfshaft bar.

13. Secure the small boot clamp with tool J35910, or equivalent, a breaker bar, and a torque wrench. Tighten the small boot clamp (1) to 100 ft. lbs. (136 Nm).

14. Check the gap dimension on the clamp ear. Continue tightening until the gap dimension is reached.

➡Assemble the CV joint with the convolute retainer in the correct position, as illustrated.

15. Install the convolute retainer over the inboard joint boot, being sure to capture three convolutions.

16. Install the tripot spider assembly onto the halfshaft bar with the counterbore towards the end of the halfshaft bar.

17. Install the spacer ring in the groove at the end of the halfshaft bar.

18. Push the spider assembly back toward the end of the halfshaft bar until the spacer ring is covered by the spider assembly counterbore.

19. Pack the tripot boot and the tripot housing with the grease supplied in the kit. The amount of grease supplied in this kit has been pre-measured for this application.

20. Reassemble the tripot housing and the tripot boot using the following procedure:
   a. Pinch the swage ring slightly by hand in order to distort it into an oval shape.
   b. Slide the distorted swage ring over the large diameter of the boot.
   c. Place the tripot housing over the spider assembly.
   d. Install the boot onto the tripot housing.
   e. Align the tripot boot with the swage ring in place, over the flat area on the tripot housing.

21. Mount tool J36652 in a vise. Install the bottom half of the split-plate swage clamp. For K15 models, use tool J36652-98. For K25 models, use tool J36652-1.

22. Check the inboard stroke position. Use measurement A for the K15 models. Use measurement B for the K25 models.

23. Position the inboard end (tripot end) of the halfshaft assembly in tool J36652. Install the top half of the proper size tool on the lower half of the tool. For K15 models, use tool J36652-98. For K25 models, use tool J36652-1.

24. Align the swage ring and the swage

ring clamp. Insert the bolts. Hand tighten the bolts in tool J36652 until the bolts are snug.

25. Align the following during this procedure:

    a. The tripot boot.

    b. The housing.

    c. The swage ring. Tighten each bolt 180 degrees at a time. Alternate between the bolts until both sides of the top half of J36652 touch the bottom half of the tool.

26. Loosen the bolts and remove the halfshaft assembly from J36652.

27. Remove the convolute retainer from the boot.

Outer Joint

1. Place protective covers over the vise jaws. Place the halfshaft in the vise.

2. Use a hand grinder to cut through the swage ring. Use side cutters to cut off the small boot clamp.

3. Slide the boot down the halfshaft bar and away from the CV-joint outer race. Wipe all grease away from the face of the CV joint.

4. Find the halfshaft bar retaining snap ring, which is located in the inner race.

5. Spread the snapring ears apart.

6. Pull the CV joint and the CV joint boot from the halfshaft bar. Discard the old CV joint boot.

7. Place a brass drift against the CV joint cage. Tap gently on the brass drift with a hammer in order to tilt the cage.

8. Remove the first chrome alloy ball when the CV joint cage tilts. Tilt the CV joint cage (1) in the opposite direction to remove the opposing chrome alloy ball. Repeat this process to remove all six of the balls.

9. Pivot the CV joint cage and the inner race 90 degrees to the center line of the outer race. At the same time, align the cage windows with the lands of the outer race. Lift out the cage and the inner race.

10. Remove the inner race from the cage by rotating the inner race upward. Clean the following items thoroughly with cleaning solvent. Remove all traces of old grease and any contaminates.

    a. The inner and outer race assemblies.

    b. The CV joint cage.

    c. The chrome alloy balls.

11. Dry all the parts. Check the CV joint assembly for unusual wear, cracks, or other damage. Replace any damaged parts. Clean the halfshaft bar. Use a wire brush to remove any rust in the boot mounting area (grooves).

To assemble:

12. Inspect all of the parts for unusual wear, cracks, or other damage. Replace the CV joint assembly if necessary. Put a light coat of the recommended grease on the inner and the outer race grooves.

13. Hold the inner race at 90 degrees to the centerline of the cage. Align the lands of the inner race with the windows of the cage. Insert the inner race into the cage by rotating the inner race downward.

14. Insert the cage and inner race into the outer race.

15. Place a brass drift against the CV joint cage. Tap gently on the brass drift with a hammer in order to tilt the cage. Install the first chrome alloy ball when the CV joint cage tilts. Tilt the CV joint cage in the opposite direction to install the opposing chrome alloy ball. Repeat this process in order to install all six of the balls.

16. Pack the CV joint boot and the CV joint assembly with the grease supplied in the kit. The amount of grease supplied in this kit has been pre-measured for this application.

17. Place the new small boot clamp onto the CV joint boot.

18. Slide the CV joint boot onto the halfshaft bar.

19. Position the small end of the CV joint boot into the joint boot groove on the halfshaft bar.

20. Secure the small boot clamp, a breaker bar, and a torque wrench. Tighten the small clamp (1) to 100 ft. lbs. (136 Nm).

21. Check the gap dimension on the clamp ear. Continue tightening until the gap dimension is reached.

22. Pinch the new swage ring slightly by hand to distort it into an oval shape. Slide the distorted swage ring over the large diameter of the boot.

➡ Be sure that the retaining ring side of the CV joint inner race faces the halfshaft bar (3) before installation.

23. Slide the CV joint onto the halfshaft bar. The retaining snap ring inside of the inner race engages in the halfshaft bar groove with a click when the CV joint is in the proper position.

24. Pull on the CV joint to verify engagement.

25. Slide the large diameter of the CV joint boot with the large swage ring in place, over the outside edge of the CV joint outer race.

26. Clamp the CV joint boot tightly to the CV joint outer race with the large swage ring, using the following procedure:

    a. Mount tool J36652 in a vise.

    b. Install the bottom half of the split-plate swage clamp. For K15 models, use tool J36652-98.

    c. For K25 models, use tool J36652-1.

    d. Position the CV joint end (outboard end) of the halfshaft assembly in the bottom half of tool J36652.

27. Align the following during this procedure:

    a. The CV joint boot.

    b. The CV joint assembly.

    c. The swage ring.

28. Install the top half of tool J36652 onto the lower half of the tool, over the CV joint boot and the CV joint assembly.

29. Align the swage ring and the swage ring clamp.

30. Insert the bolts into J36652. Hand tighten the bolts until the bolts are snug. Tighten each bolt 180 degrees at a time. Alternate between the bolts until both sides of the top half of the tool touch the bottom half of the tool.

31. Loosen the bolts and remove the halfshaft assembly from the tool.

Transfer Case Encoder Motor

REMOVAL & INSTALLATION

1. Remove the transfer case shield.

2. Remove the front propeller shaft.

3. Disconnect the transfer case switch electrical connector.

4. Disconnect the encoder motor electrical connector.

5. Remove the encoder motor bolts.

6. Remove the encoder motor.

7. Remove the actuator insulator gasket.

8. If replacing the encoder motor, remove the locating pins from the old motor.

To install:

➡ If the encoder motor is being replaced because it is defective, ensure that the transfer case is in the neutral position. Manually shift the transfer case at the shift shaft, using a crescent wrench if necessary. When installing the encoder motor, ensure that the encoder motor is indexed correctly and the motor is flat against the transfer case before tightening the bolts.

9. Install the locating pins to the new encoder motor.

10. Position a new actuator insulator gasket to the transfer case.

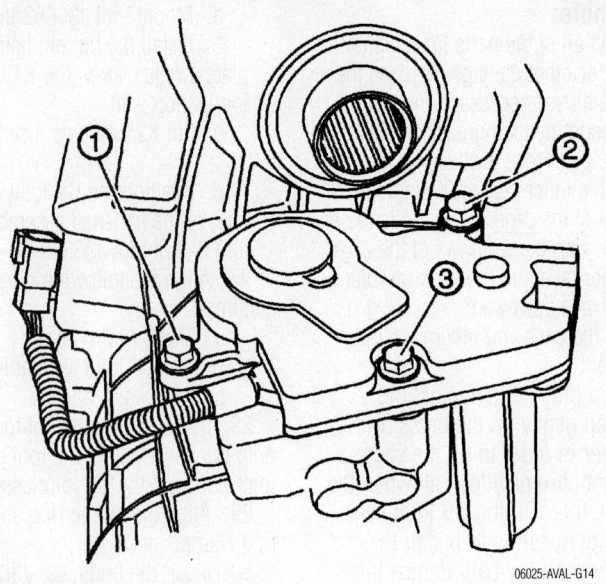

06025-AVAL-G14

Encoder motor tightening sequence

11. Install the encoder motor.
12. Install encoder motor bolts and tighten in sequence to 15 ft. lbs. (20 Nm).
13. Connect the encoder motor electrical connector.
14. Connect the transfer case switch electrical connector.
15. Install the front propeller shaft.
16. Install the transfer case shield.

Front Axle Shaft, Bearing and Seal

REMOVAL & INSTALLATION

8.25 S4WD AND 9.25 AXLES

1. Raise and support the vehicle.
2. Drain the differential carrier assembly.
3. If only replacing the right side inner shaft and/or housing, follow the steps below. If only replacing the left side inner shaft, proceed to step 18.
4. Remove the stabilizer shaft link assembly.
5. Disconnect the electrical connector from the electric motor actuator.
6. Disconnect the wire harness from the inner axle shaft housing.
7. Remove the drive shaft inboard flange bolts from the inner axle shaft.
8. Disconnect the wheel drive shaft from the inner axle shaft.
9. Remove the inner axle shaft housing nuts from the bracket.
10. For 25/35 series vehicles, remove the front axle mounting bracket to frame nuts.
11. Slide the front axle mounting bracket

towards the engine. It may be necessary to pull down on the inner axle housing and/or push up on the mounting bracket in order to gain clearance.
12. Remove the inner axle shaft housing bolts from the differential carrier case.
13. Carefully remove the inner axle shaft housing assembly from the differential carrier assembly.
14. For the 8.25 inch axle, remove the following components from the inner axle shaft housing:
    a. The clutch fork inner spring (10).
    b. The clutch fork assembly (11).
    c. The clutch shaft shim (9).
    d. The clutch sleeve (8).
    e. The clutch gear (6) by doing the following:
    f. Clamp the inner axle shaft housing (4) in a vise. Clamp only on the mounting flange.

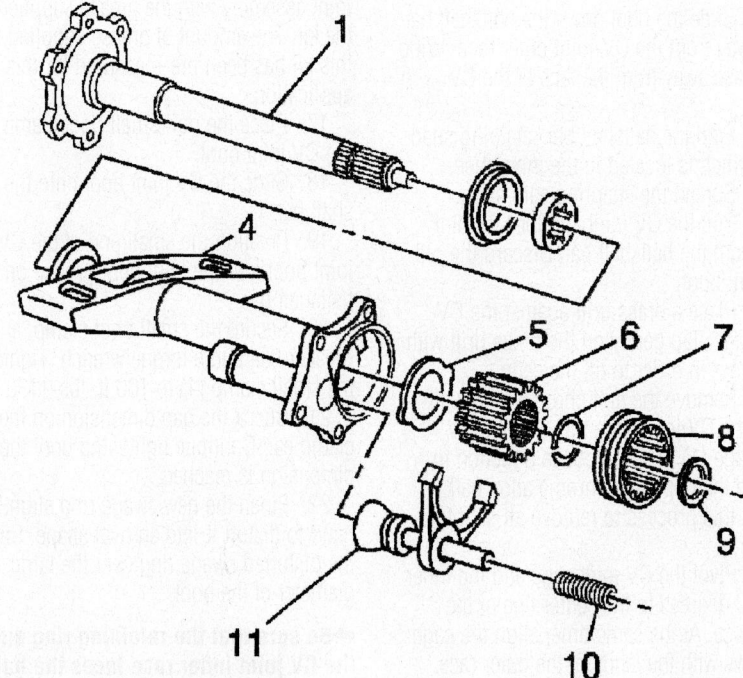

1. Inner axle shaft
4. Inner shaft housing
5. Thrust washer
6. Clutch gear
7. Washer
8. Clutch sleeve
9. Inner sleeve
10. Clutch fork inner spring
11. Clutch for assembly

06025-AVAL-G15

Exploded view of the front axle assembly—8.25 S4WD and 9.25 axles

g. Strike the inside surface of the shaft (1) flange with a hammer and a brass drift in order to dislodge the front drive axle clutch gear (6) from the inner axle shaft (1).

h. The thrust washer (5).

15. For the 9.25 inch axle, remove the following components from the inner axle shaft housing:

a. The clutch fork inner spring (10).

b. The clutch fork assembly (11).

c. The clutch shaft shim (9).

d. The clutch sleeve (8).

e. The retainer ring (7).

f. The thrust washers (5,6).

16. Remove the inner axle shaft (2). Tap out the inner axle shaft with a soft-faced mallet, if necessary.

17. Remove the inner axle seal and the bearing from the axle housing.

18. If only replacing the left side inner axle shaft, remove the wheel drive shaft inboard flange bolts from the inner axle shaft. Disconnect the wheel drive shaft from the inner axle shaft.

19. Remove the inner axle shaft using a hammer and a brass drift.

20. Install the inner axle shaft housing into a vise. Clamp only on the mounting flange of the inner axle shaft housing.

21. Install the bushing and bearing removal tool J-29369-1, 8.25 inch axle, or J-29369-2, 9.25 inch axle, behind the inner axle shaft seal or the inner axle shaft bearing as necessary.

22. Install a slide hammer to the removal tool.

23. Remove the inner axle shaft seal and/or the inner axle shaft bearing using the slide hammer.

24. If only replacing the left side seal, place an alignment mark between the inner axle shaft and the wheel drive shaft.

25. Disconnect the wheel drive shaft from the inner axle shaft.

26. Remove the inner axle shaft using a hammer and a brass drift.

27. Remove the inner axle shaft seal using a suitable seal remover tool.

To install:

28. Install the right side bearing with the square shoulder in using and axle bearing tube installer and a universal driver handle.

29. Install the new axle shaft seal using the sane tools.

30. Install the inner axle shaft into the inner axle shaft housing. Carefully tap the inner axle shaft into place with a soft-faced mallet.

31. Install the inner axle shaft and clutch fork assembly components into the inner shaft housing.

32. If only the left side inner axle shaft was removed, install the shaft by performing the following steps:

33. Install the inner axle shaft into the differential case side gear using a soft-faced mallet until the retaining ring on the inner axle shaft is fully seated within the groove in the differential case side gear.

34. Pull back on the inner axle shaft to ensure that the inner axle shaft is properly retained in the differential case side gear.

35. Connect the halfshaft to the inner axle shaft.

36. Install the halfshaft inboard flange to inner axle shaft bolts and tighten to 58 ft. lbs. (79 Nm).

37. If the right side inner axle shaft and/or housing was removed, install the shaft and/or housing using the following steps:

38. Install the new inner axle shaft bearing and the seal to the axle housing.

39. Install the inner axle shaft (2) into the inner axle shaft housing (1). Carefully tap the inner axle shaft into place with a soft-faced mallet.

40. Place the inner axle shaft housing on end so that the splines of the inner axle shaft is facing up.

41. For the 8.25 inch axle, install the following components into the inner axle shaft housing:

➡**Use chassis grease in order to hold the thrust washer in place.**

42. The thrust washer (5) Ensure the tabs on the thrust washer are aligned with the slots in the inner axle shaft housing (4).

43. The retainer ring (7) into the clutch gear (6).

44. The clutch gear (6) onto the inner axle shaft (1). Drive the clutch gear into place with a plastic hammer.

45. Install the original shim to the shaft. Use the chassis grease in order to hold the shim in place.

46. Install the inner axle housing assembly to the differential carrier case. Do not use sealer at this time.

47. Install the bolts and tighten to 30 ft. lbs. (40 Nm).

48. Install a dial indicator on the axle tube end. The plunger of the indicator must be at a right angle to the axle flange.

49. Move the shaft back and forth and read the end play. The correct end play is 0.001-0.020 in (0.03-0.51mm).

50. If the end play is incorrect, install a thicker or thinner shim as needed in order to bring the end play into the specified range.

51. Install the clutch gear shim (9).

clutch sleeve (8), clutch fork assembly (11) and clutch fork inner spring (10).

52. For the 9.25 inch axle, install the following components into the inner axle shaft housing:

53. The thrust washer (5) Ensure the tabs on the thrust washer are aligned with the slots in the inner axle shaft housing (4).

54. The second thrust washer (6).

55. The retainer ring (7) onto the inner axle shaft (1).

56. Determine the clutch gear shim thickness.

57. Install the clutch gear shim (9). clutch sleeve (8), clutch fork assembly (11) and clutch fork inner spring (10).

58. Apply sealant to the inner axle housing to differential carrier sealing surface.

59. Install the inner axle shaft housing assembly to the differential carrier assembly.

60. Install the inner axle shaft housing bolts and tighten to 30 ft. lbs. (40 Nm).

61. For 25/35 series vehicles, perform the following steps in order to install the front axle mounting bracket to the inner axle shaft housing:

62. Slide the front axle mounting bracket towards the frame. Install the front axle mounting bracket studs into the inner shaft housing mounting flange. It may be necessary to push up on the front axle mounting bracket and/or pull down on the inner axle housing in order to gain enough clearance to install the mounting bracket studs into the inner shaft housing.

63. Install the front axle mounting bracket to frame nuts.

64. Install the inner axle shaft housing washers and nuts to the bracket and tighten to 75 ft. lbs. (100 Nm).

65. Connect the wheel drive shaft inboard flange to the inner axle shaft and tighten to 30 ft. lbs. (40 Nm).

66. Install the wheel drive shaft inboard flange to the inner axle shaft bolts and tighten to 58 ft. lbs. (79 Nm).

67. Connect the wire harness to the inner axle shaft housing.

68. Connect the electrical connector to the front axle actuator.

69. Install the stabilizer shaft link assembly.

70. With either replacement procedure, fill the differential carrier assembly with axle lubricant.

71. Lower the vehicle.

8.25 F4WD AXLE

1. Raise and support the vehicle.

2. Drain the differential carrier assembly.

3. If only replacing the right side inner shaft and/or housing, follow the steps below. If only replacing the left side inner shaft, proceed to step 16.

4. Remove the stabilizer shaft link assembly.

5. Remove the wheel drive shaft inboard flange bolts from the inner axle shaft.

6. Disconnect the wheel drive shaft from the inner axle shaft.

7. Disconnect the inner axle shaft from the differential case side gear using a hammer and brass drift.

Remove the inner axle shaft housing nuts from the bracket.

8. Remove the inner axle shaft housing bolts from the differential carrier assembly.

9. Remove the inner axle shaft and inner axle shaft housing from the vehicle.

10. Remove the inner axle shaft from the inner axle shaft housing.

11. Remove the inner axle shaft seal and the bearing from the inner axle shaft housing.

12. Install the inner axle shaft housing into a vise. Clamp only on the mounting flange of the inner axle shaft housing.

13. Install the bushing and bearing removal tool J-29369-1, 8.25 inch axle, or J-29369-2, 9.25 inch axle, behind the inner axle shaft seal or the inner axle shaft bearing as necessary.

14. Install a slide hammer to the removal tool.

15. Remove the inner axle shaft seal and/or the inner axle shaft bearing using the slide hammer.

16. If only replacing the left side seal, place an alignment mark between the inner axle shaft and the wheel drive shaft.

17. Disconnect the wheel drive shaft from the inner axle shaft.

18. Remove the inner axle shaft using a hammer and a brass drift.

19. Remove the inner axle shaft seal using a suitable seal remover tool.

To install:

20. Install the right side bearing with the square shoulder in using and axle bearing tube installer and a universal driver handle.

21. Install the new axle shaft seal using the sane tools.

22. Install the inner axle shaft into the inner axle shaft housing. Carefully tap the inner axle shaft into place with a soft-faced mallet.

23. Install the inner axle shaft and clutch fork assembly components into the inner shaft housing.

24. If only the left side inner axle shaft was removed, install the shaft by performing the following steps:

25. Install the inner axle shaft into the differential case side gear using a soft-faced mallet until the retaining ring on the inner axle shaft is fully seated within the groove in the differential case side gear.

26. Pull back on the inner axle shaft to ensure that the inner axle shaft is properly retained in the differential case side gear.

27. Connect the halfshaft to the inner axle shaft.

28. Install the halfshaft inboard flange to inner axle shaft bolts and tighten to 58 ft. lbs. (79 Nm).

29. If the right side inner axle shaft and/or housing was removed, install the shaft and/or housing using the following steps.

30. Install the new inner axle shaft bearing and the new seal to the inner axle shaft housing.

31. Install the inner axle shaft into the inner axle shaft housing. Do not install the inner axle shaft completely into the inner axle shaft housing at this time.

32. Apply sealant to the inner axle housing to differential carrier sealing surface.

33. Install the inner axle shaft and the inner axle shaft housing to the differential carrier assembly.

34. Install the inner axle shaft housing bolts and tighten to 30 ft. lbs. (40 Nm).

35. Install the inner axle shaft housing nuts to the bracket and tighten to 75 ft. lbs. (100 Nm).

36. Install the inner axle shaft into the differential case side gear by doing the following:

37. Turn the inner axle shaft and align the splines of the inner axle shaft with the splines on the differential side gear.

38. Install the inner axle shaft into the differential case side gear using a soft-faced mallet until the retaining ring on the inner axle shaft is fully seated within the groove in the differential case side gear.

39. Pull back on the inner axle shaft to ensure that the inner axle shaft is properly retained in the differential case side gear.

40. Install the wheel drive shaft inboard flange to the inner axle shaft.

41. Install the wheel drive shaft inboard flange to inner axle shaft bolts and tighten to 58 ft. lbs. (79 Nm).

42. Install the stabilizer shaft link assembly.

43. Fill the differential carrier assembly with axle lubricant

44. Lower the vehicle.

Rear Axle Shaft, Bearing and Seal

REMOVAL & INSTALLATION

8.5 Inch and 9.5 Inch Rear Axle

1. Raise and support the vehicle on a hoist.

2. Remove or disconnect the following:
   • Tire and wheel assembly
   • Brake caliper
   • Rear cover and the gasket
   • Pinion shaft locking screw
   • Pinion shaft, on vehicles without a locking differential

3. On axles with a locking differential, remove the shaft part way. Rotate the case until the pinion shaft touches the housing.

4. On axles with a locking differential, use a screwdriver, or a similar tool, in order

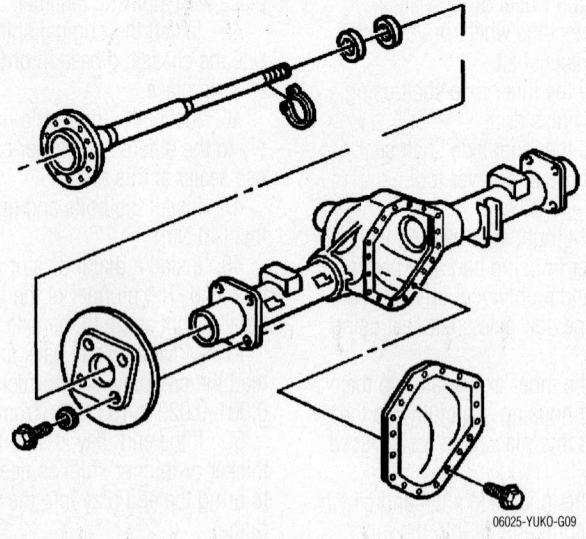

Rear axle shaft removal—8.5/9.5/9.75 inch

06025-YUKO-G09

to enter the differential case and rotate the lock until the lock aligns with the thrust block.

5. Push the flange of the axle shaft toward the differential. Remove the lock from the button end of the axle shaft.

➡When removing the axle shaft, do not rotate the shaft. Rotating the shaft will misalign the gears. Misaligning the gears will make the assembly difficult.

6. Remove the axle shaft from the housing.
7. Inspect all the parts for damage. Replace the parts as necessary.

To install:

➡Carefully insert the axle shaft in order to not damage the seal.

8. Install the axle shaft into the housing. Slide the axle shaft into place allowing the splines to engage the differential side gear.
9. On axles without a locking differential, place the lock on the button end of the axle shaft.
10. On axles with a locking differential, keep the pinion shaft partially withdrawn.
11. On axles with a locking differential, place the lock on the axle shaft so that the ends are flush with the thrust block. Pull the shaft flange outward in order to seat the lock in the differential gear.

➡Anytime you remove a differential pinion shaft locking screw, coat the screw threads with Loctite 242 before reinstalling the screws. The screw has an adhesive coating in order to prevent the screw from loosening in the case. Removing the screw removes the adhesive on the screw.

12. Align the hole in the pinion shaft with the screw hole in the differential case.
13. Install or connect the following:
- Pinion flange locking screw. Tighten the screw to 25 ft. lbs. (34 Nm).
- Rear cover and the gasket
- Brake caliper
- Tire and wheel assembly
14. Fill the rear axle.
15. Remove the supports and lower the vehicle.

8.6 Inch Rear Axle

1. Raise and support the vehicle.
2. Remove the tire and wheel assembly.
3. Remove the brake caliper bracket. It is not necessary to remove the brake caliper.

4. Remove the rear wheel speed sensor.
5. Remove the rear cover and the gasket.
6. Remve the pinion shaft locking bolt.
7. Remove the pinion shaft.
8. Push the flange of the axle shaft (1) toward the differential.

➡It may be necessary to tap the axle shaft toward the differential with a soft faced mallet to obtain the clearance needed to remove the C-lock.

9. Remove the C-lock (4) from the button end of the axle shaft (1).

➡When removing the axle shaft, do not rotate the shaft. Rotating the shaft will misalign the gears. Misaligning the gears will make the assembly difficult.

10. Remove the axle shaft (1) from the housing (5) by pulling the axle shaft away from the differential.

To install:

11. Install the axle shaft (1) into the rear axle housing (5).
12. Slide the axle shaft (1) into place allowing the splines to engage the differential side gear. It may be necessary to tap the end of the axle shaft with a soft faced mallet as it is being installed to seat the wheel speed sensor ring on the axle.
13. Place the lock (4) on the button end of the axle shaft (1).
14. Pull the shaft flange outward in order to seat the lock in the differential gear.
15. Align the hole in the pinion shaft with the bolt hole in the differential case.
16. Install the new pinion shaft locking bolt and tighten to 27 ft. lbs. (36 Nm).
17. Install the rear cover and the gasket.

18. Install the rear wheel speed sensor.
19. Install the brake caliper bracket.
20. Install the tire and wheel assembly.
21. Fill the rear axle.
22. Lower the vehicle.

9.75 Inch Rear Axles

1. Release the parking brake.
2. Raise and support the vehicle.
3. Remove the tire and wheel assembly.
4. Remove the rear steering gear assembly.
5. Remove the steering knuckle assembly.
6. Remove the lock clip from the axle shaft end. The lock clip is spring loaded and fits securely in the axle shaft slot and may need to be push off the shaft end with a screw driver or related tool. Pushing the axle shaft inwards towards the gears my help in removal of the lock clip.
7. When removing the axle shaft do not rotate the shaft. Rotating the shaft will cause the gears to move. Misalignment of the gears will make the assembly difficult.
8. Remove the axle shaft.

To install:

9. Install the axle shaft.
10. Install the spring loaded lock clip to the axle shaft end.
11. Install the steering knuckle assembly.
12. Install the rear steering gear assembly.
13. Install the tire and wheel assembly.
14. Lower the vehicle.

10.5 Inch Rear Axle

1. Remove or disconnect the following:
- Tire and wheel
- Brake caliper

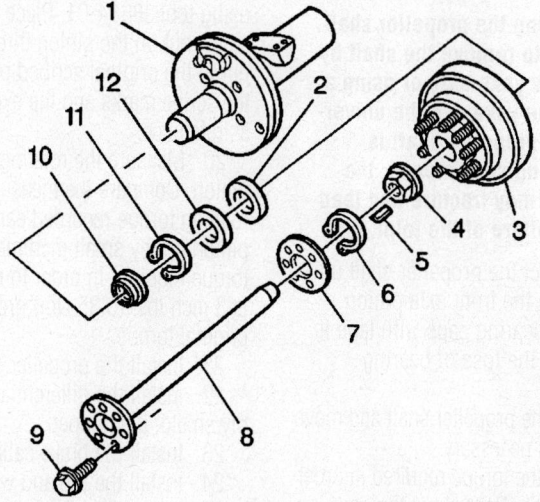

9308KG16

Rear axle shaft removal—10.5 inch 25 Series

- Brake rotor
- Flange bolts

2. Lightly rap the axle shaft with a soft-faced hammer in order to loosen the shaft. Grip the rib on the axle shaft flange with a locking pliers. Twist the axle shaft flange in order to start the axle shaft removal. Remove the axle shaft from the tube.

3. Remove the gasket.

4. Clean the axle shaft flange and the outside face of the hub assembly. Inspect all the parts. Replace the parts as necessary.

To install:

5. Install the gasket onto the axle shaft.

6. Install the gasket and the axle shaft into the tube. Ensure the shaft splines mesh into the differential side gear. Align the holes in the axle flange and the gasket with the holes in the hub.

7. Install or connect the following:
- Axle flange bolts and tighten to 110 ft. lbs. (150 Nm)
- Brake rotor
- Brake caliper
- Wheel and tire

Front Drive Axle Pinion Seal

REMOVAL & INSTALLATION

1. Raise and support the vehicle.

2. Remove the tire and wheel.

3. Remove the brake calipers.

4. Remove the differential carrier assembly shield, if equipped.

5. Reference mark the relationship of the propeller shaft to the front axle pinon yoke.

6. Remove the yoke retainer bolts and the yoke retainers from the front axle pinion yoke.

➡**When removing the propeller shaft, do not attempt to remove the shaft by pounding on the yoke ears or using a tool between the yoke and the universal joint. If the propeller shaft is removed by using such means, the injection joints may fracture and lead to premature failure of the joint.**

7. Disconnect the propeller shaft universal joint from the front axle pinion yoke. Wrap the bearing caps with tape in order to prevent the loss of bearing rollers.

8. Support the propeller shaft and move out of the way as necessary.

9. Measure the torque required in order to rotate the pinion. Record the torque value for reassembly.

10. Scribe a line on the pinion stem, the pinion nut and the companion flange. Record the number of exposed threads on the pinion stem.

11. Remove the nut.

12. Position tool J8614-01 on the flange so that the 4 notches on the tool face the flange.

13. Remove the flange. Use the special nut and the forcing screw.

➡**Carefully pry the seal from the bore. Do not distort or scratch the aluminum case.**

14. Remove the oil seal.

15. Inspect the pinion flange for a smooth oil seal surface. Inspect the pinion flange for worn drive splines. Replace the pinion flange if necessary.

16. Remove the dust deflector.

To install:

➡**Stake the new deflector at 3 new equally spaced positions. You must stake the new deflector in such a way that you do not damage the seal operating surface.**

17. Install and stake the dust deflector on the flange.

18. Position the oil seal in the bore. Then place a driver over the oil seal. Strike the driver with a hammer until the seal flange seats on the axle housing surface. Drive the seal in straight, not at an angle, as this will damage the aluminum housing.

➡**Do not hammer the pinion flange/yoke onto the pinion shaft. Pinion components may be damaged if the pinion flange/yoke is hammered onto the pinion shaft.**

19. Install the flange onto the pinion using tool J8614-01. Place the washer and a new nut on the pinion threads. Tighten the nut to the original scribed position using the scribe marks and the exposed threads as reference.

20. Measure the rotating torque of the pinion. Compare the measurement with the rotating torque recorded earlier. Tighten the pinion nut by small increments until the torque required in order to rotate the pinion is 3 inch lbs. (0.35 Nm) greater than the original torque.

21. Install the propeller shaft.

22. Install the differential carrier assembly shield, if equipped.

23. Install the brake calipers

24. Install the tire and wheel.

25. Lower the vehicle.

Rear Drive Axle Pinion Seal

REMOVAL & INSTALLATION

1. Raise the vehicle.

2. Remove the tire and wheel assemblies (8.6 inch, 9.5 inch axles).

3. Remove the rear brake calipers and rotors (8.6 inch, 9.5 inch axles).

4. Remove the axle shafts (10.5 inch, 11.5 inch axle).

5. Reference mark the rear propeller shaft to the rear axle pinion yoke.

6. Disconnect the propeller shaft from the axle.

7. Measure the torque required to turn the pinion. Record the torque number measurement which gives the combined pinion bearing, seal, carrier bearing, axle bearing and seal preload.

8. Make and accurate alignment mark on the pinion flange. Record the number of exposed threads on the pinion stem.

9. Remove the pinion flange nut and the washer. Use a container in order to catch any lubricant.

➡**Use care not to damage any of the machined surfaces.**

10. Remove the pinion flange.

➡**The pinion flange has an oil seal that is part of the pinion flange assembly. The pinion flange must be inspected to ensure that the seal is not damaged.**

11. Pry the oil seal from the bore.

12. Thoroughly clean any foreign material from the contact area. Replace any parts as necessary.

To install:

13. Lubricate the cavity between the lips of the oil seal with wheel bearing lubricant.

14. Install the oil seal into the bore using a driver.

➡**Do not hammer the pinion flange onto the pinion stem.**

15. Install the pinion flange. Use the alignment marks in the installation of the pinion flange.

16. Install the washer and a new nut. Tighten the nut on the pinion stem as close as possible to the alignment marks without going past the marks. Use the alignment marks and the thread count as a reference. Tighten the nut a little at a time. Turn the pinion flange several times after each tightening in order to seat the rollers.

17. Measure the torque required to rotate the pinion flange. Compare this to the original torque. Tighten the pinion nut, in small

increments, until the rotating torque is 3 inch lbs. (0.35 Nm) GREATER than the original torque.

18. Align the propeller shaft with the alignment marks. Connect the propeller shaft.

19. Install the axle shafts (10.5 inch, 11.5 inch axle).

20. Install the rear brake calipers and rotors (8.6 inch, 9.5 inch axles).

21. Install the tire and wheel assemblies (8.6 inch, 9.5 inch axles).

Front Drive Axle Differential Carrier

REMOVAL & INSTALLATION

1. Turn the steering wheel all the way to the left.

2. Raise and support the vehicle.

3. Place jack or utility stands at the rear end of the vehicle.

4. Remove the engine protection shield.

5. Remove the front differential carrier assembly shield, if equipped.

6. Drain the differential carrier assembly.

7. Disconnect the front propeller shaft from the differential carrier assembly.

8. Remove the relay rod.

9. Support the differential carrier assembly with a transmission jack or equivalent.

10. Remove the halfshaft inboard flange bolts from the inner axle shaft on both sides.

11. Disconnect the electrical connector from the front axle actuator, S4WD axle only.

12. Disconnect the wire harness from the inner axle shaft housing and differential, S4WD axle only.

13. Disconnect the vent hose from the differential carrier assembly.

14. Remove the inner axle housing nuts and washers from the bracket.

15. Remove the differential carrier assembly upper mounting bolt and the nut.

16. Pivot the differential carrier assembly forward and down on the lower mount bolt while it is being supported by the transmission jack.

17. Secure the differential carrier assembly to the jack.

18. Remove the differential carrier assembly lower mounting bolt and the nut.

19. Remove the differential carrier.

To install:

20. Install the differential carrier.

21. Install the differential carrier assembly lower mounting bolt and the nut. Do not tighten the bolt at this time.

22. Pivot the differential carrier assembly up and back on the lower mount bolt while it is being supported by the transmission jack.

23. Install the differential carrier assembly upper mounting bolt and the nut.

24. Install the inner axle housing washers and nuts to the bracket.

25. Tighten the inner axle housing nuts and the upper and the lower differential carrier assembly bolts to 75 ft. lbs. (100 Nm).

26. Connect the vent hose to the differential carrier assembly.

27. Remove the transmission jack.

28. Connect the electrical connector from the front axle actuator, S4WD axle only.

29. Connect the wire harness from the inner axle shaft housing and differential, S4WD axle only.

30. Install the wheel drive shaft inboard flange to inner axle shaft bolts, both sides, and tighten to 58 ft. lbs. (79 Nm).

31. Install the relay rod.

32. Install the front propeller shaft to the differential carrier assembly.

33. Fill the differential carrier assembly.

34. Install the front differential carrier assembly shield, if equipped.

35. Install the engine protection shield.

36. Lower the vehicle.

Rear Drive Axle Housing

REMOVAL & INSTALLATION

Except 9.75 Inch Axles

1. Raise and support the vehicle.

2. Drain the axle lubricant.

3. Remove the rear axle assembly.

4. Remove the brake caliper brackets.

5. Remove the wheel hub on 10.5 inch axles.

6. Remove the rear axle cover housing and gasket.

7. Remove the brake backing plates.

8. Remove the axle shafts.

9. Remove the differential.

10. Remove the drive pinion.

11. Remove the drive pinion shaft yoke and the seal.

To install:

12. Install the drive pinion.

13. Install the differential case.

14. Adjust the side bearing preload.

15. Adjust the backlash.

16. Install the brake backing plates.

17. Install the axle shafts.

18. Install the rear axle housing cover and gasket.

19. Install the brake caliper brackets.

20. Install the wheel hub on 10.5 inch axles.

21. Install the rear axle.

22. Fill the axle with lubricant.

23. Lower the vehicle.

9.75 Inch Axles

1. Raise and support the vehicle.

2. Remove the steering gear protection shield.

3. Remove the axle lubricant.

4. Remove the rear axle assembly.

5. Remove the steering knuckles.

6. Remove the axle shafts.

7. Remove the differential.

8. Remove the drive pinion shaft yoke and the pinion oil seal.

9. Remove the drive pinion and the outer pinion bearing.

10. Remove the pinion bearing cups.

To install:

11. Install the pinion bearing cups.

12. Install the drive pinion outer pinion bearing and the pinion oil seal.

13. Install the drive pinion.

14. Install the pinion yoke.

15. Install the differential case.

16. Adjust the side bearing preload.

17. Adjust the backlash.

18. Install the axle shafts.

19. Install the steering knuckles.

20. Install the rear axle.

21. Install the steering gear protection shield.

22. Fill the axle with lubricant.

23. Lower the vehicle.

STEERING AND SUSPENSION

Air Bag

✳✳ CAUTION

Some vehicles are equipped with an air bag system. The system must be disabled before performing service on or around system components, steering column, instrument panel components, wiring and sensors. Failure to follow safety and disabling procedures could result in accidental air bag deployment, possible personal injury and unnecessary system repairs.

PRECAUTIONS

Several precautions must be observed when handling the inflator module to avoid accidental deployment and possible personal injury.

• Never carry the inflator module by the wires or connector on the underside of the module

• When carrying a live inflator module, hold securely with both hands, and ensure that the bag and trim cover are pointed away

• Place the inflator module on a bench or other surface with the bag and trim cover facing up

• With the inflator module on the bench, never place anything on or close to the module that may be thrown in the event of an accidental deployment

DISARMING & ARMING

2002 Models

1. Turn the front wheels to the straight-ahead position.
2. Turn the ignition switch to the **LOCK** position and remove the key.

➡**If the key is in the RUN position when the Air Bag fuse is removed or open (blown), the Air Bag warning lamp in the dash will light up. This is normal operation, not a sign of a malfunction.**

3. Remove the Air Bag fuse from the fuse panel.
4. Remove the drivers side knee bolster and unplug the yellow 2-pin connector at the base of the steering column to disarm the driver's side Air Bag. Remove the passenger side knee bolster and unplug the

yellow 2-pin connector to disable the passenger's side Air Bag.

5. Reverse the procedure to arm the Air Bag restraint system.

2003–06 Models

1. Turn the steering wheel so that the vehicles wheels are pointing straight ahead.
2. Turn OFF the ignition.
3. Remove the key from the ignition.
4. With the SIR fuse removed and the ignition ON, the AIR BAG indicator illumi-

nates. This is normal operation and does not indicate an SIR system malfunction.

5. Remove the SIR fuse from the fuse block.
6. Raise and support the vehicle.
7. Remove the connector position assurance (CPA) from both front end sensor connectors located on the frame crossmember.
8. Disconnect both front end sensor connectors
9. Reverse the procedure to arm the system.

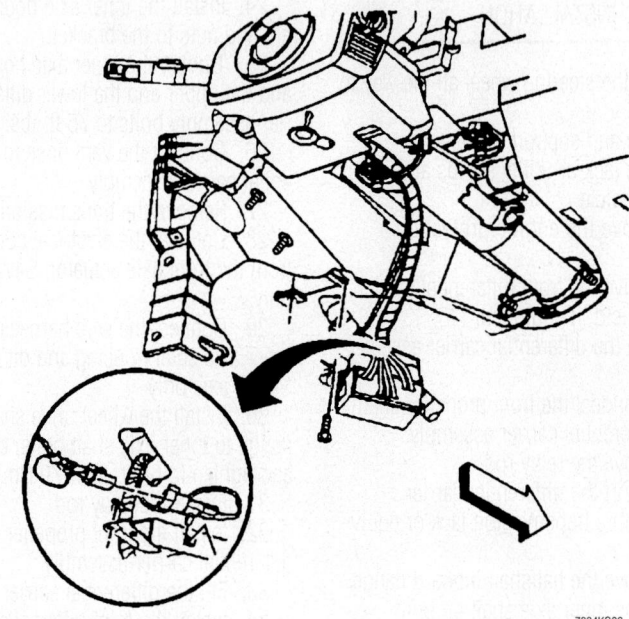

7924KG30

Typical air bag connector location—driver's side

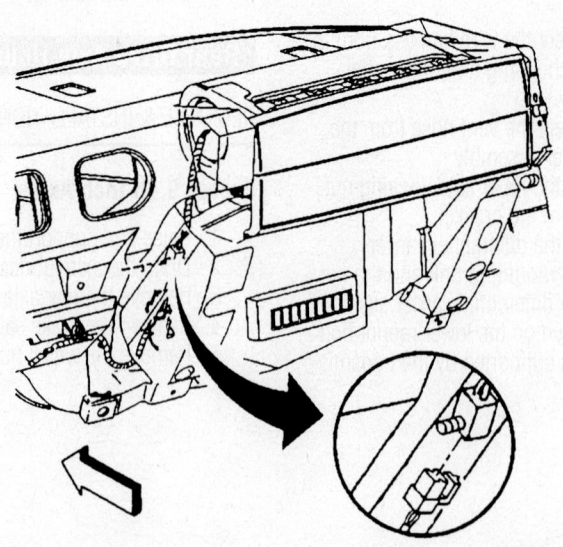

7924KG31

Typical air bag connector location—passenger's side

10. When the fuse is installed, turn ON the ignition, with the engine OFF.

11. The AIR BAG indicator will flash 7 times then turn off.

12. Perform the Diagnostic System Check if the AIR BAG indicator does not operate as described.

Power Steering Pump

REMOVAL & INSTALLATION

1. Raise and support the vehicle.
2. Remove or disconnect the following:
   • Drive belt
   • Pulley
3. Remove the oil pan skid plate.
4. Place a drain pan under the vehicle.
5. Remove the power steering pressure hose from the power steering pump from underneath the vehicle.
6. Lower the vehicle.
7. Remove the clamps retaining the power steering return hoses to the power steering pump and remove the hoses.
8. On all except 8.1L, disconnect the lower intermediate shaft from the steering gear.
9. Remove the bolt retaining the rear bracket to the engine.
10. Remove the bolts from the front of the pump.
11. Remove the pump.
   To install:
12. Install or connect the following:
   • Power steering pump
   • Bolts to the front and the rear of the pump. Tighten the bolts to 37 ft. lbs. (50 Nm).
   • Hoses to the pump. Tighten the nut to 20 ft. lbs. (28 Nm).
   • Connect the intermediate shaft to the steering gear and tighten the bolt to 37 ft. lbs. (50 Nm).
   • Pulley. Install the pulley with 0.020 in. (0.5mm) play.
   • Drive belt
13. Fill and bleed the power steering system.

Rack & Pinion Steering Gear

REMOVAL & INSTALLATION

1. Raise the vehicle.
2. Remove the shield.
3. Place a drain pan below the steering gear.
4. Remove the power steering hoses from the steering gear.
5. Cap the ends of the hoses and the power steering gear in order to prevent any entrance of dirt.
6. Disconnect the intermediate shaft from the steering gear.
7. Remove the pitman arm to relay rod nut.
8. Disconnect the pitman arm from the relay rod.
9. Remove the bolts retaining the steering gear to the frame.
10. Remove the steering gear from the vehicle.
11. Remove the pitman arm-to-power steering gear retaining nut.

12. Remove the pitman arm from the power steering gear.
   To install:
13. Install the pitman arm to the power steering gear.
14. Install the pitman arm to power steering gear retaining nut and washer and tighten to 184 ft. lbs. (250 Nm).
15. Install the steering gear to the frame and install the retaining bolts and tighten to 110 ft. lbs. (150 Nm).
16. Install the pitman arm to the relay rod.
17. Install the pitman arm to the relay rod nut and tighten to 46 ft. lbs. (62 Nm).
18. Install the intermediate shaft to the steering gear and tighten the upper and pinch bolt to 35 ft. lbs. (47 Nm)
19. Remove the caps or plugs from the steering gear and hoses.
20. Install the hoses to the steering gear and tighten the fittings to 20 ft. lbs. (28 Nm).
21. Install the engine protection shield, if equipped.
22. Remove the safety stands.
23. Lower the vehicle.
24. Bleed the power steering system.

Rear Wheel Steering Gear Motor

REMOVAL & INSTALLATION

1. Raise and support the vehicle.
2. Remove the spare tire.
3. Remove the bolts retaining the steering gear motor to the steering gear.
4. Remove the steering gear motor from the steering gear.
5. Cover the steering gear motor opening to keep contamination out of the oil in the planetary gear assembly.
6. Remove the wiring harness from the retaining clips.
7. Disconnect the wiring harness from the control module.
8. Disconnect the electrical connector from the rear position sensor.
9. Remove the steering gear motor and wiring harness from the vehicle.
   To install:
10. Install the O-ring seal to the steering gear.
11. Install the steering gear motor to the vehicle.
12. Install the bolts retaining the steering gear motor to the steering gear.
13. Tighten the 0.24 in. (6 mm) bolts to 89 inch lbs. (10 Nm) and the 0.16 in. (4mm) bolts to 35 inch lbs. (4 Nm).
14. Connect the electrical connectors to the control module and rear position sensor.

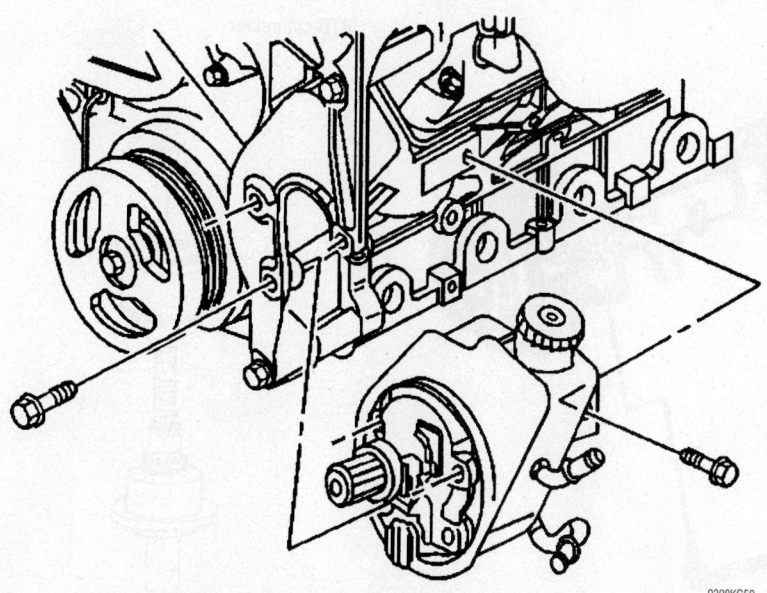

9308KG50

Power steering pump—4.8L, 5.3L and 6.0L engines shown

15. Connect the wire harness retaining clips.

16. Install the spare tire.

17. Lower the vehicle.

Rear Wheel Steering Gear Actuator

REMOVAL & INSTALLATION

1. Raise and support the vehicle.

2. Remove the tire and wheel.

3. Remove the drain plug and drain the fluid from the differential housing.

4. Remove the bolts retaining the steering gear protection shield.

5. Disconnect the electrical connectors from the control module.

6. Disconnect the electrical connector from the rear position sensor.

7. Remove the outer tie rod ends from the steering knuckle.

8. Disconnect the differential vent hose from the vent tube.

9. Remove the bolts retaining the steering gear actuator and cover to the differential.

10. Remove the steering gear actuator and cover from the vehicle.

To install:

11. Install the steering gear actuator and gasket to the differential.

12. Install the steering gear actuator and cover retaining bolts and tighten to 45 ft. lbs. (61 Nm).

13. Connect the electrical connectors to the control module.

14. Install the differential vent hose to the vent tube.

15. Install the outer tie rod ends to the steering knuckle

16. Install the drain plug and fill the differential with fluid.

17. Install the bolts retaining the steering gear protection shield and tighten to 132 ft. lbs. (180 Nm).

18. Install the tire and wheel.

19. Lower the vehicle.

20. Check the wheel alignment.

Shock Absorber

REMOVAL & INSTALLATION

2WD Front

1. Raise and support the vehicle.

2. If equipped with selectable ride, disconnect the Real Time Damping (RTD) link rod from the sensor. Grasp the connector lock tabs. Rotate the connector tabs counter clockwise until the connector is unlocked. Disengage the connector from the tennon by firmly pulling the connector up. Hold the tennon end with a wrench while removing the nut. Remove the nut.

3. Remove the upper insulator. Do not discard the plastic pilot ring.

4. Remove the shock absorber mounting bolts at the lower control arm. Remove the shock absorber through the lower control arm from below.

To install:

5. Support the lower control arm with a suitable jack in order to align the tennon with the mounting hole if equipped with selectable ride.

6. Install the shock absorber through the lower control arm from below. Insert the tennon through the mounting hole in the upper spring pocket. Align the shock absorber with the mounting holes in the lower control arm.

7. Install the shock absorber mounting bolts to the lower control arm. Tighten the bolts to 59 ft. lbs. (80 Nm).

➡The upper insulators are substantially larger that the lower insulators. The upper insulator must be installed above the shock mounting bracket on the frame. The plastic pilot ring will assist the alignment of the isolators.

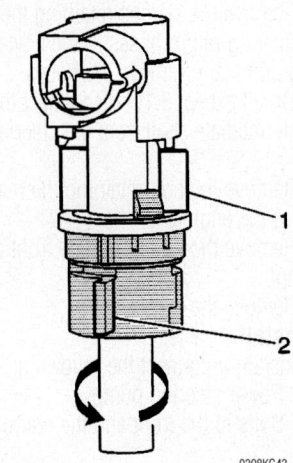

RTD connector

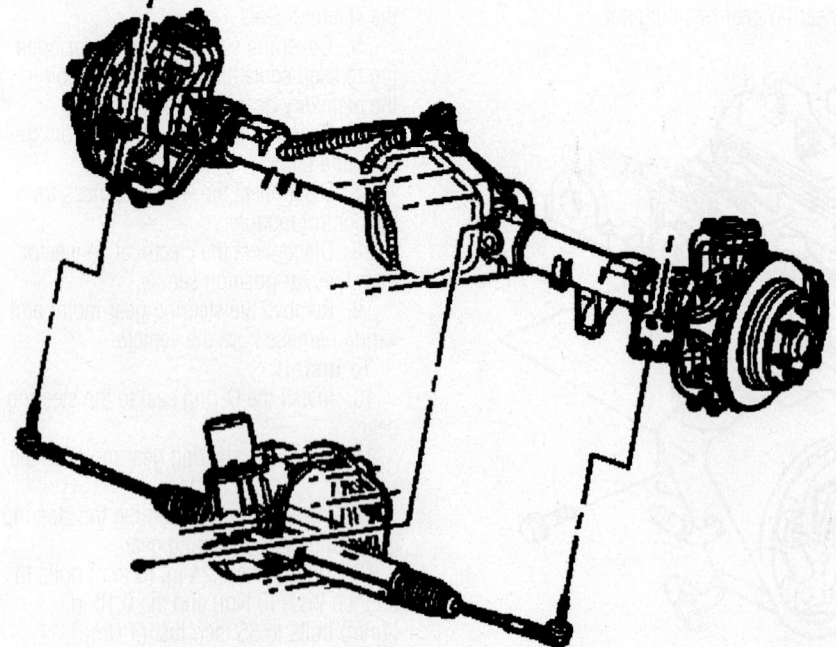

Rear wheel steering gear actuator mounting

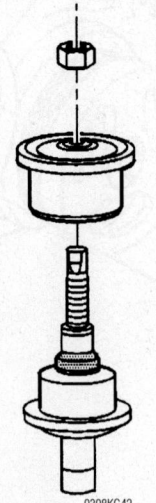

Upper shock insulator

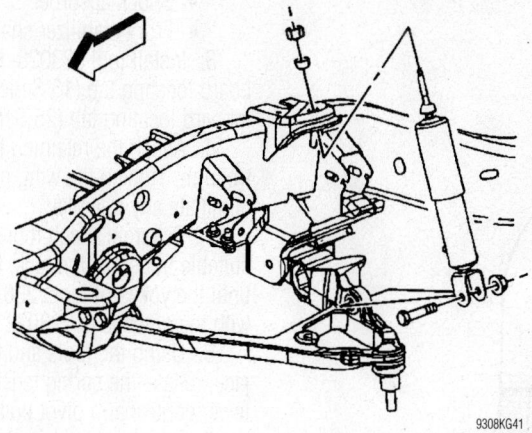

Shock absorber removal

9308KG41

8. Install the upper insulator to the shock absorber. Install the nut to the tennon end. Do not tighten the nut.

9. Connect the RTD link rod to the sensor (if equipped).

10. Remove the safety stands.

11. Lower the vehicle. Hold the tennon end with a wrench while torquing the nut. Tighten the nut to 15 ft. lbs. (20 Nm).

12. Connect the electrical connector using the following procedure:

   a. Verify that the connector is unlocked.

   b. Align the connector so that the tabs are perpendicular to the wrench flats on the tennon end.

   c. Engage the connector to the tennon by firmly pushing the connector down.

   d. Grasp the connector lock tabs. Rotate the connector counter clockwise.

13. The connector is locked into place when you hear an audible snap and the tabs are aligned.

4WD Front

1. Raise and support the vehicle.

2. Disconnect the (RTD) link rod from the sensor (if equipped).

3. Disconnect the electrical connector if equipped with selectable ride. Grasp the connector lock tabs. Rotate the connector tabs counter clockwise until the connector is unlocked. Disengage the connector from the tennon by firmly pulling the connector up. Hold the tennon end with a wrench while removing the nut. Remove the nut.

4. Remove the upper insulator. Do not discard the plastic pilot ring.

5. Remove the shock absorber mounting bolt at the lower control arm. The lower shock mounting bushing is serviceable by driving the bushing out with the appropriate tool.

6. Remove the shock absorber.

To install:

7. Install the shock absorber. Insert the stem through the hole in the shock bracket on the frame. Align the shock absorber with the mounting holes in the lower control arm.

8. Install the shock absorber through bolt to the lower control arm.

9. Install the shock absorber through bolt nut. Tighten the nut to 59 ft. lbs. (80 Nm).

➡ **The upper insulators are substantially larger that the lower insulators. The upper insulator must be installed above the shock mounting bracket on the frame. The plastic pilot ring will assist the alignment of the isolators.**

10. Install the upper insulator to the shock absorber. Install the nut to the tennon end. Do not tighten the nut. Connect the RTD link rod to the sensor (if equipped).

11. Remove the safety stands. Lower the vehicle. Hold the tennon end with a wrench while torquing the nut. Tighten the nut to 15 ft. lbs. (20 Nm).

12. Connect the electrical connector using the following procedure if equipped with selectable ride.

   a. Verify that the connector is unlocked.

   b. Align the connector so that the tabs (1) are perpendicular to the wrench flats on the tennon end.

   c. Engage the connector to the tennon by firmly pushing the connector down.

   d. Grasp the connector lock tabs (1, 2). Rotate the connector counterclockwise. The connector is locked into place when you hear an audible snap and the tabs are aligned.

Rear

1. Raise and support the vehicle.

2. Disconnect the electrical connector if equipped with Selectable Ride.

3. Remove the upper shock absorber nut and the bolt.

4. Remove the lower shock absorber nut and the bolt.

5. Remove the shock absorber.

To install:

6. Installation is the reverse of removal. Tighten the nuts to 70 ft. lbs. (95 Nm).

7. Connect the electrical connector if equipped with Selectable Ride. Remove the safety stands. Lower the vehicle.

Coil Springs

REMOVAL & INSTALLATION

Front

1. Raise and support the vehicle.

2. Remove or disconnect the following:
   • Engine protection shield
   • Frame cross bar (25 series only)
   • Tire and wheel assembly

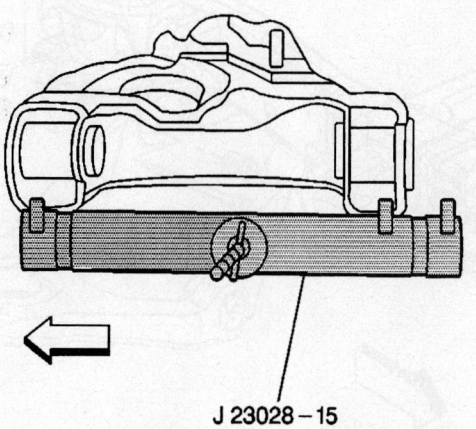

J 23028 – 15

9308KG47

Installing J23028-15 on the 25 Series

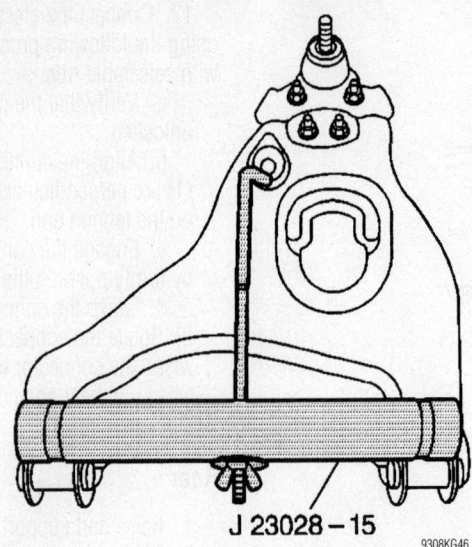

J 23028 - 15

9308KG46

Retaining hook installation

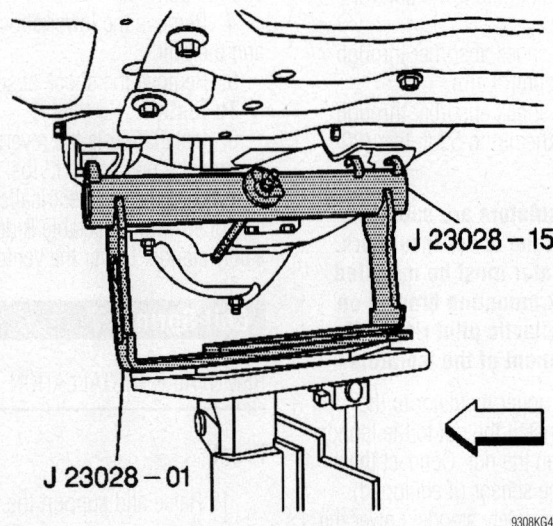

J 23028 - 15

J 23028 - 01

9308KG45

Tool attached to a jack

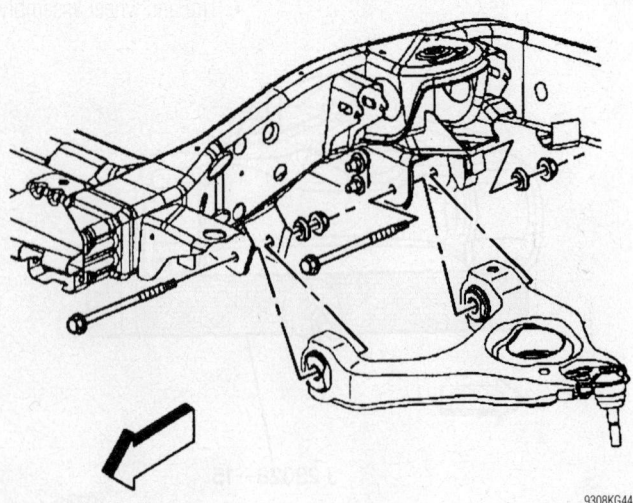

9308KG44

Lower control arm removal

- Shock absorber
- Front stabilizer shaft link

3. Install tool J23028-15 using the outboard locating tab (15 Series), or, the inboard locating tab (25 Series).

4. Attach the retaining hook to the control arm. Tighten the wing nut until you eliminate any free play.

5. Securely attach tool J23028-01 to a suitable transmission jack. Raise the jack until the yokes of tool J23028-01 line up with the notches in J23028-15.

6. Using the tools and the transmission jack, relieve the spring tension from the lower control arm pivot bolts.

7. Remove or disconnect the following:
- Lower control arm pivot bolt nuts (15 Series)
- Rear pivot bolt
- Front pivot bolt
- Lower control arm pivot bolt nuts (25 Series)
- Rear pivot bolt
- Front pivot bolt

8. Slowly lower the transmission jack in order to unload the front coil spring. It may be necessary to use a pry bar in order to guide the lower control arm out of position.
- Coil spring and the insulator

To install:

9. Install or connect the following:
- Coil spring and the insulator to the lower control arm

10. Raise the transmission jack in order to compress the front coil spring. It may be necessary to use a pry bar in order to guide the lower control arm into position.
- Front pivot bolt (15 Series)
- Rear pivot bolt
- Lower control arm pivot nuts. Tighten the pivot bolt nuts to 107 ft. lbs. (145 Nm).

11. Lower the jack. Remove the tool from the control arm.
- Front stabilizer shaft link
- Shock absorber
- Tire and wheel assembly
- Frame cross bar (25 series only). Tighten the nuts to 74 ft. lbs. (100 Nm).
- Engine protection shield

12. Remove the safety stands. Lower the vehicle.

Rear

1. Raise and support the vehicle.

2. Disconnect the Real Time Damping (RTD) sensor, if equipped.

3. Remove the lower shock absorber nuts and bolt from the rear axle.

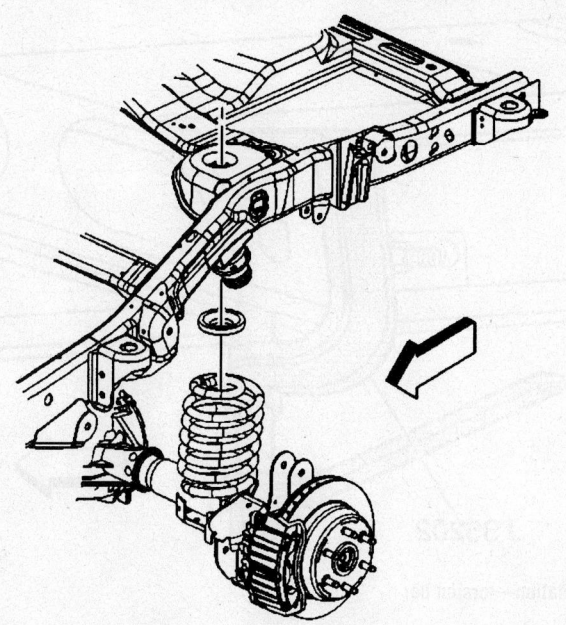

Rear coil spring removal—15 Series

4. Lower the rear axle until the springs are fully unloaded.

5. Remove the spring and the upper and lower insulators.

To install:

6. Position the spring and the upper and lower insulators.

7. Install the rear spring to the rear axle.

8. Raise the rear axle. Install the lower shock absorber nuts to the rear axle.

9. Connect the RTD sensor, if equipped.

10. Remove the rear axle support. Lower the vehicle.

Leaf Springs

REMOVAL & INSTALLATION

1. Raise and support the vehicle.

2. Support the rear axle independently in order to relieve the tension on the leaf springs.

3. Remove or disconnect the following:

- Real Time Damping (RTD) sensors, if equipped
- Trailer hitch if equipped
- Fuel tank for left side applications
- U-bolt nuts and U-bolts
- Spring spacer and anchor plate
- Shackle to the frame bracket nut and the bolt
- Front spring bracket bolt
- Leaf spring assembly from the vehicle
- Shackle from the spring

To install:

4. Loosely assemble the spring shackle bracket to the frame. Install the shackle bolt. Install the shackle nut.

5. Install the leaf spring assembly to the vehicle.

6. Loosely assemble the spring to the front hanger bracket.

7. Install or connect the following:

- Front spring hanger bracket bolt
- Front spring hanger bracket nut
- Shackle to the spring bolt
- Shackle to the spring nut

➡ **Do not reuse the U-bolts.**

- Spring spacer
- U-bolts
- Anchor plate
- U-bolt nuts

8. On 2002–04 models, tighten the U-bolt nuts as follows:

- 14mm U-bolt nuts to 59 ft. lbs. (80 Nm)

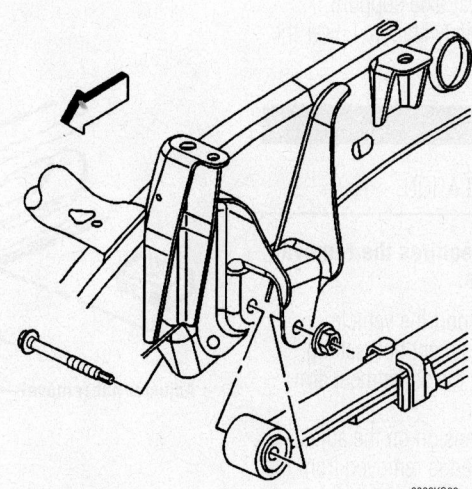

Rear leaf spring front shackle

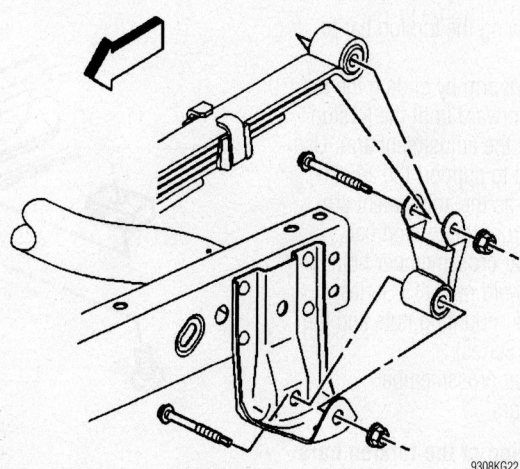

Rear leaf spring rear shackle

- 16mm U-bolt nuts to 89 ft. lbs. (120 Nm)
- Front hanger bracket nut to 92 ft. lbs. (125 Nm)
- Shackle to the frame nut to 70 ft. lbs. (95 Nm)
- Shackle to the spring nut to 70 ft. lbs. (95 Nm)

9. On 2005–06 models, tighten the U-bolt nuts as follows:
- For 25 series except 8.1L engine tighten the U-bolt nuts to 53 ft. lbs. (72 Nm)
- For all 25 series with the 8.1 liter engine tighten the U-bolts to 110 ft. lbs. (150 Nm)
- Tighten the front hanger bracket nut to 110 ft. lbs. (150 Nm)
- Tighten the rear hanger bracket nut to 70 ft. lbs. (95 Nm)

10. Install or connect the following:
- Fuel tank for left side applications
- Trailer hitch if equipped
- RTD sensors (if equipped)

11. Remove the rear axle support.

12. Remove the safety stands. Lower the vehicle.

Torsion Bars

REMOVAL & INSTALLATION

➡**This procedure requires the removal of both torsion bars.**

1. Raise and support the vehicle.

2. Mark the adjustment bolt setting. Install tool J-6202 to the adjustment arm and the crossmember.

3. Increase the tension on the adjustment arm until the load is removed from the adjustment bolt and the adjuster nut.

4. Remove or disconnect the following:
- Adjustment bolt and the adjuster nut
- Tool, allowing the torsion bar to unload
- Adjustment arm by sliding the torsion bar forward until the torsion bar clears the adjustment arm. Use your hand to support the adjustment arm as the adjustment arm releases from the torsion bar.
- Torsion bar crossmember bolts from the weld nuts (15 Series)
- Upper link mounting nuts and the bolts (25 Series)
- Torsion bar crossmember
- Torsion bars

➡**Note the position of the torsion bars as the left and right bars are different.**

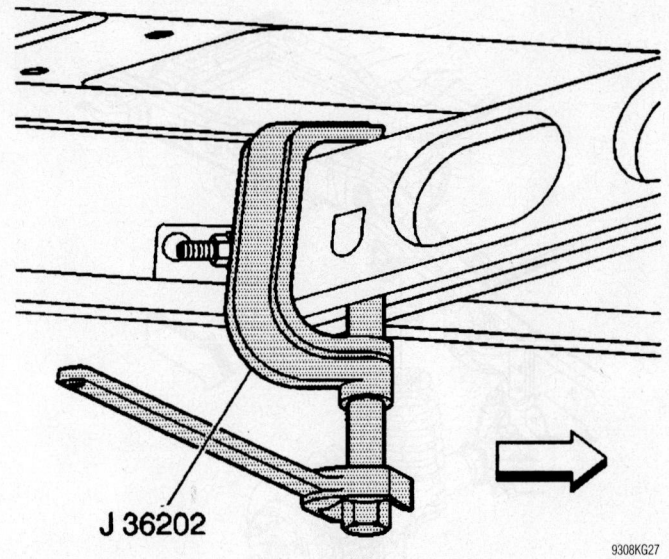

Retainer installation—torsion bar

9308KG27

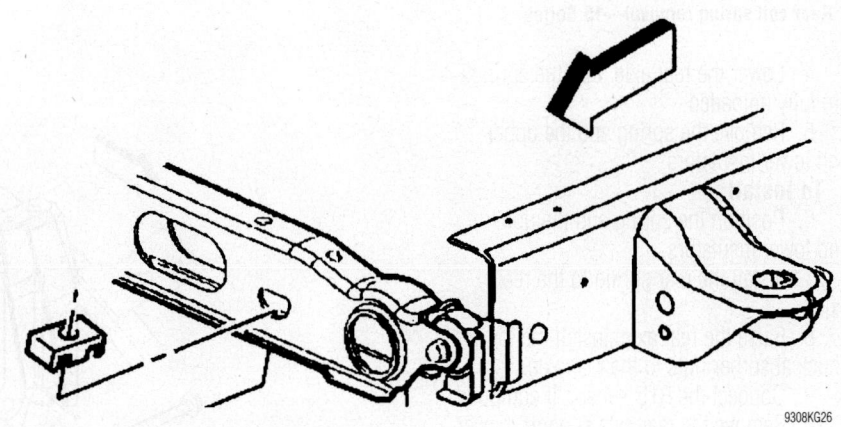

Adjuster nut removal—15 Series

9308KG26

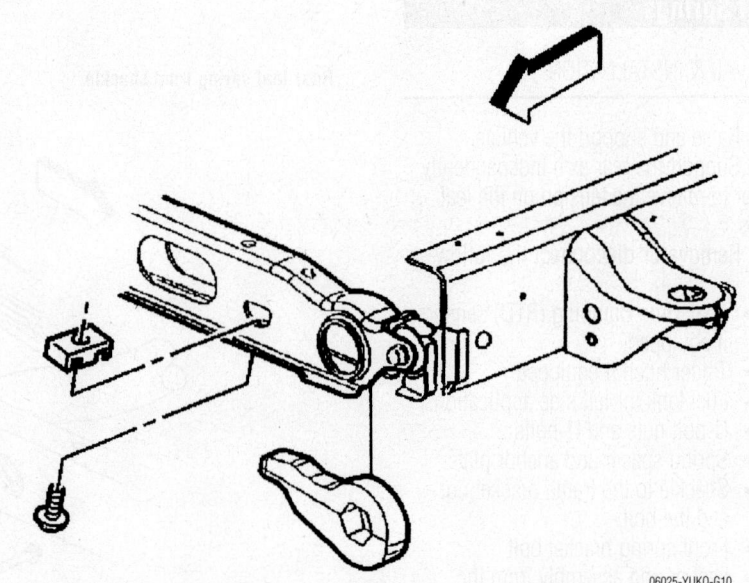

Adjuster bolt removal—15 Series

06025-YUKO-G10

To install:

5. Install or connect the following:
   - Torsion bars
   - Torsion bar crossmember
   - Torsion bar crossmember bolts to the weld nuts (15 Series). Tighten the bolt to 70 ft. lbs. (95 Nm)
   - Upper link mounting nuts and the bolts (25 Series). Tighten the nut to 70 ft. lbs. (95 Nm)

6. While supporting the adjustment arm, slide the torsion bar rearward until the torsion bar fully engages the adjustment arm. Install tool J36202 to the adjustment arm and the crossmember. Increase the tension on the adjustment arm in order to load the torsion bar.
   - Adjustment bolt and the adjuster nut

7. Remove the tool, releasing the tension on the torsion bar until the load is taken up by the adjustment bolt.

8. Remove the safety stands.
9. Lower the vehicle.
10. Measure the ride height.
11. Turn the adjustment bolt clockwise to increase the ride height and counterclockwise to decrease it.

Upper Ball Joint

REMOVAL & INSTALLATION

1. Raise and support the vehicle.
2. Remove or disconnect the following:
   - Tire and wheel assembly
   - Upper control arm
   - Upper ball joint, using a press

To install:

➡**The ball joint must be installed with the flat edges or notches in the same position as the replaced ball joint. The ball joint is directional and damage will occur if this procedure is not followed.**

3. Install or connect the following:
   - Upper ball joint, using a press
   - Upper control arm
   - Tire and wheel assembly
4. Remove the safety stands.
5. Lower the vehicle.
6. Verify the wheel alignment.

Lower Ball Joint

REMOVAL & INSTALLATION

1. Raise and support the vehicle.
2. Remove or disconnect the following:
   - Tire and wheel assembly
   - Front coil spring, if equipped

   - Torsion bar, if equipped
   - Lower control arm

3. Secure the lower control arm in a bench vise or equivalent.
4. Center punch the rivet heads.
5. Drill out the rivets.

To install:

6. Install or connect the following:
   - Ball joint to the lower control arm
   - Replacement bolts to the lower control arm
   - Nuts to the bolts. Tighten the nuts to 52 ft. lbs. (70 Nm).

7. Remove the lower control arm from the bench vise.
   - Lower control arm
   - Coil spring, if equipped
   - Torsion bas, if equipped
   - Tire and wheel tire assembly
8. Remove the safety stands.
9. Lower the vehicle.
10. Verify the wheel alignment.

Upper Control Arm

REMOVAL & INSTALLATION

1. Raise and support the vehicle.
2. Remove or disconnect the following:
   - Tire and wheel assembly
   - Real Time Damping (RTD) link rod from the sensor, if equipped
   - Retaining bolt for the brake hose and the wheel speed sensor brackets
   - Halfshaft
   - Nut at the upper ball joint. Discard the nut.
   - Upper control arm from the steering knuckle
   - Upper control arm nuts and the

adjustment cams (15 Series RWD, 4WD, and 25 Series RWD)
   - Upper control arm bolts (15 Series RWD, 4WD, and 25 Series RWD)
   - Upper control arm nuts and the adjustment cams (25 Series 4WD)
   - Upper control arm bolts (25 Series 4WD)
   - Upper control arm

To install:

3. Install or connect the following:
   - Upper control arm
   - Upper control arm bolts (25 Series 4WD)
   - Upper control arm nuts and the adjustment cams (25 Series 4WD). Tighten the nuts to 140 ft. lbs. (190 Nm).
   - Upper control arm bolts (15 Series RWD, 4WD, and 25 Series RWD)
   - Upper control arm nuts and the adjustment cams (2) (15 Series RWD, 4WD, and 25 Series RWD). Tighten the nuts to 140 ft. lbs. (190 Nm).
   - Upper control arm to the steering knuckle
   - Halfshaft
   - New nut to the upper ball joint stud. Tighten the nut to 37 ft. lbs. (50 Nm).
   - Retaining bolts for the brake hose and wheel speed sensor brackets. Tighten the bolts to 80 inch lbs. (9 Nm).
   - RTD link rod to the sensor, if equipped
   - Tire and wheel assembly
4. Remove the safety stands.
5. Lower the vehicle. Verify the wheel alignment.

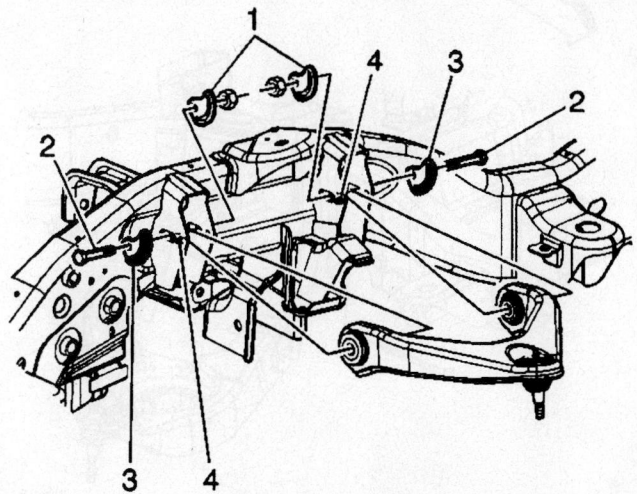

Upper control arm

9308KG31

CONTROL ARM BUSHING REPLACEMENT

The control arm bushings are removed and installed using a press.

Lower Control Arm and Bushing

REMOVAL & INSTALLATION

2WD Models

1. Raise and support the vehicle.
2. Support the lower control arm with a jack.
3. Remove or disconnect the following:
   • Tire and wheel assembly
   • Real Time Damping (RTD) link rod from the sensor, if equipped

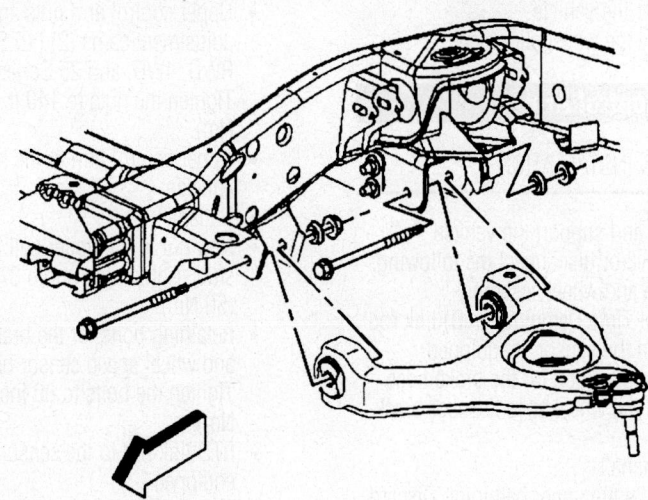

2WD lower control arm—15 Series

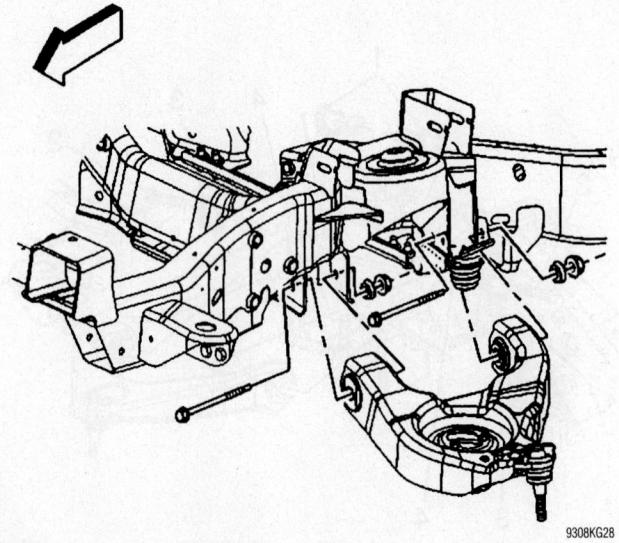

2WD lower control arm

• Shock absorber
• Front stabilizer shaft link
• Front coil spring
• Lower control arm nuts and the washers or bolts, as applicable
• Lower control arm nuts and washers and/or bolts, as applicable
• Lower ball joint stud nut
• Lower ball joint stud from the steering knuckle
• Lower control arm

To install:

4. Install or connect the following:
   • Lower control arm
   • Ball joint stud to the steering knuckle
   • Lower ball joint stud nut. Tighten the lower ball joint stud nut to 74 ft. lbs. (100 Nm)

• Front coil spring
• Lower control arm bolts (15 Series)
• Lower control arm nuts and the washers (15 Series). Tighten the lower control arm nuts to 107 ft. lbs. (145 Nm).
• Lower control arm bolt (25 Series)
• Lower control arm nuts and the washers (25 Series). Tighten the lower control arm nuts to 107 ft. lbs. (145 Nm).
• Front stabilizer shaft link.
• Shock absorber
• RTD sensor, if equipped
• Tire and wheel assembly

5. Remove the safety stands. Lower the vehicle. Verify the wheel alignment.

4WD Models

1. Raise and support the vehicle.
2. Remove or disconnect the following:
   • Tire and wheel assembly
   • Real Time Damping (RTD) link rod from the sensor, if equipped
   • Stabilizer shaft links from the lower control arm
   • Shock absorber nut and the bolt
   • Torsion bars
   • Halfshaft
   • Lower ball joint stud nut
   • Lower ball joint stud from the steering knuckle
   • Lower control arm nuts and the washers (15 Series)
   • Lower control arm bolts
   • Lower control arm nuts and the washers (25 Series)
   • Lower control arm bolts
   • Lower control arm

To install:
   • Lower control arm
   • Lower control arm bolts (15 Series)
   • Washers with the shoulder facing the arm
   • Nuts. Tighten the nuts to 129 ft. lbs. (175 Nm).
   • Halfshaft
   • Lower ball joint stud to the steering knuckle. Install the nut to the ball joint stud. Tighten the nut to 74 ft. lbs. (100 Nm).
   • Torsion bars
   • Shock absorber through nut and bolt
   • Stabilizer shaft links to the lower control arm
   • RTD link rod to the sensor, if equipped
   • Tire and wheel assembly

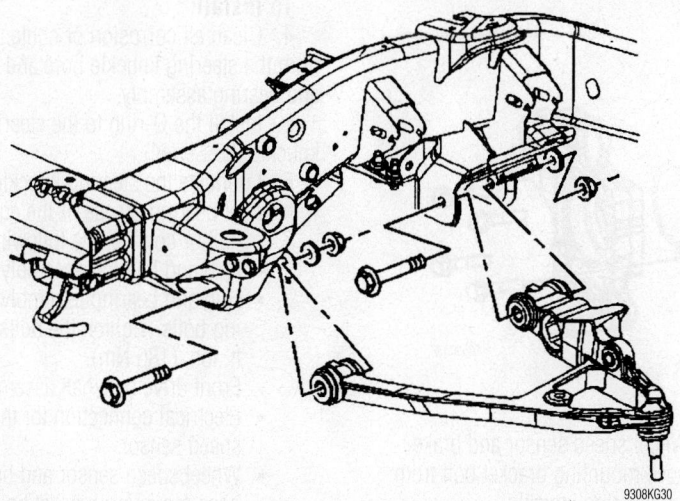

4WD lower control arm—15 Series

3. Remove the safety stands. Lower the vehicle. Verify the wheel alignment.

CONTROL ARM BUSHING REPLACEMENT

Front Bushing

1. On 15 and 25 Series, the bushings are not replaceable. If they are damaged, the control arm will have to be replaced.

2. Remove or disconnect the following:
- Wheel
- Lower control arm
- Unbend the crimps with a punch on the front bushing

3. Press out the bushings with tools J–36618–2, J–9519–23, J–36618–4 and 36618–1.

To install:

4. Lubricate the outer case of the bushing.

5. Install or connect the following:
- Bushing into control arm

6. Press in the bushings with tools J–36618–2, J–9519–23, J–36618–4 and 36618–1 until the bushing is seated in.

7. After bushing is installed crimp it in place.
- Control arm and mounting bolts. Torque the front nut first then the rear to 140 ft. lbs. (190 Nm).
- Wheel

Rear Bushing

➡On 15 and 25 Series, the bushings are not replaceable. If they are damaged, the control arm will have to be replaced.

1. Before servicing the vehicle, refer to the precautions in the beginning of this section.

2. Remove or disconnect the following:
- Wheel
- Lower control arm

3. Press out the bushings with tools. J–36618–5, J–9519–23, J–36618–3 and J–36618–2. There are no crimps.

To install:

4. Lubricate the outer case of the bushing.

5. Install or connect the following:
- Bushing into control arm

6. Press in the bushings with tools J–36618–5, J–9519–23, J–36618–3 and J–36618–2. There are no crimps.
- Control arm and mounting bolts. Torque the front nut first then the rear to 140 ft. lbs. (190 Nm).
- Wheel

Wheel Hub, Bearings and Seal

REMOVAL & INSTALLATION

2WD Front

1. Raise and support the vehicle.
2. Remove the tire and wheel.
3. Remove the caliper and rotor.
4. Remove the wheel speed sensor and brake hose mounting bracket bolt from the steering knuckle.
5. Remove the wheel hub and bearing mounting bolts.
6. Remove the wheel hub and bearing and splash shield from the vehicle.
7. Remove the O-ring seal from the steering knuckle bore on 25/35 series.
8. Remove the wheel speed sensor mounting bolt.
9. Clean and inspect the O-ring seal. Replace it if it is nicked, cut or dry.

To install:

10. Clean all corrosion or contaminates from the steering knuckle bore and the hub and bearing.
11. Lubricate the steering knuckle bore with wheel bearing grease or the equivalent.
12. Install the O-ring to the steering knuckle on 25/35 series.
13. Install the wheel speed sensor mounting bolt.
14. Install the wheel hub and bearing and splash shield.
15. Install the wheel hub and the bearing mounting bolts and tighten to 133 ft. lbs. (180 Nm).

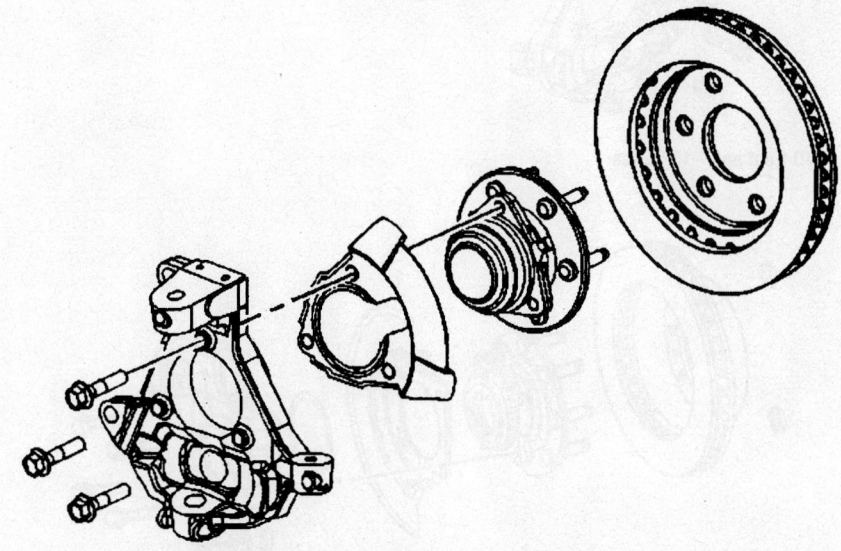

2WD/4WD front hub—15 Series

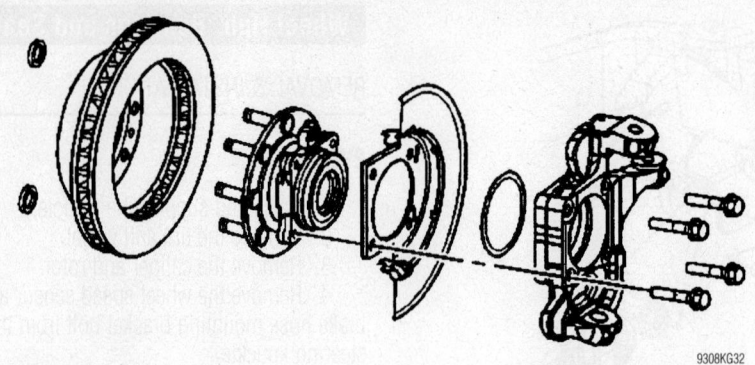

2WD/4WD front hub —25 Series

9308KG32

16. Install the wheel speed sensor and brake hose mounting bracket bolt to the steering knuckle.

17. Install the rotor and caliper.

18. Install the tire and wheel.

19. Lower the vehicle.

4WD Front

1. Raise and support the vehicle.

2. Remove or disconnect the following:
   • Tire and wheel assembly
   • Caliper and rotor

• Wheel speed sensor and brake hose mounting bracket bolt from the steering knuckle

• Electrical connection for the wheel speed sensor

• Front drive halfshaft assembly

• Hub and bearing assembly mounting bolts

• Hub and bearing assembly

• O-ring seal from the steering knuckle bore (25 Series)

3. Clean and inspect the O-ring seal (25 Series).

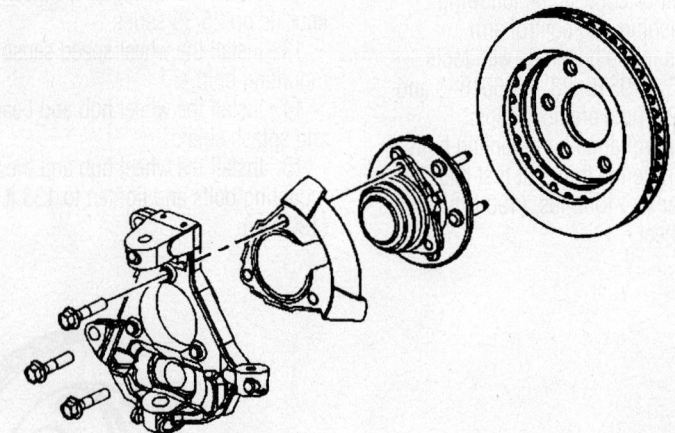

4WD front hub—15 Series

9308KG35

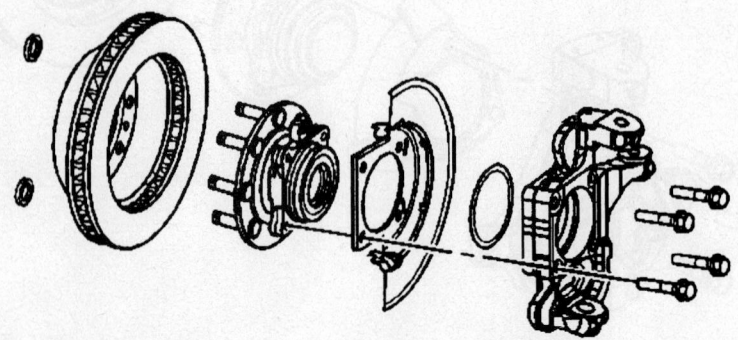

4WD front hub—25 Series

9308KG34

To install:

4. Clean all corrosion or contaminates from the steering knuckle bore and the hub and bearing assembly.

5. Install the O-ring to the steering knuckle (25 Series).

6. Lubricate the steering knuckle bore with wheel bearing grease or the equivalent.

7. Install or connect the following:
   • Hub and bearing assembly
   • Hub and bearing assembly mounting bolts. Tighten the bolts to 133 ft. lbs. (180 Nm).
   • Front drive halfshaft assembly
   • Electrical connection for the wheel speed sensor
   • Wheel speed sensor and brake hose mounting bracket bolt to the steering knuckle. Tighten to 106 inch lbs. (12 Nm).
   • Rotor
   • Tire and wheel assembly.

Rear

See the Rear Axle Shaft, Bearing and Seal procedure for bearing replacement

Stabilizer Bar

REMOVAL & INSTALLATION

1. Raise and support the vehicle.

2. Remove the tire and wheel.

3. Remove the stabilizer shaft nut from the link bolt.

4. Remove the stabilizer shaft link bolt.

5. Remove the stabilizer shaft link insulators and spacers.

6. Remove the oil pan skid plate, if equipped.

7. Remove the stabilizer shaft insulator bracket bolts.

8. Remove the stabilizer shaft bracket.

9. Remove the stabilizer shaft.

10. Remove the stabilizer shaft insulators.

11. Inspect all of the parts for wear and damage.

To install:

12. Install the insulators to the stabilizer shaft.

13. Install the stabilizer shaft.

14. Install the brackets over the insulators and the stabilizer shaft.

15. Install insulator bracket bolts and tighten to 37 ft. lbs. (50 Nm).

16. Install the stabilizer shaft link insulators and spacers.

17. Apply Loctite on the threads of the stabilizer link bolts then install the bolts.

18. Install the stabilizer shaft nut to the link bolt and tighten to 89 inch lbs. (10 Nm), and continue to tighten the nut until 2-4 threads protrude above the nut.

19. Install the oil pan skid plate, if equipped.

20. Install the tire and wheel assembly.

21. Remove the safety stands

22. Lower the vehicle.

Steering Knuckle

REMOVAL & INSTALLATION

1. Raise and support the vehicle.
2. Remove the tire and wheel.
3. Remove the wheel hub and bearing.

4. Support the lower control arm with a suitable jack.

5. Disconnect the outer tie rod from the knuckle.

6. Remove the brake hose bracket retaining bolt from the knuckle.

7. Remove the retaining nut and separate the upper and lower ball joints from the steering knuckle using a ball joint remover and adapters.

8. Remove the steering knuckle.

To install:

9. Clean all grease and contaminants from the tapered section and the threads of the upper ball joint, the lower ball joint, and the tie rod end.

10. Clean and inspect the taper holes and the mounting surfaces of the steering knuckle. If any of the tapered holes are elongated, out of round, or damaged, the replace the steering knuckle.

11. Install the steering knuckle.

12. Connect the lower ball joint to the steering knuckle and install the retaining nut and tighten to 74 ft. lbs. (100 Nm).

13. Connect the upper ball joint to the steering knuckle and install the retaining nut and tighten to 37 ft. lbs. (50 Nm).

14. Install the brake hose bracket retaining bolt to the knuckle.

15. Connect the outer tie rod to the steering knuckle.

16. Install the wheel hub and bearing.

17. Install the tire and wheel.

18. Remove the safety stands.

19. Lower the vehicle.

BRAKES

Brake Caliper

REMOVAL & INSTALLATION

Front

1. Remove or disconnect the following:
   - ⅔ of the brake fluid from the master cylinder
   - Tire and wheel assembly

2. Using a C-clamp or the equivalent, compress the caliper piston until the caliper piston bottoms in the bore.
   - Brake hose at caliper by removing the inlet fitting bolt. Plug the line.
   - Caliper mounting bolts
   - Caliper

3. Inspect the caliper assembly.

To install:

4. Install or connect the following:
   - Caliper
   - Caliper mounting bolts. Tighten the caliper guide pin bolts to 80 ft. lbs. (108 Nm).
   - Brake hose at caliper by installing the inlet fitting bolt. Tighten the inlet fitting bolt to 33 ft. lbs. (45 Nm).

5. Bleed the brakes.
   - Tire and wheel assembly

Rear

1. Remove or disconnect the following:
   - ⅔ of the brake fluid from the master cylinder
   - Tire and wheel assembly

2. Using a C-clamp or the equivalent, compress the caliper piston until the caliper piston bottoms in the bore.

- Brake hose at caliper by removing the inlet fitting bolt. Plug the line.
- Caliper mounting bolts
- Caliper

3. Inspect the caliper assembly.

To install:

4. Install or connect the following:
   - Caliper
   - Caliper mounting bolts. Tighten the caliper guide pin bolts to 31 ft. lbs. (42 Nm) on the 15 series; 80 ft. lbs. (108 Nm) on the 25 series.
   - Brake hose at the caliper by installing the inlet fitting bolt. Tighten the bolt to 33 ft. lbs. (45 Nm).

5. Bleed the brakes.
   - Tire and wheel assembly

6. Refill the brake master cylinder to the proper level with fresh brake fluid.

Disc Brake Pads

REMOVAL & INSTALLATION

Front

1. Remove ⅔ of the brake fluid from the master cylinder.

2. Remove or disconnect the following:
   - Wheel
   - Caliper. Suspend the caliper from the frame with mechanic's wire. Do not allow the caliper to hang from the brake hose.
   - Caliper mounting bracket bolts
   - Caliper mounting bracket from the steering knuckle assembly
   - Brake pads from the caliper mounting bracket

- Clips from the inside ends of the caliper mounting bracket and discard

To install:

3. Install or connect the following:
   - Clips to the inside ends of the caliper mounting bracket
   - Brake pads to the caliper mounting bracket
   - Inner pad (1 wear indicator)
   - Outer pad (2 wear indicators)
   - Caliper mounting bracket to the steering knuckle assembly

4. Perform the following procedure before installing the caliper mounting bracket bolts:

   a. Remove all traces of the original adhesive patch.

   b. Clean the threads of the bolt with brake parts cleaner or the equivalent and allow to dry.

   c. Apply red Loctite® 272 to the threads of the bolt.

5. Install or connect the following:
   - Caliper mounting bracket bolts to the steering knuckle. Tighten the caliper guide pin bolts to 74 ft. lbs. (100 Nm) on 15 series or 80 ft. lbs. (108 Nm) on 25/35 series.
   - Caliper
   - Tire and wheel assembly

6. Refill the master cylinder to the proper level with fresh brake fluid. Pump the brake pedal slowly and firmly in order to seat the brake pads. Burnish the brakes as needed.

Rear

1. Remove or disconnect the following:
   - ⅔ of the brake fluid from the master cylinder

- Tire and wheel assembly
- Caliper. Suspend the caliper from the frame with mechanic's wire. Do not allow the caliper to hang from the brake hose.
- Caliper mounting bracket bolts from the backing plate
- Brake pads from the caliper mounting bracket
- Clips from the inside ends of the caliper mounting bracket and discard

To install:

2. Install or connect the following:
- Clips to the inside ends of the caliper mounting bracket
- Brake pads to the caliper mounting bracket

- Inner pad (1 wear indicator)
- Outer pad (2 wear indicators)
- Clips to the inside ends of the caliper mounting bracket
- Caliper mounting bracket to the backing plate assembly (15 series).

3. Install the caliper mounting bracket to the backing plate assembly (25 series). Perform the following procedure before installing the caliper mounting bracket bolts.

   a. Remove all traces of the original adhesive patch.

   b. Clean the threads of the bolt with brake parts cleaner or the equivalent and allow to dry.

   c. Apply red Loctite® 272 to the threads of the bolt.

4. Install or connect the following:
- Caliper mounting bracket bolts to the steering knuckle. Tighten to 148 ft. lbs. (200 Nm) on the 15 series; 122 ft. lbs. (165 Nm) on the 25 series.
- Caliper and tighten the bolts to 80 ft. lbs. (108 Nm).
- Tire and wheel assembly

5. Refill the master cylinder to the proper level with fresh brake fluid. Pump the brake pedal slowly and firmly in order to seat the brake pads. Burnish the brakes as needed.

PONTIAC

G6

19

BRAKES **19-42**
DRIVE TRAIN **19-31**
ENGINE REPAIR **19-9**
FUEL SYSTEM **19-30**
SPECIFICATIONS AND
  MAINTENANCE CHARTS......19-2
Engine and Vehicle Identification
  Chart..............................19-2
General Engine Specifications 19-2
Gasoline Engine Tune-Up
  Specifications 19-2
Accessory Drive Belt Routing 19-3
Capacities 19-4
Valve Specifications..................19-4
Crankshaft and Connecting Rod
  Specifications 19-4
Piston and Ring Specifications 19-5
Torque Specifications 19-5
Wheel Alignment 19-6
Tire, Wheel and Ball Joint
  Specifications 19-6
Brake Specifications 19-6
Scheduled Maintenance Intervals ..19-7
STEERING AND
  SUSPENSION 19-35

A
Air Bag................................19-35
  Disarming.........................19-35
  Precautions.......................19-35
  Rearming 19-35
Alternator 19-9
  Installation 19-9
  Removal 19-9

B
Brake Caliper 19-42
  Removal & Installation............19-42
Brake Pedal Position Sensor 19-42
  Calibration 19-42

C
Camshaft and Valve Lifters 19-22
  Removal & Installation............19-22
Clutch 19-32
  Adjustment.......................19-32
  Removal & Installation............19-33
Coil Spring 19-39
  Removal & Installation............19-39
CV-Joints............................19-33
  Overhaul 19-33

Cylinder Head.......................19-16
  Removal & Installation............19-16

D
Disc Brake Pads.....................19-42
  Removal & Installation............19-42
Distributor...........................19-9
  Removal 19-9

E
Electronic Steering Module..........19-37
  Programming & Setup 19-37
Electronic Steering Motor 19-36
  Removal & Installation............19-36
Engine Assembly 19-9
  Removal & Installation............19-9
Exhaust Manifold 19-21
  Removal & Installation............19-21

F
Front Suspension Frame.............19-38
  Removal & Installation............19-38
Fuel Filter 19-30
  Removal & Installation............19-30
Fuel Pump Module 19-30
  Removal & Installation............19-30
Fuel Rail and Injector.................19-30
  Removal & Installation............19-30
Fuel System Pressure 19-30
  Relieving & Pressurizing 19-30
Fuel System Service
  Precautions.......................19-30

H
Halfshaft.............................19-33
  Removal & Installation............19-33
Heater Core.........................19-13
  Removal & Installation............19-13
Hydraulic Clutch System 19-33
  Bleeding..........................19-33
Hydraulic Steering System...........19-36
  Bleeding..........................19-36

I
Ignition Timing 19-9
Intake Manifold 19-18
  Removal & Installation............19-18

L
Lower Ball Joint......................19-40
  Removal & Installation............19-40
Lower Control Arm 19-40
  Control Arm Bushing
    Replacement 19-41

Removal & Installation............19-40

O
Oil Pan..............................19-24
  Removal & Installation............19-24
Oil Pump 19-25
  Removal & Installation............19-25

P
Piston and Ring 19-29
  Positioning 19-29
Power Rack and Pinion Steering
  Gear..............................19-35
  Removal & Installation............19-35

R
Rear Main Seal 19-25
  Removal & Installation............19-25
Rocker Arms/Shafts 19-18
  Removal & Installation............19-18

S
Shock Absorber 19-39
  Removal & Installation............19-39
Stabilizer Bar 19-39
  Removal & Installation............19-39
Starter Motor 19-23
  Removal & Installation............19-23
Steering Knuckle....................19-40
  Removal & Installation............19-40
Strut and Coil Spring.................19-38
  Removal & Installation............19-38

T
Timing Chain, Sprockets, Front
  Cover and Seal 19-25
  Removal & Installation............19-25
Trailing Arm 19-39
  Removal & Installation............19-39
Transaxle Assembly 19-31
  Removal & Installation............19-31

U
Upper Control Arm 19-41
  Removal & Installation............19-41

V
Valve Lash 19-23
  Adjustment.......................19-23

W
Water Pump.........................19-12
  Removal & Installation............19-12
Wheel Hub and Bearing.............19-41
  Removal & Installation............19-41

SPECIFICATIONS AND MAINTENANCE CHARTS

VEHICLE AND ENGINE IDENTIFICATION CHART

		Engine					Model Year	
Code ①	Liters	Cu. In.	Cyl.	Fuel Sys.	Engine Type	Eng. Mfg.	Code ②	Year
B	2.4	146	4	MFI	OHC	General Motors	5	2005
8	3.5	214	6	SEFI	OHV	General Motors	6	2006
1	3.9	238	6	SEFI	OHV	General Motors		

MFI: Multi-port Fuel Injection

SEFI: Sequential Multi-port Fuel Injection

06025-G6-C01

GENERAL ENGINE SPECIFICATIONS

All measurements are given in inches.

Year	Model	Engine Displacement Liters	Engine Series (ID/VIN)	Net Horsepower @ rpm	Net Torque @ rpm (ft. lbs.)	Bore x Stroke (in.)	Compression Ratio	Oil Pressure @ rpm
2005	G6	3.5	8	201@5600	222@3200	3.81x3.39	9.8:1	50-80@1000
2006	G6	2.4	B	167@6300	162@4500	3.81x3.39	10.0:1	30-45@1850
		3.5	8	201@5600	222@3200	3.55x3.54	9.8:1	30-45@1850
		3.9	1	240@6000	241@2800	3.55x3.54	9.8:1	30-45@1850

06025-G6-C02

GASOLINE ENGINE TUNE-UP SPECIFICATIONS

Year	Engine Displacement Liters	Engine VIN	Spark Plugs Gap (in.)	Ignition Timing (deg.) MT	Ignition Timing (deg.) AT	Fuel Pump (psi)	Idle Speed (rpm) MT	Idle Speed (rpm) AT	Valve Clearance In.	Valve Clearance Ex.
2005	3.5	8	0.060	①	①	50-60	①	①	HYD	HYD
2006	2.4	B	0.042	①	①	50-60	①	①	HYD	HYD
	3.5	8	0.060	①	①	50-60	①	①	HYD	HYD
	3.9	1	0.040	①	①	50-60	①	①	HYD	HYD

NOTE: The Vehicle Emission Control Information label often reflects specification changes changes made during production.

The label figures must be used if they differ from those in this chart.

HYD: Hydraulic

① Controlled by the Powertrain Control Module (PCM) and cannot be manually adjusted.

06025-G6-C03

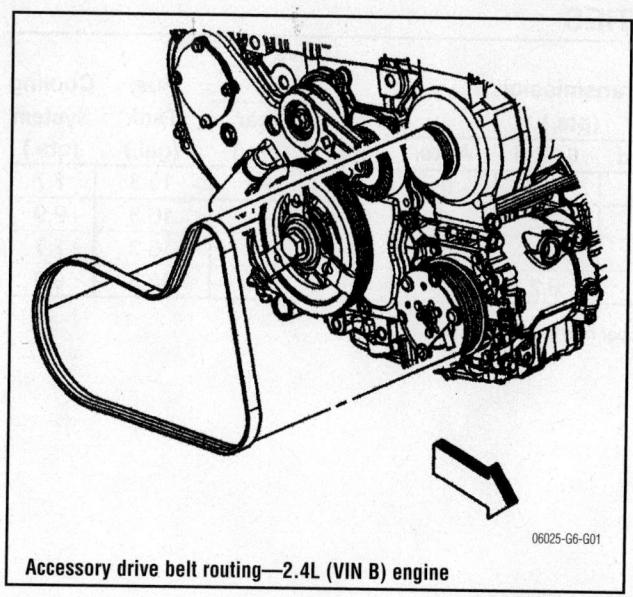

Accessory drive belt routing—2.4L (VIN B) engine

06025-G6-G01

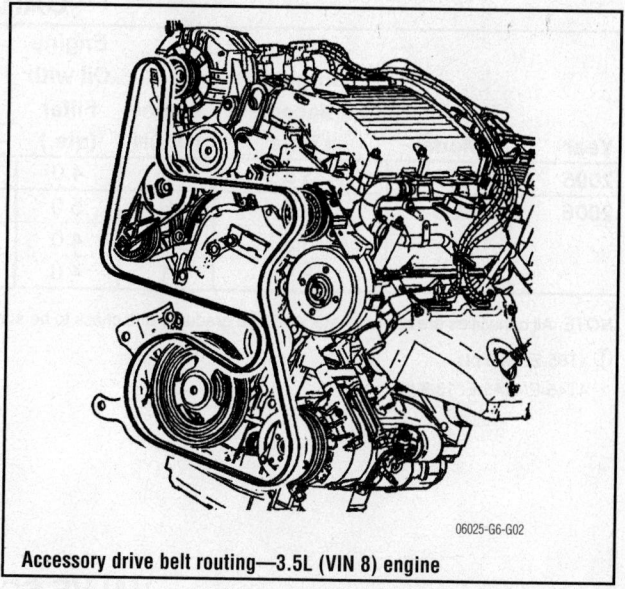

Accessory drive belt routing—3.5L (VIN 8) engine

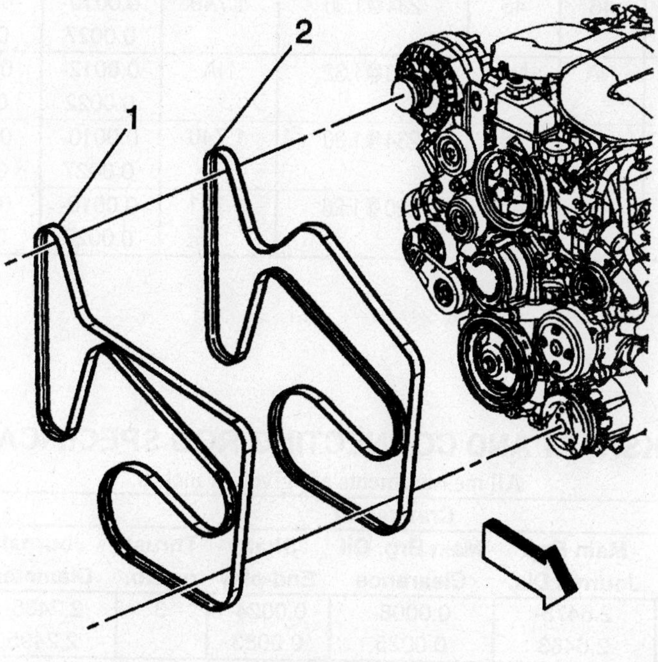

06025-G6-G02

1. With electronic power steering
2. With hydraulic power steering

06025-G6-G03

Air conditioning drive belt routing—3.9L (VIN 1) engine

CAPACITIES

Year	Model	Engine Displacement Liters	Engine ID/VIN	Engine Oil with Filter (qts.)	Transmission (pts.)			Drive Axle		Fuel Tank (gal.)	Cooling System (qts.)
					4-Spd	6-Spd	Auto.	Front (pts.)	Rear (pts.)		
2005	G6	3.5	8	4.0	—	6.2	①	—	—	16.3	7.7
2006	G6	2.4	B	5.0	—	6.2	①	—	—	16.3	9.9
		3.5	8	4.0	—	6.2	①	—	—	16.3	7.7
		3.9	1	4.0	—	6.2	①	—	—	16.3	9.8

NOTE: All capacities are approximate. Add fluid gradually and check to be sure a proper fluid level is obtained.

① 4T65-E: 14.8 pts.
4T45-E/4T45-E: 13.8 pts.

06025-G6-C04

VALVE SPECIFICATIONS

Year	Engine VIN	Engine Displacement Liters	Seat Angle (deg.)	Face Angle (deg.)	Spring Test Pressure (lbs. @ in.)	Spring Installed Height (in.)	Stem-to-Guide Clearance (in.)		Stem Diameter (in.)	
							Intake	Exhaust	Intake	Exhaust
2005	8	3.5	46	45	234@1.30	1.740	0.0010-0.0027	0.0010-0.0027	N/A	N/A
2005	B	2.4	NA	NA	220@1.32	NA	0.0012-0.0022	0.0020-0.0026	0.2344-0.3140	0.2337-0.2343
	8	3.5	46	45	234@1.30	1.740	0.0010-0.0027	0.0010-0.0027	N/A	N/A
	1	3.9	46	45	230@1.26	1.701	0.0010-0.0027	0.0010-0.0027	N/A	N/A

06025-G6-C05

CRANKSHAFT AND CONNECTING ROD SPECIFICATIONS

All measurements are given in inches.

Year	Engine Displ. Liters	Engine VIN	Crankshaft				Connecting Rod		
			Main Brg. Journal Dia.	Main Brg. Oil Clearance	Shaft End-play	Thrust on No.	Journal Diameter	Oil Clearance	Side Clearance
2005	3.5	8	2.6473-2.6483	0.0008-0.0025	0.0024-0.0083	3	2.2488-2.2495	0.0007-0.0024	0.0078-0.0094
2006	2.4	B	2.2045-2.2050	0.012-0.0026	0.0012-0.0150	NA	1.9291-1.9297	0.0011-0.0027	0.0028-0.0146
	3.5	8	2.6473-2.6483	0.0008-0.0025	0.0024-0.0083	3	2.2498-2.2495	0.0007-0.0024	0.0078-0.0094
	3.9	1	2.6473-2.6483	①	0.0024-0.0083	3	2.2488-2.2495	0.0007-0.0024	0.0078-0.0094

① 0.0008-0.0025 except no. 3
0.0012-0.0030 on no. 3

06025-G6-C06

PISTON AND RING SPECIFICATIONS

All measurements are given in inches.

Year	Engine Displ. Liters	Engine VIN	Piston Clearance	Ring Gap			Ring Side Clearance		
				Top Comp.	Bottom Comp.	Oil Control	Top Comp.	Bottom Comp.	Oil Control
2005	3.5	8	①	0.0007-0.0153	0.0188-0.0291	0.0098-0.0291	0.0011-0.0299	0.0015-0.0307	0.0035
2006	2.4	B	0.0004-0.0016	0.0078-0.0157	0.0137-0.0216	0.0098-0.0299	0.0015-0.0031	0.0012-0.0027	0.0035-0.0042
	3.5	8	①	0.0007-0.0153	0.0188-0.0291	0.0098-0.0291	0.0011-0.0299	0.0015-0.0307	0.0035
	3.9	1	②	0.0059-0.0118	0.0098-0.0177	0.0059-0.0255	0.0011-0.0025	0.0007-0.0021	0.0004

① -0.0011-+0.0011

② -0.0003-+0.0018

06025-G6-C07

TORQUE SPECIFICATIONS

All readings in ft. lbs.

Year	Engine VIN	Engine Displacement Liters	Cylinder Head Bolts	Main Bearing Bolts	Rod Bearing Bolts	Crankshaft Damper Bolts	Flywheel Bolts	Manifold		Spark Plugs	Oil Pan Drain Plug
								Intake	Exhaust		
2005	3.5	8	①	②	③	118	52	④	12	11	18
2006	2.4	B	⑤	⑥	③	⑦	⑧	⑨	⑩	15	18
	3.5	8	①	②	③	118	52	④	12	11	18
	3.9	1	①	②	③	⑪	52	④	15	11	18

① Step 1: 44 ft. lbs.
Step 2: plus 90 degrees

② Step 1: 37 ft. lbs.
Step 2: plus 77 degrees

③ Step 1: 18 ft. lbs.
Step 2: plus 110 degrees

④ Lower manifold center bolt: 15 ft. lbs.
Lower manifold corner bolt: 18 ft. lbs.
Upper manifold: 18 ft. lbs.

⑤ Step 1: 22 ft. lbs.
Step 2: plus 155 degrees

⑥ Step 1: 15 ft. lbs.
Step 2: plus 70 degrees

⑦ Step 1: 74 ft. lbs.
Step 2: plus 125 degrees

⑧ Step 1: 39 ft. lbs.
Step 2: plus 25 degrees

⑨ Stud 53 INCH lbs.
Nut/bolts: 89 INCH lbs.

⑩ Step 1: 52 ft. lbs.
Step 2: plus 72 degrees

06025-G6-C08

WHEEL ALIGNMENT

Year	Model			Caster Range (+/-Deg.)	Caster Preferred Setting (Deg.)	Camber Range (+/-Deg.)	Camber Preferred Setting (Deg.)	Toe-in (Deg.)
2005	G6	Front	Left	0.75	+3.10	0.75	-0.90	0.20+/-0.20
			Right	0.75	+3.10	0.75	-0.70	0.20+/-0.20
		Rear		—	—	0.50	-0.80	0.20+/-0.20
2006	G6	Front	Left	0.75	+3.10	0.75	-0.90	0.20+/-0.20
			Right	0.75	+3.10	0.75	-0.70	0.20+/-0.20
		Rear		—	—	0.50	-0.80	0.20+/-0.20

06025-G6-C09

TIRE AND WHEEL SPECIFICATIONS

Year	Model	OEM Tires Standard	OEM Tires Optional	Tire Pressures (psi) Front	Tire Pressures (psi) Rear	Wheel Size	Wheel Lug Nut Torque (Ft. Lbs.)
2005	G6	P215/60R16	None	35	35	①	100
	G6 GT	P225/50R17	None	35	35	①	100
	G6 GTP	P225/50R18	None	35	35	①	100
2006	G6	P215/60R16	None	35	35	①	100
	G6 GT	P225/50R17	None	35	35	①	100
	G6 GTP	P225/50R18	None	35	35	①	100

OEM: Original Equipment Manufacturer

PSI: Pounds Per Square Inch

STD: Standard

OPT: Optional

① Not available

06025-G6-C10

BRAKE SPECIFICATIONS
All measurements in inches unless noted

Year	Model		Brake Disc Original Thickness	Brake Disc Minimum Thickness	Brake Disc Maximum Runout	Minimum Lining Thickness Front	Minimum Lining Thickness Rear	Brake Caliper Bracket Bolts (ft. lbs.)	Brake Caliper Mounting Bolts (ft. lbs.)
2005	G6	F	1.023	0.898	0.002	①	—	①	26
		R	0.551	0.465	0.002	—	①	①	26
2006	G6	F	1.023	0.898	0.002	①	—	①	26
		R	0.551	0.465	0.002	—	①	①	26

① Not available

06025-G6-C11

MAINTENANCE I AND II SERVICE SCHEDULES
2005-06 Pontiac G6

When the CHANGE ENGINE OIL light appears, certain services and inspections are required. Required services are described as Maintenance I and Maintenance II.

The first service on a vehicle should be Maintenance I, and the second service should be Maintenance II. Alternate between the 2 thereafter. However, in some cases, Maintenance II may be required more often.

Maintenance I: Use Maintenance I if the CHANGE ENGINE OIL light comes on within 10 months since vehicle was purchased or, if Maintenance II was performed.

Maintenance II: Use Maintenance II if the previous service performed was Maintenance I.

Always use Maintenance II whenever the CHANGE ENGINE OIL light comes on 10 months or more since the last service, or, if the CHANGE ENGINE OIL light has not come on at all for one year.

Service	Maintenance I	Maintenance II
Change the engine oil and filter. Reset the oil life system.	✓	✓
Visually inspect the vehicle for leaks or damage. A fluid loss in the vehicle system could indicate a problem. Inspected, repair and add fluid to the system if necessary.	✓	✓
Inspect the engine air cleaner filter. If necessary, replace the filter.	✓	✓
Rotate the tires. Inspect the tire inflation pressures and the tire wear.	✓	✓
Visually inspect the brake lines and hoses for proper hook-up, binding, leaks, cracks, chafing, etc. Inspect the disc brake pads for wear and the rotors for surface condition. Inspect the drum brake linings for wear or cracks. Inspect other brake parts, including drums, wheel cylinders, calipers, parking brake, etc. Inspect the parking brake adjustment.	✓	✓
Inspect the engine coolant and the windshield washer fluid levels. Add fluid as needed.	✓	✓
Inspect the suspension and steering components. Inspect the front and rear suspension and the steering system for damaged, loose or missing parts, or signs of wear. Inspect the power steering lines and the hoses for proper hook-up, binding, leaks, cracks, chafing, etc.	--	✓
Visually inspect the coolant hoses and replace the hoses if they are cracked, swollen or deteriorated. Inspect all pipes, fittings and clamps; replace with GM parts as needed. To help ensure proper operation, a pressure test of the cooling system and pressure cap and cleaning the outside of the radiator and air conditioning condenser is recommended at least once a year.		✓
Inspect the wiper blades for wear or cracking.	--	✓
Inspect the restraint system components. Ensure the safety belt reminder light and all the belts, buckles, latch plates, retractors and anchorages are working properly. Look for any other loose or damaged safety belt system parts. If you see anything that might keep a safety belt system from working correctly, repair or replaced the damaged part. Replace torn or frayed safety belts, refer to Operational and Functional Checks in Seat Belts. Inspect for any opened or broken air bag coverings, and repair or replace as needed. The air bag system does require regular maintenance.	--	✓

06025-G6-C12

MAINTENANCE I AND II SERVICE SCHEDULES
2005-06 Pontiac G6

Lubricate the body components.Lubricate all key lock cylinders, hood latch assemblies, secondary latches, pivots, spring anchor and release pawl, hood and door hinges, rear folding seats and liftgate hinges. Frequent lubrication may be required when exposed to a corrosive environment, refer to Fluid and Lubricant Recommendations . Applying dielectric silicone grease GM P/N 12345579 (Canadian P/N 1974984) or equivalent on the weatherstrips with a clean cloth.	--	✓
Inspect the transaxle fluid level and add fluid as needed.	--	✓
Inspect the suspension and steering components.Inspect the front and rear suspension and the steering system for damaged, loose or missing parts, or signs of wear. Inspect power steering lines and hoses for proper hook-up, binding, leaks, cracks, chafing, etc.	--	✓
Inspect the throttle system for interference or binding and for damaged or missing parts. Replace the parts as needed. Replace any components that have high effort or excessive wear. Do not lubricate the accelerator or the cruise control cables.	--	✓
Replace the passenger compartment air filter.	--	✓

06025-G6-C13

ENGINE REPAIR

Distributor

REMOVAL

All engines are equipped with a distributorless ignition.

Alternator

REMOVAL

2.4L Engine

1. Remove or disconnect the following:
   - Negative battery cable
   - Air cleaner resonator
   - Drive belt
   - Alternator harness connectors
   - Alternator

3.5L Engine

1. Remove or disconnect the following:
   - Negative battery cable
   - Drive belt
   - Alternator harness connector
   - Alternator bolts and nuts
   - Alternator

3.9L Engine

1. Remove or disconnect the following:
   - Negative battery cable
   - Drive belt
   - Engine harness connector
   - Alternator harness connector
   - Upper and lower alternator bolts and stud
   - Alternator

INSTALLATION

2.4L Engine

1. Install or connect the following:

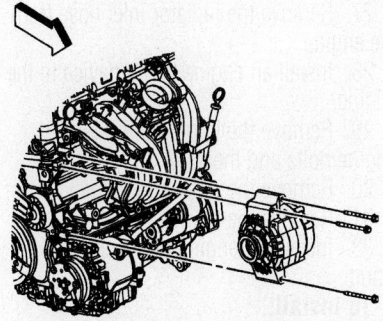

Alternator mounting—2.4L engine

- Alternator and tighten the bolts to 16 ft. lbs. (22 Nm).
- Alternator harness connectors
- Drive belt
- Air cleaner resonator
- Negative battery cable

3.5L Engine

1. Install or connect the following:
   - Alternator and tighten the bolts to 37 ft. lbs. (50 Nm), and the nuts to 22 ft. lbs. (30 Nm).
   - Alternator harness connectors
   - Drive belt
   - Negative battery cable

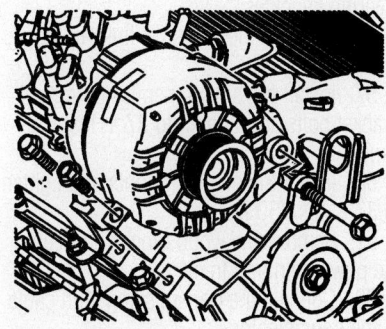

06025-G6-G05
Alternator mounting—3.5L engine

3.9L Engine

1. Install or connect the following:
   - Alternator
   - Lower bolt and stud and tighten until snug.
   - Upper bolt and tighten to 37 ft. lbs. (50 Nm)
   - Alternator harness connectors
   - Engine harness connectors
   - Drive belt
   - Negative battery cable

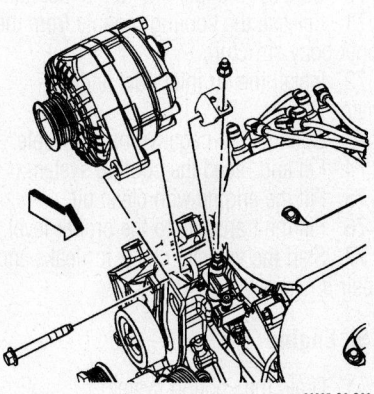

06025-G6-G06
Alternator mounting—3.9L engine

Ignition Timing

The ignition timing on all engines is controlled by the PCM and is not adjustable.

Engine Assembly

REMOVAL & INSTALLATION

2.4L Engine

1. Place the wheels in the straight ahead position.
2. Disconnect the negative battery cable.
3. Remove the air inlet duct and resonator.
4. Secure the cooling module to the upper body structure.
5. Relieve the fuel system pressure.
6. Disconnect the fuel line from the fuel rail.
7. Drain the cooling system.
8. Remove the radiator inlet hose.
9. Remove the surge tank to cylinder head hose.
10. Remove the radiator outlet hose.
11. Remove the heater hoses.
12. Disconnect the following harness connectors:
    - Electronic throttle control
    - Manifold Absolute Pressure (MAP) sensor
    - Crankshaft Position (CKP) sensor
    - Oil pressure sensor
    - Purge solenoid
    - Ignition coils
    - Heated oxygen sensor
    - Vehicle Speed Sensor (VSS)
    - Engine Coolant Temperature (ECT) sensor
    - Back-up light switch
    - Camshaft position actuator solenoid control valves
    - Fuel rail
13. Raise and support the vehicle.
14. Remove the drive belt.
15. Remove the AC compressor bolts and set the compressor aside.
16. Disconnect the starter and alternator connectors.
17. Disconnect the front exhaust pipe from the exhaust manifold.
18. Lower the vehicle.
19. Disconnect the transmission harness connectors.
20. Disconnect the transmission shift cable from the transmission.

21. Use blocks of wood to support the powertrain assembly between the frame and the powertrain.

22. Remove the engine mount.

23. Remove the side transmission mount bracket bolts.

24. Raise the vehicle.

25. Disconnect the stabilizer links from the stabilizer bar.

26. Disconnect the outer tie rod ends from the steering knuckles.

27. Disconnect the intermediate shaft from the steering gear.

28. Disconnect the lower control arms from the steering knuckles.

29. Disconnect the drive axles from the transaxle and support with wire or bungee cords.

30. Use a paint pen or magic marker in order to mark the frame to body position.

31. Lower the vehicle to about 3 feet off the ground in order to position the lift table under the frame.

32. Use wood blocks as necessary between the lift table and the frame to support the assembly.

33. Remove the front frame bolts and then the rear frame bolts.

34. Slowly raise the vehicle off of the frame and powertrain.

35. Attach an engine lift hoist to the engine lift hooks.

36. Remove the starter.

37. Remove the torque converter-to-flywheel bolts.

38. Remove the transaxle to engine bolts.

39. Separate the engine from the transaxle.

40. Remove the following components:
- Exhaust manifold
- Engine mount bracket.
- Engine block heater
- Thermostat housing and feed pipe
- Alternator
- Fuel rail
- Drive belt tensioner.

41. Install the engine to a suitable engine stand.

To install:

42. Attach a lifting device to the lifting hooks.

43. Install the following components:
- Exhaust manifold
- Engine mount bracket.
- Engine block heater
- Thermostat housing and feed pipe
- Alternator
- Fuel rail
- Drive belt tensioner.

44. Lower the engine into the vehicle and install the engine to the transaxle bolts and tighten to 55 ft. lbs. (75 Nm).

45. Install the torque converter bolts and tighten to 46 ft. lbs. (62 Nm).

46. Install the starter.

47. Remove the engine lift from the engine.

48. Lower vehicle slowly over frame and powertrain.

49. Hand start all the frame bolts while aligning the frame to the paint marks.

50. Tighten the frame bolts to 74 ft. lbs. (100 Nm) plus an additional 180 degrees.

51. Remove the lift table.

52. Connect the drive axles to the transaxle.

53. Connect the lower control arm to the steering knuckle.

54. Connect the intermediate steering shaft to the steering gear.

55. Connect the outer tie rod ends to the steering knuckles.

56. Connect the stabilizer links to the stabilizer bar.

57. Install the side transmission mount bracket bolts and tighten to 37 ft. lbs. (50 Nm).

58. Install the engine mounts and tighten to 37 ft. lbs. (50 Nm).

59. Remove the wood blocks between the powertrain and frame.

60. Connect the transmission shift cable to the transmission.

61. Connect the transmission harness connector.

62. Install the catalytic converter to the exhaust manifold and tighten to 22 ft. lbs. (30 Nm).

63. Lower the vehicle.

64. Install the alternator and starter connections.

65. Install the AC compressor to the engine.

66. Install the engine drive belt.

67. Connect all electrical connectors disconnect previously.

68. Install the heater hoses.

69. Install the radiator outlet hose.

70. Connect the fuel line to the fuel rail.

71. Release the cooling module from the upper body structure.

72. Install the air inlet duct and resonator.

73. Connect the negative battery cable.

74. Fill and bleed the cooling system.

75. Fill the engine with clean oil.

76. Fill the transaxle to the proper level.

77. Start the vehicle, check for leaks and repair if necessary.

3.5L Engine

1. Drain the cooling system.
2. Drain the engine oil.

3. Drain the transaxle fluid.

4. Properly relieve the fuel system pressure.

5. Remove the intake manifold cover.

6. Remove the air cleaner assembly.

7. Remove the hood.

8. Remove the engine mount strut.

9. Remove the drive belt.

10. Disconnect the following electrical connectors:
- Knock Sensor (KS)
- Camshaft Position Sensor (CPS)
- Crankshaft Position (CKP) sensor
- Heated oxygen sensor
- Manifold Absolute Pressure (MAP) sensor
- EGR valve
- EVAP canister purge solenoid
- Electronic throttle control
- Ignition coils
- Body wiring harness-to-engine harness

11. Raise and support the vehicle.

12. Remove the catalytic converters.

13. Remove the engine wiring harness grounds from the transaxle.

14. Remove the engine mount lower nuts.

15. Remove the torque converter covers.

16. Remove the starter.

17. Remove the air conditioning (A/C) compressor. DO NOT discharge the A/C system. Support the compressor.

18. Remove the torque converter bolts.

19. Remove the transaxle brace.

20. Remove the 6 lower transaxle-to-engine bolts and the stud.

21. Remove the radiator outlet hose from the engine.

22. Lower the vehicle and support the transaxle.

23. Remove the heater outlet and inlet hoses from the engine.

24. Remove the vacuum hoses from the upper intake manifold.

25. Remove the brake booster vacuum hose from the upper intake manifold.

26. Remove the fuel lines from the fuel rail.

27. Remove the radiator inlet hose from the engine.

28. Install an engine lifting device to the engine.

29. Remove the upper transaxle-to-engine bolts and the stud.

30. Remove the engine from the vehicle.

31. Remove the flywheel.

32. Install the engine to the engine stand.

To install:

33. Remove the engine from the engine stand.

34. Install the flywheel.
35. Install the engine to the vehicle.
36. Install the upper transaxle-to-engine bolts and the stud and tighten to 55 ft. lbs. (75 Nm).
37. Remove the engine lifting device.
38. Install the radiator inlet hose to the engine.
39. Install the fuel lines to the fuel rail.
40. Install the brake booster vacuum hose to the upper intake manifold.
41. Install the vacuum hoses to the upper intake manifold.
42. Install the heater inlet and outlet hoses to the engine.
43. Raise the vehicle and remove the transaxle support.
44. Install the radiator outlet hose to the engine.
45. Install the lower transaxle-to-engine bolt and the stud and tighten to 55 ft. lbs. (75 Nm).
46. Install the transaxle brace and tighten to 53 ft. lbs. (72 Nm).
47. Install the torque converter bolts.
48. Install the A/C compressor.
49. Install the starter motor.
50. Install the torque converter covers.
51. Install the engine mount lower nuts and tighten to 32 ft. lbs.
52. Install the engine wiring harness grounds to the transaxle.
53. Install the engine wiring harness ground nut to the transaxle stud.
54. Install the catalytic converters.
55. Lower the vehicle.
56. Connect the electrical connectors disconnected previously.
57. Install the engine mount lower nuts and tighten to 32 ft. lbs. (43 Nm).
58. Install the engine wiring harness grounds to the transaxle.
59. Install the engine wiring harness ground nut to the transaxle stud.
60. Install the catalytic converters.
61. Lower the vehicle.
62. Connect the negative battery cable.
63. Fill and bleed the cooling system.
64. Fill the engine with clean oil.
65. Fill the transaxle to the proper level.
66. Start the vehicle, check for leaks and repair if necessary.

3.9L Engine

1. Disconnect the negative battery cable.
2. Drain the engine coolant.
3. Drain the engine oil.
4. Remove the air cleaner assembly.
5. Remove the hood.
6. Remove the intake manifold cover.

7. Remove the engine mount strut.
8. Remove the drive belt.
9. Remove the power steering pump and disconnect the power steering lines, if equipped.
10. Remove the oil pressure sensor heat shield nuts and shield.
11. Disconnect the oil pressure sensor connector.
12. Disconnect the knock sensor connector.
13. Disconnect the air conditioning compressor electrical connector.
14. Lower the vehicle.
15. Disconnect the EVAP canister purge solenoid connector.
16. Disconnect the electronic throttle control connector.
17. Remove the connector position assurance retainer.
18. Disconnect the heated oxygen sensor electrical connector.
19. Disconnect the Manifold Absolute Pressure (MAP) sensor connector.
20. Disconnect the ignition control module connector.
21. Disconnect the inlet manifold valve connector.
22. Disconnect the fuel injector inline connector.
23. Disconnect the camshaft phaser sensor connector.
24. Disconnect the rear upper HO2S electrical connector.
25. Disconnect the Crankshaft Position (CKP) sensor connector.
26. Disconnect the engine harness connector from the body harness connector.
27. Disconnect the body harness electrical connector from the powertrain control module (PCM).
28. Disconnect the engine harness electrical connectors from the PCM.
29. Disconnect the engine harness electrical connector from the transmission control module (TCM).
30. Remove the engine harness attachments from the transmission stud.
31. Remove the catalytic converters.
32. Remove the engine mount.
33. Remove the torque converter cover.
34. Remove the starter.
35. Remove the torque converter bolts.
36. Unbolt and reposition the A/C compressor of to the side. DO NOT discharge the A/C system.
37. Remove the transaxle brace.
38. Remove the lower transaxle-to-engine bolt and stud.
39. Reposition the radiator outlet hose clamp at the thermostat housing.

40. Remove the radiator outlet hose from the thermostat housing.
41. Lower the vehicle and support the transaxle.
42. Reposition the radiator surge tank hose clamp at the surge tank pipe.
43. Remove the radiator surge tank hose from the surge tank pipe.
44. Reposition the brake booster vacuum hose clamp at the intake manifold.
45. Remove the brake booster vacuum hose from the intake manifold.
46. Reposition the heater inlet and outlet hose clamps at the engine.
47. Remove the heater hoses.
48. Disconnect the fuel feed line from the fuel rail.
49. Disconnect the EVAP purge line from the canister purge solenoid.
50. Reposition the radiator inlet hose clamp at the engine.
51. Remove the radiator inlet hose from the engine.
52. Install a engine lifting device to the engine.
53. Remove the remaining transaxle-to-engine bolts/studs.
54. Remove the engine from the vehicle.
55. Remove the flywheel.
56. Install the engine to the engine stand.

To install:
57. Remove the engine from the engine stand.
58. Install the flywheel and tighten the bolts to 52 ft. lbs .(70 Nm).
59. Install the engine to the vehicle.
60. Install the transaxle-to-engine bolts/studs and tighten to 55 ft. lbs. (75 Nm).
61. Remove the engine lifting device from the engine.
62. Install the radiator inlet hose to the engine.
63. Position the radiator inlet hose clamp at the engine.
64. Connect the EVAP purge line to the canister purge solenoid.
65. Connect the fuel feed line to the fuel rail.
66. Install the heater hoses.
67. Install the brake booster vacuum hose to the intake manifold.
68. Install the radiator surge tank hose to the surge tank pipe.
69. Raise and support the vehicle.
70. Install the radiator outlet hose to the thermostat housing.
71. Install the lower transaxle-to-engine bolt and stud and tighten to 55 ft. lbs. (75 Nm).

72. Install the transaxle brace and tighten to 53 ft. lbs. (72 Nm).

73. Position the A/C compressor and tighten to 37 ft. lbs. (50 Nm).

74. Install the torque converter bolts and tighten to 37 ft. lbs. (50 Nm).

75. Install the starter.

76. Install the torque converter cover.

77. Install the engine mount and tighten to 37 ft. lbs. (50 Nm).

78. Install the catalytic converters.

79. Install the engine harness attachments to the transmission stud.

80. Connect the engine harness electrical connector to the TCM and PCM.

81. Connect the engine harness connector to the body harness connector and PCM.

82. Connect the CKP sensor connector.

83. Connect the knock sensor connector.

84. Connect the rear upper HO2S connector.

85. Install the CPA retainer.

86. Connect the camshaft phaser sensor connector.

87. Connect the fuel injector inline connector.

88. Connect the inlet manifold valve connector.

89. Connect the ignition control module connector.

90. Connect the MAP sensor connector.

91. Connect the HO2S connector.

92. Install the CPA retainer.

93. Connect the ETC connector.

94. Connect the EVAP canister purge solenoid connector.

95. Raise and support the vehicle.

96. Connect the A/C compressor electrical connector.

97. Connect the knock sensor connector.

98. Connect the oil pressure sensor connector.

99. Install the oil pressure sensor heat shield and nuts.

100. Lower the vehicle.

101. Connect the power steering lines and install the power steering pump, if equipped.

102. Install the drive belt.

103. Install the engine mount strut.

104. Install the intake manifold cover.

105. Install the hood.

106. Install the air cleaner assembly.

107. Fill the engine with oil.

108. Fill the cooling system.

109. Connect the negative battery cable.

110. Start the engine and check for leaks.

Water Pump

REMOVAL & INSTALLATION

2.4L Engine

1. Before servicing the vehicle, refer to the precautions in the beginning of this section.

2. Drain the coolant from the radiator and engine block.

3. Remove the drive belt.

4. Disconnect the water pump and thermostat hoses.

5. Remove the thermostat housing.

6. Remove the coolant heater.

7. Remove the water pump access plate from the front cover.

➡ **The water pump holding tool J-43651 supports the sprocket and chain during water pump service. The tool must be used or the balance shaft must be re-timed.**

8. Install the tool into position.

9. Tighten the bolts on the water pump holding tool into the threads on the water pump sprocket.

10. Install the access cover bolts that were removed earlier to secure the water pump holding tool to the front cover assembly.

11. Remove the 3 inner water pump sprocket to water pump bolts through the holes in the water pump holding tool.

12. Remove the 2 front and 2 rear water pump bolts.

13. Remove the engine wiring harness clip nut from the water pump stud.

14. Remove the engine wiring harness clip from the stud.

15. Remove the water pump.

To install:

16. Apply sealant to the water pump drain plug.

17. Install the water pump drain plug and tighten to 15 ft. lbs. (20 Nm).

➡ **A guide pin can be created to aid in water pump alignment. Use a M6 m x 6 mm stud. Thread the pin into the water pump sprocket.**

18. Align the guide pin with the water pump holding tool.

19. Position the water pump against the engine block and hand tighten the water pump bolts.

20. Install the inner water pump sprocket bolts. After 2 are snug, remove the guide pin and install the 3rd bolt and tighten all bolts to 18 ft. lbs. (25 Nm).

21. Tighten the water pump sprocket bolts last to 89 inch lbs. (10 Nm).

22. Remove the holding tool and install the coolant heater.

23. Install the water pump access plate and bolts and tighten to 89 inch lbs. (10 Nm).

24. Install the thermostat housing.

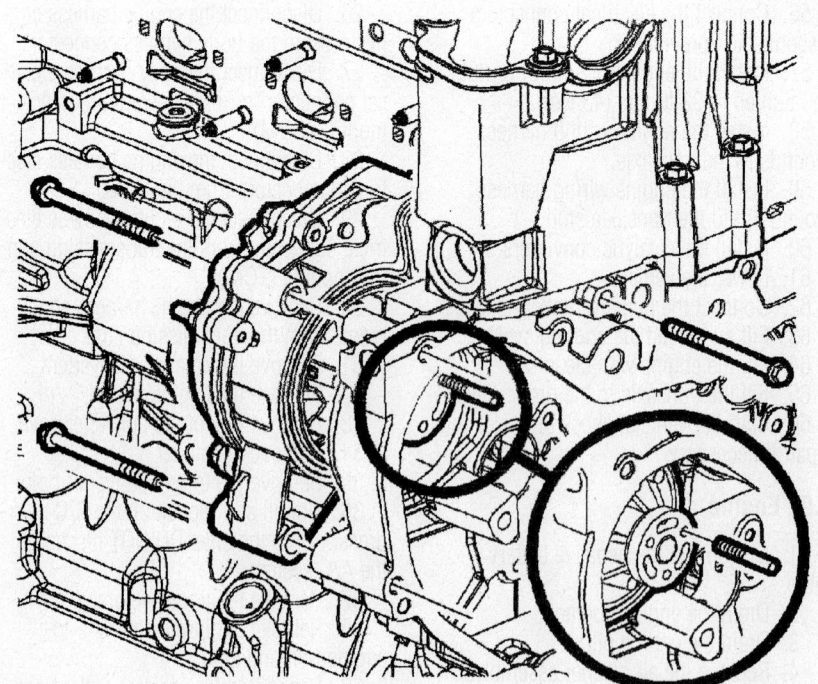

06025-G6-G07

Exploded view of the water pump mounting and guide pin installation—2.4L engine

25. Install the drive belt.
26. Connect the water pump and thermostat hoses.
27. Fill and bleed the cooling system.
28. Start the vehicle, check for leaks and repair if necessary.

3.5L Engine

1. Drain the cooling system.
2. Disconnect the radiator hoses.
3. Remove the water pump pulley bolts.
4. Remove the drive belt.
5. Remove the water pump pulley bolts and pulley.
6. Remove the water pump bolts, pump, and gasket.
7. Clean the water pump mating surfaces.

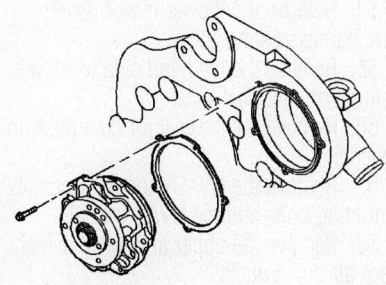

06025-G6-G08

Exploded view of the water pump mounting—3.5L engine

To install:

8. Install the water pump gasket and pump.
9. Install the water pump bolts and tighten to 89 inch lbs. (10 Nm).
10. Install the water pump pulley and bolt until snug.
11. Install the drive belt.
12. Tighten the pulley bolts to 18 ft. lbs. (25 Nm).
13. Connect the radiator hoses.
14. Connect the negative battery cable
15. Fill and bleed the cooling system.
16. Start the vehicle, check for leaks and repair if necessary.

3.9L Engine

1. Drain the cooling system.
2. Disconnect the radiator hoses.
3. Remove the water pump pulley bolts.
4. Remove the drive belt.
5. Remove the water pump pulley bolts and pulley.
6. Remove the water pump bolts, pump, and gasket.
7. Clean the water pump mating surfaces.

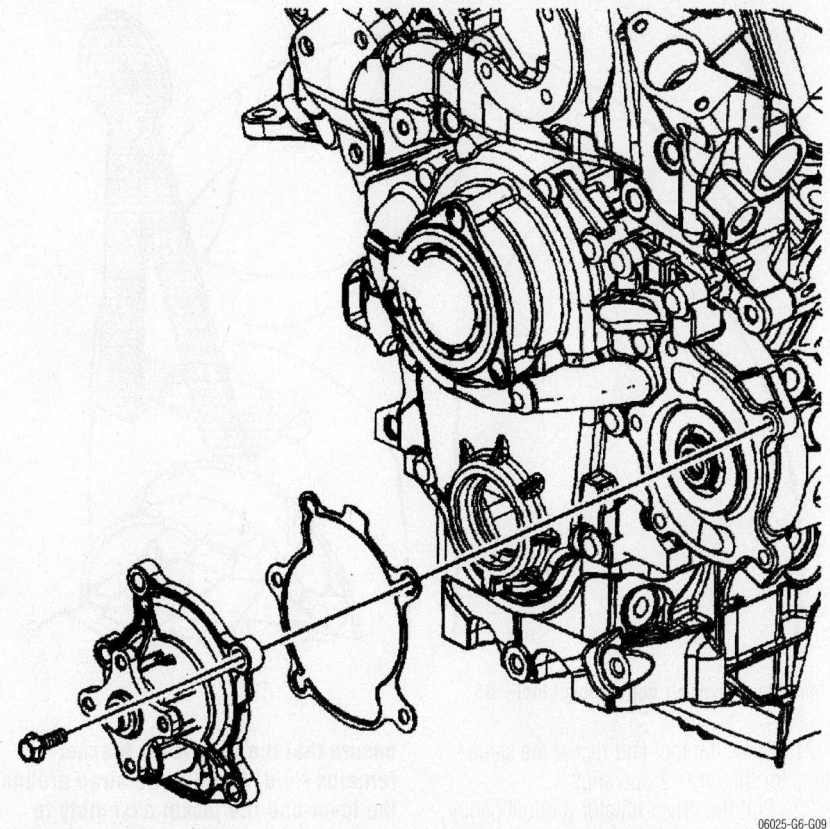

06025-G6-G09

Exploded view of the water pump mounting—3.9L engine

To install:

8. Install the water pump gasket and pump.
9. Install the water pump bolts and tighten to 89 inch lbs. (10 Nm).
10. Install the water pump pulley and bolt until snug.
11. Install the drive belt.
12. Tighten the pulley bolts to 18 ft. lbs. (25 Nm).
13. Connect the radiator hoses.
14. Connect the negative battery cable
15. Fill and bleed the cooling system.
16. Start the vehicle, check for leaks and repair if necessary.

Heater Core

REMOVAL & INSTALLATION

1. Place the front wheels in the straight ahead position.
2. Disable the air bag system.
3. Disconnect the negative battery cable.
4. Drain the engine coolant.
5. Disconnect the heater hoses at the heater core.
6. Recover the air conditioning refrigerant.

7. Remove the hose at the thermal expansion valve and plug the hose.
8. Disconnect the HVAC module to front of dash plate bolts.
9. Remove the right and left side console trim panels.
10. Remove the front floor console.
11. Remove the left closeout panel.
12. Remove both instrument panel outer trim panels.
13. Remove the instrument panel-to-body wire harness and antenna left connectors.
14. Remove the instrument panel-to-wire harness right side connectors.
15. Remove the knee bolster.
16. Remove the upper steering column trim cover.
17. On the back side of the steering wheel are 4 openings for removing the driver inflator module. Place the steering wheel so that 2 of the openings are on top.
18. Install Driver Air Bag Removal tool J-44298 into 2 of the holes so it is fully inserted.
19. Pull the handle toward the back of the steering wheel, releasing the 2 spring-loaded fasteners at the same time.
20. Turn the steering wheel and open the tool.

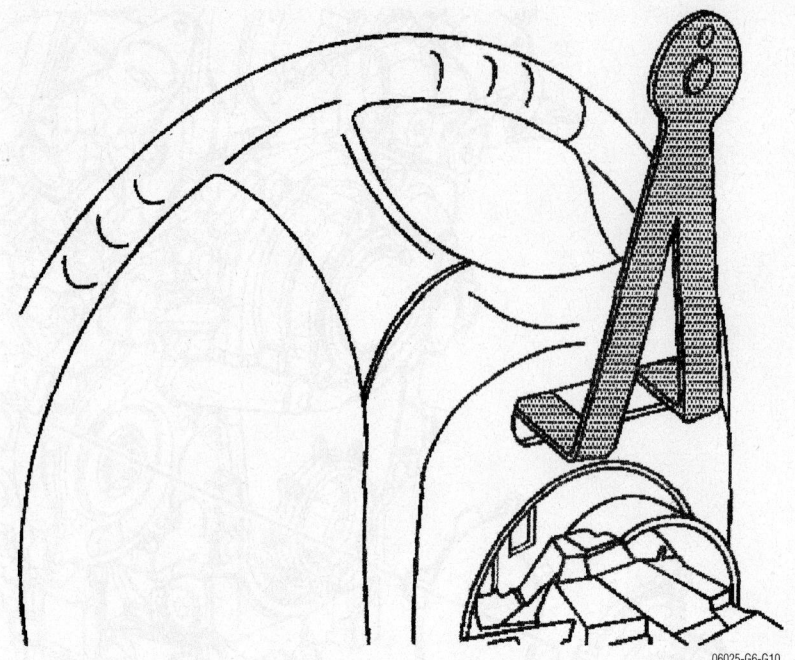

Installing driver air bag removal tool—G6

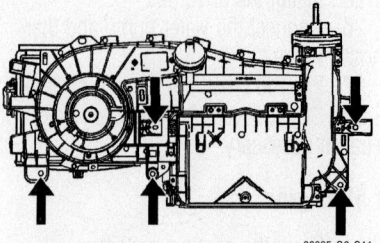

06025-G6-G11

HVAC assembly mounting—G6

21. Place the tool and repeat the same steps for the other 2 openings.

22. Pull the driver inflator module gently away from the steering wheel.

23. Remove the connector position assurance (CPA) and the electrical connector for the driver inflator module.

24. Remove the horn grounded lead from the steering wheel.

25. Remove the inflatable restraint module.

26. Disconnect the steering wheel control electrical connector.

27. Remove the steering wheel nut.

28. Remove the steering wheel using a puller.

29. Remove the controls and the control harness from the steering wheel.

30. Remove the steering column trim covers.

31. Disconnect the restraint module coil connectors and remove the coil.

32. Remove the multifunction turn signal and housing.

33. Remove the steering column knee bolster.

34. Remove the steering column shaft pinch bolt from the intermediate steering shaft.

➡**Install tie straps between the rake bracket assembly and the neck of the assist mechanism housing to prevent jacket assembly pull-apart. The steering column rake lever MUST be in the LOCK (FULL UP) position during steering column removal and installation to** ensure that the rake lever bracket remains rigid. Install a tie strap around the lever and the jacket assembly to keep the lever in the LOCK position. Do not bend the steering column energy absorbing straps located on the upper steering column mounting bracket.

35. Disconnect the steering column electrical connectors and open the steering column wire harness retainer clip.

36. Without disconnecting the adjustable brake pedal cable, remove the accelerator pedal.

37. Position the adjustable brake pedal cable aside.

38. Remove the accelerator pedal bracket from the cowl.

39. Remove the upper and lower steering column mounting bolts.

40. Remove the steering column.

41. Remove the brake pedal assembly.

42. Remove the body control module (BCM).

43. Remove the center support bracket floor bolts.

44. Remove the automatic transmission shifter assembly.

45. Remove both windshield pillar garnish moldings.

46. Remove the instrument panel upper trim panel.

47. Remove the instrument panel-to-body bolts on both sides.

48. Remove the right and left floor heater ducts at the center floor heater duct.

49. Remove the instrument panel.

50. Remove the recirculation actuator wire harness connector.

51. Remove the air temperature actuator wire harness connector.

52. Remove the mode actuator wire harness connector.

53. Remove the blower motor wire harness connector.

54. Remove the blower motor resistor wire harness connector.

55. Remove the left hand side window defogger outlet duct.

56. Remove the lower floor duct push-in fastener.

57. Remove the HVAC module assembly mounting bolts and the HVAC module.

58. Remove the upper and lower center floor air outlet ducts

59. Drill out the heater core cover heat stakes.

60. Remove the heater core cover.

61. Remove the heater core.

To install:

62. Install the heater core.

63. Install the heater core cover.

64. Install the upper and lower center floor air outlet ducts.

65. Install the HVAC module and tighten the bolts to 44 inch lbs. (5 Nm).

66. Install the lower floor duct push-in fastener.

67. Install the left hand side window defogger outlet duct.

68. Install the blower motor resistor wire harness connector.

69. Install the blower motor wire harness connector.

70. Install the mode actuator wire harness connector.

71. Install the air temperature actuator wire harness connector.

72. Install the recirculation actuator wire harness connector.

73. Install a new HVAC module drain seal and dash seal.

74. Install the instrument panel assembly.

75. Install the left and right floor heater ducts at the center floor heater duct.

76. Install the instrument panel-to-body

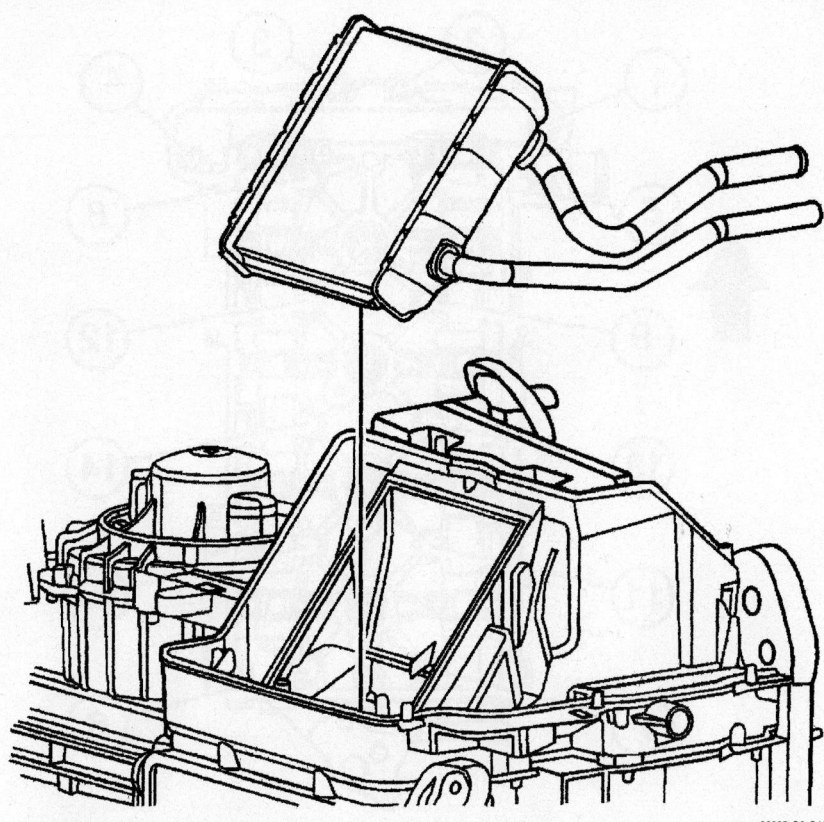

Heater core mounting—G6

06025-G6-G12

bolts on both sides and tighten to 19 ft. Lbs. (26 Nm).

77. Install the I/P upper trim panel.

78. Install both windshield pillar garnish moldings.

79. Install the shifter assembly on automatic transmissions.

80. Install the center support bracket floor bolts and tighten to 89 inch lbs. (10 Nm).

81. Install the BCM.

82. Install the brake pedal assembly.

83. Tighten the 3 brake pedal assembly to I/P carrier bolts to 18 ft. lbs. (25 Nm).

84. Install the 2 brake pedal assembly to cowl mounting nuts and tighten to 11 ft. lbs. (15 Nm).

85. Connect the brake pedal pushrod retaining clip, the wave washers and the brake booster pushrod to the brake pedal pin.

86. On electric power steering models, install the motor/module assembly to the steering column.

87. Position the steering column in the vehicle.

88. Slide the lower end of the steering column shaft into the intermediate steering shaft,

89. Loosely install the lower steering

column mounting bolt . Loosely install the upper steering column mounting bolts.

➡**Do not bend the steering column energy absorbing straps located on the upper steering column mounting bracket during installation.**

90. Align the energy absorbing straps with the bolt holes in the steering column. Loosely install the upper steering column bolts.

91. Tighten the steering column fasteners in the following sequence:

   a. Tighten the lower bolt to 18 ft. lbs. (25 Nm).

   b. Tighten the left side upper bolt to 18 ft. lbs. (25 Nm).

   c. Tighten the right side upper bolt to 18 ft. lbs. (25 Nm).

92. Connect the steering column electrical connectors and close the wire harness retainer clip on the steering column.

93. Install the adjustable accelerator bracket to the vehicle.

94. Install the 3 adjustable accelerator bracket nuts and tighten to 89 inch lbs. (10 Nm).

➡**Ensure that the adjustable accelerator pedal and adjustable brake pedal are synchronized in the full rearward position.**

95. Install a new adjustable brake pedal cable.

96. Install the accelerator pedal to the vehicle.

97. Install the steering column knee bolster.

98. Install the multifunction turn signal switch housing.

99. Install the multifunction turn signal.

100. Aim the wheels straight ahead.

101. Align the block tooth and the centering mark on the race and upper shaft assembly at the 12 o'clock position.

102. Slide the new SIR coil onto the steering shaft assembly.

103. Connect the SIR harness.

104. Remove and discard the centering tab from the new SIR coil.

105. Install the steering column trim covers.

106. Install the controls and the control harness to the steering wheel.

107. Route the steering column wiring through the steering wheel.

108. Install the steering wheel.

109. Connect the steering wheel control electrical connector.

110. Install the steering wheel nut and tighten to 24 ft. lbs. (32 Nm).

111. Connect the horn ground lead onto the steering wheel.

112. Connect the inflator module electrical connector and the CPA.

➡**This vehicle is equipped with dual stage frontal air bags. Match the right color connector to the right color opening in the module. Route the driver inflator wires, the redundant control wires, and the horn wires correctly.**

113. Align the driver inflator module fasteners to the steering column fastener holes.

114. Push the driver inflator module firmly into the steering column in order to engage the fasteners.

115. Install the steering column trim covers.

116. Install the crush bracket to the front of dash plate nuts and tighten to 89 inch lbs. (10 Nm).

117. Install the steering column to the I/P wire harness connector.

118. Install the knee bolster.

119. Install the I/P to the body wire harness RH connectors.

120. Install the I/P to the body wire harness and antenna LH connectors.

121. Install both I/P outer trim panels.

122. Install the left closeout panel.

123. Install the console.

124. Install the left and right console trim panels

125. Install the right console trim panel.

126. Install the HVAC module to the MOD plate bolts and tighten to 35 inch lbs. (4 Nm).

127. Unplug the A/C components.

128. Install new sealing washers.

129. Install the liquid and suction lines at the TXV.

130. Evacuate and charge the refrigerant system.

131. Leak test the fittings.

132. Install the heater hoses at the heater core.

133. Fill the coolant.

134. Connect the negative battery cable.

135. Enable the SIR system.

136. Perform the power steering control module programming. See STEERING and SUSPENSION.

137. Perform the brake pedal position sensor calibration. See BRAKES.

Cylinder Head

REMOVAL & INSTALLATION

2.4L Engine

1. Relieve the fuel system pressure.
2. Drain the cooling system.
3. Remove or disconnect the following:

- Intake manifold.
- Exhaust manifold
- Timing chain
- Valve cover.
- Cylinder head bolts in the sequence shown.

To install:

4. Clean the gasket surfaces and the bolt holes.

5. Install the cylinder head gasket.

6. Install the cylinder head.

7. Lightly apply clean engine oil to the threads and the bottom side flange of the head bolts and allow the oil to drain before installing.

8. Install new cylinder head bolts.

9. Install and tighten the cylinder head bolts in the sequence shown to 22 ft. lbs. (30 Nm), plus an additional 155 degrees.

10. Install and tighten the 4 front cylinder head bolts and tighten to 26 ft. lbs. (35 Nm).

11. Install the timing chain.

12. Install the exhaust manifold.

13. Install the valve cover.

14. Install the intake manifold.

15. Connect the negative battery cable.

16. Fill and bleed the cooling system.

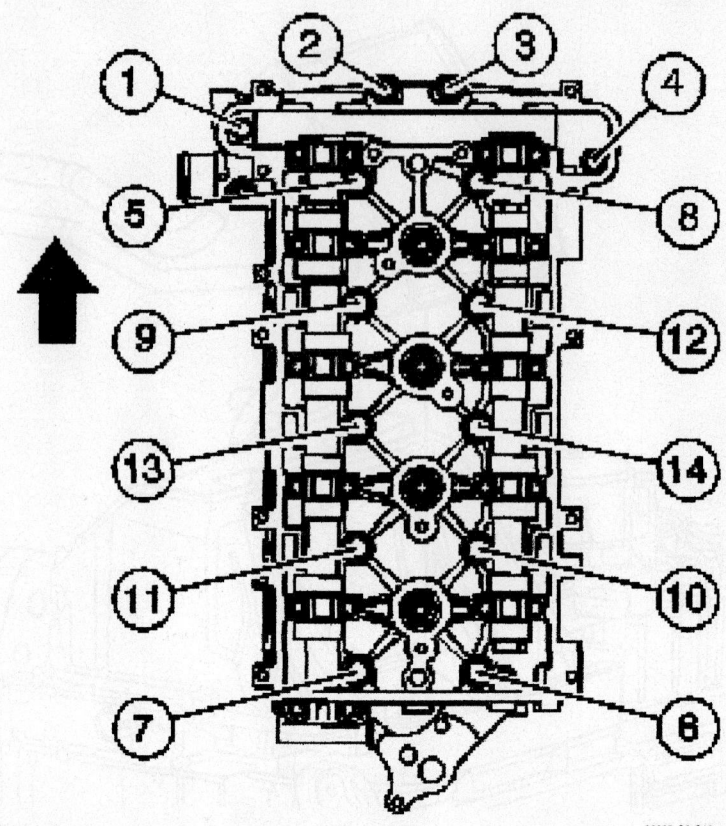

06025-G6-G13

Cylinder head bolt removal sequence—2.4L engine

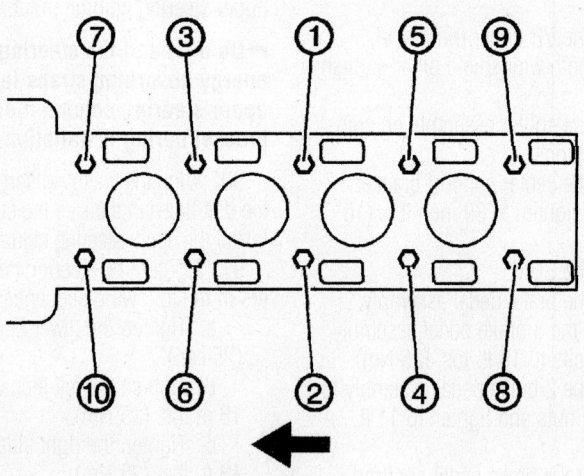

06025-G6-G14

Cylinder head bolt tightening sequence—2.4L engine

17. Start the vehicle, check for leaks and repair if necessary.

3.5L Engine

1. Relieve the fuel system pressure.
2. Drain the engine oil.
3. Drain the cooling system.
4. Remove the upper and lower intake manifolds.

5. Remove the valve covers.

6. Remove the valve rocker arms and the pushrods.

7. Remove the exhaust manifolds.

8. Remove the spark plug wires and spark plugs.

9. On the left side, remove the oil dipstick and tube.

10. On the right side, remove the fuel line bracket and alternator.

11. On both sides, remove the cylinder head bolts and the heads.

12. Clean the gasket mating surfaces.

To install:

13. Using new head gaskets, install the cylinder heads over the locator pins.

14. Install new cylinder head bolts and tighten in sequence to 44 ft. lbs. (60 Nm), plus and additional 95 degrees.

15. On the right side, install the fuel line bracket and alternator.

16. On the left side, install the oil dipstick and tube.

17. Install the spark plug wires and spark plugs.

18. Install the exhaust manifolds.

19. Install the valve rocker arms and the pushrods.

20. Install the valve covers.

21. Install the upper and lower intake manifolds.

22. Connect the negative battery cable.

23. Fill the engine with clean oil.

24. Fill and bleed the cooling system.

25. Connect the negative battery cable.

26. Start the engine and check for leaks and repair if necessary.

3.9L Engine

1. Relieve the fuel system pressure.

2. Drain the engine oil.

3. Drain the cooling system.

4. Remove the upper and lower intake manifolds.

5. Remove the valve covers.

6. Remove the valve rocker arms and pushrods.

7. Remove the exhaust manifolds.

8. Remove the spark plug wires and spark plugs.

9. On the right side, remove the fuel line bracket and the alternator.

10. On the left side, remove the oil dipstick and tube.

11. On both sides, remove and discard the cylinder head bolts and head.

To install:

12. Clean the gasket mating surfaces.

13. Install the cylinder head locator dowel pins, if necessary.

14. Inspect the cylinder head locator dowel pins for proper installation.

15. Install a new cylinder head gasket, noting the identification marks for each side.

16. Install the cylinder head onto the locator pins and the engine.

17. Install new cylinder head bolts finger tight.

18. Tighten the cylinder head bolts in sequence to 44 ft. lbs. (60 Nm), plus an additional 95 degrees.

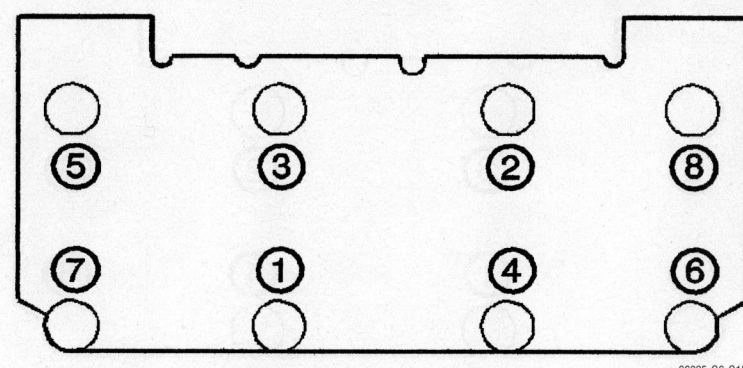

06025-G6-G15

Cylinder head bolt tightening sequence—3.5L engine

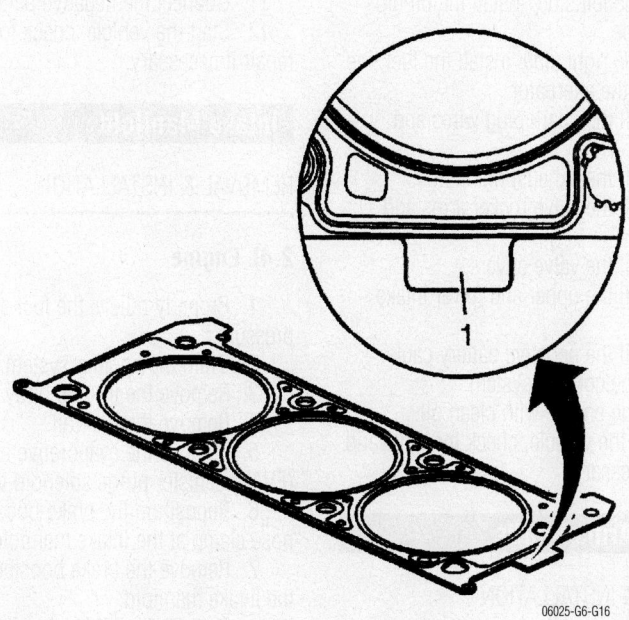

06025-G6-G16

Right side cylinder head gasket identification—3.9L engine

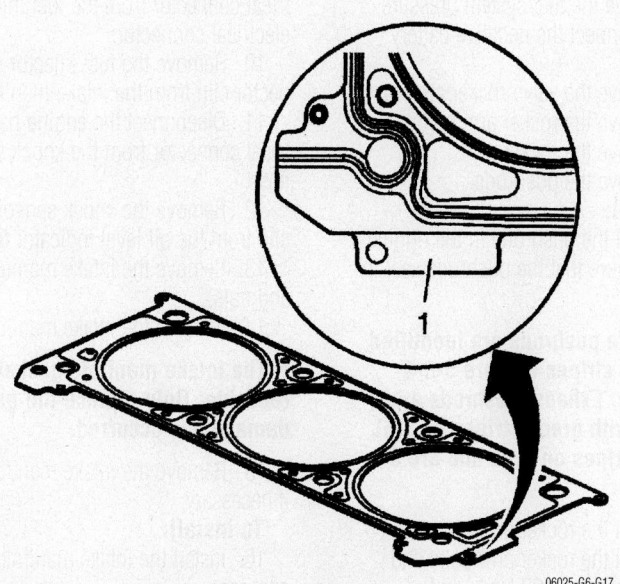

06025-G6-G17

Left side cylinder head gasket identification—3.9L engine

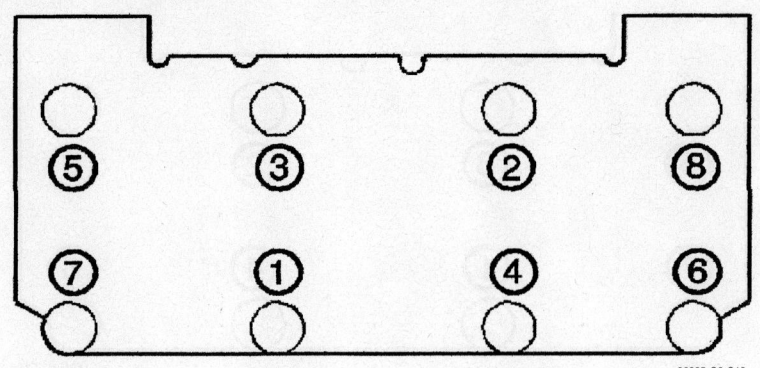

06025-G6-G18

Cylinder head bolt tightening sequence—3.9L engine

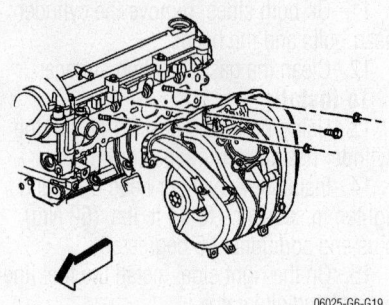

06025-G6-G19

Intake manifold mounting—2.4L engine

19. On the left side, install the oil dipstick and tube.

20. On the right side, install the fuel line bracket and the alternator.

21. Install the spark plug wires and spark plugs.

22. Install the exhaust manifolds.

23. Install the valve rocker arms and pushrods.

24. Install the valve covers.

25. Install the upper and lower intake manifolds.

26. Install the negative battery cable.

27. Fill the cooling system.

28. Fill the engine with clean oil.

29. Start the vehicle, check for leaks and repair if necessary.

Rocker Arms/Shafts

REMOVAL & INSTALLATION

3.5L and 3.9L Engines

1. Relieve the fuel system pressure.

2. Disconnect the negative battery cable.

3. Remove the valve rocker arm covers.

4. Remove the rocker arm bolts.

5. Remove the rocker arms.

6. Remove the pushrods.

To install:

7. Install the pushrods in the original location. Ensure that the pushrods seat in the lifter.

➡**The intake pushrods are identified with yellow stripes and are 5-3/4 inches long. Exhaust pushrods are identified with green stripes on 3.5L and blue stripes on 3.9L and are 6 inches long.**

8. Install the rocker arms.

9. Install the rocker arm bolts and tighten to 24 ft. lbs. (32 Nm).

10. Install the valve rocker covers.

11. Connect the negative battery cable.

12. Start the vehicle, check for leaks and repair if necessary.

Intake Manifold

REMOVAL & INSTALLATION

2.4L Engine

1. Properly relieve the fuel system pressure.

2. Drain the cooling system.

3. Remove the throttle body.

4. Remove the fuel rail.

5. Remove the evaporative emission (EVAP) canister purge solenoid valve tube.

6. Reposition the brake booster vacuum hose clamp at the intake manifold.

7. Remove the brake booster hose from the intake manifold.

8. Remove the oil level indicator tube bolt.

9. Disconnect the engine harness electrical connector from the fuel injector inline electrical connector.

10. Remove the fuel injector inline connector clip from the intake manifold.

11. Disconnect the engine harness electrical connector from the knock sensor harness.

12. Remove the knock sensor connector clip from the oil level indicator tube.

13. Remove the intake manifold bolts and nuts.

14. Remove the intake manifold.

➡**The intake manifold gasket is reusable. Only replace the gasket if damage has occurred.**

15. Remove the intake manifold gasket, if necessary.

To install:

16. Install the intake manifold gasket, if necessary.

17. Install the intake manifold.

18. Install the intake manifold bolts and nuts and tighten to 89 inch lbs. (10 Nm).

19. Connect the engine harness electrical connector to the knock sensor harness.

20. Install the knock sensor connector clip to the oil level indicator tube.

21. Connect the engine harness electrical connector to the fuel injector inline electrical connector.

22. Install the fuel injector inline connector clip to the intake manifold.

23. Install the oil level indicator tube bolt.

24. Install the brake booster hose to the intake manifold.

25. Position the brake booster vacuum hose clamp at the intake manifold.

26. Install the EVAP canister purge solenoid valve tube.

27. Install the fuel rail.

28. Install the throttle body.

29. Connect the negative battery cable

30. Fill and bleed the cooling system.

31. Start the vehicle, check for leaks and repair if necessary.

3.5L Engine

UPPER MANIFOLD

1. Release the fuel system pressure.

2. Drain the cooling system.

3. Remove the intake manifold cover.

4. Remove the vacuum hoses from the following:

  a. Evaporative emissions (EVAP) canister purge valve

  b. Manifold vacuum source

  c. Brake booster

  d. Heater and air conditioning (A/C) source

5. Disconnect the electrical connectors from the following:

  a. Exhaust Gas Recirculation (EGR) valve

  b. Mass Air Flow (MAF) sensor

  c. Intake Air Temperature (IAT) sensor

  d. Electronic throttle control

  e. EVAP canister purge valve

6. Remove the air cleaner intake duct.

7. Remove the left side spark plug wires from the spark plugs.

8. Remove the following wiring harnesses from the retainers:

   a. Camshaft position (CMP) sensor wiring harness

   b. Left side spark plug wire harness

   c. Engine wiring harness

9. Remove the ignition coil bracket with the coils.

10. Remove the EVAP canister purge solenoid valve.

11. Remove the Manifold Absolute Pressure (MAP) sensor and the bracket.

12. Remove the EGR valve.

13. Remove the upper intake manifold bolts and the stud.

14. Remove the upper intake manifold.

15. Remove the upper intake manifold gaskets.

To install:

16. Install the throttle body.

17. Install the upper intake manifold gaskets.

18. Install the upper intake manifold.

19. Install the right upper intake manifold bolts and the stud and tighten to 18 ft. lbs. (25 Nm).

20. Install the EGR valve.

21. Install the MAP sensor bracket and the sensor.

22. Install the EVAP canister purge solenoid valve.

23. Install the ignition coil bracket with the coils.

24. Install the following wiring harnesses to the retainers:

   a. Engine wiring harness

   b. Left side spark plug wire harness

   c. CMP sensor wiring harness

25. Install the left side spark plug wires to the spark plugs.

26. Install the air cleaner intake duct.

27. Connect the electrical connectors to the following:

   a. EVAP canister purge valve

   b. Electronic throttle control

   c. IAT sensor

   d. MAF sensor

   e. EGR valve

28. Install the vacuum hoses to the following: Heater and A/C source, brake booster, manifold vacuum, and EVAP canister purge valve.

29. Connect the negative battery cable.

30. Fill the cooling system.

31. Install the intake manifold cover.

LOWER MANIFOLD

1. Release the fuel system pressure.

2. Drain the cooling system.

3. Remove the upper intake manifold.

4. Remove the valve covers.

5. Disconnect the engine coolant temperature (ECT) wiring harness.

6. Disconnect and remove the fuel injector and manifold air pressure (MAP) wiring harness.

7. Remove the fuel injector rail.

8. Disconnect the heater inlet pipe with heater hose from the lower intake manifold and reposition.

9. Disconnect the radiator inlet hose from the engine.

10. Remove the thermostat housing and thermostat.

11. Remove the lower intake manifold bolts.

12. Remove the lower intake manifold.

13. Remove the valve rocker arms and pushrods.

14. Remove the lower intake manifold gaskets and seals.

15. Clean the lower intake manifold gasket and seal surfaces on the cylinder heads and the engine block.

16. Clean the gasket and seal surfaces on the lower intake manifold with degreaser.

17. Remove all the loose RTV sealer.

To install:

18. Install the lower intake manifold gaskets.

19. Install the valve rocker arms and pushrods.

20. With gaskets and seals in place apply a small drop 8-10 mm (0.31-0.39 in) of RTV sealer to the 4 corners of the intake manifold to block joints.

21. Install the lower intake manifold.

22. Apply sealer to the lower intake manifold bolt threads.

23. Install the lower intake manifold bolts and tighten in sequence as follows:

   a. Tighten the bolts in sequence to 115 inch lbs. (13 Nm),

   b. Tighten bolts (1, 2, 3, 4) in sequence to 15 ft. lbs. (20 Nm on the final pass.

   c. Tighten bolts (5, 6, 7, 8) in sequence to 18 ft. lbs. (25 Nm (18 lb ft) on the final pass.

24. Install the heater inlet pipe and tighten to 18 ft. lbs. (25 Nm).

25. Install the thermostat and housing.

26. Install the ECT sensor.

27. Connect the thermostat bypass hose to the thermostat bypass pipe and lower intake manifold pipe.

28. Connect the radiator inlet hose to the engine.

29. Connect the heater inlet pipe and heater hose to the lower intake manifold.

30. Install the power steering pump to the front engine cover.

31. Install the fuel injector rail.

32. Connect the fuel feed pipe to the fuel injector rail.

33. Connect the fuel injector and MAP wiring harness.

34. Connect the ECT wiring harness.

35. Install the valve rocker arm covers.

36. Install the upper intake manifold.

37. Connect the negative battery cable.

3.9L Engine

UPPER MANIFOLD

1. Drain the cooling system.

2. Relieve the fuel system pressure.

3. Remove the intake manifold cover.

4. Disconnect the fuel feed pipe quick connect fitting from the fuel rail.

5. Disconnect the evaporative (EVAP) emission pipe from the purge solenoid.

6. Open the retaining clip, and remove the fuel and EVAP pipes from the clip.

7. Remove the positive crankcase ventilation (PCV) air tubes.

8. Reposition the brake booster vacuum hose clamp at the intake manifold.

9. Remove the vacuum hose from the intake manifold.

10. Reposition the radiator surge tank inlet hose clamp.

11. Remove the radiator surge tank inlet hose from the inlet pipe.

12. Remove the oil fill neck.

13. Remove the radiator surge tank inlet pipe bolts.

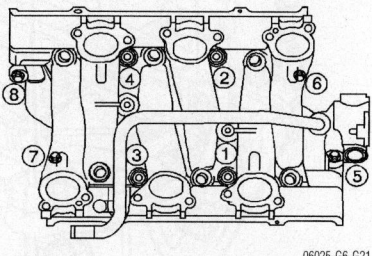

06025-G6-G20

Upper intake manifold mounting—3.5L engine

06025-G6-G21

Lower intake manifold tightening sequence —3.5L engine

14. Remove the radiator surge tank inlet pipe.

15. Disconnect the manifold absolute pressure (MAP) sensor electrical connector.

16. Disconnect the evaporative emission (EVAP) canister purge solenoid electrical connector.

17. Disconnect the electronic throttle control (ETC) electrical connector.

18. Disconnect the inlet manifold valve electrical connector>

19. Remove the air cleaner outlet duct.

20. Disconnect the left side spark plug wires from the spark plugs.

21. Disconnect the left side spark plug wires from the ignition coil.

22. Disengage the spark plug wire retainer clips from the intake manifold bracket and the heater inlet/outlet pipe.

23. Remove the left side spark plug wires.

24. Remove the heater inlet and outlet pipe nuts from the throttle body studs.

25. Remove the inlet and outlet pipe from the studs.

26. Remove the 2 ignition coil bolts.

27. Remove the generator upper bolt, ball stud and rear brace.

28. Remove the upper intake manifold bolts and stud.

29. Separate and remove the upper intake manifold from the lower intake manifold.

30. Remove the upper to lower intake manifold gaskets.

31. Remove the inlet manifold tuning valve bolts and valve.

32. Remove the throttle body bolts/studs and throttle body.

33. Remove the MAP sensor bracket and sensor.

34. Remove the EVAP canister purge solenoid valve bolt and valve.

35. Clean the upper intake to lower intake gasket mating surfaces.

36. Inspect the intake manifold tuning valve seal for damage. The tuning valve

blade attachment to the motor should be tight, with no looseness or slack present, Replace as necessary.

37. Apply lubricant to the nose of the tuning valve blade.

To install:

38. Inspect the EVAP canister purge solenoid valve seal for damage, replace as necessary.

39. Install the EVAP canister purge solenoid valve and tighten to 12 ft. lbs. (16 Nm).

40. Install the MAP sensor and bracket.

41. Apply threadlock to the throttle body bolts/studs threads and install the throttle body.

42. Install the inlet manifold tuning valve and tighten to 89 inch lbs. (10 Nm).

43. Install new upper-to-lower intake manifold gaskets.

44. Apply threadlock to the upper intake manifold bolts/stud threads.

45. Install the upper intake manifold and tighten the bolts to 18 ft. lbs. (25 Nm).

46. Install the alternator.

47. Install the ignition coils.

48. Install the heater inlet and outlet pipes to the throttle body studs.

49. Install the left side spark plug wires.

50. Connect the left side spark plug wires to the spark plugs.

51. Connect the left side spark plug wires to the ignition coil.

52. Engage the spark plug wire retainer clips to the intake manifold bracket and the heater inlet/outlet pipe.

53. Install the air cleaner outlet duct.

54. Connect the inlet manifold valve electrical connector.

55. Connect the electronic throttle control (ETC) electrical connector.

56. Connect the evaporative emission (EVAP) canister purge solenoid electrical connector.

57. Connect the manifold absolute pressure (MAP) sensor electrical connector.

58. Install the radiator surge tank inlet pipe.

59. Install the radiator surge tank inlet pipe bolts.

60. Install the oil fill neck.

61. Install the radiator surge tank inlet hose to the inlet pipe.

62. Position the radiator surge tank inlet hose clamp.

63. Install the vacuum hose to the intake manifold.

64. Position the brake booster vacuum hose clamp at the intake manifold.

65. Install the positive crankcase ventilation (PCV) air tubes.

66. Install the fuel and EVAP pipes to the retaining clip.

67. Connect the evaporative (EVAP) emission pipe to the purge solenoid.

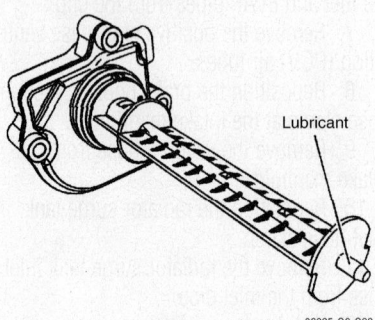

06025-G6-G22

Applying lubricant to intake manifold tuning blade—3.9L engine

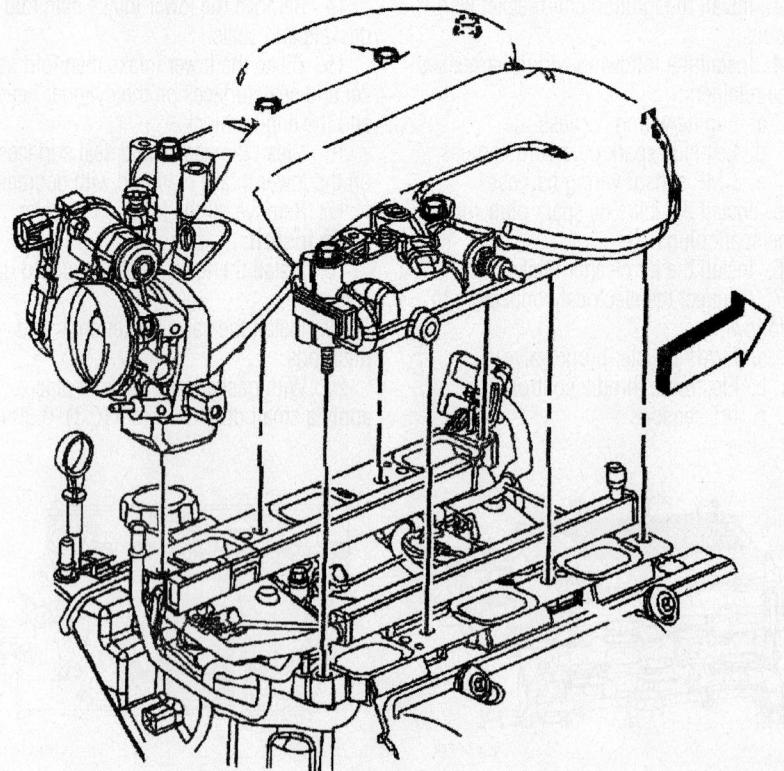

06025-G6-G23

Upper intake manifold mounting—3.9L engine

68. Connect the fuel feed pipe quick connect fitting to the fuel rail.

69. Remove the intake manifold cover.

70. Fill the cooling system.

LOWER MANIFOLD

1. Drain the cooling system.
2. Relieve the fuel system pressure.
3. Remove the coolant crossover pipe.
4. Remove the upper intake manifold.
5. Remove the valve covers.
6. Disconnect the engine coolant temperature electrical connector.
7. Disconnect the fuel feed line from the fuel rail.
8. Disconnect the fuel injector inline connector.
9. Remove the fuel injector harness connector bracket bolt from the intake manifold.
10. Disconnect the Camshaft Position (CMP) sensor electrical connector.
11. Remove the fuel rail.
12. Remove the lower intake manifold.
13. Loosen the valve rocker arm bolts.

➡**Place the valve train components in a rack in order to ensure that the components are installed in the same location from which they where removed.**

14. Remove the valve rocker arms.
15. Remove the push rods.

➡**The intake push rods measure 5.81 inches (147.51mm). The exhaust push rods measure 6.1 inches (154.87mm).**

16. Remove the lower intake manifold gaskets and seals.
17. Clean the lower intake manifold gasket and seal surfaces on the cylinder heads and the engine block.
18. Clean the gasket and seal surfaces on the lower intake manifold with degreaser.
19. Remove all the loose RTV sealer.

To install:

➡**Do not use RTV sealer under the lower intake manifold gaskets.**

20. Install the lower intake manifold gaskets and seals.
21. Coat the ends of the push rods using suitable prelube.
22. Install the push rods in their original location.
23. Coat the rocker arm friction surfaces using suitable prelube.
24. Install the valve rocker arms in their original positions.
25. Install the valve rocker arm bolts and tighten to 25 ft. lbs. (34 Nm).
26. With new gaskets and seals in place, apply a small drop of RTV sealer to the 4

corners of the intake manifold to engine block joints.

27. Install the lower intake manifold.

28. Apply sealer to the lower intake manifold bolt threads.

29. Tighten the bolts in sequence as follows:

   a. Tighten bolts 1, 2, 3, and 4 in sequence to 12 ft. lbs. (16 Nm).

   b. Tighten bolts 5, 6, 7 and 8 in sequence to 18 ft. lbs. (25 Nm).

30. Inspect the fuel rail, fuel injectors and fuel injector O-rings for damage and replace as necessary.

31. Lubricate the fuel injector O-rings.

32. Install the injector nozzles into the lower intake manifold injector bores.

33. Press on the injector rail using the palms of both hands until the injector are fully seated.

34. Install the fuel injector rail bolts and tighten to 89 inch lbs. (10 Nm).

35. Connect the CMP sensor electrical connector.

36. Position the fuel injector harness connector bracket to the intake manifold.

37. Install the fuel injector harness connector bracket bolt.

38. Connect the fuel injector inline connector.

39. Connect the fuel feed line to the fuel rail.

40. Connect the ECT electrical connector.

41. Install the valve covers.

42. Install the upper intake manifold.

43. Install the coolant crossover pipe.

44. Fill the cooling system.

Exhaust Manifold

REMOVAL & INSTALLATION

2.4L Engine

1. Remove the exhaust manifold heat shield.
2. Remove the block heater, if equipped.
3. Remove the oxygen sensor.
4. Remove the exhaust manifold.

To install:

5. Clean the gasket mating surface and install a new exhaust manifold gasket.
6. Install new exhaust manifold studs and tighten to 89 inch lbs. (10 Nm).
7. Install the exhaust manifold and finger tighten the bolts.
8. Tighten the bolts in sequence to 10 ft. lbs. (14 Nm).
9. Coat the threads of the oxygen sensor with anti-seize.
10. Install the oxygen sensor and tighten to 31 ft. lbs. (42 Nm).
11. Install the exhaust manifold heat shield.

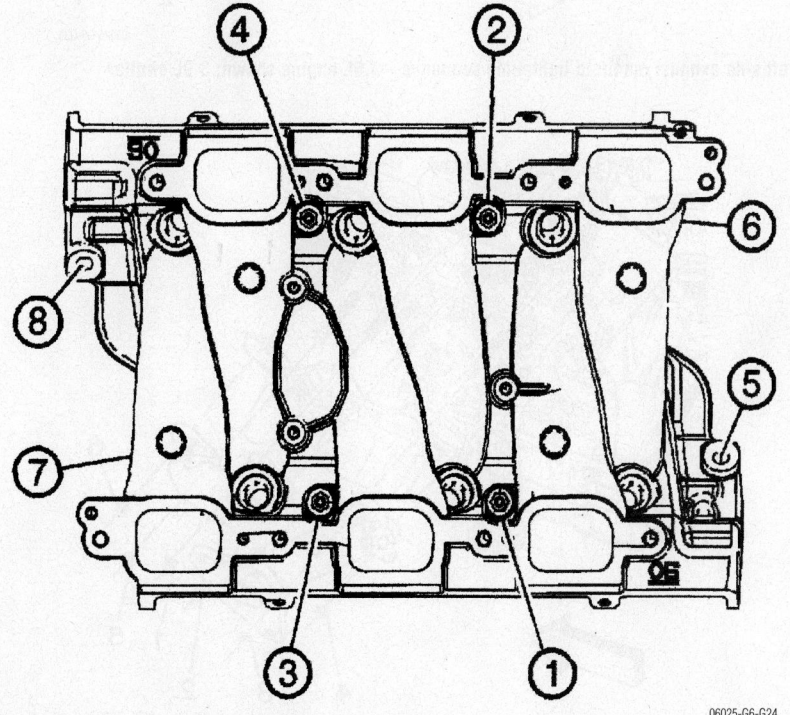

06025-G6-G24

Lower intake manifold tightening sequence—3.9L engine

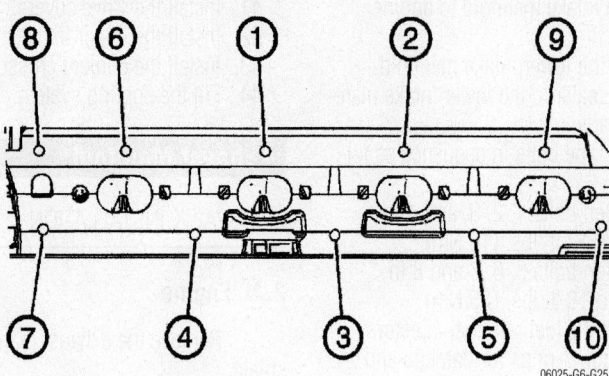

Exhaust manifold tightening sequence—2.4L engine

06025-G6-G25

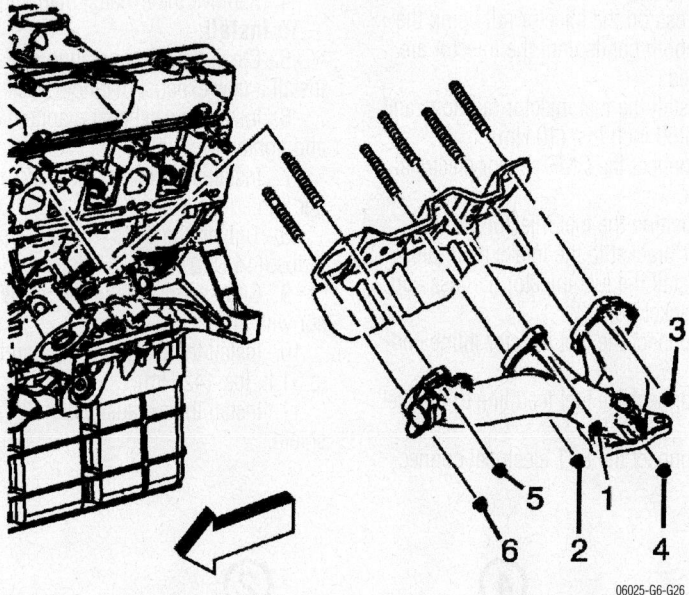

Left side exhaust manifold tightening sequence—3.5L engine shown; 3.9L similar

06025-G6-G26

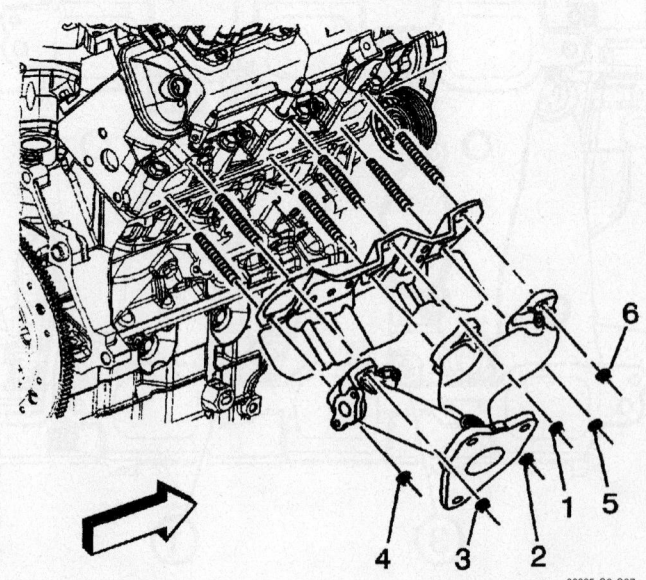

Right side exhaust manifold tightening sequence—3.5L engine shown; 3.9L similar

06025-G6-G27

3.5L and 3.9L Engines

➡️ **If necessary, soak the exhaust pipe retaining nuts with penetrating oil to loosen them.**

1. Remove the heated oxygen sensor.
2. Remove the spark plugs.
3. On the left side, remove the EGR pipe from the exhaust manifold, if equipped.
4. Remove the exhaust manifold heat shield.
5. Remove the exhaust manifold and gasket.

To install:

6. Install the exhaust manifold gasket.
7. Install the exhaust manifold and tighten the nuts in sequence to 12 ft. lbs. (16 Nm) on 3.5L or 15 ft. lbs. (20 Nm) on 3.9L.
8. Install the exhaust manifold heat shield.
9. On the left side, install the EGR pipe to the exhaust manifold, if equipped.
10. Install the spark plugs.
11. Install the heated oxygen sensor.

Camshaft and Valve Lifters

REMOVAL & INSTALLATION

2.4L Engine

1. Relieve the fuel system pressure.
2. Disconnect the negative battery cable.
3. Remove the valve cover.
4. Remove the intake and/or exhaust camshaft position actuator.

➡️ **Remove each bolt on each cap one turn at a time until there is no spring tension pushing on the camshaft.**

5. Mark the bearing caps to ensure they are installed in the original position.
6. Remove the bearing cap bolts and the bearing caps.
7. Remove the exhaust and/or intake camshaft.

➡️ **Keep all of the roller followers and hydraulic adjusters in order so that they can be reinstalled in their respective locations.**

8. Remove the camshaft roller followers.
9. Remove the lash adjusters.

To install:

10. Install the hydraulic element adjusters into their bores in the cylinder head.
11. Lubricate the hydraulic lash adjusters and valve tips.

12. Position the roller followers on the tip of the valve stem and on the lash adjuster.

13. Lubricate the camshaft(s) and install.

14. Install the camshaft bearing caps. Hand tighten the cap bolts.

15. Tighten the bearing cap bolts in increments of 3 turns to 89 inch lbs. (10 Nm) until they are seated.

16. Install the camshaft actuator(s).

17. Install the valve cover.

3.5L and 3.9L Engines

1. Drain the engine oil and cooling system.

2. Relieve the fuel system pressure.

3. To remove the valve lifters, remove the valve covers.

4. Remove the intake manifolds.

5. Remove the rocker arm bolts.

6. Remove the rocker arms.

7. Remove the pushrods.

8. Remove the intake manifold oil splash shield.

9. Remove the lifter guide bolts.

10. Remove the valve lifter guides.

11. Remove the valve lifters.

12. Clean all gasket surfaces with degreaser.

13. Clean the valve train parts.

14. Inspect the valve lifters and the cam lobes for wear.

15. To remove the camshaft, remove the engine front cover, timing chain and sprockets.

16. Remove the camshaft position sensor.

17. Remove the camshaft thrust plate.

18. Install the camshaft sprocket bolt into the camshaft. Tighten finger tight only.

19. Carefully rotate and remove the camshaft from the engine block.

   To install:

20. Coat the camshaft journals with clean engine oil.

21. Coat the camshaft lobes with pre-lube.

22. Install the camshaft sprocket bolt into the camshaft. Tighten finger tight only.

23. Carefully rotate and install the camshaft to the engine block.

24. Install the camshaft thrust plate and tighten to 89 inch lbs. (10 Nm).

25. Install the camshaft position sensor and tighten to 89 inch lbs. (10 Nm).

26. Install the timing chain, sprockets and front cover.

27. Coat the valve lifters with prelube and install them.

28. Install the valve lifter guides and tighten to 89 inch lbs. (10 Nm).

29. Install the intake manifold oil splash shield.

30. Install the pushrods in their original locations.

➡**The intake pushrods are identified with yellow stripes and are 5-3/4 inches long. Exhaust pushrods are identified with green stripes and are 6 inches long.**

31. Install the rocker arms and tighten to 24 ft. lbs. (32 Nm).

32. Install the intake manifolds.

33. Install the valve covers.

34. Connect the negative battery cable.

35. Fill the cooling system.

36. Start the vehicle, check for leaks and repair if necessary.

Valve Lash

ADJUSTMENT

Hydraulic lash adjusters are used on all engines and no adjustment is necessary.

Starter Motor

REMOVAL & INSTALLATION

2.4L Engine

1. Remove or disconnect the following:
   - Negative battery cable
   - Solenoid terminal nut
   - Starter engine harness lead
   - Positive battery cable
   - Starter mounting bolts
   - Starter.

   To install:

2. Install or connect the following:
   - Starter. Torque the bolts to 30 ft. lbs. (40 Nm).
   - Positive battery cable
   - Starter engine harness lead
   - Solenoid terminal nut
   - Negative battery cable

3.5L and 3.9L Engines

1. Remove or disconnect the following:
   - Negative battery cable

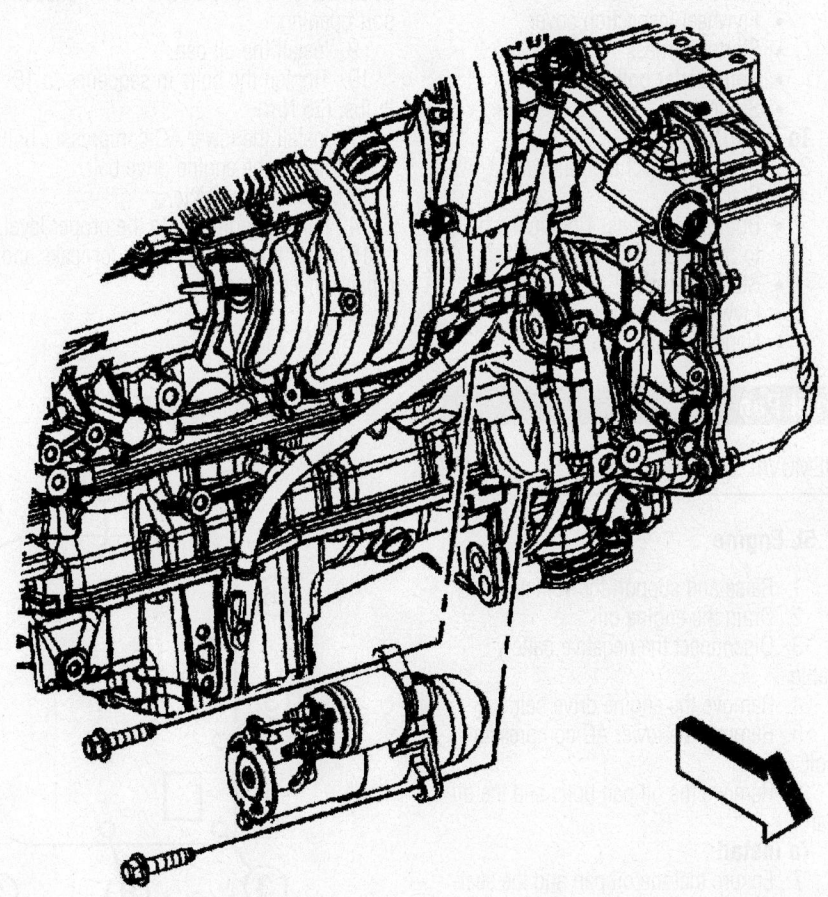

Starter mounting—2.4L engine

06025-G6-G28

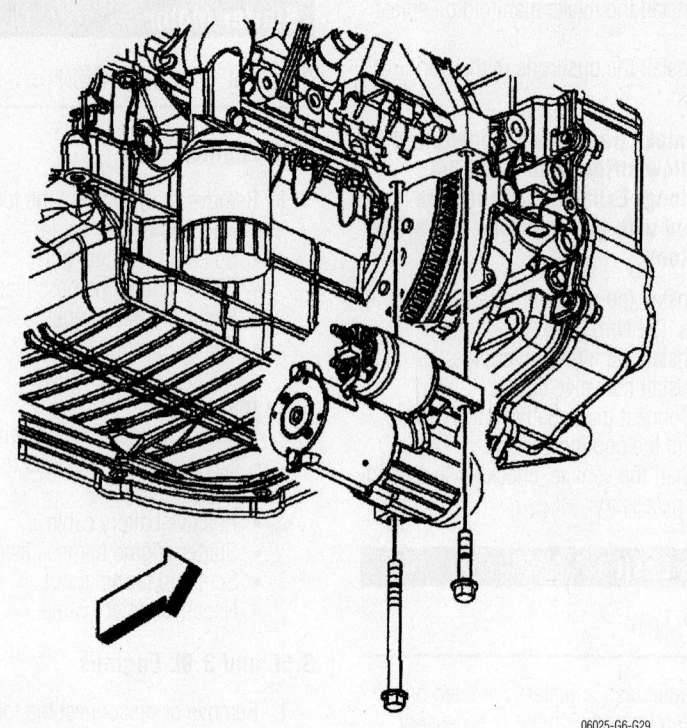

Starter mounting—3.5L and 3.9L engines

- Raise the vehicle
- Flywheel inspection cover
- Starter harness
- Both starter bolts
- Starter.

To install:

2. Install or connect the following:
- Starter
- Both starter bolts. Tighten the bolts to 30 ft. lbs. (40 Nm).
- Starter harness
- Flywheel inspection cover
- Negative battery cable

Oil Pan

REMOVAL & INSTALLATION

2.5L Engine

1. Raise and support the vehicle safely.
2. Drain the engine oil.
3. Disconnect the negative battery cable.
4. Remove the engine drive belt.
5. Remove the lower AC compressor bolt.
6. Remove the oil pan bolts and the oil pan.

To install:

7. Ensure that the oil pan and the sealing surface on the lower crankcase are free of all oil and debris.
8. Apply a bead of sealant around the perimeter of the oil pan and the oil suction port opening.
9. Install the oil pan.
10. Tighten the bolts in sequence to 18 ft. lbs. (25 Nm).
11. Install the lower AC compressor bolt.
12. Install the engine drive belt.
13. Lower the vehicle.
14. Fill the engine oil to the proper level.
15. Start the vehicle, check for leaks and repair if necessary.

3.5L and 3.9L Engines

1. Disconnect the negative battery cable.
2. Raise and support the vehicle.
3. Drain the crankcase.
4. Remove the right front tire and wheel.
5. Remove the right front splash shield.
6. Remove the oil filter and filter adapter.
7. Remove the catalytic converter.
8. Remove the wheel speed sensor harness from the right suspension support.
9. Remove the right front ball joint, bolt, and nut. Separate the ball joint from the steering knuckle.
10. Remove lower closeout panel.
11. Remove the air conditioning (A/C) compressor bolts and position the compressor aside.
12. Remove the braces that support the engine to the transmission.
13. Disconnect the oil level sensor.
14. Remove the retainers that secure the brake line to the frame.
15. Support the engine with a tall jack stand and a block of wood.
16. Remove the right side engine mount nuts and bolts.
17. Loosen the left side cradle bolts.
18. Remove the cradle bolts from the right front and the right rear.
19. Remove the starter.
20. Remove the oil pan side bolts.
21. Remove the oil pan bolts and oil pan.

To install:

22. Apply sealer to both sides of the crankshaft rear main bearing cap. Press the sealer into the gap using a putty knife.
23. Install a new oil pan gasket.

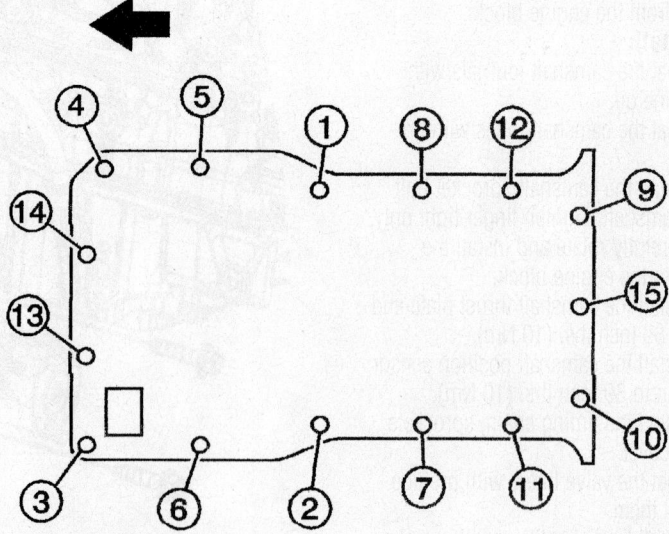

Oil pan tightening sequence—2.4L engine

24. Position the oil pan to the engine, install the bolts and tighten to 18 ft. lbs. (25 Nm).

25. Install the oil pan side bolts and tighten to 37 ft. lbs. (50 Nm).

26. Install the starter.

27. Install the flywheel inspection cover.

28. Install the cradle bolts to the right front and the right rear.

29. Tighten the left side cradle bolts to 74 ft. lbs. (100 Nm), plus an additional 90 degrees.

30. Install the right side engine mount and tighten to 37 ft. lbs. (50 Nm).

31. Remove the support from the engine.

32. Install the retainers that secure the brake line to the frame.

33. Connect the oil level sensor electrical connector.

34. Install the braces that support the engine to the transmission.

35. Position and install the A/C compressor bolts.

36. Install lower closeout panel.

37. Install the right front ball joint, bolt, and nut.

38. Install the wheel speed sensor harness to the right suspension support.

39. Install the right front splash shield.

40. Install the right front tire and wheel.

41. Lower the vehicle.

42. Fill the engine with clean oil.

43. Start the vehicle and check for leaks.

Oil Pump

REMOVAL & INSTALLATION

2.4L Engine

➡The oil pump is mounted into the engine front cover.

1. Relieve the fuel system pressure.

2. Drain the engine oil.

3. Remove the engine front cover.

4. Remove the oil pump cover.

5. Remove the oil pump from the inside of the front cover.

To install:

6. Coat the oil pump gears with oil and fit the pump to the cover.

7. Install the oil pump cover.

8. Install the engine front cover.

9. Fill the engine with clean oil.

10. Start the vehicle, check for leaks and repair if necessary.

3.5L and 3.9L Engines

1. Remove the oil pan.

2. Remove the oil pump bolt.

3. Remove the oil pump and the oil pump drive shaft.

To install:

4. Install the oil pump drive shaft and the oil pump.

5. Install the oil pump bolt and tighten to 30 ft. lbs. (40 Nm).

6. Install the oil pan.

7. Fill the engine with clean oil.

8. Start the vehicle, check for leaks and repair if necessary.

Rear Main Seal

REMOVAL & INSTALLATION

2.4L, 3.5L and 3.9L Engines

1. Remove the transmission.

2. Hold the crankshaft balancer and remove the flywheel.

3. Carefully pry the seal out of the retainer without damaging the crankshaft or the seal retainer.

To install:

4. Lubricate the seal with clean engine oil.

5. Install the seal into the retainer using the appropriate seal installer.

6. Install the flywheel and tighten the bolts to 39 ft. lbs. (53 Nm) plus an additional 25 degrees on 2.5L, or 52 ft. lbs. (71 Nm) on 3.5L and 3.9L.

7. Install the transmission.

Timing Chain, Sprockets, Front Cover and Seal

REMOVAL & INSTALLATION

2.4L Engine

1. Relieve the fuel system pressure.

2. Raise and support the vehicle safely.

3. Remove the drive belts.

➡If only the crankshaft front seal is being replaced, go to steps 9, 10 and 11.

4. Drain the cooling system.

5. Drain the engine oil.

6. Remove the no. 1 cylinder spark plug.

7. Rotate the crankshaft in the engine rotational direction clockwise, until the no. 1 piston is at TDC on the compression stroke.

8. Remove the camshaft cover.

9. Hold the crankshaft from turning and remove and discard the crankshaft balancer bolt.

10. Remove the crankshaft balancer.

11. If only the front seal is being replaced, use a flat-bladed too and pry the oil seal from the front cover. Press in a new seal with a seal installer. Install the crankshaft balancer using a new bolt and tighten to 74 ft. lbs. (100 Nm) plus an additional 125 degrees.

12. Remove the drive belt tensioner.

13. Remove the engine front cover-to-water pump bolt.

14. Remove the remaining engine front cover bolts.

15. Remove the engine front cover.

16. Remove the upper timing chain guide bolts and guide.

17. Remove the timing chain tensioner.

18. Install a 24 mm wrench on the hex on the exhaust camshaft in order to hold the camshaft.

19. Remove and discard the exhaust camshaft actuator bolt.

20. Remove the exhaust camshaft actuator from the camshaft and timing chain.

21. Remove the timing chain tensioner guide bolt and guide.

22. Remove the fixed timing chain guide access plug.

23. Remove the fixed timing chain guide bolts and guide.

24. Install a 24 mm wrench on the hex on the intake camshaft in order to hold the camshaft.

25. Remove and discard the intake camshaft actuator bolt.

26. Remove the intake camshaft actuator, and the timing chain through the top of the cylinder head.

27. Remove the timing chain crankshaft sprocket.

28. Remove the balance shaft drive chain tensioner bolts and tensioner.

29. Remove the adjustable balance shaft chain guide bolt and guide.

30. Remove the small balance shaft drive chain guide bolts and guide.

31. Remove the upper balance shaft drive chain guide bolts and guide.

32. Remove the balance shaft drive chain.

33. Remove the balance shaft drive sprocket.

To install:

34. Be sure all sealing surfaces are clean and prepared for assembly.

✳✳ CAUTION

If the balance shafts are not properly timed to the engine, the engine may vibrate or make noise.

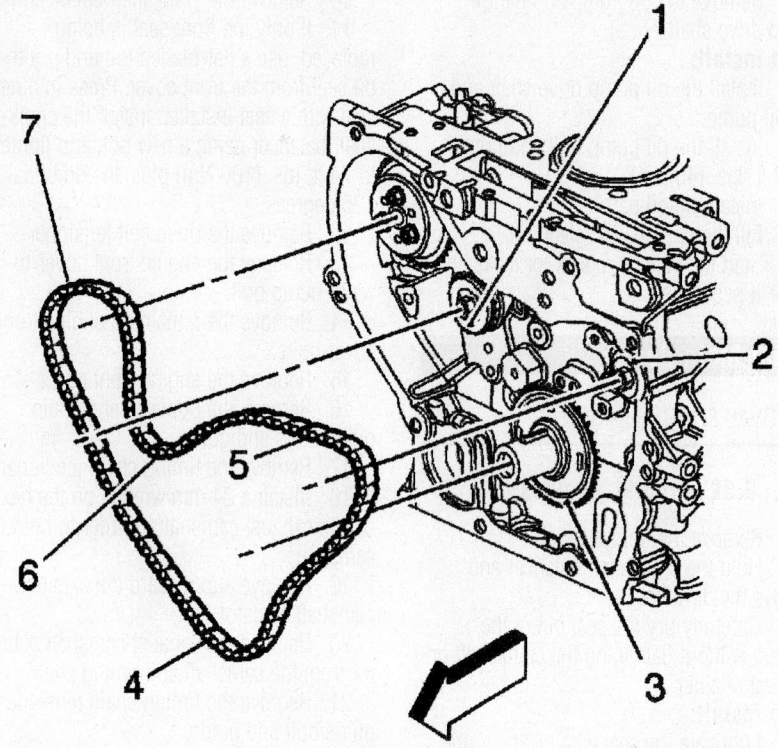

Installing balance shaft timing chain—2.4L engine

06025-G6-G31

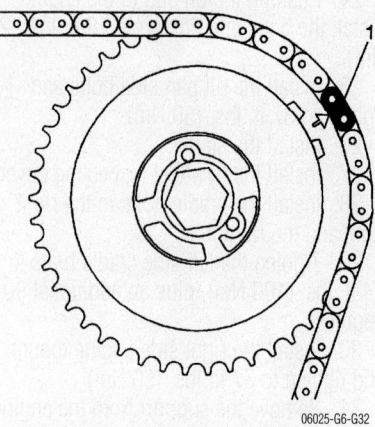

Aligning timing chain link with the intake camshaft actuator—2.4L engine

06025-G6-G32

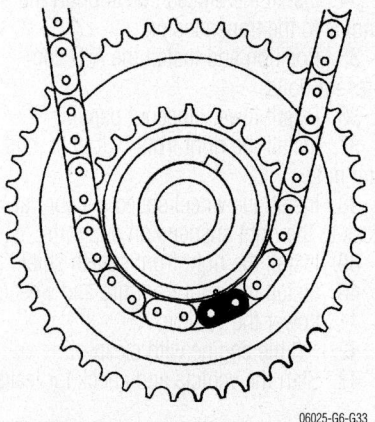

Aligning timing chain link with the crank-shaft sprocket—2.4L engine

06025-G6-G33

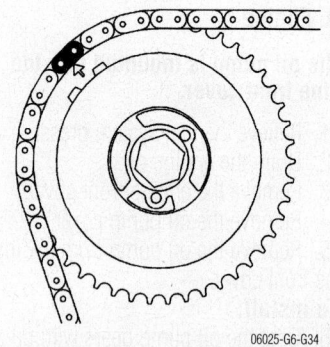

Aligning timing chain link with the exhaust camshaft actuator—2.4L engine

06025-G6-G34

35. Install the balance shaft drive chain (1) with the colored link lined up with the marks on the balance shaft sprockets and the balance shaft drive sprocket. There are three colored links on the chain. Two are chrome and one is copper. Use the following steps in order to line up the links with the sprockets.

a. Place the copper link (5) so that it lines up with the timing mark (2) on the intake side balance shaft sprocket.

b. Working clockwise around the chain, place the chrome link (4) in line with the timing mark (3) on the balance shaft drive sprocket. (approximately 6 o'clock position on the sprocket).

c. Place the chain (7) on the water pump drive sprocket. The alignment is not critical.

d. Align the last chrome link (6) with the timing mark (1) on the exhaust side balance shaft drive sprocket.

36. Install the upper balance shaft drive chain guide and bolts and tighten to 11 ft. lbs. (15 Nm).

37. Install the small balance shaft drive chain guide and bolts and tighten to 11 ft. lbs. (15 Nm).

38. Install the adjustable balance shaft chain guide and bolt and tighten to 89 inch lbs. (10 Nm).

39. Reset the timing chain tensioner by performing the following steps:.

a. Rotate the tensioner plunger 90 degrees in its bore and compress the plunger.

b. Rotate the tensioner back to the original 12 o' clock position and insert a paper clip through the hole in the plunger body and into the hose in the tensioner plunger.

40. Install the balance shaft drive chain tensioner and bolts and tighten to 89 inch lbs. (10 Nm).

41. Remove the paper clip from the balance shaft drive chain tensioner.

42. Install the timing chain crankshaft sprocket to the crankshaft with the timing mark in the 5 o'clock position,

➡There are 3 colored links on the timing chain. Two links are pink in color and one link is blue in color. Use the following procedure to line up the links with the sprockets. Orient the chain so that the colored links are visible.

43. Assemble the intake camshaft actuator to the timing chain with the timing mark lined up with the blue colored link (1). Install and hand tighten a new intake camshaft actuator bolt.

44. Lower the timing chain through the opening in the cylinder head. Use care to ensure that the chain goes around both sides of the cylinder block bosses.

45. Route the timing chain around the

crankshaft sprocket and line up the first pink link with the timing mark on the crankshaft sprocket, in approximately the 5 o'clock position.

46. Install the adjustable timing chain guide through the opening in the cylinder head. Install the adjustable timing chain bolt and tighten to 89 inch lbs. (10 Nm).

47. Install the exhaust camshaft actuator and a new bolt loosely onto the exhaust camshaft.

48. Align the timing mark on the actuator with the last pink colored link. Tighten the bolt finger tight.

49. If the camshaft is 180 degrees out of time, use a 24-mm wrench, first turn the intake camshaft until the alignment feature on the back of the camshaft actuator seats in the notch in the front of the intake camshaft.

50. Turn the crankshaft 45 degrees in either direction.

51. Turn the intake cam to the appropriate location.

52. Turn the crankshaft back to TDC.

53. When the actuator seats on the cam, tighten the actuator bolt hand tight.

54. Verify that all of the colored links and the appropriate timing marks are still aligned. If they are not, repeat the portion of the procedure necessary to align the timing marks.

55. Install the fixed timing chain guide and bolts and tighten to 89 inch lbs. (10 Nm).

56. Install the upper timing chain guide and bolts and tighten to 89 inch lbs. (10 Nm).

57. Install a 24 mm wrench onto the hex on the intake camshaft. Using a torque wrench, tighten the camshaft actuator bolt to 63 ft. lbs. (85 Nm), plus an additional 30 degrees.

58. Install a 24 mm wrench onto the hex on the exhaust camshaft. Using a torque wrench, tighten the camshaft actuator bolt to 63 ft. lbs. (85 Nm), plus an additional 30 degrees.

59. Remove the old oil from the timing chain tensioner.

60. Inspect the timing chain tensioner. If the timing tensioner, O-ring seal, or washer is damaged, replace the timing chain tensioner.

61. Measure the timing chain tensioner assembly from end to end. A new tensioner should be supplied in the fully compressed non-active state. A tensioner in the compressed state will measure 2.83 inch (72mm) from end to end. A tensioner in the active state will measure 3.35 inch (85mm) from end to end.

62. If the timing chain tensioner is not in the compressed state, perform the following steps:

　　a. Remove the piston assembly from the body of the timing chain tensioner by pulling it out.

　　b. Install the J 45027-2 (2) into a vise.

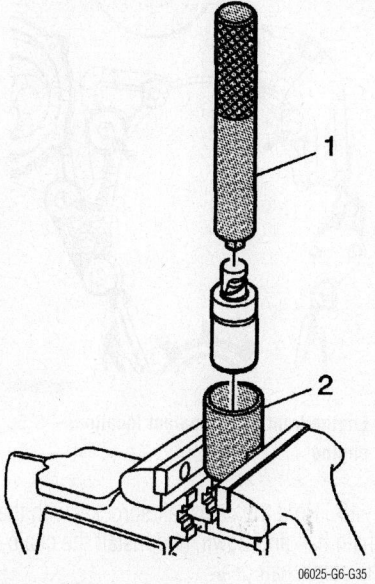

Compressing the timing chain tensioner— 2.4L engine

　　c. Install the notch end of the piston assembly into the J 45027-2 (2).

　　d. Using the J 45027-1 (1), turn the ratchet cylinder into the piston.

63. Install the compressed piston assembly back into the timing chain tensioner body until the assembly stops at the bottom of the bore. Do not compress the piston assembly against the bottom of the bore. If the piston assembly is compressed against the bottom of the bore, the assembly will activate the tensioner, which will then need to be reset again.

64. At this point the tensioner should measure approximately 2.83 inch (72mm) from end to end.

65. Ensure that all dirt and debris are removed from the timing chain tensioner threaded hole in the cylinder head.

66. Install the timing chain tensioner and tighten to 66 ft. lbs. (75 Nm).

67. The timing chain tensioner is released by compressing the tensioner 0.079 inch (2mm) which will release the locking mechanism in the ratchet. To release the timing chain tensioner, use a suitable tool with a rubber tip on the end. Feed the tool down through the cam drive chest to rest on the cam chain. Then give a sharp jolt diagonally downwards to release the tensioner.

68. Apply sealant to the threads and install the timing chain guide bolt access hole plug and tighten to 59 ft. lbs. (80 Nm).

69. Position and install the engine front cover.

70. Install the engine front cover-to-water pump bolt and tighten to 18 ft. lbs. (25 Nm).

71. Install the engine front cover bolts and tighten to 18 ft. lbs. (25 Nm).

72. Install the drive belt tensioner bolt and tighten to 33 ft. lbs. (45 Nm).

73. Install the crankshaft balancer using a new bolt and tighten to 74 ft. lbs. (100 Nm) plus an additional 125 degrees.

74. Install the camshaft cover.

75. Install the no. 1 cylinder spark plug.

76. Fill the engine oil.

77. Install the drive belts.

78. Connect the negative battery cable

79. Fill and bleed the cooling system.

80. Start the vehicle, check for leaks and repair if necessary.

3.5L Engine

➡**If only the crankshaft front seal is being replaced, perform steps 1 through 18.**

1. Properly relieve the fuel system pressure.

2. Drain the engine oil and cooling system.

3. Disconnect the negative battery cable.

4. Remove the drive belt.

5. Raise and support the vehicle.

6. Remove the right front tire and wheel.

7. Remove the right engine splash shield.

8. Install the jack stands to the frame.

9. Loosen the left frame bolts and remove the right side frame bolts.

10. Using the jack stands, lower the right side of the frame to access the crankshaft balancer.

11. Remove the torque converter covers.

12. Remove the oil pan.

13. Lock flywheel to prevent rotation.

14. Remove the crankshaft balancer bolt and the washer.

15. Remove the crankshaft balancer using a 3-jaw puller.

16. Remove the crankshaft key from the keyway.

17. Pry out the oil seal using a large screwdriver or the equivalent.

18. If only the oil seal is being replaced, lubricate the new oil seal using clean engine oil. Insert the oil seal into the front cover with the lip facing the engine and seat the seal. Install the crankshaft key into the keyway. Install the crankshaft balancer and tighten the bolt to 118 ft. lbs. (160 Nm). Reverse the removal procedure.

19. Remove the thermostat bypass pipe from the engine front cover.

20. Remove the radiator outlet hose from the engine front cover.

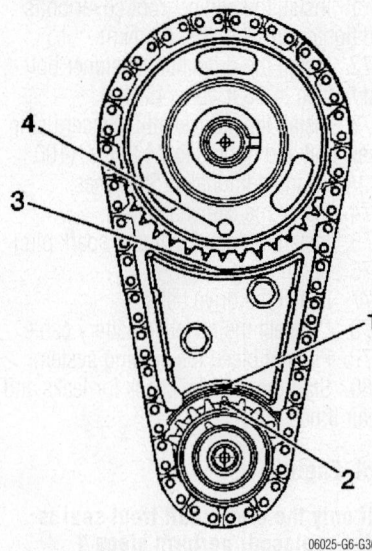

Aligning the timing chain marks—3.5L engine

21. Remove the water pump from the engine front cover.

22. Remove the CKP sensor wiring harness bracket.

23. Remove the engine front cover bolts.

24. Remove the engine front cover.

25. Remove the engine front cover gasket.

26. Clean and inspect the engine front cover.

27. If replacing the engine front cover, remove the drive belt shield bolt and the drive belt shield and the water pump.

28. Rotate the crankshaft until the timing marks in the following locations are aligned: The camshaft alignment pin (4) to the timing chain dampener (3). The timing chain dampener (1) to the crankshaft sprocket (2).

29. Remove the camshaft sprocket bolt.

30. Remove the camshaft sprocket with the timing chain.

31. Remove the crankshaft sprocket.

32. Remove the timing chain dampener bolts and dampener.

33. If necessary, remove the camshaft thrust plate.

34. Clean and inspect the timing chain and the gears.

To install:

35. If removed, install the camshaft thrust plate and tighten to 89 inch lbs. (10 Nm).

36. Install the crankshaft sprocket.

37. Apply engine oil supplement to the sprocket thrust surface.

38. Install the timing chain dampener and tighten the bolts to 15 ft. lbs. (21 Nm).

39. Install the timing chain onto the camshaft gear.

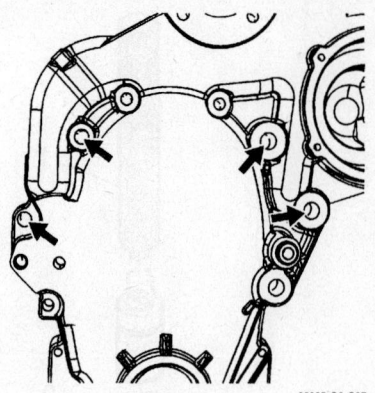

Engine front cover sealant locations—3.5L engine

40. Hold the camshaft sprocket with the chain hanging down, and install the chain to the crankshaft gear.

41. Align the chain and sprocket timing marks.

42. Install the camshaft sprocket bolt and tighten to 103 ft. lbs. (140 Nm) to draw the sprocket onto the camshaft.

43. Coat the crankshaft and camshaft sprocket with engine oil.

44. If removed install the water pump and drive belt shield to the front cover.

45. Apply sealant to both sides of the lower tabs of the engine front cover gasket.

46. Install the engine front cover gasket.

47. Install the front cover.

48. Apply sealant to the bolt hole locations shown.

49. Install the engine front cover bolts and tighten the small engine front cover bolts (1) to 20 ft. lbs. (27 Nm). Tighten the large engine front cover bolts (2, 3) to 41 ft. lbs. (55 Nm).

50. Install the radiator outlet hose to the engine front cover.

51. Install the thermostat bypass pipe to the engine front cover.

52. Apply sealant to the keyway of the crankshaft balancer.

53. Install the crankshaft balancer using Balancer Installer J-29113.

54. Install the crankshaft balancer washer and the bolt and tighten to 118 ft. lbs. (160 Nm).

55. Remove the lock from the flywheel.

56. Install the torque converter covers.

57. Raise the frame to the original position.

58. Install and tighten the frame bolts to 73 ft. lbs. (100 Nm), plus an additional 180 degrees.

59. Install the oil pan.

60. Install the right engine splash shield.

61. Install the right front tire and wheel.

62. Lower the vehicle.

63. Install the drive belt.

64. Fill the engine with clean oil.

65. Fill and bleed the cooling system.

66. Start the engine and check for proper operation.

3.9L Engine

➡ **If only the crankshaft front seal is being replaced, perform steps 1 through 18.**

1. Properly relieve the fuel system pressure.

2. Drain the engine oil and cooling system.

3. Disconnect the negative battery cable.

4. Remove the drive belt and drive belt tensioner.

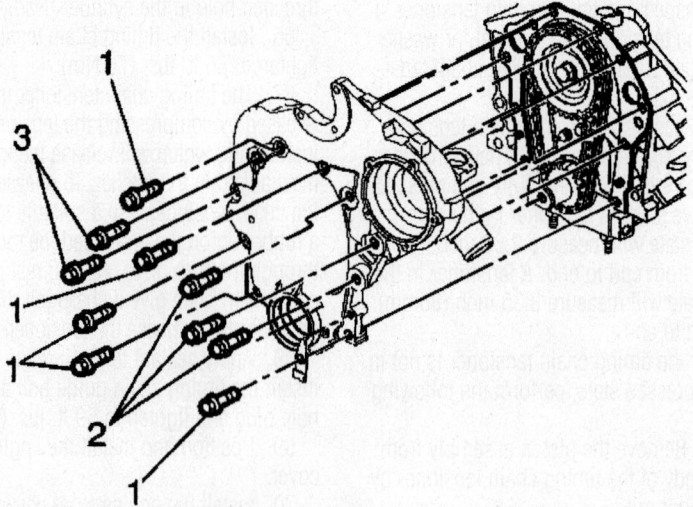

Engine front cover bolt identification—3.5L engine

5. Raise and support the vehicle.

6. Remove the right front tire and wheel.

7. Remove the right engine splash shield.

8. Install the jack stands to the frame.

9. Loosen the left frame bolts and remove the right side frame bolts.

10. Using the jack stands, lower the right side of the frame to access the crankshaft balancer.

11. Remove the torque converter covers.

12. Remove the oil pan.

13. Lock flywheel to prevent rotation.

14. Remove the crankshaft balancer bolt and the washer.

15. Remove the crankshaft balancer using a 3-jaw puller.

16. Remove the crankshaft key from the keyway.

17. Pry out the oil seal using a large screwdriver or the equivalent.

18. If only the oil seal is being replaced, lubricate the new oil seal using clean engine oil. Insert the oil seal into the front cover with the lip facing the engine and seat the seal. Install the crankshaft key into the keyway. Install the crankshaft balancer and tighten the bolt to 52 ft. lbs. (70 Nm), plus an additional 70 degrees. Reverse the removal procedure.

19. Remove the intake manifold cover.

20. Remove the air cleaner assembly.

21. Remove the engine mount strut bracket.

22. Disconnect the camshaft position actuator magnet electrical connector.

23. Remove the crankshaft position actuator magnet.

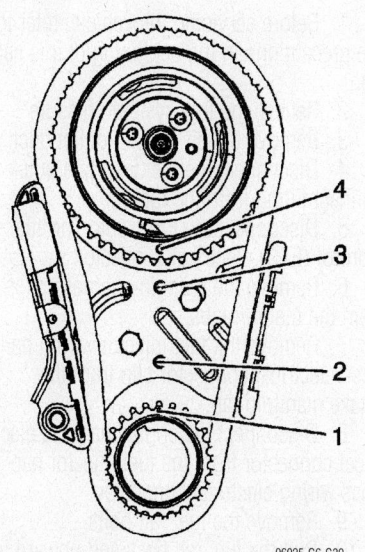

06025-G6-G39

Aligning the timing chain marks—3.9L engine

24. Remove the thermostat housing.

25. Remove the water pump.

26. Remove the engine front cover bolts.

27. Remove the engine front cover and gasket.

28. Align the crankshaft timing mark (1) to the timing mark on the bottom of the timing chain tensioner (2). Align the timing mark on the camshaft gear (3) with the timing mark on top of the timing chain tensioner (4).

29. Remove the camshaft sprocket bolts.

30. Remove the timing chain, camshaft, and crankshaft sprockets.

31. Remove the timing chain tensioner.

32. Remove and discard the camshaft position actuator filter from the end of the camshaft.

To install:

33. Install a new camshaft position actuator filter to the end of the camshaft.

34. Install the crankshaft sprocket.

35. Apply prelube to the crankshaft sprocket thrust surface.

36. Install the timing chain tensioner and tighten the bolts to 15 ft. lbs. (20 Nm).

37. Using Tensioner Compressor EN-47719, fully collapse the tensioner, and place the tensioner retaining pin into the retaining hole.

38. Align the timing marks and hold the camshaft sprocket with the timing chain hanging down and install the timing chain to the crankshaft gear.

39. Align the dowel in the camshaft sprocket with the dowel hole in the camshaft.

40. Draw the camshaft sprocket onto the camshaft using the mounting bolts and tighten to 12 ft. lbs. (16 Nm).

41. Remove the retaining pin from the timing chain tensioner in order to make the tensioner active.

42. Coat the crankshaft and camshaft sprockets with clean engine oil.

43. Install the engine front cover gasket.

44. Install the engine front cover and tighten the bolts to 18 ft. lbs. (25 Nm).

45. Install the water pump.

46. Install the thermostat housing.

47. Install the crankshaft position actuator magnet.

48. Install the camshaft position actuator magnet O-ring seal and magnet.

49. Install the camshaft position actuator magnet bolts and tighten to 89 inch lbs. (10 Nm).

50. Connect the camshaft position actuator magnet electrical connector.

51. Install the engine mount strut bracket and tighten to 18 ft. lbs. (25 Nm).

52. Install the air cleaner assembly.

53. Install the intake manifold cover.

54. Lubricate the new crankshaft front oil seal using clean engine oil.

55. Insert the oil seal into the front cover with the lip facing the engine and seat the seal.

56. Install the crankshaft key into the keyway.

57. Install the crankshaft balancer and tighten the bolt to 52 ft. lbs. (70 Nm), plus an additional 70 degrees.

58. Unlock the flywheel.

59. Install the torque converter covers.

60. Install the oil pan.

61. Raise the frame into the original position.

62. Install and tighten the right and left side frame bolts to 74 ft. lbs. (100 Nm) plus an additional 90 degrees.

63. Install the right engine splash shield.

64. Install the right front tire and wheel.

65. Lower the vehicle.

66. Install the drive belt tensioner and drive belt.

67. Fill the engine with clean oil.

68. Fill and bleed the cooling system.

69. Start the engine and check for proper operation.

Piston and Ring

POSITIONING

Install the piston rings so the oil control ring end gaps are staggered 90 degrees, and the compression ring end gaps are staggered at least 1 inch (25mm).

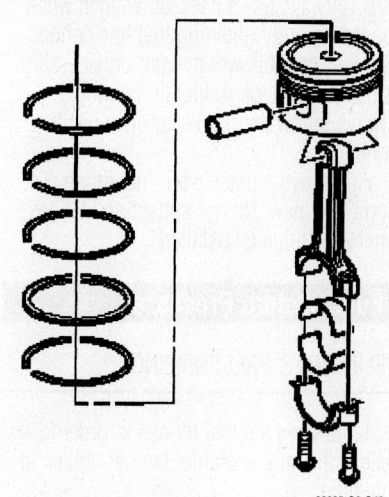

06025-G6-G40

Piston ring gap positioning—G6 engine

FUEL SYSTEM

Fuel System Service Precautions

Safety is the most important factor when performing not only fuel system maintenance but also any type of maintenance. Failure to conduct maintenance and repairs in a safe manner may result in serious personal injury or death. Maintenance and testing of the vehicle's fuel system components can be accomplished safely and effectively by adhering to the following rules and guidelines.

1. To avoid the possibility of fire and personal injury, always disconnect the negative battery cable unless the repair or test procedure requires that battery voltage be applied.

2. Always relieve the fuel system pressure prior to disconnecting any fuel system component (injector, fuel rail, pressure regulator, etc.), fitting or fuel line connection. Exercise extreme caution whenever relieving fuel system pressure, to avoid exposing skin, face and eyes to fuel spray. Please be advised that fuel under pressure may penetrate the skin or any part of the body that it contacts.

3. Always place a shop towel or cloth around the fitting or connection prior to loosening to absorb any excess fuel due to spillage. Ensure that all fuel spillage (should it occur) is quickly removed from engine surfaces. Ensure that all fuel soaked cloths or towels are deposited into a suitable waste container.

4. Always keep a dry chemical (Class B) fire extinguisher near the work area.

5. Do not allow fuel spray or fuel vapors to come into contact with a spark or open flame.

6. Always use a back-up wrench when loosening and tightening fuel line connection fittings. This will prevent unnecessary stress and torsion to fuel line piping. Always follow the proper torque specifications.

7. Always replace worn fuel fitting O-rings with new. Do not substitute fuel hose where fuel pipe is installed.

Fuel System Pressure

RELIEVING & PRESSURIZING

1. Loosen the fuel fill cap in order to relieve the tank pressure. Do not tighten at this time.

2. Raise and support the vehicle.

3. Disconnect the fuel pump electrical connector.

4. Lower the vehicle.

5. Start and run the engine until the fuel supply remaining in the fuel pipes is consumed. Engage the starter for 3.0 seconds in order to assure relief of any remaining pressure.

6. Disconnect the negative battery cable.

7. Connect the fuel pump electrical connector.

8. Lower the vehicle.

Fuel Filter

REMOVAL & INSTALLATION

The fuel filter is integral with the fuel pump module.

Fuel Pump Module

REMOVAL & INSTALLATION

All Models

1. Before servicing the vehicle, refer to the precautions in the beginning of this section.

2. Release the fuel system pressure.

3. Remove the fuel tank.

4. Disconnect the fuel pressure sensor and sender electrical connections.

5. Disconnect the EVAP vapor line quick connect fittings.

6. Disengage the fuel feed line from the retaining features built into the fuel tank.

7. Use Fuel Lock Ring Remover J-45722 and a long breaker-bar in order to unlock the fuel lock ring.

8. Raise the fuel pump assembly out of the tank far enough to access the vapor line quick connect fitting on the underside of the cover.

9. Disconnect the vapor line quick connect fitting.

10. Remove the fuel pump assembly from the fuel tank.

➡**Some lock rings were manufactured with "DO NOT REUSE" stamped into them. These lock rings may be reused if they are not damaged or warped.**

11. Place the lock ring on a flat surface. Measure the clearance between the lock ring and the flat surface using a feeler gage at 7 points. If warpage is less than 0.016 inch (0.41mm), the lock ring does not require replacement. If warpage is more than specified, replace the lock ring.

To install:

12. Install a new o-ring seal onto the fuel pump.

13. Install the pump into the fuel tank far enough to connect the vapor line quick connect fitting on the underside of the cover.

14. Connect the vapor line quick connect fitting.

15. Align the cover paddle or anti-rotation feature with the corresponding feature in the top of the fuel tank.

16. Slowly apply pressure to the top of the spring loaded sender cover until the sender aligns flush with the surface of the tank.

17. Use the remover tool in order to install the fuel sender lock ring. Turn the fuel lock ring in a clockwise direction.

18. Turn the lock ring until the ring seats on the second detent.

19. Engage the fuel feed line to the retaining features built into the fuel tank.

20. Connect the EVAP vapor line quick connect fittings.

21. Connect the fuel pressure sensor and sender electrical connections.

22. Install the fuel tank.

23. Lower the vehicle.

24. Refill the tank.

25. Connect the negative battery cable.

26. Start the vehicle, check for leaks and repair if necessary.

Fuel Rail and Injector

REMOVAL & INSTALLATION

2.4L Engine

1. Before servicing the vehicle, refer to the precautions in the beginning of this section.

2. Relieve the fuel system pressure

3. Remove the air cleaner outlet duct.

4. Disconnect the fuel feed line quick connect fitting from the fuel rail.

5. Disconnect the Electronic Throttle Control (ETC) electrical connector.

6. Remove the 2 engine harness clips from the fuel rail tabs.

7. Remove the fuel injector wiring harness electrical connector clip from the intake manifold tab.

8. Disconnect the engine harness electrical connector from the fuel injector harness wiring electrical connector.

9. Remove the fuel rail bolts.

10. Pull the fuel rail back and upward in order to release the fuel injectors from the cylinder head ports.

11. Remove the fuel rail.

12. Remove the fuel injector retainer.

13. Remove the fuel injector from the fuel rail.

14. Remove and discard the fuel injector O-rings

To install:

15. Lubricate the new O-rings with clean engine oil and install the O-rings on the injector(s).

16. Install the fuel injector to the fuel rail.

17. Install the fuel injector retainer.

18. Install the fuel injector wiring harness clips to the fuel rail.

19. Connect the fuel injector wiring harness electrical connectors to the fuel injectors.

20. Lubricate the new fuel injector tip insulators with clean engine oil.

21. Install the new fuel injector tip insulators to the cylinder head.

22. With the fuel injectors positioned downward, lower the fuel injectors into the cylinder head ports.

23. Carefully push down on the fuel rail in order to insert the injectors into the cylinder head ports.

24. Install the fuel rail bolts and tighten to 89 inch lbs. (10 Nm).

25. Connect the engine harness electrical connector to the fuel injector wiring harness electrical connector.

26. Install the fuel injector wiring har-

ness electrical connector clip to the intake manifold tab.

27. Install the 2 engine harness clips to the fuel rail tabs.

28. Connect the ETC electrical connector.

29. Connect the fuel feed line quick connect fitting to the fuel rail.

30. Install the air cleaner outlet duct.

31. Connect the negative battery cable.

32. Start the vehicle, check for leaks and repair if necessary.

3.5L and 3.9L Engines

1. Before servicing the vehicle, refer to the precautions in the beginning of this section.

2. Relieve the fuel system pressure.

3. Remove the engine fuel feed pipe from the fuel rail.

4. Remove the upper intake manifold.

5. Disconnect the main injector harness electrical connector.

6. Disconnect the main fuel injector electrical harness connector.

7. On 3.9L, disconnect the CMP sensor electrical connector.

8. On all engines, depress the lock tabs and disconnect the fuel injector electrical connectors.

9. Remove the fuel injector electrical wiring harness from the fuel rail.

10. Remove the fuel rail assembly.

11. Remove the fuel injector O-ring seal from the spray tip end of each injector.

12. Remove and discard O-rings if damaged.

13. Remove the fuel injector retainer.

14. Remove the fuel injector from the fuel rail.

15. Remove and discard the fuel injector O-rings

To install:

16. Lubricate the new O-rings with clean engine oil and install the O-rings on the injector(s).

17. Install the fuel rail assembly into the intake manifold. Tilt the fuel rail assembly slightly to install the injectors.

18. Install the fuel rail bolts and tighten to 89 inch lbs. (10 Nm).

19. Install the fuel injector electrical wiring harness to the fuel rail.

20. Connect the fuel injector electrical connectors.

21. Push the slide locks into position.

22. Connect the main fuel injector electrical harness connector.

23. On 3.9L, connect the CMP sensor electrical connector.

24. Install the fuel feed pipe to the fuel rail.

25. Install the upper intake manifold.

26. Connect the negative battery cable.

27. Start the vehicle, check for leaks and repair if necessary.

DRIVE TRAIN

Transaxle Assembly

REMOVAL & INSTALLATION

Manual

1. Drain the transaxle fluid.

2. Disconnect the negative battery cable.

3. Remove the air cleaner assembly.

4. Disconnect the clutch actuator line from the clutch actuator.

5. Disconnect the select cable from the transaxle select lever.

6. Disconnect the shift cable from the transaxle shift lever.

7. Remove both cables from the cable bracket.

8. Remove the cable bracket bolt, nuts and bracket.

9. Disconnect the vehicle speed sensor (VSS) electrical connector and position the harness out of the way.

10. Disconnect the backup lamp switch

electrical connector and position the harness out of the way.

11. Tie the radiator to the upper hood latch panel.

12. Install an engine support fixture.

13. Remove the upper transaxle mounting studs and bolt.

14. Raise and support the vehicle.

15. Remove the front wheels and tires.

16. Remove the left front fender liner.

17. Disconnect the 2 front wheel speed sensor connectors.

18. Unclip the wheel speed sensor wire harness and position the harness out of the way.

19. Remove the suspension frame.

20. Remove the starter.

21. Remove the transaxle front and rear mounts.

22. Remove the wheel driveshaft.

23. Lower the engine and transaxle assembly with the engine support fixture enough to clear the left side inner body panel.

24. Remove the front transaxle brace bolts and brace.

25. Remove the rear transaxle brace bolt and brace.

26. Remove the lower transaxle mounting bolts.

27. Support and remove the transaxle from the vehicle.

To install:

28. Support and raise the transaxle into the vehicle.

29. Install the lower transaxle mounting bolts. Hand-tighten only.

30. Lower the vehicle.

31. Install the upper transaxle mounting bolts. Hand-tighten only.

32. Tighten bolts (1, 4) in to 89 inch lbs. (10 Nm), then to 44 ft. lbs. (60 Nm).

33. Tighten the remaining bolts (2, 3, 5) to 44 ft. lbs. (60 Nm).

34. Install the rear transaxle brace and tighten the bolt to 22 ft. lbs. (30 Nm).

35. Install the front transaxle brace and bolts and tighten to 37 ft. lbs. (50 Nm).

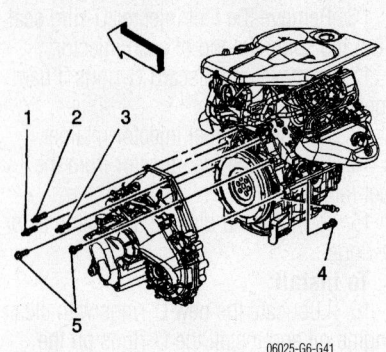

06025-G6-G41

Manual transaxle mounting bolt identification—G6

36. Raise the engine and transaxle assembly with the engine support fixture into position.

37. Install the wheel driveshaft to the vehicle.

38. Install the transaxle front and rear mounts to the transaxle and tighten to 66 ft. lbs. (90 Nm).

39. Install the starter.

40. Install the suspension frame.

41. Route the wheel speed sensor wire harness to the speed sensor.

42. Connect the wheel speed sensor electrical connectors.

43. Install the left front fender liner.

44. Install the front wheels and tires.

45. Lower the vehicle.

46. Remove the engine support fixture.

47. Connect the backup lamp switch electrical connector.

48. Connect the VSS electrical connector.

49. Install the cable bracket and tighten the fasteners to 18 ft. lbs. (25 Nm).

50. Install both cables into the cable bracket.

51. Connect the shift cable to the transaxle shift lever.

52. Connect the select cable to the transaxle select lever.

53. Connect the clutch actuator line to the clutch actuator.

54. Install the air cleaner assembly.

55. Connect the negative battery cable.

56. Fill the transaxle to the proper level with DEXRON®VI transaxle fluid.

57. Start the vehicle, check for leaks and repair if necessary.

Automatic

1. Before servicing the vehicle, refer to the precautions in the beginning of this section.

2. Drain the transaxle fluid.

3. Disconnect the negative battery cable.

4. Raise and support the vehicle.

5. Spray penetrating oil on the exposed threads of both lower ball joint bolt to facilitate their removal.

6. Lower the vehicle.

7. Remove the air intake duct.

8. Disconnect the transaxle wiring harness from the transaxle and the Park Neutral Position (PNP) switch.

9. Remove the shift cable bracket and shift cable from the lever.

10. Remove the transmission wiring harness from the retainer on the transmission.

11. Remove the upper transmission to engine bolts and stud.

12. Install an engine support fixture.

13. Support the radiator and condenser from above using the condenser tabs on each side.

14. Raise the vehicle.

15. Remove the front wheels and tires.

16. Remove the steering gear intermediate shaft.

17. Separate the control arm from the frame and the outer tie-rod end from the steering knuckles. Support the lower control arm with mechanic's wire.

18. Remove the suspension frame.

19. Remove the 3 bolts from the transmission brace near the right axle shaft.

20. Remove the 3 oil pan-to-bellhousing bracket bolts.

21. Remove the flywheel inspection cover.

22. Remove the starter.

23. Mark the relationship of the flywheel to the torque converter for reassembly.

24. Remove the torque converter-to-flywheel bolts.

25. Remove the transmission oil cooler lines by removing the nut holding the bracket to the transaxle case.

26. Disconnect the Vehicle Speed Sensor (VSS) wiring harness from the sensor.

27. Disconnect the rear Heated Oxygen Sensor (HO2S) harness from the bracket on the steering gear.

28. Disconnect the wheel drive shafts from the transaxle.

29. Remove the front transmission mount bracket from the transmission.

30. Use a transmission jack in order to support the transmission.

31. Remove the remaining bellhousing bolts and studs and separate the transmission from the engine.

32. Lower the transmission with the transmission jack far enough to remove the transmission.

33. If the transmission is being replace or installed in a holding fixture, remove the rear transmission mount bracket from the transmission.

34. Remove the transaxle to engine nut.

35. Separate the engine and the transaxle.

36. Remove the transaxle from the vehicle.

To install:

37. Position the transaxle in the vehicle.

38. Install the lower transmission-to-engine bolts and nuts and tighten to 66 ft. lbs. (90 Nm).

39. Install the front transmission mount bracket to the transmission.

40. Install the wheel drive shafts to the transaxle.

41. Connect the wiring harness to the VSS.

42. Install the torque converter-to-flywheel bolts and tighten to 46 ft. lbs. (62 Nm).

43. Install the starter.

44. Install the flywheel inspection cover bolts.

45. Connect the transaxle oil cooler pipes to the transaxle.

46. Install the 3 oil pan-to-bellhousing bracket bolts and tighten to 53 ft. lbs. (72 Nm).

47. Install the 3 bolts to the transmission brace at the final drive area and tighten.

48. Remove the transmission jack.

49. Install the suspension frame.

50. Install the engine splash shields.

51. Install the front wheels and tires.

52. Lower the vehicle.

53. Remove the radiator and condenser support and the engine support fixture.

54. Install the upper transmission to engine bolts and stud and tighten to 66 ft. lbs. (90 Nm).

55. Install the shift cable bracket and shift cable to the lever.

56. Install the exhaust pipe upper bolts and heat shield.

57. Connect the electrical connectors to the PNP switch and transaxle.

58. Connect the negative battery cable.

59. Install the air intake duct.

60. Add DEXRON® III automatic transmission fluid and verify the proper fluid level of the transaxle.

61. Road test the vehicle and check for leaks.

Clutch

ADJUSTMENT

All models are equipped with a hydraulic clutch, which is self-adjusting.

REMOVAL & INSTALLATION

1. Disconnect the negative battery cable.
2. Raise and support the vehicle safely.
3. Remove the transaxle.
4. Remove the clutch cover bolts one turn at a time, until spring pressure is relieved.
5. Remove the clutch cover and the clutch disc.

To install:

6. Align the heavy side of the flywheel assembly, stamped with an X, with the clutch cover light side, marked with paint.
7. Install an alignment tool in order to support the clutch cover to the flywheel assembly.
8. Install the clutch cover to flywheel bolts and tighten in sequence to 18 ft. lbs. (25 Nm).
9. Remove the alignment tool.
10. Lubricate the inside diameter of the bearing grease.
11. Install the transmission.
12. Bleed the hydraulic system.
13. Connect the negative battery cable.

Hydraulic Clutch System

BLEEDING

1. Clean dirt and grease from the cap in order to ensure that no foreign substances enter the system.
2. Attach a hose to the bleeder screw on the clutch actuator assembly. Submerge the other end of the hose in a container of clutch fluid.
3. Depress the clutch pedal slowly and hold.
4. Loosen the bleeder screw to purge air.
5. Tighten the bleeder screw.
6. Repeat until air is purged.
7. Fill the reservoir to the top with fluid.
8. Check the clutch operation and repeat if necessary.

Halfshaft

REMOVAL & INSTALLATION

Front

1. Raise and support the vehicle safely.
2. Remove the front wheel.

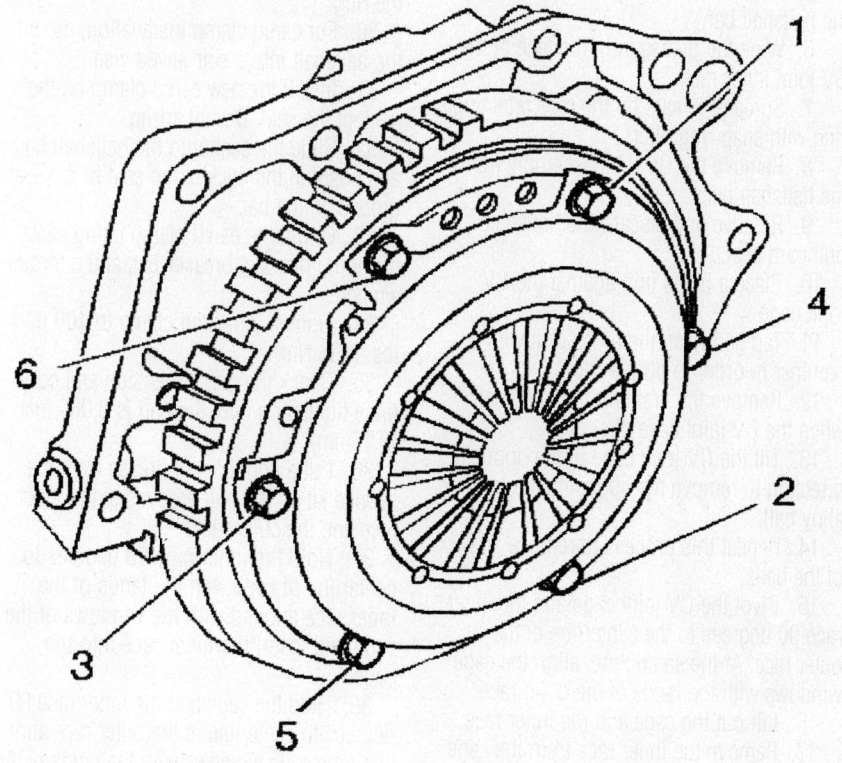

Clutch bolt tightening sequence—G6

06025-G6-G42

3. Remove the halfshaft nut.
4. Disconnect the outer tie rod assembly from the steering knuckle.
5. Separate the ball joint from the steering knuckle.
6. Separate the front wheel drive axle from the drive shaft bearing using a slide hammer and appropriate puller. The nut can be partially re-installed to protect the threads.
7. Remove the halfshaft from the transaxle.

To install:

8. Install the halfshaft into the transaxle.
9. Verify that the front wheel drive shaft retaining ring is properly seated by grasping the inner housing and pull the inner housing outward. Do not pull on the front wheel drive axle shaft.
10. The front wheel drive axle will remain in place when the front wheel drive shaft retaining ring is properly seated.
11. Install the front wheel drive shaft into the front wheel bearing.
12. Connect the ball joint to the steering knuckle.
13. Connect the outer tie rod assembly to the steering knuckle.
14. Install a new wheel drive shaft nut. Insert a drift or a flat-bladed tool into the caliper and the rotor to prevent the rotor from turning. Tighten the nut to 159 ft. lbs. (215 Nm).
15. Install the wheel and tire.
16. Lower the vehicle.
17. Inspect the transaxle fluid level.
18. Inspect the wheel alignment.

CV-Joints

OVERHAUL

Inner Joint

1. Disassemble the joint as follows:
2. Disconnect the swage ring from the halfshaft bar using a hand grinder to cut through the ring, taking care not to damage the halfshaft bar.
3. Remove the large seal retaining clamp from the tripot joint with side cutters. Discard the large seal retaining clamp.
4. Separate the inboard seal from the trilobal tripot bushing at the large diameter.
5. Slide the seal away from the joint along the halfshaft bar.
6. Remove the housing from the tripot joint spider and the halfshaft bar.
7. Spread the spacer ring.
8. Remove the spacer ring, spider

assembly, spacer ring and tripot boot. Discard the boot and rings.

9. Clean the halfshaft bar Use a wire brush in order to remove any rust in the boot mounting area (grooves).

10. Inspect the needle rollers, needle bearings, and trunnion. Check the tripot housing for unusual wear, cracks, or other damage. Replace any damaged parts with the appropriate kit.

To install:

11. Place the new small swage ring or eared clamp onto the small end of the joint seal. Slide the joint seal and the small swage ring or eared clamp onto the halfshaft bar.

12. Position the small end of the joint seal into the joint seal groove on the halfshaft bar.

13. For swage ring installation, mount Drive Axle Swage Ring Clamp J-41048 in a vise.

14. Position the inboard end of the halfshaft assembly in the tool.

15. Align the top of the seal neck on the bottom die using the indicator.

16. Place the top half of the tool on the lower half.

17. Before proceeding, ensure there are no pinch points on the halfshaft inboard seal. This could cause damage to the halfshaft inboard seal.

18. Insert the bolts and tighten the bolts by hand until snug.

19. Align the halfshaft inboard seal, the halfshaft bar and the swage ring.

20. Tighten each bolt of the tool 180 degrees at a time using a ratchet wrench. Alternate between each bolt until both sides are bottomed.

21. For eared clamp installation, mount the halfshaft into a vise.

22. Slide the tripot seal to the corresponding groove on the halfshaft bar.

23. Crimp the eared clamp using clamping pliers, a torque wrench, and a breaker bar.

24. If equipped, install the spacer ring into the groove of the halfshaft bar.

25. Slide the tripot joint spider assembly as far as it will go on the halfshaft bar.

26. Install the spacer ring into the groove of the halfshaft bar.

27. Place approximately half of the grease from the service kit in the halfshaft inboard seal. Use the remainder of the grease to repack the housing.

28. Ensure the trilobal tripot bushing is flush with the face of the housing.

29. Install the trilobal tripot bushing to the housing.

30. Position the larger new seal retaining clamp on the halfshaft inboard seal.

31. Slide the housing over the tripot joint spider assembly on the halfshaft bar.

32. Slide the large diameter of the halfshaft inboard seal, with larger clamp in place, over the outside of the trilobal tripot bushing and locate the lip of the seal in the groove.

33. Align the halfshaft inboard seal, tripot housing and large seal retaining clamp.

34. Crimp the seal retaining clamp with to 130 ft. lbs. (176 Nm). Add the breaker bar and the torque wrench if needed.

35. Check the gap dimension on the clamp ear. If the gap dimension is larger than 0.102 inch (2.6mm), continue tightening until the dimension is reached.

36. Fully stroke the joint several times to disperse the grease.

Outer Joint

1. Disassemble the joint as follows:

2. Remove the large seal retaining clamp from the CV joint with a side cutter. Discard the seal retaining clamp.

3. Use a hand grinder to cut through the swage ring in order to remove the swage ring.

4. Separate the halfshaft outboard seal from CV joint outer race at the large diameter.

5. Slide the seal away from joint along the halfshaft bar.

6. Wipe the grease from the face of the CV joint inner race.

7. Spread the ears on the race retaining ring with snap ring pliers.

8. Remove the CV joint assembly from the halfshaft bar.

9. Remove and discard the halfshaft outboard seal.

10. Place a brass drift against the CV joint cage.

11. Tap gently on the brass drift with a hammer in order to tilt the cage.

12. Remove the first chrome alloy ball when the CV joint cage tilts.

13. Tilt the CV joint cage in the opposite direction to remove the opposing chrome alloy ball.

14. Repeat this process to remove all 6 of the balls.

15. Pivot the CV joint cage and the inner race 90 degrees to the centerline of the outer race. At the same time, align the cage windows with the lands of the outer race.

16. Lift out the cage and the inner race.

17. Remove the inner race from the cage by rotating the inner race upward.

18. Clean all parts with solvent and replace any damaged parts.

To install:

19. Assemble the joint as follows:

20. Install the new swage ring or eared clamp on the neck of the outboard seal. Do not swage or crimp.

21. Slide the outboard seal onto the halfshaft bar and position the neck of the outboard seal in the seal groove on the halfshaft bar. The largest groove below the sight groove on the halfshaft bar is the seal groove.

22. For swage ring installation, position the outboard end of the halfshaft assembly Drive Axle Swage Ring Clamp J-41048.

23. Align the swage ring.

24. Place the top half of the clamp tool onto the lower half.

25. Align the outboard seal, halfshaft bar and swage ring.

26. Insert the bolts and tighten by hand until snug.

27. Tighten each bolt 180 degrees at a time using a ratchet wrench. Alternate between each bolt until both sides are bottomed.

28. Loosen the bolts and separate the dies.

29. Check swaged ring for any lip deformities. If present, place the ring back into the tool making sure the ring covers the whole swaging area. If necessary, re-swage the ring.

30. For eared clamp installation, mount the halfshaft into a soft-jawed vise.

31. Install the new eared clamp on the neck of the seal. Do not crimp.

32. Slide the seal onto the halfshaft bar and position the neck of the seal in the seal groove on the bar.

33. Crimp the eared clamp using seal clamping pliers, a breaker bar, and a torque wrench.

34. Tighten the eared clamp to 100 ft. lbs. (136 Nm).

35. Check the gap dimension and continue tightening until the gap is 0.085 inch (2.15mm).

36. Put a light coat of grease from the service kit on the ball grooves of the inner race and the outer race.

37. Hold the inner race 90 degrees to centerline of cage with the lands of the inner race aligned with the windows of the cage and insert the inner race into the cage.

38. Hold the cage and the inner race 90 degrees to centerline of the outer race and align the cage windows with the lands of the outer race.

39. Ensure that the retaining ring side of the inner race faces the halfshaft bar.

40. Place the cage and the inner race into the outer race.

41. Insert the first chrome ball then tilt the cage in the opposite direction to insert the opposing ball.

42. Repeat this process until all 6 balls are in place.

43. Place approximately half the grease

from the service kit inside the outboard seal and pack the CV joint with the remaining grease.

44. Push the CV joint onto the halfshaft bar until the retaining ring is seated in the groove on the halfshaft bar.

45. Slide large diameter of the outboard seal with the large seal retaining clamp in

place over the outside of the CV joint outer race and locate the seal lip in the groove on the CV joint outer race.

46. Crimp the seal retaining clamp to 130 ft. lbs. (174 Nm).

47. Check the gap dimension on the clamp ear. Continue tightening until the gap dimension is 0.074 inch (1.9mm).

STEERING AND SUSPENSION

Air Bag

PRECAUTIONS

Several precautions must be observed when handling the inflator module to avoid accidental deployment and possible personal injury.

1. Never carry the inflator module by the wires or connector on the underside of the module.

2. When carrying a live inflator module, hold securely with both hands, and ensure that the bag and trim cover are pointed away.

3. Place the inflator module on a bench or other surface with the bag and trim cover facing up.

4. With the inflator module on the bench, never place anything on or close to the module that may be thrown in the event of an accidental deployment.

DISARMING

➡**All Air Bag electrical wiring harnesses and connectors are covered with YELLOW outer insulation. Do not use electrical test equipment on any circuit related to the Air Bag sensors. When installing Air Bag components, always install with the arrow marks facing the front of the vehicle.**

1. Before servicing the vehicle, refer to the precautions in the beginning of this section.

2. Place the front wheels in the straight ahead position.

3. Turn the ignition switch to the **OFF** position and remove the key.

4. Remove the Body Control Module (BCM) fuse cover and remove the AIR BAG (IGN) and AIR BAG (BATT) fuses from the fuse block.

5. Open the hood and locate the front end sensor, also known as the Electronic

Frontal Sensor (EFS) on the radiator support.

6. Remove the Connector Position Assurance (CPA) from the EFS connector.

7. Remove the EFS connector from the EFS.

REARMING

1. Turn the ignition switch to the **OFF** position.

2. Connect the EFS connector to the EFS.

3. Connect the CPA to the EFS connector.

4. Install the AIR BAG (IGN) and AIR BAG (BATT) fuses into the BCM fuse center.

5. Use caution while reaching in and turn the ignition switch to the ON position. The AIR BAG indicator will flash then turn OFF.

6. Perform the SIR Diagnostic System Check if the AIR BAG warning indicator does not operate as described.

Power Rack and Pinion Steering Gear

REMOVAL & INSTALLATION

Hydraulic Steering Models

1. Raise and support the vehicle safely and remove the front wheels.

2. Remove the tie rod ends from the steering knuckle.

3. Remove the intermediate shaft to steering gear pinch bolt. Discard the bolt.

4. Disconnect the intermediate shaft from the steering gear.

5. Lower the vehicle.

6. Remove the intermediate shaft pinch bolt at the steering column shaft. Discard the pinch bolt.

7. Slide the intermediate shaft off the steering column shaft.

8. To unseat the intermediate shaft seal from the dash, squeeze the 4 tabs on the seal individually, then pull toward the inside of the vehicle.

9. Remove the intermediate shaft/seal from the vehicle.

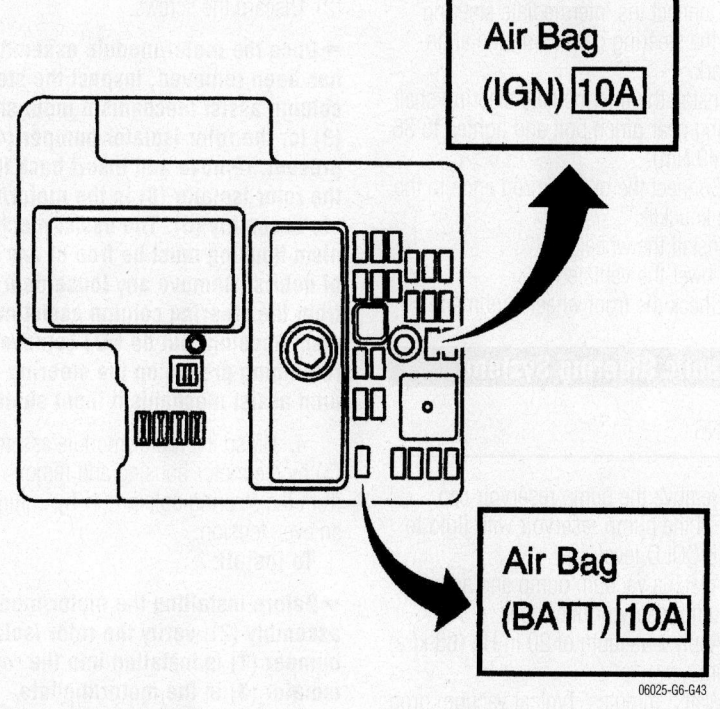

06025-G6-G43

Identifying air bag fuses—G6

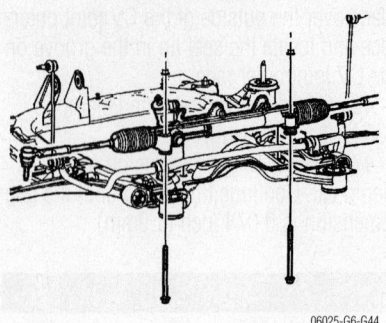

06025-G6-G44

Hydraulic power steering gear mounting— G6

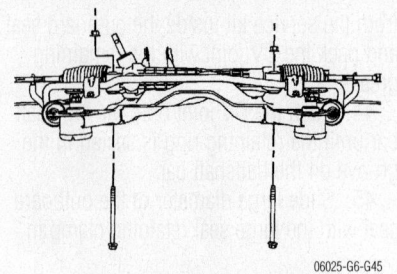

06025-G6-G45

Electronic power steering gear mounting—G6

10. Remove the power steering hoses from the from the power steering gear and cap the openings.

11. Remove the power steering gear bolts from the gear.

12. Remove the power steering gear through the left wheel opening.

To install:

13. Position the power steering gear to the vehicle.

14. Install the power steering gear bolts and tighten to 81 ft. lbs. (110 Nm).

15. Install the power steering hoses to the steering gear.

16. Install the intermediate shaft/seal through the front of dash.

➡**To ensure seal is properly seated, inspect the 4 tabs on the seal from the engine side of the seal. Do not twist or squeeze the seal during installation.**

17. Seat the intermediate shaft seal by pushing forward from the inside of the vehicle, until all four tabs of the seal are fully seated.

18. Connect the intermediate shaft to the steering column shaft.

19. Install the new steering column pinch bolt to the intermediate shaft and tighten to 46 ft. lbs. (62 Nm).

20. Connect the intermediate shaft to the steering gear shaft.

21. Install a new steering gear pinch bolt to the intermediate shaft and tighten to 36 ft. lbs. (49 Nm).

22. Install the tie rod ends to the steering knuckle.

23. Install the wheels.

24. Lower the vehicle.

25. Bleed the air from the power steering system.

26. Check the front wheel alignment.

Electronic Steering Models

1. Raise and support the vehicle safely and remove the front wheels.

2. Remove the intermediate shaft to steering gear pinch bolt. Discard the bolt.

3. Disconnect the intermediate shaft from the steering gear. Note the alignment for installation.

4. Disconnect the outer tie rod end from the steering knuckle.

5. Remove the steering gear-to-frame bolts.

6. Remove the steering gear through the left side of the vehicle. Rotate the gear 90 degrees in order to clear the rear transmission mount.

7. If replacing the steering gear, remove the outer tie rod ends.

To install:

8. Install the outer tie rod ends to the steering gear, if removed.

9. Install the steering gear to the vehicle. Rotate the gear as necessary to clear the rear transmission mount.

10. Install the steering gear-to-frame bolts and tighten to 52 ft. lbs. (70 Nm), plus an additional 90 degrees.

11. Connect the intermediate steering shaft to the steering gear using the alignment mark.

12. Install the intermediate steering shaft to steering gear pinch bolt and tighten to 36 ft. lbs. (49 Nm).

13. Connect the outer tie rod ends to the steering knuckle.

14. Install the wheels.

15. Lower the vehicle.

16. Check the front wheel alignment.

Hydraulic Steering System

BLEEDING

1. Remove the pump reservoir cap.

2. Fill the pump reservoir with fluid to the FULL COLD level.

3. Attach a vacuum pump and adapter to the pump reservoir filler neck.

4. Apply a vacuum of 20 in Hg (68 kPa) maximum.

5. Wait 5 minutes. Typical vacuum drop is 2-3 in Hg (7-10 kPa). If the vacuum does not remain steady, check the pump, lines and o-ring seals for leaks.

6. Remove the vacuum tools.

7. Reinstall the pump reservoir cap.

8. Start the engine. Allow the engine to idle.

9. Turn off the engine.

10. Verify the fluid level and fill as needed.

11. Start the engine. Allow the engine to idle.

12. Turn the steering wheel 180-360 degrees in both directions 5 times.

13. Switch the ignition off.

Electronic Steering Motor

REMOVAL & INSTALLATION

1. Disconnect the sensor wire harness (6) from the motor/module assembly (5).

➡**If replacing the motor/module assembly, you will need the sensor wire harness strap clip (7) for the new motor/module assembly installation. If replacing the steering column, a new sensor wire harness strap clip (7) will come with the column service kit. Keep the existing wire strap clip attached to the steering column sensor wire harness (6).**

2. Use needle nose pliers to remove the wire strap clip (7) from the motor/module assembly (5).

3. Use an M6x1 head bit to remove the 2 motor/module assembly TORX® screws (2). Discard the screws.

➡**Once the motor/module assembly has been removed, inspect the steering column assist mechanism input shaft (3) for the rotor isolator bumper (4). If present, remove and insert back into the rotor isolator (8) in the motor/module assembly (5). The assist mechanism housing must be free of any type of debris. Remove any loose debris from the steering column assist mechanism housing, but do NOT remove the remaining grease on the steering column assist mechanism input shaft (3).**

4. Grasp the motor/module assembly (5) by the motor housing and remove it from the steering column (1) by pulling with an even tension.

To install:

➡**Before installing the motor/module assembly (2), verify the rotor isolator bumper (1) is installed into the rotor isolator (3) in the motor/module assembly (2).**

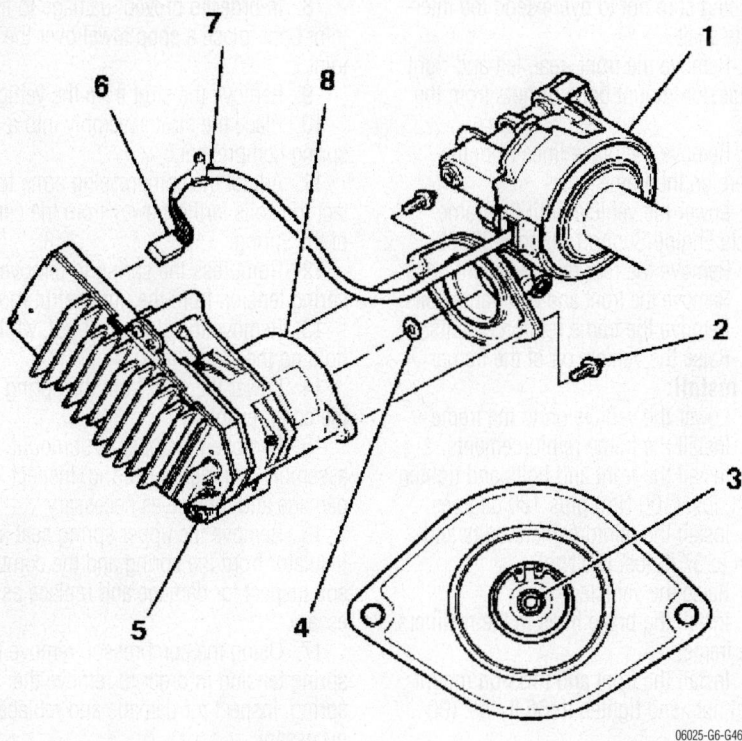

Electronic steering motor mounting—G6

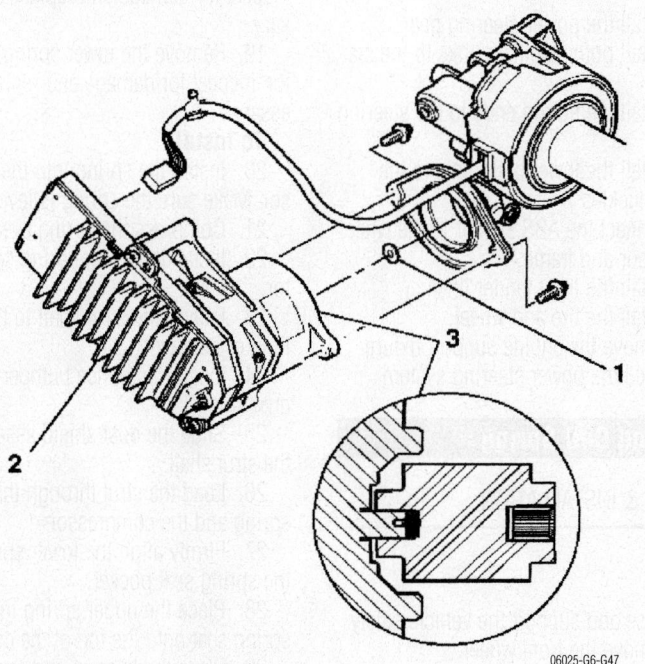

06025-G6-G47

Electronic steering rotor isolator installation—G6

5. Fit the motor/module assembly rotor isolator over the steering column assist mechanism input shaft.

6. Use an M6x1 TORX® head bit to attach the motor/module assembly (2) to the steering column with the 2 new TORX® bolts (1).

7. Tighten the steering column bolts to 80 inch lbs. (10 Nm).

8. Connect the sensor wire harness (3) to the motor/module assembly (2).

9. Install the sensor wire harness strap clip (4) into the motor/module assembly (2).

10. Perform the control module setup procedure.

Electronic Steering Module

PROGRAMMING & SETUP

Control Module Setup

After replacing the power steering motor and module assembly the following 3 procedures must be performed. After replacing the steering column assembly only the steering position sensor and the torque sensor calibration procedures must be performed. The steering position sensor and torque sensor calibration procedures should also be performed after a suspension alignment.

➡The power steering control module (PSCM) must be setup using the service programming system (SPS). There are 8 different tuning profile calibrations stored in the PSCM. The SPS will select the correct tuning profile calibration for this vehicle. Setup the PSCM to use that profile. If the PSCM has not been setup, the scan tool will display the Tuning Profile is 0 in the EPS data list, and DTC C0551 will set.

Steering Position Sensor Calibration

1. Install the scan tool.

2. Turn on the ignition, with the engine off.

3. Center the steering wheel.

4. With the scan tool select Special Functions.

5. Select Steering Position Sensor Calibration and press the Enter key. The scan tool screen will flash Calibration in Progress.

6. When done, the scan tool will display Calibration Complete.

7. Press the exit key.

8. Use the scan tool in order to clear the DTCs.

9. When done, the scan tool will display Calibration Complete.

10. Press the exit key.

11. Use the scan tool in order to clear the DTCs.

12. After turning off the ignition, allow 25 seconds of wait time before performing any procedures that require the vehicles battery to be disconnected, or module memory loss may occur.

13. Turn off the ignition.

Torque Sensor Calibration

1. Install the scan tool.

2. Turn on the ignition, with the engine off.

3. Center the steering wheel.

4. After centering the steering wheel, remove hands and other objects from the steering wheel and ensure the suspension is relaxed and no bias, or uneven force is being applied to the steering system.

5. With the scan tool select Special Functions.

6. Select Torque Sensor Calibration and press the Enter key. The scan tool screen will flash Calibration in Progress.

7. When done, the scan tool will display Calibration Complete.

8. Press the exit key

9. Use the scan tool in order to clear the DTCs

10. After turning off the ignition, allow 25 seconds of wait time before performing any procedures that require the vehicles battery to be disconnected, or module memory loss may occur.

11. Turn off the ignition.

Steering Tuning Selection

1. Install the scan tool.

2. Turn on the ignition, with the engine off.

3. With the scan tool select Special Functions

4. Select Steering Tuning Selection and press the enter key. The scan tool screen will flash Selection in Progress.

5. When done the scan tool will display Selection Complete.

6. Press the exit key.

7. Use the scan tool in order to clear the DTCs.

8. After turning OFF the ignition, allow 25 seconds of wait time before performing any procedures that require the vehicles battery to be disconnected, or module memory loss may occur.

9. Turn off the ignition.

Front Suspension Frame

REMOVAL & INSTALLATION

1. Install an engine support fixture.

2. Raise the vehicle on a hoist.

3. Remove the tire and wheel.

4. Remove the front fender liner.

5. Disconnect the ABS sensor from the wheel speed sensor and frame.

6. Remove the tie rod ends from the steering knuckles.

7. Remove both stabilizer links from the stabilizer bar.

8. Remove the power steering gear mounting bolts and secure the gear out of the way using mechanic's wire or equiva-

lent, being sure not to overextend the intermediate shaft.

9. Remove the front, rear, left and right transmission mount bolts or nuts from the frame.

10. Remove the brake lines from the retainers on the frame.

11. Lower the vehicle until the frame contacts Engine Support Stand J-39580,

12. Remove the rear reinforcement bolts.

13. Remove the front and rear frame bolts.

14. Remove the frame reinforcements.

15. Raise the vehicle off of the frame.

To install:

16. Lower the vehicle on to the frame.

17. Install the frame reinforcements.

18. Install the front and bolts and tighten to 73 ft. lbs. (100 Nm) plus 180 degrees.

19. Install the reinforcement bolts and tighten to 37 ft. lbs. (50 Nm).

20. Raise the vehicle.

21. Install the brake lines to the retainers on the frame.

22. Install the front and mission mount bracket nuts and tighten to 66 ft. lbs. (90 Nm).

23. Install the left and right transmisson mount bracket bolts and tighten to 37 (50 Nm).

24. Install the power steering gear.

25. Install both stabilizer links to the stabilizer bar.

26. Install the tie rod ends to the steering knuckles.

27. Install the lower ball joints to the steering knuckles.

28. Connect the ABS sensor to the wheel speed sensor and frame.

29. Install the front fender liner.

30. Install the tire and wheel.

31. Remove the engine support fixture.

32. Bleed the power steering system

Strut and Coil Spring

REMOVAL & INSTALLATION

Front

1. Raise and support the vehicle safely.

2. Remove the front wheel.

3. Disconnect the stabilizer link from the strut.

4. Remove the strut-to-steering knuckle nuts.

5. Reposition the wheel speed sensor/ABS harness and bracket.

6. Remove the strut-to-steering knuckle bolts.

7. Remove the upper strut cap-to-body nuts.

8. In order to prevent damage to the CV joint boot, place a shop towel over the CV joint.

9. Remove the strut from the vehicle.

10. Place the strut assembly into a spring compressor.

11. Adjust the compressing arms to contact the coils farthest away from the center of the spring.

12. Compress the spring to remove the spring tension from the upper strut mount.

13. Remove the strut shaft nut, while holding the strut shaft.

14. Lower the strut from the spring and the compressor.

15. Remove the upper strut mount assembly and mount bearing. Inspect for damage and replace as necessary.

16. Remove the upper spring seat and insulator from the spring and the compressor. Inspect for damage and replace as necessary.

17. Using the compressor, remove the spring tension in order to remove the spring. Inspect for damage and replace as necessary.

18. Remove the dust shield and jounce bumper assembly from the strut shaft. Inspect for damage and replace as necessary.

19. Remove the lower spring seat insulator. Inspect for damage and replace as necessary.

To install:

20. Install the spring into the compressor. Make sure the spring is level.

21. Compress the spring evenly.

22. Install the lower spring seat insulator.

23. Extend the strut shaft to the upper limit of its travel.

24. Insert the jounce bumper into the dust shield.

25. Slide the dust shield assembly onto the strut shaft.

26. Load the strut through the coil spring and the compressor.

27. Firmly align the lower spring coil in the spring seat pocket.

28. Place the upper spring insulator and spring seat onto the top of the coil spring.

29. Place the bearing and strut mount on the top of the spring seat.

30. Install the upper strut shaft nut and tighten to 52 ft. lbs. (70 Nm).

31. Using the compressor, remove the spring tension.

32. Remove the strut assembly from the compressor.

33. Position the strut to the vehicles strut tower, using the alignment pin as a guide.

34. Install the upper strut cap-to-body nuts and tighten to 18 ft. lbs. (25 Nm).

35. Install the strut-to-steering knuckle bolts leaving the nuts off.

36. Place the wheel speed sensor harness and bracket to the bolt end.

37. Install the strut to steering knuckle nuts and tighten to 89 ft. lbs. (120 Nm).

38. Connect the stabilizer link to the strut and tighten to 48 ft. lbs. (65 Nm).

39. Install the front wheel.

40. Lower the vehicle.

Shock Absorber

REMOVAL & INSTALLATION

Rear

1. Raise and support the vehicle.
2. Remove the rear wheel.
3. Using a suitable jack stand, raise the rear knuckle to release the spring tension.
4. Remove the lower shock bolt.
5. Remove the upper shock nuts and remove the shock absorber.

To install:

6. Install the shock in the vehicle.
7. Install the shock absorber-to-body bolts and tighten to 66 ft. lbs. (90 Nm).
8. Install the shock absorber-to-knuckle bolt and tighten to 133 ft. lbs. (180 Nm).
9. Remove the jack stand from the rear knuckle.
10. Install the tire and wheel.
11. Lower the vehicle.

Coil Spring

REMOVAL & INSTALLATION

Rear

1. Raise and support the vehicle.
2. Remove the wheel and tire.
3. Use an OTC 204-167 compressor and OTC 204-167-01 adapter shoes or equivalent on-vehicle spring compressor to compress the coil spring.
4. Using a suitable jack stand, support the lower control arm.
5. Remove the lower control arm-to-knuckle bolt, nut, and inboard fastener.
6. Use the jack stand in order to lower the lower control arm with the coil spring attached.
7. Remove the coil spring from the lower control arm.
8. Inspect the coil spring insulators for damage and replace as necessary.

To install:

9. Position the spring on the lower control arm.
10. Use the jack stand to raise the lower control arm into position.
11. Install the lower control arm-to-knuckle bolt, nut, and inboard fastener. And tighten the bolt and nut to 44 ft. lbs. (60 Nm), plus an additional 60 degrees.
12. Remove the spring compressor.
13. Remove the jack stand.
14. Install the rear tire and wheel.
15. Lower the vehicle.

Stabilizer Bar

REMOVAL & INSTALLATION

Front

1. Raise and support the vehicle.
2. Remove the wheels.
3. Disconnect the stabilizer links from the stabilizer shaft.
4. Using a suitable jack stand, support the rear of the frame assembly.
5. Remove the frame support-to-body bolts.
6. Remove the rear frame assembly mounting bolts.
7. Lower the rear of the cradle in order to gain clearance to the stabilizer shaft.
8. Remove the stabilizer bar clamps and insulators.
9. Remove the stabilizer shaft through the opening between the frame and body.

To install:

10. Position the stabilizer shaft to the frame.
11. Install the stabilizer bar clamps and insulators and tighten to 18 ft. lbs. (25 Nm).
12. Raise the rear of the cradle.
13. Install the frame support brackets. Loosely install the bracket and frame-to-body bolts.
14. Tighten the rear frame to body bolt-to-74 ft. lbs. (100 Nm).
15. Tighten the frame support bracket-to-body bolt 74 ft. lbs. (100).
16. Tighten all the bolts 90 degrees plus an additional 15 degrees.
17. Remove the jack stand.
18. Connect the stabilizer link to the stabilizer bar and tighten to 48 ft. lbs. (65 Nm).
19. Lower the vehicle.

Rear

1. Raise and support the vehicle.
2. Remove the wheels.
3. Remove the muffler assembly.
4. Disconnect the fuel pump module

electrical connector from the underbody wiring harness.

5. Disconnect the evaporative emissions (EVAP) vent valve solenoid harness electrical connector from the underbody wiring harness.

6. Suitably support the gas tank.

7. Remove the front fuel tank strap bolts.

8. Remove the fuel tank heat shield.

9. Lower the fuel tank as necessary for access.

10. Place a index mark on the rear camber adjusting bolts.

11. Suitably support the rear control arms.

12. Remove the right and left toe links.

13. Remove the stabilizer shaft bracket bolts.

14. Remove the stabilizer shaft insulators from the stabilizer shaft.

15. Remove the stabilizer link-to-knuckle bolts.

16. Remove the stabilizer shaft.

To install:

17. Position the stabilizer shaft.

18. Install the stabilizer shaft insulators and shaft brackets.

19. Install the stabilizer shaft bracket bolts and tighten to 26 ft. lbs. (35 Nm).

20. Install the stabilizer shaft links-to-knuckle bolts and tighten to 37 ft. lbs. (50 Nm).

21. Install the right and left toe links. Line up the index marks on the camber bolts.

22. Tighten the toe link bolts to 81 ft. lbs. (110 Nm) plus 70 degrees.

23. Remove the supports from the rear control arms.

24. Raise the fuel tank into position.

25. Install the fuel tank heat shield.

26. Install the front fuel tank strap bolts and tighten to 15 ft. lbs. (20 Nm).

27. Remove support from the fuel tank.

28. Connect the evaporative emissions (EVAP) vent valve solenoid harness electrical connector to the underbody wiring harness.

29. Connect the fuel pump module electrical connector to the underbody wiring harness.

30. Install the muffler assembly.

31. Install the rear tire and wheel.

32. Lower the vehicle.

Trailing Arm

REMOVAL & INSTALLATION

1. Raise and support the vehicle.
2. Remove the wheels.

3. Remove the trailing arm bracket-to-body bolts.

4. Remove the trailing arm-to-knuckle through bolt.

5. Disconnect the parking brake cable from the trailing arm.

6. Remove the trailing arm.

7. Remove the trailing arm-to-bracket bolt and nut.

8. Remove the trailing arm from the bracket.

To install:

9. Assemble the trailing arm and bracket.

10. Tighten the trailing arm-to-bracket through bolt to 44 ft. lbs. (60 Nm) plus 60 degrees.

11. Position the trailing arm to the vehicle.

12. Install the trailing arm to knuckle bolts and tighten to 133 ft. lbs. (180 Nm).

13. Connect the parking brake cable to the trailing arm.

14. Install the trailing arm bracket-to-body bolts and tighten to 66 ft. lbs. (90 Nm) plus 30 degrees plus an additional 15 degrees.

15. Install the tire and wheel.

16. Lower the vehicle.

Steering Knuckle

REMOVAL & INSTALLATION

Front

1. Raise and support the vehicle.

2. Remove the wheels.

3. Remove the brake caliper and rotor.

4. Remove the wheel hub/ bearing assembly.

5. Remove the outer tie rod-to-knuckle nut.

6. Separate the tie rod from the steering knuckle.

7. Remove the lower control arm.

8. Remove the strut-to-steering knuckle nuts and bolts.

9. Remove the steering knuckle.

To install:

10. Install the steering knuckle to the strut assembly and the ABS harness bracket. Tighten the bolts and nuts to 89 ft. lbs. (120 Nm).

11. Guide the axle through the steering knuckle.

12. Install the lower control arm.

13. Install the outer tie rod to the steering knuckle and tighten the nut to 15 ft. lbs. (20 Nm), plus an additional 180 degrees.

14. Verify the torque is 37 ft. lbs. (50 Nm).

15. Install the wheel bearing, brake rotor, brake caliper and front wheels.

16. Lower the vehicle.

17. Road test the vehicle in order to verify alignment.

Rear

1. Raise and support the vehicle.

2. Remove the wheels.

3. Remove the brake caliper and rotor.

4. Remove the wheel hub/ bearing assembly.

5. Using a suitable jack, raise the knuckle in order to relieve tension from the shock.

6. Remove the lower shock-to-knuckle bolt.

7. Remove the coil spring.

8. Remove the toe link.

9. Remove the upper control arm-to-knuckle bolt and nut.

10. Remove the trailing arm-to-knuckle bolts.

11. Remove the stabilizer shaft link-to-knuckle bolt.

12. Remove the knuckle from the vehicle.

To install:

13. Install the trailing arm-to-knuckle bolts and tighten to 133 ft. lbs. (180 Nm).

14. Install the upper control arm-to-knuckle bolt and nut and tighten to 81 ft. lbs. (110 Nm), plus 70 degrees.

15. Install the toe link.

16. Install the coil spring.

17. Install the lower shock absorber-to-knuckle bolt and tighten to 133 ft. lbs. (180 Nm).

18. Install the stabilizer link to the knuckle and tighten to 37 ft. lbs. (50 Nm).

19. Install the rear wheel bearing and brake components.

20. Install the tire and wheel.

21. Lower the vehicle.

Lower Ball Joint

REMOVAL & INSTALLATION

The lower ball joint assembly is part of the lower control arm. If replacement of the ball joint is required, the lower control arm needs to be replaced.

Lower Control Arm

REMOVAL & INSTALLATION

Front

1. Raise and support the vehicle.

2. Remove the wheel.

3. If equipped with a 3.5L engine, If removing the left lower control arm, remove the side transmission mount. If removing the right lower control arm, remove the engine mount.

4. Remove the lower control arm front bushing-to-frame bolt.

5. Remove the lower control arm rear bushing-to-frame bolts and nuts.

6. Prior to removal, note the orientation of the lower control arm ball stud to steering knuckle pinch bolt and remove the pinch bolt and discard.

7. Separate the ball stud from the steering knuckle.

8. Remove the control arm from the vehicle.

To install:

9. Position the lower control arm to the frame assembly and steering knuckle.

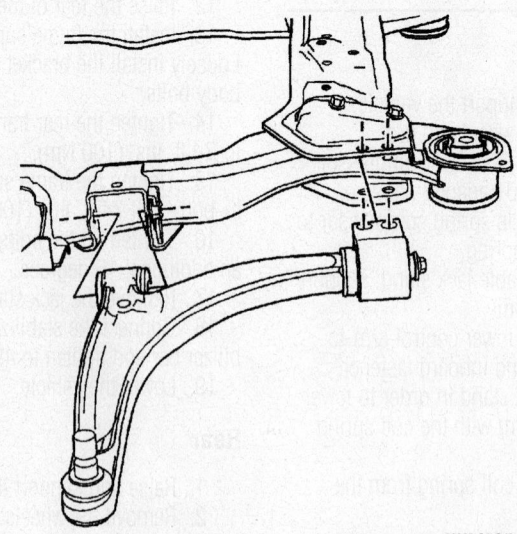

Front lower control arm mounting—G6

06025-G6-G48

10. Note the previous orientation and install the lower control arm-to-steering knuckle pinch bolt and hand tighten only.

11. Install the lower control arm front bushing to frame bolt and hand tighten only.

12. Install the lower control arm rear bushing to frame bolts and nuts.

13. Tighten the nuts with the lower control arm at the proper trim height.

14. Install a new ball stud-to-steering knuckle pinch bolt. Tighten the ball pinch nut to 37 ft. lbs. (50 Nm). Reverse the nut 3/4 of a turn . Tighten to 37 ft. lbs. (50 Nm) plus an additional 60 degrees.

15. Tighten the bolts with the front suspension loaded by using the proper jack stand.

16. Tighten the front bushing-to-frame bolt to 37 ft. lbs. (50 Nm) plus 90 degrees.

17. Tighten the rear bushing-to-frame bolts to 37 ft. lbs. (50 Nm) plus 90 degrees.

18. If equipped with a 3.5L engine, install the engine mount or transmission mount.

19. Install the tire and wheel.

20. Lower the vehicle.

Rear

1. Raise and support the vehicle.

2. Remove the wheel.

3. Remove the rear spring.

4. Remove the lower control arm-to-support assembly bolt and nut.

5. Remove the lower control arm from the vehicle.

6. Remove the lower control arm spring insulators. Inspect the lower spring insulator for damage and replace as necessary.

To install:

7. Install the lower control arm spring insulators.

8. Position the lower control arm to the support assembly.

9. Install the lower control arm-to-support assembly bolt and nut and tighten to 81 ft. lbs. (110 Nm).

10. Install the rear spring.

11. Install the tire and wheel.

12. Lower the vehicle.

13. Check the rear alignment.

CONTROL ARM BUSHING REPLACEMENT

1. Remove the lower control arm-to-rear bushing bolt.

2. Note the position of the bushing during removal. Remove the bushing off the lower control arm.

3. Install the bushing on the lower control arm as previously noted.

4. Using Loctite on the bolt threads, install the lower control arm-to-bushing bolt.

5. Hold the rear bushing inner sleeve when tightening the rear bushings to control arm bolt. Tighten the bolt to 32 ft. lbs. (44 Nm).

Wheel Hub and Bearing

REMOVAL & INSTALLATION

Front

1. Raise and support the vehicle.

2. Remove the wheel.

3. Remove the wheel drive shaft nut.

4. Remove the brake caliper and wire aside.

5. Remove the brake rotor.

6. Disconnect the electrical connector from the wheel speed sensor.

7. Remove the wheel speed sensor connector from the bracket by depressing the locking tabs.

8. Remove the 3 hub and bearing assembly bolts.

9. Install Wheel Hub Removal tool J-42129 to the hub and bearing assembly in order to remove the hub and bearing assembly from the wheel drive shaft.

10. Remove the hub and bearing assembly from the steering knuckle.

To install:

11. Install the hub and bearing assembly to the steering knuckle.

12. Install the 3 hub and bearing assembly bolts and tighten to 85 ft. lbs. (115 Nm).

13. Install the wheel speed sensor connector into the bracket until the locking tabs click into place.

14. Connect the electrical connector to the wheel speed sensor.

15. Install the axle nut to the wheel drive shaft and hand tighten.

16. Install the brake rotor.

17. Use a screw driver or similar tool to stop the rotation of the brake rotor.

18. Tighten the wheel drive shaft nut to 159 ft. lbs. (215 Nm).

19. Install the brake caliper.

20. Install the tire and wheel.

Rear

1. Raise and support the vehicle.

2. Remove the wheel.

3. Remove the brake caliper and rotor.

4. Disconnect the electrical connector from the wheel speed sensor.

5. Remove the stabilizer link bolt at the knuckle and position the stabilizer link out of the way in order to provide access to the wheel bearing/hub nuts.

6. Remove the 4 wheel bearing/hub assembly nuts.

7. Remove the wheel bearing/hub assembly from the knuckle.

To install:

8. Install the wheel bearing/hub assembly to the knuckle.

9. Install the 4 wheel bearing/hub assembly nuts and tighten to 47 ft. lbs. (63 Nm).

10. Connect the stabilizer link bolt at the knuckle.

11. Connect the electrical connector to the wheel speed sensor.

12. Install the brake rotor and caliper.

13. Install the tire and wheel.

14. Lower the vehicle.

Upper Control Arm

REMOVAL & INSTALLATION

Rear

1. Raise and support the vehicle.

2. Remove the wheel.

3. Disconnect the ABS harness connector and route the harness aside.

4. Remove the upper control arm to support assembly bolt.

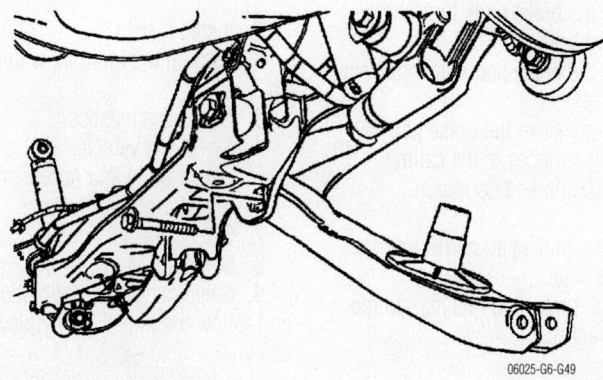

Rear lower control arm mounting—G6

06025-G6-G49

5. Remove the upper control arm-to-knuckle bolt and nut.

6. Remove the upper control arm from the vehicle through the wheel well opening.

To install:

7. Position the upper control arm to the support assembly and knuckle.

8. Install the upper control arm-to-knuckle bolt and nut and hand tighten only.

9. Install the upper control arm-to-support assembly bolt.

10. Tighten the upper control arm-to-support assembly bolt to 44 ft. lbs. (60 Nm) plus 60 degrees.

11. Tighten the upper control arm-to-knuckle bolt to 81 ft. lbs. (110 Nm) plus 70 degrees.

12. Connect the ABS harness connector. Route the harness as previously noted.

13. Install the tire and wheel.

14. Lower the vehicle.

BRAKES

Brake Caliper

REMOVAL & INSTALLATION

Front and Rear

1. Remove enough brake fluid from the reservoir to reach the half fill mark.

2. Raise and support the vehicle.

3. Remove the tire and wheel.

4. Install and firmly hand tighten 2 wheel nuts to opposite wheel studs in order to retain the rotor to the hub.

5. Install a large C-clamp over the body of the brake caliper with the C-clamp ends against the rear of the caliper body and against the outer brake pad.

6. Tighten the C-clamp until the caliper piston is compressed into the caliper bore enough to allow the caliper to slide past the brake rotor.

7. Remove the C-clamp from the caliper.

8. Remove the brake hose from the brake caliper.

9. Cap or plug the opening in the brake caliper and the brake hose to prevent fluid loss and contamination.

10. On rear brakes, disconnect the parking brake cable from the caliper.

11. Remove the brake caliper guide pin bolts.

12. Remove the brake caliper from the caliper bracket.

To install:

13. Install the brake caliper to the brake caliper bracket.

14. Install the brake caliper guide pin bolts and tighten to 26 ft. lbs. (35 Nm).

15. Remove the caps or plugs from the brake caliper opening and the brake hose.

16. Install new copper brake hose gaskets to the brake hose-to-caliper bolt and to the brake hose.

17. Install the brake hose and the brake hose-to-brake caliper bolt to the brake caliper and tighten to 37 ft. lbs. (50 Nm).

18. On rear brakes, connect the parking brake cable to the caliper.

19. Bleed the hydraulic brake system.

20. Remove the wheel nuts retaining the brake rotor to the wheel hub.

21. Install the tire and wheel assembly.

22. Lower the vehicle.

23. With the engine off, gradually apply the brake pedal to approximately 2/3 of its travel distance.

24. Slowly release the brake pedal.

25. Wait 15 seconds, then repeat the steps until a firm brake pedal apply is obtained; this will properly seat the brake caliper piston and brake pads.

Disc Brake Pads

REMOVAL & INSTALLATION

Front and Rear

1. Remove enough brake fluid from the reservoir to reach the half fill mark.

2. Raise and support the vehicle.

3. Remove the tire and wheel.

4. Install and firmly hand tighten 2 wheel nuts to opposite wheel studs in order to retain the rotor to the hub.

5. Remove the brake caliper lower guide pin bolt.

6. Push the disc brake caliper piston into the caliper bore using an old inner disc brake pad and a disc brake piston installation tool.

7. Without disconnecting the hydraulic brake flexible hose, pivot the caliper upward and secure the caliper with heavy mechanics wire.

8. Remove the brake pads from the caliper mounting bracket.

9. Remove the brake pad retainers from the caliper bracket.

10. Thoroughly clean the brake pad hardware mating surfaces of the caliper bracket, of any debris and corrosion.

To install:

11. Ensure the brake pad hardware mating surfaces are clean.

12. Install the brake pad retainers to the brake caliper bracket.

→ The wear sensor equipped disc brake pad must be mounted inboard of the rotor with the leading edge of the sensor facing the brake rotor during forward wheel rotation, or at the top of the pad when installed in vehicle position.

13. Install the brake pads to the caliper bracket.

14. Remove the support, and rotate the brake caliper into position over the disc brake pads and to the caliper mounting bracket.

15. Install the lower brake caliper guide pin bolt and tighten to 26 ft. lbs. (35 Nm).

16. Remove the wheel nuts retaining the brake rotor to the hub.

17. Install the tire and wheel.

18. Lower the vehicle.

19. With the engine OFF, gradually apply the brake pedal approximately 2/3 of its travel distance.

20. Slowly release the brake pedal.

21. Wait 15 seconds, then gradually apply the brake pedal approximately 2/3 of its travel distance again until a firm brake pedal apply is obtained. This will properly seat the brake caliper pistons and brake pads.

22. Fill the master cylinder auxiliary reservoir to the proper level.

23. Burnish the pads and rotors.

Brake Pedal Position Sensor

CALIBRATION

1. Install a scan tool.

2. Turn on the ignition, with the engine off.

3. Select Diagnostics.

4. Select the vehicle.

5. Select Body and Accessories.

6. Select Lighting Systems.

7. Select Special Functions.

8. Select BCM.

9. Select the BPPS calibration procedure and follow the directions displayed on the screen.

BRAKES 20-31
DRIVE TRAIN 20-21
ENGINE REPAIR 20-9
FUEL SYSTEM 20-19
SPECIFICATIONS AND
  MAINTENANCE CHARTS......20-2
Engine and Vehicle Identification
  Chart.....................................20-2
General Engine Specifications 20-2
Engine Tune-Up Specifications 20-2
Firing Order 20-3
Accessory Drive Belt Routing 20-3
Capacities 20-4
Valve Specifications......................20-4
Crankshaft and Connecting Rod
  Specifications 20-4
Piston and Ring Specifications 20-5
Torque Specifications 20-5
Wheel Alignment 20-6
Tire, Wheel and Ball Joint
  Specifications..............................20-6
Brake Specifications 20-6
Scheduled Maintenance
  Intervals......................................20-7
STEERING AND
  SUSPENSION 20-23
A
Air Bag...20-23
  Arming The System 20-24
  Disarming The System............20-23
  Precautions...............................20-23
Alternator 20-9
  Removal & Installation.............20-9
Axle Shaft 20-22
  Removal & Installation............20-22
B
Brake Caliper 20-31
  Removal & Installation.............20-31
Brake Pads...................................20-31
  Removal & Installation.............20-31
C
Camshaft 20-15
  Removal & Installation............20-15

Clutch..20-22
  Removal & Installation............20-22
Coil Spring 20-25
  Removal & Installation............20-25
Constant Velocity Joint...............20-22
  Overhaul 20-22
Cylinder Head 20-12
  Removal & Installation............20-12
D
Distributor....................................20-9
E
Engine Assembly 20-9
  Removal & Installation.............20-9
Engine Valley Cover.....................20-15
  Removal & Installation............20-15
Exhaust Manifold 20-15
  Removal & Installation............20-15
F
Front Lower Control Arm
  Rod..20-27
  Removal & Installation............20-27
Front Suspension Frame.............20-27
  Removal & Installation............20-27
Fuel Filter and Fuel Pump 20-19
  Removal & Installation............20-19
Fuel Rail and Injector...................20-20
  Removal & Installation............20-20
Fuel System Pressure 20-19
  Relieving...................................20-19
Fuel System Service
  Precautions...............................20-19
H
Heater Core..................................20-10
  Removal And Installation 20-10
Hydraulic Clutch System 20-22
  Bleeding....................................20-22
I
Ignition Timing 20-9
  Adjustment................................20-9
Intake Manifold 20-14
  Removal & Installation............20-14
L
Lower Control Arm 20-27
  Removal & Installation............20-27

O
Oil Pan...20-16
  Removal & Installation............20-16
Oil Pump 20-16
  Removal & Installation............20-16
P
Pinion Flange and Seal...............20-23
  Removal & Installation............20-23
Piston and Ring 20-18
  Positioning 20-18
Power Rack and Pinion Steering
  Gear...20-24
  Removal & Installation............20-24
R
Rear Inner Drive Axle Shaft and
  Seal 20-23
  Removal & Installation............20-23
Rear Main Seal 20-17
  Removal & Installation............20-17
Rocker Arms 20-13
  Removal & Installation............20-13
S
Shock Absorber 20-25
  Removal & Installation............20-25
Stabilizer Bar 20-26
  Removal & Installation............20-26
Starter Motor 20-16
  Removal & Installation............20-16
Strut and Coil Spring...................20-24
  Removal & Installation............20-24
T
Timing Chain, Sprockets,
  Front Cover and Seal................20-17
  Removal & Installation............20-17
Transmission 20-21
  Removal & Installation............20-21
V
Valve Lash 20-16
  Adjustment................................20-16
W
Water Pump 20-10
  Removal & Installation............20-10
Wheel Hub and Bearing..............20-28
  Removal & Installation............20-28

SPECIFICATION AND MAINTENANCE CHARTS

VEHICLE AND ENGINE IDENTIFICATION CHART

Engine								Model Year	
Code	Liters	Cu. In.	Cyl.	Fuel Sys.	Engine Type	Eng. Mfg.		Code	Year
G	5.7	346	8	SEFI	OHV	General Motors		4	2004
U	6.0	364	8	SEFI	OHV	General Motors		5	2005
								6	2006

SEFI: Sequential Multi-port Fuel Injection

06025-GTO-C01

GENERAL ENGINE SPECIFICATIONS

Year	Engine Displacement Liters	Engine VIN	Net Horsepower @ rpm	Net Torque @ rpm (ft. lbs.)	Bore x Stroke (in.)	Compression Ratio	Oil Pressure @ rpm
2004	5.7	G	350@5200	365@4000	3.89x3.62	10.1:1	24@4000
2005	6.0	U	400@5200	400@4000	4.00x3.62	10.9:1	24@4000

06025-GTO-C02

GASOLINE ENGINE TUNE-UP SPECIFICATIONS

Year	Engine Displacement Liters	Engine VIN	Spark Plugs Gap (in.)	Ignition Timing (deg.) MT	Ignition Timing (deg.) AT	Fuel Pump (psi)	Idle Speed (rpm) MT	Idle Speed (rpm) AT	Valve Clearance In.	Valve Clearance Ex.
2004	5.7	G	0.040	①	①	55-64	①	①	HYD	HYD
2005	6.0	U	0.040	①	①	55-64	①	①	HYD	HYD

NOTE: The Vehicle Emission Control Information label often reflects specification changes changes made during production.

The label figures must be used if they differ from those in this chart.

HYD: Hydraulic

① Controlled by the Powertrain Control Module (PCM) and cannot be manually adjusted.

06025-GTO-C03

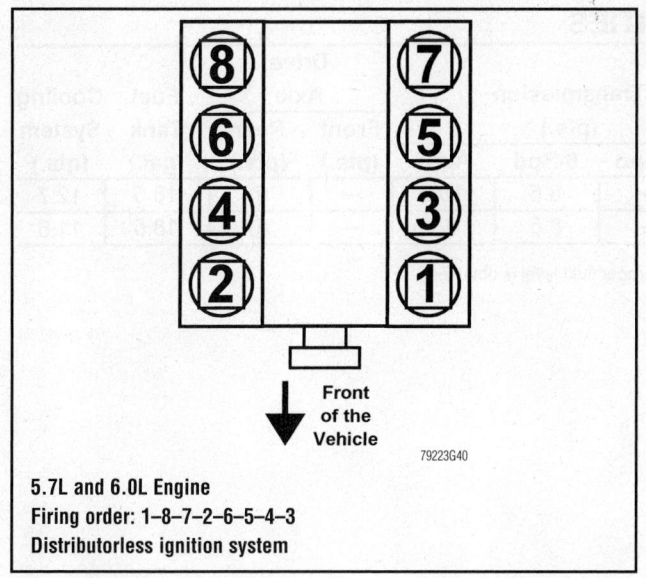

5.7L and 6.0L Engine
Firing order: 1–8–7–2–6–5–4–3
Distributorless ignition system

Front
of the
Vehicle

79223G40

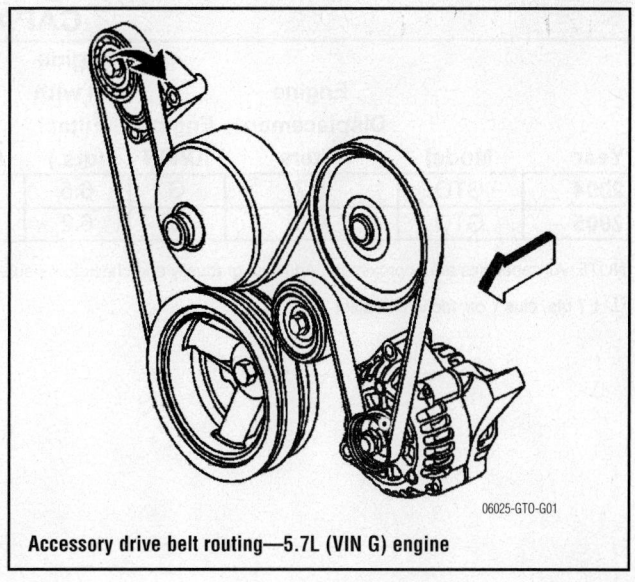

Accessory drive belt routing—5.7L (VIN G) engine

06025-GTO-G01

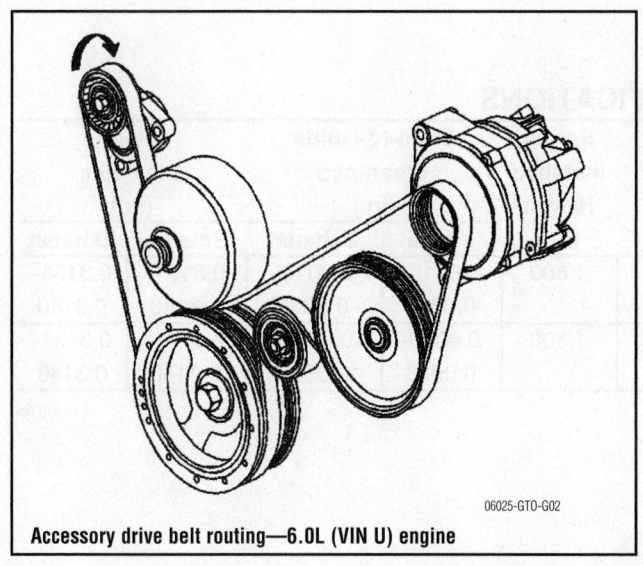

Accessory drive belt routing—6.0L (VIN U) engine

06025-GTO-G02

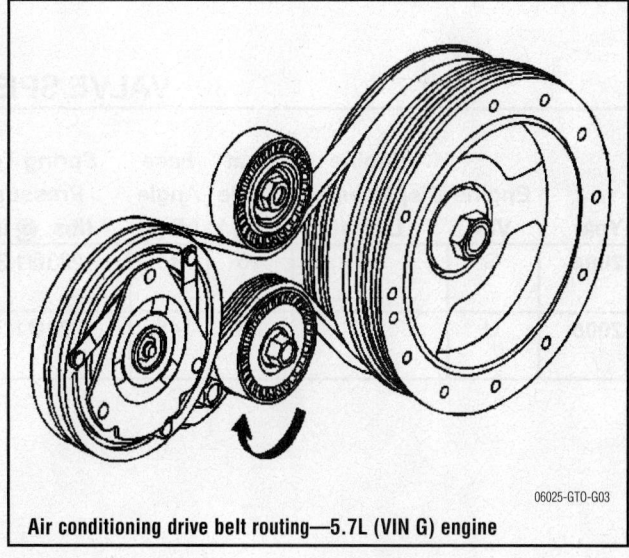

Air conditioning drive belt routing—5.7L (VIN G) engine

06025-GTO-G03

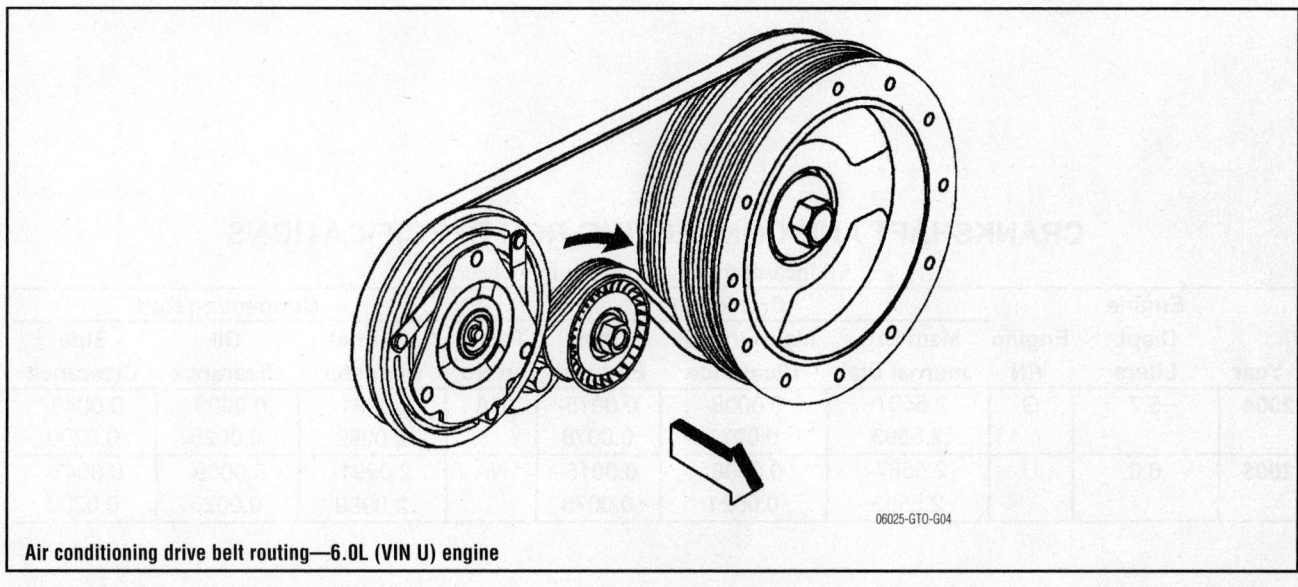

Air conditioning drive belt routing—6.0L (VIN U) engine

06025-GTO-G04

CAPACITIES

Year	Model	Engine Displacement Liters	Engine ID/VIN	Engine Oil with Filter (qts.)	Transmission (pts.)			Drive Axle		Fuel Tank (gal.)	Cooling System (qts.)
					4-Spd	6-Spd	Auto.	Front (pts.)	Rear (pts.)		
2004	GTO	5.7	G	6.5	—	8.5	10.2	—	①	18.5	12.7
2005	GTO	6.0	U	6.2	—	8.5	8.7	—	①	18.5	11.6

NOTE: All capacities are approximate. Add fluid gradually and check to be sure a proper fluid level is obtained.

① 1.7 qts. plus 1 oz. friction modifier

06025-GTO-C04

VALVE SPECIFICATIONS

Year	Engine VIN	Engine Displacement Liters	Seat Angle (deg.)	Face Angle (deg.)	Spring Test Pressure (lbs. @ in.)	Spring Installed Height (in.)	Stem-to-Guide Clearance (in.)		Stem Diameter (in.)	
							Intake	Exhaust	Intake	Exhaust
2004	G	5.7	46	45	220@1.32	1.800	0.0010-0.0026	0.0010-0.0026	0.3131-0.3140	0.3131-0.3140
2005	U	6.0	46	45	220@1.32	1.800	0.0010-0.0026	0.0010-0.0026	0.3131-0.3140	0.3131-0.3140

06025-GTO-C05

CRANKSHAFT AND CONNECTING ROD SPECIFICATIONS

All measurements are given in inches.

Year	Engine Displ. Liters	Engine VIN	Crankshaft				Connecting Rod		
			Main Brg. Journal Dia.	Main Brg. Oil Clearance	Shaft End-play	Thrust on No.	Journal Diameter	Oil Clearance	Side Clearance
2004	5.7	G	2.5587-2.5593	0.0008-0.0021	0.0015-0.0078	NA	2.0991-2.0999	0.0009-0.0025	0.0043-0.0200
2005	6.0	U	2.5587-2.5593	0.0008-0.0021	0.0015-0.0078	NA	2.0991-2.0999	0.0009-0.0025	0.0043-0.0200

06025-GTO-C06

PISTON AND RING SPECIFICATIONS

All measurements are given in inches.

Year	Engine Displ. Liters	Engine VIN	Piston Clearance	Ring Gap			Ring Side Clearance		
				Top Comp.	Bottom Comp.	Oil Control	Top Comp.	Bottom Comp.	Oil Control
2004	5.7	G	①	0.0090-0.0173	0.0173-0.0275	0.0070-0.0295	0.0015-0.0033	0.0020-0.0034	-0.0003 0.0069
2005	6.0	U	②	0.0078-0.0161	0.0145-0.0271	0.0086-0.0311	0.0012-0.0040	0.0014-0.0031	0.0005-0.0079

① Non coated skirt piston: 0.0005-0.0019
   Coated skirt piston: -0.0005-+0.0011

② -0.0009-0.0012

06025-GTO-C07

TORQUE SPECIFICATIONS

All readings in ft. lbs.

Year	Engine VIN	Engine Displacement Liters	Cylinder Head Bolts	Main Bearing Bolts	Rod Bearing Bolts	Crankshaft Damper Bolts	Flywheel Bolts	Manifold		Spark Plugs	Oil Pan Drain Plug
								Intake	Exhaust		
2004	G	5.7	①	②	③	④	⑤	⑥	⑦	11	18
2005	U	6.0	⑧	②	③	④	⑤	⑥	⑦	11	18

① Step 1: M11 bolts 22 ft. lbs.
  Step 2: M11 bolts plus 90 degrees
  Step 3: M11 bolts plus 90 degrees exc. medium bolts at front/rear of head
  Step 4: Medium bolts plus 50 degrees
  Step 5: M8 bolts 22 ft. lbs.

② Step 1: M10 bolts 15 ft. lbs.
  Step 2: M10 bolts plus 80 degrees
  Step 3: M10 studs 15 ft. lbs.
  Step 4: M10 studs plus 51 degrees
  Step 5: M8 bolts 18 ft. lbs.

③ Step 1: 15 ft. lbs.
  Step 2: plus 75 degrees

④ Step 1: 37 ft. lbs.
  Step 2: plus 140 degrees

⑤ Step 1: 15 ft. lbs.
  Step 2: 37 ft. lbs.
  Step 3: 74 15 ft. lbs.

⑥ Step 1: 44 INCH lbs.
  Step 2: 89 INCH lbs.

⑦ Step 1: 11 ft. lbs.
  Step 2: 18 ft. lbs.

⑧ Step 1: M11 bolts 22 ft. lbs.
  Step 2: M11 bolts plus 90 degrees
  Step 3: M11 bolts plus 70 degrees
  Step 4: M8 bolts 22 ft. lbs.

06025-GTO-C08

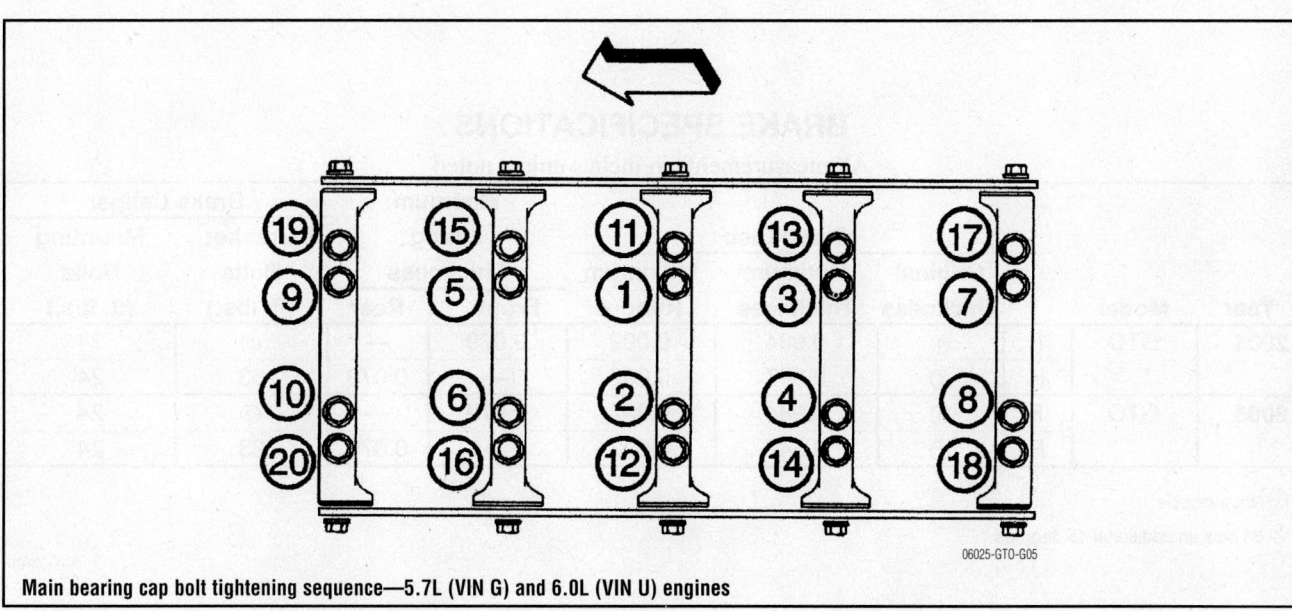

06025-GTO-G05

Main bearing cap bolt tightening sequence—5.7L (VIN G) and 6.0L (VIN U) engines

WHEEL ALIGNMENT

Year	Model			Caster Range (+/-Deg.)	Caster Preferred Setting (Deg.)	Camber Range (+/-Deg.)	Camber Preferred Setting (Deg.)	Toe-in (Deg.)
2004	GTO	Front	Left	0.80	+7.7	0.60	-0.20	-0.17+/-0.17
			Right	0.80	+7.7	0.60	-0.20	-0.17+/-0.17
		Rear		—	—	0.58	-1.05	-0.40+/-0.34
2005	GTO	Front	Left	0.80	+7.7	0.60	-0.20	-0.17+/-0.17
			Right	0.80	+7.7	0.60	-0.20	-0.17+/-0.17
		Rear		—	—	0.58	-1.05	-0.40+/-0.34

06025-GTO-C09

TIRE AND WHEEL SPECIFICATIONS

Year	Model	OEM Tires Standard	OEM Tires Optional	Tire Pressures (psi) Front	Tire Pressures (psi) Rear	Wheel Size	Wheel Lug Nut Torque (Ft. Lbs.)
2004	GTO	P245/50R16	—	35	35	①	②
2005	GTO	P245/50R16	—	35	35	①	②

OEM: Original Equipment Manufacturer

PSI: Pounds Per Square Inch

STD: Standard

OPT: Optional

① Not available

② Step 1: 50 ft. lbs.

    Step 2: 100 ft. lbs.

06025-GTO-C10

BRAKE SPECIFICATIONS
All measurements in inches unless noted

Year	Model		Brake Disc Original Thickness	Brake Disc Minimum Thickness	Brake Disc Maximum Runout	Minimum Lining Thickness Front	Minimum Lining Thickness Rear	Brake Caliper Bracket Bolts (ft. lbs.)	Brake Caliper Mounting Bolts (ft. lbs.)
2004	GTO	F	①	0.984	0.002	0.079	—	②	24
		R	①	0.547	0.002	—	0.079	63	24
2005	GTO	F	①	0.984	0.002	0.079	—	②	24
		R	①	0.547	0.002	—	0.079	63	24

① Not available

② 63 plus an additional 45 degrees

06025-GTO-C11

MAINTENANCE I AND II SERVICE SCHEDULES
2004-05 Pontiac GTO

When the CHANGE ENGINE OIL light appears, certain services and inspections are required. Required services are described as Maintenance I and Maintenance II.

The first service on a vehicle should be Maintenance I, and the second service should be Maintenance II. Alternate between the 2 thereafter. However, in some cases, Maintenance II may be required more often.

Maintenance I: Use Maintenance I if the CHANGE ENGINE OIL light comes on within 10 months since vehicle was purchased or, if Maintenance II was performed.

Maintenance II: Use Maintenance II if the previous service performed was Maintenance I.

Always use Maintenance II whenever the CHANGE ENGINE OIL light comes on 10 months or more since the last service, or, if the CHANGE ENGINE OIL light has not come on at all for one year.

Service	Maintenance I	Maintenance II
Change the engine oil and filter. Reset the oil life system.	✓	✓
Visually inspect the vehicle for leaks or damage. A fluid loss in the vehicle system could indicate a problem. Inspected, repair and add fluid to the system if necessary.	✓	✓
Inspect the engine air cleaner filter. If necessary, replace the filter.	✓	✓
Rotate the tires. Inspect the tire inflation pressures and the tire wear.	✓	✓
Visually inspect the brake lines and hoses for proper hook-up, binding, leaks, cracks, chafing, etc. Inspect the disc brake pads for wear and the rotors for surface condition. Inspect the drum brake linings for wear or cracks. Inspect other brake parts, including drums, wheel cylinders, calipers, parking brake, etc. Inspect the parking brake adjustment.	✓	✓
Inspect the engine coolant and the windshield washer fluid levels. Add fluid as needed.	✓	✓
Inspect the suspension and steering components. Inspect the front and rear suspension and the steering system for damaged, loose or missing parts, or signs of wear. Inspect the power steering lines and the hoses for proper hook-up, binding, leaks, cracks, chafing, etc.	--	✓
Visually inspect the coolant hoses and replace the hoses if they are cracked, swollen or deteriorated. Inspect all pipes, fittings and clamps; replace with GM parts as needed. To help ensure proper operation, a pressure test of the cooling system and pressure cap and cleaning the outside of the radiator and air conditioning condenser is recommended at least once a year.		✓
Inspect the wiper blades for wear or cracking.	--	✓
Inspect the restraint system components. Ensure the safety belt reminder light and all the belts, buckles, latch plates, retractors and anchorages are working properly. Look for any other loose or damaged safety belt system parts. If you see anything that might keep a safety belt system from working correctly, repair or replaced the damaged part. Replace torn or frayed safety belts, refer to Operational and Functional Checks in Seat Belts. Inspect for any opened or broken air bag coverings, and repair or replace as needed. The air bag system does require regular maintenance.	--	✓

06025-GTO-C12

MAINTENANCE I AND II SERVICE SCHEDULES
2004-05 Pontiac GTO

Lubricate the body components.Lubricate all key lock cylinders, hood latch assemblies, secondary latches, pivots, spring anchor and release pawl, hood and door hinges, rear folding seats and liftgate hinges. Frequent lubrication may be required when exposed to a corrosive environment, refer to Fluid and Lubricant Recommendations . Applying dielectric silicone grease GM P/N 12345579 (Canadian P/N 1974984) or equivalent on the weatherstrips with a clean cloth.	--	✓
Inspect the transaxle fluid level and add fluid as needed.	--	✓
Inspect the suspension and steering components.Inspect the front and rear suspension and the steering system for damaged, loose or missing parts, or signs of wear. Inspect power steering lines and hoses for proper hook-up, binding, leaks, cracks, chafing, etc.	--	✓
Inspect the throttle system for interference or binding and for damaged or missing parts. Replace the parts as needed. Replace any components that have high effort or excessive wear. Do not lubricate the accelerator or the cruise control cables.	--	✓
Replace the passenger compartment air filter.	--	✓

06025-GTO-C13

ENGINE REPAIR

➡**Disconnecting the negative battery cable on some vehicles may interfere with the functions of the on board computer system. The computer may undergo a relearning process once the negative battery cable is reconnected.**

Distributor

The 5.7L and 6.0L engines use Distributorless Ignition Systems (DIS).

Alternator

REMOVAL & INSTALLATION

1. Disconnect the negative battery cable.
2. Remove the accessory drive belt.
3. Raise and support the vehicle.
4. Disconnect the positive alternator cable.
5. Remove the alternator rear mounting bracket.
6. Disconnect the oil cooler lines from the retainer.
7. Disconnect the alternator electrical connectors.
8. Remove the alternator front mounting bracket.
9. Remove the alternator from the vehicle.
   To install:
10. Position the alternator onto engine. Tighten the front mounting bolts to 37 ft. lbs. (50 Nm).
11. Connect the alternator electrical connectors.
12. Connect the oil cooler lines to the retainer.
13. Install the alternator rear mounting bracket and tighten the bolt to 18 ft. lbs. (25 Nm).
14. Connect the positive alternator cable.

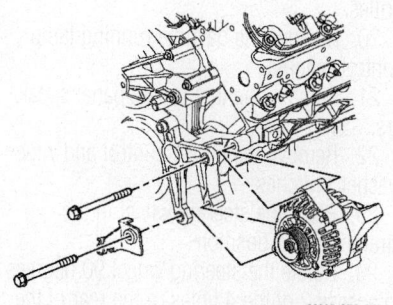

06025-GTO-G06

Alternator mounting—5.7L (VIN G) and 6.0L (VIN U) engines

15. Lower the vehicle.
16. Install the accessory drive belt.
17. Connect the negative battery cable.

Ignition Timing

ADJUSTMENT

The ignition timing is controlled by the Powertrain Control Module (PCM). No adjustment is necessary.

Engine Assembly

REMOVAL & INSTALLATION

1. Discharge and recover the air conditioning refrigerant.
2. Drain the engine cooling system.
3. Drain the engine oil and remove the oil filter.
4. Disconnect the negative battery cable.
5. Relieve the fuel system pressure.
6. Remove the hood.
7. Remove the front suspension support brace.
8. Remove the fuel rail covers.
9. Disconnect the air intake sensor connector
10. Remove the intake duct.
11. Remove the radiator.
12. Disconnect the heater hoses from the water pump.
13. Remove the heater and radiator hoses from the water pump.
14. Separate the A/C compressor and condenser hose from the compressor.
15. Remove the ground lead screw from the block and left hand engine mount.
16. Remove the battery harness ground terminals from the ABS/TCS control module.
17. Disconnect the positive battery lead and lay the harness on the engine.
18. Disconnect the A/C harness connector.
19. Disconnect the wiring harness retaining clips from the engine compartment.
20. Disconnect the connector from the theft deterrent horn.
21. Remove the coolant surge tank.
22. Remove the PCM harness connector cover.
23. Loosen the connector retaining screws at each PCM connector, then remove the connectors from the PCM.

24. Disconnect the PCM wiring harness retaining clip.
25. Remove the engine wiring harness retaining clip from the power steering pipe bracket.
26. Remove the Powertrain Interface Module (PIM) and the Throttle Relaxer Module (TRM) as an assembly.
27. Disconnect the powertrain to main wiring harness connector.
28. Remove the harness to dash panel grommet and feed the harness and connectors out into the engine bay.
29. Place the powertrain wiring harness on top of the engine.
30. Lift the throttle cable up at the throttle body mounting bracket, then remove the cable from the throttle body cam lever. Set the throttle cable aside.
31. Disconnect the fuel line from the fuel rail.
32. Disconnect the purge line from the purge valve.
33. Loosen the hose clamp on the return hose at the power steering reservoir.
34. Place a suitable container under the reservoir and remove the hose and drain the reservoir fluid.
35. Remove the high pressure line flare nut and O-ring from the pump outlet fitting at the rear of the power steering pump.
36. Disconnect the brake booster vacuum hose and heater control vacuum hose from the rear of the intake manifold.
37. Remove the four bolts securing the undertray to the crossmember.
38. Remove the tray from the crossmember.
39. Remove the 2 bolts securing the power steering high pressure line brackets to the oil pan.
40. Remove both exhaust manifolds.
41. Remove the transmission.
42. Remove the right and left engine mount to engine bracket nuts.
43. Attach a suitable lifting chain and hooks to the 2 engine lifting brackets.
44. Using a suitable lifting crane, slightly raise the engine to clear the engine mount stud.
45. Slowly lift the engine out of the vehicle.
   To install:
46. Lower the engine into the vehicle and crossmember using a suitable hoist and lifting brackets.
47. Install left and right engine mount nuts and tighten the nuts to 59 ft. lbs. (80 Nm).

48. Remove the lifting brackets.
49. Install the transmission.
50. Install the exhaust manifolds.
51. Install the power steering high pressure line brackets to the oil pan and tighten to 18 ft. lbs. (25 Nm).
52. Install the tray to the crossmember.
53. Connect the brake booster vacuum hose and heater control vacuum hose to the rear of the intake manifold.
54. Install the high pressure line flare nut and O-ring to the pump outlet fitting at the rear of the power steering pump.
55. Install the hose to the reservoir fluid. Secure the hose with the hose clamp
56. Connect the fuel line to the fuel rail.
57. Connect the purge line to the purge valve.
58. Install the cable to the throttle body cam lever then install the throttle cable to the throttle body mounting bracket.
59. Install the cable to the retaining clip.
60. Feed the harness and connectors into the passenger compartment and install the dash panel grommet.
61. Connect the powertrain to main wiring harness connector.
62. Install harness straps to retain the wiring harness.
63. Install the Powertrain Interface Module (PIM) and the Throttle Relaxer Module (TRM).
64. Install the engine wiring harness retaining clip to the power steering pipe bracket.
65. Connect the PCM wiring harness retaining clip.
66. Connect the connectors to the PCM.
67. Install the PCM harness connector cover.
68. Install the coolant surge tank.
69. Connect the connector to the theft deterrent horn.
70. Connect the wiring harness retaining clips to the engine compartment.
71. Connect the A/C wiring harness connector.
72. Connect the positive lead terminal to the battery.
73. Install the nut securing the battery harness ground terminals to the ABS/TCS control module bracket stud.
74. Install the ground lead screw to the engine block and left hand engine mount.
75. Install new O-rings onto the A/C compressor hose block fitting and install the A/C compressor and condenser hose to the A/C compressor.
76. Install the heater hoses to the water pump. Secure with the clamps.
77. Install the radiator hoses to the water pump. Secure with the clamps.

78. Install the radiator.
79. Install the intake duct and secure the intake duct with the hose clamps.
80. Connect the air intake sensor connector to the air intake sensor.
81. Install the fuel rail covers.
82. Install the front suspension support brace.
83. Install the hood.
84. Recharge the A/C system.

✳✳ WARNING

Be sure to check engine to see that all electrical connectors, hoses and cables are properly connected and secure.

85. Install new oil filter and fill engine with fresh oil.
86. Connect the negative battery cable.
87. Fill the cooling system.
88. Recharge the A/C system.
89. Start the engine and check for leaks.

Water Pump

REMOVAL & INSTALLATION

1. Disconnect the negative battery cable.
2. Drain the engine cooling system.
3. Disconnect the Mass Air Flow (MAF) sensor electrical connector.
4. Disconnect the Intake Air Temperature (IAT) sensor electrical connector.
5. Loosen two intake duct clamps and remove the duct.
6. Remove the accessory drive belt.
7. Disconnect the heater hoses from the water pump.
8. Remove the drive belt tensioner.
9. Remove the 6 attaching bolts and remove the water pump. Clean and inspect the gasket mating surfaces.

To install:
10. Install the water pump. Use a new

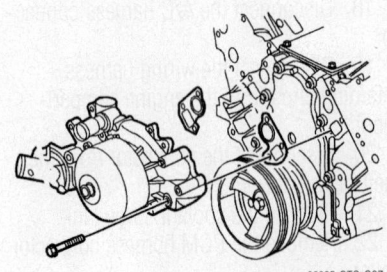

06025-GTO-G07

Exploded view of the water pump mounting—5.7L (VIN G) and 6.0L (VIN U) engines

gasket and tighten the bolts in 2 steps 11 ft. lbs. (15 Nm), then 18 ft. lbs. (25 Nm).
11. Install the drive belt tensioner.
12. Connect the heater hoses to the water pump.
13. Install the accessory drive belt.
14. Install the MAF and IAT connectors.
15. Install the intake duct and the clamps.
16. Refill the cooling system.
17. Connect the negative battery cable.
18. Start the engine and check for coolant leaks.

Heater Core

REMOVAL AND INSTALLATION

1. Disable the air bag system.
2. Disconnect the negative battery cable.
3. Recover the A/C refrigerant, if equipped.
4. Drain the cooling system.
5. Remove the floor console.
6. Remove the closeout insulator panels.
7. Remove the instrument panel compartment door.
8. Remove the instrument panel compartment lamp.
9. Remove the instrument panel compartment lamp switch.
10. Remove the instrument panel lower trim retainer panel.
11. Remove the instrument panel center trim.
12. Remove the instrument panel lower extension.
13. Remove the HVAC manual controls.
14. Remove the radio housing bracket.
15. Remove the instrument cluster.
16. Remove the headlamp switch.
17. Remove the driver information center switch.
18. Remove the instrument panel outer covers.
19. Remove the windshield defroster grilles.
20. Remove the daytime running lamp control module.
21. Remove the instrument panel speakers.
22. Remove the cruise control and wiper washer switches.
23. Place the steering wheel in the straight ahead position
24. Rotate the steering wheel 90 degrees to access 2 of the 4 holes in the rear of the steering column.
25. Using locating tool EL-46844,

relieve the tension on the 2 spring loaded retaining clips. When the tension is released the steering wheel module will away from the wheel.

26. Rotate the steering wheel 180 degrees and repeat the procedure on the other 2 access holes.

27. Disconnect the air bag module connector.

28. Remove the air bag.

29. Lock the steering wheel with it in the straight ahead position.

30. Remove the steering column trim covers.

31. Disconnect the horn connector.

32. Match mark the steering wheel-to-steering column position and remove the steering wheel bolt.

33. Remove the steering wheel.

34. Remove the 2 inner screws attaching the instrument panel inflatable restraint and instrument panel passenger's bracket and brace assembly to the instrument panel inflatable restraint bracket.

35. Remove the 12 lower screws and 5 upper screws from the instrument panel pad, and remove the pad.

36. From the rear side of the instrument panel pad, remove the 4 screws attaching the instrument panel inflatable restraint opening trim cover to the instrument panel inflatable restraint bracket.

37. While pushing on the trim cover, detach the 4 tabs and remove the trim cover.

38. From the rear of the instrument pad, remove the 4 screws, attaching the instrument panel inflatable restraint bracket to the pad assembly.

39. Remove the 6 screws, and remove the bracket from the pad assembly.

40. Disconnect the instrument panel passenger air bag module connector, the module retaining nuts and bolts and the passenger air bag module.

41. Remove the right side HVAC duct from the HVAC unit.

42. Remove the lower radio bracket.

43. Remove the instrument panel right end and center support bracket.

44. Remove the radio antenna lead from the three retaining clips on the HVAC unit.

45. Mark and remove the heater hoses from the heater core pipes.

46. Disconnect the water valve vacuum hose.

47. Disconnect the vacuum supply hose from the check valve.

48. Remove the Thermal Expansion Valve (TXV).

49. Remove the fuel line retaining bracket and nut from the HVAC mounting stud.

50. Remove the HVAC unit mounting stud and screws.

51. Remove the HVAC unit from the vehicle.

52. Remove the two heater core pipe clamp screws.

53. Remove the two heater core pipe clamps.

54. Remove the heater core retaining strap mounting screw.

55. Remove the heater core retaining strap.

56. Remove the heater pipe bracket mounting screws.

57. Remove the heater pipe bracket

58. Remove the heater core from the HVAC module.

59. Remove the heater pipe retaining screws.

60. Remove the heater pipe retaining clamps.

61. Remove the heater pipes from the heater core.

62. Remove and discard the O-rings.

To install:

63. Install new O-rings on the heater pipes.

64. Install the heater pipes to the heater core.

65. Secure the heater pipes with the retaining clamps and screws.

66. Install the heater core to the HVAC module.

67. Install the heater pipe bracket.

68. Secure the heater pipe bracket with the mounting screws

69. Install the heater core retaining strap.

70. Secure the heater core retaining strap with the mounting screw.

71. Install the two heater core pipe clamps.

72. Secure the two heater core pipe clamps with the screws.

73. Install the HVAC module assembly.

74. Install the screws from inside the vehicle and the engine compartment which secure the HVAC module assembly to the dash.

75. Install the fuel line retaining bracket and nut to the HVAC mounting stud.

76. Install the Thermal Expansion Valve (TXV).

77. Connect the vacuum supply hose to the check valve.

78. Connect the water valve vacuum hose.

79. Install the heater hoses to the heater core pipes.

80. Install the radio antenna lead to the three retaining clips on the HVAC unit.

81. Install the instrument panel center support bracket.

82. Install the instrument panel right end bracket.

83. Install the lower radio bracket.

84. Install the right side HVAC duct to the HVAC unit.

85. Install the inflatable restraint I/P module and retaining nuts and tighten to 33 ft. lbs. (45 Nm).

86. Install the I/P module retaining bolts and tighten to 84 inch lbs. (10 Nm).

87. Connect the I/P module connector.

88. Place the passenger restraint bracket in position ensuring the lower screw hole tabs are positioned in the pad assembly.

89. Install the 6 screws attaching the bracket to the pad assembly.

90. Install the 4 screws, attaching the instrument panel inflatable restraint bracket to the pad assembly.

91. Locate the trim cover in the pad assembly opening, and align the trim cover with the 4 tabs.

92. Install the 4 screws attaching the instrument panel inflatable restraint trim cover to the inflatable restraint bracket.

93. Install the instrument panel pad to the vehicle and tighten the upper screws to 80 inch lbs. (9 Nm).

94. Install the 12 lower instrument pad screws and tighten to 18 inch lbs. (2 Nm).

95. Install the 2 inner screws attaching the instrument panel inflatable restraint and instrument panel passenger's bracket and brace assembly to the instrument panel inflatable restraint bracket.

96. If replacing the steering wheel, copy the match marks from the old steering wheel to the new steering wheel.

97. Verify the front tires have not moved since the removal of the steering wheel.

98. Verify the green indexing tab on the SIR coil is aligned with the window in the coil casing. The green tab indicates the coil is locked in the centralized position.

99. Align the match marks on the steering wheel and on the steering shaft.

100. Install the steering wheel to the steering shaft.

101. Clean the threads on the steering wheel bolt and on the steering shaft and apply Loctite 242 or the equivalent to the steering wheel bolt.

102. Install the bolt to the steering wheel and tighten to 33 ft. lbs. (45 Nm).

103. Connect the horn connector.

104. Connect the driver air bag wiring harness connectors.

105. Connect the air bag electrical connector.

106. Install the air bag module. Align the module to the 4 retaining clips on the steer-

ing wheel and press the steering wheel module toward the steering wheel until the retaining clips lock.

107. Install the steering column trim covers.

108. Install the wiper washer and cruise control switches.

109. Install the instrument panel speakers.

110. Install the daytime running lamp control module.

111. Install the windshield defroster grilles.

112. Install the instrument panel outer covers.

113. Install the instrument cluster.

114. Install the driver information center switch.

115. Install the headlamp switch.

116. Install the radio housing bracket.

117. Install the HVAC manual controls.

118. Install the instrument panel lower extension.

119. Install the instrument panel center trim.

120. Install the instrument panel lower trim retainer panel.

121. Install the instrument panel compartment lamp switch and light.

122. Install the instrument panel compartment door.

123. Install the closeout insulator panels.

124. Install the floor console.

125. Enable the air bag system.

126. Connect the negative battery cable.

127. Fill the cooling system.

128. Recharge the A/C system.

129. Start the engine and check for leaks.

Cylinder Head

REMOVAL & INSTALLATION

Left Side

1. Disconnect the negative battery cable from the battery.

2. Remove the front suspension support brace.

3. Drain the cooling system.

4. Relieve the fuel system pressure.

5. Drain the engine oil.

6. Remove the fuel rail covers.

7. Disconnect the air intake sensor connector from the air intake sensor.

8. Loosen the hose clamps securing the intake duct and remove the intake duct.

9. Disconnect the fuel line from the fuel rail.

10. Disconnect the purge line from the purge valve.

11. Disconnect the spark plug wires from the ignition coils.

12. Disconnect the ignition coil wire harness main electrical connector.

13. Remove the ignition coil bracket bolts and screw from the rocker arm cover.

14. Remove the ignition coil bracket from the rocker arm cover.

15. Remove the valve cover.

16. Remove the valve rocker arm bolts and the rocker arms.

17. Remove the valve rocker arm pivot support.

18. Remove the pushrods.

19. Remove the engine coolant air bleed pipe.

20. Remove the two power steering pump bolts and power steering pump from the cylinder head.

21. Remove the exhaust manifold.

22. Remove the engine wiring harness ground bolts from the rear of the cylinder head. Reposition the ground wires.

23. Remove the cylinder head and discard the bolts.

24. Remove the cylinder head gasket and discard.

25. Clean and inspect the cylinder head.

To install:

26. Install the new cylinder head gasket on the locating pins and install the cylinder head.

➡**The cylinder head bolts are a torque-to-yield design and cannot be reused.**

27. Tighten the new cylinder head bolts on 5.7L engines in sequence as follows:

a. Tighten the M11 cylinder head bolts (1-10) a first pass in sequence to 22 ft. lbs. (30 Nm).

b. Tighten the M11 cylinder head bolts (1-10) a second pass in sequence an additional 90 degrees.

c. Tighten the M11 cylinder head bolts (1-8) an additional 90 degrees and the M11 cylinder head bolts (9 and 10) an additional 50 degrees.

d. Tighten the M8 cylinder head bolts (11-15) to 22 ft. lbs. (30 Nm). Begin with the center bolt (11) and alternating side-to-side, work outward tightening all of the bolts.

06025-GTO-G08

Left side cylinder head bolt tightening sequence—5.7L (VIN G) and 6.0L (VIN U) engines

to-side, work outward tightening all of the bolts.

28. Tighten the new cylinder head bolts on 6.0L engines in sequence as follows:

a. Tighten the M11 cylinder head bolts (1-10) a first pass in sequence to 22 ft. lbs. (30 Nm).

b. Tighten the M11 cylinder head bolts (1-10) a second pass in sequence an additional 90 degrees.

c. Tighten the M11 cylinder head bolts (1-10) an additional 70 degrees.

d. Tighten the M8 cylinder head bolts (11-15) to 22 ft. lbs. (30 Nm). Begin with the center bolt (11) and alternating side-to-side, work outward tightening all of the bolts.

29. Install the engine wiring harness ground bolts from the rear of the cylinder head. Reposition the ground wires.

30. Install the exhaust manifold.

31. Install the two power steering pump bolts and power steering pump to the cylinder head and tighten the bolts to 21 ft. lbs. (28 Nm).

32. Install the engine coolant air bleed pipe.

33. Install the pushrods.

34. Install the valve rocker arm pivot support.

35. Install the valve rocker arm bolts and the rocker arms.

36. Install a NEW cover gasket into the valve cover.

37. Install the valve cover and tighten the bolts to 106 inch lbs. (12 Nm).

38. Install the ignition coils, wire harness, and bolts onto the mounting bracket and tighten to 106 inch lbs. (12 Nm).

39. Install the ignition coils and bracket assembly and bolts onto the rocker cover and tighten to 106 inch lbs.

40. Install the crankcase ventilation hose.

41. Connect the ignition coil wire harness main electrical connector.

42. Connect the spark plug wires to the ignition coils.

43. Connect the purge line to the purge valve.

44. Connect the fuel line to the fuel rail.

45. Install the intake duct and hose clamps.

46. Connect the air intake sensor connector to the air intake sensor.

47. Install the fuel rail covers.

48. Fill the engine with oil.

49. Fill the cooling system.

50. Install the front suspension support brace.

51. Connect the negative battery cable.

52. Start the engine and check for leaks.

Right Side

1. Disconnect the negative battery cable from the battery.
2. Drain the cooling system.
3. Relieve the fuel system pressure.
4. Drain the engine oil.
5. Remove the heater hoses.
6. Remove the fuel rail covers.
7. Remove the front suspension support brace if necessary.
8. Disconnect the air intake sensor connector from the air intake sensor.
9. Loosen the hose clamps securing the intake duct and remove the duct.
10. Remove the spark plug wires from the ignition coils.
11. Disconnect the ignition coil wire harness main electrical connector.
12. Remove the ignition coil bracket from the rocker arm cover.
13. Remove the valve cover.
14. Remove the valve rocker arms and pushrods.
15. Remove the engine coolant air bleed pipe.
16. Remove the exhaust manifold.
17. Remove the cylinder head and discard the bolts.
18. Remove the cylinder head gasket and discard.
19. Clean and inspect the cylinder head.

To install:

20. Install the new cylinder head gasket on the locating pins and install the cylinder head.

➡**The cylinder head bolts are a torque-to-yield design and cannot be reused.**

21. Tighten the new cylinder head bolts on 5.7L engines in sequence as follows:
    a. Tighten the M11 cylinder head bolts (1-10) a first pass in sequence to 22 ft. lbs. (30 Nm).
    b. Tighten the M11 cylinder head bolts (1-10) a second pass in sequence an additional 90 degrees.
    c. Tighten the M11 cylinder head bolts (1-8) an additional 90 degrees and

the M11 cylinder head bolts (9 and 10) an additional 50 degrees.
    d. Tighten the M8 cylinder head bolts (11-15) to 22 ft. lbs. (30 Nm). Begin with the center bolt (11) and alternating side-to-side, work outward tightening all of the bolts.

22. Tighten the new cylinder head bolts on 6.0L engines in sequence as follows:
    a. Tighten the M11 cylinder head bolts (1-10) a first pass in sequence to 22 ft. lbs. (30 Nm).
    b. Tighten the M11 cylinder head bolts (1-10) a second pass in sequence an additional 90 degrees.
    c. Tighten the M11 cylinder head bolts (1-10) an additional 70 degrees.
    d. Tighten the M8 cylinder head bolts (11-15) to 22 ft. lbs. (30 Nm). Begin with the center bolt (11) and alternating side-to-side, work outward tightening all of the bolts.

23. Install the exhaust manifold.
24. Install the engine coolant air bleed pipe.
25. Install the valve rocker arms and pushrods.
26. Install the valve cover and tighten the bolts to 106 inch lbs. (12 Nm).
27. Install the ignition coils, wire harness, and bolts onto the mounting bracket and tighten to 106 inch lbs. (12 Nm).
28. Install the ignition coils and bracket assembly and bolts onto the rocker cover and tighten to 106 inch lbs.
29. Install the spark plug wires to the ignition coils.
30. Install the intake duct and hose clamps securing the intake duct.
31. Connect the air intake sensor connector to the air intake sensor.
32. Install the front suspension support brace if removed.
33. Install the fuel rail covers.
34. Install the heater hoses.
35. Fill the engine with oil.
36. Fill the cooling system.
37. Connect the negative battery cable.
38. Start the engine and check for leaks.

Rocker Arms

REMOVAL & INSTALLATION

➡**Place valve rocker arms, valve pushrods, and pivot support, in a rack so that they can be installed in the same location from which they were removed.**

1. Disconnect the negative battery cable.

2. Remove the valve covers. See the procedure under Cylinder Head.
3. Remove the valve rocker arm bolts.
4. Remove the valve rocker arms.
5. Remove the valve rocker arm pivot support.
6. Remove the push rods.
7. Remove the valve lifter guide bolts (211).
8. Remove the valve lifters and guide.
9. Remove the valve lifters from the guide.
10. Organize or mark the components so that they can be installed in the same location from which they were removed.

To install:

➡**If the camshaft is being replaced, the valve lifters must also be replaced.**

11. Lubricate the valve lifters and engine block valve lifter bores with clean engine oil.
12. Insert the valve lifters into the lifter guides. Align the flat area on the top of the lifter with the flat area in the lifter guide bore. Push the lifter completely into the guide bore
13. Install the valve lifters and guide assembly to the engine block.
14. Install the valve lifter guide bolt and tighten to 106 inch lbs. (12 Nm) on 5.7L engine, or 89 inch lbs. (10 Nm) on 6.0L engines.
15. Lubricate the valve rocker arms and pushrods with clean engine oil.
16. Lubricate the flange of the valve rocker arm bolts with clean engine oil. Lubricate the flange or washer surface of the bolt that will contact the valve rocker arm.
17. Install the valve rocker arm pivot support.
18. Install the pushrods ensuring that they seat properly to the ends of the rocker arms.

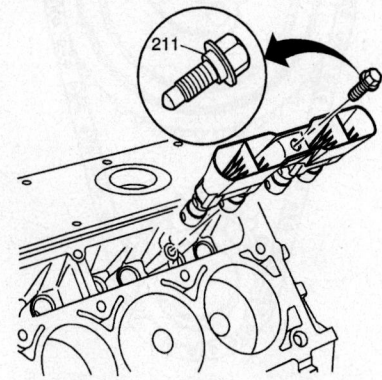

06025-GTO-G09

Removing valve lifter guide bolts—5.7L (VIN G) and 6.0L (VIN U) engines

06025-GTO-G08

Right side cylinder head bolt tightening sequence—5.7L (VIN G) and 6.0L (VIN U) engines

19. Install the rocker arms and bolts but do not tighten.

20. Rotate the crankshaft until number one piston is at top dead center of compression stroke. In this position, cylinder number one rocker arms will be off lobe lift, and the crankshaft sprocket key will be at the 1:30 position. If viewing from the rear of the engine, the additional crankshaft pilot hole, non-threaded, will be in the 10:30 position. The engine firing order is 1, 8, 7, 2, 6, 5, 4,

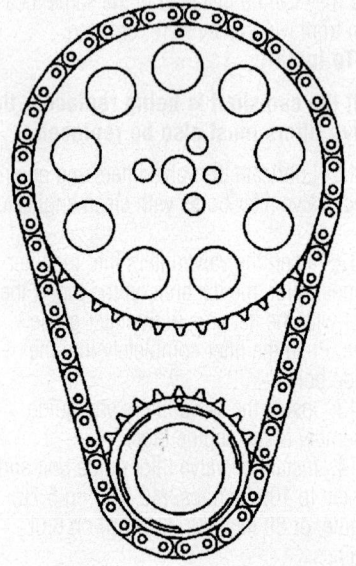

06025-GTO-G10

Positioning crankshaft sprocket keyway at the 1:30 position—5.7L (VIN G) engine

06025-GTO-G11

Positioning crankshaft sprocket keyway at the 1:30 position—6.0L (VIN U) engine

3. Cylinders 1, 3, 5 and 7 are left bank. Cylinders 2, 4, 6, and 8 are right bank.

21. With the engine in the number one firing position, tighten the following valve rocker arm bolts as follows:

   a. Tighten exhaust valve rocker arm bolts 1, 2, 7, and 8 to 22 ft. lbs. (30 Nm).

   b. Tighten intake valve rocker arm bolts 1, 3, 4, and 5 to 22 ft. lbs. (30 Nm).

22. Rotate the crankshaft 360 degrees and tighten the following valve rocker arm bolts as follows:

   a. Tighten exhaust valve rocker arm bolts 3, 4, 5, and 6 to 22 ft. lbs. (30 Nm).

   b. Tighten intake valve rocker arm bolts 2, 6, 7, and 8 to 22 ft. lbs. (30 Nm).

23. Install the valve covers.

24. Connect the negative battery cable.

25. Start the engine and check for proper operation.

Intake Manifold

REMOVAL & INSTALLATION

➡The intake manifold, throttle body, fuel injection rail, and fuel injectors may be removed as an assembly. If not servicing the individual components, remove the manifold as a complete assembly.

1. Relieve the fuel system pressure.
2. Disconnect the negative battery cable.
3. Drain the cooling system.
4. Remove the front suspension support brace.
5. Remove the fuel rail covers.
6. Disconnect the air intake sensor connector from the air intake sensor.
7. Loosen the hose clamps securing the intake duct and remove the duct.
8. Disconnect the fuel line from the fuel rail.
9. Disconnect the purge line from the purge valve.
10. Disconnect the connector from the Intake Air Control (IAC) motor.
11. Disconnect the connector from the Throttle Position (TP) sensor at the throttle body.
12. Lift the throttle cable up at the throttle body mounting bracket, then remove the cable from the throttle body cam lever.
13. Position the throttle cable aside.
14. Disconnect the fuel injector connectors from all the fuel injectors.
15. Remove the locks from the right and left side ignition coil main connectors and disconnect the connectors.
16. Disconnect the EVAP canister purge solenoid valve electrical connector.

17. Remove the harness securing clips from the fuel rail brackets and position the harnesses aside.

18. Disconnect the manifold absolute pressure (MAP) sensor connector.

19. Disconnect the knock sensor connector.

20. Disconnect the oil pressure sensor connector.

21. Remove the fresh air hose from the rocker cover and the throttle body.

22. Remove the foul air hose from the throttle body and restricting orifice external connector.

23. Remove the engine coolant vent hose from the throttle body and vent pipe.

24. Remove the coolant vent outlet hose from the throttle body and left-hand radiator tank.

25. Remove the EVAP canister purge tube from the throttle body and EVAP canister purge solenoid valve.

26. Remove the EVAP canister purge solenoid valve and bracket from the intake manifold.

27. Remove the intake manifold bolts and fuel rail stop bracket.

28. Remove the intake manifold and discard the gaskets.

To install:

29. Install the intake manifold. Use new gaskets and apply threadlock to the manifold bolts.

30. Install the fuel rail stop bracket.

31. Tighten the manifold bolts, in sequence first to 44 inch lbs. (5 Nm), then to 89 inch lbs. (10 Nm).

32. Install the EVAP canister purge solenoid valve and bracket to the intake manifold.

33. Install the EVAP canister purge tube to the throttle body and EVAP canister purge solenoid valve.

34. Install the engine coolant vent hose to the throttle body and vent pipe.

35. Install the coolant vent outlet hose to the throttle body and left-hand radiator tank.

36. Install the fresh air hose to the rocker cover and the throttle body.

37. Install the foul air hose to the throttle body and restricting orifice external connector.

38. Connect the MAP sensor connector.

39. Connect the knock sensor connector.

40. Connect the oil pressure sensor connector.

41. Connect the fuel injector connectors to all the fuel injectors.

42. Install the locks to the right and left side ignition coil main connectors and connect the connectors.

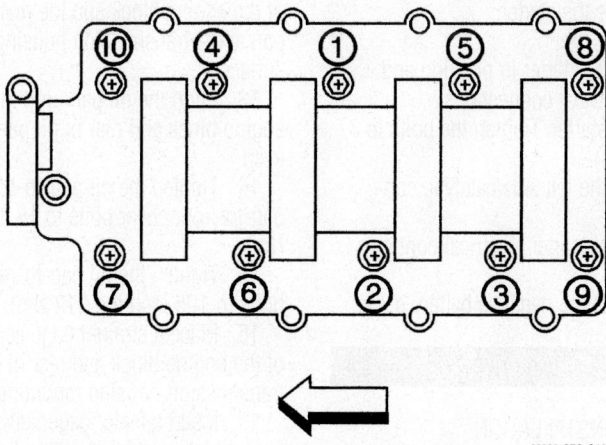

06025-GTO-G12

Intake manifold tightening sequence—5.7L (VIN G) and 6.0L (VIN U) engines

43. Connect the EVAP canister purge solenoid valve electrical connector.

44. Install the harness securing clips to the fuel rail brackets.

45. Install the cable to the throttle body cam lever then install the throttle cable to the throttle body mounting bracket.

46. Install the cable to the retaining clip.

47. Connect the connector to IAC motor.

48. Connect the connector to the TP sensor at the throttle body.

49. Connect the fuel line to the fuel rail.

50. Connect the purge line to the purge valve.

51. Install the intake duct and secure with the hose clamps.

52. Connect the air intake sensor connector to the air intake sensor.

53. Install the fuel rail covers.

54. Install the front suspension support brace.

55. Fill the cooling system.

56. Connect the negative battery cable.

57. Start the engine and check for leaks.

Engine Valley Cover

REMOVAL & INSTALLATION

1. Remove the intake manifold.

2. Remove the oil pressure switch.

3. Remove the 11 valley cover bolts.

4. Remove the valley cover and gasket.

5. Remove and discard the o-rings on the bottom side of the cover.

To install:

6. Install new o-rings to the bottom of the cover.

7. Install a new gasket and the cover.

8. Install the bolts and tighten to 18 ft. lbs. (25 Nm).

9. Install the oil pressure switch.

10. Install the intake manifold.

Exhaust Manifold

REMOVAL & INSTALLATION

1. Disconnect the negative battery cable.

2. Remove the spark plug wires from the spark plugs.

3. Remove the exhaust pipe from the manifold.

4. Remove the heat shield.

5. Remove the exhaust manifold.

To install:

6. Apply threadlock to the manifold bolts.

7. Install the manifold and tighten the bolts from the center out first to 12 ft. lbs. (15 Nm), and then to 18 ft. lbs. (25 Nm).

8. Install the heat shield.

9. Connect the exhaust pipe.

10. Connect the spark plug wires.

11. Connect the negative battery cable.

Camshaft

REMOVAL & INSTALLATION

1. Drain the cooling system.

2. Disconnect the negative battery cable.

3. Remove the engine.

4. Remove the crankshaft balancer.

5. Remove the oil level indicator tube.

6. Remove the left and right exhaust manifolds.

7. Remove the water pump.

8. Remove the intake manifold.

9. Remove the coolant air bleed pipe.

10. Remove the valve covers.

11. Remove the valve rocker arms and push rods.

12. Remove the right cylinder heads.

13. Remove the valve lifters.

14. Remove the oil pan-to-front cover bolts.

15. Remove the engine front cover.

16. Rotate the crankshaft until number one piston is at top dead center of compression stroke. In this position, cylinder number one rocker arms will be off lobe lift, and the crankshaft sprocket key will be at the 1:30 position. If viewing from the rear of the engine, the additional crankshaft pilot hole, non-threaded, will be in the 10:30 position. The engine firing order is 1, 8, 7, 2, 6, 5, 4, 3. Cylinders 1, 3, 5 and 7 are left bank. Cylinders 2, 4, 6, and 8 are right bank.

17. Remove the camshaft sensor.

18. Remove the camshaft retainer.

19. Install 3 M8-1.25 x 100 mm bolts in the camshaft front bolt holes.

20. Using the bolts as a handle, carefully rotate and pull the camshaft out of the engine block.

21. Remove the bolts from the front of the camshaft.

22. If the camshaft bearings need to be replaced, remove the oil pan-to rear cover bolts.

23. Remove the rear cover bolts.

24. Remove the rear cover and gasket. Discard the gasket.

25. Remove the camshaft bearings.

To install:

26. Install the camshaft bearings, if replaced.

27. Install the rear cover using a new gasket.

28. Install the oil pan-to-rear cover bolts.

29. Lubricate the camshaft journals and the bearings with clean engine oil.

30. Install 3 M8-1.25 x 100 mm bolts into the camshaft front bolt holes.

31. Using the bolts as a handle, carefully install the camshaft into the engine block.

32. Remove the 3 bolts from the front of the camshaft.

33. Install the camshaft retainer and the bolts and tighten to 18 ft. lbs. (25 Nm).

34. Inspect the camshaft sensor O-ring seal. If the O-ring seal is not cut or damaged, it may be used again.

35. Lubricate the O-ring seal with clean engine oil.

36. Install the camshaft sensor and tighten the bolts to 18 ft. lbs. (25 Nm).

37. Align the camshaft sprocket alignment mark in the 6 o'clock position.

38. Install the camshaft sprocket and timing chain.

39. Install the camshaft sprocket bolts and tighten to 26 ft. lbs. (35 Nm).

40. Install the engine front cover.
41. Install the valve lifters.
42. Install the cylinder heads.
43. Install the valve rocker arms and push rods.
44. Install the valve covers.
45. Install the coolant air bleed pipe.
46. Install the intake manifold.
47. Install the water pump.
48. Install the exhaust manifolds.
49. Install the oil level indicator tube.
50. Install the crankshaft balancer.
51. Install the engine assembly.
52. Connect the negative battery cable.
53. Start the engine and check for leaks.

Valve Lash

ADJUSTMENT

The 5.7L and 6.0L engines are equipped with hydraulic lash adjusters. Valve clearance is not adjustable.

Starter Motor

REMOVAL & INSTALLATION

1. Disconnect the negative battery cable.
2. Raise and support the vehicle.
3. Remove the left side catalytic converter.
4. Disconnect starter electrical connections.
5. Remove starter bolts.

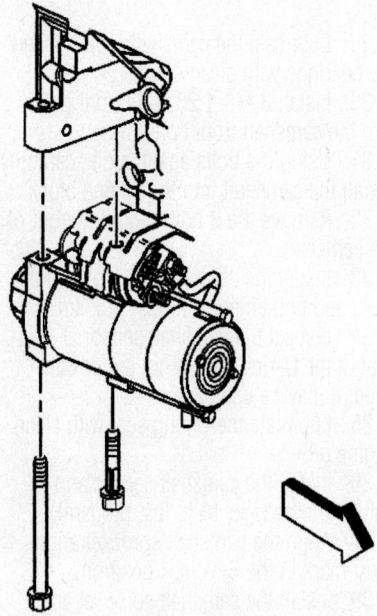

Starter mounting—5.7L (VIN G) and 6.0L (VIN U) engines

06025-GTO-G13

6. Remove the starter.

To install:

7. Place the starter in position and attach the electrical connections.
8. Install starter. Tighten the bolts to 37 ft. lbs. (50 Nm).
9. Install the left side catalytic converter.
10. Connect starter electrical connections.
11. Connect the negative battery cable.

Oil Pan

REMOVAL & INSTALLATION

➡The original oil pan gasket is retained and aligned to the oil pan by rivets. When installing a new gasket, it is not necessary to install new oil pan gasket rivets.

1. Drain the engine oil.
2. Disconnect the negative battery cable.
3. Remove the engine.
4. Remove the left and right closeout covers from the lower engine block.
5. Remove the oil pan bolts and the oil pan.
6. Drill out the oil pan rivets.
7. Remove the gasket and discard it and the rivets.

To install:

➡The alignment of the structural oil pan is critical. The rear bolt hole locations of the oil pan provide mounting points for the transmission housing. To ensure the rigidity of the powertrain and correct transmission alignment, it is important that the rear of the block and the rear of the oil pan are flush, or even. The rear of the oil pan must NEVER protrude beyond the engine block and transmission housing plane.

8. Apply a 5 mm (0.2 in) bead of sealant, 20 mm (0.8 in) long to the engine block. Apply the sealant directly onto the tabs of the front cover gasket that protrude into the oil pan surface. Apply a bead of sealant at the crankshaft rear seal retainer plate-to-cylinder block surface and the engine front cover-to-cylinder block surface.
9. Install the gasket onto the oil pan. Install the oil pan bolts to the pan and through the gasket.
10. Install the oil pan, gasket and bolts to the engine block.
11. Tighten bolts finger tight. Do not overtighten.
12. Place a straight edge across the rear

of the engine block and the rear of the oil pan at the transmission housing mounting surfaces.
13. Align the oil pan until the rear of engine block and rear of oil pan are flush or even.
14. Tighten the oil pan-to-block and oil pan-to-front cover bolts to 18 ft. lbs. (25 Nm).
15. Tighten the oil pan-to-rear cover bolts to 106 inch lbs. (12 Nm).
16. Place a straight edge across the rear of the engine block and rear of oil pan at the transmission housing mounting surfaces.
17. Insert a feeler gage between the straight edge and the oil pan transmission housing mounting surface and check to make sure that there is no more than a 0.25 mm (0.01 in) gap between the pan and straight edge.
18. If the oil pan alignment is not within specifications, remove the oil pan and repeat the above procedure.
19. Install the left and right closeout covers to the lower engine block.
20. Install the engine.
21. Connect the negative battery cable.
22. Fill the engine with oil.
23. Start the engine and check for leaks.

Oil Pump

REMOVAL & INSTALLATION

1. Drain the cooling system and the engine oil.
2. Disconnect the negative battery cable.
3. Disconnect the radiator hoses.
4. Remove the water pump.
5. Remove the radiator.
6. Remove the air conditioning (A/C) drive belt.
7. Remove the starter motor.
8. Remove the right transmission cover and bolt.
9. Lock the flywheel from turning.
10. Remove the crankshaft pulley bolt and the crankshaft damper using a puller.
11. Carefully pry out the front seal.
12. Remove the drive belt idler pulley bolt and pulley.
13. Remove the oil pan-to-front cover bolts.
14. Remove the front cover bolts.
15. Remove the front cover and gasket.
16. Discard the front cover gasket.
17. Remove the oil pan.
18. Remove the oil pump screen and discard the o-ring.
19. Remove the crankshaft oil deflector nuts.

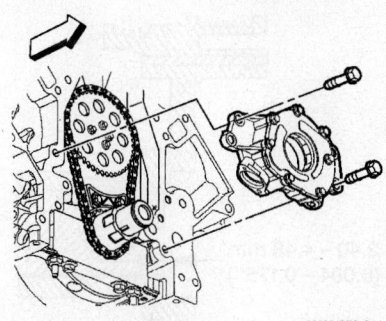

Oil pump mounting—5.7L (VIN G) and 6.0L (VIN U) engines

20. Remove the crankshaft oil deflector.
21. Remove the oil pump.

To install:

➡Inspect the oil pump and engine block oil gallery passages. These surfaces must be clear and free of debris and restrictions.

22. Align the splined surfaces of the crankshaft sprocket and the oil pump drive gear and install the oil pump.
23. Install the oil pump onto the crankshaft sprocket until the pump housing contacts the face of the engine block.
24. Tighten the bolts to 18 ft. lbs. (25 Nm).
25. Install the crankshaft oil deflector.
26. Lubricate a NEW oil pump screen O-ring seal with clean engine oil.
27. Install the NEW O-ring seal onto the oil pump screen.

➡Push the oil pump screen tube completely into the oil pump prior to tightening the bolt. Do not allow the bolt to pull the tube into the pump. Align the oil pump screen mounting brackets with the correct crankshaft bearing cap studs.

28. Install the oil pump screen and tighten the bolt to 106 inch lbs. (12 Nm).
29. Install the oil deflector and tighten the bolts to 18 ft. lbs. (25 Nm).
30. Install the oil pan.
31. Apply a 5 mm (0.20 in) bead of sealant 20 mm (0.80 in) long to the oil pan-to-engine block junction.
32. Install a new front cover gasket.
33. Install the front cover.
34. Install the crankshaft balancer bolt until snug.
35. Install the front cover bolts until snug.
36. Install the oil pan-to-front cover bolts until snug.
37. Install the drive belt idler pulley and tighten the bolt to 37 ft. lbs. (50 Nm).

38. Remove the crankshaft balancer bolt.
39. Use tool J-41665 and install the crankshaft balancer.
40. Install the used crankshaft balancer bolt and tighten to 240 ft. lbs. to position the balancer.
41. Remove the used bolt.
42. Check that the nose of the crankshaft is recessed at least 0.094-0.176 inch (2.40-4.48) into the balancer bore.
43. Install a new crankshaft balancer bolt and tighten first to 37 ft. lbs. (50 Nm) and then an additional 140 degrees.
44. Unlock the flywheel.
45. Install the right transmission cover and bolt.
46. Install the starter motor.
47. Install the air conditioning (A/C) drive belt.
48. Install the radiator.
49. Install the water pump.
50. Connect the radiator hoses.
51. Connect the negative battery cable.
52. Fill the cooling system.
53. Fill the crankcase with clean engine oil.
54. Start the engine. Check for leaks and proper operation.

Rear Main Seal

REMOVAL & INSTALLATION

1. Disconnect the negative battery cable.
2. Remove the transmission.
3. Remove the flywheel.
4. Carefully pry the oil seal from the rear cover.

To install:

5. Lubricate the outside diameter of the oil seal with clean engine oil.
6. Lubricate the rear cover oil seal bore with clean engine oil.
7. Install the J-41479 cone (2) and bolts onto the rear of the crankshaft and tighten until snug.
8. Install the rear oil seal onto the tapered cone and push the seal to the rear cover bore.
9. Thread the J-41479 threaded rod into the tapered cone until the tool (1) contacts the oil seal.
10. Align the oil seal onto the tool (1).
11. Rotate the handle of the tool (1) clockwise until the seal enters the rear cover and bottoms into the cover bore.
12. Remove the tool.
13. Apply threadlock to the flywheel bolts.

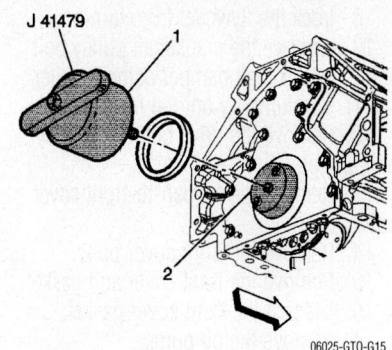

Installing crankshaft rear oil seal—5.7L (VIN G) and 6.0L (VIN U) engines

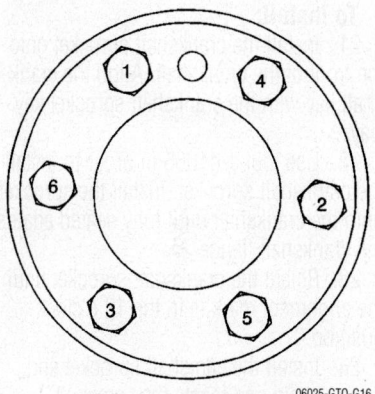

Flywheel bolt installation sequence—5.7L (VIN G) and 6.0L (VIN U) engines

14. Install the flywheel and tighten the bolts in sequence as follows:
   a. First step to 15 ft. lbs. (20 Nm).
   b. Second step to 37 ft. lbs. (50 Nm).
   c. Final step to 74 ft. lbs. (100 Nm).
15. Install the transmission.
16. Connect the negative battery cable.

Timing Chain, Sprockets, Front Cover and Seal

REMOVAL & INSTALLATION

1. Drain the cooling system and the engine oil.
2. Disconnect the negative battery cable.
3. Disconnect the radiator hoses.
4. Remove the water pump.
5. Remove the radiator.
6. Remove the air conditioning (A/C) drive belt.
7. Remove the starter motor.
8. Remove the right transmission cover and bolt.

9. Lock the flywheel from turning.

10. Remove the crankshaft pulley bolt and the crankshaft damper using a puller.

11. Carefully pry out the front seal.

12. Remove the drive belt idler pulley bolt and pulley.

13. Remove the oil pan-to-front cover bolts.

14. Remove the front cover bolts.

15. Remove the front cover and gasket.

16. Discard the front cover gasket.

17. Remove the oil pump.

18. Remove the camshaft sprocket bolts.

19. Remove the camshaft sprocket and timing chain.

20. Use tools J-8433, J-41816-2 and J-41558 and remove the crankshaft sprocket.

To install:

21. Install the crankshaft sprocket onto the front of the crankshaft. Align the crankshaft key with the crankshaft sprocket keyway.

22. Use tool J-41665 in order to install the crankshaft sprocket. Install the sprocket onto the crankshaft until fully seated against the crankshaft flange.

23. Rotate the crankshaft sprocket until the alignment mark is in the 12 o'clock position.

24. Install the camshaft sprocket and timing chain and locate the camshaft sprocket locating pin with the camshaft sprocket alignment hole.

➡**The camshaft and the crankshaft sprocket alignment marks MUST be aligned correctly. Locate the camshaft sprocket alignment mark in the 6 o'clock position.**

25. Rotate the crankshaft until number one piston is at top dead center of compression stroke.

26. Install the camshaft sprocket bolts and tighten to 26 ft. lbs. (35 Nm).

27. Install the oil pump.

28. Apply a 5 mm (0.20 in) bead of sealant 20 mm (0.80 in) long to the oil pan-to-engine block junction.

29. Install a new front cover gasket.

30. Install the front cover and hand tighten the bolts.

31. Install Front Cover Alignment tool no. J-41480 to the lower engine block.

32. Install Front Cover Alignment tool no. J-41476 to the oil seal opening.

33. Install the crankshaft balancer bolt until snug.

34. Install the front cover bolts and tighten to 18 ft. lbs. (25 Nm).

35. Remove the tools and ensure the front cover-to-oil pan surface is flush.

36. Install the oil pan-to-front cover bolts until snug.

37. Lubricate the outer edge of the front oil seal and seal bore with clean engine oil.

38. Use tool J-41478 and install the oil seal into the cover bore.

39. Remove the tool. Inspect the oil seal for proper installation. The oil seal should be installed evenly and completely into the front cover bore.

40. Install the drive belt idler pulley and tighten the bolt to 37 ft. lbs. (50 Nm).

41. Remove the crankshaft balancer bolt.

42. Use tool J-41665 and install the crankshaft balancer.

43. Install the used crankshaft balancer bolt and tighten to 240 ft. lbs. to position the balancer.

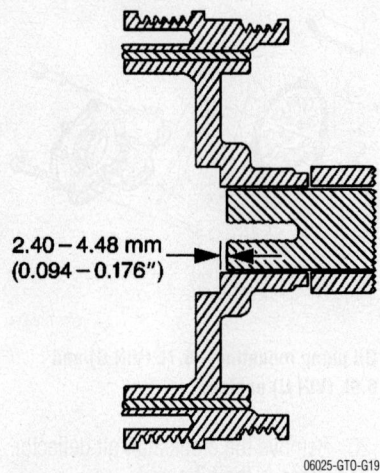

2.40 – 4.48 mm
(0.094 – 0.176")

06025-GTO-G19

Measuring crankshaft recession depth in the crankshaft balancer—5.7L (VIN G) and 6.0L (VIN U) engines

44. Remove the used bolt.

45. Check that the nose of the crankshaft is recessed at least 0.094-0.176 inch (2.40-4.48) into the balancer bore.

46. Install a new crankshaft balancer bolt and tighten first to 37 ft. lbs. (50 Nm) and then an additional 140 degrees.

47. Unlock the flywheel.

48. Install the right transmission cover and bolt.

49. Install the starter motor.

50. Install the air conditioning (A/C) drive belt.

51. Install the radiator.

52. Install the water pump.

53. Connect the radiator hoses.

54. Connect the negative battery cable.

55. Fill the cooling system.

56. Fill the crankcase with clean engine oil.

57. Start the engine. Check for leaks and proper operation.

Piston and Ring

POSITIONING

Position the oil control ring end gaps a minimum of 25 mm (1.0 in) from each other. Position the compression ring end gaps 180 degrees opposite each other.

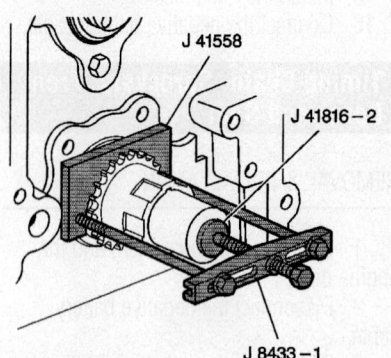

J 41558

J 41816 – 2

J 8433 – 1

06025-GTO-G17

Removing the crankshaft sprocket—5.7L (VIN G) and 6.0L (VIN U) engines

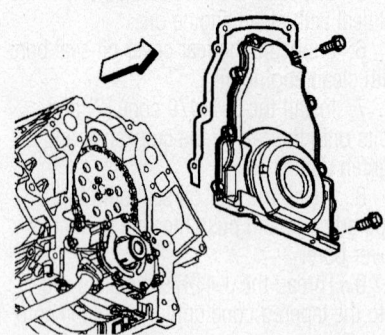

06025-GTO-G18

Front cover mounting—5.7L (VIN G) and 6.0L (VIN U) engines

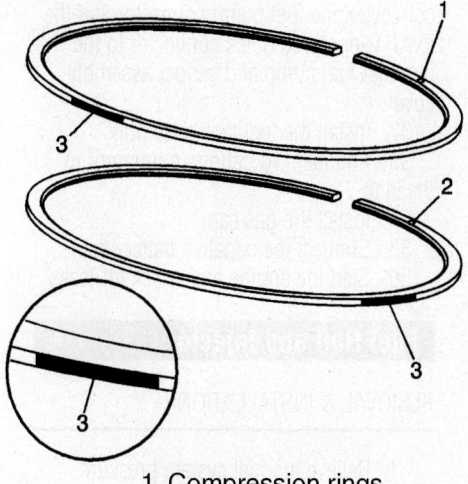

1. Compression rings
2. Oil ring
3. Paint marks

06025-GTO-G20

Piston ring identification—5.7L (VIN G) and 6.0L (VIN U) engines

FUEL SYSTEM

Fuel System Service Precautions

Safety is the most important factor when performing not only fuel system maintenance, but any type of maintenance. Failure to conduct maintenance and repairs in a safe manner may result in serious personal injury or death. Work on a vehicle's fuel system components can be accomplished safely and effectively by adhering to the following rules and guidelines.

• To avoid the possibility of fire and personal injury, always disconnect the negative battery cable unless the repair or test procedure requires that battery voltage be applied.

• Always relieve the fuel system pressure prior to disconnecting any fuel system component (injector, fuel rail, pressure regulator, etc.) fitting or fuel line connection. Exercise extreme caution whenever relieving fuel system pressure to avoid exposing skin, face and eyes to fuel spray. Please be advised that fuel under pressure may penetrate the skin or any part of the body that it contacts.

• Always place a shop towel or cloth around the fitting or connection prior to loosening to absorb any excess fuel due to spillage. Ensure that all fuel spillage is quickly removed from engine surfaces. Ensure that all fuel-soaked cloths or towels are deposited into a flame-proof waste container with a lid.

• Always keep a dry chemical (Class B) fire extinguisher near the work area.

• Do not allow fuel spray or fuel vapors to come into contact with a spark or open flame.

• Always use a second wrench when loosening or tightening fuel line connection fittings. This will prevent unnecessary stress and torsion on fuel piping. Always follow the proper torque specifications.

• Always replace worn fuel fitting O-rings with new ones. Do not substitute fuel hose where rigid pipe is installed.

Fuel System Pressure

RELIEVING

1. Disconnect the negative battery cable in order to avoid possible fuel discharge if an accidental attempt is made to start the engine. Loosen the fuel filler cap in order to relieve the fuel tank vapor pressure.
2. Remove the left fuel rail cover.
3. Connect the J 34730-1A fuel pressure gauge to the fuel pressure connection. Wrap a shop towel around the fitting while connecting the gauge in order to avoid spillage.
4. Install the bleed hose of the gauge into an approved container.
5. Open the valve on the gauge to bleed the system pressure. The fuel connections are now safe for servicing.
6. Drain any fuel remaining in the gauge into an approved container.

Fuel Filter and Fuel Pump

REMOVAL & INSTALLATION

1. Relieve the fuel system pressure.
2. Disconnect the negative battery cable.
3. Drain the fuel tank.
4. Remove the fuel tank.
5. Disconnect the electrical connections from the fuel pump module.
6. Using the lock ring removal tool J45722 and a half-inch breaker bar, remove the fuel pump cover retainer lock ring by turning in a counter-clockwise direction.
7. Partially lift the modular fuel pump and sender assembly away from the fuel tank, taking care not to damage the fuel level sender assembly.
8. Disconnect the fuel tank EVAP vapor line quick connector from the underside of the modular fuel pump and sender assembly cover.
9. Insert hand into the fuel tank opening and disconnect the fuel feed line quick connector.
10. Remove the modular fuel pump and sender assembly and discard the gasket.
11. Remove the fuel strainer assembly.
12. Clamp the protruding end of the modular fuel pump in a soft-jawed vise to support the fuel filter and pump assembly in place.
13. Insert a pair of medium sized flat bladed screwdrivers through each of the service holes in the fuel filter assembly.

14. Firmly slide the blade between the fuel pump end cap and the internal fuel filter clips that hold the fuel pump in place.

15. Push the screwdrivers in far enough so that the internal fuel filter clips are deflected just free of each of the fuel pump end cap retainer shoulders.

16. Hold the screwdrivers in place with one hand, and move the fuel filter assembly in an upward direction to separate it from the fuel pump.

17. Remove the fuel pump.

To install:

18. Install the fuel pump.

19. Ensure the washer is firmly installed along the fuel pump end cap post and firmly pressed up against the suction filter molding.

20. Using hands only, locate the fuel pump in its correct orientation into the fuel filter.

21. Push the fuel pump firmly into place and lock the fuel pump into the fuel filter.

22. Install the fuel strainer assembly.

23. Before installation into the fuel tank, check the fuel sender float position.

24. Stand the assembly upright on a flat surface. Measure the distance between the middle of the fuel sender float and the flat surface. If required, the float position should be corrected, through careful adjustment of the float arm. The float should have a position of 0.27–0.55 inch (7–14mm).

25. Clean any dirt and foreign materials from the fuel tank seal recess and position a new seal in the recess.

26. Install the modular fuel pump and sender assembly into the fuel tank, taking care not to damage the fuel level sender float and arm in the process.

27. Insert hand into the fuel tank opening and connect the fuel feed line quick connector.

28. Connect the fuel tank EVAP vapor line quick connector from the underside of the modular fuel pump and sender assembly cover.

29. Ensure the locator in the cover engages in the slot in the tank opening.

30. Install the cover retainer lock ring. Use the half-inch breaker bar with special tool J45722 and rotate the retainer in a clockwise direction until the tangs are engaged.

31. Install the fuel tank pressure sensor connector, the fuel pump connector and the EVAP vapor hose quick connector to the modular fuel pump and sender assembly cover.

32. Install the fuel tank assembly.

33. Add fuel (10 gallons minimum) to the tank.

34. Install the gas cap.

35. Connect the negative battery cable.

36. Start the engine and check for leaks.

Fuel Rail and Injector

REMOVAL & INSTALLATION

1. Relieve the fuel system pressure.

2. Disconnect the negative battery cable.

3. Disconnect the fuel feed hose from the fuel rail.

4. Disconnect the accelerator cable from the throttle body.

5. Move the accelerator cable aside.

6. Disconnect the electrical connectors from the fuel injectors. Identify the connectors to their corresponding injectors to ensure correct sequential injector firing order after reassembly.

7. Disconnect the electrical harness from the fuel rail brackets.

8. Remove the fuel rail attaching bolts.

9. Remove the fuel rail assembly.

10. Remove the injector lower O-ring seal from the spray tip end of each injector.

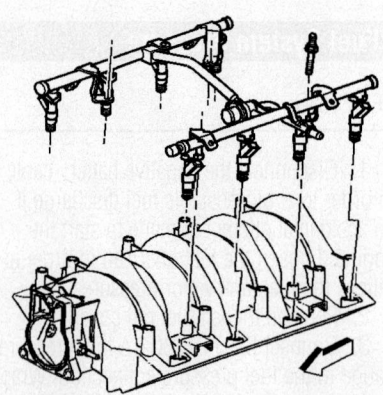

Fuel rail and injector mounting—5.7L engine

06025-GTO-G21

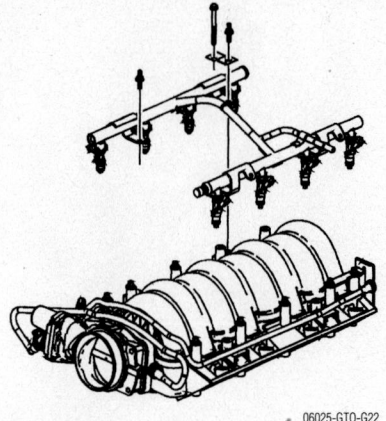

Fuel rail and injector mounting—6.0L engine

06025-GTO-G22

11. Discard the O-ring seals.

12. On both sides, remove the fuel injectors from the fuel rail.

To install:

13. Install the injectors to the fuel rail.

14. Lubricate the new lower injector O-ring seals with clean engine oil.

15. Install the new O-ring seals on the spray tip end of each injector.

16. Install the fuel rail assembly to the intake manifold.

17. Apply a 0.5 mm (0.020 in) of thread-lock or equivalent to the threads of the fuel rail bolts.

18. Install the fuel rail attaching bolts and tighten to 89 inch lbs. (10 Nm).

19. Connect the injector electrical connectors.

20. Connect the electrical harness to the fuel rail brackets.

21. Connect the accelerator cable to the throttle body and the accelerator cable bracket, if the vehicle does not have traction control.

22. Connect the fuel feed hose to the fuel rail fuel pipe.

23. Tighten the fuel filler cap.

24. Connect the negative battery cable.

25. Turn ON the ignition for 2 seconds. Turn OFF the ignition for 10 seconds.

26. Turn ON the ignition. Inspect for fuel leaks

DRIVE TRAIN

Transmission

REMOVAL & INSTALLATION

Manual

1. Disconnect the negative battery cable.
2. From inside the vehicle, remove the shift lever assembly.
3. Drain the oil from the transmission.
4. Support the rear of the vehicle with safety stands.
5. Remove the catalytic converters.
6. Remove the propeller shaft.
7. Remove the starter motor.
8. Remove the clip from the transmission.
9. Disconnect the backup lamp switch electrical connector
10. Disconnect the skip shift solenoid, reverse lockout solenoid and VSS sensor electrical connectors.
11. Remove the push in retainer and wiring harness bracket.
12. Support the engine with safety stands.
13. Remove the left and right transmission close out covers.
14. Remove the retaining clip, then disconnect the clutch hydraulic line from the slave cylinder connector.
15. Remove the transmission crossmember support.
16. Support the transmission with a transmission jack.
17. Slowly lower the transmission to access the transmission bolts.

➡ **Leave the 2 clutch housing to engine block retaining bolts located at the 10 and 2 o'clock positions tight at this time.**

18. Remove the lower 6 transmission retaining bolts.
19. Remove the 2 remaining transmission retaining bolts.
20. Pull the transmission rearward to just clear the locating dowels, then rotate the transmission 90 degrees to the right to gain clearance from the vehicle floor pan.
21. When the transmission input shaft clears the clutch pressure plate, lower the transmission from the vehicle.

To install:
22. Place the transmission in the 3rd gear position.
23. With the transmission positioned 90 degrees to the right side, position the trans-

06025-GTO-G23

Manual transmission tightening sequence—5.7L and 6.0L engines

mission in the vehicle up to the locating dowels on the engine.
24. Slowly raise the transmission until in place.
25. Install the transmission to the vehicle, then rotate the transmission to the upright position.
26. Install the transmission bolts.
27. Tighten the transmission retaining bolts in sequence to 37 ft. lbs. (50 Nm).
28. Install the transmission crossmember support and tighten the bolts to 43 ft. lbs. (58 Nm).
29. Remove the transmission jack from the transmission.
30. Install the left and right transmission close out covers.
31. Connect the clutch actuator cylinder pipe to the clutch master cylinder hose.
32. Push together the clutch hydraulic hose fittings, then install the retaining clip.
33. Check the clutch hydraulic hoses for kinks or twists.
34. Remove the engine safety stands.
35. Install the wiring harness bracket and push in retainer.
36. Connect the VSS, reverse lockout solenoid and skip shift solenoid connectors.
37. Connect the backup lamp switch electrical connector.
38. Install the clip to the transmission.
39. Install the starter motor.
40. Install the catalytic converters.
41. Install the propeller shaft.
42. Remove the rear axle safety stands.
43. Refill the transmission with fluid.
44. Bleed the clutch hydraulic system.
45. Lower the vehicle.
46. Install the transmission shift control assembly.
47. Connect the negative battery cable.

Automatic

1. Disconnect the negative battery cable.
2. Drain the transmission fluid.
3. Remove the transmission fluid filler tube.
4. Raise and suitably support the vehicle.
5. Remove the catalytic converter assembly from the vehicle.
6. Remove the starter motor.
7. Disconnect the selector lever linkage from the transmission.
8. Remove the propeller shaft.
9. Support the rear of the vehicle with a suitable jack.
10. Remove the torque converter covers.
11. Remove the torque converter bolts, then discard the bolts.
12. Remove the transmission crossmember from the vehicle.
13. Lower the tail section of the transmission slightly.
14. Disconnect the transmission oil cooler pipes from the transmission.
15. Remove the wiring harness clamp bolt attaching the clamp to the transmission.
16. Disconnect the transmission 20-way electrical connector. Compress both tabs on the connector and pull straight up; do not pry the connector.
17. Disconnect the vehicle speed sensor electrical connector.
18. Support the transmission using a transmission jack.
19. Remove the transmission bolts.
20. Separate the transmission from the engine.
21. Attach torque converter holding strap J-21366 to the transmission.
22. Lower the transmission from the vehicle.

To install:
23. Raise the transmission up to the vehicle.
24. Remove the holding strap from the transmission.
25. Align and install the transmission to the engine.
26. Install the transmission bolts and tighten to 28 ft. lbs. (38 Nm).
27. Connect the vehicle speed sensor electrical connector.
28. Connect the 20-way electrical connector.
29. Install the wiring harness clamp bolt attaching the clamp to the transmission.

30. Connect the transmission oil cooler pipes to the transmission.

31. Align the torque converter and flywheel mating marks, then install the new transmission torque converter bolts and tighten to 48 ft. lbs. (65 Nm).

32. Install the transmission crossmember support and tighten the bolts to 43 ft. lbs. (58 Nm).

33. Install and adjust the selector lever linkage.

34. Remove the transmission jack.

35. Install the propeller shaft.

36. Install the starter motor from the engine.

37. Install the catalytic converter assembly on the vehicle.

38. Remove the jack supporting the rear of the vehicle.

39. Lower the vehicle.

40. Install the transmission fluid filler tube.

41. Flush the transmission oil cooler, oil cooler pipes and the hoses.

42. Connect the negative battery cable.

43. Fill the transmission to the proper level with DEXRON® III transmission fluid.

44. Start the engine. Check for leaks and proper operation.

Clutch

REMOVAL & INSTALLATION

1. Disconnect the negative battery cable.
2. Remove the transmission.
3. Remove the clutch pressure plate bolts.
4. Remove the clutch pressure plate and driven disc from the dowel pins on the flywheel.
5. Install the clutch pressure plate and driven plate to the dowel pins on the flywheel.
6. Install the clutch pressure plate bolts finger tight.
7. Install a suitable clutch alignment arbor in order to align the clutch driven plate to the clutch pilot bearing.
8. Tighten the clutch pressure plate bolts in star pattern to 37 ft. lbs. (50 Nm).
9. Install the manual transmission.
10. Connect the negative battery cable.

Hydraulic Clutch System

BLEEDING

1. Ensure the reservoir is filled to the fill line with new hydraulic clutch fluid.

2. Press the clutch pedal all the way down to the floor.

3. Open the bleeder on the actuator cylinder to purge the air.

4. Close the bleeder and release the clutch pedal.

5. Repeat steps 2, 3 and 4 until all air is out of the clutch system.

6. Check and refill the reservoir as needed while bleeding.

7. After bleeding, pump the clutch pedal several times. If the clutch engagement is not satisfactory, repeat the bleed procedure.

8. If the previous procedures are unsuccessful, perform the following steps.

   a. Pump the clutch pedal very fast for 30 seconds.

   b. Stop pumping and let the air escape into the reservoir.

   c. Repeat this procedure as necessary.

Axle Shaft

REMOVAL & INSTALLATION

1. Place the transmission in neutral (M/T) or Park (A/T).

2. Raise the vehicle on a hoist.

3. Remove the wheel.

4. Support the axle shaft until it is removed.

5. Mark the relationship of the inner constant velocity joint and the inner axle.

6. Remove the bolts and retaining plates from the inner constant velocity joint.

7. Remove the bolts and retaining plates from the outer constant velocity joint.

8. Remove the axle shaft.

To install:

9. Align the marks on the inner constant velocity joint and the inner axle.

10. Install the inner mount bolts and retainer plates and tighten to 37 ft. lbs. (50 Nm), plus an additional 68 degrees of torque.

11. Install the outer constant velocity joint mount bolts and tighten to 37 ft. lbs. (50 Nm).

12. Remove the support stand.

13. Lower the vehicle.

Constant Velocity Joint

OVERHAUL

Inner and Outer Joint

1. Place the wheel axle shaft horizontally in a bench vise.

2. Using a side cutter or other suitable tool, remove the large and small boot retaining clamps from the constant velocity joint boot and discard the clamp.

3. Remove the dust shield and end cap from the constant velocity joint by tapping with hammer and punch.

4. Slide the boot toward the center of the drive shaft.

5. Remove the constant velocity joint retainer clip from the drive shaft.

6. Support the inner race when removing the constant velocity joint.

7. Press the drive shaft from the constant velocity joint.

8. Remove the constant velocity joint boot from the drive shaft.

9. Tilt the cage and inner race and remove one ball.

10. Repeat the process to remove the remaining five balls.

11. Remove the inner race and cage.

12. Thoroughly clean the constant velocity (CV) joint components

13. Inspect the constant velocity joint components for pitting, galling, excessive play between ball and cage, damage or cracking of the cage or cracking, galling, or chips of the races.

14. If damaged, replace the constant velocity joint.

15. Thoroughly clean and inspect the drive shaft boot for tears, cracking and deterioration.

16. If the drive shaft seal is damaged, replace the boot.

17. Install the inner race and cage into the outer race ensuring the inner race step is opposite of the outer race groove.

18. Align the thick sections with the outer race with the narrow sections on the inner race

19. Tilt the cage and inner race and fit one ball.

20. Repeat the process for the other five balls.

21. Check the plunge of the inner parts. If there is no movement, the constant velocity joint has been assembled incorrectly. Reassemble the constant velocity joint.

22. Remove the old sealing bead from the dust shields, end caps, and the constant velocity joint.

23. Install the large boot clamp over the drive shaft boot.

24. Install the dust shield onto the drive shaft boot.

25. Using crimping pliers, crimp the large clamp ensuring the crimp is positioned between 2 bolt holes.

26. Install a new small boot retaining clamp onto the wheel driveshaft.

27. Install the boot onto the driveshaft.

28. Press the constant velocity joint onto the drive shaft with the step on the inner race facing the shoulder on the driveshaft.

29. Install the constant velocity retaining clip.

30. Pack one tube of grease into the inner side of the constant velocity joint and the boot.

31. Pack a half tube of grease into the outside of the constant velocity joint.

32. Apply an 8 mm bead of RTV sealant to the inside of the end cap and the dust shield. Allow to cure.

33. Install the dust shield and end cap by gently tapping with a hammer and punch. Ensure the bolt holes are aligned.

34. Position the small boot retaining clamp onto the neck of the boot.

35. Position the boot and small retaining clamp into the boot groove on the drive shaft.

36. Using a small screwdriver, pry up the small end of the boot to equalize the air pressure.

37. Measure the distance (a) between the edge of the boot and the edge of the last

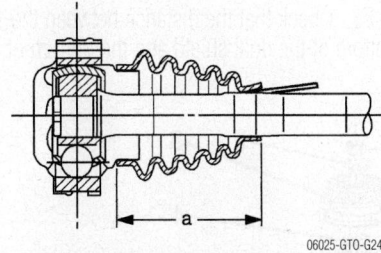

06025-GTO-G24

Measuring constant velocity joint boot length—GTO

axle shaft groove closing edge. The length should be 3.19 inch (81mm).

38. Using crimping pliers crimp the small end clamp.

39. Remove the wheel drive shaft from the bench vise.

Rear Inner Drive Axle Shaft and Seal

REMOVAL & INSTALLATION

1. Raise and support the vehicle.
2. Remove the rear wheel.
3. Remove the axle shaft.
4. Place a drain pan under the axle.
5. Remove the inner axle using a slide hammer and puller plate.
6. Remove the axle shaft seal with a seal removal tool.

To install:

7. Clean the axle seal housing.
8. Lubricate the seal bore and the seal lip with lithium grease.
9. Install the seal using an installer until the seal is fully seated in the bore.
10. Install the inner axle shaft.
11. Remove the drain pan.
12. Install the axle shaft.
13. Install the wheel.

Pinion Flange and Seal

REMOVAL & INSTALLATION

1. Place the transmission in neutral.
2. Raise the vehicle on a hoist.
3. Remove both mufflers.
4. Index mark the driveshaft, then dis-

connect driveshaft and position it out of the way.

5. Use an inch lb. torque wrench and measure the amount of torque required to maintain pinion rotation through several revolutions.

6. Hold the pinion flange from turning, then remove the pinion nut.

7. Index mark the pinion flange, then use a puller and remove the flange.

8. Remove the pinion seal.

To install:

9. Inspect the pinion shaft seal surface for nicks and burrs. If damaged, replace the pinion flange.

10. Clean the threads on the pinion shaft and pinion nut.

11. Coat the pinion shaft splines and seal with gear oil.

12. Using a pinion seal installer, install the new pinion shaft seal until the seal is flush with the differential housing.

13. Install the pinion flange on to the pinion shaft aligning the punch marks.

14. Apply thread locking compound onto the threads of the pinion shaft.

15. Align the index marks and install the pinion flange.

16. Install new pinion flange nut loosely.

17. Rotate the pinion flange occasionally while tightening the flange nut to make sure the pinion bearings seat correctly.

18. Take frequent bearing preload torque readings.

19. The pinion bearing preload specification is 2–4 inch lbs. (0.2–0.43 Nm).

20. Connect the driveshaft and install new bolts.

21. Install both mufflers.

22. Fill the differential with gear lubricant and check for leaks.

STEERING AND SUSPENSION

Air Bag

✳✳ CAUTION

Vehicles are equipped with an air bag system. The system must be disarmed before performing service on, or around, system components, the steering column, instrument panel components, wiring and sensors. Failure to follow the safety precautions and the disarming procedure could result in accidental air bag deployment, possible injury and unnecessary system repairs.

PRECAUTIONS

Several precautions must be observed when handling the inflator module to avoid accidental deployment and possible personal injury.

• Never carry the inflator module by the wires or connector on the underside of the module.

• When carrying a live inflator module, hold securely with both hands, and ensure that the bag and trim cover are pointed away.

• Place the inflator module on a bench or other surface with the bag and trim cover facing up.

• With the inflator module on the bench, never place anything on or close to the module that may be thrown in the event of an accidental deployment.

DISARMING THE SYSTEM

1. Ensure the ignition and all accessories are off.
2. Place the steering wheel in the straight-ahead position.
3. Remove the SIR fuse from the instrument panel fuse box at the base of the steering column.
4. Disconnect the negative battery cable.
5. Remove the upper steering column trim cover.

6. Release the inflatable restraint steering wheel module coil connector locking mechanism.

7. Disconnect the inflatable restraint steering wheel module coil connector.

ARMING THE SYSTEM

1. Remove the key from the ignition switch.

2. Connect the inflatable restraint steering wheel module coil connector.

3. Lock the connector with lock lever.

4. Install the upper steering column trim cover.

5. Connect the negative battery cable.

6. Install the SIR fuse.

7. Turn the ignition on.

8. The air bag indicator will flash and then turn off.

Power Rack and Pinion Steering Gear

REMOVAL & INSTALLATION

1. Lock the steering column with the steering wheel in the straight ahead position.

2. Disconnect the negative battery cable.

3. Place a drain pan under the vehicle in order to collect the fluid from the power steering system.

4. Remove front wheels.

5. Remove the pressure and return pipes from the steering gear.

6. Remove and discard the nut that retains the steering shaft coupling to the steering gear pinion.

7. Remove the bolt from the coupling.

8. Separate the retainer from the coupling.

9. Separate the coupling from the pinion.

10. Remove the 2 outer tie rod ends.

11. Remove the 2 nuts and the 2 bolts that retain the steering gear to the front frame.

12. Remove the steering gear.

To install:

13. Slide the steering gear pinion into the steering shaft coupling.

14. Install the steering gear to the front frame and tighten the bolts to 44 ft. lbs. (60 Nm), plus an additional 45 degrees.

15. Ensure the O-rings on the pipes and the hoses are in excellent condition.

16. Install the pressure and return pipes to the steering gear and tighten to 15 lbs. (20 Nm).

17. Position the steering shaft coupling on the steering gear pinion.

18. Install the retainer on the coupling.

19. Install the bolt to the coupling. Install a NEW nut to the bolt and tighten to 20 ft. lbs. (27 Nm).

20. Install the 2 outer tie rod ends.

21. Install the front tire and wheel assemblies.

22. Remove the drain pan.

23. Lower the vehicle.

24. Connect the negative battery cable.

25. Fill the power steering system with the proper type and quantity of fluid.

26. Check the front end alignment and adjust as necessary.

Strut and Coil Spring

REMOVAL & INSTALLATION

Front

1. Raise and support the vehicle.

2. Remove the front wheel.

3. Hold the stabilizer bar link upper stud with a wrench, then remove the upper nut, washer, insulator and retainer.

4. Disconnect the speed sensor harness, insulator and connector.

5. Separate the brake hose from the strut bracket.

6. Position a jack and a block of wood below the control arm ball joint and raise the jack to support the control arm.

7. Remove the lower strut attaching nuts, washers and bolts.

8. Separate the strut from the knuckle.

9. Remove the strut upper mounting nut and bumper stop.

10. Lower the strut.

11. Remove the stabilizer link from the strut bracket.

12. Remove the strut/spring assembly.

13. Use a spring compressor to compress the spring.

14. Remove the strut mount nut and washers.

15. Remove the strut bearing from the upper spring seat.

16. Remove the strut shield strap, upper spring seat, insulator and bumper.

17. Release the spring compressor.

18. Remove the spring.

19. Remove and discard the strut shield clamp.

20. Remove the dust shield and filter.

To install:

21. Hold the strut rod and housing and pull the strut rod out to its maximum length.

22. Install the dust shield and ensure the filter is seated in the housing.

23. Check that the distance between the bottom of the dust shield and the bottom of

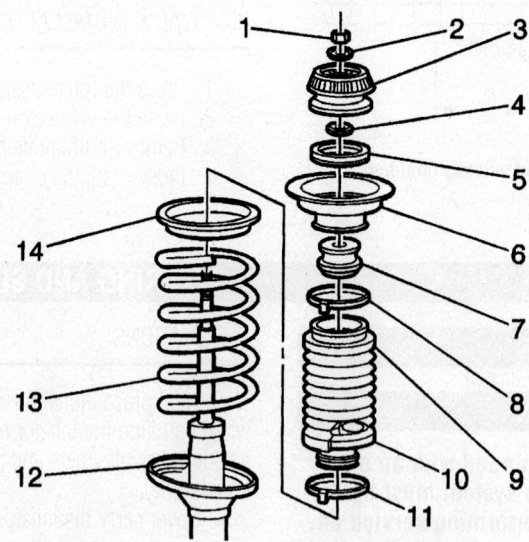

1. Nut
2. Washer
3. Mount
4. Washer
5. Strut bearing
6. Upper spring seat
7. Strut bumper
8. Shield clamp
9. Dust shield
10. Filter
11. Shield clamp
12. Lower spring seat
13. Spring
14. Upper insulator

06025-GTO-G25

Exploded view of front strut and coil spring assembly—GTO

the lower spring seat is 1.20–2.0 inch (30–50mm).

24. Install a new clamp to hold the dust shield.

25. Install the coil spring to the housing.

26. Compress the spring using a spring compressor.

27. Install the bumper.

28. Install the upper insulator with the step on the insulator to the straight projecting end of the spring.

29. Install the upper spring seat so the double notch in the upper flange of the seat in on the inboard side of the strut.

30. Pull the strut rod out to its maximum length.

31. Install the strut bearing so the narrow outer section faces toward the upper spring seat collar.

32. Install the lower washer so the cupped side faces upward.

33. Install the upper washer so the cupped side faces downward.

34. Install the strut mount nut and tighten to 58 ft. lbs. (78 Nm).

35. Install the dust shield to the upper spring shield so the shield fits over the lower flange of the spring seat collar.

36. Install a new dust shield strap.

37. Remove the spring compressor.

38. Install the stabilizer link to the strut bracket.

39. Position the strut in the strut tower.

40. Install the bumper stop, and new strut nut but do not tighten.

41. Install the lower strut to the steering knuckle and tighten the new bolts and nuts to 63 ft. lbs. (85 Nm).

42. Hold the end of the strut rod and tighten the upper nut to 41 ft. lbs. (55 Nm).

43. Install the strut cap.

44. Remove the jack

45. Attach the brake hose to the strut bracket.

46. Connect the speed sensor harness, insulator and connector.

47. Hold the stabilizer bar link upper stud with a wrench and install the washer, insulator, retainer, and upper nut and tighten to 12 ft. lbs. (16 Nm).

48. Install the front wheel.

49. Lower the vehicle and bounce the suspension 3 times to stabilize the suspension.

50. Check the wheel alignment.

51. Tighten the lower strut nuts to 74 ft. lbs. (100 Nm), plus an additional 90 degrees.

Shock Absorber

REMOVAL & INSTALLATION

Rear

1. Raise and support the vehicle.

2. Remove the bolt and the washer from the shock absorber .

3. Separate the shock absorber from the control arm.

4. Lower the vehicle to the appropriate height for accessing the rear compartment.

5. In the rear compartment, remove the rear center trim panel carpet.

6. If necessary in order to remove the right rear shock absorber, remove the fuel filler tube.

7. Remove the cap from the shock absorber.

8. Remove the upper mounting nut.

9. Remove the upper mounting upper washer.

10. Remove the upper mounting upper bushing.

11. Remove the shock absorber from the vehicle.

12. If necessary, remove the upper mounting lower bushing.

13. If necessary, remove the upper mounting lower washer.

14. Remove the shock absorber.

To install:

15. If removed, install the upper mounting lower washer to the shock absorber.

16. If removed, install the upper mounting lower bushing to the shock absorber.

17. Install the shock absorber to the vehicle.

18. Install the upper mounting upper bushing.

19. Install the upper mounting upper washer.

20. Install the upper mounting nut and tighten to 10 ft. lbs. (14 Nm).

21. Install the cap to the shock absorber.

22. Install the shock, washer and the bolt to the control arm.

23. Lower the vehicle.

24. With the weight of the vehicle on the tire and wheel assemblies, bounce the rear of the vehicle several times in order to stabilize the rear suspension.

25. Tighten the bolt to 85 ft. lbs. (115 Nm).

26. If removed, install the fuel filler tube.

27. Reinstall the carpet.

Coil Spring

REMOVAL & INSTALLATION

Rear

1. Raise and support the vehicle.

2. Remove the tire and wheel assembly.

3. Prevent the wheel hub from turning.

4. Remove the bolts and the retaining plates that retain the outer constant velocity joint to the drive shaft flange.

5. Separate the wheel drive shaft from the drive shaft flange and wire aside.

6. Loosen the nut that retains the stabilizer shaft to the shaft link.

7. Remove the nut and the bolt that retain the stabilizer shaft link to the lower control arm.

8. Loosen the nut that retains the outer adjustment link to the control arm. Position the top of the nut with the top of the outer adjustment link stud.

9. Use a tie-rod puller to separate the stud from the control arm.

10. Remove and discard the nut from the stud.

11. Position the adjustment link assembly away from the control arm.

12. Position a jack with a block of wood under the control arm.

13. Raise the jack slightly in order to reduce the spring load on the control arm.

14. Remove the bolt and washer from the shock absorber.

15. Separate the shock absorber from the control arm.

16. Lower the jack and the control arm.

17. Push down gently on the control arm and remove the spring and the 2 insulators.

18. Remove the insulators from the spring.

To install:

19. Install the lower insulator (3) to the control arm.

20. Install the upper insulator (1) to the spring (2).

21. Push down gently on the control arm. Install the spring and the upper insulator to the lower insulator.

22. Adjust the position of jack in order to align the control arm and the shock absorber.

➡**Do not tighten the bolt yet. The weight of the vehicle must be on the tire and wheel assemblies before tightening the suspension fasteners.**

23. Install the washer and the bolt in order to retain the shock absorber to the control arm.

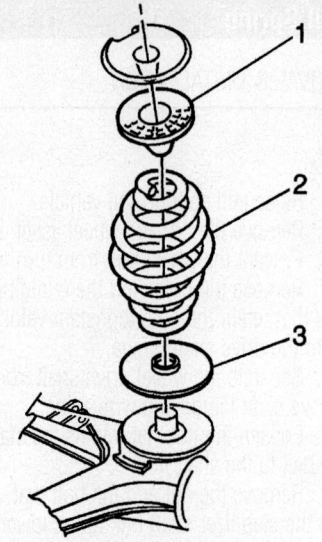

1. Upper insulator
2. Spring
3. Lower insulator

06025-GTO-G26

Exploded view of the rear coil spring assembly—GTO

➡️**Do not tighten the nut or the bolt yet.**

24. Install the nut and the bolt in order to retain the stabilizer shaft link to the control arm.

➡️**Do not tighten the nut yet.**

25. Install the outer adjustment link stud and a NEW nut to the control arm.
26. Remove the jack and the wood from the control arm.
27. Lock the wheel hub from turning.
28. Install the retainers and the bolts in order to retain the outer constant velocity joint to the drive shaft flange and tighten to 37 ft. lbs. (50 Nm), plus an additional 68 degrees.
29. Remove the wire supporting the wheel drive shaft .
30. Install the driveshaft.
31. Install the tire and wheel assembly.
32. Lower the vehicle.
33. With the weight of the vehicle on the tire and wheel assemblies, bounce the rear of the vehicle several times in order to stabilize the rear suspension.
34. Tighten the stabilizer shaft link nuts to 72 ft. lbs. (98 Nm).
35. Tighten the shock absorber-to-control arm bolt and tighten to 85 ft. lbs. (115 Nm).
36. Tighten the outer adjustment link-to-control arm to 46 ft. lbs. (63 Nm).
37. Check the wheel alignment.

Stabilizer Bar

REMOVAL & INSTALLATION

Front

1. Raise and support the vehicle.
2. Remove the front wheels.
3. Lower the front suspension frame. See Front Suspension Frame.
4. Remove the stabilizer bar nuts, brackets and insulators.
5. Use a wrench to hold the link lower studs, then remove the nuts and stabilizer bar from the links.
6. Remove the stabilizer bar from the suspension frame.

To install:

7. Install the insulators to the stabilizer bar with the slots in the insulators facing forward.

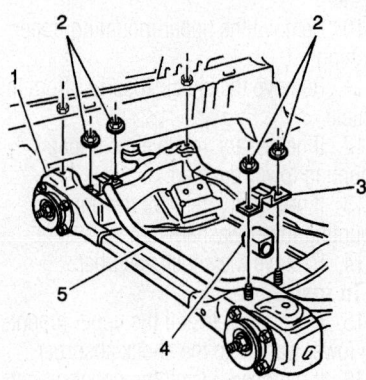

1. Suspension frame
2. Nuts
3. Insulator brackets
4. Insulator
5. Stabilizer bar

06025-GTO-G27

Front stabilizer bar mounting to the front suspension frame—GTO

8. Install the bar to the frame.
9. Install the brackets to the insulators and tighten the nuts to 20 ft. lbs. (27 Nm).
10. Install the bar to the links.
11. Use a wrench to hold the link lower studs, then install the nuts and tighten to 37 ft. lbs. (50 Nm).
12. Raise the front suspension frame.
13. Install the wheels and lower the vehicle.

Rear

1. Raise the vehicle on a hoist.
2. Use paint or a scribe in order to mark the outline of the differential carrier assembly mount on the body.
3. Use a jack in order to support the mount.
4. Remove and discard the 4 bolts that retain the mount to the body.
5. Lower the jack with the mount and the rear suspension support in order to access the bolts that retain the stabilizer shaft insulator brackets to the support.
6. Remove the 2 stabilizer shaft link bolts from the control arms.
7. Remove the stabilizer shaft links from the control arms.
8. Remove the bolts that retain the stabilizer shaft insulator brackets to the rear suspension support.
9. Use a flat blade tool as a lever in order to remove the brackets from the rear suspension support.
10. Remove the stabilizer shaft assembly from the vehicle.
11. If necessary, remove the following components from the stabilizer shaft:
    a. The 2 stabilizer shaft link nuts.
    b. The 2 stabilizer shaft links.
    c. The 2 stabilizer shaft insulator brackets.
    d. The 2 stabilizer shaft insulators.

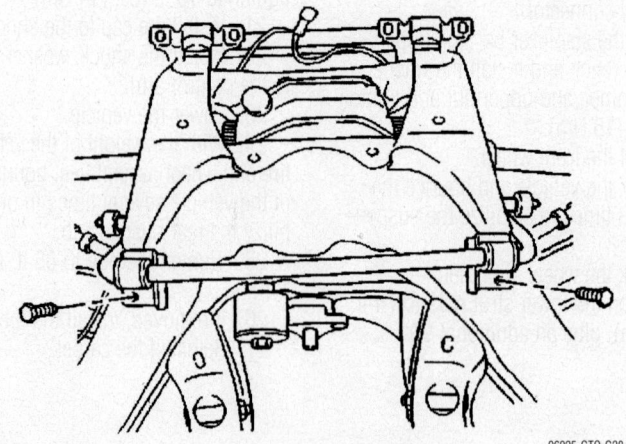

06025-GTO-G28

Rear stabilizer bar mounting to the rear suspension support—GTO

To install:

12. Install the components to the stabilizer bar, if removed.

13. Position the stabilizer shaft on the rear suspension support.

14. Install the insulator brackets to the slots on the rear suspension support.

15. Install the bolts to the stabilizer shaft insulator brackets and tighten to 16 ft. lbs. (22 Nm).

16. Raise the rear suspension support.

17. Install the stabilizer shaft link to the lower control arm.

18. Install the stabilizer shaft link bolts and the nuts to the lower control arm, but do not tighten.

19. Raise the jack with the differential carrier assembly mount and the rear suspension support.

20. Align the mount with the match marks.

21. Install 4 new bolts in order to retain the mount to the body, but do not tighten.

22. Install Crossmember Centering Tool no. CH-46839 to the underbody. The tool locates into 19 mm diameter body datum holes positioned forward of the rear suspension support.

23. With the aid of an assistant, position the rear suspension support. The locating pins of the tool engage the alignment holes in the rear suspension support.

24. Tighten the bolts that retain the differential mount to the body to 26 ft. lbs. (35 Nm), plus an additional 60 degrees.

25. Remove the centering tool.

26. Remove the jack from the vehicle.

27. Lower the vehicle.

28. With the weight of the vehicle on the tire and wheel assemblies, bounce the rear of the vehicle several times in order to stabilize the suspension.

29. Tighten the stabilizer shaft link nuts and the bolts to 72 ft. lbs. (98 Nm).

Front Suspension Frame

REMOVAL & INSTALLATION

1. Raise and support the vehicle.

2. Remove the front wheels.

3. Remove the power steering gear.

4. Remove the lower control arms and control arm rods and insulators.

5. Remove the hood.

6. Using an engine crane, support and raise the engine.

7. Remove the engine mount brackets from the frame.

8. Support the front frame with a jack.

9. Remove the 4 attaching bolts and remove the frame.

To install:

10. Install the frame to the side members.

11. Install the silver bolts to the rear of the frame and the black bolts to the front and tighten to 92 ft. lbs. (125 Nm).

12. Install the engine mount brackets and tighten to 37 ft. lbs. (50 Nm).

13. Lower the engine.

14. Install the hood.

15. Install the lower control arms and control arm rods and insulators.

16. Install the power steering gear.

17. Install the front wheels.

18. Lower the vehicle.

Front Lower Control Arm Rod

REMOVAL & INSTALLATION

1. Raise and support the vehicle.

2. Remove the front tire and wheel assembly.

3. Loosen the nut that retains the front lower control arm rod to the front lower control arm rod insulating bushing assembly.

4. Remove and discard the nut that retains the rod to the front lower control arm.

5. Remove the washer from the rod.

6. Remove the 4 nuts that retain the insulating bushing to the front frame.

7. Pull the insulating bushing away from the front frame.

8. Remove the rod and the rod retainer from the lower control arm.

9. Remove the rod retainer from the rod.

10. Pull the rod through the hole in the front frame. Remove the rod and the insulating bushing as an assembly.

11. Place the rod in a soft jaw vise.

12. Remove and discard the nut that retains the rod to the insulating bushing.

13. Remove the insulating bushing from the rod.

14. If a washer is on the rod, remove the washer.

To install:

➡ **The front of the rod has more distance between the shoulder and the end of the threads. The washer for the front of the rod is for reducing caster.**

15. If equipped, install the washer to the front of the rod.

16. Install the rod, rear end first, through the hole in the frame.

17. Install the rod retainer to the rod with the convex side toward the control arm.

18. Install the rod to the control arm.

19. Install the washer to the rear of the rod.

20. Install the new nut in order to retain the rod to the control arm.

21. Install the insulating bushing assembly over the front of the rod.

22. Install the insulating bushing to the studs on the frame.

23. Install the 4 nuts in order to retain the insulating bushing to the frame and tighten to 17 ft. lbs. (23 Nm).

24. Install the new nut in order to retain the rod to the insulating bushing but do not tighten.

25. Install the front tire and wheel assembly.

26. Lower the vehicle.

27. With the weight of the vehicle on the tire and wheel assemblies, push down on the front bumper 3 times in order to stabilize the suspension.

28. Tighten the nut that retains the the rod to the inulating bushing to 109 ft. lbs. (148 Nm).

29. Tighten the nut that retains the rod to the control arm to 76 ft. lbs. (103 Nm).

30. Check the wheel alignment.

Lower Control Arm

REMOVAL & INSTALLATION

Front

1. Raise and support the vehicle.

2. Remove the tire and wheel assembly.

3. Turn the wheel to gain access to the ball joint stud.

4. Remove the ball joint stud nut and separate the stud from the steering knuckle.

5. Push the control arm away from the knuckle.

6. Remove the control arm rod nut and washer.

7. Remove the control arm-to-frame nut and discard.

8. Remove the control arm-to-frame bolt and remove the control arm from the rod.

To install:

9. Install the retainer on the control arm rod.

10. Install the control arm to the rod.

11. Install the control arm and bolt to the frame but do not tighten.

12. Install a new nut and washer to the rod but do not tighten.

13. Place a jack and a block of wood under the control arm and install the ball joint stud to the knuckle.

14. Raise the jack to seat the stud.

15. Install a new nut to the stud and tighten to 55 ft. lbs. (60 Nm).

16. Remove the jack.

17. Lower the vehicle.

18. With the weight of the vehicle on the tires, bounce the front bumper 3 times.

19. Tighten the control arm-to-frame nut to 72 ft. lbs. (98 Nm).

20. Tighten the control arm-to-control arm nut to 76 ft. lbs. (103 Nm).

21. Check the wheel alignment.

Rear

1. Raise and support the vehicle.

2. Remove the tire and wheel assembly.

3. Loosen the nut that retains the stabilizer shaft to the shaft link.

4. Remove the nut and the bolt that retain the stabilizer shaft link to the lower control arm.

5. Loosen the nut that retains the outer adjustment link to the control arm. Position the top of the nut with the top of the outer adjustment link stud.

6. Separate the stud from the control arm.

7. Remove and discard the nut from the stud.

8. Position the adjustment link assembly away from the control arm.

9. Remove the clip from the brake hose.

10. Pull the brake pipe and hose forward from the bracket.

11. Lift the brake pipe up through the slot in the bracket in order to separate the brake pipe and hose from the control arm.

12. Loosen the brake pipe flare nut.

13. Remove the clip from the backing plate bracket.

14. Remove the brake pipe from the brake hose.

15. Plug the brake hose in order to minimize fluid loss and contamination.

16. Remove the brake pipe from the backing plate bracket.

17. Plug the brake pipe.

18. Remove the 2 bolts and the brake caliper and bracket assembly from the control arm.

19. Remove the rear parking brake cable.

20. If necessary for sufficient access, remove the propeller shaft.

21. Mark the position of the brake rotor to the hub.

22. Remove the brake rotor from the hub.

23. Lock the wheel hub from turning.

24. Separate the axle shaft from the drive shaft flange and wire it aside.

25. Position a jack with a block of wood under the control arm.

26. Raise the jack slightly in order to reduce the spring load on the control arm.

27. Remove the coil spring.

28. Pull the differential carrier breather hose out of the hole in the vehicle underbody rear suspension support.

29. Disconnect the wheel speed sensor connectors from the body harness.

30. Use paint or a scribe in order to mark the outline of the differential carrier assembly mount on the body.

31. Use a jack to support the mount.

32. Remove and discard the 4 bolts that retain the mount to the body.

33. Lower the mount and the rear suspension support in order to access the control arm fasteners.

34. Remove and discard the 2 nuts that retain the control arm to the rear suspension support.

35. Remove the 2 bolts and the control arm.

To install:

36. Install the control arm and the 2 bolts to the rear suspension support, but do not tighten.

37. Install the 2 new nuts to the control arm.

38. Position a jack with a block of wood under the control arm.

39. Install the rotor to the hub.

40. Install the brake caliper and tighten to 63 ft. lbs. (85 Nm).

41. Install the coil spring.

42. Install the lower shock absorber bolt but do not tighten.

43. Install the nut and the bolt in order to retain the stabilizer shaft link to the lower control arm, but do not tighten.

44. Install the outer adjustment link stud and a nut to the control arm.

45. Remove the jack and the wood from the control arm.

46. Lock the wheel hub from turning.

47. Install the retainers and the bolts to retain the outer constant velocity joint to the drive shaft flange and tighten to 37 ft. lbs. (50 Nm) plus an additional 68 degrees.

48. Raise the jack and the differential carrier assembly mount.

49. Align the mount with the match marks.

50. Install the new bolts that retain the mount to the body and tighten to 26 ft. lbs. (35 Nm), plus an additional 68 degrees.

51. Lower and remove the jack from under the vehicle.

52. Install the rear parking brake cable.

53. Install the brake hose and pipe.

54. Install the drive shaft, if removed.

55. Install the tire and wheel assembly.

56. Lower the vehicle.

57. With the weight of the vehicle on the tire and wheel assemblies, bounce the rear of the vehicle several times in order to stabilize the rear suspension.

58. Tighten the nuts that retain the control arm to the rear suspension support to 74 ft. lbs. (100 Nm).

59. Tighten the stabilizer shaft link nuts to 72 ft. lbs. (98 Nm).

60. Tighten the lower shock bolt to 85 ft. lbs. (115 Nm).

61. Tighten the nut that retains the outer adjustment link to the control arm to 46 ft. lbs. (63 Nm).

62. Check the wheel alignment.

Wheel Hub and Bearing

REMOVAL & INSTALLATION

Front

1. Raise and support the vehicle.

2. Remove the front wheel.

3. Remove the brake caliper and rotor.

4. Turn the steering wheel to the left or right, depending upon which is being serviced.

5. Disconnect the wheel speed sensor.

6. Remove the strut-to-knuckle bolts.

7. Remove 3 Allen bolts and remove the hub from the knuckle.

To install:

8. Clean the steering knuckle/wheel hub mating surfaces.

9. Align the wheel speed sensor connection on the hub with the hole in the knuckle.

10. Install the hub to the knuckle.

11. Install the 3 Allen bolts to the hub and tighten to 80 ft. lbs. (108 Nm).

12. Connect the wheel speed sensor connector.

13. Install the knuckle to the strut and tighten the new bolts, washers and nuts to 64 ft. lbs. (85 Nm), then to 74 ft. lbs. (100 Nm), plus an additional 90 degrees.

14. Install the brake rotor and caliper.

15. Install the wheel.

16. Lower the vehicle.

Rear

1. Raise and support the vehicle.

2. Remove the lower control arm.

3. Install the special Wheel Hub remover tool as shown, with the holes in the drive shaft flange. Use the outer CV joint bolts to hold the tool in place.

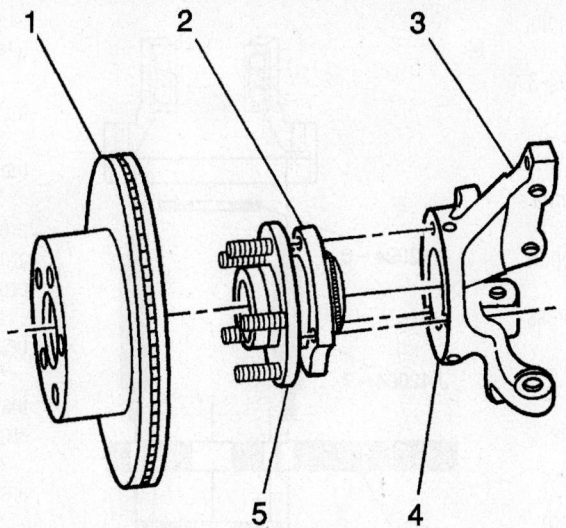

1. Brake rotor
2. Wheel hub
3. Steering knuckle

06025-GTO-G29

Exploded view of front wheel hub and steering knuckle—GTO

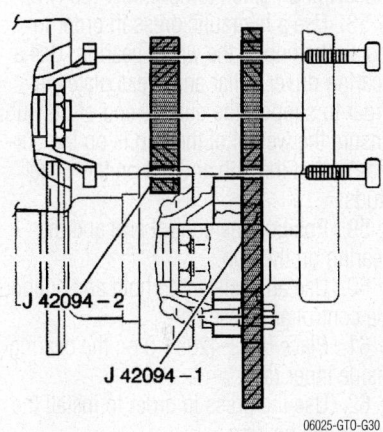

06025-GTO-G30

Installing wheel hub remover tool—GTO

4. Remove and discard the hub nut retainer and hub nut.

5. Lubricate the threads of the J-42094-3 with Extreme Press Lubricant " Ounce Tube no. J-23444-A.

6. Install the J-42094-3 to the J-42094-4.

7. Lubricate the ball end of the J-42094-5.

8. Install the J-42094-5 to the end of the J-42094-3.

9. Install the J-42094-10 to the J-42094-4.

10. Install the J-42094-4 and the 3 bolts to the J-42094-1.

11. Adjust the position of the J-42094-3 in the J-42094-4 in order to allow the J-42094-4 to be in full contact with the J-42094-1.

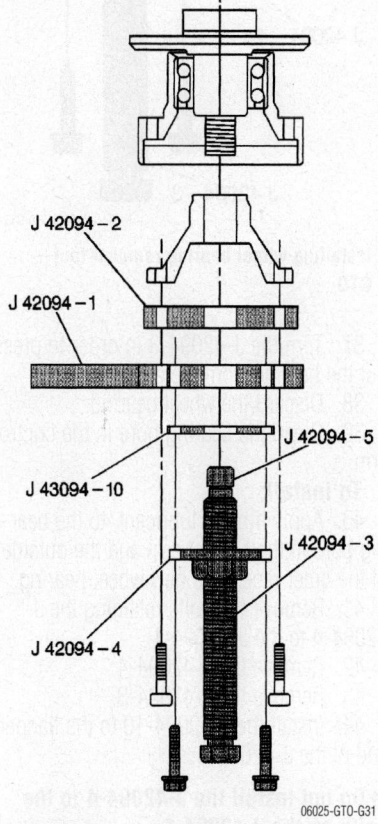

06025-GTO-G31

Installing drive shaft flange remover tool—GTO

12. Ensure the J-42094-1 is secure in a vise.

13. Use an assistant to hold and support the control arm assembly.

14. Turn the J-42094-3 in order to

remove the rear wheel drive shaft flange from the hub assembly.

15. Remove the tools from the flange.

16. Remove the 4 bolts and the 2 washers from the brake backing plate shield.

➡**Ensure the park brake adjuster anchor bracket assembly remains on the shield.**

17. Clean the threads in control arm. From the inboard side, clean the threads for the control arm to shield bolt holes. Repair the threads if necessary.

18. Install 2 J-42094-7 supports to the control arm, in the 2 shallowest control arm to shield bolt holes, near the caliper mounting holes.

19. Install the J-42094-7 support to the control arm, in the deepest control arm to shield bolt hole.

20. Install the J-42094-7 support to the control arm, in the remaining control arm to shield bolt hole.

21. Ensure the 4 supports are in the correct positions in order to attach to the flat surface of the J-42094-1.

22. Attach the J-42094-1 to the 4 supports.

23. Install the 4 nuts in order to retain the J-42094-1 to the supports.

24. Install the J-42094-3 to the J-42094-4.

25. Lubricate the ball end of the J-42094-5 with press lubricant.

26. Install the J-42094-5 to the end of the J-42094-3.

27. Install the J-42094-4 and the 3 bolts to the J-42094-1.

28. Adjust the position of the J-42094-3 in the J-42094-4 in order to allow the J-42094-4 to be in full contact with the J-42094-1.

29. Ensure the J-42094-1 is secure in a vise.

30. Use an assistant to hold and support the control arm assembly.

31. Turn the J-42094-3 in order to press out the rear wheel hub.

32. Remove the hub and the bearing outside inner race from the control arm.

33. Turn the J-42094-3 away from the bearing.

34. Remove the shield and the park brake anchor bracket as an assembly from the control arm.

35. If you are NOT replacing the hub, remove the bearing outside inner race from the hub. Use the J-22912-01, or equivalent, and a press in order to remove the race from the hub.

36. Use snap ring pliers in order to remove the retainer from the wheel bearing in the control arm.

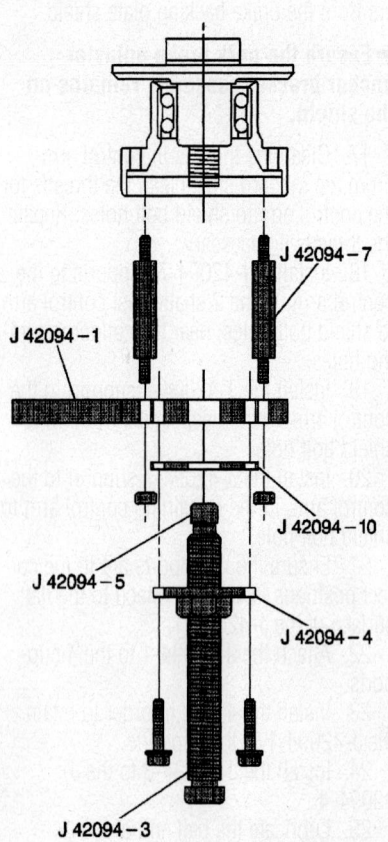

Installing hub and bearing race remover tool—GTO

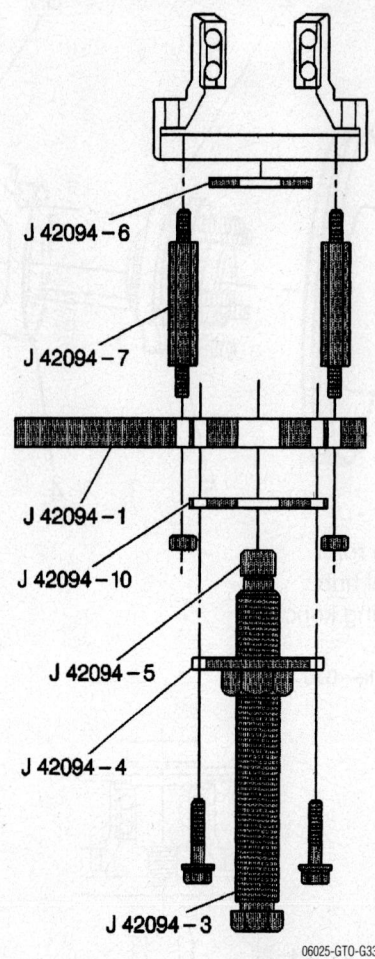

06025-GTO-G33

Installing wheel bearing remover tool—GTO

37. Turn the J-42094-3 in order to press out the bearing from the control arm.

38. Discard the wheel bearing.

39. Clean the bearing bore in the control arm.

To install:

40. Apply lithium lubricant, to the bearing bore in the control arm and the outside of the outer races of a new wheel bearing.

41. Remove the bolts retaining the J-42094-4 to the J-42094-1.

42. Remove the J-42094-4.

43. Remove the J-42094-3.

44. Install the J-42094-10 to the flanged end of the J-42094-4.

➡**Do not install the J-42094-4 to the bolts on the J-42094-1.**

45. Install the J-42094-10 and the J-42094-4 to the J-42094-1.

46. Install the J-42094-3 to the J-42094-1.

47. Install a new wheel bearing to the J-42094-8.

48. Position the wheel bearing on the wheel bearing bore in the control arm.

49. Engage a minimum of 8 threads of the J-42094-3 with the J-42094-8.

50. Use a breaker bar in order to hold the J-42094-3.

51. Rotate the J-42094-4 in order to press the wheel bearing into the control arm. Ensure the wheel bearing is seated properly in the control arm.

52. Remove the J-42094-8 from the bearing.

53. Use snap ring pliers in order to install the bearing retainer to the control arm.

54. Remove the tools from the control arm.

55. Install the rear disc brake backing plate shield and the parking brake anchor bracket assembly to the control arm.

56. Install the 2 Torx bolts and washers to the shield and tighten to 55 ft. lbs. (75 Nm).

57. Install the 2 bolts to the anchor bracket and tighten to 65 ft. lbs. (88 Nm).

58. Use a hydraulic press in order to install the hub to the wheel bearing. Use a bearing driver collar and press plates in order to support the outside end of the hub. Ensure the weight of the hub is on the outside end of the hub and not on the wheel studs.

59. Position the control arm and the bearing on the hub.

60. Use an assistant to hold and support the control arm assembly.

61. Place the J-42094-9 on the bearing inside inner race.

62. Use the press in order to install the hub to the bearing.

63. Remove the J-42094-9 from the bearing.

64. Position the J-42094-2 on the rear wheel drive shaft flange.

65. Position the J-42094-1-B on the J-42094-2.

66. Install the outer constant velocity joint bolts in order to retain the J-42094-2 and the J-42094-1 to the flange.

67. Apply lithium lubricant, to the splines on the flange and the threads on the inside end of the hub.

68. With the outside end of the hub on the collar and the press plates, position the flange on the inside end of the hub.

69. Place the J-42094-9 on the flange.

70. Use the press in order to install the flange to the hub and bearing assembly.

71. Remove the J-42094-9 from the flange.

72. With the J-42094-2 and the J-

42094-1 on the flange, remove the control arm from the press.

73. Use a vise in order to hold the J-42094-1.

74. Install a new hub nut to the inside

end of the hub and tighten to 221 ft. lbs. (300 Nm).

75. Remove the J-42094-1 and the control arm assembly from the vise.

76. Remove the bolts from the J-42094-1.

77. Remove the J-42094-2 and the J-42094-1 from the control arm assembly.

78. Install the new hub nut retainer.

79. Stake the retainer to the nut.

80. Install the control arm to the vehicle.

BRAKES

Brake Caliper

REMOVAL & INSTALLATION

Front and Rear

1. Use a siphon in order to remove half of the brake fluid from the reservoir.

2. Raise and support the vehicle.

3. Remove the tire and wheel.

4. Remove the caliper bolts.

5. Remove the caliper and the bracket from the brake rotor.

To install:

6. Ensure the mounting surfaces on the shield and on the caliper bracket are clean.

7. Ensure the caliper pistons are fully seated in the cylinder bores.

8. Position the caliper and the bracket over the rotor.

9. Install the caliper bolts and tighten to 63 ft. lbs. (85 Nm), plus an additional 45 degrees.

10. Install the tire and wheel assembly.

11. Lower the vehicle.

12. With the engine off, gradually apply and release the brake pedal several times in order to position the caliper pistons and the brake pads.

13. Fill the master cylinder reservoir.

14. Road test the vehicle and check for proper brake system operation.

Brake Pads

REMOVAL & INSTALLATION

Front and Rear

1. Use a siphon in order to remove half of the brake fluid from the master cylinder reservoir.

2. Raise and support the vehicle.

3. Remove the tire and wheel.

4. Install a large C-clamp over the brake caliper. Position 1 end of the clamp on the head of the brake hose fitting. Position the other end of the clamp against the outer pad.

5. Install a second C-clamp over the brake caliper. Position 1 end of the clamp against the caliper body adjacent to the

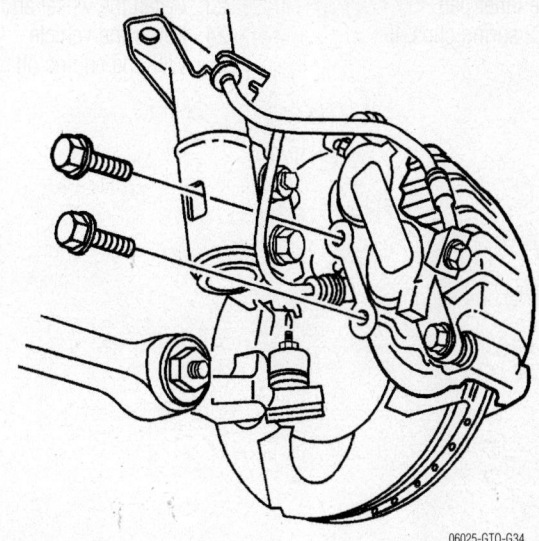

06025-GTO-G34

Front disc brake caliper mounting—GTO

brake hose fitting. Position the other end of the clamp against the outer pad.

6. Tighten the 2 clamps in order to compress the 2 caliper pistons into the bores.

7. Remove the clamps.

8. Use a wrench in order to prevent the lower guide pin from turning.

9. Remove and discard the lower guide pin bolt.

10. Rotate the caliper assembly up and support the caliper with heavy mechanic's wire, or equivalent. Verify there is no tension on the brake hose.

11. Remove the 2 brake pads and the 2 spring clips, if equipped.

To install:

12. Use a large C-clamp and a wood block to push the caliper pistons back into their bores.

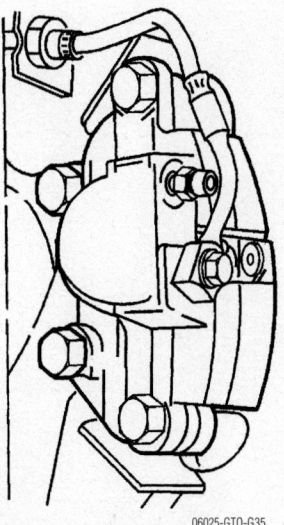

06025-GTO-G35

Rear disc brake caliper mounting—GTO

13. Clean the surfaces of the 2 pistons which contact the inner brake pad.

14. Clean the surface of the caliper which contacts the outer brake pad.

15. Install the inner brake pad to the bracket.

16. Install the outer brake pad to the caliper.

17. Position a leg of the steel spring onto the top of the inner pad.

18. Install the 2 spring clips, if equipped.

19. Rotate the caliper down over the brake rotor.

20. Ensure the outer spring clip engages the middle flange of the caliper housing.

21. Use a wrench in order to prevent the guide pin from rotating.

22. Install a new guide pin bolt and tighten to 24 ft. lbs. (32 Nm).

23. Install the wheel and tire assembly.

24. Lower the vehicle.

25. With the engine off, gradually apply and release the brake pedal several times in order to position the caliper piston.

26. Apply pressure to the brake pedal for at least 5 seconds in order to position the brake pads.

27. Fill the master cylinder fluid reservoir

28. If the disc brake calipers were replaced or repaired, be sure to bleed the system.

29. Road test the vehicle and check the brake system for proper operation.

SATURN

ION L & S Series

ENGINE REPAIR..............21-16
BRAKES 21-106
DRIVE TRAIN 21-72
FUEL SYSTEM 21-68
SPECIFICATION AND
  MAINTENANCE CHARTS......21-2
Engine and Vehicle
  Identification..................21-2
General Engine Specifications 21-3
Engine Tune-Up Specifications......21-4
Firing Order 21-4
Accessory Drive Belt Routing 21-5
Capacities.............................21-6
Valve Specifications..................21-7
Crankshaft and Connecting
  Rod Specifications...................21-8
Piston and Ring
  Specifications.......................21-9
Torque Specifications 21-10
Wheel Alignment 21-11
Tire, Wheel and Ball Joint
  Specifications......................21-12
Brake Specifications 21-13
Scheduled Maintenance
  Intervals...........................21-15
STEERING AND
  SUSPENSION.................21-91

A
Air Bag...............................21-91
  Arming..............................21-91
  Disarming...........................21-91
  Precautions.........................21-91
Alternator............................21-16
  Installation........................21-16
  Removal.............................21-16
Automatic Transaxle Assembly 21-76
  Removal & Installation.............21-76

B
Brake Caliper 21-106
  Removal & Installation.............21-106
Brake Drums..........................21-108
  Removal & Installation............21-108
Brake Shoes..........................21-109
  Removal & Installation............21-109

C
Camshaft and Lifters.................21-42
  Removal & Installation.............21-42

Clutch...............................21-82
  Adjustment..........................21-82
  Removal & Installation.............21-82
Coil Spring..........................21-95
  Removal & Installation.............21-95
CV-Joints............................21-87
  Overhaul 21-87
Cylinder Head........................21-30
  Removal & Installation.............21-30

D
Disc Brake Pads.....................21-107
  Removal & Installation...........21-107

E
Engine Assembly 21-16
  Removal & Installation.............21-16
Exhaust Manifold.....................21-40
  Removal & Installation.............21-40

F
Front Crankshaft Seal 21-41
  Replacement 21-41
Fuel Filter 21-68
  Removal & Installation.............21-68
Fuel Injector........................21-71
  Removal And Installation 21-71
Fuel Pump............................21-69
  Removal & Installation.............21-69
Fuel System Pressure 21-68
  Relieving..........................21-68
Fuel System Service
  Precautions........................21-68

H
Halfshafts...........................21-85
  Removal & Installation............21-85
Heater Core..........................21-27
  Removal & Installation.............21-27
Hydraulic Clutch System 21-84
  Bleeding...........................21-84

I
Ignition Timing 21-16
  Adjustment..........................21-16
Intake Manifold......................21-37
  Removal & Installation.............21-37

L
Lower Ball Joint.....................21-98
  Removal & Installation.............21-98
Lower Control Arm 21-98

Lower Control Arm Bushing
  Replacement........................21-99
  Removal & Installation.............21-98

M
Manual Transaxle Assembly 21-72
  Removal & Installation.............21-72

O
Oil Pan..............................21-48
  Removal & Installation.............21-48
Oil Pump.............................21-49
  Removal & Installation.............21-49

P
Piston and Ring 21-66
  Positioning........................21-66

R
Rack and Pinion Steering Gear 21-91
  Power Steering System
    Bleeding.........................21-93
  Removal & Installation.............21-91
Rear Main Seal 21-52
  Removal & Installation.............21-52
Rocker Arms/Shafts 21-36
  Removal & Installation.............21-36

S
Shock Absorber 21-95
  Removal & Installation.............21-95
Starter Motor 21-47
  Removal & Installation.............21-47
Strut................................21-93
  Removal & Installation.............21-93
Supercharger.........................21-47
  Removal & Installation.............21-47

T
Timing Belt..........................21-63
  Removal & Installation.............21-63
Timing Chain, Sprockets, Front
  Cover and Seal 21-53
  Removal & Installation.............21-53

V
Valve Lash 21-47
  Adjustment.........................21-47

W
Water Pump...........................21-25
  Removal & Installation.............21-25
Wheel Bearings.......................21-102
  Adjustment.........................21-102
  Removal & Installation..........21-102

SPECIFICATION CHARTS

ENGINE AND VEHICLE IDENTIFICATION

Engine								Model Year	
Code ①	Liters (cc)	Cu. In.	Cyl.	Fuel Sys.	Engine Type	Eng. Mfg.		Code ②	Year
7	1.9 (1901)	116	4	MFI	DOHC	Saturn		2	2002
8	1.9 (1901)	116	4	MFI	SOHC	Saturn		3	2003
F	2.0 (1998)	112	4	SFI	DOHC	Saturn		4	2004
F	2.2 (2199)	134	4	SFI	DOHC	Saturn		5	2005
R	3.0 (3000)	183	6	SFI	DOHC	Saturn		6	2006

MFI: Multi-point Fuel Injection

SFI: Sequential Fuel Injection

DOHC: Double Overhead Camshafts

SOHC: Single Overhead Camshaft

① 8th digit of VIN

② 10th digit of VIN

06025-SCAR-C01

GENERAL ENGINE SPECIFICATIONS

Year	Model	Engine Displacement Liters (VIN)	Net Horsepower @ rpm	Net Torque @ rpm (ft. lbs.)	Bore x Stroke (in.)	Com-pression Ratio	Oil Pressure @ rpm
2002	Sedan	1.9 (7)	124@5600	122@4800	3.23x3.54	9.5:1	29@2000
	Sedan	1.9 (8)	100@5000	114@2400	3.23x3.54	9.3:1	36@2000
	Wagon	1.9 (7)	124@5600	122@4800	3.23x3.54	9.5:1	29@2000
	Wagon	1.9 (8)	100@5000	114@2400	3.23x3.54	9.3:1	36@2000
	Sedan	2.2 (F)	137@5800	147@4400	3.38x3.5	9.5:1	50-80@1000
	Wagon	2.2 (F)	137@5800	147@4400	3.38x3.50	9.5:1	50-80@1000
	Sedan	3.0 (R)	182@6000	184@3600	3.38x3.50	10.0:1	22@1000
	Wagon	3.0 (R)	182@6000	184@3600	3.38x3.50	10.0:1	22@1000
2003	Sedan	1.9 (7)	124@5600	122@4800	3.23x3.54	9.5:1	29@2000
	Sedan	1.9 (8)	100@5000	114@2400	3.23x3.54	9.3:1	36@2000
	Wagon	1.9 (7)	124@5600	122@4800	3.23x3.54	9.5:1	29@2000
	Wagon	1.9 (8)	100@5000	114@2400	3.23x3.54	9.3:1	36@2000
	Sedan	2.2 (F)	137@5800	147@4400	3.38x3.5	9.5:1	50-80@1000
	Wagon	2.2 (F)	137@5800	147@4400	3.38x3.50	9.5:1	50-80@1000
	Sedan	3.0 (R)	182@6000	184@3600	3.38x3.50	10.0:1	22@1000
	Wagon	3.0 (R)	182@6000	184@3600	3.38x3.50	10.0:1	22@1000
	ION	2.2 (F)	137@5800	147@4400	3.38x3.50	9.5:1	50-80@1000
2004	Sedan	1.9 (7)	124@5600	122@4800	3.23x3.54	9.5:1	29@2000
	Sedan	1.9 (8)	100@5000	114@2400	3.23x3.54	9.3:1	36@2000
	Wagon	1.9 (7)	124@5600	122@4800	3.23x3.54	9.5:1	29@2000
	Wagon	1.9 (8)	100@5000	114@2400	3.23x3.54	9.3:1	36@2000
	Sedan	2.2 (F)	137@5800	147@4400	3.38x3.5	9.5:1	50-80@1000
	Wagon	2.2 (F)	137@5800	147@4400	3.38x3.50	9.5:1	50-80@1000
	Sedan	3.0 (R)	182@6000	184@3600	3.38x3.50	10.0:1	22@1000
	Wagon	3.0 (R)	182@6000	184@3600	3.38x3.50	10.0:1	22@1000
	ION	2.2 (F)	137@5800	147@4400	3.38x3.50	9.5:1	50-80@1000
2005	Sedan	1.9 (7)	124@5600	122@4800	3.23x3.54	9.5:1	29@2000
	Sedan	1.9 (8)	100@5000	114@2400	3.23x3.54	9.3:1	36@2000
	Wagon	1.9 (7)	124@5600	122@4800	3.23x3.54	9.5:1	29@2000
	Wagon	1.9 (8)	100@5000	114@2400	3.23x3.54	9.3:1	36@2000
	Sedan	2.2 (F)	137@5800	147@4400	3.38x3.5	9.5:1	50-80@1000
	Wagon	2.2 (F)	137@5800	147@4400	3.38x3.50	9.5:1	50-80@1000
	Sedan	3.0 (R)	182@6000	184@3600	3.38x3.50	10.0:1	22@1000
	Wagon	3.0 (R)	182@6000	184@3600	3.38x3.50	10.0:1	22@1000
	ION	2.0 (F)	105@5600	200@4400	3.38x3.38	9.5:1	50-80@1000
		2.2 (F)	137@5800	147@4400	3.38x3.50	9.5:1	50-80@1000

06025-SCAR-C02

ENGINE TUNE-UP SPECIFICATIONS

Year	Engine Displacement Liters (VIN)	Spark Plug Gap (in.)	Ignition Timing (deg.)		Fuel Pump (psi) ①	Idle Speed (rpm)		Valve Clearance	
			MT	AT		MT ②	AT ②	In.	Ex.
2002	1.9 (7)	0.040	③	③	40-55	850	750	HYD	HYD
	1.9 (8)	0.040	③	③	40-55	750	650	HYD	HYD
	2.2 (F)	0.045	③	③	55-65	④	④	HYD	HYD
	3.0 (R)	0.043	③	③	39-49	④	④	HYD	HYD
2003	1.9 (7)	0.040	③	③	40-55	850	750	HYD	HYD
	1.9 (8)	0.040	③	③	40-55	750	650	HYD	HYD
	2.2 (F)	0.045	③	③	55-65	④	④	HYD	HYD
	3.0 (R)	0.043	③	③	39-49	④	④	HYD	HYD
2004	1.9 (7)	0.040	③	③	40-55	850	750	HYD	HYD
	1.9 (8)	0.040	③	③	40-55	750	650	HYD	HYD
	2.2 (F)	0.045	③	③	55-65	④	④	HYD	HYD
	3.0 (R)	0.043	③	③	39-49	④	④	HYD	HYD
2005	1.9 (7)	0.040	③	③	40-55	850	750	HYD	HYD
	1.9 (8)	0.040	③	③	40-55	750	650	HYD	HYD
	2.0 (F)	0.040	③	③	50-60	④	④	HYD	HYD
	2.2 (F)	0.045	③	③	55-65	④	④	HYD	HYD
	3.0 (R)	0.043	③	③	39-49	④	④	HYD	HYD

NOTE: The Vehicle Emission Control Information label often reflects specification changes made during production. The label figures must be used if they differ from those in this chart.

HYD: Hydraulic

① Pressure measured at idle

② Idle speed measured with manual transmission in Neutral; automatic transmission in D (drive)

③ Engines equipped with Distributorless Ignition System (DIS). Ignition timing is not adjustable

④ Refer to the Vehicle Emission Control Information label

06025-SCAR-C03

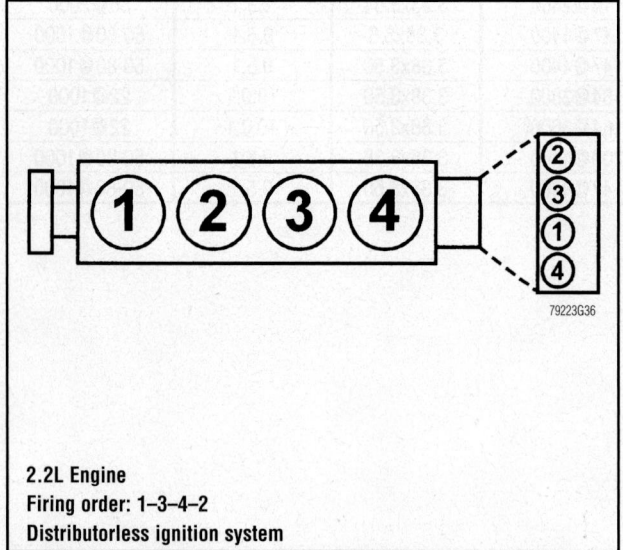

2.2L Engine
Firing order: 1–3–4–2
Distributorless ignition system

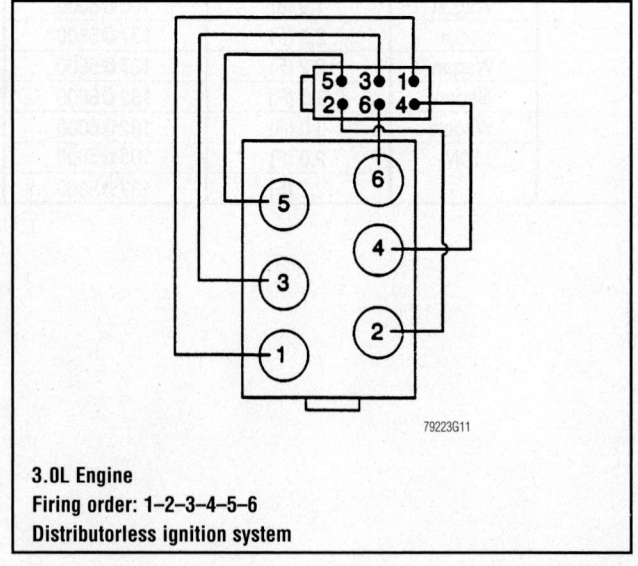

3.0L Engine
Firing order: 1–2–3–4–5–6
Distributorless ignition system

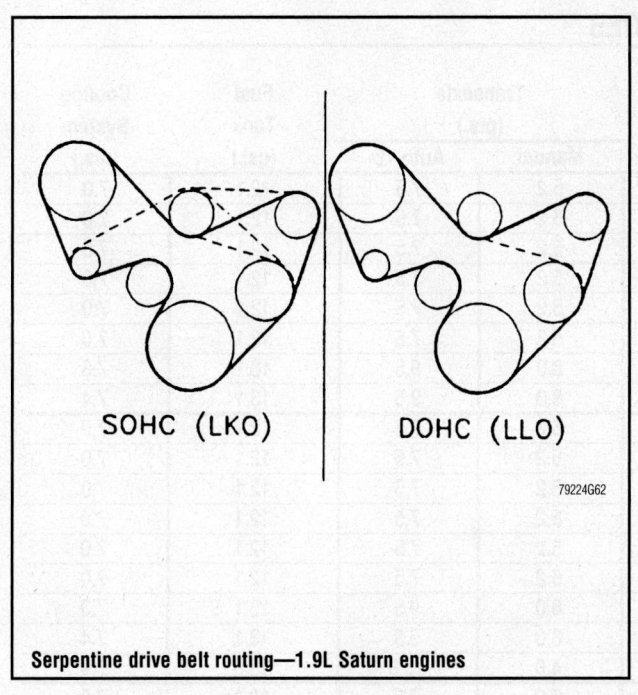

Serpentine drive belt routing—1.9L Saturn engines

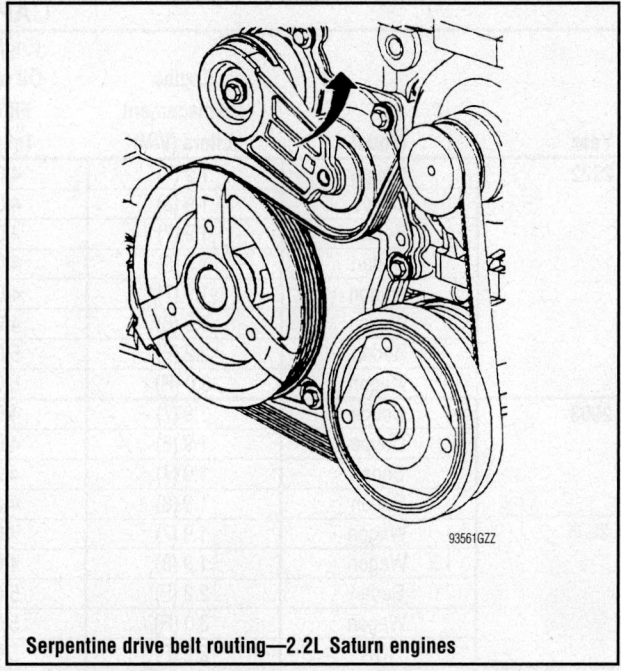

Serpentine drive belt routing—2.2L Saturn engines

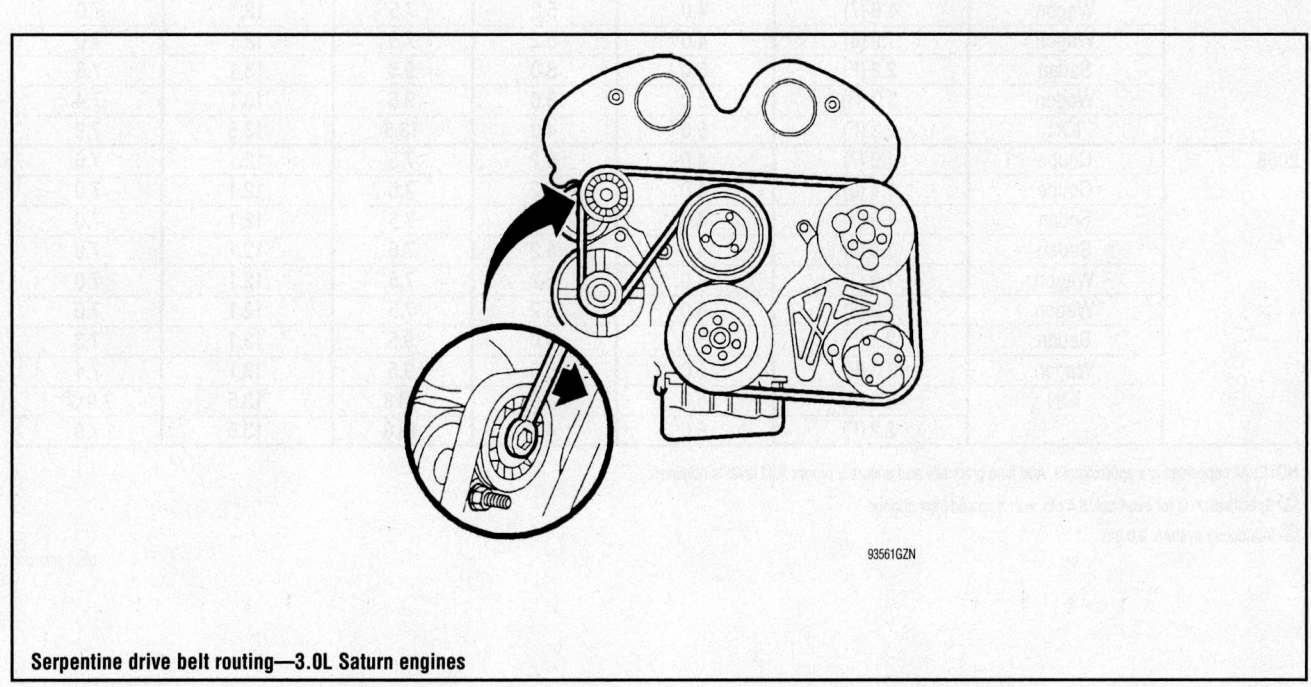

Serpentine drive belt routing—3.0L Saturn engines

CAPACITIES

Year	Model	Engine Displacement Liters (VIN)	Engine Oil with Filter (qts.)	Transaxle (pts.)		Fuel Tank (gal.)	Cooling System (qts.)
				Manual	Auto. ①		
2002	Coupe	1.9 (7)	4.0	5.2	7.5	12.1	7.0
	Coupe	1.9 (8)	4.0	5.2	7.5	12.1	7.0
	Sedan	1.9 (7)	4.0	5.2	7.5	12.1	7.0
	Sedan	1.9 (8)	4.0	5.2	7.5	12.1	7.0
	Wagon	1.9 (7)	4.0	5.2	7.5	12.1	7.0
	Wagon	1.9 (8)	4.0	5.2	7.5	12.1	7.0
	Sedan	2.2 (F)	5.0	8.0	9.5	13.1	7.3
	Wagon	3.0 (R)	5.0	8.0	9.5	13.1	7.4
2003	Coupe	1.9 (7)	4.0	5.2	7.5	12.1	7.0
	Coupe	1.9 (8)	4.0	5.2	7.5	12.1	7.0
	Sedan	1.9 (7)	4.0	5.2	7.5	12.1	7.0
	Sedan	1.9 (8)	4.0	5.2	7.5	12.1	7.0
	Wagon	1.9 (7)	4.0	5.2	7.5	12.1	7.0
	Wagon	1.9 (8)	4.0	5.2	7.5	12.1	7.0
	Sedan	2.2 (F)	5.0	8.0	9.5	13.1	7.3
	Wagon	3.0 (R)	5.0	8.0	9.5	13.1	7.4
	ION	2.2 (F)	5.0	4.0	13.8	13.5	7.9
2004	Coupe	1.9 (7)	4.0	5.2	7.5	12.1	7.0
	Coupe	1.9 (8)	4.0	5.2	7.5	12.1	7.0
	Sedan	1.9 (7)	4.0	5.2	7.5	12.1	7.0
	Sedan	1.9 (8)	4.0	5.2	7.5	12.1	7.0
	Wagon	1.9 (7)	4.0	5.2	7.5	12.1	7.0
	Wagon	1.9 (8)	4.0	5.2	7.5	12.1	7.0
	Sedan	2.2 (F)	5.0	8.0	9.5	13.1	7.3
	Wagon	3.0 (R)	5.0	8.0	9.5	13.1	7.4
	ION	2.2 (F)	5.0	4.0	13.8	13.5	7.9
2005	Coupe	1.9 (7)	4.0	5.2	7.5	12.1	7.0
	Coupe	1.9 (8)	4.0	5.2	7.5	12.1	7.0
	Sedan	1.9 (7)	4.0	5.2	7.5	12.1	7.0
	Sedan	1.9 (8)	4.0	5.2	7.5	12.1	7.0
	Wagon	1.9 (7)	4.0	5.2	7.5	12.1	7.0
	Wagon	1.9 (8)	4.0	5.2	7.5	12.1	7.0
	Sedan	2.2 (F)	5.0	8.0	9.5	13.1	7.3
	Wagon	3.0 (R)	5.0	8.0	9.5	13.1	7.4
	ION	2.0 (F)	6.0	4.0	13.8	13.5	7.9 ②
		2.2 (F)	5.0	4.0	13.8	13.5	7.9

NOTE: All capacities are approximate. Add fluid gradually and ensure a proper fluid level is obtained.

① Specification is for overhaul. 8.4 pts. with fluid and filter change

② Intercooler system. 2.0 qts.

06025-SCAR-C04

VALVE SPECIFICATIONS

Year	Engine Displacement Liters (VIN)	Seat Angle (deg.)	Face Angle (deg.)	Spring Test Pressure (lbs. @ in.)	Spring Free-Length (in.)	Stem-to-Guide Clearance (in.)		Stem Diameter (in.)	
						Intake	Exhaust	Intake	Exhaust
2002	1.9 (7)	44.5-45.4	45-45.5	163-180@ 0.984	1.5600	0.0010-0.0025	0.0015-0.0032	0.2736-0.2740	0.2729-0.2736
	1.9 (8)	44.5-45.4	45-45.25	202-211@ 1.280	1.8898-1.9134	0.0010-0.0025	0.0015-0.0032	0.2736-0.2741	0.2736-0.2740
	2.2 (F)	44.5-45.4	45-45.5	① ②	1.6100	0.0012 0.0022	0.0020 0.0026	0.2344 0.2355	0.2337 0.2343
	3.0 (R)	45	45	56.6@1.338	NA	0.0012 0.0022	0.0016 0.0026	0.2344 0.2350	0.2341 0.2346
2003	1.9 (7)	44.5-45.4	45-45.5	163-180@ 0.984	1.5600	0.0010-0.0025	0.0015-0.0032	0.2736-0.2740	0.2729-0.2736
	1.9 (8)	44.5-45.4	45-45.25	202-211@ 1.280	1.8898-1.9134	0.0010-0.0025	0.0015-0.0032	0.2736-0.2741	0.2736-0.2740
	2.2 (F)	44.5-45.4	45-45.5	① ②	1.6100	0.0012 0.0022	0.0020 0.0026	0.2344 0.2355	0.2337 0.2343
	3.0 (R)	45	45	56.6@1.338	NA	0.0012 0.0022	0.0016 0.0026	0.2344 0.2350	0.2341 0.2346
2004	1.9 (7)	44.5-45.4	45-45.5	163-180@ 0.984	1.5600	0.0010-0.0025	0.0015-0.0032	0.2736-0.2740	0.2729-0.2736
	1.9 (8)	44.5-45.4	45-45.25	202-211@ 1.280	1.8898-1.9134	0.0010-0.0025	0.0015-0.0032	0.2736-0.2741	0.2736-0.2740
	2.2 (F)	44.5-45.4	45-45.5	① ②	1.6100	0.0012 0.0022	0.0020 0.0026	0.2344 0.2355	0.2337 0.2343
	3.0 (R)	45	45	56.6@1.338	NA	0.0012 0.0022	0.0016 0.0026	0.2344 0.2350	0.2341 0.2346
2005	1.9 (7)	44.5-45.4	45-45.5	163-180@ 0.984	1.5600	0.0010-0.0025	0.0015-0.0032	0.2736-0.2740	0.2729-0.2736
	1.9 (8)	44.5-45.4	45-45.25	202-211@ 1.280	1.8898-1.9134	0.0010-0.0025	0.0015-0.0032	0.2736-0.2741	0.2736-0.2740
	2.0 (F)	44.5-45.4	45-45.5	① ②	1.6100	0.0012 0.0022	0.0020 0.0026	0.2344 0.2355	0.2337 0.2343
	2.2 (F)	44.5-45.4	45-45.5	① ②	1.6100	0.0012 0.0022	0.0020 0.0026	0.2344 0.2355	0.2337 0.2343
	3.0 (R)	45	45	56.6@1.338	NA	0.0012 0.0022	0.0016 0.0026	0.2344 0.2350	0.2341 0.2346

NA: Not available

① Valve spring load closed: 245-271 N

② Valve spring load open: 525-575 N

06025-SCAR-C05

CRANKSHAFT AND CONNECTING ROD SPECIFICATIONS
All measurements are given in inches.

Year	Engine Displacement Liters (VIN)	Crankshaft				Connecting Rod		
		Main Brg. Journal Dia.	Main Brg. Oil Clearance	Shaft End-play	Thrust on No.	Journal Diameter	Oil Clearance	Side Clearance
2002	1.9 (7)	2.2438-2.2444	0.0002-0.0020	0.0020-0.0079	3	1.9761-1.9767	0.0001-0.0021	0.0065-0.0171
	1.9 (8)	2.2438-2.2444	0.0002-0.0020	0.0020-0.0079	3	1.9761-1.9767	0.0001-0.0021	0.0065-0.0171
	2.2 (F)	2.2045-2.2050	0.0012-0.0026	0.0012-0.0150	3	1.9291-1.9297	0.0001-0.0021	0.0028-0.0146
	3.0 (R)	2.6763-2.6766	0.0060-0.0017	0.0004-0.0300	3	1.927-1.9280	0.0001-0.0021	0.0027-0.0110
2003	1.9 (7)	2.2438-2.2444	0.0002-0.0020	0.0020-0.0079	3	1.9761-1.9767	0.0001-0.0021	0.0065-0.0171
	1.9 (8)	2.2438-2.2444	0.0002-0.0020	0.0020-0.0079	3	1.9761-1.9767	0.0001-0.0021	0.0065-0.0171
	2.2 (F)	2.2045-2.2050	0.0012-0.0026	0.0012-0.0150	3	1.9291-1.9297	0.0001-0.0021	0.0028-0.0146
	3.0 (R)	2.6763-2.6766	0.0060-0.0017	0.0004-0.0300	3	1.927-1.9280	0.0001-0.0021	0.0027-0.0110
2004	1.9 (7)	2.2438-2.2444	0.0002-0.0020	0.0020-0.0079	3	1.9761-1.9767	0.0001-0.0021	0.0065-0.0171
	1.9 (8)	2.2438-2.2444	0.0002-0.0020	0.0020-0.0079	3	1.9761-1.9767	0.0001-0.0021	0.0065-0.0171
	2.2 (F)	2.2045-2.2050	0.0012-0.0026	0.0012-0.0150	3	1.9291-1.9297	0.0001-0.0021	0.0028-0.0146
	3.0 (R)	2.6763-2.6766	0.0060-0.0017	0.0004-0.0300	3	1.927-1.9280	0.0001-0.0021	0.0027-0.0110
2005	1.9 (7)	2.2438-2.2444	0.0002-0.0020	0.0020-0.0079	3	1.9761-1.9767	0.0001-0.0021	0.0065-0.0171
	1.9 (8)	2.2438-2.2444	0.0002-0.0020	0.0020-0.0079	3	1.9761-1.9767	0.0001-0.0021	0.0065-0.0171
	2.0 (F)	2.2045-2.2050	0.0012-0.0026	0.0012-0.0150	3	1.9291-1.9297	0.0001-0.0027	0.0028-0.0146
	2.2 (F)	2.2045-2.2050	0.0012-0.0026	0.0012-0.0150	3	1.9291-1.9297	0.0001-0.0027	0.0028-0.0146
	3.0 (R)	2.6763-2.6766	0.0060-0.0017	0.0004-0.0300	3	1.927-1.9280	0.0001-0.0021	0.0027-0.0110

NA: Not available

06025-SCAR-C06

PISTON AND RING SPECIFICATIONS
All measurements are given in inches.

Year	Engine Displacement Liters (VIN)	Piston Clearance	Ring Gap			Ring Side Clearance		
			Top Compression	Bottom Compression	Oil Control	Top Compression	Bottom Compression	Oil Control
2002	1.9 (7)	①	0.0071-0.0130	0.0138-0.0216	0.0039-0.0197	0.0016-0.0032	0.0012-0.0031	SNUG
	1.9 (8)	①	0.0071-0.0130	0.0138-0.0216	0.0039-0.0197	0.0016-0.0032	0.0012-0.0031	SNUG
	2.2 (F)	0.0004-0.0016	0.008-0.016	0.0014-0.0022	0.0010-0.0030	0.0028-0.0146	0.0005-0.0024	SNUG
	3.0 (R)	0.0010-0.0018	0.0008-0.0015	0.0118-0.0196	0.0157-0.0551	0.0027-0.0110	0.0005-0.0024	SNUG
2003	1.9 (7)	①	0.0071-0.0130	0.0138-0.0216	0.0039-0.0197	0.0016-0.0032	0.0012-0.0031	SNUG
	1.9 (8)	①	0.0071-0.0130	0.0138-0.0216	0.0039-0.0197	0.0016-0.0032	0.0012-0.0031	SNUG
	2.2 (F)	0.0004-0.0016	0.008-0.016	0.0014-0.0022	0.0010-0.0030	0.0028-0.0146	0.0005-0.0024	SNUG
	3.0 (R)	0.0010-0.0018	0.0008-0.0015	0.0118-0.0196	0.0157-0.0551	0.0027-0.0110	0.0005-0.0024	SNUG
2004	1.9 (7)	①	0.0071-0.0130	0.0138-0.0216	0.0039-0.0197	0.0016-0.0032	0.0012-0.0031	SNUG
	1.9 (8)	①	0.0071-0.0130	0.0138-0.0216	0.0039-0.0197	0.0016-0.0032	0.0012-0.0031	SNUG
	2.2 (F)	0.0004-0.0016	0.008-0.016	0.0014-0.0022	0.0010-0.0030	0.0028-0.0146	0.0005-0.0024	SNUG
	3.0 (R)	0.0010-0.0018	0.0008-0.0015	0.0118-0.0196	0.0157-0.0551	0.0027-0.0110	0.0005-0.0024	SNUG
2005	1.9 (7)	①	0.0071-0.0130	0.0138-0.0216	0.0039-0.0197	0.0016-0.0032	0.0012-0.0031	SNUG
	1.9 (8)	①	0.0071-0.0130	0.0138-0.0216	0.0039-0.0197	0.0016-0.0032	0.0012-0.0031	SNUG
	2.0 (F)	0.0004-0.0016	0.008-0.016	0.0014-0.0022	0.0010-0.0030	0.0015-0.0031	0.0012-0.0027	SNUG
	2.2 (F)	0.0004-0.0016	0.008-0.016	0.0014-0.0022	0.0010-0.0030	0.0015-0.0031	0.0012-0.0027	SNUG
	3.0 (R)	0.0010-0.0018	0.0008-0.0015	0.0118-0.0196	0.0157-0.0551	0.0027-0.0110	0.0005-0.0024	SNUG

NA: Not available

① Piston No. 2 and 3: 0.0002-0.0017
Piston No. 1, 4: 0.0003-0.0021

06025-SCAR-C07

TORQUE SPECIFICATIONS
All readings in ft. lbs.

Year	Engine Displacement Liters (VIN)	Cylinder Head Bolts	Main Bearing Bolts	Rod Bearing Bolts	Crankshaft Damper Bolts	Flywheel Bolts	Manifold Intake	Manifold Exhaust	Spark Plugs	Oil Pan Drain Plug
2002	1.9 (7)	①	37	33	159	59②	22③	13③	20	27
	1.9 (8)	①	37	33	159	59②	22③	16③	20	27
	2.2 (F)	④	⑤	18	⑥	②	⑦	13	15	18
	3.0 (R)	⑧	⑨	26	15	②	15	15	18	18
2003	1.9 (7)	①	37	33	159	59②	22③	13③	20	27
	1.9 (8)	①	37	33	159	59②	22③	16③	20	27
	2.2 (F)	④	⑤	18	⑥	②	⑦	13	15	18
	3.0 (R)	⑧	⑨	26	15	②	15	15	18	18
2004	1.9 (7)	①	37	33	159	59②	22③	13③	20	27
	1.9 (8)	①	37	33	159	59②	22③	16③	20	27
	2.2 (F)	④	⑤	18	⑥	②	⑦	13	15	18
	3.0 (R)	⑧	⑨	26	15	②	15	15	18	18
2005	1.9 (7)	①	37	33	159	59②	22③	13③	20	27
	1.9 (8)	①	37	33	159	59②	22③	16③	20	27
	2.0 (F)	④	⑩	⑪	⑥	②	⑫	⑬	15	18
	2.2 (F)	④	⑤	18	⑥	②	⑫	⑬	15	18
	3.0 (R)	⑧	⑨	26	15	②	15	15	18	18

① Step 1: 22 ft. lbs.
   Step 2: 33 ft. lbs.
   Step 3: 90 degrees

② Flexplate specification: 39 ft. lbs. Plus 25 degrees

③ Studs: 106 inch lbs.

④ Step 1: 22 ft. lbs.
   Step 2: 155 degrees

⑤ 15 ft. lbs. Plus 70 degrees Plus 20 degrees

⑥ 74 ft. lbs. Plus 75 degrees

⑦ 89 inch lbs.

⑧ Step 1: 22 ft. lbs.
   Step 2: plus 90 degrees
   Step 3: plus 90 degrees
   Step 4: plus 90 degrees
   Step 5: plus 15 degrees

⑨ Step 1: 37 ft. lbs.
   Step 2: plus 60 degrees
   Step 3: plus 15 degrees

⑩ 15 ft. lbs. Plus 70 degrees

⑪ 18 ft. lbs. Plus 100 degrees

⑫ Intake manifold to head nut and bolt: 18 inch lbs.
   Intake manifold to head stud: 53 inch lbs.

⑬ Exhaust manifold to head nut: 106 inch lbs.
   Exhaust manifold to head stud: 89 inch lbs.

06025-SCAR-C08

WHEEL ALIGNMENT

Year	Model		Caster Range (+/-Deg.)	Caster Preferred Setting (Deg.)	Camber Range (+/-Deg.)	Camber Preferred Setting (Deg.)	Toe-in (in.)
2002	L Series	F	1.00	+3.70	0.50	-1.00	0.20 +/- 0.15
		R	—	—	0.60	-1.00	0.15 +/- 0.07
	S Series	F	0.60	+1.70	0.70	-0.50	0.20 +/- 0.10
		R	—	—	0.70	-0.70	0.20 +/- 0.10
2003	L Series	F	1.00	+3.70	0.50	-1.00	0.20 +/- 0.15
		R	—	—	0.60	-1.00	0.15 +/- 0.07
	S Series	F	0.60	+1.70	0.70	-0.50	0.20 +/- 0.10
		R	—	—	0.70	-0.70	0.20 +/- 0.10
	ION	F	0.75	3.25	0.75	1.00	0.10 +/- 0.20
		R	—	—	0.75	-1.40	0.06 +/- 0.35
2004	L Series	F	1.00	+3.70	0.50	-1.00	0.20 +/- 0.15
		R	—	—	0.60	-1.00	0.15 +/- 0.07
	S Series	F	0.60	+1.70	0.70	-0.50	0.20 +/- 0.10
		R	—	—	0.70	-0.70	0.20 +/- 0.10
	ION	F	0.75	3.25	0.75	1.00	0.10 +/- 0.20
		R	—	—	0.75	-1.40	0.06 +/- 0.35
2005	L Series	F	1.00	+3.70	0.50	-1.00	0.20 +/- 0.15
		R	—	—	0.60	-1.00	0.15 +/- 0.07
	S Series	F	0.60	+1.70	0.70	-0.50	0.20 +/- 0.10
		R	—	—	0.70	-0.70	0.20 +/- 0.10
	ION	F	0.75	3.25	0.75	1.00	0.10 +/- 0.20
		R	—	—	0.75	-1.40	0.06 +/- 0.35

06025-SCAR-C09

TIRE, WHEEL AND BALL JOINT SPECIFICATIONS

| Year | Model | OEM Tires | | Tire Pressures (psi) | | Wheel Size | Ball Joint Inspection | Lug Nuts |
		Standard	Optional	Front	Rear			
2002	LS, LS1	P195/65R15	None	30	26	6J	NS	92
	LS2	P205/65R15	None	30	30	6J	NS	92
	LW1	P195/65R15	None	30	26	6J	NS	92
	LW2	P205/65R15	None	30	30	6J	NS	92
	SC1, SL, SL1	P185/65R14	None	30	26	6J	NS	103
	SC2	P195/60R15	None	30	26	6J	NS	103
	SL2, SW2	P185/65R15	None	30	26	6J	NS	103
2003	LS, LS1	P195/65R15	None	30	26	6J	NS	92
	LS2	P205/65R15	None	30	30	6J	NS	92
	LW1	P195/65R15	None	30	26	6J	NS	92
	LW2	P205/65R15	None	30	30	6J	NS	92
	SC1, SL, SL1	P185/65R14	None	30	26	6J	NS	103
	SC2	P195/60R15	None	30	26	6J	NS	103
	SL2, SW2	P185/65R15	None	30	26	6J	NS	103
	ION	①	①	①	①	①	①	100
2004	LS, LS1	P195/65R15	None	30	26	6J	NS	92
	LS2	P205/65R15	None	30	30	6J	NS	92
	LW1	P195/65R15	None	30	26	6J	NS	92
	LW2	P205/65R15	None	30	30	6J	NS	92
	SC1, SL, SL1	P185/65R14	None	30	26	6J	NS	103
	SC2	P195/60R15	None	30	26	6J	NS	103
	SL2, SW2	P185/65R15	None	30	26	6J	NS	103
	ION	①	①	①	①	①	①	100
2005	LS, LS1	P195/65R15	None	30	26	6J	NS	92
	LS2	P205/65R15	None	30	30	6J	NS	92
	LW1	P195/65R15	None	30	26	6J	NS	92
	LW2	P205/65R15	None	30	30	6J	NS	92
	SC1, SL, SL1	P185/65R14	None	30	26	6J	NS	103
	SC2	P195/60R15	None	30	26	6J	NS	103
	SL2, SW2	P185/65R15	None	30	26	6J	NS	103
	ION	①	①	①	①	①	①	100

OEM: Original Equipment Manufacturer

PSI: Pounds Per Square Inch

STD: Standard

OPT: Optional

NS: Not specified by manufacturer

① For tire size and information, check the label located inside the glove compartment door.

06025-SCAR-C10

BRAKE SPECIFICATIONS
All measurements in inches unless noted

| Year | Model | | Brake Disc | | | Brake Drum Diameter | | | Minimum Lining Thickness | Brake Caliper | |
			Original Thickness	Minimum Thickness	Maximum Runout	Original Inside Diameter	Max. Wear Limit	Maximum Machine Diameter		Bracket Bolt (ft. lbs.)	Mounting Bolt (ft. lbs.)
2002	Coupe ①	F	0.710	0.633	0.0024	—	—	—	0.080	81	27
		R	0.430	0.370	0.0024	7.87	7.93	7.91	0.040	63	27
	Sedan ①	F	0.710	0.633	0.0024	—	—	—	0.080	81	27
		R	0.430	0.370	0.0024	7.87	7.93	7.91	0.040	63	27
	Wagon ①	F	0.710	0.633	0.0024	—	—	—	0.080	81	27
		R	0.430	0.370	0.0024	7.87	7.93	7.91	0.040	63	27
	Coupe ②	F	0.980	0.900	0.001	—	—	—	0.080	70	22
		R	0.390	0.350	0.001	9.05	9.09	9.08	0.080	59	27
	Sedan ②	F	0.980	0.900	0.001	—	—	—	0.080	70	22
		R	0.390	0.350	0.001	9.05	9.09	9.08	0.080	59	27
	Wagon ②	F	0.980	0.900	0.001	—	—	—	0.080	70	22
		R	0.390	0.350	0.001	9.05	9.09	9.08	0.080	59	27
2003	Coupe ①	F	0.710	0.633	0.0024	—	—	—	0.080	81	27
		R	0.430	0.370	0.0024	7.87	7.93	7.91	0.040	63	27
	Sedan ①	F	0.710	0.633	0.0024	—	—	—	0.080	81	27
		R	0.430	0.370	0.0024	7.87	7.93	7.91	0.040	63	27
	Wagon ①	F	0.710	0.633	0.0024	—	—	—	0.080	81	27
		R	0.430	0.370	0.0024	7.87	7.93	7.91	0.040	63	27
	Coupe ②	F	0.980	0.900	0.001	—	—	—	0.080	70	22
		R	0.390	0.350	0.001	9.05	9.09	9.08	0.080	59	27
	Sedan ②	F	0.980	0.900	0.001	—	—	—	0.080	70	22
		R	0.390	0.350	0.001	9.05	9.09	9.08	0.080	59	27
	Wagon ②	F	0.980	0.900	0.001	—	—	—	0.080	70	22
		R	0.390	0.350	0.001	9.05	9.09	9.08	0.080	59	27
	ION	F	0.933	0.896	0.001	—	—	—	0.039	85	25
		R	—	—	—	9.06	9.09	9.08	0.020	—	—
2004	Coupe ①	F	0.710	0.633	0.0024	—	—	—	0.080	81	27
		R	0.430	0.370	0.0024	7.87	7.93	7.91	0.040	63	27
	Sedan ①	F	0.710	0.633	0.0024	—	—	—	0.080	81	27
		R	0.430	0.370	0.0024	7.87	7.93	7.91	0.040	63	27
	Wagon ①	F	0.710	0.633	0.0024	—	—	—	0.080	81	27
		R	0.430	0.370	0.0024	7.87	7.93	7.91	0.040	63	27
	Coupe ②	F	0.980	0.900	0.001	—	—	—	0.080	70	22
		R	0.390	0.350	0.001	9.05	9.09	9.08	0.080	59	27
	Sedan ②	F	0.980	0.900	0.001	—	—	—	0.080	70	22
		R	0.390	0.350	0.001	9.05	9.09	9.08	0.080	59	27
	Wagon ②	F	0.980	0.900	0.001	—	—	—	0.080	70	22
		R	0.390	0.350	0.001	9.05	9.09	9.08	0.080	59	27
	ION	F	0.933	0.896	0.001	—	—	—	0.039	85	25
		R	—	—	—	9.06	9.09	9.08	0.020	—	—

06025-SCAR-C11

BRAKE SPECIFICATIONS (Cont.)
All measurements in inches unless noted

Year	Model		Brake Disc			Brake Drum Diameter			Minimum Lining Thickness	Brake Caliper	
			Original Thickness	Minimum Thickness	Maximum Runout	Original Inside Diameter	Max. Wear Limit	Maximum Machine Diameter		Bracket Bolt (ft. lbs.)	Mounting Bolt (ft. lbs.)
2005	Coupe ①	F	0.710	0.633	0.0024	—	—	—	0.080	81	27
		R	0.430	0.370	0.0024	7.87	7.93	7.91	0.040	63	27
	Sedan ①	F	0.710	0.633	0.0024	—	—	—	0.080	81	27
		R	0.430	0.370	0.0024	7.87	7.93	7.91	0.040	63	27
	Wagon ①	F	0.710	0.633	0.0024	—	—	—	0.080	81	27
		R	0.430	0.370	0.0024	7.87	7.93	7.91	0.040	63	27
	Coupe ②	F	0.980	0.900	0.001	—	—	—	0.080	70	22
		R	0.390	0.350	0.001	9.05	9.09	9.08	0.080	59	27
	Sedan ②	F	0.980	0.900	0.001	—	—	—	0.080	70	22
		R	0.390	0.350	0.001	9.05	9.09	9.08	0.080	59	27
	Wagon ②	F	0.980	0.900	0.001	—	—	—	0.080	70	22
		R	0.390	0.350	0.001	9.05	9.09	9.08	0.080	59	27
	ION	F	③	④	0.002	—	—	—	0.039	85	25
		R	—	—	—	9.06	9.09	9.08	0.020	85	25

NA: Not Available

F: Front

R: Rear

① S series

② L series

③ Front brakes J41/JM4: 0.933 inch
   Front brakes JL9: 0.551 inch

④ Front brakes J41/JM4: 0.87
   Front brakes JL9: 0.465 inch

06025-SCAR-C12

SCHEDULED MAINTENANCE INTERVALS

SATURN—ION, L & S SERIES

TO BE SERVICED	TYPE OF SERVICE	VEHICLE MILEAGE INTERVAL (x1000)												
		3	6	9	12	15	18	21	24	27	30	33	36	39
Engine oil & filter ①	R		✓		✓		✓		✓		✓		✓	
Lubricate chassis, suspension and steering linkage	S/I		✓		✓		✓		✓		✓		✓	
Lubricate transaxle shift linkage and parking brake cable guides	S/I		✓		✓		✓		✓		✓		✓	
Lubricate underbody contact points & linkage	S/I		✓		✓		✓		✓		✓		✓	
Driveshaft boots, suspension bushings & ball joint seals	S/I		✓				✓		✓		✓		✓	
Exhaust system & throttle linkage	S/I		✓		✓		✓		✓		✓		✓	
Rotate tires	S/I		✓		✓		✓				✓			
Brake hoses & brake lining	S/I		✓				✓				✓			
Accessory drive belt(s)	S/I						✓						✓	
Engine coolant level, hoses & clamps	S/I						✓						✓	
Air filter element	R										✓			
Engine coolant	R												✓	
Manual transaxle oil	R		✓											
Spark plugs ②	R										✓			
Automatic transaxle fluid & filter	S/I										✓			
Ignition cables & fuel systems	S/I										✓			
Vacuum line/hose	S/I										✓			
Fuel filter ③	R													

S/I: Service or Inspect

R: Replace

① Newer models are equipped with an engine life oil system. The engine oil life system calculates when to change your engine oil and filter based on vehicle use.

Anytime your oil is changed, reset the system so it can calculate when the next oil change is required. If a situation occurs where you change your oil prior to the CHG OIL message being turned on reset the system as follows:

Press and release the trip/reset button until the OIL LIFE message is displayed.

Then press and hold the trip/reset button until a chime sounds five times, and RESET is displayed in the message center. When the system is reset, the odometer will again be displayedcenter.

in the message center. Turn the key to the lock position.

If the CHG OIL message comes back on when you start your vehicle, the engine oil life system has not reset. Repeat the procedure.

Your vehicle has a unique oil filter element. When installing the filler cap do not exceed 18 ft. lbs. Inspect the condition of the O-ring and replace if damaged.

② Platinum tip spark plugs: replace every 100,000 miles

③ Replace every 60,000 miles

FREQUENT OPERATION MAINTENANCE (SEVERE SERVICE)

If a vehicle is operated under any of the following conditions it is considered severe service:

- Extremely dusty areas

- 50% or more of the vehicle operation is in 32°C (90°F) or higher temperatures, or constant operation in temperatures below 0°C (32°F)

#NAME?

- Frequent short running periods (engine does not warm to normal operating temperatures)

- Police, taxi, delivery usage or trailer towing usage

Engine oil & oil filter: change every 3000 miles

06025-SCAR-C13

ENGINE REPAIR

Ignition Timing

ADJUSTMENT

The engines covered in this section utilize a Distributorless Ignition System (DIS), no adjustment is possible.

Alternator

REMOVAL

1.9L Engines

1. Before servicing the vehicle, refer to the precautions section.
2. Remove or disconnect the following:
   • Negative battery cable
   • Accessory drive belt
   • Right front wheel
   • Right front wheel well splash shield
   • Alternator electrical connectors
   • Alternator bolts
   • Alternator through the wheel well opening

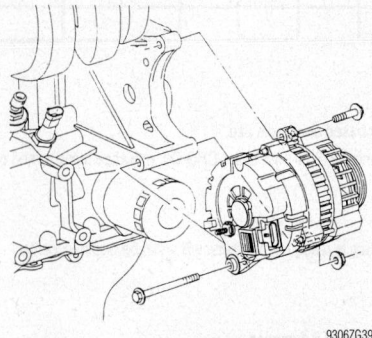

9306ZG39

Exploded view of the alternator—1.9L engines

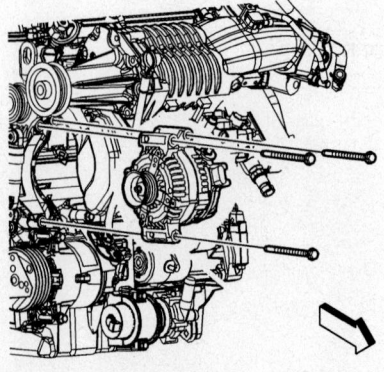

06025-SCAR-G01

Exploded view of the alternator mounting—2.0L engine

2.0L Engine

1. Before servicing the vehicle, refer to the precautions section.
2. Disconnect the negative battery cable.
3. Remove the drive belt.
4. Remove the supercharger.
5. Disconnect the alternator connectors.
6. Remove the alternator bolts.
7. Remove the alternator from the vehicle.

2.2L Engine

1. Before servicing the vehicle, refer to the precautions section.
2. Remove or disconnect the following:
   • Negative battery cable
   • Throttle body air duct
   • Accessory drive belt
   • Alternator electrical connectors
   • Alternator bolts
   • Alternator

3.0L Engine

1. Before servicing the vehicle, refer to the precautions section.
2. Remove or disconnect the following:
   • Negative battery cable
   • Accessory drive belt and tensioner
   • Upper alternator bolts
   • Alternator lower bolts, turn the steering wheel to the right to access the bolt
   • Alternator electrical connectors
   • Alternator

INSTALLATION

1.9L Engines

➡️**Make certain the alternator shield is in place before installation.**

1. Install or connect the following:
   • Alternator to the cylinder block

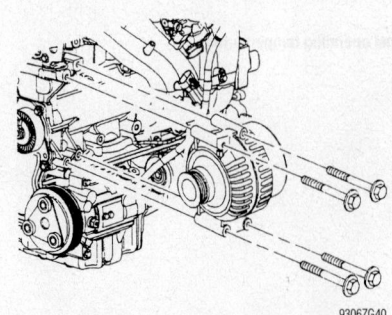

9306ZG40

Exploded view of the alternator—2.2L engine

bracket and torque the bolts to 24 ft. lbs. (32 Nm)
   • Alternator electrical connectors
   • Splash shield
   • Right front wheel
   • Accessory drive belt
   • Negative battery cable

2.0L Engine

1. Before servicing the vehicle, refer to the precautions section.
2. Install the alternator on the engine.
3. Install the alternator bolts. Tighten to 18 ft. lbs. (25 Nm).
4. Connect the positive battery harness to the alternator battery terminal. Tighten the nut to 15 ft. lbs. (25 Nm).
5. Connect the alternator harness connectors.
6. Install the supercharger.
7. Install the drive belt.
8. Connect the negative battery cable.

2.2L Engine

Install or connect the following:
   • Alternator and torque the bolts to 15 ft. lbs. (20 Nm)
   • Alternator electrical connectors
   • Accessory drive belt
   • Throttle body air duct
   • Negative battery cable

3.0L Engine

Install or connect the following:
   • Alternator electrical connectors
   • Alternator and torque the bolts to 30 ft. lbs. (40 Nm)
   • Drive belt tensioner and torque the bolts to 30 ft lbs. (40 Nm)
   • Accessory drive belt
   • Negative battery cable

Engine Assembly

REMOVAL & INSTALLATION

1.9L Engines

➡️**The manufacturer recommends that the engine and transaxle be removed as a complete unit. Disconnect the cradle and lower the entire assembly instead of lifting the assembly out of the vehicle. Both the Single Over Head Camshaft (SOHC) and Dual Over Head Camshaft (DOHC) engines are removed or installed in the same manner. We have found that it is often possible, however, to remove the engine or**

transaxle alone on some models with automatic transaxles by raising it up and out of the engine compartment.

1. Before servicing the vehicle, refer to the precautions section.

2. Properly disable the Supplemental Inflatable Restraint (SIR) system.

3. Properly relieve the fuel system pressure.

4. Drain the engine coolant.

5. Drain the engine oil.

6. Remove or disconnect the following:
- Both battery cables
- Air cleaner and intake duct assembly
- Engine Coolant Temperature (ECT) sensor electrical connector
- Oxygen Sensor (O_2S) and clip
- Idle Air Control (IAC) valve
- Ignition coil connectors
- Throttle Position Sensor (TPS)
- Manifold Absolute Pressure (MAP) sensor
- Exhaust Gas Recirculation (EGR) solenoid
- Brake booster hose
- Ground connectors at the rear of the cylinder block
- Fuel injector connectors
- Neutral Safety Selector (NSS) switch, if equipped
- Valve body actuator connector, if equipped
- Transaxle temperature sensor, if equipped
- Turbine speed sensor, if equipped
- Back-up light switch on manual transaxles
- Accelerator cable
- Fuel pressure and return lines
- Drive belt
- Upper radiator hose and de-aeration hose
- A/C compressor without removing the hoses
- Transaxle cooler lines, if equipped
- Automatic transaxle shifter cable, if equipped
- Manual transaxle shifter cables
- Hydraulic clutch actuator, if equipped
- Tie the radiator, condenser and fan module to the crossbar
- Front wheels
- Splash shields
- Brake caliper brackets and secure the calipers to the shock tower
- Struts from the knuckles
- Lower radiator hose
- Heater inlet and outlet hoses
- Steering shaft

- Power steering pressure switch electrical connector, if equipped
- Front exhaust pipe from the exhaust manifold
- Powertrain stiffening bracket
- Flywheel cover
- Torque converter bolts, if equipped
- Starter motor electrical connectors
- Alternator electrical connector
- Oil pressure sensor electrical connector
- Knock Sensor (KS) electrical connector
- Crankshaft Position (CKP) sensor electrical connector
- Electronic Variable Orifice (EVO) solenoid electrical connector, if equipped
- Vehicle Speed Sensor (VSS) electrical connector
- Evaporative Emissions (EVAP) purge solenoid electrical connector
- Powertrain Control Module (PCM) electrical connector
- Antilock Brake System (ABS) connectors, if equipped
- Brake lines from the rear of the cradle
- Electrical harness from the engine and position it out of the way

7. Place a block of wood between the torque strut and cradle.

8. Remove the 3 right side upper engine torque axis-to-front cover nuts and the 2 mount-to-midrail bracket nuts, allowing the powertrain to rest on the block of wood.

9. Properly support the engine/transaxle assembly.

10. Remove or disconnect the following:
- 2 right side front engine mount torque strut bracket-to-cradle nuts
- 4 cradle attaching bolts
- Powertrain/cradle assembly from the vehicle
- Spark plug wires from the ignition module
- Power steering pump and bracket, if equipped

11. Install an engine lifting device to the service support brackets.
- Front mount assembly
- Intake manifold support brace, DOHC engines only
- 3 axle shaft bracket support bolts and allow the bracket to rotate rearward
- Transaxle attaching bolts
- Engine from the transaxle

9346ZG04

Right hand upper engine torque axis—1.9L engines

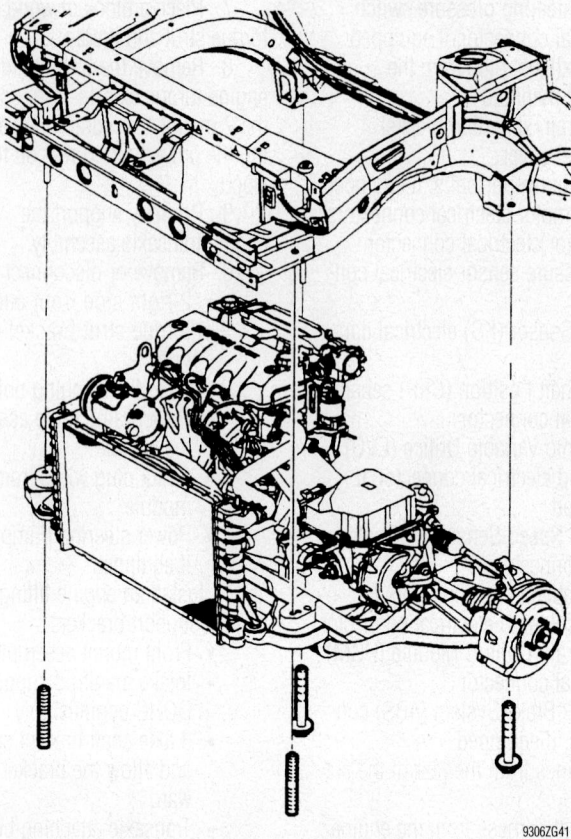

93062G41

Exploded view of the cradle to body bolts—1.9L engines

To install:

12. Install or connect the following:
- Engine to the transaxle and torque the lower transaxle bolts to 96 ft. lbs. (130 Nm) and the upper transaxle bolts to 66 ft. lbs. (90 Nm)
- Front engine mount and torque the bolts to 41 ft. lbs. (55 Nm)
- Engine mount torque strut bracket and torque the bolts to 52 ft. lbs. (70 Nm)
- Engine strut-to-engine bracket and torque the bolts to 52 ft. lbs. (70 Nm)
- Axle shaft bracket and torque the bolts to 41 ft. lbs. (55 Nm)
- Starter bracket and torque the bolts to 80 inch lbs. (9 Nm)
- Engine/transaxle assembly to the chassis and torque the cradle-to-body bolts to 151 ft. lbs. (205 Nm)
- Brake lines to the cradle
- Steering shaft and torque the bolts to 35 ft. lbs. (47 Nm)
- Starter motor electrical connectors
- Alternator electrical connectors
- Oil pressure sensor electrical connector
- KS sensor electrical connector
- CKP sensor electrical connector

- EVAP canister purge valve electrical connector
- Wiring harness PCM ground
- Wiring harness to the transaxle/engine block and torque the fasteners to 18 ft. lbs. (25 Nm)
- EVO solenoid electrical connector, if equipped
- VSS electrical connector
- ABS wheel sensor electrical connectors, if equipped
- Front exhaust pipe to the manifold and torque the bolts to 23 ft. lbs. (31 Nm)
- Exhaust pipe-to-stiffening bracket and torque the bolts to 35 ft. lbs. (47 Nm)
- Exhaust pipe-to-support bracket and torque the bolts to 23 ft. lbs. (31 Nm)
- Heater hoses
- Lower radiator hose
- Steering knuckle to the strut and torque the bolts to 148 ft. lbs. (200 Nm)
- Brake caliper and torque the bolts to 81 ft. lbs. (110 Nm)
- Hydraulic clutch slave cylinder and torque the fasteners to 19 ft. lbs. (25 Nm), if equipped

- Shift cables
- Transaxle oil cooler lines, if equipped
- A/C compressor bracket and torque the bolts to 22 ft. lbs. (30 Nm)
- A/C compressor and torque the front bolt to 40 ft. lbs. (54 Nm) and the rear bolt to 22 ft. lbs. (30 Nm)
- Accessory drive belt
- Engine mount-to-midrail bracket and torque the bolts to 37 ft. lbs. (50 Nm)
- Engine mount to the front cover and torque the bolts to 37 ft. lbs. (50 Nm)
- Torque converter-to-flexplate bolts and torque them to 52 ft. lbs. (70 Nm)
- Dust cover and torque the fasteners to 89 inch lbs. (10 Nm)
- Splash shields
- Both front wheels
- ECT sensor electrical connector
- O_2S electrical connector
- IAC valve electrical connector
- Fuel injector connectors
- TPS electrical connectors
- MAP sensor electrical connector
- EGR valve electrical connector
- A/C compressor electrical connector, if equipped
- Brake booster vacuum hose
- Ground connections
- NSS switch connectors
- PNP switch connectors
- Valve body actuator connector
- Back-up light switch connector, if equipped
- Temperature sensor electrical connector
- Accelerator cable
- Fuel feed and return lines
- Upper radiator and de-aeration hoses

➡ **Check the upper cooling module grommets for misalignment. The cooling module must be able to move freely.**

- Air cleaner and intake duct assembly
- Both battery cables
13. Fill the engine with new oil.
14. Fill the cooling system.
15. Prime the fuel system by cycling the ignition **ON** for 5 seconds and **OFF** for 10 seconds a few times without cranking the engine. Start the engine and check for leaks.
16. Perform a short road test and check the engine again for leaks. Be sure the cooling system is filled to the surge tank FULL COLD line.

L–Series With 2.2L Engine

AUTOMATIC TRANSAXLE

1. Before servicing the vehicle, refer to the precautions section.

2. Properly relieve the fuel system pressure.

3. Drain the engine coolant.

4. Drain the engine oil.

5. Drain the power steering fluid.

6. Remove or disconnect the following:
   - Both battery cables
   - Battery
   - Main wire feed from the fuse block
   - Intake Air Temperature (IAT) sensor electrical connector
   - Air cleaner and intake duct assembly
   - Purge hose from the throttle body
   - Purge solenoid electrical connector
   - Rear Heated Oxygen Sensor (HO2S) electrical connector
   - Main harness connector by the master cylinder
   - Alternator electrical connector
   - A/C pressure switch electrical connector
   - Transaxle electrical connectors
   - Right front Wheel Speed Sensor (WSS) electrical connector
   - Cowl cover
   - Powertrain Control Module (PCM) boot from the cowl
   - PCM electrical connectors
   - Engine harness connectors from the fuse block
   - Battery tray and fuse block tray
   - Cruise control and throttle cables
   - Brake assist vacuum line from the throttle body
   - Fuel feed and return
   - Evaporative Emissions (EVAP) purge hose at the solenoid
   - Starter motor
   - Torque converter bolts
   - Exhaust system from the catalytic converter to the exhaust manifold
   - A/C compressor without removing the hoses
   - Lower engine-to-transaxle bolts
   - Heater hoses at the lower cowl
   - Coolant reservoir hose
   - Upper and lower radiator hoses
   - Metal power steering line

7. Attach an engine lifting hoist to the lifting eyes.
   - Right front engine mount while supporting the weight of the transaxle with a jack
   - Engine mount bracket
   - Upper transaxle-to-engine bolts
   - Engine from the vehicle

To install:

8. Lower the engine into the vehicle and align it on the pins on the transaxle.

9. Install or connect the following:
   - Upper transaxle bolts and hand-tighten them
   - Engine mount bracket and torque the bolts to 66 ft. lbs. (90 Nm)
   - Engine mount and torque the bolts to 41 ft. lbs. (55 Nm)

10. Remove the engine lifting device.
    - Metal power steering line
    - Upper and lower radiator hoses
    - Coolant reservoir hose
    - Lower transaxle bolts and torque them to 48 ft. lbs. (65 Nm)
    - Torque converter bolts and torque them to 33 ft. lbs. (45 Nm)
    - Starter motor and torque the bolts to 37 ft. lbs. (50 Nm)
    - Heater hoses
    - A/C compressor and torque the bolts to 18 ft. lbs. (25 Nm)
    - Accessory drive belt
    - Exhaust from the catalytic converter to the exhaust manifold and torque the bolts to 12 ft. lbs. (16 Nm)

11. Torque the upper bell housing bolts to 48 ft. lbs. (65 Nm).

   - EVAP purge solenoid electrical connector and hose
   - Fuel feed and return lines
   - Brake booster vacuum hose
   - Cruise control and throttle cables
   - Fuse block and battery trays
   - Engine harness to the fuse block
   - Main engine harness under the fuse block panel
   - PCM electrical connector
   - PCM boot to the cowl
   - Cowl cover
   - Main harness connector near the master cylinder
   - Rear HO2S electrical connector
   - Purge solenoid electrical connector
   - Purge hose to the throttle body
   - Air cleaner and intake duct assembly
   - IAT sensor electrical connector
   - Ground wire to the left fender well
   - Main wire feed to the fuse block
   - Battery and both cables

12. Fill the engine with coolant.

13. Fill the engine with new oil.

14. Fill the power steering fluid.

15. Prime the fuel system by cycling the ignition **ON** for 5 seconds and **OFF** for 10 seconds a few times without cranking the engine.

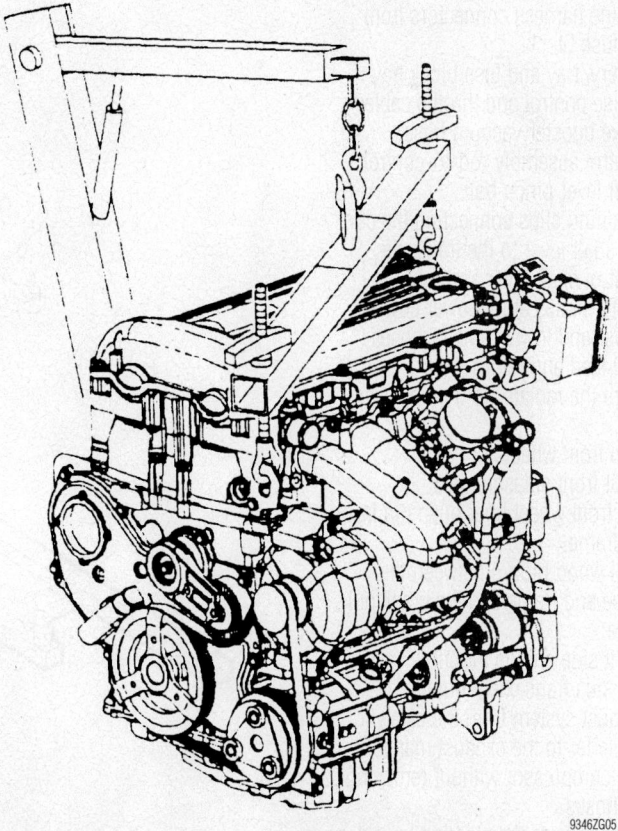

9346ZG05

Remove the engine with the lifting hoist—L–Series With 2.2L engine

16. Start the engine, check for leaks, and repair if necessary.

MANUAL TRANSAXLE

1. Before servicing the vehicle, refer to the precautions section.

2. Properly relieve the fuel system pressure.

3. Drain the engine coolant.

4. Drain the engine oil.

5. Drain the power steering fluid.

6. Remove or disconnect the following:
- Both battery cables
- Steering gear-to-intermediate shaft pinch bolt
- Main wire feed to the fuse block
- Ground wire on the left fender well
- Intake Air Temperature (IAT) sensor electrical connector
- Air cleaner and intake duct assembly
- Purge hose from the throttle body
- Purge solenoid electrical connector
- Rear Heated Oxygen Sensor (HO2S) electrical connector
- Back-up switch electrical connector
- Front dash cover
- Powertrain Control Module (PCM) boot from the front of the dash
- Main harness connector by the master cylinder
- PCM electrical connectors
- Engine harness connectors from the fuse block
- Battery, tray and fuse block tray
- Cruise control and throttle cables
- Brake booster vacuum hose
- Control assembly rod-to-control shaft lever pinch bolt
- Retaining clips connecting the control shaft lever to the transaxle
- Control shaft lever assembly
- Heater hoses at the lower cowl
- Upper and lower radiator hoses
- Fuel feed and return lines

7. Secure the radiator to the upper radiator support.
- Both front wheels
- Right front splash shield
- Left front wheel liner push pin from the frame

8. Install wood blocks between the transaxle case and frame and the crank pulley and frame.
- Right side engine mount
- Left side transaxle mount
- Exhaust system from the catalytic converter to the exhaust manifold
- A/C compressor without removing the hoses
- Tie rods from the steering knuckles
- Stabilizer bar links from the struts

- Lower ball joints from the steering knuckles
- Suspension support assemblies
- Suspension support cage nuts from the body

9. Properly support the powertrain and frame assembly.
- Frame-to-body attachment fasteners
- Powertrain/frame assembly from the vehicle

To install:

10. Raise the powertrain into the vehicle.

11. Install or connect the following:
- Powertrain/frame assembly
- New frame-to-body bolts do not tighten at this time
- Suspension supports with new bolts and torque the suspension support and frame bolts to 74 ft. lbs. (100 Nm) plus 45 degrees
- Ball joint to the steering knuckle and torque the nuts to 74 ft. lbs. (100 Nm)
- Stabilizer links to the struts and torque the fasteners to 50 ft. lbs. (65 Nm)
- Tie rods to the steering knuckles and torque the new nuts to 45 ft. lbs. (65 Nm)

- Exhaust system to the exhaust manifold and torque the bolts to 12 ft. lbs. (16 Nm)
- A/C compressor and torque the bolts to 18 ft. lbs. (25 Nm)
- Rear HO2S electrical connector
- Right side engine mount and torque the bolts to 41 ft. lbs. (55 Nm)
- Left side transaxle mount and torque the bolts to 41 ft. lbs. (55 Nm)

12. Remove the wood blocks.
- Left front wheel liner push pins
- Right front splash shield
- Both front wheels
- Fuel feed and return lines
- Upper and lower radiator hoses
- Heater hoses
- Brake booster vacuum hose
- Shift control rod

13. Adjust the shift control rod as follows:

a. Install the shift rod into the shift linkage assembly.

b. Rotate the transaxle shift control shaft clockwise and depress the spring loaded transaxle shift linkage lock pin on the case.

c. Pull up the shift lever boot.

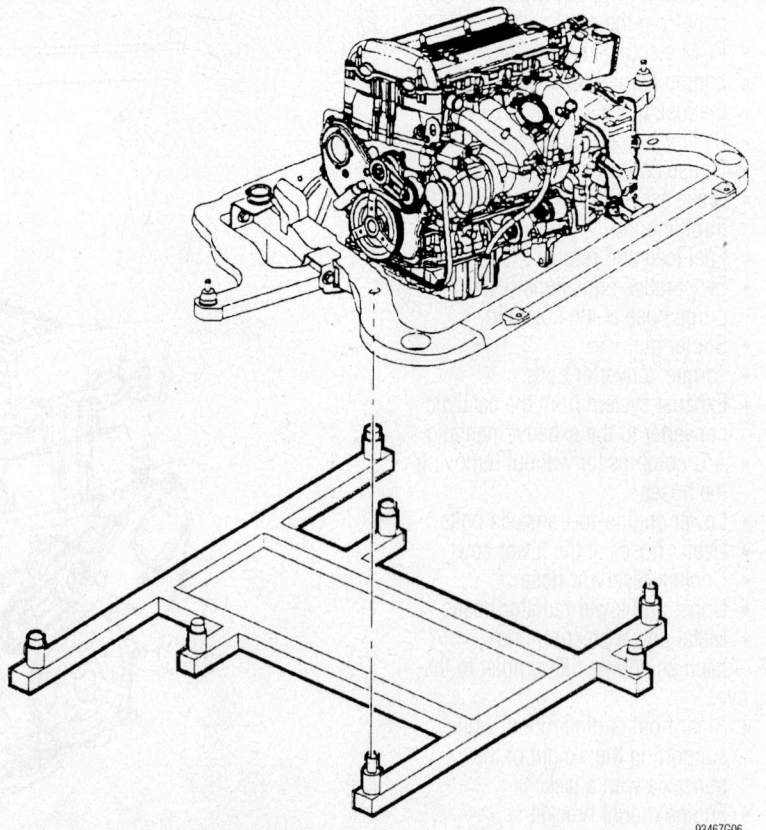

Powertrain/frame assembly with lifting table—L-Series With 2.2L engine

9346ZG06

d. Move the shift lever to the 5 (reverse gate) and use a ⅜inch punch to hold it in place.

e. Tighten the pinch bolt to 9 ft. lbs. (12 Nm) plus 180 degrees

14. Install or connect the following:
- Cruise control and throttle cables
- Fuse block and battery trays
- Fuse block
- Engine harness connector to the bottom of the fuse block
- Fuse block
- PCM electrical connector
- PCM boot to the cowl
- Cowl cover
- EVAP purge solenoid electrical connector and hose
- Main harness connector near the master cylinder
- Transaxle back-up switch electrical connector
- Purge hose to the throttle body
- Air cleaner and intake duct assembly
- IAT sensor electrical connector
- Ground wire to the left fender well
- Main wire feed to the fuse block
- Battery and both cables
- Steering gear-to-intermediate shaft and torque the pinch bolt to 22 ft. lbs. (30 Nm)

15. Fill the engine with coolant.
16. Fill the engine with new oil.
17. Fill the power steering fluid.
18. Prime the fuel system by cycling the ignition **ON** for 5 seconds and **OFF** for 10 seconds a few times without cranking the engine.
19. Start the engine, check for leaks, and repair if necessary.

ION

2.0L ENGINE

1. Before servicing the vehicle, refer to the precautions section.
2. With the tires in the straight forward position, remove the key from the ignition.
3. Disconnect the negative battery cable.
4. Remove the air outlet duct .
5. Secure the cooling module to the upper body structure.
6. Relieve the fuel system pressure.
7. Disconnect the fuel line from the fuel rail.
8. Remove the supercharger.
9. Drain the cooling system.
10. Remove the radiator inlet hose.
11. Remove the surge tank to cylinder head pipe.
12. Remove the radiator outlet hose.
13. Remove the inlet and outlet heater hoses.

14. Disconnect the following harness connectors:
- Tmap sensor
- Electronic Temperature Control (ETC) sensor
- Manifold Absolute Pressure (MAP) sensor
- Barometric Pressure (BARO) sensor
- Crankshaft Position (CKP) sensor
- Oil pressure sensor
- Purge solenoid
- Ignition coil modules
- Oxygen (O2s) sensor
- Vehicle speed sensor
- Engine temperature sensor
- Boost solenoid
- Back up lamp switch

15. Remove the drive belt.
16. Disconnect the charge air cooler hoses.
17. Recover the refrigerant.
18. Disconnect the compressor and condenser hose assembly from the compressor.
19. Remove the AC compressor bolts and set the compressor aside.
20. Disconnect the starter harness connectors.
21. Disconnect the alternator harness connectors.
22. Drain the engine oil.
23. Disconnect the front exhaust pipe from the exhaust manifold.
24. Disconnect the transmission shift cable from the transmission.
25. Use blocks of wood to support the powertrain assembly between the frame and the powertrain.
26. Remove the engine mount.
27. Remove the left transmission mount as follows:

a. Remove the underhood electrical center bracket from the vehicle and reposition the electrical center.

b. Support the transmission with a floor jack. Use a piece of wood between the jack and the transmission.

c. Remove the transmission mount-to-transmission bracket bolts.

d. Remove the transmission mount to mid-rail bolts.

e. Using a floor jack, slowly lower the transmission enough to remove the transmission mount from the vehicle

28. Disconnect the stabilizer links from the stabilizer bar.
29. Disconnect the outer tie rod ends from the steering knuckles.

➡**In order to prevent possible SIR system deployment, do not attempt to rotate the steering shaft.**

30. Disconnect the intermediate shaft from the steering gear.
31. Disconnect the lower control arms from the steering knuckles.
32. Disconnect the drive axles from the steering knuckle.
33. Use a paint pen or magic marker in order to mark the frame to body position.
34. Lower the vehicle to about 3 feet off the ground in order to position the lift table under the frame.
35. Use wood blocks as necessary between the lift table and the frame to support the assembly.
36. Slowly remove the frame bolts using the following sequence:

a. Remove the front frame bolts.

b. Partially unscrew the rear frame bolts until 1.5 inches of bolt shank is exposed.

c. Slowly lower the table to the floor.

d. Attach the engine lift hoist to the engine lift hooks.

37. Remove the starter.
38. Remove the transmission to engine bolts.
39. Separate the engine from the transmission.
40. Remove the clutch pressure plate and disk.
41. Remove the following components:
- Exhaust manifold
- Exhaust manifold studs
- Engine mount bracket
- Fuel rail
- Thermostat housing and feed pipe
- Alternator

42. Remove the engine from the engine lift.

To install:
43. Attach the engine lift hoist to the engine lift hooks.
44. Install the exhaust manifold
45. Install the engine mount bracket as follows:
46. Install the intermediate bracket to the engine.
47. Hand tighten the engine mount intermediate bracket bolts with the long bolts in the forward and front lower holes and the short bolt in the rear upper hole.
48. Tighten the intermediate bracket bolts to 74 ft. lbs. (100 Nm).
49. Install the fuel rail.
50. Install the Idler pulley.
51. Install the drive belt tensioner.
52. Install the thermostat housing and feed pipe.
53. Install the alternator.
54. Install the flywheel.
55. Install the clutch and pressure plate if equipped.

56. Install the A/C compressor.

57. Align the engine to the transmission. Tighten the transmission to engine bolts to 55 ft. lbs. (75 Nm).

58. Install the starter.

59. Remove the engine lift from the engine.

60. Raise and position the frame and powertrain assembly to the vehicle.

61. Hand start all the frame bolts while aligning the frame to the paint marks.

62. Tighten the frame bolts to 74 ft. lbs. (100 Nm) plus an additional 100 degree turn.

63. Remove the lift table.

64. Connect the drive axles to the steering knuckles. Tighten the nut to 155 ft. lbs. (210 Nm).

65. Connect the lower control arm to the steering knuckle and tighten the retainer as follows:

    a. Install the ball stud pinch bolt and nut.

    b. First Pass: Tighten the nut to 37 ft. lbs. (50 Nm), the back off ¾turn.

    c. Second Pass: Tighten the nut to 37 ft. lbs. (50 Nm plus an additional 30 degrees.

66. Connect the intermediate steering shaft to the steering gear. Tighten the bolt to 25 ft. lbs. (35 Nm).

67. Connect the outer tie rod ends to the steering knuckles. Tighten the nut to 44 ft. lbs. (60 Nm).

68. Connect the stabilizer links to the stabilizer bar. Tighten the retainers to 81 ft. lbs. (110 Nm).

69. Install the left transmission mount as follows:

    a. Install the transmission mount to the mid-rail.

    b. Install the transmission mount to mid-rail bolts and tighten to 20 ft. lbs. (27 Nm).

70. Using a floor jack, raise the transmission until it contacts the transmission mount.

➡The transmission mount to transmission bolts must be hand started. Do not pry the transmission or mount to align the holes.

71. Hand start the transmission mount to bracket bolts using the following sequence:
- Rear bolt
- Middle bolt
- Front bolt

    a. Using the previous sequence, tighten the transmission mount bolts to 37 ft. lbs. (50 Nm).

    b. Install the underhood electrical center bracket to the vehicle and install the electrical center into position on the bracket

72. Install the engine mount as follows:

    a. Place the engine mount onto the midrail and hand start the nuts.

    b. Tighten the engine mount to midrail nuts to 74 ft. lbs. (100 Nm).

➡The engine mount to intermediate bracket bolts must be hand started. Do not pry the engine mount to align the holes.

    c. Hand start the engine mount to intermediate bracket bolts.

    d. Tighten the engine mount to intermediate bracket bolts, starting with the center bolt to 37 ft. lbs. (50 Nm).

73. Remove the wood blocks between the powertrain and frame.

74. Connect the transmission shift cable to the transmission.

75. Connect the exhaust takedown pipe to the exhaust manifold and tighten the nuts to 22 ft. lbs. (30 Nm).

76. Connect the alternator harness connectors. Tighten to 15 ft. lbs. (20 Nm).

77. Connect the starter harness connectors.

78. Install the compressor and condenser hose assembly to the compressor.

79. Evacuate and charge the refrigerant system.

80. Install the supercharger.

81. Install the engine drive belt.

82. Connect the charge air cooler hoses.

83. Connect the following harness connectors:
- Tmap sensor
- Electronic Temperature Control (ETC) sensor
- Manifold Absolute Pressure (MAP) sensor
- Barometric Pressure (BARO) sensor
- Crankshaft Position (CKP) sensor
- Oil pressure sensor
- Purge solenoid
- Ignition coil modules
- O_2s sensor
- Vehicle speed sensor
- Engine temperature sensor
- Boost solenoid
- Back up lamp switch

84. Install the inlet heater hose and outlet heater hose.

85. Install the radiator outlet hose.

86. Connect the fuel line to the fuel rail.

87. Install the surge tank to the cylinder head pipe.

88. Release the cooling module from the upper body structure.

89. Install the air outlet duct .

90. Connect the negative battery cable.

91. Fill the engine with engine oil to the proper level.

92. Fill the cooling system.

93. Road test the vehicle.

2.2L ENGINE

1. Before servicing the vehicle, refer to the precautions section.

2. Properly relieve the fuel system pressure.

3. Drain the engine coolant.

4. Drain the engine oil.

5. Drain the power steering fluid.

6. Remove or disconnect the following:
- Negative battery cable
- Air cleaner and intake duct assembly

7. Secure the cooling module to the upper body.
- Throttle cable
- Fuel feed and return lines from the rail
- Radiator upper hose
- Surge tank-to-cylinder head hose
- Radiator lower hose
- Inlet and outlet heater hoses

8. Disconnect the following electrical connections:
- Intake Air Control (IAC)
- Throttle Position Sensor (TPS)
- Manifold Absolute Pressure (MAP) sensor
- Crankshaft Position (CKP) sensor
- Oil pressure sensor
- Purge solenoid
- Ignition coil and module
- Oxygen Sensor (O_2S)
- Vehicle Speed Sensor (VSS)
- Engine temperature sensor
- Back–up lamp switch

9. Remove or disconnect the following:
- Drive belt
- A/C compressor bolts and set the compressor aside without disconnecting the lines
- Starter and alternator wiring
- Front exhaust pipe from the manifold
- Transmission harness connectors
- Transmission shift cable

10. Use suitable blocks of wood to support the powertrain assembly between the assembly and the frame.

11. Support the engine with a jack using a piece of wood between the jack and the engine to avoid damage.
- Engine mount, nuts and bolts.

12. Remove the side transmission mount on models with a manual transmission as follows:

    a. Remove the under hood electrical tray.

b. Disconnect the wiring harness from the tray bracket.

c. Lift the electrical center up and swing it back out of the way.

d. Remove the mount-to-transmission bolts and mount-to-rail bolts.

e. Remove the mount.

13. Remove the side transmission mount on models with a automatic 5-speed transmission as follows:

a. Support the transmission with a jack.

b. Remove the front and rear transmission mount through bolts.

c. Remove the under hood electrical center cover.

d. Disconnect the engine control module harness connector.

e. Disconnect the positive battery cable from the under hood electrical center.

f. Make sure the surge tank hose is disconnected.

g. Remove the electrical center tray.

h. Disconnect the harness from the tray bracket.

i. Lift the electrical center up and swing it back out of the way.

j. Remove the mount-to-transmission bolts and mount-to-rail bolts.

k. Remove the mount.

14. Remove the side transmission mount on models with a automatic 5-speed continuously variable transmission as follows:

a. Remove the engine cradle.

b. Remove the under hood electrical center cover.

c. Disconnect the engine control module harness connector.

d. Disconnect the positive battery cable from the under hood electrical center.

e. Make sure the surge tank hose is disconnected.

f. Remove the electrical center tray.

g. Disconnect the harness from the tray bracket.

h. Lift the electrical center up and swing it back out of the way.

i. Install an engine support fixture. If not already removed, remove the engine mount.

j. Disconnect the upper transmission cooler line from the radiator.

k. Remove the mount-to-transmission bolts

➡Make sure the A/C clutch does not contact the inner metal rail of the vehicle.

l. Lower and tilt the powertrain assembly down approximately 4 inches.

m. Remove the mount-to-rail bolts.

n. Remove the mount.

15. Remove or disconnect the following:
• Stabilizer link from the stabilizer bar
• Outer tie rod ends from knuckles

• Intermeditae shaft from the steering gear
• Lower control arms from the knuckles
• Half shafts from the knuckles

16. Remove the front frame as follows:

a. Mark the frame-to-body position.

b. Lower the vehicle until it is 3 feet off the ground so that a lift table may be placed in position. As necessary, use blocks of wood between the frame and table to support the assembly.

c. Slowly remove the front frame bolts, the partially unscrew the bolts until 1 ½ inch of the bolt shank is exposed.

d. Slowly lower the table to the floor

17. Attach the engine hooks to a hoist.

18. Remove or disconnect the following:
• Starter
• Torque converter-to-flywheel bolts, if equipped
• Engine-to-transmission bolts and the engine from the transmission.

To install:

19. Install or connect the following:
• Engine to the transmission and tighten the bolts to 55 ft. lbs. (75 Nm)
• Torque converter-to-flywheel bolts, if equipped and tighten the bolts to 46 ft. lbs. (62 Nm)
• Starter

20. Install the front frame as follows:

a. Slowly raise the table to the into position and align the frame to the marks made during removal.

b. Hand tighten the frame bolts

c. Tighten the frame bolts to 74 ft. lbs. (100 Nm) plus an additional torque of 180 degrees using a torque angle meter.

d. Remove the lift table.

21. Install or connect the following:
• Half shafts to the knuckles
• Lower control arms to the knuckles

• Intermeditae shaft to the steering gear
• Outer tie rod ends to knuckles
• Stabilizer link to the stabilizer bar

22. Install the side transmission mount on models with a manual transmission as follows:

a. Place the mount in position and tighten the mount-to-rail bolts to 25 ft. lbs. (34 Nm).

23. Use a jack to align the mount when it is in position.

a. Install and hand tighten the mount-to-transmission bolts. Do not pry the transmission or mount to align the holes. Hand start the bolts in the following sequence; rear bolt, middle and then front bolt. Using the same sequence, tighten the bolts to 33 ft. lbs. (45 Nm).

b. Place the electrical center in position.

c. Connect the wiring harness to the tray bracket.

d. Install the under hood electrical tray.

24. Install the side transmission mount on models with a automatic 5-speed transmission as follows:

a. Place the mount in position and tighten the mount-to-rail bolts to 25 ft. lbs. (34 Nm).

25. Use a jack to align the mount when it is in position.

a. Install and hand tighten the mount-to-transmission bolts. Do not pry the transmission or mount to align the holes. Hand start the bolts in the following sequence; rear bolt, middle and then front bolt. Using the same sequence, tighten the bolts to 33 ft. lbs. (45 Nm).

b. Place the electrical center in position.

c. Connect the wiring harness to the tray bracket.

d. Install the under hood electrical tray.

e. Connect the positive battery cable to the under hood electrical center.

f. Connect the engine control module harness connector.

g. Install the under hood electrical center cover.

h. Install the front and rear transmission mount through bolts. Tighten to 74 ft. lbs. (100 Nm).

26. Install the side transmission mount on models with a automatic 5-speed continuously variable transmission as follows:

a. Place the mount in position and tighten the mount-to-rail bolts to 25 ft. lbs. (34 Nm).

27. Use a jack raise the engine back into

position and to align the mount when it is in position.

a. Install and hand tighten the mount-to-transmission bolts. Do not pry the transmission or mount to align the holes. Hand start the bolts in the following sequence; rear bolt, middle and then front bolt. Using the same sequence, tighten the bolts to 33 ft. lbs. (45 Nm).

b. If installed, install the engine mount and remove the engine support fixture.

c. Connect the upper transmission cooler line to the radiator.

d. Place the electrical center in position.

e. Connect the wiring harness to the tray bracket.

f. Install the under hood electrical tray.

g. Connect the positive battery cable to the under hood electrical center.

h. Connect the engine control module harness connector.

i. Install the under hood electrical center cover.

j. Install the front and rear transmission mount through bolts. Tighten to 74 ft. lbs. (100 Nm).

➡**The engine mount bolts must be hand started. Do not pry on the engine mount to align the holes.**

28. Install or connect the following:
- Engine mount nuts and bolts. Tighten the nuts to 74 ft. lbs. (100 Nm) and the bolts to 37 ft. lbs. (50 Nm).

29. Remove the blocks of wood used to support the powertrain assembly between the assembly and the frame.
- Transmission shift cable
- Transmission harness connectors
- Front exhaust pipe to the manifold
- Starter and alternator wiring
- A/C compressor
- Drive belt

30. Connect the following electrical connections:
- Back–up lamp switch
- Engine temperature sensor
- VSS
- O$_2$S
- Ignition coil and module
- Purge solenoid
- Oil pressure sensor
- CKP
- MAP sensor
- TPS
- IAC

31. Install or connect the following:
- Inlet and outlet heater hoses

- Radiator lower hose
- Surge tank-to-cylinder head hose
- Radiator upper hose
- Fuel feed and return lines to the rail
- Throttle cable
- Cooling module
- Air cleaner and intake duct assembly
- Negative battery cable

32. Fill the engine with coolant.

33. Fill the engine with new oil.

34. Prime the fuel system by cycling the ignition **ON** for 5 seconds and **OFF** for 10 seconds a few times without cranking the engine.

35. Start the engine, check for leaks, and repair if necessary.

3.0L Engine

1. Before servicing the vehicle, refer to the precautions section.

2. Properly relieve the fuel system pressure.

3. Drain the engine coolant.

4. Drain the engine oil.

5. Drain the power steering fluid.

6. Remove or disconnect the following:
- Both battery cables
- Air cleaner assembly
- Mass Air Flow (MAF) sensor electrical connector
- Battery ground cable from the fuse block
- Transaxle Control Module (TCM) main connector from under the cowl cover
- TCM inline connector by the brake master cylinder
- A/C pressure connector
- Black engine harness connector from under the fuse block panel
- Lower weather pack connector from inside the fuse block and secure the engine harness to the engine
- Fuse block
- Battery tray
- Evaporative Emissions (EVAP) purge connector
- Right front speed sensor connector
- Front Oxygen Sensor (O$_2$S) from the down pipe
- Transaxle ground electrical connector
- Transaxle main electrical connector
- Transaxle shift control electrical connector
- Rear connector on the Engine Control Module (ECM)
- Brake booster vacuum hose
- Fuel lines
- EVAP purge hose and solenoid

- Starter motor
- Torque converter bolts
- Exhaust system from the catalytic converter to the exhaust manifold
- Transaxle nose bracket
- A/C compressor without removing the hoses
- Heater hoses at the lower cowl
- Lower engine-to-transaxle bolts
- Upper and lower radiator hoses
- Coolant reservoir hose
- Metal power steering line
- Power steering reservoir from the radiator support

7. Attach an engine lifting device to the engine lifting eyes.
- Right front engine mount and bracket
- Upper transaxle bolts
- Engine from the vehicle

To install:

8. Lower the engine into the vehicle and align it on the pins on the transaxle.

9. Install or connect the following:
- Upper transaxle bolts and hand-tighten them
- Engine mount bracket and torque the bolt to 30 ft. lbs. (40 Nm)
- Engine mount and torque the upper bolt to 30 ft. lbs. (40 Nm) and the lower bolt to 41 ft. lbs. (55 Nm)
- Transaxle nose bracket and torque the bolts to 30 ft. lbs. (40 Nm)

10. Remove the engine lifting device.
- Metal power steering line
- Power steering reservoir to the radiator support
- Upper and lower radiator hoses
- Coolant reservoir hose
- Lower transaxle bolts and torque them to 48 ft. lbs. (65 Nm)
- Torque converter and torque them to 48 ft. lbs. (65 Nm)
- Starter motor and torque the bolts to 30 ft. lbs. (40 Nm)
- Heater hoses at the lower cowl
- A/C compressor and torque the bolts to 30 ft. lbs. (40 Nm)
- Exhaust system to the exhaust manifold and torque the nuts to 15 ft. lbs. (20 Nm)
- EVAP purge solenoid and hose
- Fuel feed and return lines
- Brake booster vacuum hose
- ECM rear connector
- Transaxle shift control electrical connector
- Transaxle ground connector
- Transaxle main connector
- Right front speed sensor electrical connector
- Front O$_2$S electrical connector

- Right front speed sensor
- EVAP purge solenoid electrical connector

11. Torque the upper bell housing bolts to 48 ft. lbs. (65 Nm).

- Battery tray
- Fuse block and route the engine harness to the block
- Lower weather pack connector
- Main wire feed to the fuse block
- A/C pressure connector
- TCM inline connector
- TCM main connector under the cowl cover and secure it to the engine
- Positive main feed cable at the fuse block
- Battery ground cable at the wheel housing
- Battery and both cables
- Air cleaner and intake duct assembly
- MAF sensor electrical connector

12. Fill the engine with coolant.
13. Fill the engine with new oil.
14. Fill the power steering fluid.
15. Prime the fuel system by cycling the ignition **ON** for 5 seconds and **OFF** for 10 seconds a few times without cranking the engine.
16. Start the engine, check for leaks, and repair if necessary.

Water Pump

REMOVAL & INSTALLATION

1.9L Engines

1. Before servicing the vehicle, refer to the precautions section.
2. Drain the cooling system.
3. Remove or disconnect the following:

- Negative battery cable
- Accessory drive belt
- Right front wheel
- Inner wheel well splash shield
- Water pump pulley bolts and allow the pulley to hang freely on the water pump hub
- Water pump flange bolts
- Water pump

To install:

4. Install or connect the following:

- Water pump with a new gasket and torque the bolts in a crisscross sequence to 22 ft. lbs. (30 Nm)
- Water pump pulley to the hub and torque the bolts to 19 ft. lbs. (25 Nm)
- Accessory drive belt

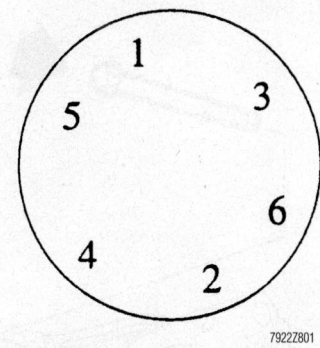

79222801

Water pump bolt torque sequence—1.9L engines

- Right front wheel well splash shield
- Right front wheel
- Negative battery cable

5. Fill the cooling system.
6. Start the vehicle and check for leaks, repair if necessary.

L–Series With 2.2L Engine

1. Before servicing the vehicle, refer to the precautions section.
2. Drain the cooling system.
3. Remove or disconnect the following:

- Negative battery cable
- Air cleaner assembly
- Exhaust manifold heat shield
- Thermostat housing
- Water pump feed pipe
- Water pump sprocket access plate from the front cover

4. Install a Water Pump Holding Tool J43651.

- Water pump retaining bolts
- Water pump

To install:

5. Install or connect the following:

- Water pump and torque the bolts to 18 ft. lbs. (25 Nm)
- Water pump sprocket and torque the bolts to 89 inch lbs. (10 Nm)
- Water pump sprocket access plate
- Water feed tube after lubricating the O-ring
- Thermostat housing and torque the bolts to 89 inch lbs. (10 Nm)
- Exhaust manifold heat shield
- Air cleaner assembly
- Negative battery cable

6. Fill the cooling system.
7. Start the vehicle and check for leaks, repair if necessary.

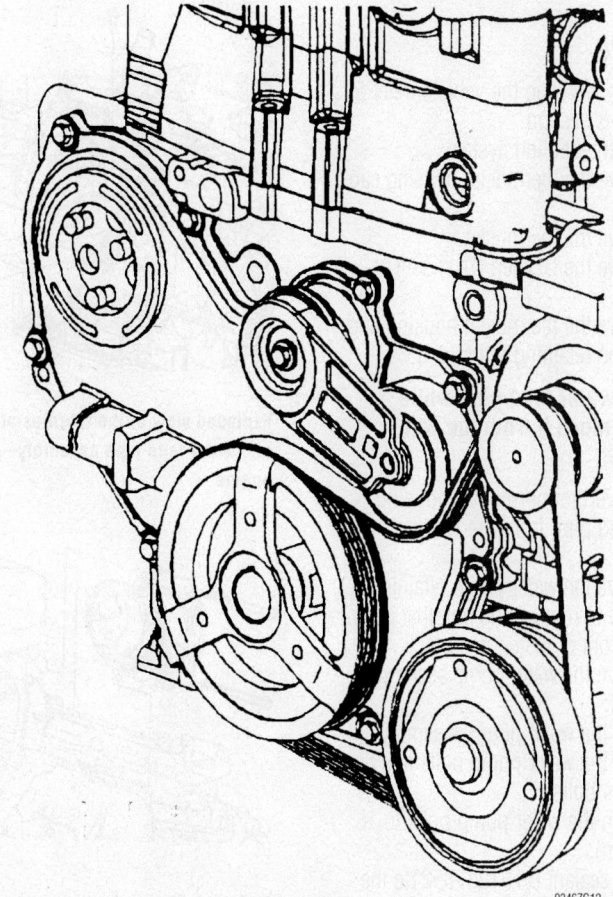

9346ZG12

Water pump holding tool J43651—all 2.2L engines

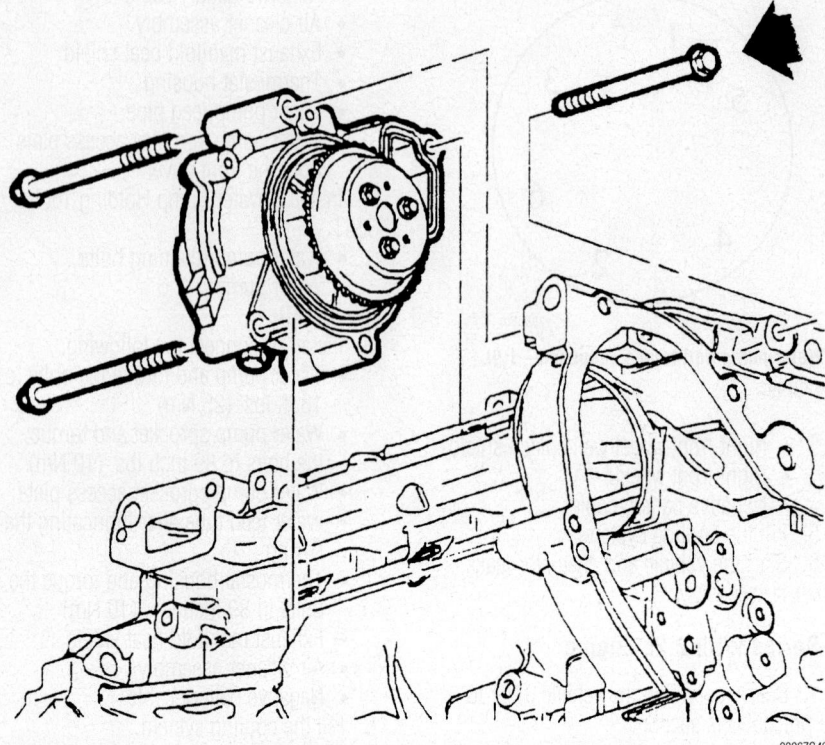

Exploded view of the water pump—all 2.2L engines

93062G42

ION

2.0L ENGINE

1. Before servicing the vehicle, refer to the precautions section.
2. Drain the cooling system.
3. Remove the thermostat housing cap and bolts.
4. Remove the thermostat.
5. Remove the Oxygen (O_2s) sensor clip.
6. Remove the thermostat housing and water feed pipe retaining bolts.

➡ **Twist the water feed pipe while pulling to remove it from the water pump cover.**

7. Remove the thermostat housing and water feed pipe from the water pump cover.
8. Remove the water pump retaining bolts. Be sure to remove the bolt that goes through the front of the engine block.
9. Remove the water pump assembly.

To install:

10. Install the water pump assembly.
11. Install the water pump bolts and finger tighten the bolts.
12. Tighten the water pump bolts to 18 ft. lbs. (25 Nm).
13. Apply sealant P/N 12378521 to the water pump drain plug.
14. Install the water pump drain plug, if

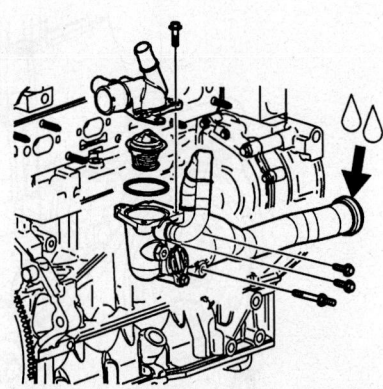

06025-SCAR-G02

Exploded view of the thermostat housing and water feed pipe assembly—2.0L engine

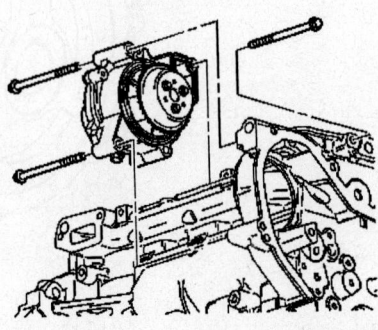

06025-SCAR-G03

Exploded view of the water pump assembly—2.0L engine

necessary and tighten to 15 ft. lbs. (20 Nm).

15. Install the water feed tube.
16. Lubricate the water feed tube O-ring with anti-freeze.
17. Install the water feed tube by twisting and pushing toward the water pump. Take care not to tear or damage the O-ring.
18. Install the thermostat housing to block bolts.
19. Install the thermostat.
20. Install the thermostat housing cap and bolts.
21. Fill the cooling system.
22. Start the vehicle and check for leaks, repair if necessary.

2.2L ENGINE

1. Before servicing the vehicle, refer to the precautions section.
2. Drain the cooling system.
3. Remove or disconnect the following:
   • Negative battery cable
   • Exhaust manifold, if equipped with an automatic transmission
   • Right front wheel
   • Front fender liner
   • Access plate on the water pump sprocket from the timing cover
4. Install a Water Pump Holding Tool J43651.
   • Water pump-to-block bolt
   • Water pump-to-engine front cover bolts
   • Feed pipe from the thermostat to the water pump
   • Water pump-to-block bolts
   • Water pump

To install:

5. Install or connect the following:
   • Water pump
   • Water pump-to-block bolts and tighten to 15 ft. lbs. (20 Nm)
   • Feed pipe from the thermostat to the water pump
   • Water pump-to-engine front cover bolts and tighten to 15 ft. lbs. (20 Nm)
   • Water pump-to-block bolt and tighten to 15 ft. lbs. (20 Nm)
   • Water pump sprocket and torque the bolts to 89 inch lbs. (10 Nm)
   • Water pump sprocket access plate
   • Front fender liner
   • Right front wheel
   • Exhaust manifold, if equipped with an automatic transmission
   • Negative battery cable
6. Fill the cooling system.
7. Start the vehicle and check for leaks, repair if necessary.

3.0L Engine

1. Before servicing the vehicle, refer to the precautions section.
2. Drain the cooling system.
3. Remove or disconnect the following:
 • Negative battery cable
 • Air cleaner assembly
 • Left front wheel
 • Wheel well splash shield
4. Loosen the water pump pulley and power steering pulley bolts, but do not remove them.
5. Install an engine support fixture.
 • Right front engine mount
 • Accessory drive belt
6. Release the retaining tabs on the wiring harness channel and remove the front cover.
 • Wiring harness from the channel
 • Water pump pulley
 • Power steering pump pulley
 • Drive belt tensioner
 • Front timing belt cover
 • Water pump

To install:

7. Install or connect the following:
 • Water pump with a new O-ring and torque the bolts to 18 ft. lbs. (25 Nm)
 • Front timing belt cover and torque the bolts to 71 inch lbs. (8 Nm)

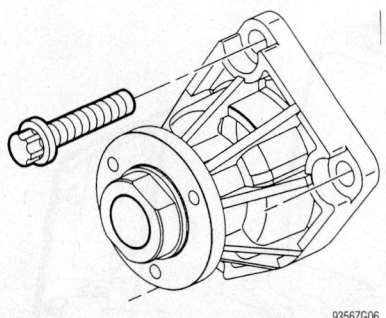

9356ZG06

Water pump mounting—3.0L engine

 • Drive belt tensioner and torque the bolts to 30 ft. lbs. (40 Nm)
 • Water pump pulley's and torque the bolts to 71 inch lbs. (8 Nm)
 • Power steering pump and torque the bolts to 15 ft. lbs. (20 Nm)
 • Wiring harness into the channel
 • Wiring harness channel front cover
 • Accessory drive belt
8. Remove the engine support fixture.
 • Wheel well splash shield and torque the bolts to 44 inch lbs. (5 Nm)
 • Front wheel
 • Air cleaner and intake duct
 • Negative battery cable

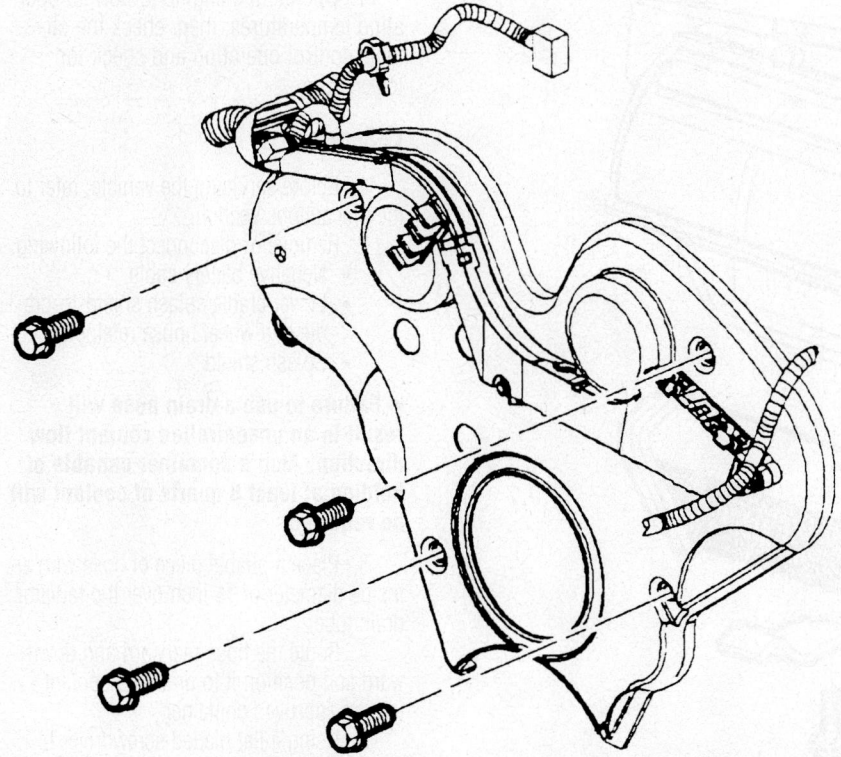

9346ZG13

Timing belt cover bolt locations—3.0L engine

9. Fill the cooling system.
10. Start the vehicle and check for leaks, repair if necessary.

Heater Core

REMOVAL & INSTALLATION

S–Series

1. Before servicing the vehicle, refer to the precautions section.
2. Disconnect the negative battery cable.
3. Drain the cooling system into a clean container for reuse.
4. Raise and safely support the vehicle.
5. Heater hoses from the heater core

➡**Carefully, use compressed air to blow the remainder of the coolant from the heater core to prevent spillage of coolant in the interior when removing the heater core.**

6. Remove or disconnect the following
 • Lower trim panel extensions (located at both sides of the console-to-center instrument panel), by pulling outward at the dual lock locations; then, rotate the panels outward to disengage the hinges at the console
 • Lower instrument panel closeout panel (located on the right side of the heater/air conditioning housing), by pulling it out at the top edge and rotating the top downward
 • Lower heater duct-to-heater/air conditioning housing 2 screws and 2 clips (located at the bottom of the heater/air conditioning housing); then, drop the duct down and carefully slide it out sideways

❊❊ WARNING

Be careful not to damage the heater duct-to-rear floor heater duct seal.

 • Release the retaining clip and disconnect the temperature control cable from the heater door hook
 • Heater core side cover
 • 4 lower heater core cover-to-heater/air conditioning housing assembly screws and the cover
 • Heater core clamp and the heater core

To install:

7. Install or connect the following:
 • Heater core, the clamp and the

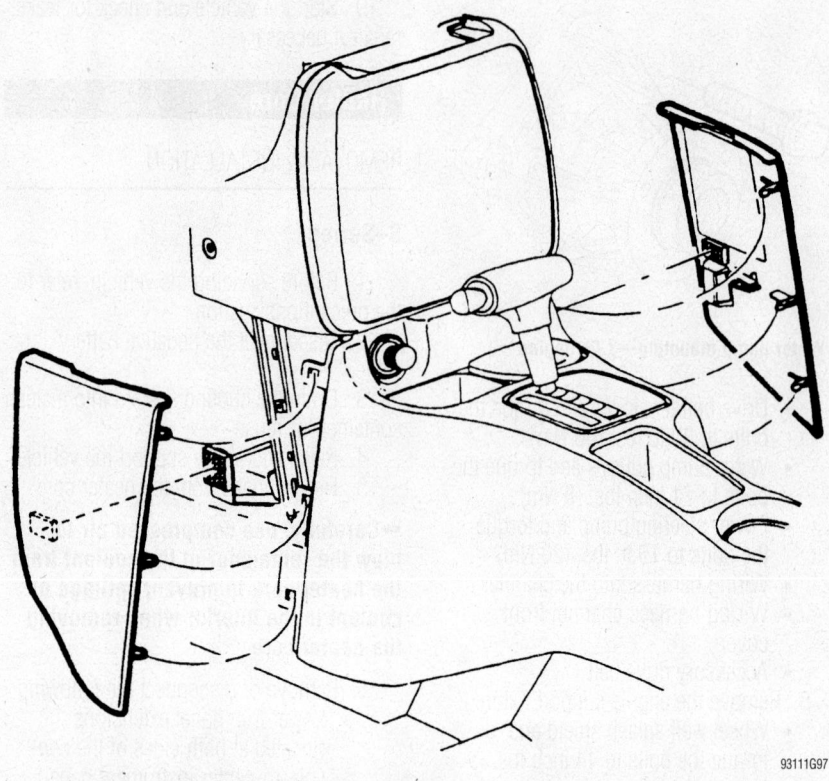

View of the lower trim panel extensions—S–Series

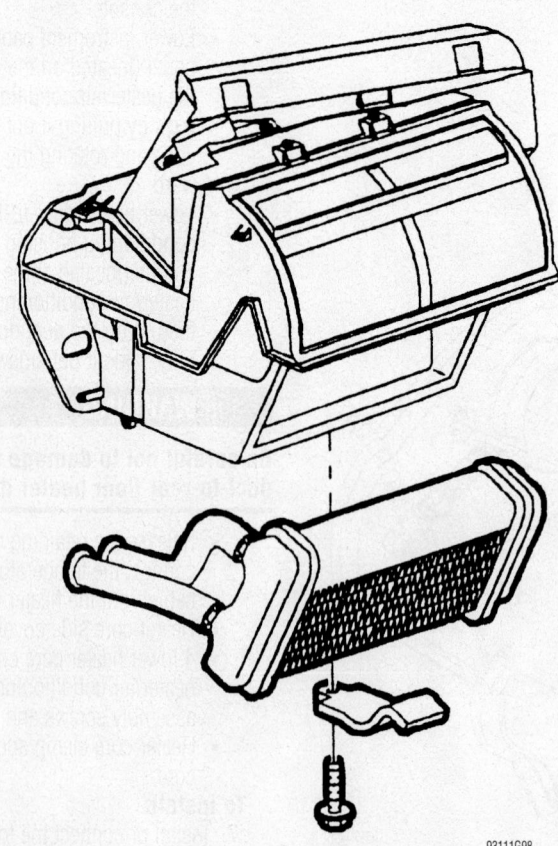

93111G98

Exploded view of the heater core and heater case—S–Series

heater core-to-heater/air conditioning housing assembly screw

• Lower heater core cover and the 4 cover-to-heater/air conditioning housing assembly screws

• Heater core side cover and the cover screw, located at the left side of the heater/air conditioning housing assembly

• Temperature control cable to the heater door hook and secure with the retaining clip

✳✳ WARNING

Be careful not to damage the heater duct-to-rear floor heater duct seal

• Lower heater duct; then, the lower heater duct-to-heater/air conditioning housing 2 screws and 2 clips, located at the bottom of the heater/air conditioning housing

• Lower instrument panel closeout panel, located at the right side of the heater/air conditioning housing

• Console-to-center instrument panel location (on both sides), install the lower trim panel extensions

• Heater hose-to-heater core clamps

8. Lower the vehicle.

9. Refill the cooling system.

10. Connect the negative battery cable.

11. Operate the engine to normal operating temperatures; then, check the climate control operation and check for leaks.

L–Series

1. Before servicing the vehicle, refer to the precautions section.

2. Remove or disconnect the following:
• Negative battery cable
• Lower cradle splash shield-to-cradle and wheel house retainers
• Splash shield

➥**Failure to use a drain hose will result in an uncontrolled coolant flow direction. Also a container capable of holding at least 8 quarts of coolant will be required.**

3. Place a pliable piece of hose with an inside diameter of ⅜ inch over the radiator drain tube.

4. Route the hose rearward and downward and position it to drain the coolant into an approved container.

5. Using a flat bladed screwdriver to open the radiator drain valve and drain the coolant.

➥**Approximately 4 quarts of coolant can be drained from the radiator. More coolant will remain in the system. When completely drained, close the drain valve.**

6. Slide the lower cradle splash shield between the wheelhouse linen and the wheelhouse. Install the fasteners and tighten to 44 inch lbs. (5 Nm).

7. Remove or disconnect the following:
- Heater inlet hose from the water pump outlet on 3.0L engines or thermostat on 2.2L engines
- Heater return hose from head bridge on 3.0L engines or block inlet on 2.2L engines

➥**Carefully, use compressed air to blow the remainder of the coolant from the heater core to prevent spillage of coolant in the interior when removing the heater core.**

- Right console extension
- Heater core-to-pipe clamps
- Heater core from the end tank by gently pulling the pipes forward. Place rags on the floor to absorb any spilled coolant.
- Heater core screw and strap
- Heater core from module. The center console forward edge may need to be pulled outward to allow core removal.

To install:

8. Install or connect the following:
- Heater core into module, being careful not to damage the foam seals
- Heater core strap and screw and tighten the screw to 9 inch lbs. (1 Nm)
- New bushings and O–rings onto the core pipes
- Heater pipes to the core and install the plastic clamps. The clamps can only be installed one way, the lower clamp hinge fits into the tab of the core retaining strap
- Right console extension
- Heater inlet hose to the water pump outlet on 3.0L engines or thermostat on 2.2L engines
- Heater return hose to head bridge on 3.0L engines or block inlet on 2.2L engines

9. Rinse off any residual coolant.
10. Refill the cooling system.
11. Connect the negative battery cable.
12. Operate the engine to normal operating temperatures; then, check the climate control operation and check for leaks.

ION

1. Before servicing the vehicle, refer to the precautions section.

2. Remove or disconnect the following:
- Surge tank cap
- Water pump drain bolt and drain the coolant from the pump. Once drained, close the bolt and torque to 88 inch lbs. (10 Nm).
- Heater outlet and inlet hose clamps
- Heater outlet and inlet hoses

3. Remove the Body Control Module (BCM) as follows:

a. Disconnect the negative battery cable.

b. Apply the parking brake and position the transmission into Neutral.

c. Remove the console shift lever bezel by lifting up around the edge carefully, if equipped with an automatic transmission.

d. Unsnap the shift boot from the console cup holder, if equipped with a manual transmission.

e. Lift the front console cup holder to disengage the retainers

f. Disconnect the cigar lighter connection.

g. Remove the cup holder, if equipped with a manual transmission, slide the boot through the cup holder.

h. Remove the console extension retainers from both sides by turning them counterclockwise.

i. Remove the console extensions by pulling the extension rearwards.

j. Remove the center extension screws, pull out the center extension to access any remaining fasteners and remove the extension.

k. Unsnap the parking brake boot.

l. If equipped with an armrest, remove the console compartment screws.

m. Lift up on the rear on the console to release the retainers.

n. Lift the console compartment and push the parking brake through the opening.

o. Slide the compartment over the parking brake lever.

p. Disconnect the rear power supply connector.

q. Remove the console screws and the console.

r. Remove the wiring harness rosebud from the right center support bracket.

s. Pull back the carpet from the right of the left Instrument Panel (IP) support bracket and remove the lower nuts.

t. Remove the center support bracket.

u. Disconnect the small then large harness connectors from the BCM.

v. Disconnect the small then large IP wiring harness connectors from the BCM.

w. Disconnect the Onstar connector, if equipped.

x. Remove the BCM retainers and BCM.

4. Remove or disconnect the following:
- Pull back the carpet at the bottom of the left IP support bracket
- Left IP support bracket nuts
- Accelerator pedal from the front of the dash and position aside
- Center floor outlet ducts by raising the duct and then pushing them to disengage them
- Rotate the center floor duct and pull down to disconnect it from the HVAC module
- Heater core cover screws
- Heater core cover down enough to clear the locating pins from the HVAC module, slide the cover rearwards until the drain tube clears the front of the dash. Slide the cover down, rearwards and to the right and remove it.
- Heater core

To install:

5. Inspect the heater core foam and if damaged use Kent Industries adhesive black foam 46480 to replace.

6. Install or connect the following:
- Heater core
- New drain tube seal on the drain tube

7. Spray the heater core seal and dashmat with soapy water to ease installation.
- Heater core cover from the right side. Slide the cover up, forward and into position. Align the drain tube with the hole in the dash, raise the cover into position while aligning the holes with the locating pins on the HVAC module. Tighten the cover screws to 15 inch lbs. (1.8 Nm).
- Center floor ducts.
- Accelerator pedal
- Carpet at the bottom of the left IP support bracket
- Left IP center bracket nuts and tighten to 88 inch lbs. (10 Nm)

8. Install the BCM as follows:

a. Install the BCM and tighten the retainers to 88 inch lbs. (10 Nm).

b. Connect the small then large IP wiring harness connectors to the BCM.

c. Connect the small then large harness connectors to the BCM.

d. Connect the Onstar connector, if equipped.

e. Pull back the carpet from the right of the left IP support bracket and the center support bracket. Tighten the nuts to 88 inch lbs. (10 Nm).

f. Connect the wiring harness rosebud to the right center support bracket.

- Install the front floor console over parking lever and shift lever. Push the console down on the carpet and tighten the screws to 22 inch lbs. (2.5 Nm).

g. Connect the rear power supply connector.

h. Slide the compartment over the parking brake lever.

i. Pull the parking brake boot through the opening.

j. Align the retainers with the console and push to engage.

k. If equipped with an armrest, install the console compartment screws and tighten the screws to 22 inch lbs. (2.5 Nm).

l. Snap the parking brake boot to the console.

m. Align the center extension to the IP and push the engage the retainers. Tighten the screws to 22 inch lbs. (2.5 Nm).

n. Install the cup holder, if equipped with a manual transmission, slide the boot through the cup holder.

o. Connect the cigar lighter connection.

p. Align the front console cup holder retainers and push to engage.

q. Snap the shift boot to the console cup holder, if equipped with a manual transmission.

r. Install the console shift lever bezel by aligning the fingers on the lever bezel with the slots on the lever base and push to engage, if equipped with an automatic transmission.

s. Apply the parking brake and position the transmission into park.

t. Install the console extensions. Turn the retainers clockwise to secure.

9. Install or connect the following:
- Heater outlet and inlet hose clamps
- Heater outlet and inlet hoses
- Surge tank cap

10. Refill the cooling system.

11. Connect the negative battery cable.

12. Operate the engine to normal operating temperatures; then, check the climate control operation and check for leaks.

Cylinder Head

REMOVAL & INSTALLATION

1.9L Engines

> ※※ **WARNING**
>
> **Only remove the cylinder head when the engine is cold. Warpage may result if the cylinder head is removed while the engine is hot.**

1. Before servicing the vehicle, refer to the precautions section.
2. Drain the cooling system.
3. Drain the engine oil.
4. Properly relieve the fuel system pressure.
5. Remove or disconnect the following:
- Negative battery cable
- Air cleaner assembly
- Rocker cover fresh air hose
- Accelerator cable from the throttle body and intake manifold bracket
- Coolant temperature gauge electrical connector
- Fuel injector connectors
- Idle Air Control (IAC) valve electrical connector
- Throttle Position Sensor (TPS) electrical connector
- A/C compressor electrical connector
- Manifold Absolute Pressure (MAP) sensor electrical connector
- Oxygen Sensor (O$_2$S) electrical connector
- Exhaust Gas Recirculation (EGR) solenoid

6. Reposition the wire harness out of the way.
- Spark plug wires
- Evaporative Emissions (EVAP) purge valve vacuum hose
- Positive Crankcase Ventilation (PCV) vacuum hose
- Throttle body connector
- Brake booster vacuum hose
- Upper radiator hose from the cylinder head outlet
- De-aeration hose from the intake manifold
- Heater hose from the intake manifold
- Fuel supply line
- Engine torque axis mount
- Accessory drive belt
- Drive belt tensioner
- Drive belt idler pulley
- Rocker arm cover
- De-aeration line from the cylinder head water outlet
- A/C compressor and front bracket without removing the hoses
- Power steering pump without removing the lines
- Right front wheel
- Wheel well splash shield
- Crankshaft pulley
- Front exhaust pipe from the exhaust manifold
- Front cover

7. Install a crankshaft gear retainer tool with the flat side toward the sprocket.
- Front four oil pan bolts and cut the oil pan seal away from the front cover
- Front cover

➡**Position the crankshaft 90 degrees off of Top Dead Center (TDC) to make certain that the pistons do not touch the valves during assembly.**

- Timing chain, tensioner and guides
- Camshaft sprocket
- Loosen and uniformly remove the 10 cylinder head bolts in several passes
- Cylinder head from the dowels on the engine block

To install:

➡**Before installing the cylinder head, it should be cleaned and inspected for excessive wear or damage.**

8. Clean the gasket mating surfaces. Be careful not to damage the aluminum components and be sure the block bolt holes are clean of any residual sealer, oil or foreign matter.

9. Using a dial gauge, check that the cylinder liners are flush or do not deviate more than 0.0005 in. (0.013mm).

10. Be sure the crankshaft is still 90 degrees past TDC and that the camshaft(s)

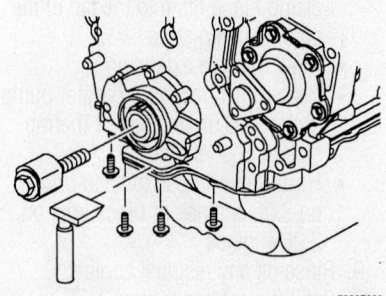

79222Z802

Crankshaft gear retaining and oil pan removal tool shown—be sure to install the crankshaft tool with the flat side toward the gear—1.9L engines

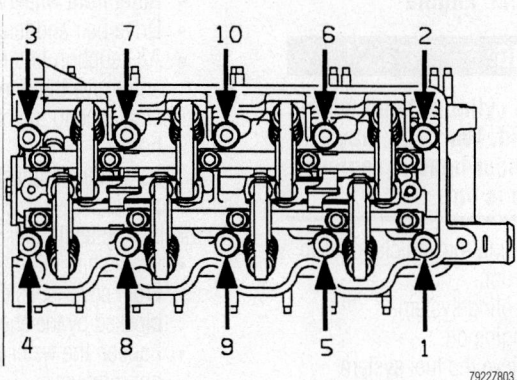

Gradually remove the cylinder head bolts in the sequence shown to prevent warping the head—1.9L engines

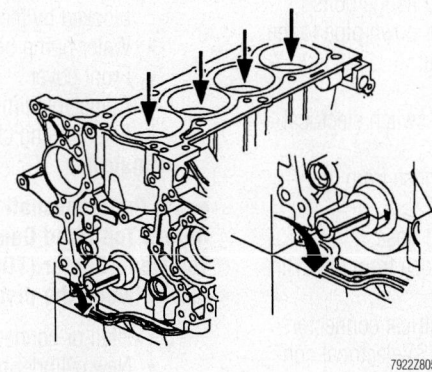

Rotate the crankshaft clockwise to 90 degrees past TDC (the crankshaft sprocket timing mark will be at 3 o'clock) to prevent valve damage during assembly—1.9L engines

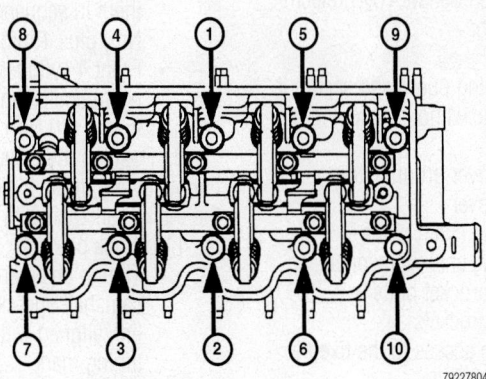

Cylinder head bolt tightening sequence—1.9L engine

are properly positioned with the dowel pin(s) at the 12 o'clock position to prevent valve damage.

11. Install the cylinder head gasket and carefully guide the head into place over the dowels.

12. If the head bolts and/or the block were replaced, install the bolts and tighten in sequence to 48 ft. lbs. (65 Nm) to insure proper clamp load; then, remove the bolts.

13. Coat the cylinder head bolts with clean engine oil and thread the bolts by hand until finger-tight.

14. Install the cylinder head bolts, in sequence, as follows:

  a. Tighten the bolts to 22 ft. lbs. (30 Nm).

  b. Tighten the bolts to 33 ft. lbs. (45 Nm).

  c. Tighten each bolt an additional 90 degree turn.

15. Install the timing chain, sprockets,

guides and tensioners, using the following procedure :

  a. Verify that the crankshaft is positioned 90 degrees clockwise past TDC. The crankshaft key way sprocket timing mark must be aligned with the cylinder block main bearing cap split line to prevent piston and valve damage.

  b. If required, rotate the camshaft up to No. 1 TDC position.

  c. Rotate the crankshaft 90 degrees counterclockwise to No. 1 TDC position. The crankshaft sprocket timing mark must align with the cylinder block timing mark.

16. Install or connect the following:

- Timing chain over the camshaft sprocket and under the crankshaft sprocket. Slide the camshaft sprocket onto the camshaft with the letters **FRT** facing away from the cylinder head
- Camshaft sprocket timing pin
- Camshaft washer and bolt and torque the bolt to 74 ft. lbs. (100 Nm)
- Fixed chain guide and torque the fastener to 19 ft. lbs. (26 Nm). The timing chain should be snug against the fixed guide.
- Pivoting chain guide making certain there is clearance between the block and head and torque the bolts to 19 ft. lbs. (26 Nm)
- Timing chain tensioner and torque the bolts to 168 inch lbs. (19 Nm)
- Crankshaft timing gear retainer tool to align the gerotor oil pump to the front cover
- Front cover and torque the perimeter bolts to 22 ft. lbs. (30 Nm) and the lower center bolt to 89 inch lbs. (10 Nm)
- Oil pan and torque the bolts to 80 inch lbs. (9 Nm)
- Water pump pulley and torque the bolts to 19 ft. lbs. (25 Nm), if removed

17. Remove the crankshaft timing gear retaining tool.

18. Apply RTV across the cylinder head and front cover T-joints and install a new rocker cover gasket.

- Crankshaft pulley and torque the bolt to 159 ft. lbs. (215 Nm)
- Rocker cover and torque the bolts to 22 ft. lbs. (30 Nm)
- Drive belt tensioner and torque the fastener to 22 ft. lbs. (30 Nm)
- Drive belt idler pulley and torque the fastener to 20 ft. lbs. (27 Nm)

- Accelerator linkage bracket and the throttle cable. Make certain that the cable is routed properly and is not binding and torque the bolts to 18 ft. lbs. (25 Nm).
- Alternator and torque the bolts to 24 ft. lbs. (32 Nm), if removed
- Power steering pump and torque the bolts to 22 ft. lbs. (30 Nm)
- A/C compressor and torque the rear bracket bolts to 18 ft. lbs. (25 Nm) and the front bracket bolts to 35 ft. lbs. (47 Nm)
- Accessory drive belt
- Upper mount and torque the bracket nuts to 37 ft. lbs. (50 Nm) and then torque the mount to cover bolts to 37 ft. lbs. (50 Nm)
- Wheel well splash shield
- Wheel
- EVAP canister purge hose
- Throttle body vacuum harness hose
- PCV hose
- Brake booster vacuum hose
- Upper radiator hose
- Heater hose
- Ground wire electrical connectors
- O2S electrical connector
- Fuel injector electrical connectors
- IAC valve electrical connector
- EGR valve electrical connector
- TPS electrical connector
- EVAP canister purge valve connectors
- MAP sensor connectors
- Alternator electrical connectors
- Starter motor electrical connectors
- A/C compressor electrical connectors
- Intake manifold support bracket and torque the bolts to 22 ft. lbs. (30 Nm)
- De-aeration line and torque the cylinder head nut to 80 inch lbs. (9 Nm) and the rear lift bracket nut to 97 inch lbs. (11 Nm)
- Fuel feed and return lines with new retainers and torque the fastener to 53 inch lbs. (6 Nm)
- Air cleaner assembly
- Fresh air hose
- Negative battery cable

19. Fill the cooling system.
20. Fill the engine with new oil.
21. Prime the fuel system by cycling the ignition **ON** for 5 seconds and **OFF** for 10 seconds a few times without cranking the engine.
22. Start the engine, check for leaks, and repair if necessary.

L–Series With 2.2L Engine

✳✳ WARNING

Only remove the cylinder head when the engine is cold. Warpage may result if the cylinder head is removed while the engine is hot.

1. Before servicing the vehicle, refer to the precautions section.
2. Drain the cooling system.
3. Drain the engine oil.
4. Properly relieve the fuel system pressure.
5. Remove or disconnect the following:
   - Negative battery cable
   - Intake manifold
   - Exhaust manifold flange bolts
   - Exhaust manifold down pipe to the catalytic converter
   - Fuel lines
   - A/C compressor switch electrical connector
   - Crankcase vent hose from the camshaft cover
   - Coolant air bleed hose
   - Upper radiator hose from the cylinder head
   - Ignition coil electrical connectors
   - Knock Sensor (KS) electrical connector
   - Engine Coolant Temperature (ECT) sensor electrical connector
   - Fuel injector jumper harness
   - Front Oxygen Sensor (O2S) electrical connector
   - Ignition coil
   - Power steering pump and secure it to the engine without removing the lines
   - Camshaft cover ground strap
   - Camshaft cover
   - Upper timing chain guide
   - Upper timing chain tensioner
   - Camshaft sprocket bolts
   - Camshaft sprockets
   - Plug to gain access to the fixed timing chain guide
   - Fixed timing chain guide upper bolt only
   - Cylinder head bolts using the proper sequence
   - Cylinder head

➡**The manufacturer recommends to perform the remaining steps to complete this procedure. It may be not be necessary to perform the remaining steps if the timing chain can be supported without slipping or moving on the crankshaft pulley.**

- Right front wheel and splash shield
- Drive belt and tensioner
- A/C suction line retaining clips and move the line away to allow clearance to the crankshaft pulley bolt

6. Position the crankshaft 60 degrees Before Top Dead Center (BTDC) and install a crankshaft pulley holder tool.
   - Crankshaft pulley
   - Front cover bolts except for the one blocked by the engine mount bracket
   - Loosen the water pump bolt but do not remove it

7. Support the engine and remove the right side engine mount.
   - Engine mount bracket
   - Upper front cover bolt which was blocked by the mount bracket
   - Water pump bolt
   - Front cover
   - Adjustable timing chain guide
   - Fixed timing chain guide lower bolt

To install:

➡**Set the crankshaft to 60 degrees Before Top Dead Center (BTDC) or after Top Dead Center (TDC) to prevent contact between the pistons and valves.**

8. Install or connect the following:
   - New cylinder head gasket with the side imprinted **OBEN** facing up
   - Cylinder head and align it on the dowels
   - New cylinder head bolts and torque them in sequence to 22 ft. lbs. (30 Nm) plus 155 degrees
   - Front 4 cylinder head bolts coated with Loctite® and torque them to 15 ft. lbs. (20 Nm)

9. Position the exhaust camshaft with the offset slot in the 2 o'clock position and the intake camshaft with the offset slot in the 11 o'clock position.
   - Timing chain around the intake camshaft sprocket with the copper link aligned with the **INT** diamond timing mark
   - Sprocket to the camshaft and align it with the offset slot
   - New camshaft sprocket bolt, but do not tighten
   - Timing chain around the crankshaft sprocket and align the silver link to the timing mark
   - Adjustable timing chain guide through the opening on top of the cylinder head and torque the bolt to 89 inch lbs. (10 Nm)
   - Timing chain around the exhaust camshaft sprocket with the silver link aligned with the offset slot

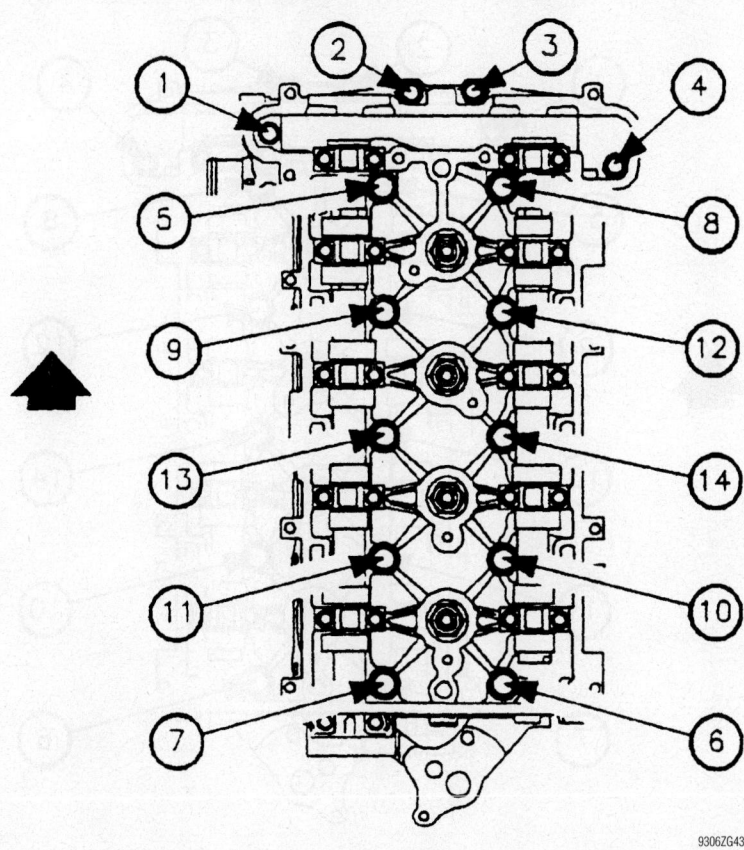

Cylinder head bolt loosening sequence—L–Series With 2.2L engine

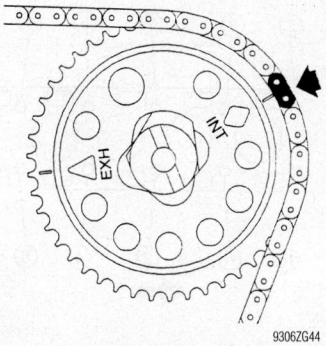

Align the copper link on the timing chain with the INT diamond timing mark

Install but do not tighten a new sprocket bolt

❊❊ WARNING

Make certain that all timing marks and colored links are aligned properly before proceeding to the next step. If the timing chain is not aligned properly, severe engine damage may occur.

10. Torque the intake and exhaust camshaft bolts to 63 ft. lbs. (85 Nm) plus a 30 degree turn.

11. Install or connect the following:
- Fixed timing guide and torque the bolt to 89 inch lbs. (10 Nm)
- Fixed timing guide bolt access plug after applying Loctite®to the threads and torque it to 30 ft. lbs. (40 Nm)
- Timing chain tensioner and torque the bolts 44 ft. lbs. (60 Nm)

12. Tap the top of the timing chain between the camshaft sprockets to engage the tensioner.
- Upper timing chain guide and torque the bolts to 89 inch lbs. (10 Nm)
- Front cover with a new gasket and torque the bolts to 18 ft. lbs. (25 Nm)
- Water pump bolt and torque it to 18 ft. lbs. (25 Nm)
- Drive belt tensioner and torque the bolts to 30 ft. lbs. (40 Nm)
- Engine mount bracket and torque the bolts to 66 ft. lbs. (90 Nm)
- Engine mount to the side rail and torque the fasteners to 41 ft. lbs. (55 Nm)
- Engine mount to the bracket and torque the bolts to 41 ft. lbs. (55 Nm)
- Crankshaft damper and torque the bolt to 74 ft. lbs. (100 Nm)
- Accessory drive belt
- A/C line clamps
- Right wheel splash shield
- Right front wheel
- Camshaft cover and torque the bolts to 89 inch lbs. (10 Nm)
- Camshaft cover ground strap and torque the bolt to 89 inch lbs. (10 Nm)
- Power steering pump and torque the bolts to 18 ft. lbs. (25 Nm)
- Ignition coil/module assembly and torque the bolts to 71 inch lbs. (8 Nm)
- Intake manifold with a new gasket and torque the bolts to 89 inch lbs. (10 Nm)
- Wiring harness to the intake manifold bracket
- Engine wiring harness and torque the fasteners to 89 inch lbs. (10 Nm)
- All component electrical connections
- Upper radiator hose
- Crankcase ventilation hose
- Fuel pressure regulator hose
- Brake booster vacuum hose
- EVAP purge hose
- Fuel lines
- Fuel lines to the bracket and torque

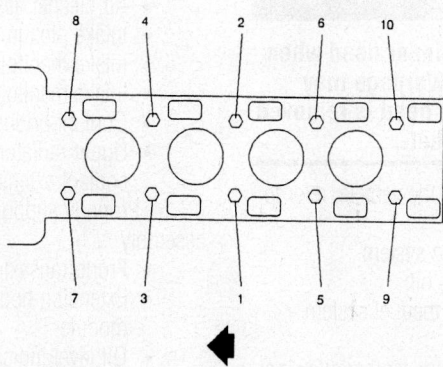

Cylinder head bolt tightening sequence—L–Series With 2.2L engine

the bracket bolt to 89 inch lbs. (10 Nm)

- Exhaust down pipe to the exhaust manifold and torque the bolts to 25 ft. lbs. (30 Nm)
- Exhaust pipe to the resonator and torque the bolts to 15 ft. lbs. (20 Nm)
- Exhaust manifold heat shield and torque the bolts to 18 ft. lbs. (25 Nm)
- Air cleaner assembly
- Negative battery cable

13. Fill the engine with clean oil.
14. Fill the cooling system.
15. Prime the fuel system by cycling the ignition **ON** for 5 seconds and **OFF** for 10 seconds a few times without cranking the engine.
16. Start the engine, check for leaks, and repair if necessary.

ION

> ❄❄ **WARNING**
>
> **Only remove the cylinder head when the engine is cold. Warpage may result if the cylinder head is removed while the engine is hot.**

1. Before servicing the vehicle, refer to the precautions section.
2. Drain the cooling system.
3. Drain the engine oil.
4. Properly relieve the fuel system pressure.
5. Remove or disconnect the following:
- Negative battery cable
- Intake manifold
- Exhaust manifold
- Timing chain
- Cylinder head bolts using the proper sequence
- Cylinder head

To install:

➡**Set the crankshaft to 60 degrees Before Top Dead Center (BTDC) or after Top Dead Center (TDC) to prevent contact between the pistons and valves.**

6. Install or connect the following:
- New cylinder head gasket
- Cylinder head and align it on the dowels
- New cylinder head bolts and torque them in sequence to 22 ft. lbs. (30 Nm) plus 155 degrees
- Front 4 cylinder head bolts coated with Loctite® and torque them to 26 ft. lbs. (35 Nm)
- Timing chain

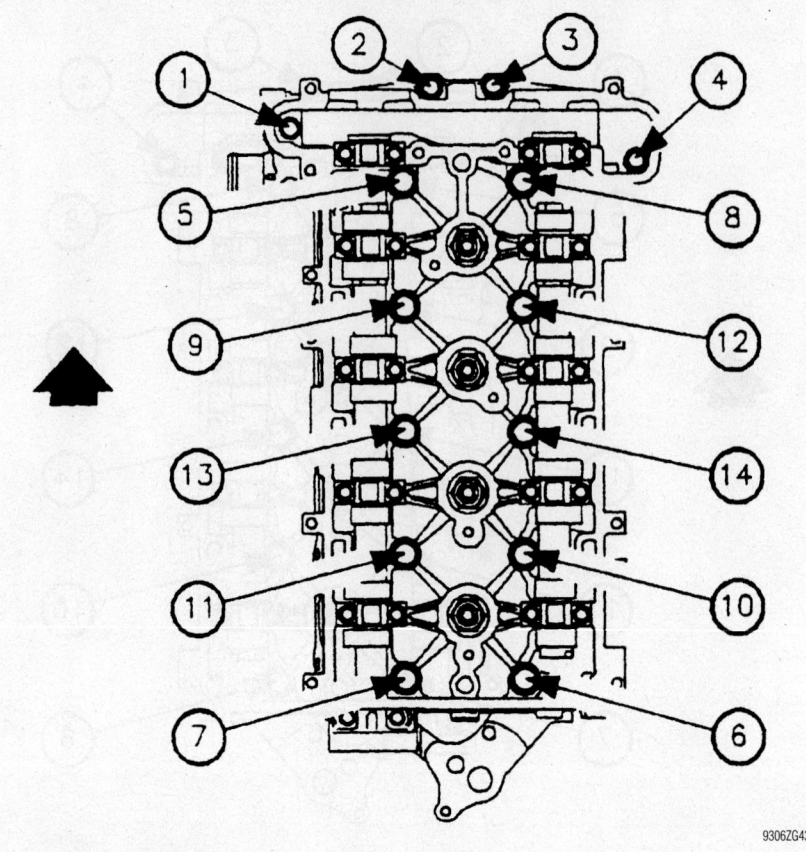

Cylinder head bolt loosening sequence—ION

- Exhaust manifold
- Intake manifold
- Negative battery cable

7. Fill the engine with clean oil.
8. Fill the cooling system.
9. Prime the fuel system by cycling the ignition **ON** for 5 seconds and **OFF** for 10 seconds a few times without cranking the engine.
10. Start the engine, check for leaks, and repair if necessary.

3.0L Engine

FRONT

> ❄❄ **WARNING**
>
> **Only remove the cylinder head when the engine is cold. Warpage may result if the cylinder head is removed while the engine is hot.**

1. Before servicing the vehicle, refer to the precautions section.
2. Drain the cooling system.
3. Drain the engine oil.
4. Properly relieve the fuel system pressure.
5. Remove or disconnect the following:
- Negative battery cable

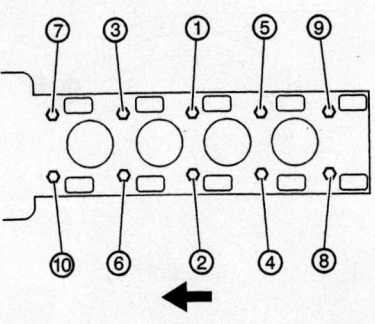

Cylinder head bolt tightening sequence—ION

- Air cleaner assembly
- Intake plenum
- Intake manifold
- Intake manifold spacer
- Coolant bridge
- Upper radiator hose from the coolant extension housing

6. Properly support the powertrain assembly.

- Front transaxle mount through bolt
- Extension housing over the coolant module
- Oil level indicator tube
- Coolant extension housing by twisting it off

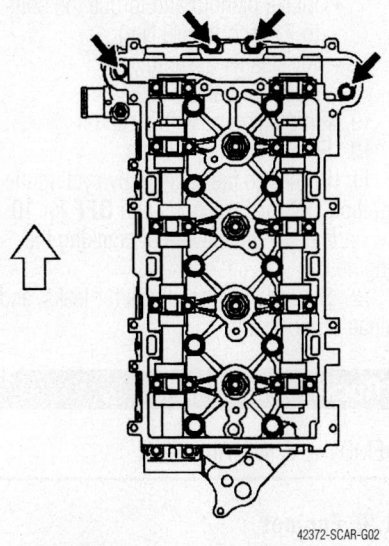

Location of the 4 front cylinder head bolts—ION

- Front camshaft cover
- Ground wires from the lift bracket
- Oxygen Sensor (O_2S) electrical connector
- Down pipe from the exhaust manifold
- Front timing belt cover
- Timing belt
- Timing belt tensioner bracket
- Rear timing belt cover
- Camshaft Position (CMP) sensor electrical connector
- Exhaust camshaft
- Loosen the cylinder head bolts in stages as shown
- Cylinder head
- Exhaust manifold from the cylinder head (if necessary)

To install:

7. Install or connect the following:
   - Exhaust manifold with a new gasket and torque the bolts to 15 ft. lbs. (20 Nm)
   - New cylinder head gasket with the part number imprint facing the top of the engine
   - Cylinder head

8. Torque the new cylinder head bolts, in sequence, as follows:
   a. 18 ft. lbs. (25 Nm)
   b. 90 degree turn
   c. 90 degree turn
   d. 90 degree turn
   e. 15 degree turn

9. Install or connect the following:
   - Coolant pipe with new sealing rings
   - Engine lift bracket bolt and torque it to 15 ft. lbs. (20 Nm)

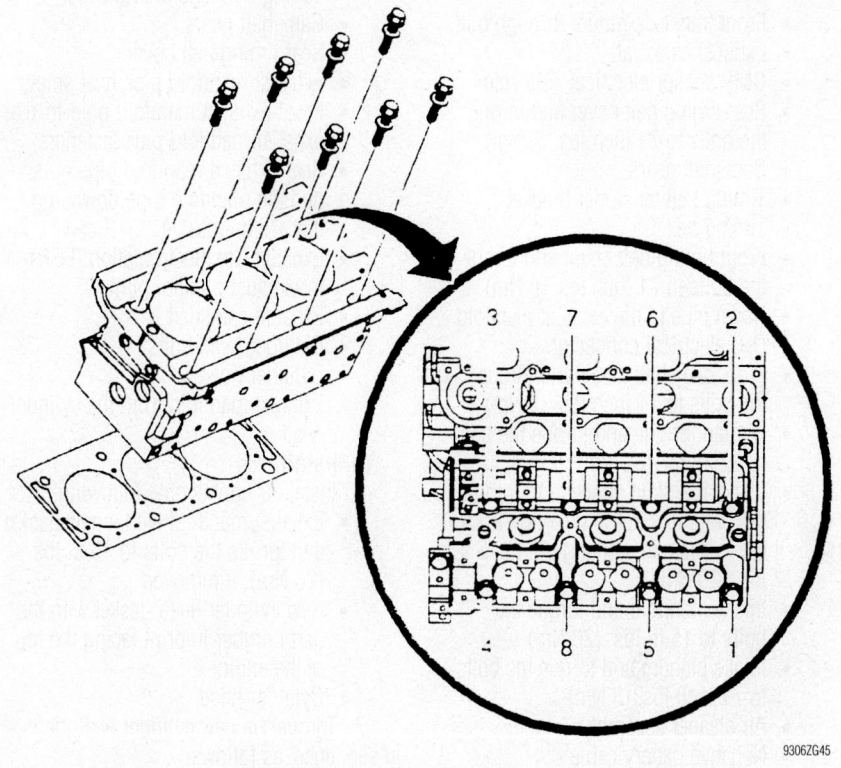

Cylinder head bolt removal for the front and rear cylinder heads—3.0L engine

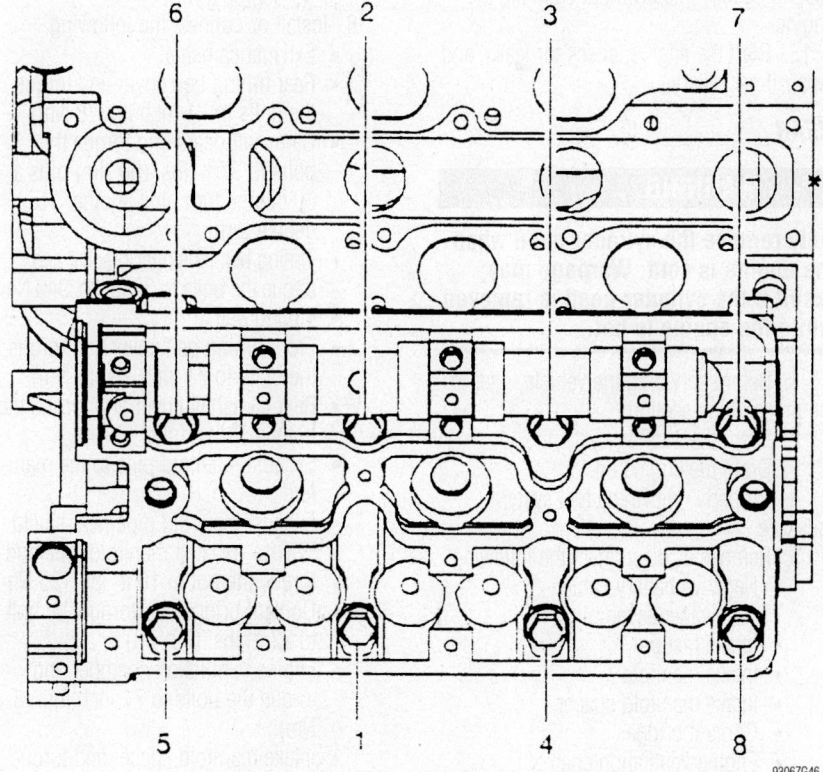

Front and rear cylinder head bolt tightening sequence—3.0L engine

- Upper radiator hose to the coolant pipe
- Front transaxle mount through bolt
- Exhaust camshaft
- CMP sensor electrical connector
- Rear timing belt cover and torque the bolts to 71 inch lbs. (8 Nm)
- Camshaft gears
- Timing belt tensioner bracket
- Timing belt
- Front timing belt cover and torque the bolts to 71 inch lbs. (8 Nm)
- Down pipe to the exhaust manifold
- O$_2$S electrical connector
- Front camshaft cover and torque the bolts to 71 inch lbs. (8 Nm)
- Coolant bridge and torque the bolt to 22 ft. lbs. (33 Nm)
- Intake manifold spacer and torque the bolts in a spiral direction from the inside and working out to 15 ft. lbs. (20 Nm)
- Intake manifold and torque the bolts to 15 ft. lbs. (20 Nm)
- Intake plenum and torque the bolts to 71 inch lbs. (8 Nm)
- Air cleaner assembly
- Negative battery cable

10. Fill the engine with clean oil.
11. Fill the cooling system.
12. Prime the fuel system by cycling the ignition **ON** for 5 seconds and **OFF** for 10 seconds a few times without cranking the engine.
13. Start the engine, check for leaks, and repair if necessary.

REAR

✻✻ WARNING

Only remove the cylinder head when the engine is cold. Warpage may result if the cylinder head is removed while the engine is hot.

1. Before servicing the vehicle, refer to the precautions section.
2. Drain the cooling system.
3. Drain the engine oil.
4. Properly relieve the fuel system pressure.
5. Remove or disconnect the following:
   - Negative battery cable
   - Air cleaner assembly
   - Intake plenum
   - Intake manifold
   - Intake manifold spacer
   - Coolant bridge
   - Engine ventilation chamber
   - Rear camshaft cover
   - Front timing belt cover

- Timing belt
- Timing belt tensioner bracket
- Camshaft gears
- Rear timing belt cover
- Exhaust manifold pipe heat shield
- Front exhaust manifold pipe-to-rear exhaust manifold pipe fasteners
- Rear exhaust manifold pipe nuts, pull the manifold pipe down and discard the gasket
- Exhaust Gas Recirculation (EGR)-to-exhaust manifold pipe
- Exhaust camshaft
- Cylinder head bolts
- Cylinder head
- Exhaust manifold from the cylinder head

To install:
6. Install or connect the following:
   - Exhaust manifold with a new gasket and torque the bolts to 15 ft. lbs. (20 Nm), if removed
   - New cylinder head gasket with the part number imprint facing the top of the engine
   - Cylinder head
7. Torque the new cylinder head bolts, in sequence, as follows:
   a. 18 ft. lbs. (25 Nm)
   b. 90 degree turn
   c. 90 degree turn
   d. 90 degree turn
   e. 15 degree turn
8. Install or connect the following:
   - Exhaust camshaft
   - Rear timing belt cover and torque the bolts to 71 inch lbs. (8 Nm)
   - Camshaft gears and torque the bolts to 37 ft. lbs. (50 Nm) plus a 60 degree turn plus another 15 degree turn
   - Timing belt tensioner bracket and torque the bolts to 30 ft. lbs. 940 Nm)
   - Timing belt
   - Front timing belt cover and torque the bolts to 71 inch lbs. (8 Nm)
   - Rear camshaft cover and torque the bolts to 71 inch lbs. (8 Nm)
   - Exhaust manifold pipe to the manifold
   - Exhaust manifold pipe heat shield
   - EGR-to-exhaust manifold pipe and torque the nut to 18 ft. lbs. (25 Nm)
   - Coolant bridge and torque the bolt to 22 ft. lbs. (33 Nm)
   - Engine ventilation chamber and torque the bolts to 71 inch lbs. (8 Nm)
   - Intake manifold spacer and torque the bolts to 15 ft. lbs. (20 Nm)
   - Intake manifold and torque the bolts to 15 ft. lbs. (20 Nm)

- Intake plenum and torque the bolts to 71 inch lbs. (8 Nm)
- Air cleaner assembly
- Negative battery cable

9. Fill the engine with clean oil.
10. Fill the cooling system.
11. Prime the fuel system by cycling the ignition **ON** for 5 seconds and **OFF** for 10 seconds a few times without cranking the engine.
12. Start the engine, check for leaks, and repair if necessary.

Rocker Arms/Shafts

REMOVAL & INSTALLATION

1.9L Engines

1. Before servicing the vehicle, refer to the precautions section.
2. Remove or disconnect the following:
   - Negative battery cable
   - Spark plug wires
   - Accessory drive belt
   - Fresh air hose
   - Camshaft cover
   - Fuel injector electrical connectors
3. Install a rocker arm removal tool J43223 to the cylinder head rail.
4. Position the crankshaft at Top Dead Center (TDC) with the pip marks on the camshaft sprockets at the 12 o'clock position.
5. Rotate the crankshaft 90 degrees past TDC and remove the following rocker arm assemblies:
   - No. 1 Intake
   - No. 2 Intake
   - No. 1 Exhaust
   - No. 3 Exhaust
6. Rotate the crankshaft 360 degrees to remove the following rocker arm assemblies:
   - No. 2 Exhaust
   - No. 4 Intake
   - No. 3 Intake
   - No. 4 Exhaust
7. Compress the valve springs.
8. Remove the rocker arm and/or valve lifters.

To install:
9. Rotate the crankshaft 90 degrees past TDC and install the following rocker arm assemblies:
   - No. 1 Intake
   - No. 2 Intake
   - No. 1 Exhaust
   - No. 3 Exhaust
10. Rotate the crankshaft 360 degrees to install the following rocker arm assemblies:
   - No. 2 Exhaust

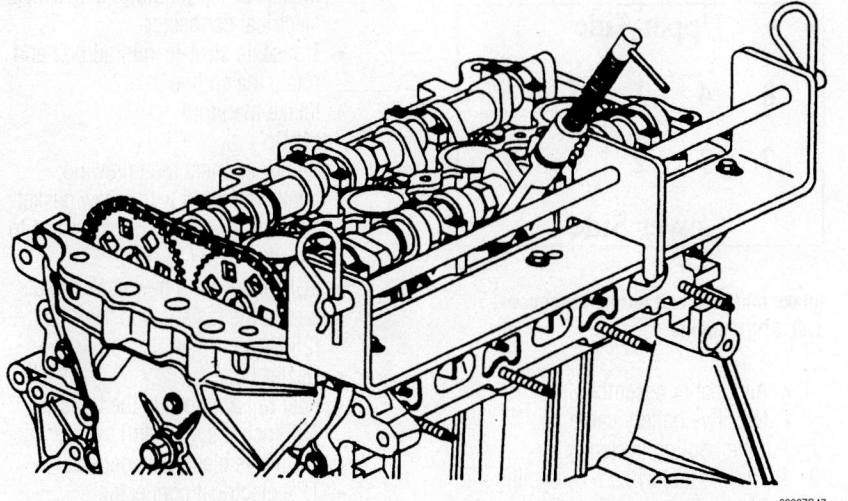

93062G47

Exploded view of the valve springs compressed—1.9L DOHC engines

- No. 4 Intake
- No. 3 Intake
- No. 4 Exhaust

11. Remove the rocker arm tool.
12. Install or connect the following:
- Camshaft cover and torque the bolts to 71 inch lbs. (8 Nm)
- Fresh air hose

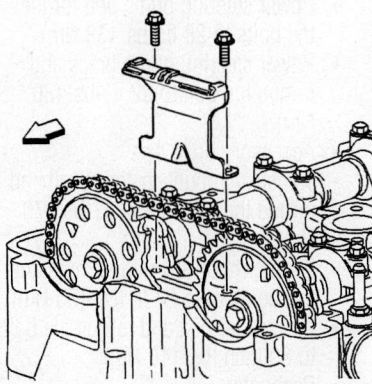

42372-SCAR-G03

Remove the upper timing chain guide—ION

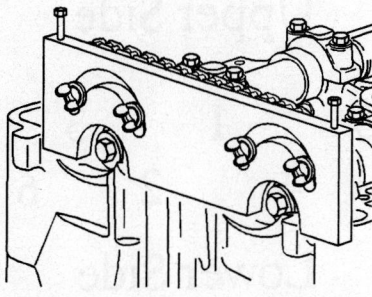

42372-SCAR-G04

Install camshaft sprocket holding tool J43655

- Accessory drive belt
- Fuel injector electrical connectors
- Spark plug wires
- Negative battery cable

2.2L Engines

1. Before servicing the vehicle, refer to the precautions section.
2. Remove or disconnect the following:
- Negative battery cable
- Camshaft cover
- Upper timing chain guide
3. Install camshaft sprocket holding tool J43655 as illustrated.
- Intake and exhaust camshaft sprocket bolts and discard the bolts
4. Slide the sprockets forward and mark the camshaft bearing caps to ensure they are installed in their correct positions.
5. Remove each bearing cap bolt one turn at a time until there is no tension.
6. Remove the bearing cap and the camshaft
- Roller followers
- Lash adjusters

To install:

7. Lubricate the valve tips.
8. Install or connect the following:
- Lash adjusters
- Roller followers
9. Make sure the alignment notches are aligned with the camshaft sprocket as illustrated.
- Camshaft
- Bearing caps except the rear camshaft cap
10. Tighten the caps in increments of 3 turns until they are seated, then tighten to 89 inch lbs. (10 Nm).
11. Apply 0.197 inch (5mm) of Permatex

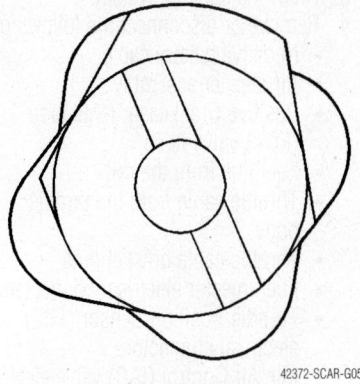

42372-SCAR-G05

Make sure the alignment notches are aligned with the camshaft sprocket

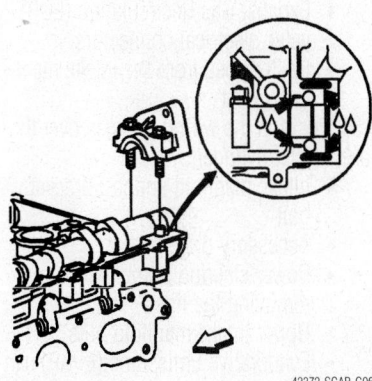

42372-SCAR-G06

Apply 0.197 inch (5mm) of anaerobic Gasket maker to the rear camshaft bearing cap

Anaerobic Gasket maker® 51813 to the rear camshaft bearing cap. Install the cap and tighten the bolts to 18 ft. lbs. (25 Nm).
- Camshaft sprockets and hand tighten the bolts
12. Remove camshaft sprocket holding tool J43655
- Camshaft sprocket bolts to 63 ft. lbs. (85 Nm) plus and additional 30 degrees
13. Install or connect the following:
- Upper timing chain guide and tighten to 89 inch lbs. (10 Nm)
- Camshaft cover
- Negative battery cable

Intake Manifold

REMOVAL & INSTALLATION

1.9L SOHC Engines

1. Before servicing the vehicle, refer to the precautions section.
2. Properly relieve the fuel system pressure.

3. Drain the cooling system.
4. Remove or disconnect the following:
- Negative battery cable
- Air cleaner assembly
- Positive Crankcase Ventilation (PCV) valve hose
- Fuel line from the rail
- Throttle cable from the throttle body
- Throttle cable bracket nuts
- Fuel injector electrical connectors
- Throttle Position Sensor (TPS) electrical connectors
- Idle Air Control (IAC) valve electrical connectors
- Manifold Absolute Pressure (MAP) sensor electrical connectors
- Exhaust Gas Recirculation (EGR) valve electrical connectors
- Heater hose from the intake manifold outlet

5. Position the wiring harness over the brake master cylinder.
- Intake manifold support bracket bolt
- Accessory drive belt
- Power steering pump without removing the lines
- Upper intake manifold nuts
- Evaporative Emissions (EVAP) canister purge valve solenoid vacuum hose
- Brake booster vacuum hose
- Lower intake manifold nuts
- Intake manifold

To install:
6. Install or connect the following:
- Intake manifold with a new gasket and torque the nuts in sequence to 22 ft. lbs. (30 Nm)
- Power steering pump and torque the fasteners to 27 ft. lbs. (35 Nm)
- Accessory drive belt
- Heater hose
- Intake manifold support bracket bolt and torque it to 22 ft. lbs. (30 Nm)
- Fuel supply and return lines
- Fuel line(s) in the retaining bracket and torque the mounting screw to 36 inch lbs. (4 Nm)
- Throttle cable to the throttle body and accelerator cable bracket and torque the bolts to 22 ft. lbs. (30 Nm)
- IAC valve electrical connector
- TPS electrical connector
- MAP sensor electrical connector
- Fuel injector electrical connectors
- EGR valve electrical connector
- EVAP purge solenoid vacuum hose
- PCV valve hose

```
        Upper Side

   8    4    1    5

   7    3    2    6    9

        Lower Side
```
79222Z808

Intake manifold bolt torque sequence—1.9L SOHC engine

- Air cleaner assembly
- Negative battery cable
7. Fill the cooling system.
8. Prime the fuel system by cycling the ignition **ON** for 5 seconds and **OFF** for 10 seconds a few times without cranking the engine.
9. Start the engine, check for leaks, and repair if necessary.

1.9L DOHC Engines

1. Before servicing the vehicle, refer to the precautions section.
2. Properly relieve the fuel system pressure.
3. Drain the cooling system.
4. Remove or disconnect the following:
- Negative battery cable
- Air cleaner fresh air hose
- Positive Crankcase Ventilation (PCV) valve hose
- Fuel line from the rail
- Throttle cable from the throttle body
- Throttle cable bracket nuts
- Fuel injector electrical connectors
- Throttle Position Sensor (TPS) electrical connector
- Idle Air Control (IAC) valve electrical connector
- Manifold Absolute Pressure (MAP) sensor electrical connector
- Exhaust Gas Recirculation (EGR) valve electrical connector
- Heater hose from the intake manifold outlet
- Power steering support brace
- Fuel rail
- EGR pipe from the cylinder head
- Brake booster vacuum hose
- Accessory drive belt
- Power steering pump without removing the lines
- Upper axis torque mount nuts and midrail bracket nuts. Allow the powertrain assembly to rest on a support fixture.
- Resonator and air cleaner box

- Intake Air Temperature (IAT) sensor electrical connector
- Transaxle strut-to-midrail bolt and rotate the engine
- Intake manifold

To install:
5. Install or connect the following:
- Intake manifold with a new gasket and torque the nuts in sequence to 115 inch lbs. (13 Nm)
- EGR pipe with a new gasket and torque the fasteners to 18 ft. lbs. (25 Nm)
- Heater hose
- Fuel rail and torque the bolts to 106 inch lbs. (12 Nm)
- IAC valve electrical connector
- TPS electrical connector
- MAP sensor electrical connector
- Fuel injector electrical connector
- EGR valve electrical connector
- EVAP canister purge solenoid vacuum hose
- PCV valve hose
- Brake booster vacuum hose
- Fuel rail to the manifold and torque the nut to 36 inch lbs. (4.5 Nm)
- Fuel supply line to the rail and torque the bolt to 36 inch lbs. (4 Nm)
- Power steering pump and torque the bolts to 28 ft. lbs. (38 Nm)
- Power steering pump brace and torque the bolt to 22 ft. lbs. (30 Nm)
- Accessory drive belt
- Transaxle mount-to-frame rail and torque the bolts to 52 ft. lbs. (70 Nm)
- Engine mount-to-midrail nuts and torque them to 37 ft. lbs. (50 Nm)
- Air cleaner box and torque the bolts to 89 inch lbs. (10 Nm)
- Resonator
- IAT sensor electrical connector
- Throttle cable to the throttle body

```
   Upper Side

 5      1      3

   4      2      6

   Lower Side
```
9306ZG49

Intake manifold bolt torque sequence—1.9L DOHC engine

and support bracket and torque the bolts to 19 ft. lbs. (25 Nm)

- Intake manifold brace and torque the bolts to 58 inch lbs. (6.5 Nm)
- Negative battery cable

6. Fill the cooling system.

7. Prime the fuel system by cycling the ignition **ON** for 5 seconds and **OFF** for 10 seconds a few times without cranking the engine.

8. Start the engine, check for leaks, and repair if necessary.

L–Series With 2.2L Engine

1. Before servicing the vehicle, refer to the precautions section.

2. Remove or disconnect the following:

- Negative battery cable
- Air cleaner assembly
- Intake Air Temperature (IAT) sensor electrical connector
- Throttle Position Sensor (TPS) electrical connectors
- Idle Air Control (IAC) valve electrical connectors
- Manifold Absolute Pressure (MAP) sensor electrical connectors
- Fuel pressure regulator vacuum pipe
- Evaporative Emissions (EVAP) purge solenoid hose from the throttle body
- Throttle cable and automatic transaxle downshift cable from the throttle body
- Throttle cable bracket nuts
- Brake booster vacuum hose
- Oil level indicator tube
- Throttle body
- Intake manifold

To install:

3. Install or connect the following:

- Intake manifold with a new gasket and torque the nuts to 89 inch lbs.

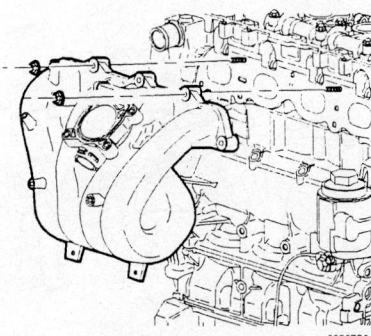

Intake manifold mounting—all 2.2L engines

9356ZG04

(10 Nm) starting from the center and working outward

- Throttle body to the intake manifold and torque the bolts to 19 ft. lbs. (25 Nm)
- Throttle cable bracket to the throttle body studs
- Oil level indicator tube and torque the fastener to 89 inch lbs. (10 Nm)
- Brake booster vacuum pipe
- Throttle cable to the throttle body and install the support bracket. Torque the bolts to 19 ft. lbs. (25 Nm)
- EVAP canister purge solenoid vacuum hose
- Fuel pressure regulator vacuum pipe to the throttle body
- IAC valve electrical connector
- TPS electrical connector
- MAP sensor electrical connector
- Air inlet hose to the throttle body
- Crankcase vent hose to the camshaft cover
- IAT sensor electrical connector
- Air cleaner assembly
- Negative battery cable

4. Start the engine, check for leaks, and repair if necessary.

ION

2.0L ENGINE

1. Before servicing the vehicle, refer to the precautions section.

2. Disconnect the negative battery cable.

3. Remove the supercharger.

4. Remove the alternator.

5. Drain the charged air cooling system.

6. Disconnect the charged air cooling system inlet and outlet hoses.

7. Remove the charged air coolant pump.

8. Remove the cooling fan assembly.

9. Remove the AC compressor bolts and reposition the compressor to access the oil level indicator tube.

10. Remove the oil level indicator tube bolt.

➡**Be sure to remove all fasteners before attempting to remove the intake manifold.**

11. Remove the intake manifold nuts and bolts.

12. Remove the intake manifold.

➡**The intake manifold gasket is reusable, only replace the gasket if damage has occurred.**

13. If applicable, remove the intake manifold gasket.

To install:

14. If applicable, install the intake manifold gasket.

15. Install the intake manifold.

16. Install the intake manifold nuts and bolts. Tighten to 89 inch lbs. (10 Nm).

17. Install the oil level indicator tube and bolt.

18. Reposition the AC compressor and install the bolts and tighten to 16 ft. lbs. (22 Nm).

19. Install the cooling fan assembly.

20. Install the charged air coolant pump.

21. Install the alternator.

22. Connect the charged air cooling system inlet and outlet hoses.

23. Fill the charged air cooling system.

24. Install the supercharger.

25. Connect the negative battery cable.

2.2L ENGINE

1. Before servicing the vehicle, refer to the precautions section.

2. Remove or disconnect the following:

- Negative battery cable
- Air cleaner assembly
- Idle Air Control (IAC) valve electrical connectors
- Throttle Position Sensor (TPS) electrical connectors
- Vacuum hoses from the throttle body
- Throttle control cable and bracket
- Throttle body
- Positive Crankcase Valve (PCV) hose
- Purge solenoid tube
- Brake booster hose
- Oil dipstick tube bolt
- Fuel rail
- Knock Sensor (KS) connector
- Intake manifold bolts and nuts
- Intake manifold

To install:

3. Install or connect the following:

- Intake manifold with a new gasket and torque the nuts to 89 inch lbs. (10 Nm) starting from the center and working outward
- Knock Sensor (KS) connector
- Fuel rail
- Oil dipstick tube bolt
- Brake booster hose
- Purge solenoid tube
- PCV hose
- Throttle body to the intake manifold and torque the bolts to 89 inch lbs. (10 Nm)
- Throttle cable bracket and torque the bolts to 89 inch lbs. (10 Nm)

- Throttle cable
- IAC sensor electrical connector
- TPS electrical connectors
- Vacuum hoses from the throttle body
- Air cleaner assembly
- Negative battery cable

4. Start the engine, check for leaks, and repair if necessary.

3.0L Engine

1. Before servicing the vehicle, refer to the precautions section.

2. Properly relieve the fuel system pressure.

3. Remove or disconnect the following:
- Negative battery cable
- Air cleaner assembly
- Throttle body electrical connector
- Fuel pressure regulator vacuum hose
- Both intake runners
- Intake plenum
- Fuel supply and return lines
- Fuel injector electrical connectors
- Fuel rail
- Intake manifold
- Intake manifold spacer and O-ring gaskets

To install:

4. Install or connect the following:
- Intake manifold spacer with new seals. Apply Loctite®242 to the bolts and torque the bolts to 15 ft. lbs. (20 Nm) in sequence.
- Intake manifold with a new gasket and torque the bolts to 15 ft. lbs. (20 Nm)
- Fuel rail
- Fuel injector electrical connectors
- Fuel supply and return hoses and torque the fastener to 11 ft. lbs. (15 Nm)
- Intake plenum and torque the bolts to 71 inch lbs. (8 Nm)
- Intake manifold runners and torque the bolts to 71 inch lbs. (8 Nm)
- Negative battery cable

5. Prime the fuel system by cycling the ignition **ON** for 5 seconds and **OFF** for 10

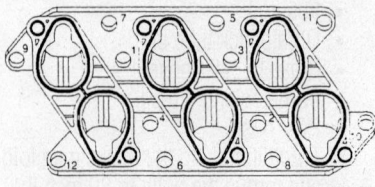

Tighten the intake manifold spacer bolts as shown—3.0L engines

9356ZG07

seconds a few times without cranking the engine.

6. Start the engine, check for leaks, and repair if necessary.

Exhaust Manifold

REMOVAL & INSTALLATION

1.9L Engines

1. Before servicing the vehicle, refer to the precautions section.

2. Remove or disconnect the following:
- Negative battery cable
- Front exhaust pipe-to-engine support bracket mounting fasteners
- Pipe-to-manifold nuts and lower the pipe
- A/C compressor and bracket from the engine without removing the lines
- Oxygen Sensor (O2S) electrical connector
- Exhaust manifold

To install:

3. Install or connect the following:
- Exhaust manifold with a new gasket

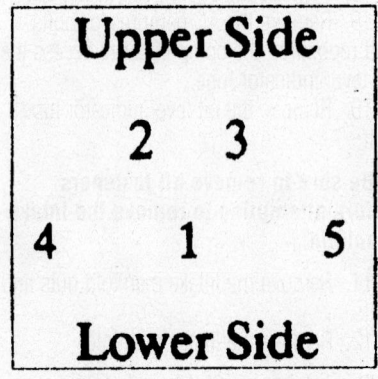

Exhaust manifold bolt torque sequence—1.9L SOHC engines

79222810

Exhaust manifold bolt torque sequence—1.9L DOHC engines

79222811

and torque the nuts in sequence to 16 ft. lbs. (22 Nm) for the SOHC engine or to 150 inch lbs. (17 Nm) for the DOHC engine
- O2S electrical connector
- A/C compressor and bracket and torque the front bracket-to-compressor to 40 ft. lbs. (54 Nm) and all remaining fasteners to 19 ft. lbs. (25 Nm)
- Pipe-to-manifold with a new gasket and torque the fasteners in a crosswise pattern to 23 ft. lbs. (31 Nm)
- Exhaust pipe-to-engine support bracket and torque the mounting fasteners to 23 ft. lbs. (31 Nm)
- Negative battery cable

4. Start the vehicle and check for leaks, repair if necessary.

L–Series With 2.2L Engine

1. Before servicing the vehicle, refer to the precautions section.

2. Remove or disconnect the following:
- Negative battery cable
- Exhaust pipe from the manifold
- Exhaust manifold heat shield
- Oxygen Sensor (O2S) from the manifold
- Exhaust manifold

To install:

3. Install or connect the following:
- Exhaust manifold with a new gasket and torque the bolts, starting from the center and working outward, to 13 ft. lbs. (18 Nm)
- O2S and torque it to 33 ft. lbs. (45 Nm)
- Exhaust manifold heat shield and torque the bolts to 18 ft. lbs. (25 Nm)
- Exhaust pipe to the manifold with a new gasket and torque the nuts to 22 ft. lbs. (30 Nm)
- Negative battery cable

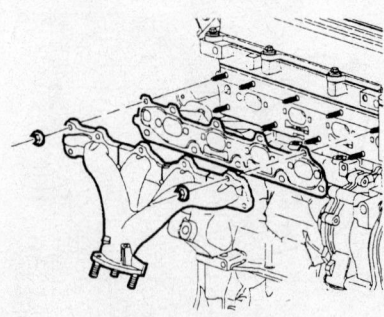

Remove the exhaust manifold and gasket.—L–Series With 2.2L engine

9346ZG16

4. Start the vehicle and check for leaks, repair if necessary.

ION

1. Before servicing the vehicle, refer to the precautions section.
2. Remove or disconnect the following:
   - Negative battery cable
   - Exhaust manifold heat shield
   - Oxygen Sensor (O$_2$S) from the manifold

➡**Do not bend the flex coupling more than 3 degrees in any direction to avoid damage.**

   - Pipe-to-manifold nuts, pull down and back on the pipe in order to separate the pipe from the manifold
   - Exhaust pipe from the manifold
   - Exhaust manifold-to-head nuts
   - Exhaust manifold

To install:

3. Install or connect the following:
   - Exhaust manifold with a new gasket

42372-SCAR-G07

Remove the exhaust manifold and gasket—ION models

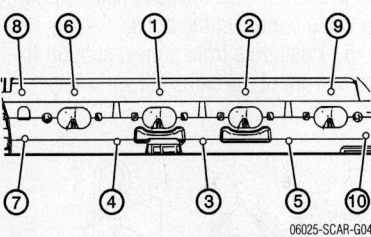

06025-SCAR-G04

Tighten the exhaust manifold nuts in the sequence shown—2.0L ION models

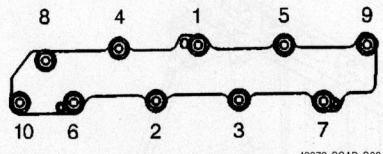

42372-SCAR-G08

Tighten the exhaust manifold nuts in the sequence shown—2.2L ION models

and torque the bolts in the sequence illustrated, to 115 inch lbs. (13 Nm)
   - New manifold-to-flex pipe gasket, and place the pipe in position. Tighten the nuts to 32 ft. lbs. (43 Nm) on 2.2L engines or 37 ft. lbs. (50 Nm) on 2.0L engines.
   - O$_2$S
   - Exhaust manifold heat shield and torque the bolts to 18 ft. lbs. (25 Nm)
   - Negative battery cable
4. Start the vehicle and check for leaks, repair if necessary.

3.0L Engine

FRONT MANIFOLD

1. Before servicing the vehicle, refer to the precautions section.
2. Drain the cooling system.
3. Remove or disconnect the following:
   - Negative battery cable
   - Exhaust manifold Oxygen Sensor (O$_2$S)
   - Upper radiator hose from the coolant extension housing
4. Install and engine support fixture.
   - Front transaxle through bolt and raise the powertrain assembly
   - Oil level indicator tube and coolant extension tube bolt
   - Coolant extension housing
   - Power steering pipe bracket bolt
   - Upper exhaust manifold nuts
   - Front exhaust manifold pipe from the manifold
   - Font exhaust manifold pipe from the rear exhaust manifold pipe
   - Oil filter housing
   - Lower exhaust manifold nuts
   - Exhaust manifold

To install:

5. Install or connect the following:
   - Exhaust manifold with a new gasket and torque the bolts to 15 ft. lbs. (20 Nm)
   - Oil filter housing and torque the filter cartridge to 33 ft. lbs. (45 Nm)
   - Front exhaust manifold pipe and gaskets, do not tighten the bolts
   - Front exhaust pipe to the rear exhaust pipe and torque the bolts to 15 ft. lbs. (20 Nm)
   - Front manifold pipe and torque the bolts to 15 ft. lbs. (20 Nm)
   - Front exhaust manifold O$_2$S and torque it to 73 ft. lbs. (45 Nm)
   - Power steering pipe bracket and torque the bolt to 71 inch lbs. (8 Nm)
   - New O-rings to the coolant extension housing

   - Coolant extension housing and torque the bolt to 15 ft. lbs. (20 Nm)
6. Lower the powertrain assembly.
   - Front transaxle mount through bolt and torque it to 41 ft. lbs. (55 Nm)
   - Upper radiator hose
   - Negative battery cable
7. Fill the cooling system.
8. Start the vehicle and check for leaks, repair if necessary.

REAR MANIFOLD

1. Before servicing the vehicle, refer to the precautions section.
2. Drain the cooling system.
3. Remove or disconnect the following:
   - Negative battery cable
   - Oxygen Sensor (O$_2$S)
   - Exhaust Gas Recirculation (EGR) pipe
   - Exhaust manifold pipe heat shield
   - Rear exhaust manifold pipe
   - Exhaust manifold

To install:

4. Install or connect the following:
   - Exhaust manifold with a new gasket and torque the bolts to 15 ft. lbs. (20 Nm)
   - Rear exhaust manifold pipe and torque the nuts to 25 ft. lbs. (30 Nm)
   - Front exhaust manifold pipe to the rear exhaust manifold pipe and torque the bolts to 15 ft. lbs. (20 Nm)
   - Rear exhaust manifold pipe to the resonator and torque the bolts to 15 ft. lbs. (20 Nm)
   - EGR pipe and torque the fasteners to 19 ft. lbs. (25 Nm)
   - O$_2$S and torque it to 37 ft. lbs. (50 Nm)
   - Exhaust manifold heat shield and torque the bolts to 71 inch lbs. (8 Nm)
   - Negative battery cable
5. Start the vehicle and check for leaks, repair if necessary.

Front Crankshaft Seal

REPLACEMENT

On the 1.9L and 2.2L engines (Except ION) the front crankshaft seal is located in the timing chain front cover. Refer to the timing chain procedure for information about removing the front cover and replacing the seal or if replacing the seal on the ION model refer to the procedure below.

ION

1. Before servicing the vehicle, refer to the precautions section.
2. Remove or disconnect the following:
   - Negative battery cable
   - Drive belt
   - Using harmonic balancer holder J 38122-A to prevent the crankshaft from rotating and remove the balancer bolt and discard
   - Crankshaft balancer
   - Front seal using a suitable prytool

To install:
3. Install or connect the following:
   - Front seal using driver J 35268-A to drive the seal into the position on the front cover
   - Crankshaft balancer
   - Balancer bolt and using tool J 38122-A to prevent the crankshaft from rotating, install a new balancer bolt and tighten to 74 ft. lbs. (100 Nm) plus an additional 75 degrees
   - Drive belt
   - Negative battery cable

3.0L Engine

1. Before servicing the vehicle, refer to the precautions section.
2. Remove or disconnect the following:
   - Negative battery cable
   - Timing belt
   - Crankshaft gear
3. Drill a small pilot hole into the steel ring of the seal.
4. Screw in a self-tapping screw.
5. Use pliers to pull out the oil seal.

To install:
6. Coat the lip of the new oil seal with engine oil.
7. Install the oil seal using a suitable seal installer.

8. Install the crankshaft gear and torque the bolt to 184 ft. lbs. (250 Nm) plus an additional 45 degrees then another 15 degrees.
9. Install the timing belt.

Camshaft and Lifters

REMOVAL & INSTALLATION

1.9L (SOHC) Engines

1. Before servicing the vehicle, refer to the precautions section.
2. Remove or disconnect the following:
   - Both battery cables
   - Battery and tray
   - Timing chain front cover
   - Timing chain and camshaft sprocket
   - Rocker arm/shaft assemblies
   - Lifters and label or position them for assembly in their original locations
3. Drive the camshaft plug inward, then remove it from the cylinder head with a magnet.
4. Carefully pull the camshaft from the rear of the cylinder head through the oversized camshaft plug hole. Turn the camshaft back and forth slowly while withdrawing to help prevent journal or bearing damage.

To install:
5. Clean and inspect all parts prior to installation. Lubricate the camshaft and carefully insert it through the hole at the rear of the cylinder head.
6. Install or connect the following:
   - New rear cylinder head plug coated with Loctite® 242
   - Valve lifters into their original bores, or if the camshaft has been replaced, install new lifters
   - Rocker arm/shaft assemblies

- Timing chain and camshaft sprocket
- Timing chain front cover
- Battery and tray and torque the battery hold-down nut and screw to 80 inch lbs. (9 Nm)
- Both battery cables

7. Start the engine and check for leaks, repair if necessary.

1.9L (DOHC) Engines

➡ **Be careful when working around the camshaft sprockets and timing chain cover during this procedure. If a bolt or washer is accidentally dropped between the front cover and engine assembly, the cover will have to be removed for retrieval.**

1. Before servicing the vehicle, refer to the precautions section.
2. Remove or disconnect the following:
   - Negative battery cable
   - Accessory drive belt
   - Spark plug wires
   - Positive Crankcase Ventilation (PCV) fresh air hose
   - Cam cover
3. Turn the crankshaft clockwise until the mark on the crankshaft pulley is in alignment with the pointer on the front cover and the No. 1 cylinder is at Top Dead Center (TDC) of the compression stroke. Both camshaft dowel pins will be at the 12 o'clock position and the timing pin holes will be aligned when the No. 1 cylinder is at TDC. If necessary, the right wheel and splash shield can be removed to help observe the timing marks.
4. Remove the camshaft sprocket retaining bolts. Use a ⅞in. (21mm) open-end wrench to hold the camshaft from turning while removing the bolts.
5. Position a front angled support fixture in front of the camshaft sprockets.

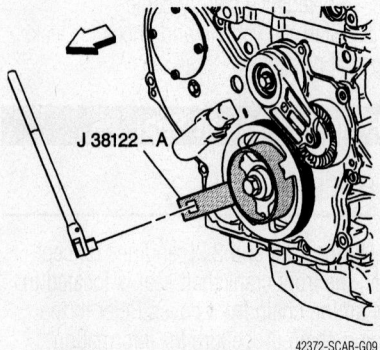

42372-SCAR-G09

Using harmonic balancer holder J 38122-A to prevent the crankshaft from rotating—ION models engine

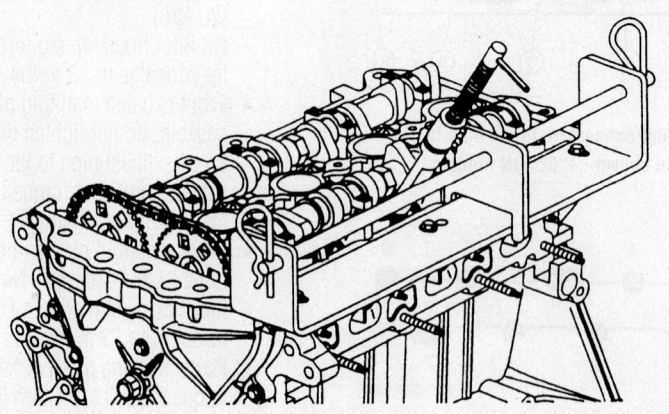

9346ZG17

Remove the rocker arm/shaft assemblies—1.9L (SOHC) engines

6. Attach the camshaft sprocket adapters to the end of each camshaft using the pilot bolts, but do not tighten the bolts. The front angled support should come between the sprocket adapters and camshaft sprockets.

7. Remove or disconnect the following:
- Upper timing chain guide
- Both front camshaft bearing caps

8. Secure the support fixture using ⅞ in. bolts/blocks and align the 2 holes in each camshaft sprocket, adapter and the front support fixture. Install the 4 nuts, but do not tighten.

➡**The steel blocks should be installed against the rearward side of the camshaft sprocket.**

9. Tighten the sprocket pilot bolts to 19 ft. lbs. (25 Nm) while holding the camshafts from turning with an open end wrench.

10. Move each camshaft sprocket off the end of the camshaft by rocking the sprocket forward or by carefully prying between the end of the camshaft and the sprocket. Then, tighten the 4 nuts and bolts with blocks from the side of the support fixture to 19 ft. lbs. (25 Nm).

11. Install the 2 bolts retaining the support fixture to the engine front cover and tighten the bolts to 89 inch lbs. (10 Nm).

Then, remove each camshaft sprocket pilot bolt while holding the camshafts with a wrench.

12. Carefully pry between the sprocket and the end of the camshaft to move the camshaft rearward. Pry only enough to remove its end from inside the sprocket pilot otherwise camshaft or lifter damage may occur.

13. Remove or disconnect the following:
- Uniformly loosen and remove the remaining camshaft bearing cap bolts. To prevent bolt/cap damage, do not use power tools and make several passes.
- Camshafts
- Pull the lifters out to remove them, always place them in the order in which they were removed and with the camshaft side facing down. Oil will drain out of the lifter if it is placed valve side down.

To install:

14. Clean and inspect all parts prior to installation. Oil the camshaft and install with the **IN** camshaft on the intake side and **EX** camshaft on the exhaust side.

➡**The dowel pin in each camshaft must be located at the 12 o'clock position during installation to prevent valve and piston damage.**

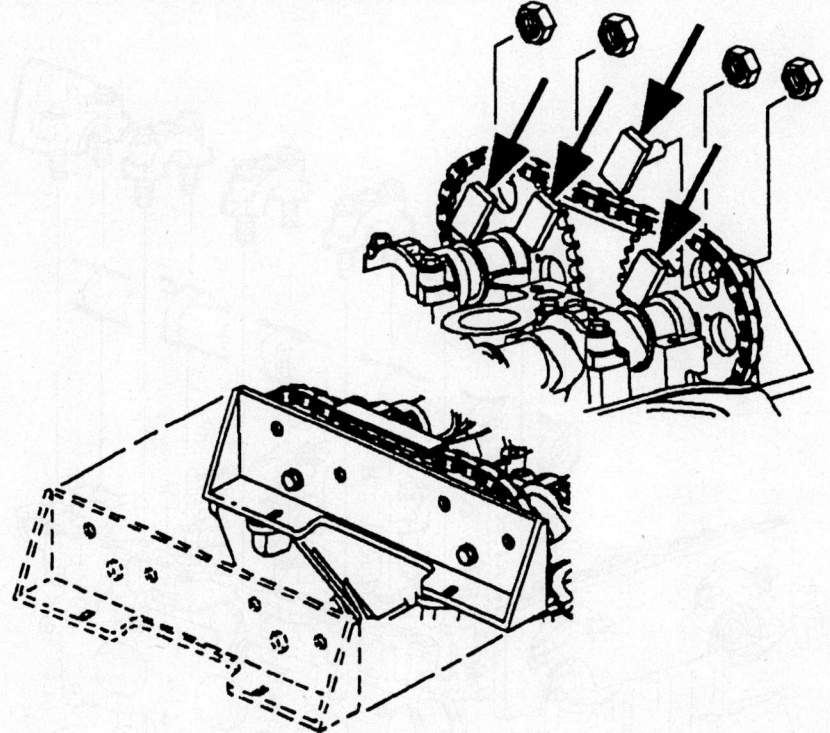

15. Install or connect the following:
- Lifters in their original locations
- Bearing caps, except for the forward pair, in their original positions, making sure the arrows on the caps are pointing forward toward the camshaft sprockets. Lightly oil each of the cap bolts, then install and uniformly torque the bolts to 124 inch lbs. (14 Nm).
- Camshaft sprocket pilot bolt in each camshaft and torque the bolts to 124 inch lbs. (14 Nm) in order to pull the camshaft fully forward and align the sprocket support for installation of the sprocket onto the camshaft

16. Remove the 4 sprocket support bolt/blocks and nuts.

17. Install the forward bearing caps and the upper chain guide. Torque the cap bolts to 124 inch lbs. (14 Nm).

18. Be sure the camshaft dowel pin aligns with the slot in each camshaft sprocket. If necessary, rotate the camshaft slightly (1–2 degrees) and move each sprocket from the adapter onto the end of the camshaft. Fully seat each sprocket on the end of each camshaft.

19. Remove the 2 sprocket pilot bolts and adapters while using a wrench on the camshaft flats to assure the camshaft cannot move.

20. Remove the support angle fixture.

21. Install the camshaft sprocket retaining bolts and washers. Hold the camshafts and torque the bolts to 76 ft. lbs. (103 Nm).

22. Verify all visible timing marks and holes are in alignment. Turn the crankshaft clockwise until the mark on the crankshaft pulley aligns with the mark on the front cover. Check timing by inserting ³⁄₁₆ in. drill bits through the camshaft sprocket alignment holes, into the cylinder head. If the alignment pins cannot be inserted, turn the

79222812

Install the camshaft support fixture to securely hold the camshafts in position—1.9L (DOHC) engines

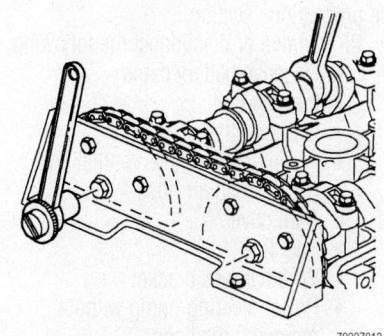

79222813

Hold the camshaft with a wrench while tightening the sprocket bolts—1.9L (DOHC) engine shown

```
Intake Side
9    5      6    10
      3   1   2
8    4    7    11
Exhaust Side
```

9356ZG01

Camshaft cover bolt tightening sequence—1.9L DOHC engines

crankshaft 360 degrees clockwise and repeat. If the pins cannot be inserted within 1–2 degrees of either TDC position, the camshafts are not properly timed. Do not start the engine until the camshafts are timed.

23. Apply a small drop of RTV across the cylinder head and front cover T-joints. Inspect the old camshaft cover gasket and replace if damaged. Install the gasket and the camshaft cover. Torque the fasteners in proper sequence to 89 inch lbs. (10 Nm).

24. Install or connect the following:
- Right splash shield and wheel, if removed to observe the timing marks
- PCV and fresh air hoses and the spark plug wires
- Accessory drive belt
- Negative battery cable

25. Start the engine and check for leaks, repair if necessary.

L–Series With 2.2L Engine

➡ Be very careful when working around the camshaft sprockets and timing chain cover during this procedure. If a bolt or washer is accidentally dropped between the front cover and engine assembly, the cover will have to be removed for retrieval.

1. Before servicing the vehicle, refer to the precautions section.

2. Remove or disconnect the following:
- Negative battery cable
- Ignition coil
- Ground strap
- Positive Crankcase Ventilation (PCV) fresh air hose
- Cam cover
- Fuel line
- Wire harness bracket
- Power steering pump without removing the lines

➡ To avoid valve piston contact, the No. 1 cylinder piston must be positioned at 60 degrees Before Top Dead

Center (BTDC). The pistons are properly aligned when the diamond shaped hole on the intake camshaft sprocket is located at 12 o'clock.

3. Remove the upper timing chain guide and front camshaft caps.

4. Install a camshaft sprocket holding tool through the sprocket holes from the timing chain side. Align the guide pins into the slot on the support head. Torque the pins to 89 inch lbs. (10 Nm).

5. Hold each camshaft in place with a 24mm open end wrench and remove the camshaft timing sprocket retaining bolts and washers. Discard the bolts.

6. Uniformly loosen and remove the remaining camshaft bearing caps.

7. Slide the camshaft sprockets away from the camshafts and remove the camshaft.

To install:

8. Lubricate the camshaft bearing journals with clean engine oil.

9. Install both camshafts and all bearing caps except the front cap on each camshaft.

10. Torque the bearing caps uniformly, except for the front caps and the rear intake cap, to 89 inch lbs. (10 Nm).

➡ Make certain that the alignment notches are properly positioned with the notches in the camshaft sprockets

before final torque is applied. Also, be sure that the timing chain is properly aligned on the fixed guide.

11. Slide the camshaft sprockets and timing chain on the guide pins toward the camshafts. Rotate the camshafts with a 24mm open end wrench to align the camshaft and sprocket.

12. Install new camshaft sprocket bolts and torque them to 63 ft. lbs. (85 Nm) plus 30 degrees.

13. Remove the camshaft sprocket holding tool.

14. Install or connect the following:
- Front camshaft bearing caps and torque the bolts to 89 inch lbs. (10 Nm)
- Upper timing chain guide and apply Loctite® to the bolts
- Rear intake camshaft bearing cap and torque the bolts 19 ft. lbs. (25 Nm)
- Power steering pump and torque the bolts to 19 ft. lbs. (25 Nm)
- Ignition coil
- PCV fresh air hose
- Ground strap
- Fuel line
- Wire harness bracket
- Cam cover
- Negative battery cable

15. Start the vehicle and check for leaks, repair if necessary.

9346ZG18

Remove the camshaft and bearing caps—L–Series With 2.2L engine

ION

2.0L ENGINE

1. Before servicing the vehicle, refer to the precautions section.

2. Disconnect the negative battery cable.

3. Remove the ignition coils.

4. Remove the ground strap and stud.

5. Disconnect the Positive Crankcase Ventilation (PCV) hose from the cam cover.

6. Disconnect the fuel feed pipe from the fuel rail.

7. Remove the camshaft cover bolts.

8. Remove the camshaft cover.

9. Remove the upper timing chain guide.

10. Install camshaft sprocket holding tool J 43655.

11. Remove both the intake and exhaust camshaft sprocket bolts and discard.

12. Slide the camshaft sprockets forward.

13. Mark the caps to ensure they are installed in the original position.

➡**Remove each bolt on each cap one turn at a time until there is no spring tension on the camshaft.**

14. Remove the caps.

15. Remove the camshaft.

16. Remove the camshaft roller followers.

17. Remove the hydraulic lash adjusters.

To install:

18. Lubricate the valve tips.

19. Install the hydraulic lash adjusters.

20. Install the camshaft roller followers.

21. Ensure that the alignment notches are aligned with the camshaft sprocket.

22. Install the intake camshaft.

23. Install the camshaft caps.

24. Tighten the caps in increments of 3 turns until they are seated, then tighten to 89 inch lbs. (10 Nm).

25. Apply 0.138 inch (3.55mm) of Permatex Anaerobic Gasket maker® 51813 to the rear camshaft bearing cap. Install the cap and tighten the bolts to 18 ft. lbs. (25 Nm).

26. Install camshaft sprockets onto the camshafts.

27. Hand tighten NEW camshaft sprocket bolts.

28. Remove camshaft sprocket holding tool J43655

29. Tighten the camshaft sprocket bolts and tighten to 63 ft. lbs. (85 Nm) plus and additional 30 degrees

30. Install the upper timing chain guide and tighten to 89 inch lbs. (10 Nm).

31. Install the camshaft cover and bolts and tighten to 89 inch lbs. (10 Nm).

32. Install the ground strap to the camshaft cover and tighten to 89 inch lbs. (10 Nm).

33. Install the ignition coils .

34. Connect the fuel feed pipe to the fuel rail.

35. Install the fuel pipe bracket and tighten to 89 inch lbs. (10 Nm).

36. Connect the PCV hose to the cam cover

37. Connect the negative battery cable.

2.2L ENGINE

1. Before servicing the vehicle, refer to the precautions section.

2. Remove or disconnect the following:

- Negative battery cable
- Air cleaner assembly
- Accelerator cable from the throttle body and bracket
- Accelerator bracket bolts and the bracket
- Positive Crankcase Ventilation (PCV) hose
- Fuel line bracket
- Ignition Coil Module (ICM) connector

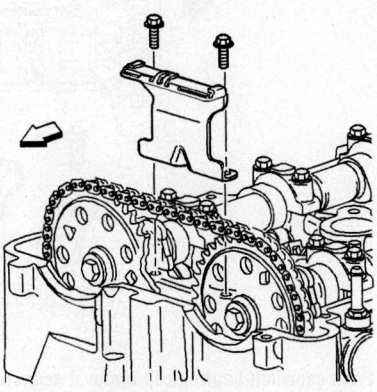

Remove the upper timing chain guide—ION

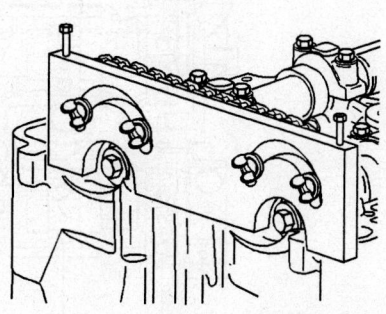

Install camshaft sprocket holding tool J43655–ION

- ICM screws and the ICM
- Ignition coil housing bolts and the housing
- Ground strap stud from the camshaft cover
- Ground strap
- Camshaft cover bolts
- Camshaft cover
- Upper timing chain guide

3. Install camshaft sprocket holding tool J43655 as illustrated.

- Intake and exhaust camshaft sprocket bolts and discard the bolts

4. Slide the sprockets forward and mark the camshaft bearing caps to ensure they are installed in their correct positions.

5. Remove each bearing cap bolt one turn at a time until there is no tension.

6. Remove the bearing cap and the camshaft

- Roller followers
- Lash adjusters

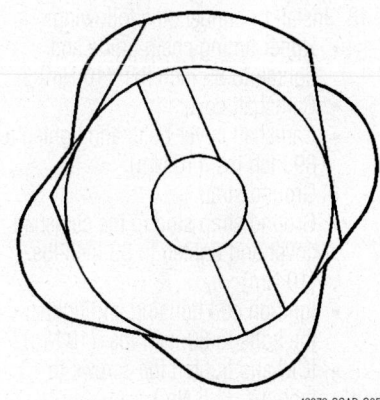

Make sure the alignment notches are aligned with the camshaft sprocket–ION

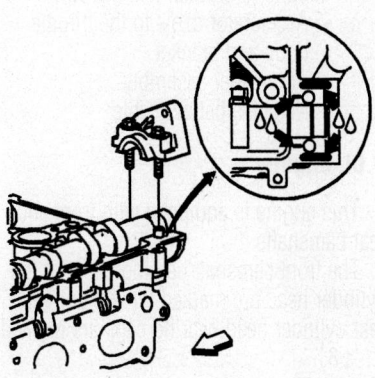

Apply 0.197 inch (5mm) on 2.2L engines or 0.138 inch (3.55mm) of anaerobic Gasket maker to the rear camshaft bearing cap—ION

To install:

7. Lubricate the valve tips.
8. Install or connect the following:
   - Lash adjusters
   - Roller followers
9. Make sure the alignment notches are aligned with the camshaft sprocket as illustrated.
   - Camshaft
   - Bearing caps except the rear camshaft cap
10. Tighten the caps in increments of 3 turns until they are seated, then tighten to 89 inch lbs. (10 Nm).
11. Apply 0.197 inch (5mm) of Permatex Anaerobic Gasket maker® 51813 to the rear camshaft bearing cap. Install the cap and tighten the bolts to 18 ft. lbs. (25 Nm).
   - Camshaft sprockets and hand tighten the bolts
12. Remove camshaft sprocket holding tool J43655
   - Camshaft sprocket bolts to 63 ft. lbs. (85 Nm) plus and additional 30 degrees
13. Install or connect the following:
   - Upper timing chain guide and tighten to 89 inch lbs. (10 Nm)
   - Camshaft cover
   - Camshaft cover bolts and tighten to 89 inch lbs. (10 Nm)
   - Ground strap
   - Ground strap stud to the camshaft cover and tighten to 89 inch lbs. (10 Nm)
   - Ignition coil housing and tighten the bolts to 89 inch lbs. (10 Nm)
   - ICM and tighten the screws to 13 inch lbs. (1.5 Nm)
   - ICM connector
   - Fuel line bracket
   - PCV hose
   - Accelerator bracket and tighten the bolts to 89 inch lbs. (10 Nm)
   - Accelerator cable to the throttle body and bracket
   - Air cleaner assembly
   - Negative battery cable

3.0L Engine

This engine is equipped with front and rear camshafts.

The front camshaft bearing caps for the cylinder head are marked R1–R8 and the rear cylinder head bearing caps are marked L1–L8.

➡Be very careful when working around the camshaft sprockets and timing chain cover during this procedure. If a bolt or washer is accidentally dropped between the front cover and engine assembly, the cover will have to be removed for retrieval.

1. Before servicing the vehicle, refer to the precautions section.
2. Remove or disconnect the following:
   - Negative battery cable
   - Intake plenum
   - Air cleaner assembly
   - Camshaft cover
   - Front timing belt cover
   - Timing belt

➡Rotate the crankshaft counterclockwise to 60 degrees Before Top Dead Center (BTDC) to prevent valve to piston contact.

3. Install a Camshaft Gear Locking Tool into the camshaft gears.
4. Remove or disconnect the following:
   - Loosen the camshaft gear bolt, remove the holding tool
   - Camshaft gear bolt and discard it
   - Camshaft gear

   - Loosen the camshaft bearing caps sequentially starting in the center and working outward in a spiral direction
   - Camshaft bearing caps
   - Camshaft

➡The bearing caps for the front camshaft bearing caps are marked with an R followed by a number and the rear caps marked with an L followed by a number.

To install:

5. Lubricate all bearing surfaces with clean engine oil.
6. Install or connect the following:
   - Camshaft to the cylinder head and make sure that the pin on the exhaust camshaft is in the 12 o'clock position or that the pin on the intake camshaft is in the 7 o'clock position
   - Camshaft bearing caps in their proper position and torque the

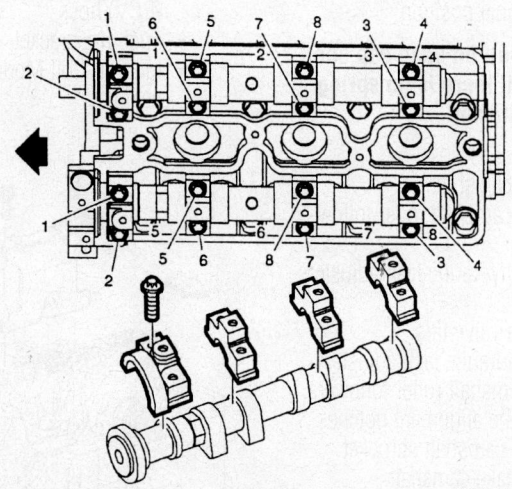

Front camshaft bearing cap removal sequence—3.0L engine

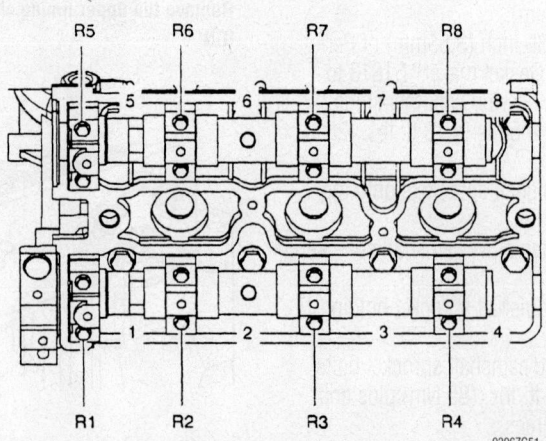

The rear camshaft bearing caps are marked to ensure proper installation—3.0L engine

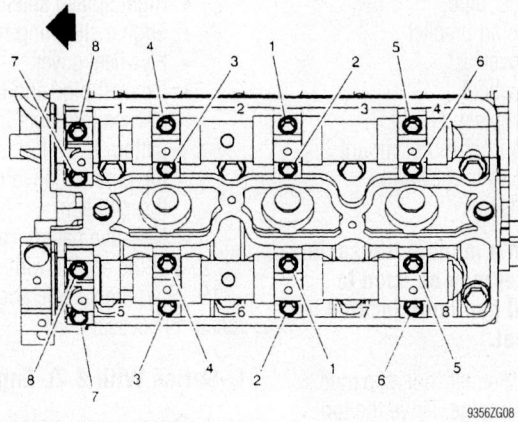

Front and rear camshaft bearing cap installation sequence—3.0L engine

bolts, starting in the center and working outward, to 71 inch lbs. (8 Nm)

- New camshaft seal lubricated with clean engine oil
- Camshaft gear with a new bolt and torque the bolt to 27 ft. lbs. (50 Nm) plus 60 degrees and an additional 15 degrees
- Timing belt and adjust as needed
- Timing belt cover
- Camshaft cover
- Intake plenum
- Air cleaner assembly
- Negative battery cable

Supercharger

REMOVAL & INSTALLATION

ION

2.0L ENGINE

1. Before servicing the vehicle, refer to the precautions section.
2. Disconnect the negative battery cable.
3. Remove the drive belt.
4. Remove the Evaporative Emission (EVAP) tube and EVAP valve.
5. Remove the throttle body and gasket.
6. Remove the Supercharger Inlet Pressure (SCIP) sensor.
7. Disconnect the vacuum brake booster hose.
8. Remove the intercooler fill neck bracket bolts.
9. Remove the supercharger.
10. Remove the gasket if damaged.

To install:

11. Install the supercharger gasket.
12. Install the supercharger and bolts. Tighten the bolts to 18 ft. lbs. (25 Nm).
13. Install the intercooler fill neck

bracket bolts and tighten to 89 inch lbs. (10 Nm).

14. Connect the vacuum brake booster hose.
15. Install the SCIP sensor and bolt and tighten to 89 inch lbs. (10 Nm).
16. Install the throttle body.
17. Install the EVAP tube and EVAP valve.
18. Install the drive belt.
19. Connect the negative battery cable.

Valve Lash

ADJUSTMENT

All engines utilize hydraulic lash adjusters; no adjustment is necessary.

Starter Motor

REMOVAL & INSTALLATION

1.9L and L–Series With 2.2L Engines

1. Before servicing the vehicle, refer to the precautions section.
2. Remove or disconnect the following:

- Negative battery cable

Starter assembly removal—1.9L engines

➡Spray the starter solenoid electrical connectors with penetrating oil before removal.

- Starter electrical connections
- Starter bolts
- Starter assembly by pulling it toward the left side of the vehicle

To install:

3. Install or connect the following:

- Starter to the flywheel housing and torque the bolts to 30 ft. lbs. (40 Nm)
- Starter electrical connectors and torque the solenoid ignition wire to 44 inch lbs. (5 Nm) and the positive battery cable to 89 inch lbs. (10 Nm)
- Negative battery cable

ION

2.0L ENGINE

1. Before servicing the vehicle, refer to the precautions section.
2. Disconnect the negative battery cable.
3. Remove the intercooler pump outer bracket.
4. Disconnect the electrical connectors from the starter.
5. Remove the starter bolts.
6. Remove the starter.

To install:

7. Install the starter.
8. Install the starter bolts and tighten to 37 ft. lbs. (50 Nm).
9. Connect the electrical connectors to the starter.
10. Install the intercooler pump outer bracket and tighten to 15 ft. lbs. (20 Nm).
11. Connect the negative battery cable

2.2L ENGINE

1. Before servicing the vehicle, refer to the precautions section.
2. Remove or disconnect the following:

- Negative battery cable

➡Spray the starter solenoid electrical connectors with penetrating oil before removal.

- Starter electrical connections
- Starter bolts
- Starter assembly

To install:

3. Install or connect the following:

- Starter to the flywheel housing and torque the bolts to 30 ft. lbs. (40 Nm)
- Starter electrical connectors and torque the solenoid ignition wire to

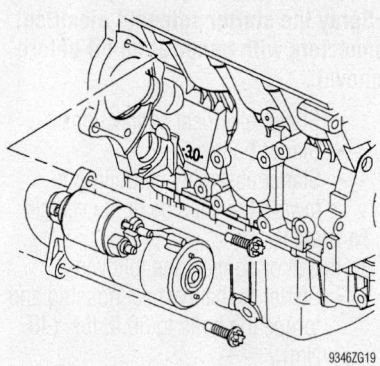

Starter motor mounting—3.0L engine

44 inch lbs. (5 Nm) and the positive battery cable to 89 inch lbs. (10 Nm), on all models except ION. On ION models, attach the electrical connectors to the starter and tighten the battery terminal nut to 13 ft. lbs. (17 Nm) and the S terminal connector to 27 inch lbs. (3 Nm).

- Negative battery cable

3.0L Engine

1. Before servicing the vehicle, refer to the precautions section.
2. Remove or disconnect the following:
   - Negative battery cable
   - Right front wheel
   - Starter electrical connections
   - Loosen the fastener securing the electrical harness bracket to the engine
   - Starter bolts
   - Starter assembly by pulling it toward the left side of the vehicle

To install:

3. Install or connect the following:
   - Starter to the flywheel housing and torque the bolts to 30 ft. lbs. (40 Nm)
   - Starter electrical connections and tighten the electrical harness bracket bolt
   - Right front wheel
   - Negative battery cable

Oil Pan

REMOVAL & INSTALLATION

1.9L Engines

1. Before servicing the vehicle, refer to the precautions section.
2. Drain the oil from the engine.
3. Remove or disconnect the following:
   - Negative battery cable

- Front exhaust pipe
- Front stiffening bracket
- Flywheel cover
- Right front wheel
- Wheel well splash shield

4. Loosen the 4 front motor mount bolts approximately ½ in. (12mm).
   - Oil pan bolts

➡**If equipped with a manual transaxle, an 8mm flex socket may be used to access the rear oil pan bolts located next to the flywheel.**

5. Using an RTV cutter tool, separate the oil pan from the engine. Drive the tool around the pan to shear the RTV seam, then tap the pan sideways with a rubber mallet to loosen.
6. Pry the engine mount away from the engine as necessary and remove the oil pan. Be careful not to damage or score component surfaces when prying.

To install:

7. Apply a 0.16 in. (4mm) bead of RTV sealer to the pan flange. Be sure the RTV is applied to the inner side of the flange from the bolt holes.
8. Install or connect the following:
   - Oil pan and torque the bolts to 80 inch lbs. (9 Nm)
   - Front engine mount bolts and torque them to 37 ft. lbs. (50 Nm)

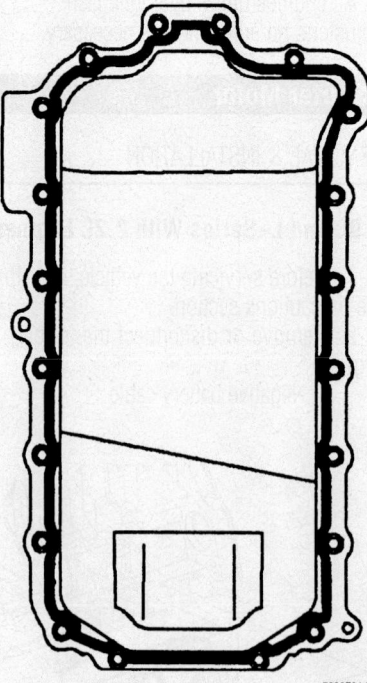

Apply a 0.16 in (4mm) bead of RTV to the oil pan flange to the inner side of the bolt holes—1.9L engines

- Right splash shield and wheel
- Engine stiffening bracket
- Flywheel cover
- Exhaust pipe and torque the pipe-to-manifold nuts in a crosswise pattern to 23 ft. lbs. (31 Nm) and the pipe to converter bolts to 33 ft. lbs. (45 Nm)
- Negative battery cable

9. Fill the engine with clean oil.
10. Start the vehicle and check for leaks, repair if necessary.

L–Series With 2.2L Engine

1. Before servicing the vehicle, refer to the precautions section.
2. Drain the engine oil.
3. Remove or disconnect the following:
   - Negative battery cable
   - Oil pan bolts

4. Using a flat-bladed tool, pry the oil pan from the engine block.

To install:

5. Apply a 0.08 in. (2mm) bead of RTV sealer to the pan flange. Be sure the RTV is applied to the inner side of the flange.
6. Install or connect the following:
   - Oil pan and torque the bolts in the proper sequence to 18 ft. lbs. (25 Nm)
   - Negative battery cable

7. Fill the engine with clean oil.
8. Start the vehicle and check for leaks, repair if necessary.

ION

2.0L ENGINE

1. Before servicing the vehicle, refer to the precautions section.
2. Disconnect the negative battery cable.
3. Drain the engine oil.
4. Remove the engine drive belt.
5. Remove the intercooler pump bracket bolts from the oil pan.
6. Remove the lower AC compressor bolt from the oil pan.
7. Remove the oil pan bolts.
8. Remove the oil pan.

To install:

9. Make sure that the oil pan and mounting surface on the lower crankcase are free of all oil and debris.
10. Apply a 0.138 inch (3.5 mm) bead of RTV sealer around the perimeter of the oil pan and the oil suction port opening.
11. Install the oil pan.
12. Install the oil pan bolts and tighten in sequence to 18 ft. lbs. (25 Nm).
13. Install the AC compressor bolts.

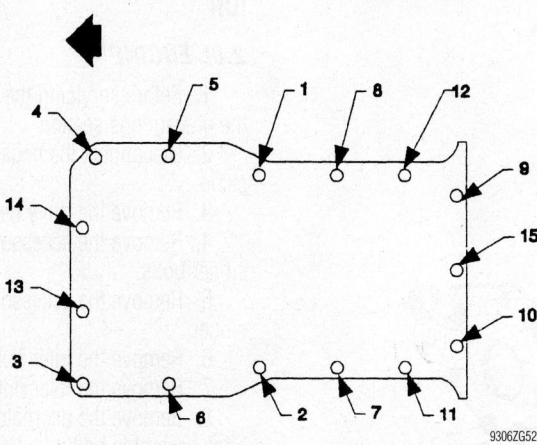

Oil pan bolts torque sequence—2.0L and 2.2L engines

14. Install the intercooler pump bracket bolts and tighten to 18 ft. lbs. (25 Nm).
15. Install the engine drive belt.
16. Fill the engine oil to the proper level.
17. Connect the negative battery cable.

2.2L ENGINE

1. Before servicing the vehicle, refer to the precautions section.
2. Drain the engine oil.
3. Remove or disconnect the following:
   • Negative battery cable
   • Drive belt
   • Lower A/C compressor bolt
   • Oil pan bolts
4. Using a flat-bladed tool, pry the oil pan from the engine block.
   To install:
5. Apply a 0.08 in. (2mm) bead of RTV sealer to the pan flange. Be sure the RTV is applied to the inner side of the flange.
6. Install or connect the following:
   • Oil pan and torque the bolts in the proper sequence to 18 ft. lbs. (25 Nm)
   • Lower A/C compressor bolt
   • Drive belt
   • Negative battery cable
7. Fill the engine with clean oil.
8. Start the vehicle and check for leaks, repair if necessary.

3.0L Engine

1. Before servicing the vehicle, refer to the precautions section.
2. Drain the engine oil.
3. Remove or disconnect the following:
   • Negative battery cable
   • Nose cone bracket bolts from the oil pan
   • Lower transaxle flange-to-oil pan bolts
   • Oil pan bolts

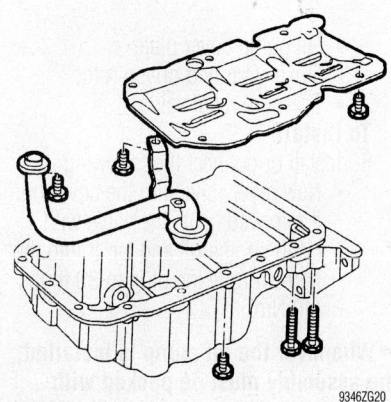

Oil pan and related components—3.0L engine

➡**Separate the oil pan from the engine with an RTV cutter tool. Drive the tool around the pan to shear the RTV seam, then tap the pan sideways with a rubber mallet to loosen.**

   • Oil pan
   To install:
4. Apply a 0.10 in. (2mm) bead of RTV sealer to the pan flange. Be sure the RTV is applied to the inner side of the flange.
5. Apply Loctite® 242 to the oil pan bolts prior to installation.
6. Install or connect the following:
   • Oil pan and torque the bolts in the proper sequence to 11 ft. lbs. (15 Nm)
   • Transaxle nose cone bracket bolts and torque them to 30 ft. lbs. (40 Nm)
   • Transaxle-to-oil pan bolts and torque them to 48 ft. lbs. (65 Nm)
7. Connect the negative battery cable.
8. Fill the engine with clean oil.
9. Start the vehicle and check for leaks, repair if necessary.

Oil Pump

REMOVAL & INSTALLATION

1.9L Engines

1. Before servicing the vehicle, refer to the precautions section.
2. Drain the engine oil.
3. Remove or disconnect the following:
   • Negative battery cable
   • Right front wheel and splash shield
   • Accessory drive belt
   • Front crankshaft vibration damper/pulley
   • Drive belt idler pulley
   • Power steering pump, SOHC only
   • Belt tensioner
   • Camshaft cover, DOHC only
   • Right side engine mount to front cover nuts and the engine mount to midrail bracket nuts
   • Front 4 oil pan bolts
   • Front cover bolts
   • Drive rotor and driven rotor
4. If necessary, remove the relief valve. Because the puller jaws will damage the relief valve sealing seat, the valve cannot be used again when removed.
   To install:
5. Install or connect the following:
   • New relief valve into the cover bore, if removed. Coat the valve with clean engine oil and tap it into the bore.

➡**Whenever the oil pump is installed, the assembly must be packed with petroleum jelly in order to prime the pump.**

   • Drive and driven rotors into the pump with the chamfer toward the front oil seal
   • Oil pump body cover using new bolts and torque them to 97 inch lbs. (11 Nm)
   • New oil pressure and suction seals in the cylinder block
   • Front oil pan bolts and torque them to 80 inch lbs. (9 Nm)
   • Front cover and torque the perimeter and center bolts to 19 ft. lbs. (25 Nm) and the lower center bolt to 89 inch lbs. (10 Nm)
   • Front oil pan bolts and torque them to 80 inch lbs. (9 Nm)
   • Camshaft cover
   • Drive belt tensioner and torque the fasteners to 26 ft. lbs. (35 Nm)
   • Power steering pump, if removed
   • Drive belt and torque the pulley bolts to 33 ft. lbs. (45 Nm)

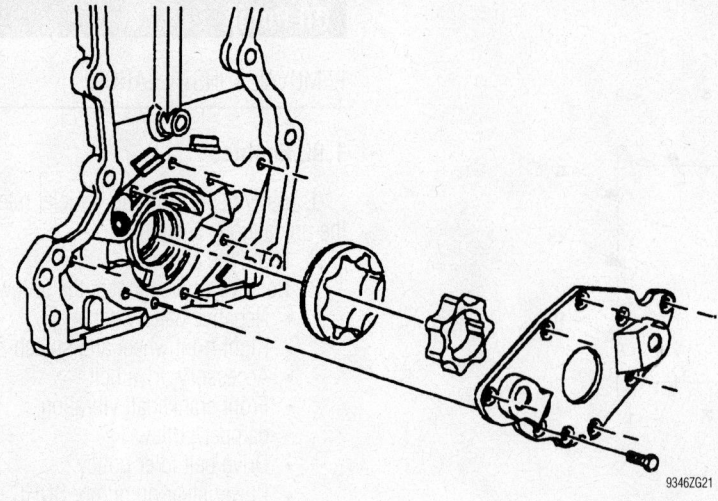

Oil pump drive rotor and driven rotor—1.9L engines

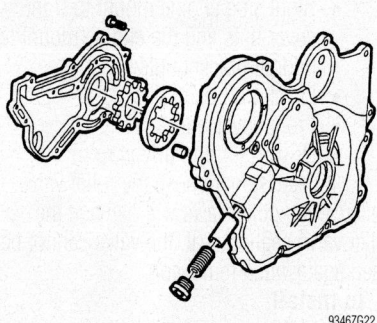

Front cover and oil pump assembly–2.2L engines

- Engine mount to midrail bracket nuts first and the engine mount to front cover nuts last and torque all nuts to 37 ft. lbs. (50 Nm)
- Right front splash shield and wheel
- Negative battery cable

6. Fill the engine with clean oil and replace the oil filter.

7. Start the vehicle and check for leaks, repair if necessary.

L–Series With 2.2L Engine

1. Before servicing the vehicle, refer to the precautions section.

2. Drain the engine oil.

3. Remove or disconnect the following:
- Negative battery cable
- Air cleaner assembly
- Right front wheel and splash shield
- Accessory drive belt
- Crankshaft damper pulley
- Belt tensioner

4. Install an engine support fixture.
- Right front engine mount
- Front cover bolts and the 13mm bolt under the water pump
- Front cover

- Oil pump cover plate
- Drive rotor and driven rotor
- Pressure relief valve

To install:

5. Install or connect the following:
- New relief valve into the cover bore, if removed. Coat the valve with clean engine oil and tap it into the bore. Torque the plug to 30 ft. lbs. (40 Nm).

➡**Whenever the oil pump is installed, the assembly must be packed with petroleum jelly in order to prime the pump.**

- Drive and driven rotors into the pump with the chamfer toward the front oil seal
- Oil pump body cover using new bolts and torque the bolts to 53 inch lbs. (6 Nm)
- Front cover with a new oil seal and torque the perimeter and center bolts to 19 ft. lbs. (25 Nm) and the lower center bolt to 89 inch lbs. (10 Nm)
- Right side engine mount and torque the bolts to 41 ft. lbs. (55 Nm)

6. Remove the engine support fixture.
- Drive belt tensioner and torque the bolts 37 ft. lbs. (50 Nm)
- Crankshaft damper pulley and torque the bolt to 74 ft. lbs. (100 Nm) plus 75 degrees
- Accessory drive belt
- Right front splash shield and wheel
- Air cleaner assembly
- Negative battery cable

7. Fill the engine with clean oil and replace the oil filter.

8. Start the vehicle and check for leaks, repair if necessary.

ION

2.0L ENGINE

1. Before servicing the vehicle, refer to the precautions section.

2. Disconnect the negative battery cable.

3. Remove the drive belt.

4. Remove the accessory drive belt tensioner bolts.

5. Remove the accessory drive belt tensioner.

6. Remove the idler pulley bolts.

7. Remove the idler pulley.

8. Remove the alternator bracket and lift hook assembly bolts.

9. Remove the alternator bracket and lift hook assembly.

10. Remove the engine front cover bolts.

11. Remove the long water pump bolt.

12. Remove the engine front cover and gaskets.

13. Remove the crankshaft front cover oil seal with an appropriate tool.

14. Disassemble the pressure relief valve.

15. Remove the oil pump gerotor cover and bolts.

16. Clean all of the parts in cleaning solvent. Remove varnish, sludge and dirt.

17. Inspect the oil pump for wear and scoring.

To install:

18. Lubricate all oil pump parts with engine oil.

19. Install the inner gear into the outer gear.

➡**If gears are improperly installed in the front cover, the gerotor cover will not bolt on.**

20. Install the gears together into the front cover with the hub of the center gear facing the front cover.

21. Install the oil pump gerotor cover and bolts. Tighten to 53 inch lbs. (6 Nm).

22. Install the pressure relief valve piston.

23. Install the pressure relief valve spring. Tighten the pressure relief valve plug to 30 ft. lbs. (40 Nm).

24. Install the engine front cover with a new gasket.

➡**Use the correct fastener in the correct location. Replacement fasteners must be the correct part number for that application. Fasteners requiring replacement or fasteners requiring the use of thread locking compound or sealant are identified in the service procedure. Do not use paints, lubri-**

cants, or corrosion inhibitors on fasteners or fastener joint surfaces unless specified. These coatings affect fastener torque and joint clamping force and may damage the fastener. Use the correct tightening sequence and specifications when installing fasteners in order to avoid damage to parts and systems.

25. Install the long water pump bolt. Tighten to 18 ft. lbs. (25 Nm).

26. Install the engine front cover bolts. Tighten to 18 ft. lbs. (25 Nm).

27. Install the generator bracket and lift hook assembly. Tighten the bolts to 31 ft. lbs. (42 Nm).

28. Install the idler pulley. Tighten the bolts to 16 ft. lbs. (22 Nm).

29. Install the accessory drive belt tensioner. Tighten the bolts to 24 ft. lbs. (32 Nm).

30. Install the drive belt.

31. Connect the negative battery cable.

2.2L ENGINE

1. Before servicing the vehicle, refer to the precautions section.

2. Drain the engine oil.

3. Remove or disconnect the following:
   - Negative battery cable
   - Air cleaner assembly
   - Right front wheel and splash shield
   - Accessory drive belt
   - Crankshaft damper pulley
   - Belt tensioner

4. Install an engine support fixture.
   - Right front engine mount
   - Front cover bolts and the 13mm bolt under the water pump
   - Front cover
   - Oil pump cover plate
   - Drive rotor and driven rotor
   - Pressure relief valve

To install:

5. Install or connect the following:
   - New relief valve into the cover bore, if removed. Coat the valve with clean engine oil and tap it into the bore. Torque the plug to 30 ft. lbs. (40 Nm).

➡Whenever the oil pump is installed, the assembly must be packed with petroleum jelly in order to prime the pump.

   - Drive and driven rotors into the pump with the chamfer toward the front oil seal
   - Oil pump body cover using new bolts and torque the bolts to 53 inch lbs. (6 Nm)

- Front cover with a new oil seal and torque the perimeter and center bolts to 19 ft. lbs. (25 Nm) and the lower center bolt to 89 inch lbs. (10 Nm)
- Right side engine mount and torque the bolts to 41 ft. lbs. (55 Nm)

6. Remove the engine support fixture.
   - Drive belt tensioner and torque the bolts 37 ft. lbs. (50 Nm)
   - Crankshaft damper pulley and torque the bolt to 74 ft. lbs. (100 Nm) plus 75 degrees
   - Accessory drive belt
   - Right front splash shield and wheel
   - Air cleaner assembly
   - Negative battery cable

7. Fill the engine with clean oil and replace the oil filter.

8. Start the vehicle and check for leaks, repair if necessary.

3.0L Engine

1. Before servicing the vehicle, refer to the precautions section.

2. Drain the engine oil.

3. Drain the cooling system.

4. Remove or disconnect the following:
   - Negative battery cable
   - Air cleaner assembly
   - Front timing belt cover

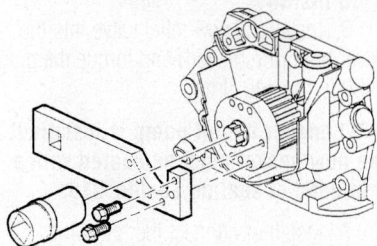

9356ZG09

Mount a crank hub holding tool to the crankshaft drive gear—3.0L engine

- Timing belt
- Rear timing belt cover
- A/C compressor and power steering pump bracket and move them away from the oil pump housing
- Alternator bolts and move the alternator out of the way
- Oil pan

5. Mount a crank hub holding tool to the crankshaft drive gear and remove the drive gear.
   - Oil pump bolts
   - Oil pan housing bolts that thread into the oil pump
   - Oil pump
   - Front main oil seal and collar
   - Pressure relief valve
   - Oil pump cover
   - Drive rotor and driven rotor

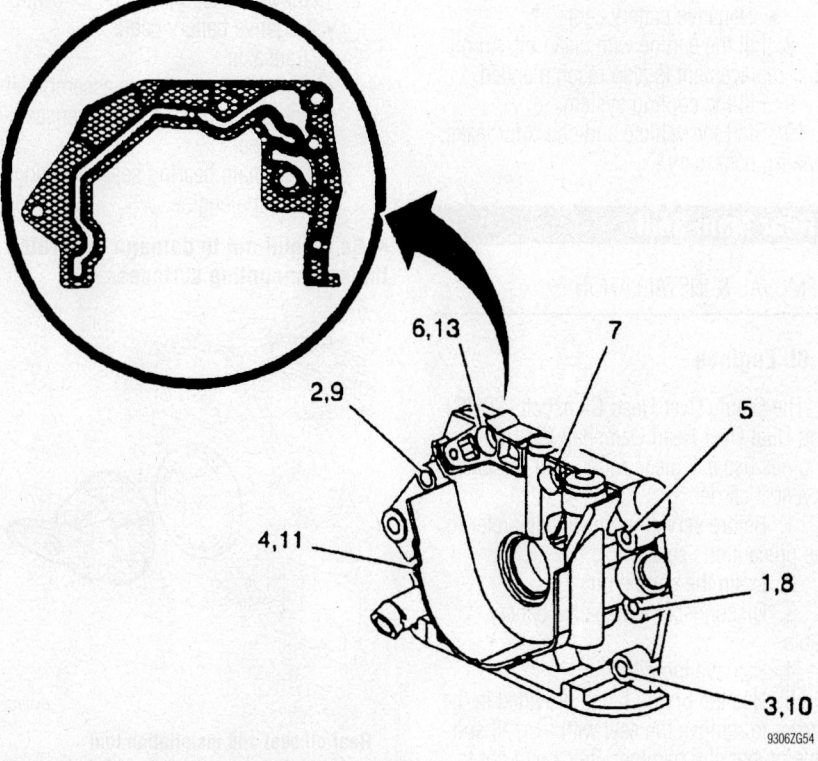

9306ZG54

Oil pump bolt tightening sequence—3.0L engine

To install:

6. Install the new relief valve into the cover bore (if removed) and torque the plug to 30 ft. lbs. (40 Nm).

➡**Whenever the oil pump is installed, the new gasket must be coated with a thin bead of sealing Loctite®518.**

7. Install or connect the following:
- Drive and driven rotors into the pump with the chamfer toward the front oil seal
- Oil pump body cover using new bolts and torque them to 89 inch lbs. (10 Nm)
- Drive gear and torque the new bolt to 184 ft. lbs. (250 Nm) plus 45 degrees then an additional 15 degrees
- Oil pan and torque the bolts to 11 ft. lbs. (15 Nm)
- Alternator and torque the bolts to 30 ft. lbs. (40 Nm)
- A/C compressor and power steering pump bracket. Torque the bolts to 30 ft. lbs. (40 Nm).
- Rear timing belt cover—Refer to section 4 for the timing belt procedure.
- Drive belt idler pulley—Refer to section 4 for the timing belt procedure.
- Timing belt
- Front cover with a new oil seal
- Air cleaner assembly
- Negative battery cable

8. Fill the engine with clean oil. An oil filter replacement is also recommended.
9. Fill the cooling system.
10. Start the vehicle and check for leaks, repair if necessary.

Rear Main Seal

REMOVAL & INSTALLATION

1.9L Engines

The Single Over Head Camshaft (SOHC) and Dual Over Head Camshaft (DOHC) engines use a 1-piece round seal mounted in a seal carrier.

1. Before servicing the vehicle, refer to the precautions section.
2. Drain the engine oil.
3. Disconnect the negative battery cable.
4. Remove the oil pan.
5. Use the prying tangs provided in the carrier to remove the seal with a small suitable prybar and hammer. Be careful not to

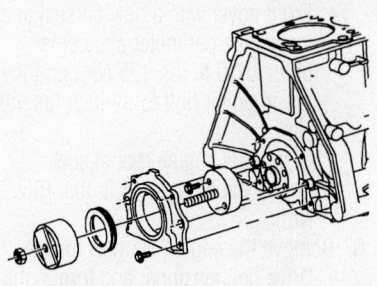

Exploded view of the rear main seal installation—1.9L engines

damage the crankshaft oil seal lip contact surface.

To install:

6. Clean the carrier and crankshaft with solvent and a rag to prevent seal lip damage during installation. Check for scores or damage to the sealing surfaces.
7. Apply a light coat of clean engine oil to the seal lip and the carrier inner diameter.
8. Install a new rear main seal, using a Seal Installer tool.
9. Install the oil pan.
10. Connect the negative battery cable.
11. Fill the engine with clean oil.
12. Start the vehicle and check for leaks, repair if necessary.

L–Series With 2.2L Engine

1. Before servicing the vehicle, refer to the precautions section.
2. Remove or disconnect the following:
- Negative battery cable
- Transaxle
- Clutch/pressure plate assembly, if equipped with a manual transaxle
- Flywheel
- Rear main bearing seal by prying it from the engine

➡**Be careful not to damage or scratch the seal mounting surfaces.**

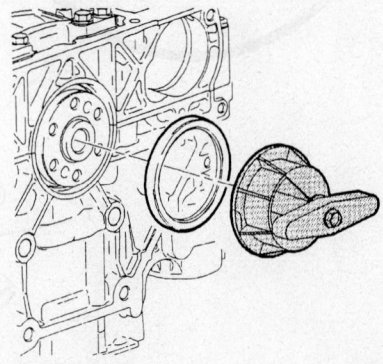

Rear oil seal and installation tool J42067–2.2L engine

To install:

3. Lubricate the new rear main bearing seal with engine oil.
4. Install or connect the following:
- New rear main seal using a Rear Main Bearing Oil Seal Installer Tool J42067 until it is flush with the block
- Flywheel
- Clutch/pressure plate assembly, if equipped with a manual transaxle
- Transaxle
- Negative battery cable

5. Start the engine and check for leaks, repair if necessary.

ION

1. Before servicing the vehicle, refer to the precautions section.
2. Remove or disconnect the following:
- Negative battery cable
- Transaxle
- Clutch/pressure plate assembly, if equipped with a manual transaxle
- Flywheel
- Rear main bearing seal by prying it from the engine

➡**Be careful not to damage or scratch the seal mounting surfaces.**

To install:

3. Lubricate the new rear main bearing seal with engine oil.
4. Install or connect the following:
- New rear main seal using a Rear Main Bearing Oil Seal Installer Tool J42067 until it is flush with the block
- Flywheel
- Clutch/pressure plate assembly, if equipped with a manual transaxle
- Transaxle
- Negative battery cable

5. Start the engine and check for leaks, repair if necessary.

3.0L Engine

1. Before servicing the vehicle, refer to the precautions section.
2. Remove or disconnect the following:
- Negative battery cable
- Transaxle
- Clutch/pressure plate assembly, if equipped with a manual transaxle
- Flywheel

3. Center punch the steel ring of the oil seal.
4. Drill a small hole into the steel ring.
5. Install a self-tapping screw and using pliers, pull out the rear main oil seal.

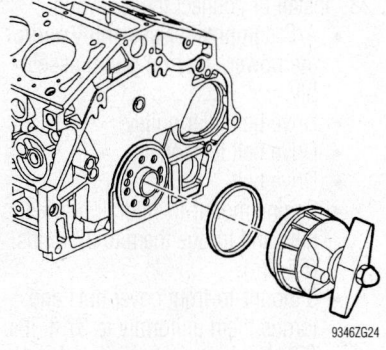

Rear main oil seal and installation tool
J42067—3.0L engine

➡ **Be careful not to damage or scratch the seal mounting surfaces.**

To install:

6. Lubricate the new rear main oil seal with engine oil.
7. Install or connect the following:
   - New rear main seal using a Rear Main Bearing Oil Seal Installer Tool SA9121E until it is flush with the block
   - Flywheel
   - Clutch/pressure plate assembly, if equipped with a manual transaxle
   - Transaxle
   - Negative battery cable
8. Start the engine and check for leaks, repair if necessary.

Timing Chain, Sprockets, Front Cover and Seal

REMOVAL & INSTALLATION

1.9L SOHC Engines

1. Before servicing the vehicle, refer to the precautions section.
2. Drain the engine oil.
3. Remove or disconnect the following:
   - Negative battery cable
   - Right front wheel and splash shield

➡ **Place a 1 x 1 x 2 in. long block of wood between the torque strut and cradle to ease removal and installation of the torque engine mount.**

   - 3 right side upper engine torque axis-to-front cover nuts and the 2 mount to midrail bracket nuts, allowing the powertrain to rest on the block of wood
   - Drive belt, tensioner and pulley
   - Power steering pump attaching bolts and set the pump to the side with the lines still attached

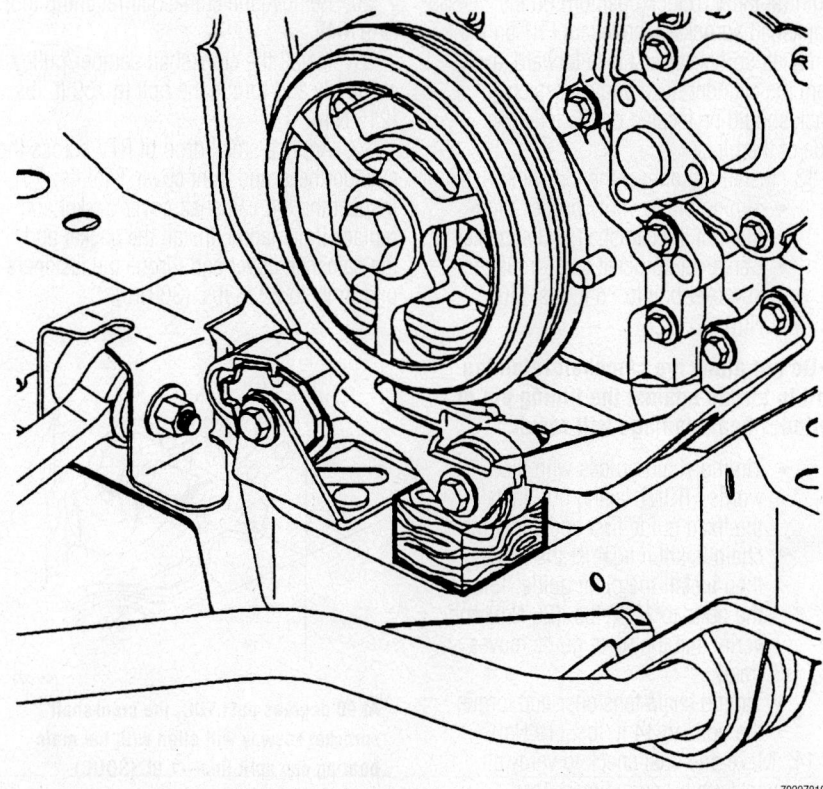

Place a 1 x 1 x 2 in. (25 x 25 x 51mm) piece of wood between the torque strut and cradle before removal of the torque engine mount—1.9L (SOHC) engines

   - A/C compressor from the bracket and set aside with the lines attached
   - Camshaft cover
   - Crankshaft damper/pulley assembly from the crankshaft
4. Install the special oil seal replacement tool SA9104E, to be sure the front crankshaft timing sprocket is held firmly in place and prevent guide damage. Install with the flat side towards the crankshaft sprocket.
5. Remove or disconnect the following:
   - Front 4 oil pan bolts and cut the seal away from the front cover
   - Front cover bolts and carefully pry the cover away from the cylinder block
   - Front cover oil seal from the cover
6. Carefully rotate the crankshaft clockwise so the timing mark on the crankshaft sprocket and keyway align with the main bearing cap split line (90 degrees past Top Dead Center [TDC]).
7. Remove or disconnect the following:
   - Timing chain guides and tensioner
   - Camshaft sprocket bolt
   - Timing chain and camshaft sprocket
   - Crankshaft sprocket

To install:

8. Inspect the chain for wear and damage. Check the inside diameter of the chain, it should be no more than 16.77 in. (426mm). Inspect the chain guides for wear or cracks and the timing sprockets for teeth or key wear. Replace components as necessary.
9. Verify that the crankshaft keyway is positioned 90 degrees clockwise past TDC (keyway at 3 o'clock). The keyway should align with the split between the bearing cap and engine block.
10. Bring the camshaft up to No. 1 TDC by loosely installing the sprocket and rotating the sprocket until the timing pin can be inserted. The camshaft contains wrench flats to assist in turning the shaft. The dowel pin should be at 12 o'clock when the camshaft is at TDC.
11. Install the crankshaft sprocket, then rotate the crankshaft counterclockwise 90 degrees up to No. 1 TDC (keyway at 12 o'clock).
12. Position the chain under the crankshaft sprocket and over the camshaft sprocket. The timing chain should be positioned so that 1 silver link plate aligns with the reference mark on the camshaft sprocket and the other aligns with the downward

tooth (at the 6 o'clock position) on the crankshaft sprocket. The letters FRT on the camshaft sprocket must face forward, away from the cylinder head and excess chain slack should be located on the tensioner side of the block.

13. Install or connect the following:
- Timing pin to verify proper alignment of the camshaft and sprocket
- Camshaft sprocket and torque the sprocket bolt to 75 ft. lbs. (102 Nm)

➡ **Do not allow the camshaft retaining bolt to torque against the timing pin or cylinder head damage will result.**

- Timing chain guides with the words FRONT facing out. Install the fixed guide first and verify the chain is snug against the guide, then install the pivot guide. Torque the bolts to 19 ft. lbs. (26 Nm) and verify that the pivot guide moves freely.
- Timing chain tensioner and torque the bolts to 14 ft. lbs. (19 Nm)

14. Make one final check to verify all components are properly timed, then remove all timing pins.

15. Install a new front cover oil seal using the installation tool with a press.

16. Apply a 0.08 in. (2mm) bead of RTV sealer along the vertical sealing surfaces of the front cover to the inside of the bolt holes and to the front of the oil pan. Extra sealer is necessary at the oil pan and cylinder head joints. Be sure to assemble the front cover to the engine within 3 minutes of RTV application.

17. Install or connect the following:
- Crankshaft sprocket retaining tool SA9104E, to align the oil pump and crankshaft during cover installation
- Front cover to the engine and torque the perimeter bolts starting at the center and working outwards on both sides to 19 ft. lbs. (25 Nm)
- Front cover center or inner bolts and torque them to 89 inch lbs. (10 Nm), except for the upper inside bolt, which should be tightened to 22 ft. lbs. (30 Nm)
- Oil pan front bolts and torque them to 80 inch lbs. (9 Nm)

18. After front cover installation, spray 6–12 squirts of oil through the front oil seal drain back hole to verify that it is not plugged.

19. Apply a thin film of RTV between the damper/pulley assembly flange and washer only; the washer and bolt head flange are designed to prevent oil leakage.

20. Remove the crankshaft retaining tool SA9104E.

21. Install the crankshaft damper/pulley assembly and torque the bolt to 159 ft. lbs. (215 Nm).

22. Apply a small drop of RTV across the cylinder head and front cover T-joints. Inspect the old camshaft cover gasket and replace if damaged. Install the gasket and the camshaft cover and torque the fasteners uniformly to 22 ft. lbs. (30 Nm).

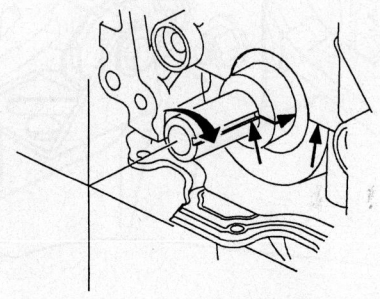

79227822

At 90 degrees past TDC, the crankshaft sprocket keyway will align with the main bearing cap split line—1.9L (SOHC) engines

23. Install or connect the following:
- A/C compressor assembly and/or the power steering pump assembly
- Drive belt idler pulley
- Drive belt tensioner
- Drive belt
- Engine mount-to-midrail bracket nuts and torque them to 37 ft. lbs. (50 Nm)
- 3 mount-to-front cover nuts and torque them uniformly to 37 ft. lbs. (50 Nm) in order to prevent front cover damage
- Splash shield and the wheel assembly
- Negative battery cable

24. Fill the engine with clean oil.

25. Start the engine and check for leaks, repair if necessary.

1.9L (DOHC) Engines

1. Before servicing the vehicle, refer to the precautions section.
2. drain the engine oil.
3. Remove or disconnect the following:
- Negative battery cable
- Right front wheel and splash shield

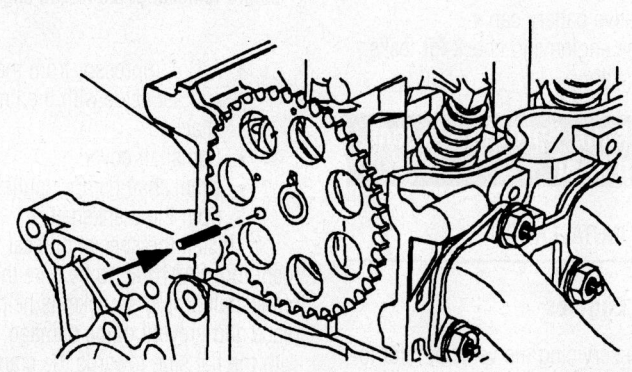

79227823

Insert a timing pin to ensure that the camshaft is at No. 1 TDC—1.9L (SOHC) engines

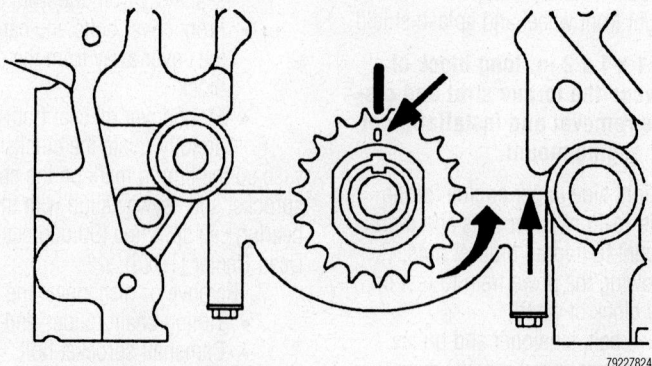

79227824

When the camshaft is at TDC, rotate the crankshaft counterclockwise 90 degrees to achieve TDC—1.9L (SOHC) engines

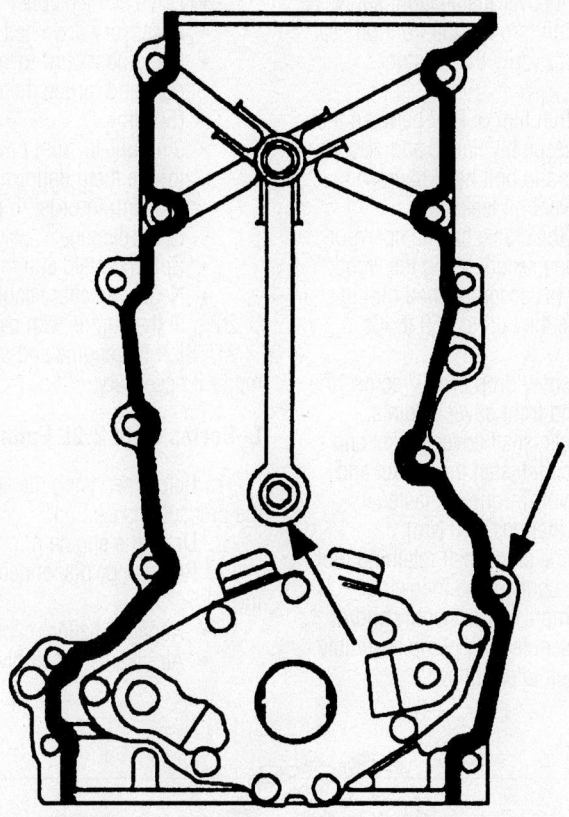

Apply a 0.08 in. (2mm) bead of RTV sealer along the vertical sealing surfaces of the front cover—1.9L (SOHC) engines

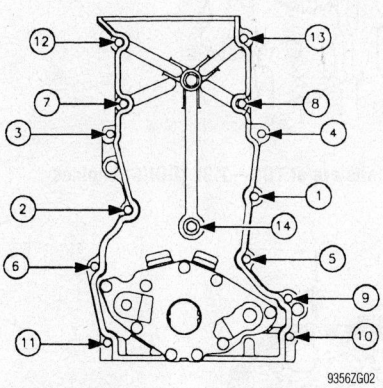

Front cover torque sequence—1.9L (SOHC) engines

➡**Place a 1 x 1 x 2 in. (25 x 25 x 51mm) block of wood between the torque strut and cradle to ease removal and installation of the torque engine mount.**

- 3 right side upper engine torque axis-to-front cover nuts and the 2 mount-to-midrail bracket nuts, allowing the powertrain to rest on the block of wood
- Accessory drive belt

- Drive belt tensioner and pulley
- Power steering pump attaching bolts and set the pump to the side with the lines still attached
- A/C compressor from the bracket and set it to the side with the lines attached
- Camshaft cover
- Crankshaft damper/pulley

4. Install an Oil Seal Replacement Tool SA9104E to be sure the front crankshaft timing sprocket is held firmly in place and prevent guide damage. Install with the flat side towards the crankshaft sprocket.

- Front 4 oil pan bolts and cut the seal away from the front cover
- Front cover bolts and carefully pry the cover away from the cylinder block
- Front cover oil seal from the cover

5. Carefully rotate the crankshaft clockwise so the timing mark on the crankshaft sprocket and keyway align with the main bearing cap split line (90 degrees past Top Dead Center [TDC]).

- Timing guides and tensioner
- Camshaft sprocket bolt

- Timing chain and camshaft sprocket
- Crankshaft sprocket

To install:

6. Inspect the chain for wear and damage. Check the inside diameter of the chain, it should be no more than 23.15 in. (588mm). Inspect the chain guides for wear or cracks and the timing sprockets for teeth or key wear. Replace components as necessary.

7. Verify that the crankshaft keyway is positioned 90 degrees clockwise past TDC (keyway at 3 o'clock). The keyway should align with the split between the bearing cap and engine block.

8. Install the crankshaft sprocket.

9. Rotate the crankshaft counterclockwise 90 degrees up to No. 1 TDC (keyway at 12 o'clock).

10. Position the chain under the crankshaft sprocket and over the camshaft sprocket. The timing chain should be positioned so that 1 silver link plate aligns with the reference mark on the camshaft sprocket and the other aligns with the downward tooth (at the 6 o'clock position) on the crankshaft sprocket. The letters FRT on the camshaft sprocket must face forward, away from the cylinder head and excess chain slack should be located on the tensioner side of the block. Torque the camshaft sprocket bolt to 75 ft. lbs. (102 Nm).

11. Verify that the crankshaft reference mark aligns with the cylinder block mark at 12 o'clock.

12. Install or connect the following:

- Timing chain fixed guide and torque the bolts to 21 ft. lbs. (28 Nm)
- Pivoting chain guide and torque the bolt to 19 ft. lbs. (26 Nm) and make certain the guide moves freely
- Two forward bearing caps and the upper timing chain guide and torque the retaining bolts to 124 inch lbs. (14 Nm)
- Timing chain tensioner and torque the bolts to 14 ft. lbs. (19 Nm)

13. Make one final check to verify all components are properly timed, then remove all timing pins.

- Seat a new front cover oil seal using the installation tool with a press
- If the engine front cover casting or assembly is replaced, the 3 torque axis mount studs should also be replaced. Torque the new studs to 19 ft. lbs. (25 Nm).

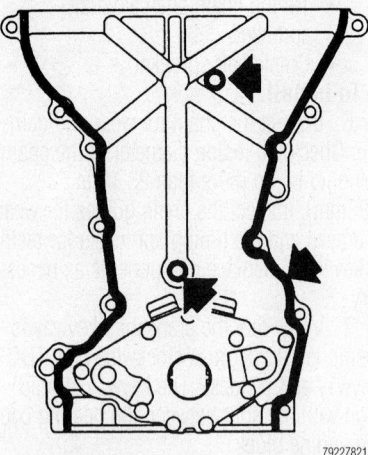

Apply a 0.08 in. (2mm) bead of RTV sealer along the vertical sealing surfaces of the front cover and the front of the oil pan—1.9L (DOHC) engines

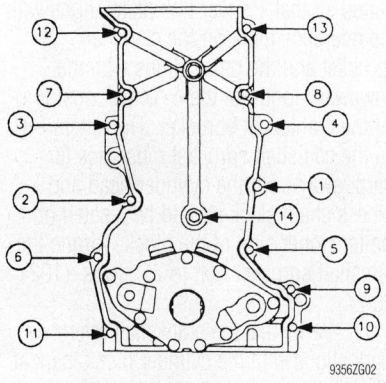

Front cover torque sequence—1.9L (DOHC) engines

➡️Apply a 0.08 in. (2mm) bead of RTV sealer along the vertical sealing surfaces of the front cover to the inside of the bolt holes and to the front of the oil pan. Extra sealer is necessary at the oil pan and cylinder head joints. Be sure to assemble the front cover to the engine within 3 minutes of RTV application.

- Crankshaft sprocket retaining tool to align the oil pump and crankshaft during cover installation
- Front cover to the engine and torque the perimeter bolts starting at the center and working outwards on both sides to 22 ft. lbs. (30 Nm)
- Front cover center or inner bolts to 89 inch lbs. (10 Nm) except for the upper inside bolt which should be tightened to 22 ft. lbs. (30 Nm)
- 4 oil pan front bolts and torque them to 80 inch lbs. (9 Nm)

14. After front cover installation, spray 6–12 squirts of oil through the front oil seal drain back hole to verify that it is not plugged.

15. Apply a thin film of RTV between the damper/pulley assembly flange and washer only; the washer and bolt head flange are designed to prevent oil leakage.

16. Position the crankshaft damper/pulley assembly, then secure using the wood or strap wrench (as accomplished during removal). Torque the bolt to 159 ft. lbs. (215 Nm).

17. Apply a small drop of RTV across the cylinder head and front cover T-joints. Inspect the old camshaft cover gasket and replace if damaged. Install the gasket and the camshaft cover. Torque the fasteners uniformly to 89 inch lbs. (10 Nm).

18. Remove the crankshaft retaining tool.

19. Install or connect the following:

- A/C compressor assembly and/or the power steering pump assembly
- Drive belt idler pulley
- Drive belt tensioner
- Accessory drive belt
- 2 engine mount-to-midrail bracket nuts and torque them to 37 ft. lbs. (50 Nm)
- 3 mount-to-front cover nuts and torque them uniformly to 37 ft. lbs. (50 Nm) in order to prevent front cover damage
- Splash shield and the wheel
- Negative battery cable

20. Fill the engine with clean oil.

21. Start the engine and check for leaks, repair if necessary.

L–Series With 2.2L Engine

1. Before servicing the vehicle, refer to the precautions section.

2. Drain the engine oil.

3. Remove or disconnect the following:

- Negative battery cable
- Air cleaner assembly

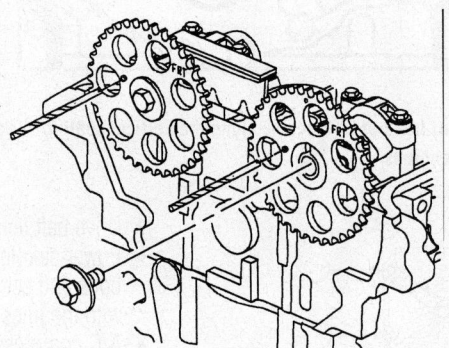

Use drill bits as timing pins to verify that the camshafts are at TDC—1.9L (DOHC) engines

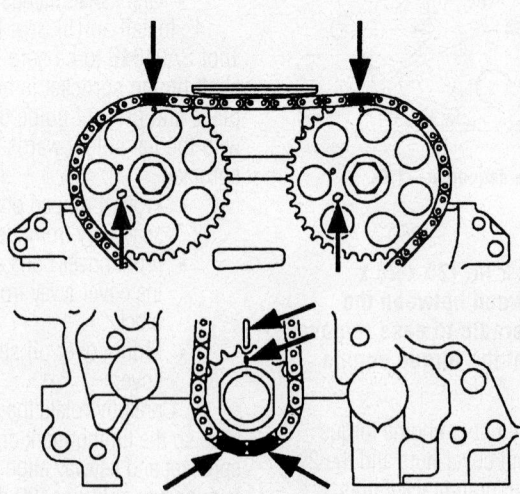

Be sure that the silver link plates and reference marks are all in alignment as shown—1.9L (DOHC) engines

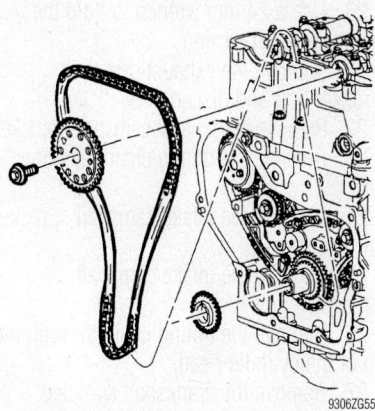

Remove the timing chain through the top of the cylinder head—L–Series with 2.2L engine

- Right front wheel and splash shield
- Accessory drive belt
- Crankshaft damper pulley
- Belt tensioner
4. Install an engine support fixture.
- Right front engine mount
- Front cover bolts and the 13mm bolt under the water pump
- Front cover
- Camshaft cover
- Timing chain tensioner
- Upper timing chain guide
- Exhaust camshaft sprocket
- Adjustable timing chain guide
- Fixed timing chain guide access plug and guide
- Intake camshaft sprocket
- Timing chain through the top of the cylinder head
- Timing chain sprocket from the crankshaft
- Timing chain oiling nozzle

To install:

5. Inspect the chain guides for wear and damage. Replace the guides if wear exceeds 0.045 inch (1.12mm). Inspect the timing chain shoe. Replace the shoe if wear exceeds 0.045 inch (1.12mm).
6. Install or connect the following:
- Timing chain sprocket to the crankshaft. Rotate the crankshaft so that the mark on the sprocket is at the 5 o'clock position.
- Timing chain to the intake camshaft sprocket alignment copper link to the "INT" diamond timing mark on the camshaft sprocket

➡**When lowering the timing chain, rotate the assembly 90 degrees to allow the chain to fall between the cylinder block bosses. Rotate the chain back so that the camshaft sprocket is facing forward.**

- Chain through the housing opening on top of the cylinder head. Make certain that the chain goes around both sides of the bosses
- Timing chain around the crankshaft sprocket and align the silver link to the timing mark
- Intake camshaft sprocket loosely on the camshaft and hand-tighten the new bolt
- Adjustable timing chain guide and torque the bolt to 89 inch lbs. (10 Nm)
- Exhaust camshaft sprocket loosely on the camshaft with the timing mark on the sprocket aligned with the silver link on the chain and hand-tighten the new bolt at this time
7. Align the camshaft sprocket to camshaft and tighten the bolt using the following procedure:
   a. Make certain that the sprocket timing mark is at the 5 o'clock position.
   b. Rotate the intake camshaft using a 24mm wrench on flats of the camshaft until the sprocket to camshaft alignment notch seats.
   c. When seated properly hand-tighten the bolt.
   d. Rotate the exhaust camshaft using a 24mm wrench on flats of the camshaft until the sprocket to camshaft alignment notch seats.
   e. When seated properly hand-tighten the bolt.
8. Verify that all colored links are aligned with the proper marks on the camshaft and crankshaft sprockets.
9. Install or connect the following:
- Fixed timing chain guide bolt and torque it to 89 inch lbs. (10 Nm)
- Fixed timing chain guide bolt access plug and torque it to 30 ft. lbs. (40 Nm)
- Upper timing chain guide and torque the bolts to 89 inch lbs. (10 Nm)
- Intake and exhaust camshaft sprocket bolts and torque them to 63 ft. lbs. (85 Nm) plus 30 degrees
- Sealing ring and tensioner assembly and torque the tensioner bolts to 44 ft. lbs. (60 Nm)
- Timing chain oiling nozzle and torque the bolt to 89 inch lbs. (10 Nm)
- Camshaft cover
- Front cover and torque the perimeter and center bolts to 19 ft. lbs. (25 Nm) and the lower center bolt to 89 inch lbs. (10 Nm)

- Front oil pan bolts and torque them to 80 inch lbs. (9 Nm)
- Camshaft cover
- Drive belt tensioner and torque the fasteners to 26 ft. lbs. (35 Nm)
- Power steering pump, if removed
- Drive belt and torque the pulley bolts to 33 ft. lbs. (45 Nm)
- Engine mount to midrail bracket nuts first and the engine mount to front cover nuts last and torque all nuts to 37 ft. lbs. (50 Nm)
- Right front splash shield and wheel
- Negative battery cable
- Negative battery cable
10. Fill the engine with clean oil.
11. Start the vehicle and check for leaks, repair if necessary.

ION

2.0L ENGINE

1. Before servicing the vehicle, refer to the precautions section.
2. Disconnect the negative battery cable.

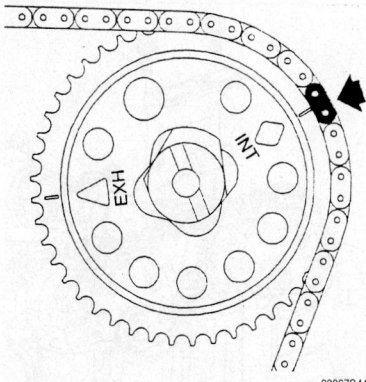

Align the copper link on the timing chain with the INT diamond timing mark—L–Series with 2.2L engine

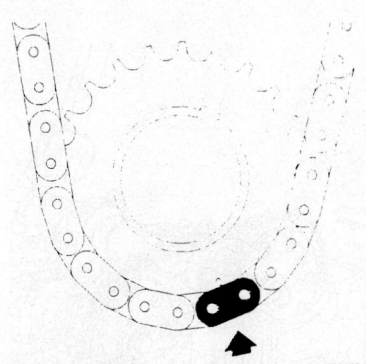

Route the timing chain around the crankshaft sprocket and align the silver link to the timing mark (5 o'clock position)—L–Series with 2.2L engines

06025-SCAR-G05

Locate the No. 1 piston to approximately 60 degrees before top dead center (diamond shaped hole on intake camshaft sprocket at 12 o'clock position)—2.0L engine

3. Remove the camshaft cover.

4. Remove the engine front cover as follows:

5. Remove the drive belt.

6. Using harmonic balancer holder J 38122-A to prevent the crankshaft from rotating and remove the balancer bolt and discard

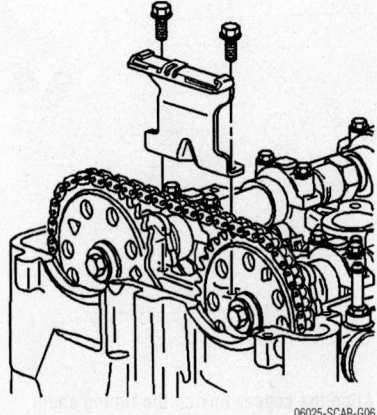

06025-SCAR-G06

Remove the fixed timing chain guide—2.0L engine

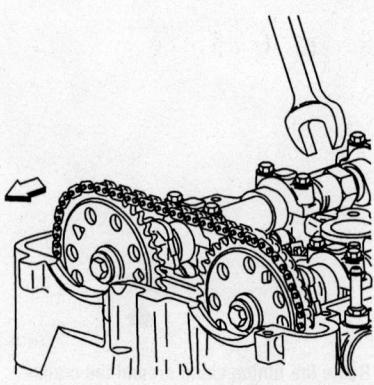

06025-SCAR-G07

Use a 24 mm wrench to hold the camshafts from turning—2.0L engine

7. Remove the crankshaft balancer.

8. Remove the drive belt tensioner.

9. Remove the idler pulley.

10. Remove the engine front cover to water pump bolt.

11. Remove the remaining engine front cover bolts.

12. Remove the engine front cover.

13. If the engine front cover gasket is damaged, remove the front engine mount.

14. Remove the front engine mount bracket

➡**To rotate the camshaft, use a 24 mm open-end wrench on the camshaft flats. Camshaft should be rotated in a clockwise direction only, facing camshaft sprockets from the passenger side of the vehicle.**

15. Locate the No. 1 piston to approximately 60 degrees before top dead center (diamond shaped hole on intake camshaft sprocket at 12 o'clock position). Remove the spark plugs. This will ease the rotation effort.

16. Remove the timing chain tensioner.

17. Remove the fixed timing chain guide access plug.

18. Remove the fixed timing chain guide.

19. Remove the upper timing chain guide.

20. Use a 24 mm wrench to hold the camshafts from turning.

21. Remove the exhaust camshaft sprocket bolt and discard.

22. Remove the exhaust camshaft sprocket.

23. Remove the timing chain tensioner guide.

24. Remove the intake camshaft sprocket bolt and discard.

25. Remove the intake camshaft sprocket.

26. Remove the timing chain through the top of the cylinder head.

27. Remove the crankshaft sprocket.

28. Remove the oil nozzle and bolt.

29. Remove the balance shaft drive chain tensioner.

30. Remove the adjustable balance shaft chain guide.

31. Remove the small balance shaft drive chain guide.

32. Remove the upper balance shaft drive chain guide.

➡**It may ease removal of the balance shaft drive chain to get all of the slack in the chain between the crankshaft and water pump sprockets.**

33. Remove the balance shaft drive chain.

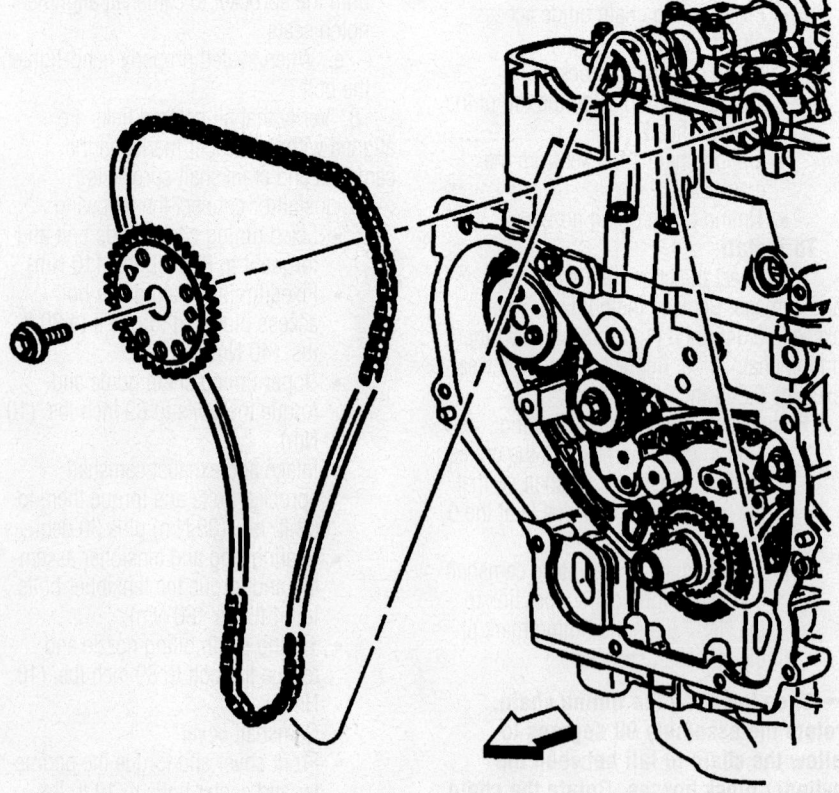

06025-SCAR-G08

Remove the timing chain through the top of the cylinder head—2.0L engine

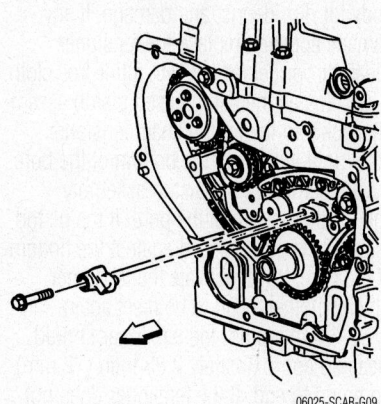

Remove the timing chain through the top of the cylinder head—2.0L engine

06025-SCAR-G09

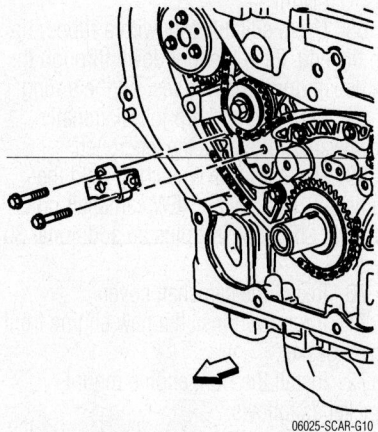

Remove the balance shaft drive chain tensioner—2.0L engine

06025-SCAR-G10

To install:

➡**If the balance shafts are not properly timed to the engine, the engine may vibrate and make noise.**

34. Install the upper balance shaft chain guide. Tighten the upper balance shaft chain guide bolts to 133 inch lbs. (15 Nm).

35. Install the balance shaft drive chain

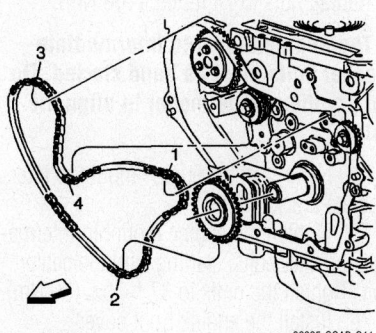

Refer to the text for the proper timing chain routing for installation—2.0L engine

06025-SCAR-G11

with the colored links lined up on with the marks on the balance shaft drive sprockets and the crankshaft sprocket. Use the following procedure to line up the links with the sprockets: Orient the chain so that the copper colored and chrome links are visible.

36. Place the uniquely colored link (1) so that it lines up with the timing mark on the intake side balance shaft sprocket.

37. Working clockwise around the chain, place the first matching colored link (2) in line with the timing mark on the crankshaft drive sprocket. (approximately 5 o'clock position on the crank sprocket).

38. Place the chain (3) on the water pump drive sprocket (alignment is not critical).

39. Align the last matching colored link (4) with the timing mark on the exhaust side balance shaft drive sprocket.

40. Install the small balance shaft chain guide. Tighten the balance shaft chain guide bolts to 89 inch lbs. (10 Nm).

41. Install the adjustable balance shaft drive chain guide. Tighten the chain guide bolts to 89 inch lbs. (10 Nm).

42. Turn the tensioner plunger 90 degrees in its bore and compress the plunger until a paper clip can be inserted through the hole in the plunger body and into hole in the tensioner plunger.

43. Install the timing chain tensioner. Tighten the chain tensioner bolts to 89 inch lbs. (10 Nm).

44. Remove the paper clip from the balance shaft drive chain tensioner.

45. Install the oil nozzle and bolt. Tighten the oil nozzle bolt to 89 inch lbs. (10 Nm).

46. Install the crankshaft sprocket with timing mark at the 5 o'clock position.

47. Lower the timing chain through the opening in the top of the cylinder head. Carefully ensure that the chain goes around both sides of the cylinder block bosses (1 and 2).

48. Install the intake camshaft sprocket with the INT diamond at the 2 o'clock position.

➡**Always install NEW sprocket bolts.**

49. Hand tighten a NEW intake camshaft sprocket bolt.

50. Route the timing chain around the crankshaft sprocket with the matching colored link aligning with the timing mark.

51. Route the timing chain around the intake camshaft sprocket with the uniquely colored link (1) aligning with the INT diamond.

52. Install the timing chain tensioner guide through the opening in the top of the

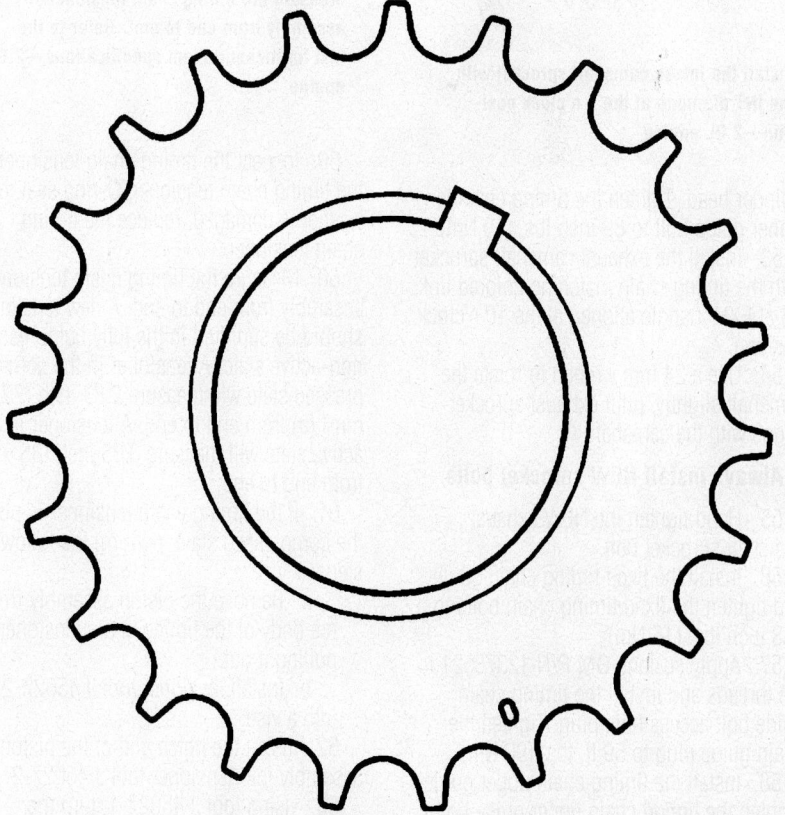

Install the crankshaft sprocket with timing mark at the 5 o'clock position—2.0L engine

06025-SCAR-G12

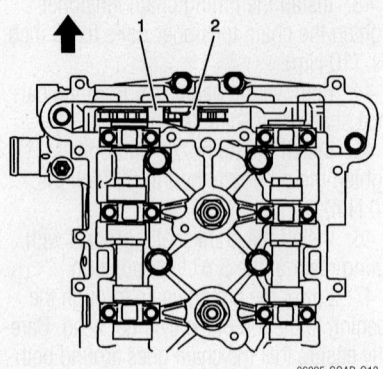

06025-SCAR-G13

Lower the timing chain through the opening in the top of the cylinder head. Carefully ensure that the chain goes around both sides of the cylinder block bosses (1 and 2)—2.0L engine

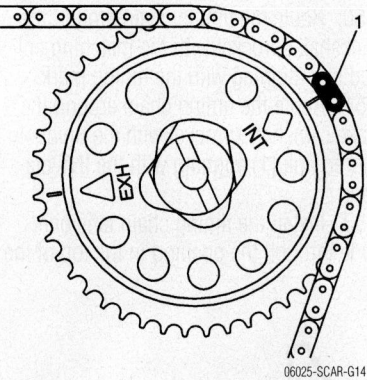

06025-SCAR-G14

Install the intake camshaft sprocket with the INT diamond at the 2 o'clock position—2.0L engine

cylinder head. Tighten the timing chain tensioner guide bolt to 89 inch lbs. (10 Nm).

53. Install the exhaust camshaft sprocket with the timing chain matching colored link (3) at EXH triangle aligned at the 10 o'clock position.

54. Use a 24 mm wrench to rotate the camshaft slightly, until exhaust sprocket aligns with the camshaft.

➡**Always install NEW sprocket bolts.**

55. Hand tighten the NEW exhaust camshaft sprocket bolt.

56. Install the fixed timing chain guide and tighten the fixed timing chain bolts to 133 inch lbs. (15 Nm)

57. Apply sealant, GM P/N 12378521 to the threads and install the timing chain guide bolt access hole plug. Tighten the chain guide plug to 59 ft. lbs. (90 Nm).

58. Install the timing chain upper guide. Tighten the timing chain upper guide bolts to 89 inch lbs. (10 Nm).

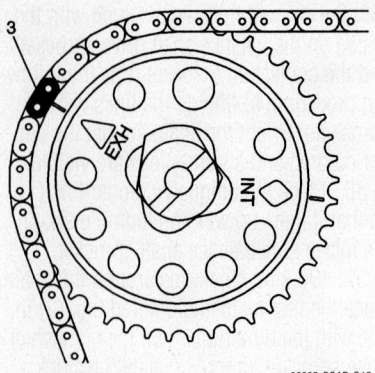

06025-SCAR-G15

Install the exhaust camshaft sprocket with the timing chain matching colored link (3) at EXH triangle aligned at the 10 o'clock position—2.0L engine

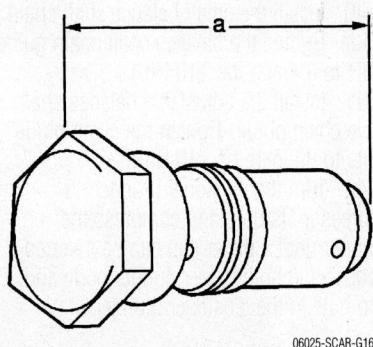

06025-SCAR-G16

Measure the timing chain tensioner assembly from end to end. Refer to the text for measurement specifications—2.0L engine

59. Inspect the timing chain tensioner. If the timing chain tensioner, O-ring seal, or washer is damaged, replace the timing chain tensioner.

60. Measure the timing chain tensioner assembly from end to end. A new tensioner should be supplied in the fully compressed non-active state. A tensioner in the compressed state will measure 2.83 inch (72 mm) (a) from end to end. A tensioner in the active state will measure 3.35 inch (85 mm) from end to end.

61. If the timing chain tensioner is not in the compressed state, perform the following steps:

a. Remove the piston assembly from the body of the timing chain tensioner by pulling it out.

b. Install tensioner tool J 45027-2 into a vise.

62. Install the notch end of the piston assembly into tensioner tool J 45027-2.

63. Using tool J 45027-1, turn the ratchet cylinder into the piston.

64. Inspect the bore of the tensioner

body for dirt, debris, and damage. If any damage appears, replace the tensioner. Clean dirt or debris out with a lint-free cloth.

65. Install the compressed piston assembly back into the timing chain tensioner body until it stops at the bottom of the bore. Do not compress the piston assembly against the bottom of the bore. If the piston assembly is compressed against the bottom of the bore, it will activate the tensioner, which will then need to be reset again.

66. At this point the tensioner should measure approximately 2.83 inch (72 mm) from end to end. If the tensioner does not read 2.83 inch (72 mm) from end to end repeat the compression steps.

67. Install the timing chain tensioner. Tighten the timing chain tensioner to 55 ft. lbs. (75 Nm).

68. Use a suitable tool with a rubber tip on the end. Feed the tool down through the camshaft drive chain to rest on the timing chain. Then give a sharp jolt diagonally downwards to release the tensioner.

69. Use a 24 mm wrench to hold the camshaft. Tighten the NEW camshaft bolts to 63 ft. Lbs. (85 Nm) plus an additional 30 degrees.

70. Install the camshaft cover.

71. If removed install a new engine front cover gasket.

72. Install the front engine mount bracket as follows:

a. Install the intermediate bracket to the engine.

b. Hand tighten the engine mount intermediate bracket bolts. Position the long bolts in the forward and front lower holes and the short bolt in the rear upper hole.

c. Tighten the intermediate bracket bolts to 74 ft. lbs. (100 Nm).

73. Install the front engine mount as follows:

a. Place the engine mount onto the midrail and hand start the nuts.

b. Tighten the engine mount to midrail nuts to 74 ft. lbs. (100 Nm).

➡**The engine mount to intermediate bracket bolts must be hand started. Do not pry the engine mount to align the holes.**

74. Hand start the engine mount to intermediate bracket bolts.

75. Tighten the engine mount to intermediate bracket bolts, starting with the center bolt. Tighten the bolts to 37 ft. lbs. (50 Nm).

76. Install the engine front cover.

77. Install the engine front cover bolts. Tighten the engine front cover bolts to 18 ft. lbs. (25 Nm).

78. Install the water pump bolt. Tighten the water pump bolt to 18 ft. lbs. (25 Nm).

79. Install the idler pulley. Tighten the bolts to 18 ft. lbs. (25 Nm).

80. Install the drive belt tensioner. Tighten the drive belt tensioner bolts to 33 ft. lbs. (45 Nm).

81. Install the crankshaft balancer.

82. Install a NEW crankshaft balancer bolt.

83. Using tool J 38122-A to prevent the crankshaft from rotating, install a new balancer bolt and tighten to 74 ft. lbs. (100 Nm) plus an additional 75 degrees

84. Install the engine drive belt.

85. Connect the negative battery cable.

2.2L ENGINE

1. Before servicing the vehicle, refer to the precautions section.

2. Drain the engine oil.

3. Remove or disconnect the following:
- Negative battery cable
- Air cleaner assembly
- Accelerator cable from the throttle body and bracket
- Accelerator bracket bolts and the bracket
- Positive Crankcase Ventilation (PCV) hose
- Fuel line bracket
- Ignition Coil Module (ICM) connector
- ICM screws and the ICM
- Ignition coil housing bolts and the housing
- Ground strap stud from the camshaft cover
- Ground strap
- Camshaft cover bolts
- Camshaft cover
- Accessory drive belt
- Using harmonic balancer holder J 38122-A to prevent the crankshaft from rotating and remove the balancer bolt and discard
- Crankshaft balancer
- Belt tensioner

4. Remove the front engine mount as follows:
   a. Support the engine with a jack and a piece of wood.
   b. Remove the mount-to-intermediate bracket bolts
   c. Remove the mount-to-midrail nuts
   d. Remove the mount.
   - Front cover bolts and the 13mm bolt under the water pump
   - Front cover

➡ **The timing chain has 2 matching colored links and 1 unique colored link.**

5. Rotate the engine until the crankshaft sprocket mark aligns with the matching colored link at the 5 o' clock position.

6. Make sure the timing chain to the intake camshaft sprocket alignment copper link to the "INT" diamond timing mark on the camshaft sprocket.

7. Make sure the "EXH" triangle on the exhaust camshaft sprocket is aligned with the colored link.

8. Remove or disconnect the following:
- Timing chain tensioner
- Fixed timing chain guide access plug and the guide
- Upper timing chain guide
- Exhaust camshaft sprocket
- Timing chain tensioner guide
- Intake camshaft sprocket
- Timing chain through the top of the cylinder head
- Timing chain sprocket from the crankshaft
- Balance shaft drive chain tensioner
- Adjustable timing chain guide
- Small balance shaft drive chain guide
- Upper balance shaft drive chain guide

➡ **To make it easier to remove the balance shaft chain, get all the slack in the chain between the crankshaft and the water pump sprockets.**

- Balance shaft drive chain

To install:

➡ **If the balance shafts are not properly aligned engine noise will occur.**

9. Install or connect the following:
- Upper balance shaft chain guide and tighten the bolts to 89 inch lbs. (10 Nm)

10. Install the balance shaft chain with the colored links lined up on the marks on

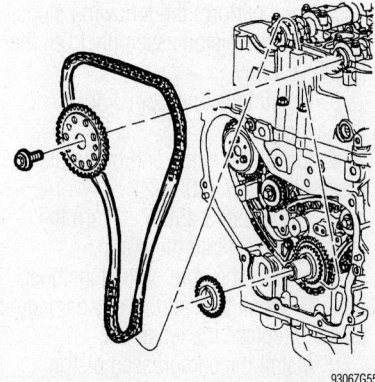

Remove the timing chain through the top of the cylinder head–Ion with 2.2L engine

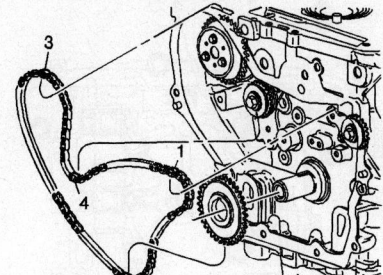

42372-SCAR-G11

Install the balance shaft chain with the colored links lined up on the marks on the balance shaft drive sprockets and crankshaft sprocket –ION with 2.2L engine

the balance shaft drive sprockets and crankshaft sprocket. Use the following to align the links with the sprockets while referring to the accompanying illustration for link location:
   a. Place the chain so that the copper colored and chrome links can be clearly seen
   b. Place link 1 (see illustration) so that it aligns with the timing mark on the intake side balance shaft sprocket.
   c. Working in a clockwise direction around the chain, place the link 2 (see illustration) in line with the timing mark on the crankshaft sprocket, this will be approximately at the 5 o'clock position.
   d. Place link 3 (see illustration) on the water pump sprocket (alignment is not critical).
   e. Align link 4 (see illustration) with the timing mark on the exhaust side balance shaft sprocket.
   f. Install the small balance shaft chain guide and tighten the bolts to 89 inch lbs. (10 Nm).
   g. Install the adjustable balance shaft drive chain guide and tighten the bolts to 89 inch lbs. (10 Nm).
   h. Turn the tensioner plunger 90 degrees in its bore and compress the plunger until a paper clip can be inserted through the hole in the plunger body and the hole in the plunger.
   i. Install the timing chain tensioner and tighten the bolts to 89 inch lbs. (10 Nm).
   j. Remove the clip from the balance shaft tensioner.

11. Inspect the chain guides for wear and damage. Replace the guides if wear exceeds 0.045 inch (1.12mm). Inspect the timing chain shoe. Replace the shoe if wear exceeds 0.045 inch (1.12mm).

12. Install or connect the following:

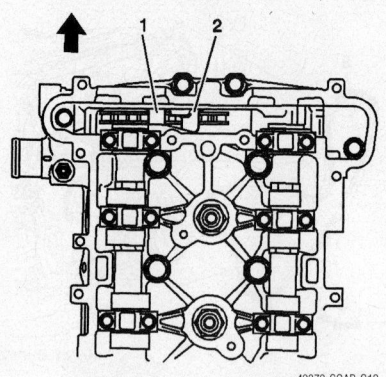

Lower the timing chain through the opening in the cylinder head. Make sure the chain goes around both sides of the cylinder block bosses–ION with 2.2L engine

- Timing chain sprocket to the crankshaft. Rotate the crankshaft so that the mark on the sprocket is at the 5 o'clock position.

13. Lower the timing chain through the opening in the cylinder head. make sure the chain goes around both sides of the cylinder block bosses (see illustration).

- Intake camshaft sprocket with the "INT" diamond at the 2 o'clock position and hand-tighten a new intake camshaft sprocket bolt
- Timing chain around the crankshaft sprocket and align the silver link to the timing mark
- Timing chain to the intake camshaft sprocket alignment copper link to the "INT" diamond timing mark on the camshaft sprocket
- Timing chain tensioner guide and torque the bolt to 89 inch lbs. (10 Nm)
- Exhaust camshaft sprocket loosely on the camshaft with the timing mark on the sprocket aligned with

Align the timing mark on the exhaust camshaft sprocket with the silver link (3) on the chain–ION with 2.2L engine

the silver link (3) on the chain at the 10 o'clock position and hand-tighten the new bolt at this time

14. Align the camshaft sprocket to camshaft and tighten the bolt using the following procedure:

a. Make certain that the sprocket timing mark is at the 5 o'clock position.

b. Rotate the intake camshaft using a 24mm wrench on flats of the camshaft until the sprocket to camshaft alignment notch seats.

c. When seated properly hand-tighten the bolt.

d. Rotate the exhaust camshaft using a 24mm wrench on flats of the camshaft until the sprocket to camshaft alignment notch seats.

e. When seated properly hand-tighten the bolt.

15. Verify that all colored links are aligned with the proper marks on the camshaft and crankshaft sprockets.

16. Install or connect the following:

- Fixed timing chain guide and tighten the bolt to 89 inch lbs. (10 Nm)
- Apply Saturn compound 21485277 to the fixed timing chain guide bolt access plug threads, install the bolt and torque it to 30 ft. lbs. (40 Nm)
- Upper timing chain guide and torque the bolts to 89 inch lbs. (10 Nm)

17. Inspect the timing chain tensioner. If the tensioner O–ring or washer is damaged the tensioner must be replaced.

18. If replacing the tensioner, measure the tensioner assembly from end-to-end. A new tensioner should be in the fully compressed non-active state. A tensioner in the non-active state should measure 2.83 inches (72mm) from end-to-end. The tensioner in the active state will measure 3.35 inches (85mm) from end-to-end.

19. If the tensioner is not in the compressed state, perform the following steps:

a. Pull the piston assembly from the tensioner body.

b. Install tensioner tool J 45027-2 into a vise.

c. Install the notch end of the piston assembly into J 45027-2.

d. Using tool J 45027-1, turn the ratchet cylinder into the pump.

e. Inspect the bore of the tensioner body for damage or dirt. If any damage is present, replace the tensioner.

f. Install the compressed piston assembly back into the tensioner body until it bottoms out in the bore. Do not compress the piston against the bottom

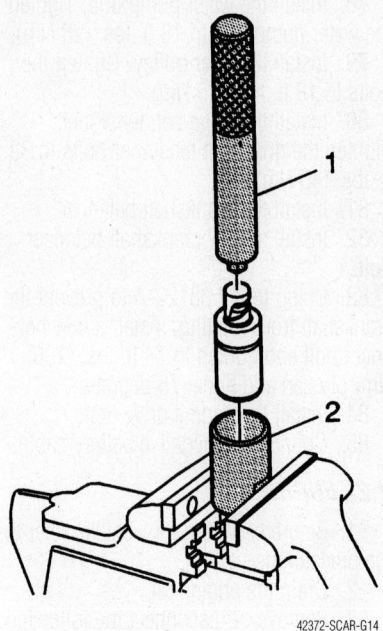

Resetting the timing chain tensioner piston–ION with 2.2L engine

of the bore. If the piston is compressed in the bore it will activate the tensioner and the piston will have to be reset.

g. Check that the tensioner measures 2.83 inches (72mm) from end-to-end. If this measurement is incorrect, reset the piston.

h. Install the tensioner assembly and tighten to 55 ft. lbs. (75 Nm)

i. Using a suitable tool with a rubber tip on the end, feed the tool down through the camshaft drive chant to rest on the timing chain. Give a sharp jolt diagonally downwards to release the tensioner.

20. Install or connect the following:
- Intake and exhaust camshaft

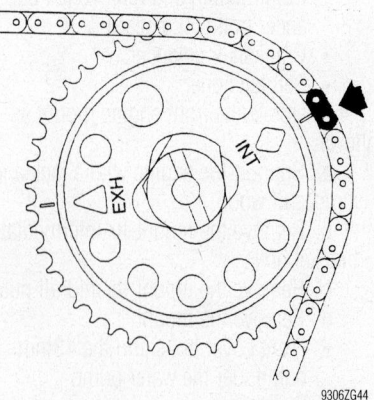

Align the copper link on the timing chain with the INT diamond timing mark—ION with 2.2L engine

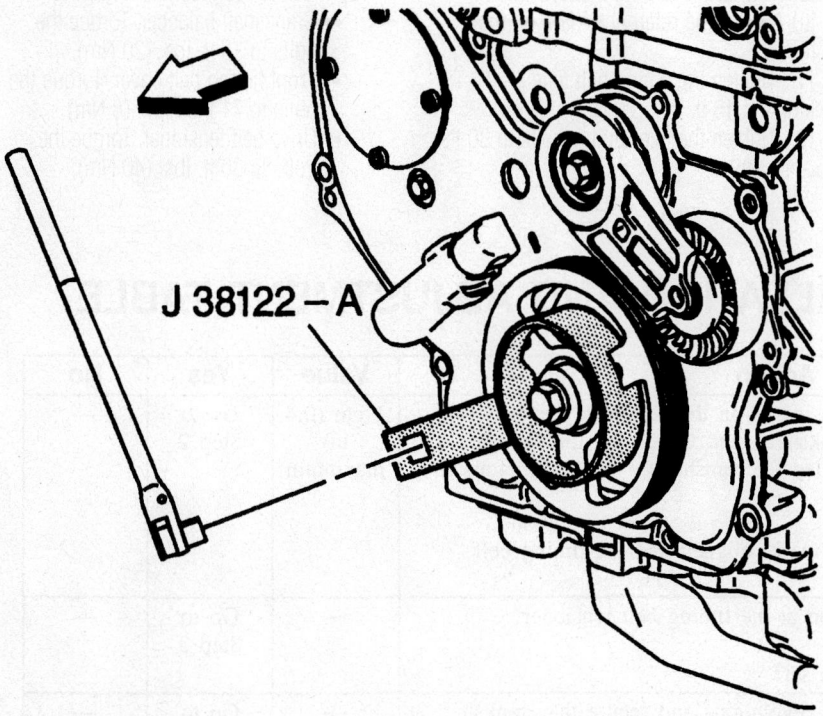

42372-SCAR-G09

Make sure the "EXH" triangle on the exhaust camshaft sprocket is aligned with the colored link—ION with 2.2L engine

sprocket bolts and torque them to 63 ft. lbs. (85 Nm) plus 30 degrees. Use a 24mm wrench as a back-up when tightening the bolts.

- Sealing ring and tensioner assembly and torque the tensioner bolts to 44 ft. lbs. (60 Nm)
- Camshaft cover
- Camshaft cover bolts and tighten to 89 inch lbs. (10 Nm)
- Ground strap
- Ground strap stud to the camshaft

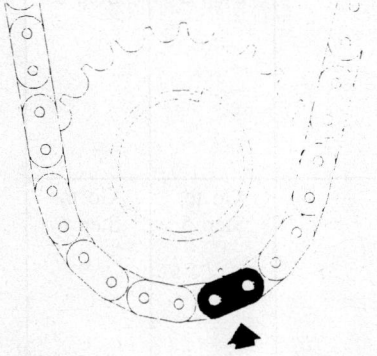

9356ZG05

Route the timing chain around the crankshaft sprocket and align the silver link to the timing mark (5 o'clock position)—ION with 2.2L engines

cover and tighten to 89 inch lbs. (10 Nm)
- Ignition coil housing and tighten the bolts to 89 inch lbs. (10 Nm)
- ICM and tighten the screws to 13 inch lbs. (1.5 Nm)
- ICM connector
- Fuel line bracket
- PCV hose
- Accelerator bracket and tighten the bolts to 89 inch lbs. (10 Nm)
- Accelerator cable to the throttle body and bracket
- Front cover gasket

21. Install the front engine mount as follows:

a. Install the mount.

b. Install the mount-to-midrail nuts and torque them to 74 ft. lbs. (100 Nm).

c. Install the mount-to-intermediate bracket bolts and torque to 37 ft. lbs. (50 Nm). Never pry the mount to align the holes, raise or lower the jack for alignment purposes.

d. Remove the jack.

22. Install or connect the following:
- Front cover and torque the bolts to 18 ft. lbs. (25 Nm) and the water pump bolt to 15 ft. lbs. (20 Nm)
- Drive belt tensioner and torque the fasteners to 33 ft. lbs. (45 Nm)
- Crankshaft balancer

- Balancer bolt and using tool J 38122-A to prevent the crankshaft from rotating, install a new balancer bolt and tighten to 74 ft. lbs. (100 Nm) plus an additional 75 degrees
- Drive belt
- Negative battery cable

23. Fill the engine with clean oil.

24. Start the vehicle and check for leaks, repair if necessary.

Timing Belt

REMOVAL & INSTALLATION

3.0L Engine

1. Before servicing the vehicle, refer to the precautions section.

2. Remove or disconnect the following:
- Negative battery cable
- Intake air resonator
- Intake plenum
- Splash shield
- Secondary air injection (AIR) pipe
- Accessory drive belt
- Water pump and power steering pump pulleys
- Drive belt tensioner
- Front timing belt cover
- Crankshaft balancer

3. Rotate the crankshaft clockwise to 60 degrees before top dead center (BTDC).

4. Loosen the timing belt tensioner and idler pulleys.

5. Remove the timing belt.

To install:

6. Start at the crankshaft sprocket and install the timing belt with the double dash (TDC) aligned with the marks on the oil pump and on the belt drive gear.

7. Set the initial timing belt tension as follows:

a. Turn the tensioner nut COUNTERCLOCKWISE to full stop, then, turn the nut back until the reference mark is 1 mm (0.003 in) over the flange.

b. Tighten the timing belt tensioner locking nut until snug, the locking nut will be tightened to specifications after all final adjustments are made.

8. Rotate the engine in the clockwise direction two revolutions stopping at 60 degrees BTDC.

9. Inspect the alignment of the reference marks on the camshaft gears with the notches on the rear timing belt cover, as well as, the mark on the crankshaft sprocket and oil pump housing.

10. If timing belt adjustment IS NOT required, set the final timing belt tension:

  a. Loosen the timing belt tensioner locking nut.

  b. Turn the locking nut counterclockwise to full stop, then back until the reference mark is 2-4 mm (0.078-0.157 in) ABOVE the reference mark on the flange.

11. Tighten the timing belt tensioner locking nut 15 ft. lbs. (20 Nm).

12. Tighten the idler pulley bolts to 30 ft. lbs. (40 Nm).

13. Install or connect the following:
  • Crankshaft balancer. Torque the bolts to 15 ft. lbs. (20 Nm).
  • Front timing belt cover. Torque the bolts to 71 inch lbs. (8 Nm).
  • Drive belt tensioner. Torque the bolts to 30 ft. lbs. (40 Nm).

TIMING BELT INSTALLATION AND ADJUSTMENT TABLE

Step	Action	Value	Yes	No
1	Install the timing belt and align marks on the belt with the marks on the camshaft gears and the crankshaft gear. Check the timing belt deflection between the idler pulley for camshafts 3 & 4 and camshaft number 4. Is the timing belt installed, the marks aligned and the timing belt deflection adjusted?	1 cm (0.4 in) maximum	Go to Step 2	—
2	Set the initial timing belt tension at the timing belt tensioner. Is the initial timing belt tension set?	—	Go to Step 3	—
3	Rotate the engine two complete revolutions and secure the crankshaft at Top Dead Center (TDC) with the J 42069-10. Has the engine been rotated and the crankshaft secured to TDC?	—	Go to Step 4	—
4	Starting with camshafts 3 and 4, check the alignment of the marks on the camshaft gears with the marks on the J 42069-20 checking gauge. Do the marks on the camshaft gears align exactly with the marks on J 42069-20?	—	Go to Step 5	Go to Step 6
5	Check the alignment of the marks on camshafts gears 1 and 2 with the marks on the J 42069-20 checking gauge. Do the marks on the camshaft gears align exactly with the marks on J 42069-20?	—	Go to Step 14	Go to Step 10
6	Do the camshaft gear marks line up to the left (BTDC) of the marks on the J 42069-20 checking gauge?	—	Go to Step 8	Go to Step 7
7	Do the camshaft gear marks line up to the right (ATDC) of the marks on the J 42069-20 checking gauge?	—	Go to Step 9	—
8	Turn the idler pulley eccentric, for camshafts 3 and 4, counterclockwise until the marks on the camshaft gear align exactly with the marks on J 42069-20. Rotate the engine two complete revolutions, lock the crankshaft at TDC with J 42069-10 and recheck the alignment of the camshaft gear marks to the marks on J 42069-20. Do the marks on the camshaft gears align exactly with the marks on J 42069-20?	—	Go to Step 5	Go to Step 6
9	Turn the idler pulley eccentric, for camshafts 3 and 4, clockwise until the marks on the camshaft gear align exactly with the marks on J 42069-20. Rotate the engine two complete revolutions, lock the crankshaft at TDC with J 42069-10 and recheck the alignment of the camshaft gear marks to the marks on J 42069-20. Do the marks on the camshaft gears align exactly with the marks on J 42069-20?	—	Go to Step 5	Go to Step 6

Timing belt installation and adjustment table—3.0L engine

79225G35

Step	Action	Value	Yes	No
10	Do the camshaft gear marks line up to the left (BTDC) of the marks on the J 42069-20 checking gauge?	—	Go to Step 12	Go to Step 11
11	Do the camshaft gear marks line up to the right (ATDC) of the marks on the J 42069-20 checking gauge?	—	Go to Step 13	—
12	Turn the idler pulley eccentric, for camshafts 1 and 2, counterclockwise until the marks on the camshaft gear align exactly with the marks on J 42069-20. Rotate the engine two complete revolutions, lock the crankshaft at TDC with J 42069-10 and recheck the alignment of the camshaft gear marks to the marks on J 42069-20. Do the marks on the camshaft gears align exactly with the marks on J 42069-20?	—	Go to Step 14	Go to Step 10
13	Turn the idler pulley eccentric, for camshafts 1 and 2, clockwise until the marks on the camshaft gear align exactly with the marks on J 42069-20. Rotate the engine two complete revolutions, lock the crankshaft at TDC with J 42069-10 and recheck the alignment of the camshaft gear marks to the marks on J 42069-20. Do the marks on the camshaft gears align exactly with the marks on J 42069-20?	—	Go to Step 14	Go to Step 10
14	Set the final timing belt tension at the timing belt tensioner. Is the final timing belt tension set?	—	Go to Step 15	—
15	Again, rotate the engine two complete revolutions and lock the crankshaft at TDC. Do a final inspection of the camshaft gear marks' relationship to the J 42069-20 marks. The marks must align exactly. Do the marks on the camshaft gears align exactly with the marks on the J 42069-20?	—	Go to Step 16	Go to Step 2
16	Remove all checking tools and ensure all idler pulleys and the tensioner locking nut are tightened to specifications. Continue with re-assembly of the engine.	—	—	—

79225G36

Timing belt installation and adjustment table (continued)—3.0L engine

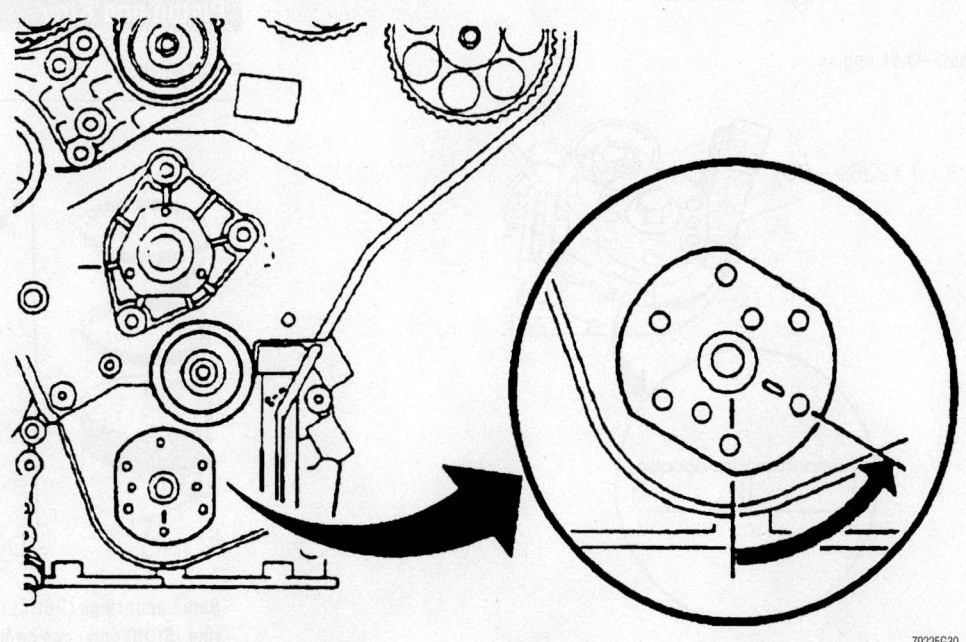

79225G30

Crankshaft alignment to 60 degrees BTDC—3.0L engine

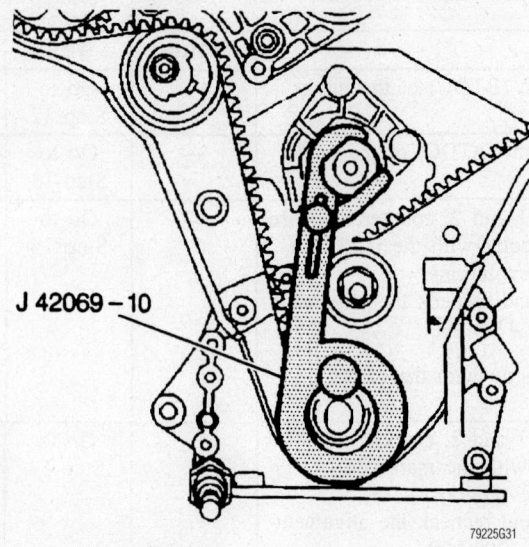

J 42069-10

Securing the crankshaft—3.0L engine

79225G31

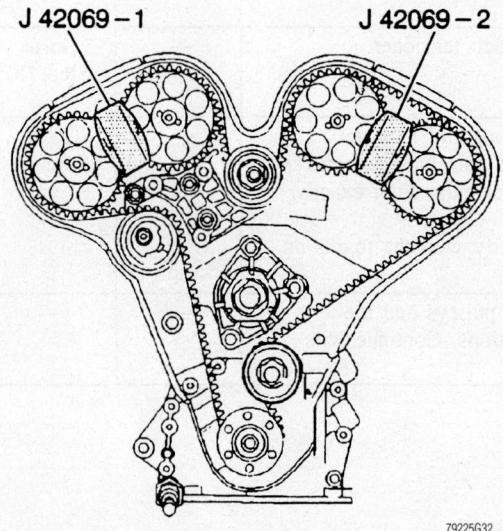

J 42069-1   J 42069-2

Locking the camshaft—3.0L engine

79225G32

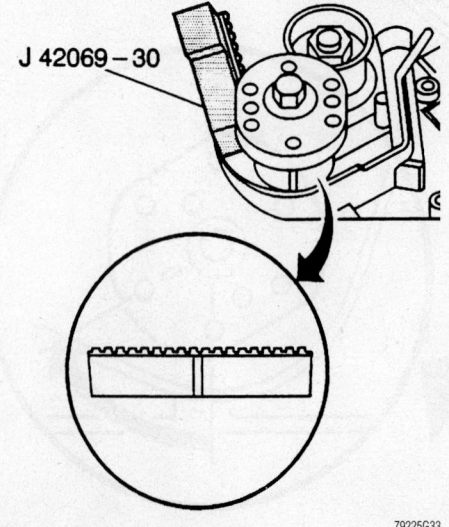

J 42069-30

Using the tool to pin the timing belt—3.0L engine

79225G33

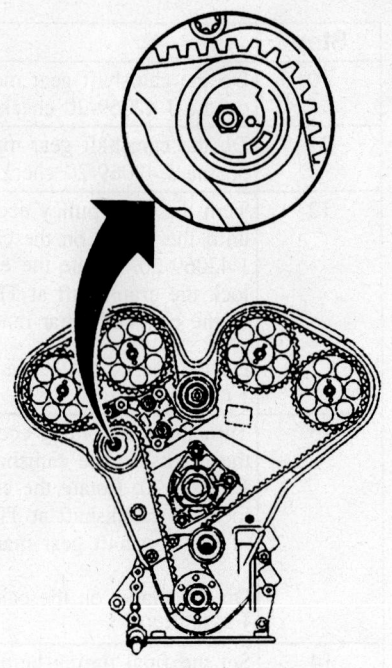

79225G34

Initial timing belt tension adjustment—3.0L engine

- Water pump pulley. Torque the bolts to 71 inch lbs. (8 Nm).
- Power steering pump pulley. Torque the bolts to 15 ft. lbs. (20 Nm).
- Accessory drive belt
- AIR injection pipe
- Splash shield
- Intake plenum
- Intake air resonator

Piston and Ring

POSITIONING

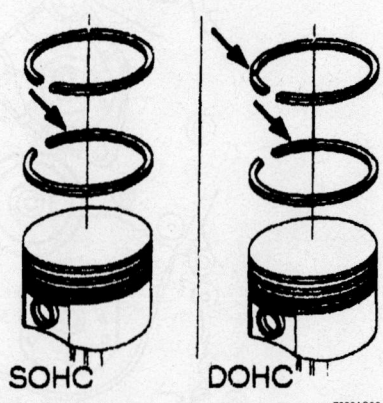

SOHC   DOHC

7922AG38

Both upper rings (DOHC) or the second ring (SOHC only) can be installed either way—1.9L engines

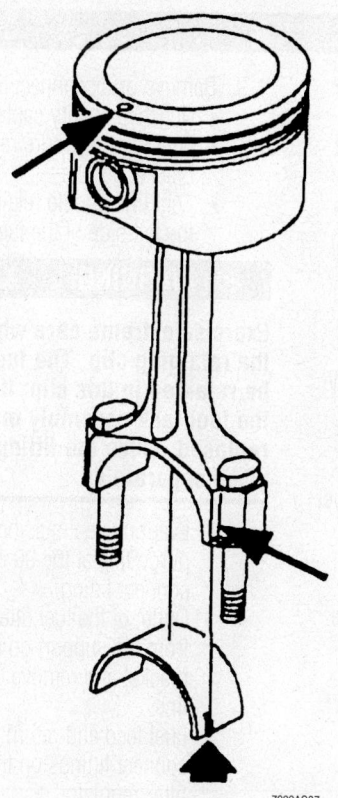

Piston and connecting rod assembly positioning mark locations. The mark on the piston and rod bearing tang slots must face the front of the engine—1.9L engines

7922AG37

Piston ring positioning—2.0L and 2.2L engines

9306ZG77

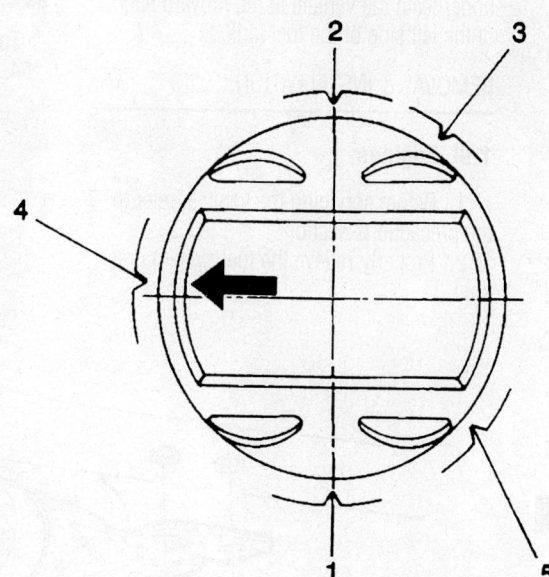

(1) 1st Compression Ring End Gap Location
(2) 2nd Compression Ring End Gap Location
(3) Oil Control Ring Upper Ring End Gap Location
(4) Oil Control Ring Spacer End Gap Location
(5) Oil Control Ring Lower Ring End Gap Location

7922AG55

Piston ring positioning—3.0L engine

FUEL SYSTEM

Fuel System Service Precautions

Safety is the most important factor when performing not only fuel system maintenance but any type of maintenance. Failure to conduct maintenance and repairs in a safe manner may result in serious personal injury or death. Maintenance and testing of the vehicle's fuel system components can be accomplished safely and effectively by adhering to the following rules and guidelines.

• To avoid the possibility of fire and personal injury, always disconnect the negative battery cable unless the repair or test procedure requires that battery voltage be applied

• Always relieve the fuel system pressure prior to disconnecting any fuel system component (injector, fuel rail, pressure regulator, etc.), fitting or fuel line connection. Exercise extreme caution whenever relieving fuel system pressure, to avoid exposing skin, face and eyes to fuel spray. Please be advised that fuel under pressure may penetrate the skin or any part of the body that it contacts

• Always place a shop towel or cloth around the fitting or connection prior to loosening to absorb any excess fuel due to spillage. Ensure that all fuel spillage (should it occur) is quickly removed from engine surfaces. Ensure that all fuel soaked cloths or towels are deposited into a suitable waste container

• Always keep a dry chemical (Class B) fire extinguisher near the work area

• Do not allow fuel spray or fuel vapors to come into contact with a spark or open flame

• Always use a back-up wrench when loosening and tightening fuel line connection fittings. This will prevent unnecessary stress and torsion to fuel line piping

• Always replace worn fuel fitting O-rings with new. Do not substitute fuel hose or equivalent, where fuel pipe is installed

Fuel System Pressure

RELIEVING

1. Before servicing the vehicle, refer to the precautions section.
2. Unless battery voltage is necessary for testing, disconnect the negative battery cable. This will prevent the fuel pump from running and causing a fuel spill through the disconnected components if the ignition key is accidentally turned **ON**.

3. Remove the air cleaner assembly, for access.
4. Wrap a shop rag around the fuel test port fitting, located at the lower rear of the engine, then remove the cap and connect a fuel pressure gauge.
5. Install the bleed hose from the pressure gauge into an approved container and open the valve to bleed the system pressure.
6. After the pressure is bled, remove the gauge from the test port and recap it.
7. Install the air cleaner assembly.
8. After servicing the vehicle, connect the negative battery cable and prime the fuel system as follows:

    a. Turn the ignition **ON** for 5 seconds, then **OFF** for 10 seconds.

    b. Repeat the **ON/OFF** cycle 2 more times.

    c. Crank the engine until it starts.

    d. If the engine does not readily start, repeat sub-steps A–C.

9. Run the engine and check for leaks.

Fuel Filter

The fuel filter and fuel pressure regulator are one integral component of the new anti-return fuel injection system, and is located underneath the vehicle at the forward edge of the left side of the fuel tank.

REMOVAL & INSTALLATION

1.9L Engines

1. Before servicing the vehicle, refer to the precautions section.
2. Properly relieve the fuel system pressure.

3. Remove or disconnect the following:
• Negative battery cable
• Fuel filter/pressure regulator bracket screws
• Fuel line bundle retaining clip on the left side of the fuel tank

✳✳ WARNING

Exercise extreme care when opening the retaining clip. The fuel lines must be retained in this clip; if damaged, the fuel tank assembly must be replaced, since the fitting is not serviced separately.

• Evaporative Emissions (EVAP) purge line at the 90 degree quick connect fitting
• Outlet of the fuel filter/regulator from the support on the fuel tank bracket and remove the fuel feed line
• Fuel feed and return line quick-connect fittings on the fuel filter/regulator
• Fuel filter

➡**It is not necessary to separate the fuel filter/regulator from the mounting bracket, since both items are serviced as an assembly.**

To install:

4. Install or connect the following:
• Fuel filter
• New fuel line retainers (3) into the female portion of the quick-connect fuel line fittings
• Fuel feed and return lines onto the fuel filter/regulator and snap them closed

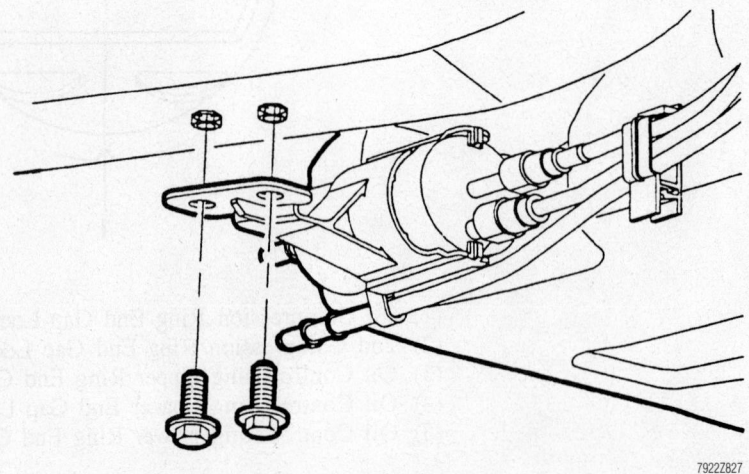

7922Z827

The fuel filter/regulator bracket is held to the frame with 2 bolts—1.9L engines

• Fuel feed, return, and EVAP purge lines into the fuel tank's retaining clip

Be sure to route the chassis fuel feed and purge lines above the parking brake cable. The parking brake cable must be firmly secured to the underbody to support the fuel and purge lines.

• 90 degree EVAP purge line quick-connect fitting to the purge line and snap it closed
• Fuel feed outlet pipe of the fuel filter/regulator into the retaining clip on the fuel tank bracket
• Fuel filter/regulator bracket mounting screws and torque them to 71 inch lbs. (8 Nm)
• Negative battery cable

5. Prime the fuel system as follows:
  a. Turn the ignition **ON** for 5 seconds, then **OFF** for 10 seconds.
  b. Repeat the **ON/OFF** cycle 2 more times.
  c. Crank the engine until it starts.
  d. If it does not start, repeat the 3 above steps.

6. Start the engine and check for leaks, repair if necessary.

L–Series

1. Before servicing the vehicle, refer to the precautions section.
2. Properly relieve the fuel system pressure.
3. Remove or disconnect the following:
  • Negative battery cable
  • Fuel filter bracket screw

• Fuel feed and return lines from the filter
• Fuel filter from the bracket

To install:

4. Install or connect the following:
  • New fuel filter into the bracket
  • New fuel line retainers to the female portion of the quick connect fittings
  • Fuel feed and return lines
  • Fuel filter bracket attaching screw and torque it to 35 inch lbs. (4 Nm)
  • Negative battery cable

5. Prime the fuel system as follows:
  a. Turn the ignition **ON** for 5 seconds, then **OFF** for 10 seconds.
  b. Repeat the **ON/OFF** cycle 2 more times.
  c. Crank the engine until it starts.
  d. If it does not start, repeat the 3 above steps.

6. Start the engine and check for leaks, repair if necessary.

ION

1. Before servicing the vehicle, refer to the precautions section.
2. Properly relieve the fuel system pressure.
3. Remove or disconnect the following:
  • Negative battery cable
  • Fuel filter bracket bolt
  • Fuel feed and return lines from the filter
  • Fuel filter from the bracket

To install:

4. Install or connect the following:
  • New fuel filter into the bracket
  • Fuel feed and return lines
  • Fuel filter bracket attaching bolt and torque it to 89 inch lbs. (10 Nm)

• Negative battery cable
5. Prime the fuel system as follows:
  a. Turn the ignition **ON** for 2 seconds, then **OFF** for 10 seconds.
  b. Turn the ignition key to the **ON** position but do not start the engine.
  c. Check for leaks.
  d. Crank the engine until it starts.

6. Start the engine and check for leaks, repair if necessary.

REMOVAL & INSTALLATION

1.9L Engines

To prevent excessive fuel spillage, whenever the tank is removed from the vehicle it should be no more than ½ full. Removal of the fuel pump module assembly requires the removal of the fuel tank.

1. Before servicing the vehicle, refer to the precautions section.
2. Properly relieve the fuel system pressure.
3. Remove or disconnect the following:
  • Negative battery cable
  • Fuel pump
  • Fuel feed and return lines from the filter/pressure regulator
  • Fuel pump vapor line from the fuel tank vent pipe
  • Fuel pump module retaining ring with service Tool SA9156E. A ½inch breaker bar will loosen the lockring

➡**To prevent bending of the sending unit float arm, lift the pump module up slightly to disengage the orientation tabs in the tank. Rotate the module 90 degrees clockwise until the fuel lines are facing the 1 o'clock position.**

• Fuel pump module from the fuel tank until the bottom of the pump module is close to the fuel tank opening
• Tilt the pump module approximately 45 degrees to the right hand side of the tank and lift the pump from the fuel tank
• Fuel pump to tank seal and discard it

To install:

4. Install or connect the following:
  • Fuel pump module into the fuel tank by orientating the float to face the right side and tilting the module approximately 45 degrees
  • Fuel pump into the tank and rotate

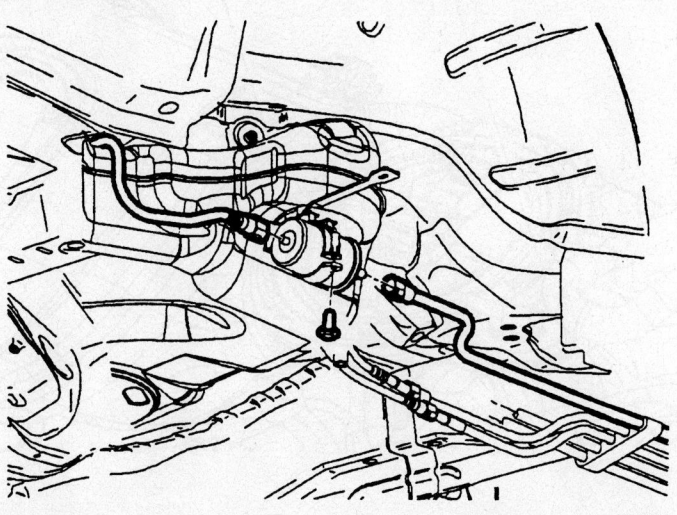

9306ZG56

Remove the fuel filter bracket attaching screw—L–Series shown, others similar

the assembly 90 degrees counter-clockwise to align the module tabs with the slots in the tank

➡ **The fuel pump retaining ring cannot be properly installed if the flange locator tabs are not aligned with the slots in the fuel tank.**

- Retaining lockring with Tool SA9156E
- Vapor line to the fuel tank vent pipe
- Fuel feed and return lines to the filter/pressure regulator
- Fuel feed, return and EVAP canister purge line in the fuel tank retaining clip
- Fuel tank
- Negative battery cable

5. Start the vehicle and check for leaks, repair if necessary.

L–Series

1. Before servicing the vehicle, refer to the precautions section.
2. Properly relieve the fuel system pressure.
3. Remove or disconnect the following:
- Negative battery cable
- Fuel tank
- Fuel lines from the fuel pump module cover
- Fuel pump module retaining ring with a Sending Unit Wrench, J43827
- Pull the retaining clip toward the float arm and lift up
- Fuel pump straight up from the fuel tank
- Fuel pump tank seal and discard the seal
- Fuel feed line from the bottom of the fuel pump cover with Clamp Pliers J43914
- Fuel pump electrical connector

To install:
4. Install or connect the following:
- Fuel pump feed line to the cover
- Fuel pump electrical connector
- Fuel pump to the new seal
- Fuel pump cover lockring with Tool J-43827
- Fuel lines and wiring harness
- Fuel tank
- Negative battery cable

5. Start the vehicle and check for leaks, repair if necessary.

ION

1. Before servicing the vehicle, refer to the precautions section.
2. Properly relieve the fuel system pressure.

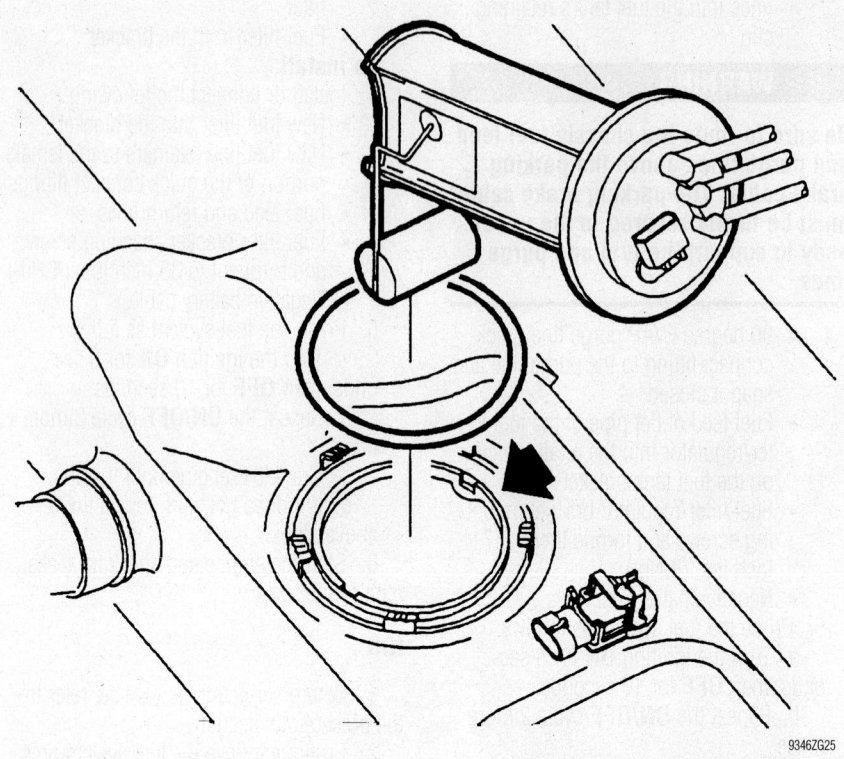

Installing the fuel sending unit and float—1.9L engines

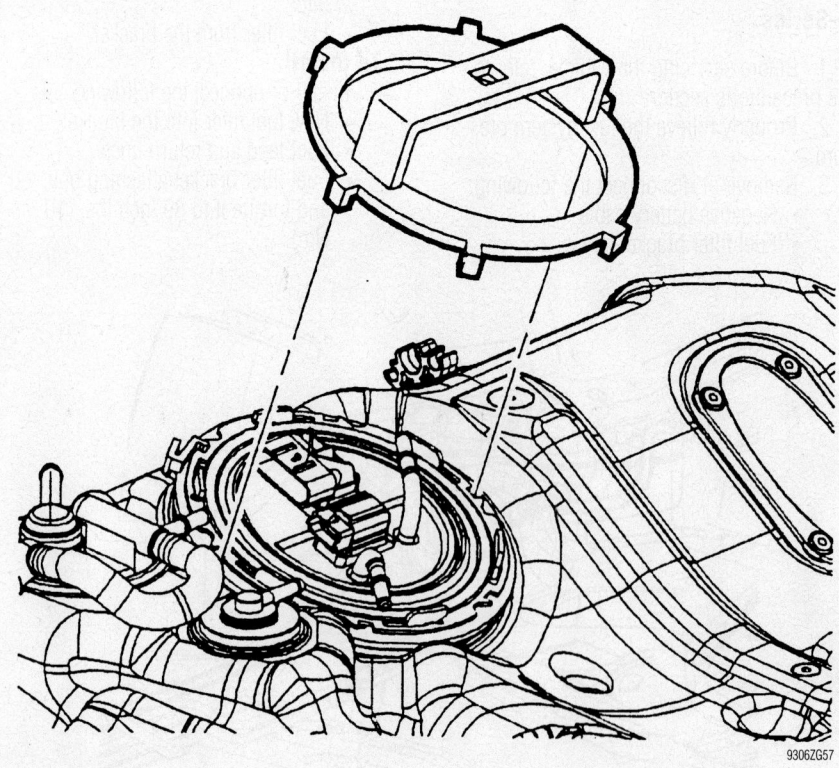

Remove the fuel pump cover lockring—L-Series

3. Remove or disconnect the following:
- Negative battery cable
- Fuel tank
- Fuel lines from the fuel pump module cover
- Fuel pump electrical connections
- Fuel pump module retaining ring with a Sending Unit Wrench, J 39765
- Pull the pump up and tilt the assembly towards the rear of the tank to allow clearance for the float arm and lift up
- Fuel pump straight up from the fuel tank
- Fuel pump tank seal and discard the seal

To install:

4. Install or connect the following:
- Fuel pump to the new seal
- Fuel pump
- Fuel pump cover lockring with Tool J 39765
- Fuel lines and wiring harness
- Fuel tank. Tighten the strap bolts to 18 ft. lbs. (25 Nm).
- Negative battery cable

5. Start the vehicle and check for leaks, repair if necessary.

Fuel Injector

REMOVAL AND INSTALLATION

1.9L Engines

1. Before servicing the vehicle, refer to the precautions section.
2. Properly relieve the fuel system pressure.
3. Remove or disconnect the following:

- Negative battery cable
- Fuel rail
- Fuel injector retaining clip off the injector
- Fuel injector
- Fuel injector O-rings and discard them

To install:

➡**The SOHC and DOHC fuel injectors are not interchangeable. The injectors are identified by a color coded plastic ring near the bottom of the injector. The SOHC injector has a blue ring near the O-ring that mates to the intake manifold and the DOHC has a pink ring.**

4. Lubricate the new fuel injector O-ring with clean engine oil.
5. Install or connect the following:

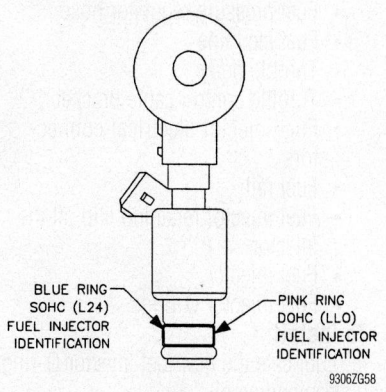

BLUE RING
SOHC (L24)
FUEL INJECTOR
IDENTIFICATION

PINK RING
DOHC (LL0)
FUEL INJECTOR
IDENTIFICATION

9306ZG58

Fuel injector identification—1.9L engine

- New O-ring seals on the fuel injector
- Retaining clip to the fuel injector
- Fuel injector to the fuel rail
- Fuel rail
- Negative battery cable

6. Start the vehicle and check for leaks, repair if necessary.

2.0L Engines

1. Before servicing the vehicle, refer to the precautions section.
2. Disconnect the negative battery cable.
3. Remove the coolant overflow pipe.
4. Relieve the fuel system pressure.
5. Use a back up wrench on the fuel rail and disconnect the fuel supply pipe.
6. Remove the fuel rail attaching studs.

➡**Use care when removing the fuel rail assembly in order to prevent damage to the fuel injectors electrical connector terminals and spray tips.**

7. Remove the fuel rail using the following procedure:
    a. Pull the fuel rail back and upward to remove the fuel injectors from the cylinder head ports.
    b. Rotate the fuel rail in order to position the injectors downward.
    c. Remove the fuel rail.
8. Disconnect the fuel injector harness connectors.
9. Remove the fuel injector retainer clip.
10. Remove the fuel injectors from the fuel rail.
11. Visually inspect the fuel injector in order to determine if the upper O-ring was also removed. If the upper O-ring is not removed, remove the O-ring from the fuel rail assembly.
12. Remove and discard the fuel injector O-rings.

To install:

➡**Always install new injector O-rings when servicing the fuel injectors. Lubricate the new injector O-rings with clean engine oil**

13. Install the O-rings on the fuel injector.
14. Install the fuel injector clip on the fuel injector.

➡**The fuel injector will click when the injector is installed correctly.**

15. Install the fuel injector in the fuel rail with the connector facing upward.
16. Connect the fuel injector harness connectors. Pull back to insure the connectors are locked in place.

➡**Install new lower O-rings when reusing fuel injectors. Lubricate the injector tip O-rings with clean engine oil prior to installing the injectors into the intake manifold.**

17. Install the fuel rail using the following procedure:
    a. With the fuel injectors positioned downward, lower the fuel injectors into the cylinder head ports.
    b. Align the injectors by rotating the fuel rail forward.
    c. Carefully push the fuel injectors into the cylinder head ports.
    d. Install the fuel rail attaching studs. Tighten the fuel rail studs to 89 inch lbs. (10 Nm).
18. Install the fuel supply pipe.
19. Using a backup wrench on the fuel rail tighten the fuel supply pipe to 10 ft .lbs. (14 Nm).
20. Install the coolant overflow pipe. Tighten the pipe bolt to 71 inch lbs. (8 Nm).
21. Connect the negative battery cable.
22. Inspect for fuel leaks using the following procedure:
    a. Turn ON the ignition, with the engine OFF for 2 seconds.
    b. Turn OFF the ignition for 10 seconds.
    c. Turn ON the ignition.
    d. Inspect for fuel leaks.

2.2L and 3.0L Engines

1. Before servicing the vehicle, refer to the precautions section.
2. Properly relieve the fuel system pressure.
3. Remove or disconnect the following:
- Negative battery cable
- Air intake tube

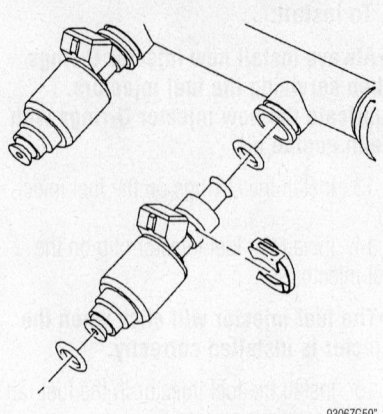

Remove the retainer clip from the fuel injector—2.2L and 3.0L engines

- Fuel pressure regulator hose
- Fuel feed line
- Throttle body
- Throttle control cable bracket
- Fuel injector electrical connectors
- Fuel rail
- Fuel injector retaining clip off the injector
- Fuel injector
- Fuel injector O-rings

To install:

4. Lubricate the new fuel injector O-ring with clean engine oil.

5. Install or connect the following:
- New O-ring seals on the fuel injector

- Retaining clip to the fuel injector
- Fuel injector to the fuel rail
- Fuel rail and torque the bolts to 89 inch lbs. (10 Nm) on 2.2L engines and 71 inch lbs. (8 Nm) on 3.0L engines
- Fuel injector electrical connectors
- Throttle control cable bracket
- Throttle body and torque the bolts to 89 inch lbs. (10 Nm)
- Fuel feed line
- Fuel pressure regulator hose
- Air intake tube
- Negative battery cable

6. Start the vehicle and check for leaks, repair if necessary.

DRIVE TRAIN

Manual Transaxle Assembly

REMOVAL & INSTALLATION

1.9L Engines

1. Before servicing the vehicle, refer to the precautions section.
2. Drain the transaxle fluid.
3. Remove or disconnect the following:
- Both battery cables
- Air cleaner assembly
- Battery and tray
- Powertrain Control Module (PCM) **J2** (black 28-way) harness connector. Do not disconnect the **J1** (80-way) connector.
- PCM attaching bolts and move it aside
- Transaxle strut to midrail bracket fastener and loosen the strut to transaxle fastener and move the strut aside
- Back-up lamp electrical connector
- Vehicle Speed Sensor (VSS) electrical connector, if equipped
- Ground terminals from the clutch housing bolts
- Vent tube retaining clip
- Top 2 clutch housing to the engine bolts
- Spark plug wires from the coil towers
- Electronic ignition module electrical connectors
- Electronic ignition module
- Shifter cables from the arms and clutch housing
- Clutch slave cylinder by rotating it ¼-turn counterclockwise and pushing it into the housing

- 2 clutch hydraulic damper to housing fasteners and wire the actuator cylinder and damper to the upper radiator hose
4. Wire the radiator to the upper support.
5. Install an engine support bar and place the feet on the outer edge of the shock tower.
6. Connect the bar hooks to the engine support bracket.
7. Position the stabilizer foot on the engine to the right of the dipstick tube.

8. Adjust the hooks and stabilizer to remove any looseness.
9. Remove or disconnect the following:
- Both front wheels and inner splash shields
- Engine splash shield
- Engine strut cradle bracket fasteners
- Transaxle mount-to-cradle fasteners
- Front exhaust pipe-to-manifold and catalytic converter fasteners
- Front exhaust pipe
- Steering gear to cradle fasteners

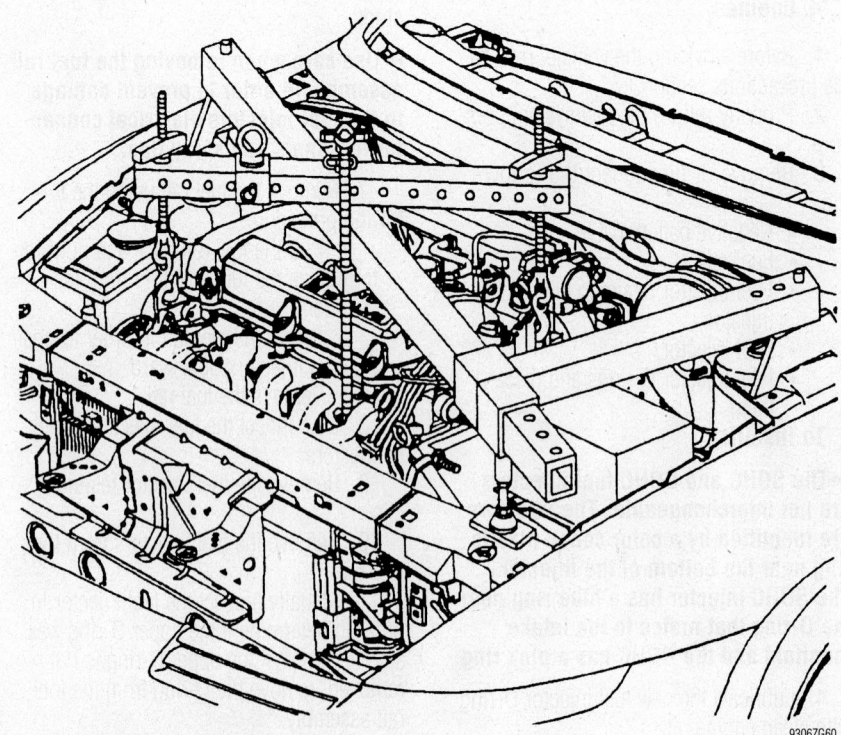

Install an Engine Support tool

- Brake pipe bracket push pin
- Engine-to-transaxle stiffening brace
- Dust cover
- Cotter pin from the lower ball joints
- Ball joint from the lower control arm with Separator Tool SA9132S
- Separate the left axle from the transaxle and install an axle seal protector
- Separate the right axle from the intermediate shaft
- Intake bracket-to-intake manifold bolt on DOHC engines
- Intake bracket to intermediate shaft support bolt
- Support bracket
- Intermediate shaft from the transaxle
- 4 cradle-to-body bolts and lower the cradle on a support dolly
- Bottom clutch housing bolts and install a guide bolt into the rear housing bolt hole

10. Separate the transaxle from the engine.

To install:

11. Place the transaxle assembly securely onto a jack and position under the vehicle. Install axle seal protectors into seals on both sides.

12. Raise the transaxle into the vehicle to align the input shaft to the center of the clutch. Rotate the transaxle back and forth to align the input shaft splines to the clutch disc.

13. Install or connect the following:

- 2 lower clutch housing-to-engine bolts and torque them to 103 ft. lbs. (140 Nm)
- Axle Seal Protector tool to the inside seal and remove the transaxle jack
- Left side axle to the transaxle. When the splines clear the seal protector, remove the tool.
- Intermediate shaft to the transaxle
- Intermediate shaft and torque the bolts to 40 ft. lbs. (54 Nm)
- Starter bracket to the intermediate shaft and torque the bolts to 22 ft. lbs. (30 Nm)
- Intake manifold support bracket (on DOHC) and torque the bolts to 22 ft. lbs. (30 Nm)
- Intake bracket to the intermediate shaft support and torque the bolt to 40 ft. lbs. (54 Nm)
- Right side axle to the intermediate shaft

14. Raise the cradle up on a support dolly and place the ball joints into the knuckles. Verify the correct positioning of

the lower control arm bar studs to the knuckles, the cooling module support bushings, the engine strut bracket and the transaxle mount.

15. Insert 9/16 in. round steel rods into the cradle-to-body alignment holes near the front cradle to body fastener holes. Guide the cradle into position making sure all mount studs are properly guided into their holes.

16. Be sure the washers are in place, then install the 2 rear cradle to body bolts. Verify proper cradle positioning and install the 2 front cradle bolts, then torque the 4 cradle bolts to 155 ft. lbs. (210 Nm).

17. Remove the support dolly and lower the vehicle sufficiently for underhood access. Remove the engine support bar assembly.

18. Install or connect the following:

- Transaxle strut-to-cradle bracket through-bolt and nut and torque them to 40 ft. lbs. (54 Nm)
- Strut cradle bracket-to-cradle bolt and torque the it to 52 ft. lbs. (70 Nm)
- Ignition module and torque the bolts to 71 inch lbs. (8 Nm)
- 2 top transaxle housing-to-engine studs and torque them to 74 ft. lbs. (100 Nm)
- VSS electrical connector
- Back-up lamp electrical connector
- Ground terminals to the clutch housing studs
- Vent hose clip to the housing
- Shifter cables to the shift arms and clutch housing
- Hydraulic clutch system to the clutch housing and torque the fasteners to 18 ft. lbs. (25 Nm)
- Actuator to the clutch housing. Push in and rotate the actuator ¼ turn
- Battery tray and torque the bolts to 89 inch lbs. (10 Nm)
- PCM and torque the bolts to 53 inch lbs. (6 Nm)
- Connect the PCM **J2** (28-way) harness
- Battery and hold down retainer and torque the fastener to 80 inch lbs. (9 Nm)
- Air cleaner assembly
- Transaxle lower mount-to-cradle fastener and torque it to 37 ft. lbs. (50 Nm)
- 2 engine strut cradle bracket-to-cradle fasteners and torque them to 37 ft. lbs. (50 Nm)
- Steering gear-to-cradle fasteners and torque them to 37 ft. lbs. (50 Nm)

- Brake pipe-to-cradle retainer
- Dust cover and torque the bolts to 8 ft. lbs. (11 Nm)
- Engine-to-transaxle stiffening brace and torque the bolts to 40 ft. lbs. (54 Nm)
- Exhaust manifold pipe and torque the bolts to 23 ft. lbs. (31 Nm)
- Intermediate exhaust pipe to the catalytic converter and torque the bolts to 18 ft. lbs. (25 Nm)
- Front exhaust pipe support to the stiffening bracket and torque the fasteners to 89 inch lbs. (10 Nm)
- Both front wheels
- Both battery cables

19. Fill the transaxle to the proper level.

20. Warm the engine and check the transaxle fluid. Check and adjust vehicle alignment, as necessary.

L-Series

1. Before servicing the vehicle, refer to the precautions section.

2. Drain the transaxle fluid.

3. Remove or disconnect the following:

- Both battery cables
- Battery feed to the underhood fuse block
- Release the retaining tabs on the fuse block cover and remove it
- Engine 68-way electrical connector
- Forward lamp 68-way connector
- Forward lamp 2-way (white) connector
- Black, green and brown instrument panel 2-way connectors
- Fuse block from the case
- Engine, forward lamp and instrument panel harness from the fuse block case
- Case from the battery tray
- Battery tray and battery
- Loosen the pinch bolt securing the control assembly rod to the control shaft lever

➡**The control shaft lever retaining pin has a spring loaded locking feature to keep the pin in place. The spring must be depressed before attempting to remove it.**

- Pin retaining the control shaft lever to shift control shaft
- Retaining clips holding the control shaft lever to the transaxle and frame brackets
- Control shaft lever
- Back-up lamp switch electrical connector
- Wire harness from the transaxle

- Spring loaded locking pin from the hole on top of the transaxle, if equipped
- Clutch hydraulic fitting from the transaxle
- Upper transaxle to engine bolts and install and engine support bar, SA9105E, and Adapter Kit J43405
- Matchmark the position of the left transaxle mount and remove the mount
- Frame assembly
- Front transaxle mount
- Rear transaxle mount
- Axle shafts from the transaxle
- A/C line retaining clips and drop the A/C line down from the body

4. Lower the left side of the powertrain assembly to allow the transaxle to clear the engine compartment rail.

5. Remove or disconnect the following:
- Left side transaxle mount bracket
- Control shaft lever assembly pivot pin bracket

6. Properly support the powertrain assembly
- Remaining engine-to-transaxle bolts
- Transaxle from the engine

To install:

7. Raise the transaxle into the vehicle and align the input shaft to the center of the clutch disc.

8. Install or connect the following:
- Lower transaxle to engine bolts and torque them to 48 ft. lbs. (65 Nm)
- Control shaft lever assembly pivot pin bracket and torque the fastener to 18 ft. lbs. (24 Nm)
- Axle shafts to the transaxle. After the splines clear the tool, remove the seal protector and snap the axle into place.

9. Remove the transaxle fill plug, fill the transaxle with 2 quarts of manual fluid and install the plug. Torque the plug to 22 ft. lbs. (30 Nm).

10. Install or connect the following:
- Left side transaxle mount bracket and torque the bolts to 41 ft. lbs. (55 Nm)
- A/C line against the body and install the retaining clips
- Rear transaxle mount and torque the bolts to 41 ft. lbs. (55 Nm)
- Front transaxle mount and torque the bolts to 41 ft. lbs. (55 Nm)
- Frame assembly
- Upper transaxle to engine bolts and torque them to 48 ft. lbs. (65 Nm)
- Left side transaxle mount bolts using the matchmarks made during removal and torque them to 41 ft. lbs. (55 Nm)
- Back-up lamp switch electrical connector

- Control shaft lever assembly onto the pivot pin brackets
- Retaining clips holding the control shaft lever to the transaxle and frame brackets
- Pin to retain the control shaft lever to the shift control shaft
- Rotate the shift control shaft clockwise and install the shift linkage lock pin
- Move the shift lever to the 5th gear/reverse gate and install a ⅜ inch punch to hold it in place
- Pinch bolt securing the control assembly rod to the control shaft lever and torque it to 9 ft. lbs. (12 Nm) plus 180 degrees
- Shift lever hole plug
- Clutch hydraulic fitting to the transaxle and bleed the hydraulic system
- Battery tray and torque the fasteners to 11 ft. lbs. (15 Nm)
- Underhood fuse block to the battery tray and torque the fasteners to 80 inch lbs. (9 Nm)
- Wire harness to the fuse case
- Snap the fuse block onto the hinges
- Engine 68-way connectors
- Forward lamp 68-way connectors
- Instrument panel 68-way connectors
- Forward 2-way (white) connector
- Black, green and brown 2-way connectors
- Snap the fuse block to the case
- Coolant hose to the fuse block
- Battery feed to the fuse block and torque the fastener to 12 inch lbs. (16 Nm)
- Battery tray and hold-down bracket and torque the fastener to 15 ft. lbs. (20 Nm)
- Battery
- Fan control module to the battery hold down bracket
- Both battery cables

11. Fill the transaxle to the proper level.

12. Warm the engine and check the transaxle fluid. Check and adjust vehicle alignment, as necessary.

ION

GETRAG 5-SPEED

1. Before servicing the vehicle, refer to the precautions section.

2. Drain the transaxle fluid.

3. Remove or disconnect the following:
- Negative battery cable
- Positive battery post from the underhood junction block

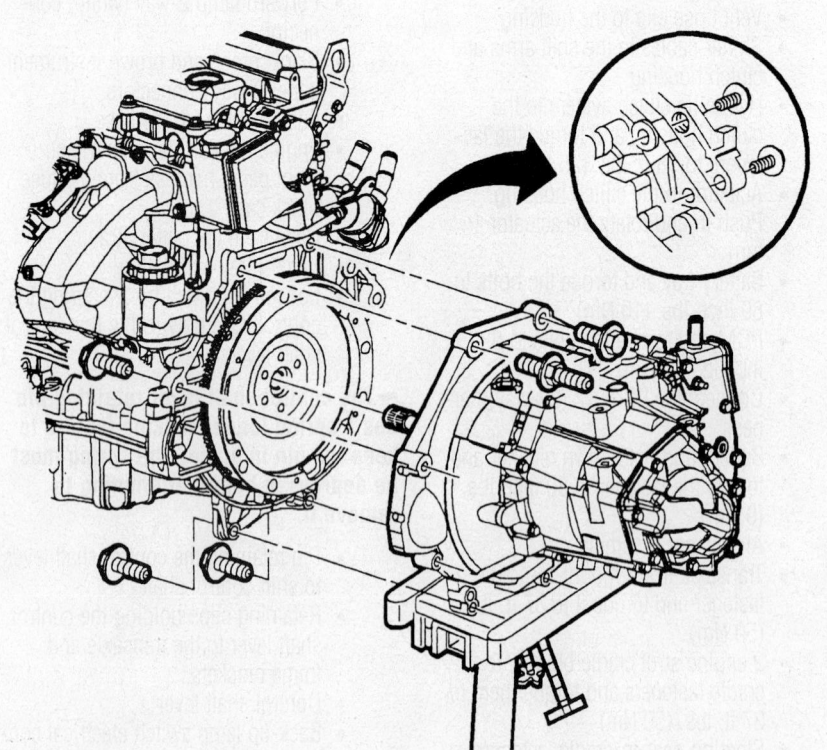

9306ZG62

Remove the remaining engine to transaxle bolts—L-Series

- Positive cables from the underhood junction block
- Surge tank inlet hose from the surge tank
- Underhood junction block bracket nuts and bolt
- Front wiring harness from the junction block bracket
- Move the junction block aside
- Clutch hose from the actuator cylinder

4. Install an engine support fixture.

5. Secure the cooling module to the upper body structure.
- Upper transmission-to-mount bolts
- Upper transmission-to-engine bolt

6. Remove the front frame as follows:

a. Mark the frame-to-body position.

b. Lower the vehicle until it is 3 feet off the ground so that a lift table may be placed in position. As necessary, use blocks of wood between the frame and table to support the assembly.

c. Slowly remove the front frame bolts, the partially unscrew the bolts until 1 ½ inch of the bolt shank is exposed.

d. Slowly lower the table to the floor

7. Remove or disconnect the following:
- Drive shafts from the transmission and support with wire
- Starter
- Shift cable from the transmission as follows:

a. Apply the parking brake.

b. Shift the transmission to **Neutral**.

c. Remove the console.

d. Lift up the cable retainers.

e. Disconnect the cable ends from the shifter assembly.

f. Disconnect the cable ends from the transmission shift levers, make sure to mark the cable locations prior to removal, this will aid during assembly.

8. Remove or disconnect the following:
- Back-up lamp switch and Vehicle Speed Sensor connections

9. Use the engine support fixture to lower the powertrain assembly to allow clearance between the side rail and powertrain.

10. Attach a transmission jack and remove the transmission-to-engine bolts.

11. Remove the transmission.

To install:

➡**The number 3 bolt position does not require a bolt. Refer to the accompanying illustration for bolt locations.**

12. Use the transmission jack to align the transmission, install the transmission-to-engine bolts and torque to 55 ft. lbs. (75 Nm).

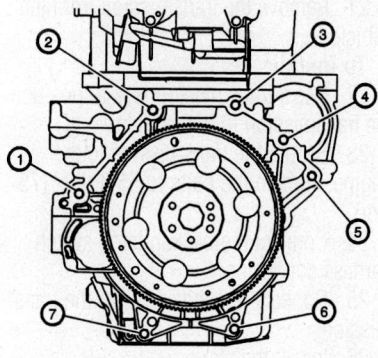

42372-SCAR-G15

The number 3 bolt position does not require a bolt–ION

13. Install or connect the following:
- Back-up lamp switch and Vehicle Speed Sensor connections

14. Connect the shift cable to the transmission as follows:

a. Connect the cable ends to the transmission shift levers as noted during removal.

b. Connect the cable ends to the shifter assembly.

c. Push the shifter neutral lock clip. Move the shifter slightly to center the clip and make sure the transmission is in **Neutral**.

d. Press down on the locking tabs.

e. Pull the shifter neutral lock into its original position.

f. Install the console.

15. Install or connect the following:
- Starter
- Drive shafts

16. Use the engine support fixture to raise the powertrain assembly into position.

17. Install the side transmission mount as follows:

a. Place the mount in position and tighten the mount-to-rail bolts to 25 ft. lbs. (34 Nm).

18. Use a jack to align the mount when it is in position.

a. Install and hand tighten the mount-to-transmission bolts. Do not pry the transmission or mount to align the holes. Hand start the bolts in the following sequence; rear bolt, middle and then front bolt. Using the same sequence, tighten the bolts to 33 ft. lbs. (45 Nm).

19. Install or connect the following:
- Underhood fuse block
- Surge tank inlet hose
- Front wiring harness to the junction block
- Positive battery cables to the junction block bracket

- Positive battery post to the junction block bracket
- Juntion block bracket bolt and tighten to 18 ft. lbs. (25 Nm) and the nuts to 89 inch lbs. (10 Nm)

20. Install the front frame as follows:

a. Slowly raise the table to the into position and align the frame to the marks made during removal.

b. Hand tighten the frame bolts

c. Tighten the frame bolts to 74 ft. lbs. (100 Nm) plus an additional torque of 180 degrees using a torque angle meter.

21. Remove the engine support fixture.

22. Install or connect the following:
- Top engine-to-transmission bolt and tighten to 55 ft. lbs. (75 Nm)
- Clutch hose to the actuator cylinder and bleed the hydraulic clutch system.
- Cooling module
- Negative battery cable.

23. Fill the transmission to the proper level.

24. Warm the engine and check the transmission fluid. Check and adjust vehicle alignment, as necessary.

MU–3

1. Before servicing the vehicle, refer to the precautions section.

2. Disconnect the negative battery cable.

3. Remove the cover from the underhood electrical center.

4. Release the forward lamp harness retainer from the ABS modulator bracket, or the proportioning valve bracket, to allow the underhood electrical center to be repositioned adequately.

5. Remove the underhood electrical center bracket from the vehicle and reposition the electrical center to access the bracket .

6. Release the wiring harness retainers above the brake booster, to allow the electrical center to be repositioned adequately.

7. Disconnect the hydraulic clutch hose from the clutch actuator cylinder and the clutch master cylinder.

8. Install an engine support fixture.

9. Secure the cooling module to the upper body structure.

10. Remove the upper transmission to mount bolts.

11. Disconnect the wiring harness retainer from the transmission stud.

12. Remove the upper transmission to engine stud and bolt.

13. Remove the frame as follows:

a. Secure the radiator and condenser assembly to the radiator support.

b. Remove the front wheels from the vehicle.

c. Remove the splash shields.

d. Remove the front transaxle mount to cradle through bolt.

e. Remove the rear transaxle mount to frame bolts.

f. Remove both stabilizer link to stabilizer shaft nuts.

g. Remove both tie rod to steering knuckle nuts.

h. Separate the outer tie rods from the steering knuckles.

i. Remove the intermediate steering shaft to steering gear pinch bolt and discard.

➡**DO NOT rotate the intermediate shaft once separated from the gear. Possible damage or a malfunction could occur.**

j. Disconnect the intermediate steering shaft from the steering gear.

k. Remove both lower control arm ball stud to steering knuckle pinch bolts.

➡**Do not free the ball stud by using a pickle fork or a wedge-type tool. Damage to the seal or bushing may result.**

l. Lower the lower control arms in order to disengage the steering knuckle. If necessary, use ball joint removal tool J 43631.

m. Mark the frame to body position with a paint pen or permanent marker.

n. Lower the vehicle to about 3 feet off the ground in order to place a hydraulic lift table under the frame.

o. Use two 2 x 4 pieces of wood between the lift table and the frame and lift the table to the frame.

p. Slowly remove the frame bolts using the following sequence:
- Front frame bolts
- Rear frame bolts

q. Slowly lower the lift table to the floor.

14. Drain the transaxle fluid.

15. Disconnect the drive axle and intermediate shaft from the transmission and secure out of the way.

16. Remove the starter.

17. Disconnect the shift cables from the transmission.

18. Disconnect the backup lamp switch harness connector.

19. Use the engine support fixture rear hook to lower the powertrain enough to allow clearance between the side rail and powertrain.

20. Use a transmission jack to secure the transmission, and remove the transmission to engine bolts.

21. Remove the transmission from the vehicle.

To install:

22. Use a transmission jack to position the transmission to the vehicle.

23. Secure the transmission to the engine. Tighten the bolts to 55 ft. lbs. (75 Nm).

24. Connect the backup lamp switch harness connector.

25. Connect the shift cable to the transmission.

26. Install the starter.

27. Connect the drive axle and intermediate shaft to the transmission.

28. Use the engine support fixture in order to raise the powertrain assembly.

29. Install the left transmission mount. Tighten the bolts to 20 ft. lbs. (27 Nm).

30. Install the frame as follows:

a. With the frame on the lift table, raise the frame to the vehicle.

b. Hand start all the frame bolts while aligning the frame to the paint marks.

c. Tighten the frame bolts. Tighten the bolts 74 ft. lbs. (100 Nm) plus and additional 180.

31. Lower and remove the hydraulic table.

a. Connect the lower control arm to the steering knuckle.

➡**The torque sequence must be followed in the order that is listed.**

b. Install the ball joint pinch bolt and nut. Tighten the nut to 37 ft. lbs. (50 Nm), back of the nut ¾ turn then retighten the nut to 37 ft. lbs. (50 Nm) plus an additional 30 degree turn.

➡**The front and rear transmission mounts must be allowed to settle with the through bolts loosened.**

c. Hand start the front transaxle mount through bolt.

d. Loosen the rear transmission mount through bolt.

e. Tighten the rear transaxle mount to frame bolts to 37 ft. lbs. (50 Nm)

f. Tighten the rear transaxle mount through bolts to 74 ft. lbs. (100 Nm), then tighten the front transaxle mount through bolts to 74 ft. lbs. (100 Nm).

g. Install the outer tie rods to the steering knuckles. Install new outer tie rod to the knuckle nuts and tighten to 15 ft. lbs. (20 Nm) plus an additional 180 degree turn.

h. Connect the stabilizer links to the stabilizer shaft. Tighten to 81 ft. lbs. (100 Nm).

i. Connect the intermediate shaft to

the steering gear. Install a new intermediate shaft pinch bolt and tighten to 25 ft. lbs. (34 Nm).

j. Install the splash shields.

k. Install the front wheels.

32. Remove the engine support fixture.

33. Install the top engine to transmission bolt. Tighten the bolts to 55 ft. lbs. (75 Nm).

34. Install the top engine to transmission stud. Tighten to 55 ft. lbs. (75 Nm).

35. Connect the wiring harness retainer to the transmission stud.

36. Connect the hydraulic clutch hose to the clutch actuator cylinder.

37. Bleed the clutch hydraulic system.

38. Install the underhood electrical center bracket to the vehicle and install the electrical center (1) into position on the bracket .

39. Secure the forward lamp harness retainer to the ABS modulator bracket, or the proportioning valve bracket .

40. Connect the electrical connector to the brake fluid level sensor, then press forward on the Connector Position Assurance (CPA) tab of the connector to secure.

41. Install the cover to the underhood electrical center.

42. Release the cooling module from the upper body structure.

43. Connect the negative battery cable.

44. Fill the transmission to the proper level.

Automatic Transaxle Assembly

REMOVAL & INSTALLATION

1.9L Engines

1. Before servicing the vehicle, refer to the precautions section.

2. Drain the transaxle fluid.

3. Remove or disconnect the following:
- Both battery cables
- Air cleaner assembly
- Battery and tray
- Powertrain Control Module (PCM) **J2** (black 28-way) harness connector. Do not disconnect the **J1** (80-way) connector.
- PCM attaching bolts and move aside
- Transaxle strut to midrail bracket fastener and loosen the strut to transaxle fastener and move the strut aside
- Transaxle solenoid harness connector
- Vehicle speed and input sensor harness connectors

- Transaxle fluid temperature sensor connector
- Range switch harness connector
- Ground terminals from the converter housing
- Ground wire from the range switch
- Spark plug wires from the coil towers
- Electronic ignition module electrical connectors
- Electronic ignition module

4. Wire the radiator to the upper support.

5. Install an engine support bar and place the feet on the outer edge of the shock tower.

6. Connect the bar hooks to the engine support bracket.

7. Position the stabilizer foot on the engine to the right of the dipstick tube.

8. Adjust the hooks and stabilizer to remove any looseness.

9. Remove or disconnect the following:
- Both front wheels and inner splash shields
- Engine splash shield
- Engine strut cradle bracket fasteners
- Transaxle mount to cradle fasteners
- Front exhaust pipe to manifold and catalytic converter fasteners
- Front exhaust pipe
- Steering gear to cradle fasteners
- Brake pipe bracket push pin
- Engine-to-transaxle stiffening brace
- Dust cover
- Torque converter-to-flywheel bolts
- Ball joint from the lower control arm
- Separate the left axle from the transaxle and install an axle seal protector
- Separate the right axle from the intermediate shaft
- Intake bracket-to-intake manifold bolt (on DOHC engines)
- Intake bracket-to-intermediate shaft support bolt
- Support bracket
- Intermediate shaft from the transaxle
- Transaxle cooler lines
- 4 cradle-to-body bolts and lower the cradle on a support dolly
- Lower torque converter bolts
- Separate the transaxle from the engine
- Shifter cable from the converter housing
- Transaxle from the vehicle

To install:

10. Flush the transaxle oil cooler lines.

11. Place the transaxle assembly securely onto a jack and position under the

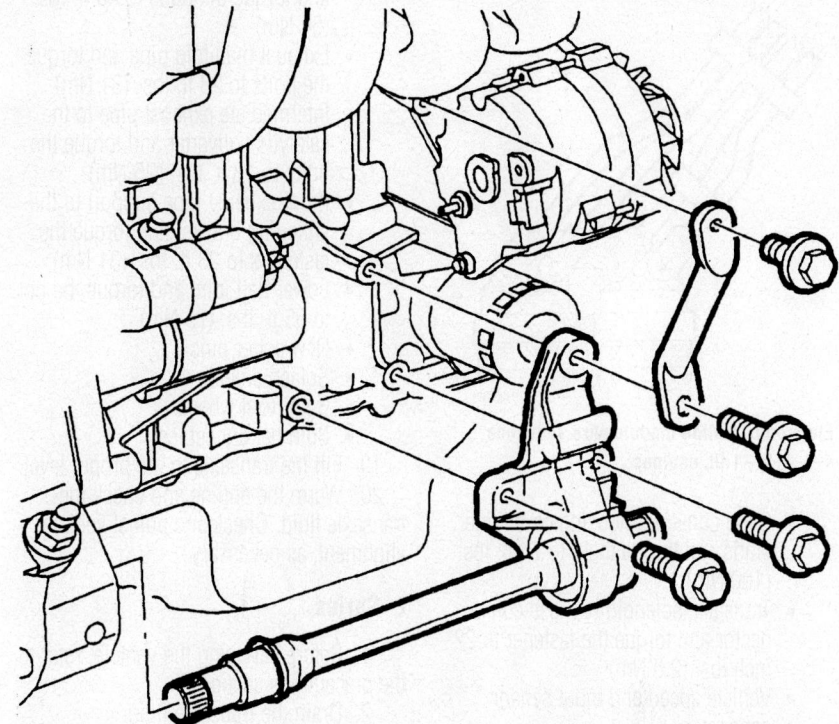

Remove the intake-to-intermediate shaft support bracket and the intermediate shaft—1.9L

9346ZG26

vehicle. Install axle seal protectors into seals on both sides.

12. Raise the transaxle high enough to connect the shifter cable.

13. Install or connect the following:
- Shifter cable to the transaxle gear selector lever and insert it in the converter housing
- 2 lower converter housing-to-engine bolts and torque them to 96 ft. lbs. (130 Nm)
- Axle Seal Protector tool to the inside seal and remove the transaxle jack
- Left side axle to the transaxle. When the splines clear the seal protector, remove the tool.
- Intermediate shaft to the transaxle
- Intermediate shaft and torque the bolts to 40 ft. lbs. (54 Nm)
- Intake manifold support bracket (on DOHC) and torque the bolts to 22 ft. lbs. (30 Nm)
- Intake bracket to the intermediate shaft support and torque the bolt to 40 ft. lbs. (54 Nm)
- Right side axle to the intermediate shaft
- Cooler lines to the transaxle

14. Raise the cradle up on a support dolly and place the ball joints into the knuckles. Verify the correct positioning of the lower control arm bar studs to the

knuckles, the cooling module support bushings, the engine strut bracket and the transaxle mount.

15. Insert ⁹⁄₁₆ in. round steel rods into the cradle-to-body alignment holes near the front cradle to body fastener holes. Guide the cradle into position making sure all mount studs are properly guided into their holes.

16. Be sure the washers are in place, then install the 2 rear cradle to body bolts. Verify proper cradle positioning and install the 2 front cradle bolts, then torque the 4 cradle bolts to 155 ft. lbs. (210 Nm).

17. Remove the support dolly and lower the vehicle sufficiently for underhood access. Remove the engine support bar assembly.

18. Install or connect the following:
- Transaxle strut-to-cradle bracket through-bolt and nut and torque them to 40 ft. lbs. (54 Nm)
- Strut midrail bracket and torque the bolt to 52 ft. lbs. (70 Nm)
- Ignition module and torque the bolts to 71 inch lbs. (8 Nm)
- Ignition module wire connector
- Spark plug wires to the coil towers

➡**Proper orientation of the ignition module is critical to the operation of the module and the on board diagnostic system. The proper sequence is 4–1–2–3 from left to right.**

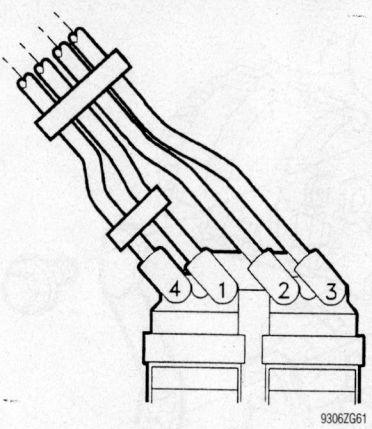

Electronic ignition module wire sequence 4–1–2–3—1.9L engines

- 2 top converter housing-to-engine studs and torque them to 74 ft. lbs. (100 Nm)
- Transaxle solenoid harness connector and torque the fastener to 22 inch lbs. (2.5 Nm)
- Vehicle speed and input sensor connectors
- Transaxle fluid temperature sensor connectors
- Transaxle range switch harness connectors
- Ground terminals to the top 2 converter housing bolts and torque the bolts to 19 ft. lbs. (25 Nm)
- Ground wire to the neutral selector switch
- Adjust the control cable
- Air cleaner assembly
- Battery tray and torque the bolts to 89 inch lbs. (10 Nm)
- PCM and torque the bolts to 53 inch lbs. (6 Nm)
- Connect the PCM **J2**(28-way) harness
- Battery and hold down retainer and torque the fastener to 80 inch lbs. (9 Nm)
- Transaxle lower mount to cradle fastener and torque it to 37 ft. lbs. (50 Nm)
- 2 engine strut cradle bracket-to-cradle fasteners and torque them to 37 ft. lbs. (50 Nm)
- Steering gear to cradle fasteners and torque them to 37 ft. lbs. (50 Nm)
- Brake pipe to cradle retainer
- Torque converter to flexplate and torque the bolts to 52 ft. lbs. (70 Nm)
- Dust cover and torque the bolts to 8 ft. lbs. (11 Nm)
- Engine to transaxle stiffening brace

and torque the bolts to 40 ft. lbs. (54 Nm)
- Exhaust manifold pipe and torque the bolts to 23 ft. lbs. (31 Nm)
- Intermediate exhaust pipe to the catalytic converter and torque the bolts to 18 ft. lbs. (25 Nm)
- Front exhaust pipe support to the stiffening bracket and torque the fasteners to 23 ft. lbs. (31 Nm)
- Lower ball joint and torque the nut to 55 ft. lbs. (75 Nm)
- New cotter pins
- Splash shields
- Both front wheels
- Both battery cables

19. Fill the transaxle to the proper level.

20. Warm the engine and check the transaxle fluid. Check and adjust vehicle alignment, as necessary.

L–Series

1. Before servicing the vehicle, refer to the precautions section.

2. Drain the transaxle fluid.

3. Remove or disconnect the following:
- Both battery cables
- Move the fan control module up and out of the way
- Battery feed to the underhood fuse block
- Coolant hose from the underhood fuse block
- Release the retaining tabs on the fuse block cover and remove it
- Engine 68-way electrical connector
- Forward lamp 68-way connector
- Forward lamp 2-way (white) connector
- Black, green and brown instrument panel 2-way connectors
- Fuse block from the case
- Engine, forward lamp and instrument panel harness from the fuse block case
- Case from the battery tray
- Battery tray and battery
- Control cable from the range switch lever
- Control cable bracket from the rear powertrain mount
- Transaxle electrical connector and ground wire
- Remaining wire harness connectors from the transaxle
- Transaxle range switch electrical connector
- Upper engine to transaxle bolts
- Drive belt from the alternator, for 3.0L engine
- Drive belt tensioner, for 3.0L engine

- Alternator, for 3.0L engine
- Output speed sensor electrical connector

4. Install an Engine Support Fixture.
- Frame
- Axle shafts from the transaxle
- Engine to transaxle bracket
- Starter solenoid electrical connector
- Starter motor

➡**The wire harness mounting bracket on the rear of the engine must be removed on the 3.0L.**

- Torque converter-to-flexplate bolts through the starter opening
- Transaxle oil cooler lines
- 2 lower engine to transaxle bolts
- Left side transaxle mount bolts

5. Slide a hydraulic lifting table under the transaxle and remove the remaining mounting bolts.

6. Separate the transaxle from the engine.

To install:

7. With the transaxle on a hydraulic lift, raise the transaxle to the engine.

8. Install or connect the following:
- Rear wire harness bracket, for 3.0L engine
- Engine to transaxle mounting bolts and torque them to 48 ft. lbs. (60 Nm)
- Left side transaxle mount bolts and torque them to 41 ft. lbs. (55 Nm)
- Bottom 2 engine to transaxle mounting bolts and torque them to 48 ft. lbs. (65 Nm)
- Transaxle oil cooler lines after lubricating them with automatic transaxle fluid and torque the fasteners to 71 inch lbs. (8 Nm)
- Loosely install the flex plate-to-torque converter bolts. When aligned properly, torque the bolts to 33 ft. lbs. (45 Nm).
- Starter motor and torque the bolts to 30 ft. lbs. (40 Nm)
- Starter solenoid electrical connectors
- Engine to transaxle bracket and torque the bolts to 26 ft. lbs. (35 Nm)
- New axle shaft retaining rings on the end of the output shafts and install the axle shafts to the transaxle
- Frame assembly
- Output speed sensor electrical connector and secure it to the mounting stud
- Alternator and torque the bolts to 30 ft. lbs. (40 Nm), for 3.0L engine

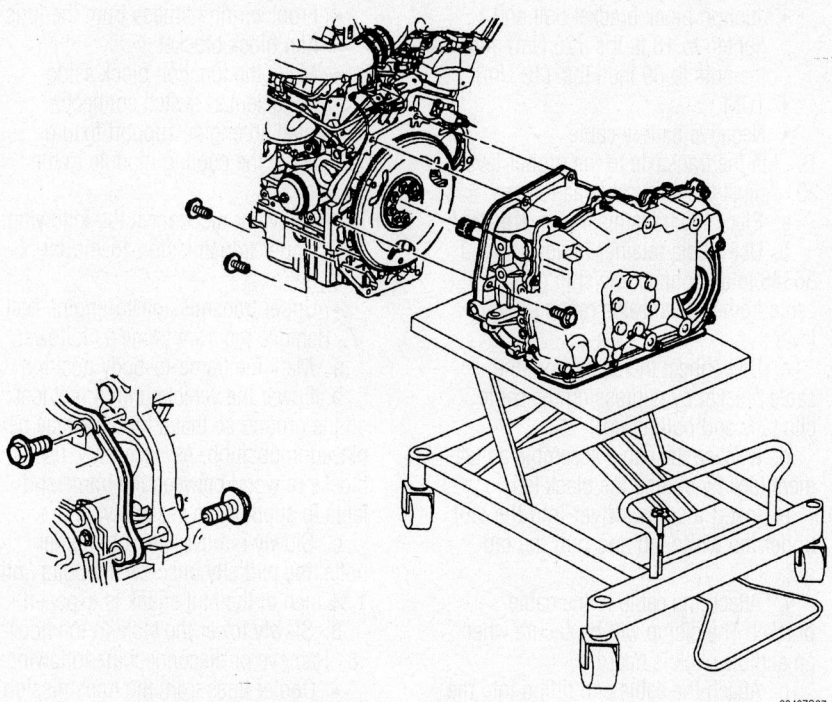

Lifting table mounting points—L–Series

- Drive belt tensioner and torque the bolts to 30 ft. lbs. (40 Nm), for 3.0L engine
- Drive belt, for 3.0L engine
- Remaining transaxle to engine mounting bolts and torque them to 48 ft. lbs. (65 Nm)
- Transaxle electrical connector and ground wire and torque the fasteners to 18 ft. lbs. (25 Nm)
- Wire harness attachments to the transaxle side cover and torque the fasteners to 15 ft. lbs. (20 Nm)
- Control cable mounting bracket and torque the bolts to 15 ft. lbs. (20 Nm)
- Control cable to the range switch lever
- Battery tray and torque the fasteners to 11 ft. lbs. (15 Nm)
- Underhood fuse block case to the battery tray. Torque the bolt to 80 inch lbs. (9 Nm) and snap the block onto the hinges.
- Engine 68-way connectors
- Forward lamp 68-way connectors
- Instrument panel 68-way connectors
- Forward 2-way (white) connector
- Black, green and brown 2-way connectors
- Snap the fuse block to the case
- Coolant hose to the fuse block
- Battery feed to the fuse block and torque the fastener to 12 inch lbs. (16 Nm)

- Battery tray and hold-down bracket and torque the fastener to 15 ft. lbs. (20 Nm)
- Battery
- Fan control module to the battery hold down bracket
- Both battery cables

9. Fill the transaxle to the proper level.

10. Warm the engine and check the transaxle fluid. Check and adjust vehicle alignment, as necessary.

ION

2003–04 AUTOMATIC 5-SPEED TRANSMISSION

1. Before servicing the vehicle, refer to the precautions section.

2. Drain the transaxle fluid.

3. Remove or disconnect the following:

- Negatibe battery cable
- Transmission Control Module (TCM) connections noting their locations to aid during installation
- TCM by releasing the tab on the junction block bracket
- Positive battery post from the underhood junction block
- Positive cables from the underhood junction block
- Surge tank inlet hose from the surge tank
- Underhood junction block bracket nuts and bolt

- Front wiring harness from the junction block bracket
- Move the junction block aside
- Park/Neutarl switch connector
- Main transmission harness connector

4. Install an engine support fixture.

5. Remove or disconnect the following:
- Upper transmission-to-mount bolts
- Upper transmission-to-engine bolt

6. Remove the front frame as follows:
a. Mark the frame-to-body position.
b. Lower the vehicle until it is 3 feet off the ground so that a lift table may be placed in position. As necessary, use blocks of wood between the frame and table to support the assembly.
c. Slowly remove the front frame bolts, the partially unscrew the bolts until 1 ½ inch of the bolt shank is exposed.
d. Slowly lower the table to the floor

7. Remove or disconnect the following:
- Cooler lines from the transmission
- Drive shafts from the transmission and support with wire
- Starter
- Torque converter-to-flywheel bolts
- Shift cable from the transmission as follows:
a. Remove the shift cable retaining nut near the exhaust heat shield.
b. Remove the shift cable from the transmission cable bracket.
c. Disconnect the shift cable from the park/neutral switch.

8. Use the engine support fixture to lower the powertrain assembly to allow clearance between the side rail and powertrain.

9. Attach a transmission jack and remove the transmission-to-engine bolts.

10. Remove the transmission.

To install:

➡ **The number 3 bolt position does not require a bolt. refer to the accompanying illustration for bolt locations.**

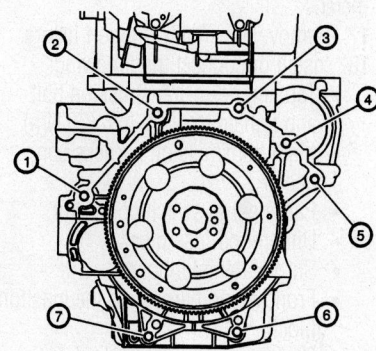

The number 3 bolt position does not require a bolt–ION

11. Use the transmission jack to align the transmission, install the transmission-to-engine bolts and torque to 55 ft. lbs. (75 Nm).

- Shift cable to the transmission as follows:

a. Connect the shift cable from the park/neutral switch.

b. Connect the shift cable to the transmission cable bracket.

c. Install the shift cable retaining nut near the exhaust heat shield and tighten to 89 inch lbs. (10 Nm).

12. Install or connect the following:

- Torque converter-to-flywheel bolts and tighten to 48 ft. lbs. (65 Nm)
- Starter
- Drive shafts
- Transmission cooler lines to the transmission. Lubricate the cooler pipe seals with transmission fluid prior to connecting the lines.

13. Use the engine support fixture to raise the powertrain assembly into position.

14. Install the side transmission mount as follows:

a. Place the mount in position and tighten the mount-to-rail bolts to 25 ft. lbs. (34 Nm).

15. Use a jack to align the mount when it is in position.

a. Install and hand tighten the mount-to-transmission bolts. Do not pry the transmission or mount to align the holes. Hand start the bolts in the following sequence; rear bolt, middle and then front bolt. Using the same sequence, tighten the bolts to 33 ft. lbs. (45 Nm).

16. Install the front frame as follows:

a. Slowly raise the table to the into position and align the frame to the marks made during removal.

b. Hand tighten the frame bolts

c. Tighten the frame bolts to 74 ft. lbs. (100 Nm) plus an additional torque of 180 degrees using a torque angle meter.

17. Remove the engine support fixture.

18. Install or connect the following:

- Top engine-to-transmission bolt and tighten to 55 ft. lbs. (75 Nm)
- Main transmission harness connector
- Park/neutral switch connector
- Underhood fuse block
- Surge tank inlet hose
- Front wiring harness to the junction block
- Positive battery cables to the junction block bracket
- Positive battery post to the junction block bracket

- Juntion block bracket bolt and tighten to 18 ft. lbs. (25 Nm) and the nuts to 89 inch lbs. (10 Nm)
- TCM
- Negative battery cable

19. Fill the transaxle to the proper level.

20. Adjust the shift cable as follows:

a. Place the transmission in **Neutral**.

b. Use fascia retainer remover tool J 36346 to disconnect the shift control cable from the transaxle range switch lever.

c. Disconnect the cable from the cable bracket by depressing the cable clip tabs and pulling up.

d. Release the cable assemble adjustment lock by sliding the black tab.

e. Insert a screw driver into the slot under the white tab and pull the tab out.

f. Attach the cable to the cable bracket. The clamp will be secure when an audible click is heard.

g. Attach the cable end fitting into the ball stud of the transaxle range switch selector. The clamp will be secure when an audible click is heard.

h. Attach the cable to the cable bracket. The clamp will be secure when an audible click is heard.

i. Lock the adjustment in place by pushing down on the white tab, slide the black tab up to lock the white tab in place and verify cable operation.

21. Warm the engine and check the transaxle fluid. Check and adjust vehicle alignment, as necessary.

2003–04 AUTOMATIC 5-SPEED CONTINUOUSLY VARIABLE TRANSMISSION

1. Before servicing the vehicle, refer to the precautions section.

2. Drain the transaxle fluid.

3. Remove or disconnect the following:

- Negatibe battery cable
- Electronic Control Module (ECM)
- Transmission Control Module (TCM) connections noting their locations to aid during installation
- TCM by releasing the tab on the junction block bracket, lift it half way, then turn it clockwise to remove
- Positive battery post from the underhood junction block
- Positive cables from the underhood junction block
- Surge tank inlet hose from the surge tank
- Underhood junction block bracket nuts and bolt

- Front wiring harness from the junction block bracket
- Move the junction block aside
- Park/Neutarl switch connector

4. Install an engine support fixture.

5. Secure the cooling module to the upper body

6. Remove or disconnect the following:

- Upper transmission-to-mount bolts
- Upper transmission-to-engine bolt

7. Remove the front frame as follows:

a. Mark the frame-to-body position.

b. Lower the vehicle until it is 3 feet off the ground so that a lift table may be placed in position. As necessary, use blocks of wood between the frame and table to support the assembly.

c. Slowly remove the front frame bolts, the partially unscrew the bolts until 1 ½ inch of the bolt shank is exposed.

d. Slowly lower the table to the floor

8. Remove or disconnect the following:

- Cooler lines from the transmission
- Drive shafts from the transmission and support with wire
- Starter
- Torque converter-to-flywheel bolts
- Shift cable from the transmission as follows:

a. Remove the shift cable retaining nut near the exhaust heat shield.

b. Remove the shift cable from the transmission cable bracket.

c. Disconnect the shift cable from the park/neutral switch.

9. Use the engine support fixture to lower the powertrain assembly 2-to-3 inches to allow clearance between the side rail and powertrain.

10. Attach a transmission jack and remove the transmission-to-engine bolts.

11. Remove the transmission.

To install:

➡**The number 3 bolt position does not require a bolt. refer to the accompanying illustration for bolt locations.**

12. Use the transmission jack to align the transmission, install the transmission-to-engine bolts and torque to 55 ft. lbs. (75 Nm).

- Shift cable to the transmission as follows:

a. Connect the shift cable from the park/neutral switch.

b. Connect the shift cable to the transmission cable bracket.

c. Install the shift cable retaining nut near the exhaust heat shield and tighten to 89 inch lbs. (10 Nm).

13. Install or connect the following:

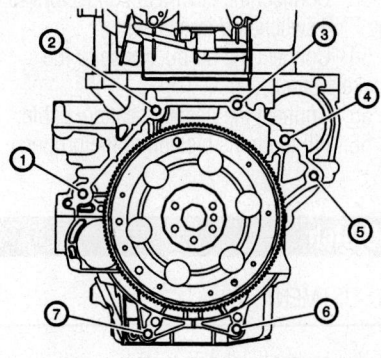

The number 3 bolt position does not require a bolt–ION

- Torque converter-to-flywheel bolts and tighten to 48 ft. lbs. (65 Nm)
- Starter
- Drive shafts
- Transmission cooler lines to the transmission. Lubricate the cooler pipe seals with transmission fluid prior to connecting the lines.

14. Use the engine support fixture to raise the powertrain assembly into position.

a. Place the mount in position and tighten the mount-to-rail bolts to 25 ft. lbs. (34 Nm).

b. Use a jack raise the engine back into position and to align the mount when it is in position.

c. Install and hand tighten the mount-to-transmission bolts. Do not pry the transmission or mount to align the holes. Hand start the bolts in the following sequence; rear bolt, middle and then front bolt. Using the same sequence, tighten the bolts to 33 ft. lbs. (45 Nm).

15. Install the front frame as follows:

a. Slowly raise the table to the into position and align the frame to the marks made during removal.

b. Hand tighten the frame bolts

c. Tighten the frame bolts to 74 ft. lbs. (100 Nm) plus an additional torque of 180 degrees using a torque angle meter.

16. Install or connect the following:

- Top engine-to-transmission bolt and tighten to 55 ft. lbs. (75 Nm)
- Park/neutral switch connector
- Cooling module

17. Remove the engine support fixture.

- Underhood fuse block
- Surge tank inlet hose
- Front wiring harness to the junction block
- Positive battery cables to the junction block bracket
- Positive battery post to the junction block bracket

- Juntion block bracket bolt and tighten to 18 ft. lbs. (25 Nm) and the nuts to 89 inch lbs. (10 Nm)
- TCM and ECM
- Negative battery cable

18. Fill the transaxle to the proper level.

19. Adjust the shift cable as follows:

a. Place the transmission in **Neutral**.

b. Use fascia retainer remover tool J 36346 to disconnect the shift control cable from the transaxle range switch lever.

c. Disconnect the cable from the cable bracket by depressing the cable clip tabs and pulling up.

d. Release the cable assemble adjustment lock by sliding the black tab.

e. Insert a screw driver into the slot under the white tab and pull the tab out.

f. Attach the cable to the cable bracket. The clamp will be secure when an audible click is heard.

g. Attach the cable end fitting into the ball stud of the transaxle range switch selector. The clamp will be secure when an audible click is heard.

h. Attach the cable to the cable bracket. The clamp will be secure when an audible click is heard.

i. Lock the adjustment in place by pushing down on the white tab, slide the black tab up to lock the white tab in place and verify cable operation.

20. Warm the engine and check the transaxle fluid. Check and adjust vehicle alignment, as necessary.

2005 AUTOMATIC TRANSMISSION

1. Before servicing the vehicle, refer to the precautions section.

2. Disconnect the negative battery cable.

3. Disconnect the air inlet duct hose from the intake plenum.

4. Disconnect the transaxle wiring harness from the transaxle and the PNP switch.

5. Remove the upper transmission to engine bolts and stud.

6. Install an engine support fixture.

7. Remove the front wheel assemblies.

8. Remove the both front fender liners.

9. Remove the steering gear mounting bolts and secure the steering gear with mechanics wire.

10. Disconnect the wheel speed sensor wires from the both front wheels and unclip from the frame.

11. Separate the ball joints from the steering knuckles.

12. Remove the frame as follows:

a. Secure the radiator and condenser assembly to the radiator support.

b. Remove the front transaxle mount to cradle through bolt.

c. Remove the rear transaxle mount to frame bolts.

d. Remove both stabilizer link to stabilizer shaft nuts.

e. Remove both tie rod to steering knuckle nuts.

f. Separate the outer tie rods from the steering knuckles.

g. Remove the intermediate steering shaft to steering gear pinch bolt and discard.

➡ **DO NOT rotate the intermediate shaft once separated from the gear. Possible damage or a malfunction could occur.**

h. Disconnect the intermediate steering shaft from the steering gear.

i. Remove both lower control arm ball stud to steering knuckle pinch bolts.

➡ **Do not free the ball stud by using a pickle fork or a wedge-type tool. Damage to the seal or bushing may result.**

j. Lower the lower control arms in order to disengage the steering knuckle. If necessary, use ball joint removal tool J 43631.

k. Mark the frame to body position with a paint pen or permanent marker.

l. Lower the vehicle to about 3 feet off the ground in order to place a hydraulic lift table under the frame.

m. Use two 2 x 4 pieces of wood between the lift table and the frame and lift the table to the frame.

n. Slowly remove the frame bolts using the following sequence:

- Front frame bolts
- Rear frame bolts

o. Slowly lower the lift table to the floor.

13. Remove the 2 bolts from the transmission brace.

14. Disconnect the shift cable from the shift linkage.

15. Disconnect the cable from the bracket.

16. Remove the flywheel inspection cover.

17. Remove the starter

18. Mark the relationship of the flywheel to the torque converter for reassembly.

19. Use the a flywheel holding tool to prevent the crankshaft from rotating.

20. Remove the torque converter to flywheel bolts.

21. Remove the transmission cooler lines by removing the nut holding the bracket to the transaxle case.

22. Disconnect the VSS wiring harness from the sensor.

23. Disconnect the wheel drive shafts from the transaxle.

24. Remove the body to transmission mount bolts.

25. Lower the transmission with the engine support fixture enough to remove the transmission.

26. Raise the vehicle.

27. Remove the oil pan to drain the transmission.

28. Support the transmission with a suitable jack.

29. Remove the transaxle to engine nut.

30. Separate the engine and the transaxle.

31. Remove the transaxle from the vehicle.

To install:

32. Position the transaxle in the vehicle.

➡**Use the correct fastener in the correct location. Replacement fasteners must be the correct part number for that application. Fasteners requiring replacement or fasteners requiring the use of thread locking compound or sealant are identified in the service procedure. Do not use paints, lubricants, or corrosion inhibitors on fasteners or fastener joint surfaces unless specified. These coatings affect fastener torque and joint clamping force and may damage the fastener. Use the correct tightening sequence and specifications when installing fasteners in order to avoid damage to parts and systems.**

33. Install the lower transmission to engine bolts and nuts. Tighten to 66 ft. lbs. (90 Nm).

34. Connect the transaxle cooler pipes to the transaxle. Tighten to 71 inch lbs. (8 Nm).

35. Use a flywheel holding tool to prevent the crankshaft from rotating.

36. Install the torque converter to flywheel bolts. Tighten to 46 ft. lbs. (62 Nm).

37. Install the wheel drive shafts to the transaxle.

38. Connect the wiring harness to the VSS.

39. Install the starter.

40. Install the flywheel inspection cover bolts.

41. Use the engine support fixture to raise the engine and transmission assembly.

42. Install the transaxle mount to body bolts. Tighten to 66 ft. lbs. (90 Nm).

43. Install the frame as follows:

a. With the frame on the lift table, raise the frame to the vehicle.

b. Hand start all the frame bolts while aligning the frame to the paint marks.

c. Tighten the frame bolts. Tighten the bolts 74 ft. lbs. (100 Nm) plus and additional 180.

44. Lower and remove the hydraulic table.

a. Connect the lower control arm to the steering knuckle.

➡**The torque sequence must be followed in the order that is listed.**

b. Install the ball joint pinch bolt and nut. Tighten the nut to 37 ft. lbs. (50 Nm), back of the nut ¾ turn then retighten the nut to 37 ft. lbs. (50 Nm) plus an additional 30 degree turn.

➡**The front and rear transmission mounts must be allowed to settle with the through bolts loosened.**

c. Hand start the front transaxle mount through bolt.

d. Loosen the rear transmission mount through bolt.

e. Tighten the rear transaxle mount to frame bolts to 37 ft. lbs. (50 Nm)

f. Tighten the rear transaxle mount through bolts to 74 ft. lbs. (100 Nm), then tighten the front transaxle mount through bolts to 74 ft. lbs. (100 Nm).

g. Install the outer tie rods to the steering knuckles. Install new outer tie rod to the knuckle nuts and tighten to 15 ft. lbs. (20 Nm) plus an additional 180 degree turn.

h. Connect the stabilizer links to the stabilizer shaft. Tighten to 81 ft. lbs. (100 Nm).

i. Connect the intermediate shaft to the steering gear. Install a new intermediate shaft pinch bolt and tighten to 25 ft. lbs. (34 Nm).

j. Install the splash shields.

k. Install the front wheels.

45. Route and clip the wheel speed sensor wiring into the proper position on both sides.

46. Install the steering gear to the front suspension crossmember.

47. Install the steering gear bolts to the front suspension crossmember. Tighten to 81 ft. lbs. (100 Nm).

48. Install the transmission to engine brace bolts. Tighten to 53 ft. lbs. (72 Nm).

49. Install the upper transmission to engine bolts and stud. Tighten to 66 ft. lbs. (90 Nm).

50. Install the engine wiring harness grounds to the transaxle to engine mount stud and nut.

51. Remove the engine support fixture

52. Install the shift linkage to the transmission.

53. Connect the electrical connectors to the PNP switch and transaxle.

54. Connect the air duct hose to the intake plenum.

55. Connect the negative battery cable.

56. Fill the transmission to the proper level.

Clutch

ADJUSTMENT

The hydraulic clutch system is self-adjusting.

REMOVAL & INSTALLATION

1.9L Engines

1. Before servicing the vehicle, refer to the precautions section.

2. Remove the transaxle from the vehicle.

3. Unsnap the release fork from the ball stud, then remove the fork and bearing from the vehicle. Slide the bearing from the fork. The bearing should be checked for excessive play and for minimal bearing drag. It should be replaced if no/little drag or excessive play is found.

➡**The release bearing is packed with grease and should not be washed with solvent.**

4. Using a feeler gauge, measure the distance between the pressure plate and flywheel surfaces in order to determine clutch face thickness. Replace the clutch disc if it is not within 0.18 –0.28 in. (4.65mm– 7.10mm).

5. Remove the pressure plate-to-flywheel bolts in a progressive crisscross pattern to prevent warping the cover, then remove the pressure plate and clutch disc.

6. Inspect the pressure plate, as follows:

a. Check for excessive wear, chatter marks, cracks or overheating (indicated by a blue discoloration). Black random spots on the friction surface of the pressure plate is normal.

b. Check the plate for warpage using a straightedge and a feeler gauge; the maximum allowable warpage is 0.006 in. (0.15mm).

c. Replace the plate, if necessary.

7. Inspect the clutch disc, as follows:

a. Check the disc face for oil or burnt spots.

b. Check the disc for loose damper springs, hub or rivets.

c. Replace the disc, if necessary.

8. Check the flywheel, as follows:

a. Check the ring gear for wear or damage.

b. Check the friction surface for excessive wear, chatter marks, cracks or overheating.

c. Check flywheel thickness; the minimum allowable is 1.102 in. (28mm).

d. Measure flywheel run-out using a dial indicator, positioned for at least 2 flywheel revolutions. Push the crankshaft forward to take up thrust bearing clearance. Maximum flywheel run-out is 0.006 in. (0.15mm).

e. Check the flywheel for warpage using a straight-edge and a feeler gauge; the maximum allowable warpage is 0.006 in. (0.15mm).

f. Replace the flywheel, if necessary.

9. If necessary, remove the flywheel retaining bolts and remove the flywheel from the crankshaft.

To install:

10. If removed, install the flywheel and torque the bolts in a crisscross sequence to 59 ft. lbs. (80 Nm).

11. Install or connect the following:

- Clutch disc and pressure plate with the yellow dot on the pressure plate aligned as close as possible to the mark on the flywheel. The clutch disc is labeled FLYWHEEL SIDE in order to help correctly position the disc. Loosely install the pressure plate bolts.
- Clutch alignment tool in the clutch disc and install the release bearing to the fork and torque the pressure plate bolts to 18 ft. lbs. (25 Nm)
- Snap the release bearing and fork onto the ball stud
- Lubricate the splines of the input

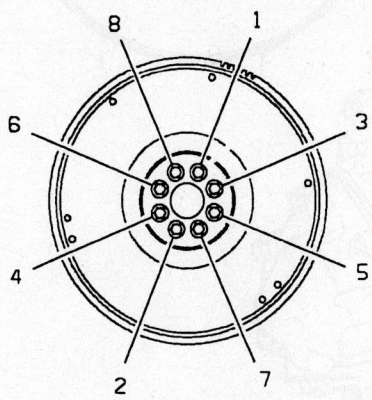

Flywheel mounting bolt tightening sequence—1.9L engines

79222831

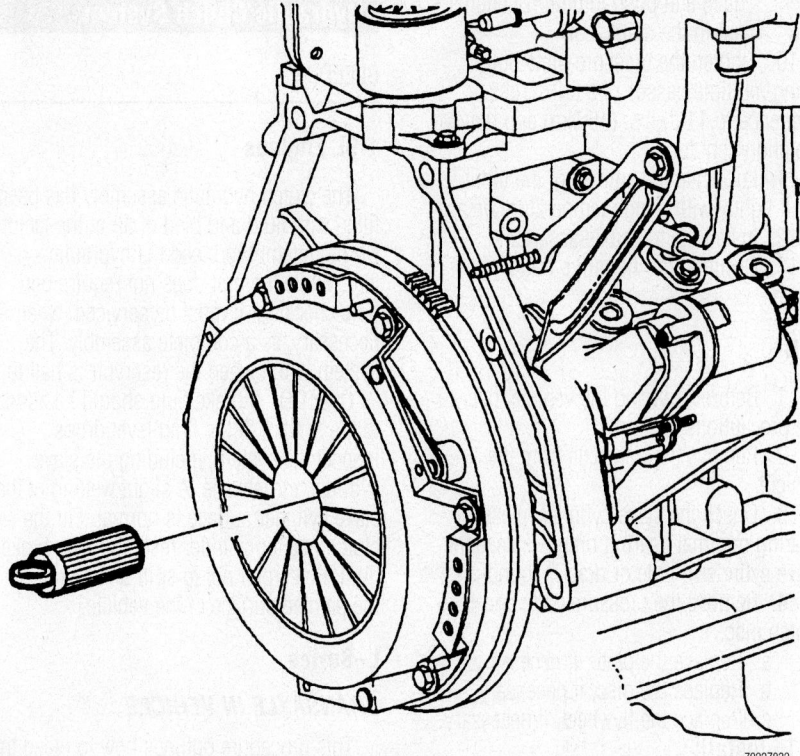

79222832

Use a proper clutch alignment tool before tightening the pressure plate retaining bolts—1.9L engines

shaft lightly with a high temperature grease
- Transaxle assembly
- Negative battery cable

L–Series

1. Before servicing the vehicle, refer to the precautions section.

2. Remove the transaxle from the vehicle.

3. Check the slave cylinder release bearing minimal bearing drag. Replace the slave cylinder if little or drag is found.

4. Remove the pressure plate and clutch disc.

5. Inspect the pressure plate, as follows:

a. Check for excessive wear, chatter marks, cracks or overheating (indicated by a blue discoloration). Black random spots on the friction surface of the pressure plate is normal.

b. Check the plate for warpage using a straightedge and a feeler gauge; the maximum allowable warpage is 0.006 in. (0.15mm).

c. Replace the plate, if necessary.

6. Inspect the clutch disc, as follows:

a. Check the disc face for oil or burnt spots.

b. Check the disc for loose damper springs, hub or rivets.

c. Replace the disc, if necessary.

7. Check the flywheel, as follows:

a. Check the ring gear for wear or damage.

b. Check the friction surface for excessive wear, chatter marks, cracks or overheating.

c. Check flywheel thickness; the minimum allowable is 1.102 in. (28mm).

d. Measure flywheel run-out using a dial indicator, positioned for at least 2 flywheel revolutions. Push the crankshaft forward to take up thrust bearing clearance. Maximum flywheel run-out is 0.006 in. (0.15mm).

e. Check the flywheel for warpage using a straight-edge and a feeler gauge; the maximum allowable warpage is 0.006 in. (0.15mm).

f. Replace the flywheel, if necessary.

8. If necessary, remove the flywheel retaining bolts and remove the flywheel from the crankshaft.

To install:

9. Install or connect the following:

- Flywheel (if removed) and torque the bolts in a crisscross pattern to 39 ft. lbs. (53 Nm) plus 25 degrees
- Clutch disc and pressure plate and loosely install the pressure plate bolts
- Clutch alignment tool in the clutch

disc, and push in until it bottoms out in the crankshaft

10. Tighten the pressure plate bolts using multiple passes of a crisscross sequence to 11 ft. lbs. (15 Nm) and remove the alignment tool.

11. Lubricate the splines of the input shaft lightly with a high temperature grease.

12. Install the transaxle assembly.

13. Connect the negative battery cable.

ION

1. Before servicing the vehicle, refer to the precautions section.

2. Remove the transaxle from the vehicle.

3. Check the slave cylinder release bearing minimal bearing drag. Replace the slave cylinder if little or drag is found.

4. Remove the pressure plate and clutch disc.

   a. Replace the plate, if necessary.

   b. Replace the disc, if necessary.

   c. Replace the flywheel, if necessary.

To install:

5. Install or connect the following:

- Clutch disc and pressure plate and loosely install the pressure plate bolts
- Clutch alignment tool in the clutch disc, and push in until it bottoms out in the crankshaft

6. Tighten the pressure plate bolts in sequence to 18 ft. lbs. (24 Nm) on all except the MU–3 transmission and tighten to 22 ft. lbs. (30 Nm) on the MU–3 transmission.

7. Remove the alignment tool.

8. Lubricate the splines of the input shaft lightly with a high temperature grease.

9. Install the transaxle assembly.

10. Connect the negative battery cable.

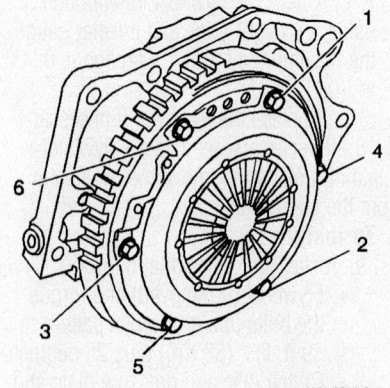

Tighten the pressure plate bolts in sequence —ION

42372-SCAR-G18

Hydraulic Clutch System

BLEEDING

1.9L Engines

The clutch hydraulic assembly has been filled with fluid and bled of air at the factory. Do not attempt to bleed the hydraulic system. While the unit does not require periodic checking, it must be serviced, when necessary, as a complete assembly. The system is full when the reservoir is half full.

Only DOT 3 brake fluid should be added to the system. If the fluid level drops, inspect the system, including the slave cylinder, for leakage. A slight wetting of the slave cylinder surface is normal. Fill the clutch master cylinder reservoir with brake fluid. Be careful not to spill brake fluid on the painted surface of the vehicle.

L–Series

TRANSAXLE IN VEHICLE

This procedure outlines how to bleed the hydraulic clutch with the transaxle in the vehicle. Only **DOT 3** brake fluid should be added to the system.

1. Before servicing the vehicle, refer to the precautions section.

2. Remove or disconnect the following:

3. Remove the reservoir cap and fill the reservoir with new brake fluid.

4. Install a Bleeder Adapter Tool J43915, to the reservoir and connect a pressure bleeder to the adapter.

5. Charge the pressure bleeder to 20–25 psi (138–172 kPa).

6. Attach a transparent hose over the clutch bleeder screw nipple and submerge the opposite end of the hose in a container of brake fluid.

7. Loosen the bleeder screw on the transaxle hydraulic fitting.

8. Bleed the system until no air bubbles are seen in the hose.

9. Tighten the bleeder screw.

10. Check the clutch pedal for a spongy feel. If the pedal feels soft, repeat the bleeding procedure.

11. Remove the bleeder tools and top off the fluid level if necessary.

TRANSAXLE OUT OF VEHICLE

This procedure outlines how to bleed the clutch slave cylinder while the transaxle is out of the vehicle. Only **DOT 3** brake fluid should be added to the system.

1. Before servicing the vehicle, refer to the precautions section.

2. Remove or disconnect the following:

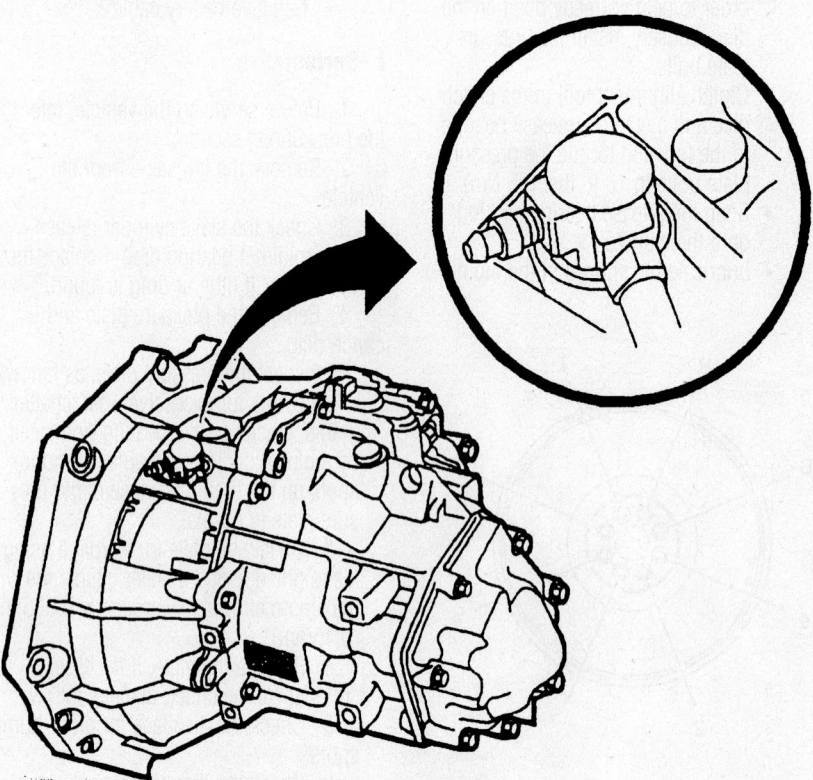

Loosen the clutch bleeder screw on the hydraulic fitting—L-Series

9306ZG63

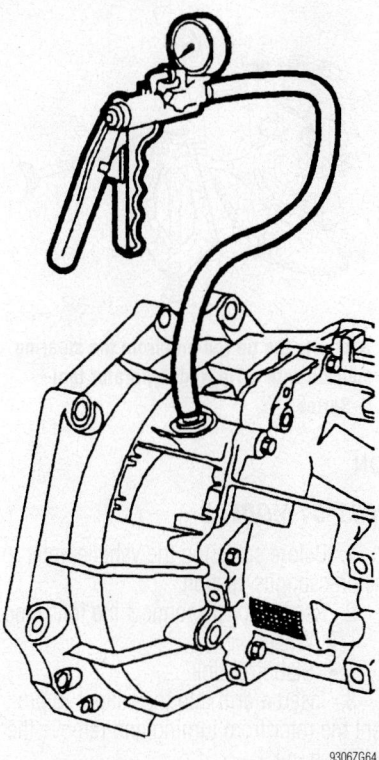

9306ZG64

Install a vacuum pump and pressure adapter to bleed the slave cylinder—L–Series

3. Connect a long piece of transparent hose to the slave cylinder fitting.

4. Depress the release bearing on the slave cylinder towards the clutch housing and release it.

5. The release bearing will spring back, but the piston will remain depressed next to the clutch housing.

6. Fill the hose with 13.8 inches (350mm) of DOT 3 brake fluid.

7. Connect a Vacuum Pump SA9180NE with a Pressure Adapter J35555-92, to the top of the hose.

8. Apply pressure to the hose to force the brake fluid into the slave cylinder. Pressure in the gauge will increase when the slave cylinder piston reaches the end of its travel.

9. Remove the vacuum pressure pump from the hose and depress the release bearing. Air bubbles should appear in the hose.

10. Repeat this procedure until no air bubbles are visible in the hose.

11. Drain and remove the hose.

ION

VACUUM BLEEDING

This procedure outlines how to bleed the hydraulic clutch with the transaxle in the vehicle. Only **DOT 3** brake fluid should be added to the system.

1. Before servicing the vehicle, refer to the precautions section.

2. Blead the system as follows:

 a. Remove the reservoir cap and fill the reservoir with new brake fluid.

 b. Install a metal mityvac too J 3555 Tool and adapter J43485.

 c. Hold J 43485 in position and apply 15–20 hg (51–68 kPa) of vacuum.

 d. Remove the adapter and refill the reservoir.

 e. Depress the clutch pedal and cycle it for 30 seconds.

 f. Lift the pedal up to the stop position and hold for 30 seconds.

 g. Repeat steps A through F until all air is removed.

 h. Place the transmission in Neutral.

 i. Start the engine and pump the clutch pedal until firm.

 j. Pump the brake pedal until firm.

 k. Add any additional fluid if needed.

MANUAL BLEEDING

This procedure outlines how to bleed the hydraulic clutch with the transaxle in the vehicle. Only **DOT 3** brake fluid should be added to the system.

1. Before servicing the vehicle, refer to the precautions section.

2. Remove the reservoir cap and fill the reservoir with new brake fluid.

3. Attach a transparent hose over the clutch bleeder screw nipple and submerge the opposite end of the hose in a container of brake fluid.

4. Depress the clutch pedal quickly to the fully depressed position.

5. Push the clip in order to move the clutch line into the bleed position.

6. Move the clutch line into normal position making sure the clip returns.

7. Lift the clutch pedal to the up position and hold for 5 seconds.

8. Bleed the system until no air bubbles are seen in the hose.

9. Check the clutch pedal for a spongy feel. If the pedal feels soft, repeat the bleeding procedure.

10. Remove the bleeder tools and top off the fluid level if necessary.

Halfshafts

REMOVAL & INSTALLATION

S–Series

1. Before servicing the vehicle, refer to the precautions section.

2. Remove the wheel cover or the center cap for access to the halfshaft nut. Have an assistant depress the brake pedal and loosen the front halfshaft nut.

3. Remove or disconnect the following:
- Wheel and splash shields
- 2 push pins
- Lower shield to cradle molded-in fasteners at the cradle
- Drain the transaxle fluid if replacing the left side halfshaft
- Drive axle nut and washer and discard them
- Lower control arm to steering knuckle cotter pin and discard it
- Loosen the lower control arm to steering knuckle castle nut so that the top of the nut is even with the top of the ball stud

✷✷ WARNING

The outer CV-joint, if equipped with Anti-lock Brake System (ABS), contains a speed sensor ring. Use of an incorrect tool to separate the control arm from the knuckle may result in damage and loss of the ABS system.

- Lower control arm from the steering knuckle by using a Lower Control Arm Separator Tool SA9132S
- Cotter pin and castle nut and discard them, if equipped
- Torque prevailing nut and discard it, if equipped
- Separate the tie rod end from the steering knuckle using a Tie Rod Separator Tool, SA91100C
- Lower control arm to steering knuckle castle nut
- Position a prybar at the proper cradle and front stabilizer and place a cloth over the ball stud boot
- Separate the lower control arm from the steering knuckle and pull the knuckle away from the ball stud
- While pulling the knuckle/strut away, pull the outer end of the halfshaft out of the wheel hub and properly support the halfshaft

4. For right side halfshafts, remove the halfshaft from the intermediate shaft by tapping the axle with a block of wood and hammer. Separate the halfshaft from the intermediate shaft.

5. For left side halfshafts, remove the halfshaft by installing a prybar and prying the halfshaft from the transaxle.

To install:

➡**Be careful not to damage the halfshaft oil seal when installing the halfshaft into the transaxle.**

6. Install or connect the following:

- Transaxle Seal Protector Tool SA91112T, when installing the left side halfshaft
- Halfshaft into the transaxle. After the splines have safely passed the oil seal, remove the protector tool
- Insert the inner end of the halfshaft onto the outer end of the intermediate shaft for the right side halfshaft. Push on the axle firmly to engage the retaining ring
- Outer end of the halfshaft into the wheel hub. Do not install any hardware at this time
- Lower control arm ball stud into the steering knuckle. Hand-tighten the castle nut at this time
- Attach the tie rod end to the steering knuckle, if equipped. Torque the nut to 33 ft. lbs. (45 Nm) and install a new cotter pin
- If equipped with a torque prevailing nut, attach the tie rod to steering knuckle with a new nut. Do not allow the ball stud to turn while tightening the prevailing nut. Torque the nut to 41 ft. lbs. (55 Nm).
- Axle to hub washer and new nut. Torque the nut to 148 ft. lbs. (200 Nm).

➡**When tightening the axle nut, have an assistant depress the brake pedal to prevent axle rotation.**

- Splash shields
- Align the molded in shield fasteners with the holes in the cradle and install push pins
- Wheel

7. Top off the transaxle with the proper fluid.

8. Check and adjust the front end alignment as necessary.

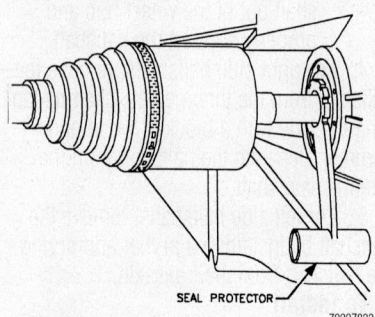

Failure to use a seal protector may allow the halfshaft splines to damage the transaxle seal

L-Series

1. Before servicing the vehicle, refer to the precautions section.

2. Remove the wheel cover or the center cap for access to the halfshaft nut. Have an assistant depress the brake pedal and loosen the front halfshaft nut.

3. Remove or disconnect the following:

- Wheel
- Tie rod end torque prevailing nut and discard it
- Tie rod end from the steering knuckle, using a Tie Rod Separator Tool SA91100C
- Lower control arm to steering knuckle

➡**Do not allow the steering knuckle to contact the ball stud seal. Contact may cause the seal to rip and the ball stud will need replacement.**

- Pull the steering knuckle/strut assembly away from the vehicle and pull the halfshaft out of the hub
- Properly support the halfshaft and remove the halfshaft from the transaxle
- Shaft retaining ring and discard it

To install:

4. Apply Output Shaft Lubricant 7847638, to the splines of the output shaft, if equipped with an automatic transaxle.

5. Install or connect the following:

- New stub shaft retaining ring
- Halfshaft to the transaxle after installing a Seal Protector Tool SA91112T

6. Remove the seal protector tool after the splines have passed the oil seal.

7. Install or connect the following:

- Fully seat the halfshaft into the transaxle
- Outer end of the halfshaft to the wheel hub with a new washer and nut
- Lower control arm ball stud to the steering knuckle. Torque the fastener to 75 ft. lbs. (100 Nm).
- Tie rod end to the steering knuckle. When seated properly, torque the fastener to 45 ft. lbs. (60 Nm).
- Wheel
- Halfshaft to wheel nut. Torque the nut to 74–118 ft. lbs. (100–160 Nm), release the nut until it turns freely and torque the nut to 15 ft. lbs. (20 Nm) plus 90 degrees.
- Cotter pin

8. Top off the transaxle with the proper fluid.

9. Check and adjust the front end alignment as necessary.

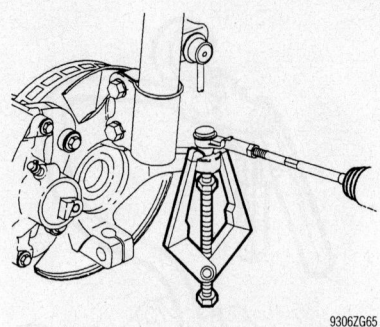

Separate the tie rod end from the steering knuckle with a Tie Rod Separator tool—L-Series

ION

2003–04 MODELS

1. Before servicing the vehicle, refer to the precautions section.

2. Remove or disconnect the following:

- Wheel
- Stabilizer link

3. Insert a drift into the caliper to prevent the rotor from turning and remove the half shaft nut.

- Tie rod end from the steering knuckle
- Wheel speed sensor connection. Make sure to position the harness away from the ball joint after disconnecting.
- Lower ball joint from the steering knuckle
- Drive shaft from the knuckle using a hammer and a block of wood. Partially install the nut before striking the wood with the hammer to prevent thread damage.

4. Using slide hammer SA 9173G and drive shaft removal tool J 45341, remove the shaft from the transmission.

To install:

5. Apply Output Shaft Lubricant 7847638, to the splines of the output shaft, if equipped with an automatic transaxle.

6. Install or connect the following:

- New stub shaft retaining ring
- Halfshaft to the transaxle after installing a Seal Protector Tool SA91112T

7. Remove the seal protector tool after the splines have passed the oil seal.

8. Install or connect the following:

- Fully seat the halfshaft into the transaxle
- Outer end of the halfshaft to the wheel hub with a new washer and nut
- Ball joint to the steering knuckle. Torque the fastener to 44 ft. lbs. (60

Nm), then an additional 30 degree turn.

- Tie rod end to the steering knuckle. When seated properly, torque the fastener to 15 ft. lbs. (20 Nm) plus an additional 180 degree turn.
- Halfshaft to wheel nut. Torque the nut to 81 ft. lbs. (110 Nm
- Cotter pin
- Stabilizer link and tighten the nut to 63 ft. lbs. (85 Nm)
- Wheel

9. Top off the transaxle with the proper fluid.

10. Check and adjust the front end alignment as necessary.

2005 MODELS

1. Before servicing the vehicle, refer to the precautions section.

2. Remove the wheel assembly.

3. Remove the wheel drive shaft nut by inserting a drift or a flat-bladed tool into the caliper and the rotor to prevent the rotor from turning.

4. Carefully loosen the wheel drive shaft splines from the wheel bearing/hub assembly using a wood block and a hammer. The nut can be partially reinstalled to protect the threads.

➡**Be sure that the wheel speed sensor wiring harness is repositioned away from the ball joint after disconnecting the electrical connector from the sensor.**

5. Disconnect the ball joint from the steering knuckle.

6. Separate the wheel drive shaft from the wheel bearing/hub assembly, then support the shaft assembly.

7. Using slide hammer SA 9173G and drive shaft removal tool J 45341, remove the shaft from the transmission.

To install:

8. Inspect the transaxle output shaft seal for damage and/or contamination and replace if necessary.

9. Install the a Seal Protector Tool SA91112T.

10. Install the wheel drive shaft into the transaxle until the drive shaft splines are past the seal, remove the seal protector tool.

11. Verify that the wheel drive shaft is properly engaged by grasping the inner tri-pod housing and pull the inner housing outward. Do NOT pull on the wheel drive axle shaft. The wheel drive shaft will remain firmly in place when properly engaged.

12. Install the wheel drive shaft to the wheel bearing/hub assembly.

13. Connect the ball joint to the steering knuckle and tighten as follows:

   a. First Pass: Tighten the nut to 37 ft. lbs. (50 Nm).

   b. Reverse the nut ¾ of a turn.

   c. Second Pass: Tighten the nut to 37 ft. lbs. (50 Nm) plus an additional 30 degrees.

14. Install the wheel drive shaft nut to the wheel drive shaft assembly and tighten to 155 ft. lbs. (210 Nm).

15. Install the wheel assembly.

16. Inspect the transaxle fluid level and wheel alignment.

CV-Joints

OVERHAUL

L and S–Series

INNER

1. Before servicing the vehicle, refer to the precautions section.

2. Remove or disconnect the following:

- Front wheel
- Halfshaft
- Swage ring using a hand grinder
- Large CV-joint boot clamp
- CV-joint boot by sliding it away from the tri-pod joint
- Tri-pod housing from the tri-pod spider
- Inboard spacer ring slide it rearward on the shaft
- Outboard retaining ring
- Tri-pod joint spider assembly
- Inboard spacer ring and CV-joint boot

To install:

3. Install or connect the following:

- Swage ring clamp on the CV-joint boot
- CV-joint boot

4. Position the CV-joint boot seal into the axle shaft's joint seal groove and align the swage ring clamp on the boot.

✳✳ WARNING

Make sure that there are no pinch points on the inboard seal.

5. Crimp the swage ring

6. Install or connect the following:

- Inboard spacer ring, slide it rearward on the shaft
- Tri-pod joint spider assembly onto the shaft
- Outboard retaining ring into the axle shaft groove

- Tri-pod joint spider assembly, slide it against the outboard retaining ring
- Inboard spacer ring, seat it in the groove
- ½ kit grease into the boot
- ½ kit grease into the tri-pod housing
- New large seal clamp onto the CV-joint boot
- Tri-pod housing, slide it over the tri-pod joint spider assembly
- CV-joint boot/clamp, slide it into place, over the trilobal tri-pod bushing with the seal lip in the groove

➡**Make sure the boot lies flat against the trilobal bushing.**

7. Position the CV-joint boot so it measures 4.9 in. (125mm).

8. Using a Crimp tool, a torque wrench and a breaker bar, crimp the large CV-joint boot clamp to 130 ft. lbs. (176 Nm).

9. Install the halfshaft and the front wheel.

OUTER

1. Before servicing the vehicle, refer to the precautions section.

2. Remove or disconnect the following:

- Axle shaft from the vehicle
- Large CV boot retaining clamp
- Small CV boot retaining clamp
- CV boot from the joint
- Axle shaft retaining ring
- Outer joint from the axle shaft
- CV boot

3. Disassemble the chrome alloy balls from the CV-joint cage as follows:

   a. Position a brass drift against the CV-joint cage and tap it with a hammer to tilt the cage.

   b. Remove the 1st chrome alloy ball from the cage.

   c. Tilt the cage in the opposite direction.

   d. Remove the opposite chrome alloy ball.

   e. Repeat the procedure until all 6 balls are removed.

4. Disassemble the CV-joint cage and inner race as follows:

   a. Pivot the cage and race 90 degrees to the center line of the outer race.

   b. Align the cage windows with outer race lands.

   c. Remove the cage from the outer race.

   d. Rotate the inner race upward and remove it from the cage.

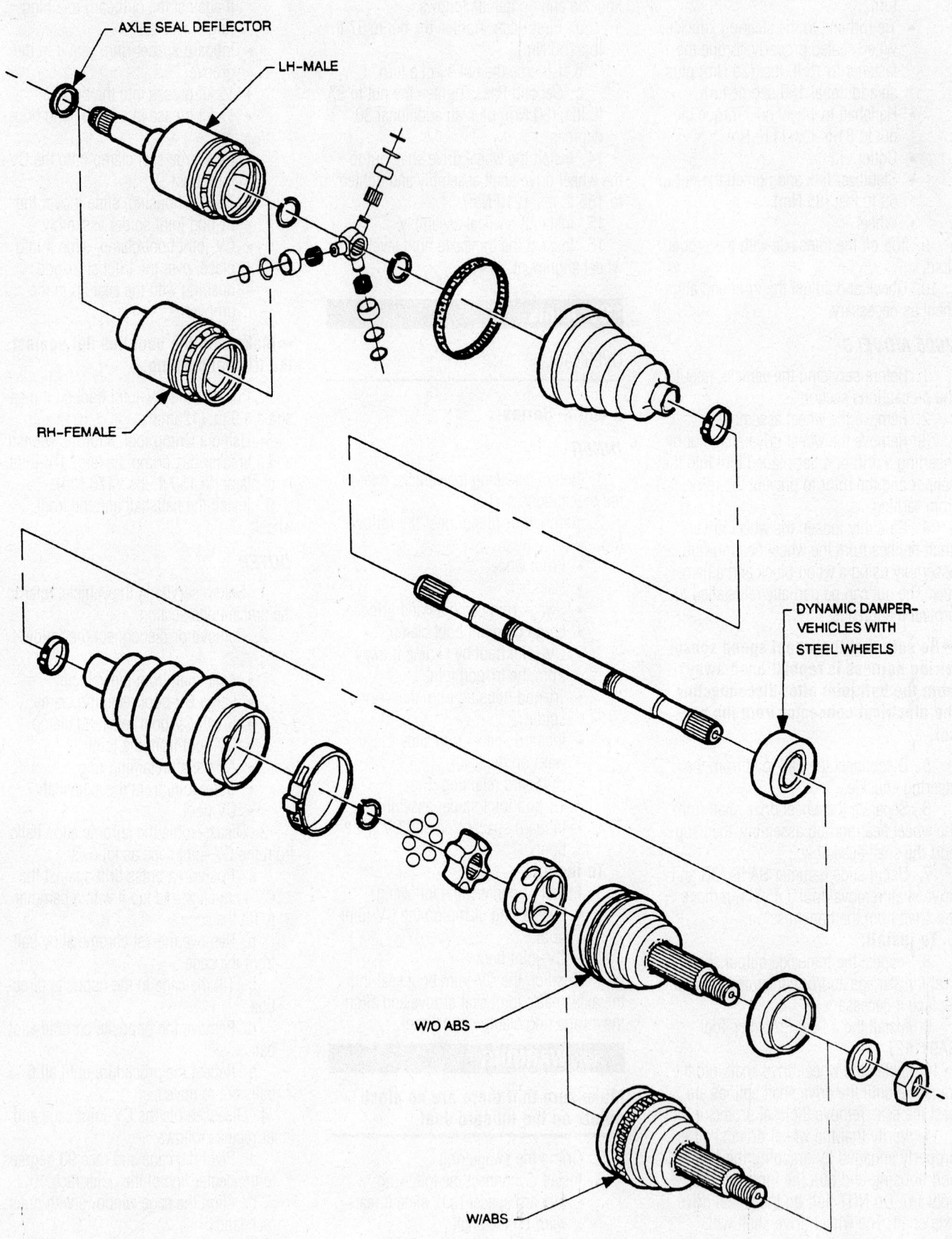

AXLE SEAL DEFLECTOR

LH-MALE

RH-FEMALE

DYNAMIC DAMPER-
VEHICLES WITH
STEEL WHEELS

W/O ABS

W/ABS

9346ZG28

Exploded view of axle shaft—S series shown, L series and ION are similar

To install:

5. Lubricate the parts with a light coat of grease.

6. Assemble the CV-joint cage and inner race, as follows:

   a. Rotate the inner race 90 degrees to the cage centerline.

   b. Align the cage windows with inner race lands.

   c. Insert the inner race into the cage by rotating the inner race downward.

   d. Insert the cage/inner race into the outer race.

7. Assemble the chrome alloy balls into the CV-joint cage, as follows:

   a. Position a brass drift against the CV-joint cage and tap it with a hammer to tilt the cage.

   b. Insert the 1st chrome alloy ball into the cage.

   c. Tilt the cage in the opposite direction.

   d. Insert the opposite chrome alloy ball.

   e. Repeat the procedure until all 6 balls are inserted.

8. Install ½ of the grease provided, into the CV-joint.

9. Install or connect the following:
- Small CV boot retaining ring
- CV boot on the halfshaft
- New retaining ring on the halfshaft
- Large ring clamp on the CV boot
- Outer joint onto the axle shaft
- Retaining ring into the outer race

10. Install the remaining grease into the CV boot.

11. Position the CV boot and the small boot clamp.

12. Crimp the small boot clamp.

13. Position and crimp in place the large boot clamp.

14. Install the Halfshaft in the vehicle.

ION

INNER

1. Before servicing the vehicle, refer to the precautions section.

2. Remove or disconnect the following:
- Front wheel
- Halfshaft
- Small boot clamp ring using a side cutter and discard
- Earless clamp and discard
- Large CV-joint boot clamp using a side cutter and discard
- CV-joint boot by sliding it away from the tri-pod joint
- Tri-pod housing from the tri-pod spider
- Inboard spacer ring slide it rearward on the shaft

- Outboard retaining ring
- Tri-pod joint spider assembly
- Inboard spacer ring and CV-joint boot

To install:

3. Install or connect the following:
- Small boot clamp onto the neck of the inboard boot but do not tighten
- Inboard boot onto the shaft

➡**The clamp must be positioned correctly during crimping to ensure a correct seal.**

4. Use boot clamp installer SA9203C to tighten the clamp making sure the end gap on the clamp does not exceed 0.118 inch (3mm)
- Spider assembly past the retaining ring groove until it is seated against the shoulder
- Spider retaining ring
- ½ kit grease into the boot
- ½ kit grease into the tri-pod housing

➡**The joint must be assembled with the convoluted retainer in position to avoid boot damage.**

- Tri-lobal housing bushing making sure it is flush with the housing
- New large seal clamp onto the CV-joint boot and slide the housing over the tri-pod joint spider assembly
- CV-joint boot/clamp, slide it into place, over the trilobal tri-pod bushing with the seal lip in the groove

➡**Make sure the boot lies flat against the trilobal bushing.**

5. Position the CV-joint boot so it measures 4.9 in. (125mm).

6. Using a boot clamp installer SA9161C tighten the large clamp.

7. Install the halfshaft and the front wheel.

OUTER

1. Before servicing the vehicle, refer to the precautions section.

2. Remove or disconnect the following:
- Axle shaft from the vehicle
- Large CV boot retaining clamp
- Small CV boot retaining clamp
- CV boot from the joint
- Axle shaft retaining ring
- Outer joint from the axle shaft
- CV boot

3. Disassemble the chrome alloy balls from the CV-joint cage as follows:

   a. Position a brass drift against the CV-joint cage and tap it with a hammer to tilt the cage.

   b. Remove the 1st chrome alloy ball from the cage.

   c. Tilt the cage in the opposite direction.

   d. Remove the opposite chrome alloy ball.

   e. Repeat the procedure until all 6 balls are removed.

4. Disassemble the CV-joint cage and inner race as follows:

   a. Pivot the cage and race 90 degrees to the center line of the outer race.

   b. Align the cage windows with outer race lands.

   c. Remove the cage from the outer race.

   d. Rotate the inner race upward and remove it from the cage.

To install:

5. Lubricate the parts with a light coat of grease.

6. Assemble the CV-joint cage and inner race, as follows:

   a. Rotate the inner race 90 degrees to the cage centerline.

   b. Align the cage windows with inner race lands.

   c. Insert the inner race into the cage by rotating the inner race downward.

   d. Insert the cage/inner race into the outer race.

7. Assemble the chrome alloy balls into the CV-joint cage, as follows:

   a. Position a brass drift against the CV-joint cage and tap it with a hammer to tilt the cage.

   b. Insert the 1st chrome alloy ball into the cage.

   c. Tilt the cage in the opposite direction.

   d. Insert the opposite chrome alloy ball.

   e. Repeat the procedure until all 6 balls are inserted.

8. Install ½ of the grease provided, into the CV-joint.

9. Install or connect the following:
- Small CV boot retaining ring
- CV boot on the halfshaft
- New retaining ring on the halfshaft
- Large ring clamp on the CV boot
- Outer joint onto the axle shaft
- Retaining ring into the outer race

10. Install the remaining grease into the CV boot.

11. Position the CV boot and the small boot clamp.

12. Crimp the small boot clamp making sure not exceed an end gap of 0.118 inch (3mm).

13. Position and crimp in place the large boot clamp making sure not exceed an end gap of 0.118 inch (3mm)

14. Install the Halfshaft in the vehicle.

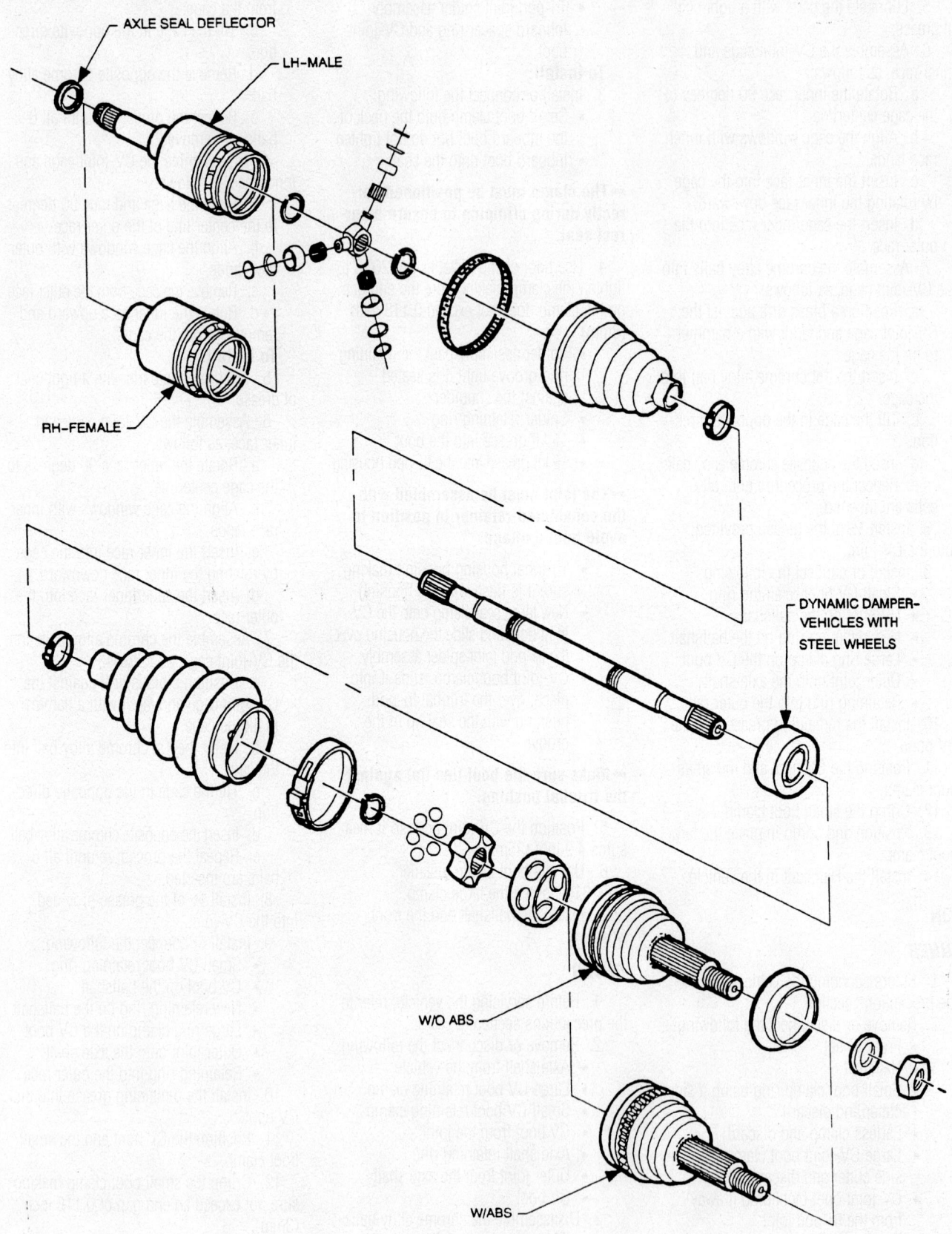

AXLE SEAL DEFLECTOR

LH-MALE

RH-FEMALE

DYNAMIC DAMPER-
VEHICLES WITH
STEEL WHEELS

W/O ABS

W/ABS

9346ZG28

Exploded view of axle shaft—S series shown, L series and ION are similar

Air Bag

✳✳ CAUTION

All vehicles are equipped with an air bag system. The system must be disabled before performing service on or around system components, steering column, instrument panel components, wiring and sensors. Failure to follow safety and disabling procedures could result in accidental air bag deployment, possible personal injury and unnecessary system repairs.

PRECAUTIONS

Several precautions must be observed when handling the inflator module to avoid accidental deployment and possible personal injury.

1. Never carry the inflator module by the wires or connector on the underside of the module.

2. When carrying a live inflator module, hold securely with both hands, and ensure that the bag and trim cover are pointed away.

3. Place the inflator module on a bench or other surface with the bag and trim cover facing up.

4. With the inflator module on the bench, never place anything on or close to the module which may be thrown in the event of an accidental deployment.

DISARMING

Except ION

1. Before servicing the vehicle, refer to the precautions section.

2. Align the steering wheel so the vehicle wheels are pointing in the straight-ahead position.

3. Turn the ignition switch to the **LOCK** position.

4. Remove the SIR or AIR BAG fuse from the fuse block.

5. Remove the Connector Position Assurance (CPA) device, then disengage the yellow 2-way SIR wiring harness connector at the base of the steering column.

ION

1. Before servicing the vehicle, refer to the precautions section.

2. Align the steering wheel so the vehicle wheels are pointing in the straight-ahead position.

3. Turn the ignition switch to the **LOCK** position.

4. Remove the SIR or AIR BAG fuse from the fuse block.

5. To disable the roof rail module left and seat belt pre-tensioner (left front), remove the garnish molding from the upper lock pillar. Remove the Connector Position Assurance (CPA) device from the roof rail module. Disconnect the roof rail module from the vehicle harness. Lower the headliner and remove the CPA from the seat belt pre-tensioner connector and disconnect the pre-tensioner connector.

ARMING

Except ION

After the repairs, enable the system as follows:

1. Turn the ignition switch to the **LOCK** position.

2. Engage the yellow 2-way connector at the base of the steering column, then install the CPA device.

3. Reinstall the Supplemental Inflatable Restraint (SIR) or AIR BAG fuse.

4. Turn the ignition switch to the **RUN** position.

5. Verify the SIR indicator light flashes 7–9 times, if not, inspect the system for malfunction.

ION

After the repairs, enable the system as follows:

1. Turn the ignition switch to the **LOCK** position and remove the key.

2. Connect the pre-tensioner and install the CPA.

3. Install the headliner.

4. Connect the roof rail module and install the CPA.

5. Install the garnish molding.

6. Reinstall the Supplemental Inflatable Restraint (SIR) or AIR BAG fuse.

7. Turn the ignition switch to the **RUN** position.

8. Verify the SIR indicator light flashes 7 times, if not, inspect the system for malfunction.

Rack and Pinion Steering Gear

REMOVAL & INSTALLATION

1.9L Engines

1. Before servicing the vehicle, refer to the precautions section.

2. Remove or disconnect the following:
- Negative battery cable
- Both front wheels

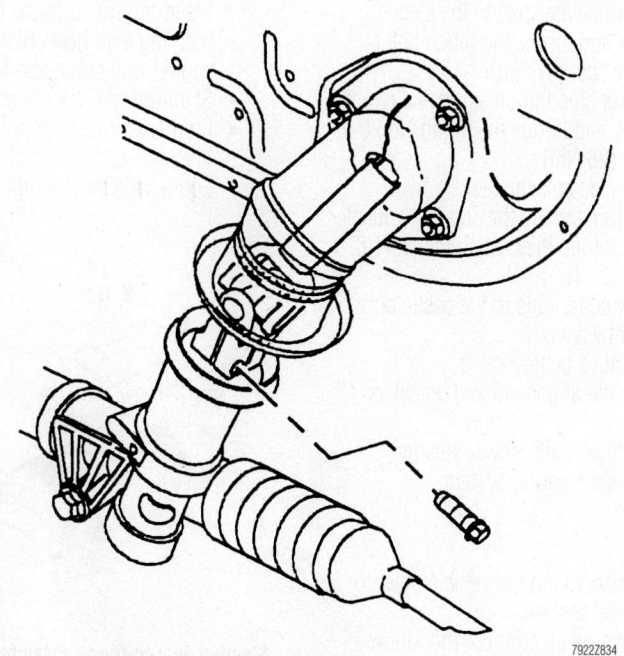

The pinch bolt is located under the intermediate shaft cover—1.9L engines

79222834

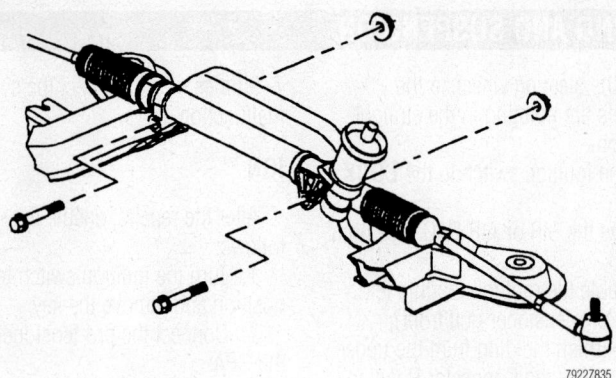

Remove the steering gear-to-cradle fasteners, then remove the gear through the left fender well—1.9L engines

7922Z835

- Outer tie rod from the steering knuckle
- Left side splash shield
- Loosen the intermediate shaft cover from the steering gear and move it aside
- Intermediate shaft pinch bolt

3. On vehicles with power steering, place a suitable container under the steering assembly. Disconnect the pressure and return lines at the steering gear and allow the system to drain.

4. Remove the steering gear to cradle fasteners.

5. Remove the steering gear through the left fender well.

To install:

6. Install or connect the following:
- Steering gear to cradle with new nuts and torque the nuts to 37 ft. lbs. (50 Nm)
- Intermediate shaft to the steering gear and torque the pinch bolt to 35 ft. lbs. (47 Nm)
- Power steering pressure and return lines and torque the fittings to 20 ft. lbs. (28 Nm)
- Left side splash shield
- Tie rod ends to the steering knuckle and torque the castle nuts to 41 ft. lbs. (55 Nm)
- New cotter pins to the castle nuts
- Front wheels
- Negative battery cable

7. Check the alignment and adjust as necessary.

8. On vehicles with power steering, bleed the power steering system.

L–Series

1. Before servicing the vehicle, refer to the precautions section.

2. Remove or disconnect the following:

- Negative battery cable
- Both front wheels
- Intermediate shaft from the steering gear pinch bolt from inside the vehicle,
- Shaft from the gear
- Rear exhaust manifold pipe heat shield (on 3.0L engine)
- Rear transaxle mount through bolt
- Transaxle mount-to-frame bolt
- Power steering pressure and return hoses
- Front wheels
- Right front lower splash shield
- Exhaust manifold pipe (on 2.2L engine with manual transaxle and all 3.0L engines)
- Remaining rear transaxle mount-to-frame bolts
- Tie rods from the steering knuckles
- Steering gear-to-frame bolts
- Steering gear heat shield (on 2.2L engine with automatic transaxle)
- Stabilizer bar links from the struts
- Front suspension support assemblies
- Loosen the remaining frame to

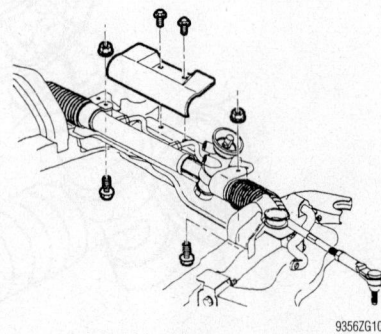

Steering gear-to-frame assembly bolts— L-series models

9356ZG10

body fasteners until there is clearance to remove the steering gear
- Steering gear through the left side wheel opening

To install:

3. One at a time remove and replace the frame bolts and nuts. The torque retention of the old fasteners may not be sufficient.

4. Install or connect the following:
- Steering gear through the left wheel opening
- Raise the frame assembly to the undercarriage and torque the fasteners to 66 ft. lbs. (90 Nm) plus a 45–60 degree turn
- Suspension supports with new bolts and torque the bolts to 66 ft. lbs. (90 Nm) plus a 45–60 degree turn
- Steering gear to the frame with new bolts and torque the bolts to 35 ft. lbs. (45 Nm) plus 45–60 degrees
- Steering gear heat shield (on 2.2L engine with an automatic transaxle) and torque the bolts to 35 inch lbs. (4 Nm)
- Tie rod ends to the steering knuckle and torque the nut to 44 ft. lbs. (60 Nm)
- Stabilizer bar links to the strut and torque the nuts to 48 ft. lbs. (65 Nm)
- Rear transaxle mount-to-frame bolts and torque them to 44 ft. lbs. (60 Nm)
- Power steering pressure and return lines and torque the fittings to 20 ft. lbs. (27 Nm)
- Exhaust manifold pipe (2.2L engines with a manual transaxle and all 3.0L engines) to the manifold and torque the nuts to 22 ft. lbs. (30 Nm)
- Exhaust manifold pipe to the exhaust pipe and torque the bolts to 15 ft. lbs. (20 Nm)
- Front wheels
- Rear transaxle mount-to-frame bolt and torque it to 45 ft. lbs. (60 Nm)
- Rear transaxle mount through bolt and torque it to 66 ft. lbs. (90 Nm)
- Exhaust manifold pipe heat shield (on 3.0L engine) and torque the bolts to 71 inch lbs. (8 Nm)
- Intermediate shaft to the steering gear and torque the new pinch bolt to 21 ft. lbs. (28 Nm)
- Negative battery cable

5. Check the alignment and adjust if necessary.

6. Bleed the power steering system.

ION

1. Before servicing the vehicle, refer to the precautions section.

2. Turn the steering wheel to the straight ahead position, remove the key and lock the steering wheel.

3. Remove or disconnect the following:
- Negative battery cable
- Both front wheels

✳✳ WARNING

Do not rotate the intermediate shaft once it is disconnected from the gear as damage may occur.

- Intermediate shaft from the steering gear pinch bolt
- Shaft from the gear
- Tie rods from the knuckles
- Steering gear bolts
- Steering gear through the left side wheel opening

To install:

4. Install or connect the following:
- Steering gear through the left wheel opening
- Steering gear bolts to torque the fasteners to 81 ft. lbs. (110 Nm)
- Intermediate shaft to the steering gear and torque the new pinch bolt to 18 ft. lbs. (25 Nm) on 2003–04 models, or 25 ft. lbs. (34 Nm) on 2005 models
- Tie rod ends to the steering knuckle and torque the nut to 15 ft. lbs. (20 Nm) plus an additional 180 degrees on 2003–04 models, or 44 ft. lbs. (60 Nm) on 2005 models
- Wheels
- Negative battery cable

5. Check the alignment and adjust if necessary.

POWER STEERING SYSTEM BLEEDING

1. Fill the power steering fluid reservoir.

2. Raise the front of the vehicle just sufficiently so that the drive wheels are off the ground, then safely support the vehicle.

3. Bleed the system by turning the wheels from side-to-side without hitting the stops. It may take several cycles to bleed the system.

➡**Maintain the reservoir at the FULL mark during this procedure.**

4. Lower the vehicle.

5. Start the engine and, with the transaxle in **P** (automatic transaxle) or **N** (manual transaxle), check the fluid level with the engine idling. If necessary, add fluid to bring the level to the FULL mark.

6. Road test the vehicle and check for proper operation. Recheck the fluid level and make sure it is at or slightly above the FULL mark after the system has stabilized at normal operating temperature.

Strut

REMOVAL & INSTALLATION

Front

1.9L ENGINES

1. Before servicing the vehicle, refer to the precautions section.

2. If equipped with an Anti-lock Brake System (ABS), disconnect the negative battery cable, then raise and support the vehicle safely. Be sure the vehicle is at a height where underhood access is still possible.

3. Remove or disconnect the following:

- Front wheel
- Unplug and disconnect the ABS wire from the strut wiring bracket. Note the wiring position for assembly purposes, then place the ABS wiring out of the way to prevent damage. If the strut is being replaced, drill the rivet head retaining the ABS wiring bracket to the strut and remove the bracket
- Loosen the 2 steering knuckle-to-strut housing bolts, but do not remove them at this time. For reassembly purposes, matchmark the strut position to the steering knuckle
- 3 upper strut-to-body nuts and discard them
- Place a rag over the CV-joint seal to protect it from damage, then remove the 2 steering knuckle-to-strut housing bolts
- Strut assembly from the vehicle

To install:

4. Install or connect the following:
- Position the strut in the vehicle and install 3 new upper mount nuts. New nuts must be used because the torque retention of the old nuts may be insufficient. Torque the nuts to 21 ft. lbs. (29 Nm).
- Knuckle bolts, also using new nuts. Push the bottom of the strut inward while tightening the fasteners to 126 ft. lbs. (170 Nm)
- ABS wiring bracket to the strut

using a new rivet. Connect the ABS wiring to the bracket and install the wiring to the speed sensor connector. Be sure the wiring is positioned as noted during removal
- Wheel assembly, remove the supports and lower the vehicle.
- Negative battery cable

5. Check and adjust the alignment if necessary.

L–SERIES

1. Before servicing the vehicle, refer to the precautions section.

2. If equipped with an Anti-lock Brake System (ABS), disconnect the negative battery cable, then raise and support the vehicle safely. Be sure the vehicle is at a height where underhood access is still possible.

3. Remove or disconnect the following:
- Front wheel
- Brake hose and ABS harness from the strut assembly
- Loosen the steering knuckle to strut fasteners, but do not remove them
- Stabilizer bar link to the strut

Remove the 3 nuts to detach the top of the strut from the body—1.9L engine

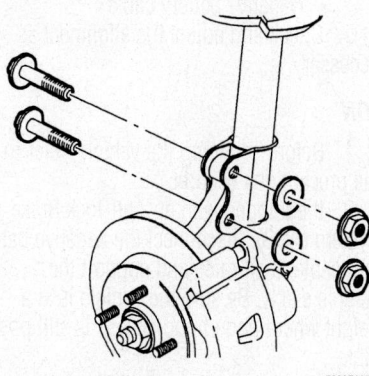

Matchmark, then remove the 2 bolts that secure the strut to the knuckle assembly—1.9L engine

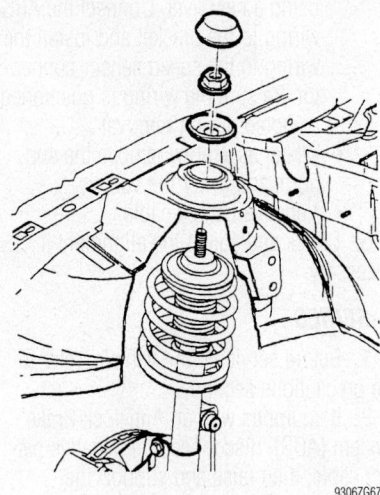

Remove the strut to body attaching nut—L–Series

assembly attaching nut and move it toward the rear of the vehicle
- Strut to body attaching nuts
- Place a rag over the CV-joint seal to protect it from damage, then remove the 2 steering knuckle-to-strut housing bolts
- Steering knuckle to strut fasteners
- Strut assembly from the vehicle

To install:
4. Install or connect the following:
- Strut to the body and torque the new attaching nut to 40 ft. lbs. (55 Nm)
- Strut to the steering knuckle and torque the new fasteners to 37 ft. lbs. (50 Nm) then to 66 ft. lbs. (90 Nm) plus a 45 degree turn
- Stabilizer bar link to the strut and torque the fastener to 50 ft. lbs. (65 Nm)
- Brake hose and ABS wiring to the strut
- Front wheel
- Negative battery cable

5. Check and adjust the alignment as necessary.

ION

1. Before servicing the vehicle, refer to the precautions section.
2. If equipped with an Anti-lock Brake System (ABS), disconnect the negative battery cable, then raise and support the vehicle safely. Be sure the vehicle is at a height where underhood access is still possible.
3. Remove or disconnect the following:

- Front wheel
- Stabilizer bar link to the strut

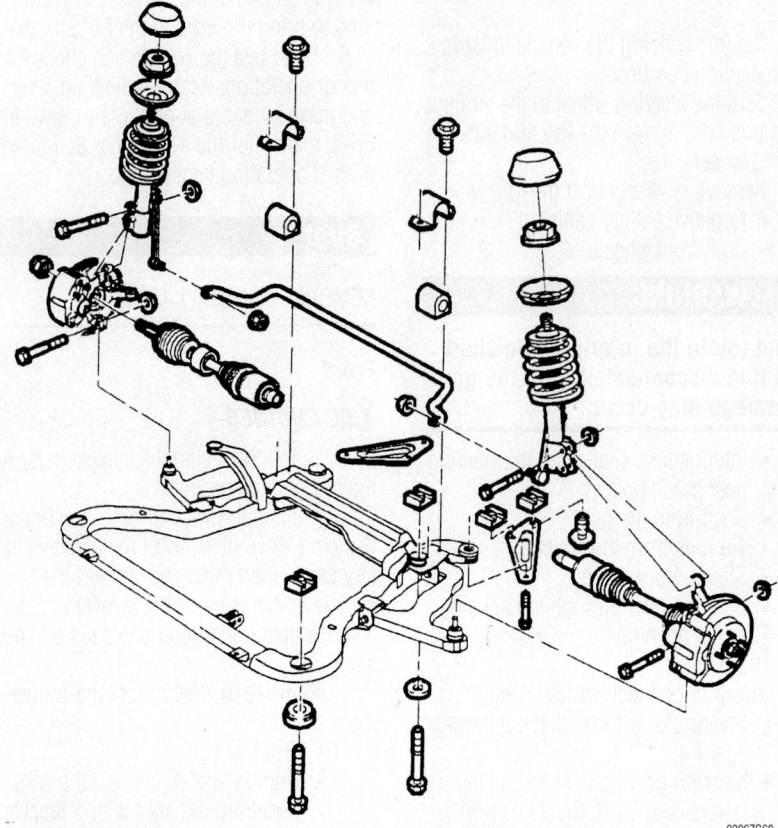

Exploded view of the front suspension—L–Series

assembly attaching nut and separate it from the strut
- Loosen the steering knuckle to strut nuts
- Wheel Speed Sensor (WSS) harness and bracket, if equipped
- Strut cap and body nut
- Place a rag over the CV-joint seal to protect it from damage, then remove the 2 steering knuckle-to-strut housing bolts
- Steering knuckle to strut fasteners
- Strut assembly from the vehicle

To install:
4. Install or connect the following:
- Strut to the body and torque the new attaching nut to 81 ft. lbs. (110 Nm)
- Strut to the steering knuckle bolts but leave the nuts off
- WSS bracket and harness, if equipped
- Strut to the steering knuckle nuts and torque to 89 ft. lbs. (120 Nm)
- Stabilizer bar link to the strut and torque the fastener to 48 ft. lbs. (65 Nm)
- Front wheel
- Negative battery cable

5. Check and adjust the alignment as necessary.

Rear

1.9L ENGINES

1. Before servicing the vehicle, refer to the precautions section.
2. Remove the rear window trim finish panel.
3. Remove or disconnect the following:
- Wheel
- Loosen the 2 strut to steering knuckle fasteners, but do not remove them
- 3 strut to body fasteners

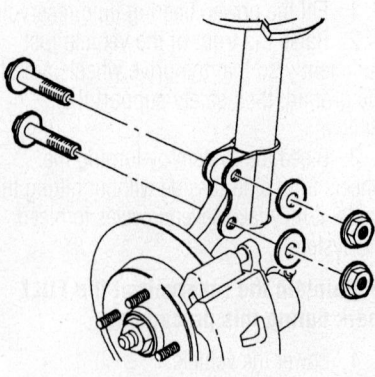

Remove the two strut to steering knuckle fasteners—1.9L engine

- Slowly raise a hoist to support the steering knuckle and lower the strut assembly
- 2 strut to steering knuckle fasteners
- Strut

To install:

4. Install the strut to the vehicle and hold it in place with strut to knuckle bolts.

5. Install or connect the following:
- Strut to the body with new upper support bolts and torque the bolts to 21 ft. lbs. (29 Nm)
- Strut to steering knuckle fasteners and push the bottom of the strut inward while tightening the bolts. Torque the new bolts to 126 ft. lbs. (170 Nm).
- Front wheel
- Rear window trim panel

6. Check and adjust the alignment if necessary.

Shock Absorber

REMOVAL & INSTALLATION

L–Series

1. Before servicing the vehicle, refer to the precautions section.

2. Remove or disconnect the following:
- Wheel
- Inner wheelhouse liner
- Shock to steering knuckle attaching bolt and discard it
- Upper carrier to body bolts
- Loosen the lower carrier bolts until the shock can be removed
- Shock

To install:

3. Install or connect the following:
- Carrier to body. Torque the bolts to 40 ft. lbs. (55 Nm).
- Carrier to steering knuckle with a

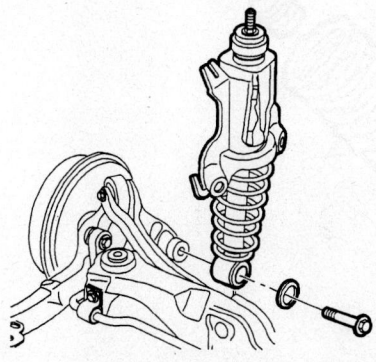

9306ZG69

Remove the shock to steering knuckle attaching bolt—L–Series

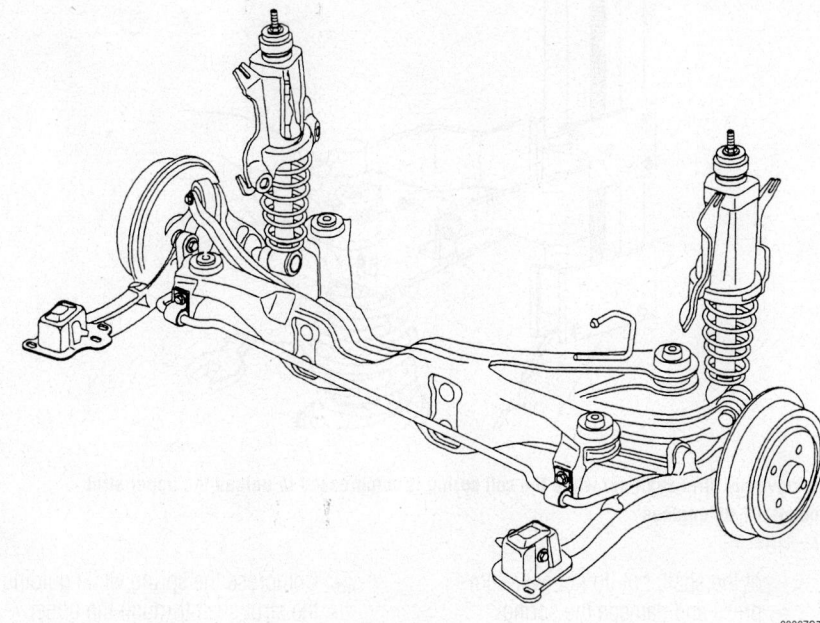

9306ZG70

Exploded view of the rear suspension—L–Series

new bolt. Torque the bolt to 110 ft. lbs. (150 Nm) plus 30 degrees
- Inner wheelhouse liner
- Wheel

ION

1. Before servicing the vehicle, refer to the precautions section.

2. Remove or disconnect the following:
- Wheel

3. Support the rear axle with a jackstand positioned near the shock absorber being removed
- Lower and upper shock absorber bolts
- Shock

To install:

4. Install or connect the following:
- Strut. Using new bolts, torque the upper bolt to 66 ft. lbs. (90 Nm) and the lower bolt to 81 ft. lbs. (110 Nm). Remove the jackstand.
- Wheel

Coil Spring

REMOVAL & INSTALLATION

Front

1.9L ENGINES

1. Before servicing the vehicle, refer to the precautions section.

2. Remove or disconnect the following:
- Strut from the vehicle
- Mount the strut in a bench vise, then

attach a spring compressor/holding fixture; be sure that the strut component is firmly secured
- Compress the spring sufficiently to completely unload the upper strut mount
- Shaft nut while holding the strut stationary with a Torx®head socket wrench
- Upper spring support and inspect the rubber for cracks or deterioration
- Spring from the strut and inspect the spring for damage
- Dust shield assembly and inspect for cracks or deterioration
- Strut from the vise or applicable holding fixture and retract the strut shaft, checking for smooth, even resistance

3. If replacing the coil spring, carefully release the spring compressor.

To install:

4. Secure the strut in the bench vise or applicable holding fixture.

5. Extend the strut shaft to the limit of its travel.

6. Install or connect the following:
- Dust shield assembly onto the strut, then install the spring with the compressor tool installed
- Spring isolator and the strut mount to the top of the assembly
- Guide the strut shaft through the upper strut mount assembly. Compress the coil until the washer and shaft nut can be installed to the end

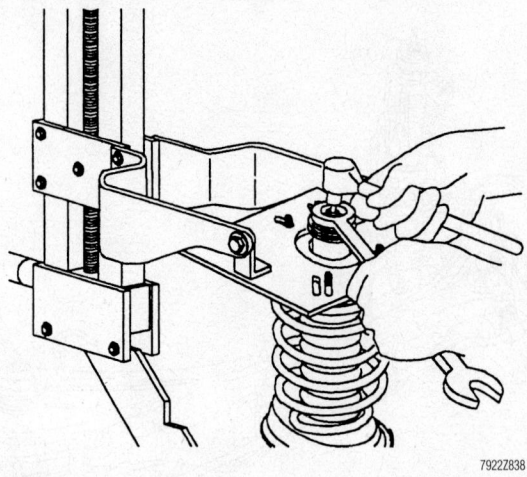

Remove the strut shaft nut while the coil spring is compressed to unload the upper strut mount—1.9L engines

of the shaft, but do not over-compress and damage the spring
- Tighten the shaft to the nut using a Torx® head socket wrench and a torque wrench, while holding the nut steady with an open end wrench. Tighten the fastener to 37 ft. lbs. (50 Nm).
- Release the spring compressor tool and remove the strut from the fixture
- Strut assembly in the vehicle.

L-SERIES

1. Before servicing the vehicle, refer to the precautions section.
2. Remove or disconnect the following:

- Strut from the vehicle
- Place the strut into a spring compressor. Fasten the assembly with a strut to steering knuckle bolt through the lower mounting hole
- Compress the spring enough to completely unload the upper strut mount
- Strut shaft nut while holding the shaft stationary with a Torx® socket
- Release the spring compressor and tilt the strut outward
- Upper strut mount assembly
- Spring

To install:

3. Install or connect the following:
- Strut into spring compressor
- Extend the strut shaft to its full travel
- Hollow bumper onto the strut
- Spring to the strut and make certain that it is properly positioned in the seat and isolator
- Upper spring seat and strut mount

- Compress the spring while guiding the strut shaft through the upper strut mount assembly. Do not over compress the spring

4. Torque the strut shaft nut to 40 ft. lbs. (55 Nm).
5. Release the spring compressor tool.
6. Strut to the vehicle.

ION

1. Before servicing the vehicle, refer to the precautions section.
2. Remove or disconnect the following:
- Strut from the vehicle

- Upper strut cap retainers and the cap
- Place the strut into a spring compressor. Fasten the assembly with a strut to steering knuckle bolt through the lower mounting hole
- Compress the spring enough to completely unload the upper strut mount
- Strut shaft nut while holding the shaft stationary with a Torx® socket
- Release the spring compressor and tilt the strut outward
- Upper strut mount assembly
- Spring

To install:

3. Install or connect the following:
- Strut into spring compressor
- Lower spring seat insulator
- Extend the strut shaft to its full travel
- Jounce bumper into the dust shield
- Dust shield assembly
- Spring to the strut and make certain that it is properly positioned in the seat and isolator
- Upper spring seat and insulator
- Bearing and strut mount
- Compress the spring while guiding the strut shaft through the upper strut mount assembly. Do not over compress the spring
- Strut shaft nut and torque to 52 ft. lbs. (70 Nm).

4. Release the spring compressor tool.
- Upper strut cap and retainers.

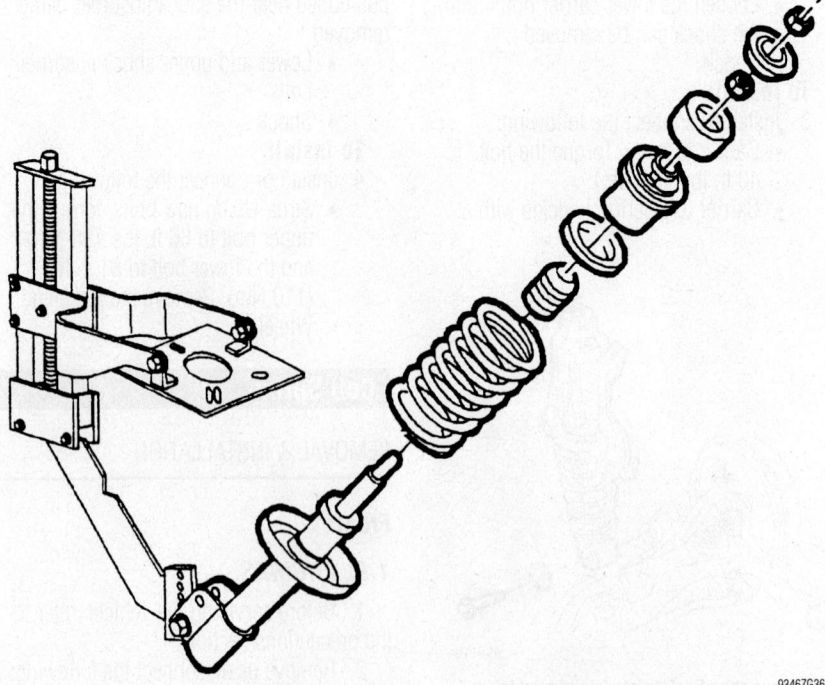

Exploded view of the coil spring and related components—L-Series

Tighten the retainers to 124 inch lbs. (14 Nm).
5. Strut to the vehicle.

Rear

1.9L ENGINES

1. Before servicing the vehicle, refer to the precautions section.
2. Remove or disconnect the following:
   - Strut assembly from the vehicle
   - Mount the strut in a bench vise, then attach a suitable spring compressor/holding fixture; be sure that the strut component is firmly secured
   - Compress the spring sufficiently to completely unload the upper strut mount
   - Strut shaft nut while holding the strut stationary with a Torx® head socket wrench
   - Upper spring support and inspect the rubber for cracks or deterioration
   - Spring from the strut and inspect the spring for damage
   - Dust shield assembly and inspect for cracks or deterioration
   - Strut from the vise or applicable holding fixture and retract the strut shaft, checking for smooth, even resistance
3. If replacing the coil spring, carefully release the spring compressor.

To assemble:

4. Secure the strut in the bench vise, or applicable holding fixture.
5. Extend the strut shaft to the limit of its travel.
6. Install or connect the following:
   - Dust shield assembly onto the strut, then install the spring with the compressor tool installed
   - Spring isolator and the strut mount to the top of the assembly
   - Guide the strut shaft through the upper strut mount assembly. Compress the coil until the washer and shaft nut can be installed to the end of the shaft, but do not over-compress and damage the spring
   - Tighten the shaft to the nut using a Torx® head socket wrench and a torque wrench, while holding the nut steady with an open-end wrench. Tighten the fastener to 37 ft. lbs. (50 Nm).
   - Release the spring compressor tool and remove the strut from the fixture
   - Strut to the vehicle

L–SERIES

1. Before servicing the vehicle, refer to the precautions section.
2. Remove or disconnect the following:
   - Shock absorber assembly from the vehicle
   - Mount the carrier assembly Spring Compressor SA9155S, in a holding fixture
   - Mount the shock assembly to the spring compressor using an adapter, SA9155–3
   - Compress the spring enough to unload the upper spring supports
   - Shock absorber shaft nut while holding the shaft securely with a Torx® socket
   - Release the spring compressor and remove the carrier assembly
   - Rear suspension support and inspect the inner and outer bumpers for cracks or deterioration, replace if necessary
   - Rear spring upper insulator
   - Extend and retract the shock absorber and check for smooth resistance

To install:

3. Install or connect the following:
   - Shock into a spring compressor

   - Extend the shock to its full limit of travel
   - Inner and outer bumpers and make certain that the springs are properly positioned in their support seats
   - Compress the spring while guiding the shock absorber shaft through the upper mount until the washer and nut can be installed. Torque the shaft nut to 15 ft. lbs. (20 Nm) while holding the shaft nut with a Torx® socket.
   - Release the compressor tool and remove the carrier from the compressor
   - Shock absorber to the vehicle

ION

1. Before servicing the vehicle, refer to the precautions section.
2. Support the rear axle with jackstands positioned near each shock absorber.
3. Remove or disconnect the following:
   - Wheel
   - U–clips from the rear brake hose brackets at the axle
   - Lower shock bolts
4. Lower the jackstands slowly to remove rear spring tension and remove the spring.

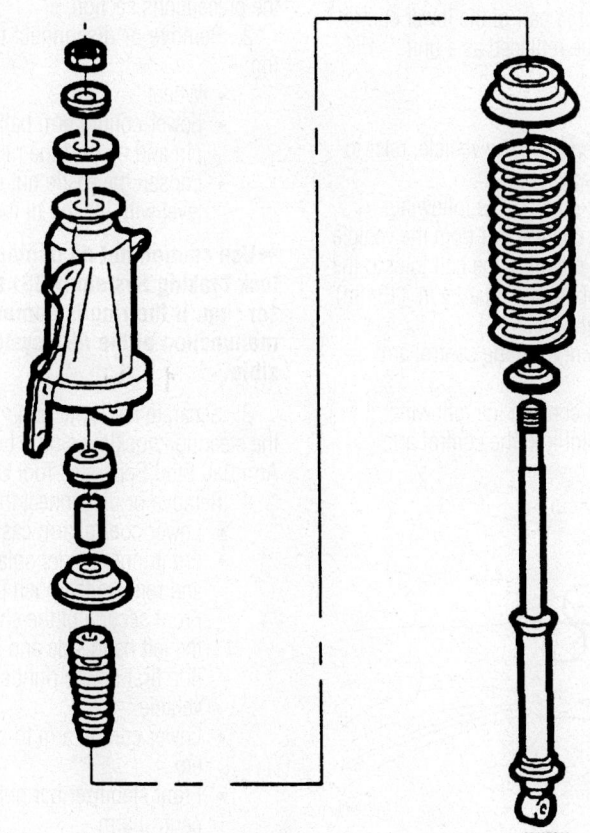

9346ZG03

Exploded view of rear shock assembly—L-Series

- Upper spring seat/jounce bumper from the spring while leaving the lower spring seat on the axle

To install:

> ### ❋❋ CAUTION
>
> **The rear springs are indexed with the colored tag towards the rear of the vehicle. No orientation for up or down is required.**

5. Install or connect the following:
   - Upper spring seat/jounce bumper to the spring
   - Spring with the spring tag towards the rear of the vehicle and the lower coil is seated into the lower spring seat. Use the jackstands to raise the rear axle into position.
   - Lower shock absorber bolt and tighten to 81 ft. lbs. (110 Nm)
   - U–clips to the rear brake hose brackets at the axle
   - Wheel

Lower Ball Joint

REMOVAL & INSTALLATION

1.9L Engines

The ball joint is part of the lower control arm and must be replaced as a unit.

L–Series

1. Before servicing the vehicle, refer to the precautions section.
2. Install or connect the following:
   - Lower control arm from the vehicle
   - Rivets retaining the ball joint to the control arm using a ½ in. (13mm) drill bit
   - Ball joint from the control arm

To install:

3. Install or connect the following:
   - Ball joint into the control arm

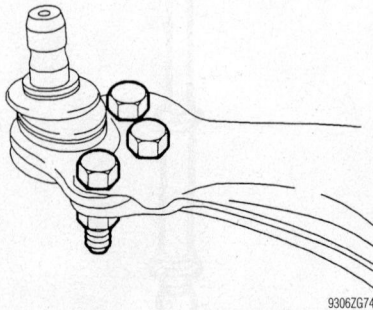

9306ZG74

The new ball stud is bolted into the control arm—L-series shown, other similar

- Nuts and bolts (included with new ball joint kit) as shown and torque them to 25 ft. lbs. (35 Nm)
- Control arm to the vehicle

ION

1. Before servicing the vehicle, refer to the precautions section.
2. Install or connect the following:
   - Lower control arm from the vehicle
   - Rivets retaining the ball joint to the control arm using a ½in. (13mm) drill bit
   - Ball joint from the control arm

To install:

3. Install or connect the following:
   - Ball joint into the control arm
   - Nuts and bolts (included with new ball joint kit) and torque them to 50 ft. lbs. (68 Nm)
   - Control arm to the vehicle

Lower Control Arm

REMOVAL & INSTALLATION

Front

1.9L ENGINES

1. Before servicing the vehicle, refer to the precautions section.
2. Remove or disconnect the following:
   - Wheel
   - Lower control arm ball stud cotter pin and discard the pin
   - Loosen the castle nut until it is level with the top of the ball stud

➡️**Use caution not do damage the Antilock Braking System (ABS) speed sensor ring. If the ring is damaged a malfunction of the ABS system is possible.**

3. Separate the lower control arm from the steering knuckle, using a Lower Control Arm Ball Stud Separator Tool SA9132S
4. Remove or disconnect the following:
   - Lower control arm castle nut
   - Front inner fender splash shield and remove the push pins
   - Front section of the shield first on the left hand side and the rear section first on the right side of the vehicle
   - Lower control arm to cradle fasteners
   - Front stabilizer bar nut at the lower control arm
   - Lower control arm

To install:

5. Install or connect the following:
   - Control arm to the front stabilizer bar
6. Position the end of the control arm into the cradle. Torque the bolt to 92 ft. lbs. (125 Nm) and the nut to 74 ft. lbs. (100 Nm).
   - Control arm to front stabilizer bar nut and torque the nut to 106 ft. lbs. (144 Nm)
   - Ball stud to the steering knuckle and torque the castle nut to 55 ft. lbs. (75 Nm) and install a new cotter pin
   - Rear section of the shield first on the left side and the front section first on the right side of the vehicle
   - Inner fender splash shield
   - Align the molded in fasteners with the cradle holes and push them straight in
   - Wheel
7. Check and adjust the alignment if necessary.

L–SERIES

1. Before servicing the vehicle, refer to the precautions section.
2. Remove or disconnect the following:
   - Wheel
   - Ball stud bolt

➡️**Use caution not do damage the Antilock Braking System (ABS) speed sensor ring. If the ring is damaged a malfunction of the ABS system is possible.**

- Separate the ball stud from the steering knuckle by using a prybar
- Lower control arm-to-frame bolts
- Lower control arm

To install:

3. Install or connect the following:
   - Control arm to the frame and

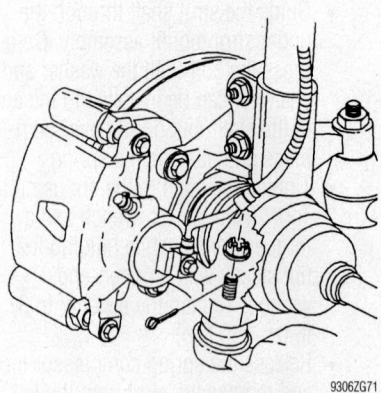

9306ZG71

Connect the ball stud to the steering knuckle—1.9L engines

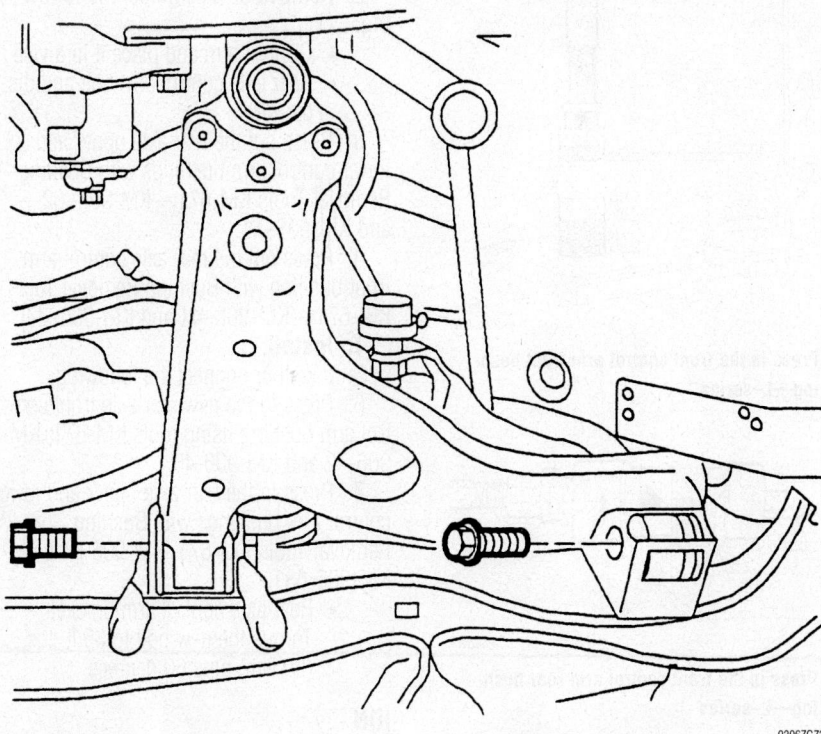

Remove the lower control arm to frame bolts—L-Series models shown others similar

torque the new bolts to 65 ft. lbs. (90 Nm) plus 75 degrees
- Ball stud to steering knuckle and torque the bolt to 75 ft. lbs. (100 Nm)
- Wheel
- Check and adjust the alignment, if necessary.

ION

1. Before servicing the vehicle, refer to the precautions section.
2. Remove or disconnect the following:
   - Wheel
   - Ball stud bolt
   - Separate the ball stud from the steering knuckle. Do not use a pickle fork as this may cause damage.
   - Rear frame bolt
   - Lower control arm-to-frame bolts
   - Lower control arm

To install:

3. Install or connect the following:
   - Control arm to the frame and loosely install the rear frame bolt

➡**The control arm contains two fore/aft movement limiting brackets. Failure to install these brackets will affect handling.**

   - Ball joint stud to the steering knuckle

- Fore/aft movement limiting brackets onto the control arm forward bushing
- Control arm-to-frame-bolts and torque the bolts to 41 ft. lbs. (55 Nm) on 2003—04 models or 49 ft. lbs. (66 Nm) on 2005 models
- Tighten the rear frame bolt to 74 ft. lbs. (100 Nm) plus an additional 180 degree turn
- Ball joint pinch bolt and nut to 44 ft. lbs. (60 Nm) plus an additional 30 degree turn on 2003–04 models. On 2005 models, tighten the bolt to 37 ft. lbs. (50 Nm), back the bolt off ¾ of a turn, retighten the bolt to 37 ft. lbs. (50 Nm) plus an additional 30 degree turn.
- Wheel
- Check and adjust the alignment, if necessary.

Rear

L-SERIES

1. Before servicing the vehicle, refer to the precautions section.
2. Remove or disconnect the following:
   - Rear wheel
   - Rear brake hose and bracket from the control arm
   - Caliper, if equipped
   - Brake drum/rotor

- Antilock Braking System (ABS) sensor harness, if equipped
- Hub
- Parking brake cable and support
- Separate the backing plate from the control arm
- Shock absorber to control arm bolt
- Rear stabilizer bar link to rear control arm bolt
- Loosen the rear suspension upper and lower control arm to rear axle fasteners
- Rear axle to control arm and discard the bolts
- Rear upper and lower suspensions control arm bolts and discard them
- Rear control arm

To install:

3. Install or connect the following:
   - Control arm and install new rear axle control arm bolts. Do not tighten them at this time
   - Upper and lower rear suspension control arm bolts
   - Torque the rear axle control arm bolts and the upper and lower rear suspension bolts to 65 ft. lbs. (90 Nm) plus 30 degrees
   - Shock absorber to rear axle control arm and torque the new bolt to 110 ft. lbs. (150 Nm) plus 30 degrees
   - Stabilizer bar link to the control arm and torque the bolt 41 ft. lbs. (55 Nm)
   - Backing plate and hub assembly and torque the bolts to 37 ft. lbs. (50 Nm) plus 30 degrees
   - Rear drum/disc and torque the bolts to 35 inch lbs. (4 Nm)
   - Rear brake caliper and torque the bolts to 59 ft. lbs. (80 Nm), if equipped
   - Parking brake cable and support bracket and torque the bolts to 71 inch lbs. (8 Nm)
   - Brake line bracket to the control arm and torque the bolts to 71 inch lbs. (8 Nm)
   - Rear wheel

4. Check and adjust the alignment, as needed.

LOWER CONTROL ARM BUSHING REPLACEMENT

L-Series

FRONT

1. Before servicing the vehicle, refer to the precautions section.
2. Remove or disconnect the following:

• Control arm

3. Press out the front bushing by using Tools KM-508–3 and KM508–1.

4. Drill out the rivets retaining the ball stud to the control arm.

5. Remove the ball stud from the control.

To install:

➡**The new ball stud is bolted to the control arm.**

6. Install or connect the following:
   • Ball stud to the control arm and torque the new bolts to 25 ft. lbs. (35 Nm)

7. Press in the new control arm bushing using Tools KM508–1, KM508–2 and KM508–3.
   • Control arm to the frame and

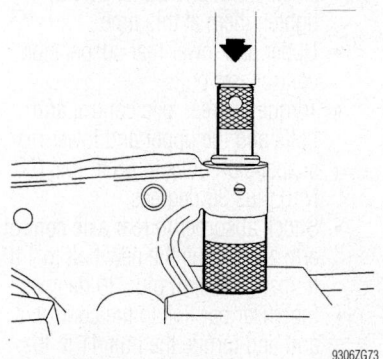

Press out the front control arm front bushing—L–series

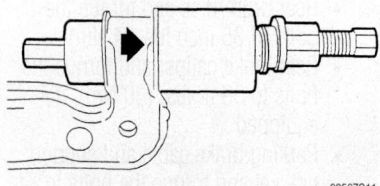

Press out the front control arm rear bushing—L–series

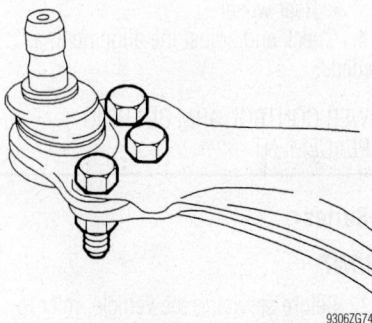

The new ball stud is bolted into the control arm—L–Series

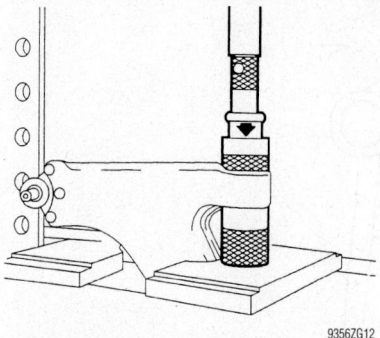

Press in the front control arm front bushing—L–series

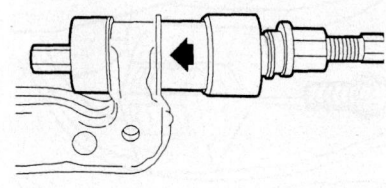

Press in the front control arm rear bushing—L–series

torque the new bolts to 65 ft. lbs. (90 Nm) plus 75 degrees
   • Ball stud to the steering knuckle and torque the bolts to 75 ft. lbs. (100 Nm)

REAR

1. Before servicing the vehicle, refer to the precautions section.

2. Remove or disconnect the following:
   • Control arm and place it in a vise
   • Rear axle control arm bolt and discard it

3. Press out the rear axle upper and lower control arm bushings with Bushing Removal Tools KM-671—KM-906–62—and KM-906–61.

4. Press out the rear axle control arm front bushing with Bushing Removal Tools KM-671—KM-906–41 and KM-906–44.

To install:

5. Install or connect the following:

6. Press in the new rear axle front control arm bushing using tools KM-671, KM-906–42 and KM-906–43.

7. Press in the rear axle upper and lower control arm bushings with Bushing Removal Tools KM-671, KM-906-64 and KM-906-631.
   • Rear axle control arm bracket. Torque the new bolt to 65 ft. lbs. (90 Nm) plus 60 degrees

ION

FRONT

1. Before servicing the vehicle, refer to the precautions section.

2. Remove or disconnect the following:
   • Control arm

➡**Make sure to note the depth and positioning of the old bushing prior to removal.**

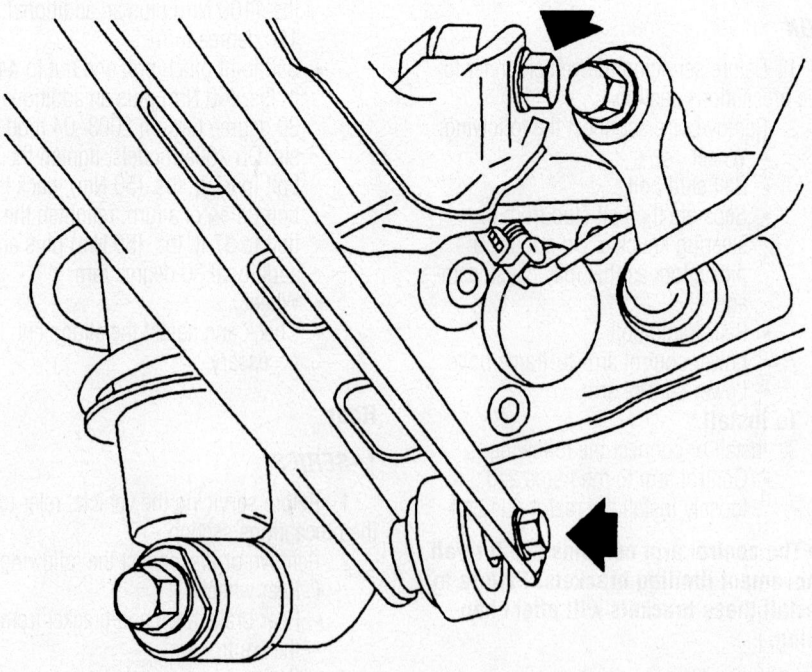

Remove the rear axle control arm lower fasteners—L–Series

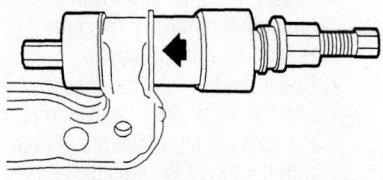

Remove the rear axle control arm upper and lower bushings—L–series

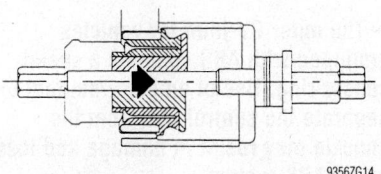

Remove the rear axle control arm front bushings—L–series

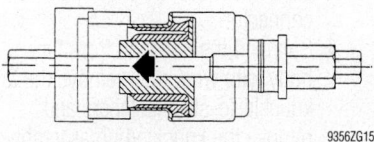

Install the rear axle control arm front bushings—L–series

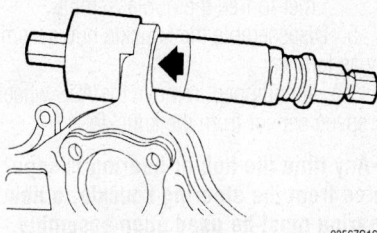

Install the rear axle control arm upper and lower bushings—L–series

3. Using bushing remover/installer KM-906B install KM-906-70, 906–42, 906-41 and 906-62 onto the bushing.

4. Hold the hex end of the threaded shaft while turning the large nut to pull the bushing through the arm.

5. Remove the tools and the bushing.

To install:

6. Install the new bushing into the tapered side of the arm.

7. Using bushing remover/installer KM-906B install KM-906-70, 906–42, 906-41 and 906-62 onto the bushing

8. Hold the hex end of the threaded shaft while turning the large nut to pull the bushing into the arm. Install the bushing to the same depth and positioning as the old one.

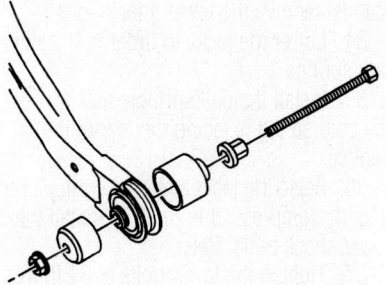

Press out the front control arm front bushing—ION

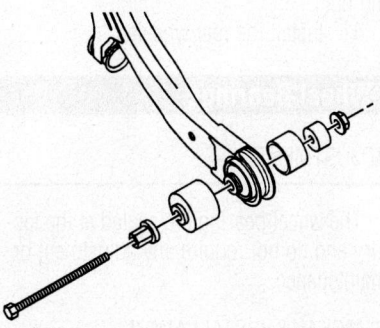

Press in the front control arm rear bushing—ION

9. Remove the removal/installer tool.
10. Install the control arm.

REAR

1. Before servicing the vehicle, refer to the precautions section.

2. Remove the rear wheels.

3. Place 2 screw type jack stands under both ends of the rear axle.

4. Remove the rear brake hose bracket attaching nuts from the body.

5. Detach the rear brake hose brackets from the body allowing the lines to hang free.

6. Remove the lower shock bolts.

➡**Do not kink the brake pipes while lowering the axle.**

7. Lower the jacks in order to remove the coil springs.

8. Temporarily re-install the lower shock bolts to support the axle.

9. Remove the bushing bracket to body bolts from both ends of the rear axle.

10. Using the jack stands, raise the rear of the axle until the bushing brackets pivot away from the body.

11. Remove the axle bushing through bolts and remove the bushing brackets.

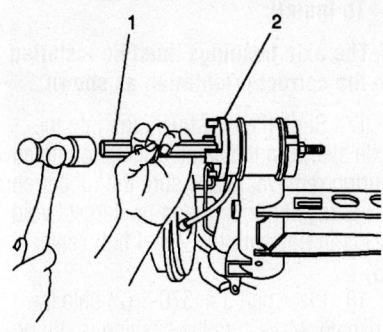

Removing the rear control arm bushings—ION

➡**Note the depth and orientation of the old bushing before removal.**

12. Using bushing removal/installation kit J44570, install the part J 44570-1 (number 2 in the accompanying illustration)with the lip between the axle sleeve and bushing flange. It may require tapping with a hammer to fully seat the tool.

13. Insert part J 44570-3 (number 1 in the accompanying illustration) through the J 44570-1 and the axle bushing.

14. Install the washer and nut by hand, tightening until the tool is snug.

15. Using a hammer, drive the bushing from the axle sleeve.

16. Disassemble the tool and remove the bushing.

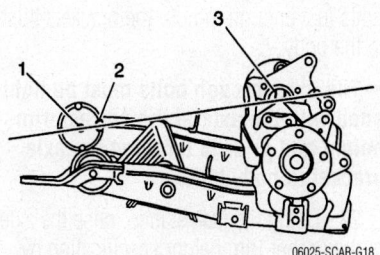

Install the rear control arm bushing in the correct orientation as illustrated—ION

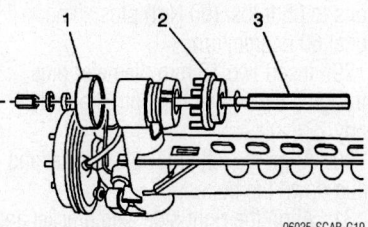

Exploded view of the rear control arm bushing and installation tool prior to installing the bushing—ION

To install:

➡ **The axle bushings must be installed in the correct orientation as shown.**

17. Slide the new bushings into the axle sleeve in the same orientation noted during removal. Make sure the rubber end (1) is facing inboard and the largest void (2) is in line with the wheel hub center (3).

18. Place part J 44570-1 (2) onto the bushing. Make sure the bushing is still oriented correctly. Refer to the exploded view illustration for component location.

19. Insert part J 44570-3 (3) through part J 44570-1 (2) and the axle bushing.

20. Install part J 44570-2 (1) bearing, washer, and nut.

21. Pull the bushing into the axle sleeve by holding the hex end of the threaded shaft while turning the nut.

22. Disassemble and remove the bushing installation tool from the axle.

23. Install the axle brackets to the axle bushings with the alignment slot on the outboard side.

➡ **The axle bushing through bolts must be installed with the bolt head facing inboard.**

24. Loosely install the bushing bolts, park brake cable brackets and nuts .

25. Using the jack stands, lower the rear of the axle until the bushing brackets contact the body.

26. Hand tighten the axle bracket to body bolts just enough to hold the brackets flush to the body.

➡ **The axle through bolts must be tightened with the axle at the correct trim height and prior to torquing the axle bracket to body bolts.**

27. Using the jack stands, raise the axle to the proper trim height specification by measuring the vertical distance between the bottom edge of the upper spring seat and the bottom of the notch in the lower spring seat.

28. Tighten the axle bushing through bolts to 66 ft lbs. (90 Nm) plus an additional 60 degree turn.

29. Insert two 12 mm diameter pins through the axle brackets into the underbody.

30. Align the left side axle bracket and snug down the bolts.

31. Align the right side axle bracket and snug down the bolts.

32. Tighten all the bracket to body bolts to 66 ft lbs. (90 Nm) plus an additional 30 degree turn.

33. With the axle supported by the jack stands, remove the lower shock bolts.

34. Lower the jacks in order to install the coil springs.

35. Install the coil springs, making sure the colored tag is facing the rear of the vehicle.

36. Raise the jacks until the springs are slightly compressed in order to install the lower shock bolts. Refer

37. Tighten the lower bolts to 81 ft. lbs. (110 Nm).

38. Remove the jack stands.

39. Reposition the rear brake hose brackets to the body.

40. Install the brake hose bracket attaching nuts.

41. Install the rear wheels.

Wheel Bearings

ADJUSTMENT

The wheel bearing are sealed at the factory and do not require any adjustment or maintenance.

REMOVAL & INSTALLATION

Front

1.9L ENGINES

1. Before servicing the vehicle, refer to the precautions section.

2. If equipped with an Antilock Braking System (ABS), disconnect the negative battery cable.

3. Loosen the front halfshaft nut, while an assistant depresses the brake pedal, then raise and support the vehicle safely.

4. Remove or disconnect the following:
- Wheel
- Brake caliper mounting bracket bolts and suspend the assembly from the strut spring with wire

- Loosen the strut-to-knuckle bolts, but do not remove at this time
- Rotor, axle nut and washer
- Cotter pin from the lower control arm ball joint. Back the ball joint nut until the top of the nut is even with the top of the threads
- Separate the lower control arm from the steering knuckle, then remove the nut. Do not use a wedge tool or seal damage may occur

➡ **The outer CV-joint for vehicles equipped with ABS contains a speed sensor ring. Use of an incorrect tool to separate the control arm from the knuckle may result in damage and loss of the ABS system.**

- Tie rod cotter pin and castle nut, then separate the tie rod end from the knuckle
- ABS wheel speed sensor electrical connector
- Suspend the halfshaft from the body with wire, then remove the 2 knuckle-to-strut fasteners and remove the knuckle/hub assembly from the vehicle. If difficulty is encountered, position a block of wood on the end of the halfshaft and tap on the wood with a hammer to free the hub assembly

5. Disassemble the knuckle hub assembly as follows:

a. If equipped, remove the ABS wheel speed sensor from the knuckle.

➡ **Any time the hub or bearing is separated from the steering knuckle, a new bearing must be used upon assembly.**

6. Install Wheel Bearing Removing Tools SA9159S, to the knuckle and secure the assembly in a vise.

7. Hold the hub driver with a wrench and tighten the hub driver screw to remove

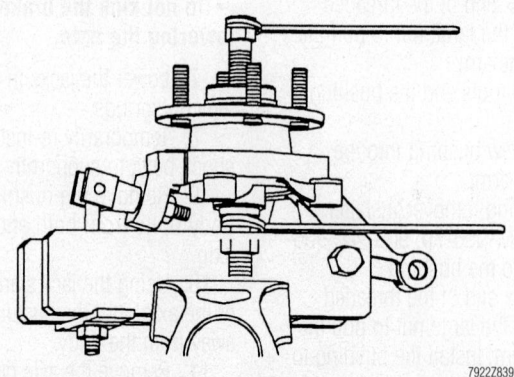

Tighten the hub driver screw to remove the hub while the assembly is held firmly in a vise

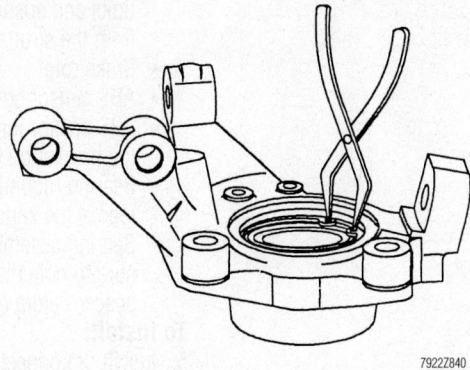

Remove the bearing retainer snapring before pressing out the bearing

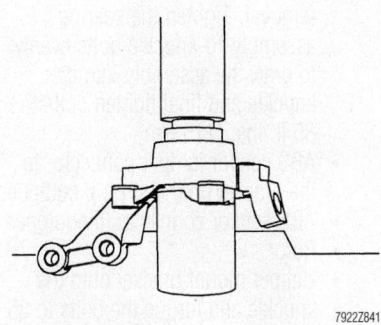

Carefully press the bearing out of the knuckle after removing the snapring

the hub. If the inner bearing race is pulled out with the hub, remove the race with a bearing race remover.

8. Remove the assembly from the vise and separate the wheel hub removal tool from the knuckle.

9. Remove the bearing retainer snapring.

10. Position the knuckle in a shop press on a knuckle support tube and press the bearing from the knuckle with a small driver.

To install:

11. If necessary, assemble the knuckle hub assembly as follows:

a. Use a large driver and press in the new bearing until seated.

b. Use the small driver and the knuckle support tube to press in the hub assembly. The small driver must be used to support the bearing inner race with its small (pilot) side facing towards the press and away from the bearing.

c. Install the bearing retainer snapring.

12. Install the ABS wheel speed sensor into the knuckle. Torque the fastener to 72 inch lbs. (8 Nm), if equipped.

➡**Service knuckles may not have holes for brake dust shield mounting. The dust shield is no longer required and does not have to be reinstalled. Also,**

should the shield become damaged, it may be removed; there is no need to repair or replace it. But, should a shield be removed and discarded, the shield should also be removed from the opposite side to maintain balance/symmetry.

13. Thoroughly clean and lubricate the ball joint stud threads of the lower control arm and tie rod end. Install the knuckle/hub assembly onto the axle shaft. Then, install the washer with a new nut, but do not tighten the nut at this time.

14. Install or connect the following:

- Lower control arm ball stud through the knuckle bore and install the nut, but do not tighten at this time
- Steering knuckle-to-strut fasteners, but do not tighten at this time
- Tie rod end and nut, then torque the nut to 33 ft. lbs. (45 Nm) and install a new cotter pin. If necessary, tighten the nut additionally, do not back off to insert the cotter pin
- Push inward on the bottom of the strut and torque the knuckle fasteners to 126 ft. lbs. (170 Nm).
- Torque the lower control arm ball stud nut to 55 ft. lbs. (75 Nm), tighten additionally, if necessary, and install a new cotter pin
- Rotor onto the hub, then install the caliper mount bracket onto the knuckle. Tighten the mount bracket assembly bolts to 81 ft. lbs. (110 Nm).
- While an assistant depresses the brake pedal, tighten the halfshaft nut to 103 ft. lbs. (140 Nm).
- Wheel assembly and lower the vehicle
- Negative battery cable

15. Check and adjust the alignment, if necessary.

L-SERIES

1. Before servicing the vehicle, refer to the precautions section.

2. If equipped with an Antilock Braking System (ABS), disconnect the negative battery cable.

3. Loosen the front halfshaft nut, while an assistant depresses the brake pedal, then raise and support the vehicle safely.

4. Remove or disconnect the following:

- Wheel
- Brake caliper mounting bracket bolts and suspend the assembly from the strut spring with wire
- Brake rotor and dust shield
- ABS sensor and bracket, if equipped
- Loosen the strut-to-knuckle bolts, but do not remove at this time
- Axle to hub nut and discard it
- Tie rod end nut and discard it

5. If difficulty is encountered, position a block of wood on the end of the halfshaft and tap on the wood with a hammer to free the hub assembly.

- Steering knuckle/hub assembly

6. To remove the wheel bearing proceed, as follows:

a. Install Wheel Bearing Removing Tools SA9159S, to the knuckle and secure the assembly in a vise.

b. Hold the hub driver with a wrench and tighten the hub driver screw to remove the hub. If the inner bearing race is pulled out with the hub, remove the race with a bearing race remover.

c. Remove the assembly from the vise and separate the wheel hub removal tool from the knuckle.

d. Remove the bearing retainer snapring.

e. Position the knuckle in a shop press on a knuckle support tube and press the bearing from the knuckle with a small driver.

To install:

7. If necessary, assemble the knuckle hub assembly, as follows:

a. Use a suitable large driver and press in the new bearing until seated.

b. Use the small driver and the knuckle support tube to press in the hub assembly. The small driver must be used to support the bearing inner race with its small (pilot) side facing towards the press and away from the bearing.

c. Install the bearing retainer snap ring.

8. Thoroughly clean and lubricate the ball joint stud threads of the lower control arm and tie rod end.

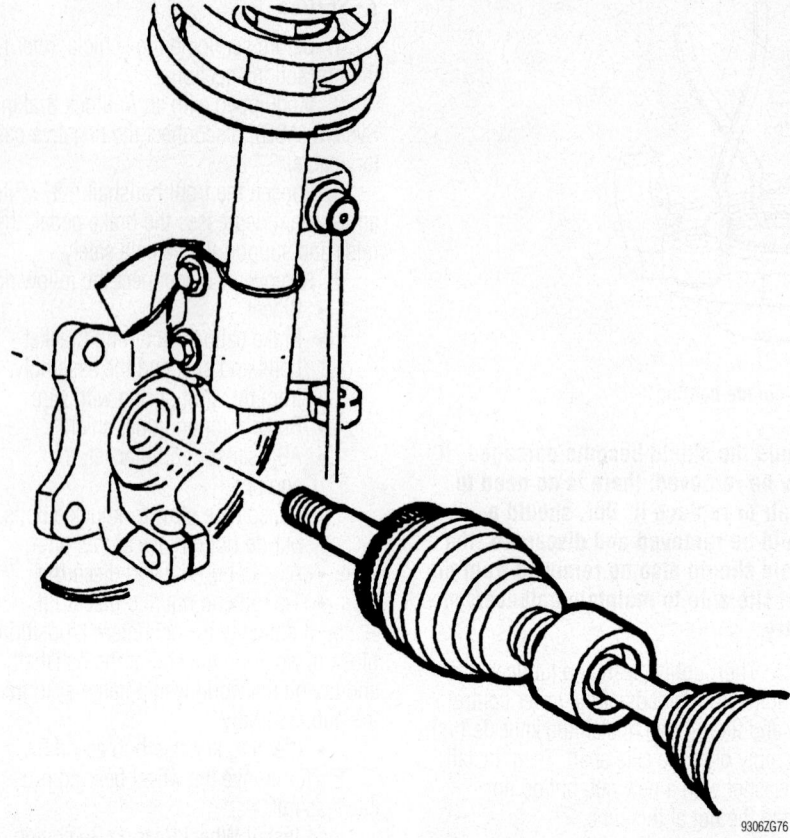

Separate the axle from the wheel hub—L–Series shown, others similar

9306ZG76

9. Install or connect the following:
- Knuckle/hub assembly onto the axle shaft. Then, install the washer with a new nut, but do not tighten the nut at this time
- Lower control arm ball stud through the knuckle bore and install the nut, but do not tighten at this time
- Steering knuckle-to-strut fasteners. Torque the fasteners to 40 ft. lbs. (50 Nm); then, to 65 ft. lbs. (90 Nm) plus 45 degrees

10. Torque the ball stud fastener to 75 ft. lbs. (100 Nm).
- Tie rod end into the steering knuckle using a Linkage Installer Tool J44015. Torque the fastener to 35 ft. lbs. (45 Nm).
- ABS sensor and mounting bracket, if equipped and torque the bolts to 71 inch lbs. (8 Nm).
- Dust shield and torque the bolts to 35 in lbs. (4 Nm)
- Rotor and screw and torque the screw to 27 inch lbs. (3.5 Nm)
- Caliper mount bracket onto the knuckle and torque the bolts to 70 ft. lbs. (90 Nm)
- Stabilizer bar link to the strut and

torque the bolts to 50 ft. lbs. (65 Nm)
- Wheel
- Negative battery cable

11. Depress the brake pedal and torque the axle to hub nut, in the following sequence:
- Step 1: 85 ft. lbs. (115 Nm) to seat the bearing.
- Step 2: Back the nut off .
- Step 3: 15 ft. lbs. (20 Nm).
- Step 4: Turn the nut an additional 90 degrees, using a torque angle gauge.

12. Install the cotter pin.

13. Check and adjust the alignment

ION

1. Before servicing the vehicle, refer to the precautions section.

2. If equipped with an Antilock Braking System (ABS), disconnect the negative battery cable.

3. Loosen the front halfshaft nut, while an assistant depresses the brake pedal, then raise and support the vehicle safely.

4. Remove or disconnect the following:
- Wheel
- Axle to hub nut and discard it
- Brake caliper mounting bracket

bolts and suspend the assembly from the strut spring with wire
- Brake rotor
- ABS sensor connector, if equipped
- ABS sensor jumper connector from the bracket on the strut, if equipped
- Bearing mounting bolts from the rear of the knuckle
- Bearing assembly and spacer. Make sure to note the positioning of the spacer before removal.

To install:

5. Install or connect the following:
- Bearing/hub assembly and spacer to the knuckle making sure to position the spacer as it was before removal. Tighten the bearing assembly-to-knuckle bolts evenly to draw the assembly into the knuckle and final tighten bolts to 85 ft. lbs. (115 Nm).
- ABS sensor jumper connector to the bracket on the strut, if equipped
- ABS sensor connector, if equipped
- Rotor
- Caliper mount bracket onto the knuckle and torque the bolts to 85 ft. lbs. (115 Nm)
- Caliper and tighten the bolts to 25 ft. lbs. (34 Nm)
- Negative battery cable

6. Depress the brake pedal and torque the axle to hub nut to 81 ft. lbs. (110 Nm) on 2003–04 models, or 155 ft. lbs. (210 Nm) on 2005 models
- Wheel

7. Check and adjust the alignment

Rear

1.9L ENGINES

Unlike the front wheel bearings, which may be removed from the hub for replacement, the rear wheel hub and bearing assembly is not serviceable. If damaged or worn, the hub and bearing assembly must be replaced as a unit.

1. Before servicing the vehicle, refer to the precautions section.

2. Remove or disconnect the following:

- Negative battery cable, if equipped with an Antilock Braking System (ABS)
- Rear wheel
- ABS speed sensor connector, if equipped
- Caliper assembly-to-knuckle mounting bolts and support it with a wire from the strut, then remove the rotor, if equipped with disc brakes

- Brake drum, if equipped with drum brakes
- 4 hub/bearing-to-knuckle bolts, then remove the assembly from the vehicle

To install:

3. Install or connect the following:
   - Brake backing plate, hub/bearing assembly and retaining bolts and torque the bolts to 63 ft. lbs. (85 Nm)
   - Brake drum or rotor and caliper assembly, if equipped
   - ABS speed sensor connector, if equipped
   - Brake drum, if equipped
   - Brake rotor, if equipped
   - Caliper to the steering knuckle, if equipped. Torque the bolts to 63 ft. lbs. (85 Nm).
   - Rear wheel
   - Negative battery cable

L–SERIES

1. Before servicing the vehicle, refer to the precautions section.
2. Remove or disconnect the following:
   - Negative battery cable, if equipped with an Antilock Braking System (ABS)

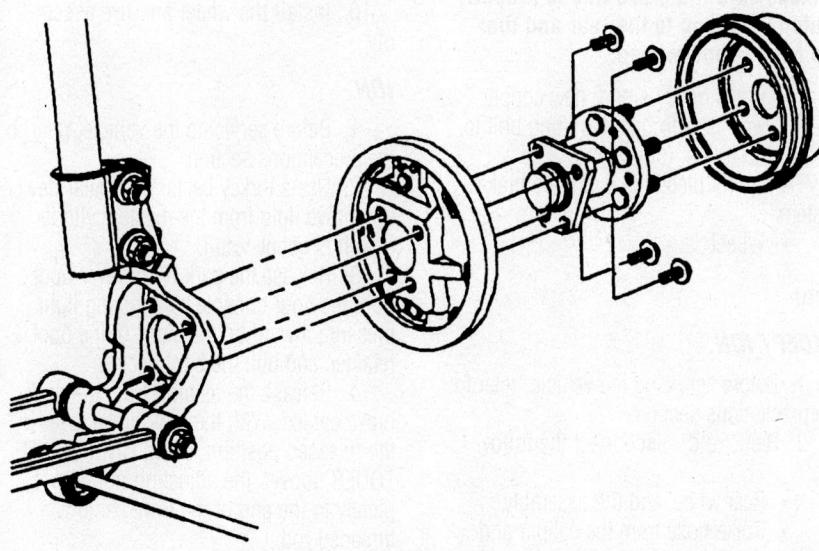

Exploded view of the rear hub/bearing assembly—drum brake set-up shown—1.9L engine

- Rear wheel
- Brake line to rear axle control arm attaching clip, if equipped with rear disc brakes
- Caliper to steering knuckle bolts and move the caliper aside, if equipped with rear disc brakes
- Brake drum/rotor
- ABS electrical connector, if equipped
- Hub to control arm nuts and discard them
- Hub

To install:

3. Install or connect the following:
   - Hub to control arm and torque the new nuts to 35 ft. lbs. (50 Nm) plus 30 degrees
   - ABS electrical connector, if equipped
   - Brake drum/rotor attaching screw and torque the screw to 35 inch lbs. (4 Nm)
   - Brake caliper, if equipped and torque the bolts to 59 ft. lbs. (80 Nm)
   - Brake line to the rear axle control arm bracket, if equipped. Torque the bolt to 71 inch lbs. (8 Nm).
   - Rear wheel
   - Negative battery cable

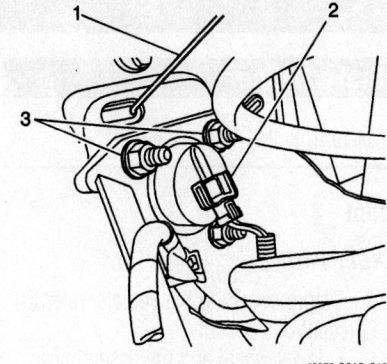

Install a support (1) for the backing plate, disconnect the electrical connector (2) and remove the hub nuts (3)—ION

ION

1. Before servicing the vehicle, refer to the precautions section.
2. On models with drum brakes, remove the brake drum. Remove the access hole plug from the brake backing plate and install a support for the backing plate
3. On models with disc brakes, remove and support the rear brake caliper and bracket as an assembly, and remove the rear brake rotor.
4. Remove or disconnect the following:
   - ABS electrical connector, if equipped
   - Hub retaining arm nuts and discard them
   - Hub

To install:

5. Install or connect the following:
   - Hub and torque the new nuts to 37 ft. lbs. (50 Nm) plus 30 degrees
   - ABS electrical connector, if equipped
6. Remove the support from the backing plate.
   - Access hole plug on the backing plate
   - Brake drum, if equipped
7. On models with drum brakes, install the rotor and bracket. Tighten the bracket bolts to 85 ft. lbs. (115 Nm) and the caliper bolts to 25 ft. lbs. (34 Nm).
8. Install or connect the following:
   - Rear wheel
   - Negative battery cable

BRAKES

Brake Caliper

REMOVAL & INSTALLATION

Front

EXCEPT ION

1. Before servicing the vehicle, refer to the precautions section.
   - Front wheel and tire assembly
   - Brake hose from the caliper and discard the 2 copper washers. Plug the openings to prevent system contamination or excessive fluid loss.
   - Lock pin and guide pin from the caliper
   - Caliper from the support, being careful not to damage the pin boots
   - Pin boots from the caliper support and inspect for damage

To install:

2. If necessary, bottom the caliper piston by hand, or by using a C-clamp.
3. If removed, install the brake pads and clips to the caliper support.
4. Lubricate the pin boots and guide pins with silicone grease.
5. Install or connect the following:
   - Pin boots into the caliper support, using the pin to assure that the boot passes all the way through the support
   - Cliper onto the support and over the brake pads
6. Lubricate the non-threaded portion of the guide and lock pins with silicone grease.
   - Pins through the caliper and torque to 27 ft. lbs. (36 Nm)

➡**Make sure the brake line is properly routed with loop to the rear and that the hose is not twisted.**

   - Brake hose using 2 new copper washers. Torque the fitting bolt to 36 ft. lbs. (49 Nm).
7. Properly bleed the hydraulic brake system.
8. Install the wheel and tire assembly.

ION

1. Before servicing the vehicle, refer to the precautions section.
2. Use a turkey baster or similar device to remove fluid from the master cylinder until it is about ½ full.
3. Remove or disconnect the following:

   - Front wheel. Hand-tighten the lug nuts on the studs to prevent the rotor from falling off.
4. Install a C-clamp over the body of the caliper with the clamp ends against the rear of the caliper body and against the outboard pad. Tighten the clamp until the caliper piston is compressed enough to allow the caliper to be removed.
   - Brake hose from the caliper and discard the 2 copper washers. Plug the openings to prevent system contamination or excessive fluid loss.
   - Caliper bolts
   - Caliper from the bracket
   - Pin boots from the caliper support and inspect for damage

To install:

5. Install or connect the following:
   - Brake pads and clips to the caliper bracket, if removed
6. Lubricate the pin boots and guide pins with silicone grease.
   - Pin boots, if removed
   - Caliper onto the bracket and over the brake pads
   - Caliper bolts through the caliper and torque to 27 ft. lbs. (34 Nm)

➡**Make sure the brake line is properly routed with loop to the rear and that the hose is not twisted.**

   - Brake hose using 2 new copper washers. Torque the fitting bolt to 35 ft. lbs. (50 Nm).
7. Properly bleed the hydraulic brake system.
   - Wheel

Rear

EXCEPT ION

1. Before servicing the vehicle, refer to the precautions section.
2. Remove or disconnect the following:
   - Rear wheel and tire assembly
   - Brake hose from the caliper and discard the 2 copper washers. Plug the openings to prevent system contamination or excessive fluid loss.
3. Slip the end of the parking cable off the parking brake lever.
   - Cable outer housing from the cable bracket with SA9151BR cable release tool
   - Lock pin and guide pin
   - Caliper from the support, being

careful not to damage the pin boots.
   - Pin boots from the caliper support for inspection and lubrication

To install:

4. Make sure the piston is bottomed in the bore. Do not compress the piston using a C-clamp; instead the piston must be rotated into the caliper on its threads using a piston driver tool.
5. If removed, install the brake pads and clips to the caliper support.
6. Lubricate the pin boots and guide pins with silicone grease.
7. Install or connect the following:
   - Pin boots into the caliper support, using the pin to assure that the boot passes all the way through the support
   - Caliper onto the caliper support
8. Lubricate the non-threaded portion of the guide and lock pins.
   - Pins and torque to 27 ft. lbs. (36 Nm)
   - Brake hose using new copper washers and torque the fitting bolt to 36 ft. lbs. (49 Nm)
   - Parking brake cable
9. Properly bleed the hydraulic brake system.
10. Install the wheel and tire assembly.

ION

1. Before servicing the vehicle, refer to the precautions section.
2. Use a turkey baster or similar device to remove fluid from the master cylinder until it is about ½ full.
3. Release the park brake lever boot from the floor console by applying light pressure inward on the sides of the boot retainer, and pull the boot back.
4. Release the tension from the park brake cables. With the park brake lever in the released position, using ONLY HAND TOOLS, loosen the adjusting nut completely to the end of the front cable threaded rod.
5. Remove the wheel assembly.
6. Install and firmly hand tighten 2 wheel nuts to opposite wheel studs in order to retain the rotor to the hub.
7. Release the park brake cable end from the lever on the caliper.
8. Release the retaining tabs securing the park brake cable to the bracket on the caliper.
9. Install a large C-clamp, over the

body of the brake caliper with the C-clamp ends against the rear of the caliper body and against the outer brake pad.

➡**When using a large C-clamp to compress a caliper piston into a caliper bore of a caliper equipped with an integral park brake mechanism, do not exceed more than 1 mm (0.039 in) of piston travel. Exceeding this amount of piston travel will cause damage to the internal adjusting mechanism and/or the integral park brake mechanism.**

10. Tighten the C-clamp just enough to compress the caliper piston 1 mm (0.039 in) of travel only.

11. Remove the C-clamp from the caliper.

12. Remove the brake hose-to-caliper bolt from the brake caliper.

13. Remove the brake hose from the brake caliper.

14. Remove and discard the 2 copper brake hose gaskets. These gaskets may be stuck to the brake caliper and/or the brake hose end.

15. Cap or plug the opening in the brake caliper and the brake hose to prevent fluid loss and contamination.

16. While using a wrench on the flats of the caliper guide pins, remove the brake caliper guide pin bolts.

17. Remove the brake caliper from the caliper bracket.

18. Inspect the brake caliper guide pins for freedom of movement, and inspect the condition of the guide pin boots. Move the guide pins inboard and outboard within the bracket bores, without disengaging the slides from the boots, and observe for the following:

- Restricted caliper guide pin movement
- Looseness in the brake caliper mounting bracket
- Seized or binding caliper guide pins
- Split or torn boots

19. If any of the conditions listed are found, the brake caliper guide pins and/or boots require replacement.

To install:

20. Install the brake caliper to the caliper bracket.

21. While using a wrench on the flats of the caliper guide pins, install the brake caliper guide pin bolts and tighten to 25 ft. lbs. (34 Nm).

22. Press the park brake cable end fitting into the bracket on the caliper to secure the retaining tabs.

23. Secure the park brake cable end to the lever on the caliper.

24. Remove the caps or plugs from the brake caliper opening and the brake hose.

➡**Do not reuse the copper brake hose gaskets.**

25. Install NEW copper brake hose gaskets to the brake hose-to-caliper bolt and to the brake hose.

26. Install the brake hose and the brake hose-to-brake caliper bolt to the caliper. Tighten to 35 ft. lbs. (48 Nm).

27. Bleed the hydraulic brake system.

28. Remove the wheel nuts retaining the brake rotor to the wheel hub.

29. Install the wheel assembly.

30. With the engine OFF, gradually apply the brake pedal to approximately ⅔ of its travel distance.

31. Slowly release the brake pedal.

32. Wait 15 seconds, then gradually apply the brake pedal approximately ⅔ of its travel distance again until a firm brake pedal apply is obtained. This will properly seat the brake caliper pistons and brake pads.

33. Adjust the park brake cable tension.

34. Position the park brake lever boot to the floor console and press the boot retainer into place to secure.

Disc Brake Pads

REMOVAL & INSTALLATION

Front

EXCEPT ION

1. Before servicing the vehicle, refer to the precautions section.

2. Remove or disconnect the following:

- Front wheels
- Caliper lower lock pins

3. Either pivot the caliper up on the guide pin or remove the upper guide pin and support the caliper from the strut using a coat hanger or length of wire.

- 2 brake pads and the pad clips from the caliper support. Discard the old pad clips.

4. Check the caliper pins, pin boots and the piston boot for deterioration or damage.

To install:

5. By hand or using a C-clamp, bottom the piston all the way into the caliper bore.

6. Carefully lift the inner edge of the piston boot by hand to release any trapped air.

7. Install or connect the following:

- New pad clips into the caliper support
- Inner and outer brake pads into the support. If installed, remove the temporary support wire from the caliper.
- Caliper body on the support and upper guide pin into position. Compress the boots by hand as the caliper is positioned onto the support.

8. Lubricate the smooth ends of the removed pin(s) with silicone grease.

- Pin(s) and torque to 27 ft. lbs. (36 Nm). Do not get grease on the pin threads.
- Wheels

9. Prior to operating the vehicle, depress the brake pedal a few times until the brake pads are seated against the rotor.

ION

1. Before servicing the vehicle, refer to the precautions section.

2. Use a turkey baster or similar device to remove fluid from the master cylinder until it is about ½ full.

3. Remove or disconnect the following:

- Front wheel. Hand-tighten the lug nuts on the studs to prevent the rotor from falling off.

4. Install a C-clamp over the body of the caliper with the clamp ends against the rear of the caliper body and against the outboard pad. Tighten the clamp until the caliper piston is compressed enough to allow the caliper to be removed.

- Caliper lower bolt
- 2 brake pads and the pad clips from the caliper support. Discard the old pad clips.

5. Check the caliper pins, pin boots and the piston boot for deterioration or damage.

To install:

6. Install or connect the following:

- Brake pads and clips to the caliper bracket

7. Lubricate the pin boots and guide pins with silicone grease.

- Pin boots, if removed
- Caliper over the brake pads
- Caliper lower bolt through the caliper and torque to 27 ft. lbs. (34 Nm)

➡**Make sure the brake line is properly routed with loop to the rear and that the hose is not twisted.**

- Wheel

8. Prior to operating the vehicle, depress

the brake pedal a few times until the brake pads are seated against the rotor.

Rear

EXCEPT ION

1. Before servicing the vehicle, refer to the precautions section.

2. Remove or disconnect the following:
   • Rear wheels
   • Caliper lock and guide pins
   • Caliper from the support, being careful not to damage the pin boots and suspend the caliper from a wire
   • Brake pads from the support

To install:

3. Using SA91110NE piston driver tool, bottom the piston by rotating it clockwise into the caliper bore; do not use a C-clamp to press the piston into the bore.

4. Align the piston slots so they are perpendicular to the brake pads.

5. Carefully lift the inner edge of the piston boot to release any trapped air. The boot must lie flat below the level of the piston face.

6. Install or connect the following:
   • New pad clips into the caliper support
   • Inner and outer brake pads into the clips on the support. The pad with the wear sensor should be located outboard. The piston indentation slots should be positioned to correctly accept the brake pads.
   • Caliper body onto the support. Lubricate the non-threaded portion of the guide and lock pins, then install the pins and torque to 27 ft. lbs. (36 Nm).

7. Check the position of the pad clips. If necessary, use a small suitable tool to re-seat or center the pad clips on the support. Repeat the procedure for the opposite side brake pads.

8. Install the rear wheel assemblies.

9. Prior to operating the vehicle, depress the brake pedal a few times until the brake pads are seated against the rotor.

ION

1. Before servicing the vehicle, refer to the precautions section.

2. Use a turkey baster or similar device to remove fluid from the master cylinder until it is about ½ full.

3. Remove the wheel assembly.

4. Install and firmly hand tighten 2 wheel nuts to opposite wheel studs in order to retain the rotor to the hub.

5. Install a large C-clamp, over the body of the brake caliper with the C-clamp

ends against the rear of the caliper body and against the outer brake pad.

➡**When using a large C-clamp to compress a caliper piston into a caliper bore of a caliper equipped with an integral park brake mechanism, do not exceed more than 1 mm (0.039 in) of piston travel. Exceeding this amount of piston travel will cause damage to the internal adjusting mechanism and/or the integral park brake mechanism.**

6. Tighten the C-clamp just enough to compress the caliper piston 1 mm (0.039 in) of travel only.

7. Remove the C-clamp from the caliper.

8. While using a wrench on the flats of the caliper guide pins, remove the brake caliper guide pin bolts.

➡**Support the brake caliper with heavy mechanic's wire, or equivalent, whenever it is separated from its mount and the hydraulic flexible brake hose is still connected. Failure to support the caliper in this manner will cause the flexible brake hose to bear the weight of the caliper, which may cause damage to the brake hose and in turn may cause a brake fluid leak.**

9. Without disconnecting the hydraulic brake flexible hose, remove the caliper from the mounting bracket and secure the caliper with heavy mechanics wire, or equivalent.

10. Remove the brake pads from the caliper mounting bracket.

11. Remove the brake pad retainers from the caliper bracket.

12. Thoroughly clean the brake pad hardware mating surfaces of the caliper bracket, of any debris and corrosion.

13. Inspect the brake caliper guide pins for freedom of movement, and inspect the condition of the guide pin boots. Move the guide pins inboard and outboard within the bracket bores, without disengaging the slides from the boots, and observe for the following:
   • Restricted caliper guide pin movement
   • Looseness in the brake caliper mounting bracket
   • Seized or binding caliper guide pins
   • Split or torn boots

14. If any of the conditions listed are found, the brake caliper guide pins and/or boots require replacement.

15. Using a spanner wrench type caliper piston installer, fully retract the piston into the caliper bore.

To install:

16. Apply a very thin coating of high temperature silicone brake lubricant to the pad hardware mating surfaces of the caliper bracket only.

17. Install the brake pad retainers to the brake caliper bracket.

➡**The wear sensor equipped disc brake pad must be mounted inboard of the rotor with the leading edge of the sensor facing the brake rotor during forward wheel rotation, or at the bottom of the pad when installed in vehicle position.**

18. Install the brake pads to the caliper bracket.

19. Remove the support, and install the caliper into position over the disc brake pads and to the caliper mounting bracket.

20. While using a wrench on the flats of the caliper guide pins, install the brake caliper guide pin bolts to 25 ft. lbs. (35 Nm).

21. Remove the wheel nuts retaining the brake rotor to the hub.

22. Install the wheel assembly.

23. With the engine OFF, gradually apply the brake pedal approximately ⅔ of its travel distance.

24. Slowly release the brake pedal.

25. Wait 15 seconds, then gradually apply the brake pedal approximately ⅔ of its travel distance again until a firm brake pedal apply is obtained. This will properly seat the brake caliper pistons and brake pads.

26. Fill the master cylinder auxiliary reservoir to the proper level.

Brake Drums

REMOVAL & INSTALLATION

1. Before servicing the vehicle, refer to the precautions section.

2. Remove or disconnect the following:
   • Rear wheel and tire assembly
   • Brake drum. If necessary, turn the starwheel of the brake adjuster assembly to loosen the brake shoes and allow for drum removal.

To install:

3. Install or connect the following:
   • Brake drum over brake shoes and onto hub
   • Tire and wheel assembly. Torque to the proper specification.

4. Adjust brakes following the proper procedure.

5. Road test for braking operation.

Brake Shoes

REMOVAL & INSTALLATION

EXCEPT ION

1. Before servicing the vehicle, refer to the precautions section.

2. Remove or disconnect the following:

  • Wheels and brake drums
  • Lower return and adjuster springs using a universal brake spring remover. Do not over extend the springs or they will damaged and will need to be replaced.

3. Compress the leading brake shoe hold-down cup and spring while removing the pin from the rear of the backing plate. Release spring compression, then remove the hold-down cup and spring.

4. Pull the leading shoe towards the front of the vehicle and remove the adjuster assembly and lever. It may be necessary to turn the adjuster starwheel to shorten the adjuster's length.

  • Leading shoe by twisting the shoe out of engagement with the upper return spring
  • Upper return spring from the park brake shoe, then the park brake shoe hold-down cup, spring and pin assembly

5. Push the park brake shoe lever into the cable spring while disengaging the cable from the end lever and remove the parking brake shoe, lever and cable spring from the vehicle.

  • Retainer and wave washer, then the park brake lever from the shoe

6. Disassemble the brake adjuster socket, screw and nut, then clean the components in denatured alcohol. Inspect the assembly, making sure the screw threads

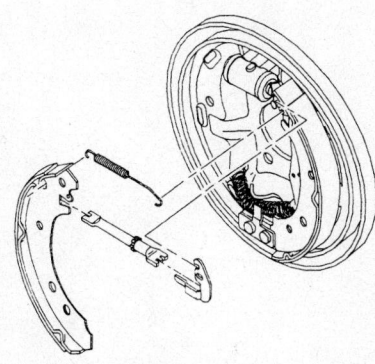

93006G81

Rear brake shoe and adjuster installation—I-Series shown

smoothly into the adjusting nut over the full threaded length.

7. Inspect the wheel cylinder for signs of leakage and for cut or damaged boots. Do not attempt to repair a damaged cylinder, the assembly must be replaced.

To install:

8. Lubricate the adjuster assembly, the 6 backing plate raised shoe contact pads, the brake lever pin and surfaces which contact brake shoe webs with brake lubricant.

9. Install or connect the following:

  • Park brake lever onto the pin on the brake shoe and secure with the wave washer and retainer clip. Crimp the ends of the retainer to secure the brake lever.
  • Cable spring into the cage on the park brake lever, then install the cable through the spring and onto the lever
  • Park brake shoe using the hold-down cup assembly; use a universal spring cup remover/installer tool. Make sure the shoe is correctly engaged into the wheel cylinder (top) and the anchor (bottom).
  • Long straight end of the upper return spring into the back hole in the park brake shoe, position the other brake shoe and install the other end of the spring into the back of the leading shoe

10. Pull the lead shoe toward the front of the vehicle and install the adjuster between the park and leading brake shoes. Verify that the adjuster notches properly engage the brake shoe notches and that the shoe is properly aligned in the wheel cylinder and anchor.

  • Adjuster lever and adjuster spring. Make sure the notch on the lever engages the pin on the park shoe and the notch on the adjusting socket. The lower leg of the lever should engage the teeth of the starwheel adjuster assembly.
  • Leading brake shoe using the hold-down cup assembly
  • Adjuster spring to the upper side of the brake shoes with the short end to the lead shoe and the long end to the adjuster lever
  • Lower return spring into the lower holes of the shoes

11. Verify the correct location of all brake components, if necessary, use the other side brake assembly for comparison.

12. Using a suitable drum clearance gauge, measure the inner diameter of the brake drum and adjust the outside diameter of the brake shoes to 0.02 inch (0.50mm) less than the inner diameter of the drum.

13. Repeat the procedure for the opposite brake shoes and install the brake drums.

14. If the wheel cylinders have been replaced, bleed the hydraulic brake system.

15. Install the rear wheels.

16. Apply and release the brake pedal 20 times to allow the adjuster to properly position the brake shoes.

17. Check and adjust the parking brake cable, as necessary.

ION

1. Before servicing the vehicle, refer to the precautions section.

2. Remove or disconnect the following:

  • Wheels and brake drums
  • Adjuster spring using a universal brake spring remover. Do not over extend the springs or they will damaged and will need to be replaced.
  • Brake adjuster lever from the pivot
  • Spread the top of the shoes apart using brake shoe spanner and spring removal tool J 38400
  • Adjuster assembly
  • Lightly pull the universal spring end out of the shoe web hole using the hook end of J 38400. Hold the universal spring while removing the trailing shoe.
  • Park brake cable from the lever
  • Lightly pull the universal spring end out of the shoe web hole using the hook end of J 38400. Hold the universal spring while removing the leading shoe.

3. Inspect the wheel cylinder for signs of leakage and for cut or damaged boots. Do not attempt to repair a damaged cylinder, the assembly must be replaced.

To install:

4. Lubricate the adjuster assembly, the 6 backing plate raised shoe contact pads, the brake lever pin and surfaces which contact brake shoe webs with brake lubricant.

5. Install or connect the following:

  • Position the hook end of tool J 38400 under the universal spring and lightly pull the spring end out while installing the leading shoe. Make sure the spring is properly engaged in the shoe web hole.
  • Park brake cable to the lever
  • Position the hook end of tool J 38400 under the universal spring and lightly pull the spring end out while installing the trailing shoe. Make sure the spring is properly engaged in the shoe web hole.

- Spread the top of the shoes apart using tool J 38400
- Adjuster assembly
- Adjuster actuator lever to the shoe and adjuster assembly. Make sure the lever is engaged properly between the adjuster and the shoe.
- Adjuster spring. Make sure the loop end of the spring engages properly to the tab on the actuator lever

6. Release the parking brake.

7. Pull the parking lever boot away from the console after applying light pressure inwards on the boot retainer, this will allow access to the cable adjusting nut.

8. Release the tension on the cable. Make sure to use hand tools only when adjusting the nut.

9. Using a suitable drum clearance gauge, measure the inner diameter of the brake drum and adjust the outside diameter of the brake shoes using the adjuster screw to 0.025 inch (0.50mm) less than the inner diameter of the drum.

10. Repeat the procedure for the opposite brake shoes and install the brake drums.

11. Adjust the parking park and install the lever boot.

12. If the wheel cylinders have been replaced, bleed the hydraulic brake system.

13. Install the rear wheels.

BUICK

22

LaCrosse

BRAKE HYDRAULIC
SYSTEM 22-64
DRIVE TRAIN 22-41
ENGINE REPAIR 22-10
FRONT BRAKES 22-59
FRONT SUSPENSION 22-50
FUEL SYSTEM 22-37
REAR BRAKES 22-61
REAR SUSPENSION 22-56
SPECIFICATIONS AND
MAINTENANCE CHARTS......22-3
Engine and Vehicle Identification
 Chart...........................22-3
General Engine Specifications 22-3
Engine Tune-Up Specifications 22-3
Firing Order 22-4
Accessory Drive Belt Routing 22-4
Capacities 22-4
Valve Specifications................22-4
Crankshaft and Connecting Rod
 Specifications...................22-5
Piston and Ring Specifications 22-5
Torque Specifications 22-5
Wheel Alignment 22-7
Tire, Wheel and Ball Joint
 Specifications...................22-7
Brake Specifications 22-7
Scheduled Maintenance
 Intervals.........................22-8
STEERING 22-45
A
Air Bag...........................22-45
 Disabling And Enabling 22-45
 Precautions....................22-45
Alternator.........................22-10
 Removal & Installation............22-10
B
Brake Caliper (Front Brakes)........22-59
 Removal & Installation............22-59
Brake Caliper (Rear Brakes) 22-61
 Removal & Installation............22-61
Brake Pads (Front Brakes) 22-60
 Removal & Installation............22-60

Brake Pads (Rear Brakes) 22-62
 Removal & Installation............22-62
Brake System Bleeding 22-64
C
Caliper Bracket (Front Brakes) 22-61
 Removal & Installation............22-61
Caliper Bracket (Rear Brakes) 22-63
 Removal & Installation............22-63
Camshaft and Valve Lifters 22-23
 Removal & Installation............22-23
Crankshaft Balancer 22-33
 Removal & Installation............22-33
CV-Joints...........................22-43
 Overhaul 22-43
Cylinder Head 22-16
 Removal & Installation............22-16
E
Engine Assembly 22-10
 Removal & Installation............22-10
Exhaust Manifold 22-22
 Removal & Installation............22-22
F
Fuel Filter 22-37
 Removal & Installation............22-37
Fuel Injectors......................22-40
 Removal & Installation............22-40
Fuel Pump 22-37
 Removal & Installation............22-37
Fuel System Pressure 22-37
 Relieving 22-37
Fuel System Service
 Precautions.....................22-37
H
Halfshaft...........................22-42
 Removal & Installation............22-42
Heater Core.........................22-15
 Removal And Installation 22-15
I
Ignition Timing 22-10
Intake Manifold.....................22-19
 Removal & Installation............22-19
K
Knuckle (Front Suspension) 22-55
 Removal & Installation............22-55

Knuckle (Rear Suspension)..........22-58
 Removal & Installation............22-58
L
Lower Ball Joint....................22-52
 Removal & Installation............22-52
Lower Control Arm 22-52
 Lower Control Arm Bushing
  Replacement..................22-53
 Removal & Installation............22-52
O
Oil Pan............................22-29
 Removal & Installation............22-29
Oil Pump 22-31
 Removal & Installation............22-31
P
Parking Brake 22-63
 Adjustment.....................22-63
Piston and Ring 22-36
 Positioning 22-36
Power Rack and Pinion Steering
 Gear...........................22-48
 Removal & Installation............22-48
R
Rear Main Seal 22-32
 Removal & Installation............22-32
Rear Suspension Support 22-57
 Removal & Installation............22-57
Rocker Arms 22-18
 Removal & Installation............22-18
S
Spindle Rod 22-57
 Removal & Installation............22-57
Stabilizer (Sway) Bar................22-56
 Removal & Installation............22-56
Stabilizer (Sway) Bar and Links ...22-51
 Removal & Installation............22-51
Starter Motor 22-29
 Removal & Installation............22-29
Strut (Front Suspension) 22-50
 Disassembly & Assembly 22-51
 Removal & Installation............22-51
Strut (Rear Suspension)..............22-56
 Disassembly And Assembly....22-56
 Removal & Installation............22-56

T

Timing Chain, Sprockets, Front
Cover and Seal 22-34
Removal & Installation............22-34
Trailing Arm 22-58
Removal & Installation............22-58
Transaxle Assembly 22-41
Removal & Installation............22-41

V

Valve Lash 22-29
Adjustment..............................22-29

W

Water Pump 22-14
Removal & Installation............22-14

Wheel Bearings (Front
Suspension) 22-55
Adjustment..............................22-55
Removal & Installation............22-55
Wheel Bearings (Rear
Suspension) 22-58
Removal & Installation............22-58

SPECIFICATIONS AND MAINTENANCE CHARTS

ENGINE AND VEHICLE IDENTIFICATION

		Engine							Model Year	
Code ①	Liters (cc)	Cu. In.	Cyl.	Fuel Sys.	Engine Type	Eng. Mfg.		Code ②		Year
2, 4	3.8 (3785)	231	6	MFI	OHV	GM		5		2005
7	3.6 (3564)	217	6	MFI	DOHC	GM		6		2006

MFI: Multi-point Fuel Injection

DOHC: Dual overhead camshafts

OHV: Overhead Valves

① 8th position of VIN

② 10th position of VIN

06025-LACR-C01

GENERAL ENGINE SPECIFICATIONS

Model	Engine Displacement Liters	Engine Series VIN	Years	Net Horsepower @ rpm	Net Torque @ rpm (ft. lbs.)	Bore x Stroke (in.)	Com- pression Ratio	Oil Pressure @ rpm
LaCrosse	3.8	2, 4	2005	205@5200	230@4000	3.80x3.40	9.4:1	60@1850
	3.6	7	2005	240@6000	225@2000	3.70x3.37	10.2:1	20@2000

06025-LACR-C02

ENGINE TUNE-UP SPECIFICATIONS

Engine Displacement Liters	Years	Engine VIN	Spark Plug Gap (in.)	Ignition Timing (deg.)	Fuel Pump (psi)	Idle Speed (rpm)	Valve Clearance Intake	Valve Clearance Exhaust
3.8	2005	2, 4	0.060	①	41-47②	③	HYD	HYD
3.6	2005	7	0.044	①	55-60	③	HYD	HYD

NOTE: The Vehicle Emission Control Information label often reflects specification changes made during production.

The label figures must be used if they differ from those in this chart.

HYD: Hydraulic

NA: Information not available

① DIS Ignition System timing not adjustable

② Pressure at fuel pump

③ Idle speed maintained by ECM. There is no recommended adjustment procedure

06025-LACR-C03

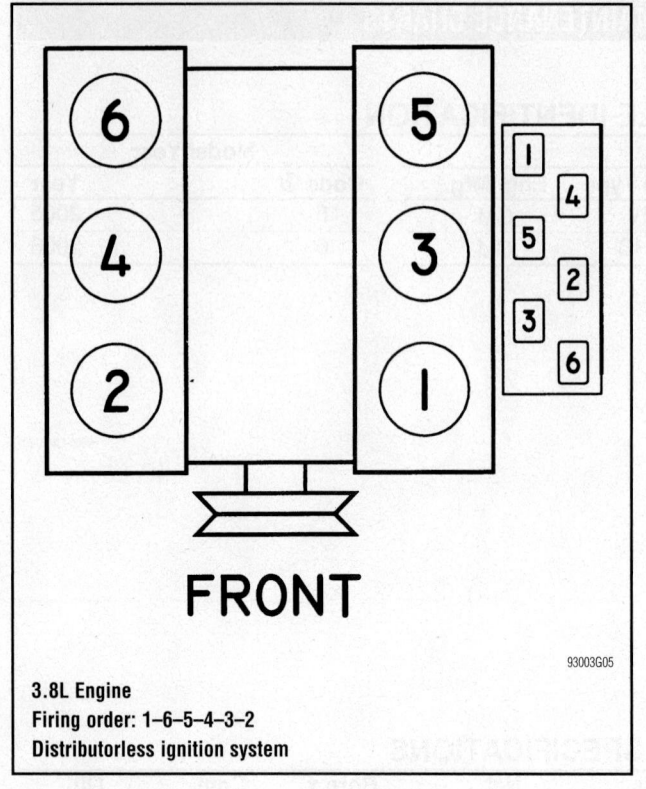

3.8L Engine
Firing order: 1–6–5–4–3–2
Distributorless ignition system

93003G05

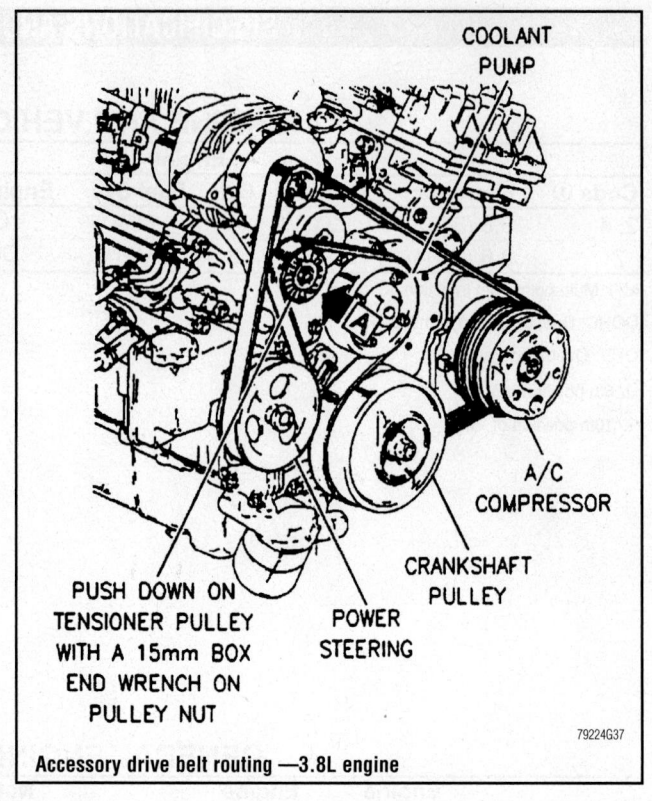

PUSH DOWN ON TENSIONER PULLEY WITH A 15mm BOX END WRENCH ON PULLEY NUT

POWER STEERING

CRANKSHAFT PULLEY

A/C COMPRESSOR

COOLANT PUMP

79224G37

Accessory drive belt routing —3.8L engine

CAPACITIES

Model	Engine Displacement Liters	Engine VIN	Years	Engine Oil with Filter (qts.)	Transmission (pts.)*	Fuel Tank (gal.)	Cooling System (qts.)
LaCrosse	3.8	2, 4	2005	4.5	14.8	17.5	11.7
	3.6	7	2005	5.5	14.8	17.5	11.0

NOTE: All capacities are approximate. Add fluid gradually and ensure a proper fluid level is obtained.

* Drain and refill

06025-LACR-C04

VALVE SPECIFICATIONS

Engine Displacement Liters	Engine VIN	Years	Seat Angle (deg.)	Face Angle (deg.)	Spring Test Pressure (lbs. @ in.)	Spring Installed Height (in.)	Stem-to-Guide Clearance (in.)		Stem Diameter (in.)	
							Intake	Exhaust	Intake	Exhaust
3.8	2, 4	2005	45	45	80@1.750	1.690-1.720	0.0015-0.0032	0.0015-0.0032	0.3129-0.3136	0.3129-0.3136
3.6	7	2005	45	44.25	134-149@0.9449	1.378	0.0010-0.0026	0.0014-0.0030	0.2344-0.2352	0.2341-0.2348

06025-LACR-C05

CRANKSHAFT AND CONNECTING ROD SPECIFICATIONS

All measurements given in inches

Engine Displacement Liters	Engine VIN	Years	Crankshaft				Connecting Rod		
			Main Brg. Journal Dia.	Main Brg. Oil Clearance	Shaft End-play	Thrust on No.	Journal Diameter	Oil Clearance	Side Clearance
3.8	2, 4	2005	2.4988-2.4998	①	0.0030-0.0110	2	2.2487-2.2499	0.0005-0.0026	0.0040-0.0200
3.6	7	2005	2.6768-2.6775	0.0004-0.0024	0.0039-0.0130	3	2.2044-2.2050	0.0004-0.0028	0.0037-0.0140

① Journal 1: 0.0007 - 0.0016
   Journals 2 and 3: 0.0010 - 0.0020
   Journal 4: 0.0009 - 0.0018

06025-LACR-C06

PISTON AND RING SPECIFICATIONS

All measurements given in inches

Engine Displacement Liters	Engine VIN	Years	Piston Clearance	Ring Gap			Ring Side Clearance		
				Top Compression	Bottom Compression	Oil Control	Top Compression	Bottom Compression	Oil Control
3.8	2, 4	2005	0.0004-0.0020	0.012-0.022	0.030-0.040	0.010-0.030	0.0013-0.0031	0.0013-0.0031	0.0009-0.0079
3.6	7	2005	0.0010-0.0021	0.0059-0.0118	0.0110-0.0189	0.0059-0.0236	0.0012-0.0026	0.0006-0.0024	0.0012-0.0067

NA: Information not available

06025-LACR-C07

TORQUE SPECIFICATIONS

All measurements given in ft. lbs. unless otherwise noted

Engine Displacement Liters	Engine VIN	Years	Cylinder Head Bolts	Main Bearing Bolts	Rod Bearing Bolts	Crankshaft Damper Bolts	Flywheel Bolts	Manifold		Spark Plugs	Oil Pan Drain Plug
								Intake	Exhaust		
3.8	2, 4	2005	①	②	③	④	⑤	⑥	22	⑦	22
3.6	7	2005	⑧	⑨	⑩	⑪	⑫	17	15	13	18

① Step 1: Tighten all bolts to 37 ft. lbs.
   Step 2: Turn all bolts 120 degrees
② Cap bolts: 30 ft. lbs. plus 110 degrees
   Side bolts: 11 ft. lbs. plus 45 degrees
③ 20 ft. lbs. plus 50 degrees
④ 111 ft. lbs. plus 76 degrees
⑤ 11 ft. lbs. plus 50 degrees
⑥ Upper manifold: 8 ft. lbs.
   Lower manifold: 11 ft. lbs.
⑦ Initial installation: 20 ft. lbs.
   Re-installation: 11 ft. lbs.
⑧ M8 bolts:
   Step 1: 10 ft. lbs.
   Step2: plus 60 degrees
   M 11 bolts:
   Step 1: 33 ft. lbs.
   Step 2: plus 120 degrees

⑨ Inner:
   Step 1: 15 ft. lbs.
   Step 2: plus 80 degrees
   Outer:
   Step 1: 10 ft. lbs.
   Step 2: plus 110 degrees
   Side:
   Step 1: 22 ft. lbs.
   Step 2: plus 60 degrees
⑩ Step 1: 22 ft. lbs.
   Step 2: back off to zero
   Step 3: 18 ft. lbs.
   Step 4: plus 110 degrees
⑪ Step 1: 74 ft. lbs.
   Step 2: plus 150 degrees
⑫ Step 1: 22 ft. lbs.
   Step 2: plus 45 degrees

06025-LACR-C08

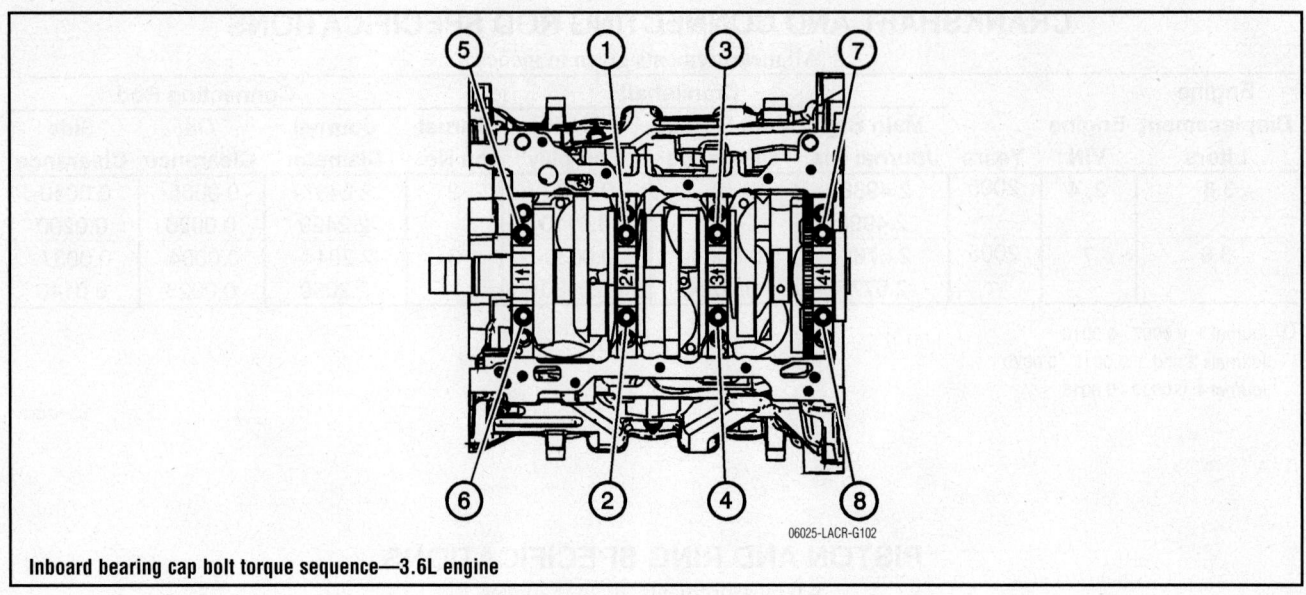

Inboard bearing cap bolt torque sequence—3.6L engine

06025-LACR-G102

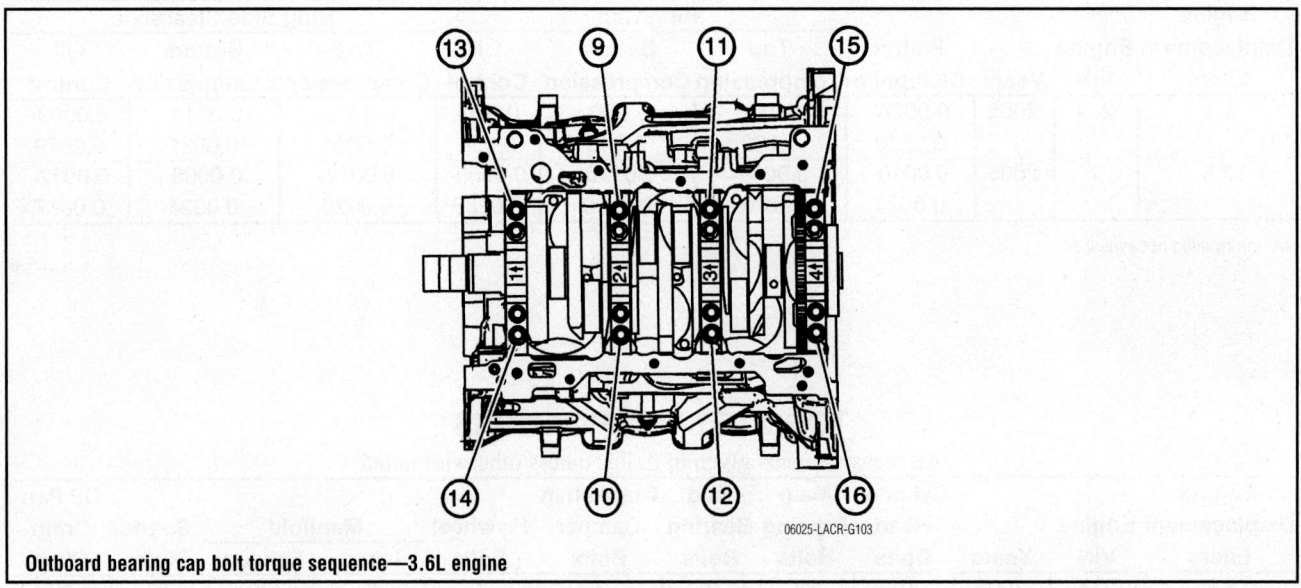

Outboard bearing cap bolt torque sequence—3.6L engine

06025-LACR-G103

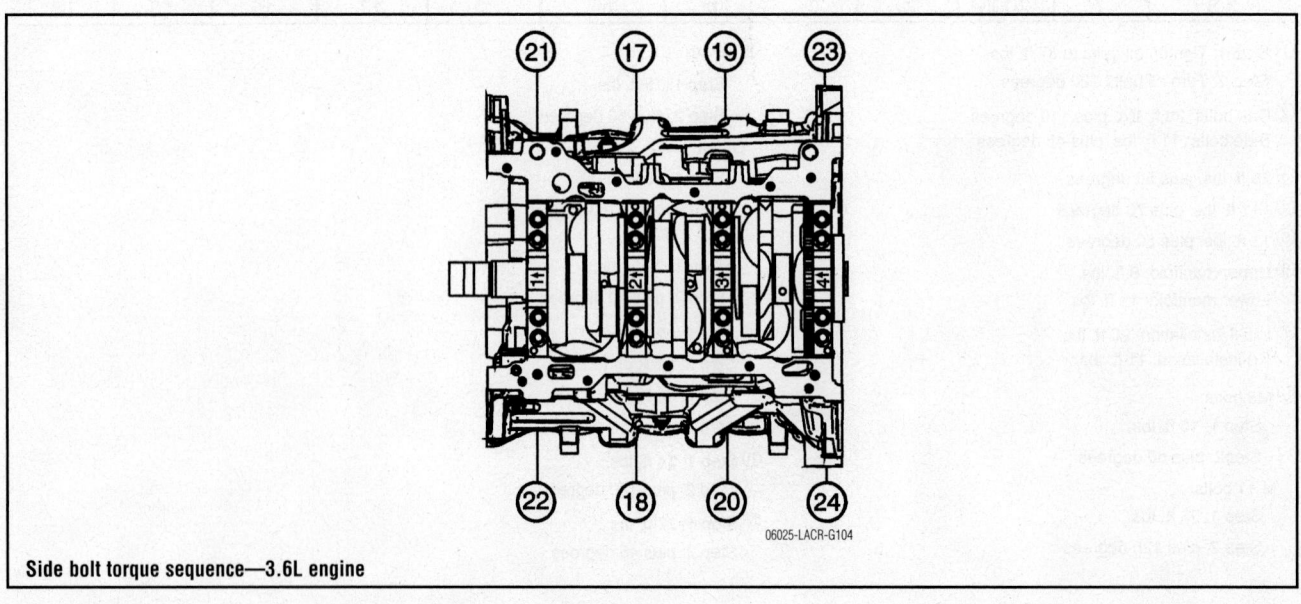

Side bolt torque sequence—3.6L engine

06025-LACR-G104

WHEEL ALIGNMENT SPECIFICATIONS

Model	Year		Caster Range (Deg.)	Caster Preferred Setting (Deg.)	Camber Range (Deg.)	Camber Preferred Setting (Deg.)	Toe-in (in.)
All	2005	F	0.75	+3.40	0.75	-0.80	0.10 +/- 0.20
		R	—	—	0.50	-0.80	0.10 +/- 0.20

06025-LACR-C09

TIRE, WHEEL AND BALL JOINT SPECIFICATIONS

Model	Years	OEM Tires Standard	OEM Tires Optional	Tire Pressures (psi) Front	Tire Pressures (psi) Rear	Wheel Size	Ball Joint Inspection	Wheel lug Torque (ft. lbs.)
CX	2005	P225/60R16	None	①	①	NA	0.125 in. ②	100
CXL	2005	P225/60R16	None	①	①	NA	0.125 in. ②	100
CXS	2005	P225/55R17	None	①	①	NA	0.125 in. ②	100

NA: Information not available

OEM: Original Equipment Manufacturer

PSI: Pounds Per Square Inch

① See placard on vehicle

② Remove tension from the joint. Measurement is max. horizontal or vertical.

06025-LACR-C10

BRAKE SPECIFICATIONS
All measurements given in inches unless otherwise noted

Model	Years		Brake Disc Original Thickness	Brake Disc Minimum Thickness	Brake Disc Maximum Runout	Minimum Lining Thickness	Caliper Bracket Bolts (ft. lbs.)	Caliper Mounting Bolts (ft. lbs.)
All	2005	F	1.270	1.210	0.002	NA	133	70
		R	0.550	0.490	0.002	NA	89	25

NA: Information not available

06025-LACR-C11

MAINTENANCE I AND II SERVICE SCHEDULES
2005 Buick LaCrosse

When the CHANGE ENGINE OIL light appears, certain services and inspections are required.

Required services are described as Maintenance I and Maintenance II.

The first service on a vehicle should be Maintenance I, and the second service should be Maintenance II.

Alternate between the 2 thereafter. However, in some cases, Maintenance II may be required more often.

Maintenance I: Use Maintenance I if the CHANGE ENGINE OIL light comes on within 10 months since vehicle was purchased or, if Maintenance II was performed.

Maintenance II: Use Maintenance II if the previous service performed was Maintenance I. Always use Maintenance II whenever the CHANGE ENGINE OIL light comes on 10 months or more since the last service, or, if the CHANGE ENGINE OIL light has not come on at all for one year.

Service	Maintenance I	Maintenance II
Change the engine oil and filter. Reset the oil life system.	✓	✓
Visually inspect the vehicle for leaks or damage. A fluid loss in the vehicle system could indicate a problem. Inspected, repair and add fluid to the system if necessary.	✓	✓
Inspect the engine air cleaner filter. If necessary, replace the filter.	--	✓
Rotate the tires. Inspect the tire inflation pressures and the tire wear.	✓	✓
Visually inspect the brake lines and hoses for proper hook-up, binding, leaks, cracks, chafing, etc. Inspect the disc brake pads for wear and the rotors for surface condition. Inspect the drum brake linings for wear or cracks. Inspect other brake parts, including drums, wheel cylinders, calipers, parking brake, etc. Inspect the parking brake adjustment.	✓	✓
Inspect the engine coolant and the windshield washer fluid levels. Add fluid as needed.	✓	✓
Inspect the suspension and steering components. Inspect the front and rear suspension and the steering system for damaged, loose or missing parts, or signs of wear. Inspect the power steering lines and the hoses for proper hook-up, binding, leaks, cracks, chafing, etc.	--	✓
Visually inspect the coolant hoses and replace the hoses if they are cracked, swollen or deteriorated. Inspect all pipes, fittings and clamps; replace with GM parts as needed. To help ensure proper operation, a pressure test of the cooling system and pressure cap and cleaning the outside of the radiator and air conditioning condenser is recommended at least once a year.	--	✓
Inspect the front and rear suspension and the steering system for damaged, loose or missing parts, or signs of wear. Inspect power steering lines and hoses for proper hook-up, binding, leaks, cracks, chafing, etc.	--	✓
Inspect the throttle system for interference or binding and for damaged or missing parts. Replace the parts as needed. Replace any components that have high effort or excessive wear. Do not lubricate the accelerator or the cruise control cables.	--	✓
Replace the passenger compartment air filter.	--	✓

To reset the CHANGE ENGINE OIL LIGHT:

1. Press the option button on the DIC until ENGINE OIL MONITOR appears on the DIC screen.
2. Press the set/reset button to reset the system. The next screen indicates that the CHANGE OIL SOON message message has been reset.If the vehicle has the uplevel DIC, when the gages button is pressed and the OIL LIFE REMAINING mode appears, it should read 100 percent OIL LIFE REMAINING
3. Turn the key to OFF.

Vehicles without Driver Information Center (DIC)

1. With the engine off, turn the ignition key to RUN.
2. Fully press and release the accelerator pedal slowly three times within five seconds.
3. Turn the key to OFF, then start the vehicle.

If the light or message comes back on when you start your vehicle, the oil life system has not reset. Repeat the procedure.

06025-LACR-C12

ADDITIONAL MAINTENANCE SERVICES
2005 Buick LaCrosse

TO BE SERVICED	TYPE OF SERVICE	VEHICLE MILEAGE INTERVAL (x1000)					
		25	50	75	100	125	150
Air cleaner filter	R	✓	✓	✓	✓	✓	✓
Accessory drive belt	I						✓
Auto. Trans. Fluid ①	R		✓		✓		✓
Cooling system hoses and clamps	S/I						✓
Engine coolant	R						✓
Fuel system	I	✓	✓	✓	✓	✓	✓
Exhaust system & heat shields	S/I	✓	✓	✓	✓	✓	✓
Spark plugs	R				✓		

R: Replace   S/I: Inspect and service, if necessary

① Replace if any of the following conditions are met:

    Heavy city traffic where the outside temperature regularly reaches 32°C (90°F) or higher

    Hilly or mountainous terrain

    Frequent trailer towing

    Taxi, police or delivery service

    Otherwise, change every 100,000 miles

06025-LACR-C13

ENGINE REPAIR

➡**Disconnecting the negative battery cable on some vehicles may interfere with the operation of the onboard computer system. The computer may undergo a relearning process once the negative battery cable is reconnected.**

Alternator

REMOVAL & INSTALLATION

3.6L Engine

1. Before servicing the vehicle, refer to the Precautions Section.
2. Disconnect the battery ground (negative) cable from the battery.
3. Remove the bolt and the nut from the engine mount strut at the left engine mount strut bracket on the engine. Remove the bolt and the nut from the engine mount strut at the engine mount strut bracket on the upper radiator support.
4. Remove the engine mount strut.
5. Remove the bolt and the nut from the engine mount strut at the right engine mount strut bracket on the engine.
6. Remove the bolt and the nut from the engine mount strut at the engine mount strut bracket on the upper radiator support.
7. Remove the engine mount strut.
8. Remove the generator B+ terminal nut and the battery cable from the generator.
9. Disconnect the generator electrical connector.
10. Remove the drive belt from the generator.
11. Remove the idler pulley.
12. Remove the generator bolts.

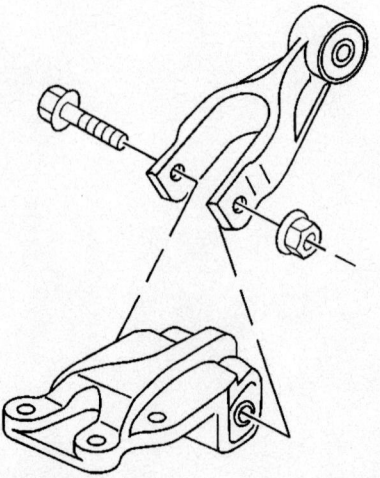

06025-LACR-G01

Engine mount strut—3.6L engine

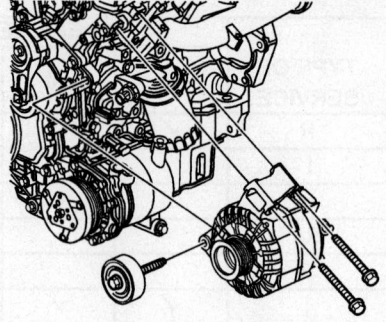

06025-LACR-G02

Alternator mounting—3.6L engine

13. Remove the generator from the vehicle.

To install:

14. Install the generator to the vehicle.
15. Install the generator bolts. Tighten the generator bolts to 50 Nm (37 ft. lbs.).
16. Install the idler pulley. Tighten the drive belt idler pulley bolt to 50 Nm (37 ft. lbs.).
17. Install the drive belt.
18. Connect the generator electrical connector.
19. Install the battery cable and the generator B+ terminal nut to the generator. Tighten the generator B+ terminal nut 20 Nm (15 ft. lbs.).
20. Install the engine mount strut.
21. Install the bolt and the nut to the engine mount strut at the engine mount strut bracket on the upper radiator support. HAND TIGHTEN ONLY.
22. Install the bolt and the nut to the engine mount strut at the right engine mount strut bracket on the engine. Tighten both engine mount strut nuts to 48 Nm (35 ft. lbs.).
23. Install the engine mount strut.
24. Install the bolt and the nut to the engine mount strut at the engine mount strut bracket on the upper radiator support. HAND TIGHTEN ONLY.
25. Install the bolt and the nut to the engine mount strut at the left engine mount strut bracket on the engine. Tighten bolt engine mount strut nuts to 48 Nm (35 ft. lbs.).
26. Connect the battery ground (negative) cable to the battery.

3.8L Engine

1. Before servicing the vehicle, refer to the Precautions Section.
2. Remove or disconnect the following:
   - Negative battery cable
   - Accessory drive belt

- Fuel injector sight shield
- Alternator brace
- Electrical connections
- Alternator bolts and the alternator

To install:

3. Install or connect the following:
   - Alternator and torque the bolts to 37 ft. lbs. (50 Nm)
   - Electrical connections and torque the nut to 111 inch lbs. (12.5 Nm)
   - Alternator brace. Torque the nut to 37 ft. lbs. (50 Nm) and the bolt to 22 ft. lbs. (30 Nm).
   - Fuel injector sight shield
   - Accessory drive belt
   - Negative battery cable

Ignition Timing

The ignition timing is not adjustable, and is set according to engine demand electronically. The Powertrain Control Module (PCM) controls the ignition timing for all driving conditions.

Engine Assembly

REMOVAL & INSTALLATION

3.6L Engine

1. Before servicing the vehicle, refer to the Precautions Section.
2. Disconnect the battery negative cable.
3. Remove the throttle body air inlet duct.

➡**Do not disconnect the battery negative cable from the vehicle.**

4. Disconnect the battery negative cable from the engine block.

➡**Do not disconnect the battery positive cable from the vehicle, underhood electrical center or the battery.**

5. Disconnect the battery positive cable from the generator and the starter.
6. Drain the cooling system.
7. Disconnect the radiator hoses from the engine.
8. Disconnect the heater hoses from the engine.
9. Remove the bolt and the nut from the engine mount strut at the left engine mount strut bracket on the engine. Remove the bolt and the nut from the engine mount strut at the engine mount strut bracket on the upper radiator support.

10. Remove the engine mount strut.

11. Remove the bolt and the nut from the engine mount strut at the right engine mount strut bracket on the engine.

12. Remove the bolt and the nut from the engine mount strut at the engine mount strut bracket on the upper radiator support.

13. Remove the engine mount strut.

➡**Relieve the fuel pressure.**

14. Disconnect the fuel pressure and evaporative emission (EVAP) pipes from the engine.

15. Remove the ECM chassis (outboard) side electrical connector from the ECM.

16. Remove the wiring harness ground from the transmission.

17. Remove the vacuum brake booster hose from the intake manifold.

18. Evacuate the air conditioning system.

19. Remove the A/C compressor hose from the A/C compressor.

20. Relocate the compressor hose to the side.

21. Remove the transmission electrical connector.

22. Raise and support the vehicle.

23. If you will be separating the engine from the transmission, remove the torque converter bolts.

24. Drain the engine oil.

25. Remove the catalytic converter.

26. Remove the front tires and wheels.

27. Remove lower radiator air baffle.

28. Remove the engine splash shields.

29. Disconnect the vehicle speed sensor (VSS) electrical connector and secure the wiring harness to the vehicle.

30. Remove the front wheel speed sensor wiring harnesses from the lower control arms and the frame.

31. Remove the tie rod ends from the steering knuckles.

32. Remove the lower ball joints from the knuckles.

33. Disconnect the drive axles from the transaxle.

34. Rotate the struts and reposition the drive axles toward the rear of the vehicle in order to provide clearance for the powertrain to be removed.

✳✳ CAUTION

Failure to disconnect the intermediate shaft from the rack and pinion steering gear stub shaft can result in damage to the steering gear and/or intermediate shaft. This damage may cause loss of steering control which could result in an accident and possible personal injury

35. Separate the intermediate steering shaft from the steering gear.

36. Remove the engine mount lower nuts.

37. Remove the transmission mount lower nuts.

38. Position a powertrain lift table below the powertrain.

39. Lower the vehicle until the powertrain is supported by the powertrain lift table.

40. Remove the fuel injector sight shield.

41. Remove the engine mount struts.

42. One engine mount strut and nut can be reinstalled to the engine mount strut bracket on engine to be used as an attachment point for front engine attachment.

43. Disconnect the electrical harness clip at the UHJB and at coolant housing.

44. Disconnect the electrical connector at the engine control module (ECM).

45. Remove the coolant recovery reservoir.

46. Remove the power steering reservoir and bracket.

47. Remove the attachment screw from the power steering pump reservoir to bracket.

48. Install the power steering reservoir

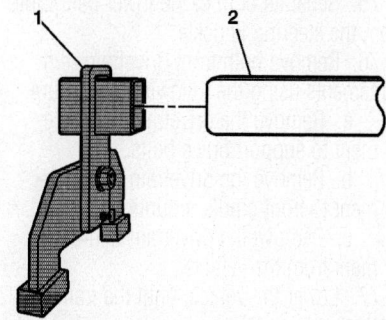

06025-LACR-G03

Install a J 28467-B and retention pin to each end of the J 28467-500 beam (2) — 3.6L engine

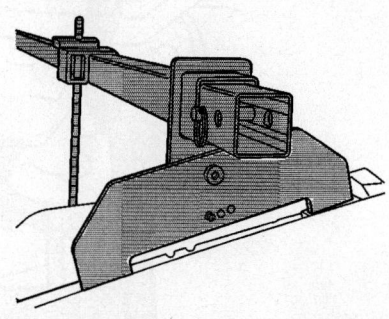

06025-LACR-G04

Position the J 28467-B to the left and right side inner fender rails in order to install the J 28467-500—3.6L engine

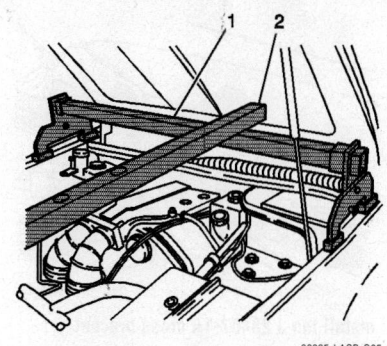

06025-LACR-G05

Install the radiator shelf tube J 28467-2A (2) on top of the strut tower tube J 28467-3 (1) above the engine front (right bank) lift bracket—3.6L engine

bracket to powertrain. Tighten the power steering reservoir bracket bolts: M6 bolt to 9 Nm (7 ft. lbs.) and M8 bolt to M8 Bolt to 25 Nm (18 ft. lbs.)

49. Install a J 28467-B and retention pin to each end of the J 28467-500 beam (2).

50. Position the J 28467-B to the left and right side inner fender rails in order to install the J 28467-500.

51. Install the radiator shelf tube J 28467-2A (2) on top of the strut tower tube J 28467-3 (1) above the engine front (right bank) lift bracket.

52. Install the round tube of the front support assembly J 28467-4A through the large hole in the radiator shelf tube J 28467-2A.

53. Locate the J 28467-4A front support assembly to the upper tie bar.

54. Install the J 28467-9 7/16 in x 2.0 in quick-release pin through the top hole in the J 28467-4A front support assembly.

55. Install the J 28467-1A cross bracket assembly.

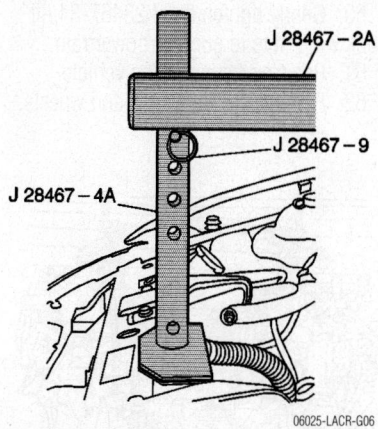

06025-LACR-G06

Install the round tube of the front support assembly J 28467-4A through the large hole in the radiator shelf tube J 28467-2A—3.6L engine

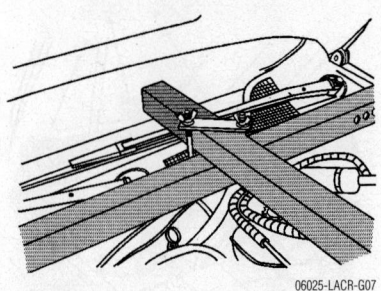

Install the J 28467-1A cross bracket assembly—3.6L engine

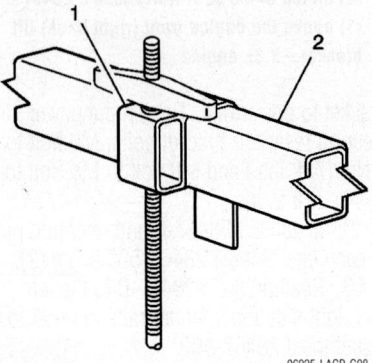

Create 2 lift hook assemblies as shown—3.6L engine

56. Assemble the following to create 2 lift hook assemblies as shown:
- J 28467-34
- J 28467-6A
- J 28467-7A

57. Install the 2 lift hook assemblies to the J 45057-2.

58. Install the LH engine mount strut nuts and bolts to the strut brackets.

59. Install the RH engine lift hook through the power steering reservoir bracket.

60. Gently tighten the J 28467-34 lift hook wing nuts to support powertrain.

61. Raise and support the vehicle.

62. Remove the front tires and wheels.

63. Remove and discard the two plastic braces from the front of the radiator lower air deflector. The plastic braces are directly below the front cradle bolts.

64. Remove the positive battery cable and the retainers from the frame and position aside.

65. Disconnect the power steering return hose from the frame.

66. Secure the power steering return hose.

67. Remove the stabilizer shaft links and rotate the stabilizer shaft upward to gain access to the mounting bolts in the power steering gear.

68. Remove the mounting bolts from the power steering gear.

69. Secure the power steering gear.

70. Remove the nuts that secure the engine mount to the frame.

71. Remove the nuts which secure the transaxle mount to the frame.

72. If applicable, disconnect the front wheel speed sensor harness connectors.

73. If applicable, disconnect the wheel speed sensor harness from the frame and lower control arms.

74. If applicable, remove the retainers at the front wheel speed harness from the frame and from the lower control arms.

75. Separate both of the lower ball joints from the steering knuckle.

76. Remove both front drivetrain reinforcements using the following procedure.

a. Remove the drivetrain reinforcement to support brace bolts.

b. Remove the drivetrain reinforcement to front cradle mounting stud nut.

c. Remove the drivetrain reinforcement from the vehicle.

77. Lower the vehicle until the frame contacts the J 39580.

78. Remove the bolts which secure the front frame to the body.

79. Remove the bolts which secure the rear frame to the body.

80. Raise the vehicle in order to separate the frame from the body.

81. Carefully raise the vehicle or lower the powertrain table in order to remove the powertrain from the vehicle.

82. Remove the exhaust crossover pipe.

83. Remove the coolant inlet pipe.

84. Remove the power steering pressure pipe/hose from the pump.

85. Remove the power steering cooler lines from the reservoir.

86. Remove the accessory drive belt.

87. Disconnect the electrical harness clip at the underhood junction block and at the coolant reservoir.

88. Disconnect the engine control module (ECM) wire harness from the ECM wire harness clip on the reservoir.

89. Disconnect the power steering hoses from the power steering reservoir.

90. Remove the power steering reservoir and bracket bolts, and remove the reservoir and bracket assembly from the vehicle.

91. Cap off the power steering reservoir and hoses to prevent contamination.

92. Remove the transmission lower brace.

93. Remove the transmission upper brace nut located behind the power steering pump.

94. Remove the engine to transmission (bell housing) bolts.

95. Use 4 M1 0x1.5x40 GM P/N 11519182, or equivalent bolts to install the engine lift brackets to the left rear and right front cylinder heads. Tighten the lift bracket bolts to 65 Nm (48 ft. lbs.).

96. Use an engine hoist in order to sep-

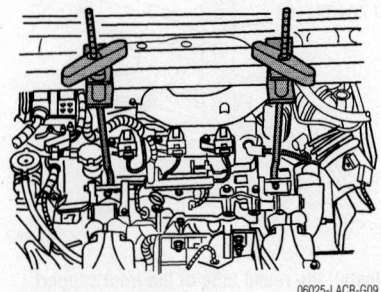

Install the 2 lift hook assemblies—3.6L engine

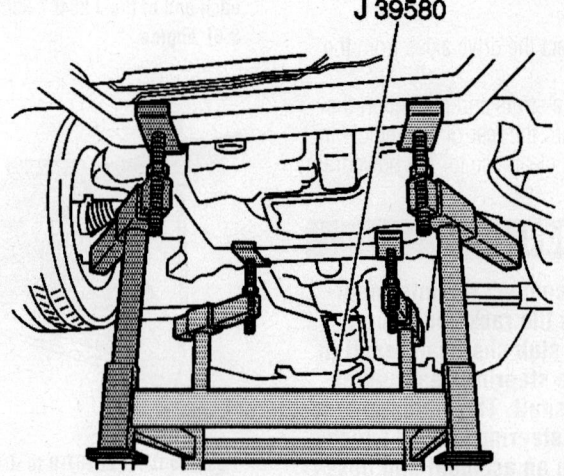

J 39580

Lower the vehicle until the frame contacts the J 39580—3.6L engine

arate the engine from the transmission and the frame.

97. Install the engine to a suitable engine stand.

To install:

98. Use an engine hoist in order to remove the engine from the engine stand.

99. Install the engine to the transmission and the frame at the powertrain lift table.

100. Remove the engine lift brackets.

101. Install the engine to transmission (bell housing) bolts. Install the torque converter bolts. Tighten the bolts to 63 Nm (46 ft. lbs.). Install the transaxle bolts and the stud. Tighten the bolts and the stud to 75 Nm (55 ft. lbs.).

102. Install the transmission upper brace nut located behind the power steering pump.

103. Tighten the transmission upper brace nut to 50 Nm (37 ft. lbs.).

104. Install the transmission lower brace. Install the transaxle brace bolts to the transaxle. Tighten the transaxle brace bolts to the transaxle to 43 Nm (32 ft. lbs.).

105. Install the transaxle brace bolts to the engine. Tighten the transaxle brace bolts to the engine to 63 Nm (46 ft. lbs.).

106. Uncap the power steering reservoir and hoses .

107. Install the power steering reservoir and bracket to the vehicle. Tighten the M6 bolt to 9 Nm (7 ft. lbs.). Tighten the M8 bolt to 25 Nm (18 ft. lbs.).

108. Connect the power steering reservoir hoses to the power steering reservoir.

109. Install the coolant recovery reservoir.

110. Connect the ECM wire harness to the ECM wire harness clip on the reservoir.

111. Connect the electrical harness clip at the underhood junction block and coolant reservoir.

112. Install the accessory drive belt.

113. Install the coolant inlet pipe.

114. Install the exhaust crossover pipe. Torque all fasteners to 18 ft. lbs. (44 Nm).

115. Carefully lower the vehicle or raise the powertrain table in order to install the powertrain to the vehicle.

116. Position the engine support table with the frame under the vehicle.

117. Lower the vehicle to the frame.

118. Loosely install the bolts to secure the rear frame to the body.

119. Loosely install the bolts to secure the front frame to the body.

120. Align the frame to the body by inserting two 19 X 203 mm (0.74 X 8 in) pins in the alignment holes on the right side of the frame.

121. Install the front and rear frame bolts. Tighten the front bolts to 145 Nm (107 ft. lbs.). Tighten the rear bolts to 160 Nm (118 ft. lbs.).

122. Install the drivetrain reinforcements using the following procedure:

a. Position the drivetrain reinforcements to the font cradle mount stud to the support brace.

b. Loosely install the drivetrain reinforcement to support brace bolts.

123. Install the drivetrain reinforcement to cradle mount nut. Tighten the drivetrain reinforcement brace nut to 50 Nm (37 ft. lbs.). Tighten the drivetrain reinforcement brace bolts to 25 Nm (18 ft. lbs.).

124. Connect both the lower ball joints to the steering knuckle.

125. Install the nuts that secure the engine mount to the frame. Tighten the engine mount upper nuts to 53 Nm (39 ft. lbs.).

126. Install the nuts which secure the transaxle mount to the frame. Tighten the transaxle mount lower nuts to 63 Nm (46 ft. lbs.). Tighten the transaxle mount upper nuts to 47 Nm (35 ft. lbs.).

127. Install the steering gear mounting bolts.

128. Install the stabilizer shaft links.

129. If applicable, connect the wheel speed sensor wiring harness to the frame and lower control arm.

130. If applicable, connect the front wheel speed sensor connectors.

131. If applicable, install the front wheel speed harness retainers to the frame and to the lower control arm.

132. Install the positive battery cable and retainers to the frame.

133. Install the power steering cooler pipe.

134. Connect the fog lamp harness connectors.

135. Install the front tires and wheels.

136. Lower the vehicle.

137. Raise the vehicle and remove the powertrain lift table.

✳✳ CAUTION

Failure to disconnect the intermediate shaft from the rack and pinion steering gear stub shaft can result in damage to the steering gear and/or intermediate shaft. This damage may cause loss of steering control which could result in an accident and possible personal injury

138. Install the intermediate steering shaft to the steering gear.

139. Rotate the struts and install the drive axles to the transaxle.

140. Install the lower ball joints to the knuckles.

141. Install the tie rod ends to the steering knuckles.

142. Install the front wheel speed sensor wiring harnesses to the lower control arms and the frame.

143. Connect the VSS electrical connector and secure the wiring harness to the vehicle.

144. Install the engine splash shields.

145. Install lower radiator air baffle.

146. Install the front tires and wheels.

147. Install the catalytic converter. Tighten the nuts to 60 Nm (44 ft. lbs.).

148. Install the torque converter bolts as necessary.

149. Lower the vehicle.

150. Fill the engine oil as necessary.

151. Install the transmission electrical connector.

152. Install the A/C compressor hose to the A/C compressor. Tighten the nut to 17 Nm (13 ft. lbs.).

153. Recharge the air conditioning system.

154. Install the brake booster vacuum hose to the intake manifold.

155. Install the transmission ground wire and the bolt. Tighten the transmission ground bolt 75 Nm (55 ft. lbs.).

156. Install the ECM chassis (outboard) side electrical connector to the ECM.

157. Connect the fuel pressure and EVAP pipes to the engine.

158. Install the engine mount struts. Tighten engine mount strut nuts to 48 Nm (35 ft. lbs.).

159. Connect the heater hoses to the engine.

160. Connect the radiator hoses to the engine.

161. Fill the cooling system.

162. Connect the battery positive cable to the generator and the starter.

163. Connect the battery negative cable to the engine block.

164. Install the throttle body air inlet duct.

165. Connect the battery negative cable to the battery.

166. Fill the power steering reservoir with power steering fluid.

167. Bleed the power steering system.

168. Inspect the system for leaks.

3.8L Engine

1. Before servicing the vehicle, refer to the Precautions Section.

2. Disconnect the negative battery cable.

3. Remove the hood.

4. Relieve the fuel system pressure.

5. Drain the coolant system and crankcase.

6. Remove or disconnect the following:
- Negative cable
- Fuel injector sight shield
- Vacuum brake booster hose from the vacuum connections
- Fuel feed and return lines from the fuel rail
- Evaporative emission canister purge valve
- Cruise control cable from the throttle body bracket and lever
- Electrical connector from the cruise control module
- Cruise control module from the mounting studs

➥**Always replace the accelerator control cable with a NEW cable whenever you remove the engine from the vehicle.**

- Accelerator control cable
- Drive belt
- Bolt securing both the battery negative cable and the engine harness ground lead to the engine block

7. Disconnect the wiring harness connectors from the following components:
- A/C compressor clutch
- A/C pressure sensor
- Knock (KS) sensor
- Engine coolant block heater
- Oil level sensor

8. Remove or disconnect the following:
- Wiring harness from the harness clip at the rear of the A/C compressor
- Torque converter cover
- Starter motor
- Bolts securing the flywheel to the torque converter

9. Disconnect then secure the following wiring harness electrical connectors to the cowl panel:
- Knock (KS) sensor number 2 which can be found behind the right exhaust manifold
- Oil pressure sensor
- Vehicle Speed Sensor (VSS)

10. Remove or disconnect the following:
- Bolts securing the transaxle brace to the transaxle
- Nuts attaching the exhaust manifold pipe to the right exhaust manifold
- Exhaust manifold pipe from the right exhaust manifold studs,

allowing it to rest on top of the power steering gear heat shield
- Exhaust manifold pipe gasket and discard the gasket
- Right wheelhouse extension
- Front A/C compressor mounting nuts
- Rear A/C compressor mounting bolt
- Compressor off of the mounting studs and rest on top of the engine frame
- Bolt securing the Powertrain Control Module (PCM) ground located at the left front cylinder head

11. Disconnect the wiring harness electrical connectors from the following components on the left side of the engine:
- Fuel injectors
- Ignition harness
- Boost control solenoid (VIN I only)
- Engine Coolant Temperature (ECT) sensor
- Throttle Position (TP) sensor
- Idle Air Control (IAC) valve
- Mass Air Flow (MAF) sensor

12. Disconnect the wiring harness connectors from the following components on the right side of the engine:
- Fuel injectors
- Exhaust Gas Recirculation (EGR) valve
- Manifold Absolute Pressure (MAP) sensor
- Heated Oxygen (O_2S) sensor
- AIR solenoid, if equipped
- Alternator

13. Remove or disconnect the following:
- Alternator
- Air cleaner intake duct

14. Attach an engine support fixture.
- Front power steering pump mounting bolts
- Side power steering pump mounting bolt and piston the power steering pump against the cowl, allowing it to rest on top of the transaxle housing
- Right engine mount bracket
- Right lower engine-to-transaxle mounting bolt
- Coolant and heater hoses

15. Use a block of wood between a floor jack and the transaxle, support the transaxle at the pan.

16. Remove the engine support fixture.

17. Attach an engine lift chain to the engine lift brackets and attach to an engine lift device.

18. Remove all remaining engine-to-transaxle bolts

19. Remove the engine from the vehicle.

To install:

20. Installation is the reverse of removal, please note the following torques:
- 5 upper engine-to-transaxle mounting bolts to 55 ft. lbs. (75 Nm)
- Right lower engine-to-transaxle mounting bolt to 55 ft. lbs. (75 Nm)
- Power steering pump bolts to 20 ft. lbs. (27 Nm)
- Bolt attaching the PCM ground to the left front cylinder head and tighten to 37 ft. lbs. (50 Nm)
- A/C compressor bolts 37 ft. lbs. (50 Nm)
- Transaxle brace bolts to 48 ft. lbs.(65 Nm)
- Flywheel-to-torque converter bolts to 46 ft. lbs. (63 Nm)

21. Refill the crankcase.

22. Refill and bleed the engine cooling system.

23. Start the engine and check for leaks.

24. Road test the vehicle and check operation.

Water Pump

REMOVAL & INSTALLATION

3.6L Engine

1. Before servicing the vehicle, refer to the Precautions Section.

2. Drain the cooling system.

3. Remove the generator drive belt.

4. Use tool EN 46104 in order to retain the water pump pulley.

5. Remove the water pump pulley bolts.

6. Remove the water pump pulley.

7. Remove the water pump bolts.

8. Remove the water pump.

9. Remove and DISCARD the water pump seal.

10. Carefully clean the water pump sealing surfaces.

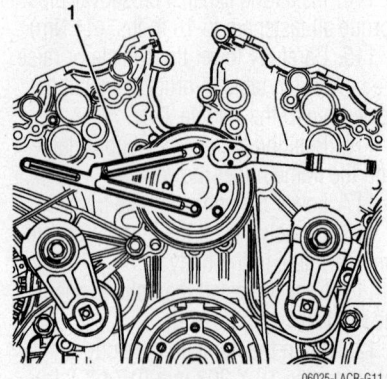

06025-LACR-G11

Use tool EN 46104 (1) in order to retain the water pump pulley

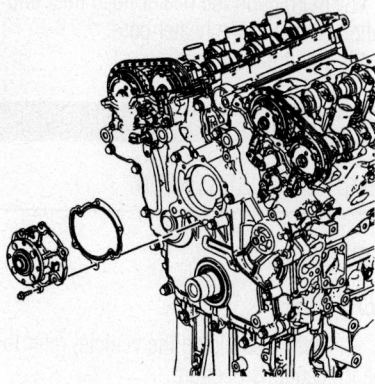

06025-LACR-G12

Water pump—3.6L engine

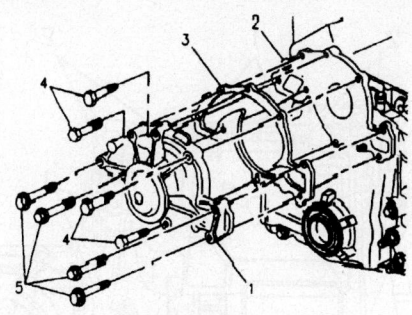

1. Coolant pump
2. Engine front cover
3. Gasket
4. 11 ft. lb. (15 Nm)
5. 22 ft. lb. (30 Nm)

7922UG01

Exploded view of the water pump—3.8L engines

To install:

11. Install a NEW water pump seal.

12. Install the water pump.

13. Install the water pump bolts. Tighten the water pump bolts to 10 Nm (89 inch lbs.).

14. Install the water pump pulley and the water pump pulley bolts.

15. Install the water pump pulley bolts. Tighten the water pump pulley bolts to 12 Nm (106 inch lbs.).

16. Install the generator drive belt.

17. Fill the cooling system.

3.8L Engine

1. Before servicing the vehicle, refer to the Precautions Section.

2. Drain the cooling system.

3. Remove or disconnect the following:
- Negative battery cable
- Accessory drive belt
- Coolant hoses from the water pump
- Water pump pulley bolts

➡**The long bolt can be removed by aligning the bolt head up with the hole in the frame rail.**

- Pulley
- Water pump bolts
- Water pump

To install:

4. Apply a thin bead of sealer around the outside edge of the water pump.

5. Install or connect the following:
- Water pump with new gasket. Torque the water pump short bolts to 11 ft. lbs. (15 Nm) and the long bolts to 22 ft. lbs. (30 Nm).
- Water pump pulley. Torque the bolts to 115 inch lbs. (13 Nm).
- Coolant hoses to the water pump
- Drive belt

6. Refill and bleed the cooling system.

7. Run the engine and check for leaks.

8. Recheck the coolant level when the engine has cooled.

Heater Core

REMOVAL AND INSTALLATION

1. Before servicing the vehicle, refer to the Precautions Section.

2. Drain the coolant.

3. Position aside the heater hose inlet and outlet clamps at the heater core.

4. Disconnect the inlet and outlet heater hose from the heater core.

5. Remove the RH instrument panel closeout/insulator panel.

6. Remove the LH instrument panel closeout/insulator panel.

7. Remove the floor carpet.

8. Remove the center console, if equipped.

9. Remove the HVAC control.

➡**The vehicle communication interface module (VCIM) has a specific set of unique numbers that tie the module to each vehicle. These numbers, the 10-digit station identification and the 11-digit electronic serial number, are used by the National Cellular Network and OnStar® to identify the specific vehicle. Because these numbers are tied to the vehicle identification number of the vehicle, you must never exchange these parts with those of another vehicle.**

10. Remove the communication interface module screws.

11. Disconnect the mobile telephone antenna cable from the communication interface module by pulling outward on the square plastic housing.

12. Disconnect the electrical connectors.

13. Disconnect the navigation antenna coaxial antenna cable from the module.

14. Remove the communication interface module.

15. Remove the rear floor air outlet duct from the holes in the floor reinforcement.

16. Disconnect the rear floor air outlet duct from the heater core outlet cover.

17. Remove the rear floor air outlet duct.

18. Remove the heater core outlet cover screws.

19. Remove the heater core outlet cover heat stakes with a small chisel.

20. Remove the heater core outlet cover from the HVAC module assembly.

21. Remove the heater core cover screws.

22. Remove the heater core cover heat stakes with a small chisel.

23. Remove the heater core cover from the HVAC module assembly.

24. Remove the heater core from the HVAC module assembly.

25. Remove the heater core foam seal from the HVAC module assembly.

To install:

26. Install a new heater core foam seal to the HVAC module assembly.

27. Install the heater core to the HVAC module assembly.

28. From the inside of the heater core cover, drill the dimples adjacent to the heat stakes using a 5.5 mm (7/32 in.) drill bit.

29. Install the heater core cover.

30. Install the heater core cover screws to the heater core cover. Tighten all the screws to 1.5 Nm (13 inch lbs.).

31. From the inside of the heater core outlet cover, drill the dimples adjacent to the heat stakes using a 5.5 mm (7/32 in.) drill bit.

32. Install the heater core outlet cover.

33. Install the heater core outlet cover

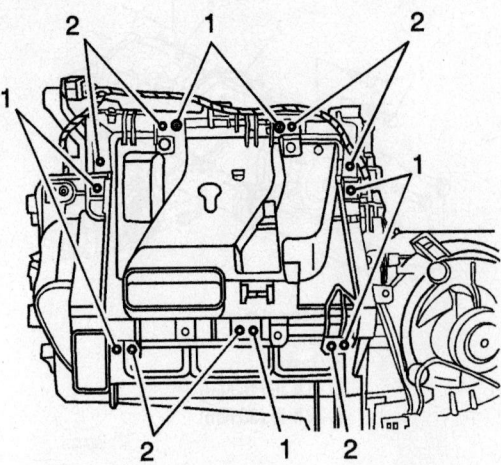

1. Outlet cover heat stakes
2. Heater core outlet cover screws

06025-LACR-G13

HVAC core outlet side

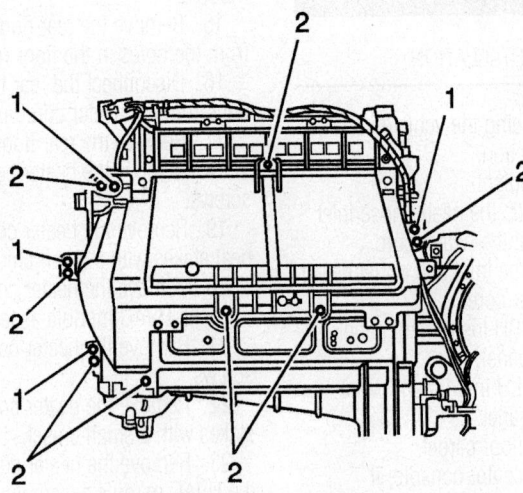

1. Heater core cover heat stakes
2. Heater core cover screws

06025-LACR-G14

HVAC cover side

screws. Tighten the screws to 1.5 Nm (13 inch lbs.).

34. Connect the rear floor air outlet duct to the heater core outlet cover.

35. Install the rear floor air outlet duct to the holes in the floor reinforcement.

36. Install the HVAC module.

37. Install the center console.

❊❊ WARNING

Before you install the antenna cable connector of the global positioning system (GPS) to the vehicle communication interface module (VCM), align the connector properly in order to avoid damaging the connector.

38. Connect the navigation antenna coaxial cable to the module.

39. Connect the electrical connectors.

40. Connect the mobile telephone antenna cable to the module by pushing inward on the square plastic housing.

41. Align the module to the vehicle antenna module bracket.

42. Install the communication interface module screws. Tighten the screws to 2 Nm (18 inch lbs.).

43. Install the floor carpet.

44. Install the RH instrument panel closeout/insulator panel.

45. Install the RH instrument panel closeout/insulator panel.

46. Connect the inlet and outlet heater hose to the heater core.

47. Reposition the heater hose inlet and outlet clamps to the heater core.

48. Refill the coolant.

Cylinder Head

REMOVAL & INSTALLATION

3.6L Engine

LEFT SIDE

1. Before servicing the vehicle, refer to the Precautions Section.

2. Remove the engine/transaxle from the vehicle.

3. Remove the lower intake manifold.

4. Remove the generator.

5. Remove the left bank secondary timing chain.

6. Remove the oil level indicator.

7. Remove the heat shield from the coolant temperature sensor and disconnect the coolant temperature sensor electrical connector.

➡ **Do not remove the exhaust crossover pipe.**

8. Disconnect the exhaust crossover pipe from the left bank exhaust manifold.

9. Remove the two front M8 left cylinder head bolts.

10. Remove the left cylinder head bolts.

11. Remove the cylinder head with the exhaust manifold.

12. Remove and discard the cylinder head gasket.

13. Clean and inspect the cylinder head and the engine block sealing surfaces.

To install:

➡ **Ensure that the crankshaft is in the stage one timing drive assembly position.**

14. Ensure the cylinder head locating pins are securely mounted in the cylinder block deck face.

15. Install a NEW left cylinder head gasket using the deck face locating pins for retention.

16. Align the left cylinder head with the deck face locating pins.

17. Place the left cylinder head in position on the deck face.

➡ **DO NOT allow oil on the cylinder head bolt bosses.**

➡ **DO NOT reuse the old M11 cylinder head bolts.**

18. Install new M11 cylinder head bolts.

a. Tighten the M11 cylinder head bolts a first pass in sequence to 45 Nm (33 ft. lbs.).

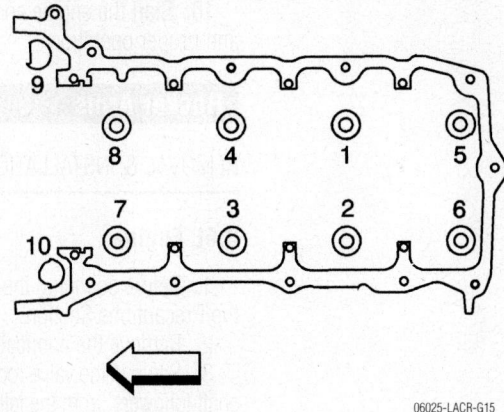

06025-LACR-G15

Left cylinder head bolt torque sequence—3.6L engine

b. Tighten the M11 cylinder head bolts a second pass in sequence an additional 120 degrees.

19. Install the 2 front M8 left cylinder head bolts.

a. Tighten the M8 cylinder head bolts a first pass to 15 Nm (11 ft. lbs.).

b. Tighten the M8 cylinder head bolts a second pass in sequence an additional 60 degrees.

➡**Do not remove the exhaust crossover pipe.**

20. Connect the exhaust crossover pipe to the left bank exhaust manifold. Torque the fasteners to 18 ft. lbs. (25 Nm).

21. Install the lower intake manifold.

22. Install the oil level indicator.

23. Install the generator.

24. Install the engine/transaxle assembly.

RIGHT SIDE

1. Before servicing the vehicle, refer to the Precautions Section.

2. Remove the powertrain module from the vehicle.

3. Remove the lower intake manifold.

4. Remove the exhaust crossover pipe.

5. Remove the right bank secondary timing chain.

6. Remove the wiring harness ground from the rear of the cylinder head.

7. Remove the wiring harness ground from the front of the cylinder head.

8. Remove the wiring harness conduit upper bolt from the cylinder head and reposition the wiring harness to provide access.

9. Remove the right cylinder head bolts.

10. Remove the right cylinder head.

11. Remove and discard the cylinder head gasket.

12. Clean and inspect the cylinder head and the engine block sealing surfaces.

To install:

13. Install a NEW cylinder head gasket.

➡**Ensure that the crankshaft is in the timing drive assembly position.**

14. Ensure the cylinder head locating pins are securely mounted in the cylinder block deck face.

15. Align the right cylinder head with the deck face locating pins.

16. Place the right cylinder head in position on the deck face.

➡**DO NOT allow oil on the cylinder head bolt bosses.**

➡**DO NOT reuse the old M11 cylinder head bolts.**

17. Install new M11 cylinder head bolts.

a. Tighten the M11 cylinder head bolts a first pass in sequence to 45 Nm (33 ft. lbs.).

b. Tighten the M11 cylinder head bolts a second pass in sequence an additional 120 degrees.

18. Install the wiring harness conduit to the cylinder head. Tighten the wiring harness upper bolt to 10 Nm (89 inch lbs.).

19. Install the wiring harness ground to the front of the cylinder head. Tighten the cylinder head front ground bolt to 10 Nm (89 inch lbs.).

20. Install the wiring harness ground to the rear of the cylinder head. Tighten the cylinder head rear ground bolt to 10 Nm (89 inch lbs.).

21. Install the right bank secondary timing chain.

22. Install the exhaust crossover pipe. Torque the fasteners to 18 ft. lbs. (25 Nm).

23. Install the engine/transaxle assembly to the vehicle.

24. Install the lower intake manifold.

3.8L Engine

1. Before servicing the vehicle, refer to the Precautions Section.

2. Disconnect the negative battery cable.

3. Relieve the fuel system pressure.

4. Drain the cooling system.

5. Remove or disconnect the following:

- Intake manifold
- Exhaust manifold
- Valve covers
- Ignition wires and ignition coil/module assembly
- Alternator front mounting bracket and alternator
- Air conditioning bracket-to-cylinder head bolt
- Power steering pump
- Accessory drive belt tensioner
- Fuel pipe heat shield
- Rocker arm assemblies, note their original position
- Pushrods and guide plate

06025-LACR-G16

Right cylinder head torque sequence—3.6L engine

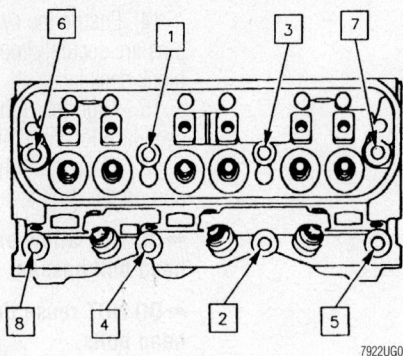

Cylinder head bolt torque sequence—3.8L engine

7922UG02

- Cylinder head bolts
- Cylinder head

To install:

6. Place the new cylinder head gasket on the engine block dowels with the note **THIS SIDE UP** facing the cylinder head and the arrow facing the front of the engine. Position the cylinder head on the engine block.

➡**The head gasket is identified by either a L or a R stamped on it next to the arrow.**

➡**This engine uses special torque-to-yield head bolts. The procedure must be followed carefully and new bolts must be used whenever the head is removed. Total bolt torque should not exceed 60 ft. lbs. (81 Nm).**

7. Install new cylinder head bolts and torque them in sequence as follows:
   a. Step 1: 37 ft. lbs. (50 Nm).
   b. Step 2: Plus 120 degrees.

8. Install or connect the following:
   - Pushrods and guide plate
   - Rocker arm assemblies into their original location

➡**Apply a thread lock compound to the rocker arm pedestal bolts before assembly.**

   - Valve covers
   - Fuel pipe heat shield
   - Accessory drive belt tensioner
   - Power steering pump
   - Air conditioning compressor bracket bolt. Torque it to 52 ft. lbs. (70 Nm).
   - Alternator front mounting bracket, and alternator
   - Ignition coil/module assembly and spark plug wires
   - Exhaust manifold. Torque the bolts to 22 ft. lbs. (30 Nm).
   - Intake manifold
   - Negative battery cable

9. Refill and bleed the cooling system.

10. Start the engine and check for leaks and proper operation.

Rocker Arms

REMOVAL & INSTALLATION

3.6L Engine

1. Before servicing the vehicle, refer to the Precautions Section.
2. Remove the applicable camshaft(s).
3. Remove the valve rocker arms, camshaft followers, from the left cylinder head.
4. Installation is the reverse of removal. Replace all parts in their original positions.

3.8L Engine

➡**When removing valvetrain components, it is very important that they are marked for installation reference, so that they can be reinstalled in their original location.**

06025-LACR-G17

Rocker arm removal—3.6L engine

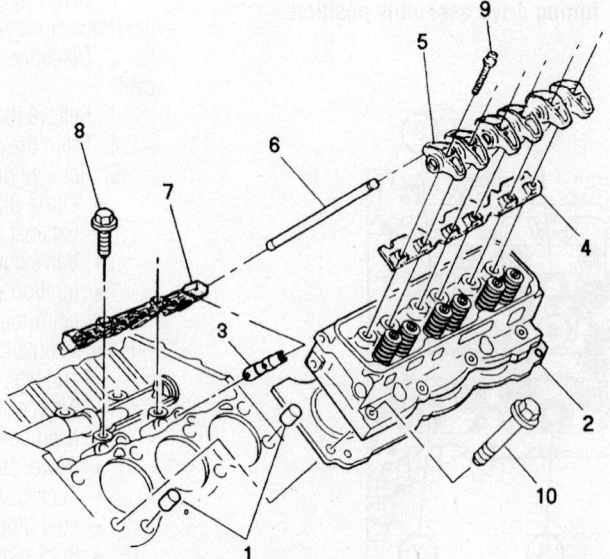

1. Dowel pin
2. Head gasket
3. Valve lifter
4. Pivot retainer
5. Rocker arm
6. Pushrod
7. Lifter guide
8. Bolt
9. Bolt
10. Head bolt

7922UG03

Exploded view of the rocker arms and related components mounting—3.8L engines

1. Before servicing the vehicle, refer to the Precautions Section.
2. Remove or disconnect the following:
   - Negative battery cable
   - Spark plug wires
   - Rocker arm cover(s)
   - Rocker arm pedestal bolts and assemblies
   - Pushrods

To install:

3. Lubricate the pushrod tips and put them in their proper locations.
4. Lubricate the rocker arms and pedestals and install them. Be sure the pushrod tips are properly seated in the rocker arms.
5. Apply thread locking compound to the rocker arm pedestal bolt threads. Torque the bolts to 11 ft. lbs. (15 Nm) plus an additional 90 degree turn.
6. Apply suitable thread locking compound to the rocker arm cover bolts.
7. Install or connect the following:
   - Rocker arm cover using a new gasket. Torque the bolts to 89 inch lbs. (10 Nm).
   - Spark plug wires
   - Accessory drive belt
   - Negative battery cable
8. Run the engine and check for leaks and proper engine operation.

Intake Manifold

REMOVAL & INSTALLATION

3.6L Engine

UPPER MANIFOLD

1. Before servicing the vehicle, refer to the Precautions Section.
2. Turn the ignition OFF.
3. Remove the air inlet duct.
4. Relieve the fuel system pressure.
5. Remove the fuel pressure and evaporative emission (EVAP) hoses from the engine.
6. Remove the purge line from the purge line retainer.
7. Remove the fuel feed hose bracket bolt and reposition the fuel feed hose.

➡**Do not disconnect the engine control module (ECM) electrical connectors. Do not remove the ECM from the ECM bracket.**

8. Remove the ECM bracket with the ECM and position it aside.
9. Disconnect the purge solenoid electrical connector.

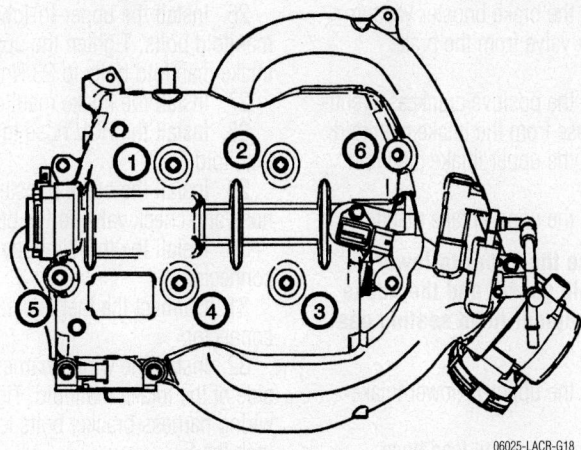

Upper intake manifold bolt loosening/tightening sequence—3.6L engine

06025-LACR-G18

10. Remove the wiring harness from the right side of the intake manifold.
11. Disconnect the fuel injector electrical connector.
12. Remove the fuel injector electrical connector from the fuel injector electrical connector bracket.
13. Remove the throttle body electrical connector.
14. Remove the brake booster vacuum hose and check valve from the brake booster.
15. Remove the electrical connector for the baro sensor.
16. Remove the positive crankcase ventilation (PCV) hose from the intake manifold.
17. Remove the intake manifold bolts.
18. Remove the upper intake manifold.
19. Remove and discard the upper intake manifold gasket.
20. Clean and inspect the intake manifold and the sealing surfaces.

To install:

21. Install the intake manifold bolts. Tighten the intake manifold bolts in the order shown to 23 Nm (17 ft. lbs.).
22. Install the PCV hose to the intake manifold.
23. Install the brake booster vacuum hose and check valve to the brake booster.
24. Install the throttle body electrical connector.
25. Reconnect the electrical connector for the baro sensor.
26. Install the fuel injector electrical connector to the fuel injector electrical connector bracket.
27. Install the ECM bracket.
28. Connect the fuel injector electrical connector.
29. Install the wiring harness and bracket to the right side of the intake manifold. Tighten the wiring harness bracket bolts to 10 Nm (89 inch lbs.).

30. Connect the purge solenoid electrical connector.

❋❋ **WARNING**

In order to prevent any possible electrostatic discharge damage to the ECM, do not touch the connector pins.

31. Install the fuel feed hose bracket and the fuel feed hose bracket bolt. Tighten the fuel feed hose bracket bolt to 10 Nm (89 inch lbs.).
32. Install the purge line to the purge line retainer.
33. Install the fuel pressure and EVAP hoses to the engine.
34. Install the air inlet duct.

LOWER MANIFOLD

1. Before servicing the vehicle, refer to the Precautions Section.
2. Turn the ignition OFF.
3. Remove the air inlet duct.
4. Remove the fuel pressure and evaporative emission (EVAP) hoses from the engine.
5. Disconnect the BARO sensor electrical connector.

➡**Do not disconnect the engine control module (ECM) electrical connectors.**

6. Remove the ECM bracket with the ECM and reposition aside.
7. Disconnect the purge solenoid electrical connector.
8. Disconnect the purge solenoid electrical connector.
9. Remove the wiring harness from the right side of the intake manifold.
10. Disconnect the fuel injector electrical connector.
11. Remove the throttle body electrical connector.

12. Remove the brake booster vacuum hose and check valve from the brake booster.

13. Remove the positive crankcase ventilation (PCV) hose from the intake manifold.

14. Remove the upper intake manifold bolts.

15. Remove the upper intake manifold.

➡ **Do not reuse the upper-to-lower intake manifold gasket and the intake manifold-to-cylinder head sealing gaskets.**

16. Remove the upper-to-lower intake manifold bolts.

17. Remove the fuel rail feed hose bracket bolts from the upper intake manifold.

18. Remove the fuel injector wiring harness bracket bolt from the upper intake manifold.

19. Remove the upper intake manifold from the lower intake manifold.

20. Remove and discard the upper-to-lower intake manifold gaskets.

21. Clean and inspect the intake manifold and the sealing surfaces.

To install:

➡ **Do not reuse the upper-to-lower intake manifold gasket and the intake manifold-to-cylinder head sealing gaskets.**

22. Install the NEW upper-to-lower intake manifold gaskets.

23. Install the upper intake manifold to the lower intake manifold.

24. Install the fuel injector wiring harness bracket bolts to the upper intake manifold. Tighten the fuel injector wiring harness bracket bolts to 10 Nm (89 inch lbs.).

25. Install the fuel rail feed hose bracket bolt to the upper intake manifold. Tighten the fuel rail feed hose bracket bolt to 10 Nm (89 inch lbs.).

26. Install the upper-to-lower intake manifold bolts. Tighten the upper-to-lower intake manifold bolts to 23 Nm (17 ft. lbs.).

27. Install the intake manifold.

28. Install the PCV hose to the intake manifold.

29. Install the brake booster vacuum hose and check valve to the brake booster.

30. Install the throttle body electrical connector.

31. Connect the fuel injector electrical connector.

32. Install the wiring harness to the right side of the intake manifold. Tighten the wiring harness bracket bolts to 10 Nm (89 inch lbs.).

33. Connect the purge solenoid electrical connector.

34. Install the ECM bracket with the ECM.

35. Connect the BARO sensor electrical connector.

36. Install the fuel pressure and EVAP hoses to the engine.

37. Install the air inlet duct.

3.8L Engine

1. Before servicing the vehicle, refer to the Precautions Section.

2. Disconnect the negative battery cable.

3. Drain the cooling system.

4. Relieve the fuel system pressure.

5. Remove or disconnect the following:
   - Fuel injector sight shield
   - Air inlet duct
   - Spark plug wires from the right side
   - Manifold Absolute Pressure (MAP) sensor
   - Vacuum lines from the intake manifold
   - Fuel lines
   - Fuel injector electrical connectors
   - Fuel regulator vacuum line
   - Fuel rail from the intake manifold
   - Exhaust Gas Recirculation (EGR) heat shield
   - Throttle cable bracket from the cylinder head mounting bracket and the throttle body cables
   - Throttle body support bracket
   - Upper intake plenum and gasket
   - Thermostat housing
   - Electrical connector from the Engine Coolant Temperature (ECT) sensor
   - Drive belt tensioner assembly
   - EGR valve outlet pipe
   - Lower intake manifold

To install:

6. Install or connect the following:
   - Intake manifold using new manifold gaskets. Torque the bolts in sequence to 11 ft. lbs. (15 Nm); then, re-torque to 11 ft. lbs. (15 Nm).
   - EGR valve outlet pipe
   - Drive belt tensioner assembly. Torque the tensioner bolts to 37 ft. lbs. (50 Nm).
   - Electrical connector to the ECT sensor
   - Thermostat housing
   - Upper intake plenum. Torque the intake plenum bolts to 88 inch. lbs. (10 Nm).
   - Throttle body support bracket
   - Throttle cable bracket to the cylinder head mounting bracket and the cables to the throttle body lever
   - EGR heat shield
   - Fuel rail. Torque the fuel rail bolts to 88 inch. lbs. (10 Nm).
   - Fuel lines
   - Fuel regulator vacuum line
   - Fuel injector electrical connectors

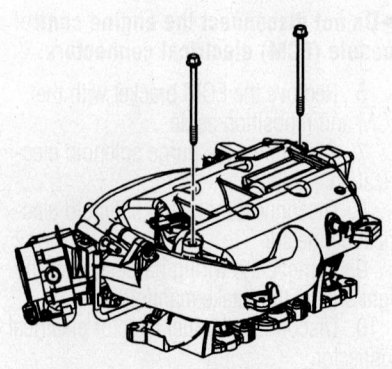

06025-LACR-G19

Upper-to-lower manifold bolts—3.6L engine

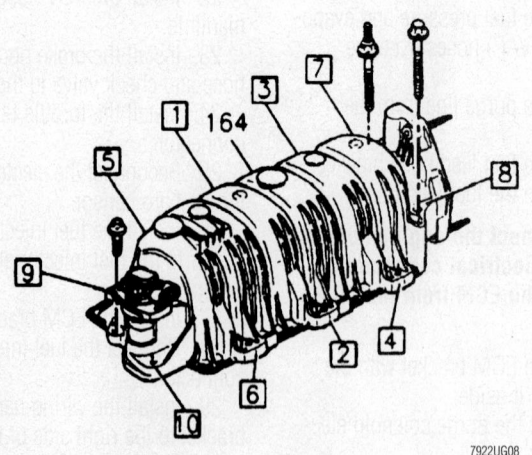

7922UG08

Upper intake manifold torque sequence—3.8L engine

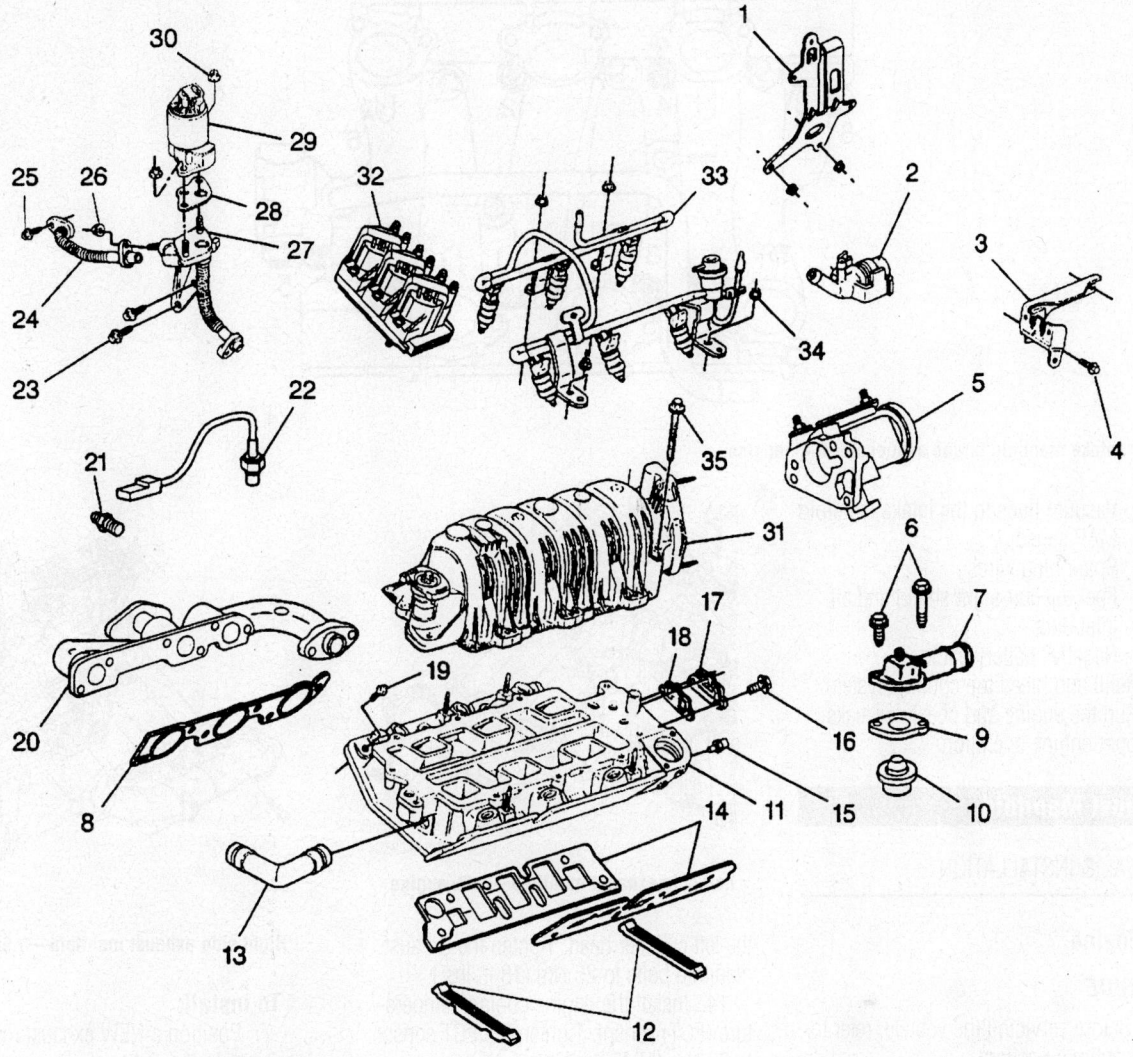

1   Fuel Injector Sight Shield Bracket
2   Vacuum Source Manifold
3   Accelerator Cable Control Bracket
4   Throttle Body Support Bolt
5   Throttle Body
6   Water Outlet Bolt
7   Water Outlet
8   Exhaust Manifold Gasket
9   Water Outlet Gasket
10 Thermostat
11 Lower Intake Manifold
12 Intake Manifold Seal
13 Heater Water Inlet Pipe
14 Lower Intake Manifold Gasket
15 Coolant Temperature Sensor
16 Engine Coolant Manifold Bolt
17 Engine Coolant Manifold
18 Engine Coolant Manifold Gasket

19 Lower Intake Manifold Bolt
20 Exhaust Manifold (Right)
21 Exhaust Manifold Bolt/Stud
22 Exhaust Oxygen Sensor
23 EGR Valve Adapter Bolt
24 EGR Valve Outlet Pipe
25 EGR Valve Outlet Pipe Bolt
26 EGR Valve Outlet Pipe Nut
27 EGR Valve Adapter
28 EGR Valve Gasket
29 EGR Valve
30 EGR Valve Nut
31 Upper Intake Manifold
32 ICM
33 Fuel Injection Rail
34 Fuel Injector Rail Nut
35 Upper Intake Manifold Bolt

9300UG02

Exploded view of the intake manifold and related components—3.8L engine

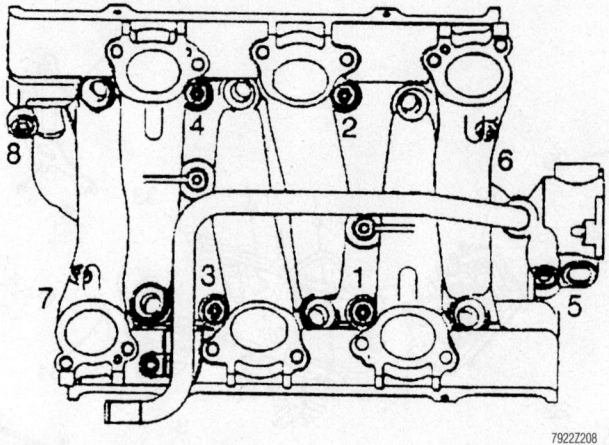

Lower intake manifold torque sequence—3.8L engine

- Vacuum lines to the intake manifold
- MAP sensor
- Spark plug wires
- Fuel injector sight shield and air inlet duct
- Negative battery cable
7. Refill and bleed the cooling system.
8. Run the engine and check for leaks and proper engine operation.

Exhaust Manifold

REMOVAL & INSTALLATION

3.6L Engine

LEFT SIDE

1. Before servicing the vehicle, refer to the Precautions Section.
2. Remove the left torque strut bracket bolts.
3. Remove the left torque strut bracket.
4. Remove the left exhaust manifold heat shield bolts.
5. Remove the left exhaust manifold heat shield.
6. Remove the engine coolant temperature (ECT) sensor.
7. Remove the exhaust manifold bolts from the left cylinder head.
8. Remove the left exhaust manifold.
9. Remove and discard the exhaust manifold gasket.
   To install:
10. Position a NEW exhaust manifold gasket onto the left exhaust manifold.
11. Install the exhaust manifold bolts into the left exhaust manifold.
12. Place the left exhaust manifold, exhaust manifold gasket and bolts as an assembly in position on the left cylinder head.
13. Install the exhaust manifold bolts into

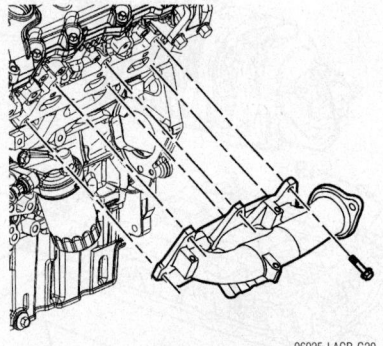

Left side exhaust manifold—3.6L engine

the left cylinder head. Tighten the exhaust manifold bolts to 25 Nm (18 ft. lbs.).
14. Install the engine coolant temperature (ECT) sensor. Tighten the ECT sensor to 22 Nm (16 ft. lbs.).
15. Install NEW O-rings on the crankshaft position sensor.
16. Place the left exhaust manifold heat shield in position.
17. Install the exhaust manifold heat shield bolts. Tighten the exhaust manifold heat shield bolts to 10 Nm (89 inch lbs.).
18. Install the left torque strut bracket.
19. Install the left torque strut bracket bolts. Tighten the left torque strut bracket bolts to 50 Nm (37 ft. lbs.).

RIGHT SIDE

1. Before servicing the vehicle, refer to the Precautions Section.
2. Remove the right exhaust manifold heat shield bolts.
3. Remove the right exhaust manifold heat shield.
4. Remove the exhaust manifold bolts from the right cylinder head.
5. Remove the right exhaust manifold.
6. Remove and discard the exhaust manifold gasket.

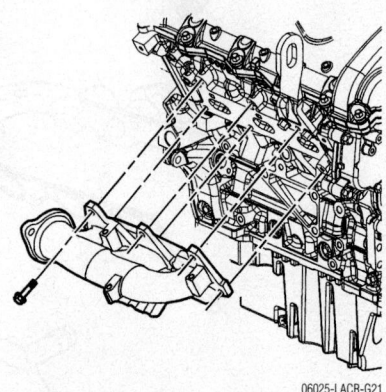

Right side exhaust manifold—3.6L engine

To install:
7. Position a NEW exhaust manifold gasket onto the right exhaust manifold.
8. Install the exhaust manifold bolts into the right exhaust manifold.
9. Place the right exhaust manifold, exhaust manifold gasket and bolts as an assembly in position on the right cylinder head.
10. Install the exhaust manifold bolts into the right cylinder head. Tighten the exhaust manifold bolts to 25 Nm (18 ft. lbs.).
11. Place the right exhaust manifold heat shield in position.
12. Install the exhaust manifold heat shield bolts. Tighten the exhaust manifold heat shield bolts to 10 Nm (89 inch lbs.).

3.8L Engine

LEFT SIDE (FRONT) MANIFOLD

1. Before servicing the vehicle, refer to the Precautions Section.
2. Remove or disconnect the following:
   - Negative battery cable
   - Spark plug wires

7922UG31

Exploded view of the left exhaust manifold mounting—3.8L engine

- Engine oil dipstick and tube
- Left side lift bracket, if necessary
- 2 bolts attaching the left exhaust manifold to the crossover pipe
- Exhaust manifold

To install:

3. Install or connect the following:
- Exhaust manifold with a new gasket. Torque the studs and bolts gradually and evenly to 22 ft. lbs. (30 Nm).
- 2 bolts attaching the left exhaust manifold to the crossover pipe. Torque the bolts to 15 ft. lbs. (20 Nm).
- Left side lift bracket, if removed
- Engine oil dipstick and tube. Torque the bolts to 15 ft. lbs. (20 Nm).
- Spark plug wires
- Negative battery cable

4. Run the engine and check for exhaust leaks.

RIGHT SIDE (REAR) MANIFOLD

1. Before servicing the vehicle, refer to the Precautions Section.

2. Remove or disconnect the following:
- Negative battery cable
- Fuel injector sight shield
- Air cleaner assembly
- Spark plug wires
- Brake booster heat shield
- Crossover pipe
- Engine harness from the right hand engine lift hook bracket
- Transaxle fluid dipstick and tube
- Oxygen (O₂S) sensor
- Exhaust Gas Recirculation (EGR) feed pipe bolt from the manifold
- Transmission oil level tube and seal
- Exhaust manifold flange nuts
- Front exhaust pipe

- Engine lift bracket
- Exhaust manifold

To install:

3. Install or connect the following:
- Manifold to the cylinder head and crossover pipe using new gaskets
- Manifold mounting studs. Torque the studs and bolts to 22 ft. lbs. (30 Nm), beginning at the center and working outwards.
- Engine lift bracket
- Front exhaust pipe
- Front exhaust pipe-to-manifold nuts. Torque the nuts to 22 ft. lbs. (30 Nm).
- Transmission dipstick tube seal and the tube
- EGR feed pipe to the manifold
- O₂S sensor
- Spark plug wires to the spark plugs
- Engine harness to the right hand engine lift hook bracket
- Crossover pipe
- Brake booster heat shield
- Air cleaner assembly
- Fuel injector sight shield
- Negative battery cable

4. Run the engine and check for exhaust leaks.

Camshaft and Valve Lifters

REMOVAL & INSTALLATION

3.6L Engine

LEFT

1. Before servicing the vehicle, refer to the Precautions Section.

2. Remove the upper intake manifold with the lower intake manifold.

3. Disconnect the ignition coil electrical connectors.

4. Remove the wiring harness from the side of the camshaft cover by sliding the conduit down and outboard.

5. Remove the wiring conduit retainers from the camshaft cover by rotating the wiring harness conduit retainers counterclockwise.

➡ **It is not necessary to disconnect the engine front cover electrical connectors.**

6. Remove the wiring harness from the front of the camshaft cover.

7. Reposition and secure the wiring harnesses away from the camshaft cover in order to provide clearance.

8. Remove the ignition coils.

9. Loosen the left engine strut bracket.

10. Loosen the left engine strut bracket-to-cylinder head bolts.

11. Remove the camshaft cover bolts and camshaft cover.

12. Remove and discard the camshaft cover seal and grommets. DO NOT reuse.

13. Remove the camshaft sensors.

➡ **Do not disconnect the power steering fluid lines/hoses from the reservoir.**

14. Remove the power steering fluid reservoir bolts and reposition the power steering fluid reservoir in order to provide access.

15. Remove the camshaft position (CMP) actuator valve electrical connector.

16. Remove the camshaft position actuator solenoid.

17. Remove the crankshaft balancer.

18. Rotate the crankshaft until the camshafts are in a neutral (low tension) position. The camshaft flats will be parallel with the camshaft cover rail.

❋❋ WARNING

A wrench must be used on the hex of the camshaft when loosening or tightening in order to prevent component damage. Failure to prevent the torque reaction against the timing drive chain can lead to timing drive chain failure.

➡ **Use an open-end wrench at the camshaft hex to prevent camshaft/engine rotation. DO NOT remove the camshaft position actuator bolt at this time.**

19. Loosen the camshaft position actuator bolt.

➡ **Ensure that the tips of tool EN 46108 are fully engaged into the timing chain (3 and 4).**

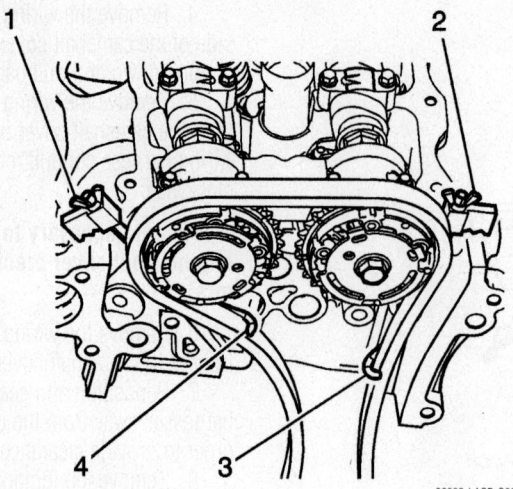

06025-LACR-G22

EN 46108 installed—3.6L engine

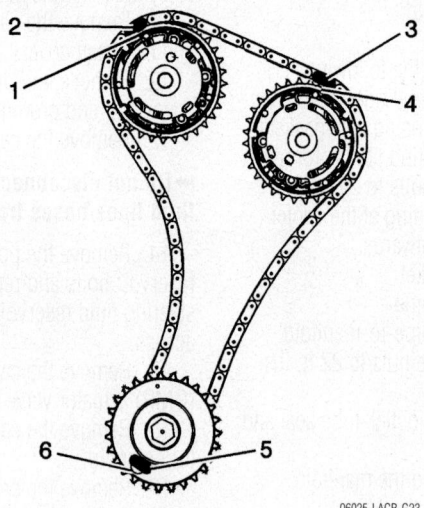

06025-LACR-G23

Mark the timing chain and the respective locations on the camshaft position actuators—3.6L engine

20. Install tool EN 46108 (1 and 2) in order to retain the timing chain. Firmly tighten the tool nuts.

➡**Ensure that the camshaft timing chain and the camshaft position actuators are marked for proper assembly.**

21. Mark the timing chain and the respective locations on the camshaft position actuators.

22. Remove the camshaft position actuator bolt.

23. Remove the timing chain from the sprockets.

24. Position the camshaft lobes in a neutral position.

25. Observe the markings on the bearing caps. Each bearing cap is marked in order to identify its location. The markings have the following meanings:
- The raised feature must always be

oriented toward the center of the cylinder head
- The E indicates the exhaust camshaft
- The I indicates the intake camshaft
- The number indicates the journal position from the front of the engine

26. Remove the camshaft bearing cap bolts.

27. Remove the camshaft bearing caps.

28. Remove the camshafts.

29. Replace the camshaft bearing caps and bolts.

To install:

➡**Ensure that the marks on the camshaft position actuator and the timing chain (1-4) are aligned. DO NOT tighten the camshaft position actuator bolt at this time.**

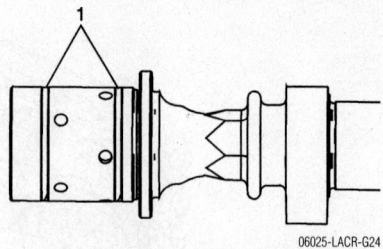

06025-LACR-G24

Ensure that the camshaft sealing rings (1) are in place in the camshaft grooves— 3.6L engine

30. Locate the camshafts to the cylinder head and assemble the camshaft actuators to the camshafts.

31. Ensure that the crankshaft is in the stage one timing drive assembly position. See the Timing Chain Removal & Installation procedure.

32. Ensure that the camshaft sealing rings are in place in the camshaft grooves.

33. Select the proper camshaft for the particular installation location. The ring placement is defined as follows:

    a. The number 4 identification ring for the left intake camshaft is machined off (1).

    b. The number 5 identification ring for the left exhaust camshaft is machined off (2).

34. Apply a liberal amount of lubricant GM P/N 12345501 (Canadian P/N 992704), or equivalent, to the camshaft journals and the left cylinder head camshaft carriers.

35. Place the left intake and left exhaust camshafts in position in the left cylinder head.

36. Position the camshaft lobes in a neutral position with the flats on the back of the camshafts up and parallel (1) with the left cylinder head camshaft cover rail.

37. Observe the markings on the left cylinder head camshaft bearing caps. Each bearing cap is marked in order to identify its location. The markings have the following meanings:
- The raised feature must always be oriented toward the center of the cylinder head.
- The E indicates the exhaust camshaft.
- The I indicates the intake camshaft.
- The number 2, 4, 6 indicates the cylinder position from the front of the engine.

38. Apply a liberal amount of lubricant GM P/N 12345501 (Canadian P/N 992704) or equivalent to the camshaft bearing caps.

39. Install the camshaft bearing thrust cap in the first journal of the left cylinder head.

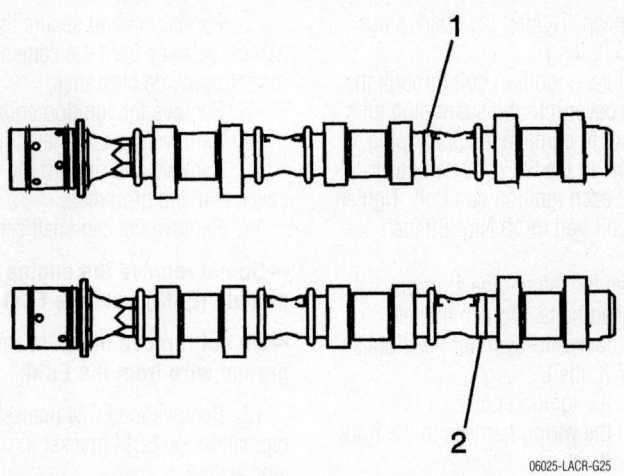

Select the proper camshaft for the particular installation location—3.6L engine

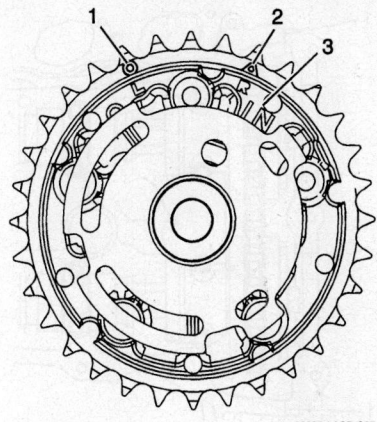

Left intake CMP actuator position—3.6L engine

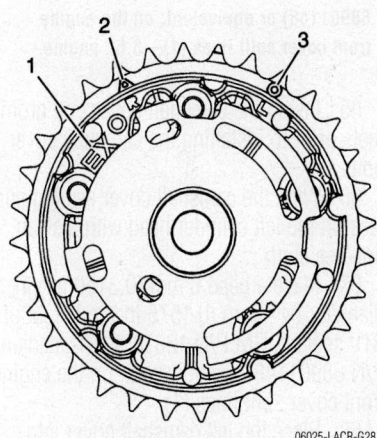

Left exhaust CMP actuator position—3.6L engine

40. Install the remaining bearing caps with their orientation mark toward the center of the cylinder head.

41. Hand start all the camshaft bearing cap bolts.

42. Tighten the camshaft bearing cap bolts in the sequence shown. Tighten the camshaft bearing cap bolts in sequence to 10 Nm (89 inch lbs.).

43. Loosen the center intake camshaft bearing cap bolts 1, 2 and the center exhaust camshaft bearing cap bolts 3, 4.

44. Retighten the center camshaft bearing cap bolts 1, 2, 3, 4. Retighten the camshaft bearing cap bolts to 10 Nm (89 inch lbs.).

✳✳ WARNING

Notice: A wrench must be used on the hex of the camshaft when loosening or tightening in order to prevent component damage. Failure to prevent the torque reaction against the timing drive chain can lead to timing drive chain failure.

➡ **Use an open-end wrench at the camshaft hex to prevent camshaft/engine rotation.**

45. Install and tighten the camshaft position actuators. Tighten the camshaft position actuator bolt to 58 Nm (43 ft. lbs.).

46. Install the CMP sensor.

47. Install the CMP sensor bolt. Tighten the CMP sensor bolt to 10 Nm (89 inch lbs.).

48. Install the CMP sensor electrical connector.

49. Install the power steering fluid reservoir. Tighten the M6 bolt to 9 Nm (7 ft. lbs.). Tighten the M8 bolt to 25 Nm (18 ft. lbs.).

50. Install the CMP actuator valve.

51. Install the CMP actuator valve bolt. Tighten the CMP actuator valve bolt to 10 Nm (89 inch lbs.).

52. Install the CMP actuator valve electrical connector.

53. Install the camshaft sensors.

54. Install the crankshaft balancer.

55. Install a NEW camshaft cover seal and NEW grommets.

56. Install a NEW camshaft cover seal and NEW grommets.

57. Install the camshaft cover.

58. Tighten the left engine strut bracket-to-cylinder head bolts. Tighten the left engine strut bracket-to-cylinder head bolts to 50 Nm (37 ft. lbs.).

59. Install the ignition coils.

60. Install the wiring harness to the front of the camshaft cover.

61. Install the wiring harness conduit retainers to the wiring harness conduit.

62. Install the wiring harness to the side of the camshaft cover.

63. Connect the ignition coil electrical connectors.

64. Install tool EN 46101 onto the spark plug tubes of the left cylinder head.

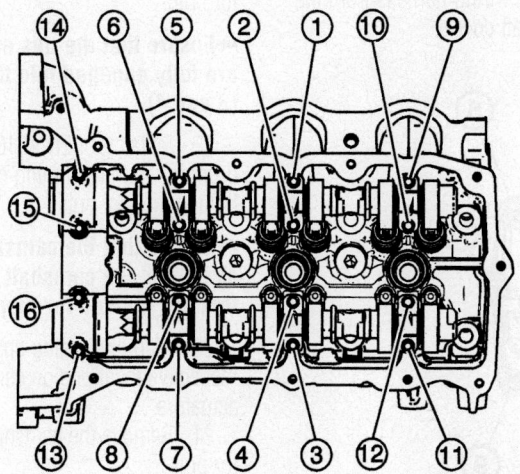

Left side camshaft bearing torque sequence—3.6L engine

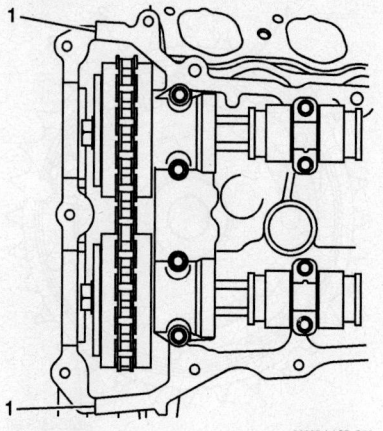

06025-LACR-G29

Place a bead 8 mm (0.3150 in.) in diameter by 4 mm (0.1575 in.) in height of RTV sealant, GM P/N 12378521 (Canadian P/N 88901148) or equivalent, on the engine front cover split lines (1)—3.6L engine

65. Install the camshaft cover bolt grommets prior to installing the camshaft cover bolts.

66. Wipe the camshaft cover sealing surface on the left cylinder head with a clean, lint-free cloth.

67. Place a bead 8 mm (0.3150 in.) in diameter by 4 mm (0.1575 in.) in height of RTV sealant, GM P/N 12378521 (Canadian P/N 88901148) or equivalent, on the engine front cover split lines (1).

68. Place the left camshaft cover into position onto the left cylinder head.

69. Loosely install the left camshaft cover bolts.

70. Tighten the left camshaft cover bolts in the sequence shown. Tighten the left camshaft cover bolts in the sequence to 10 Nm (89 inch lbs.).

71. Remove the tool from the spark plug tubes of the left cylinder head.

72. Install the NEW spark plugs into the

left cylinder head. Tighten the spark plugs to 20 Nm (15 ft. lbs.).

73. Install each ignition coil through the left camshaft cover into the spark plug tube taking care not to damage the spark plug and/or the seal in the left camshaft cover.

74. Install each ignition coil bolt. Tighten the ignition coil bolt to 10 Nm (89 inch lbs.).

75. Tighten the left engine strut bracket-to-cylinder head bolts. Tighten the left engine strut bracket-to-cylinder head bolts to 50 Nm (37 ft. lbs.).

76. Install the ignition coils.

77. Install the wiring harness to the front of the camshaft cover.

78. Install the wiring harness conduit retainers to the wiring harness conduit.

79. Install the wiring harness to the side of the camshaft cover.

80. Connect the ignition coil electrical connectors.

81. Install the upper intake manifold with the lower intake manifold.

RIGHT

1. Before servicing the vehicle, refer to the Precautions Section.

2. Remove the upper intake manifold with the lower intake manifold.

3. Disconnect the ignition coil electrical connectors.

4. Remove the wiring harness from the side of the camshaft cover by sliding the conduit down and outboard.

5. Remove the wiring conduit retainers from the camshaft cover by rotating the wiring harness conduit retainers counter-clockwise.

➡ **It is not necessary to disconnect the engine front cover electrical connectors.**

6. Remove the wiring harness from the front of the camshaft cover.

7. Reposition and secure the wiring harnesses away from the camshaft cover in order to provide clearance.

8. Remove the ignition coils.

9. Remove the camshaft cover.

10. Remove and discard the camshaft cover seal and grommets.

11. Remove the camshaft sensors.

➡ **Do not remove the engine control module (ECM) from the ECM bracket.**

➡ **Do not remove the ECM redundant ground wire from the ECM.**

12. Remove the ECM bracket bolts and reposition the ECM bracket in order to provide access.

13. Remove the camshaft position (CMP) actuator valve electrical connector.

14. Remove the CMP actuator valve bolt.

15. Remove the CMP actuator valve.

16. Remove the crankshaft balancer.

17. Rotate the crankshaft until the camshafts are in a neutral (low tension) position. The camshaft flats will be parallel with the camshaft cover rail.

✳✳ WARNING

A wrench must be used on the hex of the camshaft when loosening or tightening in order to prevent component damage. Failure to prevent the torque reaction against the timing drive chain can lead to timing drive chain failure.

➡ **Use an open-end wrench at the camshaft hex to prevent camshaft/engine rotation. DO NOT remove the camshaft position actuator bolt at this time.**

18. Loosen the camshaft position actuator bolt.

➡ **Ensure that the tips of tool EN 46108 are fully engaged into the timing chain (3 and 4).**

19. Install tool EN 46108 (1 and 2) in order to retain the timing chain. Firmly tighten the tool nuts.

➡ **Ensure that the camshaft timing chain and the camshaft position actuators are marked for proper assembly.**

20. Mark the timing chain and the respective locations on camshaft position actuators.

21. Remove the camshaft position actuator bolt.

22. Position the camshaft lobes in a neutral position.

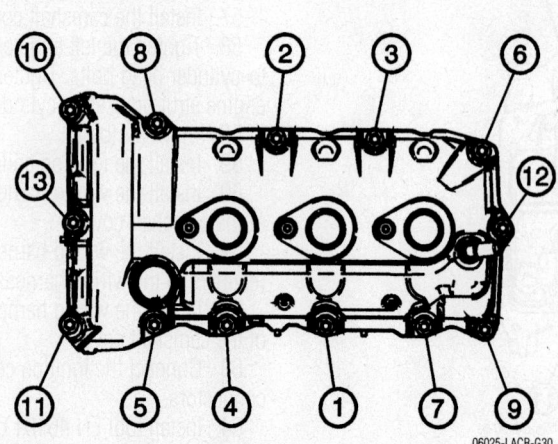

06025-LACR-G30

Left side camshaft cover torque sequence—3.6L engine

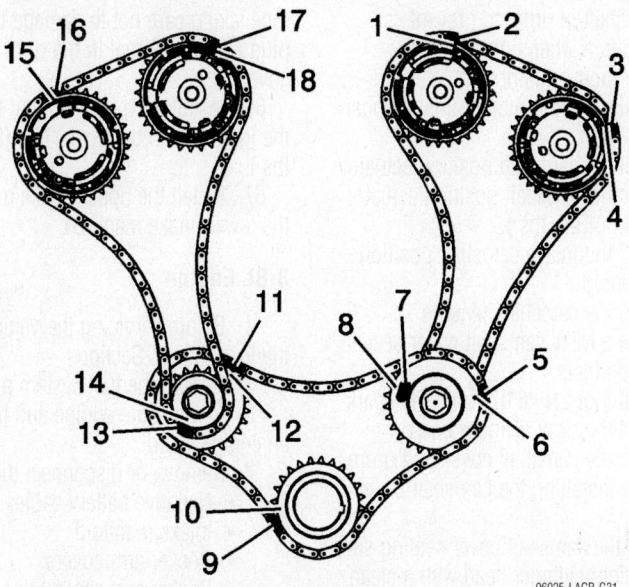

Ensure that the camshaft timing chain and the camshaft position actuators are marked for proper assembly—3.6L engine right side

06025-LACR-G31

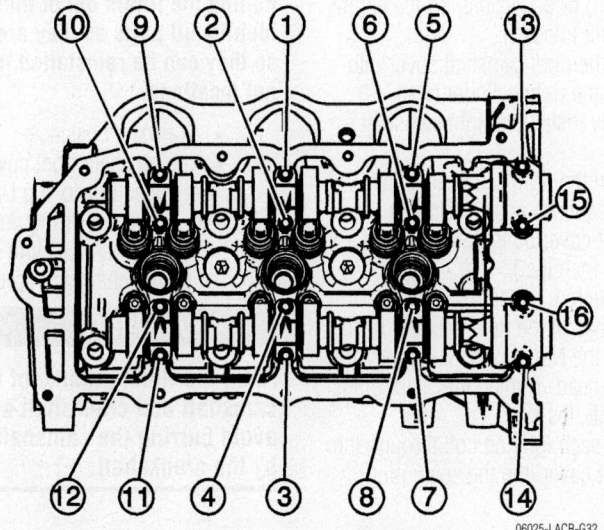

Right side camshaft bearing cap torque sequence—3.6L engine

06025-LACR-G32

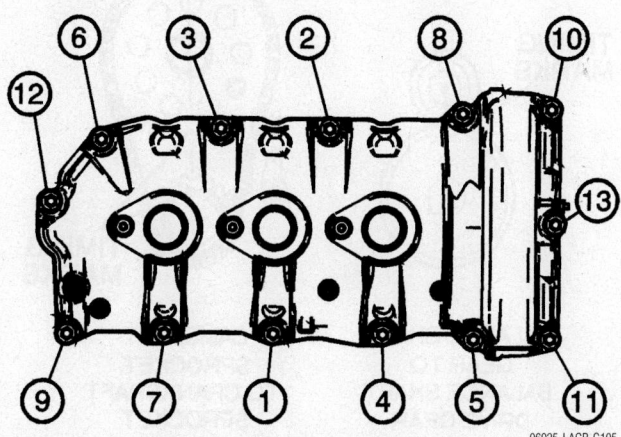

Right side camshaft cover torque sequence—3.6L engine

06025-LACR-G105

23. Observe the markings on the bearing caps. Each bearing cap is marked in order to identify its location. The markings have the following meanings:

- The raised feature must always be oriented toward the center of the cylinder head
- The I indicates the intake camshaft
- The E indicates the exhaust camshaft
- The number indicates the journal position from the front of the engine

24. Remove the camshaft bearing cap bolts.

25. Remove the camshaft bearing caps.

26. Remove the camshafts.

27. Replace the camshaft bearing caps and bolts.

To install:

➡Ensure that the marks on the camshaft position actuators and the timing chain are aligned. DO NOT tighten the camshaft position actuator bolt at this time.

28. Locate the camshafts to the cylinder head and assemble the camshaft actuators to the camshafts.

29. Ensure that the crankshaft is in the stage one timing drive assembly position. Refer to the Timing Chain Removal & Installation procedure.

30. Ensure that the camshaft sealing rings are in place in the camshaft grooves.

31. Select the proper camshaft for the particular installation location. The ring placement is defined as follows:

a. The number 2 identification ring for the right exhaust camshaft is machined off (1).

b. The number 3 identification ring for the right intake camshaft is machined off (2).

32. Apply a liberal amount of lubricant GM P/N 12345501 (Canadian P/N 992704) or equivalent to the camshaft journals and the right cylinder head camshaft carriers.

33. Place the right intake and right exhaust camshafts in position in the right cylinder head.

34. Position the camshaft lobes in a neutral position with the flats on the back of the camshafts up and parallel with the right cylinder head camshaft cover rail.

35. Observe the markings on the right cylinder head camshaft bearing caps. Each bearing cap is marked in order to identify its location. The markings have the following meanings:

- The raised feature must always be oriented toward the center of the cylinder head

- The I indicates the intake camshaft
- The E indicates the exhaust camshaft
- The number 1, 3, 5 indicates the cylinder position from the front of the engine

36. Apply a liberal amount of lubricant GM P/N 12345501 (Canadian P/N 992704) or equivalent to the camshaft bearing caps.

37. Install the camshaft bearing thrust caps in the first journal of the right cylinder head.

38. Install the remaining bearing caps with their orientation mark toward the center of the cylinder head.

39. Hand start all the camshaft bearing cap bolts.

40. Tighten the camshaft bearing cap bolts in the sequence shown. Tighten the camshaft bearing cap bolts in sequence to 10 Nm (89 inch lbs.).

41. Loosen the center intake camshaft bearing cap bolts (1, 2) and the center exhaust camshaft bearing cap bolts (3, 4).

42. Retighten the center camshaft bearing cap bolts (1, 2, 3, 4). Retighten the camshaft bearing cap bolts to 10 Nm (89 inch lbs.).

43. Install the crankshaft balancer.

❋❋ WARNING

A wrench must be used on the hex of the camshaft when loosening or tightening in order to prevent component damage. Failure to prevent the torque reaction against the timing drive chain can lead to timing drive chain failure.

➡**Use an open-end wrench at the camshaft hex to prevent camshaft/engine rotation.**

44. Install the CMP actuator valve.

45. Install the CMP actuator valve bolt. Tighten the CMP actuator valve bolt to 10 Nm (89 inch lbs.).

46. Install the CMP actuator valve electrical connector.

47. Install the ECM bracket with the ECM.

48. Ensure the proper camshaft position actuator is installed. Observe the body of the camshaft position actuator for the "IN" or "EX" marking (3).

49. Ensure the proper timing mark is used. Observe the outer ring of the camshaft position actuator for the "R" and triangle marking (2). The marking is for alignment to the highlighted timing chain link on the right side of the engine.

50. Use an open wrench on the hex cast

into the camshaft in order to prevent camshaft rotation when tightening the camshaft position actuator bolt.

51. Install the right intake camshaft position actuator.

52. Install the camshaft position actuator bolt. Tighten the camshaft position actuator bolt to 58 Nm (43 ft. lbs.).

53. Install the intake camshaft position actuator solenoid.

54. Install the camshaft sensors.

55. Install a NEW camshaft cover seal and NEW grommets.

56. Install tool EN 46101 onto the spark plug tubes of the right cylinder head.

57. Install the camshaft cover bolt grommets prior to installing the camshaft cover bolts.

58. Wipe the camshaft cover sealing surface on the right cylinder head with a clean, lint-free cloth.

59. Place a bead 8 mm (0.3150 in.) in diameter by 4 mm (0.1575 in.) in height of RTV sealant, GM P/N 12378521 (Canadian P/N 88901148) or equivalent, on the engine front cover split lines (1).

60. Place the right camshaft cover into position onto the right cylinder head.

61. Loosely install the right camshaft cover bolts.

62. Tighten the right camshaft cover bolts in the sequence shown. Tighten the right camshaft cover bolts in the sequence to 10 Nm (89 inch lbs.).

63. Remove tool EN 46101 from the spark plug tubes of the right cylinder head.

64. Install the NEW spark plugs into the right cylinder head. Tighten the spark plugs to 20 Nm (15 ft. lbs.).

65. Install each ignition coil through the right camshaft cover into the spark plug

tube taking care not to damage the spark plug and/or the seal in the right camshaft cover.

66. Install each ignition coil bolt. Tighten the ignition coil bolt to 10 Nm (89 inch lbs.).

67. Install the upper intake manifold with the lower intake manifold.

3.8L Engine

1. Before servicing the vehicle, refer to the Precautions Section.

2. Relieve the fuel system pressure.

3. Remove the engine and mount it on an engine stand.

4. Remove or disconnect the following:
- Negative battery cable
- Intake manifold
- Rocker arm covers
- Rocker arm assemblies
- Pushrods
- Lifters and guides

➡**A magnet may be helpful when pulling the lifters out of their bores. Identify all parts as they are removed, so they can be reinstalled in their original locations.**

- Crankshaft balancer
- Timing chain front cover

5. Set the engine to Top Dead Center (TDC) No. 1 cylinder (firing position) to align the timing marks, before disassembling the timing chain and sprockets.

❋❋ WARNING

Align the timing marks of the camshaft and crankshaft sprockets to avoid burring the camshaft journals by the crankshaft.

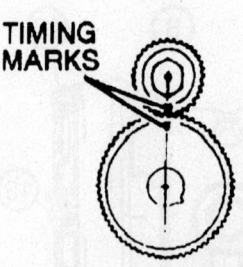

TIMING MARKS

BALANCE SHAFT GEAR TO BALANCE SHAFT DRIVE GEAR

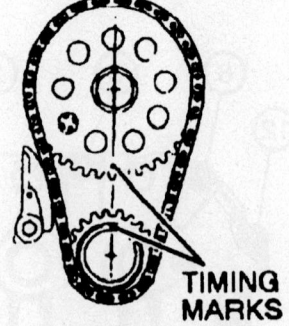

TIMING MARKS

CAMSHAFT SPROCKET TO CRANKSHAFT SPROCKET

7922UG09

The timing marks should face each other when the chain and gears are installed properly

6. Remove or disconnect the following:
- Camshaft sprocket and timing chain
- Camshaft thrust plate
- Camshaft

✴✴ WARNING

If the camshaft was replaced the lifters must also be replaced. The old lifters have developed a wear pattern and will cause the new camshaft to wear prematurely.

To install:

7. Coat the camshaft lobes and bearings with camshaft break-in prelube prior to installation.

8. Install or connect the following:
- Camshaft
- Camshaft thrust plate. Torque the bolts to 10 ft. lbs. (14 Nm).
- Camshaft sprocket and timing chain with timing marks aligned. Torque the camshaft sprocket bolt to 74 ft. lbs. (100 Nm) plus an additional 90 degree (¼) turn.
- Timing chain front cover
- Crankshaft balancer. Torque the mounting bolt to 111 ft. lbs. (150 Nm). plus an additional 76 degree turn.

9. Coat the valve lifters with camshaft break-in prelube.

10. Install or connect the following:
- Valve lifters
- Lifter guides and lifter guide retainer. Torque the retainer mounting bolts to 22 ft. lbs. (30 Nm).
- Pushrods and rocker arms. Torque the rocker arm bolts to 11 ft. lbs. (15 Nm) plus an additional 90 degree turn.
- Rocker arm covers
- Intake manifold
- Engine
- Negative battery cable

11. Verify that all fluid levels are full and correct.

12. Start the engine and check for leaks. Check engine operation.

Valve Lash

ADJUSTMENT

The valve clearance cannot be adjusted on these engines. The engine is equipped with hydraulic lifters, and adjustment is not necessary.

Starter Motor

REMOVAL & INSTALLATION

3.6L Engine

1. Before servicing the vehicle, refer to the Precautions Section.

2. Disconnect the battery ground (negative) cable from the battery.

3. Raise and support the vehicle.

4. Remove the radiator air baffle.

5. Remove the starter motor BAT terminal nut and electrical leads.

6. Remove the starter motor bolts.

7. Remove the starter motor.

To install:

8. Install the starter motor.

9. Install the starter motor bolts. Tighten the starter motor bolts to 50 Nm (37 ft. lbs.).

10. Install the starter motor S terminal electrical connector.

11. Install the battery positive cable and the BAT terminal nut to the starter motor BAT terminal. Tighten the starter motor BAT terminal nut to 13 Nm (115 inch lbs.).

12. Install the radiator air baffle.

13. Lower the vehicle.

14. Install the battery ground (negative) cable to the battery.

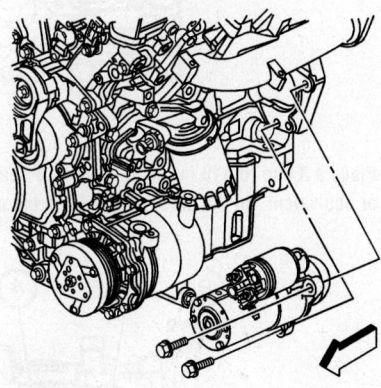

Starter installation—3.6L engine

3.8L Engine

1. Before servicing the vehicle, refer to the Precautions Section.

2. Remove or disconnect the following:
- Negative battery cable
- Flexplate inspection cover
- Splash shield, if equipped
- Electrical connectors
- Transmission cooler line clip from the transmission, if necessary
- Starter motor wiring
- Starter motor bolts

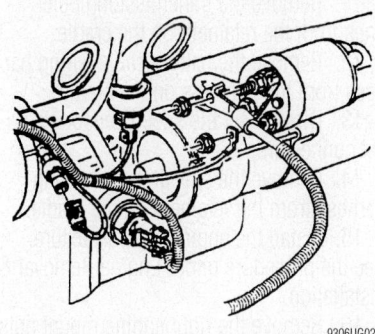

9306UG02

Starter in place with wiring—3.8L engine

- Starter

To install:

3. Install or connect the following:
- Starter and torque the bolts to 32 ft. lbs. (43 Nm)
- Wiring and torque the "B" terminal nut to 89 inch lbs. (10 Nm) and the "S" terminal nut to 22 inch lbs. (3 Nm).
- Flexplate inspection cover and torque the bolts to 62 inch lbs. (7 Nm)
- Splash shield
- Negative battery cable

Oil Pan

REMOVAL & INSTALLATION

3.6L Engine

1. Before servicing the vehicle, refer to the Precautions Section.

2. Remove the tires and wheels.

3. Remove the oil level indicator.

4. Remove the engine splash shields.

5. Drain the engine of oil.

➡**In the following service procedure, DO NOT remove the brake calipers. Relocate them to the side and properly support them.**

6. Remove the brake calipers and relocate to the side.

7. Remove the retaining bolts for the struts.

➡**In the following service procedure, DO NOT remove the steering rack from the vehicle. Relocate the steering rack and support the steering rack.**

8. Remove the steering rack from the cradle.

9. Remove the tie rods from the steering knuckle.

10. Remove the power steering cooler lines from the retainers on the cradle.

11. Remove the transmission cooler lines from the retainers on the cradle.

12. Remove the battery cable wiring harness from the retainers on the cradle.

13. Disconnect the speed sensor electrical connectors.

14. Remove the speed sensor wiring harness from the retainers on the cradle.

15. Install the engine support fixture. See the procedure under Engine Removal & Installation.

16. Remove the right engine mount nuts.

17. Remove the transmission mount bolts.

18. Remove the engine mount struts. See the procedure under Engine Removal & Installation.

19. Remove the cradle mounting bolts.

➡️Before lowering the frame, make sure that all transmission lines, power steering lines, and electrical harness have been disconnected and relocated to ensure that nothing will be damaged while lowering the frame.

20. Lower the frame enough to remove the oil pan assembly.

21. Lower the frame. See the procedure under Engine Removal & Installation.

22. Disconnect the oil level senor electrical connector.

23. Remove the oil pan bolts.

24. Using the pry points on the oil pan, separate the oil pan from the engine block.

25. Remove the oil pan from the vehicle.

26. Disassemble the oil pan.

To install:

27. Install the 8 mm (0.315 in) guides from tool set EN 46109 into the center oil pan rail bolt hole on each side of the engine block.

28. Place a 3 mm (0.118 in) bead (1) of RTV sealant, GM P/N 12378521 (Canadian P/N 88901148) or equivalent, on the block pan rail and the crankshaft rear oil seal housing.

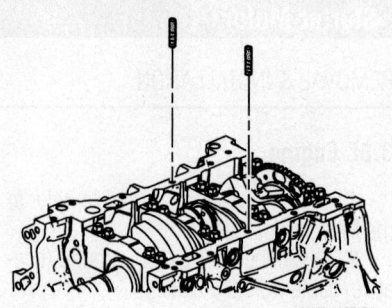

Oil pan guide studs—3.6L engine

29. Position the oil pan onto the block.

30. Remove the 8 mm (0.315 in) guides from the engine block.

31. Loosely install the oil pan bolts.

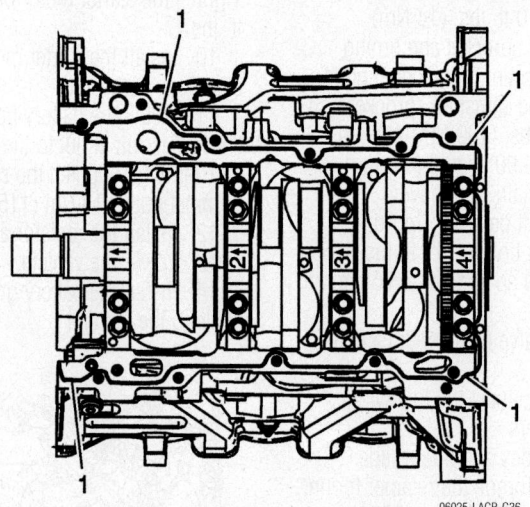

06025-LACR-G36

Place a 3 mm (0.118 in) bead (1) of RTV sealant, GM P/N 12378521 (Canadian P/N 88901148) or equivalent, on the block pan rail and the crankshaft rear oil seal housing—3.6L engine

32. Tighten the oil pan bolts in sequence shown.

- Tighten the 8 mm bolts (1–11) to 23 Nm (17 ft. lbs.).
- Tighten the 6 mm bolts (12, 13) to 10 Nm (89 inch lbs.).

33. Connect the oil level senor electrical connector.

34. Raise the frame. Refer to Engine Removal & Installation

35. Install the left engine mount strut.

36. Install the bolt and the nut to the engine mount strut at the engine mount strut bracket on the upper radiator support. HAND TIGHTEN ONLY.

37. Install the bolt and the nut to the engine mount strut at the left engine mount strut bracket on the engine. Tighten bolt

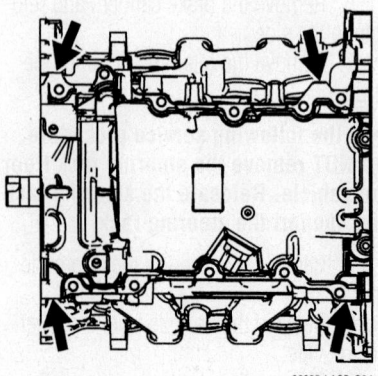

06025-LACR-G34

Oil pan pry points—3.6L engine

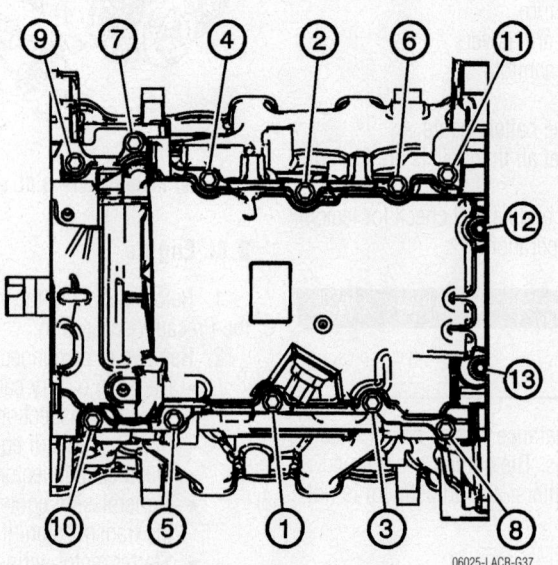

06025-LACR-G37

Oil pan bolt torque sequence—3.6L engine

engine mount strut nuts to 48 Nm (35 ft. lbs.).

38. Install the right engine mount strut.

39. Install the bolt and the nut to the engine mount strut at the engine mount strut bracket on the upper radiator support. HAND TIGHTEN ONLY.

40. Install the bolt and the nut to the engine mount strut at the right engine mount strut bracket on the engine. Tighten both engine mount strut nuts to 48 Nm (35 ft. lbs.).

41. Install the engine mount struts.

42. Install the transmission mount bolts. Tighten the transaxle mount lower nuts to 63 Nm (46 ft. lbs.). Tighten the transaxle mount upper nuts to 47 Nm (35 ft. lbs.).

43. Install the right engine mount nuts. Tighten the engine mount upper nuts to 53 Nm (39 ft. lbs.).

44. Remove the engine support fixture. Refer to Engine Removal & Installation.

45. Install the battery cable wiring harness in the retainers on the cradle.

46. Install the speed sensor wiring harness in the retainers on the cradle.

47. Install the steering rack on the cradle.

48. Install the tie rods in the steering knuckle.

49. Install the retaining bolts for the struts.

50. Install the brakes calipers.

51. Install the engine splash shields.

52. Install the oil level indicator.

53. Install the tires and wheels.

54. Fill the engine with oil.

3.8L Engine

❈❈ WARNING

The oil level sensor, located in the oil pan, must be removed prior to removal of the oil pan. If the oil pan is removed first, damage to the oil level sensor may occur.

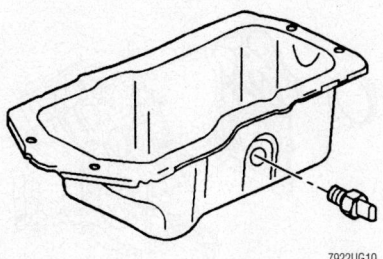

7922UG10

If equipped, be sure to remove the oil level sensor before removing the pan—3.8L engine

1. Before servicing the vehicle, refer to the Precautions Section.

2. Drain oil into an approved container.

3. Remove or disconnect the following:
   - Negative battery cable
   - Right engine mount bracket, if necessary
   - Flexplate cover
   - Oil level sensor
   - Oil filter
   - Torque axis mount bracket bolts, if necessary
   - Oil pan bolts
   - Oil pan
   - Oil pan gasket

4. Clean the oil pan and cylinder block mating surfaces.

To install:

5. Install or connect the following:
   - Oil pan with a new gasket and torque the bolts to 125 inch lbs. (14 Nm)
   - Torque axis mount bracket bolts, if removed
   - Oil filter
   - Flexplate cover
   - Oil level sensor
   - Oil drain plug and torque the plug to 30 ft. lbs. (40 Nm)
   - Right engine mount bracket, if necessary
   - Negative battery cable

6. Refill the crankcase.

7. Run the engine and check for leaks.

Oil Pump

REMOVAL & INSTALLATION

3.6L Engine

1. Before servicing the vehicle, refer to the Precautions Section.

➡**Do not remove the left bank idler sprocket.**

2. Remove the primary timing chain.

3. Remove the crankshaft sprocket.

➡**There are no serviceable components within the oil pump. Disassemble the pump only to diagnose an oiling concern. A disassembled oil pump must not be reused. A disassembled oil pump must be replaced.**

4. Remove the oil pump bolts and the oil pump.

To install:

5. Align the oil pump gerotor with the crankshaft flats and install the oil pump to the engine block.

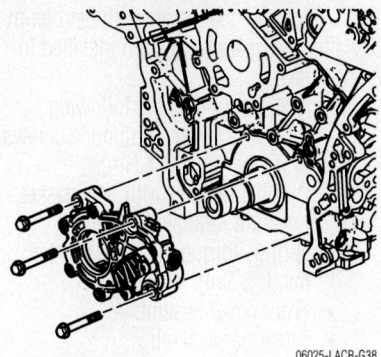

06025-LACR-G38

Oil pump removal—3.6L engine

6. Align the pump body with the mounting holes in the cylinder block.

7. Install the oil pump bolts. Tighten the oil pump bolts to 23 Nm (17 ft. lbs.).

8. Install the crankshaft sprocket.

9. Install the primary timing chain.

3.8L Engine

1. Before servicing the vehicle, refer to the Precautions Section.

2. Support the engine using an engine support fixture.

3. Remove or disconnect the following:
   - Negative battery cable
   - Engine drive belts and tensioner assembly
   - Drive belt idler pulley and bracket

4. Remove or disconnect the following:
   - Torque axis mount bracket, if necessary
   - Engine front cover assembly
   - Oil filter adapter with pressure regulator valve and spring
   - Oil pump cover
   - Inner and outer pump gears

To install:

5. Lubricate the oil pump gears with petroleum jelly.

6. Install the gears into the oil pump housing.

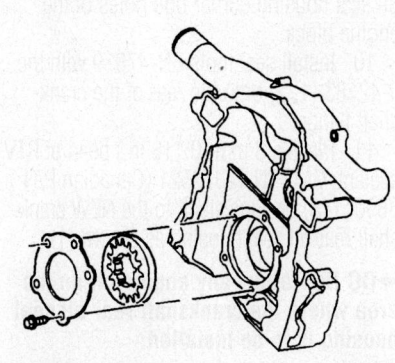

7922UG11

The oil pump is located inside the front engine cover—3.8L engines

7. Pack the gear cavity with petroleum jelly after the gears have been installed in the housing.

8. Install or connect the following:
- Oil pump cover. Torque the screws to 97 inch lbs. (11 Nm).
- Oil filter adapter with new gasket, pressure regulator valve and spring. Torque the bolts to 11 ft. lbs. (15 Nm).
- Front cover assembly
- Tensioner assembly
- Drive belt idler pulley and bracket, if removed
- Drive belts
- Torque axis mount bracket
- Negative battery cable

9. Remove the engine support fixture.
10. Verify the correct engine oil level.
11. Start the vehicle and verify no leaks and proper oil pressure.

Rear Main Seal

REMOVAL & INSTALLATION

3.6L Engine

1. Before servicing the vehicle, refer to the Precautions Section.
2. Remove the transmission.
3. Remove the engine flywheel bolts and discard.
4. Remove the engine flywheel from the crankshaft.
5. Remove the oil pan.
6. Remove the crankshaft rear oil seal housing bolts.
7. Using the pry points located at the edge of the crankshaft rear oil seal housing shear the RTV sealant.
8. Remove and discard the crankshaft rear oil seal housing.

To install:

9. Install the 6 mm (0.236 in.) guides from kit EN 46109 into the 2 crankshaft rear oil seal housing corner bolt holes of the engine block.
10. Install seal tools EN-47839 with the J-42183 (1, 2) onto the rear of the crankshaft flange.
11. Place a 3 mm (0.118 in.) bead of RTV sealant, GM P/N 12378521 (Canadian P/N 88901148) or equivalent, to the NEW crankshaft rear oil seal housing as shown (1).

→DO NOT allow any engine oil on the area where the crankshaft rear oil seal housing is to be installed.

12. Install the crankshaft rear oil seal housing to the engine block.

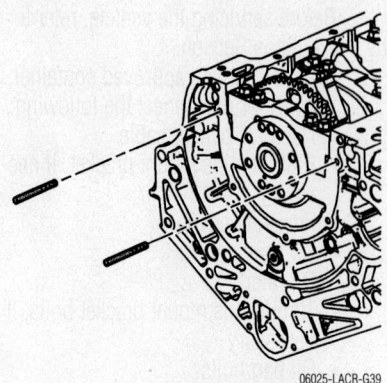

06025-LACR-G39

Install the 6 mm (0.236 in.) guides from kit EN 46109 into the 2 crankshaft rear oil seal housing corner bolt holes of the engine block—3.6L engine

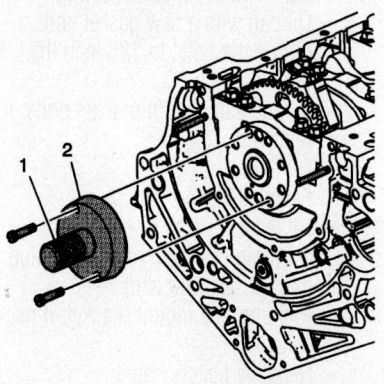

06025-LACR-G40

Install seal tools EN-47839 with the J-42183 (1, 2) onto the rear of the crankshaft flange—3.6L engine

06025-LACR-G41

Place a 3 mm (0.118 in.) bead of RTV sealant, GM P/N 12378521 (Canadian P/N 88901148) or equivalent, to the NEW crankshaft rear oil seal housing as shown (1) —3.6L engine

13. Remove the guides from the engine block.
14. Install the crankshaft rear oil seal housing bolts. Tighten the crankshaft rear oil seal housing bolts to 10 Nm (89 inch lbs.).

15. Remove the seal tool from the crankshaft flange.
16. Install the oil pan.
17. Place the engine flywheel in position on the crankshaft.
18. Install 2 NEW bolts in location at the top and bottom of the engine flywheel bolt pattern allowing the engine flywheel to hang in position.
19. Install the remaining NEW engine flywheel bolts.
- Tighten the NEW engine flywheel bolts to 30 Nm (22 ft. lbs.)
- Tighten the NEW engine flywheel bolts an additional 45 degrees
20. Install the transmission.

3.8L Engine

1. Before servicing the vehicle, refer to the Precautions Section.
2. Remove or disconnect the following:
- Transaxle assembly
- Flexplate from the crankshaft
- Rear main seal from engine block by inserting a small flat-bladed prytool through the dust lip at an angle, then pry out the crankshaft rear oil seal. Repeat as necessary around the seal until it is removed.

❊❊ WARNING

Do not damage or scratch the sealing surface of the crankshaft or the seal bore.

To install:

3. Lubricate new rear main with clean engine oil prior to installation.
4. Slide the oil seal on the mandrel of seal installer tool J-38196 until the back of the seal is seated squarely against the collar of the tool.
5. Attach the seal installer to the rear of the crankshaft with the 2 mounting bolts, then turn the T-handle until the oil seal is fully seated into the rear of the engine.
6. Loosen the T-handle of the tool completely.

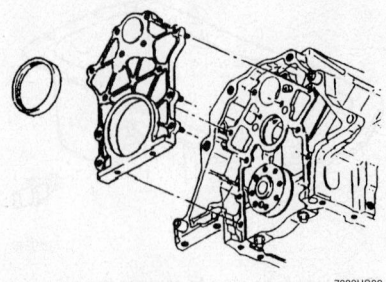

7922UG32

Rear main oil seal and rear cover—3.8L engine

7. Remove both bolts and the tool.
8. Install or connect the following:
- Flexplate. Torque the bolts to 11 ft. lbs. (15 Nm), plus an additional 50 degrees.
- Transaxle

Crankshaft Balancer

REMOVAL & INSTALLATION

3.6L Engine

1. Before servicing the vehicle, refer to the Precautions Section.
2. Remove the accessory drive belt.
3. Raise and support the vehicle.
4. Remove the starter.
5. Install flywheel holding tool EN 46106 through the starter mounting hole.
6. Remove the crankshaft balancer bolt.
7. Remove the crankshaft balancer with a 3-jawed puller.

To install:

8. Use tool J 41998-B (nut, bearing and washer) to install the crankshaft balancer.

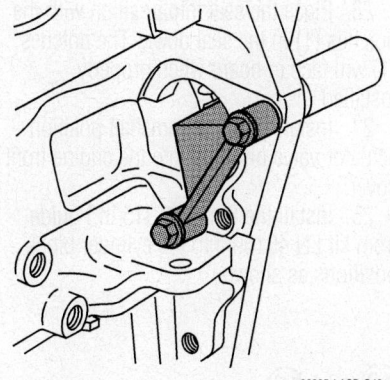

06025-LACR-G42

Flywheel holding tool—3.6L engine

06025-LACR-G43

Tool J 41998-B—3.6L engine

➡ **Do not lubricate the crankshaft front oil seal or crankshaft balancer sealing surfaces. The crankshaft balancer is installed into a dry seal.**

9. Apply lubricant to the inside of the crankshaft balancer hub bore.
10. Place the crankshaft balancer in position on the crankshaft.
11. Thread tool J 41998-B in the crankshaft. Ensure you engage at least 10 threads of the tool before pressing the crankshaft balancer in place.
12. Push the crankshaft balancer into position by tightening the nut on the tool until the large washer bottoms out on the crankshaft end.
13. Remove the tool.
14. Install the crankshaft balancer bolt.
15. Tighten the crankshaft balancer bolt.
   a. Tighten the crankshaft balancer bolt to 100 Nm (74 ft. lbs.).
   b. Tighten the crankshaft balancer bolt an additional 150 degrees.
16. Remove the holding tool.
17. Install the starter.
18. Install the accessory drive belt.
19. Lower the vehicle.

3.8L Engine

1. Before servicing the vehicle, refer to the Precautions Section.
2. Disconnect the negative battery cable.
3. Remove the drive belt.
4. Raise and support the vehicle.
5. Remove the right front tire and wheel.
6. Remove the right engine splash shield retainers and the engine splash shield.
7. Remove the torque converter covers.
8. Use tool J 37096 to secure the flywheel in order to prevent the crankshaft from rotating.
9. Remove the crankshaft balancer bolt and discard the balancer bolt.

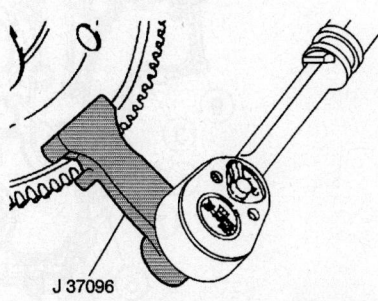

06025-LACR-G44

Use tool J 37096 to secure the flywheel in order to prevent the crankshaft from rotating

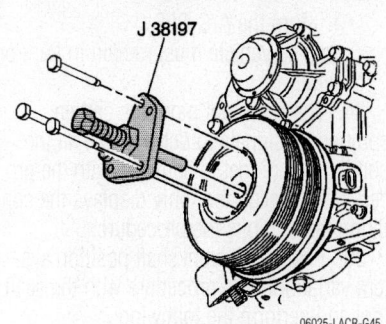

06025-LACR-G45

Remove the crankshaft balancer using tool J 38197-A

➡ **Do not separate the crankshaft pulley from the crankshaft balancer. Service the crankshaft pulley and the crankshaft balancer as an assembly.**

10. Remove the crankshaft balancer using tool J 38197-A.

To install:

11. Coat the engine front cover seal contact area on the crankshaft balancer, and the seal surface with engine oil.
12. Install the crankshaft balancer.
13. Prevent the crankshaft from rotating.
14. Install the new crankshaft balancer bolt. Tighten the bolt to 150 Nm (111 ft. lbs.), plus 76 degrees.
15. Install the torque converter covers.
16. Install the right engine splash shield and the engine splash shield retainers.
17. Install the right front tire and wheel.
18. Lower the vehicle.
19. Install the drive belt.

➡ **The following CKP System Variation Learn Procedure must be performed.**

20. Install a scan tool.
21. Monitor the powertrain control module (PCM) for DTCs with a scan tool.
22. Select the crankshaft position variation learn procedure with a scan tool.
23. The scan tool instructs you to perform the following:
   a. Accelerate to wide open throttle (WOT).
   b. Release the throttle when fuel cut-off occurs.
   c. Observe the fuel cut-off specifications for the applicable engine.
   d. The engine should not accelerate beyond the calibrated RPM value.
   e. Release the throttle immediately if the value is exceeded.
   f. Block the drive wheels.
   g. Set the parking brake.
   h. DO NOT apply the brake pedal.
   i. Cycle the ignition from OFF to ON.
   j. Apply and hold the brake pedal.
   k. Start and idle the engine.

l. Turn the A/C OFF.

m. The vehicle must remain in Park or Neutral.

24. The scan tool monitors certain component signals to determine if all the conditions are met to continue with the procedure. The scan tool only displays the condition that inhibits the procedure.

25. Enable the crankshaft position system variation learn procedure with the scan tool and perform the following:

➡ While the learn procedure is in progress, release the throttle immediately when the engine starts to decelerate. The engine control is returned to the operator and the engine responds to throttle position after the learn procedure is complete.

26. Accelerate to WOT.

a. Release when fuel cut-off occurs.

b. Test in progress.

27. The scan tool displays Learn Status: Learned this ignition. If the scan tool indicates that DTC P0315 ran and passed, the CKP variation learn procedure is complete. If the scan tool indicates DTC P0315 failed or did not run, refer to DTC P0315.

28. Turn OFF the ignition for 30 seconds after the learn procedure is completed successfully.

Timing Chain, Sprockets, Front Cover and Seal

REMOVAL & INSTALLATION

3.6L Engine

FRONT COVER AND SEAL

1. Before servicing the vehicle, refer to the Precautions Section.

2. Disconnect the negative battery cable.

3. Remove the ECU module bracket. Position it out of the way. DO NOT remove the ECU or disconnect it.

4. Remove the generator assembly.

5. Remove the coolant overflow reservoir.

6. Remove the power steering reservoir and position it out of the way.

7. Remove the camshaft covers.

8. Remove the upper radiator hose.

9. Remove the water pump.

10. Remove the engine splash shield.

11. Remove the crankshaft balancer.

12. Remove the drive belt tensioner.

13. Remove the camshaft position actuator valve bolts.

14. Remove the camshaft position actuator valves from the front cover.

15. Remove the engine front cover bolts.

❋❋ WARNING

Do not pry between the engine front cover and the camshaft position sensors or the camshaft position actuators in order to shear the RTV. Use the pry points and a bolt in the jackscrew hole in order to remove the engine front cover. Damage to the camshaft position sensors or the camshaft position actuators may occur if the camshaft position sensors or the camshaft position actuators are used to pry against in order to remove the engine front cover.

16. Loosely install a 10 x 1.5 mm bolt in the "jackscrew" hole (1).

17. Using the pry points (2) located at the edge of the front cover and the "jackscrew", shear the RTV sealant.

18. Remove the engine front cover.

19. Rotate the crankshaft until the left cylinder head camshafts align with the EN 46105-2 and the right cylinder head camshafts align with the EN 46105-1.

20. Install the EN 46105-1 to the right camshafts.

21. Install the EN 46105-2 to the left camshafts.

To install:

➡ **Do not lubricate the crankshaft front oil seal or crankshaft balancer sealing surfaces. The crankshaft balancer is installed into a dry seal.**

22. Install the NEW crankshaft front oil seal into the engine front cover using a seal driver.

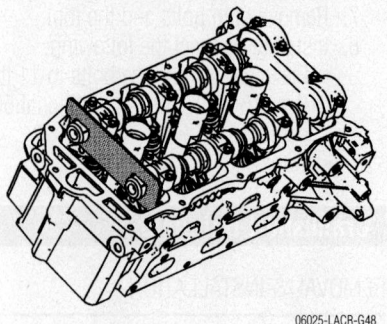

Install the EN 46105-1 to the right camshafts—3.6L engine

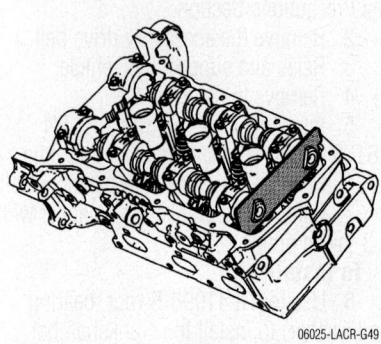

Install the EN 46105-2 to the left camshafts—3.6L engine

23. Place the seal into position with the notches (1) in the seal down. The notches (1) will face in board when properly installed.

24. Install the NEW camshaft position actuator valve oil seals into the engine front cover.

25. Install the 8 mm (0.315 in.) guide from kit EN 46109 into the cylinder block positions as shown.

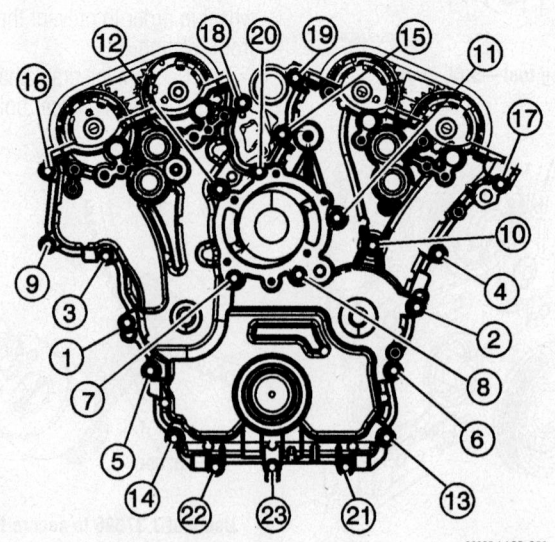

Front cover torque sequence—3.6L engine

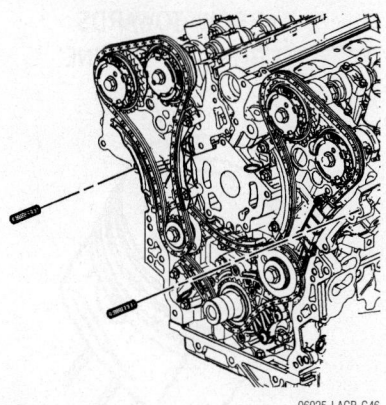

Install the 8 mm (0.315 in.) guide from kit EN 46109 into the cylinder block positions as shown—3.6L engine

26. Install the NEW engine front cover to cylinder block seal.

27. Place a 3 mm (0.118 in.) bead of RTV sealant, GM P/N 12378521 (Canadian P/N 88901148) or equivalent, on the engine front cover as shown (1).

28. Place the engine front cover onto the guide pins and slide into position.

29. Remove the guide pins from the cylinder block.

30. Hand start all the front cover bolts.

31. Tighten the engine front cover bolts in the sequence shown. Tighten the engine front cover bolts in sequence to 23 Nm (17 ft. lbs.).

32. Install NEW O-rings on the camshaft position sensor.

33. Place the camshaft position sensors in position on the front cover.

34. Install the camshaft position sensor bolts. Tighten the camshaft position sensor bolts to 10 Nm (89 inch lbs.).

35. Place the camshaft position actuator valves in position on the front cover.

36. Install the camshaft position actuator valve bolts. Tighten the camshaft position actuator valve bolts to 10 Nm (89 inch lbs.).

37. Install the water pump.

38. Install the camshaft covers.

39. Install the drive belt tensioner. Install the drive belt tensioner bolts. Tighten the drive belt tensioner center bolts to 50 Nm (37 ft. lbs.). Tighten the drive belt tensioner outer bolts to 23 Nm (17 ft. lbs.).

40. Install the crankshaft balancer.

41. Install the coolant overflow reservoir.

42. Install the power steering reservoir. Tighten the M6 bolt to 9 Nm (7 ft. lbs.). Tighten the M8 bolt to 25 Nm (18 ft. lbs.).

43. Install the upper radiator hose.

44. Install the generator assembly.

45. Install engine splash shield.

46. Install the ECU module bracket. Tighten the ECM bracket bolts to 10 Nm (89 inch lbs.).

47. Connect the negative battery cable.

3.8L Engine

1. Before servicing the vehicle, refer to the Precautions Section.

2. Drain the cooling system.

3. Support the engine.

4. Remove or disconnect the following:
- Negative battery cable
- Torque axis mount and bracket
- Drive belt
- Drive belt tensioner
- Crankshaft balancer
- Crankshaft Position (CKP) sensor shield and the CKP sensor
- Oil pan-to-front cover bolts
- Timing chain front cover

5. Align the timing marks on the camshaft and crankshaft sprockets so they are as close together as possible.
- Timing chain damper
- Camshaft sprocket bolt, the camshaft sprocket and timing chain
- Crankshaft sprocket

⁂ WARNING

Do not rotate the camshaft or crankshaft while the timing chain and sprockets are removed.

To install:
6. Install or connect the following:
- Timing chain and sprockets with the timing marks aligned
- Camshaft sprocket bolt. Torque the bolt to 74 ft. lbs. (100 Nm) plus an additional 90 degree turn.
- Timing chain damper. Torque the bolts to 16 ft. lbs. (22 Nm).

⁂ WARNING

The oil pump is built into the front cover. When the cover is removed, oil drains from the pump. Since the pump "loses its prime" it may not establish oil pressure as soon as the engine starts. Therefore, it is important to remove the oil pump cover from the back of the timing chain front cover and pack the space around the oil pump gears completely full of petroleum jelly. If this is not done, the oil pump may not pump engine oil when the engine is started, resulting in severe engine damage.

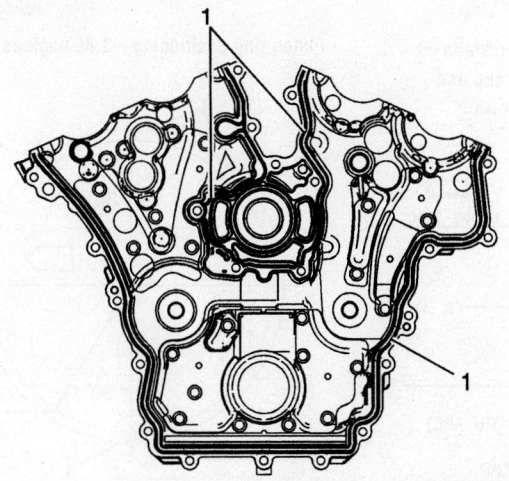

Place a 3 mm (0.118 in.) bead of RTV sealant, GM P/N 12378521 (Canadian P/N 88901148) or equivalent, on the engine front cover as shown (1)—3.6L engine

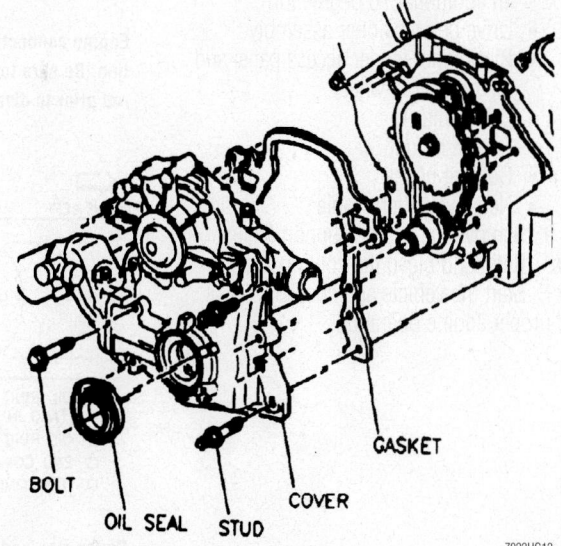

Timing chain front cover—3.8L engines

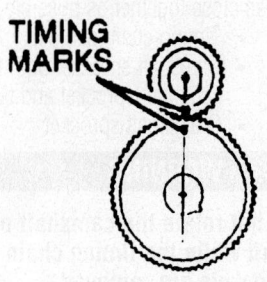

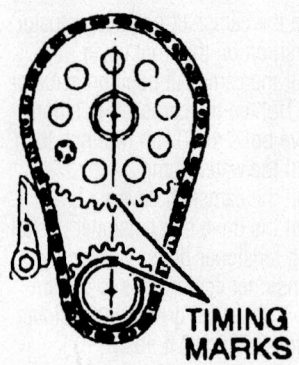

BALANCE SHAFT GEAR TO BALANCE SHAFT DRIVE GEAR

CAMSHAFT SPROCKET TO CRANKSHAFT SPROCKET

7922UG09

Timing chain sprocket and balance shaft gear alignment—3.8L engines

7. Remove the screws and the oil pump cover from the back of the timing chain front cover. Pack the space around the oil pump gears completely full of petroleum jelly. There must be no air space left inside the pump.

8. Install or connect the following:
- Pump cover with new gaskets. Torque the screws to 97 inch lbs. (11 Nm).
- Timing chain front cover. Torque the front cover-to-engine bolts to 11 ft. lbs. (15 Nm) plus an additional 40 degrees.
- Oil pan-to-front cover bolts. Torque the bolts to 125 inch lbs. (14 Nm).
- CKP sensor. Torque the bolts to 14–28 ft. lbs. (20–40 Nm).
- CKP sensor shield
- Crankshaft balancer. Torque the bolt to 111 ft. lbs. (150 Nm) plus an additional 76 degree turn.
- Drive belt tensioner assembly
- Right inner fender access panel and the right front wheel
- Drive belt(s)
- Engine mount
- Coolant hoses
- Negative battery cable

9. Remove the engine support fixture.
10. Refill and bleed the cooling system.
11. Start the vehicle and check for leaks and proper engine operation.

Piston and Ring

POSITIONING

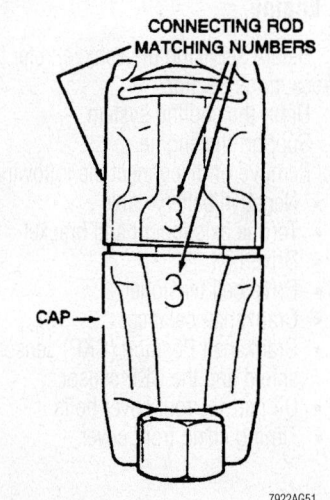

7922AG51

Engine connecting rod and cap installation. Be sure to matchmark the cap and rod prior to disassembly, as shown.

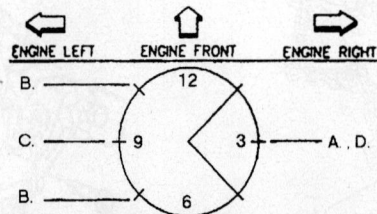

A. OIL RING SPACER GAP (TANG IN HOLE OR SLOT WITH ARC)
B. OIL RING RAIL GAPS
C. 2ND COMPRESSION RING GAP
D. TOP COMPRESSION RING GAP

7922AG46

Piston ring end–gap spacing—3.8L engines

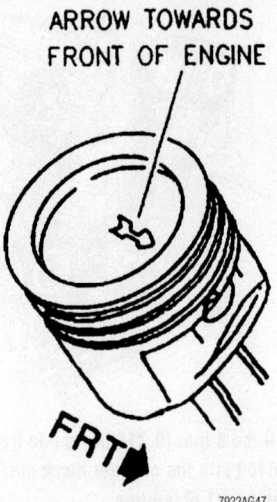

7922AG47

Piston positioning. Often the arrow is replaced by a notch, which also must face toward the front of the engine—3.8L engines

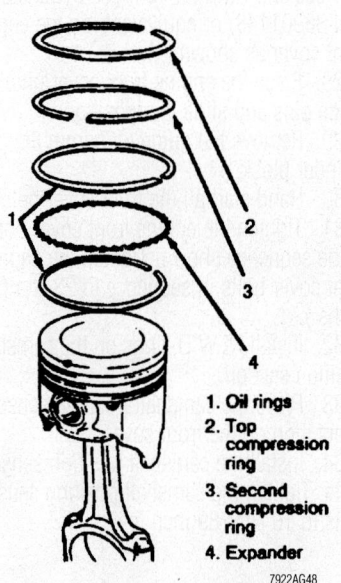

1. Oil rings
2. Top compression ring
3. Second compression ring
4. Expander

7922AG48

Piston ring positioning—3.8L engines

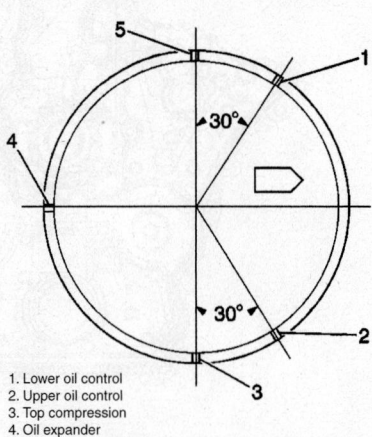

1. Lower oil control
2. Upper oil control
3. Top compression
4. Oil expander
5. Bottom compression

06025-LACR-G51

Piston ring gap placement—3.6L engine

FUEL SYSTEM

Fuel System Service Precautions

Safety is the most important factor when performing not only fuel system maintenance but any type of maintenance. Failure to conduct maintenance and repairs in a safe manner may result in serious personal injury or death. Maintenance and testing of the vehicle's fuel system components can be accomplished safely and effectively by adhering to the following rules and guidelines.

• To avoid the possibility of fire and personal injury, always disconnect the negative battery cable unless the repair or test procedure requires that battery voltage be applied.

• Always relieve the fuel system pressure prior to disconnecting any fuel system component (injector, fuel rail, pressure regulator, etc.), fitting or fuel line connection. Exercise extreme caution whenever relieving fuel system pressure, to avoid exposing skin, face and eyes to fuel spray. Please be advised that fuel under pressure may penetrate the skin or any part of the body that it contacts.

• Always place a shop towel or cloth around the fitting or connection prior to loosening to absorb any excess fuel due to spillage. Ensure that all fuel spillage (should it occur) is quickly removed from engine surfaces. Ensure that all fuel soaked cloths or towels are deposited into a suitable waste container.

• Always keep a dry chemical (Class B) fire extinguisher near the work area.

• Do not allow fuel spray or fuel vapors to come into contact with a spark or open flame.

• Always use a back-up wrench when loosening and tightening the fuel line connection fittings. This will prevent unnecessary stress and torsion to fuel line piping.

• Always replace worn fuel fitting O-rings with new. Do not substitute fuel hose or equivalent, where fuel pipe is installed.

Fuel System Pressure

RELIEVING

3.6L Engine

1. Before servicing the vehicle, refer to the Precautions Section.

❊❊ CAUTION

Remove the fuel tank cap and relieve the fuel system pressure before servicing the fuel system in order to reduce the risk of personal injury. After you relieve the fuel system pressure, a small amount of fuel may be released when servicing the fuel lines, the fuel injection pump, or the connections. In order to reduce the risk of personal injury, cover the fuel system components with a shop towel before disconnection. This will catch any fuel that may leak out. Place the towel in an approved container when the disconnection is complete.

2. Turn the ignition OFF.
3. Remove the fuel pump fuse and the fuel pump relay.
4. Loosen the fuel filler cap to relieve the fuel tank vapor pressure.
5. Attempt to start the engine and allow the engine to run until it stops.
6. Remove the fuel pressure test port cap.

❊❊ CAUTION

Wrap a shop towel around the fuel pressure connection in order to reduce the risk of fire and personal injury. The towel will absorb any fuel leakage that occurs during the connection of the fuel pressure gage. Place the towel in an approved container when the connection of the fuel pressure gage is complete.

7. Wrap a shop towel around the fuel pressure test port and use a small flat-bladed tool in order to depress (open) the fuel pressure test port valve.
8. Place the shop towel in an approved container.
9. Install the fuel pressure test port cap.
10. Tighten the fuel filler cap.

3.8L Engine

1. Before servicing the vehicle, refer to the Precautions Section.
2. Disconnect the negative battery cable to avoid possible fuel discharge if an accidental attempt is made to start the engine.
3. Remove the fuel tank cap to relieve tank pressure. Do not tighten until the service procedure has been completed.

4. Connect a fuel pressure gauge with bleed valve to the fuel pressure test port. Wrap a shop towel around the fitting while connecting the gauge to catch any spilled fuel.
5. Install the bleed hose into an approved container and open the valve to bleed off the fuel system pressure.
6. Drain any fuel remaining in the gauge into an approved container.

❊❊ CAUTION

There may still be residual fuel in the system, and a small amount of fuel may be released when servicing fuel lines or connections. In order to reduce the chance of personal injury, cover the fuel line fittings with a shop towel before disconnecting to catch any fuel that may leak out.

Fuel Filter

REMOVAL & INSTALLATION

The fuel filter is located in the fuel tank and is not normally serviced. Vehicles with a 3.6L engine have a paper filter located in the fuel pump module. Vehicles with a 3.8L engine have a fuel strainer located on the bottom of the fuel pump.

Fuel Pump

REMOVAL & INSTALLATION

With a 3.6L Engine

➡Clean the fuel and evaporative emission (EVAP) connections and surrounding areas prior to disconnecting the lines in order to avoid possible system contamination.

1. Before servicing the vehicle, refer to the Precautions Section.
2. Relieve the fuel system fuel pressure.
3. Drain the fuel tank.
4. Raise and support the vehicle.
5. Loosen the fuel fill hose clamp at the fuel tank.
6. Remove the fuel tank fill hose from the fuel tank.
7. Disconnect the EVAP vent solenoid hose on the tank from the EVAP vent valve solenoid hose.
8. Disconnect the EVAP vent pipe quick-connect fitting from the fill pipe EVAP vent pipe quick-connect fitting.

9. Disconnect the fuel feed, and the EVAP lines from the fuel tank lines.

10. Support the exhaust system.

11. Remove the rubber exhaust pipe hangers in order to allow the exhaust system to drop slightly.

12. Remove the fuel tank shield retainers.

13. Remove the fuel tank shield.

❋❋ WARNING

Do not bend the fuel tank straps as this may damage the straps.

14. Support the fuel tank with a suitable adjustable jack.

15. Remove the fuel tank strap bolts.

16. Using the jack lower the fuel tank.

17. Disconnect the fuel sender jumper harness electrical connector.

18. Remove the fuel tank and place the tank in a suitable work area.

19. Disconnect and remove the fuel pressure sensor and fuel sender jumper harness electrical connectors.

➡**Note the routing of the lines for installation.**

20. Disconnect and remove the fuel feed line, and the EVAP lines.

21. Remove the EVAP canister.

22. Remove the insulator pads from the fuel tank. Note the location of the insulator pads for installation.

23. Disconnect the fuel sender module electrical connectors.

❋❋ WARNING

Do Not handle the fuel sender assembly by the fuel pipes. The amount of leverage generated by handling the fuel pipes could damage the joints.

24. Disconnect the fuel pipes from the fuel sender.

❋❋ WARNING

Avoid damaging the lockring. Use only J-45722 to prevent damage to the lockring.

❋❋ WARNING

Do Not handle the fuel sender assembly by the fuel pipes. The amount of leverage generated by handling the fuel pipes could damage the joints.

➡**Do NOT use impact tools. Significant**

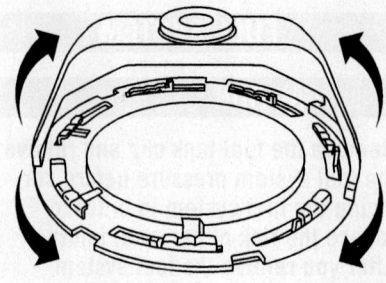

06025-LACR-G52

Fuel tank lockring removal

force will be required to release the lockring. The use of a hammer and screwdriver is not recommended. Secure the fuel tank in order to prevent fuel tank rotation.

25. Use tool J 45722 and a long breaker-bar in order to unlock the fuel sender lockring. Turn the fuel sender lockring in a counterclockwise direction.

26. Remove the fuel sender lockring and the fuel sender from the fuel tank.

27. Remove and discard the fuel sender seal.

28. Remove the fuel level sensor from the fuel sender module.

➡**Some lockring were manufactured with DO NOT REUSE stamped into them. These lockrings may be reused if they are not damaged or warped.**

➡**Inspect the lockring for damage due to improper removal or installation procedures. If damage is found, install a NEW lockring.**

➡**Check the lockring for flatness.**

29. Place the lockring on a flat surface. Measure the clearance between to lockring and the flat surface using a feeler gage at 7 points.

30. If the warpage is less than 0.41 mm (0.016 in.), the lockring does not require replacement.

31. If the warpage is greater than 0.41 mm (0.016 in.), the lockring must be replaced.

To install:

32. Install the fuel level sensor to the fuel sender module.

33. Clean the fuel sender sealing flange.

➡**Always replace the fuel sender seal when installing the fuel sender assembly.**

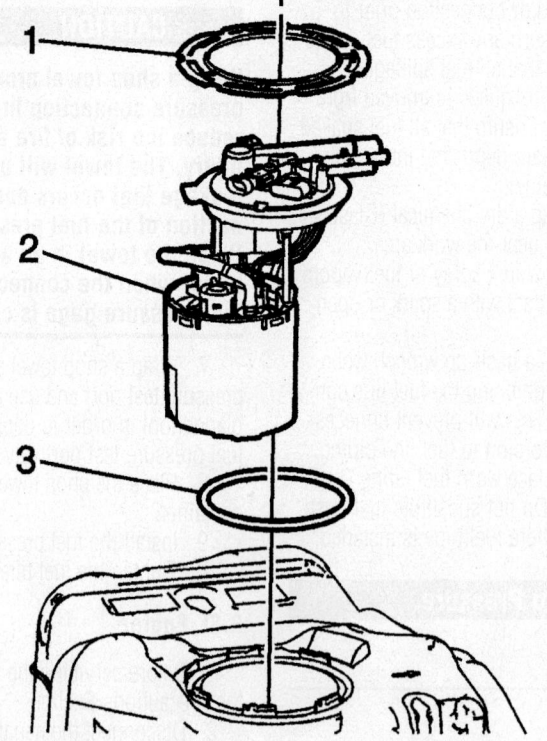

1. Lockring
2. Fuel module
3. Seal

06025-LACR-G53

Fuel module—with 3.6L engine

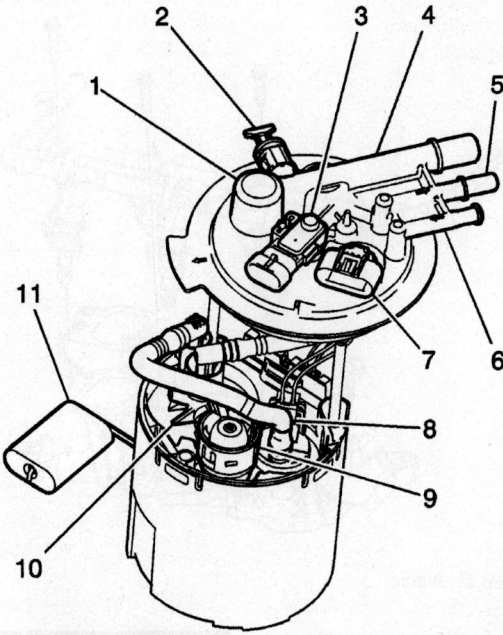

1. Fill Limit Vent Valve (FLVV)
2. T-connector for vapor hose/pipes to vent/rollover valve and fill tube
3. Fuel Tank Pressure (FTP) sensor
4. FLVV outlet to Evaporative Emission (EVAP) canister
5. Fuel feed outlet
6. Fuel return inlet-not used
7. Fuel sender assembly connector
8. Fuel pump
9. Fuel pressure regulator
10. Fuel filter assembly
11. Fuel level sensor float

06025-LACR-G54

Fuel module components—with 3.6L engine

34. Install the NEW fuel sender seal to the fuel tank seal groove.

35. Install the fuel sender and the fuel sender lockring.

➡**Always replace the fuel sender seal when installing the fuel sender assembly. Replace the lockring if necessary. Do not apply any type of lubrication in the seal groove. Ensure the lockring is installed with the correct side facing upward. A correctly installed lockring will only turn in a clockwise direction.**

36. Use tool J 45722 in order to install the fuel sender lockring. Turn the fuel sender lockring in a clockwise direction.

37. Install the fuel pipes to the fuel sender.

38. Install the fuel sender sensor electrical connectors.

39. Install the insulator pads to the fuel tank.

40. Install the EVAP canister.
- Do not attempt to straighten kinked nylon pipes. Replace any kinked nylon pipes in order to prevent damage to the vehicle.
- Do not attempt to repair sections of nylon pipes. Replace damaged nylon pipes.

- Replace the vapor pipes with original equipment or parts that meet GM specifications.
- Replace the vapor hoses with original equipment or parts meeting GM specifications. Use only reinforced fuel-resistant hose identified with the word Fluoroelastomer or GM 6163M on the hose.

41. Install and connect the fuel feed line, and the EVAP lines.

42. Install and connect the fuel pressure sensor and fuel sender jumper harness electrical connectors.

43. Install the fuel tank onto a suitable jack.

44. Partially raise the fuel tank until the electrical connection can be made.

45. Connect the fuel sender jumper harness electrical connector.

46. Completely raise the tank.

47. Install the fuel tank strap bolts. Tighten the bolts to 48 Nm (35 ft. lbs.).

48. Remove the jack from the fuel tank.

49. Position the fuel tank shield to the fuel tank.

50. Install the shield retainers.

51. Install the rubber exhaust pipe hangers.

52. Remove the support from the exhaust system.

53. Connect the fuel feed, and EVAP lines to the fuel tank lines.

54. Connect the EVAP vent pipe quick-connect fitting to the fill pipe EVAP vent pipe quick-connect fitting.

55. Connect the EVAP vent pipe quick-connect fitting to the fill pipe EVAP vent pipe quick-connect fitting.

56. Install the fuel tank fill hose onto the fuel tank. Install the hose over the orientation feature on the tank until fully seated to the tank.

57. Tighten the fuel fill hose clamp at the fuel tank. Tighten the clamp to 2.5 Nm (22 inch lbs.).

58. Lower the vehicle.

59. Add fuel and install the fuel fill cap.

60. Connect the negative battery cable.

61. Inspect the fuel system for leaks by performing the following steps:
   a. Turn ON the ignition for 2 seconds.
   b. Turn OFF the ignition for 10 seconds.
   c. Turn ON the ignition.
   d. Inspect for fuel leaks.

62. Install engine sight shield.

With a 3.8L Engine

1. Before servicing the vehicle, refer to the Precautions Section.

2. Relieve the fuel system pressure.

3. Drain the fuel tank.

4. Remove or disconnect the following:
- Negative battery cable
- Spare tire and jack
- Trunk lining
- Fuel sender access panel
- Sender and quick connect fittings from the sender
- Electrical connector from the sender and position harness and hoses aside

✳✳ CAUTION

When removing the fuel sender from the tank, the reservoir bucket is full of fuel. Use caution in containing the fuel.

- Sender retaining ring
- Sender and take note of its position
- Fuel sender O-ring and discard it

➡**Note the direction the strainer is pointing.**

- Strainer from the pump by pulling it down and twisting
- Pump electrical wires and hoses
- Pump assembly out of the rubber connectors

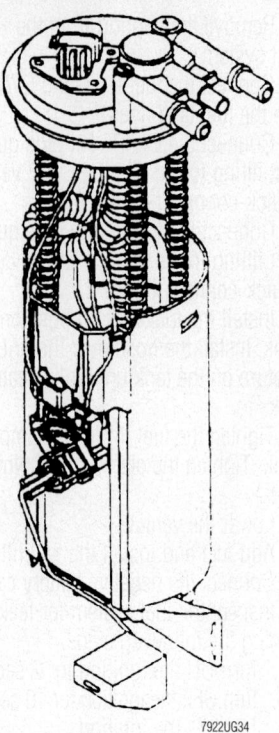

Fuel pump and level sender assembly—
3.8L engine

To install:

5. Transfer any insulators and grommets from the old pump to the new one.

6. Connect the pump to the fuel hose and tilt the bottom of the pump into the mounting bracket.

7. Install or connect the following:

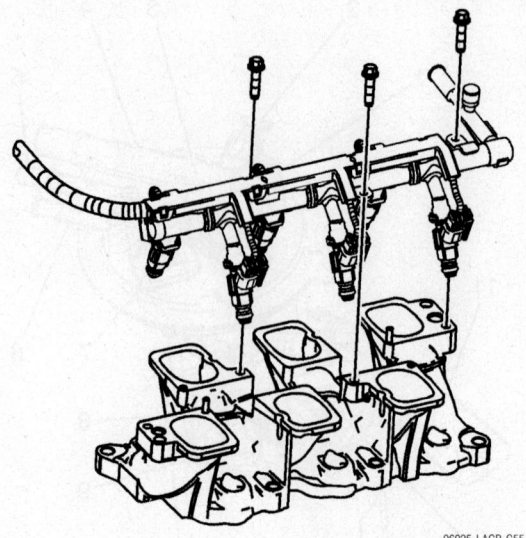

Fuel rail removal—3.6L engine

- New strainer on the pump so it points in the same direction as noted during removal
- Electrical connectors and fuel lines to the pump
- New O-ring on top of the fuel tank
- Fuel sender assembly into the tank
- Lockring
- Fuel line quick connectors
- Sender electrical connector
- Fuel sender access cover
- Trunk liner
- Spare tire and jack

8. Refill with fuel and check for leaks.

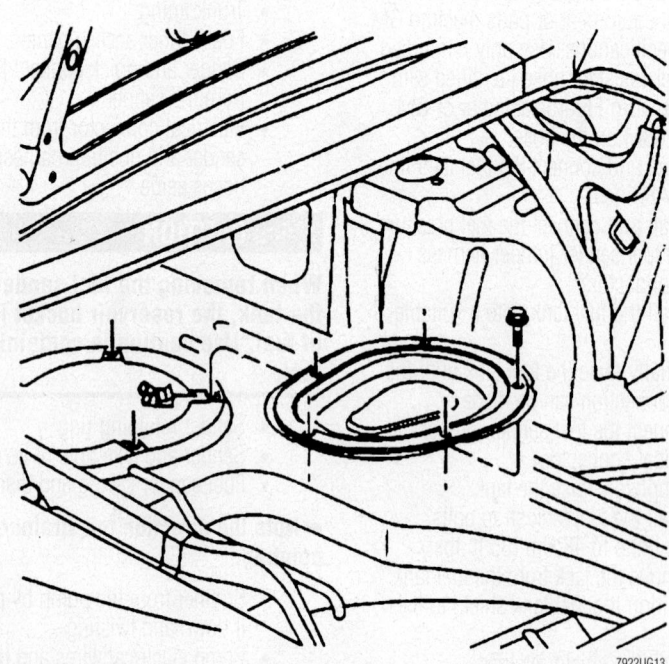

The fuel pump service cover is located in the luggage compartment under the spare tire—Park Avenue shown others similar

Fuel Injectors

REMOVAL & INSTALLATION

3.6L Engine

1. Before servicing the vehicle, refer to the Precautions Section.
2. Relieve the fuel system pressure.
3. Remove the upper intake manifold.
4. Remove the fuel pipe retaining clip.
5. Disconnect the fuel feed pipe from the fuel injector rail.

✳✳ CAUTION

Wear safety glasses.

6. Use compressed air in order to remove debris from the area where the fuel injectors enter the intake manifold.
7. Remove the fuel rail bolts.

✳✳ WARNING

Remove the fuel rail assembly carefully in order to prevent damage to the injector electrical connector terminals and the injector spray tips. Support the fuel rail after the fuel rail is removed in order to avoid damaging the fuel rail components. Cap the fittings and plug the holes when servicing the fuel system in order to prevent dirt and other contaminants from entering open pipes and passages.

8. Remove the fuel rail with the fuel injectors.
9. Disengage the fuel injector electrical connector lock.

10. Disconnect the fuel injector electrical connector.

11. Remove the fuel injector retainer clip.

12. Remove the fuel injector.

13. Remove and discard the fuel injector seals.

To install:

14. Install NEW fuel injector seals.

15. Install the fuel injector.

16. Install the fuel injector retainer clip.

17. Install the fuel injector electrical connector.

18. Engage the fuel injector electrical connector lock.

19. Install the fuel rail with the fuel injectors.

20. Install the fuel rail bolts. Tighten the fuel rail bolts to 10 Nm (89 inch lbs.).

21. Connect the fuel feed pipe to the fuel rail.

22. Install the fuel pipe retaining clip.

23. Install the upper intake manifold.

3.8L Engine

1. Before servicing the vehicle, refer to the Precautions Section.

2. Relieve the fuel system pressure.

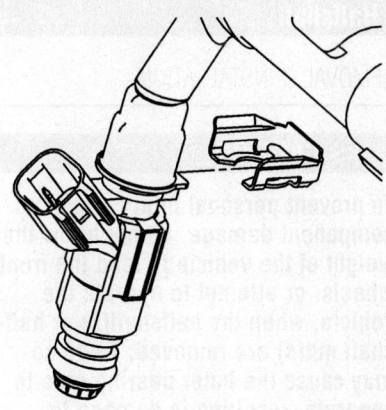

Fuel injector clip—3.6L engine

3. Remove or disconnect the following:
- Negative battery cable
- Fuel lines
- Fuel rail
- Injector retaining clips
- Fuel injector

To install:

4. Install or connect the following:
- Fuel injector with new O-rings, coat the o-rings with clean engine oil prior to installation
- Injector retaining clips

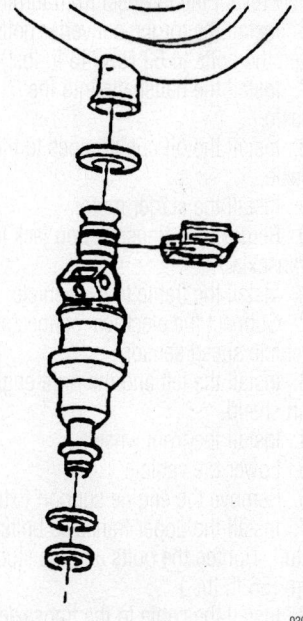

Exploded view of the fuel injector assembly—3.8L engine

- Fuel rail and torque the bolts to 89 inch lbs. (10 Nm).
- Fuel lines
- Negative battery cable

DRIVE TRAIN

Transaxle Assembly

REMOVAL & INSTALLATION

1. Before servicing the vehicle, refer to the Precautions Section.

2. Disconnect the negative battery cable.

3. Remove the air cleaner intake duct.

4. Disconnect the transaxle wiring harness electrical connector.

5. Remove the wiring harness ground nut from the transaxle.

6. Remove the wiring harness grounds from the transaxle.

7. Remove the wiring harness ground bolt from the transaxle.

8. Remove the wiring harness grounds from the transaxle.

9. Remove the wiring harness bracket bolt and reposition the wiring harness.

10. Remove the range selector cable from the transaxle shift lever.

11. Remove the range selector retainer from the cable.

12. Remove the range selector cable from the transaxle.

13. Remove the upper transaxle bolts and the stud.

14. Install the engine support fixture. Refer to Engine Removal & Installation.

15. Raise and support the vehicle.

16. Remove the front wheels.

17. Remove the left and the right engine splash shields.

18. Remove the frame from the vehicle. Refer to Engine Removal & Installation.

19. Remove the starter motor.

20. Install flywheel holding tool J 37096 in order to gain access to the torque converter bolts and to prevent the flywheel from turning.

21. Remove the torque converter bolts.

22. Remove the oil cooler pipes from the transaxle.

➡**Position and secure the halfshafts out of the way.**

23. Remove the left and the right halfshafts from the transaxle.

24. Secure the drive shafts to the steering knuckles.

25. Disconnect the electrical connector from the vehicle speed sensor.

➡**Ensure that the transmission jack is properly secured to the transaxle.**

26. Position a transmission jack under the transaxle and secure the jack firmly to the transaxle.

27. Remove the transaxle brace.

28. Remove the lower transaxle bolt and the stud.

29. Remove the transaxle from the vehicle.

30. Transfer all necessary parts as needed.

To install:

➡**Ensure that the transaxle is secured properly to the transmission jack.**

31. Position the transaxle onto a transmission jack and secure the transaxle to the jack.

32. Install the transaxle into the vehicle.

33. Install the lower transaxle bolt and the stud. Tighten the bolt and the stud to 75 Nm (55 ft. lbs.).

34. Install the transaxle brace. Tighten the transaxle brace bolts to the transaxle to 43 Nm (32 ft. lbs.). Tighten the transaxle brace bolts to the engine to 63 Nm (46 ft. lbs.).

35. Prevent the flywheel from turning.
36. Install the torque converter bolts. Tighten the bolts to 63 Nm (46 ft. lbs.).
37. Install the halfshafts into the transaxle.
38. Install the oil cooler pipes to the transaxle.
39. Install the starter motor.
40. Remove the transmission jack from the transaxle.
41. Install the frame to the vehicle.
42. Connect the electrical connector to the vehicle speed sensor.
43. Install the left and the right engine splash shield.
44. Install the front wheels.
45. Lower the vehicle.
46. Remove the engine support fixture.
47. Install the upper transaxle bolts and the stud. Tighten the bolts and the stud to 75 Nm (55 ft. lbs.).
48. Install the cable to the transaxle range selector cable bracket.
49. Install the A/T range selector cable retainer to the cable.
50. Install the transaxle range selector cable to the transaxle shift lever.
51. Reposition the wiring harness and install the wiring harness bracket and the bolt. Tighten the bolt to 25 Nm (18 ft. lbs.).
52. Install the wiring harness grounds to the transaxle.
53. Install the transaxle wiring harness ground bolt to the transaxle. Tighten the bolt to 25 Nm (18 ft. lbs.).
54. Install the wiring harness grounds to the transaxle.
55. Install the wiring harness ground nut to the transaxle. Tighten the nut to 45 Nm (33 ft. lbs.).
56. Connect the transaxle wiring harness electrical connector.
57. Install the air cleaner intake duct.
58. Connect the negative battery cable.
59. Fill the transaxle with transmission fluid.
60. Inspect the transaxle for fluid leaks.

➡It is recommended that transmission adaptive pressure (TAP) information be reset. Resetting the TAP values using a scan tool will erase all learned values in all cells. As a result, The ECM, PCM or TCM will need to relearn TAP values. Transmission performance may be affected as new TAP values are learned.

Halfshaft

REMOVAL & INSTALLATION

✻✻ CAUTION

To prevent personal injury and/or component damage, do not allow the weight of the vehicle to load the front wheels, or attempt to operate the vehicle, when the halfshaft(s) or halfshaft nut(s) are removed. To do so may cause the inner bearing race to separate, resulting in damage to brake and suspension components and loss of vehicle control.

✻✻ WARNING

Wheel drive shaft boots, seals and clamps should be protected from sharp objects any time service is performed on or near the halfshaft(s). Damage to the boot(s), the seal(s) or the clamp(s) may cause lubricant to leak from the joint and lead to increased noise and possible failure of the halfshaft.

1. Before servicing the vehicle, refer to the Precautions Section.
2. Raise and suitably support the vehicle.
3. Remove the wheel and the tire.
4. Remove the stabilizer shaft link.
5. Remove the front halfshaft nut. Insert a drift or a flat-bladed tool into the caliper and the rotor to prevent the rotor from turning.
6. Disconnect the outer tie rod assembly from the steering knuckle.
7. Separate the ball joint from the steering knuckle.
8. Separate the front halfshaft from the front halfshaft bearing using a slidehammer and adapter.

To install

9. Install the front wheel drive axle into the transaxle.
10. Verify that the front halfshaft retaining ring is properly seated:
    - Grasp the inner housing and pull the inner housing outward. Do not pull on the front wheel drive axle shaft.
    - The front wheel drive axle will remain in place when the front halfshaft retaining ring is properly seated.
11. Install the front wheel drive axle into the front halfshaft bearing.
12. Connect the ball joint to the steering knuckle.
13. Connect the outer tie rod assembly to the steering knuckle.
14. Install a new front halfshaft nut. Insert a drift or a flat-bladed tool into the caliper and the rotor to prevent the rotor from turning. Tighten the nut to 160 Nm (118 ft. lbs.).
15. Install the stabilizer shaft link.
16. Install the wheel and the tire.
17. Lower the vehicle.
18. Inspect the transaxle fluid level.
19. Inspect the wheel alignment.

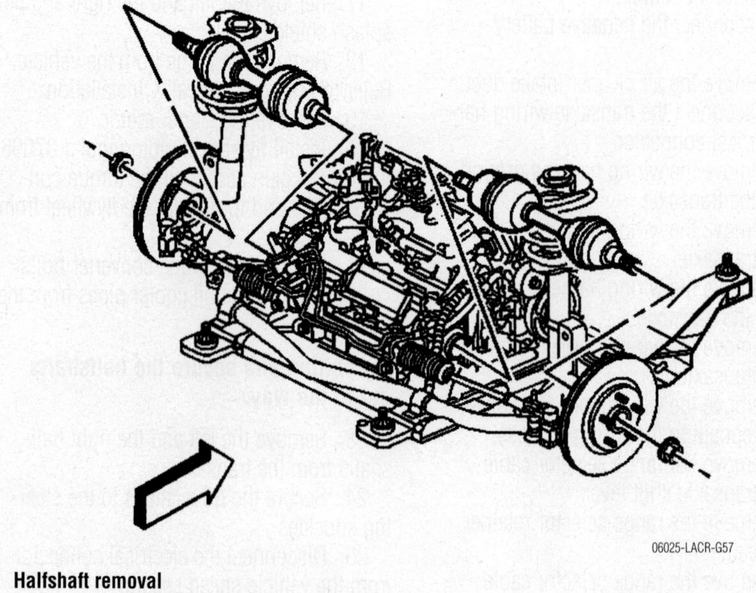
06025-LACR-G57

Halfshaft removal

CV-Joints

OVERHAUL

Inner (Tripod) Joint

1. Before servicing the vehicle, refer to the Precautions Section.
2. Raise and safely support the vehicle.
3. Remove or disconnect the following:
- Front wheel
- Halfshaft and place it in a vise
- Small CV-joint boot clamp, cut and discard it
- Large CV-joint boot clamp, cut and discard it
- CV-joint boot by sliding it away from the tripod joint
- Tripod housing from the tripod spider
- Inboard spacer ring and slide it rearward on the shaft
- Outboard retaining ring
- Tripod joint spider assembly
- Inboard spacer ring and discard it
- Tripod joint spider assembly by tapping it from the halfshaft with a brass drift
- Tripod spider retaining ring and discard it
- Trilobal tripod bushing from the housing
- CV-joint boot
4. Thoroughly clean and inspect all parts.

To install:
5. Install or connect the following:
- Small boot clamp
- CV-joint boot
- New inboard spacer ring. Slide it rearward on the shaft past the 2nd groove
- Tripod joint spider assembly onto the shaft until it passes the 2nd groove
6. Assemble the tripod spider assembly onto the halfshaft as follows:
   a. Position the tripod spider assembly onto the shop press plate.
   b. Position the halfshaft onto the tripod spider assembly, in the shop press.
   c. Press the halfshaft into the tripod spider assembly until the spider assembly passes the 2nd groove.

❊❊ WARNING

When assembling the tripod assembly onto the halfshaft, do not exceed 4,000 lbs. pressure.

7. Remove the halfshaft from the shop press and place it in vise.

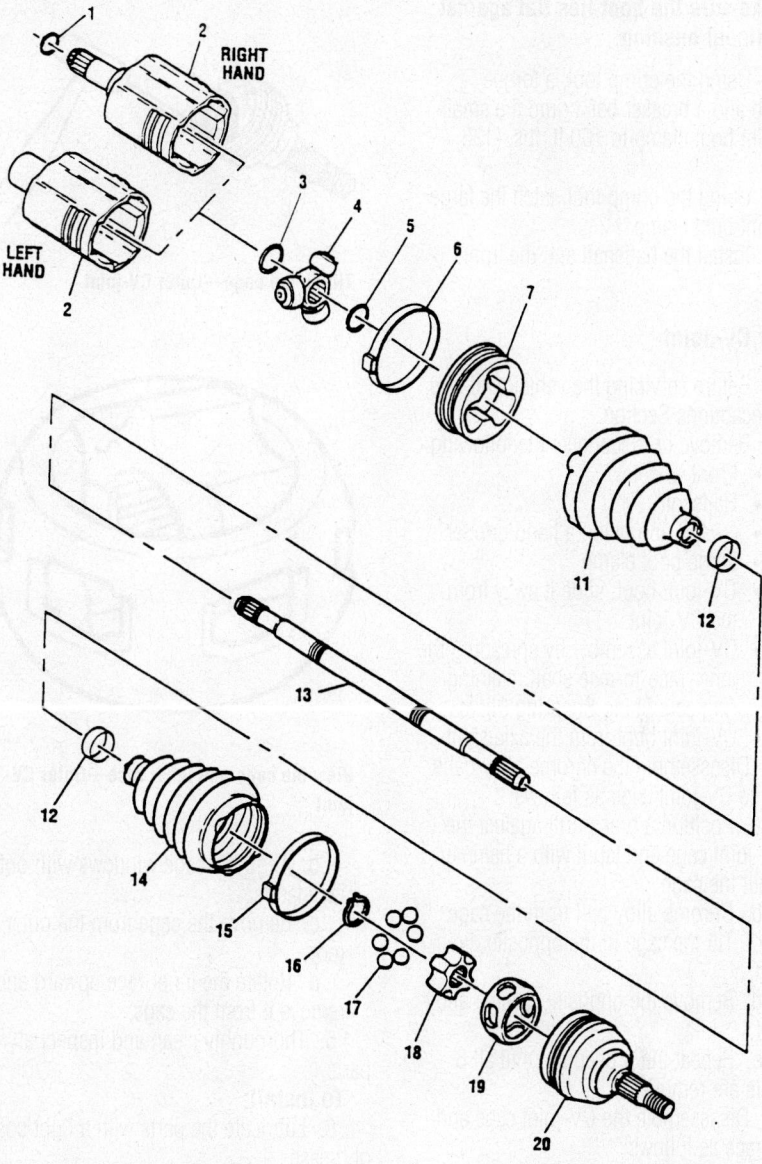

1 - RING, RETAINING
2 - HOUSING ASM, RETAINER &
3 - RING, SHAFT RETAINING
4 - SPIDER, TRIPOT JOINT
5 - RING, SPACER
6 - CLAMP, SEAL RETAINING
7 - BUSHING, TRILOBAL TRIPOT
11 - SEAL, DRIVE AXLE INBOARD
12 - RING, SWAGE

13 - SHAFT, AXLE (RH SHOWN, LH SIMILAR)
14 - SEAL, DRIVE AXLE OUTBOARD
15 - CLAMP, SEAL RETAINING
16 - RING, RACE RETAINING
17 - BALL, CHROME ALLOY
18 - RACE, C/V JOINT INNER
19 - CAGE, C/V JOINT
20 - RACE, C/V JOINT OUTER

9306UG06

Exploded view of the halfshaft assembly

8. Install or connect the following:
- New outboard retaining ring into the axle shaft groove
- Tripod joint spider assembly, slide it against the outboard retaining ring using a brass drift
- Inboard spacer ring, seat it in the groove
9. Use ½ of the grease supplied in the kit into the boot and the other ½ into the tripod housing.

- Trilobal tripod bushing flush with the tripod housing face
- New large seal clamp onto the CV-joint boot
- Tripod housing, slide it over the tripod joint spider assembly
- CV-joint boot/clamp, slide it into place, over the trilobal tripod bushing with the seal lip in the groove

➡**Make sure the boot lies flat against the trilobal bushing.**

10. Using the crimp tool, a torque wrench and a breaker bar, crimp the small CV-joint boot clamp to 100 ft. lbs. (136 Nm).

11. Using the crimp tool, latch the large CV-joint boot clamp.

12. Install the halfshaft and the front wheel.

Outer CV-Joint

1. Before servicing the vehicle, refer to the Precautions Section.

2. Remove or disconnect the following:
- Front wheel
- Halfshaft
- Swage ring using a hand grinder
- Large boot clamp
- CV-joint boot, slide it away from the CV-joint
- CV-joint assembly by spreading the inner race-to-axle shaft retaining ring ears using Snapring Pliers
- CV-joint boot from the axle shaft

3. Disassemble the chrome alloy balls from the CV-joint cage as follows:
   a. Position a brass drift against the CV-joint cage and tap it with a hammer to tilt the cage.
   b. Chrome alloy ball from the cage.
   c. Tilt the cage in the opposite direction.
   d. Remove the opposite chrome alloy ball.
   e. Repeat the procedure until all 6 balls are removed.

4. Disassemble the CV-joint cage and inner race as follows:
   a. Pivot the cage and race 90 degrees to the center line of the outer race.

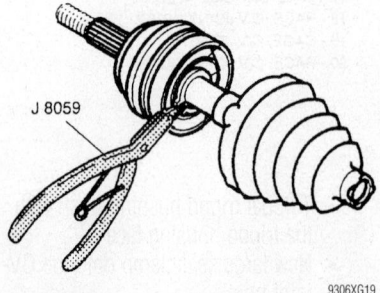

Disconnecting the outer CV-joint from the axle shaft

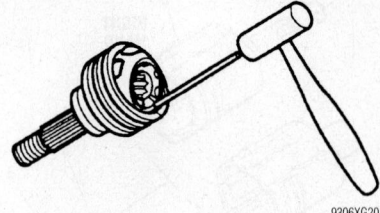

Tilting the cage—Outer CV-joint

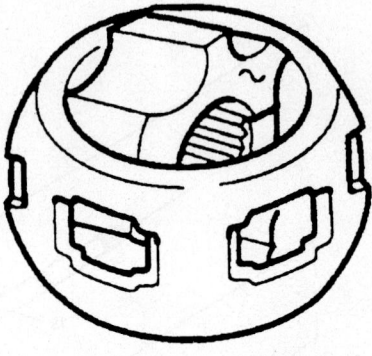

View the cage and inner race—Outer CV-joint

   b. Align the cage windows with outer race lands.
   c. Remove the cage from the outer race.
   d. Rotate the inner race upward and remove it from the cage.

5. Thoroughly clean and inspect all parts.

To install:

6. Lubricate the parts with a light coat of grease.

7. Assemble the CV-joint cage and inner race, as follows:

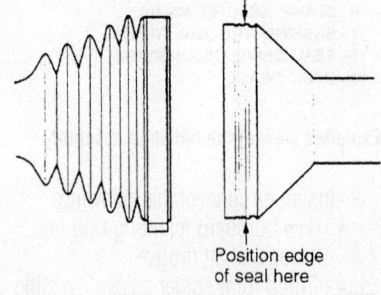

Position edge of seal here

Positioning the boot—Outer CV-joint

   a. Rotate the inner race 90 degrees to the cage centerline.
   b. Align the cage windows with inner race lands.
   c. Insert the inner race into the cage by rotating the inner race downward.
   d. Insert the cage/inner race into the outer race.

8. Assemble the chrome alloy balls into the CV-joint cage, as follows:
   a. Position a brass drift against the CV-joint cage and tap it with a hammer to tilt the cage.
   b. Insert the 1st chrome alloy ball into the cage.
   c. Tilt the cage in the opposite direction.
   d. Insert the opposite chrome alloy ball.
   e. Repeat the procedure until all 6 balls are inserted.

9. Install or connect the following:
- Swage ring clamp
- CV-joint boot
- CV-joint onto the axle shaft until the retaining ring seats into the groove

10. Position the CV-joint boot seal into the axle shaft's joint seal groove and align the swage ring clamp on the boot.

11. Secure the swage ring clamp using appropriate crimping tool.

✳✳ WARNING

Make sure that there are no pinch points on the inboard seal.

12. Install or connect the following:
- ½ kit grease into the CV-joint boot
- ½ kit grease into the CV-joint
- New large seal clamp onto the CV-joint boot
- CV-joint boot/clamp, slide it into place, over the outer race with the seal lip in the groove

➡**Make sure the boot lies flat against the outer race.**

13. Using a Crimp tool, a torque wrench and a breaker bar, crimp the large CV-joint boot clamp to 130 ft. lbs. (176 Nm).

14. Install the halfshaft and the front wheel.

STEERING

Air Bag

✳✳ CAUTION

Some vehicles are equipped with an air bag system. The system must be disabled before performing service on or around system components, the steering column, instrument panel components, wiring and sensors. Failure to follow safety precautions and the disarming procedures could result in accidental air bag deployment, possible personal injury and unnecessary system repairs.

PRECAUTIONS

Several precautions must be observed when handling the inflator module to avoid accidental deployment and possible personal injury.

• Never carry the inflator module by the wires or connector on the underside of the module.

• When carrying a live inflator module, hold it securely with both hands, and ensure that the bag and trim cover are pointed away.

• Place the inflator module on a bench or other surface with the bag and trim cover facing up.

• With the inflator module on the bench, never place anything on or close to the module which may be thrown in the event of an accidental deployment.

DISABLING AND ENABLING

Zone 1

1. Before servicing the vehicle, refer to the Precautions Section.

2. Turn the steering wheel so that the vehicles wheels are pointing straight ahead.

3. Turn the ignition switch to the OFF position.

4. Remove the key from the ignition switch.

5. Open the hood and locate the underhood fuse center on right/passenger shock tower.

➡**With the SIR Fuse removed and the ignition ON, the AIR BAG indicator illuminates. This is normal operation, and does not indicate an SIR system malfunction.**

6. Lift the cover for the underhood fuse center.

7. Locate and remove the SIR Fuse from the underhood fuse center.

8. Remove the radiator upper air baffle and deflector and locate the left and/or right front end sensor also known as electronic frontal sensor (EFS) that needs servicing.

9. Remove the connector position assurance (CPA) from both front end sensors connector.

10. Remove both front end sensor connectors from each front end sensor.

Enabling Procedure:

11. Remove the key from the ignition switch.

12. Connect both front end sensor connectors to each front end sensor.

13. Install both CPA's into each front end sensor connector.

14. Install the radiator upper air baffle and deflector.

15. Install the SIR Fuse.

16. Close the underhood fuse center cover.

17. Use caution while reaching in and turn the ignition switch to the ON position. The AIR BAG indicator will flash then turn OFF.

Zone 2

1. Before servicing the vehicle, refer to the Precautions Section.

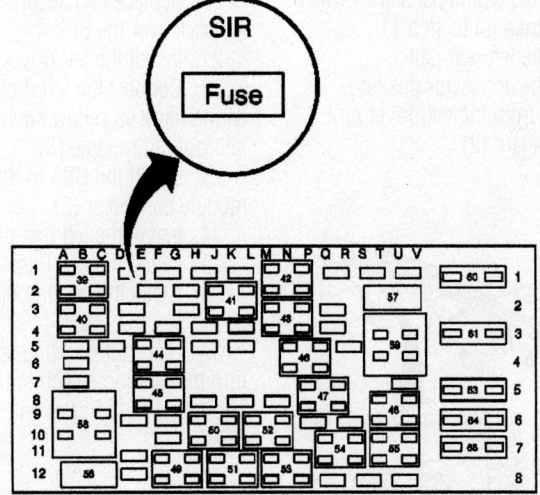

SIR fuse

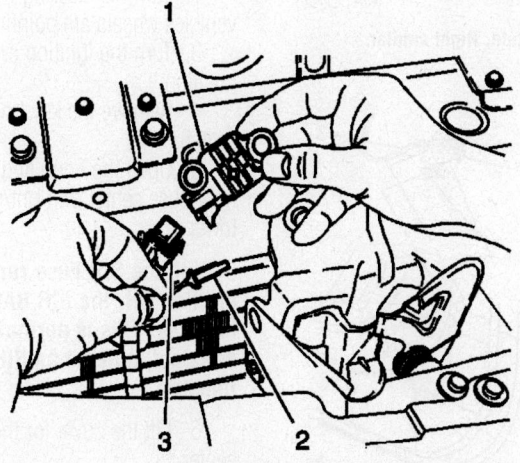

1. Front end sensor
2. Connector position assurance
3. Front end sensor connector

06025-LACR-G59

Front end sensor

2. Turn the steering wheel so that the vehicles wheels are pointing straight ahead.

3. Turn the ignition switch to the OFF position.

4. Remove the key from the ignition switch.

5. Open the hood and locate the underhood fuse center on right/passenger shock tower.

➡ **With the SIR Fuse removed and the ignition ON, the AIR BAG indicator illuminates. This is normal operation, and does not indicate an SIR system malfunction.**

6. Lift the cover for the underhood fuse center.

7. Locate and remove the SIR Fuse from the underhood fuse center.

8. When disabling the roof rail module go to step 8, if the side impact sensor (SIS) needs disabling then go to step 11.

9. Remove the left rear panel.

10. Remove the connector position assurance (CPA) from the left/driver roof rail module connector (2).

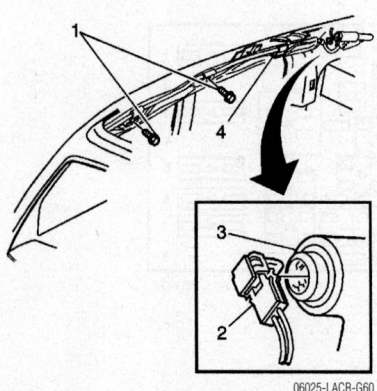

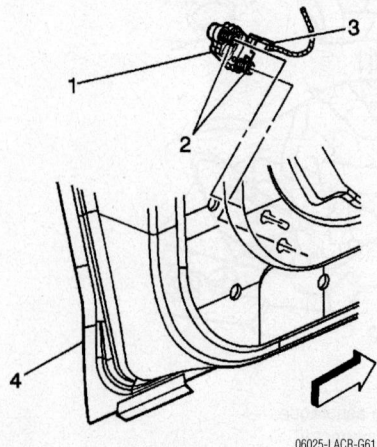

06025-LACR-G60

Left roof rail module. Right similar.

06025-LACR-G61

Left side impact lensor. Right similar.

11. Disconnect the left roof rail module wiring harness yellow connector (2) from the left roof rail module (3).

12. Remove the left/driver door trim panel.

13. Remove enough of the water deflector to access the SIS.

14. Remove the SIS CPA from the left SIS connector (3).

15. Remove the SIS connector (3) from the SIS (1).

Enabling Procedure

16. Remove the key from the ignition switch.

17. When enabling the SIS proceed to step 3, if the roof rail module needs enabling then go to step 7.

18. Install the left SIS connector (3) to the SIS (1).

19. Install the SIS CPA to the SIS connector (3).

20. Replace and secure the water deflector back over the SIS.

21. Install the left/driver door trim panel.

22. Connect the left roof rail module wiring harness yellow connector (2) to the left roof rail module (3).

23. Install the CPA to the left roof rail module connector (2).

24. Install the left rear panel.

25. Install the SIR Fuse.

26. Close the underhood fuse center cover.

27. Use caution while reaching in and turn the ignition switch to the ON position. The AIR BAG indicator will flash then turn OFF.

Zone 3

1. Before servicing the vehicle, refer to the Precautions Section.

2. Turn the steering wheel so that the vehicles wheels are pointing straight ahead.

3. Turn the ignition switch to the OFF position.

4. Remove the key from the ignition switch.

5. Open the hood and locate the underhood fuse center on right/passenger shock tower.

➡ **With the SIR Fuse removed and the ignition ON, the AIR BAG indicator illuminates. This is normal operation, and does not indicate an SIR system malfunction.**

6. Lift the cover for the underhood fuse center.

7. Locate and remove the SIR fuse from the underhood fuse center.

8. Remove the left/driver sound insulator from the instrument panel (I/P) (2).

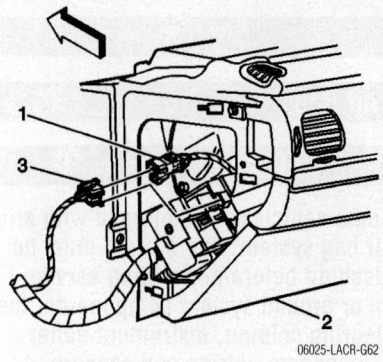

06025-LACR-G62

Steering wheel module coil yellow connector

9. Remove the connector position assurance (CPA) from the steering wheel module coil yellow connector (1).

10. Disconnect the steering wheel module coil yellow connector (1) from the vehicle harness yellow connector (3).

Enabling Procedure

11. Remove the key from the ignition switch.

12. Connect the steering wheel module coil yellow connector (3) to the vehicle harness yellow connector (1).

13. Install the CPA to the steering wheel module coil yellow connector (1).

14. Install the left sound insulator to the I/P (2).

15. Install the SIR Fuse.

16. Close the underhood fuse center cover.

17. Use caution while reaching in and turn the ignition switch to the ON position. The AIR BAG indicator will flash then turn OFF.

Zone 5

1. Before servicing the vehicle, refer to the Precautions Section.

2. Turn the steering wheel so that the vehicle wheels are pointing straight ahead.

3. Turn the ignition switch to the OFF position.

4. Remove the key from the ignition switch.

5. Open the hood and locate the underhood fuse center on right/passenger shock tower.

➡ **With the SIR Fuse removed and the ignition ON, the AIR BAG indicator illuminates. This is normal operation, and does not indicate an SIR system malfunction.**

6. Lift the cover for the underhood fuse center.

7. Locate and remove the SIR Fuse from the underhood fuse center.

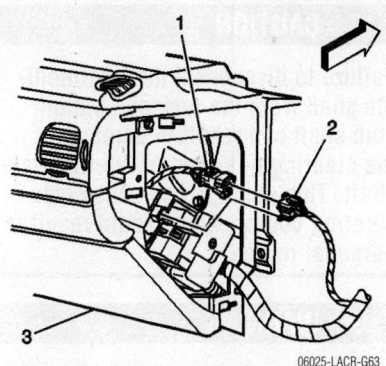

I/P module yellow connector

06025-LACR-G63

8. Remove the right/passenger sound insulator from the instrument panel (I/P) (3).

9. Remove the connector position assurance (CPA) from the I/P module yellow connector (1).

10. Disconnect the I/P module yellow connector (1) from the vehicle harness yellow connector (2).

Enabling Procedure

11. Remove the key from the ignition switch.

12. Connect the I/P module yellow connector (1) to the vehicle harness yellow connector (2).

13. Install the CPA to the I/P module yellow connector (1).

14. Install the right sound insulator to the I/P (3).

15. Install the SIR Fuse.

16. Close the underhood fuse center cover.

17. Use caution while reaching in and turn the ignition switch to the ON position. The AIR BAG indicator will flash then turn OFF.

Zone 6

1. Before servicing the vehicle, refer to the Precautions Section.

2. Turn the steering wheel so that the vehicles wheels are pointing straight ahead.

3. Turn the ignition switch to the OFF position.

4. Remove the key from the ignition switch.

5. Open the hood and locate the underhood fuse center on right/passenger shock tower.

➡**With the SIR Fuse removed and the ignition ON, the AIR BAG indicator illuminates. This is normal operation, and does not indicate an SIR system malfunction.**

6. Lift the cover for the underhood fuse center.

7. Locate and remove the SIR Fuse from the underhood fuse center.

8. When disabling the roof rail module go to step 8, if the side impact sensor (SIS) needs disabling then go to step 11.

9. Remove the right rear panel.

10. Remove the connector position assurance (CPA) from the right/passenger roof rail module connector.

11. Disconnect the right roof rail module wiring harness yellow connector from the right roof rail module.

12. Remove the right/passenger door trim panel.

13. Remove enough of the water deflector to access the SIS.

14. Remove the SIS CPA from the right SIS connector.

15. Remove the SIS connector from the SIS.

Enabling Procedure

16. Remove the key from the ignition switch.

17. When enabling the SIS proceed to step 3, if the roof rail module needs enabling then go to step 7.

18. Install the right SIS connector to the SIS.

19. Install the SIS CPA to the SIS connector.

20. Replace and secure the water deflector back over the SIS.

21. Install the right/passenger door trim panel.

22. Connect the right roof rail module wiring harness yellow connector to the right roof rail module.

23. Install the CPA to the right roof rail module connector.

24. Install the right rear panel.

25. Install the SIR Fuse.

26. Close the underhood fuse center cover.

27. Use caution while reaching in and turn the ignition switch to the ON position. The AIR BAG indicator will flash then turn OFF.

Zone 7

1. Before servicing the vehicle, refer to the Precautions Section.

2. Turn the steering wheel so that the vehicles wheels are pointing straight ahead.

3. Turn the ignition switch to the OFF position.

4. Remove the key from the ignition switch.

5. Open the hood and locate the underhood fuse center on right/passenger shock tower.

➡**With the SIR Fuse removed and the ignition ON, the AIR BAG indicator illuminates. This is normal operation, and does not indicate an SIR system malfunction.**

6. Lift the cover for the underhood fuse center.

7. Locate and remove the SIR Fuse from the underhood fuse center.

8. Remove the connector position assurance (CPA) from the left/driver seat belt pretensioner connector (1) located under the driver seat.

9. Disconnect the left seat belt pretensioner connector from vehicle wiring harness connector (1).

Enabling Procedure

10. Remove the key from the ignition switch.

11. Connect the left seat belt pretensioner connector to the vehicle wiring harness connector (1).

12. Install the CPA to the seat belt pretensioner connector (1).

13. Install the SIR Fuse.

14. Close the underhood fuse center cover.

15. Use caution while reaching in and turn the ignition switch to the ON position. The AIR BAG indicator will flash then turn OFF.

Zone 9

1. Before servicing the vehicle, refer to the Precautions Section.

2. Turn the steering wheel so that the vehicle wheels are pointing straight ahead.

3. Turn the ignition switch to the OFF position.

4. Remove the key from the ignition switch.

5. Open the hood and locate the underhood fuse center on right/passenger shock tower.

➡**With the SIR Fuse removed and the ignition ON, the AIR BAG indicator illuminates. This is normal operation, and does not indicate an SIR system malfunction.**

6. Lift the cover for the underhood fuse center.

7. Locate and remove the SIR Fuse from the underhood fuse center.

8. When disabling the right/passenger seat belt pretensioner perform step 8, if the sensing and diagnostic module (SDM) needs disabling then use entire procedure.

9. Remove the connector position assurance (CPA) from the right/passenger seat belt pretensioner connector located under the passenger seat.

10. Disconnect the seat belt preten-sioner—right connector from vehicle wiring harness connector.

11. Remove the right rear panel.

12. Remove the CPA from the right/pas-senger roof rail module connector.

13. Disconnect the right roof rail module wiring harness yellow connector from the right roof rail module.

14. Remove the right/passenger sound insulator from the instrument panel (I/P).

15. Remove the CPA from the I/P mod-ule yellow connector.

16. Disconnect the I/P module yellow connector from the vehicle harness yellow connector.

17. Remove the left/driver sound insula-tor from the I/P.

18. Remove the CPA from the steering wheel module coil yellow connector.

19. Disconnect the steering wheel mod-ule coil yellow connector from the vehicle harness yellow connector.

20. Remove the CPA from the left/driver seat belt pretensioner connector located under the driver seat.

21. Disconnect the seat belt preten-sioner-left connector from vehicle wiring harness connector.

22. Remove the left rear panel.

23. Remove the CPA from the left/driver roof rail module connector.

24. Disconnect the left roof rail module wiring harness yellow connector from the left roof rail module.

Enabling Procedure

25. Remove the key from the ignition switch.

26. When enabling the right/passenger seat belt pretensioner proceed to step 3, if the SDM needs enabling then use entire procedure.

27. Connect the seat belt pretensioner-right connector to the vehicle wiring har-ness connector.

28. Install the CPA to the seat belt pre-tensioner connector.

29. Connect the right roof rail module wiring harness yellow connector to the right roof rail module.

30. Install the CPA to the right roof rail module connector.

31. Install the right rear panel.

32. Connect the I/P module yellow con-nector to the vehicle harness yellow con-nector.

33. Install the CPA to the I/P module yellow connector.

34. Install the right sound insulator to the I/P.

35. Connect the steering wheel module coil yellow connector to the vehicle harness yellow connector.

36. Install the CPA to the steering wheel module coil yellow connector.

37. Install the left sound insulator to the I/P.

38. Connect the seat belt pretensioner-left connector to the vehicle wiring harness connector.

39. Install the CPA to the seat belt pre-tensioner connector.

40. Connect the left roof rail module wiring harness yellow connector to the left roof rail module.

41. Install the CPA to the left roof rail module connector.

42. Install the left rear panel.

43. Install the SIR Fuse.

44. Close the underhood fuse center cover.

45. Use caution while reaching in and turn the ignition switch to the ON position. The AIR BAG indicator will flash then turn OFF.

Power Rack and Pinion Steering Gear

REMOVAL & INSTALLATION

1. Before servicing the vehicle, refer to the Precautions Section.

2. Raise and support the vehicle.

3. Place a drain pan under the vehicle.

4. Remove the tire and wheel assem-blies.

✳✳ CAUTION

Failure to disconnect the intermedi-ate shaft from the rack and pinion stub shaft can result in damage to the steering gear and/or intermediate shaft. This damage can cause loss of steering control which could result in personal injury.

✳✳ WARNING

Set steering shaft so the block tooth on the upper steering shaft is at the 12 o'clock position, the wheels on the vehicle are straight ahead and set the ignition switch to the LOCK position. Failure to follow these pro-cedures could result in damage to the coil.

✳✳ WARNING

The wheels of the vehicle must be straight ahead and the steering col-umn in the LOCK position before dis-connecting the steering column or intermediate shaft from the steering gear. Failure to do so will cause the SIR coil assembly to become un-cen-tered, which may cause damage to the coil assembly.

5. Remove the lower pinch bolt from the power steering gear stub shaft.

6. Insert tool J 42640 into the steering column access hole in order to lock the steering column. This will maintain the cor-rect orientation.

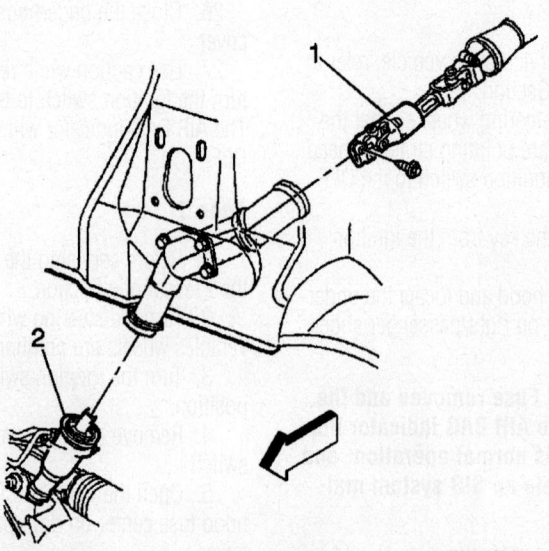

06025-LACR-G64

Disconnect the intermediate steering shaft (1) from the power steering gear stub shaft (2)

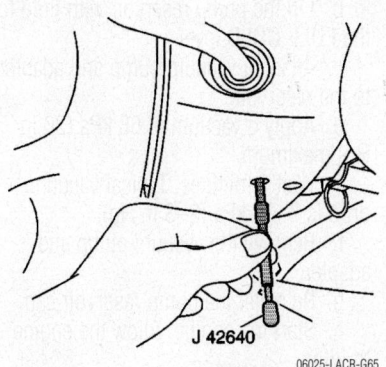

06025-LACR-G65

Insert tool J 42640 into the steering column access hole in order to lock the steering column

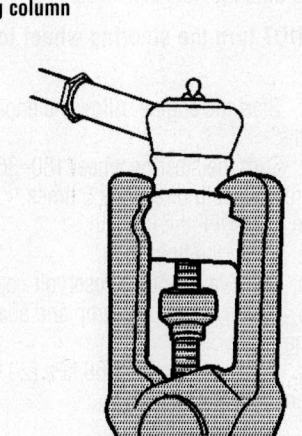

06025-LACR-G66

Separating the tie rod end from the knuckle

7. Remove the intermediate steering shaft from the power steering gear stub shaft.

8. Remove both of the tie rod ends from the steering knuckles.

9. Support the rear of the frame using jackstands.

10. Remove the frame bolts from the rear of the frame. Refer to Engine Removal & Installation.

❊❊ WARNING

Do not lower the rear of the frame too far as damage to the engine components nearest to the cowl may result.

11. Lower the rear of the frame.

12. Remove the power steering pressure hose from the power steering gear.

13. Remove the power steering return hose from the power steering gear.

14. Remove the Magnasteer Variable Assist electrical connector from the power steering gear assembly, if equipped with variable effort steering.

15. Remove the power steering gear mounting bolts and nuts.

16. Remove the power steering gear through the left wheel opening.

To install:

17. Install the power steering gear through the left wheel opening.

18. Install the power steering gear mounting bolts and nuts. Tighten the power steering gear mounting bolts to 90 Nm (66 ft. lbs.).

19. Inspect the threads on the power steering pressure hose and the power steering return hose.

20. Inspect the O-ring seals on the power steering hoses.

21. Replace the seals if damaged, lubricate the seals before installation.

22. Install the clamp that holds the power steering hoses to the power steering gear.

23. Install the Magnasteer Variable Assist electrical connector to the power steering gear assembly, if equipped with variable effort steering.

24. Install the power steering pressure

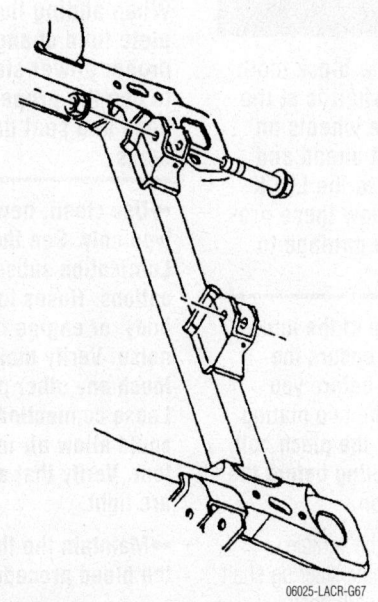

06025-LACR-G67

Remove the frame bolts from the rear of the frame

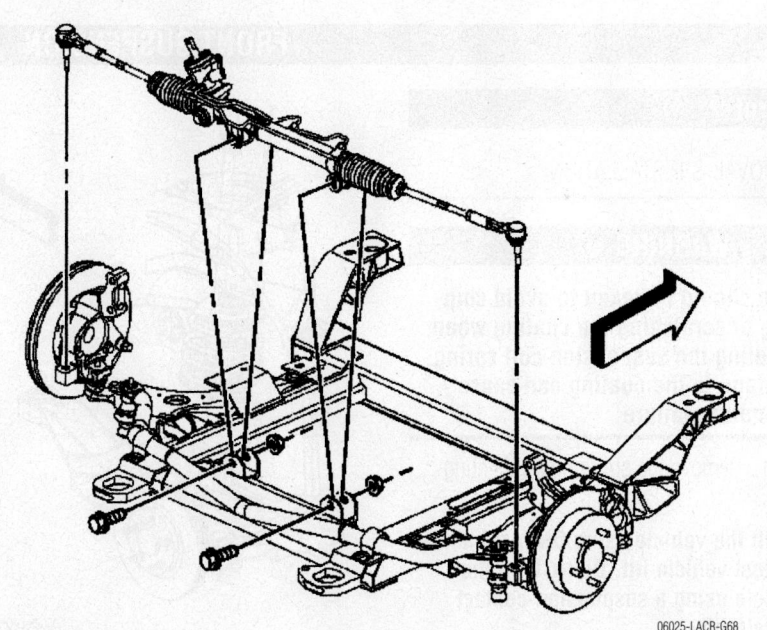

06025-LACR-G68

Steering gear mounting

hose to the power steering gear. Tighten the power steering pressure hose to power steering gear. Tighten the pressure hose fittings to 27 Nm (20 ft. lbs.).

25. Install the power steering return hose to the power steering gear. Tighten the fitting to 27 Nm (20 ft. lbs.).

26. Raise the frame into position.

27. Install rear frame bolts. Tighten the rear bolts to 160 Nm (118 ft. lbs.).

28. Remove the jackstands.

29. Install the tie rod ends to the steering knuckles. Tighten the nut to 30 Nm, plus 120 degrees (34 ft. lbs. plus 120 degrees). Inspect to ensure that 2½–4½ threads are visible above the nylon washer.

✳✳ WARNING

Set steering shaft so the block tooth on the upper steering shaft is at the 12 o'clock position, the wheels on the vehicle are straight ahead and set the ignition switch to the LOCK position. Failure to follow these procedures could result in damage to the coil.

➡ **During the installation of the intermediate steering shaft, ensure the steering shaft is seated before you install the pinch bolt. The two mating shafts may disengage if the pinch bolt is inserted into the coupling before the steering shaft installation.**

30. Raise and support the vehicle.

31. Install the intermediate steering shaft to the power steering gear stub shaft.

32. Install the lower pinch bolt to the intermediate steering shaft at the power steering gear stub shaft. Tighten the pinch bolt to 48 Nm (35 ft. lbs.).

33. Install the intermediate steering shaft seal onto the power steering gear.

34. Install the tire and wheel assemblies.

35. Remove the drain pan from under the vehicle.

36. Lower the vehicle.

37. Remove tool J 42640 from the steering column.

38. Fill the power steering system with power steering fluid.

39. Bleed the power steering system:

✳✳ WARNING

When adding fluid or making a complete fluid change, always use the proper power steering fluid. Failure to use the proper fluid will cause hose and seal damage and fluid leaks.

➡ **Use clean, new power steering fluid type only. See the Maintenance and Lubrication subsection for fluid specifications. Hoses touching the frame, body, or engine may cause system noise. Verify that the hoses do not touch any other part of the vehicle. Loose connections may not leak, but could allow air into the steering system. Verify that all hose connections are tight.**

➡ **Maintain the fluid level throughout the bleed procedure**

a. Remove the pump reservoir cap.

b. Fill the pump reservoir with fluid to the FULL COLD level.

c. Attach a vacuum pump and adapter to the reservoir.

d. Apply a vacuum of 68 kPa (20 in Hg) maximum.

e. Wait 5 minutes. Typical vacuum drop is 7–10 kPa (2–3 in Hg).

f. Remove the vacuum pump and adapter.

g. Reinstall the pump reservoir cap.

h. Start the engine. Allow the engine to idle.

i. Turn OFF the engine.

j. Verify the fluid level. Repeat steps 8-10 until the fluid stabilizes.

➡ **Do NOT turn the steering wheel to LOCK.**

k. Start the engine. Allow the engine to idle.

l. Turn the steering wheel 180–360 degrees in both directions 5 times.

m. Turn OFF the ignition.

n. Verify the fluid level.

o. Remove the pump reservoir cap.

p. Attach a vacuum pump and adapter to the reservoir.

q. Apply a vacuum of 68 kPa (20 in. Hg) maximum.

r. Wait 5 minutes.

s. Remove the vacuum pump and adapter.

t. Verify the fluid level.

u. Reinstall the pump reservoir cap.

40. Inspect the power steering system for leaks.

41. Perform a front end alignment.

FRONT SUSPENSION

Strut

REMOVAL & INSTALLATION

✳✳ WARNING

Care should be taken to avoid chipping or scratching the coating when handling the suspension coil spring. Damage to the coating can cause premature failure.

1. Remove the strut upper mounting nuts.

➡ **Lift the vehicle using ONLY a frame-contact vehicle lift. Do NOT lift the vehicle using a suspension-contact vehicle lift.**

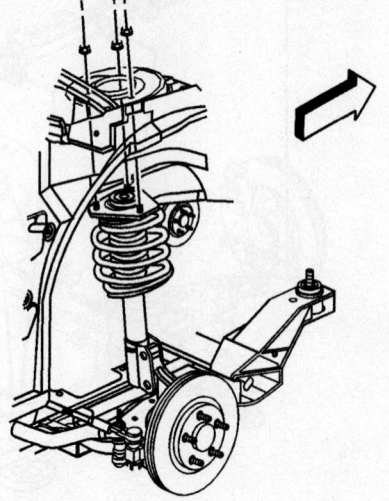

Strut upper mounting nuts

06025-LACR-G69

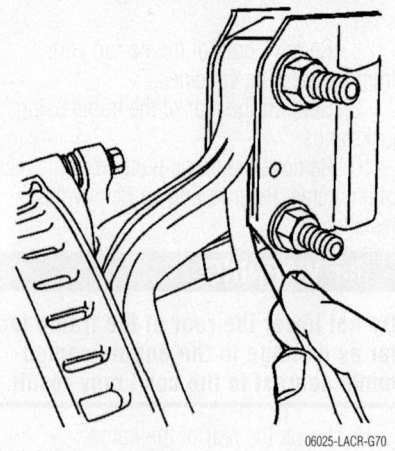

Matchmark the strut to the knuckle

06025-LACR-G70

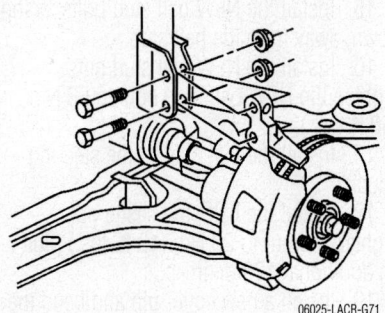

06025-LACR-G71

Strut lower bolts/nuts

2. Before servicing the vehicle, refer to the Precautions Section.

3. Raise and support the vehicle.

4. Remove the tire and wheel.

5. Matchmark the strut to the knuckle.

6. Remove the strut lower bolts and nuts.

7. Remove the strut.

To install:

8. Install the strut.

9. Install the strut upper mounting nuts. Tighten the nuts to 33 Nm (24 ft. lbs.).

10. Install the strut lower bolts and nuts.

※※ WARNING

This is a prevailing torque type fastener. This fastener may be reused ONLY if the fastener and its counterpart are clean and free from rust and the fastener develops 3 Nm (27 inch lbs.) of torque/drag against its counterpart prior to the fastener seating. If the fastener does not meet these criteria, REPLACE the fastener.

11. Align the strut to the mark on the knuckle. Tighten the strut lower nuts to 120 Nm (89 ft. lbs.).

12. Install the tire and wheel.

13. Lower the vehicle.

14. Align the front wheels.

DISASSEMBLY AND ASSEMBLY

1. Before servicing the vehicle, refer to the Precautions Section.

2. Remove the strut from the vehicle.

3. Install the strut (2) in a strut compressor tool, such as J 45400 (1).

➡**The spring is compressed when the strut moves freely.**

4. Turn the spring compressor forcing screw until the coil spring is compressed.

5. Use a 45 TORX® socket in order to hold the strut shaft. Remove the upper strut mount nut.

6. Remove the strut from the compressor.

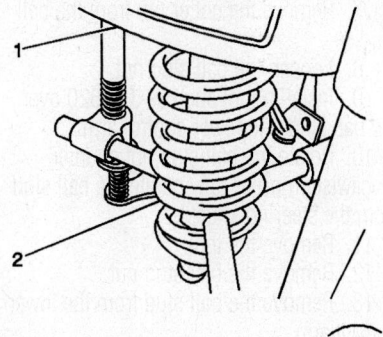

06025-LACR-G72

Strut installed in a compressor

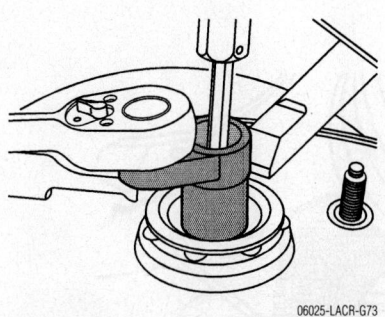

06025-LACR-G73

Use a 45 TORX® socket in order to hold the strut shaft

7. Loosen the compressor forcing screw until the upper strut mount and coil spring may be removed.

8. Remove the upper strut mount and the coil spring from the compressor.

To assemble

9. Install the coil spring and upper strut mount in the compressor.

10. Turn the spring compressor forcing screw (1) until the coil spring is compressed.

11. Install the strut to the coil spring and upper strut mount.

12. Install the strut retaining nut. Install the strut mount nut. Tighten the strut mount nut to 75 Nm (55 ft. lbs.).

13. Remove the strut from the compressor.

14. Install the strut.

Stabilizer (Sway) Bar and Links

REMOVAL & INSTALLATION

1. Before servicing the vehicle, refer to the Precautions Section.

2. Raise and support the vehicle.

3. Remove the tire and wheel assembly.

4. Remove the left and right side stabilizer shaft insulator clamp bolts.

5. Remove the left and right side stabilizer shaft insulator clamp.

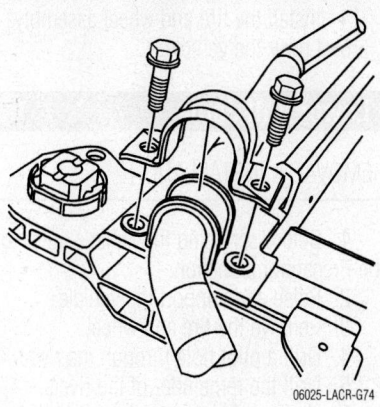

06025-LACR-G74

Stabilizer bar mounting bolts

6. Remove the left and right side stabilizer shaft insulators from the stabilizer shaft.

7. Remove the stabilizer shaft link bolt and nut.

8. Remove the stabilizer shaft link from the vehicle.

To install:

9. Install the stabilizer link into the vehicle.

10. Install the stabilizer shaft link bolt and nut. Tighten the stabilizer shaft link nut to 23 Nm (17 ft. lbs.).

11. Install the left and right side stabilizer shaft insulators to the stabilizer shaft.

12. Install the left and right side stabilizer shaft insulator clamps.

13. Install the left and right side stabilizer shaft insulator clamp bolts. Tighten the bolts to 42 Nm (31 ft. lbs.).

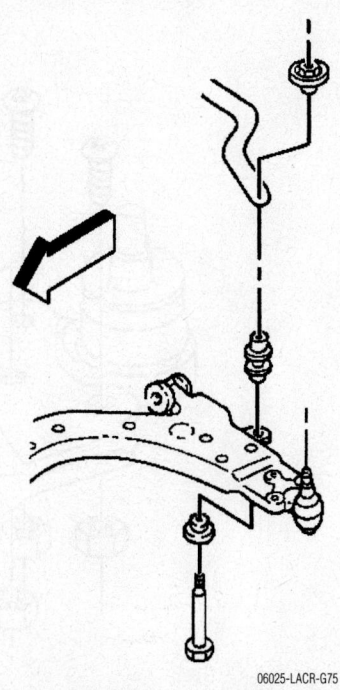

06025-LACR-G75

Stabilizer shaft link bolt and nut

14. Install the tire and wheel assembly.
15. Lower the vehicle.

Lower Ball Joint

REMOVAL & INSTALLATION

1. Before servicing the vehicle, refer to the Precautions Section.
2. Raise and support the vehicle.
3. Remove the tire and wheel.
4. Drill a pilot hole through the rivets.
5. Drill the remainder of the rivets.
6. Use a hammer and a chisel in order to remove the remainder of the rivet heads.

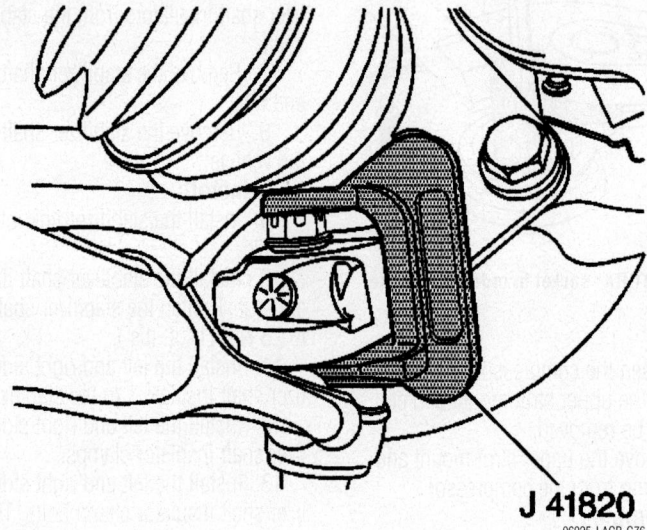

Tool J41820 installed

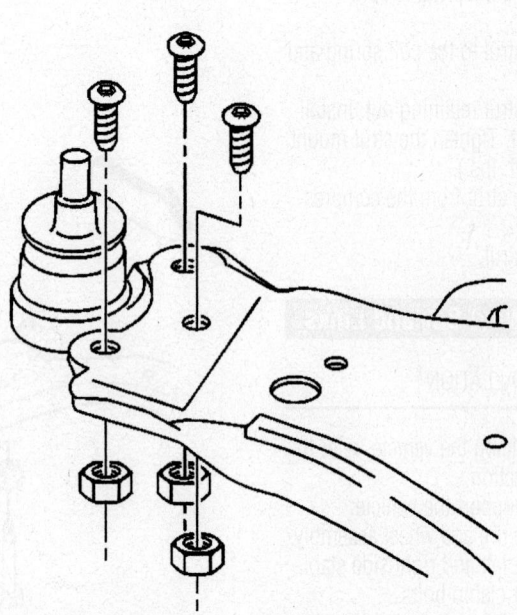

06025-LACR-G77

Install the NEW ball stud bolts facing down, away from the ball stud

7. Remove the cotter pin from the ball stud.
8. Loosen the ball stud nut.
9. Install a tool such as J 41820 over the ball stud and lower control arm.
10. Rotate the ball stud nut counter-clockwise in order to separate the ball stud from the steering knuckle.
11. Remove the tool.
12. Remove the ball stud nut.
13. Remove the ball stud from the lower control arm.
To install:
14. Install the ball stud to the lower control arm.

15. Install the NEW ball stud bolts facing down, away from the ball stud.
16. Install the NEW ball stud nuts. Tighten the NEW ball stud nuts to 68 Nm (50 ft. lbs.).
17. Install the ball stud to the steering knuckle.
18. Install the ball stud castle nut. Tighten the nut to 20 Nm (15 ft. lbs.) plus an additional 120 degrees.
19. Install a new cotter pin and bend the ends.
20. Install the tire and wheels.
21. Lower the vehicle.
22. Check the wheel alignment.

Lower Control Arm

REMOVAL & INSTALLATION

1. Before servicing the vehicle, refer to the Precautions Section.

✳✳ WARNING

Use only the recommended tools for separating the ball joint from the knuckle. Do NOT hammer or pry the ball joint from the knuckle. Failure to use the recommended tools may cause damage to the ball joint and seal.

➡**Use the ignition key in order to unlock the steering column.**

2. Turn the steering wheel in order to move the front of the applicable wheel to the outboard most position.

➡**Use ONLY a frame-contact type vehicle lift or a floor jack at the recommended lift points. Do NOT use a suspension-contact type vehicle lift. Do NOT lift the vehicle by the lower control arms.**

3. Raise and support the vehicle.
4. Remove the tire and wheel.
5. If applicable, disconnect the ABS wheel speed sensor connector.
6. If applicable, disconnect the ABS wheel speed sensor jumper harness from the harness retainer clips.
7. Remove the cotter pin from the ball stud.
8. Loosen the ball stud nut.
9. Install tool J 41820 over the ball stud and lower control arm as shown.
10. Rotate the ball stud nut counter-clockwise in order to separate the ball stud from the steering knuckle.
11. Remove the tool.

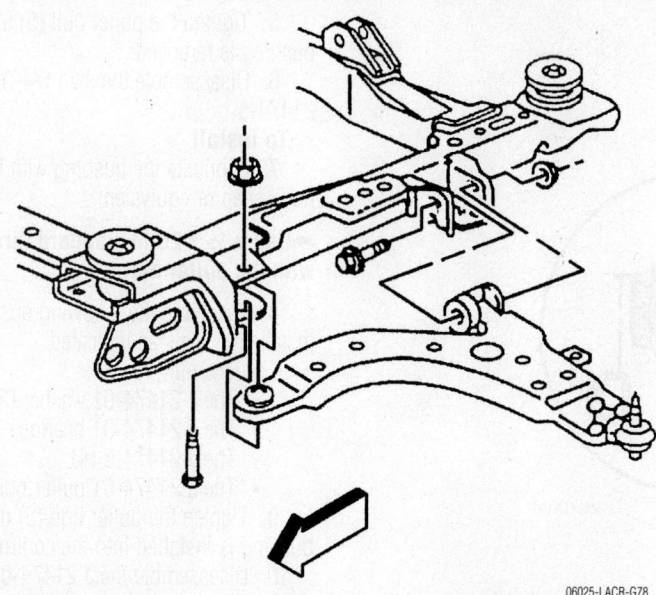

06025-LACR-G78

Lower control arm mounting

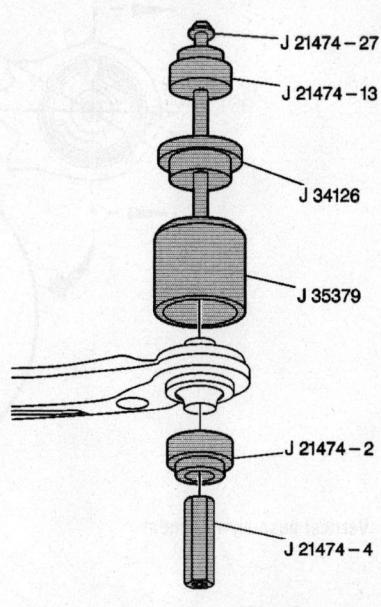

06025-LACR-G80

For vertical bushing removal, assemble the tool as shown

12. Remove the ball stud nut.
13. Remove the lower control arm bolts and nuts.
14. Remove the lower control arm.

To install:

15. Install the lower control arm.
16. Install the control arm bolts and nuts. Do not tighten at this time.

➡**Align the ball stud cotter pin hole parallel to the knuckle in order to ease the cotter pin installation.**

17. Install the ball stud to the knuckle.
18. Install the ball stud castle nut. Tighten the nut to 20 Nm plus 120 degrees (15 ft. lbs. plus 120 degrees).

➡**Do NOT loosen the ball stud nut in order to align the ball stud nut slots to the ball stud cotter pin hole.**

19. If necessary, tighten the ball stud castle nut in order to align the ball stud castle nut slot to the ball stud cotter pin hole.

➡**If applicable, ensure that the cotter pin ends do NOT contact the ABS wheel speed sensor, the ABS sensor connector or the drive axle.**

20. Install a NEW cotter pin and bend the ends as shown in either example.
21. If applicable, connect the ABS wheel speed sensor jumper harness to the harness retainer clips.
22. If applicable, connect the ABS wheel speed sensor connector.

➡**This is a prevailing torque type fastener. This fastener may be reused ONLY if the fastener and its counterpart are clean and free from rust and the**

fastener develops 3 Nm (27 inch lbs.) of torque against its counterpart prior to the fastener seating. If the fastener does not meet these criteria, REPLACE the fastener.

23. Install the lower control arm nuts. Tighten the lower control arm nuts to 125 Nm (92 ft. lbs.).
24. Install the tire and wheel.
25. Lower the vehicle.

LOWER CONTROL ARM BUSHING REPLACEMENT

Vertical

1. Before servicing the vehicle, refer to the Precautions Section.
2. Remove the lower control arm.
3. Secure the lower control arm in a vise.
4. Mark the lower control arm along the flat edge of the bushing flange.

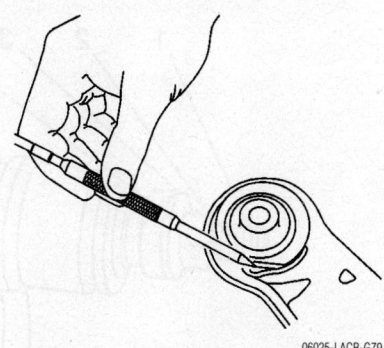

06025-LACR-G79

Mark the lower control arm along the flat edge of the bushing flange

➡**Apply high pressure lubricant such as J 23444-A, or equivalent, to the threads of J 21474-27.**

5. Assemble the following bushing removal tools, or equivalent, as shown:
   - J 21474-27
   - J 21474-13
   - J 34126
   - J 35379
   - J 21474-2
   - J 21474-4
6. Tighten J 21474-4.
7. Disassemble the bushing removal tools.

To install:

➡**You MUST install the lower control arm vertical bushing in the same position in order to maintain the original vehicle ride, handling, and road feel.**

8. Align the flat edge (1) of the bushing flange to the mark in the control arm. Ensure that the flat edge of the bushing flange is 30 degrees (2) from the centerline of the lower control arm. Ensure that the thin slot in the bushing is facing outboard.
9. Insert the bushing into the control arm.

➡**Apply J 23444-A or equivalent to the threads of J 21474-27.**

10. Assemble the following bushing installation tools as shown:
   - J 21474-27
   - J 21474-13
   - J 21474-5
   - J 21474-4

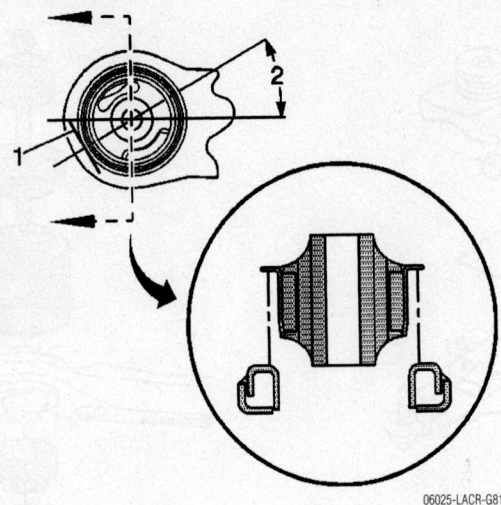

06025-LACR-G81

Vertical bushing alignment

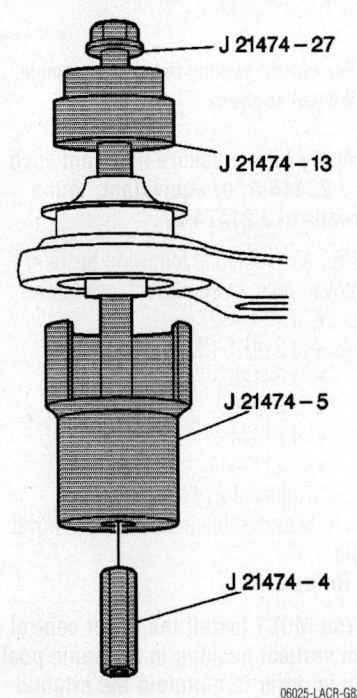

- J 21474 – 27
- J 21474 – 13
- J 21474 – 5
- J 21474 – 4

06025-LACR-G82

For vertical bushing installation, assemble the tool as shown

11. Tighten J 21474-4.
12. Disassemble the bushing installation tools.
13. Install the lower control arm.

Horizontal

1. Before servicing the vehicle, refer to the Precautions Section.
2. Remove the lower control arm.
3. Secure the lower control arm in a vise.

➡ **Use a ½ x 20 in standard thread nut with the puller bolt.**

4. Assemble the bushing removal tools as indicated:
- The nut (1)
- The J 21474-01 washer (2)
- The J 21474-01 bearing (3)
- The J 21474-5 (4)
- The J 21474-01 puller bolt (5)

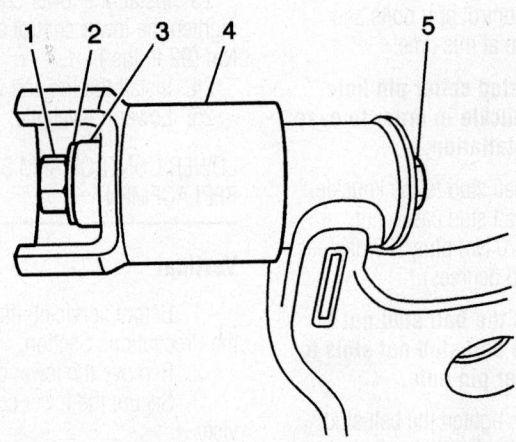

06025-LACR-G83

Tool assembly for horizontal bushing removal

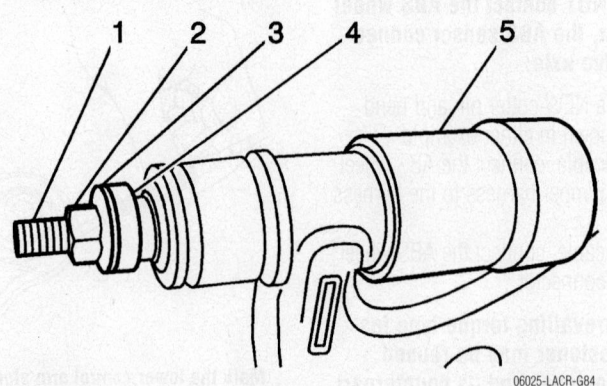

06025-LACR-G84

Tool assembly for horizontal bushing installation

5. Tighten the puller bolt (5) until the bushing is removed.
6. Disassemble the J 21474-01 and J 21474-5.

To install

7. Lubricate the bushing with liquid hand soap or equivalent.

➡ **Use a ½ x 20 in standard thread nut with the puller bolt.**

8. Assemble the following bushing installation tools as indicated:
- The nut (1)
- The J 21474-01 washer (2)
- The J 21474-01 bearing (3)
- The J 21474-5 (4)
- The J 21474-01 puller bolt (5)
9. Tighten the puller bolt (5) until the bushing is installed into the control arm.
10. Disassemble the J 21474-01 and J 21474-5.
11. Remove the control arm from the vice.
12. Install the lower control arm.

Knuckle

REMOVAL & INSTALLATION

1. Before servicing the vehicle, refer to the Precautions Section.
2. Raise and support the vehicle.
3. Remove the bearing/hub assembly.
4. Disconnect the front lower control arm ball stud.
5. Disconnect the outer tie rod end from the steering knuckle.
6. Matchmark the strut to the knuckle.
7. Remove the bolts and nuts attaching the strut to the knuckle.
8. Remove the knuckle from the vehicle.

To install:
9. Install the knuckle to the vehicle.
10. Install the through bolts and nuts attaching the strut to the knuckle. Tighten the through bolts and nuts to 120 Nm (89 ft. lbs.).
11. Connect the outer tie rod to the steering knuckle.
12. Connect the front lower control arm ball stud to the knuckle.
13. Install the front wheel drive shaft bearing.
14. Lower the vehicle.
15. Inspect the front wheel alignment and adjust if necessary.

Wheel Bearings

ADJUSTMENT

No adjustment is possible.

REMOVAL & INSTALLATION

1. Before servicing the vehicle, refer to the Precautions Section.
2. Remove the tire and wheel.
3. Disconnect the wheel speed sensor electrical connector, if equipped.
4. Remove the wheel speed sensor electrical connector from the bracket, if equipped.
5. Remove the front halfshaft nut. Insert a drift or flat-bladed tool into the caliper and rotor to prevent from turning.

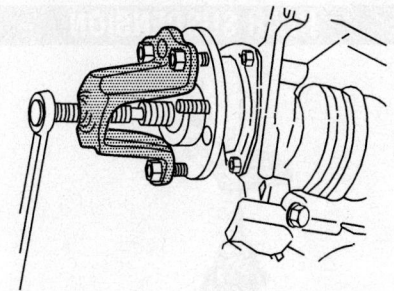

06025-LACR-G85

Front hub removal

6. Remove the brake rotor.
7. Use 3 wheel nuts in order to attach a puller to the wheel bearing/hub.
8. Use the puller to push the halfshaft out of the wheel bearing/hub.
9. Remove and DISCARD the wheel bearing/hub bolts. Remove the puller from the hub.

➡**Ensure that the halfshaft outer seal/boot is not damaged.**

10. Remove the wheel bearing/hub and splash shield-noting the position of the shield for re-installation.

To install:
11. If necessary, remove the shipping bracket from the wheel bearing/hub.

12. Install the wheel bearing/hub with the splash shield as noted during removal.

13. Install NEW wheel bearing/hub bolts. Tighten the NEW wheel bearing/hub bolts to 130 Nm (96 ft. lbs.).
14. Install the brake rotor and caliper.
15. Install the front halfshaft nut. Insert a drift on a flat-bladed tool into caliper and rotor to prevent the rotor from turning. Tighten the front halfshaft nut to 160 Nm (118 ft. lbs.).

➡**Ensure that the connector clip engages the bracket properly.**

16. Install the wheel speed sensor electrical connector to the bracket, if equipped.
17. Connect the wheel speed sensor electrical connector, if equipped.
18. Install the tire and wheel.

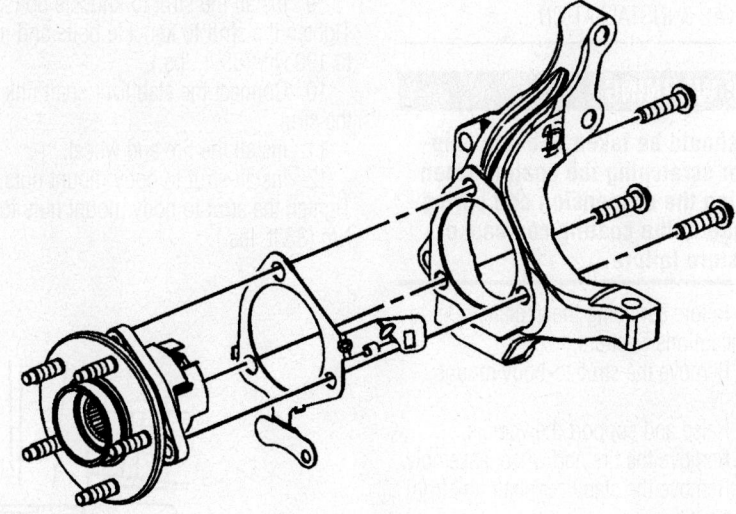

06025-LACR-G86

Front hub installation

REAR SUSPENSION

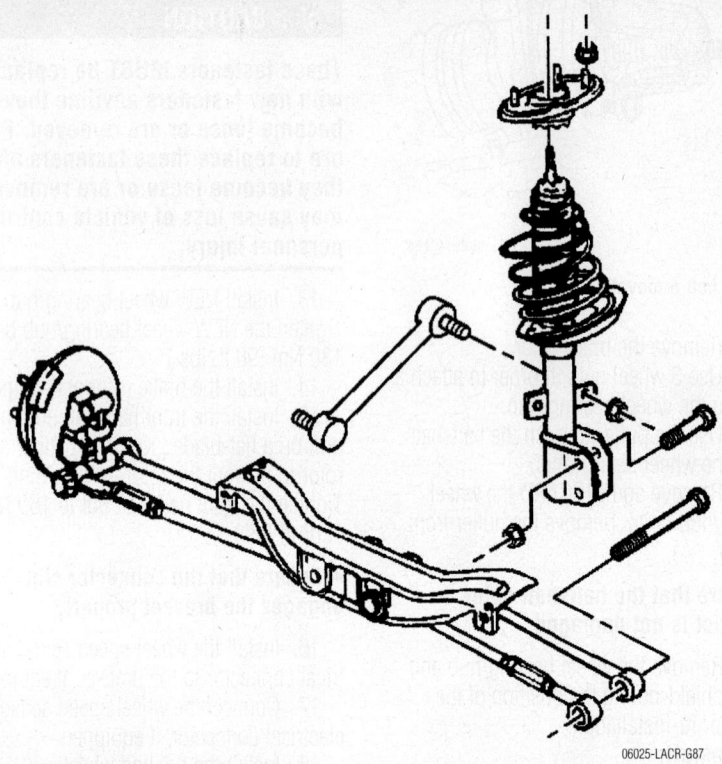

Rear strut mounting

Strut

REMOVAL & INSTALLATION

> ✳✳ **WARNING**
>
> **Care should be taken to avoid chipping or scratching the coating when handling the suspension coil spring. Damage to the coating can cause premature failure.**

1. Before servicing the vehicle, refer to the Precautions Section.
2. Remove the strut-to-body mount nuts.
3. Raise and support the vehicle.
4. Remove the tire and wheel assembly.
5. Remove the stabilizer shaft link from the strut.

> ✳✳ **WARNING**
>
> **The knuckle must be held in place after the strut-to-knuckle bolts have been removed. Failure to observe this may cause ball joint and/or wheel drive shaft damage.**

6. Remove the strut to knuckle bolts.
7. Remove the lstrut from the vehicle.

To install:

8. Install the strut into place.
9. Install the strut to knuckle bolts. Tighten the strut to knuckle bolts and nuts to 120 Nm (89 ft. lbs.).
10. Connect the stabilizer shaft link to the strut.
11. Install the tire and wheel.
12. Install strut to body mount nuts. Tighten the strut to body mount nuts to 45 Nm (33 ft. lbs.).

13. Lower the vehicle.
14. Adjust the rear wheel alignment.

DISASSEMBLY & ASSEMBLY

See the procedure in Front Suspension.

Stabilizer (Sway) Bar

REMOVAL & INSTALLATION

1. Before servicing the vehicle, refer to the Precautions Section.
2. Raise and support the vehicle.
3. Remove the tires and wheels.
4. Remove the left and right stabilizer shaft link lower nuts from the stabilizer shaft links.
5. Remove the clamp bolts from the stabilizer shaft.
6. Remove the stabilizer shaft from the rear suspension support.
7. Remove the stabilizer shaft link nut (1) from the stabilizer shaft link (2) and the strut.

To install:

8. Install the stabilizer shaft to the rear suspension support.
9. Install the stabilizer shaft clamps and bolts. Do not tighten the bolts at this time.
10. Install the right and left stabilizer shaft link nuts to the stabilizer shaft links. Tighten the stabilizer shaft link nuts to 45 Nm (33 ft. lbs.). Tighten the stabilizer shaft insulator bracket nuts to 51 Nm (38 ft. lbs.). Install the stabilizer shaft link nut to the stabilizer shaft link and the strut. Tighten the nut to 52 Nm (38 ft. lbs.).
11. Install the tires and wheels.

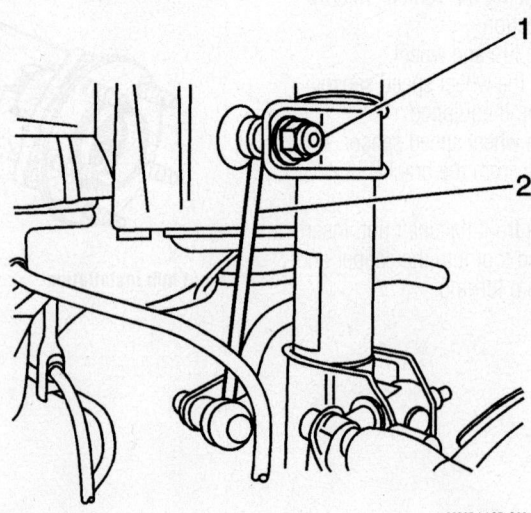

Rear stabilizer shaft end link

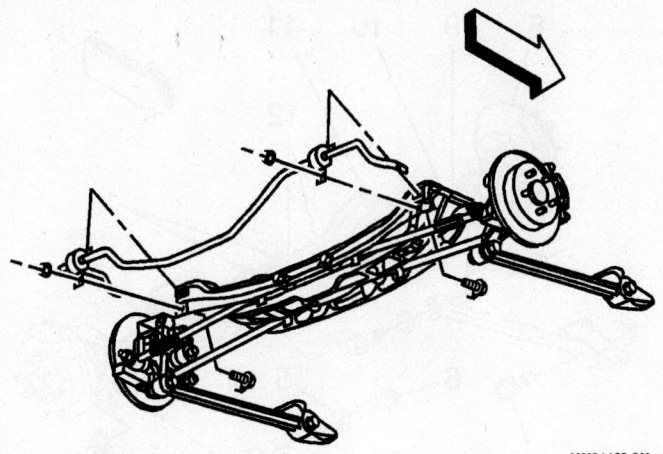

06025-LACR-G89

Rear stabilizer bar mounting

12. Lower the vehicle.

Rear Suspension Support

REMOVAL & INSTALLATION

1. Before servicing the vehicle, refer to the Precautions Section.
2. Raise and support the vehicle.
3. Remove the rear tires and wheels.
4. Remove the exhaust pipe.
5. Disconnect the brake lines from the rear suspension support.
6. Disconnect the parking brake cables and tensioner from the suspension support.

7. Remove the stabilizer shaft from the rear suspension support.

➡**Support the rear suspension support with jack stands before removing the mounting bolts.**

8. Remove the rear wheel spindle rod bolts from the knuckle.
9. Remove the rear suspension support mounting bolts.
10. Remove the rear suspension support.
To install:
11. Position the rear suspension support in place.
12. Install 2 locating pins in the suspen-

sion support alignment holes, one on each side of the suspension support.
13. Install the rear suspension support mounting bolts. Tighten the bolts to 110 Nm (81 ft. lbs.).
14. Remove the locating pins from the suspension support alignment holes.
15. Position the rear wheel spindle rod to the knuckle. Install the retaining bolts and nuts to the knuckle. Tighten the bolts and nuts to 150 Nm (110 ft. lbs.).
16. Install the stabilizer shaft to the rear suspension support.
17. Connect the brake lines to the rear suspension support.
18. Install the parking brake cables and the tensioner.
19. Install the exhaust pipe.
20. Install the tires and wheels.
21. Lower the vehicle.
22. Adjust the rear wheel alignment.

Spindle Rod

REMOVAL & INSTALLATION

1. Before servicing the vehicle, refer to the Precautions Section.
2. Raise and support the vehicle.
3. Remove the tire and wheel assembly.

➡**Use a transmission jack or suitable hoist stands to prop the rear suspension support.**

4. Lower the rear suspension support to gain clearance to the spindle rod to suspension support bolt.
5. Remove the bolt and nut from the spindle rod and the rear suspension support.
6. Remove the spindle rod to knuckle bolt and nut.
7. Remove the spindle rod from the vehicle.
To install:
8. Install the spindle rod to the vehicle.
9. Install the spindle rod to the knuckle.
10. Install the spindle rod to knuckle bolt and nut. Do not tighten at this time.
11. Install the spindle rod to the rear suspension support.
12. Install the bolt and nut to the rear suspension support. Tighten the bolt to 135 Nm (100 ft. lbs.).
13. Install the rear suspension support. Tighten the bolt to 150 Nm (110 ft. lbs.).
14. Install the tire and wheel assembly.
15. Lower the vehicle.
16. Adjust the wheel toe angle.

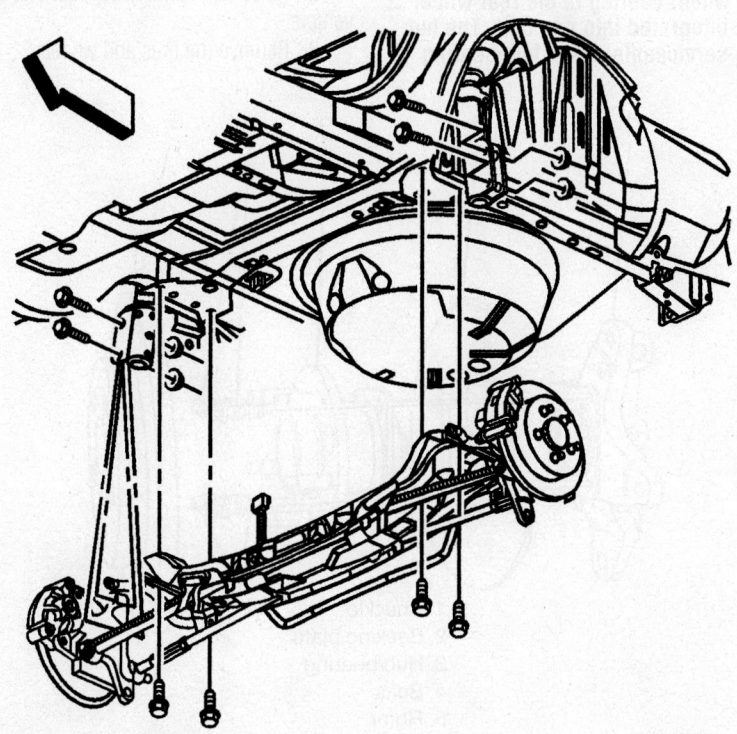

06025-LACR-G90

Rear suspension support

Knuckle

REMOVAL & INSTALLATION

1. Before servicing the vehicle, refer to the Precautions Section.
2. Raise and suitably support the vehicle.
3. Remove the tire and wheel assembly.
4. Scribe the strut to the knuckle.
5. Remove the bearing/hub assembly.
6. Remove the rear wheel spindle rods from the knuckle.
7. Remove the trailing arm from the knuckle.
8. Remove the rear strut to knuckle bolts and the nuts.
9. Remove the knuckle from the vehicle.

To install:

10. Install the knuckle to the vehicle.
11. Hand start the strut to knuckle bolts and nuts. Do not tighten the nuts at this time.
12. Connect the rear suspension trailing arm to the knuckle.
13. Connect the rear wheel spindle rods to the knuckle.
14. Install the bearing/hub assembly.
15. Tighten the strut to knuckle bolts and nuts. Tighten bolts to 120 Nm (89 ft. lbs.).
16. Install the tire and wheel assembly.
17. Lower the vehicle.
18. Inspect the rear wheel alignment, adjust if necessary.

Trailing Arm

REMOVAL & INSTALLATION

1. Before servicing the vehicle, refer to the Precautions Section.
2. Raise and support the vehicle.
3. Remove the bolt and nut from the trailing arm and the knuckle.
4. Remove the nut and bolt from the trailing arm and the trailing arm bracket.
5. Remove the trailing arm from the vehicle.

To install:

6. Install the trailing arm to the trailing arm bracket.
7. Install the bolt and nut to the trailing arm and the trailing arm bracket. Tighten the bolt and nut to 105 Nm (77 ft. lbs.).
8. Install the trailing arm to the knuckle.
9. Install the bolt and nut to the trailing arm and the knuckle. Tighten the nut to 240 Nm (177 ft. lbs.).
10. Lower the vehicle.

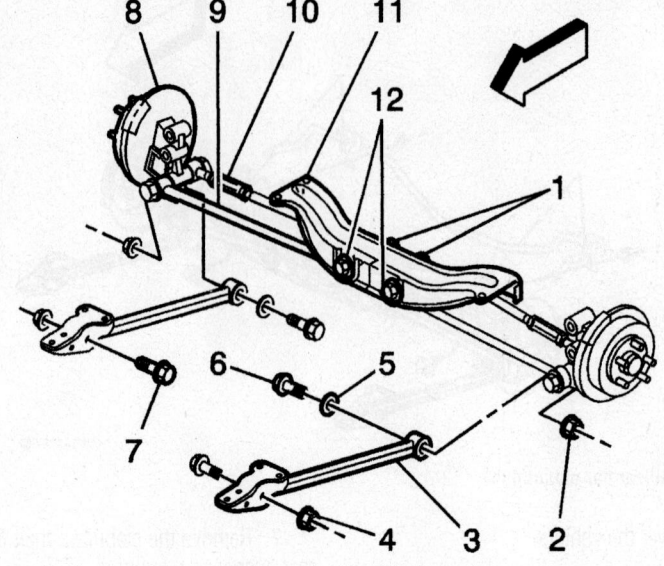

1. Nut		7. Bolt
2. Nut		8. Knuckle
3. Trailing arm		9. Stabilizer bar
4. Nut		10. Spindle rod
5. Washer		11. Suspension support
6. Bolt		12. Bolt

06025-LACR-G91

Rear suspension components

Wheel Bearings

REMOVAL & INSTALLATION

➡ **The wheel bearing in the rear wheel hub is integrated into one unit. The hub is non-serviceable. If the hub/bearing is damaged, replace the complete hub and bearing assembly.**

1. Before servicing the vehicle, refer to the Precautions Section.
2. Raise and suitably support the vehicle.
3. Remove the tires and wheels.

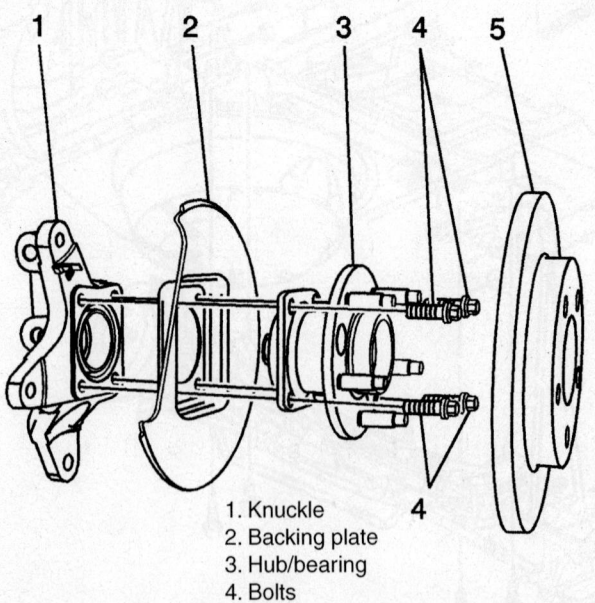

1. Knuckle
2. Backing plate
3. Hub/bearing
4. Bolts
5. Rotor

06025-LACR-G92

Rear hub/bearing and related parts

4. Remove the brake rotor.

5. Remove the ABS electrical connector from the wheel speed sensor, if equipped.

6. Remove the mounting bolts from the rear bearing/hub.

7. Remove the wheel bearing/hub and the backing plate.

To install:

8. Install the backing plate and the wheel bearing hub.

9. Install the wheel bearing/hub-to-knuckle bolts. Tighten the bolts to 75 Nm (55 ft. lbs.).

10. Install the ABS electrical connector to the wheel speed sensor, if equipped.

11. Install the brake rotor and caliper.

12. Install the tires and wheels.

13. Lower the vehicle.

FRONT BRAKES

Brake Caliper

REMOVAL & INSTALLATION

1. Before servicing the vehicle, refer to the Precautions Section.

2. Inspect the fluid level in the brake master cylinder reservoir.

3. If the brake fluid level is midway between the maximum-full point and the minimum allowable level, then no brake fluid needs to be removed from the reservoir before proceeding. If the brake fluid level is higher than midway between the maximum-full point and the minimum allowable level, then remove brake fluid to the midway point before proceeding.

4. Raise and suitably support the vehicle.

5. Remove the front tire and the wheel assembly.

6. Hand tighten 2 wheel lug nuts to retain the rotor to the hub.

7. Install a large C-clamp over the top of the brake caliper and against the back of the outboard brake pad.

8. Tighten the C-clamp until the caliper piston is pushed into the caliper bore enough to slide the caliper off the rotor.

9. Remove the C-clamp from the caliper.

10. Remove the brake hose-to-caliper bolt from the caliper. Discard the 2 copper brake hose gaskets. These gaskets may be stuck to the brake caliper and/or the brake hose end.

11. Plug the opening in the front brake hose to prevent excessive brake fluid loss and contamination.

➡**Note the location of the caliper pin bolts. The leading caliper pin bolt, or top bolt, has a bushing as part of the assembly. The trailing caliper pin bolt, or bottom bolt, is a solid design.**

12. Remove the caliper pin bolts. Note the location of the caliper pin bolts. The leading caliper pin, or top bolt, has a bushing as part of the assembly. The trailing caliper pin, or bottom bolt, is a solid design.

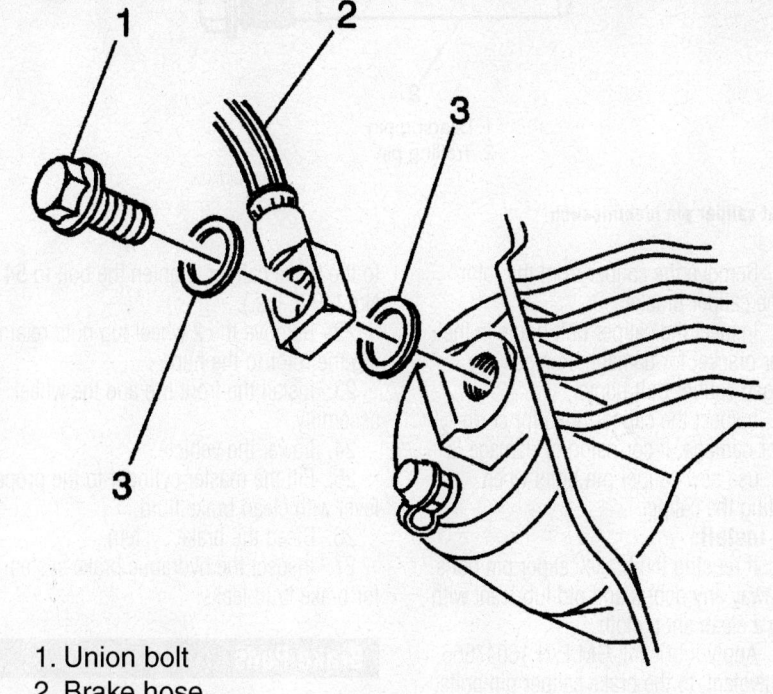

1. Union bolt
2. Brake hose
3. Copper washers

06025-LACR-G93

Brake hose attachment

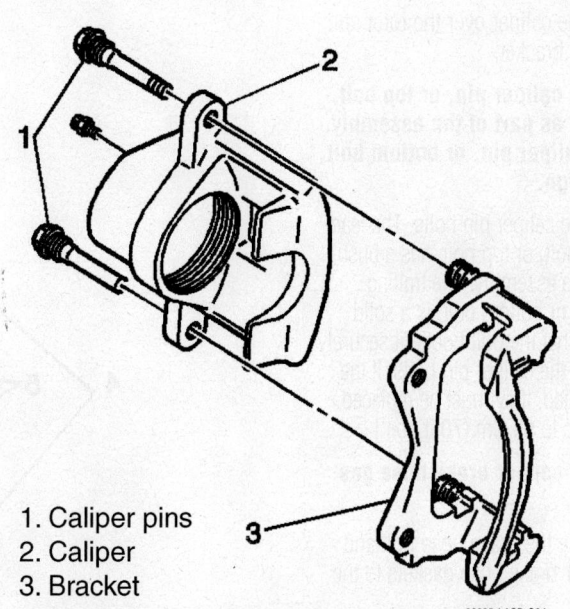

1. Caliper pins
2. Caliper
3. Bracket

06025-LACR-G94

Front caliper-to-bracket attachment

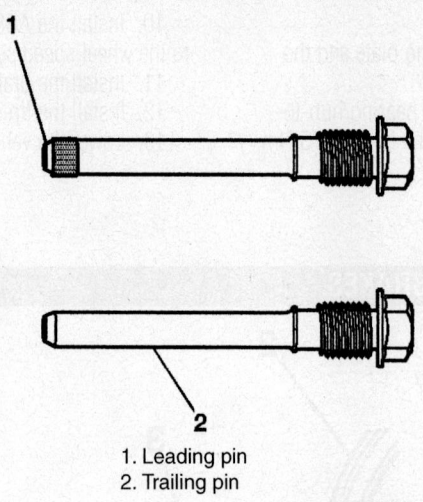

1. Leading pin
2. Trailing pin

06025-LACR-G95

Front caliper pin identification

13. Remove the caliper from the rotor and the caliper bracket.

14. Inspect the caliper bolt boots in the caliper bracket for damage. Replace any damaged caliper bolt boots.

15. Inspect the caliper bolts for corrosion or damage. If corrosion or damage is found, use new caliper pin bolts when installing the caliper.

To install:

16. If reusing the brake caliper pin bolts, wipe away any debris and old lubricant with a with a clean shop cloth.

17. Apply lubricant, GM P/N 18047666, or equivalent, to the brake caliper pin bolts. Apply a thin layer to the pin bushing and to the caliper pin bolt shank. Ensure that there is not a buildup of excess lubricant at the end of the leading caliper pin, in front of the bushing.

18. Install the caliper over the rotor and onto the caliper bracket.

➡**The leading caliper pin, or top bolt, has a bushing as part of the assembly. The trailing caliper pin, or bottom bolt, is a solid design.**

19. Install the caliper pin bolts. The leading caliper pin bolt, or top bolt, has a bushing as part of the assembly. The trailing caliper pin bolt, or bottom bolt, is a solid design. Ensure that the bolt boots fit securely in the groove of the caliper pin bolts. If the boots are damaged, they must be replaced. Tighten the bolts to 95 Nm (70 ft. lbs.).

➡**Install NEW copper brake hose gaskets.**

20. Assemble the brake hose bolt and the NEW copper brake hose gaskets to the brake hose.

21. Install the brake hose-to-caliper bolt

to the brake caliper. Tighten the bolt to 54 Nm (40 ft. lbs.).

22. Remove the 2 wheel lug nuts retaining the rotor to the hub.

23. Install the front tire and the wheel assembly.

24. Lower the vehicle.

25. Fill the master cylinder to the proper level with clean brake fluid.

26. Bleed the brake system.

27. Inspect the hydraulic brake system for brake fluid leaks.

Brake Pads

REMOVAL & INSTALLATION

1. Before servicing the vehicle, refer to the Precautions Section.

2. Inspect the fluid level in the brake master cylinder reservoir.

3. If the brake fluid level is midway between the maximum-full point and the minimum allowable level, then no brake fluid needs to be removed from the reservoir before proceeding. If the brake fluid level is higher than midway between the maximum-full point and the minimum allowable level, then remove brake fluid to the midway point before proceeding.

4. Raise and suitably support the vehicle.

5. Remove the tire and the wheel assembly.

6. Hand-tighten 2 wheel lug nuts in order to retain the rotor to the hub.

7. Install a large C-clamp over the top of the caliper housing and against the back of the outboard pad.

8. Slowly tighten the C-clamp until the piston pushes into the caliper bore enough to slide the caliper off the rotor.

9. Remove the C-clamp from the caliper.

10. Remove the lower caliper bolt (1).

✱✱ WARNING

Use care to avoid damaging pin boot when rotating caliper.

11. In order to access the pads, rotate the caliper upward and suitably support it.

12. Remove the pads (5) from the caliper bracket (3).

13. Remove the 2 retainer slides (4) from the caliper bracket (3).

14. Inspect all parts for cuts, tears, or deterioration. Replace any damaged parts.

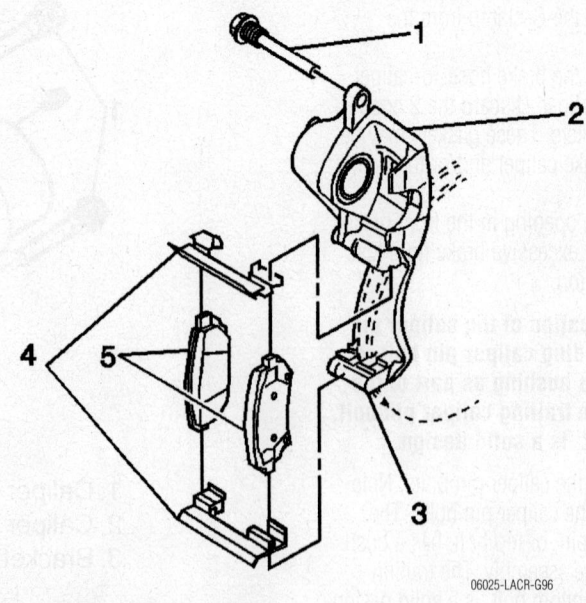

06025-LACR-G96

Front brake pads and related parts

15. Inspect the caliper bolts for corrosion or damage. If corrosion is found, use new caliper bolts when installing the caliper.

To install:

16. Using a C-clamp, bottom the piston into the caliper bore. Use an old brake pad or wooden block across the face of the piston. Do not damage the piston or the caliper boot.

17. Install the 2 retainers to the caliper bracket.

➡**The wear sensor is on the outside pad. The sensor is positioned at the leading or upward edge of the pad during forward wheel rotation.**

18. Install the pads to the caliper anchor bracket.

✳✳ **WARNING**

Use care to avoid damaging pin boot when rotating caliper.

19. Remove the support and reposition the caliper (2) back down over the front pads.

20. Lubricate the pin bolt and the inner diameter of the bolt boot with GM P/N 18047666, or equivalent. Do not lubricate the threads of the pin bolt.

21. Install the lower caliper bolt. Tighten the caliper bolts to 95 Nm (70 ft. lbs.).

22. Remove the 2 wheel lug nuts retaining the rotor to the hub.

23. Install the tire and the wheel assembly.

24. Lower the vehicle.

25. With the engine OFF, gradually apply the brake pedal to approximately ⅔ of its travel distance.

26. Slowly release the brake pedal.

27. Wait 15 seconds, then, repeat steps 10 and 11 until a firm brake pedal is obtained. This will properly seat the brake caliper pistons and brake pads.

28. Fill the brake master cylinder reservoir to the proper level.

29. Burnish the pads and rotors.

Caliper Bracket

REMOVAL & INSTALLATION

1. Before servicing the vehicle, refer to the Precautions Section.

✳✳ **WARNING**

Support the brake caliper with heavy mechanic's wire, or equivalent,

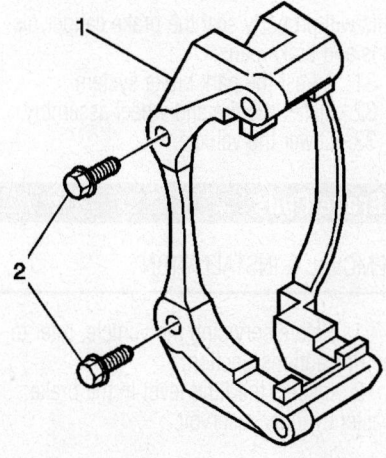

06025-LACR-G97

Front caliper bracket

whenever it is separated from its mount and the hydraulic flexible brake hose is still connected. Failure to support the caliper in this manner will cause the flexible brake hose to bear the weight of the caliper, which may cause damage to the brake hose and in turn may cause a brake fluid leak.

2. Remove the caliper from the mounting bracket and support the caliper with heavy mechanics wire or equivalent. It is not necessary to disconnect the hydraulic brake flexible hose from the caliper.

3. Remove the front brake pads.

To install:

➡**To ensure that the proper clamp load will be present when installed. It is imperative that the threads on the caliper bracket bolts, as well as the mounting holes in the knuckle, be cleaned of all debris and inspected before proceeding with installation.**

4. Clean and visually inspect threads of the caliper bracket bolts and mounting holes in the knuckle.

5. Apply THREADLOCKER, GM P/N 12345493 (Canadian P/N 10953488) or equivalent, to the threads of the brake caliper bracket bolts.

6. Install the caliper bracket with the bracket bolts. Tighten the caliper bracket bolts to 180 Nm (133 ft. lbs.).

7. Install the front brake pads.

8. Install the caliper.

REAR BRAKES

Brake Caliper

REMOVAL & INSTALLATION

1. Before servicing the vehicle, refer to the Precautions Section.

2. Inspect the fluid level in the brake master cylinder reservoir.

3. If the brake fluid level is midway between the maximum-full point and the minimum allowable level, no brake fluid needs to be removed from the reservoir before proceeding.

4. If the brake fluid level is higher than midway between the maximum-full point and the minimum allowable level, remove brake fluid to the midway point before proceeding.

5. Raise and support the vehicle.

6. Remove the tire and wheel assembly.

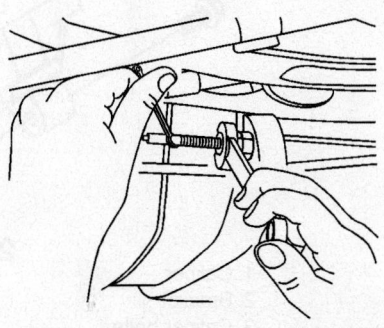

06025-LACR-G98

Release tension from park brake system at the equalizer

7. Release tension from park brake system at the equalizer.

8. Disconnect the front and rear cables from one another at the connector clip.

9. Disconnect the park brake cable from the park brake lever on the brake caliper.

10. Remove the park brake cable from the caliper bracket.

11. Remove brake hose to caliper bolt.

12. Remove the brake hose from the brake caliper.

13. Remove and discard the 2 copper brake hose gaskets. These gaskets may be stuck to the brake caliper and/or the brake hose end.

14. Plug the opening in the brake caliper and brake hose to prevent fluid loss and/or contamination.

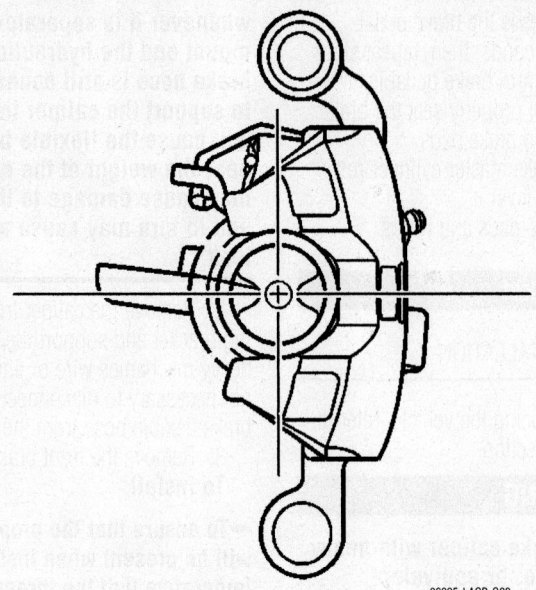

06025-LACR-G99

Align the indents on the piston face to match the pin on the brake pad

15. Remove the brake caliper bolts.
16. Remove the brake caliper.
To install:
17. Align the indents on the piston face to match the pin on the brake pad.
18. Inspect the bracket bolt guide assembly.
19. Inspect the brake pad hardware and replace if necessary.
20. Install the brake caliper onto the caliper bracket insuring that the guide boots are not damaged.
21. Install the brake caliper bolts. Tighten the brake caliper bolts to 34 Nm (25 ft. lbs.).
22. Remove the plugs in the brake hose end.

➡**Install NEW copper brake hose gaskets.**

23. Assemble the brake hose bolt and the NEW copper brake hose gaskets to the brake hose.
24. Install the brake hose to caliper bolt to the brake caliper. When installing the right rear brake hose to caliper, hold hose up while tightening. Tighten the brake hose to caliper bolt to 44 Nm (33 ft. lbs.).
25. Install the park brake cable into the park brake bracket on the caliper.
26. Connect the park brake cable to the park brake lever on the brake caliper.
27. Bleed the brake system.
28. With the engine OFF, gradually apply the brake pedal to approximately ⅔ of its travel distance.
29. Slowly release the brake pedal.
30. Wait 15 seconds, then repeat steps 12-13 until a firm brake pedal is obtained.

This will properly seat the brake caliper pistons and brake pads.
31. Adjust the park brake system.
32. Install the tire and wheel assembly.
33. Lower the vehicle.

Brake Pads

REMOVAL & INSTALLATION

1. Before servicing the vehicle, refer to the Precautions Section.
2. Inspect the fluid level in the brake master cylinder reservoir.

3. If the brake fluid level is midway between the maximum-full point and the minimum allowable level, no brake fluid needs to be removed from the reservoir before proceeding.
4. If the brake fluid level is higher than midway between the maximum-full point and the minimum allowable level, remove brake fluid to the midway point before proceeding.
5. Raise and support the vehicle.
6. Remove the tire and wheel assembly.
7. Unclamp the wheel speed sensor (WSS) harness from the lower control arm.
8. Remove both upper and lower caliper bolts from the caliper.

✳✳ WARNING

Support the brake caliper with heavy mechanic's wire, or equivalent, whenever it is separated from its mount and the hydraulic flexible brake hose is still connected. Failure to support the caliper in this manner will cause the flexible brake hose to bear the weight of the caliper, which may cause damage to the brake hose and in turn may cause a brake fluid leak.

9. Pull the caliper straight off of the bracket and secure out of the way with heavy mechanic's wire. DO NOT disconnect the hydraulic brake flexible hose from the caliper.
10. Remove the inboard and outboard pads from the brake caliper bracket.

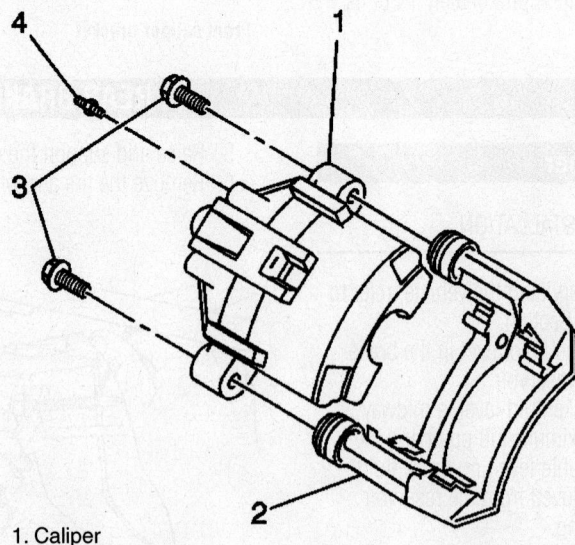

1. Caliper
2. Bracket
3. Caliper bolts
4. Bleeder screw

06025-LACR-G100

Rear caliper mounting

To install:

11. Inspect the brake caliper bracket guide boot assembly for condition.

12. Clean the brake pad hardware mating surfaces on the caliper bracket of any debris or corrosion.

13. Inspect the brake pad retainer clips for condition and replace, if necessary.

14. Inspect the piston boot for condition. Replace if damaged.

15. Retract the brake caliper piston into the brake caliper bore. Use a suitable spanner type wrench and turn the piston clockwise until it bottoms out fully in the brake caliper (Mac Tools DBC 25 C 2500 MA Disc Brake Caliper Tool Set, or equivalent).

16. Align the indents on the piston face to match the pin on the back of the inboard brake pads.

17. Install the brake pad retainers into the brake caliper bracket.

18. Install the inboard and outboard brake pads into the brake caliper bracket insuring that the pad with the metallic wear sensor is placed on the inboard side of the bracket.

19. Slide the caliper onto the bracket insuring that the bracket guide boots are not damaged.

20. Insert the brake caliper bolts. Tighten the brake caliper bolts to 34 Nm (25 ft. lbs.).

21. Re-clamp the WSS harness onto the lower control arm.

22. Install the tire and wheel assembly.

23. Lower the vehicle.

24. With the engine OFF, gradually apply the brake pedal to approximately ⅔ of its travel distance.

25. Slowly release the brake pedal.

26. Wait 15 seconds, then repeat steps 10-11 until a firm brake pedal apply is obtained. This will properly seat the brake caliper pistons and brake pads.

27. Fill the brake master cylinder reservoir to the proper level.

Caliper Bracket

REMOVAL & INSTALLATION

1. Before servicing the vehicle, refer to the Precautions Section.

✳✳ WARNING

Support the brake caliper with heavy mechanic's wire, or equivalent, whenever it is separated from its mount and the hydraulic flexible brake hose is still connected. Failure to support the caliper in this manner

will cause the flexible brake hose to bear the weight of the caliper, which may cause damage to the brake hose and in turn may cause a brake fluid leak.

2. Remove the brake caliper from the mounting bracket and support the brake caliper with heavy mechanic's wire, or equivalent. Do NOT disconnect the hydraulic brake flexible hose from the caliper.

3. Remove the brake pads from the brake caliper bracket.

4. Remove the brake pad retainers from the brake caliper bracket.

5. Remove the brake caliper bracket bolts (1).

6. Remove the rear brake caliper bracket.

7. Use a wire brush to remove any rust and debris from the brake caliper bracket.

8. Inspect the brake mounting and hardware.

To install:

➡**The caliper bracket bolts threads and the threaded holes of the caliper bracket must be free of threadlocker residue and debris prior to re-application of threadlocker in order to ensure proper adhesion and fastener retention.**

9. Prepare the bolts and the threaded holes for assembly.

10. Thoroughly clean the residue from the fastener threads using denatured alcohol or equivalent and allow to dry.

11. Thoroughly clean the residue from the threaded holes using denatured alcohol or equivalent and allow to dry.

12. If reusing the old caliper bracket bolts, apply threadlocker GM P/N 12345493 (Canadian P/N 10953488), or equivalent to

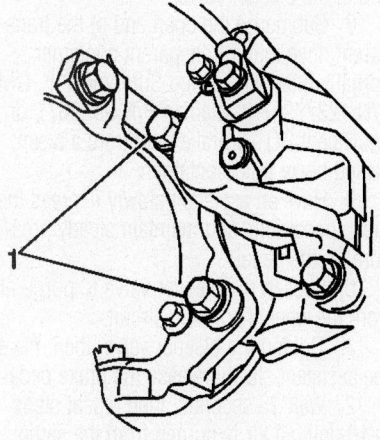

06025-LACR-G101

Rear caliper bracket

⅔ of the threaded length of the fastener. Ensure that there are no gaps in the threadlocker along the length of the filled area of the fastener.

13. Allow the threadlocker to cure approximately ten minutes before installation.

14. Install the brake caliper bracket.

15. Install the brake caliper bracket bolts. Tighten the rear brake caliper bracket bolts to 130 Nm (96 ft. lbs.).

16. Lubricate the brake caliper pin bolts with a thin coat of high temperature silicone lube.

17. Install the brake pad retainers to the brake caliper bracket.

18. Install the brake pads to the brake caliper bracket.

19. Install the brake caliper.

Parking Brake

ADJUSTMENT

Adjustment of manual adjustment park brake cable system is necessary whenever the rear brake cables have been disconnected. A need for park brake cable adjustment is indicated if the hydraulic brake system operates with good reserve, but a firm park brake pedal feel cannot be achieved with less than one full stroke of the park brake pedal.

1. Before servicing the vehicle, refer to the Precautions Section.

2. Apply and release the park brake four times.

3. Park brake light should be illuminated after park brake has been depressed slightly.

4. Check parking brake pedal assembly for full release by turning the ignition on and inspecting PARK BRAKE warning light. Light should be off. If PARK BRAKE warning light is on and park brake appears to be fully released, pull the pedal back by hand and continue with the adjustment procedure.

5. Raise the vehicle and suitably support.

6. Check park brake levers on rear calipers. Levers should be against the stops on the caliper housing. If levers are not against stops, check for binding in rear brake cables and position levers against stops.

7. Tighten park brake cable at equalizer until either the left or right lever begins to move off stop.

8. Loosen adjustment at equalizer until the lever which has moved off the stop, as

in step 5, is again resting fully against stops. Loosen tension at equalizer until the cables feel slightly loose to the touch. Cables should not sag under their own weight.

9. Operate park brake several times to check adjustment. A firm pedal feel should be obtained by depressing the pedal less than one full stroke.

10. Inspect left and right caliper levers. Both levers must be resting on stops after adjustment of parking brake.

11. Check the operation of the park brake.

12. To achieve optimal performance ensure cables are not over-tensioned at the equalizer.

BRAKE HYDRAULIC SYSTEM

Brake System Bleeding

✳✳ WARNING

When adding fluid to the brake master cylinder reservoir, use only Delco Supreme 11®, GM P/N 12377967 (Canadian P/N 992667), or equivalent DOT-3 brake fluid from a clean, sealed brake fluid container. The use of any type of fluid other than the recommended type of brake fluid may cause contamination which could result in damage to the internal rubber seals and/or rubber linings of hydraulic brake system components.

1. Before servicing the vehicle, refer to the Precautions Section.

2. Place a clean shop cloth beneath the brake master cylinder to prevent brake fluid spills.

3. With the ignition OFF and the brakes cool, apply the brakes 3–5 times, or until the brake pedal effort increases significantly, in order to deplete the brake booster power reserve.

4. If you have performed a brake master cylinder bench bleeding on this vehicle, or if you disconnected the brake pipes from the master cylinder, you must perform the following steps:

a. Ensure that the brake master cylinder reservoir is full to the maximum-fill level. If necessary, add Delco Supreme 11®, GM P/N 12377967 (Canadian P/N 992667), or equivalent DOT-3 brake fluid from a clean, sealed brake fluid container. If removal of the reservoir cap and diaphragm is necessary, clean the outside of the reservoir on and around the cap prior to removal.

b. With the rear brake pipe installed securely to the master cylinder, loosen and separate the front brake pipe from the front port of the brake master cylinder.

c. Allow a small amount of brake fluid to gravity bleed from the open port of the master cylinder.

d. Reconnect the brake pipe to the master cylinder port and tighten securely.

e. Have an assistant slowly depress the brake pedal fully and maintain steady pressure on the pedal.

f. Loosen the same brake pipe to purge air from the open port of the master cylinder.

g. Tighten the brake pipe, then have the assistant slowly release the brake pedal.

h. Wait 15 seconds, then repeat steps 3-7 until all air is purged from the same port of the master cylinder.

i. With the front brake pipe installed securely to the master cylinder, after all air has been purged from the front port of the master cylinder, loosen and separate the rear brake pipe from the master cylinder, then repeat steps 3-8.

j. After completing the final master cylinder port bleeding procedure, ensure that both of the brake pipe-to-master cylinder fittings are properly tightened.

5. Fill the brake master cylinder reservoir with Delco Supreme 11®, GM P/N 12377967 (Canadian P/N 992667), or equivalent DOT-3 brake fluid from a clean, sealed brake fluid container. Ensure that the brake master cylinder reservoir remains at least half-full during this bleeding procedure. Add fluid as needed to maintain the proper level. Clean the outside of the reservoir on and around the reservoir cap prior to removing the cap and diaphragm.

6. Install a proper box-end wrench onto the RIGHT REAR wheel hydraulic circuit bleeder valve.

7. Install a transparent hose over the end of the bleeder valve.

8. Submerge the open end of the transparent hose into a transparent container partially filled with Delco Supreme 11®, GM P/N 12377967 (Canadian P/N 992667), or equivalent DOT-3 brake fluid from a clean, sealed brake fluid container.

9. Have an assistant slowly depress the brake pedal fully and maintain steady pressure on the pedal.

10. Loosen the bleeder valve to purge air from the wheel hydraulic circuit.

11. Tighten the bleeder valve, then, have the assistant slowly release the brake pedal.

12. Wait 15 seconds, then repeat steps 8-10 until all air is purged from the same wheel hydraulic circuit.

13. With the right rear wheel hydraulic circuit bleeder valve tightened securely, after all air has been purged from the right rear hydraulic circuit, install a proper box-end wrench onto the LEFT FRONT wheel hydraulic circuit bleeder valve.

14. Install a transparent hose over the end of the bleeder valve, then, repeat steps 7-11.

15. With the left front wheel hydraulic circuit bleeder valve tightened securely, after all air has been purged from the left front hydraulic circuit, install a proper box-end wrench onto the LEFT REAR wheel hydraulic circuit bleeder valve.

16. Install a transparent hose over the end of the bleeder valve, then, repeat steps 7-11.

17. With the left rear wheel hydraulic circuit bleeder valve tightened securely, after all air has been purged from the left rear hydraulic circuit, install a proper box-end wrench onto the RIGHT FRONT wheel hydraulic circuit bleeder valve.

18. Install a transparent hose over the end of the bleeder valve, then, repeat steps 7-11.

19. After completing the final wheel hydraulic circuit bleeding procedure, ensure that each of the 4 wheel hydraulic circuit bleeder valves is properly tightened.

20. Fill the brake master cylinder reservoir to the maximum-fill level with Delco Supreme 11®, GM P/N 12377967 (Canadian P/N 992667), or equivalent DOT-3 brake fluid from a clean, sealed brake fluid container.

21. Slowly depress and release the brake pedal. Observe the feel of the brake pedal.

22. If the brake pedal feels spongy, repeat the bleeding procedure again. If the brake pedal still feels spongy after repeating the bleeding procedure, perform the following steps:

a. Inspect the brake system for external leaks.

b. Pressure bleed the hydraulic brake system in order to purge any air that may still be trapped in the system.

23. Turn the ignition key ON, with the engine OFF. Check to see if the brake system warning lamp remains illuminated.

CHEVROLET

23

Malibu • Malibu Maxx

BRAKES 23-43
DRIVE TRAIN 23-33
ENGINE REPAIR 23-9
FUEL SYSTEM 23-29
**SPECIFICATIONS AND
MAINTENANCE CHARTS**......23-2
Engine and Vehicle Identification ..23-2
General Engine Specifications 23-2
Engine Tune-Up Specifications 23-2
Firing Orders 23-3
Accessory Drive Belt Routing 23-3
Capacities 23-3
Valve Specifications 23-4
Crankshaft and Connecting Rod
  Specifications 23-4
Piston and Ring Specifications 23-5
Torque Specifications 23-5
Wheel Alignment 23-6
Tire, Wheel and Ball Joint
  Specifications 23-6
Brake Specifications 23-6
Scheduled Maintenance
  Intervals................................23-7
Maintenance I and II Service
  Schedules..............................23-8
**STEERING AND
SUSPENSION**.................23-36

A
Air Bag...................................23-36
  Arming..................................23-36
  Disarming 23-36
Alternator 23-9
  Removal & Installation 23-9

B
Brake Caliper 23-43
  Removal & Installation............23-43
Brake Drums............................23-43
  Removal & Installation............23-43
Brake Shoes............................23-43
  Removal & Installation............23-43

C
Camshaft and Valve Lifters 23-23
  Removal & Installation............23-23
Coil Spring 23-39
  Removal & Installation............23-39
CV-Joint 23-34
  Overhaul 23-34
Cylinder Head 23-15
  Removal & Installation............23-15

D
Disc Brake Pads.......................23-43
  Removal & Installation............23-43

E
Engine Assembly 23-9
  Removal & Installation...............23-9
Exhaust Manifold......................23-20
  Removal & Installation............23-20

F
Fuel Filter 23-29
  Removal & Installation............23-29
Fuel Injector............................23-31
  Removal & Installation............23-31
Fuel Pump 23-29
  Removal & Installation............23-29
Fuel System Pressure 23-29
  Relieving...............................23-29

H
Halfshaft.................................23-34
  Removal & Installation............23-34
Heater Core.............................23-12
  Removal & Installation............23-12

I
Ignition Timing 23-9
  Adjustment.............................23-9
Intake Manifold 23-18
  Removal & Installation............23-18

L
Lower Ball Joint 23-40
  Removal & Installation............23-40

Lower Control Arm 23-40
  Control Arm Bushing
    Replacement 23-41
  Removal & Installation............23-40

O
Oil Pan...................................23-24
  Removal & Installation............23-24
Oil Pump 23-25
  Removal & Installation............23-25

P
Piston and Ring 23-28
  Positioning.............................23-28
Power Rack and Pinion Steering
  Gear...................................23-36
  Removal & Installation............23-36

R
Rear Main Seal 23-26
  Removal & Installation............23-26
Rocker Arms 23-17
  Removal & Installation............23-17

S
Starter Motor 23-23
  Removal & Installation............23-23
Strut......................................23-37
  Removal & Installation............23-37

T
Timing Chain, Sprockets, Front
  Cover and Seal 23-26
  Removal & Installation............23-26
Transmission Assembly 23-33
  Removal & Installation............23-33

V
Valve Lash 23-23
  Adjustment.............................23-23

W
Water Pump..............................23-11
  Removal & Installation............23-11
Wheel Bearings.........................23-42
  Adjustment.............................23-42
  Removal & Installation............23-42

SPECIFICATIONS AND MAINTENANCE CHARTS

ENGINE AND VEHICLE IDENTIFICATION

			Engine					Model Year	
Code ①	Liters (cc)	Cu. In.	Cyl.	Fuel Sys.	Engine Type	Eng. Mfg.		Code ②	Year
M/J	3.1 (3130)	191	6	SFI	OHV	BOC		2	2002
F	2.2 (2189)	134	4	MPI	DOHC	BOC		3	2003
8	3.5 (3500)	214	6	SFI	OHV	BOC		4	2004
								5	2005

BOC: Buick/Oldsmobile/Cadillac

SFI: Sequential Fuel Injection

MPI: Multi-Point Fuel Injection

① 8th position of VIN

② 10th position of VIN

06025-MALI-C01

GENERAL ENGINE SPECIFICATIONS

Year	Model	Engine Displacement Liters	Engine Series VIN	Net Horsepower @ rpm	Net Torque @ rpm (ft. lbs.)	Bore x Stroke (in.)	Com-pression Ratio	Oil Pressure @ rpm
2002	Malibu	3.1	J	170@5200	190@4000	3.50x3.31	9.6:1	15@1100
2003	Malibu	3.1	J	170@5200	190@4000	3.50x3.31	9.6:1	15@1100
2004	Malibu	2.2	F	145@5600	155@4000	3.39x3.72	10.0:1	50-80@1000
		3.5	8	200@5400	220@3200	3.70x3.31	9.8:1	30-45@1850
2005	Malibu	2.2	F	145@5600	155@4000	3.50x3.31	10.0:1	50-80@1000
		3.5	8	200@5400	220@3200	3.70x3.31	9.8:1	30-45@1850

06025-MALI-C02

ENGINE TUNE-UP SPECIFICATIONS

Year	Engine Displacement Liters	Engine VIN	Spark Plug Gap (in.)	Ignition Timing (deg.)	Fuel Pump (psi)	Idle Speed (rpm)	Valve Clearance In.	Valve Clearance Ex.
2002	3.1	J	0.060	①	41-47	①	HYD	HYD
2003	3.1	J	0.060	①	41-47	①	HYD	HYD
2004	2.2	F	0.042	①	NA	①	HYD	HYD
	3.5	8	0.060	①	NA	①	HYD	HYD
2005	2.2	F	0.042	①	NA	①	HYD	HYD
	3.5	8	0.060	①	NA	①	HYD	HYD

NA: Information not available

NOTE: The Vehicle Emission Control Information label often reflects specification changes made during production.

The label figures must be used if they differ from those in this chart.

HYD: Hydraulic

① Refer to Vehicle Emission Control Information label

06025-MALI-C03

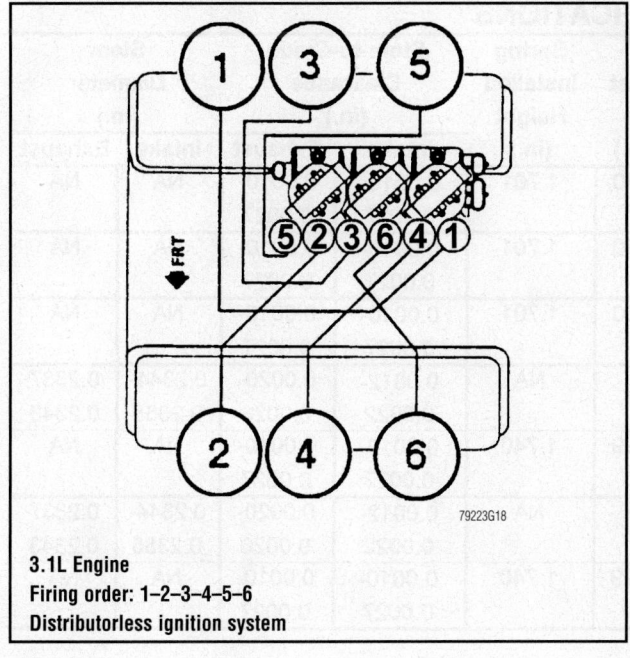

3.1L Engine
Firing order: 1–2–3–4–5–6
Distributorless ignition system

79223G18

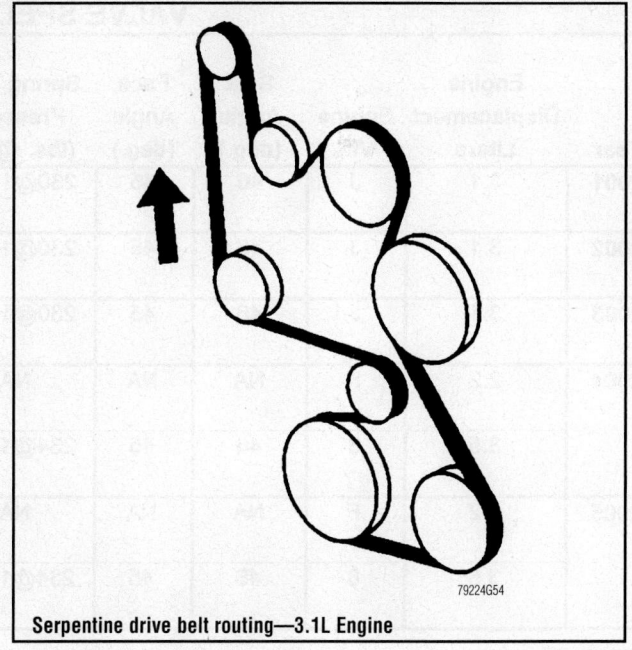

Serpentine drive belt routing—3.1L Engine

79224G54

CAPACITIES

Year	Model	Engine Displacement Liters	Engine VIN	Engine Oil with Filter (qts.)	Transmission (pts.)	Fuel Tank (gal.)	Cooling System (qts.)
2002	Malibu	3.1	J	4.5	13.8	14.3	13.6
2003	Malibu	3.1	J	4.5	13.8	14.3	13.6
2004	Malibu	2.2	F	5.0	19.0	16.1	6.9
	Malibu	3.5	8	4.0	19.0	16.1	10.1
2005	Malibu	2.2	F	5.0	19.0	16.1	6.9
	Malibu	3.5	8	4.0	19.0	16.1	10.1

NOTE: All capacities are approximate. Add fluid gradually and ensure a proper fluid level is obtained.

06025-MALI-C04

VALVE SPECIFICATIONS

Year	Engine Displacement Liters	Engine VIN	Seat Angle (deg.)	Face Angle (deg.)	Spring Test Pressure (lbs. @ in.)	Spring Installed Height (in.)	Stem-to-Guide Clearance (in.)		Stem Diameter (in.)	
							Intake	Exhaust	Intake	Exhaust
2001	3.1	J	46	45	230@1.260	1.701	0.0010-0.0027	0.0010-0.0027	NA	NA
2002	3.1	J	46	45	230@1.260	1.701	0.0010-0.0027	0.0010-0.0027	NA	NA
2003	3.1	J	46	45	230@1.260	1.701	0.0010-0.0027	0.0010-0.0027	NA	NA
2004	2.2	F	NA	NA	NA	NA	0.0012-0.0022	0.0020-0.0026	0.2344-0.2355	0.2337-0.2343
	3.5	8	46	45	234@1.299	1.740	0.0010-0.0027	0.0010-0.0027	NA	NA
2005	2.2	F	NA	NA	NA	NA	0.0012-0.0022	0.0020-0.0026	0.2344-0.2355	0.2337-0.2343
	3.5	8	46	45	234@1.299	1.740	0.0010-0.0027	0.0010-0.0027	NA	NA

NA: Information not available

06025-MALI-C05

CRANKSHAFT AND CONNECTING ROD SPECIFICATIONS
All measurements are given in inches.

Year	Engine Displacement Liters	Engine VIN	Crankshaft				Connecting Rod		
			Main Brg. Journal Dia.	Main Brg. Oil Clearance	Shaft End-play	Thrust on No.	Journal Diameter	Oil Clearance	Side Clearance
2002	3.1	J	2.6473-2.6383	0.0008-① 0.0025	0.0024-0.0083	3	1.9987-1.9994	0.0007-0.0024	0.0070-0.0170
2003	3.1	J	2.6473-2.6383	0.0008-① 0.0025	0.0024-0.0083	3	1.9987-1.9994	0.0007-0.0024	0.0070-0.0170
2004	2.2	F	2.2045-2.2050	0.0012-0.0026	0.0012-0.0150	3	2.0519-2.0525	0.0011-0.0027	0.0028-0.0146
	3.5	8	2.6473-2.6483	0.0008-0.0025	0.0024-0.0083	3	2.2490-2.2500	0.0007-0.0170	0.0080-0.0090
2005	2.2	F	2.2045-2.2050	0.0012-0.0026	0.0012-0.0150	3	2.0519-2.0525	0.0011-0.0027	0.0028-0.0146
	3.5	8	2.6473-2.6483	0.0008-0.0025	0.0024-0.0083	3	2.2490-2.2500	0.0007-0.0170	0.0080-0.0090

① Thrust bearing: 0.0012 - 0.0030

06025-MALI-C06

PISTON AND RING SPECIFICATIONS
All measurements are given in inches.

Year	Engine Displacement Liters	Engine VIN	Piston Clearance	Ring Gap			Ring Side Clearance		
				Top Compression	Bottom Compression	Oil Control	Top Compression	Bottom Compression	Oil Control
2001	3.1	J	0.0013-0.0027	0.006-0.014	0.020-0.028	0.010-0.050	0.0020-0.0033	0.0020-0.0035	0.0080
2002	3.1	J	0.0013-0.0027	0.006-0.014	0.020-0.028	0.010-0.050	0.0020-0.0033	0.0020-0.0035	0.0080
2003	3.1	J	0.0013-0.0027	0.006-0.014	0.020-0.028	0.010-0.050	0.0020-0.0033	0.0020-0.0035	0.0080
2004	2.2	F	0.0004-0.0016	0.008-0.016	0.014-0.022	0.010-0.030	0.0015-0.0031	0.0012-0.0027	0.0035-0.0042
	3.5	8	0.003 max	0.007-0.015	0.019-0.029	0.010-0.029	0.001-0.003	0.002-0.003	0.0040
2005	2.2	F	0.0004-0.0016	0.008-0.016	0.014-0.022	0.010-0.030	0.0015-0.0031	0.0012-0.0027	0.0035-0.0042
	3.5	8	0.003 max	0.007-0.015	0.019-0.029	0.010-0.029	0.001-0.003	0.002-0.003	0.0040

06025-MALI-C07

TORQUE SPECIFICATIONS
All readings in ft. lbs.

Year	Engine Displacement Liters	Engine VIN	Cylinder Head Bolts	Main Bearing Bolts	Rod Bearing Bolts	Crankshaft Damper Bolts	Flywheel Bolts	Manifold		Spark Plug	Oil Pan Drain Plug
								Intake	Exhaust		
2002	3.1	J	①	②	③	76	52	④	12	⑤	18
2003	3.1	J	①	②	③	76	52	④	12	⑤	18
2004	2.2	F	⑥	⑦	⑧	⑨	⑩	10	10	15	18
	3.5	8	⑪	⑫	⑬	⑭	52	⑮	12	11	18
2005	2.2	F	⑥	⑦	⑧	⑨	⑩	10	10	15	18
	3.5	8	⑪	⑫	⑬	⑭	52	⑮	12	11	18

NA: Not Available

① Coat threads with sealer and torque to 37 ft. lbs., then turn an additional 95 degrees

② 37 ft. lbs. plus 75 degrees

③ 15 ft. lbs. plus 75 degrees

④ Lower intake manifold bolts: 10 ft. lbs.
Upper intake manifold bolts: 18 ft. lbs.

⑤ New cylinder first-time installation: 20 ft. lbs.
All others: 11 ft. lbs.

⑥ Step 1: 22 ft. lbs.
Step 2: Plus 155 degrees

⑦ Step 1: 15 ft. lbs.
Step 2: Plus 70 degrees

⑧ Step 1: 18 ft. lbs.
Step 2: Plus 100 degrees

⑨ Step 1: 74 ft. lbs.
Step 2: Plus 125 degrees

⑩ Step 1: 39 ft. lbs.
Step 2: Plus 25 degrees

⑪ Step 1: 44 ft. lbs.
Step 2: Plus 95 degrees

⑫ Step 1: 37 ft. lbs.
Step 2: Plus 77 degrees

⑬ Step 1: 18 ft. lbs.
Step 2: Plus 110 degrees

⑭ Step 1: 52 ft. lbs.
Step 2: Plus 70 degrees

⑮ Lower intake manifold bolts: 11 ft. lbs.
Upper intake manifold bolts: 18 ft. lbs.

06025-MALI-C08

WHEEL ALIGNMENT

Year	Model		Caster Range (+/-Deg.)	Caster Preferred Setting (Deg.)	Camber Range (+/-Deg.)	Camber Preferred Setting (Deg.)	Toe-in (in.)
2002	Malibu	F	1.00	+4.10	0.80	-0.20	0.05 +/- 0.12
		R	—	—	0.50	-0.20	0.06 +/- 0.08
2003	Malibu	F	1.00	+4.10	0.80	-0.20	0.05 +/- 0.12
		R	—	—	0.50	-0.20	0.06 +/- 0.08
2004	Malibu	F	0.75	+3.25	0.50	-1.00	0.20 +/- 0.20
		R	—	—	0.50	-0.80	0.20 +/- 0.20
2005	Malibu	F	0.75	+3.00	0.75	-0.40	0.20 +/- 0.20
		R	—	—	0.50	-0.80	0.20 +/- 0.20

06025-MALI-C09

TIRE, WHEEL AND BALL JOINT SPECIFICATIONS

Year	Model	OEM Tires Standard	OEM Tires Optional	Tire Pressures (psi) Front	Tire Pressures (psi) Rear	Wheel Size	Ball Joint Inspection	Wheel Lug Torque (ft. lbs.)
2002	Malibu	P215/60R15	P215/60HR15	30	26	6-JJ	①	100
2003	Malibu	P215/60R15	P215/60HR15	30	26	6-JJ	①	100
2004	Malibu	P205/65R15	P215/60R15	②	②	6.5	①	100
2005	Malibu	P205/65R15	P215/60R15	②	②	6.5	①	100

OEM: Original Equipment Manufacturer

PSI: Pounds Per Square Inch

① Replace if any measurable movement is found.

② See placard on vehicle

06025-MALI-C10

BRAKE SPECIFICATIONS
All measurements in inches unless noted

Year	Model		Brake Disc Original Thickness	Brake Disc Minimum Thickness	Brake Disc Maximum Runout	Brake Drum Diameter Original Inside Diameter	Brake Drum Diameter Max. Wear Limit	Brake Drum Diameter Maximum Machine Diameter	Minimum Lining Thickness Front	Minimum Lining Thickness Rear	Brake Caliper Bracket Bolts (ft. lbs.)	Brake Caliper Mounting Bolts (ft. lbs.)
2002	Malibu		1.031	0.980	0.003	8.900	8.920	8.909	0.030	①	85	23
2003	Malibu		1.031	0.980	0.002	8.868	8.909	8.889	0.030	①	85	23
2004	Malibu	F	1.023	0.898	0.002	—	—	—	NA	—	85	26
		R	0.551	0.465	0.002	—	—	—	—	NA	85	26
2005	Malibu	F	1.023	0.898	0.002	—	—	—	NA	—	85	26
		R	0.551	0.465	0.002	9.060	9.075	9.094	—	NA	85	26

① 0.030 over rivet head; If bonded lining, use 0.062 from shoe

06025-MALI-C11

SCHEDULED MAINTENANCE INTERVALS
2002-03 Chevrolet Malibu

TO BE SERVICED	TYPE OF SERVICE	3	6	9	12	15	18	21	24	27	30	33	36	39	42	45	48
		**VEHICLE MILEAGE INTERVAL (x1000)**															
Accessory drive belt	S/I	Every 150,000 miles															
Air cleaner filter	S/I				✓						✓					✓	
Automatic transmission fluid	R	Every 50,000 miles															
Brake system ①	S/I	✓	✓	✓	✓	✓	✓	✓	✓	✓	✓	✓	✓	✓	✓	✓	✓
Chassis & suspension grease points	L	✓	✓	✓	✓	✓	✓	✓	✓	✓	✓	✓	✓	✓	✓	✓	✓
CV-joint boots & axle seals	S/I	✓	✓	✓	✓	✓	✓	✓	✓	✓	✓	✓	✓	✓	✓	✓	✓
Engine coolant system ②	S/I	Every 150,000 miles or 60 months															
Engine oil & filter	R	✓	✓	✓	✓	✓	✓	✓	✓	✓	✓	✓	✓	✓	✓	✓	✓
Front wheel bearings	S/I & L				✓						✓					✓	
Fuel filter	R	Every 30,000 miles															
Fuel tank, cap & lines	S/I								✓								✓
PCV valve	S/I	Every 100,000 miles															
Rear/front axle fluid level	S/I	✓	✓	✓	✓	✓	✓	✓	✓	✓	✓	✓	✓	✓	✓	✓	✓
Rotate tires	S/I		✓		✓		✓		✓		✓		✓		✓		✓
Spark plug wires	S/I	Every 100,000 miles															
Spark plugs	R	Every 100,000 miles															

R: Replace   S/I: Inspect and service, if necessary   L: Lubricate

① This should be performed when the tires are removed for rotation.

② Drain, flush and refill the cooling system, inspect the system hoses, and clean the radiator and condenser.

06025-MALI-C12

MAINTENANCE I AND II SERVICE SCHEDULES
2004-05 Chevrolet Malibu

When the CHANGE ENGINE OIL light appears, certain services and inspections are required.
Required services are described as Maintenance I and Maintenance II.
The first service on a vehicle should be Maintenance I, and the second service should be Maintenance II.
Alternate between the 2 thereafter. However, in some cases, Maintenance II may be required more often.
Maintenance I: Use Maintenance I if the CHANGE ENGINE OIL light comes on within 10 months since vehicle was purchased or, if Maintenance II was performed.
Maintenance II: Use Maintenance II if the previous service performed was Maintenance I. Always use Maintenance II whenever the CHANGE ENGINE OIL light comes on 10 months or more since the last service, or, if the CHANGE ENGINE OIL light has not come on at all for one year.

Service	Maintenance I	Maintenance II
Change the engine oil and filter. Reset the oil life system.	✓	✓
Visually inspect the vehicle for leaks or damage. A fluid loss in the vehicle system could indicate a problem. Inspected, repair and add fluid to the system if necessary.	✓	✓
Inspect the engine air cleaner filter. If necessary, replace the filter.	✓	✓
Rotate the tires. Inspect the tire inflation pressures and the tire wear.	✓	✓
Visually inspect the brake lines and hoses for proper hook-up, binding, leaks, cracks, chafing, etc. Inspect the disc brake pads for wear and the rotors for surface condition. Inspect the drum brake linings for wear or cracks. Inspect other brake parts, including drums, wheel cylinders, calipers, parking brake, etc. Inspect the parking brake adjustment.	✓	✓
Inspect the engine coolant and the windshield washer fluid levels. Add fluid as needed.	✓	✓
Inspect the suspension and steering components. Inspect the front and rear suspension and the steering system for damaged, loose or missing parts, or signs of wear. Inspect the power steering lines and the hoses for proper hook-up, binding, leaks, cracks,	--	✓
Visually inspect the coolant hoses and replace the hoses if they are cracked, swollen or deteriorated. Inspect all pipes, fittings and clamps; replace with GM parts as needed. To help ensure proper operation, a pressure test of the cooling system and pressure cap and cleaning the outside of the radiator and air conditioning condenser is recommended at least once a year.	--	✓
Inspect the front and rear suspension and the steering system for damaged, loose or missing parts, or signs of wear. Inspect power steering lines and hoses for proper hook-up, binding, leaks, cracks, chafing, etc.	--	✓
Inspect the throttle system for interference or binding and for damaged or missing parts. Replace the parts as needed. Replace any components that have high effort or excessive wear. Do not lubricate the accelerator or the cruise control cables.	--	✓
Replace the passenger compartment air filter.	--	✓

Reset the oil life system:
1. Display OIL LIFE RESET on the DIC.
2. Press and hold the ENTER button for at least one second. An ACKNOWLEDGED display message will appear for three seconds or until the next button is pressed. This will tell you the system has been reset.
3. Turn the key to OFF.

If the Change Oil Soon message comes back on when you start your vehicle, the engine oil life system has not reset, repeat the procedure.

06025-MALI-C13

ENGINE REPAIR

Alternator

REMOVAL & INSTALLATION

2.2L Engine

1. Before servicing the vehicle, refer to the Precautions Section.
2. Remove or disconnect the following:
   - Negative battery cable
   - Air intake assembly
   - Oil dipstick assembly mounting bolt
   - Accessory drive belt
   - Alternator electrical connectors
   - Alternator

To install:

3. Install or connect the following:
   - Alternator. Tighten bolts to 16 ft. lbs. (22 Nm)
   - Alternator electrical connections. Tighten nut to 15 ft. lbs. (20 Nm).
   - Accessory drive belt
   - Oil dipstick assembly mounting bolt
   - Air intake assembly
   - Negative battery cable

3.1L and 3.5L Engines

1. Before servicing the vehicle, refer to the Precautions Section.
2. Remove or disconnect the following:
   - Negative battery cable
   - Accessory drive belt
   - Alternator electrical connectors
   - Power steering line clip
   - Alternator

To install:

3. Install or connect the following:
   - Alternator. Torque the nuts to 22 ft. lbs. (30 Nm) and the bolts to 37 ft. lbs. (50 Nm).
   - Alternator electrical connector
   - Power steering line clip
   - Accessory drive belt
   - Negative battery cable

Ignition Timing

ADJUSTMENT

Ignition timing is controlled by the Powertrain Control Module (PCM). No adjustment is necessary or possible.

Engine Assembly

REMOVAL & INSTALLATION

2.2L Engine

1. Before servicing the vehicle, refer to the Precautions Section.
2. Relieve the fuel system pressure.
3. Drain the engine oil.
4. Drain the engine coolant.
5. Remove or disconnect the following:
   - Negative battery cable
   - Air intake assembly
   - Fuel lines from the fuel rail
   - Coolant hoses
   - Heater hoses
   - Throttle control electrical connector
   - MAP sensor electrical connector
   - Crankshaft sensor electrical connector
   - Oil pressure electrical connector
   - Purge solenoid electrical connector
   - Ignition coil electrical connector
   - O_2 sensor electrical connector
   - Vehicle speed sensor (VSS) electrical connector
   - Coolant temperature sensor electrical connector
   - Back-up lamp switch electrical connector
6. Raise and support the vehicle.
7. Remove or disconnect the following:
   - Accessory drive belt
   - A/C compressor mounting bolts and set compress aside
   - Starter electrical connectors
   - Alternator electrical connector
   - Front exhaust down pipe from the exhaust manifold
   - Transmission harness connectors
   - Transmission shift cable from the transmission
8. Support the powertrain assembly between the frame and the powertain using a block of wood or suitable equivalent.
9. Remove or disconnect the following:
   - Engine mount
   - Side transmission mount, if equipped with automatic transmission
   - Stabilizer links
   - Outer tie rods from steering knuckles
   - Intermediate shaft from the steering gear

✻✻ WARNING

In order to prevent possible SIR system deployment, do not attempt to rotate the steering shaft.

   - Lower control arms from the steering knuckles
   - Halfshafts from the steering knuckles
10. Matchmark the frame to body position.
11. Position a lift table under the frame to support the frame assembly.
12. Slowly remove the front frame bolts.
13. Partially unscrew the rear frame bolts until 1.5 inches of bolt shank is exposed.
14. Slowly lower the lift table.
15. Attach an engine lift hoist to the engine.
16. Remove the starter.
17. Remove the torque converter to flywheel bolts, if equipped.
18. Remove the transmission to engine bolts.
19. Separate the engine from the transmission.
20. Remove the clutch pressure plate and disc, if equipped.
21. Remove or disconnect the following:
    - Exhaust manifold and studs
    - Catalytic converter assembly
    - Engine mount bracket
    - Engine block heater, if equipped
    - Thermostat housing and feed pipe
    - Alternator

To install:

22. If removed, install or connect the following:
    - Exhaust manifold
    - Engine mount bracket
    - Fuel rail
    - Engine block heater, if equipped
    - Thermostat housing and feed pipe
    - Alternator
    - Flywheel
    - Clutch pressure plate and disc, if equipped
23. Align the engine to the transmission. Tighten to the engine-to-transmission bolts to 55 ft. lbs. (75 Nm).
24. Install the torque converter bolts, if equipped. Tighten bolts to 46 ft. lbs. (62 Nm).
25. Install the starter.
26. Reinstall the powertrain assembly

to the frame assembly. Raise and position the frame/powertrain assembly to the vehicle.

27. Hand start all the frame bolts while aligning the frame to the body matchmarks. Tighten the frame bolts to 74 ft. lbs. (100 Nm) plus 180 degrees.

28. Remove the lift table.

29. Install or connect the following:
- Halfshafts to the steering knuckles
- Lower control arms to the steering knuckles
- Intermediate steering shaft to the steering gear
- Outer tie rods to the steering knuckles
- Stabilizer links
- Side transmission mount, if equipped with automatic transmission
- Engine mount
- Transmission shift cable to the transmission
- Transmission wiring harness connector
- Front exhaust down pipe to the exhaust manifold. Tighten nuts to 22 ft. lbs. (30 Nm).
- Alternator electrical connectors
- Starter harness connectors
- A/C compressor
- Accessory drive belt
- Throttle control electrical connector
- MAP sensor electrical connector
- Oil pressure electrical connector
- Purge solenoid electrical connector
- Ignition coil electrical connector
- O$_2$ sensor electrical connector
- VSS electrical connector
- Coolant temperature sensor electrical connector
- Heater hoses
- Coolant hoses
- Fuel line to fuel rail
- Air intake assembly
- Negative battery cable

30. Fill with engine oil to the correct level.

31. Fill the cooling system to the correct level.

32. Start the engine and check for leaks.

3.1L Engine

1. Before servicing the vehicle, refer to the Precautions Section.

2. Relieve the fuel system pressure.

3. Drain the engine coolant.

4. Remove or disconnect the following:
- Air cleaner assembly
- Hood
- Accessory drive belt

- Coolant inlet line from the coolant surge tank
- Cruise control module
- Upper engine wiring harness
- Throttle and cruise control cables
- Starter motor
- A/C compressor. DO NOT remove the lines.
- Lower engine wiring harness and position aside
- Catalytic converter
- Torque converter cover
- Torque converter bolts
- Engine splash shields
- Transaxle-to-engine brace
- 2 outer transmission mounting bolts
- Upper and lower radiator hoses
- Fuel lines
- Brake booster vacuum line
- Heater hoses

5. Install an engine support fixture.

6. Remove or disconnect the following:
- Engine mount
- Power steering pump
- Transmission-to-engine bolts
- Engine from the vehicle

To install:

7. Install or connect the following:
- Transaxle to the engine. Torque the bolts to 55 ft. lbs. (75 Nm).
- Power steering pump. Torque the bolts to 25 ft. lbs. (34 Nm).
- Engine mount. Torque the bolts to 35 ft. lbs. (47 Nm).

8. Remove the engine support fixture.

9. Install or connect the following:
- Heater hoses
- Brake booster vacuum line
- Fuel lines
- Radiator hoses
- Engine splash shields
- Torque converter bolts. Torque the bolts to 46 ft. lbs. (62 Nm).
- Torque converter cover. Torque the bolts to 89 inch lbs. (10 Nm).
- Catalytic converter
- Lower engine wiring harness
- A/C compressor. Torque the bolts to 37 ft. lbs. (50 Nm).
- Starter motor
- Throttle and cruise control cables
- Upper engine wiring harness
- Cruise control module
- Coolant surge tank hose
- Accessory drive belt
- Hood
- Air cleaner assembly

10. Refill the coolant.

11. Start the engine and check for proper operation.

3.5L Engine

1. Before servicing the vehicle, refer to the Precautions Section.

2. Relieve the fuel system pressure.

3. Drain the engine oil.

4. Drain the engine coolant.

5. Remove or disconnect the following:
- Negative battery cable
- Air intake assembly
- Hood
- Engine mount strut
- Accessory drive belt
- Knock sensor (KS) electrical connector
- Camshaft position (CMP) sensor electrical connector
- Crankshaft position (CKP) sensor electrical connector
- Heated oxygen sensor (HO2S) electrical connector
- Manifold absolute pressure (MAP) sensor electrical connector
- Exhaust gas recirculation (EGR) valve electrical connector
- Evaporative emission (EVAP) canister purge solenoid electrical connector
- Throttle control electrical connector
- Ignition coil electrical connector
- Body wiring harness

6. Raise and support the vehicle.

7. Remove or disconnect the following:
- Catalytic converters
- Engine wiring harness grounds
- Engine mount lower nuts
- Torque converter covers
- Starter
- A/C compressor mounting bolts and support the compress aside
- Torque converter bolts
- Transmission support brace
- Lower transmission-to-engine bolt
- Radiator hoses

8. Lower the vehicle and support the transmission.

9. Remove or disconnect the following:
- Heater hoses
- Vacuum hoses from upper intake manifold
- Fuel lines from the fuel rail

10. Install a suitable engine lifting device to the engine.

11. Remove the remaining transmission-to-engine bolts.

12. Remove the engine from the vehicle.

To install:

13. Install the engine to the transmission. Tighten the upper bolts to 55 ft. lbs. (75 Nm).

14. Remove the engine lifting device.

15. Install or connect the following:

- Radiator hoses
- Fuel lines to the fuel rail
- Brake booster vacuum hose to the upper intake manifold
- Heater hoses
- Lower transmission bolt and tighten to 55 ft. lbs. (75 Nm)
- Transmission support brace. Tighten bolts to 53 ft. lbs. (72 Nm)
- Torque converter bolts
- A/C compressor mounting bolts
- Starter
- Torque converter covers
- Engine mount lower nuts. Tighten nuts to 32 ft. lbs. (43 Nm).
- Engine wiring grounds. Tighten nut to 26 ft. lbs. (35 Nm).
- Catalytic converters
- Body wiring harness
- Ignition coil electrical connector
- Throttle control electrical connector
- EVAP canister purge solenoid electrical connector
- EGR valve electrical connector
- MAP sensor electrical connector
- HO2S electrical connector
- CKP sensor electrical connector
- CMP sensor electrical connector
- KS electrical connector
- Accessory drive belt
- Engine mount strut
- Hood
- Air intake assembly
- Negative battery cable

16. Fill the engine with oil to the correct level.

17. Fill the cooling system to the correct level.

18. Start the engine and check for leaks.

Water Pump

REMOVAL & INSTALLATION

2.2L Engine

1. Before servicing the vehicle, refer to the Precautions Section.
2. Drain the cooling system.

➡**A drain plug is located at the bottom of the water pump assembly for additional coolant drainage from the engine block and water pump.**

3. Remove or disconnect the following:
- Negative battery cable
- Thermostat housing pipe-to-cylinder head bolt near the front of the engine
- Exhaust manifold heat shield
- Water pump access plate

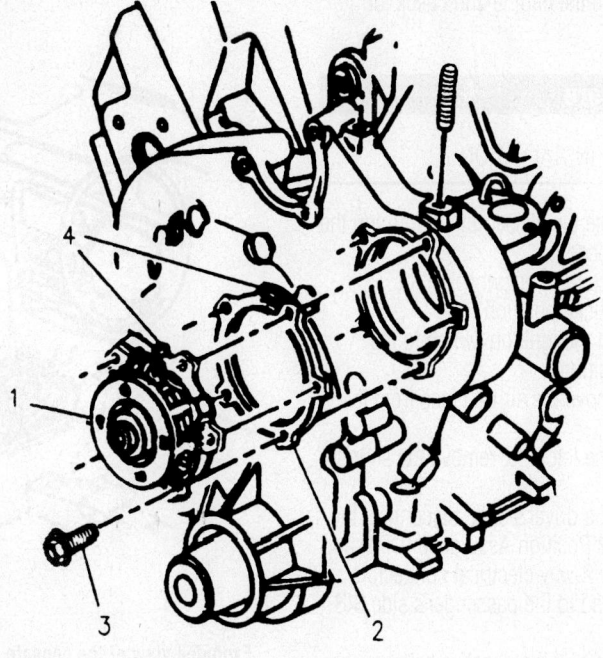

1 COOLANT PUMP
2 GASKET
3 BOLT – 10 N·m (89 LBS. IN.)
4 LOCATOR (MUST BE VERTICAL)

7922Z202

Exploded view of the water pump mounting

- Coolant temperature sensor
- Thermostat housing

➡**Leave the coolant hoses connected.**

- Coolant inlet hose

4. Install special tool J 43651 to prevent movement of the camshaft sprocket and timing chain.

5. Remove 3 inner sprocket-to-water pump mounting bolts.

6. Remove water pump assembly bolts

7. Remove the water pump.

To install:

8. Install or connect the following:
- Water pump with a new gasket. Tighten the assembly bolts to 18 ft. lbs. (25 Nm).
- Sprocket-to-water pump bolts and tighten to 89 inch lbs. (10 Nm)

9. Remove special tool J 43651.

10. Install or connect the following:
- Coolant inlet hose with new gasket
- Thermostat housing and tighten to 89 inch lbs. (10 Nm)
- Coolant temperature sensor
- Water pump access plate
- Thermostat housing pipe-to-cylinder head bolt to 71 inch lbs. (8 Nm)
- Exhaust manifold heat shield
- Negative battery cable

11. Fill the cooling system to the correct level.

12. Start the engine and check for leaks.

3.1L and 3.5L Engine

1. Before servicing the vehicle, refer to the Precautions Section.
2. Drain the cooling system.
3. Remove or disconnect the following:
- Negative battery cable
- Water pump pulley bolts, loosen
- Accessory drive belt
- Water pump pulley
- Water pump bolts, pump and gasket

To install:

4. Apply a thin bead of sealer around the outside edge of the water pump along the gasket sealing area.

5. Install or connect the following:
- Water pump with a new gasket. Torque the bolts to 89 inch lbs. (10 Nm).
- Water pump pulley and finger-tighten the bolts
- Accessory drive belt
- Water pump pulley bolts and torque to 18 ft. lbs. (24 Nm)
- Negative battery cable

6. Refill and bleed the cooling system.

7. Operate the engine and check for leaks.

Heater Core

REMOVAL & INSTALLATION

1. Disable the air bag by performing the following procedure:

a. Place the front wheel in the straight-ahead position.

b. Turn the ignition switch to the LOCK position.

c. Remove the air bag fuse from the fuse block.

d. At the left side, remove the sound insulator.

e. At the driver's side, disconnect the Connector Position Assurance (CPA) and the yellow 2-way electrical connectors and the lead to the passenger's side SIR module.

2. Disconnect the negative battery cable.

3. Drain the cooling system into a clean container for reuse.

4. Disconnect the heater hoses from the heater core and remove the drain tube from the heater/air conditioning housing.

5. Remove the steering wheel by removing or disconnecting the following:

- 2 SIR module-to-steering wheel screws (located at the rear of the steering wheel), and the module
- SIR module and disconnect the electrical connector and remove the SIR module

❊❊ CAUTION

Place the SIR module in a safe place with the front facing upward.

- Horn and cruise control electrical connectors
- Steering wheel-to-steering column nut
- Steering wheel from the steering column

6. Remove the console by removing or disconnecting the following:

- Front seats
- Fold the console compartment upward
- Console trim plate
- Rear cup holder
- Console-to-chassis screws and the console

7. Remove the instrument panel by removing or disconnecting the following:

- Defroster grille
- Instrument panel valance screws and the valance

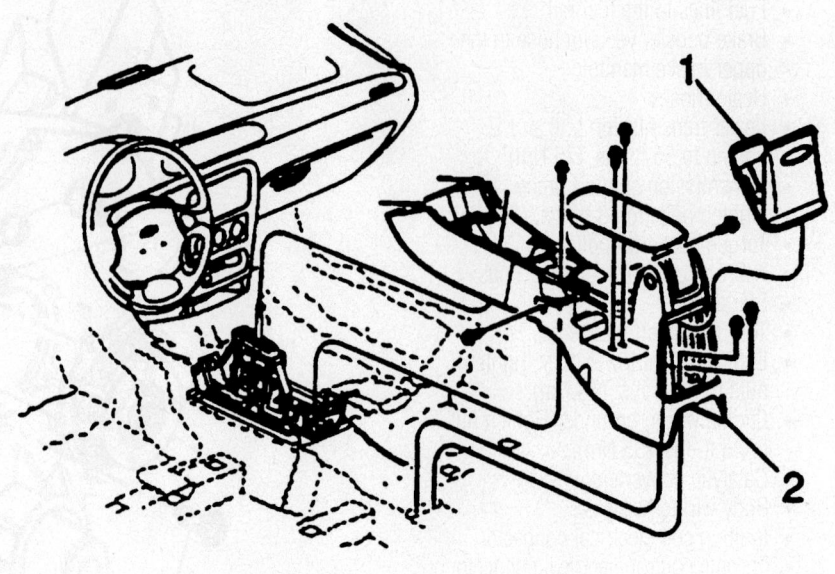

93111GB9

Exploded view of the console and related components—Malibu

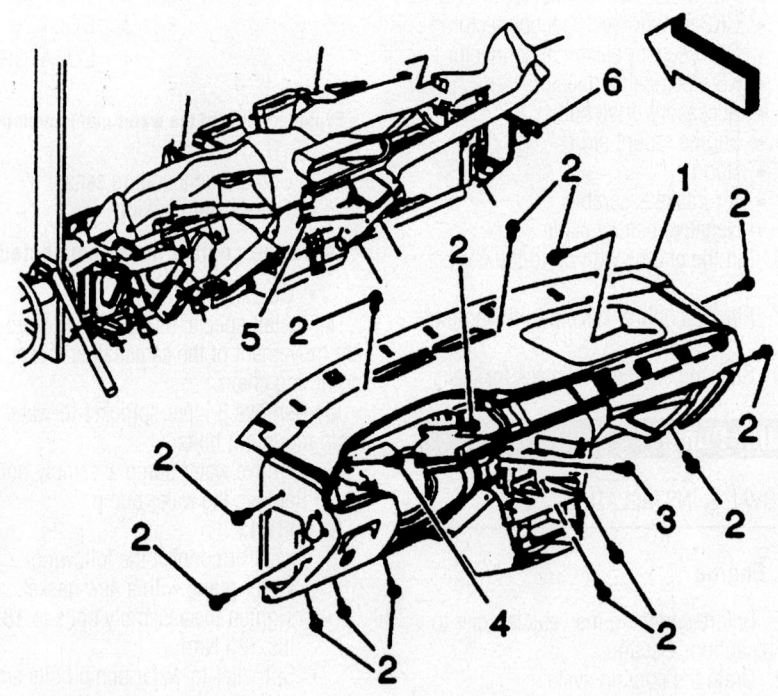

93111GB0

View of instrument panel—Malibu

- Instrument panel end caps from both sides
- Screws located under the instrument panel end caps
- Instrument panel compartment
- Steering column covers
- Steering column stalks
- Instrument cluster and accessory trim plates

- Instrument cluster fasteners, disconnect the electrical connectors and remove the cluster
- Heater/air conditioning control assembly
- Stereo/tape deck assembly
- Ignition switch
- Upper windshield side garnish molding

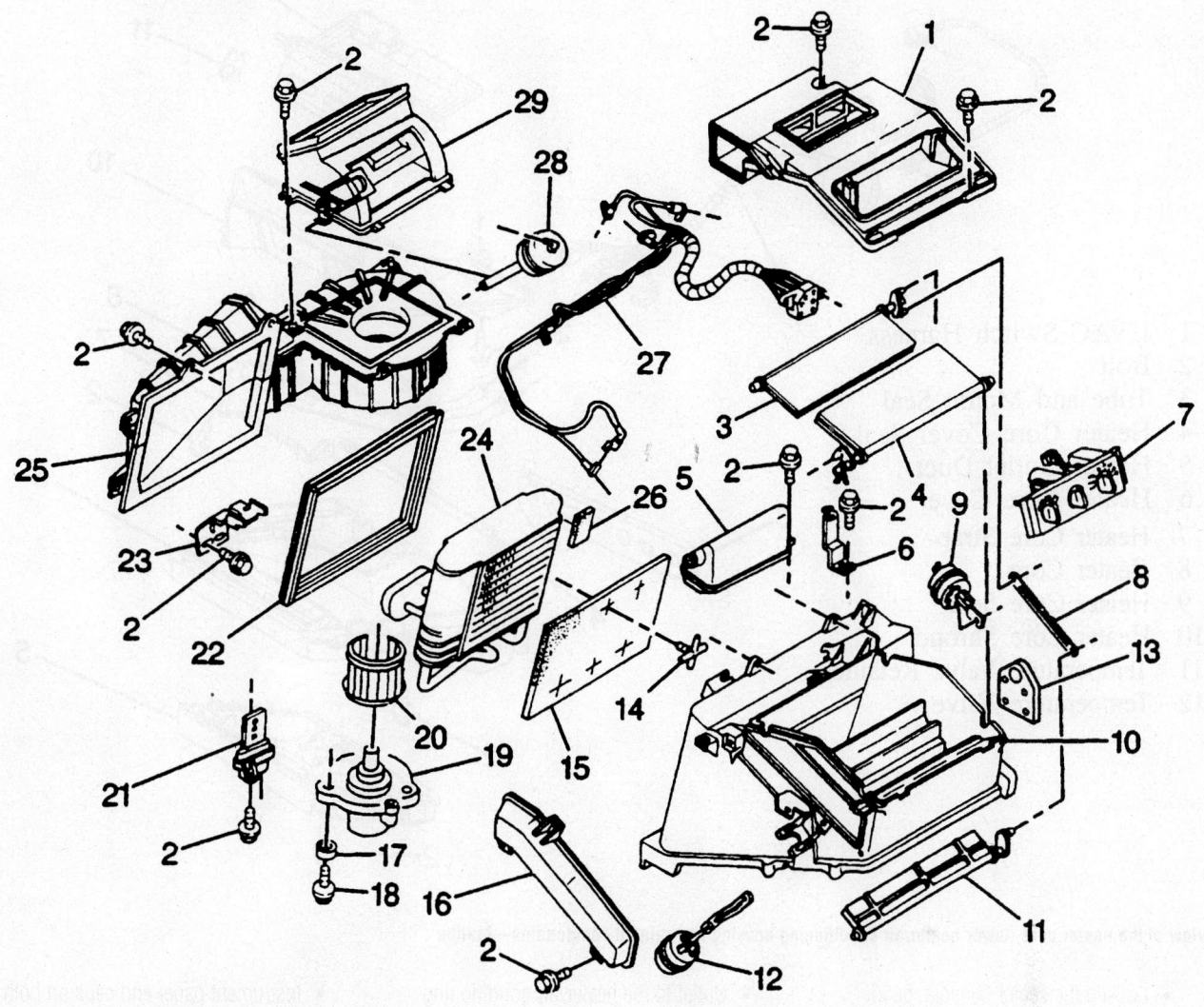

1	Valve Housing Cover	16	Defroster Case
2	Bolt	17	Blower Motor Bolt Insulator
3	Defroster Valve	18	Blower Motor Bolt
4	Mode Valve	19	Blower Motor
5	Vacuum Tank	20	Blower Motor Fan
6	HVAC Module Bracket	21	Blower Motor Resistor
7	HVAC Control Assembly	22	Evaporator Core Seal
8	Heater Valve Link	23	Evaporator Core Bracket
9	Defroster Valve Actuator	24	Evaporator Core
10	Evaporator Case	25	Blower and Air Inlet Case
11	Heater Valve	26	Evaporator Core Spacer
12	Mode Valve Actuator	27	HVAC Vacuum Harness
13	Temperature Control Motor	28	Air Inlet Valve Actuator
14	Water Filter Retainer	29	Air Inlet Case
15	Water Filter		

93111GC1

View of the evaporator core, upper heater/air conditioning housing and related components—Malibu

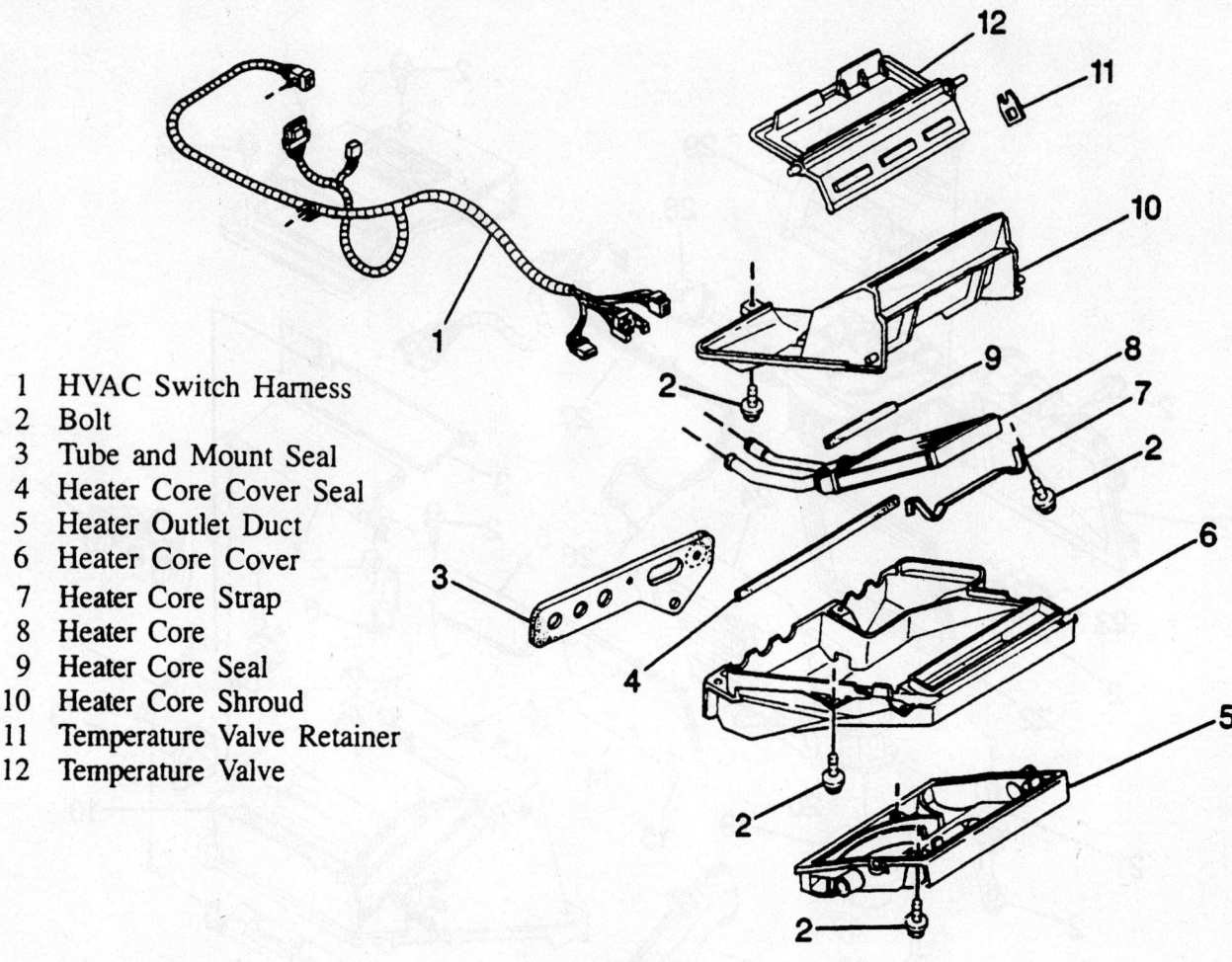

1 HVAC Switch Harness
2 Bolt
3 Tube and Mount Seal
4 Heater Core Cover Seal
5 Heater Outlet Duct
6 Heater Core Cover
7 Heater Core Strap
8 Heater Core
9 Heater Core Seal
10 Heater Core Shroud
11 Temperature Valve Retainer
12 Temperature Valve

93111GC2

View of the heater core, lower heater/air conditioning housing and related components—Malibu

- Loosen the center console, pull it rearward to disengage it from the instrument panel
- Instrument panel-to-tie bar screws
- Instrument panel
- Outlet from the heater/air conditioning housing
- Heater core cover-to-heater/air conditioning housing

➡**There is a screw located in the recess in the center of the cover.**

- Heater core-to-heater/air conditioning housing clamps
- Heater core

To install:

8. Install or connect the following:
- Heater core
- Heater core-to-heater/air conditioning housing clamps
- Heater core cover-to-heater/air conditioning housing

➡**There is a screw located in the recess in the center of the cover.**

- Outlet to the heater/air conditioning housing

9. Install the instrument panel by installing or connecting the following:
- Instrument panel
- Instrument panel-to-tie bar screws
- Move the center console forward and engage it to the instrument panel
- Upper windshield side garnish molding
- Ignition switch
- Stereo/tape deck assembly
- Heater/air conditioning control assembly
- Instrument cluster, connect the electrical connectors and install the cluster fasteners
- Instrument cluster accessory trim plates
- Steering column stalks
- Steering column covers
- Instrument panel compartment
- Instrument panel end caps, install the screws

- Instrument panel end caps on both sides
- Instrument panel valance and the valance screws
- Defroster grille

10. Install the console by installing or connecting the following:
- Console and the console-to-chassis screws
- Rear cup holder
- Console trim plate
- Front seats

11. Install the steering wheel by installing or connecting the following:
- Steering wheel to the steering column
- Steering wheel-to-steering column nut and torque to 30 ft. lbs. (41 Nm)
- Horn and cruise control electrical connectors
- SIR module and connect the electrical connector
- SIR module and torque the 2 mod-

ule-to-steering wheel screws to 89 inch lbs. (10 Nm)

- Heater hoses to the heater core and install the drain tube to the heater/air conditioning housing

12. Refill the cooling system.

13. Connect the negative battery cable.

14. Enable the SIR or air bag by installing or connecting the following:

- Connector Position Assurance (CPA), the yellow 2-way electrical connectors and the lead to the passenger's side SIR module (located on the drivers' side)
- Sound insulator, located on the left side
- Air bag fuse to the fuse block

a. Turn the ignition switch to RUN and verify that the Air Bag Warning light flashes 7–9 times and turns OFF.

➡**If the SIR system does not operate as described, perform the SIR diagnostic system check.**

15. Operate the engine to normal operating temperatures; then, check the climate control operation and check for leaks.

Cylinder Head

REMOVAL & INSTALLATION

2.2L Engine

1. Before servicing the vehicle, refer to the Precautions Section.

2. Drain the cooling system.

3. Relieve the fuel system pressure.

4. Remove or disconnect the following:

- Intake manifold
- Exhaust manifold
- Timing chain
- Cylinder head bolts in sequence
- Cylinder head

To install:

5. Install the cylinder head using new gaskets. Tighten the bolts in sequence to 22 ft. lbs. (30 Nm) plus 155 degrees.

6. Using new bolts, install the front cylinder head bolts. Tighten to 26 ft. lbs. (35 Nm).

7. Install or connect the following:

- Timing chain
- Exhaust manifold
- Intake manifold

8. Fill the cooling system to the correct level.

9. Start the engine and check for leaks.

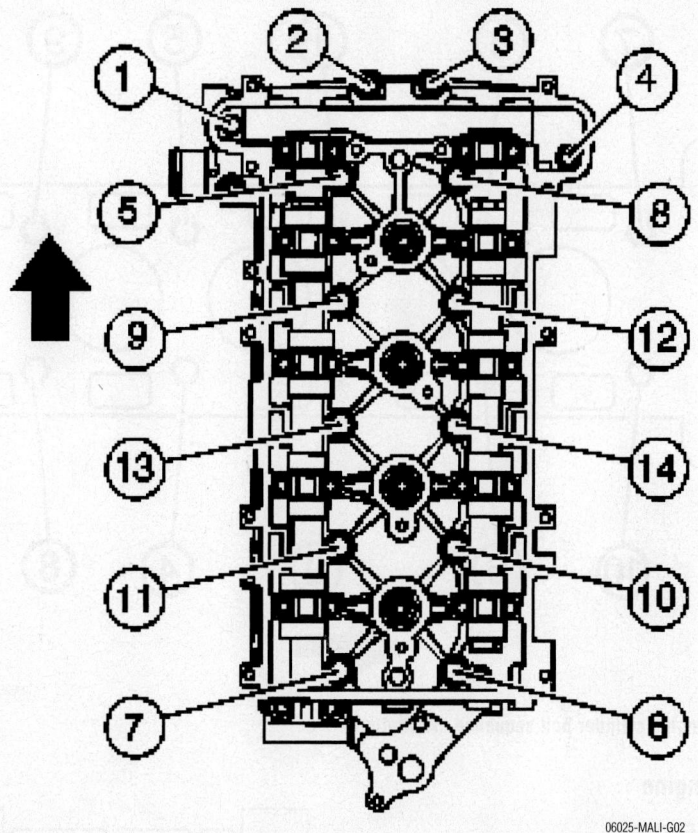

2.2L engine cylinder head bolt removal sequence.

06025-MALI-G02

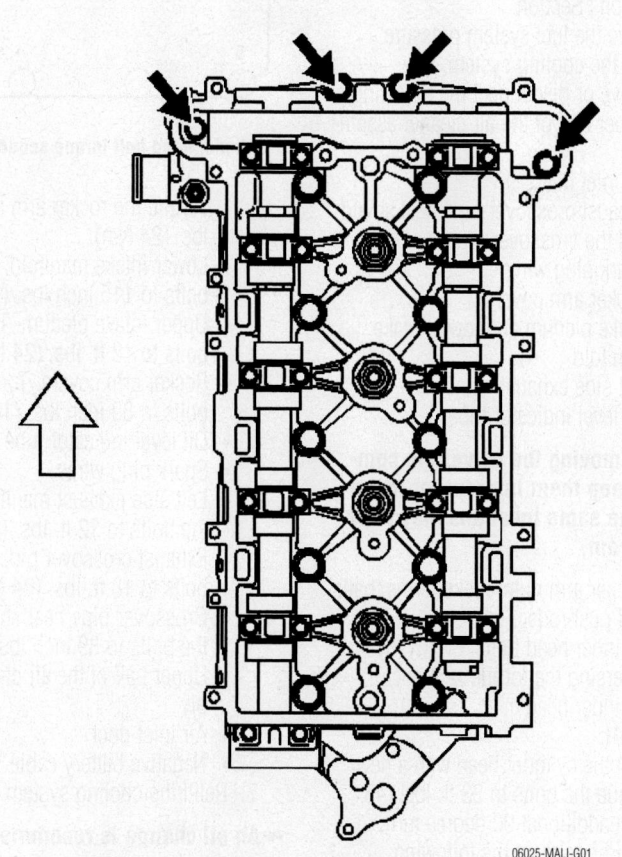

06025-MALI-G01

Location of the front four cylinder head bolts–2.2L engine

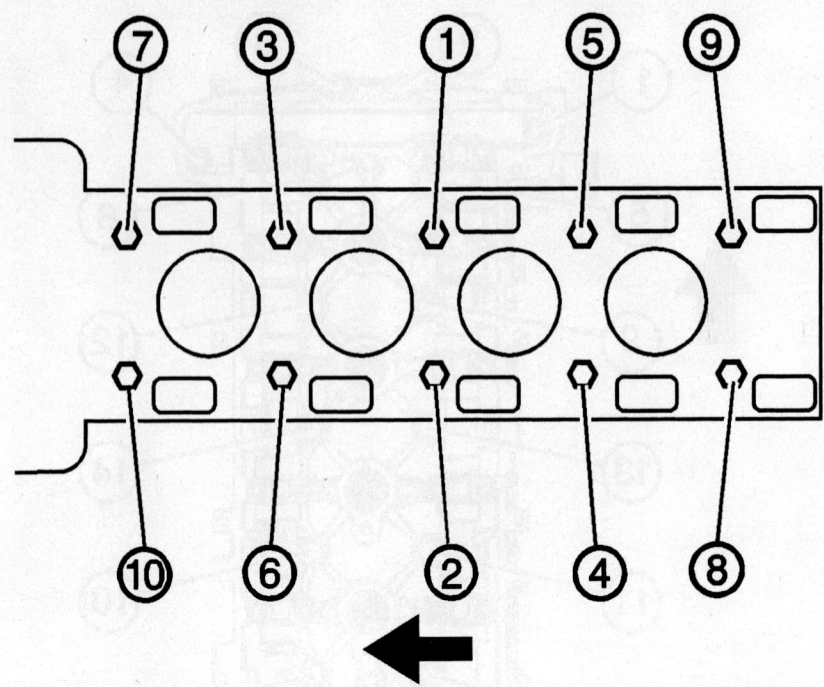

2.2L engine cylinder bolt sequence installation.

06025-MALI-G03

3.1L Engine

LEFT (FRONT)

1. Before servicing the vehicle, refer to the Precautions Section.
2. Relieve the fuel system pressure.
3. Drain the cooling system.
4. Remove or disconnect the following:
   - Upper half of the air cleaner assembly
   - Air inlet duct
   - Exhaust crossover pipe heat shield and the crossover pipe
   - Spark plug wires
   - Rocker arm covers
   - Intake plenum and lower intake manifold
   - Left side exhaust manifold
   - Oil level indicator tube

➡ **When removing the valvetrain components, keep them in order for installation in the same locations they were removed from.**

   - Rocker arm nuts, rocker arms, balls and pushrods
   - Cylinder head bolts, evenly, by reversing the torque sequence
   - Cylinder head

To install:

5. Install the cylinder head with a new gasket. Torque the bolts to 33 ft. lbs. (45 Nm) plus an additional 90 degree turn.
6. Install or connect the following:
   - Pushrods, rocker arms and balls.

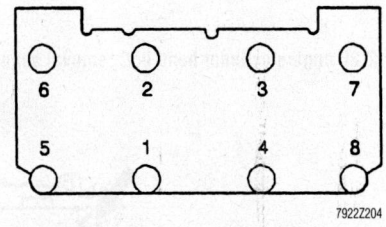

Cylinder head bolt torque sequence

79222204

Torque the rocker arm nuts to 18 ft. lbs. (24 Nm).
   - Lower intake manifold. Torque the bolts to 115 inch lbs. (13 Nm).
   - Upper intake plenum. Torque the bolts to 18 ft. lbs. (24 Nm).
   - Rocker arm covers. Torque the bolts to 89 inch lbs. (10 Nm).
   - Oil level indicator tube
   - Spark plug wires
   - Left side exhaust manifold. Torque the bolts to 12 ft. lbs. (16 Nm).
   - Exhaust crossover pipe. Torque the bolts to 18 ft. lbs. (24 Nm).
   - Crossover pipe heat shield. Torque the bolts to 89 inch lbs. (10 Nm).
   - Upper half of the air cleaner assembly
   - Air inlet duct
   - Negative battery cable
7. Refill the cooling system.

➡ **An oil change is recommended.**

8. Start the vehicle and check for leaks.

RIGHT (REAR)

1. Before servicing the vehicle, refer to the Precautions Section.
2. Relieve the fuel system pressure.
3. Drain the cooling system.
4. Remove or disconnect the following:
   - Upper half of the air cleaner assembly
   - Throttle body air inlet duct
   - Exhaust crossover pipe heat shield and the crossover pipe
   - Oxygen (O2S) sensor
   - Exhaust pipe from the exhaust manifold
   - Right side exhaust manifold
   - Spark plug wires
   - Rocker arm cover
   - Upper intake plenum
   - Lower intake manifold

➡ **When removing the valvetrain components, keep them in order for installation in the same locations they were removed from.**

   - Rocker arms nut, rocker arms, balls and pushrods
   - Cylinder head bolts, evenly, by reversing the torque sequence
   - Cylinder head

To install:

5. Install or connect the following:
   - Cylinder head with a new gasket. Torque the bolts to 37 ft. lbs. (50 Nm) plus an additional 90 degree turn.
   - Pushrods, rocker arms and balls. Torque the rocker arm nuts to 18 ft. lbs. (24 Nm).
   - Lower intake manifold. Torque the bolts to 115 inch lbs. (13 Nm).
   - Upper intake plenum. Torque the bolts to 18 ft. lbs. (24 Nm).
   - Rocker arm covers. Torque the bolts to 89 inch lbs. (10 Nm).
   - Spark plug wires
   - Exhaust manifold. Torque the bolts to 12 ft. lbs. (16 Nm).
   - Exhaust pipe to the exhaust manifold. Torque the bolts to 26 ft. lbs. (35 Nm).
   - O2S sensor
   - Exhaust crossover pipe. Torque the bolts to 18 ft. lbs. (24 Nm).
   - Heat shield. Torque the bolts to 89 inch lbs. (10 Nm).
   - Upper half of the air cleaner assembly
   - Throttle body air inlet duct
   - Negative battery cable
6. Refill the cooling system.

7. An oil and filter change is recommended.

8. Start the vehicle and check for leaks.

3.5L Engine

LEFT (FRONT)

1. Before servicing the vehicle, refer to the Precautions Section.
2. Drain the engine oil.
3. Drain the cooling system.
4. Relieve the fuel system pressure.
5. Remove or disconnect the following:
   - Lower intake manifold
   - Rocker arms and pushrods
   - Exhaust crossover pipe
   - Oil level indicator tube
   - Spark plug wires and spark plugs
   - Exhaust manifold
   - Cylinder head bolts
   - Cylinder head

To install:

6. Install or connect the following:
   - Cylinder head using new gasket. Tighten the bolts in sequence to 44 ft. lbs. (60 Nm) plus 95 degrees.
   - Exhaust manifold
   - Spark plugs and spark plug wires
   - Oil level indicator tube
   - Exhaust crossover pipe
   - Rocker arms and pushrods
   - Lower intake manifold
7. Fill the engine with oil to the correct level.
8. Fill the cooling system to the correct level.
9. Start the engine and check for leaks.

RIGHT (REAR)

1. Before servicing the vehicle, refer to the Precautions Section.
2. Drain the cooling system.
3. Drain the engine oil.
4. Relieve the fuel system pressure.
5. Remove or disconnect the following:
   - Negative battery cable
   - Lower intake manifold
   - Rocker arms and pushrods
   - Exhaust crossover pipe

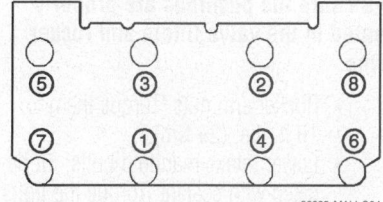

3.5L engine cylinder head bolt torque sequence

06025-MALI-G04

- Right side spark plug wires and spark plugs
- Fuel line bracket
- Alternator
- Right exhaust manifold
- Cylinder head bolts, in reverse order of the torque sequence
- Cylinder head

To install:

6. Install or connect the following:
   - Cylinder head with new gasket. Torque the bolts to 44 ft. lbs. (60 Nm) plus and additional 95 degrees.
   - Right exhaust manifold
   - Alternator
   - Fuel line bracket. Tighten bolt to 37 ft. lbs. (50 Nm).
   - Right side spark plugs and spark plug wires
   - Exhaust crossover pipe
   - Pushrods and rocker arms
   - Lower intake manifold
   - Negative battery cable
7. Fill engine with oil to the correct level.
8. Fill the cooling system to the correct level.
9. Start the engine and check for leaks.

Rocker Arms

REMOVAL & INSTALLATION

Left Side (FRONT)

1. Before servicing the vehicle, refer to the Precautions Section.
2. Disconnect the negative battery cable.
3. Partially drain the cooling system to a level below the coolant pipe.
4. Remove or disconnect the following:
   - Rear ignition wire harness at the upper intake manifold and at the spark plugs
   - Coolant bypass hose clamp at the coolant tube
   - Coolant tube from the cylinder head and move it aside
   - Positive Crankcase Ventilation (PCV) valve from the rocker arm cover
   - Rocker arm cover
   - Rocker arms and pushrods

To install:

➡ **The intake pushrods are identified with yellow stripes and are 5 ̣ inches long. The exhaust pushrods are identified with green stripes and are 6 inches long.**

5. Coat all valvetrain components with engine oil.
6. Install the pushrods and rocker arms. Torque the rocker arm as follows:
   a. 3.1L Engines: Tighten to 89 inch lbs. (10 Nm) plus an additional 30 degree turn.
   b. 3.5L Engines: Tighten to 24 ft. lbs. (32 Nm)
7. Install or connect the following:
   - Rocker arm cover with a new gasket. Torque the bolts to 89 inch lbs. (10 Nm).
   - PCV valve
   - Coolant tube and thermostat bypass hose. Torque the screw at the water pump to 106 inch lbs. (12 Nm) and the nut/bolt at the cylinder head to 18 ft. lbs. (24 Nm).
   - Rear ignition wire harness to the upper intake manifold and the spark plugs
   - Negative battery cable
8. Refill the cooling system.
9. Start the vehicle and verify no leaks.

Right Side (REAR)

1. Before servicing the vehicle, refer to the Precautions Section.
2. Remove or disconnect the following:
   - Negative battery cable
   - Spark plug wires
   - Power brake booster vacuum pipe from the intake plenum
   - Accessory drive belt
   - Alternator
   - Ignition coil assembly
   - Evaporative Emission (EVAP) canister purge solenoid
   - Rocker arm cover
   - Rocker arms and pushrods

To install:

➡ **The intake pushrods are identified with yellow stripes and are 5¾ inches long. The exhaust pushrods are identified with green stripes and are 6 inches long.**

3. Coat all valvetrain components with engine oil.
4. Install the pushrods and rocker arms. Torque the rocker arm as follows:
   a. 3.1L Engines: Tighten to 89 inch lbs. (10 Nm) plus an additional 30 degree turn.
   b. 3.5L Engines: Tighten to 24 ft. lbs. (32 Nm)
5. Install or connect the following:
   - Rocker arm cover with a new gasket. Torque the bolts to 89 inch lbs. (10 Nm)

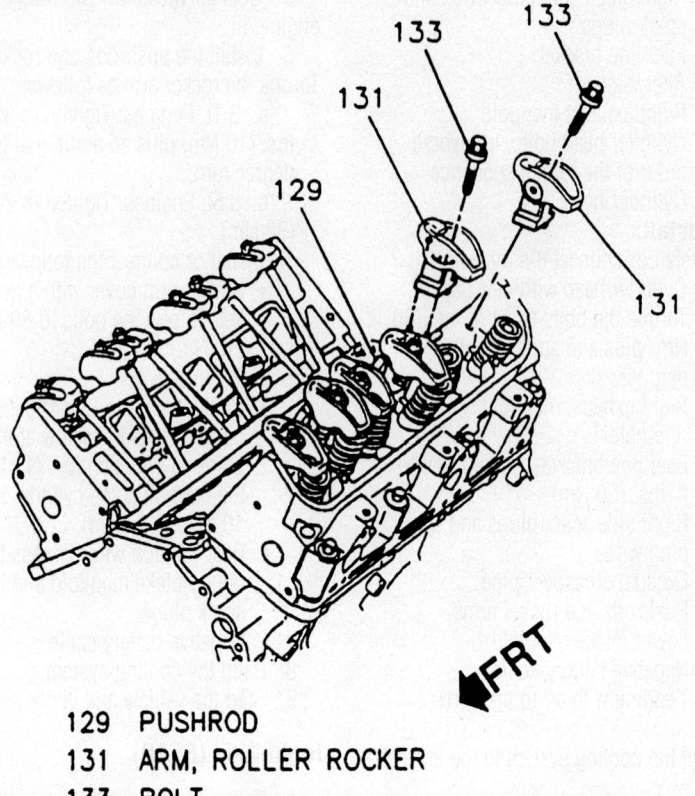

FRT

129 PUSHROD
131 ARM, ROLLER ROCKER
133 BOLT

7922Z205

Rocker arm components

- Ignition coil assembly
- EVAP canister purge solenoid
- Alternator. Torque the bolts to 18 ft. lbs. (24 Nm).
- Accessory drive belt
- Power brake booster vacuum pipe to the plenum
- Spark plug wires
- Negative battery cable

Intake Manifold

REMOVAL & INSTALLATION

2.2L Engine

1. Before servicing the vehicle, refer to the Precautions Section.
2. Drain the cooling system.
3. Relieve the fuel system pressure.
4. Remove or disconnect the following:
   - Negative battery cable
   - Air intake assembly
   - Throttle body
   - PCV hose
   - Purge solenoid tube
   - Brake booster vacuum hose
   - Oil level indicator tube mounting bolt
   - Fuel rail

- Knock Sensor (KS) electrical connector and wiring harness
- Intake manifold mounting bolts
- Intake manifold

To install:

5. Install or connect the following:
   - Intake manifold and gasket. Tighten bolts to 89 inch lbs. (10 Nm).
   - KS sensor electrical connector
   - Fuel rail
   - Oil level indicator tube mounting bolt. Tighten to 89 inch lbs. (10 Nm).
   - Brake booster vacuum hose
   - Purge solenoid rube
   - PCV hose
   - Throttle body
   - Air intake assembly
   - Negative battery cable
6. Refill the cooling system.
7. Start the engine and check for leaks.

3.1L Engine

1. Before servicing the vehicle, refer to the Precautions Section.
2. Drain the cooling system.
3. Relieve the fuel system pressure.
4. Remove or disconnect the following:
   - Upper half of the air cleaner assembly

- Brake vacuum pipe at the intake plenum
- Fuel pressure regulator vacuum line
- Spark plug wires
- Ignition coil assembly
- Evaporative Emission (EVAP) canister purge solenoid
- Throttle Position (TPS) sensor electrical connector
- Idle Air Control (IAC) valve electrical connector
- Fuel Injector electrical connectors
- Engine Coolant Temperature (ECT) sensor electrical connector
- Manifold Absolute Pressure (MAP) sensor electrical connector and vacuum line
- Camshaft Position (CMP) sensor electrical connector
- Vacuum modulator supply hose
- Control cables from the throttle body and intake plenum bracket
- Upper intake manifold
- Left and right rocker arm covers
- Fuel injector electrical connectors
- Fuel feed and return line
- Fuel rail with the injectors
- Power steering pump and move it aside without disconnecting the lines
- Heater inlet pipe
- Thermostat bypass hose
- Upper radiator hose at the thermostat housing
- Lower intake manifold bolts

➡ **When removing the valvetrain components, keep them in order for installation in their original locations.**

- Rocker arm nuts, rocker arms and pushrods
- Intake manifold

To install:

5. Place a 0.08–0.11 inch (8–12mm) bead of RTV on each ridge, where the front and rear of the intake manifold contact the block.
6. Install or connect the following:
   - Lower intake manifold using a new gasket
   - Pushrods and rocker arms

➡ **Be sure the pushrods are properly seated in the valve lifters and rocker arms.**

- Rocker arm nuts. Torque them to 18 ft. lbs. (24 Nm).
- Lower intake manifold bolts, lubricated with sealant. Torque the bolts to 115 inch lbs. (13 Nm).
- Upper radiator hose at the thermostat housing

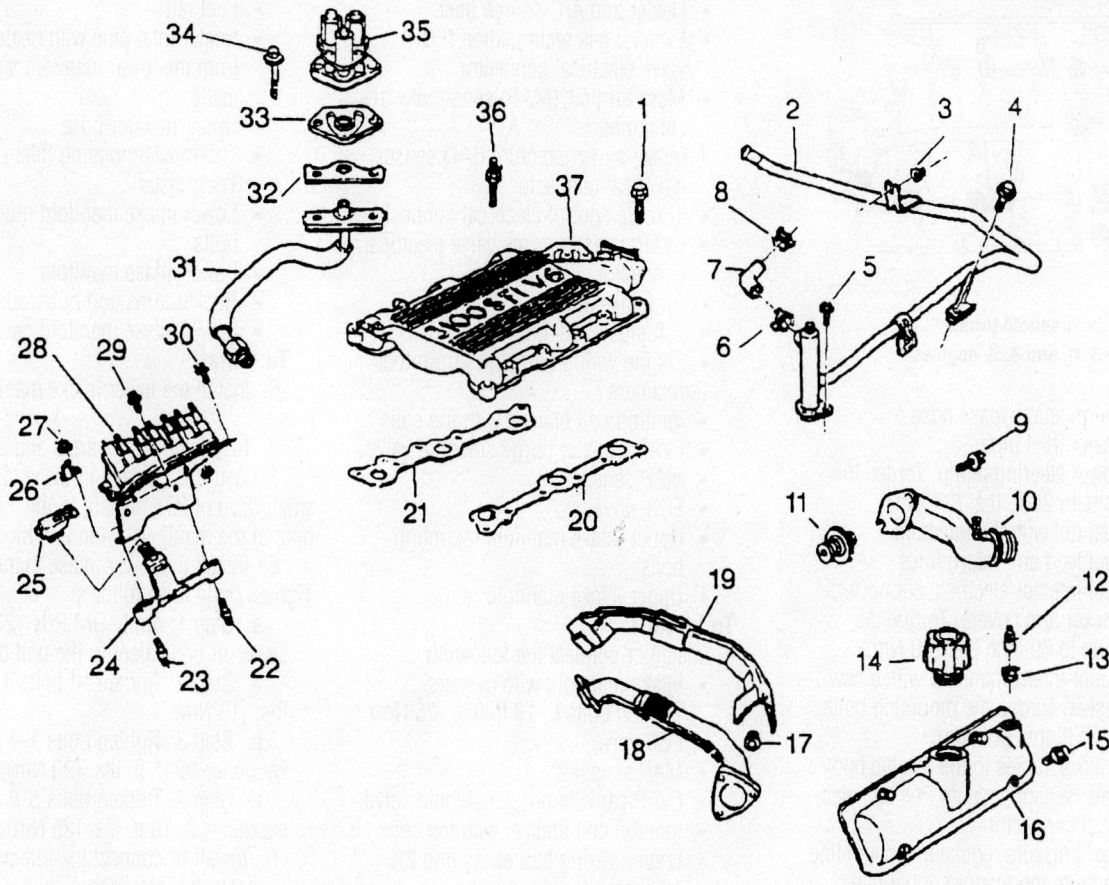

(1) Upper Intake Manifold Bolt
(2) Thermostat Bypass Pipe
(3) Thermostat Bypass Pipe Nut
(4) Thermostat Bypass Pipe Bolt
(5) Thermostat Bypass Pipe Screw
(6) Thermostat Bypass Hose Clamp
(7) Thermostat Bypass Hose
(8) Thermostat Bypass Hose Clamp
(9) Coolant Outlet Bolt
(10) Coolant Outlet Assembly
(11) Thermostat
(12) PCV Valve
(13) PCV Valve Grommet
(14) Oil Fill Cap
(15) Rocker Arm Cover Bolt
(16) Rocker Arm Cover
(17) Exhaust Crossover Nut
(18) Exhaust Crossover Pipe

(19) Exhaust Crossover Upper Heat Shield
(20) Upper Intake Manifold Gasket
(21) Upper Intake Manifold Gasket
(22) EVAP Canister Purge Valve Bracket Stud
(23) EVAP Canister Purge Valve Bracket Stud
(24) EVAP Canister Purge Valve Bracket
(25) EVAP Canister Purge Valve
(26) Spark Plug Wire Support
(27) Electronic Ignition System Nut
(28) Electronic Ignition System
(29) Electronic Ignition System Bolt
(30) Electronic Ignition System Bolt
(31) EGR Valve Pipe Assembly
(32) EGR Valve Pipe Gasket
(33) EGR Valve Gasket
(34) EGR Valve Assembly Bolt
(35) EGR Valve Assembly
(36) Upper Intake Manifold Stud
(37) Upper Intake Manifold

Exploded view of the upper intake and related components

7922Z207

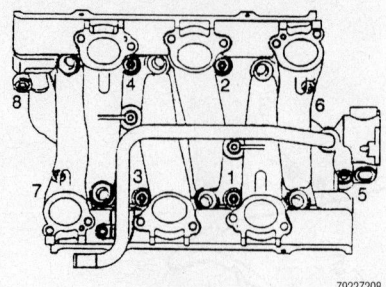

Lower intake manifold torque sequence—3.1L and 3.5L engines

79222208

- Thermostat bypass hose
- Heater inlet pipe
- Power steering pump. Torque the bolts to 25 ft. lbs. (34 Nm).
- Fuel rail with the injectors
- Fuel feed and return lines
- Fuel injector electrical connectors
- Rocker arm covers. Torque the bolts to 89 inch lbs. (10 Nm).
- Upper intake manifold with a new gasket. Torque the mounting bolts to 18 ft. lbs. (24 Nm).
- Control cables to the throttle body
- MAP sensor vacuum line and electrical connector
- Fuel pressure regulator vacuum line
- Vacuum modulator supply line
- CMP sensor electrical connector
- ECT sensor electrical connector
- IAC valve electrical connector
- TPS sensor electrical connector
- EVAP canister purge solenoid
- Ignition coil assembly
- Spark plug wires
- Brake vacuum pipe at the intake plenum
- Accessory drive belt
- Brake booster vacuum line
- Upper half of the air cleaner assembly
- Negative battery cable
7. Fill the cooling system.

➡ **An engine oil and filter change is recommended.**

8. Start the vehicle and verify no leaks.

3.5L Engine

UPPER MANIFOLD

1. Before servicing the vehicle, refer to the Precautions Section.
2. Drain the cooling system.
3. Remove or disconnect the following:
   - Negative battery cable
   - EVAP canister purge valve hose
   - Manifold vacuum source hose
   - Brake booster hose

- Heater and A/C source hose
- Exhaust gas recirculation (EGR) valve electrical connector
- Mass air flow (MAP) sensor electrical connector
- Intake air temperature (IAT) sensor electrical connector
- Throttle control electrical connector
- EVAP canister purge valve electrical connector
- Air intake assembly
- Left side spark plug wires
- Engine wiring harnesses from their retainers
- Ignition coil bracket with the coils
- EVAP canister purge solenoid valve
- MAP sensor
- EGR valve
- Upper intake manifold mounting bolts
- Upper intake manifold

To install:
4. Install or connect the following:
   - Intake manifold with gaskets. Tighten bolts to 18 ft. lbs. (25 Nm)
   - EGR valve
   - MAP sensor
   - EVAP canister purge solenoid valve
   - Ignition coil bracket with the coils
   - Engine wiring harnesses into the retainers
   - Left side spark plug wires
   - Air intake assembly
   - EVAP canister purge valve electrical connector
   - Throttle control electrical connector
   - IAT sensor electrical connector
   - MAP sensor electrical connector
   - EGR valve electrical connector
   - Heater and A/C source hose
   - Brake booster hose
   - Manifold vacuum hose
   - EVAP canister purge valve hose
   - Negative battery cable
5. Refill the cooling system to the correct level.
6. Start the engine and check for leaks.

LOWER MANIFOLD

1. Before servicing the vehicle, refer to the Precautions Section.
2. Drain the cooling system.
3. Relieve the fuel system pressure.
4. Remove or disconnect the following:
   - Negative battery cable
   - Upper intake manifold
   - Valve covers
   - Engine coolant temperature (ECT) wiring harness
   - Fuel injector wiring harness
   - Manifold air pressure (MAP) wiring harness

- Fuel rail
- Heater inlet pipe with heater hose from the lower manifold and place aside
- Upper radiator hose
- Thermostat housing inlet pipe
- Thermostat
- Lower intake manifold mounting bolts
- Lower intake manifold
- Rocker arms and pushrods
- Lower intake manifold gaskets

To install:
5. Install the lower intake manifold gaskets.
6. Install the rocker arms and pushrods.
7. With the gaskets in place, apply a small drop of RTV sealant to the four corners of the manifold-to-engine block joints.
8. Install the lower intake manifold. Tighten bolts to as follows:
   a. Step 1: Apply GM P/N 12345382 sealer or equivalent to the bolt threads.
   b. Step 2: Tighten all bolts 115 inch lbs. (13 Nm).
   c. Step 3: Tighten bolts 1-4 in sequence to 15 ft. lbs. (20 Nm).
   d. Step 4: Tighten bolts 5-8 in sequence to 18 ft. lbs. (25 Nm).
9. Install or connect the following:
   - Heater inlet pipe
   - Thermostat
   - Thermostat housing inlet pipe
   - Upper radiator hose
   - Heater inlet pipe and heater hose
   - Fuel rail
   - MAP wiring harness
   - Fuel injector and MAP wiring harness
   - ECT wiring harness
   - Valve covers
   - Upper intake manifold
   - Negative battery cable
10. Refill the cooling system to the correct level.
11. Start the engine and check for leaks.

Exhaust Manifold

REMOVAL & INSTALLATION

2.2L Engine

1. Before servicing the vehicle, refer to the Precautions Section.
2. Drain the cooling system.
3. Remove or disconnect the following:
   - Negative battery cable
   - Exhaust manifold heat shield
   - Block heater, if equipped
   - Oxygen sensor

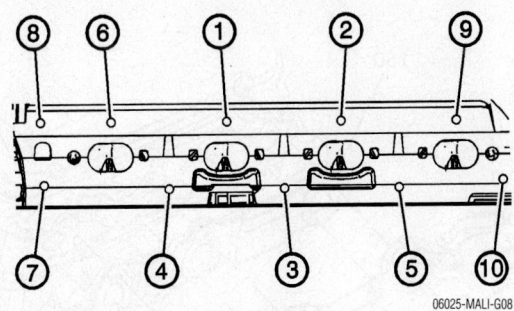

Exhaust manifold bolt torque sequence–2.2L engine

- Exhaust manifold retaining nuts
- Exhaust manifold and catalytic converter assembly

To install:

4. Install or connect the following:
- New manifold studs if necessary. Tighten to 89 inch lbs. (10 Nm).
- Exhaust manifold gasket
- Exhaust manifold and catalytic converter assembly. Tighten new nuts to 124 inch lbs. (14 Nm)
- Oxygen sensor. Tighten to 22 ft. lbs. (30 Nm)
- Block heater, if equipped
- Exhaust manifold heat shield
- Negative battery cable

5. Refill the cooling system to the correct level.

6. Start the engine and check for leaks.

3.1L Engine

LEFT SIDE (FRONT)

1. Before servicing the vehicle, refer to the Precautions Section.
2. Partially drain the cooling system.
3. Remove or disconnect the following:
- Negative battery cable
- Air cleaner assembly
- Throttle body duct
- Exhaust crossover heat shield
- Exhaust crossover pipe from the manifold
- Upper radiator hose from the thermostat housing
- Spark plug wires
- Exhaust manifold heat shield
- Exhaust manifold

To install:

4. Install or connect the following:
- Exhaust manifold using a new gasket. Torque the nuts to 12 ft. lbs. (16 Nm).
- Exhaust manifold heat shield. Torque the bolts to 89 inch lbs. (10 Nm).
- Spark plug wires
- Upper radiator hose
- Exhaust crossover pipe to the manifold. Torque the bolts to 18 ft. lbs. (24 Nm).

- Exhaust crossover pipe heat shield. Torque the bolts to 89 inch lbs. (10 Nm).
- Throttle body duct
- Air cleaner assembly
- Negative battery cable

5. Fill the cooling system.

RIGHT SIDE (REAR)

1. Before servicing the vehicle, refer to the Precautions Section.
2. Remove or disconnect the following:

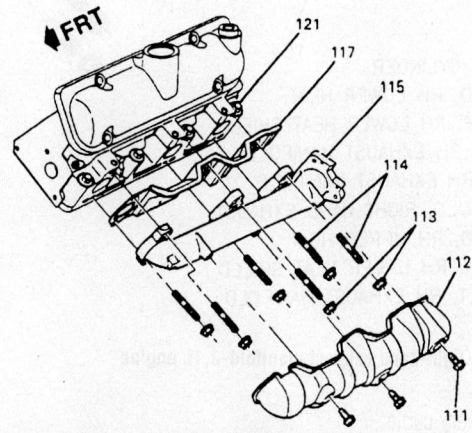

111	SCREW, LH EXHAUST MANIFOLD HEAT SHIELD
112	SHIELD LH EXHAUST MANIFOLD
113	NUT, LH EXHAUST MANIFOLD
114	STUD, LH EXHAUST MANIFOLD
115	MANIFOLD, LH EXHAUST
117	GASKET, LH EXHAUST MANIFOLD
121	HEAD, LH CYLINDER

Exploded view of the left-hand exhaust manifold–3.1L engine

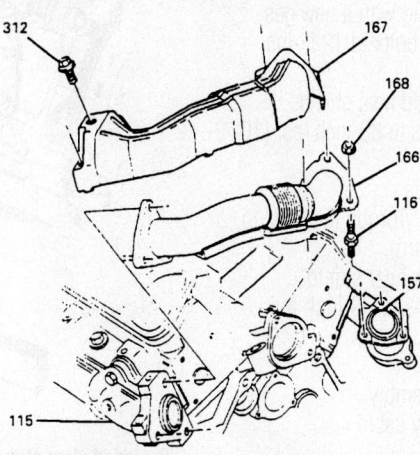

115	MANIFOLD, LEFT HAND EXHAUST
116	STUD, EXHAUST CROSSOVER
157	MANIFOLD, RIGHT HAND EXHAUST
166	CROSSOVER PIPE, EXHAUST
167	SHIELD, EXHAUST CROSSOVER UPPER HEAT
168	NUT, EXHAUST CROSSOVER
312	BOLT/SCREW, EXHAUST CROSSOVER UPPER HEAT SHIELD

Exploded view of the exhaust crossover pipe–3.1L engine

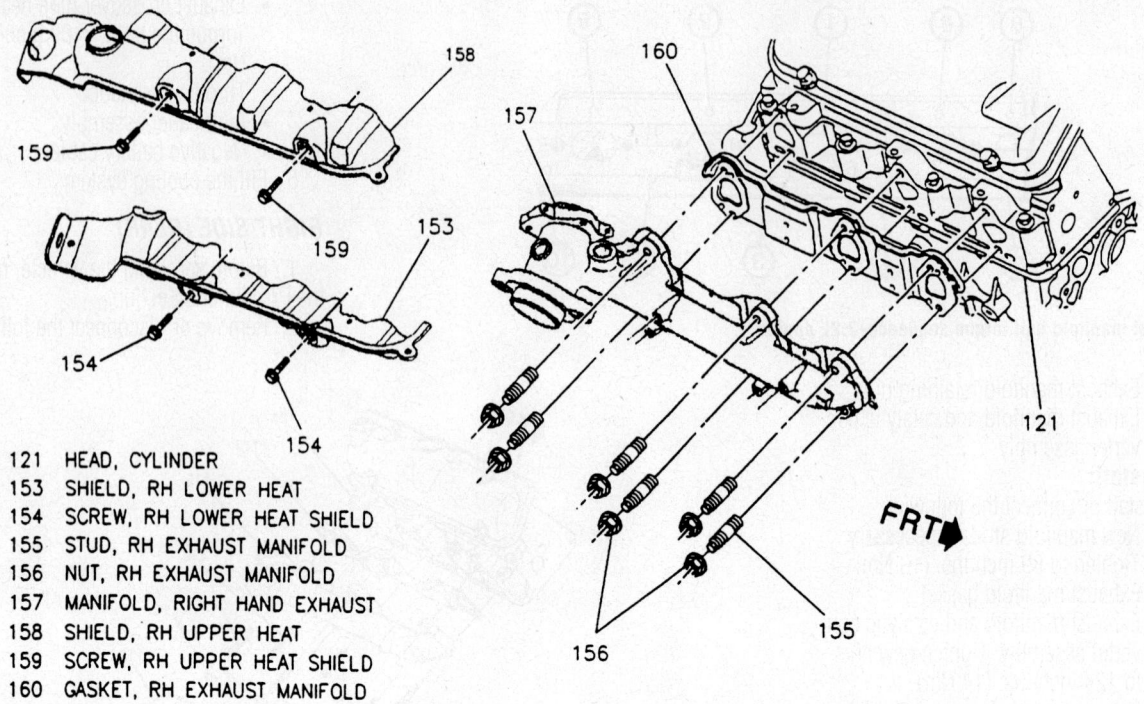

121 HEAD, CYLINDER
153 SHIELD, RH LOWER HEAT
154 SCREW, RH LOWER HEAT SHIELD
155 STUD, RH EXHAUST MANIFOLD
156 NUT, RH EXHAUST MANIFOLD
157 MANIFOLD, RIGHT HAND EXHAUST
158 SHIELD, RH UPPER HEAT
159 SCREW, RH UPPER HEAT SHIELD
160 GASKET, RH EXHAUST MANIFOLD

79222213

Exploded view of the right-hand exhaust manifold–3.1L engine

- Negative battery cable
- Air cleaner assembly
- Heated Oxygen (HO2S) sensor
- Crossover pipe heat shield
- Crossover pipe
- Spark plug wires
- Exhaust manifold heat shield
- Exhaust manifold

To install:

3. Install or connect the following:
   - Exhaust manifold with a new gasket. Torque the bolts to 12 ft. lbs. (16 Nm).
   - Exhaust manifold heat shield. Torque the bolts to 89 inch lbs. (10 Nm).
   - Spark plug wires
   - Crossover pipe. Torque the bolts to 18 ft. lbs. (24 Nm).
   - Crossover pipe heat shield. Torque the bolts to 89 inch lbs. (10 Nm).
   - HO2S sensor
   - Air cleaner assembly
   - Negative battery cable

3.5L Engine

LEFT SIDE

1. Before servicing the vehicle, refer to the Precautions Section.
2. Remove or disconnect the following:
   - Negative battery cable
   - Heated oxygen (HO2S) sensor

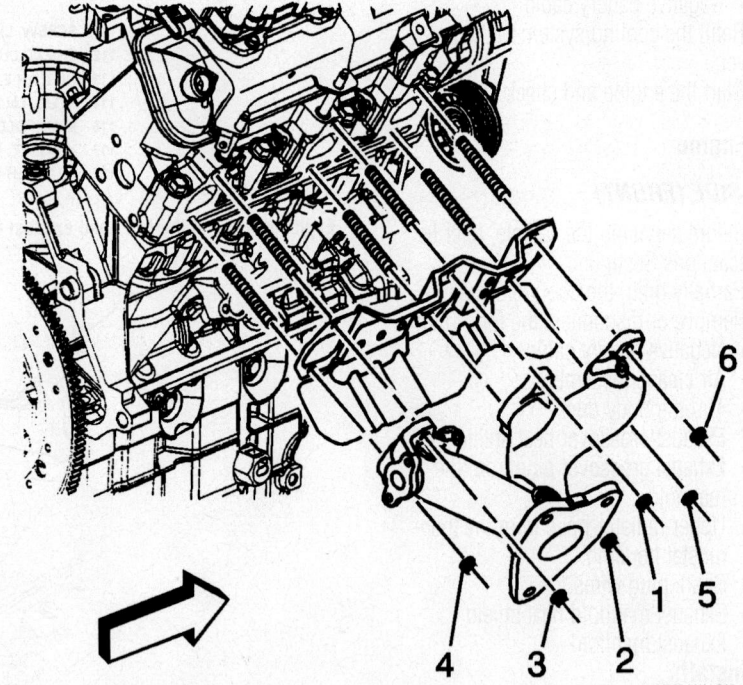

06025-MALI-G05

Exploded view of the left-hand exhaust manifold–3.5L engine

- Spark plugs
- Exhaust manifold heat shield
- Exhaust manifold

To install:

3. Install new exhaust manifold if necessary. Tighten to 13 ft. lbs. (18 Nm).
4. Install or connect the following:

- Exhaust manifold gasket
- Exhaust manifold. Tighten nuts in sequence to 12 ft. lbs. (16 Nm).
- Exhaust manifold heat shield
- Spark plugs
- HO2S sensor
- Negative battery cable

RIGHT SIDE

1. Before servicing the vehicle, refer to the Precautions Section.
2. Remove or disconnect the following:
   - Negative battery cable
   - Heated oxygen (HO2S) sensor
   - Spark plugs
   - Exhaust gas circulation (EGR) pipe
   - Exhaust manifold heat shield
   - Exhaust manifold and gasket

To install:

3. Install or connect the following:
   - New exhaust manifold studs, if necessary. Tighten to 13 ft. lbs. (18 Nm).
   - Exhaust manifold and gasket. Tighten nuts in sequence to 12 ft. lbs. (16 Nm).
   - Exhaust manifold heat shield. Tighten bolts to 89 inch lbs. (10 Nm).
   - EGR pipe
   - Spark plugs
   - HO2S sensor
   - Negative battery cable

Camshaft and Valve Lifters

REMOVAL & INSTALLATION

2.2L Engine

1. Before servicing the vehicle, refer to the Precautions Section.
2. Remove or disconnect the following:
   - Negative battery cable
   - Air intake assembly
   - PCV hose
   - Fuel line bracket
   - Ignition coil and module assembly
   - Ground strap and ground strap retaining stud from camshaft cover
   - Camshaft cover
   - Upper timing chain guide
3. Install Camshaft Sprocket Holding Tool J-43655 and remove both the intake and exhaust camshaft sprocket bolts.
4. Slide the camshaft sprockets forward.
5. Matchmark the caps and remove.
6. Remove the camshaft.

To install:

7. Install the camshaft.
8. Install the camshaft caps as follows:
   a. Apply a 0.13 inch (3.5mm) bead of GM P/N 12378521 sealant to the rear intake camshaft cap.
   b. Install camshaft caps and tighten bolts in increments of 3 turns until they are seated.
   c. Tighten cap bolts to 89 inch lbs. (10 Nm).

d. Tighten intake camshaft rear cap bolts to 18 ft. lbs. (25 Nm).
9. Install the camshaft sprockets onto the camshafts and hand tighten new camshaft sprocket bolts.
10. Remove Camshaft Sprocket Holding Tool J-43655.
11. Tighten the camshaft sprocket bolts to 63 ft. lbs. (85 Nm) plus 30 degrees.
12. Install or connect the following:
   - Upper timing chain guide. Tighten bolts to 89 inch lbs. (10 Nm).
   - Camshaft cover. Tighten bolts to 89 inch lbs. (10 Nm).
   - Ground strap
   - Ignition coil and module assembly
   - Fuel line bracket
   - PCV hose
   - Air intake assembly
   - Negative battery cable
13. Start the engine and check for leaks.

3.1L Engine and 3.5L Engine

1. Before servicing the vehicle, refer to the Precautions Section.
2. Relieve the fuel system pressure.

➡ **When removing valvetrain components, marked them for installation in the same location they are removed from.**

3. Remove or disconnect the following:
   - Rocker arm covers
   - Intake manifold
   - Rocker arms and push rods
   - Lifter guide bolts and the guide
   - Valve lifter(s) from the lifter bores
   - Crankshaft balancer
   - Front cover
   - Timing chain and sprockets
   - Oil pump driven gear bolt and gear
   - Camshaft thrust plate
   - Camshaft

To install:

4. Install or connect the following:
   - Camshaft, lubricated with camshaft lubricant
   - Camshaft thrust plate. Torque bolts to 89 inch lbs. (10 Nm).
   - Oil pump driven gear. Torque the bolt to 27 ft. lbs. (36 Nm).
   - Timing chain and sprocket. Torque the bolt to 103 ft. lbs. (140 Nm).
   - Front cover. Torque the bolts to 35 ft. lbs. (47 Nm).
   - Crankshaft balancer. Torque the bolt to 76 ft. lbs. (103 Nm).
   - Valve lifters in their original locations
   - Lifter guide. Torque the bolts to 89 inch lbs. (10 Nm).
   - Pushrods, rocker arms, rocker balls

and rocker arm nuts. Torque the rocker arm nuts to 89 inch lbs. (10 Nm) plus an additional 30 degrees.
   - Intake manifold. Torque the upper and lower intake bolts to 115 inch lbs. (13 Nm).
   - Rocker arm covers. Torque the bolts to 89 inch lbs. (10 Nm).
   - Negative battery cable
5. Start the engine and verify no oil leaks.

Valve Lash

ADJUSTMENT

All engines are equipped with hydraulic valve lifters; no adjustment is necessary.

Starter Motor

REMOVAL & INSTALLATION

2.2L Engine

1. Before servicing the vehicle, refer to the Precautions Section.
2. Remove or disconnect the following:
   - Negative battery cable
   - Starter motor electrical connections
   - Starter motor mounting bolts
   - Starter motor

To install:

3. Install or connect the following:
   - Starter motor. Tighten the mounting bolts to 30 ft. lbs. (40 Nm).
   - Starter motor electrical connections. Tighten the battery cable to 7 ft. lbs. (10 Nm). Tighten the solenoid cable to 4 ft. lbs. (5 Nm).
   - Negative battery cable

3.1L Engine

1. Before servicing the vehicle, refer to the Precautions Section.
2. Remove or disconnect the following:
   - Negative battery cable
   - Lower closeout panel
   - Starter bolts
   - Starter electrical connectors
   - Starter motor

To install:

3. Install or connect the following:
   - Starter electrical connectors. Torque the solenoid cable to 106 inch lbs. (12 Nm).
   - Starter motor. Torque the bolts to 37 ft. lbs. (50 Nm).
   - Lower closeout panel
   - Negative battery cable

3.5L Engine

1. Before servicing the vehicle, refer to the Precautions Section.
2. Remove or disconnect the following:
   - Negative battery cable
   - Flywheel inspection cover
   - Starter motor electrical connections
   - Starter motor mounting bolts
   - Starter motor

To install:

3. Install or connect the following:
   - Starter motor. Tighten mounting bolts to 30 ft. lbs. (40 Nm).
   - Starter motor electrical connections. Tighten battery cable to 13 ft. lbs. (17 Nm). Tighten the solenoid cable to 27 inch lbs. (3 Nm).
   - Flywheel inspection cover
   - Negative battery cable

Oil Pan

REMOVAL & INSTALLATION

2.2L Engine

1. Before servicing the vehicle, refer to the Precautions Section.
2. Drain the engine oil.
3. Remove or disconnect the following:
   - Negative battery cable
   - Accessory drive belt
   - Lower A/C compressor mounting bolt
   - Oil pan mounting bolts
   - Oil pan

To install:

4. Apply a 0.13 inch (3.5mm) bead of the GM P/N 12378521 sealant around the perimeter of the oil pan and oil suction port opening.
5. Install or connect the following:
   - Oil pan. Tighten the bolts in sequence to 18 ft. lbs. (25 Nm).
   - Lower A/C compressor mounting bolt. Tighten the bolt to 37 ft. lbs. (50 Nm).
   - Accessory drive belt
   - Negative battery cable
6. Fill the engine with oil to the correct level.

3.1L Engine

1. Before servicing the vehicle, refer to the Precautions Section.
2. Evacuate and recover the A/C system.
3. Drain the engine oil.
4. Remove or disconnect the following:
   - Negative battery cable
   - Accessory drive belt
   - Upper and lower radiator hoses
   - Right front wheel
   - Inner fender splash shield
   - Antilock Brake System (ABS) Wheel Speed Sensor (WSS) electrical connector
   - Right front ball joint from the steering knuckle
   - Right side outer tie rod end
   - A/C compressor bolts and move the compressor aside

❄❄ WARNING

DO NOT disconnect the refrigerant lines or allow the compressor to hang unsupported.

   - Evaporator-to-accumulator A/C line
   - Flywheel cover

   - Engine cradle bolts
   - Crankshaft balancer
   - Starter motor
   - Oil pan

To install:

5. Install or connect the following:
   - Oil pan using a new gasket

➡ **Apply silicone sealer to the portion of the pan that contacts the rear of the block.**

   - Oil pan bolts. Torque the bolts to 18 ft. lbs. (24 Nm) and the side bolts to 37 ft. lbs. (50 Nm).
   - Starter motor. Torque the bolts to 37 ft. lbs. (50 Nm).
   - Flywheel cover
   - Crankshaft balancer. Torque the bolt to 76 ft. lbs. (103 Nm).
   - A/C compressor. Torque the bolts to 37 ft. lbs. (50 Nm).
   - Evaporator-to-accumulator line
   - Engine cradle bolts. Torque the bolts to 89 ft. lbs. (120 Nm).
   - Ball joint to the steering knuckle using a new cotter pin. Torque the nut 48 ft. lbs. (60 Nm).
   - Outer tie rod end. Torque the nut to 33 ft. lbs. (45 Nm).
   - ABS wheel speed sensor electrical connector
   - Inner fender splash shield
   - Right front wheel
   - Accessory drive belt
   - Upper and lower radiator hoses
   - Negative battery cable
6. Refill the crankcase.
7. Evacuate and recharge the A/C system.

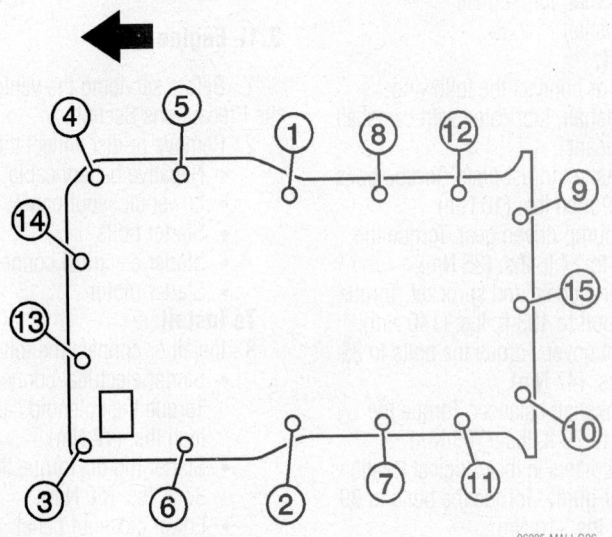

Oil pan bolt sequence–2.2L engine

06025-MALI-G06

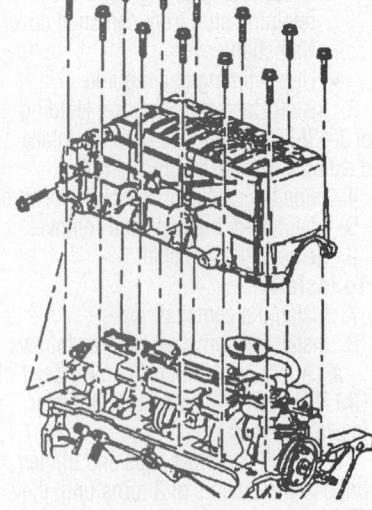

79222216

Exploded view of the oil pan mounting–3.1L and 3.5L engines

8. Start the engine and check for leaks.

➡**Whenever the vehicle subframe is removed or lowered, the wheel alignment should be checked.**

3.5L Engine

1. Before servicing the vehicle, refer to the Precautions Section.
2. Drain the engine oil.
3. Remove or disconnect the following:
- Negative battery cable
- Right front wheel
- Inner fender splash shield
- Oil filter and oil filter adapter
- Catalytic converter
- ABS wheel speed sensor
- Right front ball joint from the steering knuckle
- Lower closeout panel
- A/C compress mounting bolts and position compress aside
- Engine-to-transmission support braces
- Oil level sensor

4. Support the engine using a jackstand or equivalent and remove the right side engine mounting bolts.
5. Loosen the left side cradle bolts.
6. Remove the cradle bolts from the right front and right rear.
7. Remove the starter motor.
8. Remove the oil pan mounting bolts.
9. Remove the oil pan.

To install:

10. Install or connect the following:
- Oil pan with a new gasket.

➡**Apply silicone sealant on the tabs that insert into the gasket groove on the outer surface of the mean bearing cap.**

- Oil pan bolts. Tighten the bolts to 18 ft. lbs. (25 Nm) and the side bolts to 37 ft. lbs. (50 Nm).
- Starter motor
- Flywheel inspection cover
- A/C compressor. Tighten the bolts to 37 ft. lbs. (50 Nm).
- Engine cradle bolts. Tighten the bolts to 37 ft. lbs. (50 Nm).
- Ball joint to the steering knuckle
- ABS wheel speed sensor
- Engine-to-transmission support braces
- Catalytic converter
- Oil filter adapter and oil filter
- Inner fender splash shield
- Right front wheel
- Lower closeout panel
- Negative battery cable

11. Fill the engine with oil to the correct level.
12. Start the engine and check for leaks.

Oil Pump

REMOVAL & INSTALLATION

2.2L Engine

1. Before servicing the vehicle, refer to the Precautions Section.

➡**The oil pump is part of the engine front cover assembly.**

2. Remove or disconnect the following:
- Front cover
- Pressure relief valve from front cover
- Oil pump cover
- Oil pump

To install:

➡**If the oil pump is to be replaced, the entire front cover must also be replaced.**

➡**Lubricate all oil pump parts with clean engine oil.**

3. Install the inner gear into the outer gear.
4. Install the assembly together into the front cover the hub of the center gear facing the front cover.
5. Install the oil pump cover. Tighten the bolts to 53 inch lbs. (6 Nm).
6. Install the pressure relief valve. Tighten the plug to 30 ft. lbs. (40 Nm).
7. Install the front cover.

3.1L Engine

1. Before servicing the vehicle, refer to the Precautions Section.
2. Disconnect the negative battery cable.
3. Drain the engine oil.
4. Remove or disconnect the following:

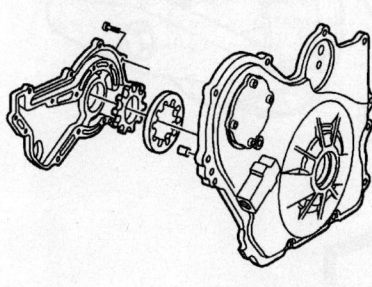

Exploded view of the oil pump assembly–2.2L engine

06025-MALI-G07

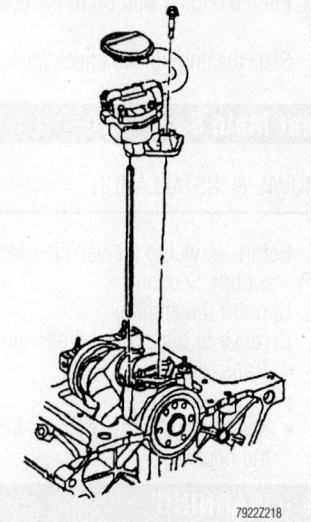

79222218

Exploded view of the oil pump mounting

- Oil pan
- Crankshaft oil deflector
- Oil pump and pump driveshaft

To install:

5. Install or connect the following:
- Oil pump and pump driveshaft. Torque the bolts to 30 ft. lbs. (41 Nm).
- Crankshaft oil deflector. Torque the nuts to 18 ft. lbs. (24 Nm).
- Oil pan. Torque the bolts to 18 ft. lbs. (24 Nm) and the side bolts to 37 ft. lbs. (50 Nm).
- Negative battery cable

6. Refill the crankcase.

➡**A filter change is recommended.**

7. Start the engine, check the oil pressure and check for leaks.

3.5L Engine

1. Before servicing the vehicle, refer to the Precautions Section.
2. Drain the engine oil.
3. Remove or disconnect the following:
- Negative battery cable
- Oil pan
- Oil pump bolt
- Oil pump and the oil pump drive shaft

To install:

➡**Rotate the oil pump drive shaft as necessary to mate with the oil pump drive unit.**

4. Install the oil pump drive shaft and oil pump.
5. Tighten the oil pump bolt to 30 ft. lbs (41 Nm).
6. Install the oil pan.
7. Connect the negative battery cable.

8. Fill the engine with oil to the correct level.

9. Start the engine and check for leaks.

Rear Main Seal

REMOVAL & INSTALLATION

1. Before servicing the vehicle, refer to the Precautions Section.

2. Support the engine.

3. Remove or disconnect the following:
- Transaxle
- Flywheel
- Rear main seal by prying it from the housing

✳✳ WARNING

Use care not to damage the crankshaft seal surface with a prytool.

To install:

4. Lubricate the seal bore and new seal with engine oil.

5. Install the new seal by performing the following procedure:

a. Slide the new seal over the mandrel until the dust lip bottoms squarely against the tool collar.

b. Align the dowel pin of the tool with the dowel pin hole in the crankshaft and attach the tool to the crankshaft. Tighten the attaching screws to 24–60 inch lbs. (2.7–6.8 Nm).

c. Tighten the T-handle of the tool to push the seal into the bore. Continue until the tool collar is flush against the block.

d. Loosen the T-handle completely. Remove the attaching screws and the tool.

➡**Check to see that the seal is squarely seated in the bore.**

6. Install or connect the following:
- Flywheel
- Transaxle

7. Start the engine and check for leaks.

Timing Chain, Sprockets, Front Cover and Seal

REMOVAL & INSTALLATION

2.2L Engine

1. Before servicing the vehicle, refer to the Precautions Section.

2. Drain the cooling system.

3. Drain the engine oil.

4. Remove or disconnect the following:
- Negative battery cable
- PCV hose
- Fuel line bracket
- Ignition coil and module assembly
- Ground strap from camshaft cover
- Camshaft cover
- Front fender liner
- Accessory drive belt
- Crankshaft balancer pulley
- Accessory drive belt tensioner
- Front cover-to-water pump bolt
- Remaining front cover bolts
- Front cover

➡**The timing chain has 2 matching colored links and 1 uniquely colored link.**

5. Rotate the engine until the crankshaft sprocket mark aligns with the matching colored link (2) at the 5 o'clock position.

6. Confirm that the INT diamond on the intake camshaft sprocket is aligned with the uniquely colored link at (1) the 2 o'clock position.

7. Confirm that the EXH triangle on the exhaust camshaft sprocket is aligned with the matching colored link (3).

8. Remove or disconnect the following:
- Timing chain tensioner
- Fixed timing chain guide access plug

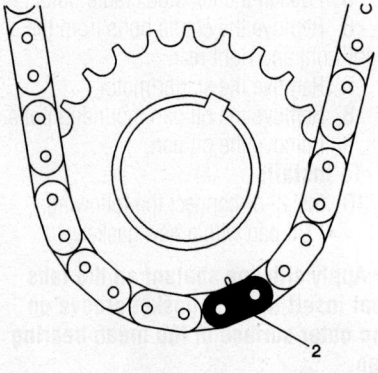

06025-MALI-G10

Lining up the crankshaft sprocket mark with the colored link in the 5 o'clock position (2).

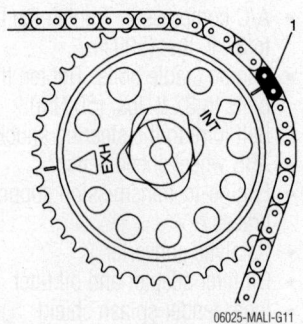

06025-MALI-G11

Aligning the intake camshaft sprocket mark with the colored link in the 2 o'clock position (1).

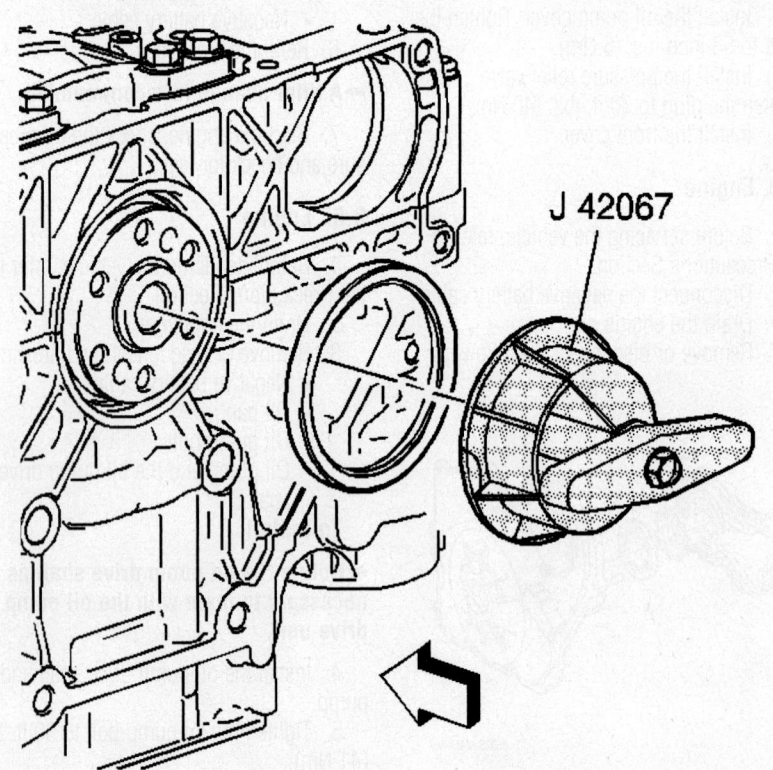

J 42067

06025-MALI-G09

Using the seal installer to install the rear main seal.

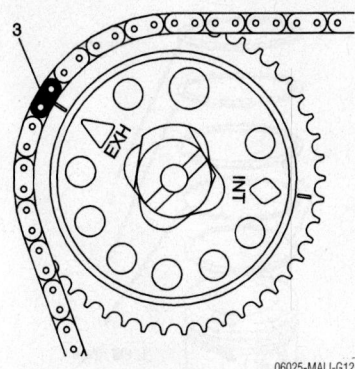

Aligning the exhaust camshaft sprocket

- Fixed timing chain guide
- Upper timing chain guide

➡**Use a 24 mm wrench to hold the camshafts to prevent them from turning.**

- Exhaust camshaft sprocket
- Timing chain tensioner guide
- Intake camshaft sprocket
- Timing chain through the top of the cylinder head
- Crankshaft sprocket

To install:

9. Install the crankshaft sprocket with the timing mark in the 5 o'clock position.

10. Assemble the intake camshaft sprocket to the timing chain with the timing mark lined up with the uniquely colored link (1). Hand tighten a new intake camshaft sprocket bolt.

11. Lower the timing chain through the opening in the cylinder head.

12. Route the timing chain around the crankshaft sprocket and line up the first marching colored link (2) with the timing mark on the crankshaft sprocket.

13. Install the exhaust camshaft sprocket with a new bolt loosely onto the exhaust camshaft.

14. Align the timing mark on the sprocket with the last matching colored (3).

15. If necessary, align the camshaft as follows:

    a. Using a 24 mm wrench, first turn the intake camshaft until the alignment feature on the back of the camshaft sprocket seats in the notch in the front of the intake camshaft.

    b. Turn the crankshaft 45 degrees in either direction.

    c. Turn the intake camshaft to the appropriate location.

    d. Turn the crankshaft back to top dead center (TDC).

16. When the sprocket seats in on the camshaft, tighten the sprocket bolt hand tight.

17. Verify all of the colored links and the appropriate timing marks are still aligned.

18. Install the fixed timing chain guide. Tighten the bolts to 133 inch lbs. (15 Nm).

19. Install the upper timing chain guide. Tighten the bolts to 89 inch lbs. (10 Nm).

20. Using a 24 mm wrench to hold the camshafts, tighten the camshaft sprocket bolts 63 ft. lbs. (85 Nm) plus 30 degrees.

21. Measure the timing chain tensioner from end to end. A new tensioner should be supplied in the fully compressed non-active state. A tensioner in the compressed state will measure 2.83 inches (72 mm) from end to end. A tensioner in the active state will measure 3.35 inches (85 mm) from end to end.

22. If the timing chain tensioner is not in the compressed state, perform the following steps:

    a. Remove the piston assembly from the body of the timing chain tensioner by pulling it out.

    b. Install the bottom half of the Tensioner tool J-45027-2 into a vise.

    c. Install the notch end of the piston assembly into the bottom half of the tensioner tool.

    d. Using the top half of the Tensioner tool J-45027-1, turn the ratchet cylinder into the piston.

    e. Install the compressed piston assembly back into the timing chain tensioner body until it stops at the bottom of the bore. Do not compress the piston assembly against the bottom of the bore.

23. Install the timing chain tensioner assembly. Tighten to 66 ft. lbs. (75 Nm).

24. Release the timing chain tensioner by compressing it approximately 0.08 inches (2mm). Feed a rubber-tipped tool down through the cam drive chest to reset on the cam chain. Give the tool a sharp jolt diagonally downwards to release the tensioner.

25. Install the timing chain oiling nozzle.

26. Install the timing chain guide bolt access hold plug with silicone sealant on the threads. Tighten the plug to 59 ft. lbs. (90 Nm).

27. Install or connect the following:

- Camshaft cover. Tighten the bolts to 89 inch lbs. (10 Nm).
- Front cover with new gasket. Tighten the bolts to 18 ft. lbs. (25 Nm).
- Accessory drive belt tensioner. Tighten the bolts to 33 ft. lbs. (45 Nm).
- Crankshaft balancer pulley using a new bolt. Tighten the bolt to 74 ft. lbs. (100 Nm) plus 75 degrees.

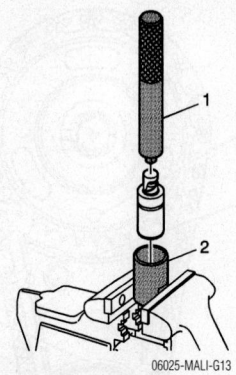

Compressing the timing chain tensioner– 2.2L engine.

- Accessory drive belt
- Front fender liner
- Ground strap to camshaft cover
- Ignition coil and module assembly
- Fuel line bracket
- Air intake assembly
- Negative battery cable

28. Refill the cooling system to the correct level.

29. Refill the engine with oil to the correct level.

30. Start the engine and check for leaks.

3.1L Engine and 3.5L Engine

1. Before servicing the vehicle, refer to the Precautions Section.

2. Disconnect the negative battery cable.

3. Drain the cooling system.

4. Drain the engine oil.

5. Recover the A/C system refrigerant.

6. Install an engine support fixture.

7. Remove or disconnect the following:

- Right engine mount assembly
- Accessory drive belt
- Air cleaner assembly
- Throttle body tube
- Power steering line at the pump
- Alternator and bracket
- Right front wheel
- Right inner fender well splash shield
- Right engine mount bracket
- Crankshaft balancer
- Drive belt tensioner
- Oil pan
- Crankshaft Position (CKP) sensor
- Coolant bypass pipe from the water pump and the intake manifold
- Lower radiator hose from the front cover outlet
- Front cover

8. Rotate the crankshaft until the timing marks on the camshaft and crankshaft sprockets are in alignment.

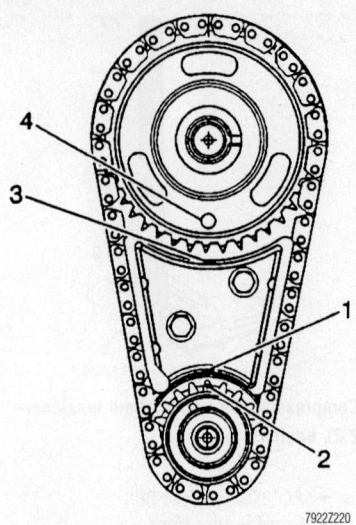

Timing chain and sprocket timing mark alignment

9. Remove or disconnect the following:
   - Camshaft sprocket and timing chain
   - Crankshaft sprocket
   - Timing chain damper

To install:

10. Install or connect the following:
    - Timing chain damper. Torque the bolts to 15 ft. lbs. (21 Nm).
    - Crankshaft sprocket

11. Be sure the crankshaft sprocket timing mark is pointing straight up.

12. Install the timing chain over the camshaft sprocket and hold the sprocket in such a way, that the timing mark is pointing down, and the timing chain is hanging down off the sprocket.

13. Loop the timing chain under the crankshaft sprocket and install the camshaft sprocket on the camshaft. The sprocket will only fit on the camshaft if the dowel on the camshaft aligns with the hole in the sprocket.

14. Verify that the marks are aligned (the camshaft sprocket will be at the 6 o'clock position and the crankshaft sprocket will be in the 12 o'clock position).

15. Torque the camshaft sprocket mounting bolt to 103 ft. lbs. (140 Nm).

16. Lubricate the timing chain components with engine oil.

17. Apply a thin bead of sealer around the gasket sealing area of the front cover.

18. Install or connect the following:
    - Front cover with a new seal and gasket. Torque the small bolts to 15 ft. lbs. (21 Nm) and the large bolts to 35 ft. lbs. (47 Nm).

- Radiator hose to the coolant outlet
- Coolant bypass pipe to the water pump and intake manifold
- CKP sensor
- Oil pan. Torque the bolts to 18 ft. lbs. (24 Nm) and the side bolts to 37 ft. lbs. (50 Nm).
- Crankshaft balancer. Torque the bolt to 76 ft. lbs. (103 Nm).
- Accessory drive belt tensioner. Torque the bolt to 40 ft. lbs. (54 Nm).
- Right engine mount bracket. Torque the bolts to 96 ft. lbs. (130 Nm).
- Right inner fender well splash shield
- Wheel

19. Remove the engine support fixture.

20. Install or connect the following:
    - Alternator. Torque the front bolt to 37 ft. lbs. (50 Nm) and the rear bolt to 18 ft. lbs. (24 Nm).
    - Power steering line
    - Throttle body tube
    - Air cleaner assembly
    - Accessory drive belt
    - Negative battery cable

21. Refill the cooling system.

22. Check the engine oil level.

➡ **An oil and filter change is recommended.**

23. Start the engine and verify that there are no leaks.

Piston and Ring

POSITIONING

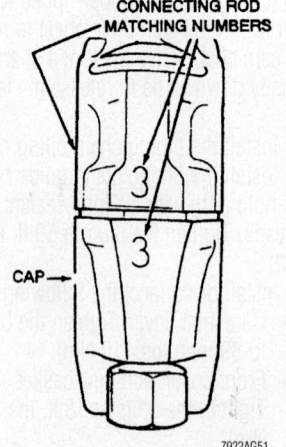

Connecting rod and cap installation. Be sure to matchmark the cap and rod prior to disassembly, as shown

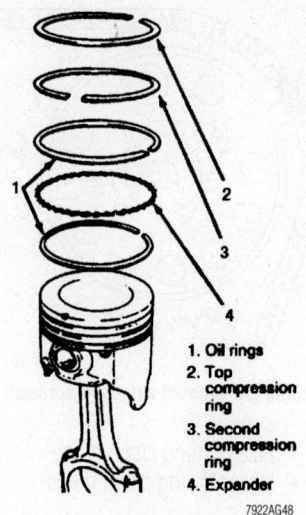

1. Oil rings
2. Top compression ring
3. Second compression ring
4. Expander

Piston ring positioning

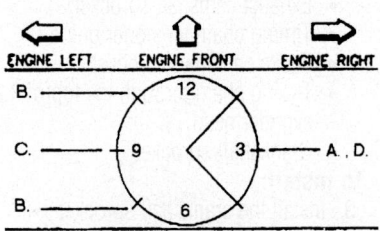

A. OIL RING SPACER GAP (TANG IN HOLE OR SLOT WITH ARC)
B. OIL RING RAIL GAPS
C. 2ND COMPRESSION RING GAP
D. TOP COMPRESSION RING GAP

Piston ring end-gap spacing

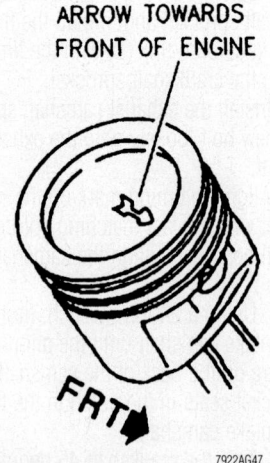

ARROW TOWARDS FRONT OF ENGINE

FRT

Piston positioning. Often the arrow is replaced by a notch, which also must face toward the front of the engine

FUEL SYSTEM

Fuel System Pressure

RELIEVING

2.2L and 3.5L Engines

1. Loosen the fuel filler cap in order to relieve the tank pressure.

2. Disconnect the fuel pump electrical connector.

3. Start and run the engine until the fuel supply remaining in the fuel lines is consumed. Engage the starter for 3 seconds in order to assure relief of any remaining pressure.

4. Disconnect the negative battery cable in order to avoid possible fuel discharge if an accidental attempt is made to start the engine.

3.1L Engine

1. Before servicing the vehicle, refer to the Precautions Section.

2. Disconnect the negative battery cable in order to avoid possible fuel discharge if an accidental attempt is made to start the engine.

3. Loosen the fuel tank filler cap in order to relieve fuel tank pressure.

4. Connect a fuel pressure gauge to the fuel pressure test port connection. Wrap a towel around the fuel pressure connection when installing the fuel pressure gauge in order to avoid fuel spillage.

5. Install the bleed hose into an approved container and open the valve in order to bleed the fuel system pressure. The fuel pipe connections are now safe for servicing.

6. Drain any fuel remaining in the fuel pressure gauge into an approved container.

Fuel Filter

REMOVAL & INSTALLATION

1. Before servicing the vehicle, refer to the Precautions Section.

2. Relieve the fuel system pressure.

3. Remove or disconnect the following:
- Fuel line from the filter using a back-up wrench
- Quick-connect fitting from the fuel filter
- Fuel filter from the mounting bracket

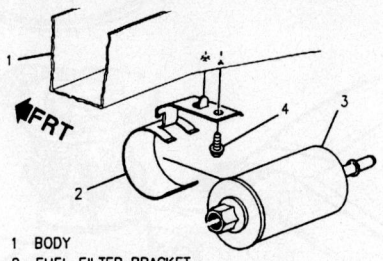

1 BODY
2 FUEL FILTER BRACKET
3 FUEL FILTER
4 SCREW – FULLY DRIVEN, SEATED AND NOT STRIPPED

7922Z221

Exploded view of the fuel filter mounting

To install:

4. Install or connect the following:
- Fuel filter to the mounting bracket
- Fuel line. Torque the fuel line fitting to 20 ft. lbs. (27 Nm).
- Quick-connect fitting to the fuel filter
- Negative battery cable

5. Pressurize the fuel system and verify no leaks.

Fuel Pump

REMOVAL & INSTALLATION

2002-2003 Models

1. Before servicing the vehicle, refer to the Precautions Section.

2. Relieve the fuel system pressure.

3. Drain the fuel tank.

4. Remove or disconnect the following:
- Fuel tank
- Modular fuel sender assembly-to-fuel tank snapring by using fuel sender lock nut Tool J-39765, press down and rotate the cam lock ring until free of the fuel sender retaining tabs.

➡ **When removing the modular fuel sender from the tank, be aware that it may spring upward.**

- Modular fuel sender assembly

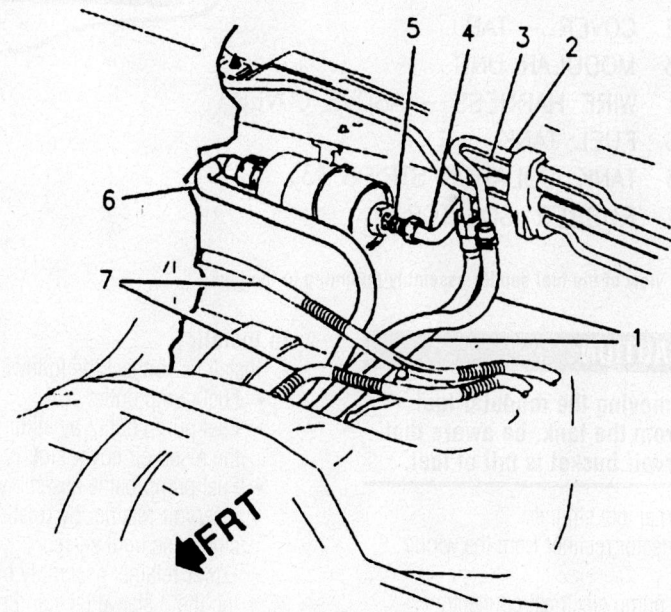

1 HOSE, PART OF FUEL SENDER
2 FUEL VAPOR PIPE
3 FUEL RETURN PIPE
4 FUEL FEED PIPE
5 FUEL FEED PIPE NUT
　27 N•m (20 LBS. FT.)
6 HOSE, PART OF FUEL SENDER
7 ABS AND FUEL SENDER HARNESS

7922Z222

Fuel filter mounting location and component identification

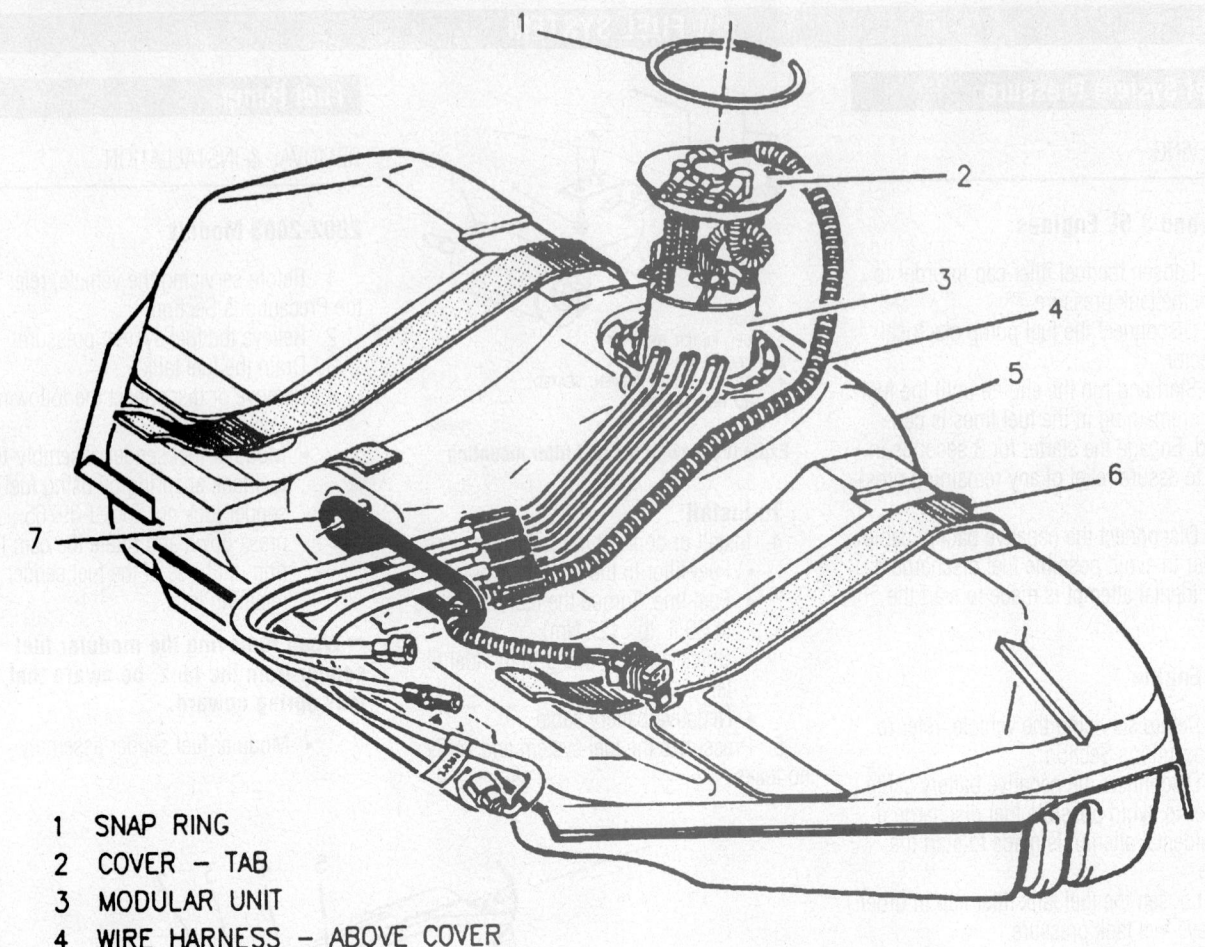

1 SNAP RING
2 COVER – TAB
3 MODULAR UNIT
4 WIRE HARNESS – ABOVE COVER
5 FUEL TANK
6 TANK ISOLATION STRIPS (3)
7 RUBBER ISOLATOR

79222223

Exploded view of the fuel sender assembly mounting to the tank

---

✳✳ CAUTION

When removing the modular fuel sender from the tank, be aware that the reservoir bucket is full of fuel.

- External fuel strainer
- Connector retainer from the wiring harness
- Fuel pump electrical connector

5. Gently release the tabs on the sides of the fuel sender at the cover assembly. Begin by squeezing the sides of the reservoir and releasing the tab opposite the fuel level sensor. Move clockwise to release the second and third tab in the same manner.

6. Remove or disconnect the following:
- Fuel pump/baffle assembly by rotating it counterclockwise
- Fuel pump by sliding the outlet away from the cover slots
- Fuel pump outlet seal

To install:

7. Install or connect the following:
- Fuel pump outlet seal
- Fuel pump outlet by sliding it into the reservoir cover slots
- Fuel pump/baffle assembly onto the reservoir retainer by rotating it clockwise until seated
- Lower retainer assembly by aligning the 3 sleeve tabs and pressing the retainer onto the reservoir until tabs are firmly seated
- Fuel pump electrical connector
- Connector retainer to the wiring harness
- External fuel strainer
- Modular fuel sender assembly
- Modular fuel sender assembly-to-fuel tank snapring using Tool J-39765
- Fuel tank
- Negative battery cable

8. Refill the fuel tank.
9. Pressurize the fuel system and verify no leaks.

2004-2005 Models

1. Before servicing the vehicle, refer to the Precautions Section.
2. Relieve the fuel system pressure.
3. Drain the fuel tank.
4. Remove or disconnect the following:
- Negative battery cable
- Fuel tank
- Electrical and ventilation connections on the modular fuel sender assembly

5. Unlock the fuel sender lock ring using Special Tool J-45722 and a long breaker-bar.

6. Raise the fuel sender assembly out of the tank far enough to access the ventilation harness connection on the underside of the module.

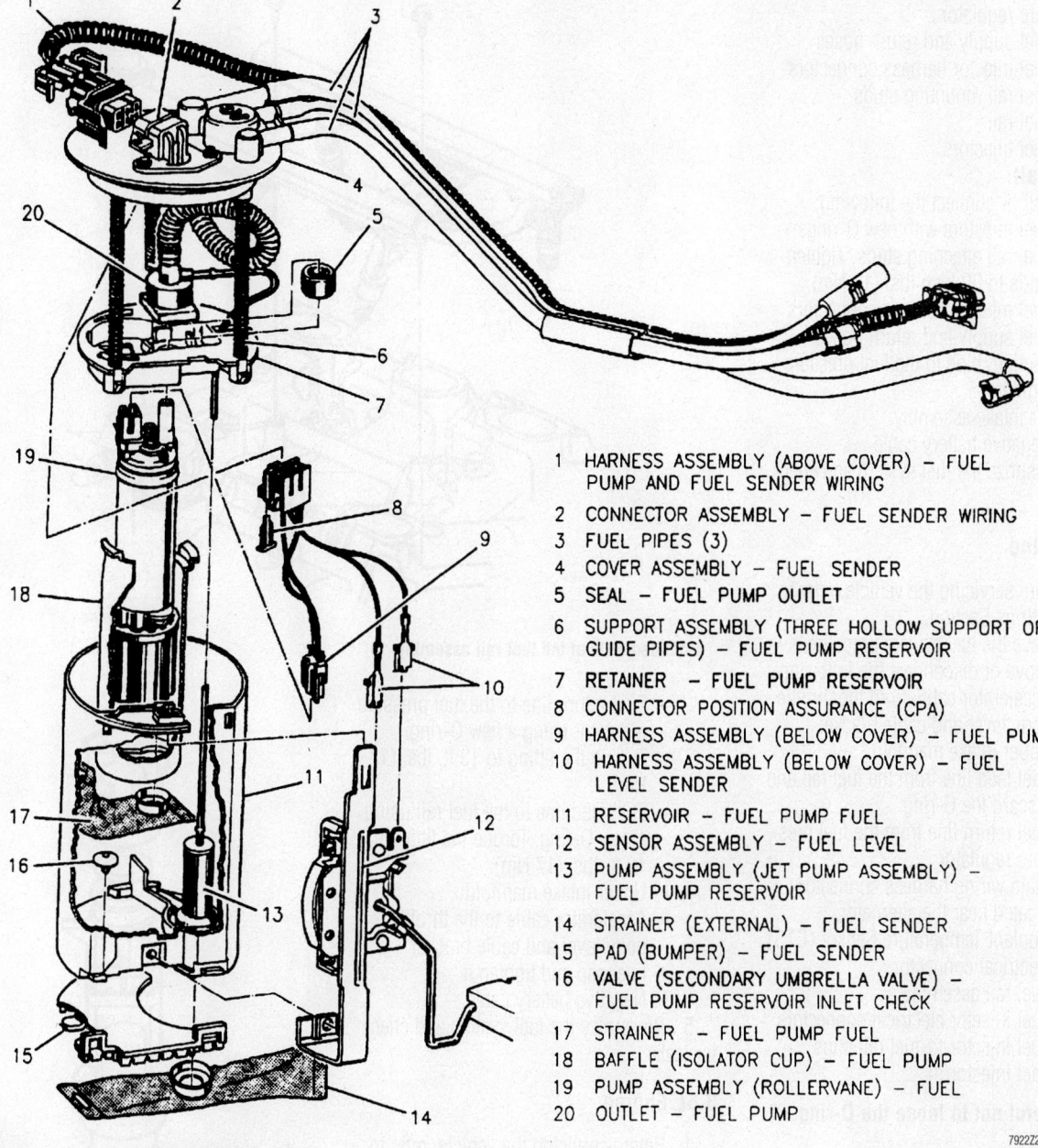

1 HARNESS ASSEMBLY (ABOVE COVER) – FUEL
  PUMP AND FUEL SENDER WIRING
2 CONNECTOR ASSEMBLY – FUEL SENDER WIRING
3 FUEL PIPES (3)
4 COVER ASSEMBLY – FUEL SENDER
5 SEAL – FUEL PUMP OUTLET
6 SUPPORT ASSEMBLY (THREE HOLLOW SUPPORT OR
  GUIDE PIPES) – FUEL PUMP RESERVOIR
7 RETAINER – FUEL PUMP RESERVOIR
8 CONNECTOR POSITION ASSURANCE (CPA)
9 HARNESS ASSEMBLY (BELOW COVER) – FUEL PUMP
10 HARNESS ASSEMBLY (BELOW COVER) – FUEL
   LEVEL SENDER
11 RESERVOIR – FUEL PUMP FUEL
12 SENSOR ASSEMBLY – FUEL LEVEL
13 PUMP ASSEMBLY (JET PUMP ASSEMBLY) –
   FUEL PUMP RESERVOIR
14 STRAINER (EXTERNAL) – FUEL SENDER
15 PAD (BUMPER) – FUEL SENDER
16 VALVE (SECONDARY UMBRELLA VALVE) –
   FUEL PUMP RESERVOIR INLET CHECK
17 STRAINER – FUEL PUMP FUEL
18 BAFFLE (ISOLATOR CUP) – FUEL PUMP
19 PUMP ASSEMBLY (ROLLERVANE) – FUEL
20 OUTLET – FUEL PUMP

7922Z224

Exploded view of the fuel pump assembly

✳✳ WARNING

Be aware that the reservoir bucket on the sender assembly is full of fuel. It must be tipped slightly during removal to avoid damage to the float.

7. Disconnect the ventilation harness quick connector and remove the sender assembly.

To install:

➡**If the lock ring is damaged, it must be replaced.**

8. Install a new O-ring on the modular fuel sender.

9. Connect the ventilation harness inside the tank to the bottom of the modular fuel sender.

10. Install the modular sender assembly and using Special Tool J-45722 install the fuel sender lock ring.

➡**Turn the lock ring until it seats on the second detent.**

11. Install or connect the following:
   • Electrical and ventilation connections on the modular fuel sender assembly
   • Fuel tank.
   • Negative battery cable
12. Refill the tank.

13. Pressurize the system and check for leaks.

Fuel Injector

REMOVAL & INSTALLATION

2.2L Engine

1. Before servicing the vehicle, refer to the Precautions Section.
2. Relieve the fuel system pressure.
3. Remove or disconnect the following:
   • Negative battery cable
   • Air intake assembly

- Vacuum hose from the fuel pressure regulator
- Fuel supply and return hoses
- Fuel injector harness connectors
- Fuel rail mounting studs
- Fuel rail
- Fuel injectors

To install:

4. Install or connect the following:
- Fuel injectors with new O-rings
- Fuel rail attaching studs. Tighten studs to 89 inch lbs. (10 Nm).
- Fuel injector harness connectors
- Fuel supply and return hoses
- Vacuum pipe to the fuel pressure regulator
- Air intake assembly
- Negative battery cable

5. Pressurize the fuel system and check for leaks.

3.1L Engine

1. Before servicing the vehicle, refer to the Precautions Section.
2. Relieve the fuel system pressure.
3. Remove or disconnect the following:
- Accelerator cable from the throttle body lever and cable bracket
- Upper intake manifold
- Fuel feed line from the fuel rail and discard the O-ring
- Fuel return line from the fuel pressure regulator
- Main wiring harness connectors located near the alternator
- Coolant Temperature Sensor (CTS) electrical connector
- Fuel rail assembly
- Fuel injector electrical connectors
- Fuel injector-to-fuel rail clips
- Fuel injector(s)

➡Be careful not to loose the O-ring backups.

To install:

➡When installing new O-rings on the fuel injector, the lower position is color coded brown and the upper position is color coded black. Be sure to install the nylon O-ring backup to properly position the O-ring on the fuel injector so it doesn't move when installing the fuel rail.

4. Install or connect the following:
- Fuel injector with new O-rings
- Fuel injector electrical connectors
- Fuel rail assembly. Torque the bolts to 7 ft. lbs. (10 Nm).
- CTS electrical connector
- Main wiring harness connectors

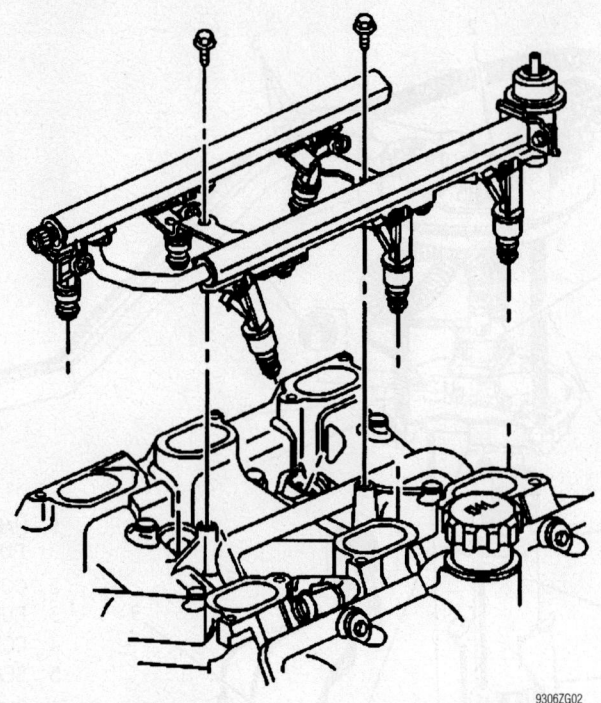

9306ZG02

Exploded view of the fuel rail assembly

- Fuel return line to the fuel pressure regulator using a new O-ring. Torque the fitting to 13 ft. lbs. (17 Nm).
- Fuel feed line to the fuel rail using a new O-ring. Torque the fitting to 13 ft. lbs. (17 Nm).
- Upper intake manifold
- Accelerator cable to the throttle body lever and cable bracket
- Fuel cap and tighten it
- Negative battery cable

5. Pressurize the fuel system and check for leaks.

3.5L Engine

1. Before servicing the vehicle, refer to the Precautions Section.
2. Relieve the fuel system pressure.
3. Remove or disconnect the following:
- Negative battery cable
- Fuel supply hose from fuel rail
- Upper intake manifold
- Main injector harness electrical connector
- Fuel injector electrical connectors
- Injector wiring harness from the fuel rail
- Fuel injector retaining clip
- Fuel injector(s)

To install:

4. Install or connect the following:
- Fuel injector(s) with new O-rings
- Fuel injector retaining clip

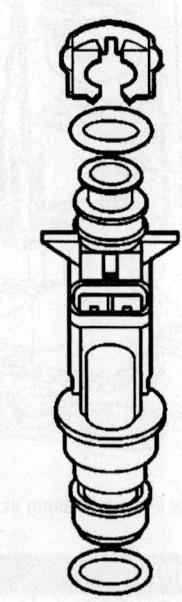

9306ZG03

Exploded view of the fuel injector

- Fuel rail assembly. Tighten the bolts to 89 inch lbs. (10 Nm).
- Injector wiring harness to the fuel rail
- Fuel injector electrical connectors
- Main fuel injector harness electrical connector
- Upper intake manifold
- Fuel supply hose to the fuel rail
- Negative battery cable

DRIVE TRAIN

Transmission Assembly

REMOVAL & INSTALLATION

2002-2003 Models

1. Before servicing the vehicle, refer to the Precautions Section.
2. Install an Engine Support Fixture.
3. Drain the transmission.
4. Remove or disconnect the following:
   - Negative battery cable
   - Air cleaner assembly
   - Front transmission mount bolts
   - Front wheels
   - Left and right splash shields
   - Shift linkage from the transaxle
   - Wiring harness connection from the transaxle
   - Ground cables from the engine block
   - Park Neutral Position (PNP) switch electrical connector
   - Lower radiator and condenser support
   - Front transmission mount bracket
   - Anti-lock Brake System (ABS) wheel speed sensors and electrical harnesses from the suspension supports
   - Brake modulator assembly
   - Vehicle Speed Sensor (VSS)
   - Torque converter cover
   - Torque converter-to-flywheel bolts
   - Ball joints from the steering knuckles
   - Halfshafts from the transaxle
   - ABS module
   - Outer tie rod ends from the steering knuckles
   - Pressure line from the rack and pinion
   - Transmission fluid cooler lines
   - Brake hose bracket from the body
   - Intermediate shaft
   - Transmission-to-engine bolts
   - Transaxle from the engine

To install:

5. Apply a thin film of grease on the torque converter pilot hub.

❋❋ WARNING

Be sure to properly seat the torque converter in the pump.

6. Install or connect the following:
   - Transaxle. Torque the transmission-to-engine bolts to 66 ft. lbs. (90 Nm).

   - Intermediate shaft. Torque the pinch bolt to 15 ft. lbs. (20 Nm).
   - Brake hose bracket to the body
   - Transaxle fluid cooler lines
   - Pressure line to the rack and pinion
   - Outer tie rod ends. Torque the nuts to 14 ft. lbs. (20 Nm) plus an additional 180 degree turn.
   - ABS module
   - Halfshafts to the transaxle
   - Ball joints. Torque the nuts to 48 ft. lbs. (65 Nm).
   - Torque converter. Torque the bolts to 46 ft. lbs. (62 Nm).
   - Torque converter cover
   - VSS electrical connector
   - Brake modulator assembly
   - Front ABS wheel speed sensors and harnesses to the suspension support
   - Front transmission mount bracket
   - Lower radiator and condenser support
   - PNP switch electrical connector
   - Ground cables to the engine block
   - Transmission electrical connections
   - Shift cable bracket. Torque the bolt to 18 ft. lbs. (24 Nm) and the nut to 37 ft. lbs. (50 Nm).
   - Air cleaner assembly
   - Left and right splash shields
   - Front wheels
7. Remove the engine support fixture.
8. Install or connect the following:
   - Shift linkage
   - Negative battery cable
9. Refill the transmission.
10. Apply the brakes and start the engine.
11. Shift the transaxle from **R** to **D** and back to **P**.
12. Recheck the fluid level.

2004-2005 Models

1. Before servicing the vehicle, refer to the Precautions Section.
2. Drain the transmission fluid.
3. Install an engine support fixture.
4. Remove or disconnect the following:
   - Negative battery cable
   - Air intake assembly
   - Transaxle wiring harness from the transaxle and Park/Neutral Position (PNP) switch
   - Shift linkage from the transaxle
   - Transmission wiring harness from the retainer on the transmission
   - Upper transmission-to-engine bolts
   - Front wheels

   - Splash shields
   - Steering gear intermediate shaft
   - Control arm from the frame. Support with mechanics wire or equivalent.
   - Outer tie rod ends from the steering knuckles.
   - Frame
   - Transmission brace mounting bolts near the right axle shaft
   - Oil pan to bell housing bolts
   - Flywheel inspection cover
   - Starter

➡**Matchmark the flywheel to the torque converter.**

   - Torque converter-to-flywheel bolts
   - Transmission cooler lines
   - Vehicle speed sensor (VSS)
   - Rear O$_2$ sensor
   - Halfshalfts from the transaxle
   - Front transmission bracket
5. Support the transmission using a transmission jack or suitable equivalent.
6. Remove the remaining bell housing bolts and separate the transmission from the engine.
7. Lower the transmission and remove.

To install:

8. Install or connect the following:
   - Transaxle. Tighten the transmission-to-engine bolts to 66 ft. lbs. (90 Nm).
   - Front transmission mount bracket
   - Halfshafts
   - Rear O$_2$ sensor
   - VSS
   - Torque converter to flywheel bolts. Tighten to 46 ft. lbs. (62 Nm).
   - Starter
   - Flywheel inspection cover
   - Transmission cooler lines
   - Oil pan-to-bell housing bolts
   - Transmission brace mounting bolts near the right axle shaft. Tighten to 53 ft. lbs. (72 Nm).
   - Frame
   - Outer tie rod ends to steering knuckles
   - Control arms to the frame
   - Steering gear intermediate shaft
   - Splash shields
   - Front wheels
   - Upper transmission-to-engine bolts. Tighten to 66 ft. lbs. (90 Nm).
   - Shift linkage
   - PNP switch and transaxle electrical connectors

- Air intake assembly
- Negative battery cable

9. Refill transmission with ATF to the proper level.

Halfshaft

REMOVAL & INSTALLATION

2002-2003 Models

1. Before servicing the vehicle, refer to the Precautions Section.

2. Remove or disconnect the following:

- Wheel
- Tie rod from the steering knuckle
- Halfshaft hub nut and washer
- Stabilizer link
- Lower ball joint from the steering knuckle
- Halfshaft from the hub/bearing assembly using a puller
- Halfshaft from the transaxle

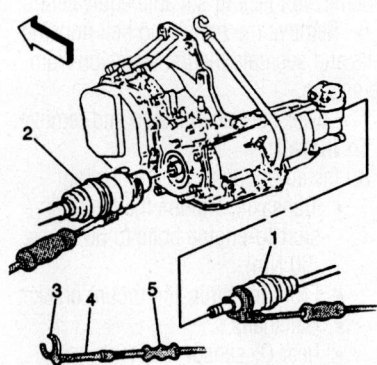

(1) Drive Axle, Right Side   (4) J 29794
(2) Drive Axle, Left Side   (5) J 2619-01
(3) J 28468 or J 33008

79222225

Removing the left and right halfshafts

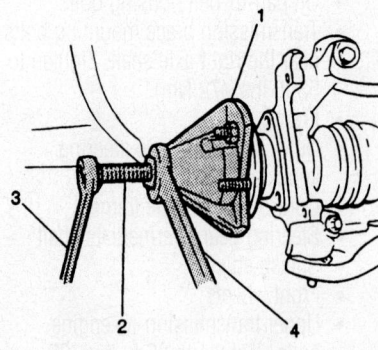

(1) J 28733A   (3) Turn Box Wrench
(2) Forcing Screw   (4) Hold Wrench

79222226

Removing the halfshaft from the hub utilizing the appropriate tools

To install:

3. Install the halfshaft into the transaxle by placing a tool in the joint housing groove and tapping until seated.

※ WARNING

Be careful not to damage the axle seal or dislodge the transaxle seal garter spring when installing the halfshaft.

4. Verify that the halfshaft is seated in the transaxle by grasping on the housing and pulling outward.

5. Install or connect the following:

- Halfshaft into the hub/bearing assembly
- Tie rod to the steering knuckle. Torque the nut to 33 ft. lbs. (45 Nm).
- Lower ball joint to the steering knuckle. Torque the nut to 48 ft. lbs. (65 Nm).
- New cotter pin to the lower ball joint
- Stabilizer link. Torque the nut to 13 ft. lbs. (17 Nm).
- New halfshaft nut. Torque it to 284 ft. lbs. (385 Nm).
- Wheel

2004-2005 Models

1. Before servicing the vehicle, refer to the Precautions Section.

2. Remove or disconnect the following:

- Wheel
- Halfshaft hub nut

3. Separate the outer tie rod assembly from the steering knuckle.

4. Separate the ball joint from the steering knuckle.

5. Separate the halfshaft from the hub bearing using Puller J-42129 or equivalent.

6. Remove the halfshaft from the transaxle.

To install:

7. Install the halfshaft into the transaxle.

➡Verify the halfshaft retaining ring is properly seated.

8. Install the halfshaft into the hub bearing.

9. Connect the ball joint to the steering knuckle. Tighten the pinch bolt to 37 ft. lbs. (50 Nm). Reverse the nut ¾ of a turn. Tighten to 37 ft. lbs. (50 Nm) plus 60 degrees.

10. Connect the outer tie rod assembly to the steering knuckle. Tighten the nut to 44 ft. lbs. (60 Nm).

11. Install a new halfshaft hub nut. Tighten to 159 ft. lbs. (215 Nm).

12. Install the wheel.

CV-Joint

OVERHAUL

Outer CV-Joint

1. Before servicing the vehicle, refer to the Precautions Section.

2. Remove or disconnect the following:

- Front wheel
- Halfshaft and position it in a vise
- Large CV-joint boot clamp
- Small CV-joint boot clamp
- CV-joint boot and slide it back on the shaft
- Outer race from the halfshaft by spreading the outer race-to-halfshaft retaining ring
- Retaining ring from the halfshaft
- CV-joint boot from the halfshaft

3. Disassemble the chrome alloy balls from the CV-joint cage as follows:

   a. Position a brass drift against the CV-joint cage and tap it with a hammer to tilt the cage.

   b. Remove the 1st chrome alloy ball from the cage.

   c. Tilt the cage in the opposite direction.

   d. Remove the opposite chrome alloy ball.

   e. Repeat the procedure until all 6 balls are removed.

4. Disassemble the CV-joint cage and inner race as follows:

   a. Pivot the cage and race 90 degrees to the center line of the outer race.

   b. Align the cage windows with outer race lands.

   c. Remove the cage from the outer race.

   d. Rotate the inner race upward and remove it from the cage.

To install:

5. Lubricate the parts with a light coat of grease.

6. Assemble the CV-joint cage and inner race, as follows:

   a. Rotate the inner race 90 degrees to the cage centerline.

   b. Align the cage windows with inner race lands.

   c. Insert the inner race into the cage by rotating the inner race downward.

   d. Insert the cage/inner race into the outer race.

7. Assemble the chrome alloy balls into the CV-joint cage, as follows:

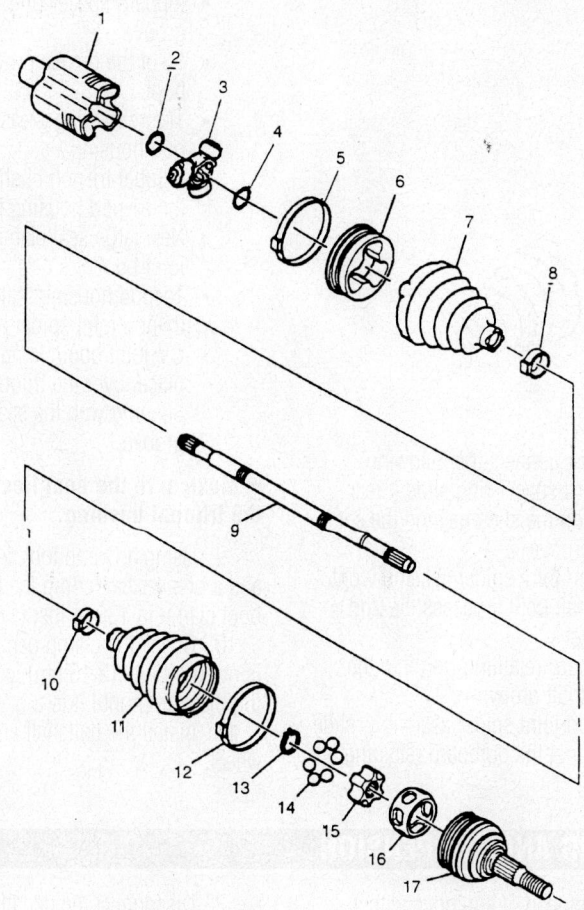

Exploded view of the halfshaft assembly

1	Retainer and Housing Assembly	10	Seal Retaining Clamp
2	Shaft Retaining Ring	11	Drive Axle Outboard Seal
3	Tripot Joint Spider Assembly	12	Seal Retaining Clamp
4	Spacer Ring	13	Race Retaining Ring
5	Seal Retaining Clamp	14	Chrome Alloy Ball
6	Tripot Trilobal Bushing	15	CV Joint Inner Race
7	Drive Axle Inboard Seal	16	CV Joint Cage
8	Seal Retaining Clamp	17	CV Joint Outer Race
9	Axle Shaft		

9306ZG01

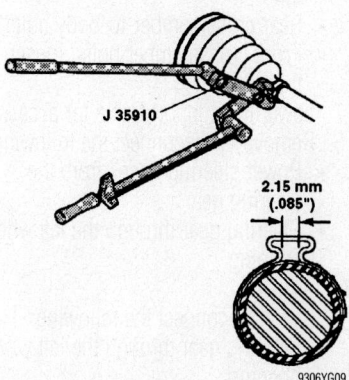

Crimping the small boot clamp—Outer CV-joint

2.15 mm (.085")

9306YG09

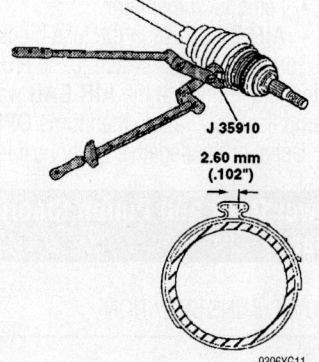

Crimping the large boot clamp—Outer CV-joint

2.60 mm (.102")

9306YG11

a. Position a brass drift against the CV-joint cage and tap it with a hammer to tilt the cage.

b. Insert the 1st chrome alloy ball into the cage.

c. Tilt the cage in the opposite direction.

d. Insert the opposite chrome alloy ball.

e. Repeat the procedure until all 6 balls are inserted.

8. Install ½ of the grease provided into the CV-joint.

9. Install or connect the following:
- Small ring clamp on the CV boot
- CV boot onto the halfshaft

10. Slide the small end of the CV-joint boot/clamp into place, with the seal lip in the halfshaft groove.

➡**Make sure the boot lies flat against the halfshaft.**

11. Using a Crimp tool, a torque wrench and a breaker bar, crimp the small CV-joint boot clamp to 100 ft. lbs. (136 Nm).

12. Install the remainder of the grease into the CV-joint boot.

13. Install or connect the following:
- New retaining ring on the halfshaft
- Outer race assembly onto the half-shaft until the ring engages the halfshaft groove
- Large ring clamp on the CV boot

14. Using a Crimp tool, a torque wrench and a breaker bar, crimp the large CV-joint boot clamp to 130 ft. lbs. (176 Nm).

15. Install the halfshaft and the front wheel.

Inner (Tri-Pod) Joint

1. Before servicing the vehicle, refer to the Precautions Section.

2. Remove or disconnect the following:
- Front wheel
- Halfshaft and place it in a vise
- Snapring from the stub shaft
- Small CV-joint boot clamp
- Large CV-joint boot clamp
- CV-joint boot by sliding it away from the tri-pod joint

3. Install a Stub Shaft Removal Tool J-38868-A to the stub shaft snapring groove.

4. Using a slide hammer puller, pull the stub shaft from the tri-pod housing.

5. Remove or disconnect the following:
- Tri-pod housing from the tri-pod spider
- Inboard spacer ring slide it rearward on the shaft
- Outboard retaining ring
- Tri-pod joint spider assembly

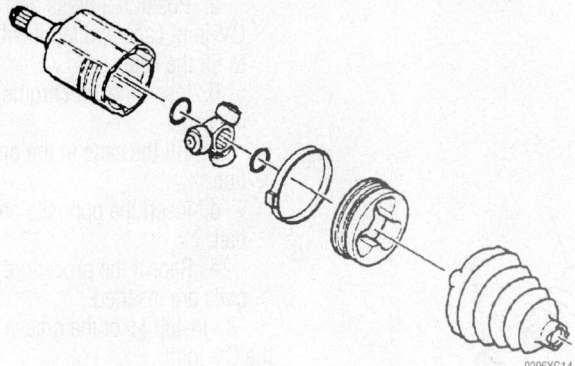

9306XG14

Exploded view of the inner (tri-pod) joint

- Inboard spacer ring and discard it
- CV-joint boot
- Trilobal tri-pod bushing from the housing

To install:

6. Install or connect the following:
   - New snapring onto the stub shaft
   - Small boot clamp
   - CV-joint boot

7. Using a Crimp tool, a torque wrench and a breaker bar, crimp the small CV-joint boot clamp to 100 ft. lbs. (136 Nm).

8. Install or connect the following:
   - Inboard spacer ring slide it rearward on the shaft beyond the second groove
   - Tri-pod joint spider assembly onto the shaft until it passes the 2nd groove
   - Outboard retaining ring into the axle shaft groove
   - Tri-pod joint spider assembly, slide it against the outboard retaining ring

- Inboard spacer ring, seat it in the groove
- ½ of the grease provided into the boot
- The remaining grease into the tri-pod housing
- Trilobal tri-pod bushing flush with the tri-pod housing face
- New large seal clamp onto the CV-joint boot
- Tri-pod housing, slide it over the tri-pod joint spider assembly
- CV-joint boot/clamp, slide it into place, over the trilobal tri-pod bushing with the seal lip in the groove

➡ **Make sure the boot lies flat against the trilobal bushing.**

9. Using a Crimp tool, a torque wrench and a breaker bar, crimp the large CV-joint boot clamp to 130 ft. lbs. (176 Nm).

10. Check the clamp gap dimension; if it is not 0.085 in. (2.16mm), continue tightening the clamp until it is.

11. Install the halfshaft and the front wheel.

STEERING AND SUSPENSION

Air Bag

DISARMING

※※ CAUTION

The Supplemental Restraint System (SRS) must be disarmed before performing service procedures around the air bag or SRS wiring. Failure to do so may cause accidental deployment of the air bag, resulting in unnecessary SRS repairs and/or personal injury.

1. Disconnect the negative battery cable.
2. Turn the steering wheel so the vehicle's wheels are pointing straight-ahead.
3. Turn the ignition switch to the **LOCK** position.
4. Remove or disconnect the following:
   - Key
   - **AIR BAG** fuse from the fuse block
   - Left sound insulator
   - Connector retainer clip from the yellow 2-way connector at the base of the steering column
   - Connector retainer and yellow 2-way connector from the passenger

air bag lead, if equipped with a passenger's side air bag

ARMING

1. Turn the ignition switch to the **LOCK** position and remove the key.
2. Install or connect the following:
   - Yellow 2-way connector at the base of steering column and the connector retainer clip
   - Yellow 2-way connector at the passenger air bag lead and secure it with the connector retainer clip, if equipped with a passenger's side air bag
   - Left sound insulator
   - **AIR BAG** fuse in the fuse block
3. Turn the ignition switch to the **RUN** position and verify that the **AIR BAG** warning lamp flashes 7 times, then turns **OFF**.
4. Connect the negative battery cable.

Power Rack and Pinion Steering Gear

REMOVAL & INSTALLATION

1. Before servicing the vehicle, refer to the Precautions Section.

2. Disconnect the negative battery cable.
3. Siphon the power steering fluid from the reservoir.
4. Remove or disconnect the following:
   - Left front wheel
   - Intermediate shaft assembly
   - Tie rod ends from the steering knuckles
   - Stabilizer shaft links
   - Steering gear mounting bolts
5. Support the rear of the support frame.
6. Remove or disconnect the following:
   - Transaxle-to-crossmember mounting bolt
   - Rear crossmember-to-body bolts
   - Front crossmember bolts, loosen them
7. Lower the support frame for access.
8. Remove or disconnect the following:
   - Power steering hoses from the steering gear
   - Steering gear through the left wheel opening

To install:

9. Install or connect the following:
   - Steering gear through the left wheel opening
   - Power steering hoses to the steering gear

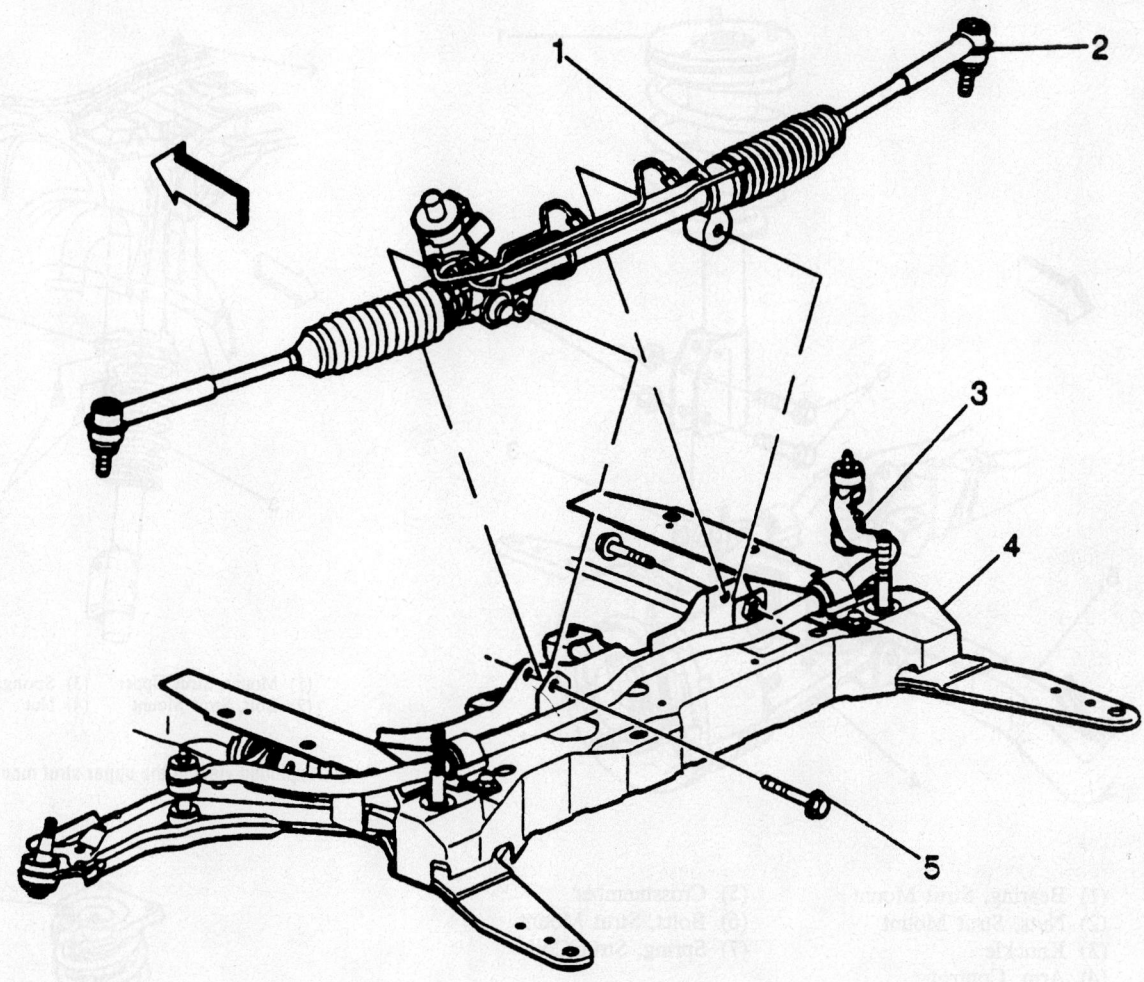

(1) Gear Assembly, Power Steering
(2) Tie Rod
(3) Shaft, Front Stabilizer
(4) Frame Assembly, Drivetrain and Front Crossmember
(5) Bolt, Power Steering Gear Assembly

79222227

Exploded view of the power rack and pinion steering gear mounting

10. Raise the support frame.
11. Install or connect the following:
- Front crossmember bolts. Torque them to 71 ft. lbs. (110 Nm) plus an additional 90 degree turn.
- Rear crossmember bolts. Torque them to 71 ft. lbs. (110 Nm) plus an additional 90 degree turn.
- Transaxle mount-to-crossmember bolt. Torque it to 49 ft. lbs. (66 Nm).
- Steering gear bolts. Torque them to 89 ft. lbs. (120 Nm).
- Stabilizer shaft links. Torque the nuts to 13 ft. lbs. (17 Nm).
- Tie rod ends to the steering knuckle. Torque the nut to 15 ft. lbs. (20 Nm) plus an additional 180 degree turn.

- Intermediate shaft. Torque the pinch bolt to 16 ft. lbs. (22 Nm).
- Left front wheel
- Negative battery cable
12. Refill and bleed the power steering system.
13. Check and/or adjust the front end alignment.

Strut

REMOVAL & INSTALLATION

Front

2002-2003 MODELS

1. Before servicing the vehicle, refer to the Precautions Section.

2. Support the front crossmember.
3. Matchmark the front strut-to-steering knuckle location.
4. Remove or disconnect the following:
- Front wheel
- Brake line bracket from the strut
- Strut lower mounting bracket nut and through-bolts
- Upper strut-to-chassis nuts
- Strut
To install:
5. Install or connect the following:
- Strut and finger-tighten the upper strut-to-chassis nuts
- Strut to the steering knuckle. Torque the through-bolts/nuts to 133 ft. lbs. (180 Nm) with the reference marks aligned.
- Brake line bracket to the strut.

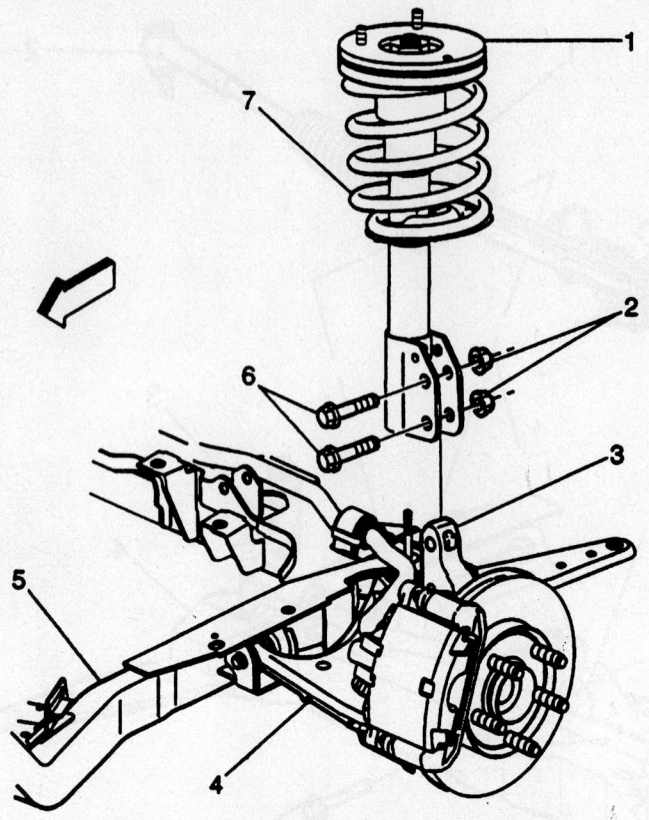

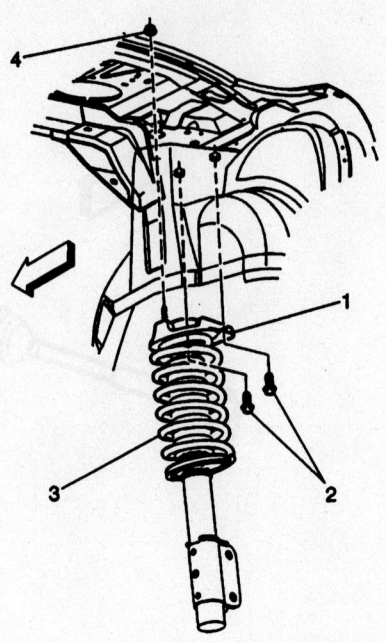

(1) Mount, Strut Upper   (3) Spring, Coil
(2) Bolt, Strut Mount   (4) Nut

79222229

Exploded view of the upper strut mounting

(1) Bearing, Strut Mount   (5) Crossmember
(2) Nuts, Strut Mount   (6) Bolts, Strut Mount
(3) Knuckle   (7) Spring, Strut Coil
(4) Arm, Control

79222228

Exploded view and component identification of the front strut

Torque the bolt to 10 ft. lbs. (14 Nm).
• Wheel

6. Torque the upper strut fasteners to 18 ft. lbs. (24 Nm).

7. Check and/or adjust the front end alignment.

2004-2005 MODELS

1. Before servicing the vehicle, refer to the Precautions Section.

2. Remove or disconnect the following:
• Front wheel
• Stabilizer link from the strut
• Strut-to-steering knuckle nuts
• Wheel speed sensor, if equipped
• Strut-to-steering knuckle bolts
• Upper strut-to-chassis nuts
• Strut

To install:

3. Position the strut in the strut tower, using the alignment pin as a guide.

4. Install or connect the following:
• Upper strut-to-chassis nuts. Tighten to 18 ft. lbs. (25 Nm).

• Strut-to-steering knuckle bolts, leaving the nuts off
• Wheel speed sensor, if equipped
• Strut-to-steering knuckle nuts and tighten to 89 ft. lbs. (120 Nm)
• Stabilizer link. Tighten to 48 ft. lbs. (65 Nm)
• Front wheel

5. Check and/or adjust the front end alignment.

Rear

2002-2003 MODELS

1. Before servicing the vehicle, refer to the Precautions Section.

2. Matchmark the strut-to-knuckle position.

3. Remove or disconnect the following:
• Rear wheels
• Upper strut-to-chassis nuts
• Strut-to-knuckle bolts
• Strut

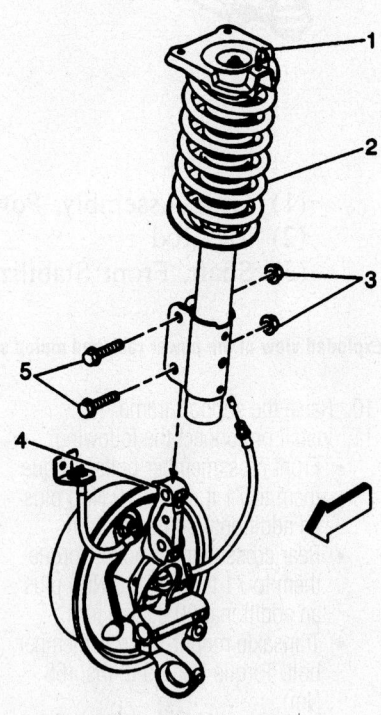

(1) Mount, Strut Upper
(2) Spring, Coil
(3) Nuts, Strut Bolt
(4) Knuckle
(5) Bolt, Strut Mount

79222230

Exploded view of the lower strut mounting

To install:

4. Install or connect the following:
   - Strut and finger-tighten the strut-to-knuckle bolts
   - Upper strut-to-chassis nuts. Torque the nuts to 18 ft. lbs. (24 Nm).

➡**Align the matchmarks to ensure proper alignment.**

   - Strut-to-knuckle bolts. Torque them to 89 ft. lbs. (120 Nm).
   - Rear wheels
5. Check and/or adjust the alignment.

Coil Spring

REMOVAL & INSTALLATION

2002-2003 Models

FRONT AND REAR

1. Before servicing the vehicle, refer to the Precautions Section.
2. Remove the strut.
3. Mount the strut in a strut compressor.

➡**The strut compressor has strut mounting holes drilled for specific vehicle lines.**

4. Compress the strut approximately ½ of its height after initial contact with the top cap.

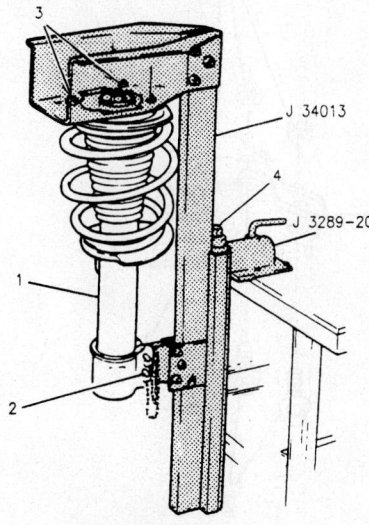

1 STRUT ASSEMBLY
2 INSTALL LOCKING PINS THROUGH STRUT ASSEMBLY
3 TIGHTEN NUTS UNTIL FLUSH WITH STRUT COMPRESSOR
4 COMPRESSOR FORCING SCREW

7922Z231

View of the strut mounted in the compressor

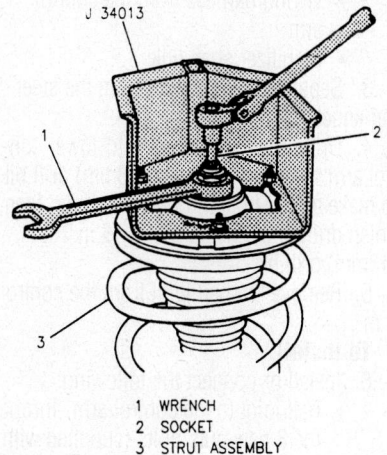

1 WRENCH
2 SOCKET
3 STRUT ASSEMBLY

7922Z232

Use a socket and a wrench to remove the damper shaft nut spring cap while compressing the spring

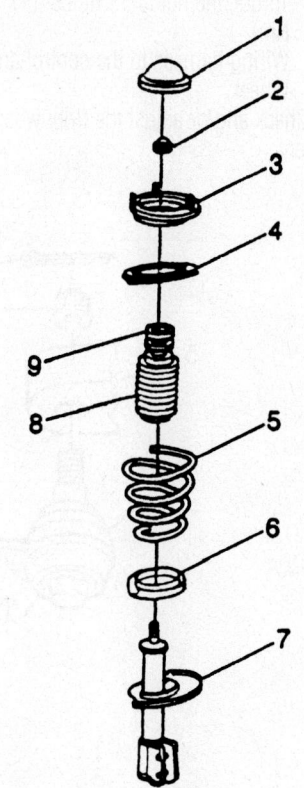

(1) Cap
(2) Nut, Strut Mount
(3) Bearing, Strut Mount
(4) Insulator, Spring (Upper)
(5) Spring, Strut Coil
(6) Insulator, Spring (Lower)
(7) Strut
(8) Bumper, Jounce
(9) Shield, Strut Dust

7922Z233

Exploded view of the front strut

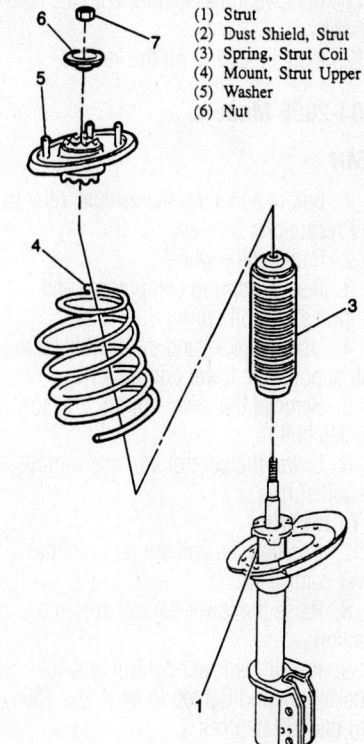

(1) Strut
(2) Dust Shield, Strut
(3) Spring, Strut Coil
(4) Mount, Strut Upper
(5) Washer
(6) Nut

7922Z234

Exploded view of the rear strut

❊❊ WARNING

Never bottom the spring or damper rod.

5. Remove the strut damper shaft nut and place a guiding rod on top of the damper shaft. Use the rod to guide the damper shaft straight down through the spring cap while compressing the spring. Remove the components.

To install:

6. Install the bearing cap onto the strut compressor, if removed.
7. Mount the strut in the strut compressor using the bottom locking pin only.
8. Install the spring over the damper and swing the assembly up so the upper locking pin can be installed.
9. Install all shields, bumpers and insulators on the spring seat. Install the spring seat on top of the spring. Be sure the flat on the upper spring seat is facing in the proper direction. The spring seat flat should be facing the same direction as the centerline of the strut assembly spindle.
10. Install the guiding rod and turn the forcing screw while the guiding rod centers the assembly. When the threads on the damper shaft are visible, remove the guiding rod and install the nut. Torque the nut to 34 ft. lbs. (47 Nm).

11. Remove the assembly from the compressor.

12. Install the strut on the vehicle.

2004-2005 Models

REAR

1. Before servicing the vehicle, refer to the Precautions Section.

2. Remove the wheel.

3. Install a spring compressor and compress the coil spring.

4. Using a jackstand or suitable equivalent, support the lower control arm.

5. Remove the lower control arm-to-knuckle bolt.

6. Lower the control arm and remove the coil spring.

To install:

7. Position the coil spring onto the lower control arm.

8. Raise the lower control arm into position.

9. Install the lower control arm-to-knuckle bolt and tighten to 44 ft. lbs. (60 Nm) plus 60 degrees.

10. Remove the spring compressor.

11. Install the wheel.

Lower Ball Joint

REMOVAL & INSTALLATION

1. Before servicing the vehicle, refer to the Precautions Section.

2. Remove or disconnect the following:

- Wheel

➡**Care must be exercised to prevent the halfshaft joints from being over-extended.**

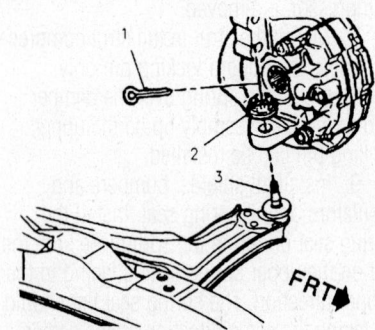

1 PIN
2 NUT — 55 N·m (41 LBS. FT.) MINIMUM TORQUE
  65 N·m (48 LBS. FT.) MAXIMUM TORQUE TO
  INSTALL PIN
3 LOWER BALL JOINT

79ZZZ235

Exploded view of the ball joint-to-knuckle mounting

- Wiring harness from the control arm
- Stabilizer shaft link

3. Separate the ball joint from the steering knuckle.

4. Drill out the 3 ball joint-to-lower control arm rivets. Use an 1/8 in. (3mm) drill bit to make a pilot hole through the rivets; then, finish drilling the rivets with a 1/2 in. (13mm) drill bit.

5. Remove the ball joint from the control arm.

To install:

6. Install or connect the following:

- Ball joint to the control arm. Torque the 3 new nuts/bolts (supplied with new ball joint) to the specifications included with the kit.
- Ball joint to the steering knuckle. Torque the nut to 48 ft. lbs. (65 Nm) and install a new cotter pin.
- Stabilizer link to the stabilizer shaft. Torque the nut to 13 ft. lbs. (17 Nm).
- Wiring harness to the control arm
- Wheel

7. Check and/or adjust the front wheel alignment.

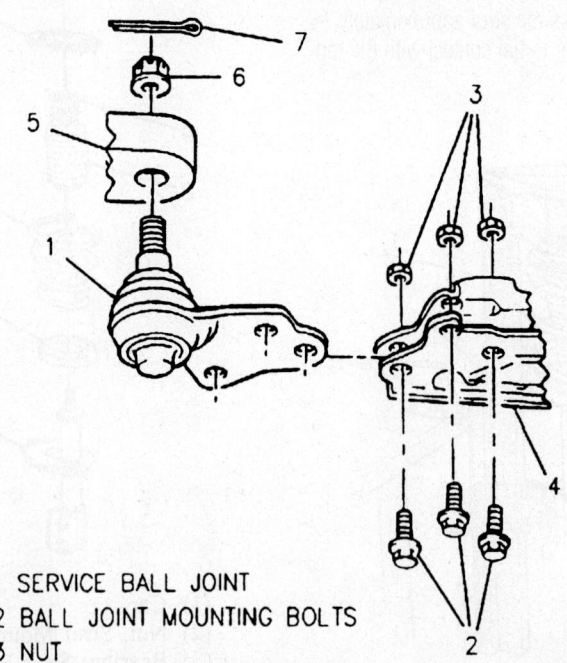

1 SERVICE BALL JOINT
2 BALL JOINT MOUNTING BOLTS
3 NUT
4 LOWER CONTROL ARM
5 STEERING KNUCKLE
6 NUT — 55 N·m (41 LBS. FT.) MINIMUM TORQUE
      65 N·m (48 LBS. FT.) MAXIMUM TORQUE
      TO INSTALL PIN
7 PIN

79ZZZ236

Exploded view of the ball joint mounting

Lower Control Arm

REMOVAL & INSTALLATION

2002-2003 Models

1. Before servicing the vehicle, refer to the Precautions Section.

2. Remove or disconnect the following:

- Front wheel
- Stabilizer shaft link

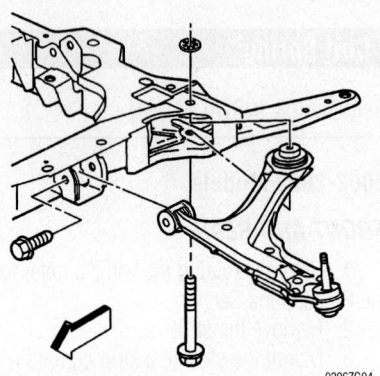

9306ZG04

Exploded view of the lower control arm and related components

- Lower ball joint from the steering knuckle
- Lower control arm-to-suspension crossmember bolts
- Lower control arm

To install:

3. Install or connect the following:

- Lower control arm-to-suspension crossmember and hand-tighten the bolts
- Lower ball joint to the steering knuckle. Torque the nut to 48 ft. lbs. (65 Nm).
- Stabilizer shaft link. Torque the nut to 13 ft. lbs. (17 Nm).
- Front wheel

4. Position the vehicle at curb height.

5. Torque the bolts as follows:

a. Front lower control arm-to-suspension crossmember bolt to 84 ft. lbs. (115 Nm) plus an additional 120 degrees.

b. Rear lower control arm-to-suspension crossmember bolt to 180 ft. lbs. (245 Nm) plus an additional 180 degrees.

6. Check and/or adjust the front alignment.

2004-2005 Models

1. Before servicing the vehicle, refer to the Precautions Section.

2. Remove the front wheel.

3. If equipped with a 3.5L engine:

a. If removing the left lower control arm, remove the side transmission mount.

b. If removing the right lower control arm, remove the engine mount.

4. Remove or disconnect the following:

- Lower control arm front bushing-to-frame bolt
- Lower control arm rear bushing-to-frame bolts
- Lower control arm-to-steering knuckle pinch bolt, noting its original position.
- Ball stud from the steering knuckle
- Control arm

To install:

5. Position the lower control arm to the frame assembly and steering knuckle.

6. Install or connect the following:

7. Install a new lower control arm-to-steering knuckle pinch bolt to its original position. Hand tighten only.

8. Install the lower control arm front bushing-to-frame bolt. Hand tighten only.

9. Install the lower control arm rear bushing-to-frame bolt.

10. Tighten components as follows:

a. Tighten the ball stud-to-steering

knuckle pinch bolt to 37 ft. lbs. (50 Nm). Reverse the nut ¾ of a turn and tighten to 37 ft. lbs. (50 Nm) plus 60 degrees.

b. Tighten the bolts with the front suspension loaded using a jackstand.

c. Tighten the front bushing bolt to 37 ft. lbs. (50 Nm).

d. Tighten the rear bushing bolt to 37 ft. lbs. (50 Nm).

11. If equipped with a 3.5L engine:

a. Install the side transmission mount if removed. Tighten bolt to 37 ft. lbs. (50 Nm).

b. Install the engine mount if removed. Tighten bolt to 37 ft. lbs. (50 Nm).

12. Install the front wheel.

CONTROL ARM BUSHING REPLACEMENT

2002-2003 Models

FRONT BUSHING

1. Before servicing the vehicle, refer to the Precautions Section.

2. Remove the lower control arm and place it in a vise.

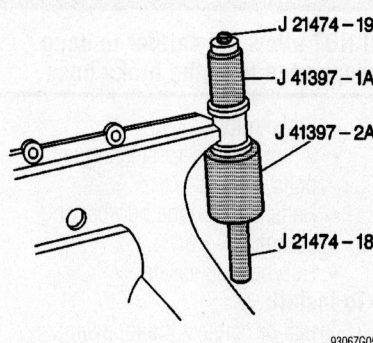

J 21474–19
J 41397–1A
J 41397–2A
J 21474–18

9306ZG05

Removing the front bushing from the lower control arm

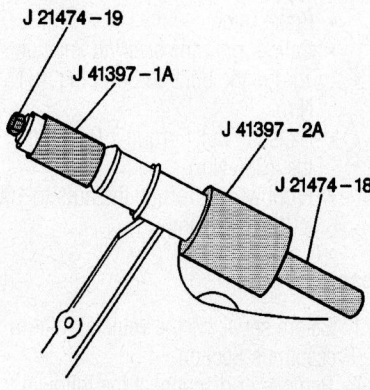

J 21474–19
J 41397–1A
J 41397–2A
J 21474–18

9306ZG06

Installing the front bushing to the lower control arm

3. Lubricate the threads of Screw J-21474-19 with high pressure lubricant.

4. Assemble Tools Screw J-21474-19, Remover/Installer J-41397-1A, Receiver J-41397-2A and J-21474-18 onto the front control arm bushing.

5. Tighten Tool J-21474-18 until the front bushing is pressed from the control arm.

6. Disassemble the tools.

To install:

7. Lubricate the new front bushing outer casing.

8. Insert the new bushing into the control arm.

9. Assemble Tools Screw J-21474-19, Remover/Installer J-41397-1A, Receiver J-41397-2A and J-21474-18 onto the front control arm bushing.

10. Tighten Screw J-21474-19 until the front bushing is pressed into the control arm.

11. Disassemble the tools.

12. Install the lower control arm.

REAR BUSHING

1. Before servicing the vehicle, refer to the Precautions Section.

2. Remove the lower control arm and place it in a vise.

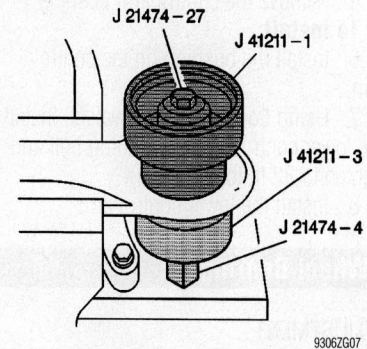

J 21474–27
J 41211–1
J 41211–3
J 21474–4

9306ZG07

Removing the rear bushing from the lower control arm

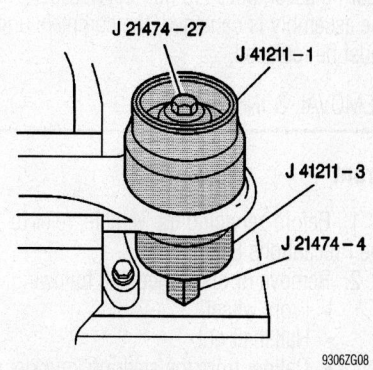

J 21474–27
J 41211–1
J 41211–3
J 21474–4

9306ZG08

Installing the rear bushing to the lower control arm

3. Assemble Tools Screw J-21474-27, Remover/Installer J-41211-1, Receiver J-41211-3 and J-21474-4 onto the rear control arm bushing.

4. Tighten Tool J-21474-27 until the rear bushing is pressed from the control arm.

5. Disassemble the tools.

To install:

6. Insert the new bushing into the control arm.

7. Assemble Tools Screw J-21474-27, Remover/Installer J-41211-1, Receiver J-41211-3 and J-21474-4 onto the rear control arm bushing.

8. Tighten Screw J-21474-4 until the rear bushing is pressed into the control arm.

9. Disassemble the tools.

10. Install the lower control arm.

2004-2005 Models

REAR BUSHING

1. Before servicing the vehicle, refer to the Precautions Section.

2. Remove the lower control arm.

3. Remove the lower control arm rear bushing bolt.

4. Matchmark the position of the bushing on the control arm.

5. Remove the control arm bushing.

To install:

6. Install the bushing on the control arm.

7. Using Locktite® or equivalent, install the lower control arm-to-bushing bolt and tighten to 32 ft. lbs. (44 Nm).

8. Install the lower control arm.

Wheel Bearings

ADJUSTMENT

These vehicles are equipped with sealed hub and bearing assemblies. The hub and bearing assemblies are non-serviceable. If the assembly is damaged, the complete unit must be replaced.

REMOVAL & INSTALLATION

Front

1. Before servicing the vehicle, refer to the Precautions Section.

2. Remove or disconnect the following:
- Front wheel
- Halfshaft nut
- Caliper from the steering knuckle and support it aside

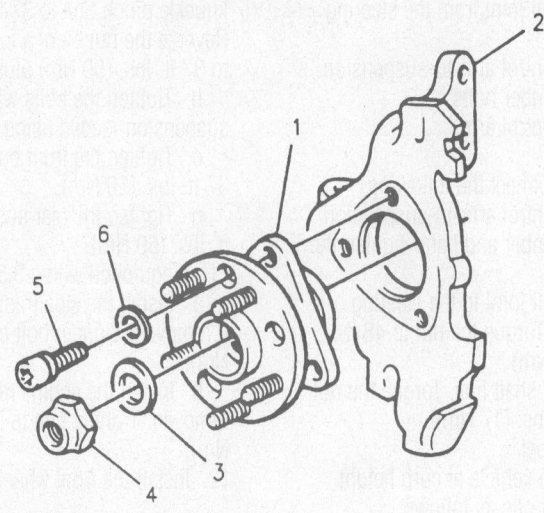

1 HUB AND BEARING ASSEMBLY
2 STEERING KNUCKLE
3 WASHER
4 DRIVE AXLE NUT — 260 N·m (192 LBS. FT.)
5 HUB AND BEARING RETAINING BOLT
6 WASHER

79222237

Exploded view of the front hub and bearing assembly

❋❋ WARNING

DO NOT allow the caliper to hang unsupported from the brake hose.

- Brake rotor
- 3 hub bearing-to-steering knuckle bolts
- Halfshaft from the hub/bearing assembly
- Hub/bearing assembly

To install:

3. Install or connect the following:
- Hub/bearing assembly on the halfshaft
- 3 hub/bearing-to-steering knuckle bolts. Torque the bolts to 70 ft. lbs. (95 Nm).
- Brake rotor
- Caliper onto the steering knuckle. Torque the bolts to 38 ft. lbs. (51 Nm).
- Halfshaft nut. Torque it to 284 ft. lbs. (385 Nm).
- Front wheel. Torque the nuts to 100 ft. lbs. (140 Nm).

Rear

1. Before servicing the vehicle, refer to the Precautions Section.

2. Remove or disconnect the following:
- Wheel
- Brake drum, if equipped

- Brake rotor, if equipped
- 4 hub/bearing assembly-to-knuckle nuts
- Anti-lock Brake System (ABS) speed sensor wire from the hub/bearing assembly
- Hub/bearing assembly

To install:

3. Install or connect the following:
- Hub/bearing assembly
- ABS wheel speed sensor
- Hub/bearing-to-knuckle nuts. Torque the nuts 47 ft. lbs. (63 Nm).
- Brake drum, if equipped
- Brake rotor, if equipped
- Wheel. Torque the bolts to 100 ft. lbs. (140 Nm).

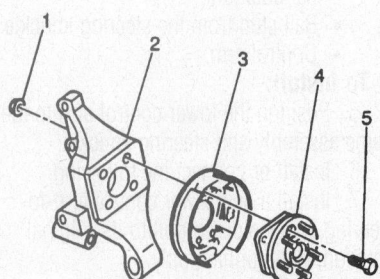

(1) Nut
(2) Knuckle, Rear
(3) Brake Assembly
(4) Hub and Bearing Assembly
(5) Bolt

79222238

Hub and bearing components

BRAKES

Brake Caliper

REMOVAL & INSTALLATION

1. Before servicing the vehicle, refer to the Precautions Section.
2. Siphon ⅔ of the brake fluid out of the master cylinder.
3. Remove the tire and wheel assembly.
4. Compress the caliper piston back into the caliper bore using a large pair of pliers, C-clamp or special piston retracting tool.
5. Remove the brake hose from the caliper and discard the copper washers.
6. Plug the hose to prevent excessive fluid loss and possible fluid contamination.
7. Remove the caliper mounting bolts
8. If removing the rear caliper, remove the park brake cable from the caliper.
9. Remove the caliper from the knuckle.
10. Remove the brake pads from the caliper, if the caliper is being replaced.

To install:
11. Inspect the condition of the caliper support for rust and corrosion that will hinder the travel of the caliper.
12. Inspect the caliper mounting hardware. New bolts are usually recommended.
13. Lubricate the mounting bushings and sleeves with silicone grease as required.
14. Install the brake pads in the caliper.
15. Install the caliper on the steering knuckle.
16. Install the mounting bolts and torque to 26 ft. lbs. (35 Nm).
17. If installing the rear caliper, install the park brake cable to the caliper.
18. Connect the brake hose to the caliper using new copper washers and torque the mounting bolt to 37 ft. lbs. (50 Nm).
19. Refill the master cylinder and bleed the brake system.
20. Install the tire and wheel assembly.
21. Verify correct brake operation.

Disc Brake Pads

REMOVAL & INSTALLATION

1. Before servicing the vehicle, refer to the Precautions Section.
2. Siphon ⅔ of the brake fluid out of the master cylinder reservoir.
3. Remove the tire and wheel assembly.
4. Remove the caliper from the steering knuckle without disconnecting the brake

hose. DO NOT allow the caliper to hang from the brake hose. Support the caliper with a piece of wire.
5. Remove the outboard pad by pushing in on the outside edge of the pad to release the mounting dowel from the hole in the caliper. When both dowels are unseated push the pad out the bottom of the caliper.
6. Remove the inboard pad from the caliper by pulling it out of the caliper.

To install:
7. Install the inboard pad in the caliper so the spring clip on the pad back engages in the caliper piston.
8. Install the outboard pad over the caliper end until the mounting dowels snap into the mounting holes in the caliper.
9. Install the caliper over the rotor onto the steering knuckle. Install the mounting bolts and sleeves and torque to 40 ft. lbs. (51 Nm).
10. Install the tire and wheel assembly.
11. Pump the brake pedal several times to seat the pads against the rotor before attempting to move the vehicle.
12. Check the master cylinder level and add fluid as necessary.

Brake Drums

REMOVAL & INSTALLATION

1. Before servicing the vehicle, refer to the Precautions Section.
2. Remove the wheel.
3. Pull the brake drum off. It may be necessary to gently tap the rear edge of the drum to start it off the studs. If extreme resistance to removal is encountered, it will be necessary to retract the brake shoe self-adjuster screw, sometimes called the star wheel. Knock out the access hole in the brake drum and turn the adjuster to retract

the linings from the drum. Install a replacement hole cover before reinstalling the drum.

➡**DO NOT hammer on the brake drum to remove it.**

To install:
4. Inspect the inside of the brake drum. If worn, heavily grooved or if the opening is distorted, the drum should be refinished or replaced. If refinishing, observe the maximum drum diameter specification.
5. Inspect the wheel cylinder for signs of brake fluid leakage. Inspect the brake shoe springs and self-adjuster mechanism. The adjuster should usually be disassembled, cleaned and lubricated when the drum is removed for brake service.
6. Install the drum over the brake shoes.
7. Install the wheel. Adjust the brakes.
8. Check brake operation.

Brake Shoes

REMOVAL & INSTALLATION

1. Before servicing the vehicle, refer to the Precautions Section.
2. Remove the tire and wheel assembly.
3. Remove the brake drum.
4. Remove the upper return springs from the shoes using tool J-8057 brake spring tool.
5. Remove the hold-down springs using J-8049 brake spring tool.
6. Remove the shoe hold-down pins from behind the brake backing plate.
7. Lift up the actuator lever for the self-adjusting mechanism and remove the actuating link. Remove the actuator lever, pivot, and the pivot return spring.
8. Spread the shoes apart to clear the wheel cylinder pistons and remove the parking brake strut and spring.

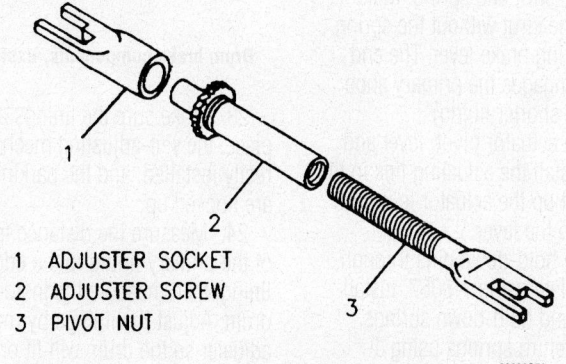

1	ADJUSTER SOCKET
2	ADJUSTER SCREW
3	PIVOT NUT

93006G73

Brake adjuster—Malibu

9. Disconnect the parking brake cable from the lever. Remove the shoes, still connected by their adjusting screw spring.

10. With the shoes removed, note the position of the adjusting spring and remove the spring and adjusting screw.

11. Remove the C-clip from the parking brake lever and the lever from the secondary shoe.

12. Use a damp cloth to remove all dirt and dust from the backing plate and brake parts.

To install:

13. Check the backing plate attaching bolts to make sure they are tight. Use fine emery cloth to clean all rust and dirt from the shoe contact surfaces on the plate and lubricate with brake grease. Check the wheel cylinder for signs of leakage.

14. Clean all parts completely in brake solvent and air dry. Clean the backing plate shoe contact points.

15. Inspect the inside of the brake drum. If worn, heavily grooved or if the opening is distorted, the drum should be refinished or replaced.

16. Inspect the brake shoe springs and self-adjuster mechanism. Disassemble the adjuster mechanism and clean the threads and coat with grease. Make sure the adjuster assembly turns freely before installing in the vehicle.

17. Install the parking brake lever on the secondary shoe and secure with C-clip.

18. Install the adjusting screw and spring on the shoes, connecting them together. The coils of the spring must not be over the star wheel on the adjuster. The left and right hand springs are not interchangeable.

19. Spread the shoe assemblies and connect the parking brake cable. Install the shoes on the backing plate, engaging the shoes at the top temporarily with the wheel cylinder pistons. Make sure the star wheel on the adjuster is lined up with the adjusting hole in the backing plate, if equipped.

20. Spread the shoes slightly and install the parking brake strut and spring. Make sure the end of the strut without the spring engages the parking brake lever. The end with the spring engages the primary shoe (the one with the shorter lining).

21. Install the actuator pivot, lever and return spring. Install the actuating link in the shoe retainer. Lift up the actuator lever and hook the link into the lever.

22. Install the hold-down pins through the back of the plate using J-8057. Install the lever pivots and hold-down springs. Install the shoe return springs using J-8049. Be careful not to stretch or distort the springs.

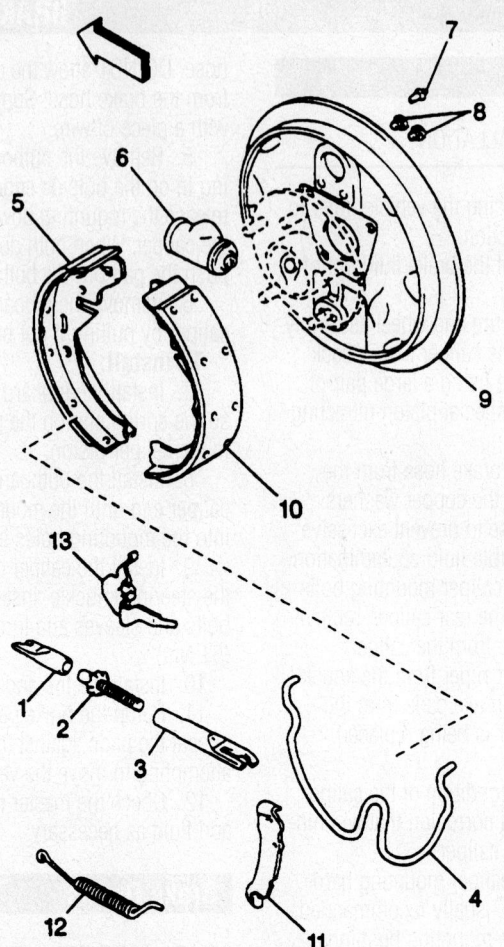

(1) Socket, Brake Adjuster
(2) Screw, Brake Adjuster
(3) Nut, Brake Pivot
(4) Spring, Retractor
(5) Brake Shoe and Lining
(6) Cylinder, Wheel Brake
(7) Valve, Bleeder
(8) Bolts, Wheel Cylinder
(9) Plate, Brake Backing
(10) Brake Shoe and Lining
(11) Lever, Park Brake
(12) Spring
(13) Adjuster Actuator

93006G78

Drum brake components, exploded view—Malibu

23. Make sure the linings are in the right place, the self-adjusting mechanism is correctly installed, and the parking brake parts are hooked up.

24. Measure the distance from the edge of the primary lining to the edge secondary lining, then measure the inside width of the drum. Adjust the linings by means of the adjuster so the drum will fit onto the linings.

25. Install the hub and bearing assembly onto the axle if removed. Torque the retaining bolts to 35 ft. lbs. (55 Nm).

26. Install the drum and the wheels.

27. Adjust the brakes. Install a rubber hole cover in the adjustment knock-out hole after the adjustment is complete. Adjust the parking brake.

28. Road test the vehicle and verify proper brake operation.

CHEVROLET, OLDSMOBILE AND PONTIAC

Montana • Silhouette • Venture

24

BRAKES **24-39**
DRIVE TRAIN **24-28**
ENGINE REPAIR...............**24-10**
FUEL SYSTEM **24-27**
HEATER CORE.................**24-11**
**STEERING AND
SUSPENSION** **24-33**
**SPECIFICATIONS AND
MAINTENANCE CHARTS**......**24-2**
Engine and Vehicle
 Identification 24-2
General Engine Specifications 24-2
Engine Tune-Up Specifications 24-2
Firing Orders 24-3
Accessory Drive Belt Routing 24-3
Capacities 24-3
Valve Specifications....................24-4
Crankshaft and Connecting
 Rod Specifications...................24-4
Piston and Ring Specifications 24-4
Torque Specifications 24-5
Brake Specifications 24-5
Wheel Alignment 24-6
Tire, Wheel and Ball Joint
 Specifications 24-6
Scheduled Maintenance
 Intervals................................24-7

A

Air Bag..................................24-33
 Disarming 24-33
 Precautions...........................24-33
Alternator...............................24-10
 Removal & Installation............24-10

B

Brake Caliper 24-39
 Removal & Installation............24-39
Brake Drums 24-40
 Removal & Installation............24-40
Brake Shoes............................24-41
 Removal & Installation............24-41

C

Camshaft and Valve Lifters 24-23
 Removal & Installation............24-23

Chevrolet Venture 24-11
 Removal & Installation............24-11
Coil Spring 24-35
 Removal & Installation............24-35
CV-Joints................................24-32
 Overhaul 24-32
Cylinder Head 24-17
 Removal & Installation............24-17

D

Disc Brake Pads........................24-40
 Removal & Installation............24-40
Distributor...............................24-10

E

Engine Assembly 24-10
 Removal & Installation............24-10
Exhaust Manifold 24-21
 Removal & Installation............24-21

F

Front Halfshaft 24-31
 Removal & Installation............24-31
Fuel Filter 24-27
 Removal & Installation............24-27
Fuel Injector 24-28
 Removal & Installation............24-28
Fuel Pump 24-27
 Removal & Installation............24-27
Fuel System Pressure 24-27
 Relieving 24-27
Fuel System Service
 Precaution 24-27

I

Intake Manifold 24-18
 Removal & Installation............24-18

L

Lower Ball Joint........................24-36
 Removal & Installation............24-36
Lower Control Arm 24-37
 Control Arm Bushing
  Replacement 24-37
 Removal & Installation............24-37

O

Oil Pan..................................24-24
 Removal & Installation............24-24

Oil Pump 24-24
 Removal & Installation............24-24
Oldsmobile Silhouette 24-13
 Removal & Installation............24-13

P

Pistons and Ring 24-26
 Postioning 24-26
Pontiac Montana........................24-15
 Removal & Installation............24-15
Power Steering Rack and
 Pinion..................................24-33
 Removal & Installation............24-33

R

Rear Halfshaft 24-31
 Removal & Installation............24-31
Rear Main Seal 24-25
 Removal & Installation............24-25
Rocker Arms 24-18
 Removal & Installation............24-18

S

Shock Absorber 24-35
 Removal & Installation............24-35
Starter Motor 24-23
 Removal & Installation............24-23
Strut.....................................24-34
 Removal & Installation............24-34

T

Timing Chain, Sprockets,
 Front Cover and Seal................24-25
 Removal & Installation............24-25
Transaxle Assembly 24-28
 Removal & Installation............24-28
Transfer Case 24-29
 Removal & Installation............24-29

V

Valve Lash 24-23
 Adjustment...........................24-23

W

Water Pump 24-11
 Removal & Installation............24-11
Wheel Bearings.........................24-37
 Adjustment...........................24-37
 Removal & Installation............24-37

SPECIFICATION AND MAINTENANCE CHARTS

ENGINE AND VEHICLE IDENTIFICATION

Engine							Model Year	
Code ①	Liters (cc)	Cu. In.	Cyl.	Fuel Sys.	Engine Type	Eng. Mfg.	Code ②	Year
E	3.4 (3350)	207	6	MFI/SFI	OHV	GM	2	2002

MFI: Multi-port Fuel Injection

OHV: Overhead Valves

SFI: Sequential Fuel Injection

① 8th position of VIN

② 10th position of VIN

Code ②	Year
2	2002
3	2003
4	2004
5	2005
6	2006

06025-VENT-C01

GENERAL ENGINE SPECIFICATIONS

All measurements are given in inches.

Year	Model	Engine Displacement Liters	Engine Series VIN	Net Horsepower @ rpm	Net Torque @ rpm (ft. lbs.)	Bore x Stroke (in.)	Compression Ratio	Oil Pressure @ rpm
2002	Silhouette	3.4	E	180@5200	205@4000	3.62x3.31	9.5:1	15@1100
	Montana	3.4	E	180@5200	205@4000	3.62x3.31	9.5:1	15@1100
	Venture	3.4	E	180@5200	205@4000	3.62x3.31	9.5:1	15@1100
2003	Silhouette	3.4	E	180@5200	205@4000	3.62x3.31	9.5:1	15@1100
	Montana	3.4	E	180@5200	205@4000	3.62x3.31	9.5:1	15@1100
	Venture	3.4	E	180@5200	205@4000	3.62x3.31	9.5:1	15@1100
2004	Silhouette	3.4	E	180@5200	205@4000	3.62x3.31	9.5:1	15@1100
	Montana	3.4	E	180@5200	205@4000	3.62x3.31	9.5:1	15@1100
	Venture	3.4	E	180@5200	205@4000	3.62x3.31	9.5:1	15@1100
2005	Silhouette	3.4	E	180@5200	205@4000	3.62x3.31	9.5:1	15@1100
	Montana	3.4	E	180@5200	205@4000	3.62x3.31	9.5:1	15@1100
	Venture	3.4	E	180@5200	205@4000	3.62x3.31	9.5:1	15@1100

06025-VENT-C02

ENGINE TUNE-UP SPECIFICATIONS

Year	Engine Displacement Liters	Engine VIN	Spark Plug Gap (in.)	Ignition Timing (deg.)	Fuel Pump (psi)	Idle Speed (rpm)	Valve Clearance	
							Intake	Exhaust
2002	3.4	E	0.060	①	41-47	②	HYD	HYD
2003	3.4	E	0.060	①	41-47	②	HYD	HYD
2004	3.4	E	0.060	①	41-47	②	HYD	HYD
2005	3.4	E	0.060	①	41-47	②	HYD	HYD

NOTE: The Vehicle Emissions Control Information label often reflects specification changes made during production.

The label figures must be used if they differ from those in this chart.

HYD: Hydraulic

① Refer to underhood label for exact setting.

② Idle speed is maintained by the PCM.

06025-VENT-C03

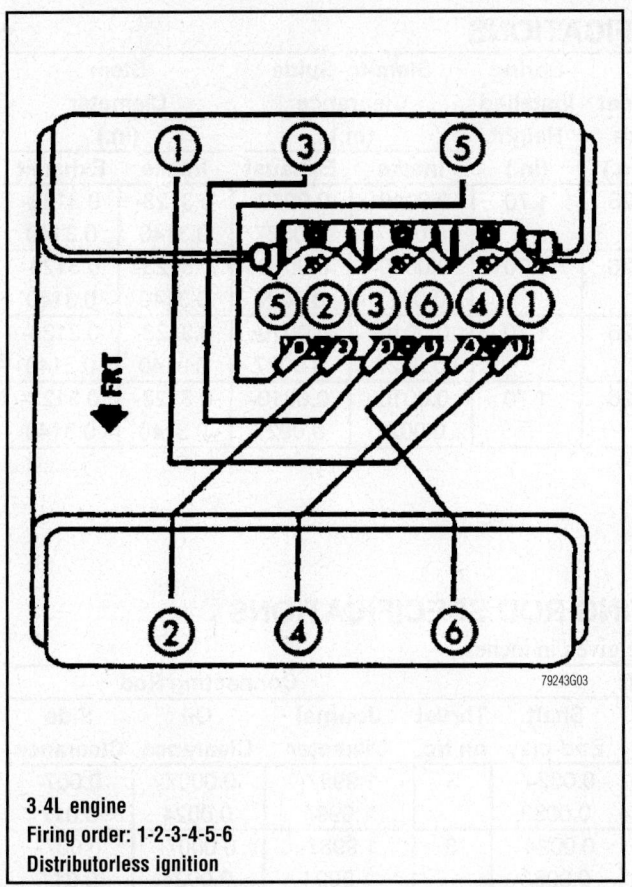

3.4L engine
Firing order: 1-2-3-4-5-6
Distributorless ignition

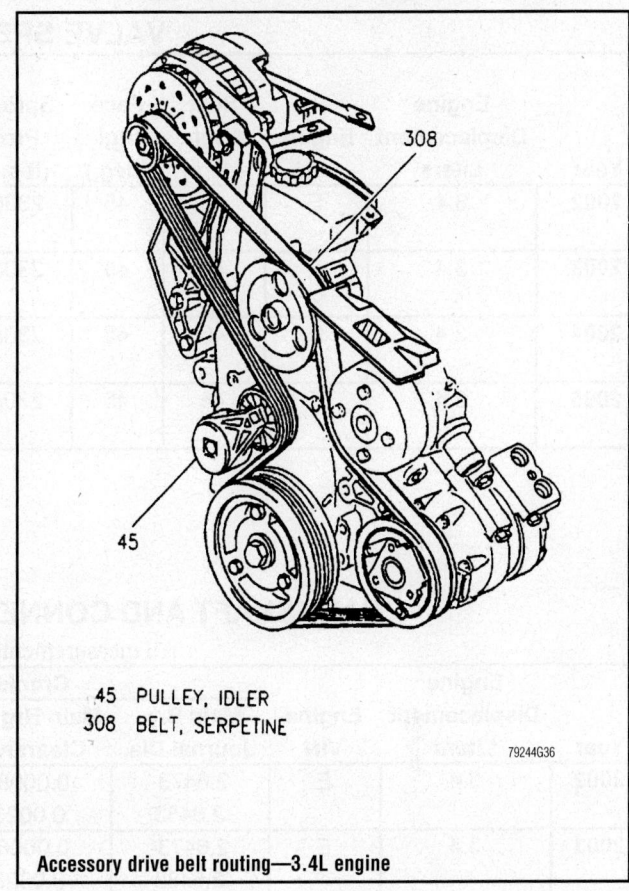

45 PULLEY, IDLER
308 BELT, SERPETINE

Accessory drive belt routing—3.4L engine

CAPACITIES

Year	Model	Engine Displacement Liters	Engine VIN	Engine Oil with Filter (qts.)	Transmission (pts.)*	Fuel Tank (gal.)	Cooling System (qts.)
2002	Montana	3.4	E	4.5	①	20 ②	③
	Silhouette	3.4	E	4.5	①	20 ②	③
	Venture	3.4	E	4.5	①	20 ②	③
2003	Montana	3.4	E	4.5	①	20 ②	③
	Silhouette	3.4	E	4.5	①	20 ②	③
	Venture	3.4	E	4.5	①	20 ②	③
2004	Montana	3.4	E	4.5	①	20 ②	③
	Silhouette	3.4	E	4.5	①	20 ②	③
	Venture	3.4	E	4.5	①	20 ②	③
2005	Montana	3.4	E	4.5	①	20 ②	③
	Silhouette	3.4	E	4.5	①	20 ②	③
	Venture	3.4	E	4.5	①	20 ②	③

NOTE: All capacities are approximate. Add fluid gradually and check to be sure a proper fluid level is obtained.

* Drain and refill
① FWD: 14.8 pts.; AWD: 15.6 pts.
② Extended van: 25.5
③ Std.: 9.9
    F/R w/o HD: 11.9
    F only w/HD: 10.5
    F/R w/HD: 13.2

VALVE SPECIFICATIONS

Year	Engine Displacement Liters	Engine VIN	Seat Angle (deg.)	Face Angle (deg.)	Spring Test Pressure (lbs. @ in.)	Spring Installed Height (in.)	Stem-to-Guide Clearance (in.)		Stem Diameter (in.)	
							Intake	Exhaust	Intake	Exhaust
2002	3.4	E	46	45	230@1.26	1.70	0.0010-0.0027	0.0010-0.0027	0.3123-0.3140	0.3123-0.3140
2003	3.4	E	46	45	230@1.26	1.70	0.0010-0.0027	0.0010-0.0027	0.3123-0.3140	0.3123-0.3140
2004	3.4	E	46	45	230@1.26	1.70	0.0010-0.0027	0.0010-0.0027	0.3123-0.3140	0.3123-0.3140
2005	3.4	E	46	45	230@1.26	1.70	0.0010-0.0027	0.0010-0.0027	0.3123-0.3140	0.3123-0.3140

06025-VENT-C05

CRANKSHAFT AND CONNECTING ROD SPECIFICATIONS

All measurements are given in inches.

Year	Engine Displacement Liters	Engine VIN	Crankshaft				Connecting Rod		
			Main Brg. Journal Dia.	Main Brg. Oil Clearance	Shaft End-play	Thrust on No.	Journal Diameter	Oil Clearance	Side Clearance
2002	3.4	E	2.6473-2.6483	0.0008-0.0023	0.0024-0.0083	3	1.9987-1.9994	0.0007-0.0024	0.007-0.017
2003	3.4	E	2.6473-2.6483	0.0008-0.0023	0.0024-0.0083	3	1.9987-1.9994	0.0007-0.0024	0.007-0.017
2004	3.4	E	2.6473-2.6483	0.0008-0.0023	0.0024-0.0083	3	1.9987-1.9994	0.0007-0.0024	0.007-0.017
2005	3.4	E	2.6473-2.6483	0.0008-0.0023	0.0024-0.0083	3	1.9987-1.9994	0.0007-0.0024	0.007-0.017

06025-VENT-C06

PISTON AND RING SPECIFICATIONS

All measurements are given in inches.

Year	Engine Displ. Liters	Engine VIN	Piston Clearance	Ring Gap			Ring Side Clearance		
				Top Compression	Bottom Compression	Oil Control	Top Compression	Bottom Compression	Oil Control
2002	3.4	E	0.0013-0.0027	0.006-0.014	0.020-0.028	0.0098-0.0303	0.0020-0.0033	0.0020-0.0035	0.0028-0.0037
2003	3.4	E	0.0013-0.0027	0.006-0.014	0.020-0.028	0.0098-0.0303	0.0020-0.0033	0.0020-0.0035	0.0028-0.0037
2004	3.4	E	0.0013-0.0027	0.006-0.014	0.020-0.028	0.0098-0.0303	0.0020-0.0033	0.0020-0.0035	0.0028-0.0037
2005	3.4	E	0.0013-0.0027	0.006-0.014	0.020-0.028	0.0098-0.0303	0.0020-0.0033	0.0020-0.0035	0.0028-0.0037

06025-VENT-C07

TORQUE SPECIFICATIONS
All readings in ft. lbs.

Year	Engine Displacement Liters	Engine VIN	Cylinder Head Bolts	Main Bearing Bolts	Rod Bearing Bolts	Crankshaft Damper Bolts	Flywheel Bolts	Manifold Intake	Manifold Exhaust	Spark Plugs	Oil Pan Drain Plug
2002	3.4	E	①	②	③	76	52	④	12	11	18
2003	3.4	E	①	②	③	76	52	④	12	11	18
2004	3.4	E	①	②	⑤	⑥	52	⑦	12	13	18
2005	3.4	E	①	②	⑤	⑥	52	⑦	12	13	18

① 44 ft. lbs. + 95 degrees
② 37 ft. lbs. plus 77 degrees
③ 18 ft. lbs. + 100 degrees
④ Lower manifold: 10 ft. lbs.
   Upper manifold: 18 ft. lbs.
⑤ 15 ft. lbs. +75 degrees
⑥ 52 ft. lbs. +72 degrees

⑦ Lower
Center bolts:
Step 1: 62 inch lbs.
Step 2: 115 inch lbs.
Corner bolts:
Step 1: 62 inch lbs.
Step 2: 18 ft. lbs.

06025-VENT-C08

BRAKE SPECIFICATIONS
All measurements in inches unless noted

Year	Model		Brake Disc Original Thickness	Brake Disc Minimum Thickness	Brake Disc Maximum Runout	Brake Drum Diameter Original Inside Diameter	Brake Drum Diameter Max. Wear Limit	Brake Drum Diameter Maximum Machine Diameter	Minimum Lining Thickness Front	Minimum Lining Thickness Rear	Brake Caliper Bracket Bolts (ft. lbs.)	Brake Caliper Mounting Bolts (ft. lbs.)
2002	FWD	F	1.270	1.250	0.002	—	—	—	NA	NA	137	40
		R	0.433	0.354	0.002	8.86	8.92	8.91	NA	NA	92	32
	AWD	F	1.181	1.063	0.002	—	—	—	NA	NA	137	40
		R	0.433	0.354	0.002	8.86	8.92	8.91	NA	NA	92	32
2003	FWD	F	1.270	1.250	0.002	—	—	—	NA	NA	137	40
		R	0.433	0.354	0.002	8.86	8.92	8.91	NA	NA	92	32
	AWD	F	1.181	1.063	0.002	—	—	—	NA	NA	137	40
		R	0.433	0.354	0.002	8.86	8.92	8.91	NA	NA	92	32
2004	FWD	F	1.270	1.250	0.002	—	—	—	NA	NA	137	40
		R	0.433	0.354	0.002	8.86	8.92	8.91	NA	NA	92	32
	AWD	F	1.181	1.063	0.002	—	—	—	NA	NA	137	40
		R	0.433	0.354	0.002	8.86	8.92	8.91	NA	NA	92	32
2005	FWD	F	1.270	1.220	0.002	—	—	—	NA	NA	137	40
		R	0.433	0.354	0.002	8.86	8.92	8.91	NA	NA	92	32
	AWD	F	1.181	1.118	0.002	—	—	—	NA	NA	137	40
		R	0.433	0.354	0.002	8.86	8.92	8.91	NA	NA	92	32

NA: Not Available

06025-VENT-C09

WHEEL ALIGNMENT

Year	Model		Caster Range (+/-Deg.)	Caster Preferred Setting (Deg.)	Camber Range (+/-Deg.)	Camber Preferred Setting (Deg.)	Toe-in (in.)
2002	All	F	0.75	+2.90	0.50	-0.70	0 +/-0.20
		R	—	—	0.25	-1.00	0 +/-0.30
2003	All	F	0.75	+2.90	0.50	-0.70	0 +/-0.20
		R	—	—	0.25	-1.00	0 +/-0.30
2004	FWD w/Twist Axle	F	0.50	+2.90	0.50	-0.67	0 +/-0.20
		R	—	—	0.25	-1.00	0 +/-0.30
	AWD and I.R.S.	F	0.50	+2.90	0.50	-0.67	0 +/-0.20
		R	—	—	0.60	-0.25	0 +/-0.20
2005	FWD w/Twist Axle	F	0.75	+3.10	0.75	-0.70	0 +/-0.20
		R	—	—	0.50	-1.00	0 +/-0.30
	AWD and I.R.S.	F	0.75	+3.10	0.75	-0.70	0 +/-0.20
		R	—	—	0.60	-0.25	0 +/-0.20

All alignment figures based on nominal ride height and standard tires

I.R.S.: Independent rear suspension

06025-VENT-C10

TIRE, WHEEL AND BALL JOINT SPECIFICATIONS

Year	Model	OEM Tires Standard	OEM Tires Optional	Tire Pressures (psi) Front	Tire Pressures (psi) Rear	Wheel Size	Ball Joint Inspection	Lugnut Torque (ft. lbs.)
2002	All	P215/70R15	P225/60R16	std: 35 opt: 32	std.: 35 opt.: 32	6-JJ	U ① L: 0.090 in.	100
2003	All	P215/70R15	P225/60R16	std: 35 opt: 32	std.: 35 opt.: 32	6-JJ	U ① L: 0.090 in.	100
2004	All	P215/70R15	P225/60R16	②	②	6-JJ	U ① L: 0.090 in.	100
2005	All	P215/70R15	P225/60R16	②	②	6-JJ	U ① L: 0.090 in.	100

OEM: Original Equipment Manufacturer

PSI: Pounds Per Square Inch

STD: Standard

OPT: Optional

L: Lower

U: Upper

① Replace if any movement is noted or if stud can be moved by hand

② See placard on vehicle

06025-VENT-C11

SCHEDULED MAINTENANCE INTERVALS
2002-03 Chevrolet Venture, Oldsmobile Silhouette & Pontiac Montana

TO BE SERVICED	TYPE OF SERVICE	VEHICLE MILEAGE INTERVAL (x1000)														
		7.5	15	22.5	30	37.5	45	52.5	60	67.5	75	82.5	90	97.5	105	112.5	120
Accessory drive belt	I								✓								✓
Air cleaner filter	R								✓								✓
Air distributor air filter	R		✓		✓		✓		✓		✓		✓		✓		✓
Brake system	I	✓	✓	✓	✓	✓	✓	✓	✓	✓	✓	✓	✓	✓	✓	✓	✓
Engine coolant	R	Every 150,000 miles															
Engine oil & filter ①	S/I	✓	✓	✓	✓	✓	✓	✓	✓	✓	✓	✓	✓	✓	✓	✓	✓
Fuel system tank, cap & lines	I								✓								✓
Rotate tires	S/I	✓	✓	✓	✓	✓	✓	✓	✓	✓	✓	✓	✓	✓	✓	✓	✓
Spark plug wires	S/I	Every 100,000 miles															
Spark plugs	R	Every 100,000 miles															

R: Replace   I: Inspect   S: Service

① Perform this at the mileage indicated or every 12 months, whichever occurs first.

FREQUENT OPERATION MAINTENANCE (SEVERE SERVICE)

If a vehicle is operated under any of the following conditions it is considered severe service:

- Towing a trailer or using a camper or car-top carrier.

- Repeated short trips of less than 5 miles in temperatures below freezing, or trips of less than 10 miles in any temperature.

- Extensive idling or low-speed driving for long distances as in heavy commercial use, such as delivery, taxi or police cars.

- Operating on rough, muddy or salt-covered roads.

- Operating on unpaved or dusty roads.

- Driving in extremely hot (over 90°) conditions.

Automatic transaxle fluid and filter: replace every 50,000 miles.

Tires: rotate every 6000 miles.

Brake system: inspect every 6000 miles.

Air distributor air filter: replace every 12,000 miles.

06025-VENT-C12

MAINTENANCE I AND II SERVICE SCHEDULES
2004-05 Chevrolet Venture, Oldsmobile Sillouette, Pontiac Montana

When the CHANGE ENGINE OIL light appears, certain services and inspections are required. Required services are described as Maintenance I and Maintenance II.

The first service on a vehicle should be Maintenance I, and the second service should be Maintenance II. Alternate between the 2 thereafter. However, in some cases, Maintenance II may be required more often.

Maintenance I: Use Maintenance I if the CHANGE ENGINE OIL light comes on within 10 months since vehicle was purchased or, if Maintenance II was performed.

Maintenance II: Use Maintenance II if the previous service performed was Maintenance I. Always use Maintenance II whenever the CHANGE ENGINE OIL light comes on 10 months or more since the last service, or, if the CHANGE ENGINE OIL light has not come on at all for one year.

Service	Maintenance I	Maintenance II
Change the engine oil and filter. Reset the oil life system.	✓	✓
Visually inspect the vehicle for leaks or damage. A fluid loss in the vehicle system could indicate a problem. Inspected, repair and add fluid to the system if necessary.	✓	✓
Inspect the engine air cleaner filter. If necessary, replace the filter.	--	✓
Rotate the tires. Inspect the tire inflation pressures and the tire wear.	✓	✓
Visually inspect the brake lines and hoses for proper hook-up, binding, leaks, cracks, chafing, etc. Inspect the disc brake pads for wear and the rotors for surface condition. Inspect the drum brake linings for wear or cracks. Inspect other brake parts, including drums, wheel cylinders, calipers, parking brake, etc. Inspect the parking brake adjustment.	✓	✓
Inspect the engine coolant and the windshield washer fluid levels. Add fluid as needed.	✓	✓
Inspect the suspension and steering components. Inspect the front and rear suspension and the steering system for damaged, loose or missing parts, or signs of wear. Inspect the power steering lines and the hoses for proper hook-up, binding, leaks, cracks, chafing, etc.	--	✓
Visually inspect the coolant hoses and replace the hoses if they are cracked, swollen or deteriorated. Inspect all pipes, fittings and clamps; replace with GM parts as needed. To help ensure proper operation, a pressure test of the cooling system and pressure cap and cleaning the outside of the radiator and air conditioning condenser is recommended at least once a year.	--	✓
Inspect the wiper blades for wear or cracking.	--	✓
Inspect all the restraint belts, buckles, latch plates, retractors.		
Lubricate all key lock cylinders, latch assemblies, and hinges.		
Inspect the transaxle fluid level and add fluid as needed.	--	✓
Lubricate the suspension, steering linkage, and transaxle shift linkage.	✓	✓
Inspect the throttle system for interference or binding and for damaged or missing parts. Replace the parts as needed. Replace any components that have high effort or excessive wear. Do not lubricate the accelerator or the cruise control cables.	--	✓
Replace the passenger compartment air filter.	--	✓

06025-VENT-C13

ADDITIONAL MAINTENANCE SERVICES
2004-05 Chevrolet Venture, Oldsmobile Sillouette, Pontiac Montana

TO BE SERVICED	TYPE OF SERVICE	VEHICLE MILEAGE INTERVAL (x1000)					
		25	50	75	100	125	150
Air cleaner filter	R		✓		✓		✓
Accessory drive belt	I						✓
Auto. Trans. Fluid ①	R		✓		✓		✓
Cooling system hoses and clamps	S/I						✓
Engine coolant	R						✓
Fuel system	I	✓	✓	✓	✓	✓	✓
Exhaust system & heat shields	S/I	✓	✓	✓	✓	✓	✓
Spark plugs and wires	R				✓		

R: Replace   S/I: Inspect and service, if necessary

① Replace if any of the following conditions are met:

   Heavy city traffic where the outside temperature regularly reaches 32°C (90°F) or higher

   Hilly or mountainous terrain

   Frequent trailer towing

   Taxi, police or delivery service

   Otherwise, change every 100,000 miles

To reset the CHANGE ENGINE OIL LIGHT:

   1. With the ignition key in RUN but the engine off, repeatedly push the trip/reset button until OIL is displayed on the Driver Information Center (DIC).

   2. Once OIL is displayed, push and hold the trip/reset button for five seconds. The number will disappear and be replaced by 100 (indicating 100% oil life remaining).

   3. Turn the key to OFF.

If the CHANGE ENGINE OIL message comes back on when you start your vehicle, the engine oil life system has not reset, repeat the procedure.

06025-VENT-C14

ENGINE REPAIR

Distributor

This engine utilizes a Distributorless ignition system (DIS). There is no distributor to remove and no provision for adjustment.

Alternator

REMOVAL & INSTALLATION

1. Before servicing the vehicle, refer to the Precautions Section.
2. Remove or disconnect the following:
   - Negative battery cable
   - Wiper system module cover
   - Fuel injector sight shield
3. Rotate the engine forward.
4. Remove or disconnect the following:
   - Alternator terminal nut, lead and electrical connector
   - Serpentine belt
   - Front bolts and two rear bolts
   - Alternator from the bracket
   - Serpentine belt tensioner
   - Bracket
   - Power steering pipes from the retainer
   - Fuel pressure test port cap from the injector rail

➡**Do not disconnect the power steering pipes from the pump**

   - Power steering pump and reposition it to gain access to the alternator
   - Alternator

To install:
5. Install or connect the following:
   - Alternator
   - Power steering pump. Torque the bolts to 25 ft. lbs. (34 Nm).
   - Fuel pressure test port cap to the fuel rail
   - Power steering pipes to the retainer. Torque the fastener to 54 inch lbs. (6 Nm).
   - Alternator bracket. Torque the bolt to 37 ft. lbs. (50 Nm).
   - Serpentine belt tensioner
   - Alternator to the bracket. Torque the bolts to 37 ft. lbs. (50 Nm).
   - Serpentine belt
   - Alternator electrical connector, lead and nut. Torque the nut to 115 inch lbs. (13 Nm).
6. Rotate the engine to its original position.
7. Install or connect the following:
   - Fuel injector sight shield. Torque the nut to 54 inch lbs. (6 Nm).
   - Wiper system module cover

   - Negative battery cable
8. Perform a charging system test and verify the proper operation of the system.

Engine Assembly

REMOVAL & INSTALLATION

1. Before servicing the vehicle, refer to the Precautions Section.
2. Drain the cooling system.
3. Drain the engine oil.
4. Relieve the fuel system pressure.
5. Remove or disconnect the following:
   - Negative battery cable
   - Fuel injector sight shield
   - Throttle body air inlet duct
   - Cruise control cable
   - Accelerator control cable
   - Radiator hoses from the engine
   - Heater hoses from the engine
   - Engine mount struts
   - Fuel lines from the fuel rail
   - Engine wiring harness connectors
   - Vacuum hoses
   - Brake booster vacuum hose
   - Automatic transaxle range selector cable
   - Wiring harness grounds
   - Catalytic converter three-way pipe from the right side exhaust manifold
   - Front wheels
   - Splash shields
   - Stabilizer shaft links from the lower control arms
   - Tie rod ends from the steering knuckles
   - Lower ball joints from the steering knuckles
   - Cooler lines and bracket from the transaxle
   - A/C compressor bolts and position compressor aside
   - Axles from the transaxle and secure them to the steering knuckle/struts

✳✳ CAUTION

Failure to remove the intermediate shaft from the steering gear may result in damage to the gear or intermediate shaft and may cause a loss of steering control.

6. Remove or disconnect the following:
   - Intermediate shaft from the steering gear
   - Frame bolts and make certain that an engine stand (such as J 39580) is aligned below the engine

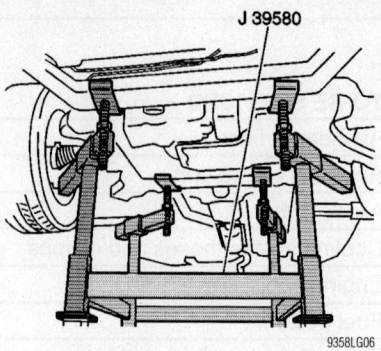

J 39580

9358LG06

Use a engine stand to support the assembly during frame removal

   - Engine to transaxle bolts and studs
   - Engine flywheel to torque converter bolts
   - Engine from the transaxle and place it on the engine stand

To install:
7. Install or connect the following:
   - Engine to the transaxle/frame and install the bolts. Torque the bolts to 55 ft. lbs. (75 Nm).
   - Torque converter to flywheel bolts. Torque the bolts to 47 ft. lbs. (63 Nm).
   - New frame to body bolts. Torque the front bolts to 111 ft. lbs. (150 Nm) and the rear bolts to 122 ft. lbs. (165 Nm).
   - Intermediate shaft to the steering gear

✳✳ CAUTION

When installing the intermediate shaft be certain that the shaft is seated properly before installing the pinch bolt. If the pinch bolt is inserted into the coupling before the shaft, the mating surfaces disengage. Disengagement of the two shafts may lead to a loss of steering control.

   - Pinch bolt at the intermediate shaft. Torque the bolt 35 ft. lbs. (48 Nm)
   - Drive axles to the transaxle.
   - Cooler lines and bracket to the transaxle. Torque the fasteners to 17 ft. lbs. (23 Nm).
   - Lower ball joints to the steering knuckles. Torque to 40 ft. lbs. (55 Nm).
   - Tie rod ends to the steering knuckles
   - Stabilizer shaft links to the lower control arms. Torque the bolts 17 ft. lbs. (23 Nm).

- Inner fender splash shield. Torque the fasteners to 18 inch lbs. (2 Nm).
- Front wheels
- Catalytic converter pipe to the right side exhaust manifold. Torque the nuts to 25 ft. lbs. (34 Nm).
- Wiring harness grounds
- Brake booster vacuum hose
- Vacuum hoses to the engine
- Range selector cable. Torque the screw to 14 ft. lbs. (20 Nm).
- Engine wiring harness connectors
- Fuel lines to the fuel rail. Torque the fasteners to 13 ft. lbs. (17 Nm).
- Throttle body brackets and cables. Torque the fasteners to 18 ft. lbs. (25 Nm).
- Engine mount strut. Torque the bolt to 35 ft. lbs. (48 Nm).
- Heater hoses
- Radiator hoses

✸✸ CAUTION

Whenever the engine has been removed from the vehicle it is neces-sary to install a new accelerator con-trol cable to avoid damage or per-sonal injury.

- New accelerator control cable
- Cruise control cable
- Throttle body air inlet duct
- Fuel injector sight shield
- Negative battery cable

8. Fill the engine with oil.
9. Fill the engine with coolant.
10. Inspect the transmission fluid level and top off, if necessary.
11. Turn the ignition to the **ON** position several times to pressurize the fuel system.
12. Start the engine and inspect for leaks, repair if necessary.
13. Check and top off the fluid levels if required.

Water Pump

REMOVAL & INSTALLATION

1. Before servicing the vehicle, refer to the Precautions Section.

2. Drain the coolant from the engine.
3. Remove or disconnect the following:
   - Negative battery cable
   - Serpentine drive belt guard
   - Loosen the water pump pulley bolts
   - Serpentine drive belt
   - Water pump pulley
   - Water pump
   - Water pump gasket

To install:
4. Clean the gasket mounting surfaces.
5. Install or connect the following:
   - Gasket
   - Water pump. Torque the bolts to 89 inch lbs. (10 Nm).
   - Water pump pulley and hand-tighten the bolts at this time.
   - Serpentine drive belt
   - Torque the water pump pulley bolts to 18 ft. lbs. (25 Nm).
   - Serpentine drive belt guard
   - Fill the cooling system

6. Start the engine and check for leaks, repair if necessary.
7. Road test the vehicle and verify there is no air in the cooling system.

HEATER CORE

Chevrolet Venture

REMOVAL & INSTALLATION

Front System

1. Disconnect the negative battery cable.
2. Drain the cooling system into a clean container for reuse.
3. Remove the air cleaner and duct assembly.
4. Remove the wiper linkage.
5. Remove the heater hoses from the heater core.

6. Remove the lower center console.
7. Remove both instrument panel insulators.
8. Remove the heater outlet module screws.
9. Remove the heater cover.
10. Remove the heater core mounting clip and line clamp screws.
11. Remove the heater core.

To install:
12. Install the heater core.
13. Install the heater core mounting clip and line clamp screws.
14. Install the heater cover.

15. Install the heater outlet module screws.
16. Install both instrument panel insulators.
17. Install the lower center console.
18. Install the heater hoses from the heater core.
19. Install the wiper linkage.
20. Install the air cleaner and duct assembly.
21. Refill the cooling system.
22. Connect the negative battery cable.
23. Run the engine to normal operating temperatures; then, check the climate control operation and check for leaks.

REAR AUXILIARY SYSTEM

1. Disconnect the negative battery cable.
2. Drain the cooling system into a clean container for reuse.
3. Remove the left rear side quarter trim.
4. Remove the rear heater hoses from the rear heater core.
5. Remove the bracket bolt securing the hoses to the heater core.
6. Release the rear heater core-to-rear heater housing tabs.
7. Remove the heater core.

To install:
8. Install the heater core.
9. Connect the rear heater core-to-rear heater housing tabs.

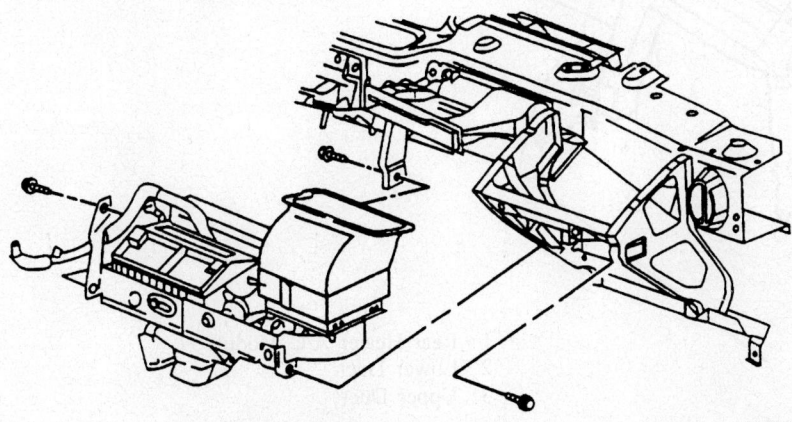

93113GK9

View of the front heater assembly—Chevrolet Venture

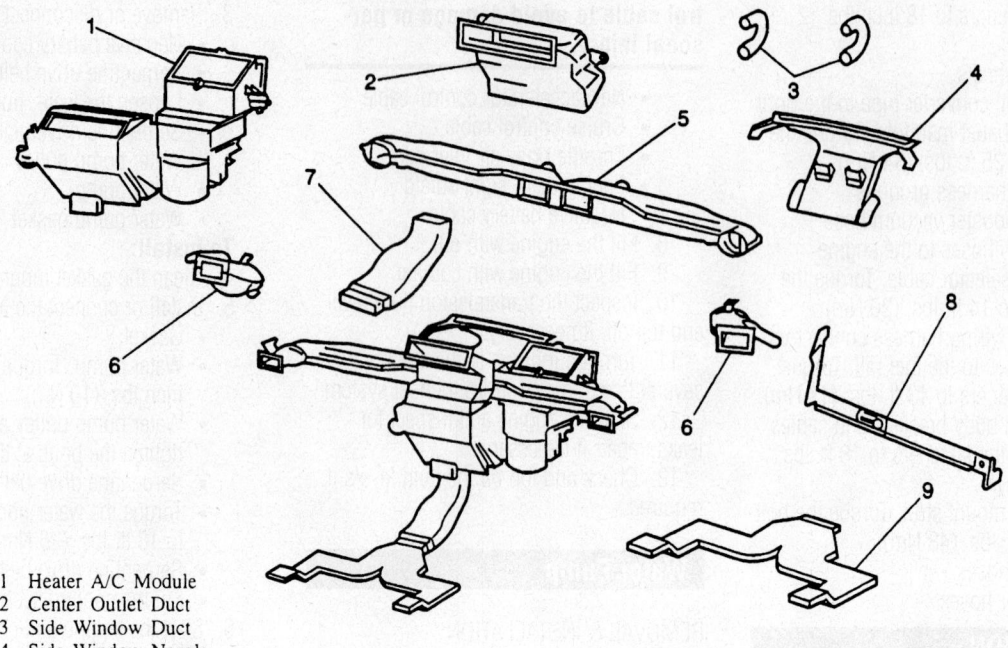

1 Heater A/C Module
2 Center Outlet Duct
3 Side Window Duct
4 Side Window Nozzle
5 Main Duct
6 Side Window Outlet Duct
7 Floor Duct
8 Module Bracket
9 Lower Floor Duct

93113GK0

Exploded view of the front heater ventilation system—Chevrolet Venture

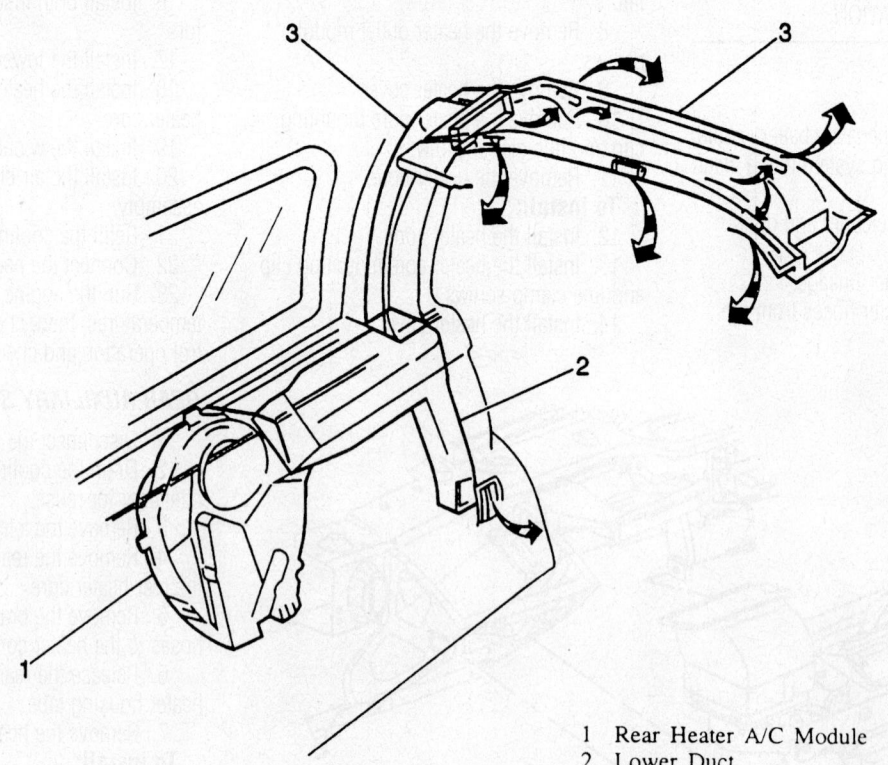

1 Rear Heater A/C Module
2 Lower Duct
3 Upper Duct

93113GL1

View of the rear auxiliary heater assembly ventilation system—Chevrolet Venture

15. Install the heater outlet module screws.

16. Install both instrument panel insulators.

17. Install the lower center console.

18. Install the heater hoses from the heater core.

19. Install the wiper linkage.

20. Install the air cleaner and duct assembly.

21. Refill the cooling system.

22. Connect the negative battery cable.

23. Run the engine to normal operating temperatures; then, check the climate control operation and check for leaks.

REAR AUXILIARY SYSTEM

1. Disconnect the negative battery cable.

2. Drain the cooling system into a clean container for reuse.

3. Remove the left rear side quarter trim.

4. Remove the rear heater hoses from the rear heater core.

5. Remove the bracket bolt securing the hoses to the heater core.

6. Release the rear heater core-to-rear heater housing tabs.

7. Remove the heater core.

To install:

8. Install the heater core.

9. Connect the rear heater core-to-rear heater housing tabs.

10. Install the bracket bolt securing the hoses to the heater core.

11. Install the rear heater hoses from the rear heater core.

12. Install the left rear side quarter trim.

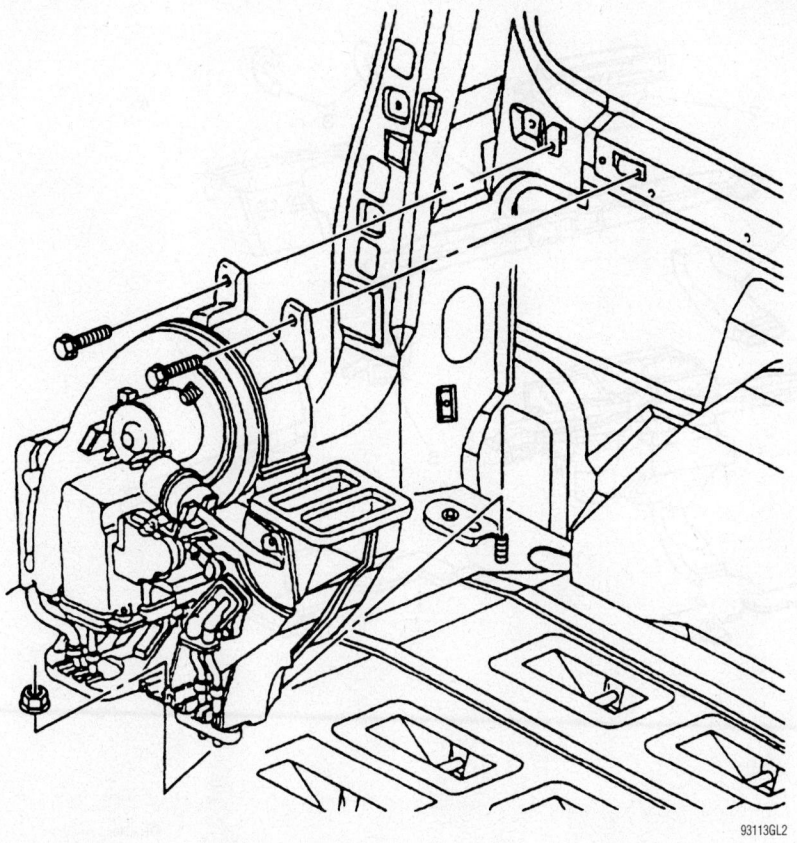

View of the rear auxiliary heater assembly—Chevrolet Venture

93113GL2

10. Install the bracket bolt securing the hoses to the heater core.

11. Install the rear heater hoses from the rear heater core.

12. Install the left rear side quarter trim.

13. Refill the cooling system.

14. Connect the negative battery cable.

Oldsmobile Silhouette

REMOVAL & INSTALLATION

Front System

1. Disconnect the negative battery cable.

2. Drain the cooling system into a clean container for reuse.

3. Remove the air cleaner and duct assembly.

4. Remove the wiper linkage.

5. Remove the heater hoses from the heater core.

6. Remove the lower center console.

7. Remove both instrument panel insulators.

8. Remove the heater outlet module screws.

9. Remove the heater cover.

10. Remove the heater core mounting clip and line clamp screws.

11. Remove the heater core.

To install:

12. Install the heater core.

13. Install the heater core mounting clip and line clamp screws.

14. Install the heater cover.

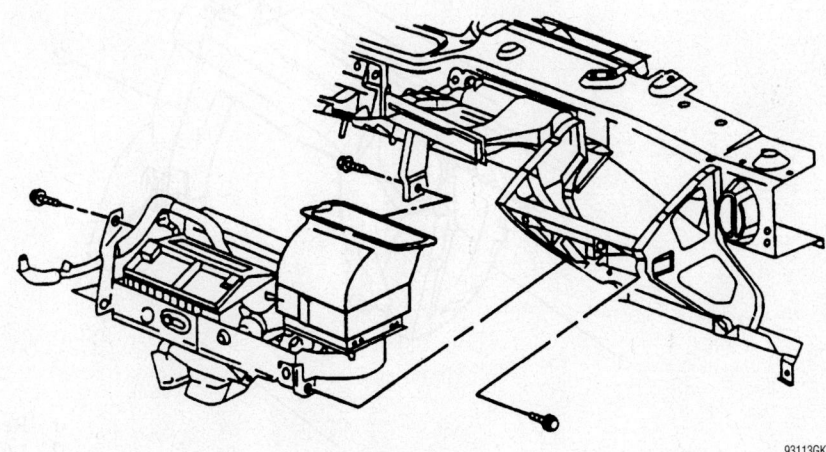

View of the front heater assembly—Oldsmobile Silhouette

93113GK9

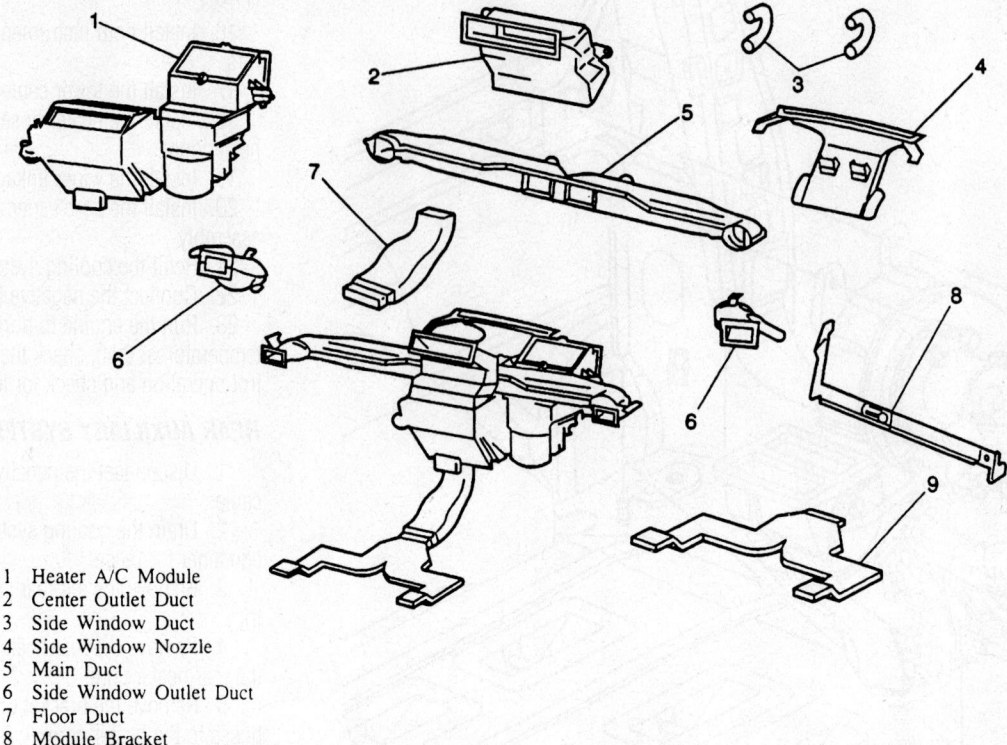

1 Heater A/C Module
2 Center Outlet Duct
3 Side Window Duct
4 Side Window Nozzle
5 Main Duct
6 Side Window Outlet Duct
7 Floor Duct
8 Module Bracket
9 Lower Floor Duct

93113GK0

Exploded view of the front heater ventilation system—Oldsmobile Silhouette

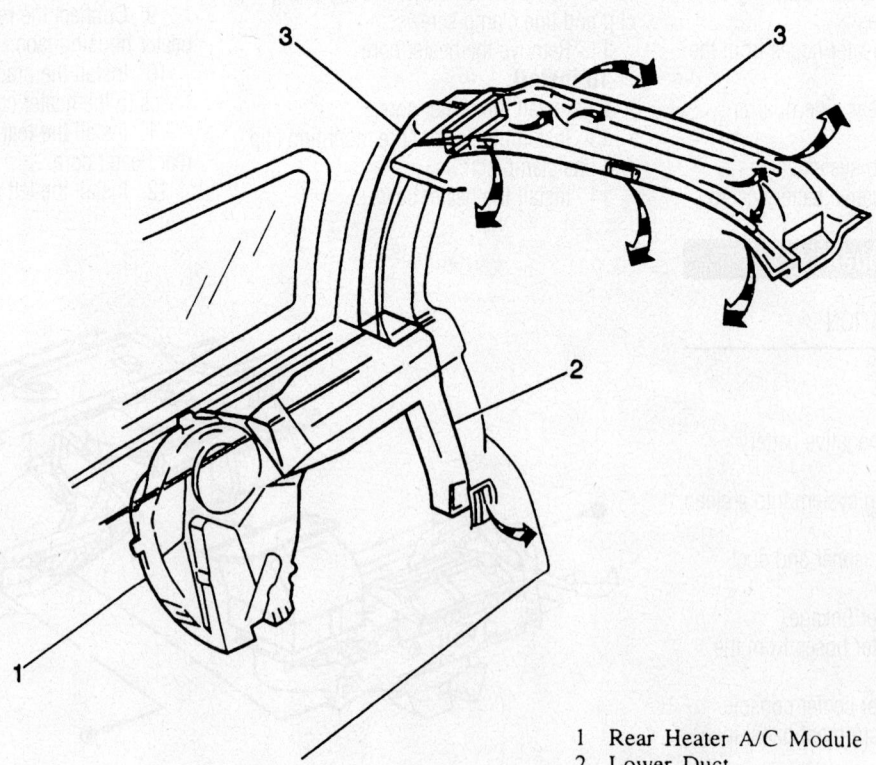

1 Rear Heater A/C Module
2 Lower Duct
3 Upper Duct

93113GL1

View of the rear auxiliary heater assembly ventilation system—Oldsmobile Silhouette

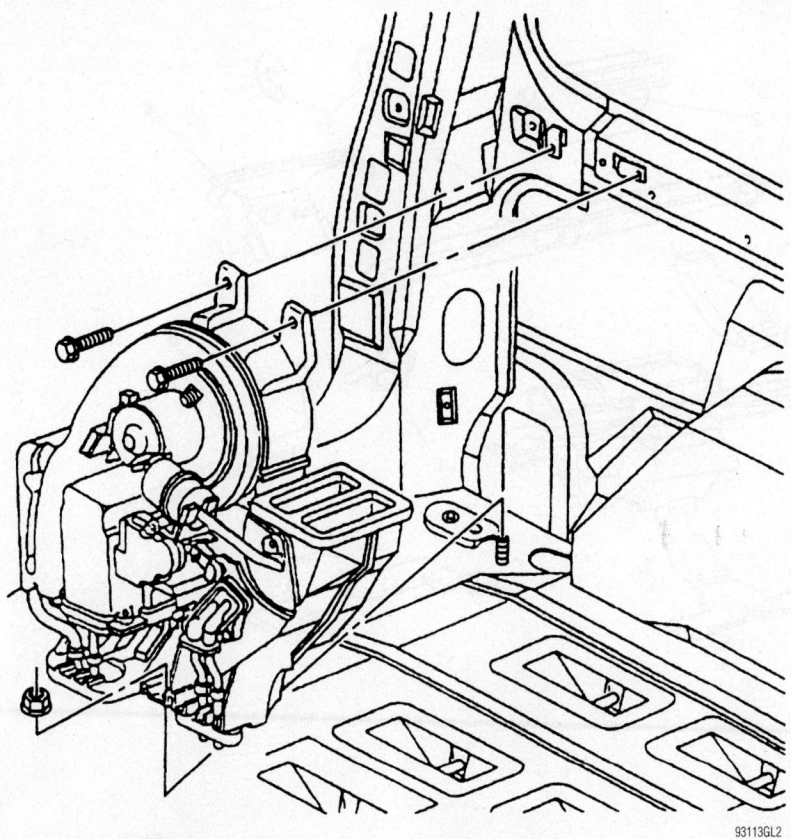

View of the rear auxiliary heater assembly—Oldsmobile Silhouette

13. Refill the cooling system.
14. Connect the negative battery cable.

Pontiac Montana

REMOVAL & INSTALLATION

Front System

1. Disconnect the negative battery cable.
2. Drain the cooling system into a clean container for reuse.
3. Remove the air cleaner and duct assembly.
4. Remove the wiper linkage.
5. Remove the heater hoses from the heater core.
6. Remove the lower center console.
7. Remove both instrument panel insulators.
8. Remove the heater outlet module screws.
9. Remove the heater cover.
10. Remove the heater core mounting clip and line clamp screws.
11. Remove the heater core.
To install:
12. Install the heater core.
13. Install the heater core mounting clip and line clamp screws.

14. Install the heater cover.
15. Install the heater outlet module screws.
16. Install both instrument panel insulators.

17. Install the lower center console.
18. Install the heater hoses from the heater core.
19. Install the wiper linkage.
20. Install the air cleaner and duct assembly.
21. Refill the cooling system.
22. Connect the negative battery cable.
23. Run the engine to normal operating temperatures; then, check the climate control operation and check for leaks.

REAR AUXILIARY SYSTEM

1. Disconnect the negative battery cable.
2. Drain the cooling system into a clean container for reuse.
3. Remove the left rear side quarter trim.
4. Remove the rear heater hoses from the rear heater core.
5. Remove the bracket bolt securing the hoses to the heater core.
6. Release the rear heater core-to-rear heater housing tabs.
7. Remove the heater core.
To install:
8. Install the heater core.
9. Connect the rear heater core-to-rear heater housing tabs.
10. Install the bracket bolt securing the hoses to the heater core.
11. Install the rear heater hoses from the rear heater core.
12. Install the left rear side quarter trim.
13. Refill the cooling system.
14. Connect the negative battery cable.

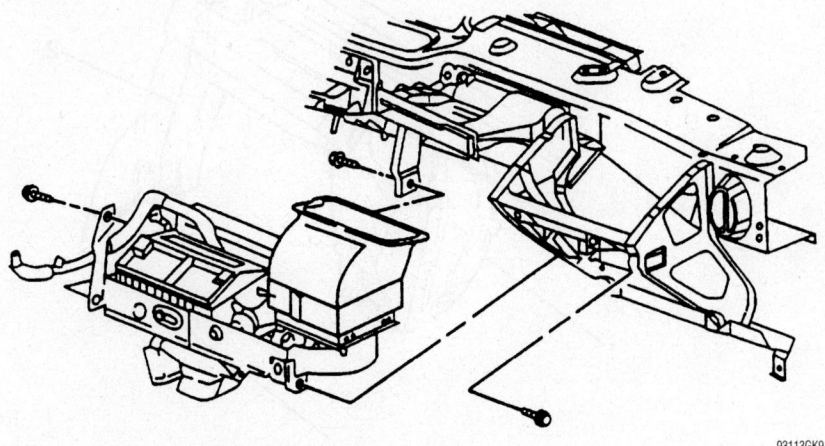

View of the front heater assembly—Pontiac Montana

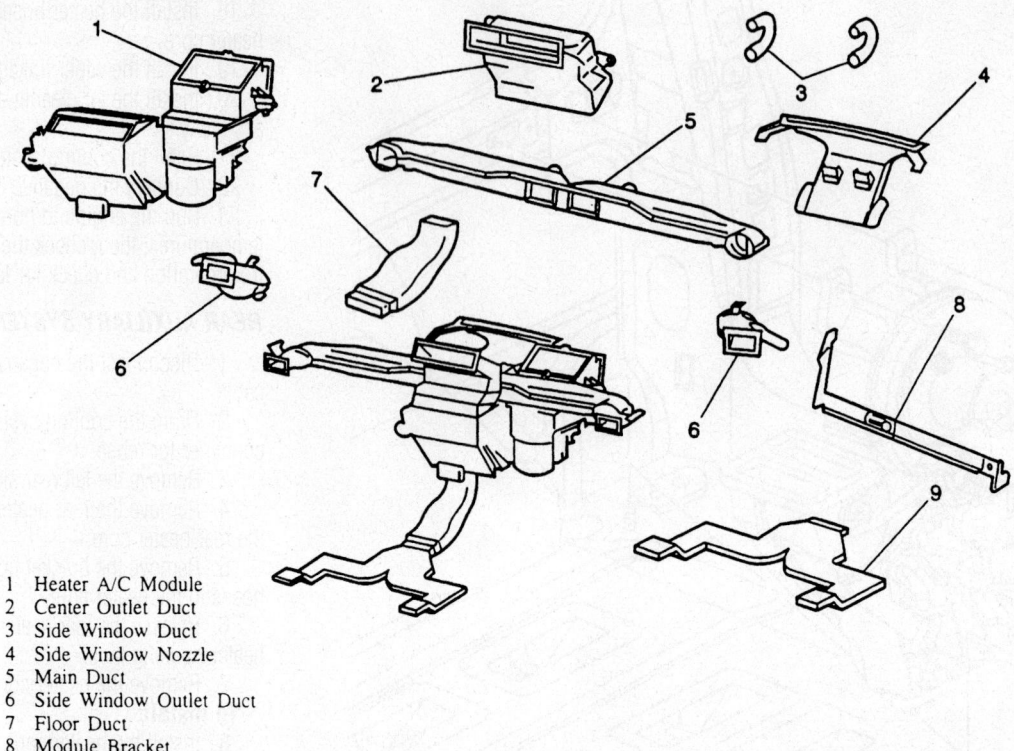

1 Heater A/C Module
2 Center Outlet Duct
3 Side Window Duct
4 Side Window Nozzle
5 Main Duct
6 Side Window Outlet Duct
7 Floor Duct
8 Module Bracket
9 Lower Floor Duct

93113GK0

Exploded view of the front heater ventilation system—Pontiac Montana

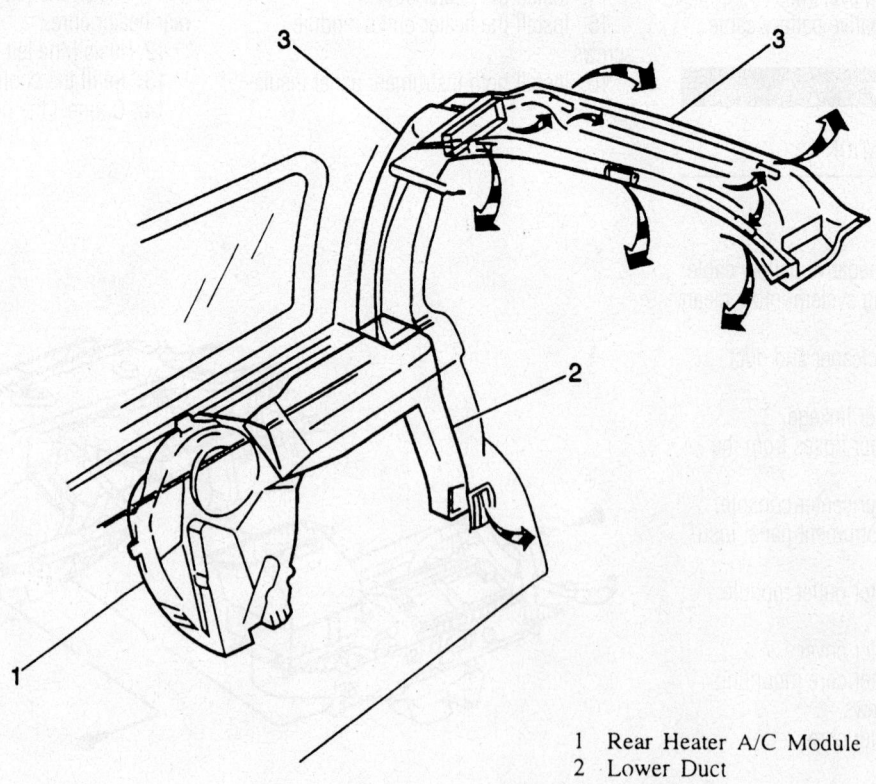

1 Rear Heater A/C Module
2 Lower Duct
3 Upper Duct

93113GL1

View of the rear auxiliary heater assembly ventilation system—Pontiac Montana

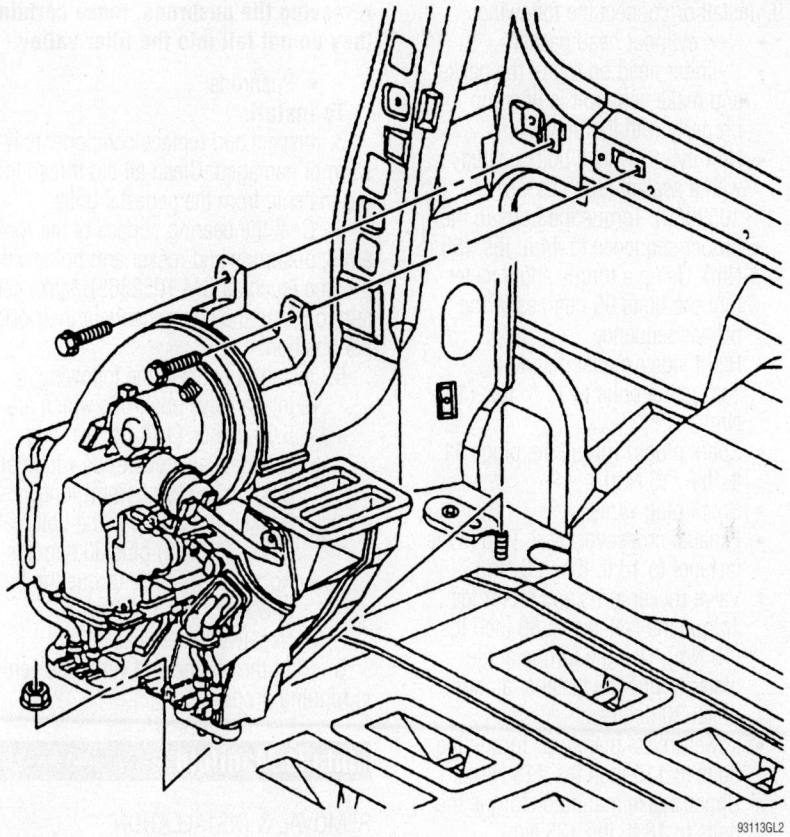

View of the rear auxiliary heater assembly—Pontiac Montana

93113GL2

Cylinder Head

REMOVAL & INSTALLATION

This engine uses aluminum cylinder heads. Use care when working with light alloy parts. Valve guides are pressed in. Roller rocker arms are located on a pedestal in a slot in the cylinder head and are retained on individual threaded bolts.

The cylinder heads are retained by torque-to-yield bolts. A torque angle meter is required for proper torque during assembly. New replacement head bolts are recommended.

Before removing the cylinder head(s) from the engine and before disassembling the valve mechanism, perform a compression test and note the results. During disassembly, be sure that the valvetrain components are kept together and identified so that they can be installed in their original locations.

Left (Front) Side

1. Before servicing the vehicle, refer to the Precautions Section.
2. Relieve the fuel system pressure using the recommended procedure.

3. Drain the cooling system.
4. Drain the oil.
5. Remove or disconnect the following:
   - Negative battery cable
   - Upper intake manifold
   - Lower intake manifold
   - Valve rocker arms and pushrods
   - Exhaust crossover pipe
   - Thermostat bypass pipe
   - Right side engine mount strut bracket
   - Oil level indicator tube
   - Left side spark plug wires and spark plugs
   - Left side exhaust manifold
   - Left side cylinder head and gasket

To install:

6. Clean all parts well. Clean all gasket surfaces. Carefully remove all varnish soot and carbon to the bare metal. DO NOT use a motorized wire brush on any gasket surface since the soft aluminum will be damaged. If necessary, the head can be disassembled for thorough inspection and reconditioning.

7. Inspect the cylinder head for cracks. Do not attempt to weld the cylinder head. If cracked, replace it. Check the cylinder head deck, intake and exhaust manifold mating surfaces for flatness. These surfaces may be reconditioned by milling. If the surfaces are warped more than 0.005 in. (0.127mm), the surface should be milled. If more than 0.010 in. (0.251mm) of metal must be removed from the head, the head should be replaced.

8. Clean the cylinder head bolts and the bolt holes. Check the head bolts for damaged threads or stretching. New replacement head bolts are recommended.

9. Install or connect the following:
   - New cylinder head gasket which is marked which side is **UP**
   - Cylinder head by aligning it with the dowel pins
   - New cylinder head bolts coated with a sealant (such as GM 1052080). Torque the bolts in the proper sequence to 44 ft. lbs. (60 Nm). Using a torque angle meter turn the bolts 95 degrees in the proper sequence.
   - Left side exhaust manifold. Torque the bolts to 12 ft. lbs. (16 Nm).
   - Spark plugs. Torque the plugs 11 ft. lbs. (15 Nm).
   - Spark plug wires
   - Oil level indicator tube. Torque the fastener to 18 ft. lbs. (25 Nm).
   - Right side mount strut bracket. Torque the fastener to 37 ft. lbs. (50 Nm).
   - Thermostat bypass pipe. Torque it to 18 ft. lbs. (25 Nm).

Cylinder head bolt torque sequence—3.4L engine

7924LG03

- Exhaust crossover pipe. Torque the fastener to 18 ft. lbs. (25 Nm).
- Valve rocker arms and pushrods. Torque the fastener to 89 inch lbs. (10 Nm). Using a torque angle meter torque the fastener an additional 30 degrees.
- Lower intake manifold. Torque the bolts to 115 inch lbs. (13 Nm).
- Upper intake manifold. Torque the bolts to 18 ft. lbs. (25 Nm).
- Negative battery cable

10. Refill the coolant system.

11. Change the oil filter and fill the engine with clean oil.

12. Turn the ignition to the **ON** position several times to pressurize the fuel system. Start the engine and inspect for any leaks, repair if necessary. Check and top off the fluid levels if required.

Right (Rear) Side

1. Before servicing the vehicle, refer to the Precautions Section.

2. Relieve the fuel system pressure using the recommended procedure.

3. Drain the coolant system.

4. Drain the oil from the engine.

5. Remove or disconnect the following:
- Negative battery cable
- Upper intake manifold
- Lower intake manifold
- Valve rocker arms and pushrods
- Exhaust crossover pipe
- Right side spark plug wires
- Right side exhaust manifold
- Right side cylinder head and gasket
- Right side spark plugs from the cylinder head

To install:

6. Clean all parts well. Clean all gasket surfaces. Carefully remove all varnish soot and carbon to the bare metal. DO NOT use a motorized wire brush on any gasket surface since the soft aluminum will be damaged. If necessary, the head can be disassembled for thorough inspection and reconditioning.

7. Inspect the cylinder head for cracks. Do not attempt to weld the cylinder head. If cracked, replace it. Check the cylinder head deck, intake and exhaust manifold mating surfaces for flatness. These surfaces may be reconditioned by milling. If the surfaces are "out of flat" by more than 0.005 inch, the surface should be milled. If more than 0.010 inch of metal must be removed from the head, the head should be replaced.

8. Clean the cylinder head bolts and the bolt holes. Check the head bolts for damaged threads or stretching. New replacement head bolts are recommended.

9. Install or connect the following:
- New cylinder head gasket
- Cylinder head on top of the gasket and make certain it is lined up properly with the dowel pins
- New cylinder head bolts coated with a sealant (such as GM 1052080). Torque the bolts in the proper sequence to 44 ft. lbs. (50 Nm). Using a torque angle meter turn the bolts 95 degrees in the proper sequence.
- Right side exhaust manifold. Torque the bolts to 12 ft. lbs. (16 Nm).
- Spark plugs. Torque the plugs 11 ft. lbs. (15 Nm).
- Spark plug wires
- Exhaust crossover pipe. Torque the fastener to 18 ft. lbs. (25 Nm).
- Valve rocker arms and pushrods. Torque the fastener to 89 inch lbs. (10 Nm). Using a torque angle meter torque the fastener an additional 30 degrees.
- Lower intake manifold. Torque the bolts to 115 inch lbs. (13 Nm).
- Upper intake manifold. Torque the bolts to 18 ft. lbs. (25 Nm).
- Negative battery cable

10. Refill the coolant system.

11. Change the oil filter and fill the engine with clean oil.

12. Start the vehicle and verify no leaks, abnormal noises and correct engine operation.

13. Check the fluid levels and top off if necessary.

Rocker Arms

REMOVAL & INSTALLATION

➡️**Valve train components which are to be reused must be installed in their original positions. If removed, be sure to tag or arrange all rocker arms and pushrods to assure proper installation.**

1. Before servicing the vehicle, refer to the Precautions Section.

2. Remove or disconnect the following:
- Negative battery cable
- Rocker arm cover
- Rocker arm bolts
- Rocker arms

➡️**Place the valve train parts in order to ensure they are installed in the proper location. Intake pushrods are yellow and measure 5.68 inches (144.18mm). Exhaust pushrods are green and measure 6.0 inches (152.51mm). When**

removing the pushrods, make certain they do not fall into the lifter valley.

- Pushrods

To install:

3. Inspect and replace components if worn or damaged. Clean all old thread locking material from the pedestal bolts.

4. Coat the bearing surface of the rocker arms, pushrods and rocker arm bolts with a prelube (such as GM 1052365). Make certain to install the components in their original position.

5. Install or connect the following:
- Intake valve pushrods which are 5.68 inches (144.18mm) long
- Exhaust valve pushrods which are 6.0 inches (152.51mm) long
- Rocker arms. Torque the bolt to 14 ft. lbs. (19 Nm) plus 30 degrees
- Rocker arm cover. Torque the bolt to 89 inch lbs. (10 Nm).
- Negative battery cable

6. Start the engine and verify the vehicle is running properly.

Intake Manifold

REMOVAL & INSTALLATION

Upper

This engine uses a 2-piece intake manifold. The upper half (often called a plenum) mounts the throttle body. The lower half of the manifold bolts to the engine and contains the fuel injectors. Please note that this engine uses a sequential multi-port fuel injection system. Injector connectors must be connected to their appropriate fuel injector assembly or engine emissions and engine performance will be seriously affected. Identify and tag for identification all wiring connectors as well as vacuum and other components as required to assure correct assembly.

1. Before servicing the vehicle, refer to the Precautions Section.

2. Drain the engine coolant. Remove the coolant recovery bottle.

3. Relieve the fuel system pressure using the recommended procedure.

4. Remove or disconnect the following:
- Negative battery cable
- Throttle body air inlet duct
- Accelerator and cruise control cables and bracket from the throttle body
- Throttle Position (TP) sensor connector from the throttle body
- Idle Air Control (IAC) valve connector from the throttle body

- Left side spark plug wires
- Left side spark plug wire harness clip and harness
- Throttle body heater hoses
- Evaporative emissions (EVAP) canister purge solenoid valve vacuum hoses
- EVAP canister purge solenoid valve
- Ignition coil bracket and coils
- Wire harness for the Manifold Air Pressure (MAP) sensor and upper intake manifold
- Emissions control vacuum harness
- Brake booster vacuum hose from the upper intake manifold
- Vacuum hose connection for the Heater Vent Air Conditioning (HVAC) source hose
- Vacuum hose connection for the fuel pressure regulator
- Exhaust Gas Recirculation (EGR) valve
- MAP sensor and bracket
- Upper intake manifold
- Upper intake manifold gasket
- Throttle body, if replacing the manifold

To install:

5. Clean all parts well. Use care in cleaning old gasket material from the machined aluminum surfaces on the

plenum and manifold as sharp tools may damage sealing surfaces.

6. Clean the mating surfaces to the upper intake manifold and engine block. Remove any loose pieces of RTV sealer.

7. Install or connect the following:
- Throttle body to the upper intake manifold (if removed). Torque the bolts to 18 ft. lbs. (25 Nm).
- Upper intake manifold gasket
- Upper intake manifold
- MAP sensor and bracket. Torque the bolt to 44 inch lbs. (5 Nm).
- Upper intake manifold bolts. Torque the bolts to 18 ft. lbs. (25 Nm).
- EGR valve. Torque the fastener to 18 ft. lbs. (25 Nm).
- HVAC vacuum source hose to the upper intake manifold
- Fuel pressure regulator vacuum hose to the upper intake manifold
- Brake booster vacuum hose
- MAP sensor and bracket. Torque the bolt to 44 inch lbs. (5 Nm).
- Emissions control vacuum harness
- Vacuum hose for the MAP sensor
- Wiring harness to the MAP sensor
- Ignition coil bracket and coils. Torque the fasteners to 18 ft. lbs. (25 Nm).
- EVAP canister purge solenoid valve

- Vacuum hoses to the EVAP canister purge solenoid valve
- Throttle body heater hose
- Left side spark plug wire harness clip
- Spark plugs wires
- TP sensor wire harness connector to the throttle body
- IAC valve wire harness connector to the throttle body
- Accelerator and cruise control cables and bracket to the throttle body. Torque the fasteners to 106 inch lbs. (12 Nm).
- Throttle body air inlet duct
- Negative battery cable

8. Fill the coolant system.
9. Fill the engine with new oil.
10. Turn the ignition to the **ON** position several times to pressurize the fuel system.
11. Start the engine and check for any leakage and repair if necessary.
12. Check and top off all fluid levels if needed.

Lower

This engine uses a 2-piece intake manifold. The upper half (often called a plenum) mounts the throttle body. The lower half of the manifold bolts to the engine and contains the fuel injectors. Please note that this engine uses a sequential multi-port fuel injection system. Injector connectors must be connected to their appropriate fuel injector assembly or engine emissions and engine performance will be seriously affected. Identify and tag for identification all wiring connectors as well as vacuum and other components as required to assure correct assembly.

1. Before servicing the vehicle, refer to the Precautions Section.
2. Drain the engine coolant.
3. Relieve the fuel system pressure using the recommended procedure.
4. Remove or disconnect the following:
- Negative battery cable
- Upper intake manifold
- Left side valve rocker arm cover
- Right side valve rocker arm cover
- Wire harness from the Engine Coolant Temperature (ECT) sensor
- Fuel injector, Manifold Absolute Pressure (MAP) and ECT wire harness
- Fuel feed and return pipe from the injector rail
- Fuel injector rail
- Power steering pump from the front cover

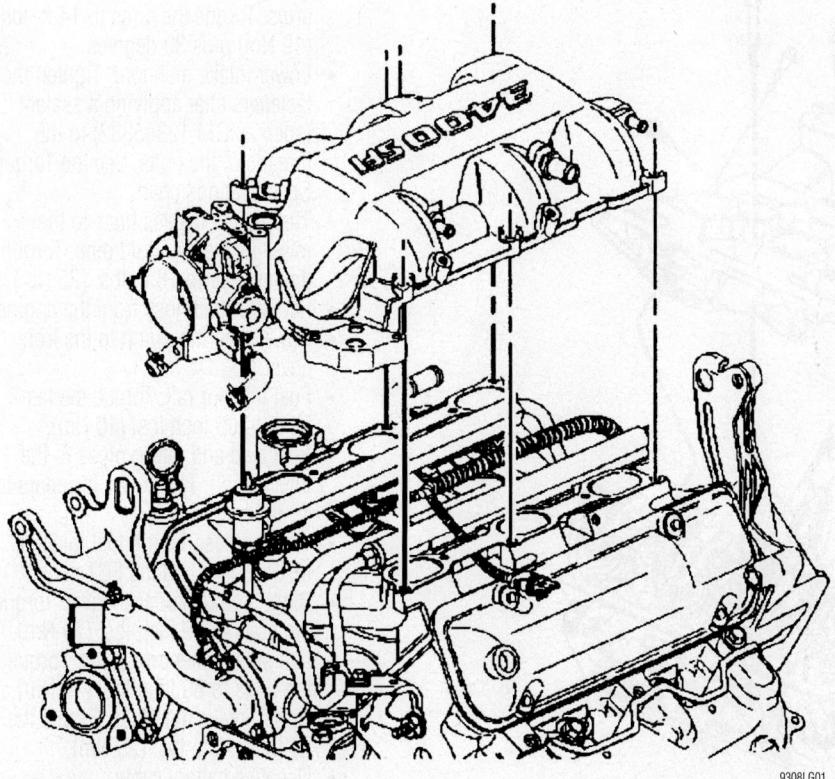

9308LG01

Remove upper intake manifold

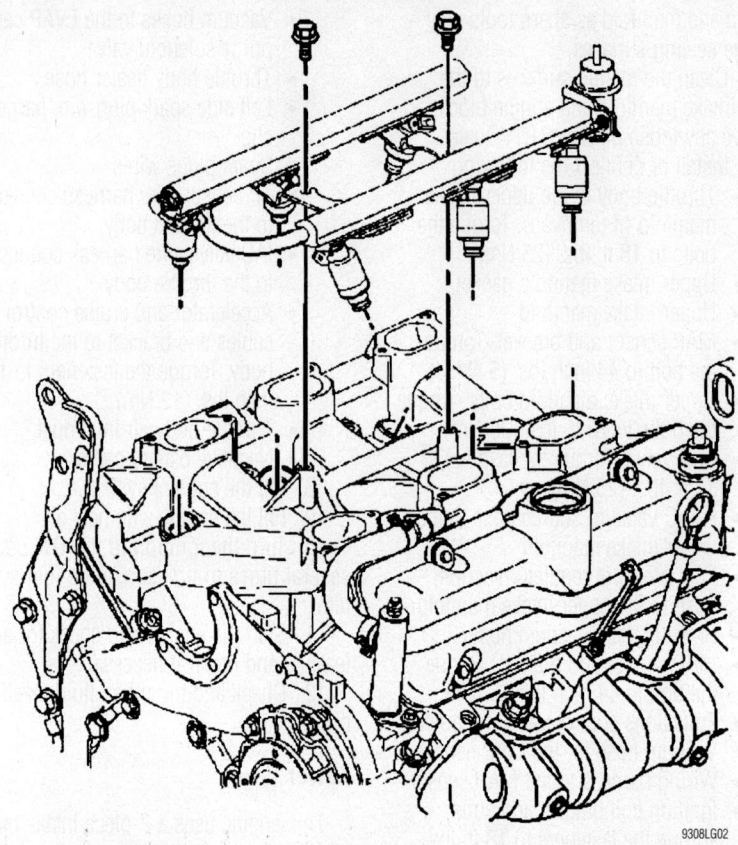

Remove the fuel injector rail

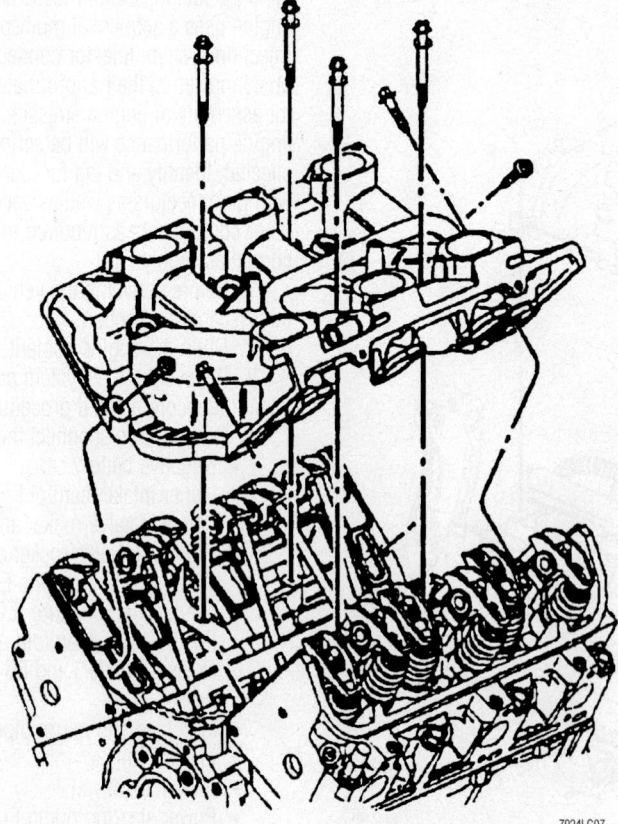

Lower intake manifold assembly

➡ **Do not disconnect the power steering pipes or hoses from the steering pump.**

- Heater inlet pipe with the heater hose from the lower intake manifold
- Radiator inlet hose
- Thermostat bypass hose from the manifold
- Lower intake manifold
- Pushrods after loosening the rocker arms
- Lower intake manifold gasket and seals
- ECT sensor, if replacing the manifold
- Thermostat and housing, if replacing the manifold

To install:

5. Clean the gasket mounting surfaces.

6. Inspect the intake manifold for cracks or damage, replace if necessary.

7. Install or connect the following:

- ECT sensor, if removed. Torque the sensor to 17 ft. lbs. (23 Nm).
- Thermostat and housing, if removed. Torque the fastener to 18 ft. lbs. (25 Nm).
- Thin bead of RTV sealer (such as GM 12345739) on the ridge of the engine block where the lower intake manifold makes contact
- Lower intake manifold gaskets
- Pushrods and tighten the rocker arms. Torque the arms to 14 ft. lbs. (19 Nm) plus 30 degrees.
- Lower intake manifold. Tighten the fasteners after applying a sealant (such as GM 12345382) to the threads of the bolts. See the Torque Specifications chart.
- Thermostat bypass hose to the lower intake manifold pipe. Torque the fastener to 18 ft. lbs. (25 Nm).
- Radiator inlet hose from the engine
- Power steering pump to the front cover
- Fuel injector rail. Torque the fastener to 89 inch lbs. (10 Nm).
- Fuel feed and return pipes to the injector rail. Torque the fasteners to 13 ft. lbs. (17 Nm).
- Wire harness for the fuel injector, MAP sensor and the ECT sensor
- Right side rocker arm cover. Torque the bolts to 89 inch lbs. (10 Nm).
- Left side rocker arm cover. Torque the bolts to 89 inch lbs. (10 Nm).
- Upper intake manifold. Torque the bolts to 18 ft. lbs. (25 Nm).
- Negative battery cable

8. Fill the coolant system.

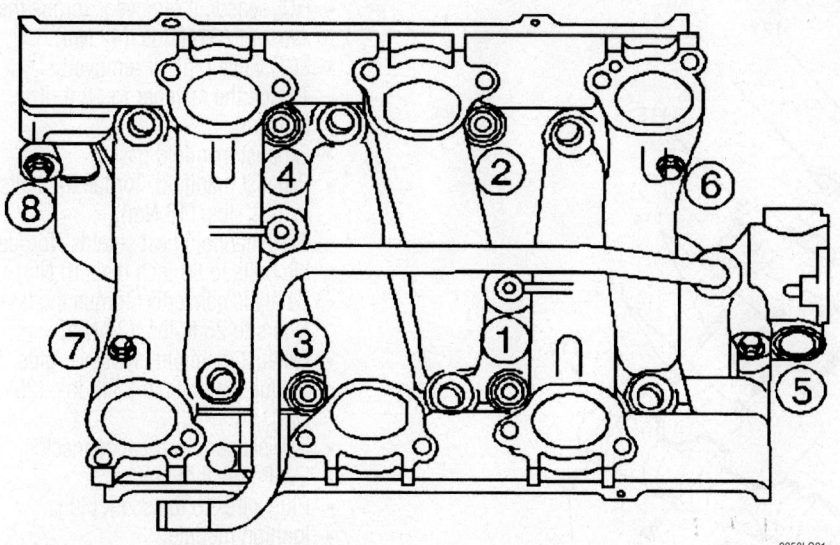

9358LG01

Lower intake manifold assembly torque sequence

9. Turn the ignition to the **ON** position several times to pressurize the fuel system.

10. Start the engine and check for any leakage and repair if necessary.

11. Check and top off all fluid levels if needed.

Exhaust Manifold

REMOVAL & INSTALLATION

The exhaust manifolds are conventional iron castings. The left and right manifolds are connected by a crossover pipe. Use care with the exhaust manifold-to-cylinder head fasteners. The cylinder heads are aluminum.

Left (Front) Side

1. Before servicing the vehicle, refer to the Precautions Section.
2. Drain the cooling system.
3. Remove or disconnect the following:
   - Negative battery cable
   - Throttle body air inlet duct
   - Right side engine mount strut bracket
   - Radiator inlet hose
   - Thermostat bypass pipe
   - Exhaust crossover pipe heat shield
   - Exhaust crossover pipe
   - Left side exhaust manifold heat shield
   - Left side exhaust manifold and discard the gasket

To install:
4. Clean the gasket mounting surfaces.
5. Install or connect the following:
   - Left side exhaust manifold gasket
   - Left side exhaust manifold. Torque the nuts to 12 ft. lbs. (16 Nm).

- Left side exhaust manifold heat shield. Torque the fasteners to 89 inch lbs. (10 Nm).
- Exhaust crossover pipe. Torque the bolts to 18 ft. lbs. (25 Nm).

- Exhaust crossover pipe heat shield. Torque the bolts to 89 inch lbs. (10 Nm).
- Thermostat bypass pipe. Torque the fastener to 18 ft. lbs. (25 Nm).
- Radiator inlet hose to the engine
- Right side engine mount strut bracket. Torque the bolts to 35 ft. lbs. (48 Nm).
- Throttle body air inlet duct
- Negative battery cable

6. Fill the cooling system
7. Start the vehicle and check for leaks, repair if necessary.
8. Check and top off all fluid levels if necessary.

Right (Rear) Side

1. Before servicing the vehicle, refer to the Precautions Section.
2. Remove or disconnect the following:
   - Negative battery cable
   - Windshield wiper motor cover
   - Fuel injector sight shield
   - Throttle body air inlet duct

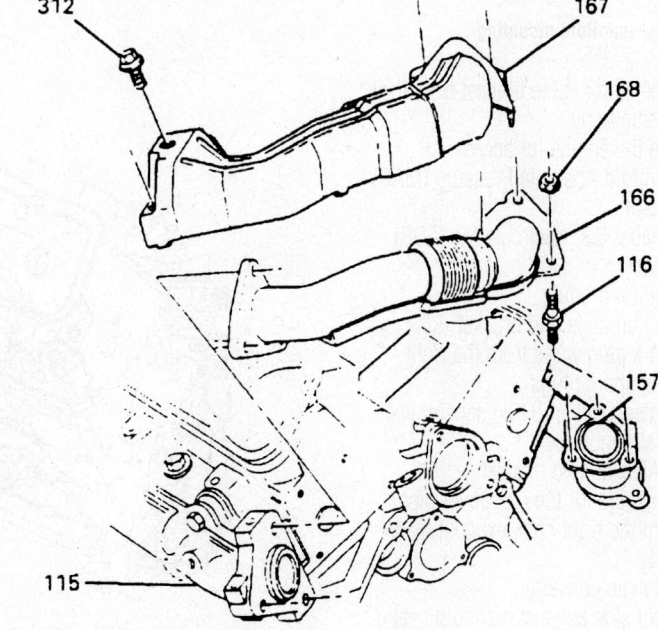

115	MANIFOLD, LEFT HAND EXHAUST
116	STUD, EXHAUST CROSSOVER
157	MANIFOLD, RIGHT HAND EXHAUST
166	CROSSOVER PIPE, EXHAUST
167	SHIELD, EXHAUST CROSSOVER UPPER HEAT
168	NUT, EXHAUST CROSSOVER
312	BOLT/SCREW, EXHAUST CROSSOVER UPPER HEAT SHIELD

7924LG28

Exploded view of the exhaust crossover and heat shield mounting

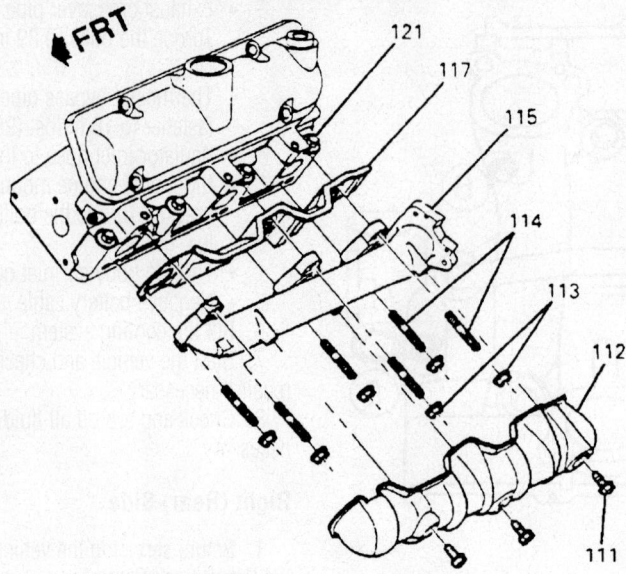

111 SCREW, LH EXHAUST MANIFOLD HEAT SHIELD
112 SHIELD LH EXHAUST MANIFOLD
113 NUT, LH EXHAUST MANIFOLD
114 STUD, LH EXHAUST MANIFOLD
115 MANIFOLD, LH EXHAUST
117 GASKET, LH EXHAUST MANIFOLD
121 HEAD, LH CYLINDER

7924LG29

Left exhaust manifold mounting

- Accelerator cable bracket from the throttle body
3. Rotate the engine for access.
  - Manifold Absolute Pressure (MAP) sensor
  - Exhaust Gas Recirculation (EGR) valve
  - Ignition module
  - Ignition coils and bracket
  - Spark plug wires from the right bank spark plugs
  - Heated Oxygen (HO_2) sensor electrical connector
  - EVAP solenoid bracket
  - Fasteners for the crossover pipe from the right bank exhaust manifold
  - Catalytic converter
  - Right side exhaust manifold heat shields
  - Right side exhaust manifold
  - Right side exhaust manifold gasket
  - EGR valve pipe, if replacing the exhaust manifold
  - HO_2 sensor, if replacing the exhaust manifold

To install:
4. Clean the gasket mounting surfaces.
5. Install or connect the following:

- HO_2 sensor, if removed. Torque the sensor to 31 ft. lbs. (42 Nm).
- EGR valve pipe, if removed. Torque the fastener to 18 ft. lbs. (25 Nm).
- Exhaust manifold gasket
- Exhaust manifold. Torque the bolts to 12 ft. lbs. (16 Nm).
- Both manifold heat shields. Torque the bolts to 89 inch lbs. (10 Nm).
- Catalytic converter. Torque the fasteners to 25 ft. lbs. (34 Nm).
- Exhaust manifold crossover pipe. Torque the bolts to 18 ft. lbs. (25 Nm).
- HO_2 sensor electrical connector
- EVAP solenoid bracket
- Plug wires to the spark plugs
- Ignition module
- Ignition coils and bracket. Torque the fastener to 18 ft. lbs. (25 Nm).
6. Rotate the engine to its original position.
7. Install or connect the following:
   - EGR valve to the intake manifold
   - MAP sensor
   - Accelerator cable bracket to the throttle body. Torque the bolt to 89 inch lbs. (10 Nm).
   - Throttle body air inlet duct

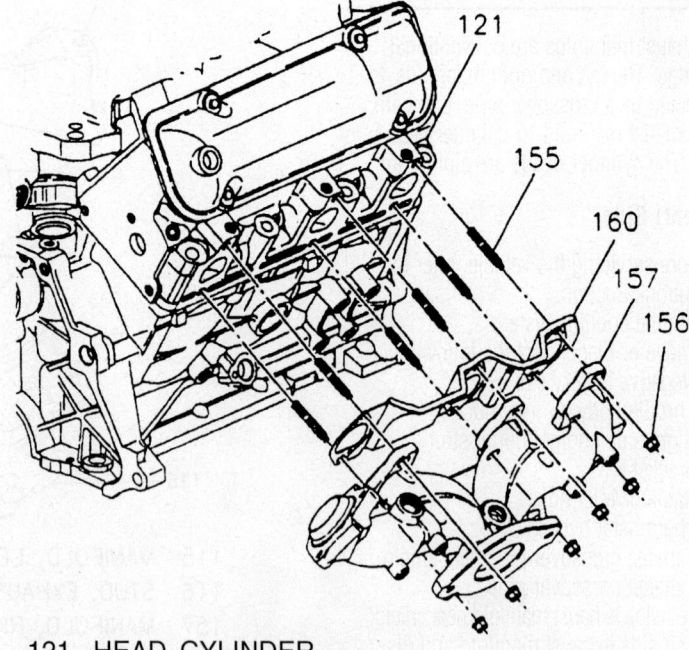

121 HEAD, CYLINDER
155 STUD, EXHAUST MANIFOLD
156 NUT, EXHAUST MANIFOLD
157 MANIFOLD RIGHT EXHAUST
160 GASKET, RIGHT EXHAUST MANIFOLD

7924LG30

Exploded view of the right exhaust manifold mounting

- Windshield wiper motor cover
- Fuel injector sight shield
- Negative battery cable

8. Start the engine and inspect for leaks, repair if necessary.

Camshaft and Valve Lifters

REMOVAL & INSTALLATION

1. Before servicing the vehicle, refer to the Precautions Section.
2. Relieve the fuel system pressure.
3. Remove or disconnect the following:
   - Engine assembly

❊❊ WARNING

When removing valvetrain components they must be marked for installation in their original location. When the camshaft is being replaced, the valve lifters must also be replaced.

- Rocker arm covers
- Intake manifold
- Rocker arm bolts, balls, rocker arms and pushrods
- Lifter guide bolts and the guide
- Valve lifter(s) from the bores
- Crankshaft balancer and front cover
- Timing chain and sprockets
- Oil pump driven gear bolt and gear
- Camshaft thrust plate
- Camshaft

❊❊ WARNING

Avoid damaging the camshaft bearing surfaces.

To install:
4. Coat the camshaft with prelube.
5. Install or connect the following:
   - Camshaft
   - Camshaft thrust plate. Tighten the bolts to 89 inch lbs. (10 Nm).
   - Oil pump driven gear. Tighten the bolt to 27 ft. lbs. (36 Nm).
   - Timing chain and sprocket
   - Camshaft thrust button and front cover
   - Crankshaft balancer
6. Lubricate the bearing surfaces with Molykote®.

➡**Installation of a new camshaft or a wear pattern on the old valve lifter will require the replacement of the camshaft and lifters together. If camshaft replacement is not necessary, be sure to install the used valve lifters in their original position.**

7. Install or connect the following:
   - Lifters in their original locations
   - Lifter guide. Tighten the guide bolts to 89 inch lbs. (10 Nm).
   - Pushrods, rocker arms, balls and bolts. Tighten the nuts to 89 inch lbs. (10 Nm) plus an additional 30 degree turn.
   - Intake manifold
   - Rocker arm covers
   - Engine assembly
   - Negative battery cable

8. Adjust the valves, as required. Start the engine and verify no oil leaks.

Valve Lash

ADJUSTMENT

Because the rocker arm fasteners are secured and tightened, valve lash is not adjustable. If a valve train problem is suspected, check that the rocker arm pedestals bolts are tightened to specification. During initial installation the bolts are coated with thread locking compound. If they are sufficiently loosened to cause valvetrain noise, they should be removed and thoroughly cleaned. Apply thread locking compound to the rocker arm pedestal bolts. Tighten the bolts to 14 ft. lbs. (19 Nm) plus 30 degrees.

When valve lash falls out of specification (valve tap is heard) and tightening the bolts does not solve the problem, replace the rocker arm, pushrod and hydraulic lifter on the offending cylinder.

Starter Motor

REMOVAL & INSTALLATION

1. Before servicing the vehicle, refer to the Precautions Section.
2. Remove or disconnect the following:
   - Negative battery cable
   - Radiator air baffle assembly
   - Electrical connections
   - Torque converter cover
   - Starter

To install:
3. Install or connect the following:
   - Starter. Torque the bolts to 35 ft. lbs. (47 Nm).

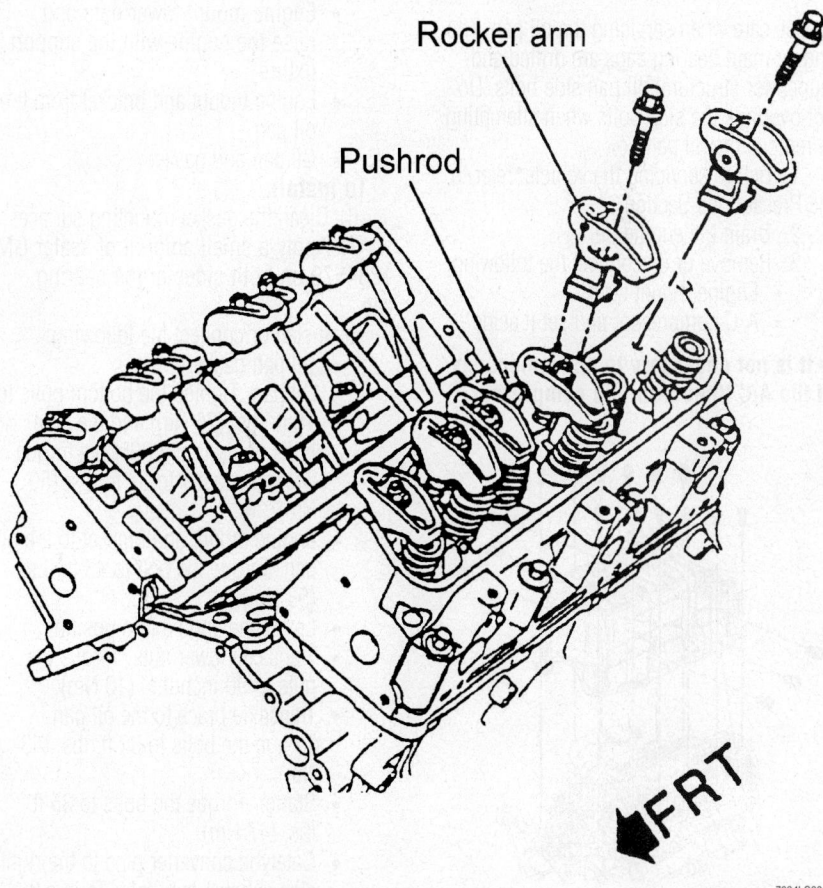

Valve rocker arm and related components

7924LG33

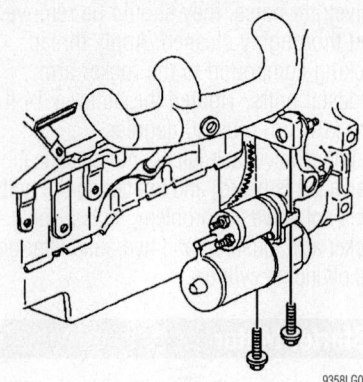

Starter motor mounting

- Torque converter cover
- Solenoid "BAT" terminal. Torque the nut to 89 inch lbs. (10 Nm).
- Solenoid "S" terminal. Torque the nut to 27 inch lbs. (3 Nm).
- Radiator air baffle assembly
- Negative battery cable

4. Perform a charging system test and verify the starter is operating properly

Oil Pan

REMOVAL & INSTALLATION

Use care when servicing the oil pan. The engine main bearing caps are drilled and tapped for structural oil pan side bolts. Do not overlook the side bolts when attempting to remove the oil pan.

1. Before servicing the vehicle, refer to the Precautions Section.
2. Drain the engine oil.
3. Remove or disconnect the following:
   - Engine mount struts
   - A/C compressor and set it aside

➡ **It is not necessary to disconnect any of the A/C lines from the compressor.**

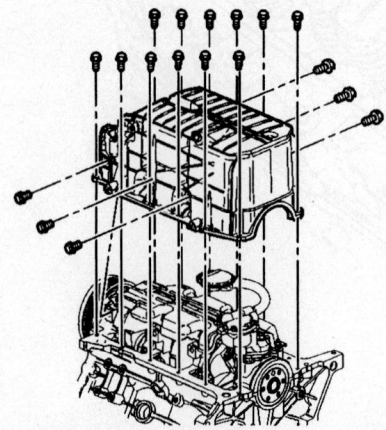

Install the oil pan bolts

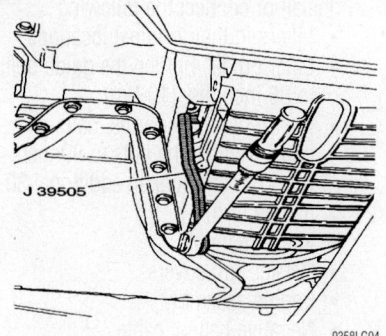

Use a torque wrench adapter to tighten the oil pan side bolts

4. Install an engine support fixture.
5. Remove or disconnect the following:

   - Catalytic converter pipe from the right side exhaust manifold
   - Frame bolts and make certain that an engine stand (such as J 39580) is aligned below the engine, lower the frame
   - Oil level sensor wiring harness connector
   - Starter
   - Transaxle brace from the oil pan
   - Transaxle mount lower nuts
   - Engine mount lower nuts and raise the engine with the support fixture
   - Engine mount and bracket from the oil pan
   - Oil pan and gasket

To install:
6. Clean the gasket mounting surfaces.
7. Apply a small amount of sealer GM 1234579 on both sides of the bearing cap.
8. Install or connect the following:
   - Oil pan gasket
   - Oil pan. Tighten the bottom bolts to 18 ft. lbs. (25 Nm) and the side bolts using tool J 39505 a torque wrench adapter to 37 ft. lbs. (50 Nm).
   - Engine mount and bracket to the oil pan. Torque the bolt to 43 ft. lbs. (58 Nm).
   - Lower the engine into position
   - Transaxle lower nuts. Torque the nuts to 90 inch lbs. (10 Nm).
   - Transaxle brace to the oil pan. Torque the bolts to 32 ft. lbs. (43 Nm).
   - Starter. Torque the bolts to 35 ft. lbs. (47 Nm).
   - Catalytic converter pipe to the right side exhaust manifold. Torque the bolts to 26 ft. lbs. (35 Nm).

- Frame and use new frame to body bolts. Torque the front bolts to 111 ft. lbs. (150 Nm) and the rear bolts to 122 ft. lbs. (165 Nm).
9. Remove the engine support fixture.
10. Install or connect the following:
   - A/C compressor. Torque the bolts to 37 ft. lbs. (50 Nm).
   - Engine mount struts. Torque the fasteners to 52 ft. lbs. (70 Nm).
   - Negative battery cable
11. Fill the engine with oil.
12. Start the vehicle and inspect for leaks, repair if necessary.

Oil Pump

REMOVAL & INSTALLATION

1. Before servicing the vehicle, refer to the Precautions Section.
2. Drain the engine oil
3. Remove or disconnect the following:

   - Negative battery cable
   - Oil pan
   - Bolt attaching the oil pump to the rear crankshaft bearing cap
   - Oil pump and driveshaft

To install:
   - Oil pump and driveshaft
   - Oil pump to the rear crankshaft

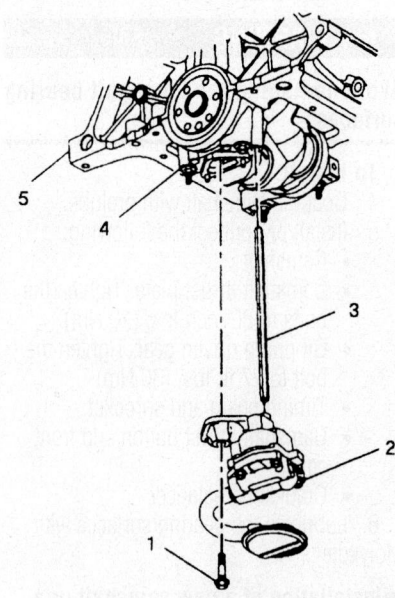

1. Oil pump bolt
2. Oil pump
3. Oil pump drive rod
4. Main bearing cap
5. Engine block

Exploded view of the oil pump mounting

bearing cap. Torque the bolt to 30 ft. lbs. (41 Nm).
- Oil pan
- Negative battery cable

4. Fill the engine with oil.

5. Start the vehicle and inspect for leaks, repair if necessary.

Rear Main Seal

REMOVAL & INSTALLATION

The transaxle assembly must be removed to perform this service. This requires special tooling to support the engine assembly while the transaxle and sub-frame are lowered from under the vehicle.

1. Before servicing the vehicle, refer to the Precautions Section.

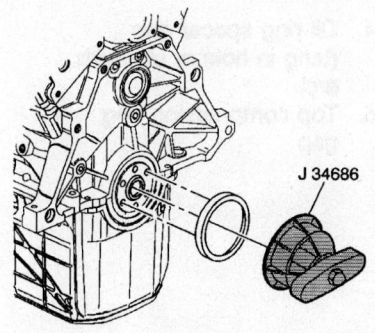

9358LG05

Use the correct installation tool when replacing the rear seal

2. Remove or disconnect the following:
- Negative battery cable
- Transmission assembly
- Engine flywheel
- Oil seal

✳✳ WARNING

When removing the seal, use care so that no damage occurs to the crankshaft. Once the seal is removed, inspect the crankshaft surface for any nicks or burrs. Repair or replace crankshaft as necessary.

To install:

3. Install or connect the following:
- New oil seal lubricated with engine oil, using an Oil Seal Installer tool J 34686 until it is seated properly over the crankshaft
- Flywheel
- Transmission assembly
- Negative battery cable

4. Start the vehicle and check for leaks, repair if necessary.

Timing Chain, Sprockets, Front Cover and Seal

REMOVAL & INSTALLATION

1. Before servicing the vehicle, refer to the Precautions Section.

2. Drain the engine oil.

3. Drain the coolant.

4. Remove or disconnect the following:
- Negative battery cable
- Crankshaft balancer
- Drive belt tensioner
- Power steering pump and lines. Do not disconnect the lines from the pump
- Thermostat bypass pipe from the front cover
- Radiator outlet hose from the water pump
- Water pump
- Upper and lower Crankshaft Position (CKP) sensor wire harness bracket from the front cover
- CKP sensor from the front cover
- Front cover and gasket

5. Rotate the crankshaft until the timing marks are aligned in the following locations:
- Camshaft alignment pin (1)
- Timing chain damper (2) to the crankshaft sprocket (3)
- Crankshaft key (4)
- Timing chain damper (5) to the camshaft sprocket locator hole (6)

6. Remove or disconnect the following:
- Camshaft sprocket bolt
- Timing chain, timing chain sprockets and damper
- Front oil seal

To install:

7. Install or connect the following:
- New front oil seal by making certain the seal is fully seated

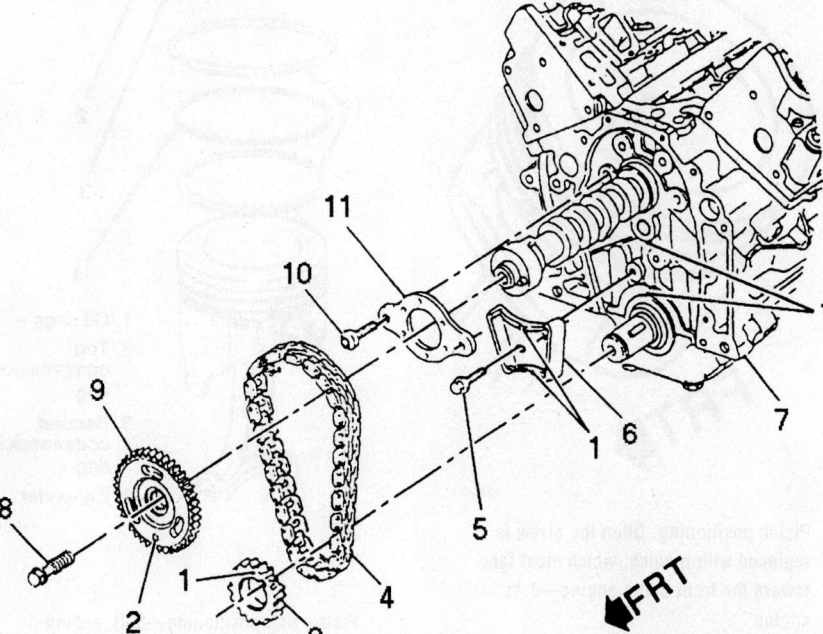

1. Timing alignment marks
2. Locator hole
3. Crankshaft sprocket
4. Timing chain
5. Timing chain dampener bolt
6. Timing chain dampener
7. Engine block
8. Camshaft sprocket bolt
9. Camshaft sprocket
10. Thrust plate bolt
11. Thrust plate

Exploded view of the timing chain assembly

7924LG14

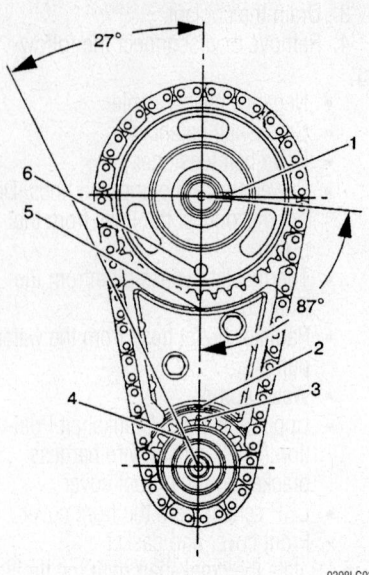

Crankshaft timing mark locations

- Timing chain damper. Torque the bolts to 15 ft. lbs. (21 Nm).
- Timing chain to the camshaft sprocket
- Crankshaft sprocket
- Timing chain to the crankshaft sprocket by making certain the chain is fully seated

8. Align the crankshaft timing mark to the bottom mark on the damper.

9. Align the timing mark on the camshaft gear center line of the locator hole with the timing mark on the top of the damper.

10. Align the dowel in the camshaft with the dowel hole in the camshaft sprocket.

11. Install or connect the following:
- Camshaft sprocket bolt. Torque the bolt to 103 ft. lbs. (140 Nm).

12. Apply a 0.20 inch (5mm) bead of sealer to both sides of the lower tabs of the engine front cover gasket.
- Front cover. Torque the 5 small bolts to 15 ft. lbs. (21 Nm), the 3 large bolts to 41 ft. lbs. (55 Nm) and the 2 remaining bolts to 35 ft. lbs. (47 Nm).
- Water pump to the front cover. Torque the bolts to 89 inch lbs. (10 Nm).
- Water pump pulley. Torque the bolt to 18 ft. lbs. (25 Nm).
- CKP sensor to the front cover
- Upper/lower CKP wire harness brackets to the front cover
- Radiator outlet hose to the water pump

- Thermostat bypass pipe to the front cover
- Power steering pump and lines
- Drive belt tensioner
- Crankshaft balancer
- Negative battery cable

13. Fill the engine with oil.

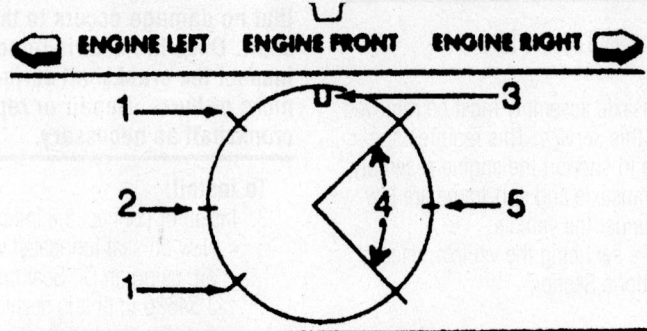

1. Oil ring rail gaps
2. 2nd Compression ring gap
3. Notch in piston
4. Oil ring spacer gap (tang in hole or slot with arc)
5. Top compression ring gap

Piston ring end-gap spacing—3.4L engine

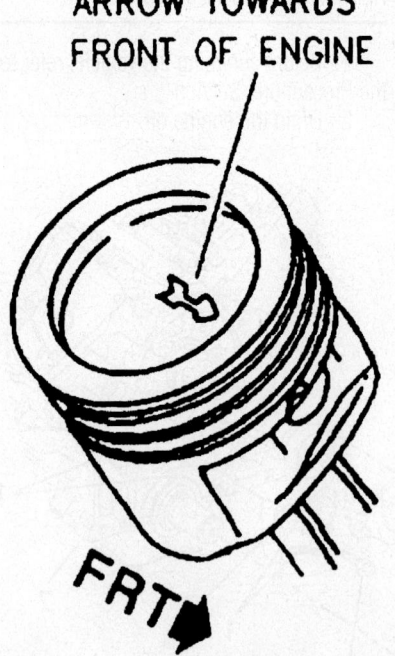

Piston positioning. Often the arrow is replaced with a notch, which must face toward the front of the engine—3.4L engine

14. Fill the coolant system.
15. Start the vehicle and verify that the engine is running properly.

Pistons and Ring

POSTIONING

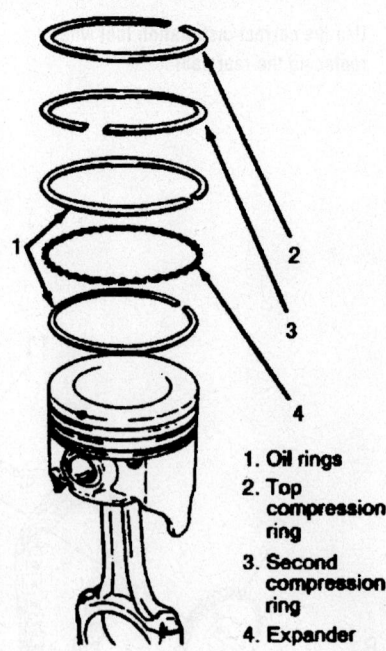

1. Oil rings
2. Top compression ring
3. Second compression ring
4. Expander

Piston ring positioning—3.4L engine

FUEL SYSTEM

Fuel System Service Precaution

Safety is the most important factor when performing not only fuel system maintenance but any type of maintenance. Failure to conduct maintenance and repairs in a safe manner may result in serious personal injury or death. Maintenance and testing of the vehicle's fuel system components can be accomplished safely and effectively by adhering to the following rules and guidelines.

• To avoid the possibility of fire and personal injury, always disconnect the negative battery cable unless the repair or test procedure requires that battery voltage be applied.

• Always relieve the fuel system pressure prior to disconnecting any fuel system component (injector, fuel rail, pressure regulator, etc.), fitting or fuel line connection. Exercise extreme caution whenever relieving fuel system pressure, to avoid exposing skin, face and eyes to fuel spray. Please be advised that fuel under pressure may penetrate the skin or any part of the body that it contacts.

• Always place a shop towel or cloth around the fitting or connection prior to loosening to absorb any excess fuel due to spillage. Ensure that all fuel spillage (should it occur) is quickly removed from engine surfaces. Ensure that all fuel soaked cloths or towels are deposited into a suitable waste container.

• Always keep a dry chemical (Class B) fire extinguisher near the work area.

• Do not allow fuel spray or fuel vapors to come into contact with a spark or open flame.

• Always use a back-up wrench when loosening and tightening fuel line connection fittings. This will prevent unnecessary stress and torsion to fuel line piping. Always follow the proper torque specifications.

• Always replace worn fuel fitting O-rings with new ones. Do not substitute fuel hose, or equivalent, where fuel pipe is installed.

Fuel System Pressure

RELIEVING

A Schrader valve is provided on these fuel systems to conveniently test or release the system pressure. A fuel pressure gauge and adapter will be necessary to connect the gauge to the fitting. Most of the MFI systems utilize a service valve on one end of the fuel rail assembly.

1. Before servicing the vehicle, refer to the Precautions Section.
2. Disconnect the negative battery cable
3. Loosen the fuel filler cap to relieve tank vapor pressure.
4. Connect a fuel pressure gauge to the connector. Wrap a shop towel around the fittings to prevent spillage.
5. Install the bleed hose into an approved container and open the valve.
6. Drain any remaining fuel from the pressure gauge.
7. When fuel service is finished, tighten the fuel filler cap and connect the negative battery cable.

Fuel Filter

REMOVAL & INSTALLATION

1. Before servicing the vehicle, refer to the Precautions Section.
2. Relieve fuel system pressure.
3. Remove or disconnect the following:

• Negative battery cable
• Quick-connect fittings at the inlet/outlet sides of the in-pipe fuel filter
• Fuel filter and drain any remaining fuel

To install:
4. Install or connect the following:
• Fuel filter to the bracket
• Fuel filter assembly to the side rail near the fuel tank. Torque the nut to 89 inch lbs. (10 Nm).
• Inlet/outlet quick connectors to the fuel filter
• Negative battery cable
5. Start the vehicle and checks for leaks, repair if necessary.

Fuel Pump

REMOVAL & INSTALLATION

1. Before servicing the vehicle, refer to the Precautions Section.
2. Properly relieve the fuel system pressure.
3. Drain and remove the fuel tank from the vehicle
4. Remove or disconnect the following:
• Negative battery cable
• Quick-connect fittings at the fuel pump
• Fuel pump locking nut with a Fuel Pump Spanner Wrench J 39348
• Fuel pump assembly from the fuel tank and discard the O-ring

To install:
5. Install or connect the following:
• New O-ring on the fuel tank
• Fuel pump into the fuel tank mak-

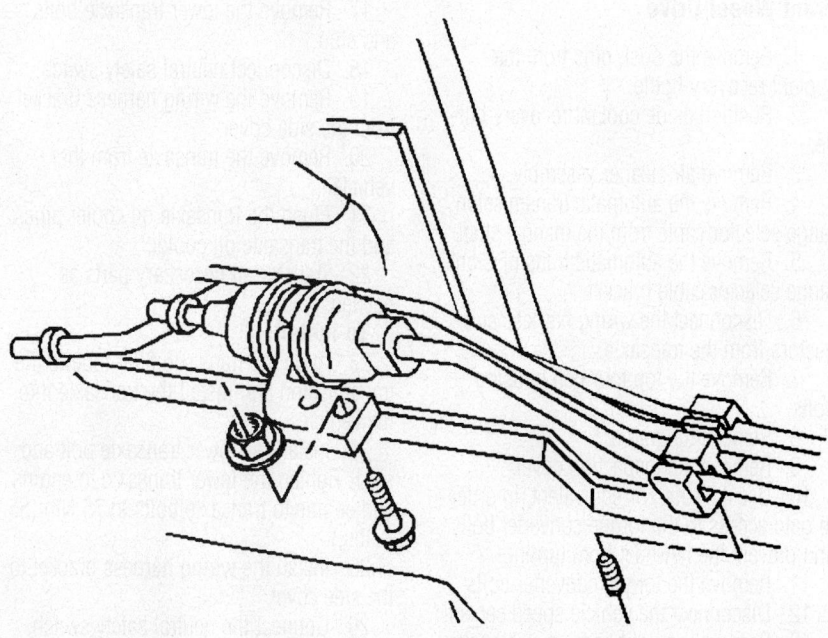

9308LG04

Fuel filter component identification—the fuel filter is located on the frame rail near the tank

ing certain not to fold or twist the strainer and that it does not interfere with the full travel of the float arm
- Fuel pump locking nut with the scanner tool
- Quick-connect fittings at the fuel pump
- Fuel tank and fill the tank
- Negative battery cable

6. Prime the fuel system as follows:

a. Turn the ignition switch **ON** for two seconds.

b. Turn the ignition switch **OFF** for 10 seconds.

c. Turn the ignition switch **ON** and checks for leaks. Repair if necessary.

Fuel Injector

REMOVAL & INSTALLATION

1. Before servicing the vehicle, refer to the Precautions Section.
2. Relieve the fuel system pressure.
3. Remove or disconnect the following:
   - Negative battery cable
   - Upper intake manifold
   - Fuel rail

- Fuel injector retaining clips and injectors
- O-rings and discard them

To install:

➡**When replacing the fuel injector O-rings install the brown O-ring in the lower position. The lower O-ring uses a nylon collar to properly position it on the injector. Be sure to install the O-ring backup or the sealing O-ring may move when the injector is installed to the fuel rail. If the sealing ring is not seated properly, a vacuum leak is possible thus causing driveability complaints.**

4. Install or connect the following:
   - Upper O-ring to the fuel injector
   - Lower O-ring backup to the injector
   - Lower O-ring to the injector
   - Fuel injector to the fuel rail
   - Fuel rail with the retaining clips
   - Upper intake manifold
   - Negative battery cable

5. Prime the fuel system as follows:

a. Turn the ignition switch **ON** for two seconds.

b. Turn the ignition switch **OFF** for 10 seconds.

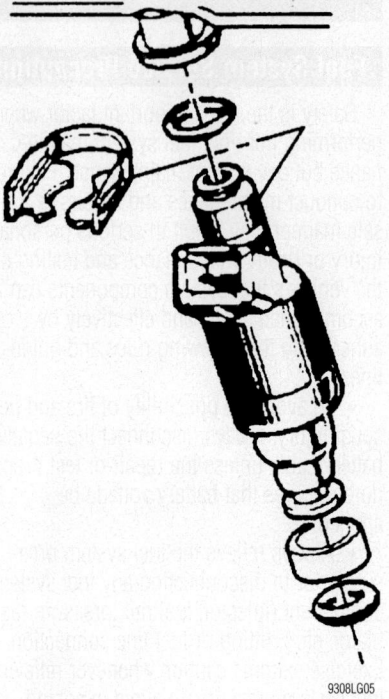

9308LG05

Fuel injector

c. Turn the ignition switch **ON** and checks for leaks. Repair if necessary.

DRIVE TRAIN

Transaxle Assembly

REMOVAL & INSTALLATION

Front Wheel Drive

1. Remove the push pins from the coolant recovery bottle.
2. Position aside coolant recovery bottle.
3. Remove air cleaner assembly.
4. Remove the automatic transmission range selector cable from the manual shaft.
5. Remove the automatic transmission range selector cable bracket.
6. Disconnect the wiring harness connectors from the transaxle.
7. Remove the top four bell housing bolts.
8. Remove the frame.
9. Remove the inspection cover.
10. Use J 37096, or equivalent, in order to gain access to the torque converter bolts and prevent the flywheel from turning.
11. Remove the torque converter bolts.
12. Disconnect the vehicle speed sensor.
13. Remove the right and left axle shafts from the transmission.

14. Disconnect the transmission cooler lines.
15. Position the transmission jack under the vehicle.
16. Remove the transaxle brace.
17. Remove the lower transaxle bolts and stud.
18. Disconnect neutral safety switch.
19. Remove the wiring harness bracket from the side cover.
20. Remove the transaxle from the vehicle.
21. Flush the transaxle oil cooler pipes and the transaxle oil cooler.
22. Transfer all necessary parts as needed.

To install:

23. Align the transaxle filler tube to the transmission and install the transaxle into the vehicle.
24. Install the lower transaxle bolt and stud. Tighten the lower transaxle to engine and engine to transaxle bolts to 75 Nm (55 ft. lbs.).
25. Install the wiring harness bracket to the side cover.
26. Connect the neutral safety switch.
27. Install the transaxle brace.

28. Connect the transmission cooler lines.
29. Install the transaxle brace.
30. Install the right and left axle shafts into the transaxle.
31. Connect the vehicle speed sensor.
32. Install the torque converter bolts. Tighten the bolts to 63 Nm (46 ft. lbs.).
33. Install the torque converter cover.
34. Install the filler tube bracket retaining nut on the transmission.
35. Install the frame.
36. Install the front wheels.
37. Lower vehicle.
38. Remove the Engine Support Fixture.
39. Install the upper transaxle bolts and stud. Tighten the bolts and stud to 75 Nm (55 ft. lbs.).
40. Connect the wiring harness to the transaxle.
41. Install transmission range selector cable bracket.
42. Install the automatic transmission range selector cable on the manual shaft.
43. Install the air cleaner assembly.
44. Install the coolant recovery bottle.
45. Adjust the fluid level.
46. Inspect for fluid leaks.

All Wheel Drive

➡ **Transmission oil circulates between the transmission assembly and the transfer case. In situations where transmission related failures circulate debris into the transfer case, the transfer case must be disassembled, cleaned, and inspected for damage.**

➡ **The transmission-to-transfer case end play check procedure must be performed each time the transmission, transfer case, or the internal components (excluding gaskets and seals) are replaced.**

1. Remove the push pins from the coolant recovery bottle.
2. Position aside coolant recovery bottle.
3. Remove air cleaner assembly.
4. Remove the automatic transmission range selector cable from the manual shaft.
5. Remove the automatic transmission range selector cable bracket.
6. Disconnect the wiring harness connectors from the transaxle.
7. Remove the wiring harness bracket from the side cover.
8. Remove the top four bell housing bolts.
9. Remove the prop shaft.
10. Remove the engine to transfer case brace.
11. Remove the frame.
12. Remove the filler tube bracket retaining bolt.
13. Remove the inspection cover.
14. Remove the torque converter bolts.
15. Disconnect the vehicle speed sensor.
16. Remove the right and left axle shafts into the transaxle.
17. Disconnect the transmission cooler lines.
18. Position the transmission jack under the transmission.
19. Remove the transfer case to engine mount bolts.
20. Remove the lower transaxle bolts and stud.
21. Disconnect neutral safety switch.
22. Remove the transaxle from the vehicle.
23. Remove the transfer case from the transaxle.
24. The transmission-to-transfer case end play check procedure must be performed each time the transmission, transfer case, or the internal components (excluding gaskets and seals) are replaced.

To install:

25. The transmission-to-transfer case end play check procedure must be performed each time the transmission, transfer case, or the internal components (excluding gaskets and seals) are replaced.
26. Install the transfer case to the transaxle.
27. Position the flex plate alignment hole to the seven o'clock position.
28. Align the transaxle filler tube to the transmission and install the transaxle into the vehicle.
29. Install the lower transaxle bolt and stud. Tighten the lower transaxle to engine and engine to transaxle bolts to 75 Nm (55 ft. lbs.).
30. Install transfer case to transmission bracket.
31. Install the wiring harness bracket to the side cover.
32. Connect the neutral safety switch.
33. Install the transaxle brace.
34. Install the right and left axle shafts into the transaxle.
35. Connect the vehicle speed sensor.
36. Install the torque converter bolts. Tighten the bolts to 63 Nm (46 ft. lbs.).
37. Install the torque converter cover.
38. Install the filler tube bracket retaining bolt on the transmission.
39. Install the prop shaft.

➡ **Thoroughly clean and apply LOCTITE® DRI-LOC 201® (GM P/N 12345493, or equivalent) to the bolt threads prior to assembly.**

40. Install the propeller shaft-to-transfer case bolts. Ensure the special washer is in place on each pair of bolts. Tighten the bolts to 33 Nm (24 ft. lbs.).
41. Install the propeller shaft-to-torque tube bolts. Ensure the special washer is in place on each pair of bolts. Tighten the bolts to 33 Nm (24 ft. lbs.).
42. Install the frame.
43. Install the front wheels.
44. Remove engine support fixture.
45. Install the right side engine strut.
46. Install the upper transaxle bolts and stud. Tighten the bolts and stud to 75 Nm (55 ft. lbs.).
47. Connect the wiring harness the transaxle.
48. Install transmission range selector cable bracket.
49. Install the automatic transmission range selector cable on the manual shaft.
50. Install the air cleaner assembly.
51. Install the coolant recovery bottle.
52. Adjust the fluid level.
53. Inspect for fluid leaks.

Transfer Case

REMOVAL & INSTALLATION

1. Raise and support the vehicle.
2. Remove propeller shaft.

➡ **Transmission oil circulates between the transmission assembly and the transfer case differential carrier and three helical gear set. In situations where transmission related failures circulate fiber type debris, such as clutch friction disc material into the transfer case, the helical gear portion of the transfer case must be disassembled, cleaned, and inspected for damage. In situations where transmission or transfer case related failures circulate metal type debris, such as bearing material, repair or overhaul of both components may be required.**

3. Remove the drain plug and washer from the extension housing. Allow the gear oil to drain.
4. Remove the drain plug and washer from the case. Allow the transmission fluid to drain.
5. Disconnect the electrical connector from the speed sensor.
6. Remove the transfer case lower brace bolts.
7. Remove the transfer case lower brace.
8. Disconnect the clamp from the extension housing vent hose coupling.
9. Remove the bolt retaining the vent hose bracket to the transfer case.
10. Remove the vent hose coupling, vent hose, and bracket from the transfer case.
11. Remove the transaxle.
12. Remove the output shaft retaining ring.
13. Rotate the transaxle 90 degrees.
14. Remove the transfer case lower brace bolt.
15. Rotate the transaxle 90 degrees to the installed position.
16. Remove the transfer case side brace bolts. Remove the side brace.
17. Remove the transfer case-to-transaxle bolts.

❋❋ CAUTION

This component weighs approximately 60 lbs. Personal injury may result if you lift the component improperly.

18. Remove the transfer case assembly from the transaxle. Note that the transaxle

output shaft withdraws from the transaxle with the transfer case assembly.

19. Remove the output shaft.

20. Remove the transfer case O-ring seal from the transfer case.

To install:

➡ **The end play clearance between the transmission and transfer case internal components is adjusted by selection of the proper size selective washer (located behind the transmission sun gear). The transmission-to-transfer case washer selection procedure must be performed each time a transmission and/or transfer case assembly or the internal components (excluding gaskets and seals) are replaced.**

21. Perform the transmission-to-transfer case end play check.

22. Install the O-ring seal to the transfer case.

23. Rotate the transaxle so that the bottom pan is facing the floor.

24. Install the transfer case assembly to the transaxle.

25. Install the transfer case-to-transaxle bolts and tighten the bolts in the following sequence:

   a. Tighten the transfer case-to-transaxle bolts (1) and (2).
   First pass: 35 Nm (26 ft. lbs.).
   Final pass: an additional 160 degrees.
   b. Tighten the transfer case-to-transaxle bolt (3).
   First pass: 35 Nm (26 ft. lbs.).
   Final pass: an additional 70 degrees.
   c. Tighten the transfer case-to-transaxle bolts (4) and (5) to 40 Nm (30 ft. lbs.).

26. Install the transfer case side brace.

27. Install the transfer case side brace-to-transaxle bolts. Hand tighten bolts.

28. Tighten the bolts in the following sequence (1, 2, 3, 4, 5) to 47 Nm (35 ft. lbs.).

29. Rotate the transaxle 90 degrees.

30. Install the transfer case lower brace-to-transaxle bolt. Tighten transfer case lower brace bolt to 47 Nm (35 ft. lbs.).

31. Install the output shaft to the transmission.

32. Install a new output shaft retaining ring.

33. Install the transaxle.

34. Install the vent hose and coupling to the extension housing.

35. Connect the clamp to the vent hose coupling.

36. Install the vent hose bracket and bolt/stud to the transfer case. Tighten the bolt/stud to 10 Nm (106 inch lbs.).

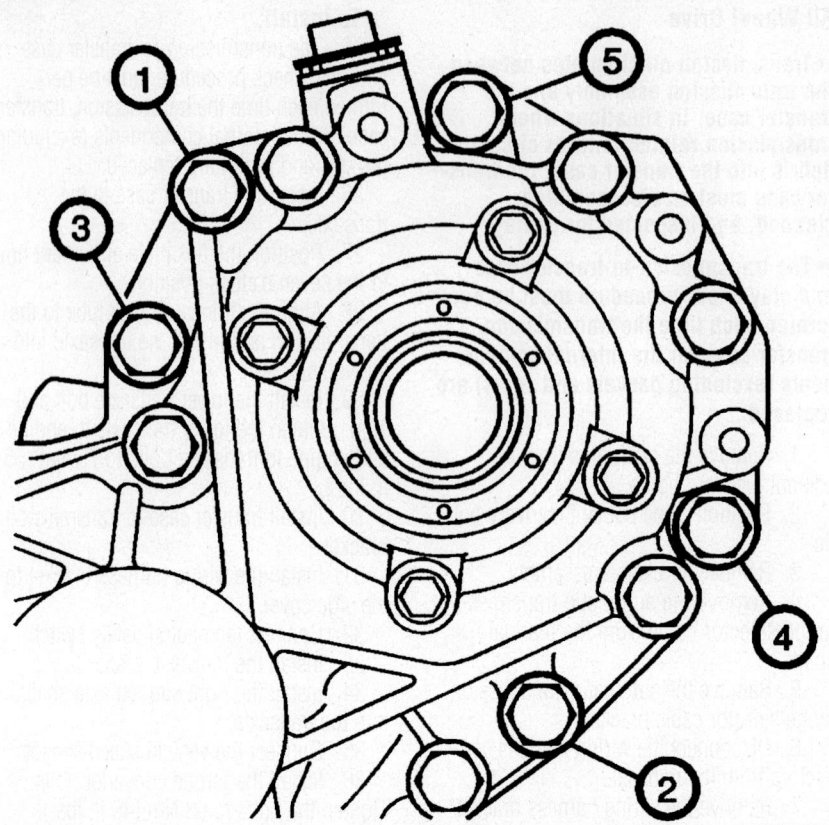

71461-VENT-G01

Install the transfer case-to-transaxle bolts and tighten the bolts in this sequence

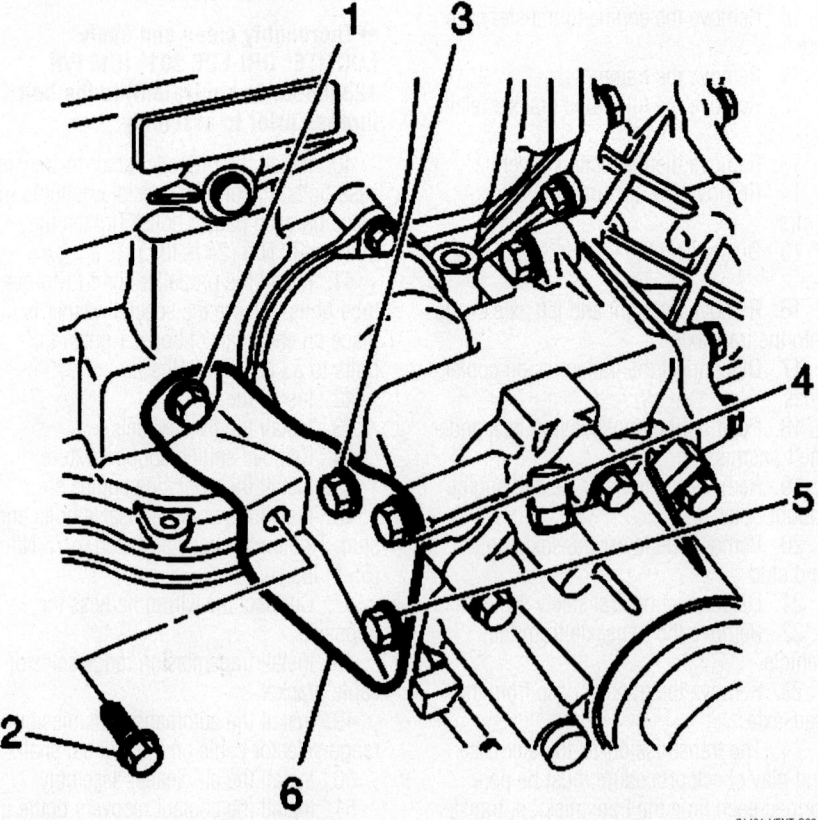

71461-VENT-G02

Transfer case side brace-to-transaxle bolts torque sequence

37. Connect the electrical connector to the speed sensor.

38. Install propeller shaft.

39. Install the transfer case lower brace.

40. Install the transfer case lower brace bolts. Tighten the brace bolts to 47 Nm (35 ft. lbs.).

41. Install the lubricant drain plug and gasket to the case. Tighten the drain plug to 32 Nm (24 ft. lbs.).

42. Inspect the transmission fluid level.

43. Install the lubricant drain plug and gasket to the extension housing. Tighten the drain plug to 32 Nm (24 ft. lbs.).

44. Remove the extension housing fill plug and gasket.

45. Fill the extension housing with synthetic gear oil GM P/N United States 12378514 GM P/N Canada 88901045.

46. Install the fill plug and gasket to the extension housing. Tighten the extension housing fill plug to 32 Nm (24 ft. lbs.).

47. Lower the vehicle.

Front Halfshaft

REMOVAL & INSTALLATION

1. Before servicing the vehicle, refer to the Precautions Section.

2. Remove or disconnect the following:
   • Front wheel
   • Stabilizer shaft link
   • Tie rod end from the steering knuckle
   • Lower ball joint from the steering knuckle by inserting a flat blade tool through the caliper hole and into the rotor to prevent the rotor from rotating
   • Halfshaft nut
   • Halfshaft from the wheel bearing/hub with a front hub spindle remover tool
   • Right side halfshaft from the transaxle with an axle shaft remover tool
   • Left side halfshaft by installing a flat blade tool between the lower control arm and the frame. Insert the tip of the blade into the groove of the tri-pot joint
   • Left side halfshaft

To install:

3. Install or connect the following:
   • Left side halfshaft and make certain that it is fully seated
   • Push the halfshaft into place and pull on the tri-pot joint and verify that the retaining ring in fully engaged

☀ WARNING

Do not pull on the halfshaft.

   • Right side halfshaft and make certain that it is fully seated
   • Push the halfshaft into place and pull on the tri-pot joint and verify that the retaining ring in fully engaged

☀ WARNING

Do not pull on the halfshaft.

   • Halfshaft into the wheel bearing/hub by inserting a flat blade tool through the caliper and into the rotor so that the rotor does not move
   • New halfshaft nut. Torque it to 118 ft. lbs. (160 Nm).
   • Ball joint to the steering knuckle
   • Tie rod end to the steering knuckle
   • Stabilizer shaft link. Torque the nut to 17 ft. lbs. (23 Nm).
   • Front wheel

4. Road test the vehicle and check for any abnormal noise.

5. Check and adjust the alignment as needed.

Rear Halfshaft

REMOVAL & INSTALLATION

1. Apply the parking brake.
2. Raise and support the vehicle.
3. Remove the tire and wheel assembly.

➡ **The wheel drive shaft nut must not be reused. Replace the wheel drive shaft nut with new nut whenever it is removed.**

4. Remove and discard the wheel drive shaft nut.

5. Release the parking brake.

6. Remove and support the brake caliper bracket.

7. Remove the nut securing the park brake cable routing bracket, if necessary.

8. Remove the bolt retaining the rear tie rod end from the rear suspension knuckle; do NOT loosen the tie rod end jam nut.

9. Loosen, but do not remove, the bolts securing the park brake cable bracket to the suspension knuckle.

10. Disconnect the wheel speed sensor electrical connector.

11. Install a puller such as J 42129 onto the wheel hub and secure with wheel nuts.

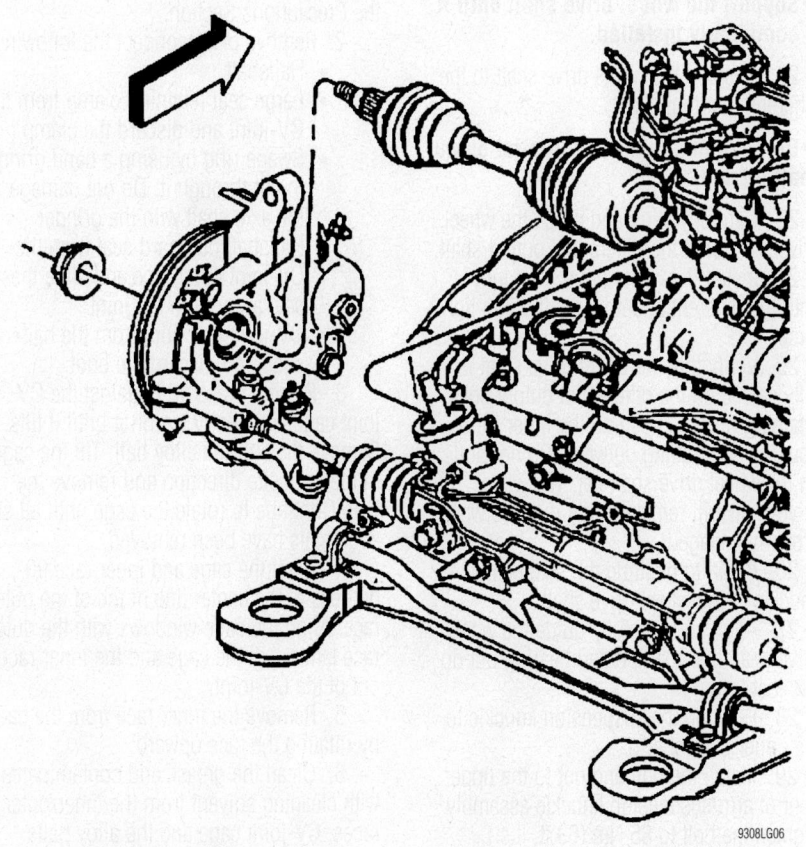

9308LG06

Install the left side halfshaft

12. Begin to disengage the wheel drive shaft from the wheel hub and bearing.

13. Remove the bolt and nut securing the upper control arm to the suspension knuckle.

14. Disengage the wheel drive shaft completely from the wheel hub and bearing.

15. Reposition the suspension knuckle toward the rear of the vehicle.

16. Remove the puller from the wheel hub.

➡**Support the wheel drive shaft until it is removed.**

17. Assemble an appropriate slidehammer.

18. Install the slidehammer evenly onto the rear beveled surface of the wheel drive shaft inner joint housing.

19. Disengage the wheel drive shaft from the rear axle differential, then remove the tool assembly.

20. Remove the wheel drive shaft from the vehicle.

➡**The differential output shaft oil seal must be replaced when removing the rear wheel drive shaft.**

21. Replace the rear wheel drive shaft oil seal.

To install:

➡**Support the wheel drive shaft until it is completely installed.**

22. Position the wheel drive shaft to the differential output shaft.

➡**Do not damage the differential output shaft oil seal.**

23. Carefully align and guide the wheel drive shaft onto the differential output shaft.

24. Install the wheel drive shaft fully onto the differential output shaft using light force.

25. Verify that the wheel drive shaft is fully seated on the differential output shaft retaining ring by grasping the inner tripot housing and pulling outward. Do not pull on the wheel drive shaft bar. The wheel drive shaft will remain firmly in place when properly engaged.

26. Begin to position the suspension knuckle to the wheel drive shaft.

27. Align and carefully guide the wheel drive shaft into the hub and bearing but do not seat fully.

28. Position the suspension knuckle to the upper control arm.

29. Install the bolt and nut to the upper control arm/suspension knuckle assembly. Tighten the bolt to 85 Nm (63 ft. lbs.).

30. Secure the park brake cable bracket.

31. Install the rear tie rod to the rear suspension knuckle.

32. Install the tie rod-to-knuckle bolt. Tighten the bolt to 85 Nm (63 ft. lbs.).

33. Install the brake caliper bracket.

34. Connect the wheel speed sensor electrical connector.

35. Install the park brake cable routing bracket, if removed.

36. Install the park brake cable routing bracket nut, if removed. Tighten the nut to 10 Nm (89 lb in).

37. Set the park brake.

38. Install a NEW wheel drive shaft spindle nut.

39. Slowly tighten the nut in order to draw the wheel drive shaft spindle into the wheel hub and bearing. Tighten the wheel drive shaft spindle nut to 260 Nm (192 ft. lbs.).

40. Install the tire and wheel assembly.

41. Lower the vehicle.

42. Release the park brake.

CV-Joints

OVERHAUL

1. Before servicing the vehicle, refer to the Precautions Section.

2. Remove or disconnect the following:
   • Halfshaft
   • Large seal retaining clamp from the CV-joint and discard the clamp
   • Swage ring by using a hand grinder to cut through it. Do not damage the axle shaft with the grinder
   • Halfshaft outboard seal from the CV-joint outer race and slide the seal away from the joint
   • CV-joint and boot from the halfshaft and discard the boot

3. Place a brass drift against the CV-joint cage and gently tap on it until it tilts. Remove the chrome alloy ball. Tilt the cage in the opposite direction and remove the ball. Continue to rotate the cage until all six alloy balls have been removed.

4. Pivot the cage and inner race 90 degrees to the center line of the of the outer race. Align the cage windows with the outer race lands. Lift the cage and the inner race out of the CV-joint.

5. Remove the inner race from the cage by rotating the race upward.

6. Clean the grease and contaminates with cleaning solvent from the inner/outer races; CV-joint cage and the alloy balls.

7. Remove any rust from the boot mounting area and clean the halfshaft bar.

To assemble:

8. Coat the inner and outer race grooves with grease and align the inner race with the windows of the cage.

9. Insert the inner race to the cage by rotating the race downward.

10. Insert the cage and inner race into the outer race.

11. Install the six alloy balls into the cage by tilting the cage. Repeat this process until all the balls are in place.

12. Pack the CV boot and joint with the grease supplied in the kit.

13. Install or connect the following:
    • New boot clamp onto the boot
    • CV boot on to the halfshaft bar and position the small end of the boot into the groove on the halfshaft bar. Secure the clamp to the boot with a Seal Clamp tool J 35910. Torque the clamp to 100 ft. lbs. (136 Nm).
    • Swage ring over the large diameter of the boot by pinching the ring into an oval shape

➡**Make certain that the retaining ring side of the inner race faces the halfshaft bar before installation.**

14. Slide the joint onto the halfshaft with the retaining snapring inside of the inner race. The race is properly seated when it snaps into place. Pull on the CV-joint to verify full engagement.

15. Install the large diameter of the boot with the large swage ring in place over the outside edge of the joint outer race.

16. Clamp the boot tightly to the outer race with the large swage ring by mounting Split Plate Swage Clamp tool J 36652 in a vise.

17. Position the outboard end of the halfshaft in the bottom of the tool.

18. Align the CV boot, joint and swage ring.

19. Install the top half of the tool and align the swage ring and clamp. Install the bolts to the top of the tool and tighten snugly. Tighten each bolt an additional 180 degrees. Alternate between the bolts until both sides of the top portion of the tool touch the bottom half.

20. Loosen the bolts and remove the split plate swage clamp tool.

21. Install the halfshaft.

22. Road test the vehicle and make certain there are no abnormal noises in the front end.

23. Check and adjust the alignment if necessary.

STEERING AND SUSPENSION

Air Bag

✳✳ CAUTION

All models are equipped with a Supplemental Inflatable Restraint (SIR) system. Before attempting any work on or near the steering column, ALWAYS disarm the air bag to prevent a costly and possibly dangerous accidental deployment.

PRECAUTIONS

Several precautions must be observed when handling the inflator module to avoid accidental deployment and possible personal injury.

• Never carry the inflator module by the wires or connector on the underside of the module

• When carrying a live inflator module, hold securely with both hands, and ensure that the bag and trim cover are pointed away from your body

• Place the inflator module on a bench or other surface with the bag and trim cover facing up

• With the inflator module on the bench, never place anything on or close to the module that may be thrown in the event of an accidental deployment

DISARMING

1. Turn the wheels to the straight-ahead position, then turn the ignition switch to **LOCK**.

2. Remove the instrument panel lower extension for access to the fuse block.

3. Remove the "AIR BAG" or "SIR" fuse from the block, as applicable.

4. Remove the steering column filler

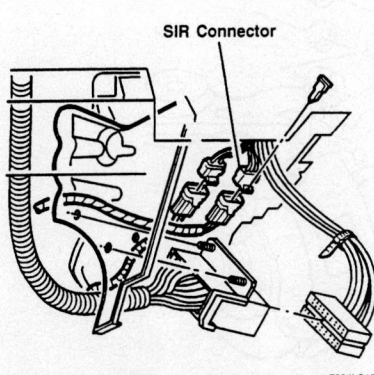

7924LG17

Driver's side air bag connector location

panel or left-hand sound insulator, as applicable, for access to the SIR wiring harness.

5. Remove the Connector Position Assurance (CPA) device, then disengage the yellow 2-way connector at the base of the steering column.

6. Remove the right hand insulator panel.

• Remove the CPA from the inflatable restraint instrument panel module connector located behind the RH insulator panel.

7. Disconnect the instrument panel module connector

8. If equipped with side air bags, remove the CPA from the inflatable restraint side impact module—left front connector located under the driver seat.

9. Disconnect the side impact module—left front connector.

10. Remove the CPA from the seat belt pretensioner —left front connector located under the driver seat.

11. Disconnect the seat belt pretensioner—left front connector.

12. Remove the CPA from the inflatable restraint side impact module—right front connector located under the front passenger seat.

13. Disconnect the side impact module—right front connector.

14. Remove the CPA from the seat belt pretensioner—right front connector located under the front passenger seat.

15. Disconnect the seat belt pretensioner—right front connector.

➡ **With the fuse removed, the AIR BAG or SIR light will illuminate if the ignition switch is turned ON at any time. This is normal and does not indicate a problem when the system is disarmed.**

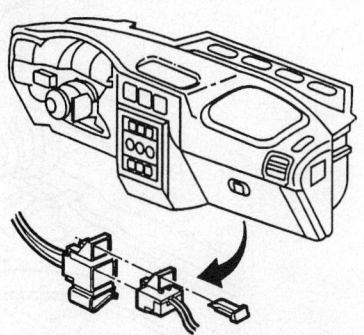

7924LG18

Passenger's side air bag connector location

To enable:

16. Be sure the ignition is in the **LOCK** position.

17. Connect the seat belt pretensioner—right front connector located under the front passenger seat.

18. Install the CPA to the seat belt pretensioner—right front connector.

19. Connect the inflatable restraint side impact module—right front connector located under the front passenger seat.

20. Install the CPA to the side impact module—right front connector.

21. Connect the seat belt pretensioner—left front connector located under the driver seat.

22. Install the CPA to the seat belt pretensioner—left front connector.

23. Connect the inflatable restraint side impact module—left front connector located under the driver seat.

24. Install the CPA to the side impact module—left front connector.

25. Engage the yellow SIR connector, then secure using the CPA device for both the drivers and passenger sides.

26. Install the steering column filler or sound insulator panel, as applicable.

27. Install the SIR system fuse to the fuse block.

28. Turn the ignition switch to the **ON** position and verify that the AIR BAG indicator light flashes 7 times, then extinguishes. If it does not go out, troubleshoot the SIR system fault.

29. Install the instrument panel lower extension.

Power Steering Rack and Pinion

REMOVAL & INSTALLATION

1. Before servicing the vehicle, refer to the Precautions Section.

2. Remove or disconnect the following:

• Negative battery cable
• Left front wheel
• Stabilizer shaft
• Tie rod ends from the steering knuckle
• Intermediate shaft from the steering gear
• Frame rear bolts and discard them. Properly support the frame using engine support stand J39580.
• Power steering gear heat shield
• Cooler pipe from the power steering gear
• Pressure hose from the power steering gear

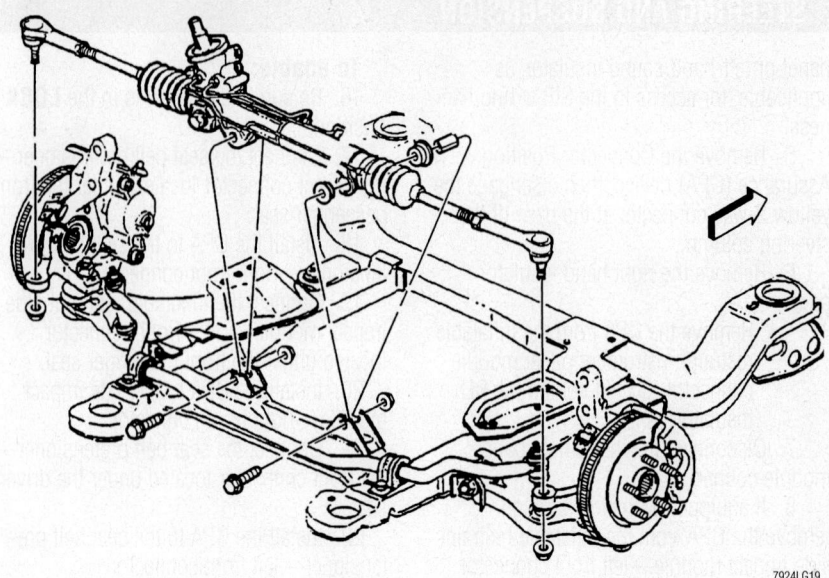

The rack and pinion steering gear is bolted to the rear of the subframe, as shown

- Power steering gear through the left wheel opening

To install:

3. Install or connect the following:
   - Power steering gear through the left wheel opening
   - New power steering gear bolts. Torque them to 59 ft. lbs. (80 Nm).
   - Pressure hose and cooler pipe to the power steering gear. Torque the fasteners to 20 ft. lbs. (27 Nm).
   - Heat shield. Torque the bolts to 54 inch lbs. (6 Nm).
   - Utility stand to support the frame
   - New rear frame bolts. Torque them to 122 ft. lbs. (165 Nm).
4. Remove the utility stand.
5. Install or connect the following:
   - Intermediate shaft to the steering gear. Torque the bolt to 35 ft. lbs. (47 Nm).
   - Tie rod ends to the steering knuckle
   - Stabilizer shaft. Torque the bolt to 17 ft. lbs. (23 Nm).
   - Left front wheel
6. Fill and bleed the power steering system and check for leaks.
7. Road test the vehicle and adjust the toe as necessary.

Strut

REMOVAL & INSTALLATION

❊❊ CAUTION

Do not remove the top center nut from the strut assembly. This nut

should only be removed when the strut assembly is out of the vehicle, mounted in a holding fixture and the coil spring is in a compressed position using the proper coil spring compressor.

1. Before servicing the vehicle, refer to the Precautions Section.
2. Remove or disconnect the following:
   - Front wheel
   - Three upper strut nuts
   - Lower strut bolts after marking the position of the strut to the knuckle

➡ **The strut to steering knuckle position must be marked so that the camber angle will not change. If the angle is change, the wheel alignment will also be affected.**

   - Strut
   - Nut from the top of the strut by placing the assembly in a Strut Compressor tool J 34013-B and Damper Rod Clamp J 34013-20. Turn the compressor forcing screw until the spring compresses slightly
   - Strut mount
   - Spring from the strut assembly

To install:

3. Install or connect the following:
   - Spring over the strut in the proper position
   - Strut mount
   - Compressor screw and start turning the screw clockwise until the strut

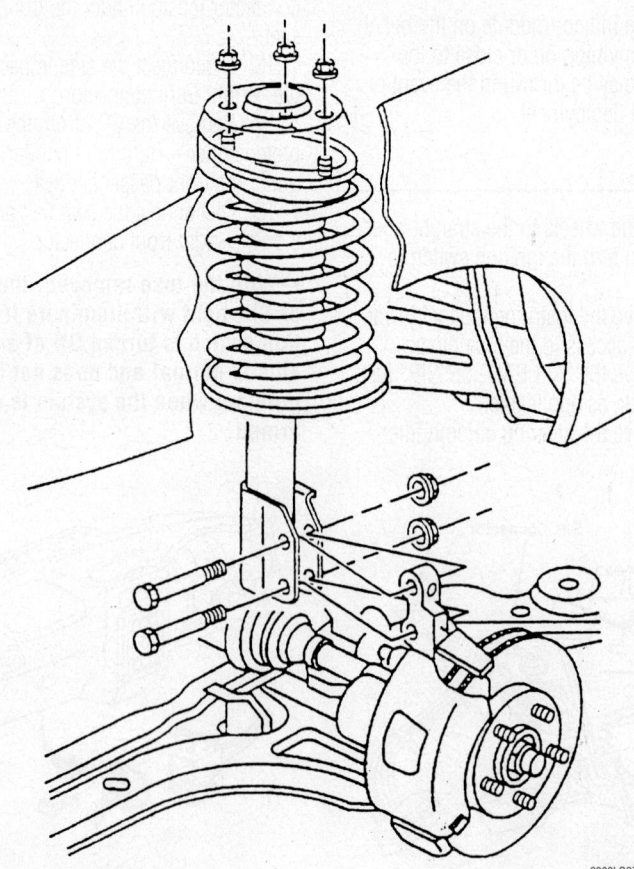

Strut assembly mounting

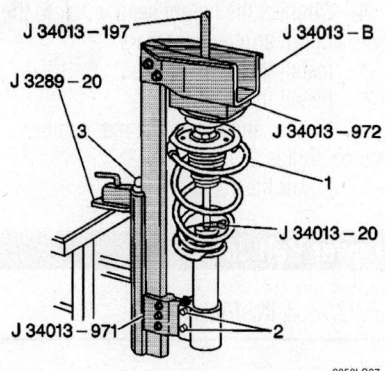

Disassembling the strut assembly

shaft threads are visible through the top of the strut. Torque the nut to 63 ft. lbs. (85 Nm).
- Strut and the upper nuts. Torque the nuts to 30 ft. lbs. (41 Nm).
- Lower strut bolts by aligning the strut to the steering knuckle. Torque the bolts 90 ft. lbs. (123 Nm).
- Front wheel

4. Road test the vehicle and check the front end alignment and adjust as needed.

Shock Absorber

REMOVAL & INSTALLATION

1. Before servicing the vehicle, refer to the Precautions Section.
2. Set the parking brake and chock the wheels.
3. Support the rear axle.
4. If the vehicle is equipped with automatic level control, disconnect the air tube connector from the shock absorber.
5. Remove or disconnect the following:
- Shock absorber at the lower bracket
- Shock absorber at the upper bracket
- Shock absorber from the brackets by compressing it slightly
- Remaining hardware from the shock absorber

To install:

6. Inspect the shock absorber, upper and lower mounting brackets and the frame mounting hole for cracks excessive wear and burrs.
7. Install or connect the following:
- Flat washer and insulator on the lower stem of the shock absorber
- Lower stem of the shock into the lower mounting bracket
- Flat washer and insulator on the upper stem of the shock absorber
- Upper stem of the shock in to the upper mounting bracket

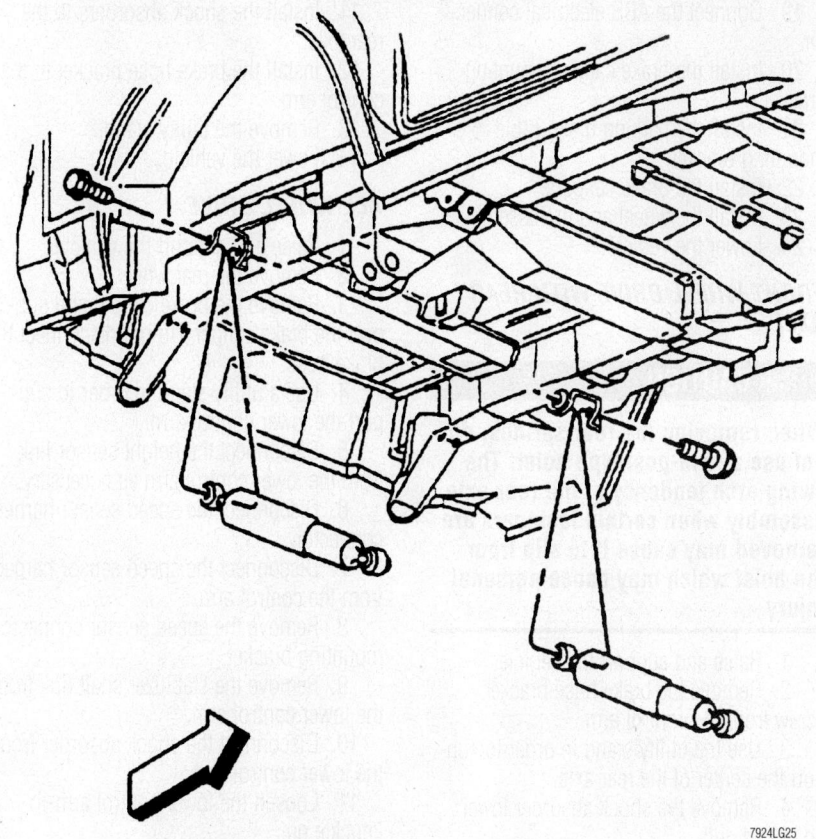

Rear shock absorber upper mounting

- Lower shock absorber insulator, flat washer and nut. Torque the nut to 62 ft. lbs. (84 Nm).
- Upper shock absorber insulator, flat washer and nut. Torque the nut to 62 ft. lbs. (84 Nm).

8. If the vehicle is equipped with automatic level control, install the air tube connector to the shock absorber.
9. Remove the support from the rear axle.
10. Road test the vehicle.

Coil Spring

REMOVAL & INSTALLATION

Front

The service procedure for the front coil springs is covered under MacPherson Strut removal and installation.

Rear

FRONT WHEEL DRIVE WITHOUT REAR AXLE

1. Raise and support the vehicle.
2. Remove the wheel and tire assemblies.

3. Remove the brake calipers.
4. Remove the parking brake cable mounting bracket.
5. Remove the brake caliper mounting bracket.
6. Disconnect the ABS electrical connector.
7. Remove the ABS electrical connector mounting bracket.
8. Use the utility jack in order to support the lower control arm.
9. Remove the stabilizer link from the lower control arm.
10. Disconnect the ball joint from the knuckle.
11. Use the utility jack in order to lower the lower control arm and relieve the coil spring tension.
12. Remove the coil spring.

To install:

13. Install the coil spring.
14. Use the utility jack in order to raise the lower control arm and load the suspension.
15. Connect the ball joint to the knuckle.
16. Install the stabilizer link to the lower control arm.
17. Use the utility jack in order to support the lower control arm.
18. Install the ABS electrical connector mounting bracket.

19. Connect the ABS electrical connector.

20. Install the brake caliper mounting bracket.

21. Install the parking brake cable mounting bracket.

22. Install the brake calipers.

23. Install the wheel and tire assemblies.

24. Lower the vehicle.

FRONT WHEEL DRIVE WITH REAR AXLE

> ❋❋ **CAUTION**
>
> **When removing the rear springs, do not use a twin-post type hoist. The swing arch tendency of the rear axle assembly when certain fasteners are removed may cause it to slip from the hoist which may cause personal injury.**

1. Raise and support the vehicle.

2. Remove the brake hose bracket screw from the control arm.

3. Use the utility stand in order to support the center of the rear axle.

4. Remove the shock absorber lower bolts and nuts.

5. Remove the rear axle tie rod from the rear axle.

➡ **Do not suspend the rear axle by the brake hoses. Damage to the brake hoses may result.**

6. Use the utility stand in order to lower the rear axle.

➡ **Care should be taken to avoid chipping or scratching the spring coating when handling the rear suspension coil spring. Damage to the coating can cause premature failure.**

7. Remove the springs and the insulators.

To install:

➡ **Care should be taken to avoid chipping or scratching the spring coating when handling the rear suspension coil spring. Damage to the coating can cause premature failure.**

➡ **Position the spring so that the paint stripe is facing rearward and centered to the shock absorber.**

8. Install the springs and the insulators to the rear axle.

9. Use the utility stand in order to raise the rear axle.

10. Install the rear axle tie rod to the rear axle.

11. Install the shock absorbers to the rear axle.

12. Install the brake hose bracket to the control arm.

13. Remove the utility stand.

14. Lower the vehicle.

ALL WHEEL DRIVE

1. Raise and support the vehicle.

2. Remove the rear wheel.

3. Remove the brake caliper and support the brake caliper. Do not disconnect the brake hose.

4. Use a utility stand in order to support the lower control arm.

5. Disconnect the height sensor link from the lower control arm as necessary.

6. Disconnect the speed sensor harness connector.

7. Disconnect the speed sensor harness from the control arm.

8. Remove the speed sensor connector mounting bracket.

9. Remove the stabilizer shaft link from the lower control arm.

10. Disconnect the shock absorber from the lower control arm.

11. Loosen the lower control arm-to-knuckle nut.

12. Disconnect the ball joint from the knuckle.

13. Remove the lower ball joint nut.

14. Use the utility stand in order to lower the lower control arm and relieve the coil spring tension.

15. Remove the utility stand.

16. Carefully remove the coil spring.

To install:

17. Install the coil spring.

18. Install the spring jounce bumper to the spring.

19. Index and install the spring insulator to the spring.

20. Index and install the spring assembly to the control arms.

21. Use the utility stand in order to raise the lower control arm.

22. Install the lower ball joint to the knuckle.

23. Install the lower ball joint nut. Tighten the lower ball joint nut to 60 Nm (44 ft. lbs.).

24. Install the stabilizer shaft link to the lower control arm.

25. Install the speed sensor connector mounting bracket.

26. Connect the speed sensor harness to the control arm.

27. Connect the speed sensor harnesses.

28. Connect the shock absorber to the lower control arm.

29. Connect the height sensor link to the lower control arm as necessary.

30. Install the brake caliper.

31. Install the rear wheel.

32. Inspect and adjust the rear camber and toe angles as necessary.

33. Lower the vehicle.

Lower Ball Joint

REMOVAL & INSTALLATION

1. Before servicing the vehicle, refer to the Precautions Section.

2. Remove the lower control arm.

3. Place the control arm in a vise.

4. Drill or grind off the ball stud rivet heads and use a punch to remove the rivets.

5. Remove the ball joint from the lower control arm

To install:

6. Install the ball joint using new fasteners facing down away from the joint and tighten to 50 ft. lbs. (63 Nm)

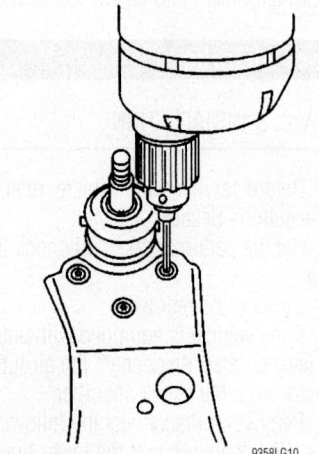

Drill off the rivet heads

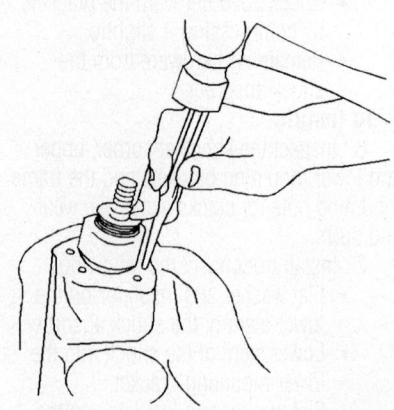

Use a punch and hammer to remove the rivets

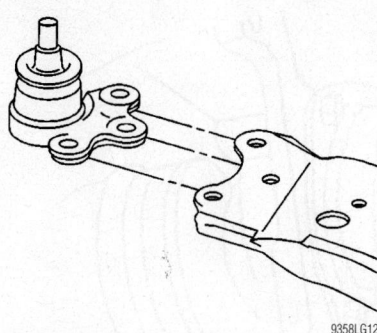

Remove the ball joint from the control arm

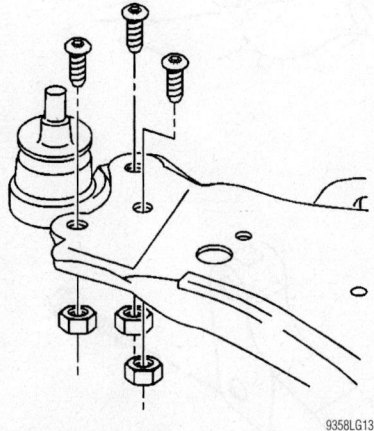

Install the ball joint fasteners facing down away from the joint

7. Install the lower control arm.
8. Road test the vehicle and check the front wheel alignment and adjust if necessary.

Lower Control Arm

REMOVAL & INSTALLATION

1. Before servicing the vehicle, refer to the Precautions Section.
2. Remove or disconnect the following:
   - Front wheel
   - Anti-lock Brake System (ABS) wheel speed sensor connector and jumper harness
   - Stabilizer shaft link
   - Cotter pin from the ball joint stud and loosen the nut
   - Ball Joint from the steering knuckle
   - Lower control arm

To install:
3. Install or connect the following:
   - Lower control arm
   - Ball joint stud to the knuckle

➡**Align the ball stud cotter pin hole parallel to the knuckle to ease the pin installation.**

- Ball joint stud castle nut. Torque it to 40 ft. lbs. (55 Nm).
- New cotter pin
- Stabilizer shaft link. Torque the nut to 17 ft. lbs. (23 Nm).
- ABS jumper harness to the retainer clips
- ABS sensor connector
- Lower control arm nuts. Torque them to 83 ft. lbs. (113 Nm).
- Front wheel

4. Road test the vehicle and check the front end alignment, adjust if necessary.

CONTROL ARM BUSHING REPLACEMENT

1. Before servicing the vehicle, refer to the Precautions Section.
2. Remove the lower control arm and secure it in a vise and mark the control arm along the flat edge of the bushing flange.
3. Assemble the bushing removal tool.
4. Tighten the assembly until the bushing is removed.

To install:
5. Install the bushing into the control arm by align the flat edge of the bushing to the mark in the control arm.
6. Make certain that the flat edge of the bushing is 30 degrees from the centerline of the control arm and the thin slot in the bushing is facing outboard.
7. Fully seat the bushing in the control arm.

8. Install the lower control arm.
9. Road test the vehicle and adjust the alignment, if necessary.

Wheel Bearings

ADJUSTMENT

Both front and rear wheel bearings are integral to the hub assembly and are not adjustable. If the bearings are found to be defective, the hub assembly must be replaced.

REMOVAL & INSTALLATION

Front

1. Before servicing the vehicle, refer to the Precautions Section.
2. Remove or disconnect the following:
   - Front wheel
   - Wheel speed sensor electrical connector and the connector from the bracket
   - Brake caliper and bracket
   - Brake rotor
   - Halfshaft nut
3. Attach a front hub spindle removal tool to the wheel bearing/hub.
4. Push the halfshaft out of the wheel bearing hub assembly.
5. Remove or disconnect the following:

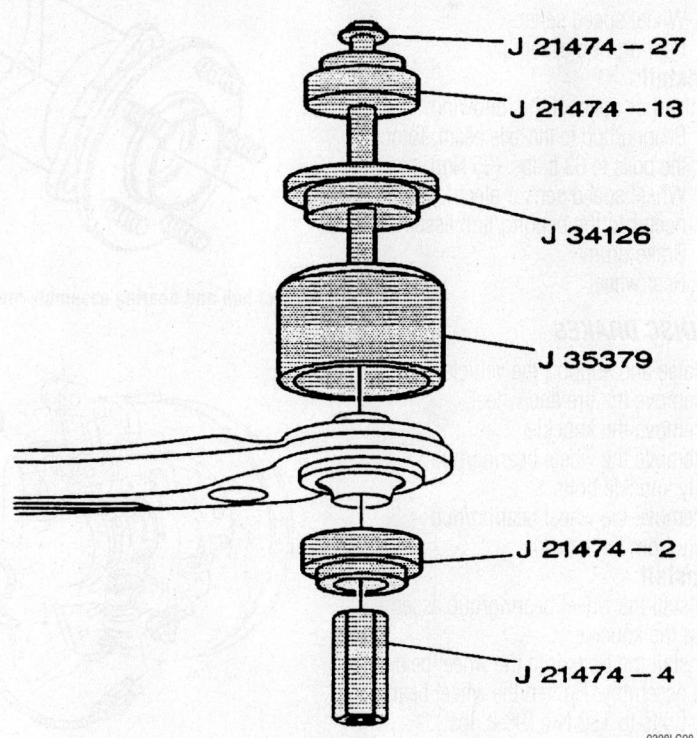

J 21474 – 27
J 21474 – 13
J 34126
J 35379
J 21474 – 2
J 21474 – 4

View of the lower control arm bushing

- Wheel bearing/hub bolts and discard them
- Wheel bearing/hub assembly

To install:

6. Install or connect the following:
- Wheel bearing/hub assembly

✷✷ CAUTION

The wheel bearing/hub bolts must be replaced whenever they are loosened or removed.

- New wheel bearing/hub bolts. Torque them to 96 ft. lbs. (130 Nm).
- Halfshaft nut
- Brake rotor
- Brake caliper
- Wheel speed sensor electrical connector to the bracket
- Wheel speed sensor electrical connector
- Front wheel

7. Road test the vehicle and check the front alignment, adjust if necessary.

Rear

WITH DRUM BRAKES

1. Before servicing the vehicle, refer to the Precautions Section.
2. Remove or disconnect the following:
- Rear wheel
- Brake drum
- Bearing/hub assembly from the axle beam
- Wheel speed sensor
- Bearing/hub assembly

To install:

3. Install or connect the following:
- Bearing/hub to the axle beam. Torque the bolts to 63 ft. lbs. (85 Nm).
- Wheel speed sensor electrical connector to the bearing/hub assembly
- Brake drum
- Rear wheel

WITH DISC BRAKES

1. Raise and support the vehicle.
2. Remove the tire and wheel.
3. Remove the knuckle.
4. Remove the wheel bearing/hub assembly knuckle bolts.
5. Remove the wheel bearing/hub assembly from the knuckle.

To install:

6. Install the wheel bearing/hub assembly on to the knuckle.
7. Install the bolts into the wheel bearing/hub assembly. Tighten the wheel bearing/hub bolts to 130 Nm (96 ft. lbs.).
8. Install the tire and wheel.
9. Lower the vehicle.

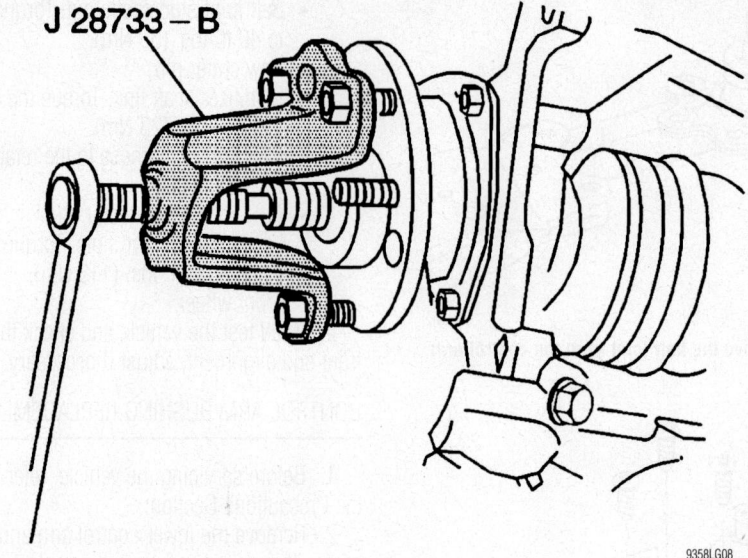

Use a puller to separate the hub from the halfshaft

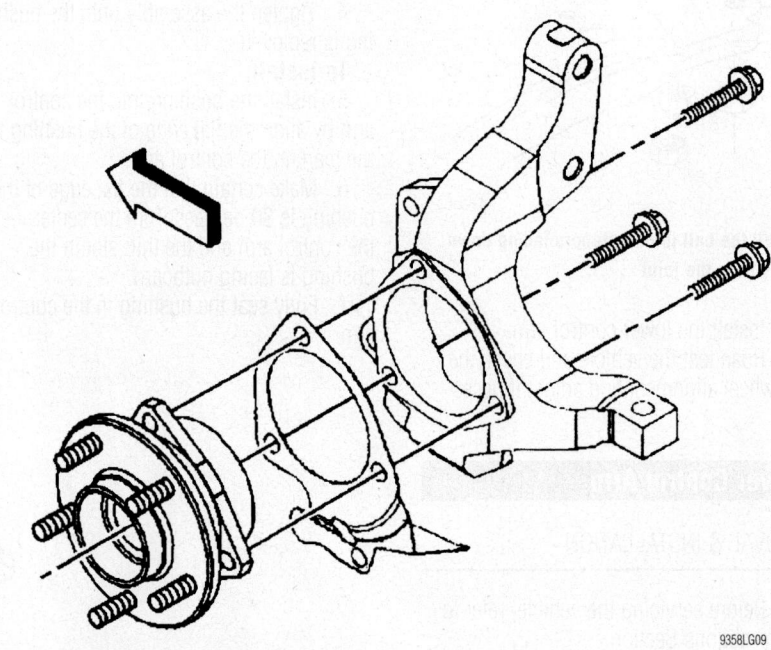

Front hub and bearing assembly mounting

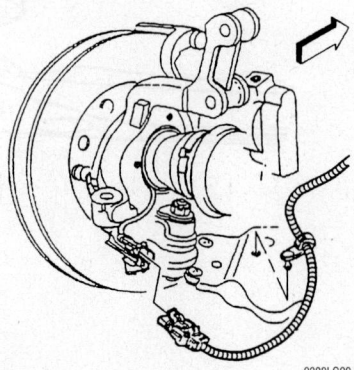

View of the front steering assembly

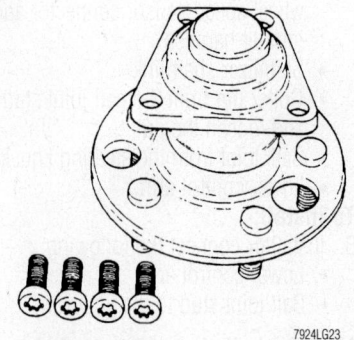

The rear wheel hub is mounted with 4 Torx® head bolts

BRAKES

Brake Caliper

REMOVAL & INSTALLATION

Front

1. Inspect the fluid level in the brake master cylinder reservoir.

2. If the brake fluid level is midway between the maximum-full point and the minimum allowable level, then no brake fluid needs to be removed from the reservoir before proceeding. If the brake fluid level is higher than midway between the maximum-full point and the minimum allowable level, then remove brake fluid to the midway point before proceeding.

3. Raise and support the vehicle.

4. Remove the tire and wheel.

5. Install two wheel lug nuts to retain the rotor to the hub.

6. Install a large C-clamp over the top of the brake caliper and against the back of the outboard brake pad.

7. Tighten the C-clamp until the caliper piston is pushed into the caliper bore enough to slide the caliper off the rotor.

8. Remove the brake hose bolt.

9. Remove the brake hose from the brake caliper.

10. Remove and discard the two copper brake hose gaskets. These gaskets may be stuck to the brake caliper and/or the brake hose end.

11. Plug the openings in the brake caliper and the brake hose in order to prevent brake fluid loss and contamination.

12. Clean off any dirt or corrosion on the brake caliper near the brake hose fitting.

13. Remove the brake caliper bolts.

14. Remove the brake caliper from the brake caliper bracket.

15. Inspect the brake caliper pin boots.

16. If the caliper pin boots are damaged, inspect the caliper pins for corrosion or damage. If corrosion is found on the brake caliper pin shaft, replace the brake caliper pin and the brake caliper pin boot. Do not attempt to polish away the corrosion.

To install:

17. Ensure that the caliper bolt boots are properly installed.

18. Install the caliper to the caliper bracket.

19. Install the brake caliper bolts. Tighten the bolts to 54 Nm (40 ft. lbs.).

20. Remove the plugs in the brake hose end.

➡**Install NEW copper brake hose gaskets.**

21. Assemble the brake hose bolt and the NEW copper brake hose gaskets to the brake hose.

22. Install the brake hose bolt to the brake caliper. Tighten the bolt to 54 Nm (40 ft. lbs.).

23. Remove the nuts securing the rotor to the hub.

24. Install the tire and wheel.

25. Lower the vehicle.

26. Fill the master cylinder to the proper level.

27. Bleed the brake system.

28. Apply approximately 778 Nm (175 ft. lbs.) of force to the brake pedal for ten seconds.

29. Inspect the brake caliper and hydraulic brake system for brake fluid leaks.

Rear

1. Inspect the fluid level in the brake master cylinder reservoir.

2. If the brake fluid level is midway between the maximum-full point and the minimum allowable level, then no brake fluid needs to be removed from the reservoir before proceeding. If the brake fluid level is higher than midway between the maximum-full point and the minimum allowable level, then remove brake fluid to the midway point before proceeding.

3. Raise and support the vehicle.

4. Remove the tire and wheel.

5. Install two nuts to retain the rotor to the hub.

6. Install a large C-clamp over the top of the brake caliper and against the back of the outboard brake pad.

7. Tighten the C-clamp until the caliper piston is pushed into the caliper bore enough to slide the caliper off the rotor.

8. Remove the brake hose bolt.

9. Remove the brake hose from the brake caliper.

10. Remove and discard the two copper brake hose gaskets. These gaskets may be stuck to the brake caliper and/or the brake hose end.

11. Plug the openings in the brake caliper and the brake hose in order to prevent brake fluid loss and contamination.

12. Clean off any dirt or corrosion on the brake caliper near the brake hose fitting.

13. Remove the brake caliper bolts.

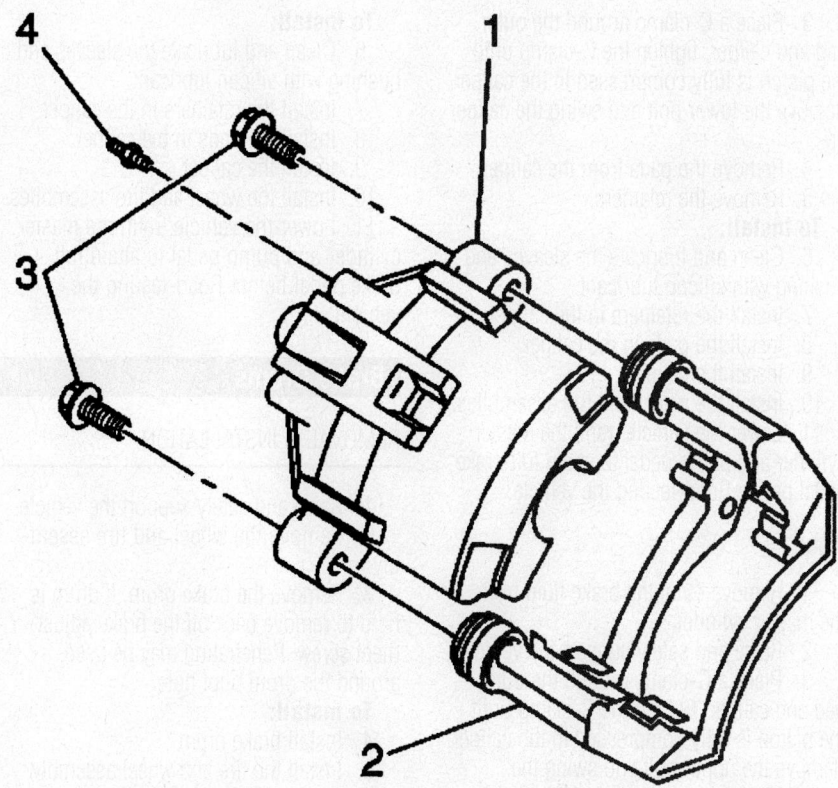

71461-VENT-G03

Front caliper (1), bracket (2), caliper bolts (3), bleeder screw (4)—rear is similar

14. Remove the brake caliper from the brake caliper bracket.

15. Inspect the brake caliper piston boot.

16. Inspect the brake caliper bolt boots.

17. Inspect the caliper bolts for corrosion or damage.

➡ **Do not attempt to polish away the corrosion.**

To install:

18. Ensure that the caliper bolt bushings and the caliper bolt boots are properly installed.

19. Install the brake caliper to the brake caliper bracket.

20. Lubricate the brake caliper bolt shaft. Do not lubricate the threads.

21. Lubricate the brake caliper bolt boots in the brake caliper bracket.

22. Install the brake caliper bolts. Tighten the bolts to 44 Nm (32 ft. lbs).

23. Use a flat-bladed tool in order to install the brake caliper bracket boot over the shoulder of the brake caliper bolt.

24. Remove the plugs in the brake hose end.

➡ **Install NEW copper brake hose gaskets.**

25. Assemble the brake hose bolt and the NEW copper brake hose gaskets to the brake hose.

26. Install the brake hose bolt to the brake caliper.

27. Remove the nuts securing the rotor to the hub.

28. Install the tire and wheel.

29. Lower the vehicle.

30. Fill the master cylinder to the proper level with clean brake fluid.

31. Bleed the brake system.

32. Pump the brake pedal (¾ of a full stroke) as many times as necessary to obtain a firm brake pedal.

33. Apply approximately 778 Nm (175 ft. lbs.) of force to the brake pedal for ten seconds.

34. Inspect the brake caliper and hydraulic brake system for brake fluid leaks.

Disc Brake Pads

REMOVAL & INSTALLATION

Front

1. Remove ⅔ of the brake fluid from the master cylinder.

2. Raise and safely support the vehicle.

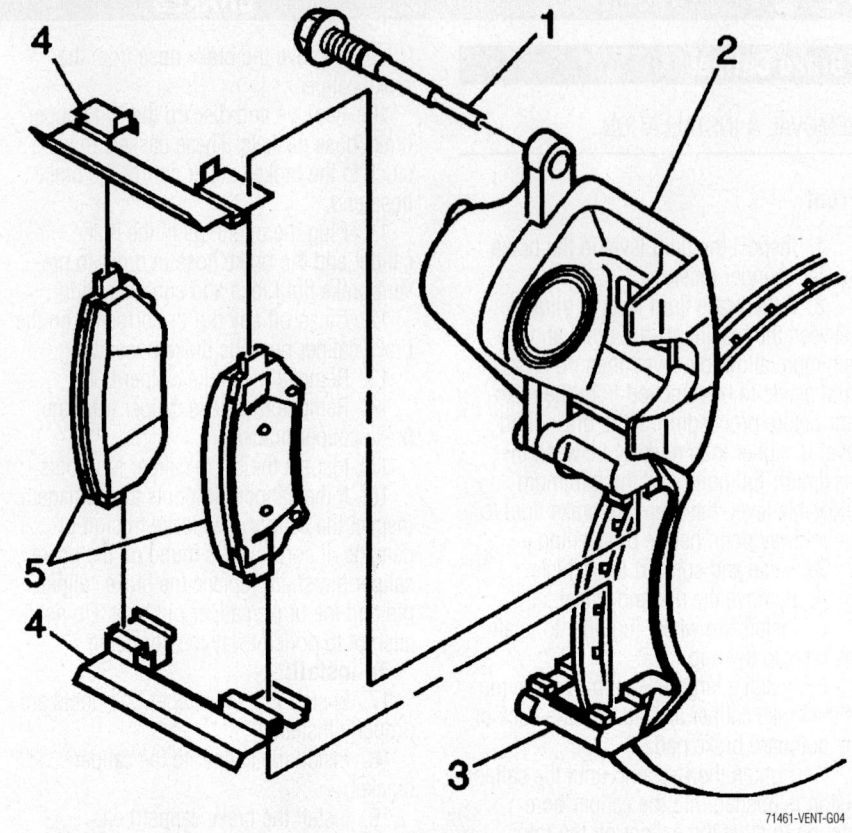

Brake pad replacement: lower caliper bolt (1), caliper (2), bracket (3), pad retainers (4) and pads (5)

3. Place a C-clamp around the outer pad and caliper; tighten the C-clamp until the piston is fully compressed in the caliper. Remove the lower bolt and swing the caliper up.

4. Remove the pads from the caliper.

5. Remove the retainers.

To install:

6. Clean and lubricate the sleeves and bushing with silicon lubricant.

7. Install the retainers in the caliper.

8. Install the pads in the caliper.

9. Install the caliper.

10. Install the wheel and tire assemblies.

11. Lower the vehicle, refill the master cylinder and pump pedal to attain full brake pedal before Road-testing the vehicle.

Rear

1. Remove ⅔ of the brake fluid from the master cylinder.

2. Raise and safely support the vehicle.

3. Place a C-clamp around the outer pad and caliper; tighten the C-clamp until the piston is fully compressed in the caliper. Remove the upper bolt and swing the caliper down.

4. Remove the pads from the caliper.

5. Remove the retainers.

To install:

6. Clean and lubricate the sleeves and bushing with silicon lubricant.

7. Install the retainers in the caliper.

8. Install the pads in the caliper.

9. Install the caliper.

10. Install the wheel and tire assemblies.

11. Lower the vehicle, refill the master cylinder and pump pedal to attain full brake pedal before Road-testing the vehicle.

Brake Drums

REMOVAL & INSTALLATION

1. Raise and safely support the vehicle.

2. Remove the wheel and tire assembly.

3. Remove the brake drum. If drum is hard to remove back off the brake adjustment screw. Penetrating may be used around the drum pilot hole.

To install:

4. Install brake drum

5. Install the tire and wheel assembly and adjust brake lining.

6. Lower the vehicle.

Brake Shoes

REMOVAL & INSTALLATION

1. Raise and safely support the vehicle.
2. Remove the wheel and tire assembly.
3. Remove the brake drum.
4. Remove the actuator spring from the brake shoes. Remove the retractor spring from the shoe web, being careful not to over stretch the spring.

5. Remove the adjuster shoe, adjuster actuator and adjusting screw assembly.
6. Do not remove the parking brake cable from the parking brake lever, unless the lever is being replaced. Remove the parking brake shoe.
7. Remove the retractor spring, as required.

To install:

8. Lubricate the contact points on the backing plate with lithium grease. Clean and lubricate the adjuster with lithium grease.

9. Install the retractor spring, if removed.
10. Install the parking brake shoe against the backing plate and snap the retractor spring into the slot on the brake shoe. Install the parking brake lever onto the parking brake shoe.
11. Install the adjuster shoe and adjusting screw assembly. Install the retractor spring into the slot on the adjuster shoe web.
12. Lubricate and install the adjuster

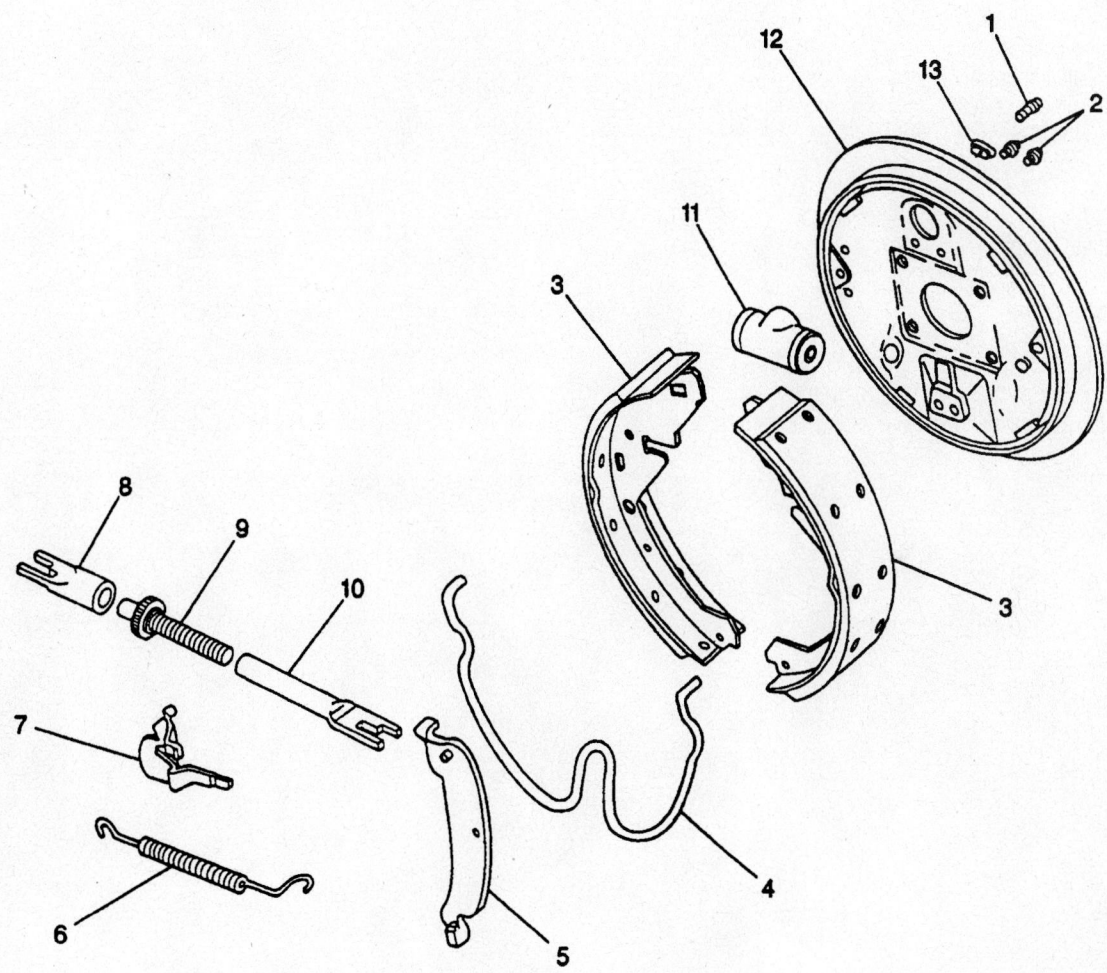

(1)	Bleeder Valve	(8)	Socket Adjuster
(2)	Wheel Cylinder Mounting Bolts	(9)	Adjuster Screw
(3)	Shoe and Lining	(10)	Pivot Nut
(4)	Return Spring	(11)	Wheel Cylinder
(5)	Parking Brake Lever	(12)	Backing Plate
(6)	Adjuster Spring	(13)	Access Hole Plug
(7)	Adjuster Actuator		

93026G96

Exploded view of the rear drum brake components

actuator onto the adjuster shoe. Install the actuator spring.

13. Ensure the parking brake system is adjusted properly with no tension on the cables or parking brake lever. The tops of the shoes should be firmly seated against the upper spring retaining anchor, if not as specified, loosen the parking brake cables.

14. Adjust the brakes using J-21177-A or equivalent. Turn the adjuster screw until the brake lining diameter is 0.050 in. (1.27mm) less than the inside diameter of the brake drum. Install the brake drum.

15. Install the wheel and tire assembly.

16. Lower the vehicle.

BRAKES 25-51
DRIVE TRAIN 25-25
ENGINE REPAIR 25-9
FUEL SYSTEM 25-22
**SPECIFICATIONS AND
  MAINTENANCE CHARTS**......25-2
Engine and Vehicle
  Identification 25-2
General Engine Specifications 25-2
Engine Tune-Up Specifications......25-2
Accessory Drive Belt Routing 25-3
Capacities 25-3
Valve Specifications.....................25-4
Crankshaft and Connecting
  Rod Specifications.....................25-4
Piston and Ring Specifications......25-4
Torque Specifications 25-5
Wheel Alignment 25-5
Tire, Wheel and Ball Joint
  Specifications 25-5
Brake Specifications 25-6
Scheduled Maintenance
  Intervals.............................25-7
**STEERING AND
  SUSPENSION**.................25-35

A

Air Bag..................................25-35
  Arming.................................25-40
  Diagnostic System Checks 25-41
  Disarming.............................25-36
  Sir Disabling And
    Enabling Zones....................25-35
  Sir Service Precautions...........25-35
Alternator.................................25-9
  Removal & Installation...............25-9
Automatic Transmission
  Assembly.............................25-25
  Removal & Installation..............25-25
Auxiliary Heater Core...............25-12
  Removal & Installation............25-12
Axle Shaft, Bearing and Seal........25-33
  Removal & Installation............25-33

B

Ball Joints..................................25-46
  Removal & Installation............25-46
Brake Caliper 25-51
  Removal & Installation............25-51

C

Camshafts................................25-18
  Removal & Installation............25-18
Coil Springs 25-44
  Removal & Installation............25-44
CV-Joints.................................25-30
  Overhaul 25-30
Cylinder Head 25-14
  Removal & Installation............25-14

D

Differential Carrier 25-35
  Removal & Installation............25-35
Disc Brake Pads........................25-53
  Removal & Installation............25-53

E

Engine Assembly 25-10
  Removal & Installation............25-10
Exhaust Manifold 25-18
  Removal & Installation............25-18

F

Fuel Pump/Sender 25-23
  Removal & Installation............25-23
Fuel Rail and Injectors 25-24
  Removal & Installation............25-24
Fuel System Pressure 25-22
  Relieving 25-22
Fuel System Service
  Precautions...........................25-22

H

Halfshaft..................................25-28
  Removal & Installation............25-28
Heater Core.............................25-12
  Removal & Installation............25-12

I

Ignition Timing 25-10
  Adjustment...........................25-10
Intake Manifold 25-15
  Removal & Installation............25-15

L

Lower Control Arm 25-47
  Removal & Installation............25-47

O

Oil Pan/Pump 25-19
  Removal & Installation............25-19

P

Pinion Seal 25-33
  Removal & Installation............25-33
Piston and Ring 25-22
  Positioning 25-22
Power Steering Gear 25-42
  Removal & Installation............25-42

R

Rear Main Seal 25-20
  Removal & Installation............25-20
Rocker Arms/Cover.....................25-15
  Removal & Installation............25-15

S

Shocks/Struts 25-42
  Removal & Installation............25-42
Stabilizer Bar 25-50
  Removal & Installation............25-50
Starter Motor 25-19
  Removal & Installation............25-19
Steering Knuckle.......................25-49
  Removal & Installation............25-49

T

Timing Chain, Sprockets, Front
  Cover and Seal 25-21
  Removal & Installation............25-21
Transfer Case Assembly...............25-26
  Removal & Installation............25-26

U

Upper Control Arm 25-46
  Removal & Installation............25-46

W

Water Pump 25-11
  Removal & Installation............25-11
Wheel Hub/Bearing Assembly 25-48
  Removal & Installation............25-48

SPECIFICATION AND MAINTENANCE CHARTS

ENGINE AND VEHICLE IDENTIFICATION

	Engine						Model Year	
Code ①	Liters (cc)	Cu. In.	Cyl.	Fuel Sys.	Engine Type	Eng. Mfg.	Code ②	Year
L	3.5 (3497)	214	6	SFI	DOHC	CPC	5	2005
							6	2006

CPC: Chevrolet/Pontiac/Canada

SFI: Sequential Fuel Injection

① 8th position of VIN

② 10th position of VIN

06025-MONT-C01

GENERAL ENGINE SPECIFICATIONS
All measurements are given in inches.

Year	Model	Engine Displacement Liters	Engine Series VIN	Net Horsepower @ rpm	Net Torque @ rpm (ft. lbs.)	Bore x Stroke (in.)	Compression Ratio	Oil Pressure @ rpm
2005	Montana SV6	3.5	L	200@5200	200@5200	3.70x3.31	9.8:1	30-45@1850
	Terraza	3.5	L	200@5200	200@5200	3.70x3.31	9.8:1	30-45@1850
	Relay	3.5	L	200@5200	200@5200	3.70x3.31	9.8:1	30-45@1850
	Uplander	3.5	L	200@5200	200@5200	3.70x3.31	9.8:1	30-45@1850
2006	Montana SV6	3.5	L	200@5200	200@5200	3.70x3.31	9.8:1	30-45@1850
	Terraza	3.5	L	200@5200	200@5200	3.70x3.31	9.8:1	30-45@1850
	Relay	3.5	L	200@5200	200@5200	3.70x3.31	9.8:1	30-45@1850
	Uplander	3.5	L	200@5200	200@5200	3.70x3.31	9.8:1	30-45@1850

SFI: Sequential Fuel Injection

06025-MONT-C02

GASOLINE ENGINE TUNE-UP SPECIFICATIONS

Year	Engine Displacement Liters	Engine VIN	Spark Plug Gap (in.)	Ignition Timing (deg) AT	Fuel Pump (psi)	Idle Speed (rpm) AT	Valve Clearance In.	Valve Clearance Ex.
2005	3.5	L	0.060	①	50-60	②	HYD	HYD
2006	3.5	L	0.060	①	50-60	②	HYD	HYD

NOTE: The Vehicle Emission Control Information label often reflects specification changes made during production.

The label figures must be used if they differ from those in this chart.

HYD: Hydraulic

① Ignition timing is preset and cannot be adjusted

② Idle speed is maintained by the PCM

06025-MONT-C03

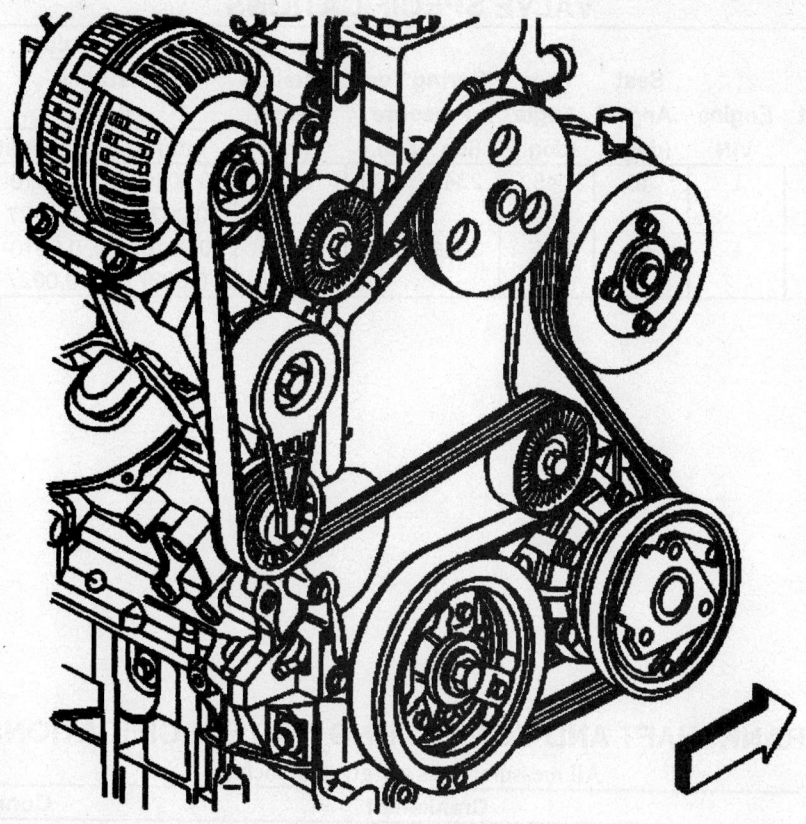

06025-MONT-G05

Accessory drive belt routing on the 3.5L engine

CAPACITIES

Year	Model	Engine Displacement Liters	Engine VIN	Engine Oil with Filter (qts.)	Transmission (pts.) * AWD	Transmission (pts.) * 2WD	Transfer Case (pts.)	Drive Axle Rear (pts.)	Fuel Tank (gal.)	Cooling System (qts.)
2005	Montana SV6	3.5	L	4	27.6	26.8	NA	4.2	20 ①	②
	Terraza	3.5	L	4	27.6	26.8	NA	4.2	20 ①	②
	Relay	3.5	L	4	27.6	26.8	NA	4.2	20 ①	②
	Uplander	3.5	L	4	27.6	26.8	NA	4.2	20 ①	②
2006	Montana SV6	3.5	L	4	27.6	26.8	NA	4.2	20 ①	②
	Terraza	3.5	L	4	27.6	26.8	NA	4.2	20 ①	②
	Relay	3.5	L	4	27.6	26.8	NA	4.2	20 ①	②
	Uplander	3.5	L	4	27.6	26.8	NA	4.2	20 ①	②

NOTE: All capacities are approximate. Add fluid gradually and check to be sure a proper fluid level is obtained.

NA: Not available

* Dry fill.

① Extended: 25 gal.

② Without rear A/C: 11.3 qts.
   With rear A/C: 12.8 qts.

VALVE SPECIFICATIONS

Year	Engine Displacement Liters	Engine VIN	Seat Angle (deg.)	Face Angle (deg.)	Spring Test Pressure (lbs. @ in.)	Spring Installed Height (in.)	Stem-to-Guide Clearance (in.)		Stem Diameter (in.)	
							Intake	Exhaust	Intake	Exhaust
2005	3.5	L	46	45	234@1.299	1.740	0.0010-0.0027	0.0010-0.0027	NA	NA
2006	3.5	L	46	45	234@1.299	1.740	0.0010-0.0027	0.0010-0.0027	NA	NA

NA: Not Available

06025-MONT-C05

CRANKSHAFT AND CONNECTING ROD SPECIFICATIONS

All measurements are given in inches.

Year	Engine Displacement Liters	Engine VIN	Crankshaft				Connecting Rod		
			Main Brg. Journal Dia.	Main Brg. Oil Clearance	Shaft End-play	Thrust on No.	Journal Diameter	Oil Clearance	Side Clearance
2005	3.5	L	2.6473-2.6483	0.0008-0.0025	0.0024-0.0083	3	2.2480-2.2490	0.0007-0.0170	0.008-0.009
2006	3.5	L	2.6473-2.6483	0.0008-0.0025	0.0024-0.0083	3	2.2480-2.2490	0.0007-0.0170	0.008-0.009

06025-MONT-C06

PISTON AND RING SPECIFICATIONS

All measurements are given in inches.

Year	Engine Disp. Liters	Engine VIN	Piston Clearance	Ring Gap			Ring Side Clearance		
				Top Compression	Bottom Compression	Oil Control	Top Compression	Bottom Compression	Oil Control
2005	3.5	L	0.0011-0.0110	0.007-0.015	0.019-0.029	0.010-0.029	0.001-0.030	0.002-0.003	0.004
2006	3.5	L	0.0011-0.0110	0.007-0.015	0.019-0.029	0.010-0.029	0.001-0.030	0.002-0.003	0.004

06025-MONT-C07

TORQUE SPECIFICATIONS
All readings in ft. lbs.

Year	Engine Displ. Liters	Engine VIN	Cylinder Head Bolts	Main Bearing Bolts	Rod Bearing Bolts	Crankshaft Damper Bolts	Flywheel Bolts	Manifold Intake	Manifold Exhaust	Spark Plugs	Oil Pan Drain Plug
2005	3.5	L	①	②	③	118	52	④	13	11	18
2006	3.5	L	①	②	③	118	52	④	13	11	18

① 1st pass: 44 ft. lbs.
  2nd pass: Plus 95 degrees

② 1st pass: 37 ft. lbs.
  2nd pass: plus 77 degrees

③ 1st pass: 18 ft. lbs.
  2nd pass: plus 110 degrees

④ 1st pass: 115 in. lbs.
  2nd pass: Center bolts 15 ft. lbs.
  2nd pass: Corner bolts 18 ft. lbs.

06025-MONT-C08

WHEEL ALIGNMENT

Year	All Models		Caster Range (+/-Deg.)	Caster Preferred Setting (Deg.)	Camber Range (+/-Deg.)	Camber Preferred Setting (Deg.)	Toe-in (Deg.)
2005	Twist Axle	Front	0.75	2.70	0.75	-0.65	0+/-0.20
		Rear	--	--	0.50	-1.00	0+/-0.30
	IRS	Front	0.75	2.70	0.75	-0.65	0+/-0.20
		Rear	--	--	0.50	-1.00	0+/-0.30
2006	Twist Axle	Front	0.75	2.70	0.75	-0.65	0+/-0.20
		Rear	--	--	0.50	-1.00	0+/-0.30
	IRS	Front	0.75	2.70	0.75	-0.65	0+/-0.20
		Rear	--	--	0.50	-1.00	0+/-0.30

(IRS) - Independent Rear Suspension

(--) No adjustment provided

06025-MONT-C09

TIRE, WHEEL AND BALL JOINT SPECIFICATIONS

Year	Model	OEM Tires Standard	OEM Tires Optional	Tire Pressures (psi) Front	Tire Pressures (psi) Rear	Wheel Size	Ball Joint Inspection	Wheel Lug Nut Torque (Ft. Lbs.)
2005	All	P225/60R17	None	①	①	NA	②	100
2006	All	P225/60R17	None	①	①	NA	②	100

NA: Not Available

OEM: Original Equipment Manufacturer

PSI: Pounds Per Square Inch

STD: Standard

OPT: Optional

① A tire and loading Information label is attached to the vehicle's center pillar (B-pillar), below the driver's door latch. This label shows your vehicle's original equipment tires and the correct inflation pressures for your tires when they are cold. The recommended cold tire inflation pressure, shown on the label, is the minimum amount of air pressure needed to support your vehicle's maximum load carrying capacity.

② Horizontal and vertical play, unloaded: 0.125 in. max.

06025-MONT-C10

BRAKE SPECIFICATIONS
All measurements in inches unless noted

Year	Model		Brake Disc Original Thickness	Brake Disc Minimum Thickness	Brake Disc Maximum Runout	Minimum Lining Thickness	Brake Caliper Bracket Bolts (ft. lbs.)	Brake Caliper Mounting Bolts (ft. lbs.)
2005	All	F	1.270	1.210	0.002	NA	137	40
		R	0.472	0.413	0.002	NA	96	25
2006	All	F	1.270	1.210	0.002	NA	137	40
		R	0.472	0.413	0.002	NA	96	25

06025-MONT-C11

MAINTENANCE I AND II SERVICE SCHEDULES
2005-06 Buick Terraza, Chevrolet Uplander, Pontiac Montana SV6, Saturn Relay

When the CHANGE ENGINE OIL light appears, certain services and inspections are required. Services are described below. Generally, it is recommended that the first service be Maintenance I, second service be Maintenance II, and that services are then alternated from Maintenance I and Maintenance II thereafter. In some cases, Maintenance II may be Required services are described as Maintenance I and Maintenance II.

The first service of a vehicle should be Maintance I, and the second service should be Maintenance II.

Alternate between the 2 services thereafter. However, in some cases, Maintenance II may be required more often.

Maintenance I: Use Maintenance I if the CHANGE ENGINE OIL light comes on within 10 months since the vehicle was purcahses or, if Maintenance II was performed.

Maintenance II: Use Maintenance II if the previous service performed was Maintenance I. Always used Maintenance II whenever the CHANGE ENGINE OIL light comes on 10 months or more since the last service, or, if the CHANGE ENGINE OIL light has not come on at all for one year.

Service	Maintenance I	Maintenance II
Change engine oil and filter. Reset oil life system.	✓	✓
Visually check for any leaks or damage. A fluid loss in the vehicle system could indicate a problem. Inspect, repair and add fluid to the system, if necessary.	✓	✓
Inspect engine air cleaner filter. If necessary, replace filter.	—	✓
Rotate tires and check inflation pressures and wear.	✓	✓
Visually inspect brake lines and hoses for proper hook-up, binding, leaks, cracks, chafing, etc. Inspect the disc brake pads for wear and the rotors for surface condition. Inspect the drum brake lings for wear or cracks. Inspect other brake parts, including drums, wheel cylinders, calipers, parking brake, etc. Inspect parking brake adjustment.	✓	✓
Check engine coolant and windshield washer fluid levels and add fluid as needed.	✓	✓
Inspect the suspension and steering components. Inspect the front and rear suspension systems and steering system for damaged, loose, or missing parts, or signs of wear. Inspect the power steering lines and the hoses for proper hook-up, binding, leaks, cracks, chafing, etc.	—	✓
Inspect the coolant hoses and replace the hoses if they are crackes, swollen or deteriorated. Inspect all pipes, fittings and clamps; replace with OEM parts as needed. To help ensure proper operation, a pressure test of the cooling system and pressure cap, and cleaning the outside of the radiator and A/C condesnser is recommended at least once a year.	—	✓
Inspect wiper blades for wear or cracking		✓
Inspect restraint system components.	—	✓
Lubricate all key lock cylinders, latch assemblies and hinges		✓
Inspect the transmission and transaxle fluid level and add fluid as needed.	—	✓
Replace passenger compartment air filter.		✓
Inspect throttle system	—	✓

To reset the CHANGE ENGINE OIL LIGHT:

1. Press the up or down arrow to scroll the DIC to show OIL LIFE.
2. Once the XXX% ENGINE OIL LIFE menu item is highlighted, press and hold the RESET button until the percentage shows 100%. If the percentage does not return to 100% or if the CHANGE ENGINE OIL SOON message comes back on when the vehicle is started, the engine oil life system was not properly reset. Repeat the procedure.

06025-MONT-C12

ADDITIONAL MAINTENANCE SERVICES
2005-06 Buick Terraza, Chevrolet Uplander, Pontiac Montana SV6, Saturn Relay

TO BE SERVICED	TYPE OF SERVICE	VEHICLE MILEAGE INTERVAL (x1000)					
		25	50	75	100	125	150
Air cleaner filter	R		✓		✓		✓
Accessory drive belt	I						✓
Auto. Trans. Fluid ①	R		✓		✓		✓
Cooling system hoses and clamps	S/I						✓
Transfer case fluid	R		✓		✓		✓
Throttle body	I	✓	✓	✓	✓	✓	✓
Engine coolant	R						✓
Fuel system	I	✓	✓	✓	✓	✓	✓
Exhaust system & heat shields	S/I	✓	✓	✓	✓	✓	✓
Spark plugs	R				✓		

R: Replace

S/I: Inspect and service, if necessary

① Replace if any of the following condition are met:

Heavy city traffic where the outside temperature regularly reaches 90°F (32°C) or higher.

Hilly or mountainous terrain

Frequent trailer towing

Taxi, police or delivery service

Otherwise, change every 100,000 miles

06025-MONT-C13

ENGINE REPAIR

Alternator

REMOVAL & INSTALLATION

1. Before servicing the vehicle, refer to the Precautions Section.

2. Disconnect the negative battery cable.

3. Disconnect the windshield wiper transmission link in front of the wiper motor and position out of the way as follows:

    a. Remove the driver side and passenger side wiper arms.

    b. Remove the passenger side wiper arm (3).

    c. Remove the air inlet grille panel by removing the retaining screws after the wiper arms are removed.

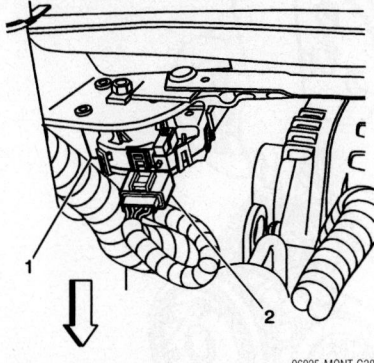

06025-MONT-G38

Detaching the electrical connector from the wiper/washer motor

d. Lower the washer container into the engine compartment.

e. Disconnect the electrical connector (2) from the wiper motor (1).

f. Remove the 4 bolts, in sequence as shown, from the wiper module (5).

Carefully guide the wiper modules out of the way to access the alternator.

4. Rotate the engine forward as follows:

    a. Remove the throttle body air inlet duct.

    b. Set the park brake.

    c. Shift the transaxle into Neutral.

    d. Remove the engine mount strut bolts. Swing the engine mount struts aside as follows:

- Remove the bolt and the nut from the engine mount strut at the left engine mount strut bracket on the engine.
- Remove the bolt and the nut from the engine mount strut at the engine mount strut bracket on the upper radiator support.
- Swing the engine mount strut out of the way.

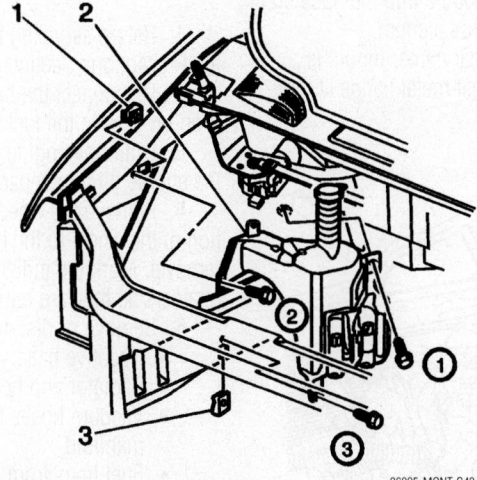

06025-MONT-G43

Disconnect washer container to lower it into the engine compartment

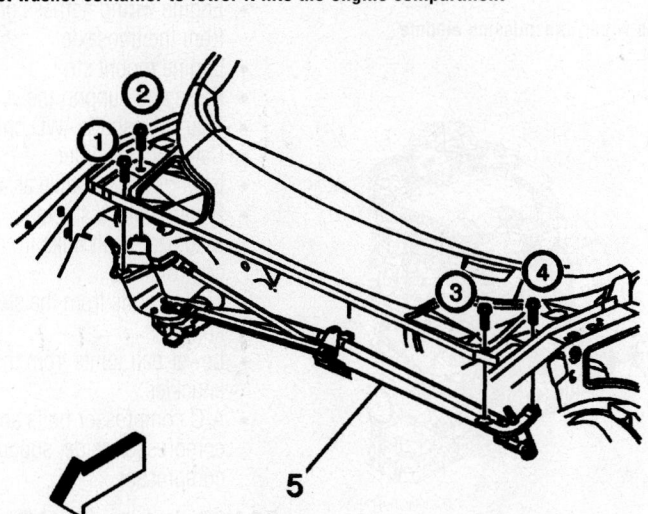

06025-MONT-G44

Removing the wiper module bolts

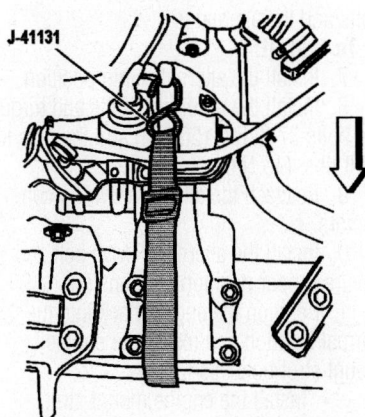

06025-MONT-G01

Identifying the J 41131 engine strap

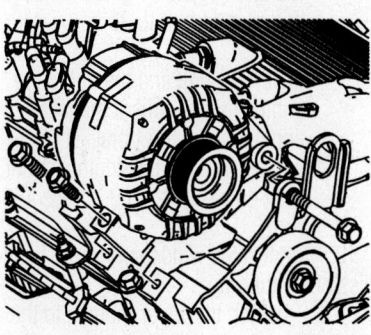

06025-MONT-G02

Indicating the alternator mounting bolts

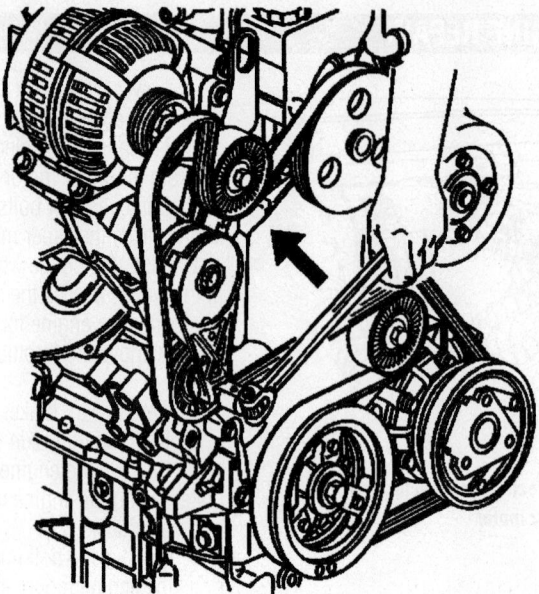

06025-MONT-G06

Removing the drive belt from the 3.5L engine

e. Install the engine strap (J 41131) and pull on the engine in order to rotate the engine forward.

f. Tighten the engine strap to hold in this position.

5. Remove the drive belt.

6. Remove the alternator mounting bolts and the alternator.

To install:

7. Install the alternator into position.

8. Install the mounting bolts and torque the bolts 37 ft. lbs. (50 Nm) and the nuts to 22 ft. lbs. (15 Nm).

9. Reattach the alternator wiring connectors.

10. Install the alternator drive belt and ensure proper position and tension.

11. Position the engine back into the normal location and install the engine mount strut bolts, as follows:

- Install the engine mount strut.
- Install the bolt and the nut to the engine mount strut at the engine mount strut bracket on the upper radiator support. Tighten the engine mount strut bolt to 36 ft. lbs. (48 Nm).
- Install the bolt and the nut to the engine mount strut at the left engine mount strut bracket on the engine. Tighten the engine mount strut bolt and nut to 36 ft. lbs. (48 Nm).

12. Connect the windshield wiper transmission as follows:

a. Position the wiper module to the lower plenum in the driver side fender flange opening first.

b. Rotate the module into the opening at the passenger side plenum.

c. Ensure the rear center mount is secure into the sheet metal flange at the rear of the module.

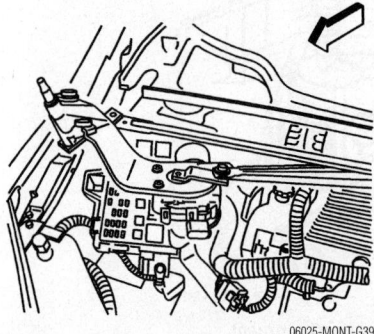

06025-MONT-G39

Moving the wiper transmission module aside

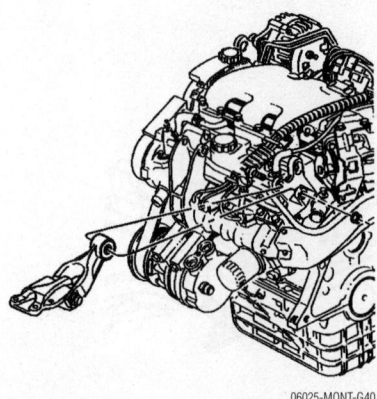

06025-MONT-G40

Showing the engine strut for removal to reposition the engine

d. Install the wiper module bolts. Tighten the bolts in sequence, as shown, to 10 Nm (89 inch lbs.).

e. Connect the electrical connector to the motor.

f. Install the air inlet grille panel.

g. Install the driver side wiper arm and the passenger side wiper arm.

h. Secure the washer container.

i. Close the hood. Inspect the wiper system for proper operation.

13. Connect the negative battery cable.

Ignition Timing

ADJUSTMENT

The ignition timing is preset and cannot be adjusted.

Engine Assembly

REMOVAL & INSTALLATION

1. Before servicing the vehicle, refer to the Precautions Section.

2. Disconnect the battery cables and properly relieve the fuel system pressure.

3. Drain the engine cooling system and the engine oil into separate drain pans.

4. Remove the wheel, marking the location of the wheel to the hub prior to removal. Mark the individual location of all retainers as they are removed.

5. Remove or disconnect the following:

- Negative battery cable
- Radiator and heater hoses
- Vacuum hoses from upper intake manifold
- Fuel lines from the fuel rail
- Electrical connectors
- Engine wiring harness grounds from the transaxle
- Engine mount strut
- Raise and support the vehicle
- Rear driveshaft (AWD only)
- Catalytic converter
- Lower radiator baffle assembly
- Engine splash shields
- Stabilizer shaft links from the lower control arms
- Tie rod ends from the steering knuckles
- Lower ball joints from the steering knuckles
- A/C compressor bolts and position compressor aside, support the compressor

➡ **DO NOT discharge the A/C system**

- Drive axles from the transaxle, secure the drive axles
- Intermediate shaft pinch bolt from the steering gear

✴✴ CAUTION

Failure to disconnect the intermediate shaft from the rack and pinion steering gear stub shaft can result in damage to the steering gear and/or intermediate shaft.

6. Lower the vehicle until the frame contacts the transaxle table J39580.

7. Remove the frame bolts.

8. Raise the vehicle to separate the powertrain/frame assembly from the vehicle.

9. Remove or disconnect the following:
- Starter motor
- Torque converter covers
- Torque converter bolts
- Engine mount lower nuts
- Transaxle brace
- Exhaust crossover pipe

10. Install engine hoist to engine.

11. Remove the transaxle to engine bolts (3, 4, 5, 6) and the studs (1, 2)

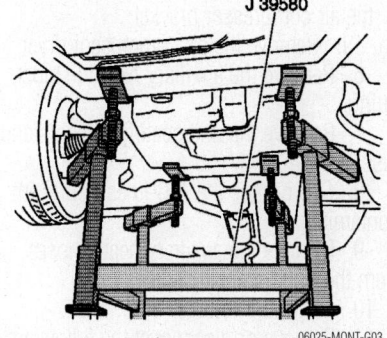

J 39580

Showing the transaxle table J39580

12. Separate and remove the engine from the transaxle/frame

13. Remove the flywheel.

14. Remove the drive belt.

To install:

15. Install or connect the following:
- Drive belt
- Flywheel
- Engine to the transaxle/frame
- Transaxle to engine bolts and tighten the bolts to 55 ft. lbs. (75 Nm)

16. Remove the engine hoist from engine.

17. Install or connect the following:
- Exhaust crossover pipe
- Transaxle brace
- Engine mount lower nuts and tighten to 32 ft. lbs. (43 Nm)
- Torque converter bolts
- Torque converter covers
- Starter motor

18. Position the transaxle table with the powertrain/frame under the vehicle.

19. Lower the vehicle until the frame contacts the transaxle table.

20. Install NEW frame bolts.

21. Raise and support the vehicle.

22. Remove the transaxle table.

23. Install or connect the following:
- Intermediate shaft pinch bolt to the steering gear
- Drive axles to transaxle
- A/C compressor to engine and install bolts
- Lower ball joints to steering knuckles
- Stabilizer shaft links to lower control arms
- Engine splash shields
- Lower radiator baffle assembly
- Catalytic converter
- Rear driveshaft (AWD only)
- Lower vehicle

- Fill engine with oil
- Engine mount strut
- Electrical connectors
- Engine wiring harness ground nut to transaxle stud and tighten to 18 ft. lbs. (2 Nm)
- Fuels lines to fuel rail
- Vacuum hoses to upper intake manifold
- Radiator and heater hoses

24. Install the tire and wheel assembly. Tighten the lug nuts to 100 ft. lbs. (140 Nm) in a criss—cross pattern, after aligning the wheel hub with the reference mark and holes as shown in appropriate illustration.

25. Check all powertrain fluid levels and add, as necessary. Be sure to properly fill the engine crankcase with clean engine oil.

26. Connect the battery cables and properly fill the engine cooling system.

27. Start and run the engine, then check for leaks.

Water Pump

REMOVAL & INSTALLATION

1. Before servicing the vehicle, refer to the Precautions Section.

2. Disconnect the negative battery cable.

3. Drain the engine cooling system.

4. Relieve the belt tension and remove the accessory drive belt.

5. Remove or disconnect the following:
- Water pump pulley
- Water pump bolts
- Water pump and gasket

To install:

6. Clean the gasket mating surfaces.

7. Install or connect the following:
- Water pump using a new gasket. Tighten the water pump bolts to 89 inch lbs. (10 Nm)

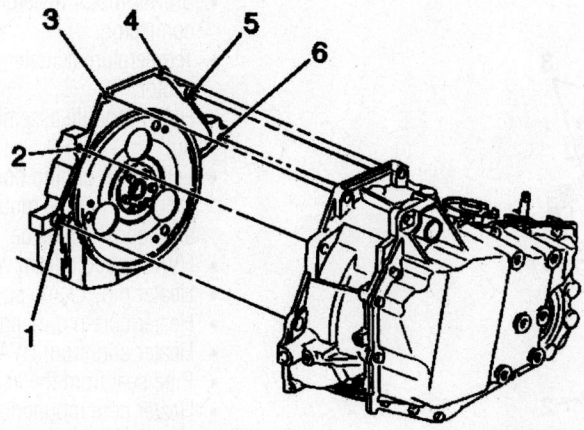

06025-MONT-G04

Transaxle to engine mounting bolts

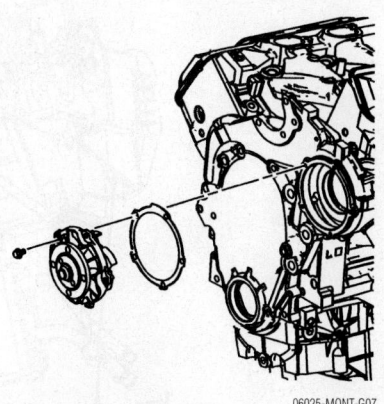

06025-MONT-G07

Showing the water pump and gasket on the 3.5L engine

- Water pump pulley and tighten the bolts to 18 ft. lbs. (25 Nm).
- Drive belt
- Negative battery cable

8. Refill the engine cooling system.
9. Run the engine and check for leaks.

Heater Core

REMOVAL & INSTALLATION

1. Before servicing the vehicle, refer to the Precautions Section.
2. Drain the engine cooling system.
3. Remove or disconnect the following:
   - Heater hoses from the heater core

➡ **Cap off the heater core inlet and outlet pipes to prevent coolant spilling inside the vehicle**

- Floor air outlet duct

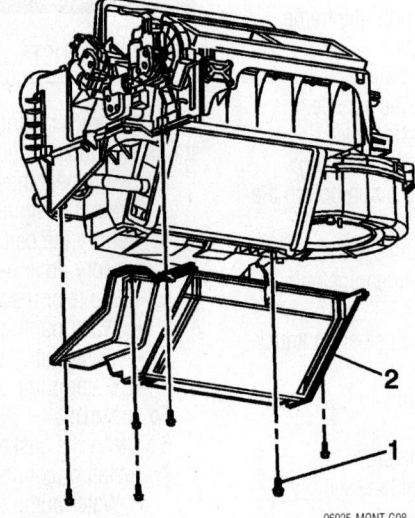

06025-MONT-G08

Showing the heater core cover screws

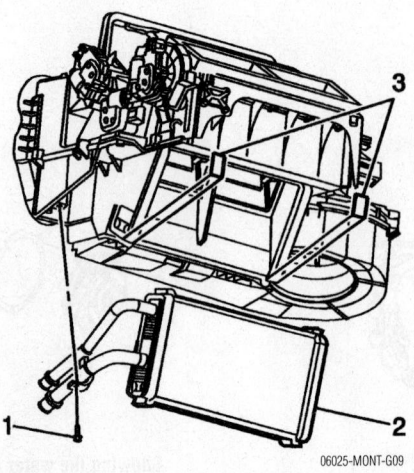

06025-MONT-G09

Identifying the heater core

- Heater core cover screws and cover
- Heater core pipe clamp screw
- Heater core

To install:

4. Install or connect the following:
   - Heater core
   - Heater core pipe clamp screw
   - Heater core cover and screws
   - Floor air outlet duct
   - Heater hoses and clamps
5. Refill the cooling system.
6. Run the engine to normal operating temperatures; then, check the climate control operation and check for leaks.

Auxiliary Heater Core

REMOVAL & INSTALLATION

1. Before servicing the vehicle, refer to the Precautions Section.
2. Drain the engine cooling system.
3. Properly evacuate and recover the A/C system refrigerant.
4. Remove or disconnect the following:
   - Left rear quarter trim panel
   - Raise and support vehicle
5. Remove the level ride air compressor as follows:
   a. Remove the electronic level control ELC/TRAILER fuse.
   b. Raise and support the vehicle.
   c. Remove the A/C hoses for access.
   d. Remove the air compressor
   e. Squeeze the clips on the side of the air compressor filter in order to remove the filter from the underbody rail.
   f. Remove the compressor bracket bolt.
   g. Remove the compressor bracket nuts. Allow the compressor to hang from the bracket hook.
   h. Remove the air line from the air dryer.
   i. Disconnect the air compressor electrical connector.
   j. Rotate the air compressor and the bracket 90 degrees in order to disengage the compressor bracket hook from the vehicle.
   k. Remove the air compressor with the air compressor bracket.
   l. Remove the air compressor dryer.
6. Remove the auxiliary A/C line block fitting.
7. Remove the unit sealing washers and discard them.
8. Cap or tape the A/C lines to prevent contamination.
9. Remove the auxiliary heater hoses from the heater core pipes.
10. Lower the vehicle.
11. Remove or disconnect the following:
    - A/C outlet pillar duct
    - A/C outlet duct
    - Blower motor electrical connector
    - Blower motor resistor electrical connector
    - Temperature actuator electrical connector
    - HVAC module assembly bolts and nut
    - Left power sliding door (PSD) module from the mounting bracket and reposition aside, if equipped
    - HVAC module from vehicle
    - Heater pipe clamp screw
    - Heater core mounting screws
    - Heater core from HVAC module
    - Pipe seal from the heater core
    - Heater core mounting bracket

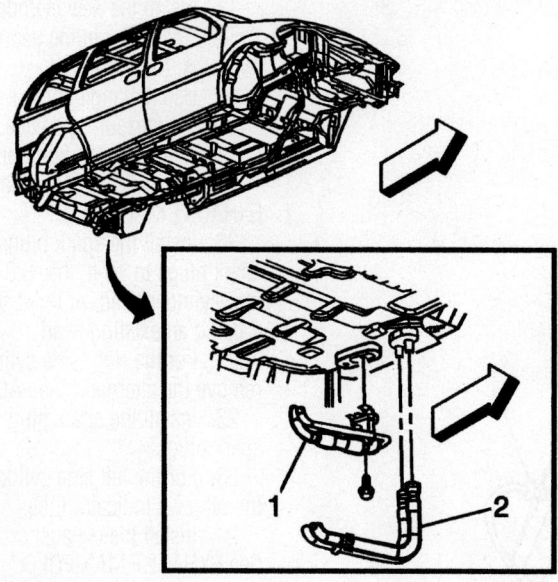

06205-MONT-G10

Location of the auxiliary heater hoses

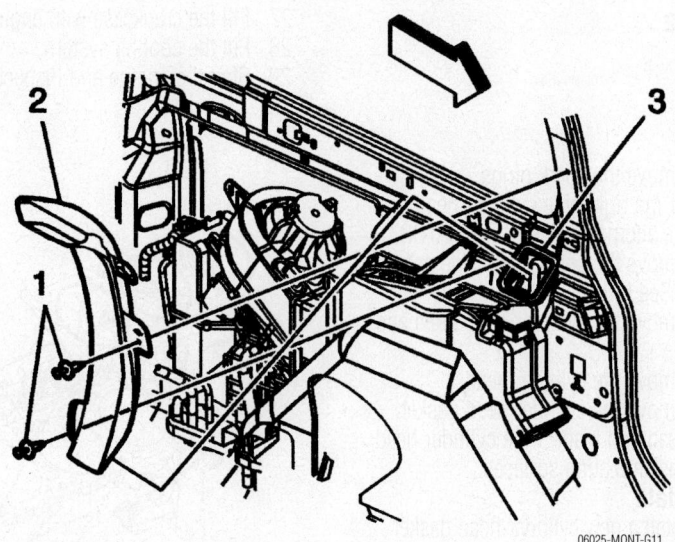

06025-MONT-G11

Indicating the pillar duct and bolts

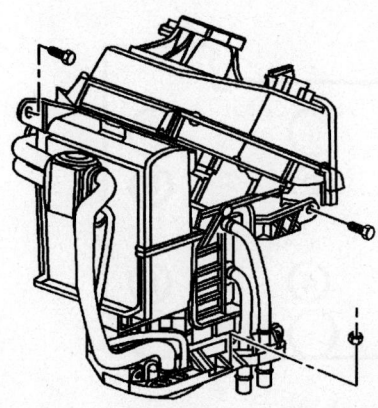

06025-MONT-G12

Showing the auxiliary HVAC module

- PSD module
- Temperature actuator electrical connector
- Blower motor resistor electrical connector
- Blower motor electrical connector
- A/C outlet duct; tighten bolts to 89 inch lbs. (10 Nm)
- A/C outlet pillar duct; tighten bolts to 89 inch lbs. (10 Nm)
- Raise and support vehicle
- Heater hoses to heater core pipes
- New sealing washers onto the A/C line
- A/C line block fitting bolt; tighten bolts to 15 ft. lbs. (20 Nm)

13. Install the level ride air compressor as follows:

   a. If removed, install the air compressor filter hose to the air compressor at 2 places.

   b. Install the air compressor to the air compressor bracket. Tighten the bracket to 37 inch lbs. (50 Nm).

   c. Install the air compressor bolts/screws. Tighten the air compressor bolts/screws to 35 inch lbs. (4 Nm).

   d. Install the air compressor shield. Tighten the bolts to 27 inch lbs. (3 Nm).

   e. Install the air compressor relay to the air compressor bracket.

   f. Install the air compressor dryer.

   g. Install the air compressor bracket to the underbody rail and rotate the air compressor bracket 90 degrees in order to allow the air compressor to hang from the air compressor bracket hook.

   h. Connect the air compressor electrical connector.

   i. Install the air line to the air dryer.

   j. Install the air compressor bracket nuts. Tighten the air compressor bracket nuts to 89 inch lbs. (10 Nm).

   k. Install the air compressor bracket bolt. Tighten the air compressor bracket bolt/screw to 37 ft. lbs. (50 Nm).

   l. Install the air compressor filter to the underbody rail.

   m. Install the A/C hoses if previously removed.

   n. Lower the vehicle.

   o. Install the ELC/TRAILER fuse.

14. Install the left rear quarter trim panel.
15. Refill the cooling system.
16. Evacuate and recharge the A/C system.
17. Run the engine to normal operating temperatures, and then check the climate control operation. Check the system for leaks.

To install:

12. Install or connect the following:
- New seals onto the heater core
- Heater core mounting bracket
- New pipe seal onto the heater core
- Heater core into the HVAC module
- Heater core mounting screws; tighten screws to 14 inch lbs. (1.6 Nm)
- Heater pipe clamp screw; tighten screw to 14 inch lbs. (1.6 Nm)
- HVAC module body seals
- HVAC module assembly to the vehicle
- HVAC module assembly bolts and the nut, tighten bolts/nut to 89 inch lbs. (10 Nm)

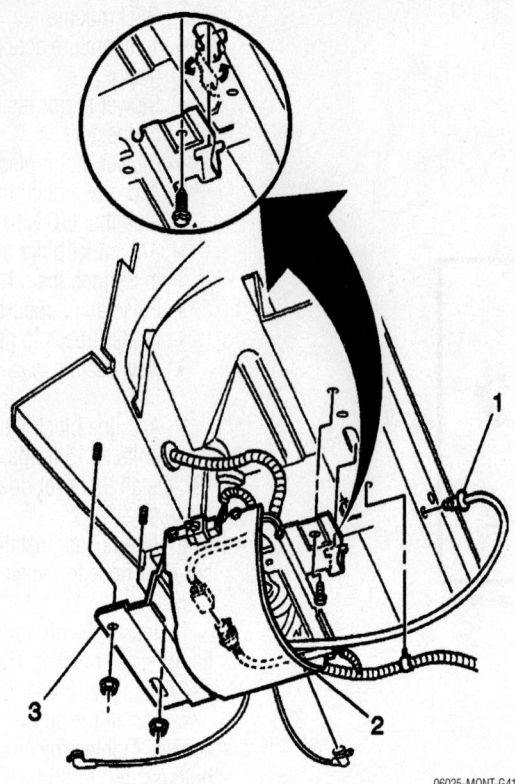

Removing components to access air compressor assembly

17. Install the new cylinder head bolts. Tighten the bolts, in the sequence shown, to 44 ft. lbs. (60 Nm).

18. Use the angle tool, J 45059, or suitable angle measuring device, to rotate the bolts in sequence an additional 95 degrees.

19. Install the exhaust manifold. See EXHAUST MANIFOLD.

20. Install the spark plugs. Torque the spark plugs to 15 ft. lbs. (20 Nm), if using a new cylinder head, or to 11 ft. lbs. (15 Nm), if using an existing head.

21. For the right side cylinder head, remove the alternator. See ALTERNATOR.

22. Install the spark plug wires to the spark plugs.

23. For the left side cylinder head, install the oil level indicator tube.

24. Install the exhaust crossover pipe. See EXHAUST MANIFOLD.

25. Install the valve rocker arms and pushrods. See ROCKER ARMS/COVERS.

26. Install the lower intake manifold. See INTAKE MANIFOLD.

27. Fill the crankcase with engine oil.

28. Fill the cooling system.

29. Start the engine and inspect for leaks.

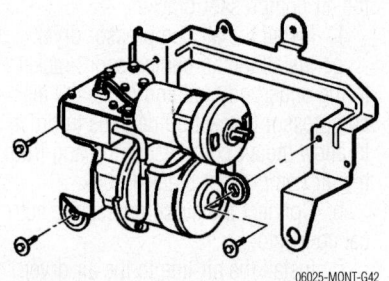

Identifying the air compressor assembly

Cylinder Head

REMOVAL & INSTALLATION

1. Drain the cooling system.
2. Drain the engine oil.
3. Remove the lower intake manifold. See INTAKE MANIFOLD.
4. Remove the valve rocker arms and the pushrods. See ROCKER ARMS/COVERS.
5. Remove the exhaust crossover pipe. See EXHAUST MANIFOLD.
6. For the left cylinder head, remove the oil level indicator tube.
7. Remove the spark plug wires from the spark plugs.

8. Remove the spark plugs.
9. For the right side cylinder head, remove the alternator. See ALTERNATOR.
10. Remove the appropriate side exhaust manifold. See EXHAUST MANIFOLD.
11. Remove the cylinder head bolts and discard.
12. Remove the cylinder head.
13. Remove the cylinder head gasket.
14. Clean and inspect the cylinder head and the gasket mating surfaces.

To install:
15. Install a new cylinder head gasket over the alignment pins.
16. Install the cylinder head over the locator pins and the gasket.

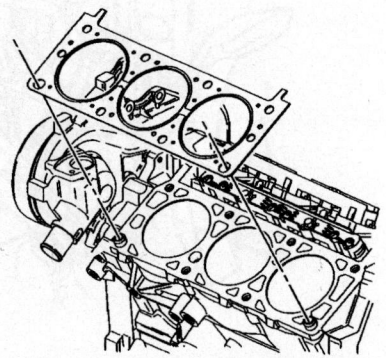

Cylinder head gasket installation

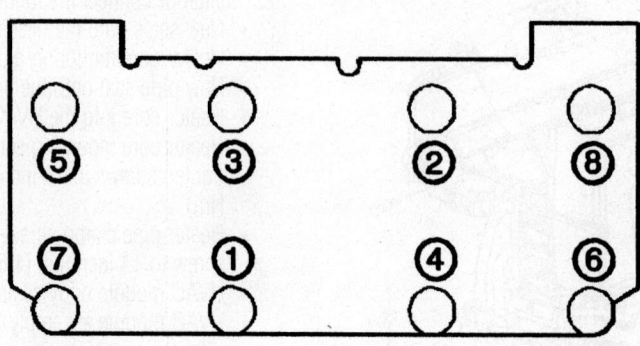

Cylinder head bolt torque sequence

Rocker Arms/Cover

REMOVAL & INSTALLATION

1. Before servicing the vehicle, refer to the Precautions Section.
2. Partially drain the cooling system (left side only)
3. Remove or disconnect the following:
   - Negative battery cable
   - Drive belt (Right)
   - Generator (Right)
   - Generator bracket (Right)
   - Spark plug wires
   - Vacuum hoses from the EVAP purge valve (Right)
   - EVAP purge valve (Right)
   - Ignition coil bracket and coils (Right)
   - Thermostat bypass pipe (Left)
   - PCV vacuum hose (Left)
   - Rocker arm cover bolts and cover
   - Trim the valve cover sealant away from the lower intake manifold gasket at the cylinder head to the lower intake manifold joints

☀ CAUTION

Rocker arm cover gasket and sealant must be carefully trimmed away from the lower intake manifold gasket. Failure to do so will damage the lower intake manifold gasket and cause a severe oil leak.

   - Rocker arm bolts
   - Rocker arms

☀ CAUTION

Keep components separated in order to reinstall in the same location.

To install:
4. Install or connect the following:
   - Rocker arms

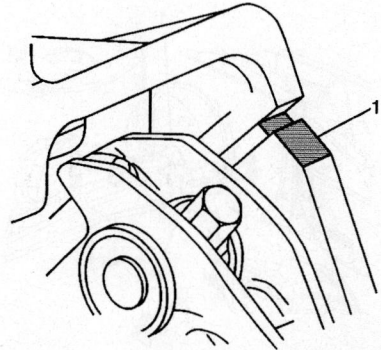

06025-MONT-G13

View of the area to trim the gasket sealant

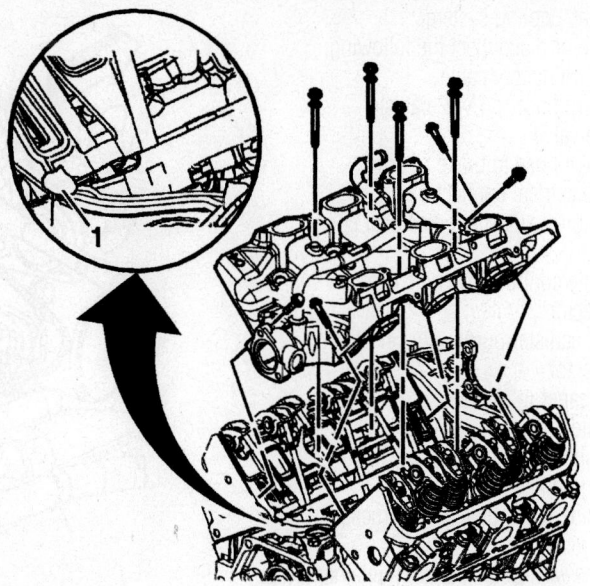

06025-MONT-G14

Lower intake manifold joint

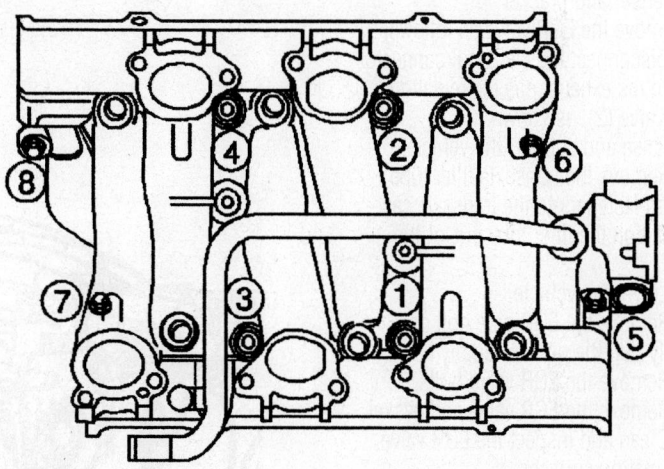

06025-MONT-G15

Lower intake manifold torque sequence

   - Rocker arm bolts and tighten to 24 ft. lbs. (32 Nm)
   - Apply sealant (GM P/N12378521 or equivalent) at the cylinder head to lower intake manifold joint (1)
   - New gasket and rocker arm cover, tighten bolts to 89 inch lbs. (10 Nm)

➡Use an alternating criss—cross pattern when tightening the rocker cover bolts. Failure to do so may cause an oil leak due to improper seating of the gasket.

   - PCV vacuum line (Left)
   - Thermostat bypass pipe (Left)
   - Ignition coil bracket and coils (Right)
   - EVAP purge valve (Right)

   - Vacuum hose to EVAP purge valve (Right)
   - Spark plug wires
   - Generator bracket (Right)
   - Generator (Right)
   - Drive belt (Right)
   - Negative battery cable
   - Refill cooling system (Left)
5. Start and run the engine to check for leaks.

Intake Manifold

REMOVAL & INSTALLATION

Upper manifold

1. Before servicing the vehicle, refer to the Precautions Section.

2. Drain the cooling system
3. Remove or disconnect the following:
   - Negative battery cable
   - Vacuum hose to EVAP canister purge valve
   - Vacuum hose to brake booster
   - EGR electrical connector
   - Mass air flow sensor electrical connector
   - Throttle control valve electrical connector
   - EVAP canister purge valve electrical connector
   - Air cleaner intake duct
   - Left side spark plug wires
   - Camshaft position sensor (CMP) wiring harness from retainer
   - Left side spark plug wire harness from retainer
   - Engine wiring harness from retainer
   - Ignition coil bracket and coils
   - EVAP canister purge solenoid valve
   - Manifold absolute pressure (MAP) sensor and bracket

4. Remove the EGR valve as follows:
   a. Disconnect the electrical connector (1) from the exhaust gas recirculation (EGR) valve (2), as shown.
   b. Raise and support the vehicle.
   c. Remove the transaxle filler tube retaining fasteners to the transaxle case and position the filler tube out-of-the-way.
   d. Lower the vehicle.
   e. Remove the EGR pipe bolt and carefully pull the pipe assembly back.
   f. Remove the EGR valve bolts.
   g. Remove the EGR valve and gasket.
   h. Clean and inspect the EGR valve gasket mating surfaces.

5. Remove the upper intake manifold bolts and the stud.
6. Remove the alternator bracket after removing the alternator and drive belt tensioner.
7. Remove the upper intake manifold and gasket.
8. If replacing the upper intake manifold, transfer the throttle body.
9. If necessary to transfer the throttle body, remove it as follows:
   a. Disconnect the electronic throttle control (ETC) electrical connector (1) from the throttle body (2).
   b. Remove the heater pipe nut at the throttle body.
   c. Remove only the heater inlet pipe.
   d. Remove the nuts and the bolts from the throttle body.
   e. Remove the throttle body assembly.
   f. Remove the throttle body gasket.

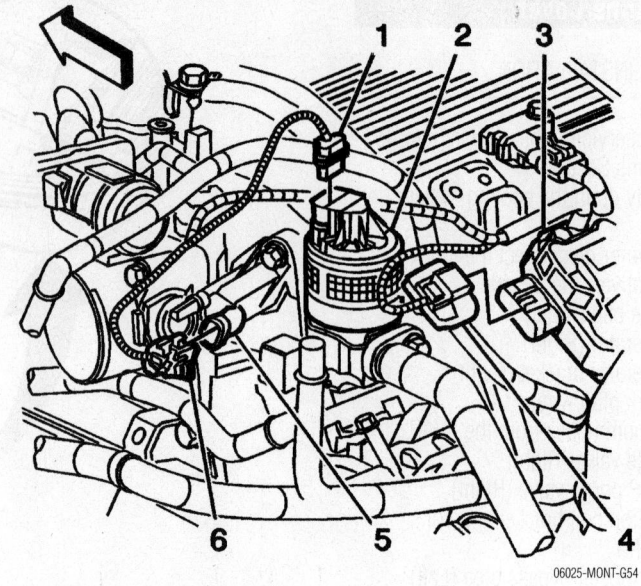

06025-MONT-G54

Detaching EGR Valve electrical connector

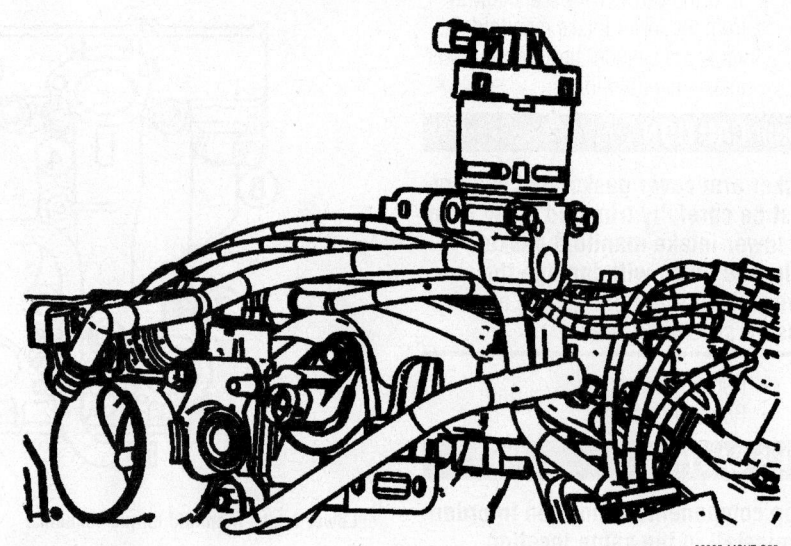

06025-MONT-G55

Removing the EGR Valve

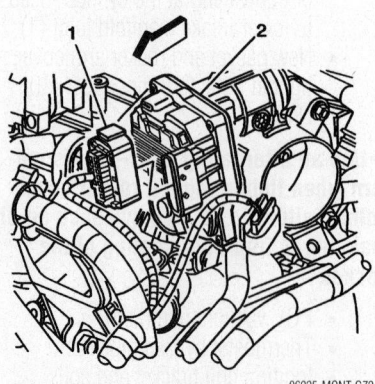

06025-MONT-G72

Disconnect ETC connector (1) from the throttle body (2)

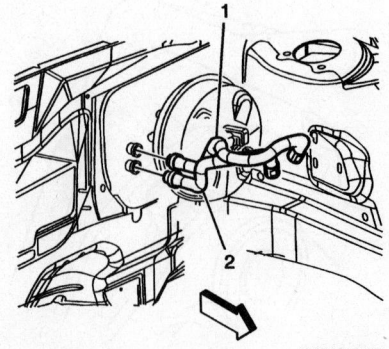

06025-MONT-G73

Remove the heater inlet pipe (1) only

✳✳ CAUTION

Do not use solvent of any type when cleaning the gasket surfaces on the intake manifold and the throttle body assembly, as damage to the gasket surfaces and throttle body assembly may result.Use care in cleaning the gasket surfaces on the intake manifold and the throttle body assembly, as sharp tools may damage the gasket surfaces.

  g. Clean and inspect the throttle body gasket mating surfaces.

To install:

10. If removed, install the throttle body to the upper intake manifold, as follows:

  a. Install a new gasket, if necessary.

  b. Install the throttle body assembly.

  c. Install the throttle body nuts and the bolts. Tighten the nuts and the bolts to 89 inch lbs. (10 Nm).

  d. Install the heater inlet pipe.

  e. Install the heater pipe nut to the throttle body. Tighten the nut to 18 ft. lbs. (25 Nm).

  f. Connect the ETC electrical connector (1) to the throttle body (2).

11. Install or connect the following:

- Intake manifold with new gasket
- Apply threadlock (GM P/N 12345382, or equivalent) to the intake manifold bolts
- Upper intake manifold bolts and the stud; tighten the bolts to 18 ft. lbs. (25 Nm)
- Generator bracket: torque bracket bolts to 37 ft. lbs. (50 Nm).

12. Install the EGR valve as follows:

  a. Install a new EGR valve gasket.

  b. Install the EGR valve.

  c. Install the EGR valve bolts. Tighten the bolt to 22 ft. lbs. (30 Nm).

  d. Install the EGR pipe to the EGR valve.

  e. Install the EGR pipe. Tighten the bolt to 18 ft. lbs. (25 Nm).

  f. Connect the electrical connector (1) to the EGR valve (2).

  g. Raise and support the vehicle.

  h. Position the transaxle filler tube to the normal installed position and install the filler tube to transaxle case fasteners. Tighten the fasteners to 115 ft. lbs. (130 Nm).

  i. Lower the vehicle.

13. Install or connect the following:

- MAP sensor bracket and sensor; tighten retaining bolt to 89 inch lbs. (10 Nm)

- EVAP canister purge solenoid valve; tighten bracket bolt to 89 inch lbs. (10 Nm)
- Ignition coil bracket and coils; tighten bolts to 40 inch lbs. (4.5 Nm)
- Camshaft position sensor (CMP) wiring harness to retainer
- Left side spark plug wire harness to retainer
- Engine wiring harness to retainer
- Left side spark plug wires
- Air cleaner intake duct
- EGR electrical connector
- Mass air flow sensor electrical connector
- Throttle control valve electrical connector
- EVAP canister purge valve electrical connector
- Vacuum hose to EVAP canister purge valve
- Vacuum hose to brake booster
- Negative battery cable

14. Fill the cooling system.

15. Start the engine and check for leaks.

Lower Manifold

✳✳ CAUTION

This engine uses a sequential multiport fuel injection system. Injector wiring harness connectors must be connected to the appropriate fuel injector.

1. Before servicing the vehicle, refer to the Precautions Section.

2. Drain the cooling system.

3. Remove or disconnect the following:

- Upper intake manifold; see Upper Manifold
- Both rocker arm covers; see Rocker Arms/Covers
- Engine coolant temperature (ECT) wiring harness
- Fuel injector and manifold air pressure (MAP) wiring harness

4. Remove the fuel injector rail after removing 2 retaining bolts.

5. Remove the heater inlet pipe, with heater hose, from the lower intake manifold and reposition aside.

6. Remove the radiator inlet hose from the engine.

7. Remove or disconnect the following:

- Water outlet
- Thermostat
- Lower intake manifold bolts
- Lower intake manifold
- Valve rocker arms and pushrods

✳✳ CAUTION

Keep components separated in order to reinstall in the same location.

- Lower intake manifold gaskets and seals

To install:

8. Install or connect the following:

- Lower intake manifold gaskets
- Valve rocker arms and pushrods

➡The intake pushrods are identified with yellow stripes and are 5–¾ inches long. Exhaust pushrods are identified with green stripes and are 6 inches long.

➡With gaskets and seals in place apply a small drop of RTV sealer or equivalent, to the 4 corners of the intake manifold to block joints (1).

- Lower intake manifold

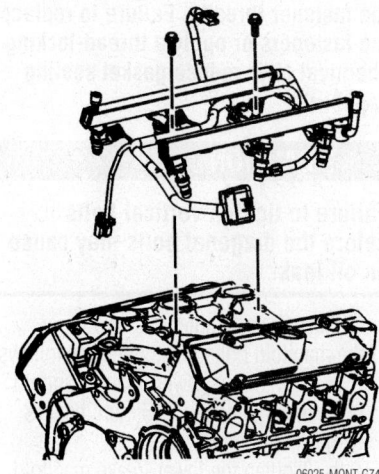

06025-MONT-G74

Removing/installing the fuel rail assembly

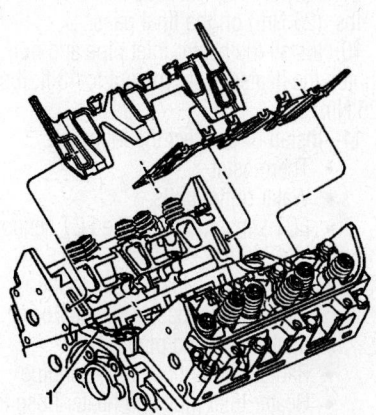

06025-MONT-G57

Removing lower intake manifold and gaskets

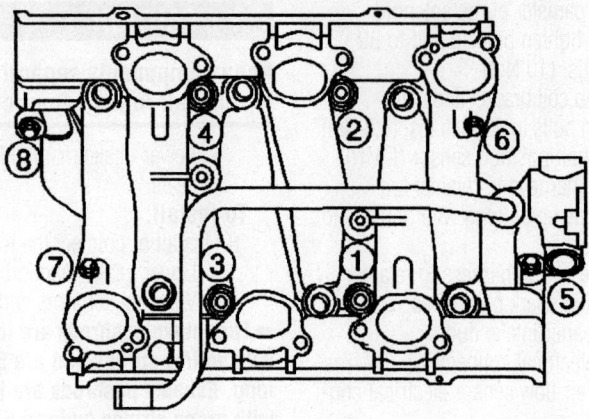

06025-MONT-G15

Lower intake manifold torque sequence

➡**Maximum gasket performance is achieved when using new fasteners that contain a thread locking patch. If the fasteners are not replaced, a thread locking chemical must be applied to the fastener threads. Failure to replace the fasteners or apply a thread-locking chemical MAY reduce gasket sealing capability.**

❋❋ CAUTION

Failure to tighten vertical bolts before the diagonal bolts may cause an oil leak.

9. Install and tighten the new lower intake manifold bolts in sequence as follows:

  a. Tighten the lower intake manifold bolts in sequence to 115 inch lbs. (13 Nm) on the first pass.

  b. Tighten the lower intake manifold bolts (1, 2, 3, 4) in sequence to 15 ft. lbs. (20 Nm) on the final pass.

  c. Tighten the lower intake manifold bolts (5, 6, 7, 8) in sequence to 18 ft. lbs. (25 Nm) on the final pass.

10. Install the heater inlet pipe and nut; tighten the heater inlet pipe nut to 18 ft. lbs. (25 Nm).

11. Install or connect the following:
  • Thermostat
  • Water outlet bolts
  • ECT sensor, tighten the ECT sensor to 15 ft. lbs. (20 Nm)
  • Thermostat bypass hose to the thermostat bypass pipe and lower intake manifold pipe
  • Radiator inlet hose to the engine
  • Heater inlet pipe and heater hose to the lower intake manifold
  • Fuel injector rail
  • Fuel injector and MAP wiring harness
  • ECT wiring harness

  • Rocker arm covers; see Rocker Arms/Covers.
  • Upper intake manifold; see Upper Manifold.
  • Negative battery cable

12. Refill the cooling system.

13. Start and run the engine to check for leaks.

Exhaust Manifold

REMOVAL & INSTALLATION

1. Before servicing the vehicle, refer to the Precautions Section.

2. Remove the negative battery cable.

3. Remove the spark plug wires.

4. Remove the spark plugs.

5. For right side exhaust manifold, remove the following:
  • EGR pipe from the exhaust manifold
  • Heated oxygen sensor

6. Remove the exhaust manifold heat shield bolts.

7. Remove the exhaust manifold heat shields.

8. Remove the exhaust manifold nuts.

9. Remove the exhaust manifold and gasket.

To install:

10. Install the exhaust manifold studs. Tighten the exhaust manifold studs to 13 ft. lbs. (18 Nm).

11. Install the exhaust manifold gasket.

12. Install the exhaust manifold.

13. Install the exhaust manifold nuts. Tighten the exhaust manifold nuts working from the center out to 12 ft. lbs. (16 Nm).

14. Install the right side spark plugs. Tighten the spark plugs to 11 ft. lbs. (15 Nm).

15. Install the spark plug wires onto the spark plugs.

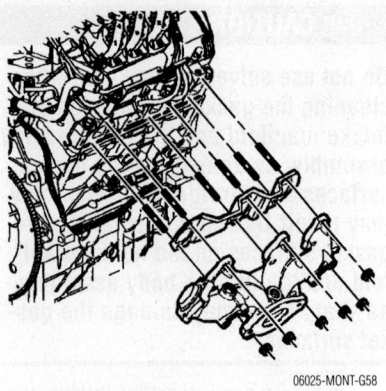

06025-MONT-G58

Showing the RH exhaust manifold mounting position

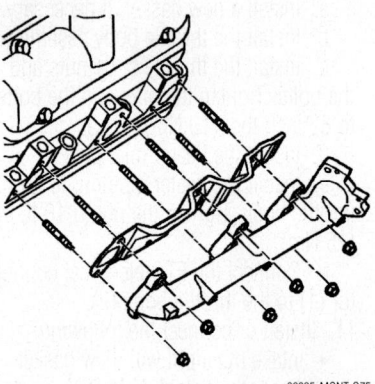

06025-MONT-G75

Showing the LH exhaust manifold mounting position

16. Install the lower exhaust manifold heat shield.

17. Install the upper exhaust manifold heat shield.

18. Install the exhaust manifold heat shield bolts. Tighten the exhaust manifold heat shield bolts to 89 inch lbs. (10 Nm).

19. For right exhaust manifold, coat the threads of the heated oxygen sensor with anti-seize compound.

20. For right exhaust manifold, install the heated oxygen sensor. Tighten the heated oxygen sensor to 31 ft. lbs. (42 Nm).

21. Install the spark plug wires.

22. Reconnect the negative battery cable.

Camshafts

REMOVAL & INSTALLATION

1. Before servicing the vehicle, refer to the Precautions Section.

2. Properly relieve the fuel system pressure.

3. Disconnect the negative battery cable.

4. Drain the engine cooling system and the engine oil.

5. Remove timing chain and front cover. See Timing Chain, Sprockets, Front Cover and Seal.

6. Remove or disconnect the following:
- Intake manifold; see Intake Manifold
- Rocker arms and pushrods
- Timing chain and sprockets; see Timing Chain
- Camshaft position sensor bolt
- Camshaft position sensor
- Camshaft thrust plate screws
- Camshaft thrust plate

✳✳ CAUTION

All camshaft journals are the same diameter, so care must be used in removing or installing the camshaft to avoid damage to the camshaft bearings.

7. Install the camshaft sprocket bolt into the camshaft and tighten finger tight only

8. Carefully rotate and remove the camshaft from the engine block.

To install:

9. Coat the camshaft journals with clean engine oil.

10. Coat the camshaft lobes with pre-lube.

11. Install the camshaft sprocket bolt into the camshaft and tighten finger tight only.

12. Carefully rotate the camshaft while installing the camshaft into the camshaft bearings.

13. Install or connect the following:
- Camshaft thrust plate
- Camshaft thrust plate screws and tighten the screws to 89 inch lbs. (10 Nm)
- Camshaft position sensor
- Camshaft position sensor bolt and

tighten the bolt to 89 inch lbs. (10 Nm)
- Timing chain and sprockets; see Timing Chain
- Rocker arms and push rods; see Rocker Arms/Covers
- Intake manifold; see Intake Manifold

14. Install timing chain and front cover. See Timing Chain, Sprockets, Front Cover and Seal.
- Negative battery cable

15. Refill the engine cooling system and engine oil.

Starter Motor

REMOVAL & INSTALLATION

1. Before servicing the vehicle, refer to the Precautions Section.

2. Remove or disconnect the following:
- Negative battery cable
- Flywheel inspection cover bolts
- Flywheel inspection cover
- Electrical connections from the starter motor
- Starter motor mounting bolts
- Starter motor

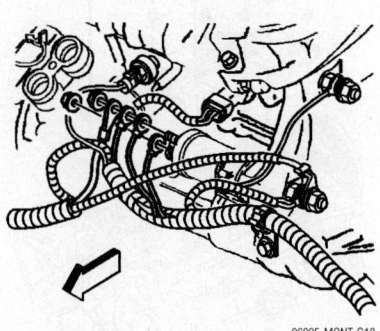

06025-MONT-G18

Showing the starter motor location and electrical connections

To install:

3. Install or connect the following:

➡**Before installing the starter motor to the engine, tighten the nut next to the cap on the solenoid BAT terminal. If this terminal is not tight in the solenoid cap, the cap may be damaged during installation of electrical connections and cause the starter motor to fail later.**

- Starter motor
- Starter motor mounting bolts and tighten the bolts to 30 ft. lbs. (40 Nm)
- Electrical connection to the battery terminal on the solenoid and tighten the battery terminal nut to 13 ft. lbs. (17 Nm)
- Electrical connections to the S terminal on the solenoid and tighten solenoid S terminal nut to 27 inch lbs. (3 Nm)
- Flywheel inspection cover
- Flywheel inspection cover bolts

4. Connect the negative battery cable

Oil Pan/Pump

REMOVAL & INSTALLATION

1. Before servicing the vehicle, refer to the Precautions Section.

2. Drain the engine oil.

3. Remove the oil pan support bracket bolts and brackets as needed

4. Remove the oil pan side bolts.

5. Remove the oil pan bolts.

6. Remove the oil pan.

7. Remove the oil pan gasket.

8. Remove the oil pump bolt.

9. Remove the oil pump and oil pump drive shaft.

10. Remove the crankshaft oil deflector nuts.

11. Remove the crankshaft oil deflector.

To install:

12. Install the crankshaft oil deflector.

13. Install the crankshaft oil deflector nuts. Tighten the crankshaft oil deflector nuts to 18 ft. lbs. (25 Nm).

✳✳ CAUTION

Do not reuse the oil pump driveshaft retainer. During assembly, install a NEW oil pump driveshaft retainer.

14. Install the oil pump.

15. Position the oil pump onto the pins.

16. Install the oil pump bolt attaching the oil pump to the rear crankshaft bearing

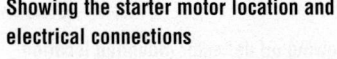

06025-MONT-G59

Showing camshaft mounting in block

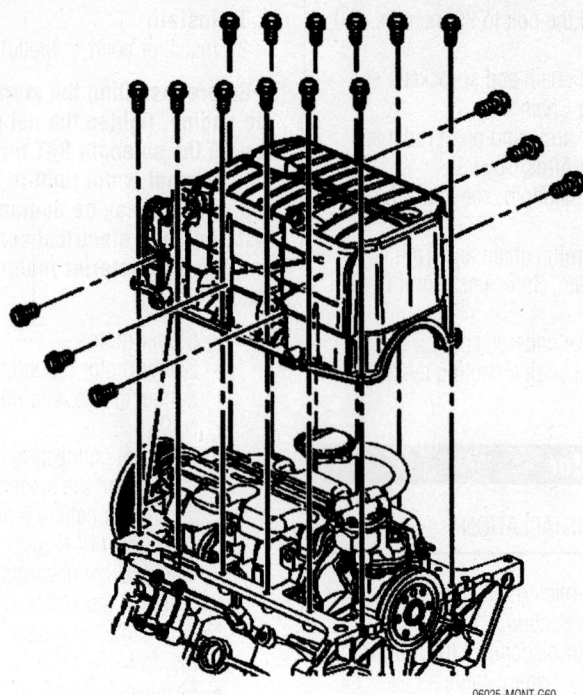

06025-MONT-G60

Showing oil pan mounting bolt locations

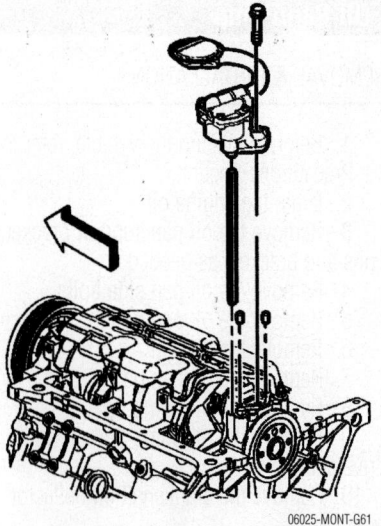

06025-MONT-G61

Showing oil pump mounting location

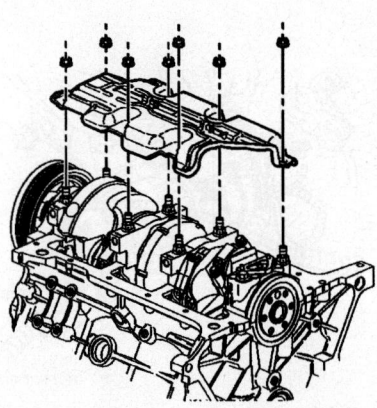

06025-MONT-G62

Showing oil deflector mounting location

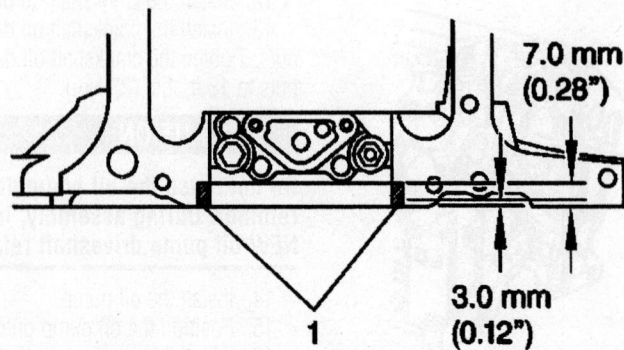

06025-MONT-G63

Applying sealant to rear main bearing cap

cap. Tighten the oil pump bolt to 30 ft. lbs. (41 Nm).

17. Apply sealer GM P/N 12378521, (Canadian P/N 88901148) or the equivalent to both sides of the crankshaft rear main bearing cap (1). Press sealer into gap using a putty knife.

18. Install the oil pan gasket.

19. Install the oil pan.

20. Install the oil pan bolts. Tighten the oil pan bolts to 18 ft. lbs. (25 Nm).

21. Install the oil pan side bolts. Tighten the oil pan side bolts to 37 ft. lbs. (50 Nm).

22. Install the oil pan drain plug. Tighten the oil pan drain plug to 18 ft. lbs. (25 Nm). Bottom of Form

Rear Main Seal

REMOVAL & INSTALLATION

1. Before servicing the vehicle, refer to the Precautions Section.

2. Remove or disconnect the following:
- Negative battery cable
- Transmission assembly
- Flywheel
- Crankshaft seal by prying it from out oil seal housing

➡**Be careful not to damage the crankshaft seal surface with the prying tool.**

To install:

❋ CAUTION

Note the direction of the rear oil seal. The new design seal is a reverse style as opposed to what has been used in the past. "THIS SIDE OUT" has been stamped into the seal.

➡**Do not apply or use any oil lubrication on the crankshaft rear oil seal or the seal installer. Do not touch the sealing lip of the oil seal once the protective sleeve is removed. Doing so will damage or deform the seal. Clean the crankshaft sealing surface with a clean, lint free towel. Inspect the edge of crankshaft for burrs or sharp edges that could damage the rear main oil seal. Remove burrs or sharp edges with a crocus cloth.**

3. Install the new rear seal by using a seal installer.

4. Install or connect the following:
- Flywheel
- Transmission assembly
- Negative battery cable

5. Start the engine and check for leaks.

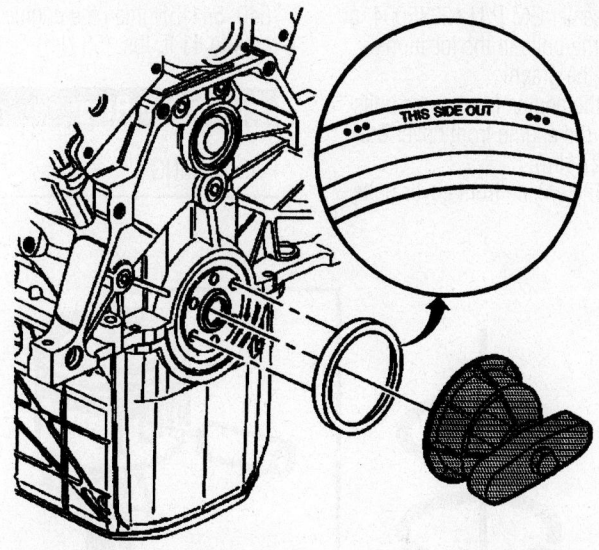

06205-MONT-G19

Indicating the proper orientation of the rear main seal during installation

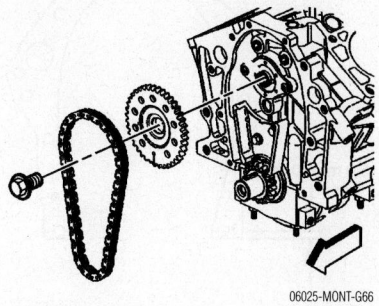

06025-MONT-G66

Removing camshaft sprocket and timing chain

Timing Chain, Sprockets, Front Cover and Seal

REMOVAL & INSTALLATION

1. Before servicing the vehicle, refer to the Precautions Section.

2. Drain the engine cooling system.

3. Drain the engine oil

4. Pry out the crankshaft front oil seal using a suitable tool. Use care not to damage the engine front cover or the crankshaft.

5. Remove the engine front cover bolts (540, 541, 545).

6. Remove the engine front cover.

7. Remove the engine front cover gasket.

8. Remove the camshaft sprocket bolt.

9. Remove the camshaft sprocket and timing chain.

10. Remove the crankshaft sprocket.

11. Remove the timing chain dampener bolts.

12. Remove the timing chain dampener.

To install:

13. Install the crankshaft sprocket.

14. Apply prelube GM P/N 12345501, or equivalent, to the crankshaft sprocket thrust surface. Install the timing chain dampener.

15. Install the timing chain dampener bolts. Tighten the timing chain dampener bolt to 15 ft. lbs. (21 Nm).

16. Align the crankshaft timing mark (2) to the timing mark on the bottom of the timing chain dampener (1).

17. Hold the camshaft sprocket with the timing chain hanging down and install the timing chain to the crankshaft gear.

18. Align the timing mark on the

camshaft gear (4) with the timing mark on top of the timing chain dampener (3).

19. Align the dowel in the camshaft with the dowel hole in the camshaft sprocket.

20. Draw the camshaft sprocket onto the camshaft using the mounting bolt.

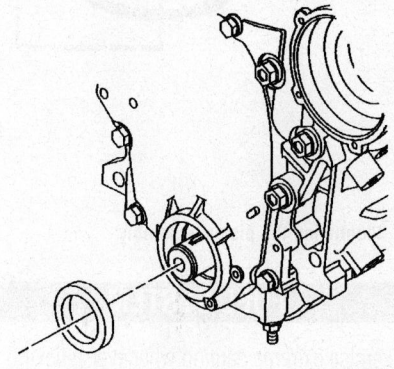

06025-MONT-G64

Removing front oil seal

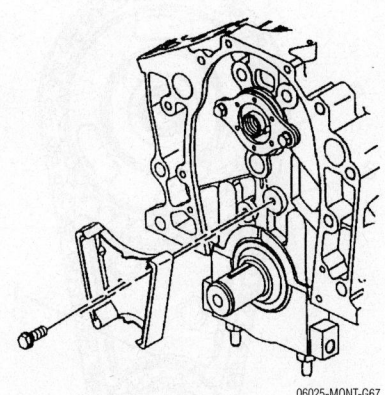

06025-MONT-G67

Removing timing chain damper

21. Coat the crankshaft and camshaft sprocket with engine oil. Tighten the bolt to 103 ft. lbs. (140 Nm).

22. Install the engine front cover gasket.

23. Apply sealer GM P/N 12346004, or equivalent, to both sides of the lower tabs of the engine front cover gasket. Apply the sealer no less than 5.0 mm (0.20 in) wide.

24. Install the engine front cover.

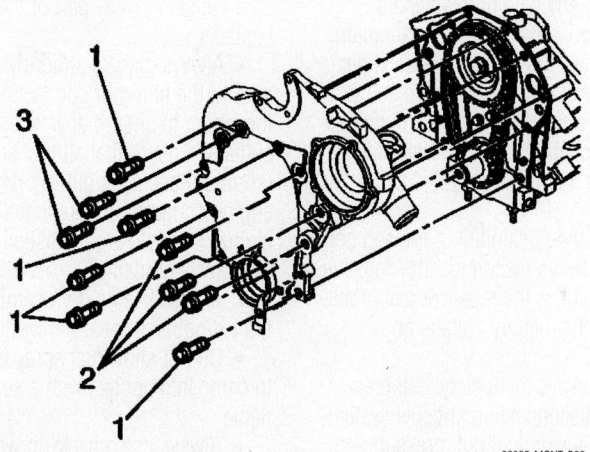

06025-MONT-G22

Removing/installing the front cover

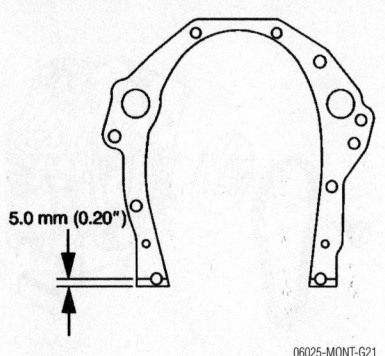

5.0 mm (0.20")

06025-MONT-G21

Applying sealant to the cover gasket

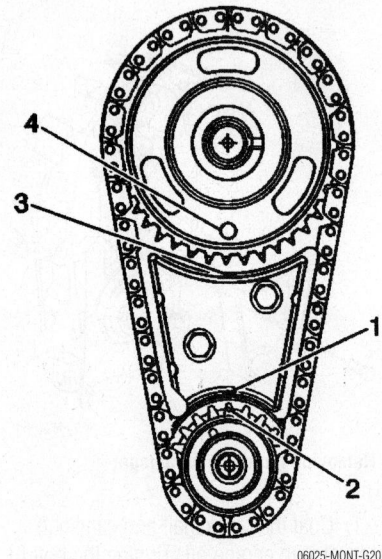

06025-MONT-G20

Showing the proper timing chain-to-sprocket alignment

25. Apply sealer GM P/N 12346004, or equivalent, to the bolts in the locations pointed out in the graphic.

26. Install the engine front cover bolts (545). Tighten the engine front cover bolts to 20 ft. lbs. (27 Nm).

27. Install the engine front cover bolts (540, 541). Tighten the engine front cover bolts to 41 ft. lbs. (55 Nm).

Piston and Ring

POSITIONING

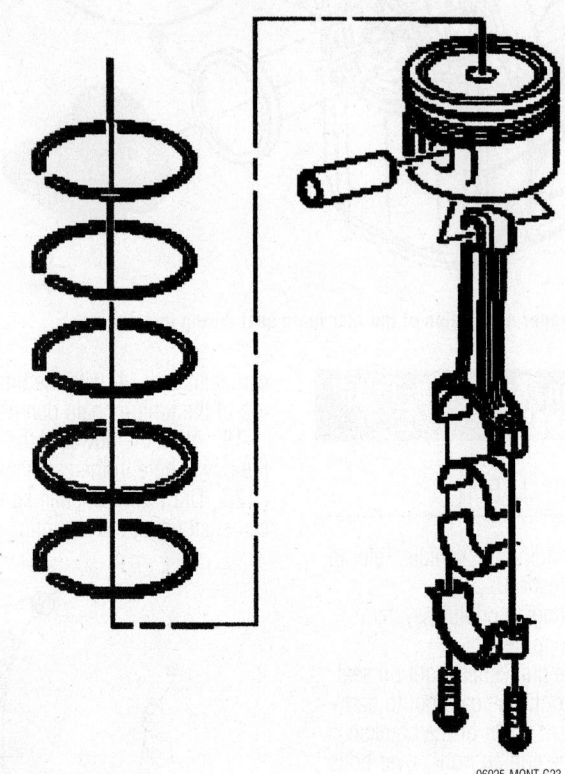

06025-MONT-G23

Identifying the piston assembly

FUEL SYSTEM

Fuel System Service Precautions

Safety is the most important factor when performing not only fuel system maintenance but also any type of maintenance. Failure to conduct maintenance and repairs in a safe manner may result in serious personal injury or death. Maintenance and testing of the vehicle's fuel system components can be accomplished safely and effectively by adhering to the following rules and guidelines.

• To avoid the possibility of fire and personal injury, always disconnect the negative battery cable unless the repair or test procedure requires that battery voltage be applied.

• Always relieve the fuel system pressure prior to disconnecting any fuel system component (injector, fuel rail, pressure regulator, etc.), fitting or fuel line connection.

Exercise extreme caution whenever relieving fuel system pressure, to avoid exposing skin, face and eyes to fuel spray. Please be advised that fuel under pressure may penetrate the skin or any part of the body that it contacts.

• Always place a shop towel or cloth around the fitting or connection prior to loosening to absorb any excess fuel due to spillage. Ensure that all fuel spillage (should it occur) is quickly removed from engine surfaces. Ensure that all fuel soaked cloths or towels are deposited into a suitable waste container.

• Always keep a dry chemical (Class B) fire extinguisher near the work area.

• Do not allow fuel spray or fuel vapors to come into contact with a spark or open flame.

• Always use a back-up wrench when loosening and tightening fuel line connection fittings. This will prevent unnecessary stress and torsion to fuel line piping. Always follow the proper torque specifications.

• Always replace worn fuel fitting O-rings with new. Do not substitute fuel hose or equivalent where fuel pipe is installed.

Fuel System Pressure

RELIEVING

The fuel systems operate under high fuel pressures. It is very important that the pressure be properly relieved prior to servicing the system or any of its components.

1. Before servicing the vehicle, refer to the Precautions Section.

2. Loosen the fuel filler cap in order to relieve the tank pressure. Do not tighten at this time.

3. Raise the vehicle.

4. Disconnect the fuel pump electrical connector.

5. Lower the vehicle.

6. Start and run the engine until the fuel supply remaining in the fuel pipes is consumed. Engage the starter for 3.0 seconds in order to assure relief of any remaining pressure.

7. Raise the vehicle.

8. Connect the fuel pump electrical connector.

9. Lower the vehicle.

10. Disconnect the negative battery cable in order to avoid possible fuel discharge if an accidental attempt is made to start the engine.

11. When fuel service is finished, tighten the fuel filler cap and connect the negative battery cable.

Fuel Pump/Sender

REMOVAL & INSTALLATION

1. Before servicing the vehicle, refer to the Precautions Section.

2. Properly relieve the fuel system pressure.

3. Disconnect the negative battery cable.

4. Drain the fuel tank into an approved container.

✳✳ WARNING

Do not attempt to straighten any kinked nylon fuel lines. Replace any kinked nylon fuel feed or return pipes in order to prevent damage to the vehicle.

✳✳ WARNING

Do not attempt to repair sections of nylon fuel pipes. If the nylon fuel pipes are damaged, replace the pipes.

5. Raise the vehicle.

6. Disconnect the fuel supply pipe (5) quick-connect fitting.

7. Disconnect the evaporative emission (EVAP) purge pipe (6) quick connect fitting at the EVAP canister.

8. Loosen the fuel tank filler pipe hose clamp.

9. Disconnect the fuel tank filler from the fuel tank.

10. Disconnect the vapor recirculation pipe from the filler tube.

11. Remove the fuel tank shield.

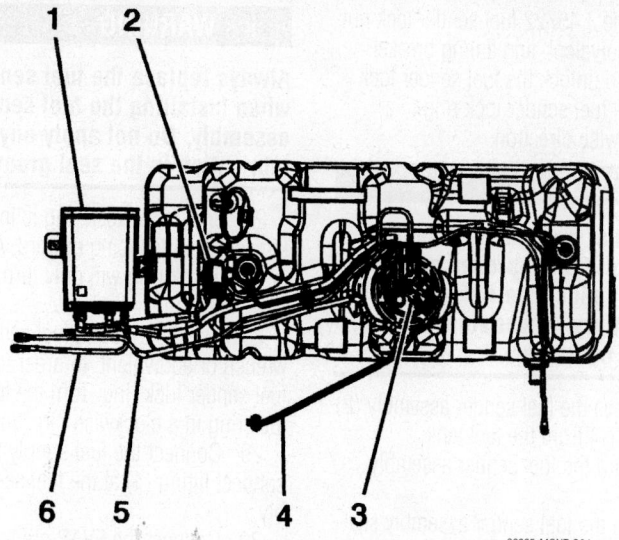

Showing the fuel tank assembly and related components

12. With the aid of an assistant, support the fuel tank.

13. Remove the fuel tank strap attaching bolts.

14. Disconnect the fuel sender and the fuel tank pressure (FTP) sensor electrical connectors at the body pass through.

15. Remove the fuel tank from the vehicle.

16. Remove the fuel sender assembly (3).

17. Disconnect the fuel sender electrical connections (3).

18. Clean all of the fuel pipe connections, all of the hose connections, and all of the areas surrounding the connections before disconnecting the connections in order to avoid possible contamination of the fuel system.

19. Disconnect the fuel supply pipe quick-connect fitting (4) at the fuel sender assembly.

20. Disconnect the evaporative emission (EVAP) pipe quick-connect fittings (5, 6) at the fuel sender assembly.

✳✳ CAUTION

Do NOT use impact tools. Significant force will be required to release the lock ring. The use of a hammer and screwdriver is not recommended. Secure the fuel tank in order to prevent fuel tank rotation.

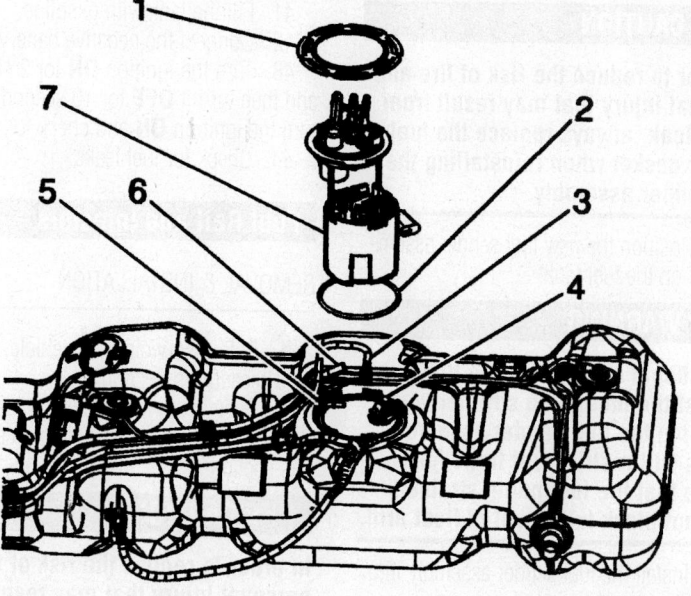

Identifying the fuel sender assembly

21. Use the J 45722 fuel sender lock nut wrench or equivalent, and a long breaker bar in order to unlock the fuel sender lock ring. Turn the fuel sender lock ring in a counterclockwise direction.

❋❋ CAUTION

Drain the fuel from the fuel sender assembly into an approved container in order to reduce the risk of fire and personal injury. Never store the fuel in an open container.

22. Remove the fuel sender assembly (2) and the seal (7) from the fuel tank.
23. Discard the fuel sender assembly seal.
24. Clean the fuel sender assembly sealing surfaces.

➡Some lock ring were manufactured with DO NOT REUSE stamped into them. These lock rings may be reused if they are not damaged or warped. Inspect the lock ring for damage due to improper removal or installation procedures. If damage is found, install a NEW lock ring.

➡Check the lock ring for flatness. Place the lock ring on a flat surface. Measure the clearance between to lock ring and the flat surface using a feeler gage. If the warpage is less than 0.41mm (0.016 in), the lock ring does not require replacement. If the warpage is greater than 0.41mm (0.016 in), the lock ring must be replaced.

To install:

❋❋ CAUTION

In order to reduce the risk of fire and personal injury that may result from a fuel leak, always replace the fuel sender gasket when reinstalling the fuel sender assembly.

25. Position the new fuel sender assembly seal on the fuel tank.

❋❋ WARNING

Care should be taken not to fold over or twist the fuel pump strainer when installing the fuel sender assembly, as this will restrict fuel flow. Also, ensure that the fuel pump strainer does not block full travel of float arm.

26. Install the fuel sender assembly into the fuel tank.

❋❋ WARNING

Always replace the fuel sender seal when installing the fuel sender assembly. Do not apply any type of lubrication in the seal groove.

27. Ensure the lock ring is installed with the correct side facing upward. A correctly installed lock ring will only turn in a clockwise direction.
28. Use the J 45722 fuel sender lock nut wrench or equivalent, in order to install the fuel sender lock ring. Turn the fuel sender lock ring in a clockwise direction.
29. Connect the fuel supply pipe quick-connect fitting (4) at the fuel sender assembly.
30. Connect the EVAP quick-connect fittings at the fuel sender assembly.
31. Connect the fuel sender electrical connections.
32. With the aid of an assistant, position and support the fuel tank.
33. Connect the fuel sender and the FTP sensor electrical connectors at the body pass through.
34. Install the fuel tank retaining strap attaching bolts and tighten the bolts to 35 ft. lbs. (47.5 Nm).
35. Install the fuel tank shield.
36. Connect the vapor recirculation pipe to the fuel fill pipe.
37. Connect the fuel tank filler pipe to the fuel tank.
38. Connect the fuel feed pipe quick connect fitting.
39. Connect the EVAP purge pipe quick connect fitting at the EVAP canister.
40. Lower the vehicle.
41. Fill the tank with gasoline.
42. Connect the negative battery cable.
43. Turn the ignition **ON** for 2 seconds and then turn it **OFF** for 10 seconds. Again turn the ignition **ON** and check for leaks.
44. Check for fuel leaks.

Fuel Rail and Injectors

REMOVAL & INSTALLATION

1. Before servicing the vehicle, refer to the Precautions Section.
2. Relieve the fuel system pressure. Refer to the fuel system relief procedure in this section.

❋❋ CAUTION

In order to reduce the risk of fire and personal injury that may result from a fuel leak, always install the fuel

injector O-rings in the proper position. If the upper and lower O-rings are different colors (black and brown), be sure to install the black O-ring in the upper position and the brown O-ring in the lower position on the fuel injector. The O-rings are the same size but are made of different materials.

3. Clean all engine fuel pipe connections and areas surrounding the engine fuel pipe connections before disconnecting the engine fuel pipe connections to avoid possible contamination of the fuel system.
4. Remove or disconnect the following:
   - Fuel feed pipe retaining clip
   - Fuel feed pipe from the fuel rail
   - Evaporative emission (EVAP) pipe from the EVAP canister purge solenoid valve
   - Upper intake manifold
   - Fuel injector harness from the fuel rail
   - Coolant temperature sensor and camshaft position sensor electrical connectors
   - Main fuel injector electrical connector
   - Fuel rail retaining bolts
   - Fuel rail assembly

❋❋ WARNING

Use care in removing the fuel injectors in order to prevent damage to the fuel injector electrical connector pins or the fuel injector nozzles. Do not immerse the fuel injector in any type of cleaner. The fuel injector is an electrical component and may be damaged by this cleaning method.

❋❋ WARNING

If the fuel injectors are found to be leaking, the engine oil may be contaminated with fuel.

   - Fuel injector retaining clip
   - Fuel injector from the fuel rail
   - Fuel injector upper O-ring
   - Fuel injector lower O-ring

To install:

➡Be sure to use the correct part number when ordering replacement fuel injectors. The fuel injector assembly is stamped with a part number identification

5. Lubricate the new injector O-ring seats with engine oil.

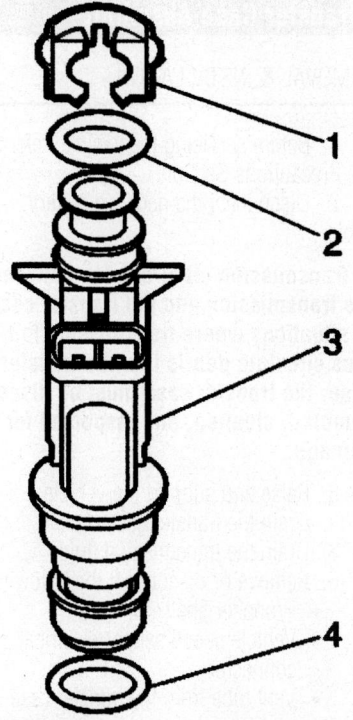

Exploded view of the fuel injector

6. Install or connect the following:
- Fuel injector upper O-ring
- Fuel injector lower O-ring
- Fuel injector to the fuel rail
- Fuel injector retaining clip

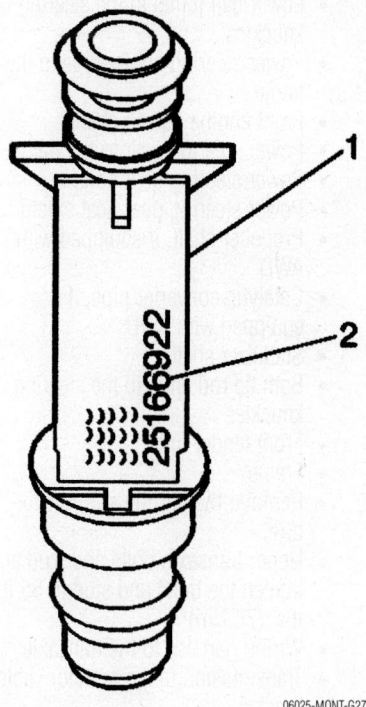

Indicating the fuel injector ID number

✴✴ WARNING

Plug the inlet and the outlet ports of the fuel rail in order to prevent contamination. Do not use compressed air to clean the fuel rail assembly as this may damage the fuel rail components. Do not immerse the fuel rail assembly in a solvent bath in order to prevent damage to the fuel rail assembly.

- Fuel injector nozzles into the lower intake manifold injector bores
- Press on the fuel rail using the palms of both hands until the fuel injectors are fully seated
- Install the fuel rail attaching bolts and tighten the bolt to 89 inch lbs. (10Nm)
- Injector electrical harness to the fuel rail
- Fuel injector electrical connectors
- Coolant temperature sensor and camshaft position sensor electrical connectors
- Main fuel injector electrical harness connector
- Engine fuel feed pipe to the fuel rail
- EVAP pipe to the EVAP canister purge solenoid valve
- Upper intake manifold
- Negative battery cable

7. Turn the ignition **ON** for 2 seconds and then turn it **OFF** for 10 seconds. Again turn the ignition **ON** and check for leaks.

DRIVE TRAIN

Automatic Transmission Assembly

REMOVAL & INSTALLATION

1. Before servicing the vehicle, refer to the Precautions Section.
2. Remove or disconnect the following:
- Negative battery cable
- Position aside coolant recovery bottle
- Air cleaner assembly
- Automatic transmission range selector cable from the manual shaft
- Automatic transmission range selector cable bracket
- Wiring harness connector from the transaxle
- Wiring harness retainer from the side cover
- The top 4 bell housing bolts and stud

- Install the engine support fixture
- Front tires and wheels

➡ **When removing the wheel, mark the location of the wheel to the hub prior to removal. Mark the individual location of all retainers as they are removed.**

- Front fender liner enough to gain access to the front frame bolts
- Both outer tie rod ends from the steering knuckles
- Stabilizer shaft
- Propeller shaft, if equipped with all wheel drive (AWD).

➡ **It is NOT necessary to remove the catalytic converter pipe on FWD models.**

- Catalytic converter pipe, if equipped with AWD
- Power steering gear heat shield
- Power steering gear bolts and suspend steering gear from the frame
- Front engine splash shield
- Power steering cooler pipe from the frame, use mechanics wire to secure the power steering cooler line out of the way
- Front wheel speed sensor connectors
- Front wheel speed sensor wiring harnesses from the lower control arms
- Lower ball joints from the steering knuckles
- Engine mount nuts
- Transaxle mount nuts

3. Lower the vehicle. until the frame contacts the J 39580, or equivalent
- Frame front bolts
- Frame rear bolts
- Frame strap bolts and straps
- Raise the vehicle in order to separate the frame from the vehicle
- Torque converter inspection cover
- Torque converter bolts
- Vehicle speed sensor
- Right and left axle shafts from the transmission

- Transmission cooler lines
- Position the transmission jack under the transaxle
- Transaxle brace
- Lower transaxle bolt and stud
- Transaxle from the vehicle

To install:

4. Install or connect the following:
   - Align the transaxle filler tube to the transmission and Install the transaxle into the vehicle.
   - Lower transaxle bolt and stud and tighten the bolt and stud to 55 ft. lbs. (75 Nm)
   - Transaxle brace
   - Transmission cooler lines
   - Right and left axle shafts into the transaxle
   - Vehicle speed sensor
   - Torque converter bolts and tighten the bolts to 46 ft. lbs. (63 Nm)
   - Torque converter inspection cover
   - Position the transaxle table with the frame under the vehicle

❄❄ WARNING

Ensure that the power steering cooler line does not become trapped by the engine mount during this step

5. Lower the vehicle. until the frame is close to the vehicle
   - Adjust the utility straps as necessary in order to align the powertrain mounts with the frame

❄❄ WARNING

Ensure that the alignment pins remain installed during the frame installation.

- Insert two 19mm (0.75 inch) diameter X 203mm (8.0 inches) long guide pins or drill bits into the frame right side alignment holes in order to align the frame
- Frame front bolts and tighten the bolts to 96 ft. lbs. (130 Nm)
- Frame straps and bolts and tighten the bolts to 37 ft. lbs. (50 Nm)
- Frame rear bolts and tighten the bolts to 177 ft. lbs. (240 Nm)
- Remove the alignment pins from the frame
- Transaxle mount nuts
- Engine mount nuts
- Wheel speed sensor wiring harnesses to the lower control arms
- Wheel speed sensor electrical connectors.

- Lower ball joints to the steering knuckles
- Power steering cooler pipe to the frame
- Front engine splash shield
- Power steering gear to the frame
- Power steering gear bolts
- Power steering gear heat shield
- Propeller shaft, if equipped with AWD
- Catalytic converter pipe, if equipped with AWD
- Stabilizer shaft
- Both tie rod ends to the steering knuckles
- Front fender liner
- Frame
- Remove the engine support fixture.
- Upper transaxle bolts and stud and tighten the bolts and stud to 55 ft. lbs. (75 Nm)
- Wiring harness to the transaxle
- Transmission range selector cable bracket
- Transmission range selector cable on the manual shaft
- Air cleaner assembly
- Coolant recovery bottle

6. Install the tire and wheel assembly. Tighten the lug nuts to 100 ft. lbs. (140 Nm) in a criss—cross pattern, after aligning the wheel hub with the reference mark and holes as shown in appropriate illustration.

❄❄ WARNING

Do NOT overfill the transaxle. The overfilling of the transaxle causes foaming, loss of fluid, shift complaints, and possible damage to the transaxle. Adjust the fluid level.

➡ **It is recommended the ate transmission adaptive pressure (TAP) information be reset. Resetting the TAP values using a scan tool will erase all learned values in all cells. As a result the ECM, PCM, or TCM will need to relearn TAP values. Transmission performance may be affected as new TAP values are learned.**

7. Reset the TAP values.
8. Refill the transmission with the proper amount and type of fluid.
9. Connect the negative battery cable. Start the vehicle and allow to warm while checking for leaks. Road test the vehicle to check for shift quality.

Transfer Case Assembly

REMOVAL & INSTALLATION

1. Before servicing the vehicle, refer to the Precautions Section.
2. Disconnect the negative battery cable.

➡ **Transmission oil circulates between the transmission and the transfer case. In situations where transmission failures circulate debris into the transfer case, the transfer case must be disassembled, cleaned, and inspected for damage.**

3. Raise and support the vehicle.
4. Drain the transfer case oil.
5. Drain the transmission fluid.
6. Remove or disconnect the following:
   - Propeller shaft
   - Vehicle speed sensor electrical connector
   - Vent tube from the transfer case

❄❄ WARNING

Removal and installation of the transfer case while the transmission is in vehicle may cause improper positioning of the park gear thrust bearing. If the transfer case fasteners are tightened while the park gear thrust bearing is out of position, the park gear thrust bearing, transmission, and/or the transfer case will be damaged.

   - Transfer case with the transaxle
7. Install the transaxle to the transmission support fixture.
8. Remove and discard the left output shaft retaining ring.
9. Rotate the transaxle 90 degrees so that the transmission side cover is facing down.
10. Remove the transfer case side brace bolts, and the transfer case side brace.
11. Remove the transfer case bolts.

❄❄ WARNING

During removal of the transfer case/output shaft, do not use excessive force or damage to the bushings may occur.

12. Remove the transfer case with the output shaft from the transaxle.
13. Install the transfer case assembly to the J 44755, or equivalent.

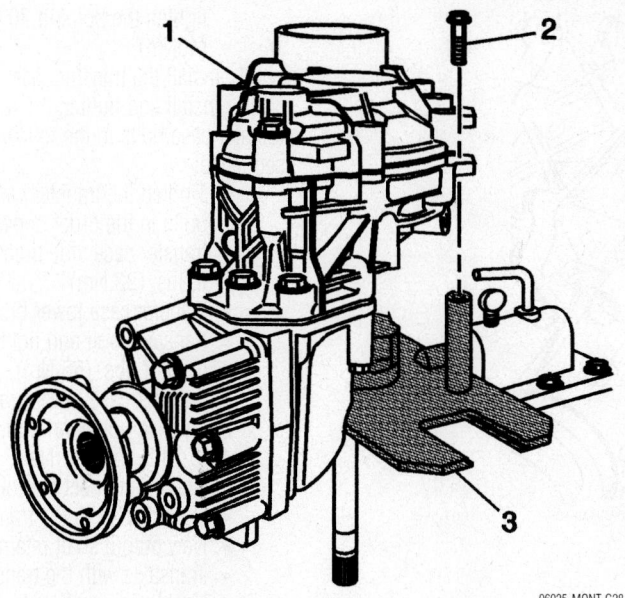

06025-MONT-G28

Showing the J 44755 transfer case holding fixture

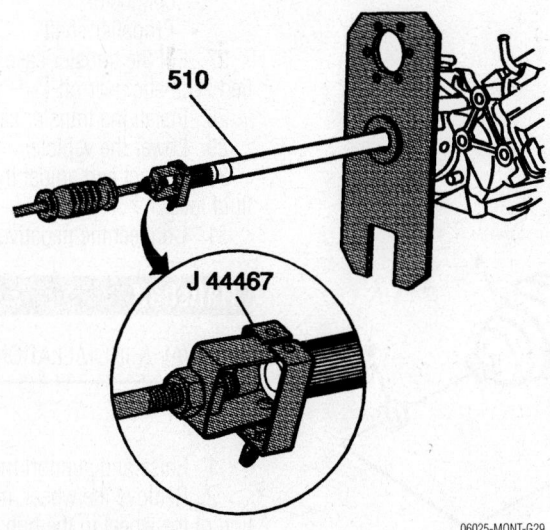

06025-MONT-G29

Indicating the J 44467 output shaft remover

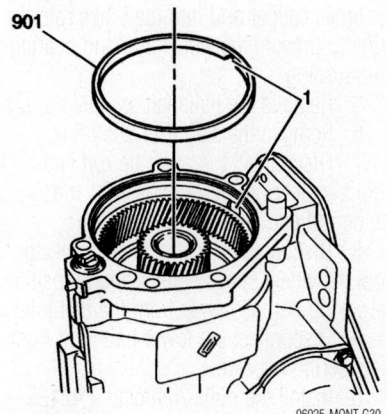

06025-MONT-G30

Removing the oil dam

14. Install the retaining bolts and tighten the bolts to 37 ft. lbs. (50 Nm).

15. Attach the J 6125-1B, or equivalent slide hammer to the J 44467.

16. Install the J 44467 into the snap ring groove on the output shaft (510) and tighten securely.

17. Use the J 6125-1B and the J 44467 to remove the output shaft.

18. Remove the output shaft from the transfer case.

19. Remove the transfer case seal from the transfer case.

20. Remove the oil dam (901) from the transaxle.

To install:

➡**If you are replacing anything other than gaskets or seals, the transmission to transfer case end play check must be performed.**

21. Install or connect the following:
- New transfer case seal

✳✳ WARNING

The oil dam must be installed with the notch aligned to the oil passage in the transaxle case.

- Oil dam to the transaxle

22. Rotate the transaxle 90 degrees, then position the transfer case to the transaxle.

23. Install and tighten the transfer case bolts in the following sequence:
- Transfer case bolts (1) and (2) and tighten the bolts to 26 ft. lbs. (35 Nm), then rotate the bolts 160 degrees
- Transfer case bolt (3) and tighten the bolt to 26 ft. lbs. (35 Nm), then rotate the bolts 70 degrees
- Transfer case bolts (4) and (5) and

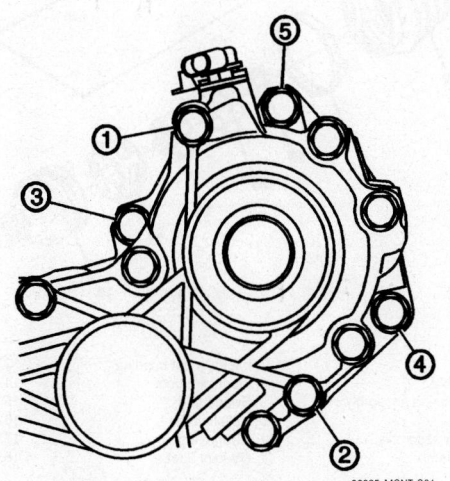

06025-MONT-G31

Showing the transfer case bolt torque sequence

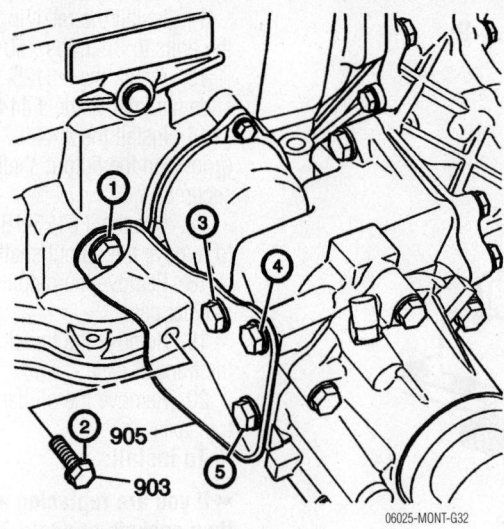

Showing the transfer case side bolts torque sequence

06025-MONT-G32

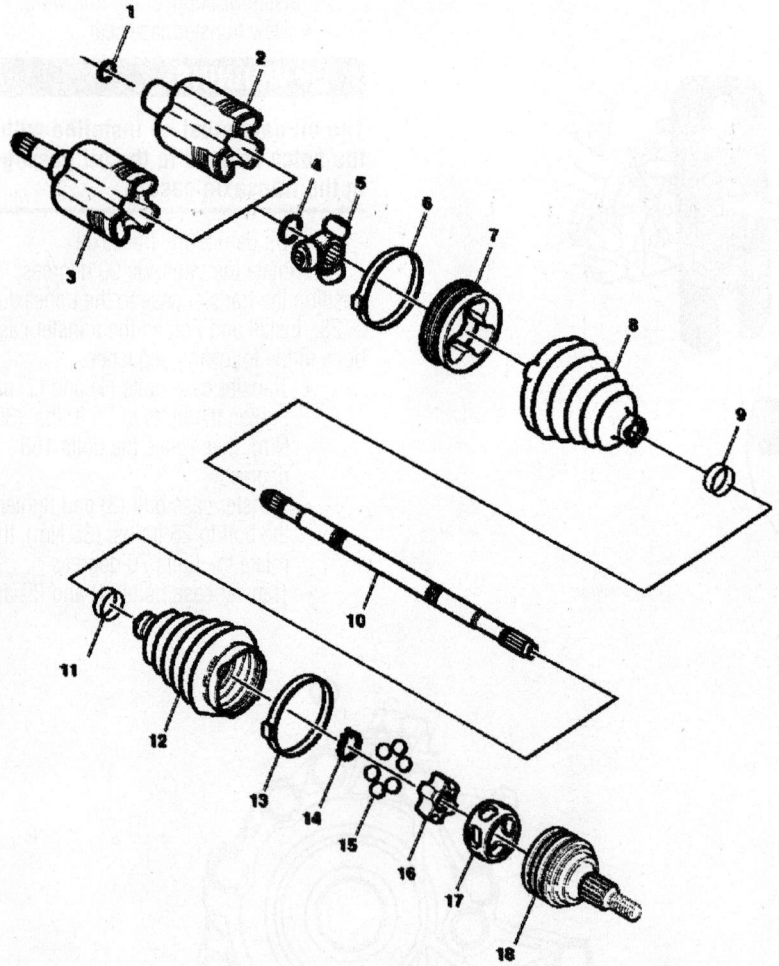

1. Retaining ring	7. trilobal tripot bushing	13. Boot retaining clamp
2. Housing assembly	8. tripot joint boot	14. Race retaining ring
3. Retainer and housing assembly	9. Swage ring	15. Chrome alloy ball
4. Spacer ring	10. Halfshaft bar	16. CV joint inner race
5. tripot joint spider assembly	11. Swage ring	17. CV joint cage
6. Boot retaining clamp	12. CV joint boot	18. CV joint outer race

06025-MONT-G68

Exploded view of the front halfshaft assembly

tighten the bolts to 30 ft. lbs. (40Nm)

24. Install the transfer case side brace.

25. Install and tighten the remaining transfer case bolts in the following sequence:

- Tighten the transfer case side brace bolts in the order shown. Tighten transfer case side brace bolts to 24 ft. lbs. (32 Nm)
- Transfer case lower brace to transaxle bolt and tighten the bolt to 42 ft. lbs. (56 Nm)
- Transfer case lower brace to transfer case bolts and tighten the bolts to 24 ft. lbs. (32 Nm)

26. Install or connect the following:
- Output shaft to the transmission
- New output shaft retaining ring
- Transaxle with the transfer case
- Vent hose and the clamp to the transfer case
- Vehicle speed sensor electrical connector
- Propeller shaft

27. Fill the transfer case with the specified synthetic gear oil.

28. Install the transfer case lower brace.

29. Lower the vehicle.

30. Inspect and adjust the transmission fluid level.

31. Connect the negative battery cable.

Halfshaft

REMOVAL & INSTALLATION

Front

1. Raise and support the vehicle.

2. Remove the wheel, marking the location of the wheel to the hub prior to removal. Mark the individual location of all retainers as they are removed.

3. Remove the engine splash shield.

4. Insert a drift or punch (1) through the brake caliper and into the brake rotor in order to prevent the wheel hub and bearing from turning.

5. Remove the halfshaft spindle nut (2).

6. Remove the stabilizer shaft link.

7. Disconnect the outer tie rod end from the steering knuckle; do NOT loosen the tie rod end jam nut.

8. Disconnect the electrical connector from the wheel speed sensor and reposition the wiring harness away from the ball joint.

9. Disconnect the lower ball joint from the steering knuckle.

10. Install the puller/remover J 42129 onto the wheel hub and secure with wheel nuts.

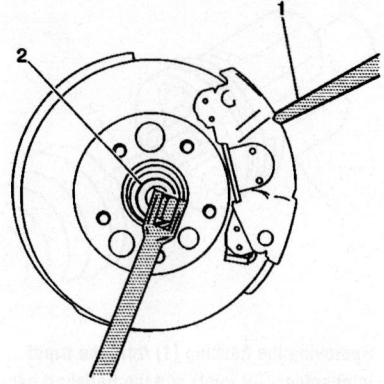

Using drift through brake caliper to hold wheel hub and bearing from turning

06025-MONT-G70

❄ CAUTION

Be sure to support the halfshaft until it is fully removed from the vehicle.

11. Using the puller/remover tool, disengage the halfshaft from the wheel hub and bearing.

12. Assemble the special slide hammer and tool attachments (J 33008-A, J 29794, J 2619-01) and disengage the halfshaft from the transaxle.

13. Remove the halfshaft from the vehicle

To install:

14. Install the halfshaft to the transaxle

➡**Verify that the wheel halfshaft is properly engaged to the transaxle by grasping the inner tripot housing and pulling outward. Do not pull on the halfshaft bar. The wheel halfshaft will remain firmly in place when properly engaged.**

15. Install or connect the following:
- Halfshaft to the hub and bearing
- Ball joint to the steering knuckle. See Lower Control Arm

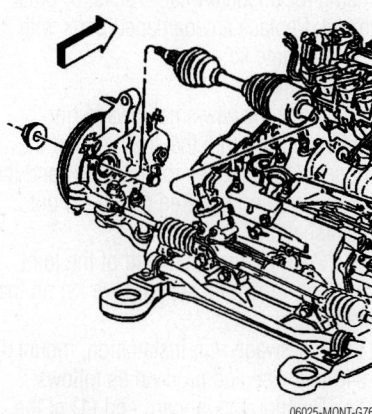

06025-MONT-G76

Removing/installing the front halfshaft

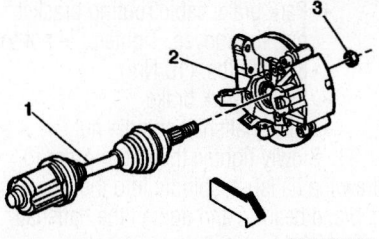

06025-MONT-G77

Installing the hub and bearing (2) to the halfshaft (1), with retaining nut (3)

- Wheel speed sensor electrical connector
- Stabilizer shaft link. See Stabilizer Bar under FRONT SUSPENSION.

➡**Insert a drift or punch through the brake caliper and into the brake rotor in order to prevent the hub and bearing from turning.**

- Install the nut to the halfshaft spindle and tighten the nut to 118 ft. lbs. (160 Nm)
- Outer tie rod end to the steering knuckle. See Tie Rod Ends
- Engine splash shield

16. Install the tire and wheel assembly. Tighten the lug nuts to 100 ft. lbs. (140 Nm) in a criss—cross pattern, after aligning the wheel hub with the reference mark and holes as shown in appropriate illustration.

17. Lower the vehicle.

Rear

1. Apply the parking brake.
2. Raise and support the vehicle.
3. Remove the tire and wheel assembly.

➡**When removing the wheel, mark the location of the wheel to the hub prior to removal. Mark the individual location of all retainers as they are removed.**

➡**The halfshaft nut must not be reused. Replace the halfshaft nut with new nut whenever it is removed.**

4. Remove the halfshaft nut and discard it.

5. Release the parking brake.

6. Remove and support the brake caliper bracket as follows:

a. Remove the brake caliper from the mounting bracket and support the brake caliper (2) with heavy mechanics wire (1), or equivalent.

❄ CAUTION

Do NOT disconnect the hydraulic brake flexible hose from the caliper.

b. Remove the brake pads from the brake caliper bracket (3).

c. Remove the brake pad retainers from the brake caliper bracket (3).

7. Remove the nut securing the park brake cable routing bracket, if necessary.

8. Remove the bolt retaining the rear tie rod end from the rear suspension knuckle; do NOT loosen the tie rod end jam nut.

9. Loosen, but do not remove, the bolts securing the park brake cable bracket to the suspension knuckle.

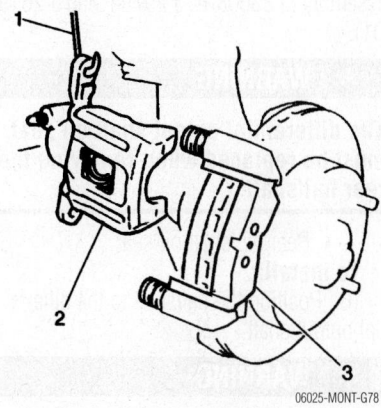

06025-MONT-G78

Suspending the brake caliper bracket (2) out of the way on mechanic's wire (1) after detaching it from the brake caliper (3)

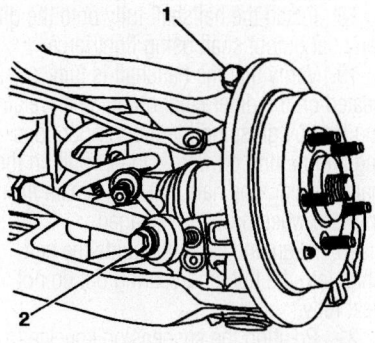

06025-MONT-G79

Identifying the rear tie rod end retaining bolt (2) and the upper control arm bolt (1)

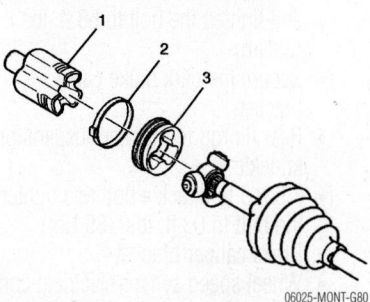

06025-MONT-G80

Separating the halfshaft from the wheel hub and bearing, using the special puller (J 42129)

10. Detach the wheel speed sensor electrical connector.

11. Install the puller, J 42129, onto the wheel hub and secure with wheel nuts.

12. Begin to disengage the halfshaft from the wheel hub and bearing.

13. Reposition the suspension knuckle toward the rear of the vehicle.

14. Remove the special puller from the wheel hub.

15. Remove the halfshaft from the rear axle differential and remove the halfshaft from the vehicle, using special removal tool assembly (J 33008-A, J 29794, and J 2619-01).

✳✳ WARNING

The differential output shaft oil seal must be replaced when removing the rear halfshaft.

- Rear halfshaft oil seal

To install:

16. Position the halfshaft to the differential output shaft.

✳✳ WARNING

Do not damage the differential output shaft oil seal.

17. Carefully align and guide the halfshaft onto the differential output shaft.

18. Install the halfshaft fully onto the differential output shaft using light force.

19. Verify that the halfshaft is fully seated on the differential output shaft retaining ring by grasping the inner tripot housing and pulling outward. Do not pull on the halfshaft bar. The halfshaft will remain firmly in place when properly engaged.

20. Align and carefully guide the halfshaft into the hub and bearing but do not seat fully.

21. Position the suspension knuckle to the upper control arm.

22. Install or connect the following:

- Bolt and nut to the upper control arm/suspension knuckle assembly and tighten the bolt to 63 ft. lbs. (85Nm)
- Secure the park brake cable bracket.
- Rear tie rod to the rear suspension knuckle
- Tie rod to knuckle bolt and tighten the bolt to 63 ft. lbs. (85 Nm)
- Brake caliper bracket
- Wheel speed sensor electrical connector
- Park brake cable routing bracket, if removed

- Park brake cable routing bracket nut, if removed. Tighten the nut to 89 inch lbs. (10 Nm)
- Set the park brake
- NEW halfshaft spindle nut

23. Slowly tighten the nut in order to draw the halfshaft spindle into the wheel hub and bearing and tighten the halfshaft spindle nut to 192 ft. lbs. (260 Nm).

24. Install the tire and wheel assembly.

25. Lower the vehicle.

26. Release the parking brake.

CV-Joints

OVERHAUL

Front CV-Joint

1. Before servicing the vehicle, refer to the Precautions Section.

✳✳ CAUTION

Do not cut through the wheel drive shaft inboard seal during service. Cutting through the seal may damage the sealing surface of the housing and the tripot bushing. Damage to the sealing surface may lead to water and dirt intrusion and premature wear of the constant velocity joint.

2. Disconnect the swage ring from the halfshaft bar using a hand grinder to cut through the ring, taking care not to damage the halfshaft bar.

3. Remove the large seal retaining clamp (2) from the tripot joint with side cutters. Discard the large seal retaining clamp.

4. Separate the inboard seal from the tri-lobal tripot bushing (3) at the large diameter.

5. Slide the seal away from the joint along the halfshaft bar.

6. Remove the housing (1) from the tripot joint spider and the halfshaft bar (2).

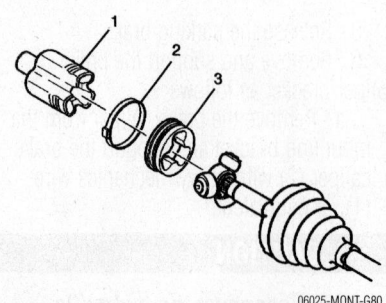

Exploded view of the front CV-Joint assembly

06025-MONT-G80

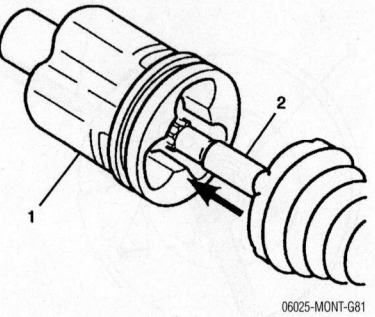

Removing the housing (1) from the tripot joint spider (CV joint) and the halfshaft bar (2)

06025-MONT-G81

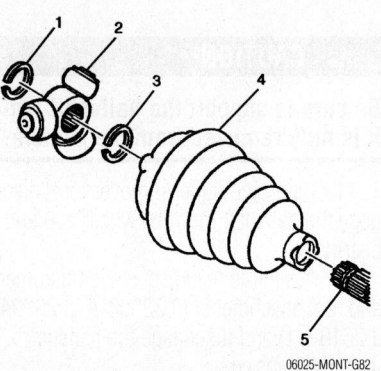

Disassembling the CV joint components: spacer ring (1), spider assembly (2), spacer ring (3), tripot boot (4), halfshaft bar (5)

06025-MONT-G82

7. Spread the spacer ring (1) using J 8059 (or equivalent).

8. Remove the spacer ring (1), spider assembly (2), spacer ring (3) (if equipped) using J 8059, and tripot boot (4). Discard the boot and rings.

9. Clean the halfshaft bar (5). Use a wire brush in order to remove any rust in the boot mounting area (grooves).

10. Inspect the needle rollers, needle bearings, and trunnion. Check the tripot housing for unusual wear, cracks, or other damage. Replace any damaged parts with the appropriate kit.

To install:

11. Place the new small swage ring or eared clamp (2) onto the small end of the joint seal (1). Slide the joint seal (1) and the small swage ring or eared clamp (2) onto the halfshaft bar.

12. Position the small end of the joint seal (1) into the joint seal groove (3) on the halfshaft bar.

13. For swage ring installation, mount J 41048 in a vise and proceed as follows:

14. Position the inboard end (1) of the halfshaft assembly in tool J 41048.

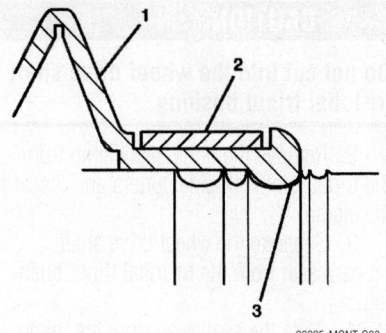

06025-MONT-G83

Assembling the CV joint components: joint seal (1), swage ring (2), joint seal groove (3)

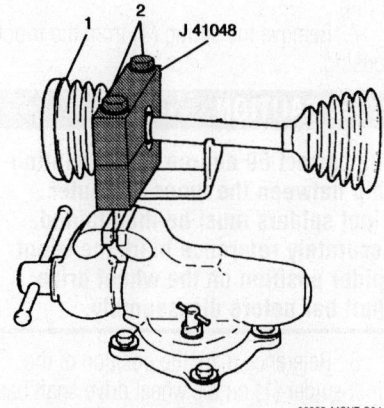

06025-MONT-G84

Installing the swage ring, using a J 41048 holding tool. Position inboard end (1) into tool, assemble J 41048, tighten tool bolts (2) and align components

15. Align the top of the seal neck on the bottom die using the indicator.

16. Place the top half of the J 41048 on the lower half.

✳✳ CAUTION

Before proceeding, ensure there are no pinch points on the halfshaft inboard seal. This could cause damage to the halfshaft inboard seal.

17. Insert the bolts (2).

18. Tighten the bolts by hand until snug.

19. Align the following items:
- Halfshaft inboard seal (1)
- Halfshaft bar
- Swage ring (2)

20. Tighten each bolt of the clamping tool (J 41048) 180 degrees at a time using a ratchet wrench.

21. Alternate between each bolt until both sides are bottomed.

22. For eared clamp installation, mount the halfshaft into a vise.

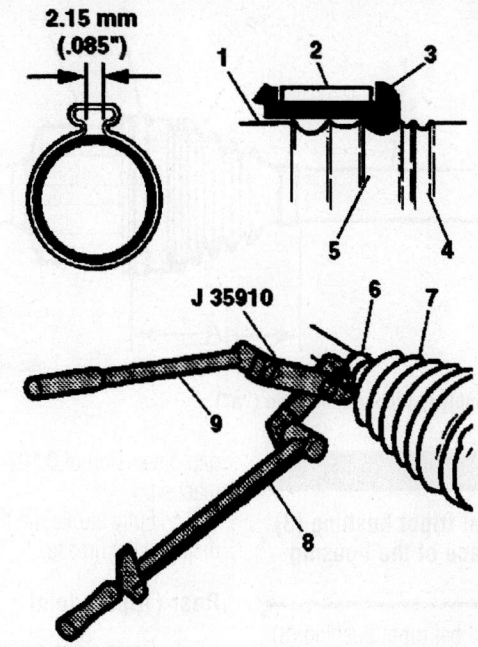

1. Halfshaft
2. Level bubble
3. Level tool
4. Clamp
5. Boot
6. Clamp
7. Boot
8. Torque wrench
9. Breaker bar

06025-MONT-G85

Assembling the eared clamp with special tool J 35910 to properly crimp clamp

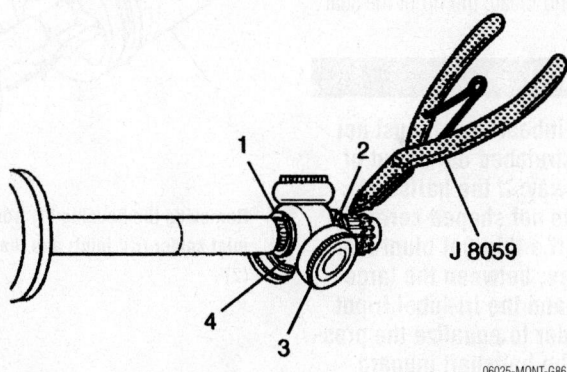

06025-MONT-G86

Installing spacer ring into halfshaft bar groove, using J 8059 ring pliers

23. Slide the tripot seal (7) to the corresponding groove on the halfshaft bar.

24. Crimp the eared clamp (6) using J 35910, a torque wrench (8), and a breaker bar (9).

25. If equipped, install the spacer ring (2) into the groove of the halfshaft bar using J 8059.

26. Slide the tripot joint spider assembly (4) as far as it will go on the halfshaft bar.

27. Install the spacer ring (2) into the groove of the halfshaft bar J 8059.

28. Place approximately half of the grease from the service kit in the halfshaft inboard seal. Use the remainder of the grease to repack the housing.

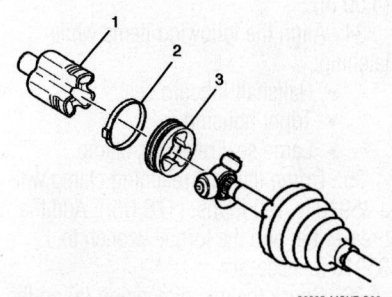

06025-MONT-G80

Exploded view of the CV-Joint assembly

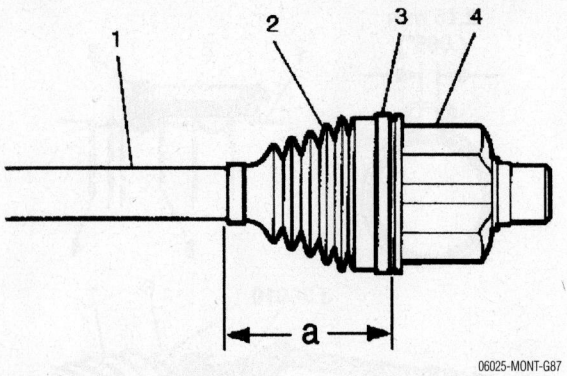

06025-MONT-G87

Installing joint assembly to proper dimension ("a")

> **✳✳ CAUTION**
>
> **Ensure the tri-lobal tripot bushing (3) is flush with the face of the housing (1).**

29. Install the tri-lobal tripot bushing (3) to housing (1).

30. Position the larger new seal retaining clamp (2) on the halfshaft inboard seal.

31. Slide the housing (1) over the tripot joint spider assembly on the halfshaft bar.

32. Slide the large diameter of the halfshaft inboard seal (2), with larger clamp (3) in place, over the outside of the tri-lobal tripot bushing and locate the lip of the seal in the groove.

> **✳✳ CAUTION**
>
> **The halfshaft inboard seal must not be dimpled, stretched out or out of shape in any way. If the halfshaft inboard seal is not shaped correctly, carefully insert a thin flat blunt tool, no sharp edges, between the large seal opening and the tri-lobal tripot bushing in order to equalize the pressure. Shape the halfshaft inboard seal properly by hand. Remove the tool.**

33. Position the joint assembly at the proper vehicle dimension, a = 106 mm (4.00 in).

34. Align the following items while latching:
- Halfshaft inboard seal
- Tripot housing
- Large seal retaining clamp

35. Crimp the seal retaining clamp with J 35910 to 130 ft. lbs. (176 Nm). Add the breaker bar and the torque wrench to J 35910 if necessary.

36. Check the gap dimension (a) on the clamp ear. If the gap dimension is larger than shown, continue tightening until the gap dimension of 0.102 inch (2.6mm) is reached.

37. Fully stroke the joint several times to disperse the grease.

Rear (Tripot) Joint

1. Remove the small seal clamp from the wheel drive shaft bar using side cutters and discard the clamp.

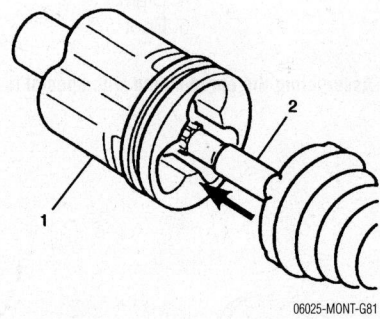

06025-MONT-G81

Removing the housing (1) from the tripot joint spider (CV joint) and the halfshaft bar (2)

> **✳✳ CAUTION**
>
> **Do not cut into the wheel drive shaft tri-lobal tripot bushing.**

2. Remove the large seal clamp from the tripot joint with side cutters and discard the clamp.

3. Separate the wheel drive shaft inboard seal from the tri-lobal tripot bushing.

4. Slide the seal away from the joint along the wheel drive shaft bar.

5. Remove the housing (1) from the tripot joint spider and the wheel drive shaft bar (2).

6. Remove the guide (3) from the spring.

7. Remove the spring (4) from the tripot housing.

> **✳✳ CAUTION**
>
> **The correct 60 degree offset relationship between the inner and outer tripot spiders must be maintained. Accurately reference mark the tripot spider position on the wheel drive shaft bar before disassembly.**

8. Reference mark the position of the tripot spider (1) on the wheel drive shaft bar (2).

9. Using a brass drift and hammer, carefully tap around the tripot spider face in order to compress the barrel retaining ring on the wheel drive shaft bar.

10. Remove the tripot spider from the wheel drive shaft bar.

11. Remove and discard the barrel retaining ring from the wheel drive shaft bar.

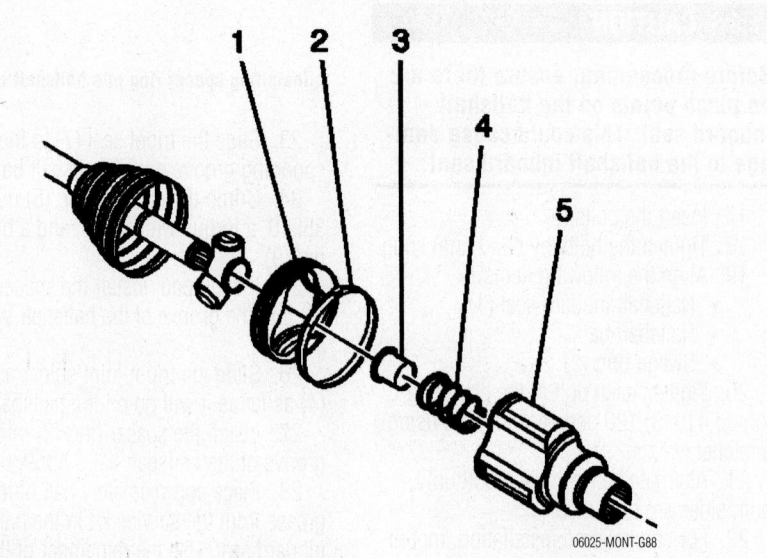

06025-MONT-G88

Exploded view of the rear (tripot) joint assembly

12. Remove the joint seal from the wheel drive shaft bar.

13. Inspect the following parts for damage or wear:
- Wheel drive shaft inboard seal
- Tripot joint spider assembly
- Housing
- Tri-lobal tripot bushing

To install:

14. Place the new small swage ring or eared clamp onto the small end of the joint seal. Slide the joint seal and the small swage ring or eared clamp onto the halfshaft bar.

15. Position the small end of the joint seal into the joint seal groove on the halfshaft bar.

16. For swage ring installation, use the J 41048.

17. Position the outboard end of the halfshaft assembly into the J 41048 holding tool.

18. Align the swage ring. Insert the bolts and tighten by hand until snug. Tighten each bolt 180 degrees at a time using a ratchet wrench. Alternate between each bolt until both sides are bottomed.

19. Loosen the bolts and separate the dies.

20. Check swaged ring for any lip deformities.

21. For eared clamp installation, mount the halfshaft into a vise.

22. Slide the tripot seal to the corresponding groove on the halfshaft bar.

23. If equipped, install the spacer ring into the groove of the halfshaft bar.

24. Slide the tripot joint spider assembly as far as it will go on the halfshaft bar.

25. Install the spacer ring into the groove of the halfshaft bar.

26. Place approximately half of the grease from the service kit in the halfshaft inboard seal. Use the remainder of the grease to repack the housing.

➡**Ensure the tri-lobal tripot bushing is flush with the face of the housing.**

27. Install the tri-lobal tripot bushing to housing.

28. Position the larger new seal retaining clamp on the halfshaft inboard seal.

29. Slide the housing over the tripot joint spider assembly on the halfshaft bar.

30. Slide the large diameter of the halfshaft inboard seal, with larger clamp in place, over the outside of the tri-lobal tripot bushing and locate the lip of the seal in the groove.

✳✳ WARNING

The halfshaft inboard seal must not be dimpled, stretched out or out of shape in any way. If the halfshaft inboard seal is not shaped correctly, carefully insert a thin flat blunt tool, no sharp edges, between the large seal opening and the tri-lobal tripot bushing in order to equalize the pressure. Shape the halfshaft inboard seal properly by hand.

31. Position the joint assembly at the proper vehicle dimension ("a"—4.0 inches (106mm).

32. Check the gap dimension (a) on the clamp ear. If the gap dimension is larger than shown, continue tightening until the gap dimension of 0.102 inch (2.6mm) is reached.

33. Fully stroke the joint several times to disperse the grease.

34. Reinstall the halfshaft

Axle Shaft, Bearing and Seal

REMOVAL & INSTALLATION

Rear

1. Raise and support the vehicle.
2. Remove the rear tire and wheel assembly.

➡**When removing the wheel, mark the location of the wheel to the hub prior to removal. Mark the individual location of all retainers as they are removed.**

3. Remove the axle shaft.
4. Remove the differential output shaft oil seal. Do not damage the differential sealing surfaces.

To install:

5. Install the axle shaft oil seal.

➡**Inspect the sealing surface of the halfshaft inner tri-pot housing to ensure it is free of corrosion. Use a crocus cloth in order to remove any light corrosion and clean the sealing surface with denatured alcohol, or equivalent.**

6. Lubricate the halfshaft sealing surface of the oil seal with synthetic gear oil.

7. Install the halfshaft.
8. Install the tire and wheel assembly.
9. Inspect the differential lubricant level.
10. Lower the vehicle.

Pinion Seal

REMOVAL & INSTALLATION

1. Before servicing the vehicle, refer to the Precautions Section.

2. Raise and support the vehicle.

3. Position a jack stand under the rear differential and firmly secure the differential to the jack.

4. Remove the front propeller shaft.

5. Remove the torque tube to bracket through bolt and nut.

6. Remove the torque tube to differential bolts.

7. Pull the torque tube toward the front of the vehicle in order to disengage the torque tube from the differential pinion shaft.

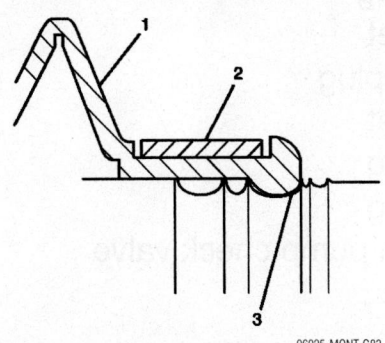

Assembling the CV joint components: joint seal (1), swage ring (2), joint seal groove (3)

06025-MONT-G83

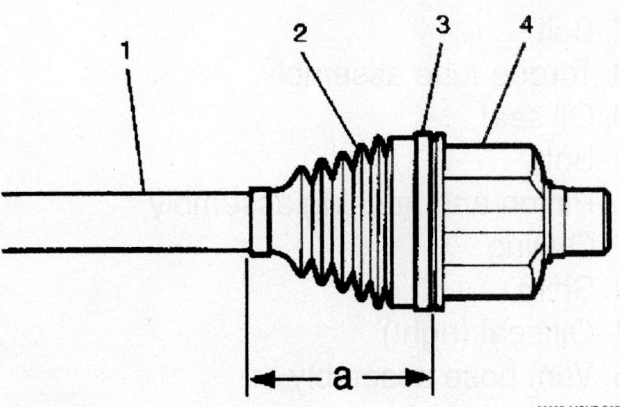

Installing joint assembly to proper dimension ("a")

06025-MONT-G87

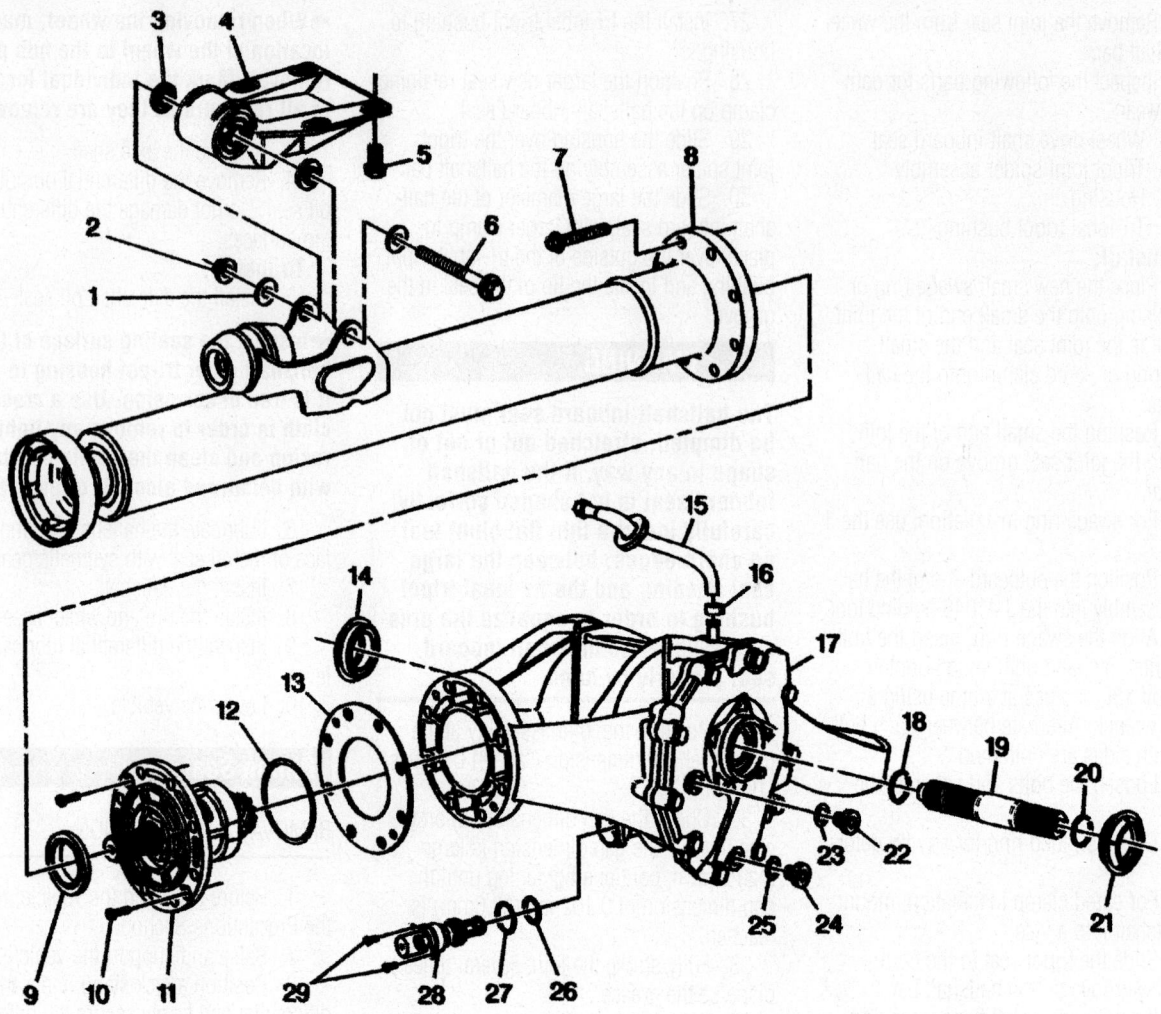

1. Washer
2. Nut
3. Spacer
4. Bracket
5. Bolt
6. Bolt
7. Bolt
8. Torque tube assembly
9. Oil seal
10. Bolt
11. Pinion and housing assembly
12. O-Ring
13. Shim
14. Oil seal (right)
15. Vent hose assembly
16. Clamp
17. Differential assembly
18. Snap ring
19. Axle shaft
20. Snap ring
21. Oil seal (left)
22. Fill plug
23. Gasket
24. Drain plug
25. Gasket
26. O-Ring
27. O-Ring
28. Clutch pump check valve
29. Bolts

Exploded view of differential assembly, showing pinion components

06025-MONT-G89

8. Remove the torque tube from the differential.

9. Remove the drive pinion housing oil seal and discard.

➡ **The internal components of the torque tube assembly cannot be serviced separately. If corrosion exists that cannot be polished off, or there is excessive scoring or wear, the torque tube must be replaced as a complete assembly.**

10. Inspect the sealing surface of the propeller shaft contained inside the torque tube, polish corrosion off with a crocus cloth.

To install:

11. Lubricate the new pinion housing assembly oil seal with synthetic gear oil.

12. Use the seal installer, J 44915, in order to install a new oil seal to the drive pinion housing.

13. Install the torque tube assembly.

14. Inspect the differential carrier lubricant level.

15. Align the splines on the torque tube shaft to the differential pinion shaft.

16. Install the torque tube to the differential.

17. Install the torque tube to bracket through bolt and tighten the torque tube to bracket through bolt nut to 47 ft. lbs. (64 Nm).

18. Install the torque tube-to-differential carrier bolts and tighten the bolts to 18 ft. lbs. (25 Nm).

19. Install the front propeller shaft.

20. Remove the jack stand from under the rear differential.

21. Lower the vehicle.

Differential Carrier

REMOVAL & INSTALLATION

1. Before servicing the vehicle, refer to the Precautions Section.

2. Set the parking brake.

3. Drain the rear differential.

4. Remove or disconnect the following:
   - Right rear tire and wheel assembly

➡ **When removing the wheel, mark the location of the wheel to the hub prior to removal. Mark the individual location of all retainers as they are removed.**

   - Electrical connector from the clutch pump check valve
   - Right rear halfshaft
   - Front propeller shaft
   - Place an adjustable support beneath the torque tube
   - Loosen, but do not remove the torque tube to bracket through bolt and nut
   - Bolts from the torque tube bracket
   - Differential carrier to cradle mounting bolts, nuts, washers, and mounts from the differential

✵✵ WARNING

During the removal of the halfshaft, the differential output shaft may become disengaged from the differential. If this occurs, firmly grasp and separate the output shaft from the halfshaft. Align the splines on the output shaft to the differential and reposition the output shaft to the differential.

   - While simultaneously moving the differential assembly to the right side of the vehicle, disengage the left halfshaft from the differential.
   - Rear differential and torque tube as an assembly
   - Torque tube from the differential

To install:

5. Install the torque tube to the differential.

6. Install the rear differential and torque tube assembly to the suspension cradle; at the same time, guide the left halfshaft onto the differential output shaft while positioning the differential assembly to the suspension cradle.

7. Place an adjustable support under the torque tube.

8. Ensure that the left halfshaft is fully engaged to the differential output shaft.

9. Install the differential carrier mounts, washers, bolts, and nuts to the differential. Tighten the differential carrier to cradle mounting bolts to 37 ft. lbs. (50 Nm).

10. Torque tube bracket to body bolts to 41 ft. lbs. (55 Nm). Tighten the torque tube to bracket through bolt and nut to 47 ft. lbs. (64 Nm).

11. Install the front propeller shaft.

12. Install the right rear halfshaft.

13. Install the right rear tire and wheel assembly.

14. Fill the axle with synthetic gear oil.

15. Inspect the differential oil level to ensure it is even with, but not lower than, 6mm (0.25 in) below the opening of the fill hole.

16. Attach the electrical connector to the clutch pump check valve.

17. Lower the vehicle.

STEERING AND SUSPENSION

Air Bag

SIR SERVICE PRECAUTIONS

✵✵ CAUTION

When performing service on or near the SIR components or the SIR wiring, the SIR system must be disabled. Refer to SIR Disabling and Enabling Zones. Failure to observe the correct procedure could cause deployment of the SIR components, personal injury, or unnecessary SIR system repairs.

The following are general service instructions which must be followed in order to properly repair the vehicle and return it to its original integrity:

- Do not expose inflator modules to temperatures above 65°C (150°F).
- Verify the correct replacement part number. Do not substitute a component from a different vehicle.
- Use only original GM replacement parts available from your authorized GM dealer. Do not use salvaged parts for repairs to the supplemental inflatable restraint (SIR) system.
- Discard any SIR component if it has

been dropped from a height of 3 ft. (91cm) or greater.

SIR DISABLING AND ENABLING ZONES

✵✵ CAUTION

Before disabling the SIR system, refer to SIR Service Precautions.

The supplemental inflatable restraint (SIR) system has been divided into Disabling and Enabling Zones. When performing service on or near SIR components or SIR wiring, it may be necessary to disable the SIR components in that zone. It may be

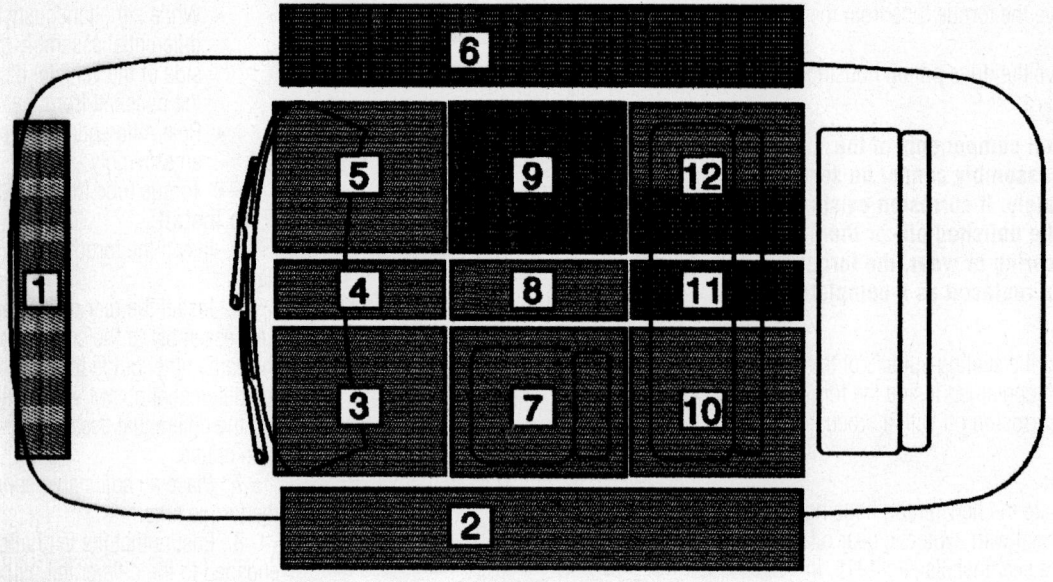

Zone	Description
1	Left and right front end
2	Driver/left side impact sensor (SIS) and seat belt retractor pretensioner
3	Inflatable restraint steering wheel module and coil
4	Not used
5	Inflatable restraint instrument panel (I/P) module
6	Passenger/right side impact module
7	Driver seat with LF side impact module
8	Not used
9	Passenger seat with RF side impact module, passenger presence system (PPS), and inflatable restraint sensing and diagnostic module (SDM)
10-12	Not used

06025-MONT-G36

Identifying the SIR System Zones

necessary to disable more than one zone depending on the location of other SIR components and the area being serviced. See the illustration to identify the specific zone or zones in which service will be performed. After identifying the zone or zones, proceed to the disabling and enabling procedures for that particular zone or zones.

DISARMING

Zone 1

✳✳ CAUTION

Before disabling the SIR system, refer to SIR Service Precautions.

1. Turn the steering wheel so that the vehicle's wheels are pointing straight ahead.
2. Turn OFF the ignition.
3. Remove the key from the ignition switch.

4. Open the hood and locate the under-hood fuse center.
5. Lift the cover for the underhood fuse center.

✳✳ WARNING

With the Air Bag Fuse removed and the ignition ON, the AIR BAG indicator illuminates. This is normal operation, and does not indicate an SIR System malfunction.

6. Locate and remove the air bag fuse from the underhood fuse center.
7. Open the front hood and locate both right and left front-end sensors (1), also known as the electronic front sensor (EFS).
8. Remove both connector position assurances (CPAs) from the right and left front-end sensor.
9. Disconnect both front-end sensor

wiring harness connectors from the left and right front-end sensor (1).
10. Open the front hood and locate both right and left front-end sensors, also known as the electronic front sensor (EFS).
11. Remove both connector position assurances (CPAs) from the right and left front-end sensor.

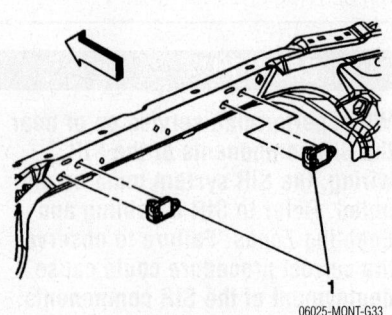

06025-MONT-G33

Showing the location of the electronic front sensors—Zone 1

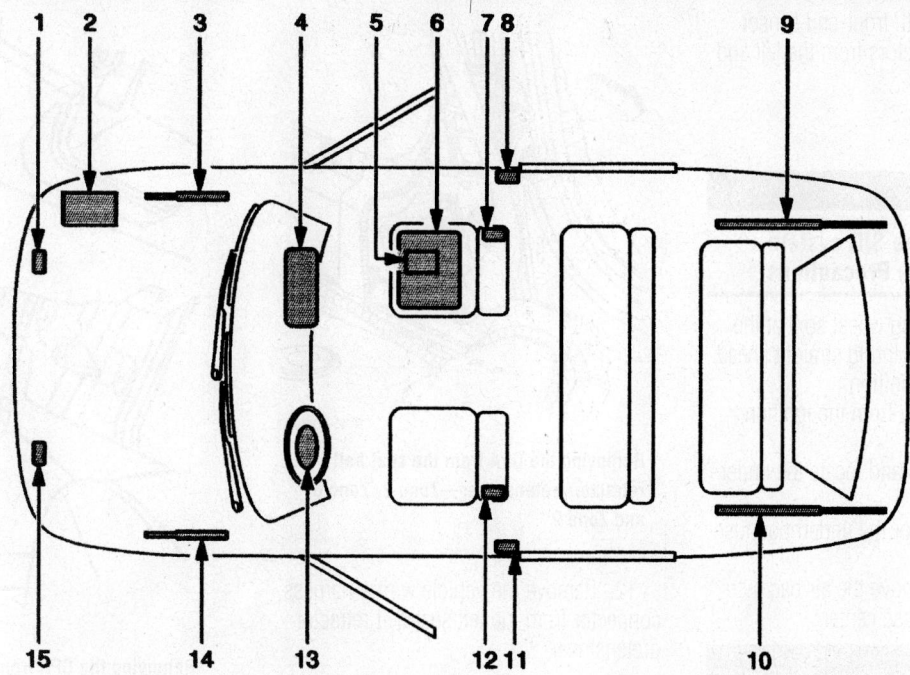

1- Right front end sensor - located on the front of the vehicle in the engine compartment

2- Vehicle battery- located under the hood on the right side

3- Front hood assist rod- a gas shock located under the front hood on the passenger side

4- I/P air bag- located at the top right under the instrument panel

5- Sensing and diagnostic module (SDM)- located underneath the passenger front seat

6- Passenger presence system (PPS)- located on the passenger front seat underneath the seat bottom trim

7- RF side impact air bag- located on the seat back of the passenger front seat

8- Right seat belt retractor pretensioner and right side impact sensor (SIS)- the right SIS is located above the right seat belt retractor pretensioner under the center pillar trim near the bottom on passenger side of vehicle

9- Rear compartment lid assist rod- a gas shock is located under the rear trunk lid on the passenger side

10- Rear compartment lid assist rod- a gas shock is locate under the rear trunk lid on the driver side

11- Left seat belt retractor pretensioner and left side impact sensor (SIS)- the left sis is located above the left seat belt retractor pretensioner under the center pillar trim near the bottom on the driver side of vehicle

12- LF side impact air bag- located on the seat back of the driver front seat

13- Steering wheel air bag- located on the steering wheel

14- Front hood assist rod- a gas shock located under the front hood on the driver side

15- Left front end sensor- located on the front of the vehicle in the engine compartment

06025-MONT-G37

Identifying components and their locations in the SIR System

12. Disconnect both front-end sensor wiring harness connectors from the left and right front-end sensor.

Zone 2

> ✱✱ **CAUTION**
>
> **Before disabling the SIR system, refer to SIR Service Precautions.**

1. Turn the steering wheel so that the vehicle's wheels are pointing straight ahead.
2. Turn OFF the ignition.
3. Remove the key from the ignition switch.
4. Open the hood and locate the underhood fuse center.
5. Lift the cover for the underhood fuse center.
6. Locate and remove the air bag fuse from the underhood fuse center.

> ✱✱ **CAUTION**
>
> **With the Air Bag Fuse removed and the ignition ON, the AIR BAG indicator illuminates. This is normal operation, and does not indicate an SIR System malfunction.**

7. Remove the driver/left lower center pillar trim cover.
8. Loosen the left side impact sensor (SIS) fasteners, then slide the left SIS (1) up and remove the sensor from the center pillar.
9. Remove the connector position assurance (CPA) (3) from the SIS connector (2).
10. Disconnect the SIS wiring harness connector (2) from the SIS (1).
11. Remove the CPA from the driver/left seat belt retractor pretensioner connector.

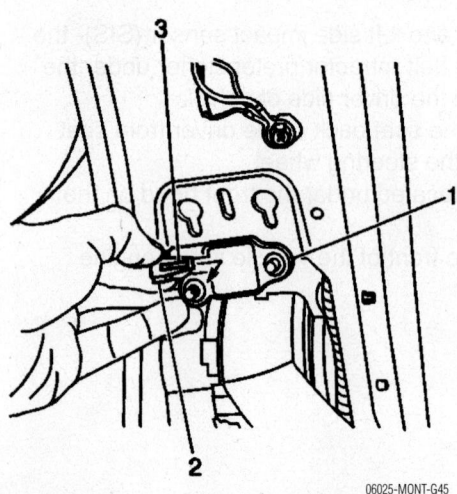

Removing the left side impact sensor from the driver's side pillar—Zone 2

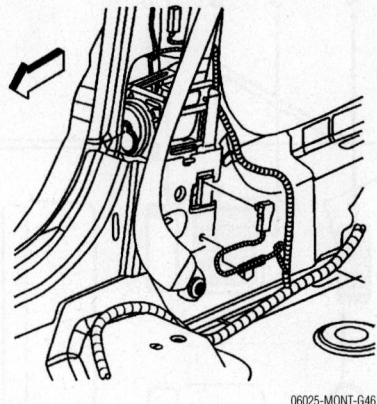

06025-MONT-G46

Removing the CPA from the seat belt retractor/pretensioner—Zone 2, Zone 6 and Zone 9

12. Remove the vehicle wiring harness connector from the left seat belt retractor pretensioner.

Zone 3

> ✱✱ **CAUTION**
>
> **Before disabling the SIR system, refer to SIR Service Precautions.**

1. Turn the steering wheel so that the vehicle's wheels are pointing straight ahead.
2. Turn OFF the ignition.
3. Remove the key from the ignition switch.
4. Open the hood and locate the underhood fuse center.
5. Lift the cover for the underhood fuse center.
6. Locate and remove the air bag fuse from the underhood fuse center.

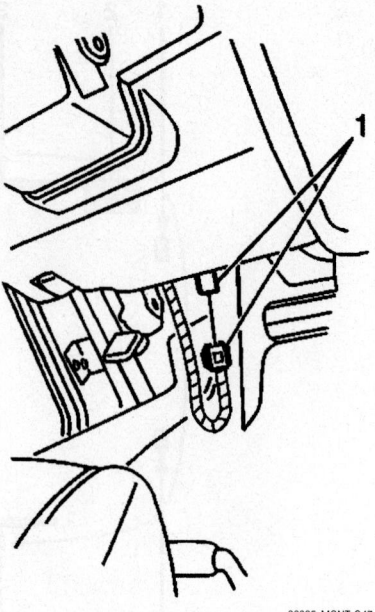

06025-MONT-G47

Removing the CPA from the steering wheel module coil yellow connector—Zone 3 and Zone 9

> ✱✱ **CAUTION**
>
> **With the Air Bag Fuse removed and the ignition ON, the AIR BAG indicator illuminates. This is normal operation, and does not indicate an SIR System malfunction.**

7. Remove the driver/left instrument panel (I/P) insulator panel.
8. Remove the connector position assurance (CPA) from the steering wheel module coil yellow connector (1) located at the base of the steering column.
9. Disconnect the steering wheel module coil connector (1).

➡**Zone 4 not used on these vehicles.**

Zone 5

> ✱✱ **CAUTION**
>
> **Before disabling the SIR system, refer to SIR Service Precautions.**

1. Turn the steering wheel so that the vehicle's wheels are pointing straight ahead.
2. Turn OFF the ignition.
3. Remove the key from the ignition switch.
4. Open the hood and locate the underhood fuse center.
5. Lift the cover for the underhood fuse center.
6. Locate and remove the air bag fuse from the underhood fuse center.

✳✳ CAUTION

With the Air Bag Fuse removed and the ignition ON, the AIR BAG indicator illuminates. This is normal operation, and does not indicate an SIR System malfunction.

7. Remove the passenger/right instrument panel (I/P) insulator panel.

8. Remove the connector position assurance (CPA) (2) from the I/P module yellow connector (1).

9. Disconnect the I/P module connector (1).

Zone 6

✳✳ CAUTION

Before disabling the SIR system, refer to SIR Service Precautions.

1. Turn the steering wheel so that the vehicle's wheels are pointing straight ahead.

2. Turn OFF the ignition.

3. Remove the key from the ignition switch.

4. Open the hood and locate the underhood fuse center.

5. Lift the cover for the underhood fuse center.

6. Locate and remove the air bag fuse from the underhood fuse center.

✳✳ CAUTION

With the Air Bag Fuse removed and the ignition ON, the AIR BAG indicator illuminates. This is normal operation, and does not indicate an SIR system malfunction.

7. Remove the passenger/right lower center pillar trim cover.

8. Loosen the right side impact sensor (SIS) fasteners, then slide the right SIS (1) up and remove from the center pillar.

9. Remove the connector position assurance (CPA) (3) from the SIS connector (2).

10. Disconnect the SIS wiring harness connector (2) from the SIS (1).

11. Remove the CPA from the passenger/right seat belt retractor pretensioner connector.

12. Remove the vehicle wiring harness connector from the right seat belt retractor pretensioner.

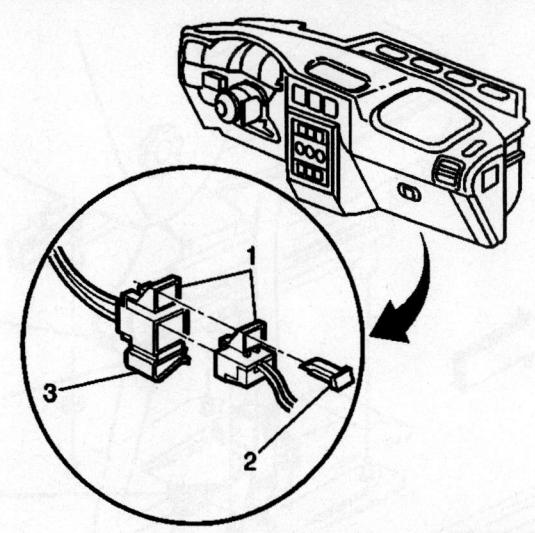

Removing the CPA from the instrument panel module yellow connector—Zone 5

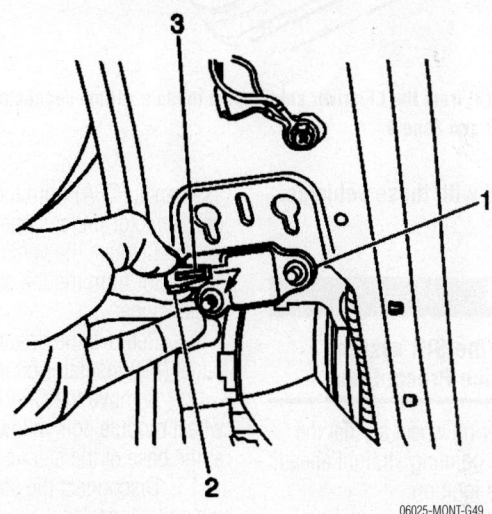

Removing the right side impact sensor from the center pillar—Zone 6

Zone 7

✳✳ CAUTION

Before disabling the SIR system, refer to SIR Service Precautions.

1. Turn the steering wheel so that the vehicle's wheels are pointing straight ahead.

2. Turn OFF the ignition.

3. Remove the key from the ignition switch.

4. Open the hood and locate the underhood fuse center.

5. Lift the cover for the underhood fuse center.

✳✳ CAUTION

With the Air Bag Fuse removed and the ignition ON, the AIR BAG indicator illuminates. This is normal operation, and does not indicate an SIR system malfunction.

6. Locate and remove the air bag fuse from the underhood fuse center

7. Remove the connector position assurance (CPA) (3) from the LF/driver side impact module yellow connector (3), which is located under the driver seat.

8. Disconnect the vehicle harness connector (4) from the LF side impact module connector (1).

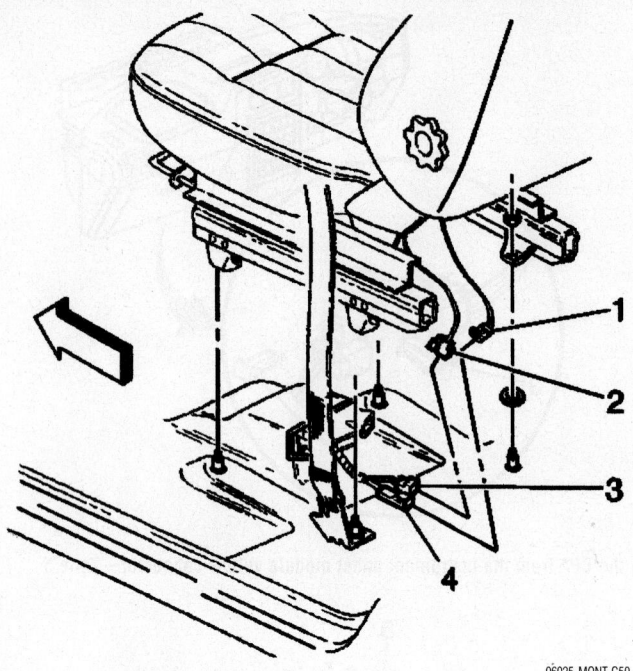

06025-MONT-G50

Removing the CPA (3) from the LF/driver side impact module yellow connector (3) under the driver seat—Zone 7 and Zone 9

➡**Zone 8 not used with these vehicles.**

Zone 9

✳ CAUTION

Before disabling the SIR system, refer to SIR Service Precautions.

1. Turn the steering wheel so that the vehicle's wheels are pointing straight ahead.
2. Turn OFF the ignition.
3. Remove the key from the ignition switch.
4. Open the hood and locate the underhood fuse center.
5. Lift the cover for the underhood fuse center.

✳ CAUTION

With the Air Bag Fuse removed and the ignition ON, the AIR BAG indicator illuminates. This is normal operation, and does not indicate an SIR system malfunction.

6. Locate and remove the air bag fuse from the underhood fuse center.
7. When disabling only the side impact module—RF proceed to step 21. If the entire SIR system needs to be disabled, then go to step 7.
8. Remove the driver/left lower center pillar trim cover.
9. Remove the connector position assurance (CPA) from the driver/left seat belt retractor pretensioner connector.
10. Remove the vehicle wiring harness connector from the left seat belt retractor pretensioner.
11. Remove the driver/left instrument panel (I/P) insulator panel.
12. Remove the CPA from the steering wheel module coil yellow connector, located at the base of the steering column.
13. Disconnect the steering wheel module coil connector.
14. Remove the CPA (3) from the LF/driver side impact module yellow connector (3), which is located under the driver seat.
15. Disconnect the vehicle harness connector (4) from the LF side impact module connector (1).
16. Remove the passenger/right instrument panel (I/P) insulator panel.
17. Remove the CPA (2) from the I/P module yellow connector (1).
18. Disconnect the I/P module connector (1).
19. Remove the passenger/right lower center pillar trim cover.
20. Remove the CPA from the right seat belt retractor pretensioner connector.
21. Remove the vehicle wiring harness connector from the right seat belt retractor pretensioner.
22. Remove the CPA (3) from the RF/passenger side impact module yellow connector (3), which is located under the passenger seat.
23. Disconnect the vehicle harness connector (4) from the RF side impact module connector (1).

ARMING

Zone 1

1. Remove the key from the ignition.
2. Connect both front end sensor wiring harness connectors to the left and right front end sensor (1).
3. Install the CPAs to the left and right front end sensor connector.
4. Install the Air Bag Fuse.
5. Install the cover for the underhood fuse center.
6. Use caution while reaching in and turning the ignition switch to the ON posi-

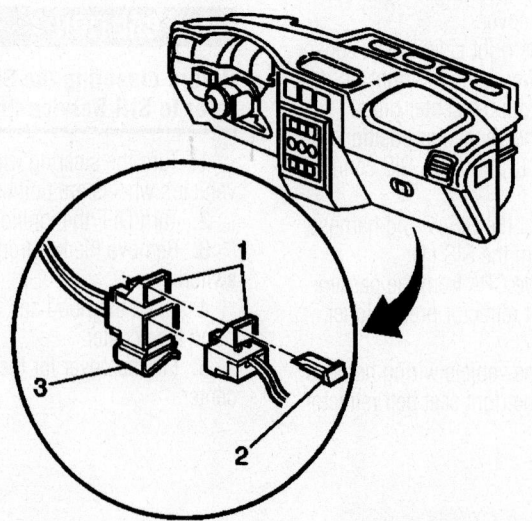

06025-MONT-G48

Removing the CPA (2) from the instrument panel module yellow connector (1)—Zone 9

tion. The AIR BAG indicator will flash, then turn OFF.

7. Perform the Diagnostic System Checks. Ensure the Air Bag indicator operates properly.

Zone 2

1. Remove the key from the ignition.
2. Connect the vehicle wiring harness connector to the left seat belt retractor pretensioner.
3. Install the CPA to the left seat belt retractor pretensioner connector.
4. Connect the left SIS wiring harness connector (2) to the left SIS (1).
5. Install the CPA (3) to the SIS connector (1).
6. Position the SIS (1) back inside the center pillar and slide back into place and tighten the SIS fasteners.
7. Install the left lower center pillar trim cover.
8. Install the Air Bag Fuse.
9. Install the cover for the underhood fuse center.
10. Use caution while reaching in and turning the ignition switch to the ON position. The AIR BAG indicator will flash, then turn OFF.
11. Perform the Diagnostic System Check, if the AIR BAG warning indicator does not operate as described.

Zone 3

1. Remove the key from the ignition.
2. Connect the steering wheel module coil yellow connector (1).
3. Install the CPA to the steering wheel module coil connector (1) located at the base of the steering column.
4. Install the left I/P insulator panel.
5. Install the Air Bag Fuse.
6. Install the cover for the underhood fuse center.
7. Use caution while reaching in and turning the ignition switch to the ON position. The AIR BAG indicator will flash, then turn OFF.
8. Perform the Diagnostic System Check, if the AIR BAG warning indicator does not operate as described.

Zone 5

1. Remove the key from the ignition.
2. Connect the I/P module yellow connector (1).
3. Install the CPA (2) to the I/P module connector (1).
4. Install the right I/P insulator panel.
5. Install the Air Bag Fuse.

6. Install the cover for the underhood fuse center.
7. Use caution while reaching in and turning the ignition switch to the ON position. The AIR BAG indicator will flash, then turn OFF.
8. Perform the Diagnostic System Check, if the AIR BAG warning indicator does not operate as described.

Zone 6

1. Remove the key from the ignition.
2. Connect the vehicle wiring harness connector to the right seat belt retractor pretensioner.
3. Install the CPA to the right seat belt retractor pretensioner connector.
4. Connect the right SIS wiring harness connector (2) to the right SIS (1).
5. Install the CPA (3) to the SIS connector (1).
6. Position SIS (1) back inside the center pillar and slide back into place and tighten the SIS fasteners.
7. Install the right lower center pillar trim cover.
8. Install the Air Bag Fuse.
9. Install the cover for the underhood fuse center.
10. Use caution while reaching in and turning the ignition switch to the ON position. The AIR BAG indicator will flash, then turn OFF.
11. Perform the Diagnostic System Check, if the AIR BAG warning indicator does not operate as described.

Zone 7

1. Remove the key from the ignition.
2. Connect the vehicle harness connector (4) to the LF side impact module yellow connector (1).
3. Install the CPA to the LF side impact module connector (1).
4. Install the Air Bag Fuse.
5. Install the cover for the underhood fuse center.
6. Use caution while reaching in and turning the ignition switch to the ON position. The AIR BAG indicator will flash, then turn OFF.
7. Perform the Diagnostic System Check, if the AIR BAG warning indicator does not operate as described.

Zone 9

1. Remove the key from the ignition.
2. When enabling the side impact module—RF proceed to step 17. If the entire SIR system needs to be enabled, then go to step 3.

3. Connect the vehicle wiring harness connector to the left seat belt retractor pretensioner.
4. Install the CPA to the left seat belt retractor pretensioner connector.
5. Install the left lower center pillar trim cover.
6. Connect the steering wheel module coil yellow connector (1).
7. Install the CPA to the steering wheel module coil connector (1), located at the base of the steering column.
8. Install the left I/P insulator panel.
9. Connect the vehicle harness connector (4) to the LF side impact module yellow connector (1).
10. Install the CPA to the LF side impact module connector (1).
11. Connect the I/P module yellow connector (1).
12. Install the CPA (2) to the I/P module connector (1).
13. Install the right I/P insulator panel.
14. Connect the vehicle wiring harness connector to the right seat belt retractor pretensioner.
15. Install the CPA to the right seat belt retractor pretensioner connector.
16. Install the right lower center pillar trim cover.
17. Connect the vehicle harness connector (4) to the RF side impact module yellow connector (1).
18. Install the CPA to the RF side impact module connector (1).
19. Install the Air Bag Fuse.
20. Install the cover for the underhood fuse center.
21. Use caution while reaching in and turning the ignition switch to the ON position. The AIR BAG indicator will flash, then turn OFF.
22. Perform the Diagnostic System Check, if the AIR BAG warning indicator does not operate as described.

DIAGNOSTIC SYSTEM CHECKS

1. Ensure that the battery, and the vehicle primary power and ground systems are functioning correctly.
2. With scan tool attached, check for proper communication. Lack of communication may be due to a particular malfunction of a serial data circuit. Further scan tool or communications diagnosis may be required.
3. With the scan tool, check that all indicated engine electronic modules are operating in the incorrect power mode, based on key position. If not, this may cause other vehicle symptoms and/or DTCs to set.
4. With the scan tool, check for any

Power Mode Mismatch and correct the condition before checking for module DTCs or symptoms.

5. Ensure that all data link communication DTCs are diagnosed before system level DTCs.

6. Ensure that all electronic control unit (ECU) internal DTCs are diagnosed before other system level DTCs.

7. Ensure that all device voltage DTCs are diagnosed before other system level DTCs.

Power Steering Gear

REMOVAL & INSTALLATION

1. Before servicing the vehicle, refer to the Precautions Section.
2. Raise and support the vehicle.
3. Remove or disconnect the following:
   - Stabilizer shaft
   - Tie rod ends from the steering knuckles
   - Steering intermediate shaft from the steering gear
   - Rear frame bolts, use a utility stand in order to support the frame

✳✳ WARNING

Do not lower the rear of the frame too far as damage to the engine components nearest to the cowl may result.

   - Use the utility stand in order to lower the frame and the powertrain
   - Power steering gear heat shield

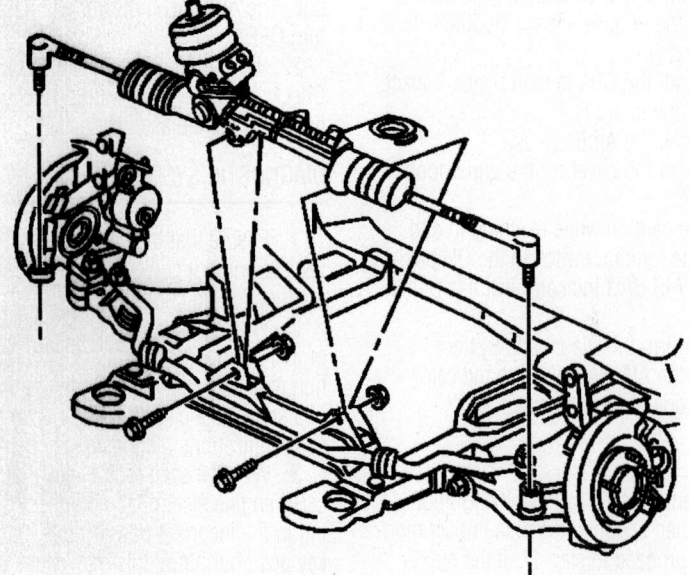

06025-MONT-G91

Installed position of the power steering gear

   - Power steering gear cooler pipe from the steering gear
   - Power steering gear pressure hose from the power steering gear
   - Power steering gear bolts and nuts
   - Power steering gear through the vehicle left side wheel opening

To install:
4. Install or connect the following:
   - Power steering gear through the vehicle left side wheel opening to the frame
   - Power steering gear bolts

➡ **This is a prevailing torque type fastener. This fastener may be reused ONLY if: The fastener and its counterpart are clean and free from rust. The fastener develops 18 inch lbs. (2 Nm) of torque/drag against its counterpart prior to the fastener seating. If the fastener does not meet these criteria, REPLACE the fastener.**

   - Power steering gear nuts and tighten nuts to 44 ft. lbs. (60 Nm) plus 60 degrees
   - Power steering gear heat shield
   - Power steering gear pressure hose to the power steering gear and tighten the fitting to 20 ft. lbs. (27 Nm)
   - Power steering gear cooler pipe to the power steering gear and tighten the fitting to 20 ft. lbs. (27 Nm)
   - Power steering gear cooler pipe retaining clip

➡ **Use the utility stand in order to raise the frame and the powertrain.**

   - Rear frame bolts and tighten the bolts to 122 ft. lbs. (165 Nm)
   - Remove the utility stand from the frame
   - Steering intermediate shaft to the steering gear
   - Tie rod ends to the steering knuckles
   - Stabilizer shaft
5. Lower the vehicle.
6. Fill the power steering system.
7. Bleed the power steering system.

Shocks/Struts

REMOVAL & INSTALLATION

Front

1. Before servicing the vehicle, refer to the Precautions Section.

✳✳ WARNING

Lift the vehicle using ONLY a frame contact vehicle lift. Do NOT lift the vehicle using a suspension contact vehicle lift.

2. Raise and support the vehicle.
3. Remove the wiper module as follows:
   a. Open and support the hood.
   b. Remove the driver side wiper arm
   c. Remove the passenger side wiper arm.
   d. Remove the air inlet grille panel.
   e. Lower the washer container into the engine compartment.
   f. Disconnect the electrical connector from the wiper motor.
   g. Remove the 4 bolts from the wiper module.
   h. Carefully guide the passenger side module out first from the opening in the front plenum.
   i. Pull the module from the drive side fender flange opening.
   j. Remove the module from the vehicle.
4. Remove the tire and wheel assembly, marking the location to the hub assembly. Mark the location of each hub retainer before removal.
5. Lower the vehicle.
6. Remove the strut upper nuts.
7. Raise the vehicle.
8. Scribe the strut to the knuckle position for installation reference.
9. Remove the strut lower clamp retaining bolts and nuts.
10. Remove the strut assembly from the vehicle.

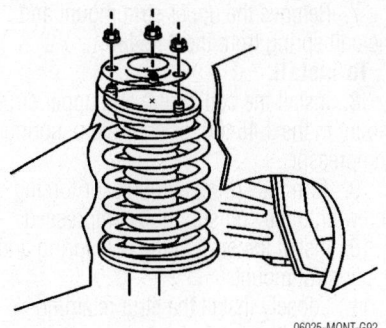

Removing the strut assembly upper nuts

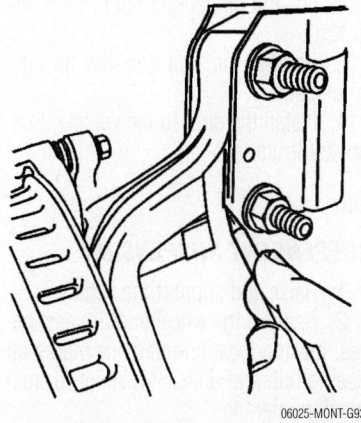

Scribing a reference line on the knuckle-to-strut position

To install:

➡**Care should be taken to avoid chipping or scratching the coating when handling the suspension coil spring. Damage to the coating can cause premature failure.**

11. Install the strut assembly into position in the vehicle. Ensure the upper studs align with the mounting holes and the lower clamp retaining portion aligns with the steering knuckle holes.

12. Lower the vehicle to

13. Install the strut upper nuts; tighten the strut upper nuts to 30 ft. lbs. (41 Nm)

14. Raise the vehicle.

15. Install the strut lower bolts and nuts.

➡**This is a prevailing torque type fastener. This fastener may be reused ONLY if: The fastener and its counterpart are clean and free from rust. The fastener develops 27 inch lbs. (3 Nm) of torque/drag against its counterpart prior to the fastener seating. If the fastener does not meet these criteria, REPLACE the fastener.**

16. Align the strut to the mark on the knuckle; tighten the strut lower nuts to 90 ft. lbs. (123 Nm).

17. Install the tire and wheel assembly. Tighten the lug nuts to 100 ft. lbs. (140 Nm) in a criss—cross pattern, after aligning the wheel hub with the reference mark and holes as shown.

✳✳ CAUTION

Be sure to use the correct fastener in the proper location, according to the illustration provided.

18. Install the wiper motor module as follows:

a. Position the wiper module to the lower plenum in the driver side fender flange opening first.

b. Rotate the module into the opening at the passenger side plenum.

c. Ensure the rear center mount is secure into the sheet metal flange at the rear of the module.

d. Install the wiper module bolts. Tighten the bolts in sequence, as shown, to 10 Nm (89 inch lbs.).

e. Connect the electrical connector to the motor.

f. Install the air inlet grille panel.

g. Install the driver side wiper arm and the passenger side wiper arm.

h. Secure the washer container.

i. Close the hood. Inspect the wiper system for proper operation.

19. Inspect the alignment.

Rear

1. Before servicing the vehicle, refer to the Precautions Section.

2. Support the rear axle.

➡**On independent rear suspension vehicles, use a utility stand to slightly raise the rear suspension at the shock absorber to compress the coil spring and relieve some spring tension prior to shock absorber removal.**

3. Remove or disconnect the following:
- If the vehicle is equipped with automatic level control, disconnect the air tube connector from the shock absorber
- Remove the shock absorber upper bolt and nut
- Remove the shock absorber lower bolt and nut
- Remove the shock absorber

To install:

4. Install the shock absorber.

5. Install the shock absorber bolts and nuts and tighten the shock absorber bolts and nuts to 63 ft. lbs. (85 Nm).

6. If the vehicle is equipped with automatic level control, install the air tube connector to the shock absorber.

7. Remove the support from the rear axle.

8. Lower the vehicle.

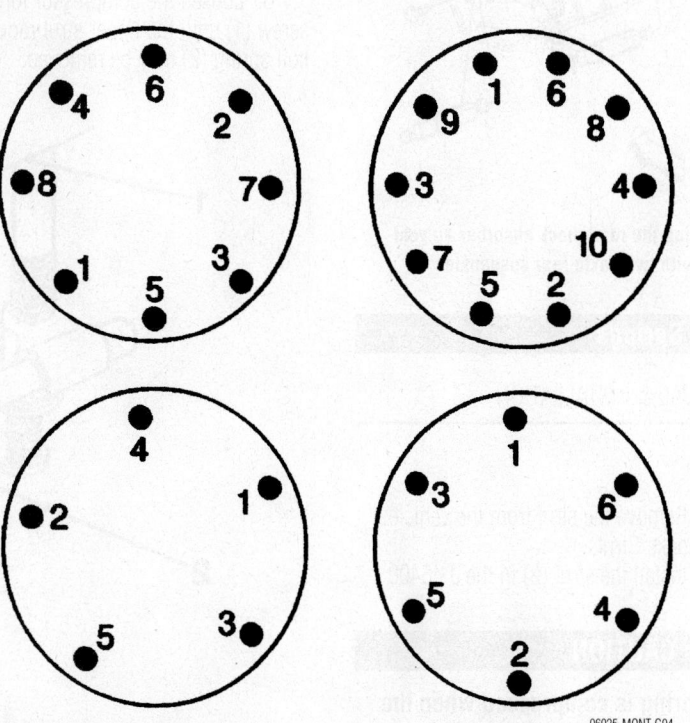

Showing wheel hub hole patterns, for proper alignment and installation

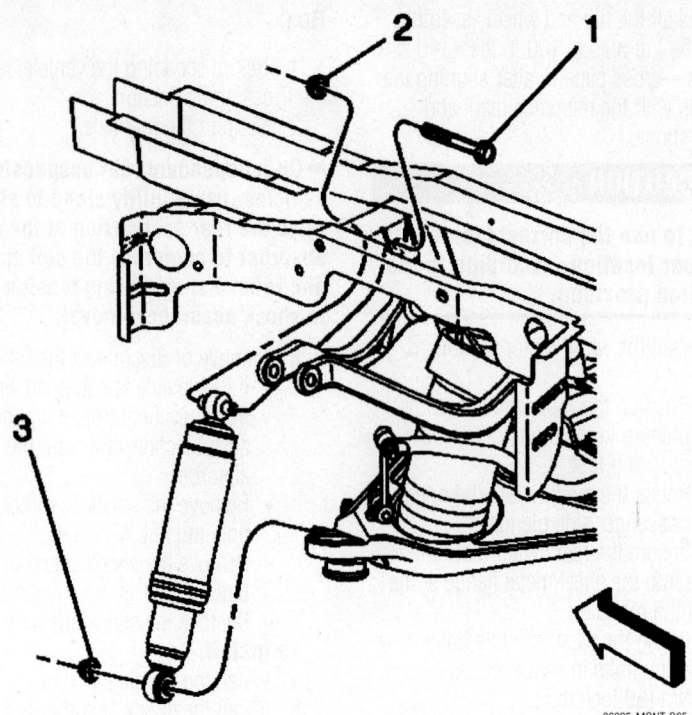

06025-MONT-G95

Removing the rear shock absorber on vehicles with independent rear suspension, showing upper nut and bolt (1, 2) and lower nut (3)

3. Turn the spring compressor forcing screw (1) until the coil spring (2) is compressed.

4. Use a 45 TORX® socket in order to hold the strut shaft. Use the J 42991 to remove the upper strut mount nut.

5. Remove the strut from the J 45400.

6. Loosen the compressor forcing screw (1) until the upper strut mount and coil spring (2) may be removed.

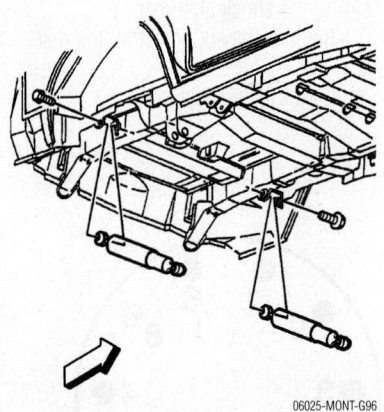

06025-MONT-G96

Showing the rear shock absorber on vehicles with twist axle rear suspension

Coil Springs

REMOVAL & INSTALLATION

Front

1. Remove the strut from the vehicle. See Shocks/Struts.

2. Install the strut (2) in the J 45400 (1).

✳✳ CAUTION

The spring is compressed when the strut moves freely.

7. Remove the upper strut mount and the coil spring from the J 45400.

To install:

8. Install the coil spring and upper strut mount to the J 45400, or equivalent, spring compressor.

9. Turn the spring compressor forcing screw until the coil spring is compressed.

10. Install the strut to the coil spring and upper strut mount.

11. Loosely install the strut retaining nut.

12. Use a 45 TORX socket in order to hold the strut shaft. Install the upper strut mount nut and tighten the nut to 63 ft. lbs. (85 Nm).

13. Remove the strut from the spring compressor.

14. Install the strut to the vehicle. See Shocks/Struts.

Rear

INDEPENDENT SUSPENSION

1. Raise and support the vehicle.

2. Remove the wheel and tire assemblies, making location reference marks on wheel to hub and on each retainer for reinstallation reference.

3. Remove the brake calipers. See Disc Brakes.

4. Remove the parking brake cable mounting bracket. See Parking Brake.

5. Remove the brake caliper mounting bracket. See Disc Brakes.

6. Disconnect the ABS electrical connector from the wheel speed sensor.

7. Remove the ABS electrical connector mounting bracket.

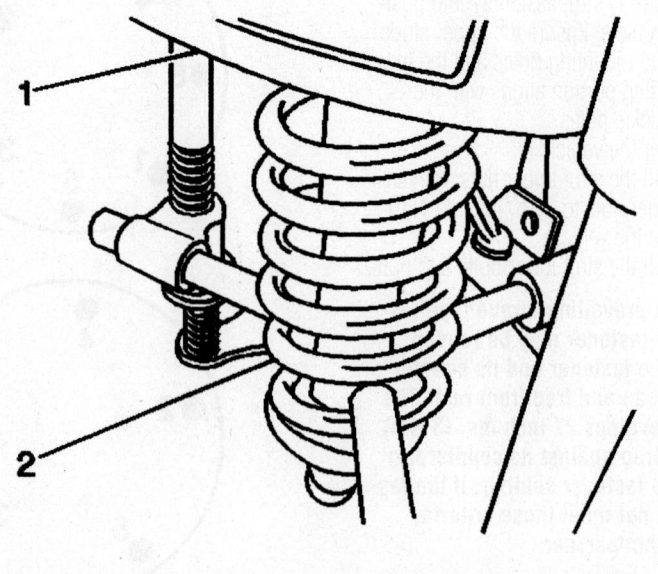

06025-MONT-G98

Showing strut (2) installed in J 45400 coil spring retaining/compressing device (1)

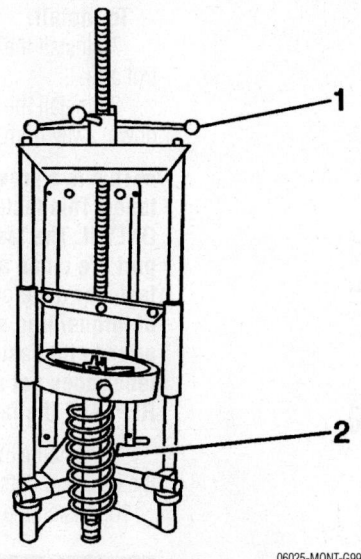

06025-MONT-G99

Turning the forcing screw (1) until the upper strut mount and coil spring (2) can be removed

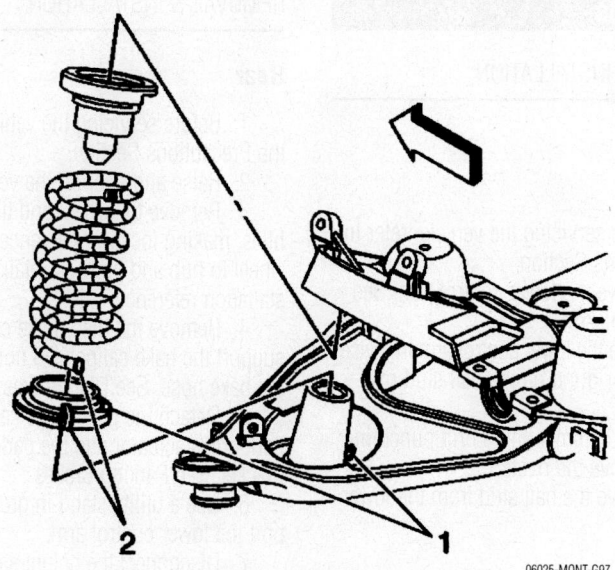

06025-MONT-G97

Showing coil spring position, showing spring alignment marks (1, 2)

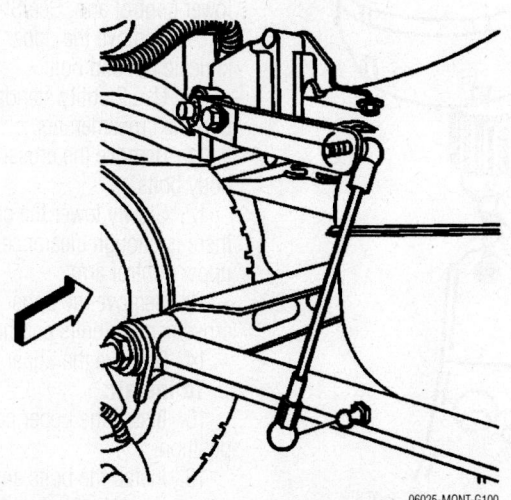

06025-MONT-G100

Disconnecting height sensor link from the lower control arm

8. Use the utility jack in order to support the lower control arm.

9. Remove the stabilizer link from the lower control arm.

10. Use the J 41820 in order to disconnect the ball joint from the knuckle.

11. Use the utility jack in order to lower the lower control arm and relieve the coil spring tension.

12. Remove the coil spring.

To install:

13. Install the coil spring.

14. Use the utility jack in order to raise the lower control arm and load the suspension.

15. Use the J 41820 in order to connect the ball joint to the knuckle.

16. Install the stabilizer link to the lower control arm.

17. Use the utility jack in order to support the lower control arm.

18. Install the ABS electrical connector mounting bracket.

19. Connect the ABS electrical connector.

20. Install the brake caliper mounting bracket. See Disc Brakes.

21. Install the parking brake cable mounting bracket.

22. Install the brake calipers. See Disc Brakes.

23. Install the tire and wheel assembly. Tighten the lug nuts to 100 ft. lbs. (140 Nm) in a criss—cross pattern, after aligning the wheel hub with the reference mark and holes as shown in appropriate illustration.

24. Lower the vehicle.

WITH ALL WHEEL DRIVE

1. Raise and support the vehicle.

2. Remove the wheel and tire assemblies, making location reference marks on wheel to hub and on each retainer for reinstallation reference.

3. Remove the brake caliper and support the brake caliper. Do not disconnect the brake hose. See Disc Brakes.

4. Disconnect the tie rod from the knuckle. See Tie Rods.

5. Use a utility stand in order to support the lower control arm.

6. Disconnect the height sensor link from the lower control arm, as shown.

7. Disconnect the shock absorber from the lower control arm. See Shocks/Struts.

8. Use the J 41820 in order to disconnect the ball joint from the knuckle. See Ball Joints.

9. Remove the lower ball joint nut.

10. Use the utility stand in order to lower the control arm and relieve the coil spring tension.

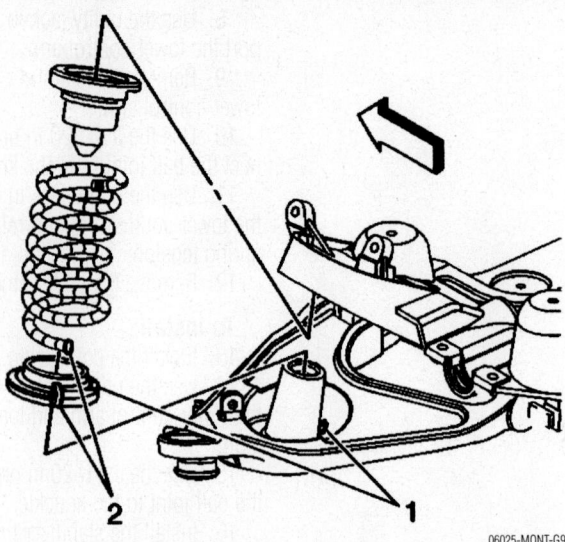

06025-MONT-G97

Showing coil spring position, showing spring alignment marks (1, 2)

11. Carefully remove the coil spring and jounce bumper.

12. Remove the spring insulator (2).

13. Remove the lower control arm.

To install:

14. Install the spring jounce bumper to the spring.

15. Index and install the spring insulator to the spring (2).

16. Index and install the spring insulator with the spring and the jounce bumper to the lower control arm (1).

17. Use the utility stand in order to raise the lower control arm.

18. Install the lower ball joint to the knuckle.

19. Install the lower ball joint nut. Tighten the ball joint nut to 26 ft. lbs. (35 Nm), plus 130 degrees.

20. Connect the shock absorber to the lower control arm. See Shocks/Struts.

21. Connect the height sensor link to the lower control arm.

22. Connect the tie rod to the knuckle. See Tie Rods.

23. Install the brake caliper. See Disc Brakes.

24. Install the tire and wheel assembly. Tighten the lug nuts to 100 ft. lbs. (140 Nm) in a criss—cross pattern, after aligning the wheel hub with the reference mark and holes as shown in appropriate illustration.

25. Lower the vehicle.

Ball Joints

REMOVAL & INSTALLATION

Front

LOWER

1. Before servicing the vehicle, refer to the Precautions Section.

2. Remove the lower control arm. See Lower Control Arm.

3. Secure the lower control arm in a vice.

4. Drill or grind off the ball stud rivet heads.

5. Use a hammer and a drift punch in order to remove the rivets.

6. Remove the ball stud from the lower control arm.

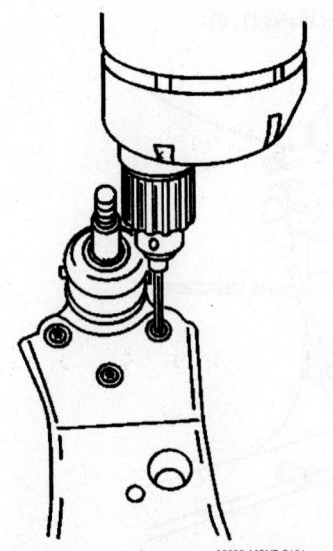

06025-MONT-G101

Drilling rivets to remove lower ball joint

To install:

7. Install the ball stud to the lower control arm.

8. Install the NEW ball stud bolts facing down, away from the ball stud.

➡**This is a prevailing torque type fastener. This fastener may be reused ONLY if: The fastener and its counterpart are clean and free from rust. The fastener develops 18 inch lbs. (2 Nm) of torque/drag against its counterpart prior to the fastener seating. If the fastener does not meet these criteria, REPLACE the fastener.**

9. Install NEW ball stud nuts, tighten the ball stud nuts to 50 ft. lbs. (68 Nm).

10. Install the lower control arm.

Upper Control Arm

REMOVAL & INSTALLATION

Rear

1. Before servicing the vehicle, refer to the Precautions Section.

2. Raise and support the vehicle.

3. Remove the wheel and tire assemblies, making location reference marks on wheel to hub and on each retainer for reinstallation reference.

4. Remove the disc brake caliper and support the bake caliper. Do not disconnect the bake hose. See Disc Brakes.

5. Detach the park brake cable from the park brake actuator and the park brake cable bracket. See Parking Brakes.

6. Use a utility stand in order to support the lower control arm.

7. Disconnect the height sensor link from the lower control arm, if equipped.

8. Remove the shock absorber from the lower control arm. See Shocks/Struts.

9. Remove the upper control arm to knuckle nut and bolt.

10. Use 2 utility stands in order to support the crossmember.

11. Remove the crossmember to underbody bolts.

12. Slowly lower the crossmember until there is enough clearance to remove the upper control arm.

13. Remove the upper control arm to crossmember bolts and nuts.

14. Remove the upper control arm.

To install:

15. Install the upper control arm into position.

16. Install the bolts and the nuts which secure the upper control arm to the cross-

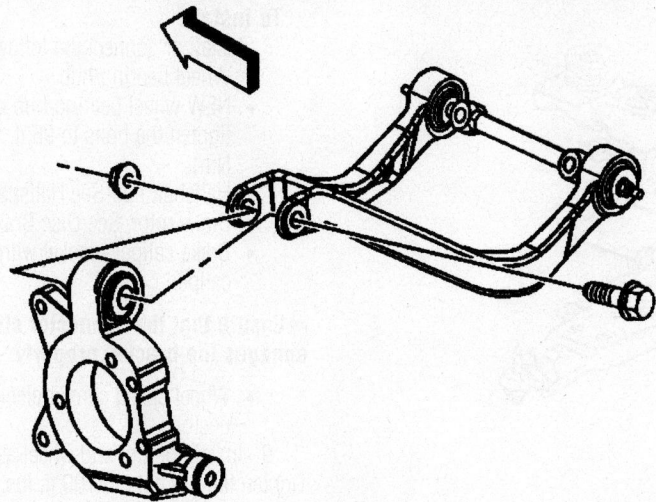

06025-MONT-G102

Removing nut and bolt retaining upper control arm to knuckle

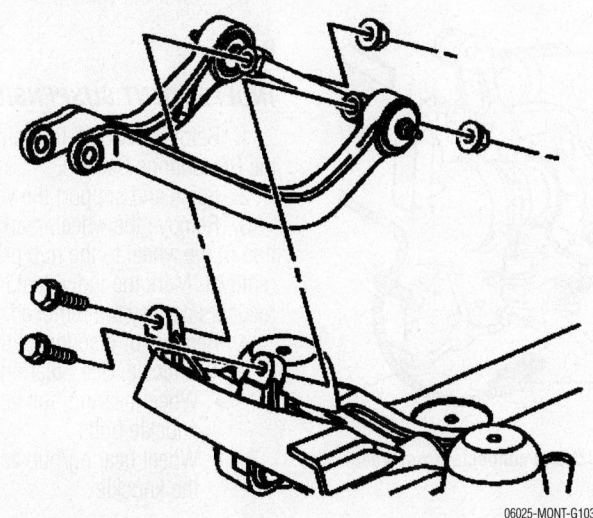

06025-MONT-G103

Removing nut and bolt retaining upper control arm to crossmember

member. Tighten the nuts to 55 ft. lbs. (75 Nm).

17. Use 2 utility stands in order to raise the crossmember. Install and tighten the crossmember to underbody bolts to 96 ft. lbs. (130 Nm).

18. Install or connect the following:
- Upper control arm to the knuckle, tighten the nut which secures the upper control arm to 74 ft. lbs. (100 Nm)
- Shock absorber to the lower control arm
- Height sensor link to the lower control arm as necessary
- Park brake cable to the park brake actuator and the park brake cable bracket
- Brake caliper

19. Install the tire and wheel assembly. Tighten the lug nuts to 100 ft. lbs. (140 Nm)

in a criss—cross pattern, after aligning the wheel hub with the reference mark and holes as shown in appropriate illustration.

20. Lower the vehicle.

Lower Control Arm

REMOVAL & INSTALLATION

Front

1. Before servicing the vehicle, refer to the Precautions Section.

2. Raise and support the vehicle.

3. Remove the wheel, marking the location of the wheel to the hub prior to removal. Mark the individual location of all retainers as they are removed.

4. Detach the ABS wheel speed sensor connector. See Disc Brakes.

5. Remove the ABS wheel speed sensor jumper harness from the harness retainer clips. See Disc Brakes.

6. Remove the stabilizer shaft link. See Stabilizer Bar.

7. Remove the cotter pin from the ball stud and loosen the ball stud nut.

8. Install the ball joint separator tool, J 41820, over the ball stud and lower control arm as shown.

9. Rotate the ball stud nut counter-clockwise in order to separate the ball stud from the steering knuckle.

10. Remove the lower control arm bolts and nuts, and then the lower control arm.

To install:

11. Install or connect the following:
- Lower control arm
- Control arm bolts and nuts
- Ball stud to the knuckle
- Ball stud castle nut, tighten the ball stud castle nut to 22 ft. lbs. (30 Nm), plus 135 degrees

➥**Do NOT loosen the ball stud nut in order to align the ball stud nut slots to the ball stud cotter pin hole. If necessary, tighten the ball stud castle nut in order to align the ball stud castle nut slot to the ball stud cotter pin hole. Ensure that the cotter pin ends do NOT contact the ABS wheel speed sensor, the ABS sensor connector or the drive axle.**

- NEW cotter pin and bend the ends
- Stabilizer shaft link
- ABS wheel speed sensor jumper harness to the harness retainer clips
- ABS wheel speed sensor connector

➥**This is a prevailing torque type fastener. This fastener may be reused ONLY if: The fastener and its counterpart are clean and free from rust. The fastener develops 18 inch lbs. (2 Nm) of torque/drag against its counterpart prior to the fastener seating. If the fastener does not meet these criteria, REPLACE the fastener.**

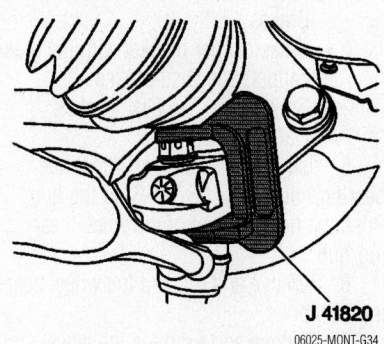

J 41820

06025-MONT-G34

Using the J 41820 removal tool

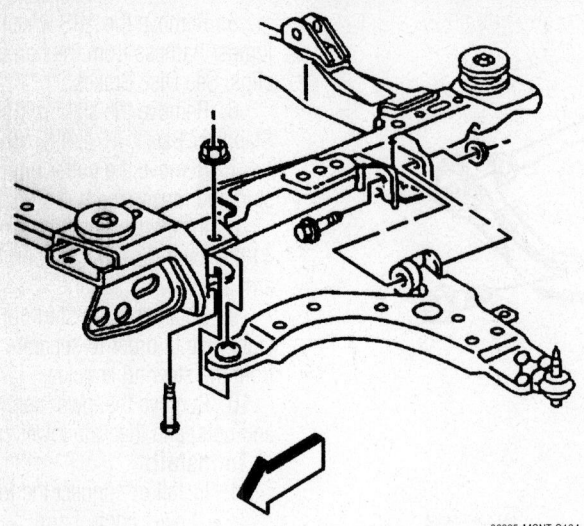

06025-MONT-G104

Showing the lower control arm

• Lower control arm nuts, tighten the lower control arm nuts to 71 ft. lbs. (96 Nm)

12. Install the tire and wheel assembly. Tighten the lug nuts to 100 ft. lbs. (140 Nm) in a criss—cross pattern, after aligning the wheel hub with the reference mark and holes as shown in appropriate illustration.

13. Lower the vehicle

Wheel Hub/Bearing Assembly

REMOVAL & INSTALLATION

Front

1. Before servicing the vehicle, refer to the Precautions Section.

2. Raise and support the vehicle.

3. Remove the wheel, marking the location of the wheel to the hub prior to removal. Mark the individual location of all retainers as they are removed.

4. Remove or disconnect the following:
• Wheel speed sensor electrical connector. See Disc Brakes.
• Wheel speed sensor electrical connector from the bracket. See Disc Brakes
• Brake caliper bracket with the brake caliper. See Disc Brakes
• Brake rotor. See Disc Brakes.
• Halfshaft nut

5. Use 3 wheel nuts and attach the bearing/hub tool, J 28733-B, to the hub. Push the halfshaft out of the wheel bearing/hub.

6. Remove and discard the wheel bearing/hub bolts.

7. Remove and examine the wheel bearing/hub.

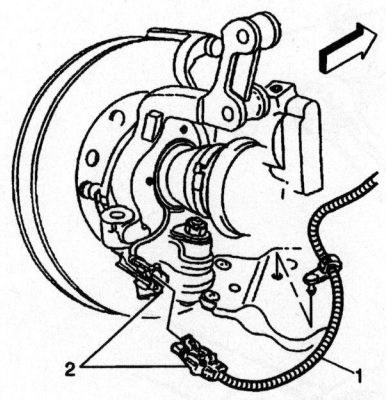

06025-MONT-G105

Removing electrical connector from wheel speed sensor

To install:

8. Install or connect the following:
• Wheel bearing/hub
• NEW wheel bearing/hub bolts and tighten the bolts to 96 ft. lbs. (130 Nm)
• Halfshaft nut. See Halfshaft
• Brake rotor. See Disc Brakes
• Brake caliper bracket with the brake caliper

➡**Ensure that the connector clip engages the bracket properly.**

• Wheel speed sensor electrical connector

9. Install the tire and wheel assembly. Tighten the lug nuts to 100 ft. lbs. (140 Nm) in a criss—cross pattern, after aligning the wheel hub with the reference mark and holes as shown in appropriate illustration.

10. Lower the vehicle.

Rear

INDEPENDENT SUSPENSION

1. Before servicing the vehicle, refer to the Precautions Section.

2. Raise and support the vehicle.

3. Remove the wheel, marking the location of the wheel to the hub prior to removal. Mark the individual location of all retainers as they are removed.

4. Remove or disconnect the following:
• Knuckle. See Steering Knuckle
• Wheel bearing/hub assembly knuckle bolts
• Wheel bearing/hub assembly from the knuckle

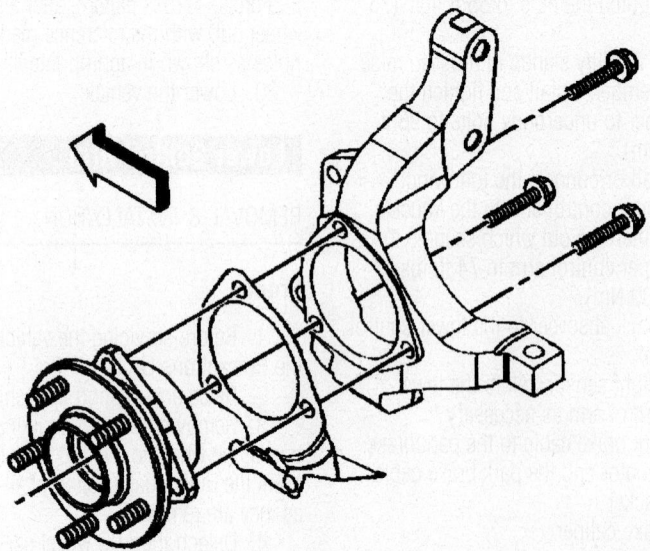

06025-MONT-G106

Exploded view of the wheel bearing/hub assembly

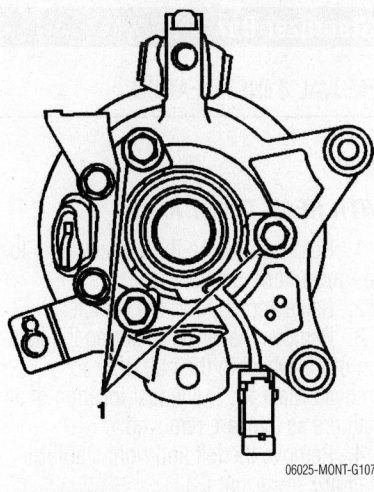

06025-MONT-G107

Identifying wheel bearing/hub-to-knuckle retaining bolts

To install:

5. Install or connect the following:
   - Wheel bearing/hub assembly on to the knuckle
   - Bolts into the wheel bearing/hub assembly, tighten the wheel bearing/hub bolts to 96 ft. lbs. (130 Nm)

6. Install the tire and wheel assembly. Tighten the lug nuts to 100 ft. lbs. (140 Nm) in a criss—cross pattern, after aligning the wheel hub with the reference mark and holes as shown in appropriate illustration.

7. Lower the vehicle.

Rear

WITHOUT INDEPENDENT SUSPENSION

1. Before servicing the vehicle, refer to the Precautions Section.
2. Raise and support the vehicle.
3. Remove the wheel, marking the location of the wheel to the hub prior to removal. Mark the individual location of all retainers as they are removed.
4. Remove or disconnect the following:
   - Brake caliper

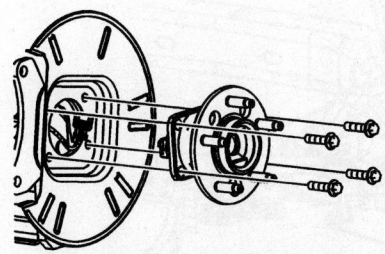

06025-MONT-G108

Showing mounting view of rear bearing/hub assembly mounting

- Wheel bearing/hub bolts, discard bolts
- Wheel bearing/hub
- Wheel speed sensor electrical connector

To install:

5. Install or connect the following:
   - Wheel speed sensor electrical connector
   - Wheel bearing/hub

❈❈ CAUTION

These fasteners MUST be replaced with new fasteners anytime they become loose or are removed. Failure to replace these fasteners after they become loose or are removed may cause loss of vehicle control and personal injury.

- NEW wheel bearing/hub bolts, tighten the bolts to 59 ft. lbs. (80 Nm)
- Brake caliper. See Disc Brakes

6. Install the tire and wheel assembly. Tighten the lug nuts to 100 ft. lbs. (140 Nm) in a criss—cross pattern, after aligning the wheel hub with the reference mark and holes as shown in appropriate illustration.

7. Lower the vehicle.

Steering Knuckle

REMOVAL & INSTALLATION

Front

1. Before servicing the vehicle, refer to the Precautions Section.
2. Raise and support the vehicle.
3. Remove the wheel, marking the location of the wheel to the hub prior to

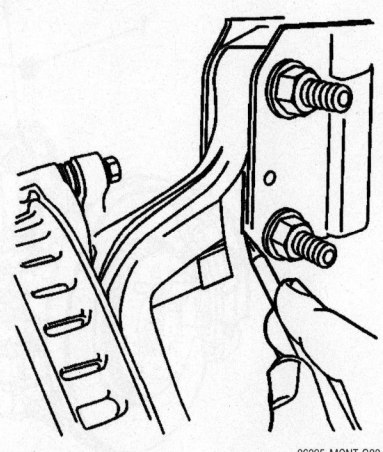

06025-MONT-G93

Scribing a reference line on the knuckle-to-strut position

removal. Mark the individual location of all retainers as they are removed.

4. Remove or disconnect the following:
   - Front halfshaft bearing. See Wheel Hub/Bearing Assembly.
   - Front lower control arm ball stud. See Lower Control Arm.
   - Outer tie rod end from the steering knuckle
   - Scribe the strut to the knuckle for reinstallation reference
   - Bolts connecting the strut to the knuckle
   - Steering knuckle

To install:

5. Install or connect the following:
   - Steering knuckle
   - Bolts which connect the strut to the knuckle, tighten the strut to knuckle bolts to 83 ft. lbs. (112 Nm)
   - Outer tie rod to the steering knuckle
   - Front lower control arm ball stud to knuckle; tighten to 22 ft. lbs. (30 Nm), plus 135 degrees.
   - Front halfshaft bearing. See Wheel Hub/Bearing Assembly.

6. Install the tire and wheel assembly. Tighten the lug nuts to 100 ft. lbs. (140 Nm) in a criss—cross pattern, after aligning the wheel hub with the reference mark and holes as shown in appropriate illustration.

7. Lower the vehicle

Rear

1. Before servicing the vehicle, refer to the Precautions Section.
2. Raise and support the vehicle.
3. Remove the wheel, marking the location of the wheel to the hub prior to removal. Mark the individual location of all retainers as they are removed.
4. Remove or disconnect the following:
   - Brake caliper and support the bake caliper. Do not disconnect the bake hose. See Disc Brakes.
   - Brake caliper bracket
   - Brake rotor. See Disc Brakes.
   - Tie rod from the knuckle
   - Stabilizer shaft link at the lower control arm
   - Halfshaft from the wheel bearing/hub. See Wheel Hub/Bearing Assembly.
   - Park brake cable and the park brake cable bracket aside
   - Wheel bearing and the backing plate
   - Use a utility stand in order to support the lower control arm
   - Height sensor link from the lower control arm as necessary. See Lower Control Arm.

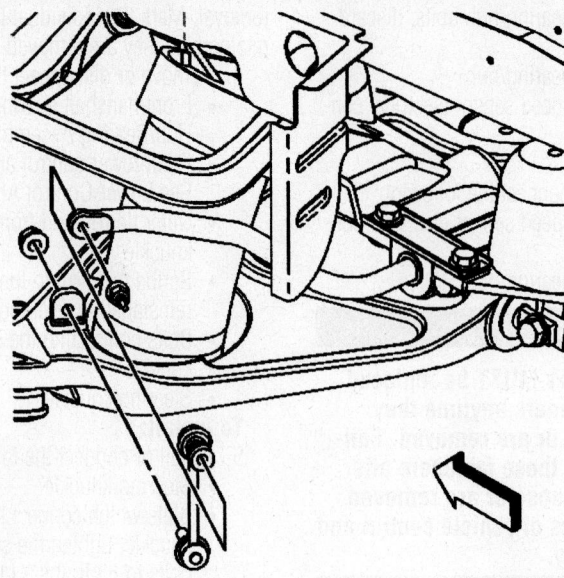

06025-MONT-G109

Removing stabilizer shaft link from lower control arm

- Shock absorber from the lower control arm. See Shocks/Struts.
- Upper control arm from the knuckle. See Upper Control Arm.

5. Using the J 41820 separator, disconnect the lower ball joint from the knuckle.

6. Remove the knuckle from vehicle.

7. Detach the wheel speed sensor connector bracket from the knuckle.

To install:

8. Install or connect the following:

- Wheel speed sensor connector bracket to the knuckle; tighten the bolt and nut to 106 inch lbs. (12 Nm)
- Knuckle to the vehicle
- Lower ball joint to the knuckle; tighten the ball joint nut to 26 ft. lbs. (35 Nm) plus 130 degrees
- Upper control arm to the knuckle. Tighten to 74 ft. lbs. (100 Nm).
- Shock absorber to the lower control arm. Tighten to 63 ft. lbs. (85 Nm) without IRS or to 66 ft. lbs. (90 Nm) with IRS.
- Height sensor link to the lower control arm as necessary. See Lower Control Arm.
- Backing plate and the wheel bearing; tighten bolts to 96 ft. lbs. (130 Nm) with IRS or to 59 ft. lbs. (80 Nm) on new bolts for models without IRS
- Halfshaft to the wheel bearing/hub; tighten halfshaft nut to 192 ft. lbs. (260 Nm)
- Stabilizer shaft link to the lower control arm; tighten nuts to 30 ft. lbs. (40 Nm)

- Tie rod to the knuckle; with IRS, tighten tie rod nut to 59 ft. lbs. (80 Nm) and bolt to 66 ft. lbs. (90 Nm); with twist axle, tighten tie rod bolt to body to 100 ft. lbs. (135 Nm) and tie rod bolt to axle to 92 ft. lbs. (125 Nm)
- Brake rotor. See Disc Brakes
- Brake caliper bracket
- Brake caliper. See Disc Brakes

9. Install the tire and wheel assembly. Tighten the lug nuts to 100 ft. lbs. (140 Nm) in a criss—cross pattern, after aligning the wheel hub with the reference mark and holes as shown in appropriate illustration.

10. Lower the vehicle.

Stabilizer Bar

REMOVAL & INSTALLATION

Front

WITH REAR TWIST AXLE

1. Before servicing the vehicle, refer to the Precautions Section.

2. Raise and support the vehicle.

3. Remove the wheel, marking the location of the wheel to the hub prior to removal. Mark the individual location of all retainers as they are removed.

4. Remove the left and right stabilizer shaft link lower nut (3).

5. Remove the stabilizer bar (link) (2) from the suspension strut and stabilizer shaft.

To install:

6. Install the stabilizer link (2) to the suspension strut and stabilizer shaft.

7. Install the stabilizer shaft link nuts. Tighten the nut to 33 ft. lbs. (45 Nm).

8. Install the tire and wheel assembly. Tighten the lug nuts to 100 ft. lbs. (140 Nm) in a criss—cross pattern, after aligning the wheel hub with the reference mark and holes as shown in appropriate illustration.

9. Lower the vehicle.

WITH INDEPENDENT REAR SUSPENSION

1. Before servicing the vehicle, refer to the Precautions Section.

2. Raise and support the vehicle.

3. Remove the wheel, marking the loca-

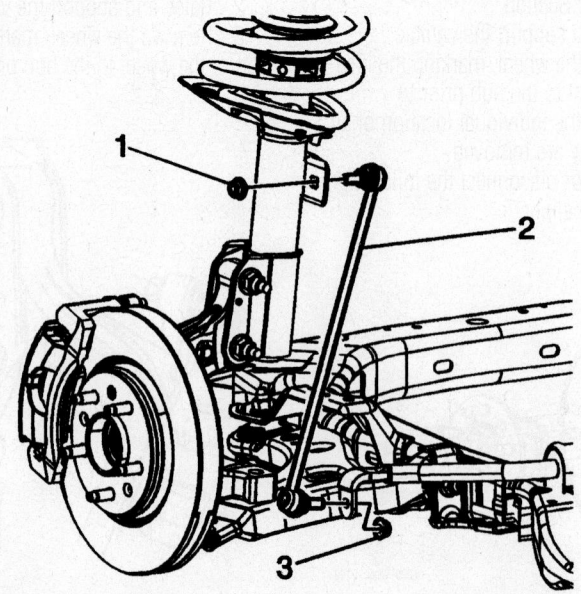

06025-MONT-G110

Identifying front stabilizer bar components: upper nut (1); shaft bar (link) (2); lower nut (3)

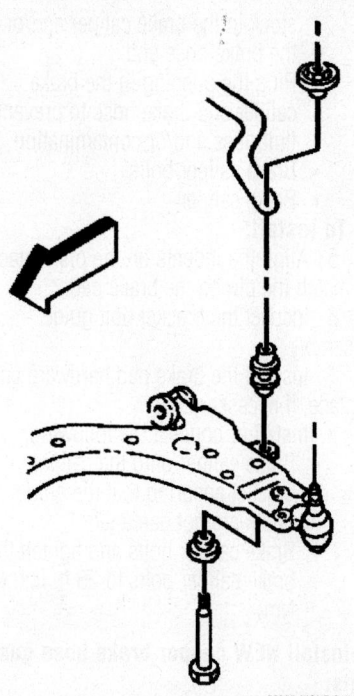

06025-MONT-G111

Showing stabilizer bar links to control arm

tion of the wheel to the hub prior to removal. Mark the individual location of all retainers as they are removed.

4. Remove or disconnect the following:
   - Left and right stabilizer shaft links
   - Left and right stabilizer shaft insulators and brackets
   - Stabilizer shaft from the left side of the vehicle

To install:
5. Install or connect the following:
   - Stabilizer shaft from the left side of the vehicle
   - Stabilizer shaft link bolt and nut, tighten the stabilizer shaft link nut to 14 ft. lbs. (19 Nm)
   - Left and right stabilizer shaft insulators and brackets
   - Left front wheel
6. Lower the vehicle.

Rear

1. Before servicing the vehicle, refer to the Precautions Section.
2. Raise and support the vehicle.
3. Remove or disconnect the following:
4. Install the tire and wheel assembly. Tighten the lug nuts to 100 ft. lbs. (140 Nm) in a criss—cross pattern, after aligning the wheel hub with the reference mark and holes as shown in appropriate illustration.
   - Load level sensor
   - Clamp holding the right park brake cable to the stabilizer shaft
   - Left and right side stabilizer shaft insulators and brackets
   - Left and right stabilizer shaft links
   - Spare tire from the spare tire hoist
   - Stabilizer shaft

To install:
5. Install or connect the following:
   - Stabilizer shaft
   - Spare tire to the spare tire hoist
   - Left and right stabilizer shaft links

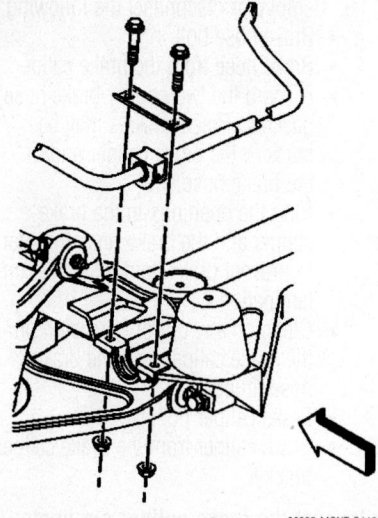

06025-MONT-G112

Showing rear stabilizer bar removal

   - Left and right stabilizer shaft insulators and brackets
   - Clamp holding the right park brake cable to the stabilizer shaft
   - Load level sensor
6. Install the tire and wheel assembly. Tighten the lug nuts to 100 ft. lbs. (140 Nm) in a criss—cross pattern, after aligning the wheel hub with the reference mark and holes as shown in appropriate illustration.
7. Lower the vehicle

BRAKES

Brake Caliper

REMOVAL & INSTALLATION

Front

1. Before servicing the vehicle, refer to the Precautions Section.

➡**Inspect the fluid level in the brake master cylinder reservoir. If the brake fluid level is midway between the maximum-full point and the minimum allowable level, then no brake fluid needs to be removed from the reservoir before proceeding. If the brake fluid level is higher than midway between the maximum full point and the minimum allowable level, then remove brake fluid to the midway point before proceeding.**

2. Remove the wheel, marking the location of the wheel to the hub prior to

removal. Mark the individual location of all retainers as they are removed.

3. Install two wheel lug nuts to retain the rotor to the hub
4. Install a large C-clamp over the top

of the brake caliper and against the back of the outboard brake pad

5. Tighten the C-clamp until the caliper piston is pushed into the caliper bore enough to slide the caliper off the rotor

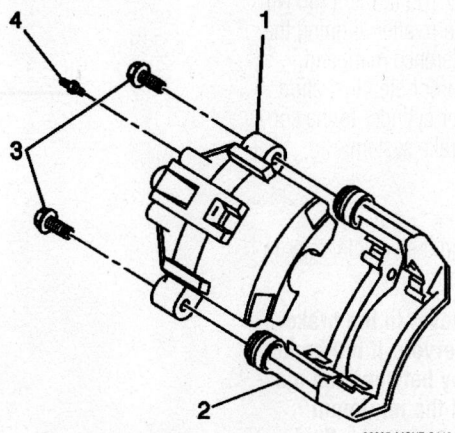

06025-MONT-G113

Exploded view of front brake caliper assembly

6. Remove or disconnect the following:
- Brake hose bolt
- Brake hose from the brake caliper
- Discard the two copper brake hose gaskets. These gaskets may be stuck to the brake caliper and/or the brake hose end
- Plug the openings in the brake caliper and the brake hose in order to prevent brake fluid loss and contamination.
- Clean off any dirt or corrosion on the brake caliper near the brake hose fitting
- Brake caliper bolts
- Brake caliper from the brake caliper bracket

➡Inspect the brake caliper pin boots, if the caliper pin boots are damaged, inspect the caliper pins for corrosion or damage. If corrosion is found on the brake caliper pin shaft, replace the brake caliper pin and the brake caliper pin boot. Do not attempt to polish away the corrosion.

To install:

7. Ensure that the caliper bolt boots are properly installed.
8. Install or connect the following:
- Caliper to the caliper bracket
- Caliper bolts and tighten the bolts to 40 ft. lbs. (54 Nm)

❄❄ WARNING

Install NEW copper brake hose gaskets

- Brake hose bolt and the NEW copper brake hose gaskets to the brake hose
- Brake hose bolt to the brake caliper and tighten the bolt to 40 ft. lbs. (54 Nm)

9. Install the tire and wheel assembly. Tighten the lug nuts to 100 ft. lbs. (140 Nm) in a criss—cross pattern, after aligning the wheel hub with the reference mark and holes as shown in appropriate illustration.
10. Refill the master cylinder to the correct level. Bleed the brake system.

Rear

1. Before servicing the vehicle, refer to the Precautions Section.

➡Inspect the fluid level in the brake master cylinder reservoir. If the brake fluid level is midway between the maximum-full point and the minimum allowable level, then no brake fluid needs to be removed from the reservoir

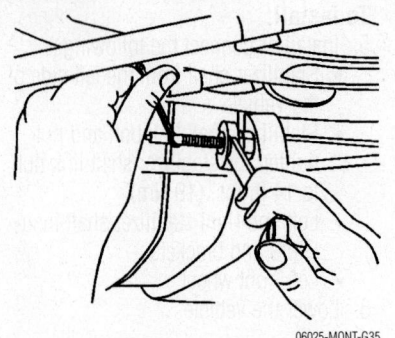

Showing the park brake system equalizer

before proceeding. If the brake fluid level is higher than midway between the maximum full point and the minimum allowable level, then remove brake fluid to the midway point before proceeding.

2. Remove the tire and wheel assembly
3. Release tension from the park brake system at the equalizer
4. Remove or disconnect the following:
- Front and rear cables from one another at the connector clip
- Park brake cable from the park brake lever on the brake caliper
- Park brake cable from the caliper bracket
- Brake hose to caliper bolt
- Brake hose from the brake caliper
- Discard the 2 copper brake hose gaskets. These gaskets may be

stuck to the brake caliper and/or the brake hose end
- Plug the opening in the brake caliper and brake hose to prevent fluid loss and/or contamination
- Brake caliper bolts
- Brake caliper

To install:

5. Align the indents on the piston face to match the pin on the brake pad.
6. Inspect the bracket bolt guide assembly.
7. Inspect the brake pad hardware and replace, if necessary.
8. Install or connect the following:
- Brake caliper onto the caliper bracket ensuring that the guide boots are not damaged
- Brake caliper bolts and tighten the brake caliper bolts to 25 ft. lbs. (34 Nm)

➡Install NEW copper brake hose gaskets.

- Brake hose bolt and the NEW copper brake hose gaskets to the brake hose
- Brake hose to caliper bolt to the brake caliper. When installing the rear brake hose to caliper, hold the hose up while tightening. Tighten the brake hose to caliper bolt to 30 ft. lbs. (40 Nm)
- Park brake cable into the park brake bracket on the caliper

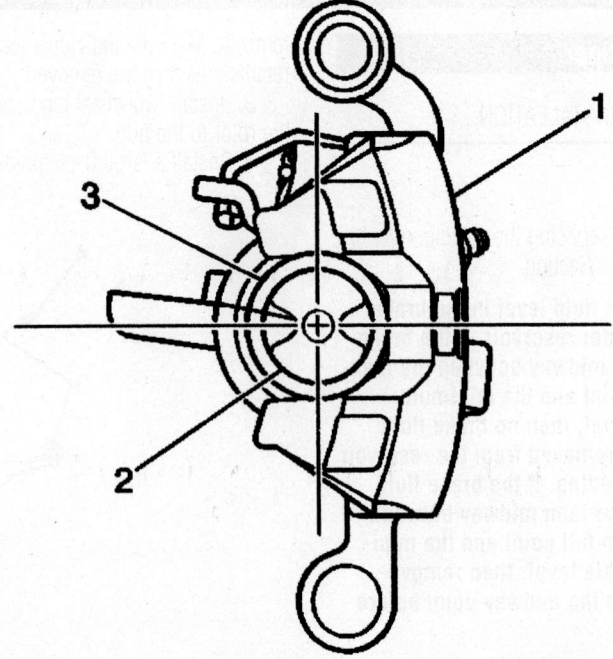

Aligning indents on piston face during caliper installation

- Park brake cable to the park brake lever on the brake caliper

9. Bleed the brake system.

10. With the engine OFF, gradually apply the brake pedal to approximately 2/3 of its travel distance. Slowly release the brake pedal.

11. Wait 15 seconds, then repeat steps until a firm brake pedal is obtained. This will properly seat the brake caliper pistons and brake pads.

12. Adjust the park brake system.

13. Install the tire and wheel assembly.

14. Lower the vehicle.

Disc Brake Pads

REMOVAL & INSTALLATION

Front

1. Before servicing the vehicle, refer to the Precautions Section.

➡**Inspect the fluid level in the brake master cylinder reservoir. If the brake fluid level is midway between the maximum-full point and the minimum allowable level, then no brake fluid needs to be removed from the reservoir before proceeding. If the brake fluid level is higher than midway between the maximum full point and the minimum allowable level, then remove brake fluid to the midway point before proceeding.**

2. Remove the wheel, marking the location of the wheel to the hub prior to removal. Mark the individual location of all retainers as they are removed.

3. Remove or disconnect the following:
- Install a C-clamp over the top of the brake caliper and against the back of the outboard brake pad
- Tighten the C-clamp until the brake caliper piston pushes into the brake caliper bore enough to slide the brake caliper off of the rotor
- C-clamp from the brake caliper
- Lower brake caliper guide point
- Rotate the brake caliper upward in order to access the brake pads
- Brake pads from the brake caliper bracket
- Brake pad retainers from the brake caliper bracket

❋❋ WARNING

Insert a block of wood or the old brake pad brake between the C-clamp and the brake caliper piston in order to prevent damage to the brake caliper piston and the brake caliper piston boot.

- Install a C-clamp over the brake caliper and against the block of wood or the old brake pad
- Tighten the C-clamp until the brake caliper piston pushes completely into the brake caliper bore

- Remove the C-clamp from the brake caliper
- Inspect the brake caliper guide pin boots, repair or replace as necessary
- Inspect the brake caliper piston boot, repair or replace as necessary
- Inspect the brake caliper guide pins for corrosion or damage. If corrosion is found, use new parts, including bushings, when installing the brake caliper. Do not attempt to polish away corrosion.

To install:

4. Install or connect the following:

➡**Use denatured alcohol to clean the outside surface of caliper boot before installing new brake pads.**

❋❋ CAUTION

If you are installing new brake pads, use a C-clamp in order to compress the piston to the bottom of the caliper bore. Use the old brake pad, a metal plate or a wooden block across the face of the piston in order to protect the piston and the caliper boot.

- Brake pad retainers to the caliper bracket

➡**Ensure that the wear sensor is positioned at the trailing edge (upward) of the outer pad during forward wheel rotation.**

- Brake pads over the brake pad retainers and onto the caliper bracket
- Lubricate the brake caliper guide pin and the guide pin boot with high temp brake silicone lube, or equivalent.
- Pivot the caliper down onto the pads
- Lower caliper guide pin and tighten the caliper guide pin to 40 ft. lbs. (54 Nm)

5. Install the tire and wheel assembly. Tighten the lug nuts to 100 ft. lbs. (140 Nm) in a criss—cross pattern, after aligning the wheel hub with the reference mark and holes as shown in appropriate illustration.

6. Lower the vehicle.

7. Fill the master cylinder to the proper level with new clean brake fluid.

8. Pump the brake pedal as many times as necessary to obtain a firm brake pedal.

Rear

1. Before servicing the vehicle, refer to the Precautions Section.

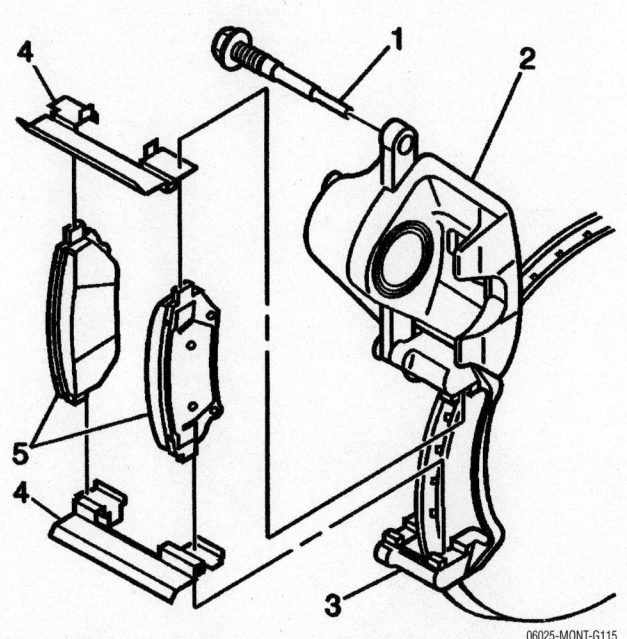

06025-MONT-G115

Exploded view of caliper and pads on front rotor: caliper bolt (1), caliper (2), caliper bracket (3), pad retainers (4), brake pads (5)

→**Inspect the fluid level in the brake master cylinder reservoir. If the brake fluid level is midway between the maximum-full point and the minimum allowable level, then no brake fluid needs to be removed from the reservoir before proceeding. If the brake fluid level is higher than midway between the maximum full point and the minimum allowable level, then remove brake fluid to the midway point before proceeding.**

2. Remove the wheel, marking the location of the wheel to the hub prior to removal. Mark the individual location of all retainers as they are removed.

3. Remove or disconnect the following:
  • Unclamp the wheel speed sensor (WSS) harness from the lower control arm
  • Upper and lower caliper bolts from the caliper
  • Pull the caliper straight off of the bracket and secure out of the way with heavy mechanics wire. DO NOT disconnect the hydraulic brake flexible hose from the caliper
  • Inboard and outboard pads from the brake caliper bracket

To install:

4. Clean the brake pad hardware mating surfaces on the caliper bracket of any debris or corrosion.

5. Inspect the brake pad retainer clips and replace, if necessary.

6. Inspect the piston boot. Replace if damaged.

7. Retract the brake caliper piston into the brake caliper bore. Use a suitable spanner type wrench and turn the piston clockwise until it bottoms out fully in the brake caliper.

8. Align the indents on the piston face to match the pin on the back of the inboard brake pads.

9. Install or connect the following:
  • Brake pad retainers into the brake caliper bracket
  • Inboard and outboard brake pads into the brake caliper bracket insuring that the pad with the metallic wear sensor is placed on the inboard side of the bracket
  • Slide the caliper onto the bracket insuring that the bracket guide boots are not damaged
  • Brake caliper bolts and tighten the bolts to 25 ft. lbs. (34 Nm)
  • WSS harness onto the lower control arm

10. Install the tire and wheel assembly. Tighten the lug nuts to 100 ft. lbs. (140 Nm) in a criss—cross pattern, after aligning the wheel hub with the reference mark and holes as shown in appropriate illustration.

11. Lower the vehicle.

12. With the engine OFF, gradually apply the brake pedal to approximately 2/3 of its travel distance. Slowly release the brake pedal.

13. Wait 15 seconds, then repeat steps until a firm brake pedal apply is obtained. This will properly seat the brake caliper pistons and brake pads.

14. Fill the brake master cylinder reservoir to the proper level.

BRAKE HYDRAULIC
  SYSTEM 26-53
DRIVE TRAIN 26-32
ENGINE REPAIR 26-8
FRONT BRAKES...............26-50
FRONT SUSPENSION 26-42
FUEL SYSTEM 26-27
REAR BRAKES 26-51
REAR SUSPENSION 26-46
SPECIFICATIONS AND
  MAINTENANCE CHARTS......26-3
Engine and Vehicle
  Identification.........................26-3
General Engine Specifications 26-3
Engine Tune-Up Specifications......26-3
Accessory Drive Belt Routing 26-4
Capacities..............................26-4
Valve Specifications....................26-4
Crankshaft and Connecting Rod
  Specifications 26-5
Piston and Ring Specifications......26-5
Torque Specifications 26-5
Wheel Alignment 26-6
Tire, Wheel and Ball Joint
  Specifications 26-6
Brake Specifications 26-7
Scheduled Maintenance
  Intervals.................................26-7
STEERING 26-38
A
Air Bag.....................................26-38
  Disarming & Enabling 26-38
  Precautions............................26-38
Alternator 26-8
  Removal................................26-8
B
Balance Shaft 26-20
  Removal & Installation............26-20
Balance Shaft Chain....................26-25
  Removal & Installation............26-25
Brake Bleeding...........................26-53
  Manual Bleeding....................26-53
Brake Caliper Front Brakes 26-50
  Removal & Installation............26-50
Brake Caliper Rear Brakes 26-51
  Removal & Installation............26-51

C
Caliper Anchor Plate Front
  Brakes.................................26-51
  Removal & Installation............26-51
Caliper Anchor Plate Rear
  Brakes.................................26-52
  Removal & Installation............26-52
Camshaft and Lifters...................26-19
  Removal & Installation............26-19
Camshaft Cover 26-18
  Removal & Installation............26-18
Clutch 26-32
  Removal & Installation............26-32
Crankshaft Damper 26-22
  Removal & Installation............26-22
Crankshaft Front Seal 26-22
  Removal & Installation............26-22
CV-Joints.................................26-34
  Overhaul 26-34
Cylinder Head 26-15
  Removal & Installation............26-15
D
Differential Pinion Seal 26-37
  Removal & Installation............26-37
Disc Brake Pads Front Brakes......26-50
  Removal & Installation............26-50
Disc Brake Pads Rear Brakes......26-52
  Removal & Installation............26-52
Driveshaft 26-36
  Removal & Installation............26-36
E
Engine......................................26-8
  Removal & Installation..............26-8
Exhaust Manifold.......................26-17
  Removal & Installation............26-17
F
Front Cover...............................26-22
  Removal & Installation............26-22
Front Hub/Bearing/Speed
  Sensor.................................26-45
  Removal & Installation............26-45
Fuel Filter 26-27
  Removal & Installation............26-27
Fuel Injectors 26-28
  Removal & Installation............26-28
Fuel Pump Module 26-27
  Removal & Installation............26-27

Fuel System Service
  Precautions...........................26-27
H
Halfshaft..................................26-33
  Removal & Installation............26-33
Heater Core..............................26-12
  Removal & Installation............26-12
Hydraulic Clutch System 26-33
  Bleeding................................26-33
I
Intake Manifold 26-16
  Removal & Installation............26-16
K
Knuckle....................................26-47
  Removal & Installation............26-47
L
Lower Control Arm Front
  Suspension...........................26-43
  Removal & Installation............26-43
Lower Control Arm Rear
  Suspension...........................26-46
  Removal & Installation............26-46
Manual Transmission 26-32
  Removal & Installation............26-32
O
Oil Pan....................................26-21
  Removal & Installation............26-21
P
Piston and Ring 26-26
  Positioning 26-26
Power Steering Gear 26-41
  Removal & Installation............26-41
Q
Quick-Connect Fittings 26-29
  Opening & Closing.................26-29
R
Rear Hub/Bearing 26-49
  Removal & Installation............26-49
Rear Main Seal 26-22
  Removal & Installation............26-22
Relieving Fuel System
  Pressure 26-27
Rotor Front Brakes.....................26-51
  Removal & Installation............26-51
Rotor Rear Brakes 26-53
  Removal & Installation............26-53

S

Stabilizer Bar & End Links
 Front Suspension 26-44
  Removal & Installation............26-44
Stabilizer Bar and End Links
 Rear Suspension 26-48
  Removal & Installation............26-48
Steering Knuckle..........................26-43
  Removal & Installation............26-43
Strut Front Suspension................26-42
  Disassembly & Assembly 26-42
  Removal & Installation............26-42

Strut Rear Suspension 26-46
 Disassembly & Assembly 26-46
 Removal & Installation............26-46

T

Timing Chain and Sprockets.......26-22
 Removal & Installation............26-22
Toe Links 26-47
 Removal & Installation............26-47

U

Upper Control Arm Front
 Suspension...............................26-43
  Removal & Installation............26-43

Upper Control Arm Rear
 Suspension...............................26-46
  Removal & Installation............26-46

W

Water Pump 26-10
 Removal & Installation............26-10

SPECIFICATIONS AND MAINTENANCE CHARTS

ENGINE AND VEHICLE IDENTIFICATION

			Engine					Model Year	
Code ①	Liters	Cu. In.	Cyl.	Fuel Sys.	Engine Type	Eng. Mfg.		Code ②	Year
B	2.4	146	4	MFI	DOHC	GM		6	2006

MFI: Multi-port Fuel Injection

DOHC: Double Overhead Camshafts

NA: Information not available

① 8th digit of VIN

② 10th digit of VIN

06025-SOLS-C01

GENERAL ENGINE SPECIFICATIONS

Engine Displacement Liters	Engine VIN	Year	Net Horsepower @ rpm	Net Torque @ rpm (ft. lbs.)	Bore x Stroke (in.)	Com- pression Ratio	Oil Pressure @ rpm
2.4	B	2006	177@6600	166@4800	3.467x3.861	10:1	50-80@1000

06025-SOLS-C02

ENGINE TUNE-UP SPECIFICATIONS

Engine Displacement Liters	Engine VIN	Year	Spark Plug Gap (in.)	Ignition Timing (deg.)		Fuel Pump (psi)	Idle Speed (rpm)		Valve Clearance	
				MT	AT		MT	AT	Intake	Exhaust
2.4	B	2006	0.042	NA	NA	50-60	NA	NA	Hyd.	Hyd.

Hyd. Hydraulic

NA: Information not available

06025-SOLS-C03

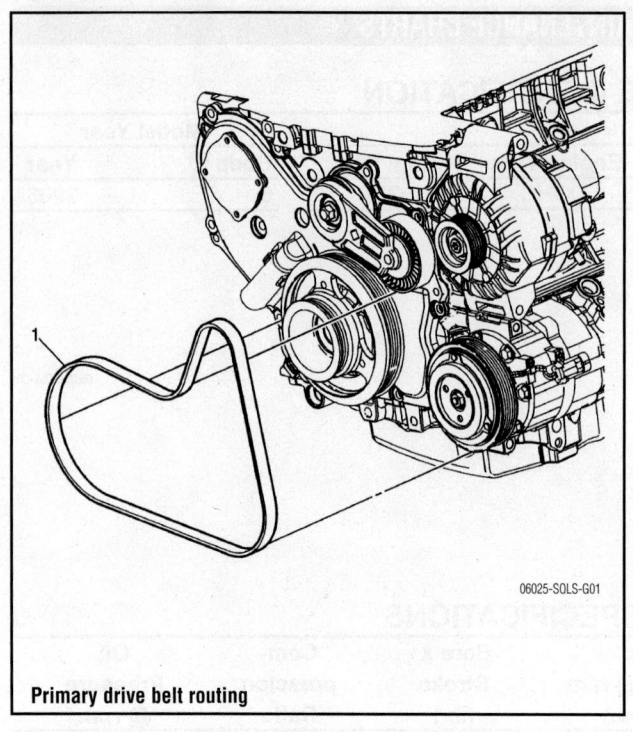

Primary drive belt routing

06025-SOLS-G01

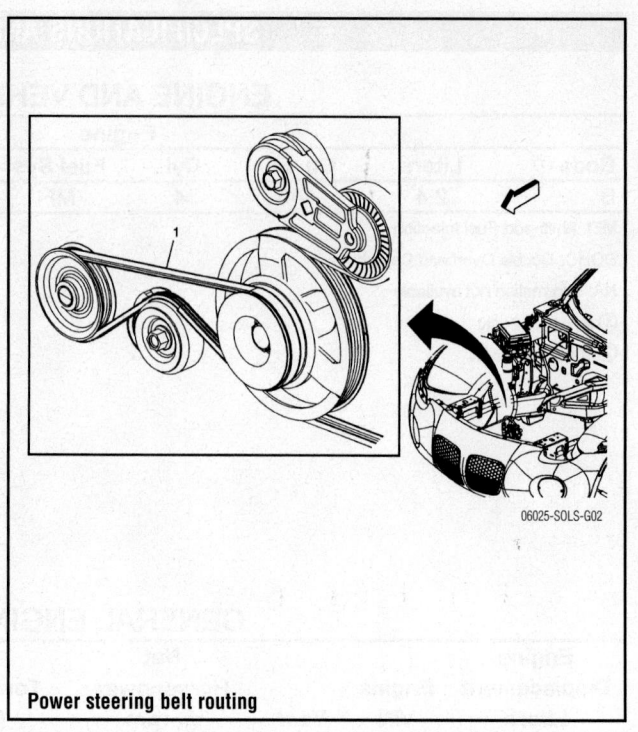

Power steering belt routing

06025-SOLS-G02

CAPACITIES

Engine Displacement Liters	Model	Engine VIN	Year	Engine Oil with Filter (qts.)	Transmission (pts.)		Rear Axle (pts.)	Fuel Tank (gal.)	Cooling System (qts.)
					Manual	Auto. *			
2.4	Solstice	B	2006	5.0	5.5	18	2.74	13.8	7.9

* Dry fill

06025-SOLS-C04

VALVE SPECIFICATIONS

Engine Displacement Liters	Engine VIN	Year	Seat Angle (deg.)	Face Angle (deg.)	Spring Test Pressure (lbs. @ in.)	Spring Installed Height (in.)	Stem-to-Guide Clearance (in.)		Stem Diameter (in.)	
							Intake	Exhaust	Intake	Exhaust
2.4	B	2006	NA	NA	NA	1.28	0.0012-0.0022	0.0020-0.0026	0.2344-0.2355	0.2337-0.2343

NA: Information not available

06025-SOLS-C05

CRANKSHAFT AND CONNECTING ROD SPECIFICATIONS

All values shown in inches

Engine Displacement (Liters)	Engine VIN	Year	Crankshaft				Connecting Rod		
			Main Brg. Journal Dia.	Main Brg. Oil Clearance	Shaft End-play	Thrust on No.	Journal Diameter	Oil Clearance	Side Clearance
2.4	B	2006	2.2045-2.2050	0.0012-0.0026	0.0012-0.0150	2	1.19291-1.9297	0.0011-0.0027	0.0028-0.0146

06025-SOLS-C06

PISTON AND RING SPECIFICATIONS

All measurements given in inches

Engine Displacement Liters	Engine VIN	Year	Piston Clearance	Ring Gap			Ring Side Clearance		
				Top Compression	Bottom Compression	Oil Control	Top Compression	Bottom Compression	Oil Control
2.4	B	2006	0.0004-0.0016	0.008-0.016	0.014-0.022	0.010-0.030	0.0015-0.0031	0.0012-0.0027	0.0035-0.0042

06025-SOLS-C07

TORQUE SPECIFICATIONS

All values given in ft. lbs. unless otherwise noted

Engine Displacement Liters	Engine VIN	Year	Cylinder Head Bolts	Main Bearing Bolts	Rod Bearing Bolts	Crankshaft Damper Bolts	Flywheel Bolts	Manifold		Spark Plugs	Oil Pan Drain Plug
								Intake	Exhaust		
2.4	B	2006	①	②	③	④	⑤	⑥	10	15	18

① Step 1: 22 ft. lbs.
　Step 2: plus 155 degrees
The 4 front bolts: 26 ft. lbs. (see text)

② Bedplate-to-block
　Cap bolts
　Step 1: 15 ft. lbs.
　Step 2: plus 70 degrees
　Perimeter bolts: 18 ft. lbs.

③ Step 1: 18 ft. lbs.
　Step 2: plus 100 degrees

④ Step 1: 74 ft. lbs.
　Step 2: plus 125 degrees

⑤ Step 1: 39 ft. lbs.
　Step 2: plus 25 degrees

⑥ Bolts and nuts: 89 inch lbs.; studs: 59 inch lbs.

06025-SOLS-C08

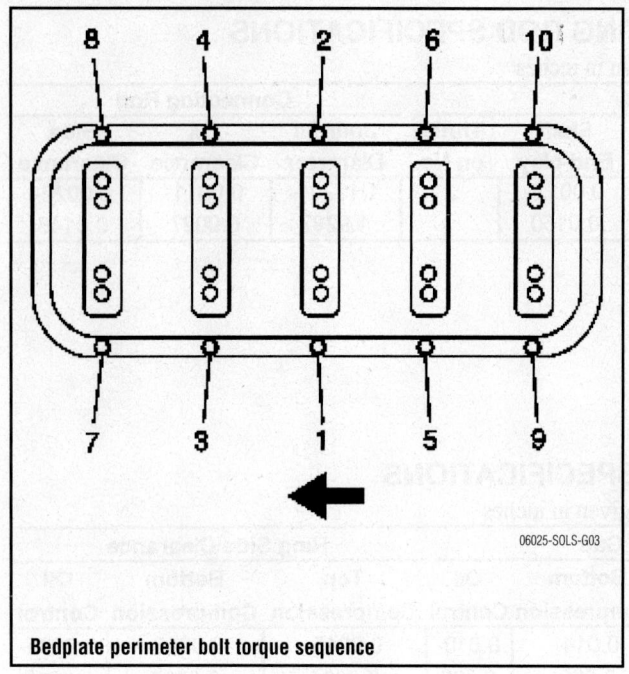

Bedplate perimeter bolt torque sequence

06025-SOLS-G03

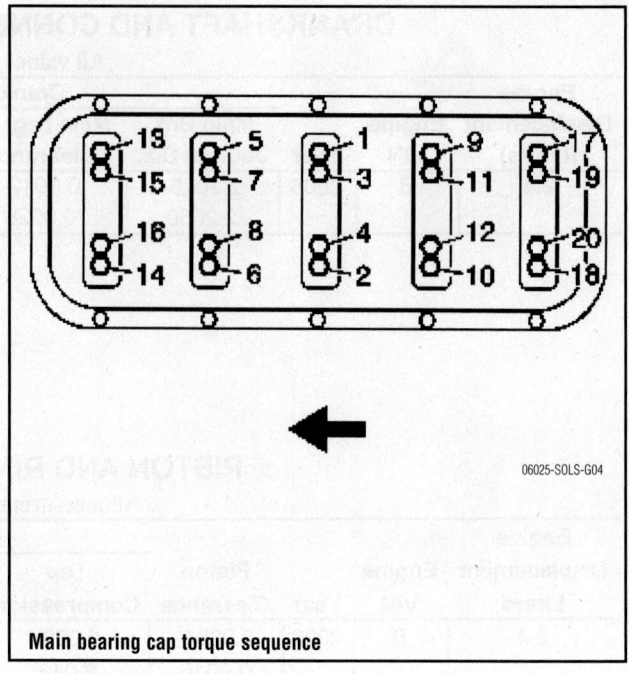

Main bearing cap torque sequence

06025-SOLS-G04

WHEEL ALIGNMENT

Model	Year		Caster Range (+/-Deg.)	Caster Preferred Setting (Deg.)	Camber Range (+/-Deg.)	Camber Preferred Setting (Deg.)	Toe-in (in.)
Solstice	2006	F	0.60	+8.00	0.60	-0.50	0.10+/-0.20
		R	0.75	-4.00	0.50	-0.50	0.10+/-0.20

06025-SOLS-C09

TIRE, WHEEL AND BALL JOINT SPECIFICATIONS

Model	Year	OEM Tires Standard	OEM Tires Optional	Tire Pressures (psi) Front	Tire Pressures (psi) Rear	Wheel Size	Ball Joint Inspection	Lug Nuts (ft. lbs.)
Solstice	2006	P245/45R18	none	①	①	8.5	②	100

OEM: Original Equipment Manufacturer

PSI: Pounds Per Square Inch

① See placard on vehicle

② Replace if any movement is noted

06025-SOLS-C10

BRAKE SPECIFICATIONS

All measurements in inches unless noted

Model	Year		Brake Disc Original Thickness	Minimum Thickness	Maximum Run-out	Minimum Lining Thickness	Brake Caliper Bracket Bolts (ft. lbs.)	Mounting Bolts (ft. lbs.)
Solstice	2006	F	1.023	0.906	0.002	NA	85	25
		R	0.465	0.394	0.002	NA	85	20

NA: Information not available

06025-SOLS-C11

SCHEDULED MAINTENANCE INTERVALS
PONTIAC Solstice

TO BE SERVICED	TYPE OF SERVICE	5	10	15	20	25	30	35	40	45	50	55	60	65	70	75	80
Accessory drive belt	S/I	Every 150,000 miles															
Air cleaner filter ①	S/I			✓			✓			✓			✓			✓	
Automatic transmission fluid and filter ②	R										✓						
Brake fluid level	I	Every 6 months															
Clutch fluid level	I	Every 6 months															
Engine coolant system ③	S/I	Every 150,000 miles or 60 months															
Engine oil & filter ④	R	✓	✓	✓	✓	✓	✓	✓	✓	✓	✓	✓	✓	✓	✓	✓	✓
Hinges and latches	L	Once a year															
Restraint system	I	Every 6 months															
Tires ⑤	Rotate		✓		✓		✓		✓		✓		✓		✓		✓
Spark plugs	R	Every 100,000 miles															
Wiper blades	S/I	Every 6 months															

R: Replace   S/I: Inspect and service, if necessary   L: Lubricate

① Replace as necessary, but, replace every 45,000 miles if not previously done.

② Replace at this interval if driven under any of these conditons:

    Heavy city traffic where the outside temperature regularly reaches 32°C (90°F) or   higher

    Hilly or mountainous terrain

    Frequent trailer towing

    Taxi, police or delivery service

    Otherwise, change every 100,000 miles

③ Drain an flush

④ For vehicle with a Driver Information Center, change the oil and filter when specified by the DIC.

⑤ Check wear and inflation every month

After each oil change, reset the Engine Oil Life System as follows:

1. Turn the ignition switch to RUN with the engine OFF.

2. Press the Information and Reset buttons on the DIC at the same time to enter the personalization menu.

3. Press the Information button to scroll through the menu to Oil Life Reset.

4. Press and hold the Reset button until the display shows Acknowledged.

5. Turn the key to OFF.

06025-SOLS-C12

ENGINE REPAIR

Alternator

REMOVAL

1. Before servicing the vehicle, refer to the Precautions Section.
2. Disconnect the negative battery cable.
3. Remove the drive belt.
4. Disconnect the alternator wiring.
5. Remove the bolts.
6. Installation is the reverse of removal. Torque the mounting bolts to 16 ft. lbs. (22 Nm).

Engine

REMOVAL & INSTALLATION

1. Before servicing the vehicle, refer to the Precautions Section.
2. Disconnect the negative battery cable.
3. Remove the headlamps.
4. Remove the hood.
5. Drain the cooling system.
6. Remove the upper radiator air baffle.
7. Remove the fan shroud assembly.
8. Remove the radiator inlet hose from the radiator.
9. Remove the radiator outlet hose from the radiator.
10. Remove the radiator support brackets.
11. Remove the radiator.
12. Relieve the fuel system pressure.
13. Disconnect the evaporative emission (EVAP) canister purge solenoid tube from the valve.
14. Disconnect the fuel feed pipe from the fuel rail.
15. Reposition the radiator inlet hose clamp at the engine.
16. Remove the radiator inlet hose.
17. Reposition the radiator outlet hose clamps at the thermostat housing and oil cooler.
18. Remove the radiator outlet hose from the thermostat housing and oil cooler.
19. Reposition the outlet hose out of the way.
20. Reposition the surge tank outlet hose clamps at the thermostat housing and oil cooler.
21. Remove the surge tank outlet hose from the thermostat housing and oil cooler.
22. Reposition the surge tank outlet hose clamp at the surge tank.
23. Remove the surge tank clip from the oil level indicator tube bracket.

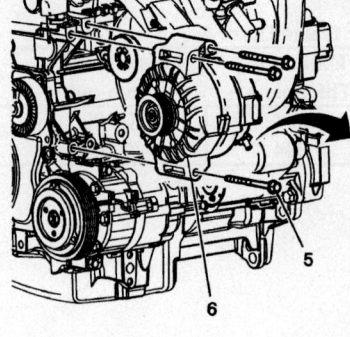
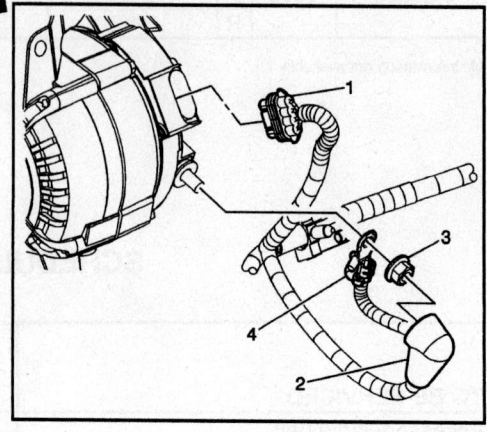

06025-SOLS-G05

Alternator mounting

24. Remove the surge tank outlet hose.
25. Reposition the surge tank air bleed hose clamp at the engine.
26. Remove the surge tank air bleed hose from the engine.
27. Reposition the air bleed hose out of the way.
28. Reposition the heater inlet and outlet hose clamps at the thermostat.
29. Remove the heater inlet and outlet hoses from the thermostat.
30. Reposition brake booster vacuum hose clamp at the intake manifold.
31. Remove the brake booster vacuum hose from the intake manifold.
32. Remove the fuel feed and EVAP line clip from the oil level indicator tube bracket.
33. Remove the engine harness clips from the oil level indicator tube and the radiator outlet hose bracket.
34. Raise and suitably support the vehicle.
35. Drain the engine oil.
36. Remove the catalytic converter to bracket bolt.
37. Disconnect the crankshaft position (CKP) sensor electrical connector.
38. Disconnect the oil pressure sensor electrical connector.
39. Remove the starter solenoid "S" terminal nut.
40. Remove the engine harness terminal from the starter.
41. Remove the engine harness clip from the intake manifold.
42. Remove the starter solenoid terminal nut.

43. Remove the positive battery cable clips from the intake manifold.
44. Remove the positive battery cable terminal from the starter.
45. Lower the vehicle.
46. Remove the positive battery cable clip from the bracket.
47. Cut the engine harness tie straps.
48. Remove the negative battery cable ground bolt.
49. Disconnect the generator electrical connector.
50. Reposition the positive battery cable terminal boot.
51. Remove the generator terminal nut.
52. Remove the positive battery cable terminal from the generator.
53. Disconnect the air conditioning compressor electrical connector.
54. Disconnect the knock sensor electrical connector.
55. Reposition the negative/positive battery cable.
56. Disconnect the A/C refrigerant pressure sensor electrical connector.
57. Disconnect the windshield wiper motor electrical connector.
58. Remove the engine harness clip from the wiper motor hole.
59. Remove the engine harness clip from the surge tank air bleed hose.
60. Disconnect the fuel injector electrical connector.
61. Disconnect the electronic throttle control (ETC) electrical connector.
62. Remove the engine harness clips.

63. Disconnect the intake camshaft position (CMP) actuator electrical connector.

64. Disconnect the exhaust CMP actuator electrical connector.

65. Remove the engine harness clip from the camshaft cover.

66. Disconnect the ignition coil electrical connectors.

67. Remove the connector position assurance (CPA) retainers from the heated oxygen sensor (HO2S) electrical connectors.

68. Disconnect the front HO2S electrical connector.

69. Disconnect the rear HO2S electrical connector.

70. Disconnect the engine harness clips from the junction block bracket.

71. Disconnect the HO2S electrical connector clip from the strut.

72. Disconnect the intake CMP sensor electrical connector.

73. Disconnect the exhaust CMP sensor electrical connector.

74. Remove the engine harness ground terminal bolt.

75. Disconnect the engine coolant temperature (ECT) sensor electrical connector.

76. Disconnect the EVAP canister purge solenoid electrical connector.

77. Disconnect the engine harness clip from the EVAP canister purge solenoid valve bracket.

78. Disconnect the engine harness clips from the camshaft cover.

79. Remove the EVAP canister purge solenoid valve.

80. Remove the engine harness bracket bolt.

81. Disconnect the engine harness clip from the bracket.

82. Gather all branches of the engine harness and lay off to the side.

83. Remove the A/C compressor bolts. Position the compressor off to the side.

84. Remove the catalytic converter nuts.

85. Remove the catalytic converter gasket and converter.

86. Raise and support the vehicle.

87. Remove the muffler.

88. Remove the starter motor bolts and starter.

89. Remove the torque converter to flywheel bolts, if equipped with a automatic transmission.

90. Remove the lower bellhousing bolts.

91. Remove the bellhousing bolts.

92. Lower the vehicle.

93. Remove the left engine mount upper nut.

94. Remove the right engine mount upper nut.

95. Install a suitable engine lifting devise to the engine.

96. Separate the engine from the transmission.

97. Remove the engine from the vehicle.

98. Remove the clutch pressure plate and disc and the flywheel.

99. Install the engine to an engine stand.

To install:

100. Install a suitable engine lifting devise to the engine.

101. Remove the engine from the stand.

102. Install the flywheel and clutch.

103. Install the engine to the vehicle.

104. Install the right engine mount upper nut. Tighten the nut to 50 Nm (37 ft. lbs.).

105. Install the left engine mount upper nut. Tighten the nut to 50 Nm (37 ft. lbs.).

106. Raise and support the vehicle.

107. Install the bellhousing bolts. Tighten the bolts to 50 Nm (37 ft. lbs.).

108. Install the lower bellhousing bolts. Tighten the bolts to 50 Nm (37 ft. lbs.).

109. Install the torque converter to flywheel bolts, if equipped with a automatic transmission. Tighten the bolts to 50 Nm (37 ft. lbs.).

110. Install the starter motor and bolts. Tighten the bolts to 40 Nm (30 ft. lbs.).

111. Install the muffler.

112. Install the catalytic converter and gasket.

113. Install the catalytic converter nuts. Tighten the nuts to 25 Nm (18 ft. lbs.).

114. Position the compressor and install the A/C compressor bolts. Tighten the bolts to 22 Nm (16 ft. lbs.).

115. Position the branches of the engine harness to the engine.

116. Connect the engine harness clip to the bracket.

117. Install the engine harness bracket bolt. Tighten the bolt to 10 Nm (89 inch lbs.).

118. Install the EVAP canister purge solenoid valve.

119. Connect the engine harness clips to the camshaft cover.

120. Connect the engine harness clip to the EVAP canister purge solenoid valve bracket.

121. Connect the EVAP canister purge solenoid electrical connector.

122. Connect the ECT sensor electrical connector.

123. Install the engine harness ground terminal bolt. Tighten the bolt to 10 Nm (89 inch lbs.).

124. Connect the exhaust CMP sensor electrical connector.

125. Connect the intake CMP sensor electrical connector.

126. Connect the HO2S electrical connector clip to the strut.

127. Connect the engine harness clips to the junction block bracket.

128. Connect the rear HO2S electrical connector.

129. Connect the front HO2S electrical connector.

130. Install the CPA retainers to the HO2S electrical connectors.

131. Connect the ignition coil electrical connectors.

132. Install the engine harness clip to the camshaft cover.

133. Connect the exhaust CMP actuator electrical connector.

134. Connect the intake CMP actuator electrical connector.

135. Install the engine harness clips.

136. Connect the ETC electrical connector.

137. Connect the fuel injector electrical connector.

138. Install the engine harness clip to the surge tank air bleed hose.

139. Connect the windshield wiper motor electrical connector.

140. Install the engine harness clip to the wiper motor hole.

141. Connect the A/C refrigerant pressure sensor electrical connector.

142. Position the negative/positive battery cable.

143. Connect knock sensor electrical connector.

144. Connect the air conditioning compressor electrical connector.

145. Install the positive battery cable terminal to the generator.

146. Install the generator terminal nut. Tighten the nut to 20 Nm (15 ft. lbs.).

147. Position the positive battery cable terminal boot.

148. Connect the generator electrical connector.

149. Install the negative battery cable ground bolt. Tighten the bolt to 20 Nm (15 ft. lbs.).

150. Install NEW tie straps to the engine harness.

151. Install the positive battery cable clip to the bracket.

152. Raise and support the vehicle.

153. Install the positive battery cable terminal to the starter.

154. Install the positive battery cable clips to the intake manifold.

155. Install the starter solenoid terminal nut. Tighten the nut to 10 Nm (89 inch lbs.).

156. Install the engine harness clip to the intake manifold.

157. Install the engine harness terminal to the starter.

158. Install the starter solenoid "S" terminal nut. Tighten the nut to 3 Nm (27 inch lbs.).

159. Connect the oil pressure sensor electrical connector.

160. Connect the CKP sensor electrical connector.

161. Install the catalytic converter to bracket bolt. Tighten the bolt to 25 Nm (18 ft. lbs.).

162. Lower the vehicle.

163. Install the engine harness clips to the oil level indicator tube and the radiator outlet hose bracket.

164. Install the fuel feed and EVAP line clip from the oil level indicator tube bracket.

165. Install the brake booster vacuum hose to the intake manifold.

166. Position brake booster vacuum hose clamp at the intake manifold.

167. Install the heater inlet and outlet hoses to the thermostat.

168. Position the heater inlet and outlet hose clamps at the thermostat.

169. Position the air bleed hose.

170. Install the surge tank air bleed hose to the engine.

171. Position the surge tank air bleed hose clamp at the engine.

172. Install the surge tank outlet hose.

173. Install the surge tank clip to the oil level indicator tube bracket.

174. Position the surge tank outlet hose clamp at the surge tank.

175. Install the surge tank outlet hose to the thermostat housing and oil cooler.

176. Position the surge tank outlet hose clamps at the thermostat housing and oil cooler.

177. Position the outlet hose.

178. Install the radiator outlet hose to the thermostat housing and oil cooler.

179. Position the radiator outlet hose clamps at the thermostat housing and oil cooler.

180. Install the radiator inlet hose.

181. Position the radiator inlet hose clamp at the engine.

182. Connect the fuel feed pipe to the fuel rail.

183. Connect the EVAP canister purge solenoid tube to the valve.

184. Install the radiator.

185. Install the headlamps.

186. Install the hood. Torque to 19 ft. lbs. (25 Nm).

187. Connect the negative battery cable.

188. Fill the engine oil.

Water Pump

REMOVAL & INSTALLATION

1. Before servicing the vehicle, refer to the Precautions Section.

2. Drain the cooling system.

➠**A drain has been provided at the bottom of the water pump for engine block coolant drainage.**

3. Drain the coolant from the engine block at the water pump drain. After the coolant has drained, tighten the drain bolt.

4. Lower the vehicle.

5. Disconnect the engine coolant temperature (ECT) sensor electrical connector.

6. Remove the ECT sensor, if necessary.

7. Reposition the surge tank outlet hose clamp at the thermostat housing.

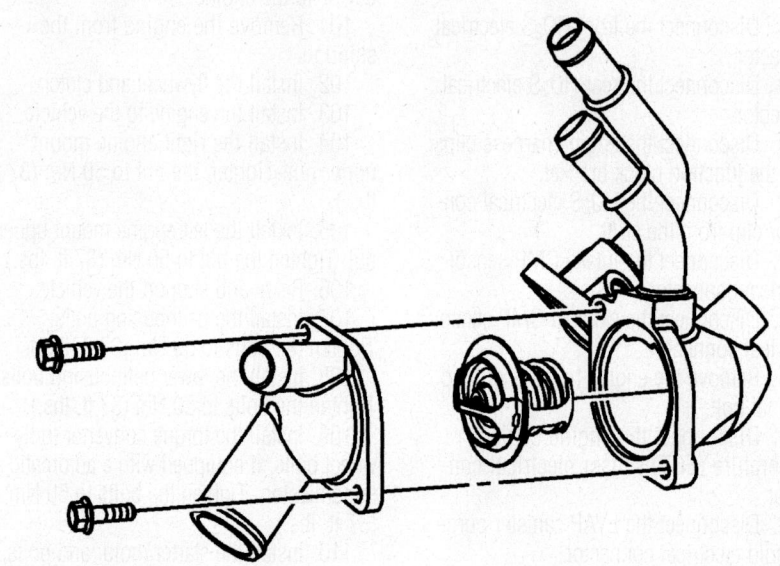

06025-SOLS-G06

Thermostat housing

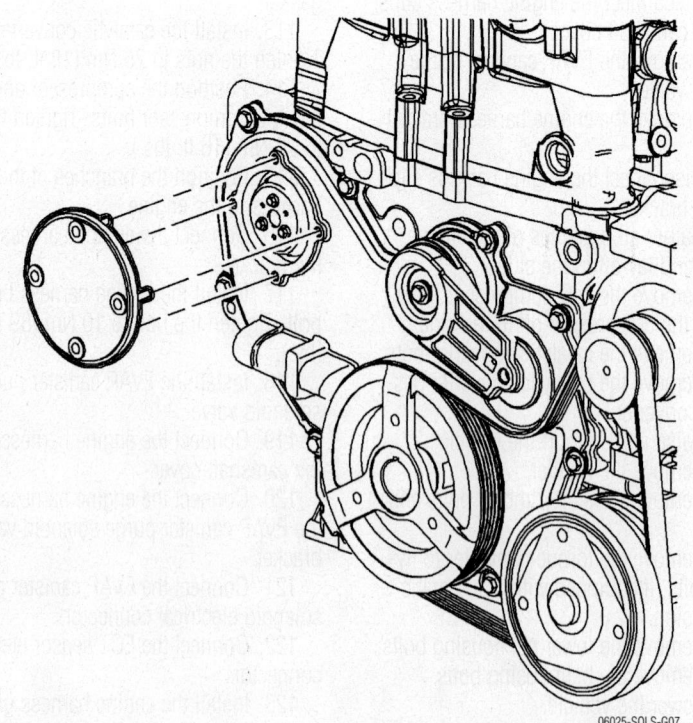

06025-SOLS-G07

Water pump access cover

8. Remove the surge tank outlet hose from the thermostat housing.

9. Reposition the radiator outlet hose clamp at the thermostat housing.

10. Remove the radiator outlet hose from the thermostat housing.

11. Remove the exhaust heat shield bolts.

12. Remove the exhaust heat shield.

13. Reposition the heater inlet and outlet hose clamps at the thermostat housing pipes.

14. Disconnect the heater inlet and outlet hoses from the thermostat housing pipes.

15. Remove the thermostat housing bolts.

➡**Twist the water transfer pipe while pulling in order to remove it from the water pump.**

16. Remove the thermostat housing from the vehicle.

17. Remove the water transfer pipe from the thermostat housing, if necessary.

18. Remove and discard the water transfer pipe O-ring seals, if necessary.

19. Remove the thermostat housing cover bolts and cover, if necessary.

20. Remove the thermostat, if necessary.

21. Remove and discard the thermostat housing O-ring seal, if necessary.

22. Remove all debris and thread sealant from the engine coolant temperature sensor and bolt holes if the housing is being re-used.

23. Remove the water pump access plate from the front cover.

24. Remove the right hand fender liner.

➡**A drain plug has been provided at the bottom of the water pump assembly for additional coolant drainage from the engine block and water pump.**

25. Drain the coolant from the water pump using the plug at the bottom of the pump.

➡**The water pump holding tool supports the sprocket and chain during water pump service. The tool must be used or the balance shaft must be re-timed.**

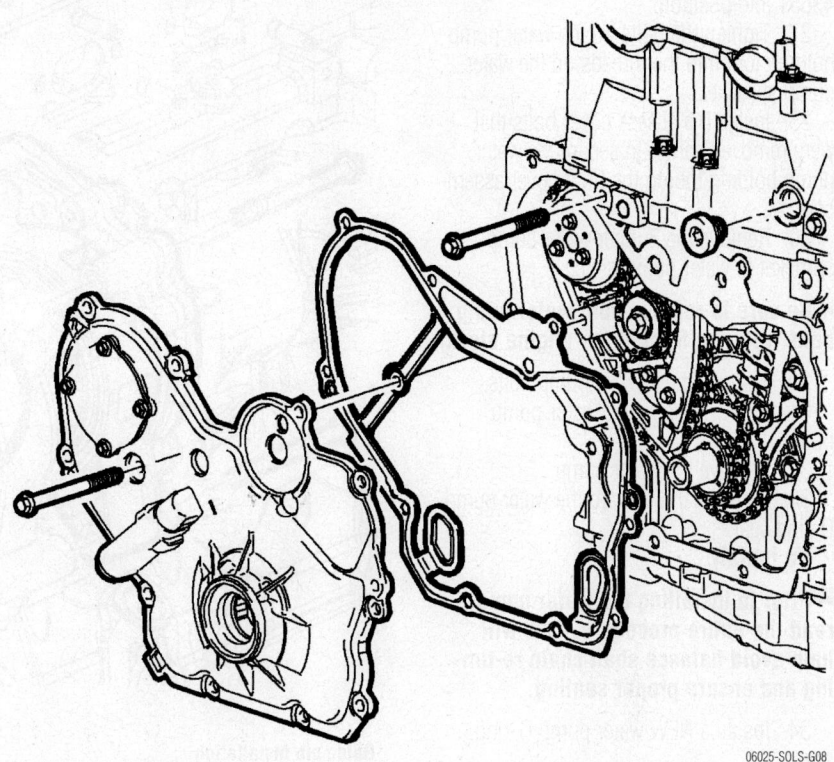

06025-SOLS-G08

Front water pump bolts

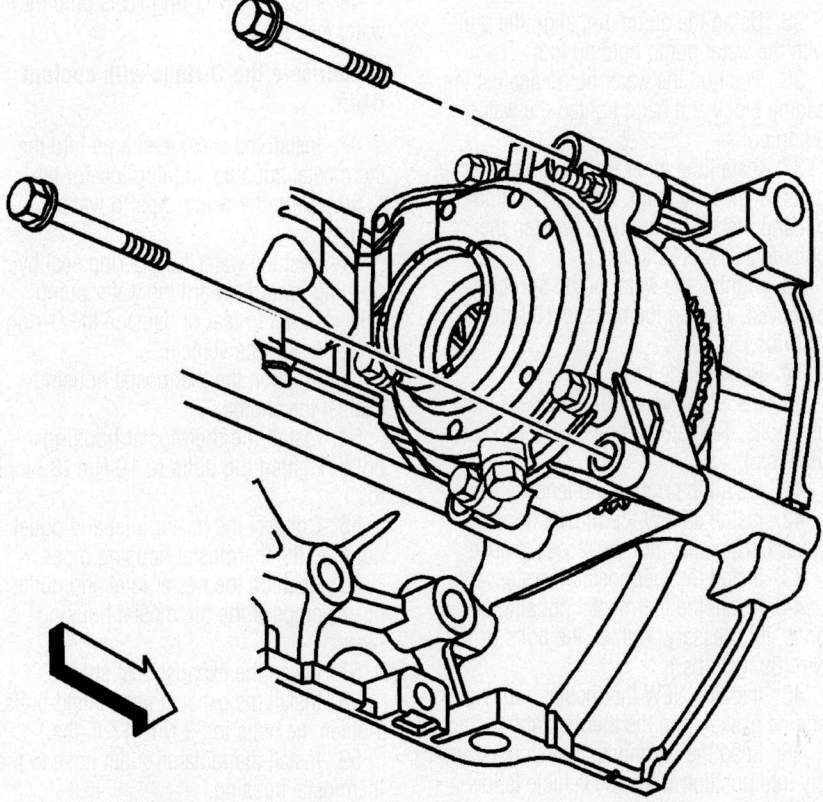

06025-SOLS-G10

Rear water pump bolts

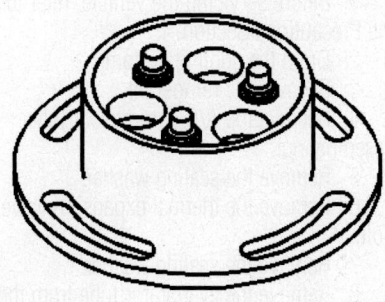

06025-SOLS-G09

Water pump holding tool

26. Install water pump holding tool J 43651 into position.

27. Tighten the bolts on the water pump holding tool into the threads on the water pump sprocket.

28. Install the access cover bolts that were removed earlier to secure the water pump holding tool to the front cover assembly.

29. Remove the 3 inner water pump sprocket to water pump blots.

➡ **Be sure to remove both water pump bolts from the front of the engine block.**

30. Remove the 2 water pump bolts.

31. Remove the rear 2 water pump bolts.

32. Remove the water pump.

33. Remove and discard the water pump O-ring seal.

To install:

➡ **Prior to installing the water pump, read the entire procedure. This will help avoid balance shaft chain re-timing and ensure proper sealing.**

34. Install a NEW water pump O-ring seal.

➡ **A guide pin can be created to aid in water pump alignment. Use an M 6 m x 6 mm stud. Thread the pin into the water pump sprocket.**

35. Using the guide pin, align the pin with the water pump holding tool.

36. Position the water pump against the engine block and hand tighten the water pump bolts.

37. Install the inner water pump sprocket bolts. After 2 are snug, remove the guide pin and install the 3rd bolt. Tighten the bolts to 25 Nm (18 ft. lbs.).

38. Tighten the water pump sprocket bolts last. Tighten the bolts to 10 Nm (89 inch lbs.).

39. Remove the tool.

40. Install the water pump access plate and bolts. Tighten the bolts to 10 Nm (89 inch lbs.).

41. Install the right hand fender liner.

42. Install a NEW thermostat housing cover O-ring seal into the recess groove.

43. Install the thermostat, if necessary.

44. Install the thermostat housing cover bolts, if necessary. Tighten the bolts to 10 Nm (89 inch lbs.).

45. Install a NEW thermostat housing to engine gasket onto the thermostat housing.

46. Load the thermostat housing assembly into position while the vehicle is lowered.

47. Raise and support the vehicle.

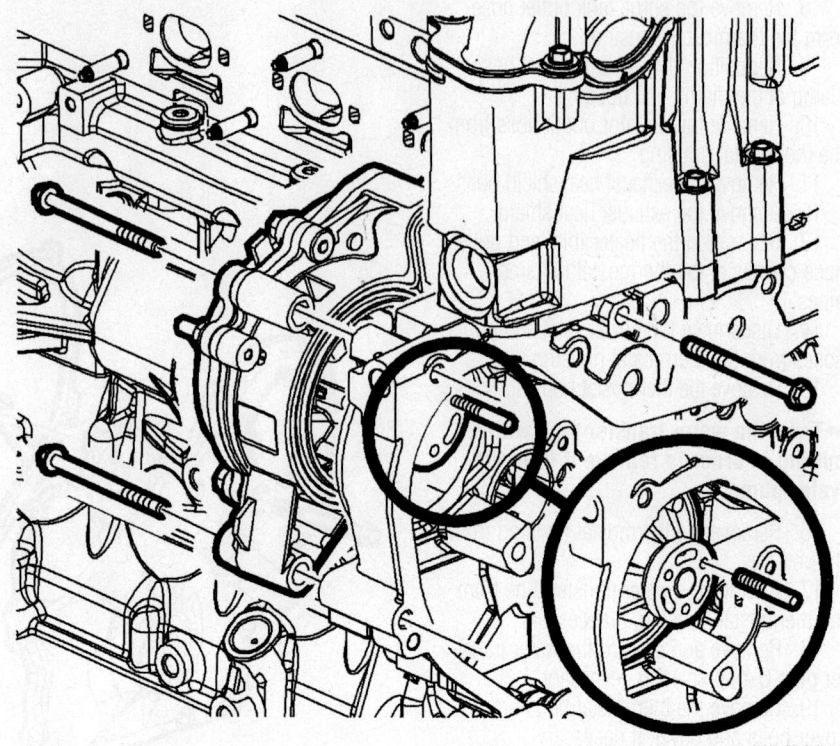

06025-SOLS-G11

Guide pin installation

➡ **The water feed pipe seals can be lightly lubricated with coolant to aid during installation.**

48. Install NEW O-ring seals onto the water feed pipe.

➡ **Lubricate the O-rings with coolant ONLY.**

49. Install the water feed pipe into the thermostat housing aligning locator tab.

50. Align the water pipe to water pump.

51. Seat the water feed O-ring seal by pushing inward toward the water pump. Take care not to tear or damage the O-ring.

52. Lower the vehicle.

53. Position the thermostat housing against the engine.

54. Install the thermostat housing bolts. Tighten the bolts to 10 Nm (89 inch lbs.).

55. Connect the heater inlet and outlet hoses to the thermostat housing pipes.

56. Position the heater inlet and outlet hose clamps at the thermostat housing pipes.

57. Install the exhaust heat shield.

58. Install the exhaust heat shield bolts. Tighten the bolts to 23 Nm (17 ft. lbs.).

59. Install the radiator outlet hose to the thermostat housing.

60. Position the radiator outlet hose clamp at the thermostat housing.

61. Install the surge tank outlet hose to the thermostat housing.

62. Position the surge tank outlet hose clamp at the thermostat housing.

63. Install the ECT sensor, if necessary.

64. Connect the ECT sensor electrical connector.

➡ **The vehicle must be level when filling the cooling system.**

65. Verify the drain valves at the radiator and water pump are closed.

66. Fill the cooling system.

67. Lower the vehicle.

68. Check for any leaks.

Heater Core

REMOVAL & INSTALLATION

1. Before servicing the vehicle, refer to the Precautions Section.

2. Drain the cooling system.

3. Recover the refrigerant.

4. Remove the A/C compressor tube assembly nut.

5. Remove the sealing washer.

6. Remove the thermal expansion valve bolt.

7. Remove the sealing washer.

8. Remove the evaporator tube from the thermal expansion valve.

9. Remove the air inlet grille panel.

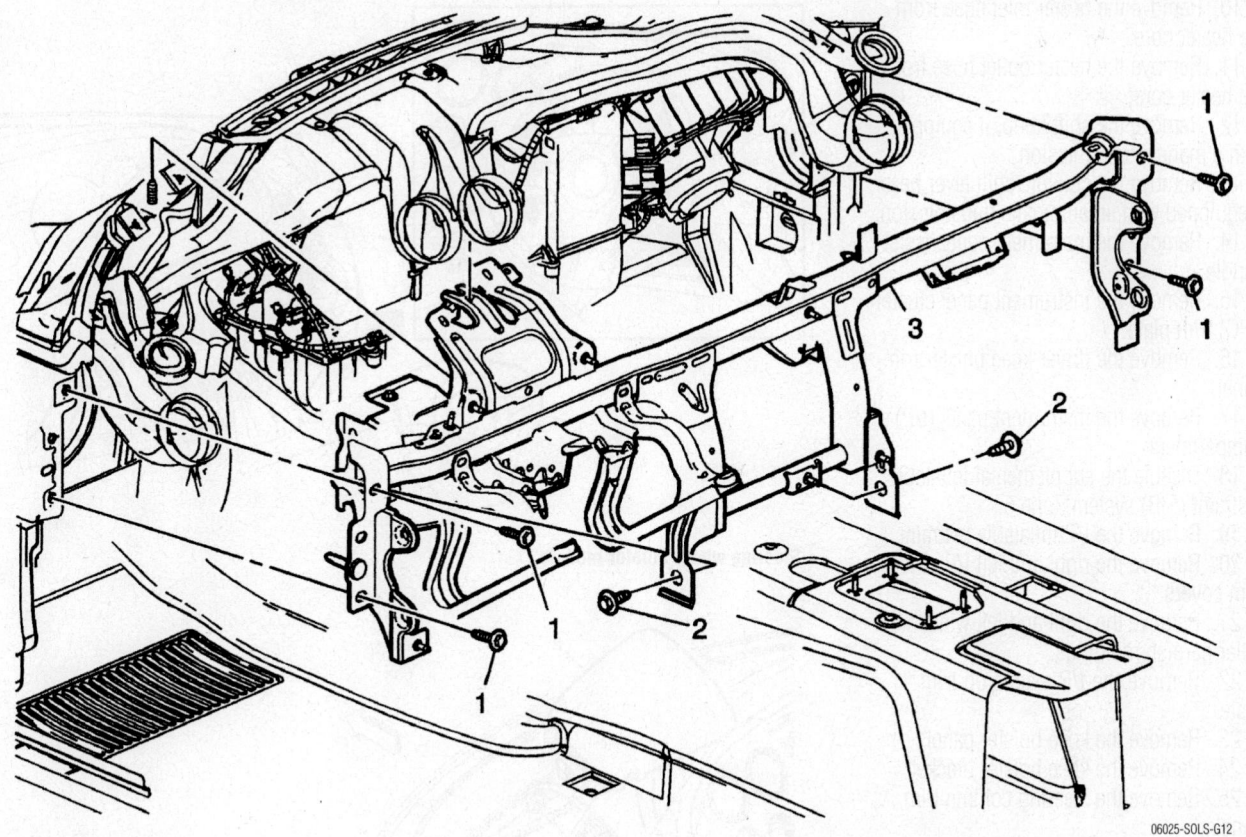

06025-SOLS-G12

IP tie bar

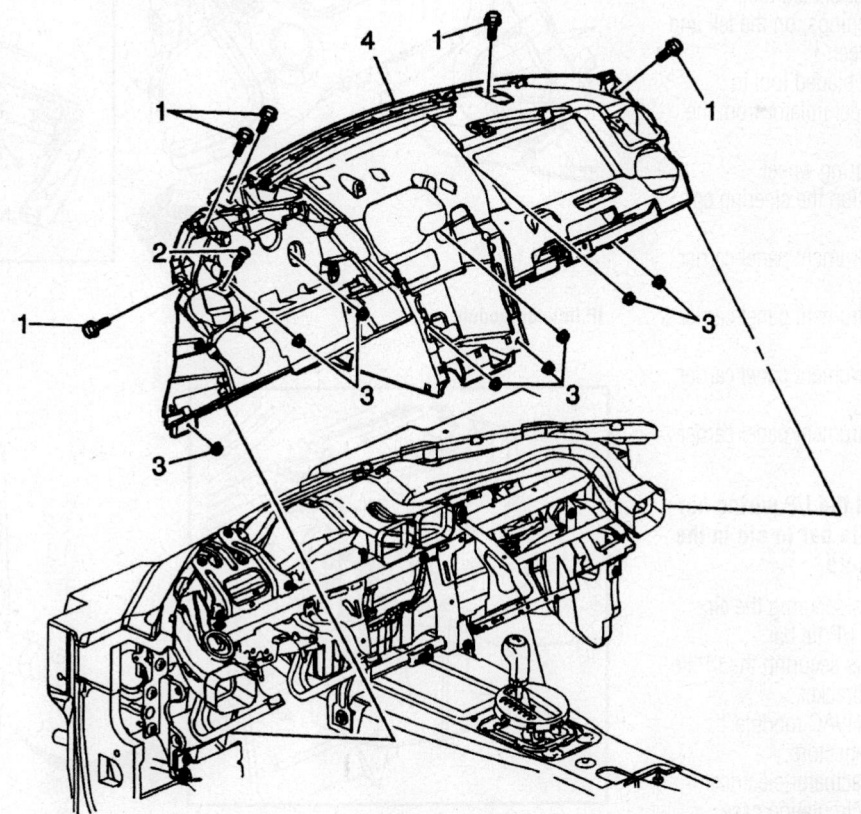

06025-SOLS-G13

IP carrier

10. Remove the heater inlet hose from the heater core.

11. Remove the heater outlet hose from the heater core.

12. Remove the shift knob, if equipped with a manual transmission.

13. Remove the console shift lever bezel, if equipped with an automatic transmission.

14. Remove the instrument panel assist handle.

15. Remove the instrument panel cluster (IPC) trim plate.

16. Remove the driver knee bolster trim panel.

17. Remove the instrument panel (I/P) compartment.

18. Disable the supplemental inflatable restraint (SIR) system Zone 5.

19. Remove the I/P inflatable restraint.

20. Remove the right and left I/P outer trim covers.

21. Remove the right and left windshield pillar garnish moldings.

22. Remove the I/P extension trim plate.

23. Remove the knee bolster panel.

24. Remove the knee bolster bracket.

25. Remove the steering column trim panels.

26. Disable the supplemental inflatable restraint (SIR) system Zone 3.

27. Insert a small flat bladed tool through the access openings, on the left and right of the steering wheel.

28. Push on the flat bladed tool to release the steering wheel inflator from the steering wheel.

29. Remove the steering wheel.

30. Lower and position the steering column out of the way.

31. Remove the instrument panel carrier bolt.

32. Remove the instrument panel carrier screw.

33. Remove the instrument panel carrier nut.

34. Remove the instrument panel carrier assembly.

➡**Note the routing of the I/P wiring harness around the I/P tie bar to aid in the reinstallation procedure**

35. Remove the nuts securing the air distribution duct to the I/P tie bar.

36. Remove the bolts securing the I/P tie bar to the brake pedal bracket.

37. Disconnect the HVAC module assembly electrical connectors.

38. Disconnect the actuator electrical connectors from the recirculation case.

39. Remove the air outlet duct screw.

40. Remove the air outlet duct.

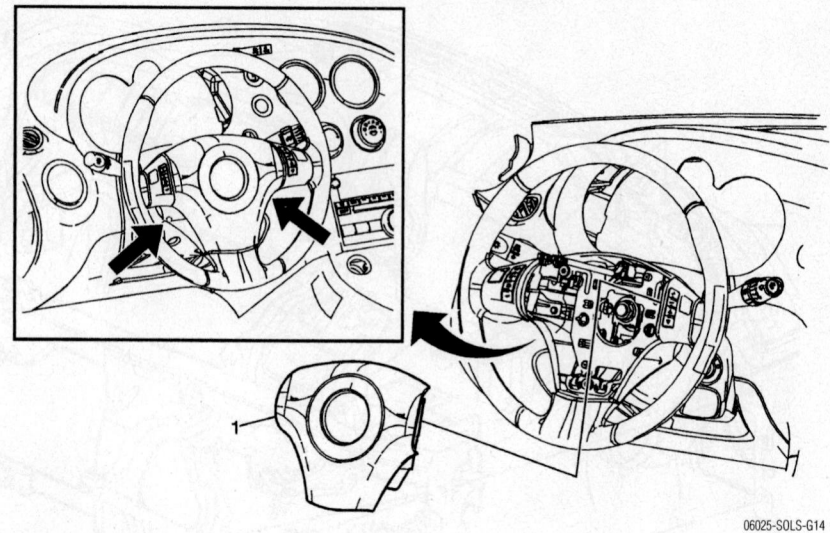

06025-SOLS-G14

Steering wheel inflator module

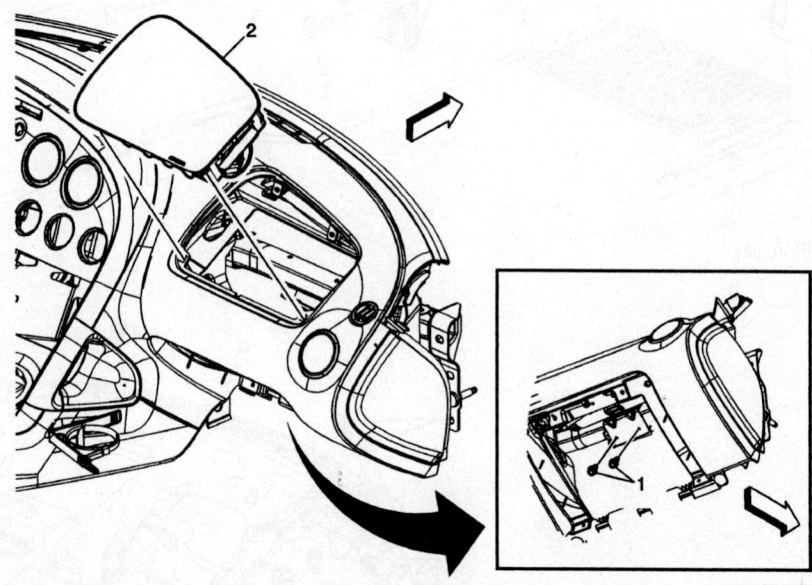

06025-SOLS-G17

IP inflator module

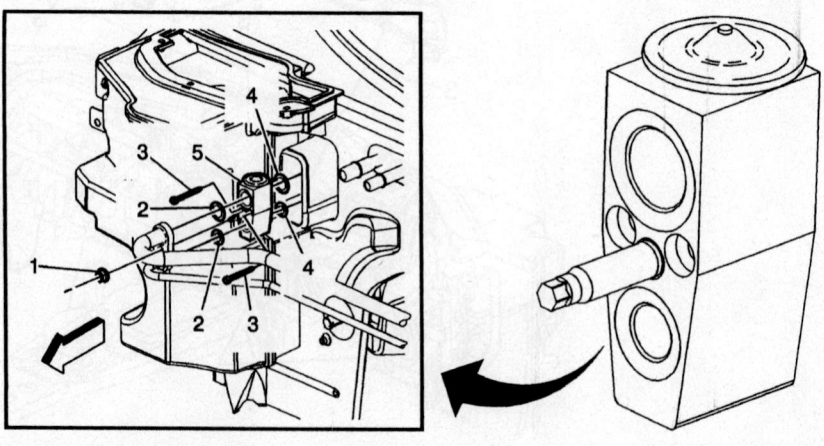

06025-SOLS-G18

Thermal expansion valve

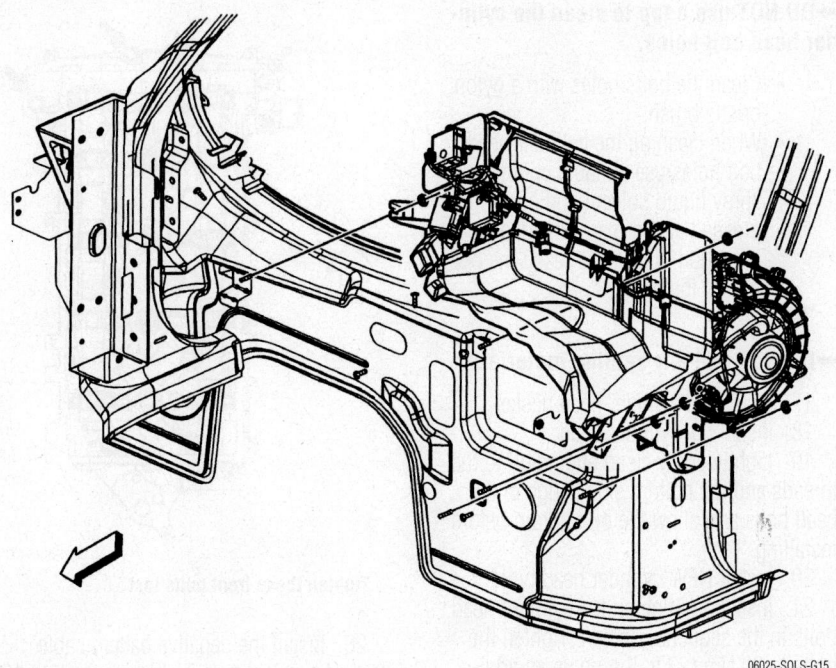

HVAC unit

06025-SOLS-G15

- A/C compressor tube nut: 12 ft. lbs. (16 Nm)
- Thermal expansion valve bolt: 12 ft. lbs. (16 Nm)
- IP tie bar bolts (1): 18 ft. lbs. (25 Nm)
- IP tie bar bolts (2) 106 inch lbs. (12 Nm)
- IP carrier fasteners: 80 inch lbs. (9 Nm)

48. Replace all seals.

Cylinder Head

REMOVAL & INSTALLATION

1. Before servicing the vehicle, refer to the Precautions Section.
2. Remove the intake manifold.
3. Remove the exhaust manifold.
4. Remove the timing chain.
5. Drain the cooling system.
6. Remove the negative battery cable ground terminal bolt.

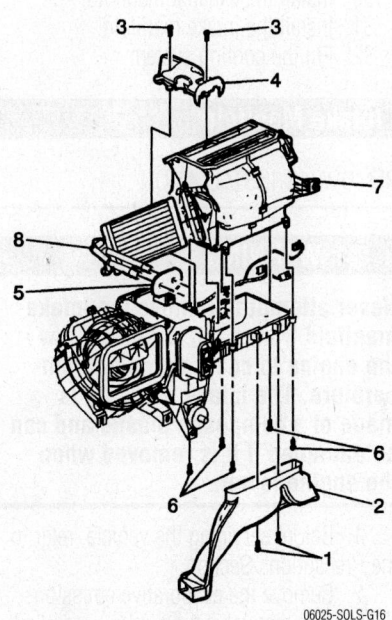

Heater core removal

06025-SOLS-G16

41. Remove the heater core cover screw.
42. Remove the heater core cover.
43. Remove the heater core pass-through seal.
44. Remove the air distribution case screw.
45. Remove the air distribution case.
46. Remove the heater core.
47. Installation is the reverse of removal. Observe the following torques:
- HVAC module nuts: 89 inch lbs. (9 Nm)

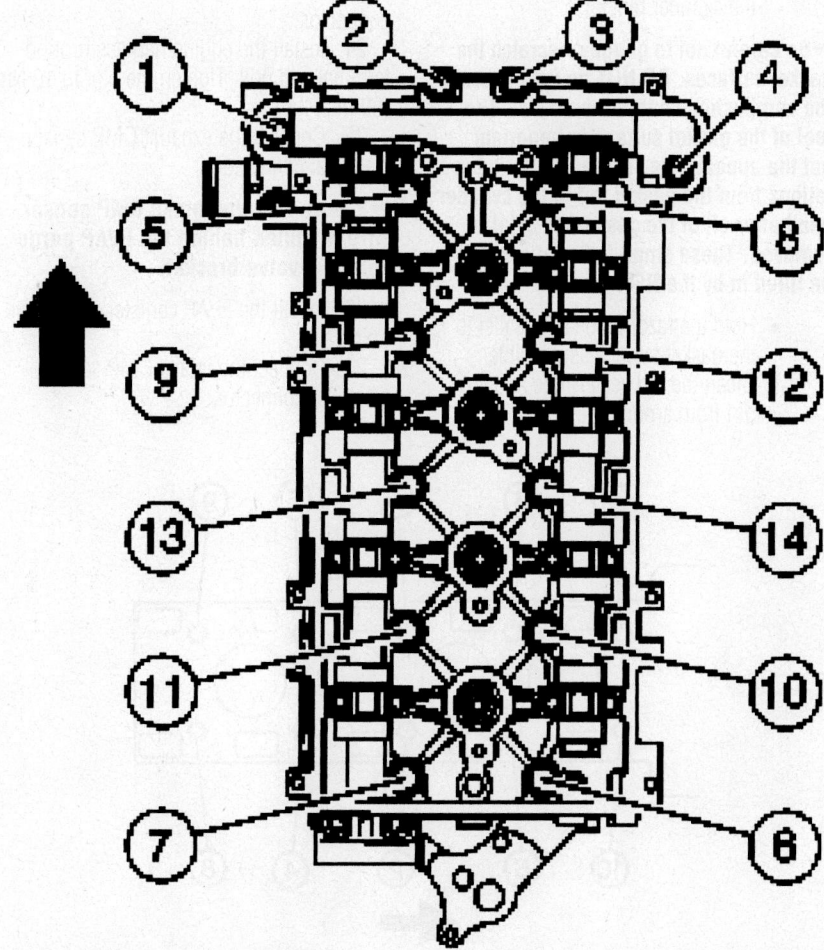

Head bolt removal sequence

06025-SOLS-G19

7. Disconnect the intake camshaft position (CMP) sensor electrical connector.

8. Remove the evaporative emission (EVAP) canister purge solenoid valve.

9. Disconnect the exhaust CMP sensor electrical connector.

10. Remove the engine harness ground terminal bolt and terminal.

11. Disconnect the engine coolant temperature (ECT) sensor electrical connector.

12. Remove the cylinder head bolts in the sequence shown. Discard the bolts.

13. Remove the cylinder head.

14. Remove the cylinder head gasket.

15. Clean all of the gasket surfaces.

16. Use the following steps when cleaning the cylinder head and cylinder block surfaces:

- Use a razor blade gasket scraper to clean the cylinder head and cylinder block gasket surfaces. Do not scratch or gouge either surface

➡**DO NOT use any other method or technique to clean these gasket surfaces.**

- Use a NEW razor blade on the cylinder head and a NEW blade on the cylinder block.

➡**Be careful not to gouge or scratch the gasket surfaces. DO NOT gouge or scrape the combustion chamber surfaces. The feel of the gasket surface is important, not the appearance. There will be indentations from the gasket left in the cylinder head after all of the gasket material is removed. These small indentations will be filled in by the NEW gasket.**

- Hold the razor blade as parallel to the gasket surface as possible
- Clean the old sealer/lube and any dirt from around the bolt holes.

➡**DO NOT use a tap to clean the cylinder head bolt holes.**

- Clean the bolts holes with a nylon bristle brush.
- When cleaning the cylinder head bolt holes use suitable commercial spray liquid solvent and compressed air from an extended-tip blow gun in order to reach the bottom of the holes.

To install:

➡**DO NOT use any sealing material.**

17. Install the cylinder head gasket.

18. Install the cylinder head.

19. Lightly apply clean engine oil to the threads and the bottom side flange of the head bolts and allow the oil to drain before installing.

20. Install NEW cylinder head bolts.

21. Install and tighten the cylinder head bolts in the sequence shown. Tighten the bolts to 30 Nm (22 ft. lbs.) plus an additional 155 degrees.

22. Install the NEW front cylinder head bolts. Tighten the bolts to 35 Nm (26 ft. lbs.).

23. Connect the ECT sensor electrical connector.

24. Install the engine harness ground terminal and bolt. Tighten the bolt to 10 Nm (89 inch lbs.).

25. Connect the exhaust CMP sensor electrical connector.

➡**Ensure that the intake CMP sensor wire is routed behind the EVAP purge solenoid valve bracket.**

26. Install the EVAP canister purge solenoid valve.

27. Connect the intake CMP sensor electrical connector.

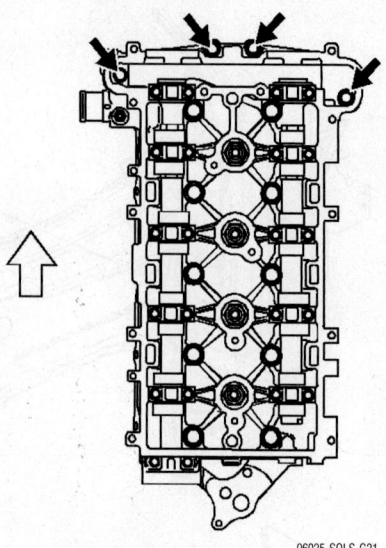

06025-SOLS-G21

Tighten these front bolts last

28. Install the negative battery cable ground terminal bolt. Tighten the bolt to 10 Nm (89 inch lbs.).

29. Install the timing chain.

30. Install the exhaust manifold.

31. Install the intake manifold.

32. Fill the cooling system.

Intake Manifold

REMOVAL & INSTALLATION

✳✳ WARNING

Never attempt to remove the intake manifold from a hot engine, allow the engine to cool to ambient temperature. The intake manifold is made of a composite plastic and can be damaged if it is removed when the engine is hot.

1. Before servicing the vehicle, refer to the Precautions Section.

2. Remove the evaporative emission (EVAP) canister valve tube.

3. Remove the EVAP canister valve.

4. Remove the throttle body bolts.

5. Remove the throttle body.

6. Remove fuel pipes and clips.

7. Remove the fuel rail bolts.

8. Remove the fuel rail.

9. Remove the fuel injector tip insulators and discard.

10. Remove the intake manifold retaining nuts and bolts.

11. Remove the intake manifold.

12. Remove the intake manifold gasket, if necessary. The gasket can be used again if it is not damaged.

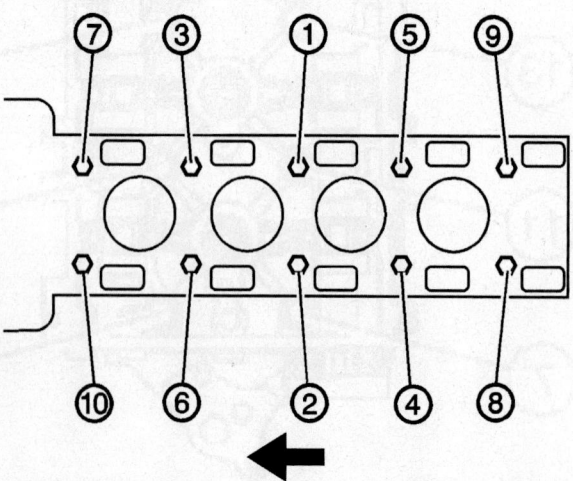

06025-SOLS-G20

Head bolt torque sequence

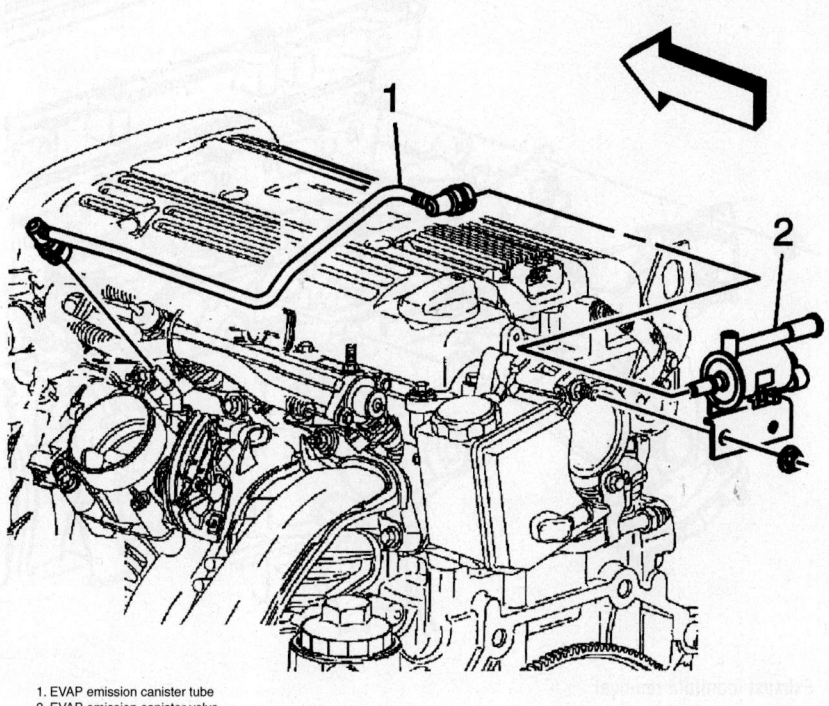

1. EVAP emission canister tube
2. EVAP emission canister valve

06025-SOLS-G22

EVAP emission canister valve

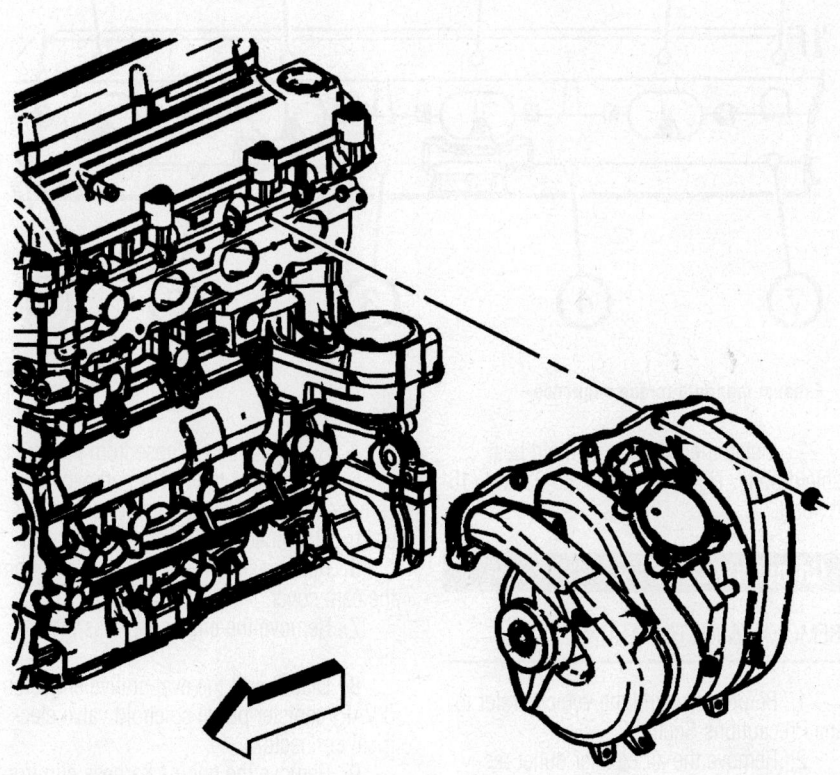

06025-SOLS-G23

Intake manifold

To install:

13. Clean the intake manifold mating surfaces.

14. Inspect the intake manifold for damage.

15. Inspect the intake manifold for cracks near metallic inserts.

16. Inspect the crankcase ventilation passages in the intake manifold face for blockage.

❋❋ CAUTION

Wear safety glasses in order to avoid eye damage.

17. Clean the crankcase ventilation passages with compressed air if necessary. Use a maximum of 172 kPa (25 psi) of air pressure.

18. Replace the intake manifold as necessary.

19. Install the intake manifold studs in the manifold face. Tighten the intake manifold studs to 6 Nm (53 inch lbs.).

20. Install a new intake manifold gasket on the intake manifold.

21. Install the intake manifold.

22. Install the intake manifold bolts and nuts. Tighten the bolts and nuts to 10 Nm (89 inch lbs.).

23. Lubricate the fuel injector tip insulators with engine oil.

24. Install new fuel injector tip insulators.

25. Lubricate the fuel injector oil rings with engine oil.

26. Install the fuel rail assembly.

27. Install the fuel rail bolts. Tighten the bolts to 10 Nm (89 inch lbs.).

28. Install a new throttle body gasket.

29. Install the throttle body. Tighten the bolts and nuts to 10 Nm (89 inch lbs.).

❋❋ WARNING

Ensure that the rear metal tab of the EVAP emission canister valve is resting on the power steering pump metal body.

30. Install the EVAP emission canister valve. Tighten the EVAP canister valve bolt to 22 Nm (16 ft. lbs.).

31. Install the EVAP emission canister valve tube.

Exhaust Manifold

REMOVAL & INSTALLATION

1. Before servicing the vehicle, refer to the Precautions Section.

2. Remove the exhaust manifold heat shield bolts.

3. Remove the exhaust manifold heat shield.

4. Remove the block heater if equipped.

5. Remove the oxygen sensor.

6. Remove and discard the exhaust manifold to cylinder head retaining nuts.

7. Remove the exhaust manifold.

8. If the exhaust manifold is being replaced, transfer the following parts:
- The exhaust manifold heat shield
- The oxygen sensor

To install:

➡**Do not reuse the exhaust manifold-to-cylinder head gaskets. Upon installation of the exhaust manifold, install a NEW gasket. An improperly installed gasket or leaking exhaust system may effect On-Board Diagnostics (OBD) II system performance. Remove the oxygen sensor prior to cleaning the manifold, do not submerge the oxygen sensor in cleaning solvent.**

9. Remove the oxygen sensor from the manifold.

10. Clean the exhaust manifold in solvent.

✳✳ CAUTION

Wear safety glasses in order to avoid eye damage.

11. Dry the exhaust manifold with compressed air.

12. Inspect the heat shield for damage.

13. Use a straightedge and a feeler gage and measure the exhaust manifold mounting face for warpage. An exhaust manifold face with warpage in excess of 0.25 mm (0.010 in.) may cause an exhaust leak and may affect OBD II system performance. Exhaust manifolds not within specifications must be replaced.

14. Install new exhaust manifold studs. Tighten the studs to 10 Nm (89 inch lbs.).

15. Install the exhaust manifold gasket.

16. Install the exhaust manifold to the cylinder head.

17. Install NEW exhaust manifold to cylinder head retaining nuts finger tight.

18. Tighten the NEW exhaust manifold to cylinder head retaining nuts in sequence. Tighten the nuts to 14 Nm (10 ft. lbs.).

19. Coat the threads of the oxygen sensor with anti-seize P/N 12397953, or equivalent.

20. Install the oxygen sensor. Tighten the oxygen sensor to 42 Nm (31 ft. lbs.).

21. Install the exhaust manifold heat shield.

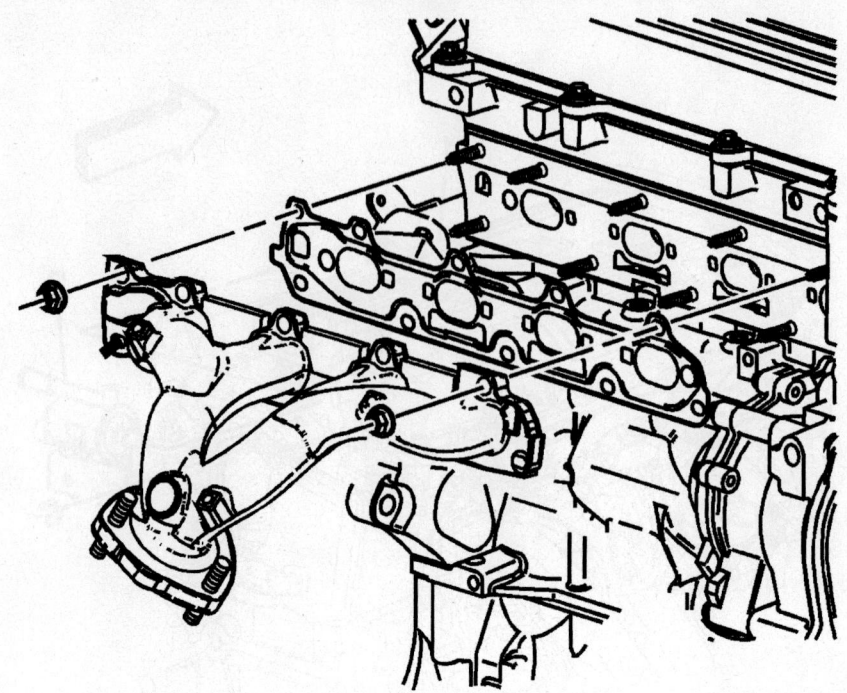

Exhaust manifold removal

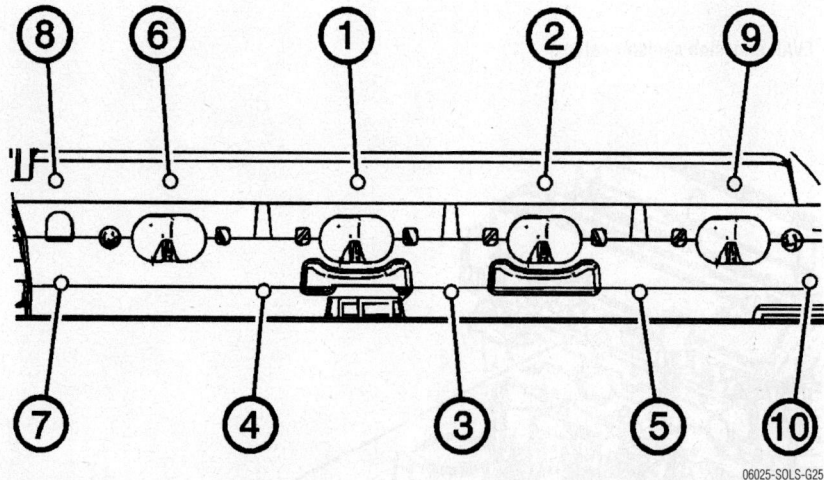

Exhaust manifold torque sequence

22. Install the exhaust manifold heat shield bolts. Tighten the bolts to 22 Nm (16 ft. lbs.).

Camshaft Cover

REMOVAL & INSTALLATION

1. Before servicing the vehicle, refer to the Precautions Section.

2. Remove the air cleaner outlet resonator.

3. Reposition the positive crankcase ventilation (PCV) hose clamp.

4. Remove the PCV hose from the cover.

5. Disconnect the intake and exhaust camshaft position actuator solenoid valve electrical connectors.

6. Remove the engine harness clip from the cam cover.

7. Remove the engine harness bracket bolt.

8. Disconnect the evaporative emission (EVAP) canister purge solenoid valve electrical connector.

9. Remove the engine harness clip from the EVAP purge solenoid bracket.

10. Remove the engine harness clips from the cam cover.

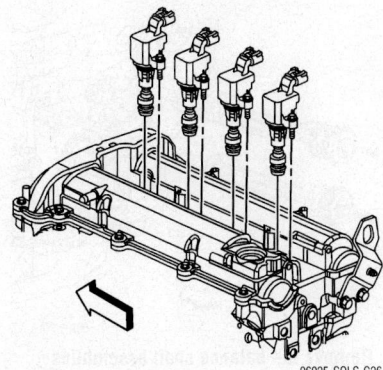

Ignition coils

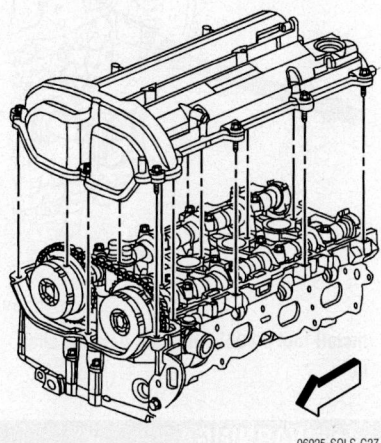

Camshaft cover

11. Remove the ignition coils.
12. Remove the camshaft cover bolts.
13. Remove the camshaft cover.
To install:
14. Install the camshaft cover and bolts. Tighten the bolts to 10 Nm (89 inch lbs.).
15. Install the ignition coils. Tighten the bolts to 10 Nm (89 inch lbs.).
16. Install the engine harness clips to the cam cover.
17. Install the engine harness clip to the EVAP purge solenoid bracket.
18. Connect the EVAP canister purge solenoid valve electrical connector.
19. Install the engine harness bracket bolt. Tighten the bolt to 5 Nm (44 inch lbs.).
20. Install the engine harness clip to the cam cover.
21. Connect the intake and exhaust camshaft position actuator solenoid valve electrical connectors.
22. Install the PCV hose from the cover.
23. Position the PCV hose clamp.
24. Install the air cleaner outlet resonator.

Camshaft and Lifters

REMOVAL & INSTALLATION

Intake

1. Before servicing the vehicle, refer to the Precautions Section.
2. Remove the camshaft cover.
3. Remove the upper timing chain guide bolts and guide.
4. Install tool J 43655, or equivalent.
5. Remove and discard both the intake and exhaust camshaft sprocket bolts.
6. Slide the camshaft sprockets forward.

➡**Remove each bolt on each cap one turn at a time until there is no spring tension pushing on the camshaft.**

7. Mark the bearing caps to ensure they are installed in the original position.
8. Remove the bearing cap bolts.
9. Remove the bearing caps.
10. Remove the intake camshaft.

➡**Keep all of the roller followers and hydraulic adjusters in order so that they can be reinstalled in their respective locations.**

11. Remove the camshaft roller followers.
12. Remove the hydraulic adjusters.
To install:
13. Install the hydraulic element lash adjusters into their bores in the cylinder head.
14. Lubricate the hydraulic lash adjusters with GM PN 12345501 (Canadian PN 992704) or equivalent.
15. Lubricate the valve tips.

➡**Used roller followers MUST be returned to their original position on the camshaft. If the camshaft is being replaced, the roller followers actuated by the camshaft must also be replaced.**

16. Position the camshaft roller followers on the tip of the valve stem and on the lash adjuster. Lubricate the roller followers with GM PN 12345501 (Canadian PN 992704) or equivalent.
17. Install the intake camshaft. Lubricate with GM PN 12345501 (Canadian PN 992704) or equivalent.
18. Install the camshaft bearing caps. Hand-tighten the cap bolts.
19. Ensure that the alignment notches are aligned with the camshaft sprocket.
20. Tighten the camshaft bearing cap bolts in increments of 3 turns until they are seated. Tighten the bolts to 10 Nm (89 inch lbs.).

21. Install the camshaft sprockets onto the camshafts.
22. Install and hand tighten NEW camshaft sprocket bolts.
23. Remove the tool.
24. Tighten the camshaft sprocket bolts. Tighten the bolts to 85 Nm (63 ft. lbs.) plus 30 degrees.
25. Install the upper timing chain guide and bolts. Tighten the bolts to 10 Nm (89 inch lbs.).
26. Install the camshaft cover.

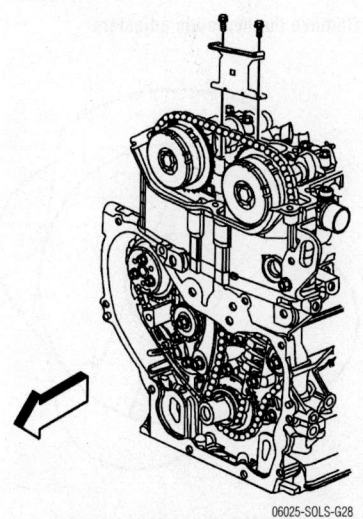

Remove the upper timing chain guide bolts and guide

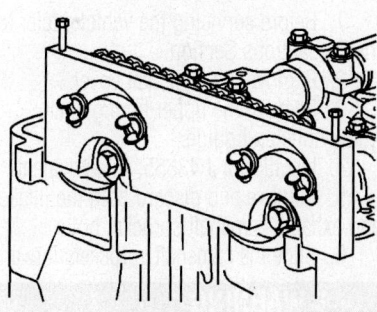

Install tool J 43655, or equivalent

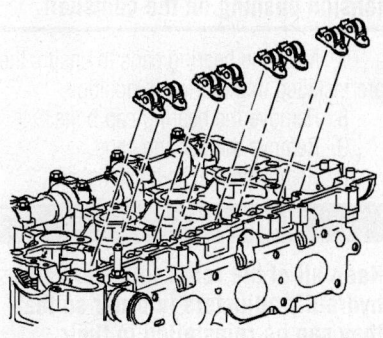

Remove the camshaft roller followers

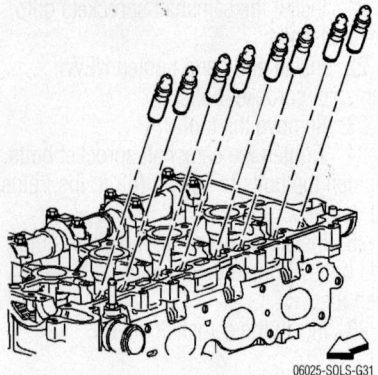

06025-SOLS-G31

Remove the hydraulic adjusters

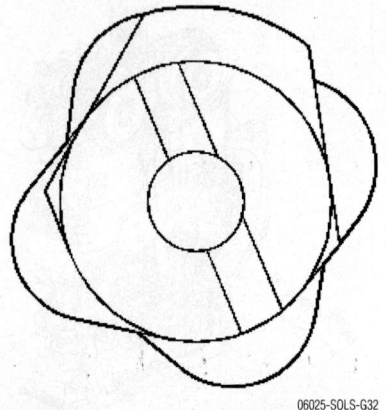

06025-SOLS-G32

Ensure that the alignment notches are aligned with the camshaft sprocket

Exhaust

1. Before servicing the vehicle, refer to the Precautions Section.

2. Remove the camshaft cover.

3. Remove the upper timing chain guide bolts and guide.

4. Install tool J 43655, or equivalent.

5. Remove and discard both the intake and exhaust camshaft sprocket bolts.

6. Slide the camshaft sprockets forward.

✳✳ WARNING

Remove each bolt on each cap one turn at a time until there is no spring tension pushing on the camshaft.

7. Mark the bearing caps to ensure they are installed in the original position.

8. Remove the bearing cap bolts.

9. Remove the bearing caps.

10. Remove the exhaust camshaft.

✳✳ WARNING

Keep all of the roller followers and hydraulic adjusters in order so that they can be reinstalled in their respective locations.

11. Remove the camshaft roller followers.

12. Remove the hydraulic element adjusters.

To install:

13. Install the hydraulic element adjusters into their bores in the cylinder head. Lubricate the hydraulic lash adjusters with GM PN 12345501 (Canadian PN 992704) or equivalent.

14. Lubricate the valve tips with GM PN 12345501 (Canadian PN 992704) or equivalent.

✳✳ WARNING

Used roller followers MUST be returned to the original position on the camshaft. If the camshaft is being replaced, the roller followers actuated by the camshaft must also be replaced.

15. Position the roller followers on the tip of the valve stem and on the lash adjuster. Apply lubricate GM PN 12345501 (Canadian PN 992704) or equivalent.

16. Install the exhaust camshaft. Lubricate with GM PN 12345501 (Canadian PN 992704) or equivalent.

17. Install the camshaft bearing caps.

18. Install and hand tighten the camshaft bearing cap bolts.

19. Ensure that the alignment notches are aligned with the camshaft sprocket.

20. Install and hand tighten the camshaft bearing cap bolts in increments of 3 turns until they are seated.

21. Tighten the camshaft bearing cap bolts to 10 Nm (89 lb in).

22. Install camshaft sprockets onto the camshafts.

23. Hand tighten the NEW camshaft sprocket bolts.

24. Remove the tool.

25. Tighten the camshaft sprocket bolts. Tighten the bolts to 85 Nm (63 lb ft) plus 30 degrees.

26. Install the upper timing chain guide and bolts. Tighten the bolts to 10 Nm (89 lb in).

27. Install the camshaft cover.

Balance Shaft

REMOVAL & INSTALLATION

1. Before servicing the vehicle, refer to the Precautions Section.

2. Remove the balance shaft bearing carrier bolts.

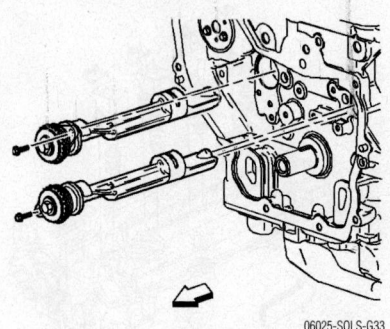

06025-SOLS-G33

Remove the balance shaft assemblies

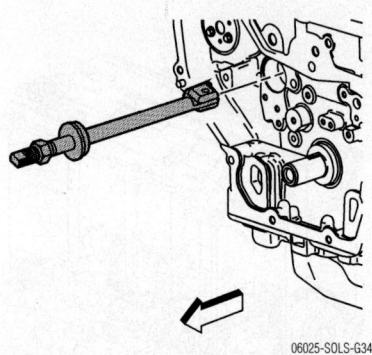

06025-SOLS-G34

Install tool J 43650 into the balance shaft hole

✳✳ WARNING

It is possible to install the intake side balance shaft into the exhaust side and vice versa. Please use care not to install the balance shafts into the wrong bores. Engine vibration will result. Do not remove the bolt holding the sprocket.

3. Remove the balance shaft assemblies.

✳✳ WARNING

Proper centering of the tool is required on the balance shaft bushing. If the tool is not properly centered then damage to the bearing bore and block will occur.

4. Install tool J 43650 into the balance shaft hole. Insert the tool with the foot parallel to the shaft.

5. When the tool is inserted in the block turn the tool so that the foot becomes perpendicular to the shaft. Center the foot of the tool on the balance shaft bushing.

6. Once the tool is centered on the balance shaft bushing, then insert the centering guide into the front balance shaft bore and tighten the nut with an appropriate wrench.

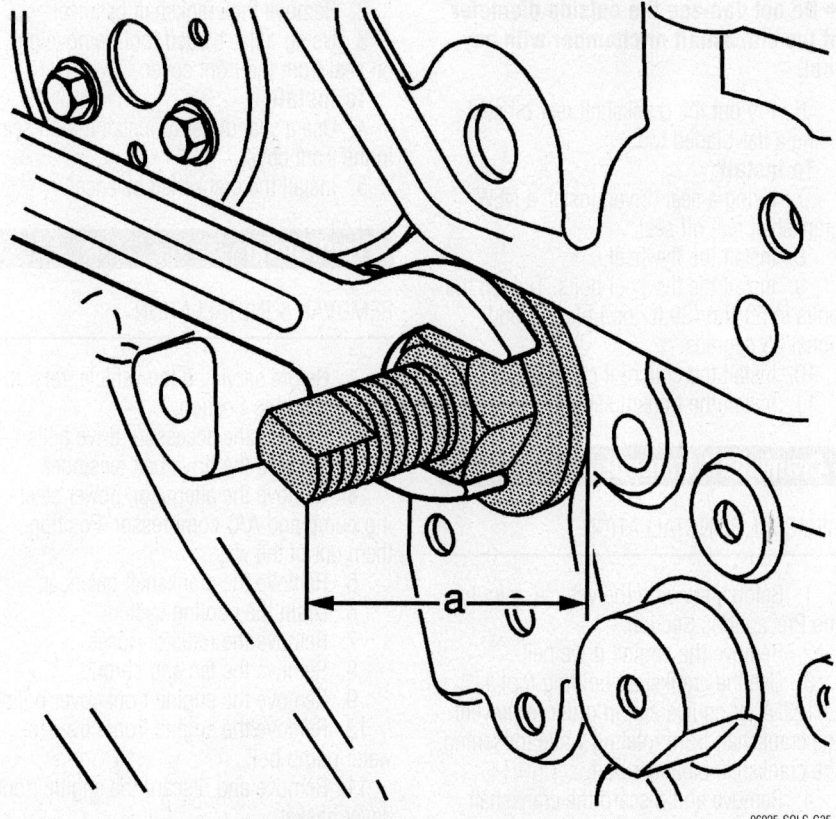

06025-SOLS-G35

When tool J 43650 is properly installed, before removing the bushing, the end of the tool should be 116 mm (4.6 in) (a) from the block face

When tool J 43650 is properly installed, before removing the bushing, the end of the tool should be 116 mm (4.6 in) (a) from the block face. If the tool is less than approximately 114 mm (4.5 in) (a), recheck the tool alignment.

7. Tighten the nut on tool J 43650 until the tension releases. When the tension releases, remove the tool and the balance shaft bushing.

To install:

8. Install the balance shaft bushing using tool J 43650.

9. Seat the balance shaft bushing into the bore using tool J 43650 and a wrench.

10. When tool J 43650 is fully seated in the engine block remove it with a wrench.

➡**If the balance shafts are not properly timed to the engine, the engine may vibrate or make noise.**

11. Place the number one piston at top dead center (TDC).

12. Lubricate the balance shaft lobes with engine oil.

13. Install the balance shafts into their bores.

14. Install the balance shaft retaining bolts. Tighten the balance shaft retaining bolts to 10 Nm (89 inch lbs.).

Oil Pan

REMOVAL & INSTALLATION

1. Before servicing the vehicle, refer to the Precautions Section.

2. Raise and safely support the car.

3. Remove the oil pan bolts.

4. Remove the oil pan at pry points.

5. Clean the oil pan mating surface.

6. Clean the oil pan. Remove all the sludge and the oil deposits.

7. Inspect the threads for the engine oil drain plug.

8. Inspect the oil pan for cracking near the pan rail and the transmission mounting points.

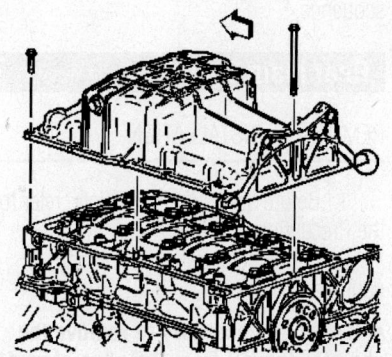

06025-SOLS-G36

Oil pan pry points

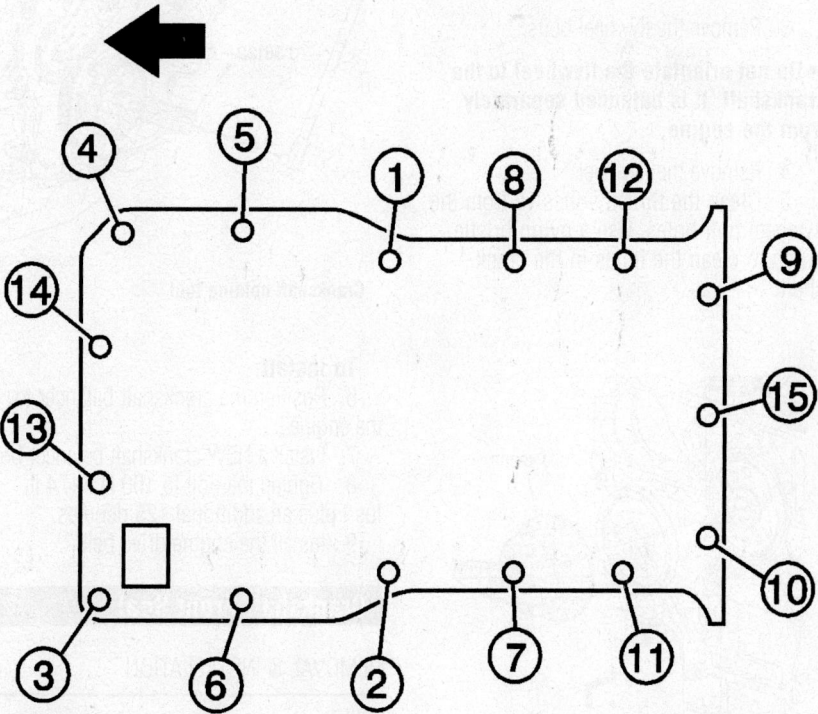

06025-SOLS-G37

Oil pan bolt torque sequence

9. Inspect the oil pan for cracking resulting from impact or flying road debris.

➡**The oil pan baffle and pickup screen are not removable from the oil pan.**

10. Inspect the oil pan baffle and pickup screen.

11. Repair or replace the oil pan as necessary.

12. Make sure that the oil pan and mounting surface on the lower crankcase are free of all oil and debris.

13. Apply a 3.5 mm bead of GM P/N 12378521 (Canadian P/N 88901148) or equivalent around the perimeter of the oil pan and the oil suction port opening.

14. Install the oil pan.

15. Install the oil pan bolts. Tighten the oil pan bolts to 25 Nm (18 ft. lbs.) in sequence.

Rear Main Seal

REMOVAL & INSTALLATION

1. Before servicing the vehicle, refer to the Precautions Section.

2. Remove the transmission and, if equipped, the clutch.

➡**It may be necessary to remove the chamfer (bevel) from the edge of an 18 mm socket in order to get full engagement on the thin-headed flywheel bolts.**

3. Remove the flywheel bolts.

➡**Do not orientate the flywheel to the crankshaft. It is balanced separately from the engine.**

4. Remove the flywheel.

5. Clean the thread adhesive from the flywheel bolt holes. Use a nylon bristle brush to clean the holes in the crankshaft.

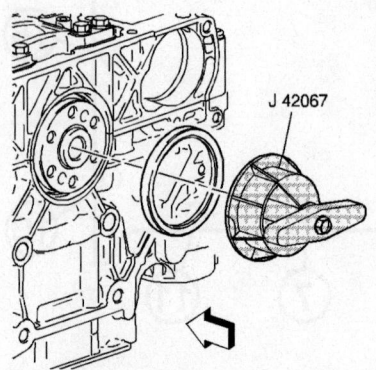

J 42067

Rear main seal installation

06025-SOLS-G38

➡**Do not damage the outside diameter of the crankshaft or chamber with any tool.**

6. Pry out the crankshaft rear oil seal using a flat-bladed tool.
To install:
7. Using a seal driver, install a NEW crankshaft real oil seal.
8. Install the flywheel.
9. Install the flywheel bolts. Tighten the bolts to 53 Nm (39 ft. lbs.) plus an additional 25 degrees.
10. Install the clutch, if equipped.
11. Install the transmission.

Crankshaft Damper

REMOVAL & INSTALLATION

1. Before servicing the vehicle, refer to the Precautions Section.
2. Remove the engine drive belt.
3. Use the crankshaft holding tool J 38122-A, or equivalent, in order to prevent the crankshaft from rotating while loosening the crankshaft balancer bolt.
4. Remove and discard the crankshaft balancer bolt.
5. Remove the crankshaft balancer.

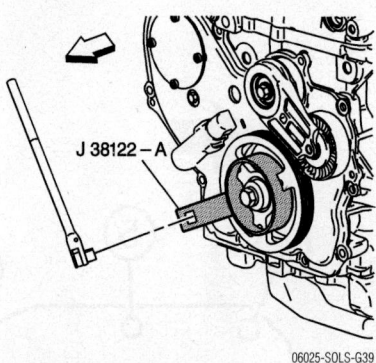

J 38122 – A

Crankshaft holding tool

06025-SOLS-G39

To install:
6. Position the crankshaft balancer to the engine.
7. Install a NEW crankshaft balancer bolt.
8. Tighten the bolt to 100 Nm (74 ft. lbs.) plus an additional 125 degrees.
9. Install the engine drive belt.

Crankshaft Front Seal

REMOVAL & INSTALLATION

1. Before servicing the vehicle, refer to the Precautions Section.

2. Remove the crankshaft balancer.
3. Using a flat-bladed tool, remove the oil seal from the front cover.
To install:
4. Use a seal driver to install the oil seal in the front cover.
5. Install the crankshaft balancer.

Front Cover

REMOVAL & INSTALLATION

1. Before servicing the vehicle, refer to the Precautions Section.
2. Remove the accessory drive belts.
3. Remove the drive belt tensioner.
4. Remove the alternator, power steering pump and A/C compressor. Position them out of the way.
5. Remove the crankshaft balancer.
6. Drain the cooling system.
7. Remove the radiator hoses.
8. Remove the fan and shroud.
9. Remove the engine front cover bolts.
10. Remove the engine front cover-to-water pump bolt.
11. Remove and discard the engine front cover gasket.
To install:
12. Position and install the engine front cover.
13. Install the engine front cover to water pump bolt. Tighten the bolt to 25 Nm (18 ft. lbs.).
14. Install the engine front cover bolts. Tighten the bolts to 25 Nm (18 ft. lbs.).
15. Install the crankshaft balancer.
16. Install the drive belt tensioner. Torque to 33 ft. lbs. (45 Nm).
17. The remainder of installation is the reverse of removal.

Timing Chain and Sprockets

REMOVAL & INSTALLATION

1. Before servicing the vehicle, refer to the Precautions Section.
2. Remove the camshaft cover.
3. Remove the engine front cover.
4. Rotate the engine until the crankshaft sprocket mark aligns with the matching colored link (2) at the 5 o'clock position.
5. Lower the vehicle.
6. Ensure that the timing mark on the intake camshaft sprocket is aligned with the blue colored link at the 2 o'clock position.
7. Ensure that the timing mark on the exhaust camshaft sprocket is aligned with the pink colored link.

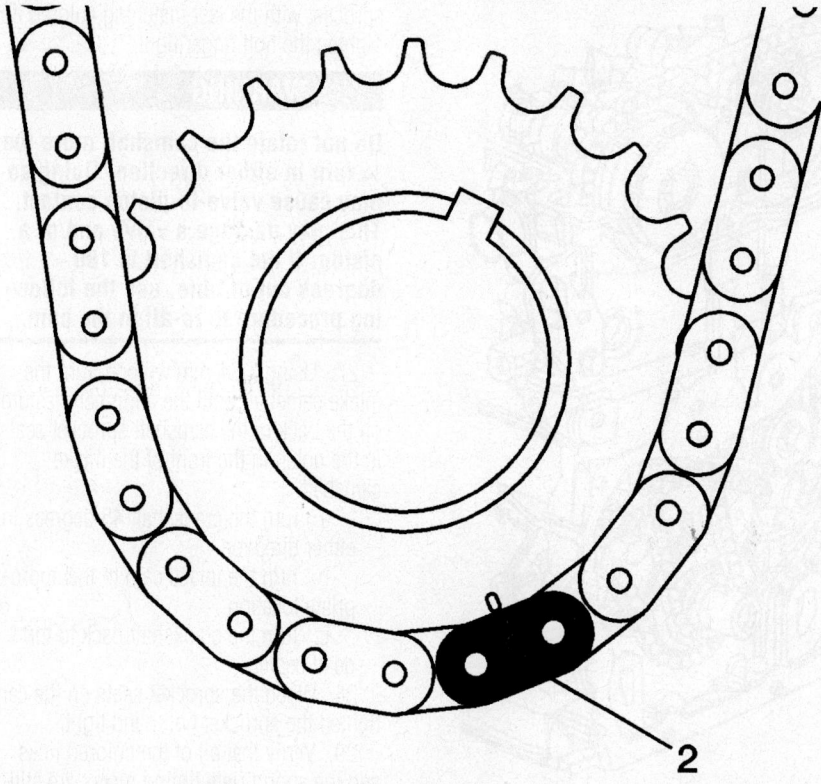

06025-SOLS-G40

Rotate the engine until the crankshaft sprocket mark aligns with the matching colored link (2) at the 5 o'clock position

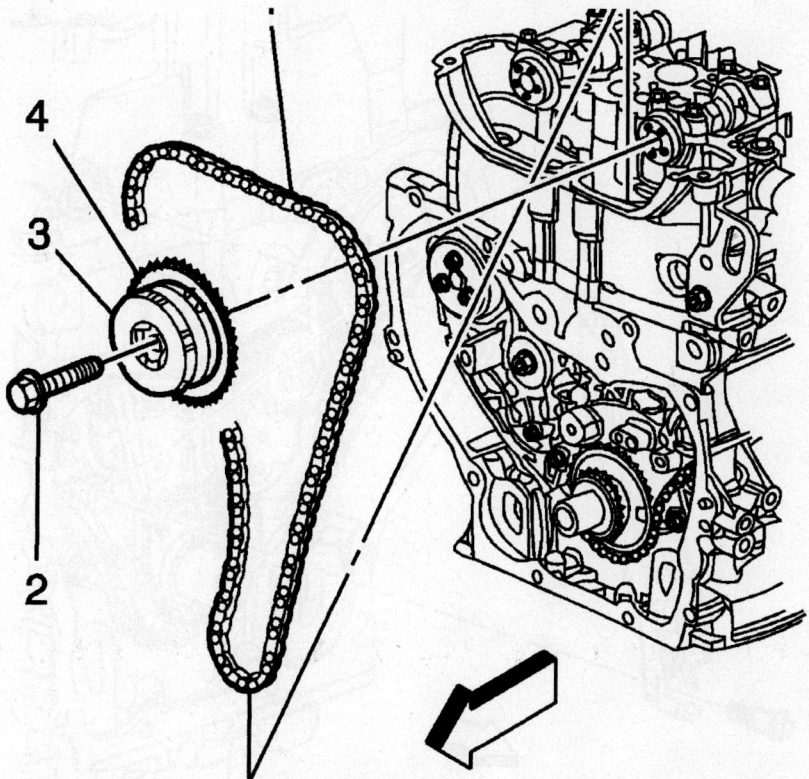

06025-SOLS-G41

Ensure that the timing mark (4) on the intake camshaft sprocket is aligned with the blue colored link (1) at the 2 o'clock position

8. Remove the timing chain tensioner.

9. Remove the fixed timing chain guide access plug.

10. Remove the fixed timing chain guide bolts and guide.

11. Remove the upper timing chain guide bolts and guide.

12. Using a 24-mm wrench, hold the camshaft in order to prevent the camshaft from turning.

13. Remove and discard the exhaust camshaft sprocket bolt.

14. Remove the exhaust camshaft sprocket.

15. Remove the timing chain tensioner guide bolt and guide.

16. Remove and discard the intake camshaft sprocket bolt.

17. Remove the intake camshaft sprocket.

18. Remove the timing chain out through the top of the cylinder head.

19. Remove the crankshaft sprocket.

To install:

20. Install the timing chain drive sprocket to the crankshaft with the timing mark in the 5 o'clock position.

➡**There are 3 colored links on the timing chain. Two links are of matching color and one link is of a unique color. Use the following procedure to line up the links with the sprockets. Orient the chain so that the colored links are visible. Always use NEW sprocket bolts.**

21. Assemble the intake camshaft sprocket (3) to the timing chain with the timing mark (4) lined up with the blue colored link (1). Install and hand tighten a NEW intake camshaft sprocket bolt (2).

22. Lower the timing chain through the opening in the cylinder head. Use care to ensure that the chain goes around both sides of the cylinder block bosses.

23. Route the timing chain around the crankshaft sprocket and line up the first matching colored link with the timing mark on the crankshaft sprocket, in approximately the 5 o'clock position.

24. Install the adjustable timing chain guide through the opening in the cylinder head. Install the adjustable timing chain bolt. Tighten the bolt to 10 Nm (89 inch lbs.).

➡**Always install NEW sprocket bolts.**

25. Install the exhaust camshaft sprocket and a NEW bolt loosely onto the exhaust camshaft.

26. Align the timing mark on the

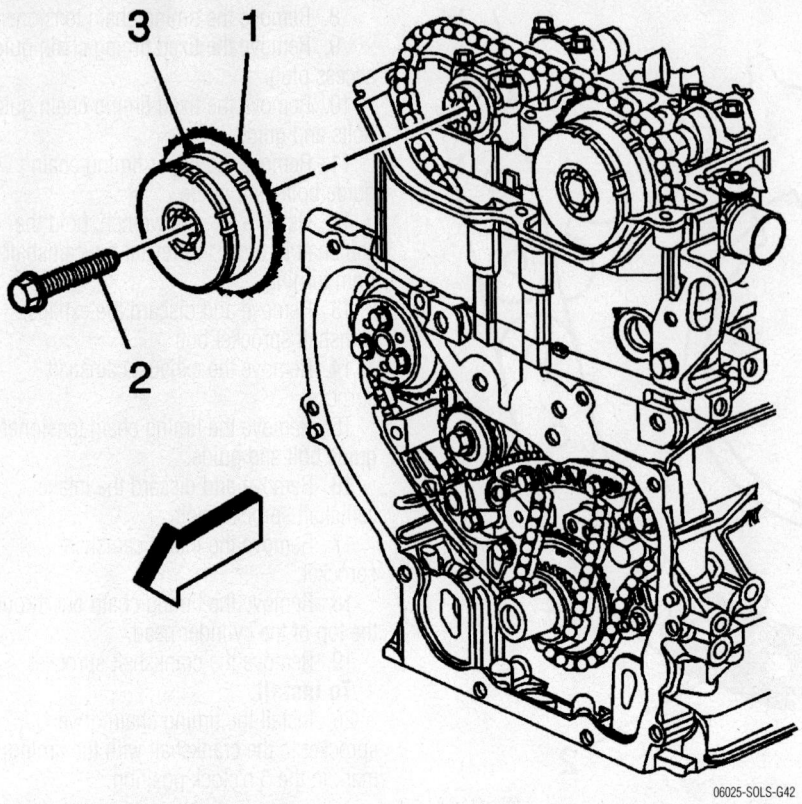

06025-SOLS-G42

Ensure that the timing mark (3) on the exhaust camshaft sprocket is aligned with the pink colored link

sprocket with the last matching colored link. Tighten the bolt finger tight.

❋❋ WARNING

Do not rotate the camshaft more than ½ turn in either direction. Doing so may cause valve-to-piston contact. This may damage a valve and/or a piston. If the camshaft is 180 degrees out of time, use the following procedure to re-align the cam.

27. Using a 24-mm wrench, turn the intake camshaft until the alignment feature on the back of the camshaft sprocket seats in the notch in the front of the intake camshaft.

   a. Turn the crankshaft 45 degrees in either direction.

   b. Turn the intake cam to the appropriate location.

   c. Turn the crankshaft back to top dead center.

28. When the sprocket seats on the cam, tighten the sprocket bolt hand tight.

29. Verify that all of the colored links and the appropriate timing marks are still aligned. If they are not, repeat the portion of the procedure necessary to align the timing marks.

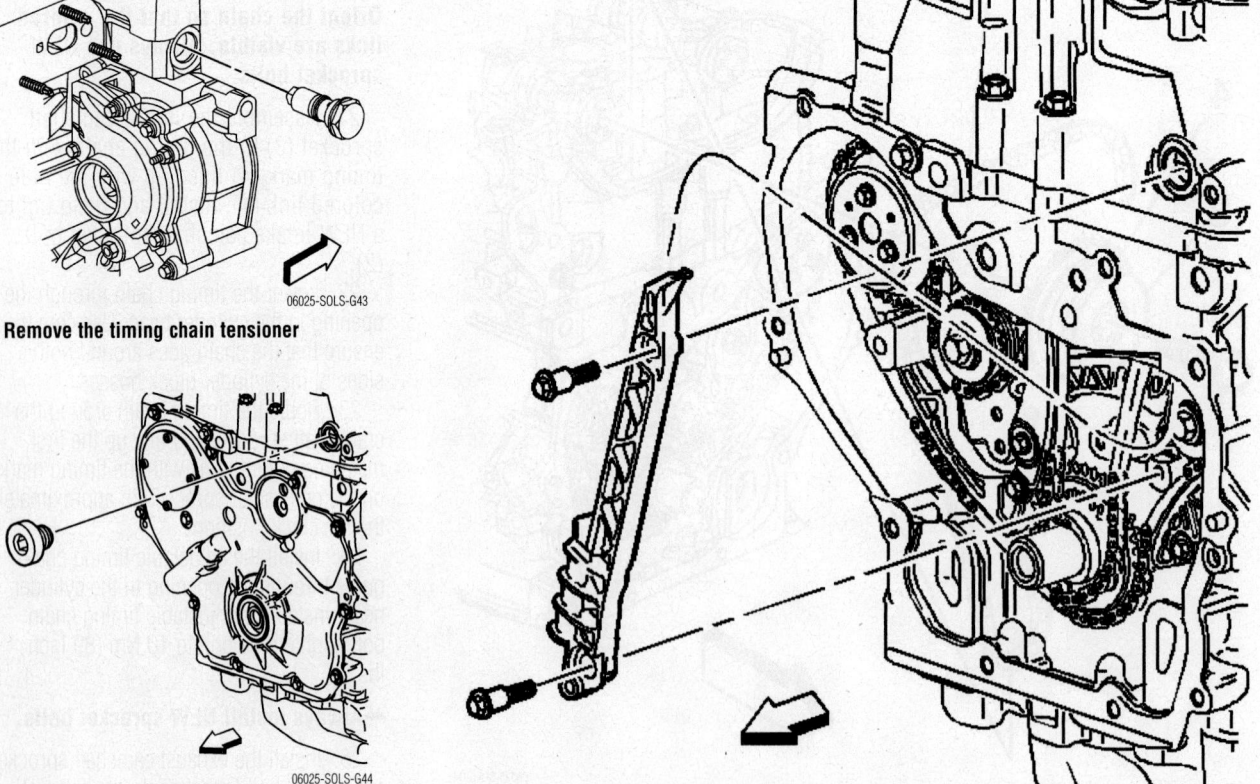

06025-SOLS-G43

Remove the timing chain tensioner

06025-SOLS-G44

Remove the fixed timing chain guide access plug

06025-SOLS-G45

Remove the fixed timing chain guide bolts and guide

06025-SOLS-G47

Install the timing chain drive sprocket to the crankshaft with the timing mark in the 5 o'clock position

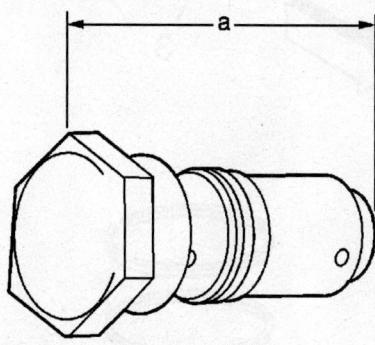

06025-SOLS-G48

Measure the timing chain tensioner

30. Install the fixed timing chain guide and bolts. Tighten the bolts to 10 Nm (89 inch lbs.).

31. Install the upper timing chain guide and bolts. Tighten the bolts to 10 Nm (89 inch lbs.).

32. Using a 24-mm wrench, engage the hex on the intake camshaft. Using a torque wrench, tighten the camshaft sprocket bolt. Tighten the bolt to 85 Nm (63 ft. lbs.), plus 30 degrees.

33. Using a 24-mm wrench, engage the hex on the exhaust camshaft. Using a torque wrench, tighten the camshaft sprocket bolt. Tighten the bolt to 85 Nm (63 ft. lbs.), plus 30 degrees.

34. Remove the old oil from the timing chain tensioner.

35. Inspect the timing chain tensioner for scoring or free movement.

36. Inspect the timing chain washer and O-ring for damage. If damaged, replace the timing chain tensioner.

37. Measure the timing chain tensioner assembly from end to end. A NEW tensioner should be supplied in the fully compressed

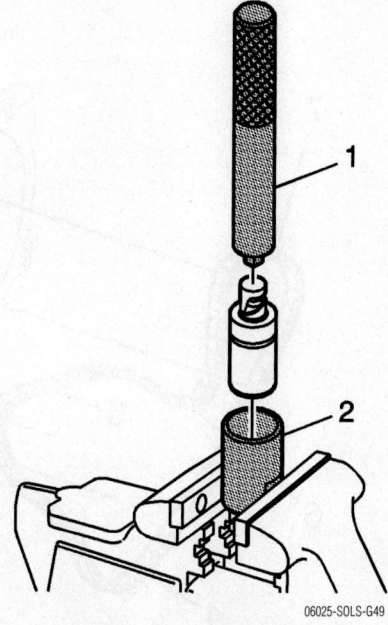

06025-SOLS-G49

Timing chain tensioner compression

non-active state. A tensioner in the compressed state will measure 72 mm (2.83 in.) (a) from end to end. A tensioner in the active state will measure 85 mm (3.35 in.) (a) from end to end.

38. If the timing chain tensioner is not in the compressed state, perform the following steps:

 a. Remove the piston assembly from the body of the timing chain tensioner by pulling it out.

 b. Install the J 45027-2 (2) into a vise.

 c. Install the notch end of the piston assembly into the J 45027-2.

 d. Using the J 45027-1 (1), turn the ratchet cylinder into the piston.

39. Inspect the bore of the tensioner body for dirt, debris, and damage. If any damage appears, replace the tensioner. Clean dirt or debris with a lint free cloth.

40. Install the compressed piston assembly back into the timing chain tensioner body until the assembly stops at the bottom of the bore. Do not compress the piston assembly against the bottom of the bore. If the piston assembly is compressed against the bottom of the bore, the assembly will activate the tensioner, which will then need to be reset again.

41. At this point the tensioner should measure approximately 72 mm (2.83 in.) (a) from end to end. If the tensioner does not read 72 mm (2.83 in.) (a) from end, to end repeat steps 1-4.

42. Ensure that all dirt and debris are removed from the timing chain tensioner threaded hole in the cylinder head.

43. Install the timing chain tensioner. Tighten the tensioner to 75 Nm (66 ft. lbs.).

44. The timing chain tensioner is released by compressing the tensioner 2 mm (0.079 in.) which will release the locking mechanism in the ratchet. To release the timing chain tensioner, use a suitable tool with a rubber tip on the end. Feed the tool down through the cam drive chest to rest on the cam chain. Then give a sharp jolt diagonally downwards to release the tensioner.

45. Ensure that all dirt and debris are removed from the timing chain oiling nozzle hole in the engine block.

46. Install the timing chain oiling nozzle and bolt. Tighten the bolt to 10 Nm (89 inch lbs.).

47. Apply sealant GM PN 12345382 (Canadian PN 10953489) or equivalent, to the threads and install the timing chain guide bolt access hole plug. Tighten the plug to 90 Nm (59 ft. lbs.).

48. Install the engine front cover.

49. Install the camshaft cover.

Balance Shaft Chain

REMOVAL & INSTALLATION

1. Before servicing the vehicle, refer to the Precautions Section.

2. Remove the balance shaft drive chain tensioner bolts and tensioner.

3. Remove the adjustable balance shaft chain guide bolt and guide.

4. Remove the small balance shaft drive chain guide bolts and guide.

5. Remove the upper balance shaft drive chain guide bolts and guide.

➡**The balance shaft drive chain will be easier to remove if you gather all of the slack in the chain between the crankshaft and water pump sprockets.**

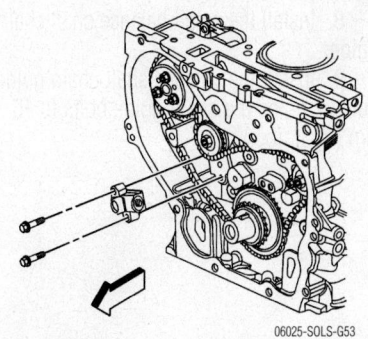

06025-SOLS-G53

Remove the balance shaft drive chain tensioner bolts and tensioner

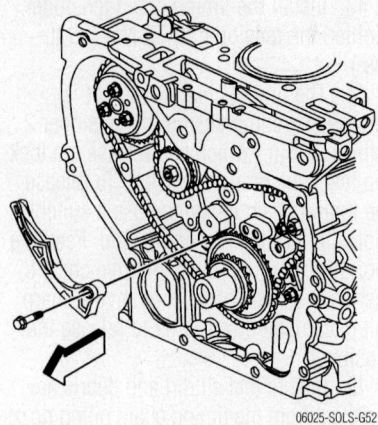

Remove the adjustable balance shaft chain guide bolt and guide

06025-SOLS-G52

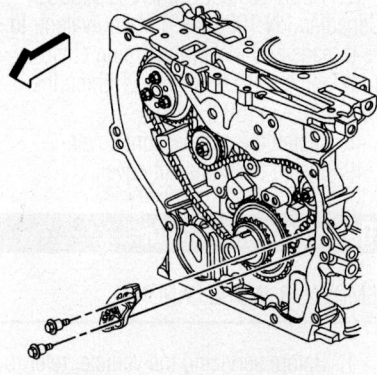

Remove the small balance shaft drive chain guide bolts and guide

06025-SOLS-G51

6. Remove the balance shaft drive chain.

To install:

➡**If the balance shafts are not properly timed to the engine, the engine may vibrate and make noise.**

7. Install the upper balance shaft chain guide and bolts. Tighten the upper balance shaft chain guide bolts to 15 Nm (11 ft. lbs.).

8. Install the small balance shaft chain guide.

9. Install the balance shaft chain guide bolts. Tighten the chain guide bolts to 15 Nm (11 ft. lbs.).

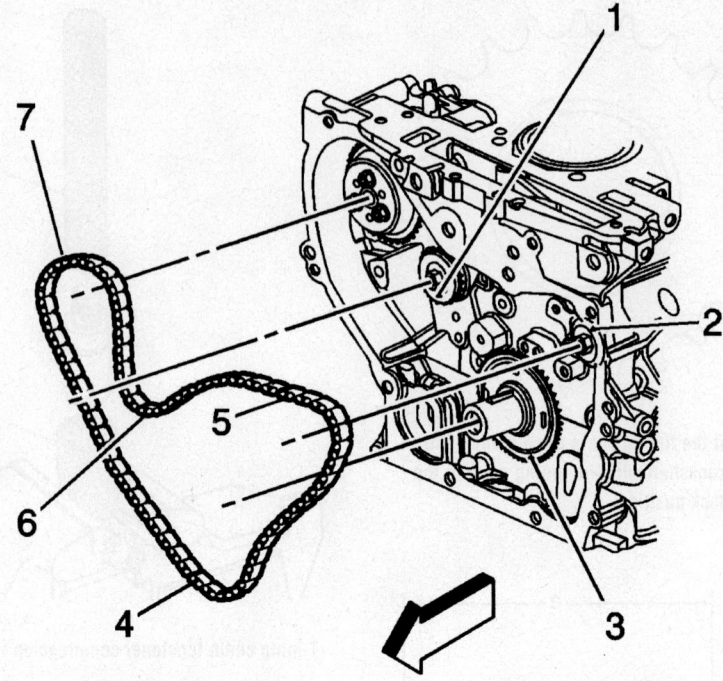

06025-SOLS-G50

Balance shaft drive chain

10. Install the adjustable balance shaft drive chain guide.

11. Install the adjustable balance shaft drive chain guide bolts. Tighten the chain guide bolts to 10 Nm (89 inch lbs.).

12. Reset the timing chain tensioner by performing the following steps:

   a. Turn the tensioner plunger 90 degrees in its bore and compress the plunger.

   b. Turn the tensioner back to the original 12 o'clock position and insert a paper clip through the hole in the plunger body and into the hole in the tensioner plunger.

13. Install the timing chain tensioner.

14. Install the chain tensioner bolts. Tighten the chain tensioner bolts to 10 Nm (89 inch lbs.).

15. Remove the paper clip from the balance shaft drive chain tensioner.

Piston and Ring

POSITIONING

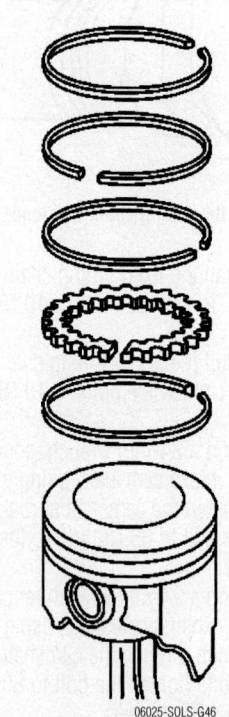

06025-SOLS-G46

Assembling the piston and rings

FUEL SYSTEM

Fuel System Service Precautions

Safety is the most important factor when performing not only fuel system maintenance but any type of maintenance. Failure to conduct maintenance and repairs in a safe manner may result in serious personal injury or death. Maintenance and testing of the vehicle's fuel system components can be accomplished safely and effectively by adhering to the following rules and guidelines.

• To avoid the possibility of fire and personal injury, always disconnect the negative battery cable unless the repair or test procedure requires that battery voltage be applied.

• Always relieve the fuel system pressure prior to disconnecting any fuel system component (injector, fuel rail, pressure regulator, etc.), fitting or fuel line connection. Exercise extreme caution whenever relieving fuel system pressure, to avoid exposing skin, face and eyes to fuel spray. Please be advised that fuel under pressure may penetrate the skin or any part of the body that it contacts.

• Always place a shop towel or cloth around the fitting or connection prior to loosening to absorb any excess fuel due to spillage. Ensure that all fuel spillage (should it occur) is quickly removed from engine surfaces. Ensure that all fuel soaked cloths or towels are deposited into a suitable waste container.

• Always keep a dry chemical (Class B) fire extinguisher near the work area.

• Do not allow fuel spray or fuel vapors to come into contact with a spark or open flame.

• Always use a back-up wrench when loosening and tightening fuel line connection fittings. This will prevent unnecessary stress and torsion to fuel line piping. Always follow the proper torque specifications.

• Always replace worn fuel fitting O-rings with new ones. Do not substitute fuel hose, or equivalent, where fuel pipe is installed.

Relieving Fuel System Pressure

1. Before servicing the vehicle, refer to the Precautions Section.

✳✳ CAUTION

Relieve the fuel system pressure before servicing fuel system compo-nents in order to reduce the risk of fire and personal injury. After relieving the system pressure, a small amount of fuel may be released when servicing the fuel lines or connections. In order to reduce the chance of personal injury, cover the regulator and the fuel line fittings with a shop towel before disconnecting. This will catch any fuel that may leak out. Place the towel in an approved container when the disconnection is complete.

2. Disconnect the negative battery cable.
3. Unscrew the cap on the fuel pressure service port.
4. Install fuel pressure gauge J 34730-1A, or equivalent, to the fuel pressure service port, located on the fuel rail.
5. Loosen the fuel fill cap in order to relieve fuel tank vapor pressure.
6. Open the valve on the pressure gauge in order to bleed the system pressure. The fuel connections are now safe for servicing.
7. Drain any fuel remaining in the gage into an approved container.
8. Once the system pressure is completely relieved, remove the gauge.

Fuel Filter

REMOVAL & INSTALLATION

There is no routinely replaced fuel filter. A plastic mesh strainer is part of the fuel pump module located in the fuel tank.

Fuel Pump Module

REMOVAL & INSTALLATION

1. Before servicing the vehicle, refer to the Precautions Section.

✳✳ CAUTION

In order to reduce the risk of fire and personal injury that may result from a fuel leak, always replace the fuel sender gasket when reinstalling the fuel sender assembly.

✳✳ WARNING

Cap the fittings and plug the holes when servicing the fuel system in order to prevent dirt and other conta-minants from entering the open pipes and passages.

2. Relieve the fuel system pressure.
3. Remove the rear compartment trim panel.
4. Remove the fuel sending unit access cover bolts.
5. Remove the access cover.
6. Disconnect the fuel sender electrical connector.
7. Disconnect the fuel pressure sensor electrical connector.
8. Disconnect the fuel fill pipe evaporative emission (EVAP) pipe quick connect fitting.
9. Disconnect the fuel feed pipe quick connect fitting.

✳✳ WARNING

Avoid damaging the lockring. Use only tool J-45722 to prevent damage to the lockring.

✳✳ WARNING

Do Not handle the fuel sender assembly by the fuel pipes. The amount of leverage generated by handling the fuel pipes could damage the joints.

➡The fuel sender assembly may spring up from its position. When removing the fuel sender assembly from the fuel tank, be aware that the reservoir bucket is full of fuel. It must be tipped slightly during removal to avoid damage to the float. Discard the fuel sender assembly O-ring and replace it with a new one. Carefully discard the fuel in the reservoir bucket into an approved container.

➡Do NOT use impact tools. Significant force will be required to release the lockring. The use of a hammer and screwdriver is not recommended. Secure the fuel tank in order to prevent fuel tank rotation.

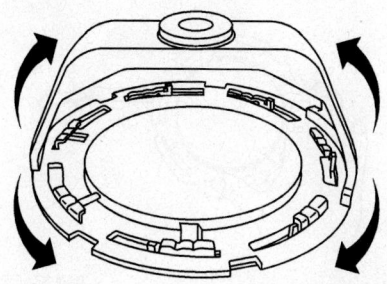

06025-SOLS-G54

Tool J-45722

10. Use tool J 45722 and a long breaker-bar in order to unlock the fuel sender lockring. Turn the fuel sender lockring in a counterclockwise direction.

11. Raise the fuel sender up slightly.

12. Connect the large EVAP canister quick connect fitting.

13. Remove the fuel sender assembly.

14. Remove and discard the fuel sender O-ring.

➡️**Some lockrings were manufactured with DO NOT REUSE stamped into them. These lockrings may be reused if they are not damaged or warped.**

➡️**Inspect the lockring for damage due to improper removal or installation procedures. If damage is found, install a NEW lockring.**

➡️**Check the lockring for flatness.**

15. Place the lockring on a flat surface. Measure the clearance between to lockring and the flat surface using a feeler gage at 7 points.

16. If the warpage is less than 0.41 mm (0.016 in.), the lockring does not require replacement.

17. If the warpage is greater than 0.41 mm (0.016 in.), the lockring must be replaced.

To install:

18. Install a NEW fuel sender O-ring.

19. Install the fuel sender assembly.

➡️**Always replace the fuel sender seal when installing the fuel sender assembly. Replace the lockring if necessary. Do not apply any type of lubrication in the seal groove. Ensure the lockring is installed with the correct side facing upward. A correctly installed lockring will only turn in a clockwise direction.**

20. Using the tool, rotate the fuel sender assembly lockring clockwise until the ring is locked into place on the fuel tank.

21. Connect the large EVAP canister quick connect fitting.

22. Connect the fuel feed pipe quick connect fitting.

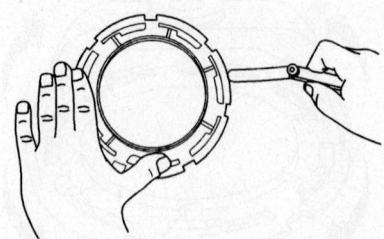

06025-SOLS-G55

Check the lockring for flatness

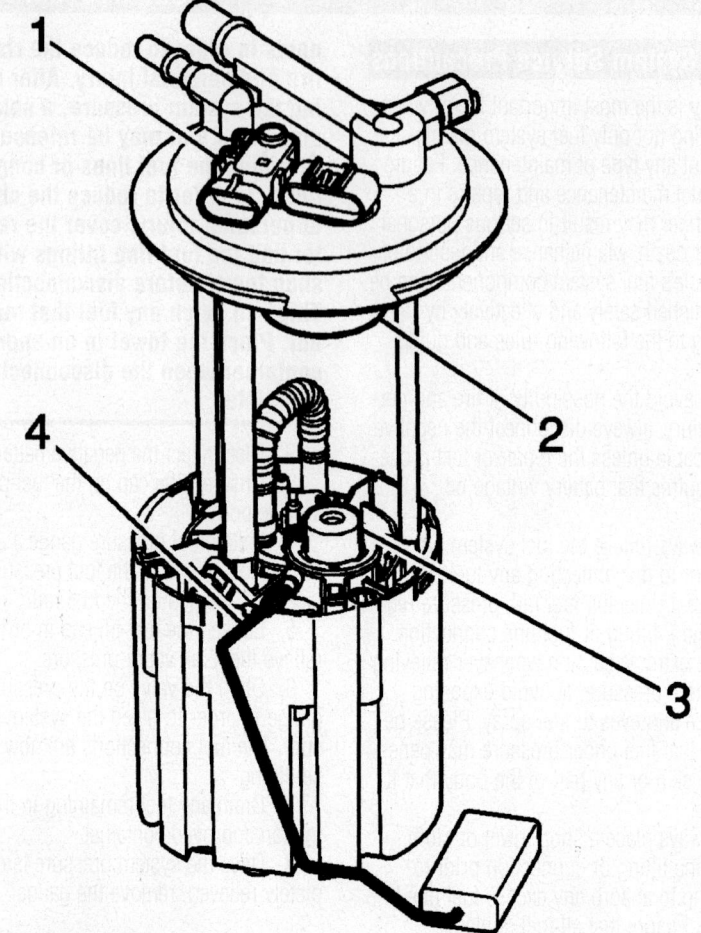

1. Fuel tank pressure sensor
2. Fill limit valve
3. Fuel pressure regulator
4. Fuel level sensor

06025-SOLS-G56

Fuel pump module

23. Connect the fuel fill pipe EVAP pipe quick connect fitting.

24. Connect the fuel pressure sensor electrical connector.

25. Connect the fuel sender electrical connector.

26. Install the access cover.

27. Install the fuel sending unit access cover bolts.

28. Install the rear compartment trim panel.

29. Refill the tank.

30. Connect the negative battery cable.

31. Inspect for fuel leaks through the following steps:

    a. Turn the ignition to the ON position for 2 seconds.

    b. Turn the ignition to the OFF position for 10 seconds.

    c. Turn the ignition to the ON position for 2 seconds.

    d. Check for fuel leaks.

Fuel Injectors

REMOVAL & INSTALLATION

1. Before servicing the vehicle, refer to the Precautions Section.

2. Relieve the fuel system pressure.

3. Remove the air cleaner outlet duct.

4. Disconnect the fuel injector inline electrical connector.

5. Disconnect the electronic throttle control (ETC) electrical connector.

6. Remove the 2 engine harness clips from the fuel rail tabs.

7. Remove the fuel rail bolts.

➡️**Use care when removing the fuel rail assembly in order to prevent damage to the fuel injectors electrical connector terminals and spray tips.**

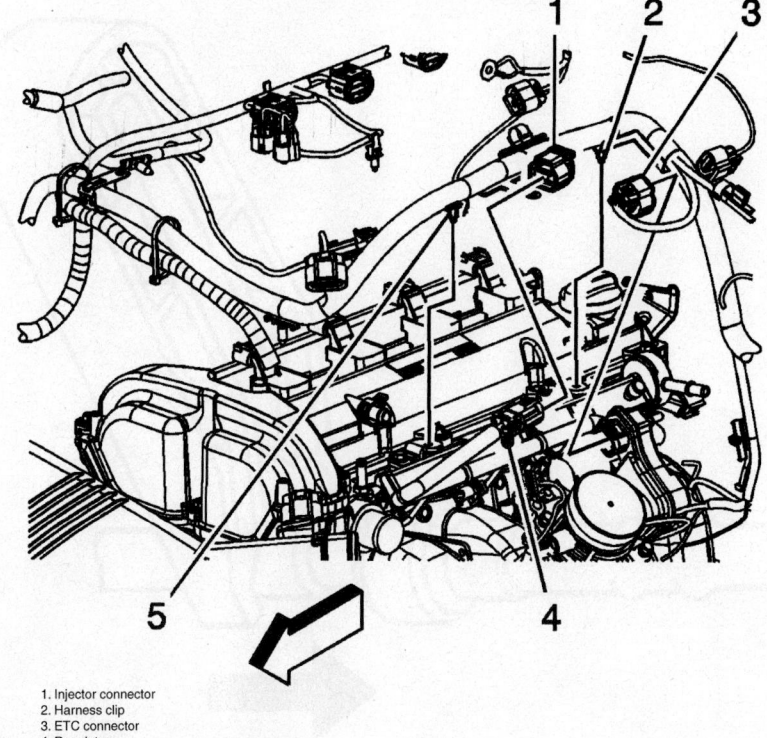

1. Injector connector
2. Harness clip
3. ETC connector
4. Regulator
5. Harness clip

06025-SOLS-G57

Fuel rail

8. Pull the fuel rail back and upward in order to release the fuel injectors from the cylinder head ports.

9. Remove the fuel rail.

✳✳ WARNING

Use care in removing the fuel injectors in order to prevent damage to the fuel injector electrical connector pins or the fuel injector nozzles. Do not immerse the fuel injector in any type of cleaner. The fuel injector is an electrical component and may be damaged by this cleaning method.

➡**If the fuel injectors are found to be leaking, the engine oil may be contaminated with fuel.**

10. Remove the fuel injector retaining clip.

11. Remove the fuel injector from the fuel rail.

12. Remove the fuel injector upper O-ring.

13. Remove the fuel injector lower O-ring.

To install:

➡**Be sure to use the correct part number when ordering replacement fuel injectors.**

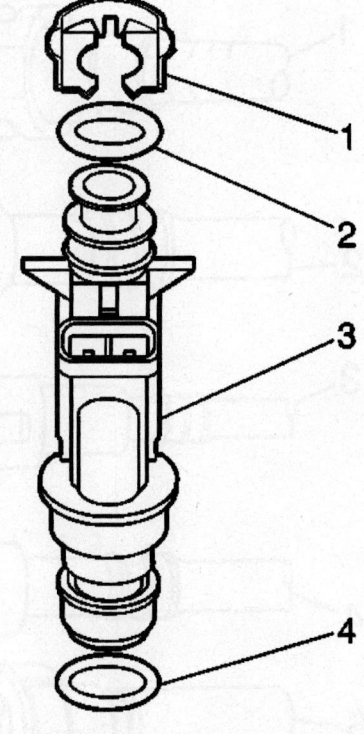

1. Retaining clip
2. Upper O-ring
3. Injector
4. Lower O-ring

06025-SOLS-G58

Fuel injector components

14. The fuel injector assembly is stamped with a part number identification.

15. Lubricate the new injector O-rings with clean engine oil.

16. Install the fuel injector upper O-ring.

17. Install the fuel injector lower O-ring.

18. Install the fuel injector to the fuel rail.

19. Install the fuel injector retaining clip.

➡**Install NEW lower O-rings when reusing fuel injectors. Lubricate the lower O-rings prior to installing the injectors into the intake manifold.**

20. With the fuel injectors positioned downward, lower the fuel injectors into the cylinder head ports.

21. Carefully push the fuel injectors into the cylinder head ports.

22. Install the fuel rail bolts. Tighten the bolts to 10 Nm (89 inch lbs.).

23. Install the 2 engine harness clips to the fuel rail tabs.

24. Connect the ETC electrical connector.

25. Connect the fuel injector inline electrical connector.

26. Install the air cleaner outlet duct.

27. Connect the negative battery cable.

28. Inspect for fuel leaks using the following procedure:

   a. Turn ON the ignition, with the engine OFF for 2 seconds.

   b. Turn OFF the ignition for 10 seconds.

   c. Turn ON the ignition.

   d. Inspect for fuel leaks.

Quick-Connect Fittings

OPENING & CLOSING

Metal Collar

1. Before servicing the vehicle, refer to the Precautions Section.

➡**Tool Required: J 37088-A Fuel Line Disconnect Tool Set, or equivalent.**

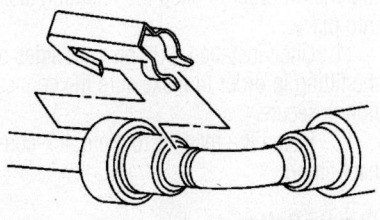

06025-SOLS-G59

Remove the retainer from the quick-connect fitting—metal collar type

2. Relieve the fuel system pressure before servicing any fuel system connection.

3. Remove the retainer from the quick-connect fitting.

❊❊ CAUTION

Wear safety glasses when using compressed air, as flying dirt particles may cause eye injury.

4. Blow dirt out of the fitting using compressed air.

5. Choose the correct tool from the set for the size of the fitting. Insert the tool into the female connector, then push inward in order to release the locking tabs.

6. Pull the connection apart.

➡ **If necessary, remove rust or burrs from the fuel pipes with an emery cloth. Use a radial motion with the fuel pipe end in order to prevent damage to the O-ring sealing surface. Use a clean shop towel in order to wipe off the male tube ends. Inspect all the connections for dirt and burrs. Clean or replace the components and assemblies as required.**

7. Use a clean shop towel in order to wipe off the male pipe end.

8. Inspect both ends of the fitting for dirt and burrs. Clean or replace the components as required.

To connect

❊❊ CAUTION

In order to reduce the risk of fire and personal injury, before connecting fuel pipe fittings, always apply a few drops of clean engine oil to the male pipe ends. This will ensure proper reconnection and prevent a possible fuel leak. During normal operation, the O-rings located in the female connector will swell and may prevent proper reconnection if not lubricated.

9. Apply a few drops of clean engine oil to the male pipe end.

10. Push both sides of the fitting together in order to snap the retaining tabs into place.

11. Once installed, pull on both sides of the fitting in order to make sure the connection is secure.

12. Install the retainer to the quick-connect fitting.

Plastic Collar

1. Before servicing the vehicle, refer to the Precautions Section.

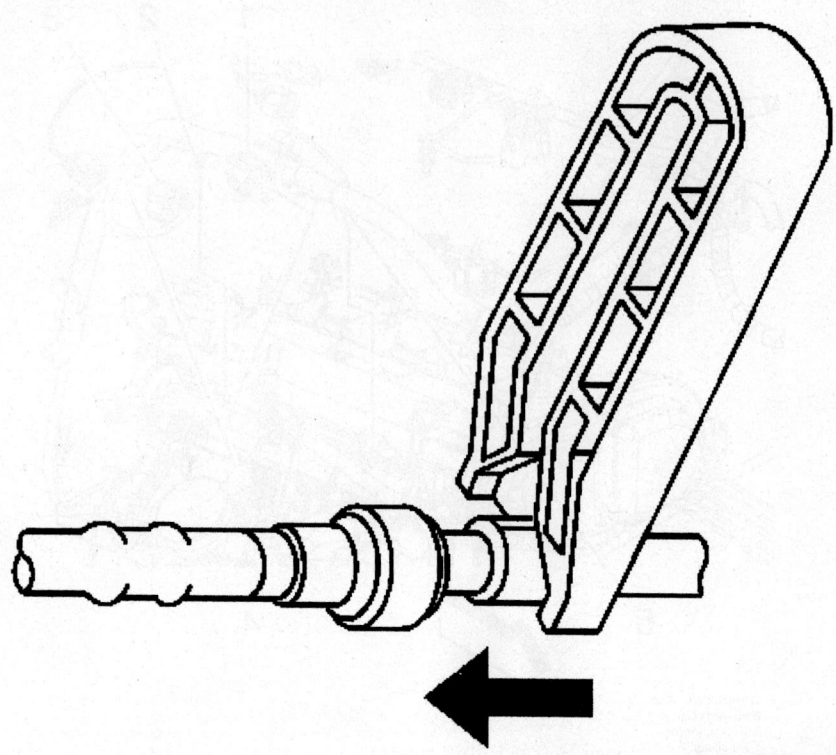

06025-SOLS-G60

Insert the tool into the female connector, then push inward in order to release the locking tabs—metal collar type

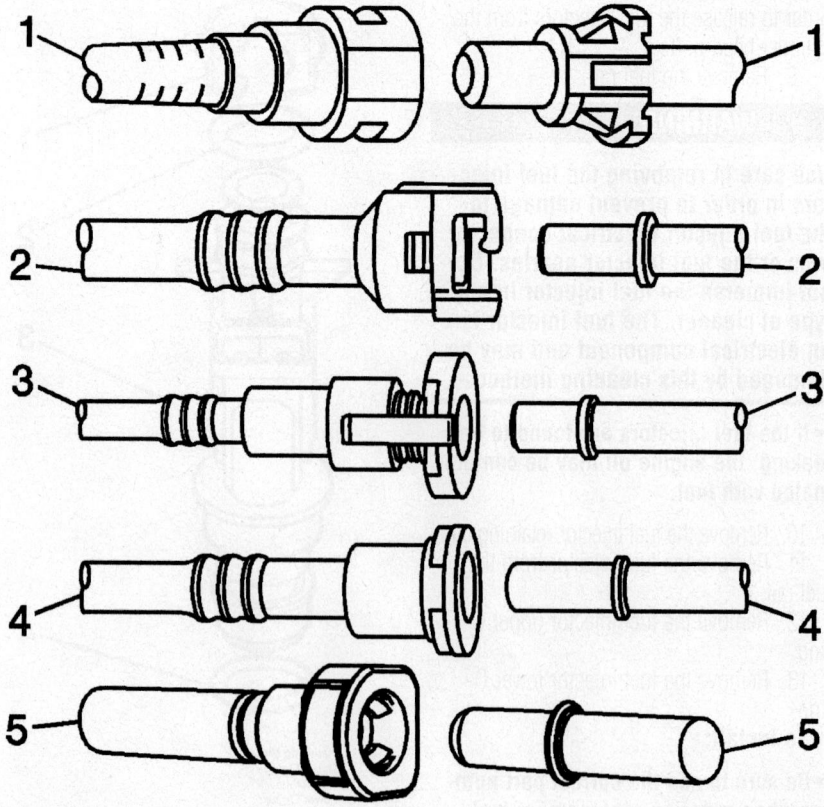

06025-SOLS-G61

Plastic collar type fittings

There are several types of plastic collar fuel and evaporative emission quick connect fittings that may be used on this vehicle.

- Bartholomew (1)
- Q release (2)
- Squeeze to release (3)
- Sliding retainer (4)
- Push down TI (5)

The following instructions apply to all of these types of fittings except where indicated.

2. Relieve the fuel system pressure.

❋ CAUTION

Wear safety glasses when using compressed air in order to prevent eye injury.

3. Using compressed air, blow any dirt out of the quick-connect fitting.

- Bartholomew style connectors ONLY: Squeeze the plastic quick-connect fitting release tabs.
- Q release style connector ONLY: Release the fitting by pushing the tab toward the other side of the slot in the fitting.
- Squeeze to release style connectors ONLY: Squeeze where indicated by arrows on both sides of the plastic ring surrounding the quick connect fitting.

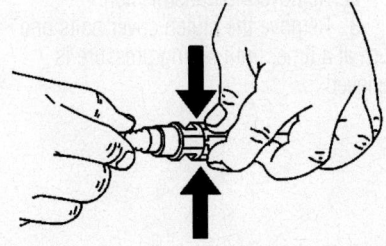

Opening Bartholomew connector

06025-SOLS-G62

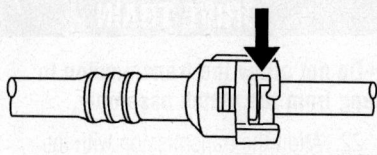

Opening Q release connector

06025-SOLS-G63

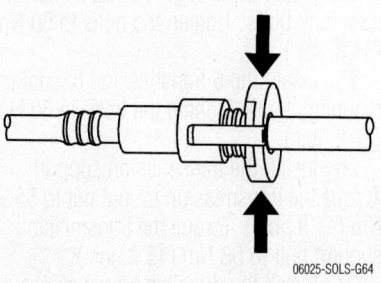

Opening straight Squeeze to release connector

06025-SOLS-G64

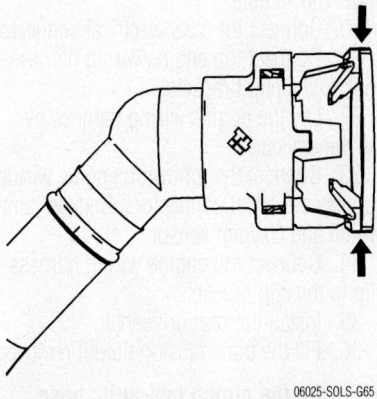

Opening angled Squeeze to release connector

06025-SOLS-G65

- Sliding retainer style connectors ONLY: Release the fitting by pressing on one side of the release tab causing it to push in slightly. If the tab does not move, try pressing the

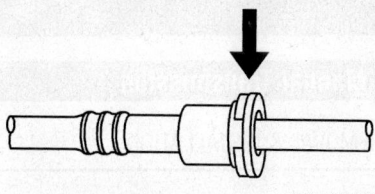

Opening Sliding retainer connector

06025-SOLS-G66

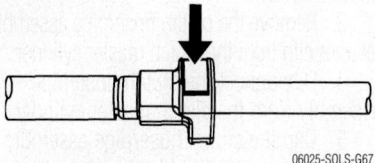

Opening TI connector

06025-SOLS-G67

tab in from the opposite side. The tab will only move in one direction.

- Push down TI style connectors ONLY: Release the fitting by pressing on the tab indicated by the arrow.

4. Pull the connection apart.

Use an emery cloth in order to remove rust or burrs from the fuel pipe. Use a radial motion with the fuel pipe end in order to prevent damage to the O-ring sealing surface.

5. Wipe off the male pipe end using a clean shop towel.

6. Inspect both ends of the fitting for dirt and burrs.

7. Clean or replace the components as required.

To connect:

8. Apply a few drops of clean engine oil to the male connection end.

9. Push both sides of the quick-connect fitting together in order to cause the retaining feature to snap into place.

10. Once installed, pull on both sides of the quick connect fitting in order to make sure the connection is secure.

DRIVE TRAIN

Manual Transmission

REMOVAL & INSTALLATION

1. Before servicing the vehicle, refer to the Precautions Section.

2. Remove the control lever knob and boot assembly.

3. Remove the clutch hose/pipe assembly retainer clip from the clutch master cylinder.

4. Disconnect the clutch hose/pipe assembly from the clutch master cylinder.

5. Cap the clutch hose/pipe assembly in order to prevent fluid loss and contamination. It is not necessary to plug the lower hose end or slave cylinder fitting as they are equipped with check valves, only minimal fluid loss may be experienced.

6. Drain the transmission fluid if necessary.

7. Remove the driveshaft.

8. Disconnect the following engine wiring harness electrical connectors: Backup lamp switch; Oxygen sensor.

9. Disconnect the engine wiring harness clip from the clip bracket, and position the harness over the transmission.

10. Disconnect the vehicle speed sensor (VSS) electrical connector.

11. Disconnect the engine wiring harness clips from the clip brackets, and position the harness aside.

12. Remove the front floor closeout panel.

13. Remove the driveline tunnel closeout panel.

14. Support the transmission.

15. Remove the transmission support.

16. Remove the 5 transmission to engine mounting bolts.

17. Remove the 2 engine to transmission mounting bolts.

18. Remove the 2 remaining transmission mounting bolts.

➡ **Do not allow the transmission to hang from the clutch assembly.**

19. Pull the transmission straight back off the clutch hub splines.

20. Using the transmission jack, carefully lower the transmission from the vehicle.

To install:

➡ **Ensure clearance is maintained between the transmission and the following: The catalytic converter; The clutch assembly; The engine wiring harness.**

21. Using the transmission jack, carefully raise the transmission to the engine.

➡ **Do not allow the transmission to hang from the clutch assembly.**

22. Align the transmission with the engine dowels.

23. Install the 2 transmission mounting bolts. Tighten the bolts to 50 Nm (37 ft. lbs.).

24. Install the 2 engine to transmission mounting bolts. Tighten the bolts to 50 Nm (37 ft. lbs.).

25. Install the 5 transmission to engine mounting bolts. Tighten the bolts to 50 Nm (37 ft. lbs.).

26. Install the transmission support. Torque the transmission mount nut to 55 Nm (41 ft. lbs.). Torque the transmission support bolt to 58 Nm (43 ft. lbs.).

27. Install the driveline tunnel closeout panel.

28. Install the front floor closeout panel.

29. Remove the transmission jack from under the vehicle.

30. Connect the VSS electrical connector.

31. Connect the engine wiring harness clips to the clip brackets.

32. Lay the engine wiring harness over the transmission.

33. Connect the following engine wiring harness electrical connectors: Backup lamp switch and Oxygen sensor.

34. Connect the engine wiring harness clip to the clip bracket.

35. Install the rear driveshaft.

36. Fill the transmission fluid if removed.

➡ **Ensure the clutch hydraulic hose does not come in contact with any sharp or potentially hot surfaces.**

37. Install the clutch hose/pipe assembly retainer clip to the clutch master cylinder.

38. Connect the clutch hose/pipe assembly to the clutch master cylinder.

39. Tug gently on the clutch hose/pipe assembly to ensure proper retention into the clutch master cylinder.

40. Install the control lever knob and boot assembly. Refer to shift control knob replacement.

41. Complete the following procedure after the transmission is installed in the vehicle:

a. With the ignition OFF or disconnected and clutch pedal depressed, crank the engine several times. Listen for any unusual noises or evidence that any parts are binding.

b. Place the transmission in neutral, start the engine and listen for any unusual noises or evidence that any parts are binding.

c. Turn OFF the ignition.

d. Perform a final inspection for the proper fluid level.

e. Road test the vehicle.

Clutch

REMOVAL & INSTALLATION

1. Before servicing the vehicle, refer to the Precautions Section.

2. Remove the transmission.

3. Remove the clutch cover bolts one turn at a time, until spring pressure is relieved.

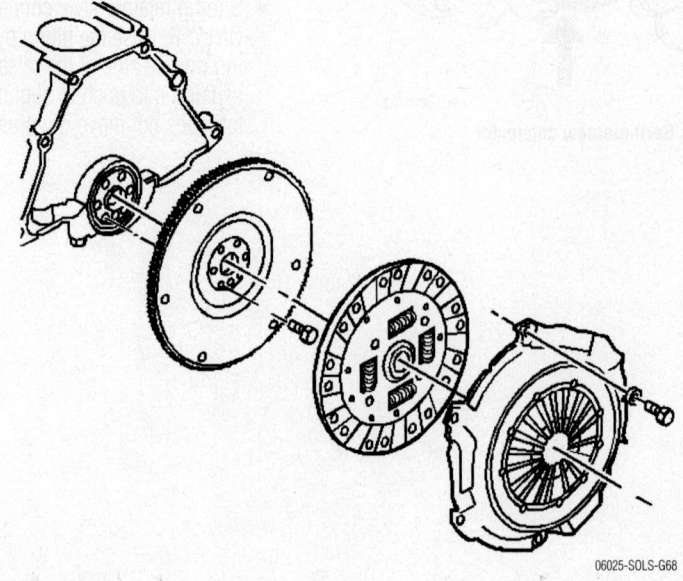

06025-SOLS-G68

Clutch components

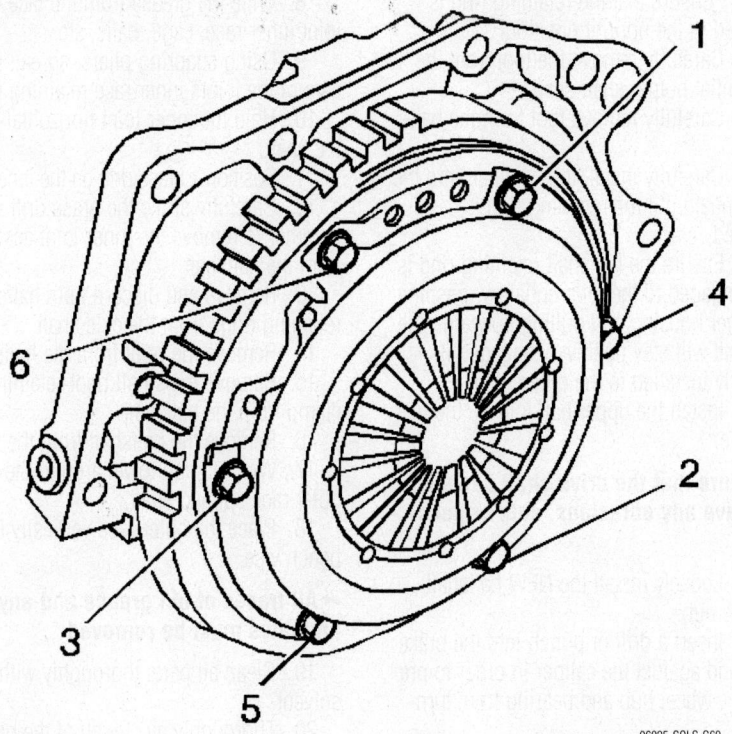

Clutch pressure plate torque sequence

06025-SOLS-G69

4. Remove the clutch cover and the clutch disc.

To install:

5. Install the clutch disc and the clutch cover.

6. Hand-start the clutch cover to flywheel bolts, leaving the clutch cover loose enough to reposition for alignment.

7. Install a clutch alignment tool to center the disc.

8. Tighten the clutch cover to flywheel bolts in the sequence shown. Tighten the bolts to 30 Nm (22 ft. lbs.).

9. Recheck each bolt torque using the tightening sequence.

➡**Excessive amounts of lubricant on the input shaft splines may contaminate the clutch disc and cause clutch shudder.**

10. Lubricate the inside diameter of the bearing with Saturn P/N 21005995, or equivalent.

11. Install the transmission.

12. Bleed the hydraulic system.

13. Connect the negative battery cable.

Hydraulic Clutch System

BLEEDING

1. Before servicing the vehicle, refer to the Precautions Section.

❋❋ WARNING

Do not reuse the fluid that has been bled from a system in order to fill the clutch master cylinder reservoir.

➡**Maintain the fluid level in the clutch reservoir to the top step with DOT 3 hydraulic fluid.**

2. Clean dirt and grease from the cap in order to ensure that no foreign substances enter the system.

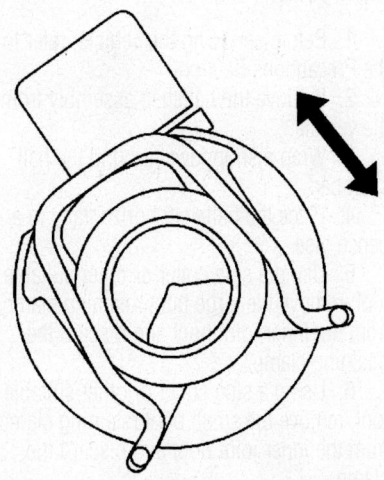

06025-SOLS-G70

Push the clip in order to move the clutch line into the bleed position

3. Attach a hose to the bleeder port on the clutch actuator assembly. Submerge the other end of the hose in a container of DOT 3 hydraulic fluid.

4. Depress the clutch pedal quickly to the full depressed position.

5. Push the clip in order to move the clutch line into the bleed position.

6. Move the clutch line into the normal position ensuring that the clip returns.

7. Lift the clutch pedal to the up stop position and hold for 5 seconds.

8. Repeat steps 3-6 until air is purged from the clutch system.

Halfshaft

REMOVAL & INSTALLATION

1. Before servicing the vehicle, refer to the Precautions Section.

2. Raise and support the vehicle.

3. Remove the tire and wheel assembly.

❋❋ WARNING

The halfshaft spindle nut must not be reused. Replace the halfshaft spindle nut with a new nut whenever it is removed.

4. Insert a drift or punch into the brake rotor and against the caliper in order to prevent the wheel hub and bearing from turning.

5. Remove and discard the halfshaft spindle nut.

6. Remove the drift or punch from the rotor.

7. Separate the upper ball joint from the rear knuckle.

8. Using a prybar, carefully release the halfshaft from the rear differential enough to install tool J 44394.

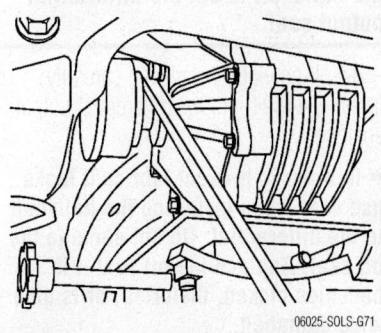

06025-SOLS-G71

Using a prybar, carefully release the halfshaft from the rear differential enough to install tool J 44394

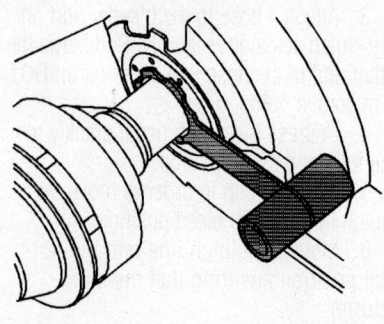

J 44394 installed over the halfshaft and into the seal

06025-SOLS-G72

✳✳ WARNING

J-44394 must be installed into the differential output shaft seal prior to removing and installing the halfshaft. Failure to install J-44394 as indicated may cause the splines of the halfshaft to cut the differential output seal.

9. Carefully install the tool over the halfshaft.

10. Carefully slide the tool into the differential output shaft seal.

11. Remove the halfshaft from the vehicle.

12. If reusing the halfshaft remove and discard the halfshaft retaining ring. The halfshaft retaining ring is on the splined shaft of the cross groove joint.

To install:

13. Install the new halfshaft retaining ring. The halfshaft retaining ring is on the splined shaft of the cross groove joint.

✳✳ WARNING

Tool J-44394 must be installed into the differential output shaft seal prior to removing and installing the halfshaft. Failure to install J-44394 as indicated may cause the splines of the halfshaft to cut the differential output seal.

14. If previously removed carefully install J 44394 into the differential output shaft seal.

➡**In order to prevent lubricant leaks, use care when installing the halfshaft to the differential. Do not damage the oil seal. Replace the oil seal if it becomes nicked, distorted, or is otherwise damaged.**

15. Carefully install the halfshaft into the differential until the splines are past the J

44394. Ensure that the retaining ring is installed in the upright position.

16. Carefully remove the tool from the differential output shaft seal.

17. Carefully remove tool from the halfshaft.

18. Carefully install the halfshaft into the differential until the retaining ring is engaged.

19. Ensure the halfshaft retaining ring is fully engaged to the differential by grasping the inner housing and pulling outward. The halfshaft will stay positively engaged if properly installed to the differential.

20. Install the upper ball joint to the rear knuckle.

➡**Ensure that the drive shaft seals do not have any abrasions, cuts or punctures.**

21. Loosely install the NEW halfshaft spindle nut.

22. Insert a drift or punch into the brake rotor and against the caliper in order to prevent the wheel hub and bearing from turning.

23. Use the New halfshaft spindle nut to slowly pull the spindle to the wheel hub and bearing assembly. Tighten the halfshaft spindle nut to 160 Nm (118 ft. lbs.).

24. Remove the drift or punch from the rotor.

25. Install the tire and wheel assembly.

26. Inspect the differential lubricant level.

27. Lower the vehicle.

CV-Joints

OVERHAUL

Inner Joint

1. Before servicing the vehicle, refer to the Precautions Section.

2. Remove the halfshaft assembly from the vehicle.

3. Wrap a shop towel around the halfshaft bar.

4. Place the halfshaft horizontally in a bench vise.

5. Using a side cutter or other suitable tool, remove the large boot retaining clamp from the inner joint boot and discard the retaining clamp.

6. Using a side cutter or other suitable tool, remove the small boot retaining clamp from the inner joint boot and discard the clamp.

7. Slide the boot along the halfshaft bar away from the joint face.

8. Wipe the grease from the face of the joint inner race, cage, balls, etc.

9. Using snapring pliers, spread the ears of the joints inner race retaining ring.

10. Hold the inner joint horizontally to the shaft.

11. Position a brass drift on the inner race.

12. Carefully strike the brass drift with a hammer to remove the inner joint assembly from the halfshaft.

13. Remove and discard both halfshaft retaining rings from the axle shaft.

14. Remove the boot from the halfshaft.

15. Remove the small boot retaining clamp from the halfshaft.

16. Remove the halfshaft from the vise.

17. Wrap a shop towel around the joint outer race splined shaft.

18. Place the outer race vertically in a bench vise.

➡**All traces of old grease and any contaminates must be removed.**

19. Clean all parts thoroughly with a safe solvent.

20. Thoroughly air dry all of the parts.

21. Inspect all parts for damage and/or wear.

To install:

22. Insert approximately 60 percent of the grease from the service kit into the inner joint. Spread the most of that 60 percent onto the ball tracks, the balls, the cage and the inner race. Spread the remainder of the grease into the bottom of the outer race.

23. Remove the inner joint from the bench vise.

24. Wrap a shop towel around the halfshaft.

25. Place the halfshaft horizontally in a bench vise.

26. Install a new small boot retaining clamp onto the halfshaft.

27. Slide the boot onto the halfshaft.

28. Place the new large retaining clamp over the large end of the boot.

29. Install the new halfshaft large retaining ring onto drive shaft.

30. Seat the new halfshaft large retaining ring into the large ring groove.

31. Install the new halfshaft small retaining ring onto drive shaft.

32. Seat the new halfshaft small retaining ring into the small ring groove.

33. Push the inner joint assembly onto the halfshaft until the small retaining ring seats itself in the retaining groove on the halfshaft.

34. Insert the remaining grease from the service kit into the boot.

35. Position the small boot retaining clamp onto the neck of the boot.

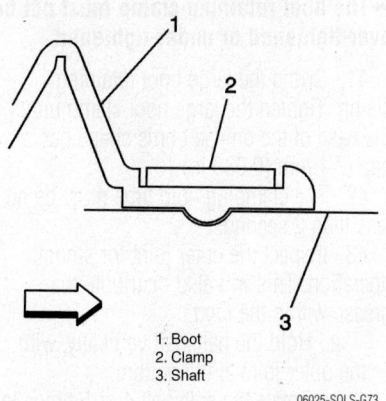

1. Boot
2. Clamp
3. Shaft

06025-SOLS-G73

**Position the boot and small retaining
clamp to the halfshaft as shown**

36. Position the boot and small retaining
clamp to the axle shaft as shown.

➡**The boot retaining clamp must not be
over-tightened or under-tightened.**

37. Using the crimping pliers, crimp the
small boot retaining clamp. Tighten the
small boot clamp until the base of the
omega shape has a gap of 1 mm (0.039 in).

38. The clamping hold time must be no
less than 2 seconds.

➡**The inner boot must not be dimpled,
stretched, or out of shape in any way. If**

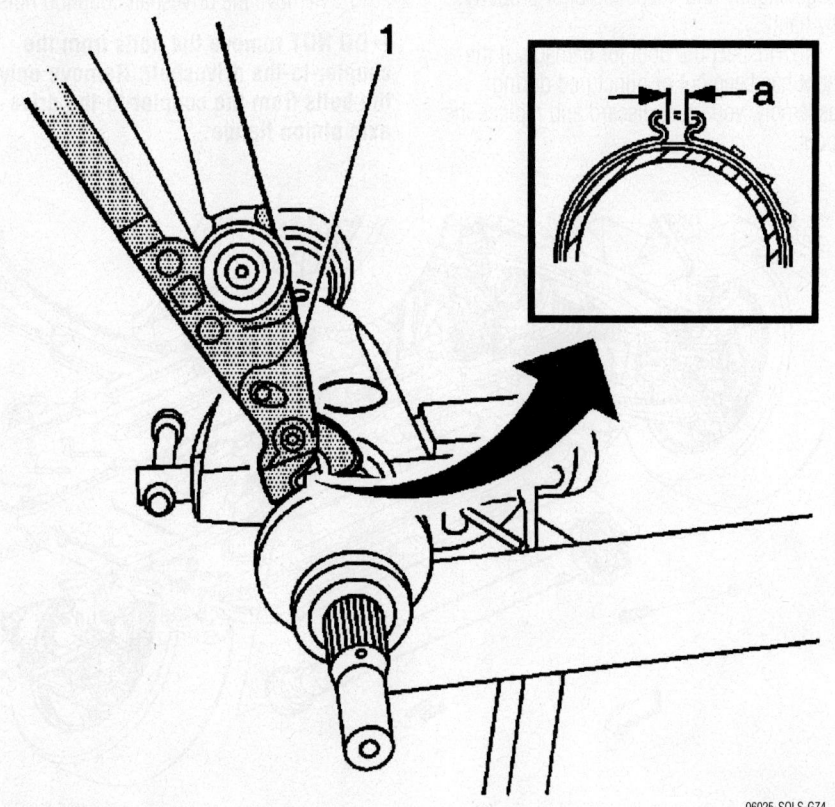

06025-SOLS-G74

Measuring the clamp gap

**necessary, equalize the air pressure in
the inner boot and shape properly by
hand.**

39. Position the large boot retaining
clamp onto the boot.

40. Position the boot and large retaining
clamp to the joint's boot groove as shown.

41. Measure the distance (1) between
the boot edges. The gap should be 88.6 mm
(3.50 in.)

➡**The boot must not be dimpled,
stretched or otherwise deformed.**

42. Inspect the boot for proper shape. If
the boot is not shaped correctly, equalize
the pressure in the boot by lifting the boot
edge slightly and shape the boot properly
by hand.

43. Inspect the boot for damage. If the boot
has been cut or punctured during assembly,
you must discard and replace the boot.

➡**The boot retaining clamp must not be
over-tightened or under-tightened.**

44. Crimp the large boot retaining
clamp. Tighten the large boot clamp until
the base of the omega shape has a gap of 1
mm (0.039 in.).

45. Inspect the inner joint for smooth
operation. This will also distribute the
grease within the joint.

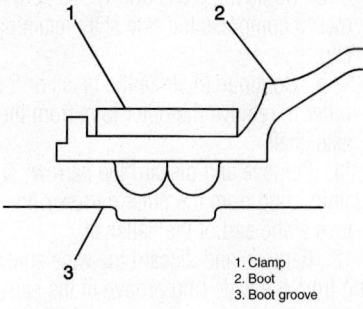

1. Clamp
2. Boot
3. Boot groove

06025-SOLS-G75

**Position the boot and large retaining
clamp to the joint's boot groove as shown**

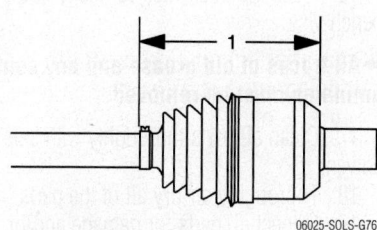

06025-SOLS-G76

Measure the boot length

a. Hold the halfshaft vertically, with
the inner joint at the bottom.

b. Rotate the halfshaft 4 or 5 times in
a circular motion.

c. Install the halfshaft.

Outer Joint

1. Before servicing the vehicle, refer to
the Precautions Section.

2. Remove the halfshaft from the
vehicle.

3. Wrap a shop towel around the shaft.

4. Place the halfshaft horizontally in a
bench vise.

5. Using a side cutter or other suitable
tool, remove the large boot retaining clamp
from the outer joint boot and discard the
clamp.

6. Using a side cutter or other suitable
tool, remove the small boot retaining clamp
from the outer joint boot and discard the
clamp.

7. Separate the boot from the joint
outer race at the large diameter end.

8. Slide the boot away from the joint
face.

9. Wipe the grease from the face of the
joint inner race, cage, balls, etc.

10. Remove the outer joint from the axle
shaft using the following steps:

a. Hold the outer joint housing horizontally to the shaft.

b. Position a brass drift on the inner
race.

c. Strike the brass drift with a hammer to compress the axle shaft retaining clip.

d. Continue to strike the brass drift in order to remove the outer joint from the axle shaft.

11. Remove and discard the narrow retaining ring from the outer narrow ring groove at the end of the halfshaft.

12. Remove and discard the wide spacer ring from the wide ring groove of the halfshaft.

13. Remove the boot from the axle shaft.

14. Remove the halfshaft from the vise.

15. Wrap a shop towel around the joint outer race splined shaft.

16. Place the outer race vertically in a bench vise.

➡**All traces of old grease and any contaminates must be removed.**

17. Clean all parts thoroughly with safe solvent.

18. Thoroughly air dry all of the parts.

19. Inspect all parts for damage and/or wear.

To install:

20. Insert approximately 60 percent of the grease from the service kit into the outer joint. Spread most of the grease onto the ball tracks, the balls, the cage, and the inner race. Spread the remainder of the grease into the bottom of the outer race.

21. Remove the inner joint from the bench vise.

22. Wrap a shop towel around the axle shaft.

23. Place the halfshaft horizontally in a bench vise.

24. Install a new small boot retaining clamp onto the axle shaft.

25. Install the boot onto the axle shaft.

26. Install the New narrow retaining ring to the outer narrow ring groove at the end of the halfshaft.

27. Install the New wide spacer ring to the wide ring groove of the halfshaft.

28. Position the outer joint horizontally.

29. Engage the inner race splines onto the axle shaft splines.

30. Compress the axle shaft ring spacer.

a. Press the end of the retaining ring, using a flat-bladed tool, into the axle shaft groove while firmly pressing the inner joint onto the axle shaft.

b. Continue to work around the retaining ring, until it is compressed.

➡**The axle shaft and inner race must be fully seated to each other.**

31. Install the outer joint to the axle shaft.

a. Position a wood block over the end of the outer joint threaded shaft.

b. Use a hammer to drive the outer joint onto the shaft.

c. Continue to drive the outer joint until the inner joint seats fully onto the axle shaft.

32. Position the small boot retaining clamp into the boot groove.

33. Position the boot and small retaining clamp to the shaft boot groove as shown.

➡**The boot retaining clamp must not be over-tightened or under-tightened.**

34. Crimp the small boot retaining clamp. Tighten the small boot clamp until the base of the omega ohms shape has a gap of 1 mm (0.039 in).

35. The clamping hold time must be no less than 2 seconds.

36. Insert the remaining grease from the service kit into the boot.

37. Position the large boot retaining clamp onto the boot.

38. Position the boot and large retaining clamp to the joint outer race as shown.

➡**The boot must not be dimpled, stretched, or otherwise deformed.**

39. Inspect the boot for proper shape. If the boot is not shaped correctly, equalize the pressure in the boot by lifting the boot edge slightly and shape the boot properly by hand.

40. Inspect the boot for damage. If the boot has been cut or punctured during assembly, you must discard and replace the boot.

➡**The boot retaining clamp must not be over-tightened or under-tightened.**

41. Crimp the large boot retaining clamp. Tighten the large boot clamp until the base of the omega ohms shape has a gap of 1 mm (0.039 in).

42. The clamping hold time must be no less than 2 seconds.

43. Inspect the outer joint for smooth operation. This will also distribute the grease within the joint.

a. Hold the halfshaft vertically, with the outer joint at the bottom.

b. Rotate the halfshaft 4 or 5 times in a circular motion.

c. Install the halfshaft.

Driveshaft

REMOVAL & INSTALLATION

1. Before servicing the vehicle, refer to the Precautions Section.

2. Raise and safely support the vehicle.

3. Remove the driveline tunnel closeout panel.

4. Support the rear axle.

5. Matchmark the driveshaft flanges.

6. Remove the bolt from the differential case bracket assembly to body.

7. Remove the driveshaft coupling nuts.

➡**DO NOT remove the bolts from the coupler-to-the driveshaft. Remove only the bolts from the coupler-to-the drive axle pinion flange.**

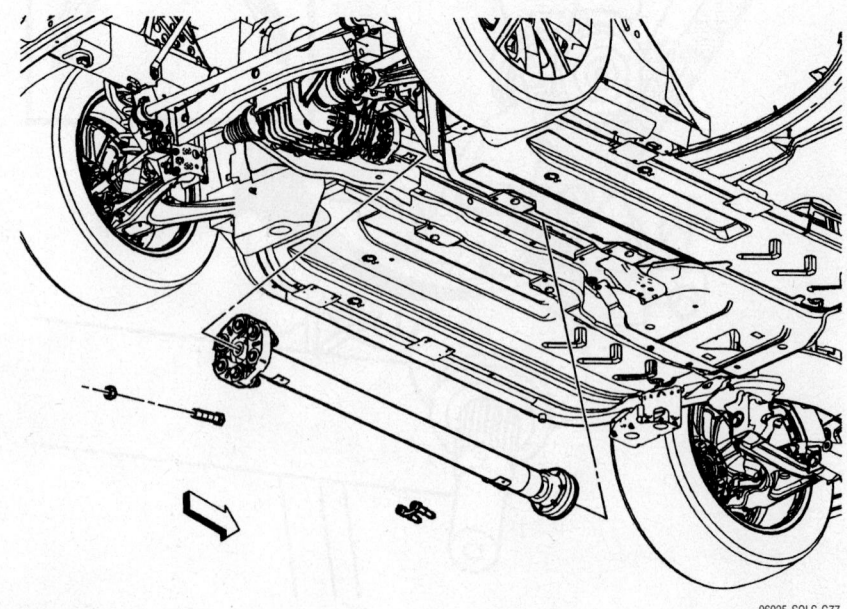

06025-SOLS-G77

Driveshaft removal

8. Remove the driveshaft coupler bolt.

9. Remove the CV-Joint mounting bolts.

10. Remove the CV-Joint mounting bolt spacers.

11. Remove the driveshaft.

➡**Lower the front of the differential to remove the driveshaft from the pinion flange.**

12. Installation is the reverse of removal. Note the following torques:
- Driveshaft attaching bolts: 63 ft. lbs. (85 Nm)
- Tunnel closeout: 11 ft. lbs. (15 Nm)

Differential Pinion Seal

REMOVAL & INSTALLATION

1. Before servicing the vehicle, refer to the Precautions Section.

2. Raise and support vehicle.

3. Remove the driveline tunnel closeout panel.

4. Support the rear drive module (RDM).

5. Remove the bolt from the differential case bracket assembly to body.

➡**Remove only the driveshaft coupler-to-differential flange bolts. Do NOT remove the coupler from the driveshaft.**

6. Remove the driveshaft coupler-to-differential flange bolts, nuts, and washers.

7. Push the rear driveshaft toward the front of the vehicle in order to release the driveshaft coupler from the differential pinion flange.

8. Lower the driveshaft and the front of the RDM until disconnected.

9. Carefully position the driveshaft out of the way and support the driveshaft using a suitable jack.

10. Install a holding tool on the flange.

11. Remove the drive pinion nut.

12. Using a puller, remove the flange.

13. Using a flat-bladed tool, remove the drive pinion seal. Take care not to damage any sealing surfaces.

To install:

14. Lubricate the drive pinion flange sealing surface of the drive pinion seal with

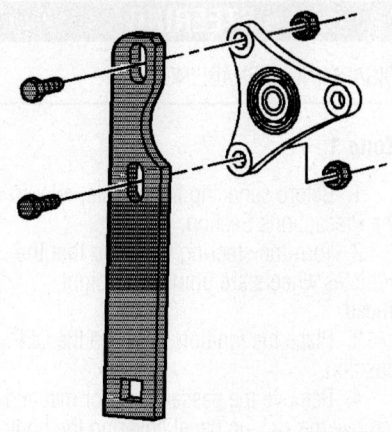

06025-SOLS-G78

Install a holding tool on the flange

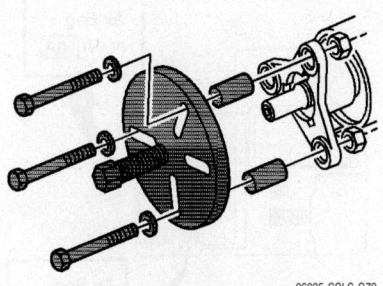

06025-SOLS-G79

Using a puller, remove the flange

synthetic gear oil GM P/N 12378514 (Canadian P/N 88901045), or equivalent.

15. Using a seal driver, install the pinion seal to the differential.

16. Install the pinion flange on the drive pinion shaft.

➡**The pinion shaft threads and the pinion flange nut must be free of residue and debris prior to application of threadlocker in order to ensure proper adhesion and fastener retention.**

17. Prepare the pinion shaft threads and the pinion flange nut for assembly:

a. Thoroughly clean the residue from the pinion shaft threads by using denatured alcohol or equivalent and allow to dry.

b. Thoroughly clean the residue from the pinion flange nut by using denatured alcohol or equivalent and allow to dry.

18. Apply threadlocker GM P/N

12345382 (Canadian P/N 10953489), or equivalent to ⅔ of the threaded length of the pinion shaft threads. Ensure that there are no gaps in the threadlocker along the length of the filled area of the pinion shaft threads.

19. Allow the threadlocker to cure approximately 10 minutes before installation.

20. Install the drive pinion flange nut to the pinion shaft. Tighten the drive pinion flange nut to 285 Nm (210 ft. lbs.).

21. Raise the driveshaft and the front of the RDM until the driveshaft is installed to the pinion flange.

22. Install the driveshaft coupler to the differential flange.

23. Remove the jack.

24. Inspect the driveshaft to flange nuts, bolts, and washers. Replace if damaged or worn.

➡**If reusing the driveshaft to flange nuts and bolts, to ensure proper adhesion and fastener retention, the threads must be free of debris prior to the application of threadlocker.**

25. Thoroughly clean the threads using denatured alcohol or equivalent and allow to dry. Apply threadlocker GM P/N 12345493 (Canadian P/N 10953488), or equivalent to the driveshaft to the flange bolt. Ensure that there are no gaps in the threadlocker along the length of the filled area of the bolt. Allow the threadlocker to cure approximately 10 minutes before installation.

26. Install the driveshaft coupler-to-differential flange washers to the driveshaft coupler-to-differential flange bolts.

27. Install the driveshaft coupler-to-differential flange bolts and washers to the differential flange and driveshaft coupler.

28. Install the driveshaft coupler-to-differential flange nuts. Tighten the driveshaft coupler-to-differential flange bolts to 85 Nm (63 ft. lbs.).

29. Install the bolt from the differential case bracket assembly to body.

30. Remove the RDM support.

31. Install driveline tunnel closeout panel.

32. Inspect the fluid level.

33. Lower the vehicle.

STEERING

Air Bag

PRECAUTIONS

☀☀ CAUTION

When performing service on or near the SIR components or the SIR wiring, the SIR system must be disabled. Refer to SIR Disabling and Enabling Zones. Failure to observe the correct procedure could cause deployment of the SIR components, personal injury, or unnecessary SIR system repairs.

The inflatable restraint sensing and diagnostic module (SDM) maintains a reserved energy supply. The reserved energy supply provides deployment power for the air bags. Deployment power is available for as much as 1 minute after disconnecting the vehicle power. Disabling the SIR system prevents deployment of the air bags from the reserved energy supply.

The following are general service instructions which must be followed in order to properly repair the vehicle and return it to its original integrity:

• Do not expose inflator modules to temperatures above 65°C (150°F).

• Verify the correct replacement part number. Do not substitute a component from a different vehicle.

• Use only original GM replacement parts available from your authorized GM dealer. Do not use salvaged parts for repairs to the SIR system.

Discard any of the following components if it has been dropped from a height of 91 cm (3 feet) or greater:

• Inflatable restraint sensing and diagnostic module (SDM)

• Inflatable restraint I/P module

• Inflatable restraint steering wheel module

• Inflatable restraint steering wheel module coil

• Inflatable restraint passenger presence system (PPS)

• Inflatable restraint seat belt retractor pretensioners

• Inflatable restraint front end sensors

Use caution when handling or storing a live (undeployed) inflator module. An inflator module deployment produces a rapid generation of gas. This may cause the inflator module, or an object in front of the inflator module, to project through the air in the event of an unlikely deployment.

DISARMING & ENABLING

Zone 1

1. Before servicing the vehicle, refer to the Precautions Section.

2. Turn the steering wheel so that the vehicles wheels are pointing straight ahead.

3. Place the ignition switch to the OFF position.

4. Remove the passenger floor mat and remove the kick up panel covering the body control module (BCM) fuse center.

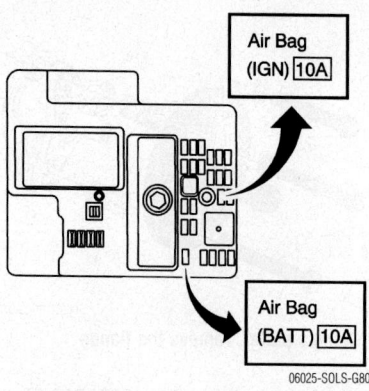

BCM fuse center

☀☀ WARNING

This sensing and diagnostic module (SDM) has two fused power inputs. To ensure there is no unwanted SIR deployment, personal injury, or unnecessary SIR system repairs, remove both AIR BAG (IGN) and AIR BAG (BATT) fuses from the BCM fuse center. With the AIR BAG fuses removed and the ignition switch in the ON position, the AIR BAG warning indicator illuminates. This is normal operation, and does not indicate an SIR system malfunction.

5. Locate and remove the AIR BAG (IGN) and AIR BAG (BATT) fuses from the BCM fuse center.

6. Open front hood and locate both the right and left front end sensors.

7. Remove both the right and left connector position assurance (CPA) from the right and left front end sensor connector.

8. Remove both connectors from left and right front end sensors.

Enabling Procedure

9. Place the ignition in the OFF position.

10. Connect both connectors to the left and right front end sensors.

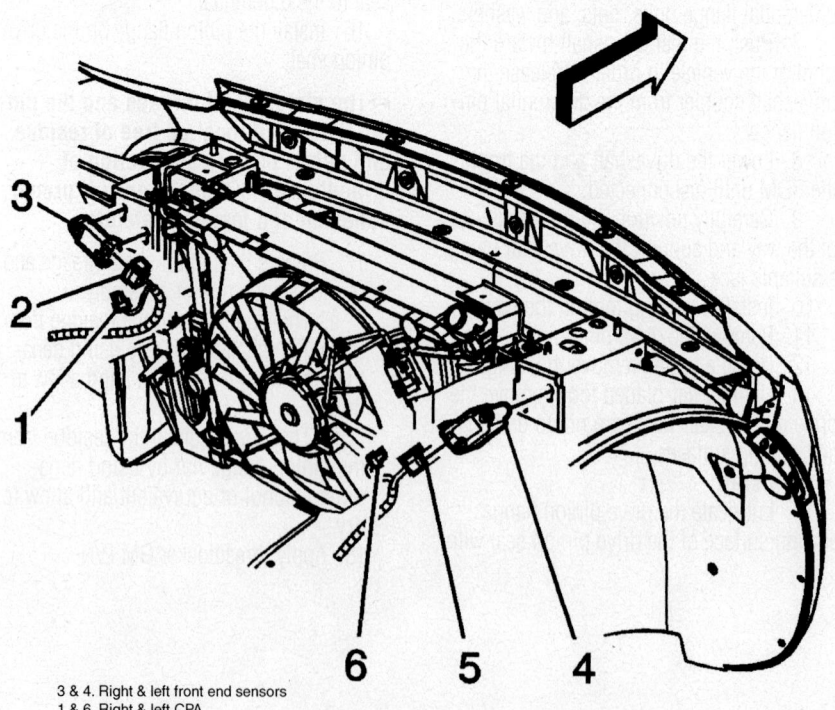

3 & 4. Right & left front end sensors
1 & 6. Right & left CPA
2 & 5. Connectors

Front end sensors

06025-SOLS-G81

11. Connect both CPAs to the left and right front sensor connectors.

12. Install both AIR BAG (IGN) fuse and the AIR BAG (BATT) fuse into the BCM fuse center.

13. Install the kick up panel to cover the BCM fuse center then replace the passenger floor mat.

14. Use caution while reaching in and turn the ignition switch to the ON position. The AIR BAG indicator will flash then turn OFF.

Zone 3

1. Before servicing the vehicle, refer to the Precautions Section.

2. Turn the steering wheel so that the vehicles wheels are pointing straight ahead.

3. Place the ignition switch to the OFF position.

4. Remove the passenger floor mat and remove the kick up panel covering the body control module (BCM) fuse center.

❋❋ WARNING

This sensing and diagnostic module (SDM) has two fused power inputs. To ensure there is no unwanted SIR deployment, personal injury, or unnecessary SIR system repairs, remove both AIR BAG (IGN) and AIR BAG (BATT) fuses from the BCM fuse center. With the AIR BAG fuses removed and the ignition switch in the ON position, the AIR BAG warning indicator illuminates. This is normal operation, and does not indicate an SIR system malfunction.

5. Locate and remove the AIR BAG (IGN) and AIR BAG (BATT) fuses from the BCM fuse center.

6. On the back side of the steering wheel there are 2 openings to access the fasteners securing the steering wheel module.

7. Insert a 3.175 mm ($\frac{1}{8}$ inch) diameter blunt ended punch or equivalent tool into each of the holes in order to release the fastener securing the module to the steering wheel.

8. Pull the steering wheel module gently away from the steering wheel.

9. Remove both connector position assurance (CPA) from the steering wheel module connectors.

10. Remove both connectors from the steering wheel module.

11. Remove the steering wheel module.
Enabling Procedure

12. Place the ignition in the OFF position.

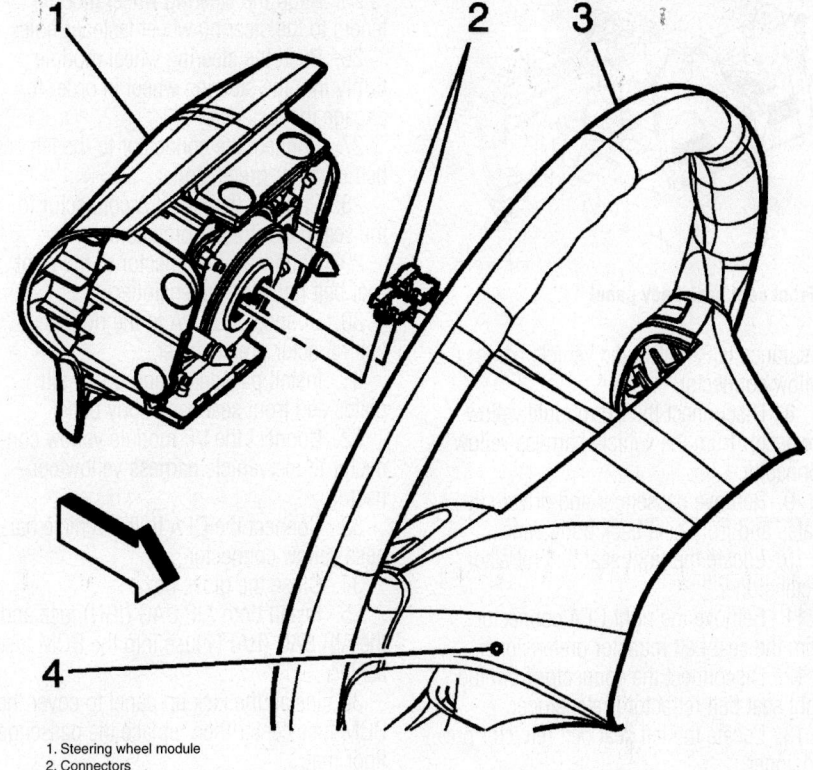

1. Steering wheel module
2. Connectors
3. Steering wheel
4. Housing

06025-SOLS-G82

Steering wheel air bag module

➡This vehicle is equipped with dual stage air bags, you will find 2 connectors. Match the right color connector to the right color opening in the module. Route the steering wheel module wires, the redundant control wires, and the horn wires correctly.

13. Connect both connectors to the steering wheel module.

14. Connect both CPA's to the steering wheel module connectors.

15. Align the steering wheel module fasteners to the steering wheel fastener holes.

16. Push the steering wheel module firmly into the steering wheel in order to engage the fasteners.

17. Install both AIR BAG (IGN) fuse and the AIR BAG (BATT) fuse into the BCM fuse center.

18. Install the kick up panel to cover the BCM fuse center then replace the passenger floor mat.

19. Use caution while reaching in and turn the ignition switch to the ON position. The AIR BAG indicator will flash then turn OFF.

Zone 4

1. Before servicing the vehicle, refer to the Precautions Section.

2. Turn the steering wheel so that the vehicles wheels are pointing straight ahead.

3. Place the ignition switch to the OFF position.

4. Remove the passenger floor mat and remove the kick up panel covering the body control module (BCM) fuse center.

❋❋ WARNING

This sensing and diagnostic module (SDM) has two fused power inputs. To ensure there is no unwanted SIR deployment, personal injury, or unnecessary SIR system repairs, remove both AIR BAG (IGN) and AIR BAG (BATT) fuses from the BCM fuse center. With the AIR BAG fuses removed and the ignition switch in the ON position, the AIR BAG warning indicator illuminates. This is normal operation, and does not indicate an SIR system malfunction.

5. Locate and remove the AIR BAG (IGN) and AIR BAG (BATT) fuses from the BCM fuse center.

6. Open and lower glove box door fully.

7. Locate the I/P module yellow connector and remove the connector position

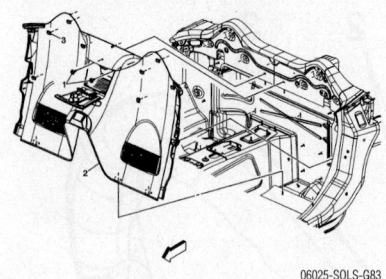

06025-SOLS-G83

Front seat back body panel

assurance (CPA) from the vehicle harness yellow connector.

8. Disconnect the I/P module yellow connector from the vehicle harness yellow connector.

9. Remove passenger and driver sill plates and front seat back body panel.

10. Locate the right seat belt retractor pretensioner.

11. Remove the right CPA connector from the seat belt retractor pretensioner.

12. Disconnect the connector from the right seat belt retractor pretensioner.

13. Locate the left seat belt retractor pretensioner.

14. Remove the left CPA connector from the seat belt retractor pretensioner.

15. Disconnect the connector from the left seat belt retractor pretensioner.

16. On the back side of the steering wheel there are 2 openings to access the fasteners securing the steering wheel module.

17. Insert a 3.175 mm (⅛ inch) diameter blunt ended punch or equivalent tool into each of the holes in order to release the fastener securing the module to the steering wheel.

18. Pull the steering wheel module gently away from the steering wheel.

19. Remove both CPAs from the steering wheel module connectors.

20. Remove both connectors from the steering wheel module.

21. Remove the steering wheel module.

Enabling Procedure

22. Place the ignition in the OFF position.

➡This vehicle is equipped with dual stage air bags. You will find 2 connectors. Match the right color connector to the right color opening in the module. Route the steering wheel module wires, the redundant control wires, and the horn wires correctly.

23. Connect both connectors to the steering wheel module.

24. Connect both CPAs to the steering wheel module connectors.

25. Align the steering wheel module fasteners to the steering wheel fastener holes.

26. Push the steering wheel module firmly into the steering wheel in order to engage the fasteners.

27. Connect the connector to the left seat belt retractor pretensioner.

28. Connect the left CPA connector to the seat belt retractor pretensioner.

29. Connect the connector to the right seat belt retractor pretensioner.

30. Connect the CPA to the right seat belt retractor pretensioner.

31. Install passenger and driver sill plates and front seat back body panel.

32. Connect the I/P module yellow connector to the vehicle harness yellow connector.

33. Connect the CPA to the vehicle harness yellow connector.

34. Close the glove box.

35. Install both AIR BAG (IGN) fuse and the AIR BAG (BATT) fuse into the BCM fuse center.

36. Install the kick up panel to cover the BCM fuse center then replace the passenger floor mat.

37. Use caution while reaching in and turn the ignition switch to the ON position. The AIR BAG indicator will flash then turn OFF.

Zone 5

1. Before servicing the vehicle, refer to the Precautions Section.

2. Turn the steering wheel so that the vehicles wheels are pointing straight ahead.

3. Place the ignition switch to the OFF position.

4. Remove the passenger floor mat and remove the kick up panel covering the body control module (BCM) fuse center.

✷✷ WARNING

This sensing and diagnostic module (SDM) has two fused power inputs. To ensure there is no unwanted SIR deployment, personal injury, or unnecessary SIR system repairs, remove both AIR BAG (IGN) and AIR BAG (BATT) fuses from the BCM fuse center. With the AIR BAG fuses removed and the ignition switch in the ON position, the AIR BAG warning indicator illuminates. This is normal operation, and does not indicate an SIR system malfunction.

5. Locate and remove the AIR BAG (IGN) and AIR BAG (BATT) fuses from the BCM fuse center.

6. Open and lower glove box door fully.

7. Locate the I/P module yellow connector and remove the connector position assurance (CPA) from the vehicle harness yellow connector.

8. Disconnect the I/P module yellow connector from the vehicle harness yellow connector.

Enabling Procedure

9. Place the ignition in the OFF position.

10. Connect the I/P module yellow connector to the vehicle harness yellow connector.

11. Connect the CPA to the vehicle harness yellow connector.

12. Close the glove box.

13. Install both AIR BAG (IGN) fuse and the AIR BAG (BATT) fuse into the BCM fuse center.

14. Install the kick up panel to cover the BCM fuse center then replace the passenger floor mat.

15. Use caution while reaching in and turn the ignition switch to the ON position. The AIR BAG indicator will flash then turn OFF.

Zone 7

1. Before servicing the vehicle, refer to the Precautions Section.

2. Turn the steering wheel so that the vehicles wheels are pointing straight ahead.

3. Place the ignition switch to the OFF position.

4. Remove the passenger floor mat and remove the kick up panel covering the body control module (BCM) fuse center.

✷✷ WARNING

This sensing and diagnostic module (SDM) has two fused power inputs. To ensure there is no unwanted SIR deployment, personal injury, or unnecessary SIR system repairs, remove both AIR BAG (IGN) and AIR BAG (BATT) fuses from the BCM fuse center. With the AIR BAG fuses removed and the ignition switch in the ON position, the AIR BAG warning indicator illuminates. This is normal operation, and does not indicate an SIR system malfunction.

5. Locate and remove the AIR BAG (IGN) and AIR BAG (BATT) fuses from the BCM fuse center.

6. Remove driver sill plate and front seat back body panel.

7. Locate the left seat belt retractor pretensioner.

8. Remove the connector position assurance (CPA) from the left seat belt retractor pretensioner.

9. Disconnect the connector from the left seat belt retractor pretensioner.

Enabling Procedure

10. Place the ignition in the OFF position.

11. Connect the connector to the left seat belt retractor pretensioner.

12. Connect the CPA to the left seat belt retractor pretensioner.

13. Install driver sill plate and front seat back body panel.

14. Install both AIR BAG (IGN) fuse and the AIR BAG (BATT) fuse into the BCM fuse center.

15. Install the kick up panel to cover the BCM fuse center then replace the passenger floor mat.

16. Use caution while reaching in and turn the ignition switch to the ON position. The AIR BAG indicator will flash then turn OFF.

Zone 9

1. Before servicing the vehicle, refer to the Precautions Section.

2. Turn the steering wheel so that the vehicles wheels are pointing straight ahead.

3. Place the ignition switch to the OFF position.

4. Remove the passenger floor mat and remove the kick up panel covering the body control module (BCM) fuse center.

✳✳ WARNING

This sensing and diagnostic module (SDM) has two fused power inputs. To ensure there is no unwanted SIR deployment, personal injury, or unnecessary SIR system repairs, remove both AIR BAG (IGN) and AIR BAG (BATT) fuses from the BCM fuse center. With the AIR BAG fuses removed and the ignition switch in the ON position, the AIR BAG warning indicator illuminates. This is normal operation, and does not indicate an SIR system malfunction.

5. Locate and remove the AIR BAG (IGN) and AIR BAG (BATT) fuses from the BCM fuse center.

6. Remove passenger sill plate and front seat back body panel.

7. Locate the right seat belt retractor pretensioner.

8. Remove the connector position assurance (CPA) from the right seat belt retractor pretensioner.

9. Disconnect the connector from the right seat belt retractor pretensioner.

Enabling Procedure

10. Place the ignition in the OFF position.

11. Connect the connector to the right seat belt retractor pretensioner.

12. Connect the CPA to the right seat belt retractor pretensioner.

13. Install passenger sill plate and front seat back body panel.

14. Install both AIR BAG (IGN) fuse and the AIR BAG (BATT) fuse into the BCM fuse center.

15. Install the kick up panel to cover the BCM fuse center then replace the passenger floor mat.

16. Use caution while reaching in and turn the ignition switch to the ON position. The AIR BAG indicator will flash then turn OFF.

Power Steering Gear

REMOVAL & INSTALLATION

1. Before servicing the vehicle, refer to the Precautions Section.

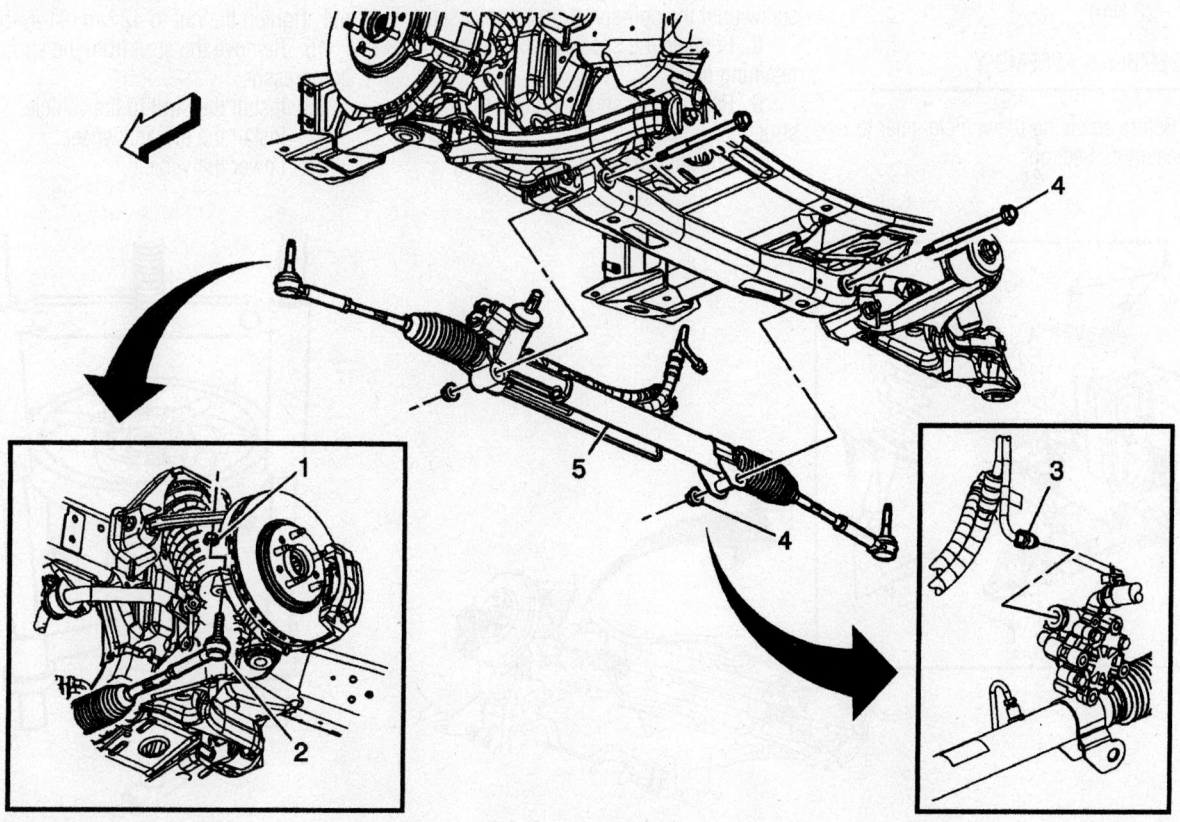

06025-SOLS-G84

Power steering gear removal

2. Raise and safely support the vehicle.
3. Remove the wheels.
4. Remove the tie rod end outer nut.
5. Using a 2-jawed tool, remove the outer tie rod end.
6. Drain the power steering system.
7. Disconnect the power steering outlet pipe/hose fitting.

8. Using tool J 42640 Steering Column Anti-Rotation Pin, or equivalent, lock the steering wheel in place.
9. Remove the intermediate shaft bolt from the steering gear.
10. Remove the power steering gear mounting nuts/bolts.
11. Remove the power steering gear.

12. Installation is the reverse of removal. Observe the following torques:
- Steering gear mount bolts/nuts: 81 ft. lbs. (110 Nm).
- Power steering outlet pipe/hose fitting: 20 ft. lbs. (27 Nm)
- Tie rod end ball stud nut: 44 ft. lbs. (60 Nm)

FRONT SUSPENSION

Strut

REMOVAL & INSTALLATION

1. Before servicing the vehicle, refer to the Precautions Section.
2. Raise and safely support the vehicle.
3. Remove the wheels.
4. Remove the lower strut mounting bolts.
5. Remove the upper strut mounting nuts.
6. Remove the strut.
7. Installation is the reverse of removal. Observe the following torques:
- Upper strut nuts: 35 ft. lbs. (47 Nm)
- Lower strut bolts/nuts: 21 ft. lbs. (28 Nm)

DISASSEMBLY & ASSEMBLY

1. Before servicing the vehicle, refer to the Precautions Section.

✳✳ WARNING

Use care when handling the coil springs in order to avoid chipping or scratching the coating. Damage to the coating will result in premature failure of the coil springs.

2. Raise and support the vehicle.
3. Remove the tire and wheel.
4. Remove the strut from the vehicle.
5. Install the strut into the spring compressor.
6. Mark the upper control arm assembly and insulator for proper installation.

➡**The spring is compressed when the shock absorber moves freely.**

7. Turn the spring compressor forcing screw until the coil spring is compressed.
8. Remove the shock absorber upper retaining nut.
9. Remove the shock absorber from the strut.
10. Loosen the compressor forcing

screw until the upper mounting plate and the coil spring may be removed.
11. Remove the upper control arm bracket assembly, the insulator, and the coil spring from the spring compressor.

To assemble

➡**Ensure the alignment pins in the upper control arm bracket are orientated 90 degrees with the shock absorber lower mounting holes.**

12. Install the coil spring, the insulator, the upper control arm bracket assembly, and the shock absorber to the spring compressor, aligning all marks made in disassembly procedure.
13. Turn the spring compressor forcing screw until the coil spring is compressed.
14. Install the shock absorber retaining nut. Tighten the nut to 42 Nm (31 ft. lbs.).
15. Remove the strut from the spring compressor.
16. Install the strut to the vehicle.
17. Install the tire and wheel.
18. Lower the vehicle.

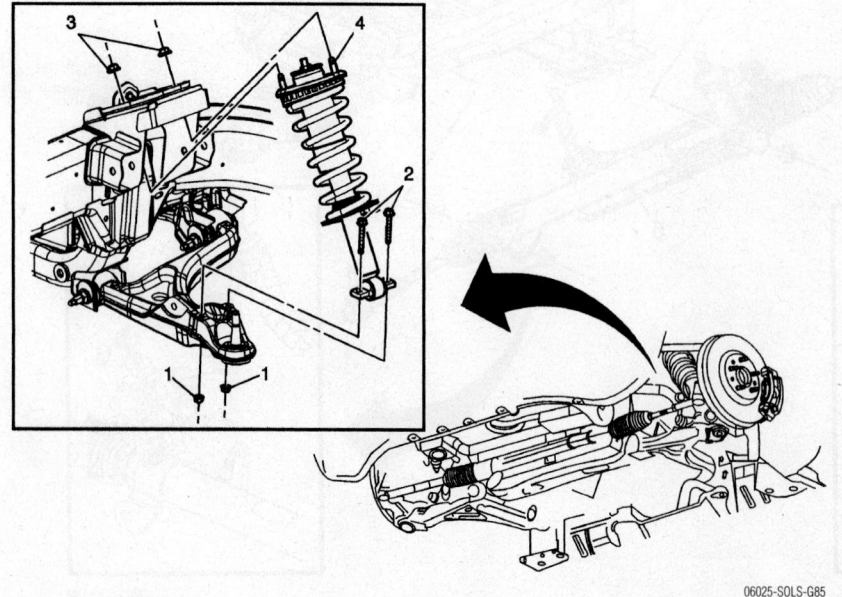

06025-SOLS-G85

Front strut mounting

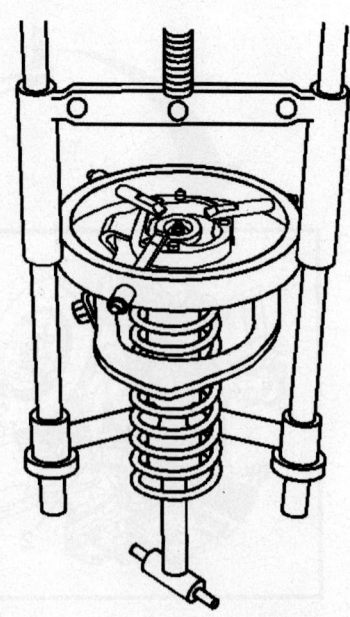

06025-SOLS-G86

Strut mounted in the compressor

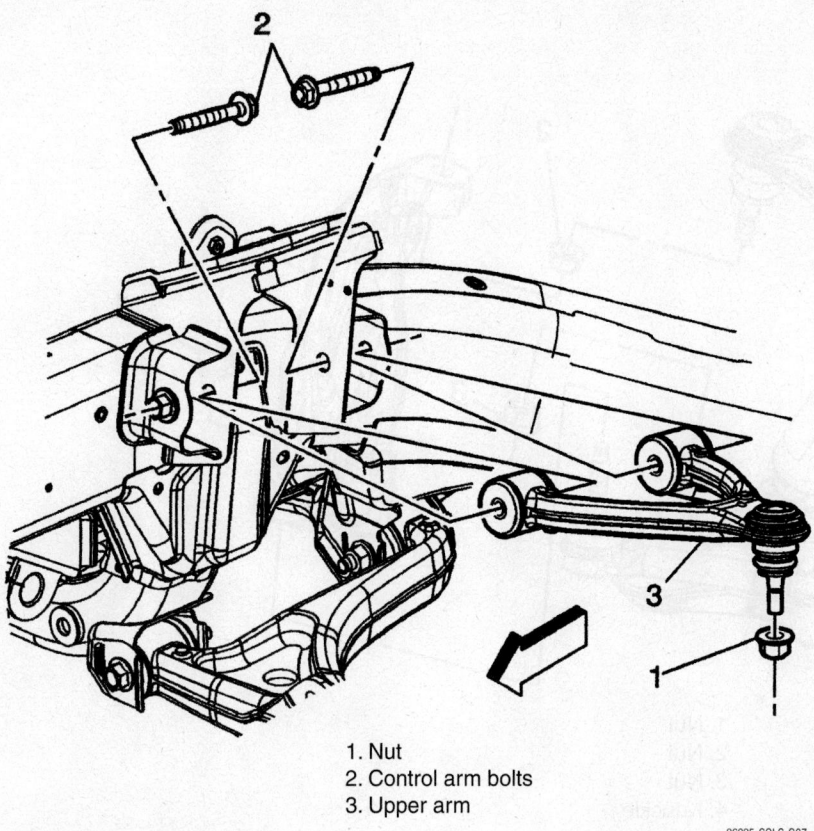

1. Nut
2. Control arm bolts
3. Upper arm

06025-SOLS-G87

Front upper control arm mounting

Upper Control Arm

REMOVAL & INSTALLATION

1. Before servicing the vehicle, refer to the Precautions Section.
2. Remove the tire and wheel.
3. Remove the shock module.
4. Remove the upper ball joint stud nut.
5. Remove the upper control arm mounting bolts.
6. Remove the upper control arm.
7. Installation is the reverse of removal. Observe the following torques:
- Control arm mounting bolts: 110 Nm (81 ft. lbs.)
- Ball stud nut: 30 Nm (22 ft. lbs.) + 150 degrees

Lower Control Arm

REMOVAL & INSTALLATION

1. Before servicing the vehicle, refer to the Precautions Section.
2. Remove the strut.
3. Remove the steering knuckle.
4. Remove the lower control arm to frame nuts.

5. Remove the lower control arm alignment cam.
6. Remove the lower control arm alignment bolt.
7. Remove the lower control arm.
8. Installation is the reverse of removal. Torque the lower arm-to-frame nuts to 165 Nm (122 ft. lbs.). Torque the ball stud nut to 40 Nm (30 ft. lbs.) + 135 degrees.
9. Adjust the alignment.

Steering Knuckle

REMOVAL & INSTALLATION

1. Before servicing the vehicle, refer to the Precautions Section.
2. Remove the tire and wheel.
3. Remove the tie rod end nut.
4. Using the appropriate tool, remove the tie rod end from the steering knuckle.
5. Remove the upper ball joint nut.
6. Use the appropriate tool to remove the upper ball joint from the steering knuckle.
7. Remove the lower ball joint nut.
8. Use the appropriate tool to remove the lower ball joint from the steering knuckle.

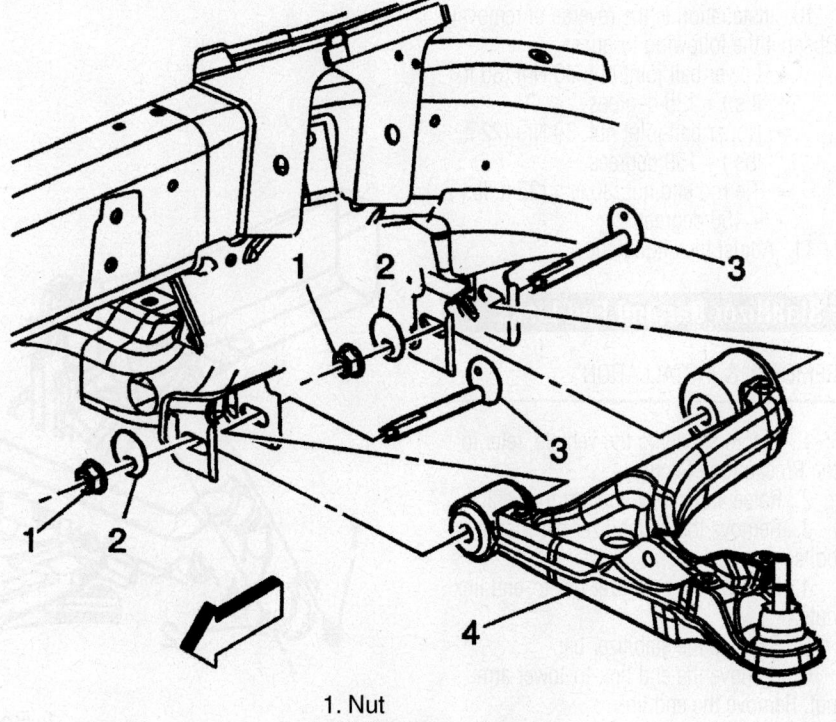

1. Nut
2. Alignment cam
3. Alignment bolt
4. Lower arm

06025-SOLS-G88

Front lower control arm mounting

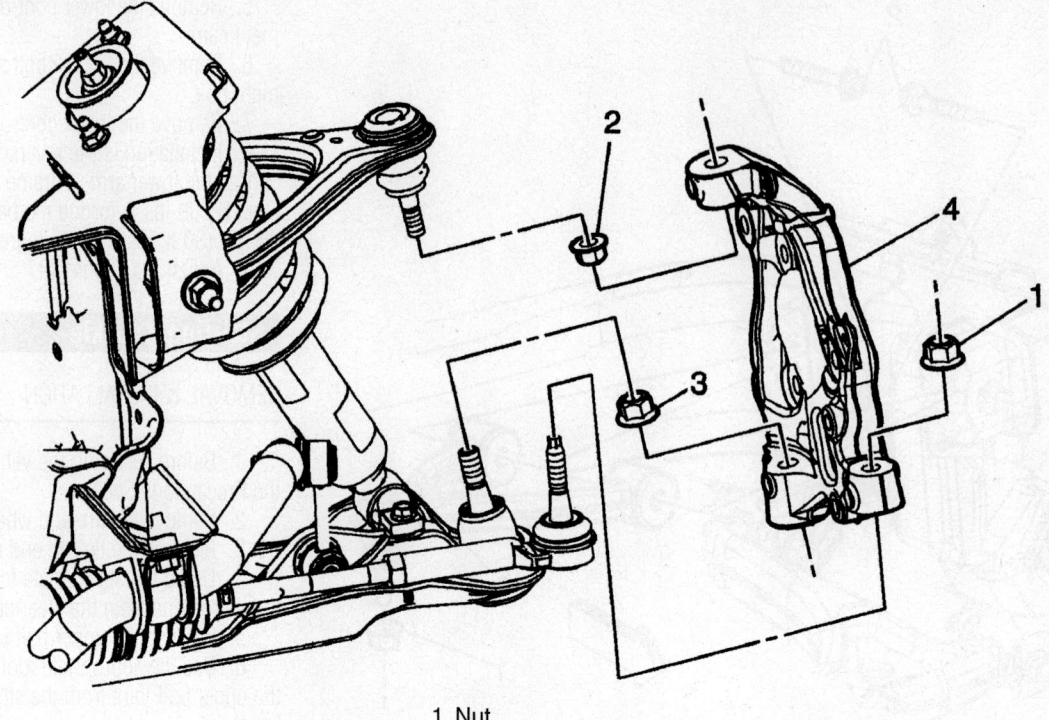

1. Nut
2. Nut
3. Nut
4. Knuckle

06025-SOLS-G89

Steering knuckle removal

9. Remove the steering knuckle.
10. Installation is the reverse of removal. Observe the following torques:
- Lower ball joint nut: 40 Nm (30 ft. lbs.) + 135 degrees
- Upper ball joint nut: 30 Nm (22 ft. lbs.) + 150 degrees
- Tie rod end nut: 30 Nm (22 ft. lbs.) + 150 degrees
11. Adjust the alignment.

Stabilizer Bar and End Links

REMOVAL & INSTALLATION

1. Before servicing the vehicle, refer to the Precautions Section.
2. Raise and safely support the vehicle.
3. Remove the 4 stabilizer bar mounting bolts.
4. Remove the stabilizer bar-to-end link nuts.
5. Remove the stabilizer bar.
6. Remove the end link-to-lower arm nut. Remove the end link.
7. Installation is the reverse of removal. Observe the following torques:
- End link nuts: 72 Nm (53 ft. lbs.)
- Stabilizer bar mounting bolts: 55 Nm (41 ft. lbs.)

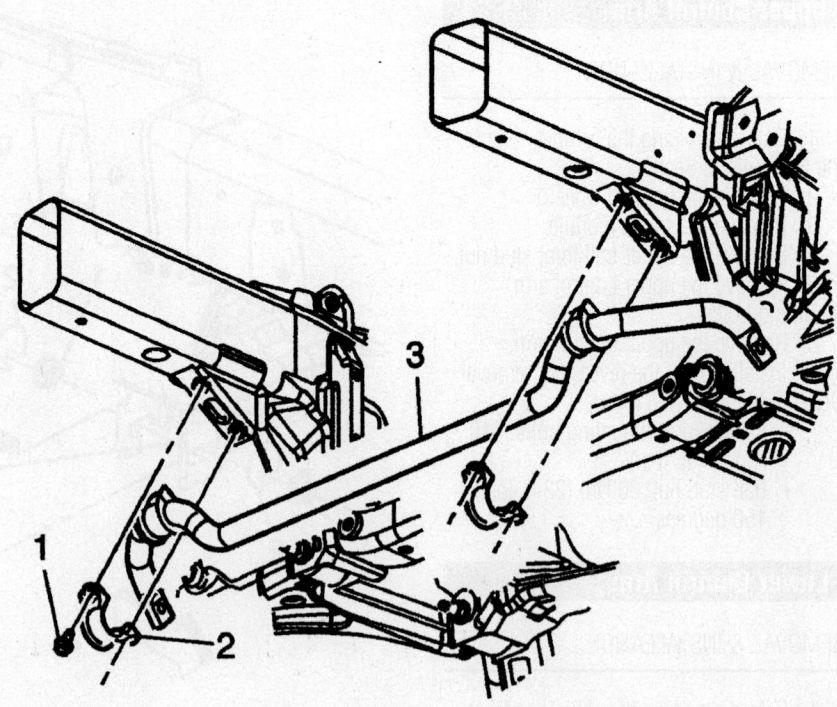

1. Bolt
2. Bushing bracket
3. Stabilizer bar

06025-SOLS-G90

Front stabilizer bar mounting

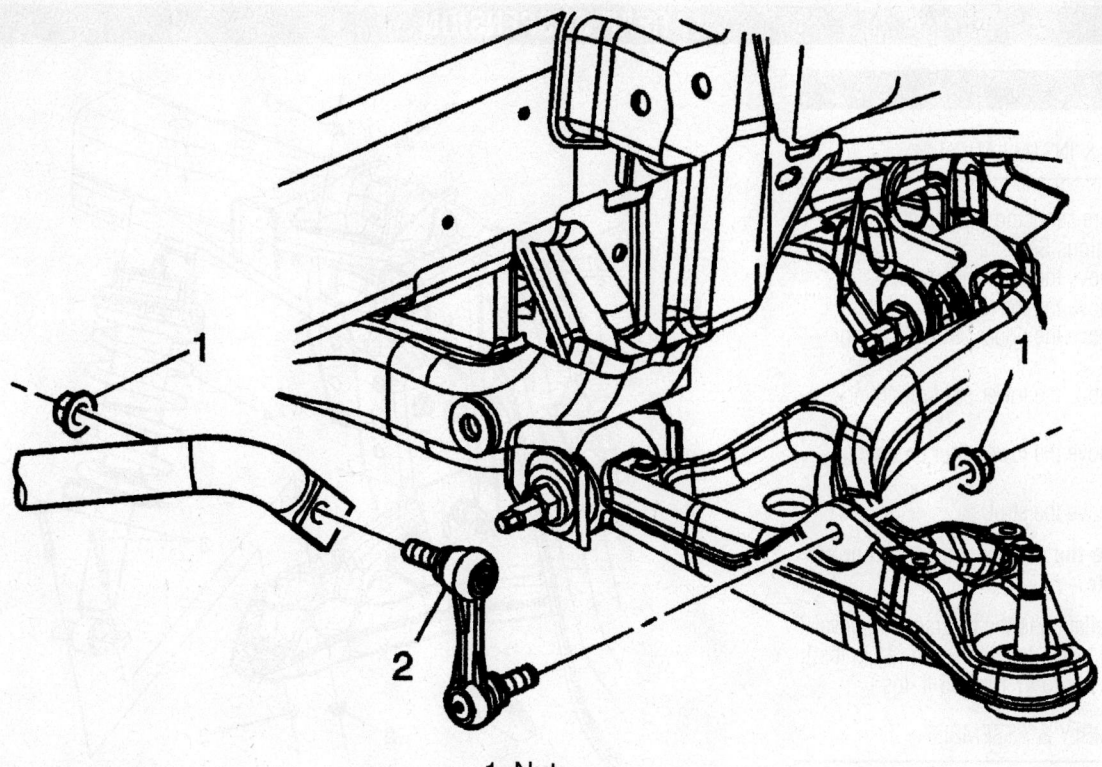

1. Nut
2. End link

06025-SOLS-G91

Front stabilizer bar end link

Front Hub/Bearing/Speed Sensor

REMOVAL & INSTALLATION

1. Before servicing the vehicle, refer to the Precautions Section.

2. Remove the tire and wheel.

3. Remove the brake caliper mounting bracket.

4. Remove the brake rotor.

5. Disconnect the speed sensor electrical connector and the wiring harness from the retainers on the steering knuckle.

6. Remove the wheel hub mounting bolts.

7. Remove the wheel hub/bearing/speed sensor assembly.

8. Installation is the reverse of removal. Torque the mounting bolts to 115 Nm (85 ft. lbs.).

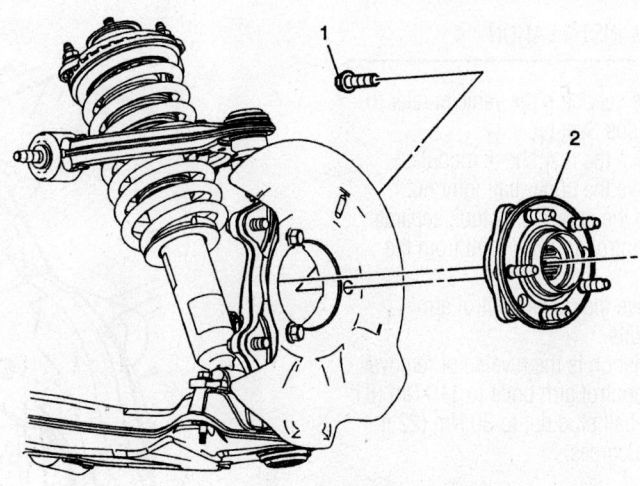

1. Bolt
2. Front hub/bearing/speed sensor assembly

06025-SOLS-G92

Front hub/bearing/speed sensor

REAR SUSPENSION

Strut

REMOVAL & INSTALLATION

1. Before servicing the vehicle, refer to the Precautions Section.
2. Remove the tires and wheels.
3. Remove the toe links.
4. Remove the upper strut mounting nuts.
5. Remove the lower strut mounting nuts.
6. Remove the lower strut mounting bolts.
7. Remove the strut.

➡**Remove the strut toward the rear of the vehicle.**

8. Installation is the reverse of removal. Torque the lower nuts to 28 Nm (21 ft. lbs.); the upper nuts to 47 Nm (35 ft. lbs.)

DISASSEMBLY & ASSEMBLY

See the procedure under Front Strut Disassembly and Assembly.

Upper Control Arm

REMOVAL & INSTALLATION

1. Before servicing the vehicle, refer to the Precautions Section.
2. Remove the rear shock module.
3. Remove the upper ball joint nut.
4. Using the appropriate tool, separate the upper control arm ball joint from the knuckle.
5. Remove the upper control arm mounting bolts.
6. Installation is the reverse of removal. Torque the control arm bolts to 110 Nm (81 ft. lbs.); the ball stud nut to 30 Nm (22 ft. lbs.) + 150 degrees.

Lower Control Arm

REMOVAL & INSTALLATION

1. Before servicing the vehicle, refer to the Precautions Section.
2. Remove the tire and wheel.
3. Remove the toe link.
4. Remove the lower control arm ball joint nut.
5. Using the appropriate tool, separate the lower control arm ball joint from the knuckle.

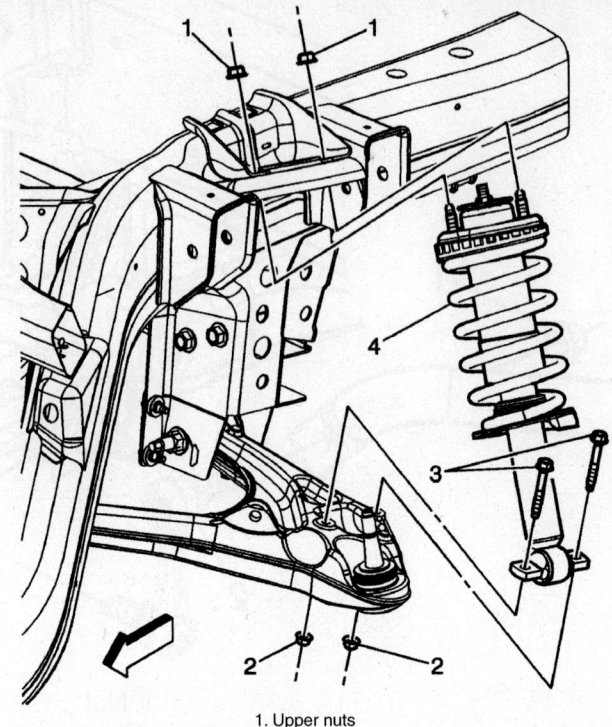

1. Upper nuts
2. Lower nuts
3. Lower nuts
4. Strut

06025-SOLS-G93

Rear strut mounting

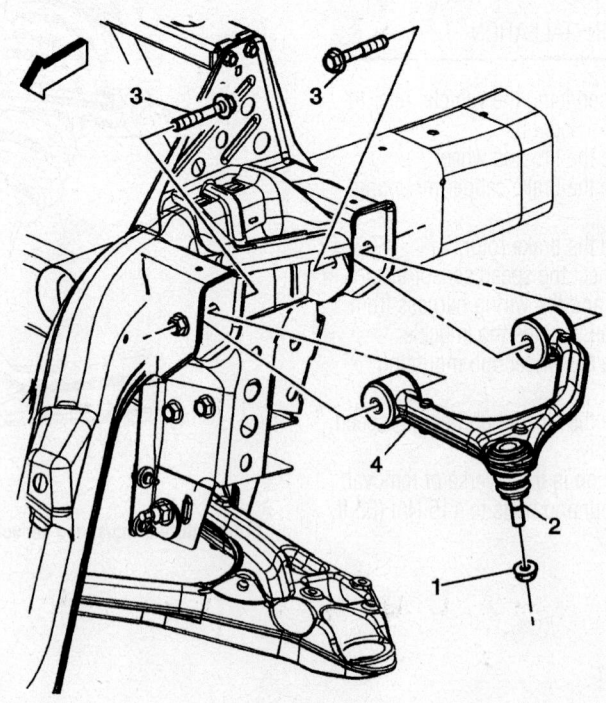

1. Ball stud nut
2. Ball joint
3. Mounting bolts
4. Control arm

06025-SOLS-G94

Rear upper control arm removal

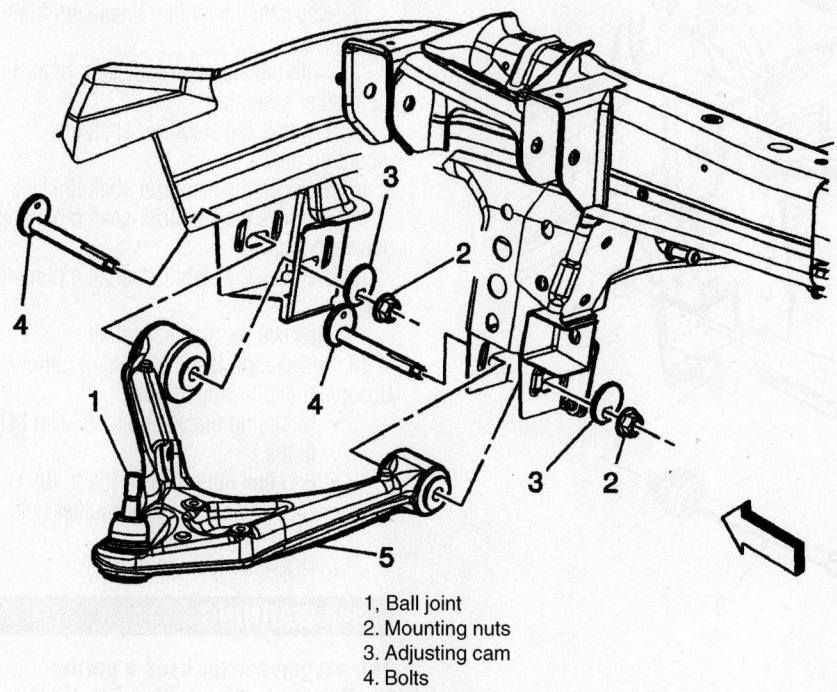

1. Ball joint
2. Mounting nuts
3. Adjusting cam
4. Bolts
5. Control arm

06025-SOLS-G95

Rear lower control arm mounting

6. Remove the lower control arm nuts.
7. Remove the adjusting cams.
8. Remove the lower control arm-to-frame bolts.
9. Remove the lower control arm.
10. Installation is the reverse of removal. Observe the following torques:
   - Mounting nuts: 165 Nm (122 ft. lbs.)
   - Ball stud nut: 40 Nm (30 ft. lbs.) + 135 degrees
11. Adjust the alignment.

Knuckle

REMOVAL & INSTALLATION

1. Before servicing the vehicle, refer to the Precautions Section.
2. Remove the tire and wheel.
3. Remove the wheel bearing/ hub assembly.
4. Remove the outer tie rod end retaining nut.
5. Use the appropriate tool to remove the tie rod end from the steering knuckle.
6. Remove the upper ball joint retaining nut.
7. Use the appropriate tool to remove the upper ball joint from the knuckle.

8. Remove the lower ball joint retaining nut.
9. Use the appropriate tool to remove the lower ball joint ball joint from the knuckle.
10. Remove the steering knuckle.
11. Installation is the reverse of removal. Observe the following torques:
   - Tie rod end stud nut: 30 Nm (22 ft. lbs.) + 150 degrees
   - Upper ball stud nut: 30 Nm (22 ft. lbs.) + 150 degrees
   - Lower ball stud nut: 40 Nm (30 ft. lbs.) + 135 degrees
12. Check the alignment.

Toe Links

REMOVAL & INSTALLATION

1. Before servicing the vehicle, refer to the Precautions Section.
2. Remove the tire and wheel.
3. Remove the toe link retaining nut.
4. Use the appropriate tool to remove the toe link ball joint from the rear knuckle.
5. Clean off threads and apply a small amount of penetrating oil to the thread of the link and allow to sit for a very minutes.

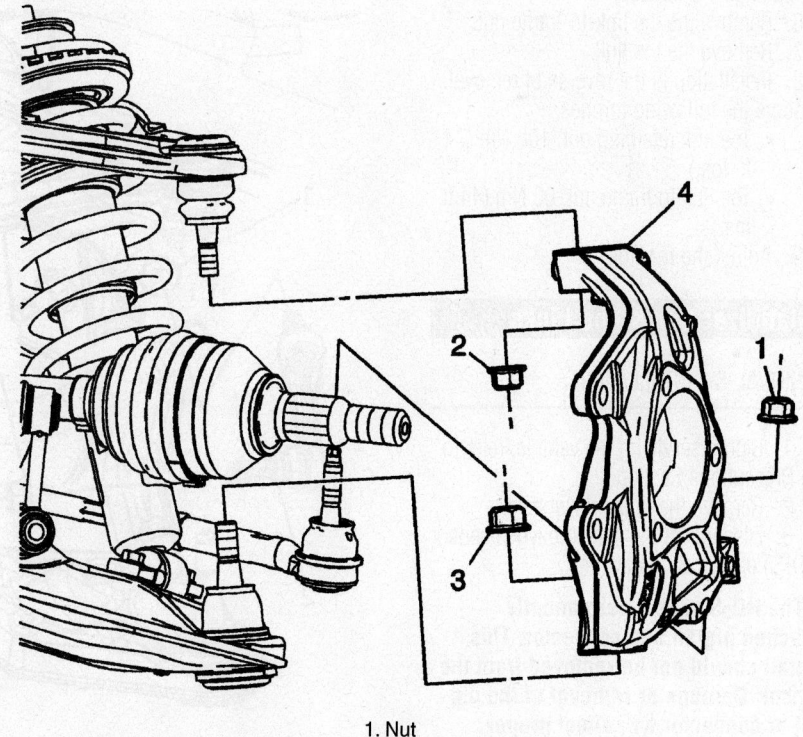

1. Nut
2. Nut
3. Nut
4. Knuckle

06025-SOLS-G96

Rear knuckle

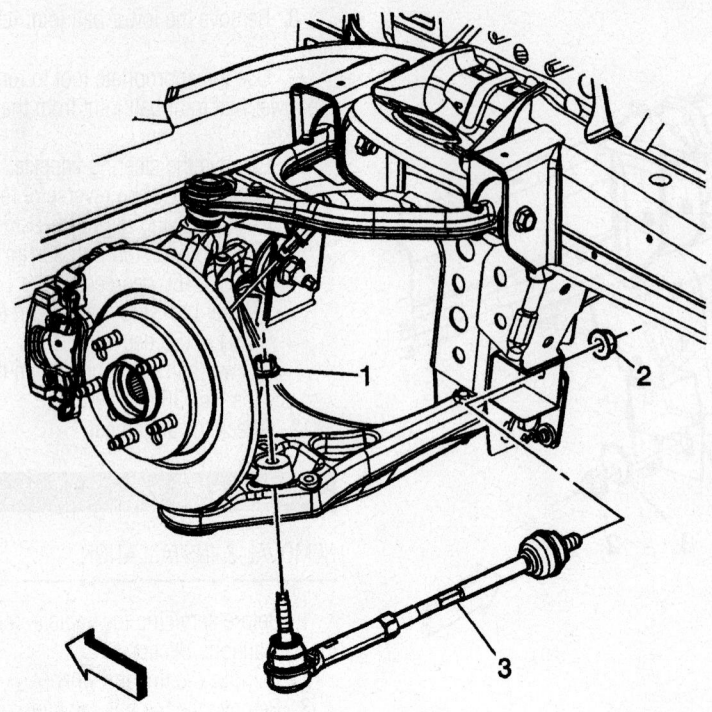

1. Ball stud nut
2. Link nut
3. Toe link

06025-SOLS-G97

Toe link mounting

This will aid in the removal of the nut and not damage the threads.

6. Remove the toe link-to-frame nut.
7. Remove the toe link.
8. Installation is the reverse of removal. Observe the following torques:
   - Toe link retaining nut: 100 Nm (74 ft. lbs.)
   - Toe link to frame nut: 60 Nm (44 ft. lbs.)
9. Adjust the rear toe.

Stabilizer Bar & End Links

REMOVAL & INSTALLATION

1. Before servicing the vehicle, refer to the Precautions Section.
2. Remove the tires and wheels.
3. Disconnect the heated oxygen sensor (HO_2S) electrical connector.

➡ **The HO_2S uses a permanently attached pigtail and connector. This pigtail should not be removed from the sensor. Damage or removal of the pigtail or connector will affect proper operation of the sensor.**

4. Remove the HO_2S.
5. Remove the catalytic converter to muffler nuts.

6. Have an assistant support the muffler assembly.
7. Separate the muffler insulators from the hangers.
8. With the aid of an assistant, remove the muffler assembly.
9. Remove the stabilizer shaft link mounting nuts.
10. Remove the stabilizer shaft links.
11. Remove the stabilizer shaft mounting bracket bolts.
12. Remove the stabilizer shaft mounting brackets.
13. Remove the stabilizer shaft.
14. Installation is the reverse of removal. Observe the following torques:
    - Mounting bracket bolts: 55 Nm (41 ft. lbs.)
    - End link nuts: 72 Nm (53 ft. lbs.)
    - Catalytic converter to muffler nuts: 17 Nm (13 ft. lbs.).
    - HO_2S: 41 Nm (30 ft. lbs.).

✳✳ WARNING

The oxygen sensor uses a permanently attached pigtail and connector. Do not remove the pigtail from the oxygen sensor. Damage to or removal of the pigtail connector could affect proper operation of the oxygen sensor.

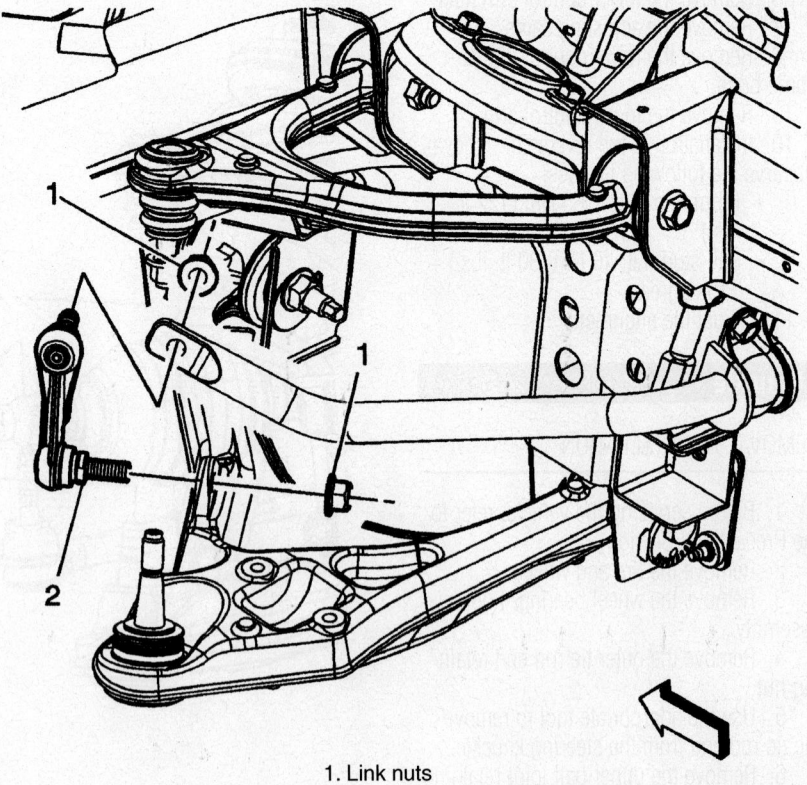

1. Link nuts
2. End link

06025-SOLS-G98

Rear stabilizer bar end links

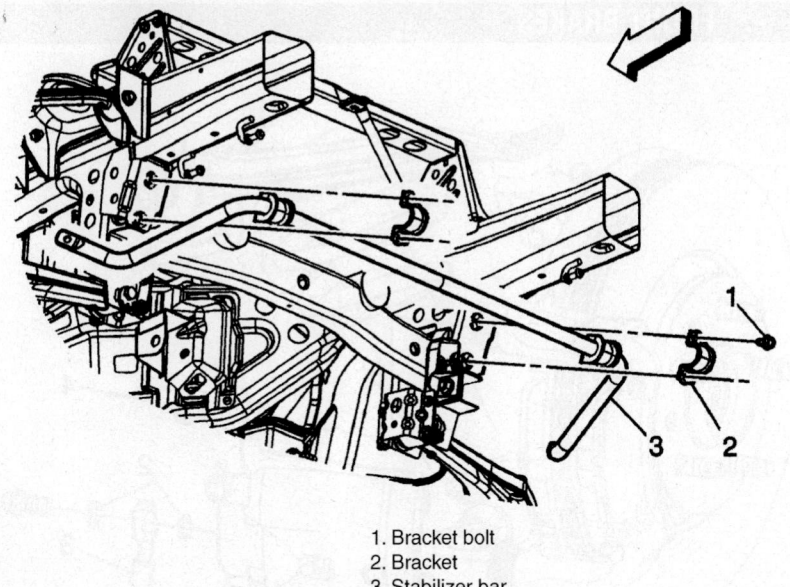

1. Bracket bolt
2. Bracket
3. Stabilizer bar

06025-SOLS-G99

Rear stabilizer bar

☀ WARNING

The use of excessive force may damage the threads in the exhaust manifold/pipe.

➡The in-line connector and louvered end must be kept clear of grease, dirt or other contaminants. Avoid using cleaning solvents of any type. DO NOT drop or roughly handle the heated oxygen sensor (HO$_2$S).

➡The HO$_2$S may be difficult to remove when the engine temperature is less than 48°C (120°F).

➡A special anti-seize compound is used on the HO$_2$S threads. The compound consists of a liquid graphite and glass beads. The graphite will burn away, but the glass beads will remain, making the sensor easier to remove. New or service sensors will have the compound applied to the threads. If a sensor is removed and is to be reinstalled, the threads must have an anti-seize compound applied before installation.

Rear Hub/Bearing

REMOVAL & INSTALLATION

1. Before servicing the vehicle, refer to the Precautions Section.
2. Remove the tire and wheel.
3. Remove the brake caliper mounting bracket.
4. Remove the drive axle retaining nut.
5. Remove the wheel/hub mounting bolt.
6. Using a small flat-blade screw driver, remove the speed sensor wiring harness from the upper control arm.
7. Disconnect the speed sensor electrical connector.
8. Remove the wheel hub.
9. Installation is the reverse of removal. Torque the NEW halfshaft nut to 160 Nm (118 ft. lbs.).

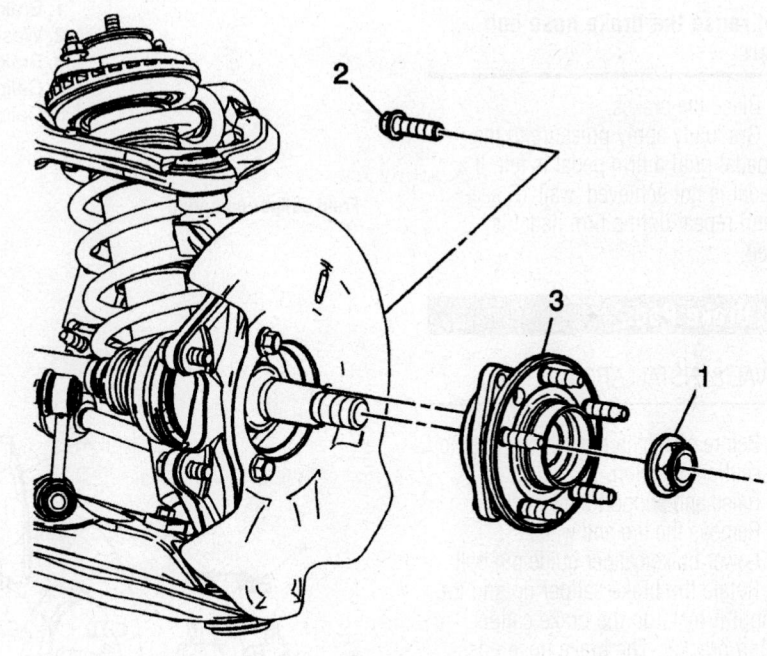

1. Halfshaft nut
2. Hub bolt
3. Hub/bearing assembly

06025-SOLS-G100

Rear hub/bearing assembly

FRONT BRAKES

Brake Caliper

REMOVAL & INSTALLATION

1. Before servicing the vehicle, refer to the Precautions Section.
2. Raise and support the vehicle.
3. Remove the tire and wheel.
4. Remove the brake hose bolt.
5. Remove the brake hose bolt washers.
6. Remove the front brake hose.
7. Cap or plug the brake hose to prevent fluid loss and contamination.
8. Remove the brake caliper bolt.
9. Remove the brake caliper.
10. Installation is the reverse of removal. Torque the caliper mounting bolts to 34 Nm (25 ft. lbs.). Torque the brake hose bolt to 40 Nm (30 ft. lbs.)

✳✳ WARNING

Do not reuse the brake hose bolt washers.

11. Bleed the brakes.
12. Gradually apply pressure to the brake pedal until a firm pedal is felt. If a firm pedal is not achieved, wait 15 seconds and repeat until a firm pedal is obtained.

Disc Brake Pads

REMOVAL & INSTALLATION

1. Before servicing the vehicle, refer to the Precautions Section.
2. Raise and support the vehicle.
3. Remove the tire and wheel.
4. Lower brake caliper guide pin bolt.
5. Rotate the brake caliper up and forward until it rests on the brake caliper mounting bracket. The brake hose does not have to be removed from the brake caliper.
6. Remove the brake pads.
7. Remove the brake pad springs.
8. Remove about half of the fluid from the master cylinder.
9. Using an appropriate tool, force the caliper piston back into the caliper.
10. Installation is the reverse of removal. Torque the caliper bolt to 34 Nm (25 ft. lbs.). Apply brake lubricant to the front caliper guide pins ensuring adequate lubrication to both pin areas of the front brake caliper bracket with GM P/N 22688644 or equivalent.

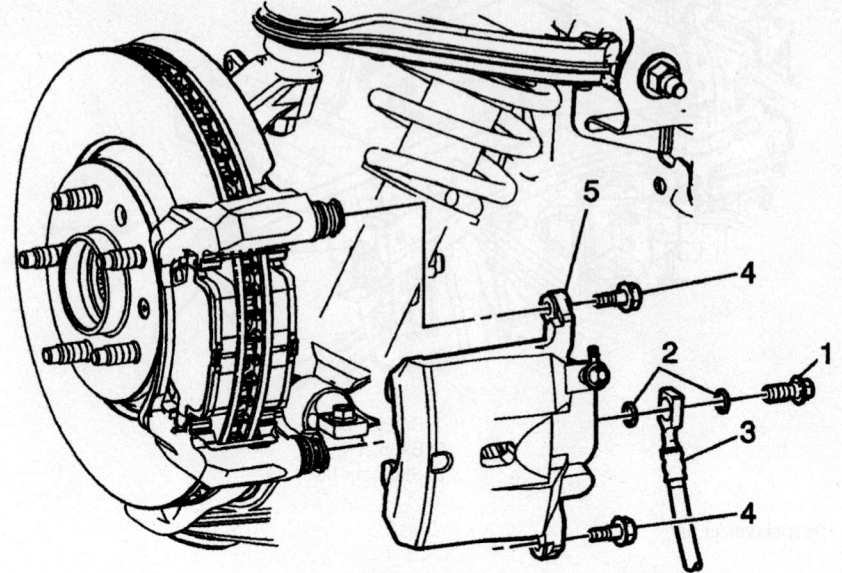

1. Brake hose bolt
2. Washers
3. Brake hose
4. Caliper bolts
5. Caliper

06025-SOLS-G101

Front caliper mounting

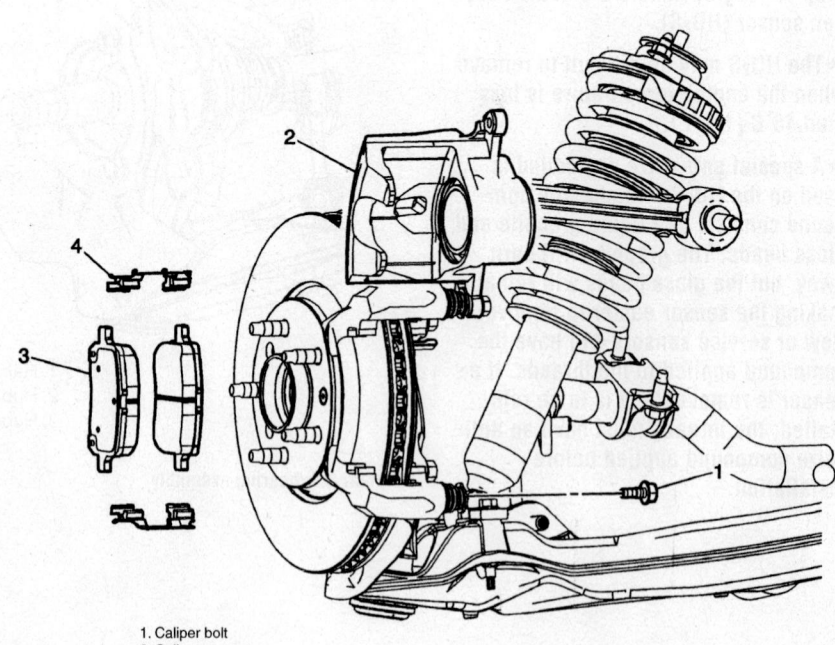

1. Caliper bolt
2. Caliper
3. Brake pads
4. Anti-rattle springs

06025-SOLS-G102

Front brake pads and related parts

➡**DO NOT reuse the old brake pad springs if replacing the brake pads, use new springs.**

11. Gradually apply pressure to the brake pedal until a firm pedal is felt. If a firm pedal is not achieved, wait 15 seconds and repeat until a firm pedal is obtained.

Caliper Anchor Plate

REMOVAL & INSTALLATION

1. Before servicing the vehicle, refer to the Precautions Section.
2. Raise and support the vehicle.
3. Remove the tire and wheel.
4. Remove the front caliper and support it out of the way. Do not disconnect the brake hose. Do not let the caliper hang by the hose.
5. Remove the brake caliper bracket bolts.
6. Remove the brake caliper bracket.
7. Installation is the reverse of removal. Torque the bracket bolts to 85 ft. lbs. (115 Nm). Apply lubricant to the front caliper guide pins ensuring adequate lubrication to both pin areas of the front brake caliper bracket with GM P/N 22688644 or equivalent.
8. Gradually apply pressure to the brake pedal until a firm pedal is felt. If a firm pedal is not achieved, wait 15 seconds and repeat until a firm pedal is obtained.

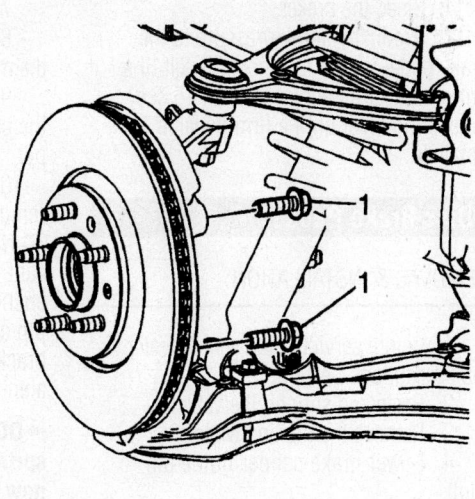

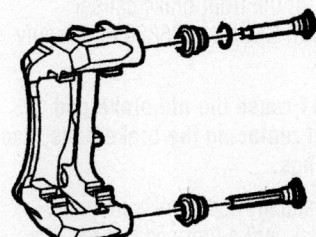

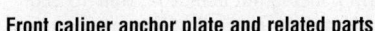

Front caliper anchor plate and related parts

06025-SOLS-G103

Rotor

REMOVAL & INSTALLATION

1. Before servicing the vehicle, refer to the Precautions Section.
2. Raise and support the vehicle.
3. Remove the tire and wheel.
4. Remove the anchor plate.
5. Remove the rotor retaining bolt.

➡**If the rotor is to be re-used, mark the position of the brake rotor to the wheel stud.**

6. Installation is the reverse of removal. Torque the rotor bolt to 89 inch lbs. (10 Nm).
7. Gradually apply pressure to the brake pedal until a firm pedal is felt. If a firm pedal is not achieved, wait 15 seconds and repeat until a firm pedal is obtained.

REAR BRAKES

Brake Caliper

REMOVAL & INSTALLATION

1. Before servicing the vehicle, refer to the Precautions Section.
2. Raise and support the vehicle.
3. Remove the tire and wheel.
4. Remove the brake hose bolt.
5. Remove the brake hose bolt washers.
6. Remove the brake hose.
7. Cap or plug the brake hose to prevent fluid loss and contamination.
8. Remove the brake caliper bolt.
9. Remove the brake caliper.
10. Installation is the reverse of removal. Torque the caliper mounting bolts to 30 Nm (22 ft. lbs.). Torque the brake hose bolt to 40 Nm (30 ft. lbs.)

❋❋ WARNING

Do not reuse the brake hose bolt washers.

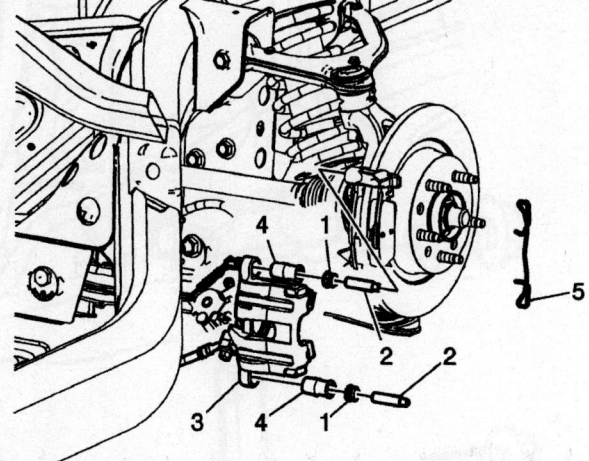

1. Guide pin seal
2. guide pin
3. Caliper
4. Bushing
5. Anti-rattle spring

06025-SOLS-G104

Rear brake caliper and related parts

11. Bleed the brakes.

12. Gradually apply pressure to the brake pedal until a firm pedal is felt. If a firm pedal is not achieved, wait 15 seconds and repeat until a firm pedal is obtained.

Disc Brake Pads

REMOVAL & INSTALLATION

1. Before servicing the vehicle, refer to the Precautions Section.

2. Raise and support the vehicle.

3. Remove the tire and wheel.

4. Lower brake caliper guide pin bolt.

5. Rotate the brake caliper up and forward until it rests on the brake caliper mounting bracket. The brake hose does not have to be removed from the brake caliper.

6. Remove the brake pads.

7. Remove the brake pad springs.

8. Remove about half of the fluid from the master cylinder.

9. Using an appropriate tool, force the caliper piston back into the caliper.

10. Installation is the reverse of removal. Torque the caliper bolt to 30 Nm (22 ft. lbs.). Apply brake lubricant to the front caliper guide pins ensuring adequate lubrication to both pin areas of the front brake caliper bracket with GM P/N 22688644 or equivalent.

➡**DO NOT reuse the old brake pad springs if replacing the brake pads, use new springs.**

11. Gradually apply pressure to the brake pedal until a firm pedal is felt. If a firm pedal is not achieved, wait 15 seconds and repeat until a firm pedal is obtained.

Caliper Anchor Plate

REMOVAL & INSTALLATION

1. Before servicing the vehicle, refer to the Precautions Section.

2. Raise and support the vehicle.

3. Remove the tire and wheel.

4. Remove the caliper and support it out of the way. Do not disconnect the brake hose. Do not let the caliper hang by the hose.

5. Remove the brake caliper bracket bolts.

6. Remove the brake caliper bracket.

7. Installation is the reverse of removal. Torque the bracket bolts to 85 ft. lbs. (115 Nm). Apply lubricant to the front caliper guide pins ensuring adequate lubrication to both pin areas of the front brake caliper bracket with GM P/N 22688644 or equivalent.

8. Gradually apply pressure to the brake pedal until a firm pedal is felt. If a firm pedal is not achieved, wait 15 seconds and repeat until a firm pedal is obtained.

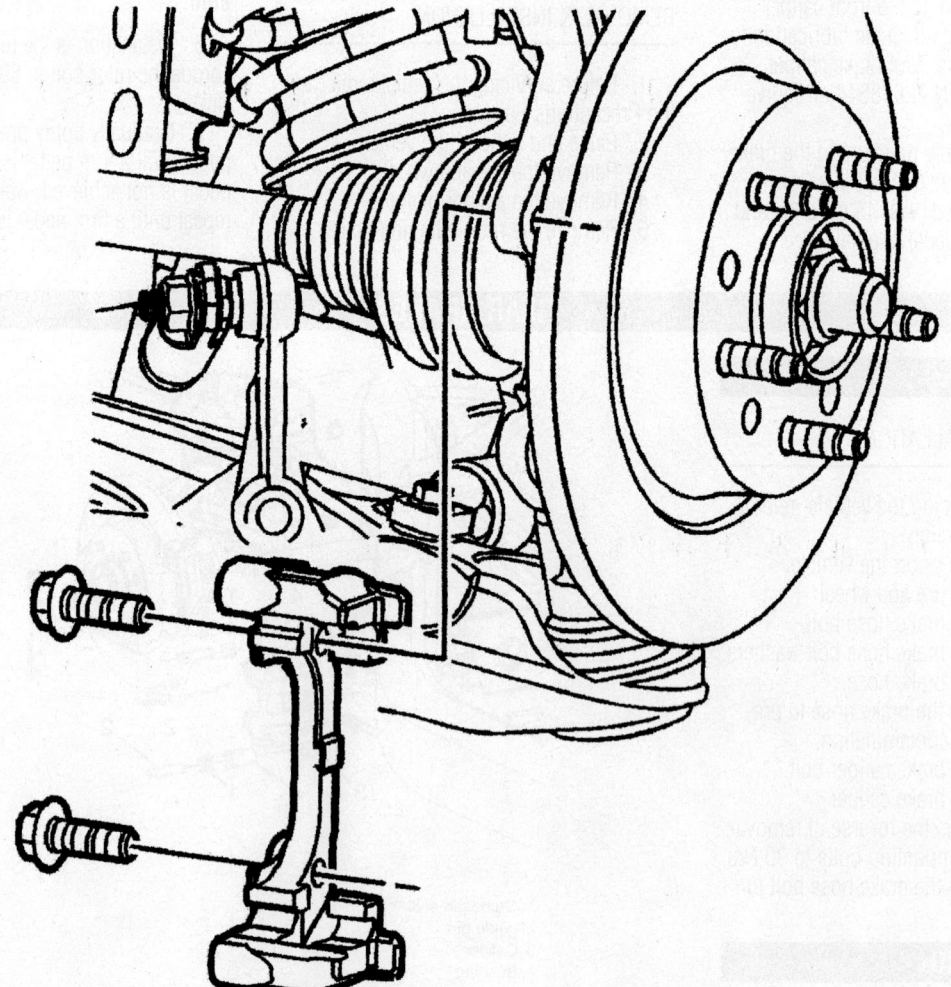

Rear caliper anchor plate

06025-SOLS-G105

Rotor

REMOVAL & INSTALLATION

1. Before servicing the vehicle, refer to the Precautions Section.

2. Raise and support the vehicle.
3. Remove the tire and wheel.
4. Remove the anchor plate.
5. Remove the rotor retaining bolt.

➥If the rotor is to be re-used, mark the position of the brake rotor to the wheel stud.

6. Installation is the reverse of removal. Torque the rotor bolt to 89 inch lbs. (10 Nm).

7. Gradually apply pressure to the brake pedal until a firm pedal is felt. If a firm pedal is not achieved, wait 15 seconds and repeat until a firm pedal is obtained.

BRAKE HYDRAULIC SYSTEM

Brake Bleeding

MANUAL BLEEDING

1. Before servicing the vehicle, refer to the Precautions Section.

✳✳ CAUTION

When adding fluid to the brake master cylinder reservoir, use only Delco Supreme 11®, GM P/N 12377967 (Canadian P/N 992667), or equivalent DOT-3 brake fluid from a clean, sealed brake fluid container. The use of any type of fluid other than the recommended type of brake fluid, may cause contamination which could result in damage to the internal rubber seals and/or rubber linings of hydraulic brake system components.

2. Place a clean shop cloth beneath the brake master cylinder to prevent brake fluid spills.

3. With the ignition OFF and the brakes cool, apply the brakes 3-5 times, or until the brake pedal effort increases significantly, in order to deplete the brake booster power reserve.

4. If you have performed a brake master cylinder bench bleeding on this vehicle, or if you disconnected the brake pipes from the master cylinder, or disconnected the brake pipes from the proportioning valve assembly or the brake modulator assembly, you must perform the following steps to bleed air at the ports of the hydraulic component:

   a. Ensure that the brake master cylinder reservoir is full to the maximum-fill level. If necessary, add GM approved, or equivalent DOT-3 brake fluid from a clean, sealed brake fluid container. If removal of the reservoir cap and diaphragm is necessary, clean the outside of the reservoir on and around the cap prior to removal.

   b. With the brake pipes installed securely to the master cylinder, proportioning valve assembly, or brake modulator assembly, loosen and separate one of the brake pipes from the port of the component. For the proportioning valve assembly or the brake modulator assembly perform these steps in the sequence of system flow; begin with the fluid feed pipes from the master cylinder.

   c. Allow a small amount of brake fluid to gravity bleed from the open port of the component.

   d. Reconnect the brake pipe to the component port and tighten securely.

   e. Have an assistant slowly depress the brake pedal fully and maintain steady pressure on the pedal.

   f. Loosen the same brake pipe to purge air from the open port of the component.

   g. Tighten the brake pipe, then have the assistant slowly release the brake pedal.

   h. Wait 15 seconds, then repeat steps 3-7 until all air is purged from the same port of the component.

   i. With the brake pipe installed securely to the master cylinder, proportioning valve assembly, or brake modulator assembly-after all air has been purged from the first port of the component that was bled-loosen and separate the next brake pipe from the component, then repeat steps 3-8 until each of the ports on the component have been bled.

   j. After completing the final component port bleeding procedure, ensure that each of the brake pipe-to-component fittings are properly tightened.

5. Fill the brake master cylinder reservoir with GM approved, or equivalent DOT-3 brake fluid from a clean, sealed brake fluid container. Ensure that the brake master cylinder reservoir remains at least half-full during this bleeding procedure. Add fluid as needed to maintain the proper level. Clean the outside of the reservoir on and around the reservoir cap prior to removing the cap and diaphragm.

6. Install a proper box-end wrench onto the RIGHT REAR wheel hydraulic circuit bleeder valve.

7. Install a transparent hose over the end of the bleeder valve.

8. Submerge the open end of the transparent hose into a transparent container partially filled with GM approved, or equivalent DOT-3 brake fluid from a clean, sealed brake fluid container.

9. Have an assistant slowly depress the brake pedal fully and maintain steady pressure on the pedal.

10. Loosen the bleeder valve to purge air from the wheel hydraulic circuit.

11. Tighten the bleeder valve, then have the assistant slowly release the brake pedal.

12. Wait 15 seconds, then repeat steps 8-10 until all air is purged from the same wheel hydraulic circuit.

13. With the right rear wheel hydraulic circuit bleeder valve tightened securely, after all air has been purged from the right rear hydraulic circuit, install a proper box-end wrench onto the LEFT FRONT wheel hydraulic circuit bleeder valve.

14. Install a transparent hose over the end of the bleeder valve, then repeat steps 7-11.

15. With the left front wheel hydraulic circuit bleeder valve tightened securely, after all air has been purged from the left front hydraulic circuit, install a proper box-end wrench onto the LEFT REAR wheel hydraulic circuit bleeder valve.

16. Install a transparent hose over the end of the bleeder valve, then repeat steps 7-11.

17. With the left rear wheel hydraulic circuit bleeder valve tightened securely, after all air has been purged from the left rear hydraulic circuit, install a proper box-end wrench onto the RIGHT FRONT wheel hydraulic circuit bleeder valve.

18. Install a transparent hose over the end of the bleeder valve, then repeat steps 7-11.

19. After completing the final wheel hydraulic circuit bleeding procedure, ensure that each of the 4 wheel hydraulic circuit bleeder valves are properly tightened.

20. Fill the brake master cylinder reservoir to the maximum-fill level with GM approved, or equivalent DOT-3 brake fluid from a clean, sealed brake fluid container.

21. Slowly depress and release the brake pedal. Observe the feel of the brake pedal.

➡ **If it is determined that air was induced into the system upstream of the ABS modulator prior to servicing, the ABS Automated Bleed Procedure must be performed.**

22. If the brake pedal feels spongy, repeat the bleeding procedure again. If the brake pedal still feels spongy after repeating

the bleeding procedure, perform the following steps:

    a. Inspect the brake system for external leaks.

    b. Pressure bleed the hydraulic brake system in order to purge any air that may still be trapped in the system.

23. Turn the ignition key ON, with the

engine OFF. Check to see if the brake system warning lamp remains illuminated.

✳✳ WARNING

DO NOT allow the vehicle to be driven until it is diagnosed and repaired.

CADILLAC

SRX

27

BRAKES **27-52**
DRIVE TRAIN **27-36**
ENGINE REPAIR **27-9**
FUEL SYSTEM **27-35**
SPECIFICATIONS AND
  MAINTENANCE CHARTS......27-2
Engine and Vehicle
  Identification 27-2
General Engine Specifications 27-2
Gasoline Engine Tune-Up
  Specifications 27-2
Accessory Drive Belt Routing 27-3
Capacities 27-4
Valve Specifications...............27-4
Crankshaft and Connecting
  Rod Specifications.................27-4
Piston and Ring Specifications 27-5
Torque Specifications 27-5
Wheel Alignment 27-6
Tire, Wheel and Wheel
  Specifications 27-6
Brake Specifications 27-6
Scheduled Maintenance
  Intervals.........................27-7
STEERING AND
  SUSPENSION **27-42**
A
Air Bag (Supplemental Restraint)
System.........................27-42
  Disabling And Enabling The
    System 27-43
Alternator.........................27-9
  Removal & Installation............27-9
B
Ball Joints and Steering Knuckle .27-50
  Removal & Installation............27-50
Brake Caliper 27-52
  Removal & Installation............27-52
C
Camshafts.........................27-19
  Removal & Installation............27-19
Coil Springs 27-48
  Removal & Installation............27-48

CV-Joint 27-38
  Overhaul 27-38
Cylinder Head 27-14
  Removal & Installation............27-14
D
Disc Brake Pads.....................27-52
  Removal & Installation............27-52
E
Engine Assembly 27-10
  Removal & Installation............27-10
Exhaust Manifold 27-18
  Removal & Installation............27-18
F
Front Differential 27-41
  Removal & Installation............27-41
Front Output Shaft Seal 27-41
  Removal & Installation............27-41
Front Pinion Seal 27-41
  Removal & Installation............27-41
Front Suspension Frame.............27-47
  Removal & Installation............27-47
Fuel Filter 27-35
  Removal & Installation............27-35
Fuel Injectors 27-35
  Removal & Installation............27-35
Fuel Pump 27-35
  Removal & Installation............27-35
Fuel System Pressure 27-35
  Relieving 27-35
H
Halfshafts.........................27-37
  Removal & Installation............27-37
Heater Core 27-13
  Removal & Installation............27-13
Hub and Wheel Bearing 27-51
  Removal & Installation............27-51
I
Intake Manifold.....................27-17
  Removal & Installation............27-17
L
Lower Control Arm 27-49
  Removal & Installation............27-49

O
Oil Pan.........................27-25
  Removal & Installation............27-25
Oil Pump 27-26
  Removal & Installation............27-26
P
Piston and Ring 27-34
  Positioning 27-34
R
Rack and Pinion Steering Gear 27-47
  Removal & Installation............27-47
Rear Differential 27-41
  Removal & Installation............27-41
Rear Main Seal 27-27
  Removal & Installation............27-27
Rear Pinion Seal 27-42
  Removal & Installation............27-42
Rocker Arms and Valve Lifters.....27-17
  Removal & Installation............27-17
S
Shock Absorber Module 27-48
  Disassembly 27-48
  Reassembly.........................27-48
  Removal & Installation............27-48
Stabilizer Bar 27-49
  Removal & Installation............27-49
Starter Motor 27-24
  Removal & Installation............27-24
T
Timing Chain, Sprockets, Front
  Cover and Seal 27-28
  Removal & Installation............27-28
Transfer Case 27-36
  Removal & Installation............27-36
Transmission 27-36
  Removal & Installation............27-36
U
Upper Control Arm 27-50
  Removal & Installation............27-50
W
Water Pump 27-12
  Removal & Installation............27-12

SPECIFICATION AND MAINTENANCE CHARTS

VEHICLE AND ENGINE IDENTIFICATION CHART

Engine							Model Year	
Code	Liters	Cu. In.	Cyl.	Fuel Sys.	Engine Type	Eng. Mfg.	Code	Year
7	3.6	217	6	SEFI	DOHC	GM	4	2004
A	4.6	279	8	SEFI	DOHC	GM	5	2005

SEFI: Sequential Electronic Fuel Injection

06025-SRX-C01

GENERAL ENGINE SPECIFICATIONS

Year	Engine Displacement Liters	Engine VIN	Net Horsepower @ rpm	Net Torque @ rpm (ft. lbs.)	Bore x Stroke (in.)	Compression Ratio	Oil Pressure @ rpm
2004	3.6	7	260@6500	252@2800	3.70x3.37	10.2:1	20@2000
	4.6	A	320@6400	315@4400	3.66x3.30	10.5:1	35@2000
2005	3.6	7	255@6500	252@2800	3.70x3.37	10.2:1	20@2000
	4.6	A	320@6400	315@4400	3.66x3.30	10.5:1	35@2000

06025-SRX-C02

GASOLINE ENGINE TUNE-UP SPECIFICATIONS

Year	Engine Displacement Liters	Engine VIN	Spark Plugs Gap (in.)	Ignition Timing (deg.) MT	Ignition Timing (deg.) AT	Fuel Pump (psi)	Idle Speed (rpm) MT	Idle Speed (rpm) AT	Valve Clearance In.	Valve Clearance Ex.
2004	3.6	7	0.044	—	①	55-60	—	②	HYD	HYD
	4.6	A	0.051	—	①	55-60	—	②	HYD	HYD
2005	3.6	7	0.044	—	①	55-60	—	②	HYD	HYD
	4.6	A	0.040	—	①	55-60	—	②	HYD	HYD

NOTE: The Vehicle Emission Control Information label often reflects specification changes changes made during production.

The label figures must be used if they differ from those in this chart.

HYD: Hydraulic

① Controlled by the Powertrain Control Module (PCM) and cannot be manually adjusted.

② 600 with A/C off, 700 with A/C on.

06025-SRX-C03

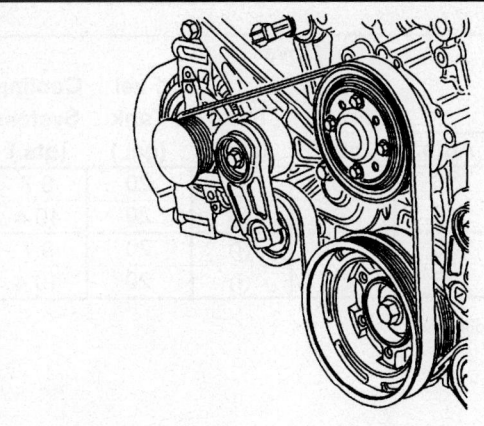

71461-SRX-G01

Engine accessory drive belt routing—3.6L crankshaft, alternator and water pump belt

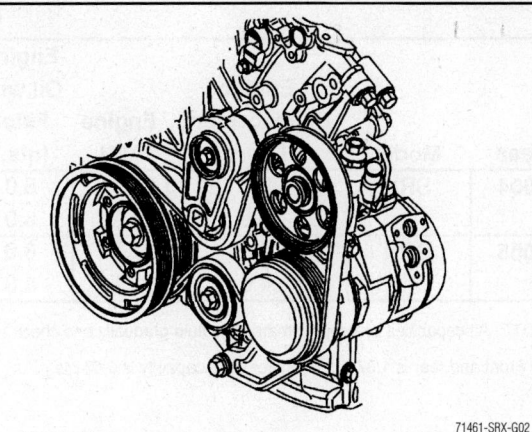

71461-SRX-G02

Engine accessory drive belt routing—3.6L crankshaft, A/C compressor, tensioner and power steering pump belt

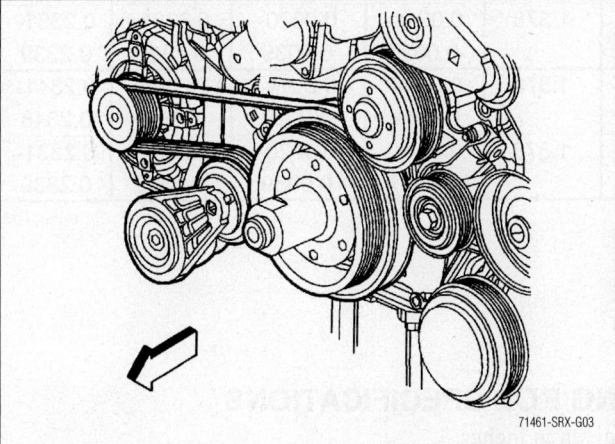

71461-SRX-G03

Engine accessory drive belt routing—4.6L crankshaft, alternator and tensioner belt

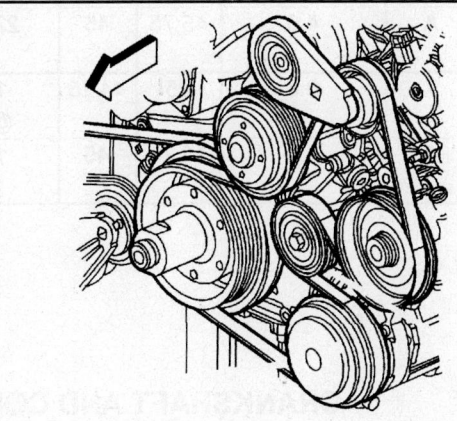

71461-SRX-G04

Engine accessory drive belt routing—4.6L crankshaft, A/C compressor, tensioner, idler and power steering pump belt

CAPACITIES

Year	Model	Engine Displacement Liters	Engine VIN	Engine Oil with Filter (qts.)	Transmission (pts.) 4-Spd	5-Spd	Auto.	Drive Axle Front (pts.)	Rear (pts.)	Fuel Tank (gal.)	Cooling System (qts.)
2004	SRX	3.6	7	6.0	—	—	18	①	①	20	9.7
		4.6	A	8.0	—	—	18	①	①	20	10.4
2005	SRX	3.6	7	6.0	—	—	18	①	①	20	9.7
		4.6	A	8.0	—	—	18	①	①	20	10.4

NOTE: All capacities are approximate. Add fluid gradually and check to be sure a proper fluid level is obtained.

① Front and rear is 1.37 qts. Transfer case capacity is 0.53 qts.

06025-SRX-C04

VALVE SPECIFICATIONS

Year	Engine VIN	Engine Displacement Liters	Seat Angle (deg.)	Face Angle (deg.)	Spring Test Pressure (lbs. @ in.)	Spring Installed Height (in.)	Stem-to-Guide Clearance (in.) Intake	Exhaust	Stem Diameter (in.) Intake	Exhaust
2004	7	3.6	45	44.25	224@1.16	1.377	0.0010-0.0026	0.0014-0.0030	0.2344-0.2352	0.2341-0.2348
	A	4.6	45.75	45	224@1.16	1.378	0.0011-0.0027	0.0020-0.0039	0.2331-0.2339	0.2331-0.2339
2005	7	3.6	45	44.25	134-139 @0.9449	1.378	0.0010-0.0026	0.0014-0.0030	0.2344-0.2352	0.2341-0.2348
	A	4.6	45.75	45	130-142 @0.965	1.378	0.0011-0.0027	0.0020-0.0039	0.2331-0.2339	0.2331-0.2339

06025-SRX-C05

CRANKSHAFT AND CONNECTING ROD SPECIFICATIONS

All measurements are given in inches.

Year	Engine Displ. Liters	Engine VIN	Crankshaft Main Brg. Journal Dia.	Main Brg. Oil Clearance	Shaft End-play	Thrust on No.	Connecting Rod Journal Diameter	Oil Clearance	Side Clearance
2004	3.6	7	2.6768-2.6775	0.0004-0.0024	0.0039-0.0130	NA	2.2044-2.2050	0.0004-0.0028	0.0374-0.0140
	4.6	A	2.5335-2.5341	0.0006-0.0022	0.0020-0.0197	NA	2.1239-2.1245	0.0010-0.0014	0.0079-0.0197
2005	3.6	7	2.6768-2.6775	0.0004-0.0024	0.0039-0.0130	NA	2.2044-2.2050	0.0004-0.0028	0.0374-0.0140
	4.6	A	2.5335-2.5341	0.0006-0.0022	0.0020-0.0197	NA	2.1239-2.1245	0.0010-0.0014	0.0079-0.0197

NA: Not available

06025-SRX-C06

PISTON AND RING SPECIFICATIONS
All measurements are given in inches.

Year	Engine Displ. Liters	Engine VIN	Piston Clearance	Ring Gap			Ring Side Clearance		
				Top Comp.	Bottom Comp.	Oil Control	Top Comp.	Bottom Comp.	Oil Control
2004	3.6	7	0.0008-0.0013	0.0059-0.0118	0.0110-0.0189	0.00989-0.0295	0.0012-0.0026	0.0002-0.0013	0.0083-0.0155
	4.6	A	0.0008-0.0020	0.0098-0.0157	0.0138-0.0200	0.0098-0.0299	0.0016-0.0037	0.0016-0.0037	Snug
2005	3.6	7	0.0010-0.0021	0.0059-0.0118	0.0110-0.0189	0.0059-0.0236	0.0012-0.0026	0.0006-0.0024	0.0012-0.0067
	4.6	A	0.0008-0.0020	0.0098-0.0157	0.0138-0.0200	0.0098-0.0299	0.0016-0.0037	0.0016-0.0037	Snug

06025-SRX-C07

TORQUE SPECIFICATIONS
All readings in ft. lbs.

Year	Engine VIN	Engine Displacement Liters	Cylinder Head Bolts	Main Bearing Bolts	Rod Bearing Bolts	Crankshaft Damper Bolts	Flywheel Bolts	Manifold		Spark Plugs	Oil Pan Drain Plug
								Intake	Exhaust		
2004	7	3.6	①	②	③	④	⑤	17	15	13	18
	A	4.6	⑥	⑦	⑧	⑨	⑩	⑪	15	11	18
2005	7	3.6	①	②	③	④	⑤	17	15	13	18
	A	4.6	⑥	⑦	⑧	⑨	⑩	⑪	15	11	18

① M8 bolt step 1: 10 ft. lbs.
   Step 2: plus 60 degrees
   M11 bolt step 1: 33 ft. lbs.
   Step 2: plus 120 degrees

② Inner bolt step 1: 15 ft. lbs.
   Step 2: plus 80 degrees
   Outer bolt step1: 10 ft. lbs.
   Step 2: plus 110 degrees

③ Step 1: 22 ft. lbs.
   Step 2: loosen to zero degrees
   Step 3: 18 ft. lbs.
   Step 4: plus 110 degrees

④ Step 1: 74 ft. lbs.
   Step 2: plus 150 degrees

⑤ Step 1: 22 ft. lbs.
   Step 2: plus 150 degrees

⑥ M6 bolt: 106 inch lbs.
   M11 bolt step 1: 22 ft. lbs.
   Step 2: plus 60 degrees
   Step 3: plus 60 degrees
   Step 4: plus 60 degrees

⑦ M8 bolt: 22 ft. lbs.
   M10 bolt step 1: 15 ft. lbs.
   Step 2: plus 65 degrees

⑧ Step 1: 22 ft. lbs.
   Step 2: loosen to zero degrees
   Step 3: 18 ft. lbs.
   Step 4: plus 100 degrees

⑨ Step 1: 37 ft. lbs.
   Step 2: plus 150 degrees

⑩ Step 1: 11 ft. lbs.
   Step 2: plus 50 degrees

⑪ 89 inch lbs.

06025-SRX-C08

WHEEL ALIGNMENT

Year	Model		Caster Range (+/-Deg.)	Caster Preferred Setting (Deg.)	Camber Range (+/-Deg.)	Camber Preferred Setting (Deg.)	Toe-in (Deg.)
2004	SRX	Front	0.50	+4.10	0.50	-0.50	0.20+/-0.20
		Rear	—	—	0.75	-1.00	-0.40+/-0.10
2005	SRX	Front	0.60	+4.10	0.60	-0.50	0.20+/-0.20
		Rear	—	—	0.50	-1.00	-0.20+/-0.20

06025-SRX-C09

TIRE AND WHEEL SPECIFICATIONS

Year	Model	OEM Tires Front	OEM Tires Rear	Tire Pressures (psi) Front	Tire Pressures (psi) Rear	Wheel Size	Lug Nut (ft. lbs.)
2004	SRX-V6	P235/65R17	P255/60R17	①	①	②	100
	SRX-V8	P235/60R18	P255/55R18	①	①	②	100
2005	SRX-V6	P235/65R17	P255/60R17	①	①	②	100
	SRX-V8	P235/60R18	P255/55R18	①	①	②	100

① See vehicle tire placard.

② Not available

OEM: Original Equipment Manufacturer

PSI: Pounds Per Square Inch

06025-SRX-C10

BRAKE SPECIFICATIONS

All measurements in inches unless noted

Year	Model		Brake Disc Original Thickness	Brake Disc Minimum Thickness	Brake Disc Maximum Runout	Minimum Lining Thickness Front	Minimum Lining Thickness Rear	Brake Caliper Bracket Bolts (ft. lbs.)	Brake Caliper Mounting Bolts (ft. lbs.)
2004	SRX	F	1.270	1.210	0.002	①	—	96	25
		R	1.020	0.944	0.002	—	①	88	44
2005	SRX	F	1.267	1.209	0.002	①	—	96	25
		R	1.020	0.944	0.002	—	①	88	44

① Not available

06025-SRX-C11

MAINTENANCE I AND II SERVICE SCHEDULES
Cadillac SRX

When the CHANGE ENGINE OIL light appears, certain services and inspections are required.

Required services are described as Maintenance I and Maintenance II.

The first service on a vehicle should be Maintenance I, and the second service should be Maintenance II.

Alternate between the 2 thereafter. However, in some cases, Maintenance II may be required more often.

Maintenance I: Use Maintenance I if the CHANGE ENGINE OIL light comes on within 10 months since vehicle was purchased or, if Maintenance II was performed.

Maintenance II: Use Maintenance II if the previous service performed was Maintenance I. Always use Maintenance I whenever the CHANGE ENGINE OIL light comes on 10 months or more since the last service, or, if the CHANGE ENGINE OIL light has notcome on at all for one year.

Service	Maintenance I	Maintenance II
Change the engine oil and filter. Reset the oil life system.	✓	✓
Visually inspect the vehicle for leaks or damage. A fluid loss in the vehicle system could indicate a problem. Inspected, repair and add fluid to the system if necessary.	✓	✓
Inspect the engine air cleaner filter. If necessary, replace the filter.	✓	✓
Rotate the tires. Inspect the tire inflation pressures and the tire wear.	✓	✓
Visually inspect the brake lines and hoses for proper hook-up, binding, leaks, cracks, chafing, etc. Inspect the disc brake pads for wear and the rotors for surface condition. Inspect the drum brake linings for wear or cracks. Inspect other brake parts, including drums, wheel cylinders, calipers, parking brake, etc. Inspect the parking brake adjustment.	✓	✓
Inspect the engine coolant and the windshield washer fluid levels. Add fluid as needed.	✓	✓
Inspect the suspension and steering components. Inspect the front and rear suspension and the steering system for damaged, loose or missing parts, or signs of wear. Inspect the power steering lines and the hoses for proper hook-up, binding, leaks, cracks,	--	✓
Visually inspect the coolant hoses and replace the hoses if they are cracked, swollen or deteriorated. Inspect all pipes, fittings and clamps; replace with GM parts as needed. To help ensure proper operation, a pressure test of the cooling system and pressure cap and cleaning the outside of the radiator and air conditioning condenser is recommended at least once a year.	--	✓
Inspect the front and rear suspension and the steering system for damaged, loose or missing parts, or signs of wear. Inspect power steering lines and hoses for proper hook-up, binding, leaks, cracks, chafing, etc.	--	✓
Inspect the throttle system for interference or binding and for damaged or missing parts. Replace the parts as needed. Replace any components that have high effort or excessive wear. Do not lubricate the accelerator or the cruise control cables.	--	✓
Replace the passenger compartment air filter.	--	✓

Press the CLR button located to the right of the DIC display to acknowledge the Change Engine Oil message. This will clear the message from the display and reset it.

To reset the oil life indicator, use the following steps:

1. Press the up or down arrow on the INFO button located to the right of the DIC display to access the DIC menu.

2. Once XXX% ENGINE OIL LIFE menu item is highlighted, press and hold the CLR button.

The percentage will return to 100, and the oil life indicator will be reset.

3. Turn the key to OFF.

06025-SRX-C12

ADDITIONAL MAINTENANCE SERVICES
Cadillac SRX

TO BE SERVICED	TYPE OF SERVICE	VEHICLE MILEAGE INTERVAL (x1000)					
		25	50	75	100	125	150
Air cleaner filter	R		✓		✓		✓
Accessory drive belt	I						✓
Auto. Trans. Fluid and Filter①	R		✓		✓		✓
Cooling system hoses and clamps	S/I						✓
Engine coolant	R						✓
Fuel system	I	✓	✓	✓	✓	✓	✓
Exhaust system & heat shields	S/I	✓	✓	✓	✓	✓	✓
Supercharger oil level	S/I	✓	✓	✓	✓	✓	✓
Spark plugs and wires	R				✓		

R: Replace   S/I: Inspect and service, if necessary

① Replace if any of the following conditions are met:

    Heavy city traffic where the outside temperature regularly reaches 32°C (90°F) or higher

    Hilly or mountainous terrain

    Frequent trailer towing

    Taxi, police or delivery service

    Otherwise, change every 100,000 miles

06025-SRX-C13

ENGINE REPAIR

Alternator

REMOVAL & INSTALLATION

3.6L Engine

1. Before servicing the vehicle, refer to the Precautions Section.
2. Disconnect the negative battery cable.
3. Remove the accessory drive belt.
4. Raise and support the vehicle.
5. Disconnect the alternator wiring connector.
6. Remove the positive cable nut.
7. Remove the alternator.

To install:

8. Position the alternator on the engine.
9. Install the alternator mounting bolts. Tighten the bolts to 37 ft. lbs. (50 Nm).
10. Tighten the positive cable nut to 89 inch lbs. (10 Nm).
11. Connect the wiring connector.
12. Install and tension the accessory drive belt.
13. Connect the negative battery cable.

4.6L Engine

2WD MODELS

1. Before servicing the vehicle, refer to the Precautions Section.
2. Disconnect the negative battery cable.

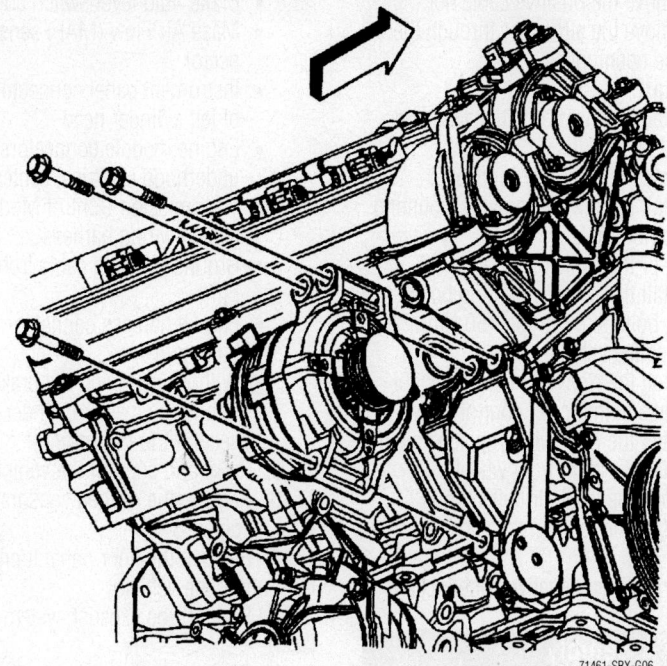

Alternator mounting—4.6L engine

3. Remove the accessory drive belt.
4. Remove the alternator upper mounting bolts.
5. Raise and support the vehicle.
6. Remove the front air deflector.
7. Remove the lower alternator mounting bolt.
8. Disconnect the alternator wiring connector.

9. Remove the positive cable nut.
10. Remove the alternator.

To install:

11. Install alternator positive lead and tighten to 111 inch lbs. (13 Nm).
12. Connect the wiring connector.
13. Install the lower alternator mounting bolt but do not tighten.
14. Lower the vehicle.
15. Install the upper mounting bolts. Tighten all bolts to 37 ft. lbs. (50 Nm).
16. Install the front air deflector.
17. Install and tension the accessory drive belt.
18. Connect the negative battery cable.

AWD MODELS

1. Before servicing the vehicle, refer to the Precautions Section.
2. Disconnect the negative battery cable.
3. Remove the accessory drive belt.
4. Remove the alternator upper mounting bolts.
5. Raise and support the vehicle.
6. Remove the front air deflector.
7. Remove the right front wheel.
8. Remove the right wheel splash shield
9. Remove the right and left front stabilizer bar links at the lower control arms.
10. Rotate the stabilizer bar down enough to access the alternator.
11. Remove the lower alternator mounting bolt.

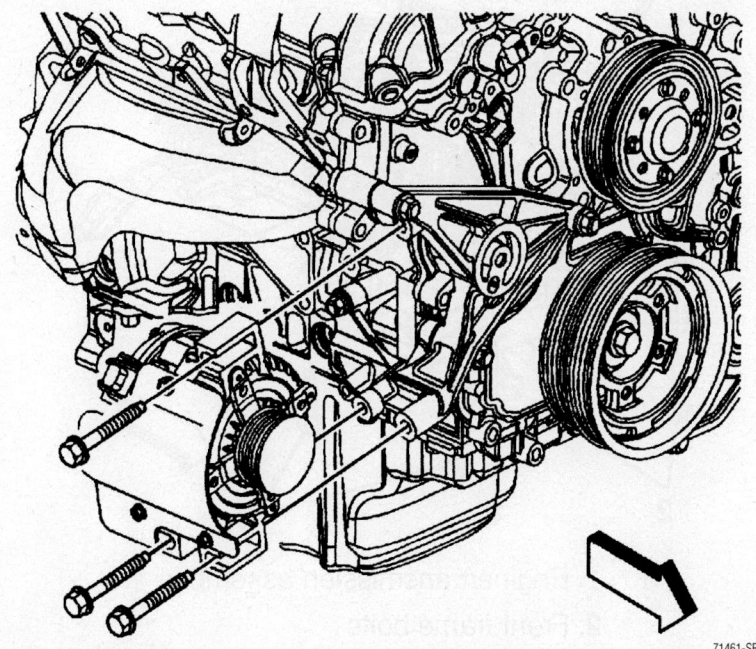

Alternator mounting—3.6L engine

12. Disconnect the alternator wiring connector.

13. Remove the positive cable nut.

14. Remove the alternator through the wheelhouse opening.

To install:

15. Install alternator positive lead and tighten to 111 inch lbs. (13 Nm).

16. Connect the wiring connector.

17. Install the lower alternator mounting bolt but do not tighten.

18. Lower the vehicle.

19. Install the upper mounting bolts. Tighten all bolts to 37 ft. lbs. (50 Nm).

20. Raise the vehicle.

21. Install the right and left front stabilizer bar links at the lower control arms.

22. Install the right wheel splash shield

23. Install the right front wheel.

24. Install the front air deflector.

25. Install and tension the accessory drive belt.

26. Connect the negative battery cable.

Engine Assembly

REMOVAL & INSTALLATION

3.6L Engine

➡The front wheels must be in the straight ahead position and the steering column locked before disconnect the intermediate shaft. Failure to do so may result in damage to the Supplemental Restraint System (SRS) coil.

1. Before servicing the vehicle, refer to the Precautions Section.

2. Relieve the fuel system pressure.

3. Drain the engine coolant.

4. Recover the air conditioning refrigerant, into a refrigerant recovery station

5. Center the steering wheel.

6. Install Steering Column Anti-rotation pin J-42640 to lock the steering column.

7. Disconnect the battery cables from the battery and the body.

8. Remove or disconnect the following:
- Battery
- Fuel injector shield
- Air cleaner duct
- Cooling fan connectors
- Surge tank hoses
- Heater hoses
- Purge solenoid line
- Fuel line from fuel rail
- Wiper module
- A/C suction hose from evaporator
- Suction hose bracket
- A/C pressure switch connector
- Radiator support brackets

- Brake booster check valve and vacuum hose
- Brake fluid level switch connector
- Mass Air Flow (MAF) sensor connector
- Instrument panel connector at rear of left cylinder head
- Engine module connectors from underhood electrical center
- Transmission Control Module (TCM) wiring harness
- Ground bolt and cable from frame rail
- Engine harness connector from frame rail
- Without removing the brake lines, unbolt the master cylinder and secure it to the engine

9. Raise and support the vehicle.

10. Remove the oxygen sensors from the exhaust pipes.

11. Remove the floor panel tunnel brace from under the vehicle.

12. Support the exhaust system with a jack.

13. Disconnect the exhaust pipes from the exhaust manifold.

14. Remove the front and rear exhaust hangars from the frame.

15. With the help of an assistant, remove the exhaust system.

16. Mark the driveshaft-to-transmission flange and differential flange locations and remove the driveshaft.

17. Remove the front air deflector.

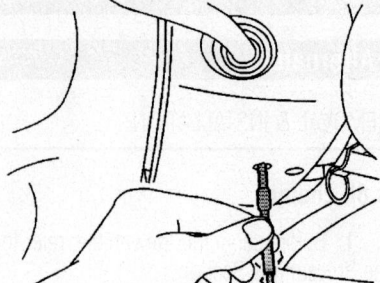

71461-SRX-G07

Installing steering column anti-rotation pin

18. Remove the washer bottle bracket, but not the washer bottle.

19. Remove the radiator side air baffles.

20. Disconnect the front brake pipe retainers.

21. Disconnect the 2 center pipes from the brake proportion modulator valve and cap the openings.

22. Remove the front wheels.

23. Remove the upper to center intermediate steering shaft bolt.

24. Remove the lower intermediate steering shaft-to-steering gear bolt.

25. Remove the center intermediate steering shaft with the lower shaft attached.

26. Remove the lower engine mount nuts.

27. Disconnect the transmission shift linkage.

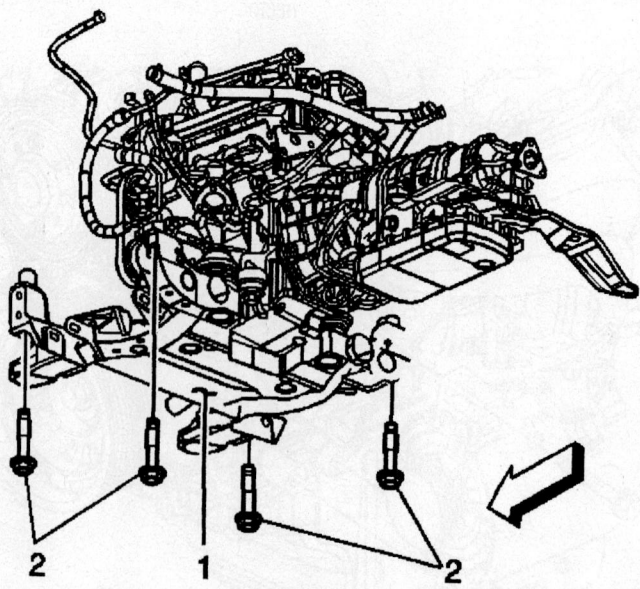

1. Engine/transmission assembly

2. Front frame bolts

71461-SRX-G08

Removing the front frame mounting bolts—3.6L engine

28. Disconnect the oil level sensor connector.

29. Remove the headlight leveling sensors.

30. Secure the shock modules to the lower control arms with a suitable strap to avoid damage to the brake lines.

31. Remove the shock yoke.

32. Remove the left and right shock module upper mounting nuts.

33. Raise the vehicle enough to place a suitable engine lift table under the engine, transmission, front frame and front suspension assembly.

34. Lower the vehicle or raise the lift until the engine assembly is supported by the lift.

35. Remove the transmission brace-to-underbody bolts.

36. Remove the 4 front frame bolts.

37. With the aid of an assistant, remove the engine, transmission, front frame and front suspension assembly from the vehicle.

38. If the engine itself is to be serviced, the engine will have to be separated from the transmission and the front frame and suspension assembly.

To install:

39. With the aid of an assistant, raise the table and/or lift the vehicle to install the engine, transmission, front frame and front suspension assembly to the vehicle.

40. Install the front frame bolts. Tighten the bolts to 141 ft. lbs. (191 Nm).

41. Install the transmission support to underbody bolts. Tighten the bolts 44 ft. lbs. (60 Nm).

42. Remove the powertrain lift/support table.

43. Install the right and left shock module upper mounting bolts. Tighten the bolts 83 ft. lbs. (112 Nm).

44. Install the headlamp leveling sensors.

45. Connect the transmission shift linkage to the transmission.

46. Connect the low oil level sensor electrical connector.

47. Install the lower engine mount nuts. Tighten the nut to 59 ft. lbs. (80 Nm).

48. Install the lower and intermediate steering shafts. Tighten the bolts to 23 ft. lbs. (80 Nm).

49. Install the front tire and wheel assemblies.

50. Install the front brake pipes and retainers to the underbody.

51. Connect the rear brake pipes (two center pipes) to the brake pressure modulator valve (BPVM).

52. Connect the radiator side air baffles to the radiator.

53. Install the washer bottle bracket.

54. Install the air deflector.

55. Install the propeller shaft using the reference marks previously made. Tighten the bolts to 63 ft. lbs. (85 Nm).

56. Install the exhaust system. Tighten the exhaust pipe-to-manifold nuts to 22 ft. lbs. (30 Nm).

57. Install the floor tunnel brace. Tighten the bolts to 18 ft. lbs. (25 Nm).

58. Install the master cylinder.

59. Connect the engine harness electrical connector to the frame rail.

60. Install the ground wire and bolt to the longitudinal rail. Tighten the bolt to 89 inch lbs. (10 Nm).

61. Connect the wiring harness to the TCM.

62. Connect the engine module wiring harness connectors to the underhood electrical center.

63. Connect and lock the instrument panel electrical connector to the engine at the rear of the left cylinder head.

64. Connect the mass air flow sensor electrical connector.

65. Connect the brake fluid level switch electrical connector from the master cylinder.

66. Connect the brake booster vacuum hose.

67. Install the radiator support brackets.

68. Connect the purge line to the purge solenoid.

69. Connect the fuel pipe to the fuel rail.

70. Connect the heater hoses to the heater core.

71. Install the air inlet duct.

72. Position the surge tank inlet hose to the vehicle.

73. Connect the surge tank inlet hose to the water outlet housing and the radiator.

74. Connect the surge tank outlet hose to the surge tank.

75. Connect the A/C pressure switch electrical connector and the liquid line to the evaporator.

76. Connect the air conditioning suction hose to the evaporator and install the suction hose bracket to the shock tower.

77. Install the cooling fan wiring harnesses to the fan shroud.

78. Install the cooling fan electrical connectors.

79. Install the wiper module.

80. Install the fuel injector sight shield.

81. Connect the battery cables.

82. Connect the battery negative cable from the battery and the body

83. Remove the locking pin from the steering column.

84. Bleed the brake rear circuits

85. Refill the engine, transaxle and cooling system with the correct amount of the appropriate fluids before starting the engine. Recharge the A/C system using approved recycling equipment.

4.6L Engine

➡**The front wheels must be in the straight ahead position and the steering column locked before disconnect the intermediate shaft. Failure to do so may result in damage to the Supplemental Restraint System (SRS) coil.**

1. Before servicing the vehicle, refer to the Precautions Section.

2. Relieve the fuel system pressure.

3. Drain the engine coolant.

4. Recover the air conditioning refrigerant, into a refrigerant recovery station

5. Center the steering wheel.

6. Install Steering Column Anti-rotation pin J-42640 to lock the steering column.

7. Disconnect the battery cables from the battery and wire them to the engine.

8. Remove or disconnect the following:
- Cross vehicle brace
- Fuel injector shield
- Air cleaner assembly
- Surge tank hoses
- A/C suction hose fitting on shock tower
- A/C liquid hose from condenser
- Brake booster vacuum hose
- Brake fluid level switch connector
- Without removing the brake lines, unbolt the master cylinder and secure it to the engine
- Fuel line retainer
- Engine harness connector at firewall
- Underhood fuse block connector near right shock tower
- Underhood electrical center cover
- Ground bolt from right shock tower
- Positive battery cable from inside electrical center
- Chassis electrical connector from right shock tower
- Transmission Control Module (TCM) wiring harness
- Engine wiring harness connector inside electrical center
- Electrical connector at right frame rail
- Cooling fans

9. Raise and support the vehicle.

10. Remove the front wheels.

11. Remove the wheel house splash shields and the fender liners.

12. From the right wheel opening, disconnect the washer reservoir brace.

13. From the left wheel opening, disconnect the transmission oil cooler lines.

14. Remove the upper to center intermediate steering shaft bolt.

15. Remove the lower intermediate steering shaft-to-steering gear bolt.

16. Remove the center intermediate steering shaft with the lower shaft attached.

17. Disconnect the power steering cooler lines from the radiator.

18. Lower the vehicle.

19. Remove the radiator, condenser and transmission oil cooler as an assembly.

20. Raise the vehicle.

21. Remove the power steering oil cooler from the bracket and tie the cooler to the engine.

22. On 4WD models, remove the transfer case.

23. Remove the transmission.

24. Remove the brake bundle clips from both frame rails.

25. Disconnect the fuel line from the filter.

26. Disconnect the EVAP hose from the rear of the fuel filter.

27. Disconnect the rear brake lines from the bracket above the rear axle assembly.

28. Remove the fuel and brake line bundle retainers from the frame rail the length of the vehicle. Do not remove the retainers from the lines.

29. Remove the fuel filter bracket to provide a removal path for the fuel and brake line bundle assembly.

30. Remove the fuel and brake line bundle bracket from the right side wheelhouse.

31. Lower the vehicle.

32. Disconnect the heater outlet hose from the heater outlet pipe at the right frame rail. Position the hose to the engine.

33. Disconnect the heater inlet hose from the water housing and position the hose to the vehicle.

34. If the vehicle is equipped with Magnaride, disconnect the electrical connectors from the top of the right and left shock modules.

35. Secure the shock modules to the lower control arms with a suitable strap to avoid damage to the brake lines.

36. Remove the shock yoke.

37. Remove the left and right shock module upper mounting nuts.

38. Raise the vehicle enough to place a suitable engine lift table under the engine, transmission, front frame and front suspension assembly.

39. Support the rear of the vehicle with jack stands.

40. Raise the lift table and/or lower the vehicle to preload the weight of the engine, front frame, and front suspension assembly.

41. Remove the 4 front frame bolts.

42. With the aid of an assistant, lower the table and/or raise the vehicle to remove the engine, front frame, fuel/brake bundle and front suspension assembly from the vehicle.

43. Ensure that all the hoses, wires, pipes and shock modules clear the vehicle during the removal process

44. If the engine itself is to be serviced, the engine will have to be separated from the transmission and the front frame and suspension assembly.

To install:

45. With the aid of an assistant, raise the table and/or lift the vehicle to install the engine, fuel/brake bundle, front frame and front suspension assembly to the vehicle.

46. Install the front frame bolts. Tighten the bolts to 141 ft. lbs. (191 Nm).

47. Install the right and left shock module upper mounting bolts. Tighten the bolts 83 ft. lbs. (112 Nm).

48. Connect the shock module connectors, if equipped with Magnaride.

49. Connect the heater inlet hose to the water housing.

50. Connect the heater outlet hose to the outlet pipe.

51. Raise and support the vehicle.

52. Install the fuel and brake line bundle to the right side wheelhouse. Tighten bundle bracket to 80 inch lbs. (9 Nm).

53. Install the fuel filter bracket. Tighten the bracket bolt to 80 inch lbs. (9 Nm).

54. Install the fuel/brake line bundle to the bundle brackets the length of the vehicle.

55. Install the rear brake line to the rear axle assembly. Tighten the bracket bolt to 80 inch lbs. (9 Nm).

56. Connect the EVAP hose to the fuel filter.

57. Connect the fuel line to the fuel filter.

58. Install the brake lines to the bundle clips on the left and right frame rails.

59. Install the transmission.

60. On 4WD models, install the transfer case.

61. Install the power steering oil cooler to the mounting bracket.

62. Lower the vehicle.

63. Install the radiator, condenser and transmission oil cooler as an assembly.

64. Raise and support the vehicle.

65. Connect the power steering cooler lines to the radiator.

66. Install the lower and intermediate steering shafts. Tighten the bolts to 23 ft. lbs. (30 Nm).

67. From the left wheel opening, connect the transmission oil cooler lines.

68. From the right wheel opening, connect the washer reservoir brace.

69. Install the wheel house splash shields and the fender liners.

70. Install the front wheels.
Install or connect the following:
- Cooling fans
- Electrical connector at right frame rail
- Engine wiring harness connector inside electrical center
- Transmission Control Module (TCM) wiring harness
- Chassis electrical connector to right shock tower
- Positive battery cable inside electrical center
- Ground bolt to right shock tower
- Underhood electrical center cover
- Underhood fuse block connector near right shock tower
- Engine harness connector at firewall
- Fuel line retainer
- Master cylinder. Tighten the retaining nuts to 18 ft. lbs (25 Nm).
- Brake fluid level switch connector
- Brake booster vacuum hose
- A/C liquid hose to condenser
- A/C suction hose fitting on shock tower
- Surge tank hoses
- Air cleaner assembly
- Fuel injector shield
- Cross vehicle brace

71. Connect the battery cables.

72. Remove the locking pin from the steering column.

73. Bleed the brake circuits.

74. Refill the engine, transaxle and cooling system with the correct amount of the appropriate fluids before starting the engine. Recharge the A/C system using approved recycling equipment.

Water Pump

REMOVAL & INSTALLATION

3.6L Engine

1. Before servicing the vehicle, refer to the Precautions Section.

2. Drain the cooling system.

3. Remove or disconnect the following:
- Negative battery cable
- Accessory drive belt

4. Use Special Tool EN-46104 to retain the water pump pulley

Using water pump pulley holding tool EN-46104 to remove the pulley bolts.

06025-SRX-G01

5. Remove or disconnect the following:
- Water pump pulley
- Water pump

To install:

➡**Clean the water pump sealing surfaces**

6. Install the water pump and new gasket. Tighten the bolts to 89 inch lbs. (10 Nm).

7. Install the water pump pulley.

8. Use Special Tool EN-46104 to retain the water pump pulley. Tighten the water pump pulley bolts to 106 inch lbs. (12 Nm).

9. Install or connect the following:
- Accessory drive belt
- Negative battery cable

10. Fill the cooling system to the correct level.

11. Start the engine and check for leaks.

4.6L Engine

1. Before servicing the vehicle, refer to the Precautions Section.

2. Drain the cooling system.

3. Remove or disconnect the following:
- Negative battery cable
- Cooling fan
- Water pump drive belt
- Drive belt tensioner
- Water pump pulley
- Water pump mounting bolts
- Water pump

To install:

4. Install or connect the following:
- Water pump and new gasket. Tighten the bolts to 89 inch lbs. (10 Nm).
- Water pump pulley. Tighten the bolts to 106 inch lbs. (12 Nm).
- Drive belt tensioner
- Water pump drive belt
- Cooling fan
- Negative battery cable

5. Fill the cooling system to the correct level.

6. Start the engine and check for leaks.

Heater Core

REMOVAL & INSTALLATION

1. Before servicing the vehicle, refer to the Precautions Section.

2. Disconnect the negative battery cable.

3. Drain the cooling system into a clean container for reuse.

4. Recover the air conditioning refrigerant, into a refrigerant recovery station

5. Disable the air bag system.

6. Disconnect the heater hoses from the heater core inlet and outlet tubes in the engine compartment.

7. Disconnect both A/C line fittings at

the cowl. Remove the quick connect fittings, then remove and discard the O-rings.

8. Remove the floor console shifter trim panel.

9. Disconnect the traction control connector and the console connectors on each side.

10. Remove 6 screws and remove the floor console.

11. Remove the center console A/C vents.

12. Remove the HVAC control panel/ashtray assembly.

13. Remove the radio.

14. Remove the glove box door and the glove box.

15. Remove the instrument cluster assembly.

16. Remove the defroster grille.

17. On each side of the instrument panel, remove the screws attaching the finish panels, release the locking tabs, disconnect the connectors and remove the finish panels.

➡**The front wheels must be in the straight ahead position and the steering column locked before disconnect the intermediate shaft. Failure to do so may result in damage to the SRS coil.**

18. Center the steering wheel.

19. Install Steering Column Anti-rotation pin J-42640 to lock the steering column.

20. Raise and support the vehicle.

21. Remove the upper to lower intermediate steering shaft bolt.

22. Remove the knee bolster.

23. Remove the steering column trim covers.

24. Disconnect the steering column electrical connectors.

25. Support the steering column.

26. Remove the 4 steering column mounting nuts and lower the column.

27. Remove the bolts at the lower instrument panel retainer at the bottom of the center console.

28. Remove the center console attaching screws.

29. Remove the screws behind the glove box opening attaching the instrument panel retainer.

30. Remove the instrument panel fascia retaining screws and the upper retaining screws and remove the instrument panel retainer.

31. Remove the nuts attaching the brake pedal bracket to the instrument panel carrier.

32. Remove the parking brake pedal assembly.

33. Disconnect all electrical wire retain-

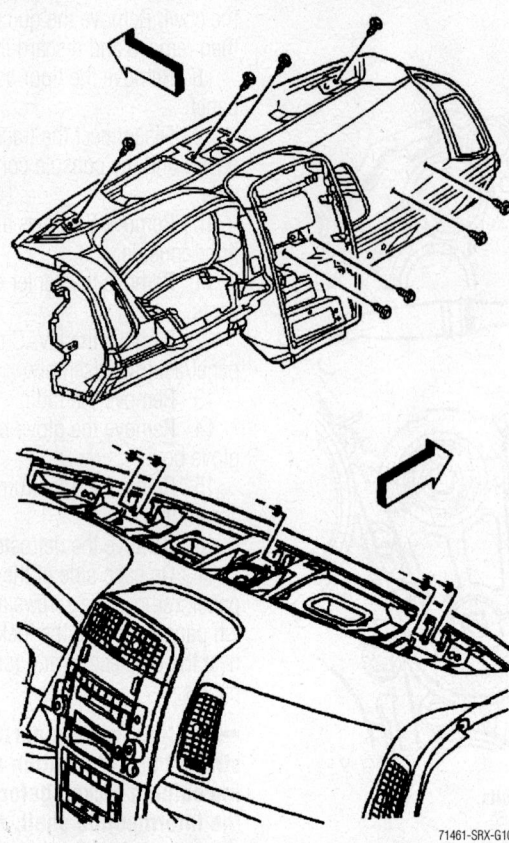

Instrument panel retainer mounting screw locations

71461-SRX-G10

ers from the instrument panel carrier to free the carrier.

34. Remove the screws attaching the carrier to the cowl.

35. Remove the instrument panel carrier.

36. Remove the air inlet assembly.

37. Disconnect the HVAC module connector.

38. Press the tabs and release the left and right rear heater ducts from the HVAC module.

39. Remove the drain tube.

40. Remove the upper and lower HVAC mounting screws and remove the HVAC module.

41. Remove the heater core pipe bracket screw and bracket.

42. Remove the heater core.

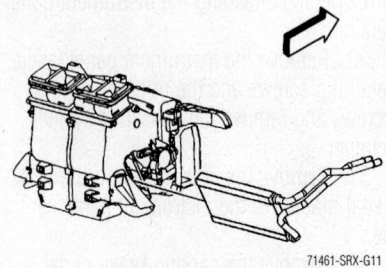

71461-SRX-G11

Removing the heater core from the HVAC module

To install:

43. Install the heater core to the HVAC module.

44. Install the heater core pipe bracket and screw.

45. Install the HAVC module. Tighten the mounting nuts to 89 inch lbs. (10 Nm).

46. Connect left and right rear heater ducts to the HVAC module

47. Install the drain tube.

48. Connect the HVAC module connector.

49. Install the air inlet assembly.

50. Install new O-rings to the A/C suction and liquid lines and connect them to the HVAC module.

51. Connect the heater hoses.

52. Install the instrument panel carrier and tighten the bolts and nuts.

53. Connect the electrical connections to the carrier.

54. Connect the brake pedal bracket to the cowl.

55. Install the parking brake pedal assembly.

56. Install the HVAC ducts.

57. Install the instrument panel fascia and the upper retaining screws and tighten to 80 inch lbs. (9 Nm).

58. Install the screws behind the glove box opening attaching the instrument panel retainer.

59. Install the center console attaching screws.

60. Install the bolts at the lower instrument panel retainer and tighten the bolts to 18 ft. lbs. (25 Nm).

61. Install the upper intermediate steering shaft-to-steering column bolt and tighten the bolt to 35 ft. lbs. (48 Nm).

62. Raise the steering column and install the 4 steering column mounting nuts. Tighten the nuts to 18 ft. lbs. (25 Nm).

63. Install the upper intermediate steering shaft-to-lower intermediate shaft bolt and tighten the bolt to 35 ft. lbs. (48 Nm).

64. Install the knee bolster.

65. Install the steering column trim covers.

66. Connect the steering column electrical connectors.

67. Remove the Steering Column Anti-rotation pin.

68. Install the instrument panel side finish panels.

69. Install the defroster grille.

70. Install the instrument cluster assembly.

71. Install the glove box door and the glove box.

72. Install the radio.

73. Install the HVAC control panel/ashtray assembly.

74. Install the center console A/C vents.

75. Install the floor console.

76. Connect the traction control connector and the console connectors on each side.

77. Install the floor console shifter trim panel.

78. Recharge the air conditioning refrigerant.

79. Fill the cooling system.

80. Connect the negative battery cable.

81. Start the vehicle and check for leaks.

Cylinder Head

REMOVAL & INSTALLATION

3.6L Engine

LEFT SIDE

1. Before servicing the vehicle, refer to the Precautions Section.

2. Relieve the fuel system pressure.

3. Drain the cooling system.

4. Remove or disconnect the following:
   - Negative battery cable
   - Left side secondary timing chain
   - Oil level indicator
   - Coolant temperature sensor heat shield

- Coolant temperature sensor electrical connector
- Wiring harness ground, connector and connector bracket
- Power steering pump pulley
- Power steering pump, but leave the fluid lines attached
- Surge tank hose
- Exhaust manifold heat shield
- Catalytic converter
- Oil filter adapter upper bolt
- Two front cylinder head M8 bolts
- Cylinder head bolts

5. Remove the cylinder head with the exhaust manifold attached.

6. Discard the head gasket.

To install:

7. The cylinder head should be cleaned and inspected prior to installation.

8. Lightly oil all bolt threads and stud bolt threads before installation.

9. Clean all gasket mating surfaces thoroughly.

10. Install or connect the following:
- Exhaust manifold, if removed.
- New head gasket on the cylinder block.

✳✳ WARNING

Always use new cylinder head bolts when installing the cylinder head or damage to the engine may occur.

- Cylinder head on the cylinder block.
- Tighten the M11 cylinder head bolts in steps following the proper torque sequence. The first step is 33 ft. lbs. (45 Nm), the second step is an additional 120°.
- Tighten the front M8 cylinder head bolts in steps following the proper torque sequence. The first step is 11 ft. lbs. (15 Nm), the second step is an additional 60°.
- Oil filter adapter upper bolt
- Catalytic converter
- Exhaust manifold heat shield
- Surge tank hose
- Power steering pump, but leave the fluid lines attached
- Power steering pump pulley
- Wiring harness ground, connector and connector bracket
- Coolant temperature sensor electrical connector
- Coolant temperature sensor heat shield
- Oil level indicator
- Left side secondary timing chain
- Negative battery cable

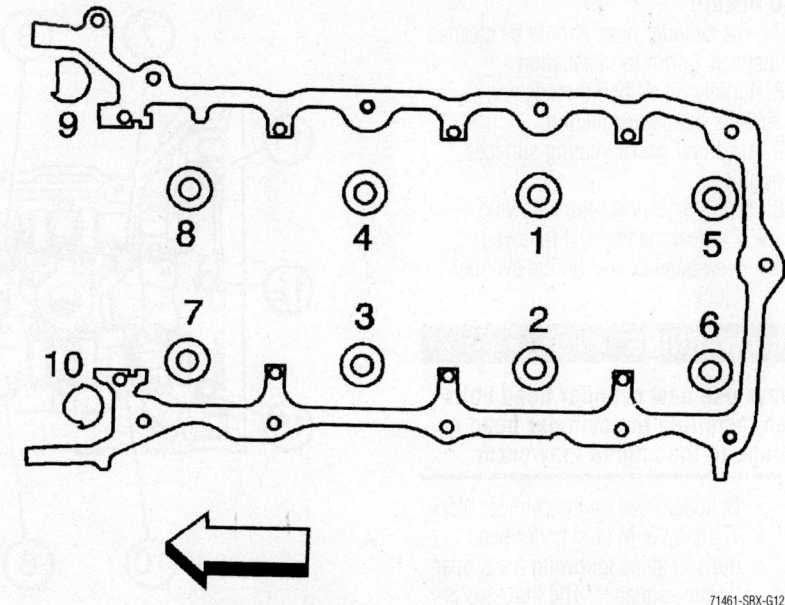

Left side cylinder head bolt torque sequence—3.6L engines

71461-SRX-G12

11. Fill and bleed the cooling system.

➡ **Engine coolant is corrosive to engine bearing material. Replace the engine oil after removal of any coolant-carrying component to help prevent potential bearing damage.**

12. Change the engine oil and filter

13. Connect the negative battery cable.

14. Start the engine and check for leaks.

RIGHT SIDE

1. Before servicing the vehicle, refer to the Precautions Section.

2. Relieve the fuel system pressure.

3. Drain the cooling system.

4. Remove or disconnect the following:
- Negative battery cable
- Right side secondary timing chain
- Coolant inlet pipe
- Wiring harness ground and harness bracket
- Negative battery cable bolt on head
- Exhaust manifold heat shield
- Catalytic converter
- Cylinder head bolts

5. Remove the cylinder head with the exhaust manifold attached.

6. Discard the head gasket.

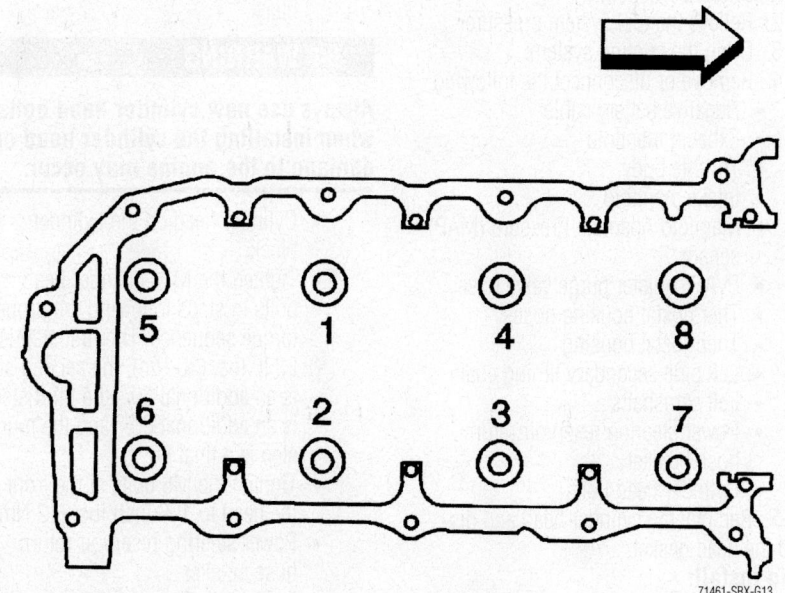

Right side cylinder head bolt torque sequence—3.6L engines

71461-SRX-G13

To install:

7. The cylinder head should be cleaned and inspected prior to installation.

8. Lightly oil all bolt threads and stud bolt threads before installation.

9. Clean all gasket mating surfaces thoroughly.

10. Install or connect the following:
- Exhaust manifold, if removed.
- New head gasket on the cylinder block.

✳✳ WARNING

Always use new cylinder head bolts when installing the cylinder head or damage to the engine may occur.

- Cylinder head on the cylinder block.
- Tighten the M11 cylinder head bolts in steps following the proper torque sequence. The first step is 33 ft. lbs. (45 Nm), the second step is an additional 120°.
- Catalytic converter
- Exhaust manifold heat shield
- Negative battery cable bolt on head
- Wiring harness ground and harness bracket. Tighten bolt to 89 inch lbs. (10 Nm).
- Coolant inlet pipe
- Right side secondary timing chain
- Negative battery cable

11. Fill and bleed the cooling system.

12. Start the engine and check for leaks.

4.6L Engine

LEFT SIDE

1. Before servicing the vehicle, refer to the Precautions Section.

2. Relieve the fuel system pressure.

3. Drain the cooling system.

4. Remove or disconnect the following:
- Negative battery cable
- Exhaust manifold
- Throttle body
- Intake manifold
- Manifold Absolute Pressure (MAP) sensor
- EVAP canister purge valve hose
- Thermostat housing hoses
- Thermostat housing
- Left side secondary timing chain
- Left camshafts
- Power steering reservoir return hose bracket
- Cylinder head bolts

5. Remove the cylinder head and discard the head gasket.

To install:

6. The cylinder head should be cleaned and inspected prior to installation.

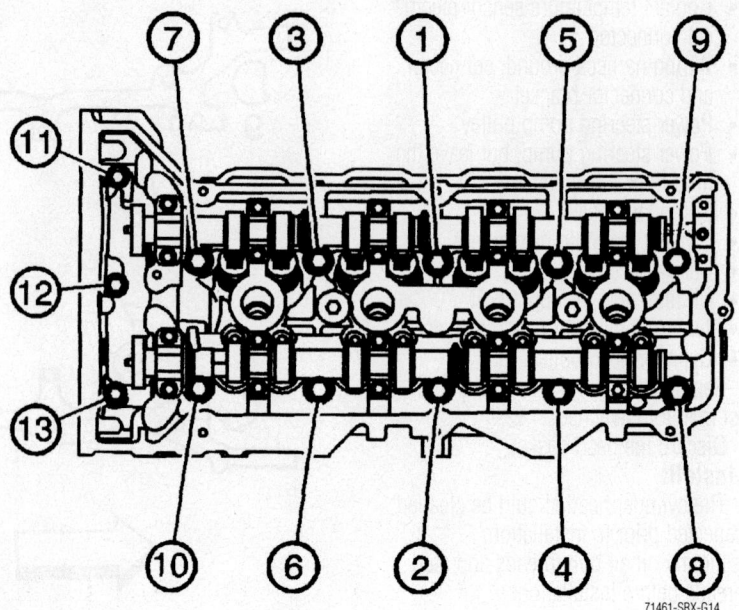

Left side cylinder head bolt torque sequence—4.6L engines

71461-SRX-G14

7. Lightly oil all bolt threads and stud bolt threads before installation.

8. Clean all gasket mating surfaces thoroughly.

➡**Ensure the M11 cylinder head bolts have the proper pitch or engine damage will occur. The bolts have been revised. Identify the bolts before installation. Bolts with a pitch of 1.5mm have a thread length of about 1.89 inches (48mm). Bolts with a pitch of 2mm have a thread length of about 2.64 inches (67mm).**

9. Install or connect the following:
- New head gasket on the cylinder block.

✳✳ WARNING

Always use new cylinder head bolts when installing the cylinder head or damage to the engine may occur.

- Cylinder head on the cylinder block.
- Tighten the M11 cylinder head bolts in steps following the proper torque sequence. The first step is 22 ft. lbs. (30 Nm), the second step is an additional 60°, the third step is an additional 60° and the fourth step is a final 60°.
- Tighten the M6 bolts at the front of the head to 106 inch lbs. (12 Nm).
- Power steering reservoir return hose bracket
- Left camshafts
- Left side secondary timing chain

- Thermostat housing and tighten the bolts to 18 ft. lbs. (25 Nm).
- Thermostat housing hoses
- EVAP canister purge valve hose
- MAP sensor
- Throttle body
- Intake manifold
- Exhaust manifold
- Coolant
- Negative battery cable

10. Fill and bleed the cooling system.

11. Start the engine and check for leaks.

RIGHT SIDE

1. Before servicing the vehicle, refer to the Precautions Section.

2. Relieve the fuel system pressure.

3. Remove or disconnect the following:
- Negative battery cable
- Coolant
- Exhaust manifold
- Throttle body
- Intake manifold
- Manifold Absolute Pressure (MAP) sensor
- EVAP canister purge valve hose
- Thermostat housing hoses
- Thermostat housing
- Right side secondary timing chain
- Right camshafts
- Cylinder head bolts

4. Remove the cylinder head and discard the head gasket.

To install:

5. The cylinder head should be cleaned and inspected prior to installation.

6. The cylinder head should be cleaned and inspected prior to installation.

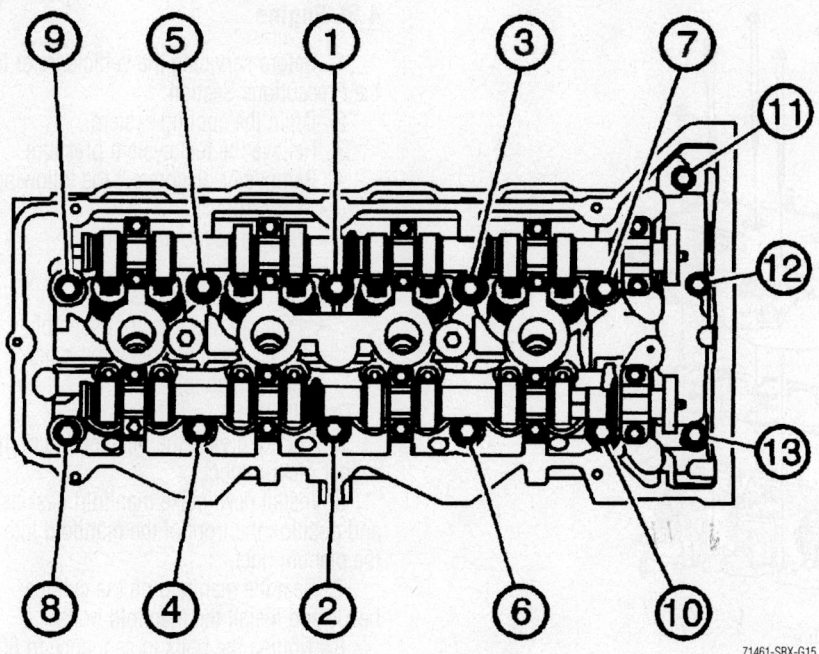

Right side cylinder head bolt torque sequence—4.6L engines

71461-SRX-G15

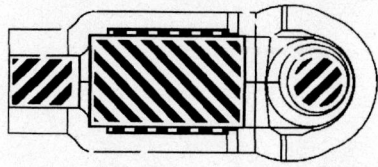

71461-SRX-G16

Rocker arm engine assembly lubricant application locations—3.6L and 4.6L engines

➡ **The camshaft follower must be positioned squarely on the valve tip so the full width of the roller contacts the camshaft lobe.**

9. Install the follower so the rounded head goes on the hydraulic lash adjuster and the flat end goes on the valve tip.

10. Install the camshafts.

Intake Manifold

REMOVAL & INSTALLATION

3.6L Engine

1. Before servicing the vehicle, refer to the Precautions Section.
2. Relieve the fuel system pressure.
3. Drain the cooling system.
4. Remove or disconnect the following:
   - Negative battery cable
   - Engine cover
   - Air intake assembly
   - Brake booster hose
   - Intake manifold brace
   - PCV tube assembly
   - EVAP hose
   - EVAP solenoid
   - All necessary electrical connectors
   - Intake manifold mounting bolts
   - Intake manifold

To install:

5. Installation is the reverse of the removal procedure, using the following torque specifications.
   - New intake manifold gasket
   - Upper-to-lower intake manifold bolts, if removed, to 17 ft. lbs. (23 Nm).
   - Intake manifold bolts and tighten all bolts in a circular manner from the center outward to 17 ft. lbs. (23 Nm).
   - Intake manifold brace—Bolt 1 to 89 inch lbs. (10 Nm). Bolt 2 to 48 ft. lbs. (65 Nm).

6. Fill and bleed the engine cooling system.
7. Connect the negative battery cable.
8. Start the engine and check for leaks.

---

7. Lightly oil all bolt threads and stud bolt threads before installation.

8. Clean all gasket mating surfaces thoroughly.

➡ **Ensure the M11 cylinder head bolts have the proper pitch or engine damage will occur. The bolts have been revised. Identify the bolts before installation. Bolts with a pitch of 1.5mm have a thread length of about 1.89 inches (48mm). Bolts with a pitch of 2mm have a thread length of about 2.64 inches (67mm).**

9. Install or connect the following:
   - New head gasket on the cylinder block.

❋❋ WARNING

Always use new cylinder head bolts when installing the cylinder head or damage to the engine may occur.

- Cylinder head on the cylinder block.
- Tighten the M11 cylinder head bolts in steps following the proper torque sequence. The first step is 22 ft. lbs. (30 Nm), the second step is an additional 60°, the third step is an additional 60° and the fourth step is a final 60°.
- Tighten the M6 bolts at the front of the head to 106 inch lbs. (12 Nm).
- Right camshafts
- Right side secondary timing chain

- Thermostat housing and tighten the bolts to 18 ft. lbs. (25 Nm).
- Thermostat housing hoses
- EVAP canister purge valve hose
- MAP sensor
- Throttle body
- Intake manifold
- Exhaust manifold
- Coolant
- Negative battery cable

10. Fill and bleed the cooling system.
11. Start the engine and check for leaks.

Rocker Arms and Valve Lifters

REMOVAL & INSTALLATION

1. Before servicing the vehicle, refer to the Precautions Section.
2. Remove the camshafts.
3. Remove the rocker arm and camshaft follower.

➡ **The arms should be installed in their original location during assembly.**

4. Remove the rocker arms. If more than 1 rocker arm is to be removed, identify each rocker arm location.

5. Remove the valve lifters.

To install:

6. Using clean engine oil, fill the lifter and lubricate the bores in the cylinder heads.

7. Install the lifters.

8. Apply engine assembly lubricant to the rocker arm as shown.

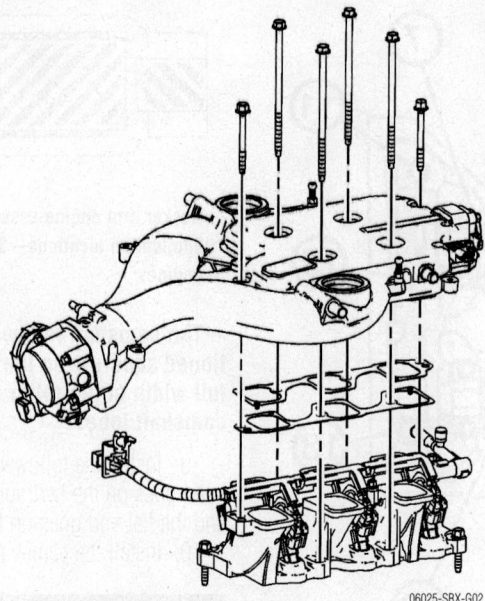

Upper to lower intake manifold mounting—3.6L engine

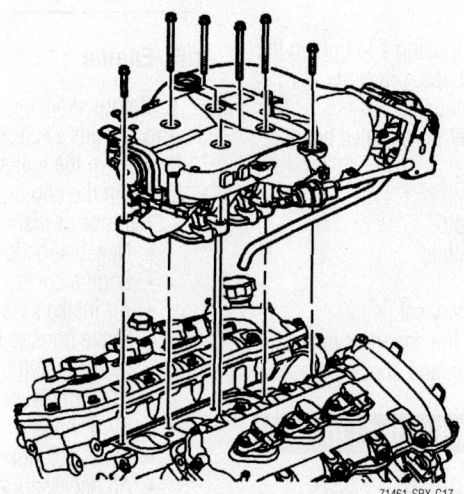

Intake manifold mounting—3.6L engine

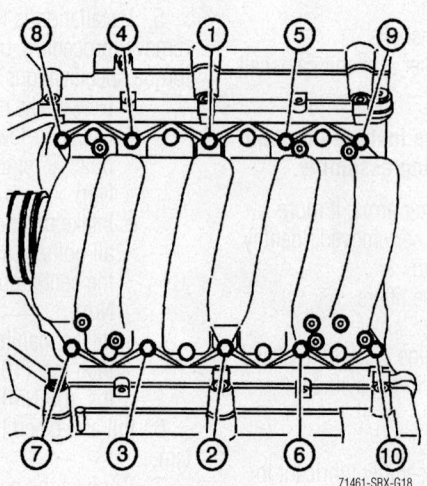

Intake manifold tightening sequence—4.6L engine

4.6L Engine

1. Before servicing the vehicle, refer to the Precautions Section.
2. Drain the cooling system.
3. Relieve the fuel system pressure.
4. Remove or disconnect the following:
   - Cross brace
   - Engine cover/sight shield
   - PCV air tubes
   - Sight shield bracket
   - Fuel rail and injector assembly
   - Intake manifold retaining bolts
   - Intake manifold

To install:

5. Lightly grease the inside edge of the rubber plenum duct.
6. Install new intake manifold gaskets and position the front of the manifold into the plenum duct.
7. Seat the manifold on the cylinder heads and install the manifold bolts.
8. Tighten the bolts in sequence to 89 inch lbs. (10 Nm).
9. Ensure the plenum duct is fully attached to the manifold, then install the plenum duct clamp.
10. Lightly lubricate the fuel injector bores with clean engine oil.
11. Install the fuel rail and injectors and tighten to 89 inch lbs. (10 Nm).
12. Install or connect the following.
    - Sight shield bracket
    - PCV air tubes
13. Fill and bleed the cooling system.
14. Start the engine and check for leaks.

Exhaust Manifold

REMOVAL & INSTALLATION

➡Spray the exhaust system fasteners with penetrating lubricant before removing them to help prevent broken studs and bolts. The use of a 6-point socket is highly recommended when removing exhaust system fasteners.

✳✳ CAUTION

To prevent serious burns, allow the exhaust manifold to cool down before attempting to remove it.

3.6L Engine

LEFT

1. Before servicing the vehicle, refer to the Precautions Section.
2. Disconnect the negative battery cable.

3. Raise and support the vehicle safely on jackstands.

4. Disconnect the catalytic converter from the exhaust manifold.

5. Remove the heat shield.

6. On the left side, remove the upper insulator from the oil dipstick tube.

7. Remove the exhaust manifold.

To install:

8. Clean all gasket mating surfaces thoroughly.

9. Install a new exhaust manifold gasket and the exhaust manifold on the cylinder head. Start 2 bolts to hold the manifold in position.

10. Install the remaining bolts. Tighten the bolts to 18 ft. lbs. (25 Nm).

11. Raise and support the vehicle safely.

12. Connect the dual converter Y-pipe.

13. Install the heat shield.

14. Install the upper insulator on the left side.

15. Connect the negative battery cable.

16. Start the engine and check for exhaust leaks.

4.6L Engine

1. Before servicing the vehicle, refer to the Precautions Section.

2. Disconnect the negative battery cable.

3. Raise and support the vehicle safely on jackstands.

4. Disconnect the catalytic converter from the exhaust manifold.

5. Remove the heat shield.

6. Remove the exhaust manifold.

To install:

7. Clean all gasket mating surfaces thoroughly.

8. Install a new gasket and the exhaust manifold on the cylinder head.

9. Install the bolts. Tighten the bolts in the sequence shown to 18 ft. lbs. (25 Nm).

10. Install the heat shield.

11. Raise and support the vehicle safely on jackstands.

12. Connect the catalytic converter.

13. Lower the vehicle.

14. Connect the negative battery cable.

15. Start the engine and check for exhaust leaks.

Camshafts

REMOVAL & INSTALLATION

3.6L Engine

LEFT SIDE

➡The camshaft position sensors, camshaft position actuators and crankshaft damper are removed in the Front Cover and Timing Chain procedure.

1. Before servicing the vehicle, refer to the Precautions Section.

2. Remove or disconnect the following:
- Engine
- Intake manifold assembly
- Ignition coil connectors
- Wiring harnesses on camshaft cover
- Ignition coils
- Camshaft cover
- Camshaft Position (CMP) sensors
- Camshaft position actuator solenoid
- Crankshaft damper

3. Rotate the crankshaft until the camshafts flats are parallel with the camshaft cover rail as shown.

4. Place an open end wrench on the camshaft flats to hold it in place, then loosen the camshaft position actuator bolt.

5. Install holding tool EN-46108 to retain the timing chain in place.

6. Mark the timing chain and camshaft position actuators for reassembly reference.

7. Remove the camshaft position actuator bolt.

8. Note the locations of the camshaft bearing caps for reassembly reference.

9. Remove the camshaft bearing caps.

10. Remove the camshafts.

To install:

11. Install the camshaft sealing rings in the camshaft grooves.

12. Ensure the camshafts are placed in the correct position by locating the identification numbers on the appropriate camshaft.

13. Apply engine lubricant to the camshaft journals and carriers and the camshaft bearing caps.

14. Install the camshafts.

15. Ensure the camshafts flats are parallel with the camshaft cover rail as shown.

16. Install thrust cap in the first camshaft journal.

17. Install the camshaft bearing caps in

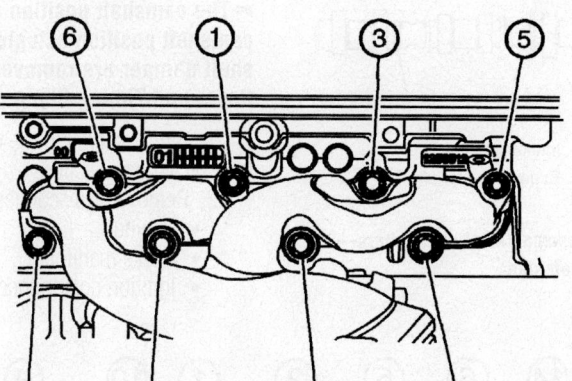

Left Manifold

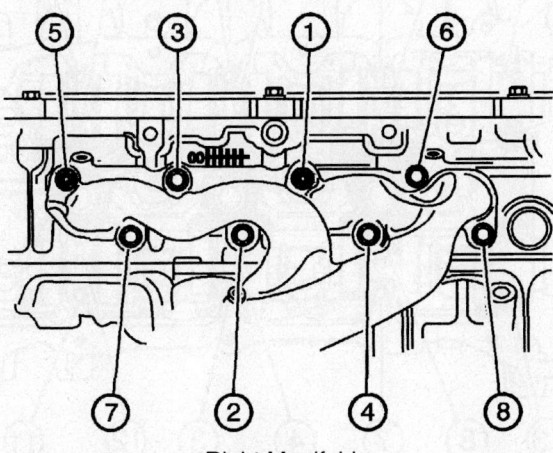

Right Manifold

71461-SRX-G19

Exhaust manifold tightening sequence—4.6L engine

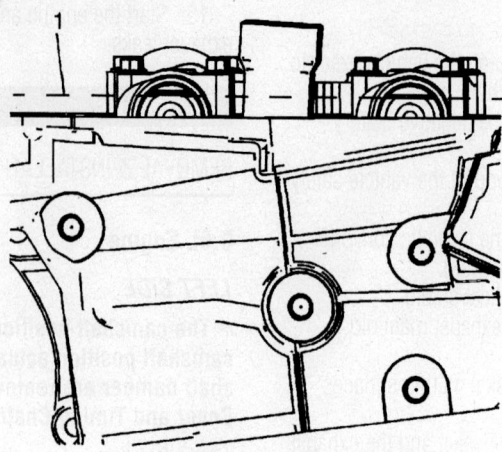

1. Camshaft cover rail location

71461-SRX-G20

Aligning the camshaft flats with the camshaft cover rail—3.6L engine left side

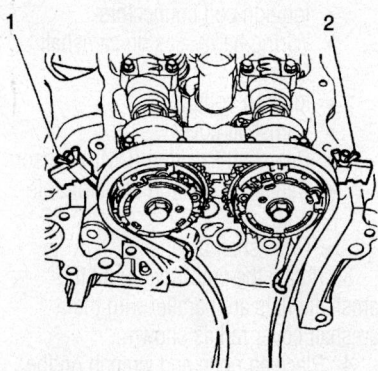

1-2. Timing chain retaining tool EN-46108

71461-SRX-G21

Installing tool EN-46108 to lock the timing chain—3.6L engine left side

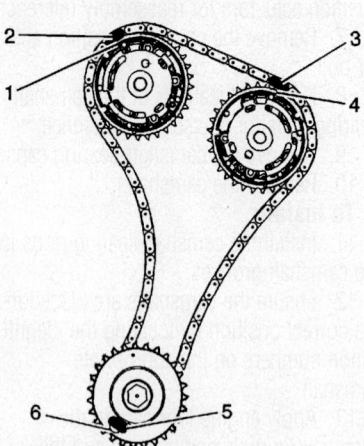

1. Intake camshaft actuator mark
2. Timng chain mark
3. Exhaust camshaft actuator mark
4. Timing chain mark

71461-SRX-G22

Aligning the camshaft position actuators and timing chain smrks—3.6L engine

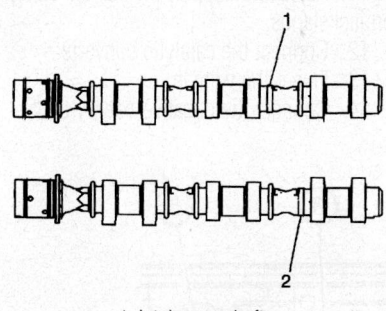

1. Intake camshaft
2. Exhaust camshaft

71461-SRX-G23

Identifying camshaft locating numbers—3.6L engine left side

the correct locations so the raised boss is toward the center of the engine.

18. Hand tighten the bearing cap bolts.

19. Tighten the bearing caps bolt in the sequence shown to 89 inch lbs. (10 Nm).

20. Loosen the bolts number 1, 2, 3 and 4, then retighten the bolts to 89 inch lbs. (10 Nm).

21. Place an open end wrench on the camshaft flats to hold it in place, then tighten the camshaft position actuator bolt to 48 ft. lbs. (58 Nm).

22. Install or connect the following:
- Crankshaft damper
- Camshaft position actuator solenoids
- CMP sensors
- Camshaft cover with a new gasket
- Ignition coils
- Wiring harnesses on camshaft cover
- Ignition coil connectors
- Intake manifold
- Engine

RIGHT SIDE

➡The camshaft position sensors, camshaft position actuators and crankshaft damper are removed in the Front Cover and Timing Chain procedure.

1. Before servicing the vehicle, refer to the Precautions Section.

2. Remove or disconnect the following:
- Engine
- Intake manifold
- Ignition coil connectors

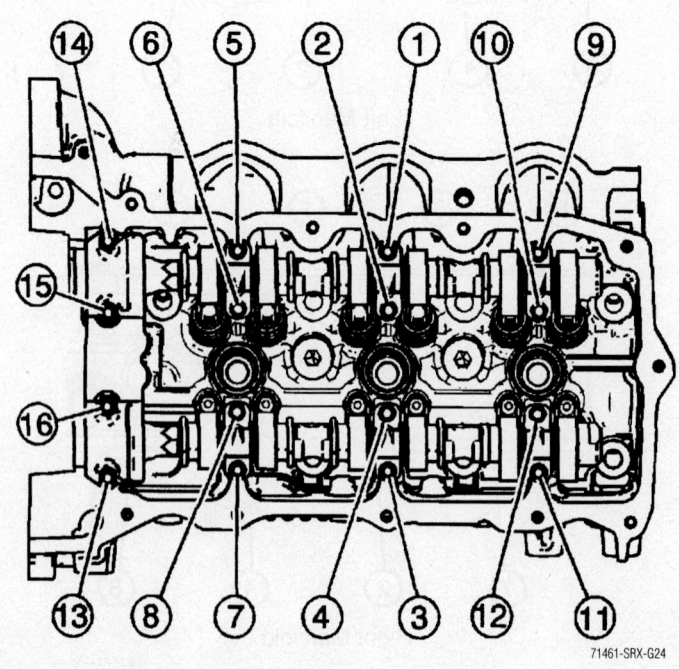

71461-SRX-G24

Camshaft bearing cap tightening sequence—3.6L engine left side

- Wiring harnesses on camshaft cover
- Ignition coils
- Camshaft cover
- Camshaft Position (CMP) sensors
- Camshaft position actuator solenoid
- Crankshaft damper

3. Rotate the crankshaft until the camshafts flats are parallel with the camshaft cover rail as shown.

4. Place an open end wrench on the camshaft flats to hold it in place, then loosen the camshaft position actuator bolt.

5. Install holding tool EN-46108 to retain the timing chain in place.

6. Mark the timing chain and camshaft position actuators for reassembly reference.

7. Remove the camshaft position actuator bolt.

8. Note the locations of the camshaft bearing caps for reassembly reference.

9. Remove the camshaft bearing caps.

10. Remove the camshafts.

To install:

11. Install the camshaft sealing rings in the camshaft grooves.

12. Ensure the camshafts are placed in the correct position by locating the identification numbers on the appropriate camshaft.

13. Apply engine lubricant to the camshaft journals and carriers and the camshaft bearing caps.

14. Install the camshafts.

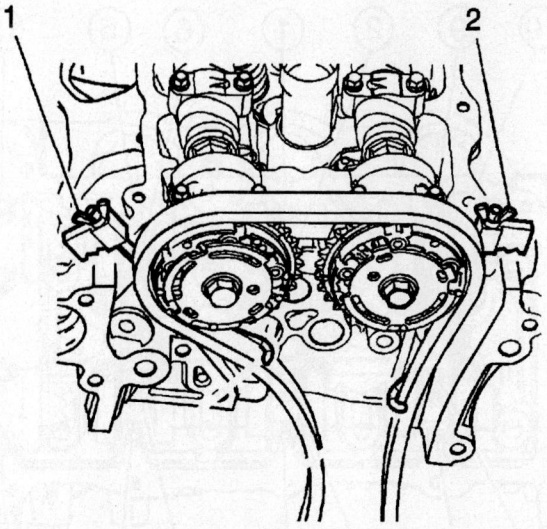

1-2. Timing chain retaining tool EN-46108

71461-SRX-G26

Installing tool EN-46108 to lock the timing chain—3.6L engine right side

15. Ensure the camshafts flats are parallel with the camshaft cover rail as shown.

16. Install thrust cap in the first camshaft journal.

17. Install the camshaft bearing caps in the correct locations so the raised boss is toward the center of the engine.

18. Hand tighten the bearing cap bolts.

19. Tighten the bearing caps bolt in the sequence shown to 89 inch lbs. (10 Nm).

20. Loosen the bolts number 1, 2, 3 and 4, then retighten the bolts to 89 inch lbs. (10 Nm).

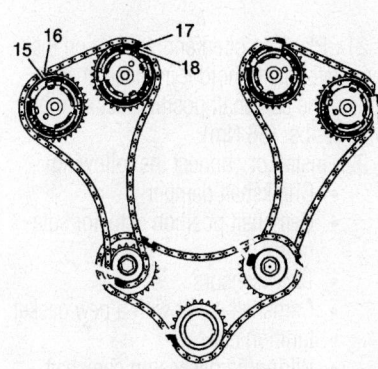

15. Exhaust camshaft timing marks
16. Timing chain marks
17. Timing chain marks
18. Intake camshaft timing marks

71461-SRX-G27

Aligning the camshaft position actuator and timing chain marks—3.6L engine

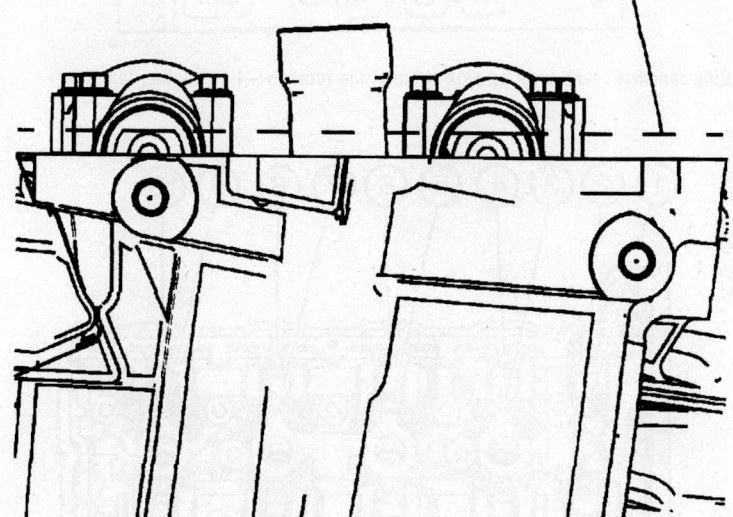

1. Camshaft cover rail

71461-SRX-G25

Aligning the camshaft flats with the camshaft cover rail—3.6L engine right side

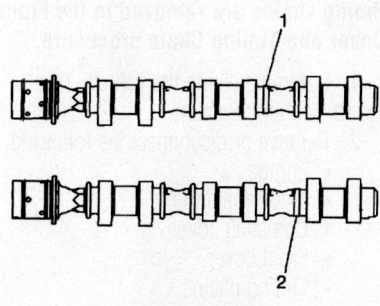

1. Intake camshaft
2. Exhaust camshaft

71461-SRX-G28

Identifying camshaft locating numbers—3.6L engine right side

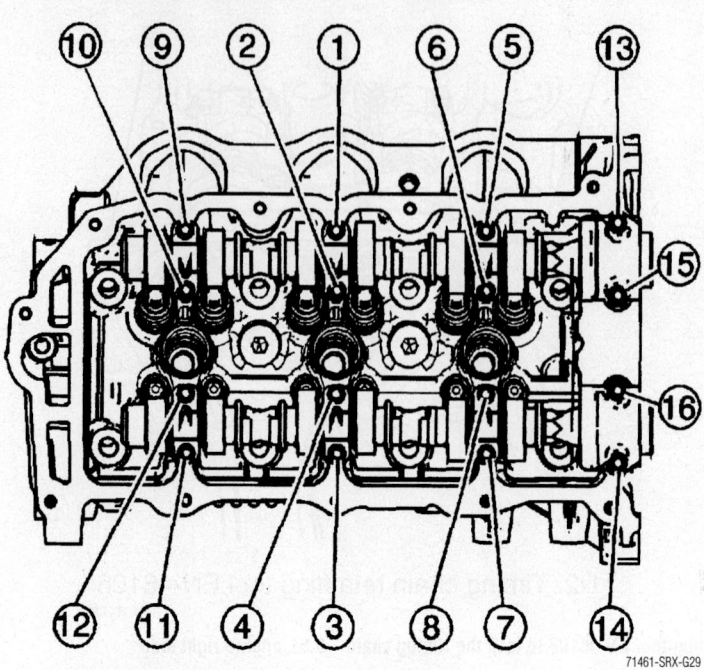

Camshaft bearing cap tightening sequence—3.6L engine right side

71461-SRX-G29

21. Place an open end wrench on the camshaft flats to hold it in place, then tighten the camshaft position actuator bolt to 48 ft. lbs. (58 Nm).

22. Install or connect the following:
- Crankshaft damper
- Camshaft position actuator solenoids
- CMP sensors
- Camshaft cover with a new gasket
- Ignition coils
- Wiring harnesses on camshaft cover
- Ignition coil connectors
- Intake manifold
- Engine

4.6L Engine

LEFT SIDE

➡The camshaft position actuators and timing chains are removed in the Front Cover and Timing Chain procedure.

1. Before servicing the vehicle, refer to the Precautions Section.

2. Remove or disconnect the following:
- Engine
- Intake manifold
- Camshaft cover
- Front cover
- Timing chains
- Camshaft position actuators
- Camshaft bearing caps

➡Observe the positions of the camshaft bearing caps. The arrow on the cap points toward the front of the engine, the I or E indicates intake or exhaust and the number indicates the journal position from the front of the engine.

3. Remove the camshafts.

To install:

4. Apply engine lubricant to the camshaft journals and carriers and the camshaft bearing caps.

5. Ensure the camshafts are placed in the correct position by locating the identification letters stamped near the rear journal. For example: L-INT indicates left intake camshaft.

6. Install the camshafts with the camshaft sprocket drive pins at the top of their rotation and the lobes in the neutral position.

7. Install the bearing caps in their correct locations and hand start the bearing cap bolts.

➡Ensure each rocker arm is properly aligned with the valve tip, the lifter and the cam lobe.

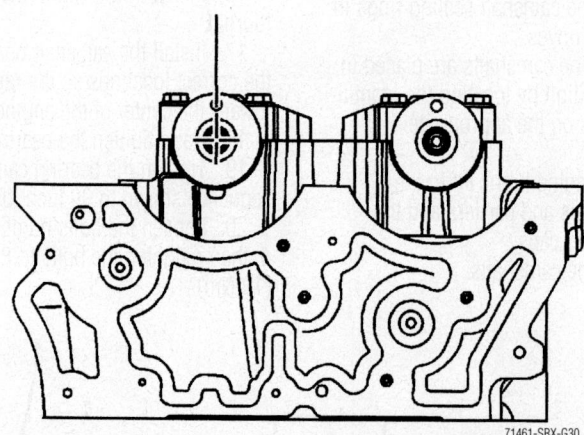

Installing camshafts with locating pins at top of the rotation—4.6L engine, left side

71461-SRX-G30

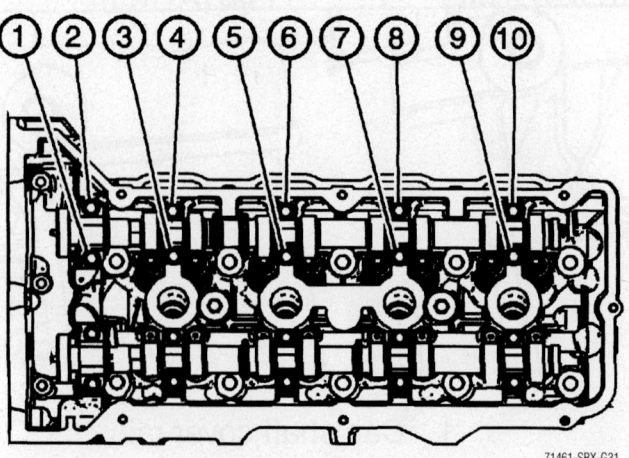

71461-SRX-G31

Camshaft bearing cap tightening sequence—4.6L engine, left side intake

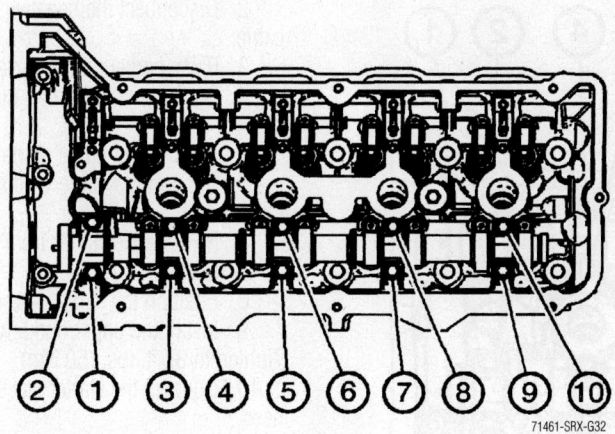

Camshaft bearing cap tightening sequence—4.6L engine, left side exhaust

8. Tighten the bearing cap bolts in sequence to 44 inch lbs. (5 Nm), plus an additional 30°.

9. Install or connect the following:
- Camshaft position actuators
- Timing chains
- Front cover
- Camshaft cover
- Ignition coils
- Wiring harnesses on camshaft cover
- Ignition coil connectors
- Intake manifold
- Engine

RIGHT SIDE

➡**The camshaft position actuators and timing chains are removed in the Front Cover and Timing Chain procedure.**

1. Before servicing the vehicle, refer to the Precautions Section.

2. Remove or disconnect the following:
- Engine
- Intake manifold
- Camshaft cover
- Front cover
- Timing chains
- Camshaft position actuators
- Camshaft bearing caps

➡**Observe the positions of the camshaft bearing caps. The arrow on the cap points toward the front of the engine, the I or E indicates intake or exhaust and the number indicates the journal position from the front of the engine.**

3. Remove the camshafts.

To install:

4. Apply engine lubricant to the camshaft journals and carriers and the camshaft bearing caps.

5. Ensure the camshafts are placed in the correct position by locating the identifi-

cation letters stamped near the rear journal. For example: R-INT indicates right intake camshaft.

6. Install the right exhaust camshaft with the camshaft sprocket drive pins at the 10 and 60 degree locations and the lobes in the neutral position.

7. Install the right intake camshaft with the camshaft sprocket drive pins at the 25 and 60 degree locations and the lobes in the neutral position.

8. Install the bearing caps in their correct locations and hand start the bearing cap bolts.

➡**Ensure each rocker arm is properly aligned with the valve tip, the lifter and the cam lobe.**

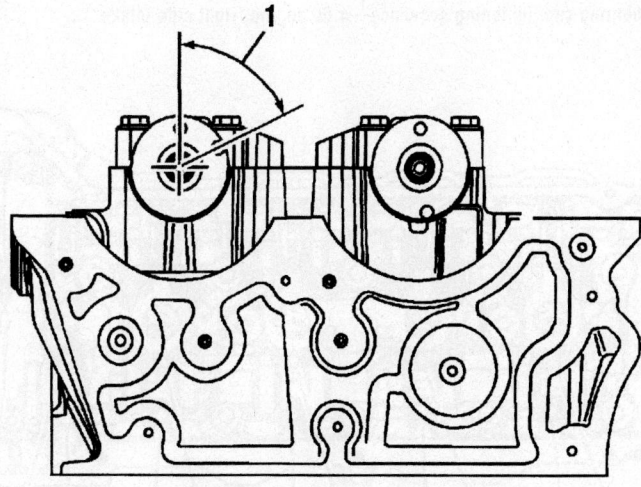

1. 60 and 10 degrees

Installing the right exhaust camshaft with locating pins at 10 and 60 degree locations—4.6L engine, right side

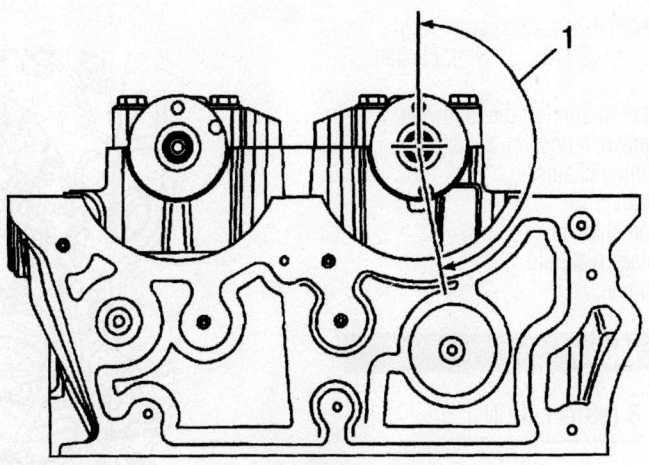

1. 25 and 60 degrees

Installing the right intake camshaft with locating pins at 25 and 60 degree locations—4.6L engine, right side

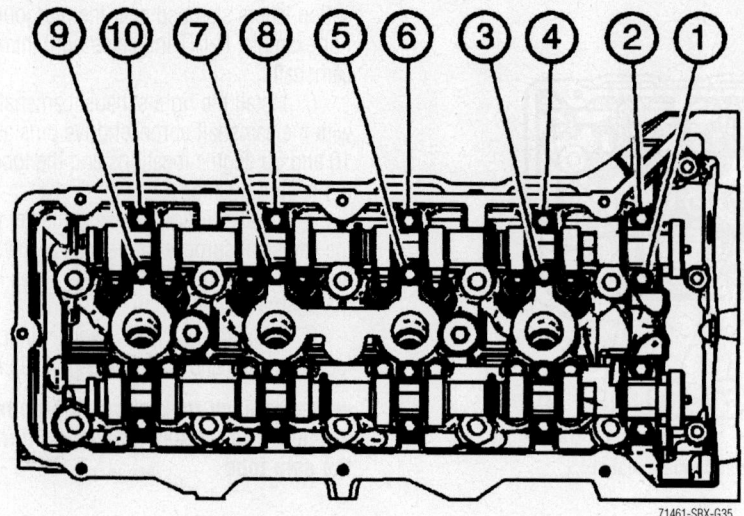

Camshaft bearing cap tightening sequence—4.6L engine, right side intake

71461-SRX-G35

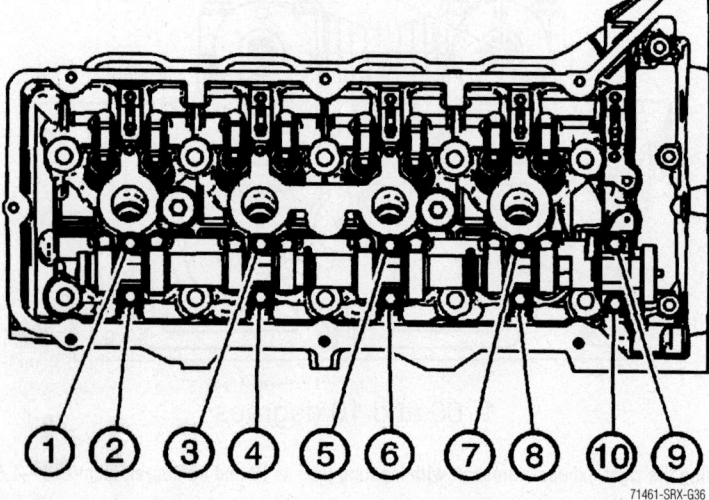

Camshaft bearing cap tightening sequence—4.6L engine, right side exhaust

71461-SRX-G36

9. Tighten the bearing cap bolts in sequence to 44 inch lbs. (5 Nm), plus an additional 30°.

10. Install or connect the following:
- Camshaft position actuators
- Timing chains
- Front cover
- Camshaft cover
- Intake manifold
- Engine

Starter Motor

REMOVAL & INSTALLATION

3.6L Engine

1. Before servicing the vehicle, refer to the Precautions Section.

2. Disconnect the negative battery cable.

3. Raise and support the vehicle safely.

4. Disconnect the starter electrical harness.

5. Remove the upper starter bolt.

6. Support the starter and remove the lower bolt.

7. Remove the starter from the vehicle.

To install:

8. Position the starter in the vehicle.

9. Install the upper and lower bolts. Tighten to 37 ft. lbs. (50 Nm).

10. Connect the starter electrical harness.

11. Lower the vehicle.

12. Connect the negative battery cable.

4.6L Engine

1. Before servicing the vehicle, refer to the Precautions Section.

2. Disconnect the negative battery cable.

3. Remove the intake manifold.

4. Disconnect the starter electrical harness.

5. Remove the starter mounting bolts.

6. Remove the starter from the engine.

To install:

7. Position the starter on the engine.

8. Install or connect the following:
- Starter mounting bolts. Tighten to 22 ft. lbs. (30 Nm).
- Starter electrical harness. Tighten the motor stud nut to 84 inch lbs. (9.5Nm) and tighten the solenoid stud nut to 30 inch lbs. (3.4Nm).
- Intake manifold
- Negative battery cable

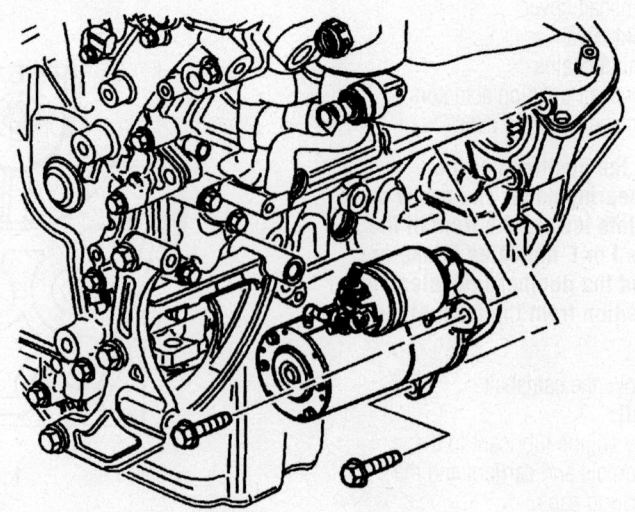

Starter motor mounting—3.6L engine

71461-SRX-G37

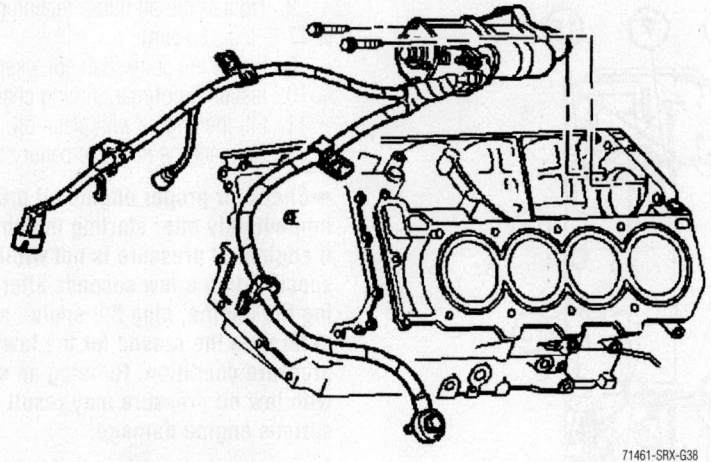

Starter motor mounting—4.6L engine

71461-SRX-G38

Oil Pan

REMOVAL & INSTALLATION

3.6L Engine

1. Before servicing the vehicle, refer to the Precautions Section.
2. Disconnect the negative battery cable.
3. Drain the engine oil.
4. Remove the engine front cover.
5. Remove the power steering hose retainer from the A/C compressor bracket.
6. Disconnect the intermediate steering shaft.
7. Remove the engine mount lower nuts.
8. Position the A/C compressor aside.
9. Remove the transmission oil cooler pipe retainer from the right side of the engine.
10. Install an engine lifting kit to raise the engine for clearance.
11. Remove the retaining bolts and remove the oil pan.

To install:

12. Clean the gasket mating surfaces thoroughly.
13. Trial fit the oil pan to the cylinder block. Ensure that enough clearance has been provided to allow the oil pan to be installed without sealant being scraped off when pan is positioned under the engine.
14. Apply a bead of silicone sealer to the oil pan flange as shown.
15. Install the oil pan and loosely install the attaching bolts.
16. Tighten the bolts in sequence.
   - Tighten bolts 1–11 to 17 ft. lbs. (23 Nm).
   - Tighten bolts 11–12 to 89 inch lbs. (10 Nm).
17. Lower the engine to engage the engine mounts.

18. Remove the engine lift.
19. Install the transmission oil cooler pipe retainer to the right side of the engine.
20. Reposition the A/C compressor.
21. Install the engine mount lower nuts. Tighten the nuts to 59 ft. lbs. (80 Nm).
22. Connect the intermediate steering shaft.
23. Install the power steering hose retainer to the A/C compressor bracket.
24. Install the engine front cover.
25. Fill the engine with oil.
26. Connect the negative battery cable.

4.6L Engine

1. Before servicing the vehicle, refer to the Precautions Section.

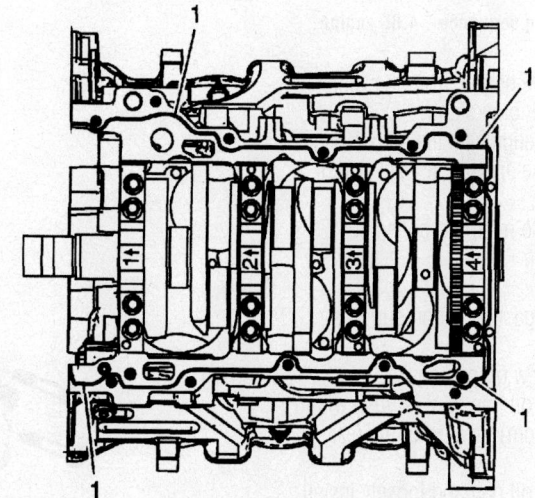

1. Sealant application areas

71461-SRX-G39

Oil pan sealant application locations—3.6L engine

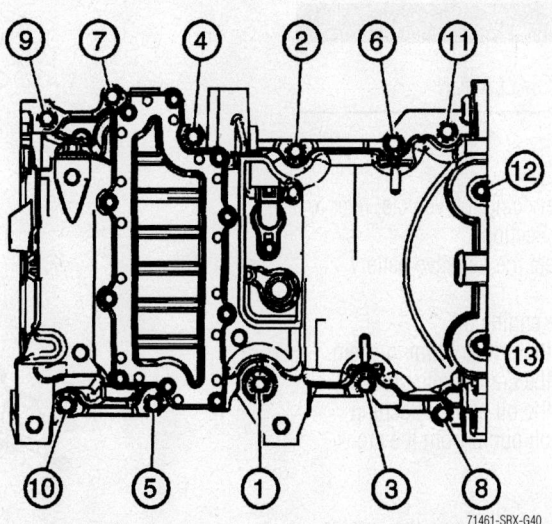

71461-SRX-G40

Oil pan tightening sequence—3.6L engine

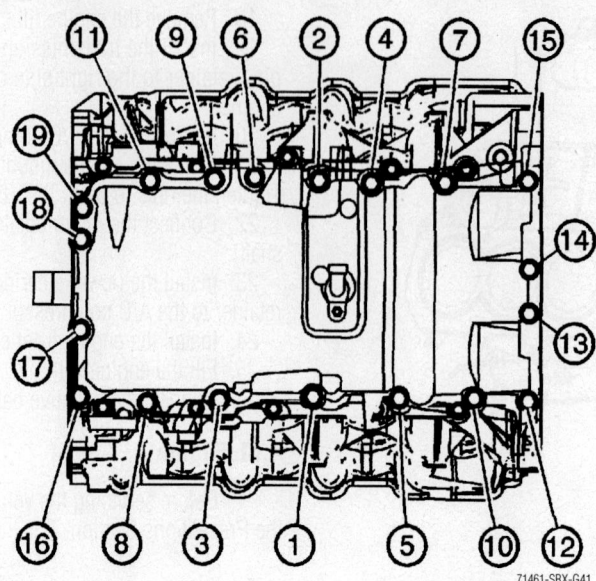

Oil pan tightening sequence—4.6L engine

71461-SRX-G41

2. Remove the engine from the vehicle and place on an engine stand.

3. Rotate the engine so the oil pan is up.

4. Remove the engine oil level sensor from the oil pan.

5. Remove the retaining bolts and remove the oil pan.

To install:

6. Clean the gasket mating surfaces thoroughly.

7. Install a new oil pan gasket.

8. Install the oil pan so it is flush within 0.020 inch (0.50mm) forward of the rear face of the block.

9. Install the oil pan and loosely install the attaching bolts.

10. Tighten the bolts in sequence as shown to 89 inch lbs. (10 Nm).

11. Install the oil level sensor.

Oil Pump

REMOVAL & INSTALLATION

3.6L Engine

1. Before servicing the vehicle, refer to the Precautions Section.

2. Disconnect the negative battery cable.

3. Drain the engine oil.

4. Remove the primary timing chain.

5. Remove the crankshaft sprocket.

6. Remove the oil pump attaching bolts. Slide the oil pump from the crankshaft.

To install:

7. Install the oil pump so the gerotor aligns with the crankshaft flats.

8. Tighten the oil pump retaining bolts to 17 ft. lbs. (23 Nm).

9. Install the crankshaft sprocket.

10. Install the primary timing chain.

11. Fill the engine with clean oil.

12. Connect the negative battery cable.

➡**Check for proper engine oil pressure immediately after starting the engine. If engine oil pressure is not within specification a few seconds after starting the engine, stop the engine and determine the reason for the low oil pressure condition. Running an engine with low oil pressure may result in serious engine damage.**

13. Start the engine and check for leaks.

4.6L Engine

1. Before servicing the vehicle, refer to the Precautions Section.

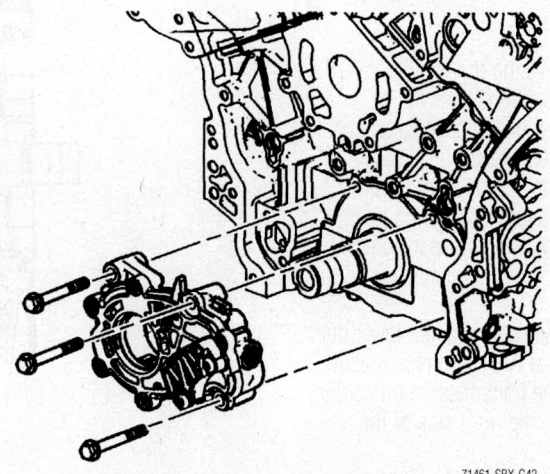

Oil pump mounting—3.6L engine

71461-SRX-G42

Oil pump bolt tightening sequence—4.6L engine

71461-SRX-G43

2. Disconnect the negative battery cable.

3. Drain the engine oil.

4. Remove the engine front cover.

5. Remove the oil pump attaching bolts. Slide the oil pump from the crankshaft.

To install:

6. Install the oil pump drive spacer so the drive flat engages the pump rotor.

7. Tighten the oil pump retaining bolts and tighten in sequence to 89 inch lbs. (10 Nm), plus an additional 35°.

8. Install the engine front cover.

9. Fill the engine with clean oil.

10. Connect the negative battery cable.

➥**Check for proper engine oil pressure immediately after starting the engine. If engine oil pressure is not within specification a few seconds after starting the engine, stop the engine and determine the reason for the low oil pressure condition. Running an engine with low oil pressure may result in serious engine damage.**

11. Start the engine and check for leaks.

Rear Main Seal

REMOVAL & INSTALLATION

3.6L Engine

1. Before servicing the vehicle, refer to the Precautions Section.

2. Disconnect the negative battery cable.

3. Raise and support the vehicle safely on jackstands.

4. Remove the transmission.

5. Remove the flywheel.

6. Remove the oil pan.

7. Remove the real seal and housing attaching bolts and remove the housing.

To install:

8. Install Crankshaft Rear Oil Seal Installation Tool EN-47839 onto the rear of the crankshaft flange.

9. Place a bead of RTV sealant around the seal housing mounting surface.

10. Install the seal housing and tighten the bolts to 89 inch lbs. (10 Nm).

11. Remove Installation Tool EN-47839

12. Install the oil pan.

13. Install the flywheel. Tighten the NEW bolts to 22 ft. lbs. (30 Nm), plus an additional 45°.

14. Install the transmission.

15. Lower the vehicle and connect the battery.

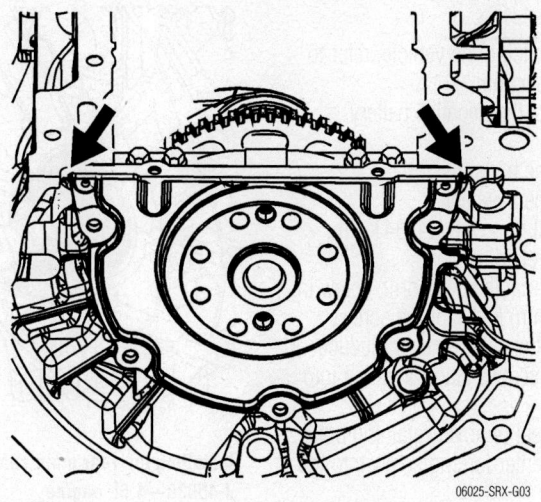

Location of the pry points to remove the rear main seal—3.6L engine

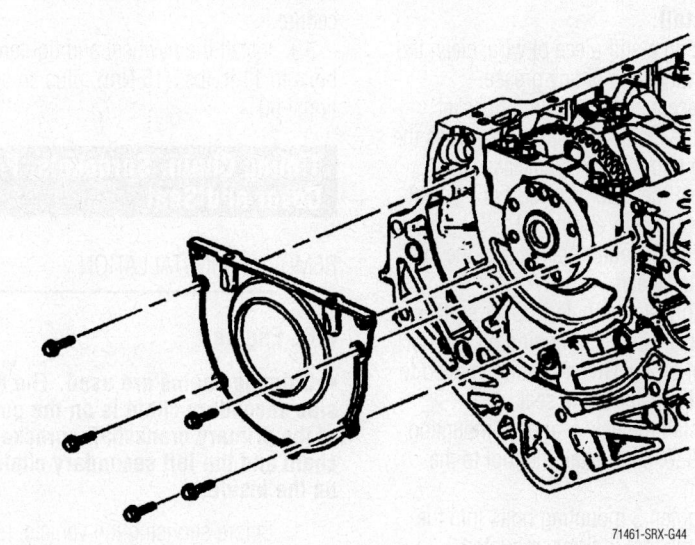

Rear main seal mounting—3.6L engine

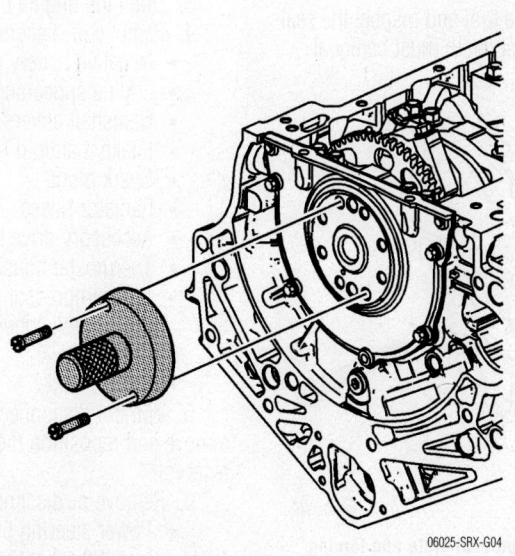

Using the rear oil seal installation tool–3.6L engine.

4.6L Engine

1. Before servicing the vehicle, refer to the Precautions Section.

2. Disconnect the negative battery cable.

3. Remove the transmission.

4. Remove the flywheel.

5. Install seal remover J-42841 onto the seal using 2 retaining bolts.

6. Using a variable speed drill, screw in eight 1 inch (25mm) self-tapping screws into the holes in the seal remover. Reduce drill speed when screws start threading into the seal.

7. Remove seal remover retaining bolts.

8. Install a center forcing screw into the remover.

9. Using a socket wrench, tighten the center screw until the seal is pulled off the crankshaft.

To install:

10. Using a stiff piece of wire, clean the seal mounting area of any grease.

11. Place a small amount of gasket maker at the crankshaft split line across the end of the upper/lower crankcase seal.

12. Coat the outer diameter of the seal opening with clean engine oil.

13. Wipe the outer diameter of the flywheel flange clean.

14. Lubricate the outer rubber surface of the new seal with clean engine oil. Do not get any oil on the Green coating applied to the inner diameter of the seal.

15. Install the new seal onto mounting tool J-45930, and install the tool to the crankshaft.

16. Tighten 3 mounting bolts into the tool until the tool is firmly mounted.

17. Tighten the tool center bolt until the tool bottoms against the crankcase.

18. Remove the tool and inspect the seal mounting. The seal depth must be equal

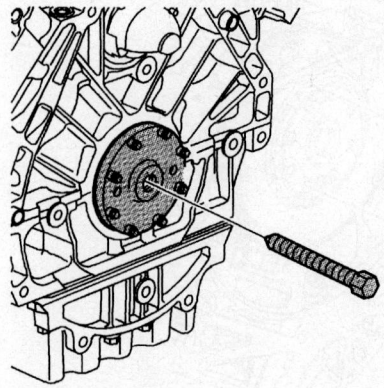

Rear main seal removal plate and forcing screw—4.6L engine

71461-SRX-G45

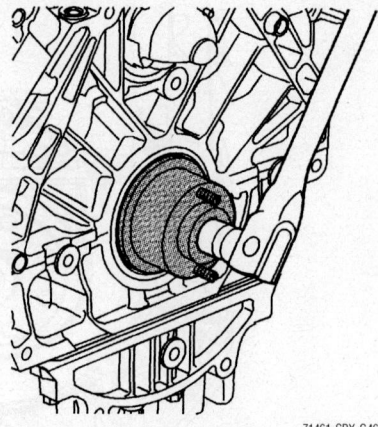

Installing the rear main seal using tool J-45930—4.6L engine

71461-SRX-G46

around the circumference. If the depth is not equal, reinstall the tool and repeat the procedure.

19. Install the flywheel and tighten the bolts to 11 ft. lbs. (15 Nm), plus an additional 50°.

Timing Chain, Sprockets, Front Cover and Seal

REMOVAL & INSTALLATION

3.6L Engine

➡3 timing chains are used. The right side secondary chain is on the outside of the primary crankshaft sprocket chain and the left secondary chain is on the inside.

1. Before servicing the vehicle, refer to the Precautions Section.

2. Drain the cooling system.

3. Drain the engine oil.

4. Remove or disconnect the following:
- Negative battery cable
- Engine appearance cover
- Camshaft covers
- Intake manifold
- Spark plugs
- Radiator hoses
- Accessory drive belts
- Thermostat housing
- A/C compressor and power steering belt tensioners
- Alternator
- Starter

5. Without disconnecting the lines, remove and reposition the power steering reservoir.

6. Remove or disconnect the following:
- Power steering pump pulley
- Install flywheel locking tool EN-

46106 in the starter mounting holes
- Crankshaft damper bolt
- Using a gear puller, remove the crankshaft damper

7. To remove the front seal only, pry the seal out of the front cover opening.

8. Remove or disconnect the following:
- 4 Camshaft Position (CMP) sensors
- 4 CMP actuator solenoids
- Water pump pulley
- Front cover bolts, cover and gasket

9. Ensure the engine is placed in the Stage 2 timing drive chain alignment position as shown.

10. Remove or disconnect the following:
- Right side secondary drive chain tensioners, guides and shoes
- Right secondary drive chain
- Primary drive chain tensioner and upper guide
- Primary drive chain
- Right side secondary drive chain idler sprockets
- Left side secondary drive chain tensioners, guides and shoes
- Left side secondary drive chain idler sprockets
- Left secondary drive chain

➡The lower primary drive chain guide is not serviceable. If the guide needs to be replaced, the guide and oil pump must be replaced as an assembly.

11. If necessary, remove the crankshaft sprocket.

➡There are 4 camshaft position actuator sprockets used. Left and right exhaust and left and right intake camshafts.

12. If the camshafts are to be removed, place an open end wrench on the appropriate camshaft hex location and hold the camshaft, while the removing the camshaft position actuator sprocket bolt and sprocket.

13. If the front seal was not removed earlier, pry the seal out of the front cover opening.

To install:

14. Clean all the gasket mating surfaces.

15. If the camshafts were removed, install the camshafts using the procedure described under Camshafts.

16. Install the camshaft position actuator sprockets and tighten the bolts to 43 ft. lbs. (58 Nm).

17. If the front cover was removed, use a seal installer and install a new front seal into the front cover opening.

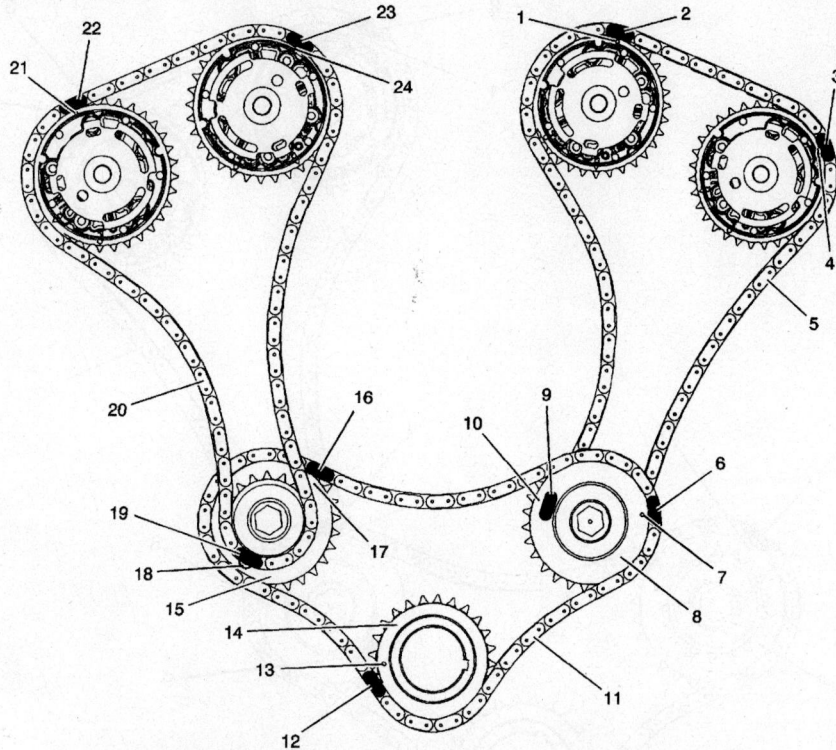

1. Left Intake Camshaft Position Actuator (CMP) Timing Mark
2. Left Intake Secondary Camshaft Timing Drive Chain Bright Plated Link
3. Left Exhaust Secondary Camshaft Timing Drive Chain Bright Plated Link
4. Left Exhaust Camshaft Position Actuator (CMP) Timing Mark
5. Left Secondary Camshaft Timing Drive Chain
6. Primary Camshaft Drive Chain Bright Plated Link for the
   Left Primary Camshaft Intermediate Drive Chain Sprocket
7. Left Primary Camshaft Intermediate Drive Chain Sprocket
   Timing Mark for the Primary Camshaft Drive Chain
8. Left Primary Camshaft Intermediate Drive Chain Sprocket
9. Left Secondary Camshaft Timing Drive Chain Bright Plated
   Link for the Left Primary Camshaft Intermediate Drive Chain Sprocket
10. Left Primary Camshaft Intermediate Drive Chain Sprocket Timing Window
11. Primary Camshaft Drive Chain
12. Primary Camshaft Drive Chain Bright Plated Link for the Crankshaft Sprocket
13. Crankshaft Sprocket Timing Mark
14. Crankshaft Sprocket
15. Right Primary Camshaft Intermediate Drive Chain Sprocket
16. Primary Camshaft Drive Chain Bright Plated Link for the Right
    Primary Camshaft Intermediate Drive Chain Sprocket
17. Right Primary Camshaft Intermediate Drive Chain Sprocket
    Timing Mark for the Primary Camshaft Drive Chain
18. Right Primary Camshaft Intermediate Drive Chain Sprocket
    Timing Mark/Window for the Right Secondary Camshaft Timing Drive Chain
19. Right Secondary Camshaft Timing Drive Chain Bright Plated Link
    for the Right Primary Camshaft Intermediate Drive Chain Sprocket
20. Right Secondary Camshaft Timing Drive Chain
21. Right Exhaust Camshaft Position Actuator (CMP) Timing Mark
22. Right Exhaust Secondary Camshaft Timing Drive Chain Bright Plated Link
23. Right Intake Camshaft Position Actuator (CMP) Timing Mark
24. Right Intake Camshaft Position Actuator (CMP) Timing Mark

71461-SRX-G47

Stage 2 timing drive chain alignment position—3.6L engine

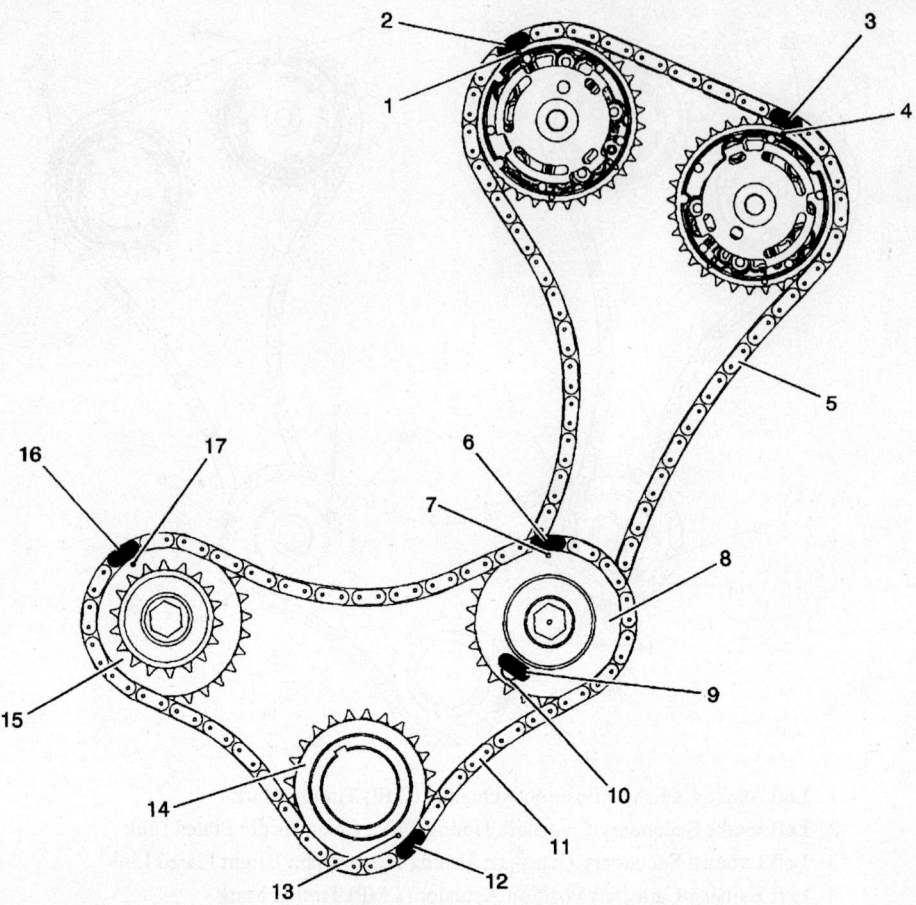

1. Left Intake Camshaft Position Actuator (CMP) Timing Mark
2. Left Intake Secondary Camshaft Timing Drive Chain Bright Plated Link
3. Left Exhaust Secondary Camshaft Timing Drive Chain Bright Plated Link
4. Left Exhaust Camshaft Position Actuator (CMP) Timing Mark
5. Left Secondary Camshaft Timing Drive Chain
6. Primary Camshaft Drive Chain Bright Plated Link for the Left Primary Camshaft Intermediate Drive Chain Sprocket
7. Left Primary Camshaft Intermediate Drive Chain Sprocket Timing Mark for the Primary Camshaft Drive Chain
8. Left Primary Camshaft Intermediate Drive Chain Sprocket
9. Left Secondary Camshaft Timing Drive Chain Bright Plated Link for the Left Primary Camshaft Intermediate Drive Chain Sprocket
10. Left Primary Camshaft Intermediate Drive Chain Sprocket Timing Window for the Left Secondary Camshaft Timing Drive Chain Bright Plated Link
11. Primary Camshaft Drive Chain
12. Primary Camshaft Drive Chain Bright Plated Link for the Crankshaft Sprocket
13. Crankshaft Sprocket Timing Mark
14. Crankshaft Sprocket
15. Right Primary Camshaft Intermediate Drive Chain Sprocket
16. Primary Camshaft Drive Chain Bright Plated Link for the Right Primary Camshaft Intermediate Drive Chain Sprocket
17. Right Primary Camshaft Intermediate Drive Chain Sprocket Timing Mark

71461-SRX-G48

Stage 1 timing drive chain alignment position—3.6L engine

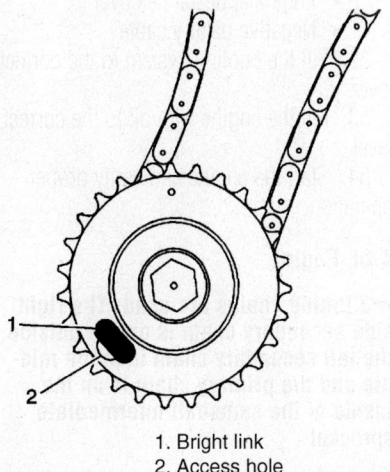

1. Bright link
2. Access hole

71461-SRX-G49

Aligning left secondary timing chain to lower idler sprocket—3.6L engine

18. If removed, install the crankshaft sprocket with the timing mark out and placed in the Stage 1 timing drive position.

19. Install camshaft holding tools EN-46105 on the camshaft rear flats to hold them in position.

20. Install the left side secondary drive chain idler sprockets with the part number outs and the larger sprocket on the outside. Tighten the bolt to 48 ft. lbs. (65 Nm).

21. Install the left secondary timing chain on the inner idler sprocket with the bright plate chain drive link aligned with the hole in the outer sprocket as shown.

22. Place the timing chain around the camshaft sprockets and ensure there are 7 links between the bright links as shown.

23. There will be 18 darkened links between the left camshaft intermediate drive chain idler bright plated secondary camshaft drive chain link and each left camshaft posi-

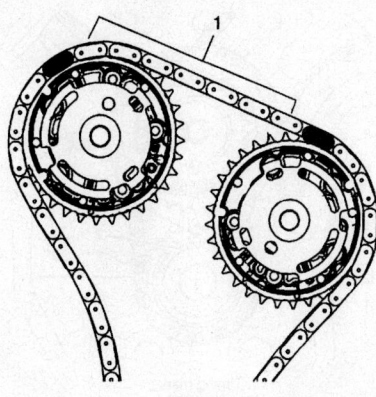

1. 7 links

71461-SRX-G50

Aligning left secondary timing chain to camshaft sprockets—3.6L engine

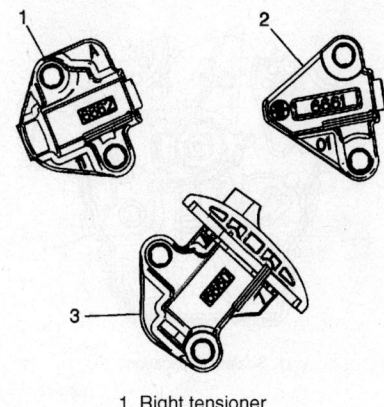

1. Right tensioner
2. Left tensioner
3. Primary tensioner

71461-SRX-G51

Identifying timing chain tensioners—3.6L engine

tion actuator sprocket bright plated secondary camshaft drive chain link.

24. Ensure that the correct tensioner is installed in the correct positions as shown.

25. Using tensioner tool J-45027 reset the left tensioner plunger.

26. Install the plunger in the tensioner body and lock it in place with a paper clip.

27. Release the pressure on the tensioner and ensure it remains compressed.

28. Install a new gasket on the tensioner, install the tensioner and tighten the bolts to 17 ft. lbs. (23 Nm).

29. Install the left side chain guide and chain shoe and tighten the bolts to 17 ft. lbs. (23 Nm).

30. Install the right side secondary drive chain idler sprockets with the part number outs and the larger sprocket on the outside. Tighten the bolt to 48 ft. lbs. (65 Nm).

31. Place the primary timing chain around the camshaft idler sprockets and crankshaft sprocket with the links and timing marks aligned as shown.

32. Using tensioner tool J-45027 reset the primary tensioner plunger.

33. Install the plunger in the tensioner body and lock it in place with a paper clip.

34. Release the pressure on the tensioner and ensure it remains compressed.

35. Install a new gasket on the tensioner, install the tensioner and tighten the bolts to 44 inch lbs. (5 Nm) and then a final pass of 17 ft. lbs. (23 Nm).

36. Release the tension on the primary tensioner, and ensure the primary and left side timing chains are in the Stage 1 position.

37. Remove the camshaft holding tool from the left camshafts.

38. Rotate the crankshaft sprocket clock-

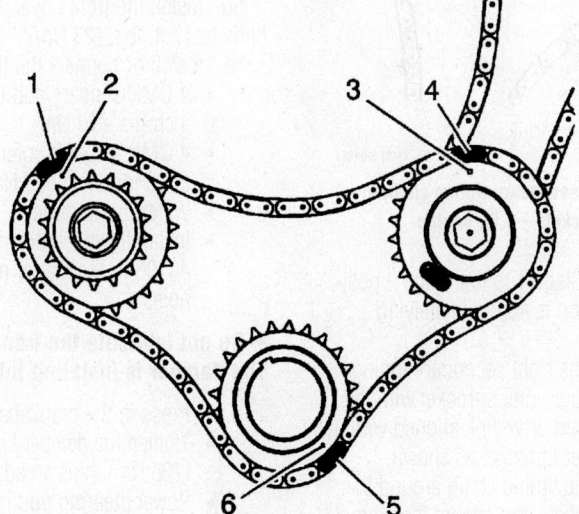

1. Bright chain link
2. Timing mark
3. Timing mark
4. Bright chain link
5. Bright chain link
6. Timing mark

71461-SRX-G52

Aligning primary timing chain to camshaft idler and crankshaft sprockets—3.6L engine

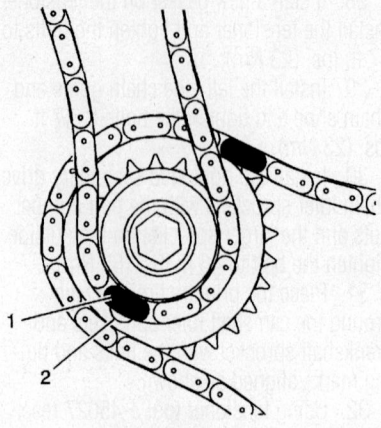

1. Bright link
2. Access hole

71461-SRX-G53

Aligning right secondary timing chain to lower idler sprocket—3.6L engine

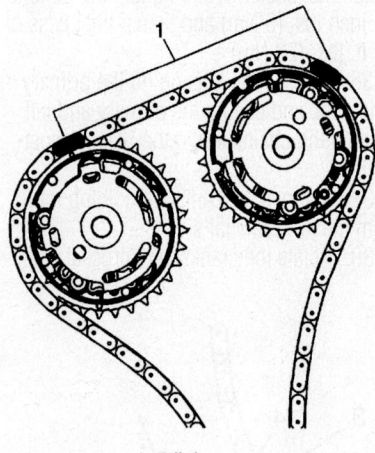

1. 7 links

71461-SRX-G54

Aligning right secondary timing chain to camshaft sprockets—3.6L engine

wise from the Stage 1 to the Stage 2 position. The rotation is approximately 15 degrees.

39. Install the right secondary timing chain on the outer idler sprocket with the bright plate chain drive link aligned with the hole in the inner sprocket as shown.

40. Place the timing chain around the camshaft sprockets and ensure there are 7 links between the bright links as shown.

41. There will be 18 darkened links between the right camshaft intermediate drive chain idler bright plated secondary camshaft drive chain link and each right camshaft position actuator sprocket bright plated secondary camshaft drive chain link.

42. Using tensioner tool J-45027 reset the right tensioner plunger.

43. Install the plunger in the tensioner body and lock it in place with a paper clip.

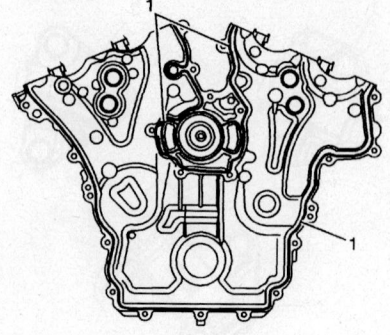

1. Sealant application

71461-SRX-G55

Front cover sealant application areas— 3.6L engine

44. Release the pressure on the tensioner and ensure it remains compressed.

45. Install a new gasket on the tensioner, install the tensioner and tighten the bolts to 17 ft. lbs. (23 Nm).

46. Install the right side chain guide and chain shoe and tighten the bolts to 17 ft. lbs. (23 Nm).

47. Verify the timing chains are aligned in the Stage 2 timing drive chain alignment position as shown.

48. Install a new front cover-to-cylinder block seal.

49. Place a bead of RTV sealant onto the front cover as shown.

50. Install the front cover and tighten the bolts to 17 ft. lbs. (23 Nm).

51. Install or connect the following:
- 4 CMP sensors and tighten to 89 inch lbs. (10 Nm).
- 4 CMP actuator solenoids tighten to 89 inch lbs. (10 Nm).
- Water pump pulley
- Install flywheel locking tool EN-46106 in the starter mounting holes

➡**Do not lubricate the front seal bore. The damper is installed into a dry bore.**

- Press in the crankshaft damper
- Tighten the damper bolt to 74 ft. lbs. (100 Nm), plus an additional 150˚.
- Power steering pump pulley
- Power steering reservoir
- Starter
- Alternator
- A/C compressor and power steering belt tensioners
- Thermostat housing
- Accessory drive belts
- Radiator hoses
- Spark plugs
- Intake manifold
- Camshaft covers

- Engine appearance cover
- Negative battery cable

52. Fill the cooling system to the correct level.

53. Fill the engine with oil to the correct level.

54. Start the engine and verify proper operation.

4.6L Engine

➡**3 timing chains are used. The right side secondary chain is on the outside, the left secondary chain is in the middle and the primary chain is on the inside of the camshaft intermediate sprocket.**

1. Before servicing the vehicle, refer to the Precautions Section.

2. Drain the cooling system.

3. Drain the engine oil.

4. Remove or disconnect the following:
- Negative battery cable
- Air outlet duct
- Radiator hoses
- Camshaft covers
- Accessory drive belts
- Drive belt tensioners
- Drive belt idler pulley
- Water pump pulley
- Fan adapter, if equipped
- Starter
- Water pump pulley

5. Install a flywheel holding tool in the starter mounting holes.

6. Remove or disconnect the following:
- Crankshaft damper bolt
- Using a puller, remove the crankshaft damper
- Engine front cover and gasket
- Oil pump

7. Rotate the crankshaft until the primary timing marks are aligned as shown.

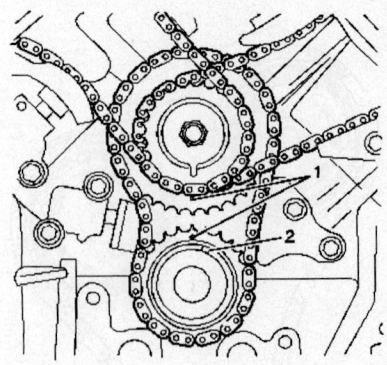

1. Timing marks
2. Crankshaft sprocket

71461-SRX-G56

Aligning primary timing marks—4.6L engine

8. Remove or disconnect the following:
- Right side Camshaft Position (CMP) sensors
- Right side CMP solenoids.
- Right side CMP actuator housings
- Install camshaft holding tool EN-46328 on the right side camshafts
- Right side timing chain tensioner

9. Place an open end wrench on the exhaust camshaft hex to hold the camshaft and remove the oil control valve.

10. Slide the right exhaust camshaft position actuator off of the camshaft and remove the secondary timing chain from the actuator.

11. Place an open end wrench on the intake camshaft hex to hold the camshaft and remove the oil control valve.

12. Slide the right intake camshaft position actuator off of the camshaft and remove the secondary timing chain from the actuator.

13. Remove the timing chain.

14. Remove the timing chain guide and shoe.

15. Remove or disconnect the following:
- Left side Camshaft Position (CMP) sensors
- Left side CMP actuator solenoids
- Left side CMP actuator housings
- Install camshaft holding tool EN-46328 on the left side camshafts
- Left side timing chain tensioner

16. Place an open end wrench on the exhaust camshaft hex to hold the camshaft and remove the oil control valve.

17. Slide the left exhaust camshaft position actuator off of the camshaft and remove the secondary timing chain from the actuator.

18. Place an open end wrench on the intake camshaft hex to hold the camshaft and remove the oil control valve.

19. Slide the left intake camshaft position actuator off of the camshaft and remove the secondary timing chain from the actuator.

20. Remove the timing chain.

21. Remove the timing chain guide and shoe.

22. Remove the primary timing chain tensioner.

23. Remove the oil outlet tube.

24. Remove the camshaft intermediate sprocket bolt.

25. Remove the primary timing drive chain guide.

26. Remove the intermediate sprocket, primary drive chain and crankshaft sprocket as an assembly.

27. Pry the front seal from the timing cover.

To install:

28. Clean all the gasket mating surfaces.

29. Using a seal installer, press in a new front crankshaft seal into the front cover.

30. Align the intermediate and crankshaft sprocket timing marks so they are vertical as shown.

31. Install the primary timing chain around the intermediate and crankshaft sprockets,

32. Verify that the no. 1 piston is at TDC and the crankshaft keyway is at about the 5° ATDC position.

33. Install the intermediate/crankshaft sprocket/timing chain assembly onto the crankshaft and camshaft driveshafts.

34. Install the intermediate sprocket bolt and tighten to 44 ft. lbs. (60 Nm).

35. Install the primary drive chain guide and tighten the bolts to 18 ft. lbs. (25 Nm).

36. Holding the primary drive chain tensioner in one hand, rotate the ratchet release lever counterclockwise with the other hand and place a paper clip to hold it.

37. Collapse the tensioner shoe and hold while slowly releasing the ratchet lever to relive tension on the shoe.

38. Install the chain tensioner and tighten the bolt to 18 ft. lbs. (25 Nm).

39. Remove the paper clip and allow the tensioner to load.

40. Install the oil outlet tube and tighten the bolts to 89 inch lbs. (10 Nm).

41. Install the left side drive chain guide and shoe and tighten the bolts to 18 ft. lbs. (25 Nm).

42. Guide the left timing chain through the cylinder head and onto the inner row of the intermediate drive chain sprocket teeth.

43. Install the left camshaft position actuators into the drive chain.

44. Install the camshaft sprockets on the camshafts.

➡**Ensure the camshaft sprockets notches marked LI or LE engage the proper camshaft pins.**

45. Loosely install the actuator oil control valves.

46. Verify that the camshaft sprocket notches and the camshaft pins are 90° perpendicular to each other.

47. Install camshaft locking tool J-46328 to the left camshafts.

48. Install the upper drive chain shoe guide bolt and then tighten the upper and lower bolts to 18 ft. lbs. (25 Nm).

49. Holding the left drive chain tensioner in one hand, rotate the ratchet release lever counterclockwise with the other hand and place a paper clip to hold it.

50. Collapse the tensioner shoe and

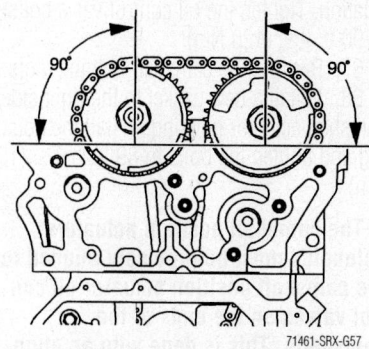

71461-SRX-G57

Aligning left camshaft sprocket marks and camshaft pin marks—4.6L engine; right side similar

hold while slowly releasing the ratchet lever to relive tension on the shoe.

51. Install the chain tensioner and tighten the bolt to 18 ft. lbs. (25 Nm).

52. Remove the paper clip and allow the tensioner to load.

53. Install the right side drive chain guide and shoe and tighten the bolts to 18 ft. lbs. (25 Nm).

54. Guide the right timing chain through the cylinder head and onto the outer row of the intermediate drive chain sprocket teeth.

55. Install the right camshaft position actuators into the drive chain.

56. Install the camshaft sprockets on the camshafts.

➡**Ensure the camshaft sprockets notches marked RI or RE engage the proper camshaft pins.**

57. Loosely install the actuator oil control valves.

58. Verify that the camshaft sprocket notches and the camshaft pins are 90° perpendicular to each other.

59. Install camshaft locking tool J-46328 to the right camshafts.

60. Install the upper drive chain shoe guide bolt and then tighten the upper and lower bolts to 18 ft. lbs. (25 Nm).

61. Holding the right drive chain tensioner in one hand, rotate the ratchet release lever counterclockwise with the other hand and place a paper clip to hold it.

62. Collapse the tensioner shoe and hold while slowly releasing the ratchet lever to relive tension on the shoe.

63. Install the chain tensioner and tighten the bolt to 18 ft. lbs. (25 Nm).

64. Remove the paper clip and allow the tensioner to load.

65. Ensure the correct alignment of all timing chain marks.

66. Tighten all 4 oil control valves, holding the camshaft flats to prevent camshaft

rotation. Tighten the oil control valve bolts to 90 ft. lbs. (120 Nm).

67. Remove the camshaft holding tools.

68. Install a new gasket to the right side camshaft actuator housing, install the housing and tighten the bolts to 89 inch lbs. (10 Nm).

➡ **The camshaft position actuators solenoids must be precisely aligned to the camshaft position actuator oil control valves on the ends of the camshafts. This is done with an alignment pin. Failure to align the solenoids correctly can lead to poor engine performance and component damage.**

69. Fabricate an alignment pin using a 15/64 inch drill bit at least 2 inches long.

70. Verify the alignment pin will pass through intake actuator solenoid alignment hole and the control valve alignment hole.

71. Apply a bead of RTV sealant around the actuator solenoid flange.

72. Install the actuator over the oil control valve and insert the alignment pin to align the valve to the solenoid.

73. Tighten the actuator solenoid bolts to 71 inch lbs. (8 Nm).

74. Install a new actuator solenoid plug.

75. Repeat this procedure on the right exhaust actuator solenoid, and both left side actuator solenoids.

76. Install the oil pump.

77. Clean the front cover gasket surface.

78. Place a small amount of sealant to the split line of the upper and lower crankcases and the top edge of the block face.

79. Install a new front cover gasket over the crankcase dowel pins.

80. Install the front cover and hand tighten the bolts.

81. Tighten the front cover bolts in the sequence shown to 106 inch lbs. (12 Nm).

82. Install the 4 CMP sensors and tighten to 89 inch lbs. (10 Nm).

83. Install the camshaft covers

84. Press the crankshaft damper on.

85. Coat the damper bolt with clean engine oil, install the bolt and tighten to 37 ft. lbs. (50 Nm), plus an additional 120°.

86. Remove the crankshaft holding tool from the starter opening and install the starter.

87. Install the water pump pulley and tighten the bolts to 89 inch lbs. (10 Nm).

88. Install the drive belt idler pulley and tighten to 37 ft. lbs. (50 Nm).

89. Install or connect the following:
- Fan adapter, if equipped
- Drive belt tensioners
- Accessory drive belts

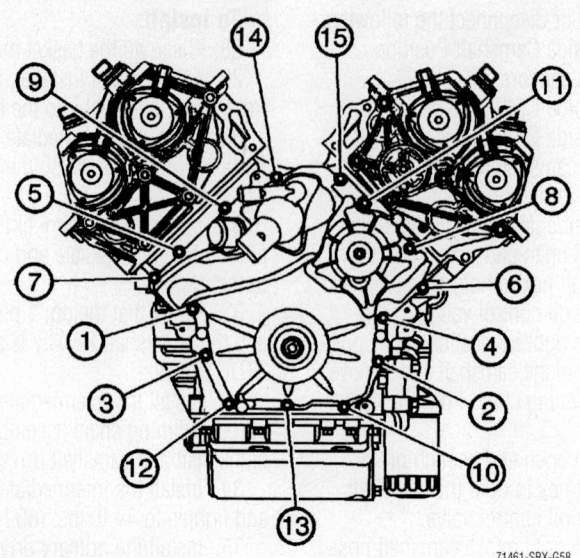

Engine front cover bolt tightening sequence—4.6L engine

- Radiator hoses
- Engine appearance cover
- Air outlet duct
- Negative battery cable

90. Fill the cooling system to the correct level.

91. Fill the engine with oil to the correct level.

92. Start the engine and verify proper operation.

Piston and Ring

POSITIONING

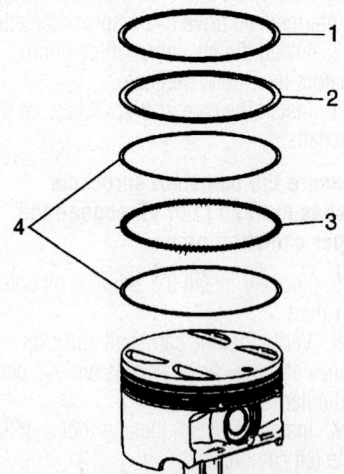

1. Top compression ring
2. Second compression ring
3. Expander ring
4. Oil scraper rings

71461-SRX-G59

Ring positioning on pistons—3.6L shown; 4.6L similar

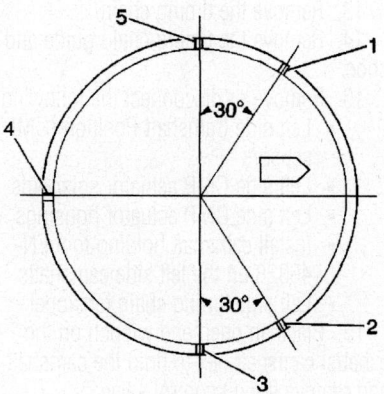

1. Lower oil scraper ring
2. Upper oil scraper ring
3. Top compression ring
4. Expander ring
5. Second compression ring

71461-SRX-G60

Ring gap positioning on pistons—3.6L engine

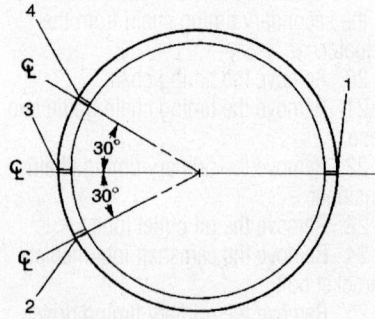

1. Expander and second compression ring
2. Upper oil scraper ring
3. Top compression ring
4. Lower oil scraper ring

71461-SRX-G61

Ring gap positioning on pistons—4.6L engine

FUEL SYSTEM

Fuel System Pressure

RELIEVING

Turn the ignition **OFF** and disconnect the negative battery cable. Remove the engine appearance cover and locate the fuel pressure test port. Remove the fuel pressure test port cap. Wrap a shop towel around the test port and use a flat bladed tool or fuel pressure gauge to depress the port valve, relieving any system pressure. Install the port cap. When vehicle service is complete, turn the ignition **ON** to pressurize the fuel system. Start the vehicle and check the system for leaks.

Fuel Filter

REMOVAL & INSTALLATION

1. Before servicing the vehicle, refer to the Precautions Section.
2. Relieve the fuel system pressure.
3. Raise and support the vehicle safely on jackstands.
4. Place a rag under the fuel filter to catch any residual fuel that may leak out when the filter is removed.
5. Remove the quick-connect fitting at the fuel filter inlet.
6. Remove the threaded fitting at the fuel outlet line.
7. Remove the discard the O-ring.
8. Remove the fuel filter.
   To install:
9. Lubricate a new O-ring with clean engine oil and install it.
10. Install the fuel filter in its bracket, ensuring proper direction of flow.
11. Connect the fuel inlet and outlet lines. Tighten the outlet fitting to 22 ft. lbs. (30 Nm).
12. Start the engine and check the filter connections for leaks by running the tip of your finger around each connection.
13. Turn the engine off and lower the vehicle.

Fuel Pump

REMOVAL & INSTALLATION

➡**The following procedure applies to both the primary and secondary fuel pump modules. To gain access to the fuel pump, it is necessary to remove the fuel tank.**

1. Before servicing the vehicle, refer to the Precautions Section.
2. Disconnect the negative battery cable.
3. Depressurize the fuel system and drain the fuel into a suitable container.
4. Remove the fuel tank.
5. Disconnect the electrical connectors and fuel line fittings.
6. Remove any dirt that has accumulated around the fuel pump module attaching flange to prevent it from entering the tank during service.
7. Turn the fuel pump module locking ring counterclockwise using a locking ring removal tool, and remove the locking ring.
8. Raise the fuel pump and disconnect the fuel transfer tube.
9. Remove the fuel pump module.
10. Remove the seal gasket and discard it.
   To install:
11. Ensure the new seal bead is facing the fuel tank.
12. Reverse the removal procedure to install the fuel pump(s).
13. Install the tank in the vehicle.
14. Install a minimum of 10 gallons (38L) of fuel and check for leaks.
15. Check for fuel leaks at the fittings.
16. Turn the ignition on for 2 seconds. Turn the ignition off for 10 seconds. Turn the ignition on and check for fuel leaks.

Fuel Injectors

REMOVAL & INSTALLATION

3.6L Engine

1. Before servicing the vehicle, refer to the Precautions Section.
2. Disconnect the negative battery cable.
3. Relieve the fuel system pressure.
4. Remove the intake manifold.
5. Remove the fuel line retaining clip.
6. Disconnect the feed line from the fuel rail.

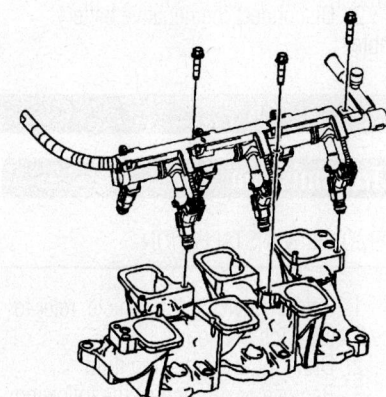

71461-SRX-G62

Fuel injectors and fuel rail—3.6L engine

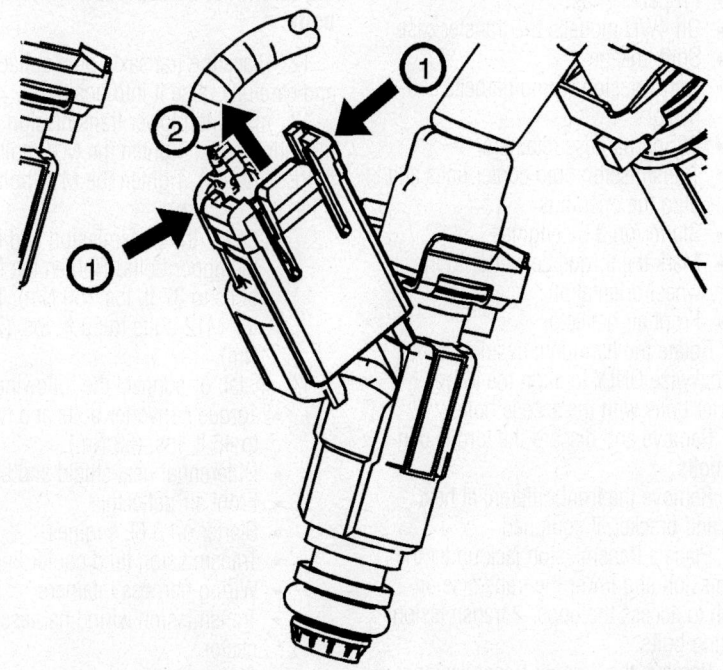

06025-SRX-G05

Disengage the connector lock—3.6L engine

7. Remove the fuel rail attaching bolts and remove the fuel rail with the injectors.

8. Disengage the fuel injector electrical connector lock.

9. Disconnect the injector electrical connectors.

10. Remove the retaining clips and remove the fuel injectors. Discard the seals.

To install:

11. Install new fuel injector seals.

12. Reverse the removal procedure to install the fuel rail and injectors. Tighten the fuel rail bolts to 89 inch lbs. (10 Nm).

13. Start the engine and check for fuel leaks.

4.6L Engine

1. Before servicing the vehicle, refer to the Precautions Section.

2. Disconnect the negative battery cable.

3. Relieve the fuel system pressure.

4. Remove the front end cross vehicle brace.

5. Disconnect the feed line from the fuel rail.

6. Remove the PCV air hose.

7. Disconnect the EVAP quick connect fitting.

8. Open the fuel line retainers at the rear of the engine.

9. Disconnect the fuel injector electrical connectors.

10. Remove the injector shield bracket.

11. Remove the fuel rail attaching bolts and remove the fuel rail with the injectors.

12. Remove the retaining clips and remove the fuel injectors. Discard the O-ring seals.

To install:

13. Install new fuel injector seals after coating them with clean engine oil.

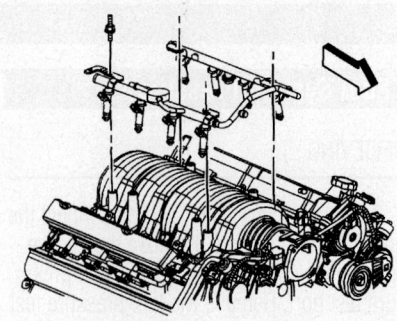

Fuel injectors and fuel rail—4.6L engine

14. Reverse the removal procedure to install the fuel rail and injectors. Tighten the fuel rail bolts to 89 inch lbs. (10 Nm).

15. Turn the ignition on for 2 seconds. Turn the ignition off for 10 seconds. Turn the ignition on and check for fuel leaks.

DRIVE TRAIN

Transmission

REMOVAL & INSTALLATION

1. Before servicing the vehicle, refer to the Precautions Section.

2. Drain the transmission fluid

3. Remove or disconnect the following:
   - Negative battery cable
   - Thermostat housing
   - Exhaust system
   - Propeller shaft
   - On 4WD models, the transfer case
   - Shift linkage
   - Transmission wiring harness connector
   - Wiring harness retainers
   - Transmission fluid cooler lines and plug the openings
   - Starter on 3.6L engines
   - Mark the torque converter to flywheel orientation
   - Front air deflector

4. Rotate the harmonic balancer center bolt clockwise ONLY to align the torque converter bolts with the access hole.

5. Remove and discard the torque converter bolts.

6. Remove the front differential heat shield and bracket, if equipped.

7. Place a transmission jack under the transmission and lower the transmission enough to access the upper 2 transmission mounting bolts.

8. Remove the 2 upper transmission mounting bolts.

9. Raise the transmission so the engine and transmission are in the normal position, then place a jack under the engine to keep it level.

10. Remove the lower transmission mounting bolts.

11. Carefully pull the transmission back off the engine dowels, then lower the transmission away from the vehicle.

To install:

➡ **Torque converter bolts are self-locking and must be replaced with new bolts.**

12. Place the transaxle on a suitable jack and carefully raise it into position.

13. Install the lower transmission mounting bolts. Tighten the M10 bolts to 37 ft. lbs. (50 Nm). Tighten the M12 bolts to 55 ft. lbs. (75 Nm).
   - Lower the transmission and install the upper bolts. Tighten the M10 bolts to 37 ft. lbs. (50 Nm). Tighten the M12 bolts to 55 ft. lbs. (75 Nm).

14. Install or connect the following:
   - Torque converter bolts and tighten to 46 ft. lbs. (63 Nm).
   - Differential heat shield and bracket.
   - Front air deflector
   - Starter on 3.6L engines
   - Transmission fluid cooler lines
   - Wiring harness retainers
   - Transmission wiring harness connector
   - Shift linkage
   - On 4WD models, the transfer case

   - Propeller shaft
   - Exhaust system
   - Thermostat housing
   - Negative battery cable

15. Fill the transmission with fluid to the correct level.

16. Adjust the shift control linkage.

17. Start the vehicle and check for leaks and check the fluid level.

Transfer Case

REMOVAL & INSTALLATION

4.6L AWD Models

1. Before servicing the vehicle, refer to the Precautions Section.

2. Drain the transfer case fluid.

3. Remove or disconnect the following:
   - Negative battery cable
   - Exhaust system
   - Rear propeller shaft

4. Support the transmission with a jack

5. Remove the rear transmission mount-to-body bolts.

6. Remove the transmission mount-to-transmission bolts.

7. Insert a flat bladed tool into the notch on the transfer case flange and carefully move the front propeller shaft forward and remove the transfer case flange.

8. Wire the propeller shaft up and out of the way.

9. Support the transfer case with a jack and remove the mounting bolts.

10. Remove the transfer case.

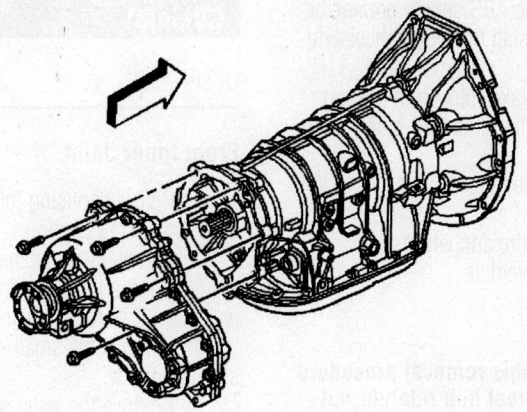

Transfer case mounting—4.6L AWD models

71461-SRX-G64

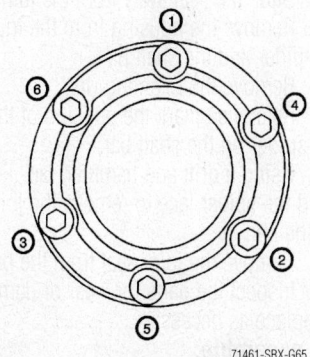

71461-SRX-G65

Transfer case flange bolt tightening sequence—4.6L AWD models

To install:

11. Install the transfer case and tighten the bolts to 44 ft. lbs. (60 Nm).

12. Remove the jack.

13. Untie the front propeller shaft and position it to the transfer case so the coned end CV joint is installed into the transfer case flange.

14. Clean the transfer case flange bolts and apply threadlock to the threads and allow it to dry for 10 minutes.

15. Install the transfer case flange bolts with the crescent washers and tighten the bolts in sequence to 22 ft. lbs. (30 Nm).

16. Install or connect the following:
- Transmission mount-to-transmission bolts. Tighten to 81 ft. lbs. (110 Nm).
- Rear transmission mount-to-body bolts. Tighten the bolts to 44 ft. lbs. (60 Nm).

17. Remove the jack.

18. Install or connect the following:
- Rear propeller shaft
- Exhaust system
- Negative battery cable

19. Fill the transfer case with fluid to the correct level.

Halfshafts

REMOVAL & INSTALLATION

Front

➡**Do not begin this removal procedure unless a new wheel hub retainer nut and a new retainer circlip are available. Once removed, these parts must not be reused during assembly. Their torque holding ability, or retention capability, is diminished during removal.**

➡**This procedure requires the use of Slide Hammer and Adapter J-2619-01, or equivalent, Extension J-29794, Axle Shaft Puller J-35341, Seal Protector J-44394 and Wheel Hub Remover J-45859.**

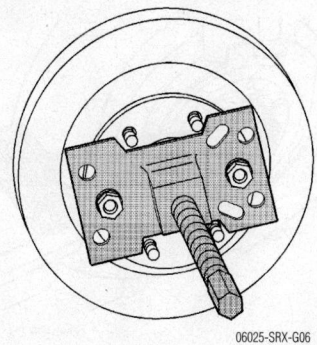

06025-SRX-G06

Install Special Tool J-45859 to disengage the halfshaft from the wheel hub assembly.

1. Before servicing the vehicle, refer to the Precautions Section.

2. Raise and support the vehicle.

3. Remove or disconnect the following:
- Front wheels
- Outer tie rod end, but DO NOT loosen the jam nut
- Axle hub nut and washer. Discard the nut.
- Anti-lock Brake System (ABS) sensor connector
- Upper ball joint

4. Install tool J-45859 onto the wheel hub and secure with 2 lug nuts.

5. Use the tool to disengage the halfshaft from the wheel hub and bearing. Support the halfshaft.

6. Remove the tool from the wheel hub.

7. Assemble tools J-2619-01, J-29794 and J-45341 and install it to the halfshaft inner joint pull groove.

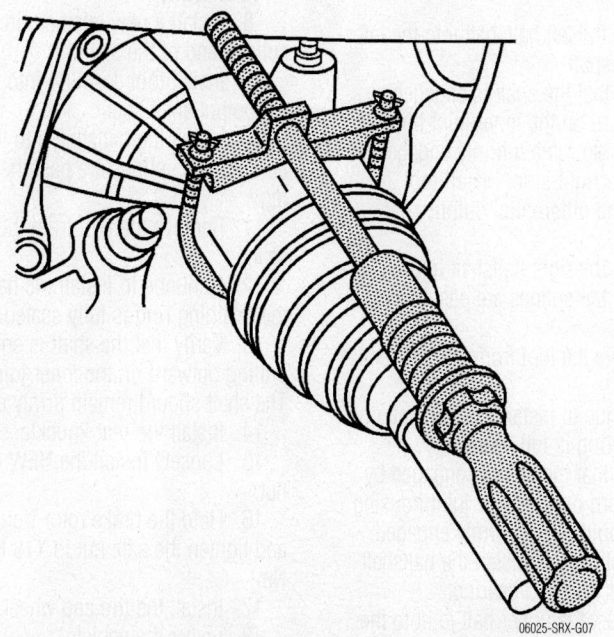

06025-SRX-G07

Special tools J-2619-01, J-29794 and J-45341 assembled on the inner joint.

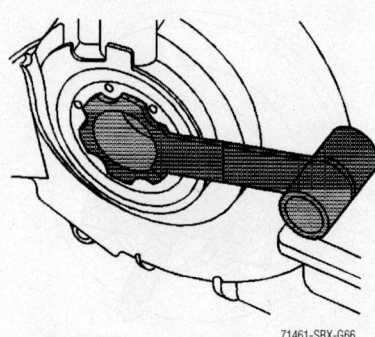

71461-SRX-G66

Using tool J-44394 to protect the right side differential output seal

8. On the left side, use the tool to separate the wheel halfshaft from the intermediate wheel driveshaft.

9. Remove the halfshaft from the left side.

10. On the right side, use the tool to disengage the halfshaft away from the differential enough to install tool J-44394.

11. Install tool J-44394 over the halfshaft and into the differential output seal to protect the seal.

12. Remove the halfshaft from the right side.

13. If reusing the halfshaft, remove and discard the retaining ring from the intermediate shaft ring groove.

To install:

14. On the left side install a new O-ring to the intermediate shaft O-ring groove.

15. On both sides, install a new retainer circlip on the splined end of the shaft

16. On the left side, apply a small amount of grease to the intermediate driveshaft splines.

17. Install the left halfshaft into the intermediate driveshaft.

18. Verify that the shaft is engaged by pulling outward on the inner joint housing. The shaft should remain firmly engaged.

19. On the right side, install tool J-44394 into the differential output shaft seal.

20. Install the right halfshaft into the differential until the splines are past the tool opening.

21. Remove the tool from the differential seal.

22. Continue to install the halfshaft until the retaining ring is fully seated.

23. Verify that the shaft is engaged by pulling outward on the inner joint housing. The shaft should remain firmly engaged.

24. On both sides, install the halfshaft into the wheel hub and bearing.

25. Connect the upper ball joint to the steering knuckle.

26. Connect the ABS sensor connector.

27. Loosely install the NEW wheel axle nut.

28. Hold the brake rotor from turning and tighten the axle nut to 118 ft. lbs. (160 Nm).

29. Connect the outer tie rod end to the steering knuckle.

30. Install the tire and wheel.

31. Lower the vehicle.

Rear

➡**Do not begin this removal procedure unless a new wheel hub retainer nut and a new retainer circlip are available. Once removed, these parts must not be reused during assembly. Their torque holding ability, or retention capability, is diminished during removal.**

1. Before servicing the vehicle, refer to the Precautions Section.

2. Raise and support the vehicle.

3. Remove or disconnect the following:
   • Rear wheels
   • Axle hub nut and washer. Discard the nut.
   • Rear knuckle assembly

4. Pry the halfshaft away from the differential enough to install tool J-44394.

5. Install tool J-44394 over the halfshaft and into the differential output seal to protect the seal.

6. Remove the halfshaft from the vehicle.

7. If reusing the halfshaft, remove and discard the retaining ring from the intermediate shaft ring groove.

To install:

8. Install a new retainer circlip on the splined end of the shaft.

9. Install tool J-44394 into the differential output shaft seal.

10. Install the halfshaft into the differential until the splines are past the tool opening.

11. Remove the tool from the differential seal.

12. Continue to install the halfshaft until the retaining ring is fully seated.

13. Verify that the shaft is engaged by pulling outward on the inner joint housing. The shaft should remain firmly engaged.

14. Install the rear knuckle.

15. Loosely install the NEW wheel axle nut.

16. Hold the brake rotor from turning and tighten the axle nut to 118 ft. lbs. (160 Nm).

17. Install the tire and wheel.

18. Lower the vehicle.

CV-Joint

OVERHAUL

Front Inner Joint

1. Before servicing the vehicle, refer to the Precautions Section.

2. Remove the halfshaft.

3. Wrap a towel around the halfshaft and place it horizontally in a vise.

4. Remove the small seal clamp using side cutters.

5. Remove the large seal clamp from the tripot joint.

6. Separate the inboard joint seal from the tripot joint.

7. Slide the seal away from the joint.

8. Remove the housing from the tripot joint spider and the shaft bar.

9. Remove the retaining ring.

10. Reference mark the position of the tripot spider on the shaft bar.

11. Using a drift and hammer, tap around the spider face to remove the joint from the bar.

12. Remove the joint seal from the bar.

13. Inspect the parts for wear or damage and replace as necessary.

To assemble:

14. Place a new small seal clamp on the small end of the inner joint seal.

15. Slide the inner joint seal and the small seal clamp into the boot groove on the shaft bar.

16. Position the small end of the inner joint seal into the inner joint seal groove.

17. Crimp the small seal clamp until the crimping joint gap is 0.039 inch (1mm).

➡**The clamping hold time must be no less than 2 seconds.**

18. Align the reference marks and install the tripot spider to the shaft bar.

19. Install a new retaining ring.

20. Pull out on the spider to verify that the spider is fully engaged.

21. Place half of the grease from the service kit in the shaft inboard seal. Use the remainder of the grease to repack the housing.

22. Install the tripot bushing to the housing.

23. Position the large seal clamp on the inboard seal.

24. Slide the housing over the joint spider.

25. Slide the large inboard seal with the large clamp in place, over the bushing and locate the lip of the seal in the groove.

26. Ensure the seal is properly shaped around the circumference of the joint.

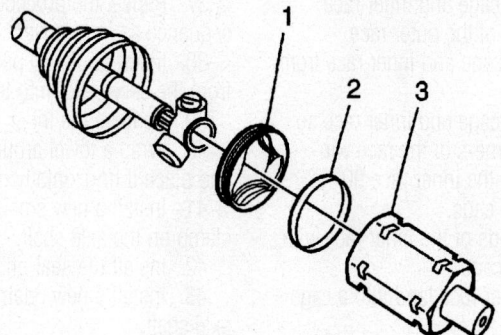

1. Tripot bushing
2. Clamp
3. Housing

71461-SRX-G67

Exploded view of front inner CV joint assembly

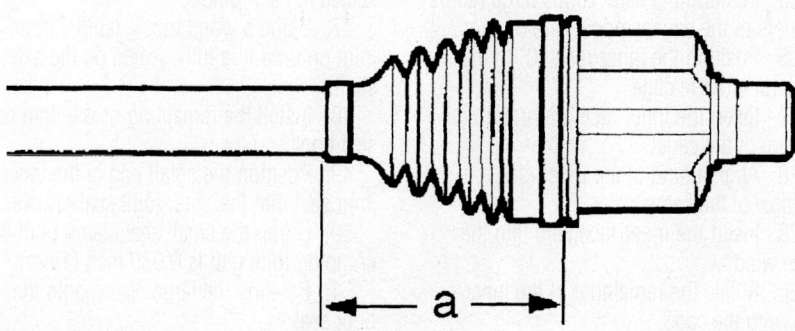

71461-SRX-G68

Measuring installed boot seal distance—front inner joint

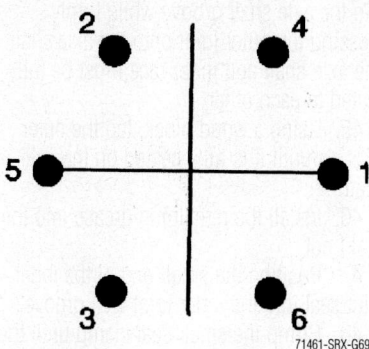

71461-SRX-G69

CV joint ball removal and installation sequence

27. Measure the distance (a) between the seal edges. The correct distance is 4.12 inches (104.7mm).

28. Remove the air from the inner joint seal.

29. Using crimping pliers, crimp the large clamp around the boot, ensuring the tangs are fully engaged.

30. Rotate the inner tripot housing a few times to fully distribute the gear into the bearings.

Front and Rear Outer Joint

1. Before servicing the vehicle, refer to the Precautions Section.

2. Remove the halfshaft.

3. Wrap a towel around the halfshaft and place it horizontally in a vise.

4. Remove the small and large seal clamps using side cutters.

5. Separate the seal from the joint outer race at the large end.

6. Wipe the grease away from the joint face.

7. Hold the outer joint housing horizontally and position a wood block between the seal and the joint on the joint face.

8. Hammer around the wood block to remove the outer joint from the axle shaft.

9. Remove and discard the shaft retaining ring.

10. Remove the seal from the shaft.

11. Remove the axle shaft from the vise.

12. Wrap a towel around the joint outer face and place the joint in a vise vertically.

13. Using a drift and hammer, tap gently on the inner cage until it tilts up enough to remove the first ball.

14. Repeat this procedure using the proper sequence until all the balls are removed.

15. Position the cage and inner race 90° to the centerline of the outer race.

16. Remove the cage and inner race from the outer race.

17. Position the cage and inner race so the larger radius corners of the race windows are up. Rotate the inner race 90° to the centerline of the cage.

18. Align the lands of the inner race with the windows of the cage.

19. Insert an inner race land into a cage window.

20. Pivot the inner down and remove it from the cage.

21. Clean all parts thoroughly with solvent and allow to air dry.

22. Inspect all parts for wear or damage and replace as necessary.

To assemble:

23. Position the cage so the large radius corners of the cage windows are up.

24. Position the inner race 90° to the centerline of the cage.

25. Insert the inner race up through the bottom of the cage.

26. Align a land of the inner race to a window of the cage.

27. Insert the inner race land into the cage window.

28. Rotate the remainder of the inner race into the cage.

29. Align the inner race ball tracks with the cage windows.

30. Wrap a towel around the splined shaft and place the shaft vertically in a vise with the joint opening up.

31. Position the cage and inner race 90° to the centerline of the outer race.

32. Align 2 cage windows 180° apart.

33. Place a cage window and inner race ball track in position to insert the ball.

34. Insert a ball through the cage window and into the ball track. Tap the ball with a plastic hammer until it is fully engaged.

35. Repeat the procedure using the same sequence as the removal.

36. Insert about 60 percent of the grease from the service kit into the outer joint.

37. Remove the outer joint from the vise.

38. Wrap a towel around the axle shaft and place it horizontally in a vise.

39. Install a new small seal retaining clamp on the axle shaft.

40. Install the seal on the axle shaft.

41. Install a new retaining ring on the axle shaft.

42. Position the outer joint horizontally.

43. Engage the inner race splines into the axle shaft splines.

44. Press the end of the retaining ring

into the axle shaft groove while firmly pressing the outer joint onto the axle shaft. The axle shaft and inner race must be fully seated to each other.

45. Using a wood block, tap the outer joint on until it is fully seated on the axle shaft.

46. Install the remaining grease into the seal boot.

47. Position the small end of the inner joint seal into the inner joint seal groove.

48. Crimp the small seal clamp until the crimping joint gap is 0.039 inch (1mm).

49. Position the large clamp onto the boot seal

50. Slide the large inboard seal with the large clamp in place, over the bushing and locate the lip of the seal in the groove.

51. Ensure the seal is properly shaped around the circumference of the joint.

52. Remove the air from the inner joint seal.

53. Using crimping pliers, crimp the large clamp around the boot, until the crimping joint gap is 0.039 inch (1mm).

54. Rotate the outer joint a few times to fully distribute the gear into the bearings.

Rear Inner Joint

1. Before servicing the vehicle, refer to the Precautions Section.

2. Remove the halfshaft:

3. Wrap a towel around the halfshaft and place it horizontally in a vise.

4. Remove the small and large seal clamps using side cutters.

5. Separate the seal from the joint outer race at the large end.

6. Slide the seal away from the joint face.

7. Wipe the grease away from the joint face.

8. Hold the inner joint housing horizontally and position a drift on the inner face.

9. Tap the drift to compress the axle shaft retaining ring.

10. Remove and discard the shaft retaining ring.

11. Use the hammer and drift again to remove the inner joint.

12. Remove the seal from the shaft.

13. Remove the axle shaft from the vise.

14. Wrap a towel around the joint outer face and place the joint in a vise vertically.

15. Using a drift and hammer, tap gently on the inner cage until it tilts up enough to remove the first ball.

16. Repeat this procedure using the proper sequence until all the balls are removed.

17. Position the cage and inner race 90°to the centerline of the outer race.

18. Remove the cage and inner race from the outer race.

19. Position the cage and inner race so the larger radius corners of the race windows are up. Rotate the inner race 90° to the centerline of the cage.

20. Align the lands of the inner race with the windows of the cage.

21. Insert an inner race land into a cage window.

22. Pivot the inner down and remove it from the cage.

23. Clean all parts thoroughly with solvent and allow to air dry.

24. Inspect all parts for wear or damage and replace as necessary.

To assemble:

25. Position the cage so the large radius corners of the cage windows are up.

26. Position the inner race 90° to the centerline of the cage.

27. Insert the inner race up through the bottom of the cage.

28. Align a land of the inner race to a window of the cage.

29. Insert the inner race land into the cage window.

30. Rotate the remainder of the inner race into the cage.

31. Align the inner race ball tracks with the cage windows.

32. Wrap a towel around the splined shaft and place the shaft vertically in a vise with the joint opening up.

33. Position the cage and inner race 90° to the centerline of the outer race.

34. Align 2 cage windows 180° apart.

35. Place a cage window and inner race ball track in position to insert the ball.

36. Insert a ball through the cage window and into the ball track. Tap the ball with a plastic hammer until it is fully engaged.

37. Repeat the procedure using the same sequence as in removal.

38. Insert about 60 percent of the grease from the service kit into the inner joint.

39. Remove the inner joint from the vise.

40. Wrap a towel around the axle shaft and place it horizontally in a vise.

41. Install a new small seal retaining clamp on the axle shaft.

42. Install the seal on the axle shaft.

43. Install a new retaining ring on the axle shaft.

44. Position the inner joint horizontally.

45. Engage the inner race splines into the axle shaft splines.

46. Press the end of the retaining ring into the axle shaft groove while firmly pressing the inner joint onto the axle shaft. The axle shaft and inner race must be fully seated to each other.

47. Using a wood block, tap the inner joint on until it is fully seated on the axle shaft.

48. Install the remaining grease into the seal boot.

49. Position the small end of the inner joint seal into the inner joint seal groove.

50. Crimp the small seal clamp until the crimping joint gap is 0.039 inch (1mm).

51. Position the large clamp onto the boot seal

52. Slide the large inboard seal with the large clamp in place, over the bushing and locate the lip of the seal in the groove.

53. Ensure the seal is properly shaped around the circumference of the joint.

54. Measure the distance (a) between the seal edges. The correct distance is 3.42 inches (86.85mm) for AWD, or 3.50 inches (88.6mm)

55. Using crimping pliers, crimp the large clamp around the boot, until the crimping joint gap is 0.039 inch (1mm).

56. Rotate the inner joint a few times to fully distribute the gear into the bearings.

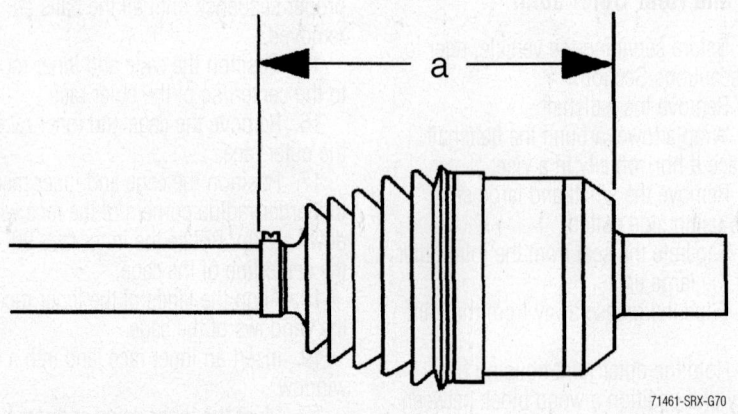

71461-SRX-G70

Measuring installed boot seal distance—Rear inner joint

Front Differential

REMOVAL & INSTALLATION

1. Before servicing the vehicle, refer to the Precautions Section.
2. Remove the front wheels.
3. Remove the front frame assembly. See the procedure in suspension.
4. Remove or disconnect the following:
   - Front propeller shaft
   - Front axle shafts
   - Intermediate drive shaft support bearing
5. Secure the differential to a transmission jack.
6. Remove the differential-to-oil pan bolts.
7. Remove the differential and the intermediate drive shaft as an assembly.

To install:

8. Installation is the reverse of the removal procedure. Tighten the differential bolts in the sequence shown to 81 ft. lbs. (110 Nm).
9. Check the differential fluid level and fill as needed.

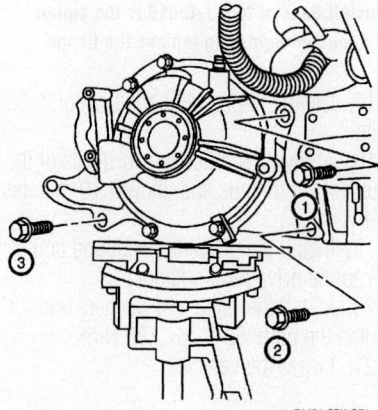

71461-SRX-G71

Front differential bolt tightening sequence—AWD models

Rear Differential

REMOVAL & INSTALLATION

1. Before servicing the vehicle, refer to the Precautions Section.
2. Remove the rear propeller shaft.
3. Raise and support the vehicle.
4. Remove the rear wheels.
5. Remove the right halfshaft.
6. Position a transmission jack under the differential.
7. Remove the front differential-to-support mounting bolt.

8. Remove the left and right rear differential-to-support mounting bolt. The left bolt cannot be completely removed due to interference with the underbody.
9. Lower the jack until the mounting ear at the front of the differential clears the support attachment point.
10. Pry the left halfshaft away from the differential enough to install tool J-44394.
11. Install seal protector J-44394 over the halfshaft and into the differential output seal to protect the seal.
12. Continue lowering the jack while disengaging the halfshaft.
13. Remove the differential.

To install:

14. Installation is the reverse of the removal procedure. Tighten the differential mounting bolts to 129 ft. lbs. (175 Nm).
15. Check the differential fluid level and fill as needed.

Front Output Shaft Seal

REMOVAL & INSTALLATION

1. Before servicing the vehicle, refer to the Precautions Section.
2. Raise and support the vehicle.
3. Remove the right front wheel.
4. Remove the right halfshaft.
5. Pry the output shaft seal from the seal opening.

To install:

6. Lubricate the seal surface with synthetic gear oil.
7. Using a seal installer, install the new seal.
8. Install the right halfshaft.
9. Install the right wheel.
10. Lower the vehicle.

Front Pinion Seal

REMOVAL & INSTALLATION

1. Before servicing the vehicle, refer to the Precautions Section.
2. Remove the front propeller shaft-to-drive pinion flange bolts and carefully slide the shaft away from the flange.
3. Wire the propeller shaft up with mechanics wire.
4. Install holding tool J-45012 to the pinion flange.
5. While holding the tool rigid, remove the pinion flange nut.
6. Remove the holding tool and install J-45019 to the pinion flange and tighten the tool to remove the pinion flange.
7. Pry the pinion seal from the seal opening.

To install:

8. Lubricate the pinion flange sealing surface with synthetic gear oil.
9. Using installer tool J-46262, install the new pinion seal.
10. Install the drive pinion flange.

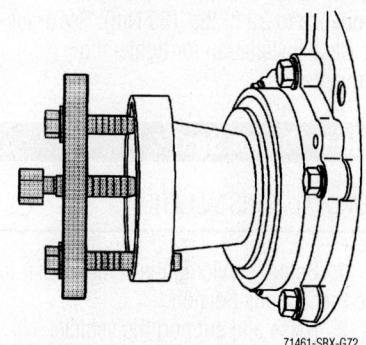

71461-SRX-G72

Remove the front drive pinion flange using tool J-45019—AWD models

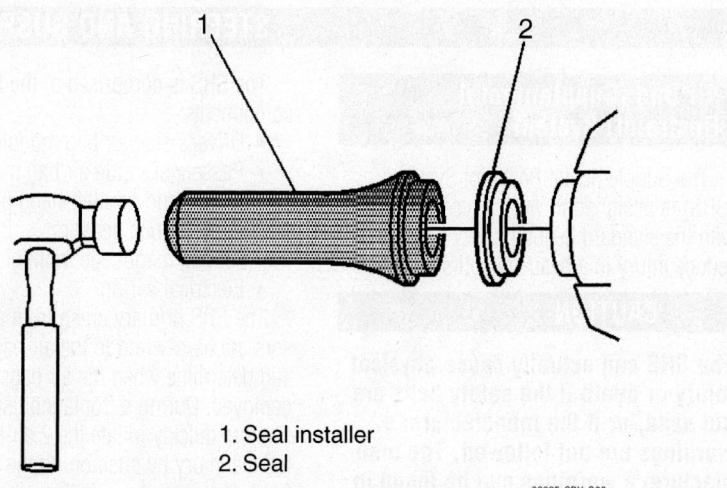

1. Seal installer
2. Seal

06025-SRX-G08

Using the Seal installer tool (1) to install the output shaft seal (2).

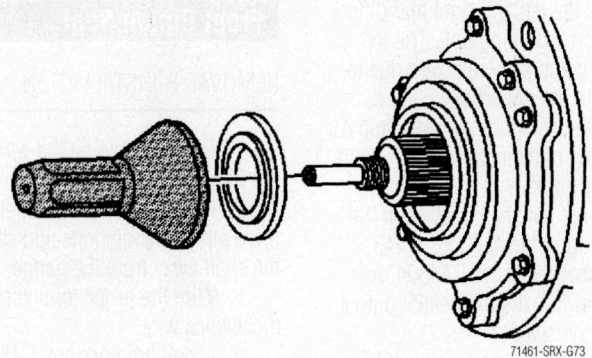

Installing the front pinion seal using tool J-46262—AWD models

71461-SRX-G73

11. Clean the pinion shaft threads and pinion nut.

12. Apply Threadlock to two-thirds of the pinion shaft threads and allow it 10 minutes to dry.

13. Install the pinion nut to the pinion flange.

14. Install the holding tool and while holding the tool, tighten the nut to 178 ft. lbs. (241 Nm).

15. Remove the holding tool.

16. Install the propeller shaft to the drive pinion flange and tighten the bolts in sequence to 22 ft. lbs. (30 Nm). See Transfer Case installation for tightening sequence.

Rear Pinion Seal

REMOVAL & INSTALLATION

1. Before servicing the vehicle, refer to the Precautions Section.

2. Raise and support the vehicle.

3. Remove the rear propeller shaft coupler-to-drive pinion flange bolts and carefully slide the shaft away from the flange.

4. Wire the propeller shaft up with mechanics wire.

5. Install holding tool J-45012 to the pinion flange.

6. While holding the tool rigid, remove the pinion flange nut.

7. Remove the holding tool and install J-45019 to the pinion flange and tighten the tool to remove the pinion flange.

8. Pry the pinion seal from the seal opening.

9. Lubricate the pinion flange sealing surface with synthetic gear oil.

10. Using a seal installer, install the new pinion seal.

11. Install the drive pinion flange.

12. Clean the pinion shaft threads and pinion nut.

13. Apply Threadlock to two-thirds of the pinion shaft threads and allow it 10 minutes to dry.

14. Install the pinion nut to the pinion flange.

15. Install the holding tool and while holding the tool, tighten the nut to 210 ft. lbs. (285 Nm).

16. Remove the holding tool.

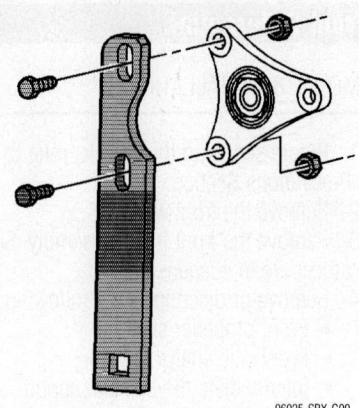

Installing Special tool J-45012 to the pinion flange.

06025-SRX-G09

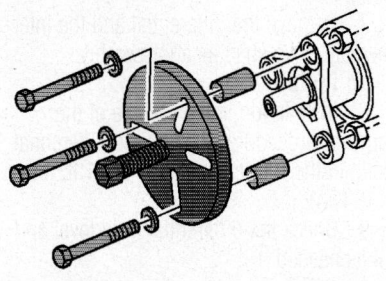

06025-SRX-G10

Install Special tool J-45019 to the pinion flange and tighten to remove the flange.

17. Clean the propeller shaft coupler bolts.

18. Apply threadlock to two-thirds of the coupler bolt threads and allow it 10 minutes to dry.

19. Install the propeller shaft and coupler to the drive pinion flange.

20. Install the bolts and washers and tighten the bolts 63 ft. lbs. (85 Nm).

21. Lower the vehicle.

STEERING AND SUSPENSION

Air Bag (Supplemental Restraint) System

The Supplemental Restraint System (SRS) is designed to work in conjunction with the standard 3-point safety belts to reduce injury in a head-on collision.

☀ CAUTION

The SRS can actually cause physical injury or death if the safety belts are not used, or if the manufacturer's warnings are not followed. The manufacturer's warnings can be found in your owner's manual, or, in some cases, on your sun visor.

The SRS is comprised of the following components:
- Driver's side air bag module
- Passenger's side air bag module
- Right-hand and left-hand primary crash front air bag sensors
- Air bag diagnostic monitor computer
- Electrical wiring

The SRS primary crash front air bag sensors are hard-wired to the air bag modules and determine when the air bags are deployed. During a frontal collision, the sensors quickly inflate the 2 air bags to reduce injury by cushioning the driver and front passenger from striking the dashboard, windshield, steering wheel and any other hard surfaces. The air bag inflates so quickly (in a fraction of a second) that in most cases it is fully inflated before you actually start to move during a collision.

Since the SRS is a complicated and essentially important system, its components are constantly being tested by a diagnostic computer. The computer illuminates the air bag indicator light on the instrument cluster for approximately 6 seconds when the ignition switch is turned to the **RUN** position when the SRS is functioning properly. After being illuminated for the 6 seconds, the indicator light should then turn off.

If the air bag light does not illuminate at all, stays on continuously, or flashes at any time, a problem has been detected by the diagnostic computer.

❋❋ CAUTION

If at any time the air bag light indicates that the computer has noted a problem, immediately diagnose the problem. A faulty SRS can cause severe physical injury or death.

DISABLING AND ENABLING THE SYSTEM

Zone 1

1. Turn the steering wheel so that the vehicles wheels are pointing straight ahead.
2. Turn the ignition switch to the OFF position.
3. Remove the key from the ignition switch.
4. Adjust the right rear seat to the rear of vehicle.
5. Pull the carpet away from under the right side of rear seat.
6. When the carpet is pulled away from the right rear seat the right rear fuse center will be exposed, then remove the fuse center top cover.

➡**With the SIR fuse removed and the ignition switch in the ON position, the AIR BAG warning indicator illuminates. This is normal operation, and does not indicate an SIR system malfunction.**

7. Locate and remove the SIR fuse from the right rear fuse center.
8. Open front hood, and locate both front end sensors (3, 4).
9. Remove both connector position assurance (CPA) (1, 6) from left and right front end sensor connectors (2, 5).
10. Remove both connectors (2, 5) from left and right front end sensors (3, 4).

Enabling Procedure:
11. Remove the key from the ignition switch.
12. Connect both connectors (2, 5) to the left and right front end sensors (3, 4).
13. Install both CPAs (1, 6) to the left and right front end sensor connectors (2, 5).
14. Close front hood.
15. Install the SIR fuse into the right rear fuse center.
16. Install the right rear fuse center cover.
17. Position the carpet back under the right rear seat.
18. Use caution while reaching in and turn the ignition switch to the ON position. The AIR BAG indicator will flash then turn OFF.

Zone 2

1. Turn the steering wheel so that the vehicle's wheels are pointing straight ahead.

2. Turn the ignition switch to the OFF position.
3. Remove the key from the ignition switch.
4. Adjust the right rear seat to the rear of vehicle.
5. Pull the carpet away from under the right side of rear seat.
6. When the carpet is pulled away from the rear seat the right rear fuse center (1) will be exposed, then remove the fuse center top cover.

➡**With the SIR fuse removed and the ignition switch in the ON position, the AIR BAG warning indicator illuminates. This is normal operation, and does not indicate an SIR system malfunction.**

7. Locate and remove the SIR fuse from the right rear fuse center.
8. When disabling the roof rail module, go to step 9. If the side impact sensor (SIS) needs disabling, go to step 12.
9. Remove the left carpet retainer trim.
10. Remove the connector position assurance (CPA) from the left/driver roof rail module yellow connector.
11. Disconnect the left roof rail module connector from the left roof rail module.
12. Remove the left center pillar trim panel.
13. Remove the SIS CPA (1) from the left SIS connector (2).
14. Remove the SIS connector (2) from the SIS (3).

Enabling Procedure:
15. Remove the key from the ignition switch.
16. When enabling the SIS, proceed to step 3. If the roof rail module needs enabling, go to step 6.
17. Connect the SIS connector to the SIS.
18. Connect the CPA to the SIS connector.
19. Install the left center pillar trim panel.
20. Connect the roof rail module yellow connector to the left roof rail module.
21. Install the CPA to the left roof rail module connector.
22. Install the left carpet retainer trim.
23. Install the SIR fuse into the right rear fuse center.
24. Install the right rear fuse center cover.
25. Position the carpet back under the right rear seat.
26. Use caution while reaching in and turn the ignition switch to the ON position. The AIR BAG indicator will flash then turn OFF.

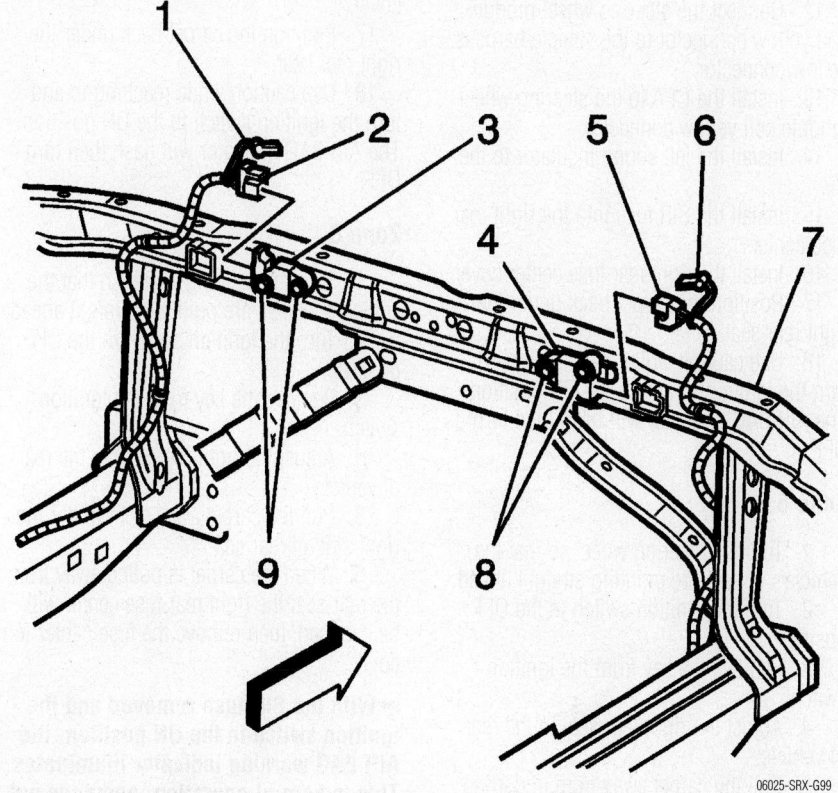

06025-SRX-G99

Front end sensors

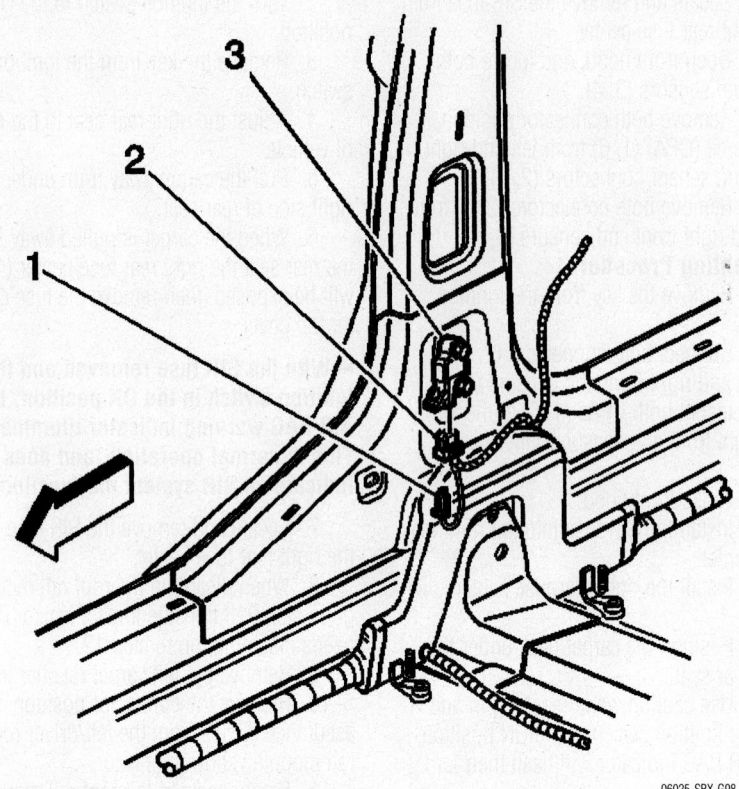

Left side impact sensor

06025-SRX-G98

Zone 3

1. Turn the steering wheel so that the vehicle's wheels are pointing straight ahead.

2. Turn the ignition switch to the OFF position.

3. Remove the key from the ignition switch.

4. Adjust the right rear seat to the rear of vehicle.

5. Pull the carpet away from under the right side of rear seat.

6. When the carpet is pulled away from the rear seat the right rear fuse center will be exposed, then remove the fuse center top cover.

➡ **With the SIR fuse removed and the ignition switch in the ON position, the AIR BAG warning indicator illuminates. This is normal operation, and does not indicate an SIR system malfunction.**

7. Locate and remove the SIR fuse from the right rear fuse center.

8. Remove the left/driver sound insulator from the instrument panel (I/P).

9. Remove the connector position assurance (CPA) from the steering wheel module coil yellow connector.

10. Disconnect the steering wheel module coil yellow connector from the vehicle harness yellow connector.

Enabling Procedure:

11. Remove the key from the ignition switch.

12. Connect the steering wheel module coil yellow connector to the vehicle harness yellow connector.

13. Install the CPA to the steering wheel module coil yellow connector.

14. Install the left sound insulator to the I/P.

15. Install the SIR fuse into the right rear fuse center.

16. Install the right rear fuse center cover.

17. Position the carpet back under the right rear seat.

18. Use caution while reaching in and turn the ignition switch to the ON position. The AIR BAG indicator will flash then turn OFF.

Zone 5

1. Turn the steering wheel so that the vehicle's wheels are pointing straight ahead.

2. Turn the ignition switch to the OFF position.

3. Remove the key from the ignition switch.

4. Adjust the right rear seat to the rear of vehicle.

5. Pull the carpet away from under the right side of rear seat.

6. When the carpet is pulled away from the rear seat the right rear fuse center will be exposed, then remove the fuse center top cover.

➡ **With the SIR fuse removed and the ignition switch in the ON position, the AIR BAG warning indicator illuminates. This is normal operation, and does not indicate an SIR system malfunction.**

7. Locate and remove the SIR fuse from the right rear fuse center.

8. Remove the right/passenger sound insulator from the instrument panel (I/P).

9. Remove the connector position assurance (CPA) from the I/P module yellow connector.

10. Disconnect the I/P module yellow connector from the vehicle harness yellow connector.

Enabling procedure:

11. Remove the key from the ignition switch.

12. Connect the I/P module yellow connector to the vehicle harness yellow connector.

13. Install the CPA to the I/P module yellow connector.

14. Install the right sound insulator to the I/P.

15. Install the SIR fuse into the right rear fuse center.

16. Install the right rear fuse center cover.

17. Position the carpet back under the right rear seat.

18. Use caution while reaching in and turn the ignition switch to the ON position. The AIR BAG indicator will flash then turn OFF.

Zone 6

1. Turn the steering wheel so that the vehicle's wheels are pointing straight ahead.

2. Turn the ignition switch to the OFF position.

3. Remove the key from the ignition switch.

4. Adjust the right rear seat to the rear of vehicle.

5. Pull the carpet away from under the right side of rear seat.

6. When the carpet is pulled away from the rear seat the right rear fuse center will be exposed, then remove the fuse center top cover.

➡ **With the SIR fuse removed and the ignition switch in the ON position, the AIR BAG warning indicator illuminates. This is normal operation, and does not indicate an SIR system malfunction.**

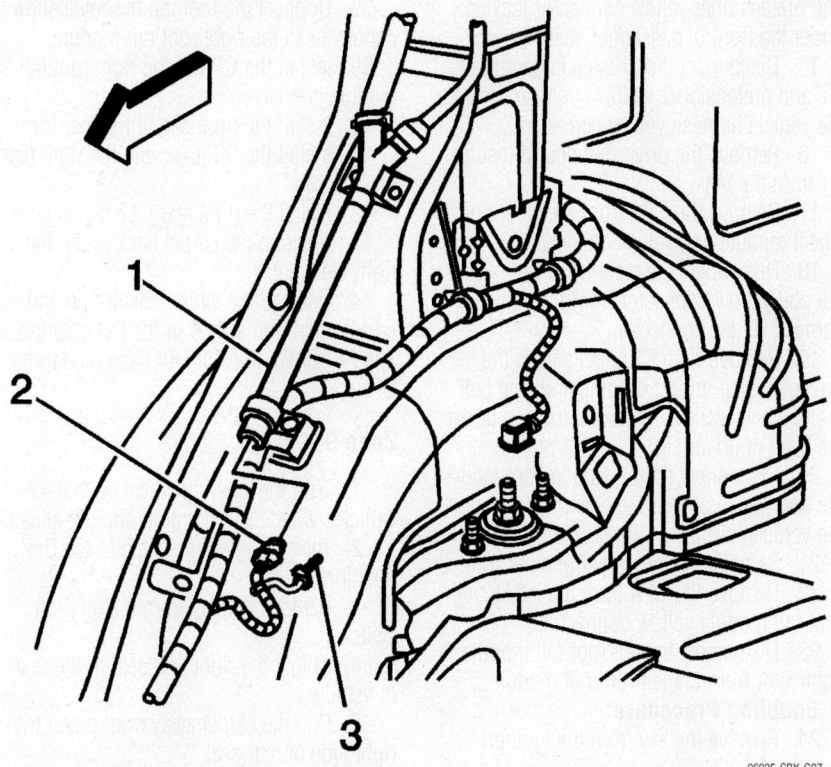

Right/passenger roof rail module

06025-SRX-G97

7. Locate and remove the SIR fuse from the right rear fuse center.

8. When disabling the roof rail module, go to step 9. If the side impact sensor (SIS) needs disabling, go to step 12.

9. Remove the right carpet retainer trim.

10. Remove the connector position assurance (CPA) (2) from the right/passenger roof rail module yellow connector (3).

11. Disconnect the right roof rail module connector from the right roof rail module (1).

12. Remove the right center pillar trim panel.

13. Remove the SIS CPA from the left SIS connector.

14. Remove the SIS connector from the SIS.

Enabling Procedure:

15. Remove the key from the ignition switch.

16. When enabling the SIS, proceed to step 3. If the roof rail module needs enabling, go to step 6.

17. Connect the SIS connector to the SIS.

18. Connect the CPA to the SIS connector.

19. Install the right center pillar trim panel.

20. Connect the roof rail module yellow connector to the right roof rail module.

21. Install the CPA to the right roof rail module connector.

22. Install the right carpet retainer trim.

23. Install the SIR fuse into the right rear fuse center.

24. Install the right rear fuse center cover.

25. Position the carpet back under the right rear seat.

26. Use caution while reaching in and turn the ignition switch to the ON position. The AIR BAG indicator will flash then turn OFF.

Zone 7

1. Turn the steering wheel so that the vehicles wheels are pointing straight ahead.

2. Turn the ignition switch to the OFF position.

3. Remove the key from the ignition switch.

4. Adjust the right rear seat to the rear of vehicle.

5. Pull the carpet away from under the right side of rear seat.

6. When the carpet is pulled away from the rear seat the right rear fuse center will be exposed, then remove the fuse center top cover.

➡ **With the SIR fuse removed and the ignition switch in the ON position, the AIR BAG warning indicator illuminates. This is normal operation, and does not indicate an SIR system malfunction.**

7. Locate and remove the SIR fuse from the right rear fuse center.

8. Remove both connector position assurance (CPA) from the LF/driver's side impact module and seat belt pretensioner yellow connector (1) which is located under the front of driver's seat.

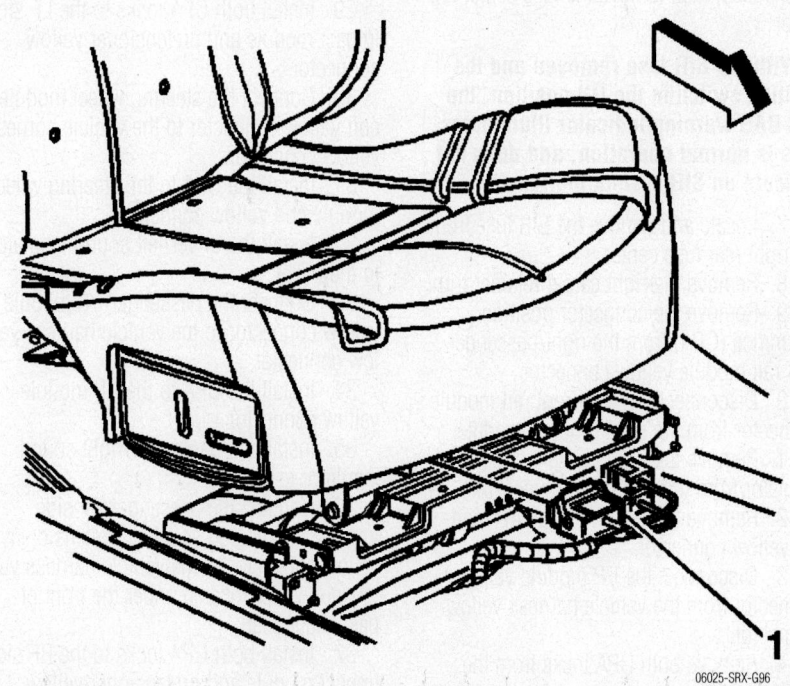

LF/driver's side impact module

06025-SRX-G96

9. Disconnect the LF side impact module and pretensioner yellow connector from the vehicle harness yellow connector.

Enabling Procedure:

10. Remove the key from the ignition switch.

11. Connect the LF side impact module and pretensioner yellow connector (1) to the vehicle harness yellow connector.

12. Install both CPA locks to the LF side impact module and pretensioner yellow connector.

13. Install the SIR fuse into the right rear fuse center.

14. Install the right rear fuse center cover.

15. Position the carpet back under the right rear seat.

16. Use caution while reaching in and turn the ignition switch to the ON position. The AIR BAG indicator will flash then turn OFF.

Zone 8

1. Turn the steering wheel so that the vehicle's wheels are pointing straight ahead.

2. Turn the ignition switch to the OFF position.

3. Remove the key from the ignition switch.

4. Adjust the right rear seat to the rear of vehicle.

5. Pull the carpet away from under the right side of rear seat.

6. When the carpet is pulled away from the rear seat the right rear fuse center will be exposed, then remove the fuse center top cover.

➡ **With the SIR fuse removed and the ignition switch in the ON position, the AIR BAG warning indicator illuminates. This is normal operation, and does not indicate an SIR system malfunction.**

7. Locate and remove the SIR fuse from the right rear fuse center.

8. Remove the right carpet retainer trim.

9. Remove the connector position assurance (CPA) from the right/passenger roof rail module yellow connector.

10. Disconnect the right roof rail module connector from the right roof rail module.

11. Remove the right/passenger sound insulator from the instrument panel (I/P).

12. Remove the CPA from the I/P module yellow connector.

13. Disconnect the I/P module yellow connector from the vehicle harness yellow connector.

14. Remove both CPA locks from the passenger/RF side impact module and seat

belt pretensioner yellow connector located under the front of passenger seat.

15. Disconnect the RF side impact module and pretensioner yellow connector from the vehicle harness yellow connector.

16. Remove the driver/left sound insulator from the I/P.

17. Remove the CPA from the steering wheel module coil yellow connector.

18. Disconnect the steering wheel module coil yellow connector from the vehicle harness yellow connector.

19. Remove both CPA locks from the driver/LF side impact module and seat belt pretensioner yellow connector located under the front of driver seat.

20. Disconnect the LF side impact module and pretensioner yellow connector from the vehicle harness yellow connector.

21. Remove the left carpet retainer trim.

22. Remove the CPA from the left/driver roof rail module yellow connector.

23. Disconnect the left roof rail module connector from the left roof rail module.

Enabling Procedure:

24. Remove the key from the ignition switch.

25. Connect the roof rail module yellow connector to the left roof rail module.

26. Install the CPA to the left roof rail module connector.

27. Install the left carpet retainer trim.

28. Connect the driver/LF side impact module and seat belt pretensioner yellow connector to the vehicle harness yellow connector located under the front of driver seat.

29. Install both CPA locks to the LF side impact module and pretensioner yellow connector.

30. Connect the steering wheel module coil yellow connector to the vehicle harness yellow connector.

31. Install the CPA to the steering wheel module coil yellow connector.

32. Install the driver/left sound insulator to the I/P.

33. Connect the passenger/I/P module yellow connector to the vehicle harness yellow connector.

34. Install the CPA to the I/P module yellow connector.

35. Install the passenger/right sound insulator to the I/P.

36. Connect the passenger/RF side impact module and seat belt pretensioner yellow connector to the vehicle harness yellow connector located under the front of passenger seat.

37. Install both CPA locks to the RF side impact module and pretensioner yellow connector.

38. Connect the roof rail module yellow connector to the right roof rail module.

39. Install the CPA to the right roof rail module connector.

40. Install the right carpet retainer trim.

41. Install the SIR fuse into the right rear fuse center.

42. Install the right rear fuse center cover.

43. Position the carpet back under the right rear seat.

44. Use caution while reaching in and turn the ignition switch to the ON position. The AIR BAG indicator will flash then turn OFF.

Zone 9

1. Turn the steering wheel so that the vehicles wheels are pointing straight ahead.

2. Turn the ignition switch to the OFF position.

3. Remove the key from the ignition switch.

4. Adjust the right rear seat to the rear of vehicle.

5. Pull the carpet away from under the right side of rear seat.

6. When the carpet is pulled away from the rear seat the right rear fuse center will be exposed, then remove the fuse center top cover.

➡ **With the SIR fuse removed and the ignition switch in the ON position, the AIR BAG warning indicator illuminates. This is normal operation, and does not indicate an SIR system malfunction.**

7. Locate and remove the SIR fuse from the right rear fuse center.

8. Remove both connector position assurance (CPA) from the RF/passenger side impact module and seat belt pretensioner yellow connector which is located under the front of passenger seat.

9. Disconnect the RF side impact module and pretensioner yellow connector from the vehicle harness yellow connector.

Enabling Procedure:

10. Remove the key from the ignition switch.

11. Connect the RF side impact module and pretensioner yellow connector to the vehicle harness yellow connector.

12. Install both CPA locks to the RF side impact module and pretensioner yellow connector.

13. Install the SIR fuse into the right rear fuse center.

14. Install the right rear fuse center cover.

15. Position the carpet back under the right rear seat.

16. Use caution while reaching in and turn the ignition switch to the ON position. The AIR BAG indicator will flash then turn OFF.

Rack and Pinion Steering Gear

REMOVAL & INSTALLATION

➡**The front wheels must be in the straight ahead position and the steering column locked before disconnect the intermediate shaft. Failure to do so may result in damage to the SRS coil.**

1. Before servicing the vehicle, refer to the Precautions Section.
2. Center the steering wheel.
3. Install Steering Column Anti-rotation pin J-42640 to lock the steering column.
4. Remove or disconnect the following:
   - Front wheels
   - Front air deflector
   - Intermediate steering shaft lower pinch bolt
   - Intermediate shaft from the steering gear
   - Variable effort steering harness connector
   - Outer tie rod retaining nuts
   - Outer tie rods from steering knuckles
   - Power steering hoses from steering gear
   - Left brake line from brake hose and plug opening
   - On AWD models, place a jack under the front differential
   - On AWD models, remove the right engine mount-to-frame nut
   - Raise the differential enough to clear the steering rack bolt
5. On all models, remove the steering gear mounting bolts.
6. Remove the left rear lower control arm-to-frame mounting bolt and nut.

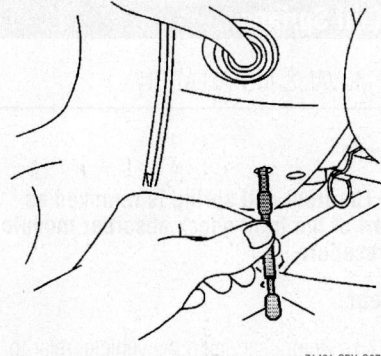

71461-SRX-G07

Installing steering column anti-rotation pin

7. Remove the steering gear through the left wheel opening.
   To install:
8. Install the steering gear through the left wheel opening.
9. Install the left rear lower control arm-to-frame mounting bolt and nut.
10. Install the steering gear mounting bolts. Tighten the bolts to 134 ft. lbs. (180 Nm).
11. On AWD models, lower the differential.
12. On AWD models, install the right engine mount-to-frame nut and tighten to 59 ft. lbs. (80 Nm).
13. On AWD models, remove the differential jack.
14. Install or connect the following:
    - Left brake line to brake hose
    - Power steering hoses
    - Outer tie rod to steering knuckles. Tighten the nuts to 52 ft. lbs. (70 Nm).
    - Variable effort steering harness connector
    - Intermediate shaft-to-steering gear
    - Intermediate shaft pinch bolt and tighten to 37 ft. lbs. (50 Nm).
    - Front air deflector
    - Front wheels
15. Lower the vehicle.
16. Remove the steering wheel anti-rotation pin.
17. Fill the power steering oil reservoir.
18. Bleed the steering system.
19. Check and adjust the front wheel toe.
20. Start the vehicle and check for leaks.

Front Suspension Frame

REMOVAL & INSTALLATION

1. Before servicing the vehicle, refer to the Precautions Section.
2. Install an engine support fixture to the top of the engine.
3. Raise and support the vehicle.
4. Remove the front wheels.
5. Remove the front air deflector.
6. Remove or disconnect the following:
   - Wheel speed sensor
   - Outer tie rod nuts
   - Separate the outer tie rods
   - Power steering return and pressure hose clamp bolts
   - Engine Control Module (ECM) and retaining bracket
   - Power steering cooler from the A/C condenser
   - Variable effort steering harness connector

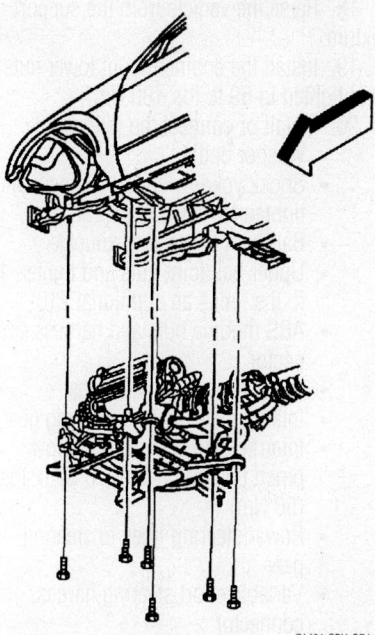

71461-SRX-G74

Removing front suspension frame mounting bolts

   - Power steering lines from steering gear
   - Intermediate steering shaft lower pinch bolt
   - Intermediate shaft from steering gear
   - Brake lines from brake hoses
   - ABS module nuts and harness connector
   - Upper ball joint nuts
   - Separate the ball joint from steering knuckle
   - Shock yoke-to-shock nuts and bolts
   - Washer bottle-to-knuckle bolts
   - Engine mount lower retaining nuts
7. Install frame support table J-39580 under the vehicle and lower the vehicle to the frame support.
8. Remove the 6 frame mounting bolts.
9. With an assistant, slowly raise the body up away from the frame.
10. Remove the lower control arms and the stabilizer bar.
11. Remove the frame from the support fixture.
    To install:
12. Install the frame to the support fixture.
13. Install the stabilizer bar.
14. Install the lower control arms.
15. Place the frame assembly under the vehicle.
16. Carefully lower the vehicle to the frame.
17. Install the frame mounting bolts and tighten to 141 ft. lbs. (191 Nm).

18. Raise the vehicle from the support fixture.

19. Install the engine mount lower nuts and tighten to 59 ft. lbs. (80 Nm).

20. Install or connect the following:
- Washer bottle
- Shock yoke-to-shock retainers and tighten to 133 ft. lbs. (180 Nm).
- Ball joint to steering knuckle
- Upper ball joint nuts and tighten 15 ft. lbs., plus an additional 210°
- ABS module nuts and harness connector
- Brake lines to brake hoses
- Intermediate shaft to steering gear
- Intermediate steering shaft lower pinch bolt and tighten to 37 ft. lbs. (50 Nm)
- Power steering lines to steering gear
- Variable effort steering harness connector
- Power steering cooler to the A/C condenser
- ECM and retaining bracket
- Power steering return and pressure hose clamp bolts
- Connect the outer tie rods and tighten the nuts to 52 ft. lbs. (70 Nm)
- Wheel speed sensor
- Front air deflector
- Front wheels
- Lower the vehicle

21. Remove the engine support fixture.

22. Fill and bleed the power steering system

Shock Absorber Module

REMOVAL & INSTALLATION

Front

1. Before servicing the vehicle, refer to the Precautions Section.

2. Raise and support the vehicle.

3. Remove or disconnect the following:
- Front wheel
- Shock-to yoke retaining nut
- Shock-to-yoke retaining bolt by pulling up on the lower control arm to relieve tension

4. Using a puller, separate the yoke from the lower arm and remove the yoke.

5. On vehicles with Magnaride or automatic headlight aiming, disconnect the sensor connector and the link from the upper control arm.

6. Remove the upper arm-to-steering knuckle nut.

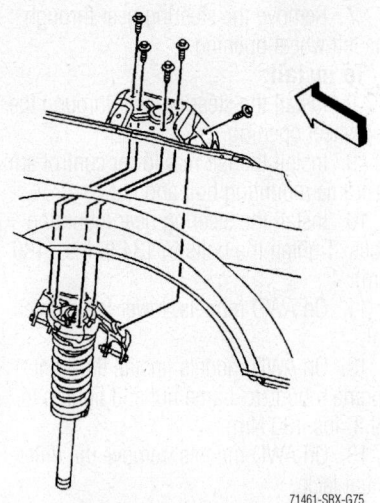

71461-SRX-G75

Front shock module assembly mounting

7. Separate the control arm from the knuckle.

8. Lower the vehicle.

9. Remove the shock module upper mounting bolts and remove the shock module.

To install:

10. Install the shock module and tighten the upper mounting nuts to 83 ft. lbs. (112 Nm).

11. Install the upper control arm to the knuckle and tighten the nut to 18 ft. lbs. (20 Nm), plus an additional 210°.

12. On vehicles with Magnaride or automatic headlight aiming, connect the sensor connector and the link to the upper control arm.

13. Connect the yoke to the lower arm and install the nut.

14. Install the yoke to the shock.

15. Tighten the shock-to-yoke nut to 81 ft. lbs. (110 Nm), and the yoke to lower control arm nut to 133 ft. lbs. (180 Nm).

16. Install the front wheel.

17. Lower the vehicle.

Rear

1. Before servicing the vehicle, refer to the Precautions Section.

2. Remove the rear interior trim panel to access the upper shock mounting.

3. Move the sound insulator away from the shock tower.

4. On vehicles with Magnaride, disconnect the electrical connector.

5. Remove the upper shock mounting nuts.

6. Raise and support the vehicle.

7. Disconnect the Magnaride connector from the shock.

8. Remove the lower shock mounting bolt.

9. Remove the shock absorber.

To install:

10. Install the shock to the vehicle.

11. Tighten the lower mounting bolt to 111 ft. lbs. (150 Nm).

12. Connect the connector to the shock.

13. Install the shock into the shock tower and tighten the upper nuts to 18 ft. lbs. (25 Nm).

14. Connect the Magnaride connector.

15. Reposition the sound insulator.

16. Install the trim panel.

DISASSEMBLY

1. Install the shock module in a spring compressor.

2. Mark the upper control arm assembly and insulator for installation reference.

3. Compress the coil spring.

4. If equipped with Magnaride, remove the sensor nut.

5. Remove the shock upper mounting nut.

6. Remove the shock absorber.

7. Loosen the spring compressor.

8. Remove the upper mounting plate, insulator, coil spring and upper control arm bracket.

REASSEMBLY

➡**Ensure the alignment pins in the upper control arm bracket are oriented 90° with the shock absorber lower mounting holes.**

1. Install the upper mounting plate, insulator, coil spring and upper control arm bracket to the spring compressor, aligning all marks.

2. Turn the compressor forcing screw until the coil spring is compressed.

3. Install the shock absorber upper nut and tighten the nut to 18 ft. lbs. (25 Nm).

4. Install the Magnaride sensor nut, if equipped.

5. Remove the shock module from the spring compressor.

Coil Springs

REMOVAL & INSTALLATION

Front

➡**The front coil spring is removed as part of the front shock absorber module procedure.**

Rear

1. Before servicing the vehicle, refer to the Precautions Section.

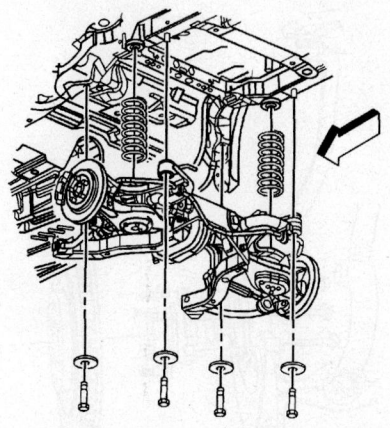

Removing the rear coil spring

2. Raise and safely support the vehicle.

3. Remove the rear wheels.

4. Disconnect the Magnaride and headlight adjustment sensor link from the upper control arm.

5. Raise and support the lower control arm with a jack.

6. Remove the shock absorber lower mounting bolt.

7. Lower the control arm and remove the jack.

8. Support the rear frame with a jack.

9. Remove the 4 rear frame to body bolts and lower the frame far enough to remove the coil spring without going past the guide pins.

10. Remove the coil spring.

To install:

11. Install the coil spring.

12. Raise the frame and install the mounting bolts. Tighten the front bolts to 195 ft. lbs. (265 Nm), and the rear bolts to 140 ft. lbs. (191 Nm).

13. Remove the frame jack.

14. Place a jack under the lower control arm.

15. Raise the jack until the shock absorber aligns with the lower control arm.

16. Install the lower shock bolt and tighten the bolt to 111 ft. lbs. (150 Nm).

17. Remove the jack.

18. Connect the Magnaride and headlight adjustment link to the upper control arm.

19. Install the rear wheels.

20. Lower the vehicle.

Stabilizer Bar

REMOVAL & INSTALLATION

1. Before servicing the vehicle, refer to the Precautions Section.

2. Raise and safely support the vehicle.

3. Remove the wheels.

4. Remove the stabilizer bar link nuts.

5. Disconnect the stabilizer shaft links from the stabilizer bar.

6. Remove the bar mounting bolts and bracket.

7. Remove the insulators from the stabilizer shaft.

8. Remove the stabilizer bar.

To install:

9. Install the stabilizer bar.

10. Install the bar insulators to the shaft with the slits facing rearward.

11. Install the stabilizer shaft bracket and bolts but do not tighten.

12. Connect the shaft links to the bar and tighten the nuts to 95 ft. lbs. (115 Nm), and the bolts to 81 ft. lbs. (110 Nm).

13. Install the wheels and lower the vehicle.

Lower Control Arm

REMOVAL & INSTALLATION

Front

1. Before servicing the vehicle, refer to the Precautions Section.

2. Remove or disconnect the following:
   - Wheel
   - Shock yoke
   - Stabilizer shaft link
   - ABS harness
   - Lower control arm-to-steering knuckle nut

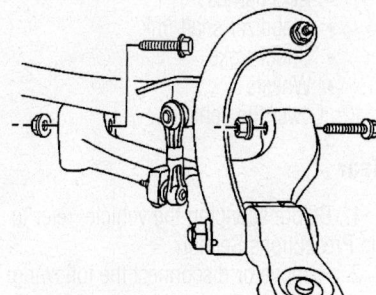

Front lower control arm-to-cradle mounting bolts

3. Separate the lower control arm from knuckle using Ball Joint Remover J-43631.

4. Loosen the steering gear mounting nuts and raise the steering gear.

5. Remove the control arm to cradle nuts and bolts.

6. Lower the control arm at the frame and move the ball joint upward.

7. Remove the lower control arm.

To install:

8. Install the lower arm on the ball joint and move the arm up to meet the cradle.

9. Install the lower arm-to-cradle nuts and bolts. Tighten the nuts to 96 ft. lbs. (135 Nm).

10. Lower the steering gear and install the bolts. Tighten the bolts to 89 ft. lbs. (120 Nm).

11. Install or connect the following:
   - Lower control arm-to-steering knuckle nut. Tighten to 15 ft. lbs. (20 Nm) plus 210°

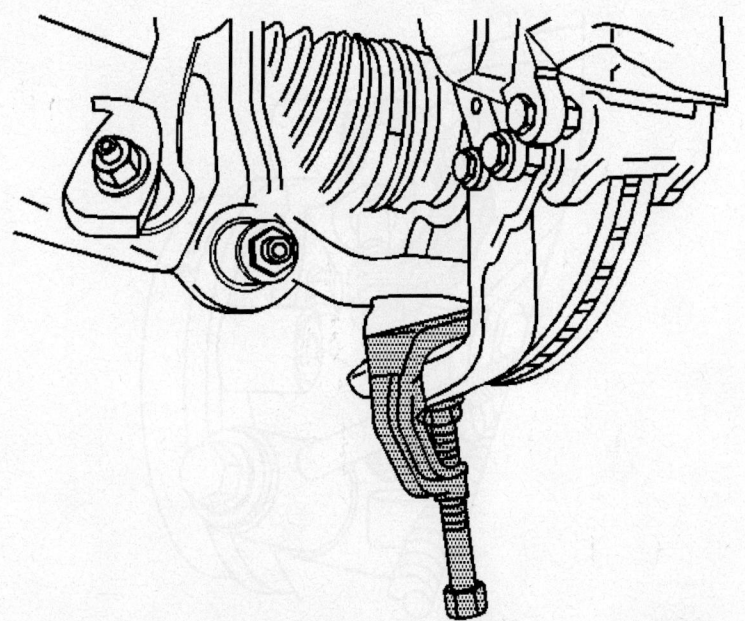

Using the ball joint remover tool to separate the lower control arm from the knuckle.

- ABS harness
- Stabilizer shaft link
- Shock yoke
- Wheels
12. Lower the vehicle.

Rear

1. Before servicing the vehicle, refer to the Precautions Section.
2. Remove or disconnect the following:
   - Wheel
   - Stabilizer shaft link
   - Coil spring
   - Lower arm-to-knuckle bolt
   - Lower arm-to-frame bolt
   - Lower control arm

To install:

3. Install the lower arm and the retaining bolts and nuts.
4. Install the coil spring.
5. Tighten the lower arm-to-frame bolt and nut to 100 ft. lbs. (135 Nm).
6. Tighten the lower arm-to-knuckle bolt to 118 ft. lbs. (160 Nm).
7. Install or connect the following:
   - Coil spring
   - Stabilizer shaft link
   - Wheel

Upper Control Arm

REMOVAL & INSTALLATION

Rear

1. Before servicing the vehicle, refer to the Precautions Section.

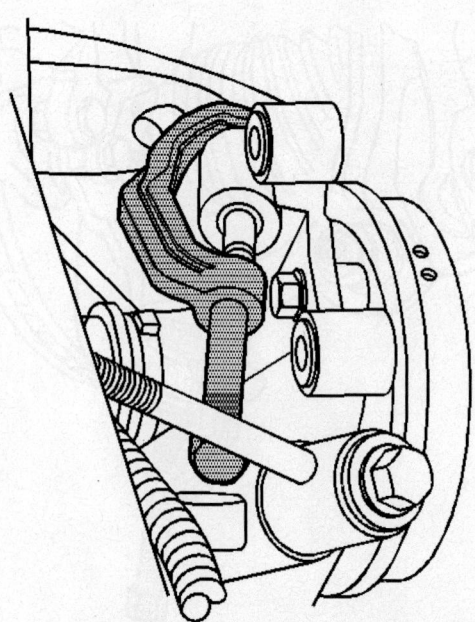

06025-SRX-G12

Using the ball joint remover to separate the upper control arm from the knuckle.

2. Remove or disconnect the following:
   - Wheel
   - Upper arm-to-knuckle nut
3. Separate the upper arm from the knuckle using Ball Joint Remover J-43631. Do not use a pickle fork or pry bar to separate the arm.
4. Remove the upper arm-to-frame mounting bolts and nuts and remove the upper control arm.

To install:

5. Install the upper control arm and loosely install the mounting nuts and bolts.
6. Connect the arm to the knuckle.
7. Tighten the arm-to-knuckle nut to 15 ft. lbs. (20 Nm), plus an additional 210°. Tighten the arm-to-frame nuts and bolts to 111 ft. lbs. (150 Nm).
8. Install the wheel.

Ball Joints and Steering Knuckle

REMOVAL & INSTALLATION

Front

1. Before servicing the vehicle, refer to the Precautions Section.
2. Raise and safely support the vehicle.
3. Remove the wheel and tire assembly.
4. Remove the wheel hub/bearing assembly.
5. Remove the outer tie rod-to-steering knuckle nut.
6. Disconnect the tie rod from the steering knuckle.
7. Remove the brake hose bracket bolts.

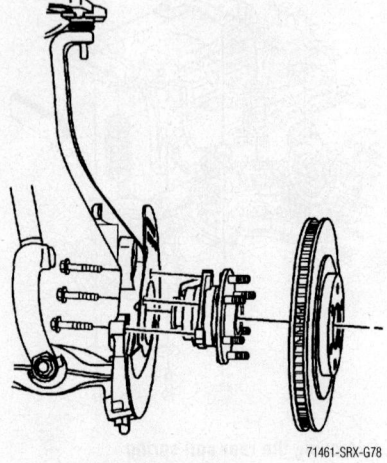

71461-SRX-G78

Front steering knuckle mounting

8. Remove the upper and lower control arm ball joint-to-knuckle nuts.
9. Separate the lower ball joint from the knuckle.
10. Remove the steering knuckle.
11. If the ball joints are to be removed, press the appropriate ball joint out the mounting.

To install:

12. If a ball joint was removed, install the appropriate ball joint.
13. Install the steering knuckle to the upper and lower ball joints.
14. Install new ball joint nuts and tighten the nuts to 15 ft. lbs. (20 Nm), plus an additional 210°.
15. Install the brake hose bracket bolts.
16. Connect the outer tie rod to the steering knuckle and tighten the retaining nut to 52 ft. lbs. (70 Nm).
17. Install the wheel hub/ bearing assembly.
18. Install the wheel and tire assembly.

Rear

1. Before servicing the vehicle, refer to the Precautions Section.
2. Raise and safely support the vehicle.
3. Remove the wheel and tire assembly.
4. Remove the brake caliper and wire it out of the way.
5. Remove or disconnect the following:
   - Brake rotor
   - ABS sensor connector
   - Axle hub nut
   - Parking brake cable bracket
   - Parking brake cable from the brake lever
   - Upper ball joint nut
6. Separate the upper arm from the knuckle. Do not use a pickle fork or pry bar to separate the arm.

7. Support the lower control arm with a jack.

8. Remove or disconnect the following:
- Lower shock mounting bolt
- Trailing arm-to-knuckle bolt and nut
- Lower control arm-to-knuckle bolt
- Adjustment link-to-knuckle nut

9. Install tool J-45859 onto the wheel hub and secure with 2 lug nuts.

10. Use the tool to disengage the half-shaft from the wheel hub and bearing. Support the halfshaft.

11. Remove the knuckle.

12. Separate the wheel hub/bearing from the knuckle and backing plate.

To install:

13. Install the wheel hub/bearing to the knuckle and backing plate.

14. Install the wheel hub/bearing and tighten the bolts to 92 ft. lbs. (125 Nm).

15. Install the knuckle.

16. Install the adjustment link to the knuckle and tighten the bolt to 118 ft. lbs. (160 Nm).

17. Install the trailing arm-to-knuckle bolt and nut and tighten to 125 ft. lbs. (170 Nm).

18. Install the lower shock mounting bolt and tighten to 111 ft. lbs. (150 Nm).

19. Install the lower control arm-to-knuckle bolt.

20. Connect the upper ball joint to the knuckle.

21. Install the upper ball joint mounting nut.

22. Install the parking brake bracket.

23. Remove the jack.

24. Connect the ABS sensor connector.

25. Install a new axle hub nut and tighten to 118 ft. lbs. (160 Nm).

26. Install the brake rotor.

27. Install the brake caliper.

28. Install the wheel and tire assembly.

Hub and Wheel Bearing

REMOVAL & INSTALLATION

The wheel bearing is integral with the wheel hub and cannot be replaced separately. If the wheel bearing is found to be defective, the wheel hub must be replaced as an assembly.

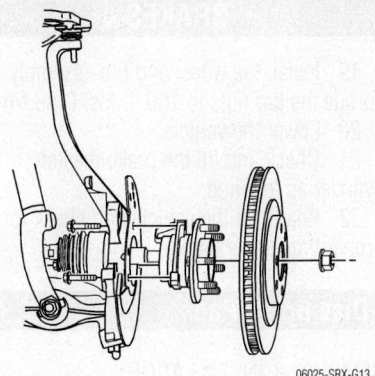

06025-SRX-G13

Front wheel hub and bearing assembly mounting.

Front

1. Before servicing the vehicle, refer to the Precautions Section.

2. Raise and safely support the vehicle.

3. Remove the wheel and tire assembly.

4. On AWD models, remove axle hub nut and discard.

5. On all models, remove the brake caliper and wire it out of the way.

6. Remove the brake rotor.

7. Disconnect the ABS sensor connector.

8. On AWD models, install tool J-45859 onto the wheel hub and secure with 2 lug nuts.

9. Use the tool to disengage the half-shaft from the wheel hub and bearing. Support the halfshaft.

10. On all models from the backside, remove the wheel hub/bearing retaining bolts.

11. Remove the wheel hub/bearing assembly.

To install:

12. Install the wheel hub/bearing and tighten the bolts to 100 ft. lbs. (135 Nm).

13. Install a new axle shaft nut and tighten the nut to 118 ft. lbs. (160 Nm).

14. Install or connect the following:
- ABS sensor connector
- Brake rotor
- Brake caliper
- Wheel and tire assembly

Rear

1. Before servicing the vehicle, refer to the Precautions Section.

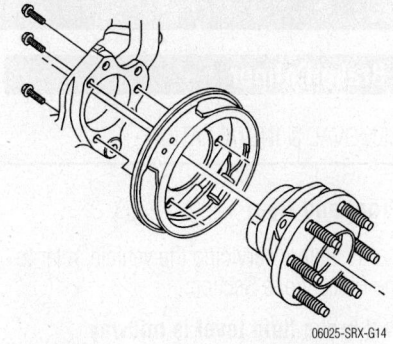

06025-SRX-G14

Rear wheel hub and bearing assembly mounting.

2. Raise and safely support the vehicle.

3. Remove the wheel and tire assembly.

4. Remove the brake caliper and wire it out of the way.

5. Remove the brake rotor.

6. Disconnect the ABS sensor connector.

7. Remove the axle hub nut.

8. Remove the upper arm-to-knuckle nut

9. Separate the upper arm from the knuckle. Do not use a pickle fork or pry bar to separate the arm.

10. From the backside, remove the wheel hub/bearing retaining bolts.

11. Install tool J-45859 onto the wheel hub and secure with 2 lug nuts.

12. Use the tool to disengage the half-shaft from the wheel hub and bearing. Support the halfshaft.

13. Remove the tool from the wheel hub and remove the wheel hub/bearing assembly.

To install:

14. Install the wheel hub/bearing and tighten the bolts to 92 ft. lbs. (125 Nm).

15. Connect the arm to the knuckle and tighten the nut to 15 ft. lbs. (20 Nm), plus an additional 210°.

16. Install a new axle shaft nut and tighten the nut to 118 ft. lbs. (160 Nm).

17. Install or connect the following:
- ABS sensor connector
- Brake rotor
- Brake caliper
- Wheel and tire assembly

BRAKES

Brake Caliper

REMOVAL & INSTALLATION

Front and Rear

1. Before servicing the vehicle, refer to the Precautions Section.

➡ **If brake fluid level is midway between the maximum-full point and the minimum allowable level, no fluid needs to be removed from the reservoir.**

2. Remove ½ of the brake fluid from the brake master cylinder reservoir. Properly dispose of the brake fluid.

3. Raise and safely support the vehicle.

4. Remove the wheel and tire assembly.

5. Install a large C-clamp over the rear of the caliper and against the outer brake pad.

6. Tighten the clamp until the caliper piston is compressed into the caliper bore enough to clear the rotor.

7. Remove the C-clamp.

8. Remove the brake line, discard the 2 copper washers and plug the line openings.

9. Remove the 2 brake pin retainer bolts.

10. Remove the disc brake caliper from the vehicle.

To install:

11. Inspect the guide pin boots for tears or cuts and replace as needed.

12. Seat the guide pin boots into the guide pin retaining seat.

13. Clean the guide pin bolt threads, then apply Threadlock to two-thirds of the lower guide pin bolt threads and allow it 10 minutes to dry.

14. Apply a thin coat of high temperature brake lubricant to the guide bolts.

15. Ensure that the disc brake pads are properly positioned and that the lining material is facing the rotor.

16. Place the disc brake caliper over the rotor and tighten the brake pin retainer bolts-to-guide pin bolts to 25 ft. lbs. (34 Nm) for front brakes, or 44 ft. lbs. (60 Nm) for rear brakes.

17. Unplug and install the brake hose and retaining bolt to the disc brake caliper using new copper sealing washers on each side of the hose fitting. Tighten the brake hose bolts to 37 ft. lbs. (50 Nm).

18. Bleed the brake system, then apply the brake pedal two-thirds down and slowly release the pedal. Wait 15 seconds and repeat until a firm pedal is obtained.

19. Install the wheel and tire assembly. Torque the lug nuts to 100 ft. lbs. (136 Nm).

20. Lower the vehicle.

21. Check and fill the brake master cylinder as required.

22. Road-test the vehicle and check for proper brake operation.

Disc Brake Pads

REMOVAL & INSTALLATION

Front and Rear

1. Before servicing the vehicle, refer to the Precautions Section.

2. Remove ½ of the brake fluid from the brake master cylinder reservoir. Properly dispose of the brake fluid.

3. Raise and safely support the vehicle.

4. Remove the wheel and tire assembly.

5. Install a large C-clamp over the rear of the caliper and against the outer brake pad.

6. Tighten the clamp until the caliper piston is compressed into the caliper bore enough to clear the rotor.

7. Remove the C-clamp.

8. Remove the lower brake pin retainer bolt.

9. Pivot the caliper upward away from the rotor.

10. Hang the disc brake caliper with a length of wire or equivalent to prevent damage to the brake hose.

11. Remove the inner and outer disc brake pads and the retainers.

12. Inspect the disc brake rotor surfaces for grooves, cracks or glazing. Resurface or replace as required. If resurfacing, observe the minimum thickness specification.

To install:

13. Inspect the guide pin boots for tears or cuts and replace as needed.

14. Seat the guide pin boot into the guide pin retaining seat.

15. Clean the guide pin bolt threads, then apply Threadlock to two-thirds of the lower guide pin bolt threads and allow it 10 minutes to dry.

16. Apply a thin coat of high temperature brake lubricant to the guide bolt.

17. Retract the caliper piston fully into the caliper bore using a C-clamp and wood block or equivalent. This will allow room for the new disc brake pads.

18. Install new inner and outer disc brake pads and the retainers. Ensure that the disc brake pads are properly positioned

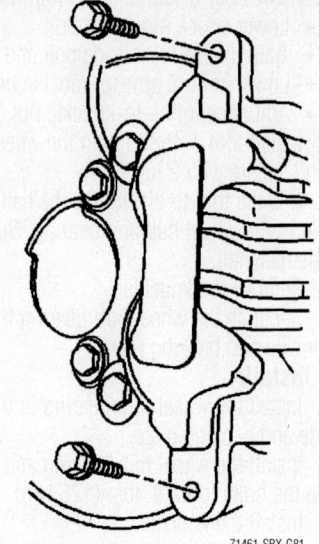

Front brake caliper mounting

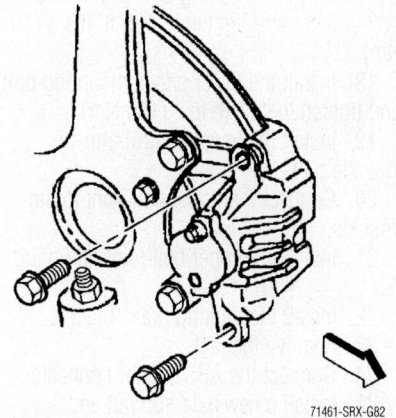

Rear brake caliper mounting

and that the lining material is facing the rotor.

19. Ensure that the disc brake pads are properly positioned and that the lining material is facing the rotor.

20. Place the disc brake caliper over the rotor and tighten the brake pin retainer bolts-to-guide pin bolts to 25 ft. lbs. (34 Nm) for front brakes, or 44 ft. lbs. (60 Nm) for rear brakes.

21. Install the wheel and tire assembly. Torque the lug nuts to 100 ft. lbs. (136 Nm).

22. Lower the vehicle.

23. Pump the brake pedal to position the brake pads before attempting to move the vehicle.

24. Check and fill the brake master cylinder reservoir, as required.

25. Road-test the vehicle and check for proper brake system operation.

CADILLAC

STS

BRAKES **28-91**
DRIVE TRAIN **28-66**
ENGINE REPAIR **28-8**
FUEL SYSTEM **28-62**
SPECIFICATION AND
  MAINTENANCE CHARTS......**28-2**
Engine and Vehicle
  Identification 28-2
General Engine Specifications 28-2
Gasoline Engine Tune-Up
  Specifications 28-2
Capacities 28-3
Valve Specifications.....................28-3
Crankshaft and Connecting
  Rod Specifications....................28-3
Piston and Ring Specifications 28-4
Torque Specifications 28-4
Wheel Alignment 28-5
Brake Specifications 28-5
Tire, Wheel and Ball Joint
  Specifications.........................28-5
Scheduled Maintenance
  Intervals.................................28-6
STEERING AND
  SUSPENSION................**28-79**

A

Air Bag.....................................28-79
  Arming.................................28-83
  Diagnostic System Checks 28-84
  Disarming 28-80
  Sir Disabling And Enabling
    Zones 28-80
  Sir Service Precautions...........28-79
Alternator 28-8
  Removal & Installation..............28-8
Automatic Transmission
  Assembly.............................28-66
  Removal & Installation...........28-66
Axle Shaft, Bearing and Seal........28-77
  Removal & Installation...........28-77

B

Ball Joints...................................28-87
  Removal & Installation...........28-87
Brake Caliper 28-91
  Removal & Installation...........28-91

C

Camshaft Timing 28-40
Camshafts..................................28-35
  Removal & Installation...........28-35
Coil Springs 28-86
  Removal & Installation...........28-86
CV-Joints....................................28-74
  Overhaul 28-74
Cylinder Head 28-22
  Removal & Installation...........28-22

D

Disc Brake Pads..........................28-92
  Removal & Installation...........28-92

E

Engine Assembly 28-10
  Removal & Installation...........28-10
Engine Mounts 28-18
Exhaust Manifold 28-32
  Removal & Installation...........28-32

F

Flywheel.....................................28-69
  Removal & Installation...........28-69
Fuel Filter 28-62
  Removal & Installation...........28-62
Fuel Pump 28-63
  Removal & Installation...........28-63
Fuel Rail and Injectors 28-64
  Removal & Installation...........28-64
Fuel System Pressure 28-62
  Relieving 28-62
Fuel System Service
  Precautions..........................28-62

H

Halfshaft....................................28-69
  Removal & Installation...........28-69
Heater Core................................28-20
  Removal & Installation...........28-20

I

Ignition Timing 28-10
  Adjustment...........................28-10
Intake Manifold 28-28
  Removal & Installation...........28-28

L

Lower Control Arm 28-88
  Control Arm Bushing
    Replacement....................28-88
  Removal & Installation...........28-88

O

Oil Pan.......................................28-42
  Removal & Installation...........28-42
Oil Pump 28-46
  Removal & Installation...........28-46

P

Pinion Seal 28-77
  Removal & Installation...........28-77
Piston and Ring 28-61
  Positioning 28-61
Power Steering Gear 28-84
  Removal & Installation...........28-84
Propeller Shaft............................28-72
  Removal & Installation...........28-72

R

Rear Main Seal 28-47
  Removal & Installation...........28-47

S

Shock Absorbers 28-85
  Removal & Installation...........28-85
Stabilizer Bar 28-90
  Removal & Installation...........28-90
Starter Motor 28-42
  Removal & Installation...........28-42
Steering Knuckle.........................28-90
  Removal & Installation...........28-90

T

Timing Chain, Sprockets, Front
  Cover and Seal 28-48
  Removal & Installation...........28-48

U

Upper Control Arm 28-87
  Removal & Installation...........28-87

W

Water Pump 28-19
  Removal & Installation...........28-19
Wheel Bearings...........................28-88
  Adjustment...........................28-88
Wheel Hub/Bearing Assembly 28-88
  Removal & Installation...........28-88

SPECIFICATION AND MAINTENANCE CHARTS

ENGINE AND VEHICLE IDENTIFICATION

	Engine						Model Year	
Code ①	Liters (cc)	Cu. In.	Cyl.	Fuel Sys.	Engine Type	Eng. Mfr.	Code ②	Year
7	3.6 (3564)	217	6	SFI	DOHC	GM	5	2005
A	4.6 (4564)	279	8	SFI	DOHC	GM	6	2006

SFI: Sequential Fuel Injection

① 8th position of VIN

② 10th position of VIN

06025-GSTS-C01

GENERAL ENGINE SPECIFICATIONS

All measurements are given in inches.

Year	Model	Engine Displacement Liters	Engine Series VIN	Net Horsepower @ rpm	Net Torque @ rpm (ft. lbs.)	Bore x Stroke (in.)	Com-pression Ratio	Oil Pressure @ rpm
2005	STS	3.6	7	260@6500	252@2800	3.701x3.370	10.2:1	20@2000
		4.6	A	320@6400	315@4400	3.661x3.307	10.5:1	35@2000

06025-GSTS-C02

GASOLINE ENGINE TUNE-UP SPECIFICATIONS

Year	Engine Displacement Liters	Engine VIN	Spark Plug Gap (in.)	Ignition Timing (deg.) MT	AT	Fuel Pump (psi)	Idle Speed (rpm) MT	AT	Valve Clearance In.	Ex.
2005	3.6	7	0.044	①	①	55-60	②	②	HYD	HYD
	4.6	A	0.040	①	①	55-60	②	②	HYD	HYD

NOTE: The Vehicle Emission Control Information label often reflects specification changes made during production.

The label figures must be used if they differ from those in this chart.

HYD: Hydraulic

① Ignition timing is preset and cannot be adjusted

② Idle speed is maintained by the PCM

06025-GSTS-C03

CAPACITIES

Year	Model	Engine Displacement Liters	Engine VIN	Engine Oil with Filter (qts.)	Transmission (pts.) * 5-Spd	Auto.	Transfer Case (pts.)	Drive Axle Front (pts.)	Rear (pts.)	Fuel Tank (gal.)	Cooling System (qts.)
2005	STS	3.6	7	6	NA	14.8	1.03	2.74	2.74	17.5	9.7
		4.6	A	8	NA	14.8	1.03	2.74	2.74	17.5	10.4

NOTE: All capacities are approximate. Add fluid gradually and check to be sure a proper fluid level is obtained.

NA: Not available

* With bottom pan removal; 19 pts. with overhaul.

06025-GSTS-C04

VALVE SPECIFICATIONS

Year	Engine Displacement Liters	Engine VIN	Seat Angle (deg.)	Face Angle (deg.)	Spring Test Pressure (lbs. @ in.)	Spring Installed Height (in.)	Stem-to-Guide Clearance (in.) Intake	Exhaust	Stem Diameter (in.) Intake	Exhaust
2005	3.6	7	45	44.25	134-149 @0.9449	1.3779	0.0010-0.0026	0.0014-0.0030	0.2344-0.2352	0.2341-0.2348
	4.6	A	45.75	45	130-142 @.9646	1.3780	0.0011-0.0027	0.0020-0.0027	0.2341-0.2348	0.2331-0.2339

06025-GSTS-C05

CRANKSHAFT AND CONNECTING ROD SPECIFICATIONS

All measurements are given in inches.

Year	Engine Displ. Liters	Engine VIN	Crankshaft Main Brg. Journal Dia.	Main Brg. Oil Clearance	Shaft End-play	Thrust on No.	Connecting Rod Journal Diameter	Oil Clearance	Side Clearance
2005	3.6	7	2.6768-2.6775	0.0004-0.0024	0.0039-0.0130	3	2.2044-2.2050	0.0004-0.0028	0.0374-0.0140
	4.6	A	2.5335-2.5341	0.0006-0.0022	0.0020-0.0197	3	2.1239-2.1245	0.0010-0.0030	0.0079-0.0197

06025-GSTS-C06

PISTON AND RING SPECIFICATIONS

All measurements are given in inches.

Year	Engine Displ. Liters	Engine VIN	Piston Clearance	Ring Gap			Ring Side Clearance		
				Top Compression	Bottom Compression	Oil Control	Top Compression	Bottom Compression	Oil Control
2005	3.6	7	0.0010-0.0021	0.0059-0.0118	0.0110-0.0190	0.0059-0.0236	0.0012-0.0026	0.0006-0.0024	0.0012-0.0067
	4.6	A	0.0008-0.0020	0.0098-0.0157	0.0138-0.0020	0.0098-0.0299	0.0016-0.0037	0.0016-0.0037	①

① Side-sealing

06025-GSTS-C07

TORQUE SPECIFICATIONS

All readings in ft. lbs.

Year	Engine Displacement Liters	Engine VIN	Cylinder Head Bolts	Main Bearing Bolts	Rod Bearing Bolts	Crankshaft Damper Bolts	Flywheel Bolts	Manifold		Spark Plugs	Oil Pan Drain Plug
								Intake	Exhaust		
2005	3.6	7	①	②	③	④	⑤	17	15	13	18
	4.6	A	⑥	⑦	③	⑧	⑨	⑩	18	11	18

① M8 bolts:
  1st pass: 10 ft. lbs.
  Final pass: Plus 60 degrees
  M11 bolts:
  1st pass: 33 ft. lbs.
  Final pass: Plus 120 degrees

② Inner:
  1st pass: 15 ft. lbs.
  Final: 80 degrees
  Outer:
  1st pass: 11 ft. lbs.
  Final: 110 degrees
  Side:
  1st pass: 22 ft. lbs.
  Final: 60 degrees

③ 1st pass: 22 ft. lbs.
  2nd pass: back off to zero
  3rd pass: 18 ft. lbs.
  Final pass: 110 degrees

④ 1st pass: 74 ft. lbs.
  Final pass: plus 150 degrees

⑤ 1st pass: 22 ft. lbs.
  Final pass: plus 45 degrees

⑥ M6 bolt: 106 inch lbs.
  M11 bolts:
  1st pass: 22 ft. lbs.
  2nd pass: 60 degrees
  3rd pass: 60 degrees
  Final pass: 60 degrees
  Total: 180 degrees

⑦ Bearing bolt (M10):
  1st pass: 15 ft. lbs.
  2nd pass: Plus 65 degrees
  Perimeter bolt: 22 ft. lbs.

⑧ 1st pass: 37 ft. lbs.
  Final pass: Plus 150 degrees

⑨ 1st pass: 22 ft. lbs.
  Final pass: Plus 50 degrees

⑩ 89 inch lbs.

06025-GSTS-C08

WHEEL ALIGNMENT

Year	Model		Caster		Camber		Toe-in (Deg.)
			Range (+/-Deg.)	Preferred Setting (Deg.)	Range (+/-Deg.)	Preferred Setting (Deg.)	
2005	STS RWD	Front	0.60	+5.85	0.60	-0.50	0.20 +/-0.20
		Rear	—	—	0.50	-0.90	0.20 +/-0.20
	STS AWD	Front	0.60	+5.5	0.60	-0.50	0.20 +/-0.20
		Rear	—	—	0.50	-0.90	0.20 +/-0.20

06025-GSTS-C09

BRAKE SPECIFICATIONS
All measurements in inches unless noted

Year	Model		Brake Disc			Minimum Lining Thickness	Brake Caliper	
			Original Thickness	Minimum Thickness	Maximum Runout		Bracket Bolts (ft. lbs.)	Mounting Bolts (ft. lbs.)
2005	Standard &	F	1.267	1.209	0.002	NA	96	46
	Luxury Package	R	1.023	0.944	0.002	NA	88	44
	Luxury Performance	F	1.259	1.181	0.002	NA	96	25
	Package	R	1.102	1.062	0.002	NA	88	44

NA: Not Available

06025-GSTS-C10

TIRE, WHEEL AND BALL JOINT SPECIFICATIONS

Year	Model	OEM Tires		Tire Pressures (psi)		Wheel Size	Ball Joint Inspection	Lug Nut Torque (ft. lbs.)
		Standard	Optional	Front	Rear			
2005	Standard & Luxury Package	①	②	③	③	17 in. x 7 in.	④	100
	Performance Luxury Package	⑤	⑥	③	③	18 in. x 8 in.	④	100

OEM: Original Equipment Manufacturer

PSI: Pounds Per Square Inch

in.: inches

① Front: P235/50R17

   Rear: P255/45R17

② Front & Rear: P235/50R17

③ Refer to tire inflation pressure label on drivers door pillar

④ Horizontal and vertical looseness: 0.125 inch unloaded.

⑤ Front: P235/50R18

   Rear: P255/45R18

⑥ Front & Rear: P255/45R18

06025-GSTS-C11

MAINTENANCE I AND II SERVICE SCHEDULES
2005 Cadillac STS

When the CHANGE ENGINE OIL light appears, certain services and inspections are required. Services are described below. Generally, it is recommended that the first service be Maintenance I, second service be Maintenance II, and that services are then alternated from Maintenance I and Maintenance II thereafter. In some cases, Maintenance II may be Required services are described as Maintenance I and Maintenance II.

The first service of a vehicle should be Maintance I, and the second service should be Maintenance II.

Alternate between the 2 services thereafter. However, in some cases, Maintenance II may be required more often.

Maintenance I: Use Maintenance I if the CHANGE ENGINE OIL light comes on within 10 months since the vehicle was purcahsed or, if Maintenance II was performed.

Maintenance II: Use Maintenance II if the previous service performed was Maintenance I. Always used Maintenance II whenever the CHANGE ENGINE OIL light comes on 10 months or more since the last service, or, if the CHANGE ENGINE OIL light has not come on at all for one year.

Service	Maintenance I	Maintenance II
Change engine oil and filter. Reset oil life system.	✓	✓
Visually check for any leaks or damage. A fluid loss in the vehicle system could indicate a problem. Inspect, repair and add fluid to the system, if necessary.	✓	✓
Inspect engine air cleaner filter. If necessary, replace filter.	—	✓
Rotate tires and check inflation pressures and wear.	✓	✓
Visually inspect brake lines and hoses for proper hook-up, binding, leaks, cracks, chafing, etc. Inspect the disc brake pads for wear and the rotors for surface condition. Inspect the drum brake lings for wear or cracks. Inspect other brake parts, including drums, wheel cylinders, calipers, parking brake, etc. Inspect parking brake adjustment.	✓	✓
Check engine coolant and windshield washer fluid levels and add fluid as needed.	✓	✓
Inspect the suspension and steering components. Inspect the front and rear suspension systems and steering system for damaged, loose, or missing parts, or signs of wear. Inspect the power steering lines and the hoses for proper hook-up, binding, leaks, cracks, chafing, etc.	—	✓
Inspect the coolant hoses and replace the hoses if they are crackes, swollen or deteriorated. Inspect all pipes, fittings and clamps; replace with OEM parts as needed. To help ensure proper operation, a pressure test of the cooling system and pressure cap, and cleaning the outside of the radiator and A/C condesnser is recommended at least once a year.	—	✓
Inspect wiper blades for wear or cracking		✓
Inspect restraint system components.	—	✓
Lubricate all key lock cylinders, latch assemblies and hinges		✓
Inspect the transmission and transaxle fluid level and add fluid as needed.	—	✓
Replace passenger compartment air filter.		✓
Inspect throttle system	—	P

To reset the CHANGE ENGINE OIL LIGHT:

1. Press the up or down arrow to scroll the DIC to show OIL LIFE.

2. Once the XXX% ENGINE OIL LIFE menu item is highlighted, press and hold the RESET button until the percentage shows 100%. If the percentage does not return to 100% or if the CHANGE ENGINE OIL SOON message comes back on when the vehicle is started, the engine oil life system was not properly reset. Repeat the procedure.

06025-GSTS-C12

ADDITIONAL MAINTENANCE SERVICES
2005 Cadillac STS

TO BE SERVICED	TYPE OF SERVICE	VEHICLE MILEAGE INTERVAL (x1000)					
		25	50	75	100	125	150
Air cleaner filter	R		✓		✓		✓
Accessory drive belt	I						✓
Auto. Trans. Fluid ①	R		✓		✓		✓
Cooling system hoses and clamps	S/I						✓
Transfer case fluid	R		✓		✓		✓
Throttle body	I	✓	✓	✓	✓	✓	✓
Engine coolant	R						✓
Fuel system	I	✓	✓	✓	✓	✓	✓
Exhaust system & heat shields	S/I	✓	✓	✓	✓	✓	✓
Spark plugs	R				✓		

R: Replace

S/I: Inspect and service, if necessary

① Replace if any of the following condition are met:

Heavy city traffic where the outside temperature regularly reaches 90deg.F (32deg.C) or higher.

Hilly or mountainous terrain

Frequent trailer towing

Taxi, police or delivery service

Otherwise, change every 100,000 miles

06025-GSTS-C13

ENGINE REPAIR

Alternator

REMOVAL & INSTALLATION

3.6L Engines

1. Before servicing the vehicle, refer to the precautions in the beginning of this section.

2. Disconnect the battery negative cable.

3. Remove the drive belt from the alternator as follows:

   a. Remove the front compartment sight shield, if necessary.

   b. Rotate the drive belt tensioner clockwise to release the drive belt tension.

   c. Slide the drive belt off of the water pump pulley.

   d. Slowly release the drive belt tensioner.

   e. Remove the drive belt from the accessory drive pulleys.

4. Raise and support the vehicle.

5. Disconnect the electrical connector (1) from the alternator.

6. Reposition the protective boot from the alternator output BAT terminal for access.

7. Remove the alternator output BAT terminal nut (3) and disconnect the battery positive lead (2) from the alternator.

8. Remove the alternator lower bolts.

9. Lower the vehicle.

10. Remove the alternator upper bolt.

11. Remove the alternator from the engine.

To install:

12. Install the alternator to the engine.

13. Install the alternator upper bolt.

14. Raise and support the vehicle.

15. Install the alternator lower bolts.

16. Tighten the bolts to 37 ft. lbs. (50 Nm).

17. Install the BAT terminal boot.

18. Connect the electrical connector (1) to the alternator.

19. Install the drive belt to the alternator as follows:

   a. Install the drive belt to the crankshaft pulley, the tensioner and the alternator.

   b. Rotate the drive belt tensioner clockwise.

   c. Install the drive belt to the water pump.

✳✳ CAUTION

Ensure the drive belt is properly aligned and seated into the grooves of the accessory drive pulleys.

   d. Slowly release the drive belt tensioner.

   e. Install the front compartment sight shield, if necessary.

20. Connect the battery negative cable.

4.6L

1. Before servicing the vehicle, refer to the precautions in the beginning of this section.

2. Disconnect the battery negative cable

3. Remove the alternator drive belt as follows:

   a. Remove the air conditioning, power steering, and water pump belt (3).

   b. Rotate the alternator drive belt tensioner (4) clockwise to release drive belt tension.

   c. Slide the alternator drive belt from the alternator pulley (1).

   d. Allow the drive belt tensioner to return to the relaxed position.

   e. Remove the alternator drive belt from the pulleys.

4. Remove the upper alternator mounting bolts (1, 3).

5. Raise and support the vehicle.

6. Remove the front air deflector.

7. Remove the front wheels after making a reference mark between the wheel and hub.

8. Remove the right side wheelhouse liner after removing 6 wheelhouse liner screws.

9. Remove the right and left side stabi-

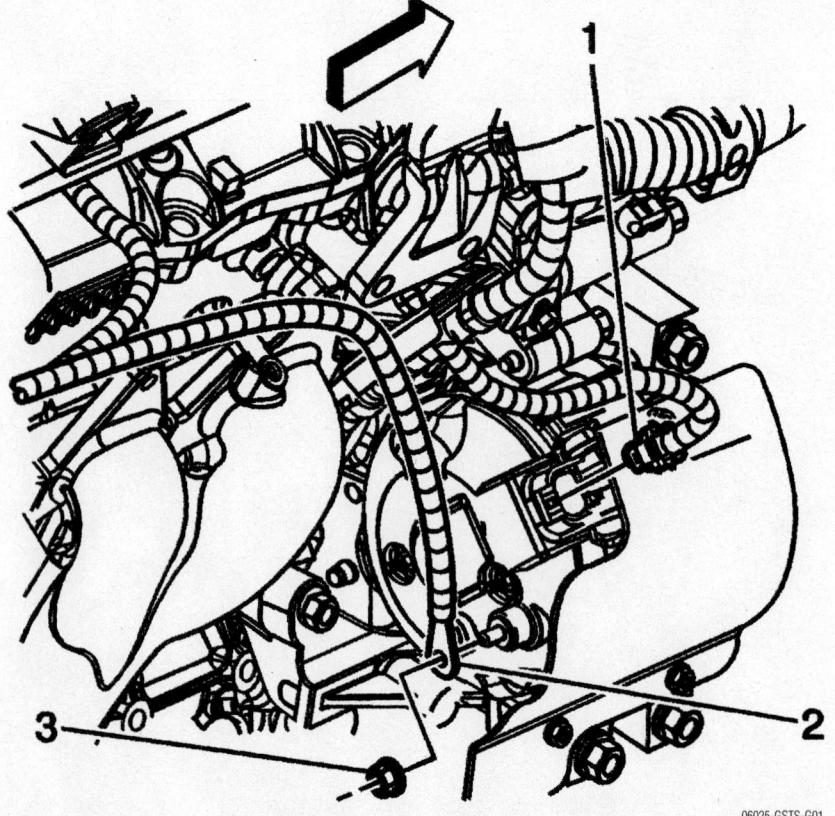

06025-GSTS-G01

Alternator wiring connection locations: 1. electrical connector; 2. battery cable; 3. nut—3.6L

06025-GSTS-G02

Alternator mounting bolt locations—3.6L

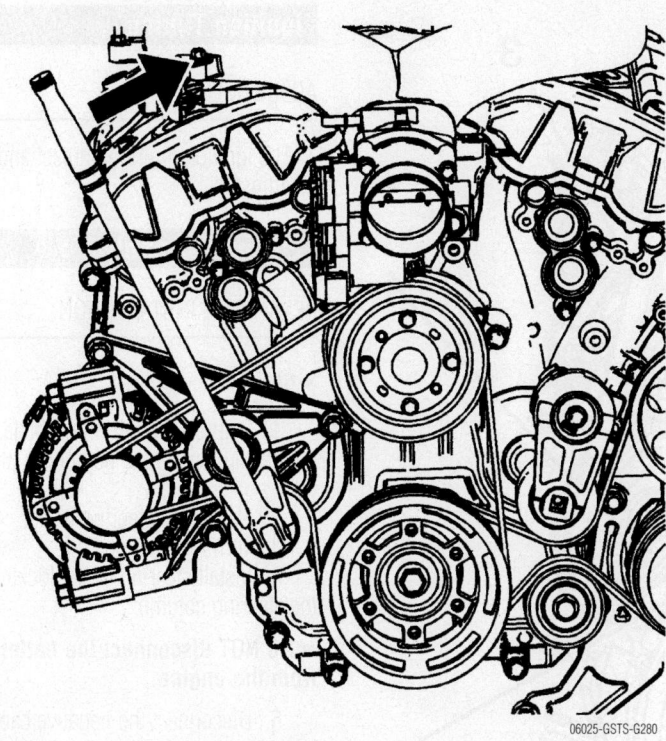

06025-GSTS-G280

Removing or installing the alternator drive belt

lizer shaft links at the lower control arm. See Stabilizer Bar.

10. Rotate the stabilizer shaft downward to gain access to the alternator and to create a removal envelop for alternator removal.

➡**Perform the remainder of the removal process through the wheel-house opening.**

11. Cut the tie strap securing the wiring harness to the alternator.

12. Remove the lower alternator mounting bolt.

13. Lift the alternator off of the mounting bracket in order to gain access to the connector and the alternator output BAT terminal nut.

14. Disconnect the wire harness electrical connector from the alternator.

15. Reposition the protective boot from the alternator output battery terminal for access.

16. Remove the alternator output battery terminal nut and disconnect the battery positive lead from the alternator.

17. Remove the alternator from the vehicle.

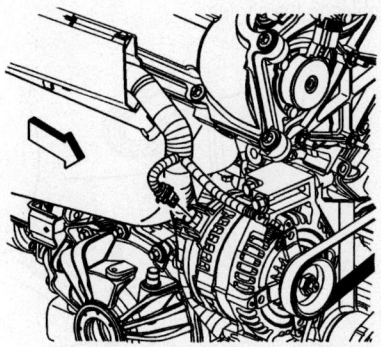

06025-GSTS-G03

Alternator mounting bolt locations—4.6L AWD

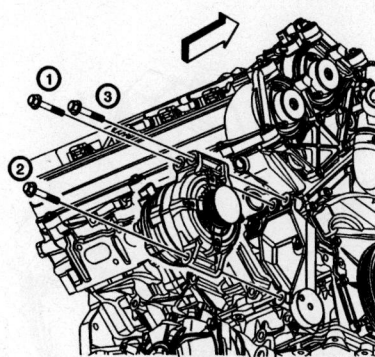

06025-GSTS-G04

Alternator wiring harness illustration— 4.6L AWD

Indicating the installed position of the alternator drive belt (2) around the alternator pulley (1), crankshaft pulley (3) and tensioner (4)

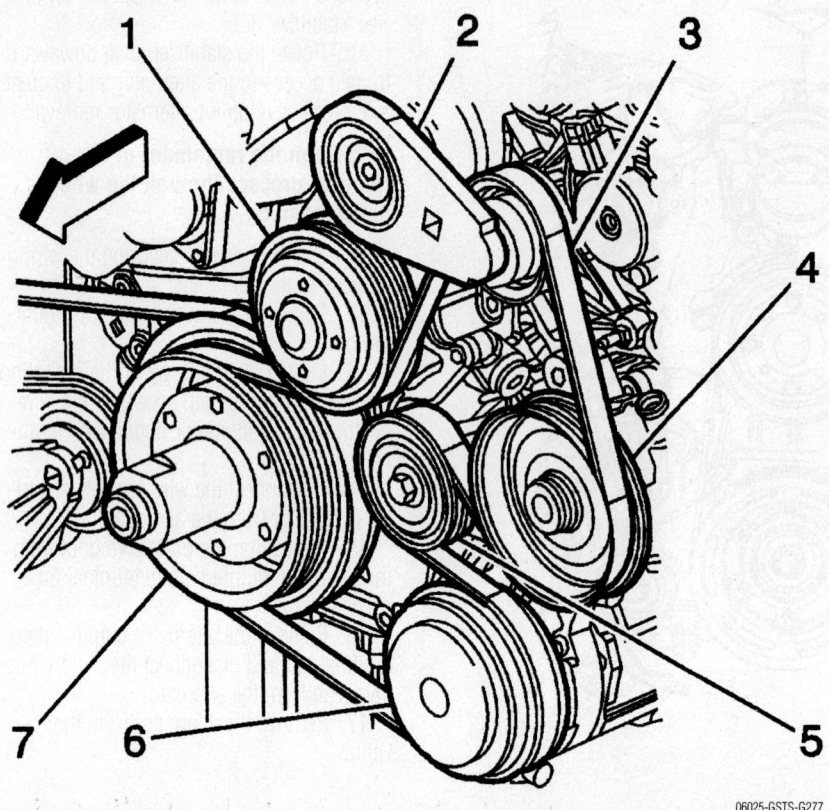

06025-GSTS-G277

Indicating the installed position of the serpentine drive belt around the: water pump pulley (1), camshaft pulley with tensioner (3, 2) AIR pump pulley (4), idler pulley (5), A/C compressor pulley (6), and crankshaft pulley (7)

To install:

18. Position the alternator near the installed position.

19. Connect the battery positive lead to the alternator and install the alternator output battery terminal nut.

20. Tighten the nut to 89 inch lbs. (10 Nm).

21. Press the protective boot on to the alternator output BAT terminal.

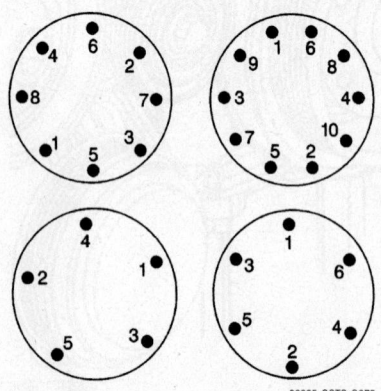

06025-GSTS-G275

Showing wheel bolt hole orientations for proper wheel and hub alignment during wheel installation

22. Connect the wiring harness connector to the alternator.

23. Position the alternator to the alternator bracket on the engine.

24. Install the lower alternator bolt, but do not tighten at this time.

25. Lower the vehicle.

26. Install the upper alternator bolts.

27. Tighten the bolts in the sequence shown to 37 ft. lbs. (50 Nm).

28. Raise and support the vehicle.

29. Install a new tie strap to secure the wiring harness to the alternator.

30. Install the right and left side stabilizer shaft links to the lower control arm. Tighten retaining bolt to 95 ft. lbs. (115 Nm).

31. Install the right side wheelhouse liner with 6 screws.

32. Install the front wheels, aligning the reference mark made and ensuring the wheel bolt holes are properly aligned as shown. Torque wheel nuts to 100 ft. lbs. (140 Nm).

33. Install the front air deflector.

34. Lower the vehicle.

35. Install the alternator drive belt, alternator belt first, then serpentine belt, as shown.

36. Connect the battery negative cable.

Ignition Timing

ADJUSTMENT

The ignition timing is preset and cannot be adjusted.

Engine Assembly

REMOVAL & INSTALLATION

3.6L

1. Before servicing the vehicle, refer to the precautions in the beginning of this section.

2. Center the steering wheel.

3. Turn the ignition OFF.

4. Install steering wheel locking tool to the steering column.

➡**Do NOT disconnect the battery cables from the engine.**

5. Disconnect the negative cable from the battery and the body.

6. Disconnect the positive cable from the battery and the underhood electrical center.

7. Position and secure the battery cables to the engine.

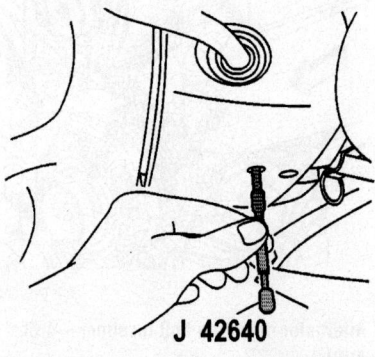

J 42640

06025-GSTS-G07

Illustrating the steering wheel locking tool

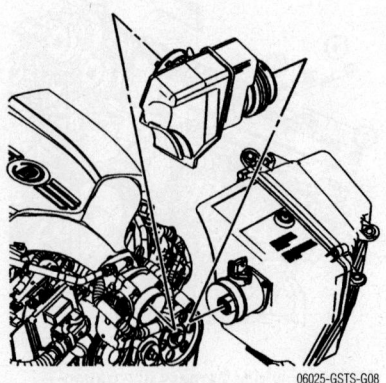

06025-GSTS-G08

Illustrating the air cleaner duct removal

8. Remove the battery.

9. Remove the fuel injector sight shield.

10. Remove the air cleaner duct.

11. Mark and disconnect the cooling fan electrical connectors from each fan.

12. Mark and remove the cooling fan wiring harnesses from the fan shroud.

13. Secure the wiring harnesses to the vehicle.

14. Drain the cooling system.

➡ **Do NOT disconnect the surge hoses from the engine or the radiator.**

15. Disconnect the surge tank outlet hose from the surge tank.

16. Position and secure the surge hose to the engine.

17. Disconnect the surge tank inlet hose from the water outlet housing and the radiator.

18. Position and secure the surge tank inlet hose to the vehicle.

19. Disconnect the heater hoses from the heater core.

20. Disconnect the purge line from the purge solenoid.

21. Disconnect the fuel pipe from the fuel rail.

22. Plug the fuel pipe and cap the fuel rail to prevent fuel loss or contamination.

23. Properly evacuate and recover the A/C system refrigerant.

24. Remove the wiper module as follows:

a. Remove the wiper motor mini 10A and 30A fuses. The wiper motor fuses are located in the underhood fuse block.

b. Remove the wiper arm assemblies.

c. Remove the air inlet grille (1,3).

d. Remove the wiper motor module mounting bolts (3,5).

e. Disconnect the wiper motor electrical connector (4).

f. Remove the wiper motor module (1) from the vehicle.

➡ **Do NOT disconnect the suction hose from the A/C compressor.**

25. Disconnect the suction hose from the evaporator and remove the suction hose bracket from the shock tower.

26. Position and secure the suction hose to the engine.

➡ **Do NOT disconnect the liquid line from the condenser.**

27. Disconnect the air conditioning pressure switch electrical connector and remove the liquid line.

28. Remove the 2 radiator support brackets from the top of the radiator.

29. Disconnect the brake booster check valve and vacuum hose from the brake booster.

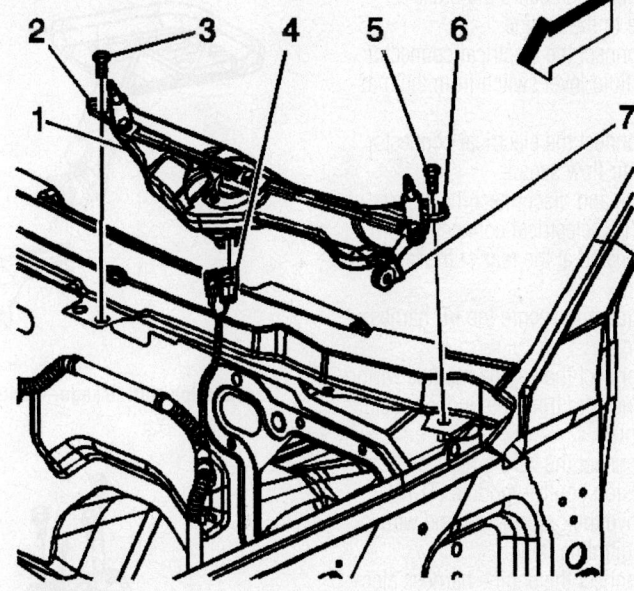

1. Wiper mechanism
2. Locator hole
3. Bolt
4. Electrical connector
5. Bolt
6. Locator hole
7. Module locator hole

06025-GSTS-G278

Removing/installing the wiper arm/motor module assembly

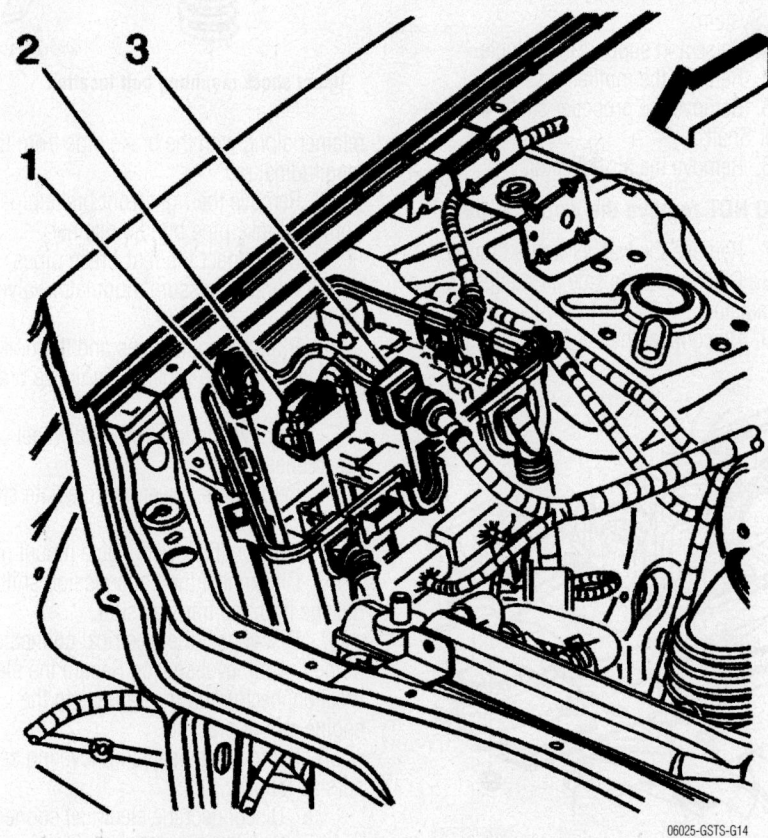

06025-GSTS-G14

Disconnecting the ECM harness connectors

30. Position and secure the brake booster hose to the engine.

31. Disconnect the electrical connector to the brake fluid level switch from the master cylinder.

32. Disconnect the electrical connector to the mass air flow sensor.

33. Unlock and disconnect the instrument panel (I/P) electrical connector from the engine located at the rear of the left bank cylinder head.

34. Position and secure the I/P harness to the vehicle.

35. Disconnect the engine module wiring harness connectors from the under hood electrical center.

36. Disconnect the wiring harness from the transmission control module (TCM).

37. Remove the ground bolt and wire from the longitudinal rail.

38. Disconnect the engine harness electrical connector at the longitudinal rail.

39. Position and secure the ground wire, the engine harness and the TCM harness to the engine.

➡ **Do NOT disconnect the brake pipes from the master cylinder.**

40. Remove the master cylinder nuts.

41. Reposition and secure the master cylinder to the engine.

42. Relieve the fuel system pressure. See Fuel System.

43. Raise and support the vehicle.

44. Remove the muffler assembly.

45. Remove the propeller shaft. See Propeller Shaft.

46. Remove the air deflector.

➡ **DO NOT remove the water bottle.**

47. Remove the washer bottle bracket.

48. Disconnect the side air baffles from the radiator.

49. Disconnect the left front brake pipe

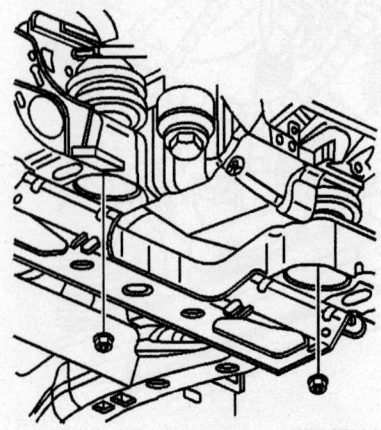

Lower engine mounting nuts—3.6L

06025-GSTS-G09

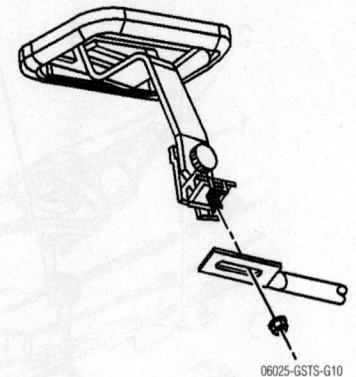

Transmission shift linkage—5L40/5L50-E

06025-GSTS-G10

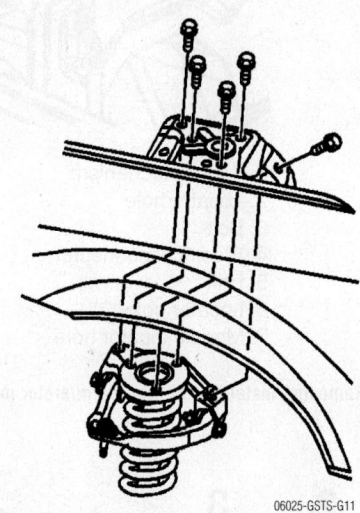

Upper shock mounting bolt location

06025-GSTS-G11

retainer along with the brake pipe from the longitudinal rail.

50. Remove the right front brake pipe from the brake pipe bundle retainer.

51. Disconnect the rear brake pipes from the brake pressure modulator valve (BPMV).

52. Plug the brake pipes and the brake modulator valve in order to minimize brake fluid loss.

53. Remove the front tire and wheel assemblies.

54. Remove the lower intermediate steering shaft.

55. Remove the lower engine mount nuts.

56. Disconnect the transmission shift linkage from the transmission.

57. Disconnect the electrical connector to the low oil level sensor. Secure the electrical connector and the harness to the engine mount bracket.

58. Remove the headlamp leveling sensors as follows:

   a. Disconnect the electrical connector from the headlamp leveling sensor.

   b. Disconnect the ball and socket from the vehicle (located near the outer edge of the brake rotor).

   c. Remove the bolt from the leveling sensor.

   d. Remove the sensor from the vehicle.

59. Secure the shock modules to the lower control arms with a suitable strap in order to prevent damage to the front brake hoses.

60. Remove the upper mounting bolts from the right and left shock module.

61. Raise the vehicle enough to place a suitable lift table under the engine, transmission, front frame and front suspension assembly.

62. Position a suitable lift table below the frame, engine and transmission.

63. Raise the lift table or lower the vehicle to support the frame, engine and transmission.

64. Remove the bolts which secure the transmission brace to the underbody.

65. Remove the front frame bolts (2).

✳✳ WARNING

Ensure that all the hoses, wires, pipes and shock modules clear the vehicle during the removal process.

66. With the aid of an assistant, lower the table or raise the vehicle to remove the engine, transmission, front frame and front suspension assembly from the vehicle.

67. Disconnect the engine control module (ECM) upper electrical connector.

➡ **Do NOT remove the oxygen sensors.**

68. Remove the catalytic converters with the oxygen sensors.

69. Remove the thermostat housing with the heater pipes, heater hoses and surge tank outlet hose.

70. Remove the starter motor.

71. Remove the flywheel bolts.

72. Remove the heated oxygen sensor (HO2S) connector bracket (1) from the left cylinder head.

73. Remove the engine wiring harness and related components.

74. Disconnect the transmission oil cooler pipes from the engine, radiator and the transmission.

75. Disconnect the radiator hoses from the water outlet housing and the coolant inlet pipe.

76. Disconnect the power steering cooler hoses from the condenser radiator and fan module (CRFM).

77. Plug the power steering hoses and

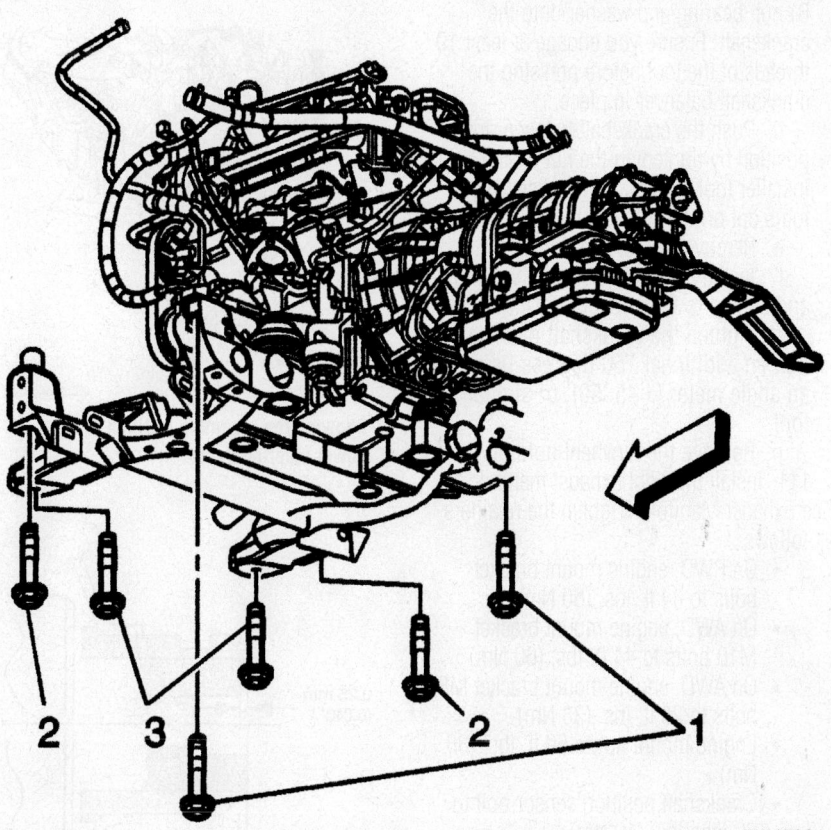

06025-GSTS-G12

Transmission brace bolt location

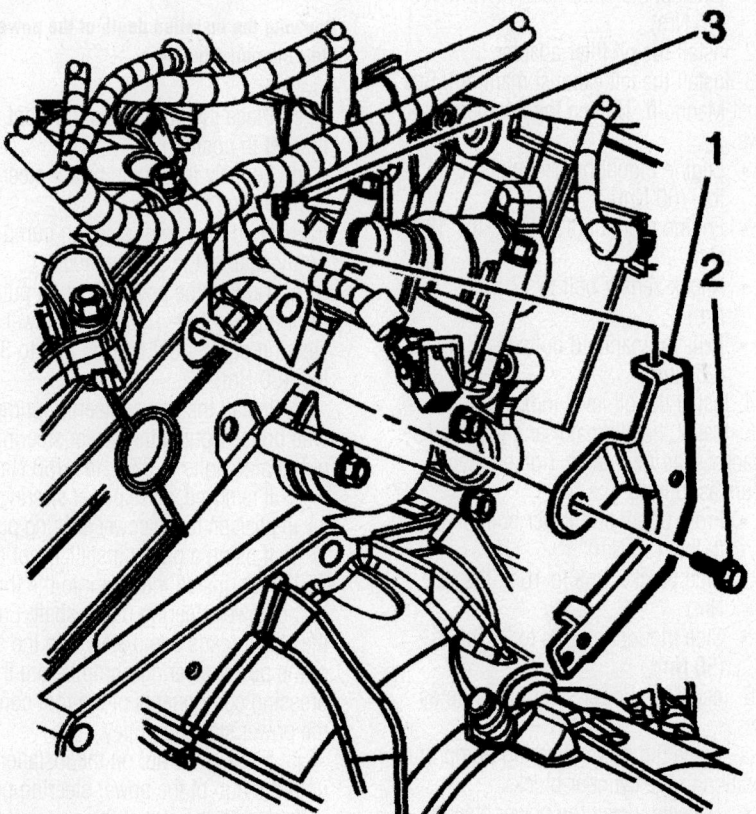

06025-GSTS-G13

Oxygen sensor bracket location

pipes in order to prevent fluid loss and contamination.

78. Remove the CRFM with the radiator hoses from the frame.

79. Install an engine support lift assembly to the engine.

80. Connect a floor crane to the engine lift brackets and raise the floor crane to partially support the engine.

81. Position a second powertrain lift table below the transmission.

82. Remove the bolts that secure the transmission to the engine.

83. Remove or disconnect the transmission from the engine.

84. Remove the drive belts.

85. Remove the alternator and water pump drive belt tensioner.

86. Remove the A/C compressor and power steering drive belt tensioner.

87. Remove the alternator and the alternator bracket.

88. Remove the A/C compressor.

89. Remove the power steering pump.

90. Remove the power steering reservoir from the engine.

91. Remove the oil level indicator.

92. Remove the left exhaust manifold. See Exhaust Manifold.

93. Remove the oil filter adapter.

94. Remove the right exhaust manifold. See Exhaust Manifold.

95. Remove the crankshaft balancer.

96. Remove the flywheel. See Flywheel.

97. Remove the intake manifold. See Intake Manifold.

98. Remove the water outlet.

99. Remove the block heater.

100. Remove the ECM with the ECM bracket.

101. Use a floor crane in order to remove the engine from the frame.

102. Remove the engine mount brackets with the engine mounts.

To install:

103. Install the engine mount brackets with the engine mounts.

104. Use a floor crane in order to install the engine to the frame.

105. Install the ECM with the ECM bracket.

106. Install the water outlet.

107. Install the block heater.

108. Install the flywheel and bolts as follows:

    a. Place the engine flywheel in position on the crankshaft.

    b. Install 2 NEW bolts in location at the top and bottom of the engine flywheel bolt pattern allowing the engine flywheel to hang in position.

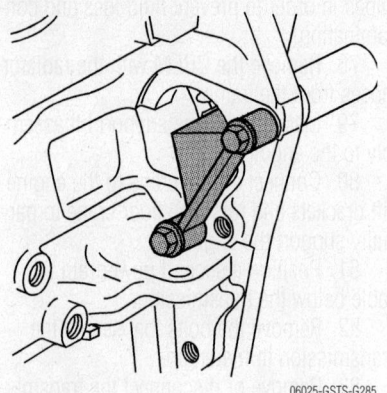

06025-GSTS-G285

Using flywheel holding tool during installation

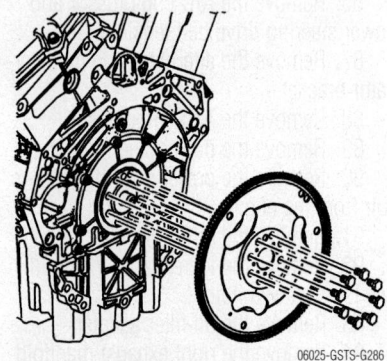

06025-GSTS-G286

Showing flywheel bolt installation

c. Install the flywheel holding tool (EN 46106).

d. Install the remaining NEW engine flywheel bolts: the NEW engine flywheel bolts to 22 ft. lbs. (30 Nm), then tighten them an additional 45 degrees using an angle meter (J 45059) or similar tool.

109. Install the intake manifold. See Intake Manifold. Tighten the intake manifold bolts to 17 ft. lbs. (23 Nm), the intake manifold brace-to-manifold bolt to 89 inch lbs. (10 Nm), and intake manifold brace-to-front cover bolt to 48 ft. lbs. (65 Nm).

110. Install the crankshaft balancer as follows:

a. Install the flywheel holding tool (EN 46106) onto the flywheel to hold it in place.

✳✳ CAUTION

Do not lubricate the crankshaft front oil seal or crankshaft balancer sealing surfaces. The crankshaft balancer is installed into a dry seal.

b. Apply lubricant to the inside of the crankshaft balancer hub bore.

c. Thread the installer tool (J 41998-

B) nut, bearing and washer onto the crankshaft. Ensure you engage at least 10 threads of the tool before pressing the crankshaft balancer in place.

d. Push the crankshaft balancer into position by tightening the nut on the installer tool until the large washer bottoms out on the crankshaft end.

e. Remove the installer tool.

f. Install the crankshaft balancer bolt and tighten to 74 ft. lbs. (100 Nm).

g. Tighten the crankshaft balancer bolt an additional 150 degrees using an angle meter (J 45059), or similar tool.

h. Remove the flywheel holding tool.

111. Install the right exhaust manifold. See Exhaust Manifold. Tighten the retainers as follows:

- On RWD, engine mount bracket bolts to 44 ft. lbs. (60 Nm)
- On AWD, engine mount bracket M10 bolts to 44 ft. lbs. (60 Nm)
- On AWD, engine mount bracket M8 bolts to 28 ft. lbs. (38 Nm)
- Engine mount nut to 59 ft. lbs. (80 Nm)
- Crankshaft position sensor bolt to 89 inch lbs. (10 Nm)
- Knock sensor bolt to 17 ft. lbs. (23 Nm)
- Exhaust manifold bolts to 18 ft. lbs. (25 Nm)

112. Install the oil filter adapter.

113. Install the left exhaust manifold. See Exhaust Manifold. Tighten the retainers as follows:

- Engine mount bracket bolts to 44 ft. lbs. (60 Nm)
- Engine mount nut to 59 ft. lbs. (80 Nm)
- Knock sensor bolt to 17 ft. lbs. (23 Nm)
- Exhaust manifold bolts to 18 ft. lbs. (25 Nm)

114. Install the oil level indicator.

115. Install the alternator and the alternator bracket. See Alternator. Tighten the retainers as follows:

- Front upper and lower bolts to 37 ft. lbs. (50 Nm)
- Side bracket bolt to 16 ft. lbs. (22 Nm)
- Side mounting bolts to 37 ft. lbs. (50 Nm)

116. Install the power steering pump as follows:

a. Place the power steering pump in position to the cylinder block.

b. Loosely install the power steering side bolts in order to hold the power steering pump in position.

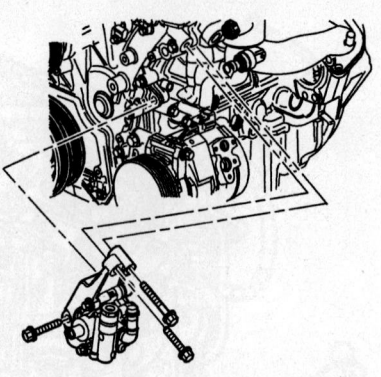

06025-GSTS-G287

Showing the mounting position of the power steering pump

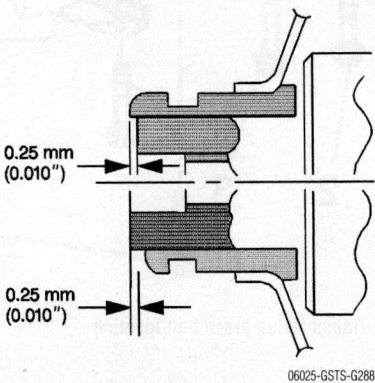

0.25 mm (0.010")

0.25 mm (0.010")

06025-GSTS-G288

Showing the installed depth of the power steering pump pulley

c. Place the power steering front bracket in position.

d. Loosely install the power steering front bracket bolts.

e. Hold the power steering pump firmly against the engine block.

f. Tighten the power steering pump front bracket bolt. Tighten the power steering pump front bracket bolt to 37 ft. lbs. (50 Nm).

g. Install the power steering pump side bolts. Tighten the power steering pump side bolts to 37 ft. lbs. (50 Nm).

h. If removed, place power steering pulley in position to the power steering pump.

i. If using a pulley installer tool (J 25033-C), install the tool into the threads of the power steering pump shaft. Ensure the tool threads completely into the shaft of the power steering pump before the pressing components of the tool contact the power steering pulley.

j. Tighten the nut on the installer tool until the hub of the power steering pulley is flush with the end of the power steering pump shaft. Remove the tool.

117. Install the power steering reservoir to the engine.

118. Install the A/C compressor. Tighten the retainers as follows:

- A/C compressor bracket center bolt: 37 ft. lbs. (50 Nm)
- A/C compressor bracket upper bolt: 37 ft. lbs. (50 Nm)
- A/C compressor bracket lower bolt: 37 ft. lbs. (50 Nm)
- A/C compressor front upper and lower bolts: 18 ft. lbs. (23 Nm)
- A/C compressor rear upper and lower bolts: 18 ft. lbs. (23 Nm)

119. Install the A/C compressor and power steering drive belt tensioner.

120. Install the alternator and water pump drive belt tensioners. Tighten the retaining bolt for the left tensioner to 37 ft. lbs. (50 Nm) to for the right tensioner to 18 ft. lbs. (23 Nm).

121. Install the drive belts.

122. Install the bolts that secure the transmission to the engine and the flywheel to the torque converter as follows:

   a. Tighten the 3 lower mounting bolts, then the 3 upper mounting bolts to 37 ft. lbs. (50 Nm). Install and tighten the rear engine cradle bolts to 141 ft. lbs. (191 Nm).

123. Install the starter motor. Tighten the starter mounting bolts to 37 ft. lbs. (50 Nm).

124. Remove the engine lift assembly from the engine.

125. Install the condenser-radiator fan module (CRFM) with the radiator hoses to the frame.

126. Connect the power steering cooler hoses to the (CRFM).

127. Connect the radiator hoses to the water outlet housing and the coolant inlet pipe.

128. Install the transmission oil cooler pipes to the engine, radiator and the transmission.

129. Install the engine wiring harness and related components.

130. Install the HO2S connector bracket (1) to the left cylinder head.

131. Install the thermostat housing with the heater pipes, heater hoses and surge tank outlet hose.

132. Install the catalytic converters with the oxygen sensors.

133. Install the fuel lines to the fuel rail and the EVAP purge solenoid.

134. Connect the ECM chassis side electrical connector.

➡**Ensure all the hoses, wires, pipes and shock modules clear the vehicle during the installation process.**

135. With the aid of an assistant, raise the table or lift the vehicle to install the engine, transmission, front frame and front suspension assembly to the vehicle.

136. Install the front frame bolts. Tighten the mounting bolts (1, 2) to 141 ft. lbs. (191 Nm) and mounting bolts (3) to 185 ft. lbs. (250 Nm).

137. Install the lower engine mount nuts and tighten to 59 ft. lbs. (80 Nm).

138. Install and tighten the washer bottle to knuckle retaining bolt.

139. Install the upper mounting bolts to the right and left shock module.

140. Secure the shock modules to the front frame with mechanics wire to avoid stretching the front brake hoses.

141. Install the headlamp leveling sensors.

142. Connect the transmission shift linkage to the transmission.

143. Connect electrical connector to the low oil level sensor.

144. Install the lower engine mount nuts. Tighten to 59 ft. lbs. (80 Nm).

145. Install the lower intermediate steering shaft as follows:

   a. Connect the lower intermediate shaft to the center intermediate shaft.

   b. Connect the lower intermediate shaft to the power steering gear.

   c. Install the lower intermediate shaft to the power steering gear retaining bolt. Tighten the bolt to 37 ft. lbs. (50 Nm).

   d. Install the center intermediate shaft to the lower intermediate shaft retaining bolt (3). Tighten the bolt to 23 ft. lbs. (30 Nm).

146. Install the front wheels, aligning the reference mark made and ensuring the wheel bolt holes are properly aligned as shown in applicable illustration. Torque wheel nuts to 100 ft. lbs. (140 Nm).

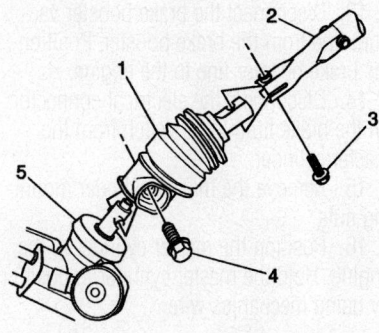

1. Steering column boot
2. Center steering column
3. Bolt
4. Bolt
5. Steering gear

06025-GSTS-G21

Center and lower intermediate shafts

147. Install the front brake pipes and retainers to the underbody.

148. Connect the rear brake pipes to the BPMV.

149. Connect the side air baffles to the radiator.

150. Install the washer bottle bracket.

151. Install the air deflector.

152. Install the propeller shaft. See Propeller Shaft. Tighten the retainers as follows:

- Coupler to flange bolts to 63 ft. lbs. (85 Nm)
- On AWD, the CV joint to transfer case bolts to 23 ft. lbs. (30 Nm)
- On RWD, the prop shaft coupler to transmission bolts to 63 ft. lbs. (85 Nm)
- The support bearing bolts to 37 ft. lbs. (50 Nm)

153. Install the muffler assembly.

154. Install the master cylinder. Tighten the master cylinder mounting nuts to 18 ft. lbs. (25 Nm). Tighten the brake pipe fittings at the master cylinder to 28 ft. lbs. (38 Nm).

155. Connect the engine harness electrical connector at the longitudinal rail.

156. Install the ground wire and bolt to the longitudinal rail. Tighten the bolt to 89 inch lbs. (10 Nm).

157. Connect the wiring harness to the transmission control module.

158. Connect the engine module wiring harness connectors to the under hood electrical center.

159. Connect and lock the I/P electrical connector to the engine at the rear of the left Bank 2 cylinder head.

160. Connect the mass air flow sensor electrical connector.

161. Connect the electrical connector for the brake fluid level switch from the master cylinder.

162. Connect the brake booster vacuum hose to the brake booster.

163. Install the radiator support brackets.

164. Connect the purge line to the purge solenoid.

165. Connect the fuel pipe to the fuel rail.

166. Connect the heater hoses to the heater core.

167. Install the air inlet duct.

168. Position and secure the surge tank inlet hose to the vehicle.

169. Connect the surge tank inlet hose to the water outlet housing and to the radiator.

170. Connect the outlet hose to the surge tank.

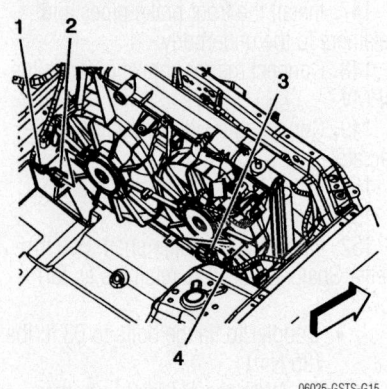

Cooling fan wiring harness and electrical connectors.

171. Connect the electrical connector for the air conditioning pressure switch and the liquid line to the evaporator.

172. Connect the air conditioning suction hose to the evaporator and install the suction hose bracket to the shock tower.

173. Install the air cleaner duct.

174. Install the cooling fan wiring harnesses (1,4) to the fan shroud.

175. Install the cooling fan (2,3) electrical connectors.

176. Install the wiper module.

177. Install the battery.

178. Install the fuel injector sight shield.

179. Connect the positive cable to the battery and the under hood electrical center.

180. Connect the negative cable from the battery and the body.

181. Remove locking tool (J 42640) from the steering column.

182. Fill the cooling system.

183. Charge the air conditioning system.

184. Bleed the brake rear circuits.

4.6L

1. Before servicing the vehicle, refer to the precautions in the beginning of this section.

2. Install steering wheel locking tool to the steering column.

3. Disconnect the positive and negative battery cables from the battery.

4. Remove the cross vehicle brace.

5. Remove the fuel injector sight shield.

6. Remove the air cleaner assembly.

7. Disconnect the surge tank inlet hose and position the hose to the engine.

8. Evacuate the air conditioning system.

9. Disconnect the air conditioning suction hose fitting at the top of the left shock tower.

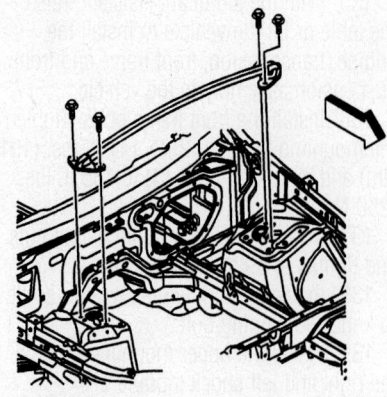

Illustrating the vehicle cross brace

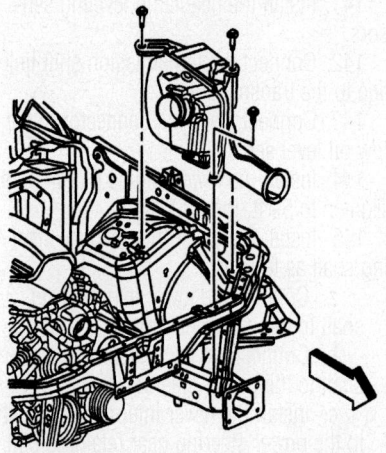

Illustrating the vehicle air cleaner assembly

10. Remove the suction hose from the retaining feature. Position the line to the engine.

11. Disconnect the air conditioning liquid line from the condenser.

12. Disconnect the air conditioning liquid line from the retaining feature on the fan shroud.

13. Disconnect the brake booster vacuum line from the brake booster. Position the brake booster line to the engine.

14. Disconnect the electrical connector on the brake fluid level switch from the master cylinder.

15. Remove the master cylinder mounting nuts.

16. Position the master cylinder to the engine. Hold the master cylinder in position by using mechanics wire.

➡**Do not disconnect the brake lines from the master cylinder.**

17. Relieve the fuel system pressure.

18. Remove the fuel line retainer from the vehicle. The retainer is located on the

bracket attached to the heater lines on the front of the dash.

19. Disconnect the engine wiring harness connector at the front of the dash.

20. Disconnect the underhood fuse block connector located between the right shock tower and the cam cover.

21. Remove the underhood electrical center cover.

22. Remove the negative battery cable bolt from the right shock tower.

23. Remove the cable retainer from the stud on the right shock tower and position the cable to the engine.

24. Remove the positive battery cable nut from inside the underhood electrical center.

25. Remove the cable from the electrical center and position the cable to the engine.

26. Disconnect the chassis electrical connector at the top of the right shock tower. Position the wire to the engine.

27. Disconnect the electrical connector for the transmission control module and position the wiring harness to the engine.

28. Disconnect the engine wiring harness connector from the inside of the underhood electrical center

29. Disconnect the electrical connector at the right frame rail.

30. Remove the engine cooling fans.

31. Raise and support the vehicle.

32. Remove the front wheels after making a reference mark between the wheel and hub.

33. Remove the right and left wheelhouse splash shields.

34. Remove the right and left wheelhouse liners.

35. Working through the right wheelhouse, remove the brace that secures the windshield washer reservoir to front frame.

36. Place a suitable drain pan under the vehicle.

37. Working through the left wheelhouse, disconnect the transmission oil

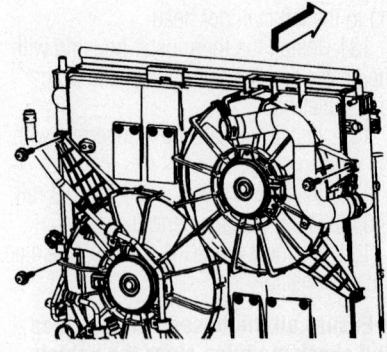

Illustrating the electric cooling fans

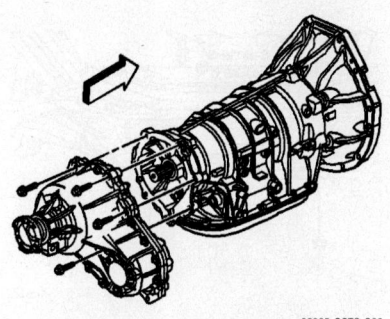

06025-GSTS-G22

Showing the transfer case mounting position

cooler lines located near the air conditioning compressor.

38. Remove the bolt (3) that secures the center intermediate shaft to the lower intermediate shaft.

39. Separate the 2 shafts.

40. Drain the cooling system.

41. Remove the bolt securing the power steering cooler lines to the radiator assembly.

42. Lower the vehicle.

43. Remove the radiator, condenser, and transmission oil cooler as an assembly.

44. Raise the vehicle.

45. Remove the bolts from the mounting bracket on the power steering oil cooler.

46. Position the oil cooler to the engine. Secure the cooler to the engine with mechanics wire.

47. Remove the transfer case, if equipped.

48. Remove the transmission assembly. See Automatic Transmission Assembly.

49. Remove the brake bundle clips from the right and left frame rails. Do not remove the clips from the brake lines.

50. Disconnect the fuel line from the fuel filter.

51. Disconnect the evaporative emission (EVAP) hose connection at the rear of the fuel filter.

52. Disconnect the rear brake lines from the bracket above the rear axle assembly.

53. Remove the fuel and brake line bundle retainers from the frame rail along the length of the vehicle. Do not remove the retainers from the lines.

54. Remove the fuel filter bracket to provide a removal path for the fuel and brake line bundle assembly.

55. Remove the fuel and brake line bundle bracket from the right side wheelhouse.

56. Lower the vehicle.

57. Disconnect the outlet hose from the heater outlet pipe at the right frame rail. Position the hose to the engine.

58. Disconnect the inlet hose from the water housing and position the hose to the vehicle.

59. If the vehicle is equipped with Magnaride® disconnect the electrical connectors from the top of the right and left shock modules.

60. Remove the upper mounting bolts from the right and left shock module.

61. Secure the shock modules to the front frame with mechanics wire to avoid stretching the front brake hoses.

62. Raise the vehicle enough to place a suitable lift table under the engine, front frame, and front suspension assembly.

✳✳ WARNING

To avoid any vehicle damage, serious personal injury or death when major components are removed from the vehicle and the vehicle is supported by a hoist, support the vehicle with jack stands at the opposite end from which the components are being removed.

63. Support the rear of the vehicle with suitable jack stands.

64. Raise the lift table and/or lower the vehicle to preload the weight of the engine, front frame, and front suspension assembly.

65. Remove the front frame bolts.

66. With the aid of an assistant, lower the table and/or raise the vehicle to remove the engine, front frame, fuel/brake bundle and front suspension assembly from the vehicle.

67. Ensure that all the hoses, wires, pipes and shock modules clear the vehicle during the removal process.

68. Install lifting bracket to the left cylinder head.

69. Install lifting bracket to the right cylinder head

70. Using a suitable engine lift, remove the engine from the front frame assembly.

To install:

71. Install the engine to the front frame.

72. Ensure that all the hoses, wires, pipes and shock modules are positioned out of the way to avoid contact during engine installation.

73. With the aid of an assistant, raise the table and/or lower the vehicle to install the engine, front frame, fuel/brake bundle and front suspension.

74. Install the frame bolts.

75. Tighten the frame bolts to 141 ft. lbs. (191 Nm).

76. Install the right and left upper shock modules bolts. Tighten the bolts to 83 ft. lbs. (112 Nm).

77. If the vehicle is equipped with Magnaride® connect the shock module electrical connectors.

78. Connect the heater inlet hose to the water housing.

79. Connect the heater outlet hose to the outlet pipe.

80. Raise the vehicle.

81. Install the fuel and brake line bundle bracket to the right side wheelhouse.

82. Tighten the bracket bolt to (80 inch lbs. (9 Nm).

83. Install the fuel filter bracket bolt.

84. Tighten the bracket bolt to 80 inch lbs. (9 Nm).

85. Install the fuel and brake line to the bundle brackets, the length of the vehicle.

86. Install the rear brake line to the bracket bolt above the rear axle assembly.

87. Tighten the bracket bolt to 80 inch lbs. (9 Nm).

88. Connect the EVAP hose to the rear of the fuel filter.

89. Connect the fuel line to the fuel filter.

90. Install the brake lines to the bundle clips on the right and left frame rails.

91. Install the transmission assembly. See Automatic Transmission Assembly. Install the 8 lower and then the upper transmission bolts and torque to 37 ft. lbs. (50 Nm).

92. Install the transfer case, if equipped.

93. Install the power steering oil cooler to the mounting bracket.

94. Lower the vehicle.

95. Install the radiator, condenser and transmission oil cooler as an assembly.

96. Raise the vehicle.

97. Install the bolt that secures the power steering to the radiator. Tighten the bolt to 80 inch lbs. (9 Nm).

98. Connect the center intermediate shaft to the lower intermediate shaft. Tighten the bolt to 23 ft. lbs. (30 Nm).

99. Install the bolt that secures the shafts.

100. Working through the left wheelhouse, connect the transmission oil cooler lines near the air conditioning compressor.

101. Working through the right hand wheelhouse, install the bolt which secures the washer fluid reservoir to the front frame brace.

102. Tighten the bolt to 53 inch lbs. (6 Nm).

103. Install the right and left front wheelhouse liners.

104. Install the right and left wheelhouse splash shields.

105. Install the right and left front tire assemblies.

106. Lower the vehicle.

107. Install the electric cooling fans.

108. Connect the electrical connector at the right hand frame rail.

109. Connect the electrical con-nector to the underhood bussed electrical center.

110. Connect the transmission control module.

111. Connect the chassis electrical connector located at the top of the right hand shock tower.

112. Install the positive battery cable to the underhood bussed electrical center.

113. Tighten the battery cable nut to 11 ft. lbs. (15 Nm).

114. Install the negative battery cable retainer and bolt to the right hand shock tower.

115. Tighten the retainer bolt to 36 Nm (27 ft. lbs.).

116. Install the underhood electrical center cover.

117. Connect the underhood fuse block connector.

118. Connect the front of the dash electrical connector.

119. Install the fuel line retainer located on the bracket attached to the heater lines on the front of the dash.

120. Install the brake master cylinder. Tighten the retaining nuts to 18 ft. lbs. (25 Nm).

121. Connect the brake fluid level switch.

122. Connect the brake booster vacuum line to the brake booster.

123. Install the air conditioning liquid line to the condenser. Always use a NEW O-ring.

124. Install the air conditioning suction hose located at the top of the left hand shock tower.

125. Install the air conditioning liquid and suction lines to the retaining feature on the fan shroud.

126. Connect the coolant surge tank hose to the radiator.

127. Install the air cleaner assembly.

128. Install the fuel injector sight shield.

129. Install the cross vehicle brace.

130. Install the positive battery cable.

131. Install the negative battery cable.

132. Remove the steering wheel locking tool

133. Fill the engine with engine oil.

134. Fill the cooling system.

135. Recharge the air conditioning system.

136. Bleed the brake system.

Engine Mounts

3.6L

1. Raise and support the vehicle.

2. Install a screw jack with a block of wood under the oil pan for support.

3. Remove the lower engine mount retaining nuts.

4. Raise the engine using the screw jack until the weight is removed from the mount.

5. Remove the upper engine mount retaining nut.

6. Remove the engine mount from the vehicle.

To install:

7. Install the engine mount to the vehicle.

8. Install the upper engine mount retaining nut. Tighten the nut to 59 ft. lbs. (80 Nm).

9. Lower the engine using the screw

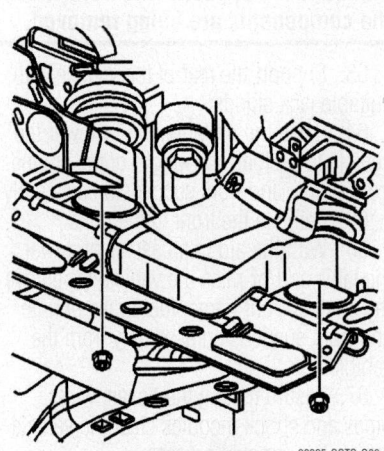

Lower engine mounting nuts—3.6L

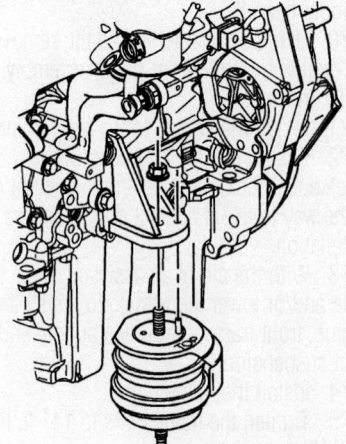

Removing the upper engine mount nut and engine mount (left mount shown)—3.6L with RWD

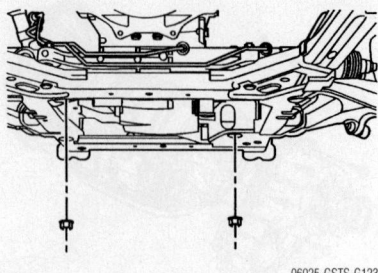

Removing the engine mount nuts—3.6L with AWD

jack until the weight is resting on the mount.

10. Install the engine mount lower retaining nut. Tighten the nut to 59 ft. lbs. (80 Nm).

11. Remove the screw jack and block of wood from under the oil pan.

12. Lower the vehicle.

4.6L

1. Raise and support the vehicle.

2. Remove the left engine mount bracket, as follows:

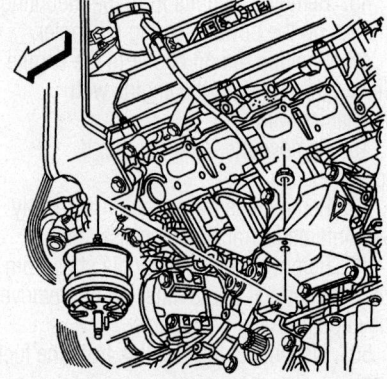

Removing the engine mount nuts—4.6L with RWD

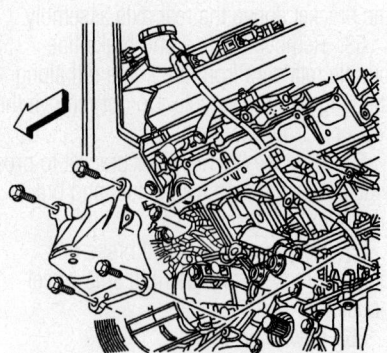

Removing the left engine mount bracket—4.6L with RWD

a. Remove the left exhaust manifold. See Exhaust Manifold.

b. Install a screw jack with a block of wood under the oil pan for support.

c. Remove the upper engine mount retaining nut.

d. Remove the left engine mount bracket retaining bolts.

e. Remove the left engine mount bracket.

3. Remove the lower engine mount retaining nut.

4. Remove the engine mount from the vehicle.

To install:

5. Install the engine mount to the vehicle.

6. Install the engine mount lower retaining nuts. Tighten the nut to 59 ft. lbs. (80 Nm).

7. Install the engine mount brackets.

8. Install the engine mount bracket retaining bolts. Tighten the bolts to 44 ft. lbs. (60 Nm).

9. Install the upper engine mount retaining nut. Tighten the nut to 59 ft. lbs. (80 Nm).

10. Remove the screw jack.

11. Install the exhaust manifold. See Exhaust Manifold.

12. Lower the vehicle.

Water Pump

REMOVAL & INSTALLATION

3.6L & 4.6L Engines

1. Before servicing the vehicle, refer to the precautions in the beginning of this section.

2. Disconnect the negative battery cable.

3. Drain the engine cooling system.

4. Remove the alternator drive belt.

5. Remove the front compartment sight shield as necessary.

6. Rotate the drive belt tensioner clockwise to release the drive belt tension.

7. Slide the drive belt off of the water pump pulley.

8. Slowly release the drive belt tensioner.

9. Remove the drive belt from the accessory drive pulleys.

10. Remove the water pump pulley bolts.

11. Remove the water pump pulley.

12. Remove the water pump bolts

13. Remove the water pump

14. Remove and DISCARD the water pump seal.

Belt and pulley illustration—3.6L

06025-GSTS-G05

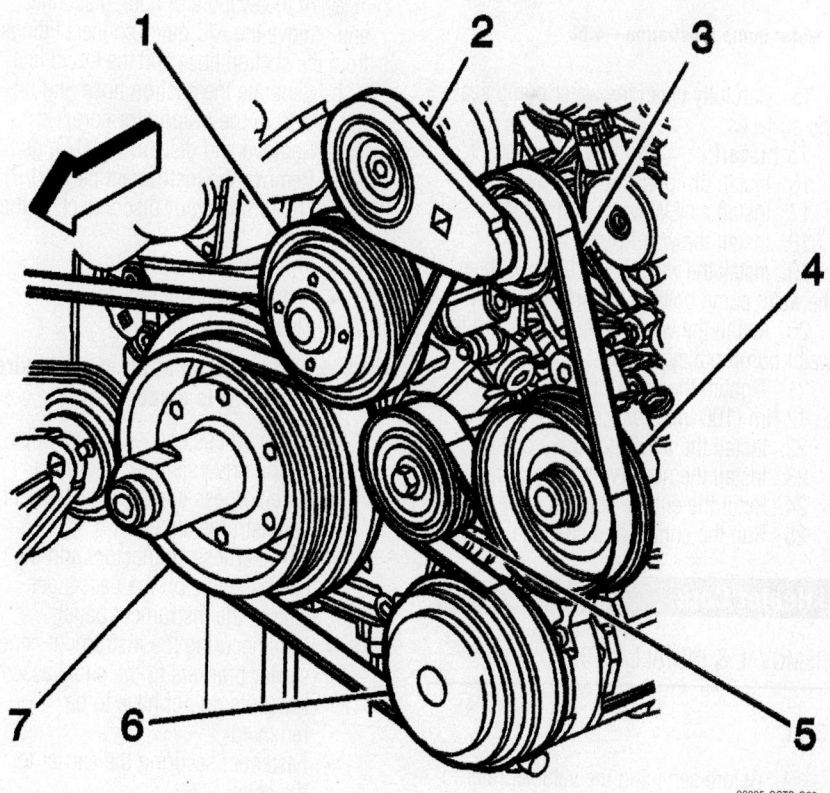

Belt and pulley illustration—4.6L

06025-GSTS-G23

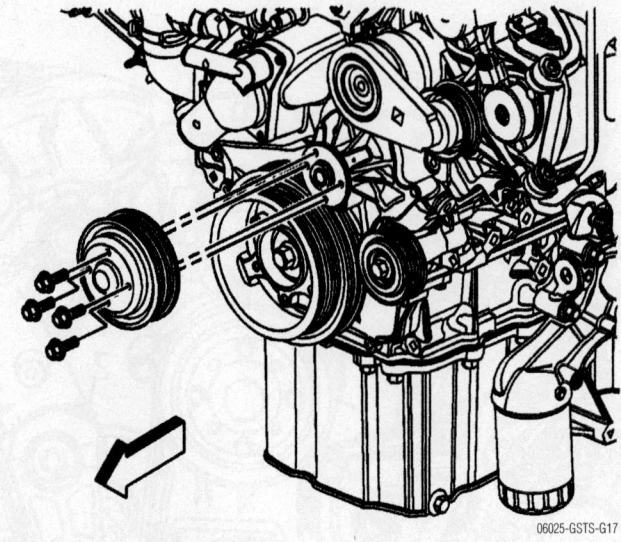

Water pump pulley illustration—4.6L

06025-GSTS-G17

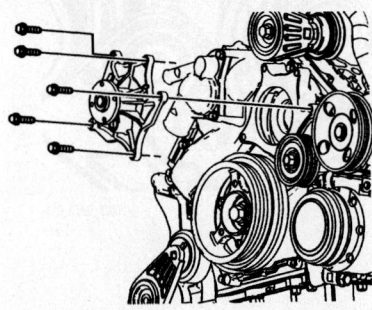

06025-GSTS-G16

Water pump illustration—4.6L

15. Carefully clean the water pump sealing surfaces

To install:

16. Install or connect the following:
17. Install a NEW water pump seal.
18. Install the water pump.
19. Install the water pump bolts. Tighten the water pump bolts to 89 inch lbs. (10 Nm).
20. Install the water pump pulley and the water pump pulley bolts
21. Tighten the water pump pulley bolts to 12 Nm (106 inch lbs.).
22. Install the alternator drive belt.
23. Install the negative battery cable
24. Refill the engine cooling system.
25. Run the engine and check for leaks.

Heater Core

REMOVAL & INSTALLATION

3.6L

1. Before servicing the vehicle, refer to the precautions in the beginning of this section.

2. Disable the air bag system. See Air Bag.
3. Disconnect the negative battery cable.
4. Properly discharge and recover the A/C refrigerant.
5. Drain the coolant.
6. Disconnect the heater inlet hose from the heater core.
7. Disconnect the heater outlet hose from the heater core.
8. Remove the HVAC module as follows:
   a. At the evaporator core, disconnect and remove the A/C quick connect fittings from the suction hose and the liquid line.
   b. Separate the suction hose and liquid line from the evaporator core.
   c. Remove and discard the O-rings.
   d. Remove the instrument panel (I/P) carrier by removing or disconnecting the following:
   • Front seats
   • Instrument panel retainer
   • Steering column

➡**Mark name and location of each wire harness before it is disconnected.**

   • Wire harness connector and clips on the drivers side
   • Wire harness clips in the center of the instrument panel
   • Wire harness connectors and the harness clips on the passenger side of the instrument panel
   • Nuts securing the instrument panel carrier brackets to the pedal assembly (nuts do not have to be removed)
   • Fasteners securing the carrier to the cowl
   • Bolt securing the carrier to the HVAC housing

   • Move the carrier back away from the cowl far enough to disconnect the wire harness retaining clips off of the carrier
   • Carrier from the vehicle
   e. Remove the air inlet assembly.
   f. Disconnect the HVAC module electrical connector.
   g. Disconnect the HVAC module drain tube from the floor.
   h. Remove the lower left HVAC module mounting nut
   i. Remove the upper left HVAC module mounting nut.
   j. Remove the HVAC module from the vehicle.
   k. Remove the foam from the A/C lines.
   l. Remove the foam from around the heater pipes.
9. Remove the battery tray.
10. Remove the cowl panel retaining bolts.
11. Remove the cowl panel.
12. Slide the clamp away from the base of the heater inlet hose.
13. Remove the inlet heater hose from the heater core.
14. Remove the heater hose bracket
15 Slide the heater core out of the HVAC module

To install:

16. Slide the heater core into the HVAC module. Install the inlet heater hose to the heater core.
17. Install the heater hose bracket.
18. Install the heater hose bracket screw.
19. Install the cowl panel retaining bolts.
20. Install the HVAC module as follows:
   a. Install the thick foam packing on the heater core pipes.
   b. Install the thin foam packing on the heater core pipes
   c. Install the thick foam packing on the A/C lines.
   d. Install the thin foam packing on the A/C lines.

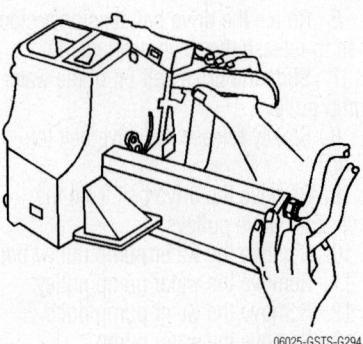

06025-GSTS-G294

Removing the heater core from the HVAC module

j. Connect the HVAC electrical connector.

k. Install the air inlet assembly.

l. Install the I/P carrier by installing or connecting the following:

- Wire harness retaining clips to the rear of the carrier
- Locate the carrier on to the pedal assembly brackets
- Tighten the nuts securing the instrument panel carrier brackets to the pedal assembly to 18 ft. lbs. (25 Nm)
- I/P carrier to the vehicle
- Wire harness back into noted position
- Fasteners securing the carrier to the cowl; tighten the bolts to 80 inch lbs. (9 Nm)
- Bolt securing the carrier to the HVAC housing; tighten the bolt to 80 inch lbs. (9 Nm)
- Wire harness connectors and the harness clips on the passenger side of the instrument panel
- Wire harness clips in the center of the instrument panel
- Wire harness connector and clips on the drivers side of the vehicle
- Steering column
- Instrument panel retainer
- Front seats

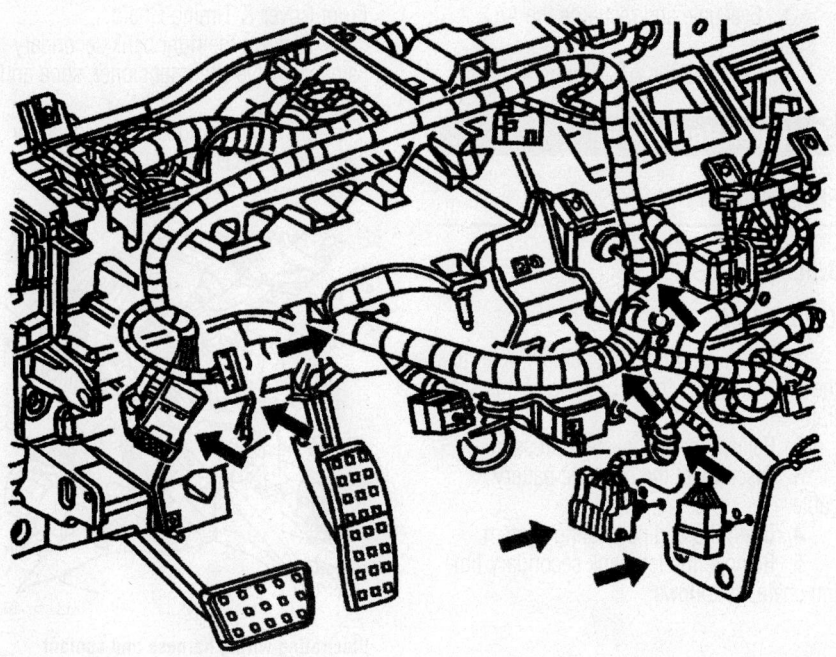

06025-GSTS-G24

Illustrating instrument panel and HVAC wiring

e. Install the HVAC module into the vehicle.

f. Install the upper left HVAC module mounting nut. Tighten the nut to 80 inch lbs. (9 Nm).

g. Install the lower left HVAC module mounting nut. Tighten the nut to 80 inch lbs. (9 Nm).

h. Connect the drain tube to the bottom of the HVAC module.

i. Connect the drain tube to the floor.

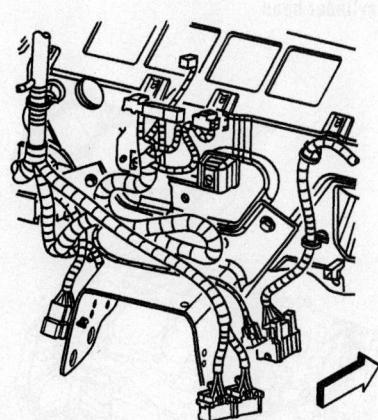

06025-GSTS-G26

Illustrating wire harness connectors in the center of the instrument panel.

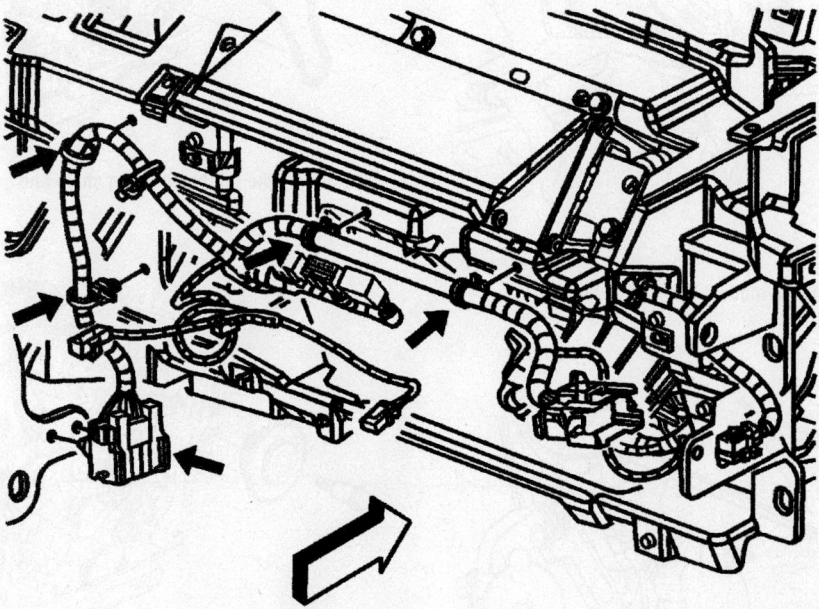

06025-GSTS-G27

Illustrating wire harness connectors on the passenger side instrument panel

m. Install new O-rings to the suction hose and liquid line.

n. Connect the suction hose and liquid line to the evaporator core.

o. Install the A/C quick connect fittings to the suction hose and liquid line.

Illustrating instrument panel carrier nuts

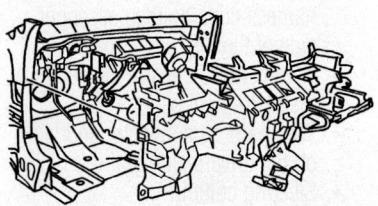

Illustrating the HVAC carrier separation

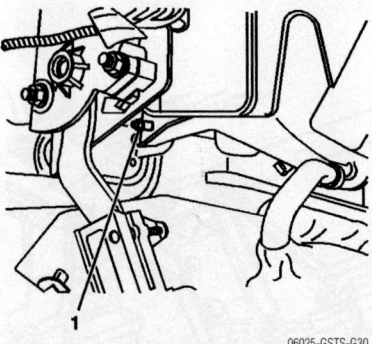

HVAC module lower mounting nut

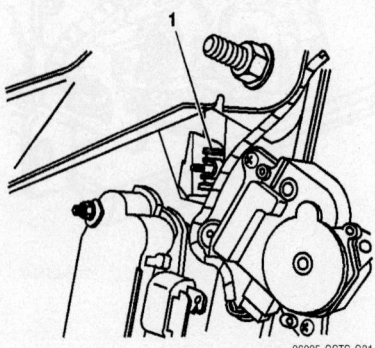

HVAC module upper mounting nut

p. Connect the heater inlet and outlet hoses to the heater core.

q. Fill the coolant.

r. Evacuate and recharge the A/C system.

s. Leak test the A/C fittings.

Cylinder Head

REMOVAL & INSTALLATION

3.6L

LEFT BANK

1. Before servicing the vehicle, refer to the precautions in the beginning of this section.

2. Relieve the fuel system pressure.

3. Disconnect the negative battery cable.

4. Drain the engine cooling system.

5. Remove the left bank secondary timing chain, as follows:

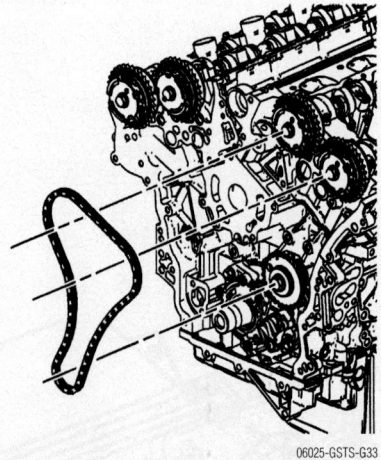

Illustrating the left bank timing chain and gears

Showing the right bank intermediate drive chain idler.

a. Remove the spark plugs in order to ease crankshaft/engine rotation.

b. Remove the engine front cover. See Front Cover & Timing Chain.

c. Remove the right bank secondary camshaft drive chain tensioner, shoe and guide.

d. Remove the right bank secondary camshaft drive chain.

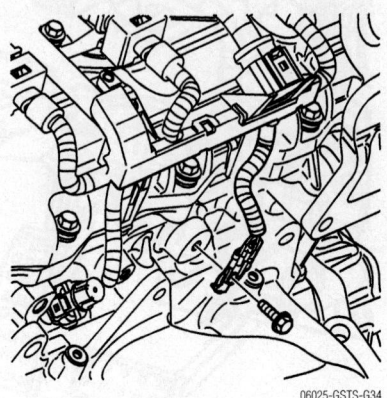

Illustrating wiring harness and coolant sensor

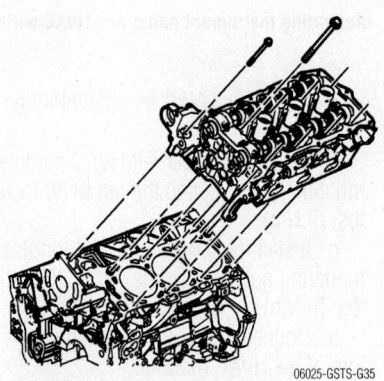

Illustrating the cylinder head bolts and cylinder head

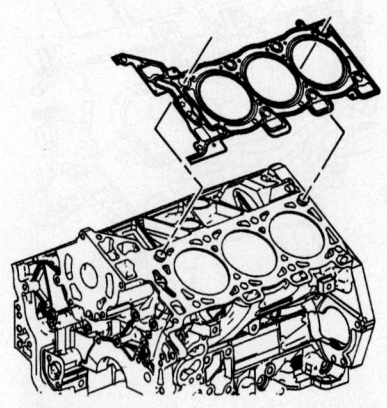

Illustrating cylinder head gasket

e. Remove the primary camshaft drive chain, upper chain guide, and chain.

f. Remove the right bank camshaft intermediate drive chain idler.

g. Remove the left bank secondary camshaft drive chain tensioner, shoe and guide.

h. Remove the left bank camshaft intermediate drive chain idler.

i. Remove the left bank secondary camshaft drive chain.

6. Remove the oil level indicator.

7. Remove the heat shield from the coolant temperature sensor and disconnect the coolant temperature sensor electrical connector.

8. Remove the wiring harness ground from the cylinder head.

9. Disconnect the wiring harness electrical connector located at the side of the cylinder head.

10. Remove the wiring harness connector bracket from the side of the cylinder head.

➡**DO NOT disconnect the power steering pipes and/or hoses.**

11. Remove the power steering pump bolts.

12. Remove the surge tank hose from the bracket at the rear of the cylinder head.

13. Remove the wiring harness bracket from the rear of the cylinder head.

14. Remove the catalytic converter.

➡**Do not remove the oil filter adapter.**

15. Remove the oil filter adapter upper bolt.

16. Remove the cylinder head with the exhaust manifold

17. Remove and discard the cylinder head gasket.

18. Clean and inspect the cylinder head and the engine block sealing surfaces.

To install:

LEFT BANK

1. If removed, install the exhaust manifold to the cylinder head. See Exhaust Manifold.

2. Install a NEW left cylinder head gasket, using the deck face locating pins for retention.

3. Align the left cylinder head with the deck face locating pins.

4. Place the left cylinder head in position on the deck face.

➡**DO NOT allow oil on the cylinder head bolt bosses.**

➡**DO NOT reuse the old M11 cylinder head bolts.**

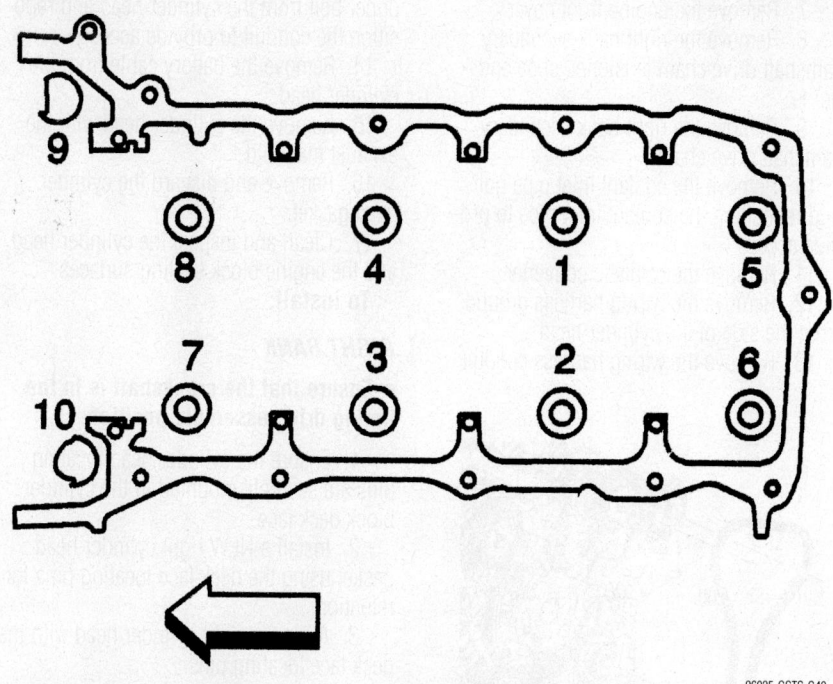

06025-GSTS-G40

Illustrating the left bank cylinder head bolt tightening sequence

5. Install NEW M11 cylinder head bolts and tighten as follows:

a. First pass, in sequence, to 33 ft. lbs. (45 Nm).

b. Second pass, in sequence, an additional 120 degrees.

6. Install the 2 front M8 left cylinder head bolts and tighten as follows:

a. First pass to 11 ft. lbs. (15 Nm).

b. Second pass an additional 60 degrees.

7. Carefully install the cylinder head with the exhaust manifold to the engine.

8. Install the oil filter adapter upper bolt.

9. Install the catalytic converter to the exhaust manifold.

10. Install the wiring harness bracket to the rear of the cylinder head.

11. Install the surge tank hose to the bracket at the rear of the cylinder head.

12. Install the power steering pump bolts.

13. Install the wiring harness connector bracket from the side of the cylinder head.

14. Install the wiring harness electrical connector located at the side of the cylinder head.

15. Install the wiring harness ground to the cylinder head.

16. Tighten the wiring harness ground bolt to 89 inch lbs. (10 Nm).

17. Install the coolant temperature sensor electrical connector heat shield.

18. Install the oil level indicator.

19. Install the left bank secondary camshaft drive chain.

20. Install the left bank camshaft intermediate drive chain idler.

21. Install the left bank camshaft guide, shoe and tensioner.

22. Install the right bank camshaft intermediate drive chain idler.

23. Install the primary camshaft drive chain.

24. Install the primary upper camshaft drive chain guide.

25. Install the primary camshaft drive chain tensioner.

26. Install the right bank secondary camshaft drive chain.

27. Install the right bank secondary camshaft drive chain guide, shoe and tensioner.

28. Properly refill the engine cooling system.

29. Run the engine to check for leaks.

RIGHT BANK

1. Before servicing the vehicle, refer to the precautions in the beginning of this section.

2. Relieve the fuel system pressure.

3. Disconnect the negative battery cable.

4. Drain the engine cooling system.

5. Remove the right bank secondary timing chain using the following sequence.

6. Remove the spark plugs in order to ease crankshaft and engine rotation.

7. Remove the engine front cover.

8. Remove the right bank secondary camshaft drive chain tensioner, shoe and guide.

9. Remove the right bank secondary camshaft drive chain.

10. Remove the coolant inlet pipe bolts and reposition the coolant inlet pipe to provide access.

11. Remove the catalytic converter.

12. Remove the wiring harness ground from the side of the cylinder head.

13. Remove the wiring harness conduit

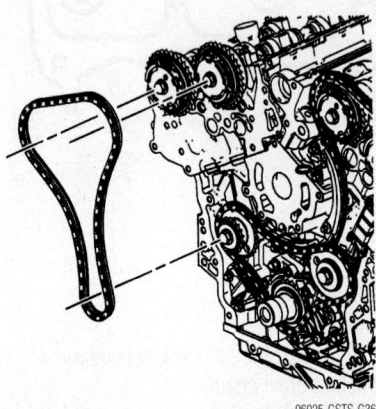

Illustrating the right bank timing chain and gears

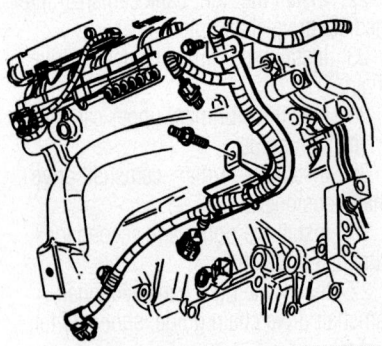

Illustrating wiring harness conduit

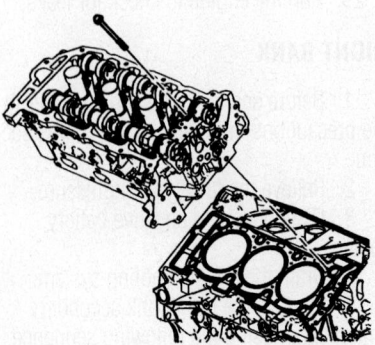

Removing the cylinder head

upper bolt from the cylinder head and reposition the conduit to provide access.

14. Remove the battery cable from the cylinder head.

15. Remove the cylinder head with the exhaust manifold.

16. Remove and discard the cylinder head gasket.

17. Clean and inspect the cylinder head and the engine block sealing surfaces.

To install:

RIGHT BANK

➡**Ensure that the crankshaft is in the timing drive assembly position.**

1. Ensure the cylinder head locating pins are securely mounted in the cylinder block deck face.

2. Install a NEW right cylinder head gasket using the deck face locating pins for retention.

3. Align the right cylinder head with the deck face locating pins.

4. Place the right cylinder head in position on the deck face.

➡**DO NOT allow oil on the cylinder head bolt bosses.**

➡**DO NOT reuse the old M11 cylinder head bolts.**

5. Install NEW M11 cylinder head bolts and tighten as follows:
- First pass, in sequence, to 33 ft. lbs. (45 Nm)
- Second pass, in sequence, an additional 120 degrees

6. Carefully install the cylinder head with the exhaust manifold to the engine.

7. Connect the catalytic converter to the exhaust manifold.

8. Install the battery negative cable to the cylinder head.

9. Install the wiring harness conduit to the cylinder head.

10. Tighten the wiring harness upper bolt to 89 inch lbs. (10 Nm).

11. Install the wiring harness ground to the side of the cylinder head.

12. Tighten the wiring harness ground bolt to 89 inch lbs. (10 Nm).

13. Install the catalytic converter.

14. Install the coolant inlet pipe.

15. Install the right bank secondary timing chain using the following sequence.

16. Install the right bank secondary camshaft drive chain.

17. Install the right bank camshaft intermediate drive chain idler.

18. Install the right bank secondary

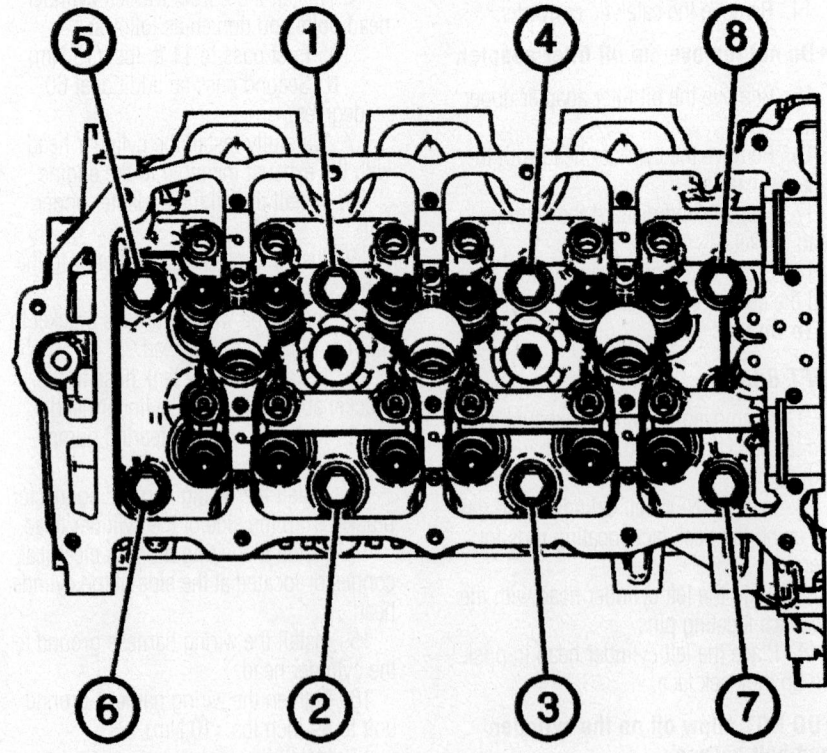

Showing the right bank cylinder head bolt tightening sequence

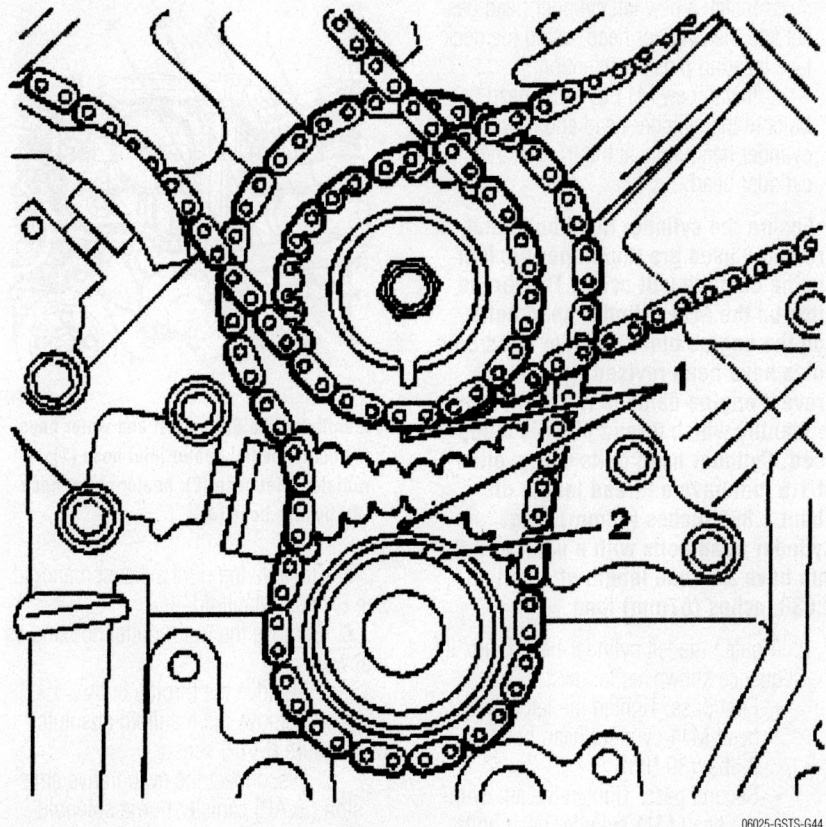

Indicating the timing gear alignment locations—4.6L

06025-GSTS-G44

Removing/installing the drive belt tensioner

06025-GSTS-G297

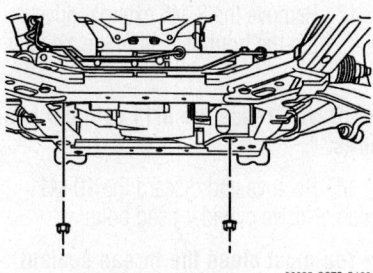

Showing the left camshaft position actuator housing

06025-GSTS-G123

Illustrating the camshaft holding tool

06025-GSTS-G45

Using a wrench on the camshaft hex cast to hold the camshaft during bolt removal

06025-GSTS-G46

camshaft drive chain guide, shoe and tensioner.

19. Install the right bank camshaft intermediate drive chain idler.

20. Properly refill the engine cooling system.

21. Run the engine to check for leaks.

4.6L

LEFT BANK

1. Before servicing the vehicle, refer to the precautions in the beginning of this section.

2. Relieve the fuel system pressure.

3. Disconnect the negative battery cable.

4. Drain the engine cooling system.

5. Remove the left exhaust manifold. See Exhaust Manifold.

6. Remove the water cross-over pipe.

7. Remove the intake manifold. See Intake Manifold.

8. Remove the left secondary camshaft drive chain using the following sequence:

   a. Rotate the crankshaft until the primary timing gear alignment marks (1) are adjacent to each other as shown.

9. Remove the left camshaft position actuator housing as follows:

✼✼ CAUTION

DO NOT remove the actuator solenoids from the housing.

   a. Remove the fuel injector sight shield.

   b. Remove the left exhaust camshaft sensor.

   c. Remove the left intake camshaft sensor.

   d. Remove the power steering reservoir.

   e. Remove the left camshaft cover. See Rocker Arms/Covers.

   f. Remove the drive belt tensioner.

   g. Remove the left intake camshaft position actuator solenoid.

   h. Remove the left intake camshaft position actuator solenoid alignment plug and discard.

   i. Remove the left exhaust camshaft position actuator solenoid.

   j. Remove the left exhaust camshaft position actuator solenoid alignment plug and discard.

   k. Remove the left camshaft position actuator housing.

10. Install a camshaft holding tool on the (left) bank camshafts.

11. Loosen and remove the left sec-

ondary timing chain tensioner bolts and tensioner.

12. Use an open-end wrench on the hex cast into the camshaft in order to prevent the camshaft from rotating when removing the camshaft oil control valve.

13. Loosen and remove the (left) exhaust camshaft position oil control valve.

14. Slide the left exhaust camshaft position actuator off of the camshaft and remove the secondary timing chain from the camshaft actuator teeth.

15. Remove the left secondary timing chain from the engine.

16. Remove the power steering reservoir return hose retaining bolts from the cylinder head.

17. Remove the 3 M6 external drive bolts from the front portion of the cylinder head.

➡ **DO NOT reuse the M11 cylinder head bolts.**

18. Remove and discard the 10 M11 internal-drive cylinder head bolts.

➡ **You must clean the thread sealant material from the cylinder head bolt holes in the cylinder block. Failure to do so could cause false torque readings during reassembly.**

19. After removing the cylinder head, remove any remaining bolt thread sealant material from the threaded cylinder block holes.

20. Remove the left cylinder head. Make sure that no dowel guide pins are stuck in the cylinder.

21. Remove the left cylinder head gasket.

22. Remove all remaining gasket material from the cylinder head and cylinder block.

23. Clean and inspect the cylinder head.

To install:

LEFT BANK

1. Install the left cylinder head as follows:

a. Ensure all the cylinder head locating pins are securely mounted in the cylinder block deck face.

❄❄ CAUTION

Failure to remove all the old thread sealant material from the cylinder block could cause false torque readings.

b. Ensure any old thread sealant material is removed from the cylinder head bolt holes in the cylinder block.

c. Install a new left cylinder head gasket and the cylinder head, using the deck face locating pins for retention.

d. Install new M11 cylinder head bolts in the cylinder head and new M6 cylinder head bolts at the front of the cylinder head.

➡ **Ensure the cylinder head bolts that are being used are the proper pitch or engine damage will occur. The thread pitch on the M11 cylinder head bolts and the engine block cylinder head bolt holes have been revised. In order to prevent engine damage it is important to identify which thread pitch is being used. Cylinder head bolts with a pitch of 1.5 mm have a thread length of about 1.890 inches (48mm) long. Cylinder head bolts with a pitch of 2.0 mm have a thread length of about 2.638 inches (67mm) long.**

2. Tighten the left cylinder head bolts, in the sequence shown, as follows:

- First pass: Tighten the left cylinder head M11 cylinder head bolts to 22 ft. lbs. (30 Nm)
- Second pass: Tighten the left cylinder head M11 cylinder head bolts an additional 60 degrees
- Third pass: Repeat the sequence turning each bolt another 60 degrees
- Final pass: Repeat the sequence again turning each bolt a final 60 degrees, total 180 degrees
- Tighten the M6 bolts at the front of the cylinder head to 106 inch lbs. (12 Nm).

3. Install the power steering reservoir and return hose retaining bolts to the cylinder head. Tighten the power steering reservoir and return hose retaining bolts to 37 ft. lbs. (50 Nm).

4. Install the left camshafts. See Camshaft.

5. Install the left secondary camshaft drive chain. See Camshaft.

6. Install the intake manifold. See Intake Manifold.

7. Install the water crossover.

8. Install the left exhaust manifold. See Exhaust Manifold.

RIGHT BANK

1. Before servicing the vehicle, refer to the precautions in the beginning of this section.

2. Relieve the fuel system pressure.

3. Disconnect the negative battery cable.

4. Drain the engine cooling system.

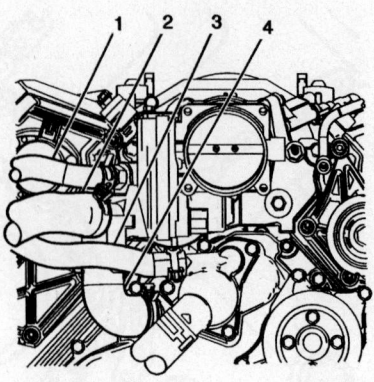

06025-GSTS-G43

Identifying the thermostat and water housing components: heater inlet hose (1); radiator inlet hose (2); heater outlet hose (3); bypass hose (4)

5. Remove the right exhaust manifold. See Exhaust Manifold.

6. Remove the water outlet housing as follows:

a. Remove the throttle body.

b. Remove the manifold absolute pressure (MAP) sensor.

c. Disconnect the evaporative emission (EVAP) canister purge solenoid valve hose and electrical connector.

d. Compress the clamps and remove the hoses from the thermostat and the water housing.

e. Loosen the intake manifold duct clamp.

f. Remove the bolts securing the water outlet housing to the cylinder heads.

7. Remove the water outlet housing from the vehicle.

8. Remove the intake manifold. See Intake Manifold.

9. Remove the right secondary camshaft drive chain, using the following sequence:

a. Rotate the crankshaft until the primary timing gear alignment marks are adjacent to each other as shown.

b. Remove the right camshaft position actuator housing. DO NOT remove the actuator solenoids from the housing.

c. Install a camshaft holding tool (EN 46328) on the (right) bank camshafts.

d. Loosen and remove the right secondary timing chain tensioner bolts and tensioner.

e. Use an open-end wrench on the hex cast into the camshaft in order to prevent the camshaft from rotating when removing the camshaft oil control valve.

f. Loosen and remove the (right)

exhaust camshaft position oil control valve.

g. Slide the right exhaust camshaft position actuator off of the camshaft and remove the secondary timing chain from the camshaft actuator teeth.

h. Remove the right secondary timing chain from the engine.

10. Remove the right camshaft. See Camshaft.

11. Remove the right cylinder head. Make sure that no dowel guide pins are stuck in the cylinder head.

12. Remove the right cylinder head gasket.

13. Remove all remaining gasket material from the cylinder head and cylinder block.

14. Clean and inspect the cylinder head.

To install:

LEFT OR RIGHT BANK

1. Ensure all the cylinder head locating pins are securely mounted in the cylinder block deck face.

➡**Failure to remove all the old thread sealant material from the cylinder block could cause false torque readings.**

2. Ensure any old thread sealant material is removed from the cylinder head bolt holes in the cylinder block.

3. Install a new left cylinder head gasket using the deck face locating pins for retention.

4. Align the cylinder head with the deck face locating pins.

5. Place the cylinder head in position on the deck face.

❋ CAUTION

DO NOT reuse the old M11 cylinder head bolts.

6. Install NEW M11 cylinder head bolts in the cylinder head.

7. Install the M6 cylinder head bolts at the front of the cylinder head.

❋ WARNING

Ensure the cylinder head bolts that are being used are the proper pitch or engine damage will occur. The thread pitch on the M11 cylinder head bolts and the engine block cylinder head bolt holes have been revised. In order to prevent engine damage it is important to identify which thread pitch is being used. Cylinder head bolts with a pitch of

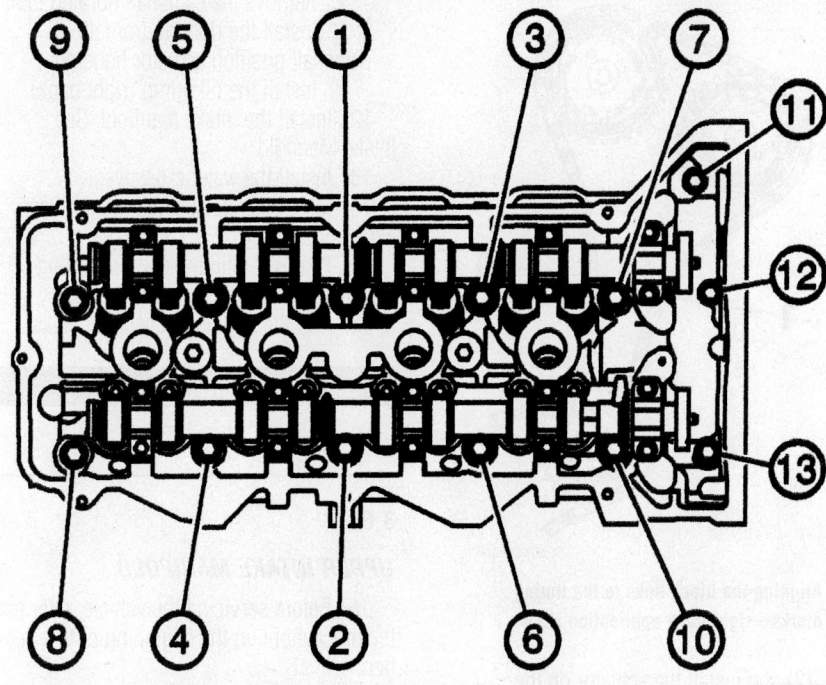

Illustrating cylinder head bolt torque sequence—4.6L

06025-GSTS-G52

1.5 mm have a thread length of about 48 mm (1.890 in) long. Cylinder head bolts with a pitch of 2.0 mm have a thread length of about 67 mm (2.638 in) long.

8. Tighten the cylinder head bolts in the sequence shown, as follows:

a. First Pass: M11 cylinder head bolts to 22 ft. lbs. (30 Nm)

b. Second Pass: M11 cylinder head bolts an additional 60 degrees

c. Third Pass: Repeat the sequence turning each bolt another 60 degrees.

d. Final Pass: Repeat the sequence again turning each bolt a final 60 degrees, total 180 degrees.

e. Tighten the M6 bolts at the front of the cylinder head.

f. Tighten the M6 cylinder head bolts to 106 inch lbs. (12 Nm).

9. Install the power steering reservoir and return hose retaining bolts to the left cylinder head.

10. Tighten the power steering reservoir and return hose retaining bolts to 37 ft. lbs. (50 Nm).

11. Install the secondary camshaft drive chain, noting 3 black links that aid in timing the camshaft position actuators to the intermediate sprocket, as follows:

• The black link (2) is aligned with the bank 1 exhaust actuator timing mark.

• The black link (3) is aligned with

the bank 1 intake actuator timing mark.

• The black link (1) is aligned with the intermediate sprocket.

a. The intermediate sprocket right bank timing mark is labeled "RB" as shown, while the left bank timing mark is labeled "LB" as shown.

Assemble the secondary timing chain to the intermediate sprocket aligning the sprocket "RB" or "LB" timing mark to the timing chain black link.

❋ CAUTION

A wrench must be used on the hex of the camshaft when loosening or tightening in order to prevent component damage. Failure to prevent the torque reaction against the timing drive chain can lead to timing drive chain failure.

b. On right bank, align the timing mark (5) of the "RB" intake camshaft position actuator with the timing chain black link (8) and install the actuator on the camshaft with the actuator timing mark perpendicular (90 degrees) to the cylinder head deck surface near the top of its rotation. Loosely install the oil control valve (2) to secure the intake actuator.

c. On left bank, align the timing mark (5) of the "LB" intake camshaft position actuator with the timing chain black link

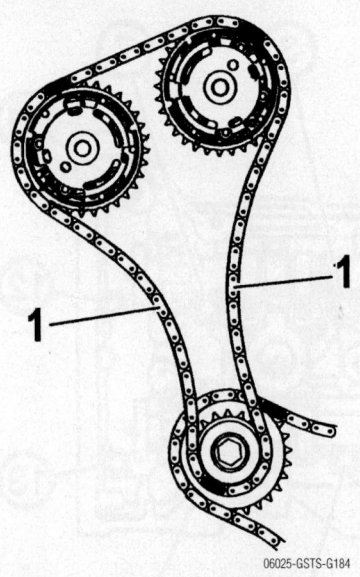

Aligning the black links to the timing marks—right bank application shown

(3) and install the actuator on the camshaft with the actuator timing mark perpendicular (90 degrees) to the cylinder head deck surface at the top of its rotation. Loosely install the oil control valve (10) to secure the intake actuator.

d. Use an open-end wrench on the hex cast into the camshaft in order to prevent the camshaft from rotating when tightening the oil control valve. Tighten the oil control valve to 89 ft. lbs. (120 Nm).

e. On right bank, align the timing mark (3) of the "RB" exhaust camshaft position actuator with the timing chain black link and install the actuator on the camshaft with the actuator timing mark perpendicular (90 degrees) to the cylinder head deck surface near the top of its rotation. Loosely install the oil control valve (1) to secure the exhaust actuator.

f. On left bank, align the timing mark (7) of the "LB" exhaust camshaft position actuator with the timing chain black link (4) and install the actuator on the camshaft with the actuator timing mark perpendicular (90 degrees) to the cylinder head deck surface near the top of its rotation. Loosely install the oil control valve (8) to secure the exhaust actuator.

g. Use an open-end wrench on the hex cast into the camshaft in order to prevent the camshaft from rotating when tightening the oil control valve. Tighten the oil control valve to 89 ft. lbs. (120 Nm).

h. Install the right secondary timing chain tensioner.

i. Remove the camshaft holding tool.

j. Install the right and/or left camshaft position actuator housing.

k. Install the oil pump (right bank).

12. Install the intake manifold. See Intake Manifold.

13. Install the water crossover.

14. Install the exhaust manifolds. See Exhaust Manifold.

15. Properly refill the engine cooling system.

16. Run the engine to check for leaks.

Intake Manifold

REMOVAL & INSTALLATION

3.6L

UPPER INTAKE MANIFOLD

1. Before servicing the vehicle, refer to the precautions in the beginning of this section.

2. Turn the ignition OFF.

3. Remove the engine cover (fuel injector sight shield).

4. Remove the air inlet duct.

5. Disconnect the brake booster vacuum hose from the intake manifold.

6. Disconnect the purge solenoid valve electrical connector.

7. Disconnect the purge line (3) from the purge solenoid valve.

8. Disconnect the fuel injector harness electrical connector.

9. Remove the wiring harness retainer.

10. Disconnect the throttle body electrical connector.

11. Remove the upper intake manifold brace stud (1), bolt (2), and brace.

12. Disconnect the positive crankcase ventilation (PCV) hose from the right bank camshaft cover.

13. Disconnect the barometric pressure sensor electrical connector.

14. Disconnect the intake manifold runner control solenoid electrical connector.

15. Remove the injector harness bracket bolt.

16. Remove the left bank ignition coil wiring harness from the bracket.

17. Remove the injector harness bracket bolt.

18. Remove the left bank ignition coil wiring harness from the bracket.

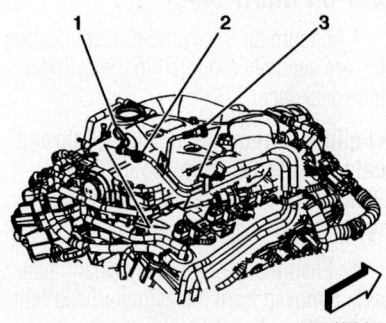

Locating the EVAP purge solenoid valve hose and electrical connections

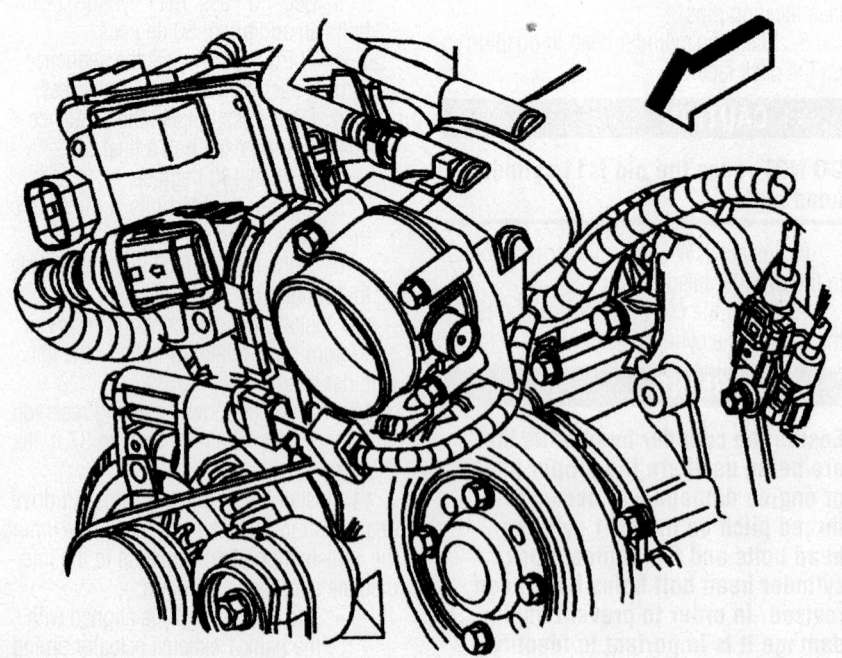

Disconnecting the throttle body electrical connector

Showing the upper intake manifold brace stud (1) and bolts (2)

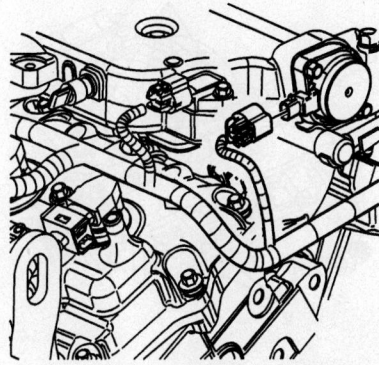

Indicating the electrical connections for the barometric sensor (left) and the intake manifold runner control solenoid (right)—3.6L

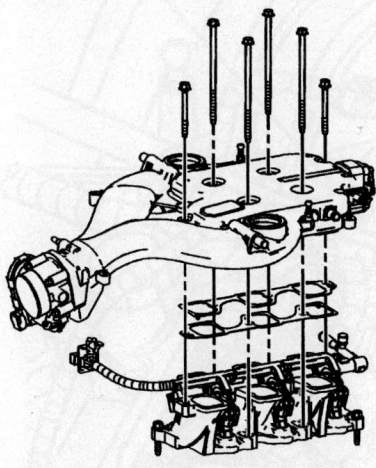

Illustrating upper intake manifold bolt locations—3.6L

19. Remove the intake manifold bolts.

20. Reposition the intake manifold to gain access to the fuel pipe connector.

21. Clean and inspect the intake manifold.

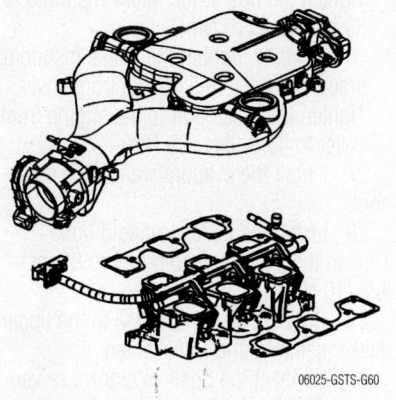

Showing the upper intake manifold gaskets—3.6L

22. Remove the upper-to-lower intake manifold bolts.

➡ **Do not reuse the upper-to-lower intake manifold gasket or the intake manifold-to-cylinder head sealing gaskets.**

23. Remove the upper intake manifold with the throttle body.

To install:

24. If the upper intake manifold is being replaced or serviced, assemble the intake manifold as follows.

25. Install a NEW upper intake manifold gasket.

26. Install the upper intake manifold.

➡ **Tighten the intake manifold bolts in a circular pattern starting at the center (long bolts) and moving outward.**

27. Install the upper intake manifold bolts. Tighten the bolts to 17 ft. lbs. (23 Nm).

28. Connect the intake manifold runner control solenoid electrical connector.

29. Connect the barometric pressure sensor electrical connector.

30. Connect the PVC hose to the right bank camshaft cover.

31. Install the upper intake manifold brace, stud (1) and bolt (2).

32. Tighten the intake manifold brace bolt (2) to 48 ft. lbs. (65 Nm).

33. Tighten the intake manifold brace stud (1) to 89 inch lbs. (10 Nm).

✳✳ WARNING

Ensure proper engagement of the wiring harness connector. The wiring harness connector must be installed straight onto the component connector and firmly seated. Visually inspect the connector to ensure that the connector latches are engaged

and locked. Any damage to the connector or wiring must be repaired. Failure to follow this procedure can lead to an intermittent electrical connection, driveability concerns, and/or wiring harness or wiring harness connector damage or failure.

34. Connect the electrical connector to the throttle body.

35. Install the wiring harness retainer over the throttle body.

36. Connect the EVAP purge line to the EVAP solenoid.

37. Connect the purge solenoid valve electrical connector.

38. Connect the brake booster vacuum hose to the intake manifold.

39. Install the air inlet duct.

40. Install the engine cover (fuel injector sight shield).

LOWER INTAKE MANIFOLD

1. Remove the fuel injector sight shield.

2. Remove the air inlet duct.

3. Note the booster hose routing. Disconnect the brake booster check valve and hose from the brake booster.

4. Reposition the brake booster hose to the intake manifold.

5. Disconnect the EVAP purge solenoid valve electrical connector.

6. Disconnect the purge line from the EVAP purge solenoid valve.

7. Disconnect the fuel injector harness electrical connector.

8. Disconnect the throttle body electrical connector.

9. Remove the wiring harness retainer.

10. Remove the upper intake manifold brace stud and bolt, reposition the brace with the wiring harness.

11. Remove the wiring harness retainer from the right side of the throttle body.

12. Disconnect the positive crankcase ventilation (PCV) hose from the right bank camshaft cover.

13. Disconnect the barometric pressure sensor electrical connector.

14. Disconnect the intake manifold runner control solenoid electrical connector.

15. Remove the left bank ignition coil wiring harness from the bracket.

✳✳ CAUTION

Do not remove the upper to lower intake manifold bolts. DO not kink or damage the fuel pipe.

16. Remove the intake manifold bolts.

17. Reposition the intake manifold to gain access to the fuel pipe connector.

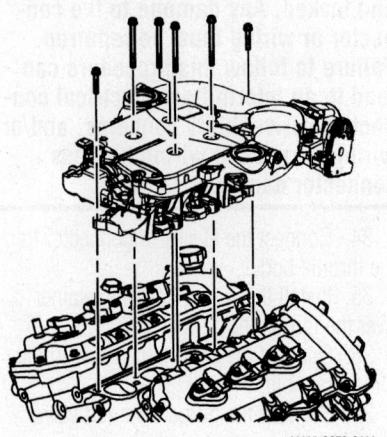

Removing/installing the lower intake manifold retaining bolts

18. Remove the fuel feed pipe retainer.
19. Disconnect the fuel feed pipe from the fuel rail.
20. Disassemble the intake manifold as necessary.
21. Clean and inspect the intake manifold and the sealing surfaces.

To install:
22. Install the NEW intake manifold gasket.
23. Install the intake manifold assembly.

✷✷ CAUTION

Tighten the intake manifold bolts in a circular pattern starting from the center and moving outward.

24. Install the intake manifold bolts. Tighten the intake manifold bolts to 17 ft. lbs. (23 Nm).
25. Place the intake manifold brace to the engine front cover and intake manifold.
26. Loosely install the bolts and tighten as follows:
  a. First tighten the intake manifold brace bolt (1) to the intake manifold.

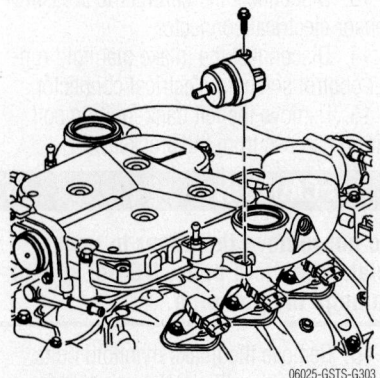

Showing the location of the EVAP solenoid

Tighten the bolt at the intake manifold to 89 inch lbs. (10 Nm).
  b. Finally, tighten the intake manifold brace bolt (2) to the engine front cover. Tighten the brace bolt to the engine front cover to 48 ft. lbs. (65 Nm).
27. Install the evaporative (EVAP) solenoid.
28. Install the EVAP solenoid bolt. Tighten the EVAP solenoid bolt to 89 inch lbs. (10 Nm).
29. Connect the EVAP hose to the upper intake manifold and EVAP solenoid.
30. Connect the positive crankcase ventilation (PCV) tube assembly to the upper

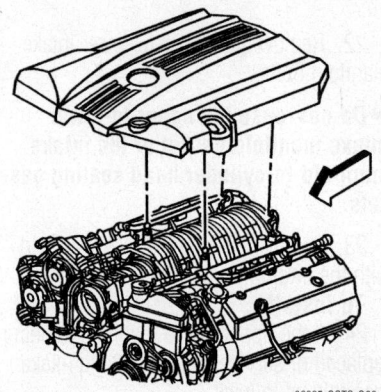

Illustrating fuel injector sight shield—4.6L

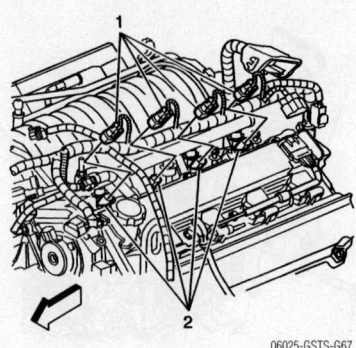

Illustrating left side fuel injector electrical connections—4.6L

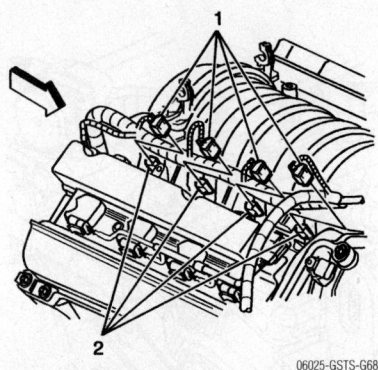

Illustrating right side fuel injector electrical connections—4.6L

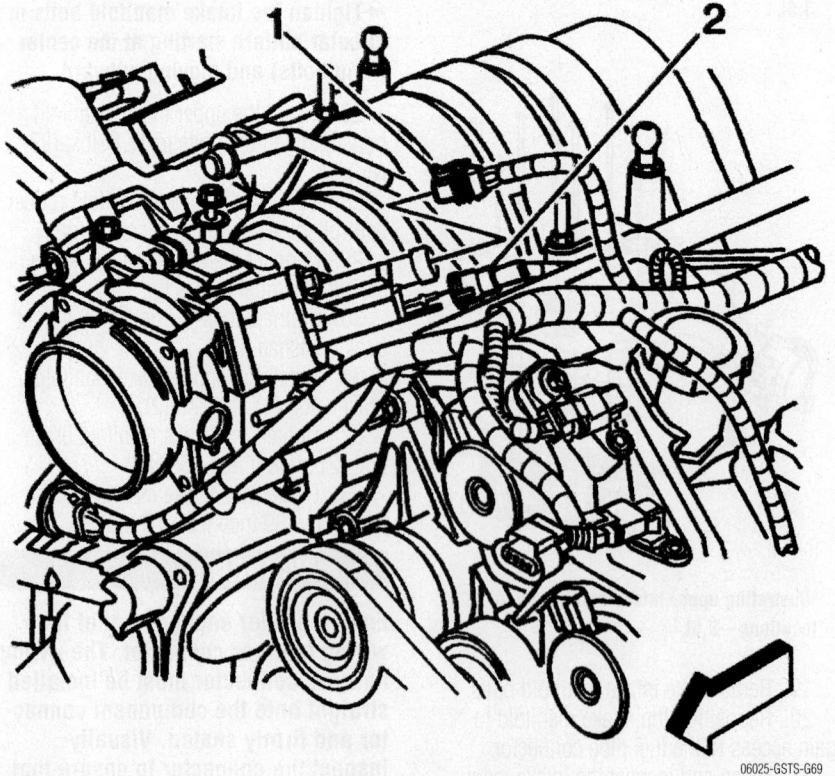

Illustrating emissions purge valve line and electrical connector

intake manifold and the right camshaft cover.

31. Install the PCV tube assembly bracket bolts. Tighten the bolts to 89 inch lbs. (10 Nm).

32. Install the brake booster hose to the upper intake manifold.

4.6L

UPPER INTAKE MANIFOLD

1. Before servicing the vehicle, refer to the precautions in the beginning of this section.

2. Remove the cross car brace.

3. Remove the fuel injector sight shield.

4. Remove the PCV dirty air tube from the camshaft cover.

5. Remove the PCV fresh air tube.

6. Remove the sight shield bracket nuts.

7. Remove the sight shield bracket.

8. Disconnect the left side fuel injector electrical connectors (1) from the fuel injectors (2).

9. Disconnect the right side fuel injector electrical connectors (1) from the fuel injectors (2).

10. Disconnect the evaporative emissions purge valve line (2) and electrical connector (1).

11. Remove the EVAP line from the retaining feature (2) at the rear of the right head. Position the line aside.

12. Relieve the fuel system pressure.

13. Disconnect the fuel line from the fuel rail

14. Remove the fuel line from the retainer at the rear of the right head.

15. Remove the fasteners attaching the fuel rail to the intake manifold.

16. Lift the entire fuel rail and injector assembly from the intake manifold.

17. Loosen the plenum duct clamp at the front of the intake manifold.

18. Loosen the bolts attaching the intake manifold to the cylinder heads.

19. Remove the intake manifold by using an upward lifting motion at the rear of the manifold assembly.

20. Inspect the intake manifold.

To install:

21. Transfer the necessary parts when replacing the intake manifold.

22. Lightly grease the inside edge of the rubber plenum duct.

23. Install NEW gaskets on the intake manifold.

24. Place the intake manifold in position.

25. Position the front of the intake manifold into the plenum duct.

26. Position the rear of the intake manifold downward onto the cylinder heads.

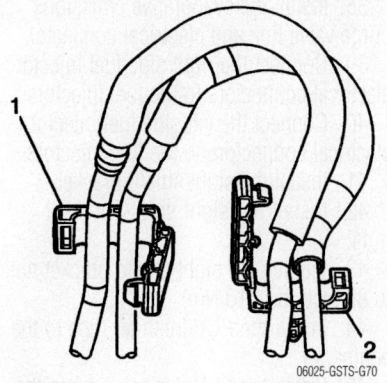

Indicating the EVAP lines and retainers

06025-GSTS-G70

Removing/installing the intake manifold—4.6L

06025-GSTS-G72

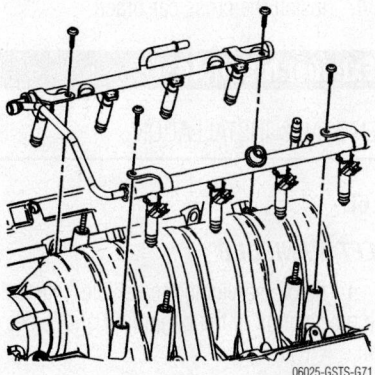

Showing the fuel rail fastener locations

06025-GSTS-G71

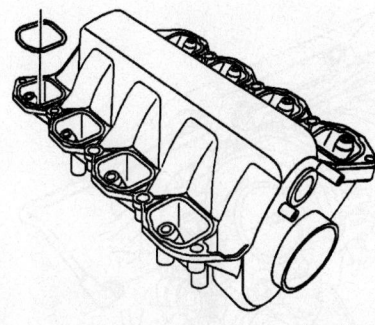

Showing the intake manifold gasket location—4.6L

06025-GSTS-G73

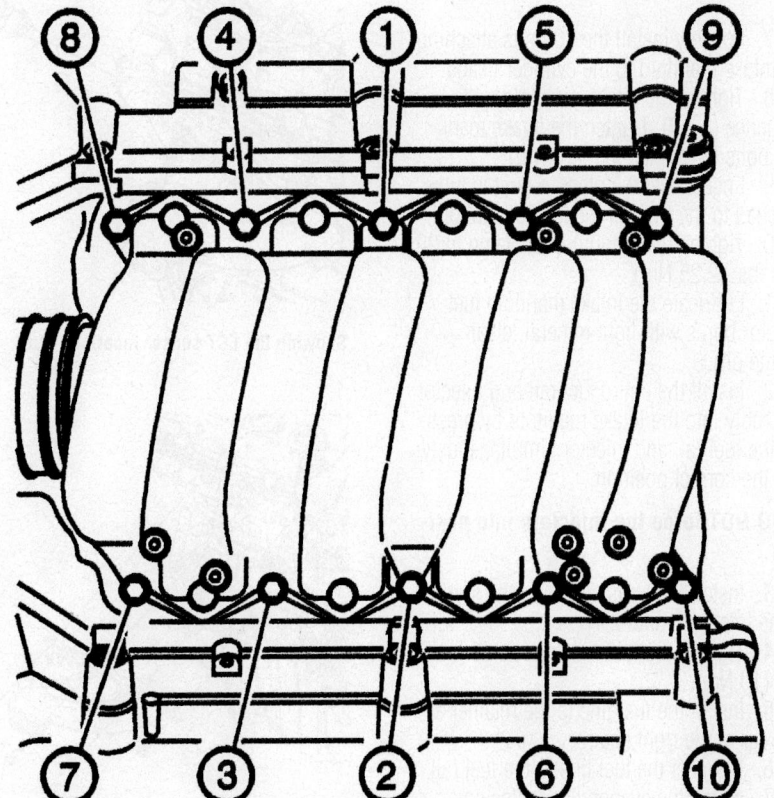

Illustrating the intake manifold bolt tightening sequence—4.6L

06025-GSTS-G74

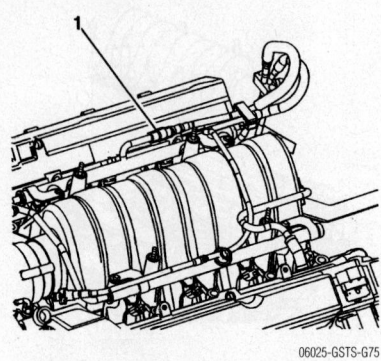

06025-GSTS-G75

Identifying the fuel line to fuel rail connector

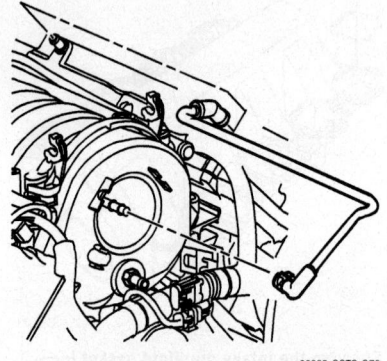

06025-GSTS-G76

Removing/installing the PCV dirty air tube at the front of the engine—4.6L

27. Loosely install the 10 bolts attaching the intake manifold to the cylinder heads.

28. Tighten the intake manifold bolts in sequence (1-10). Tighten the intake manifold bolts to 89 inch lbs. (10 Nm).

29. Ensure that the plenum duct is fully attached to the front of the intake manifold.

30. Tighten the plenum duct clamp to 20 inch lbs. (2.25 Nm).

31. Lubricate the intake manifold fuel injector bores with light mineral, clean engine oil).

32. Install the entire fuel rail and injector assembly into the intake manifold by pressing the fuel rail and injector simultaneously into the correct position.

➡ **DO NOT force the injectors into position.**

33. Install the bolts attaching the fuel rail to the intake manifold.

34. Tighten the fuel rail bolts to 89 inch lbs. (10 Nm).

35. Install the fuel line to the retainer at the rear of the right head.

36. Connect the fuel line to the fuel rail

37. Install the evaporative emissions purge valve line to the retainer at the rear of the right head.

38. Install the evaporative emissions purge valve line and electrical connector.

39. Connect the right side fuel injector electrical connectors to the fuel injectors.

40. Connect the left side fuel injector electrical connectors to the fuel injectors.

41. Install the sight shield bracket.

42. Install the sight shield bracket nuts.

43. Tighten the sight shield bracket nuts to 89 inch lbs. (10 Nm).

44. Install the PCV fresh air tube to the engine.

45. Install the PCV dirty air tube to the engine.

46. Install the fuel injector sight shield.

47. Install the cross car brace.

Exhaust Manifold

REMOVAL & INSTALLATION

3.6L

LEFT MANIFOLD

1. Before servicing the vehicle, refer to the precautions in the beginning of this section.

06025-GSTS-G77

Showing the ECT sensor location—3.6L

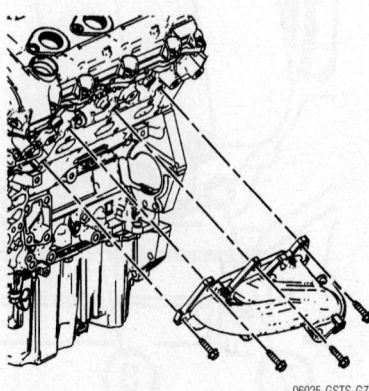

06025-GSTS-G78

Showing the installed position of the left exhaust manifold—3.6L

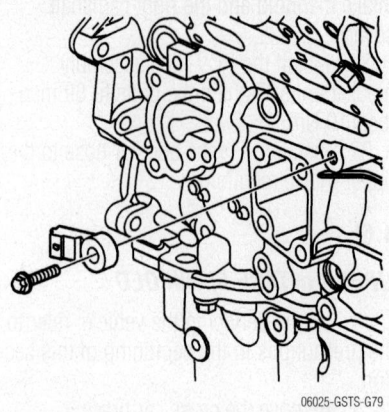

06025-GSTS-G79

Showing the location of the knock sensor—3.6L

2. Remove the engine coolant temperature (ECT) sensor.

3. Remove the exhaust manifold bolts from the left cylinder head.

4. Remove the left exhaust manifold.

5. Remove and discard the exhaust manifold gasket from the left cylinder head.

6. Remove the left knock sensor bolt.

7. Remove the left knock sensor.

8. Remove the left engine mount nut.

9. Remove the left engine mount. See Engine Mounts.

10. Remove the left engine mount bracket bolt and bracket.

To install:

11. Install the engine mount bracket.

12. Install the engine mount bracket bolts.

13. Tighten the engine mount bracket bolts to 44 ft. lbs. (60 Nm).

14. Install the engine mount.

15. Install the engine mount nut.

16. Tighten the engine mount nut to 59 ft. lbs. (80 Nm).

17. Position the left knock sensor to the cylinder block as shown.

18. Install the knock sensor bolt.

19. Tighten the knock sensor bolt to 17 ft. lbs. (23 Nm).

20. Ensure proper knock sensor orientation.

21. Position a NEW exhaust manifold gasket onto the left exhaust manifold.

22. Install the exhaust manifold bolts into the left exhaust manifold.

23. Place the left exhaust manifold, exhaust manifold gasket and bolts as an assembly in position on the left cylinder head.

24. Install the exhaust manifold bolts into the left cylinder head.

25. Tighten the exhaust manifold bolts to 18 ft. lbs. (25 Nm).

26. Install the engine coolant tempera-

ture (ECT) sensor. Tighten the ECT sensor to 16 ft. lbs. (22 Nm).

RIGHT MANIFOLD

1. Before servicing the vehicle, refer to the precautions in the beginning of this section.

2. Remove the right exhaust manifold heat shield bolts.

3. Remove the right exhaust manifold heat shield.

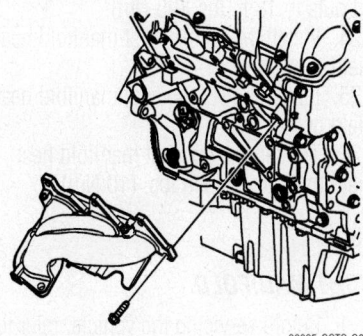

06025-GSTS-G83

Illustrating the right exhaust manifold— 3.6L

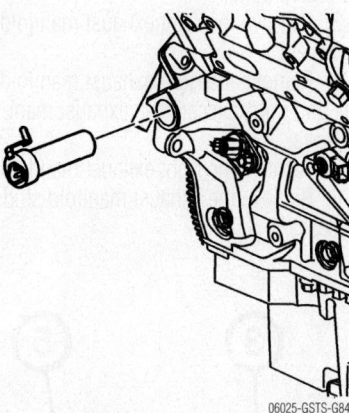

06025-GSTS-G84

Showing the block heater cartridge location

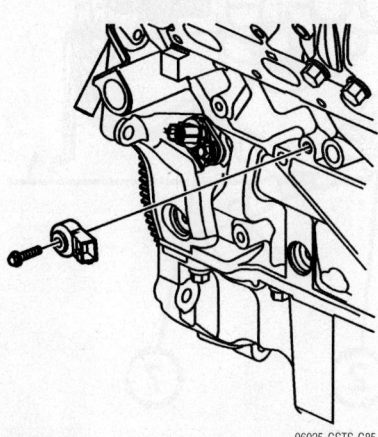

06025-GSTS-G85

Indicating the right knock sensor location

4. Remove the exhaust manifold bolts from the right cylinder head.

5. Remove the right exhaust manifold.

6. Remove and discard the exhaust manifold gasket from the right cylinder head.

7. Remove the block heater cartridge.

8. Remove the right knock sensor bolt.

9. Remove the right knock sensor.

10. Remove the crankshaft position sensor bolt.

11. Remove the crankshaft position sensor.

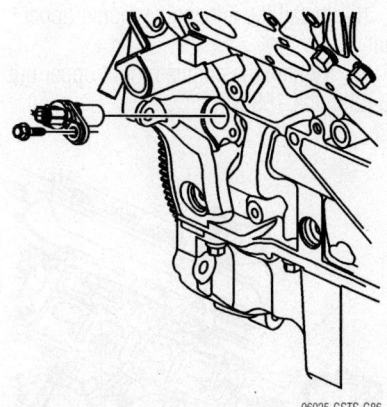

06025-GSTS-G86

Showing the CKP sensor location—3.6L

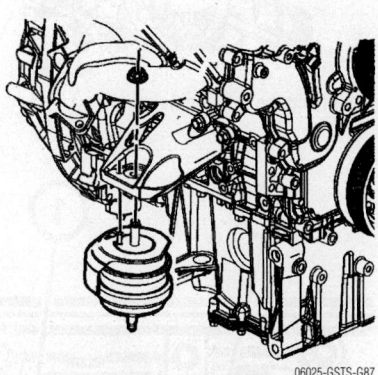

06025-GSTS-G87

Illustrating right engine mount

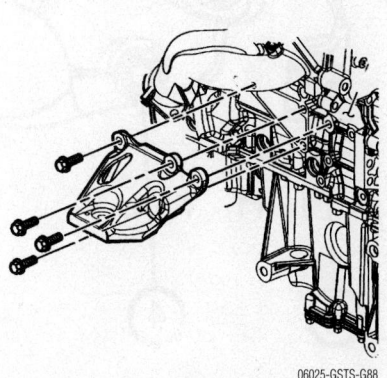

06025-GSTS-G88

Illustrating right engine bracket

12. Remove and discard the crankshaft position sensor O-ring, if damaged.

13. Remove the engine mount nut and the engine mount. See Engine Mounts.

14. Remove the right engine mount bracket bolts and bracket.

To install:

RIGHT MANIFOLD

1. Install the engine mount bracket.

2. Install the engine mount bracket bolts.

3. Tighten the engine mount bracket bolts to 44 ft. lbs. (60 Nm).

4. Install the engine mount.

5. Install the engine mount nut.

6. Tighten the engine mount nut to 59 ft. lbs. (80 Nm).

7. Install the NEW O-ring on the crankshaft position sensor, if damaged.

8. Position the crankshaft position sensor into the cylinder block.

9. Install the crankshaft position sensor bolt.

10. Tighten the crankshaft position sensor bolt to 89 inch lbs. (10 Nm).

11. Position the right knock sensor to the cylinder block as shown.

12. Install the knock sensor bolt.

13. Tighten the knock sensor bolt to 17 ft. lbs. (23 Nm).

14. Ensure proper knock sensor orientation.

15. Install the block heater cartridge, if equipped.

16. Position a NEW exhaust manifold gasket onto the right exhaust manifold.

17. Install the exhaust manifold bolts into the right exhaust manifold.

18. Place the right exhaust manifold, exhaust manifold gasket and bolts as an assembly in position on the right cylinder head.

19. Install the exhaust manifold bolts into the right cylinder head.

20. Tighten the exhaust manifold bolts to 18 ft. lbs. (25 Nm).

21. Place the right exhaust manifold heat shield in position.

22. Install the exhaust manifold heat shield bolts.

23. Tighten the exhaust manifold heat shield bolts to 89 inch lbs. (10 Nm).

LEFT MANIFOLD

1. Before servicing the vehicle, refer to the precautions in the beginning of this section.

2. Remove the left exhaust manifold heat shield bolts and heat shield.

3. Remove the left exhaust manifold

bolts and nuts. Discard the exhaust manifold bolts.

4. Remove the left exhaust manifold and gasket from the engine. Do not reuse the gasket.

5. Remove the left exhaust manifold studs from the left cylinder head.

6. Remove the left engine mount upper nut and engine mount.

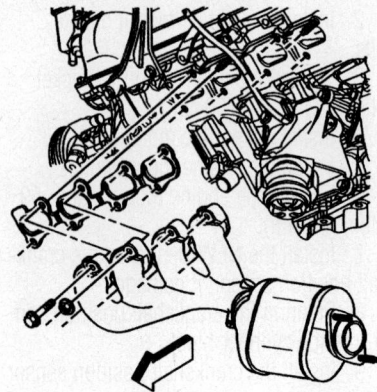

06025-GSTS-G90

Illustrating the left exhaust manifold and gasket—4.6L

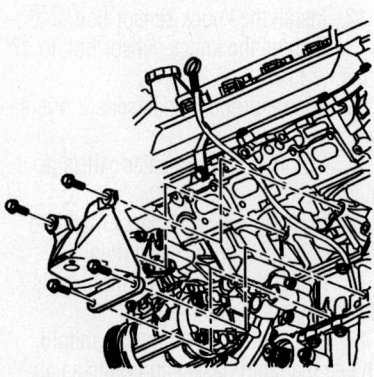

06025-GSTS-G94

Removing/installing the left engine mount bracket location

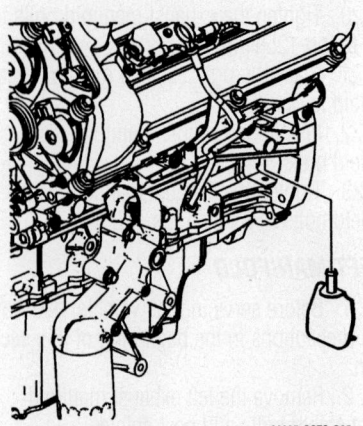

06025-GSTS-G95

Illustrating engine flywheel cover location

7. Remove the left engine mount bracket bolts and engine mount bracket.

8. Remove the engine flywheel housing left cover.

9. Clean and inspect the left exhaust manifold.

To install:

10. Install the left engine flywheel housing cover.

11. Install the left engine mount bracket.

12. Install the left engine mount bracket bolts.

13. Tighten the engine mount bracket bolts to 43 ft. lbs. (58 Nm).

14. Install the left engine mount.

15. Install the left engine mount upper nut.

16. Tighten the engine mount upper nut to 43 ft. lbs. (58 Nm).

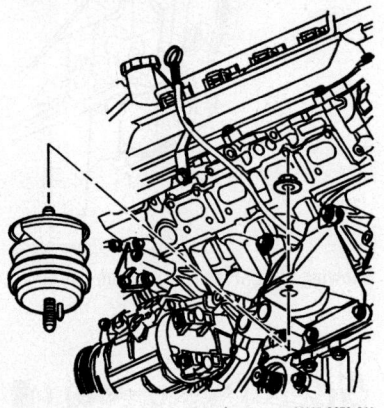

06025-GSTS-G93

Identifying the left engine mount

17. Install the left exhaust manifold studs to the left cylinder head.

18. Tighten the exhaust manifold studs to 53 inch lbs. (6 Nm).

19. Install the NEW left exhaust manifold gasket.

20. Install the left exhaust manifold.

21. Loosely install the left exhaust manifold bolts to the left cylinder head.

22. Tighten the exhaust manifold bolts and nuts in the sequence shown.

23. Tighten the exhaust manifold bolts and nuts to 18 ft. lbs. (25 Nm).

24. Install the left exhaust manifold heat shield.

25. Install the left exhaust manifold heat shield bolts.

26. Tighten the exhaust manifold heat shield bolts to 89 inch lbs. (10 Nm).

4.6L

RIGHT MANIFOLD

1. Before servicing the vehicle, refer to the precautions in the beginning of this section.

2. Remove the right exhaust manifold heat shield bolts.

3. Remove the right exhaust manifold heat shield.

4. Remove the right exhaust manifold bolts and nuts. Discard the exhaust manifold bolts.

5. Remove the right exhaust manifold.

6. Remove the exhaust manifold studs, if necessary.

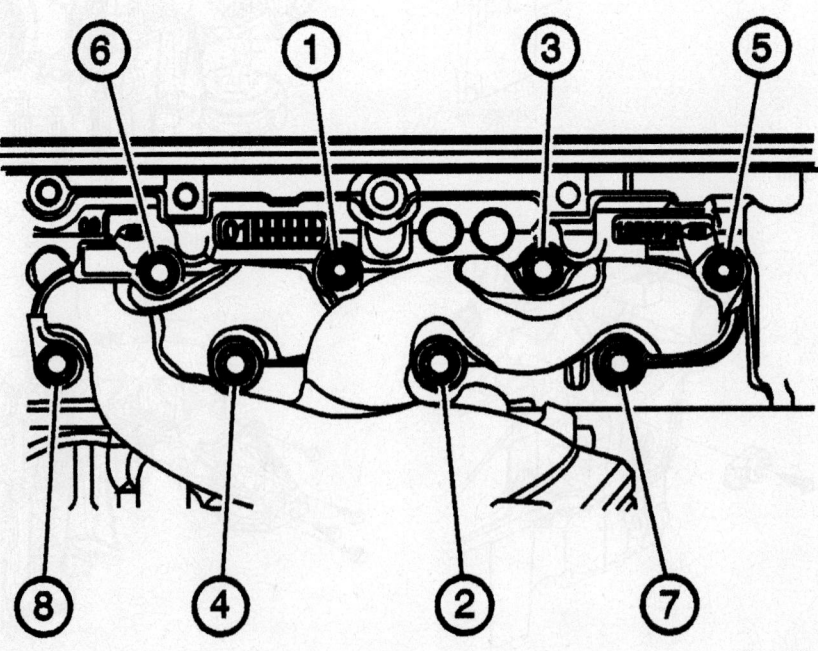

06025-GSTS-G96

Illustrating the left exhaust manifold bolt tightening sequence—4.6L

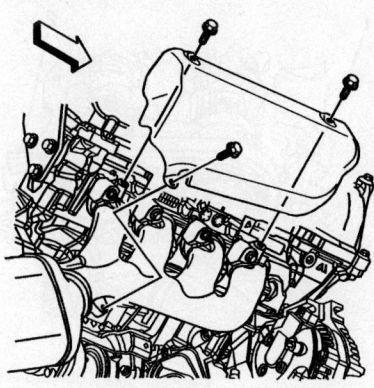

06025-GSTS-G91

Showing the position of the right exhaust heat shield—4.6L

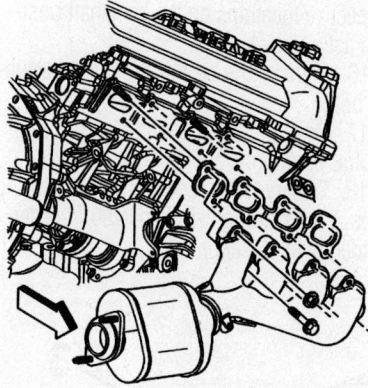

06025-GSTS-G92

Illustrating right exhaust manifold and gasket—4.6L

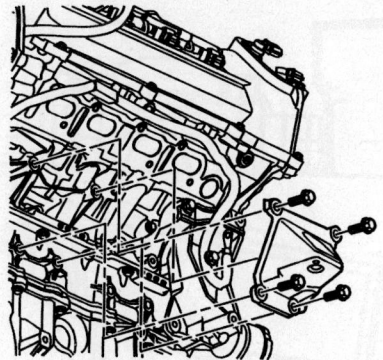

06025-GSTS-G98

Illustrating right engine bracket

7. Remove the right exhaust gasket from the engine. Do not reuse the gasket.

8. Remove the right engine mount upper nut and engine mount.

9. Remove the right engine mount bracket bolts and engine mount bracket.

10. Clean and inspect the right exhaust manifold.

To install:

11. Install the right engine mount bracket. Tighten the engine mount bracket bolts to 43 ft. lbs. (58 Nm).

12. Install the right engine mount and upper nut. Tighten the engine mount upper nut to 43 ft. lbs. (58 Nm).

13. Install the right exhaust manifold studs to the right cylinder head. Tighten the studs to 53 inch lbs. (6 Nm).

14. Install the NEW right exhaust manifold gasket and the right exhaust manifold.

15. Loosely install the right exhaust manifold bolts and nuts to the right cylinder head.

16. Tighten the exhaust manifold bolts and nuts in the sequence shown. Tighten the exhaust manifold bolts and nuts to 18 ft. lbs. (25 Nm).

17. Install the right exhaust manifold heat shield and bolts. Tighten the exhaust manifold heat shield bolts to 89 inch lbs. (10 Nm).

18. Exhaust pipe to the manifold

19. Disconnect the negative battery cable..

Camshafts

REMOVAL & INSTALLATION

3.6L

1. Before servicing the vehicle, refer to the precautions in the beginning of this section.

2. Properly relieve the fuel system pressure.

3. Disconnect the negative battery cable.

4. Drain the engine cooling system and the engine oil.

5. Remove the upper intake manifold with the lower intake manifold.

6. Remove the bank camshaft cover.

7. Remove the camshaft sensors.

8. Remove the camshaft position actuator solenoid.

9. Remove the crankshaft balancer.

10. Rotate the crankshaft with the proper socket until the camshafts are in a neutral (low tension) position.

11. The camshaft flats will be parallel with the camshaft cover rail.

✳✳ WARNING

A wrench must be used on the hex of the camshaft when loosening or tightening in order to prevent component damage. Failure to prevent the torque reaction against the timing drive chain can lead to timing drive chain failure.

➡**Use an open-end wrench at the camshaft hex to prevent camshaft/ engine rotation.**

12. DO NOT remove the camshaft position actuator bolt at this time.

13. Loosen the camshaft position actuator bolt.

14. Install the timing chain holding tools (1 and 2) in order to retain the timing chain. Firmly tighten the retainer nuts.

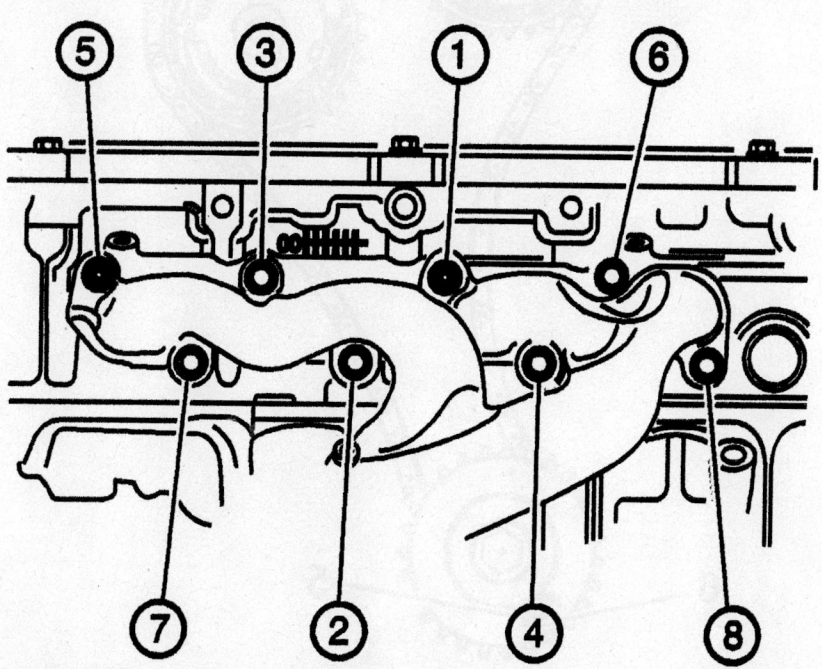

06025-GSTS-G99

Showing the right exhaust manifold bolt tightening sequence—4.6L

Illustrating the cylinder head with camshafts in neutral position—3.6L

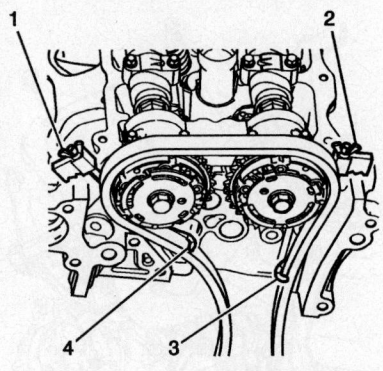

Illustrating the timing chain retention tools (1, 2) in place on the chains (positions 3, 4)

15. Mark the timing chain and the respective locations on the camshaft position actuators (1-4).

16. Remove the camshaft position actuator bolt.

17. Position the camshaft lobes in a neutral position.

18. Observe the markings on the bearing caps. Each bearing cap is marked in order to identify its location.

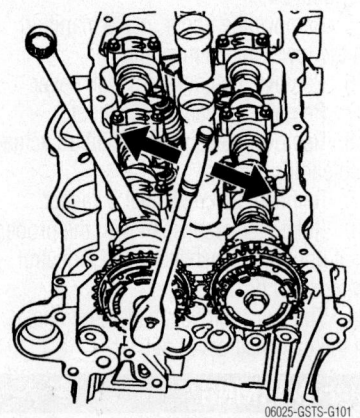

Removing the camshaft position actuator bolt

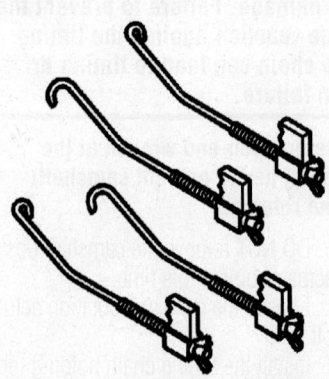

Showing the timing chain holding tools

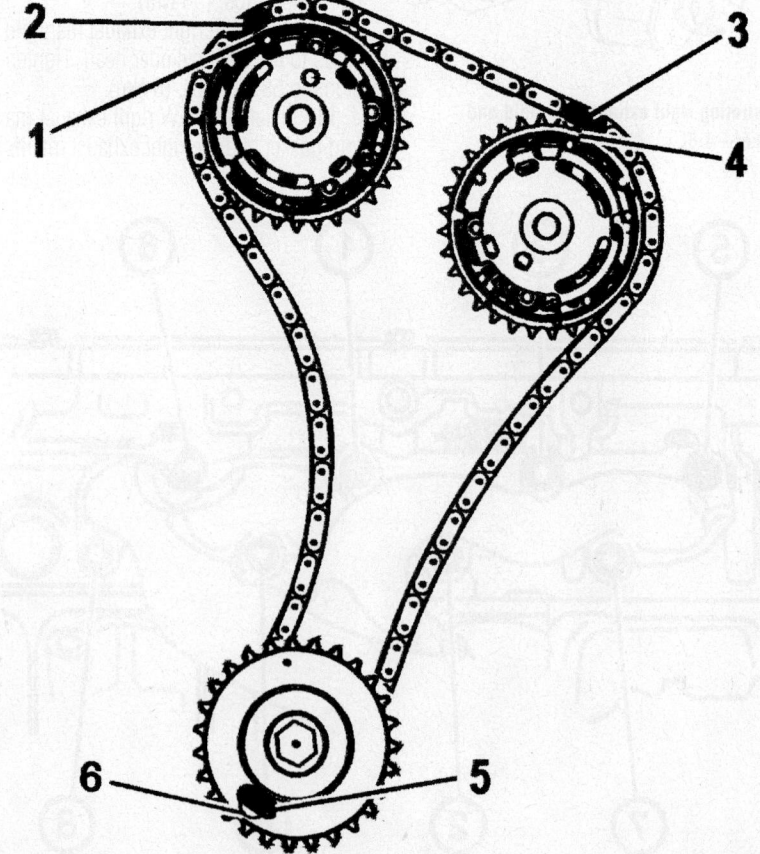

Indicating proper alignment of the timing chain marks

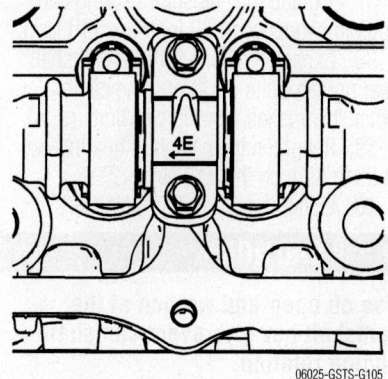

06025-GSTS-G105

Illustrating camshaft bearing cap identification markings

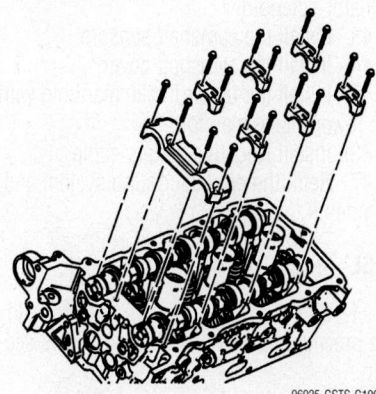

06025-GSTS-G106

Removing the camshaft bearing caps

19. The markings have the following meanings:
- The "I" indicates the intake camshaft.
- The "E" indicates the exhaust camshaft.
- The number indicates the journal position from the front of the engine.

20. Remove the camshaft bearing cap bolts and bearing caps.

21. Remove the camshafts.

22. Replace the camshaft bearing caps and bolts into their installed positions.

To install:

➡ **Ensure that the marks on the camshaft position actuators and the timing chain (15-18) are aligned. DO NOT tighten the camshaft position actuator bolt at this time.**

23. Locate the camshafts to the cylinder head and assemble the camshaft actuators to the camshafts.

24. Ensure that the crankshaft is in the stage one timing drive assembly position.

25. Ensure that the camshaft sealing rings (1) are in place in the camshaft grooves.

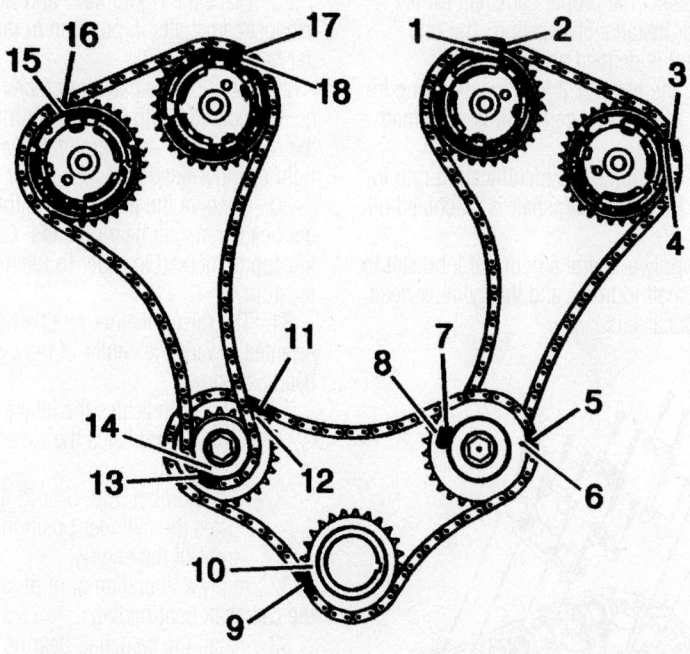

1. Left intake camshaft sprocket timing mark
2. Timing link
3. Left exhaust camshaft sprocket timing mark
4. Timing chain timing link
5. Timing link
6. Intermediate camshaft idler sprocket
7. Sprocket ID
8. Sprocket timing mark to hub
9. Timing link
10. Crankshaft sprocket
11. Timing link
12. Intermediate camshaft idler sprocket
13. Timing link
14. Idler
15. Right exhaust camshaft sprocket timing mark
16. Timing mark
17. Right intake camshaft sprocket timing mark
18. Timing link

06025-GSTS-G107

Matching the camshaft and crankshaft timing chain timing marks—3.6L

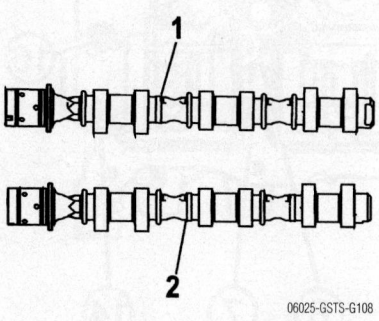

06025-GSTS-G108

Illustrating the positions of the camshaft sealing rings (1, 2)

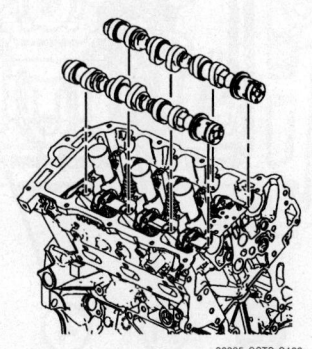

06025-GSTS-G109

Positioning the camshaft into the cylinder head

26. Select the proper camshaft for the particular installation location. The ring placement is defined as follows:

- The number 2 identification ring for the exhaust camshaft is machined off (1).
- The number 3 identification ring for the intake camshaft is machined off (2).

27. Apply a liberal amount of lubricant to the camshaft journals and the cylinder head camshaft carriers.

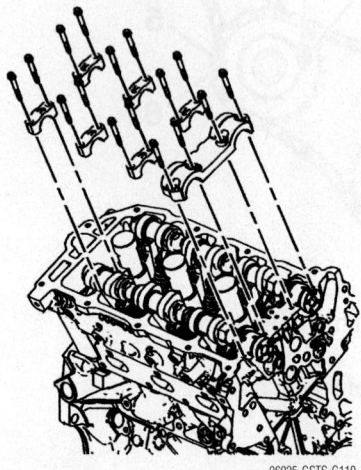

06025-GSTS-G110

Indicating the camshaft bearing cap positions—3.6L

28. Place the right intake and right exhaust camshafts in position in the cylinder head.

29. Position the camshaft lobes in a neutral position with the flats on the back of the camshafts up and parallel (1) with the right cylinder head camshaft cover rail.

30. Observe the markings on the cylinder head camshaft bearing caps. Each bearing cap is marked in order to identify its location.

31. The raised feature must always be oriented toward the center of the cylinder head as follows:

- The "I" indicates the intake camshaft.
- The "E" indicates the exhaust camshaft.
- The number 1, 3, 5 or 2, 4, 6 indicates the cylinder position from the front of the engine.

32. Apply a liberal amount of lubricant to the camshaft bearing caps.

33. Install the camshaft bearing thrust caps in the first journal of the right cylinder head.

34. Install the remaining bearing caps with their orientation mark toward the center of the cylinder head.

35. Hand start all the camshaft bearing cap bolts.

36. Tighten the camshaft bearing cap bolts in the sequence shown.

37. Tighten the camshaft bearing cap bolts in sequence to 89 inch lbs. (10 Nm).

38. Loosen the center intake camshaft bearing cap bolts (1, 2) and the center exhaust camshaft bearing cap bolts (3, 4).

39. Retighten the camshaft bearing cap bolts to 89 inch lbs. (10 Nm).

40. Install the crankshaft balancer.

✸✸ CAUTION

Use an open-end wrench at the camshaft hex to prevent camshaft/engine rotation.

41. Install and tighten the camshaft position actuators

42. Install the intake camshaft position actuator solenoid.

43. Install the camshaft sensors.

44. Install the camshaft cover.

45. Install the upper intake manifold with the lower intake manifold.

46. Install Negative battery cable

47. Refill the engine cooling system and engine oil.

4.6L

1. Before servicing the vehicle, refer to the precautions in the beginning of this section.

2. Properly relieve the fuel system pressure.

3. Disconnect the negative battery cable.

4. Drain the engine cooling system and the engine oil.

5. Remove or disconnect the following:

- Intake manifold
- Front engine cover
- Fuel injector sight shield
- Exhaust camshaft sensors
- Intake camshaft sensors
- Power steering reservoir
- Camshaft covers
- Intake camshaft position actuator solenoid bolts
- Intake camshaft position actuator solenoid
- Intake camshaft position actuator solenoid alignment plug and discard
- Exhaust camshaft position actuator solenoid bolts
- Exhaust camshaft position actuator solenoid
- Exhaust camshaft position actuator solenoid alignment plug and discard
- Camshaft position actuator housing bolts
- Camshaft position actuator housing

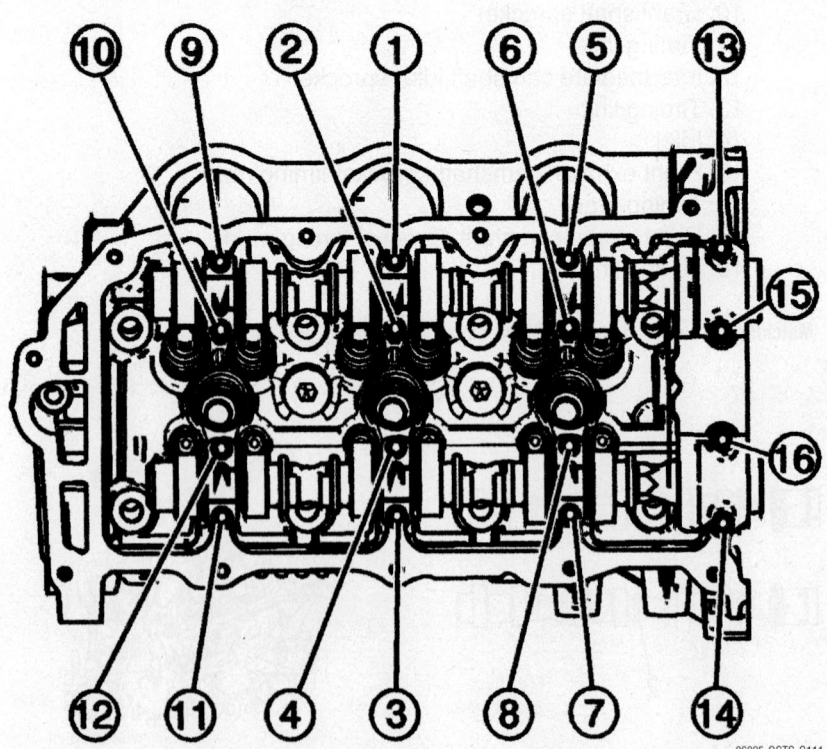

06025-GSTS-G111

Showing the camshaft bearing cap bolt tightening sequence—3.6L

- Intake camshaft position actuator oil control valve

6. Pull the actuator forward to disengage the actuator from the camshaft alignment pin.

7. Disengage the secondary timing chain from the actuator and remove the actuator.

✳✳ WARNING

Bearing caps must remain with their original cylinder head and in their original location. Do not mix bearing caps.

➡**Observe the markings on the bearing caps. Each bearing cap is marked in order to identify its location. The markings have the following meanings: The arrow points toward the front of the engine. The "E" indicates the exhaust camshaft. The "I" indicates the intake camshaft. The number indicates the journal position from the front of the engine.**

8. Remove the camshaft bearing cap bolts, bearing caps, and camshafts.

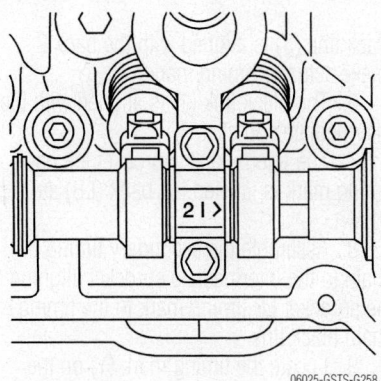

06025-GSTS-G258

Showing the identification numbering of the camshaft bearing caps

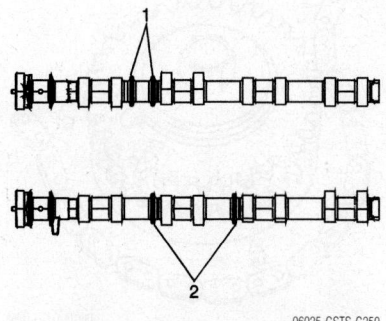

06025-GSTS-G259

Indicating the locations of the camshaft ID rings

To install:

9. Apply a liberal amount of lubricant ((GM P/N 12345001,)) to the camshaft bearing journals.

10. Select the proper camshaft by the identifying rings cast into the camshaft.

11. Apply a liberal amount of lubricant (GM P/N 12345001) to the camshaft lobes and the camshaft journals.

12. Place the camshaft in the camshaft journals with the camshaft sprocket drive pins near the top of their rotation (1) and the camshaft lobes in a neutral position. The camshaft can be identified by a stamping near the rear journal. For example: "L-EXH" is defined as Left Bank Exhaust.

13. Observe the markings on the camshaft bearing caps. Each camshaft bearing cap is marked in order to identify its location. The markings have the following meanings: The arrow should point toward the front of the engine. The number indicates the position from the front of the engine. The "E" indicates the exhaust camshaft.

14. Apply a liberal amount of lubricant (GM P/N 12345001) to the left exhaust camshaft bearing cap journals.

15. Install the left exhaust camshaft bearing caps according to the identifications marks.

16. Hand start all the left exhaust camshaft bearing cap bolts.

✳✳ WARNING

Ensure each valve rocker arm is properly aligned to the valve tip, the valve lifter and the camshaft lobe. Inspect the alignment prior to and after the camshaft caps are tightened to specifications.

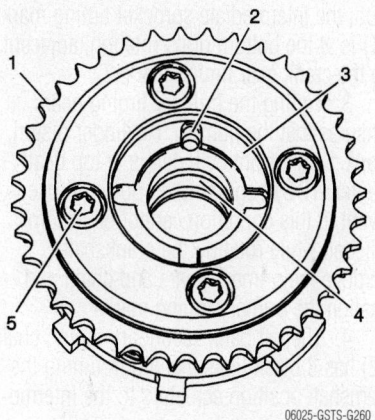

06025-GSTS-G260

Showing the camshaft actuator and the alignment pin hole (2)

17. Install the left cylinder head exhaust camshaft bearing cap bolts, and tighten as follows:

- First Pass, tighten the camshaft bearing cap bolts to 44 inch lbs. (5 Nm)
- Final Pass, tighten the camshaft bearing cap bolts an additional 30 degrees

18. Locate the camshaft alignment pin hole on the back of the actuator. This hole will mate with the camshaft alignment pin.

19. Locate the timing alignment mark on the front of the actuator.

20. Engage the actuator timing gear teeth with the secondary timing chain ensuring that the actuator alignment mark corresponds with the paint mark created on the chain link.

21. Slide the actuator over the camshaft alignment pin and install the actuator onto the camshaft.

22. Install or connect the following:

- Intake camshaft position actuator oil control valve
- NEW gasket to the camshaft position actuator housing
- Camshaft position actuator housing to the cylinder head
- Actuator housing bolts and studs, tighten the actuator housing bolts and studs to 89 lb. in. (10 Nm)

✳✳ CAUTION

The camshaft position actuator solenoids must be precisely aligned to the camshaft position actuator oil control valves on the end of the camshafts. This is accomplished with an alignment pin. Failure to align the camshaft position actuator solenoids to the camshaft position actuator oil control valves can lead to poor engine performance and engine component damage.

23. To allow for camshaft position actuator solenoid to camshaft position actuator oil control valve alignment, DO NOT install NEW camshaft position actuator solenoid alignment plugs at this time.

24. Use the following procedure for the camshaft position actuator solenoid alignment:

a. Make an alignment pin from drill rod 15/64 inch diameter and at least 50 mm (1.97 in) long.

b. Verify that the alignment pin will pass through the camshaft position actuator solenoid alignment hole.

c. Verify that the alignment pin will fit

into the alignment hole in the camshaft position actuator oil control valve.

d. Apply a 2 mm (0.079 in) bead of RTV around the flange of the camshaft position actuator solenoid.

e. Install the camshaft position actuator solenoid over the oil control valve.

f. Install the alignment pin through the solenoid alignment hole and into the oil control valve alignment hole.

g. With the alignment pin in place, install the camshaft position actuator solenoid bolts. Tighten the camshaft position actuator solenoid bolts to 71 inch lbs. (8 Nm).

25. Remove the alignment pin.

26. Install a NEW camshaft position actuator solenoid plug.

27. Repeat the above steps for the exhaust camshaft position actuator solenoid.

28. Set the camshaft timing. See Camshaft Timing.

29. Install or connect the following:
- Intake camshaft sensor
- Exhaust camshaft sensor
- Camshaft cover
- Front engine cover
- Intake manifold
- Negative battery cable

30. Refill the engine cooling system and engine oil.

Camshaft Timing

The crankshaft and camshafts are correctly timed when the crank sprocket and the intermediate shaft sprocket have their timing marks aligned and all 4 camshaft position actuator timing marks are perpendicular (90 degrees) to the cylinder head deck surface at the top of their rotation. The black timing chain links are used for assembly purposes. Do not reference these black links to determine if the engine timing components are properly aligned. The black links will only align with the actuator timing marks and the intermediate shaft sprocket, once every 126 revolutions. For this reason, it is not necessary to have the black secondary chain links aligned with the timing marks when servicing only one of the secondary timing components. Proper alignment is maintained by marking a chain link and sprocket gear tooth with a paint stick prior to disassembling individual components. When servicing individual components, refer to the appropriate procedure.

1. Remove the following components for access to the timing chains:
- Camshaft actuator housing

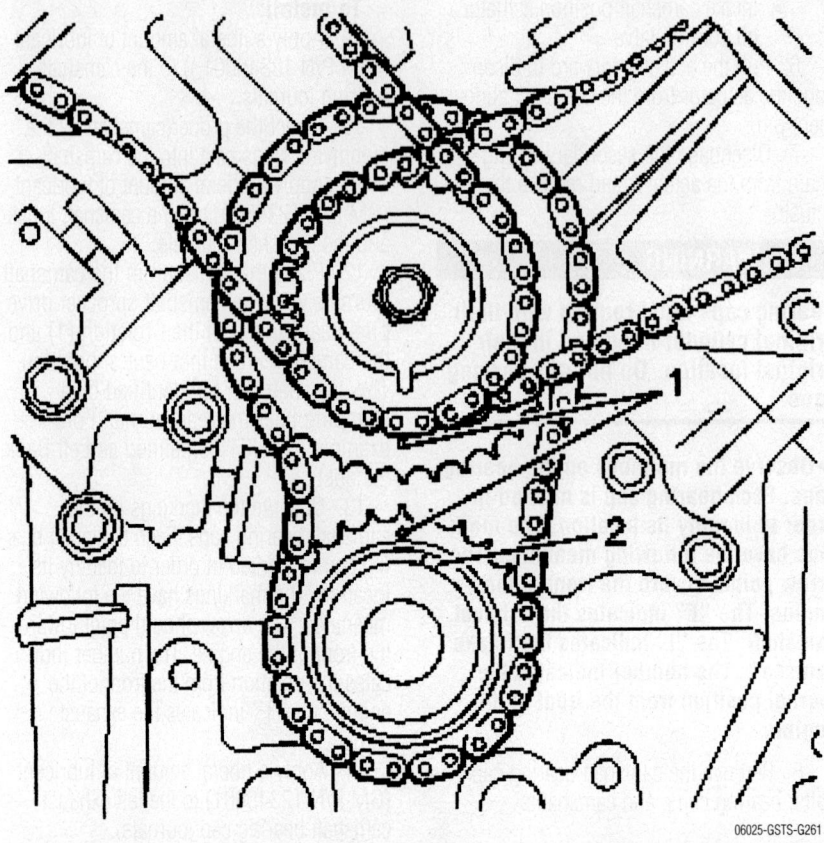

Showing the primary gear alignment, setting timing marks in proper locations

- Engine front cover
- Engine oil pump
- Secondary camshaft drive chain

❊❊ CAUTION

The primary gears and chain must be properly timed prior to setting the left or right secondary camshaft timing.

2. Rotate the crankshaft until the primary timing gear alignment mark (1) is at the top of its position, the crankshaft keyway (2) is approximately at the 1 o'clock position, the intermediate sprocket timing mark (1) is at the bottom of its rotation, adjacent to the crank gear timing mark.

3. Timing the primary timing gears ensures that the number 1 cylinder piston, bank 1 right front cylinder, is at top dead center (TDC) beginning its power stroke event. If this condition cannot be accomplished while rotating the crankshaft, remove the primary gears and chain and realign the primary timing marks.

4. The left bank secondary timing chain (2) has 3 black links that aid in timing the camshaft position actuators to the intermediate sprocket.

5. The black link (4) is aligned with the bank 2 exhaust actuator timing mark. The

black link (3) is aligned with the bank 2 intake actuator timing mark.

6. The black link (1) is aligned with the intermediate sprocket.

7. The intermediate sprocket left bank timing mark is labeled left bank (LB) as shown.

8. Assemble the secondary timing chain to the intermediate sprocket aligning the sprocket LB timing mark to the timing chain black link.

9. Locate the timing mark (1) on the front of the intake actuator marked "LI" which stands for left intake.

10. Align the timing mark of the left bank

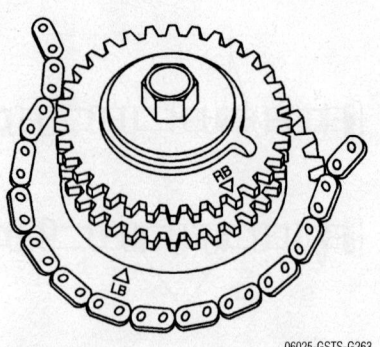

Noting the ID marks on the intermediate sprocket

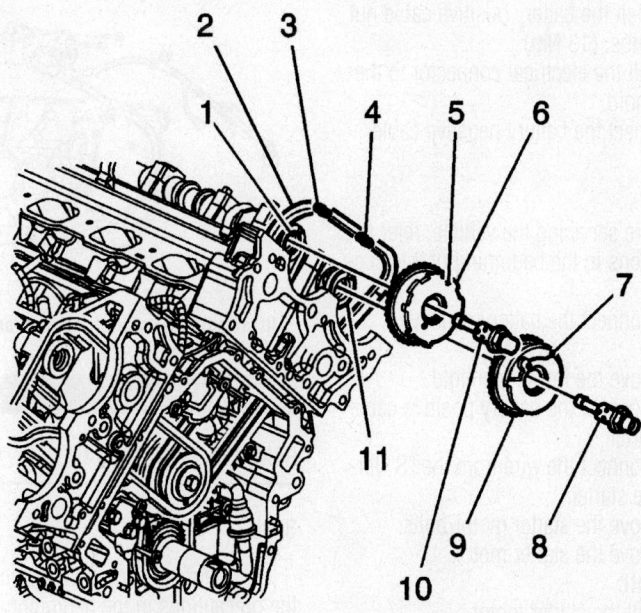

1. Camshaft
2. Backing plate
3. Timing mark
4. Timing mark
5. Actuator wheel
6. Sprocket
7. Actuator wheel
8. Oil control valve
9. Sprocket
10. Oil control valve
11. Camshaft

06025-GSTS-G264

Installing the left bank gear for proper alignment

intake camshaft position actuator with the timing chain black link and install the actuator on the camshaft with the actuator timing mark perpendicular (90 degrees) to the cylinder head deck surface at the top of its rotation.

11. Ensure that the camshaft alignment pin hole on the rear of the actuator engages with the camshaft alignment pin. If necessary, use an open-end wrench on the hex cast into the camshaft to obtain proper camshaft pin to actuator engagement.

12. Loosely install the oil control valve to secure the intake actuator.

✴✴ WARNING

A wrench must be used on the hex of the camshaft when loosening or tightening in order to prevent component damage. Failure to prevent the torque reaction against the timing drive chain can lead to timing drive chain failure.

13. Use an open-end wrench on the hex cast into the camshaft in order to prevent the camshaft from rotating when tightening the oil control valve. Tighten the oil control valve to 89 ft. lbs. (120 Nm).

14. Locate the timing mark on the front of the exhaust actuator marked "LE", which stands for left exhaust.

15. Align the timing mark of the left bank exhaust camshaft position actuator with the timing chain black link and install the actuator on the camshaft with the actuator timing mark perpendicular (90 degrees) to the cylinder head deck surface at the top of its rotation. Ensure that the camshaft alignment pin hole on the rear of the actuator engages with the camshaft alignment pin. If necessary, use an open end wrench on the hex cast into the camshaft to obtain proper camshaft pin to actuator engagement.

16. Loosely install the oil control valve to secure the exhaust actuator.

17. Use an open end wrench on the hex cast into the camshaft in order to prevent the camshaft from rotating when tightening the oil control valve. Tighten the oil control valve to 89 ft. lbs. (120 Nm).

18. Install the left secondary timing chain tensioner.

19. The right bank secondary timing chain has 3 black links that aid in timing the camshaft position actuators to the intermediate sprocket.

20. The black link is aligned with the bank 1 exhaust actuator timing mark. The

black link is aligned with the bank 1 intake actuator timing mark.

21. The black link is aligned with the intermediate sprocket.

22. The intermediate sprocket right bank timing mark is labeled right bank (RB).

23. Assemble the secondary timing chain to the intermediate sprocket aligning the sprocket RB timing mark to the timing chain black link.

24. Locate the timing mark on the intake actuator marked "RI" which stands for right intake.

25. Align the timing mark of the right bank intake camshaft position actuator with the timing chain black link and install the actuator on the camshaft with the actuator timing mark perpendicular (90 degrees) to the cylinder head deck surface near the top of its rotation. Ensure that the camshaft alignment pin hole on the rear of the actuator engages with the camshaft alignment pin. If necessary, use an open end wrench on the hex cast into the camshaft to obtain proper camshaft pin to actuator engagement.

26. Loosely install the oil control valve to secure the intake actuator.

27. Use an open end wrench on the hex cast into the camshaft in order to prevent the camshaft from rotating when tightening the oil control valve. Tighten the oil control valve to 89 ft. lbs. (120 Nm).

28. Locate the timing mark on the actuator marked RE which stands for right exhaust.

29. Align the timing mark of the right bank exhaust camshaft position actuator with the timing chain black link and install the actuator on the camshaft with the actuator timing mark perpendicular (90 degrees) to the cylinder head deck surface near the top of its rotation. Ensure that the camshaft alignment pin hole on the rear of the actuator engages with the camshaft alignment pin. If necessary, use an open end wrench on the hex cast into the camshaft to obtain proper camshaft pin to actuator engagement.

30. Loosely install the oil control valve to secure the exhaust actuator.

31. Use an open-end wrench on the hex cast into the camshaft in order to prevent the camshaft from rotating when tightening the oil control valve. Tighten the oil control valve to 89 ft. lbs. (120 Nm).

32. Install the right secondary timing chain tensioner.

33. Remove the camshaft holding tool (EN 46328).

34. Ensure the correct alignment of all secondary timing components.

35. Ensure the correct alignment of the primary timing drive components.

36. Install the engine oil pump.
37. Install the engine front cover.
38. Install the right camshaft actuator housing.
39. Install the left camshaft actuator housing.

Starter Motor

REMOVAL & INSTALLATION

3.6L

1. Before servicing the vehicle, refer to the precautions in the beginning of this section.
2. Turn the ignition OFF.
3. Disconnect the battery negative cable
4. Remove the starter solenoid electrical connector from the starter.
5. Remove the starter terminal nut and the battery positive cable from the starter.
6. Remove the starter motor bolts.
7. Remove the starter motor.

To install:

8. Install the starter motor.
9. Install the starter motor mounting bolts
10. Tighten the starter motor mounting bolts to 37 ft. lbs. (50 Nm).
11. Install the battery positive cable (2) to the starter and install the starter terminal nut (1).

12. Tighten the battery positive cable nut to 115 inch lbs. (13 Nm).
13. Install the electrical connector to the starter solenoid.
14. Connect the battery negative cable

4.6L

1. Before servicing the vehicle, refer to the precautions in the beginning of this section.
2. Disconnect the battery negative cable.
3. Remove the intake manifold.
4. Disconnect the battery positive cable from the starter.
5. Disconnect the wire from the "S" terminal on the starter.
6. Remove the starter motor bolts.
7. Remove the starter motor.

To install:

8. Install the starter motor.
9. Install the starter motor bolts.
10. Tighten the bolts to 22 ft. lbs. (30 Nm).
11. Connect the wire and nut to the "S" terminal on the starter.
12. Tighten the nut to 35 inch lbs. (4 Nm).
13. Connect the battery positive cable and nut to the starter terminal.
14. Tighten the nut to 89 inch lbs. (10 Nm).
15. Install the intake manifold.
16. Connect the battery negative cable

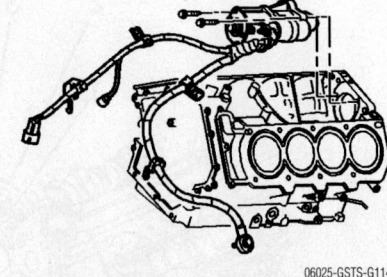

06025-GSTS-G114

Illustrating the starter location on the 4.6L

Oil Pan

REMOVAL & INSTALLATION

3.6L

1. Before servicing the vehicle, refer to the precautions in the beginning of this section.
2. Drain the engine oil.
3. Remove the engine front cover
4. Remove the upper intake manifold with the lower intake manifold.
5. Remove the camshaft covers.
6. Drain the engine coolant.
7. Disconnect the purge vent hose from the water outlet.
8. Remove the water outlet with the radiator hose and reposition aside.
9. Remove the accessory drive belts.
10. Remove the A/C compressor and power steering belt tensioner.
11. Remove the alternator bracket with the alternator and the belt tensioner.

➡**Do not disconnect the power steering pipes or drain the power steering fluid.**

12. Remove the power steering fluid reservoir and reposition the power steering fluid reservoir in order to provide access.
13. Remove the power steering pump pulley.
14. Remove the power steering pump upper front bolt and loosen the remaining two bolts.
15. Remove the crankshaft balancer.
16. Remove the oil control valves.
17. Remove the engine front cover with the water pump.
18. Remove the power steering hose retainer from the air conditioning (A/C) compressor bracket.
19. Disconnect the intermediate steering shaft.
20. Remove the engine mount lower nuts.
21. Remove the A/C compressor bracket bolts and reposition aside.

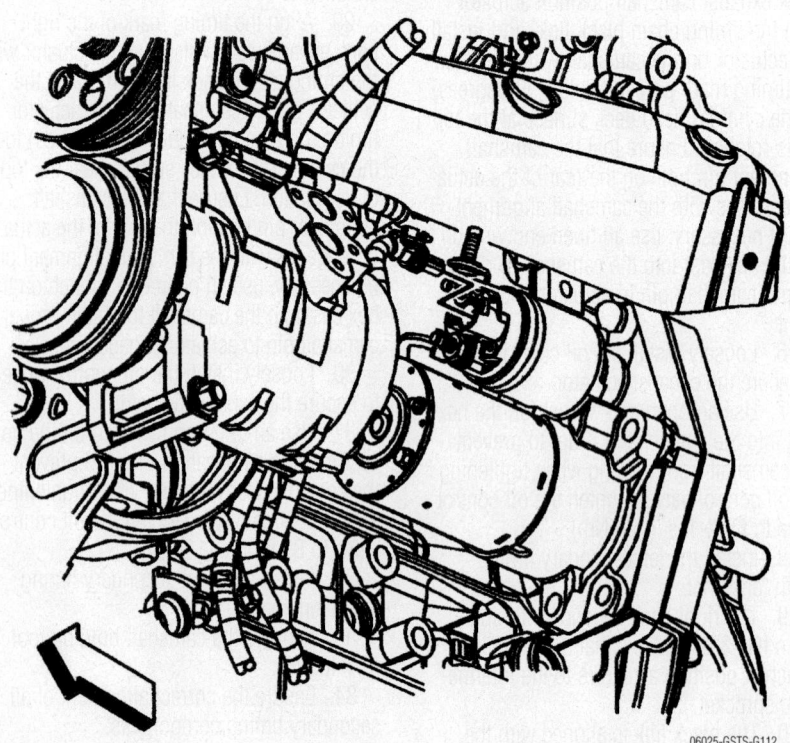

06025-GSTS-G112

Showing the starter location on the 3.6L

06025-GSTS-G116

Illustrating the front cover assembly with water pump—3.6L

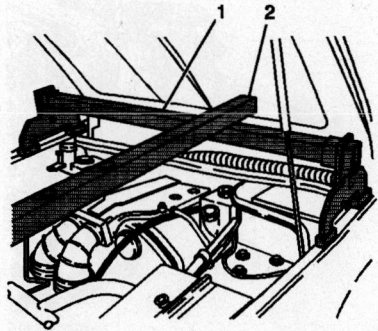

06025-GSTS-G118

Installing the engine support fixture

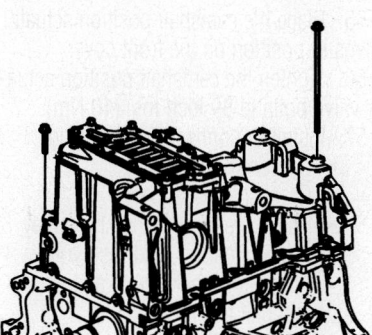

06025-GSTS-G119

Identifying the oil pan bolts for removal—3.6L

➡ **Do NOT disconnect the A/C pipes and/or hoses.**

22. Drain the engine oil.

23. Remove the transmission oil cooler pipe retainer from the engine right side.

24. Install the engine support fixture.

25. Tighten the support fixture wing nuts in order to provide clearance for the oil pan.

26. Remove the front differential carrier, if equipped (AWD).

27. Remove the oil pan bolts.

28. Using the pry points located at the edge of the oil pan shear the RTV sealant.

29. Remove the oil pan from the block.

To install:

30. Install the 0.315 inch (8mm) guides into the center oil pan rail bolt hole on each side of the engine block.

31. Place a 0.118 inch (3mm) bead of RTV sealant, on the block pan rail and the crankshaft rear oil seal housing

32. Position the oil pan onto the block.

33. Remove the guides from the engine block.

34. Loosely install the oil pan bolts. Then, tighten the oil pan bolts in sequence shown, as follows:

- 8 mm bolts (1-11) to 17 ft. lbs. (23 Nm)

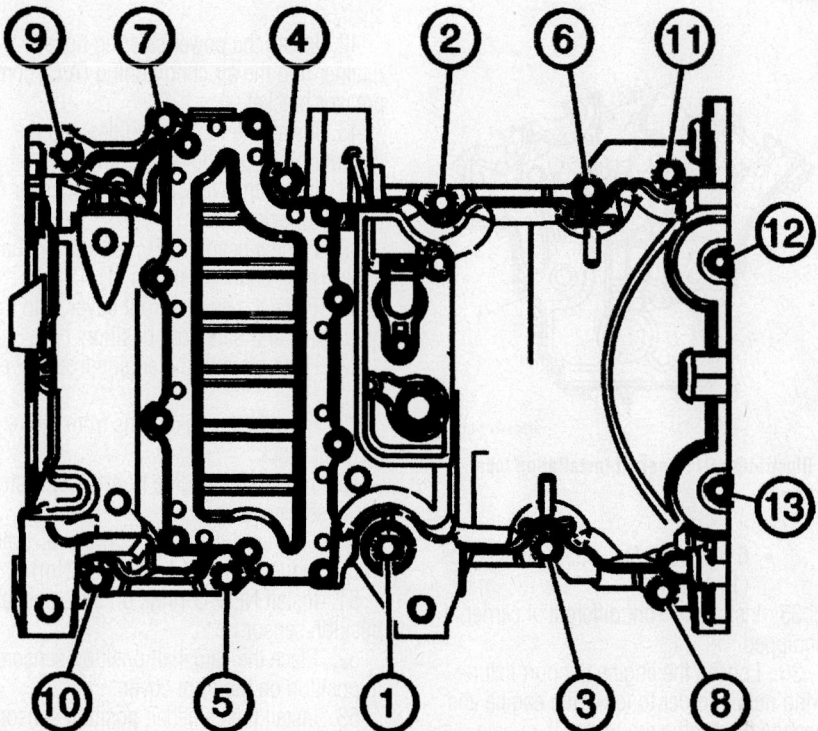

06025-GSTS-G127

Illustrating the oil pan bolt tightening sequence—3.6L

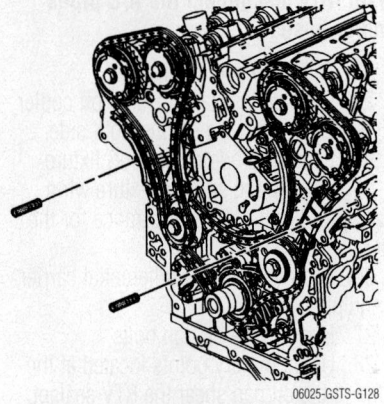

Showing the locations of the engine block guides for front cover installation

06025-GSTS-G128

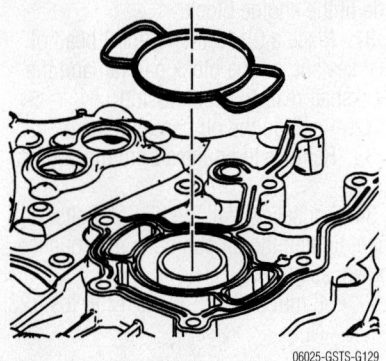

Illustrating front cover to block seal location

06025-GSTS-G129

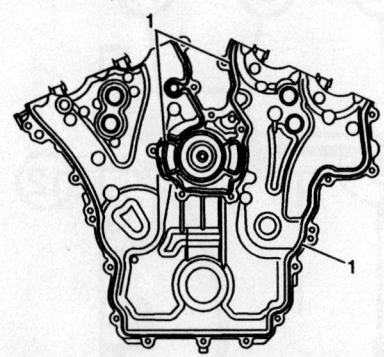

Illustrating RTV sealant installation location

06025-GSTS-G130

• 6 mm bolts (12, 13) to 89 inch lbs. (10 Nm)

35. Install the front differential carrier, if equipped.

36. Loosen the engine support fixture wing nuts in order to lower the engine and engage the engine mounts to the frame.

37. Remove the engine support fixture.

38. Install the transmission oil cooler pipe retainer on the engine right side.

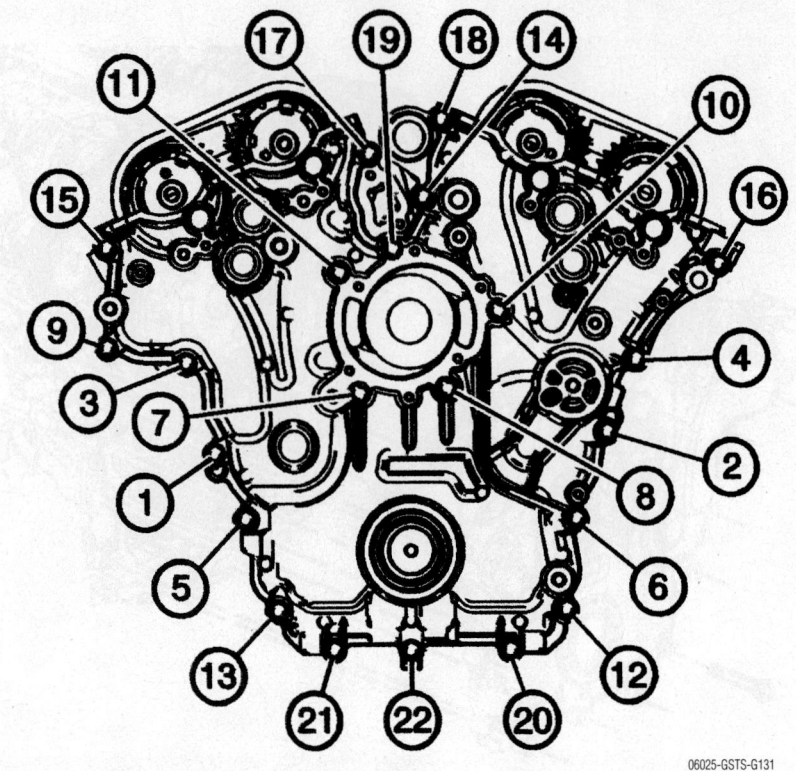

Indicating the locations of the front cover bolts

06025-GSTS-G131

39. Install the A/C compressor bracket bolts and reposition aside.

40. Install the engine mount lower nuts.

41. Install the intermediate steering shaft.

42. Install the power steering hose retainer into the air conditioning (A/C) compressor bracket.

43. Install the locating guides into the cylinder block positions as shown.

44. Install the NEW engine front cover to cylinder block seal.

45. Place a bead of RTV sealant, on the engine front cover as shown (1).

46. Place the engine front cover onto the guide pins and slide into position.

47. Hand start all the engine front cover bolts.

48. Remove the guide pins from the cylinder block.

49. Tighten the engine front cover bolts in the sequence shown.

50. Tighten the engine front cover bolts in the sequence to 17 ft. lbs. (23 Nm).

51. Install NEW O-rings on the camshaft position sensor.

52. Place the camshaft position sensors in position on the front cover.

53. Install the camshaft position sensor bolts.

54. Tighten the camshaft position sensor bolts to 89 inch lbs. (10 Nm).

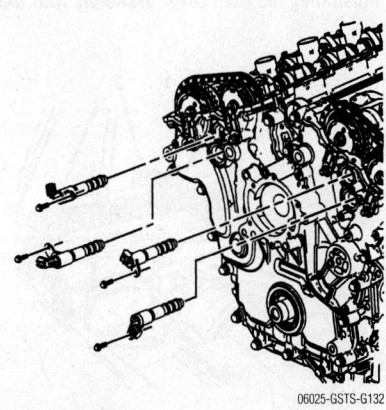

Illustrating the camshaft position actuator valve locations

06025-GSTS-G132

55. Place the camshaft position actuator valves in position on the front cover.

56. Tighten the camshaft position actuator valve bolts to 89 inch lbs. (10 Nm).

57. Install or connect the following:

• Oil control valves
• Crankshaft balancer
• Power steering pump and pulley
• Power steering fluid reservoir
• Alternator bracket with alternator and belt tensioner
• Accessory drive belts
• Water outlet
• EVAP purge vent hose to the water outlet

58. Fill the cooling system.
59. Install the camshaft covers.
60. Install the upper intake manifold with the lower intake manifold.
61. Install the engine cover.
62. Change the engine oil.

4.6L

1. Before servicing the vehicle, refer to the precautions in the beginning of this section.
2. Install the engine support fixture.

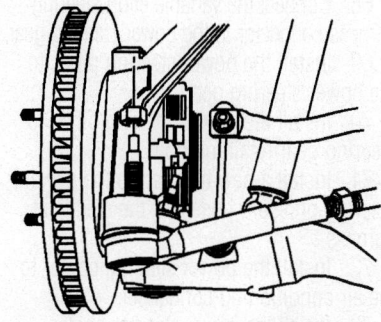

06025-GSTS-G120

Illustrating brake rotor, tie rod and steering knuckle

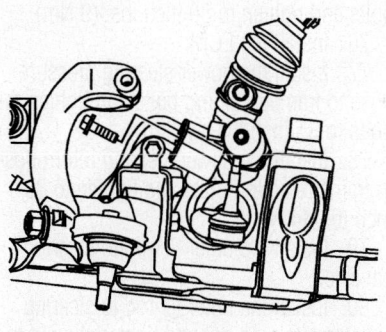

06025-GSTS-G121

Illustrating the power steering gear pipe retainer bolt

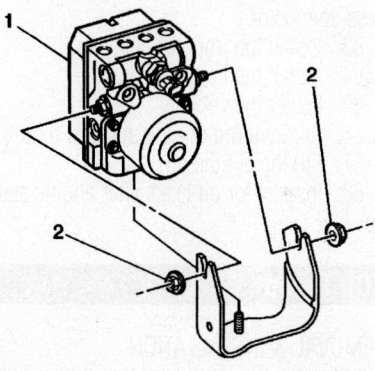

06025-GSTS-G122

Indicating the location of the ABS module retaining bolts and bracket

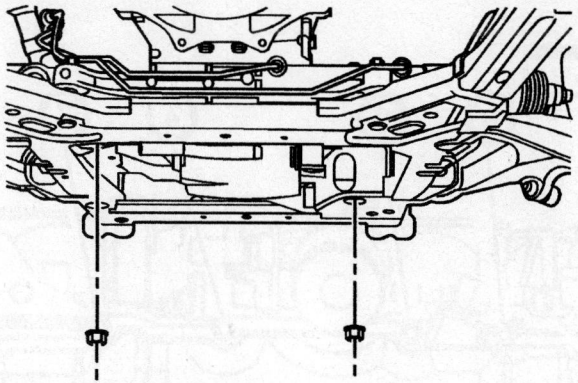

06025-GSTS-G123

Illustrating frame and engine mount retaining nuts

3. Drain the engine oil.
4. Raise and support the vehicle.
5. Remove the front wheels.
6. Remove the front air deflector.
7. Disconnect the wheel speed sensor harness connectors on the front brakes.
8. Disconnect the wheel speed sensor harness from the lower control arm.
9. Remove the outer tie rod to steering knuckle retaining nut.
10. Separate the outer tie rod from the steering knuckle. See Ball Joints.
11. Remove the power steering return hose to frame retaining bolts.
12. Remove the engine control module (ECM), bracket retaining bolts, and bracket.
13. Remove the power steering cooler bracket to air conditioning condenser retaining bolts.
14. Remove the power steering cooler from the air conditioning condenser.
15. Remove the tie strap from the power steering pressure hose to the electrical harness.
16. Remove the power steering pipes to steering gear retaining bolt.
17. Disconnect the power steering pipes from the power steering gear.
18. Disconnect the variable effort steering electrical connector from the power steering gear.
19. Remove the intermediate shaft to power steering gear pinch bolt.
20. Disconnect the intermediate shaft from the power steering gear.
21. Disconnect the brake lines from the brake hoses
22. Loosen the ABS module retaining nuts (2, 3).
23. Disconnect the ABS module harness connector.
24. Remove the upper ball joint retaining nut and separate the ball stud from the steering knuckle. See Ball Joint.

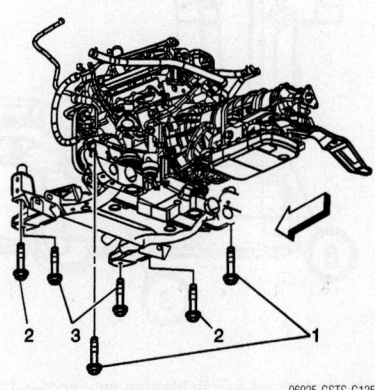

06025-GSTS-G125

Showing the location of the frame mounting bolts

> ※※ **WARNING**
>
> **Do not free the ball stud by using a pickle fork or a wedge-type tool. Damage to the seal or bushing may result.**

25. Remove the yoke to shock retaining nut.
26. Remove the washer bottle to knuckle retaining bolts.
27. Remove the engine mount lower retaining nuts.
28. Lower the vehicle to a frame support table.
29. Remove the frame mounting bolts (1-3).
30. With the aid of an assistant, carefully raise the body from the frame.
31. Ensure when raising the body the following items are clear from the frame:
- Brake pipes
- Steering knuckle assembly
- Wheel speed sensor electrical harness
32. Remove the lower control arms. See Lower Control Arms.

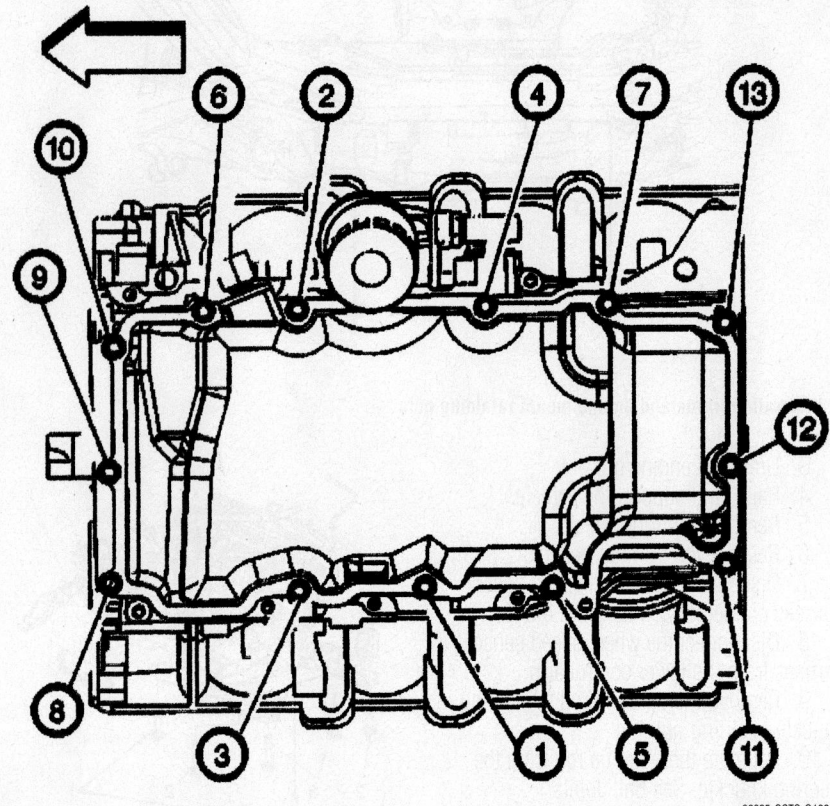

Illustrating oil pan tightening sequence—4.6L

06025-GSTS-G133

33. Remove the stabilizer shaft. See Stabilizer Bar.

34. With the aid of an assistant, remove the frame from the support fixture.

35. Remove the front differential carrier, if equipped (AWD).

36. Disconnect the electrical connector from the engine oil level sensor.

37. Remove the engine oil level sensor from the oil pan.

38. Remove the oil pan bolts.

39. Remove the oil pan.

To install:

40. Install a new oil pan seal.

41. Clean any residual oil from the seal groove.

42. Work the seal into the pan groove in both directions around the pan.

43. Position the oil pan to the crankcase.

44. Install the oil pan bolts. Tighten the bolts, in the sequence shown, as follows:

- First pass to 11 ft. lbs. (15 Nm)
- Final pass to 18 ft. lbs. (25 Nm)

45. Install the engine oil level sensor into the oil pan.

46. Tighten the sensor to 89 inch lbs. (10 Nm).

47. Connect the electrical connector to the engine oil level sensor.

48. Install the front differential carrier, if equipped.

49. Install the front frame.

50. With the aid of an assistant, install the frame to the support table.

51. Install the stabilizer shaft. See Stabilizer Bar.

52. Install the lower control arms. See Lower Control Arm.

53. With the aid of an assistant, carefully lower the body to the frame.

54. When lowering the body, ensure the brake pipes, steering knuckle and wheel speed sensor electrical harness each are clear from the frame.

55. Install the frame mounting bolts (1-2) and tighten to 141 ft. lbs. (191 Nm).

56. Install the frame mounting bolts (3) and tighten to 185 ft. lbs. (250 Nm).

57. Raise the vehicle from the support table.

58. Install the engine mount lower retaining nuts.

59. Tighten the nuts to 59 ft. lbs. (80 Nm).

60. Install the washer bottle to knuckle retaining bolts and tighten to 53 inch lbs. (6 Nm).

61. Install the yoke to shock retaining bolt and tighten the nut to 133 ft. lbs. (180 Nm).

62. Install the upper ball joint retaining

nut and tighten to 15 ft. lbs. (20 Nm), plus 210 degrees.

63. Tighten the ABS module retaining nuts to 71 inch lbs. (8 Nm).

64. Connect the ABS module harness connector.

65. Connect the brake lines to the brake hoses and tighten the fittings to 20 ft. lbs. (27 Nm).

66. Connect the intermediate shaft to the power steering gear.

67. Install the intermediate shaft to power steering gear retaining bolt and tighten to 37 ft. lbs. (50 Nm).

68. Connect the variable effort steering harness connector to the power steering gear.

69. Install the power steering pipes to the power steering gear.

70. Install the power steering pipes to steering gear retaining bolt.

71. Install a new tie strap to the power steering pressure hose and the electrical harness.

72. Install the power steering cooler to the air conditioning condenser.

73. Install the power steering cooler bracket to air conditioning condenser retaining bolts and tighten to 80 inch lbs. (9 Nm).

74. Install the ECM bracket.

75. Install the ECM bracket retaining bolts and tighten to 80 inch lbs. (9 Nm).

76. Install the ECM.

77. Install the power steering pressure hose to frame retaining bolts and tighten the bolts to 35 inch lbs. (4 Nm).

78. Install the power steering return hose to frame retaining bolts and tighten to 35 inch lbs. (4 Nm).

79. Install the outer tie rod to steering knuckle.

80. Install the outer tie rod to steering knuckle retaining nut and tighten the nut to 52 ft. lbs. (70 Nm).

81. Connect the wheel speed sensor harness to the lower control arm.

82. Connect the wheel speed sensor harness connector.

83. Install the front air deflector.

84. Install the front wheels.

85. Lower the vehicle.

86. Remove the engine support fixture.

87. Fill the engine oil.

88. Inspect for oil leaks after engine start up.

Oil Pump

REMOVAL & INSTALLATION

1. Before servicing the vehicle, refer to the precautions in the beginning of this section.

3.6L Engine

1. Remove the primary timing chain.
2. Remove the crankshaft sprocket.
3. Remove the oil pump bolts and the oil pump.
4. Clean and inspect the oil pump

➡**Do not remove the left bank idler sprocket.**

To install:

5. Install the oil pump.
6. Align the oil pump gerotor with the crankshaft flats and install the oil pump to the engine block.
7. Align the pump body with the mounting holes in the cylinder block.
8. Install the oil pump bolts.
9. Tighten the oil pump bolts to 17 ft. lbs. (23 Nm).
10. Install the crankshaft sprocket.
11. Install the primary timing chain.

4.6L Engine

1. Before servicing the vehicle, refer to the precautions in the beginning of this section.
2. Remove the engine front cover.

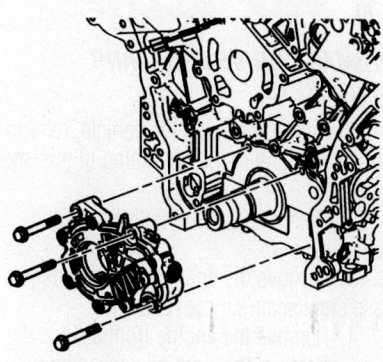

Illustrating oil pump mounting bolt location—3.6L

Illustrating oil pump mounting bolt location—4.6L

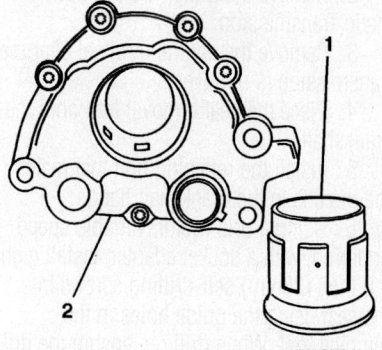

06025-GSTS-G136

Identifying the oil pump drive spacer (1) —4.6L

3. Remove the 3 oil pump assembly retaining bolts (1, 2, 3) identified by the larger head size.
4. Slide the oil pump assembly off the nose of the crankshaft with the drive collar in place.
5. Clean and inspect the oil pump.

To install:

6. Install the oil pump drive spacer (1) into the oil pump (2) so that the drive flat engages the pump rotor.
7. Position the oil pump on the crankshaft.
8. Install the retaining bolts.
9. Apply upward pressure on the pump while tightening the three retaining bolts. Tighten the bolts, in the sequence (1, 2, 3) shown, as follows:
- First Pass Tighten the oil pump mounting bolts in sequence to 89 inch lbs. (10 Nm).
- Final Pass Tighten the oil pump mounting bolts in sequence an additional 35 degrees.
- Install the engine front cover.

Rear Main Seal

REMOVAL & INSTALLATION

3.6L Engine

1. Before servicing the vehicle, refer to the precautions in the beginning of this section.
2. Remove the negative battery cable.
3. Remove the transmission. See Automatic Transmission.
4. Remove the engine flywheel after the transmission is removed.
5. Remove the oil pan.
6. Using the pry points located at the edge of the crankshaft rear oil seal housing shear the RTV sealant.
7. Remove and discard the crankshaft rear oil seal housing.

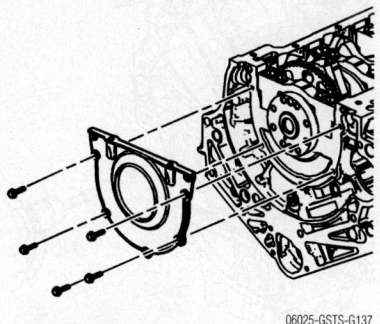

06025-GSTS-G137

Illustrating crankshaft rear oil seal and housing location—3.6L

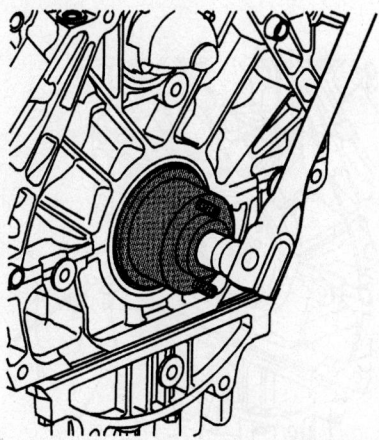

06025-GSTS-G142

Using the crankshaft rear oil seal installation tool

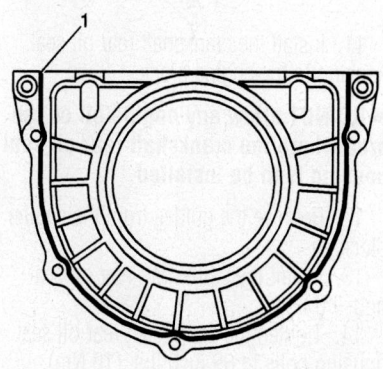

06025-GSTS-G144

Showing the area on the crankshaft rear oil seal housing for sealant application

To install:

8. Install two 0.236 inch (6mm) guides into 2 crankshaft rear oil seal housing corner bolt holes of the engine block.
9. Install the oil seal installation tool onto the rear of the crankshaft flange.
10. Place a 0.118 inch (3mm) bead of RTV sealant, to the NEW crankshaft rear oil seal housing as shown (1).

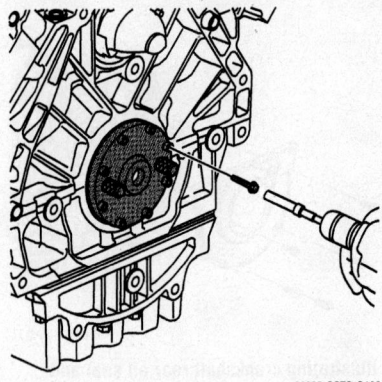

Installing the rear oil seal removal tool screws

06025-GSTS-G139

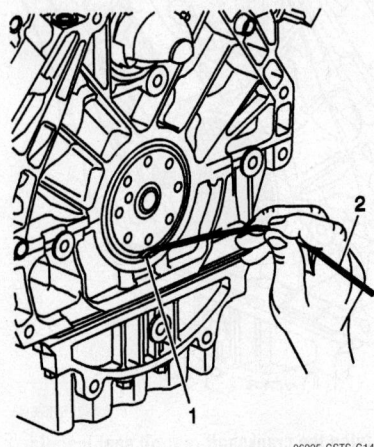

Cleaning debris from the rear oil seal drain location

06025-GSTS-G140

11. Install the crankshaft rear oil seal housing to the engine block.

➡ **DO NOT allow any engine oil on the area where the crankshaft rear oil seal housing is to be installed.**

12. Remove the guides from the engine block.

13. Install the crankshaft rear oil seal housing bolts.

14. Tighten the crankshaft rear oil seal housing bolts to 89 inch lbs. (10 Nm).

15. Remove the seal installation tool from the crankshaft flange.

16. Install the oil pan.

17. Install the engine flywheel.

18. Install the transmission.

19. Install the negative battery cable.

20. Start the engine and check for leaks.

4.6L Engine

1. Before servicing the vehicle, refer to the precautions in the beginning of this section.

2. Remove the transmission. See Automatic Transmission.

3. Remove the engine flywheel after the transmission is removed.

4. Place the seal removal tool onto the crankshaft.

5. Install the retaining bolt that match the threads in the crankshaft flange.

6. Using a drill motor, variable speed preferred, with a socket adapter, install eight 1,0 inch (25mm) self-drilling screws into the seal using the guide holes in the removal tool. When drilling, ensure the drill motor speed is reduced when the screw begins threading into the seal.

7. With all 8 removal screws installed, remove the retaining bolts.

8. Install the center forcing screw.

9. Tighten the center screw on the seal pulling tool to pull the seal assembly off the end of the crankshaft.

To install:

10. Clean any debris from the crankshaft rear oil seal drain (1) using wire or an unbound plastic tie-wrap (2).

❋❋ WARNING

Make sure the drain is clear before installing the new crankshaft rear oil seal. Failure to clear the drain could cause the crankshaft rear oil seal to leak.

11. Place a small amount of Gasket Maker, at the crankcase split line across the end of the upper/lower crankcase seal.

12. Coat the outer diameter of the cylinder block crankshaft rear oil seal area with clean engine oil.

❋❋ WARNING

DO NOT allow any engine oil on the area where the crankshaft rear oil seal is to be pressed onto the crankshaft. The green coating pre-applied to the inner diameter of the crankshaft rear oil seal must not be contaminated.

13. Wipe the outer diameter of the flywheel flange clean with a lint-free cloth.

14. Lubricate the outer rubber surface of the crankshaft rear oil seal with clean engine oil.

15. Install the proper sized bolts into the seal installation tool, that matches the threads in the crankshaft flange.

16. Install the seal and tool to the rear of the crankshaft.

17. Thread the mounting bolts into the crankshaft flange.

18. Tighten the mounting bolts until the tool is firmly mounted on the crankshaft.

19. Install the crankshaft rear oil seal by tightening the center bolt until the tool bottoms against the crankcase.

20. Loosen the center bolt to release pressure on the crankcase.

21. Loosen the mounting bolts.

22. Remove the installation tool from the crankshaft flange.

23. Inspect to ensure the installation depth is equal around the crankshaft rear oil seal's circumference. If the depth is not equal reinstall the installation tool and repeat the installation procedures.

24. Install the flywheel.

25. Install the transmission. See Automatic Transmission.

26. Install the negative battery cable.

27. Start the engine and check for leaks.

Timing Chain, Sprockets, Front Cover and Seal

REMOVAL & INSTALLATION

3.6L

PRIMARY DRIVE CHAIN AND SPROCKETS

1. Before servicing the vehicle, refer to the precautions in the beginning of this section.

2. Disconnect the negative battery cable.

3. Remove the spark plugs in order to ease crankshaft/engine rotation.

4. Remove the engine front cover

5. Remove the right bank secondary camshaft drive chain tensioner.

6. Remove the right bank secondary camshaft drive chain shoe.

7. Remove the right bank secondary camshaft drive chain guide

8. Remove the right bank secondary camshaft drive chain.

9. Remove the primary camshaft drive chain tensioner.

10. Remove the primary camshaft drive chain upper guide

11. Remove the primary camshaft timing chain.

To install:

➡ **Ensure that the crankshaft is in the stage one timing drive assembly position.**

12. Install the primary camshaft drive chain.

06025-GSTS-G145

Removing/installing the primary camshaft chain

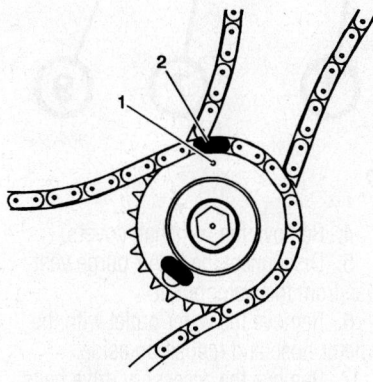

06025-GSTS-G165

Positioning the left intermediate drive chain alignment marks—3.6L

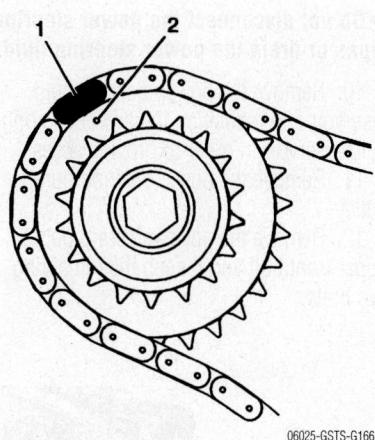

06025-GSTS-G166

Aligning the right intermediate drive chain timing marks—3.6L

13. Wrap the primary camshaft drive chain around the large sprockets of each camshaft intermediate drive chain idler and the crankshaft sprocket.

14. The left camshaft intermediate drive chain idler timing mark (1) will align with a timing camshaft drive chain link (2).

15. The right camshaft intermediate drive

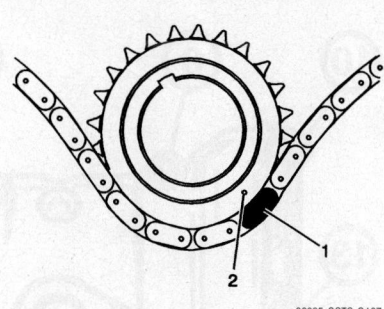

06025-GSTS-G167

Set the timing link opposite the crankshaft sprocket timing mark (2)—3.6L

chain idler timing mark (2) will align with a timing camshaft drive chain link (1).

16. The crankshaft sprocket timing mark (2) will align with a timing camshaft drive chain link (1).

17. Ensure all the timing marks (2, 3, 6) are properly aligned with the timing camshaft drive chain links (1, 4, 5).

18. Install the primary camshaft timing chain.

19. Install the primary upper camshaft drive chain guide.

20. Install the primary camshaft drive chain tensioner.

21. Install the right bank secondary camshaft drive chain.

22. Install the right bank secondary camshaft drive chain guide.

23. Install the right bank secondary camshaft drive chain shoe.

24. Install the right bank secondary camshaft drive chain tensioner.

25. Install the engine front cover.

26. Install the spark plugs.

CAMSHAFT COVER

1. Before servicing the vehicle, refer to the precautions in the beginning of this section.

2. Remove the engine cover.

3. Remove the upper intake manifold with the lower intake manifold.

4. Disconnect the ignition coil electrical connectors.

5. Remove the wiring harness from the side of the camshaft cover by sliding the conduit down and outboard.

6. Remove the wiring conduit retainers from the camshaft cover by rotating the wiring harness conduit retainers counter-clockwise.

7. Remove the wiring harness from the front of the camshaft cover.

8. Reposition and secure the wiring harnesses away from the camshaft cover in order to provide clearance.

9. Remove the ignition coils.

10. Remove the camshaft cover.

11. Remove and discard the camshaft cover seal and grommets.

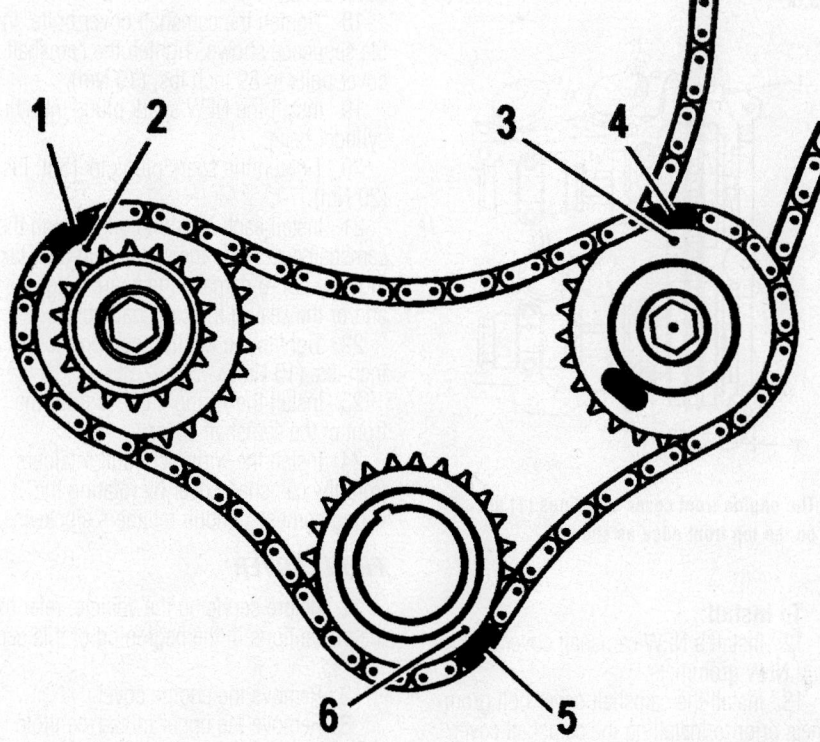

06025-GSTS-G168

Ensure all timing marks are properly aligned—3.6L

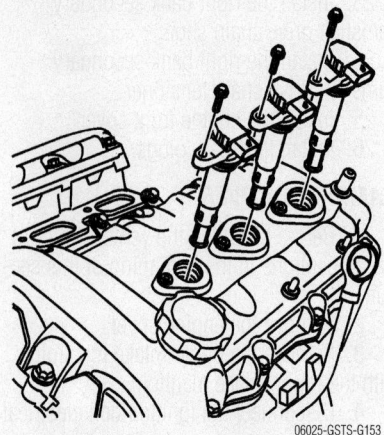

Showing the ignition coils on the left bank—3.6L

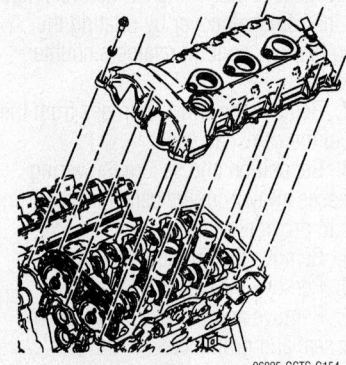

Removing/installing the camshaft cover—3.6L

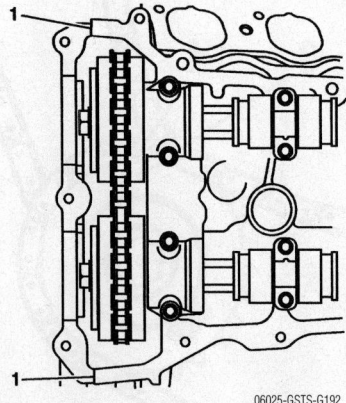

The engine front cover split lines (1) are on the top front edge as shown

To install:

12. Install a NEW camshaft cover seal and NEW grommets.

13. Install the camshaft cover bolt grommets prior to installing the camshaft cover bolts.

14. Wipe the camshaft cover sealing sur-

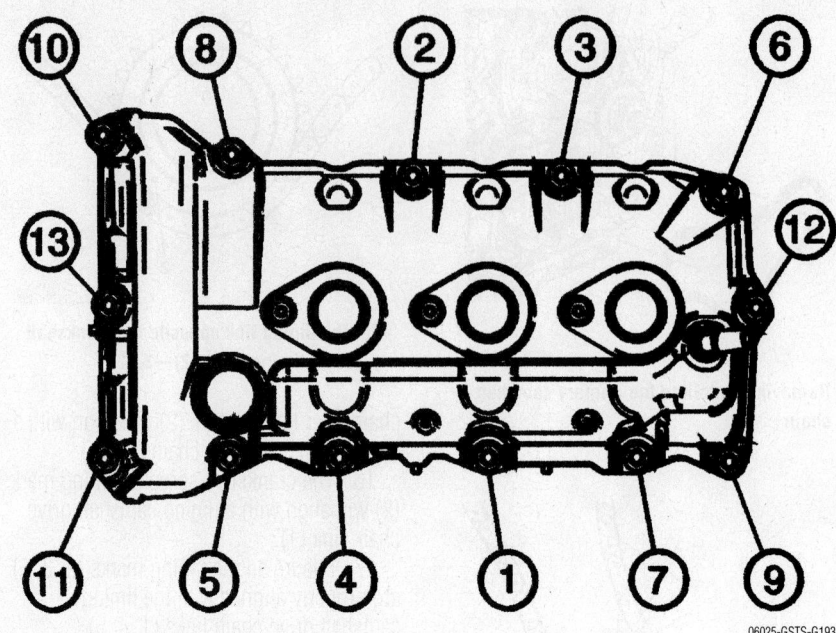

Showing the camshaft cover bolt tightening sequence

face on the cylinder head with a clean, lint-free cloth.

15. Place a bead 8 mm (0.3150 in) in diameter by 0.1575 in (4mm) in height of RTV sealant, on the engine front cover split lines (1).

16. Place the camshaft cover into position onto the cylinder head.

17. Loosely install the left camshaft cover bolts.

18. Tighten the camshaft cover bolts, in the sequence shown. Tighten the camshaft cover bolts to 89 inch lbs. (10 Nm).

19. Install the NEW spark plugs into the cylinder head.

20. Tighten the spark plugs to 15 ft. lbs. (20 Nm).

21. Install each ignition coil through the camshaft cover into the spark plug tube taking care not to damage the spark plug and/or the seal in the camshaft cover.

22. Tighten the ignition coil bolt to 89 inch lbs. (10 Nm).

23. Install the wiring harness from the front of the camshaft cover.

24. Install the wiring conduit retainers from the camshaft cover by rotating the wiring harness conduit retainers clockwise.

FRONT COVER

1. Before servicing the vehicle, refer to the precautions in the beginning of this section.

2. Remove the engine cover.

3. Remove the upper intake manifold with the lower intake manifold. See Intake Manifold.

4. Remove the camshaft covers.

5. Disconnect the EVAP purge vent hose from the water outlet.

6. Remove the water outlet with the radiator hose and reposition aside.

7. Remove the accessory drive belts.

8. Remove the A/C compressor and power steering belt tensioner.

9. Remove the alternator bracket with the alternator and the belt tensioner.

➡ **Do not disconnect the power steering pipes or drain the power steering fluid.**

10. Remove the power steering fluid reservoir and reposition the power steering fluid reservoir in order to provide access.

11. Remove the power steering pump pulley.

12. Remove the power steering pump upper front bolt and loosen the remaining two bolts.

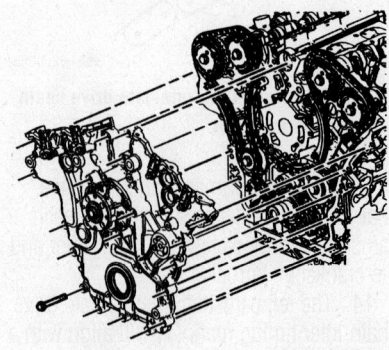

Removing the timing chain front cover

13. Remove the crankshaft balancer.

14. Remove the oil control valves.

15. Remove the camshaft position actuator valve bolts.

16. Remove the camshaft position actuator valves from the front cover.

17. Remove the engine front cover bolts.

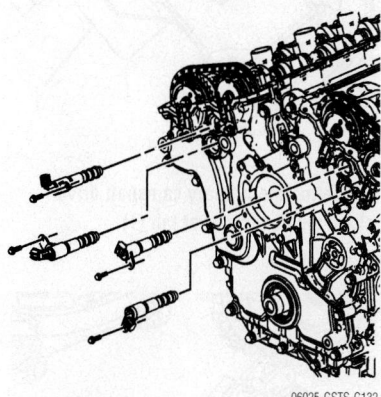

06025-GSTS-G132

Showing the camshaft position actuator valve locations

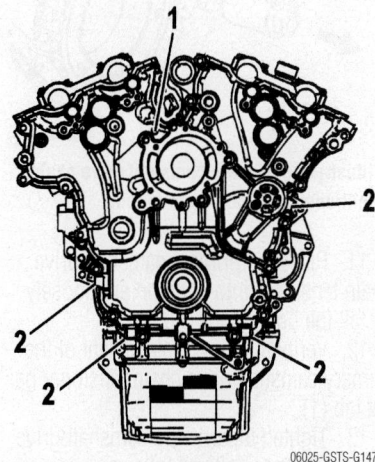

06025-GSTS-G147

Illustrating front cover prying locations (2) and jackscrew hole (1)

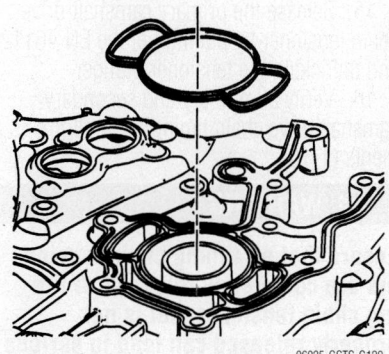

06025-GSTS-G129

Installing the front cover-to-block seal

✳✳ WARNING

Do not pry between the engine front cover and the camshaft position sensors or the camshaft position actuators in order to shear the RTV. Use the pry points and a bolt in the jackscrew hole in order to remove the engine front cover. Damage to the camshaft position sensors or the camshaft position actuators may occur if the camshaft position sensors or the camshaft position actuators are used to pry against in order to remove the engine front cover.

18. Loosely install a 10 x 1.5 mm bolt in the jackscrew hole (1).

19. Using the pry points (2) located at the edge of the front cover and the jackscrew, shear the room temperature vulcanizing (RTV) sealant.

20. Remove the engine front cover.

To install:

21. Install the 0.315 inch (8mm) guides into the front of the cylinder block.

22. Install the NEW engine front cover to cylinder block seal.

23. Place a 0.118 inch (3mm) bead of RTV sealant, on the engine front cover as shown (1).

24. Place the engine front cover onto the guide pins and slide into position.

25. Hand start all the engine front cover bolts.

26. Remove the guide pins from the cylinder block.

27. Tighten the engine front cover bolts, in the sequence shown, to 17 ft. lbs. (23 Nm).

28. Install NEW O-rings on the camshaft position sensor.

29. Place the camshaft position sensors in position on the front cover.

30. Install the camshaft position sensor

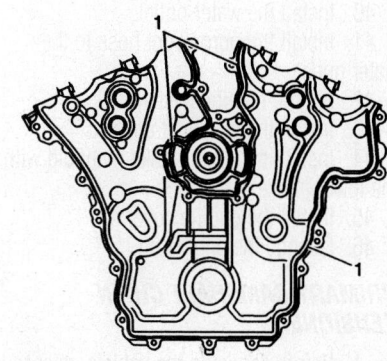

06025-GSTS-G130

Place RTV sealant on front cover in areas indicated

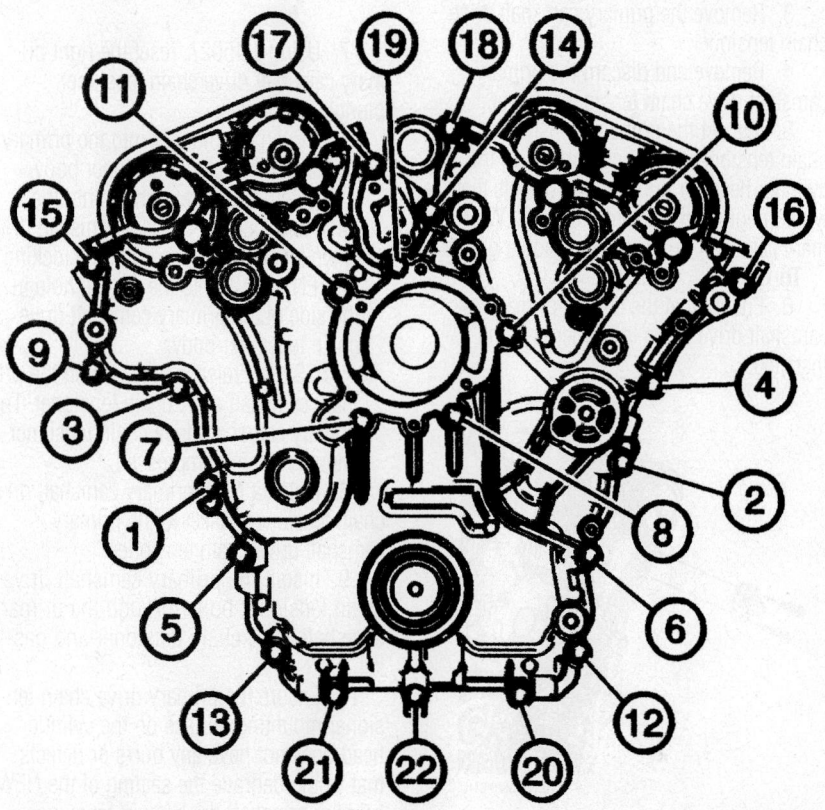

06025-GSTS-G131

Front cover bolt tightening sequence

bolts. Tighten the camshaft position sensor bolts to 89 inch lbs. (10 Nm).

31. Place the camshaft position actuator valves in position on the front cover.

32. Tighten the camshaft position actuator valve bolts to 89 inch lbs. (10 Nm).

33. Install the oil control valves.

34. Install the crankshaft balancer.

35. Install the power steering pump.

36. Install the power steering pump pulley.

37. Install the power steering fluid reservoir.

38. Install the alternator bracket with the alternator and the belt tensioner.

39. Install the accessory drive belts.

40. Install the water outlet.

41. Install the purge vent hose to the water outlet

42. Fill the cooling system.

43. Install the camshaft covers.

44. Install the upper intake manifold with the lower intake manifold.

45. Install the engine cover.

46. Change the engine oil.

PRIMARY CAMSHAFT CHAIN TENSIONER

1. Before servicing the vehicle, refer to the precautions in the beginning of this section.

2. Remove the primary camshaft drive chain tensioner bolts.

3. Remove the primary camshaft drive chain tensioner.

4. Remove and discard the primary camshaft drive chain tensioner gasket.

5. Inspect the primary camshaft drive chain tensioner mounting surface on the cylinder head for burrs or any defects that would degrade the sealing of the NEW primary camshaft drive chain tensioner gasket.

To install:

6. Ensure that the correct primary camshaft drive chain tensioner is being installed.

Removing/installing the camshaft primary drive chain tensioner

06025-GSTS-G148

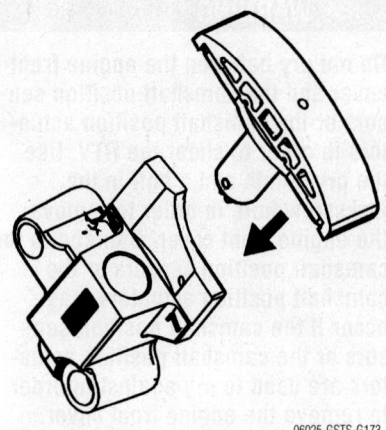

06025-GSTS-G173

Resetting the primary drive chain tensioner plunger

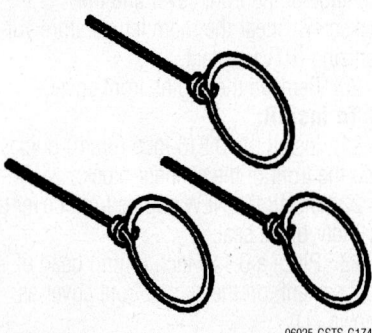

06025-GSTS-G174

Showing the locking pin tool (EN 46112)

7. Using J 45027, reset the right primary camshaft drive chain tensioner plunger as follows:

a. Push the plunger into the primary camshaft drive chain tensioner body.

b. Compress the plunger into the body and lock the primary camshaft drive chain tensioner by inserting the locking pin (EN 46112) into the access hole in the side of the primary camshaft drive chain tensioner body.

c. Slowly release pressure on the primary camshaft drive chain tensioner. The primary camshaft drive chain tensioner should remain compressed.

8. Install a NEW primary camshaft drive chain tensioner gasket to the primary camshaft drive chain tensioner.

9. Install the primary camshaft drive chain tensioner bolts through the primary camshaft drive chain tensioner and gasket.

10. Ensure the primary drive chain tensioner mounting surface on the cylinder head does not have any burrs or defects that would degrade the sealing of the NEW primary camshaft drive chain tensioner gasket.

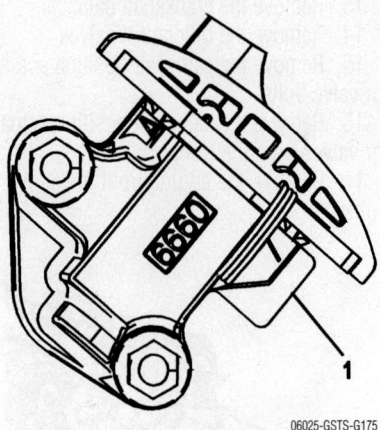

06025-GSTS-G175

Indicating the primary camshaft drive chain tensioner gasket tab (1)

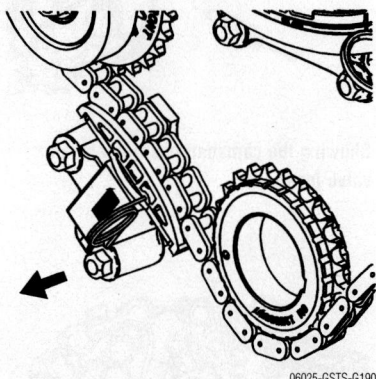

06025-GSTS-G190

Illustrating primary camshaft drive chain tentioner and shoe installation.

11. Place the primary camshaft drive chain tensioner into position and loosely install the bolts to the block.

12. Verify the proper placement of the primary camshaft drive chain tensioner gasket tab (1).

13. Tighten the primary camshaft drive chain tensioner bolts as follows:

- First Pass: 44 inch lbs. (5mm).

14. Final Pass tighten the primary camshaft drive chain tensioner bolts to 17 ft. lbs. (23 Nm).

15. Release the primary camshaft drive chain tensioner by pulling out the EN 46112 and unlocking the tensioner plunger

16. Verify all primary and secondary camshaft drive chain timing mark alignments (1-18)

❉ WARNING

Ensure that all timing chain tensioners are completely released. A timing chain tensioner that is not properly released can lead to serious engine damage.

PRIMARY UPPER CAMSHAFT CHAIN GUIDE

1. Before servicing the vehicle, refer to the precautions in the beginning of this section.

2. Remove the spark plugs in order to ease crankshaft/engine rotation.

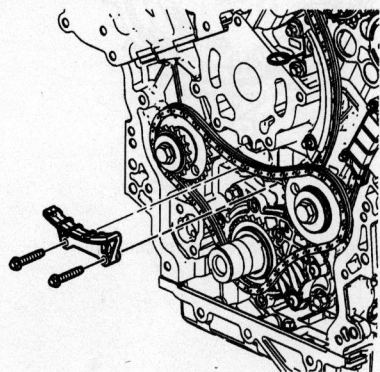

Identifying the camshaft upper drive chain guide location

06025-GSTS-G152

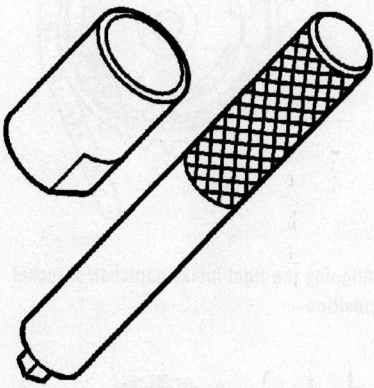

Illustrating the special tool (J45027) for resetting the tensioner plunger

06025-GSTS-G172

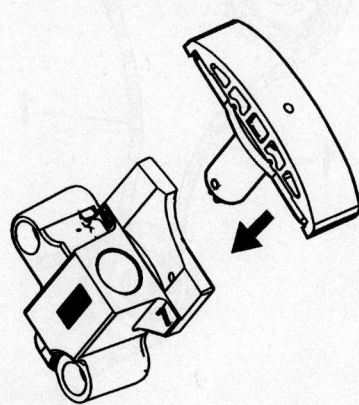

Inserting the camshaft drive chain tensioner plunger

06025-GSTS-G173

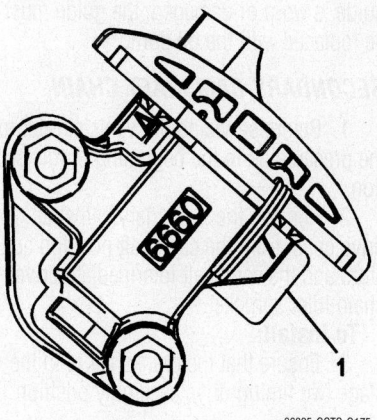

Identifying the camshaft drive chain tensioner gasket tab

06025-GSTS-G175

3. Remove the engine front cover.

4. Remove the primary camshaft drive chain tensioner.

5. Remove the primary camshaft drive chain upper guide bolts.

6. Remove the primary camshaft drive chain upper guide.

To install:

7. Using special tool J45027 reset the primary camshaft drive chain tensioner plunger.

8. Install the plunger into the primary camshaft drive chain tensioner body.

9. Compress the plunger into the body and lock the primary camshaft drive chain tensioner by inserting tool EN 46112 into the access hole in the side of the primary camshaft drive chain tensioner body.

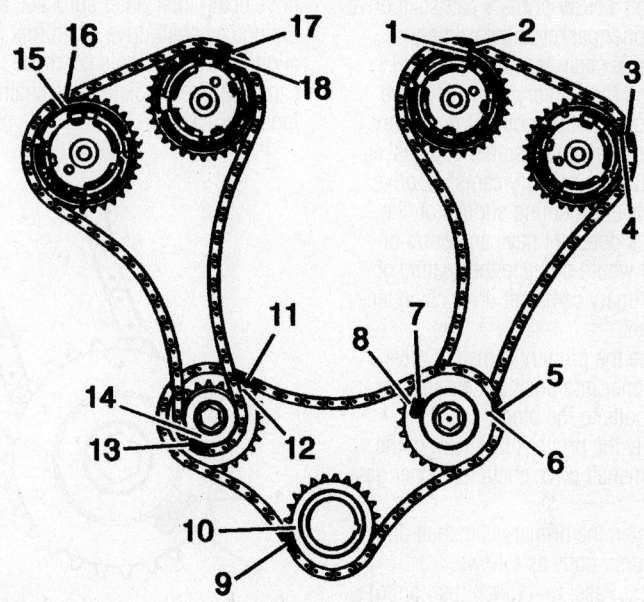

1. Left intake camshaft sprocket timing mark
2. Timing link
3. Left exhaust camshaft sprocket timing mark
4. Timing chain timing link
5. Timing link
6. Intermediate camshaft idler sprocket
7. Sprocket ID
8. Sprocket timing mark to hub
9. Timing link
10. Crankshaft sprocket
11. Timing link
12. Intermediate camshaft idler sprocket
13. Timing link
14. Idler
15. Right exhaust camshaft sprocket timing mark
16. Timing mark
17. Right intake camshaft sprocket timing mark
18. Timing link

06025-GSTS-G107

Indicating all of the camshaft drive chain timing mark alignments

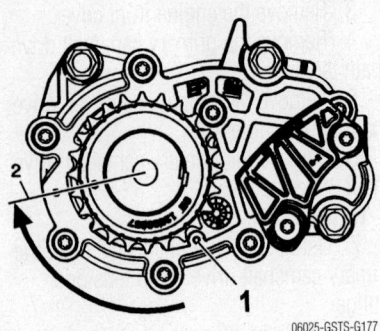

Ensure proper alignment of the stage two position

10. Slowly release pressure on the primary camshaft drive chain tensioner. The primary camshaft drive chain tensioner should remain compressed.

11. Install a NEW primary camshaft drive chain tensioner gasket to the primary camshaft drive chain tensioner.

12. Install the primary camshaft drive chain tensioner bolts through the primary camshaft drive chain tensioner and gasket.

13. Ensure the primary camshaft drive chain tensioner mounting surface on the engine block does not have any burrs or defects that would degrade the sealing of the NEW primary camshaft drive chain tensioner gasket.

14. Place the primary camshaft drive chain tensioner into position and loosely install the bolts to the block.

15. Verify the proper placement of the primary camshaft drive chain tensioner gasket tab (1).

16. Tighten the primary camshaft drive chain tensioner bolts as follows:
- First Pass: to 44 inch lbs. (5mm)
- Final Pass: to 17 ft. lbs. (23 Nm)

17. Release the primary camshaft drive chain tensioner by pulling out the locking pins (EN 46112) and unlocking the tensioner plunger.

18. Verify the primary and left secondary camshaft drive chain timing mark alignments (1-12)

19. Remove the camshaft holding tool (EN 46105-1) from the rear of the left camshafts.

20. Rotate the crankshaft and crankshaft sprocket from the stage one alignment position (1) to the stage two alignment position (2), 115 crankshaft degrees, in order to install the right secondary camshaft drive chain components.

PRIMARY LOWER CAMSHAFT CHAIN GUIDE

1. The primary camshaft drive chain guide is not serviceable separately. If the guide is worn or damaged, the guide must be replaced with the oil pump.

SECONDARY CAMSHAFT CHAIN

1. Before servicing the vehicle, refer to the precautions in the beginning of this section.

2. Remove the secondary camshaft drive chain from the camshaft position actuators and the camshaft intermediate drive chain idler sprocket.

To install:

3. Ensure that the crankshaft is in the stage two timing drive assembly position (1).

4. Install the secondary camshaft drive chain.

5. Place the secondary camshaft drive chain around the camshaft intermediate drive chain idler outer sprocket, aligning the timing camshaft drive chain link (1) with the alignment access hole (2) made in the right camshaft intermediate drive chain idler inner sprocket.

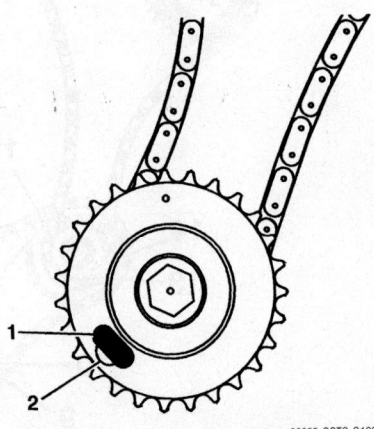

Aligning the secondary drive chain

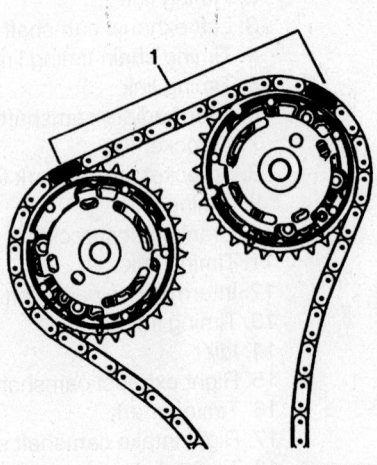

Check for 7 links (1) between the secondary drive chain actuator sprockets

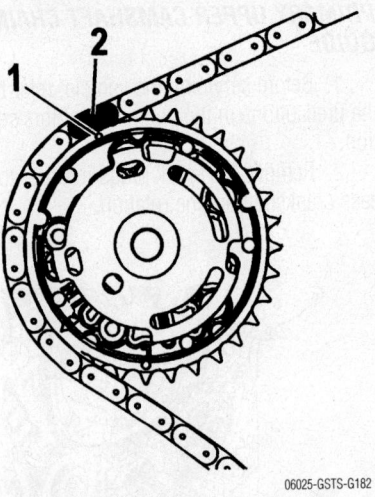

Aligning the right exhaust camshaft sprocket position

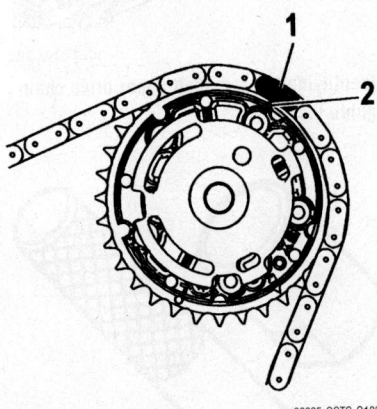

Aligning the right intake camshaft sprocket position

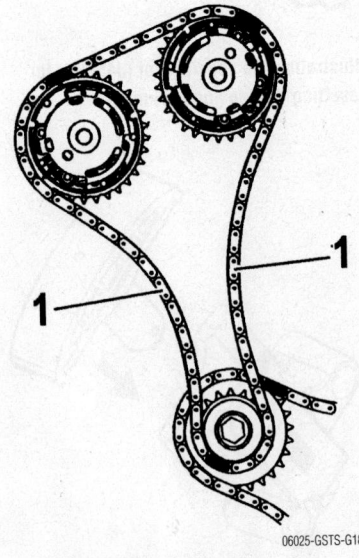

Ensure there are 18 links between the intermediate idler link and each actuator sprocket timing links

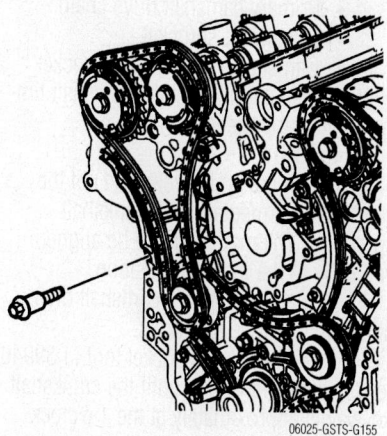

06025-GSTS-G155

Indicating the camshaft drive chain shoe location

6. Wrap the secondary camshaft drive chain around both actuator drive sprockets.

7. Ensure there are 7 links (1) between the timing camshaft drive chain links for the camshaft position actuator sprockets.

8. Align the exhaust camshaft position actuator sprocket alignment triangle mark (1) with the timing camshaft drive chain link (2).

9. Align the intake camshaft position actuator sprocket alignment triangle mark (2) with the timing camshaft drive chain link (1).

10. There will be 18 links (1) between the camshaft intermediate drive chain idler timing camshaft drive chain link and each right camshaft position actuator sprocket timing camshaft drive chain link.

SECONDARY CAMSHAFT CHAIN SHOE

1. Before servicing the vehicle, refer to the precautions in the beginning of this section.

2. Remove the right secondary camshaft drive chain shoe bolt.

3. Remove the right secondary camshaft drive chain shoe.

To install:

4. Position the right secondary camshaft drive chain shoe.

5. Install the secondary camshaft drive chain shoe bolt.

6. Tighten the secondary camshaft drive chain shoe bolt to 17 ft. lbs. (23 Nm).

7. Follow the same procedures for the left secondary chain shoe.

SECONDARY CAMSHAFT CHAIN GUIDE

1. Before servicing the vehicle, refer to the precautions in the beginning of this section.

2. Remove the right secondary camshaft drive chain guide bolts.

3. Remove the right secondary camshaft drive chain guide.

To install:

4. Position the right secondary camshaft drive chain guide.

5. Install the secondary camshaft drive chain guide bolts.

6. Tighten the secondary camshaft drive chain guide bolts to 17 ft. lbs. (23 Nm).

7. Follow the same procedures for the left secondary chain shoe.

SECONDARY CAMSHAFT CHAIN TENSIONER

1. Before servicing the vehicle, refer to the precautions in the beginning of this section.

2. Remove the left secondary camshaft drive chain tensioner bolts.

3. Remove the left secondary camshaft drive chain tensioner.

4. Remove and discard the left secondary camshaft drive chain tensioner gasket.

5. Inspect the left secondary camshaft drive chain tensioner mounting surface on the left cylinder head for burrs or any defects that would degrade the sealing of the NEW left secondary camshaft drive chain tensioner gasket.

To install:

6. Using special tool (J45027) reset the right secondary camshaft drive chain tensioner plunger.

7. Install the plunger into the right secondary camshaft drive chain tensioner body.

8. Compress the plunger into the body

06025-GSTS-G188

Assembling the secondary drive chain tensioner plunger

and lock the right secondary camshaft drive chain tensioner by inserting the locking pin (EN 46112) into the access hole in the side of the right secondary camshaft drive chain tensioner body.

9. Slowly release pressure on the right secondary camshaft drive chain tensioner. The right secondary camshaft drive chain tensioner should remain compressed.

10. Install a NEW right secondary camshaft drive chain tensioner gasket to the right secondary camshaft drive chain tensioner.

11. Install the right secondary camshaft drive chain tensioner bolts through the right secondary camshaft drive chain tensioner and gasket.

12. Ensure the right secondary camshaft drive chain tensioner mounting surface on the right cylinder head does not have any burrs or defects that would degrade the sealing of the NEW right secondary camshaft drive chain tensioner gasket.

13. Place the right secondary camshaft drive chain tensioner into position and loosely install the bolts to the block.

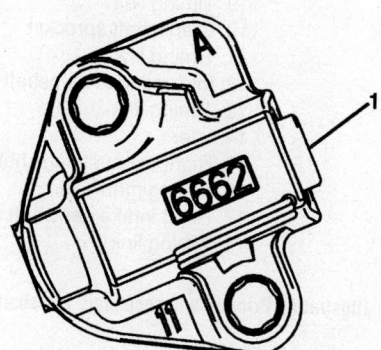

06025-GSTS-G189

Identifying the secondary camshaft drive chain tensioner gasket tab

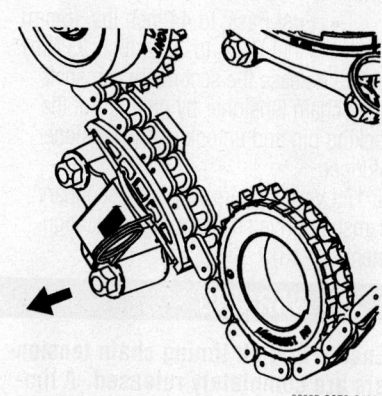

06025-GSTS-G190

Showing the proper installation of the secondary camshaft drive chain tensioner and shoe

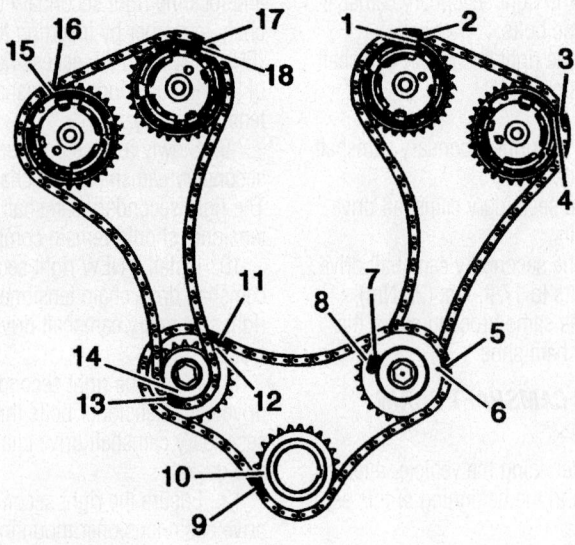

1. Left intake camshaft sprocket timing mark
2. Timing link
3. Left exhaust camshaft sprocket timing mark
4. Timing chain timing link
5. Timing link
6. Intermediate camshaft idler sprocket
7. Sprocket ID
8. Sprocket timing mark to hub
9. Timing link
10. Crankshaft sprocket
11. Timing link
12. Intermediate camshaft idler sprocket
13. Timing link
14. Idler
15. Right exhaust camshaft sprocket timing mark
16. Timing mark
17. Right intake camshaft sprocket timing mark
18. Timing link

06025-GSTS-G107

Illustrating Primary and secondary camshaft drive chain alignment locations—3.6L

14. Verify the proper placement of the right secondary camshaft drive chain tensioner gasket tab (1).

15. Tighten the tensioner bolts as follows:

- First Pass: to 44 inch lbs. (5mm)
- Final Pass: to 17 ft. lbs. (23 Nm)

16. Release the secondary camshaft drive chain tensioner by pulling out the locking pin and unlocking the tensioner plunger.

17. Verify all primary and secondary camshaft drive chain timing mark alignments (1-18)

❊❊ WARNING

Ensure that all timing chain tensioners are completely released. A timing chain tensioner that is not properly released can lead to serious engine damage.

4.6L Engine

PRIMARY DRIVE CHAIN AND SPROCKETS

1. Before servicing the vehicle, refer to the precautions in the beginning of this section.

2. Remove the oil pump.

3. Rotate the crankshaft with a special socket (J 39946) to align the primary timing marks.

4. Remove the secondary camshaft drive chains.

5. Remove the primary drive chain tensioner.

6. Remove the oil outlet tube (4).

7. Remove the primary camshaft drive chain guide bolts and guide.

8. Remove the camshaft intermediate sprocket retaining bolt.

9. Remove the following as an assembly:

- Primary camshaft drive chain
- Crankshaft sprocket
- Camshaft intermediate sprocket

10. Clean and inspect the camshaft timing drive components.

To install:

11. Align the timing marks (1) of the camshaft intermediate and crankshaft sprockets. The marks should be aligned vertically in the installed position.

12. Install the primary camshaft drive chain on the drive sprockets.

13. Use the special socket tool (J 39946) to rotate the crankshaft until the crankshaft keyway is approximately at the 1 o'clock position.

14. Install the following as an assembly:

- Primary camshaft drive chain
- Crankshaft sprocket
- Camshaft intermediate sprocket

15. Install the camshaft intermediate sprocket retaining bolt.

16. Tighten the camshaft intermediate sprocket retaining bolt to 44 ft. lbs. (60 Nm).

17. Install the primary drive chain guide.

18. Install the primary drive chain guide bolts: tighten to 18 ft. lbs. (25 Nm).

19. Install the primary drive chain tensioner.

20. Install the oil outlet tube (4) and bolts.

21. Tighten the oil outlet tube bolts to 89 inch lbs. (10 Nm).

22. Install the secondary camshaft drive chains.

23. Remove the pin from the primary timing chain tensioner release lever.

24. Ensure the primary timing marks are aligned vertically.

CAMSHAFT COVER

1. Before servicing the vehicle, refer to the precautions in the beginning of this section.

2. Remove the fuel injector sight shield.

3. Disconnect the positive crankcase ventilation (PCV) fresh air tube from the left camshaft cover.

4. Remove the left side ignition module.

5. Remove the bolt connecting the ground strap to the left camshaft cover.

6. Remove the bolt securing the oil level indicator tube to the left cylinder head and reposition the tube away from the camshaft cover.

7. Remove the camshaft cover bolts.

8. Lift the camshaft drive end of the camshaft cover up.

9. Discard the camshaft cover gasket and spark plug port seals if there is any evi-

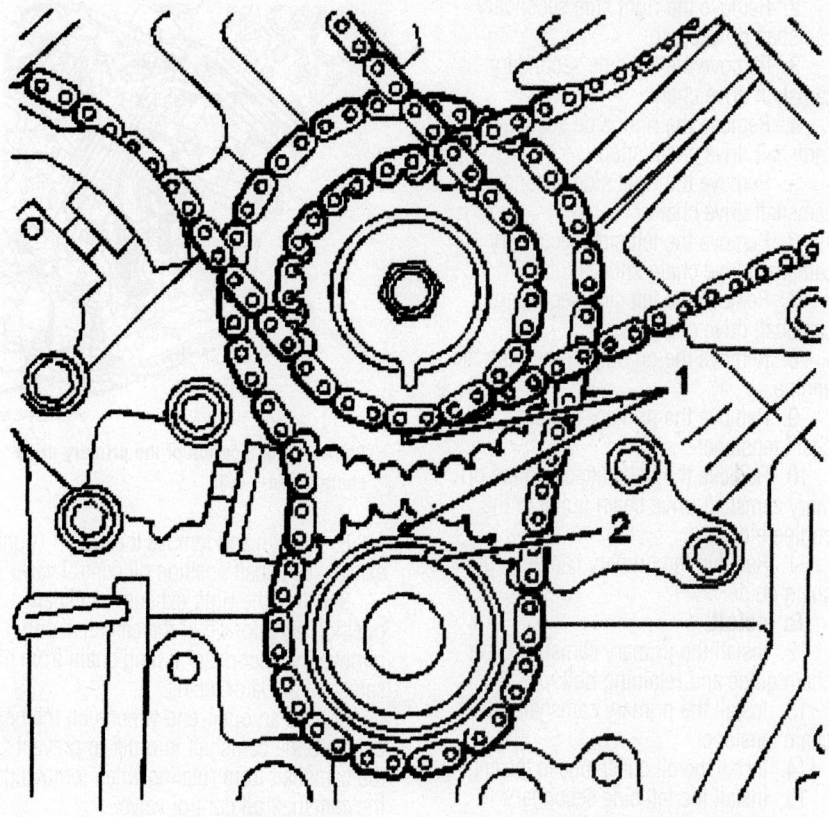

Align the primary timing marks—4.6L

06025-GSTS-G44

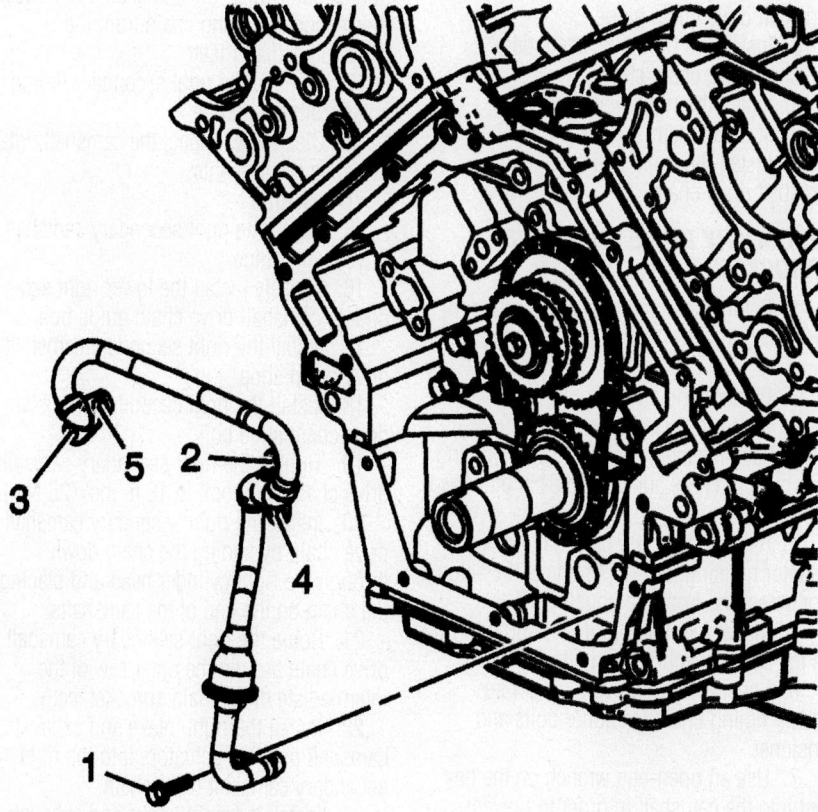

Removing the oil outlet tube—4.6L

06025-GSTS-G158

dence of damage or if the seal comes out of the groove in the cover during removal.

10. Clean the gasket mating surface on the cylinder head.

11. Clean and inspect the camshaft cover.

To install:

12. Install a new camshaft cover gasket to the camshaft cover if necessary.

13. Place a small amount of sealant (GM P/N 12345739) at the split line (1, 2) of the left cylinder head and the left camshaft position actuator housing.

➡ **Be careful to prevent the exposed section of the camshaft cover gasket from being damaged by the edge of the cylinder head casting.**

14. Work the camshaft cover into position by pivoting the cover down and aligning the bolt holes.

15. Install the camshaft cover bolts.

16. Tighten the camshaft cover bolts to 89 inch lbs. (10 Nm).

17. Install the bolt connecting the left camshaft cover ground strap to the left camshaft cover.

18. Tighten the left ground strap bolt to 89 inch lbs. (10 Nm).

19. Rotate the oil level indicator tube back into its original position and install the bolt securing the tube to the cylinder head.

20. Tighten the oil level indicator tube bolt to 89 inch lbs. (10 Nm).

21. Install the ignition module.

22. Connect the PCV fresh air tube to the left camshaft cover.

23. Install the fuel injector sight shield.

24. Follow the same procedures for the right side camshaft cover.

PRIMARY DRIVE CHAIN TENSIONER

1. Before servicing the vehicle, refer to the precautions in the beginning of this section.

2. Remove the engine oil pump.

3. Align the primary timing marks (1) using the tool J 39946

4. Remove the 2 bolts attaching the primary camshaft drive chain tensioner to the engine block.

5. Remove the primary camshaft drive chain tensioner, allowing the tensioner to expand during removal.

6. Clean and inspect the camshaft timing drive components.

To install:

7. Collapse the primary camshaft drive chain tensioner using the following procedure:

8. Rotate the ratchet release lever (2) clockwise and hold.

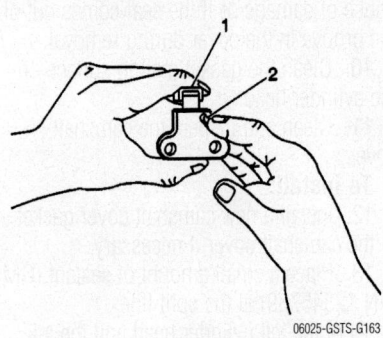

06025-GSTS-G163

Compressing the tensioner shoe—4.6L

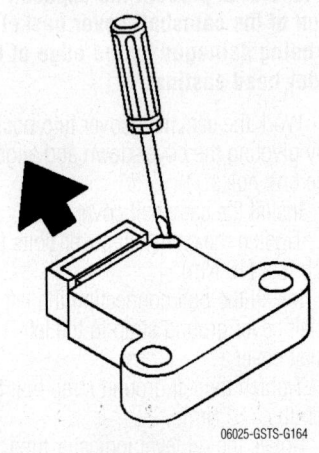

06025-GSTS-G164

Insert a pin to release the lever in the primary drive chain tensioner

9. Collapse the tensioner shoe (1) and hold.

10. Release the ratchet lever (2).

11. Slowly release the pressure on the shoe (1), until the ratchet lever (2) moves to the first detent and a click is heard and felt.

12. Collapse the tensioner shoe (1) and hold.

13. Insert a pin through the hole in the release lever in order to lock the tensioner shoe in the collapsed position.

14. Install the primary camshaft drive chain tensioner and retaining bolts.

15. Tighten the primary camshaft drive chain tensioner bolts to 18 ft. lbs. (25 Nm).

➡**Ensure the tensioner release lever is facing outward.**

16. Remove the pin holding the tensioner to tighten any slack in the timing chain.

17. Ensure the primary timing marks are aligned vertically.

18. Install the oil pump.

PRIMARY DRIVE CHAIN GUIDE

1. Before servicing the vehicle, refer to the precautions in the beginning of this section.

2. Remove the right side secondary camshaft drive chain.

3. Remove the left side secondary camshaft drive chain.

4. Remove the right side secondary camshaft drive chain shoe.

5. Remove the right side secondary camshaft drive chain

6. Remove the left side secondary camshaft drive chain shoe

7. Remove the left side secondary camshaft drive chain guide.

8. Remove the oil outlet tube from the engine.

9. Remove the primary camshaft drive chain tensioner

10. Remove the bolts attaching the primary camshaft drive chain guide to the engine block.

11. Remove the primary camshaft drive chain guide.

To install:

12. Install the primary camshaft drive chain guide and retaining bolts.

13. Install the primary camshaft drive chain tensioner.

14. Install the oil outlet tube to the engine.

15. Install the left side secondary camshaft drive chain guide

16. Install the left side secondary camshaft drive chain shoe

17. Install the right side secondary camshaft drive chain guide.

18. Install the right side secondary camshaft drive chain shoe.

19. Install the right side secondary camshaft drive chain shoe.

20. Install the right side secondary camshaft drive chain.

SECONDARY DRIVE CHAIN AND SPROCKETS

1. Before servicing the vehicle, refer to the precautions in the beginning of this section.

2. Remove the oil pump.

3. Using the special socket (J 39946) rotate the crankshaft until the primary timing gear marks (1) are adjacent to each other as shown.

4. Remove the right camshaft position actuator housing. DO NOT remove the actuator solenoids from the housing.

5. Install a holding tool (EN 46328) (1) on the bank 1 (right) camshafts (2).

6. Loosen and remove the right secondary timing chain tensioner bolts and tensioner.

7. Use an open-end wrench on the hex cast into the camshaft in order to prevent the camshaft from rotating when removing the camshaft oil control valve.

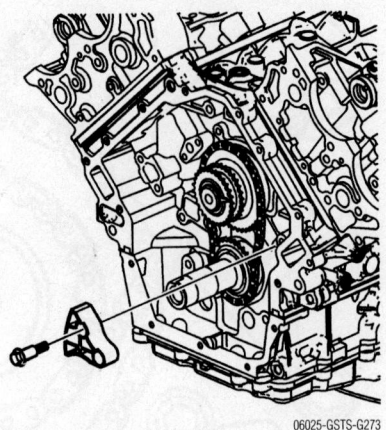

06025-GSTS-G273

Showing the location of the primary drive chain guide—4.6L

8. Loosen and remove the bank 1 (right) exhaust camshaft position oil control valve.

9. Slide the right exhaust camshaft position actuator off of the camshaft and remove the secondary timing chain from the camshaft actuator teeth.

10. Use an open-end wrench on the hex cast into the camshaft in order to prevent the camshaft from rotating when removing the camshaft oil control valve.

11. Loosen and remove the bank intake camshaft position oil control valve.

12. Slide the right intake camshaft position actuator off of the camshaft and remove the secondary timing chain from the camshaft actuator teeth.

13. Remove the right secondary timing chain from the engine.

14. Clean and inspect the camshaft timing drive components.

To install:

15. Install the right secondary camshaft drive chain guide.

16. Loosely install the lower right secondary camshaft drive chain guide bolt.

17. Install the right secondary camshaft drive chain shoe.

18. Install the right secondary camshaft drive chain shoe bolt.

19. Tighten the right secondary camshaft drive chain shoe bolt to 18 ft. lbs. (25 Nm).

20. Install the right secondary camshaft drive chain by sliding the chain down through the right cylinder head and placing the chain on the end of the camshafts.

21. Route the right secondary camshaft drive chain around the outer row of the intermediate drive chain sprocket teeth.

22. Install the right intake and exhaust camshaft position actuators into the right secondary camshaft drive chain.

23. Install the right intake and exhaust camshafts onto the camshafts. The camshaft

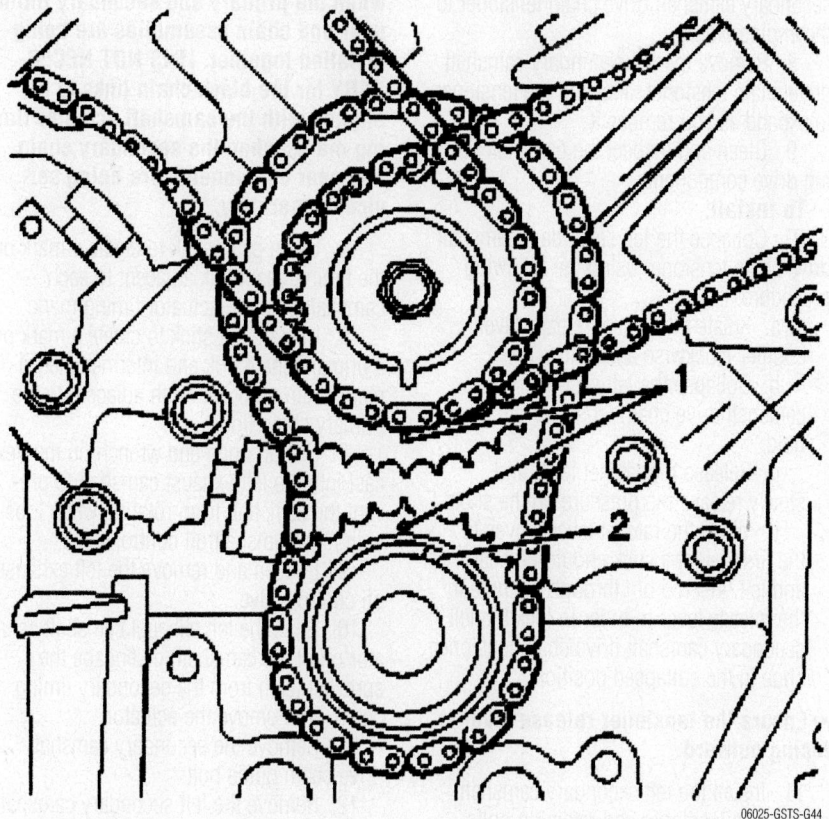

Aligning the primary timing gear marks—4.6L

06025-GSTS-G44

b. Collapse the right secondary camshaft drive chain tensioner shoe and hold.

c. Release the ratchet lever and slowly release the pressure on the shoe.

d. When the ratchet lever moves to the first detent a click should be heard and felt. Insert a pin through the hole in the release lever in order to lock the right secondary camshaft drive chain tensioner shoe in the collapsed position.

31. Install the right secondary camshaft drive chain tensioner.

➡ **Install the right secondary camshaft drive chain tensioner.**

32. Install the right secondary camshaft drive chain tensioner bolts.

33. Tighten the right secondary camshaft drive chain tensioner bolts to 18 ft. lbs. (25 Nm).

34. Remove pin from right secondary camshaft drive chain tensioner lever.

35. Ensure the correct alignment of all secondary timing components.

36. Ensure the correct alignment of all primary timing components.

37. Use an open-end wrench on the hex cast into the left intake camshaft, in order

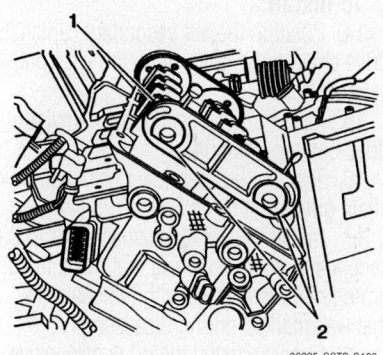

Using the holding tool (EN 46328) to secure the camshafts—4.6L

06025-GSTS-G160

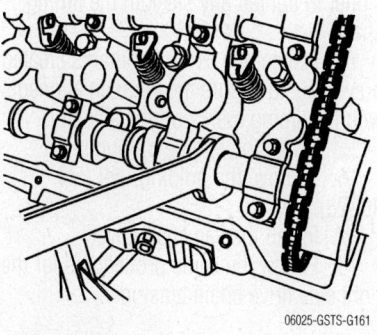

Using an open-end wrench on the camshaft hex cast—4.6L

06025-GSTS-G161

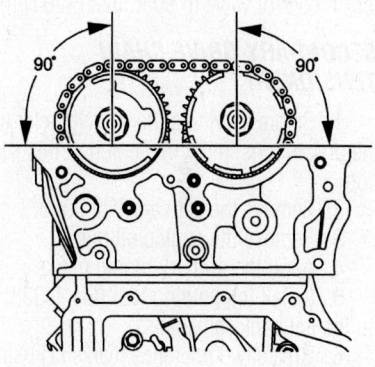

Illustrating perpendicular alignment of intake and exhaust camshafts—4.6L

06025-GSTS-G271

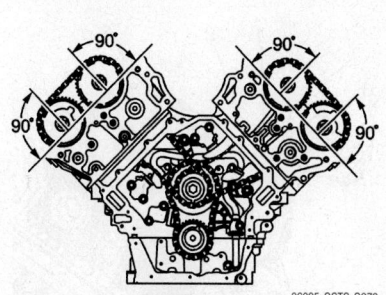

Showing the correct alignment of secondary timing components

06025-GSTS-G272

sprocket notch marked "RI" which indicates right intake, engages the intake camshaft pin and the camshaft sprocket notch marked "RE" which indicates right exhaust, engages the exhaust camshaft pin.

24. If necessary, use an open wrench on the hex cast near the front of each camshaft to help align the sprocket notch to the camshaft pin.

25. Loosely install the right intake and exhaust camshaft position actuator oil control valves.

26. Ensure the perpendicular alignment of the right intake and exhaust camshaft

sprocket notches and camshaft pins to the cylinder head.

27. Install the holding tool (EN 46328) to the right cylinder head camshafts.

28. Install the upper right secondary camshaft drive chain guide bolt.

29. Tighten BOTH the upper and lower right secondary camshaft drive chain guide bolts to 18 ft. lbs. (25 Nm).

30. Collapse the right secondary camshaft drive chain tensioner using the following procedure:

a. Rotate the ratchet release lever counter-clockwise and hold.

to prevent the camshaft from rotating, when tightening the camshaft position actuator oil control valve on the left intake camshaft.

38. Tighten the camshaft position actuator oil control valve on the left intake camshaft.

39. Tighten the camshaft position actuator oil control valve to 90 ft. lbs. (120 Nm).

40. Use an open-end wrench on the hex cast into the right intake camshaft, in order to prevent the camshaft from rotating, when tightening the camshaft position actuator oil control valve on the right intake camshaft.

41. Tighten the camshaft position actuator oil control valve on the right intake camshaft.

42. Tighten the camshaft position actuator oil control valve to 90 ft. lbs. (120 Nm).

43. Use an open end wrench on the hex cast into the right exhaust camshaft, in order to prevent the camshaft from rotating, when tightening the camshaft position actuator oil control valve on the right exhaust camshaft.

44. Tighten the camshaft position actuator oil control valve on the right exhaust camshaft.

45. Tighten the camshaft position actuator oil control valve to 90 ft. lbs. (120 Nm).

SECONDARY DRIVE CHAIN TENSIONER

1. Before servicing the vehicle, refer to the precautions in the beginning of this section.

2. Remove the left camshaft cover.

3. Remove the engine oil pump.

4. Align the primary timing marks.

5. Install the holding tool (EN 46328) on the left bank camshafts.

6. Create two reference marks (1) using a paint stick to identify the chain link adjacent to each actuator timing mark (2, 3).

7. Remove the 2 bolts attaching the left

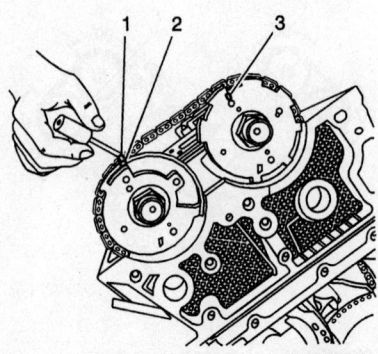

06025-GSTS-G274
Making reference marks on chain links

secondary camshaft drive chain tensioner to the engine block.

8. Remove the left secondary camshaft drive chain tensioner, allowing the tensioner to expand as you remove it.

9. Clean and inspect the camshaft timing drive components.

To install:

10. Collapse the left secondary camshaft drive chain tensioner using the following procedure:

   a. Rotate the ratchet release lever counter-clockwise and hold.

   b. Collapse the left secondary camshaft drive chain tensioner shoe and hold.

   c. Release the ratchet lever and slowly release the pressure on the shoe.

   d. When the ratchet lever moves to the first detent a click should be heard and felt. Insert a pin through the hole in the release lever in order to lock the left secondary camshaft drive chain tensioner shoe in the collapsed position.

➥**Ensure the tensioner release lever is facing outward.**

11. Install the left secondary camshaft drive chain tensioner and retaining bolts.

12. Tighten the secondary camshaft drive chain tensioner bolts to 18 ft. lbs. (25 Nm).

13. Remove the pin holding the tensioner to tighten any slack in the timing chain.

14. Inspect the two paint marks created previously to ensure they remain aligned with the timing marks on the actuator.

15. Install the engine oil pump.

16. Remove the holding tool (EN 46328).

17. Install the camshaft cover.

18. Follow the same procedures for the right side drive chain tensioner.

SECONDARY DRIVE CHAIN GUIDE

1. Before servicing the vehicle, refer to the precautions in the beginning of this section.

2. Remove the left secondary camshaft tensioner.

3. Remove the left camshaft position actuator housing. DO NOT remove the camshaft actuator solenoids from the housing.

4. Using the J 39946, rotate the crankshaft until both camshaft position actuator timing marks are at the top of their rotation.

5. Install the holding tool to the camshafts.

➥**The secondary timing chain has black links that are used for alignment**

when the primary and secondary timing gear and chain assemblies are being installed together. IT IS NOT NECESSARY for the black chain links to be aligned with the camshaft actuator timing marks when the secondary chain and gear components are being serviced separately.

6. Use a paint stick to create a mark on the timing chain link adjacent to each camshaft position actuator timing mark.

7. Use a paint stick to create a mark on a primary chain link and intermediate primary chain sprocket tooth adjacent to the primary chain link.

8. Use an open end wrench on the hex cast into the left exhaust camshaft to prevent the camshaft from rotating when loosening the camshaft oil control valve.

9. Loosen and remove the left exhaust oil control valve.

10. Slide the left exhaust camshaft actuator off of the camshaft, disengage the sprocket teeth from the secondary timing chain, and remove the actuator.

11. Remove the secondary camshaft drive chain guide bolt.

12. Remove the left secondary camshaft drive chain guide.

13. Clean and inspect the camshaft timing drive components.

To install:

14. Position the left secondary camshaft drive chain guide in through the left front cylinder head

15. Install the left secondary camshaft drive chain guide bolt.

16. Tighten the secondary camshaft drive chain guide bolt to 18 ft. lbs. (25 Nm).

17. Install the left exhaust actuator to the secondary chain, ensuring that the actuator timing mark is aligned with the chain link that was marked during disassembly.

18. Loosely install the oil control valve to secure the exhaust actuator to the camshaft.

19. Use an open-end wrench on the hex cast into the camshaft in order to prevent the camshaft from rotating when tightening the oil control valve.

20. Tighten the oil control valve to 90 ft. lbs. (120 Nm).

21. Install the left secondary camshaft drive chain tensioner.

22. Ensure that the previously created paint marks still line up with the sprocket timing marks.

23. Install the left camshaft actuator housing.

24. Follow the same procedures for the right side drive chain guide.

SECONDARY DRIVE CHAIN SHOE

1. Before servicing the vehicle, refer to the precautions in the beginning of this section.

2. Remove the left secondary camshaft tensioner.

3. Remove the left camshaft position actuator housing. DO NOT remove the camshaft actuator solenoids from the housing

4. Using the J 39946 , rotate the crankshaft until both camshaft position actuator timing marks are at the top of their rotation.

5. Install the holding tool (EN 46328) to the camshafts.

➡**The secondary timing chain has black links that are used for alignment when the primary and secondary timing gear and chain assemblies are being installed together. IT IS NOT NECESSARY for the black chain links to be aligned with the camshaft actuator timing marks when the secondary chain and gear components are being serviced separately.**

6. Use a paint stick to create a mark on the timing chain link adjacent to each camshaft position actuator timing mark

7. Use a paint stick to create a mark on a primary chain link and intermediate primary chain sprocket tooth adjacent to the primary chain link.

8. Use an open end wrench on the hex cast into the left intake camshaft to prevent the camshaft from rotating when loosening the camshaft oil control valve.

9. Loosen and remove the left intake oil control valve.

10. Slide the left intake camshaft actuator off of the camshaft, disengage the sprocket teeth from the secondary timing chain, and remove the actuator.

11. Remove the left secondary camshaft drive timing chain shoe bolt.

12. Remove the left secondary camshaft drive timing shoe.

13. Clean and inspect the camshaft timing drive components

To install:

14. Position the left secondary camshaft drive chain shoe in through the left front cylinder head

15. Install the left secondary camshaft drive chain shoe bolt.

16. Tighten the secondary camshaft drive chain shoe bolt to 18 ft. lbs. (25 Nm).

17. Install the left intake actuator to the secondary chain, ensuring that the actuator timing mark is aligned with the chain link that was marked during disassembly.

18. Loosely install the oil control valve to secure the intake actuator to the camshaft.

19. Use an open-end wrench on the hex cast into the camshaft in order to prevent the camshaft from rotating when tightening the oil control valve.

20. Tighten the oil control valve to 90 ft. lbs. (120 Nm).

21. Install the left secondary camshaft drive chain tensioner.

22. Ensure that the previously created paint marks still line up with the sprocket timing marks.

23. Install the left camshaft actuator housing.

24. Follow the same procedures for the right side drive chain shoe.

Piston and Ring

POSITIONING

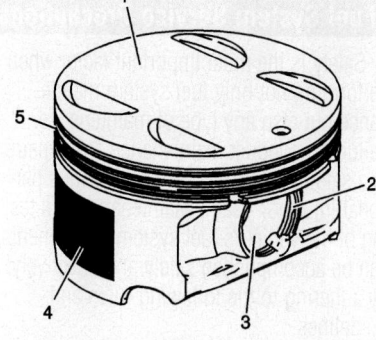

1. Top of the piston
2. Piston pin retainer grooves
3. Piston pin bores and piston pins
4. Piston skirt
5. Ring grooves and ring lands (place dimple on rings upward)

06025-GSTS-G304

Illustrating the key components of the piston assembly

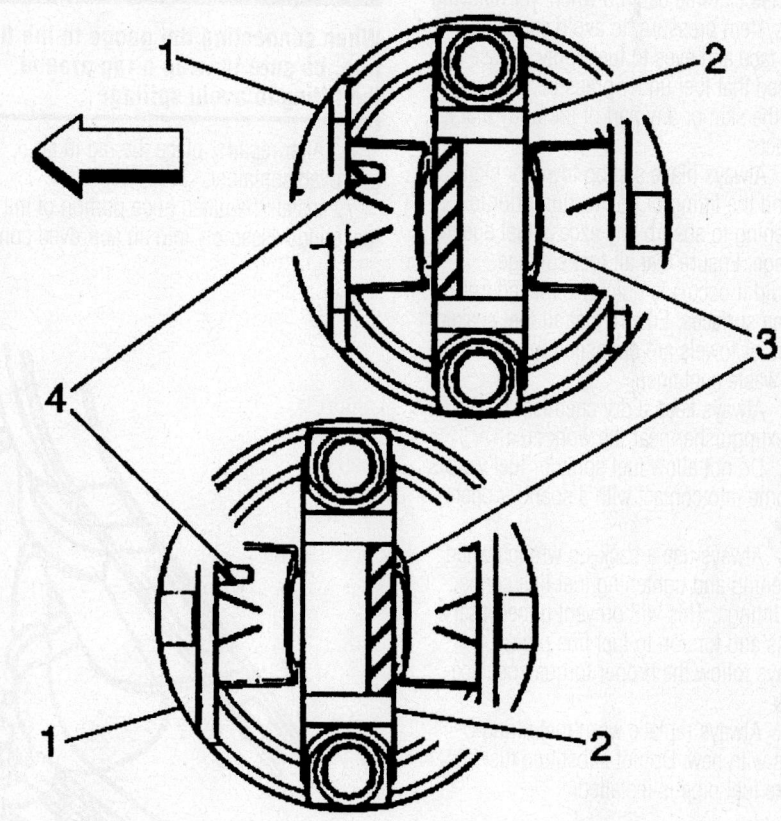

1. Piston
2. Connecting rod
3. Locating notch (toward front)
4. Locating mark (toward front)

06025-GSTS-G305

Showing the orientation of the piston and rings in the block

FUEL SYSTEM

Fuel System Service Precautions

Safety is the most important factor when performing not only fuel system maintenance but also any type of maintenance. Failure to conduct maintenance and repairs in a safe manner may result in serious personal injury or death. Maintenance and testing of the vehicle's fuel system components can be accomplished safely and effectively by adhering to the following rules and guidelines.

1. To avoid the possibility of fire and personal injury, always disconnect the negative battery cable unless the repair or test procedure requires that battery voltage be applied.

2. Always relieve the fuel system pressure prior to disconnecting any fuel system component (injector, fuel rail, pressure regulator, etc.), fitting or fuel line connection. Exercise extreme caution whenever relieving fuel system pressure, to avoid exposing skin, face and eyes to fuel spray. Please be advised that fuel under pressure may penetrate the skin or any part of the body that it contacts.

3. Always place a shop towel or cloth around the fitting or connection prior to loosening to absorb any excess fuel due to spillage. Ensure that all fuel spillage (should it occur) is quickly removed from engine surfaces. Ensure that all fuel soaked cloths or towels are deposited into a suitable waste container.

4. Always keep a dry chemical (Class B) fire extinguisher near the work area.

5. Do not allow fuel spray or fuel vapors to come into contact with a spark or open flame.

6. Always use a back-up wrench when loosening and tightening fuel line connection fittings. This will prevent unnecessary stress and torsion to fuel line piping. Always follow the proper torque specifications.

7. Always replace worn fuel fitting O-rings with new. Do not substitute fuel hose) where fuel pipe is installed.

Fuel System Pressure

RELIEVING

The fuel systems operate under high fuel pressures. It is very important that the pressure be properly relieved prior to servicing the system or any of its components.

A Schrader valve is provided on these fuel systems to conveniently test or release the system pressure. A fuel pressure gauge and adapter will be necessary to connect the gauge to the fitting. This system utilizes a service valve on one end of the fuel rail assembly.

1. Before servicing the vehicle, refer to the precautions in the beginning of this section.

2. Disconnect the negative battery cable to assure the prevention of fuel spillage if the ignition switch is accidentally turned **ON** while a fitting is still detached.

3. Loosen the fuel filler cap to release the fuel tank pressure.

4. Remove the fuel injector sight shield

5. Be sure the release valve on the fuel gauge is closed, then connect the fuel gauge to the pressure fitting located on the inlet fuel pipe fitting.

✳✳ CAUTION

When connecting the gauge to the fitting, be sure to wrap a rag around the fitting to avoid spillage.

6. After repairs, place the rag in an approved container.

7. Install the bleed hose portion of the fuel gauge assembly into an approved container, then open the gauge release valve and bleed the fuel pressure from the system.

8. When the gauge is removed, be sure to open the bleed valve and drain all fuel from the gauge assembly.

9. When fuel service is finished, tighten the fuel filler cap and connect the negative battery cable.

Fuel Filter

REMOVAL & INSTALLATION

1. Before servicing the vehicle, refer to the precautions in the beginning of this section.

2. Properly relieve the fuel system pressure.

3. Keep a shop cloth and a container ready to capture any spilled fuel.

4. Disconnect the negative battery cable.

5. Remove the fuel filler cap.

6. Quick connect fittings from the filter

7. Raise and support the vehicle.

8. Remove the rear suspension bracket.

9. Disconnect the quick-connect fitting at the fuel filter inlet.

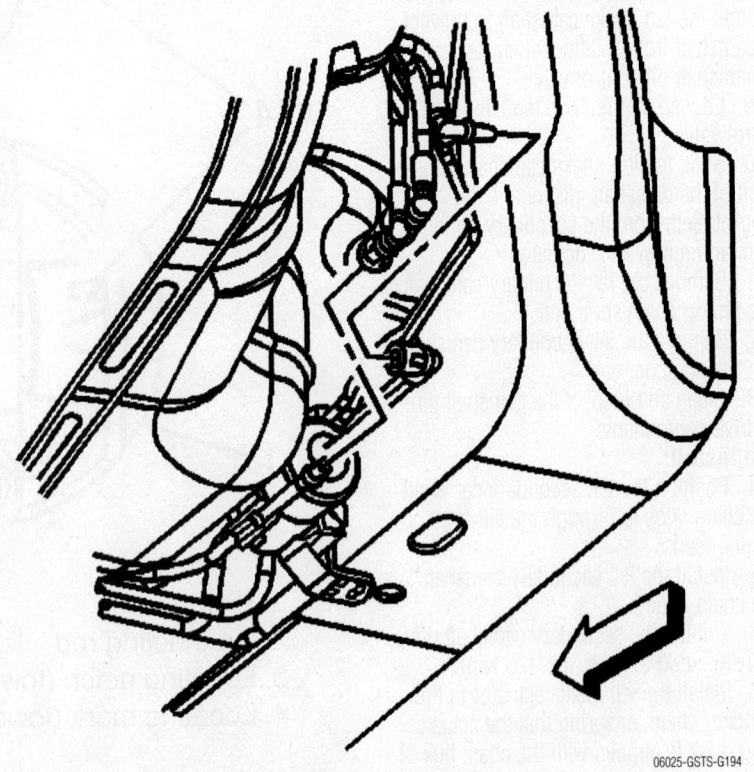

06025-GSTS-G194

Identifying the fuel filter mounting at the rear of the vehicle

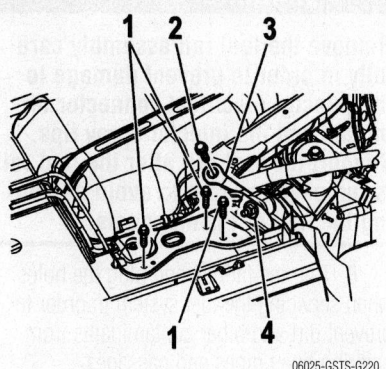

06025-GSTS-G220

Removing the rear suspension bracket

10. Using a primary and back-up wrench, remove the threaded fitting at the fuel filter outlet.

11. Slide the fuel filter rearward from the fuel filter bracket.

12. Remove the fuel pipe O-ring seal. Discard the O-ring seal if damaged.

13. Drain any remaining fuel into an approved gasoline container.

14. Discard the fuel filter into an approved container.

To install:

15. Lubricate the fuel pipe O-ring seal with clean engine oil.

16. Install the fuel pipe O-ring seal to the fuel pipe.

17. Remove the protective caps from the new fuel filter.

18. Slide the fuel filter forward into the fuel filter bracket.

19. Install the threaded fitting to the fuel filter outlet. Use a back-up wrench in order to prevent the fuel filter from turning.

20. Tighten the fuel filter outlet fitting to 22 ft. lbs. (30 Nm).

21. Connect the quick-connect fitting at the fuel filter inlet

22. Install the rear suspension bracket.

23. Lower the vehicle.

24. Start the engine and check for leaks.

Fuel Pump

REMOVAL & INSTALLATION

1. Before servicing the vehicle, refer to the precautions in the beginning of this section.

2. Properly relieve the fuel system pressure.

3. Disconnect the negative battery cable.

4. Drain the fuel tank into an approved container.

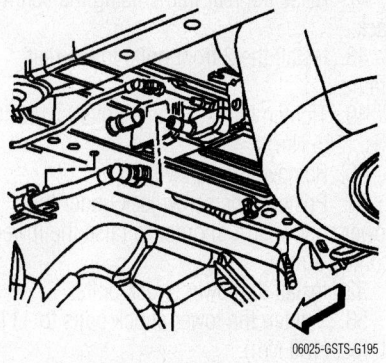

06025-GSTS-G195

Disconnecting hose from the EVAP canister

5. Raise and support the rear of the vehicle.

6. Remove or disconnect the following:
- Exhaust system
- Propeller shaft. See Propeller Shaft.
- Filler hose from the fuel tank
- Filler vent tube from the EVAP hose
- Fuel feed hoses
- Fuel EVAP hose
- Fuel tank electrical connector
- EVAP hoses from the EVAP canister
- Electrical connector from the EVAP canister

7. Pull outward on the retainer tab in order to disengage the retainer from the chassis.

8. Disconnect the electrical connector from the EVAP canister.

9. Raise the lower control arms, using a suitable screw jack, in order to remove the load from the lower shock bolts.

10. Remove the lower shock bolts.

11. Remove the screw jack.

12. Position the screw jack under the rear frame near the adjuster tie bar, in order to support the front of the rear frame.

13. Remove the 2 front bolts from the rear frame.

✳✳ CAUTION

Use care not to over extend the rear brake hoses.

14. Lower the screw jack until there is approximately 50 mm (2 in) between the front mounting surface of the rear frame and the chassis. This will allow clearance to access the fuel tank strap bolts.

15. Remove the fuel tank strap bolts.

16. Position the fuel tank straps downward around the rear frame.

➡**Ensure the following are not damaged while lowering the fuel tank:**

- Fuel tank wiring harness
- EVAP wiring harness

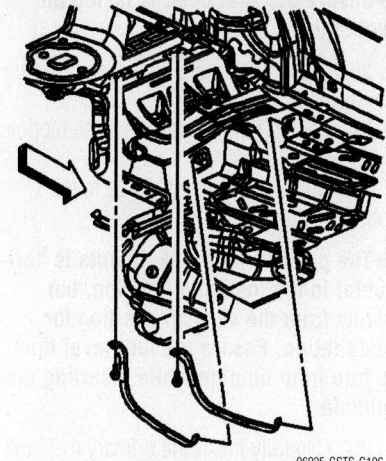

06025-GSTS-G196

Removing the fuel tank straps

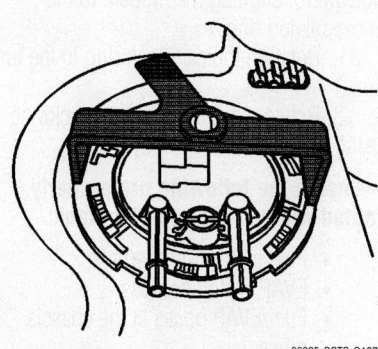

06025-GSTS-G197

Using the lock ring tool J45747 to remove the cam lock ring

- Fuel/EVAP hoses at the chassis pipes

17. With the aid of an assistant, carefully lower the fuel tank from the vehicle.

18. Rotate the cam lock ring counter-clockwise using the J 45747.

19. Remove the cam lock ring from the fuel tank.

➡**The primary fuel tank module is horizontal in the installed position, but pivots vertically for removal.**

20. Carefully lift the primary fuel tank module from the fuel tank only enough to access the transfer tube.

21. Pull the locking mechanism away from the module.

22. Remove the transfer tube from the module.

23. Remove the primary fuel tank module from the fuel tank.

24. Remove the primary fuel tank module seal from the module. Do not reuse the seal.

To install:

25. Place the new primary fuel tank module seal over the module.

➡**Ensure the seal bead is facing the fuel tank.**

26. Grasp the transfer tube from inside the fuel tank.

27. Position the module near the module opening.

28. Connect the transfer tube to the module.

➡**The primary fuel tank module is horizontal in the installed position, but pivots from the vertical position for installation. Ensure the fuel level float is free from binding while inserting the module**

29. Carefully insert the primary fuel tank module into the fuel tank.

30. Press the primary fuel tank module downward, aligning the module to the encapsulated ring.

31. Position the cam lock ring to the fuel tank.

32. Rotate the cam lock ring clockwise using the J 45747 until fully seated.

➡**Ensure the following are properly routed while raising the fuel tank:**

- Fuel tank wiring harness
- EVAP wiring harness
- Fuel/EVAP hoses at the chassis pipes

33. With the aid of an assistant, carefully raise the fuel tank to the vehicle, aligning the filler neck with the filler hose.

➡**Ensure the fuel tank straps are not pressed into the fuel tank.**

34. Carefully bend the fuel tank straps back to their original form.

35. Position the fuel tank straps around the rear frame and upward into position, aligning the holes in the straps with the threaded holes in the chassis.

36. Install the fuel tank strap bolts.

37. Tighten the fuel tank strap bolts to 37 ft. lbs. (50 Nm).

38. Connect the filler hose to the fuel tank.

39. Tighten the fuel filler tube hose clamp (3) to 31 inch lbs. (3.5 Nm).

40. Connect the filler vent tube to the EVAP hose.

41. Connect the EVAP hoses to the EVAP canister.

42. Insert the retainer into the chassis and press inward on the tab to engage.

43. Connect the electrical connector to the EVAP canister.

44. Connect the fuel tank electrical connector.

45. Connect the fuel feed hoses.

46. Connect fuel EVAP hose.

47. Raise the rear frame using the screw jack.

48. Install the 2 front bolts to the rear frame.

49. Tighten the rear frame bolts to 195 ft. lbs. (265 Nm).

50. Remove the screw jack.

51. Position the screw jack under the lower control arm in order to raise the lower control arms.

52. Install the lower shock bolts.

53. Tighten the lower shock bolts to 111 ft. lbs. (150 Nm).

54. Remove the screw jack.

55. Install the propeller shaft. See Propeller Shaft.

56. Install the exhaust system.

57. Refill the fuel tank.

58. Turn the ignition ON for 10 seconds and then turn it OFF for 10 seconds. Again turn the ignition ON and inspect for leaks.

Fuel Rail and Injectors

REMOVAL & INSTALLATION

3.6L

1. Before servicing the vehicle, refer to the precautions in the beginning of this section.

2. Relieve the fuel system pressure. Refer to the fuel system relief procedure in this section.

3. Remove the upper intake manifold with the lower intake manifold.

4. Remove the fuel pipe retaining clip.

5. Disconnect the fuel feed pipe (2) from the fuel injector rail.

6. Use compressed air in order to remove debris from the area where the fuel injectors enter the intake manifold.

7. Remove the fuel rail bolts.

06025-GSTS-G202

Showing the fuel rail bolt locations

✳✳ **WARNING**

Remove the fuel rail assembly carefully in order to prevent damage to the injector electrical connector terminals and the injector spray tips. Support the fuel rail after the fuel rail is removed in order to avoid damaging the fuel rail components.

8. Cap the fittings and plug the holes when servicing the fuel system in order to prevent dirt and other contaminants from entering open pipes and passages.

9. Remove the fuel rail with the fuel injectors.

10. Disengage the fuel injector electrical connector lock.

11. Disconnect the fuel injector electrical connector.

12. Remove the fuel injector retainer clip.

13. Remove the fuel injector.

14. Remove and discard the fuel injector seals.

To install:

15. Install NEW fuel injector seals.

16. Install the fuel injector.

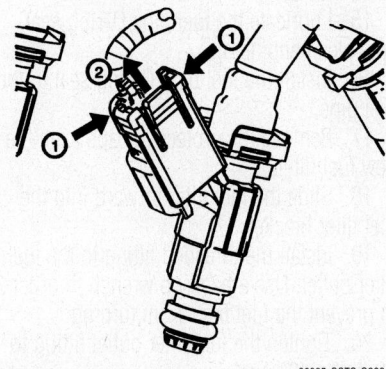

06025-GSTS-G203

Sequence for removing the fuel injector electrical connector

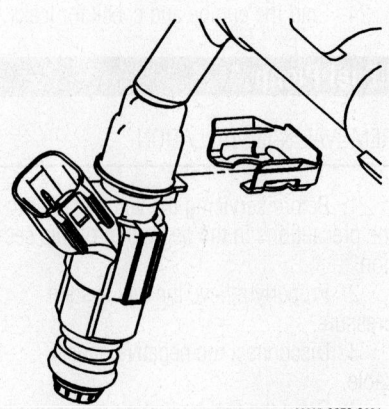

06025-GSTS-G204

Removing the fuel injector retainer clip

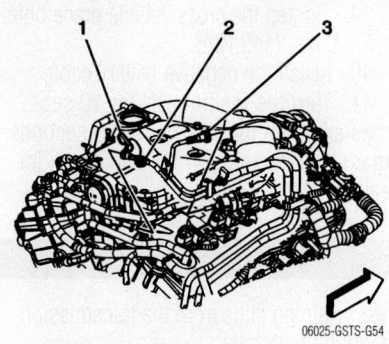

Showing the fuel feed pipe connection locations

17. Install the fuel injector retainer clip.

18. Install the fuel injector electrical connector.

19. Engage the fuel injector electrical connector lock.

20. Install the fuel rail with the fuel injectors.

21. Install the fuel rail bolts.

22. Tighten the fuel rail bolts to 89 inch lbs. (10 Nm).

23. Connect the fuel feed pipe (2) to the fuel rail.

24. Install the fuel pipe retaining clip.

25. Install the upper intake manifold.

26. Install the negative battery cable.

27. Turn the ignition ON for 10 seconds, and then turn it OFF for 10 seconds. Again, turn the ignition ON and check for leaks.

4.6L

1. Before servicing the vehicle, refer to the precautions in the beginning of this section.

2. Relieve the fuel system pressure.

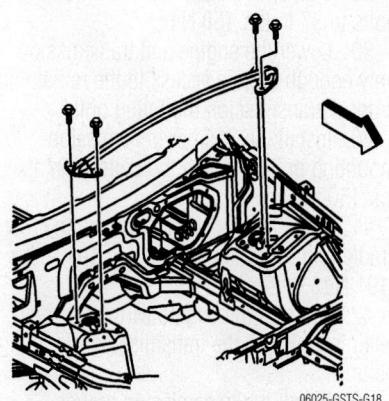

Removing the vehicle cross brace

Refer to the fuel system relief procedure in this section.

3. Disconnect the negative battery cable.

4. Remove the cross vehicle brace.

5. Clean the fuel rail assembly with a spray type engine cleaner (GM X-30A), if necessary.

❋❋ CAUTION

Follow the package instructions. Do not soak the fuel rail in liquid cleaning solvent.

6. Disconnect the fuel feed hose/pipe from the fuel rail.

7. Remove the positive crankcase ventilation (PCV) dirty air hose.

8. Disconnect the evaporative emission (EVAP) quick connect fitting

9. Open the retainers located at the right side rear of the engine and the front of dash. Position the lines aside.

10. Disconnect the left side fuel injector wiring harness connectors from the fuel injectors.

11. Disconnect the right side fuel injec-

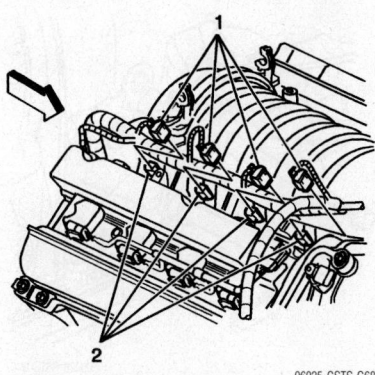

Disconnecting the fuel injector wiring harness and injectors

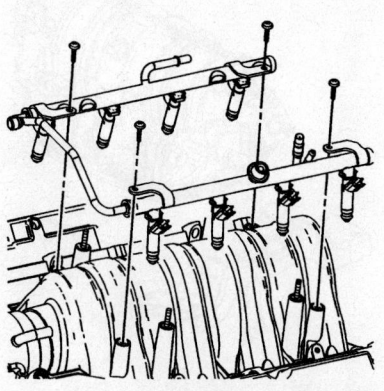

Removing the fuel injector rail assembly

tor wiring harness connectors (1) from the fuel injectors (2).

12. Remove the fuel injector sight shield bracket.

13. Remove the fuel rail attaching studs.

14. Remove the fuel rail assembly.

15. Remove the injector lower O-ring seal from the spray tip end of each injector.

16. Discard the O-ring seals.

17. Remove the retainers from the fuel injectors.

18. Remove the fuel injectors from the fuel rail.

19. Remove the O-ring seals from the fuel injectors. Discard the O-ring seals.

To install:

❋❋ WARNING

Use care when servicing the fuel system components, especially the fuel injector electrical connectors, the fuel injector tips, and the injector O-rings. Plug the inlet and the outlet ports of the fuel rail in order to prevent contamination.

- Do not use compressed air to clean the fuel rail assembly as this may damage the fuel rail components.
- Do not immerse the fuel rail assembly in a solvent bath in order to prevent damage to the fuel rail assembly.

20. Lubricate the new fuel injector O-ring seals with clean engine oil.

21. Install the new fuel injector O-ring seals on to the fuel injectors.

22. Install the fuel injectors to the fuel rail using new retainer clips.

23. Ensure that the injectors are aligned by orientating the electrical connectors perpendicular to the crankshaft centerline.

24. Lubricate and install the new O-ring seals on the spray tip end of each injector.

25. Align the fuel injectors and fuel rail to the intake manifold.

26. Carefully press the fuel rail assembly downward until fully seated against the intake manifold.

27. Install the fuel rail attaching studs.

28. Tighten the studs to 89 inch lbs. (10 Nm).

29. Install the fuel injector sight shield bracket nuts.

30. Tighten the nuts to 89 inch lbs. (10 Nm).

31. Connect the right side fuel injector electrical connectors (1) to each fuel injector (2).

32. Connect the left side fuel injector electrical connectors to each fuel injector.

33. Install the fuel feed and EVAP lines to the retaining clips at the right side rear of the engine and at the front of dash.

34. Install the EVAP quick connect fitting to the purge valve.

35. Install the positive crankcase ventilation (PCV) dirty air hose.

36. Connect the fuel feed hose/pipe to the fuel rail

37. Install the cross vehicle brace.

38. Install the bolts to the cross vehicle brace.

39. Tighten the cross vehicle brace bolts to 83 ft. lbs. (112 Nm).

40. Install the negative battery cable

41. Turn the ignition ON for 10 seconds and then turn it OFF for 10 seconds. Again turn the ignition ON and check for leaks.

DRIVE TRAIN

Automatic Transmission Assembly

REMOVAL & INSTALLATION

3.6L

1. Before servicing the vehicle, refer to the precautions in the beginning of this section.

2. Disconnect the negative battery cable.

3. Drain the transmission fluid.

4. Remove the thermostat housing to gain access to the upper transmission mounting bolts.

5. Raise and support the vehicle.

6. Remove the transmission manual shift shaft nut.

7. Disconnect the shift linkage from the transmission.

8. Place the transmission in neutral by rotating the transmission shift shaft clockwise 2 clicks.

9. Remove the exhaust system.

10. Disconnect the propeller shaft coupler (1) from the transmission flange. See Propeller Shaft.

11. Disconnect the transmission wiring harness connector from the transmission by rotating the locking latch counterclockwise

12. Disconnect the wiring harness clips from the transmission, and position the wiring harness aside.

13. Remove the starter motor to gain access to the torque converter bolts.

14. Remove the front air deflector.

15. Mark the torque converter to flexplate/flywheel orientation to ensure proper realignment.

16. Remove the torque converter bolts by rotate the harmonic balancer center bolt clockwise ONLY, in order to align the torque converter bolt with the starter motor opening in the engine block.

17. Remove and discard the torque converter bolt. The bolt is self locking and is NOT reusable.

18. Place an oil drain pan under the transmission fluid cooler pipes.

19. Remove the bolt securing the transmission fluid cooler pipes brace to the engine.

20. Disconnect the fluid cooler pipes from the transmission.

21. Plug the open outlet ports to prevent fluid loss and contamination.

22. Position a suitable transmission jack under the transmission.

23. On RWD models, remove the transmission rear mount.

24. Remove the 3 lower transmission mounting bolts.

25. Support the engine with a suitable jack. Loosen the two rear cradle bolts approximately one inch. Lower the engine and transmission just enough to gain access to the remaining 3 upper transmission mounting bolts.

26. Disconnect the engine wiring har-

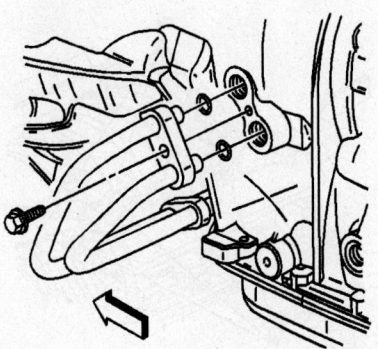

06025-GSTS-G207

Disconnecting the transmission cooler pipes

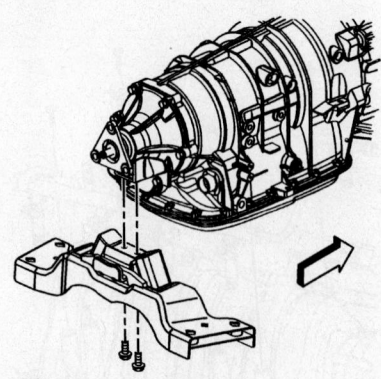

06025-GSTS-G208

Removing the transmission rear mount (RWD)

ness retaining clips from the transmission mounting bolts.

27. Remove the 3 upper transmission mounting bolts.

28. Pull the transmission free from the engine dowels

➡Ensure clearance is maintained between the transmission and the following:

- Catalytic converters
- Wiring harnesses
- Cooler pipes
- Propeller shaft

29. Carefully lower the transmission from the vehicle.

30. If the transmission is being replaced, remove the drive flange and install it on the replacement unit

To install:

➡Ensure clearance is maintained between the transmission and the following:

31. The catalytic converters

32. The wiring harnesses

33. The cooler pipes

34. The propeller shaft

35. Using the transmission jack, carefully raise the transmission to the vehicle.

36. Align the engine dowels (1) with the transmission.

37. Install the 3 lower transmission mounting bolts (2).

38. Tighten the transmission mounting bolts to 37 ft. lbs. (50 Nm).

39. Lower the engine and transmission only enough to gain access to the remaining 3 upper transmission mounting bolts.

40. Install the 3 upper transmission mounting bolts. Tighten the bolts to 37 ft. lbs. (50 Nm).

41. Raise the engine and tighten the rear cradle bolts. Tighten the bolts to 141 ft. lbs. (191 Nm).

42. Connect the engine wiring harness retaining clips to the transmission mounting bolts.

43. Install the transmission mount.

44. Remove transmission jack from under the transmission.

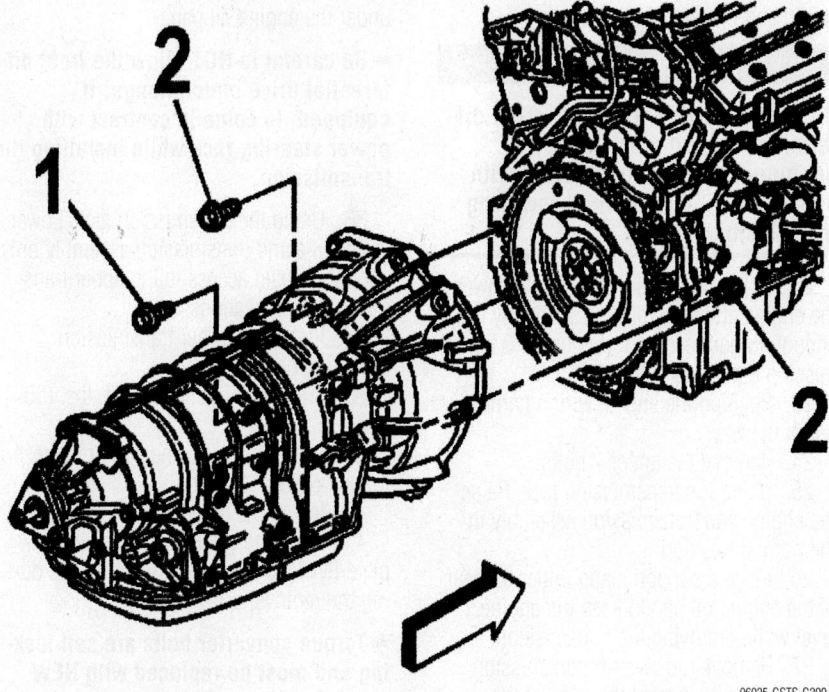

Installing the transmission

06025-GSTS-G209

45. Lubricate the O-rings with automatic transmission fluid.

➡**Replace the O-rings if cracked, cut, or distorted.**

46. Install the O-rings onto the cooler pipes prior to inserting the cooler pipes into the transmission.

47. Insert the transmission fluid cooler pipes into the transmission.

48. Install the bolt securing the transmission fluid cooler pipe retainer to the transmission. Tighten the bolt to 18 ft. lbs. (25 Nm).

49. Install the bolt securing the transmission fluid cooler pipes brace to the engine. Tighten the bolt to 37 ft. lbs. (50 Nm).

50. Align the torque converter to flexplate/flywheel orientation marks.

⁂ **CAUTION**

Torque converter bolts are self locking and must be replaced with NEW torque converter bolts every time the bolts are removed.

51. Install the NEW torque converter bolts by rotate the harmonic balancer center bolt clockwise ONLY, in order to align the torque converter bolt with the starter motor opening in the engine block.

52. To aid in alignment of the torque converter to the flexplate/flywheel. Install all

3 NEW torque converter bolts before fully tightening.

53. Tighten the torque converter bolts to 46 ft. lbs. (63 Nm).

54. Install the starter motor.

55. Install the front air deflector.

56. Connect the wiring harness clips to the transmission.

57. Connect the transmission wiring harness connector to the transmission by rotating the locking latch clockwise.

58. Remove the mechanics wire securing front propeller shaft to the shift control lever.

59. Install the propeller shaft coupler to the transmission flange. See Propeller Shaft.

60. Place the transmission in the park position by rotating the shift shaft fully counter clockwise.

61. Connect the shift linkage to the transmission.

62. Install the transmission manual shift shaft nut.

63. Tighten the transmission manual shift shaft nut to 11 ft. lbs. (15 Nm).

64. Install the exhaust system.

65. Check the transmission fluid level (fill if necessary).

66. Adjust the shift control linkage.

67. Hold the selector lever on the transmission against the rear stop in the PARK position to eliminate any play in the forward section of the linkage.

68. Grasp the linkage at the control lever.

69. Push the linkage forward and make a note of the amount of travel the rod makes going into the adjustment sleeve.

70. Pull the linkage rearward and make a note of the amount of travel the rod makes coming out of the adjustment sleeve.

71. Place the rod into the adjustment sleeve half the distance of the complete amount of travel.

72. Tighten the shift control linkage adjustment nut to 80 inch lbs. (9 Nm).

73. Lower the vehicle.

74. Inspect the operation of the starting system with the shift lever in each position. The engine should only crank when the lever is in the PARK or NEUTRAL position.

75. Ensure that the shifter detents align with the gear indicators.

76. Install the thermostat housing.

77. Connect the negative battery cable.

78. Complete the following procedure after the transmission is installed in the vehicle:

79. With the ignition OFF or disconnected, crank the engine several times. Listen for any unusual noises or evidence that any parts are binding.

80. Start the engine and listen for abnormal conditions.

81. While the engine continues to idle raise and support the vehicle.

82. Inspect for fluid leaks while the engine is idling.

83. Lower the vehicle.

84. Perform a final inspection for the proper fluid level.

85. Road test the vehicle.

4.6L

1. Before servicing the vehicle, refer to the precautions in the beginning of this section.

2. Raise and support the vehicle.

3. Remove the exhaust system.

4. Remove the rear propeller shaft. See Propeller Shaft.

5. Remove the transfer case if equipped. See Transfer Case.

6. Disconnect the shift linkage from the transmission. The linkage may be on the right or left side of the transmission depending on the application.

7. Disconnect the transmission wiring harness connector from the transmission by rotating the locking latch counterclockwise.

8. Disconnect the wiring harness retainers from the transmission, and position the wiring harness aside.

9. Remove the front air deflector.

10. Remove the transmission fluid cooler

lines form the plastic cooler line retainer bracket.

11. Place an oil drain pan under the transmission fluid cooler pipes.

12. Remove the transmission fluid cooler pipes from the transmission, and position aside.

13. Remove and discard the O-rings. Do NOT reuse the O-rings.

14. Plug the open outlet ports to prevent fluid loss and contamination.

15. Remove the transmission close out plug.

16. Mark the torque converter to flex-plate/flywheel orientation to ensure proper realignment.

17. Pull the torque converter bolt close out cover upper retaining pin downward unlocking it from the engine block.

18. Remove the torque converter bolt close out cover from the engine block.

19. Remove the torque converter bolts by rotate the harmonic balancer center bolt clockwise ONLY, in order to align the torque converter bolt with the starter motor opening in the engine block.

20. Remove and discard the torque converter bolt. The bolt is self locking and is NOT reusable

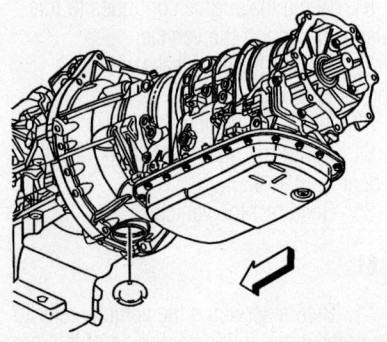

06025-GSTS-G210

Removing the transmission close out plug

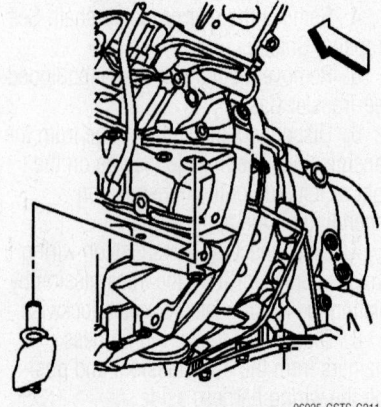

06025-GSTS-G211

Removing the close out cover retaining pin

21. Remove the front differential heat shield and bracket if equipped.

❊❊ CAUTION

Be careful to NOT allow the front differential drive pinion flange, if equipped, to come in contract with power steering rack while lowering the transmission.

22. Using the transmission jack. Lower the engine and transmission assembly only enough to gain access to the upper 2 transmission mounting bolts.

23. Position the engine wiring harness out of the way.

24. Remove the upper 2 bolts.

25. Using the transmission jack. Raise the engine and transmission assembly to the normal position.

26. Place a support stand under the rear of the engine oil pan to keep the engine level while removing the transmission.

27. Remove the lower 8 transmission mounting bolts around the edge of the housing.

28. Pull the transmission free from the engine dowels.

29. Separate the Transmission from the engine approximately 1 inch (25mm) to clear the starter drive nose cone.

➥**During transmission removal, ensure clearance is maintained between the transmission and the following components:**

- Starter drive nose cone
- Catalytic converters
- Oxygen sensors
- Wiring harnesses
- Cooler pipes

30. Carefully lower the transmission from the vehicle.

To install:

➥**Ensure clearance is maintained between the transmission and the following:**

- Starter drive nose cone
- Catalytic converters
- Oxygen sensors
- Wiring harnesses
- Cooler pipes

31. Using the transmission jack, carefully raise the transmission to meet the engine.

32. Align the transmission with the engine dowels

33. Install the lower 8 transmission mounting bolts.

34. Tighten the bolts to 37 ft. lbs. (50 Nm).

35. Remove the support stand from under the engine oil pan.

➥**Be careful to NOT allow the front differential drive pinion flange, if equipped, to come in contract with power steering rack while installing the transmission.**

36. Using the transmission jack. Lower the engine and transmission assembly only enough to gain access to the upper transmission bolt locations.

37. Install the upper transmission mounting bolts.

38. Tighten the bolts to 37 ft. lbs. (50 Nm).

39. Position the engine wiring harness to the original location.

40. Install the transfer case if equipped.

41. Align the torque converter to flexplate/flywheel orientation marks made during the removal procedure.

➥**Torque converter bolts are self locking and must be replaced with NEW torque converter bolts every time the bolts are removed.**

42. Install the NEW torque converter bolts by rotate the harmonic balancer center bolt clockwise ONLY, in order to align the torque converter bolt with the starter motor opening in the engine block.

43. To aid in alignment of the torque converter to the flexplate/flywheel. Install all 3 NEW torque converter bolts before fully tightening.

44. Tighten the torque converter bolts to 46 ft. lbs. (63 Nm).

45. Install the torque converter bolt close out cover to the engine block.

46. Push the torque converter bolt close out cover upper retaining pin upward locking it into the engine block.

47. Install the transmission close out plug.

48. Place NEW O-rings over the transmission fluid cooler pipes.

49. Install the transmission fluid cooler pipes to the transmission.

50. Clip the transmission fluid cooler lines into the plastic cooler line retainer bracket.

51. Install the front differential heat shield and bracket.

52. Install the front air deflector.

53. Connect the wiring harness retainers to the transmission.

54. Connect the transmission wiring harness connector to the transmission by rotating the locking latch clockwise.

55. Connect the shift linkage to the transmission.

56. Install the rear propeller shaft. See Propeller Shaft.

57. Install the exhaust system.

58. Check the transmission fluid level (fill if necessary).

59. Adjust the shift control linkage.

60. Hold the selector lever on the transmission against the rear stop in the PARK position to eliminate any play in the forward section of the linkage.

61. Grasp the linkage at the control lever.

62. Push the linkage forward and make a note of the amount of travel the rod makes going into the adjustment sleeve.

63. Pull the linkage rearward and make a note of the amount of travel the rod makes coming out of the adjustment sleeve.

64. Place the rod into the adjustment sleeve half the distance of the complete amount of travel.

65. Tighten the shift control linkage adjustment nut to 80 inch lbs. (9 Nm).

66. Lower the vehicle.

67. Inspect the operation of the starting system with the shift lever in each position. The engine should only crank when the lever is in the PARK or NEUTRAL position.

68. Ensure that the shifter detents align with the gear indicators.

69. Install the thermostat housing.

70. Connect the negative battery cable.

71. Complete the following procedure after the transmission is installed in the vehicle:

72. With the ignition OFF or disconnected, crank the engine several times. Listen for any unusual noises or evidence that any parts are binding.

73. Start the engine and listen for abnormal conditions.

74. While the engine continues to idle raise and support the vehicle.

75. Inspect for fluid leaks while the engine is idling.

76. Lower the vehicle.

77. Perform a final inspection for the proper fluid level.

78. Road test the vehicle.

Flywheel

REMOVAL & INSTALLATION

1. Before servicing the vehicle, refer to the precautions in the beginning of this section.

2. Disconnect the negative battery cable.

3. Remove the transmission. See Automatic Transmission.

4. Remove the engine flywheel bolts and the flywheel.

5. Clean and inspect the flywheel. Replace the starter if you find excessive wear or damage to the starter drive.

To install:

6. Place the engine flywheel in position on the crankshaft.

7. Install 2 NEW bolts in location at the top and bottom of the engine flywheel bolt pattern allowing the engine flywheel to hang in position.

8. Install the remaining NEW engine flywheel bolts as follows:

- With 3.6L engine, tighten the NEW engine flywheel bolts to 22 ft. lbs. (30 Nm), plus an additional 45 degrees.
- With 4.6L engine, tighten the engine flywheel bolts to 11 ft. lbs. (15 Nm), plus an additional 50 degrees.

9. Install the transmission. See Automatic Transmission.

10. Install the negative battery cable.

Halfshaft

REMOVAL & INSTALLATION

Front Left Halfshaft

1. Before servicing the vehicle, refer to the precautions in the beginning of this section.

2. Unlock the steering column so the steering linkage is free to move.

3. Raise and support the vehicle.

4. Remove the left front tire and wheel assembly after making a reference mark from the wheel to the hub for reinstallation positioning.

5. Disconnect the left outer tie rod end from the steering knuckle. Do NOT loosen the tie rod end jam nut.

6. Insert a drift or punch into the brake rotor and against the brake caliper mounting bracket in order to prevent the wheel hub and bearing from turning.

✳✳ CAUTION

The wheel halfshaft spindle nut must not be reused. Replace the wheel halfshaft spindle nut with a new nut whenever it is removed.

7. Remove and discard the wheel halfshaft spindle nut retaining the wheel halfshaft to the hub.

✳✳ CAUTION

Be sure that the wheel speed sensor wiring harness is repositioned away from the ball joint after disconnecting the electrical connector from the sensor.

8. Disconnect the electrical connector from the wheel speed sensor and reposition the wiring harness away from the ball joint.

9. Disconnect the left upper ball joint from the steering knuckle.

10. Install the wheel hub and removal tool (J 45859) onto the wheel hub and secure with wheel nuts.

➡Be sure to support the wheel halfshaft until it is fully removed from the vehicle.

11. With the tool, disengage the left front wheel halfshaft from the wheel hub and bearing and support the wheel halfshaft.

12. Remove tool from the wheel hub.

13. Assemble the halfshaft puller tool assembly (J2619-01, J 29794, J 45341).

14. Install tool to the wheel halfshaft inner joint pull groove.

15. Using the halfshaft puller, or an

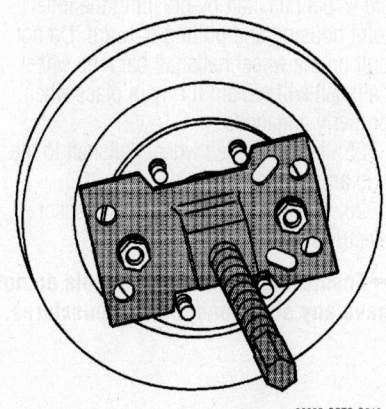

06025-GSTS-G213

Using the J 45859 to remove the hub

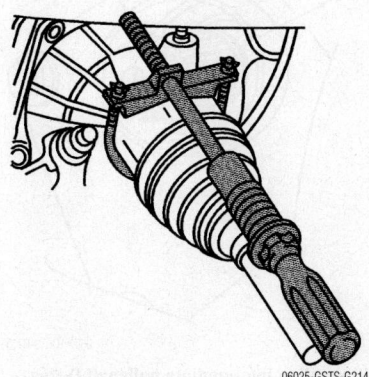

06025-GSTS-G214

Halfshaft puller tool assembled in place

equivalent tool disengage the wheel half-shaft from the intermediate wheel halfshaft.

16. Remove the left front wheel halfshaft from the vehicle.

17. Remove the halfshaft puller from the wheel halfshaft inner joint.

18. If reusing the intermediate wheel halfshaft remove and discard the wheel half-shaft retaining ring from the intermediate shaft retaining ring groove. The wheel half-shaft retaining ring is on the splined end of the shaft.

19. If reusing the intermediate wheel halfshaft remove and discard the wheel half-shaft O-ring from the O-ring groove. The O-ring is on the splined end of the shaft.

To install:

20. Install a new O-ring (1) to the inter-mediate wheel halfshaft O-ring groove.

21. Install the new wheel halfshaft retain-ing ring (2) to the retaining ring groove. The intermediate wheel halfshaft retaining ring groove is on the splined end of the shaft.

22. Apply a small amount of grease (GM P/N 01051344) to the intermediate wheel halfshaft splines.

23. Install the left front wheel halfshaft to the intermediate wheel halfshaft.

24. Verify that the left front wheel half-shaft is properly engaged to the intermedi-ate wheel halfshaft by grasping the inner joint housing and pulling outward. Do not pull on the wheel halfshaft bar. The wheel halfshaft will remain firmly in place when properly engaged.

25. Install the left wheel halfshaft to the hub and bearing.

26. Connect the wheel speed sensor electrical connector.

➡**Ensure that the halfshaft seals do not have any abrasions, cuts or punctures.**

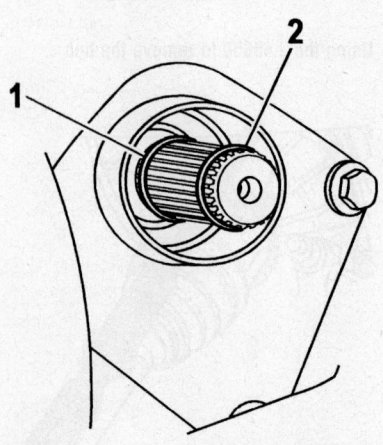

Illustrating intermediate halfshaft O-ring (1) and retaining ring (2)

06025-GSTS-G215

27. Loosely install the NEW wheel half-shaft spindle nut.

28. Insert a drift or punch into the rotor and against the caliper mounting bracket in order to prevent the hub and bearing from turning.

29. Use the New wheel halfshaft spindle nut to slowly pull the spindle to the wheel hub and bearing assembly.

30. Tighten the wheel halfshaft spindle nut to 118 ft. lbs. (160 Nm).

31. Connect the outer tie rod end to the steering knuckle.

32. Install the left tire and wheel assem-bly.

33. Lower the vehicle.

Front Right Halfshaft

1. Raise and support the vehicle.

2. Remove the right front tire and wheel assembly.

3. Disconnect the right outer tie rod end from the steering knuckle. Do NOT loosen the tie rod end jam nut.

4. Insert a drift or punch into the brake rotor and against the brake caliper mounting bracket, in order to prevent the wheel hub and bearing from turning.

➡**The wheel halfshaft spindle nut must not be reused. Replace the wheel half-shaft spindle nut with a new nut when-ever it is removed.**

5. Remove and discard the right front wheel halfshaft spindle nut retaining the wheel halfshaft to the hub.

➡**Be sure that the wheel speed sensor wiring harness is repositioned away from the ball joint after disconnecting the electrical connector from the sen-sor.**

6. Disconnect the electrical connector from the wheel speed sensor and reposition the wiring harness away from the ball joint.

7. Disconnect the right upper ball joint from the steering knuckle.

8. Install the J 45859) onto the wheel hub and secure with wheel nuts.

➡**Be sure to support the wheel half-shaft until it is fully removed from the vehicle.**

9. Using the removal tool, disengage the right front wheel halfshaft from the wheel hub and bearing and support the wheel halfshaft.

10. Remove the tool from the wheel hub.

11. Assemble the halfshaft puller tool assembly (J2619-01, J 29794, J 45341).

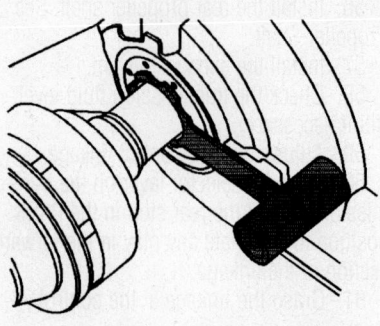

06025-GSTS-G216

Differential output shaft seal protector (J 44394) installed

12. Install puller tool assembly to the wheel halfshaft inner joint pull groove.

13. Using the halfshaft puller disengage the wheel halfshaft from the differential enough to install the differential output shaft seal protector (J 44394).

✲✲ WARNING

The protector tool (J-44394) must be installed into the differential output shaft seal prior to removing and installing the wheel halfshaft. Failure to install this tool may cause the splines of the wheel halfshaft to cut the differential output seal.

14. Carefully install the seal protector (J 44394) over the wheel halfshaft.

15. Carefully slide the tool (J 44394) into the differential output shaft seal.

16. Remove the wheel halfshaft from the vehicle.

17. Remove the halfshaft puller from the wheel halfshaft inner joint.

18. Remove and discard the wheel half-shaft retaining ring from the retaining ring groove. The wheel halfshaft retaining ring is on the splined shaft of the inner joint housing.

To install:

19. Install the new wheel halfshaft retain-ing ring to the retaining ring groove. The wheel halfshaft retaining ring is on the splined shaft of the inner joint housing.

20. If previously removed, carefully install J 44394 into the differential output shaft seal.

➡**In order to prevent lubricant leaks, use care when installing the wheel halfshaft to the differential. Do not damage the oil seal. Replace the oil seal if it becomes nicked, distorted, or is otherwise damaged.**

21. Carefully install the wheel halfshaft into the differential until the splines are past the seal protector (J 44394).

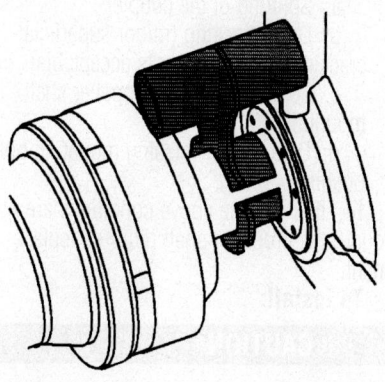

06025-GSTS-G217

Removing the differential output shaft seal protector (J 44394)

22. Carefully remove the J 44394 from the differential output shaft seal.

23. Carefully continue installing the wheel halfshaft into the differential until the retaining ring is fully seated.

24. Verify that the wheel halfshaft is properly engaged to the differential by grasping the inner joint housing and pulling outward. Do not pull on the wheel halfshaft bar. The wheel halfshaft will remain firmly in place when properly engaged.

25. Install the right front wheel halfshaft to the hub and bearing.

26. Connect the right upper ball joint to the steering knuckle.

➡**Ensure that the halfshaft seals do not have any abrasions, cuts or punctures.**

27. Loosely install the NEW wheel halfshaft spindle nut.

28. Insert a drift or punch into the rotor and against the caliper mounting bracket in order to prevent the hub and bearing from turning.

29. Use the new wheel halfshaft spindle nut to slowly pull the spindle to the wheel hub and bearing assembly. Once the hub is in place, tighten the wheel halfshaft spindle nut to 118 ft. lbs. (160 Nm).

30. Remove the drift or punch from the rotor.

31. Connect the outer tie rod end to the steering knuckle.

32. Install the tire and wheel assembly.

33. Lower the vehicle.

Rear Halfshaft

1. Raise and support the vehicle.
2. Remove the tire and wheel assembly

➡**The wheel halfshaft spindle nut must not be reused. Replace the wheel halfshaft spindle nut with a new nut whenever it is removed.**

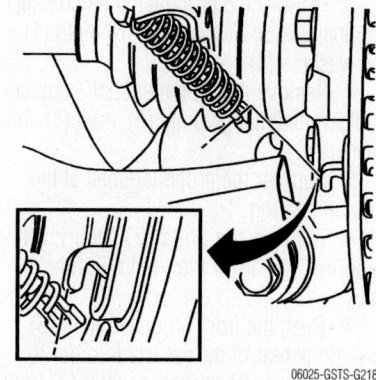

06025-GSTS-G218

Removing the parking brake cable lever

3. Insert a drift or punch into the brake rotor and against the caliper in order to prevent the wheel hub and bearing from turning.

4. Remove and discard the wheel halfshaft spindle nut.

5. Remove the drift or punch from the rotor.

6. Remove the rear knuckle assembly by:

7. Disconnect the Antilock Brake System (ABS) sensor harness connector.

8. Disconnect the ABS sensor harness connector from the backing plate.

9. Remove the parking brake cable bracket from the knuckle.

10. Remove the parking brake cable from the parking brake lever.

11. Remove the upper ball joint mounting nut.

12. Use a proper tool to separate the ball stud from the knuckle. See Lower Control Arm.

13. Support the lower control arm with a suitable jack.

14. Remove the lower shock mounting bolt.

15. Remove the trailing arm to knuckle mounting bolt and nut.

16. Remove the lower control arm to knuckle mounting bolt.

17. Remove the adjustment link to knuckle mounting bolt.

18. Using a suitable tool, carefully release the wheel halfshaft from the rear differential enough to install the differential output shaft seal protector (J 44394).

✳✳ WARNING

The output shaft seal protector (J-44394) must be installed into the differential output shaft seal prior to removing and installing the wheel halfshaft. Failure to install the protector may cause the splines of the wheel halfshaft to cut the differential output seal.

19. Carefully install the seal protector over the wheel halfshaft.

20. Carefully slide the protector tool into the differential output shaft seal.

21. Remove the wheel halfshaft from the vehicle.

22. Remove and discard the wheel halfshaft retaining ring. The wheel halfshaft retaining ring is on the splined shaft of the inner joint housing.

To install:

23. Install the new wheel halfshaft retaining ring. The wheel halfshaft retaining ring is on the splined shaft of the inner joint housing.

✳✳ WARNING

The output shaft seal protector (J-44394) must be installed into the differential output shaft seal prior to removing and installing the wheel halfshaft. Failure to install the protector may cause the splines of the wheel halfshaft to cut the differential output seal.

24. If previously removed, carefully install the seal protector (J 44394) to the differential output shaft seal.

➡**In order to prevent lubricant leaks, use care when installing the wheel halfshaft to the differential. Do not damage the oil seal. Replace the oil seal if it becomes nicked, distorted, or is otherwise damaged.**

25. Carefully install the wheel halfshaft into the differential until the splines are past the seal protector tool.

26. Carefully remove the protector tool from the differential output shaft seal.

27. Carefully install the wheel halfshaft into the differential until the retaining ring is engaged.

28. Ensure the wheel halfshaft retaining ring is fully engaged to the differential by grasping the inner joint housing and pulling outward. The wheel halfshaft will stay positively engaged if properly installed to the differential.

29. Install the rear knuckle as follows:

a. Install the knuckle to the vehicle.

b. Install the adjustment link to knuckle mounting bolt.

c. Tighten the adjustment link to knuckle bolt to 118 ft. lbs. (160 Nm).

d. Install the trailing arm to knuckle mounting bolt and nut.

e. Tighten the trailing arm to knuckle bolt to 170 Nm (125 ft. lbs.).

f. Install the lower shock and mounting bolt.

g. Tighten the lower shock mounting bolt to 111 ft. lbs. (150 Nm).

h. Install the lower control arm to knuckle mounting bolt.

i. Connect the upper ball joint to the knuckle.

j. Install the upper ball joint mounting nut.

k. Install the parking brake cable bracket to the knuckle.

30. Remove the jack.

31. Connect the ABS sensor electrical connector to the backing plate.

32. Connect the ABS sensor electrical connector.

33. Loosely install the NEW wheel half-shaft spindle nut.

34. Insert a drift or punch into the brake rotor and against the caliper in order to prevent the wheel hub and bearing from turning.

✳✳ WARNING

Do not use paints, lubricants, or corrosion inhibitors on fasteners or fastener joint surfaces unless specified. These coatings affect fastener torque and joint clamping force and may damage the fastener. Use the correct tightening sequence and specifications when installing fasteners in order to avoid damage to parts and systems.

35. Use the new wheel halfshaft spindle nut to slowly pull the spindle to the wheel hub and bearing assembly.

36. Tighten the wheel halfshaft spindle nut to 118 ft. lbs. (160 Nm).

37. Remove the drift or punch from the rotor.

38. Install the tire and wheel assembly.

39. Inspect the differential lubricant level.

40. Lower the vehicle.

Propeller Shaft

REMOVAL & INSTALLATION

Rear Wheel Drive

1. Raise and support the vehicle.
2. Remove the exhaust system

✳✳ CAUTION

Remove only the propeller shaft coupler-to-transmission flange bolts. Do NOT remove the coupler from the propeller shaft.

3. Remove the propeller shaft coupler-to-transmission flange bolts (6), nuts (4) and washers (5).

4. Remove the propeller shaft coupler-to-differential flange bolts (6), nuts (4) and washers (5).

5. Support the propeller shaft at the support bearing.

6. Remove the bolts (2) securing the support bearing to the vehicle underbody.

7. Push the front propeller shaft (3) toward the rear of the vehicle in order to release the propeller shaft coupler (2) from the transmission flange (1).

8. While holding the front propeller shaft (3), lower the support device under the propeller shaft support bearing.

9. Remove the support bearing from the mounting studs.

10. Pull the rear propeller shaft (1) forward to release the coupler (2) from the differential flange (3).

11. Note the number and location of the shim packs between the support bearing mounting bracket and the underbody to ensure proper assembly.

12. Remove the propeller shaft from the vehicle.

13. Inspect the propeller shaft coupler for the following conditions:

a. Splitting of the coupler

b. Deep cracking (minor superficial cracking of the coupler is acceptable).

c. Looseness at the propeller shaft mounting bolts

d. Distorted or missing mounting bolt bushings

14. If any of the above conditions are found, the propeller shaft requires replacement.

To install:

✳✳ CAUTION

The propeller shaft flange bolts must be installed toward the flanges. There are arrows on the coupler pointing toward the flange as a reference.

15. Using a suitable jack, support the propeller shaft close to the support bearing.

16. Apply a small amount of lubricant (GM P/N 1051344 (Canadian P/N 993037,),), to the rear propeller shaft centering bushing.

17. Inspect the propeller shaft-to-flange nuts (4), bolts (6) and washers (5). Replace if damaged or worn.

18. Install the rear propeller shaft coupler (2) to the differential flange (3).

19. Push the propeller shaft (1) to the

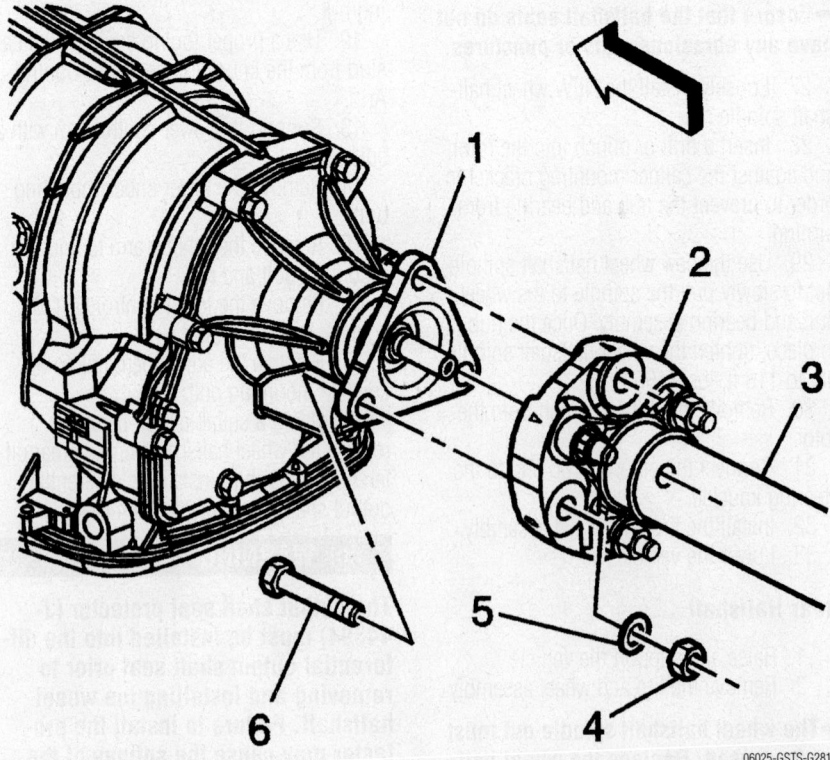

Removing front retainers (4, 5, 6) from propeller shaft (3) to separate the coupling (2) from the flange (1)

06025-GSTS-G281

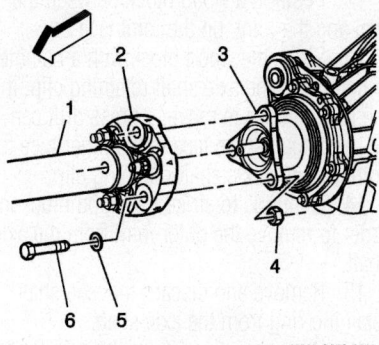

Removing rear retainers (4, 5, 6) from propeller shaft (3) to separate the coupling (2) from the flange (1)

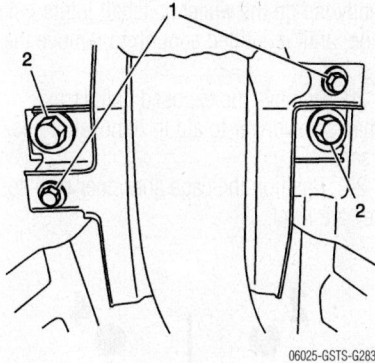

Removing center support cover bearings (2)

rear of the vehicle and install the propeller shaft coupler to the transfer case flange.

> ✳✳ **CAUTION**
>
> **If reusing the propeller shaft-to-flange nuts and bolts, to ensure proper adhesion and fastener retention, the threads must be free of debris prior to the application of thread lock sealant.**

20. Thoroughly clean the threads and allow to dry. Apply a suitable thread lock sealant to the propeller shaft to the flange bolt. Ensure that there are no gaps in the sealant along the length of the filled area of the bolt.

21. Allow the sealant to cure approximately 10 minutes before installation.

22. Install the propeller shaft coupler-to-differential flange washers (5) to the propeller shaft coupler-to-differential flange bolts (6).

23. Install the propeller shaft coupler-to-differential flange bolts (6) and washers (5) to the differential flange (3) and propeller shaft coupler (2).

24. Install the propeller shaft coupler-to-differential flange nuts (4). Tighten the pro-

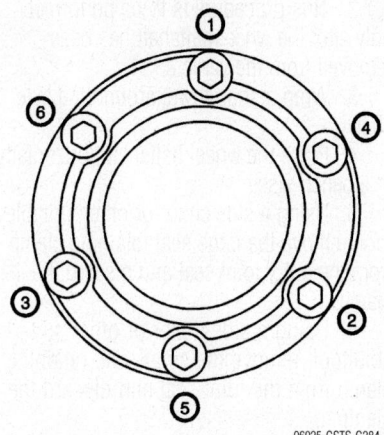

Showing propeller shaft bolt tightening sequence

peller shaft coupler to differential flange bolts to 63 ft. lbs. (85 Nm).

25. Thoroughly clean the threads and allow to dry. Apply a suitable thread lock to the propeller shaft to flange bolt. Ensure that there are no gaps in the thread lock sealant along the length of the filled area of the bolt. Allow the sealant to cure approximately 10 minutes before installation.

26. Install the propeller shaft CV joint to transfer case flange bolts. Tighten the propeller shaft CV joint to transfer case flange bolts in sequence (1-6) to 22 ft. lbs. (30 Nm).

27. Install the center support bearing to the vehicle underbody.

28. Raise the support device in order to hold the support bearing in position.

29. Install the center support bearing bolts (2) to the studs on the vehicle underbody. Tighten the support bearing bolts to 37 ft. lbs. (50 Nm).

30. Install the heat shield bolts to the mounting studs on the underbody. Tighten the heat shield mounting bolts to 71 inch lbs. (8 Nm).

31. Remove the support device from the propeller shaft.

32. Install the exhaust system.

33. Lower the vehicle.

All Wheel Drive

➡**Refer to illustrations under Rear Wheel Drive above.**

1. Raise and support the vehicle.
2. Remove the exhaust system.
3. Using a suitable jack, support the propeller shaft close to the support bearing.
4. Remove the bolts (1) securing the heat shield to the vehicle underbody.
5. Remove the bolts (2) securing the support bearing to the vehicle underbody.
6. Reference mark the location of the

propeller shaft consent velocity (CV) joint to the transfer case flange.

7. Remove the propeller shaft CV joint to transfer case flange bolts.

> ✳✳ **CAUTION**
>
> **Remove only the propeller shaft coupler-to-differential flange bolts. Do NOT remove the coupler from the propeller shaft.**

8. Remove the propeller shaft coupler-to-differential flange bolts (6), nuts (4) and washers (5).

9. Install a flat-bladed tool into the notch on the transfer case flange.

10. Using a flat-bladed tool, move the propeller shaft toward the rear of the vehicle in order to release the propeller shaft CV joint from the transfer case flange.

11. While holding the propeller shaft, lower the support device under the propeller shaft.

12. Move the propeller shaft forward to release the coupler from the differential flange.

13. Remove the propeller shaft from the vehicle.

14. Inspect the propeller shaft coupler for the following conditions:

    a. Splitting of the coupler

    b. Deep cracking (minor superficial cracking of the coupler is acceptable).

    c. Looseness at the propeller shaft mounting bolts

    d. Distorted or missing mounting bolt bushings

15. If any of the above conditions are found, the propeller shaft requires replacement.

To install:

16. Using a suitable jack, support the propeller shaft close to the support bearing.

17. Apply a small amount of lubricant (GM P/N 1051344) to the rear propeller shaft centering bushing.

18. Inspect the propeller shaft-to-flange nuts (4), bolts (6) and washers (5). Replace if damaged or worn.

19. Install the rear propeller shaft coupler (2) to the differential flange (3).

20. Push the propeller shaft (1) to the rear of the vehicle and install the propeller shaft coupler to the transfer case flange.

> ✳✳ **CAUTION**
>
> **If reusing the propeller shaft-to-flange nuts and bolts, to ensure proper adhesion and fastener retention, the threads must be free of debris prior to the application of thread lock sealant.**

21. Thoroughly clean the threads using denatured alcohol and allow to dry. Apply thread lock sealant (GM P/N 12345493), to the propeller shaft to the flange bolt. Ensure that there are no gaps in the thread lock sealant along the length of the filled area of the bolt. Allow the thread lock sealant to cure approximately 10 minutes before installation.

22. Install the propeller shaft coupler-to-differential flange washers (5) to the propeller shaft coupler-to-differential flange bolts (6).

23. Install the propeller shaft coupler-to-differential flange bolts (6) and washers (5) to the differential flange (3) and propeller shaft coupler (2).

24. Install the propeller shaft coupler-to-differential flange nuts (4). Tighten the propeller shaft coupler to differential flange bolts to 63 ft. lbs. (85 Nm).

25. Thoroughly clean the threads using denatured alcohol and allow to dry. Apply thread lock sealant (GM P/N 12345493) to the propeller shaft to flange bolt. Ensure that there are no gaps in the thread lock sealant along the length of the filled area of the bolt. Allow the thread lock sealant to cure approximately 10 minutes before installation.

26. Install the propeller shaft CV joint to transfer case flange bolts. Tighten the propeller shaft CV joint to transfer case flange bolts in sequence (1-6) to 22 ft. lbs. (30 Nm).

27. Install the center support bearing to the vehicle underbody.

28. Raise the support device in order to hold the support bearing in position.

29. Install the center support bearing bolts (2) to the studs on the vehicle underbody. Tighten the support bearing bolts to 37 ft. lbs. (50 Nm).

30. Install the heat shield bolts to the mounting studs on the underbody. Tighten the heat shield mounting bolts to 71 inch lbs. (8 Nm).

31. Remove the support device from the propeller shaft.

32. Install the exhaust system.

33. Lower the vehicle.

CV-Joints

OVERHAUL

Outer CV-Joint

1. Before servicing the vehicle, refer to the precautions in the beginning of this section.

2. This procedure is to be performed only after the wheel halfshaft has been removed from the vehicle.

3. Wrap a shop towel around the axle shaft.

4. Place the wheel halfshaft horizontally in a bench vise.

5. Using a side cutter or other suitable tool, remove the large seal retaining clamp from the outer joint seal and discard the clamp.

6. Using a side cutter or other suitable tool, remove the small seal retaining clamp from the joint seal and discard the clamp.

7. Separate the seal from the joint outer race at the large diameter end.

8. Slide the seal away from the joint face.

9. Wipe the grease from the face of the joint inner race, cage, balls, etc.

10. Remove the outer joint from the axle shaft using the following steps:

11. Hold the outer joint housing horizontally to the shaft.

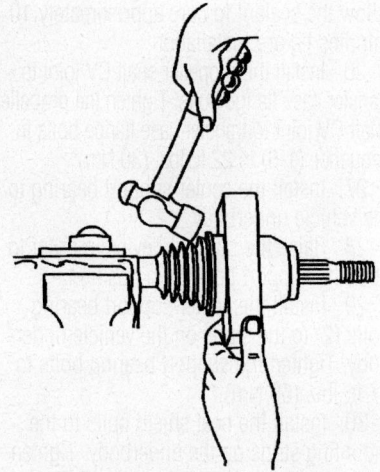

Removing the axle and outer joint

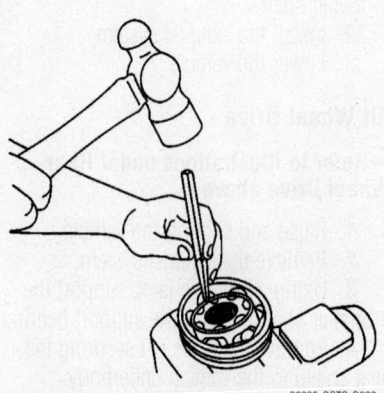

Removing/installing the wheel joint inner cage ball

12. Position a wood block between the seal and the joint, on the joint face.

13. Strike the wood block with a hammer to compress the axle shaft retaining clip. If the joint refuses to move, a brass drift can be used against the face of the inner race to compress the axle shaft retaining clip.

14. Continue to strike the wood block in order to remove the outer joint from the axle shaft.

15. Remove and discard the axle shaft retaining ring from the axle shaft.

16. Remove the seal from the axle shaft.

17. Wrap a shop towel around the joint outer race splined shaft.

18. Place the outer race vertically in a bench vise.

19. Use a brass drift and hammer to gently tap on the wheel halfshaft joints inner cage, until it is tilted enough to remove the first ball.

20. Remove the exposed ball. Use a small screwdriver to aid in removal, if necessary.

21. Position the cage and inner race so they are level.

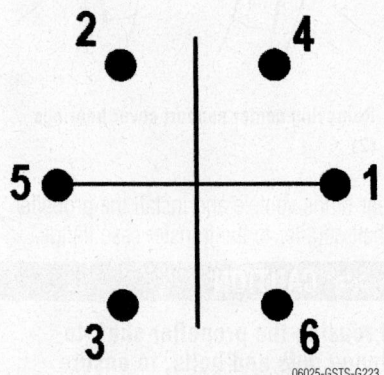

Showing the ball removal sequence

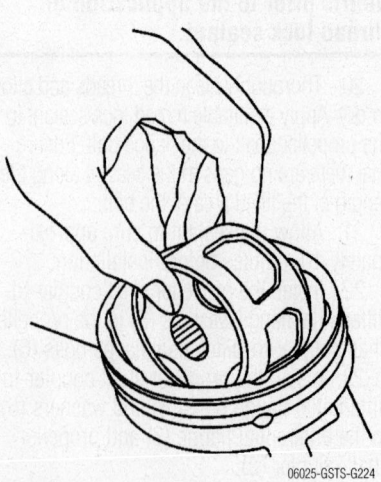

Removing the cage and inner race from the outer race

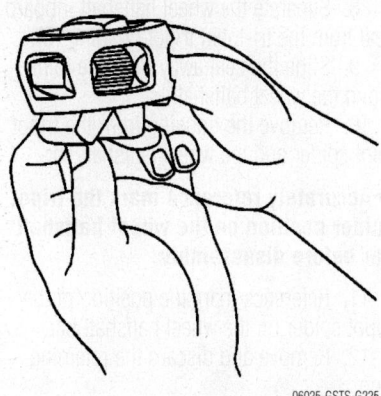

Removing or installing the inner race from the cage

22. Repeat the previous steps in the removal sequence as shown until all of the balls are removed.

23. Position the cage and the inner race 90 degrees to the centerline of the outer race.

24. Align the cage windows with the lands of the outer race.

25. Lift to remove the cage and the inner race from the outer race.

26. Position the cage and inner race so that the larger radius corners of the cage windows are up.

27. Rotate the inner race 90 degrees to the centerline of the cage.

28. Align the lands of the inner race with the windows of the cage.

29. Insert an inner race land into a cage window.

30. Pivot the inner race down and remove it from the cage.

➡**All traces of old grease and any contaminates must be removed.**

31. Inspect all parts for damaged and/or wear and replace as needed.

To install:

32. Position the cage so the larger radius corners of the cage windows are up.

33. Position the inner race 90 degrees to the centerline of the cage.

34. Begin to insert the inner race up through the bottom of the cage.

35. Align a land of the inner race to a window of the cage.

36. Insert the inner race land into the cage window.

37. Rotate the remainder of the inner race into the cage.

38. Rotate the inner race within the cage so that the grooved surface of the inner race is facing up.

➡**Ensure that the inner race is fully assembled into the cage.**

39. Align the inner race ball tracks with the cage windows.

40. Wrap a shop towel around the joint outer race splined shaft.

41. Place the outer race vertically in a bench vise.

42. Position the cage and inner race 90 degrees to the centerline of the outer race.

43. Align 2 cage windows at 0 and 180 degrees, within the outer race. Rotate the inner race and cage assembly downward in the vertical plane.

➡**The inner race ring groove must be positioned down.**

44. Position the cage and inner race so they are level.

45. Align the cage windows and inner race ball tracks with the outer race ball tracks.

46. Position a cage window and inner race ball track for ball installation.

47. Press down on the cage following one of the outer race ball tracks. The opposing cage window and inner race ball track will be accessible for ball installation.

48. After installing the first ball, use a brass drift and a hammer to tap gently on the cage, in order to drive the cage and inner race down completely.

➡**No gap should exist between the ball and the inner race ball track.**

49. Insert a ball through the cage window onto the inner race ball track. Tap the ball lightly with a plastic tipped hammer.

50. Position the cage and inner race so they are level.

51. Repeat the previous steps in the installation sequence as shown until all of the balls are installed.

➡**The ball installation sequence must be followed as shown.**

52. Insert approximately 60 percent of the grease from the service kit into the outer joint.

53. Spread 60 percent of the grease onto the ball tracks, the balls, the cage and the inner race.

54. Spread the remainder of the grease into the bottom of the outer race.

55. Remove the outer joint from the bench vise.

56. Wrap a shop towel around the axle shaft.

57. Place the wheel halfshaft horizontally in a bench vise.

58. Install a new small seal retaining clamp onto the axle shaft.

59. Install the seal onto the axle shaft.

60. Install a new retaining ring to the axle shaft.

61. Position the outer joint horizontally.

62. Engage the inner race splines onto the axle shaft splines.

63. Compress the axle shaft retaining ring.

64. Press the end of the retaining ring, using a flat-bladed tool, into the axle shaft groove while firmly pressing the outer joint onto the axle shaft.

65. Continue to work around the retaining ring, until it is compressed.

➡**The axle shaft and inner race must be fully seated to each other.**

66. Install the outer joint to the axle shaft.

67. Position a wood block over the end of the outer joint threaded shaft.

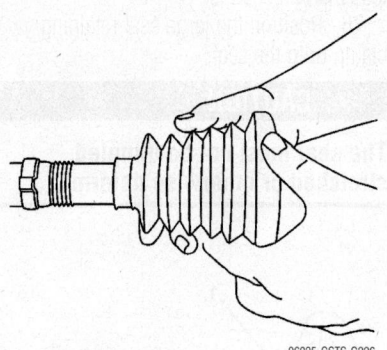

Installing the seal on the axle shaft

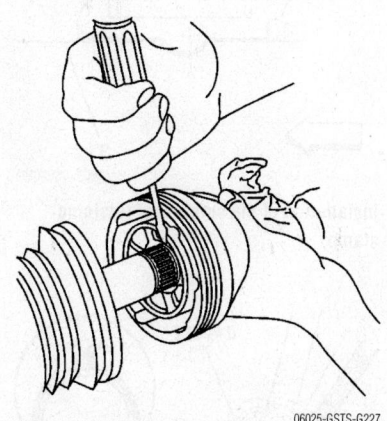

Installing the retaining ring

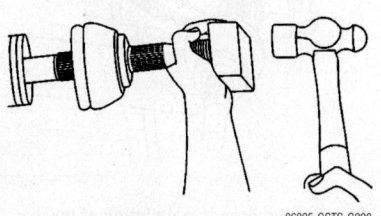

Installation of the outer joint

68. Use a hammer to drive the outer joint onto the shaft.

69. Continue to drive the outer joint until the outer joint seats fully onto the axle shaft.

70. Insert the remaining grease from the service kit into the seal.

71. Position the small seal retaining clamp (2) into the seal boot groove.

72. Position the seal and small retaining clamp to the axle shaft boot groove (3) as shown.

➡ **The seal retaining clamp must not be over-tightened or under-tightened.**

73. Using seal clamp pliers, crimp the small seal retaining clamp.

74. Tighten the small seal clamp until the base of the omega ohms shape has a gap of 0.039 inch (1mm).

75. The clamping hold time must be no less than 2 seconds.

76. Position the large seal retaining clamp onto the seal.

❋ CAUTION

The seal must not be dimpled, stretched or otherwise deformed.

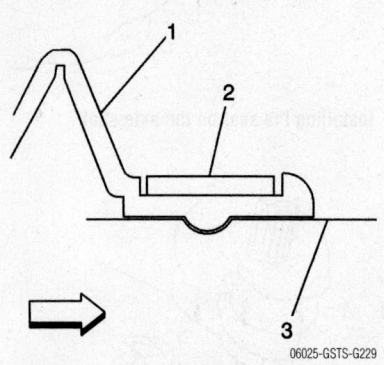

Installation of the small seal retaining clamp

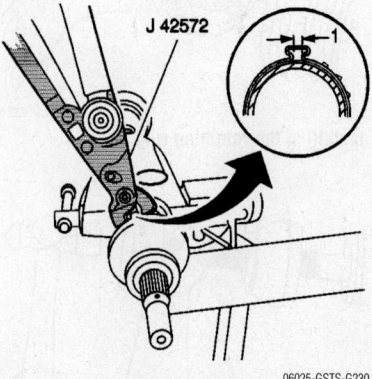

Showing the proper installation of the large seal retaining clamp

77. Inspect the seal for proper shape. If the seal is not shaped correctly, equalize the pressure in the seal by lifting the seal edge slightly and shape the seal properly by hand.

78. Inspect the seal for damage. If the seal has been cut or punctured during assembly, you must discard and replace the seal.

➡ **The seal retaining clamp must not be over-tightened or under-tightened.**

79. Using the J 42572 seal clamp pliers, crimp the large seal retaining clamp.

80. Tighten the large seal clamp until the base of the omega ohms shape has a gap of 1 mm (0.039 in).

81. The clamping hold time must be no less than 2 seconds.

82. Inspect the outer joint for smooth operation. This will also distribute the grease within the joint.

83. Hold the wheel halfshaft vertically, with the outer joint at the bottom.

84. Rotate the wheel halfshaft 4 or 5 times in a circular motion.

Inner (Tri-Pot) Joint

1. Before servicing the vehicle, refer to the precautions in the beginning of this section.

2. This procedure is to be performed only after the wheel halfshaft has been removed from the vehicle.

3. Wrap a shop towel around the axle shaft.

4. Place the wheel halfshaft horizontally in a bench vise.

5. Using side cutters, remove the small seal clamp (4) from the wheel halfshaft bar and discard the clamp.

6. Using the J35566, unlatch the large seal retaining clamp.

7. Remove the large seal clamp (2) from the tripot joint and discard the clamp.

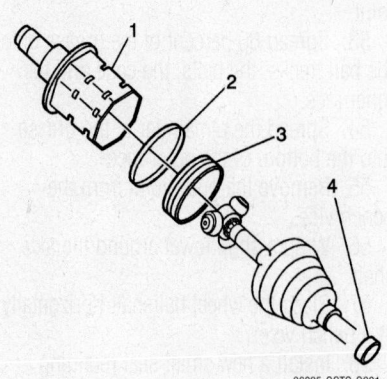

Illustrating inner tripot joint assembly.

8. Separate the wheel halfshaft inboard seal from the tri-lobal tripot bushing (3).

9. Slide the seal away from the joint along the wheel halfshaft bar.

10. Remove the housing from the tripot joint spider and the wheel halfshaft bar.

➡ **Accurately reference mark the tripot spider position on the wheel halfshaft bar before disassembly.**

11. Reference mark the position of the tripot spider on the wheel halfshaft bar.

12. Remove and discard the retaining ring.

13. Using a brass drift and hammer, carefully tap around the tripot spider face to remove the tripot spider (1) from the wheel halfshaft bar.

14. Remove the joint seal from the wheel halfshaft bar.

15. Inspect the following parts for damage and/or wear:

16. The wheel halfshaft inboard seal

17. The tripot joint spider assembly

18. The housing

19. The tri-lobal tripot bushing

To install:

20. Place the new small seal clamp (2) onto the small end of the joint seal (1).

21. Slide the joint seal (1) and the small seal clamp (2) into the boot groove on the wheel halfshaft bar.

22. Position the small end of the joint seal (1) into the joint seal groove (3) on the wheel halfshaft bar.

➡ **The seal retaining clamp must not be over-tightened or under-tightened.**

23. Using seal clamp pliers, crimp the large seal retaining clamp.

24. Tighten the large seal clamp until the base of the omega ohms shape has a gap of 0.039 inch (1 mm).

25. The clamping hold time must be no less than 2 seconds.

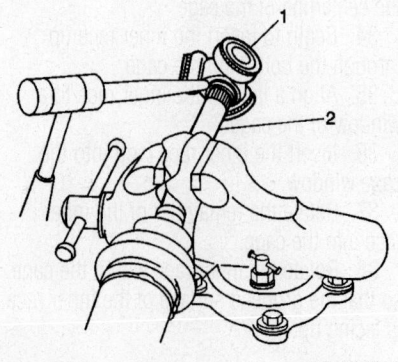

Illustrating tripot spider removal

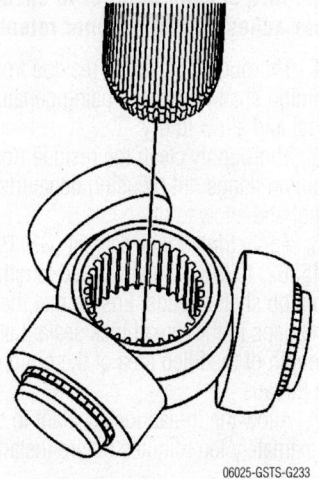

06025-GSTS-G233

Illustrating the spider to wheel bar shaft installation

➤**If reusing the old tripot spider, align the reference mark on the tripot spider to the wheel halfshaft bar. Ensure that the beveled edge of the tripot spider faces the wheel halfshaft bar during reassembly.**

26. Install the tripot spider to the wheel halfshaft bar.

27. Install a NEW retaining ring to the wheel halfshaft bar.

28. Verify positive engagement of the tripot spider to the wheel halfshaft bar by grasping the tripot spider and attempting to pull free from the wheel halfshaft bar.

29. Place approximately half of the grease from the service kit in the wheel half-shaft inboard seal. Use the remainder of the grease to repack the housing.

30. Install the tri-lobal tripot bushing (3) to the housing (1).

31. Position the larger new seal retaining clamp (2) on the wheel halfshaft inboard seal.

32. Slide the housing over the tripot joint spider assembly on the wheel halfshaft bar.

33. Slide the large diameter of the wheel halfshaft inboard seal with the large clamp in place, over the outside of the tri-lobal tripot bushing and locate the lip of the seal in the groove.

➤**The seal must not be dimpled, stretched or otherwise deformed.**

34. Inspect the seal for proper shape. If the seal is not shaped correctly, equalize the pressure in the seal by lifting the seal edge slightly and shape the seal properly by hand.

35. Position the seal to the proper vehicle dimension.

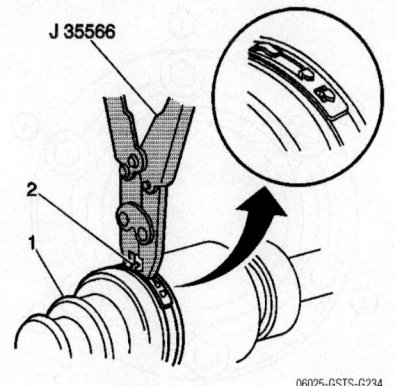

06025-GSTS-G234

Identifying proper installation of the inner tripot large clamp

36. Measure the distance (a) between the seal edges. The proper distance is 4.12 inch (104.7mm).

➤**Burp the air from the inner joint seal.**

37. Align the following items while latching the retaining clamp:
- Halfshaft inboard seal
- Tripot housing
- Large seal retaining clamp

38. Using the J 35566), latch the large seal retaining clamp.

39. Ensure that the latching tangs are fully engaged in the large clamp band.

40. Rotate the inner tripot housing 4 or 5 times in order to distribute the grease throughout the tripot spider bearings.

Axle Shaft, Bearing and Seal

REMOVAL & INSTALLATION

➤**This information is covered in Half-shafts and CV-Joint sections.**

Pinion Seal

REMOVAL & INSTALLATION

Front

1. Before servicing the vehicle, refer to the precautions in the beginning of this section.

2. Raise and support vehicle.

3. Remove the front propeller shaft constant velocity (CV) joint to drive pinion flange bolts.

4. Carefully move the front propeller shaft toward the rear of the vehicle in order to release the propeller shaft (CV) joint from the drive pinion flange.

5. Position the front propeller shaft out of the way.

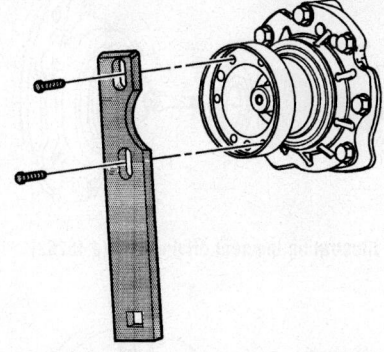

06025-GSTS-G235

Identifying the J 45012 holding tool

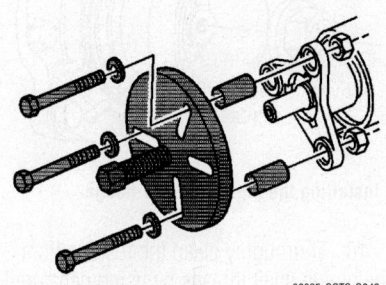

06025-GSTS-G242

Illustrating drive pinion flange tool J

6. Support the propeller shaft with heavy mechanic's wire).

7. Install the holding tool (J 45012) to the drive pinion flange.

8. While holding the tool, remove the drive pinion flange nut.

9. Remove the holding tool from the drive pinion flange.

10. Install the drive pinion flange tool (J 45019) assembly.

11. Using this tool assembly, remove the drive pinion flange.

12. Remove the pinion flange tool assembly from the drive pinion flange.

13. Using a flat bladed tool carefully remove the drive pinion seal. Take care not to damage any sealing surfaces.

To install:

14. Lubricate the drive pinion flange sealing surface of the drive pinion seal with synthetic gear oil (GM P/N 12378514).

15. Install the drive pinion seal using a seal driving tool (J 46262).

16. Remove the J 46262 from the new drive pinion seal.

17. Install the pinion flange to the drive pinion shaft.

➤**The pinion shaft threads and the pinion flange nut must be free of residue and debris prior to application of thread lock sealant in order to ensure proper adhesion and fastener retention.**

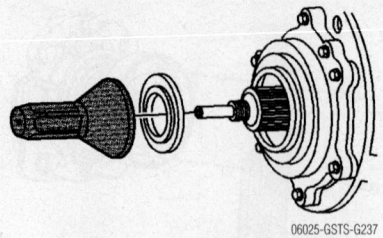

Illustrating the seal driving tool (J 45262)

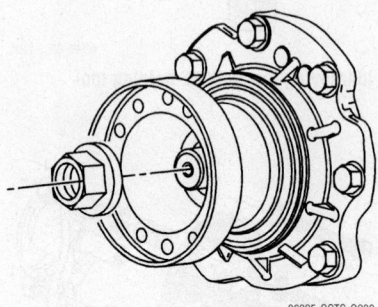

Installing the pinion nut and flange

18. Thoroughly clean the residue from the pinion shaft threads by using denatured alcohol and allow to dry.

19. Thoroughly clean the residue from the pinion flange nut by using denatured alcohol and allow to dry.

20. Apply thread lock sealant (GM P/N 12345382) to 2/3 of the threaded length of the pinion shaft threads. Ensure that there are no gaps in the thread lock sealant along the length of the filled area of the pinion shaft threads.

21. Allow the thread lock sealant to cure approximately ten minutes before installation. Install the drive pinion flange nut to the pinion shaft.

22. Install the holding tool (J 45012) to the drive pinion flange.

23. While holding the tool in place, tighten the drive pinion nut to 178 ft. lbs. (241 Nm).

24. Remove the holding tool (J 45012) from the pinion flange.

25. Install the propeller shaft to the drive pinion flange.

26. Install and tighten the propeller shaft to the drive pinion flange bolts, in the sequence shown.

27. Tighten the (6) propeller shaft coupler-to-differential flange bolts in sequence (1-6) to 22 ft. lbs. (30 Nm).

28. Inspect the fluid level.

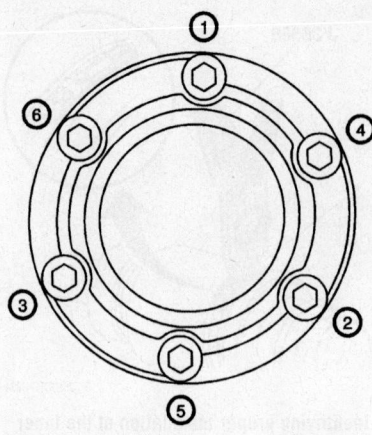

Propeller shaft bolt tightening sequence.

➡Ensure that there is no rotational movement between the constant velocity and the propeller shaft tube.

29. Lower the vehicle.

Rear

1. Raise and support vehicle.

2. Remove the propeller shaft coupler-to-differential flange bolts and propeller shaft from the pinion flange. See Propeller Shaft.

3. Carefully position the propeller shaft (1) out of the way and support the propeller shaft using a suitable jack.

4. Install the pinion flange holding tool (J 45012) to the flange.

5. While holding the tool in place, remove the drive pinion nut, using a proper sized deep socket.

6. Remove the holding tool.

7. Install the flange removal tool (J 45019) to the flange and remove the flange.

8. Using a flat-bladed tool, remove the drive pinion seal. Take care not to damage any sealing surfaces.

To install:

9. Lubricate the drive pinion flange sealing surface of the drive pinion seal with synthetic gear oil (GM P/N 12378514).

10. Install the drive pinion seal with the seal installer (J 45005).

11. Remove the flange removal tool.

12. Install the flange holding tool (J 45012).

13. Install the pinion flange to the drive pinion shaft.

➡The pinion shaft threads and the pinion flange nut must be free of residue and debris prior to application of

thread lock sealant in order to ensure proper adhesion and fastener retention.

14. Thoroughly clean the residue from the pinion shaft threads by using denatured alcohol and allow to dry.

15. Thoroughly clean the residue from the pinion flange nut by using denatured alcohol and allow to dry.

16. Apply thread lock sealant (GM P/N 12345382) to 2/3 of the threaded length of the pinion shaft threads. Ensure that there are no gaps in the thread lock sealant along the length of the filled area of the pinion shaft threads.

17. Allow the thread lock sealant to cure approximately ten minutes before installation.

18. Install the drive pinion flange nut to the pinion shaft. While holding the J 45012, use the proper deep socket to tighten the drive pinion nut.

19. Tighten the pinion flange nut to 210 ft. lbs. (285 Nm).

20. Remove the holding tool (J 45012).

21. Install the propeller shaft coupler to the differential flange.

22. Remove the jack.

23. Inspect the propeller shaft to flange nuts, bolts, and washers. Replace if damaged or worn.

➡If reusing the propeller shaft to flange nuts and bolts, to ensure proper adhesion and fastener retention, the threads must be free of debris prior to the application of thread lock sealant.

24. Thoroughly clean the threads using denatured alcohol and allow to dry. Apply thread lock sealant (GM P/N 12345493) to the propeller shaft to the flange bolt. Ensure that there are no gaps in the thread lock sealant along the length of the filled area of the bolt. Allow the thread lock sealant to cure approximately 10 minutes before installation.

25. Install the propeller shaft coupler-to-differential flange washers to the propeller shaft coupler-to-differential flange bolts.

26. Install the propeller shaft coupler-to-differential flange bolts and washers to the differential flange and propeller shaft coupler.

27. Install the propeller shaft coupler-to-differential flange nuts.

28. Tighten the propeller shaft coupler-to-differential flange bolts to 63 ft. lbs. (85 Nm).

29. Lower the vehicle.

STEERING AND SUSPENSION

Air Bag

SIR SERVICE PRECAUTIONS

❋❋ CAUTION

When performing service on or near the SIR components or the SIR wiring, the SIR system must be disabled. Refer to SIR Disabling and Enabling Zones. Failure to observe the correct procedure could cause deployment of the SIR components, personal injury, or unnecessary SIR system repairs.

The following are general service instructions which must be followed in order to properly repair the vehicle and return it to its original integrity:

• Do not expose inflator modules to temperatures above 65°C (150°F).

• Verify the correct replacement part number. Do not substitute a component from a different vehicle.

• Use only original GM replacement parts available from your authorized GM

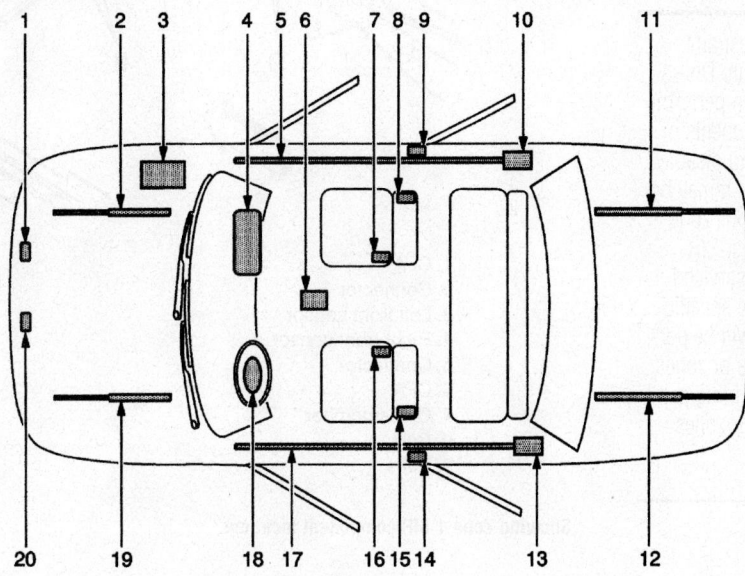

1. RF end sensor: On RF of vehicle in engine compartment
2. Front hood assit rod: A gas shock under the right side of the hood
3. Vehicle battery: Under hood on right side
4. I/P air bag: At top right under instrument panel
5. Right roof rail air bag: Under headliner, extending from passenger front windshield pillar to passenger rear windshield pillar
6. Sensing & Diagnostic Module (SDM): Under carpet under center console
7. Seat belt tensioner: On inboard side of passenger seat
8. RF side impact air bag: On seat back of passenger seat
9. Side Impact Sensor (SIS): Under center pillar trim near the bottom
10. Inflator module for right roof rail air bag: Behind reat seat back cushion near rear passenger door opening
11. Rear compartment lid assist rod: A gas shock located under the rear trunk lid on the right side
12. Rear compartment lid assist rod: A gas shock located under the rear trunk lid on the left side
13. Inflator module for left roof rail air bag: Behind rear seat back cushion near rear driver door opening
14. Side Impact Sensor (SIS): Under center pillar trim near the bottom
15. LF side impact air bag: On seat back of driver seat
16. Seat belt pretensioner: On inboard side of driver seat
17. Left roof rail air bag: Under the headliner, extending from driver front windshield pillar to driver rear windshield pillar
18. Steering wheel air bag: On steering wheel
19. Front hood assist rod: A gas shock located under the hood of the left side
20. LF end sensor: On left front side of vehicle in the engine compartment

06025-GSTS-G306

Identifying the SIR zones and related components

dealer. Do not use salvaged parts for repairs to the supplemental inflatable restraint (SIR) system.

• Discard any SIR component if it has been dropped from a height of 3 ft. (91cm) or greater.

SIR DISABLING AND ENABLING ZONES

✳✳ CAUTION

Before disabling the SIR system, refer to SIR Service Precautions.

The supplemental inflatable restraint (SIR) system has been divided into Disabling and Enabling Zones. When performing service on or near SIR components or SIR wiring, it may be necessary to disable the SIR components in that zone. It may be necessary to disable more than one zone depending on the location of other SIR components and the area being serviced. See the illustration to identify the specific zone or zones in which service will be performed. After identifying the zone or zones, proceed to the disabling and enabling procedures for that particular zone or zones.

DISARMING

Zone 1

To disarm:

✳✳ CAUTION

Before disabling the SIR system, refer to SIR Service Precautions.

1. Turn the steering wheel so that the vehicle's wheels are pointing straight ahead.
2. Place the ignition in the OFF position.
3. Remove the rear seat.

✳✳ WARNING

This sensing and diagnostic module (SDM) has 2 fused power inputs. To ensure there is no unwanted SIR deployment, personal injury, or unnecessary SIR system repairs, remove both the AIR BAG (IGN) fuse in the right fuse center and the AIR BAG (BATT) fuse from the left rear fuse center. With both AIR BAG fuses removed and the ignition switch in the ON position, the AIR BAG warning indicator illuminates. This is normal operation, and does not indicate an SIR system malfunction.

4. Locate and remove the AIR BAG (IGN) fuse in the right rear fuse center and the AIR

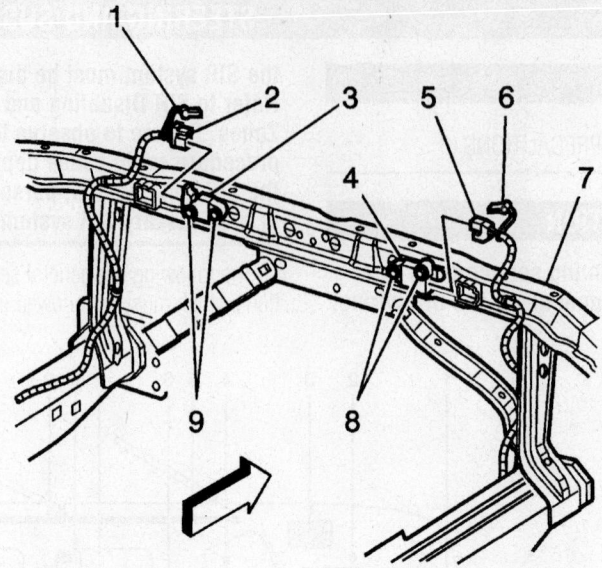

1. CPA
2. Connector
3. Left front sensor
4. Right front sensor
5. Connector
6. CPA
7. Crossmember
8. Bolts
9. Bolts

06025-GSTS-G267

Showing Zone 1 SIR component locations

BAG (BATT) fuse from the left rear fuse center.

5. Open the front hood and remove the fasteners and plastic cover for the radiator.

6. Remove both of the upper radiator support brackets and push the radiator toward the engine to gain access to both front sensors (3, 4).

7. Remove both connector position assurances (CPAs) (1, 6) from the left and right front end sensor connectors (2, 5).

8. Remove both connectors (2, 5) from the left and right front end sensors (3, 4).

Zone 2

✳✳ CAUTION

Before disabling the SIR system, refer to SIR Service Precautions.

1. Turn the steering wheel so that the vehicle's wheels are pointing straight ahead.
2. Place the ignition in the OFF position.
3. Remove the rear seat.

✳✳ WARNING

This sensing and diagnostic module (SDM) has 2 fused power inputs. To

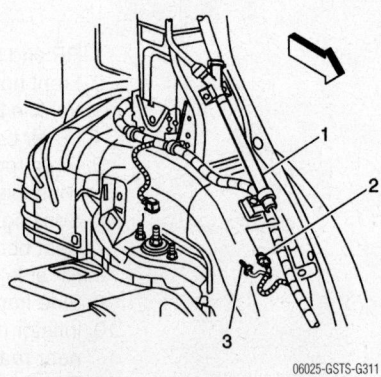

06025-GSTS-G311

Disconnecting the CPA from the yellow connector at the rear pillar

ensure there is no unwanted SIR deployment, personal injury, or unnecessary SIR system repairs, remove both the AIR BAG (IGN) fuse in the right fuse center and the AIR BAG (BATT) fuse from the left rear fuse center. With both AIR BAG fuses removed and the ignition switch in the ON position, the AIR BAG warning indicator illuminates. This is normal operation, and does not indicate an SIR system malfunction.

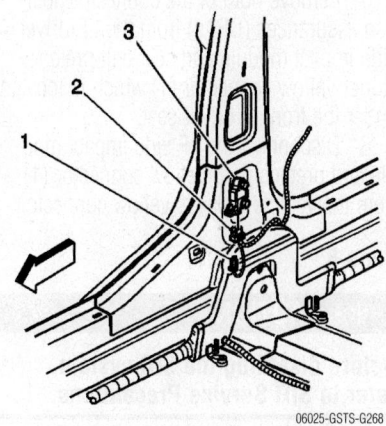

06025-GSTS-G268

Disconnecting the SIS CPA and connectors at the center pillar

4. Locate and remove the AIR BAG (IGN) fuse in the right rear fuse center and the AIR BAG (BATT) fuse from the left rear fuse center.

5. When disabling the roof rail module, go to step 6. If the side impact sensor (SIS) needs disabling, then go to step 10.

6. Remove the left carpet retainer trim.

7. Push the rear seat back away from the rear pillar to expose the left/driver roof rail module connector.

8. Remove the connector position assurance (CPA) (3) from the roof rail module yellow connector (2).

9. Disconnect the left roof rail module yellow connector (2) from the left roof rail module (1).

10. Remove the left center pillar trim panel.

11. Remove the SIS CPA (1) from the left SIS connector (2).

12. Remove the SIS connector (2) from the SIS (3).

Zone 3

✳✳ CAUTION

Before disabling the SIR system, refer to SIR Service Precautions.

1. Turn the steering wheel so that the vehicle's wheels are pointing straight ahead.

2. Place the ignition in the OFF position.

3. Remove the rear seat.

✳✳ WARNING

This sensing and diagnostic module (SDM) has 2 fused power inputs. To ensure there is no unwanted SIR deployment, personal injury, or unnecessary SIR system repairs, remove both the AIR BAG (IGN) fuse

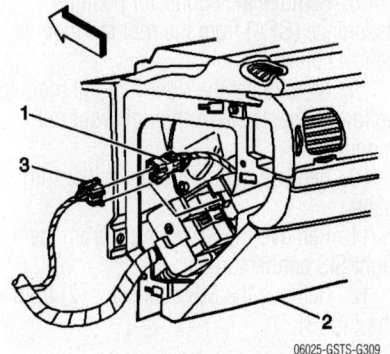

06025-GSTS-G309

Disconnecting the CPA through the left end of the I/P

in the right fuse center and the AIR BAG (BATT) fuse from the left rear fuse center. With both AIR BAG fuses removed and the ignition switch in the ON position, the AIR BAG warning indicator illuminates. This is normal operation, and does not indicate an SIR system malfunction.

4. Locate and remove the AIR BAG (IGN) fuse in the right rear fuse center and the AIR BAG (BATT) fuse from the left rear fuse center.

5. Remove the left/driver sound insulator from the instrument panel (I/P)

6. Remove the connector position assurance (CPA) from the steering wheel module coil yellow connector

7. Disconnect the steering wheel module coil yellow connector from the vehicle harness yellow connector

Zone 5

✳✳ CAUTION

Before disabling the SIR system, refer to SIR Service Precautions.

1. Turn the steering wheel so that the vehicle's wheels are pointing straight ahead.

2. Place the ignition in the OFF position.

3. Remove the rear seat.

✳✳ WARNING

This sensing and diagnostic module (SDM) has 2 fused power inputs. To ensure there is no unwanted SIR deployment, personal injury, or unnecessary SIR system repairs, remove both the AIR BAG (IGN) fuse in the right fuse center and the AIR BAG (BATT) fuse from the left rear fuse center. With both AIR BAG fuses removed and the ignition switch in the ON position, the AIR BAG warning

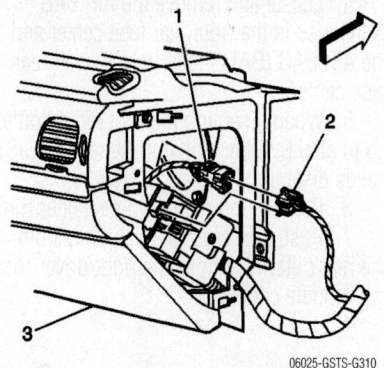

06025-GSTS-G310

Disconnecting the CPA from the I/P module yellow connector

indicator illuminates. This is normal operation, and does not indicate an SIR system malfunction.

4. Locate and remove the AIR BAG (IGN) fuse in the right rear fuse center and the AIR BAG (BATT) fuse from the left rear fuse center.

5. Remove the right/driver sound insulator from the instrument panel (I/P)

6. Remove the connector position assurance (CPA) from the steering wheel module coil yellow connector

7. Disconnect the steering wheel module coil yellow connector from the vehicle harness yellow connector.

Zone 6

✳✳ CAUTION

Before disabling the SIR system, refer to SIR Service Precautions.

1. Turn the steering wheel so that the vehicle's wheels are pointing straight ahead.

2. Place the ignition in the OFF position.

3. Remove the rear seat

✳✳ WARNING

This sensing and diagnostic module (SDM) has 2 fused power inputs. To ensure there is no unwanted SIR deployment, personal injury, or unnecessary SIR system repairs, remove both the AIR BAG (IGN) fuse in the right fuse center and the AIR BAG (BATT) fuse from the left rear fuse center. With both AIR BAG fuses removed and the ignition switch in the ON position, the AIR BAG warning indicator illuminates. This is normal operation, and does not indicate an SIR system malfunction.

4. Locate and remove the AIR BAG (IGN) fuse in the right rear fuse center and the AIR BAG (BATT) fuse from the left rear fuse center.

5. When disabling the roof rail module, go to step 6. If the side impact sensor (SIS) needs disabling, then go to step 10.

6. Remove the right carpet retainer trim.

7. Push the rear seat back away from the rear pillar to expose the right/driver roof rail module connector.

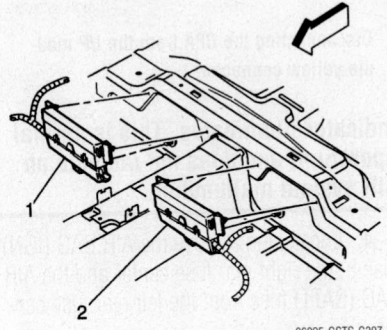

Removing the AIR BAG fuses (IGN and BATT) from the fuse centers

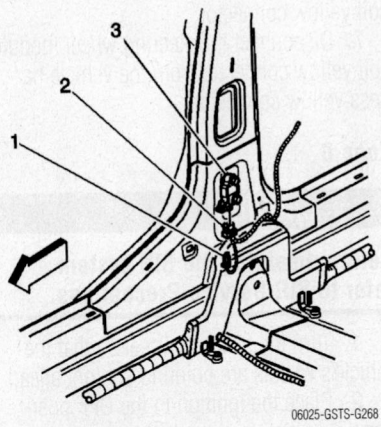

Disconnecting the SIS CPA and connectors

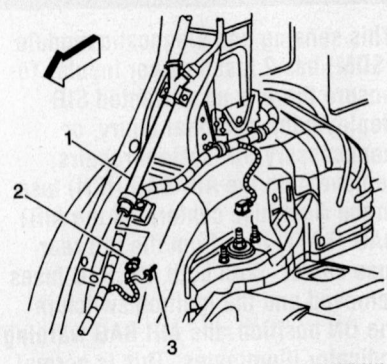

Removing the SIS CPA (1) and connector (2)

8. Remove the connector position assurance (CPA) from the roof rail module yellow connector.

9. Disconnect the right roof rail module yellow connector from the right roof rail module.

10. Remove the right center pillar trim panel.

11. Remove the SIS CPA (1) from the right SIS connector (2).

12. Remove the SIS connector (2) from the SIS (3).

Zone 7

✳✳ CAUTION

Before disabling the SIR system, refer to SIR Service Precautions.

1. Turn the steering wheel so that the vehicle's wheels are pointing straight ahead.
2. Place the ignition in the OFF position.
3. Remove the rear seat.

✳✳ WARNING

This sensing and diagnostic module (SDM) has 2 fused power inputs. To ensure there is no unwanted SIR deployment, personal injury, or unnecessary SIR system repairs, remove both the AIR BAG (IGN) fuse in the right fuse center and the AIR BAG (BATT) fuse from the left rear fuse center. With both AIR BAG fuses removed and the ignition switch in the ON position, the AIR BAG warning indicator illuminates. This is normal operation, and does not indicate an SIR system malfunction.

4. Locate and remove the AIR BAG (IGN) fuse in the right rear fuse center and the AIR BAG (BATT) fuse from the left rear fuse center.

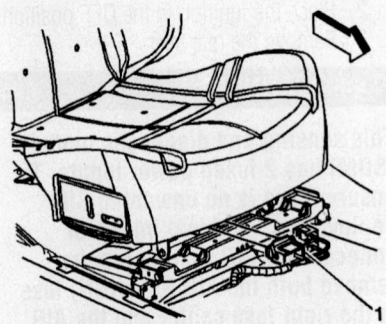

Disconnecting the Zone 7 CPAs from the impact module and pretensioner connector under the driver seat

5. Remove both of the connector position assurances (CPAs) from the LF/driver side impact module and seat belt pretensioner yellow connector (1) which is located under the front of driver seat.

6. Disconnect the LF side impact module and pretensioner yellow connector (1) from the vehicle harness yellow connector.

Zone 8

✳✳ CAUTION

Before disabling the SIR system, refer to SIR Service Precautions.

1. Turn the steering wheel so that the vehicle's wheels are pointing straight ahead.
2. Place the ignition in the OFF position.
3. Remove the rear seat.

✳✳ WARNING

This sensing and diagnostic module (SDM) has 2 fused power inputs. To ensure there is no unwanted SIR deployment, personal injury, or unnecessary SIR system repairs, remove both the AIR BAG (IGN) fuse in the right fuse center and the AIR BAG (BATT) fuse from the left rear fuse center. With both AIR BAG fuses removed and the ignition switch in the ON position, the AIR BAG warning indicator illuminates. This is normal operation, and does not indicate an SIR system malfunction.

4. Locate and remove the AIR BAG (IGN) fuse in the right rear fuse center and the AIR BAG (BATT) fuse from the left rear fuse center.

5. Remove the right carpet retainer trim.

6. Push the rear seat back away from the rear pillar to expose the right/passenger roof rail module connector.

7. Remove the connector position assurance (CPA) from the roof rail module yellow connector.

8. Disconnect the right roof rail module yellow connector from the right roof rail module.

9. Remove the passenger/right sound insulator from the instrument panel (I/P).

10. Remove the CPA from the I/P module yellow connector.

11. Disconnect the I/P module yellow connector from the vehicle harness yellow connector.

12. Remove both CPA locks from the passenger/RF side impact module and seat belt pretensioner yellow connector located under the front of passenger seat.

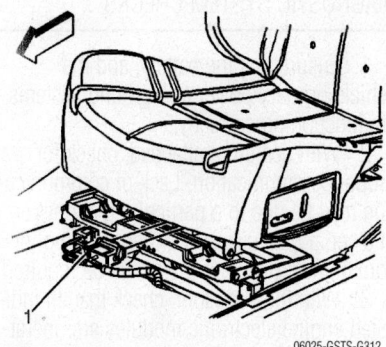

06025-GSTS-G312

Disconnecting the Zone 8 CPAs from the impact module and pretensioner connector under the driver seat

13. Disconnect the RF side impact module and pretensioner yellow connector from the vehicle harness yellow connector.

14. Remove the driver/left sound insulator from the I/P.

15. Remove the CPA from the steering wheel module coil yellow connector.

16. Disconnect the steering wheel module coil yellow connector from the vehicle harness yellow connector.

17. Remove both CPA locks from the driver/LF side impact module and seat belt pretensioner yellow connector located under the front of driver seat.

18. Disconnect the LF side impact module and pretensioner yellow connector from the vehicle harness yellow connector.

19. Remove the left carpet retainer trim

20. Push the rear seat back away from the rear pillar to expose the left/driver roof rail module connector.

21. Remove the CPA from the roof rail module yellow connector.

22. Disconnect the left roof rail module yellow connector from the left roof rail module.

Zone 9

❊❊ CAUTION

Before disabling the SIR system, refer to SIR Service Precautions.

1. Turn the steering wheel so that the vehicle's wheels are pointing straight ahead.
2. Place the ignition in the OFF position.
3. Remove the rear seat.

❊❊ WARNING

This sensing and diagnostic module (SDM) has 2 fused power inputs. To ensure there is no unwanted SIR deployment, personal injury, or unnecessary SIR system repairs,

remove both the AIR BAG (IGN) fuse in the right fuse center and the AIR BAG (BATT) fuse from the left rear fuse center. With both AIR BAG fuses removed and the ignition switch in the ON position, the AIR BAG warning indicator illuminates. This is normal operation, and does not indicate an SIR system malfunction.

4. Locate and remove the AIR BAG (IGN) fuse in the right rear fuse center and the AIR BAG (BATT) fuse from the left rear fuse center.

5. Remove both connector position assurance (CPA) from the RF/passenger side impact module and seat belt pretensioner yellow connector which is located under the front of passenger seat.

6. Disconnect the RF side impact module and pretensioner yellow connector from the vehicle harness yellow connector.

ARMING

Zone 1

1. Place the ignition in the OFF position.
2. Connect both connectors (2, 5) to the left and right front end sensors (3, 4).
3. Connect both CPAs (1, 6) to the left and right front sensor connectors (2, 5).
4. Install both upper radiator support brackets.
5. Install the plastic cover for the radiator.
6. Install both the AIR BAG (IGN) fuse in the right rear fuse center and the AIR BAG (BATT) fuse from the left rear fuse center.
7. Install the right and left rear fuse center covers.
8. Install the rear seat.
9. Use caution while reaching in and switch the ignition to the ON position. The AIR BAG indicator will flash, then turn OFF.
10. Perform the Diagnostic System Check—Vehicle if the AIR BAG warning indicator does not operate as described.

Zone 2

1. Place the ignition in the OFF position.
2. When enabling the SIS, proceed to step 3. If the roof rail module needs enabling, then go to step 6.
3. Connect the SIS connector to the SIS.
4. Install the left center pillar trim panel.
5. Connect the left roof rail module yellow connector to the roof rail module.
6. Install the CPA to the left roof rail module connector.
7. Release rear seat back.

8. Install the left carpet retainer trim.
9. Install both the AIR BAG (IGN) fuse in the right rear fuse center and the AIR BAG (BATT) fuse from the left rear fuse center.
10. Install the right and left rear fuse center covers.
11. Install the rear seat.
12. Use caution while reaching in and switch the ignition to the ON position. The AIR BAG indicator will flash, then turn OFF.
13. Perform the Diagnostic System Check—Vehicle if the AIR BAG warning indicator does not operate as described.

Zone 3

1. Place the ignition in the OFF position.
2. Connect the steering wheel module coil yellow connector to the vehicle harness yellow connector
3. Install the CPA to the steering wheel module coil yellow connector
4. Install the left sound insulator to the I/P
5. Install both the AIR BAG (IGN) fuse in the right rear fuse center and the AIR BAG (BATT) fuse from the left rear fuse center.
6. Install the right and left rear fuse center covers.
7. Install the rear seat.
8. Use caution while reaching in and switch the ignition to the ON position. The AIR BAG indicator will flash, then turn OFF.
9. Perform the Diagnostic System Check—Vehicle if the AIR BAG warning indicator does not operate as described.

Zone 5

1. Place the ignition in the OFF position.
2. Connect the steering wheel module coil yellow connector to the vehicle harness yellow connector
3. Install the CPA to the steering wheel module coil yellow connector
4. Install the right sound insulator to the I/P
5. Install both the AIR BAG (IGN) fuse in the right rear fuse center and the AIR BAG (BATT) fuse from the left rear fuse center.
6. Install the right and left rear fuse center covers.
7. Install the rear seat.
8. Use caution while reaching in and switch the ignition to the ON position. The AIR BAG indicator will flash, then turn OFF.
9. Perform the Diagnostic System Check—Vehicle if the AIR BAG warning indicator does not operate as described.

Zone 6

1. Place the ignition in the OFF position.

2. When enabling the SIS, proceed to step 3. If the roof rail module needs enabling, then go to step 6.

3. Connect the SIS connector to the SIS.

4. Install the right center pillar trim panel.

5. Connect the right roof rail module yellow connector to the roof rail module.

6. Install the CPA to the right roof rail module connector.

7. Release rear seat back.

8. Install the right carpet retainer trim.

9. Install both the AIR BAG (IGN) fuse in the right rear fuse center and the AIR BAG (BATT) fuse from the left rear fuse center.

10. Install the right and left rear fuse center covers.

11. Install the rear seat.

12. Use caution while reaching in and switch the ignition to the ON position. The AIR BAG indicator will flash, then turn OFF.

13. Perform the Diagnostic System Check—Vehicle if the AIR BAG warning indicator does not operate as described.

Zone 7

1. Place the ignition in the OFF position.

2. Connect the LF side impact module and pretensioner yellow connector to the vehicle harness yellow connector.

3. Install both CPA locks to the LF side impact module and pretensioner yellow connector.

4. Install both the AIR BAG (IGN) fuse in the right rear fuse center and the AIR BAG (BATT) fuse from the left rear fuse center.

5. Install the right and left rear fuse center covers.

6. Install the rear seat.

7. Use caution while reaching in and switch the ignition to the ON position. The AIR BAG indicator will flash, then turn OFF.

8. Perform the Diagnostic System Check—Vehicle if the AIR BAG warning indicator does not operate as described.

Zone 8

1. Place the ignition in the OFF position.

2. Connect the steering wheel module coil yellow connector to the vehicle harness yellow connector.

3. Install the CPA to the steering wheel module coil yellow connector.

4. Install the driver/left sound insulator to the I/P

5. Connect the driver/LF side impact module and seat belt pretensioner yellow connector to the vehicle harness yellow connector located under the front of driver seat.

6. Install both CPA locks to the LF side impact module and pretensioner yellow connector.

7. Connect the left roof rail module yellow connector to the roof rail module.

8. Install the CPA (3) to the left roof rail module connector.

9. Release the rear seat back.

10. Install the left carpet retainer trim.

11. Connect the passenger/I/P module yellow connector to the vehicle harness yellow connector.

12. Install the CPA to the I/P module yellow connector.

13. Install the passenger/right sound insulator to the I/P

14. Connect the passenger/RF side impact module and seat belt pretensioner yellow connector to the vehicle harness yellow connector located under the front of passenger seat.

15. Install both CPA locks to the RF side impact module and pretensioner yellow connector.

16. Connect the right roof rail module yellow connector to the roof rail module.

17. Release the rear seat back.

18. Install the right carpet retainer trim.

19. Install both the AIR BAG (IGN) fuse in the right rear fuse center and the AIR BAG (BATT) fuse from the left rear fuse center.

20. Install the right and left rear fuse center covers.

21. Install the rear seat.

22. Use caution while reaching in and switch the ignition to the ON position. The AIR BAG indicator will flash, then turn OFF.

23. Perform the Diagnostic System Check—Vehicle if the AIR BAG warning indicator does not operate as described.

Zone 9

1. Place the ignition in the OFF position.

2. Connect the RF side impact module and pretensioner yellow connector to the vehicle harness yellow connector.

3. Install both CPA locks to the RF side impact module and pretensioner yellow connector.

4. Install both the AIR BAG (IGN) fuse in the right rear fuse center and the AIR BAG (BATT) fuse from the left rear fuse center.

5. Install the right and left rear fuse center covers.

6. Install the rear seat

7. Use caution while reaching in and switch the ignition to the ON position. The AIR BAG indicator will flash, then turn OFF.

8. Perform the Diagnostic System Check—Vehicle if the AIR BAG warning indicator does not operate as described.

DIAGNOSTIC SYSTEM CHECKS

1. Ensure that the battery, and the vehicle primary power and ground systems are functioning correctly.

2. With scan tool attached, check for proper communication. Lack of communication may be due to a particular malfunction of a serial data circuit. Further scan tool or communications diagnosis may be required.

3. With the scan tool, check that all indicated engine electronic modules are operating in the incorrect power mode, based on key position. If not, this may cause other vehicle symptoms and/or DTCs to set.

4. With the scan tool, check for any Power Mode Mismatch and correct the condition before checking for module DTCs or symptoms.

5. Ensure that all data link communication DTCs are diagnosed before system level DTCs.

6. Ensure that all electronic control unit (ECU) internal DTCs are diagnosed before other system level DTCs.

7. Ensure that all device voltage DTCs are diagnosed before other system level DTCs.

Power Steering Gear

REMOVAL & INSTALLATION

1. Before servicing the vehicle, refer to the precautions in the beginning of this section.

2. Position a fluid catch pan under the power steering gear.

3. Place the front wheels in the straight ahead position.

4. Turn the ignition key to the LOCK position and remove the key.

5. Insert Steering Column Locking Pin J-42640 into the access hole in the lower steering column trim cover.

6. Raise and support the vehicle.

7. Remove the front tires and wheels, after making a reference mark between the hub and wheel.

8. Remove the front air deflector.

9. Remove the intermediate shaft lower pinch bolt.

10. Disconnect the intermediate shaft from the power steering gear.

✳✳ CAUTION

Failure to disconnect the intermediate shaft from the rack and pinion stub shaft can result in damage to the steering gear and/or intermediate shaft. This damage can cause loss of steering control which could result in personal injury.

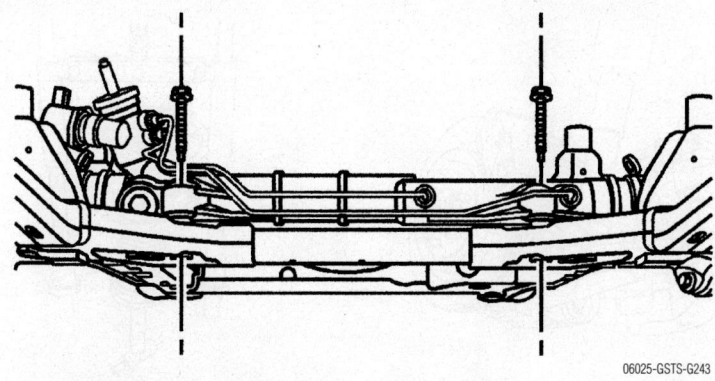

Showing the location of the power steering gear mounting bolts

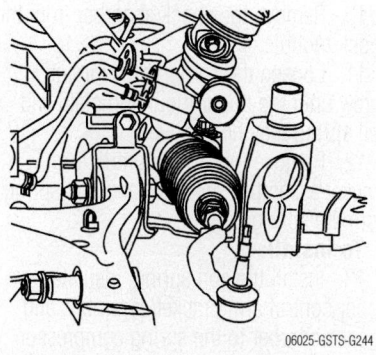

Showing the line connections on the power steering gear

11. Disconnect the variable effort steering harness connector.

12. Remove the outer tie rod retaining nuts.

13. Separate the outer tie rod from the steering knuckles.

14. Remove the power steering hoses to power steering gear retaining bolt.

15. Disconnect the power steering hoses from the power steering gear.

16. Immediately cap or plug all openings to prevent system contamination or excessive fluid loss.

17. If equipped with the 4.6L engine, perform the following steps:

   a. Disconnect the left brake line from the brake hose.

   b. Plug the brake line.

   c. Carefully position the brake line to the side.

18. If equipped with AWD, using a suitable jack, support the bottom of the front differential housing.

19. If equipped with AWD, loosen the right engine mount nut.

20. If equipped with AWD, raise the differential housing to clear the steering rack bolt.

21. Remove the power steering gear mounting bolts.

22. If equipped with the 4.6L engine, remove the left rearward lower control arm to frame mounting nut and bolt.

23. Remove the power steering gear through the left wheel opening.

To install:

24. Install the power steering gear through the left wheel opening.

25. If equipped with the 4.6L (LH2) engine, install the left rearward lower control arm to frame mounting bolt and nut.

26. Install the power steering gear retaining bolts. Tighten the bolts to 134 ft. lbs. (180 Nm).

27. If equipped with AWD, lower the differential housing.

28. If equipped with AWD, tighten the right lower engine mount to the frame nut. Tighten the nut to 59 ft. lbs. (80 Nm).

29. If equipped with AWD, remove the support from the bottom of the differential housing.

30. If equipped with the 4.6L engine, install the left brake line to the brake hose. Tighten the line to 20 ft. lbs. (27 Nm).

31. Install the power steering hoses to the power steering gear.

32. Install the power steering hoses to the power steering gear retaining bolt. Tighten the bolt to 17 ft. lbs. (23 Nm).

33. Remove the drain pan from under the vehicle.

34. Install the outer tie rod to the steering knuckles. Tighten the nuts to 52 ft. lbs. (70 Nm).

35. Connect the variable effort steering harness connector.

36. Connect the intermediate shaft to the power steering gear.

37. Install the intermediate shaft lower pinch bolt. Tighten the bolt to 37 ft. lbs. (50 Nm).

38. Install the tire and wheels.

39. Lower the vehicle.

40. Remove the holding tool (J 42640) from the steering column.

41. Bleed the power steering system.

42. Adjust the front toe.

Shock Absorbers

REMOVAL & INSTALLATION

Front

1. Before servicing the vehicle, refer to the precautions in the beginning of this section.

2. Raise and support the vehicle.

3. Remove the shock yoke.

4. Vehicles equipped with electronic suspension control (ESC), or automatic headlamp aiming, disconnect the suspension position sensor link rod from the upper control arm.

➡ **The ball stud must not rotate during disassembly/reassembly. Hand tools must be used to keep the ball stud from rotating. If air tools are used, and the stud is allowed to rotate, damage to the ball stud may occur.**

➡ **To prevent the ball stud from slipping, insert a hex-head tool while removing the upper control arm-to-steering knuckle nut.**

5. Remove the upper control arm to steering knuckle nut.

6. Use the J 24319-B to separate the upper control arm from the steering knuckle.

7. Lower the vehicle.

8. Vehicles equipped with ESC, disconnect the damper coil harness connector.

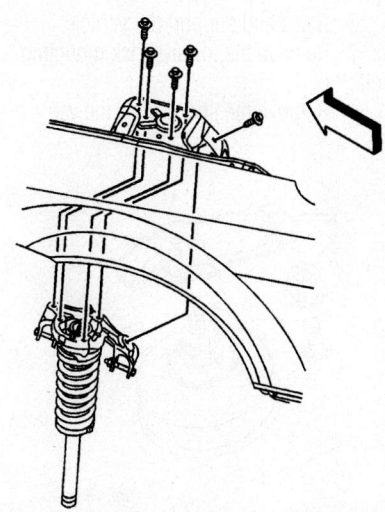

Front strut/shock absorber upper mounting bolt locations

9. Remove the shock module upper mounting bolts.

10. Remove the shock module upper mounting bolts.

11. Remove the shock module from the vehicle.

To install:

12. Install the shock module to the vehicle.

13. Tighten the bolts to 83 ft. lbs. (112 Nm).

14. Vehicles equipped with ESC, connect the damper coil harness connector.

15. Raise the vehicle.

16. Install the upper control arm to the steering knuckle and tighten the retaining nut to 18 ft. lbs. (20 Nm), plus an additional 210 degrees.

17. Vehicles equipped with ESC, or automatic headlamp aiming, connect the suspension position sensor link rod to the upper control arm.

18. Install the shock yoke.

19. Install the tire and wheel.

20. Lower the vehicle.

Rear

1. Before servicing the vehicle, refer to the precautions in the beginning of this section.

2. Remove the rear compartment trim panel.

3. Move the sound insulator away from the shock tower.

4. Disconnect the electrical connector (1) on vehicles equipped with F55 (Chassis Continuously Variable Real Time Damping Magneto Rheological) suspension.

5. Remove the upper shock mounting nuts (2).

6. Raise and support the vehicle.

7. Remove the lower shock mounting bolt.

8. Remove the shock from the vehicle.

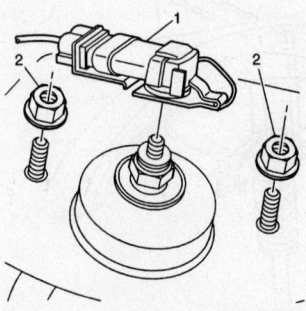

06025-GSTS-G246

On models with F55 suspension, detach electrical connection from top of shock tower

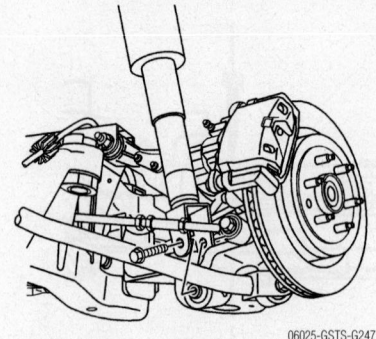

06025-GSTS-G247

Installing the rear suspension lower shock mounting bolt

To install:

9. Install the shock to the vehicle.

10. Install the lower shock mounting bolt.

11. Tighten the bolt to 111 ft. lbs. (150 Nm).

12. Connect the automatic level control air line to the shock on vehicle with F55 suspension.

13. Guide the shock to the body while lowering the vehicle.

14. Install the upper shock mounting nuts.

15. Tighten the nuts to 18 ft. lbs. (25 Nm).

16. Connect the electrical connector on vehicles equipped with F55 suspension.

17. Reposition the sound insulator around the shock tower.

18. Install the trim panel.

Coil Springs

REMOVAL & INSTALLATION

Front

1. Before servicing the vehicle, refer to the precautions in the beginning of this section.

2. Raise and support the vehicle.

3. Remove the tire and wheel after marking between the wheel and hub for reassembly reference.

4. Remove the shock module from the vehicle.

5. Install the shock module into the spring compressor.

6. Mark the upper control arm assembly and insulator for proper installation.

7. Turn the spring compressor forcing screw until the coil spring is compressed.

➡**The spring is compressed when the shock absorber moves freely.**

8. Remove the Magnaride® sensor nut.

9. Remove the shock absorber upper retaining nut.

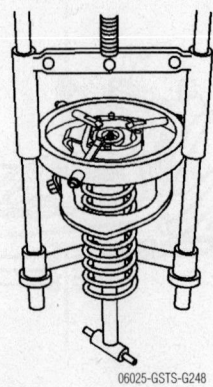

06025-GSTS-G248

Compressing the coil spring

10. Remove the shock absorber from the shock module.

11. Loosen the compressor forcing screw until the upper mounting plate and coil spring may be removed.

12. Remove the upper control arm bracket assembly, insulator and coil spring from the spring compressor.

To install:

13. Install the coil spring, insulator, upper control arm bracket assembly, and shock absorber to the spring compressor aligning all marks.

14. Turn the spring compressor forcing screw until the coil spring is compressed.

➡**Ensure the alignment pins in the upper control arm bracket is oriented 90 degrees with the shock absorber lower mounting holes.**

15. Install the shock absorber retaining nut.

16. Install the Magnaride® sensor nut.

17. Remove the shock module from the spring compressor.

18. Install the tire and wheel, matching the reference mark made during removal.

19. Lower the vehicle.

Rear

1. Before servicing the vehicle, refer to the precautions in the beginning of this section.

2. Raise and support the vehicle.

3. Remove the tire and wheel after marking between the wheel and hub for reassembly reference.

4. Disconnect the automatic level control sensor link from the upper control arm.

5. Disconnect the head lamp adjustment link from the upper control arm.

6. Support and raise the lower control arm using a suitable jack.

7. Remove the shock absorber lower mounting bolt.

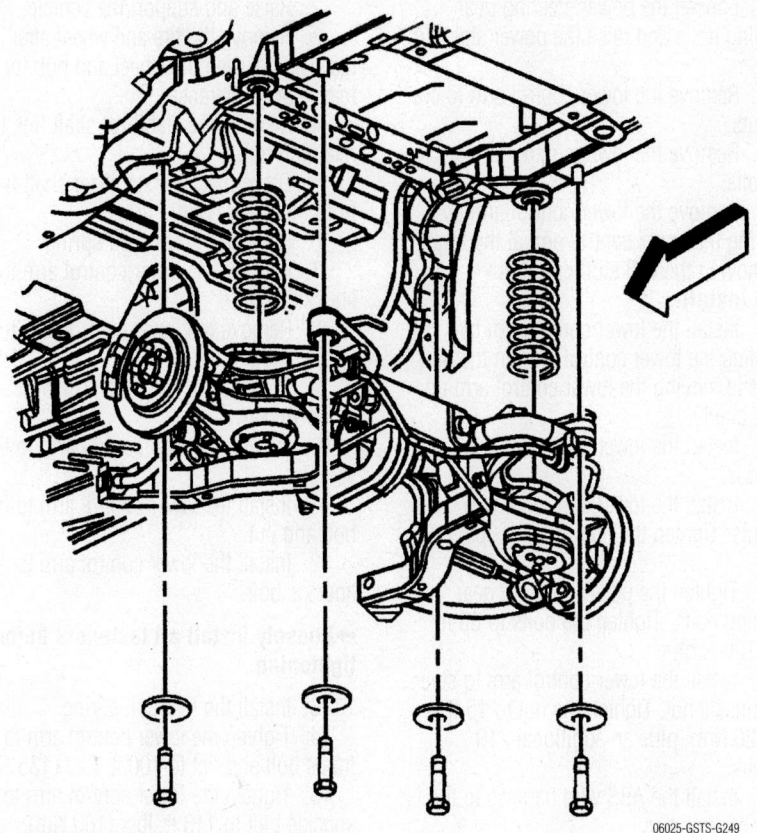

06025-GSTS-G249

Illustrating the frame to body bolt locations

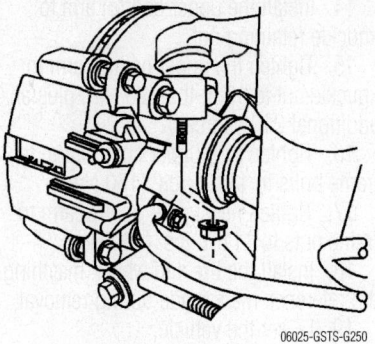

06025-GSTS-G250

Removing the upper control arm retaining nut

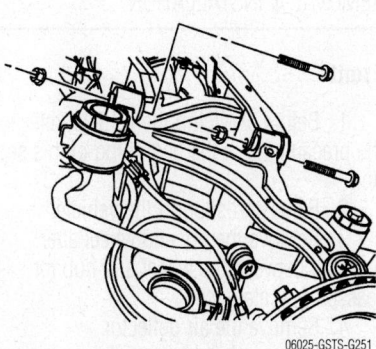

06025-GSTS-G251

Showing the rear upper control arm to frame bolt locations

8. Lower the lower control arm and remove the support.

9. Support the rear frame with a suitable jack.

10. Remove the frame to body side mounting bolts and washers.

➡**Lower the frame far enough to remove the coil spring without going past the guide pins.**

11. Lower the side of the frame.

12. Remove the coil spring from the vehicle.

To install:

13. Install the coil spring to the vehicle.

14. Raise the frame.

15. Install the frame to body mounting bolts and washers.

16. Tighten the front bolts to 195 ft. lbs. (265 Nm).

17. Tighten the rear bolts to 191 Nm (140 ft. lbs.).

18. Remove the jack from the vehicle.

19. Install a suitable jack under the lower control arm.

20. Using the jack raise the lower control arm until the shock absorber aligns with the knuckle.

21. Install the shock absorber lower retaining bolt.

22. Tighten the bolt to 111 ft. lbs. (150 Nm).

23. Remove the jack from the vehicle.

24. Connect the head lamp adjustment link to the upper control arm.

25. Connect the automatic level control sensor link to the upper control arm.

26. Install the tire and wheel, matching the reference mark made during removal.

27. Lower the vehicle.

Ball Joints

REMOVAL & INSTALLATION

➡**Ball joints are not replaceable in this application.**

Upper Control Arm

REMOVAL & INSTALLATION

Rear

1. Before servicing the vehicle, refer to the precautions in the beginning of this section.

2. Disconnect the negative battery cable.

3. Raise and support the vehicle.

4. Remove the tire and wheel after

marking between the wheel and hub for reassembly reference.

5. Remove the upper control arm to knuckle retaining nut.

6. Disconnect the upper control arm from the knuckle.

➡**Do not free the ball stud by using a pickle fork or a wedge-type tool. Damage to the seal or bushing may result.**

7. Remove the upper control arm to frame mounting nuts and discard.

8. Remove the upper control arm to frame mounting bolts and discard.

9. Remove the upper control arm from the vehicle.

To install:

10. Install the upper control arm to the vehicle.

11. Install NEW upper control arm to frame mounting bolts.

12. Install NEW upper control arm to frame mounting nuts.

13. Install the upper control arm to the knuckle.

✳✳ CAUTION

Loosely install all fasteners before tightening.

14. Install the upper control arm to knuckle retaining nut.

15. Tighten the upper control arm to knuckle nut to 15 ft. lbs. (20 Nm) plus an additional 210 degrees.

16. Tighten new upper control arm to frame bolts to 111 ft. lbs. (150 Nm).

17. Tighten new upper control arm to frame nuts to 111 ft. lbs. (150 Nm).

18. Install the tire and wheel, matching the reference mark made during removal.

19. Lower the vehicle.

Lower Control Arm

REMOVAL & INSTALLATION

Front

1. Before servicing the vehicle, refer to the precautions in the beginning of this section.

2. Raise and support the vehicle.

3. Remove the tire and wheel after marking between the wheel and hub for reassembly reference.

4. Remove the air deflector.

5. Remove the shock module yoke from the lower control arm.

6. Remove the stabilizer shaft link lower retaining nut.

7. Remove the stabilizer shaft link from the lower control arm.

8. Remove the ABS wire harness from the lower control arm.

9. Remove the lower control arm to steering knuckle nut.

10. Separate the lower control arm from the steering knuckle.

➡ **The power steering gear needs to be raised to have clearance when removing the rear lower control arm to frame retaining bolt.**

11. Loosen the power steering gear retaining bolts and raise the power steering gear.

12. Remove the lower control arm to cradle nuts.

13. Remove the lower control arm to cradle bolts.

14. Remove the lower control arm by lowering the lower control arm at the frame and moving the ball stud upwards.

To install:

15. Install the lower control arm by installing the lower control arm on the ball stud and moving the lower control arm up to the cradle.

16. Install the lower control arm to cradle bolts.

17. Install the lower control arm to cradle nuts. Tighten the nuts to 96 ft. lbs. (135 Nm).

18. Tighten the power steering gear retaining bolts. Tighten the bolts to 89 ft. lbs. (120 Nm).

19. Install the lower control arm to steering knuckle nut. Tighten the nut to 15 ft. lbs. (20 Nm), plus an additional 210 degrees.

20. Install the ABS wire harness to the lower control arm.

21. Install the stabilizer shaft link to the lower control arm.

22. Install the stabilizer shaft link lower retaining nut. Tighten the nut to 81 ft. lbs. (110 Nm).

23. Install the shock module yoke to the lower control arm.

24. Install the air deflector.

25. Install the tire and wheel, matching the reference mark made during removal.

26. Align the front end.

Rear

1. Before servicing the vehicle, refer to the precautions in the beginning of this section.

2. Raise and support the vehicle.

3. Remove the tire and wheel after marking between the wheel and hub for reassembly reference.

4. Remove the stabilizer shaft link lower retaining nut.

5. Disconnect the stabilizer shaft link from the lower control arm.

6. Remove the rear coil spring.

7. Remove the lower control arm to knuckle bolt.

8. Remove the lower control arm to frame bolt and nut.

9. Remove the lower control arm.

To install:

10. Install the lower control arm to the vehicle.

11. Install the lower control arm to frame bolt and nut

12. Install the lower control arm to knuckle bolt.

➡ **Loosely install all fasteners before tightening.**

13. Install the rear coil spring.

14. Tighten the lower control arm to frame bolt and nut to 100 ft. lbs. (135 Nm).

15. Tighten the lower control arm to knuckle bolt to 118 ft. lbs. (160 Nm).

16. Connect the stabilizer shaft link to the lower control arm.

17. Install the stabilizer shaft link lower retaining nut.

18. Install the tire and wheel, matching the reference mark made during removal.

19. Lower the vehicle.

CONTROL ARM BUSHING REPLACEMENT

➡ **Control arms are serviced as an assembly in this application. Control arm bushings are not replaceable.**

Wheel Bearings

ADJUSTMENT

➡ **All models use sealed wheel bearings that are pre-adjusted. If the bearing need replacing, replace the wheel hub/bearing assembly.**

Wheel Hub/Bearing Assembly

REMOVAL & INSTALLATION

Front (RWD Vehicles)

1. Before servicing the vehicle, refer to the precautions in the beginning of this section.

2. Raise and support the vehicle.

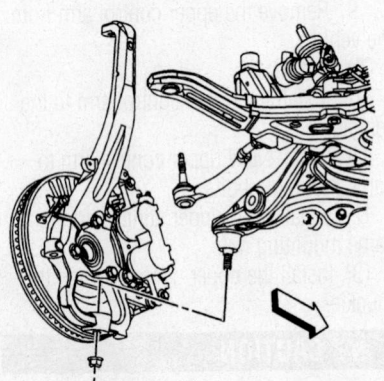

06025-GSTS-G252

Illustrating lower control arm and front suspension components.

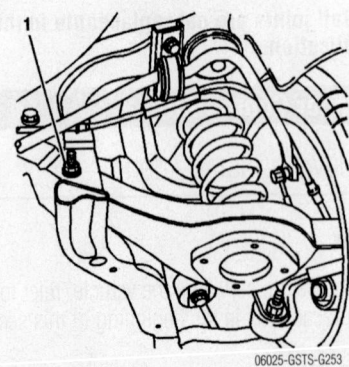

06025-GSTS-G253

Identifying the rear lower control arm retaining nut

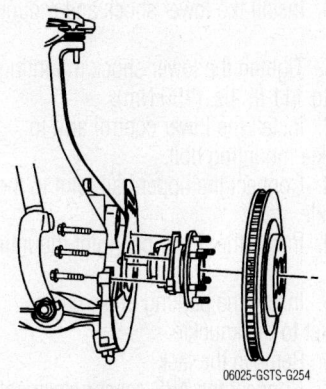

Front hub assembly exploded view

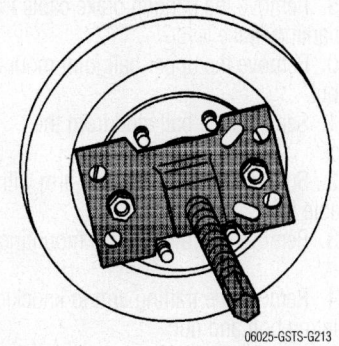

Illustrating the wheel hub and removal tool

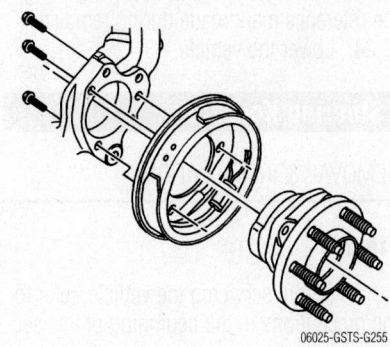

Exploded view of the rear hub assembly

3. Remove the tire and wheel after marking between the wheel and hub for reassembly reference.

4. Remove the brake rotor. See Disc Brakes.

5. Disconnect the ABS electrical connector.

6. Remove the electrical connector from the splash shield.

7. Remove the wheel hub/bearing mounting bolts.

8. Remove the wheel hub/bearing assembly.

To install:

9. Install the wheel hub/bearing.

10. Install the wheel hub/bearing mounting bolts. Tighten the bolts to 100 ft. lbs. (135 Nm).

11. Install the electrical connector to the splash shield.

12. Connect the ABS electrical connector.

13. Install the brake rotor.

14. Install the tire and wheel, matching the reference mark made during removal.

15. Lower the vehicle.

Front AWD Vehicles

1. Before servicing the vehicle, refer to the precautions in the beginning of this section.

2. Raise and support the vehicle.

3. Remove the tire and wheel after marking between the wheel and hub for reassembly reference.

4. Remove the wheel driveshaft retaining nut and discard.

5. Remove the brake rotor.

6. Disconnect the Antilock Brake System (ABS) electrical connector.

7. Remove the ABS electrical connector from the splash shield.

8. Carefully remove the wheel bearing/hub mounting bolts.

➡Avoid tool contact to the outer constant velocity boot seal when removing

the wheel bearing mounting bolts. Failure to observe this notice may result in damage to the CV boot.

9. Install the hub removal tool (J 45859) to the wheel bearing/hub.

10. Carefully disengage the wheel driveshaft from the wheel bearing/hub.

11. Remove the wheel bearing/hub assembly.

To install:

12. Install the wheel bearing/hub.

➡Avoid tool contact to the outer constant velocity boot seal when removing the wheel bearing mounting bolts. Failure to observe this notice may result in damage to the CV boot.

13. Install the wheel bearing/hub mounting bolts.

14. Tighten the wheel bearing/hub mounting bolts to 100 ft. lbs. (135 Nm).

15. Install the ABS electrical connector to the splash shield.

16. Connect the ABS electrical connector.

17. Install a NEW wheel driveshaft retaining nut.

18. Tighten the nut to 118 ft. lbs. (160 Nm).

19. Install the brake rotor.

20. Install the tire and wheel, matching the reference mark made during removal.

21. Lower the vehicle.

Rear

1. Before servicing the vehicle, refer to the precautions in the beginning of this section.

2. Raise and support the vehicle.

3. Remove the tire and wheel after marking between the wheel and hub for reassembly reference.

4. Remove the rear brake rotor

5. Disconnect the wheel speed sensor electrical connector.

6. Disconnect the wheel speed sensor electrical connector from the backing plate.

7. Remove the wheel halfshaft retaining nut and discard.

8. Carefully remove the upper control arm to knuckle retaining nut.

9. Separate the upper control arm from the knuckle.

➡Avoid tool contact to the outer constant velocity boot seal when removing the wheel bearing mounting bolts. Failure to observe this notice may result in damage to the CV boot.

10. Carefully remove the wheel bearing/hub retaining bolts.

11. Separate (with J 45859) the wheel driveshaft from the wheel bearing/hub.

12. Remove the wheel bearing/hub from the vehicle.

To install:

13. Install the wheel bearing/hub.

➡Avoid tool contact to the outer constant velocity boot seal when removing the wheel bearing mounting bolts. Failure to observe this notice may result in damage to the CV boot.

14. Install the wheel bearing/hub mounting bolts.

15. Tighten the bolts to 92 ft. lbs. (125 Nm).

16. Install the upper control arm to the knuckle.

17. Install the upper control arm to knuckle retaining nut.

18. Tighten the upper ball joint nut to 15 ft. lbs. (20 Nm), plus a additional 210 degree turn.

19. Install a NEW wheel halfshaft retaining nut. Tighten the nut to 118 ft. lbs. (160 Nm).

20. Install the brake rotor.

21. Connect the wheel speed sensor electrical connector to the backing plate.

22. Connect the wheel speed sensor electrical connector.

23. Install the tire and wheel, matching the reference mark made during removal.

24. Lower the vehicle.

Steering Knuckle

REMOVAL & INSTALLATION

Front

1. Before servicing the vehicle, refer to the precautions in the beginning of this section.

2. Raise and support the vehicle.

3. Remove the tire and wheel after marking between the wheel and hub for reassembly reference.

4. Remove the wheel hub/bearing assembly. See Wheel Hub/Bearing Assembly.

5. Remove the outer tie rod to steering knuckle.

6. Remove the brake hose bracket to steering knuckle retaining bolts.

7. Remove the upper control arm ball stud from the steering knuckle.

8. Remove the lower control arm ball stud from the steering knuckle.

9. Remove the steering knuckle.

To install:

10. Install the steering knuckle to the lower control arm ball stud.

11. Install the upper control arm ball stud to the steering knuckle.

12. Install the brake hose bracket to the steering knuckle retaining bolts. Tighten the bolts to 10 ft. lbs. (14 Nm).

13. Install the outer tie rod to the steering knuckle.

14. Install the wheel bearing/hub.

15. Install the tire and wheel, matching the reference mark made during removal.

16. Lower the vehicle.

Rear

1. Before servicing the vehicle, refer to the precautions in the beginning of this section.

2. Raise and support the vehicle.

3. Remove the tire and wheel after marking between the wheel and hub for reassembly reference.

4. Remove the rear brake rotor.

5. Remove the wheel driveshaft nut and discard.

6. Disconnect the Antilock Brake System (ABS) sensor harness connector.

7. Disconnect the ABS sensor harness connector from the backing plate.

8. Remove the parking brake cable bracket from the knuckle.

9. Remove the parking brake cable from the parking brake lever

10. Remove the upper ball joint mounting nut.

11. Separate the ball stud from the knuckle.

12. Support the lower control arm with a suitable jack.

13. Remove the lower shock mounting bolt.

14. Remove the trailing arm to knuckle mounting bolt and nut.

15. Remove the lower control arm to knuckle mounting bolt.

16. Remove the adjustment link to knuckle mounting bolt.

17. Install the removal tool (J 45859) to the wheel hub/bearing.

18. Using J 45859 carefully separate the wheel driveshaft from the wheel hub/bearing.

19. Remove the knuckle from the vehicle.

20. Remove the wheel bearing/hub bolts.

21. Remove the knuckle and backing plate from the wheel bearing/hub.

To install:

22. Install the knuckle and backing plate to the wheel bearing/hub.

23. Install the wheel bearing/hub mounting bolts.

24. Tighten the bolts to 92 ft. lbs. (125 Nm).

25. Install the knuckle to the vehicle.

26. Install the adjustment link to knuckle mounting bolt.

➡**Loosely install all fasteners before tightening.**

27. Tighten the adjustment link to knuckle bolt to 118 ft. lbs. (160 Nm).

28. Install the trailing arm to knuckle mounting bolt and nut.

29. Tighten the trailing arm to knuckle bolt to 125 ft. lbs. (170 Nm).

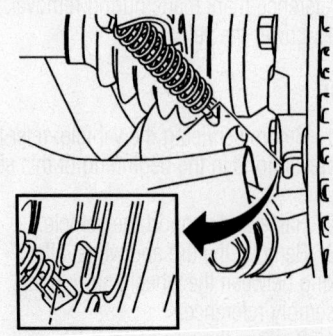

06025-GSTS-G218

Removing the parking brake cable from the lever

30. Install the lower shock and mounting bolt.

31. Tighten the lower shock mounting bolt to 111 ft. lbs. (150 Nm).

32. Install the lower control arm to knuckle mounting bolt.

33. Connect the upper ball joint to the knuckle.

34. Install the upper ball joint mounting nut.

35. Install the parking brake cable bracket to the knuckle.

36. Remove the jack.

37. Connect the ABS sensor electrical connector to the backing plate.

38. Connect the ABS sensor electrical connector.

39. Install a NEW wheel halfshaft nut. Tighten the nut to 118 ft. lbs. (160 Nm).

40. Install the tire and wheel, matching the reference mark made during removal.

41. Lower the vehicle.

Stabilizer Bar

REMOVAL & INSTALLATION

Front

1. Before servicing the vehicle, refer to the precautions in the beginning of this section.

2. Raise and support the vehicle.

3. Remove the tire and wheel after marking between the wheel and hub for reassembly reference.

4. Remove the stabilizer shaft link to stabilizer shaft retaining nuts

5. Disconnect the stabilizer shaft links from the stabilizer shaft.

6. Remove the stabilizer shaft mounting bolts and brackets.

7. Remove the stabilizer shaft insulators from the stabilizer shaft.

8. Remove the stabilizer shaft from the vehicle.

To install:

9. Install the stabilizer shaft to the vehicle.

10. Install the stabilizer shaft insulators to the stabilizer shaft. Install the insulator to the stabilizer shaft with the slit facing rearward.

➡**Do not tighten the bolts at this time.**

11. Install the stabilizer shaft brackets and mounting bolts.

12. Connect the stabilizer shaft links to the stabilizer shaft.

13. Install the stabilizer shaft link retaining nuts.

14. Tighten the link nuts to 115 Nm (95 ft. lbs.).

15. Tighten the bracket bolts to 81 ft. lbs. (110 Nm).

16. Install the tire and wheel, matching the reference mark made during removal.

17. Lower the vehicle.

Rear

1. Before servicing the vehicle, refer to the precautions in the beginning of this section.

2. Raise and support the vehicle.

3. Remove the tire and wheel after marking between the wheel and hub for reassembly reference.

4. Remove the stabilizer shaft links to the stabilizer shaft upper retaining nuts.

5. Remove the stabilizer shaft links from the stabilizer shaft.

6. Remove the stabilizer shaft brackets mounting bolts.

7. Remove the stabilizer shaft brackets.

8. Remove the stabilizer shaft from the vehicle.

9. Remove the stabilizer shaft insulator from the stabilizer shaft.

To install:

10. Install the stabilizer shaft insulator to the stabilizer shaft with the slit facing forwards.

11. Install the stabilizer shaft to the vehicle.

12. Install the stabilizer shaft brackets and mounting bolts.

➡**Loosely install all fasteners before tightening.**

13. Install the stabilizer shaft links to the stabilizer shaft.

14. Install the stabilizer shaft links to the stabilizer shaft retaining nuts.

15. Tighten the stabilizer shaft bracket bolts to 44 ft. lbs. (60 Nm).

16. Tighten the stabilizer shaft links nuts to 49 ft. lbs. (66 Nm).

17. Install the tire and wheel, matching the reference mark made during removal.

18. Lower the vehicle.

BRAKES

Brake Caliper

REMOVAL & INSTALLATION

Front and Rear

1. Before servicing the vehicle, refer to the precautions in the beginning of this section.

2. Inspect the fluid level in the brake master cylinder reservoir.

3. If the brake fluid level is midway between the maximum-full point and the minimum allowable level, no brake fluid needs to be removed from the reservoir before proceeding.

4. If the brake fluid level is higher than midway between the maximum-full point and the minimum allowable level, remove brake fluid to the midway point before proceeding.

5. Raise and suitably support the vehicle.

6. Remove the tire and wheel after marking between the wheel and hub for reassembly reference.

7. Install a large C-clamp over the body

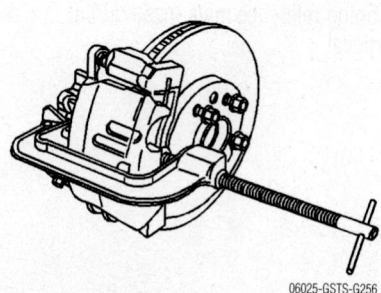

06025-GSTS-G256

Compressing the caliper piston with a C-clamp

of the brake caliper with the C-clamp ends against the rear of the caliper body and against the outer brake pad.

8. Tighten the C-clamp until the caliper piston is compressed into the caliper bore enough to allow the caliper to slide past the brake rotor.

9. Remove the C-clamp from the caliper.

10. Remove the brake hose to caliper bolt attaching the brake hose to the brake caliper.

11. Remove the brake hose from the brake caliper.

12. Remove and discard the 2 copper brake hose gaskets. These gaskets may be stuck to the brake caliper and/or the brake hose end.

13. Plug the opening in the brake caliper and the brake hose to prevent fluid loss and contamination.

14. Remove the brake caliper pin bolts.

15. Remove the brake caliper from the brake caliper bracket.

To install:

16. Inspect the caliper guide pin boots (2) for cuts, tears, or deterioration. If damaged, replace the slides and boots.

➡**Ensure that the caliper guide pin boots are fully seated to the caliper guide pin retaining seat of the caliper guide pin. Ensure that the caliper guide pin boots are fully seated to the caliper boot seal retaining seat of the brake caliper mounting bracket.**

17. Install the brake caliper to the brake caliper bracket.

18. Prepare the bolts and the threaded holes for assembly.

19. Thoroughly clean the residue from the bolt threads by using denatured alcohol and allow to dry.

20. Thoroughly clean the residue from the threaded holes by using denatured alcohol and allow to dry.

21. Apply thread lock sealant (GM P/N 12345493) to 2/3 of the threaded length of the lower caliper bracket bolts. Ensure that there are no gaps in the thread lock sealant along the length of the filled area of the bolts.

22. Allow the thread lock sealant to cure approximately 10 minutes before installation.

23. Apply a thin coat of high temperature silicone brake lubricant to the brake caliper pin bolts.

24. Install the brake caliper pin bolts.

25. Tighten the brake caliper pin bolts to 25 ft. lbs. (34 Nm).

26. Remove the plug from the brake caliper opening and the brake hose.

27. Assemble the NEW copper brake hose gaskets, and the brake caliper bolt to the brake hose.

28. Install the brake hose and the brake caliper bolt to the brake caliper.

29. Tighten the brake hose to caliper bolt to 37 ft. lbs. (50 Nm).

30. Bleed the hydraulic brake system.

31. With the engine OFF, gradually apply the brake pedal to approximately 2/3 of its travel distance.

32. Slowly release the brake pedal.

33. Wait 15 seconds, then repeat the previous steps until a firm brake pedal apply is obtained. This will properly seat the brake caliper pistons and brake pads.

34. Install the tire and wheel, matching the reference mark made during removal.

35. Lower the vehicle.

36. Fill the brake master cylinder reservoir to the proper level.

Disc Brake Pads

REMOVAL & INSTALLATION

Front and Rear

1. Before servicing the vehicle, refer to the precautions in the beginning of this section.

2. Inspect the fluid level in the brake master cylinder reservoir.

3. If the brake fluid level is midway between the maximum-full point and the minimum allowable level, no brake fluid needs to be removed from the reservoir before proceeding.

4. If the brake fluid level is higher than midway between the maximum-full point and the minimum allowable level, remove brake fluid to the midway point before proceeding.

5. Raise and suitably support the vehicle.

6. Remove the tire and wheel after marking between the wheel and hub for reassembly reference.

7. Install a large C-clamp over the body of the brake caliper with the C-clamp ends against the rear of the caliper body and against the outer brake pad.

8. Tighten the C-clamp until the caliper piston is compressed into the caliper bore enough to allow the caliper to slide past the brake rotor.

9. Remove the C-clamp from the caliper.

10. To loosen the brake caliper lower pin bolt, hold the brake caliper guide pin with a wrench.

11. Remove the brake caliper pin bolt.

❊❊ WARNING

Support the brake caliper with heavy mechanic's wire whenever it is separated from its mount and the hydraulic flexible brake hose is still connected. Failure to support the caliper in this manner will cause the flexible brake hose to bear the weight of the caliper, which may cause damage to the brake hose and in turn may cause a brake fluid leak.

12. Pivot the brake caliper body upward and secure the caliper out of the way with heavy mechanic's wire. Ensure that there is no tension on the hydraulic brake flexible hose. Do NOT disconnect the hydraulic brake flexible hose from the caliper.

13. Remove the brake pads from the caliper bracket.

14. Remove and inspect the brake pad retainers from the caliper bracket.

To install:

15. Inspect the brake caliper guide pin bolts. If damaged or corroded, replace the brake caliper guide bolts.

16. Inspect the brake caliper guide pins. If damaged or corroded, replace the brake caliper guide pin. Do not attempt to clean away any corrosion.

17. Inspect the brake caliper guide pin boots for cuts, tears, or deterioration. If damaged, replace the brake caliper guide pin boots.

18. Inspect the brake caliper piston boot for deterioration.

19. Install a large C-clamp (1) over the body of the brake caliper (3), with the C-clamp ends against the rear of the caliper body and against an old inboard brake pad (2) or a wood block installed against the caliper pistons.

20. Tighten the C-clamp (1) evenly until

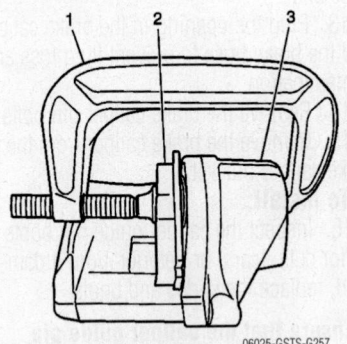

06025-GSTS-G257

Compressing the caliper piston with an old brake pad and C-clamp.

the caliper pistons are compressed completely into the caliper bores.

21. Remove the C-clamp and the old brake pad or wood block from the caliper.

22. Install the brake pad retainers to the caliper bracket.

23. Install the brake pads to the caliper bracket.

24. Pivot the brake caliper downward, over the brake pads and into the caliper bracket.

25. If reusing the caliper pin bolts, prepare the bolt and the threaded hole for assembly.

26. Thoroughly clean the residue from the bolt threads by using denatured alcohol and allow to dry.

27. Thoroughly clean the residue from the threaded holes by using denatured alcohol and allow to dry.

28. Apply thread lock sealant (GM P/N 12345493) to 2/3 of the threaded length of the lower caliper pin bolt. Ensure that there are no gaps in the thread lock sealant along the length of the filled area of the bolt.

29. Allow the thread lock sealant to cure approximately 10 minutes before installation.

30. Apply a thin coat of high temperature silicone lube to the front brake caliper guide pin.

31. Install the lower brake caliper pin bolt.

32. Hold the lower brake caliper guide pin with a wrench and tighten the lower brake caliper pin bolt to 46 ft. lbs. (63 Nm).

33. Lower the vehicle.

34. With the engine OFF, gradually apply the brake pedal to approximately 2/3 of its travel distance.

35. Slowly release the brake pedal.

36. Wait 15 seconds, then repeat the previous steps until a firm brake pedal apply is obtained. This will properly seat the brake caliper pistons and brake pads.

37. Fill the brake master cylinder reservoir to the proper level.

38. Install the tire and wheel assembly, aligning reference mark made during removal.

CHEVROLET

Tracker

BRAKES 29-36
DRIVE TRAIN 29-28
ENGINE REPAIR.............29-12
FUEL SYSTEM 29-27
SPECIFICATION AND
  MAINTENANCE CHARTS......29-2
Engine and Vehicle Identification ..29-2
General Engine Specifications 29-2
Engine Tune-Up Specifications......29-2
Firing Order 29-3
Accessory Drive Belt Routing 29-3
Capacities 29-4
Crankshaft and Connecting Rod
  Specifications 29-5
Valve Specifications...........29-5
Piston and Ring Specifications......29-6
Torque Specifications 29-6
Wheel Alignment 29-7
Tire, Wheel and Ball Joint
  Specifications 29-7
Brake Specifications 29-7
Scheduled Maintenance
  Intervals....................29-8
STEERING AND
  SUSPENSION 29-32

A
Air Bag.........................29-32
  Arming......................29-32
  Disarming...................29-32
  Precautions.................29-32
Alternator 29-12
  Installation.................29-12
  Removal.....................29-12

B
Brake Caliper 29-36
  Removal & Installation...........29-36
Brake Drums...................29-37
  Removal & Installation...........29-37
Brake Shoes...................29-38
  Removal & Installation...........29-38

C
Camshaft and Valve Lifters 29-19
  Removal & Installation...........29-19
Clutch.........................29-29
  Adjustments.................29-29
  Removal & Installation...........29-29

Coil Spring.....................29-34
  Removal & Installation...........29-34
CV-Joints......................29-31
  Overhaul 29-31
Cylinder Head..................29-16
  Removal & Installation...........29-16

D
Disc Brake Pads.................29-36
  Removal & Installation...........29-36
Distributor.....................29-12
  Removal.....................29-12

E
Engine Assembly 29-13
  Removal & Installation...........29-13
Exhaust Manifold...............29-19
  Removal & Installation...........29-19

F
Fuel Filter 29-27
  Removal & Installation...........29-27
Fuel Injector...................29-27
  Removal & Installation...........29-27
Fuel Pump.....................29-27
  Removal & Installation...........29-27
Fuel System Pressure 29-27
  Relieving...................29-27
Fuel System Service
  Precautions.................29-27

H
Halfshaft.......................29-30
  Removal & Installation...........29-30
Heater Core....................29-14
  Removal & Installation...........29-14
Hydraulic Clutch System 29-29
  Bleeding....................29-29

I
Ignition Timing 29-12
  Adjustment..................29-12
Intake Manifold.................29-18
  Removal & Installation...........29-18

L
Lower Ball Joint.................29-35
  Removal & Installation...........29-35
Lower Control Arm 29-35
  Control Arm Bushing
    Replacement................29-35
  Removal & Installation...........29-35

O
Oil Pan........................29-21
  Removal & Installation...........29-21
Oil Pump......................29-22
  Removal & Installation...........29-22

P
Pinion Seal 29-31
  Removal & Installation...........29-31
Piston and Ring 29-26
  Positioning 29-26
Power Rack and Pinion
  Steering Gear...............29-32
  Removal & Installation...........29-32

R
Rear Axle Housing Assembly.......29-31
  Removal & Installation...........29-31
Rear Axle Shaft, Bearing
  and Seal 29-31
  Removal & Installation...........29-31
Rear Main Seal 29-22
  Removal & Installation...........29-22
Rocker Arms/Shafts 29-18
  Removal & Installation...........29-18

S
Shock Absorber 29-33
  Removal & Installation...........29-33
Starter Motor 29-21
  Removal & Installation...........29-21
Strut.........................29-33
  Removal & Installation...........29-33

T
Timing Chain, Sprockets,
  Front Cover and Seal.........29-22
  Removal & Installation...........29-22
Transfer Case Assembly.........29-29
  Removal & Installation...........29-29
Transmission Assembly.........29-28
  Removal & Installation...........29-28

V
Valve Lash 29-21
  Adjustment.................29-21

W
Water Pump..................29-13
  Removal & Installation...........29-13
Wheel Bearings...............29-35
  Adjustment.................29-35
  Removal & Installation...........29-35

SPECIFICATION AND MAINTENANCE CHARTS

ENGINE AND VEHICLE IDENTIFICATION CHART

		Engine Code						Model Year	
Code	Liters (cc)	Cu. In.	Cyl.	Fuel Sys.	Engine Type	Eng. Mfg.		Code	Year
1	2.5 (2494)	152	6	MFI	DOHC	Suzuki		1	2001
C	2.0 (1997)	122	4	MFI	DOHC	Suzuki		2	2002
								3	2003
								4	2004
								5	2005

MFI: Multiport Fuel Injection

DOHC: Dual Overhead Cam

71461-TRAC-C01

GENERAL ENGINE SPECIFICATIONS

Year	Model	Engine Displacement Liters	Engine VIN	Net Horsepower @ rpm	Net Torque @ rpm (ft. lbs.)	Bore x Stroke (in.)	Compression Ratio	Oil Pressure @ rpm
2001	Tracker	2.0	C	127@6000	134@3000	3.31x3.54	9.7:1	55-67@4000
		2.5	1	140@6500	151@4000	3.31x2.95	9.5:1	55-67@4000
2002	Tracker	2.0	C	127@6000	134@3000	3.31x3.54	9.7:1	55-67@4000
		2.5	1	140@6500	151@4000	3.31x2.95	9.5:1	55-67@4000
2003	Tracker	2.0	C	127@6000	134@3000	3.31x3.54	9.7:1	55-67@4000
		2.5	1	140@6500	151@4000	3.31x2.95	9.5:1	55-67@4000
2004	Tracker	2.0	C	127@6000	134@3000	3.31x3.54	9.7:1	55-67@4000
		2.5	1	140@6500	151@4000	3.31x2.95	9.5:1	55-67@4000

MFI: Multi-port Fuel Injection

NA: Not available

71461-TRAC-C02

ENGINE TUNE-UP SPECIFICATIONS

Year	Engine Displacement Liters	Engine VIN	Spark Plugs Gap (in.)	Ignition Timing (deg.) MT	Ignition Timing (deg.) AT	Fuel Pump (psi)	Idle Speed (rpm) MT	Idle Speed (rpm) AT	Valve Clearance In.	Valve Clearance Ex.
2001	2.5	1	0.040	5B	5B	30-45	700-800	700-800	HYD	HYD
	2.0	C	0.041	5B	5B	35-43	700-800	700-800	HYD	HYD
2002	2.5	1	0.040	5B	5B	30-45	700-800	700-800	HYD	HYD
	2.0	C	0.041	5B	5B	35-43	700-800	700-800	HYD	HYD
2003	2.5	1	0.040	5B	5B	30-45	700-800	700-800	HYD	HYD
	2.0	C	0.041	5B	5B	35-43	700-800	700-800	HYD	HYD
2004	2.5	1	0.041	5B	5B	30-45	700-800	700-800	HYD	HYD
	2.0	C	0.041	5B	5B	35-43	700-800	700-800	HYD	HYD

HYD: Hydraulic

71461-TRAC-C03

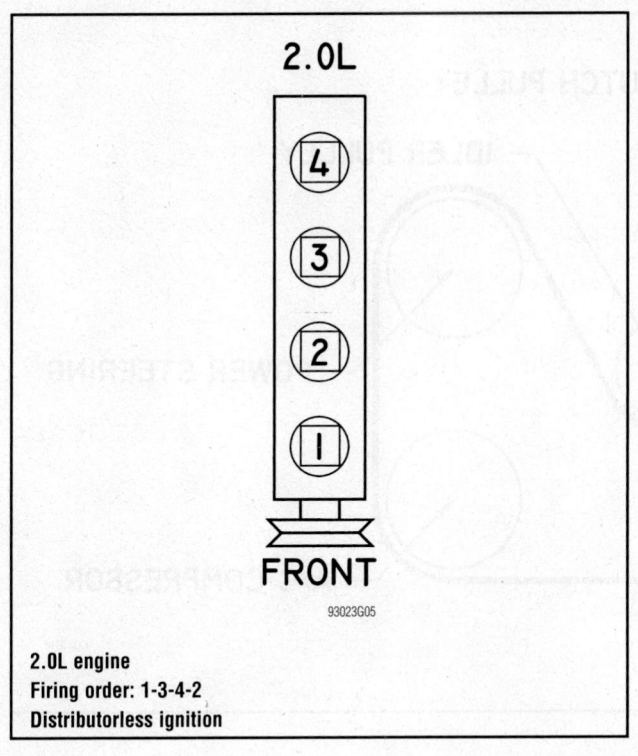

2.0L

FRONT

93023G05

2.0L engine
Firing order: 1-3-4-2
Distributorless ignition

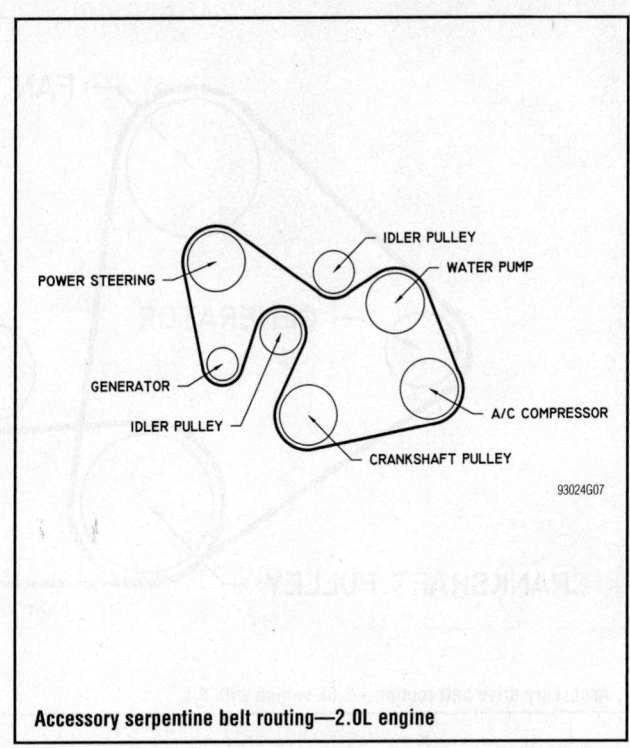

93024G07

Accessory serpentine belt routing—2.0L engine

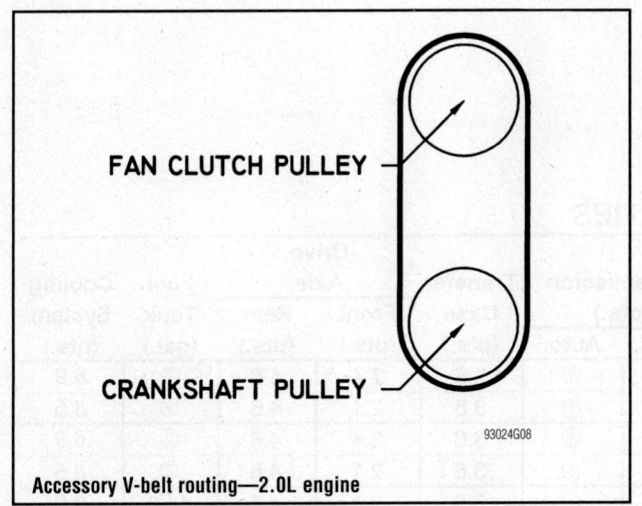

FAN CLUTCH PULLEY

CRANKSHAFT PULLEY

93024G08

Accessory V-belt routing—2.0L engine

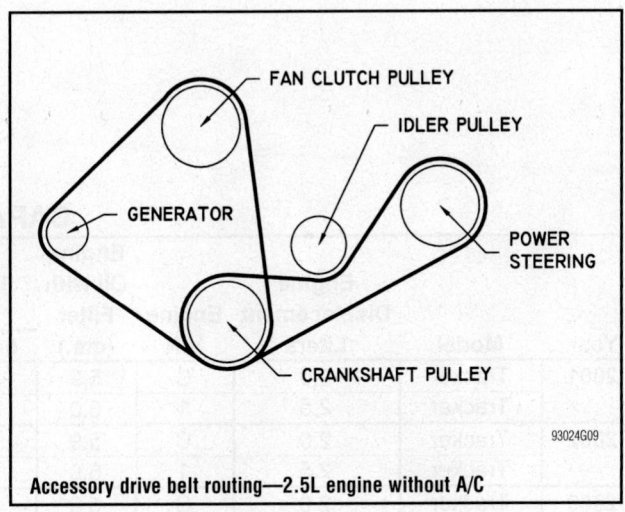

93024G09

Accessory drive belt routing—2.5L engine without A/C

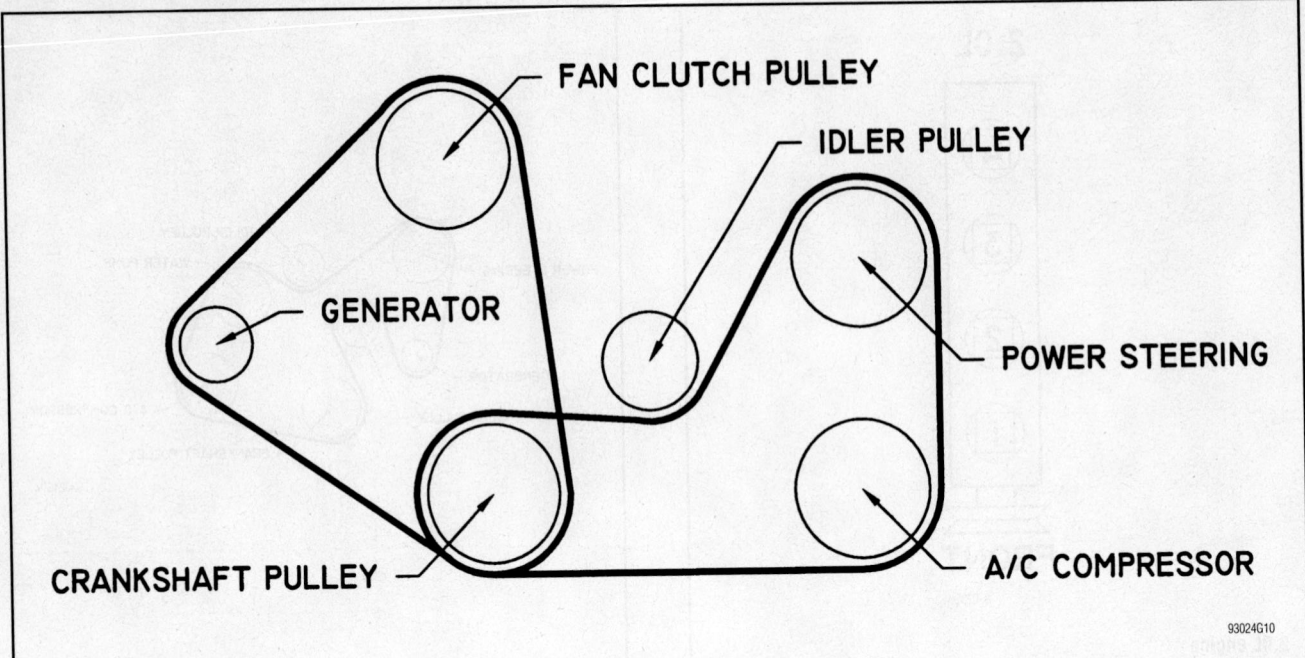

Accessory drive belt routing—2.5L engine with A/C

93024G10

CAPACITIES

Year	Model	Engine Displacement Liters	Engine VIN	Engine Oil with Filter (qts.)	Transmission (pts.) 5-Spd	Transmission (pts.) Auto.	Transfer Case (pts.)	Drive Axle Front (pts.)	Drive Axle Rear (pts.)	Fuel Tank (gal.)	Cooling System (qts.)
2001	Tracker	2.0	C	5.9	3.2	①	3.6	2.4	4.6	②	6.9
	Tracker	2.5	1	6.0	5.5	①	3.6	2.1	4.6	②	8.5
2002	Tracker	2.0	C	5.9	3.2	①	3.6	2.4	4.6	②	6.9
	Tracker	2.5	1	6.0	5.5	①	3.6	2.1	4.6	②	8.5
2003	Tracker	2.0	C	5.9	3.2	①	3.6	2.4	4.6	17.0	6.9
	Tracker	2.5	1	6.0	5.5	①	3.6	2.1	4.6	17.0	8.5
2004	Tracker	2.0	C	5.5	③	①	3.6	2.2	4.6	④	6.9
	Tracker	2.5	1	5.8	③	①	3.6	2.2	4.6	④	8.5

Note: All capacities are approximate. Add fluid gradually and check to be sure a proper fluid level is obtained.

① 2WD: 6.0 pts
   4WD: 5.2 pts.

② 2dr: 14.8
   4dr: 17.4

③ 2-wheel drive: 4.0 pts.
   4-wheel drive: 3.2 pts.

④ 2-door: 14.8
   4-door: 17.4

71461-TRAC-C04

VALVE SPECIFICATIONS

Year	Engine Displacement Liters	Engine VIN	Seat Angle (deg.)	Face Angle (deg.)	Spring Test Pressure (lbs. @ in.)	Spring Installed Height (in.)	Stem-to-Guide Clearance (in.)		Stem Diameter (in.)	
							Intake	Exhaust	Intake	Exhaust
2001	2.5	1	45	45	①	1.250	0.0008-0.0027	0.0018-0.0035	0.2348-0.2354	0.2339-0.2344
	2.0	C	45	45	①	②	0.0008-0.0027	0.0018-0.0035	0.2348-0.2354	0.2339-0.2344
2002	2.5	1	45	45	①	1.250	0.0008-0.0027	0.0018-0.0035	0.2348-0.2354	0.2339-0.2344
	2.0	C	45	45	①	②	0.0008-0.0027	0.0018-0.0035	0.2348-0.2354	0.2339-0.2344
2003	2.5	1	45	45	①	1.250	0.0008-0.0027	0.0018-0.0035	0.2348-0.2354	0.2339-0.2344
	2.0	C	45	45	①	②	0.0008-0.0027	0.0018-0.0035	0.2348-0.2354	0.2339-0.2344
2004	2.5	1	45	45	①	1.250	0.0008-0.0027	0.0018-0.0035	0.2348-0.2354	0.2339-0.2344
	2.0	C	45	45	①	②	0.0008-0.0027	0.0018-0.0035	0.2348-0.2354	0.2339-0.2344

① Inner: 13.6-17.4@1.08
Outer: 30.4-39.2@1.25

② Inner spring: 1.08
Outer spring: 1.25

71461-TRAC-C05

CRANKSHAFT AND CONNECTING ROD SPECIFICATIONS
All measurements are given in inches.

Year	Engine Displacement Liters	Engine VIN	Crankshaft				Connecting Rod		
			Main Brg. Journal Dia.	Main Brg. Oil Clearance	Shaft End-play	Thrust on No.	Journal Diameter	Oil Clearance	Side Clearance
2001	2.5	1	2.5583-2.5590	0.0008-0.0023	0.0044-0.0149	2	1.9678-1.9685	0.0016-0.0031	0.0099-0.0157
	2.0	C	2.2828-2.2834	0.0008-0.0023	0.0039-0.0165	3	1.9678-1.9685	0.0016-0.0031	0.0099-0.0157
2002	2.5	1	2.5583-2.5590	0.0008-0.0023	0.0044-0.0149	2	1.9678-1.9685	0.0016-0.0031	0.0099-0.0157
	2.0	C	2.2828-2.2834	0.0008-0.0023	0.0039-0.0165	3	1.9678-1.9685	0.0016-0.0031	0.0099-0.0157
2003	2.5	1	2.5583-2.5590	0.0008-0.0023	0.0044-0.0149	2	1.9678-1.9685	0.0016-0.0031	0.0099-0.0157
	2.0	C	2.2828-2.2834	0.0008-0.0023	0.0039-0.0165	3	1.9678-1.9685	0.0016-0.0031	0.0099-0.0157
2004	2.5	1	2.5583-2.5590	0.0008-0.0023	0.0044-0.0149	2	1.9678-1.9685	0.0016-0.0031	0.0099-0.0157
	2.0	C	2.2828-2.2834	0.0008-0.0023	0.0039-0.0165	3	1.9678-1.9685	0.0016-0.0031	0.0099-0.0157

71461-TRAC-C06

PISTON AND RING SPECIFICATIONS

All measurements are given in inches.

Year	Engine Displacement Liters	Engine VIN	Piston Clearance	Ring Gap			Ring Side Clearance		
				Top Compression	Bottom Compression	Oil Control	Top Compression	Bottom Compression	Oil Control
2001	2.5	1	0.0008-0.0015	0.0079-0.0276	0.0138-0.0276	0.0079-0.0709	0.0012-0.0027	0.0008-0.0023	NA
	2.0	C	0.0008-0.0015	0.0079-0.0276	0.0138-0.0276	0.0079-0.0709	0.0012-0.0027	0.0008-0.0023	NA
2002	2.5	1	0.0008-0.0015	0.0079-0.0276	0.0138-0.0276	0.0079-0.0709	0.0012-0.0027	0.0008-0.0023	NA
	2.0	C	0.0008-0.0015	0.0079-0.0276	0.0138-0.0276	0.0079-0.0709	0.0012-0.0027	0.0008-0.0023	NA
2003	2.5	1	0.0008-0.0015	0.0079-0.0276	0.0138-0.0276	0.0079-0.0709	0.0012-0.0027	0.0008-0.0023	NA
	2.0	C	0.0008-0.0015	0.0079-0.0276	0.0138-0.0276	0.0079-0.0709	0.0012-0.0027	0.0008-0.0023	NA
2004	2.5	1	0.0008-0.0015	0.0079-0.0276	0.0138-0.0276	0.0079-0.0709	0.0012-0.0027	0.0008-0.0023	NA
	2.0	C	0.0008-0.0015	0.0079-0.0276	0.0138-0.0276	0.0079-0.0709	0.0012-0.0027	0.0008-0.0023	NA

NA: Information Not Available

71461-TRAC-C07

TORQUE SPECIFICATIONS

All readings in ft. lbs.

Year	Engine Displacement Liters	Engine VIN	Cylinder Head Bolts	Main Bearing Bolts	Rod Bearing Bolts	Crankshaft Damper Bolts	Flywheel Bolts	Manifold		Spark Plugs	Oil Pan Drain Plug
								Intake	Exhaust		
2001	2.5	1	①	②	33	109	51	17	22	18	25
	2.0	C	①	②	33	109	51	17	17	18	25
2002	2.5	1	①	②	33	109	51	17	22	18	25
	2.0	C	①	②	33	109	51	17	17	18	25
2003	2.5	1	①	②	33	109	51	17	22	18	25
	2.0	C	①	②	33	109	51	17	17	18	25
2004	2.5	1	①	③	33	109	51	17	22	18	25
	2.0	C	①	③	33	109	51	19	37	18	25

① Step 1: 38 ft. lbs.
  Step 2: 61 ft. lbs.
  Step 3: Loosen in reverse order to 0 ft. lbs.
  Step 4: 38 ft. lbs.
  Step 5: 76 ft. lbs.
  Step 6: Tighten 6mm bolt to 8 ft. lbs.

② 10mm: 43.5 ft. lbs.
  8mm: 19.5 ft. lbs.

③ Step 1: 31 ft. lbs.
  Step 2: 42 ft. lbs.

71461-TRAC-C08

WHEEL ALIGNMENT

Year	Model	Caster* Range (+/-Deg.)	Caster* Preferred Setting (Deg.)	Camber* Range (+/-Deg.)	Camber* Preferred Setting (Deg.)	Toe-in (in.)
2001	Tracker	1.00	+2.67	1.00	0	0+/-0.16
2002	Tracker	1.00	+2.67	1.00	0	0+/-0.16
2003	Tracker	1.00	+2.67	1.00	0	0+/-0.16
2004	Tracker	1.00	+2.67	1.00	0	0+/-0.16

* Alignment is non-adjustable. Figures are for reference only.

71461-TRAC-C09

TIRE, WHEEL AND BALL JOINT SPECIFICATIONS

Year	Model	OEM Tires Standard	OEM Tires Optional	Tire Pressures (psi) Front	Tire Pressures (psi) Rear	Wheel Size	Ball Joint Inspection	Lugnut Torque (ft. lbs.)
2001	Tracker 2wd	P195/75R15	None	23	23	5.5-JJ	①	70
	Tracker 4wd	P205/75R15	None	23	23	5.5-JJ	①	70
2002	Tracker 2wd	P195/75R15	None	23	23	5.5-JJ	①	70
	Tracker 4wd	P205/75R15	None	23	23	5.5-JJ	①	70
2003	Tracker 2wd	P195/75R15	None	23	23	5.5-JJ	①	70
	Tracker 4wd	P205/75R15	None	23	23	5.5-JJ	①	70
2004	Tracker 2wd	P215/70R15	None	26	26	NA	①	70
	Tracker 4wd	P215/70R15	None	26	26	NA	①	70
	Tracker ZR2	P215/75R15	None	26	26	NA	①	70

OEM: Original Equipment Manufacturer

PSI: Pounds Per Square Inch

STD: Standard

OPT: Optional

① Replace if any measurable movement is found.

71461-TRAC-C10

BRAKE SPECIFICATIONS
All measurements in inches unless noted

Year	Model	Brake Disc Original Thickness	Brake Disc Minimum Thickness	Brake Disc Maximum Runout	Brake Drum Diameter Original Inside Diameter	Brake Drum Diameter Max. Wear Limit	Brake Drum Diameter Maximum Machine Diameter	Minimum Lining Thickness Front	Minimum Lining Thickness Rear	Brake Caliper Bracket Bolts (ft. lbs.)	Brake Caliper Mounting Bolts (ft. lbs.)
2001	Tracker ①	0.670	0.590	0.006	8.66	8.74	8.74	0.08	0.04	61.5	19-21
	Tracker ②	0.670	0.590	0.006	8.66	8.74	8.74	0.08	0.04	61.5	19-21
2002	Tracker ①	0.670	0.590	0.006	8.66	8.74	8.74	0.08	0.04	61.5	19-21
	Tracker ②	0.670	0.590	0.006	8.66	8.74	8.74	0.08	0.04	61.5	19-21
2003	Tracker ①	0.670	0.590	0.006	8.66	8.74	8.74	0.08	0.04	61.5	19-21
	Tracker ②	0.670	0.590	0.006	8.66	8.74	8.74	0.08	0.04	61.5	19-21
2004	Tracker ①	0.670	0.590	0.006	8.66	8.74	8.74	0.08	0.04	61.5	19-21
	Tracker ②	0.670	0.590	0.006	8.66	8.74	8.74	0.08	0.04	61.5	19-21

① 2-door model

② 4-door model

71461-TRAC-C11

SCHEDULED MAINTENANCE INTERVALS
2001-02 CHEVROLET TRACKER

TO BE SERVICED	TYPE OF SERVICE	VEHICLE MILEAGE INTERVAL (x1000)												
		7.5	15	22.5	30	37.5	45	52.5	60	67.5	75	82.5	90	97.5
Engine oil & filter	R	✓	✓	✓	✓	✓	✓	✓	✓	✓	✓	✓	✓	✓
Automatic transmission fluid ①	S/I	✓	✓	✓	✓	✓	✓	✓	✓	✓	✓	✓	✓	✓
Manual transmission oil ②	S/I	✓	✓	✓	✓	✓	✓	✓	✓	✓	✓	✓	✓	✓
Steering system	S/I	✓	✓	✓	✓	✓	✓	✓	✓	✓	✓	✓	✓	✓
Transfer & differential oil ②	S/I	✓	✓	✓	✓	✓	✓	✓	✓	✓	✓	✓	✓	✓
Wheel discs & free wheeling hubs	S/I	✓	✓	✓	✓	✓	✓	✓	✓	✓	✓	✓	✓	✓
Suspension system	S/I	✓	✓	✓	✓	✓	✓	✓	✓	✓	✓	✓	✓	✓
Brake discs & pads (front)	S/I		✓		✓		✓		✓		✓		✓	
Brake drums & shoes (rear)	S/I		✓		✓		✓		✓		✓		✓	
Brake fluid ③	S/I		✓		✓		✓		✓		✓		✓	
Brake hoses & pipes	S/I		✓		✓		✓		✓		✓		✓	
Brake pedal	S/I		✓		✓		✓		✓		✓		✓	
Brake lever & cable	S/I		✓		✓		✓		✓		✓		✓	
Clutch	S/I		✓		✓		✓		✓		✓		✓	
Idle speed	S/I		✓		✓		✓		✓		✓		✓	
Propeller shafts	S/I		✓		✓		✓		✓		✓		✓	
Valve lash (clearance)	S/I		✓		✓		✓		✓		✓		✓	
Wheel bearings	S/I		✓		✓		✓		✓		✓		✓	
Air cleaner filter element	R				✓				✓				✓	
Engine coolant	R				✓				✓				✓	
Fuel filter	R				✓				✓				✓	
Spark plugs	R				✓				✓				✓	
Cooling system hoses	S/I				✓				✓				✓	
Drive belt(s)	S/I				✓				✓				✓	
Exhaust pipes & mountings	S/I				✓				✓				✓	
Fuel lines & connections	S/I				✓				✓				✓	
Camshaft timing belt	R								✓				✓	
Distributor cap & rotor	S/I								✓					
Emission-related hoses	S/I								✓					
Oxygen sensor	S/I											✓		
EVAP canister	R	every 100,000 miles												
PCV valve	R							✓						

71461-TRAC-C12

SCHEDULED MAINTENANCE INTERVALS
2001-02 CHEVROLET TRACKER

TO BE SERVICED	TYPE OF SERVICE	VEHICLE MILEAGE INTERVAL (x1000)												
		7.5	15	22.5	30	37.5	45	52.5	60	67.5	75	82.5	90	97.5
EGR system	S/I							✓						
Fuel Injectors	S/I	every 100,000 miles												
TWC converter	S/I	every 100,000 miles												

R: Replace S/I: Service or Inspect

① Replace at 100,000 miles.
② Replace oil every 30,000 miles.
③ Replace every 60,000 miles.

FREQUENT OPERATION MAINTENANCE (SEVERE SERVICE)

If a vehicle is operated under any of the following conditions it is considered severe service:

- Extremely dusty areas.
- 50% or more of the vehicle operation is in 32°C (90°F) or higher temperatures, or constant operation in temperatures below 0°C (32°F).
- Prolonged idling (vehicle operation in stop and go traffic).
- Frequent short running periods (engine does not warm to normal operating temperatures).
- Police, taxi, delivery usage or trailer towing usage.

Oil & oil filter: replace every 3000 miles.

Air cleaner filter element: service or inspect every 3000 miles & replace every 15,000 miles.

Steering wheel free play, gear box oil & linkage: service or inspect every 3000 miles.

Brake & nuts on chassis: tighten every 6000 miles.

Brake discs & pads (front): service or inspect every 6000 miles.

Brake drums & shoes (rear): service or inspect every 6000 miles.

Exhaust pipes & mountings: tighten every 6000 miles.

Propeller shafts: service or inspect every 6000 miles.

Automatic transmission fluid & filter: replace every 15,000 miles.

Distributor cap & ignition wires: service or inspect every 15,000 miles.

Drive belt(s): service or inspect every 15,000 miles.

Manual transmission oil: replace every 15,000 miles.

Transfer & differential oil: replace every 15,000 miles.

71461-TRAC-C13

SHORT TRIP/CITY SCHEDULED MAINTENANCE INTERVALS
2003-04 CHEVROLET TRACKER

TO BE SERVICED	TYPE OF SERVICE	3	6	9	12	15	18	21	24	27	30	33	36	39
		VEHICLE MILEAGE INTERVAL (x1000)												
Engine oil & filter	R	✓	✓	✓	✓	✓	✓	✓	✓	✓	✓	✓	✓	✓
Passenger compartment air filter	I		✓		✓		✓		✓		✓		✓	
Passenger compartment air filter	R										✓			
Rotate tires	S/I		✓		✓		✓		✓		✓		✓	
Auto. trans. fluid and filter ①	R	every 100,000 miles												
Auto. trans. fluid lines	I	every 45,000 miles												
Manual transmission oil	R					✓					✓			
Transfer case oil	R					✓					✓			
Differential oil	R					✓					✓			
Brake fluid	R	every 60,000miles												
Ignition coil plug cap	I										✓			
Propeller shafts	S/I					✓					✓			
Air cleaner filter element	R					✓					✓			
Engine coolant	R										✓			
Fuel filter	R										✓			
Spark plugs	R	every 60,000miles												
Accessory drive belt(s)	S/I										✓			
Fuel lines & connections	S/I										✓			
Emission-related hoses	S/I	every 60,000miles												
EVAP canister	R	every 120,000 miles												

R: Replace   S/I: Service or Inspect

① Replace every 15,000 miles if driven under any of the following conditions:

  a. heavy city traffic at temps. above 90 degrees F

  b. hilly or mountainous terrain

  c. frequent trailer towing

  d. taxi, police or delivery service

71461-TRAC-C14

LONG TRIP/HIGHWAY SCHEDULED MAINTENANCE INTERVALS
2003-04 CHEVROLET TRACKER

TO BE SERVICED	TYPE OF SERVICE	VEHICLE MILEAGE INTERVAL (x1000)												
		7.5	15	22.5	30	37.5	45	52.5	60	67.5	75	82.5	90	97.5
Engine oil & filter	R	✓	✓	✓	✓	✓	✓	✓	✓	✓	✓	✓	✓	✓
Rotate tires	S/I	✓	✓	✓	✓	✓	✓	✓	✓	✓	✓	✓	✓	✓
Automatic transmission fluid ①	R	every 100,000 miles												
Brake fluid	R								✓					
Propeller shafts	S/I		✓		✓		✓		✓		✓		✓	
Accessory drive belts	I		✓		✓		✓		✓		✓		✓	
Passenger compartment air filter	I		✓		✓		✓		✓		✓		✓	
Engine coolant	R				✓				✓				✓	
Ignition coil plug cap	I				✓				✓				✓	
Engine air cleaner	R				✓				✓				✓	
Fuel filter	R				✓				✓					
Spark plugs	R								✓					
Fuel lines & connections	S/I				✓				✓				✓	
Automatic transmission fluid lines	I						✓						✓	
Emission-related hoses	S/I								✓					

R: Replace   S/I: Service or Inspect

① Replace at 15,000 miles if operated under any of the following conditions:

    a. heavy city traffic at temps. above 90 degrees F

    b. hilly or mountainous terrain

    c. frequent trailer towing

    d. taxi, police or delivery service

71461-TRAC-C15

ENGINE REPAIR

➡ **Disconnecting the negative battery cable on some vehicles may interfere with the functions of the on board computer system. The computer may undergo a relearning process once the negative battery cable is reconnected.**

Distributor

REMOVAL

All engines are equipped with a Distributorless Ignition System (DIS).

Alternator

REMOVAL

1. Before servicing the vehicle, refer to the precautions in the beginning of this section.
2. Remove or disconnect the following:
   - Negative battery cable
   - Evaporative Emission (EVAP) canister
   - Accessory drive belt
   - Alternator harness connectors
   - Alternator mounting bracket
   - Alternator

INSTALLATION

Install or connect the following:
   - Alternator
   - Alternator mounting bracket. Tighten the bolts to 20 ft. lbs. (27 Nm).
   - Alternator harness connectors
   - Accessory drive belt. Tighten the alternator bolts to 24 ft. lbs. (33 Nm).
   - EVAP canister
   - Negative battery cable

Ignition Timing

ADJUSTMENT

2.0L and 2.5L Engines

➡ **The 2.0L and 2.5L engines use a Camshaft Position (CMP) sensor that is rotated to set base timing.**

➡ **Check and adjust the ignition timing with the engine at normal operating temperature, all electrical accessories OFF and transmission in P, N for automatic transmission or neutral for manual transmission.**

1. Before servicing the vehicle, refer to the precautions in the beginning of this section.

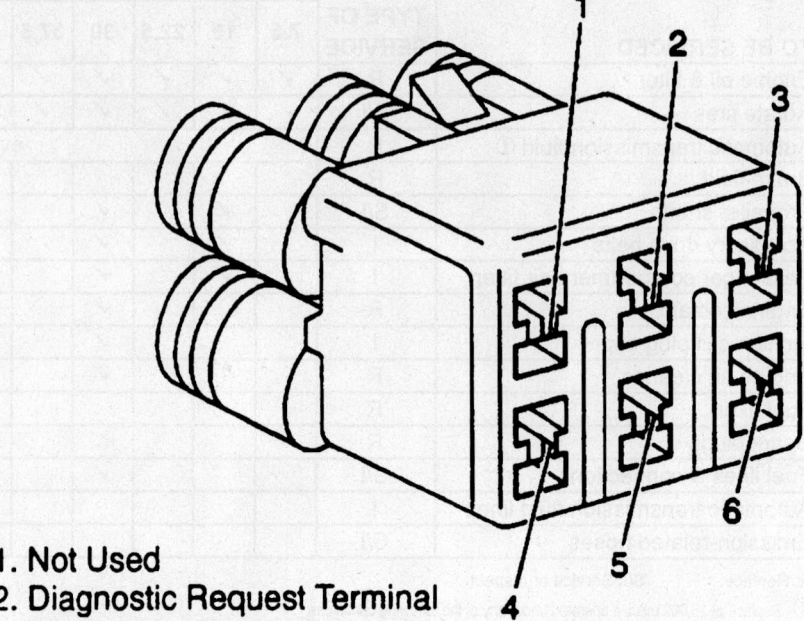

1. **Not Used**
2. **Diagnostic Request Terminal**
3. **Diagnostic Output Terminal**
4. **Ground Terminal**
5. **Test Switch Terminal**
6. **Duty Check Terminal**

7924HG01

Duty Check Data Link Connector terminal identification for ignition timing

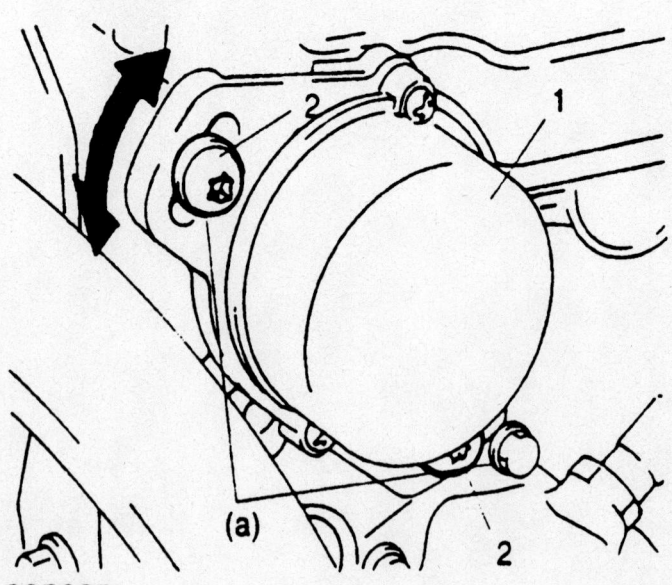

1. **CMP sensor**
2. **Bolt**

7924HG82

Camshaft position sensor

2. With the engine **OFF**, connect a jumper wire between terminals **4** and **5** of the Data Link Connector (DLC) for Tracker or between terminals **D** and **E** of the DLC for all others.

3. Connect a timing light to the No. 1 spark plug wire and start the engine.

4. Ignition timing at idle should be 4–6 degrees Before Top Dead Center (BTDC).

5. Adjust the timing as necessary, then turn the engine **OFF**.

6. Remove the jumper wire from the DLC and remove the timing light.

Engine Assembly

REMOVAL & INSTALLATION

2.0L and 2.5L Engines

1. Before servicing the vehicle, refer to the precautions in the beginning of this section.

2. Relieve the fuel system pressure.

3. Drain the cooling system.

4. Drain the engine oil.

5. Remove or disconnect the following:
- Negative battery cable
- Hood
- Heater hoses
- Radiator hoses
- Cooling fan and shroud
- Radiator overflow tank
- Radiator
- Accelerator cable
- Transmission cable, if equipped
- Strut tower bar, if equipped
- Air intake assembly
- Engine oil dipstick tube
- Transmission oil dipstick tube, if equipped
- Ignition coil covers
- Ignition coil connectors
- Injector connectors
- Camshaft Position (CMP) sensor connector
- Crankshaft Position (CKP) sensor connector
- Throttle Position (TP) sensor connector
- Mass Air Flow (MAF) sensor connector
- Idle Air Control (IAC) valve
- Intake manifold ground cable
- Evaporative Emission (EVAP) canister purge valve
- Exhaust Gas Recirculation (EGR) valve connector
- Heated Oxygen (HO2S) sensor connectors
- Engine Coolant Temperature (ECT) sensor connector
- Alternator wiring connectors
- Oil pressure gauge sender connector
- Power Steering Pressure (PSP) switch connector
- Alternator bracket ground cable
- Brake booster vacuum line
- Tank pressure control vacuum valve hose
- Fuel lines
- EVAP canister
- Power steering pump
- A/C compressor
- Steering shaft lower assembly
- Front differential housing, if equipped
- Exhaust front pipe and bracket
- Transmission oil cooler lines, if equipped
- Transmission stiffener brackets, if equipped
- Flywheel access cover
- Torque converter, if equipped
- Starter motor
- Transmission flange fasteners and support the transmission
- Left and right engine mounts
- Engine

To install:

6. Install or connect the following:
- Engine
- Left and right engine mounts. Tighten the nuts to 36 ft. lbs. (50 Nm).
- Transmission flange fasteners. Tighten them to 58 ft. lbs. (80 Nm).
- Starter motor
- Torque converter. Tighten the bolts to 47 ft. lbs. (65 Nm).
- Flywheel access cover
- Transmission stiffener brackets. Tighten the bolts to 36 ft. lbs. (50 Nm).
- Transmission oil cooler lines, if equipped
- Exhaust front pipe and bracket
- Front differential housing, if equipped
- Steering shaft lower assembly
- A/C compressor
- Power steering pump
- EVAP canister
- Fuel lines
- Tank pressure control vacuum valve hose
- Brake booster vacuum line
- Alternator bracket ground cable
- PSP switch connector
- Oil pressure gauge sender connector

- Alternator wiring connectors
- ECT sensor connector
- HO2S sensor connectors
- EGR valve connector
- EVAP canister purge valve
- Intake manifold ground cable
- IAC valve
- MAF sensor connector
- TP sensor connector
- CKP sensor connector
- CMP sensor connector
- Injector connectors
- Ignition coil connectors
- Ignition coil covers
- Transmission oil dipstick tube, if equipped
- Engine oil dipstick tube
- Air intake assembly
- Strut tower bar, if equipped
- Transmission cable, if equipped
- Accelerator cable
- Radiator
- Radiator overflow tank
- Cooling fan and shroud
- Radiator hoses
- Heater hoses
- Hood
- Negative battery cable

7. Fill the crankcase to the correct level.

8. Fill the cooling system.

9. Start the engine and check for leaks.

Water Pump

REMOVAL & INSTALLATION

2.0L Engines

1. Before servicing the vehicle, refer to the precautions in the beginning of this section.

2. Drain the cooling system.

3. Remove or disconnect the following:
- Negative battery cable
- Radiator hose at the thermostat housing
- Heater outlet pipe bolt
- Alternator belt
- Water pump

To install:

➡**Use new water pump bolts for assembly.**

4. Install or connect the following:
- Water pump with a new O-ring seal. Tighten the bolts to 19 ft. lbs. (27 Nm).
- Alternator belt
- Heater outlet pipe bolt

- Radiator hose at the thermostat housing
- Negative battery cable

5. Fill the cooling system.
6. Start the engine and check for leaks.

2.5L Engine

1. Before servicing the vehicle, refer to the precautions in the beginning of this section.
2. Drain the cooling system.
3. Remove or disconnect the following:
- Negative battery cable
- Accessory drive belts
- Front cover
- Water pump

To install:

4. Install or connect the following:
- Water pump with a new O-ring seal. Tighten the bolts to 19 ft. lbs. (27 Nm).
- Front cover
- Accessory drive belts
- Negative battery cable

5. Fill the cooling system.
6. Start the engine and check for leaks.

Heater Core

REMOVAL & INSTALLATION

1. Disconnect the negative battery cable.
2. Disable the SIR by performing the following procedure:

a. From the fuse box, located near the base of the steering column, remove the AIR BAG fuse.

b. Remove the steering wheel side cap, disconnect the Connector Positive Assurance (CPA) and the yellow 2-way driver's inflator module connectors.

c. Pull the instrument panel compartment out by pushing the right-side and left-side stoppers (located on both sides) inward.

d. Disconnect the CPA and the yellow 4-way passenger's inflator module connectors.

➡ **With the AIR BAG fuse removed and the ignition switch turned ON, the AIR BAG warning light will be ON; this is normal operation and does not indicate a SIR system malfunction.**

3. Drain the cooling system into a clean container for reuse.
4. Remove the instrument panel as follows:

a. Remove the center console.

b. Remove the lower steering column cover by loosening the mounting screws.

c. Remove the glove box.

d. Detach the wiring harness connectors from the heater unit and the blower motor assembly.

e. Detach the wiring harness connectors from the ignition switch, contact coil and combination switch.

f. Open the hood.

g. Remove the steering column shaft joint bolt, then separate the steering column shaft from the lower steering shaft.

h. Loosen all of the steering column-to-firewall and instrument panel brace bolts.

i. If equipped, remove the shift (key) interlock cable screw. Disconnect the cable from the ignition switch.

j. Remove the steering column from the vehicle.

✳✳ WARNING

Do not rest the steering column assembly on the steering wheel with the air bag module facing downward and the column vertical—personal injury may be the result.

k. Disconnect the speedometer cable from the speedometer, then remove the instrument cluster.

l. Remove the hood latch handle.

m. Remove the radio, the heater control panel and the heater control cables from the instrument panel.

n. Disconnect and label all wiring harness connectors from the instrument panel.

o. Remove the instrument panel mounting screws and bolts. Remove the side cover plates and the instrument panel mounting fasteners from the side of the assembly. Then, remove the upper cover plates and loosen the remaining mounting fasteners

p. Have an assistant help you carefully lift the instrument panel up and out of the vehicle. When separating the instrument panel from the firewall, ensure that all of the cables, wires and hoses are disconnected form the instrument panel.

5. Remove the 2 bolts and the right-side instrument panel center support.
6. If equipped with air conditioning, remove the evaporator.
7. Remove the 2 screws securing the SIR harness clip on the Sensing and Diagnostic Module (SDM) bracket.
8. Disconnect the SDM electrical connector.

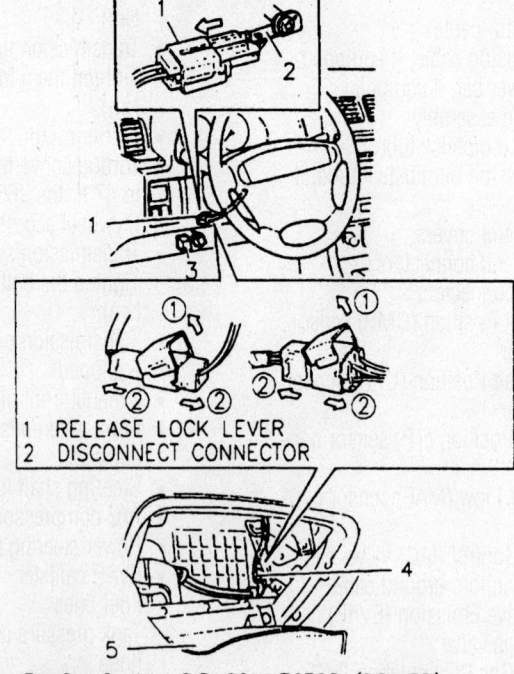

1 RELEASE LOCK LEVER
2 DISCONNECT CONNECTOR

1 YELLOW 2-WAY SIR CONNECTOR (DRIVER)
2 CONNECTOR POSITION ASSURANCE (CPA)
3 AIR BAG FUSE
4 YELLOW 4-WAY SIR CONNECTOR (PASSENGER)
5 GLOVE BOX

93113G90

Disabling the air bag system

9. Remove the 4 screws and the SDM bracket from the vehicle.

10. Remove the speedometer cable and antenna cable (if equipped) from the heater case.

11. Remove the floor duct from the heater case.

12. If equipped with air conditioning, disconnect the electrical jumper harness for the air conditioning amplifier.

13. Remove the relay bracket screws and the relay bracket.

14. From the engine compartment, remove the 2 heater assembly-to-chassis nuts and the 2 bolts.

15. Remove the heater case from the vehicle.

16. Remove the dampers and linkages from the heater case.

17. Remove the heater core bracket screw and the bracket.

18. Remove the heater core from the heater case.

To install:

19. Install the heater core to the heater case.

20. Install the heater core bracket and the bracket screw.

21. Install the dampers and linkages to the heater case.

22. Install the heater case to the vehicle.

23. In the engine compartment, install the 2 heater assembly-to-chassis nuts and the 2 bolts. Torque the nuts/bolts to 89 inch lbs. (10 Nm).

24. Install the relay bracket and the relay bracket screws.

25. If equipped with air conditioning, connect the electrical jumper harness for the air conditioning amplifier.

26. Install the floor duct to the heater case.

27. Install the speedometer cable and antenna cable (if equipped) to the heater case.

28. Install the SDM bracket and the 4 screws to the vehicle. Torque the screws to 49 inch lbs. (5.5 Nm).

29. Connect the SDM electrical connector.

30. Install the 2 screws securing the SIR harness clip on the Sensing and Diagnostic Module (SDM) bracket. Torque the screws to 49 inch lbs. (5.5 Nm)

31. If equipped with air conditioning, install the evaporator.

32. Install the 2 bolts and the right-side instrument panel center support.

33. Install the instrument panel as follows:

a. Have an assistant help you position the instrument panel in the vehicle. When installing the instrument panel on the firewall, ensure that all of the cables, wires and hoses are routed properly.

b. Install and tighten the instrument panel mounting screws and bolts.

c. Reattach all wiring harness connectors to the instrument panel.

d. Install the radio, the heater control panel and the heater control cables. Be sure to adjust the heater control cables.

e. Install the hood latch handle.

f. Install the instrument cluster, then connect the cable to the speedometer.

g. Install the steering column in the vehicle.

h. If equipped, connect the cable from the ignition switch, then install the shift (key) interlock cable screw.

i. Install and tighten all of the steering column-to-firewall and instrument panel brace bolts to 221 inch lbs. (25 Nm).

j. Install and tighten the steering column shaft joint bolt to 221 inch lbs. (25 Nm).

k. Reattach the wiring harness connectors to the ignition switch, contact coil and combination switch.

l. Reattach the wiring harness connectors to the heater unit and the blower motor assembly.

m. Install the glove box.

n. Install the lower steering column cover.

o. Install the center console.

34. Refill the cooling system.

35. Enable the SIR by performing the following procedure:

a. Turn the ignition switch to LOCK and remove the key.

100 HEATER CONTROL UNIT
106 BLOWER MOTOR CASE-TO-HEATER CASE DUCT
107 HEATER CASE
114 TEMPERATURE CONTROL CABLE
115 MODE CONTROL CABLE
116 FRESH/RECIRC CONTROL CABLE
117 HEATER CORE
118 DAMPERS

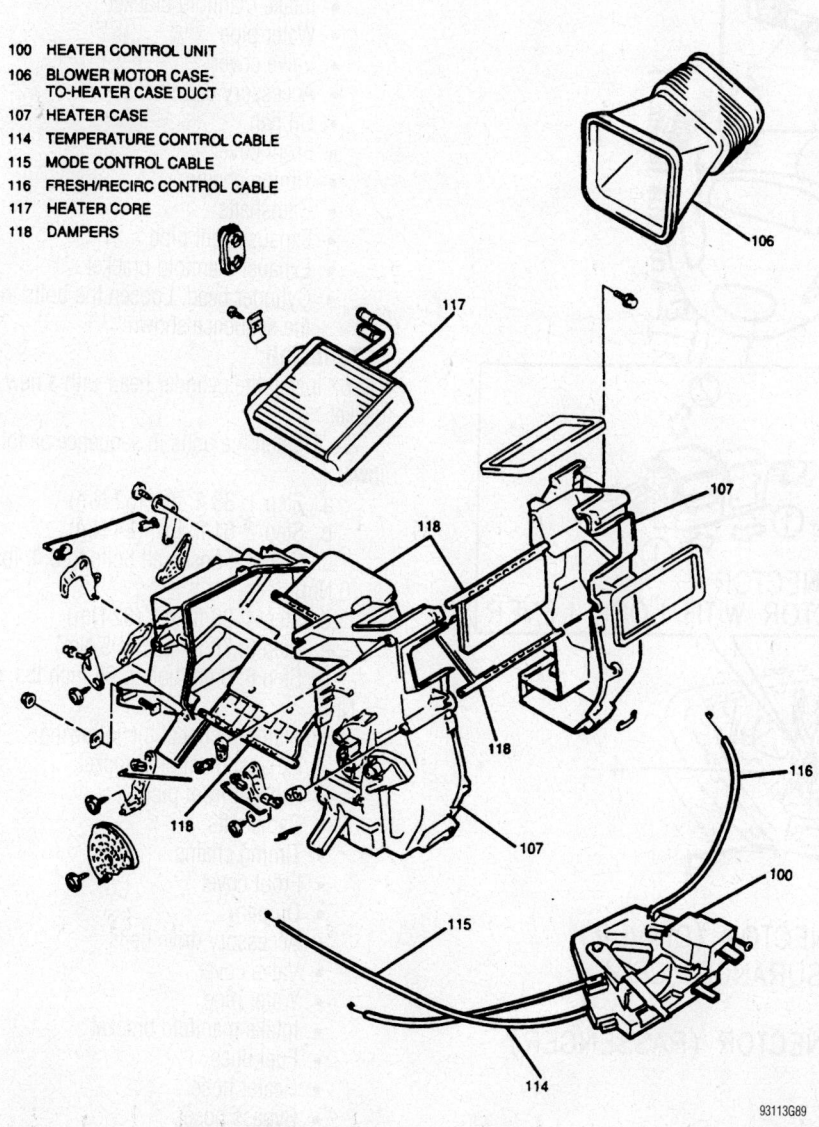

93113G89

Exploded view of the heater case assembly and related components

b. Connect the Connector Positive Assurance (CPA) and the yellow 4-way passenger's inflator module connectors.

c. Close the instrument panel compartment.

d. Connect the Connector Positive Assurance (CPA) and the yellow 2-way driver Inflator module connectors and install the steering wheel side cap.

e. At the fuse box, located near the base of the steering column, install the AIR BAG fuse.

36. Connect the negative battery cable.

37. Run the engine to normal operating temperatures; then, check the climate control operation and check for leaks.

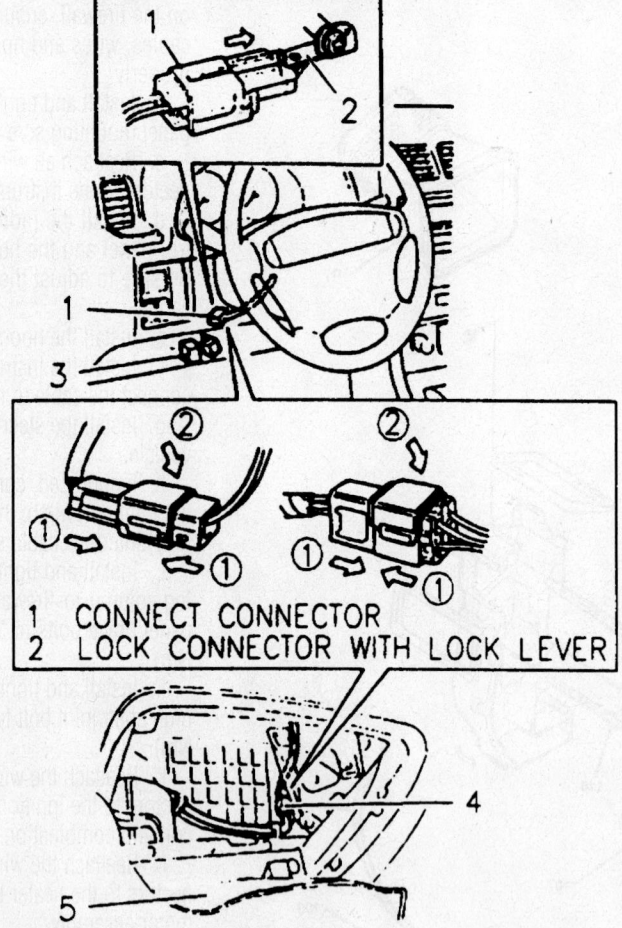

1 YELLOW 2-WAY SIR CONNECTOR (DRIVER)
2 CONNECTOR POSITION ASSURANCE (CPA)
3 AIR BAG-IG FUSE
4 YELLOW 4-WAY SIR CONNECTOR (PASSENGER)
5 GLOVE BOX

93113G91

Enabling the air bag system

Cylinder Head

REMOVAL & INSTALLATION

2.0L Engines

1. Before servicing the vehicle, refer to the precautions in the beginning of this section.
2. Relieve the fuel system pressure.
3. Drain the cooling system.
4. Drain the engine oil.
5. Remove or disconnect the following:
   - Negative battery cable
   - Strut tower brace
   - Air intake tube
   - Exhaust Gas Recirculation (EGR) valve connector
   - Idle Air Control (IAC) valve connector
   - Throttle Position (TP) sensor connector
   - Evaporative Emission (EVAP) canister purge valve connector and hose
   - Intake manifold ground cable
   - Heated Oxygen (HO$_2$S) sensor connectors
   - Camshaft Position (CMP) sensor connector
   - Engine Coolant Temperature (ECT) sensor connector
   - Fuel injector connectors
   - Ignition coils
   - Accelerator cable
   - Transmission cable, if equipped
   - Brake booster vacuum hose
   - Radiator hose
   - Bypass hose
   - Heater hose
   - Fuel lines
   - Intake manifold bracket
   - Water pipe
   - Valve cover
   - Accessory drive belts
   - Oil pan
   - Front cover
   - Timing chains
   - Camshafts
   - Exhaust front pipe
   - Exhaust manifold bracket
   - Cylinder head. Loosen the bolts in the sequence shown.

To install:

6. Install the cylinder head with a new gasket.

7. Tighten the bolts in sequence as follows:
   a. Step 1: 38 ft. lbs. (52 Nm)
   b. Step 2: 61 ft. lbs. (84 Nm)
   c. Step 3: Loosen all bolts to 0 ft. lbs. (0 Nm)
   d. Step 4: 38 ft. lbs. (52 Nm)
   e. Step 5: 76 ft. lbs. (105 Nm)
   f. Step 6: 6mm bolt to 96 inch lbs. (8 Nm)

8. Install or connect the following:
   - Exhaust manifold bracket
   - Exhaust front pipe
   - Camshafts
   - Timing chains
   - Front cover
   - Oil pan
   - Accessory drive belts
   - Valve cover
   - Water pipe
   - Intake manifold bracket
   - Fuel lines
   - Heater hose
   - Bypass hose
   - Radiator hose

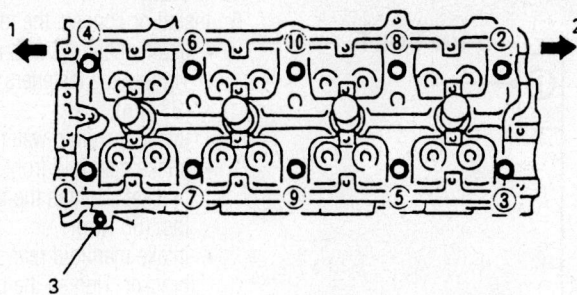

1. Crankshaft pulley side
2. Flywheel side
3. Bolt (M6)

7924HG09

Cylinder head loosening sequence—2.0L engines

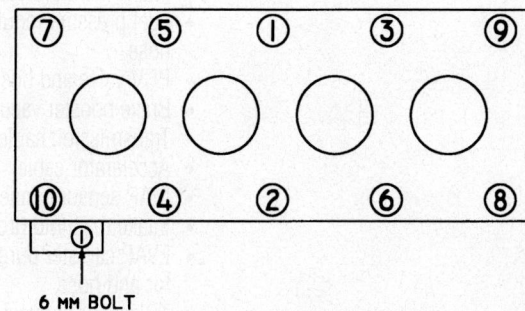

6 MM BOLT

9308HG07

Cylinder head torque sequence—2.0L engines

RH bank

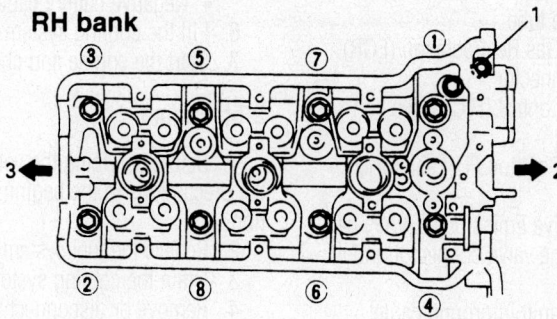

LH bank

1. Hex hole bolt
2. Timing chain side
3. Flywheel side

9302HG01

Cylinder head loosening sequence—2.5L engine

- Brake booster vacuum hose
- Transmission cable, if equipped
- Accelerator cable
- Ignition coils
- Fuel injector connectors
- ECT sensor connector
- CMP sensor connector
- HO_2S sensor connectors
- Intake manifold ground cable
- EVAP canister purge valve connector and hose
- TP sensor connector
- IAC valve connector
- EGR valve connector
- Air intake tube
- Strut tower brace
- Negative battery cable

9. Fill the crankcase to the correct level.
10. Fill the cooling system.
11. Start the engine and check for leaks.

2.5L Engine

1. Before servicing the vehicle, refer to the precautions in the beginning of this section.
2. Relieve the fuel system pressure.
3. Drain the cooling system and engine oil.
4. Remove or disconnect the following:
- Negative battery cable
- Intake manifold
- Ignition coil covers and ignition coils
- Valve covers
- Oil pan
- Timing chain cover and timing chains
- Camshaft Position (CMP) sensor
- Camshafts
- Exhaust manifolds
- Water outlet caps
- Cylinder heads. Loosen the bolts in the sequence shown.

To install:

5. Install the cylinder heads with new gaskets. Tighten the bolts in sequence as follows:
   a. Step 1: 38 ft. lbs. (52 Nm)
   b. Step 2: 61 ft. lbs. (84 Nm)
   c. Step 3: Loosen all bolts to 0 ft. lbs. (0 Nm)
   d. Step 4: 38 ft. lbs. (52 Nm)
   e. Step 5: 76 ft. lbs. (105 Nm)
   f. Step 6: 6mm bolt to 96 inch lbs. (8 Nm)
6. Install or connect the following:
- Water outlet caps
- Exhaust manifolds
- Camshafts
- CMP sensor
- Timing chain cover and timing chains
- Oil pan
- Valve covers

RIGHT BANK

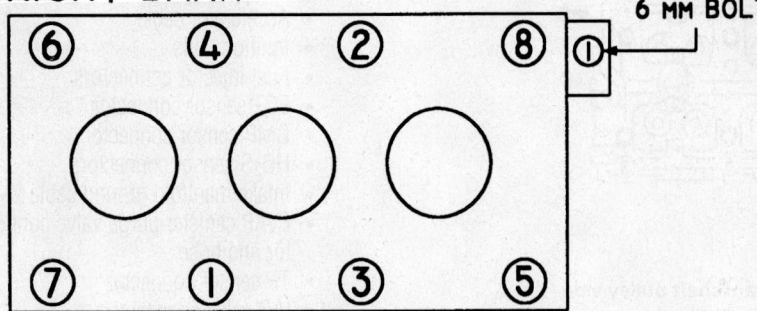

6 MM BOLT

LEFT BANK

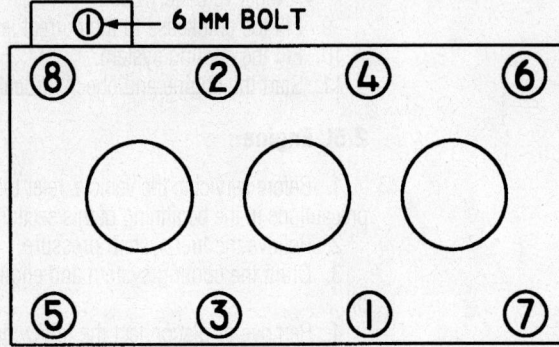

6 MM BOLT

9308HG08

Cylinder head torque sequence—2.5L engine

- Ignition coil covers and ignition coils
- Intake manifold
- Negative battery cable
7. Fill the crankcase to the correct level
8. Fill the cooling system.
9. Start the engine and check for leaks.

Rocker Arms/Shafts

REMOVAL & INSTALLATION

2.0L and 2.5L Engines

The 2.0L and 2.5L engines do not utilize rocker arms or rocker arm shafts.

Intake Manifold

REMOVAL & INSTALLATION

2.0L Engines

1. Before servicing the vehicle, refer to the precautions in the beginning of this section.
2. Relieve the fuel system pressure.
3. Drain the cooling system.
4. Remove or disconnect the following:

- Negative battery cable
- Air intake tube
- Exhaust Gas Recirculation (EGR) valve connector
- Idle Air Control (IAC) valve connector
- Throttle Position (TP) sensor connector
- Evaporative Emissions (EVAP) canister purge valve connector and hose
- Intake manifold ground cable
- Manifold Absolute Pressure (MAP) sensor connector
- Accelerator cable
- Transmission cable, if equipped
- Brake booster vacuum hose
- Positive Crankcase Ventilation (PCV) valve and hose
- Fuel pressure regulator vacuum hose
- Intake manifold vacuum hose
- Throttle body coolant hoses
- Water bypass pipe
- Fuel lines
- Fuel supply manifold with injectors attached
- Intake manifold support brackets
- Intake manifold water pipe
- Intake manifold

To install:

5. Install or connect the following:
- Intake manifold with a new gasket. Tighten the fasteners to 17 ft. lbs. (23 Nm).
- Intake manifold water pipe
- Intake manifold front support bracket. Tighten the bolts to 36 ft. lbs. (50 Nm).
- Intake manifold rear support bracket. Tighten the bolts to 18 ft. lbs. (25 Nm).
- Fuel supply manifold with injectors attached
- Fuel lines
- Water bypass pipe
- Throttle body coolant hoses
- Intake manifold vacuum hose
- Fuel pressure regulator vacuum hose
- PCV valve and hose
- Brake booster vacuum hose
- Transmission cable, if equipped
- Accelerator cable
- MAP sensor connector
- Intake manifold ground cable
- EVAP canister purge valve connector and hose
- TP sensor connector
- IAC valve connector
- EGR valve connector
- Air intake tube
- Negative battery cable
6. Fill the cooling system.
7. Start the engine and check for leaks.

2.5L Engine

1. Before servicing the vehicle, refer to the precautions in the beginning of this section.
2. Relieve the fuel system pressure.
3. Drain the cooling system.
4. Remove or disconnect the following:
- Negative battery cable
- Strut tower bar
- Intake Air Temperature (IAT) sensor connector
- Surge tank cover
- Air intake assembly
- Accelerator cable
- Transmission cable, if equipped
- Throttle body coolant hoses
- Fuel injector connectors
- Throttle Position (TP) sensor connector
- Mass Air Flow (MAF) sensor connector
- Idle Air Control (IAC) valve connector
- Intake manifold ground cables
- Brake booster vacuum hose

- Evaporative Emissions (EVAP) canister purge valve connector and hoses
- Exhaust Gas Recirculation (EGR) valve connector
- Positive Crankcase Ventilation (PCV) valve and hose
- Heater hoses
- EGR pipe
- Fuel lines
- Throttle body and intake collector
- Intake manifold

To install:

5. Install or connect the following:
- Intake manifold with new gaskets. Tighten the fasteners to 16 ft. lbs. (23 Nm).
- Throttle body and intake collector with new gaskets. Tighten the fasteners to 102 inch lbs. (12 Nm).
- Fuel lines
- EGR pipe
- Heater hoses
- PCV valve and hose
- EGR valve connector
- EVAP canister purge valve connector and hoses
- Brake booster vacuum hose
- Intake manifold ground cables
- IAC valve connector
- MAF sensor connector
- TP sensor connector
- Fuel injector connectors
- Throttle body coolant hoses
- Transmission cable, if equipped
- Accelerator cable
- Air intake assembly
- Surge tank cover
- IAT sensor connector
- Strut tower bar
- Negative battery cable
6. Fill the cooling system.
7. Start the engine and check for leaks.

Exhaust Manifold

REMOVAL & INSTALLATION

2.0L Engines

1. Before servicing the vehicle, refer to the precautions in the beginning of this section.
2. Remove or disconnect the following:
- Negative battery cable
- Strut tower bar, if equipped
- Air intake assembly and bracket
- Heated Oxygen (HO2S) sensor connector
- Exhaust front pipe
- Exhaust manifold heat shield

- Exhaust manifold bracket, if equipped
- Exhaust manifold

To install:

3. Install or connect the following:
- Exhaust manifold with a new gasket. Tighten the fasteners to 13–20 ft. lbs. (18–28 Nm).
- Exhaust manifold bracket, if equipped. Tighten the bolts to 36–43 ft. lbs. (50–60 Nm).
- Exhaust manifold heat shield
- Exhaust front pipe. Tighten the fasteners to 29–43 ft. lbs. (40–60 Nm).
- HO2S sensor connector
- Air intake assembly and bracket
- Strut tower bar, if equipped. Tighten the fasteners to 66 ft. lbs. (90 Nm).
- Negative battery cable
4. Start the engine and check for leaks.

2.5L Engine

1. Before servicing the vehicle, refer to the precautions in the beginning of this section.
2. Remove or disconnect the following:
- Negative battery cable
- Strut tower bar
- Air intake assembly
- Heated Oxygen (HO2S) sensor connectors
- Oil dipstick tube
- Exhaust Gas Recirculation (EGR) pipe
- Exhaust manifold heat shields
- Evaporative Emissions (EVAP) canister
- Front driveshaft, if equipped
- Exhaust front pipe
- Exhaust manifold brace
- Exhaust manifolds

To install:

3. Install or connect the following:
- Exhaust manifolds with new gas-

kets. Tighten the nuts to 21 ft. lbs. (30 Nm).
- Exhaust manifold brace
- Exhaust front pipe. Tighten the fasteners to 37 ft. lbs. (50 Nm).
- Front driveshaft, if equipped
- EVAP canister
- Exhaust manifold heat shields
- EGR pipe
- Oil dipstick tube
- HO2S sensor connectors
- Air intake assembly
- Strut tower bar
- Negative battery cable
4. Start the engine and check for leaks.

Camshaft and Valve Lifters

REMOVAL & INSTALLATION

2.0L Engines

1. Before servicing the vehicle, refer to the precautions in the beginning of this section.
2. Drain the engine oil.
3. Drain the cooling system.
4. Remove or disconnect the following:
- Negative battery cable
- Oil pan
- Valve cover
- Accessory drive belts
- Crankshaft pulley
- Front cover
- Secondary timing chain
- Camshaft Position (CMP) sensor

➡ **Keep all valvetrain components in order for installation.**

- Camshaft bearing caps. Loosen the bolts in several steps in reverse of the tightening sequence.
- Camshafts
- Hydraulic lash adjusters

To install:

5. Install or connect the following:

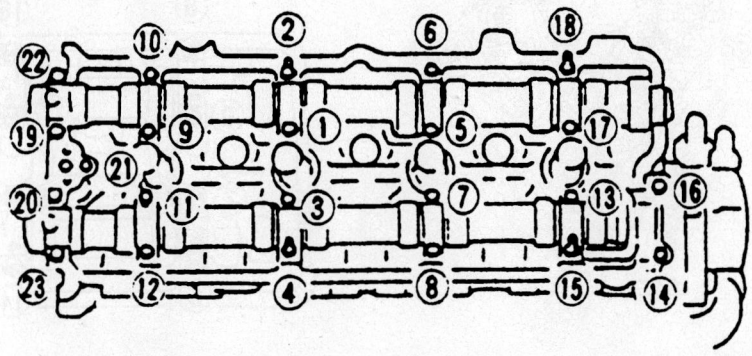

Camshaft housing torque sequence—2.0L engines

7924HG26

- Hydraulic lash adjusters in their original positions
- Camshafts
- Camshaft bearing caps in their original positions. Tighten the bolts in several steps in sequence to 96 inch lbs. (11 Nm).
- CMP sensor
- Secondary timing chain
- Front cover
- Crankshaft pulley. Tighten the bolt to 109 ft. lbs. (148 Nm).
- Accessory drive belts
- Valve cover
- Oil pan
- Negative battery cable

6. Fill the crankcase to the correct level.
7. Fill the cooling system.
8. Start the engine and check for leaks.

❋❋ WARNING

Wait ½ hour after installing the lash adjusters and camshafts before cranking or starting the engine to allow the lash adjusters to bleed down. Operating the engine before this time period may result in interference between the valves and pistons.

2.5L Engine

1. Before servicing the vehicle, refer to the precautions in the beginning of this section.
2. Drain the engine oil.
3. Drain the cooling system.
4. Remove or disconnect the following:
   - Negative battery cable

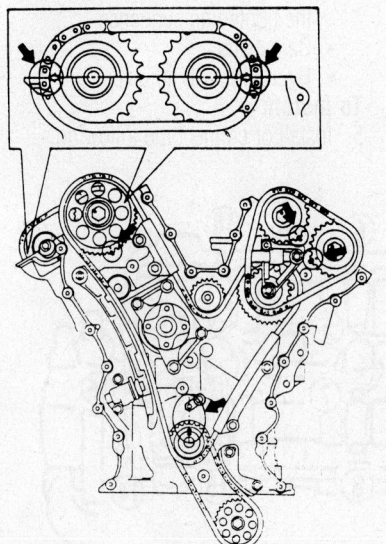

Timing mark alignment—2.5L engine

9302HG02

Left bank camshaft housing loosening sequence—2.5L engine

9302HG03

- Intake manifold
- Oil pan
- Accessory drive belts
- Water pump pulley
- Crankshaft pulley
- Timing chain cover

5. Align the timing marks as shown.

❋❋ WARNING

Do not allow the crankshaft or camshafts to rotate once the timing chains have been removed. Valve or piston damage could result.

6. Remove or disconnect the following:
   - Left bank secondary timing chain
   - Primary timing chain
   - Valve covers

➡**Keep all valvetrain components in order for assembly.**

- Right bank camshaft bearing caps. Loosen the bolts in several steps and in the sequence shown.
- Right bank secondary timing chain, exhaust and intake camshafts as an assembly
- Camshaft Position (CMP) sensor
- Left bank camshaft bearing caps. Loosen the bolts in several steps and in the sequence shown.
- Left bank camshafts
- Hydraulic lash adjusters

To install:

7. Install or connect the following:
   - Hydraulic lash adjusters in their original positions
   - Left bank camshafts
   - Left bank camshaft bearing caps. Tighten the bolts in several steps and in reverse of the loosening sequence to 102 inch lbs. (12 Nm).
   - CMP sensor
   - Right bank secondary timing chain, exhaust and intake camshafts as an assembly
   - Right bank camshaft bearing caps. Tighten the bolts in several steps and in reverse of the loosening sequence to 102 inch lbs. (12 Nm).

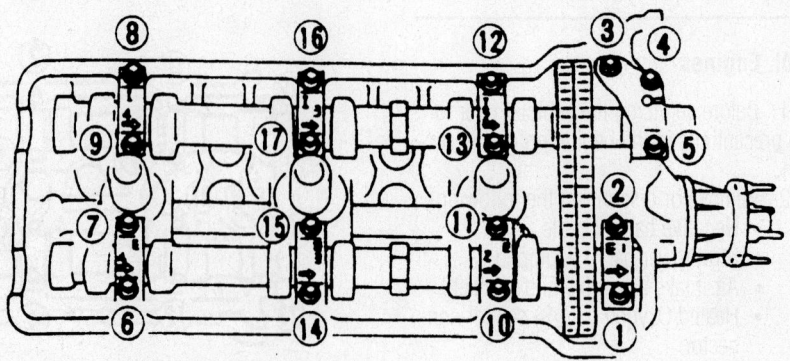

Right bank camshaft housing loosening sequence—2.5L engine

9302HG04

☀☀ WARNING

Wait ½ hour after installing the lash adjusters and camshafts before cranking or starting the engine to allow the lash adjusters to bleed down. Operating the engine before this time period may result in interference between the valves and pistons.

- Valve covers. Tighten the bolts to 90 inch lbs. (10.5 Nm).
- Primary timing chain
- Left bank secondary timing chain
- Timing chain cover
- Crankshaft pulley. Tighten the bolt to 109 ft. lbs. (148 Nm).
- Water pump pulley
- Accessory drive belts
- Oil pan
- Intake manifold
- Negative battery cable

8. Fill the crankcase to the correct level.
9. Fill the cooling system.
10. Start the engine and check for leaks.

Valve Lash

ADJUSTMENT

2.0L and 2.5L Engines

2.0L and 2.5L engines utilize automatic hydraulic lash adjusters to maintain proper valve lash at all times. Periodic valve lash inspection and adjustment is not necessary or possible.

Starter Motor

REMOVAL & INSTALLATION

1. Before servicing the vehicle, refer to the precautions in the beginning of this section.
2. Remove or disconnect the following:

- Negative battery cable
- Starter motor wiring connectors
- Starter motor

To install:

3. Install or connect the following:

- Starter motor. Tighten the bolts to 22 ft. lbs. (30 Nm).
- Starter motor wiring connectors. Tighten the solenoid nut to 11 ft. lbs. (15 Nm).
- Negative battery cable

Oil Pan

REMOVAL & INSTALLATION

2.0L Engines

1. Before servicing the vehicle, refer to the precautions in the beginning of this section.
2. Drain the engine oil.
3. Remove or disconnect the following:

- Negative battery cable
- Oil dipstick tube
- Front wheels
- Front skidplate, if equipped
- Steering gear
- Front differential, if equipped
- Left transmission stiffener bracket, if equipped
- Flywheel access panel
- Exhaust front pipe
- Left and right motor mounts and raise the engine about 1 inch (25mm) for clearance
- Oil pan and oil pump pickup tube

To install:

4. Apply a bead of silicone sealant to the oil pan flange. Install new oil pump pickup tube O-ring seals.
5. Install or connect the following:

- Oil pan and oil pump pickup tube. Tighten the fasteners to 97 inch lbs. (11 Nm).
- Left and right engine mounts. Tighten the nuts to 36 ft. lbs. (50 Nm).
- Exhaust front pipe
- Flywheel access panel
- Left transmission stiffener bracket, if equipped
- Front differential, if equipped
- Steering gear
- Front skidplate, if equipped
- Front wheels
- Oil dipstick tube
- Negative battery cable

6. Fill the crankcase to the correct level.
7. Start the engine and check for leaks.

2.5L Engine

1. Before servicing the vehicle, refer to the precautions in the beginning of this section.
2. Drain the engine oil.
3. Remove or disconnect the following:

- Negative battery cable
- Oil dipstick tube
- Front wheels
- Front skidplate, if equipped

1. O-ring

Lower crankcase O-ring seal

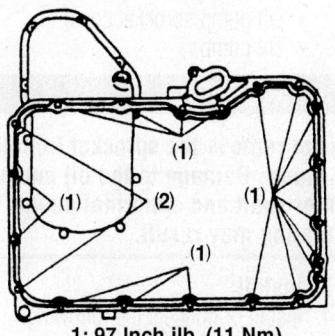

1: 97 Inch ilb. (11 Nm)
2: 20 Ft. lbs. (27 Nm)

Upper oil pan bolt torque values—2.5L engine

- Steering gear
- Front differential, if equipped
- Lower oil pan
- Oil pickup tube bracket
- Radiator outlet pipe
- Upper oil pan and oil pickup tube

To install:

4. Install a new O-ring to the lower crankcase.
5. Apply a bead of silicone sealant to the upper oil pan flange.
6. Install or connect the following:

- New oil pump pickup tube O-ring seals
- Upper oil pan and oil pump pickup tube and tighten the fasteners as shown
- Radiator outlet pipe
- Oil pickup tube bracket
- Lower oil pan. Tighten the bolts to 97 inch lbs. (11 Nm).
- Front differential, if equipped
- Steering gear
- Front skidplate, if equipped
- Front wheels
- Oil dipstick tube
- Negative battery cable

7. Fill the crankcase to the correct level.
8. Start the engine and check for leaks.

Oil Pump

REMOVAL & INSTALLATION

2.0L Engines

1. Before servicing the vehicle, refer to the precautions in the beginning of this section.
2. Drain the engine oil.
3. Remove or disconnect the following:
   - Negative battery cable
   - Oil pan and pickup tube
   - Oil pump sprocket cover
   - Oil pump

> ❊❊ **WARNING**
>
> **Do not remove the sprocket from the oil pump. Damage to the oil pump center shaft and abnormal pump operation may result.**

To install:

4. Install or connect the following:
   - Oil pump. Tighten the bolts to 20 ft. lbs. (27 Nm).
   - Oil pump sprocket cover. Tighten the bolts to 108 inch lbs. (12 Nm).
   - Oil pan and pickup tube
   - Negative battery cable
5. Fill the crankcase to the correct level.
6. Start the engine and check for leaks.

2.5L Engine

1. Before servicing the vehicle, refer to the precautions in the beginning of this section.
2. Drain the cooling system.
3. Drain the engine oil.
4. Remove or disconnect the following:
   - Negative battery cable
   - Accessory drive belts
   - Intake manifold
   - Oil pan and oil pickup tube
   - Front cover

   - Oil pump chain guide
   - Oil pump

> ❊❊ **WARNING**
>
> **Do not remove the sprocket from the oil pump. Damage to the oil pump center shaft and abnormal pump operation may result.**

To install:

5. Install or connect the following:
   - Oil pump. Tighten the bolts to 20 ft. lbs. (27 Nm).
   - Oil pump chain guide. Tighten the bolts to 97 inch lbs. (11 Nm).
   - Front cover
   - Oil pan and oil pickup tube
   - Intake manifold
   - Accessory drive belts
   - Negative battery cable
6. Fill the crankcase to the correct level.
7. Fill the cooling system.
8. Start the engine and check for leaks.

Rear Main Seal

REMOVAL & INSTALLATION

1. Before servicing the vehicle, refer to the precautions in the beginning of this section.

2. Remove or disconnect the following:
   - Negative battery cable
   - Transmission
   - Clutch assembly, if equipped
   - Flywheel
   - Rear main seal

To install:

3. Install or connect the following:
   - Rear main seal flush with the cylinder block
   - Flywheel. Tighten the bolts in a crossing pattern to 51 ft. lbs. (69 Nm) for all other engines.
   - Clutch assembly, if equipped
   - Transmission
   - Negative battery cable

Timing Chain, Sprockets, Front Cover and Seal

REMOVAL & INSTALLATION

2.0L Engines

1. Before servicing the vehicle, refer to the precautions in the beginning of this section.
2. Drain the cooling system.
3. Drain the engine oil.
4. Remove or disconnect the following:

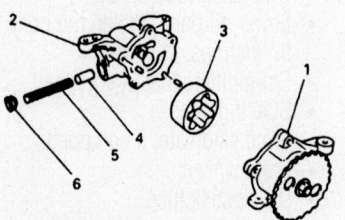

1. Oil pump case No.1
2. Oil pump case No.2
3. Outer rotor
4. Relief valve
5. Relief spring
6. Retainer

7924HG15

Exploded view of oil pump—2.0L and 2.5L engines

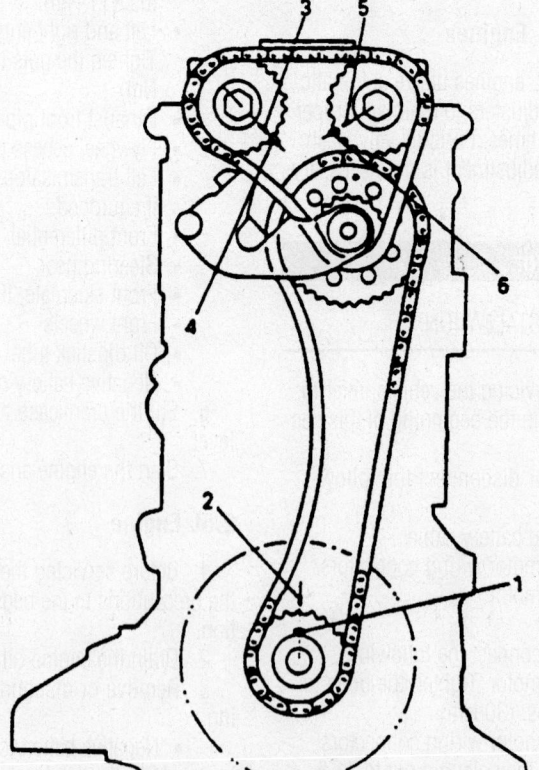

7924HG22

Timing mark alignment—2.0L engines

- Negative battery cable
- Oil pan and pickup tube
- Valve cover
- Bypass pipe and hose
- Accessory drive belts
- Cooling fan and shroud
- Water pump pulley
- Alternator belt tensioner and idler pulleys
- Upper radiator hose
- A/C compressor and bracket, if equipped
- Crankshaft pulley
- Front crankshaft seal
- Front cover

5. Rotate the crankshaft to align the timing marks as shown.

❋❋ WARNING

Do not allow the crankshaft or camshafts to rotate once the timing chains have been removed. Valve or piston damage could result.

6. Remove or disconnect the following:
- Second timing chain tensioner
- Camshaft sprockets and second timing chain
- First timing chain tensioner
- Timing chain idler sprocket and first timing chain

To install:

7. Prepare the timing chain tensioners for installation by releasing the latches,

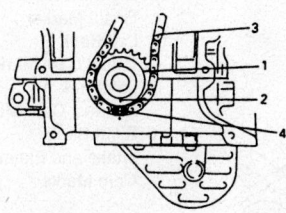

1. Crankshaft timing sprocket   3. 1st timing chain
2. Match mark   4. Yellow plate

7924HG17

Crankshaft and first timing chain alignment—2.0L engines

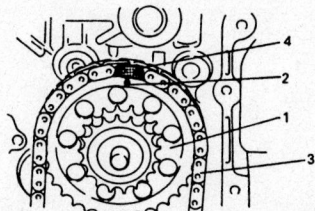

1. Idler sprocket   3. 1st timing chain
2. Match mark on idler sprocket   4. Dark blue plate

7924HG16

Idler sprocket and first timing chain alignment—2.0L engines

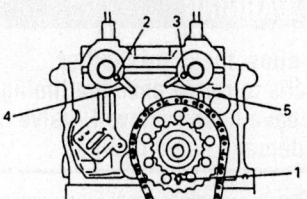

1. Arrow mark on idler sprocket
2. Knock pin of intake camshaft
3. Knock pin of exhaust camsaft
4. Timing mark of intake side
5. Timing mark of exhaust side

7924HG19

Idler sprocket and camshaft alignment— 2.0L engines

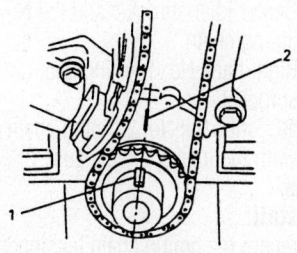

1. Crank timing sprocket key
2. Timing mark

7924HG18

Crankshaft sprocket alignment—2.0L engines

compressing the tensioner piston fully into the bore and instaing retaining pins.

8. Install or connect the following:
- Timing chain idler sprocket and first timing chain with the match-marks and colored links aligned as shown
- First timing chain tensioner. Tighten the bolts to 97 inch lbs. (11 Nm).
- Camshaft sprockets and second timing chain with the matchmarks and colored links aligned as shown
- Second timing chain tensioner. Tighten the bolts to 97 inch lbs. (11 Nm) and the nut to 33 ft. lbs. (45 Nm).

9. Tighten the camshaft sprocket bolts to 59 ft. lbs. (80 Nm).

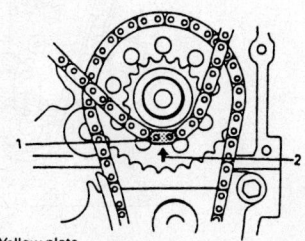

1. Yellow plate
2. Match mark of 2nd timing chain (Arrow mark)

7924HG20

Idler sprocket and second timing chain alignment—2.0L engines

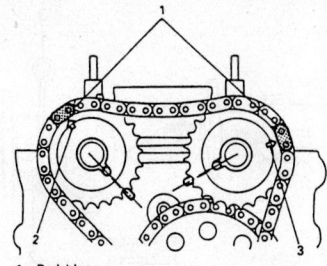

1. Dark blue
2. Arrow mark on intake camshaft timing sprocket
3. Arrow mark on exhaust camshaft timing sprocket

7924HG21

Camshaft sprocket and second timing chain alignment—2.0L engines

10. Remove the timing chain tensioner retaining pins.

11. Rotate the crankshaft two complete turns and check that the timing marks align.

12. Install or connect the following:
- Front cover. Apply sealant as shown.
- Front crankshaft seal
- Crankshaft pulley. Tighten the bolt to 109 ft. lbs. (130 Nm).
- A/C compressor and bracket, if equipped
- Upper radiator hose
- Alternator belt tensioner and idler pulleys
- Water pump pulley
- Cooling fan and shroud
- Accessory drive belts
- Bypass pipe and hose

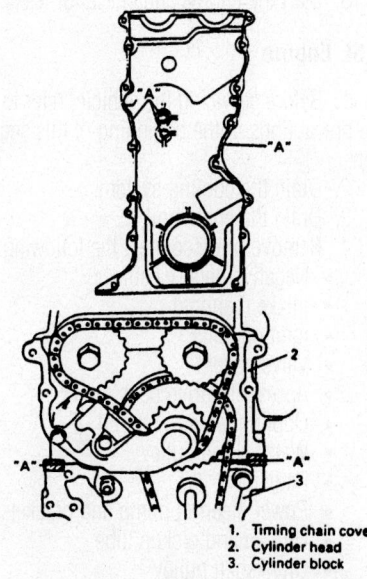

1. Timing chain cover
2. Cylinder head
3. Cylinder block

7924HG10

Prior to installing the timing chain cover on the engine block and cylinder head, apply silicone sealant to the cover as indicated (areas marked A)—2.0L and 2.5L engines

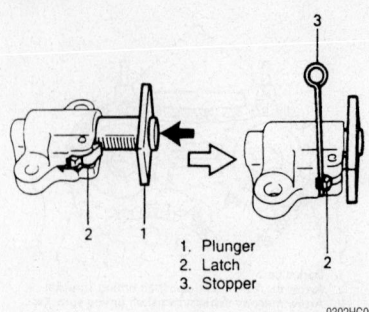

Preparing the No. 1 timing chain tensioner adjuster for installation—2.0L and 2.5L engines

1. Plunger
2. Latch
3. Stopper

9302HG09

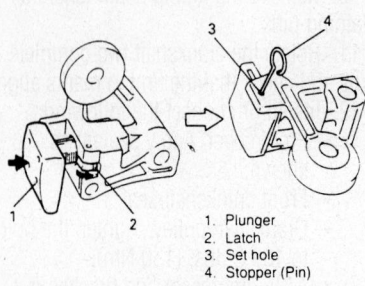

No. 2 timing chain tensioner—left bank tensioner shown—2.0L and 2.5L engines

1. Plunger
2. Latch
3. Set hole
4. Stopper (Pin)

9302HG13

- Valve cover
- Oil pan and pickup tube
- Negative battery cable

13. Fill the crankcase to the correct level.

14. Fill the cooling system.

15. Start the engine and check for leaks.

2.5L Engine

1. Before servicing the vehicle, refer to the precautions in the beginning of this section.

2. Drain the cooling system.

3. Drain the engine oil.

4. Remove or disconnect the following:
- Negative battery cable
- Intake manifold
- Ignition coils
- Valve covers
- Accessory drive belts
- Cooling fan and shroud
- Water pump pulley
- Radiator
- Power steering pump and brackets
- Oil pan and pickup tube
- Crankshaft pulley
- Front crankshaft seal
- Crankshaft Position (CKP) sensor
- Front cover

5. Rotate the crankshaft so that the timing marks are aligned as shown.

✳✳ WARNING

Do not allow the crankshaft or camshafts to rotate once the timing chains have been removed. Valve or piston damage could result.

6. Remove or disconnect the following:
- Left bank No. 2 timing chain tensioner
- Left bank intake and exhaust camshaft sprockets with the No. 2 timing chain
- No. 1 timing chain guides
- No. 1 timing chain tensioner
- Center idler sprocket and the No. 1 timing chain
- Right bank No. 1 timing chain sprocket

7. The right bank No. 2 timing chain is removed with the intake and exhaust camshafts.

To install:

8. Prepare the timing chain tensioners for installation by releasing the latches, compressing the tensioner piston fully into the bore and installing retaining pins.

9. Align the timing chain sprocket matchmarks and colored chain links as shown during assembly.

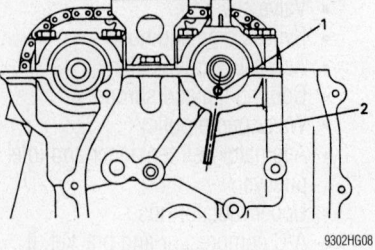

1. Knock pin of intake camshaft
2. Match mark

9302HG08

Right bank camshaft timing marks—2.5L engine

10. Install or connect the following:
- Right bank intake and exhaust camshafts with the No. 2 timing chain
- Right bank No. 1 timing chain sprocket. Tighten the bolt to 58 ft. lbs. (80 Nm).
- Center idler sprocket and the No. 1 timing chain. Tighten the fastener to 32 ft. lbs. (45 Nm).
- No. 1 timing chain tensioner and guides. Tighten the bolts to 97 inch lbs. (11 Nm).
- Left bank intake and exhaust camshaft sprockets with the No. 2 timing chain. Tighten the bolts to 57 ft. lbs. (80 Nm).

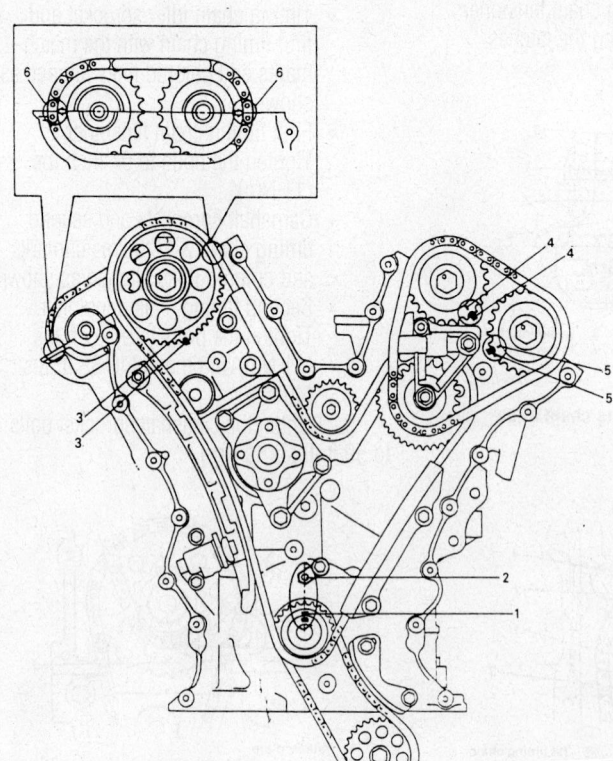

1. Crankshaft Keyway
2. Oil Jet
3. Right Bank No. 1 Chain Marks
4. Left Bank Intake Cam Marks
5. Left Bank Exhaust Cam Marks
6. Right Bank Intake and Exhaust Cam Marks

9302HG07

Timing chain alignment marks—2.5L engine

- Left bank No. 2 timing chain tensioner. Tighten the bolts to 97 inch lbs. (11 Nm).
11. Remove the retaining pins from the timing chain tensioners.
12. Rotate the crankshaft two complete turns and check that the timing marks align.
13. Install or connect the following:
  - Front cover. Tighten the bolts to 97 inch lbs. (11 Nm).
  - CKP sensor
  - Front crankshaft seal
  - Crankshaft pulley. Tighten the bolt to 109 ft. lbs. (148 Nm).

- Oil pan and pickup tube
- Power steering pump and brackets
- Radiator
- Water pump pulley
- Cooling fan and shroud
- Accessory drive belts
- Valve covers
- Ignition coils
- Intake manifold
- Negative battery cable
14. Fill the crankcase to the correct level.
15. Fill the cooling system.
16. Start the engine and check for leaks.

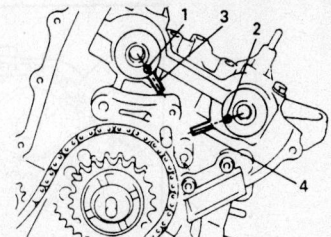

1. Knock pin of LH bank intake camshaft
2. Knock pin of LH bank exhaust camshaft
3. Match mark of intake side
4. Match mark of exhaust side

9302HG11

Left bank camshaft alignment—2.5L engine

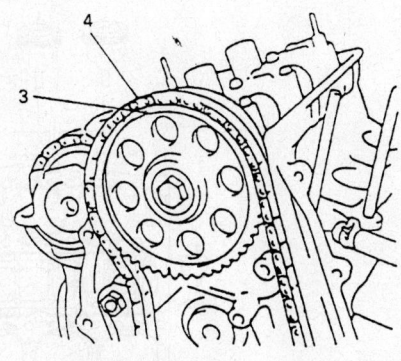

3. Match mark of RH bank 1st timing chain sprocket
4. Silver plate (LH) of 1st timing chain

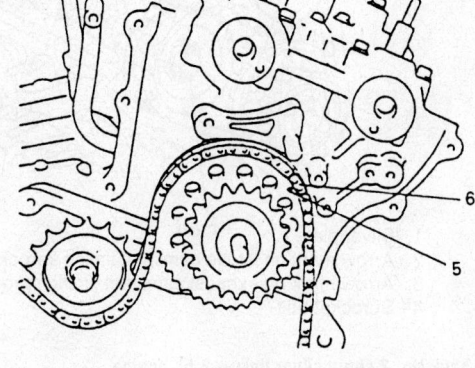

5. Match mark of idler sprocket No.2
6. Silver plate (RH) of 1st timing chain

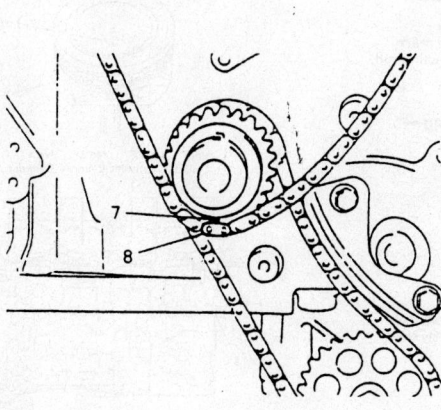

7. Match mark of crankshaft timing sprocket
8. Gold or Yellow plate of 1st timing chain

1. Crank timing pulley key
2. Oil jet

9302HG10

No. 1 timing chain alignment—2.5L engine

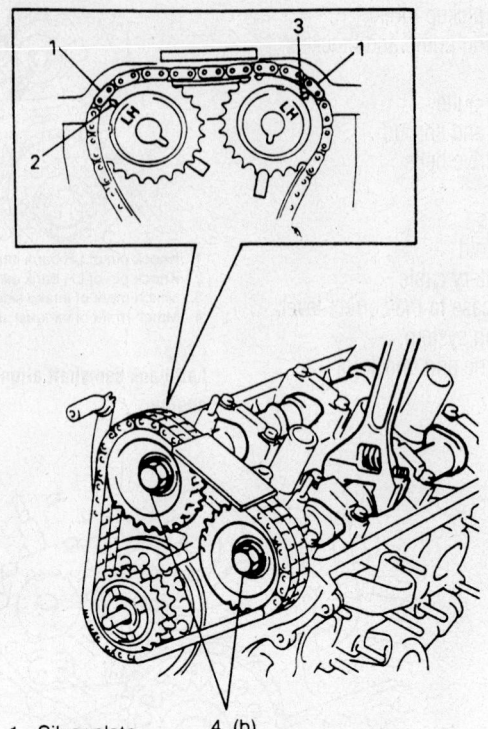

1. Silver plate
2. Arrow mark on intake camshaft timing sprocket
3. Arrow mark on exhaust camshaft timing sprocket
4. Sprocket bolt

4, (b)

9302HG12

Align the left bank No. 2 chain silver links—2.5L engine

Piston and Ring

POSITIONING

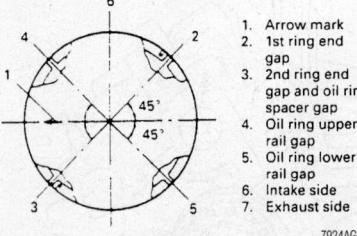

1. Arrow mark
2. 1st ring end gap
3. 2nd ring end gap and oil ring spacer gap
4. Oil ring upper rail gap
5. Oil ring lower rail gap
6. Intake side
7. Exhaust side

7924AG67

Piston ring end-gap spacing—All engines

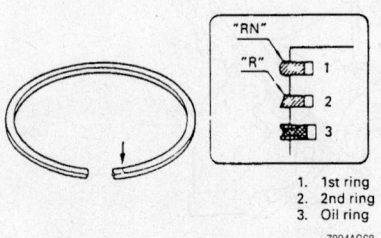

"RN"
"R"
1
2
3

1. 1st ring
2. 2nd ring
3. Oil ring

7924AG68

Compression ring identification marks—All engines

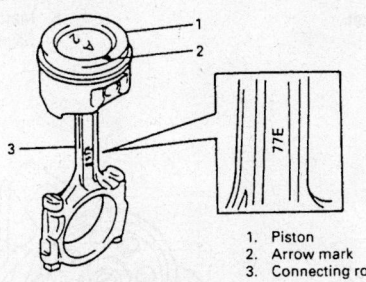

1. Piston
2. Arrow mark
3. Connecting rod

7924AG61

Piston and connecting rod positioning—2.0L and 2.5L engines

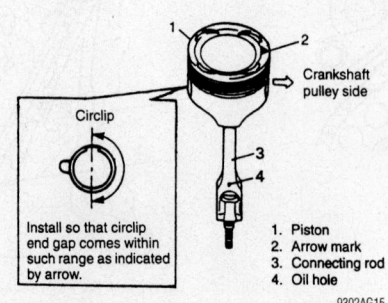

Circlip

Install so that circlip end gap comes within such range as indicated by arrow.

1. Piston
2. Arrow mark
3. Connecting rod
4. Oil hole

9302AG15

Piston pin circlip installation—2.5L engine

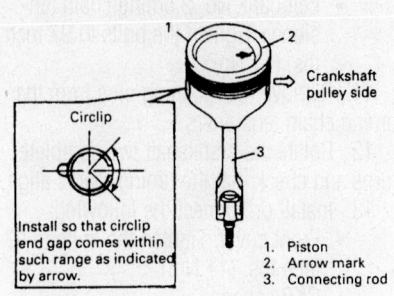

Crankshaft pulley side

Circlip

Install so that circlip end gap comes within such range as indicated by arrow.

1. Piston
2. Arrow mark
3. Connecting rod

7924AG62

Piston pin circlip installation—2.0L engines

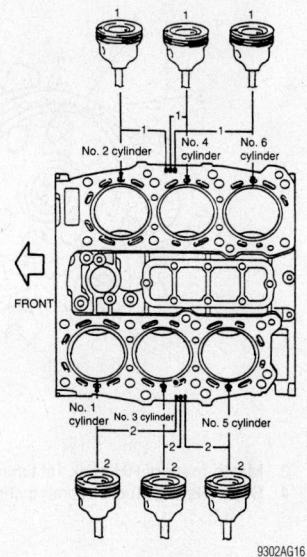

No. 2 cylinder
No. 4 cylinder
No. 6 cylinder

FRONT

No. 1 cylinder
No. 3 cylinder
No. 5 cylinder

9302AG16

Piston identification—2.5L engine

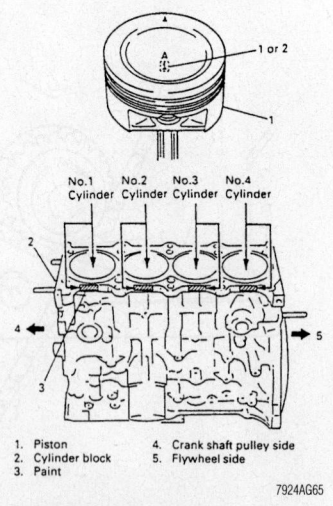

1 or 2

No.1 Cylinder
No.2 Cylinder
No.3 Cylinder
No.4 Cylinder

1. Piston
2. Cylinder block
3. Paint
4. Crank shaft pulley side
5. Flywheel side

7924AG65

Piston identification—2.0L engines
Match pistons with "1" indicators to red cylinder paint marks
Match pistons with "2" indicators to blue cylinder paint marks

FUEL SYSTEM

Fuel System Service Precautions

Safety is the most important factor when performing not only fuel system maintenance but any type of maintenance. Failure to conduct maintenance and repairs in a safe manner may result in serious personal injury or death. Maintenance and testing of the vehicle fuel system components can be accomplished safely and effectively by adhering to the following rules and guidelines.

• To avoid the possibility of fire and personal injury, always disconnect the negative battery cable unless the repair or test procedure requires that battery voltage be applied.

• Always relieve the fuel system pressure prior to disconnecting any fuel system component (injector, fuel rail, pressure regulator, etc.), fitting or fuel line connection. Exercise extreme caution whenever relieving fuel system pressure to avoid exposing skin, face and eyes to fuel spray. Please be advised that fuel under pressure may penetrate the skin or any part of the body that it contacts.

• Always place a shop towel or cloth around the fitting or connection prior to loosening to absorb any excess fuel due to spillage. Ensure that all fuel spillage (should it occur) is quickly removed from engine surfaces. Ensure that all fuel soaked cloths or towels are deposited into a suitable waste container.

• Always keep a dry chemical (Class B) fire extinguisher near the work area.

• Do not allow fuel spray or fuel vapors to come into contact with a spark or open flame.

• Always use a backup wrench when loosening and tightening fuel line connection fittings. This will prevent unnecessary stress and torsion to fuel line piping.

• Always replace worn fuel fitting O-rings with new. Do not substitute fuel hose or equivalent, where fuel pipe is installed.

Fuel System Pressure

RELIEVING

1. Before servicing the vehicle, refer to the precautions in the beginning of this section.

2. Detach the wiring harness connector from the fuel pump relay, located under the left-hand side of the instrument panel near the ECM.

3. Start the engine and run it until it stops from lack of fuel. Crank the engine 2–3 times for a 3 second period. The fuel lines should now be depressurized.

4. After servicing, reattach the wiring harness connector to the fuel pump relay.

Fuel Filter

REMOVAL & INSTALLATION

1. Before servicing the vehicle, refer to the precautions in the beginning of this section.

2. Relieve fuel system pressure.

3. Remove or disconnect the following:
   • Negative battery cable
   • Fuel lines from the fuel filter
   • Fuel filter

To install:

4. Install or connect the following:
   • Fuel filter and tighten the bracket bolt. Note the fuel flow directional arrow.
   • Fuel lines to the fuel filter.
   • Negative battery cable

5. Start the engine and inspect the fuel filter connections for leaks.

Fuel Pump

REMOVAL & INSTALLATION

1. Before servicing the vehicle, refer to the precautions in the beginning of this section.

2. Relieve the fuel system pressure.

3. Remove or disconnect the following:
   • Negative battery cable
   • Fuel filler hose and vent hose
   • Fuel tank inlet valve and drain the fuel tank
   • Fuel filter inlet hose
   • Evaporative Emissions (EVAP) vapor hose
   • Fuel return line
   • Fuel tank skidplate
   • Fuel pump connector
   • Fuel tank pressure sensor connector
   • Fuel tank
   • Fuel pump module

To install:

4. Install or connect the following:
   • Fuel pump module with a new seal. Tighten the bolts to 44 inch lbs. (5 Nm).

   • Fuel tank. Tighten the strap bolts to 37 ft. lbs. (50 Nm).
   • Fuel tank pressure sensor connector
   • Fuel pump connector
   • Fuel tank skidplate
   • Fuel return line
   • EVAP vapor hose
   • Fuel filter inlet hose
   • Fuel tank inlet valve and drain the fuel tank
   • Fuel filler hose and vent hose
   • Negative battery cable

5. Fill the fuel tank.

6. Start the engine and check for leaks.

Fuel Injector

REMOVAL & INSTALLATION

2.0L Engines

1. Before servicing the vehicle, refer to the precautions in the beginning of this section.

2. Relieve the fuel system pressure.

3. Remove or disconnect the following:
   • Negative battery cable
   • Front intake manifold bracket, if equipped
   • Positive Crankcase Ventilation (PCV) valve and hose
   • Fuel injector harness connectors
   • Fuel line bracket
   • Fuel supply manifold
   • Fuel injectors

To install:

4. Install or connect the following:
   • Fuel injectors with new O-ring seals
   • Fuel supply manifold. Tighten the bolts to 17 ft. lbs. (23 Nm).
   • Fuel line bracket
   • Fuel injector harness connectors
   • PCV valve and hose
   • Front intake manifold bracket, if equipped
   • Negative battery cable

5. Start the engine and check for leaks.

2.5L Engine

1. Before servicing the vehicle, refer to the precautions in the beginning of this section.

2. Relieve the fuel system pressure.

3. Remove or disconnect the following:

- Negative battery cable
- Air intake tube
- Throttle body intake collector
- Fuel lines
- Fuel pressure regulator vacuum line
- Fuel injector harness connectors
- Fuel supply manifold connect pipe
- Fuel supply manifolds
- Fuel injectors

To install:

4. Install or connect the following:
- Fuel injectors with new O-ring seals
- Fuel supply manifolds. Tighten the bolts to 17 ft. lbs. (23 Nm).
- Fuel supply manifold connect pipe. Tighten the bolts to 22 ft. lbs. (30 Nm).

- Fuel injector harness connectors
- Fuel pressure regulator vacuum line
- Fuel lines
- Throttle body intake collector
- Air intake tube
- Negative battery cable

5. Start the engine and check for leaks.

DRIVE TRAIN

Transmission Assembly

REMOVAL & INSTALLATION

Manual Transmission

1. Before servicing the vehicle, refer to the precautions in the beginning of this section.
2. Drain the transmission fluid.
3. Drain the transfer case fluid, if equipped.
4. Remove or disconnect the following:
- Negative battery cable
- Shift lever boots
- Gear shift lever
- Transfer case shift lever, if equipped
- 4WD switch connector, if equipped
- Reverse light switch connector
- Starter motor
- Front driveshaft, if equipped
- Rear driveshaft
- Speedometer cable, if equipped
- Vehicle Speed (VSS) sensor, if equipped
- Clutch slave cylinder or cable
- Flywheel access cover
- Transmission flange bolts and nuts
- Transmission braces, if equipped
5. Support the transmission with a jack

A	WOOD BLOCK
H	200 mm (8.0")
T	45 mm (1.8")
W	100–150 mm (4.0–6.0")
7017	DISTRIBUTOR CAP
7018	BULKHEAD

7924HG37

Support the engine with a wooden block between the cylinder head and the firewall—All models

and remove the transmission mount and crossmember.

6. Place a wooden block at the rear of the cylinder head as shown to support the engine when the transmission is removed.
7. Lower the transmission away from the vehicle.

To install:

8. Install or connect the following:
- Transmission. Tighten the flange fasteners to 62–72 ft. lbs. (85–98 Nm).
- Transmission mount and crossmember. Tighten the fasteners to 29–43 ft. lbs. (40–60 Nm).
- Transmission braces, if equipped. Tighten the bolts to 62–72 ft. lbs. (85–98 Nm).
- Flywheel access cover
- Clutch slave cylinder or cable
- VSS sensor, if equipped
- Speedometer cable, if equipped
- Rear driveshaft
- Front driveshaft, if equipped
- Starter motor
- Reverse light switch connector
- 4WD switch connector, if equipped
- Transfer case shift lever, if equipped
- Gear shift lever
- Shift lever boots
- Negative battery cable

9. Fill the transmission to the correct level.
10. Fill the transfer case, if equipped.

Automatic Transmission

1. Before servicing the vehicle, refer to the precautions in the beginning of this section.
2. Drain the transfer case oil, if equipped.
3. Remove or disconnect the following:
- Negative battery cable
- Center console and transfer case shift lever, if equipped
- Transmission dipstick tube
- Transmission wiring harness connectors
- Starter motor

- Front driveshaft, if equipped
- Rear driveshaft
- Gear select cable and bracket
- Throttle Valve (TV) cable, if equipped
- Exhaust front pipe
- Transmission oil cooler lines
- Transmission brace
- Flywheel access cover
- Torque converter
- Speedometer cable, if equipped
- Vehicle Speed (VSS) sensor connector, if equipped
- Transmission flange bolts and nuts
- Transmission braces, if equipped

4. Support the transmission with a jack and remove the transmission mount and crossmember.
5. Place a wooden block at the rear of the cylinder head as shown to support the engine when the transmission is removed.
6. Lower the transmission away from the vehicle.

To install:

7. Install or connect the following:
- Transmission. Tighten the flange fasteners to 62–72 ft. lbs. (85–98 Nm).
- Transmission mount and crossmember. Tighten the fasteners to 29–43 ft. lbs. (40–60 Nm).
- Transmission braces, if equipped. Tighten the bolts to 62–72 ft. lbs. (85–98 Nm).
- VSS sensor connector, if equipped
- Speedometer cable, if equipped
- Torque converter. Tighten the bolts to 47 ft. lbs. (65 Nm).
- Flywheel access cover
- Transmission brace
- Transmission oil cooler lines
- Exhaust front pipe
- TV cable, if equipped
- Gear select cable and bracket
- Rear driveshaft
- Front driveshaft, if equipped
- Starter motor
- Transmission wiring harness connectors
- Transmission dipstick tube

- Center console and transfer case shift lever, if equipped
- Negative battery cable

8. Fill the transmission to the correct level.

9. Fill the transfer case, if equipped.

Clutch

ADJUSTMENTS

These vehicles are equipped with a hydraulic clutch system. No adjustment is necessary.

REMOVAL & INSTALLATION

1. Before servicing the vehicle, refer to the precautions in the beginning of this section.

2. Remove the transmission.

3. Loosen the pressure plate mounting bolts in a 2-step crisscross sequence until the spring tension is relieved.

4. Remove the pressure plate and the clutch disc.

To install:

5. Using a clutch alignment tool, assemble the clutch disc and pressure plate onto the flywheel.

6. Tighten the pressure plate bolts in multiple passes to 17 ft. lbs. (23 Nm).

7. Install the transmission.

8. Check for proper clutch operation.

Hydraulic Clutch System

BLEEDING

1. Before servicing the vehicle, refer to the precautions in the beginning of this section.

2. Fill the master cylinder reservoir to the MAX line with clean brake fluid and keep it at least half full throughout the bleeding procedure.

3. From beneath the vehicle, remove the bleeder plug cap, then attach a clear vinyl tube to the slave cylinder bleeder plug. Insert the open end of the hose into a container.

4. Have an assistant depress the clutch pedal. Open the bleeder after the pedal is depressed.

5. Close the bleeder before releasing the clutch pedal.

6. Repeat until all air bubbles are gone from the hydraulic fluid.

7. Install the bleeder plug cap.

8. Fill the clutch master cylinder fluid reservoir to the specified full level.

Transfer Case Assembly

REMOVAL & INSTALLATION

✳✳ CAUTION

Before servicing any electrical component, the ignition key must be in the OFF or LOCK position and all electrical loads must be OFF, unless instructed otherwise in these procedures. If a tool or equipment could easily come in contact with a live exposed electrical terminal, also disconnect the negative battery cable. Failure to follow these precautions may cause personal injury and/or damage to the vehicle or its components.

1. Disconnect the negative battery cable.

2. Remove the 2 screws, the 2 plastic retainers and the rear console box from the floor.

3. Remove the 2 screws, the 2 plastic retainers and the front console box from the floor.

4. Remove the 1 screw and the transfer case gearshift control lever knob from the transfer case gearshift control lever.

5. Remove the 6 screws, the boot cover and the gearshift control lever boot from the gearshift control lever.

6. Remove the 1 clamp and the gearshift lever case boot from the gearshift lever case.

7. Remove the gearshift control lever from the gearshift lever case by pushing down on the gearshift control lever pivot and turning counterclockwise 90 degrees, ¼ turn, and lifting up.

8. Remove the 4-wheel drive switch electrical connector from the four wheel drive switch, if equipped.

9. Remove the four-wheel drive low switch electrical connector from the four-wheel drive low switch, if equipped.

10. Remove the 4 bolts from the fan shroud at the radiator.

11. Raise and suitably support the vehicle.

12. Place a drain pan or suitable container under the transfer case.

13. Remove the transfer case oil level/filler plug from the transfer case.

14. Remove the transfer case drain plug from the transfer case and drain the transfer case oil.

➡**An index mark should be made on the front and rear propeller shaft pinion**

flange yokes and the front and rear differential pinion flanges to ensure that the front and rear propeller shafts are installed in the same positions from which they were removed. If this precaution is not observed, a driveline imbalance may result causing vibration, premature component wear or other undesirable conditions.

15. Place an index mark on the front propeller shaft pinion flange yoke and the front deferential pinion flange, if equipped.

16. Remove the 4 bolts, the 4 nuts and the front propeller shaft from the vehicle, if equipped.

17. Place an index mark on the rear propeller shaft pinion flange yoke and the differential pinion flange.

18. Remove the 4 bolts, the 4 nuts and the rear propeller shaft from the vehicle.

19. Remove the VSS from the case.

20. Remove the sensor harness from the crossmember.

21. Support the transfer case with a suitable hydraulic jack.

22. Remove the following components:
- The 6 nuts
- The 6 bolts
- The transfer case skid plate
- The 2 transfer case braces

23. Remove the 10 bolts and the transfer case crossmember from the undercarriage.

24. Remove the 1 clamp and the breather hose from the gearshift lever case.

25. Remove the 1 clamp and the breather hose from the cover.

26. Remove the gearshift lever case from the rear case manual transmission equipped vehicles.

27. Remove the 12 transfer case-to-transmission bolts.

28. Slide the transfer case off the transmission output shaft and slowly lower the transfer case making sure there are no obstructions.

To install:

29. Raise the transfer case into position and slide it onto the transmission output shaft.

30. Install the 10 transfer case-to-transmission bolts. Tighten the 10 transfer case-to-transmission bolts to 28 Nm (21 ft. lbs.).

31. Install the breather hose to the gearshift lever case. Secure with the clamp.

32. Install the breather hose to the cover. Secure with the clamp.

33. Install the gearshift lever case to the rear case manual transmission equipped vehicles.

34. Install the transfer case crossmember to the undercarriage. Secure the crossmember with the 6 bolts. Tighten the transfer case crossmember bolts to 50 Nm (37 ft. lbs.).

35. Install the skid plate and the 2 transfer case braces.

36. Secure the skid plate and the 2 braces with the 6 nuts and the bolts. Tighten the transfer case skid plate bolts and the nuts to 50 Nm (37 ft. lbs.).

37. Remove the hydraulic jack from under the transfer case.

38. Install the sensor harness to the crossmember.

39. Install the ground wire and bolt.

40. Install the rear propeller shaft into the vehicle aligning the reference marks made during the removal. Secure the rear propeller shaft with the 4 bolts and 4 nuts. Tighten the rear propeller shaft bolts and nuts to 50 Nm (37 ft. lbs.).

41. Install the front propeller shaft into the vehicle aligning the reference marks made during removal. Secure the front propeller shaft the 4 bolts and the 4 nuts, if equipped.

42. Tighten the front propeller shaft bolts and nuts to 50 Nm (37 ft. lbs.).

43. Apply GM P/N 12346004 or the equivalent, to the threaded portion of the transfer case drain plug.

44. Connect the VSS to the case.

45. Install the transfer case drain plug into the transfer case. Tighten the transfer case drain plug to 28 Nm (21 ft. lbs.).

46. Add approximately 1.7 liters (1.8 qt.) of 75W-90 GL4 lubricant, or the equivalent, into the transfer case oil level/filler plug hole. The oil level should be even with the bottom of the oil level/filler plug.

47. Apply GM P/N 12346004 (Canadian P/N 10953480), or the equivalent, to the threaded portion of the transfer case oil level/filler plug.

48. Install the transfer case oil level/filer plug into the transfer case. Tighten the transfer case oil level/filler plug to 28 Nm (21 ft. lbs.).

49. Lower the vehicle.

50. Install the 4 bolts to the fan shroud at the radiator. Tighten the fan shroud bolt to 10 Nm (89 inch lbs.).

51. Connect the four-wheel drive switch electrical connector to the four-wheel drive switch.

52. Install the four-wheel drive low switch electrical connector to the four-wheel drive low switch.

53. Install the gearshift control lever.

54. Connect the negative battery cable.

Halfshaft

REMOVAL & INSTALLATION

Left

1. Before servicing the vehicle, refer to the precautions in the beginning of this section.

2. Remove or disconnect the following:
   - Front wheel
   - Hub drive flange or locking hub, as equipped
   - Snapring
   - Thrust washer
   - Halfshaft flange fasteners
   - Halfshaft

To install:

3. Install or connect the following:
   - Halfshaft. Tighten the flange bolts to 41 ft. lbs. (55 Nm).
   - Thrust washer
   - Snapring
   - Locking hub, if equipped. Tighten the bolts to 24 ft. lbs. (33 Nm).
   - Hub drive flange, if equipped. Tighten the bolts to 35 ft. lbs. (48 Nm).
   - Front wheel

Right

1. Raise and support the vehicle.
2. Remove the front skid plate.
3. Remove the right front tire and wheel assembly.
4. Remove the 6 bolts and the flange.
5. Remove the right drive axle shaft snap ring.

6. Remove the right drive axle shaft spindle washer from the drive axle shaft.

7. Make a reference mark on the left drive axle mating flanges and.

8. Disconnect the left drive axle from the front differential.

9. Make a reference mark on the pinion flange yoke and the differential pinion flange.

10. Disconnect the propeller shaft from the front differential

11. Support the front differential.

12. Remove the front differential bolts.

13. Lower the front differential and right drive shaft from the vehicle as an assembly.

➡The drive axle circlip may break during the removal and possibly fall into the differential case. If this occurs, the pieces must be removed from the differential.

➡The drive axle is retained with a circlip. If the circlip is off center, the drive axle is very difficult to remove. Center the circlip by rotating and lightly pulling on the drive axle. Continue rotating and pulling until the drive axle pops out about 1 mm (0.039 inch). This movement indicates that the circlip is now centered on the drive axle.

14. Using the J 45531 remove the right drive axle from the front differential.

To install:

15. Install the right drive axle shaft to the differential housing.

1. Drive shaft oil seal
2. Double off-set joint (DOJ)
3. Joint circlip
4. DOJ boot
5. Ball joint boot
6. Ball joint assembly (RH side)
7. Drive shaft assembly (LH side)
8. Left drive shaft
9. Drive shaft bearing circlip
10. Drive shaft bearing

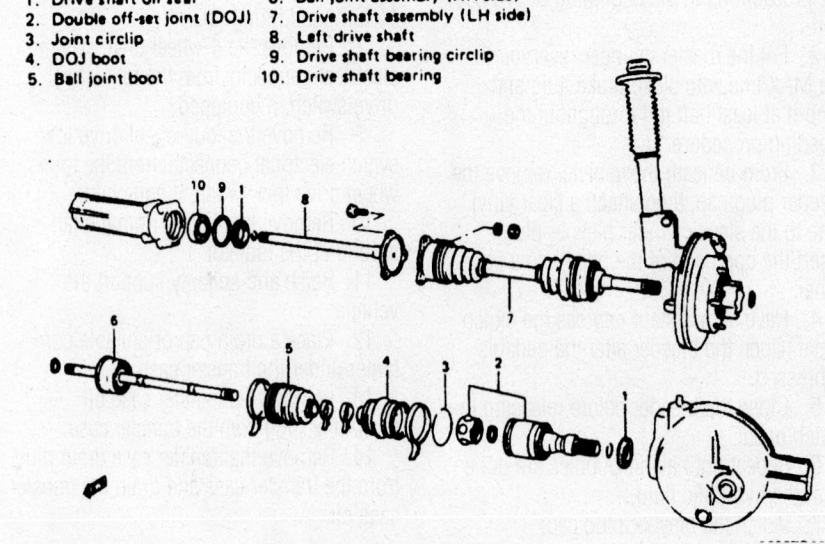

7924HG31

Exploded view of the left- and right-hand halfshaft assemblies

16. Raise the front differential and right drive shaft as an assembly into position.

17. Install the front differential retaining bolts. Tighten the bolts to 50 Nm (37 ft. lbs.).

18. Align the reference marks, then connect the left drive axle to the differential. Install the drive axle retaining bolts. Tighten the retaining bolts to 55 Nm (41 ft. lbs.).

19. Align the front propeller shaft reference mark to the differential pinion flange.

20. Install the propeller shaft retaining bolts Tighten the retaining bolts to 50 Nm (37 ft. lbs.).

21. Install the drive axle shaft spindle washer onto the drive axle shaft.

22. Install the drive axle shaft snap ring onto the end of the drive axle.

23. Apply General Motors Sealant, GM P/N 1052366 to the flange.

24. Install the flange.

25. Install the 6 flange bolts.

26. Secure the flange bolts. Tighten the flange bolts to 48 Nm (35 ft. lbs.).

27. Install the front tire and wheel assembly.

28. Lower the vehicle.

CV-Joints

OVERHAUL

Outer CV-Joint

The outer CV-joint is serviced with the axle shaft as an assembly. The outer CV-joint boot can be serviced by removing the inner CV-joint.

Inner CV-Joint

1. Before servicing the vehicle, refer to the precautions in the beginning of this section.

2. Remove or disconnect the following:

- Halfshaft from the vehicle
- Grease boot clamps
- Outer race snapring
- Outer race
- Shaft snapring
- Inner race, cage and balls

To install:

3. Install or connect the following:

- Inner race, cage and balls
- Shaft snapring
- Outer race
- Outer race snapring

4. Fill the outer race and the grease boot with CV-joint grease and tighten the boot clamps.

5. Install the axle halfshaft.

Rear Axle Shaft, Bearing and Seal

REMOVAL & INSTALLATION

1. Before servicing the vehicle, refer to the precautions in the beginning of this section.

2. Loosen the parking brake cable for clearance.

3. Remove or disconnect the following:

- Rear wheel
- Brake drum
- Wheel speed sensor, if equipped
- Bearing retainer nuts
- Axle shaft and bearing
- Axle shaft inner oil seal

4. If equipped with ABS, grind a flat spot on the wheel speed sensor tone ring, then split the ring with a chisel.

5. Grind flat spots on the bearing retainer and split it with a chisel.

6. Press the wheel bearing off the axle shaft.

7. Remove the bearing retainer and the outer oil seal.

To install:

8. Install or connect the following:

- Outer oil seal to the bearing retainer
- Bearing retainer to the axle shaft
- Bearing and retainer ring pressed onto the axle shaft
- Wheel speed sensor tone ring pressed onto the axle shaft, if equipped
- Axle shaft inner oil seal
- Axle shaft and bearing
- Bearing retainer nuts. Tighten them to 17 ft. lbs. (23 Nm).
- Wheel speed sensor, if equipped
- Brake drum
- Rear wheel

9. Fill the rear differential to the correct level.

Pinion Seal

REMOVAL & INSTALLATION

1. Before servicing the vehicle, refer to the precautions in the beginning of this section.

2. Remove or disconnect the following:

- Driveshaft
- Wheels
- Brake calipers and pads or brake drum

➡ **The brake calipers and pads or brake drum must be removed so that there is**

no additional drag when measuring pinion bearing preload.

3. Use an inch lb. torque wrench and measure and record the amount of torque required to maintain pinion rotation through several revolutions.

4. Remove or disconnect the following:

- Pinion flange
- Pinion seal
- Pinion bearing
- Collapsible spacer

To install:

➡ **Use a new collapsible spacer and flange nut for assembly.**

5. Install or connect the following:

- Collapsible spacer
- Pinion bearing
- Pinion seal
- Pinion flange

6. Rotate the pinion flange occasionally while tightening the flange nut to make sure the pinion bearings seat correctly.

7. Take frequent bearing preload torque readings. Tighten the flange nut to achieve the preload torque readings originally recorded.

✳✳ CAUTION

Never loosen the pinion nut to reduce bearing preload. If it is necessary to reduce bearing preload, install a new collapsible spacer and pinion nut.

8. Install or connect the following:

- Driveshaft
- Brake calipers and pads or brake drum
- Wheels

9. Fill the differential with gear lubricant and check for leaks.

Rear Axle Housing Assembly

REMOVAL & INSTALLATION

1. Before servicing the vehicle, refer to the precautions in the beginning of this section.

2. Drain the gear oil.

3. Support the vehicle at the frame with a hoist or jackstands.

4. Support the rear axle with a floor jack.

5. Remove or disconnect the following:

- Rear wheels
- Rear brake drums
- Rear axle shafts
- Load sensing proportioning valve linkage, if equipped
- Brake fluid hose

- Brake backing plates
- Wheel speed sensor connector, if equipped
- Axle vent tube
- Rear driveshaft
- Differential carrier assembly
- Shock absorber lower bolts
- Coil springs
- Upper rods
- Lower rods
- Lateral rod
- Axle housing

To install:

6. Install or connect the following:

- Axle housing
- Upper rods
- Lower rods
- Coil springs
- Lateral rod
- Shock absorber lower bolts
- Differential carrier assembly. Tighten the nuts to 40 ft. lbs. (55 Nm).
- Rear driveshaft
- Axle vent tube
- Wheel speed sensor connector, if equipped
- Brake backing plates

- Brake fluid hose
- Load sensing proportioning valve linkage, if equipped
- Rear axle shafts
- Rear brake drums
- Rear wheels

7. Fill the rear axle to the correct level.

8. Lower the vehicle so that the rear suspension is at curb height.

9. Tighten the upper, lower and lateral rod fasteners to 65 ft. lbs. (90 Nm).

10. Tighten the lower shock absorber fasteners to 62 ft. lbs. (85 Nm).

STEERING AND SUSPENSION

Air Bag

✳✳ CAUTION

Some vehicles are equipped with an air bag system. The system must be disarmed before performing service on, or around, system components, the steering column, instrument panel components, wiring and sensors. Failure to follow the safety precautions and the disarming procedure could result in accidental air bag deployment, possible injury and unnecessary system repairs.

PRECAUTIONS

Several precautions must be observed when handling the inflator module to avoid accidental deployment and possible personal injury.

- Never carry the inflator module by the wires or connector on the underside of the module.
- When carrying a live inflator module, hold securely with both hands and ensure that the bag/trim cover are pointed away.
- Place the inflator module on a bench or other surface with the bag and trim cover facing up.
- With the inflator module on the bench, never place anything on or close to the module which may be thrown in the event of an accidental deployment.
- Never use air bag component parts from another vehicle.
- If there is a chance of electrical shock to any of the air bag components, remove the air bag module before servicing the vehicle.

DISARMING

1. Before servicing the vehicle, refer to the precautions in the beginning of this section.

2. Remove or disconnect the following:
- Negative battery cable
- AIR BAG fuse
- Driver air bag connector
- Glove box
- Passenger air bag connector

ARMING

When repairs are complete, install or connect the following:
- Passenger air bag connector
- Glove box

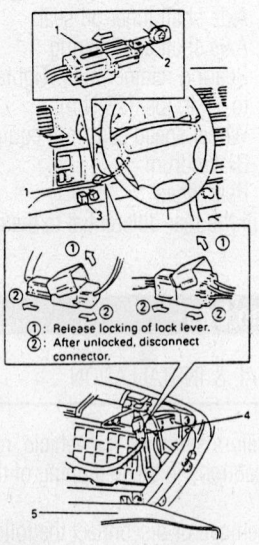

①: Release locking of lock lever.
②: After unlocked, disconnect connector.

1. Yellow connector of driver air bag (inflator) module
2. Connector stay
3. Air bag fuse box
4. Yellow connector of passenger air bag (inflator) module
5. Glove box

7924HG36

Air bag component location and identification—All models

- Driver air bag connector
- AIR BAG fuse
- Negative battery cable

Power Rack and Pinion Steering Gear

REMOVAL & INSTALLATION

1. Turn the key to the LOCK position and remove the key.

➡**The wheels of the vehicle must be straight ahead and the steering column in the LOCK position before disconnecting the steering column or intermediate shaft from the steering gear. Failure to do so will cause the coil assembly in the steering column to become un-centered which will cause damage to the coil assembly.**

2. Lock the steering column in the straight ahead position.

3. Attempt to turn the steering wheel in order to verify the steering column is locked.

4. Disable the supplemental inflatable restraint system.

5. Raise and support the vehicle.

6. Remove the skid plate, if equipped.

7. Remove the left outer tie rod nut from the knuckle.

8. Remove the tie rod end from the knuckle.

9. Remove the right outer tie rod nut from the knuckle.

10. Remove the tie rod end from the knuckle.

11. Place a drain pan or a suitable container under the steering gear.

➡**The first design banjo bolt style fitting utilizes 2 gaskets. The second design flare nut style fitting utilizes seals.**

12. Remove the power steering gear inlet hose fitting from the power steering gear.

13. Remove the power steering gear inlet hose and the 2 gaskets, if equipped, from the steering gear.

14. Loosen the hose clamp and remove the power steering fluid reservoir hose from the steering gear.

15. Plug the hoses in order to prevent the fluid from escaping.

16. Using paint, place match marks on the lower steering shaft and on the steering gear housing.

17. Remove the clamp bolt from the lower steering shaft.

18. Disconnect the lower steering shaft from the steering gear.

19. Using paint, place match marks on the pinion shaft and on the steering gear housing.

➡**Do NOT disassemble the power steering gear.**

20. Remove the 4 steering gear mounting bolts.

21. Remove the steering gear with the insulators and the brackets from the vehicle.

22. Remove the 2 brackets from the steering gear.

23. Remove the 2 insulators from the steering gear.

24. Place a match mark on the left inner tie rod thread and the right inner tie rod thread, indicating the positions of the tie rod lock nuts.

25. Loosen the 2 tie rod lock nuts and remove the 2 outer tie rod ends from the inner tie rods.

To install:

➡**If you are replacing the steering gear, measure the location of the match mark made on the original inner tie rods. Use this measurement in order to make a corresponding mark on the replacement steering gear.**

26. Using paint, place match marks on the inner tie rods in order to position the 2 lock nuts during assembly.

27. Install the 2 tie rod lock nuts to the inner tie rods. Position the lock nuts at the match marks. Do NOT tighten the lock nuts.

28. Install the 2 outer tie rod ends to the inner tie rods. Do NOT tighten the outer tie rod ends or the lock nuts.

➡**If you are replacing the steering gear, measure the location of the match mark made on the pinion shaft and the steering gear housing. Make**

corresponding marks on the replacement steering gear in order to set the rack in the straight-ahead position.

29. Place match marks on the pinion shaft and on the steering gear housing.

30. Verify the steering wheel and the front wheels are in the straight-head position.

31. Install the 2 insulators to the steering gear.

32. Install the 2 brackets to the 2 insulators.

➡**Do NOT tighten the bracket bolts before tightening the lower steering shaft clamp bolt.**

33. Install the steering gear and the 4 bolts to the vehicle.

34. Align the match mark on the pinion shaft with the match mark on the steering gear housing.

35. Connect the lower steering shaft to the steering gear. Install the clamp bolt to the lower steering shaft. Tighten the bolt to 25 Nm (18 ft. lbs.).

36. Tighten the mounting bolts on the steering gear. Tighten the bolts to 55 Nm (40 ft. lbs.).

37. Remove the plugs from the disconnected pipes and hoses.

38. Install the power steering fluid reservoir hose and the hose clamp to the steering gear.

➡**The first design banjo bolt style fitting uses 2 gaskets. The second design flare nut style fitting uses seals. The first design power steering components are not interchangeable with the second design power steering components.**

39. Install the following components to the steering gear:

- The power steering gear inlet hose
- The 2 gaskets, if equipped
- The power steering gear inlet hose fitting

40. Tighten the fitting to 35 Nm (25.5 ft. lbs.).

41. Install the left outer tie rod end to the knuckle.

42. Install a NEW nut to the ball stud. Tighten the nut to 43 Nm (31.5 ft. lbs.).

43. Install the right outer tie rod end to the knuckle.

44. Install a NEW nut to the ball stud. Tighten the nut to 43 Nm (31.5 ft. lbs.).

45. Install the skid plate, if equipped.

46. Lower the vehicle.

47. Add fluid to the reservoir.

48. Bleed the power steering system.

49. Enable the SIR.

50. Measure the wheel alignment.

51. Adjust the front toe, if necessary.

52. Tighten the tie rod lock nuts. Tighten the nuts to 65 Nm (47 ft. lbs.).

Strut

REMOVAL & INSTALLATION

1. Before servicing the vehicle, refer to the precautions in the beginning of this section.

2. Support the control arm with a stand or floor jack.

3. Remove or disconnect the following:

- Front wheel
- Brake hose bracket
- Strut bracket bolts
- Upper strut mount nuts
- Strut

To install:

4. Install or connect the following:

- Strut. Tighten the upper mount nuts to 40 ft. lbs. (55 Nm) and the bracket bolts to 70 ft. lbs. (95 Nm).
- Brake hose bracket
- Front wheel

5. Check the wheel alignment and adjust as necessary.

Shock Absorber

REMOVAL & INSTALLATION

1. Before servicing the vehicle, refer to the precautions in the beginning of this section.

2. Support the rear axle housing with a hydraulic jack or stand.

3. Remove or disconnect the following:

- Shock absorber upper locknut and retaining nut
- Lower shock absorber mounting nut and bolt
- Rear shock absorber

To install:

4. Install or connect the following:

- Rear shock absorber
- Lower mounting nut and bolt
- Upper retaining nut and locknut

5. Torque the upper mounting nuts to 21 ft. lbs. (29 Nm) and the lower mounting nut/bolt to 74 ft. lbs. (100 Nm).

6. Remove the jack or stand from the rear axle assembly.

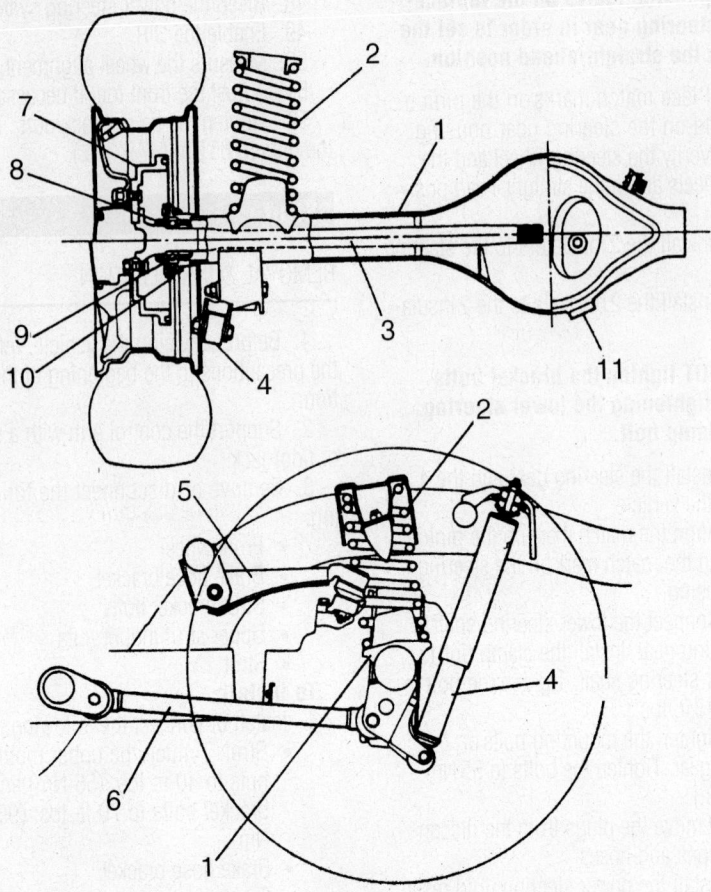

1. Rear axle housing
2. Coil spring
3. Axle shaft
4. Shock absorber
5. Upper arm
6. Trailing rod
7. Brake drum
8. Wheel bearing retainer
9. Rear wheel bearing
10. Brake back plate
11. Oil drain plug

7924HG32

Rear suspension component identification

Coil Spring

REMOVAL & INSTALLATION

Front

1. Raise and support the vehicle.
2. Remove the tire and wheel assembly.
3. Remove the stabilizer shaft links from the lower control arms.

❋❋ CAUTION

To prevent personal injury and/or component damage, use the proper tools to support the lower control arm when removing the coil spring. The coil spring is under extreme pressure and can become a projectile should the spring separate from the lower control arm before all of the tension is relieved.

4. Support the lower control arm with a jack.
5. Remove the ball stud nut.
6. Separate the lower control arm ball stud from the knuckle.
7. Use the jack in order to raise the control arm and disconnect the ball joint from the knuckle.
8. Move the knuckle and disc brake assembly away from the control arm.
9. While lowering the floor jack, remove the coil spring.
10. Remove the spring seat and the spring bumper. Inspect and replace the components if necessary.

To install:

11. Install the spring bumper and the spring seat.

➡**Care should be taken to avoid chipping or scratching the coating when handling the suspension coil spring. Damage to the coating can cause premature failure.**

➡**The top of the coil spring is narrower than the bottom of the coil spring.**

12. Install the coil spring. Install the narrower diameter end to the upper spring seat and the larger diameter end to the lower control arm.
13. Raise the lower control arm with a jack.
14. Lower the jack in order to install the ball stud to the knuckle.
15. Install a NEW ball stud nut. Tighten the nut to 60 Nm (43.5 ft. lbs.).
16. Remove the jack.

➡**DO NOT tighten the stabilizer shaft link nuts and bolts completely. The weight of the vehicle must be on the wheels and tires before tightening the nuts and bolts.**

17. Install the stabilizer shaft links to the lower control arms.
18. Install the tire and wheel assembly.
19. Lower the vehicle.
20. Verify the stabilizer shaft is centered, side-to-side. Verify that the stabilizer shaft links are set at a neutral position, as seen from above. Tighten the stabilizer shaft link nut to 29 Nm (21.0 ft. lbs.).

Rear

1. Raise and support the vehicle.
2. Use a floor jack in order to support the rear axle housing.
3. Disconnect the park brake cable hanger from the frame. Disconnect the clamp from the cross member.
4. Remove the lower nut and bolt from the shock absorber.
5. If the vehicle has ABS, remove the wheel speed sensor clamps from the upper control arm and the axle housing.
6. Remove the E-ring from the right rear brake pipe.
7. Disconnect the breather hose from the axle housing.

8. Lower the rear axle housing gradually as far down as necessary in order to remove the coil spring.

To install:

➡**When seating the coil spring, mate the spring end with the stepped part of the rear axle spring seat.**

9. Install the coil spring on the spring seat of the axle housing. Raise the axle housing.

➡**DO NOT tighten the fasteners before lowering the vehicle. The weight of the vehicle must be on the tires and wheels before tightening the fasteners.**

10. Install the shock absorber lower bolt and the nut.
11. If the vehicle has ABS, install the wheel speed sensor clamps.
12. Install the E-ring to the right rear brake pipe.
13. Connect the breather hose to the axle housing.
14. Connect the park brake cable hanger to the chassis frame and clamp the hanger to the cross member.
15. Remove the floor jack from the axle housing.
16. Lower the vehicle.
17. With the weight of the vehicle on the tires and wheels, tighten the nut and the bolt.
18. Tighten the nut and the bolt to 100 Nm (74 ft. lbs.).

Lower Ball Joint

REMOVAL & INSTALLATION

The lower ball joint is serviced with the lower control arm as an assembly.

Lower Control Arm

REMOVAL & INSTALLATION

1. Before servicing the vehicle, refer to the precautions in the beginning of this section.
2. Support the vehicle at the frame with a hoist or jackstand.
3. Support the control arm with a floor jack.

4. Remove or disconnect the following:

- Front wheel
- Brake caliper and rotor
- Locking hub or drive flange, if equipped
- Axle shaft snapring and thrust washer, if equipped
- Wheel speed sensor, if equipped
- Stabilizer bar link
- Lower ball joint
- Strut bracket bolts

5. Lower the floor jack and remove the coil spring.
6. Remove the inner control arm bolts and remove the control arm.

To install:

7. Install the inner control arm bolts.
8. Install the coil spring onto the control arm and raise the floor jack.
9. Install or connect the following:

- Strut bracket bolts. Tighten them to 70 ft. lbs. (95 Nm).
- Lower ball joint. Tighten the nut to 40 ft. lbs. (55 Nm).
- Stabilizer bar link. Tighten the nut to 21 ft. lbs. (29 Nm).
- Wheel speed sensor, if equipped
- Axle shaft snapring and thrust washer, if equipped
- Locking hub or drive flange, if equipped
- Brake caliper and rotor
- Front wheel

10. Lower the vehicle so that the front suspension is at curb height.
11. Tighten the front inner bolt to 62 ft. lbs. (85 Nm) and the rear inner bolt to 92 ft. lbs. (127 Nm).
12. Check the wheel alignment and adjust as necessary.

CONTROL ARM BUSHING REPLACEMENT

1. Before servicing the vehicle, refer to the precautions in the beginning of this section.
2. Remove the control arm from the vehicle.
3. Remove the control arm bushings with a hydraulic press.

To install:

4. Lubricate the control arm bushings with liquid soap.
5. Press the bushings into the control

arm until the bushing flange contacts the housing edge of the control arm.
6. Install the control arm to the vehicle.
7. Check the wheel alignment and adjust as necessary.

Wheel Bearings

ADJUSTMENT

The wheel bearings are not adjustable.

REMOVAL & INSTALLATION

1. Before servicing the vehicle, refer to the precautions in the beginning of this section.
2. Remove or disconnect the following:

- Front wheel
- Brake caliper and rotor
- Locking hub or hub drive flange, if equipped
- Hub grease cap, if equipped
- Wheel speed sensor, if equipped
- Wheel bearing lockwasher
- Wheel bearing locknut and inner washer
- Wheel hub and bearing assembly
- Wheel hub oil seal
- Wheel bearing oil seal
- Snapring

3. Press the wheel bearing and race out of the hub.

To install:

4. Press the wheel bearing and race into the hub so that the race is fully seated in the hub bore.
5. Install or connect the following:

- Snapring
- Wheel bearing oil seal
- Wheel hub oil seal
- Wheel hub and bearing assembly
- Wheel bearing locknut and inner washer. Tighten the nut to 157 ft. lbs. (216 Nm).
- Wheel bearing lockwasher. Tighten the retaining screws to 13 inch lbs. (1.5 Nm).
- Wheel speed sensor, if equipped
- Hub grease cap, if equipped
- Locking hub or hub drive flange, if equipped
- Brake caliper and rotor
- Front wheel

BRAKES

Brake Caliper

REMOVAL & INSTALLATION

1. Raise and safely support the vehicle.
2. Remove the wheels.
3. Disconnect and plug the brake line.
4. Remove the caliper mounting bolts (guide pins) and remove the caliper from the vehicle.
   To install:
5. Install the caliper on the vehicle. Tighten the mounting bolts to 20 ft. lbs. (27 Nm).
6. Connect the hydraulic brake line, using 2 new washers. Torque the union bolt to 17 ft. lbs. (23 Nm).
7. Replace the front wheels.

8. Lower the vehicle.
9. Fill the brake reservoir and bleed the hydraulic brake system.

Disc Brake Pads

REMOVAL & INSTALLATION

1. Siphon about ⅔ of the fluid out of the master cylinder.
2. Raise and safely support the vehicle.
3. Remove the wheels.
4. Remove the brake caliper mounting bolts and remove the caliper from the mounting bracket.
5. Support the caliper with a wire.
6. Using a large pair of plies or a C-clamp compress the caliper piston back into the bore.

7. Remove the disc brake pads and any shims from the caliper mounting bracket.
   To install:
8. Install the brake pads and any shims removed from the caliper mounting bracket.
9. Install the caliper on the mounting bracket and install the mounting bolts. Tighten the mounting bolts to 20 ft. lbs. (27 Nm).
10. Install the front wheels and lower the vehicle.

✳✳ CAUTION

Do not attempt to drive the vehicle until after the following step is performed.

11. Depress the brake pedal repeatedly until a firm pedal is obtained. Do not

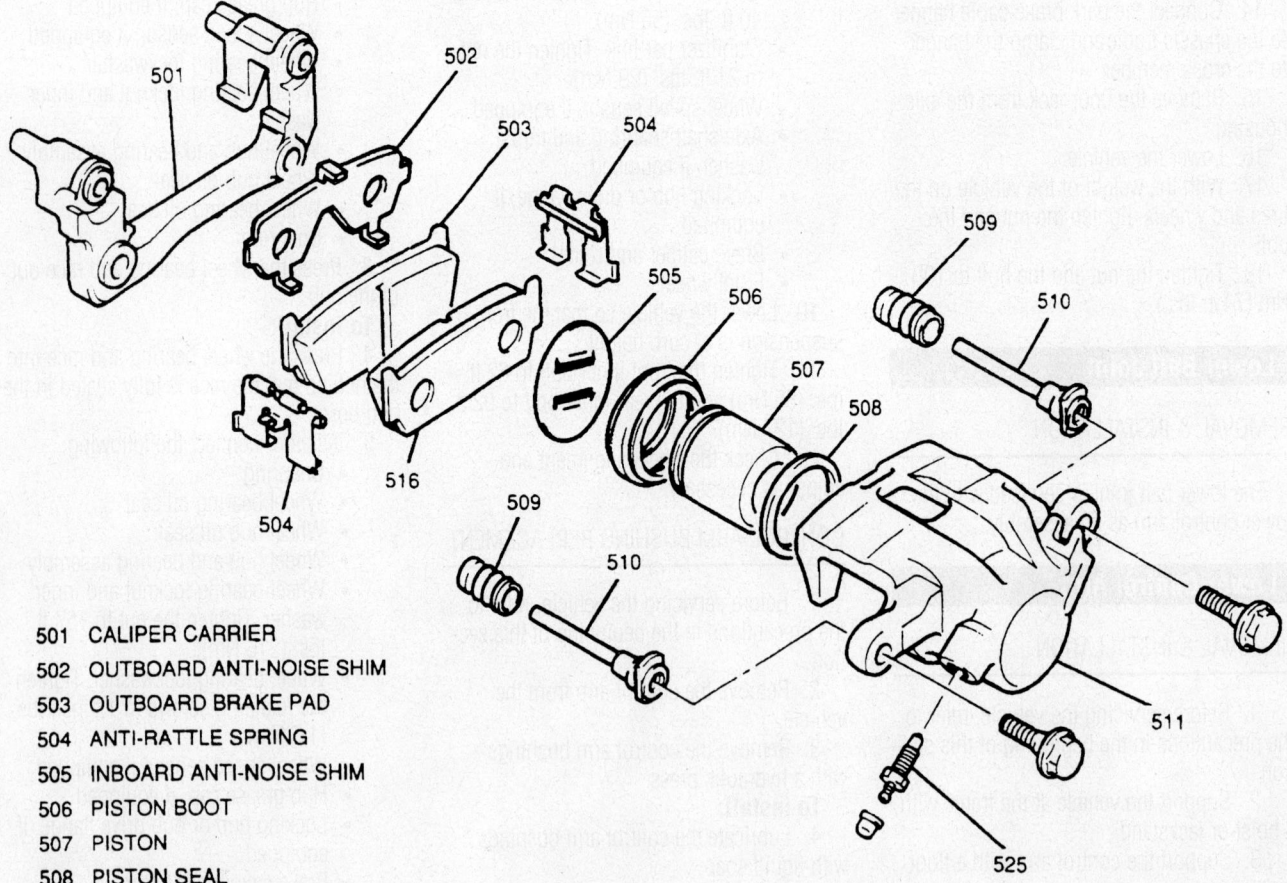

501	CALIPER CARRIER
502	OUTBOARD ANTI-NOISE SHIM
503	OUTBOARD BRAKE PAD
504	ANTI-RATTLE SPRING
505	INBOARD ANTI-NOISE SHIM
506	PISTON BOOT
507	PISTON
508	PISTON SEAL
509	CALIPER PIN BOLT BOOT
510	CALIPER PIN BOLT
511	CALIPER
516	INBOARD BRAKE PAD
525	BLEEDER VALVE

Front disc brake components

93026G37

attempt to drive the vehicle unless a firm pedal is obtained.

12. Check the fluid level in the master cylinder. Add fresh brake fluid, as necessary.

13. Road-test the vehicle.

Brake Drums

REMOVAL & INSTALLATION

1. Raise and safely support the vehicle.
2. Remove the rear wheel(s).
3. Release the parking brake.
4. Remove the parking brake lever cover screws and loosen the brake cable locking nut.
5. Install 2, 8mm bolts into the brake drum holes and uniformly tighten each bolt. Tighten each bolt until the brake drum is removed from the vehicle. If there is difficulty in removing the drum, insert a small tool through the hole in the rear of the backing plate, and hold the automatic adjusting lever away from the adjuster. Using another narrow, flat tool at the same time, reduce the

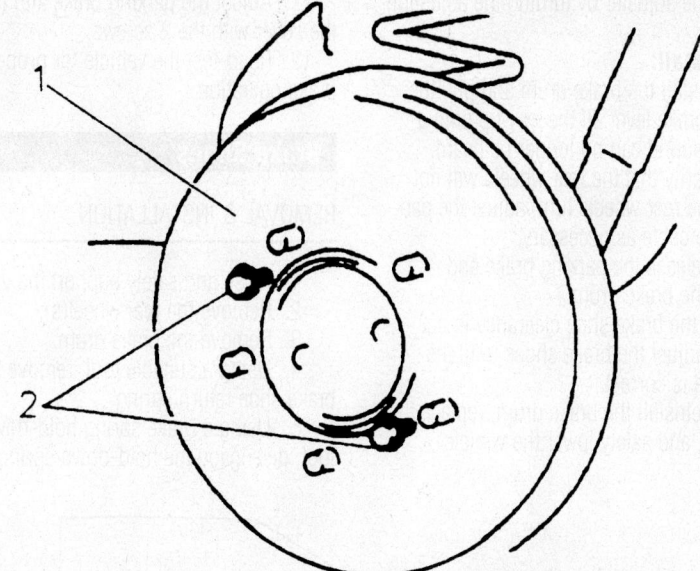

1 DRUM
2 TWO 8mm BOLTS

93026G39

Removing the brake drum with the 2, 8mm bolts

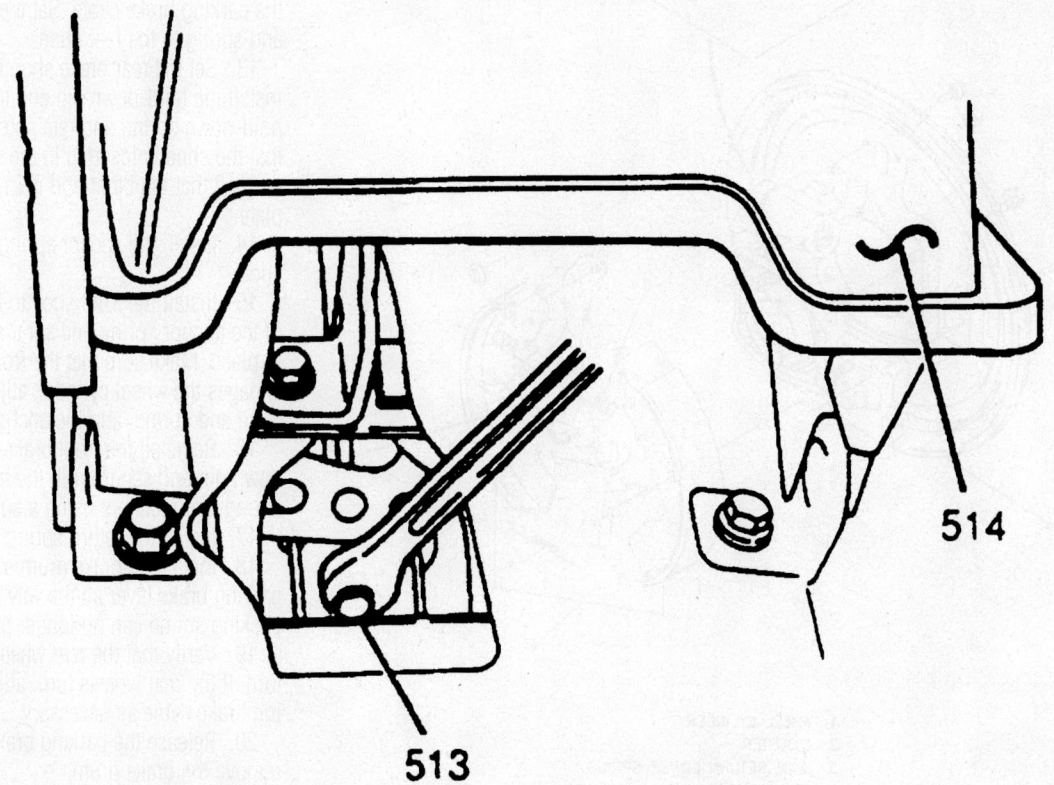

513 PARKING BRAKE CABLE LOCKNUT

514 PARKING BRAKE LEVER COVER

93026G38

Reducing the adjuster to remove the brake drum

brake shoe adjuster by turning the adjusting wheel.

To install:

6. Install the brake drum and pull the parking brake lever all the way up until a clicking sound can no longer be heard.

7. Verify that the rear wheels will not turn. If the rear wheels turn, adjust the parking brake cable as necessary.

8. Release the parking brake and remove the brake drum.

9. If the brake shoe clearance is not correct, adjust the brake shoes until the clearance is correct.

10. Reinstall the brake drum, replace the wheel(s), and safely lower the vehicle.

11. Adjust the parking brake and install the cover with the 2 screws.

12. Road-test the vehicle for proper brake operation.

Brake Shoes

REMOVAL & INSTALLATION

1. Raise and safely support the vehicle.
2. Remove the rear wheel(s).
3. Remove the brake drum.
4. Using a suitable tool, remove the brake shoe return spring.
5. Using a brake spring hold-down tool, disengage the hold-down spring and

retainers from the front shoe. Remove the hold-down retainer pinch

6. Disconnect the anchor spring from the front shoe and remove the front shoe.

7. Remove the anchor spring from the rear shoe. Using a brake spring hold-down tool, disengage the hold-down spring and retainers from the rear shoe. Remove the hold-down pinch

8. Disengage the parking brake lever from the parking brake cable and remove the rear shoe.

9. Remove the C-washer, the automatic adjuster lever and spring, the C-washer, and the parking brake lever from the rear shoe.

10. Thoroughly clean the backing plate and brake hardware with brake cleaning solvent. Apply high temperature grease to the backing plate shoe contact points, anchor plate and shoe contact points, adjusting bolt, and adjuster and brake shoe contact points.

To install:

11. Reinstall the automatic adjuster lever and the parking brake lever to the rear shoe using new C-washers.

12. Connect the parking brake lever to the parking brake cable. Set the adjuster and spring to the rear shoe.

13. Set the rear brake shoe in place, install the hold-down pin and install the hold-down spring and retainers. Make sure that the shoe is inserted in the wheel cylinder and that the other end is in the anchor plate.

14. Install the anchor spring to the rear shoe.

15. Install the front shoe to the other end of the anchor spring and set the front shoe in place. Make sure that the front shoe engages the wheel cylinder, adjuster mechanism and spring, and the anchor plate.

16. Reinstall the front brake shoe hold-down pin and secure with the hold-down spring and retainers using a suitable tool.

17. Install the return spring.

18. Install the brake drum and pull the parking brake lever all the way up until a clicking sound can no longer be heard.

19. Verify that the rear wheels will not turn. If the rear wheels turn, adjust the parking brake cable as necessary.

20. Release the parking brake and remove the brake drum.

21. If the brake shoe clearance is not correct, adjust the brake shoes until the clearance is correct.

22. Reinstall the brake drum, replace the wheel(s), and safely lower the vehicle.

23. Road-test the vehicle for proper brake operation.

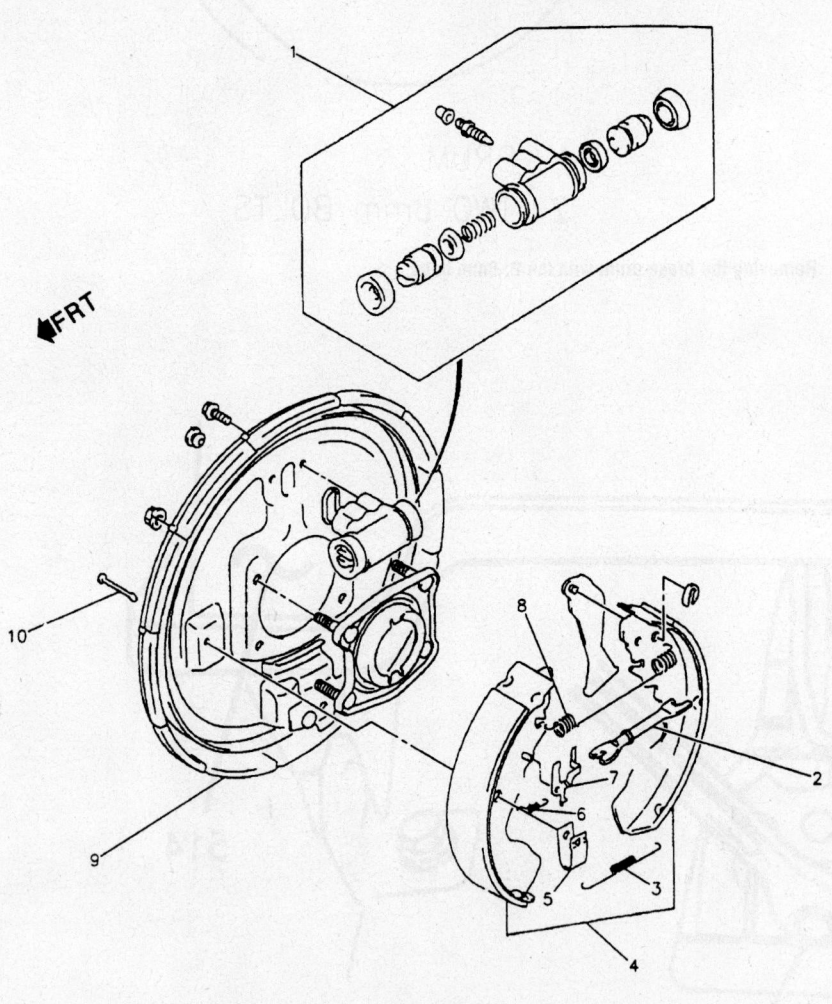

1	WHEEL CYLINDER
2	ADJUSTER
3	SHOE RETURN LOWER SPRING
4	BRAKE SHOES
5	SHOE HOLD DOWN SPRING
6	ADJUSTER SPRING
7	PAWL LEVER
8	SHOE RETURN UPPER SPRING
9	BACKING PLATE
10	SHOE HOLD DOWN PIN

93026G40

Exploded view of the rear brake components

PONTIAC

30

Vibe

BRAKES **30-59**
DRIVE TRAIN **30-37**
ENGINE REPAIR **30-8**
FUEL SYSTEM **30-32**
SPECIFICATION AND
MAINTENANCE CHARTS.....**30-2**
Engine and Vehicle Identification...30-2
General Engine Specifications 30-2
Engine Tune-Up Specifications 30-2
Capacities 30-3
Valve Specifications 30-3
Crankshaft and Connecting Rod
Specifications 30-4
Piston and Ring Specifications 30-4
Torque Specifications 30-5
Wheel Alignment 30-5
Tire, Wheel and Ball Joint
Specifications 30-6
Brake Specifications 30-6
Scheduled Maintenance
Intervals..................................30-7
STEERING AND
SUSPENSION.................**30-49**

A
Air Bag......................................30-49
Disarming 30-49
Precautions 30-49
Rearming 30-50
Alternator 30-8
Removal & Installation 30-8
Automatic Transaxle 30-41
Removal & Installation 30-41

B
Brake Caliper 30-59
Removal & Installation 30-59
Brake Drums..............................30-60
Removal & Installation 30-60
Brake Shoes...............................30-60
Removal & Installation 30-60

C
Camshaft(s) 30-17
Removal & Installation 30-17

Clutch 30-43
Adjustments 30-43
Removal & Installation 30-43
CV-Joints...................................30-47
Overhaul 30-47
Cylinder Head 30-13
Removal & Installation 30-13

D
Disc Brake Pads..........................30-59
Removal And Installation 30-59

E
Engine Assembly 30-8
Removal & Installation 30-8
Exhaust Manifold 30-17
Removal & Installation 30-17

F
Fuel Filter 30-32
Removal & Installation 30-32
Fuel Injectors 30-35
Removal & Installation 30-35
Fuel Pump 30-32
Removal & Installation 30-32
Fuel System Pressure 30-32
Relieving 30-32
Fuel System Service Precautions.30-32

H
Halfshaft....................................30-46
Removal & Installation 30-46
Heater Core 30-10
Removal & Installation 30-10
Hydraulic Clutch System 30-44
Bleeding.................................30-44

I
Ignition Timing 30-8
Adjustment..............................30-8
Intake Manifold 30-16
Removal & Installation 30-16

L
Lower Ball Joint 30-55
Removal & Installation 30-55
Lower Control Arm 30-56
Removal & Installation 30-56

M
Manual Transaxle Assembly 30-37
Removal & Installation 30-37

O
Oil Pan......................................30-28
Removal & Installation 30-28
Oil Pump 30-28
Removal & Installation 30-28

P
Piston and Ring Positioning 30-31

R
Rack and Pinion Steering Gear 30-51
Removal & Installation 30-51
Rear Main Seal 30-29
Removal & Installation 30-29

S
Starter......................................30-27
Removal & Installation 30-27
Strut and Coil Spring..................30-53
Removal & Installation 30-53

T
Timing Chain, Sprockets, Front
Cover and Seal 30-29
Removal & Installation 30-29
Transfer Case.............................30-45
Removal & Installation 30-45

U
Upper Control Arm 30-56
Removal & Installation 30-56

V
Valve Lash 30-21
Adjustment..............................30-21

W
Water Pump................................30-9
Removal & Installation 30-9
Wheel Bearings...........................30-56
Removal & Installation 30-56

SPECIFICATION AND MAINTENANCE CHARTS

ENGINE AND VEHICLE IDENTIFICATION

			Engine				Model Year	
Code ①	Liters (cc)	Cu. In.	Cyl.	Fuel Sys.	Engine Type	Eng. Mfg.	Code ②	Year
1ZZ-FE	1.8 (1794)	109	4	EFI	DOHC	Toyota	3	2003
2ZZ-GE	1.8 (1796)	109.5	4	EFI	DOHC	Toyota	4	2004
EFI: Electronic Fuel Injection							5	2005
DOHC: Double Overhead Camshaft							6	2006

① 8th digit of VIN

② 10th digit of VIN

06025-VIBE-C01

GENERAL ENGINE SPECIFICATIONS

Year	Model	Engine Displacement Liters (VIN)	Net Horsepower @ rpm	Net Torque @ rpm (ft. lbs.)	Bore x Stroke (in.)	Compression Ratio	Oil Pressure @ idle
2003	Vibe	1.8 (1ZZ-FE)	①	②	3.11x3.60	10.0:1	4.2
	Vibe	1.8 (2ZZ-GE)	180@7600	130@6800	3.23x3.35	11.5:1	2.8
2004	Vibe	1.8 (1ZZ-FE)	①	②	3.11x3.60	10.0:1	4.2
	Vibe	1.8 (2ZZ-GE)	180@7600	130@6800	3.23x3.35	11.5:1	2.8
2005	Vibe	1.8 (1ZZ-FE)	①	②	3.11x3.60	10.0:1	4.2
	Vibe	1.8 (2ZZ-GE)	170@7600	130@6800	3.23x3.35	11.5:1	2.8

EFI: Electronic Fuel Injection

① 2WD models: 130@6000
   4WD models: 123@6000

② 2WD models: 125@4400
   4WD models: 118@4400

06025-VIBE-C02

ENGINE TUNE-UP SPECIFICATIONS

Year	Engine Displacement Liters (VIN)	Spark Plug Gap (in.)	Ignition Timing (deg.)	Fuel Pump (psi)	Idle Speed (rpm) MT	Idle Speed (rpm) AT	Valve Clearance In.	Valve Clearance Ex.
2003	1.8 (1ZZ-FE)	0.043	①	44-50	650-750	650-750	0.0059-0.0098	0.0098-0.0138
	1.8 (2ZZ-GE)	0.043	②	44-50	750-850	700-800	0.0031-0.0071	0.0087-0.0126
2004	1.8 (1ZZ-FE)	0.043	①	44-50	650-750	650-750	0.0059-0.0098	0.0098-0.0138
	1.8 (2ZZ-GE)	0.043	②	44-50	750-850	700-800	0.0031-0.0071	0.0087-0.0126
2005	1.8 (1ZZ-FE)	0.040-0.048	①	44-50	650-750	650-750	0.0059-0.0098	0.0098-0.0138
	1.8 (2ZZ-GE)	0.040-0.051	②	44-50	750-850	700-800	0.0031-0.0071	0.0087-0.0126

Note: The Vehicle Emission Control Information label often reflects specification changes made during production. The label figures must be used if they differ from those in this chart.

① With terminal TC and CG of DLC3 connected: 8-12 degrees BTDC
   With terminal TC and CG of DLC3 disconnected: 10-18 degrees BTDC

② With terminal TC and CG of DLC3 connected: 8-12 degrees BTDC
   With terminal TC and CG of DLC3 disconnected:
   A/T: 10-18 degrees BTDC
   M/T: 4-12 degrees BTDC

06025-VIBE-C03

CAPACITIES

Year	Model	Engine Displacement Liters (VIN)	Engine Oil with Filter	Transmission (pts.)			Drive Axle		Fuel Tank (gal.)	Cooling System (qts.)
				5-Spd	6-Spd	Auto.	Front (pts.)	Rear (pts.)		
2003	Vibe	1.8 (1ZZ-FE)	3.9	4.0	4.0	6.0	3.0	—	12.0	6.9
	Vibe	1.8 (2ZZ-GE)	4.8	4.8	4.8	6.6	①	—	13.0	7.1
2004	Vibe	1.8 (1ZZ-FE)	3.9	4.0	4.0	6.0	3.0	—	12.0	6.9
	Vibe	1.8 (2ZZ-GE)	4.8	4.8	4.8	6.6	①	—	13.0	7.1
2005	Vibe	1.8 (1ZZ-FE)	3.9	4.0	4.0	6.0	3.0	—	12.0	6.9
	Vibe	1.8 (2ZZ-GE)	4.8	4.8	4.8	6.6	①	—	13.0	7.1

Note: All capacities are approximate. Add fluid gradually and check to be sure a proper fluid level is obtained.

① Included in transaxle capacity

06025-VIBE-C04

VALVE SPECIFICATIONS

Year	Engine Displacement Liters (VIN)	Seat Angle (deg.)	Face Angle (deg.)	Spring Test Pressure (lbs. @ in.)	Spring Installed Height (in.)	Stem-to-Guide Clearance (in.)		Stem Diameter (in.)	
						Intake	Exhaust	Intake	Exhaust
2003	1.8 (1ZZ-FE)	45	44.5	31.3-34.8 @ 1.252	1.323	0.0010-0.0024	0.0012-0.0026	0.2154-0.2159	0.2152-0.2158
	1.8 (2ZZ-GE)	45	44.5	①	1.516	0.0010-0.0023	0.0012-0.0025	0.2145-0.2156	0.2144-0.2154
2004	1.8 (1ZZ-FE)	45	44.5	31.3-34.8 @ 1.252	1.323	0.0010-0.0024	0.0012-0.0026	0.2154-0.2159	0.2152-0.2158
	1.8 (2ZZ-GE)	45	44.5	①	1.516	0.0010-0.0023	0.0012-0.0025	0.2145-0.2156	0.2144-0.2154
2005	1.8 (1ZZ-FE)	45	45	31.3-34.8 @ 1.252	②	0.0010-0.0023	0.0011-0.0028	0.2150-0.2155	0.2143-0.2153
	1.8 (2ZZ-GE)	45	45	①	②	0.0010-0.0023	0.0011-0.0028	0.2150-0.2155	0.2143-0.2153

① Intake: 49.6-55.5 @ 1.516
   Exhaust: 47.6-52.6 @ 1.516

② Intake: 1.831

06025-VIBE-C05

CRANKSHAFT AND CONNECTING ROD SPECIFICATIONS

All measurements are given in inches.

| Year | Engine Displacement Liters (VIN) | Crankshaft | | | | Connecting Rod | | |
		Main Brg. Journal Dia.	Main Brg. Oil Clearance	Shaft End-play	Thrust on No.	Journal Diameter	Oil Clearance	Side Clearance
2003	1.8 (1ZZ-FE)	1.8893-1.8898	0.0006-0.0013	0.0008-0.0087	3	1.7320-1.7323	0.0011-0.0024	0.0063-0.0135
	1.8 (2ZZ-GE)	1.8893-1.8898	0.0006-0.0013	0.0016-0.0094	3	1.7713-1.7717	0.0011-0.0020	0.0063-0.0135
2004	1.8 (1ZZ-FE)	1.8893-1.8898	0.0006-0.0013	0.0008-0.0087	3	1.7320-1.7323	0.0011-0.0024	0.0063-0.0135
	1.8 (2ZZ-GE)	1.8893-1.8898	0.0006-0.0013	0.0016-0.0094	3	1.7713-1.7717	0.0011-0.0020	0.0063-0.0135
2005	1.8 (1ZZ-FE)	1.8893-1.8898	0.0006-0.0013	0.0016-0.0094	3	1.7320-1.7323	0.0011-0.0024	0.0063-0.0135
	1.8 (2ZZ-GE)	1.8893-1.8898	0.0006-0.0013	0.0016-0.0094	3	1.7713-1.7717	0.0011-0.0020	0.0063-0.0135

06025-VIBE-C06

PISTON AND RING SPECIFICATIONS

All measurements are given in inches.

| Year | Engine Displacement Liters (VIN) | Piston Clearance | Ring Gap | | | Ring Side Clearance | | |
			Top Compression	Bottom Compression	Oil Control	Top Compression	Bottom Compression	Oil Control
2003	1.8 (1ZZ-FE)	0.0026-0.0035	0.0098-0.0138	0.0138-0.0197	0.0059-0.0197	0.0009-0.0028	0.0012-0.0028	0.0012-0.0043
	1.8 (2ZZ-GE)	0.0003-0.0015	0.0098-0.0138	0.0138-0.0197	NA	0.0009-0.0028	0.0012-0.0028	NA
2004	1.8 (1ZZ-FE)	0.0026-0.0035	0.0098-0.0138	0.0138-0.0197	0.0059-0.0197	0.0009-0.0028	0.0012-0.0028	0.0012-0.0043
	1.8 (2ZZ-GE)	0.0003-0.0015	0.0098-0.0138	0.0138-0.0197	NA	0.0009-0.0028	0.0012-0.0028	NA
2005	1.8 (1ZZ-FE)	0.0039	0.0098-0.0138	0.0138-0.0197	0.0059-0.0197	0.0009-0.0028	0.0012-0.0028	0.0012-0.0043
	1.8 (2ZZ-GE)	0.0039	0.0098-0.0138	0.0138-0.0197	0.0059-0.0197	0.0009-0.0028	0.0012-0.0028	NA

NA - Not available

06025-VIBE-C07

TORQUE SPECIFICATIONS
All readings in ft. lbs.

Year	Engine Displacement Liters (VIN)	Cylinder Head Bolts	Main Bearing Bolts	Rod Bearing Bolts	Crankshaft Damper Bolts	Flywheel Bolts	Manifold Intake	Manifold Exhaust	Spark Plugs	Oil Pan Drain Plug
2003	1.8 (1ZZ-FE)	①	②	③	102	①	22	27	18	26
	1.8 (2ZZ-GE)	④	⑤	⑥	87	①	⑦	37	13	26
2004	1.8 (1ZZ-FE)	①	②	③	102	①	22	27	18	26
	1.8 (2ZZ-GE)	④	⑤	⑥	87	①	⑦	37	13	26
2005	1.8 (1ZZ-FE)	⑧	⑨	③	105	⑩	13	37	21	26
	1.8 (2ZZ-GE)	④	⑤	⑥	87	⑪	⑦	37	21	26

① Step 1: 36 ft. lbs.
Step 2: 90 degree turn

② 12 pointed bolts:
Step 1: 33 ft. lbs.
Step 2: 90 degree turn
Hex head bolts: 14 ft. lbs.

③ Step 1: 15 ft. lbs.
Step 2: 90 degree turn

④ Step 1: 26 ft. lbs.
Step 2: 180 degree turn

⑤ 12 pointed bolts:
Step 1: 16 ft. lbs.
Step 2: 32 ft. lbs.
Step 3: 45 degree turn
Step 4: 45 degree turn
Hex head bolts: 13 ft. lbs.

⑥ Step 1: 22 ft. lbs.
Step 2: 90 degree turn

⑦ Bolt A: 25 ft. lbs.
Bolt B: 34 ft. lbs.

⑧ Step 1: 18 ft. lbs.
Step 2: 36 ft. lbs.
Step 3: 36 ft. lbs. Plus a 90 degree turn

⑨ Step 1: 16 ft. lbs.
Step 2: 32 ft. lbs.
Step 2: 32 ft. lbs. Plus a 90 degree turn

⑩ Automatic transmission: 61 ft. lbs.
Manual transmission: 36 ft. lbs. plus a 90 degree turn

⑪ Automatic transmission: 65 ft. lbs.
Manual transmission: 36 ft. lbs. plus a 90 degree turn

06025-VIBE-C08

WHEEL ALIGNMENT

Year	Model		Caster Range (+/-Deg.)	Caster Preferred Setting (Deg.)	Camber Range (+/-Deg.)	Camber Preferred Setting (Deg.)	Toe-in (in.)
2003	Vibe - 2WD	F	0.75	+2.78	0.75	-0.77	0+/-0.08
		R	—	—	0.50	-1.45	0.11+/-0.11
	Vibe - 4WD	F	0.75	+2.77	0.75	-0.48	0+/-0.08
		R	—	—	0.75	-0.73	0.08+/-0.08
2004	Vibe - 2WD	F	0.75	+2.78	0.75	-0.77	0+/-0.08
		R	—	—	0.50	-1.45	0.11+/-0.11
	Vibe - 4WD	F	0.75	+2.77	0.75	-0.48	0+/-0.08
		R	—	—	0.75	-0.73	0.08+/-0.08
2005	Vibe - 2WD	F	0.75	+2.78	0.75	-0.77	0+/-0.08
		R	—	—	0.50	-1.45	0.11+/-0.11
	Vibe - 4WD	F	0.75	+2.77	0.75	-0.48	0+/-0.08
		R	—	—	0.75	-0.73	0.08+/-0.08

06025-VIBE-C09

TIRE, WHEEL AND BALL JOINT SPECIFICATIONS

| Year | Model | OEM Tires | | Tire Pressures (psi) | | Wheel Size | Ball Joint Inspection | Lug Nuts |
		Standard	Optional	Front	Rear			
2003	Vibe	205/55R16	—	33	33	6.5-JJ	9-26 in. ①	76
2004	Vibe	205/55R16	—	33	33	6.5-JJ	9-26 in. ①	76
2005	Vibe	205/55R16	—	33	33	6.5-JJ	9-26 in. ①	76

OEM: Original Equipment Manufacturer

PSI: Pounds Per Square Inch

STD: Standard

OPT: Optional

① Torque required in inch lbs. to rotate ball joint when removed from the knuckle

06025-VIBE-C10

BRAKE SPECIFICATIONS
All measurements in inches unless noted

| Year | Model | | Brake Disc | | | Brake Drum Diameter | | | Minimum Lining Thickness | Brake Caliper | |
			Original Thickness	Minimum Thickness	Maximum Runout	Original Inside Diameter	Max. Wear Limit	Maximum Machine Diameter		Bracket Bolts (ft. lbs.)	Mounting Bolts (ft. lbs.)
2003	Vibe	F	0.984	0.906	0.0020	—	—	—	0.039	79	25
		R	0.354	0.295	0.0059	9.00	—	9.04	0.039	34	—
2004	Vibe	F	0.984	0.906	0.0020	—	—	—	0.039	79	25
		R	0.354	0.295	0.0059	9.00	—	9.04	0.039	34	—
2005	Vibe	F	0.984	0.906	0.0020	—	—	—	0.039	79	25
		R	0.354	0.295	0.0059	9.00	—	9.04	0.039	34	—

F: Front

R: Rear

06025-VIBE-C11

SCHEDULED MAINTENANCE INTERVALS
PONTIAC—VIBE

TO BE SERVICED	TYPE OF SERVICE	VEHICLE MILEAGE INTERVAL (x1000)												
		7.5	15	22.5	30	37.5	45	52.5	60	67.5	75	82.5	90	97.5
Engine oil & filter	R	✓	✓	✓	✓	✓	✓	✓	✓	✓	✓	✓	✓	✓
Drive belts	S/I								✓	✓	✓	✓	✓	✓
Automatic transaxle fluid & filter	S/I		✓		✓		✓		✓		✓		✓	
Ball joints & dust covers	S/I		✓		✓		✓		✓		✓		✓	
Bolts & nuts on body & chassis	S/I		✓		✓		✓		✓		✓		✓	
Brake line pipes & hoses	S/I		✓		✓		✓		✓		✓		✓	
Brake linings & drums	S/I		✓		✓		✓		✓		✓		✓	
Brake pads & discs (front & rear if equipped)	S/I		✓		✓		✓		✓		✓		✓	
Differential oil	S/I		✓		✓		✓		✓		✓		✓	
Drive shaft boots (except Supra)	S/I		✓		✓		✓		✓		✓		✓	
Manual transaxle oil	S/I		✓		✓		✓		✓		✓		✓	
Steering gear housing oil	S/I		✓		✓		✓		✓		✓		✓	
Steering linkage	S/I		✓		✓		✓		✓		✓		✓	
Air filter	R				✓				✓				✓	
Spark plugs	R				✓				✓				✓	
Spark plugs (platinum tip)	R								✓					
Exhaust system	S/I				✓				✓					
Fuel lines & connections	S/I				✓				✓				✓	
Valve clearance	S/I				✓				✓				✓	
Engine coolant	R						✓				✓			
Fuel tank cap gasket	R								✓					
Charcoal canister	S/I								✓					

R: Replace   S/I: Service or Inspect

FREQUENT OPERATION MAINTENANCE (SEVERE SERVICE)

If a vehicle is operated under any of the following conditions it is considered severe service:

- Extremely dusty areas.

- 50% or more of the vehicle operation is in 32°C (90°F) or higher temperatures, or constant operation in temperatures below 0°C (32°F).

- Prolonged idling (vehicle operation in stop and go traffic).

- Frequent short running periods (engine does not warm to normal operating temperatures).

- Police, taxi, delivery usage or trailer towing usage.

Oil & oil filter: change every 6000 miles.

Bolts & nuts on chassis & body: tighten every 7500 miles.

Ball joints & dust covers: service or inspect every 12,000 miles.

Brake linings & drums: service or inspect ever 12,000 miles.

Brake pads & discs (front & rear if equipped): service or inspect every 12,000 miles.

Drive shaft boots & except Supra): service or inspect every 12,000 miles.

Steering linkage: service or inspect every 12,000 miles.

Air filter: service or inspect every 15,000 miles.

Exhaust system: service or inspect every 15,000 miles.

Timing belt: replace every 60,000 miles.

06025-VIBE-C12

ENGINE REPAIR

Alternator

REMOVAL & INSTALLATION

1. Before servicing the vehicle, refer to the precautions section.
2. Remove or disconnect the following:
   - Negative battery cable
   - Drive belt
   - Wire clamp from the clip on the end frame
   - Rubber clamp and nut
   - Alternator wiring and connector
   - Alternator

To install:

3. Install or connect the following:
   - Alternator. Torque the 12mm bolt to 18 ft. lbs. (25 Nm) and the 14mm bolt to 39 ft. lbs. (54 Nm).
   - Alternator connector and wiring
   - Rubber clamp and nut
   - Wire clamp
   - Drive belt
   - Negative battery cable

Ignition Timing

ADJUSTMENT

➡ **The timing on engines equipped with Distributorless Ignition Systems (DIS) is not adjustable.**

Engine Assembly

REMOVAL & INSTALLATION

1. Before servicing the vehicle, refer to the precautions section.
2. Relieve the fuel system pressure.
3. Drain the cooling system.
4. Drain the engine oil.
5. Drain the transaxle fluid and transfer fluid, if equipped.
6. Remove or disconnect the following:
   - Negative battery cable. Wait at least 90 seconds before proceeding.
   - Battery
   - Hood
   - Undercovers
   - Radiator inlet and outlet hoses
   - Radiator hose outlet
   - Oil cooler inlet and outlet tubes
   - Upper radiator support and radiator, if equipped with A/C
   - Battery
   - Air cleaner assembly
   - Fuel pipe clamp

- Fuel tube sub-assembly
- Accelerator control cable
- Cruise control actuator, if equipped
- Union-to-connector tube hose
- Heater inlet and outlet hoses
- Transmission shift cable(s)
- Clutch release cylinder, on manual transaxle
- Glove compartment door
- Engine relay block cover
- 3 connectors from the relay block
- 2 ground cables
- Engine wire from the Engine Control Module (ECM) and junction block
- Engine wire from the cabin
- Drive belt
- Compressor and magnetic clutch, if equipped with A/C. Unbolt and position aside, DO NOT disconnect the lines.
- Vane pump oil reservoir from the bracket
- Return tube
- Right side front door scuff plate
- Right side cowl side trim plate
- Right side rear door scuff plate, AWD
- Lower right side center pillar garnish, AWD
- Right front seat, AWD
- Column hole cover silencer sheet
- Steering intermediate shaft
- Front floor panel brace, FWD
- Center exhaust pipe, AWD
- Propeller shaft with center bearing shaft, AWD
- Front exhaust pipe
- Front hub nuts
- Tie rod ends from the steering knuckles

7. Separate the front stabilizer links and lower control arm ball joints
   - Front halfshafts
8. Remove the engine from the vehicle, as follows:
   a. Set the engine lifter.
   b. Remove the bolts and nuts, then remove the engine mounting insulator.
   c. Remove the through bolt and nut, then detach the engine mounting insulator from the vehicle.
   d. Remove the 6 bolts as shown.
   e. Use a suitable tool to suspend the engine assembly, as shown in the figure.
   f. No. 1 engine hanger: P/N 12281-15040 (1ZZ-FE), 12281-88600 (2ZZ-GE).
   g. No. 2 engine hanger: P/N 12281-22021 (1ZZ-FE), 12281-88600 (2ZZ-GE).

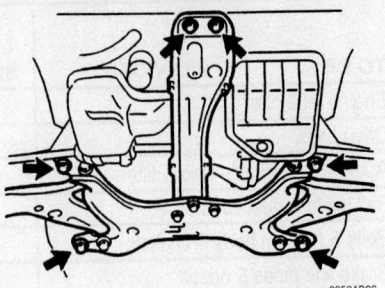

Remove the 6 bolts, as indicated by arrows

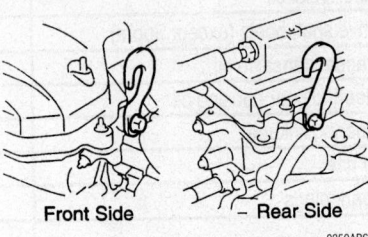

Install the engine hangers—1ZZ-FE shown, 2ZZ-GE similar

   h. Bolt: P/N 91512-B1016.
   i. Torque the bolts to 28 ft. lbs. (38 Nm).

✳✳ CAUTION

Do not try to suspend the engine by hooking the chain to any other part.

   j. Attach an engine chain hoist to the hangers.
   k. Using the chain block and sling device, suspend the engine.
   l. Remove the engine and transaxle assembly from the vehicle.
9. Remove or disconnect the following components, as necessary:
   - Vane pump
   - Steering gear
   - Crossmember
   - Manifold stay
   - Oxygen (O_2) sensor
   - Exhaust manifold
   - Starter
   - Transaxle
   - Transfer case
   - Clutch
   - Flywheel
   - Alternator
   - Ignition coil
   - Fuel delivery pipe
   - Intake manifold
   - Oil level gauge
   - Water inlet and bypass pipes

- Thermostat
- Oil pressure switch
- Crankshaft Position (CKP) sensor
- Knock Sensor (KS)
- Drive belt tensioner
- Engine mounts and brackets
- Coolant Temperature Sensor (CTS)

To install:

10. Install any removed components to the engine and transaxle assembly.

11. To install the engine:

a. Place the engine and transaxle on an engine lifter.

b. Install the engine with the transaxle to the vehicle.

c. Temporarily install the crossmember and 6 bolts.

d. Install the left engine mounting insulator. Tighten the bolts to 59 ft. lbs. (80 Nm).

e. Install the right engine mounting insulator. Tighten the bolts to 38 ft. lbs. (52 Nm).

f. Insert SST 09670-00010 to the positioning holes of the right handle crossmember and on the right handle of the vehicle. Temporarily tighten bolt A, then bolt B.

g. Insert SST 09670-00010 to the positioning holes of the left handle crossmember and on the left handle of

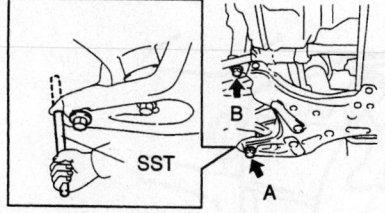

9359AB68

Insert the SST to the positioning holes of the right handle crossmember and on the right handle of the vehicle. Temporarily tighten bolt A, then bolt B

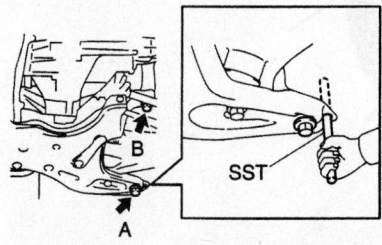

9359AB69

Insert the SST to the positioning holes of the left handle crossmember and on the left handle of the vehicle. Temporarily tighten bolt A, then bolt B

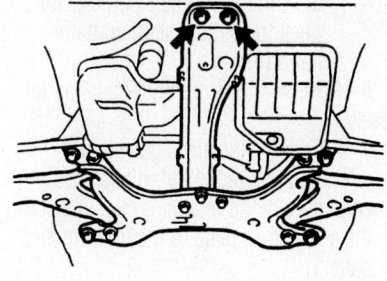

9359AB70

Tighten the 2 crossmember bolts, indicated by arrows

the vehicle. Temporarily tighten bolt A, then bolt B.

h. Insert the SST to the positioning holes on the right-handle crossmember and right handle. Tighten bolt A to 116 ft. lbs. (157 Nm) and bolt B to 83 ft. lbs. (113 Nm).

i. Insert the SST to the positioning holes on the left-handle crossmember and left handle. Tighten bolt A to 116 ft. lbs. (157 Nm) and bolt B to 83 ft. lbs. (113 Nm).

j. Tighten the 2 crossmember bolts, shown in the figure, to 29 ft. lbs. (39 Nm).

12. Installation of the remaining components is the reverse of the removal procedure.

13. Make sure all fluid levels are accurate, then start the engine check for leaks.

Water Pump

REMOVAL & INSTALLATION

1ZZ-FE Engine

1. Before servicing the vehicle, refer to the precautions section.

2. Drain the cooling system.

3. Remove or disconnect the following:
- Negative battery cable
- Right-hand engine under cover
- Drive belt
- Alternator
- Water pump

To install:

4. Install or connect the following:
- Water pump. Torque bolts marked **A** (short) to 80 inch lbs. (9 Nm) and bolts marked **B** (long) to 96 inch lbs. (11 Nm).
- Alternator
- Drive belt
- Right engine under cover
- Negative battery cable

5. Fill the cooling system to the proper level.

6. Start the vehicle, check for leaks and repair if necessary.

2ZZ-GE Engine

1. Before servicing the vehicle, refer to the precautions section.

2. Drain the cooling system.

3. Remove or disconnect the following:
- Negative battery cable
- Right-hand engine under cover

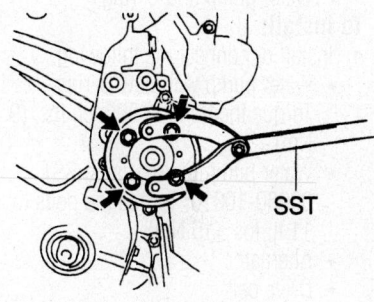

9359AB02

View of the special tool needed to remove and install the water pump pulley

7923VG06

Water pump bolt identification—1.8L (1ZZ-FE) engine

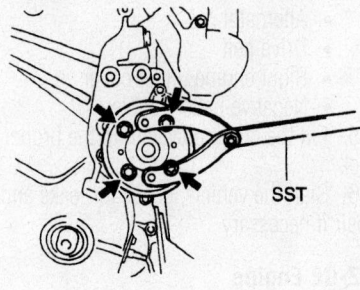

Water pump mounting and bolt locations—2ZZ-GE engine

- Drive belt
- Alternator
- Water pump pulley, using SST 09960-10010
- Water pump and O-ring

To install:

4. Install or connect the following:
- Water pump with new O-ring. Torque the bolts to 80 inch lbs. (9 Nm).
- Water pump pulley, using SST 09960-10010. Torque the bolts to 11 ft. lbs. (15 Nm).
- Alternator
- Drive belt
- Right engine under cover
- Negative battery cable

5. Fill the cooling system to the proper level.

6. Start the vehicle, check for leaks and repair if necessary.

Heater Core

REMOVAL & INSTALLATION

1. Before servicing the vehicle, refer to the precautions section.

2. Drain the cooling system.

3. Discharge and recover the A/C system refrigerant using approved equipment.

4. Remove or disconnect the following:
- Negative battery cable
- Heater hoses from the core
- Evaportor inlet and outlet tubes

from the evaporator and cap the lines to avoid system contamination

5. Remove the instrument panel as follows:

a. Disable the air bag system.

b. Using a taped flat–bladed tool, carefully pry the retaining clips attaching the center trim plate to the instrument panel.

c. Disconnect the A/C switch, hazard switch; rear defogger switch and passenger seat belt indicator switch electrical connections.

d. Remove the radio retaining screws, clamp from the radio bracket, slide the radio forward to disconnect the power and antenna connections. Remove the radio.

e. Remove the A/C switch and screw.

f. Remove the hazard switch.

g. Remove the rear defogger switch.

h. Remove the manual transmission shift knob.

i. Using a taped flat–bladed tool, carefully pry the retaining clips attaching the front floor console trim plate to the floor console assembly.

j. Disconnect the 2 cigar lighter connectors.

k. Disconnect the accessory power receptacle connectors.

l. Remove the cigar lighters and power receptacle.

m. Place both wheels in the straight ahead position.

n. Remove the bolts from the steering wheel module.

o. Release the connector Position Assurance (CPA) from the inflator module.

p. Disconnect the steering wheel module connectors.

q. Remove the steering wheel module.

r. Matchmark the steering wheel nut–to–shaft position, then remove the steering wheel nut and the wheel.

s. Remove the upper and lower steering column cover screws and the covers.

t. Disconnect the turn signal/headlamp assembly connectors.

u. Remove the turn signal/headlamp switch assembly

v. Remove the wiper switch by depressing the tab.

w. Remove the glove box.

x. Disconnect the instrument panel connector.

y. Remove the instrument panel module connectors and the passenger air bag assembly.

z. Remove the cluster trim plate by disengaging the clips.

aa. Remove the cluster screw and disengage the 2 lower clips.

bb. Disconnect the cluster electrical connectors and remove the cluster.

cc. Remove the windshield garnish moldings.

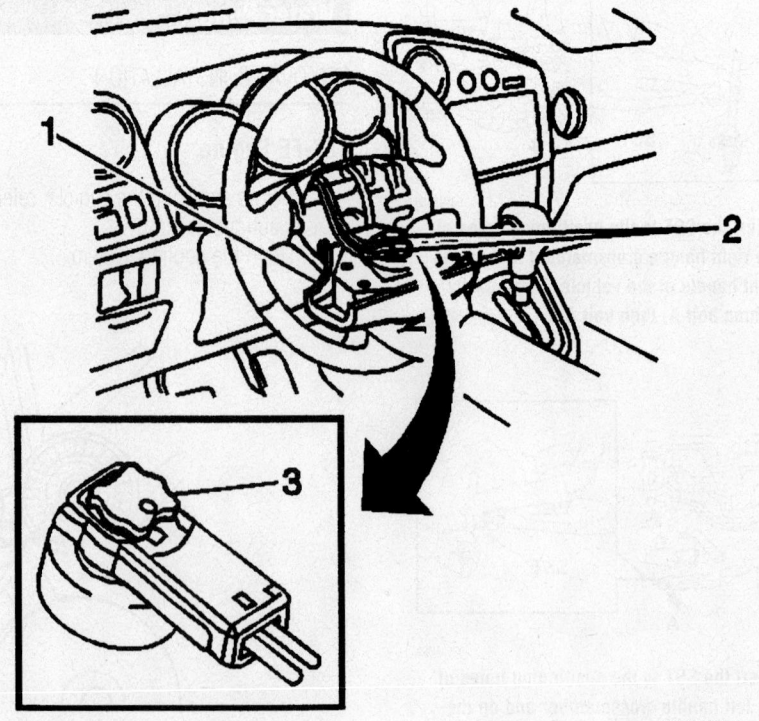

Exploded view of the CPA assembly

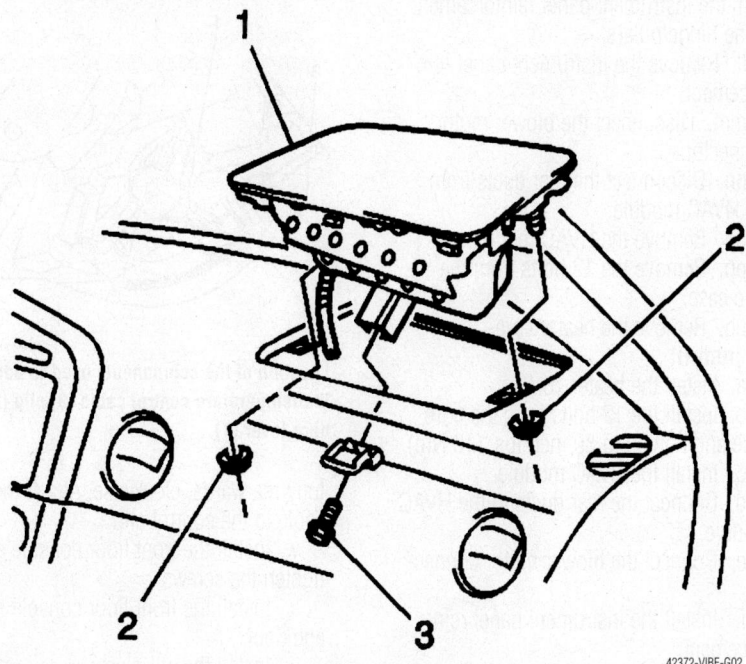

42372-VIBE-G02

Remove the instrument panel module connectors and the passenger air bag assembly

dd. Using a taped flat–bladed tool, carefully pry the retaining clips attaching the instrument panel left trim plate to the instrument panel.

ee. Disconnect the power mirror and dimmer switch connectors.

ff. Remove the power mirror and dimmer switches.

gg. Disconnect any remaining electrical connections.

hh. Remove the upper instrument panel screws and the panel by pulling towards the rear to disengage the tabs.

ii. Disconnect the steering wheel coil connector.

jj. Release the 3 claws and remove the coil assembly.

kk. If the vehicle is equipped with an automatic transmission, insert the key into the cylinder, turn to the ACC position, push in the release butt, disconnect the park lock cable, remove the key from the cylinder and lock the steering wheel.

ll. Move the silencer pad from the column.

mm. Matchmark the steering shaft coupling to the shaft.

nn. Loosen the upper bolt on the coupling.

oo. Remove the lower bolt from the coupling.

pp. Move the coupling onto the column shaft.

qq. Disconnect the wiring harness clamps from the column.

rr. Remove the 3 bolts and the column.

ss. Remove the body hinge trim panels.

tt. Remove the sill plates.

uu. Remove the front floor console storage door.

vv. Remove the screws attaching the console to the instrument panel, pull the console rewards and up and remove the front floor console.

ww. Remove the HVAC retaining screw; disconnect the electrical connectors and module control, temperature control and A/C cables. Remove the unit.

xx. Push in the clip and disconnect the cable from the manual selector shifter assembly.

yy. Using a suitable prytool, disconnect the park lock cable from the bracket.

zz. Disconnect the shift select cable from the manual selector lever.

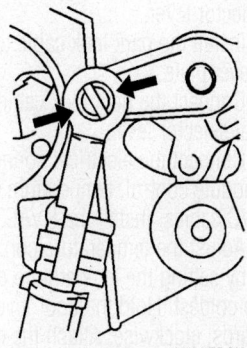

42372-VIBE-G03

Push in the clip and disconnect the cable from the manual selector shifter assembly

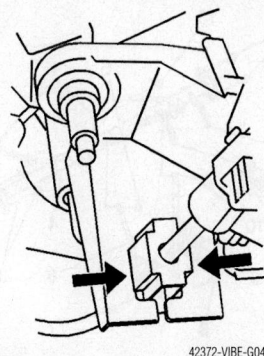

42372-VIBE-G04

Using a suitable prytool, disconnect the park lock cable from the bracket

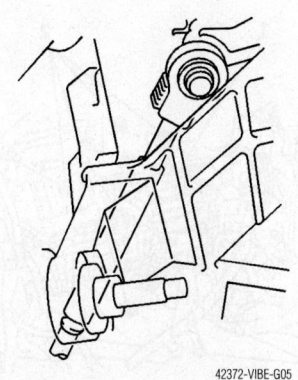

42372-VIBE-G05

Disconnect the shift select cable from the manual selector lever

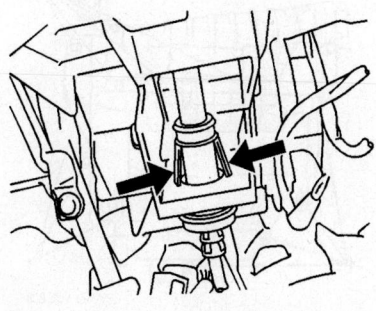

42372-VIBE-G06

Using a suitable prytool, disconnect the shift select cable from the shift lever plate

aa. Using a suitable prytool, disconnect the shift select cable from the shift lever plate.

bb. Disconnect the electrical connectors and the wire harness clip.

cc. Remove the nuts from the selector and remove the selector.

dd. Disonnect the hood release cable from the release handle.

ee. Remove the 8 bolts, 4 push retainers and the wire harness clamps from the lower instrument panel and remove the panel.

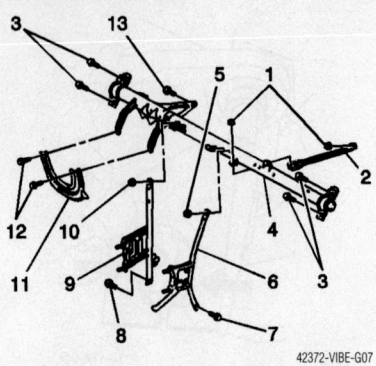

Exploded view of the instrument panel reinforcement

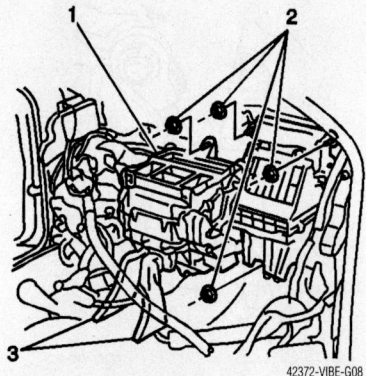

Remove the HVAC module

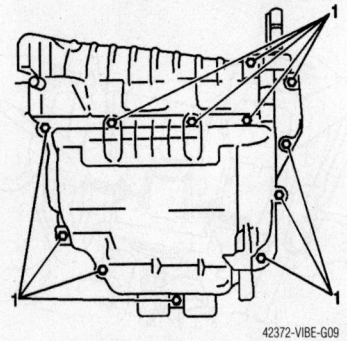

Remove the 12 bolts from the core case to access the heater core

ff. Disenage the wiring harness clips, remove the bolts retaining the lower instrument panel pad and the pad.

gg. Remove the ground cable.

hh. Remove the connector housing bracket from the right instrument panel center support brace.

ii. Remove the left brace nut, right brace nut, left brace bolt, right brace bolt, left center support brace and right center support brace.

jj. Remove the windshield defroster nozzle duct from the heater case.

kk. Remove the 5 bolts and the nuts

from the instrument panel reinforcement at the hinge pillars.

ll. Remove the instrument panel reinforcement.

mm. Disconnect the blower motor connector.

nn. Disconnect the rear ducts from the HVAC module.

oo. Remove the HVAC module.

pp. Remove the 12 bolts from the core case.

qq. Remove the heater core.

To install:

a. Install the heater core.

b. Install the 12 bolts from the core case and tighten to 89 inch lbs. (10 Nm).

c. Install the HVAC module.

d. Connect the rear ducts to the HVAC module.

e. Connect the blower motor connector.

f. Install the instrument panel reinforcement.

g. Install the 5 bolts and the nuts to the instrument panel reinforcement at the hinge pillars. Tighten to 21 ft. lbs. (28 Nm).

h. Install the windshield defroster nozzle duct to the heater case.

i. Install left center support brace and right center support brace. Tighten the nuts and bolt to 15 ft. lbs. (20 Nm).

j. Install the connector housing bracket.

k. Connect the ground cable.

l. Install the lower instrument panel pad and the bolts and attach the wiring harness clips.

m. Connect the hood release cable to the release handle.

n. Install the lever and tighten the nuts to 12 ft. lbs. (18 Nm).

o. Connect the manual selector electrical connections.

p. Attach the shift cable to the shift lever plate.

q. Connect the shift select cable to the selector lever.

r. Install the park lock cable to the shift lever plate.

s. Connect the park lock cable to the manual selector lever.

t. Connect the electrical connectors and module control, temperature control and A/C cables. Install the HVAC unit.

u. Adjust the temperature control cable by setting the temperature control dial to coldest. Hold the door lever fully rearwards, clockwise. Attach the cable to the control clip.

v. Adjust the Mode linkage by setting the dial to defrost. Hold the door lever

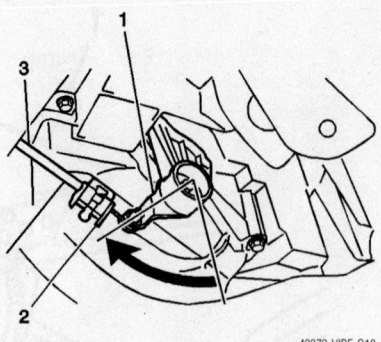

Location of the components used to adjust the temperature control cable (3) clip (2), door lever (1).

fully rearwards, clockwise. Attach the cable to the control clip.

w. Install the front floor console and tighten the screws.

x. Install the front floor console storage door.

y. Install the sill plates.

z. Install the column.

aa. Install the 3 bolts and tighten the lower bolt to 16 ft. Lbs. (21 Nm) and the 2 upper bolts to 16 ft. Lbs. (21 Nm).

bb. Align the matchmarks made prior to removal.

cc. Lower the coupling onto the shaft. Install the bolts and tighten to 26 ft. lbs. (35 Nm).

dd. Connect the wiring harness clamps to the column.

ee. Move the silencer pad to the column.

ff. If the vehicle is equipped with an automatic transmission, insert the key into the cylinder, turn to the ACC position, insert the park lock cable making sure the release button engages. Make sure the key will not rotate to the lock position unless the shifter is in the park position, remove the key from the cylinder and lock the steering wheel.

gg. Install the body hinge trim panels.

hh. Make sure the turn signal switch is in the neutral position.

ii. If installing a new coil, remove the lock pin.

jj. Install the coil making sure the 3 claws engage.

kk. While holding the coil casing, turn the coil center casing counterclockwise until the coil reaches its stop.

ll. Turn the coil center casing clockwise 2 ½ turns.

mm. Align the center casing with the arrow on the outer casing.

nn. Connect the coil electrical connector.

oo. Install the upper instrument panel and screws.

pp. Connect the electrical connections.

qq. Install the power mirror and dimmer switches.

rr. Connect the power mirror and dimmer switch connectors.

ss. Install the instrument panel left trim plate to the instrument panel.

tt. Install the windshield garnish moldings.

uu. Connect the cluster electrical connectors and install the cluster.

vv. Enage the cluster lower clips and install the screw.

ww. Install the cluster trim plate.

xx. Install the passenger air bag assembly and the instrument panel module connectors.

yy. Connect the instrument panel connector.

zz. Install the glove box.

aa. Install the wiper switch.

bb. Install the turn signal/headlamp switch assembly

cc. Connect the turn signal/headlamp assembly connectors.

dd. Install the upper and lower steering column covers and screws.

ee. Install the steering wheel and nut aligning the matchmarks made prior to removal and tighten the nut to 37 ft. lbs. (50 Nm).

ff. Connect the steering wheel module connectors.

gg. Install the CPA to the inflator module.

hh. Install the steering wheel module and tighten the retainers 78 inch lbs. (9 Nm)

ii. Install cigar lighters and power receptacle.

jj. Connect the accessory power receptacle connectors.

kk. Connect the 2 cigar lighter connectors.

ll. Install the front floor console trim plate to the floor console assembly.

mm. Install the manual transmission shift knob.

nn. Install the rear defogger switch.

oo. Install the hazard switch.

pp. Install the A/C switch and screw.

qq. Install the radio.

rr. Connect the A/C switch, hazard switch, rear defogger switch and passenger seat belt indicator switch electrical connections.

ss. Install the center trim plate to the instrument panel.

tt. Connect the evaporator inlet and outlet tubes.

uu. Connect the heater hoses to the core.

vv. Connect the negative battery cable.

ww. Recharge the A/C system and fill the cooling system.

Cylinder Head

REMOVAL & INSTALLATION

1. Before servicing the vehicle, refer to the precautions section.

2. Drain the cooling system.

3. Remove or disconnect the following:
- Right side engine under cover
- Right front wheel and tire
- Cylinder head cover
- Air cleaner assembly with hose
- Accelerator control cable
- Wire harness clamp and suction hose assembly, 2ZZ-GE engine only
- Water bypass hoses
- Fuel pipe clamp
- Fuel tube sub-assembly
- Union-to-connector tube hose
- Radiator and heater inlet hoses
- Drive belt

4. Separate the vane pipe assembly, but do not disconnect the hose, 1ZZ-FE engine.
- Alternator bracket, 2ZZ-GE
- Alternator

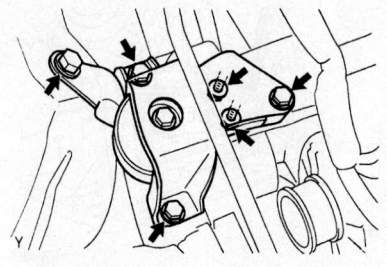

With the engine supported, remove the right side engine mount—1ZZ-FE engine shown, 2ZZ-GE similar

5. Separate the compressor and magnetic clutch, on 2ZZ-GE engines with air conditioning.
- Front exhaust pipe assembly
- Power steering pump reservoir and position it aside, 1ZZ-FE engine

6. Place a jack with a wooden block under the vehicle for support, then remove the 4 bolts and 2 nuts and remove the right side engine mount.

7. Remove the engine wire, on 1ZZ-FE engines as follows:

a. Remove the 5 clamps from the brackets.

b. Detach the connectors.

c. Remove the ignition coil connectors.

d. Bolt and nut holding the engine wire.

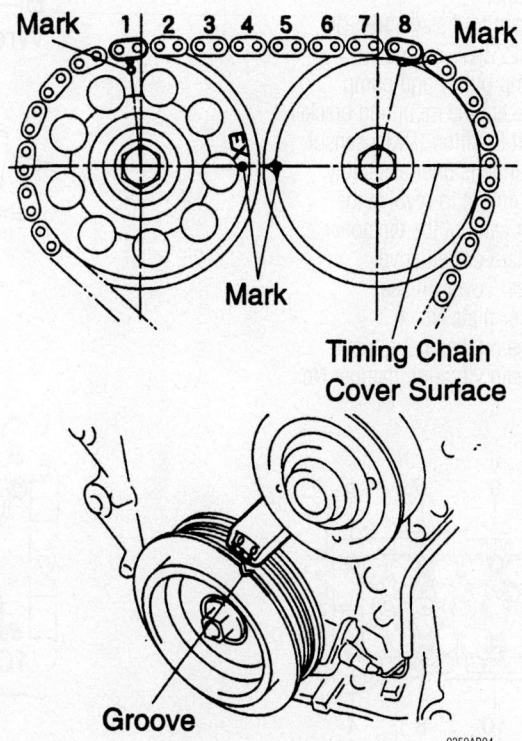

Proper timing mark alignment for TDC

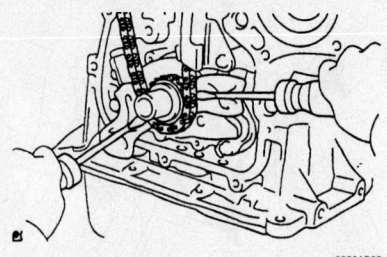

Remove the timing chain with the crankshaft gear

8. Remove or disconnect the following:
- Ignition coil assembly
- Positive Crankcase Ventilation (PCV) hoses
- Valve (cylinder head) cover sub-assembly

9. Set the No. 1 cylinder to Top Dead Center (TDC) of the compressor stroke as follows:

a. Turn the crankshaft pulley, and align its groove with the "0" timing mark of the timing chain cover.

b. Make sure the point marks of the camshaft timing sprockets and Variable Valve Timing (VVT) timing sprockets are in a straight line as shown. If not, turn the crankshaft 1 complete revolution (360°) and align the marks.

10. Remove or disconnect the following:
- Crankshaft pulley, using SST 09960-10010
- Belt tensioner
- Exhaust manifold stay and head insulator, 2ZZ-GE engine
- Water pump pulley and pump
- Transverse engine mounting bracket
- Crankshaft Position (CKP) sensor
- No. 1 chain tensioner assembly, making sure not to revolve the crankshaft without the tensioner
- Timing chain or belt cover
- Timing gear cover oil seal
- CKP sensor plate No. 1
- Timing chain tensioner slipper
- Timing chain vibration damper No. 1

➡️ **In case you turn the camshafts with the timing chain removed, turn the crankshaft ¼ turn for the valve to avoid contact with the pistons.**

- Timing chain sub-assembly. Remove the chain with the crankshaft gear, using screwdrivers as shown.
- Surge tank stay, 2ZZ-GE engine
- Intake manifold
- Oil level gauge
- Water bypass pipe bolts and pipe, 1ZZ-FE engine
- Camshafts
- Camshaft timing oil control valve, 1ZZ-FE engine
- Manifold stay, 1ZZ-FE engine
- Cylinder head bolts in sequence. To prevent damage to the cylinder head, loosen each bolt about ¼ of a turn during each pass until the bolts are loose.
- Cylinder head

To install:

11. Clean and degrease the surface of the cylinder head and engine block.

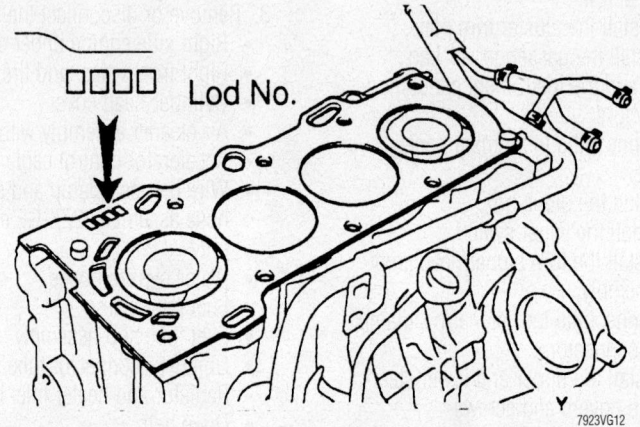

Position the head gasket correctly on the cylinder head—1.8L (1ZZ-FE) engine

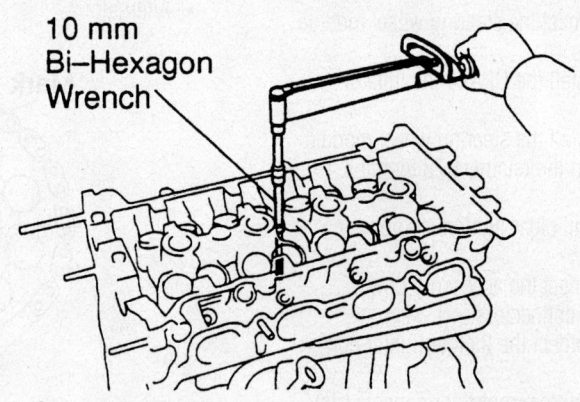

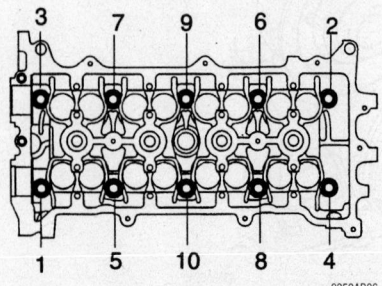

Cylinder head bolt loosening sequence

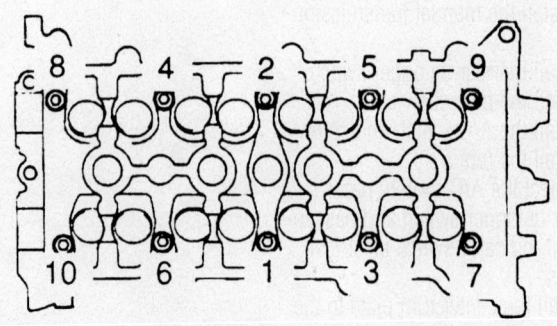

Cylinder head bolt tightening sequence—1ZZ-FE and 2ZZ-GE engines

12. Check the length of the cylinder head bolts. They should be 5.780–5.835 in. (146.8–148.2mm) long. If they are longer than 5.846 in. (148.5mm), they must be replaced.

13. Install or connect the following:
- New gasket on the engine block with the Lot No. stamp facing up.
- Cylinder head
- Apply a light coat of oil to cylinder head bolt threads;

14. On 1ZZ-FE engines tighten the bolts as follows:
    a. Step 1: 36 ft. lbs. (49 Nm)
    b. Step 2: Tighten an additional 90 degrees.

15. On 2003–04 2ZZ-GE engines tighten the bolts as follows:
    a. Step 1: 26 ft. lbs. (35 Nm)
    b. Step 2: Tighten an additional 180 degrees.

16. On 2005 2ZZ-GE engines tighten the bolts as follows:
    a. Step 1: 16 ft. lbs. (25 Nm)
    b. Step 2: 36 ft. lbs. (49 Nm)
    c. Step 3: 36 ft. lbs. (49 Nm), plus an additional 90 degrees.

17. Install or connect the following:
- Manifold stay, 1ZZ-FE engine. Tighten the bolts to 36 ft. lbs. (49 Nm).
- Camshaft timing oil control valve, on 1ZZ-FE engines, and tighten to 80 inch lbs. (9 Nm)
- Camshaft
- Water by-pass pipe, on 1ZZ-FE engines, and tighten to 80 inch lbs. (9 Nm)
- Oil level gauge
- Intake manifold
- Surge tank stay, 2ZZ-GE engine. Tighten to 18 ft. lbs. (24 Nm).
- Timing chain
- Timing chain vibration damper. Tighten the bolts to 80 inch lbs. (9 m).
- Timing chain tensioner slipper and tighten the bolt to 14 ft. lbs. (19 Nm).
- Crankshaft position sensor plate, with the "F" mark facing forward.
- Timing gear cover oil seal
- Timing cover. For 1ZZ-FE engine, tighten the "A" bolts to 10 ft. lbs. (13 Nm), the "B" bolts to 14 ft. lbs. (19 Nm) and the stud bolt to 84 inch lbs. (9.5 Nm), using a Torx® wrench. For 2ZZ-GE engines, tighten the M8 bolts to 15 ft. lbs. (21 m), the M6 bolts to 8 ft. lbs. (11 Nm) and the stud bolt to 84 inch lbs. (9.5 Nm).

➡**When installing the tensioner, make sure to set the hook again if the hook releases the plunger.**

- Timing chain tensioner. Torque the nuts to 80 inch lbs. (9 Nm).
- CKP sensor and tighten the bolts to 80 inch lbs. (9 Nm)
- Transverse engine mounting bracket. Tighten the bolts to 35 ft. lbs. (47 Nm).
- Water pump and pulley
- Exhaust manifold stay and heat insulator, 2ZZ-GE engine
- Belt tensioner. Tighten the nut to 21 ft. lbs. (29 Nm) and the bolt to 51 ft. lbs. (69 Nm) on 1ZZ-FE engines or to 74 ft. lbs. (100 Nm) on 2ZZ-GE engines.

18. Install the crankshaft pulley, as follows:
    a. Align the pulley set key with the key groove of the pulley and slide on the pulley.

    b. Use SST 09960-11010 to install the bolt and tighten to 102 ft. lbs. (138 Nm) for 1ZZ-FE engine or to 87 ft. lbs. (118 Nm) on 2ZZ-GE engines.

    c. Turn the crankshaft counterclockwise and disconnect the plunger knock pin from the hook.

    d. Turn the crankshaft clockwise and check that the slipper is pushed by the plunger. If the plunger does not spring out, press the slipper into the chain tensioner with a screwdriver so that the hook is released from the knock pin and the plunger springs out.

19. Install or connect the following:
- Cylinder head sub-assembly cover. Install seal packing into the locations shown and install within 3 minutes. Tighten the "A" bolts to 8 ft. lbs. (11 Nm) and the "B" bolts to 80 inch lbs. (9 Nm) for 1ZZ-FE engines. For 2ZZ-GE engines, tighten the bolts to 7 ft. lbs. (10 Nm).

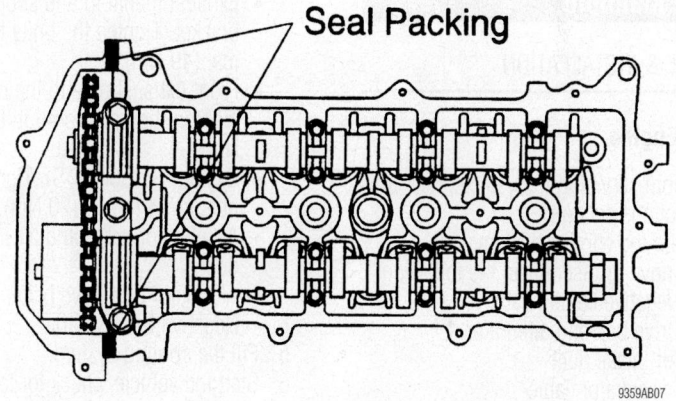

Seal packing installation locations

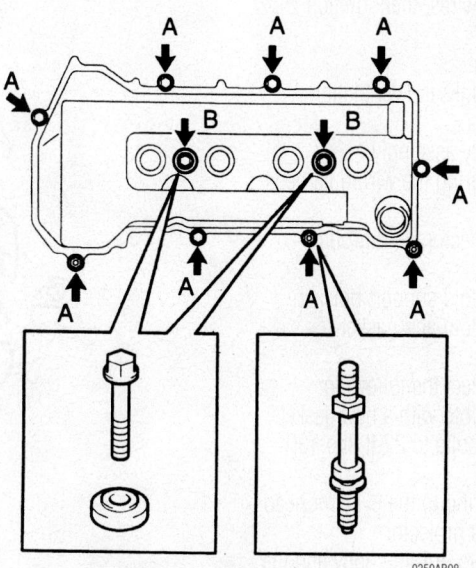

Cylinder head (valve) cover bolt locations—1ZZ-FE engine

- Ignition coil assembly. Torque the bolts to 80 inch lbs. (9 Nm).
- Engine wire and tighten to 80 inch lbs. (9 Nm), 1ZZ-FE
- Right side engine mount. Tighten to 38 ft. lbs. (52 Nm).
- Front exhaust pipe
- Vane pump, 1ZZ-FE
- Compressor and magnetic clutch, 2ZZ-GE
- Alternator bracket, 2ZZ-GE engine
- Alternator
- Suction hose and wire harness clamp, 2ZZ-GE engine
- Air cleaner and hose
- Main cylinder head cover and tighten to 62 inch lbs. (7 Nm)
- Right front wheel and tire. Tighten the lug nuts to 76 ft. lbs. (103 Nm).

20. Fill the cooling system to the proper level.

21. Start the vehicle, check for leaks and repair if necessary.

Intake Manifold

REMOVAL & INSTALLATION

1ZZ-FE Engine

1. Before servicing the vehicle, refer to the precautions section.
2. Drain the cooling system.
3. Remove or disconnect the following:
   - Negative battery cable
   - Drive belt and alternator
   - Air intake duct
   - Accelerator cable
   - Exhaust pipe from the manifold.
   - Exhaust manifold support bracket
   - Spark plug wires, then ignition coils
   - Spark plugs
   - Positive Crankcase Ventilation (PCV) hoses
   - Throttle body assembly
   - 2 bolts securing the wiring harness protector
   - Wiring connectors and ground wires
   - Intake manifold support bracket
   - Intake manifold and gasket

To install:

4. Install or connect the following:
   - Intake manifold with a new gasket. Torque the bolts to 22 ft. lbs. (30 Nm).
   - Harness wiring to the cylinder head and harness protector
   - Fuel injectors, throttle body and the PCV hoses

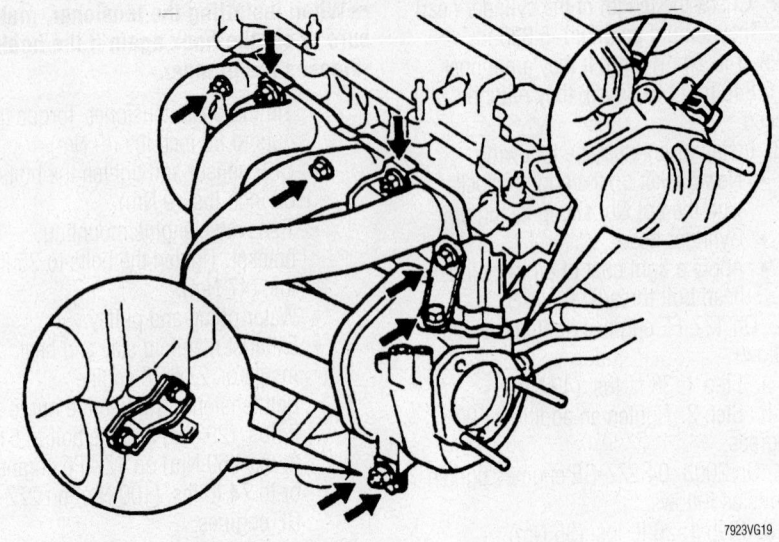

Intake manifold mounting fastener locations—1.8L (1ZZ-FE) engine

7923VG19

- Spark plugs and ignition coils. Tighten the bolts and nuts to 80 inch lbs. (9 Nm).
- Exhaust manifold and support bracket. Tighten the bolts to 37 ft. lbs. (49 Nm).
- Front exhaust pipe to the manifold. Tighten the bolts to 46 ft. lbs. (62 Nm).
- Oxygen Sensor (O_2S). Tighten the nuts to 14 ft. lbs. (20 Nm).
- Accelerator cable and air intake duct
- Alternator and drive belt
- Negative battery cable

5. Fill the cooling system.

6. Start the vehicle, check for leaks and repair if necessary.

2ZZ-GE Engine

1. Before servicing the vehicle, refer to the precautions section.
2. Drain the cooling system.
3. Remove or disconnect the following:
   - Negative battery cable
   - Drive belt and alternator
   - Air intake duct
   - Accelerator cable
   - Spark plug wires, then ignition coils
   - Spark plugs
   - Positive Crankcase Ventilation (PCV) hoses
   - Throttle body assembly
   - Wiring harness
   - Hoses and tubes connected to the head

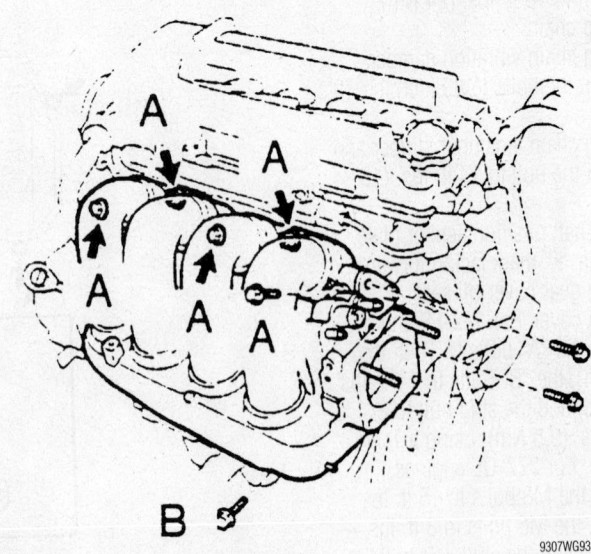

Intake manifold bolt installation—2ZZ-GE engine

9307WG93

- Intake manifold support bracket
- Intake manifold and gasket

To install:

4. Install or connect the following:
  - Intake manifold with a new gasket. Tighten bolts A to 25 ft. lbs. (34 Nm) and bolt B to 34 ft. lbs. (46 Nm).
  - Harness wiring to the cylinder head and harness protector
  - Fuel injectors, throttle body and the PCV hoses
  - Spark plugs and ignition coils. Tighten the bolts and nuts to 80 inch lbs. (9 Nm).
  - Oxygen Sensor (O_2S). Tighten the nuts to 14 ft. lbs. (20 Nm).
  - Accelerator cable and air intake duct
  - Alternator and drive belt
  - Negative battery cable

5. Fill the cooling system.

6. Start the vehicle, check for leaks and repair if necessary.

Exhaust Manifold

REMOVAL & INSTALLATION

1ZZ-FE Engine

1. Before servicing the vehicle, refer to the precautions section.

2. Drain the cooling system.

3. Remove or disconnect the following:
  - Negative battery cable
  - Drive belt and alternator
  - Air intake duct

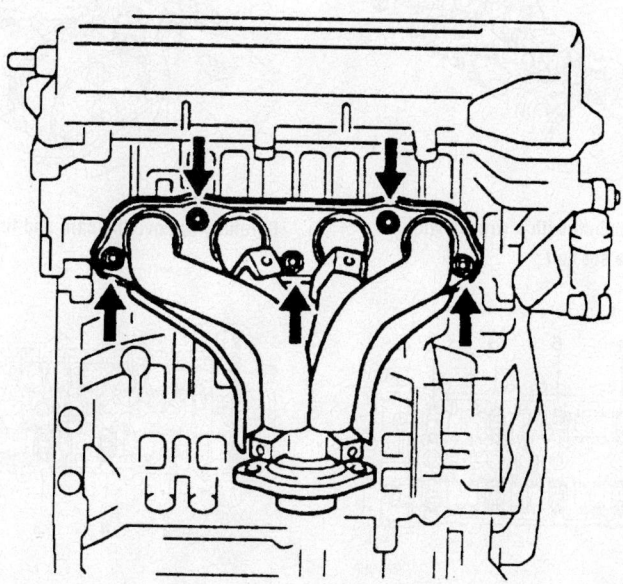

Exhaust manifold mounting nut locations—1.8L (1ZZ-FE) engine

7923VG22

- Accelerator cable
- Exhaust pipe from the manifold
- Exhaust manifold support bracket
- Heat insulator from the dash panel
- Upper heat insulator
- Exhaust manifold and gasket
- If necessary, the lower heat insulator from the exhaust manifold.

To install:

4. Install or connect the following:
  - Lower heat insulator on the exhaust manifold. Tighten the bolts to 108 inch lbs. (12 Nm).
  - Exhaust manifold using a new gasket. Tighten the nuts, in several passes, to 27 ft. lbs. (37 Nm).
  - Upper heat insulator. Tighten the bolts to 108 inch lbs. (12 Nm).
  - Heat insulator on the dash panel
  - Exhaust manifold support bracket. Tighten the bolts, in an alternating pattern, to 37 ft. lbs. (49 Nm).
  - Front exhaust pipe to the manifold. Tighten the bolts to 46 ft. lbs. (62 Nm).
  - Oxygen Sensor (O_2S). Tighten the nuts to 14 ft. lbs. (20 Nm).
  - Accelerator cable and air intake duct
  - Alternator and drive belt
  - Negative battery cable

5. Fill the cooling system.

6. Start the vehicle, check for leaks and repair if necessary.

2ZZ-GE Engine

1. Before servicing the vehicle, refer to the precautions section.

2. Drain the cooling system.

3. Remove or disconnect the following:
  - Negative battery cable
  - Drive belt and alternator
  - Air intake duct
  - Accelerator cable
  - Exhaust pipe from the manifold
  - Exhaust manifold support bracket
  - Heat insulator from the dash panel
  - Upper heat insulator
  - Exhaust manifold and gasket
  - If necessary, the lower heat insulator from the exhaust manifold.

To install:

4. Install or connect the following:
  - Lower heat insulator on the exhaust manifold. Tighten the bolts to 15 ft. lbs. (20 Nm).
  - Exhaust manifold using a new gasket. Tighten the nuts, in several passes to 37 ft. lbs. (50 Nm).
  - Upper heat insulator. Tighten the bolts to 15 ft. lbs. (20 Nm).
  - Heat insulator on the dash panel.
  - Exhaust manifold support bracket. Tighten the bolts to 37 ft. lbs. (49 Nm).
  - Front exhaust pipe to the manifold. Tighten the bolts to 46 ft. lbs. (62 Nm).
  - Oxygen Sensor (O_2S). Tighten the nuts to 14 ft. lbs. (20 Nm).
  - Accelerator cable and air intake duct.
  - Alternator and drive belt.
  - Negative battery cable

5. Fill the cooling system.

6. Start the vehicle, check for leaks and repair if necessary.

Camshaft(s)

REMOVAL & INSTALLATION

1. Before servicing the vehicle, refer to the precautions section.

2. Remove or disconnect the following:
  - Negative battery cable
  - Right side engine under cover
  - Cylinder head cover
  - Suction hose sub-assembly, 2ZZ-GE engine
  - Drive belt
  - Power steering pump reservoir and position it aside, 1ZZ-FE engine

3. Place a jack with a wooden block under the vehicle for support, then remove the 4 bolts and 2 nuts and remove the right side engine mount.

4. Remove the engine wire, on 1ZZ-FE engines:

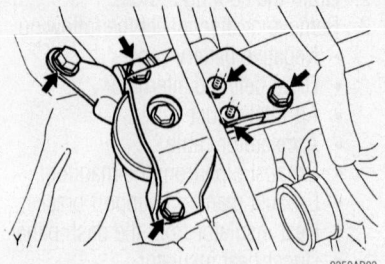

With the engine supported, remove the right side engine mount—1ZZ-FE engine shown, 2ZZ-GE similar

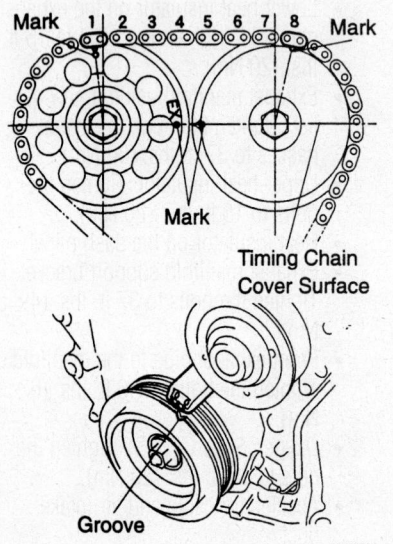

Proper timing mark alignment for TDC

a. Remove the 5 clamps from the brackets.

b. Detach the connectors.

c. Remove the ignition coil connectors.

d. Bolt and nut holding the engine wire.

5. Remove or disconnect the following:
- Ignition coil assembly
- Positive Crankcase Ventilation (PCV) hoses from the valve cover
- Valve (cylinder head) cover subassembly

6. Set the No. 1 cylinder to Top Dead Center (TDC) of the compressor stroke as follows:

a. Turn the crankshaft pulley, and align its groove with the "0" timing mark of the timing chain cover.

b. Make sure the point marks of the camshaft timing sprockets and VVT timing sprockets are in a straight line as shown. If not, turn the crankshaft 1 complete revolution (360°) and align the marks.

7. Remove the drive belt tensioner.

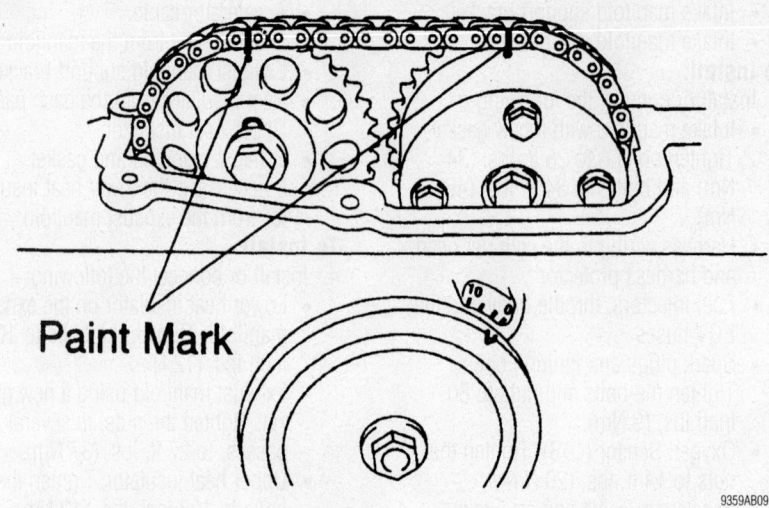

Paint Mark

Matchmark the timing chain and cam sprockets

Do not turn the crankshaft without the tensioner installed.

8. Make sure the No. 1 cylinder is at TDC of the compression stroke.

9. Matchmark the timing chain and camshaft sprockets

10. Remove the 2 nuts and chain tensioner.

11. Hold the camshafts with a wrench and loosen the camshaft set bolt.

12. Using several passes, gradually remove the bearing cap bolts from the No. 2 camshaft, in the proper sequence.

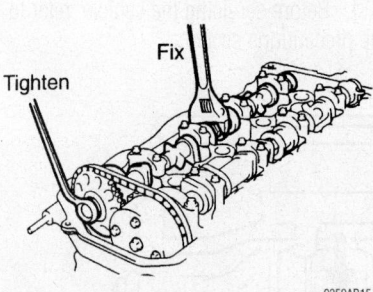

Hold the camshaft with a wrench while removing the set bolt

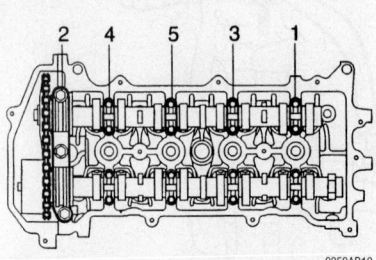

Camshaft bearing cap bolt removal sequence—1ZZ-FE engine

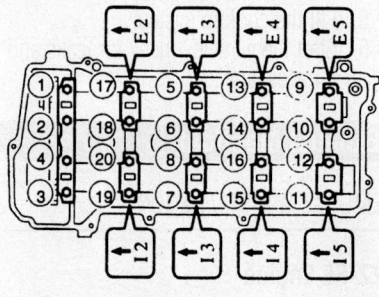

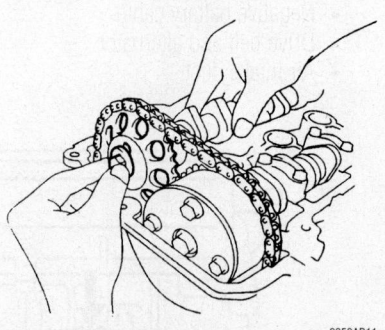

Camshaft bearing cap bolt removal sequence—2ZZ-GE engine

Carefully remove the cam and timing gear

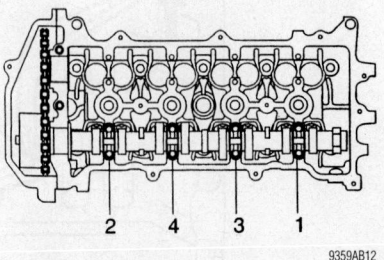

Camshaft bearing cap bolt removal sequence—1ZZ-FE engine

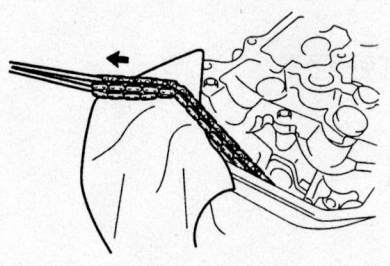

Secure the timing chain with string to prevent it from slipping down into the timing chain cover

13. Remove the camshaft and timing gear as shown.

14. Using several passes, gradually remove the bearing cap bolts from the other camshaft, in the proper sequence.

15. Remove the camshaft while holding the timing chain.

❋❋ WARNING

Do not let anything drop down into the timing chain cover while the camshafts are removed.

16. Tie the timing chain with a string as shown, to prevent it from dropping down into the timing chain cover.

To install:

17. Position the camshaft on the cylinder head, then install the timing chain on the cam timing gear, with the painted links aligned with the marks on the timing gear.

18. Check the front marks and numbers and torque the camshaft cap bolts, in sequence, to 10 ft. lbs. (13 Nm) for 1ZZ-FE engine, or to 14 ft. lbs. (19 Nm) for 2ZZ-GE engines.

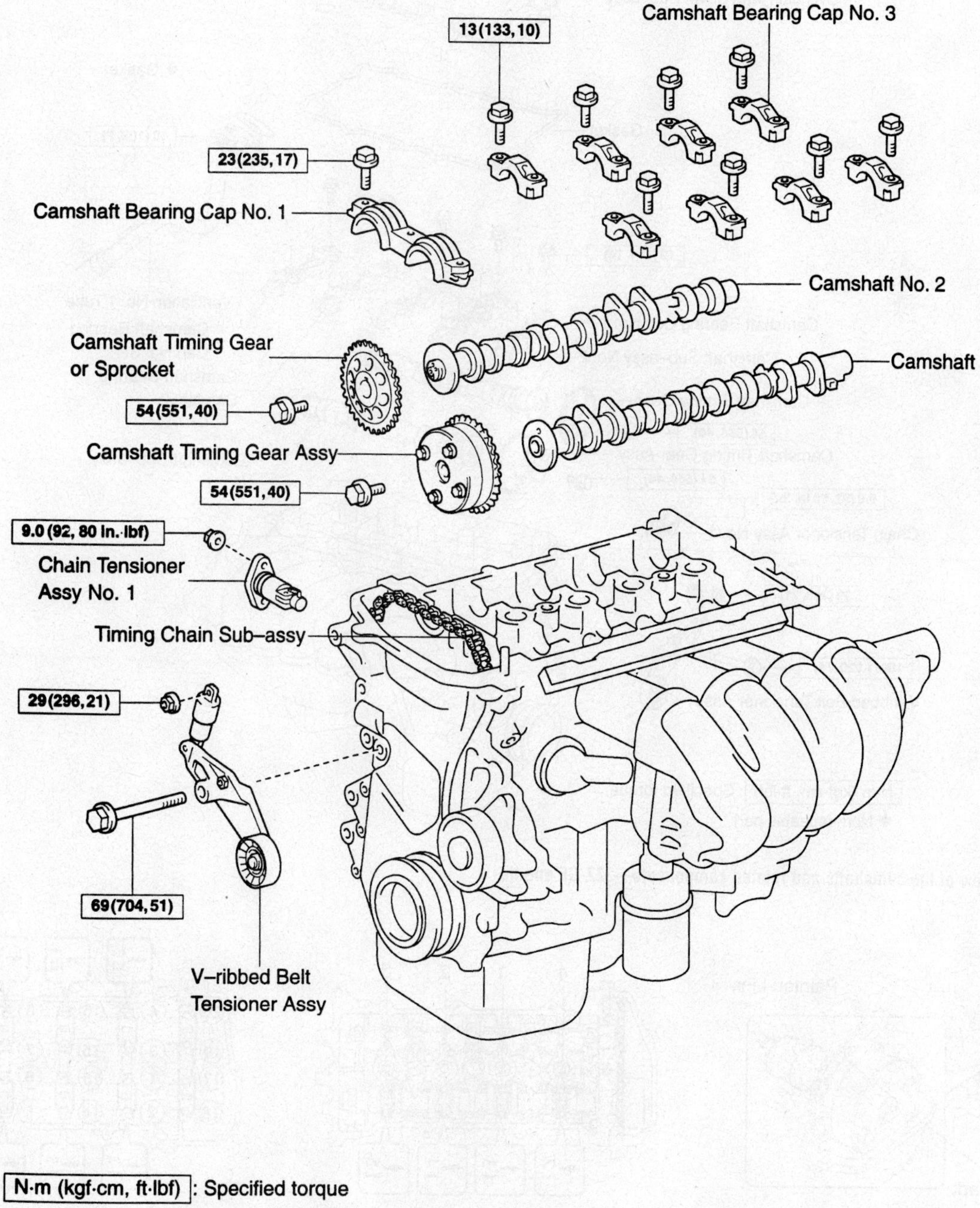

13 (133, 10)

Camshaft Bearing Cap No. 3

23 (235, 17)

Camshaft Bearing Cap No. 1

Camshaft No. 2

Camshaft Timing Gear or Sprocket

Camshaft

54 (551, 40)

Camshaft Timing Gear Assy

9.0 (92, 80 in.·lbf)

Chain Tensioner Assy No. 1

Timing Chain Sub–assy

29 (296, 21)

69 (704, 51)

V–ribbed Belt Tensioner Assy

54 (551, 40)

N·m (kgf·cm, ft·lbf) : Specified torque

Exploded view of the camshafts and related components—1ZZ-FE engine

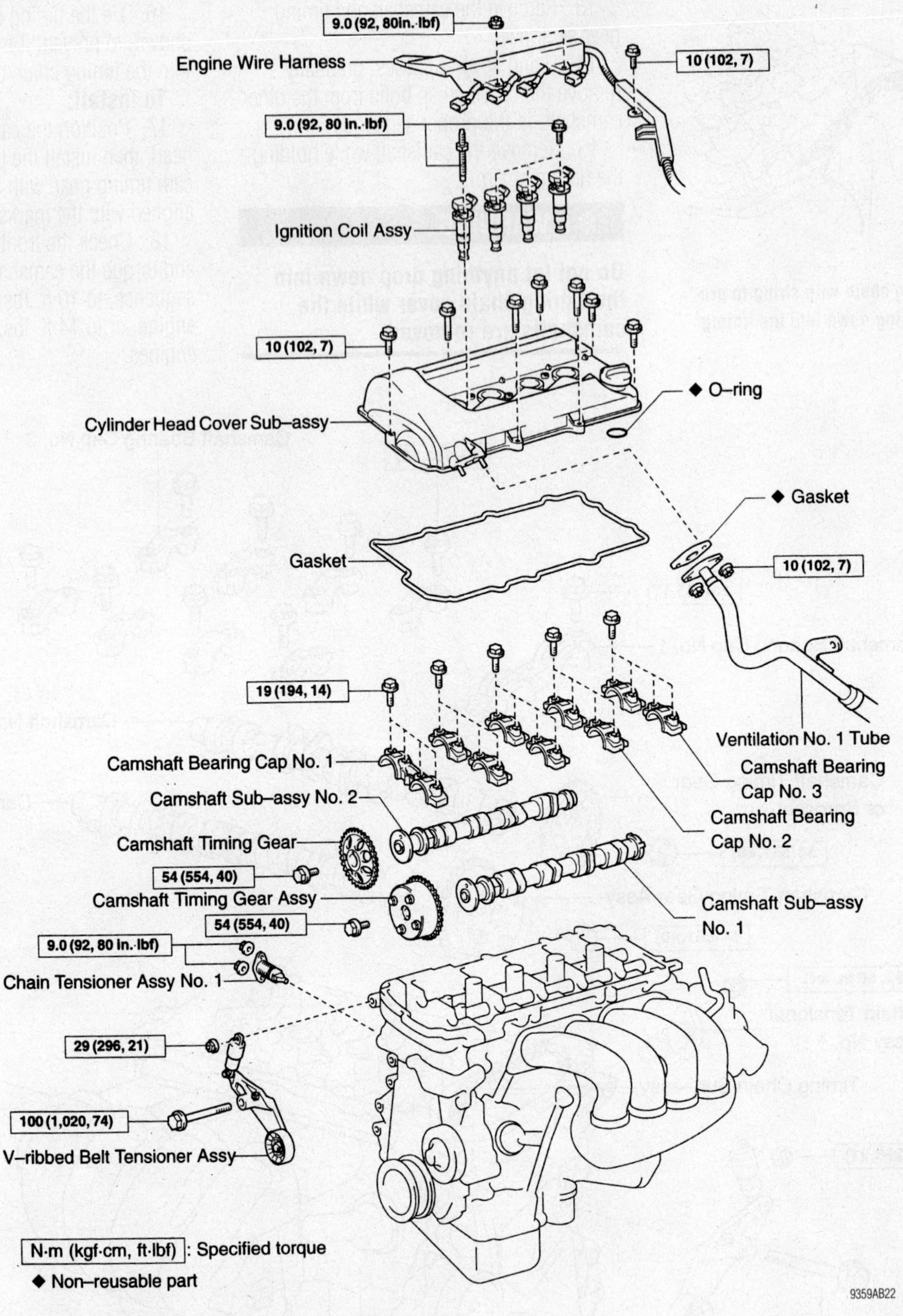

9.0 (92, 80in.·lbf)
Engine Wire Harness
10 (102, 7)
9.0 (92, 80 in.·lbf)
Ignition Coil Assy
10 (102, 7)
◆ O–ring
Cylinder Head Cover Sub–assy
◆ Gasket
Gasket
10 (102, 7)
19 (194, 14)
Ventilation No. 1 Tube
Camshaft Bearing Cap No. 1
Camshaft Bearing Cap No. 3
Camshaft Sub–assy No. 2
Camshaft Bearing Cap No. 2
Camshaft Timing Gear
54 (554, 40)
Camshaft Timing Gear Assy
54 (554, 40)
Camshaft Sub–assy No. 1
9.0 (92, 80 in.·lbf)
Chain Tensioner Assy No. 1
29 (296, 21)
100 (1,020, 74)
V–ribbed Belt Tensioner Assy
N·m (kgf·cm, ft·lbf) : Specified torque
◆ Non–reusable part
9359AB22

Exploded view of the camshafts and related components—2ZZ-GE engine

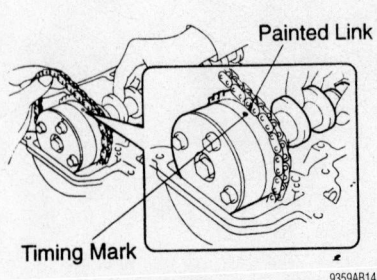

Painted Link
Timing Mark
9359AB14

Make sure the alignment marks on the timing chain and camshaft gear match up

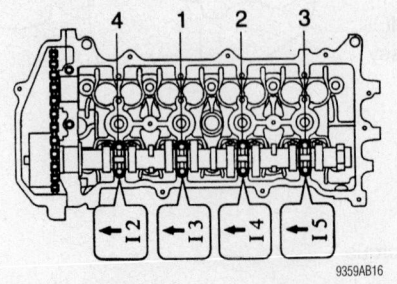

4
1
2
3
I2
I3
I4
I5
9359AB16

Camshaft cap bolt tightening sequence—1ZZ-FE engine

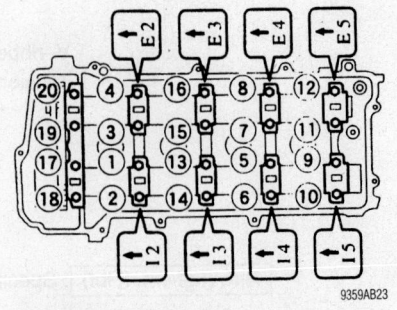

E2
E3
E4
E5
20
4
16
8
12
19
3
15
7
11
17
1
13
5
9
18
2
14
6
10
I2
I3
I4
I5
9359AB23

Camshaft cap bolt tightening sequence—2ZZ-GE engine

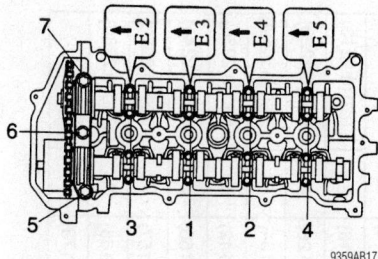

Camshaft cap bolt tightening sequence—1ZZ-FE

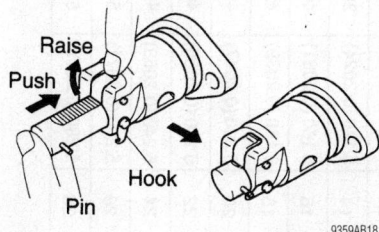

Set the timing chain tensioner hook properly

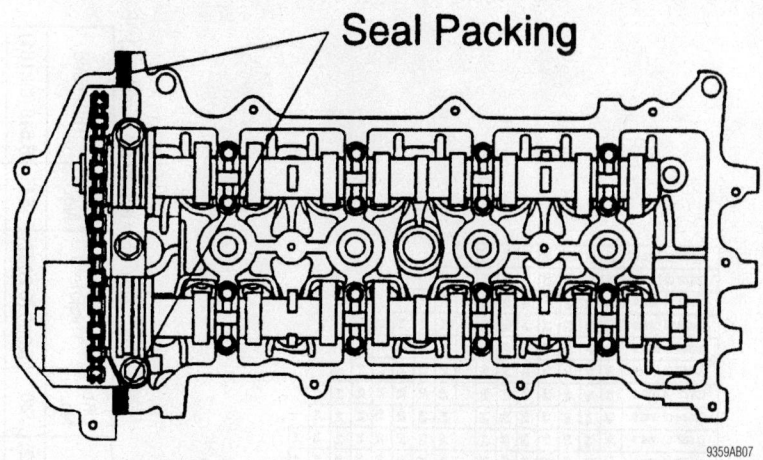

Seal packing installation locations

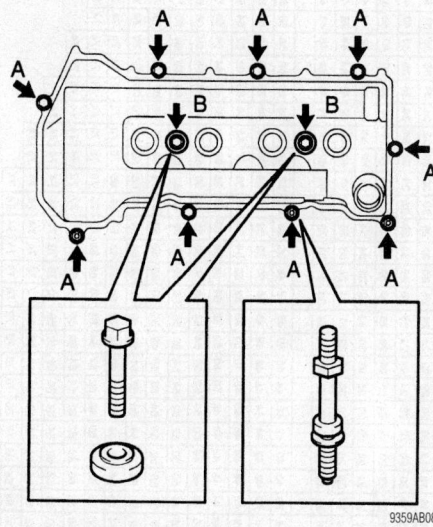

Cylinder head (valve) cover bolt locations—1ZZ-FE engine

19. Put camshaft No. 2 on the cylinder head, with the painted links of the chain aligned with the mark on the timing gear.

20. Tighten the camshaft gear set bolt temporarily.

21. Check the front marks and numbers and torque the camshaft cap bolts, in sequence, to 10 ft. lbs. (13 Nm). Install the No. 1 bearing cap and tighten to 17 ft. lbs. (23 Nm).

22. Hold the camshaft secure with a wrench and tighten the set bolt to 40 ft. lbs. (54 Nm). Be careful not the damage the lifters.

23. Check to be sure the matchmarks on the timing chain and cam sprockets, and the alignment of the pulley groove with the timing mark on the cover are still aligned.

24. Install the chain tensioner:

a. Make sure the O-ring is clean, then set the hook as shown.

b. Oil the tensioner, then install and tighten to 80 inch lbs. (9 Nm).

➥**When installing the tensioner, set the hook again if the hook releases the plunger.**

c. Turn the crankshaft counterclockwise, and disconnect the plunger knock pin from the hook.

d. Turn the crankshaft clockwise and check that the slipper is pushed by the plunger. If the plunger does not spring out, press the slipper into the chain tensioner with a screwdriver so that the hook is released from the knock pin and the plunger springs out.

25. Check the valve clearance and make adjustments as needed.

26. Install or connect the following:

- Belt tensioner. Tighten the nut to 21 ft. lbs. (29 Nm) and the bolt to 51 ft. lbs. (69 Nm).
- Cylinder head sub-assembly cover. Install seal packing into the locations shown and install within 3 minutes. Tighten the "A" bolts to 8 ft. lbs. (11 Nm) and the "B" bolts to 80 inch lbs. (9 Nm) for 1ZZ-FE engine and to 7 ft. lbs. (10 Nm) for 2ZZ-GE engines.
- Ignition coil assembly. Torque the bolts to 80 inch lbs. (9 Nm).
- Engine wire and tighten to 80 inch lbs. (9 Nm)
- Right side engine mount. Tighten to 38 ft. lbs. (52 Nm).

- Cylinder head (valve) cover
- Negative battery cable

Valve Lash

ADJUSTMENT

1ZZ-FE Engine

➥**Adjust the valve clearance when the engine is cold.**

1. Before servicing the vehicle, refer to the precautions section.

2. Remove or disconnect the following:
- Negative battery cable.
- Cylinder head covers
- Engine wire
- Ignition coil
- Positive Crankcase Ventilation (PCV) hoses
- Cylinder head cover sub-assembly

1ZZ–FE: Valve Lifter Selection Chart (Intake)

Adjusting shim chart (intake)—1ZZ-FE engine

New lifter thickness — mm (in.)

Lifter No.	Thickness	Lifter No.	Thickness	Lifter No.	Thickness
06	5.060 (0.1992)	30	5.300 (0.2087)	54	5.540 (0.2181)
08	5.080 (0.2000)	32	5.320 (0.2094)	56	5.560 (0.2189)
10	5.100 (0.2008)	34	5.340 (0.2102)	58	5.580 (0.2197)
12	5.120 (0.2016)	36	5.360 (0.2110)	60	5.600 (0.2205)
14	5.140 (0.2024)	38	5.380 (0.2118)	62	5.620 (0.2213)
16	5.160 (0.2031)	40	5.400 (0.2126)	64	5.640 (0.2220)
18	5.180 (0.2039)	42	5.420 (0.2134)	66	5.660 (0.2228)
20	5.200 (0.2047)	44	5.440 (0.2142)	68	5.680 (0.2236)
22	5.220 (0.2055)	46	5.460 (0.2150)	70	5.700 (0.2244)
24	5.240 (0.2063)	48	5.480 (0.2157)	72	5.720 (0.2252)
26	5.260 (0.2071)	50	5.500 (0.2165)	74	5.740 (0.2260)
28	5.280 (0.2079)	52	5.520 (0.2173)		

Intake valve clearance (Cold):
0.15 – 0.25 mm (0.006 – 0.010 in.)

EXAMPLE: The 5.250 mm (0.2067 in.) lifter is installed, and the measured clearance is 0.400 mm (0.0157 in.).

Replace the 5.250 mm (0.2067 in.) lifter with a new No. 48 lifter.

9307WG70

1ZZ–FE: Valve Lifter Selection Chart (Exhaust)

Adjusting shim chart (exhaust)—1ZZ-FE engine

New lifter thickness — mm (in.)

Lifter No.	Thickness	Lifter No.	Thickness	Lifter No.	Thickness
06	5.060 (0.1992)	30	5.300 (0.2087)	54	5.540 (0.2181)
08	5.080 (0.2000)	32	5.320 (0.2094)	56	5.560 (0.2189)
10	5.100 (0.2008)	34	5.340 (0.2102)	58	5.580 (0.2197)
12	5.120 (0.2016)	36	5.360 (0.2110)	60	5.600 (0.2205)
14	5.140 (0.2024)	38	5.380 (0.2118)	62	5.620 (0.2213)
16	5.160 (0.2031)	40	5.400 (0.2126)	64	5.640 (0.2220)
18	5.180 (0.2039)	42	5.420 (0.2134)	66	5.660 (0.2228)
20	5.200 (0.2047)	44	5.440 (0.2142)	68	5.680 (0.2236)
22	5.220 (0.2055)	46	5.460 (0.2150)	70	5.700 (0.2244)
24	5.240 (0.2063)	48	5.480 (0.2157)	72	5.720 (0.2252)
26	5.260 (0.2071)	50	5.500 (0.2165)	74	5.740 (0.2260)
28	5.280 (0.2079)	52	5.520 (0.2173)		

Exhaust valve clearance (Cold):
0.25 – 0.35 mm (0.010 – 0.014 in.).
EXAMPLE: The 5.340 mm (0.2102 in.) lifter is installed, and the measured clearance is 0.440 mm (0.0173 in.).
Replace the 5.340 mm (0.2102 in.) lifter with a new No. 48 lifter.

9307WG71

Adjusting shim chart (intake)—2ZZ-GE engine

New Shim thickness   mm (in.)

Shim No.	Thickness	Shim No.	Thickness	Shim No.	Thickness
00	2.000 (0.0787)	28	2.280 (0.0898)	56	2.560 (0.1008)
02	2.020 (0.0795)	30	2.300 (0.0906)	58	2.580 (0.1016)
04	2.040 (0.0803)	32	2.320 (0.0913)	60	2.600 (0.1024)
06	2.060 (0.0811)	34	2.340 (0.0921)	62	2.620 (0.1031)
08	2.080 (0.0819)	36	2.360 (0.0929)	64	2.640 (0.1039)
10	2.100 (0.0827)	38	2.380 (0.0937)	66	2.660 (0.1047)
12	2.120 (0.0835)	40	2.400 (0.0945)	68	2.680 (0.1055)
14	2.140 (0.0843)	42	2.420 (0.0953)	70	2.700 (0.1063)
16	2.160 (0.0850)	44	2.440 (0.0961)	72	2.720 (0.1071)
18	2.180 (0.0858)	46	2.460 (0.0969)	74	2.740 (0.1079)
20	2.200 (0.0866)	48	2.480 (0.0976)	76	2.760 (0.1087)
22	2.220 (0.0874)	50	2.500 (0.0984)	78	2.780 (0.1094)
24	2.240 (0.0882)	52	2.520 (0.0992)	80	2.800 (0.1102)
26	2.260 (0.0890)	54	2.540 (0.1000)		

Intake valve clearance (Cold):
0.08 – 0.18 mm (0.0031 – 0.0071 in.)

EXAMPLE: The 2.200 mm (0.0826 in.) shim is installed, and the measured clearance is 0.400 mm (0.0157 in.).

Replace the 2.600 mm (0.1024 in.) shim with a new No. 60 shim.

Measure clearance mm (in.):

Measure clearance mm (in.)
0.000 – 0.030 (0.0000 – 0.0012)
0.031 – 0.050 (0.0012 – 0.0020)
0.051 – 0.070 (0.0020 – 0.0028)
0.071 – 0.090 (0.0028 – 0.0035)
0.091 – 0.099 (0.0036 – 0.0039)
0.100 – 0.160 (0.0039 – 0.0063)
0.161 – 0.180 (0.0063 – 0.0071)
0.181 – 0.200 (0.0071 – 0.0079)
0.201 – 0.220 (0.0079 – 0.0087)
0.221 – 0.240 (0.0087 – 0.0094)
0.241 – 0.260 (0.0095 – 0.0102)
0.261 – 0.280 (0.0103 – 0.0110)
0.281 – 0.300 (0.0111 – 0.0118)
0.301 – 0.320 (0.0119 – 0.0126)
0.321 – 0.340 (0.0126 – 0.0134)
0.341 – 0.360 (0.0134 – 0.0142)
0.361 – 0.380 (0.0142 – 0.0150)
0.381 – 0.400 (0.0150 – 0.0157)
0.401 – 0.420 (0.0158 – 0.0165)
0.421 – 0.440 (0.0166 – 0.0173)
0.441 – 0.460 (0.0174 – 0.0181)
0.461 – 0.480 (0.0181 – 0.0189)
0.481 – 0.500 (0.0189 – 0.0197)
0.501 – 0.520 (0.0197 – 0.0205)
0.521 – 0.540 (0.0205 – 0.0213)
0.541 – 0.560 (0.0213 – 0.0220)
0.561 – 0.580 (0.0221 – 0.0228)
0.581 – 0.600 (0.0229 – 0.0236)
0.601 – 0.620 (0.0237 – 0.0244)
0.621 – 0.640 (0.0244 – 0.0252)
0.641 – 0.660 (0.0252 – 0.0260)
0.661 – 0.680 (0.0260 – 0.0268)

Installed shim thickness mm (in.): 2.000 (0.0787), 2.020 (0.0795), 2.040 (0.0803), 2.060 (0.0811), 2.080 (0.0819), 2.100 (0.0827), 2.120 (0.0835), 2.140 (0.0843), 2.160 (0.0850), 2.180 (0.0858), 2.200 (0.0866), 2.220 (0.0874), 2.240 (0.0882), 2.250 (0.0886), 2.260 (0.0890), 2.270 (0.0894), 2.280 (0.0898), 2.290 (0.0902), 2.300 (0.0906), 2.310 (0.0909), 2.320 (0.0913), 2.330 (0.0917), 2.340 (0.0921), 2.350 (0.0925), 2.360 (0.0929), 2.370 (0.0933), 2.380 (0.0937), 2.390 (0.0941), 2.400 (0.0945), 2.410 (0.0949), 2.420 (0.0953), 2.430 (0.0957), 2.440 (0.0961), 2.450 (0.0965), 2.460 (0.0969), 2.470 (0.0972), 2.480 (0.0976), 2.490 (0.0980), 2.500 (0.0984), 2.510 (0.0988), 2.520 (0.0992), 2.530 (0.0996), 2.540 (0.1000), 2.550 (0.1004), 2.560 (0.1008), 2.580 (0.1016), 2.600 (0.1024), 2.620 (0.1031), 2.640 (0.1039), 2.660 (0.1047), 2.680 (0.1055), 2.700 (0.1063), 2.720 (0.1071), 2.740 (0.1079), 2.760 (0.1087), 2.780 (0.1094), 2.800 (0.1102)

9359AB26

New Shim thickness mm (in.)

Shim No.	Thickness	Shim No.	Thickness	Shim No.	Thickness
00	2.000 (0.0787)	28	2.280 (0.0898)	56	2.560 (0.1008)
02	2.020 (0.0795)	30	2.300 (0.0906)	58	2.580 (0.1016)
04	2.040 (0.0803)	32	2.320 (0.0913)	60	2.600 (0.1024)
06	2.060 (0.0811)	34	2.340 (0.0921)	62	2.620 (0.1031)
08	2.080 (0.0819)	36	2.360 (0.0929)	64	2.640 (0.1039)
10	2.100 (0.0827)	38	2.380 (0.0937)	66	2.660 (0.1047)
12	2.120 (0.0835)	40	2.400 (0.0945)	68	2.680 (0.1055)
14	2.140 (0.0843)	42	2.420 (0.0953)	70	2.700 (0.1063)
16	2.160 (0.0850)	44	2.440 (0.0961)	72	2.720 (0.1071)
18	2.180 (0.0858)	46	2.460 (0.0969)	74	2.740 (0.1079)
20	2.200 (0.0866)	48	2.480 (0.0976)	76	2.760 (0.1087)
22	2.220 (0.0874)	50	2.500 (0.0984)	78	2.780 (0.1094)
24	2.240 (0.0882)	52	2.520 (0.0992)	80	2.800 (0.1102)
26	2.260 (0.0890)	54	2.540 (0.1000)		

Exhaust valve clearance (Cold):
0.22 – 0.32 mm (0.0087 – 0.0126 in.)
EXAMPLE: The 2.200 mm (0.0862 in.) shim is installed, and the measured clearance is 0.500 mm (0.0197 in.). Replace the 2.540 mm (0.1000 in.) shim with a new No. 54 shim.

Adjusting shim chart (exhaust)—2ZZ-GE engine

Measure clearance mm (in.):

Measure clearance mm	Measure clearance (in.)
0.000 – 0.030	0.0000 – 0.0012
0.031 – 0.050	0.0012 – 0.0020
0.051 – 0.070	0.0020 – 0.0028
0.071 – 0.090	0.0028 – 0.0035
0.091 – 0.110	0.0036 – 0.0043
0.111 – 0.130	0.0044 – 0.0051
0.131 – 0.150	0.0052 – 0.0059
0.151 – 0.170	0.0059 – 0.0067
0.171 – 0.190	0.0067 – 0.0075
0.191 – 0.210	0.0075 – 0.0083
0.211 – 0.230	0.0083 – 0.0091
0.231 – 0.239	0.0091 – 0.0094
0.301 – 0.320	0.0119 – 0.0126
0.321 – 0.340	0.0126 – 0.0134
0.341 – 0.360	0.0134 – 0.0142
0.361 – 0.380	0.0142 – 0.0150
0.381 – 0.400	0.0150 – 0.0157
0.401 – 0.420	0.0158 – 0.0165
0.421 – 0.440	0.0166 – 0.0173
0.441 – 0.460	0.0174 – 0.0181
0.461 – 0.480	0.0181 – 0.0189
0.481 – 0.500	0.0189 – 0.0197
0.501 – 0.520	0.0197 – 0.0205
0.521 – 0.540	0.0205 – 0.0213
0.541 – 0.560	0.0213 – 0.0220
0.561 – 0.580	0.0221 – 0.0228
0.581 – 0.600	0.0229 – 0.0236
0.601 – 0.620	0.0237 – 0.0244
0.621 – 0.640	0.0244 – 0.0252
0.641 – 0.660	0.0252 – 0.0260
0.661 – 0.680	0.0260 – 0.0268
0.681 – 0.700	0.0268 – 0.0276
0.701 – 0.720	0.0276 – 0.0283
0.721 – 0.740	0.0284 – 0.0291
0.741 – 0.760	0.0292 – 0.0299
0.761 – 0.780	0.0299 – 0.0307
0.781 – 0.800	0.0307 – 0.0315
0.801 – 0.820	0.0315 – 0.0323

Installed shim thickness mm(in.) across top of chart: 2.000 (0.0787), 2.020 (0.0795), 2.040 (0.0803), 2.060 (0.0811), 2.080 (0.0819), 2.100 (0.0827), 2.120 (0.0835), 2.140 (0.0843), 2.160 (0.0850), 2.180 (0.0858), 2.200 (0.0866), 2.210 (0.0870), 2.220 (0.0874), 2.230 (0.0878), 2.240 (0.0882), 2.250 (0.0886), 2.260 (0.0890), 2.270 (0.0894), 2.280 (0.0898), 2.290 (0.0902), 2.300 (0.0906), 2.310 (0.0909), 2.320 (0.0913), 2.330 (0.0917), 2.340 (0.0921), 2.340 (0.0921), 2.350 (0.0925), 2.360 (0.0929), 2.370 (0.0933), 2.380 (0.0937), 2.390 (0.0941), 2.400 (0.0945), 2.410 (0.0949), 2.420 (0.0953), 2.430 (0.0957), 2.440 (0.0961), 2.450 (0.0965), 2.460 (0.0969), 2.470 (0.0972), 2.480 (0.0976), 2.490 (0.0980), 2.500 (0.0984), 2.510 (0.0988), 2.520 (0.0992), 2.530 (0.0996), 2.540 (0.1000), 2.550 (0.1004), 2.560 (0.1008), 2.580 (0.1016), 2.660 (0.1047), 2.680 (0.1055), 2.700 (0.1063), 2.720 (0.1071), 2.740 (0.1079), 2.760 (0.1087), 2.780 (0.1094), 2.800 (0.1102)

9359AB27

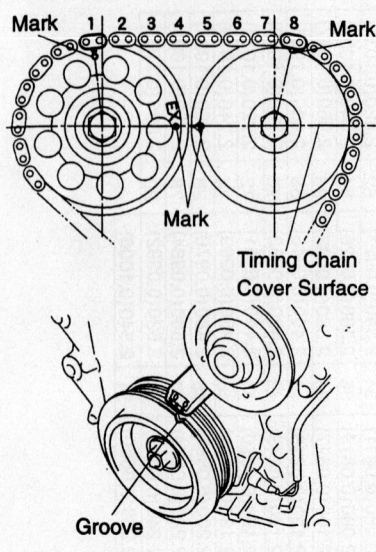

Proper timing mark alignment for TDC—1ZZ-FE and 2ZZ-GE engines

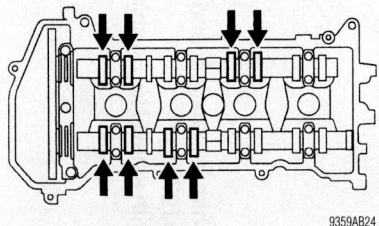

Check the clearance of the 1st set of valves–1ZZ-FE engine

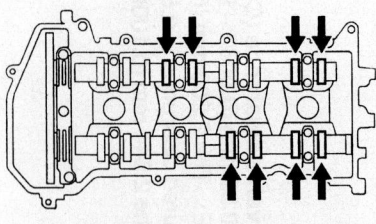

Check the clearance of the 2nd set of valves–1ZZ-FE engine

3. Set the No. 1 cylinder to Top Dead Center (TDC) of the compressor stroke as follows:

   a. Turn the crankshaft pulley, and align its groove with the "0" timing mark of the timing chain cover.

   b. Make sure the point marks of the camshaft timing sprockets and VVT timing sprockets are in a straight line as shown. If not, turn the crankshaft 1 complete revolution (360°) and align the marks.

4. Check the valve clearance of the first set of the valves shown:

   a. Use a feeler gauge to measure the clearance between the valve lifter and camshaft. The clearance of the intake valves should be 0.0059–0.0098 in. (0.15–0.25mm). The clearance of the exhaust valves should be 0.0098–0.0138 in. (0.25–0.35mm).

   b. Note the out-of-specification valve clearance measurements. You will need them later to determine the required replacement valve lifter.

   c. Turn the crankshaft 1 revolution (360°) to set the No. 4 cylinder to TDC.

5. Check the valve clearance of the second set of the valves shown:

   a. Use a feeler gauge to measure the clearance between the valve lifter and camshaft. The clearance of the intake valves should be 0.0059–0.0098 in. (0.15–0.25mm). The clearance of the exhaust valves should be 0.0098–0.0138 in. (0.25–0.35mm).

   b. Note the out-of-specification valve clearance measurements. You will need them later to determine the required replacement valve lifter.

6. Remove or disconnect the following:
   • Drive belt
   • Right side engine mount
   • Drive belt tensioner

✷✷ WARNING

DO NOT turn the crankshaft while the tensioner is removed!

7. Set the No. 1 cylinder to TDC of the compression stroke.
   • Camshafts
   • Valve lifters.

8. Use a micrometer to measure the thickness of the used lifter. Calculate the thickness of a new lifter. so the valve clearance comes within the specified value:

   a. A: Thickness of new lifter.

   b. B: Thickness of used lifter.

   c. C: Measured valve clearance.

   d. Intake valve clearance: A = B + (C—0.0079 in. (0.20mm).

   e. Exhaust valve clearance: A = B + (C—0.0118 in. (0.30mm).

   f. Select a new lifter with a thickness as close as possible to the calculated values. Lifters come in 35 sizes in increments of 0.0008 in. (0.020mm) from 0.1992–0.2260 in (5.060–5.740mm).

9. Install or connect the following:
   • Camshafts
   • Drive belt tensioner
   • Right hand engine mount
   • Cylinder head (valve) cover sub-assembly

   • Ignition coil
   • Engine wire
   • Cylinder head (valve) cover
   • Negative battery cable

2ZZ-GE Engine

➡**Adjust the valve clearance when the engine is cold.**

1. Before servicing the vehicle, refer to the precautions section.

2. Remove or disconnect the following:
   • Negative battery cable.
   • Right side engine under cover
   • Cylinder head cover
   • Ignition coil assembly
   • Wire harness clamp
   • Suction hose sub-assembly
   • Cylinder head cover sub-assembly
   • Drive belt
   • Right side engine mount

3. Set the No. 1 cylinder to Top Dead Center (TDC) of the compressor stroke as follows:

   a. Turn the crankshaft pulley, and align its groove with the "0" timing mark of the timing chain cover.

   b. Make sure the point marks of the camshaft timing sprockets and VVT timing sprockets are in a straight line as shown. If not, turn the crankshaft 1 complete revolution (360°) and align the marks.

4. Check the valve clearance of the first set of the valves shown:

   a. Use a feeler gauge to measure the clearance between the valve lifter and camshaft. The clearance of the intake valves should be 0.0031–0.0071 in. (0.08–0.18mm). The clearance of the exhaust valves should be 0.0087–0.0126 in. (0.22–0.32mm).

   b. Note the out-of-specification valve clearance measurements. You will need them later to determine the required replacement valve lifter.

   c. Turn the crankshaft 1 revolution (360°) to set the No. 4 cylinder to TDC.

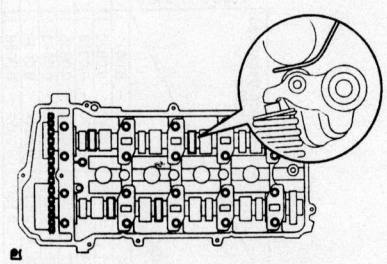

Check the clearance of the 1st set of valves–2ZZ-GE engine

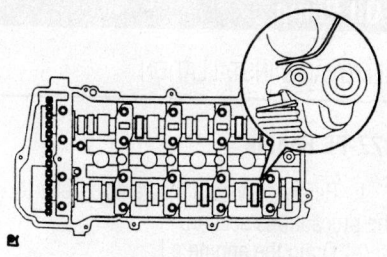

Check the clearance of the 2nd set of valves—2ZZ-GE engine

5. Check the valve clearance of the second set of the valves shown:

a. Use a feeler gauge to measure the clearance between the valve lifter and camshaft. The clearance of the intake valves should be 0.0031–0.0071 in. (0.08–0.18mm). The clearance of the exhaust valves should be 0.0087–0.0126 in. (0.22–0.32mm).

b. Note the out-of-specification valve clearance measurements. You will need them later to determine the required replacement valve lifter.

6. To adjust the intake valve clearance:

a. Set the SST. Turn the crankshaft so the related rocker arm, where the valve clearance is adjusted, is fully pushed down.

➡**Remove the spark plug and take off the compression.**

b. Insert SST 09248-77010 into the plug tube. The tool cannot be inserted unless the set screw is loosened.

c. Operate the lever so that the SST's seat surface comes to contact with the valve retainer and lock them with the set screw. Clearance between the valve retainer and SST's set surface is not allowed. Be careful not to make clearance when inserting the SST, since clearance may unlock the keeper.

d. Lock the set screw on the tube side of the SST.

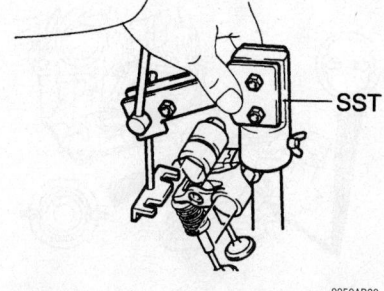

Insert the special tool into the plug tube— 2ZZ-GE

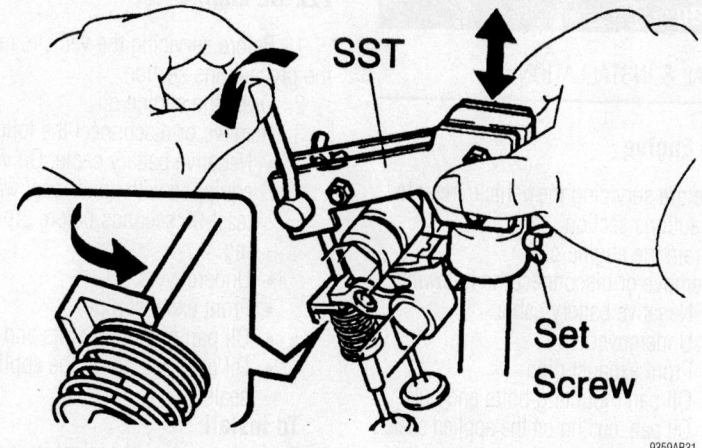

Operate the lever so that the SST's seat surface comes to contact with the valve retainer and lock them with the set screw

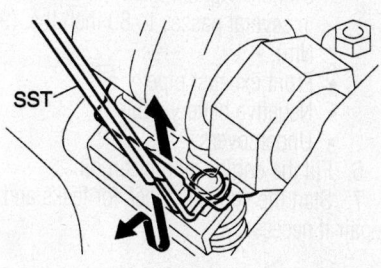

Setting the tool from the right side, makes shim removal easier—2ZZ-GE

e. Rotate the crankshaft so that the camshaft is position as shown. During rotation, pay attention to the direction, to prevent the nose of the camshaft from interfering with the SST's shaft. Do not rotate the crankshaft excessively.

f. Lift the rocker arm to make room and remove the adjusting shim using SST 09248-77010.

7. Determine the size of the replaced shim according to the chart or the following formula:

a. Use a dial indicator to measure the thickness of the removed shim.

b. Calculate the thickness of a new shim so that the valve clearance comes within the specified value.

c. A: Thickness of new shim.

d. B: Thickness of used shim.

e. C: Measured valve clearance.

f. Intake: A = B + (C—0.005 in. [0.13mm])

g. Exhaust: A = B + (C—0.011 in. [0.27mm])

h. Select a new shim with a thickness as close as possible to the calculated values. Shims come in 41 sizes in incre-

ments of 0.0008 in. (0.020mm) from 0.0787–0.1102 in (2.0–2.8mm).

8. Lift the rocker arm to make room, then install the adjusting shim using the SST. To remove the tool from the shim, push down on the rocker arm.

9. Turn the crankshaft so the related rocker arm, where the valve clearance is adjusted, is fully pushed down.

10. Loosen the 2 set-screws, then remove the SST.

11. Install all components in the reverse of the removal procedure.

Starter

REMOVAL & INSTALLATION

1. Before servicing the vehicle, refer to the precautions section.

2. Remove or disconnect the following:
- Negative battery cable
- Right side engine undercover
- Starter wiring
- Starter

3. Installation is the reverse of removal. Torque the bolts to 27 ft. lbs. (37 Nm) and the nut to 7 ft. lbs. (10 Nm).

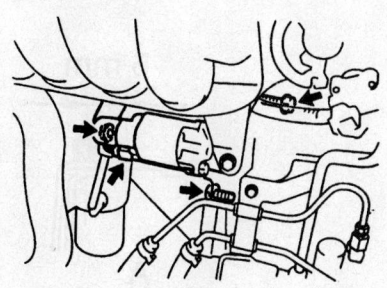

Starter mounting—Vibe

Oil Pan

REMOVAL & INSTALLATION

1ZZ-FE Engine

1. Before servicing the vehicle, refer to the precautions section.
2. Drain the engine oil.
3. Remove or disconnect the following:
   - Negative battery cable
   - Undercovers
   - Front exhaust pipe
   - Oil pan mounting bolts and nuts
   - Oil pan, cutting off the applied sealer.

To install:

4. Remove any old sealant from the oil pan flange and thoroughly clean the sealing surface.
5. Install or connect the following:
   - Oil pan. Tighten the bolts and nuts in several passes to 80 inch lbs. (9 Nm).
   - Front exhaust pipe
   - Negative battery cable
   - Undercovers
6. Fill the engine with clean oil.
7. Start the vehicle, check for leaks and repair if necessary.

2ZZ-GE Engine

1. Before servicing the vehicle, refer to the precautions section.
2. Drain the engine oil.
3. Remove or disconnect the following:
   - Negative battery cable. On vehicles equipped with an air bag, wait at least 90 seconds before proceeding.
   - Undercovers
   - Front exhaust pipe
   - Oil pan mounting bolts and nuts
   - Oil pan, cutting off the applied sealer

To install:

4. Remove any old sealant from the oil pan flange and thoroughly clean the sealing surface.
5. Install or connect the following:
   - Oil pan. Tighten the bolts and nuts in several passes to 80 inch lbs. (9 Nm).
   - Front exhaust pipe
   - Negative battery cable
   - Undercovers
6. Fill the engine with clean oil.
7. Start the vehicle, check for leaks and repair if necessary.

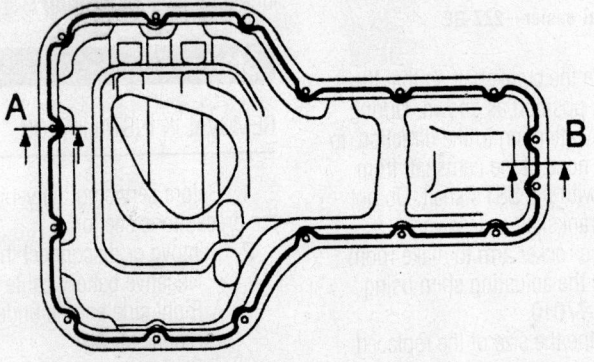

Seal Width
4 – 5 mm

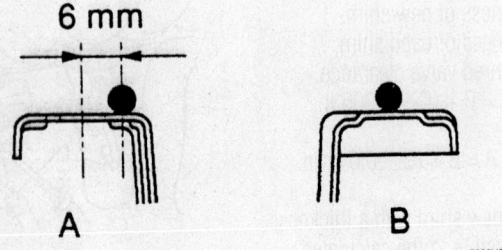

Apply sealant to the oil pan as shown—1.8L (1ZZ-FE) engine

7923VG72

Oil Pump

REMOVAL & INSTALLATION

1ZZ-FE Engine

1. Before servicing the vehicle, refer to the precautions section.
2. Drain the engine oil.
3. Remove or disconnect the following:
   - Negative battery cable
   - Timing chain and crankshaft sprocket
   - Timing chain vibration damper
   - Oil pump bolts, pump and gasket

To install:

4. Clean the mounting surface.
5. Install or connect the following:
   - Oil pump, with new gasket. Engage the spline teeth of the oil pump drive rotor with the larger teeth of the crankshaft, and slide the pump on.
   - Oil pump bolts and tighten to 97 inch lbs. (11 Nm)
   - Crankshaft vibration damper and tighten to 80 inch lbs. (9 Nm)
   - Crankshaft sprocket and timing chain
   - Negative battery cable
6. Fill the engine with clean oil.
7. Start the vehicle, check for leaks and repair if necessary.

2ZZ-GE Engine

1. Before servicing the vehicle, refer to the precautions section.
2. Drain the engine oil.
3. Remove or disconnect the following:
   - Negative battery cable
   - Timing chain and crankshaft sprocket
   - Oil pump and gasket

To install:

4. Clean the mounting surface.
5. Install or connect the following:

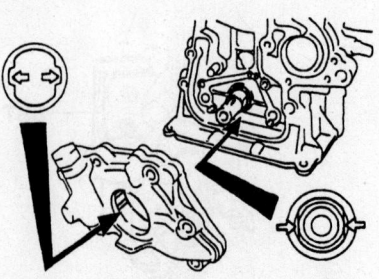

Oil pump mounting—1ZZ-FE and 2ZZ-GE engines

9359AB34

- Oil pump, with new gasket. Engage the spline teeth of the oil pump drive rotor with the larger teeth of the crankshaft, and slide the pump on.
- Oil pump bolts and tighten to 97 inch lbs. (11 Nm)
- Crankshaft sprocket and timing chain
- Negative battery cable

6. Fill the engine with clean oil.

7. Start the vehicle, check for leaks and repair if necessary.

Rear Main Seal

REMOVAL & INSTALLATION

1. Remove or disconnect the following:
   - Transaxle
   - Clutch assembly
   - Flywheel or flexplate

2. Use a small sharp knife to cut off the lip of the oil seal. Take great care not to score any metal with the knife.

3. Use a small prytool to pry the old seal from the retaining plate. Be careful not to damage the plate. Protect the tip of the tool with tape and pad the fulcrum point with cloth.

4. Inspect the crankshaft and seal lip contact surfaces for any sign of damage.

To install:

5. Apply a light coat of multi-purpose grease to the lip of a new oil seal. Loosely fit the seal into place by hand, making sure it is not crooked.

6. Use a seal driver of the correct size to install the seal. Tap it into place until the surface of the seal is flush with the edge of the housing.

Timing Chain, Sprockets, Front Cover and Seal

REMOVAL & INSTALLATION

1. Before servicing the vehicle, refer to the precautions section.

2. Drain the cooling system.

3. Remove or disconnect the following:
   - Right side engine under cover
   - Right front wheel and tire
   - Cylinder head cover
   - Wire harness clamp and suction hose assembly, 2ZZ-GE engine
   - Drive belt

4. Separate the vane pipe assembly, but do not disconnect the hose, 1ZZ-FE engine.
   - Alternator bracket, 2ZZ-GE

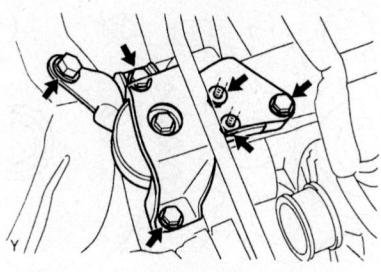

With the engine supported, remove the right side engine mount—1ZZ-FE engine shown, 2ZZ-GE similar

- Alternator
- Power steering pump reservoir and position it aside, 1ZZ-FE engine

5. Place a jack with a wooden block under the vehicle for support, then remove the 4 bolts and 2 nuts and remove the right side engine mount.

6. Remove the engine wire as follows, on 1ZZ-FE engines:

   a. Remove the 5 clamps from the brackets.

   b. Detach the connectors.

   c. Remove the ignition coil connectors.

   d. Bolt and nut holding the engine wire.

7. Remove the engine wire as follows, on 2ZZ-GE engines:

   a. Detach the ignition coil, oil control valve and Crankshaft Position Sensor (CKP) sensor electrical connectors.

   b. Bolt and nut for the engine ground, then position the engine wire aside

8. Remove or disconnect the following:

   - Ignition coil assembly
   - Positive Crankcase Ventilation (PCV) hoses from the cylinder head cover, if necessary
   - Cylinder head (valve) cover sub-assembly

9. Set the No. 1 cylinder to Top Dead Center (TDC) of the compressor stroke as follows:

   a. Turn the crankshaft pulley, and align its groove with the "0" timing mark of the timing chain cover.

   b. Make sure the point marks of the camshaft timing sprockets and VVT timing sprockets are in a straight line as shown. If not, turn the crankshaft 1 complete revolution (360°) and align the marks.

   - Crankshaft pulley, using SST 09960-10010
   - Belt tensioner

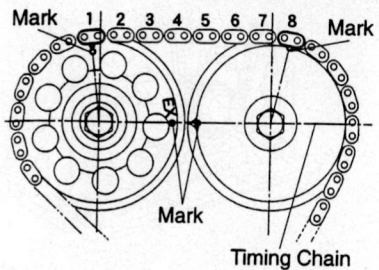

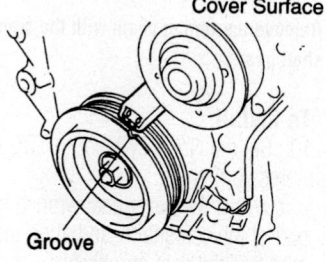

Proper timing mark alignment for TDC

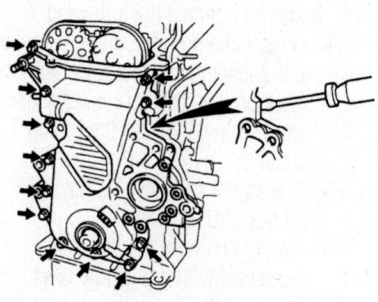

Timing chain cover mounting—1ZZ-FE engine shown, 2ZZ-GE similar

- Water pump pulley, if equipped, and pump
- Transverse engine mounting bracket
- Crankshaft Position (CKP) sensor
- No. 1 chain tensioner assembly, making sure not to revolve the crankshaft without the tensioner
- Timing chain cover. The cover is retained with 11 bolts and nuts and a Torx® stud bolt. Pry the cover between the cylinder head and block to remove it.
- Timing gear cover oil seal
- CKP sensor plate No. 1
- Timing chain tensioner slipper

➡ **In case you turn the camshafts with the timing chain removed, turn the crankshaft ¼ turn for the valve to avoid contact with the pistons.**

- Timing chain sub-assembly. Remove the chain with the crankshaft gear, using screwdrivers as shown.

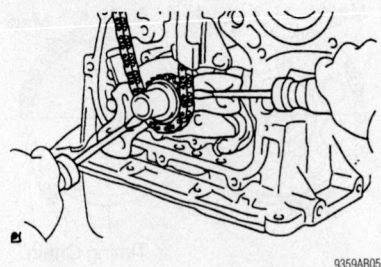

Remove the timing chain with the crankshaft gear

To install:

10. Set the No. 1 cylinder to TDC of the compression stroke:

a. Turn the hexagonal wrench head part of the camshafts, and align the point marks of the cam sprockets.

b. Using the crankshaft pulley bolt, turn the crankshaft and position the crankshaft set key upward.

11. Install or connect the following:

- Timing chain on the crank sprocket with the yellow link aligned with the mark on the crank sprocket. There are 3 yellow links on the timing chain.
- Crankshaft sprocket, using SST 09223-22010
- Timing chain on the camshaft sprockets with the yellow links aligned with the marks on the cam sprockets
- Timing chain tensioner slipper and tighten the bolt to 14 ft. lbs. (19 Nm)
- Crankshaft position sensor plate, with the "F" mark facing forward
- Timing gear cover oil seal
- Timing cover. For 1ZZ-FE engine, tighten the "A" bolts to 10 ft. lbs. (13 Nm), the "B" bolts to 14 ft. lbs. (19 Nm) and the stud bolt to 84 inch lbs. (9.5 Nm), using a Torx® wrench. For 2ZZ-GE engines, tighten the M8 bolts to 15 ft. lbs. (21 m), the M6 bolts to 8 ft. lbs. (11 Nm) and the stud bolt to 84 inch lbs. (9.5 Nm).

➡When installing the tensioner, make sure to set the hook again if the hook releases the plunger.

- Timing chain tensioner. Torque the nuts to 80 inch lbs. (9 Nm).
- CKP sensor and tighten the bolts to 80 inch lbs. (9 Nm)
- Transverse engine mounting bracket. Tighten the bolts to 35 ft. lbs. (47 Nm).

Proper alignment of the camshaft sprockets—1ZZ-FE engine

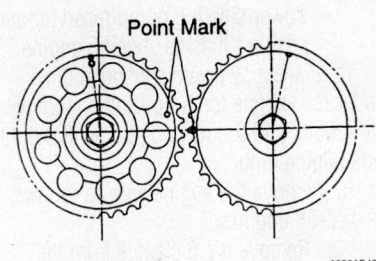

Proper alignment of the camshaft sprockets—2ZZ-GE engine

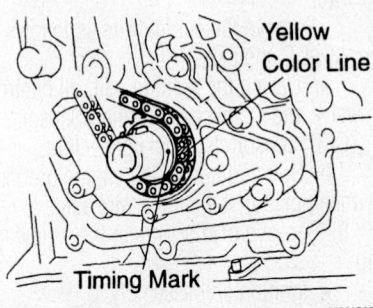

Make sure the yellow link is aligned with the crankshaft sprocket timing mark—1ZZ-FE and 2ZZ-GE engines

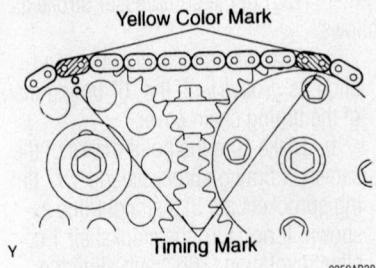

The yellow links of the timing chain must align with the camshaft sprocket timing marks—1ZZ-FE and 2ZZ-GE engines

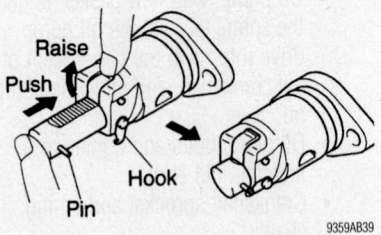

Timing chain tensioner—1ZZ-FE engine

- Water pump and pulley
- Drive belt tensioner. Tighten the nut to 21 ft. lbs. (29 Nm) and the bolt to 51 ft. lbs. (69 Nm) on 1ZZ-FE engines or to 74 ft. lbs. (100 Nm) on 2ZZ-GE engines.

12. Install the crankshaft pulley, as follows:

a. Align the pulley set key with the key groove of the pulley and slide on the pulley.

b. Use SST 09960-11010 to install the bolt and tighten to 102 ft. lbs. (138 Nm) for 1ZZ-FE engine or to 87 ft. lbs. (118 Nm) on 2ZZ-GE engines.

c. Turn the crankshaft counterclockwise and disconnect the plunger knock pin from the hook.

d. Turn the crankshaft clockwise and check that the slipper is pushed by the plunger. If the plunger does not spring out, press the slipper into the chain tensioner with a screwdriver so that the hook is released from the knock pin and the plunder springs out.

- Cylinder head sub-assembly cover. Install seal packing into the locations shown and install within 3 minutes. Tighten the "A" bolts to 8 ft. lbs. (11 Nm) and the "B" bolts to 80 inch lbs. (9 Nm) for 1ZZ-FE engines. For 2ZZ-GE engines, tighten the bolts to 7 ft. lbs. (10 Nm).
- Ignition coil assembly. Torque the bolts to 80 inch lbs. (9 Nm).
- Engine wire and tighten to 80 inch lbs. (9 Nm)
- Right side engine mount. Tighten to 38 ft. lbs. (52 Nm).
- Alternator bracket, 2ZZ-GE engine
- Alternator
- Vane pump, 1ZZ-FE
- Main cylinder head cover and tighten to 62 inch lbs. (7 Nm)
- Right front wheel and tire. Tighten the lug nuts to 76 ft. lbs. (103 Nm).

13. Fill the cooling system to the proper level.

14. Start the vehicle, check for leaks and repair if necessary.

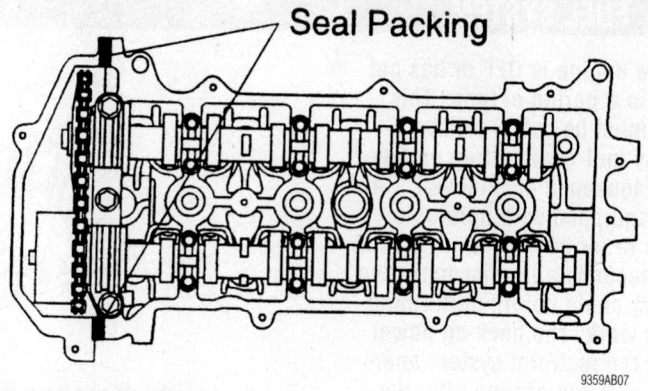

Seal packing installation locations

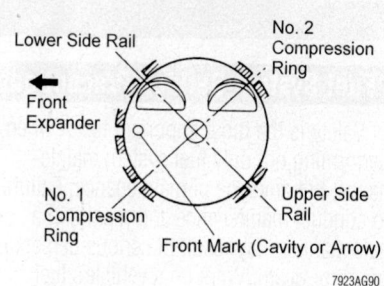

Piston ring end-gap spacing —1ZZ-FE and 2ZZ-GE engines

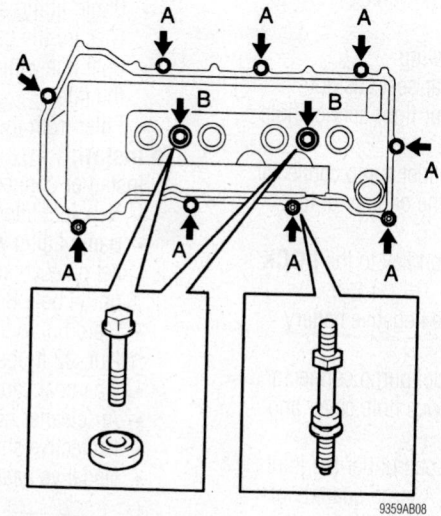

Cylinder head (valve) cover bolt locations—1ZZ-FE engine

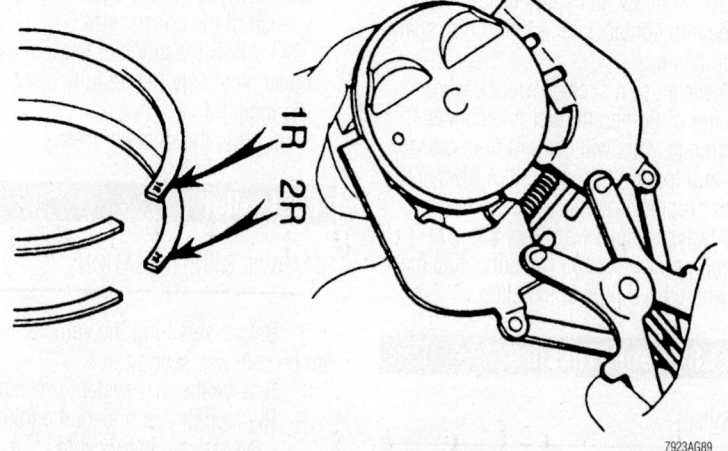

Front Mark (Cavity or Arrow)

Front Mark (Protrusion)

Piston-to-connecting rod assembly —1ZZ-FE and 2ZZ-GE engines

Piston and Ring Positioning

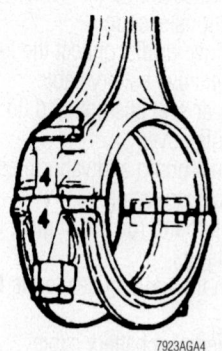

Before removing the caps from the connecting rods, be sure to matchmark them as shown

Piston ring identification mark locations—1ZZ-FE and 2ZZ-GE engines

FUEL SYSTEM

Fuel System Service Precautions

Safety is the most important factor when performing not only fuel system maintenance, but any type of maintenance. Failure to conduct maintenance and repairs in a safe manner may result in serious personal injury or death. Work on a vehicle's fuel system components can be accomplished safely and effectively by adhering to the following rules and guidelines.

• To avoid the possibility of fire and personal injury, always disconnect the negative battery cable unless the repair or test procedure requires that battery voltage by applied.

• Always relieve the fuel system pressure prior to disconnecting any fuel system component (injector, fuel rail, pressure regulator, etc.) fitting or fuel line connection. Exercise extreme caution whenever relieving fuel system pressure, to avoid exposing skin, face and eyes to fuel spray. Please be advised that fuel under pressure may penetrate the skin or any part of the body that it contacts.

• Always place a shop towel or rag around the fitting or connection prior to loosening to absorb any excess fuel due to spillage. Ensure that all fuel spillage is quickly remove from engine surfaces. Ensure that all fuel-soaked cloths or towels are deposited into a flame-proof waste container with a lid.

• Always keep a dry chemical (Class B) fire extinguisher near the work area.

• Do not allow fuel spray or fuel vapors to come into contact with a light bulb, spark or open flame.

• Always use a second wrench when loosening or tightening fuel line connections fittings. This will prevent unnecessary stress and torsion to fuel piping. Always follow the proper torque specifications.

• Always replace worn fuel fitting O-rings with new ones. Do not substitute fuel hose where rigid pipe is installed.

Fuel System Pressure

RELIEVING

❄ CAUTION

Failure to relieve fuel pressure before repairs or disassembly can cause serious personal injury and/or property damage. Fuel pressure is maintained within the fuel lines, even if the engine is OFF or has not been run in a period of time. This pressure must be safely relieved before any fuel-bearing line or component is loosened or removed. On vehicles equipped with inflatable restraints or air bag systems, wait at least 90 seconds after disconnecting the battery cable before performing any other work. The back-up power will keep the restraint system energized for a period of time after the battery is disconnected.

1. Before servicing the vehicle, refer to the precautions section.
2. Perform the following:
    a. Remove the rear seat cushion.
    b. Remove the rear floor service hole cover.
    c. Disconnect the fuel pump connector.
    d. Start and run the engine, until it stalls.
    e. Turn the ignition key to the **LOCK** position.
    f. Disconnect the negative battery cable.
    g. Connect the fuel pump connector.
    h. Install the service hole cover and rear seat cushion.
    i. Place a catch-pan under the joint to be disconnected. A large quantity of fuel may be released when the joint is opened.
    j. Wear eye or full face protection.
    k. Place a shop towel over the area and slowly release the joint using a wrench of the correct size.
    l. Allow the any fuel left in the line to bleed off slowly before fully disconnecting the joint.
    m. Plug the opened lines.

Fuel Filter

REMOVAL & INSTALLATION

1. Before servicing the vehicle, refer to the precautions section.
2. Relieve the fuel system pressure.
3. Remove or disconnect the following:
    • Negative battery cable
    • Protective shield for the fuel filter
    • Air cleaner hose and cap, if necessary
    • Charcoal canister, if necessary
    • Slowly loosen the lower flare nut fitting until all the pressure is relieved

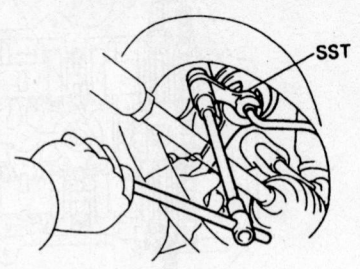

7923VG85

A line wrench with an extension may be needed to loosen the inlet line at the filter

    • Banjo fitting and 2 metal gaskets. Discard the gaskets.
    • Fuel line with the flared nut from the filter
    • Filter from the mounting bracket

To install:
4. Install or connect the following:
    • New fuel filter
    • Banjo fitting with a new metal gasket on each side and install the union bolt. Bolt: 22 ft. lbs. (30 Nm).
    • Flare nut to the lower connection. Nut: 22 ft. lbs. (30 Nm).
    • Charcoal canister
    • Air cleaner hose and cap
    • Protective shield
    • Negative battery cable

Fuel Pump

REMOVAL & INSTALLATION

1. Before servicing the vehicle, refer to the precautions section.
2. Remove or disconnect the following:
    • Negative battery cable
    • Rear seat cushion and floor service hole cover
    • Fuel pump and vapor pressure sensor connectors
    • Start and run the engine, until it stalls
3. Turn the ignition key to the **LOCK** position.
    • Negative battery cable
4. Connect the fuel pump connector.
    • Fuel tank protector, AWD vehicles
    • Fuel tank main tube sub-assembly
    • Fuel emission tube sub-assembly No. 1, FWD vehicles
    • Fuel tank vent tube set plate. The plate is secured with 8 bolts on FWD vehicles, or 5 bolts on AWD vehicles.

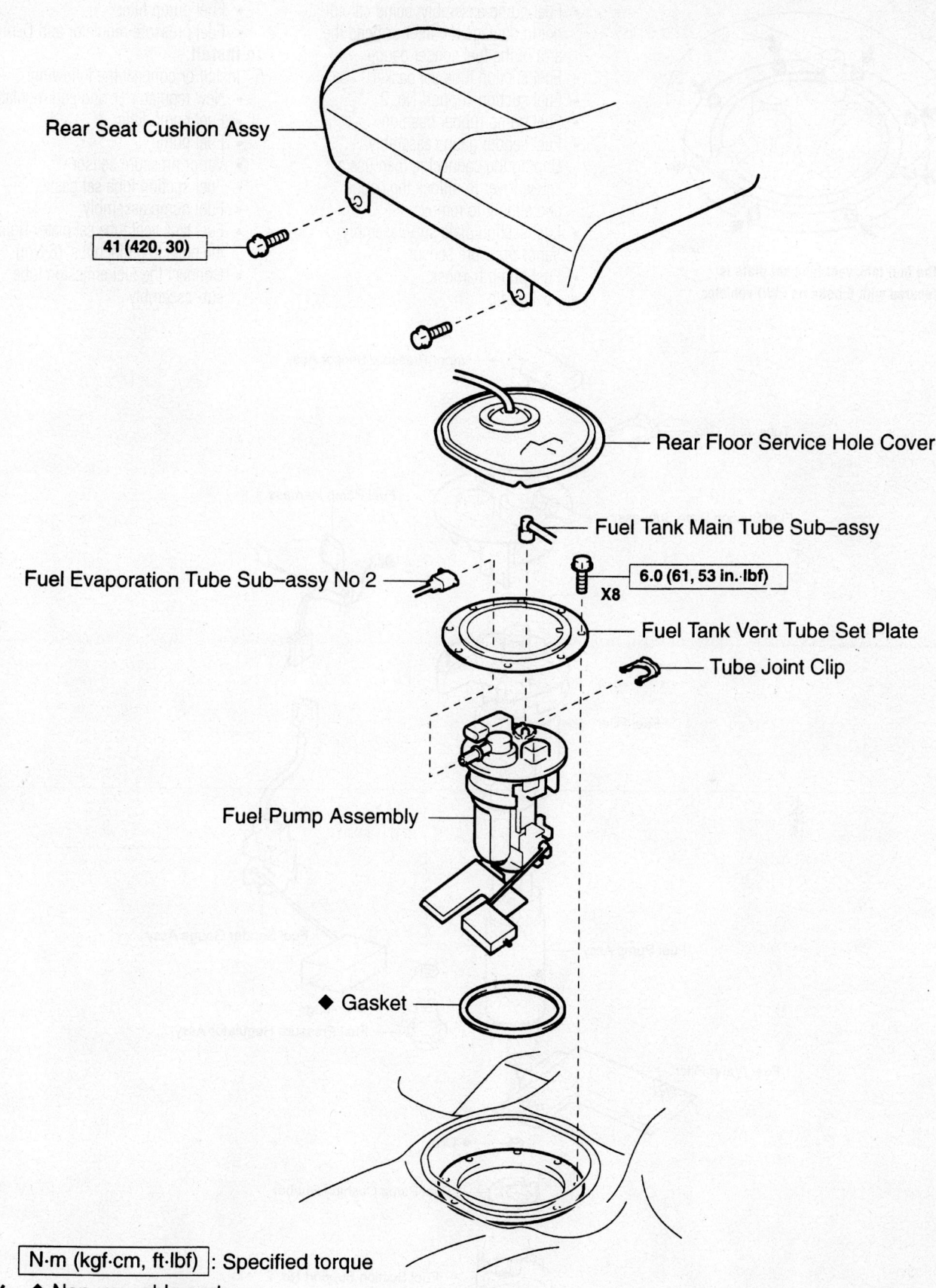

Rear Seat Cushion Assy

41 (420, 30)

Rear Floor Service Hole Cover

Fuel Tank Main Tube Sub–assy

Fuel Evaporation Tube Sub–assy No 2

6.0 (61, 53 in.·lbf)
X8

Fuel Tank Vent Tube Set Plate

Tube Joint Clip

Fuel Pump Assembly

◆ Gasket

N·m (kgf·cm, ft·lbf) : Specified torque

◆ Non–reusable part

9359AB41

Exploded view of the fuel pump mounting—FWD shown, AWD similar

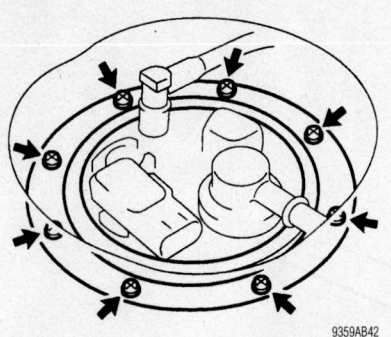

The fuel tank vent tube set plate is secured with 8 bolts on FWD vehicles

- Fuel pump assembly, being careful not to damage the filter or bend the arm of the fuel sender gauge
- Fuel suction tube set gasket
- Fuel suction support No. 2
- Fuel pump rubber cushion
- Fuel sender gauge assembly. Unplug the connector, then use a screwdriver to unlock the gauge and slide it to remove.
- Fuel section plate sub-assembly
- Vapor pressure sensor
- Fuel pump harness
- Fuel pump

- Fuel pump filter
- Fuel pressure regulator and O-ring

To install:

5. Install or connect the following:
- New regulator O-ring and regulator
- Fuel pump filter
- Fuel pump
- Vapor pressure sensor
- Fuel suction tube set gasket
- Fuel pump assembly
- Fuel tank vent tube set plate. Tighten the bolts to 53 inch lbs. (6 Nm).
- Connect the fuel emission tube sub-assembly

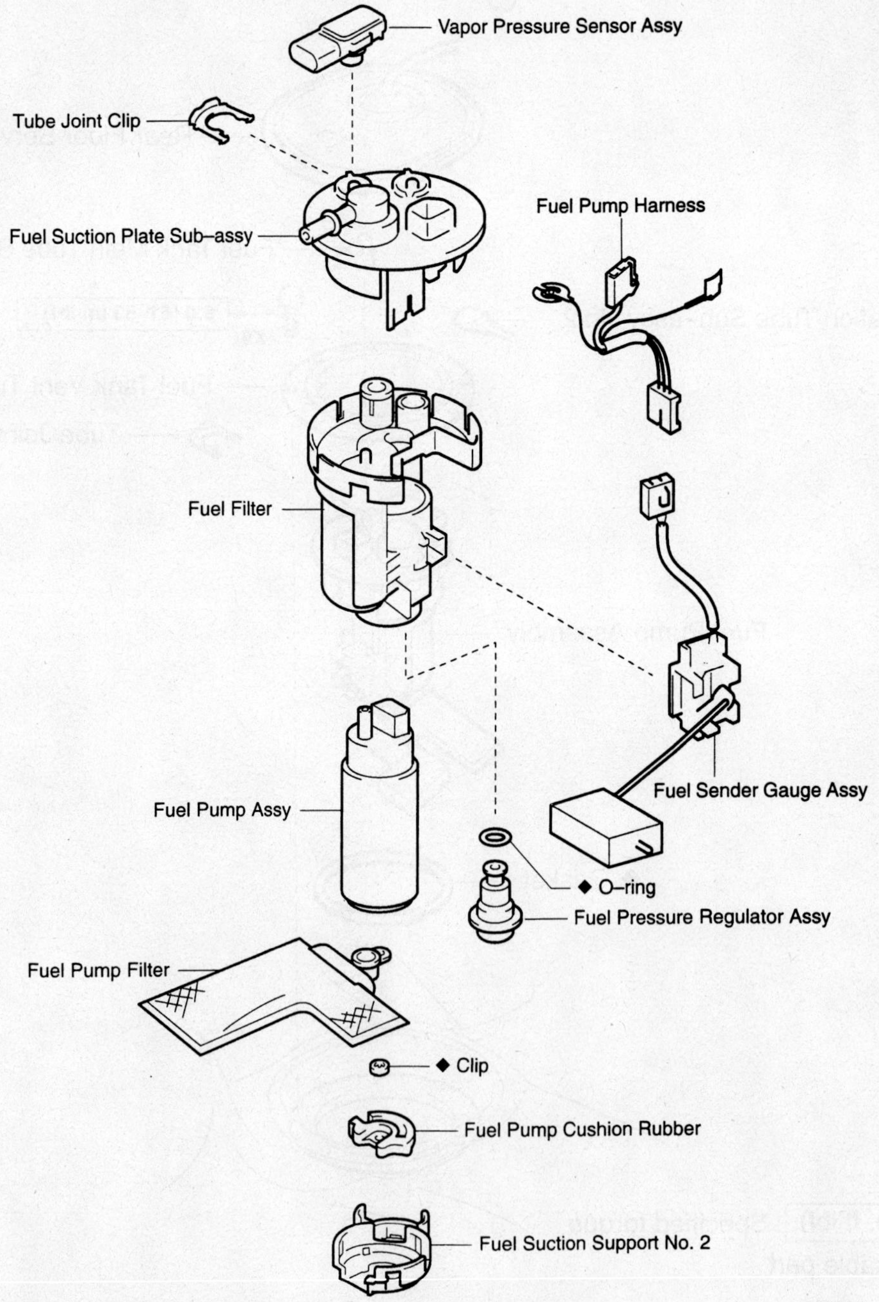

Vapor Pressure Sensor Assy

Tube Joint Clip

Fuel Suction Plate Sub–assy

Fuel Pump Harness

Fuel Filter

Fuel Pump Assy

Fuel Sender Gauge Assy

◆ O-ring

Fuel Pressure Regulator Assy

Fuel Pump Filter

◆ Clip

Fuel Pump Cushion Rubber

Fuel Suction Support No. 2

Fuel pump assembly components—FWD vehicles shown, AWD similar

- Fuel tank main tube sub-assembly
- Fuel tank protector No. 2, AWD vehicles
- Negative battery cable. Check for fuel leaks.
- Floor service hole cover. Use butyl tape to seal the cover.
- Rear seat cushion

Fuel Injectors

REMOVAL & INSTALLATION

1ZZ-FE Engine

1. Before servicing the vehicle, refer to the precautions section.

2. Properly relieve the fuel system pressure.

3. Remove or disconnect the following:
- Negative battery cable.
- No. 2 cylinder head cover
- Positive Crankcase Ventilation (PCV) hose
- Engine wire, unplugging the injector connectors and clamps

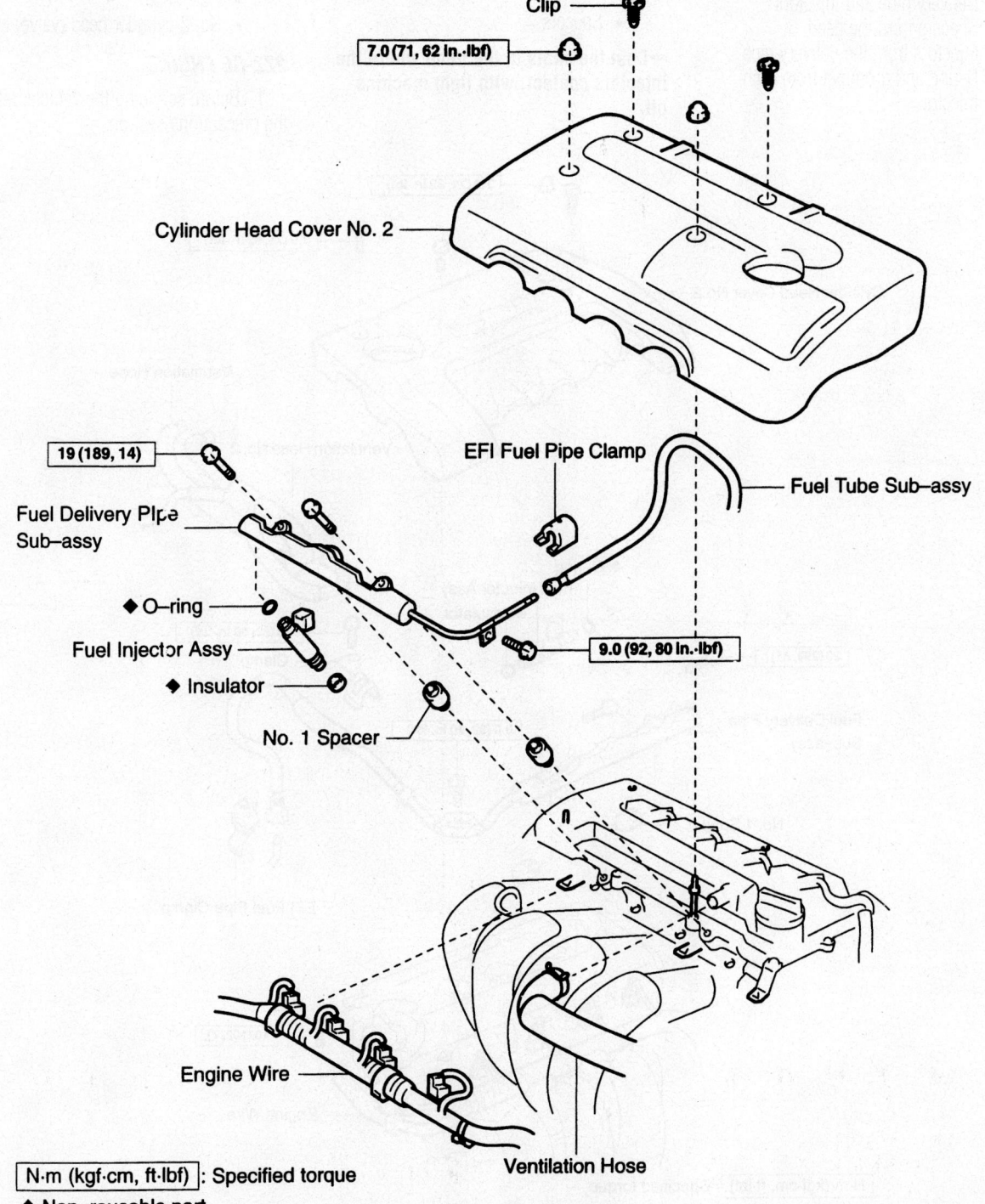

Clip

7.0 (71, 62 in.·lbf)

Cylinder Head Cover No. 2

EFI Fuel Pipe Clamp

Fuel Tube Sub–assy

19 (189, 14)

Fuel Delivery Pipe Sub–assy

◆ O–ring

Fuel Injector Assy

◆ Insulator

No. 1 Spacer

9.0 (92, 80 in.·lbf)

Engine Wire

Ventilation Hose

N·m (kgf·cm, ft·lbf) : Specified torque

◆ Non–reusable part

9359AB44

Fuel injector removal and installation—1ZZ-FE engine

- Fuel pipe clamp
- Fuel line/tube sub-assembly

✳✳ WARNING

Be careful not to drop the fuel injectors when removing the delivery pipe.

- Fuel delivery pipe sub-assembly with the injectors attached
- Delivery pipe and injectors
- Spacers from the head
- Injectors from the delivery pipe
- O-ring and grommet from each injector

To install:

4. Install or connect the following:
   - New grommets
   - New O-rings coated with light machine oil
   - Injectors on the delivery pipe

➡**Coat the contact point on the pipe with light machine oil and twist the injectors into place. The connector should face outward.**

   - Spacers

➡**Coat the seats in the head where the injectors contact, with light machine oil.**

- Delivery pipe and injectors

5. Loosely install the hold-down bolts and check that the injectors rotate smoothly. If they don't, the probable cause is incorrect O-ring installation. Torque the delivery pipe hold-down bolts to 14 ft. lbs. (19 Nm) and the fuel pipe bolt to 80 inch lbs. (9 Nm).

- Engine wire, attaching the injector connectors and clamps
- Fuel line/tube sub-assembly
- PCV hose
- No. 2 cylinder head (valve) cover

2ZZ-GE ENGINE

1. Before servicing the vehicle, refer to the precautions section.

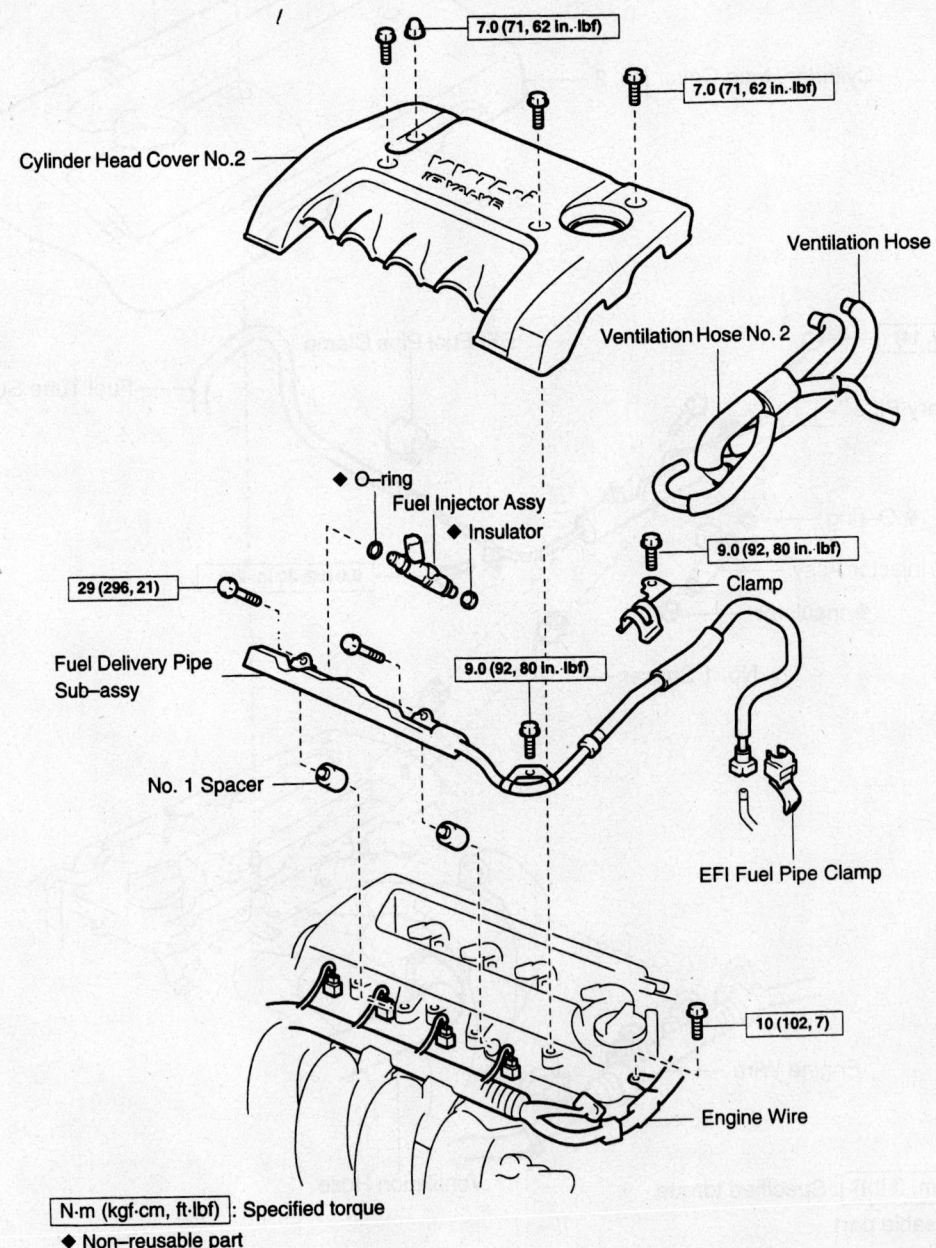

N·m (kgf·cm, ft·lbf) : Specified torque

◆ Non-reusable part

9359AB45

Fuel injector removal and installation—2ZZ-GE engine

2. Properly relieve the fuel system pressure.

3. Remove or disconnect the following:
- Negative battery cable.
- No. 2 cylinder head cover
- Positive Crankcase Ventilation (PCV) hose
- Engine wire, by removing the bolt, then unplugging the injector and Camshaft Position (CMP) sensor connectors
- Fuel pipe clamp

❋❋ WARNING

Be careful not to drop the fuel injectors when removing the delivery pipe.

- Fuel delivery pipe sub-assembly with the injectors attached
- Delivery pipe and injectors
- Spacers from the head
- Injectors from the delivery pipe
- O-ring and grommet from each injector

To install:

4. Install or connect the following:
- New grommets
- New O-rings coated with light machine oil
- Injectors on the delivery pipe

➡**Coat the contact point on the pipe with light machine oil and twist the injectors into place. The connector should face outward.**

- Spacers

➡**Coat the seats in the head where the injectors contact, with light machine oil.**

- Delivery pipe and injectors

5. Loosely install the hold-down bolts and check that the injectors rotate smoothly. If they don't, the probable cause is incorrect O-ring installation. Torque the delivery pipe hold-down bolts to 14 ft. lbs. (19 Nm) and the fuel pipe bolt to 80 inch lbs. (9 Nm).
- Fuel line/tube sub-assembly
- PCV hose
- Engine wire, by connecting the CMP sensor and injector connectors and installing the bolt. Tighten the bolt to 7 ft. lbs. (10 Nm).
- No. 2 cylinder head (valve) cover

DRIVE TRAIN

Manual Transaxle Assembly

REMOVAL & INSTALLATION

2003–04 Models

1. Before servicing the vehicle, refer to the precautions section.
2. Drain the transaxle fluid.
3. Place the front wheels in the straight-ahead position.
4. Remove or disconnect the following:
- Steering intermediate shaft
- Front wheel and tires
- Right and left side undercovers
- Exhaust pipe
- Hood
- Cylinder head (valve) cover
- Air cleaner assembly
- Battery clamp, battery, battery tray and battery carrier
- Cruise control actuator assembly, if equipped
5. Remove the wire harness as follows:
   a. Remove the wire harness clamp, 2 bolts and wire harness brackets.
   b. Remove the 2 bolts and 2 ground cables.
6. Remove or disconnect the following:
- Back-up lamp switch connector, with ABS
- Speed sensor connector, without ABS
- 5 bolts, then separate the release cylinder with the clutch pipes from the transaxle
- Shift cable clips and washer, then disconnect the cable from the transaxle and bracket

- Select cable clips and washer, then disconnect the cable from the transaxle and bracket
- Starter
- Right and left side tie rod ends
- Pressure feed tube
- Front halfshafts
7. Use a suitable tool to suspend the engine assembly, as shown in the illustration:
   a. No. 1 engine hanger: P/N 12281-22021 (5-speed M/T), 12281-88600 (6-speed M/T)
   b. No. 2 engine hanger: P/N 12281-15040 (5-speed M/T), 12281-88600 (6-speed M/T).
   c. Bolt: P/N 91512-B1016.
   d. Torque the bolts to 28 ft. lbs. (38 Nm).

❋❋ CAUTION

Do not try to suspend the engine by hooking the chain to any other part.

   e. Attach an engine chain hoist to the hangers.
8. Remove or disconnect the following:
- Front suspension crossmember
9. Support the transaxle with a floor jack.
- Transverse engine mounting insulator and brackets
- Manual transaxle assembly
- Transverse engine mounting brackets from the transaxle, if necessary

To install:
- Transverse engine mounting brackets to the transaxle, if necessary
- Manual transaxle, by aligning the

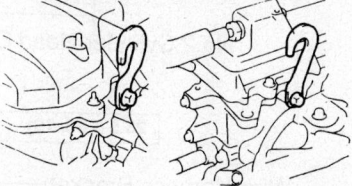

No. 1 Engine Hanger   No. 2 Engine Hanger

9359AB46

Secure the engine using the proper tools— 5-speed manual transmission shown

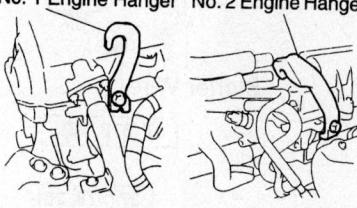

No. 1 Engine Hanger   No. 2 Engine Hanger

9359AB49

Secure the engine using the proper tools— 6-speed manual transmission shown

input shaft with the clutch disc. Torque the "A" bolts to 47 ft. lbs. (64 Nm), the "B" bolts to 35 ft. lbs. (47 Nm) and the "C" bolts to 17 ft. lbs. (23 Nm).
- Transverse engine mounting bracket. Tighten to 38 ft. lbs. (52 Nm).
- Transverse engine mounting insulator. Tighten the "A" bolts to 38 ft. lbs. (52 Nm) and the "B" bolts to 59 ft. lbs. (80 Nm).
10. The remainder of installation is the reverse of the removal procedure, noting the following specifications:

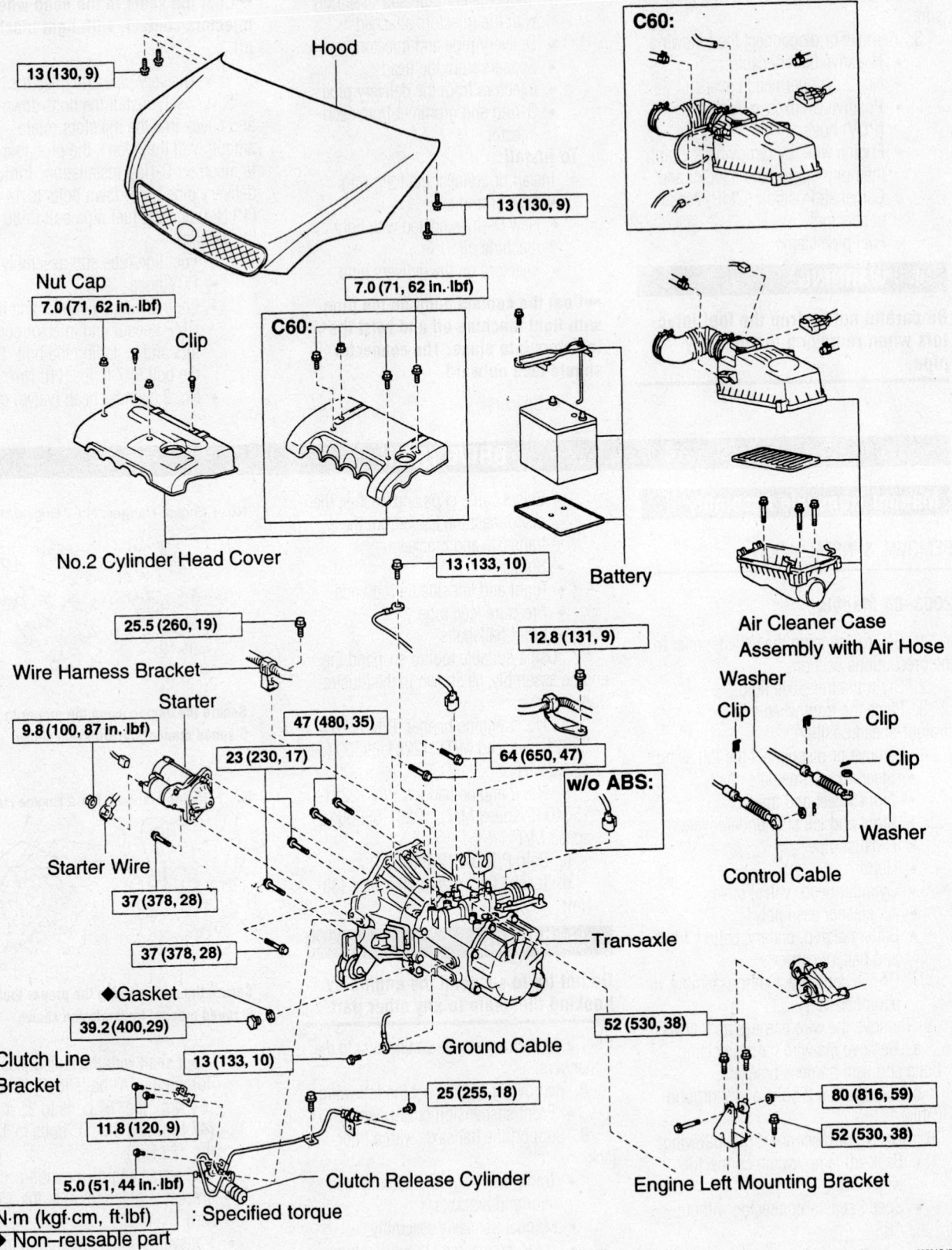

Hood

13 (130, 9)

13 (130, 9)

C60:

Nut Cap

7.0 (71, 62 in. lbf)

7.0 (71, 62 in. lbf)

Clip

C60:

Battery

No.2 Cylinder Head Cover

13 (133, 10)

Air Cleaner Case Assembly with Air Hose

25.5 (260, 19)

12.8 (131, 9)

Wire Harness Bracket

Washer

Clip

Clip

Clip

Starter

9.8 (100, 87 in. lbf)

47 (480, 35)

23 (230, 17)

64 (650, 47)

w/o ABS:

Washer

Starter Wire

Control Cable

37 (378, 28)

37 (378, 28)

Transaxle

◆Gasket

39.2 (400, 29)

52 (530, 38)

Clutch Line Bracket

13 (133, 10)

Ground Cable

11.8 (120, 9)

25 (255, 18)

80 (816, 59)

5.0 (51, 44 in. lbf)

52 (530, 38)

N·m (kgf·cm, ft·lbf) : Specified torque

Clutch Release Cylinder

Engine Left Mounting Bracket

◆ Non–reusable part

9359AB47

Exploded view of the manual transaxle (1 of 2)

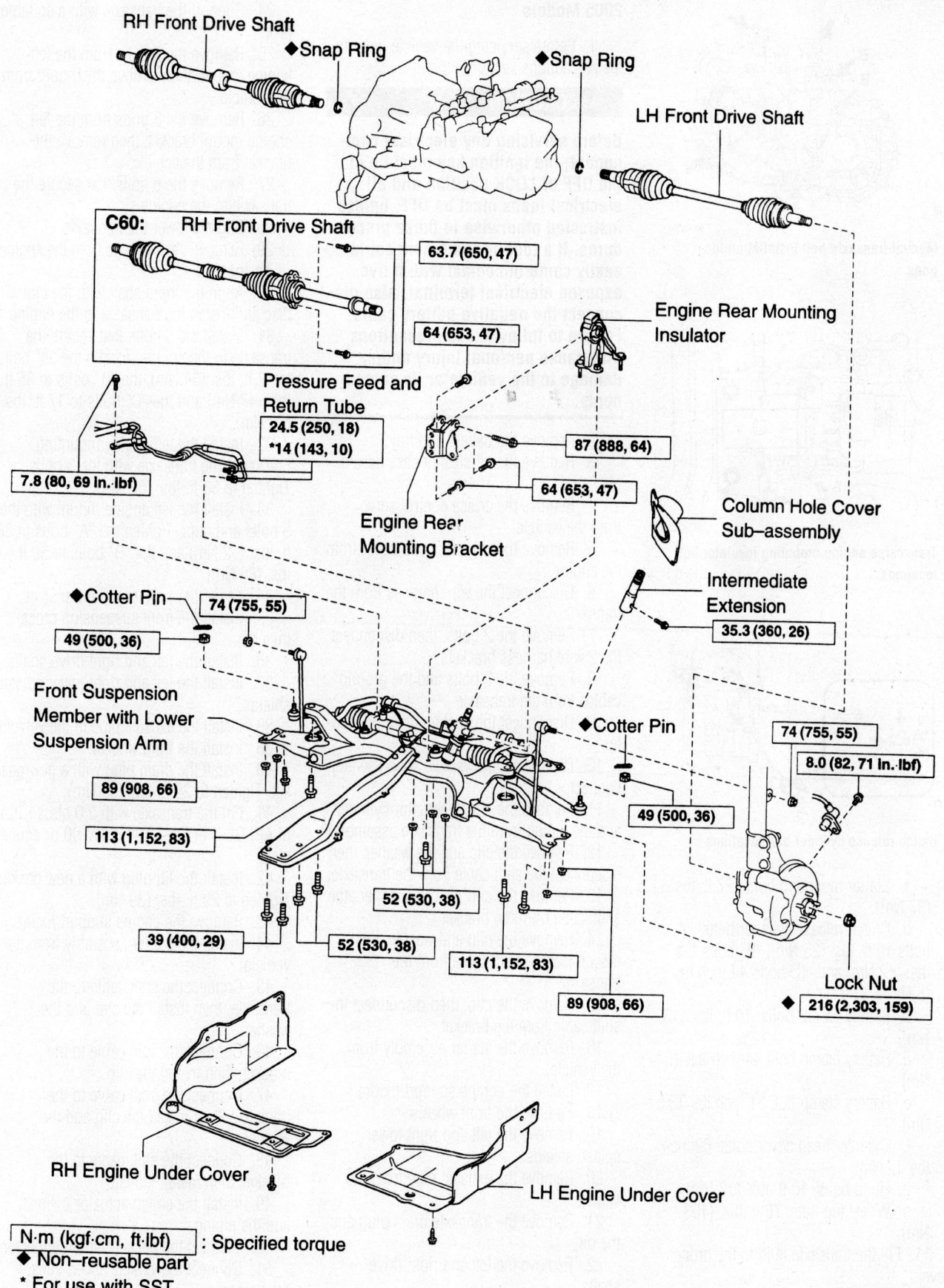

RH Front Drive Shaft

◆Snap Ring

◆Snap Ring

LH Front Drive Shaft

C60: RH Front Drive Shaft

63.7 (650, 47)

64 (653, 47)

Engine Rear Mounting Insulator

Pressure Feed and Return Tube

24.5 (250, 18)
***14 (143, 10)**

87 (888, 64)

7.8 (80, 69 In.·lbf)

64 (653, 47)

Column Hole Cover Sub–assembly

Engine Rear Mounting Bracket

◆Cotter Pin

74 (755, 55)

49 (500, 36)

Intermediate Extension

35.3 (360, 26)

Front Suspension Member with Lower Suspension Arm

◆Cotter Pin

74 (755, 55)

8.0 (82, 71 In.·lbf)

89 (908, 66)

49 (500, 36)

113 (1,152, 83)

52 (530, 38)

39 (400, 29)

52 (530, 38)

113 (1,152, 83)

89 (908, 66)

Lock Nut
◆ 216 (2,303, 159)

RH Engine Under Cover

LH Engine Under Cover

N·m (kgf·cm, ft·lbf) : Specified torque
◆ Non–reusable part
* For use with SST

9359AB48

Exploded view of the manual transaxle (2 of 2)

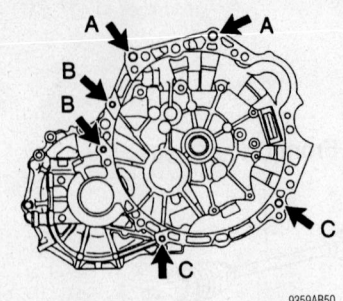

Manual transaxle bolt installation locations

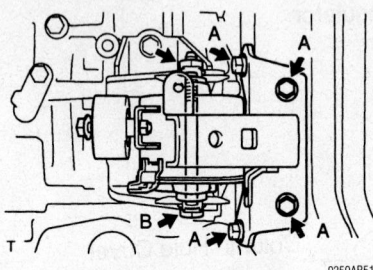

Transverse engine mounting insulator bolt locations

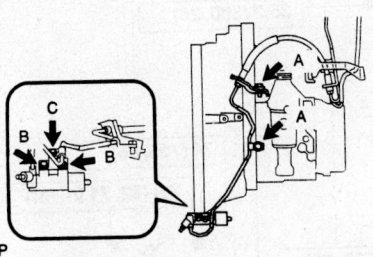

Clutch release cylinder bolt locations

a. Starter mounting bolts: 27 ft. lbs. (37 Nm).

b. Clutch release cylinder bolts: "A" bolts 19 ft. lbs. (25 Nm), "B" bolts 9 ft. lbs. (12 Nm) and "C" bolts 44 inch lbs. (5 Nm).

c. Battery carrier bolts: 10 ft. lbs. (13 Nm).

d. Battery clamp bolt: 44 inch lbs. (5 Nm).

e. Battery clamp nut: 31 inch lbs. (3.5 Nm).

f. Cylinder head cover bolts: 62 inch lbs. (7 Nm).

g. Hood bolts: 10 ft. lbs. (13 Nm).

h. Wheel lug nuts: 76 ft. lbs. (103 Nm).

11. Fill the transaxle fluid to the proper level.

12. Start the vehicle, check for leaks and repair if necessary.

2005 Models

1. Before servicing the vehicle, refer to the precautions section.

✳✳ CAUTION

Before servicing any electrical component, the ignition key must be in the OFF or LOCK position and all electrical loads must be OFF, unless instructed otherwise in these procedures. If a tool or equipment could easily come in contact with a live exposed electrical terminal, also disconnect the negative battery cable. Failure to follow these precautions may cause personal injury and/or damage to the vehicle or its components.

2. Remove the battery and tray.

3. Remove the air cleaner case assembly.

4. Remove the cruise control servo from the vehicle.

5. Remove the cylinder head cover from the engine.

6. Disconnect the wire harness from the transaxle.

7. Remove the 2 bolts, then disconnect the 2 wire harness brackets.

8. Remove the 2 bolts and the ground cables from the transaxle.

9. Disconnect the backup lamp connector.

10. Disconnect the vehicle speed sensor connector.

11. Remove the clutch actuator cylinder and the piping from the transaxle assembly.

12. Remove the clip and the washer, then disconnect the shift cable from the transaxle.

13. Remove the clip, then disconnect the shift cable from the bracket.

14. Remove the clip and washer, then disconnect the shift cable from the transaxle.

15. Remove the clip, then disconnect the shift cable from the bracket.

16. Remove the starter assembly from the vehicle.

17. Install the engine support fixture.

18. Remove the front wheels.

19. Remove the left and right lower splash shields.

20. Remove the exhaust pipe from the vehicle.

21. Remove the transaxle drain plug and the oil.

22. Remove the left and right drive shafts.

23. Remove the front suspension crossmember.

24. Support the transaxle with a suitable jack.

25. Remove the 5 bolts from the left engine mount, then remove the mount from the vehicle.

26. Remove the 3 bolts from the left engine mount bracket, then remove the bracket from the vehicle.

27. Remove the 6 bolts that secure the transaxle to the engine.

28. Slightly lower the transaxle.

29. Remove the transaxle from the engine.

To install:

30. Align the input shaft with the clutch disc and install the transaxle to the engine.

31. Install the 6 bolts that secure the transaxle to the engine. Torque the "A" bolts to 47 ft. lbs. (64 Nm), the "B" bolts to 35 ft. lbs. (47 Nm) and the "C" bolts to 17 ft. lbs. (23 Nm).

32. Install the left engine mounting bracket to the transaxle with the 3 bolts. Tighten to 38 ft. lbs. (52 Nm).

33. Install the left engine mount with the 5 bolts and nuts. Tighten the "A" bolts to 38 ft. lbs. (52 Nm) and the "B" bolts to 59 ft. lbs. (80 Nm).

34. Lower the jack from the transaxle.

35. Install the front suspension crossmember.

36. Install the left and right drive shafts.

37. Install the left and right lower splash shields.

38. Install the exhaust pipe in the vehicle.

39. Install the front wheels.

40. Install the drain plug with a new gasket. Tighten to 29 ft. lbs. (39 Nm).

41. Fill the transaxle with 2.0 qts. (1.9L) of API GL-4 or GL-5 SAE 75W-90 or equivalent.

42. Install the fill plug with a new gasket. Tighten to 29 ft. lbs. (39 Nm).

43. Remove the engine support fixture.

44. Install the starter assembly from the vehicle.

45. Connect the shift cable to the transaxle, then install the clip and the washer.

46. Connect the shift cable to the bracket, then install the clip.

47. Connect the shift cable to the transaxle, then install the clip and the washer.

48. Connect the shift cable to the bracket, then install the clip.

49. Install the clutch actuator cylinder and the piping.

50. Connect the backup lamp connector.

51. Connect the vehicle speed sensor connector.

52. Connect the wire harness to the transaxle.

53. Connect the 2 wire harness brackets, then install the 2 bolts.

54. Install the 2 bolts and the ground cables to the transaxle.

55. Install the cruise control servo in the vehicle.

56. Install the battery tray and the 4 bolts.

❋❋ CAUTION

Before servicing any electrical component, the ignition key must be in the OFF or LOCK position and all electrical loads must be OFF, unless instructed otherwise in these procedures. If a tool or equipment could easily come in contact with a live exposed electrical terminal, also disconnect the negative battery cable. Failure to follow these precautions may cause personal injury and/or damage to the vehicle or its components.

57. Install the battery.
58. Install the air cleaner case assembly.

59. Install the cylinder head cover in the engine.

Automatic Transaxle

REMOVAL AND INSTALLATION

FWD—A246E & U240E Transaxles

1. Before servicing the vehicle, refer to the precautions section.
2. Drain the transaxle fluid.
3. Remove or disconnect the following:

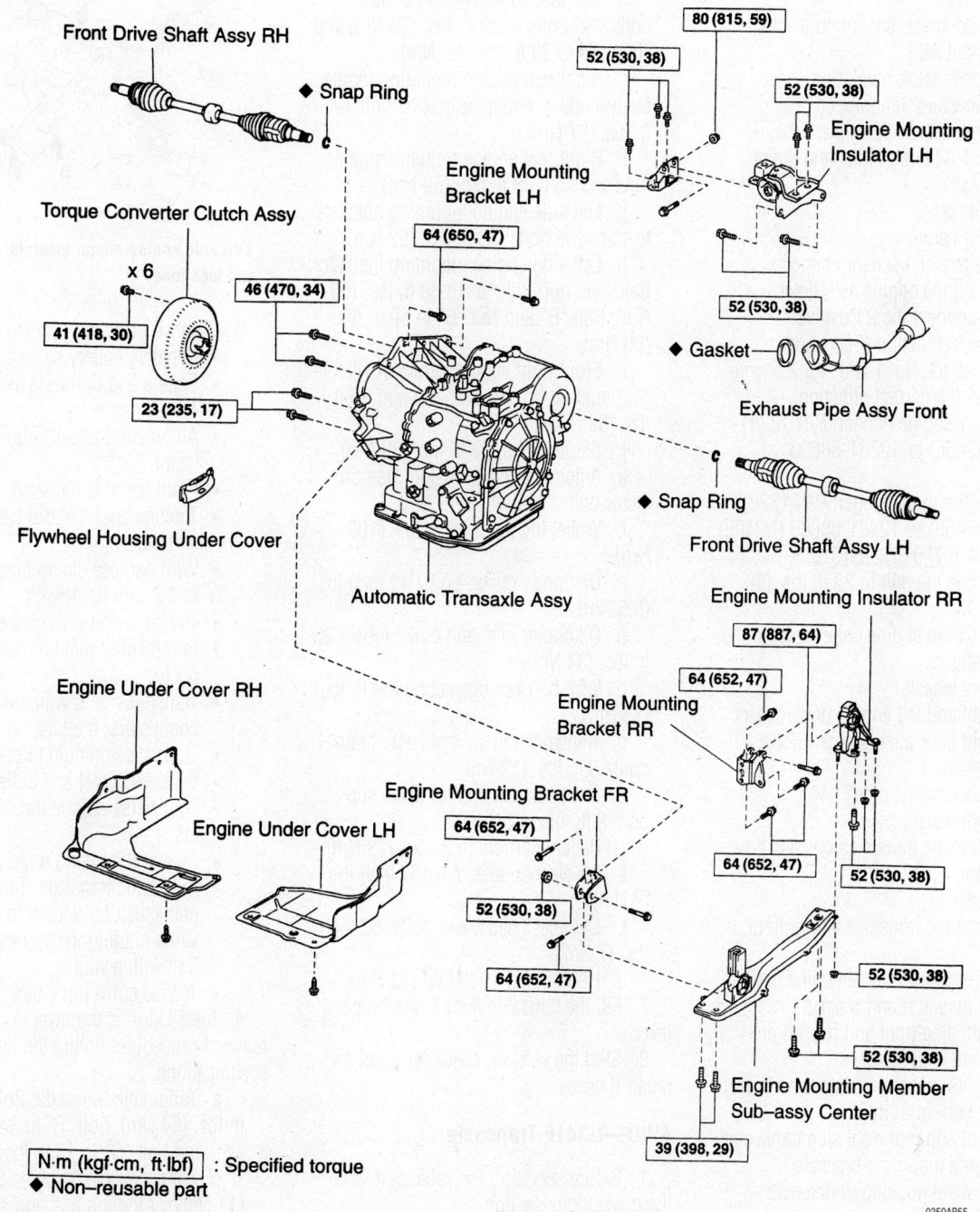

N·m (kgf·cm, ft·lbf) : Specified torque
◆ Non—reusable part

Automatic transaxle and related components—U240E transaxle shown, A246E similar

9359AB55

- Negative battery cable
- Hood
- No. 2 cylinder head cover
- Battery and battery carrier
- Air cleaner assembly with hose
- Floor shift cable transmission control shift
- Transmission control cable support
- No. 1 transmission control cable bracket
- Wiring harness and brackets
- Transmission wire connector
- Park/neutral position switch connector, with Anti-lock Brake System (ABS)
- Speedometer sensor connector, without ABS
- Transmission revolution sensor connectors, if equipped
- Transmission fluid filler tube
- No. 1 oil cooler inlet and outlet tubes
- Foot rest
- Floor carpet
- Oxygen (O_2) sensor connector

4. Suspend the engine as follows:

a. Disconnect the 2 Positive Crankcase Ventilation (PCV) hoses.

b. Install the No. 1 and No. 2 engine hangers in the correct direction.

c. No. 1 engine hanger: P/N 12281-22021 (A246E) or 12281-88600 (U240E).

d. No. 2 engine hanger: P/N 12281-15040 (A246E) or 12281-88600 (U240E)

e. Bolt: P/N 91512-B1016.

f. Torque the bolt to 28 ft. lbs. (38 Nm).

g. Attach an engine chain hoist to the engine hangers.

- Front wheels
- Right and left engine undercovers
- Front floor panel brace, U240E transaxle
- Front exhaust pipe
- Front halfshafts
- Automatic transmission case protector
- Starter

5. Support the transaxle with a floor jack

- Left side transverse engine mounting insulator and bracket
- Right side front and rear engine mount insulators
- 4 bolts, dynamic damper and member sub-assembly
- Front and rear right side transverse engine mounting brackets
- Flywheel housing undercover
- Automatic transaxle. Turn the crankshaft for access to the 6 bolts

while holding the crankshaft pulley bolt with a wrench.

- Torque converter clutch

6. Installation is the reverse of the removal procedure, noting the following specifications:

a. Automatic transaxle: Bolt "A" to 47 ft. lbs. (64 Nm), bolt "B" to 34 ft. lbs. (47 Nm) and bolt "C" to 17 ft. lbs. (23 Nm).

b. Torque converter bolts: 20 ft. lbs. (28 Nm).

c. Front and rear right transverse engine mounting bracket bolts: 47 ft. lbs. (64 Nm).

d. Member sub-assembly center bolts: "A" bolts to 29 ft. lbs. (39 Nm) and "B" bolts to 38 ft. lbs. (52 Nm).

e. Right rear engine mounting insulator-to-engine mounting bracket bolt: 64 ft. lbs. (87 Nm).

f. Right rear engine mount insulator nuts and bolt: 38 ft. lbs. (52 Nm).

g. Left side engine mounting bracket-to-transaxle bolts: 38 ft. lbs. (52 Nm).

h. Left side engine mounting insulator bolts and nut: Bolt "A" to 38 ft. lbs. (52 Nm), Bolt "B" and Nut "B" to 59 ft. lbs. (80 Nm).

i. Front right engine mount insulator-to-mounting bracket bolt and nut: 38 ft. lbs. (52 Nm).

j. Starter bolts: 29 ft. lbs. (39 Nm).

k. Automatic transmission case protector bolts: 14 ft. lbs. (18 Nm).

l. Wheel lug nuts: 76 ft. lbs. (103 Nm).

m. Oil cooler clamp bolts: 49 inch lbs. (5.5 Nm).

n. Oil cooler inlet and outlet tubes: 25 ft. lbs. (34 Nm).

o. Wire harness bracket bolt: 9 ft. lbs. (13 Nm).

p. Transmission control cable bracket bolts: 9 ft. lbs. (12 Nm).

q. Transmission control cable support: 9 ft. lbs. (12 Nm).

r. Battery carrier: 10 ft. lbs. (13 Nm).

s. Air cleaner assembly: 62 inch lbs. (7 Nm).

t. Cylinder head cover bolts: 62 inch lbs. (7 Nm).

u. Hood bolts: 10 ft. lbs. (13 Nm).

7. Fill the transaxle fluid to the proper level.

8. Start the vehicle, check for leaks and repair if necessary.

AWD—U341F Transaxle

1. Before servicing the vehicle, refer to the precautions section.

2. Drain the transaxle fluid.

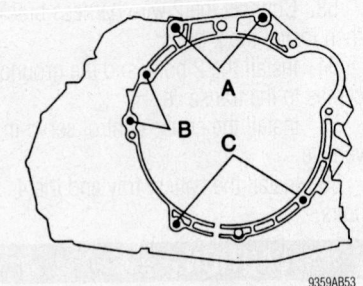

Automatic transaxle bolt locations

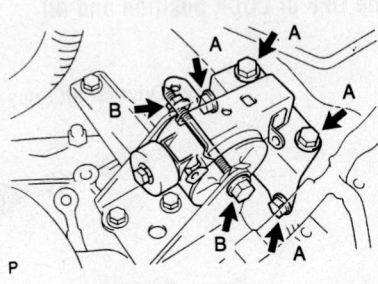

Left side engine mount insulator bolt and nut locations

3. Remove or disconnect the following:

- Negative battery cable
- Engine and transaxle assembly
- Transfer case
- Automatic transmission case protector
- Front left side halfshaft
- Transmission control cable support and bracket
- Wire harness clamp bracket, bolts and 2 wire harnesses
- Transmission wire connector
- Park/neutral position switch connector
- Transmission revolution sensor connectors, if equipped
- Transmission fluid filler tube
- Oil cooler inlet and outlet tubes
- Transverse engine mounting brackets
- Flywheel housing undercover
- Automatic transaxle. Turn the crankshaft for access to the 6 bolts while holding the crankshaft pulley bolt with a wrench.
- Torque converter clutch

4. Installation is the reverse of the removal procedure, noting the following specifications:

a. Automatic transaxle: Bolt "A" to 47 ft. lbs. (64 Nm), bolt "B" to 34 ft. lbs. (47 Nm) and bolt "C" to 17 ft. lbs. (23 Nm).

b. Oil cooler clamp bolts: 8 ft. lbs. (11 Nm) for the top bolt and 49 inch lbs. (5.5 Nm) for the bottom bolt

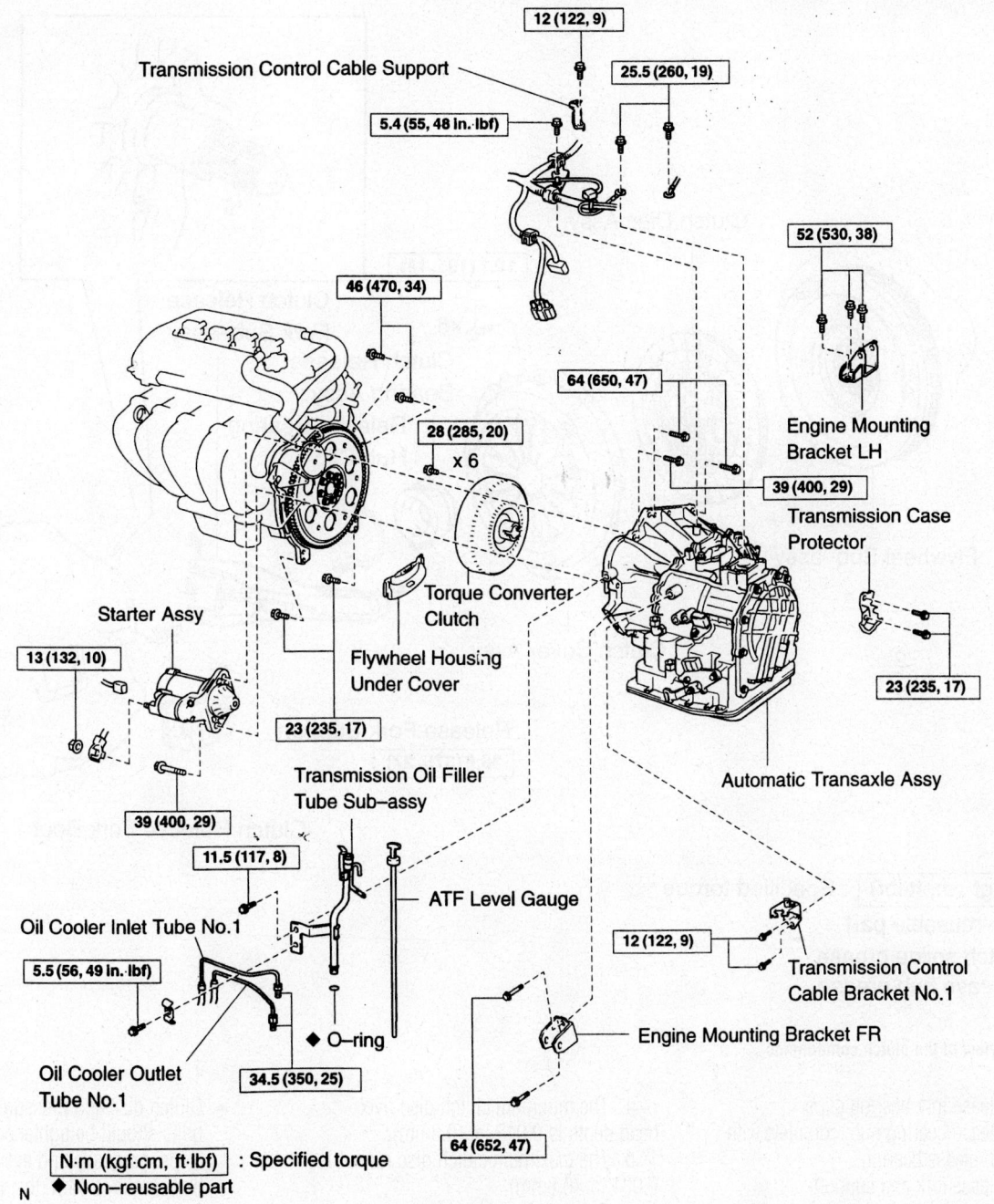

12 (122, 9)

Transmission Control Cable Support

25.5 (260, 19)

5.4 (55, 48 in. lbf)

52 (530, 38)

46 (470, 34)

64 (650, 47)

Engine Mounting
Bracket LH

28 (285, 20)
x 6

39 (400, 29)

Transmission Case
Protector

Torque Converter
Clutch

Starter Assy

Flywheel Housing
Under Cover

13 (132, 10)

23 (235, 17)

Automatic Transaxle Assy

23 (235, 17)

Transmission Oil Filler
Tube Sub–assy

39 (400, 29)

11.5 (117, 8)

ATF Level Gauge

Oil Cooler Inlet Tube No.1

5.5 (56, 49 in. lbf)

◆ O–ring

12 (122, 9)

Transmission Control
Cable Bracket No.1

Engine Mounting Bracket FR

Oil Cooler Outlet
Tube No.1

34.5 (350, 25)

64 (652, 47)

N·m (kgf·cm, ft·lbf) : Specified torque
◆ Non–reusable part
N

9359AB56

Automatic transaxle and related components—U341F transaxle

c. Oil cooler inlet and outlet tube
bolts: 25 ft. lbs. (34 Nm).

d. Wire harness clamp bracket bolt:
48 inch lbs. (5 Nm).

e. Transmission control cable bracket
and support bolts: 9 ft. lbs. (12 Nm).

f. Automatic transmission case protector bolts: 17 ft. lbs. (23 Nm).

5. Fill the transaxle fluid to the proper
level.

6. Start the vehicle, check for leaks and
repair if necessary.

Clutch

ADJUSTMENTS

Hydraulic clutch actuating systems used
in Pontiac vehicles do not require adjustment.

REMOVAL & INSTALLATION

1. Before servicing the vehicle, refer to
the precautions section.

➡**Do not allow grease or oil to get on
any part of the disc, pressure plate, or
flywheel surfaces.**

2. Remove or disconnect the following:
• Negative battery cable. On vehicles
equipped with an air bag, wait at
least 90 seconds before proceeding
• Transaxle assembly

3. Make matchmarks on the clutch cover
(pressure plate) and flywheel so that the
pressure plate can be returned to its original
position during installation.

4. Remove or disconnect the following:

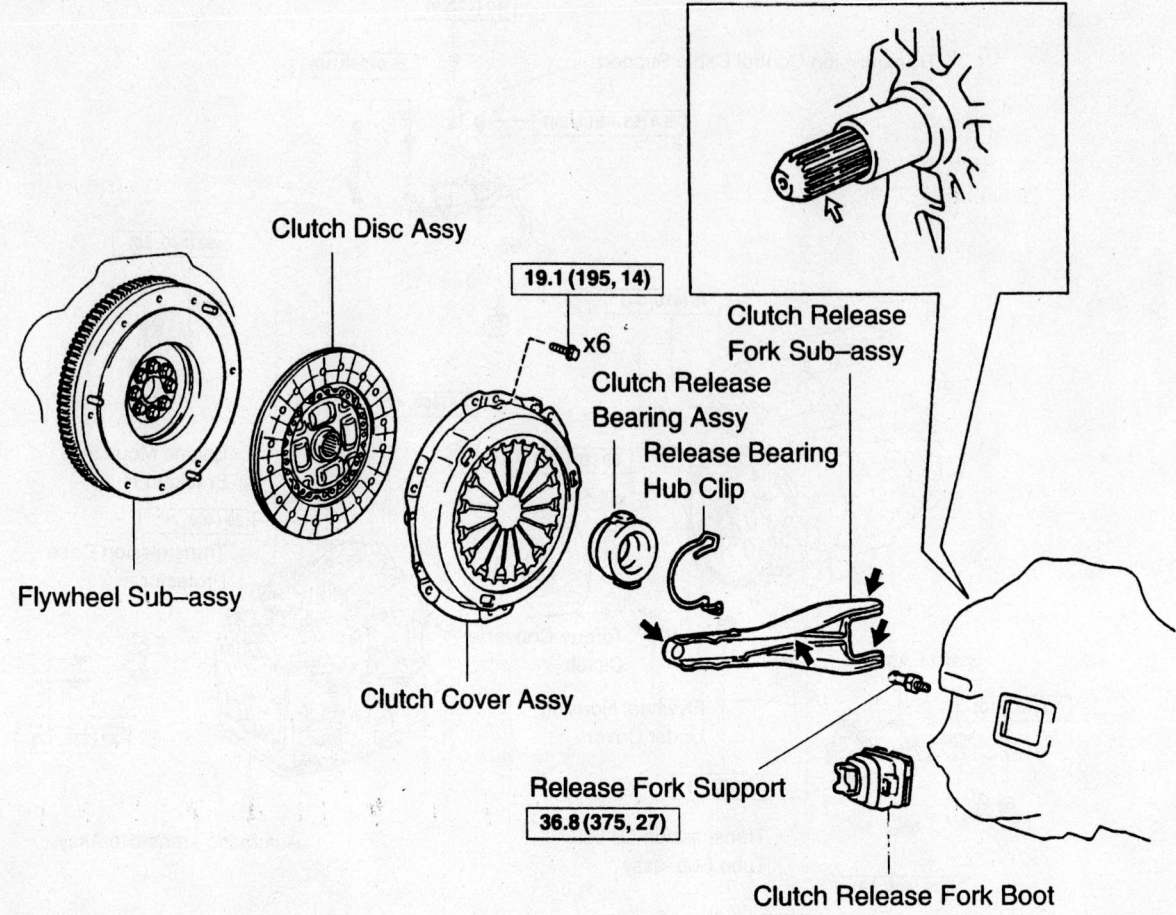

Clutch Disc Assy

19.1 (195, 14)

×6

Clutch Release Bearing Assy

Release Bearing Hub Clip

Clutch Release Fork Sub–assy

Flywheel Sub–assy

Clutch Cover Assy

Release Fork Support

36.8 (375, 27)

Clutch Release Fork Boot

N·m (kgf·cm, ft·lbf) : Specified torque

◆ Non–reusable part

⇦ Clutch spline grease

⇐ Release hub grease

9359AB57

Exploded view of the clutch components

- Release fork bearing clips
- Release bearing hub, complete with the release bearing
- Release fork and support

❊❊ CAUTION

Slowly unfasten the bolts which attach the pressure plate. Loosen each bolt 1 turn at a time until the spring tension is released. If the bolts are released improperly the clutch assembly could fly apart, causing possible injury.

- Pressure plate from the clutch cover/spring assembly

5. Inspect the disc, pressure plate and flywheel for damage and wear using a caliper to measure depth and width and a dial indicator to measure runout.

   a. The minimum clutch disc rivet head depth is 0.012 in. (0.3mm).

   b. The maximum clutch disc runout is 0.031 in. (0.8mm).

   c. The maximum pressure plate spring depth is 0.024 in. (0.6mm).

   d. The maximum pressure plate spring width is 0.197 in. (5.0mm).

   e. The maximum flywheel runout is 0.004 in. (0.1mm).

6. Replace or machine parts as necessary.

To install:

7. When reassembling, apply a thin coating of multipurpose grease to the release bearing hub and release fork contact points. Also, pack the groove inside the clutch hub with multipurpose grease and lubricate the pivot points of the release fork.

8. Install or connect the following:

- Clutch disc and pressure plate. The bolts should be tightened in 2 or 3 steps, gradually and evenly. Final bolt torque is 14 ft. lbs. (19 Nm).
- Release bearing, fork and boot
- Transaxle assembly
- Negative battery cable

Hydraulic Clutch System

BLEEDING

➡**If any maintenance on the clutch system was performed or the system is suspected of containing air, bleed the system. Use care; brake fluid will remove the paint from any surface. If the brake fluid spills onto any painted surface, wash it off immediately with soap and water.**

1. Before servicing the vehicle, refer to the precautions section.

2. Fill the clutch reservoir with brake fluid. Check the reservoir level frequently and add fluid as needed.

3. Connect one end of a vinyl tube to the bleeder plug on the slave cylinder and submerge the other end into a clear container half-filled with brake fluid.

4. Slowly pump the clutch pedal several times.

5. Have an assistant hold the clutch pedal down and loosen the bleeder plug until fluid and/or air starts to run out of the bleeder plug. Close the bleeder plug while the pedal is held to the floor.

➡**Do not allow the pedal to rise back-up while the bleeder is still open. If this happens, it will allow air to re-enter the slave cylinder and cause the clutch system not to work properly.**

6. Repeat Steps 2 and 3 until all the air bubbles are removed from the system.

7. Tighten the bleeder plug when all the air is gone.

8. Refill the master cylinder to the proper level as required.

9. Check the system for leaks.

Transfer Case

REMOVAL & INSTALLATION

1. Before servicing the vehicle, refer to the precautions section.

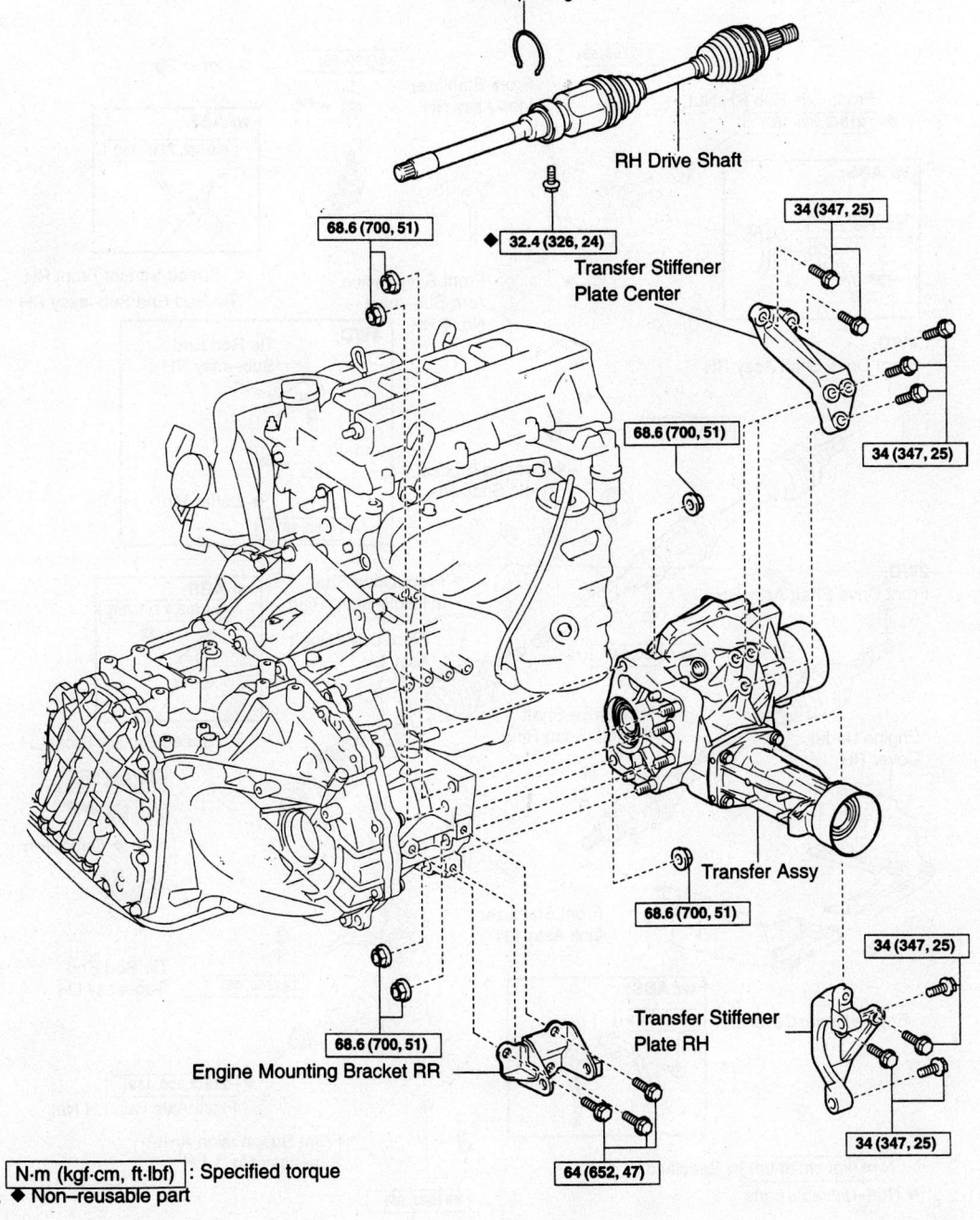

◆ Snap Ring

RH Drive Shaft

68.6 (700, 51)

◆ 32.4 (326, 24)

Transfer Stiffener Plate Center

34 (347, 25)

68.6 (700, 51)

34 (347, 25)

Transfer Assy

68.6 (700, 51)

34 (347, 25)

Transfer Stiffener Plate RH

68.6 (700, 51)

Engine Mounting Bracket RR

34 (347, 25)

64 (652, 47)

N·m (kgf·cm, ft·lbf) : Specified torque
◆ Non–reusable part

9359AB65

Exploded view of the transfer case mounting

2. Drain the transfer case fluid.
3. Remove or disconnect the following:
- Negative battery cable. Due to the air bag system, wait at least 90 seconds before proceeding
- Engine and transaxle assembly
- Separate vane pump
- Steering gear
- Crossmember
- Manifold stay
- Oxygen (O2) sensor
- Exhaust manifold heat shield
- Exhaust manifold
- Starter
- Right side halfshaft

- Transverse engine mounting bracket
- Center and right side transfer stiffener plates

✳✳ WARNING

When removing the transfer case, DO NOT touch the oil seal.

- Transfer case bolts, and transfer assembly, using a mallet to dislodge it from the transaxle
4. Installation is the reverse of the removal procedure, noting the following specifications:

a. Transfer case stiffener case bolts: 25 ft. lbs. (34 Nm).
b. Engine mounting bracket bolts: 47 ft. lbs. (64 Nm).
5. Add fluid to the transfer case, and check for leaks.

Halfshaft

REMOVAL & INSTALLATION

➡ **The hub bearing could be damaged if subjected to the full weight of the vehicle, such as if the vehicle is moved without the halfshafts. If it is absolutely**

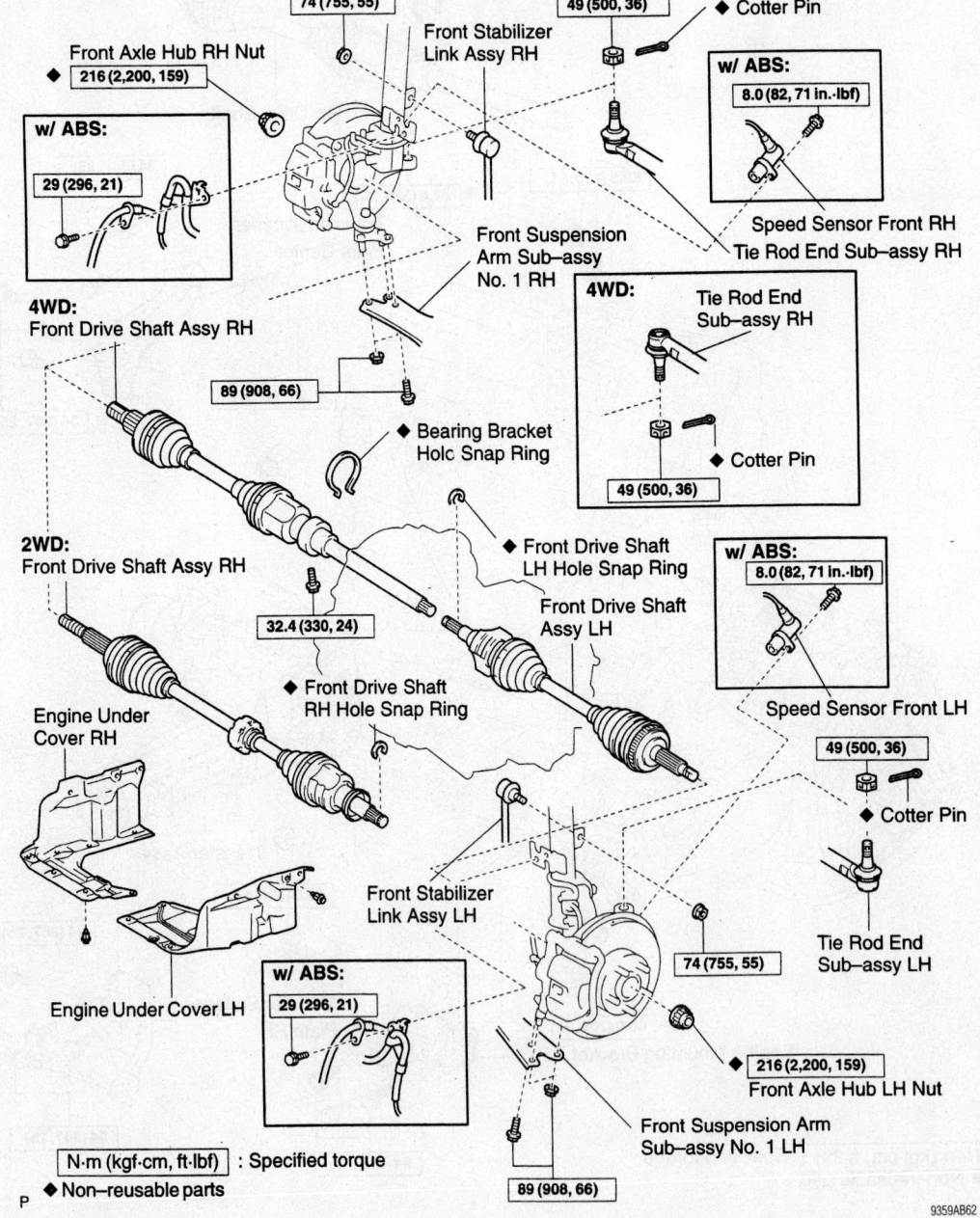

Front Axle Hub RH Nut
◆ 216 (2,200, 159)

74 (755, 55)

Front Stabilizer Link Assy RH

49 (500, 36)

◆ Cotter Pin

w/ ABS:
8.0 (82, 71 in.-lbf)

Speed Sensor Front RH
Tie Rod End Sub–assy RH

w/ ABS:
29 (296, 21)

Front Suspension Arm Sub–assy No. 1 RH

4WD:
Tie Rod End Sub–assy RH

4WD:
Front Drive Shaft Assy RH

89 (908, 66)

◆ Bearing Bracket Hole Snap Ring

◆ Cotter Pin

49 (500, 36)

2WD:
Front Drive Shaft Assy RH

◆ Front Drive Shaft LH Hole Snap Ring

Front Drive Shaft Assy LH

w/ ABS:
8.0 (82, 71 in.-lbf)

32.4 (330, 24)

◆ Front Drive Shaft RH Hole Snap Ring

Speed Sensor Front LH

Engine Under Cover RH

49 (500, 36)

◆ Cotter Pin

Tie Rod End Sub–assy LH

Engine Under Cover LH

Front Stabilizer Link Assy LH

74 (755, 55)

216 (2,200, 159)
Front Axle Hub LH Nut

w/ ABS:
29 (296, 21)

Front Suspension Arm Sub–assy No. 1 LH

N·m (kgf·cm, ft·lbf) : Specified torque
P ◆ Non-reusable parts

89 (908, 66)

9359AB62

Halfshafts and related components

necessary to place the full vehicle weight on the hub bearing, first support the bearing with SST No. 09608-16041.

1. Before servicing the vehicle, refer to the precautions section.
2. Drain the transaxle fluid.
3. Remove or disconnect the following:
- Negative battery cable. Due to the air bag system, wait at least 90 seconds before proceeding.
- Both front wheels
- Cotter pin, locknut cap, and the hub nut
- Undercovers
- Speed sensors
- Tie rod ball joint from the steering knuckle
- Stabilizer bar link from the lower suspension arm
- Lower ball joint from the lower suspension arm
- Halfshaft from the knuckle

➡**Be careful not to damage the inner oil seal or the ABS sensor rotor on the halfshaft.**

4. To remove the left side halfshaft, separate the halfshaft from the transaxle.
5. To remove the right side halfshaft perform the following steps:
- Remove the 2 bolts of the center bearing bracket
- Pull the halfshaft out together with the center bearing case and the center halfshaft.
- Remove the center shaft with the right-hand halfshaft from the transaxle through the bearing bracket.

➡**Do not damage the oil seal lip.**

To install:

6. Install or connect the following:
- Snapring opening side facing downward, on the oiled inboard joint tulip
- Left side halfshaft into the transaxle
- Right side halfshaft, with the bearing case and center shaft, into the transaxle
- Center bearing case (right side).

7. After installing either halfshaft, check that there is 0.08–0.12 in. (2–3mm) of axial play. Check that the halfshaft is making contact with the pinion shaft and that the halfshaft cannot be pulled out.

8. Install or connect the following:
- Halfshaft into the knuckle
- Lower suspension arm to the lower ball joint. Torque the bolt and nuts to 66 ft. lbs. (89 Nm).

- Tie rod end to the steering knuckle. Tighten the nut to 36 ft. lbs. (49 Nm).
- Stabilizer bar link to the lower suspension arm. Torque the nuts to 55 ft. lbs. (74 Nm).
- Front wheels
- Hub nut and washer and tighten to 159 ft. lbs. (216 Nm)
- Negative battery cable
- Locknut cap and a new cotter pin.
- Speed sensors
- Undercover

9. Fill the transaxle fluid to the proper level

10. Start the vehicle, check for leaks and repair if necessary.

2WD:

CV-Joints

OVERHAUL

1. Before servicing the vehicle, refer to the precautions section.
2. Remove or disconnect the following:
- Inboard joint boot clips
- Inboard joint tulip from the driveshaft
- Snapring
- Using a brass rod and hammer, the tri-pot joint off the driveshaft without hitting the joint roller
- Inboard joint boot
- Clamp and driveshaft damper
- Clamps and the outboard drive

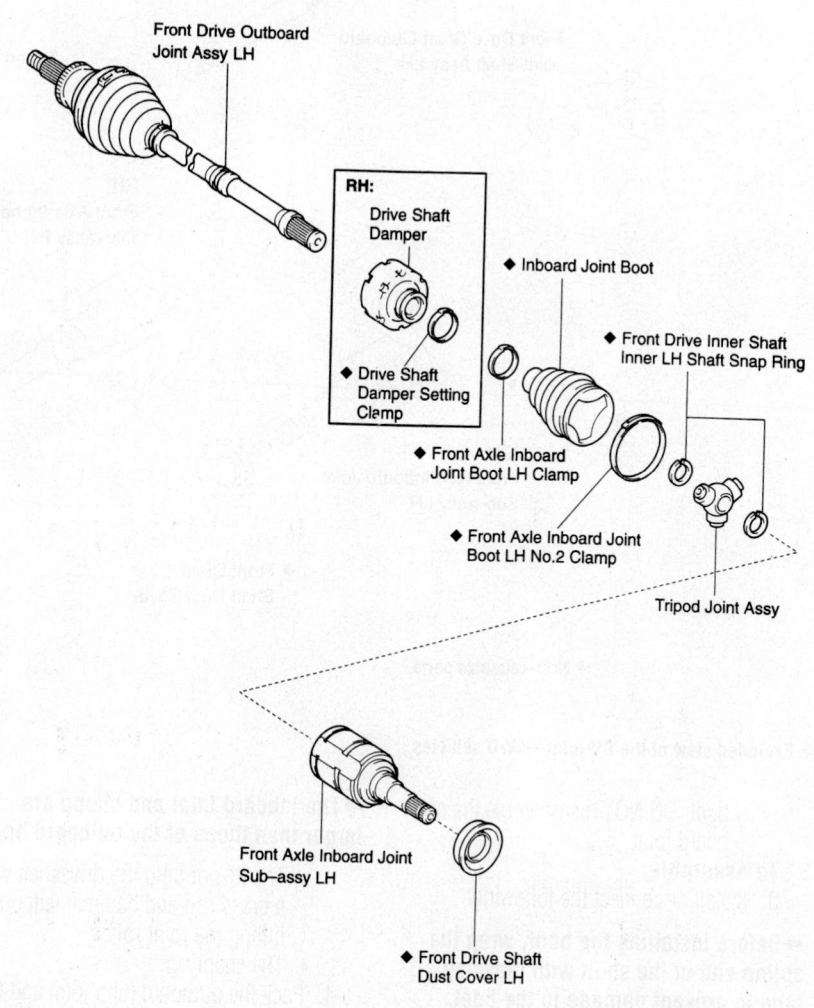

♦ Non–reusable parts

Exploded view of the CV-joint—FWD vehicles

9359AB63

4WD:

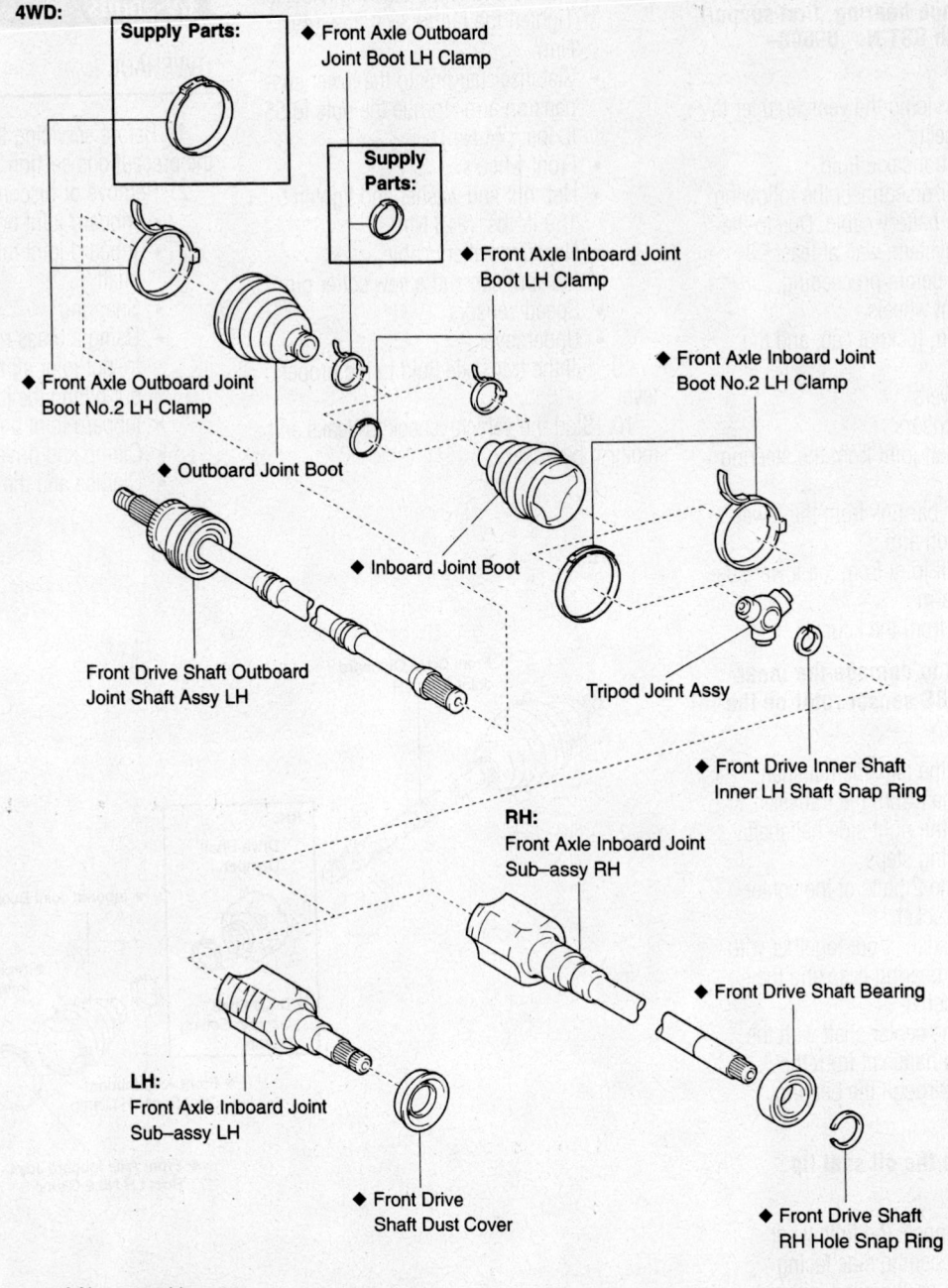

Supply Parts:

◆ Front Axle Outboard Joint Boot LH Clamp

Supply Parts:

◆ Front Axle Inboard Joint Boot LH Clamp

◆ Front Axle Inboard Joint Boot No.2 LH Clamp

◆ Front Axle Outboard Joint Boot No.2 LH Clamp

◆ Outboard Joint Boot

◆ Inboard Joint Boot

Front Drive Shaft Outboard Joint Shaft Assy LH

Tripod Joint Assy

◆ Front Drive Inner Shaft Inner LH Shaft Snap Ring

RH:
Front Axle Inboard Joint Sub–assy RH

◆ Front Drive Shaft Bearing

LH:
Front Axle Inboard Joint Sub–assy LH

◆ Front Drive Shaft Dust Cover

◆ Front Drive Shaft RH Hole Snap Ring

◆ Non–reusable parts

P

9359AB64

Exploded view of the CV-joint—AWD vehicles

boot. DO NOT disassemble the outboard joint.

To assemble:

3. Install or connect the following:

➡**Before installing the boot, wrap the spline end of the shaft with masking tape to prevent damage to the boot.**

- Driveshaft damper with a new clamp
- Temporarily, the inboard boot with new clamp to the drive joint

➡**The inboard boot and clamp are larger than those of the outboard boot.**

- The tri-pot onto the driveshaft with a brass rod and hammer without hitting the joint roller
- The snapring

4. Pack the outboard tulip joint and the outboard boot with about 0.26–0.33 lbs. ounces of grease that was supplied with the boot kit.

5. Install or connect the following:
- Boot onto the outboard joint

6. Pack the inboard tulip joint and boot with ½ lb. of grease that was supplied with the boot kit.

- Inboard tulip joint onto the driveshaft
- Boot onto the driveshaft

7. Before checking the standard length, bend the band and lock it. Make sure that the boot is not stretched or squashed when the driveshaft is at standard length. Standard driveshaft length: LH: 540.2 mm (21.268 in.); RH: 857.4 mm (33.756 in.)

STEERING AND SUSPENSION

Air Bag

PRECAUTIONS

Several precautions must be observed when handling the inflator module to avoid accidental deployment and possible personal injury.

• Never carry the inflator module by the wires or connector on the underside of the module.

• When carrying a live inflator module, hold securely with both hands, and ensure that the bag and trim cover are pointed away.

• Place the inflator module on a bench or other surface with the bag and trim cover facing up.

• With the inflator module on the bench, never place anything on or close to the module that may be thrown in the event of an accidental deployment.

DISARMING

Zone 1

1. Turn the steering wheel so that the vehicle's wheels are pointing straight ahead.
2. Turn the ignition switch to the OFF position.

➡ **With the SIR fuse removed and the ignition switch in the ON position, the AIR BAG warning indicator illuminates. This is normal operation, and does not indicate a SIR system malfunction.**

3. Locate and remove the AM2 Fuse from the junction block, which is located near the base of the steering column.
4. Disconnect the battery negative cable.
5. Locate the front end sensor—right electrical connector.
6. Remove the Connector Position Assurance (CPA) from the front end sensor—right connector.
7. Remove the front end sensor—right electrical connector from the front end sensor—right.
8. Open the hood, and locate the front end discriminating sensor—left electrical connector.
9. Remove the CPA from the front end sensor—left connector.
10. Remove the front end sensor—left electrical connector from the front end sensor—left.

Zone 2

1. Turn the steering wheel so that the vehicle wheels are pointing straight ahead.
2. Turn the ignition switch to the OFF position.
3. With the SIR fuse removed and the ignition switch in the ON position, the AIR BAG warning indicator illuminates. This is normal operation, and does not indicate an SIR system malfunction.
4. Locate and remove the AM2 Fuse from the junction block, which is located near the base of the steering column.
5. Disconnect the battery negative cable.
6. Remove the front door sill plate.
7. Remove the rear door sill plate.
8. Remove the center pillar lower trim panel.
9. Remove the front seat belt retractor.
10. Disconnect the SIS electrical connector.
11. Remove the quarter upper trim panel.
12. Remove the center pillar upper trim panel.
13. Disconnect the left roof rail module wiring harness yellow connector from the left roof rail module.

Zone 4

1. Turn the steering wheel so that the vehicle wheels are pointing straight ahead.
2. Turn the ignition switch to the OFF position.

➡ **With the SIR fuse removed and the ignition switch in the ON position, the AIR BAG warning indicator illuminates. This is normal operation, and does not indicate an SIR system malfunction.**

3. Locate and remove the AM2 Fuse from the junction block, which is located near the base of the steering column.
4. Disconnect the battery negative cable.
5. Release the inflatable restraint steering wheel module coil connector locking mechanism.
6. Disconnect the inflatable restraint steering wheel module coil connector C3.
7. Remove the Instrument Panel (I/P) glove compartment.
8. Release then unlock the I/P module connector.
9. Disconnect the I/P module pigtail.
10. Release and unlock the driver and passenger seat module connectors.

11. Disconnect the driver and passenger seat modules.
12. Remove the quarter upper trim panel.
13. Remove the center pillar upper trim panel.
14. Disconnect the left and right roof rail module wiring harness yellow connectors from the roof rail modules.

Zone 5

1. Turn the steering wheel so that the vehicle wheels are pointing straight ahead.
2. Turn the ignition switch to the OFF position.

➡ **With the SIR fuse removed and the ignition switch in the ON position, the AIR BAG warning indicator illuminates. This is normal operation, and does not indicate an SIR system malfunction.**

3. Locate and remove the AM2 fuse from the junction block, which is located near the base of the steering column.
4. Disconnect the negative battery cable.
5. Remove the glove compartment.
6. Release then unlock the Instrument Panel (I/P) module connector.
7. Disconnect the I/P module pigtail.

Zone 6

1. Turn the steering wheel so that the vehicle wheels are pointing straight ahead.
2. Turn the ignition switch to the OFF position.

➡ **With the SIR fuse removed and the ignition switch in the ON position, the AIR BAG warning indicator illuminates. This is normal operation, and does not indicate an SIR system malfunction.**

3. Locate and remove the AM2 fuse from the junction block, which is located near the base of the steering column.
4. Disconnect the negative battery cable.
5. Remove the front door sill plate.
6. Remove the rear door sill plate.
7. Remove the center pillar lower trim panel.
8. Remove the front seat belt retractor.
9. Disconnect the SIS electrical connector.
10. Remove the quarter upper trim panel.
11. Remove the center pillar upper trim panel.
12. Disconnect the right roof rail module wiring harness yellow connector from the right roof rail module.

Zone 7

1. Turn the steering wheel so that the vehicle wheels are pointing straight ahead.

2. Turn the ignition switch to the OFF position.

➡ **With the SIR fuse removed and the ignition switch in the ON position, the AIR BAG warning indicator illuminates. This is normal operation, and does not indicate an SIR system malfunction.**

3. Locate and remove the AM2 fuse from the junction block, which is located near the base of the steering column.

4. Disconnect the negative battery cable.

5. Release and unlock the Driver Seat Module (DSM) connector.

6. Disconnect the DSM connector.

Zone 9

1. Turn the steering wheel so that the vehicle wheels are pointing straight ahead.

2. Turn the ignition switch to the OFF position.

➡ **With the SIR fuse removed and the ignition switch in the ON position, the AIR BAG warning indicator illuminates. This is normal operation, and does not indicate an SIR system malfunction.**

3. Locate and remove the AM2 fuse from the junction block, which is located near the base of the steering column.

4. Disconnect the negative battery cable.

5. Release and unlock the Passenger Seat Module (PSM) connector.

6. Disconnect the PSM connector.

Zone 10

1. Turn the steering wheel so that the vehicle wheels are pointing straight ahead.

2. Turn the ignition switch to the OFF position.

➡ **With the SIR fuse removed and the ignition switch in the ON position, the AIR BAG warning indicator illuminates. This is normal operation, and does not indicate an SIR system malfunction.**

3. Locate and remove the AM2 fuse from the junction block, which is located near the base of the steering column.

4. Disconnect the negative battery cable.

5. Remove the rear door sill plate.

6. Remove the rear seat garnish molding.

7. Disconnect the Side Impact Sensor (SIS) electrical connector.

Zone 12

1. Turn the steering wheel so that the vehicle wheels are pointing straight ahead.

2. Turn the ignition switch to the OFF position.

➡ **With the SIR fuse removed and the ignition switch in the ON position, the AIR BAG warning indicator illuminates. This is normal operation, and does not indicate an SIR system malfunction.**

3. Locate and remove the AM2 fuse from the junction block, which is located near the base of the steering column.

4. Disconnect the negative battery cable.

5. Remove the rear door sill plate.

6. Remove the rear seat garnish molding.

7. Disconnect the Side Impact Sensor (SIS) electrical connector.

REARMING

Zone 1

1. Connect the front end sensor—left electrical connector from the front end sensor—left.

2. Connect the CPA from the front end sensor—left connector.

3. Connect the front end sensor—right electrical connector from the front end sensor—right.

4. Connect the CPA from the front end sensor—right connector.

5. Connect the battery negative cable.

6. Install the AM2 Fuse into the junction block.

❊❊ CAUTION

Use caution while reaching in and turn the ignition switch to the ON position.

7. The AIR BAG indicator will flash then turn OFF.

8. Perform the Diagnostic System Check—Vehicle if the AIR BAG warning indicator does not operate as described.

Zone 2

1. Connect the SIS electrical connector.

2. Install the front seat belt retractor.

3. Install the center pillar lower trim panel.

4. Interior Trim.

5. Install the rear door sill plate.

6. Install the front door sill plate.

7. Connect the left roof rail module

wiring harness yellow connector to the left roof rail module.

8. Install the center pillar upper trim panel.

9. Install the quarter upper trim panel.

10. Connect the battery negative cable.

11. Install the AM2 Fuse into the junction block.

❊❊ CAUTION

Use caution while reaching in and turn the ignition switch to the ON position.

12. The AIR BAG indicator will flash then turn OFF.

13. Perform the Diagnostic System Check—Vehicle if the AIR BAG warning indicator does not operate as described.

Zone 4

1. Remove the key from the ignition switch.

2. Connect the left and right roof rail module wiring harness yellow connectors to the roof rail modules.

3. Install the center pillar upper trim panel.

4. Install the quarter upper trim panel.

5. Install the yellow 2-way connectors to the driver and passenger seat modules.

6. Connect the connectors and lock the connectors with the connector lock levers.

7. Install the yellow 2-way connector to the inflatable restraint I/P module pigtail. Connect the connector and lock the connector with the connector lock lever.

8. Install the I/P glove compartment.

9. Install the yellow 2-way connector for the inflatable restraint steering wheel module coil. Connect the connector and lock the connector with the connector lock lever.

10. Install the lower steering column trim cover.

11. Connect the battery negative cable.

12. Install the AM2 fuse into the junction block.

❊❊ CAUTION

Use caution while reaching in and turn the ignition switch to the ON position.

13. The AIR BAG indicator will flash then turn OFF.

Perform the Diagnostic System Check—Vehicle if the AIR BAG warning indicator does not operate as described.

Zone 5

1. Install the yellow 2-way connector to the inflatable restraint I/P module pigtail. Connect the connector and lock the connector with the connector lock lever.
2. Install the glove compartment.
3. Connect the battery negative cable.
4. Install the AM2 fuse into the junction block.

✳✳ CAUTION

Use caution while reaching in and turn the ignition switch to the ON position.

5. The AIR BAG indicator will flash then turn OFF.
6. Perform the Diagnostic System Check—Vehicle if the AIR BAG warning indicator does not operate as described.

Zone 6

1. Connect the SIS electrical connector.
2. Install the front seat belt retractor.
3. Install the center pillar lower trim panel.
4. Install the rear door sill plate.
5. Install the front door sill plate.
6. Connect the right roof rail module wiring harness yellow connector to the right roof rail module.
7. Install the center pillar upper trim panel.
8. Install the quarter upper trim panel.
9. Connect the negative battery cable.
10. Install the AM2 fuse into the junction block.

✳✳ CAUTION

Use caution while reaching in and turn the ignition switch to the ON position.

11. The AIR BAG indicator will flash then turn OFF.
12. Perform the Diagnostic System Check—Vehicle if the AIR BAG warning indicator does not operate as described.

Zone 7

1. Install the yellow 2-way connector to the DSM.
2. Connect the connector and lock the connector with the connector lock lever.
3. Connect the negative battery cable.
4. Install the AM2 fuse into the junction block.

✳✳ CAUTION

Use caution while reaching in and turn the ignition switch to the ON position.

5. The AIR BAG indicator will flash then turn OFF.
6. Perform the Diagnostic System Check—Vehicle if the AIR BAG warning indicator does not operate as described.

Zone 9

1. Install the yellow 2-way connector to the PSM.
2. Connect the connector and lock the connector with the connector lock lever.
3. Connect the negative battery cable.
4. Install the AM2 fuse into the junction block.

✳✳ CAUTION

Use caution while reaching in and turn the ignition switch to the ON position.

5. The AIR BAG indicator will flash then turn OFF.
6. Perform the Diagnostic System Check—Vehicle if the AIR BAG warning indicator does not operate as described.

Zone 10

1. Connect the SIS electrical connector.
2. Install the rear seat garnish molding.
3. Connect the negative battery cable.
4. Install the AM2 fuse into the junction block.

✳✳ CAUTION

Use caution while reaching in and turn the ignition switch to the ON position.

5. The AIR BAG indicator will flash then turn OFF.
6. Perform the Diagnostic System Check—Vehicle if the AIR BAG warning indicator does not operate as described.

Zone 12

1. Connect the SIS electrical connector.
2. Install the rear seat garnish molding.
3. Install the rear door sill plate.
4. Connect the negative battery cable.
5. Install the AM2 fuse into the junction block.

✳✳ CAUTION

Use caution while reaching in and turn the ignition switch to the ON position.

6. The AIR BAG indicator will flash then turn OFF.
7. Perform the Diagnostic System Check—Vehicle if the AIR BAG warning indicator does not operate as described.

Rack and Pinion Steering Gear

REMOVAL & INSTALLATION

2003–04 Models

1. Before servicing the vehicle, refer to the precautions section.
2. Position the front wheels straight ahead.
3. Remove or disconnect the following:
   - Negative battery cable. Because these vehicles are equipped with air bags, wait at least 90 seconds before proceeding.
   - Horn button
   - Steering wheel
   - Front wheels
   - Left and right engine undercovers
   - Left and right tie rod ends
   - Column hose cover silencer sheet
   - Steering intermediate shaft
   - Pressure feed and return tubes
   - Left and right side front stabilizer links

1ZZ–FE:

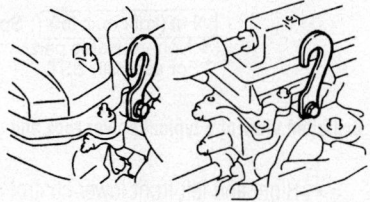

2ZZ–GE:

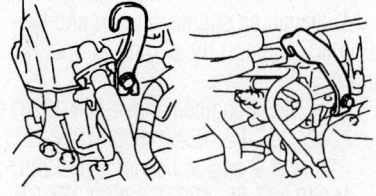

9359AB58

Proper installation of engine hangers

2WD:

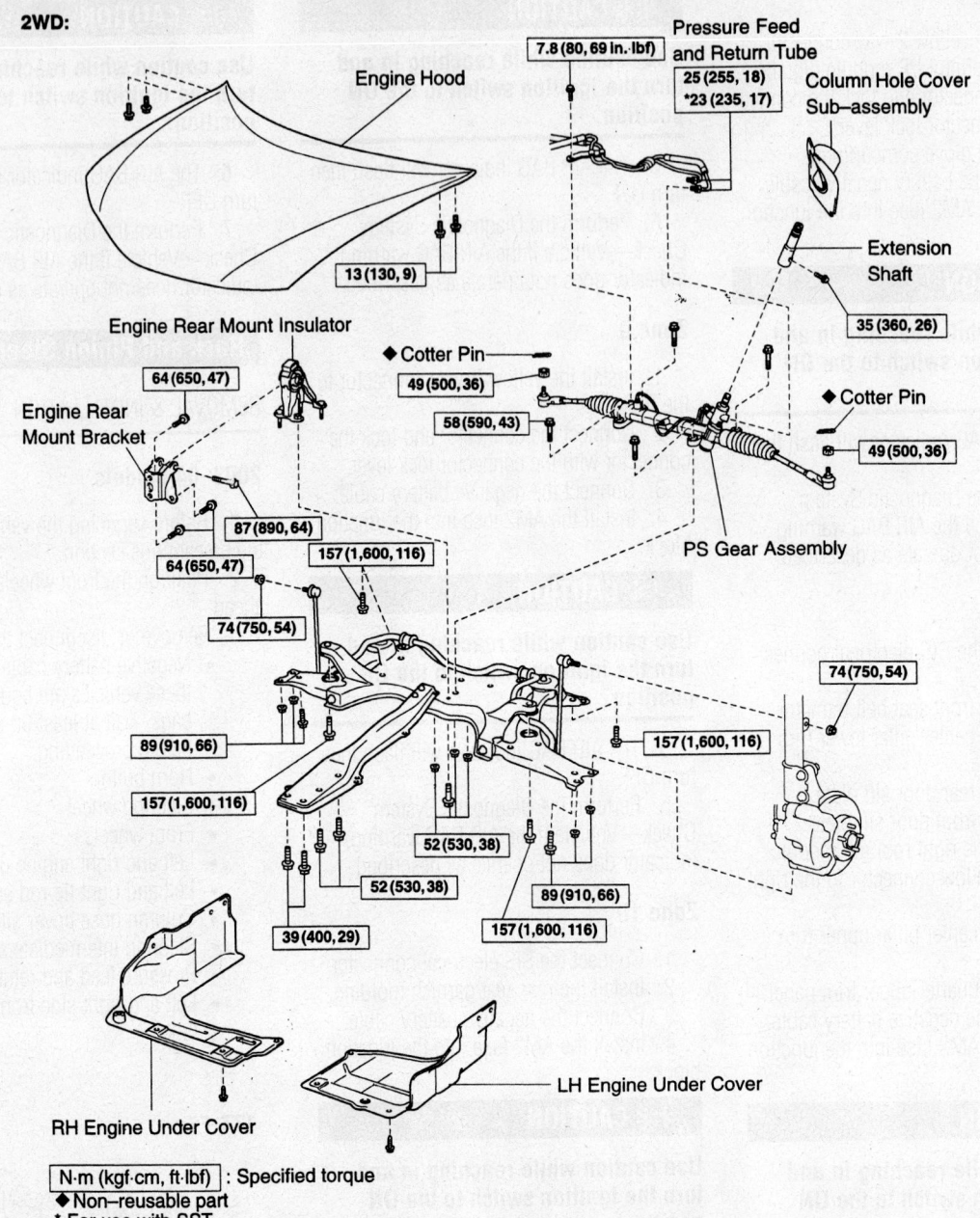

Engine Hood

7.8 (80, 69 in.·lbf)

Pressure Feed and Return Tube

25 (255, 18)
*23 (235, 17)

Column Hole Cover Sub–assembly

13 (130, 9)

Extension Shaft

35 (360, 26)

Engine Rear Mount Insulator

64 (650, 47)

◆ Cotter Pin
49 (500, 36)

Engine Rear Mount Bracket

58 (590, 43)

◆ Cotter Pin
49 (500, 36)

87 (890, 64)

157 (1,600, 116)

PS Gear Assembly

64 (650, 47)

74 (750, 54)

74 (750, 54)

89 (910, 66)

157 (1,600, 116)

157 (1,600, 116)

52 (530, 38)

52 (530, 38)

89 (910, 66)

39 (400, 29)

157 (1,600, 116)

RH Engine Under Cover

LH Engine Under Cover

N·m (kgf·cm, ft·lbf) : Specified torque
◆ Non–reusable part
* For use with SST

9359AB59

Exploded view of a typical power rack and pinion steering gear unit—FWD shown

- Right and left front lower control arms from the ball joints
- Hood
- No. 2 cylinder head (valve) cover

4. Install an engine support and tension it to support the engine without raising it.
   a. No. 1 engine hanger: P/N 12281-22021 1ZZ-FE, 12281-88600 2ZZ-GE.
   b. No. 2 engine hanger: P/N 12281-15040 1ZZ-FE, 12281-88600 2ZZ-GE.
   c. Bolt: P/N 91512-B1016.
   d. Torque the bolts to 28 ft. lbs. (38 Nm).

✳✳ CAUTION

Do not try to suspend the engine by hooking the chain to any other part.

   e. Attach an engine chain hoist to the hangers.

✳✳ CAUTION

The engine hoist is now in place and under tension. Use care when repositioning the vehicle and make necessary adjustments to the engine support.

5. Remove or disconnect the following:
   - Bolt and nuts holding in the middle of the crossmember and support the crossmember with a jack
   - Bolts from the outer side of the suspension crossmember
   - Suspension crossmember with the steering gear assembly
   - Steering intermediate shaft, after matchmarking it
   - Rack and pinion steering gear from the crossmember

6. Installation is the reverse of the removal procedure, noting the following specifications:

a. Steering gear bolts and nuts: 43 ft. lbs. (58 Nm) FWD, 60 ft. lbs. (82 Nm) AWD.

b. Steering intermediate shaft: 26 ft. lbs. (35 Nm).

c. Suspension crossmember bolts: 116 ft. lbs. (157 Nm).

d. Engine mount insulator bolts: 38 ft. lbs. (52 Nm).

e. Center member-to-frame bolts: 29 ft. lbs. (39 Nm).

f. Stabilizer bar link-to-the lower control arms nuts: 55 ft. lbs. (74 Nm).

g. Fluid return and pressure tubes: 17 ft. lbs. (23 Nm).

h. Tie rod ends: 36 ft. lbs. (49 Nm).

i. Wheel lug nuts: 76 ft. lbs. (103 Nm).

7. Check and top off the power steering fluid.

8. Check and adjust the alignment, if needed.

2005 Models

1. Before servicing the vehicle, refer to the precautions section.

➡ **The steering column must be in the LOCK position before disconnecting the following components:**

- Steering column
- Steering shaft coupling
- Intermediate shaft
- Lower steering shaft

➡ **After disconnecting these components, do not move the front tires and wheels. Failure to follow these procedures may cause improper alignment of some components during installation and result in possible damage to the SIR coil.**

2. LOCK the steering column and verify the front wheels are in the straight ahead position.

3. Move the silencer pad away from the steering column.

4. Use paint in order to place match marks on the steering shaft coupling and on the intermediate shaft.

5. Loosen the upper coupling bolt.

6. Remove the lower coupling bolt.

7. Remove the steering column hole cover from the bulkhead.

8. Install the Engine Support Fixture.

9. Remove the front tire and wheel assemblies.

10. Remove the engine splash shields.

11. Remove the 2 outer tie rod ends.

12. Place a drain pan under the vehicle in order to collect the fluid from the power steering system.

13. Remove the pressure and return pipes from the steering gear.

14. Remove the bolt and the pipe bracket from the steering gear.

15. Remove the following components together as a unit:
- Steering gear
- Intermediate steering shaft
- Front suspension crossmember
- Trans support
- Control arms
- Front stabilizer shaft

16. Remove the bolt and the rear engine mount insulator from the crossmember.

17. Remove the 3 bolts and the rear engine mount bracket from the crossmember.

18. Use paint in order to place match marks on the intermediate shaft and on the steering gear.

19. Remove the bolt and the intermediate shaft (4).

20. Remove the 4 bolts and the steering gear from the crossmember.

To install:

21. Install the rear engine mount bracket to the crossmember.

22. Install the 3 bolts to the rear engine mount bracket. Tighten to 47 ft. lbs. (64 Nm).

23. Install the rear engine mount insulator to the crossmember.

24. Install the bolt to the rear engine mount insulator. Tighten to 64 ft. lbs. (87 Nm).

25. If you are replacing the steering gear or the intermediate shaft, copy the match marks from the old parts to the same locations on the new parts.

26. Install the steering gear and the 4 bolts to the crossmember. Tighten to 42 ft. lbs. (58 Nm).

27. Install the intermediate shaft to the steering gear. Align the match marks.

28. Install the bolt to the intermediate shaft. Tighten to 26 ft. lbs. (35 Nm).

29. Install the steering column hole cover to the bulkhead.

30. Install the following components as a unit:
- Steering gear
- Intermediate steering shaft
- Front suspension crossmember
- Trans support
- Control arms
- Front stabilizer shaft

31. Install the 2 outer tie rod ends.

32. Install the pressure and return pipes to the steering gear. Tighten the fittings to 17 ft. lbs. (23 Nm).

33. Install the pipe bracket bolt. Tighten to 69 inch lbs. (8 Nm).

34. Install the splash shields.

35. Install the front tire and wheel assemblies.

36. Remove the Engine Support Fixture.

37. Align the match marks on the intermediate shaft and on the steering shaft coupling.

38. Install the lower coupling bolt. Tighten to 26 ft. lbs. (35 Nm).

39. Tighten the upper coupling bolt. Tighten to 26 ft. lbs. (35 Nm).

40. Place the silencer pad into the correct position.

41. Fill the power steering fluid reservoir.

42. Bleed the power steering system.

43. Inspect the power steering system for leaks. Repair as necessary.

44. Measure the wheel alignment. Adjust as necessary

Strut and Coil Spring

REMOVAL & INSTALLATION

Front

1. Before servicing the vehicle, refer to the precautions section.

2. Remove or disconnect the following:
- Negative battery cable. Because of the air bag system, wait at least 90 seconds before proceeding

✷✷ WARNING

Do not support the weight of the vehicle on the suspension arm; the arm will deform under its weight.

- Wheel
- Stabilizer link from the strut
- Bolt, and disconnect the brake hose from the strut
- With ABS brakes, speed sensor wiring harness from the strut
- Lower strut bolts and nuts
- Upper strut nuts
- Strut from the steering knuckle
- Strut

3. To disassemble the strut:
- Install a bolt and 2 nuts to the bracket at the lower portion of the strut shell and secure it in a vise
- Compress the coil spring
- Dust cover and hold the spring seat so that it will not turn
- Nut on the top of the strut
- Suspension support, bearing, dust seal, spring seat, spring, insulators and bumper

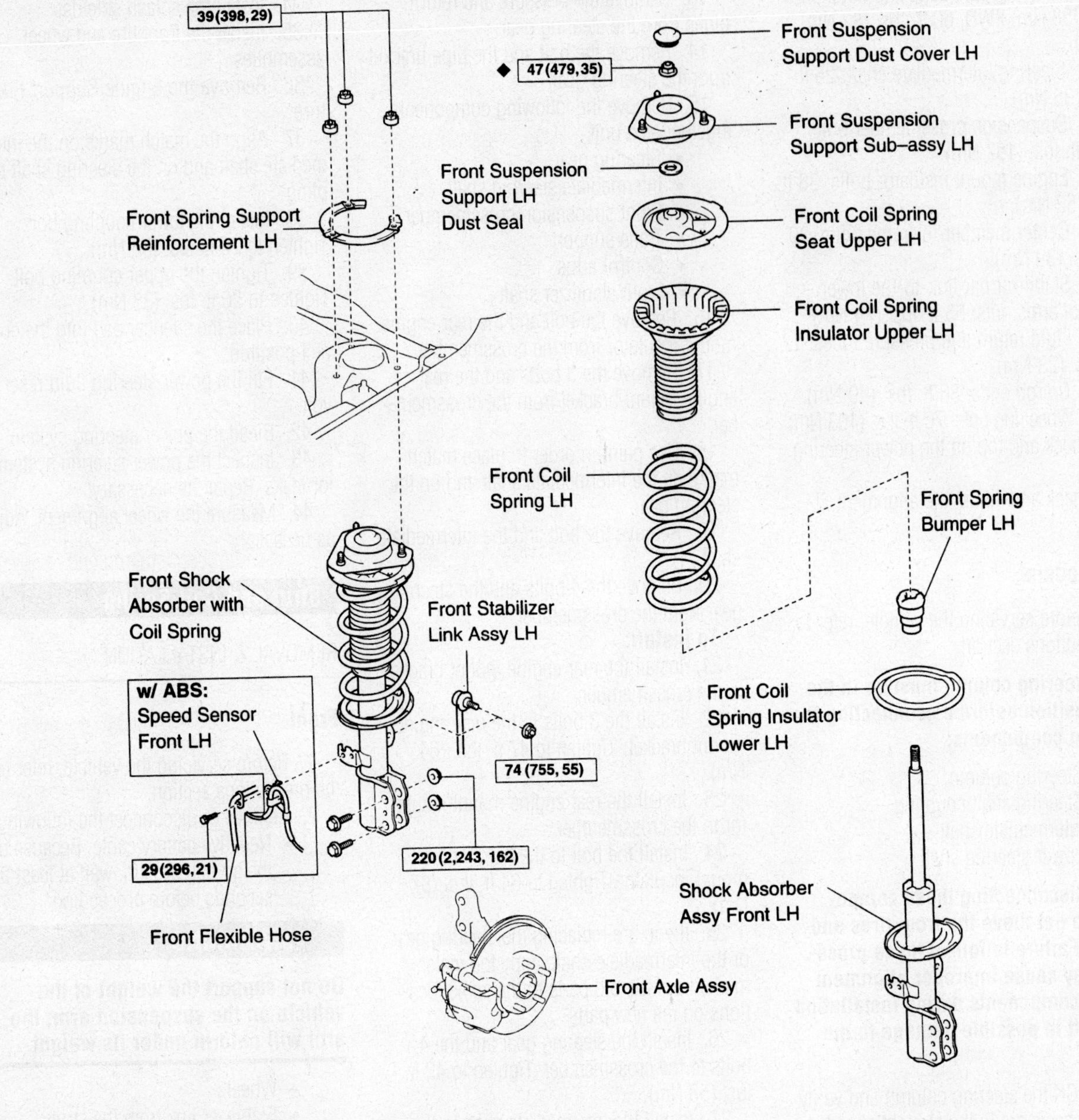

39(398,29)

Front Suspension
Support Dust Cover LH

◆ 47(479,35)

Front Suspension
Support Sub–assy LH

Front Suspension
Support LH
Dust Seal

Front Coil Spring
Seat Upper LH

Front Spring Support
Reinforcement LH

Front Coil Spring
Insulator Upper LH

Front Coil
Spring LH

Front Spring
Bumper LH

Front Shock
Absorber with
Coil Spring

Front Stabilizer
Link Assy LH

w/ ABS:
Speed Sensor
Front LH

Front Coil
Spring Insulator
Lower LH

74 (755, 55)

29(296, 21)

220 (2,243, 162)

Shock Absorber
Assy Front LH

Front Flexible Hose

Front Axle Assy

N·m (kgf·cm, ft·lbf) : Specified torque
P
◆ Non–reusable part

9359AB60

Common coil spring and strut component assembly

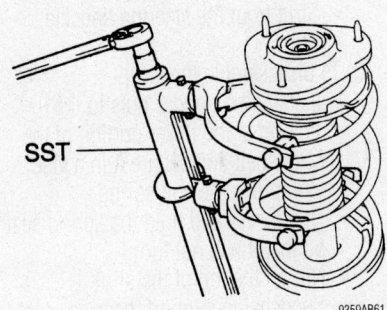

SST

9359AB61

Proper method of supporting the strut in a vise

To install:
4. To assemble the strut:
  • Install the spring bumper to piston
5. Using a spring compressor, compress the spring.
  • Coil spring to the strut. Fit the lower end of the coil spring into the gap of the lower seat.
  • Spring seat with the insulator
  • Dust seal on the spring seat
  • Suspension support and tighten 35 ft. lbs. (47 Nm). After the nut has been tighten, release the compressor tool tension.

6. Pack multipurpose grease into the suspension support.
  • Dust cover.

➡ **Do not use an impact wrench to tighten the nut. Also, check that the bearing fits into the recess in the suspension support.**

  • Strut
  • Nuts holding the strut to the strut tower. Tighten the nuts to 29 ft. lbs. (39 Nm).
  • 2 lower strut bolts and nuts. Tighten to 162 ft. lbs. (220 Nm).

- Brake line to the steering knuckle. Tighten the line bolt to 21 ft. lbs. (29 Nm).
- Secure the wiring harness, if equipped with ABS
- Stabilizer link. Tighten the nut to 55 ft. lbs. (74 Nm).
- Wheel. Tighten the lug nuts to 76 ft. lbs. (103 Nm).
- Negative battery cable

7. Check and adjust the alignment, if needed.

Rear

1. Before servicing the vehicle, refer to the precautions section.
2. Remove or disconnect the following:

- Negative battery cable. Because of the air bag system, wait at least 90 seconds before proceeding.
- Rear wheel
- Rear deck board, luggage compartment tray and any trim necessary to access the strut towers
- Shock absorber head cover

3. On AWD vehicles, separate the rear stabilizer link.
4. For FWD vehicles:
    a. Support the axle beam with a jack.
    b. Remove the strut tower nuts and bolt.
    c. Remove the lower strut nut, cushion retainer and strut .
5. For AWD vehicles:
    a. Support the rear control arm.
    b. Remove the bolt and nut from the rear control arm.
    c. Remove the strut tower nuts.
    d. Remove the 3 rear control arm bolts.
    e. Press the rear control arm down to the outside of the vehicle, then remove the strut.
6. To disassemble the strut:
    a. Place the strut assembly in a pipe vise or strut vise.

❊❊ WARNING

Do not attempt to clamp the strut assembly in a flat jaw vise as this will result in damage to the strut tube.

b. Compress the spring until the upper suspension support is free of any spring tension. Do not over-compress the spring.
c. Hold the upper support, then remove the nut on the end of the shock piston rod.
    d. Remove the support, coil spring, insulator, and bumper.
7. Inspect the strut as follows:
    a. Check the shock absorber by moving the piston shaft through its full range of travel. It should move smoothly and evenly throughout its entire travel without any trace of binding or notching.
    b. Use a small straightedge to check the piston shaft for any bending or deformation.
    c. Inspect the spring for any sign of deterioration or cracking. The waterproof coating on the coils should be intact to prevent rusting.

To install:

➡**Never reuse a self-locking nut. Always replace self-locking nuts and cotter pins as applicable.**

8. Assemble the strut as follows:
    a. Loosely assemble all components onto the strut assembly. Be sure the spring end aligns with the hollow in the lower seat.
    b. Align the upper suspension support with the piston rod and install the support.
    c. Align the suspension support with the strut lower bracket. This assures the spring will be properly seated top and bottom.
    d. Compress the spring to expose the strut piston rod threads.
    e. Install a new strut piston nut and tighten to 41 ft. lbs. (56 Nm).
    f. Remove the spring compressor. Be sure the paint mark on the upper support faces the outside of the strut.
9. Install or connect the following:

- Strut on the vehicle. Tighten the strut-to-strut tower nuts to 59 ft. lbs. (80 Nm).
- Strut to the axle carrier and install the nut and cushion retainer/bolt snug. Do not fully tighten at this time.
- Strut head cover
- Rear control arm (AWD). Tighten the bolts to 48 ft. lbs. (65 Nm).
- Rear stabilizer link (AWD)
- Trunk tray, deckboard and any other trim pieces removed
- Wheel

10. With the vehicle's weight on the suspension, tighten the bolt holding the strut to the axle carrier to 59 ft. lbs. (80 Nm) for FWD vehicles, or 103 ft. lbs. (140 Nm) for AWD vehicles.

- Negative battery cable

11. Check and adjust the rear wheel alignment.

Lower Ball Joint

REMOVAL & INSTALLATION

1. Before servicing the vehicle, refer to the precautions section.
2. Remove or disconnect the following:

- Negative battery cable. Wait at least 90 seconds before proceeding.
- Front wheel

3. Depress the brake pedal and loosen the hub nut

- ABS speed sensor, if equipped
- Cotter pin and nut from the tie rod end. Using a tie rod end removal tool, separate the tie rod end from the steering knuckle.
- Lower control arm ball joint, using a suitable puller
- Separate the front halfshaft
- Lower ball joint cotter pin and castle nut
- Lower ball joint from the steering knuckle using a puller

To install:

4. Install or connect the following:

- Lower ball joint to the lower arm. Tighten the castle nut to 76 ft. lbs. (103 Nm).
- New cotter pin
- Front halfshaft
- Lower control arm
- Tie rod end to the knuckle
- ABS speed sensor
- Hub nut
- Wheel
- Negative battery cable

5. Check and adjust the alignment, if needed.

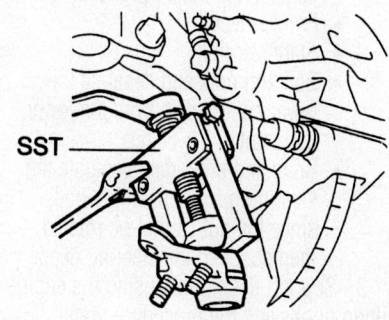

Removing the ball joint from the knuckle

Upper Control Arm

REMOVAL & INSTALLATION

Rear—AWD Only

1. Before servicing the vehicle, refer to the precautions section.
2. Remove or disconnect the following:
   - Negative battery cable. Wait at least 90 seconds before proceeding
   - Rear wheel
   - Exhaust pipe
   - Propeller shaft with center bearing shaft
   - Rear stabilizer links
   - Rear hub nuts
   - Rear brake drum
   - Speed sensor
   - Front brake shoe
   - Parking brake shoe strut set
   - Rear brake shoe
   - Parking brake cables
   - Rear brake hoses
   - Separate the rear suspension arms
   - Separate the upper control arm
   - Rear drive axle assembly
   - Rear strut nut and bolt
   - Rear strut
   - Rear suspension arm
   - Rear suspension member
   - Upper control arm assembly. Matchmark the camber adjust cams and rear suspension member prior to removal.
3. Installation is the reverse of the removal procedure.

Lower Control Arm

REMOVAL & INSTALLATION

1. Before servicing the vehicle, refer to the precautions section.
2. Remove or disconnect the following:
   - Negative battery cable. Wait at least 90 seconds before proceeding..
   - Front wheel
   - Stabilizer link
   - Bolt and nuts and separate the lower control arm from the lower ball joint
   - Bolts and nuts, then separate the steering gear. Loosen the bolt, since the nut cannot be rotated, then suspend the steering gear.
3. Support the engine, using the engine lifting hooks and the procedure under Engine Removal & Installation.
   - Crossmember

   - Lower control arm from the crossmember
4. Installation is the reverse of the removal procedure.

Wheel Bearings

REMOVAL & INSTALLATION

Front

1. Before servicing the vehicle, refer to the precautions section.
2. Remove or disconnect the following:
   - Negative battery cable. On vehicles equipped with an air bag, wait at least 90 seconds before proceeding.
   - Wheels
   - Hub nut
   - Front stabilizer link
   - Anti-lock Brake System (ABS) speed sensor
   - Brake caliper
   - Rotor
   - Tie rod end from the steering knuckle
   - Lower control arm ball joint
   - Front halfshaft from the hub, using a mallet to tap it out. Be careful not to damage the boot or speed sensor.
3. Loosen the nuts on the lower side of the strut assembly. Do not remove at this time.
   - Lower ball joint using a puller
   - Tie rod end from the steering knuckle
   - Steering knuckle from the lower control arm
   - Knuckle from the strut assembly
   - Hub

➡ **Cover the halfshaft boot with a shop rag to protect it from any damage.**

4. Clamp the steering knuckle in a vise and remove the dust deflector. Remove the nut holding the steering knuckle to the ball joint. Press the ball joint out of the steering knuckle.
5. Remove the inner axle seal.
6. Using a Torx® wrench, remove the bolts securing the dust cover.
7. Using hub puller, remove the hub and backing plate from the steering knuckle.
8. Using a proper sized driver and a press, remove the inner hub race from the axle hub.
9. Using seal removal tool, remove the outer axle seal.

10. Using snapring pliers, remove the snapring from the inner side of the steering knuckle.
11. Using a proper sized driver and a press, remove the bearing from the steering knuckle. The bearing is pressed from the front of the steering knuckle and is removed through the back of the steering knuckle.

To install:
12. Perform the following:
13. Using a proper sized driver and a press, install a new bearing to the steering knuckle.
14. Install the snapring to the steering knuckle using snapring pliers.
15. Using a seal driver and a hammer, install a new outer oil seal. Apply multipurpose grease to the oil seal lip.
16. Place the dust cover on the steering knuckle. Tighten the bolts: 78 inch lbs. (9 Nm).
17. Using a press and a proper sized driver, install the axle hub to the steering knuckle.
18. Attach the ball joint to the steering knuckle. Install a new cotter pin.
19. Using a seal driver and a hammer, install a new inner oil seal. Apply multipurpose grease to the oil seal lip.
20. Install the knuckle and hub assembly to the axle and temporarily tighten the axle nut.
21. Connect the knuckle assembly to the lower strut bracket. Temporarily insert the mounting bolts from the rear and install the nuts making sure the matchmarks made earlier are in alignment.
22. Connect the lower ball joint to lower arm.
23. Connect the tie rod end to the knuckle.
24. Tighten the bolts on the lower side of the strut assembly.
25. If equipped, install the ABS speed sensor.
26. Install the brake disc and the caliper.
27. Tighten the axle nut while someone depresses the brake pedal.
28. Install the wheels to the vehicle. Verify that the wheel turns freely.
29. Connect the negative battery cable to the battery.
30. Check alignment.

Rear

1. Before servicing the vehicle, refer to the precautions section.
2. Remove or disconnect the following:
   - Negative battery cable. On vehicles

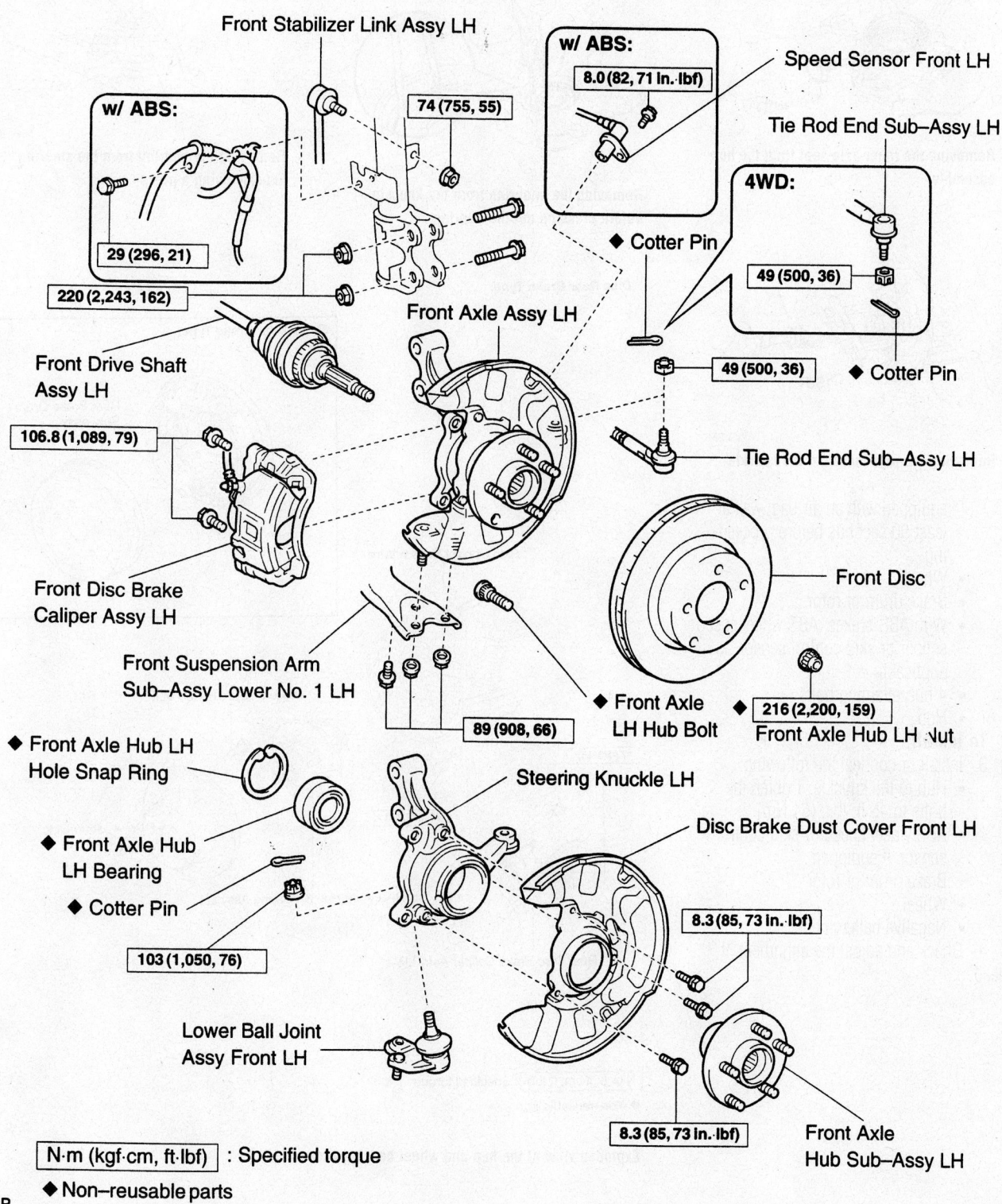

Front Stabilizer Link Assy LH

w/ ABS:
8.0 (82, 71 in. lbf)
Speed Sensor Front LH

w/ ABS:
29 (296, 21)

74 (755, 55)

Tie Rod End Sub–Assy LH

4WD:
49 (500, 36)

220 (2,243, 162)

Front Axle Assy LH

◆ Cotter Pin

◆ Cotter Pin

Front Drive Shaft
Assy LH

49 (500, 36)

106.8 (1,089, 79)

Tie Rod End Sub–Assy LH

Front Disc Brake
Caliper Assy LH

Front Disc

Front Suspension Arm
Sub–Assy Lower No. 1 LH

216 (2,200, 159)

89 (908, 66)

◆ Front Axle
LH Hub Bolt

Front Axle Hub LH Nut

◆ Front Axle Hub LH
Hole Snap Ring

Steering Knuckle LH

◆ Front Axle Hub
LH Bearing

Disc Brake Dust Cover Front LH

◆ Cotter Pin

8.3 (85, 73 in. lbf)

103 (1,050, 76)

Lower Ball Joint
Assy Front LH

8.3 (85, 73 in. lbf)

Front Axle
Hub Sub–Assy LH

N·m (kgf·cm, ft·lbf) : Specified torque

◆ Non–reusable parts

P

9359AB72

Exploded view of the front hub and bearing, and related components

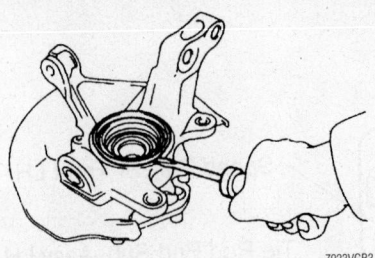

Removing the inner axle seal from the hub assembly

7923VGB3

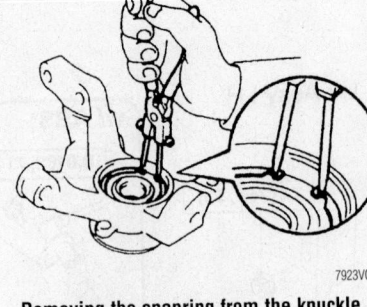

Removing the snapring from the knuckle before pressing out the bearing

7923VGB5

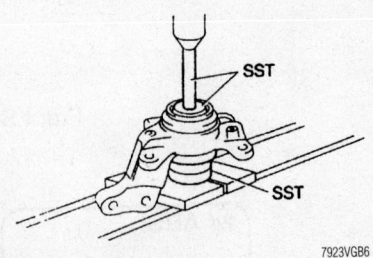

Removing the bearing from the steering knuckle using a press

7923VGB6

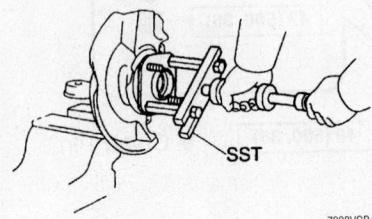

Removing the axle hub from the knuckle

7923VGB4

equipped with an air bag, wait at least 90 seconds before proceeding.

- Wheel
- Brake drum or rotor
- With ABS brakes, ABS wheel speed sensor or skid control sensor, as applicable
- 4 hub retaining bolts
- Hub

To install:

3. Install or connect the following:
- Hub to the knuckle. Tighten the bolts to 45 ft. lbs. (61 Nm).
- ABS wheel speed or skid control sensor, if equipped
- Brake drum or rotor
- Wheel
- Negative battery cable

4. Check and adjust the alignment, if needed.

Disc Rear Brake Type:

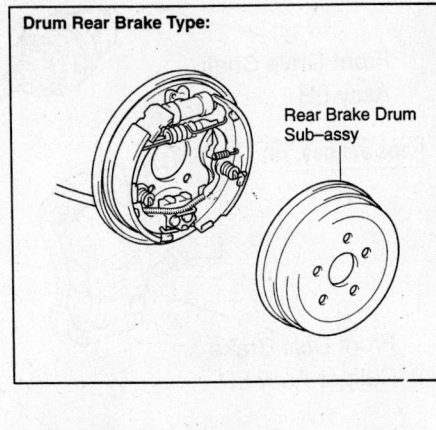

Drum Rear Brake Type:

Rear Brake Drum Sub-assy

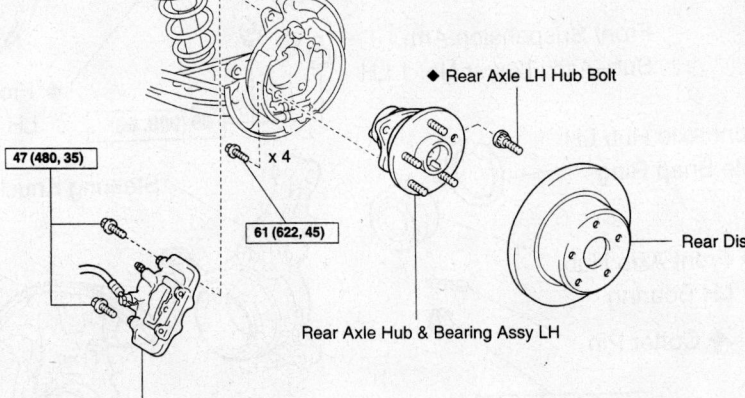

Skid Control Sensor Wire

47 (480, 35)

x 4

61 (622, 45)

◆ Rear Axle LH Hub Bolt

Rear Disc

Rear Axle Hub & Bearing Assy LH

Rear Disc Brake Caliper Assy LH

N·m (kgf·cm, ft·lbf) : Specified torque

◆ Non-reusable part

9359AB73

Exploded view of the hub and wheel bearing assembly

BRAKES

Brake Caliper

REMOVAL AND INSTALLATION

Front

1. Before servicing the vehicle, refer to the precautions section.
2. Remove some fluid from the reservoir with a suction pump.
3. Remove or disconnect the following:
   - Front wheels
   - Banjo bolt and disconnect the brake hose from the caliper. Plug the hose to prevent fluid loss and contamination.
   - Mounting bolts while holding the slide pin
   - Caliper

To Install:

4. Compress the caliper piston using a C–clamp or other suitable tool.
5. Install or connect the following:
   - Caliper
   - Mounting bolts and tighten to 25 ft. lbs. (34 Nm)
   - Brake hose to the caliper using new sealing washers. Carefully torque the banjo bolt to 21 ft. lbs. (29 Nm).
6. Fill the reservoir with fluid and bleed the brakes.
   - Front wheels

Rear

1. Before servicing the vehicle, refer to the precautions section.
2. Remove some fluid from the reservoir with a suction pump.

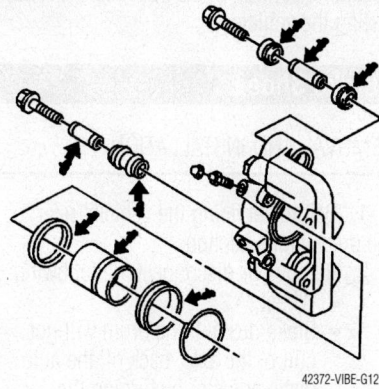

42372-VIBE-G12

Exploded view of the rear caliper components

3. Remove or disconnect the following:
   - Rear wheels
   - Clip and both anti-rattle springs
   - Two pad guide pins
   - Pads with the shims
   - Banjo bolt and disconnect the brake hose from the caliper. Plug the hose to prevent fluid loss and contamination.
   - 2 caliper mounting bolts and the caliper from its mounting bracket

To Install:

4. Compress the caliper piston using a C–clamp or other suitable tool.
5. Install or connect the following:
   - Caliper. Tighten the caliper bolts to 34 ft. lbs. (46 Nm).
   - Brake hose with new sealing washers. Tighten the banjo bolt to 21 ft. lbs. (29 Nm).
   - New anti-squeal shims, apply disc brake grease to the inside of the shim before installation

- Inner pad with the wear indicator facing upwards
- Outer pad
- Two pad guide pins
- Anti-rattle springs and the clip
6. Fill the reservoir with fluid and bleed the brake system. Adjust the parking brake if necessary.
   - Rear wheels

Disc Brake Pads

REMOVAL AND INSTALLATION

Front

1. Before servicing the vehicle, refer to the precautions section.
2. Remove some fluid from the reservoir with a suction pump.
3. Remove or disconnect the following:
   - Front wheels
4. Compress the caliper piston using a C–clamp or other suitable tool.
   - Mounting bolts while holding the slide pin
   - Caliper
   - Brake pads
   - Inside and outside pad wear indicators
   - Insulators
   - Pad insulators

To Install:

5. Compress the caliper piston using a C–clamp or other suitable tool.
6. Install or connect the following:
   - Upper, then lower brake pad retainer to the caliper bracket
   - New wear indicators
   - Pad insulators
   - Insulators
   - Brake pads
   - Caliper
   - Mounting bolts and tighten to 25 ft. lbs. (34 Nm)

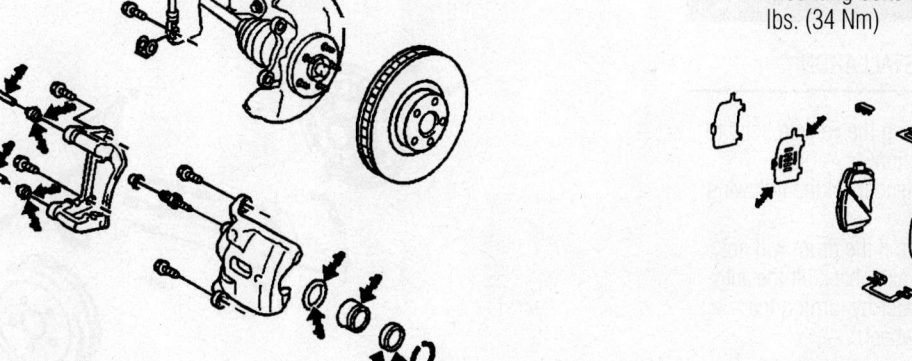

42372-VIBE-G11

Exploded view of the front caliper components

42372-VIBE-G13

Exploded view of the front pads and related components

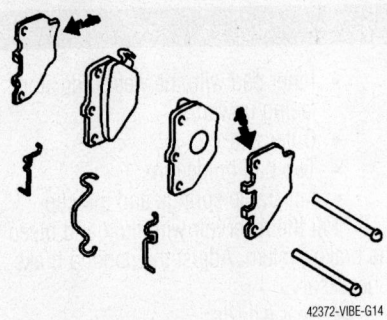

Exploded view of the front pads and related components

7. Fill the reservoir with fluid and bleed the brakes, if necessary.
- Front wheels

Rear

1. Before servicing the vehicle, refer to the precautions section.
2. Remove some fluid from the reservoir with a suction pump.
3. Remove or disconnect the following:
- Rear wheels
- Clip and both anti-rattle springs
- Two pad guide pins
- Pads with the shims

To Install:

4. Compress the caliper piston using a C–clamp or other suitable tool.
5. Install or connect the following:
- New anti-squeal shims, apply disc brake grease to the inside of the shim before installation
- Inner pad with the wear indicator facing upwards
- Outer pad
- Two pad guide pins
- Anti-rattle springs and the clip
6. Fill the reservoir with fluid and bleed the brake system. Adjust the parking brake if necessary.
- Rear wheels

Brake Drums

REMOVAL AND INSTALLATION

1. Before servicing the vehicle, refer to the precautions section.
2. Remove or disconnect the following:
- Wheel
- Brake drum. If the drum will not pull of the axle, back off the automatic adjuster by turning the adjusting wheel.

To install:

3. Install or connect the following:
- Drum on the axle

- Wheel
4. Refill the master cylinder and pump pedal to attain full brake pedal before road-testing the vehicle.

Brake Shoes

REMOVAL AND INSTALLATION

1. Before servicing the vehicle, refer to the precautions section.
2. Remove or disconnect the following:
- Wheel
- Brake drum. If the drum will not pull of the axle, back off the automatic adjuster by turning the adjusting wheel.
- Upper return spring
- Lower return spring
- Hold-down springs and pins from the front shoe
- Anchor side spring using needle nosed pliers
- Front shoe with the adjuster lever
- Adjuster lever and spring from the front shoe
- Hold-down springs and pins from the rear shoe
- Parking brake cable from the rear shoe using needle nosed pliers
- Rear brake shoe
- C–washer using a suitable pry tool from the shoe
- Parking brake lever from the shoe
- Automatic strut

To install:

3. Lubricate the contact points on the backing plate and the adjuster with lithium grease.
4. Install or connect the following:
- Adjuster strut
- Parking brake lever, and attach using a new C–washer
- Rear shoe
- Parking brake lever to the shoe lever

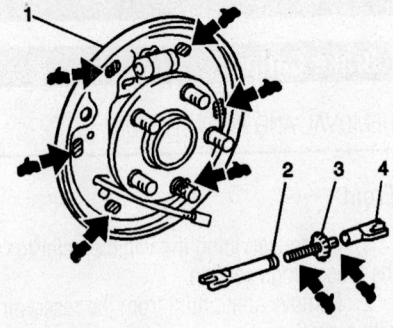

Lubricate the contact points on the backing plate and the adjuster with lithium grease

- Hold-down springs and pins to the rear shoe
- Automatic adjuster lever and spring to the front shoe
- Front shoe
- Anchor spring
- Hold-down springs and pins to the front shoe
- Lower return spring
- Upper return spring
- Drum
5. Adjust the rear brakes as follows:
   a. Temporarily install the drum and hub nuts.
   b. Remove the hole plug from the backing plate.
   c. Turn the adjuster to expand the shoe until the drum locks.
   d. Back off the adjuster eight notches using a suitable adjustment tool.
   e. Install the hole plug. into the backing plate to prevent dirt and moisture from entering.
   f. Readjust the parking brake cable as necessary.
6. Install the wheels.
7. Refill the master cylinder and pump pedal to attain full brake pedal before Road-testing the vehicle.

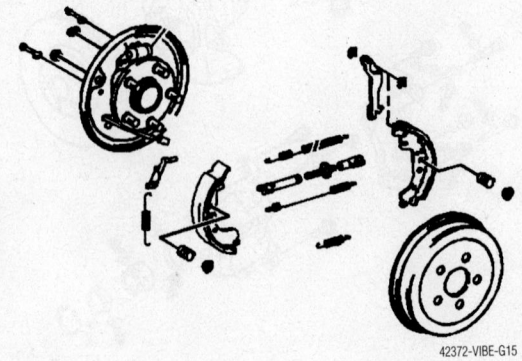

Exploded view of the drum brake assembly—2WD shown–AWD similar

BRAKES **31-52**
DRIVE TRAIN **31-41**
ENGINE REPAIR **31-9**
FUEL SYSTEM **31-38**
SPECIFICATION AND
MAINTENANCE CHARTS......**31-2**
Engine and Vehicle
Identification..........................31-2
General Engine Specifications 31-2
Engine Tune-Up Specifications 31-2
Firing Order 31-3
Accessory Drive Belt Routing 31-3
Capacities 31-4
Valve Specifications.................31-4
Crankshaft and Connecting Rod
Specifications 31-5
Piston and Ring Specifications 31-5
Torque Specifications 31-6
Wheel Alignment 31-6
Tire, Wheel and Ball Joint
Specifications.......................31-7
Brake Specifications 31-7
Scheduled Maintenance Intervals ..31-8
STEERING AND
SUSPENSION **31-47**
A
Air Bag..................................31-47
Arming..................................31-47
Disarming 31-47
Precautions 31-47
Alternator 31-9
Installation 31-9
Removal................................31-9
Automatic Transmission
Assembly..............................31-41
Removal & Installation............31-41
B
Brake Caliper 31-52
Removal & Installation............31-52
Brake Drums 31-52
Removal & Installation............31-52
Brake Shoes...........................31-53
Removal & Installation............31-53
C
Camshaft and Lifters.................31-22
Removal & Installation............31-22

Clutch...................................31-43
Adjustment............................31-43
Removal & Installation............31-43
Coil Spring 31-48
Removal & Installation............31-48
CV-Joints................................31-45
Overhaul 31-45
Cylinder Head 31-15
Removal & Installation............31-15
D
Disc Brake Pads.......................31-52
Removal & Installation............31-52
E
Engine Assembly 31-9
Removal & Installation............31-9
Exhaust Manifold.....................31-21
Removal & Installation............31-21
F
Front Crankshaft Seal 31-22
Replacement 31-22
Fuel Filter 31-39
Removal & Installation............31-39
Fuel Injector............................31-39
Removal & Installation............31-39
Fuel Pump 31-39
Removal & Installation............31-39
Fuel System Pressure 31-38
Relieving 31-38
Fuel System Service
Precautions 31-38
H
Halfshafts...............................31-44
Removal & Installation............31-44
Heater Core.............................31-13
Removal & Installation............31-13
Hydraulic Clutch System 31-43
Bleeding................................31-43
I
Ignition Timing 31-9
Adjustment............................31-9
Intake Manifold........................31-19
Removal & Installation............31-19
L
Lower Ball Joint.......................31-49
Removal & Installation............31-49
Lower Control Arm 31-49

Lower Control Arm Bushing
Replacement.........................31-50
Removal & Installation............31-49
M
Manual Transmission Assembly ..31-41
Removal & Installation............31-41
O
Oil Pan...................................31-28
Removal & Installation............31-28
Oil Pump 31-29
Removal & Installation............31-29
P
Piston and Ring 31-38
Positioning 31-38
R
Rack and Pinion Steering Gear 31-47
Removal & Installation............31-47
Rear Main Seal 31-31
Removal & Installation............31-31
S
Shock Absorber 31-48
Removal & Installation............31-48
Starter Motor 31-27
Removal & Installation............31-27
Strut......................................31-48
Removal & Installation............31-48
T
Timing Belt 31-33
Removal & Installation............31-33
Timing Chain, Sprockets, Front
Cover and Seal 31-32
Removal & Installation............31-32
U
Upper Control Arm 31-51
Removal & Installation............31-51
V
Valve Lash 31-27
Adjustment............................31-27
W
Water Pump 31-12
Removal & Installation............31-12
Wheel Bearings........................31-51
Adjustment............................31-51
Removal & Installation............31-51

SPECIFICATION CHARTS

ENGINE AND VEHICLE IDENTIFICATION

Code ①	Liters (cc)	Cu. In.	Cyl.	Fuel Sys.	Engine Type	Eng. Mfg.
D	2.2 (2199)	134	4	SFI	DOHC	Saturn
B	3.0 (3000)	183	6	SFI	DOHC	Saturn
4	3.5 (3471)	212	6	SFI	DOHC	Saturn

Code ②	Year
2	2002
3	2003
4	2004
5	2005
6	2006

SFI: Sequential Fuel Injection

DOHC: Double Overhead Camshafts

① 8th digit of VIN

② 10th digit of VIN

06025-SVUE-C01

GENERAL ENGINE SPECIFICATIONS

Year	Model	Engine Displacement Liters (cc)	Net Horsepower @ rpm	Net Torque @ rpm (ft. lbs.)	Bore x Stroke (in.)	Compression Ratio	Oil Pressure @ rpm
2002	VUE	2.2 (D)	137@5800	147@4400	3.38x3.50	9.5:1	50-80@1000
		3.0 (B)	182@6000	184@3400	3.38x3.50	10.0:1	50-80@1000
2003	VUE	2.2 (D)	137@5800	147@4400	3.38x3.50	9.5:1	50-80@1000
		3.0 (B)	182@6000	184@3400	3.38x3.50	10.0:1	50-80@1000
2004	VUE	2.2 (D)	137@5800	147@4400	3.38x3.50	9.5:1	50-80@1000
		3.0 (B)	182@6000	184@3400	3.38x3.50	10.0:1	50-80@1000
		3.5 (4)	250@5800	242@4500	3.50x3.66	10.0:1	71@3000
2005	VUE	2.2 (D)	143@5400	152@4000	3.38x3.72	10.0:1	50-80@1000
		3.5 (4)	250@5800	242@4500	3.50x3.66	10.0:1	71@3000

06025-SVUE-C02

ENGINE TUNE-UP SPECIFICATIONS

Year	Engine Displacement Liters (cc)	Spark Plug Gap (in.)	Ignition Timing (deg.) MT	Ignition Timing (deg.) AT	Fuel Pump (psi) ①	Idle Speed (rpm) MT ②	Idle Speed (rpm) AT ②	Valve Clearance In.	Valve Clearance Ex.
2002	2.2 (D)	0.045	③	③	50-60	④	④	HYD	HYD
	3.0 (B)	0.043	③	③	50-60	④	④	HYD	HYD
2003	2.2 (D)	0.045	③	③	50-60	④	④	HYD	HYD
	3.0 (B)	0.043	③	③	50-60	④	④	HYD	HYD
2004	2.2 (D)	0.045	③	③	50-60	④	④	HYD	HYD
	3.0 (B)	0.043	③	③	50-60	④	④	HYD	HYD
	3.5 (4)	0.051	③	③	48-56	④	④	HYD	HYD
2005	2.2 (D)	0.045	③	③	50-60	④	④	HYD	HYD
	3.5 (4)	39-43	③	③	48-56	④	④	HYD	HYD

NOTE: The Vehicle Emission Control Information label often reflects specification changes made during production. The label figures must be used if they differ from those in this chart.

HYD: Hydraulic

① Pressure measured at idle

② Idle speed measured with manual transmission in Neutral; automatic transmission in D (drive)

③ Engines equipped with Distributorless Ignition System (DIS). Ignition timing is not adjustable

④ Refer to the Vehicle Emission Control Information label

06025-SVUE-C03

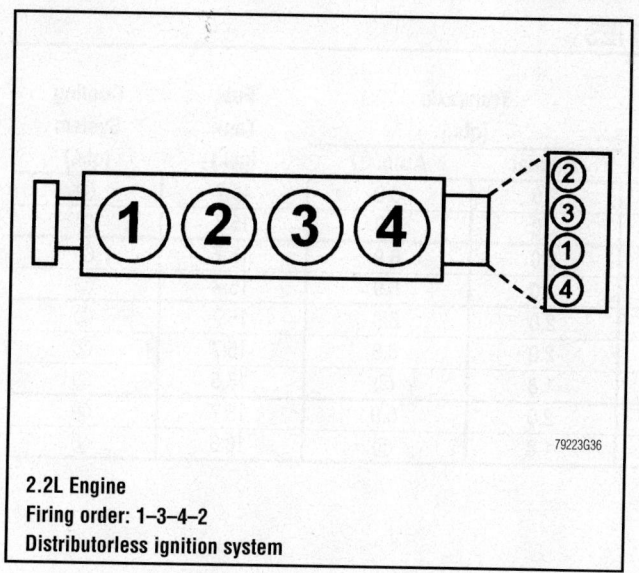

2.2L Engine
Firing order: 1–3–4–2
Distributorless ignition system

79223G36

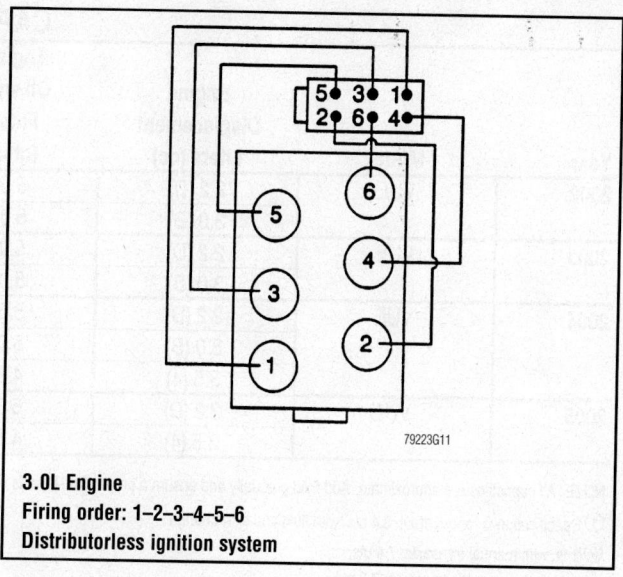

3.0L Engine
Firing order: 1–2–3–4–5–6
Distributorless ignition system

79223G11

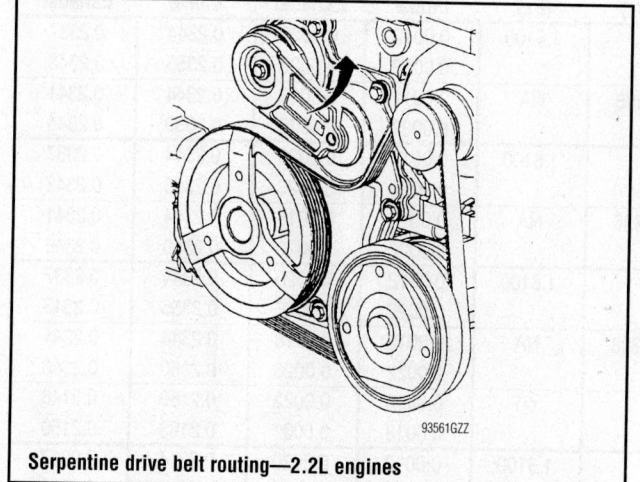

Serpentine drive belt routing—2.2L engines

93561GZZ

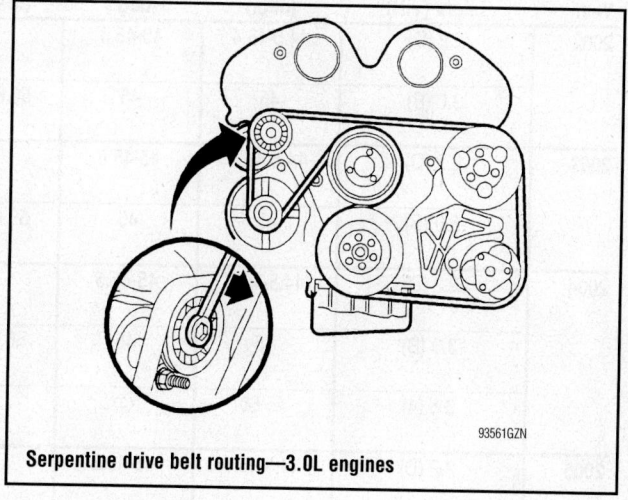

Serpentine drive belt routing—3.0L engines

93561GZN

CAPACITIES

Year	Model	Engine Displacement Liters (cc)	Engine Oil with Filter (qts.)	Transaxle (qts.) Manual	Transaxle (qts.) Auto. ①	Fuel Tank (gal.)	Cooling System (qts.)
2002	VUE	2.2 (D)	5.0	2.0	6.9	15.7	②
		3.0 (B)	5.0	2.0	6.9	15.7	②
2003	VUE	2.2 (D)	5.0	2.0	6.9	15.7	②
		3.0 (B)	5.0	2.0	6.9	15.7	②
2004	VUE	2.2 (D)	5.0	2.0	6.9	15.7	②
		3.0 (B)	5.0	2.0	6.9	15.7	②
		3.5 (4)	4.5	1.8	③	16.5	②
2005	VUE	2.2 (D)	5.0	2.0	6.9	15.7	②
		3.5 (4)	4.5	1.8	③	16.5	②

NOTE: All capacities are approximate. Add fluid gradually and ensure a proper fluid level is obtained.

① Specification is for overhaul. 8.4 pts. with fluid and filter change

② 2.2L with manual transaxle: 7.4 qts.

2.2L with automatic transaxle: 7.3 qts.

3.0L: 7.8 qts.

3.5L: 9.7 qts.

③ Except with Vti variable transmission: 2.9 qts.

With Vti variable transmission: 6.9 qts.

06025-SVUE-C04

VALVE SPECIFICATIONS

Year	Engine Displacement Liters (VIN)	Seat Angle (deg.)	Face Angle (deg.)	Spring Test Pressure (lbs. @ in.)	Spring Free-Length (in.)	Stem-to-Guide Clearance (in.) Intake	Stem-to-Guide Clearance (in.) Exhaust	Stem Diameter (in.) Intake	Stem Diameter (in.) Exhaust
2002	2.2 (D)	44.5-45.4	45-45.5	①	1.6100	0.0012	0.0020	0.2344	0.2337
				②		0.0022	0.0026	0.2355	0.2343
	3.0 (B)	45	45	56.6@1.338	NA	0.0012	0.0016	0.2344	0.2341
						0.0022	0.0026	0.2350	0.2346
2003	2.2 (D)	44.5-45.4	45-45.5	①	1.6100	0.0012	0.0020	0.2344	0.2337
				②		0.0022	0.0026	0.2355	0.2343
	3.0 (B)	45	45	56.6@1.338	NA	0.0012	0.0016	0.2344	0.2341
						0.0022	0.0026	0.2350	0.2346
2004	2.2 (D)	44.5-45.4	45-45.5	①	1.6100	0.0012	0.0020	0.2344	0.2337
				②		0.0022	0.0026	0.2355	0.2343
	3.0 (B)	45	45	56.6@1.338	NA	0.0012	0.0016	0.2344	0.2341
						0.0022	0.0026	0.2350	0.2346
	3.5 (4)	45-60	③	NA	④	0.0008	0.0022	0.2159	0.2146
						0.0018	0.0031	0.2163	0.2150
2005	2.2 (D)	44.5-45.4	45-45.5	①	1.6100	0.0012	0.0020	0.2344	0.2337
				②		0.0022	0.0026	0.2355	0.2343
	3.5 (4)	45-60	③	NA	④	0.0008	0.0022	0.2159	0.2146
						0.0018	0.0031	0.2163	0.2150

NA: Not available

① Valve spring load closed: 245-271 N

② Valve spring load open: 525-575 N

③ Intake: 45-60-70

Exhaust: 40-60

④ Intake: 2.029

Exhaust: 2.01

06025-SVUE-C05

CRANKSHAFT AND CONNECTING ROD SPECIFICATIONS

All measurements are given in inches.

| Year | Engine Displacement Liters (VIN) | Crankshaft | | | | Connecting Rod | | |
		Main Brg. Journal Dia.	Main Brg. Oil Clearance	Shaft End-play	Thrust on No.	Journal Diameter	Oil Clearance	Side Clearance
2002	2.2 (D)	2.2045-2.2050	0.0012 0.0026	0.0012-0.0150	3	1.9291-1.9297	0.0001-0.0021	0.0028-0.0146
	3.0 (B)	2.6763-2.6766	0.0060 0.0017	0.0004-0.0300	3	1.927-1.9280	0.0001-0.0021	0.0027-0.0110
2003	2.2 (D)	2.2045-2.2050	0.0012 0.0026	0.0012-0.0150	3	1.9291-1.9297	0.0001-0.0021	0.0028-0.0146
	3.0 (B)	2.6763-2.6766	0.0060 0.0017	0.0004-0.0300	3	1.927-1.9280	0.0001-0.0021	0.0027-0.0110
2004	2.2 (D)	2.2045-2.2050	0.0012 0.0026	0.0012-0.0150	3	1.9291-1.9297	0.0001-0.0021	0.0028-0.0146
	3.0 (B)	2.6763-2.6766	0.0060 0.0017	0.0004-0.0300	3	1.927-1.9280	0.0001-0.0021	0.0027-0.0110
	3.5 (4)	2.8337-2.8346	0.0008 0.0017	0.0004-0.0140	3	2.1644-2.1654	0.0008-0.0017	NA
2005	2.2 (D)	2.2045-2.2050	0.0012 0.0026	0.0012-0.0150	3	1.9291-1.9297	0.0001-0.0021	0.0028-0.0146
	3.5 (4)	2.8337-2.8346	0.0008 0.0017	0.0004-0.0140	3	2.1644-2.1654	0.0008-0.0017	NA

06025-SVUE-C06

PISTON AND RING SPECIFICATIONS

All measurements are given in inches.

| Year | Engine Displacement Liters (VIN) | Piston Clearance | Ring Gap | | | Ring Side Clearance | | |
			Top Compression	Bottom Compression	Oil Control	Top Compression	Bottom Compression	Oil Control
2002	2.2 (D)	0.0004-0.0016	0.008-0.016	0.0014 0.0022	0.0010 0.0030	0.0028-0.0146	0.0005-0.0024	SNUG
	3.0 (B)	0.0010-0.0018	0.0008-0.0015	0.0118 0.0196	0.0157 0.0551	0.0027-0.0110	0.0005-0.0024	SNUG
2003	2.2 (D)	0.0004-0.0016	0.008-0.016	0.0014 0.0022	0.0010 0.0030	0.0028-0.0146	0.0005-0.0024	SNUG
	3.0 (B)	0.0010-0.0018	0.0008-0.0015	0.0118 0.0196	0.0157 0.0551	0.0027-0.0110	0.0005-0.0024	SNUG
2004	2.2 (D)	0.0004-0.0016	0.008-0.016	0.0014 0.0022	0.0010 0.0030	0.0028-0.0146	0.0005-0.0024	SNUG
	3.0 (B)	0.0010-0.0018	0.0008-0.0015	0.0118 0.0196	0.0157 0.0551	0.0027-0.0110	0.0005-0.0024	SNUG
	3.5 (4)	0.0006-0.0016	0.0008-0.0014	0.0160 0.0220	0.008-0.0280	0.0022-0.0031	0.0012-0.0022	SNUG
2005	2.2 (D)	0.0004-0.0016	0.008-0.016	0.0014 0.0022	0.0010 0.0030	0.0028-0.0146	0.0005-0.0024	SNUG
	3.5 (4)	0.0006-0.0016	0.0008-0.0014	0.0160 0.0220	0.008-0.0280	0.0022-0.0031	0.0012-0.0022	SNUG

NA: Not available

06025-SVUE-C07

TORQUE SPECIFICATIONS
All readings in ft. lbs.

Year	Engine Displacement Liters (cc)	Cylinder Head Bolts	Main Bearing Bolts	Rod Bearing Bolts	Crankshaft Damper Bolts	Flywheel Bolts	Manifold		Spark Plugs	Oil Pan Drain Plug
							Intake	Exhaust		
2002	2.2 (D)	①	②	③	④	②	⑤	13	15	18
	3.0 (B)	⑥	⑦	26	15	②	15	15	18	18
2003	2.2 (D)	①	②	③	④	②	⑤	13	15	18
	3.0 (B)	⑥	⑦	26	15	②	15	15	18	18
2004	2.2 (D)	①	②	③	④	②	⑤	13	15	18
	3.0 (B)	⑥	⑦	26	15	②	15	15	18	18
	3.5 (4)	⑧	⑨	⑩	181	54	⑪	31	13	29
2005	2.2 (D)	①	⑫	③	④	②	⑬	13	15	18
	3.5 (4)	⑧	⑨	⑩	181	54	⑪	31	13	29

① Step 1: 22 ft. lbs.
  Step 2: 155 degrees
② 39 ft. lbs. Plus 25 degrees
③ 18 ft. lbs. Plus 100 degrees
④ 74 ft. lbs. Plus 125 degrees
⑤ 89 inch lbs.

⑥ Step 1: 18 ft. lbs.
  Step 2: plus 90 degrees
  Step 3: plus 90 degrees
  Step 4: plus 90 degrees
  Step 5: plus 15 degrees
⑦ Step 1: 37 ft. lbs.
  Step 2: plus 60 degrees
  Step 3: plus 15 degrees

⑧ Step 1: 29 ft. lbs.
  Step 2: 51 ft. lbs.
  Step 3: 72.3 ft. lbs.
⑨ Bottom M11 bolts: 54 ft. lbs.
  Side M10 bolts: 36 ft. lbs.
⑩ 14 ft. lbs. Plus 90 degrees

⑪ Intake manifold nuts and bolts:
  Step 1: 97 inch lbs.
  Step 2: 16 ft. lbs.
  Intake manifold top cover gasket:
  Step 1: 53 inch lbs.
  Step 2: 106 inch lbs.
⑫ Step 1: 15 ft. lbs.
  Step 2: plus 70 degrees
⑬ Nut: 124 inch lbs.
  Stud: 89 inch lbs.

06025-SVUE-C08

WHEEL ALIGNMENT

Year	Model		Caster		Camber		Toe-in (in.)
			Range (+/-Deg.)	Preferred Setting (Deg.)	Range (+/-Deg.)	Preferred Setting (Deg.)	
2002	VUE	F	2.60-3.40	3.00	-1.00	0.60	0.20 +/- 0.15
		R	—	—	—	-0.05	0.10 +/- 0.10
2003	VUE	F	2.60-3.40	3.00	-1.00	0.60	0.20 +/- 0.15
		R	—	—	—	-0.05	0.10 +/- 0.10
2004	VUE	F	2.60-3.40	3.00	-1.00	0.60	0.20 +/- 0.15
		R	—	—	—	-0.05	0.10 +/- 0.10
2005	VUE	F	2.60-3.40	3.00	-1.00	0.60	0.20 +/- 0.15
		R	—	—	—	-0.05	0.10 +/- 0.10

06025-SVUE-C09

TIRE, WHEEL AND BALL JOINT SPECIFICATIONS

Year	Model		OEM Tires		Tire Pressures (psi)		Wheel Size	Ball Joint Inspection	Lug Nuts
			Standard	Optional	Front	Rear			
2002	VUE	F	P235/65R16	P215/70R16	①	①	NS	NS	92
		R	P235/65R16	P215/70R16	①	①	NS	NS	92
2003	VUE	F	P235/65R16	P215/70R16	①	①	NS	NS	92
		R	P235/65R16	P215/70R16	①	①	NS	NS	92
2004	VUE	F	P235/65R16	P215/70R16	①	①	NS	NS	92
		R	P235/65R16	P215/70R16	①	①	NS	NS	92
2005	VUE	F	P235/65R16	P215/70R16	①	①	NS	NS	92
		R	P235/65R16	P215/70R16	①	①	NS	NS	92

OEM: Original Equipment Manufacturer

PSI: Pounds Per Square Inch

① Check the placard on the drivers side sill

NS: Not specified by manufacturer

06025-SVUE-C10

BRAKE SPECIFICATIONS
All measurements in inches unless noted

Year	Model		Brake Disc			Brake Drum Diameter			Minimum Lining Thickness	Brake Caliper	
			Original Thickness	Minimum Thickness	Maximum Runout	Original Inside Diameter	Max. Wear Limit	Maximum Machine Diameter		Bracket Bolt (ft. lbs.)	Mounting Bolt (ft. lbs.)
2002	VUE	F	1.020	0.960	0.001	—	—	—	0.080	136	32
		R	—	—	—	9.84	9.90	9.90	0.040	63	32
2003	VUE	F	1.020	0.960	0.001	—	—	—	0.080	136	32
		R	—	—	—	9.84	9.90	9.90	0.040	63	32
2004	VUE	F	1.020	0.960	0.001	—	—	—	0.080	136	32
		R	—	—	—	9.84	9.90	9.90	0.040	63	32
2005	VUE	F	1.024	0.960	0.002	—	—	—	0.080	136	32
		R	—	—	—	9.84	9.90	9.90	0.040	—	—

NA: Not Available

F: Front

R: Rear

06025-SVUE-C11

SCHEDULED MAINTENANCE INTERVALS
SATURN—VUE

TO BE SERVICED	TYPE OF SERVICE	VEHICLE MILEAGE INTERVAL (x1000)												
		3	6	9	12	15	18	21	24	27	30	33	36	39
Engine oil & filter	R		✓		✓		✓		✓		✓		✓	
Lubricate chassis, suspension and steering linkage	S/I		✓		✓		✓		✓		✓		✓	
Lubricate transaxle shift linkage and parking brake cable guides	S/I		✓		✓		✓		✓		✓		✓	
Lubricate underbody contact points & linkage	S/I		✓		✓		✓		✓		✓		✓	
Driveshaft boots, suspension bushings & ball joint seals	S/I		✓				✓		✓		✓		✓	
Exhaust system & throttle linkage	S/I		✓		✓		✓		✓		✓			
Rotate tires	S/I		✓		✓		✓				✓			
Brake hoses & brake lining	S/I		✓				✓						✓	
Accessory drive belt(s)	S/I						✓						✓	
Engine coolant level, hoses & clamps	S/I						✓						✓	
Air filter element	R										✓			
Engine coolant	R												✓	
Manual transaxle oil	R		✓											
Spark plugs ①	R										✓			
Automatic transaxle fluid & filter	S/I										✓			
Ignition cables & fuel systems	S/I										✓			
Vacuum line/hose	S/I										✓			
Fuel filter ②	R													

S/I: Service or Inspect

R: Replace

① Platinum tip spark plugs: replace every 100,000 miles

② Replace every 60,000 miles

FREQUENT OPERATION MAINTENANCE (SEVERE SERVICE)

If a vehicle is operated under any of the following conditions it is considered severe service:

- Extremely dusty areas

- 50% or more of the vehicle operation is in 32°C (90°F) or higher temperatures, or constant operation in temperatures below 0°C (32°F)

- Prolonged idling (vehicle operation in stop and go traffic)

- Frequent short running periods (engine does not warm to normal operating temperatures)

- Police, taxi, delivery usage or trailer towing usage

Engine oil & oil filter: change every 3000 miles

06025-SVUE-C12

ENGINE REPAIR

Ignition Timing

ADJUSTMENT

The engines covered in this section utilize a Distributorless Ignition System (DIS), no adjustment is possible.

Alternator

REMOVAL

2.2L Engine

1. Before servicing the vehicle, refer to the precautions section.
2. Remove or disconnect the following:
   - Negative battery cable
   - Throttle body air duct
   - Accessory drive belt
   - Alternator electrical connectors
   - Alternator bolts
   - Alternator

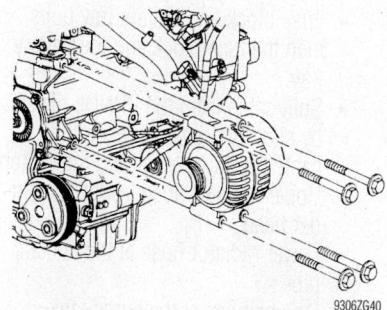

9306ZG40

Typical alternator mounting–2.2L engine

3.0L Engine

1. Before servicing the vehicle, refer to the precautions section.
2. Remove or disconnect the following:
   - Negative battery cable
   - Accessory drive belt and tensioner
   - Upper alternator bolts
   - Alternator lower bolts
   - Alternator electrical connectors
   - Alternator

3.5L Engine

1. Before servicing the vehicle, refer to the precautions section.
2. Disconnect the negative battery cable.
3. Remove the air cleaner.
4. Remove the accessory drive belt tensioner.
5. Install an engine support fixture.

6. Remove the front of the engine mount throughbolt and raise the engine for clearance.
7. Remove the alternator.

INSTALLATION

2.2L Engine

Install or connect the following:
- Alternator and torque the bolts to 26 ft. lbs. (35 Nm)
- Alternator electrical connectors
- Accessory drive belt
- Throttle body air duct
- Negative battery cable

3.0L Engine

Install or connect the following:
- Alternator electrical connectors
- Alternator and torque the bolts to 26 ft. lbs. (35 Nm)
- Drive belt tensioner and torque the bolts to 30 ft lbs. (40 Nm)
- Accessory drive belt
- Negative battery cable

3.5L Engine

1. Before servicing the vehicle, refer to the precautions section.
2. Install the alternator and handtighten the top bolts.
3. Install the bottom bolts and tighten all the bolts to 33 ft. lbs. (44 Nm).
4. Lower the engine and tighten the engine mount throughbolt to 81 ft. lbs. (110 Nm).
5. Remove the engine support fixture.
6. Install the accessory drive belt tensioner.
7. Install the air cleaner assembly.
8. Connect the negative battery cable.

Engine Assembly

REMOVAL & INSTALLATION

2.2L Engine

1. Before servicing the vehicle, refer to the precautions section.
2. Properly relieve the fuel system pressure.
3. Drain the engine coolant.
4. Drain the engine oil.
5. Drain the power steering fluid.
6. Remove or disconnect the following:
   - Both battery cables

- Battery
- Air cleaner and intake duct assembly
- Intake Air Temperature (IAT) sensor electrical connector
- Underhood fuse panel cover
- 3 connector through bolts from the fuse block
- Battery and electronic power steering feed wire nut and the wires from the stud
- Fuse block connectors from the fuse block
- Fuse block and battery tray bolts, then the fuse block and the battery tray
- Electrical connectors from the Transmission Control Module (TCM), if equipped
- Rear Heated Oxygen (HO$_2$S) sensor electrical connector
- 8-way electrical connection
- Main harness connector

➡**Do not remove the shifter cable from the bracket before removing the cable from the switch.**

- Shift cable from the transaxle range switch by slightly prying between the cable plastic retainer and the switch, if equipped
- Shift lever cables from the shift control housing using tool J36346, manual transmission only
- Shift lever cable from the bracket, manual transmission only
- Pressure line from the clutch actuator
- Back-up lamp switch
- Upper radiator hose from the cylinder head
- Lower radiator hose at the coolant pipe
- De-gas hose at the surge tank
- Surge hose at the surge tank
- Heater hoses from the core at the firewall
- Fuel lines
- Purge hoses from the solenoid
- Headlamp assemblies

7. Attach the radiator/condenser assembly to the radiator support using tie straps as this assembly stays in the vehicle.
   - Left front wheel and splash shield

8. Install a block of wood 1 inch x 2 inch x 4 inch between the transfer case and the engine cradle.
   - Right front wheel and splash shield

9. Install a block of wood 1 inch x 2

inch x 4 inch between the oil pane and the engine cradle. Do not place the wood under the oil pan and plug boss.

- Drive belt
- Electrical connections from the A/C compressor and pressure transducer
- A/C compressor bolts and position the compressor aside
- Push pins that attach the air deflector to the cradle

10. Drain the transaxle fluid.

- Transaxle lines from the transaxle and discard the seals. Replace with new seals during assembly.
- Rear HO2S sensor
- Exhaust pipe-to-manifold flange and intermediate fasteners
- Converter pipe and support the intermediate pipe
- Propshaft bolt to Power Take Off (PTO) on all wheel drive models
- Propshaft bolts, bracket and the shaft, on all wheel drive models
- Shifter cable from the bracket, on automatic transaxle models
- Power steering gear-to-intermediate shaft bolt
- Tie rod from the knuckles
- Lower control arms from the knuckles
- Stabilizer link nuts
- Left hand axle shaft from the transaxle
- Right hand shaft from the intermediate shaft using tool J45341
- Right hand engine mount and the left hand transaxle mount and let the engine rest on the wood blocks

➡ **Support the rear of the vehicle with a jackstand prior to engine removal.**

11. Place an engine support table under the engine.

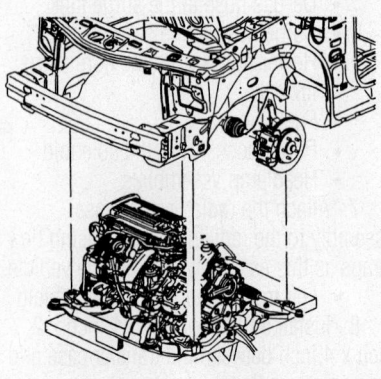

9359ZG01

Remove the engine with a support table–2.2L engine

12. Place blocks of wood to level the powertrain assembly, if necessary. The blocks can be placed between the oil pan and the cradle.

13. Raise the support table until it supports the powertrain assembly.

14. Remove the cradle-to-body bolts.

15. Check that all hoses, lines and wiring is free, then lower the engine table and raise the body on a hoist to remove the assembly

To install:

16. Installation is the reverse of removal, please note the following torque specifications.

a. Frame-to-body bolts to 114 ft. lbs. (155 Nm).

b. Right hand engine mount-to-bracket bolts to 37 ft. lbs. (50 Nm).

c. Right hand engine mount assembly and nuts to 92 ft. lbs. (125 Nm).

d. Left hand transaxle mount-to-bracket bolts to 37 ft. lbs. (50 Nm).

e. Lower control arm to knuckle bolts to 89 inch lbs. (10 Nm) plus 150 degrees.

f. Stabilizer link nuts to 48 ft. lbs. (65 Nm).

g. Tie rod-to-knuckle assembly to 37 ft. lbs. (50 Nm).

h. Steering shaft-to-rack bolt to 25 ft. lbs. (34 Nm).

i. Propshaft-to-PTU bolts to 19 ft. lbs. (25 Nm).

j. Propshaft-to-rear module bolts to 37 ft. lbs. (50 Nm).

k. Propshaft support bearing bolts to 19 ft. lbs. (25 Nm).

l. Propshaft guard strap to 19 ft. lbs. (25 Nm).

m. Transaxle cooler lines with new seals, stud to 15 ft. lbs. (20 Nm) and nut to 7 ft. lbs. (10 Nm).

n. Down pipe-to-manifold nuts to 22 ft. lbs. (30 Nm).

o. Down pipe-to-intermediate pipe nuts to 37 ft. lbs. (50 Nm).

p. A/C compressor bolts to 18 ft. lbs. (25 Nm).

q. Engine ground-to-body bolt to 15 ft. lbs. (20 Nm).

17. Fill the engine with coolant.

18. Fill the engine with new oil.

19. Prime the fuel system by cycling the ignition **ON** for 5 seconds and **OFF** for 10 seconds a few times without cranking the engine.

20. Start the engine, check for leaks, and repair if necessary.

3.0L Engine

1. Before servicing the vehicle, refer to the precautions section.

2. Properly relieve the fuel system pressure.

3. Drain the engine coolant.

4. Drain the engine oil.

5. Drain the power steering fluid.

6. Remove or disconnect the following:

- Both battery cables
- Battery
- Air cleaner and intake duct assembly
- Intake Air Temperature (IAT) sensor electrical connector
- Underhood fuse panel cover
- 3 connector through bolts from the fuse block
- Battery and electronic power steering feed wire nut and the wires from the stud
- Fuse block connectors from the fuse block
- Electrical connectors from the Transmission Control Module (TCM), if equipped
- Rear Heated Oxygen (HO2S) sensor electrical connector
- Vacuum hose with the check valve from the brake booster and lay on the engine
- Main harness connector
- Fuse block and battery tray bolts, then the fuse block and the battery tray
- Shift cable from the PRNDL switch by slightly prying between the cable plastic retainer and the switch
- Upper radiator hose from the cylinder head
- Lower radiator hose at the coolant pipe
- De-gas hose at the surge tank
- Surge hose at the surge tank
- Heater hoses from the core at the firewall
- Fuel lines
- Purge hoses from the solenoid

7. Attach the radiator/condenser assembly to the radiator support using tie straps as this assembly stays in the vehicle.

8. Evacuate the A/C system using approved equipment and disconnect the transducer and connector.

- Left front wheel and splash shield

9. Install a block of wood 1 inch x 2 inch x 4 inch between the transfer case and the engine cradle.

- Right front wheel and splash shield

10. Install a block of wood 1 inch x 2 inch x 4 inch between the oil pane and the engine cradle. Do not place the wood under the oil pan and plug boss.

- Drive belt
- A/C compressor and position the A/C line aside

- Push pins that attach the air deflector to the cradle
11. Drain the transaxle fluid.
- Transaxle lines from the transaxle and discard the seals. Replace with new seals during assembly.
- Exhaust pipe-to-manifold flange fasteners
- Exhaust pipe-to-muffler nuts
- Converter pipe and support the intermediate pipe
- Propshaft bolt-to-Power Take Off (PTO) ,on all wheel drive models
- Propshaft bolts, bracket and the shaft, on all wheel drive models
- Shifter cable from the bracket
- Power steering gear-to-intermediate shaft bolt
- Tie rod from the knuckles
- Lower control arms from the knuckles
- Stabilizer link nuts
- Left hand axle shaft from the transaxle
- Right hand shaft from the intermediate shaft using tool J45341
- Right hand engine mount and the left hand transaxle mount and let the engine rest on the wood blocks

➡**Support the rear of the vehicle with a jackstand prior to engine removal.**

12. Place an engine support table under the engine.
13. Place blocks of wood to level the powertrain assembly, if necessary. The blocks can be placed between the oil pan and the cradle.
14. Raise the support table until it supports the powertrain assembly.
15. Remove the cradle-to-body bolts.
16. Check that all hoses, lines and wiring is free, then lower the engine table and raise the body on a hoist to remove the assembly

To install:

17. Installation is the reverse of removal, please note the following torque specifications.
a. Frame-to-body bolts to 114 ft. lbs. (155 Nm).
b. Right hand engine mount-to-bracket bolts to 37 ft. lbs. (50 Nm).
c. Left hand transaxle mount-to-bracket bolts to 37 ft. lbs. (50 Nm).
d. Lower control arm to knuckle bolts to 89 inch lbs. (10 Nm) plus 150 degrees.
e. Stabilizer link nuts to 48 ft. lbs. (65 Nm).
f. Tie rod-to-knuckle assembly to 37 ft. lbs. (50 Nm).

g. Steering shaft-to-rack bolt to 25 ft. lbs. (34 Nm).
h. Propshaft-to-PTU bolts to 19 ft. lbs. (25 Nm).
i. Propshaft-to-rear module bolts to 37 ft. lbs. (50 Nm).
j. Propshaft support bearing bolts to 19 ft. lbs. (25 Nm).
k. Propshaft guard strap to 19 ft. lbs. (25 Nm).
l. Transaxle cooler lines with new seals, stud to 15 ft. lbs. (20 Nm) and nut to 7 ft. lbs. (10 Nm).
m. Down pipe-to-manifold nuts to 22 ft. lbs. (30 Nm).
n. Down pipe-to-intermediate pipe nuts to 22 ft. lbs. (30 Nm).
o. A/C compressor bolts to 18 ft. lbs. (25 Nm).
p. Engine ground-to-body bolt to 15 ft. lbs. (20 Nm).
18. Fill the engine with coolant.
19. Fill the engine with new oil.
20. Prime the fuel system by cycling the ignition **ON** for 5 seconds and **OFF** for 10 seconds a few times without cranking the engine.
21. Start the engine, check for leaks, and repair if necessary.

3.5L Engine

1. Before servicing the vehicle, refer to the precautions section.
2. Place the wheels in the straight forward position, remove the key from the ignition.
3. Disconnect the negative battery cable.
4. Remove the air cleaner assembly.
5. Secure the cooling module to the upper body structure.
6. Remove the battery and battery tray.
7. Disconnect the transmission shifter cable.
8. Disconnect the wiring harness from the underhood junction block.
9. Evacuate the A/C system.
10. Drain the cooling system.
11. Remove the Powertrain Control Module (PCM).
12. Remove the A/C low pressure tube at the front lift bracket.
13. Disconnect the alternator positive cable and the A/C high pressure switch harness.
14. Remove the A/C tube from the compressor.
15. Disconnect the A/C line from the condenser to the compressor.
16. Disconnect the coolant reservoir hose from the engine.

17. Disconnect the radiator inlet and outlet hoses at the engine.
18. Disconnect the heater hoses.
19. Remove the starter positive cable.
20. Relieve the fuel pressure.
21. Disconnect the fuel feed line.
22. Disconnect the fuel Evaporative Emission (EVAP) line.
23. Remove the lower transaxle-to-engine bolts.
24. Remove the PTU as follows:
a. Remove the propeller shaft.
b. Drain the transfer case oil.
c. Remove the exhaust cross-under pipe.
d. Remove the vent tube clamp.
e. Remove the vent tube from the transfer case.
f. Remove the transfer case.

➡**When removing the transfer case/output shaft, do not use excessive force or damage to the bushings may occur.**

g. Remove the transfer case from the transaxle.
25. Remove the torque converter inspection cover.
26. Remove the torque converter to flywheel bolts.
27. Remove the front wheels.
28. Remove the left inner liner.
29. Disconnect the transmission cooler lines from the transmission and bracket.
30. Remove the tie rod ends from the steering knuckles.
31. Remove the stabilizer bar links.
32. Disconnect the lower ball joints.
33. Remove the axle shaft nuts.
34. Disconnect the intermediate shaft from the steering gear.
35. Remove the front exhaust pipe.
36. Remove the 3 front fender pushpins to allow the front fender to flex.
37. Matchmark mark the frame to the body position for installation purposes.
38. Support the engine in the cradle with wood blocks.
39. Disconnect the front engine mount from the body.
40. Lower the vehicle to 3 feet off the ground in order to position an engine support table such as J 39580 under the frame.
41. Remove the cradle bolts.
42. Lower the table to the floor slowly.
43. Remove the starter.
44. Remove the A/C compressor.
45. Remove the alternator.
46. Remove the front covers.
47. Remove the rocker covers.
48. Remove the catalytic converters.
49. Remove the timing belt.

50. Remove the cylinder heads.
51. Remove the front engine mount from the engine.
52. Remove the right engine mount.
53. Separate the engine from the transmission.

To install:

54. Attach the engine to the transmission.
55. Install the right engine mount.
56. Install the front engine mount to the engine.
57. Install the cylinder heads.
58. Install the timing belt.
59. Install the catalytic converters.
60. Install the rocker covers.
61. Install the front covers.
62. Install the alternator.
63. Install the A/C compressor.
64. Install the starter.
65. Install the engine and transmission assembly in the vehicle.
66. Install the cradle bolts. Tighten the bolts to 114 ft. lbs. (155 Nm).
67. Remove the lift table.
68. Install the front engine mount bolts to the body. Tighten the transmission mount–to–frame bolts to 37 ft. lbs. (50 Nm) and the transmission mount–to–bracket through bolt, while aligning the transmission mount to the bracket to 81 ft. lbs. (110 Nm).
69. Remove the wood blocks from the cradle.
70. Install the lower transaxle-to-engine bolts and tighten to 47 ft. lbs. (64 Nm).
71. Install the PTU as follows:
   a. Install the transfer case to the transaxle and tighten the bolts to 38 ft. lbs. (51 Nm).
72. Install the vent hose and the clamp to the transfer case.
73. Install the exhaust cross–under pipe.
74. Remove the transfer case check plug and fill the transfer case with synthetic gear oil.
75. Install the check plug and gasket to the case and tighten to 33 ft. lbs. (44 Nm).
76. Install the torque converter–to–flywheel bolts and tighten to 9 ft. lbs. (12 Nm).
77. Install the torque converter inspection cover and tighten to 9 ft. lbs. (12 Nm).
78. Install the 3 front fender pushpins
79. Install the front exhaust pipe.
80. Connect the intermediate shaft from the steering gear.
81. Install the propeller shaft.
82. Install the axle shaft nuts.
83. Connect the lower ball joints.
84. Install the stabilizer bar links.
85. Install tie rod ends to the steering knuckles.

86. Connect the transmission cooler lines to the transmission and bracket.
87. Install the left inner liner.
88. Install the front tires.
89. Install the fuel EVAP line.
90. Connect the fuel feed line.
91. Connect the starter positive cable.
92. Install the heater hoses.
93. Connect the radiator hoses to the engine.
94. Attach the A/C tube to the A/C compressor.
95. Connect the coolant reservoir hose.
96. Connect the A/C line from the condenser to compressor.
97. Connect the A/C high pressure switch harness.
98. Connect the alternator wiring.
99. Install the A/C lower pressure tube at the front lift bracket.
100. Install the PCM.
101. Connect the radiator hoses to the engine.
102. Fill the cooling system.
103. Connect the wiring harness to the underhood junction block.
104. Connect the transmission shifter cable.
105. Remove the cooling module support.
106. Install the battery tray and battery.
107. Install the air cleaner assembly and ducts.
108. Connect the negative battery cable.

Water Pump

REMOVAL & INSTALLATION

2.2L Engine

1. Before servicing the vehicle, refer to the precautions section.
2. Drain the cooling system.
3. Remove or disconnect the following:
   - Negative battery cable
   - Air cleaner assembly
   - Thermostat housing pipe-to-cylinder head bolt (near the front of the engine)
   - Exhaust manifold heat shield
   - Water pump access plate
   - Right hand wheel and splash shield
4. Remove the drain plug from the bottom of the pump and drain the remaining coolant.
   - Engine Coolant Temperature (ECT) sensor connection
   - Thermostat housing bolts, then move the housing towards the left hand side of the vehicle while twisting the feed pipe from the rear

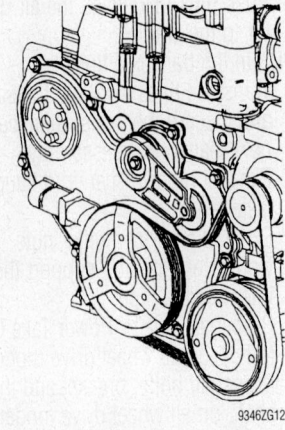

Water pump holding tool J43651–2.2L engine

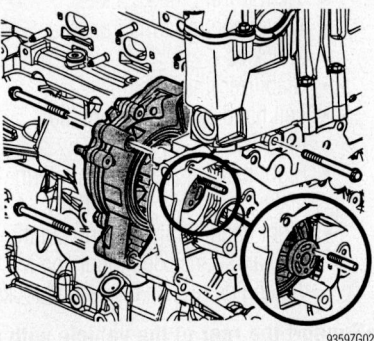

Exploded view of the water pump mounting–2.2L engine

of the pump. Leave the coolant hoses and the housing connected.
   - Water feed pipe and discard the seals
5. Install a Water Pump Holding Tool J43651. Tighten the bolts on the tool into threads on the pump sprocket, then install some of the access plate bolts to attach the tool to the front cover.
   - Water pump retaining bolts
   - Water pump

To install:

6. Install or connect the following:
   - Water pump with a new seal and torque the bolts to 18 ft. lbs. (25 Nm)
   - Water pump sprocket and torque the bolts to 89 inch lbs. (10 Nm)
   - Water pump sprocket access plate and torque the bolts to 89 inch lbs. (10 Nm)
   - Water feed tube after lubricating the O-ring
   - Thermostat housing and torque the bolts to 89 inch lbs. (10 Nm)
   - Exhaust manifold heat shield
   - Air cleaner assembly
   - Negative battery cable

7. Fill the cooling system.
8. Start the vehicle and check for leaks, repair if necessary.

3.0L Engine

1. Before servicing the vehicle, refer to the precautions section.
2. Drain the cooling system.
3. Remove or disconnect the following:
- Negative battery cable
- Air cleaner assembly
- Left front wheel
- Wheel well splash shield
- Loosen the water pump pulley and power steering pulley bolts, but do not remove them
4. Install an engine support fixture.
- Right front engine mount
- Accessory drive belt
5. Release the retaining tabs on the wiring harness channel and remove the front cover.
- Wiring harness from the channel
- Water pump pulley
- Power steering pump pulley
- Drive belt tensioner
- Front timing belt cover
- Water pump

To install:
6. Install or connect the following:
- Water pump with a new O-ring and torque the bolts to 18 ft. lbs. (25 Nm)
- Front timing belt cover and torque the bolts to 71 inch lbs. (8 Nm)
- Drive belt tensioner and torque the bolts to 30 ft. lbs. (40 Nm)
- Water pump pulley's and torque the bolts to 71 inch lbs. (8 Nm)

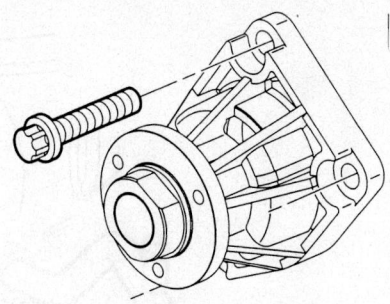

Water pump mounting–3.0L engine

- Power steering pump and torque the bolts to 15 ft. lbs. (20 Nm)
- Wiring harness into the channel
- Wiring harness channel front cover
- Accessory drive belt
7. Remove the engine support fixture.
- Wheel well splash shield and torque the bolts to 44 inch lbs. (5 Nm)
- Front wheel
- Air cleaner and intake duct
- Negative battery cable
8. Fill the cooling system.
9. Start the vehicle and check for leaks, repair if necessary.

3.5L Engine

1. Before servicing the vehicle, refer to the precautions section.
2. Drain the coolant.
3. Remove the timing belt cover.

➡ **If a vehicle is diagnosed to have coolant leaking inside the timing belt cover, a visual inspection of the timing belt should be done. If there is an indi-** cation that coolant has leaked onto the timing belt such as wetness or staining, the timing belt should be replaced. **Also when replacing the water pump, coolant may be spilled onto the timing belt. Replacement of the timing belt due to coolant spillage on the belt is not necessary.**

4. Remove the water pump assembly bolts.
5. Remove the water pump and O-ring.
To install:
6. Clean the water pump mating surfaces.
7. Install a new water pump O-seal to the water pump.
8. Install the water pump assembly and tighten the bolts to 18 inch lbs. (12 Nm.
9. Install the timing belt cover.
10. Fill the cooling system.

Heater Core

REMOVAL & INSTALLATION

1. Before servicing the vehicle, refer to the precautions section.
2. Disable the air bag system.
3. Record all preset radio stations.
4. Disconnect the negative battery cable.
5. Drain and recycle the engine coolant.
6. Recover the A/C system refrigerant using approved equipment.
7. Remove or disconnect the following:
- Suction line from the Thermal Expansion Valve (TXV) and cap the TXV and the line
- TXV thermistor connection
- Heater core outlet and inlet hose's from the core and plug the lines
- Instrument Panel (IP) right end panel by gently tugging to disengage the clip
- IP knee bolster panel
- IP left end panel by gently tugging to disengage the clip
- Door jamb switch electrical connections on both sides
8. Place the shifter in neutral.
- Horse shoe bezel at the shifter by first pulling up at the rear to disengage the clips and slide it up and over the shifter to access the electrical connections. Disengage all electrical connections.
- Glove box-to-IP screws and upper glove box-to-radio bezel screws
- Glove box
- Radio bezel by pulling the lower edge forward first and then the top to disengage the clips

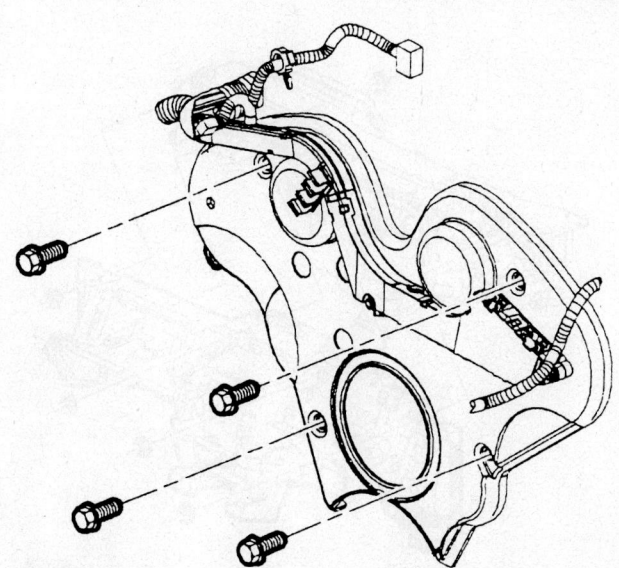

Timing belt cover bolt locations–3.0L engine

- Temperature cable, blower switch and IP 20-way connections from the controller
- Air bag telltale, foglamp switch, dimmer switch and hazard switch connections
- Upper and lower steering column shrouds
- Cluster bezel screws and the bezel
- A-pillar garnish moldings
- Right IP deflector assembly-to-intermediate duct screw
- IP cover-to-cross car beam bolt covers and the cover-to-beam bolts
- IP cover-to-IP retaining screws at the lower edge of the cover and the radio opening
- IP cover by lifting it then moving it rearwards passenger side first and walk the cluster opening around the steering wheel and out the drivers side door
- Cluster-to-IP screws and the cluster
- Radio
- Passenger side air bag
- Center, right and left shifter close-out panels
- Right and left intermediate ducts
- Cluster connector from the retainer
- IP fuse block from the bracket
- IP ground wire from the H brace
- Brake Control Module (BCM) from the retainer
- Data Link Controller (DLC) from the retainer
- Right door sill plate trim
- IP retainer fasteners and retainer
- Heater duct
- Heater core cover
- Heater core pipe cover and pipe foam seal

9. Grab the heater core at the end tanks and remove the core. If the core sticks, spray the perimeter of the core seal and the pips at the front of the dash with soapy water can aid in removal.

To install:

10. Spray the dash seal at the core pipe openings and seal with soapy water to aid in installation.

11. Install or connect the following:
- Heater core
- Pipe seal and cover. Tighten the cover retainers to 9 inch lbs. (1 Nm).
- Heater core cover and tighten the cover retainers to 9 inch lbs. (1 Nm)
- Heater duct
- IP retainer by aligning the 4-way locator (tapered boss) and outboard

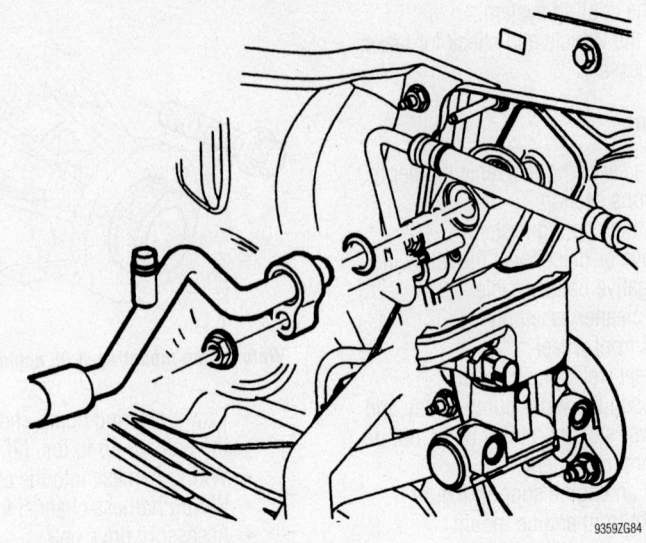

Disconnect the suction line from the Thermal Expansion Valve (TXV)

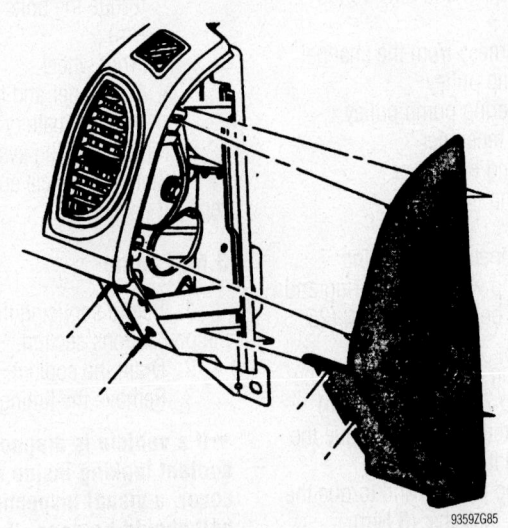

Remove the instrument Panel right end panel

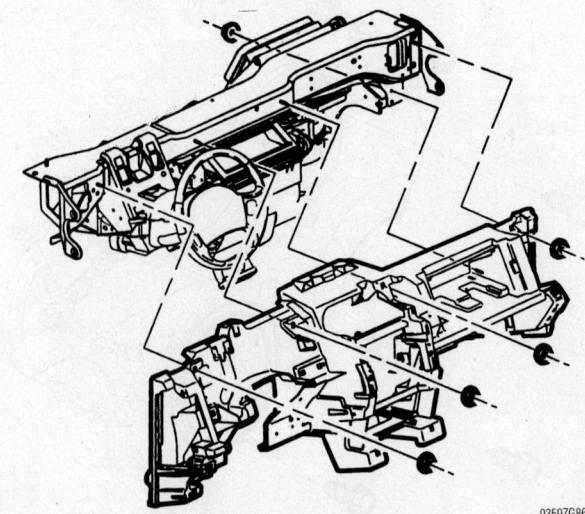

Remove the IP retainer

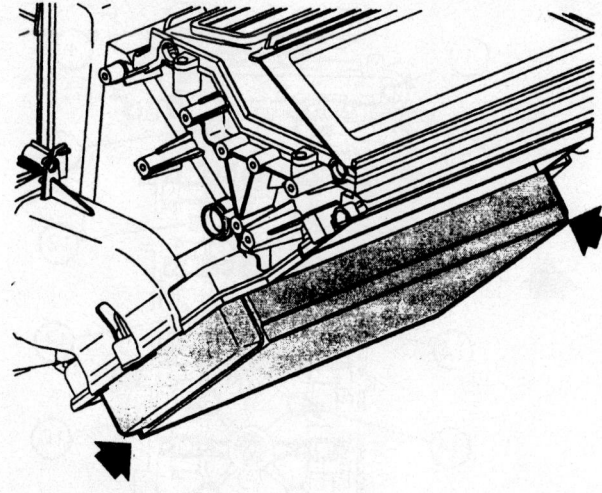

9359ZG83

Grab the heater core at the end tanks and remove the core

locators with the corresponding holes or slots in the beam, then tighten the fasteners starting from the center to 88 inch lbs. (10 Nm)
- Right door sill plate trim
- DLC to the retainer
- BCM to the retainer
- IP ground wire to the H brace
- IP fuse block to the bracket
- Cluster connector to the retainer
- Right and left intermediate ducts
- Center, right and left shifter close-out panels
- Passenger side air bag and tighten the fasteners to 88 inch lbs. (10 Nm)
- Radio
- Cluster and tighten the screws
- IP top cover and tighten the retainers to 22 inch lbs. (2.5 Nm)
- IP top cover-to-beam bolts and tighten to 88 inch lbs. (10 Nm)
- IP top cover bolt covers
- Right IP deflector assembly-to-intermediate duct screw
- A-pillar garnish moldings
- Cluster bezel and screws
- Upper and lower steering column shrouds

12. Center the temperature knob by aligning the controller housing alignment tab with the slot in the shaft.

13. Align the air temperature cable lug with the detent spring point.
- Temperature cable to the control head by aligning the retention tabs and the knob shaft and snap into place
- IP harness to the blower switch and controller
- Air bag telltale, foglamp switch, dimmer switch and hazard switch connections
- Radio bezel
- Glove box

14. Place the shifter in neutral.
- Horse shoe bezel electrical connections
- Horse shoe bezel and place the shifter in park
- Door jamb switch electrical connections on both sides
- IP left end panel
- IP knee bolster panel
- IP right end panel
- Heater core outlet and inlet hose's to the core and position the clamp at 9 o'clock
- TXV thermistor connection
- Suction line to the TXV using new seal washer and tighten to 12 ft. lbs. (16 Nm) and cap the TXV and the line

15. Recharge the A/C system refrigerant using approved equipment.
16. Refill the engine cooling system.
17. Connect the negative battery cable.
18. Reset all preset radio stations.
19. Enable the air bag system.

Cylinder Head

REMOVAL & INSTALLATION

2.2L Engine

✳✳ WARNING

Only remove the cylinder head when the engine is cold. Warpage may result if the cylinder head is removed while the engine is hot.

1. Before servicing the vehicle, refer to the precautions section.
2. Drain the cooling system.
3. Drain the engine oil.
4. Properly relieve the fuel system pressure.
5. Remove or disconnect the following:
- Negative battery cable
- Intake Air Temperature (IAT) sensor connection
- Air cleaner assembly
- Ignition module assembly
- Electronic Control Module (ECM) connections
- Oil dipstick tube bolt
- Throttle body electrical connection
- Electrical connector from the fuel injector harness and attachment at the bottom of the intake manifold
- Electrical connector at the purge solenoid and Manifold Absolute Pressure (MAP) sensor
- Vacuum hose at the brake booster
- Coolant pipe bracket bolts from the cylinder head
- De-gas hose clamp from the cylinder head and fuel rail and position aside
- Ground strap from the rear cam cover
- Fuel rail bracket from the cam cover
- Fuel lines
- Cam cover

6. Position the No. 1 piston 60 degrees Before Top dead center (TDC) using a 24mm wrench to rotate the camshafts in a clockwise motion and make sure the diamond shaped hole on the intake sprocket is at the 12 O'clock position.
- Upper timing guide

➡**Remove the timing chain tensioner to unload chain tension before removing the timing chain.**

9359ZG03

Position the No. 1 piston 60 degrees Before Top dead center (TDC) using a 24mm wrench to rotate the camshafts in a clockwise motion and make sure the diamond shaped hole on the intake sprocket is at the 12 O'clock position—2.2L engine

- Fixed timing chain guide access plug
- Upper fixed guide bolt using a magnetic socket

7. Install a three bar engine support fixture.

- Right hand engine mount
- Right hand mount bracket
- Right wheel splash shield
- Install a 1 x 2 x 4 inch block of wood between the oil pan and cradle
- Drive belt
- Drive belt tensioner assembly

8. Install crankshaft pullet holder J38122A.

- Crankshaft balancer bolt and pulley
- Front cover bolts
- Lower water pump bolt
- Front cover and gasket
- Lower fixed guide
- Upper radiator hose from the cylinder head
- Exhaust manifold pipe nuts
- Front and rear Oxygen (O_2S) sensor electrical connector
- Down pipe-to-intermediate pipe nuts
- Down pipe
- Exhaust camshaft sprocket bolts while holding the camshaft with a 24mm wrench and discard the camshaft sprocket bolts
- Exhaust sprocket
- Adjustable guide through the top of the cylinder head
- Intake camshaft sprocket bolts while holding the camshaft with a 24mm wrench and discard the camshaft sprocket bolts
- Intake sprocket
- Timing chain assembly
- Timing drive sprocket from the crankshaft

9. Install a floor jack to support the engine and remove the engine support fixture.

- Cylinder head bolts using the proper sequence
- Cylinder head

To install:

➡ **Set the crankshaft to 60 degrees Before Top Dead Center (BTDC) to prevent contact between the pistons and valves.**

10. Install or connect the following:
- New cylinder head gasket with the side imprinted **OPEN** facing up
- Cylinder head and align it on the dowels
- New cylinder head bolts and torque

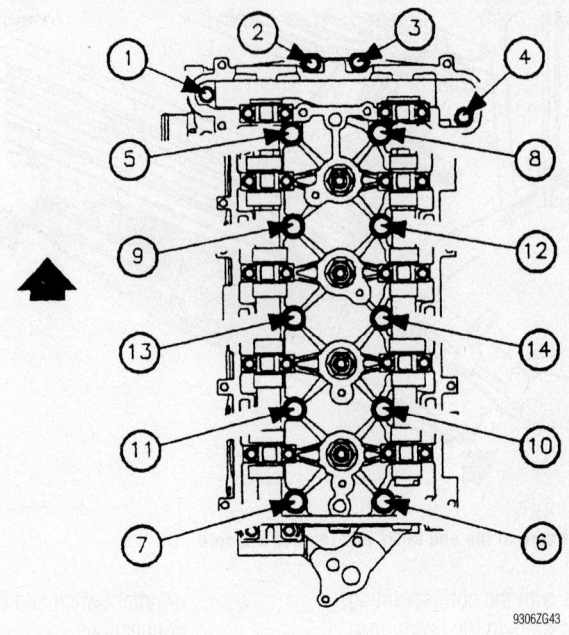

Cylinder head bolt loosening sequence—2.2L engine

them in sequence to 22 ft. lbs. (30 Nm) plus 155 degrees
- Front 4 cylinder head bolts coated with Loctite® and torque them to 26 ft. lbs. (35 Nm)

11. Position the exhaust camshaft with the offset slot in the 2 o'clock position and the intake camshaft with the offset slot in the 11 o'clock position.

- Timing chain around the intake camshaft sprocket with the copper link aligned with the **INT** diamond timing mark
- Sprocket to the camshaft and align it with the offset slot
- New camshaft sprocket bolt but do not tighten
- Timing chain around the crankshaft sprocket and align the silver link to the timing mark
- Adjustable timing chain guide through the opening on top of the cylinder head and torque the bolt to 89 inch lbs. (10 Nm)
- Timing chain around the exhaust camshaft sprocket with the silver link aligned with the offset slot. Install but do not tighten a new sprocket bolt

✷✷ CAUTION

Make certain that all timing marks and colored links are aligned properly before proceeding to the next step. If the timing chain is not aligned properly, severe engine damage may occur.

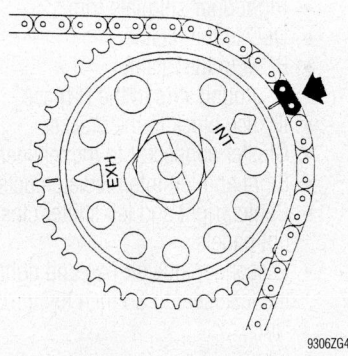

Align the copper link on the timing chain with the INT diamond timing mark

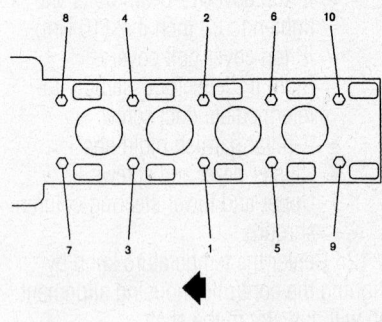

Cylinder head bolt tightening sequence—2.2L engine

12. Torque the intake and exhaust camshaft bolts to 63 ft. lbs. (85 Nm) plus a 30 degree turn.

13. Install or connect the following:
- Fixed timing guide and torque the bolt to 89 inch lbs. (10 Nm)

- Fixed timing guide bolt access plug after applying Loctite® to the threads and torque it to 30 ft. lbs. (40 Nm)
- Timing chain tensioner and torque the bolts 55 ft. lbs. (75 Nm)

14. Tap the top of the timing chain between the camshaft sprockets to engage the tensioner.

- Upper timing chain guide and torque the bolts to 89 inch lbs. (10 Nm)
- Front cover with a new gasket and torque the bolts to 18 ft. lbs. (25 Nm)
- Water pump bolt and torque it to 18 ft. lbs. (25 Nm)
- Crankshaft damper and torque the bolt to 74 ft. lbs. (100 Nm) plus 75 degrees
- Drive belt tensioner and torque the bolts to 30 ft. lbs. (40 Nm)
- Exhaust manifold pipe to the manifold and tighten the nuts to 22 ft. lbs. (30 Nm).
- Oil dipstick tube bolt to 89 inch lbs. (10 Nm).
- Engine mount bracket and torque the bolts to 66 ft. lbs. (90 Nm)
- Engine mount to body nuts to 81 ft. lbs. (110 Nm)
- Engine mount to the bracket and torque the bolts to 41 ft. lbs. (55 Nm)

15. Remaining components in the reverse order of removal.

16. Fill the engine with clean oil.

17. Fill the cooling system.

18. Prime the fuel system by cycling the ignition **ON** for 5 seconds and **OFF** for 10 seconds a few times without cranking the engine.

19. Start the engine, check for leaks, and repair if necessary.

3.0L Engine

FRONT

✳✳ WARNING

Only remove the cylinder head when the engine is cold. Warpage may result if the cylinder head is removed while the engine is hot.

1. Before servicing the vehicle, refer to the precautions section.

2. Drain the cooling system.

3. Drain the engine oil.

4. Properly relieve the fuel system pressure.

5. Remove or disconnect the following:

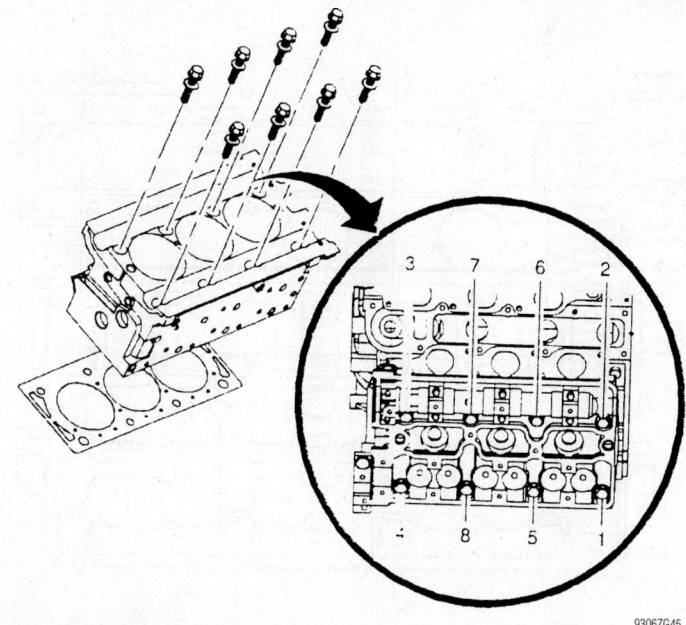

Cylinder head bolt removal for the front and rear cylinder heads—3.0L engine

9306ZG45

- Negative battery cable
- Air cleaner assembly
- Intake plenum
- Intake manifold
- Intake manifold spacer
- Coolant bridge
- Upper radiator hose from the coolant extension housing

6. Properly support the powertrain assembly.

- Front transmission mount through bolt
- Extension housing over the coolant module
- Oil level indicator tube
- Coolant extension housing by twisting it off
- Front camshaft cover
- Electronic Control Module (ECM) from the bracket
- Ground wires from the lift bracket
- Oxygen (O2S) sensor electrical connector
- Rear lift bracket
- Down pipe from the exhaust manifold
- Front timing belt cover
- Timing belt
- Timing belt tensioner bracket
- Rear timing belt cover
- Camshaft position (CMP) sensor electrical connector
- Exhaust camshaft
- Loosen the cylinder head bolts in stages as shown
- Cylinder head
- Exhaust manifold from the cylinder head (if necessary)

To install:

7. Install or connect the following:

- Exhaust manifold with a new gasket and torque the bolts to 15 ft. lbs. (20 Nm)
- New cylinder head gasket with the part number imprint facing the top of the engine
- Cylinder head

8. Torque the new cylinder head bolts, in sequence, as follows:

a. 18 ft. lbs. (25 Nm).
b. 90 degree turn.
c. 90 degree turn.
d. 90 degree turn.
e. 15 degree turn.

9. Install or connect the following:

- Coolant pipe with new sealing rings
- Engine lift bracket bolt and torque it to 15 ft. lbs. (20 Nm)
- Upper radiator hose to the coolant pipe
- Front transmission mount through bolt
- Exhaust camshaft
- CMP sensor electrical connector
- Rear timing belt cover and torque the bolts to 71 inch lbs. (8 Nm)
- Camshaft gears
- Timing belt tensioner bracket
- Timing belt
- Front timing belt cover and torque the bolts to 71 inch lbs. (8 Nm)
- Down pipe to the exhaust manifold
- O2S electrical connector
- Front camshaft cover and torque the bolts to 71 inch lbs. (8 Nm)

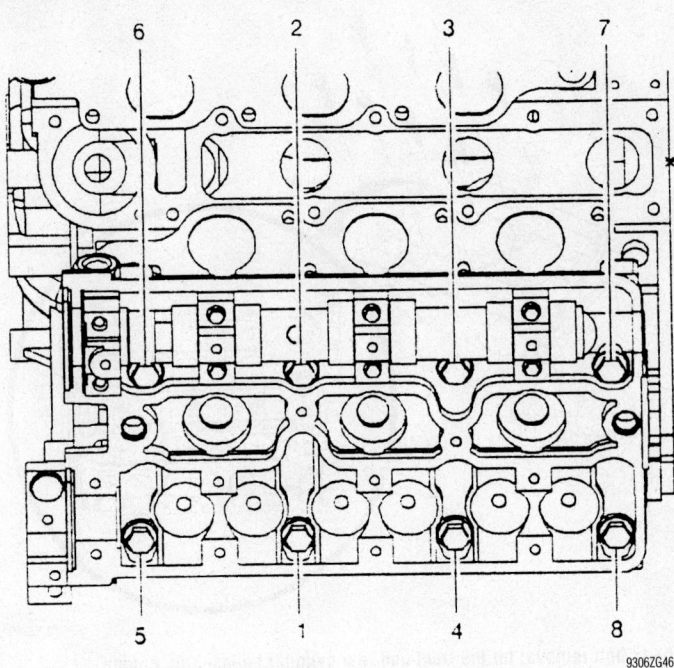

Front and rear cylinder head bolt tightening sequence—3.0L engine

- Coolant bridge and torque the bolt to 22 ft. lbs. (33 Nm)
- Intake manifold spacer and torque the bolts in a spiral direction from the inside and working out to 11 ft. lbs. (16 Nm)
- Intake manifold and torque the bolts to 15 ft. lbs. (20 Nm)
- Intake plenum and torque the bolts to 71 inch lbs. (8 Nm)
- Air cleaner assembly
- Negative battery cable

10. Fill the engine with clean oil.
11. Fill the cooling system.
12. Prime the fuel system by cycling the ignition **ON** for 5 seconds and **OFF** for 10 seconds a few times without cranking the engine.
13. Start the engine, check for leaks, and repair if necessary.

REAR

❊❊ WARNING

Only remove the cylinder head when the engine is cold. Warpage may result if the cylinder head is removed while the engine is hot.

1. Before servicing the vehicle, refer to the precautions section.
2. Drain the cooling system.
3. Drain the engine oil.
4. Properly relieve the fuel system pressure.
5. Remove or disconnect the following:

- Negative battery cable
- Air cleaner assembly
- Intake plenum
- Intake manifold
- Intake manifold spacer
- Coolant bridge
- Rear camshaft cover
- Oxygen (O_2S) sensor electrical connector
- Front timing belt cover
- Timing belt
- Timing belt tensioner bracket
- Camshaft gears
- Rear timing belt cover
- Exhaust manifold pipe heat shield
- Front exhaust manifold pipe-to-rear exhaust manifold pipe fasteners
- Rear exhaust manifold pipe nuts, pull the manifold pipe down and discard the gasket
- Exhaust Gas Recirculation (EGR)-to-exhaust manifold pipe
- Exhaust camshaft
- Cylinder head bolts
- Cylinder head
- Exhaust manifold from the cylinder head

To install:

6. Install or connect the following:
- Exhaust manifold with a new gasket and torque the bolts to 15 ft. lbs. (20 Nm), if removed
- New cylinder head gasket with the part number imprint facing the top of the engine
- Cylinder head

7. Torque the new cylinder head bolts, in sequence, as follows:
  a. 18 ft. lbs. (25 Nm).
  b. 90 degree turn.
  c. 90 degree turn.
  d. 90 degree turn.
  e. 15 degree turn.
8. Install or connect the following:
- Exhaust camshaft
- Rear timing belt cover and torque the bolts to 71 inch lbs. (8 Nm)
- Camshaft gears and torque the bolts to 37 ft. lbs. (50 Nm) plus a 60 degree turn plus another 15 degree turn
- Timing belt tensioner bracket and torque the bolts to 30 ft. lbs. 940 Nm)
- Timing belt
- Front timing belt cover and torque the bolts to 71 inch lbs. (8 Nm)
- Rear camshaft cover and torque the bolts to 71 inch lbs. (8 Nm)
- Exhaust manifold pipe to the manifold
- Exhaust manifold pipe heat shield
- EGR-to-exhaust manifold pipe and torque the nut to 18 ft. lbs. (25 Nm)
- Coolant bridge and torque the bolt to 22 ft. lbs. (33 Nm)
- Engine ventilation chamber and torque the bolts to 71 inch lbs. (8 Nm)
- Intake manifold spacer and torque the bolts to 25 ft. lbs. (20 Nm)
- Intake manifold and torque the bolts to 15 ft. lbs. (20 Nm)
- Intake plenum and torque the bolts to 71 inch lbs. (8 Nm)
- Air cleaner assembly
- Negative battery cable

9. Fill the engine with clean oil.
10. Fill the cooling system.
11. Prime the fuel system by cycling the ignition **ON** for 5 seconds and **OFF** for 10 seconds a few times without cranking the engine.
12. Start the engine, check for leaks, and repair if necessary.

3.5L Engine

LEFT

1. Before servicing the vehicle, refer to the precautions section.
2. Remove the oil dipstick tube.
3. Remove the oil fill cap.
4. Remove the Positive Crankcase Ventilation (PCV) valve and bolt.
5. Disconnect the ignition wring harness from the ignition coils and retaining bracket.

6. Remove the wiring harness bracket bolt from the cylinder head.

7. Remove the bolts the ignition coils.

8. Remove the bolts, grommets, valve cover and gasket.

9. Remove the seals, if required.

10. Remove the timing belt.

11. Remove the coolant bridge.

12. Disconnect the left bank Oxygen (O_2S) sensors.

13. Remove the cylinder head bolts.

14. Remove the cylinder head and gasket.

To install:

15. Install a new gasket and the cylinder head.

16. Install the cylinder head bolts and tighten in the following sequence:

   a. First pass: 29 ft. lbs. (39 Nm).
   b. Second pass: 51 ft. lbs. (69 Nm).
   c. Final pass: 72 ft. lbs. (98 Nm).

17. Connect the left bank O_2S sensors.

18. Install the coolant bridge.

19. Install the timing belt.

20. Install new seals, new gasket and the cover

21. Install new grommets and the bolts, tighten the bolts in sequence as follows:

   a. First pass: 53 inch lbs. (6 Nm).
   b. Final pass: 106 inch lbs. (12 Nm).

22. Install the ignition coils and tighten the bolts to 106 inch lbs. (12 Nm).

23. Install the wiring harness bracket bolt to the cylinder head and tighten to 89 inch lbs. (10 Nm).

24. Connect the ignition wiring harness to the retaining bracket and ignition coils.

25. Install the oil dipstick tube and new O-rings.

26. Install a new O-rings, the PCV valve and bolt. Tighten the bolt to 106 inch lbs. (12 Nm.

27. Install a new O-ring and the oil fill cap.

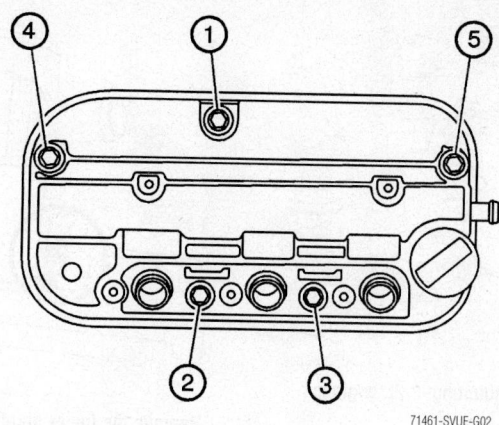

Valve cover bolt tightening sequence—3.5L engine

71461-SVUE-G02

RIGHT

1. Before servicing the vehicle, refer to the precautions section.

2. Disconnect the vacuum brake booster hose.

3. Disconnect the Manifold Absolute Pressure (MAP) sensor connector.

4. Disconnect the Intake Air Temperature (IAT) sensor connector.

5. Disconnect the Engine Coolant Temperature (ECT) sensor connector.

6. Disconnect the EVAP purge hose from the purge valve.

7. Remove the intake manifold.

8. Disconnect the fuel injector connectors.

9. Remove the wiring harness bracket bolt from the valve rocker arm cover.

10. Remove the ignition coils.

11. Remove the bolts, grommets, valve cover and gasket.

12. Remove the seals, if required.

13. Remove the timing belt.

14. Remove the coolant bridge.

15. Disconnect the right bank Oxygen (O_2S) sensors.

16. Remove the cylinder head bolts.

17. Remove the cylinder head and gasket.

To install:

18. Install a new gasket and the cylinder head.

19. Install the cylinder head bolts and tighten in the following sequence:

   a. First pass: 29 ft. lbs. (39 Nm).
   b. Second pass: 51 ft. lbs. (69 Nm).
   c. Final pass: 72 ft. lbs. (98 Nm).

20. Connect the right bank O_2S sensors.

21. Install the coolant bridge.

22. Install the timing belt.

23. Install new seals, new gasket and the cover

24. Install new grommets and the bolts, tighten the bolts in sequence as follows:

   a. First pass: 53 inch lbs. (6 Nm).
   b. Final pass: 106 inch lbs. (12 Nm).

25. Install the ignition coils and tighten the bolts to 106 inch lbs. (12 Nm).

26. Install the wiring harness bracket bolt to the cylinder head and tighten to 89 inch lbs. (10 Nm).

27. Connect the fuel injector connectors.

28. Install the intake manifold.

29. Connect the EVAP purge hose.

30. Connect the IAT sensor connector.

31. Connect the MAP sensor connector.

32. Connect the vacuum brake booster hose.

33. Connect the ECT sensor connector.

Intake Manifold

REMOVAL & INSTALLATION

2.2L Engine

1. Before servicing the vehicle, refer to the precautions section.

2. Remove or disconnect the following:

   - Negative battery cable
   - Intake Air Temperature (IAT) sensor electrical connector

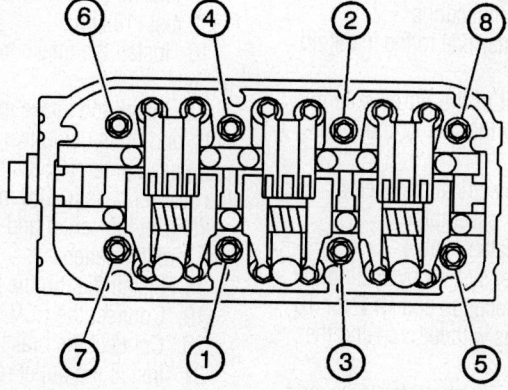

Cylinder head bolt tightening sequence—3.5L engine

71461-SVUE-G01

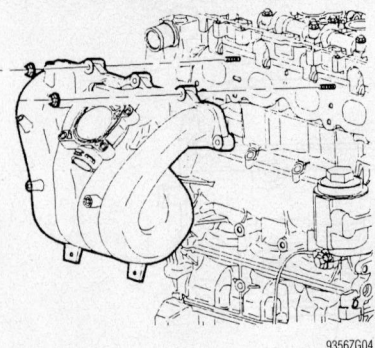

9356ZG04

Intake manifold mounting–2.2L engines

- Air cleaner assembly
- Throttle body electrical connection
- Throttle cable and automatic transmission downshift cable from the throttle body
- Throttle body
- Intake manifold

To install:

3. Install or connect the following:

- Intake manifold with a new gasket and torque the nuts to 89 inch lbs. (10 Nm) starting from the center and working outward
- Throttle body to the intake manifold and torque the bolts to 89 inch lbs. (10 Nm)
- Throttle cable and automatic transmission downshift cable from the throttle body
- Throttle body electrical connection
- Air cleaner assembly
- Intake Air Temperature (IAT) sensor electrical connector
- Negative battery cable

4. Start the engine, check for leaks, and repair if necessary.

3.0L Engine

1. Before servicing the vehicle, refer to the precautions section.

2. Properly relieve the fuel system pressure.

3. Remove or disconnect the following:

- Negative battery cable
- Air cleaner assembly
- Fuel line
- Manifold electrical connector
- Attachment of the purge solenoid and line from the rear of the manifold
- Brake booster line from the manifold
- Fuel injector and manifold control solenoid connections
- Resonator and bracket

9359ZG06

Remove the lower manifold bolt located near the throttle body–3.0L engines

- Positive Crankcase Valve (PCV) hose from the throttle body
- Throttle body bolts
- Throttle body cooling hose bracket bolts

4. Lay the throttle body over the master cylinder with the coolant hoses attached.

- Intake manifold-to-spacer bolts using a T–30 Torx® bit and a long ¼ inch extension
- Lower manifold bolt located near the throttle body
- Intake manifold
- Spacer plate bolts, spacer and seals

To install:

5. Install or connect the following:

- Intake manifold spacer with new seals. Apply Loctite® 242 to the bolts and torque the bolts to 12 ft. lbs. (16 Nm) in sequence.
- Intake manifold with a new gasket and torque the bolts to 66 inch lbs. (7.5 Nm)
- Throttle body and gasket, tighten the bolts to 66 inch lbs. (7.5 Nm)
- PCV hose to the throttle body
- Resonator and bracket
- Fuel injector and manifold control solenoid connections
- Brake booster line to the manifold
- Fuel line
- Attachment of the purge solenoid and line to the rear of the manifold
- Manifold electrical connector
- Air cleaner assembly
- Negative battery cable

6. Prime the fuel system by cycling the ignition **ON** for 5 seconds and **OFF** for 10 seconds a few times without cranking the engine.

7. Start the engine, check for leaks, and repair if necessary.

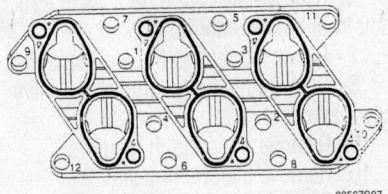

9356ZG07

Tighten the intake manifold spacer bolts as shown—3.0L engines

3.5L Engine

1. Before servicing the vehicle, refer to the precautions section.

2. Remove the outlet resonator/duct assembly.

3. Remove the throttle body.

4. Disconnect the Positive Crankcase Ventilation (PCV) hoses from the intake manifold.

5. Disconnect the brake booster hose from the intake manifold.

6. Remove the intake manifold top cover nuts and bolts.

7. Remove and discard the intake manifold top cover gasket.

8. Remove the Intake Air Temperature (IAT) sensor.

9. Remove the air outlet duct bracket.

10. Remove the intake manifold nuts and bolts.

11. Remove the intake manifold and gasket and discard the gasket.

To install:

12. Install the intake manifold with a new gasket.

13. Install the intake manifold nuts and bolts in using two passes, on the first pass tighten to 97 inch lbs. (11 Nm) and the second pass to 16 ft. lbs. (22 Nm) starting in the center and working outwards in a circular pattern.

14. Install the outlet duct bracket and nuts and tighten the nuts to 89 inch lbs. (10 Nm).

15. Install the IAT sensor and tighten to 13 ft. lbs. (18 Nm).

16. Install the intake manifold top cover gasket.

17. Install the intake manifold top cover, nuts, and bolts using two passes, on the first pass tighten to 53 inch lbs. (6 Nm) and the second pass to 106 inch. lbs. (12 Nm) starting in the center and working outwards in a circular pattern.

18. Install the throttle body.

19. Connect the PCV hoses.

20. Connect the brake booster hose.

21. Install the outlet resonator/duct assembly.

Exhaust Manifold

REMOVAL & INSTALLATION

2.2L Engine

1. Before servicing the vehicle, refer to the precautions section.
2. Remove or disconnect the following:

- Negative battery cable
- Air cleaner assembly
- Exhaust manifold heat shield
- Oxygen (O_2S) sensor from the manifold
- Exhaust pipe from the manifold
- Exhaust pipe-to-resonator pipe nuts from behind the converter
- Exhaust manifold pipe and resonator pipe
- Exhaust manifold

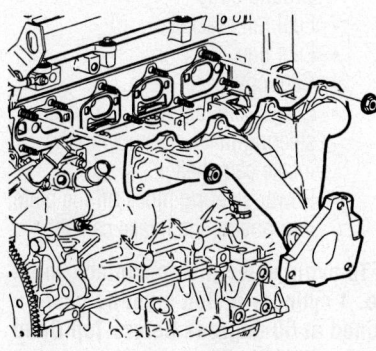

9359ZG07

Remove the exhaust manifold and gasket—2.2L engine

To install:

3. Install or connect the following:

- Exhaust manifold with a new gasket and torque the bolts, starting from the center and working outward, to 13 ft. lbs. (18 Nm)
- O_2S sensor and torque it to 33 ft. lbs. (45 Nm)
- Exhaust pipe to the manifold with a new gasket and torque the nuts to 22 ft. lbs. (30 Nm)
- Exhaust pipe-to-resonator pipe nuts and tighten to 31 ft. lbs. (42 NM)
- Exhaust manifold heat shield and torque the bolts to 18 ft. lbs. (25 Nm)
- Air cleaner assembly
- Negative battery cable

4. Start the vehicle and check for leaks, repair if necessary.

3.0L Engine

FRONT MANIFOLD

1. Before servicing the vehicle, refer to the precautions section.
2. Drain the cooling system.
3. Remove or disconnect the following:

- Negative battery cable
- Exhaust manifold Oxygen (O_2S) sensor
- Coolant extension pipe and engine lift bracket bolts
- Oil level indicator tube
- Upper exhaust manifold nuts
- Front down pipe assembly
- Right hand splash shield
- Accessory drive belt
- A/C compressor and pressure transducer connections
- A/C compressor-to-bracket bolts and position the compressor aside without disconnecting the lines
- Lower exhaust manifold nuts
- Exhaust manifold

To install:

4. Install or connect the following:

- Exhaust manifold with a new gasket and torque the bolts to 15 ft. lbs. (20 Nm)
- Front down pipe and gaskets
- A/C compressor and tighten the bolts to 16 ft. lbs. (22 Nm)
- Electrical connections to the pressure transducer and compressor
- Accessory drive belt
- Right hand splash shield
- Oil level indicator tube
- Coolant extension pipe/engine lift bracket bolt and torque to 15 ft. lbs. (20 Nm)
- O_2S sensor, tighten to 33 ft. lbs. (45 Nm) and attach the electrical connection
- Negative battery cable

5. Fill the cooling system.
6. Start the vehicle and check for leaks, repair if necessary.

REAR MANIFOLD

1. Before servicing the vehicle, refer to the precautions section.
2. Drain the cooling system.
3. Remove or disconnect the following:

- Negative battery cable
- Air cleaner assembly
- Purge solenoid from the manifold
- Middle exhaust manifold nut using a boxed end wrench
- Oxygen (O_2S) sensor connection at the rear of the engine
- Rear exhaust manifold pipe

- Right hand stabilizer bar-to-frame bolts
- Left-to-right hand stabilizer bar to lower link nut and discard the nut
- Exhaust manifold pipe heat shield
- Exhaust manifold

To install:

4. Install or connect the following:

- Exhaust manifold with a new gasket and torque the bolts to 15 ft. lbs. (20 Nm)
- Manifold heat shield and tighten the retainers to 18 ft. lbs. (25 Nm)
- Right hand stabilizer bar-to-frame bolts and tighten to 37 ft. lbs. (50 Nm)
- New left-to-right hand stabilizer bar to lower link nut to 48 ft. lbs. (65 Nm)
- Rear exhaust manifold pipe
- O_2S sensor connection at the rear of the engine
- Middle exhaust manifold nut using a boxed end wrench to 15 ft. lbs. (20 Nm)
- Purge solenoid to the manifold
- Air cleaner assembly
- Negative battery cable

5. Start the vehicle and check for leaks, repair if necessary.

3.5L Engine

1. Before servicing the vehicle, refer to the precautions section.
2. Disconnect the Intake Air Temperature (IAT) sensor connector.
3. Loosen the clasp at the air cleaner assembly.
4. Remove the push-pin attachment from the outlet resonator/duct assembly to support the bracket.
5. Loosen the clamp at the throttle body assembly.
6. Disconnect the Positive Crankcase Ventilation (PCV) hose at the cam cover.
7. Remove the outlet resonator/duct assembly.
8. Remove the exhaust manifold heat shield bolts.
9. Remove the exhaust manifold heat shield.
10. Disconnect the front and rear Oxygen (O_2S) sensor connectors.
11. Remove the O_2S sensor if the exhaust manifold is being replaced.
12. Remove the exhaust manifold pipe–to–manifold nuts.
13. Remove the exhaust manifold pipe–to–resonator pipe nuts behind the converter.

14. Disconnect the rear O2S sensor wire from the heat shield.

15. Separate the exhaust manifold pipe and resonator pipe and discard the gaskets.

16. Remove the exhaust manifold nuts.

17. Remove the exhaust manifold assembly and discard the gasket.

To install:

18. Install a new manifold gasket on the cylinder head.

19. Install the exhaust manifold and torque the nuts to 13 ft. lbs. (18 Nm).

20. If necessary, transfer the O2S sensor from the old manifold.

➡**Whenever the oxygen sensor is removed, coat the threads with a nickel-based anti-seize.**

21. Install the O2 O2S sensor into the exhaust manifold and tighten to 33 ft. lbs. (45 Nm).

22. Install the gasket and exhaust pipe to the intermediate pipe and tighten the nuts to 22 ft. lbs. (30 Nm).

23. Attach the rear O2S sensor wire to the heat shield.

24. Install a new exhaust manifold gasket onto the exhaust manifold flange studs.

25. Attach the exhaust manifold pipe to the exhaust manifold studs and install the exhaust manifold nuts. Tighten the nuts to 31 ft. lbs. (42 Nm).

26. Install the exhaust manifold heat shield and tighten the bolts to 17 ft. lbs. (23 Nm).

27. Install the outlet resonator/duct assembly.

28. Connect the PVC fresh air vent hose.

29. Tighten the clamp at the throttle body assembly.

30. Position the outlet resonator/duct assembly up with the support bracket and install the push-pin.

31. Tighten the air cleaner assembly.

32. Connect the IAT sensor connector.

Front Crankshaft Seal

REPLACEMENT

On the 2.2L engines the front crankshaft seal is located in the timing chain front cover. Refer to the timing chain procedure for information about removing the front cover and replacing the seal.

3.0L Engine

1. Before servicing the vehicle, refer to the precautions section.

2. Remove or disconnect the following:
   • Negative battery cable

   • Timing belt
   • Crankshaft gear

3. Drill a small pilot hole into the steel ring of the seal.

4. Screw in a self taping screw.

5. Use pliers to pull out the oil seal.

To install:

6. Coat the lip of the new oil seal with engine oil.

7. Install the oil seal using a suitable seal installer.

8. Install the crankshaft gear and torque the bolt to 184 ft. lbs. (250 Nm) plus an additional 45 degrees then another 15 degrees.

9. Install the timing belt.

3.5L Engine

1. Before servicing the vehicle, refer to the precautions section.

2. Before servicing the vehicle, refer to the precautions section.

3. Remove or disconnect the following:
   • Negative battery cable
   • Timing belt
   • Crankshaft gear

4. Drill a small pilot hole into the steel ring of the seal.

5. Screw in a self taping screw.

6. Use pliers to pull out the oil seal.

To install:

7. Coat the lip of the new oil seal with engine oil.

8. Install the oil seal using a suitable seal installer.

9. Install the crankshaft gear and torque the bolt to 181 ft. lbs. (245 Nm).

10. Install the timing belt.

Camshaft and Lifters

REMOVAL & INSTALLATION

2.2L Engine

➡**Be very careful when working around the camshaft sprockets and timing chain cover during this procedure. If a bolt or washer is accidentally dropped between the front cover and engine assembly, the cover will have to be removed for retrieval.**

1. Before servicing the vehicle, refer to the precautions section.

2. Relieve the fuel system pressure.

3. Remove or disconnect the following:
   • Negative battery cable
   • Air cleaner assembly
   • Ignition coil
   • Coolant degas hose clips from the fuel rail
   • Ground strap
   • Fuel rail bracket
   • Fuel line
   • Cam cover
   • Purge solenoid from the power steering plate, if removing the intake camshaft
   • Power steering block off the plate, if removing the intake camshaft

➡**To avoid valve piston contact, the No. 1 cylinder piston must be positioned at 60 degrees Before Top Dead Center (BTDC). The pistons are properly aligned when the diamond shaped hole on the intake camshaft sprocket is located at 12 o'clock.**

9346ZG18

Remove the camshaft and bearing caps–2.2L engine

4. Remove the upper timing chain guide and front camshaft caps.

5. Install a camshaft sprocket holding tool J43655 through the sprocket holes from the timing chain side. Align the guide pins into the slot on the support head. Torque the pins to 89 inch lbs. (10 Nm).

6. Hold each camshaft in place with a 24mm open end wrench and remove the camshaft timing sprocket retaining bolts and washers. Discard the bolts.

7. Uniformly loosen and remove the remaining camshaft bearing caps.

8. Slide the camshaft sprockets away from the camshafts and remove the camshaft.

To install:

9. Lubricate the camshaft bearing journals with clean engine oil.

10. Install both camshafts and all bearing caps except the front cap on each camshaft.

11. Torque the bearing caps uniformly, except for the front caps and the rear intake cap, to 89 inch lbs. (10 Nm).

➡**Make certain that the alignment notches are properly positioned with the notches in the camshaft sprockets before final torque is applied. Also, be sure that the timing chain is properly aligned on the fixed guide.**

12. Slide the camshaft sprockets and timing chain on the guide pins toward the camshafts. Rotate the camshafts with a 24mm open end wrench to align the camshaft and sprocket.

13. Install new camshaft sprocket bolts and torque them to 63 ft. lbs. (85 Nm) plus 30 degrees.

14. Remove the camshaft sprocket holding tool.

15. Install or connect the following:
- Front camshaft bearing caps and torque the bolts to 89 inch lbs. (10 Nm)

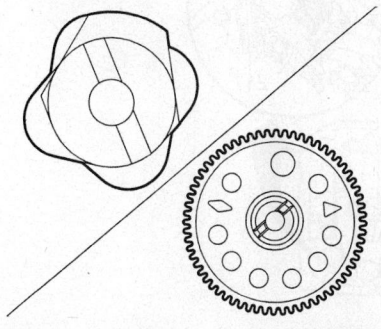

Make certain that the alignment notches are properly positioned with the notches in the camshaft sprockets—2.2L engines

9359ZG09

- Upper timing chain guide and apply Loctite® to the bolts
- Rear intake camshaft bearing cap and torque the bolts 19 ft. lbs. (25 Nm)
- Power steering block off plate and torque the bolts to 19 ft. lbs. (25 Nm)
- Cam cover
- Fuel line and tighten the fitting to 89 inch lbs. (10 Nm)
- Fuel line bracket and tighten to 89 inch lbs. (10 Nm)
- Ground strap
- Coolant degas hose
- Ignition coil
- Air cleaner assembly
- Negative battery cable

16. Start the vehicle and check for leaks, repair if necessary.

3.0L Engine

This engine is equipped with front and rear camshafts.

The front camshaft bearing caps for the cylinder head are marked R1–R8 and the rear cylinder head bearing caps are marked L1–L8.

➡**Be very careful when working around the camshaft sprockets and timing chain cover during this procedure. If a bolt or washer is accidentally dropped between the front cover and engine assembly, the cover will have to be removed for retrieval.**

1. Before servicing the vehicle, refer to the precautions section.

2. Remove or disconnect the following:
- Negative battery cable
- Air cleaner assembly
- Intake plenum
- Camshaft cover
- Front timing belt cover
- Timing belt

➡**Rotate the crankshaft counterclockwise to 60 degrees Before Top Dead Center (BTDC) to prevent valve to piston contact.**

3. Install a Camshaft Gear Locking Tool J42069-2 into the camshaft gears.

4. Remove or disconnect the following:
- Loosen the camshaft gear bolt, remove the holding tool
- Camshaft gear bolt and discard it
- Camshaft gear
- Loosen the camshaft bearing caps sequentially starting in the center and working outward in a spiral direction
- Camshaft bearing caps
- Camshaft

➡**The bearing caps for the front camshaft bearing caps are marked with an R followed by a number and the rear caps marked with an L followed by a number.**

To install:

5. Lubricate all bearing surfaces with clean engine oil.

6. Install or connect the following:
- Camshaft to the cylinder head and make sure that the pin on the exhaust camshaft is in the 12 o'clock position or that the pin on

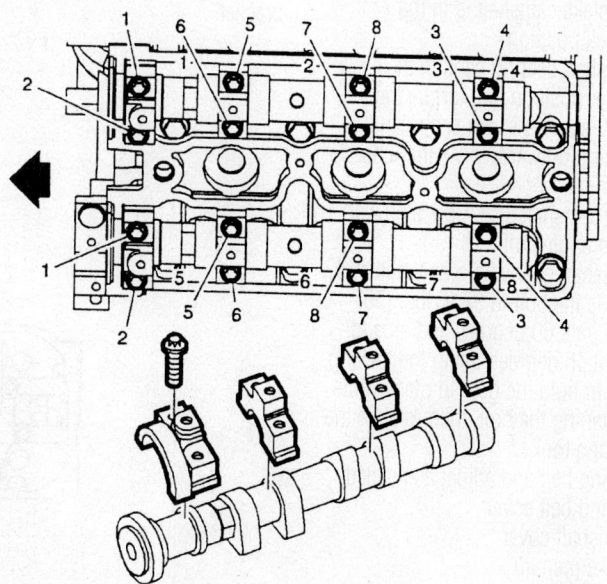

Front camshaft bearing cap removal sequence–3.0L engine

9306ZG50

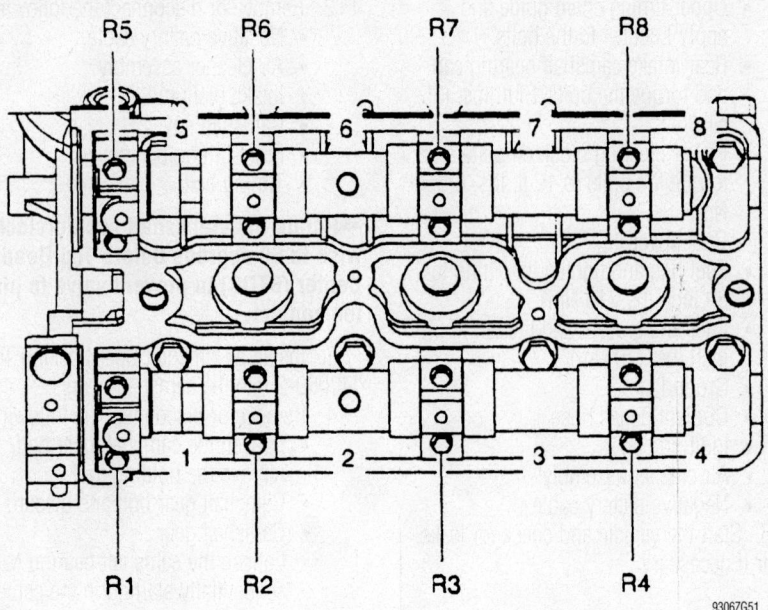

The rear camshaft bearing caps are marked to ensure proper installation–3.0L engine

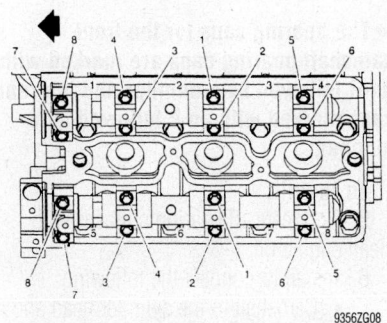

Front and rear camshaft bearing cap installation sequence–3.0L engine

the intake camshaft is in the 7 o'clock position
- Camshaft bearing caps in their proper position and torque the bolts, starting in the center and working outward, to 71 inch lbs. (8 Nm)
- New camshaft seal lubricated with clean engine oil
- Camshaft gear with a new bolt and torque the bolt to 37 ft. lbs. (50 Nm) plus 60 degrees and an additional 15 degrees using the locking tool to hold the gear in place while tightening the bolt, then remove the locking tool
- Timing belt and adjust as needed
- Timing belt cover
- Camshaft cover
- Intake plenum
- Air cleaner assembly
- Negative battery cable

3.5L Engine

LEFT

1. Before servicing the vehicle, refer to the precautions section.
2. Remove the battery and tray.
3. Drain the cooling system.
4. Remove the upper radiator hose.
5. Remove the Exhaust Gas Recirculation (EGR) valve.
6. Remove the oil dipstick tube.
7. Remove the oil fill cap.
8. Remove the Positive Crankcase Ventilation (PCV) valve and bolt.
9. Disconnect the ignition wring harness from the ignition coils and retaining bracket.

10. Remove the wiring harness bracket bolt from the cylinder head.
11. Remove the bolts the ignition coils.
12. Remove the bolts, grommets, valve cover and gasket.
13. Remove the seals, if required.
14. Remove the left camshaft drive sprocket as follows:
   a. Remove the timing belt.
   b. Use timing belt alignment kit EN 46337 to retain the camshaft sprocket.
15. Remove the camshaft bolt and sprocket.
16. Loosen the adjusting nuts and bolts.
17. Loosen the left side rocker shaft retaining bolts in the sequence illustrated .
18. Loosen the valve rocker shaft mounting bolts two turns at a time in sequence, to prevent damaging the valves or rocker arms.

➡**Make sure to not each components location as it is being removed so that it may be installed in its correction location upon assembly. When removing or installing the rocker arm shaft assembly, do not remove the rocker arm shaft mounting bolts. The bolts will retain the springs and rocker arms on the shaft.**

19. Remove the bolts, valve rocker arm and shaft assemblies.
20. Remove the lash adjusters.
21. Remove the intake valve rocker arms as an assembly from the shaft.
22. Remove the exhaust valve rocker arms and springs from the shaft.
23. Remove the nuts and bolts.
24. Note the installed location of the exhaust valve rocker arms.
25. Remove the rear camshaft cap, bolts, and O-ring.

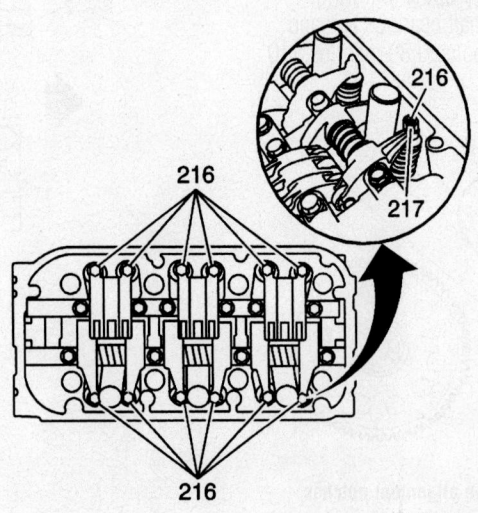

Loosen the adjusting nuts (217) and bolts (216)—3.5L engine

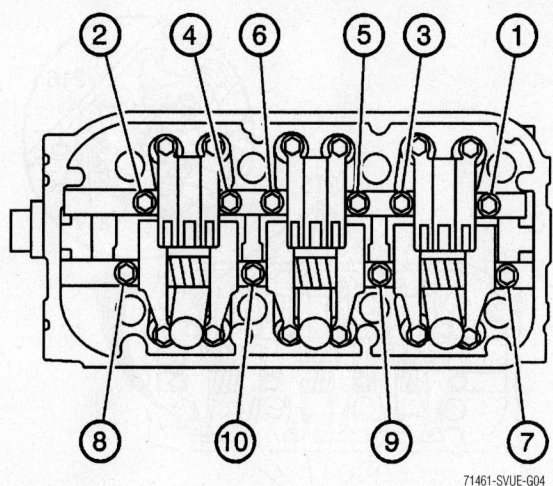

Loosen the left side rocker shaft retaining bolts in sequence—3.5L engine

71461-SVUE-G04

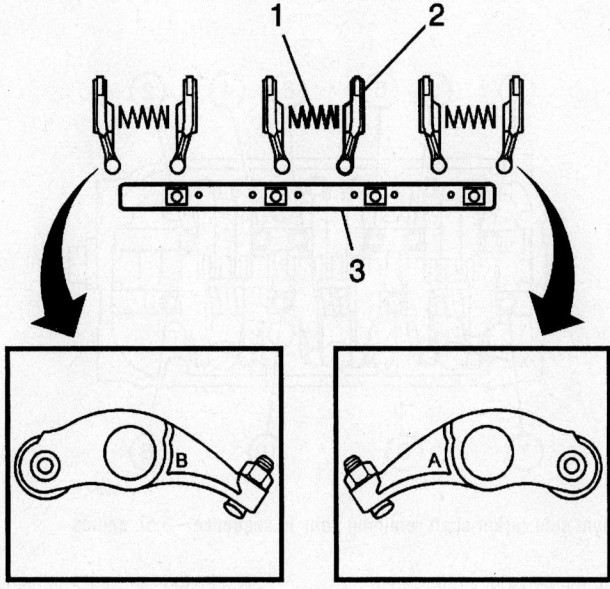

71461-SVUE-G05

Exploded view of the rocker assembly—3.5L engine

26. Remove the camshaft from the cylinder head.

27. Remove the camshaft seal.

28. Clean all of the bearing surfaces and sealing surfaces.

To install:

29. Lubricate the camshaft journals and bores with clean engine oil.

30. Install the camshaft, with new O-ring, cap, and bolts. Tighten the bolts to 16 ft. lbs. (22 Nm).

31. Install a new camshaft seal.

32. Install the bolts and nuts. Do not tighten the nuts at this time.

33. Install the intake valve rocker arm assemblies onto the shaft.

34. Install the bolts to the shaft.

35. Install the exhaust valve rocker arms and springs onto the shaft.

36. Note the installed position of the exhaust valve rocker arms.

37. Install the bolts onto the shaft.

38. Install the lash adjusters.

➡**The intake rocker arm shaft front locating pin serves as an oil passage for VTEC system operation. During assembly, use care to locate the shaft onto the pin. Replace pins that are bent or damaged.**

39. Install the rocker arm and shaft assemblies and bolts.

40. Tighten the left side rocker shaft retaining bolts in the sequence illustrated. Tighten the bolts 2 turns at a time in, to ensure that the rocker arms do not bind on the valves as follows:

    a. First pass: 106 inch lbs. (12 Nm).

    b. Final pass: 27 ft. lbs. (24 Nm).

41. Install the left camshaft drive sprocket.

42. Install new seals, new gasket and the cover

43. Install new grommets and the bolts, tighten the bolts in sequence as follows:

    a. First pass: 53 inch lbs. (6 Nm).

    b. Final pass: 106 inch lbs. (12 Nm).

44. Install the ignition coils and tighten the bolts to 106 inch lbs. (12 Nm).

45. Install the wiring harness bracket bolt to the cylinder head and tighten to 89 inch lbs. (10 Nm).

46. Connect the ignition wiring harness to the retaining bracket and ignition coils.

47. Install the oil dipstick tube and new O-rings.

48. Install a new O-rings, the PCV valve and bolt. Tighten the bolt to 106 inch lbs. (12 Nm.

49. Install a new O-ring and the oil fill cap.

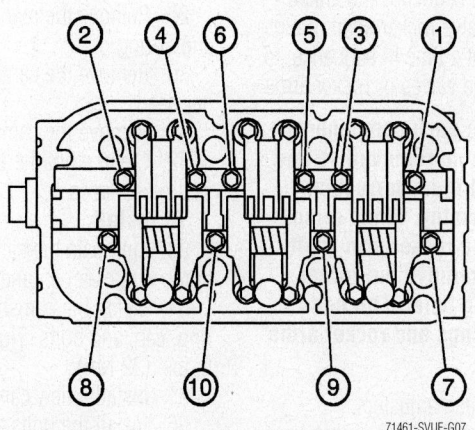

71461-SVUE-G07

Tighten the left side rocker shaft retaining bolts in sequence—3.5L engine

50. Install the EGR valve.
51. Install the upper radiator hose.
52. Fill the cooling system.
53. Install the battery tray and tray.

RIGHT

1. Before servicing the vehicle, refer to the precautions section.
2. Remove the battery and tray.
3. Remove the battery tray.
4. Remove the underhood fuse/relay box.
5. Disconnect the fuel feed line from the fuel rail.
6. Disconnect the engine harness connectors to gain access.
7. Remove the Powertrain Control Module (PCM).
8. Disconnect the vacuum brake booster hose.
9. Disconnect the Manifold Absolute Pressure (MAP) sensor connector.
10. Disconnect the Intake Air Temperature (IAT) sensor connector.
11. Disconnect the Engine Coolant Temperature (ECT) sensor connector.
12. Disconnect the EVAP purge hose from the purge valve.
13. Remove the intake manifold.
14. Disconnect the fuel injector connectors.
15. Remove the wiring harness bracket bolt from the valve rocker arm cover.
16. Remove the ignition coils.
17. Remove the bolts, grommets, valve cover and gasket.
18. Remove the seals, if required.
19. Remove the right camshaft drive sprocket as follows:
   a. Remove the timing belt.
   b. Use timing belt alignment kit EN 46337 to retain the camshaft sprocket.
20. Remove the camshaft bolt and sprocket.
21. Loosen the adjusting nuts and bolts.
22. Loosen the right side rocker shaft retaining bolts in the sequence illustrated .
23. Loosen the valve rocker shaft mounting bolts two turns at a time in sequence, to prevent damaging the valves or rocker arms.

➡**Make sure to not each components location as it is being removed so that it may be installed in its correction location upon assembly. When removing or installing the rocker arm shaft assembly, do not remove the rocker arm shaft mounting bolts. The bolts will retain the springs and rocker arms on the shaft.**

24. Remove the lash adjusters.
25. Remove the intake valve rocker arms as an assembly from the shaft.

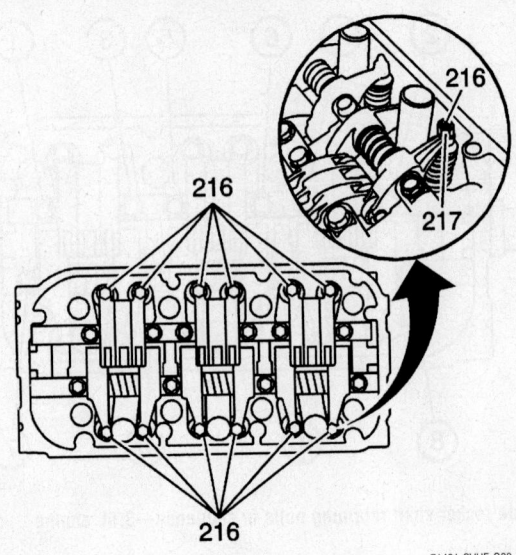

Loosen the adjusting nuts (217) and bolts (216)—3.5L engine

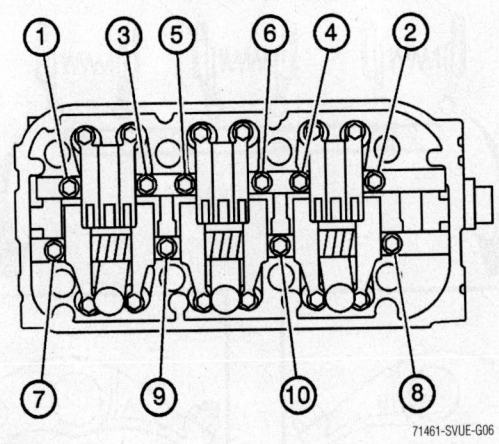

Loosen the right side rocker shaft retaining bolts in sequence—3.5L engine

26. Remove the exhaust valve rocker arms and springs from the shaft.
27. Remove the nuts and bolts.
28. Note the installed location of the exhaust valve rocker arms.
29. Remove the rear camshaft cap, bolts, and O-ring.
30. Remove the camshaft from the cylinder head.
31. Remove the camshaft seal.
32. Clean all of the bearing surfaces and sealing surfaces.
   To install:
33. Lubricate the camshaft journals and bores with clean engine oil.
34. Install the camshaft, with new O-ring, cap, and bolts. Tighten the bolts to 16 ft. lbs. (22 Nm).
35. Install a new camshaft seal.
36. Install the bolts and nuts. Do not tighten the nuts at this time.

37. Install the intake valve rocker arm assemblies onto the shaft.
38. Install the bolts to the shaft.
39. Install the exhaust valve rocker arms and springs onto the shaft.
40. Note the installed position of the exhaust valve rocker arms.
41. Install the bolts onto the shaft.
42. Install the lash adjusters.

➡**The intake rocker arm shaft front locating pin serves as an oil passage for VTEC system operation. During assembly, use care to locate the shaft onto the pin. Replace pins that are bent or damaged.**

43. Install the rocker arm and shaft assemblies and bolts.
44. Tighten the right side rocker shaft retaining bolts in the sequence illustrated. Tighten the bolts 2 turns at a time in, to

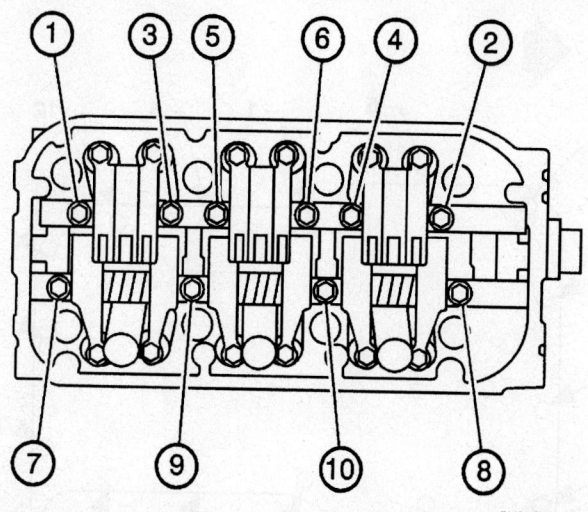

Tighten the right side rocker shaft retaining bolts in sequence—3.5L engine

71461-SVUE-G08

ensure that the rocker arms do not bind on the valves as follows:

    a. First pass: 106 inch lbs. (12 Nm).
    b. Final pass: 27 ft. lbs. (24 Nm).

45. Install the left camshaft drive sprocket.

46. Install new seals, new gasket and the cover

47. Install new grommets and the bolts, tighten the bolts in sequence as follows:

    a. First pass: 53 inch lbs. (6 Nm).
    b. Final pass: 106 inch lbs. (12 Nm).

48. Install the ignition coils and tighten the bolts to 106 inch lbs. (12 Nm).

49. Install the wiring harness bracket bolt to the cylinder head and tighten to 89 inch lbs. (10 Nm).

50. Connect the fuel injector connectors.
51. Install the intake manifold.
52. Connect the EVAP purge hose.
53. Connect the IAT sensor connector.
54. Connect the MAP sensor connector.
55. Connect the vacuum brake booster hose.
56. Connect the ECT sensor connector.
57. Install the PCM.
58. Connect the engine harness connectors .
59. Connect the fuel feed line to the fuel rail.
60. Install the underhood fuse/relay box.
61. Install the battery tray and tray.

Valve Lash

ADJUSTMENT

All engines utilize hydraulic lash adjusters; no adjustment is necessary.

Starter Motor

REMOVAL & INSTALLATION

2.2L Engines

1. Before servicing the vehicle, refer to the precautions section.
2. Remove or disconnect the following:
    • Negative battery cable

➡**Spray the starter solenoid electrical connectors with penetrating oil before removal.**

    • Starter electrical connections
    • Starter bolts
    • Starter assembly by pulling it toward the left side of the vehicle

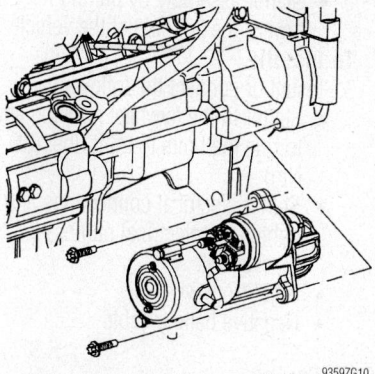

9359ZG10

Starter assembly removal–2.2L engines

To install:

3. Install or connect the following:
    • Starter to the flywheel housing and torque the bolts to 30 ft. lbs. (407 Nm)
    • Starter electrical connectors and torque the solenoid ignition wire to 44 inch lbs. (5 Nm) and the positive battery cable to 89 inch lbs. (10 Nm)
    • Negative battery cable

3.0L Engine

1. Before servicing the vehicle, refer to the precautions section.
2. Remove or disconnect the following:
    • Negative battery cable
    • Right front wheel
    • Starter electrical connections
    • Loosen the fastener securing the electrical harness bracket to the engine
    • Starter bolts

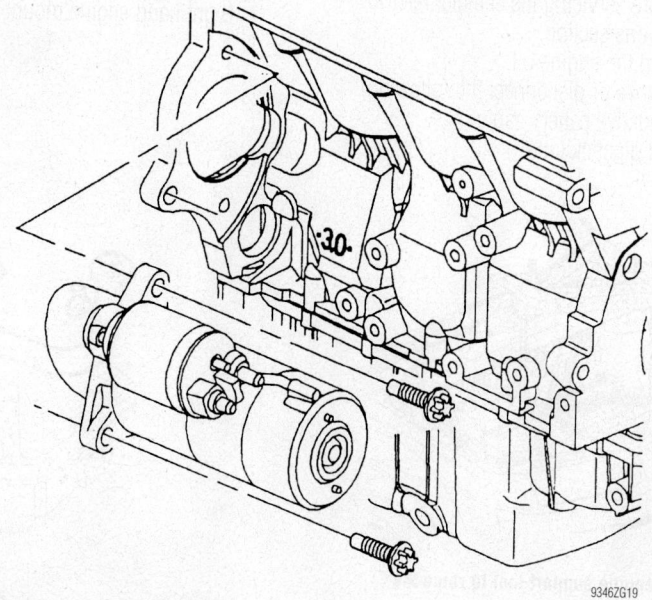

9346ZG19

Starter motor mounting–3.0L engine

- Starter assembly by pulling it toward the left side of the vehicle

To install:

3. Install or connect the following:
- Starter to the flywheel housing and torque the bolts to 26 ft. lbs. (35 Nm)
- Starter electrical connections and tighten the electrical harness bracket bolt
- Right front wheel
- Negative battery cable

3.5L Engine

1. Before servicing the vehicle, refer to the precautions section.

➡**Record all pre-set radio stations.**

2. Disconnect the negative battery cable.
3. Remove starter electrical connections.
4. Remove the lower starter–to–transmission bolt and Oxygen (O_2S) sensor connector bracket.
5. Remove the upper starter–to–transmission bolt.
6. Remove the starter.

To install:

7. Install the starter. Tighten the starter bolts to 33 ft. lbs. (44 Nm).
8. Install the starter electrical connections.
9. Connect the negative battery cable and reprogram the radio stations.

Oil Pan

REMOVAL & INSTALLATION

2.2L Engine

1. Before servicing the vehicle, refer to the precautions section.
2. Drain the engine oil.
3. Remove or disconnect the following:
- Negative battery cable
- Oil dipstick tube

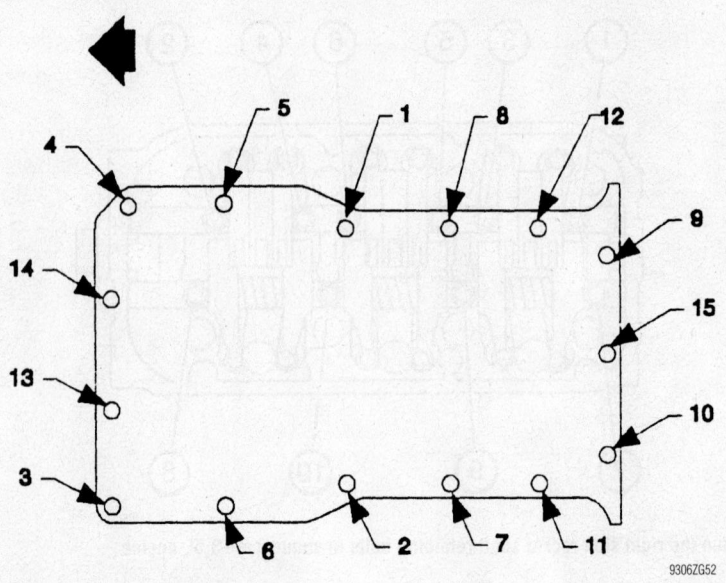

Oil pan bolts torque sequence–2.2L engine

4. Install an engine support fixture.
- Right hand engine mount and raise the engine 3 inches using the support tool
- Lower A/C compressor bolt
- Oil pan bolts
5. Using a flat blade tool, pry the oil pan from the engine block.

To install:

6. Apply a 0.08 in. (2mm) bead of RTV sealer to the pan flange. Be sure the RTV is applied to the inner side of the flange.
7. Install or connect the following:
- Oil pan and torque the bolts in the proper sequence to 11 ft. lbs. (15 Nm)
- Lower A/C compressor bolt and tighten to 15 ft. lbs. (20 Nm)
- Right hand engine mount

8. Remove the engine support fixture.
- Oil dipstick tube
- Negative battery cable
9. Fill the engine with clean oil.
10. Start the vehicle and check for leaks, repair if necessary.

3.0L Engine

1. Before servicing the vehicle, refer to the precautions section.
2. Drain the engine oil.
3. Remove or disconnect the following:
- Negative battery cable
- Nose cone bracket bolts from the oil pan
- Lower transmission flange-to-oil pan bolts
- Oil pan bolts

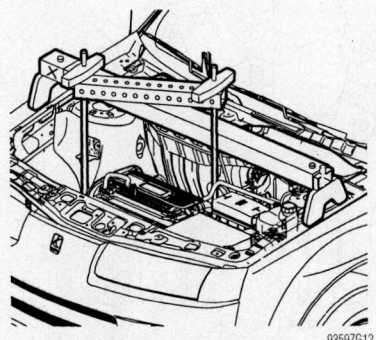

Use the engine support tool to raise the engine—2.2L engines

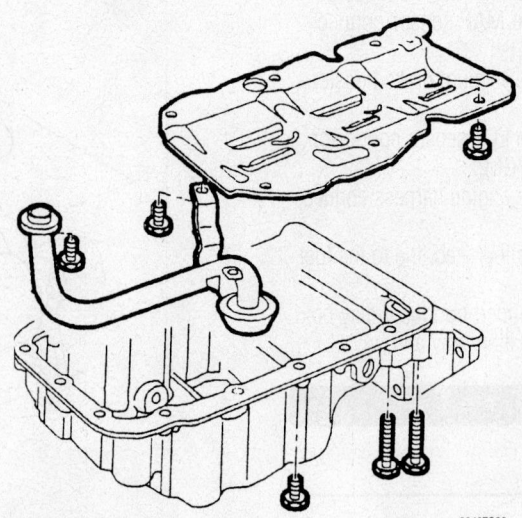

Oil pan and related components–3.0L engine

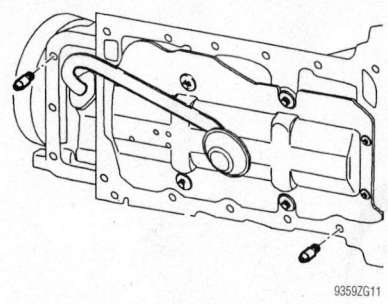

Install the oil pan alignment pins J44715 into the Datum holes—3.0L engines

➡**Separate the oil pan from the engine with an RTV cutter tool. Drive the tool around the pan to shear the RTV seam, then tap the pan sideways with a rubber mallet to loosen.**

- Oil pan

To install:

➡**The alignment of the oil pan is critical in ensuring transaxle flange-to-oil pane sealing. You must apply the proper amount of RTV to ensure a positive seal of the chamfered edge. Once the pan ahs been placed on the bloc, do not allow the pan to move as this will not allow the sealant to properly seal the chamfered edge**

4. Apply a 0.08 in. (2mm) bead of RTV sealer to the pan flange. Be sure the RTV is applied to the inner side of the flange.

5. Install oil pan alignment pins J44715 into the Datum holes shown in the accompanying illustration.

6. Apply Loctite® 242 to the oil pan bolts prior to installation.

7. Install or connect the following:
- Oil pan and finger tighten all bolts, and remove the alignment pins, then torque the oil pan bolts to 11 ft. lbs. (15 Nm) and the transaxle-to-oil pan bolts to 48 ft. lbs. (65 Nm)

8. Connect the negative battery cable.

9. Fill the engine with clean oil.

10. Start the vehicle and check for leaks, repair if necessary.

3.5L Engine

1. Before servicing the vehicle, refer to the precautions section.

2. Drain the engine oil.

3. Remove the rear engine cover.

4. Remove the cross-under exhaust pipe brace and pipe.

5. Remove the oil pan bolts and pan.

6. Clean the pan mating surfaces.

To install:

7. Apply sealant GM P/N 12346240 evenly to the oil pan mating surface of the engine block.

8. Install the oil pan to the engine. Tighten the bolts in two passes. First pass to 53 inch lbs. (6 Nm) and the second pass to 106 inch lbs. (12 Nm).

9. Install the rear engine cover.

10. Install the cross-under exhaust pipe brace and pipe.

11. Refill the engine with the proper type and amount of engine.

Oil Pump

REMOVAL & INSTALLATION

2.2L Engine

1. Before servicing the vehicle, refer to the precautions section.

2. Drain the engine oil.

3. Remove or disconnect the following:
- Negative battery cable
- Air cleaner assembly
- Right front wheel and splash shield
- Accessory drive belt
- Crankshaft damper pulley
- Belt tensioner

4. Install an engine support fixture.
- Right front engine mount
- Front cover bolts and the 13mm bolt under the water pump
- Front cover
- Oil pump cover plate
- Drive rotor and driven rotor
- Pressure relief valve

To install:

5. Install or connect the following:
- New relief valve into the cover bore, if removed. Coat the valve with clean engine oil and tap it into the

bore. Torque the plug to 30 ft. lbs. (40 Nm).

➡**Whenever the oil pump is installed, the assembly must be packed with petroleum jelly in order to prime the pump.**

- Drive and driven rotors into the pump with the chamfer toward the front oil seal
- Oil pump body cover using new bolts and torque the bolts to 53 inch lbs. (6 Nm)
- Front cover with a new oil seal and torque the perimeter and center bolts to 19 ft. lbs. (25 Nm) and the lower center bolt to 89 inch lbs. (10 Nm)
- Right side engine mount and torque the bolts to 41 ft. lbs. (55 Nm)

6. Remove the engine support fixture.
- Drive belt tensioner and torque the bolts 37 ft. lbs. (50 Nm)
- Crankshaft damper pulley and torque the bolt to 74 ft. lbs. (100 Nm) plus 75 degrees
- Accessory drive belt
- Right front splash shield and wheel
- Air cleaner assembly
- Negative battery cable

7. Fill the engine with clean oil and replace the oil filter.

8. Start the vehicle and check for leaks, repair if necessary.

3.0L Engine

1. Before servicing the vehicle, refer to the precautions section.

2. Drain the engine oil.

3. Drain the cooling system.

4. Remove or disconnect the following:

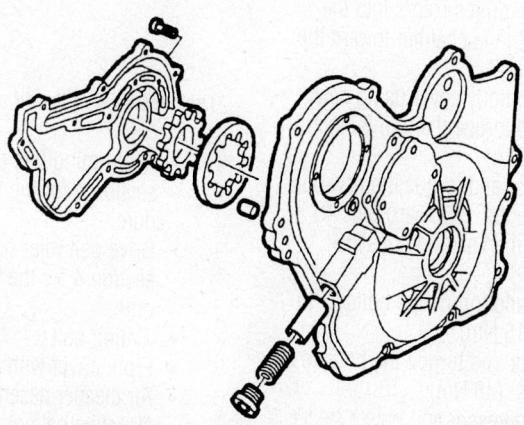

Front cover and oil pump assembly–2.2L engine

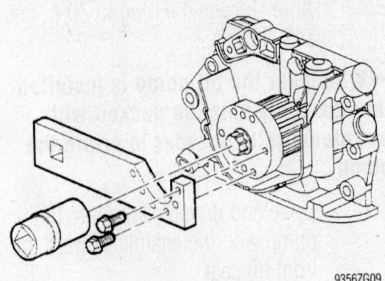

Mount a crank hub holding tool to the crankshaft drive gear—3.0L engine

- Negative battery cable
- Air cleaner assembly
- Front timing belt cover
- Timing belt
- Rear timing belt cover
- A/C compressor and power steering pump bracket and move them away from the oil pump housing
- Alternator bolts and move the alternator out of the way
- Oil pan

5. Mount a crank hub holding tool to the crankshaft drive gear and remove the drive gear.
- Oil pump bolts
- Oil pan housing bolts that thread into the oil pump
- Oil pump
- Front main oil seal and collar
- Pressure relief valve
- Oil pump cover
- Drive rotor and driven rotor

To install:

6. Install the new relief valve into the cover bore (if removed) and torque the plug to 30 ft. lbs. (40 Nm).

➡ **Whenever the oil pump is installed, the new gasket must be coated with a thin bead of sealing Loctite 518®.**

7. Install or connect the following:
- Drive and driven rotors into the pump with the chamfer toward the front oil seal
- Oil pump body cover using new bolts and torque them to 89 inch lbs. (10 Nm)
- Drive gear and torque the new bolt to 184 ft. lbs. (250 Nm) plus 45 degrees then an additional 15 degrees
- Oil pan and torque the bolts to 11 ft. lbs. (15 Nm)
- Alternator and torque the bolts to 30 ft. lbs. (40 Nm)
- A/C compressor and power steering pump bracket. Torque the bolts to 30 ft. lbs. (40 Nm).

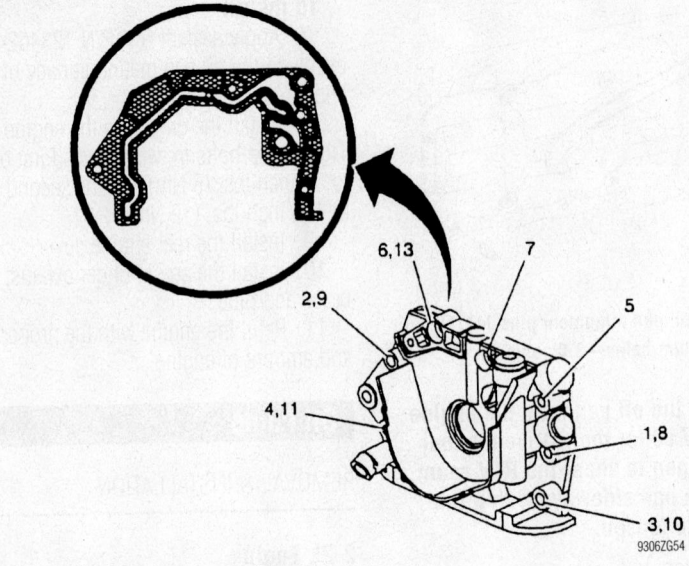

Oil pump bolt tightening sequence–3.0L engine

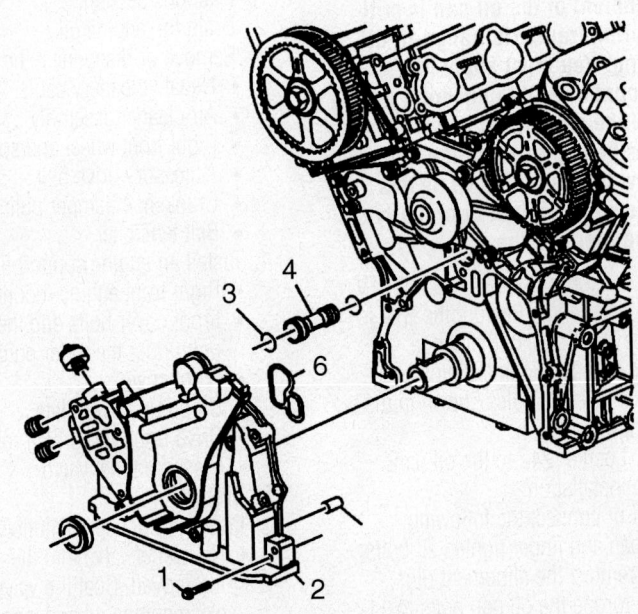

1. Oil pump bolts
2. Oil pump
3. O-ring
4. Pipe
5. Pins
6. O-ring

Exploded view of the oil pump assembly—3.5L engine

- Rear timing belt cover—Refer to section 4 for the timing belt procedure.
- Drive belt idler pulley—Refer to section 4 for the timing belt procedure.
- Timing belt
- Front cover with a new oil seal
- Air cleaner assembly
- Negative battery cable

8. Fill the engine with clean oil. An oil filter replacement is also recommended.

9. Fill the cooling system.
10. Start the vehicle and check for leaks, repair if necessary.

3.5L Engine

1. Before servicing the vehicle, refer to the precautions section.
2. Remove the timing belt.
3. Remove the timing belt idler pulley.
4. Remove the oil flow control module.
5. Remove the oil pan.

6. Remove the oil pump pickup tube.

7. Remove the oil pump bolts and pump.

8. Clean the oil pump mating surfaces.

To install:

9. Install new O-rings onto the oil transfer pipe.

10. Install the pipe into the block.

11. Apply sealant GM P/N 12346240 evenly to the block mating surface of the oil pump housing and to the inner threads of the bolt holes.

12. Install the pins

13. Install a new O-ring.

14. Install the oil pump assembly and tighten the bolts to 106 inch lbs. (12 Nm).

15. Install a new O-ring and the oil pump pickup tube. Tighten the tube bolts to 106 inch lbs. (12 Nm).

16. Install the oil pan.

17. Install the oil flow control module.

18. Install the timing belt idler pulley.

19. Install the timing belt.

Rear Main Seal

REMOVAL & INSTALLATION

2.2L Engine

1. Before servicing the vehicle, refer to the precautions section.

2. Remove or disconnect the following:
- Negative battery cable
- Transmission
- Clutch/pressure plate assembly, if equipped with a manual transmission
- Flywheel

3. Center punch the steel ring of the oil seal.

4. Drill a small hole into the steel ring.

5. Install a self-tapping screw and using pliers, pull out the rear main oil seal.

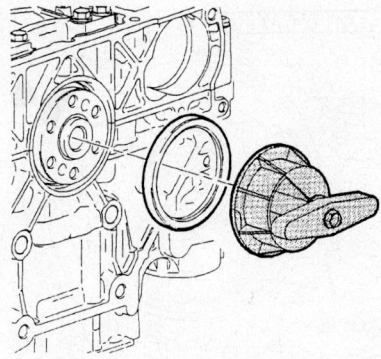

9346ZG23

Rear oil seal and installation tool J42067–2.2L engine

➡️**Be careful not to damage or scratch the seal mounting surfaces.**

To install:

6. Lubricate the new rear main bearing seal with engine oil.

7. Install or connect the following:
- New rear main seal using a Rear Main Bearing Oil Seal Installer Tool J42067 until it is flush with the block
- Flywheel
- Clutch/pressure plate assembly, if equipped with a manual transmission
- Transmission
- Negative battery cable

8. Start the engine and check for leaks, repair if necessary.

3.0L Engine

1. Before servicing the vehicle, refer to the precautions section.

2. Remove or disconnect the following:
- Negative battery cable
- Transmission
- Clutch/pressure plate assembly, if equipped with a manual transmission
- Flywheel

3. Center punch the steel ring of the oil seal.

4. Drill a small hole into the steel ring.

5. Install a self-tapping screw and using pliers, pull out the rear main oil seal.

➡️**Be careful not to damage or scratch the seal mounting surfaces.**

To install:

6. Lubricate the new rear main oil seal with engine oil.

7. Install or connect the following:
- New rear main seal using a Rear Main Seal Installer Tool J42067 until it is flush with the block

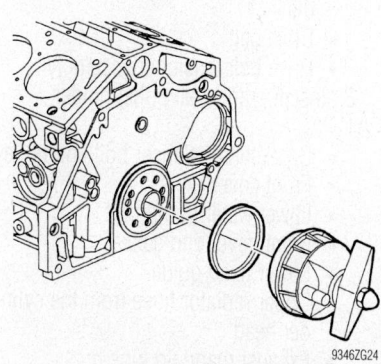

9346ZG24

Rear main oil seal and installation tool J42067–3.0L engine

- Flywheel
- Clutch/pressure plate assembly, if equipped with a manual transmission
- Transmission
- Negative battery cable

8. Start the engine and check for leaks, repair if necessary.

3.5L Engine

1. Before servicing the vehicle, refer to the precautions section.

2. Remove the flywheel.

3. Remove the seal from the rear cover.

To install:

4. Lubricate the lip of the oil seal with a light coat of grease.

5. Use the driver handle tool EN 46342 and driver tool EN 46351 to install the seal squarely into the housing.

6. Drive the new crankshaft oil seal until the tool bottoms onto the housing. A properly installed seal will be flush with the face of the housing.

7. Install the engine flywheel.

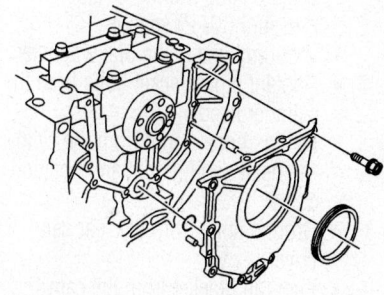

71461-SVUE-G10

Exploded view of the rear main seal—3.5L engine

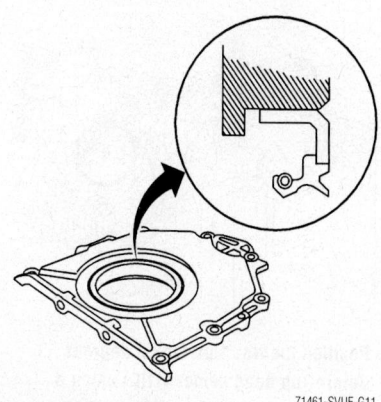

71461-SVUE-G11

A properly installed rear main seal will be flush with the face of the housing—3.5L engine

Timing Chain, Sprockets, Front Cover and Seal

REMOVAL & INSTALLATION

2.2L Engine

1. Before servicing the vehicle, refer to the precautions section.
2. Drain the cooling system.
3. Drain the engine oil.
4. Properly relieve the fuel system pressure.
5. Remove or disconnect the following:

- Negative battery cable
- Intake Air Temperature (IAT) sensor connection
- Air cleaner assembly
- Ignition module assembly
- Electronic Control Module (ECM) connections
- Oil dipstick tube bolt
- Throttle body electrical connection
- Electrical connector from the fuel injector harness and attachment at the bottom of the intake manifold
- Electrical connector at the purge solenoid and Manifold Absolute Pressure (MAP) sensor
- Vacuum hose at the brake booster
- Coolant pipe bracket bolts from the cylinder head
- De-gas hose clamp from the cylinder head and fuel rail and position aside
- Ground strap from the rear cam cover
- Fuel rail bracket from the cam cover
- Fuel lines
- Cam cover

6. Position the No. 1 piston 60 degrees Before Top dead center (TDC) using a

Position the No. 1 piston 60 degrees Before Top dead center (TDC) using a 24mm wrench to rotate the camshafts in a clockwise motion and make sure the diamond shaped hole on the intake sprocket is at the 12 O'clock position—2.2L engine

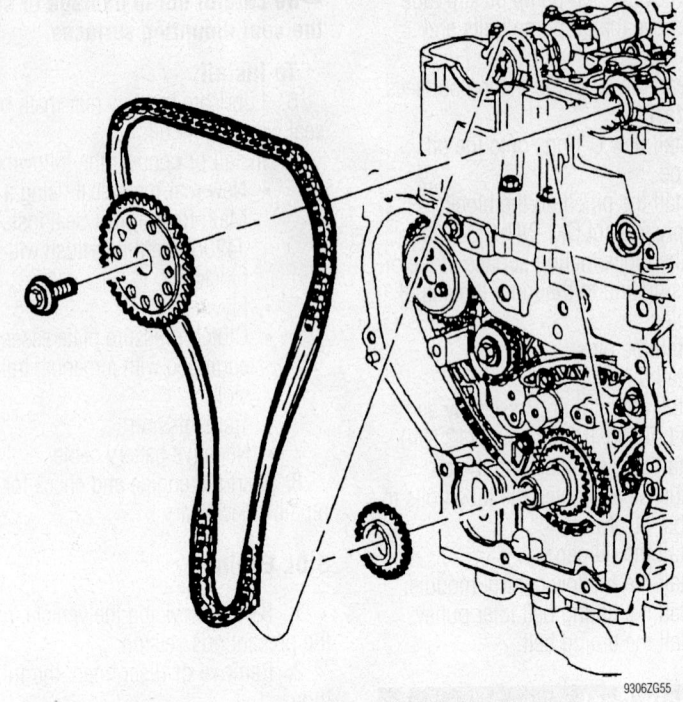

Remove the timing chain through the top of the cylinder head–2.2L engine

24mm wrench to rotate the camshafts in a clockwise motion and make sure the diamond shaped hole on the intake sprocket is at the 12 O'clock position.

- Upper timing guide

➡ **Remove the timing chain tensioner to unload chain tension before removing the timing chain.**

- Fixed timing chain guide access plug
- Upper fixed guide bolt using a magnetic socket

7. Install a three bar engine support fixture.

- Right hand engine mount
- Right hand mount bracket
- Right wheel splash shield
- Install a 1 x 2 x 4 inch block of wood between the oil pan and cradle
- Drive belt
- Drive belt tensioner assembly

8. Install crankshaft pullet holder J38122A.

- Crankshaft balancer bolt and pulley
- Front cover bolts
- Lower water pump bolt
- Front cover and gasket
- Lower fixed guide
- Upper radiator hose from the cylinder head
- Exhaust manifold pipe nuts
- Front and rear Oxygen (O₂S) sensor electrical connector

- Down pipe-to-intermediate pipe nuts
- Down pipe
- Exhaust camshaft sprocket bolts while holding the camshaft with a 24mm wrench and discard the camshaft sprocket bolts
- Exhaust sprocket
- Adjustable guide through the top of the cylinder head
- Intake camshaft sprocket bolts while holding the camshaft with a 24mm wrench and discard the camshaft sprocket bolts
- Intake sprocket
- Timing chain assembly
- Timing drive sprocket from the crankshaft

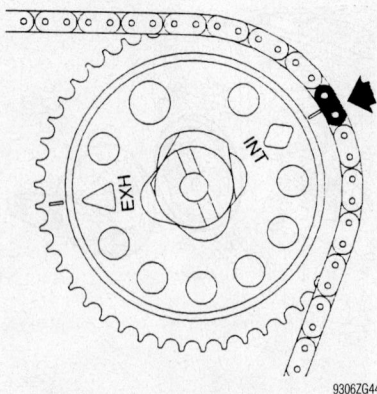

Align the copper link on the timing chain with the INT diamond timing mark

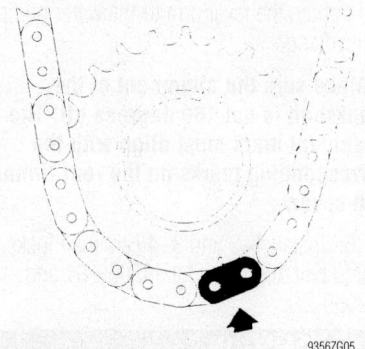

Route the timing chain around the crankshaft sprocket and align the silver link to the timing mark (5 o'clock position)—2.2L engines

To install:

➡️ **Set the crankshaft to 60 degrees Before Top Dead Center (BTDC) to prevent contact between the pistons and valves.**

9. Install or connect the following:
10. Position the exhaust camshaft with the offset slot in the 2 o'clock position and the intake camshaft with the offset slot in the 11 o'clock position.

- Timing chain around the intake camshaft sprocket with the copper link aligned with the **INT** diamond timing mark
- Sprocket to the camshaft and align it with the offset slot
- New camshaft sprocket bolt but do not tighten
- Timing chain around the crankshaft sprocket and align the silver link to the timing mark
- Adjustable timing chain guide through the opening on top of the cylinder head and torque the bolt to 89 inch lbs. (10 Nm)
- Timing chain around the exhaust camshaft sprocket with the silver link aligned with the offset slot. Install but do not tighten a new sprocket bolt

✳✳ CAUTION

Make certain that all timing marks and colored links are aligned properly before proceeding to the next step. If the timing chain is not aligned properly, severe engine damage may occur.

11. Torque the intake and exhaust camshaft bolts to 63 ft. lbs. (85 Nm) plus a 30 degree turn.
12. Install or connect the following:

- Fixed timing guide and torque the bolt to 89 inch lbs. (10 Nm)
- Fixed timing guide bolt access plug after applying Loctite® to the threads and torque it to 30 ft. lbs. (40 Nm)
- Timing chain tensioner and torque the bolts 55 ft. lbs. (75 Nm)

13. Tap the top of the timing chain between the camshaft sprockets to engage the tensioner.

- Upper timing chain guide and torque the bolts to 89 inch lbs. (10 Nm)
- Front cover with a new gasket and torque the bolts to 18 ft. lbs. (25 Nm)
- Water pump bolt and torque it to 18 ft. lbs. (25 Nm)
- Crankshaft damper and torque the bolt to 74 ft. lbs. (100 Nm) plus 75 degrees
- Drive belt tensioner and torque the bolts to 30 ft. lbs. (40 Nm)
- Exhaust manifold pipe to the manifold and tighten the nuts to 22 ft. lbs. (30 Nm).
- Oil dipstick tube bolt to 89 inch lbs. (10 Nm).
- Engine mount bracket and torque the bolts to 66 ft. lbs. (90 Nm)
- Engine mount to body nuts to 81 ft. lbs. (110 Nm)
- Engine mount to the bracket and torque the bolts to 41 ft. lbs. (55 Nm)

14. Remaining components in the reverse order of removal.
15. Fill the engine with clean oil.
16. Fill the cooling system.

17. Prime the fuel system by cycling the ignition **ON** for 5 seconds and **OFF** for 10 seconds a few times without cranking the engine.
18. Start the engine, check for leaks, and repair if necessary.

Timing Belt

REMOVAL & INSTALLATION

3.0L Engine

1. Before servicing the vehicle, refer to the precautions section.
2. Remove or disconnect the following:

- Negative battery cable
- Intake air resonator
- Right front wheel and splash shield
- Water pump and idler bolts, loosen only
- Accessory drive belt
- Right engine mount
- Water pump pulley
- Idler pulley
- Electrical harness connectors at the harness channel
- Electrical harness channel from the front cover and position aside
- Drive belt tensioner
- Front timing belt cover

3. Rotate the crankshaft clockwise to 60 degrees Before Top Dead Center (BTDC) using crank hub Torx® socket J42098.
4. Install crankshaft locking tool J42069-10.
5. Rotate the crankshaft in a clockwise direction using tool J42098 until the number one cylinder is at Top dead center (TDC)

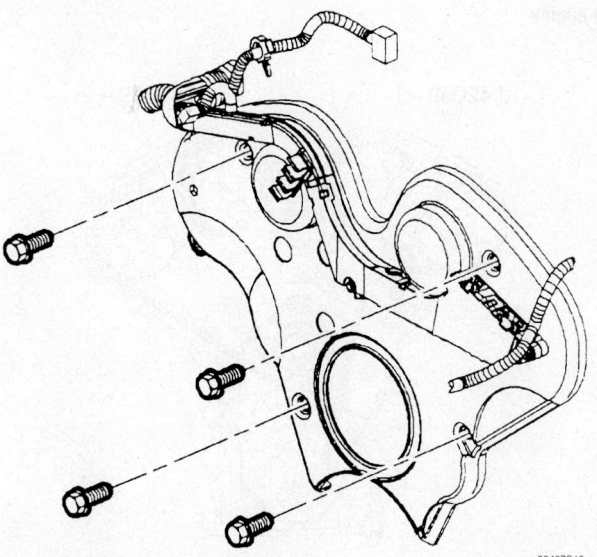

Timing belt cover bolt locations–3.0L (VIN B) engine

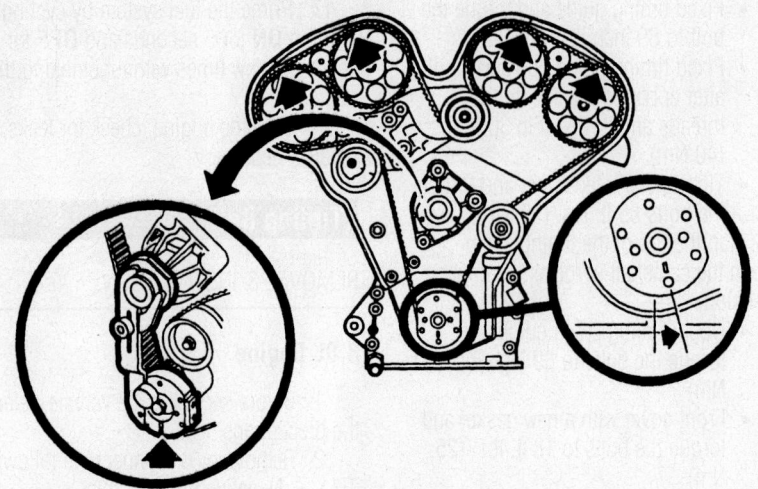

Rotate the crankshaft clockwise to 60 degrees Before Top Dead Center (BTDC) using crank hub Torx® socket J42098—GM 3.0L (VIN B) engine

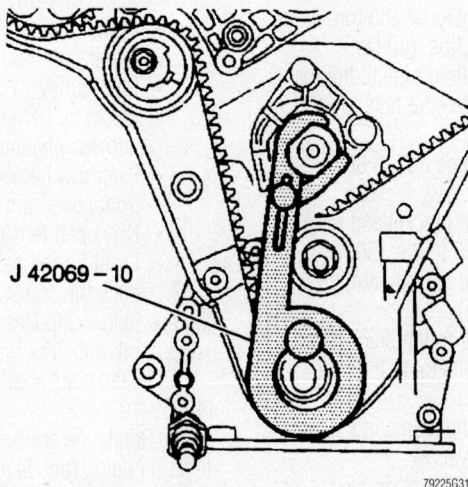

J 42069 – 10

Rotate the crankshaft in a clockwise direction using tool J42098 until the number one cylinder is at Top dead center (TDC) and tighten the lever arm to the water pump pulley flange—GM 3.0L (VIN B) engine

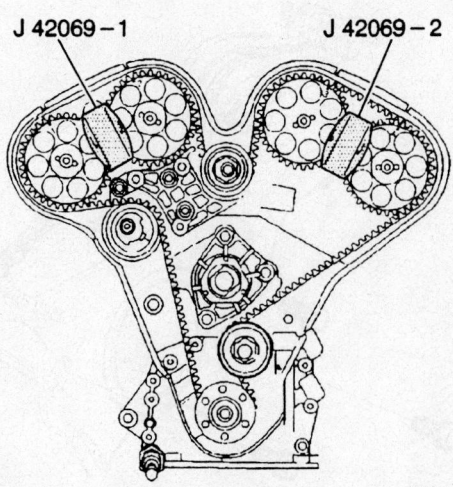

J 42069 – 1   J 42069 – 2

Locking the camshaft—GM 3.0L (VIN B) engine

and tighten the lever arm to the water pump pulley flange.

➡️Make sure the alignment of the crankshaft is not 180 degrees off. The alignment mark must align with the corresponding marks on the rear timing belt cover.

6. Install 1–2 and 3–4 camshaft locks, timing belt alignment kit J42069-01 and J42069-2.

❊❊ CAUTION

Do not rotate the crankshaft unless crankshaft is at 60 degrees BTDC or the valve could hit the crankshaft.

7. Loosen the timing belt tensioner and idler pulleys.
8. Remove the timing belt.
To install:
9. Position the crank at 60 degrees BTDC.
10. Install the lower idler pulley and tighten the bolt to 30 ft. lbs. (40 Nm).
11. Install the upper idle pulley and hand tighten the bolt.
12. Install the timing belt tensioner and hand tighten the nut.
13. With the camshaft locks installed use the green test belt from the alignment kit J42069 to make sure the distance between the camshaft sprockets.
14. Start at the number 1 and 2 sprockets, route the belt over the sprockets while aligning the belt marks to marks on the sprocket and rear cover.
15. Install the belt around the tensioner and upper idler pulley.
16. Route the belt over the 3 and 4 sprockets while aligning the belt marks to marks on the sprocket and rear cover.
17. Install the belt around the crankshaft hub sprockets while aligning marks on the belt to marks on the sprocket.
18. Install wedge tool J42069-30 to lock the belt onto the crankshaft hub sprocket.
19. Carefully rotate the crankshaft counterclockwise to get belt slack on the lower idler pulley, then route the belt around the lower idler pulley and remove the wedge tool.
20. Install the crankshaft locking tool and tighten the lever arm to the water pump pulley flange.
21. Install camshaft checking gauge to cams 3 and 4 and make sure the alignment marks are within 2mm of each other. Install 3 and 4 camshaft lock if properly aligned, then install checking gauge on cams 1 and 2. Rotate the number 1 camshaft sprocket

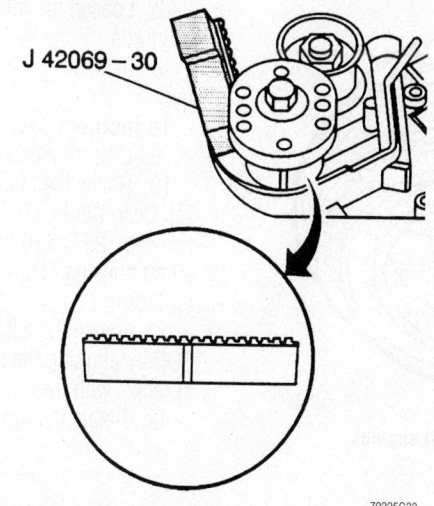

J 42069-30

Install wedge tool J42069-30 to lock the belt onto the crankshaft hub sprocket—GM 3.0L (VIN B) engine

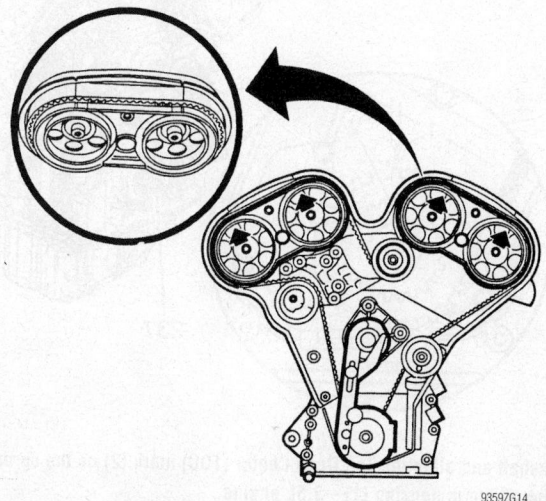

Install camshaft checking gauge to cams 3 and 4 and make sure the alignment marks are within 2mm of each other. Install 3 and 4 camshaft lock if properly aligned, then install checking gauge on cams 1 and 2. Rotate the number 1 camshaft sprocket counterclockwise to remove slack between all the cam sprockets—3.0L (VIN B) engines

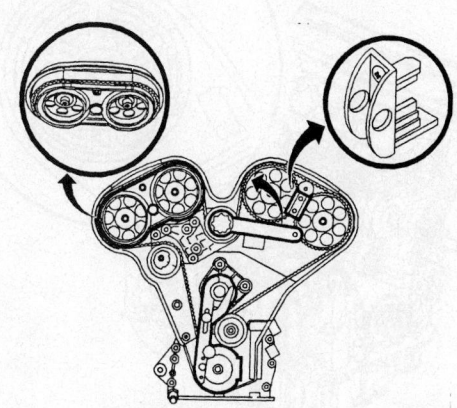

Make sure the camshaft timing marks are to the left (retard) side of the checking gauge by rotating the upper idler pulley counterclockwise until marks are properly aligned within 2mm of each other and install cam lock 1 and 2—3.0L (VIN B) engines

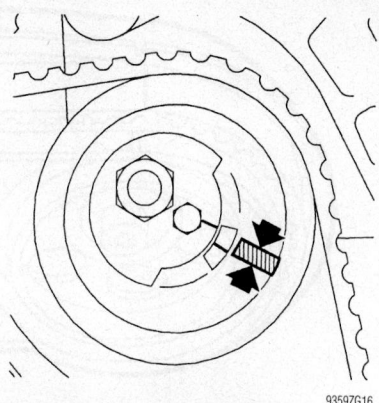

Adjust the belt tensioner so that the center alignment mark is about ⅛ inch (3mm) above the mark on the spring loader—3.0L (VIN B) engines

counterclockwise to remove slack between all the cam sprockets.

22. Make sure the camshaft timing marks are to the left (retard) side of the checking gauge as follows:

a. Rotate the upper idler pulley counterclockwise until marks are properly aligned within 2mm of each other and install cam lock 1 and 2.

23. Tighten the upper idler pulley bolt to 30 ft. lbs. (40 Nm).

24. Adjust the belt tensioner so that the center alignment mark is about ⅛ inch (3mm) above the mark on the spring loader.

25. Remove 1 and 2, 3 and 4 camshaft locks, belt alignment kit J42069-1 and J42069-2 and camshaft locking tool J42069-10.

➤If TDC is passed, do not rotate the crankshaft counterclockwise. this will not allow proper slack to be taken up between the belt tensioner and crank sprocket. To correct, rotate the crank an additional 2 turns.

26. Rotate the crankshaft 1 ¾ turns clockwise and install crankshaft locking tool J42069-10 and stop at TDC. Tighten the lever arm to the water pump pulley flange.

27. Install gauge J42069-20 and verify the marks are within 2mm.

28. Adjust the belt tensioner so that the center alignment mark is about ⅛ inch (3mm) above the mark on the spring loaded idler.

29. Tighten the belt tensioner nut to 15 ft. lbs. (20 Nm).

30. Remove the crankshaft locking tool.

31. Install or connect the following:

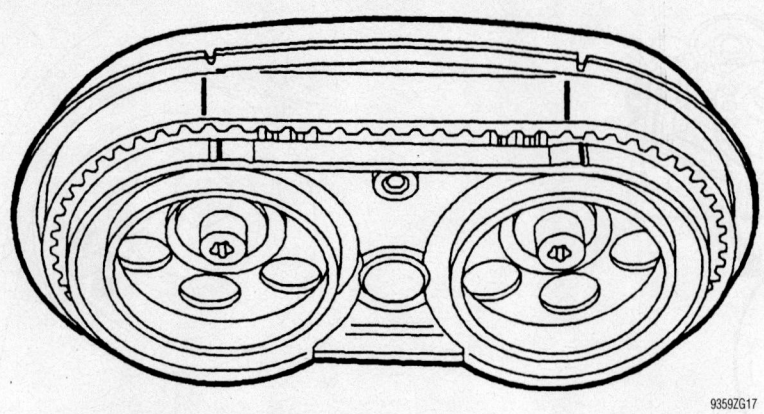

Install gauge J42069-20 and verify the marks are within 2mm—3.0L (VIN B) engines

- Front timing belt cover. Torque the bolts to 71 inch lbs. (8 Nm).
- Drive belt tensioner. Torque the bolts to 30 ft. lbs. (40 Nm).
- Idler pulley and snug the bolt.
- Electrical harness.
- Water pump pulley
- Accessory drive belt

32. Tighten the water pump pulley bolts to 71 inch lbs. (8 Nm0 and the idler puller bolt to 15 ft. lbs. (20 Nm).

- Engine mount bracket and tighten the bolt to 41 ft. lbs. (55 Nm).
- Engine mount-to-frame rail and engine mount bracket, tighten the fasteners to 37 ft. lbs. (50 Nm)

33. Remove the engine support.

- Splash shield
- Wheel
- Intake air resonator
- Negative battery cable

3.5L Engine

The Powertrain Control Module (PCM) has to perform the crankshaft position learn process under the following conditions:

   a. The Crankshaft Position (CKP) sensor is removed for replacement or during disassembly.

   b. The timing belt is removed for replacement or during disassembly.

   c. The CKP Pattern Clear is executed from the Tech II.

   d. The PCM is replaced

1. Before servicing the vehicle, refer to the precautions section.

2. Remove the accessory drive belt.

3. Remove the right splash shield.

4. Remove the crankshaft balancer using balancer holding tool EN 46337.

5. Remove the timing belt cover bolts and the cover.

6. Use timing belt tensioner pulley retaining bolt tool EN 36331 to retain the tensioner pulley.

7. Loosen the idler pulley bolt about 5 or 6 turns.

8. Remove the timing belt from the pulleys.

To install:

9. Clean the belt pulleys and covers.

10. Rotate the crankshaft and align the Top Dead Center (TDC) mark (2) on the sprocket (237) with the pointer on the oil pump housing (1). Refer to the illustration for location.

11. Rotate the left camshaft sprocket to TDC by aligning the mark (2) on the sprocket with the pointer (1) on the cover.

12. Rotate the right camshaft sprocket to

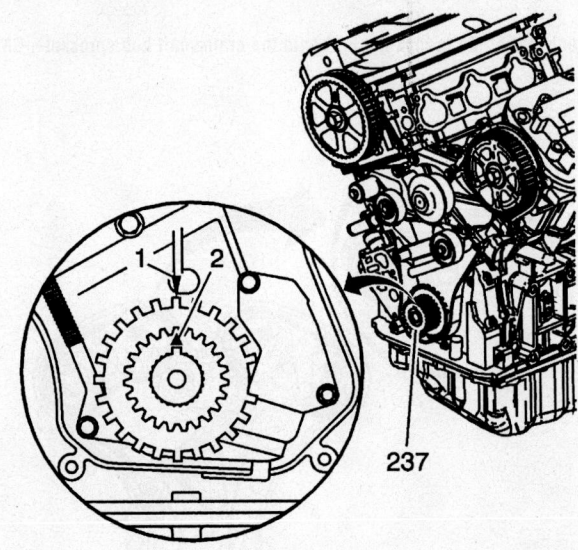

Rotate the crankshaft and align the Top Dead Center (TDC) mark (2) on the sprocket (237) with the pointer on the oil pump housing (1)—3.5L engine

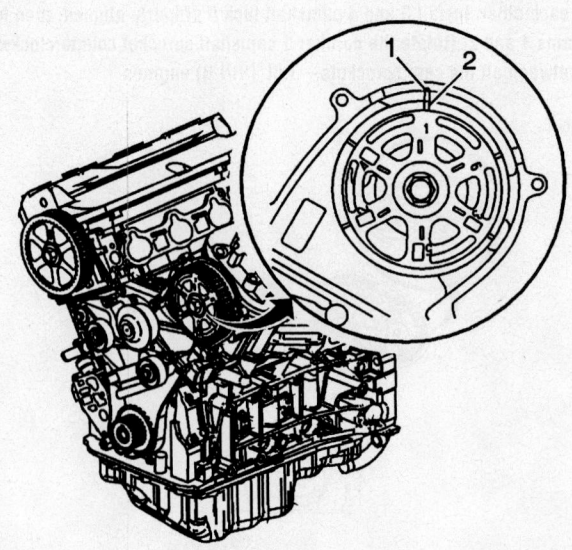

Rotate the left camshaft sprocket to TDC by aligning the mark (2) on the sprocket with the pointer (1) on the cover—3.5L engine

TDC by aligning the mark (2) on the sprocket with the pointer (1) on the cover.

13. Install the timing belt tensioner pulley retaining bolt tool EN 36331. Screw the tool all the way in by hand, until the tool contacts the timing belt tensioner.

14. Install the belt onto the sprockets. Install the timing belt in a clockwise sequence starting with the tensioner pulley in the sequence illustrated.

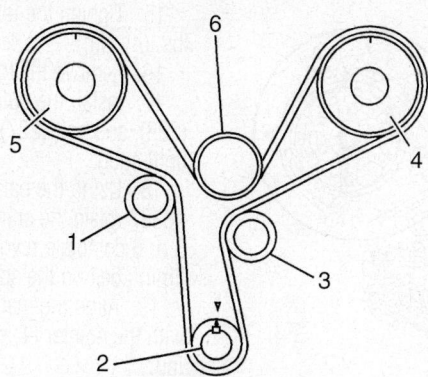

1. Tensioner pulley
2. Crankshaft sprocket
3. Idler pulley
4. Camshaft sprocket - left
5. Camshaft sprocket - right
6. Water pump

71461-SVUE-G17

Install timing belt on the pulleys in the sequence shown—3.5L engine

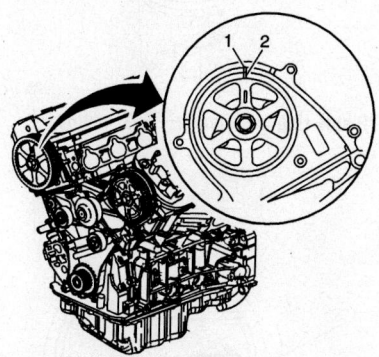

71461-SVUE-G14

Rotate the right camshaft sprocket to TDC by aligning the mark (2) on the sprocket with the pointer (1) on the cover—3.5L engine

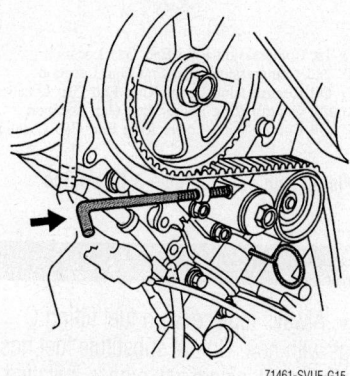

71461-SVUE-G15

Install timing belt tensioner pulley retaining bolt tool EN 36331—3.5L engine

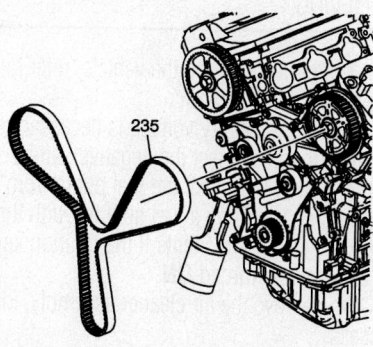

71461-SVUE-G16

Install timing belt onto the sprockets—3.5L engine

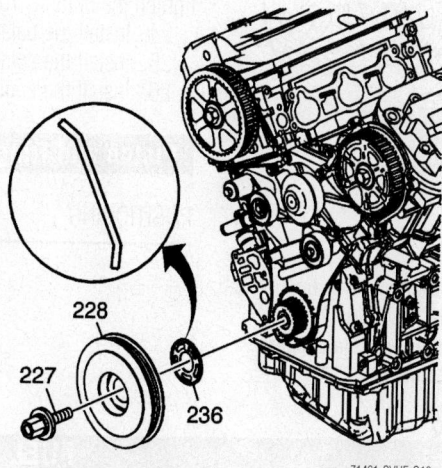

71461-SVUE-G18

Install the guide (236), balancer (228), and bolt (227)—3.5L engine

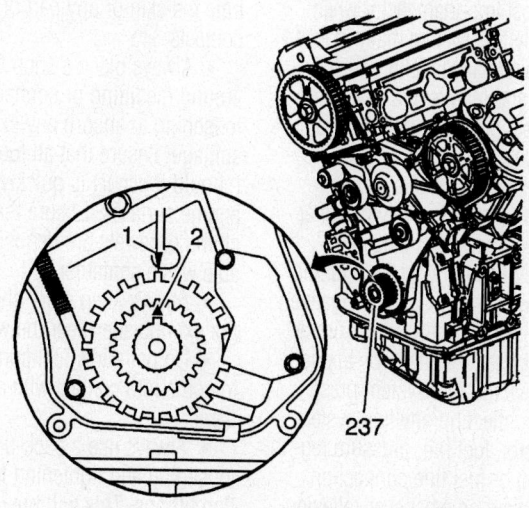

71461-SVUE-G12

Align the mark (2) on the sprocket with the pointer (1) on the oil pump housing—3.5L engine

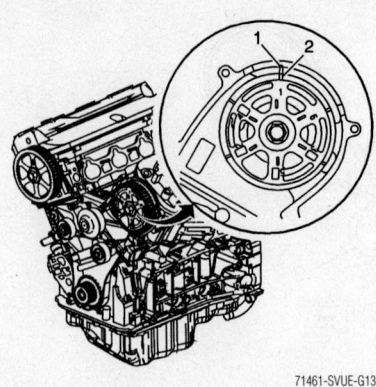

71461-SVUE-G13

Inspect the left camshaft sprocket for proper alignment. The TDC mark (2) on the sprocket should align with the pointer (1) on the cover—3.5L engine

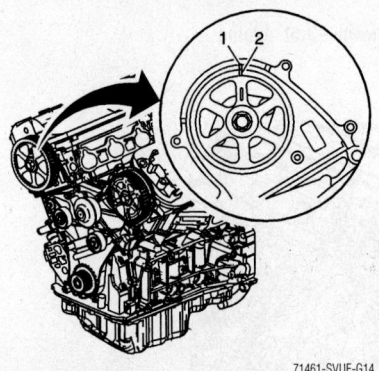

71461-SVUE-G14

Inspect the right camshaft sprocket for proper alignment. The TDC mark (2) on the sprocket should align with the pointer (1) on the cover—3.5L engine

15. Tighten the idler pulley bolt to 33 ft. lbs. (44 Nm).

16. Remove the tools.

17. Install the guide (236), balancer (228), and bolt (227) and tighten the bolt until snug.

18. Using the balancer holding tool EN 46337 rotate the crankshaft clockwise about 5 or 6 complete revolutions to position the timing belt on the sprockets.

19. Align the mark (2) on the sprocket with the pointer (1) on the oil pump housing.

20. Inspect the left camshaft sprocket for proper alignment. The TDC mark (2) on the sprocket should align with the pointer (1) on the cover.

21. Inspect the right camshaft sprocket for proper alignment. The TDC mark (2) on the sprocket should align with the pointer (1) on the cover.

22. Remove the balancer.

23. Install the timing belt cover and tighten the bolts to 106 inch lbs. (12 Nm).

24. Install the balancer.

25. Install the splash shield.

26. Install the engine drive belt.

Piston and Ring

POSITIONING

9306ZG77

Piston ring positioning—2.2L engine

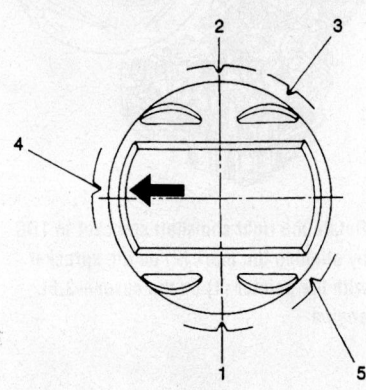

(1) 1st Compression Ring End Gap Location
(2) 2nd Compression Ring End Gap Location
(3) Oil Control Ring Upper Ring End Gap Location
(4) Oil Control Ring Spacer End Gap Location
(5) Oil Control Ring Lower Ring End Gap Location

7922AG55

Piston ring positioning—3.0L engine

FUEL SYSTEM

Fuel System Service Precautions

Safety is the most important factor when performing not only fuel system maintenance but any type of maintenance. Failure to conduct maintenance and repairs in a safe manner may result in serious personal injury or death. Maintenance and testing of the vehicle's fuel system components can be accomplished safely and effectively by adhering to the following rules and guidelines.

• To avoid the possibility of fire and personal injury, always disconnect the negative battery cable unless the repair or test procedure requires that battery voltage be applied

• Always relieve the fuel system pressure prior to disconnecting any fuel system component (injector, fuel rail, pressure regulator, etc.), fitting or fuel line connection. Exercise extreme caution whenever relieving fuel system pressure, to avoid exposing

skin, face and eyes to fuel spray. Please be advised that fuel under pressure may penetrate the skin or any part of the body that it contacts

• Always place a shop towel or cloth around the fitting or connection prior to loosening to absorb any excess fuel due to spillage. Ensure that all fuel spillage (should it occur) is quickly removed from engine surfaces. Ensure that all fuel soaked cloths or towels are deposited into a suitable waste container

• Always keep a dry chemical (Class B) fire extinguisher near the work area

• Do not allow fuel spray or fuel vapors to come into contact with a spark or open flame

• Always use a back-up wrench when loosening and tightening fuel line connection fittings. This will prevent unnecessary stress and torsion to fuel line piping

• Always replace worn fuel fitting O-rings with new. Do not substitute fuel hose or equivalent, where fuel pipe is installed

Fuel System Pressure

RELIEVING

1. Before servicing the vehicle, refer to the precautions section.

2. Unless battery voltage is necessary for testing, disconnect the negative battery cable. This will prevent the fuel pump from running and causing a fuel spill through the disconnected components if the ignition key is accidentally turned **ON**.

3. Remove the air cleaner assembly, for access.

4. Connect gauge bar 53476 to fuel gauge pressure adapter 309725 using a flexible hose from gauge bar set SA9127E.

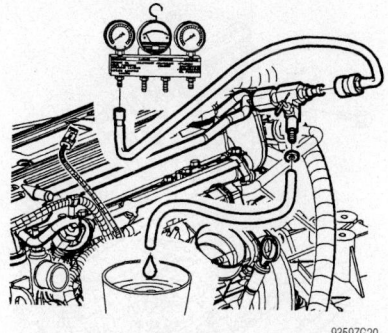

9359ZG20

Connect gauge bar 53476 to fuel gauge pressure adapter 309725 using a flexible hose from gauge bar set SA9127E

Make sure the needle valve on the pressure adapter is off.

➡**Do not use tools to tighten the adapter to the pressure port. If the adapter will not hand-tighten the seals are defective and needs to be replaced.**

5. Wrap a shop rag around the fuel test port fitting, located at the lower rear of the engine, then remove the cap and connect fuel pressure gauge.

6. Install the bleed hose from the pressure gauge into an approved container and open the valve to bleed the system pressure.

7. After the pressure is bled, remove the gauge from the test port and recap it.

8. Install the air cleaner assembly.

9. After servicing the vehicle, connect the negative battery cable and prime the fuel system as follows:

a. Turn the ignition **ON** for 5 seconds, then **OFF** for 10 seconds.

b. Repeat the **ON/OFF** cycle 2 more times.

c. Crank the engine until it starts.

d. If the engine does not readily start, repeat sub-steps A–C.

10. Run the engine and check for leaks.

Fuel Filter

REMOVAL & INSTALLATION

1. Before servicing the vehicle, refer to the precautions section.

2. Properly relieve the fuel system pressure.

3. Remove or disconnect the following:
- Negative battery cable
- Fuel filter bracket screw
- Fuel feed and return lines from the filter
- Fuel filter from the bracket

To install:

4. Install or connect the following:

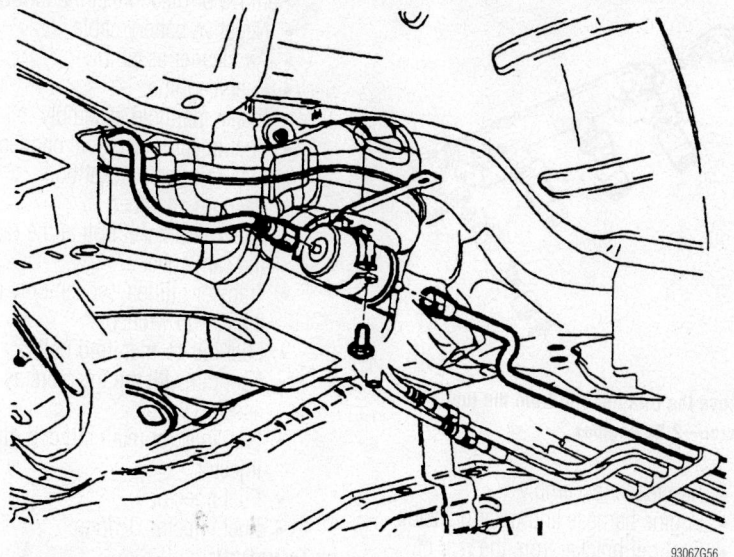

9306ZG56

Remove the fuel filter bracket attaching screw 2.2L and 3.0L engines

- New fuel filter into the bracket
- New fuel line retainers to the female portion of the quick connect fittings
- Fuel feed and return lines
- Fuel filter bracket attaching screw and torque it to 18 inch lbs. (2 Nm)
- Negative battery cable

5. Prime the fuel system as follows:

a. Turn the ignition **ON** for 5 seconds, then **OFF** for 10 seconds.

b. Repeat the **ON/OFF** cycle 2 more times.

c. Crank the engine until it starts.

d. If it does not start, repeat the 3 above steps.

6. Start the engine and check for leaks, repair if necessary.

Fuel Pump

REMOVAL & INSTALLATION

1. Before servicing the vehicle, refer to the precautions section.

2. Properly relieve the fuel system pressure.

3. Remove or disconnect the following:
- Negative battery cable
- Fuel tank
- Fuel pump electrical connector
- Fuel lines and hoses from the fuel pump module cover
- Fuel pump module retaining ring with Wrench SA9156E
- Pull the retaining clip toward the float arm and lift up
- Fuel pump straight up from the fuel tank

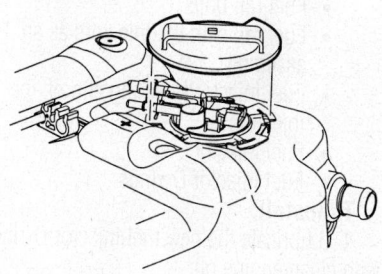

9359ZG21

Remove the fuel pump cover lockring

- Fuel pump tank seal and discard the seal

To install:

4. Install or connect the following:
- Fuel pump with new seal
- Fuel pump cover lockring with Tool SA9156E
- Fuel lines, hoses and wiring harness
- Fuel tank
- Negative battery cable

5. Start the vehicle and check for leaks, repair if necessary.

Fuel Injector

REMOVAL & INSTALLATION

2.2L Engines

1. Before servicing the vehicle, refer to the precautions section.

2. Properly relieve the fuel system pressure.

3. Remove or disconnect the following:
- Negative battery cable

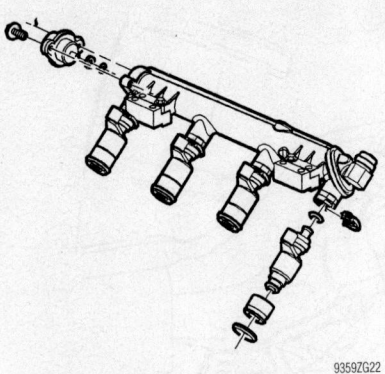

Remove the retainer clip from the fuel injector—2.2L engines

- Air cleaner assembly
- Engine harness and position aside
- Fuel rail bracket from the rear of the can cover
- Fuel feed line
- Fuel injector electrical connectors
- Fuel rail bolts
- Fuel rail with the injectors as an assembly
- Fuel injector retaining clip off the injector
- Fuel injector
- Fuel injector O-rings

To install:

4. Lubricate the new fuel injector O-ring with clean engine oil.

5. Install or connect the following:
- New O-ring seals on the fuel injector
- Fuel injector to the fuel rail
- Retaining clip to the fuel injector
- Fuel rail and torque the bolts to 89 inch lbs. (10 Nm)
- Fuel injector electrical connectors
- Fuel feed line and tighten the fitting to 89 inch lbs. (10 Nm)
- Fuel line bracket and tighten the retainer to 89 inch lbs. (10 Nm)
- Air cleaner assembly
- Negative battery cable

6. Start the vehicle and check for leaks, repair if necessary.

3.0L Engines

1. Before servicing the vehicle, refer to the precautions section.

2. Properly relieve the fuel system pressure.

3. Remove or disconnect the following:
- Negative battery cable
- Air cleaner assembly
- Fuel feed line
- Intake manifold assembly
- Fuel injector electrical connectors
- Fuel rail line bracket bolt
- Vacuum chamber
- Fuel rail bracket bolt at the rear of the cam cover
- Transfer fittings, loosen only using a back-up wrench
- Fuel rail-to-manifold bolts
- Fuel rail with the injectors as an assembly
- Fuel injector retaining clip off the injector
- Fuel injector
- Fuel injector O-rings

To install:

4. Lubricate the new fuel injector O-ring with clean engine oil.

5. Install or connect the following:
- New O-ring seals on the fuel injector
- Fuel injector to the fuel rail
- Retaining clip to the fuel injector aligning the notch in the clip to the tab on the rail
- Fuel rail and torque the bolts to 66 inch lbs. (7.5 Nm)
- Transfer fittings, and tighten to 11 ft. lbs. (15 Nm) using a back-up wrench
- Fuel rail bracket bolt at the rear of the cam cover and tighten to 39 inch lbs. (4.5 Nm)
- Vacuum chamber with a new gasket and tighten the retainers to 18 inch lbs. (2 Nm), then attach the vacuum hose
- Intake manifold top cover and tighten to 35 inch lbs. (4 Nm)
- Fuel injector electrical connectors
- Intake manifold assembly
- Air cleaner assembly
- Negative battery cable

6. Start the vehicle and check for leaks, repair if necessary.

3.5L Engine

1. Before servicing the vehicle, refer to the precautions section.

2. Relieve the fuel system pressure.

3. Remove the intake manifold.

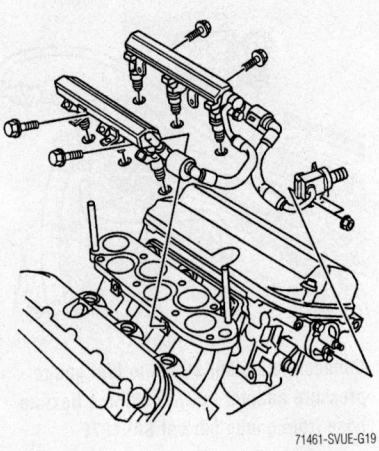

Exploded view of the fuel rail assembly— 3.5L engine

4. Disconnect the fuel injector electrical connectors.

5. Remove the rail bolts.

6. Remove the fuel rail and injectors from the manifold as an assembly.

7. Remove the fuel injector retaining clip off the injector.

8. Remove the fuel injector.

9. Fuel injector O-rings and discard

To install:

10. Lubricate the new fuel injector O-ring with clean engine oil.

11. Install new O-ring seals on the fuel injector.

12. Install the fuel injector to the fuel rail.

13. Install the retaining clip to the fuel injector.

14. Install new O-rings to the injectors.

15. Install the fuel rail and injectors as an assembly to the manifold. Torque the bolts to 87 inch lbs. (10 Nm).

16. Connect the fuel injector electrical connectors.

17. Install the intake manifold.

18. Inspect for fuel leaks using the following procedure:

a. With the engine OFF, turn ON the ignition for 2 seconds and inspect for fuel leaks.

b. Turn the ignition OFF for at least 10 seconds.

c. With the engine OFF, turn ON the ignition for 2 seconds and inspect for fuel leaks.

19. Turn the ignition OFF.

DRIVE TRAIN

Manual Transmission Assembly

REMOVAL & INSTALLATION

1. Before servicing the vehicle, refer to the precautions section.
2. Drain the transmission fluid.
3. Remove or disconnect the following:
   - Both battery cables
4. Install an engine support fixture.
5. Fasten the radiator to the upper radiator support.
   - Wheels
   - Splash shields
   - Ball joints and tie rods from the knuckle
   - Lower control arm from the knuckle
   - Lower stabilizer bar links
   - Steering gear from the steering gear assembly
   - Rear transaxle mount-to-cradle bolts
   - Rear transaxle mount bracket-to-transaxle bolts
   - Front lower mount through bolt from the cradle
   - Front air deflector from the body but leave it attached to the cradle
6. Support the cradle with a jack
   - Cradle bolts and the cradle
   - Shift lever cable from the shift control housing using tool J36346
   - Shift lever cable from the bracket
   - Pressure line from the clutch actuator by removing the C-clip and pulling it away from the actuator
   - Back-up lamp switch connector
   - Front transaxle mount from the transaxle
   - Right side drive axle from the intermediate drive shaft using removal tool J45341 and slide hammer SA9173G
7. Secure the drive axle from the intermediate drive shaft.

➡**Remove the retainer ring from the stub shaft before removing the tool and discard the ring.**

   - Intermediate drive shaft using removal tool J440177 and axle seal puller SA9133T
   - Left side drive axle drive shaft using removal tool J45341 and slide hammer SA9173G and secure the axle aside
   - Top transaxle mount bolts
8. Use the engine support fixture to

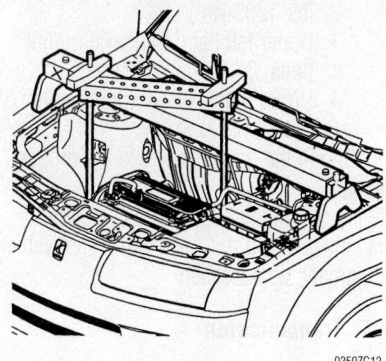

Install an engine support tool

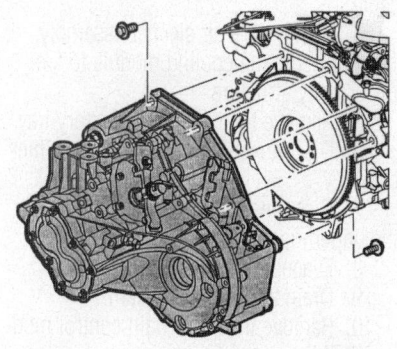

Remove the engine to manual transmission bolts

lower the transaxle enough so that the assembly can be removed.

9. Attach the transaxle to a suitable transaxle jack.
10. Remove the transaxle bolts on the engine side and the transaxle bolt on the transaxle side.

To install:

11. Installation is the reverse of removal, please note the following torque specifications:
   - Transaxle-to-engine and engine-to-transaxle bolts: 55 ft. lbs. (75 Nm)
   - Top mount-to-transaxle bolts: 37 ft. lbs. (50 Nm)
   - Front transaxle mount bolts: 37 ft. lbs. (50 Nm)
   - Rear transaxle mount-to-transaxle bolts: 37 ft. lbs. (50 Nm)
   - Cradle-to-body bolts: 2001–03 models to 140 ft. lbs. (190 Nm), or 114 ft. lbs. (155 Nm) on 2004 models
   - Front transaxle mount through bolt: 81 ft. lbs. (110 Nm)
   - Rear transaxle mount through bolt: 81 ft. lbs. (110 Nm)

   - Steering gear-to-column bolt: 25 ft. lbs. (34 Nm)
   - Lower control arm nut: 89 inch lbs. (10 Nm) plus 150 degrees
   - Stabilizer bar link: 48 ft. lbs. (65 Nm)
   - Tie rod end using installer J44015 and tighten to 30 ft. lbs. (40 Nm)
   - Tie rod-to-steering knuckle nut: 37 ft. lbs. (50 Nm)
12. Fill the transmission to the proper level.
13. Warm the engine and check the transmission fluid. Check and adjust vehicle alignment, as necessary.

Automatic Transmission Assembly

REMOVAL & INSTALLATION

Except 5AT Transmission

1. Before servicing the vehicle, refer to the precautions section.
2. Drain the transmission fluid.
3. Remove or disconnect the following:
   - Both battery cables
   - Battery and battery tray bracket
   - Control cable from the Transaxle Range Switch using prytool J36346
   - Control cable from the bracket
   - TRS electrical connection
   - Input and output speed sensor connections
   - Positive battery cable from the transaxle stud
   - Headlamp fasteners and wire radiator to the core support
4. Install an engine support fixture.
   - Transaxle fluid dipstick tube
   - Upper left hand transaxle mount
   - Wheels
   - Splash shields
   - Front air deflector
   - Transaxle cooler line nut
   - Transaxle cooler lines from the transaxle

➡**Mark the position of the Power take Off (PTO) prior to removal.**

   - Drive shaft from the PTO marking alignment locations prior to removal
   - Lower stabilizer nut from the cradle on both sides
   - Lower control arm from the knuckle
   - Ball joints and tie rods from the knuckle

- Front pitch restrictor bolts, through bolts and restrictor
- Rear restrictor through bolt from the cradle

5. Position a support table under the cradle.

- Cradle bolts and lower the cradle onto the table
- Left side drive axle drive shaft using removal tool J45341 and slide hammer SA9173G and secure the axle aside

➡ **The stub shaft may disengage from the PTO if this occurs, plug the PTO to avoid fluid loss.**

- Right side drive axle from the PTO using removal tool J45341 and slide hammer SA9173G
- PTO-to-engine bracket
- Starter
- Torque converter-to-flexplate bolts through the starter hole
- 3 lower transaxle-to-engine bolts
- 1 engine-to-transaxle bolt located above the PTO

6. Attach the transaxle to a suitable transaxle jack.

7. Separate the engine from the transaxle , lower the assembly and disconnect the PTO hose.

8. Remove the PTO, if necessary.

To install:

9. Installation is the reverse of removal, please note the following torque specifications:

- Transaxle-to-engine bolts: 55 ft. lbs. (75 Nm)
- PTO bracket bolts: 44 ft. lbs. (60 Nm)
- Torque converter-to-flexplate bolts: 44 ft. lbs. (60 Nm)
- Bracket-to-engine bolts: 26 ft. lbs. (35 Nm)
- Cradle bolts: 114 ft. lbs. (155 Nm)
- Rear pitch restrictor through bolt: 81 ft. lbs. (110 Nm)
- Front pitch restrictor bolts: 37 ft. lbs. (50 Nm)
- Front pitch restrictor through bolt: 81 ft. lbs. (110 Nm)
- Tie rod end using installer J44015 and tighten to 30 ft. lbs. (40 Nm)
- Tie rod-to-steering knuckle nut: 37 ft. lbs. (50 Nm)
- Lower control arm nut: 89 inch lbs. (10 Nm) plus 150 degrees
- Rack and pinion bolts: 81 ft. lbs. (110 Nm)
- Lower stabilizer nut: 48 ft. lbs. (65 Nm)
- Drive shaft retainers: 74 ft. lbs. (100 Nm)

- Transaxle oil cooler line assembly: 71 inch lbs. (8 Nm)
- Front axle stub shaft nut: 151 ft. lbs. (205 Nm)
- Upper left hand transaxle mount bolts: 37 ft. lbs. (50 Nm)
- Upper transaxle-to-engine bolts: 55 ft. lbs. (75 Nm)

10. Fill the transmission to the proper level.

11. Warm the engine and check the transmission fluid. Check and adjust vehicle alignment, as necessary.

5AT Transmission

1. Before servicing the vehicle, refer to the precautions section.

2. Disconnect the negative battery cable.

3. Remove the air cleaner assembly.

4. Secure the cooling module to the upper body structure.

5. Remove the battery and battery tray.

6. Disconnect the transmission shiftier cable.

7. Disconnect the wiring harness from the underhood junction block.

8. Evacuate the A/C system.

9. Drain the cooling system.

10. Remove the powertrain control module (PCM).

11. Remove the A/C low pressure tube at the front lift bracket.

12. Disconnect the alternator positive cable.

13. Disconnect the A/C high pressure switch harness.

14. Remove the A/C tube from the A/C compressor.

15. Disconnect the A/C line from the condenser to the compressor.

16. Disconnect the coolant reservoir hose from the engine.

17. Disconnect the radiator hoses from the engine.

18. Remove the heater hoses.

19. Remove the starter positive cable.

20. Relieve the fuel pressure.

21. Disconnect the fuel feed line.

22. Disconnect the fuel EVAP line.

23. Remove the lower transaxle–to–engine bolts.

24. Remove the PTU as follows:
   a. Remove the propeller shaft.
   b. Drain the transfer case oil.
   c. Remove the exhaust cross–under pipe.
   d. Remove the vent tube clamp.
   e. Remove the vent tube from the transfer case.
   f. Remove the transfer case.

➡ **When removing the transfer case/output shaft, do not use excessive force or damage to the bushings may occur.**

   g. Remove the transfer case from the transaxle.

25. Remove the torque converter inspection cover.

26. Remove the torque converter to flywheel bolts.

27. Remove the front wheels.

28. Remove the left inner liner.

29. Disconnect the transmission cooler lines from the transmission and bracket.

30. Remove the tie rod ends from the steering knuckles.

31. Remove the stabilizer bar links.

32. Disconnect lower ball joints.

33. Remove the axle shaft nuts.

➡ **In order to prevent possible SIR system deployment, do not attempt to rotate the steering shaft.**

34. Disconnect the intermediate shaft from the steering gear.

35. Remove the front exhaust pipe.

36. Remove the three front fender pushpins to allow the front fender to flex.

37. Matchmark the frame to the body position.

38. Support the engine in the cradle with wood blocks.

39. Disconnect the front engine mount from the body.

40. Lower the vehicle to 3 feet off the ground in order to position Engine support table J 39580 under the frame.

41. Remove the cradle bolts.

42. Slowly lower the table to the floor.

43. Remove the starter.

44. Remove the following components:
   a. Remove the transmission shift cable bracket.
   b. Remove the Park/Neutral position switch.
   c. Remove the transmission vent hose.
   d. Remove the transmission mount from the transmission.
   e. Remove the rear transmission mount bracket.

45. Separate the transmission from the engine.

46. Remove the transfer case, if equipped.

To install:

47. Attach the engine to the transmission.

48. Install the rear mount.

49. Install the front engine mount to the engine

50. Install the following components:

a. Install the transmission vent hose.

b. Install the Park/Neutral position switch.

c. Install the transmission shift cable bracket and bolts. Tighten the bolts to 89 inch lbs. (10 Nm).

d. Install the starter.

51. Install the engine and transmission assembly in the vehicle.

52. Install the cradle bolts. Tighten the bolts to 114 ft. lbs. (155 Nm).

53. Remove the lift table.

54. Install the front engine mount bolts to the body. Tighten the transmission mount–to–frame bolts to 37 ft. lbs. (50 Nm) and the transmission mount–to–bracket through bolt, while aligning the transmission mount to the bracket to 81 ft. lbs. (110 Nm).

55. Remove the wood blocks from the cradle.

56. Install the lower transaxle-to-engine bolts and tighten to 47 ft. lbs. (64 Nm).

57. Install the PTU as follows:

a. Install the transfer case to the transaxle and tighten the bolts to 38 ft. lbs. (51 Nm).

58. Install the vent hose and the clamp to the transfer case.

59. Install the exhaust cross–under pipe.

60. Remove the transfer case check plug and fill the transfer case with synthetic gear oil.

61. Install the check plug and gasket to the case and tighten to 33 ft. lbs. (44 Nm).

62. Install the torque converter–to–flywheel bolts and tighten to 9 ft. lbs. (12 Nm).

63. Install the torque converter inspection cover and tighten to 9 ft. lbs. (12 Nm).

64. Install the 3 front fender pushpins

65. Install the front exhaust pipe.

66. Connect the intermediate shaft from the steering gear.

67. Install the propeller shaft.

68. Install the axle shaft nuts.

69. Connect the lower ball joints.

70. Install the stabilizer bar links.

71. Install tie rod ends to the steering knuckles.

72. Connect the transmission cooler lines to the transmission and bracket.

73. Install the left inner liner.

74. Install the front tires.

75. Install the fuel EVAP line.

76. Connect the fuel feed line.

77. Connect the starter positive cable.

78. Install the heater hoses.

79. Connect the radiator hoses to the engine.

80. Attach the A/C tube to the A/C compressor.

81. Connect the coolant reservoir hose.

82. Connect the A/C line from the condenser to compressor.

83. Connect the A/C high pressure switch harness.

84. Connect the alternator wiring.

85. Install the A/C lower pressure tube at the front lift bracket.

86. Install the PCM.

87. Connect the radiator hoses to the engine.

88. Fill the cooling system.

89. Connect the wiring harness to the underhood junction block.

90. Connect the transmission shifter cable.

91. Remove the cooling module support.

92. Install the battery tray and battery.

93. Install the air cleaner assembly and ducts.

94. Connect the negative battery cable.

Clutch

ADJUSTMENT

The hydraulic clutch system is self-adjusting.

REMOVAL & INSTALLATION

1. Before servicing the vehicle, refer to the precautions section.

2. Remove the transmission from the vehicle.

3. Remove the pressure plate and clutch disc.

4. Inspect the pressure plate, as follows:

a. Check for excessive wear, chatter marks, cracks or overheating (indicated by a blue discoloration). Black random spots on the friction surface of the pressure plate is normal.

b. Check the plate for warpage using a straightedge and a feeler gauge; the maximum allowable warpage is 0.006 in. (0.15mm).

c. Replace the plate, if necessary.

5. Inspect the clutch disc, as follows:

a. Check the disc face for oil or burnt spots.

b. Check the disc for loose damper springs, hub or rivets.

c. Replace the disc, if necessary.

6. Check the flywheel, as follows:

a. Check the ring gear for wear or damage.

b. Check the friction surface for excessive wear, chatter marks, cracks or overheating.

c. Check flywheel thickness; the minimum allowable is 1.102 in. (28mm).

d. Measure flywheel run-out using a dial indicator, positioned for at least 2 flywheel revolutions. Push the crankshaft forward to take up thrust bearing clearance. Maximum flywheel run-out is 0.006 in. (0.15mm).

e. Check the flywheel for warpage using a straight-edge and a feeler gauge; the maximum allowable warpage is 0.006 in. (0.15mm).

f. Replace the flywheel, if necessary.

7. If necessary, remove the flywheel retaining bolts and remove the flywheel from the crankshaft.

To install:

8. Install or connect the following:

- Flywheel (if removed) and torque the bolts in a crisscross pattern to 39 ft. lbs. (53 Nm) plus 25 degrees
- Clutch disc and pressure plate and loosely install the pressure plate bolts
- Clutch alignment tool in the clutch disc, and push in until it bottoms out in the crankshaft

9. Tighten the pressure plate bolts using multiple passes of a crisscross sequence to 11 ft. lbs. (15 Nm) on 2002–03 models or 17 ft. lbs. (24 Nm) on 2004–05 models and remove the alignment tool.

10. Lubricate the splines of the input shaft lightly with a high temperature grease.

11. Install the transmission assembly.

12. Connect the negative battery cable.

Hydraulic Clutch System

BLEEDING

Vacuum Bleeding

This procedure outlines how to bleed the hydraulic clutch with the transmission in the vehicle. Only **DOT 3** brake fluid should be added to the system.

1. Before servicing the vehicle, refer to the precautions section.

2. Remove the reservoir cap and fill the reservoir with new brake fluid.

3. Install a Bleeder Adapter Tool J23738A, to the reservoir and connect a pressure bleeder to the adapter.

4. Charge the pressure bleeder to 15–20 psi (103–138 kPa).

5. Attach a transparent hose over the clutch bleeder screw nipple and submerge the opposite end of the hose in a container of brake fluid.

6. Loosen the bleeder screw on the transmission hydraulic fitting.

7. Bleed the system until no air bubbles are seen in the hose.

8. Tighten the bleeder screw.

9. Check the clutch pedal for a spongy feel. If the pedal feels soft, repeat the bleeding procedure.

10. Remove the bleeder tools and top off the fluid level if necessary.

Manual Bleeding

1. Before servicing the vehicle, refer to the precautions section.

   a. Fill the clutch reservoir with brake fluid. Check the reservoir level frequently and add fluid as needed.

   b. Connect one end of a vinyl tube to the bleeder plug on the slave cylinder and submerge the other end into a clear container half-filled with clean brake fluid.

   c. Slowly pump the clutch pedal 10–15 times without bring the pedal the full way up.

   d. Repeat Steps 2 and 3 until all of the air bubbles are removed from the system.

   e. Tighten the bleeder screw to 62 inch lbs. (7 Nm).

   f. Refill the master cylinder to the proper level.

   g. Check the system for leaks.

Halfshafts

REMOVAL & INSTALLATION

1. Before servicing the vehicle, refer to the precautions section.

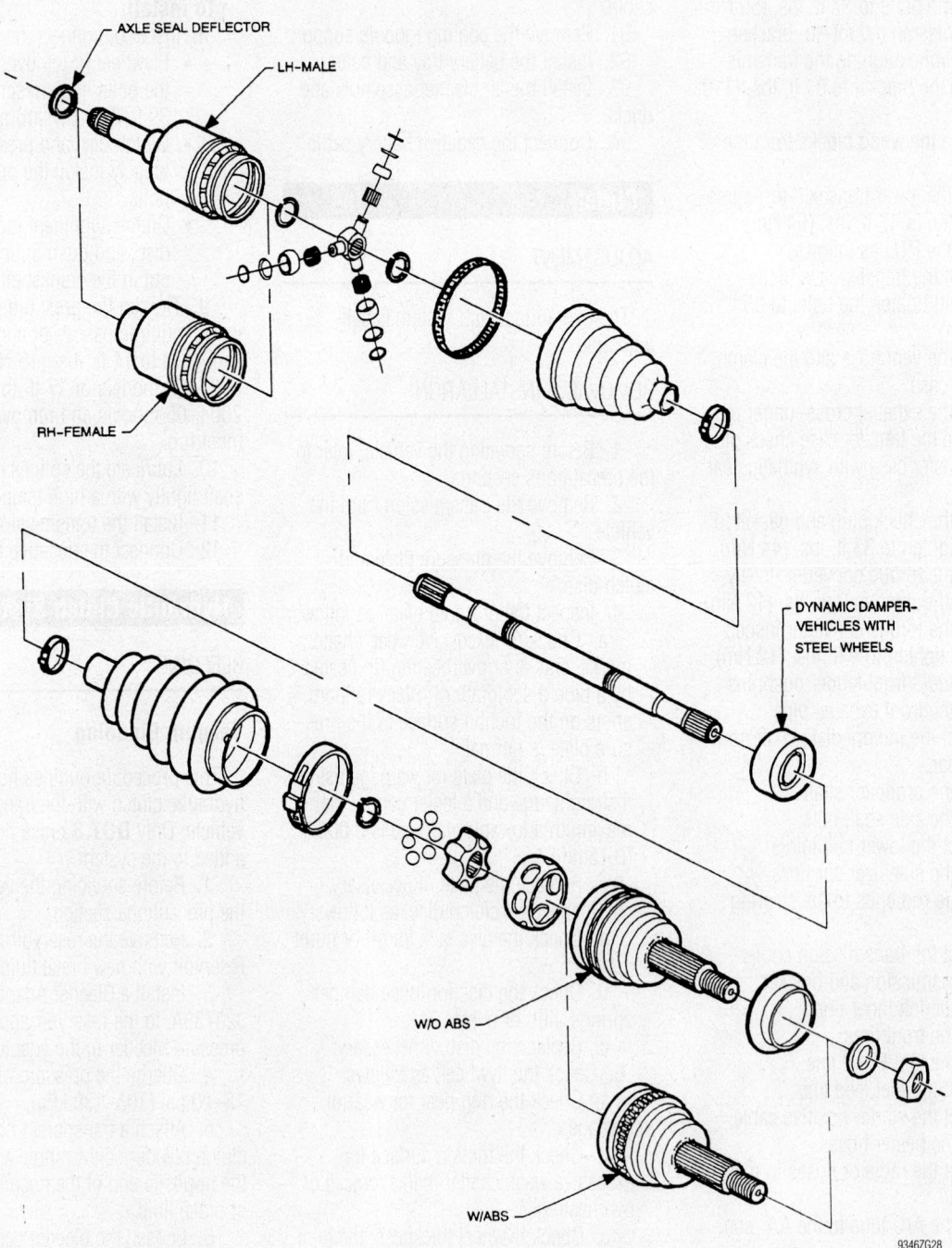

AXLE SEAL DEFLECTOR

LH-MALE

RH-FEMALE

DYNAMIC DAMPER-
VEHICLES WITH
STEEL WHEELS

W/O ABS

W/ABS

9346ZG28

Exploded view of a typical axle shaft assembly

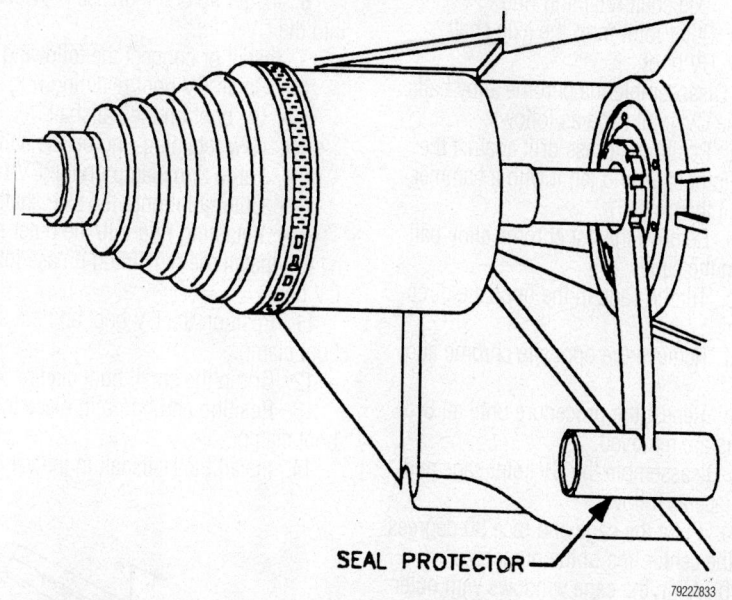

SEAL PROTECTOR —

79222Z833

Failure to use a seal protector may allow the halfshaft splines to damage the transaxle seal

2. Remove the wheel cover or the center cap for access to the halfshaft nut. Have an assistant depress the brake pedal and loosen the front halfshaft nut.

3. Remove or disconnect the following:
- Wheel
- Tie rod end torque prevailing nut and discard it
- Tie rod end from the steering knuckle, using a Tie Rod Separator Tool SA91100C
- Lower control arm to steering knuckle
- Stabilizer bar link nut

➡**Do not allow the steering knuckle to contact the ball stud seal. Contact may cause the seal to rip and the ball stud will need replacement.**

- Pull the steering knuckle/strut assembly away from the vehicle and pull the halfshaft out of the hub
- Properly support the halfshaft and remove the halfshaft from the transmission or Power Take Off (PTO) unit
- Shaft retaining ring and discard it

To install:

4. Apply Output Shaft Lubricant 7847638, to the splines of the output shaft, if equipped with an automatic transmission.

5. Install or connect the following:
- New stub shaft retaining ring

- Halfshaft to the transmission or PTO after installing a Seal Protector Tool SA91112T

6. Remove the seal protector tool after the splines have passed the oil seal.

7. Install or connect the following:
- Fully seat the halfshaft into the transmission
- Outer end of the halfshaft to the wheel hub with a new washer and nut
- Lower control arm ball stud to the steering knuckle. Torque the fastener to 89 inch lbs. (10 Nm) plus 150 degrees on 2002–04 models or 30 ft. lbs. (40 Nm) on 2005 models.
- Stabilizer bar link nut to 48 ft. lbs. (65 Nm)
- Tie rod end to the steering knuckle using tool J44015. When seated properly, torque the fastener to 37 ft. lbs. (50 Nm) on 2002–04 models or 30 ft. lbs. (40 Nm) on 2005 models.
- Wheel
- Halfshaft to wheel nut. Torque the nut to 151 ft. lbs. (205 Nm).
- Cotter pin

8. Top off the transmission with the proper fluid.

9. Check and adjust the front end alignment as necessary.

CV-Joints

OVERHAUL

Constant Velocity (Outer) Joint

1. Before servicing the vehicle, refer to the precautions section.

2. Remove or disconnect the following:
- Front wheel
- Halfshaft
- Swage ring using a hand grinder
- Large CV-joint boot clamp
- CV-joint boot by sliding it away from the tri-pod joint
- Tri-pod housing from the tri-pod spider
- Inboard spacer ring slide it rearward on the shaft
- Outboard retaining ring
- Tri-pod joint spider assembly
- Inboard spacer ring and CV-joint boot

To install:

3. Install or connect the following:
- Swage ring clamp on the CV-joint boot
- CV-joint boot

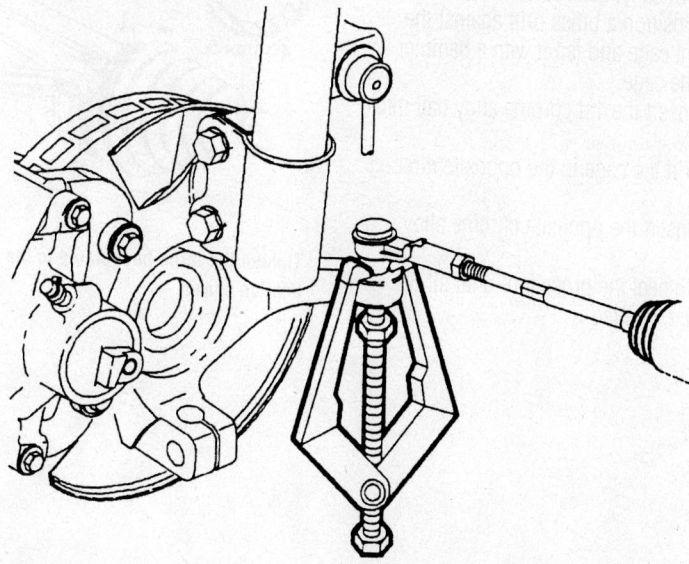

9306ZG65

Separate the tie rod end from the steering knuckle with a Tie Rod Separator tool

4. Position the CV-joint boot seal into the axle shaft's joint seal groove and align the swage ring clamp on the boot.

�֍ WARNING

Make sure that there are no pinch points on the inboard seal.

5. Crimp the swage ring
6. Install or connect the following:
   - Inboard spacer ring, slide it rearward on the shaft
   - Tri-pod joint spider assembly onto the shaft
   - Outboard retaining ring into the axle shaft groove
   - Tri-pod joint spider assembly, slide it against the outboard retaining ring
   - Inboard spacer ring, seat it in the groove
   - ½ kit grease into the boot
   - ½ kit grease into the tri-pod housing
   - New large seal clamp onto the CV-joint boot
   - Tri-pod housing, slide it over the tri-pod joint spider assembly
   - CV-joint boot/clamp, slide it into place, over the trilobal tri-pod bushing with the seal lip in the groove

➡ **Make sure the boot lies flat against the trilobal bushing.**

7. Position the CV-joint boot so it measures 4.9 in. (125mm).
8. Using a Crimp tool, a torque wrench and a breaker bar, crimp the large CV-joint boot clamp to 130 ft. lbs. (176 Nm).
9. Install the halfshaft and the front wheel.

Tri-Pot (Inner) Joint

1. Before servicing the vehicle, refer to the precautions section.
2. Remove or disconnect the following:
   - Axle shaft from the vehicle
   - Large CV boot retaining clamp
   - Small CV boot retaining clamp
   - CV boot from the joint
   - Axle shaft retaining ring
   - Outer joint from the axle shaft
   - CV boot

3. Disassemble the chrome alloy balls from the CV-joint cage as follows:
   a. Position a brass drift against the CV-joint cage and tap it with a hammer to tilt the cage.
   b. Remove the 1st chrome alloy ball from the cage.
   c. Tilt the cage in the opposite direction.
   d. Remove the opposite chrome alloy ball.
   e. Repeat the procedure until all 6 balls are removed.

4. Disassemble the CV-joint cage and inner race as follows:
   a. Pivot the cage and race 90 degrees to the center line of the outer race.
   b. Align the cage windows with outer race lands.
   c. Remove the cage from the outer race.
   d. Rotate the inner race upward and remove it from the cage.

To install:

5. Lubricate the parts with a light coat of grease.
6. Assemble the CV-joint cage and inner race, as follows:
   a. Rotate the inner race 90 degrees to the cage centerline.
   b. Align the cage windows with inner race lands.
   c. Insert the inner race into the cage by rotating the inner race downward.
   d. Insert the cage/inner race into the outer race.
7. Assemble the chrome alloy balls into the CV-joint cage, as follows:
   a. Position a brass drift against the CV-joint cage and tap it with a hammer to tilt the cage.
   b. Insert the 1st chrome alloy ball into the cage.
   c. Tilt the cage in the opposite direction.
   d. Insert the opposite chrome alloy ball.
   e. Repeat the procedure until all 6 balls are inserted.

8. Install ½ of the grease provided, into the CV-joint.
9. Install or connect the following:
   - Small CV boot retaining ring
   - CV boot on the halfshaft
   - New retaining ring on the halfshaft
   - Large ring clamp on the CV boot
   - Outer joint onto the axle shaft
   - Retaining ring into the outer race
10. Install the remaining grease into the CV boot.
11. Position the CV boot and the small boot clamp.
12. Crimp the small boot clamp.
13. Position and crimp in place the large boot clamp.
14. Install the Halfshaft in the vehicle.

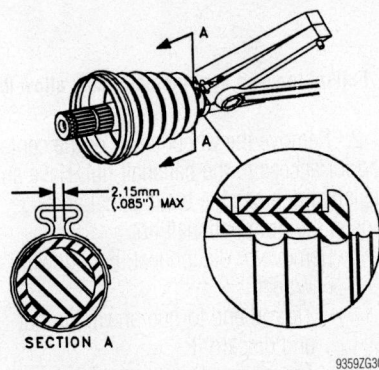

9359ZG30

Tighten the small boot clamp to the specification shown

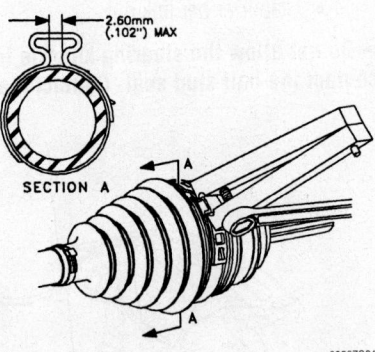

9359ZG31

Tighten the large boot clamp to the specification shown

STEERING AND SUSPENSION

Air Bag

✳✳ CAUTION

All vehicles are equipped with an air bag system. The system must be disabled before performing service on or around system components, steering column, instrument panel components, wiring and sensors. Failure to follow safety and disabling procedures could result in accidental air bag deployment, possible personal injury and unnecessary system repairs.

PRECAUTIONS

Several precautions must be observed when handling the inflator module to avoid accidental deployment and possible personal injury.

1. Never carry the inflator module by the wires or connector on the underside of the module.

2. When carrying a live inflator module, hold securely with both hands, and ensure that the bag and trim cover are pointed away.

3. Place the inflator module on a bench or other surface with the bag and trim cover facing up.

4. With the inflator module on the bench, never place anything on or close to the module which may be thrown in the event of an accidental deployment.

DISARMING

1. Before servicing the vehicle, refer to the precautions section.

2. Align the steering wheel so the vehicle wheels are pointing in the straight-ahead position.

3. Turn the ignition switch to the **LOCK** position.

4. Remove the SIR or AIR BAG fuse from the fuse block.

5. Disable the passenger side air bag as follows:

   a. Locate the SIR connector attached to the HVAC blower motor and disconnect the clip.

➡**Do not remove the upper trim panel as you are able to disable the passenger side air bag from the underside of the instrument panel.**

   b. Remove the Connector Position Assurance (CPA) device, then disengage

the yellow 2-way SIR wiring harness connector.

6. Disable the drivers side air bag as follows:

7. Remove the Connector Position Assurance (CPA) device, then disengage the yellow 2-way SIR wiring harness connector at the base of the steering column.

8. Disable the curtain air bag as follows:

   a. Remove the center push pins located in the upper headliner trim panel.

   b. Remove the left D-pillar upper trim panel.

   c. Using a flat bladed tool, partially remove the right and left coat hook center retainers, then remove the coat hooks.

   d. Pull back the headliner gently to access the yellow 2-way connectors.

   e. Remove the Connector Position Assurance (CPA) device, then disengage the yellow 2-way SIR wiring harness connector.

ARMING

✳✳ CAUTION

After the repairs, enable the system as follows:

1. Turn the ignition switch to the **LOCK** position.

2. Engage the yellow 2-way connectors for the airbags, then install the CPA device.

3. Install any removed trim pieces.

4. Reinstall the Supplemental Inflatable Restraint (SIR) or AIR BAG fuse.

5. Turn the ignition switch to the **RUN** position.

6. Verify the SIR indicator light flashes 7–9 times, if not, inspect the system for malfunction.

Rack and Pinion Steering Gear

REMOVAL & INSTALLATION

1. Before servicing the vehicle, refer to the precautions section.

2. Remove or disconnect the following:
   - Negative battery cable
   - Both front wheels
   - Tie rods from the steering knuckles
   - Intermediate shaft from the steering gear pinch bolt and discard the bolt
   - Shaft from the gear
   - Stabilizer bar link nuts, links and swing the bar upwards

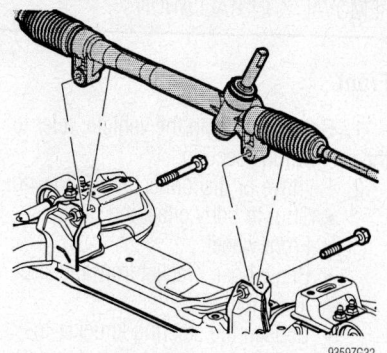

Steering gear-to-frame assembly bolts

   - Shift cable clip from the gear housing, if equipped
   - Steering gear-to-cradle bolts
   - Steering gear by sliding it out the right side of the vehicle

To install:

3. Install or connect the following:
   - Steering gear through the right wheel opening

4. Center the gear mounting bushings into the cradle supports.
   - Steering gear-to-mounting bolts and tighten to 81 ft. lbs. (110 Nm)
   - Shift cable clip to the gear housing, if equipped
   - Intermediate shaft to the steering gear and torque the new pinch bolt to 25 ft. lbs. (34 Nm)
   - Stabilizer bar into position
   - Stabilizer bar links and nuts, then tighten the nuts to 48 ft. lbs. (65 Nm)
   - Tie rods to the steering knuckles using installer tool JJ44015 and tighten to 30 ft. lbs. (40 Nm)
   - New tie rod nut and tighten to 37 ft. lbs. (50 Nm) on 2002–04 models or 44 ft. lbs. (60 Nm) on 2005 models
   - Front wheels
   - Negative battery cable

5. Check the alignment and adjust if necessary.

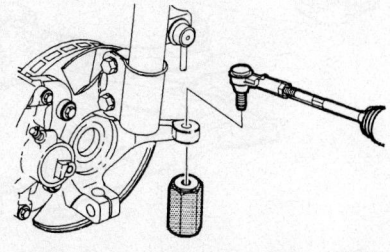

Use installer tool J44015 to attach the tie rods to the steering knuckle

Strut

REMOVAL & INSTALLATION

Front

1. Before servicing the vehicle, refer to the precautions section.
2. Remove or disconnect the following:
   - Strut to body attaching nuts
   - Front wheel
   - Brake hose bracket from the strut assembly
   - Loosen the steering knuckle-to-strut fasteners, but do not remove them
   - Stabilizer bar link to the strut assembly attaching nut and move it toward the rear of the vehicle
   - Place a rag over the CV-joint seal to protect it from damage, then remove the 2 steering knuckle-to-strut housing bolts
   - Steering knuckle-to-strut fasteners
   - Strut assembly from the vehicle

To install:

3. Install or connect the following:
   - Strut to the body and torque the new attaching nuts and bolt to 18 ft. lbs. (25 Nm)
   - Strut to the steering knuckle and torque the new fasteners to 133 ft. lbs. (180 Nm)
   - Stabilizer bar link to the strut and torque the fastener to 48 ft. lbs. (65 Nm)
   - Brake hose bracket to the strut
   - Front wheel
   - Negative battery cable

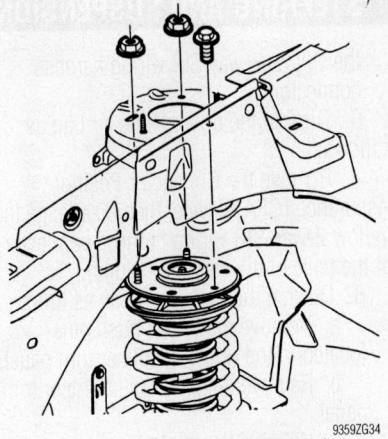

Remove the strut-to-body attaching nut

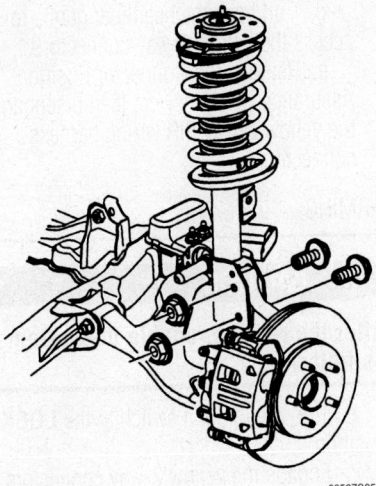

Remove the steering knuckle-to-strut fasteners

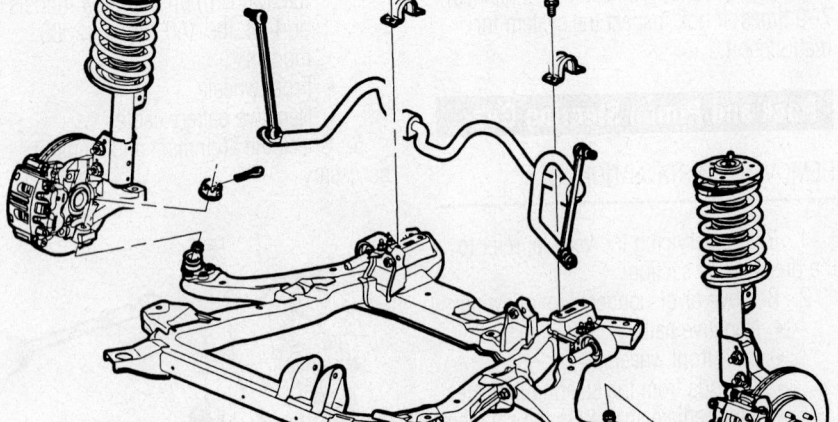

Exploded view of the front suspension

4. Check and adjust the alignment as necessary.

Shock Absorber

REMOVAL & INSTALLATION

1. Before servicing the vehicle, refer to the precautions section.
2. Remove or disconnect the following:
   - Wheel
   - Lower shock bolt

➡ **If removing the right side shock absorber, remove the splash shield.**

   - Upper shock bolt and the shock

To install:

3. Install or connect the following:
   - Shock and upper bolt. Torque the bolt to 81 ft. lbs. (110 Nm).
   - Shock lower bolt and torque the bolt to 81 ft. lbs. (110 Nm)
   - Splash shield, if removed
   - Wheel

Coil Spring

REMOVAL & INSTALLATION

Front

1. Before servicing the vehicle, refer to the precautions section.
2. Remove or disconnect the following:
   - Strut from the vehicle
   - Place the strut into spring compressor J45400 or a similar compressor.
   - Compress the spring enough to completely unload the upper strut mount

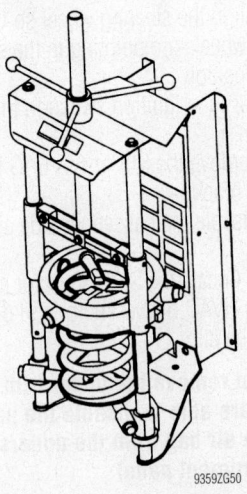

Place the strut into spring compressor J45400

Exploded view of the strut assembly

- Strut shaft nut while holding the shaft stationary with a TORX® socket
- Release the spring compressor
- Spring from the strut
- Remaining strut assembly components, examine for wear or damage and replaces as necessary

To install:

3. Install or connect the following:
- Strut into spring compressor
- Extend the strut shaft to its full travel
- Strut to the spring
- Upper spring seat onto the strut shaft and align the flat with the strut-to-knuckle bracket
- Top mount onto the strut shaft and align the flat 180 degrees from the flat on the upper spring seat
- Strut top nut and tighten with a wrench while holding the strut shaft with a socket

4. Make sure the flats are aligned on the upper spring seat and top mount before tightening the nut. Then tighten the nut to 55 ft. lbs. (75 Nm).

5. Release the spring compressor tool.

6. Install the strut to the vehicle.

Rear

1. Before servicing the vehicle, refer to the precautions section.

2. Remove or disconnect the following:
- Wheel

- Stabilizer link-to-lower control arm nut
- Trailing arm bracket-to-underbody bolts

3. Support the lower control arm with a suitable jack.
- Lower shock bolt
- Jounce jumper nut from the lower control arm
- Lower control arm-to-support frame bolt
- Lower control arm-to-knuckle retainers
- Lower the jack supporting the control arm slowly to unload the spring
- Spring

To install:
- Spring
- Jounce jumper and hand tighten the nut
- Lower control arm into position using the jack
- Lower control arm-to-knuckle fasteners and hand tighten making sure the bolt heads face the rear of the vehicle. Tighten the bolts to 81 ft. lbs. (110 Nm).
- Lower control arm-to-support bolt to 81 ft. lbs. (110 Nm)
- Shock absorber lower bolt and tighten to 81 ft. lbs. (110 Nm)
- Jounce jumper nut to 46 ft. lbs. (63 Nm)
- Stabilizer bar link nut and tighten to 11 ft. lbs. (15 Nm)

4. When installing the training arm, push upward and install the front bolt, then use a drift to align the rear holes and install the bolts. Tighten the bolts to 81 ft. lbs. (110 Nm).
- Wheel

Lower Ball Joint

REMOVAL & INSTALLATION

1. Before servicing the vehicle, refer to the precautions section.

2. Install or connect the following:
- Lower control arm from the vehicle
- Rivets retaining the ball joint to the control arm using a 5/16 in. (8 mm) drill bit
- Ball joint from the control arm

To install:

3. Install or connect the following:
- Ball joint into the control arm
- Nuts and bolts (included with new ball joint kit) as shown and torque them to 50 ft. lbs. (68 Nm)
- Control arm to the vehicle

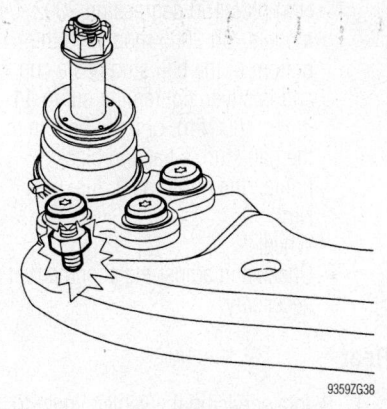

The new ball stud is bolted into the control arm

Lower Control Arm

REMOVAL & INSTALLATION

Front

1. Before servicing the vehicle, refer to the precautions section.

2. Remove or disconnect the following:
- Wheel
- Ball stud bolt
- Separate the ball stud from the steering knuckle by using ball joint removal tool J43828
- Lower control arm-to-frame retainers
- Lower control arm

To install:

3. Install or connect the following:
- Control arm to the frame and torque the rear bolts and nuts to 52 ft. lbs. (70 Nm) and the front arm-to-frame bolt and nut to 89 ft. lbs. (120 Nm) on 2002–04 models or 148 ft. lbs. (200 Nm) on 2005 models
- Ball stud to steering knuckle and torque the bolt to 89 inch. lbs. (10

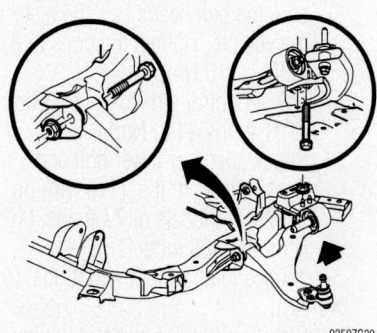

Remove the lower control arm to frame bolts

Nm) plus 150 degrees on 2002–04 models. On 2005 models, if the bottom of the ball stud has a cup and is silver, tighten the nut to 44 ft. lbs. (60 Nm). of if the bottom of the ball stud is flat and black, tighten the nut to 30 ft. lbs. (40 Nm).

- Wheel
- Check and adjust the alignment, if necessary.

Rear

1. Before servicing the vehicle, refer to the precautions section.
2. Remove or disconnect the following:
   - Wheel
   - Stabilizer link-to-lower control arm nut
   - Trailing arm bracket-to-underbody bolts
3. Support the lower control arm with a suitable jack.
   - Lower shock bolt
   - Jounce jumper nut from the lower control arm
   - Lower control arm-to-support frame bolt
   - Lower control arm-to-knuckle retainers
   - Lower the jack supporting the control arm slowly to unload the spring
   - Lower control arm-to-knuckle retainers
   - Spring
   - Lower control arm-to-support nut and bolt

To install:
- Lower control arm-to-support and hand tighten the nut
- Spring
- Jounce jumper and hand tighten the nut
- Lower control arm into position using the jack
- Lower control arm-to-knuckle fasteners and hand tighten making sure the bolt heads face the rear of the vehicle. Tighten the bolts to 81 ft. lbs. (110 Nm).
- Lower control arm-to-support bolt to 81 ft. lbs. (110 Nm)
- Shock absorber lower bolt and tighten to 81 ft. lbs. (110 Nm) on 2002–04 models or 77 ft. lbs. (105 Nm) on 2005 models
- Jounce jumper nut to 46 ft. lbs. (63 Nm)
- Stabilizer bar link nut and tighten to 11 ft. lbs. (15 Nm)
4. When installing the training arm,

push upward and install the front bolt, then use a drift to align the rear holes and install the bolts. Tighten the bolts to 81 ft. lbs. (110 Nm).
- Wheel

LOWER CONTROL ARM BUSHING REPLACEMENT

Front Control Arm

FRONT BUSHING

1. Before servicing the vehicle, refer to the precautions section.
2. Remove or disconnect the following:
   - Wheel
   - Control arm
3. Front bushing by using bearing removal/installer tools J44971 to press out the bushing.

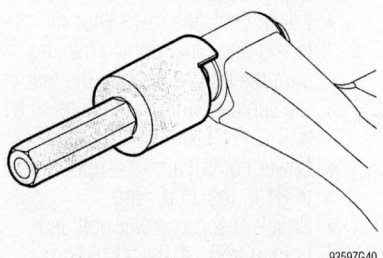

9359ZG40

Press out the front control arm front bushing

To install:
4. Install or connect the following:
5. Press in the new control arm bushing using bearing removal/installer tools J44971.
   - Control arm

FRONT BUSHING

1. Before servicing the vehicle, refer to the precautions section.
2. Remove or disconnect the following:

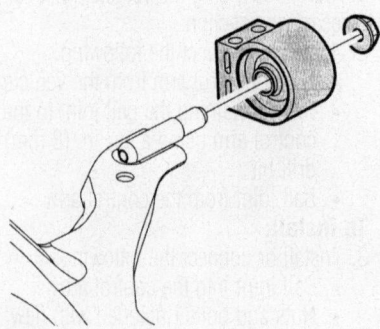

9359ZG41

Remove the rear bushing nut and the bushing

- Wheel
- Control arm
- Rear bushing nut and the bushing

To install:
3. Install or connect the following:
   - Bushing and nut. Tighten the nut to 85 ft. lbs. (115 Nm).
   - Control arm
   - Wheel

Rear Control Arm

INNER BUSHING

1. Before servicing the vehicle, refer to the precautions section.
2. Remove the control arm.
3. Remove the bushing from the arm using removal/installer J45097 in the direction shown in the accompanying illustration as follows:
   a. Place the push out socket against the bushing from the flanged side of the arm.
   b. Install the through-bolt with the washer and bearing against the push-out socket.
   c. Install a backing socket against the control arm on the opposite side of the flange.

➡**Apply high pressure lube to the tool threads.**

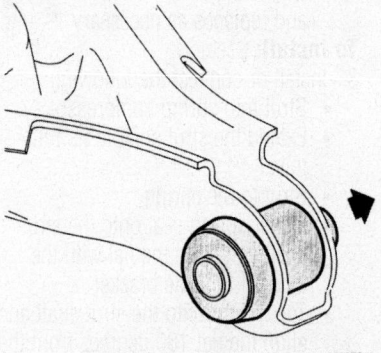

9359ZG42

Remove the inner bushing in this direction

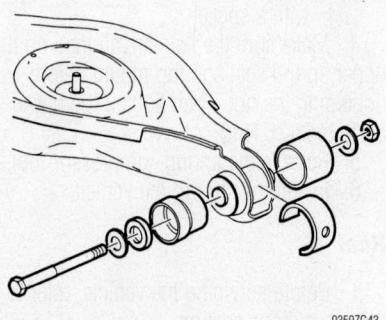

9359ZG43

Remove the bushing from the arm using removal/installer J45097

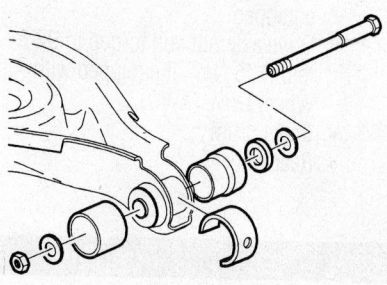

Reverse the installer removal tool and press the bearing into the control arm

d. Install the flat washer and nut, then tighten the nut to remove the bushing.
To install:
4. Install the bushing centering it to align with the center of the control arm approximately 0.079 inch (2mm) beyond the flange.
5. Reverse the installer removal tool and press the bearing into the control arm
6. Install the control arm.

Upper Control Arm

REMOVAL & INSTALLATION

Rear

1. Before servicing the vehicle, refer to the precautions section.
2. Remove or disconnect the following:
   • Trailing arm bracket-to-underbody bolts
   • Anti-lock Brake System (ABS) harness from the upper control arm, if equipped
   • Upper control arm-to-knuckle fasteners
   • Upper control arm-to-support frame retainers
   • Upper control arm
To install:
   • Upper control arm

➡**Make sure the bolt heads face the front of the vehicle.**

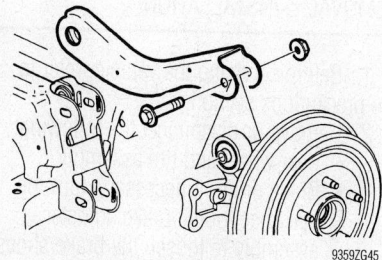

Remove the lower control arm

• Upper control arm-to-knuckle fasteners and hand tighten
• Upper control arm-to-support frame cam nut
• Tighten the control arm-to-knuckle bolts to 118 ft. lbs. (160 Nm)
• Tighten the upper control arm-to-support bolt to 118 ft. lbs. (160 Nm)
• Anti-lock Brake System (ABS) harness to the upper control arm, if equipped
3. When installing the training arm, push upward and install the front bolt, then use a drift to align the rear holes and install the bolts. Tighten the bolts to 81 ft. lbs. (110 Nm).
4. Align the vehicle.

Wheel Bearings

ADJUSTMENT

The wheel bearing are sealed at the factory and do not require any adjustment or maintenance.

REMOVAL & INSTALLATION

Front

1. Before servicing the vehicle, refer to the precautions section.
2. Remove or disconnect the following:
   • Wheel
   • Brake caliper mounting bracket bolts and suspend the assembly from the strut spring with wire
   • Brake rotor
   • ABS sensor electrical connection, if equipped
   • ABS sensor electrical connection from the bracket, if equipped
   • Drive shaft axle nut
3. Support the axle with wire
   • Drive shaft from the knuckle
   • Bearing bolts, the bearing assembly and shield

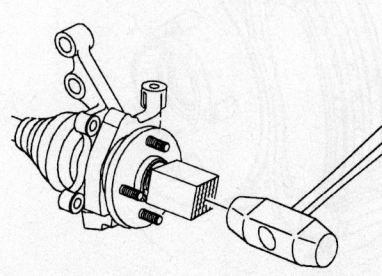

Separate the axle from the wheel hub

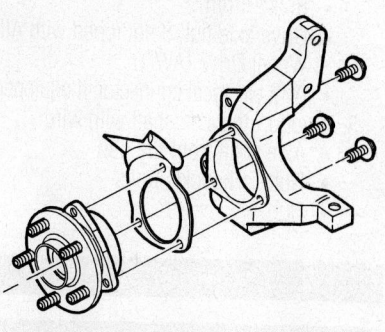

Exploded view of the front bearing/hub assembly

To install:
➡**Make sure the drive shaft splines are aligned with the wheel bearing assembly splines. If you do not align the splines you could damage the bearing or drive shaft assemblies.**

4. Install or connect the following:
   • Bearing assembly with the shield.
   • Bearing assembly bolts and tighten to 96 ft. lbs. (130 Nm)
   • Drive shaft axle nut and torque to 151 ft. lbs. (205 Nm)
   • ABS sensor electrical connection to the bracket, if equipped
   • ABS sensor electrical connection, if equipped
   • Brake rotor
   • Brake caliper
   • Wheel

Rear

1. Before servicing the vehicle, refer to the precautions section.
2. Remove or disconnect the following:
   • Negative battery cable, if equipped with an Antilock Braking System (ABS)
   • Rear wheel

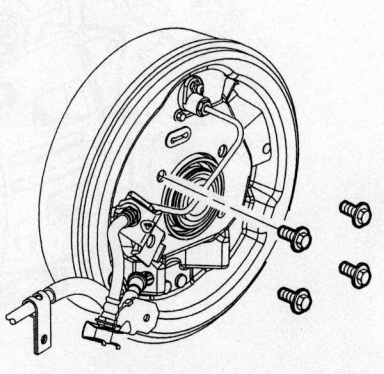

Remove the hub-to-knuckle bolts from the rear bearing/hub assembly

- Brake drum
- Drive axle nut, if equipped with All Wheel Drive (AWD)
- ABS electrical connector, if equipped
3. Support the axle shaft with wire.
- Axle shaft from the hub
- Hub-to-knuckle bolts
- Hub

To install:
4. Install or connect the following:
- Hub to the knuckle and torque the new bolts to 62 ft. lbs. (84 Nm)
- ABS electrical connector through the hole and seat the rubber grommet, if equipped

- ABS electrical connector, if equipped
- Drive axle nut and torque to 92 ft. lbs. (125 Nm), if equipped with All Wheel Drive (AWD)
- Brake drum
- Rear wheel

BRAKES

Brake Caliper

REMOVAL & INSTALLATION

1. Before servicing the vehicle, refer to the precautions section.
2. Remove or disconnect the following:
- Front wheel and tire assembly
- Brake hose from the caliper and discard the 2 copper washers. Plug the openings to prevent system contamination or excessive fluid loss.
- Lock pin and guide pin from the caliper
- Caliper from the support, being careful not to damage the pin boots
- Pin boots from the caliper support and inspect for damage

To install:
3. If necessary, bottom the caliper piston by using a C-clamp.
4. If removed, install the brake pads and clips to the caliper support.
5. Lubricate the pin boots and guide pins with silicone grease.
6. Install or connect the following:

- Pin boots into the caliper support, using the pin to assure that the boot passes all the way through the support
- Caliper onto the support and over the brake pads
7. Lubricate the non-threaded portion of the guide and lock pins with silicone grease.
- Pins through the caliper and torque to 32 ft. lbs. (44 Nm)

➥**Make sure the brake line is properly routed with loop to the rear and that the hose is not twisted.**

- Brake hose using 2 new copper washers. Torque the fitting bolt to 32 ft. lbs. (44 Nm).
8. Properly bleed the hydraulic brake system.
9. Install the wheel and tire assembly.

Disc Brake Pads

REMOVAL & INSTALLATION

1. Before servicing the vehicle, refer to the precautions section.

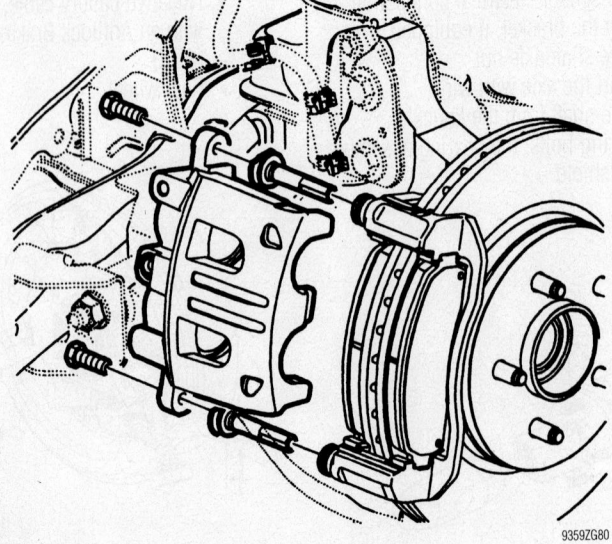

9359ZG80

Exploded view of the caliper mounting

2. Remove or disconnect the following:
- Front wheels
- Caliper lower lock pin
3. Either pivot the caliper up on the guide pin or remove the upper guide pin and support the caliper from the strut using a coat hanger or length of wire.
- 2 brake pads and the pad clips from the caliper support. Discard the old pad clips.
4. Check the caliper pins, pin boots and the piston boot for deterioration or damage.

To install:
5. Using a C-clamp, bottom the piston all the way into the caliper bore.
6. Carefully lift the inner edge of the piston boot by hand to release any trapped air.
7. Install or connect the following:
- New pad clips into the caliper support
- Inner and outer brake pads into the support. If installed, remove the temporary support wire from the caliper.
- Caliper body on the support and upper guide pin into position. Compress the boots by hand as the caliper is positioned onto the support.
8. Lubricate the smooth ends of the removed pin(s) with silicone grease
- Pin(s) and torque to 32 ft. lbs. (44 Nm). Do not get grease on the pin threads.
- Wheels
9. Prior to operating the vehicle, depress the brake pedal a few times until the brake pads are seated against the rotor.

Brake Drums

REMOVAL & INSTALLATION

1. Before servicing the vehicle, refer to the precautions section.
2. Remove or disconnect the following:
- Rear wheel and tire assembly
- Brake drum. If necessary, turn the starwheel of the brake adjuster assembly to loosen the brake shoes and allow for drum removal.

To install:

3. Install or connect the following:
- Brake drum over brake shoes and onto hub
- Tire and wheel assembly. Torque to the proper specification.

4. Adjust the brakes.

5. Road test for braking operation.

Brake Shoes

REMOVAL & INSTALLATION

1. Remove the brake drum.

➡**Do not over stretch the adjuster spring. Damage can occur if the spring is over stretched.**

2. Disengage the adjuster spring hook end from the tab on the adjuster actuator.

3. Remove the straight end of the adjuster spring from the brake shoe.

4. Remove the adjuster actuator from the brake shoe.

5. Remove the return spring from the brake shoes.

6. Remove the park brake cable from the parking brake actuator lever.

7. Remove the brake shoe hold-down springs and retainers from the brake shoes.

8. Remove the adjuster from the brake shoes and the parking brake actuator lever.

9. Remove the horseshoe clip retaining the parking brake actuator lever to the brake shoe.

10. Remove the parking brake actuator lever and wave washer from the brake shoe.

11. Clean all of the drum brake system components with denatured alcohol.

12. Inspect all of the drum brake system components.

13. Inspect the wheel cylinder for the following conditions:
   a. Brake fluid leakage
   b. Worn or damaged dust boots

14. Replace damaged or leaking wheel cylinders as necessary.

To install:

15. Lubricate the adjuster assembly, the 6 backing plate raised shoe contact pads, the brake lever pin and surfaces which contact brake shoe webs with brake lubricant.

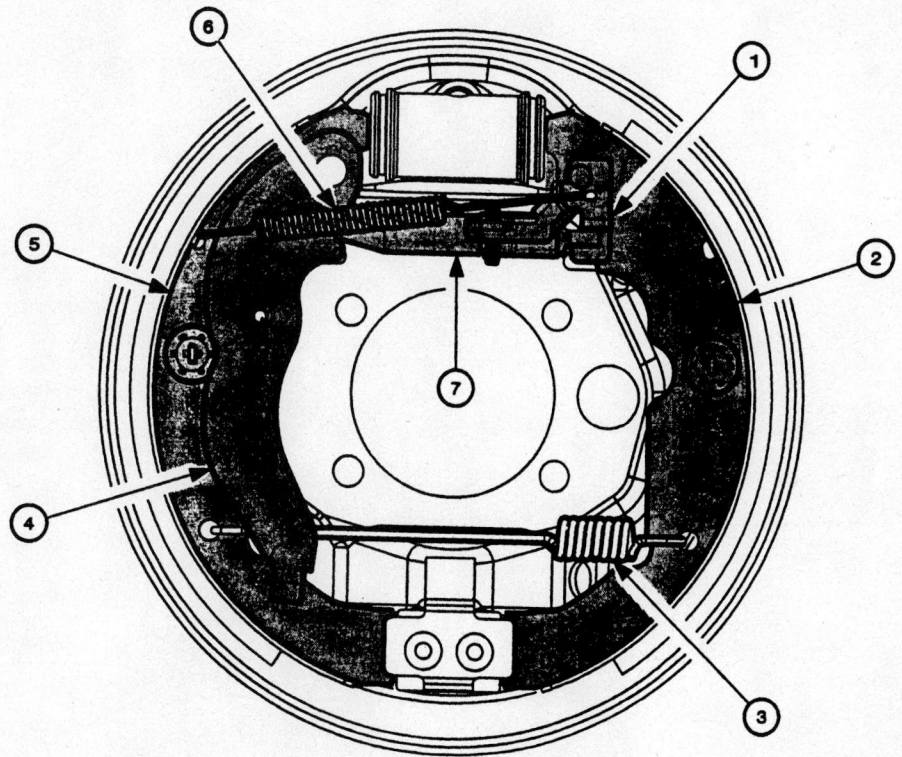

1. Adjuster lever

2. Leading (front) brake shoe

3. Lower return spring

4. Park brake lever

5. Trailing (rear) brake shoe

6. Upper return spring

7. Adjuster assembly

9359ZG81

View of the installed drum brake assembly components

16. Install the parking brake actuator lever to the lever pivot pin.

17. Install the horseshoe clip to the parking brake actuator lever pivot pin.

18. Install the brake shoes to the brake backing plate.

19. Install the brake shoe hold-down pins, springs and retainers to the brake shoes.

20. Install the parking brake cable to the park brake actuator lever.

➡ **Make sure that the adjuster engages the brake shoe and the parking brake actuator properly.**

21. Install the adjuster screw to the brake shoe and the parking brake actuator.

22. Apply a thin, light coat of high temperature, silicone brake lubricant to the adjuster actuator/brake shoe interface.

23. Install the adjuster actuator to the brake shoe.

➡ **Do not over stretch the adjuster spring. Damage can occur if the spring is over stretched.**

24. Install the straight end of the adjuster spring to the brake shoe.

25. Install the adjuster spring hook end to the tab on the adjuster actuator.

26. Install the return spring to the brake shoes.

27. Make sure that the adjuster operates properly.

28. Move the parking brake actuator lever in order to spread the brake shoes apart. The adjuster actuator lever should move downward, then upward as the park brake actuator lever is released, forcing the adjuster wheel to rotate. If the adjuster does not operate properly, remove then reinstall the adjuster.

29. Adjust the brake shoes.

30. Adjust the parking brake cable.

31. Install the brake drum.

ENGLISH TO METRIC CONVERSION: TORQUE

To convert foot-pounds (ft. lbs.) to Newton-meters (Nm), multiply the number of ft. lbs. by 1.36

To convert Newton-meters (Nm) to foot-pounds (ft. lbs.), multiply the number of Nm by 0.7376

ft. lbs.	Nm	ft. lbs.	Nm	ft. lbs.	Nm	ft. lbs.	Nm
0.1	0.1	34	46.2	76	103.4	118	160.5
0.2	0.3	35	47.6	77	104.7	119	161.8
0.3	0.4	36	49.0	78	106.1	120	163.2
0.4	0.5	37	50.3	79	107.4	121	164.6
0.5	0.7	38	51.7	80	108.8	122	165.9
0.6	0.8	39	53.0	81	110.2	123	167.3
0.7	1.0	40	54.4	82	111.5	124	168.6
0.8	1.1	41	55.8	83	112.9	125	170.0
0.9	1.2	42	57.1	84	114.2	126	171.4
1	1.4	43	58.5	85	115.6	127	172.7
2	2.7	44	59.8	86	117.0	128	174.1
3	4.1	45	61.2	87	118.3	129	175.4
4	5.4	46	62.6	88	119.7	130	176.8
5	6.8	47	63.9	89	121.0	131	178.2
6	8.2	48	65.3	90	122.4	132	179.5
7	9.5	49	66.6	91	123.8	133	180.9
8	10.9	50	68.0	92	125.1	134	182.2
9	12.2	51	69.4	93	126.5	135	183.6
10	13.6	52	70.7	94	127.8	136	185.0
11	15.0	53	72.1	95	129.2	137	186.3
12	16.3	54	73.4	96	130.6	138	187.7
13	17.7	55	74.8	97	131.9	139	189.0
14	19.0	56	76.2	98	133.3	140	190.4
15	20.4	57	77.5	99	134.6	141	191.8
16	21.8	58	78.9	100	136.0	142	193.1
17	23.1	59	80.2	101	137.4	143	194.5
18	24.5	60	81.6	102	138.7	144	195.8
19	25.8	61	83.0	103	140.1	145	197.2
20	27.2	62	84.3	104	141.4	146	198.6
21	28.6	63	85.7	105	142.8	147	199.9
22	29.9	64	87.0	106	144.2	148	201.3
23	31.3	65	88.4	107	145.5	149	202.6
24	32.6	66	89.8	108	146.9	150	204.0
25	34.0	67	91.1	109	148.2	151	205.4
26	35.4	68	92.5	110	149.6	152	206.7
27	36.7	69	93.8	111	151.0	153	208.1
28	38.1	70	95.2	112	152.3	154	209.4
29	39.4	71	96.6	113	153.7	155	210.8
30	40.8	72	97.9	114	155.0	156	212.2
31	42.2	73	99.3	115	156.4	157	213.5
32	43.5	74	100.6	116	157.8	158	214.9
33	44.9	75	102.0	117	159.1	159	216.2

METRIC TO ENGLISH CONVERSION: TORQUE

To convert foot-pounds (ft. lbs.) to Newton-meters (Nm), multiply the number of ft. lbs. by 1.36
To convert Newton-meters (Nm) to foot-pounds (ft. lbs.), multiply the number of Nm by 0.7376

Nm	ft. lbs.	Nm	ft. lbs.	Nm	ft. lbs.	Nm	ft. lbs.	Nm	ft. lbs.
0.1	0.1	34	25.0	76	55.9	118	86.8	160	117.6
0.2	0.1	35	25.7	77	56.6	119	87.5	161	118.4
0.3	0.2	36	26.5	78	57.4	120	88.2	162	119.1
0.4	0.3	37	27.2	79	58.1	121	89.0	163	119.9
0.5	0.4	38	27.9	80	58.8	122	89.7	164	120.6
0.6	0.4	39	28.7	81	59.6	123	90.4	165	121.3
0.7	0.5	40	29.4	82	60.3	124	91.2	166	122.1
0.8	0.6	41	30.1	83	61.0	125	91.9	167	122.8
0.9	0.7	42	30.9	84	61.8	126	92.6	168	123.5
1	0.7	43	31.6	85	62.5	127	93.4	169	124.3
2	1.5	44	32.4	86	63.2	128	94.1	170	125.0
3	2.2	45	33.1	87	64.0	129	94.9	171	125.7
4	2.9	46	33.8	88	64.7	130	95.6	172	126.5
5	3.7	47	34.6	89	65.4	131	96.3	173	127.2
6	4.4	48	35.3	90	66.2	132	97.1	174	127.9
7	5.1	49	36.0	91	66.9	133	97.8	175	128.7
8	5.9	50	36.8	92	67.6	134	98.5	176	129.4
9	6.6	51	37.5	93	68.4	135	99.3	177	130.1
10	7.4	52	38.2	94	69.1	136	100.0	178	130.9
11	8.1	53	39.0	95	69.9	137	100.7	179	131.6
12	8.8	54	39.7	96	70.6	138	101.5	180	132.4
13	9.6	55	40.4	97	71.3	139	102.2	181	133.1
14	10.3	56	41.2	98	72.1	140	102.9	182	133.8
15	11.0	57	41.9	99	72.8	141	103.7	183	134.6
16	11.8	58	42.6	100	73.5	142	104.4	184	135.3
17	12.5	59	43.4	101	74.3	143	105.1	185	136.0
18	13.2	60	44.1	102	75.0	144	105.9	186	136.8
19	14.0	61	44.9	103	75.7	145	106.6	187	137.5
20	14.7	62	45.6	104	76.5	146	107.4	188	138.2
21	15.4	63	46.3	105	77.2	147	108.1	189	139.0
22	16.2	64	47.1	106	77.9	148	108.8	190	139.7
23	16.9	65	47.8	107	78.7	149	109.6	191	140.4
24	17.6	66	48.5	108	79.4	150	110.3	192	141.2
25	18.4	67	49.3	109	80.1	151	111.0	193	141.9
26	19.1	68	50.0	110	80.9	152	111.8	194	142.6
27	19.9	69	50.7	111	81.6	153	112.5	195	143.4
28	20.6	70	51.5	112	82.4	154	113.2	196	144.1
29	21.3	71	52.2	113	83.1	155	114.0	197	144.9
30	22.1	72	52.9	114	83.8	156	114.7	198	145.6
31	22.8	73	53.7	115	84.6	157	115.4	199	146.3
32	23.5	74	54.4	116	85.3	158	116.2	200	147.1
33	24.3	75	55.1	117	86.0	159	116.9	201	147.8

ENGLISH/METRIC CONVERSION: TEMPERATURE

To convert Fahrenheit (F°) to Celsius (C°), take F° temperature and subtract 32, multiply the result by 5 and divide the result by 9
To convert Celsius (C°) to Fahrenheit (F°), take C° temperature and multiply it by 9, divide the result by 5 and add 32

F°	C°	F°	C°	C°	F°	C°	F°
-40	-40.0	150	65.6	-38	-36.4	46	114.8
-35	-37.2	155	68.3	-36	-32.8	48	118.4
-30	-34.4	160	71.1	-34	-29.2	50	122
-25	-31.7	165	73.9	-32	-25.6	52	125.6
-20	-28.9	170	76.7	-30	-22	54	129.2
-15	-26.1	175	79.4	-28	-18.4	56	132.8
-10	-23.3	180	82.2	-26	-14.8	58	136.4
-5	-20.6	185	85.0	-24	-11.2	60	140
0	-17.8	190	87.8	-22	-7.6	62	143.6
1	-17.2	195	90.6	-20	-4	64	147.2
2	-16.7	200	93.3	-18	-0.4	66	150.8
3	-16.1	205	96.1	-16	3.2	68	154.4
4	-15.6	210	98.9	-14	6.8	70	158
5	-15.0	212	100.0	-12	10.4	72	161.6
10	-12.2	215	101.7	-10	14	74	165.2
15	-9.4	220	104.4	-8	17.6	76	168.8
20	-6.7	225	107.2	-6	21.2	78	172.4
25	-3.9	230	110.0	-4	24.8	80	176
30	-1.1	235	112.8	-2	28.4	82	179.6
35	1.7	240	115.6	0	32	84	183.2
40	4.4	245	118.3	2	35.6	86	186.8
45	7.2	250	121.1	4	39.2	88	190.4
50	10.0	255	123.9	6	42.8	90	194
55	12.8	260	126.7	8	46.4	92	197.6
60	15.6	265	129.4	10	50	94	201.2
65	18.3	270	132.2	12	53.6	96	204.8
70	21.1	275	135.0	14	57.2	98	208.4
75	23.9	280	137.8	16	60.8	100	212
80	26.7	285	140.6	18	64.4	102	215.6
85	29.4	290	143.3	20	68	104	219.2
90	32.2	295	146.1	22	71.6	106	222.8
95	35.0	300	148.9	24	75.2	108	226.4
100	37.8	305	151.7	26	78.8	110	230
105	40.6	310	154.4	28	82.4	112	233.6
110	43.3	315	157.2	30	86	114	237.2
115	46.1	320	160.0	32	89.6	116	240.8
120	48.9	325	162.8	34	93.2	118	244.4
125	51.7	330	165.6	36	96.8	120	248
130	54.4	335	168.3	38	100.4	122	251.6
135	57.2	340	171.1	40	104	124	255.2
140	60.0	345	173.9	42	107.6	126	258.8
145	62.8	350	176.7	44	111.2	128	262.4

LENGTH CONVERSION

To convert inches (in.) to millimeters (mm), multiply the number of inches by 25.4
To convert millimeters (mm) to inches (in.), multiply the number of millimeters by 0.04

Inches	Millimeters	Inches	Millimeters	Inches	Millimeters	Inches	Millimeters
0.0001	0.00254	0.005	0.1270	0.09	2.286	4	101.6
0.0002	0.00508	0.006	0.1524	0.1	2.54	5	127.0
0.0003	0.00762	0.007	0.1778	0.2	5.08	6	152.4
0.0004	0.01016	0.008	0.2032	0.3	7.62	7	177.8
0.0005	0.01270	0.009	0.2286	0.4	10.16	8	203.2
0.0006	0.01524	0.01	0.254	0.5	12.70	9	228.6
0.0007	0.01778	0.02	0.508	0.6	15.24	10	254.0
0.0008	0.02032	0.03	0.762	0.7	17.78	11	279.4
0.0009	0.02286	0.04	1.016	0.8	20.32	12	304.8
0.001	0.0254	0.05	1.270	0.9	22.86	13	330.2
0.002	0.0508	0.06	1.524	1	25.4	14	355.6
0.003	0.0762	0.07	1.778	2	50.8	15	381.0
0.004	0.1016	0.08	2.032	3	76.2	16	406.4

ENGLISH/METRIC CONVERSION: LENGTH

To convert inches (in.) to millimeters (mm), multiply the number of inches by 25.4

To convert millimeters (mm) to inches (in.), multiply the number of millimeters by 0.04

Inches Fraction	Inches Decimal	Millimeters Decimal	Inches Fraction	Inches Decimal	Millimeters Decimal	Inches Fraction	Inches Decimal	Millimeters Decimal
1/64	0.016	0.397	11/32	0.344	8.731	11/16	0.688	17.463
1/32	0.031	0.794	23/64	0.359	9.128	45/64	0.703	17.859
3/64	0.047	1.191	3/8	0.375	9.525	23/32	0.719	18.256
1/16	0.063	1.588	25/64	0.391	9.922	47/64	0.734	18.653
5/64	0.078	1.984	13/32	0.406	10.319	3/4	0.750	19.050
3/32	0.094	2.381	27/64	0.422	10.716	49/64	0.766	19.447
7/64	0.109	2.778	7/16	0.438	11.113	25/32	0.781	19.844
1/8	0.125	3.175	29/64	0.453	11.509	51/64	0.797	20.241
9/64	0.141	3.572	15/32	0.469	11.906	13/16	0.813	20.638
5/32	0.156	3.969	31/64	0.484	12.303	53/64	0.828	21.034
11/64	0.172	4.366	1/2	0.500	12.700	27/32	0.844	21.431
3/16	0.188	4.763	33/64	0.516	13.097	55/64	0.859	21.828
13/64	0.203	5.159	17/32	0.531	13.494	7/8	0.875	22.225
7/32	0.219	5.556	35/64	0.547	13.891	57/64	0.891	22.622
15/64	0.234	5.953	9/16	0.563	14.288	29/32	0.906	23.019
1/4	0.250	6.350	37/64	0.578	14.684	59/64	0.922	23.416
17/64	0.266	6.747	19/32	0.594	15.081	15/16	0.938	23.813
9/32	0.281	7.144	39/64	0.609	15.478	61/64	0.953	24.209
19/64	0.297	7.541	5/8	0.625	15.875	31/32	0.969	24.606
5/16	0.313	7.938	41/64	0.641	16.272	63/64	0.984	25.003
21/64	0.328	8.334	21/32	0.656	16.669	1/1	1.000	25.400
			43/64	0.672	17.066			

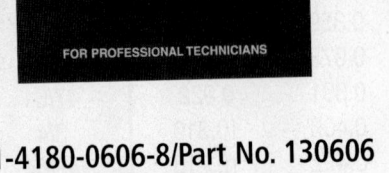

Domestic Manual ISBN 1-4180-0606-8/Part No. 130606
Import Manual ISBN 1-4180-1537-7/Part No. 131537

Chilton has added so much to its labor guide manual that we've had to put it in two volumes! We've added hundreds of new labor operations—including maintenance services and electronic system diagnosis—to the *Chilton® 2006 Import and Domestic Labor Guide* manuals. All labor times for 1981 through 2006 vehicles consider the real world environment in which technicians work: worn, rusted or dirty components, being serviced with tools commonly used in the aftermarket. Chilton labor times are accepted by most insurance and extended warranty companies. Vehicle makes and models conform to current Automotive Aftermarket Industry Association standards.

Labor Guide Manual Benefits:

- parts terminology is more standardized across different OEMs to simplify reference.
- a total of more than 2,500 pages of updated Chilton labor times appear in these two volumes
- make sure your students have this latest edition because our experts have updated hundreds of labor times for earlier models

Hardcover Manuals are 8 7/8" x 11", ©2006

Labor Guide CD-ROM Benefits:

- easy-to-use software to create and print professional-quality estimates and invoices
- three user-defined levels of labor rates correspond to different types of job scenarios, for "real-world" application
- functions as a database of aftermarket labor times for monitoring warranty and insurance claims
- software keeps track of customers and prior estimates for time-saving recall
- customizable application allows service writers to add labor operations and times, and parts companies to add labor times to existing parts ordering systems

CD ISBN 1-4180-0605-X/Part No. 130605
©2006

Previous Year Editions:
Chilton 2005 Labor Guide Manual, ISBN 1-4018-7412-6/Part No. 27412
Chilton 2005 Labor Guide CD-ROM, ISBN 1-4018-7818-0/Part No. 27818

FOR CUSTOMER SUPPORT CALL **1-800-477-3692**

The *Chilton® 2006 Mechanical Service Manuals* provides updated coverage through 2005 models and even many 2006 models, as made available from original equipment manufacturers (OEMs). Chilton is still your reliable source for fast, accurate repairs and reassembly and it still provides the lowest-priced professional repair manuals on the market! These manuals are organized by make, model and system so information gathering is easier. Now with even more illustrations and a streamlined index, it's no wonder more automotive professionals turn to Chilton Professional Manuals for their mechanical service and repair information.

Mechanical Service Manual Benefits:

- access up-to-date service and repair information covering model years 2002-2006, all logically arranged by manufacturer
- follow clear, step-by-step procedures—from drive train to chassis —to yield fast, accurate results
- service more mechanical systems, including brakes, engines, suspensions, steering and related components
- know what special tools are required for specific jobs, as Chilton editors describe and illustrate them to make repair work go more smoothly

2006 Editions

Chilton 2006 DaimlerChrysler Mechanical Service Manual—ISBN 1-4180-0600-9/Part No. 130600
Chilton 2006 Ford Mechanical Service Manual—ISBN 1-4180-0601-7/Part No. 130601
Chilton 2006 General Motors Mechanical Service Manual—ISBN 1-4180-0602-5/Part No. 130602
Chilton 2006 Asian Mechanical Service Manual—Volume I—ISBN 1-4180-0947-4/Part No. 130947
Chilton 2006 Asian Mechanical Service Manual—Volume II—ISBN 1-4180-0948-2/Part No. 130948
Chilton 2006 Asian Mechanical Service Manual—Volume III—ISBN 1-4180-0949-0/Part No. 130949
Chilton 2006 Asian Mechanical Service Manual—3 Volume Set—ISBN 1-4180-0603-3/Part No. 130603
Chilton 2006 European Mechanical Service Manual—ISBN 1-4180-0604-1/Part No. 130604

Manuals are 8 1/2" x 11", ©2006

2005 Editions

Chilton 2005 General Motors Mechanical Service Manual—ISBN 1-4018-7146-1/Part No. 27146
Chilton 2005 Chrysler Mechanical Service Manual—ISBN 1-4018-6718-9/Part No. 26718
Chilton 2005 Ford Mechanical Service Manual—ISBN 1-4018-6719-7/Part No. 26719
Chilton 2005 European Mechanical Service Manual—ISBN 1-4018-6720-0/Part No. 126720
Chilton 2005 Asian Mechanical Service Manual – Volume I—(Acura-Mazda) ISBN 1-4018-6716-2/Part No. 26716
Chilton 2005 Asian Mechanical Service Manual – Volume II—(Mitsubishi-Toyota)
    ISBN 1-4018-6717-0/Part No. 26717
Chilton 2005 Asian Mechanical Service Manual – Set of Volumes I & II—ISBN 1-4018-7180-1/Part No. 27180

Manuals are 8 1/2" x 11", ©2005

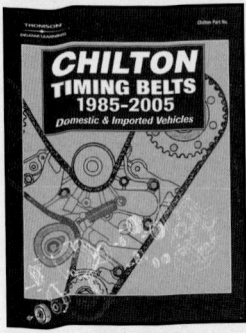

Chilton Timing Belts, 1985-2005

Chilton
ISBN 1-4018-9880-7/Part No. 129880

Timing belt procedures can represent increased profits for automotive repair shops and service stations, and this manual contains all the information automotive technicians need to properly service timing belts on domestic and imported cars, vans, and light trucks through 2005 models. Clear, straightforward procedures, illustrations, and specifications help to communicate 20 years of vehicle applications for fast, accurate inspection, replacement, and tensioning of timing belts. Readers will learn step-by-step how to perform key procedures both quickly and safely, while learning the correct labor time to charge for the service. OEM-recommended replacement intervals for proper maintenance of customer's vehicles are also featured.

BENEFITS

- detailed illustrations clearly demonstrate important concepts, such as how to correctly align camshaft and crankshaft timing marks, and how to simplify serpentine belt installation
- readers are made aware of potential hazards and time-wasting practices that can impede safe and profitable service procedures
- special tools are identified so that completing the service is as easy and quick as possible

544 pp, 8 1/2" x 11", softcover, ©2006

The *Chilton® 2006 Diagnostic Service Manuals* provide technicians with the critical diagnostic information they need to accurately identify and solve engine performance problems. Clear explanations, specifications, and illustrations help technicians diagnose second generation on-board diagnostic (OBD-II) systems. Chilton Diagnostic Service Manuals, when used with an engine analyzer, scan tool, or lab scope, allow diagnosticians to understand functions of engine performance components and systems, simplify testing procedures, and diagnose trouble codes.

Diagnostic Service Manual Benefits:

- provide training information in addition to reference material
- explain engine performance components and system operation
- function as exceptional diagnostic companions when analyzing automotive drive-train performance problems
- provide a comprehensive list of trouble code titles, conditions, and possible causes
- reduce diagnostic and repair time using expert testing procedures and trouble-shooting hints

2006 Editions
Chilton 2006 DaimlerChrysler Diagnostic Service Manual
ISBN 1-4180-2118-0/Part No. 132118
Chilton 2006 Ford Diagnostic Service Manual
ISBN 1-4180-2119-9/Part No. 132119
Chilton 2006 General Motors Diagnostic Service Manual
ISBN 1-4180-2120-2/Part No. 132120
Chilton 2006 Asian Diagnostic Service Manual, Volume 1
ISBN 1-4180-2913-0/Part No. 132913
Chilton 2006 Asian Diagnostic Service Manual, Volume 2
ISBN 1-4180-2914-9/Part No. 132914
Chilton 2006 Asian Diagnostic Service Manual, Volume 3
ISBN 1-4180-2915-7/Part No. 132915
Chilton 2006 European Diagnostic Service Manual
ISBN 1-4180-2924-6/Part No. 132924
Manuals are 8 1/2" x 11", ©2006

2005 Editions
Chilton 2005 General Motors Diagnostic Service Manual
ISBN 1-4180-0552-5/Part No. 130552
Chilton 2005 Chrysler Diagnostic Service Manual
ISBN 1-4180-0550-9/Part No. 130550
Chilton 2005 Ford Diagnostic Service Manual
ISBN 1-4180-0551-7/Part No. 130551
Chilton 2005 Asian Diagnostic Service Manual
ISBN 1-4180-0553-3/Part No. 130553
Manuals are 8 1/2" x 11", ©2005

THOMSON
DELMAR LEARNING

The *Chilton® Perennial Editions* contain repair and maintenance information for popular mechanical systems that may not be available elsewhere. They offer a wide range of repair information on cars, trucks, vans, and SUVs dating back to the early 1960s, and as current as 2002. Information for 1993 and later model years includes scheduled maintenance interval charts.

Benefits:

- covers the most common vehicle models found in the repair aftermarket today
- gain quick understanding of systems using exploded-view illustrations, diagrams, and charts
- simplify tough jobs with easy-to-follow removal and installation instructions for heater core and other components
- obtain complete coverage of repair procedures from drive train to chassis and associated components

Auto Repair Manual, 1998-2002, 1,426 pages
    ISBN 0-8019-9362-8/Part No. 9362
Auto Repair Manual, 1993-1997, 2,064 pages
    ISBN 0-8019-7919-6/Part No. 7919
Auto Repair Manual, 1988-1992, 1,284 pages
    ISBN 0-8019-7906-4/Part No. 7906
Auto Repair Manual, 1980-1987, 1,344 pages
    ISBN 0-8019-7670-7/Part No. 7670

Import Car Repair Manual, 1998-2002, 1,792 pps
    ISBN 0-8019-9363-6/Part No. 9363
Import Car Repair Manual, 1993-1997, 2,080 pps
    ISBN 0-8019-7920-X/Part No. 7920
Import Car Repair Manual, 1988-1992, 1,632 pages
    ISBN 0-8019-7907-2/Part No. 7907
Import Car Repair Manual, 1980-1987, 1,488 pages
    ISBN 0-8019-7672-3/Part No. 7672

Truck & Van Repair Manual, 1998-2002, 1,408 pages
    ISBN 0-8019-9364-4/Part No. 9364
Truck & Van Repair Manual, 1993-1997, 2,096 pages
    ISBN 0-8019-7921-8/Part No. 7921
Truck & Van Repair Manual, 1991-1995, 1,664 pages
    ISBN 0-8019-7911-0/Part No. 7911
Truck & Van Repair Manual, 1986-1990, 1,536 pages
    ISBN 0-8019-7902-1/Part No. 7902
Truck & Van Repair Manual, 1979-1986, 1,440 pages
    ISBN 0-8019-7655-3/Part No. 7655

SUV Repair Manual, 1998-2002, 1,292 pages
    ISBN 0-8019-9365-2/Part No. 9365

Hardcover manuals are 8 1/2" x 11".

Chilton Collector's Editions—*Reference Manuals for Vintage Vehicles*
Auto Repair Manual, 1964-1971, ISBN 0-8019-5974-8/Part No. 5974,
Truck & Van Repair Manual, 1961-1971, ISBN 0-8019-6198-X/Part No. 6198
Truck & Van Repair Manual, 1971-1978, ISBN 0-8019-7012-1/Part No. 7012